A Concordance to
CHARLES DARWIN'S NOTEBOOKS,
1836–1844

Edited by Paul H. Barrett and Donald J. Weinshank:

A Concordance to Darwin's "Origin of Species," First Edition
with Timothy T. Gottleber

*A Concordance to Darwin's "The Expression of the Emotions
in Man and Animals"*
with Paul Ruhlen, Stephan J. Ozminski,
and Barbara N. Berghage

*A Concordance to Darwin's "The Descent of Man, and
Selection in Relation to Sex"*
with Paul Ruhlen and Stephan J. Ozminski

A Concordance to
Charles Darwin's Notebooks,
1836–1844

Edited by
Donald J. Weinshank, Stephan J. Ozminski,
Paul Ruhlen, *and* Wilma M. Barrett

Cornell University Press

ITHACA AND LONDON

First published 1990 by Cornell University Press.

International Standard Book Number 0-8014-2352-X
Library of Congress Catalog Card Number 89-45977
Printed in the United States of America
*Librarians: Library of Congress cataloging information
appears on the last page of the book.*

⊗ The paper used in this publication meets the minimum requirements of the American National Standard for Permanence of Paper for Printed Library Materials Z39.48–1984.

Library of Congress Cataloging-in-Publication Data

A Concordance to Charles Darwin's notebooks, 1836–1844 / edited by Donald J. Weinshank . . . [et al.].
    p.   cm.
Includes bibliographical references.
ISBN 0-8014-2352-X (alk. paper)
1. Darwin, Charles, 1809–1882. Charles Darwin's notebooks, 1836–1844—Concordances. I. Weinshank, Donald J. (Donald Jerome), 1937–   . II. Darwin, Charles, 1809–1882. Charles Darwin's notebooks, 1836–1844.
QH365.Z9W45   1990
575.01'62—dc20   89-45977

To C.R.D., who started it all, and to P.H.B.,
who introduced each of us to his friend.

d.j.w.
s.j.o.
p.r.
w.m.b.

CONTENTS

FOREWORD

This concordance is a companion to *Charles Darwin's Notebooks, 1836–1844: Geology, Transmutation of Species, Metaphysical Enquiries*, edited by Paul H. Barrett, Peter J. Gautrey, Sandra Herbert, David Kohn, and Sydney Smith and published jointly in 1987 by the British Museum (Natural History), Cornell University Press, and Cambridge University Press (England). *Charles Darwin's Notebooks* contains eleven notebooks and four related manuscripts from Darwin's most creative years as a theorist. The earliest notes in this collection date from 1836, when Darwin was twenty-seven years old and returning home from his service as naturalist aboard H.M.S. *Beagle* during the years 1831–36; the latest explicitly dated notes made in regular sequence are from 1844, though later entries exist. In the notes Darwin developed many of the major ideas contained in several of his geological publications and in *On the Origin of Species by Means of Natural Selection* (1859), *The Variation of Animals and Plants under Domestication* (1868), *The Descent of Man, and Selection in Relation to Sex* (1871), and *The Expression of the Emotions in Man and Animals* (1872). The notes thus contain in outline the main program of research and publication he was to follow throughout his life.

The scholar who wishes to trace the development of Darwin's thought will find this concordance a research tool of exceptional value, both in its own right and as a complement to the index to the *Notebooks* volume. The arguments presented in the notebooks were in the early stages of formation; they display the probing and discursive logic of discovery rather than the coherent and fully articulated logic of final exposition. Thus a subject—such as "mockingbird" or "isolation" or "Galapagos"—might be treated at any number of points in various manuscripts. The concordance does a great service in pulling together these word references for quick consultation. In addition, the concordance enables one to look for patterns in Darwin's employment of words and to observe his development of a conceptual vocabulary adequate to the subjects he was treating. Further, the concordance can help one identify more readily the authors whose views were of moment to Darwin and in what context.

Paul H. Barrett planned for this concordance at the inception of the *Notebooks* project, but he died before he could begin work on it. His colleagues Donald J. Weinshank, Stephan J. Ozminski, and Paul Ruhlen at Michigan State

University, East Lansing, and his wife, Wilma M. Barrett, made his intentions a reality. While the *Notebooks* volume was in preparation, it was Paul Barrett who was most alert to the utility of computer technology for making the volume accessible to readers. For example, he suggested that the bibliography include the locations in the text where each work was cited. Record keeping of such magnitude becomes more feasible with computer assistance, and as the *Notebooks* project progressed, the techniques for gaining computer assistance expanded.

Similarly, computer techniques for producing concordances have expanded in recent years, becoming more sophisticated and efficient. For the present skillful adaptation and application of such techniques we are indebted to those at East Lansing.

SANDRA HERBERT

Washington, D.C.

PREFACE

We have produced this concordance as one of a series designed to aid in the study of Charles Darwin's writings. Already published are concordances to the *Origin of Species, First Edition*; *The Expression of the Emotions in Man and Animals*; and *The Descent of Man, and Selection in Relation to Sex* (Cornell University Press, 1981, 1986, 1987).

Our system of representing and organizing words in this concordance remains identical to that in the preceding concordances. Keywords are centered in the page, surrounded by as much of the original text as possible, with fragments of words commonly occurring at the beginnings and ends of lines. Keywords are printed in capital letters, whereas the surrounding text appears as in the original Darwin text. Each keyword is coded to the page number of the British Museum–Cornell source text, *Charles Darwin's Notebooks, 1836–1844* (1987; cited hereafter as Cornell), and to the page of Darwin's notebook in which that keyword appears. The letters used to denote the notebooks are as follows:

Identifier	Notebook
a, b, c, d, e, m, n	A, B, C, D, E, M, N
r	Red
g	Glen Roy
t	Torn Apart
s	Summer 1842
z	Zoology Notes, Edinburgh
q	Questions & Experiments
o	Old & Useless Notes
j (for John Macculloch)	Abstract of Macculloch

For example, for the keyword "meteoric," 077r172 indicates that the word can be found on Cornell page 77 and that it appears in Darwin's Red Notebook on page 172. We have also used the abbreviations "r" for recto (or right-hand page) and "v" for verso (or left-hand page), as well as BC (back cover), FC (front cover), IBC (inside back cover), and IFC (inside front cover). Thus, for "peas," 497q06v indicates that the keyword can be found on Cornell page 497 and in the Questions & Experiments Notebook on page 6 verso. The notation 331dIFC refers to the inside front cover of Notebook D. Using three characters to represent Darwin's page numbers proved adequate except for the solitary case of 167r and 167v in the Abstract of Macculloch on Cornell page 640, where all words were keyed as 640j167.

In our experience, the text surrounding the keyword is sufficient to enable the user to locate any citation in the source text. Diacritical marks such as French accents and the German umlaut appear in the text surrounding each keyword, but not in keywords themselves, in part because French accents are generally not used with capital letters. For a single keyword, occurrences appear in the same order as they do in the source text. Hyphenated words are keyed as two separate words, with both the first and second words listed as if not hyphenated.

The list of suppressed words, Appendix 1, contains all words excluded from the keywords, with their frequencies of occurrence in the notebooks. Among the items in Appendix 1 are prepositions, articles, and common verbs. Every keyword in the concordance, with its frequency, is given in Appendix 2.

We have retained all of Darwin's variant spellings (e.g., camelion / chamaeleon / chamaelion / chamelion). Although in our previous concordances it seemed entirely appropriate to correct obvious mistakes silently, here, following the lead of the editors of the notebooks, we have indexed all variants. Because these variant spellings sometimes occur far apart in the main body of the concordance, they are listed in Appendix 3. That appendix also includes non-English words, proper names, word fragments and abbreviations, and words of common British but not American spelling (e.g., neighbourhood).

Transcription of the original notebooks necessitated certain typographical conventions, described in the introduction to the Cornell volume:

> In transcribing Darwin's text all words and fragments of words are recorded. Deletions are enclosed by single angled brackets ‹ ›. . . . Insertions in the text are set off by double angled brackets « » [p. 12]. Darwin's brackets are kept with a superscript 'CD' added; his use of the ampersand and plentiful employment of the dash are unaltered; and his abbreviations 'do' for ditto and 'V.' for 'Vide' remain unexpanded. . . . Two idiosyncratic markings have also been kept: first, a backwards question mark that he borrowed from Spanish, and a cross-hatch he used on occasion for emphasis [p. 13]. Where a section of text is illegible, bracketed ellipses marks ('[. . .]') appear in the text [p. 14].

We have done our utmost to retain all of these conventions in the concordance. Fragmented words or phrases such as "cou[n]tries" and "I«ndian»" and "‹i›nfertility" are printed in the concordance just as they appear in the source text. However, for purposes of sorting, they are treated as intact words. We have represented Darwin's backwards question mark with the modern Spanish inverted question mark. We have substituted the words "degree" and "minute" for the symbols ° and ′ and have replaced ligatures (e.g., æ) with the complete letters (e.g., ae). Because a concordance is a work that reduces text to a series of characters that precede and follow keywords, we made no attempt to represent cases where Darwin's brackets (i.e., braces) span several lines. Nor did we represent the various type fonts found in the original text. Finally, we introduced two symbols not found in the original notebooks: {P}, which indicates a picture or sketch made by Darwin; and {T}, which indicates a table.

We transferred the notebooks in the 1987 Cornell edition to a computer text file using the Kurzweil Data Entry Machine model 4000, except that Darwin's page numbers were inserted with PC-WRITE 2.71. We then checked the spelling of each word with the Simon & Schuster Webster's New World Combo program. According to the documentation supplied, "The Speller dictionary contains every entry in *Webster's New World Dictionary* (Second Concise Edition)." Words rejected by that spell-checker were included, after verification, in the file that became Appendix 3.

We proofed the entire edition aloud back-to-front against our computer text file, correcting errors with SideKick Plus, and then spell-checked the corrected file a second time to correct errors that might have been introduced in the editing process. We spell-checked the entire text a third time to identify all word fragments to ensure that the concordance program was handling them correctly. Finally, we used our concordance program (described below) in an interactive mode to check occurrence of problematic words.

Stephan J. Ozminski, with help from Paul Ruhlen, wrote the program used to produce this concordance. In just thirty minutes the concordance was generated on a Zenith 159-series (8088/16 bit) microcomputer with a 20-megabyte hard disk, a roughly twentyfold gain in speed over the program used for the preceding concordance. The concordance was printed on a Hewlett Packard LaserJet + Printer.

We express our gratitude to many people at Michigan State University without whose unselfish and extraordinary contributions this project could not have been completed: Sharon Bennett, KDEM Operator, and Charles J. Gillengerten and Dean Franklin, Shift Supervisors, of the Computer Laboratory; Marsha Wise, Secretary, Department of Natural Science; Diana Marinez, former Chairperson of the Department of Natural Science, who obtained essential grant funds; Cathy M. Davison and Michelle Sidel, Secretaries, Mark Urban-Lurain, Instructional Resources Coordinator, and Anthony S. Wojcik, Chairperson, of the Department of Computer Science. John E. Cantlon, Vice President for Research and Graduate Studies, has given welcome advice and support to this project for many years.

Preparation of the concordance was partially funded by Grant Number 20066 from the College of Natural Science, Michigan State University.

THE EDITORS

East Lansing, Michigan

A Concordance to
CHARLES DARWIN'S NOTEBOOKS,
1836–1844

Page
(Key Word)
099a051 n stratum OP. let force drag particls to line {P} AB, & likewise gravity MN. Then every particle woul
634j54v d to its utmost exhaustion, & till it must be ABANDONED for another".-- What bosch!! Put it to case o
634j54v designs of an omnipotent creator, exhausted & ABANDONED. Such is Man's philosophy. when he argues abo
469t151 ered «all» came up in 1840 true. Shrewsbury.-- ABERLEY-- Early Magazine-- &c. double-blossomed «& dw
498q008 rs in frames are not artificially impregnated. ABERLEY says Ants-- Enquire (13) Do any of same speci
501q011 ect, except by very minute insects.-- (30) Get ABERLEY to plant SINGLE Peas, Kidney Bean & Bean, int
501q011 impregnated Cowcumbers will they seed.?-- (11) ABERLEY has planted seeds of pale green Cynoglossum.
505q014 e stamens.-- show crossing & ¿heredetary? (15) ABERLEY has a hooked Pea.-- intends to breed from it
505q014 ll it seed?-- (Skim through Penny Cyclopaedia) ABERLEY says that some Bees are smaller & more viciou
603o013 come corrupt, & whole classes of words «are ABBREVIATIONS» he thus derives from nouns & verbs-- so th
342d034 y well answers that nearly all F. W. Fish are ABDOMINALS. ˙.that order first converted-- is it an old
590n094 count of a monkey in a passion like Jenny.-- Dr. ABEL has given an account of an Ourang.-- see his Tr
177b028 near arrangement.-- ¿How is it that there come ABERANT species in each genus «(with well characterize
350d056 cted with locomotion.-- «Mem. Dr. Blackwell (ABERCRMBIES) comparison of sight to threads.--» Hence th
258c061 lationship, the dissenblances analogy,-- See ABERCROMBIE p. 172. for definition of Analogy. All the d
322c270 alia Walter Scotts life I & 2d & 3rd Volumes ABERCROMBIE on the Intellectual powers. Hunters Animal O
606o025 it, but other motives prevent the action see ABERCROMBIE conclusive remarks p. 205 & 206.]CD Motives
548m114 ore one may be imperfectly reason -- <In a> ABERCROMBIE'S case of «in Botanical Student» somnambulism,
554m141 ledge.-- Sept. 13th It will be good to give ABERCROMBIE'S definition of "reason" & "reasoning," & tak
201b126 tion of that axiom in Natural History that all ABERRANT & osculant groups are not only few in species
258c061 on of analogy may chiefly be looked for in the ABERRANT groups.-- It is having walking fly catcher, w
262c073 as. show how nicely things adapted--.-- These «ABERRANT» varieties will be formed in any kingdom of n
262c073 n round of chances every family will have some ABERRANT groups.-- but as for number five in each grou
264c082 ground woodpecker-- -- but tail of some ducks ABERRANT from-- habits.-- Gould I see quite recognizes
265c087 ts, & how much heredetary. The circumstance of ABERRANT groups being small it is truism. for it not s
265c087 ps being small it is truism. for it not so not ABERRANT.-- Tenioptera rufiventris, is instance of bir
283c144 not some statement about diversity of forms in ABERRANT circles.-- explained by such not having been
297c185 only kills them when urged by hunger.-- p. 65. ABERRANT groups few in numbers & vary much in characte
297c185 nces., which by causing death, makes the group ABERRANT When species more rare we infer extermination, whe
302c202 . Characters of analogy.-- last acquired,-- be ABERRANT, therefore more easily modified = = this is n
304c207 ld whether not descended from long way back.-- ABERRANT-- the two former connecting classes like Toxo
448e168 osiren-- Amblyrhyncus & Toxodon, <all> equally ABERRANT.-- its because instincts to woman is not foll
591n099 dual are same as in normal cases) are held in ABHORRENCE it is because instincts to woman is not foll
591n100 ct) think not fit, as cannabalism, is held in ABHORRENCE.-- all this makes analogy of actions with <&
320c276 yages & Quadrupeds of Paraguay Dobrizhoffer. ABIPOM<E>NES. Edinburgh New. Phil Journal. about 13 numb
240c013 scription of wild animals, nor in Dobrizhoffer ABIPONES.-- Voyage. de L'Astrolabe Zoologie. p. 60. Vo
481o010 e. Dobrizhoffer., Vol I p. 310. History of the ABIPONES, Vol II p. 125 Allusion to
226b221 chance of some one of the different orders being ABLE to survive or chance having transported them to
278c132 .-- the wonder is that the Europaean forms, were ABLE to escape to some more fitting country,-- if To
293c175 ld be disagreeable to Musquitoes We never may be ABLE to trace the steps by which the organization of
335d016 id, "an animal <acquires <th> any new> is <only> ABLE to transmit «only» those peculiarities, to its
350d055 ms common, but that perfection consists in being ABLE to reproduce Here there is some error-- Observe
355d066 e fertile]; as long as opponents will «are» not «ABLE to» tie themselves down, they can find loophole
384d162 become so related to each other, as never to be ABLE to impregnate themselves (this never happens in
407e039 reeminently equable climate. might not have been ABLE to have survived a change, (& become transmuted
424e103 s.-- CD[Camel does not vary «one ought not to be ABLE to hybridise the Camel» like ass «same way some
530m043 have had. yet during whole illness, he had been ABLE to direct about his own health.-- his complaint
583n071 ans are sufficient for hence it must some way be ABLE to measure the cell; p. 22. instincts & structu
596n184 ot move eyebrows.-- or skin of head,-- «scarcely ABLE St.-- » Cyanocephalus, macacus. Cercopithecus?
611o034 n to whole, that is enough must be present to be ABLE to exist as individual.-- [RHC] 2) In animals,
619o043 others injured these objects, without his being ABLE to prevent it, he would likewise feel pain.-- I
591n099 yed-- I often have «as a boy» wondered why all ABNORMAL sexual actions or even impulses. (where sensa
147g016 would intersect alley above the 300 ft Alluvium <ABO> by Loch Dochart-- Rivers could not have deposit
277c126 law; habits determining fertility Scheme for ABOLISHING specific names & giving subgenera. true valu
229b233 -- Dr. Smith's Information Long Horned (very) ABORIGINAL at Cape: crossed with English Bull. offsprin
229b233 English.-- Hottentots say great tailed sheep ABORIGINAL at Cape & a thinner-tailed kind further inla
355d066 e approaches to ass.» or fowls to the several ABORIGINAL species «or ducks» (here argue if it be said
400e013 at present time Uncle J. does not suppose one ABORIGINAL variety.-- for they are all made by fertiliz
495q005 of wild cabbage in VERY rich soil, will plants ABORT?, does it require successive generations to acc
304c207 uide in this opinion?-- EXCELLENT PRINCIPLE OF ABORTION ISOLATION of range « <far more prob> » tends
383d159 of Horses poenis reduced to extreme degree of ABORTION).-- Insecta.-- hermaphrodite, being not only
418e080 on from domestication of Monoecious plants, & ABORTION of others.-- ¿ in hemi-hermaphrodite insects
427e109 ous; & really dioecious plants being effect of ABORTION of one sex.-- Linnaean class Dioecia & Monoe
502q11v to sit on own eggs Flowers in short turf. for ABORTION. or for sterility Land Birds Madeira Migrator
509q017 ebrae process developed into ribs.) & does its ABORTION vary, according to Bentham's Remark. Horse or
515q021 horned, Suffolk have <abortive> «no» horns by ABORTION, but sometimes have dangling ones.-- Is there
515q021 of plants, «in» which some organ is absent by ABORTION, but appears in abortive state either in the
297c186 g of Zoolog. & Bot. Vol. II p. 125 Allusion to ABORTIVE spiracles in Hemiptera do. p. 160. soft pluma
307c215 of some moths, like glowworm <are> have «These ABORTIVE organs in some Males animals, Mammae in Men,
307c215 cies of Apterix This is important for if these ABORTIVE wings in the female are allowed to the   fully
307c215 the fully organized wings of the male rendered ABORTIVE in the womb.-- if these apparently useless or
307c215 ate such crippled, then we are bound to consider ABORTIVE organs of same tendency in species, this is c
307c216 s is capital & novel argument.-- Are there any ABORTIVE organs in neuter bee, (There is paper by Yarr
307c216 lendid argument old female, turning into cock, ABORTIVE spurs. growing.-- Are there any abortive orga
307c216 cock, abortive spurs. growing.-- Are there any ABORTIVE organs produced in domesticated animals, in p
307c216 re a tendril passes into a mere stump.-- Shall ABORTIVE organs «of very same kind» in these cases, ha
309c219 to Mr Herbert's views.-- Argue <argue> case of ABORTIVE organs to mules in their genitals & even to a
311c226 to analyse) will not this separate facts about ABORTIVE organs &c The doctrine of monsters is preemin
325c267 views very ancient? Study with profound care, ABORTIVE organs produced in domesticated plants; where
335d014 .-- Study what these monsters are:-- are they «ABORTIVE» twins.--] The fertility of first cross, as
335d014 g new species,] -- Are not dreadful monsters, ABORTIVE, just like mules. Fox's half bred Persians «c
335d015 ans «cat» favour the Persian side.-- Theory of ABORTIVE hybrids.-- If mules did breed, the offspring
352d057 rare.-- 2d Sept Those animals which have many ABORTIVE organs, might be expected to have larvae more
352d058 se them with more difficulty, «contradicted by ABORTIVE wings in the female are allowed to the fully
352d058 by abortive organs, but number of species with ABORTIVE organ of any kind few.-- » hence become EXTIN
352d059 malia As every organ is modified by use, every ABORTIVE organ must have been once changed.-- what is
352d059 e organ must have been once changed.-- what is ABORTIVE? when it does not perform that function which
352d059 ost important law.-- Penguins wing perhaps not ABORTIVE???. Apterix certainly.-- Lyell's excellent vi
373d132 . analogous to other males feeding young, & to ABORTIVE <organs> «mammae» in male Mammalia:-- ¿is not
373d132 reation.-- why. what tendency can there be for ABORTIVE organ ever disappearing??-- Have Marsupiata a
373d132 ve organ ever disappearing??-- Have Marsupiata ABORTIVE Mammae?.-- My view would make every individua
381d157 ctinibr Mollusca.-- «or Cephalopoda» are there ABORTIVE traces of other «sexual» organs; for if so, s
382d158 erfection --How came it nipples <are> «though» ABORTIVE, are so plain in Man, & yet no trace of abort
382d158 ortive, are so plain in Man, & yet no trace of ABORTIVE womb, or ovarium.-- or testicles in female.--
384d162 seen distinct. (in dioecious plants are there ABORTIVE sexual organs?): they then become so related
386d166 law acting against heredetary tendency causes ABORTIVE organs.-- the origin of this law is part of t
388d172 - The view that man & «or cock» pheasant &c is ABORTIVE hermaphrodite is supported by change which ta
388d172 on in structure = Neuter bee having both sexes ABORTIVE fact of same tendency. -- Mammae in man. havi
389d176 th capable of propagating, but one is rendered ABORTIVE as far as parturition is concerned.-- Generat
391d174 CD[The view of <In> each Man or mammalia being ABORTIVE hermaphrodite simplifys case much; & original
415e007 bimanous,, is to see, what parts of structure ABORTIVE.-- Remember my fathers remark about the Bladd
415e070 , that varieties are generally additive, & not ABORTIVE: with reference to the non-necessity of the «
434e129 nerally dioicous. yet parts only very slightly ABORTIVE, & bed of female flowers will sometimes produc
434e130 rs female flowers, the organs are most clearly ABORTIVE, so that they become so by suppression of one
434e130 ioecious & Monooecious plants have rudimentary ABORTIVE organs, even more so than Polygamia: Monooeci
466t099 1841. Maer Examined the Lemon-thyme.-- equally ABORTIVE as it was in autumn: filaments united in whol
467t099 en.-- Common Thyme growing close by is equally ABORTIVE--and both growing within Kitchen Garden.-- As
467t099 t Henslow's remark that pistil does not become ABORTIVE. Examined in microscope--some of the stigmas

Page **(Key Word)**
467t100 -- some with no division in young flowers. The ABORTIVE stamen are of useful height.-- In Lupine, Bee
470t177 s & common» On silene, many plants of wh. have ABORTIVE stamens= Many Humbles on hedge Linaria= (Plen
491q01v t other branches-- The French Apple tree «with ABORTIVE stamens» answers first question in negative.-
497q06v n Cultivated Plants, as Pentstemon, which have ABORTIVE parts, whether such vary.-- Do Bees go to Swe
504q013 3) ______ in Heartease (4) Does the Thyme bear ABORTIVE stamens every year & Spring. & within garden
505q014 impregnate it (7) History of Potato field= (8) ABORTIVE Thyme seeds weather wet--? Linum flavum put i
505q014 ium pyrenaicum. small white-flowered var. with ABORTIVE stamens.-- show crossing & ¿heredetary? (15)
505q014 germinate 16 Will plant some of the Thyme with ABORTIVE stamens by Terrace to see, whether stamens wi
509q017 ave wide range-- How is this in «Plants??» Are ABORTIVE organs as <young> teeth, more plain in young
509q017 . Has he dissected any animal often, which has ABORTIVE bone. (ask more about the lowest cervical ver
509q017 ree of soldering of tibia & fibula: in Man any ABORTIVE bones??? do. Wing in Apteryx ||| no as Os Cocc
513q020 ing varieties Amongst varieties cross one with ABORTIVE tail or horn, with another & see result, for
515q021 moner kinds-- Cattle are horned, Suffolk have <ABORTIVE> «no» horns by abortion, but sometimes have d
515q021 me organ is absent by abortion, but appears in ABORTIVE state either in the species, or in the indivi
555m143 n of head & ears,-- ∴ some men have this power ABORTIVE muscles) The black Spider Monkey, very differ
589n092 actions associated with ideas.-- A sigh, is an ABORTIVE groan.-- more power over muscles of voice tha
633j53v mental.-- NB. One limit to the transmission of ABORTIVE organs will be as long as they are not detrim
634j54v es & constantly alludes «(& at p. 312)» to the ABORTIVE bones. He explains it <"By> saying "It is the
634j54v ents in the Holocentrus ruber (a fish) Man has ABORTIVE muscles to his ears.-- p. 313 Many other good
507q016 rity in the males of a family-- Where one tooth ABORTS, do you know whether any trace in germ. (2) An
039r061 d rock, which elevates a continent> We are more ABOUND to take analogy of movements of W coast in exp
078r176 Porphyries which are destitute of quartz, & wh ABOUND both in hornblend & vitreous felspar".-- p. 21
399e010 rative of Miss. Enterprise, p. 497. Vampire bat ABOUND in the Navigators & at Manguia, but are unknow
451e177 Isld & Preparis between Andaman & Pegu. <have> ABOUND with monkeys & squirrels.-- Horsbrugh E. I. Di
486z018 ater duration of Heat. hence musquitoes & knats ABOUND during short summer far N. where this other or
470t176 oduced about '65 years ago--& soon after mules ABOUNDED--so that palmated has now nearly disappeared.
451e177 te S. adjoining it are several small islands. ABOUNDING with deer-- Horsburgs. Vol II. p. 527.-- <Sci
453e182 ountenance the theory of polymorphous plants, ABOUNDING in volcanic islds.-- <Cocks> The possibility
641j29r yet this is contradicted by continents <bri> ABOUNDING with species-- there will be a balance, conti
128a129 of an arch weighted in its centre.-- Will not ABRASION of land on one side. produce subsidence of wa
104a069 fragments fall off cliffs. but then how spread ABROAD?-- There is thus wide difference between erosi
263c076 anted!!) & whole fabric totters & falls.-- look ABROAD, study gradation. study unity of type-- Study
023r010 e surface of the ocean, before arriving at the ABROLHOS shoals ‖ -- N.B. The view of the Volcanos of
024r016 about same as last to N. of C. Frio Except at ABROLHOS. [18 degree S.] Bahia [12 degree 57 minute S.
269c100 use applicable to N. Hemisphere (NB.-- Examine ABROLHOS Flora with this view) Tristan D'Acunha, St He
115a098 e-- {P} if not course enough flat top. ended by ABRUPT slope {P} each stratum would thin out, both in
058r118 e centre de' l île, elles presentent une coupe ABRUPTE et souvent tailée a pic. Toutes ces montagnes
029r034 sures in England (Excepting Conglomerates?) «& ABSENCE of limestone?» have been collected on the open
036r051 ure, it is necessary to ascend the hill.-- The ABSENCE of Second form, except near submarine Volc: in
037r052 nts. V. Lyell. Chap XI Vol II. Urge the entire ABSENCE of any rock situated beneath low water in the
103a065 in proportion to weight of super [...] mass.-- ABSENCE of Caverns, in Plutonic rocks argument against
115a097 struction Sound in S. America The very general ABSENCE of fragments «& pebbles» in mica slate & gneis
132a139 fresh sediment added to bottom) be caused, by ABSENCE of circulating water.-- & therefore that tempe
151g041 Lauder Dicks Hypothesis impossible to explain ABSENCE of lines in certain parts.-- At the Pass of Gl
221b201 esh water.-- origin of Fresh-water genera? The ABSENCE of lime in Plutonic & Volcanic rocks. most rem
225b219 mpare to Van Diemen's land.» glorious fact. of ABSENCE of quadrupeds East India Archipelago very good
294c176 ies-- I fear argument must rest upon analogy & ABSENCE of varieties in a wild state-- it may be said
440e147 ything about habits-- no one can be shocked at ABSENCE of final cause mammae in man & wings under uni
616o39v sness. it is a point of indifference 2) In the ABSENCE of such a guide we can only <shew> point out t
641j29r d Oceans I think not.-- Does this bear on, the ABSENCE of their remains in the Wealden? In the strong
052r099 peaks elevated.-- Greywacke. as a general fact ABSENT in T. del Fuego, excepting in Port Famine Mr S
234b256 of a Harpalus. common at South end, but <rare> «ABSENT from» near London. = Dr. Smith, he says, is de
265c085 best attempt of nature, colouring matter being ABSENT.-- again dwarf plant in alpine district & dwar
308c217 go to tea chest almost unconsciously.-- why do ABSENT «Dr. Black. tea & sugar» people. reverse habit
315c241 itual actions of plants «when exciting cause is ABSENT» & memory of animals.-- (surely in plants move
406e036 og. will not attack any animal except, dog when ABSENT from its master.-- dogs when strayed hang thei
515q021 e any genus of plants, «in» which some organ is ABSENT by abortion, but appears in abortive state eit
547m113 very action, & always running on in mind, being ABSENT. one could not compare the castle with them, t
569n021 by VIVID power of conception between one or two ABSENT things.-- reason probably mere consequence of
601o009 hat is the only evidence. when consciousness is ABSENT) in fibres united with nervous filaments.-- ¿p
046r080 ny smaller prior ones might have been owing to ABSOLUTE movement of ground). Michell (Philos: Transac
047r081 ord. -- «also neighbouring sea must partake in ABSOLUTE movement» Moreover wave «with same general ch
115a098 hen {P} that form.-- All this depending not on ABSOLUTE <force> «size of» of <currents> «fragments» b
120a110 ivers.-- Excess of matter brought down Mention ABSOLUTE elevation of Patagonian blocks (1200 ft??). S
179b035 on? «if stating from same epoch certainly» The ABSOLUTE end of certain form from considering. S. Amer
195b104 this strong so many thousand miles distant.-- ABSOLUTE knowledge that species die &. others replace
281c138 ory explains that family likeness, which as in ABSOLUTE human family is undescribable, yet holds good
299c194 Many intermediate shades in these cases,-- but ABSOLUTE species formed. The Anagallis perhaps, offers
315c240 s same in Australia & Europe.-- if creation be ABSOLUTE thing, the creation must take place on[ly] wh
419e085 I have sometimes speculated might be owing to ABSOLUTE quantity of vitality «in the World»,-- the pr
441e150 hange.-- The weakest part of my theory is, the ABSOLUTE necessity, that every <animal> «organic being
526m030 cording to usual method, but what urges him,-- ABSOLUTE free will, motive may be anything ambition, a
528m036 - Analysis of pleasures of scenery.-- There is ABSOLUTE pleasure independent of imagination, (as in h
528m036 ses from (1) harmony of colours, <whi> & their ABSOLUTE beauty. (which is as real a cause as in music
528m038 s of the foreground either owe their beauty to ABSOLUTE forms or to the repetition of similar forms a
543m099 A man at Cambridge, during his time, almost an ABSOLUTE fool used to play regularly with D'Arblay of
576n047 well deteriorate. «CD[in my theory there is no ABSOLUTE tendency to progression, excepting from favou
529m043 mind, every occurrence for a day or two are ABSOLUTELEY forgotten.-- My father signed a bond, yet wh
100a054 avage.-- Phillips (113) «Lardner Encyclop.--» ABSOLUTELY considers gneiss an aqueo deposit resulting
119a106 metamorphic rocks have just floated over the ABSOLUTELY fluid pool.-- (this is shown by the softness
170b001 f generation the coeval kind, all individuals ABSOLUTELY similar; for instance fruit trees, probably
221b201 earches. facts of salt-water shells living in ABSOLUTELY fresh water.-- origin of Fresh-water genera?
243c019 en) in East Ind. Arch:-- Birds of New Zealand ABSOLUTELY different.-- --Philedon circinnatus not foun
291c167 endant in Europe<)> Toxodon in S. America) is ABSOLUTELY necessary to explain genera & classes. if ex
292c169 be great, otherwise it would be unlimited. We ABSOLUTELY know that tendency is greater in Mammalia, t
298c192 e on Guernsey & on West coast of Ireland, are ABSOLUTELY (& who better authority) similar with those
334d010 , that first cross <not se> plentiful, second ABSOLUTELY sterile.-- My case of Stallion, according to
358d075 ays accompanying us from camp to camp; it was ABSOLUTELY necessary to watch our meat, while in kettle
391d175 e.-- (many monster are really twins.)-- It is ABSOLUTELY necessary that some «but not great» differen
398e005 the result with that clearness of conviction, ABSOLUTELY necessary as the «basal» foundation stone of
398e005 ing proper ideas of these subjects. should be ABSOLUTELY necessary to arrive at right conclusion abou
410e050 . «& hence not «be» higher animals» -- it was ABSOLUTELY necessary that Physical changes should act n
442e153 ndividuals» propagated by gemmation should be ABSOLUTELY similar; [all the gorze in Norway ought to b
461t037 ny structure can be handed down without being ABSOLUTELY injurious «(or requiring nutrition)» to a ce
538m078 s, one only «little» less perfect than other, ABSOLUTELY two people. Consider this profoundly, may th
585n076 magnetic powe in birds, seeing the sun &c are ABSOLUTELY useless when applied to birds, which have be
124a119 en in making brass a piece of copper not melted ABSORB, zinc thrugout its thickness.-- this most curi
386d166 hys) can make a head; the other part may surely ABSORB a useless member.-- in fact they do it in dise
612o035 amount of ignorance [RHC] The radicle of plants ABSORB by physical laws of endosmic & exosmic juices.
028r031 ch «rain» is cause, hence at least no water is ABSORBED into the earth <I did not see one dike in the
071r158 There the Cordova earthquake in which lake was ABSORBED.--Earthquakes felt. different case from shore
124a119 still more curious to know whether it would be ABSORBED.-- if so exactly parallel to limestone & volc
386d166 of nature that any organ. which is not used is ABSORBED.-- this law acting against heredetary tendenc
547m112 m.-- It appears to me, that the mind is wholly ABSORBED with one idea (hence apparent vividness) & th
574n040 idea of a mind so limited as Birgos to become ABSORBED by one end of Cocoa nut.-- November 27th.-- S
637j57v veness, & laws of adaptation‖ p. 234. The non-ABSORBING Camel's stomach is puzzler p. do says inconve
516q23v Peat (2) Athenaeum 1840 p. 777. Decaying wood ABSORBS oxygen & forms Carbonic Acid. will this bear o
531m049 birds, this most curious instance of reason & ABSTINENCE.-- My Father remarks that things of great im

```
033r043 ian:» in the I Vol. Humb: There is rather good ABSTRACT of Humboldt. S. American Geolog. in Daubeny.
196b108 om parasitical nature of insects & worms.-- In ABSTRACT we may say that vegetables & mass of insects
255c052 f all birds, & so for birds. We thus obtain an ABSTRACT idea of a bird-- An animal with skeleton of s
267c095 gs L'Institut. 1838. p. 67. Australian rodents ABSTRACT of Infusoria. <p> do p. 62. Ehrenberg Annals
292c170 surdity of Quinary arrangement let him look at ABSTRACT of Swainson on Classification "Let anyone eve
296c185 tries & present tropical countries. p. 564. an ABSTRACT of Mr Swainsons views. which if abstract true
296c185 4. an abstract of Mr Swainsons views. which if ABSTRACT true are wonderfully absurd.-- p. 565 <breed>
303c204 monkeys clitoris wonderfully produced».-- make ABSTRACT on this subject from Lawrence. Blumenbach & P
325c267 Stewart works. & lives of Reid, Smith & giving ABSTRACT of their views Mackintosh Ethical Philos: Pro
421e091 ated.-- X.do. p. 305.-- Mr Owen says <tha> «in ABSTRACT» in his paper on the Dugong, "The generative
541m092 of sea.-- Is the effort greater if the idea is ABSTRACT as love, (or an emotion not so) than if simpl
559m155 f Adam Smith Reid, &c worth reading. as giving ABSTRACT of Smith's views «Take & pound up inflorescen
602o010 as showing "the perfection of this science of ABSTRACT form" is the source of part of the highest en
610o034 eems to doubt. [RHC] 1) Effects of Life in the ABSTRACT is matter united by certain laws different fr
627o52v king blows.--' p. 224.-- Hume's Inquiry-- good ABSTRACT of Butler & arguments of beneficial tendency
632j53r f Hutcheson unfolded by D. Hartley.-- Darwin's ABSTRACT of John Macculloch 1837 Proofs and Illustrati
085aIFC ral subjects. Feb 24th 1839 As far as p 140-- ABSTRACTED as far as concerns "Geolog Observat on Volum
193b092 aphie des Plants. I Vol in 4 degree.-- I have ABSTRACTED Mr Swainson's trash.-- at beginning of Readin
250c039 m rest of world??? Lyells Principles, must be ABSTRACTED & answered Much might be argued what is  not
320c275 Sebright's Pamphlets Wilkinson on Cattle not ABSTRACTED Scientific Memoirs. published by Taylor Maga
477zIFC down-feathers. Zoology 1856 Skimmed through & ABSTRACTED Zoology Some excellent references in L. Jeny
527m034 e a long castle in the air, is as hard work (ABSTRACTING it being done in open air, with exercise  &c
571n028 acquire  many notions unconsciously, without ABSTRACTING them & reasoning on them (as justice?? as an
580n062 cted with Nat. Theology.-- says animals have ABSTRACTION because they understand signs.-- very profou
587n089 Ethics p. 97. on Devotional feeling p. 103-- ABSTRACTION p. 152. Perception very different from emoti
535m069 ndly that these are replaced by metaphysical ABSTRACTIONS, such as plastic virtue, «&c» (Very true, n
189b074 nimals.-- & the story will be complete.-- It is ABSURD to talk of one animal being higher than anothe
225b216 t to how species are. Lamaks "willing" doctrine ABSURD. (as equally are arguments against it-- namely
259c063 iarity. or any race of plants--Lamark's willing ABSURD, ∵ not applicable to plant, Epidemics of South
262c073 groups.-- but as for number five in each group ABSURD.-- the mere fact of division of lesser &  more
263c075 effort of the mind (although this may appear an ABSURD saying) & will never be conquered by anyone (i
296c185 s views. which if abstract true are wonderfully ABSURD.-- p. 565 <breed> Scotch wild Cattle. breed fr
401e015 others--  so my experiment of strawberry not so ABSURD.-- Thinks-- that such variety as red cabbage p
403e024 t of species.-- they must deny species which is ABSURD.-- their only escape is that rule applies to w
439e144 y.-- «Elizabeth & Hensleigh. seemed to think it ABSURD. that the presence of of the Leopard & Tiger t
440e145 - fox-dog-- turnspit & two other kinds It seems ABSURD proposition, that every «budding» tree, & ever
441e151 ed. & cannot vary.-- which all facts show to be ABSURD.-- As there are plants, in northern latitudes,
566n010 e not beautiful-- is there -- anything in these ABSURD ideas.-- do they indicate mind & body retrogra
292c170 azine of Zoology & Botany p. 566 wants to see ABSURDITY of Quinary arrangement let him look at abstra
589n092 en given us to exist, is clearly seen. in the ABSURDITY of a tree having reason: or dog, having high
594n113 out of the room. really 3 months old. What is ABSURDITY, why does one laugh at it-- sensation of disg
553m137 pig-tailed baboon, shoved out its lip, looking ABSURDLY sulky «as» often as keeper spoke to it,-- but
557m147 out & flaccid, when furious «with fright» back ABSURDLY arched. & tail stiff.-- is shame, jealousy, a
271c106 tary whilst species have changed Argumentum ad ABSURDUM. The creative American halo has extended to J
116a099 res where currents very weak??-- too great an ABUNDANCE of matter would have same effect as too coars
135a144 some miles from coast-- quote passage to show ABUNDANCE Bengal Journal. Vol 4. 1835. p. 437. Tours by
190b077 an character There appears in Australia great ABUNDANCE of species of few genera or families.-- (long
528m039 kened during music.-- connection with poetry, ABUNDANCE, fertility, rustic life, virtuous happiness.-
040r063 bed as very rare Mem. St Helena; probably more ABUNDANT in this case from intersecting a mass probabl
044r074 er high pressure of gazes. especially the most ABUNDANT. Sulp. Hyd: Carb: A. Mur: A. = (& this effect
064r135 -- May I not generalize the fact glaciers most ABUNDANT in interior channels. there no outer coast.--
071r156 if  so coniferous must formerly have been most ABUNDANT tree-- Metamorphic action: <most> coming so n
074r164 Limestone & Grauwacke: Silver appears far more ABUNDANT in the upper limestone, which H. calls by sev
075r167 er, do not name native silver because not very ABUNDANT.--X muriated silver. which is so rare in Euro
076r169 in, contrary to Europe. argentiferous lead not ABUNDANT. ='considerable quantity of silver procured f
228b229 ow run together, were not both genera formerly ABUNDANT. Seed of Ribston Pippin tree <go> producing c
259c062 cted by-- for instance, fish being excessively ABUNDANT & tempting the Jaguar to use its feet much in
272c109 es «of subgenera» scattered over it.-- We have ABUNDANT instances of remarkable structure which as fa
467t103 n the Lupine flower is perfectly ripe & pollen ABUNDANT filaments & stamens all protrude «there is  a
467t105 sulphur Broccoli not many do-- pollen not very ABUNDANT. not very small-- Saw one small Bee; saw anot
468t105 » Rhubarb. pollen very minute--not excessively ABUNDANT flowers not attractive, very small--stigma ra
482z012 zara. Voyage dans l'Amerique Merid. Tatu noir. ABUNDANT from Paraguay to 27 degree, then the Mulita f
499q09v le insect?= or is pollen excessively minute or ABUNDANT? do they seed plentifully? Look for  isolated
156g072 <granite &> «veined» gneiss <unite> «occurs» ABUNDANTLY with perfectly rounded pebbles of granite  &
426e106 tween the fact that different species produce ABUNDANTLY infertile hybrids, & the fact that old varie
161g094 uarter of a mile further on, where three [...] ABUTTED Having crossed the mouth, (deep) of above vall
133a140 show  how slowly heat travels; & therefore the ABYSSES where fluid rock has been ejected must  remain
182b048 a black face, & similar to those brought from ABYSSINIA, & others dark brown, with long clotted  hair
229b233 inks several species of Rhinoceros range from ABYSSINIA to extreme South coast. Elephant he believes
267c093 » black face, & similar to those brought from ABYSSINIA; the others dark brown with long clotted hair
130a134 arked line of separation A Paper by Parrott Mem. ACAD. Peters. Scienc Math. Phy-- Nat. t. I, 1831. su
132a137 sage of the moon.-- Ask Hopkins. M. Parrot, Mem. ACAD. Imp. des Sciences. (Sc Math. Phys. et Naturell
127a126 of ice ought to be less.-- Memoir of the Irish ACADEMY Vol 8. p. 118 water no--. oil will freeze if c
128a130 h forces dilemma. Transactions of the Maryland ACADEMY (at Athenaeum.) I. Part. I Vol.-- some notices
315c241 es.-- Menoir will be published St. Petersburgh ACADEMY Imperial-- Paper read in 1837. semestre I susp
317c251 Appendix to Tuckey's Expedition Journal of the ACADEMY of Natural Sciences of Philadelphia Vol VII. P
322c270 History.  Octob 2d Transactions of Royal Irish ACADEMY. ----do Lavater's Physiognomy ---- Octob 3d Ma
385d164 varium of women (Paper in Vol I of Irish Royal ACADEMY) have contained perfect teeth & hair-- showing
447e164 cent» mollusca, than between the corresponding ACALEPHA?-- But if Acalepha do not cross there would b
447e164 n between the corresponding acalepha?-- But if ACALEPHA do not cross there would by my theory gradati
381d156 isible). Oyster. cystic Entozoa. Echinoderms. ACALEPHES. Polyps. Sponges Heautandrous, male organs fo
535m064 lder species. The earwig & a doubtful one of ACANTHOSOMA grisea described [not located] as first caus
073r163 r trap cover it to great thickness. = Coast of ACAPULCO granitic rock.--in parts of table granits & g
079r177 na.--» Humboldt. New Spain Vol. IV. «p. 58» At ACAPULCO earthquakes are recognized as coming from thr
581n066 s not first felt?-- «without «slight» flush, ACCELERATION of pulse. or rigidity of muscles.-- man can
063r132 g an animal in two. (gemmiparous. by nature or ACCIDENT). we see an individual divided either at  one
176b022 n those which have changed most. «owing to the ACCIDENT of positions» must in each state of existence
302c203 he formation of genera may sometimes be due to ACCIDENT as submersion of land containing all of inter
309c221 rt legged sheep. heredetary proceeding from an ACCIDENT. New England farmer,-- useful could not  leap
504q013 ne branch of Rhod. flowered later.-- effect of ACCIDENT?? (7) Which. Rhododendrum seeds??-- Bladder-n
540m089 of  distress.-- see how a crowd collects at an ACCIDENT,-- children with other children naughty.-- Wh
612o035 as much as growth of tissue and are subject to ACCIDENT; the sexual willing comes on period of year a
023r011 present  Volcanos have been said to be merely ACCIDENTAL apertures still open.--The fault like appear
045r076 [...]> part of a regular system can be called ACCIDENTAL; the proportional force of crust of globe  &
046r079 ed as often as Volcanic. I consider latter as ACCIDENTAL on the afflux of the former. -- Ascension. V
146g013 merate on S. Ventana, excellent instance, how ACCIDENTAL is the preservation in situ of even imperish
171b004 other hand, generation destroys the effect of ACCIDENTAL injuries, <on> which if animals lived for ev
187b064 ties, (therefore adaptation), & to obliterate ACCIDENTAL varieties, & to accomodate itself to change,
199b118 tive want into a fundamental propenity, of an ACCIDENTAL habit into an instinct." Ed. N. Phi. J. p 29
265c083 ture by crossing with other varieties»-- but <ACCIDENTAL> changes after birth do not effect progeny--
315c242 emory. very remarkably-- scenes in themselves ACCIDENTAL.-- My first thought of sea side-- Study Bell
427e111 do not become acclimatised by crossing, or by ACCIDENTAL production of seedling with hardier constitu
428e111 estruction of all the less hardy ones. & the <ACCIDENTAL> preservation of ACCIDENTAL hardy seedlings:
428e112 ardy ones. & the <accidental> preservation of ACCIDENTAL hardy seedlings: (which are confessed to  by
523m016 ver the misery of an illness at Rome, when by ACCIDENTAL <was> delay of money, he was «only» NEARLY t
607o025 varied  capability of receiving impressions-- ACCIDENTAL (so called like chance) circumstances. As ma
608o026 admonition succeeds who does not recognize an ACCIDENTAL spark falling on prepared materials. From co
633j53v are  necessary adaptations.-- May they not be ACCIDENTAL? We have good reason to know that they would
```

342d035 6th. D Israeli (Cur of Literat. Vol II p 11) ACCIDENTALLY says "--is distinctly marked as whole dynas
621o047 lmost» any taste to a young person. or it is ACCIDENTALLY acquired from some trifling circumstance.--
621o047 e taught to think almost anything nasty. (<& ACCIDENTALLY» «by odd association» comes to this conclus
045r076 in the Cordillera), they may be considered as ACCIDENTS (if <[...]> part of a regular system can be c
206b148 , heredetary disease, effects of contagions & ACCIDENTS: yet some causes are evident, as for instance
507q016 . Andrew Smith (1) Are cross-births, or other ACCIDENTS of delivery-- inheritable.?-- Bell cd ask Acc
530m044 iving directions,-- & forgetfulness after bad ACCIDENTS:-- After journey, a fit of = gout, has affect
611o034 variable, as long as not modified by external ACCIDENTS, & in such cases modifications bear fixed rel
611o034 ses modifications bear fixed relation to such ACCIDENTS. But such tissue <must> bears relation to who
633j53v on to know that they would not be detrimental ACCIDENTS, & domesticated variations show us accidents
633j53v accidents, & domesticated variations show us ACCIDENTS may become heredetary [produce some peculiari
633j56r &c)» These are reasons, just as liability to ACCIDENTS & any other cause.-- (& my theory [ALL PARTS
318c253 white hares, fitted for regions of snow.-- on ACCLIMATISATION.-- Bachman tells me in Audubon there is mo
502q11v t from AEgypt ripen in Scotland?-- to show on ACCLIMATISATION.-- July <1842> When nettle leaf. put into
427e11l ys <p 347. Amyyralidae» Plants do not become ACCLIMATISED by crossing, or by accidental production of
427e11l . says Zizenia in 16 generations did become, ACCLIMATIZED. & says Laurels have not been so. (which is
326c265 ? Sweet. Hortus Britannicus. has remarks on ACCLIMATIZING of Plants. Herbert. p. 348. gives reference
185b055 appears attempt in each dominant structure to ACCOMODATE itself to as many situations as possible.--
187b064), & to obliterate accidental varieties, & to ACCOMODATE itself to change, (for of course change even
461t037 e probably any structure would rather become ACCOMODATED to new circumstances than it would be elimin
187b064 , (for of course change even in varieties is ACCOMODATION). Now this argument applies to species.-- f
075r167 er. which is so rare in Europe. common there ACCOMPANIED by molybdated lead & «argentiferous lead»; s
076r169 of lime-- native silver in Mexico is always ACCOMPANIED by Sulp. silver sometimes by selenite.-- in
102a062 racting gravity.-- As volcanic eruptions are ACCOMPANIED by horizontal elevations, so are injection o
102a062 ations, so are injection of mountain chains. ACCOMPANIED by do.-- Give this after supposition p. 461
271c107 s of one tribe taking on structure (probably ACCOMPANIED by habits) of other, thus in Chalcididous in
332d003 en limb «in children» & other such disorders ACCOMPANIED with some fever, be attended by the transmis
522m012 rly> «slight habitual» intemperance.-- often ACCOMPANIED by extreme anger, at not being understood.--
531m053 ger-- Fear, shamming death, or running away. ACCOMPANIED with want of muscular exertion, palpitation,
532m054 tened at.-- (again diseases of the heart are ACCOMPANIED by much involuntary fear) In these cases pro
532m057 feelings of Man.-- The sensation of fear is ACCOMPANIED by «troubled» beating of heart, sweat, tremb
565n009 ative expression? Expression of affection is ACCOMPANIED by slight protrusion of lips, as if going to
570n025 ng so similar to shame after asinine.-- both ACCOMPANIED by depending head., & active vessels of skin
605o19v s an inward pride & glorying. (often however ACCOMPANIED with terror & wonderment) <which> «this» emo
629o55v nto mind in first case-- seeing how shame is ACCOMPANIED by blushing, bears some relation to others 5
531m052 uent on the violent muscular exertion» which ACCOMPANIES violent attack,-- Even the worm when trod up
532m057 which many animals put on.-- The flush which ACCOMPANIES passion & not sweat is the <state> effect of
570n023 shoulder-- or is it from inspiration, which ACCOMPANIES surprise.-- & why does one inspire, when sur
590n093 retion attends passion p. 39. The sweat that ACCOMPANIES fear is the same, as that which attends grea
079r176 lph. of do.[,]galena[,]quartz, Carb. of Lime. ACCOMPANY.-- Ulloa has said silver in the highest & gol
569n022 can say, I am very sorry so it is"-- does not ACCOMPANY I will not. I am sorry I cannot.-- Expression
358d075 . 306 The crows were amazingly bold, always ACCOMPANYING us from camp to camp; it was absolutely nec
531m052 especially about heart as of excited action, ACCOMPANYING violent movement; may not passion be the fe
556m146 838 good instance of useless muscular tricks ACCOMPANYING emotion.-- when horses fighting, they put d
569n021 ings remembered & the associated pleasure &c ACCOMPANYING such memory.-- A Melody on flute & Epic poe
581n066 ions are the heredetary effects on the mind, ACCOMPANYING certain bodily actions]CD., but what first
605o18v ty & "inward glorrying, which height. by its ACCOMPANYING & associated sensations so often gives, whe
057r114 ulders not in lower strata. only in upper. in ACCORDANCE in Europe with ice theory.-- Capt Ross found
191b082 wife being constant together for life, is in ACCORDANCE with The Male animal affecting all the Proge
265c088 loured plumage, where colours have changed in ACCORDANCE to habits.-- one is tempted to suppose from
500q10a few-- Range of mundane genera, «in Birds» in ACCORDANCE with range of species?-- Are there any fine
619o042 s own expense.-- Moreover <the> any action in ACCORDANCE to an instinct gives great pleasure, & <an>
619o043 -- If he saw another man <say go> «acting in» ACCORDANCE to his instincts, <he would know that many e
624o051 e same law effects both.-- <such> changes «in ACCORDANCE to beneficial tendency» will most readily af
624o051 eadily affect. the instincts, for they are in ACCORDANCE with it. thus a dog may be trained to hunt o
633j54r perfectly adapted to all situations, where in ACCORDANCE to certain laws they can live.-- Hence the m
302c202 ee my notes on p. 37. of Macleay. wonderfully ACCORDANT. with fact there stated, only in most discord
032r040 ly fall, (3) the case would be as at first. & ACCORDING to the greater or less time of rest. so would
070r155 ee ridges in Copiapo, as well as in latter.-- ACCORDING to Mr Brown, a person (whom I met at S. W. P.
103a065 ocks argument against great bodies of vapour. ACCORDING to Hopkins theory.-- general presence of dike
117a101 . p. 91. a classification of Europaean strata ACCORDING to composition thinks sand with vegetable rem
126a124 ter does percolate, & springs beneath sea-- → ACCORDING to this latter view the rod is reversed, uppe
132a138 n coast lines, than on continents. it ought, (ACCORDING to M. Parrots argument against central heat t
132a139 than the transmission from ocean's bottom.-- (ACCORDING to M.. Parrots own hypothesis some such expla
171b003 me permanently changed or subject to variety, ACCORDING «to» circumstance,-- seeds of plants sown in
172b007 n to intermarry who will dare say what result ACCORDING to this view animals, on separate islands, ou
173b009 surely is not Produced?-- <Granting> Species ACCORDING to Lamarck disappear as collection made perfe
177b027 rganization.-- birds-- not. {P} We may fancy, ACCORDING to shortness of life of species that in perfe
179b034 t the species (for species they certainly are ACCORDING to all common language) will keep to their ty
180b038 hanging circumstances are continued & produce ACCORDING to the adaptation of such circumstances & the
184b053 nyx mastodon, & the species now living.-- Now ACCORDING to my view. in S. America parent of all armad
188b067 , & another to Sumatra --Mem Parrots peculiar ACCORDING to Swainson to certain islets in East Indian
195b101 re simple, & sublime power let attraction act ACCORDING to certain laws such are inevitable consequen
210b161 ery easily stage) & which, follow certain laws ACCORDING to species. present an analogy to production
212b167 lls between herbivorous & zoophagous Mollusca ACCORDING to periods.-- NB. Was Europe desert (like S.
212b168 ra dead? --Examine into this «in Phillips».-- ACCORDING to this, formerly there would have been many
221b201 existed for ages as metamorphic; & therefore ACCORDING to Lyells doctrine removed?? Is the prevalenc
222b202 eeing with Senegal. whilst Crag <agrees with> ACCORDING to Beck has none recent, yet genera same.-- S
222b204 tock cannot be supposed to be «most» perfect (ACCORDING to our ideas<)> of perfection); but intermedi
223b210 ones are formed) but yet propagates varieties ACCORDING to same law with animals?? Why are species no
224b212 e is no test but generation, «(but experience ACCORDING to each group)» whether good species, & hence
232b247 so would the plants from extreme north, which ACCORDING to all analogy would have been very unlike So
233b249 alia Fauna so far. Indian all the rest. Timor ACCORDING to Mountain chain ought to be Australian.?--
235b263 nimals separate &c & Work out Quinary system ACCORDING to three elements How is Fauna of Van Diemens
247c028 d» & «even» Java. & very common on Otaheite-- ACCORDING «stated in note to p 21» to Quoy & Gaimard in
247c028 o p 21» to Quoy & Gaimard in Sandwich isld. & ACCORDING to Chamisso in Radack isld.-- p. 69. Sharks v
255c051 arks.-- such as the bands on pidgeons back.-- ACCORDING to this description of class is description o
258c062 der first becomes developed & then another-- (ACCORDING as parent types are present) must follow afte
264c080 t the other into into Crows. yet all forming, ACCORDING to Gould, good genus Gould seems to doubt how
265c084 ntries few will say it is direct effect, <of> ACCORDING to Physical laws, as sulphuric acid disorgani
278c130 uld I not give Catalogue of Mammalia arranged ACCORDING to my own methods. Dasyurus being found fossi
284c149 g p. 32 "where it (mode of generation) varies ACCORDING to the species, it is manifestly of less impo
285c152 individual were put together, they would not ACCORDING to all analogy breed together.-- The bottom o
301c199 deity to teach squirrel to kill ears of corn ACCORDING to my views, habits give structure,.. habits
302c202 ise, my premises <in di> would be disputed.-- ACCORDING to Principles of last page. <an> osculant gro
305c210 itude of forms «each having acting principle» ACCORDING to subordinate laws.-- There is one thinking
305c211 nking principle. seems to be given or assumed ACCORDING to a more extended relations of the individua
309c219 m climate & habits, & therefore less fertile. ACCORDING to Mr Herbert's views.-- Argue <argue> case o
312c232 e same principle transferable., not wonderful ACCORDING to my view <ins> beccause actions are constan
316c246 ole world Many shells at present day same (or ACCORDING to Sowerby fine species) on coasts of N. Amer
334d010 nd absolutely sterile.-- My case of Stallion, (ACCORDING to Erasmus preferring young mare to old, expl
334d010 & young mare to old, explained by Stallions, (ACCORDING to Fox) being guided entirely by their smell.
335d015 its they might return to either parent), then ACCORDING law, that in proportion as things are long in
336d019 ated. it is probable if created at once. <wd> ACCORDING to ordinary laws, the character of offspring
339d025 des animals «world into Zoological Provinces» ACCORDING to varieties of Man.? «In Australia. plants E
339d026 o animals being distributed after flood (!) ACCORDING to affinities!. confounds, like Whewell affin
349d052 -- NB. This paper worth referring to again.-- ACCORDING to my theory, every species in any sub-genus
354d062 rdering of the nerves, but in different parts ACCORDING to age of individuals-- (see Mammae of Women)

356d071 I> Mitchell p. 244. vol I) spit & throw dust <ACCORDING to my theory of generation (p. 175) if> 8th S
366d108 h could not have been persistent in nature.-- ACCORDING to my view, the domesticated animals would ce
370d116 eloped.-- Sept. 19th <Are> There is no scale, ACCORDING to importance of divisions in arrangement, of
378d148 show them.-- Anyhow not connected with habits ACCORDING as child is like parent, so is species old: H
381d157 <it> «organs» open to water? Would not Ferns ACCORDING to this doctrine be considered as really cryp
388d171 ight partake of shade of fathers character.-- ACCORDING to this view more semen to one child. more li
392d180 cks do young take most after father or Mother ACCORDING as they are crossed? & How is it with China &
397e003 ature, have been conducted almost! invariably ACCORDING to fixed laws: And since the world began, the
424e101 ry much closer, than the present ones., which ACCORDING to Beck are different.-- Subsidence of Greenl
427e109 tionary, hence all present types are ancient. ACCORDING to my views of <Dioecious p> all plants, bein
432e124 s adaptation to the surrounding circumstances ACCORDING to my theory no land animal with fluid seeds
436e136 t «may» be said, that wild animals will vary, ACCORDING to my Malthusian views, within certain limits
442e153 r flowers!!-- <How did it get there? whether> ACCORDING to the above suggestion my theory would requi
448e165 d to Asclepias, where this is always the case ACCORDING to Brown.-- Voyage of Adventure & Beagle Vol
477z001 son CXII. & CXV do Azara Voyage Vol I p. 196. ACCORDING to Charpentier de Cossigny. only 10 years ago
493q004 ith <old> female of old breed & see result.-- ACCORDING to Mr Walker the form of male ought to prepon
493q004 alker the form of male ought to preponderate; ACCORDING to Mr Yarrell the latter ought: either in fir
509q017 veloped into ribs.) & does its abortion vary, ACCORDING to Bentham's Remark. Horse or cow.-- degree o
526m030 not this free will.-- he improves the faculty ACCORDING to usual method, but what urges him,-- absolu
527m031 he uppermost.-- so in thoughts, one will rise ACCORDING to law. How strange <all> «so many» birds sin
553m135 he theological age of science in every nation ACCORDING to M. le Comte).-- Those savages who thus arg
571n029 otions, which are taste? Real taste in mouth, ACCORDING to my theory must be acquired, by certain foo
582n067 & heredetary remains of savages state.-- N B. ACCORDING to my view marrying late, will make average o
610o033 e amount of our instincts-- surely in animals ACCORDING to usual definition, there is much knowledge
611o034 rganic matter, without action of vital laws-- ACCORDING to the individual forms of living beings, mat
613o036 e is one one instinct to all animals modified ACCORDING to species. This I suppose he deduces from th
622o049 grafted.-- Origin of the instincts Hartley, (ACCORDING to Sir J) explains our love of another, as pl
623o050 nerations giving the social feelings.-- [LHC] ACCORDING to my theory, all instincts demand some expla
397e003 ines., yet very inexplicable.-- do p. 529. "It ACCORDS with the most liberal! spirit of philosophy to
507q016 nts of delivery-- inheritable.?-- Bell cd ask ACCOUCHERS Is any peculiarity in milk teeth inheritable
026r022 he bottom of sea near T. del Fuego.-- Is there ACCOUNT of Baron Roussin's voyage.-- In Europe proofs
034r044 rtillo Pass example of do? <Poor> Daubeny good ACCOUNT of ejected granitic fragments P. 386 Mem. Lyel
034r045 ast Indian Volcanos Gypsum Andes Mem. Beechey. ACCOUNT of regular change in soundings. on approaching
043r070 raccas.-- Is this mentioned by Humboldt in his ACCOUNT of extensive areas. -- P. 322 In any archipela
051r095 ction on upper tidal band, I do not see how to ACCOUNT for oceans power.--excepting when pebbles are
054r102 t. Catherine & coast Granite: P. 199; Falkland ACCOUNT of cleavage differs wonderfully with mine: phy
054r102 : refers to broken hill described by Pernetty: ACCOUNT of streams of stones agrees with mine.--At Con
067r142 mplete.= Silliman Journal. year 1835 excellent ACCOUNT of N. American geology. Conybeare Lava in Cord
074r165 lls by several secondary names «Study Hoffmans ACCOUNT of steam acting on trachytes. also Azores. We
087a010 in salt marshes Efflorescence nothing -- Study ACCOUNT.-- Alluvial plains of Mississippi -- No Vol. I
088a015 Kotzebue. Study Humboldt. Fragmens Asiatiques ACCOUNT of American Volcanic action. -- Fragments of s
105a070 Prevost.-- In Cordillera. a rush of water will ACCOUNT for filling up of valleys-- subsequent opening
115a097 ng on Falkland Islds p. 94. Von Buch's Travels ACCOUNT of Norway chain being broken through like that
129a132 t poor. Athenaeum. 1838 p. 791 -- Most curious ACCOUNT of great subsidence «20 miles long I in with.»
138a151 Berlin Transactions (1832. or 3?) there is an ACCOUNT of Sellow Geolog. Observat. in Southern Brazil
139a1BC usset would be well worth visiting really good ACCOUNT of ice.-- C. Darwin A. Glen Roy Generally rece
148q026 but more easily fatten, This man confirmed the ACCOUNT of the «YOUNG» Shepherd dogs Saturday. Before
193b090 e preserved which are well adapted, This would ACCOUNT for each tribe <being> «acting as» in vacuum t
194b096 ent. about varieties being difficult to keep on ACCOUNT of pollen from other plants because this may b
197b112 heory of Conditions of existence is thought to ACCOUNT resemblances &. quinary system, or three eleme
199b119 Journ.-- Paper by Crawford on Mission to Ava. ACCOUNT of HAIRY man. because ancestors hairy with one
201b126 ll as 2 recent See Geolog. Proc. p. 569. 1837. ACCOUNT of wonderful fossils of India.-- & p. 545 «gre
206b149 y perhaps some one single one.-- Will not this ACCOUNT for the odd genera with few species which stan
211b162 . in Zoolog. Proceedings. Jan 1837, «by Eyton» ACCOUNT of three, kinds of pigs. difference in skeleto
212b165 o be removed.-- In L.' Institut. 1837: p. 404. ACCOUNT of instinct of dogs.-- agreement & reason Some
219b196 have been said it was as gret a difficulty to ACCOUNT for movement of all, by one law. as to account
219b196 account for movement of all, by one law. as to ACCOUNT for each separate one, so to say that all Mamm
223b208 d Anatomists have said.-- ¿where is Pentland's ACCOUNT of Bolivian human species?-- Small «new» anima
232b248 yton to procure me some Get Hope to give me an ACCOUNT of parasitic animals of beasts varying in diff
233b250 neighbouring localities Institute 1838. p 38. ACCOUNT of fossils of Sewalick «India» Monkeys of old
239cIFC Instances of old Breeds taking greatest effect ACCOUNT of the [...] the world.-- Charles Darwin writt
240c004 birds & brought up by hand.-- These facts all ACCOUNT for [not located] Falkner Patagonia no descrip
246c025 - Wide space of sea, to East of America. would ACCOUNT for this.-- Coquille Voyage Says no reptiles.
246c027 c, yet there appears to be one at Botouma from ACCOUNT of natives, & probably on Oualan.-- "Mitchill
252c045 ecies-- FOX IS & Mice of America «good case on ACCOUNT of varieties in N America» «some doubt, from w
253c047 rds. Zoolog. Transact. Vol I p. 165.-- <a> "an ACCOUNT of the MANELESS lion of Guzerat by Capt. W. Sh
255c051 thers.-- if wing totally obliterated. This may ACCOUNT for permanence in many trifling marks.-- such
260c067 ist In the Zoological Journal I read a curious ACCOUNT to show that very many birds of different kind
260c069 hite, snow.-- the fine green of vegetation,-- ¿ACCOUNT for colour of bird in district. which they fre
266c092 riods.-- Arcana of Science & Art. 1831. p 160. ACCOUNT of Bulbous root from Mummy: after 2000 years,
268c096 als of Natural History. «vol I» p. 159 curious ACCOUNT of Tit mouse feeding young of redstart & actua
281c139 ixed & therefore most subject to change.-- may ACCOUNT for certain organs not being fixed, <whi> in s
289c161 follow.-- examine structure of this bird & get ACCOUNT of habits My definition <in wild-o> species.
294c178 s, its extension.-- Von Buch. Travels. p. 306. ACCOUNT of trees ceasing to grow far N. becomes stunte
306c214 dent on localities.-- Hamilton will give an ACCOUNT in his Travels in Asia Minor of the domestic a
310c223 between Man (--<a> «&» chasms» «necessary to ACCOUNT> «consequence of» for the scheme of nature) an
317c251 iences of Philadelphia Vol VII. Part II/. 1837 ACCOUNT of the various hares «some since discovered» o
325c267 man's Journal Rengger on Mammalia of Paraguay. ACCOUNT of wild cattle & Montagu on birds (facts about
327cIBC s Read Volney's travels in Syria Vol I. p. 71. ACCOUNT of Europaean plants transported.-- Crawford. E
327cIBC - Crawford. Eastern Archipelago. probably some ACCOUNT Raffles. Sir. S do. do-- Buffon Suites Cline o
338d024 ed races.-- Athenaeum. p. 505. some (very poor ACCOUNT) of plants of Nova Zenbla -- in review of Baer
346d047 ore tendency to fatten-- This man confirmed my ACCOUNT of the Shepherd dogs.-- Aug. 24th. Was struck
346d048 25th Athenaeum (1838) p. 611. Ld. Tankerville ACCOUNT of wild cattle of Chillingham,-- habits peculi
346d048 of more than 100.-- Agrees, «nearly» with. the ACCOUNT. given by Boethius of ancient Caledonian Cattl
366d112 red from Australians knocking out teeth.-- the ACCOUNT of the people on the NW. Coast blinking to kee
402e019 Kotzebue's second «1st» Voyage. Vol II p. 344. ACCOUNT of insects of St. Peter & St. Pauls in Lat' 53
403e021 ar[s]den p. 94 (1st Edit) of Sumatra has given ACCOUNT of Buffalo of the East which differs from that
403e022 mb,-- one leg, hare lip &c &c. in Vol II p 363 ACCOUNT of Flora of pacific, given in my coral paper O
405e035 -- L'Institut. 1838. p. 338.«V[ide]» Important ACCOUNT of cross of sheep & Moufflon of Corsica. <woul
411e056 can be produced. & yet sexual apparatus.-- My ACCOUNT of Circus cinereus of the Falklands Isld. is i
412e058 amm: from Poland. &c.-- Three principles, will ACCOUNT for all (1) Grandchildren. like. grandfathers
420e088 ns settled there L'Institut do. p. 419, «long» ACCOUNT of Hyaenodon, a fossil dog-- leading towards H
421e090 s.-- Annals of Natural History. (p 225. 1838.) ACCOUNT of metamorphosis in the young of Syngnathus.=
424e103 e of mind is most relied on, but Bell has some ACCOUNT of wolf in Zoolog. Gardens, which brought its
427e109 f Greenland render climate less extreme. (& so ACCOUNT for descent of snow line there «& there & ther
429e113 as. we may infer it in animals.)-- Azara gives ACCOUNT of production of hornless cattle-- ¿& others?-
430e117 -- even if extinction is denied.-- it will not ACCOUNT for all species. even if it will do.-- Va
453e182 to the Caribs.-- Vol II p. 650. Long attested ACCOUNT of fall of fish in India.-- Windsor Earl-- Eas
458t001 long preserved.-- vol VI. p. 539. Dr Cantor's ACCOUNT of fossil frog. 40 inches in length--! alludes
480z006 II, p 24 Proceedings of Zoolog Soc. Important ACCOUNT of habits of Tubularia. p 52. May 1836 dimensi
480z006 erica. D'.Orbigny. L'.Institut. No.-- 221 Good ACCOUNT of Condor by Humboldt Zoologie Recuiel-- Meyen
480z006 Humboldt. Zoologie Recuiel-- Meyen has written ACCOUNT of Guanaca. In transaction of Bonn Society M.
484z014 olog must be studied before writing my general ACCOUNT-- ¿ Do not the Penguins replace the «Auk» Guil
504q013 & pear to see if pollen naturally carried, on ACCOUNT of Van Mons views-- Also PEAS-- N.B. I think v
527m032 not:-- Sir J. Reynolds explanation may perhaps ACCOUNT for our acquiring «the instinct» our notion of
544m102 ayo Philosophy of Living. p. 140-- Dreams good ACCOUNT of «thinks» are recollected when intense, or w
544m102 nks no dreams except at this time. how does he ACCOUNT for dogs & men speaking in their sleep.-- Char
555m144 ssibility of person recovering from hanging on ACCOUNT of blood. but all these idea came one after ot

Page ***(Key Word)***
```
576n047 seems opposed to progressive. developement) on ACCOUNT of dark ages.-- «effects of external circumsta
590n094 tural. (Vol I. p. 234) Vol. II p 153. «do». an ACCOUNT of a monkey in a passion like Jenny.-- Dr. Abe
590n094  a passion like Jenny.-- Dr. Abel has given an ACCOUNT of an Ourang.-- see his Travels.-- When one se
618o41v r knowledge of matter is quite insufficient to ACCOUNT for the phenomena of thought. (The objects of
074r165 syenite» «strangling &c of veins can only be ACCOUNTED for by concretionary action, conjoined with o
115a097 late & gneiss, can only (see «supra» p 94) be ACCOUNTED for by great molecular attraction of every at
132a138 lcanicity has warmed it. Is not cold of ocean ACCOUNTED for, by the circulation being greater, than t
293c174 Now these exquisite adaptations can hardly be ACCOUNTED for by My method of breeding there must be so
403e023 n.-- now on my theory this «certainly» can be ACCOUNTED for, on any other it is the will of God.-- Oc
617o40v orces manifested would be <sat> fundamentally ACCOUNTED for, we prefer this metaphorical mode of stat
448e167 y improbable.-- yet I can see no other way of ACCOUNTING for them.-- Think over this-- The Superga be
535m069 e them.-- next step plastic <virtue> natures. ACCOUNTING for fossils). & lastly the tracing facts to
047r082 e similar fact at Concepcion? Read the various ACCOUNTS & see if fall is not the first very evident m
070r155 mooth. Sir W. P. states that in Helm's travels ACCOUNTS of travelled boulders. from the Cordovise ran
132a138 temperature of earth. at the freezing point.-- ACCOUNTS for increase on earth by volcanic action.-- <
205b143 Kirby Bridgewater Treatise There are some good ACCOUNTS of passages of legs into mouth-pieces of Crus
223b211 species & intermediate character of offsprings ACCOUNTS for uniformity of species & we Must confess.
313c233 e, when brought over on tropical animals which ACCOUNTS for the species changing which accounts for t
313c233 which accounts for the species changing which ACCOUNTS for the species changing « ∴ because mammalia
523m015 of curing it. by keeping the sum-total of his ACCOUNTS in his pocket, & studying mathematics.-- My F
091a027 wopenny periodical said so. «Campbell the Poet» ACCRA. Coast of Africa. Clay Slate & Quartz. strike S
200b121 show the possibility of common branching off.? ACCRA, Coast of Africa. Clay slate. strike. SSW. & NN
213b172 -- I think it is certain strata could not now ACCUMULATE without seal-bones & cetaceans.-- both found
261c069 ere is capital table of extent of all species ACCUMULATE instances of one family sending out structur
412e057 heir slow formation» these variations tend to ACCUMULATE. «on any structure.» L'Institut. 1838. p. 38
028r029 ocess? How does it come that all Lime is not ACCUMULATED in the Tropical.oceans detained by Organic p
073r161 ere the stalactiform masses have layers been ACCUMULATED, round knobs, or pushed where soft, or redis
114a095 other deposits.-- NB. because lowest. first ACCUMULATED in bed of ocean With the exception of sandst
248c033 animals unite, all the change that has been ACCUMULATED cannot be transmitted;.-- hence the tendency
278c131 dugong, therefore immense age since breccia ACCUMULATED-- surely ask Owen to see whether species sam
028r029 her minerals containing Alumen.--This matter ACCUMULATING in deep seas forms slates: How is the Lime
037r055 currents off the Galapagos.--strata must be ACCUMULATING which like the secondary strata of England,
028r032 where that action commenced before any great ACCUMULATION of such matter.-- Dr A. Smith says. that Bo
032r041 m Patagonian steps, because the deposition & ACCUMULATION is brought into play As in Ocean & Air; the
032r041 ation «however slow & weak.»; «(cause of not ACCUMULATION of Coral limestone in intertropical)» hence
222b206 y epoch: it is impossible to suppose such an ACCUMULATION at present day & not include Mammalian rema
153g052 opposite Glen collarig at bend & here most ACCUMULATIONS At gentler bends roads disappear The normal
538m079 half ideotic in some respects & with store of ACCURATE & even profound knowledge or other & unusual
495q005 rt?, does it require successive generations to ACCUSTOM them to such soil.-- Sow weeds in such soil.-
086a003 with peculiar character of Vegetation. -- So ACCUSTOMED to utter confusion in Europe, that the simpl
086a004 th. old doubters of what are fossil shells.-- ACCUSTOMED to such terms "fixed as the land, stable as
332d004 No houses to Eaton Mascott, where he had been ACCUSTOMED to turn down.-- -- applicable to birds migra
334d011 , by being at first beaten from her, & always ACCUSTOMED to her.-- case parallel to brothers & sister
488q007 caeous plants no other case.-- (6) Will plant ACCUSTOMED to rich soil, when placed in very poor flowe
531n053 en sound or noise, & therefore brain has been ACCUSTOMED to send a mandate to the muscles & when the
533m059 les in different way from what they have been ACCUSTOMED to, in certain actions-- the difficulty of g
570n026 eauty, that which we have been most generally ACCUSTOMED to:-- analogous case to my idea of conscienc
577n052 ost curiously shown in the sudden cures of tooth ACHE before being drawn,-- My father «even» believes
590n093 ows bile in circulation p. 75. Haller says tooth ACHE, even from carious tooth cured by sight of inst
064r137 nos differ from all others in quantity of Sulph. ACID emitted: mem: Grand gypseous formation of Cordi
096a041 lime <n> being heated without parting with Carb. ACID.-- Mr Malcolmson in Paper on India gives reason
108a080 ht be almost as well said probably much Carbonic ACID gaz here.-- [top portion page excised, not loca
138a176 [blank] Would rotting wood by yielding Carbonic ACID unite with «piece of cabbage» alklali & precipi
138a176 itate silica / or charcoal charged with carbonic ACID [blank] Many interesting experiments might be t
176b021 will of animal, but law of adaptation as much as ACID & alkali organized beings represent a tree. irr
196b109 vice versâ. ¿could plants live without carbonic ACID gaz.)-- Yet unquestionably animals most depende
265c084 t, <of> according to Physical laws, as sulphuric ACID disorganizes wood, but adaptation.-- albino how
516q23v 7. Decaying wood absorbs oxygen & forms Carbonic ACID. will this bear on Petrifaction?-- [blank] Ques
587n087 ly disagreeable to organs adapted to like sugar, ACID, &c, which may be doubted for possibly even tas
618o041 not perceive the thought attraction of sulphuric ACID for metal of another person at all, we can only
149g030 this Remember however the great Chilian valley ACONGUA, must there have deposited much-- On other han
440e145 is,-- the law of growth, that which changes the ACORN into the oak.-- In short all which «Nutrition,
058r116 ines of sea weed-- Histoire Naturelle des Indes ACOSTA. p. 125. of French «?» Edition states that the
317c250 o be distinguished from it.-- & several old ACQUAINTANCES. which grow on the lower region of the Cana
397e003 s any of the laws of nature with which we are ACQUAINTED."-- this applies to one species-- I would ap
300c198 low of, acquirement of language. heredetary & ACQUIRABLE.-- therefore mans mind not so different from
301c199 . is not squirrel hoarding, & killing grains. ACQUIRABLE through hoarding from short time.-- My theor
171b004 n.--birds rendered wild <through> generations, ACQUIRE ideas ditto. V. Zoonomia.-- There may be unkno
301c199 inct gained during life.-- do Elephants easily ACQUIRE habits is this the Key to their mental powers.
352d058 al?? Is there some law in nature an animal may ACQUIRE organs, but lose them with more difficulty, «c
390d179 g the passing through whole series of forms to ACQUIRE differences: if none are added object failed,
391d174 esent I can only say the whole object being to ACQUIRE differences «indifferently of what kind, eithe
533m058 It is known that birds learn to sing & do not ACQUIRE it instinctively. may not this be connected wi
571n028 quired by education. else why do some children ACQUIRE it soon. & why do all men. agree ultimately?--
571n028 soon. & why do all men. agree ultimately?-- We ACQUIRE many notions unconsciously, without abstractin
571n028 ter???) Why may not our heredetary nature thus ACQUIRE some general notions, which are taste? Real ta
621o047 g in action, so a child may be taught, or will ACQUIRE from seeing conduct of others, the feeling tha
149g031 ls subject to much variation which have lately ACQUIRED their peculiarities? The slope of A & B regul
224b215 mixed & physical changes (¿intellectual being ACQUIRED alters case) other species or angels. produce
226b221 presentative species. this must happen. & thus ACQUIRED will explain representative system Of this we
230b239 lace, Nature [not located] Any change suddenly ACQUIRED is with difficulty permanently transmitted.--
231b244 ees with non-blending of languages?-- Till man ACQUIRED reason, he would be limited animal in range--
259c066 rosity, if brought into cold country, «& there ACQUIRED» then adaptation.-- No Carrion Vultures in Au
263c076 to each other.-- grant that one instinct to be ACQUIRED (if the medullary point in ovum. has such org
297c186 on to nocturnal habits-- to cats &c.-- must be ACQUIRED by my theory-- else my theory not applicable
300c197 death, most difficult case to imagine how art ACQUIRED.-- They reason however on this to a degree. M
300c197 have any fear of death, only of pain. of death ACQUIRED?. The S. American dung beetles will each beco
302c202 ds same remarks. Characters of analogy.-- last ACQUIRED,-- or aberrant, therefore more easily modifie
302c202 CTER VARIABLE it is (one of analogy or) LATELY ACQUIRED. In pigs number of vertebrae. subject to vari
302c202 tebrae. subject to variation. therefore lately ACQUIRED.-- I fear «great evil» from vast opposition i
310c222 profound ignorance.-- but being such passions ACQUIRED & heredetary & such definite thoughts, I will
344d042 The peculiarities of our breeds must have been ACQUIRED & heredetary tameness.-- In comparing my theo
356d071 ery curious as pointing out difference between ACQUIRED & heredetary.-- thus Yarrel has Lark & Nightingale which b
363d102 - Å. In singing birds, part instinctive & part ACQUIRED,-- thus Yarrel has Lark & Nightingale which b
371d127 other instances are there of such changes, not ACQUIRED by parent, being handed down? Are not Loddige
391d179 y individual except by incestuous marriage has ACQUIRED from father some differences. V. Supra <v. in
416e075 n Greyhound not so species every part of newly ACQUIRED structure is fully practised & perfected Henc
441e148 t «as» in producing bud.-- Fewer of the lately ACQUIRED peculiarities are transmitted it is doubtful
445e158 ive power in chicken, yet says it is evidently ACQUIRED by experience in baby Lamarck. Vol II p. 152.
466t095 ily & many other flowers!-- My view of «variety ACQUIRED?) « <character> » of characters being inherite
522m013 t indelicate actions,-- as if «these emotions» ACQUIRED.-- this may be doubted, whether rather not go
525m023 back game, or picking up a stone, though only ACQUIRED rules by art.-- like the law of honour.-- the
534m061 ive knowledge.-- but if so, yet this knowledge ACQUIRED by senses,-- then thinking consists of sensat
569n019 he beautiful. (distinct from sexual beauty) is ACQUIRED taste.-- Whilst music extremely primitive.--
571n028 urely we have taste naturally all has not been ACQUIRED by education. else why do some children acqui
571n029 taste in mouth, according to my theory must be ACQUIRED, by certain foods being habitual-- & hence be
571n029 ; on same principle we know many tastes become ACQUIRED during life time:-- the latter correspond to
576n047 ife in doing so.-- nor would he regret «having ACQUIRED» this sense of right (& Whether wholly instin
```

```
580n062 wo years are soon lost; yet many of the habits ACQUIRED in that age are retained through life" p. 200
585n077 knowledge  of things which might be «possibly» ACQUIRED by habit. so bees in building cells, must hav
586n081 in  man is learnt by experience is in other is ACQUIRED instinctively" So with <sight> sight-- so a B
587n087 ted for possibly even taste of senna. might be ACQUIRED. as the Turks have of Rhubarb: again on other
606o20v ental power Although taste must necessarily be ACQUIRED by a long series of experiments & observation
613o036 on Will Consciousness Definite instincts being ACQUIRED, is a most important argument, to show that t
614o037 mory; but first memory in many cases cannot be ACQUIRED by experience for child sucking.-- And is  it
614o037 e.-- Must be so if Lamarck's theory true [RHC] ACQUIRED instincts analogous «(& replace)» to experien
615o36v rement or obliteration of instincts But habits ACQUIRED even by <children> «plants»! [RHC] 7) As defi
615o038 eneral-- Instincts, certainly appear a sort of ACQUIRED memory. a permanent secretion of thought, (or
615o038 general kind taking pleasure in virtue because ACQUIRED in past ages; seems to indicate that when  we
621o047 taste to a young person. or it is accidentally ACQUIRED from some trifling circumstance.-- Thus a chi
623o050 minutely our instincts extend, yet as they are ACQUIRED by social animals, living under certain condi
623o050 l tendency through <all> «many» ages. could be ACQUIRED, & we are certain from our reason, that all w
623o050 at all which <has> (as we must admit) has been ACQUIRED, does possess the beneficial tendency [RHC] 1
624o051 of  right & wrong, though, that part, which is ACQUIRED by association from education & imitation, ha
629o055 ried on to other feelings, such as temperance, ACQUIRED by education.-- CD[In similar manner our desi
293c173 tinct explains its loss ¿ if it explains its ACQUIREMENT.-- Analogy. a bird can swim without being we
300c108 instincts,-- & those powers which allow of, ACQUIREMENT of language. heredetary & acquirable.-- ther
574o043 or brain very hard to define.-- Consider the ACQUIREMENT of instinct by dogs, would show habit.-- Tak
575n046 detary ideas.-- being lower faculty than the ACQUIREMENT of new ideas.-- Walter Scott «(Antiquary)» V
584o072 which  my theory no way applies.-- it is the ACQUIREMENT of a new sense,-- bats avoiding strings «in
615o36v ts when imported & plants sleeping good show ACQUIREMENT or obliteration of instincts But habits acqu
335d016 In  last page. I should have said, "an animal <ACQUIRES <th> any new> is <only> able to transmit «onl
527m032 nolds explanation may perhaps account for our ACQUIRING «the instinct» our notion of beauty & negroes
533m058 may not this be connected with their power of ACQUIRING language.-- Hensleigh. W. says that babies kn
312c228 inst this, without «number of vertebrae» new ACQUISITION, we must [not located] Henry Thompson  tells
641j29v ickly plants or animals-- Exudation of fetid «& ACRID» secretion in Mollusca. insects «Carabids & Sta
254c049 loped <is> not surprising to find many forms in ACRITA,-- typical of other, (surely rather  parents).
254c048 ebral structure??-- «do» p. 390. All classes of ACRITE exhibit lowest stages of animal  organization,
254c048 w!!! p.  392.-- except generation & digestion in ACRITE Kingdom all organs blended together, & same or
047r081 om violence. Is it not same as swell travelling ACROSS Pacifick.--excepting in number of waves & in w
070r155 & not far from Tucama[n]. & at Chuquisaca. half ACROSS the continent.--He states plains of Mendoza sm
130a133 the chance in sounding over a continent to fall ACROSS a hot.--spring.-- Hot water would not lie.  at
154g059 cts from side of hill if line suppose continued ACROSS to {P} side removed all well & good, but how c
157g074 surement of shelf of 3d:-- granite block a yard ACROSS. On side of «that» hill, in front of which she
159g084 below it some way; several large ones (one 6 ft ACROSS) on top of spit between river & dry Corry Scar
529m041 e were admiring one in India. & a tiger stalked ACROSS the plains, how ones feelings would be excited
534m062 able in cups of water which they waded. or swam ACROSS.-- they then stretched themselves from wall to
534m062 urther, they ascended about a foot & then leapt ACROSS. (Col Sykes compares this with pidgeons findin
539m082 ciation with much pleasure immediately thrilled ACROSS me, bringing up old indistinct ideas of FitzWi
578n055 the» thought of his knowing «it», suddenly came ACROSS her, the blood rushed to her face,"-- One blus
593n111 rd of command, they all took flight & flappered ACROSS pool to bed of flags I was astonished & having
057r112 kind than the Volcano & Earthquake.--Earthquakes ACT as ploughs [,] Volcanos as Marl-pits: Consider w
068r146 amorphic Volcanos only burst out where strata in ACT of dislocation (NB. dislocation connected with f
109a082 st of England.-- Sea must always on actual beach ACT same way.-- a little further from beach action p
195b101 much more simple, & sublime power let attraction ACT according to certain laws such are inevitable co
198b115 large Mammalia not being found on all isld, (if ACT of fresh creation why not produced on New Zealan
206b148 tions consequent on climate &c-- the whole races ACT towards each other, and are acted on, just  like
206b148 amilies «no doubt a different set of causes must ACT in the two case,» May this not be extended to al
216b180 rid lillies &c &c &, (V Herbert on hybrids) thus ACT.-- Now the point will be to find whether know va
269c102 Flinders Voyage by Brown.-- great space seems to ACT per se as barrier-- Mem. Tartary & China.--, bot
293c173 brain would become webfooted & there would be no ACT of memory.-- [There is no corelation between our
332d001 2d  As a proof. what <trifling> «unknown» causes ACT upon people. My Father mention, than for ten yea
371d118 hould be provided with many contingencies how to ACT--so with the mind. the simplest transmission is
410e050 bsolutely necessary that Physical changes should ACT not on individuals, but on masses of individuals
414e063 destroyed)-- -- When two races of men meet, they ACT precisely like two species of animals.-- they fi
467t103 l height.-- In Lupine, Bees «frequent» & seem to ACT, something like on Kidney Bean, they go to necta
513q21. ious plants, when crossed R. BROWN-- will pollen ACT on any flower before stigmas expanded-- in refer
533m060 ed the idea that I was tending to make myself in ACT less grateful.-- How comes this tendency in thes
535m063 p has this much intellect. yet habit may make it ACT wrong, as I have done when taking lid off  <tea>
537m077 ry., & therefore he has these strong, & does not ACT up to them, no doubt disobeys & hurts conscience
541m090 in  saying delirium rest-- therefore dreams thus ACT.-- ∴ weak minded people are fickle & full of lev
553m136 (<by>  ¿ individually or in race?) by a separate ACT of God, & not as a necessary integrant part of h
557m147 fox--  I can conceive the opposite muscles would ACT, to when in a passion.-- dog tail curled when an
559m154 bably that as long as we consider each object an ACT of separate creation. we admire it more. because
573m036 insect, we wish to be created at once by special ACT, provided with its instincts its place in nature
573m036 ture. its range, its-- &c &c:--must be a special ACT, or result of laws. yet we placidly believe  the
578n054 er giving a beggar, & expecting admiration or an ACT of cowardice, or cheating.-- one does not  blush
582n068 lls Lecture on animal instinct. 1834: p. 15. "To ACT from instinct is to be guided to the performance
583n069 of fox, industry of bee &c &c-- p. 15. "instincts ACT with unerring precision".-- no p. 17. Contrast t
586n079 ause can not have been taught, where to go-- the ACT of crossing the sea in dark night & not  loosing
600o08v Master  and sport &c &c -- The Bitch does not so ACT, because maternal instinct gives most  pleasure.
611o034 mely «one individual» vegetables, the vital laws ACT definitely (<like> «as» chemical laws,) as  long
620o044 eels remorse.-- He reasons on it & determines to ACT more wisely other time, for he knows that the in
626o53v l dogs might be taught, but not cat, that is not ACT by gusto, though by fear it might be partly made
042r068 aved in a dry form It is clear the forces have ACTED with far more regularity in S. America: in Fran
059r123 he rude symmetry of the globe shows powers have ACTED from great depths. so changes. acting in  those
195b098 in the shells.-- The question if creative power ACTED at Galapagos it so acted that bi[r]ds with plum
195b098 tion if creative power acted at Galapagos it so ACTED that bi[r]ds with plumage «&» tone of voice par
206b148 the whole races act towards each other, and are ACTED on, just like the two fine families «no doubt a
219b195 ot know how transported.-- (Glaciers might have ACTED at Tristan d'Acunha-- Carmichael Linn. Transact
302c202 out  hypothesis, & compare it with resuts. if I ACTED otherwise, my premises <in di> would be dispute
415e065 o Man. were to be discovered. Man acts on. & is ACTED on by «the» organic and inorganic agents of thi
522m013 from delirium, a peculiar complaint stomach not ACTED upon by Emetics.-- people recognized,-- sudden
605o19v nsfer to ourselves in the same manner as we are ACTED on by sympathy. D. Stewart on taste The object
619o043 he  would feel part of that pleasure, which the ACTER received.-- If either man did not obey his inst
053r101 ountain appear to me to be effect of expansions ACTING at great depths (mem: profound earthquakes), w
059r123 owers have acted from great depths, so changes, ACTING in those lines. must now proceed from great de
074r165 econdary names «Study Hoffmans account of steam ACTING on trachytes. also Azores. We here have case o
102a062 discussion, state broadly indication of new law ACTING in certain directions predominantly, connectio
183b053 American  quadrupeds. part of some great system ACTING over whole world, the period of great quadrupe
193b090 ted, This would account for each tribe <being> «ACTING as» in vacuum to each other p. 306.--. Chamiss
305c210 which assumes a multitude of forms «each having ACTING principle» according to subordinate laws.-- Th
305c211 ing beings.-- We see thus Unity in thinking and ACTING principle in the various shades of <dif> separ
386d166 rgan. which is not used is absorbed.-- this law ACTING against heredetary tendency causes abortive or
408e043 plants,-- who can say. what <light> «colours». ACTING. by a most delicate organ, on the whole system
523m018 suddenly from <I> (in one case ipecacuahn-- not ACTING) in others from drinking cold drink.-- then br
535m064 s blind storge-- They continue till death, thus ACTING 4 to 6 weeks. The deserted broods appeared hea
549m122 o this-- though believing it to be true, & then ACTING on it, will add to happiness.-- Men having som
558m150 t moral sense arise from our enlarged capacity <ACTING> «yet being obscurely guided» or strong instin
558m151 instinct more than mere love.-- fear for others ACTING in unison.-- active assistance. &c &c. it come
564m004 moral sense.-- Notion of deity effect of reason ACTING on (<not social instinct>) but a causation.  &
567m014 blood to its breast &c &c All Science is reason ACTING «systematizing» on principles, which even anim
570m024 & hence carried on as trick) «Shrugging aroused ACTING» Octob 25. Why is modesty, mixed with triumpha
595m121 ing ugly-- a beau-ideal feeling. Same effect as ACTING on us-- <The Baby> «Effie Wedgwood» April 28th
602o11v applicable to the high idea «p. 131.» in Tragic ACTING-- CD [My idea. would make the mind have myster
619o043 se feel pain.-- If he saw another man <say go> «ACTING in» accordance to his instincts, <he would kno
```

022r006 ms commonly to occur where rocks have undergone ACTION of heat. it is so found in Anglesea. amongst t
026r021 arrange themselves in determinate planes ∴ such ACTION can take place in melted rocks The frequent co
028r032 revented by sedimentary rocks, & hence Volcanic ACTION, contradicted by Cordillera, where that action
028r032 action, contradicted by Cordillera, where that ACTION commenced before any great accumulation of suc
031r038 f great lava cliffs [Fig. I] line of high tidal ACTION {P} NB. patches of modern Conglomerates [Fig.
032r039 atches of modern Conglomerates [Fig. 2] {P} The ACTION of sea A. B. will be to eat in the land in lin
032r039 be to eat in the land in line of highest tidal ACTION. this will at length be checked by increased v
032r039 waves. by the part beneath the band of greatest ACTION not having been worn away.--If the level of th
032r039 (o) which was before beneath band. of greatest ACTION. would now by degrees be exposed to it, & the
032r040 le.-- Must first explain «top of» tidal band of ACTION. This case differs. I think. from Patagonian s
035r048 ly.-- These facts become easy if we look at the ACTION as a deep & extensive movement of viscid nucle
036r051 --In England much subsidence: hence difference; ACTION on land different Volney, P 351. Vol I. woody
037r055 graniniverous» a herbivorous lizard.-- from the ACTION of torrents. «marine» Tortoise & other species
038r058 ?--In former not so much; or no rapilli; & from ACTION of water probably not so much aluminated. As a
042r068 atches of strata in eocene lakes of France, & unequal ACTION of Earthquakes. «on Chili & delta of Indus», m
043r069 ions of period «& manner» of elevation Volcanic ACTION, must be more exclusively confined to that cou
043r070 ame time: this is contrasted to contemporaneous ACTION over larger spaces of the globes & "periods" o
044r073 anic remains.--Unequal distribution of Volcanic ACTION, Australia S. Africa-- on one side. S. America
048r084 attraction, as a blade of grass penetrating by ACTION of Organic power a lump of hard clay. -- In th
051r093 any Corals?? Breccia--Stratification? Anomalous ACTION of ocean.--at Ascension. (where occasionally
051r094 t St Helena) I have mentioned point of greatest ACTION; I now having seen Pernambuco believe much is
051r094 substances worn into bare cliffs evident); the ACTION is anomalous; It is wonderful to see Coral ree
051r094 al of large fragments by mere force of waves: & ACTION on upper tidal band, I do not see how to accou
056r111 whole world, general circulation. But Volcanic ACTION separates some sulphur (perhaps lime) salt. &
059r121 close   analogy to Obsidian, & all show chemical ACTION as well as effects of cooling [misnumbering, n
060r125 & Europe.-- If great chain of Volc. had been in ACTION during secondary period how diff. would the ro
064r135 t effect.--? Capt. FitzRoy. -- Limited Volcanic ACTION & limited earthquakes & great but local elevat
071r156 erly have been most abundant tree-- Metamorphic ACTION: <most> coming so near surface most important
072r159 onsult Dr Holland about bubbles. -- No Volcanic ACTION on coast line of Old Greenland, close to W of
074r165 eins can only be accounted for by concretionary ACTION, conjoined with other» «(state simplest case.
080r180 e world P. 14-91. gradual shoaling of coasts 93 ACTION of sea on coast. 27. Bahama Isd De Lucs travel
088a015 ragmens Asiatiques account of American Volcanic ACTION. -- Fragments of slate converted into crystals
088a016 lte brun -- Main character of Andes Metamorphic ACTION -- red sand of Europe no fossil shells --
088a016 -- Mem: red sand of Europe no fossil shells --¿ ACTION of Heat bubbles volatilized at bottom, condens
089a017 re rising?-- Mem. granite heated.-- Metamorphic ACTION in red sandstone.-- Certainly Volcanic-- CD[Mi
096a039 f shells at Iquique. <Ceylon>. Band of Volcanic ACTION in Iceland parellel to Greenland: Mem.¿ Greenl
099a049 fication evidently small scale of concretionary ACTION all fluid at once, the films vertical. Ascerta
103a065 might be considered a level.-- Dikes being last ACTION. (effect of horizontal movement) hence general
105a071 y successive torrent spread out. by sea-- beach ACTION -- no one will dispute. sea. once came to Mend
109a082 ach act same way.-- a little further from beach ACTION probably modified by form of waves & currents.
114a094 ent composition using difference in metamorphic ACTION which I give at C. of Good Hope.-- A bare hill
115a096 non cleaving beds. metamorphosed. The chemical ACTION which gives polarity to atoms in slates that c
119a105 ation is independent of spreading out matter by ACTION of the sea.-- as no sea exists there.-- But Si
120a107 qual movements of Glen Roy road. (¿ metamorphic ACTION at the bottom of the sea?) All this profoundly
120a109 n each side, such as now exist.-- caution about ACTION of rivers.-- Excess of matter brought down Men
122a114 e over the platform:-- On my view the degrading ACTION must prevent internal fluid arriving at equili
124a118 's theory offers no explanation of intermittent ACTION of elevatory force-- Erasmus says he has seen
124a119 s.-- this most curious with respect to epigmous ACTION.-- if the zinc were mixed with 90 percent of l
132a138 t.-- accounts for increase on earth by volcanic ACTION.-- <Why> now as we know volcanic action prevai
132a138 olcanic action.-- <Why> now as we know volcanic ACTION prevails more beneath the sea, <than> «&» on c
157g077 Glen Roy very little back from line 2d; little ACTION since «that shelf» formed Upper terrace near L
269c101 d currents which, may take place in Metamorphic ACTION. -- Geograph Journal. vol I. p. 17 &c excellent
270c104 ranus, Biologie referred to., as compilation of ACTION of organic nature on inorganic It is very rema
292c171 possible thing. see men walking in sleep».-- an ACTION becomes habitual is probably first stage, & an
292c171 habitual is probably first stage, & an habitual ACTION implies want of consciousness & will & therefo
293c172 s written about it]CD If we saw a child do some ACTION-- which its father. had done habitually we sho
301c198 y what habitual in men & what reasonable-- Same ACTION may be either in same individual p. 7. is not
301c199 lows of any degree in lowest animals --habitual ACTION. even which Man performs.-- child striking a p
313c236 t showing itself, not from instruction Even the ACTION of the viscera. under sympathetic nerve may be
315c242 vous from illness. <the> it must be an excited ACTION, in intestines subject to sympathetic nerves--
417e077 ion-- Looking at simple generation as being the ACTION of two organs in one body.-- or in two bodies,
431e123 view of the crossing of mosses & all others by ACTION of wind difficult.-- Cline on the breeding of
435e133 , affinity to tadpoles. p. 210. Shows. that the ACTION of light is concerned with the developement of
441e148 ibility of concourse of two «individuals» & the ACTION always of two organs-- instead of one part «as
460t019 Forms lost; if» «of this old stock (which from ACTION & reaction grew more complex)» some perhaps re
514q21. tructive principle" be better. or "constructive ACTION on germ." '=?? answered Does Mormodes (one of
521m008 ause she did not remembered, it was an habitual ACTION-- If Miss Elspeth's «in Antiquary» power of repea
521m008 t-secreting organs, brought into play by morbid ACTION.-- Old Elspeth's «in Antiquary» power of repea
524m020 n.-- affections of the thinking organs -- the ACTION of brain which gives sensation of pain, emits
526m027 things, one doubts existence of free will every ACTION determined by hereditary constitution, example
531m052 iscomfort, especially about heart as of excited ACTION, accompanying violent movement; may not passio
531m052 consequent on the injury & consequently excited ACTION of heart.-- now this is the oldest <her> inher
532m055 vous from illness. <the> it must be an excited ACTION in the involuntary mind which is startled.-- M
532m057 t is the <state> effect of short -- but violent ACTION.-- To avoid stating how far, I believe, in Mat
536m070 walks hard.-- «He cannot avoid sending will of ACTION to muscles, any more than «prevent» heart beat
537m075 no universal moral sense.-- «from difference of ACTION of approved» Yet as, I think, the opposite sid
538m081 . These facts showing what a train of though[t] ACTION &c will arise from physical action on the brai
538m081 of though[t] action &c will arise from physical ACTION on the brain, renders much less wondefful the
541m090 re fickle & full of levity (¿ Do I not confound ACTION & thought here?) The opposite extreme of this
545m106 they dance with passion, ie. nervous impulse to ACTION is sent so fast to limbs that they cannot rema
545m107 heredetary habitual movement consequent on some ACTION, which the progenitor did, when excited or dis
546m110 throw light on instinct, showing what trains of ACTION may be done unconsciously as far as the ordina
547m113 llel trains of thought necessary heirs of every ACTION, & always running on in mind, being absent. on
557m148 » passion from blood rushing in face, with less ACTION of the heart.-- tendency to muscular movement,
565n009 howing thought, no compression of mouth showing ACTION,-- sulkiness all negative expression? Expressi
566n010 «& mind» sympathetics with internal organs, as ACTION of heart‖ Malthus on Pop. p. 32, origin of Cha
567n013 y notion of cause & effect, «they have habitual ACTION. which depends on such confidence» when does s
568n017 imply» habituated «in life time» to any line of ACTION, or thought one feels pain, at not performing
568n017 in case of temperance, or real virtue, that is ACTION which experience shows will be for general goo
572n032 o not expect any bodily harm-- case of habitual ACTION.-- L'Institut. 1838. p. 340. Mr Carlyle says t
574n041 bly connected with flow of saliva, & hence with ACTION of mouth & jaws.-- Lascivious women. are descr
574n042 are fact, on my theory intelligible An habitual ACTION. must some way affect the brain in a manner whi
577n049 & hence there is great probability against free ACTION.-- on my view of free will, no one could disco
578n051 :-- Let a person have committed any «concealed» ACTION he should not, & let him be thinking over it w
578n055 suspects one of having done either good or bad ACTION, it always bear some references to thoughts of
578n056 y of the spine-- that it paralyzes all muscular ACTION -- «in man & animals» Blubbering of a child (d
578n056 fainting, sphincters are loosed is a convulsive ACTION.-- to remove disagreeable impression like true co
581n063 instinctive memory, & consequently instinctive ACTION.-- Sir. J. Sebright. has given the phrase "her
581n066 actions]CD.¿ but what first caused this bodily ACTION. if the emotion was not first felt?-- «without
586n079 to know how much of the wonder consists in the ACTION being performed on emotion felt in early child
587n089 t Sir. J. M. says of pure reason not leading to ACTION.-- p. 248. Theory of Association. owing to tim
589n092 t Sir. J. M. says of pure reason not leading to ACTION & yet our emotions being only bodily actions a
592n103 un's work.-- Shutting eyes in contempt opposite ACTION to opening eyes in fear The effect of habitual
600o08a en, when the five senses were the same-- In its ACTION-- emotions-- p 176 & 177 good passage in Frenc
601o109 when the excised heart is pricked) and certain ACTION. (only evidence. when not consciousness) are p
603o012 of sacrifices. «common to many races»-- thinks ACTION towards <man> «a king» <changed into> is carri
606o025 inctive impression 1) September 6th. 1838 Every ACTION whatever is the effect of a motive.-- [-- must
606o025 isms-- one knows it, when one wishes to do some ACTION (as jump off a bridge to save another) & yet d

Page **(Key Word)**
```
606o025 one   could do it, but other motives prevent the ACTION see Abercrombie conclusive remarks p. 205 & 20
608o026 t free will obvious.-- because man has power of ACTION, & he can seldom analyse his motives (original
609o030 ness we must look far forward «& to the general ACTION» -- certainly because it is the result of what
611o034 the union of simple non-organic matter, without ACTION of vital laws-- According to the individual fo
616o038 y become perfect & we may look back to definite ACTION or to our conscious selves.-- Such memory may
616o039 of   attraction to ordinary matter is that which ACTION bears to the agent. Matter is by a metaphor sa
616o39v n only <shew> point out the nature «of perceptive ACTION» by which we come to conceive of matter as att
616o39v to the brain. There are two modes «of perceptive ACTION by which bodily action is made known to us, re
616o39v two   modes of perceptive action by which bodily ACTION is made known to us, revealing respectively wh
617o39v jective aspect. The subjective aspect of bodily ACTION is revealed to us by the effort it costs us to
617o040 ell as we can. <The objective aspect of> bodily ACTION as recognised by our external senses  consists
617o040 raced to the body of the individual to whom the ACTION is attributed; force (be it remembered)  being
617o40v But   coming round to the <subjective> aspect of ACTION as known by the exertion of our own power & co
617o40v n trace any force in inanimate matter up to the ACTION of some animated agent Now the phenomena of gr
618o041 houghts, perception &c. are modes of subjective ACTION-- they are known only by internal consciousnes
618o041 n is perceived but they are known by courses of ACTION quite independent of each other. A person migh
618o41v that   weight, the blueness, still less between <ACTION> «things» so different as action thought & org
618o41v less   between <action> «things» so different as ACTION thought & organization: But if the weight neve
619o042 others at its own expense.-- Moreover <the> any ACTION in accordance to an instinct gives great pleas
620o044 nce is improved by attending & reasoning on its ACTION, & on the results following our conduct.--  If
620o045 instinct, even when our reason tells-- + us the ACTION was superfluous, as one man trying to save ano
621o047 there nice & nasty in taste, & right & wrong in ACTION, so a child may be taught, or will acquire fro
621o047 st (rarely if opposed to natural instincts) any ACTION is either right or wrong.-- [RHC] 7) Hence, wh
622o048 [RHC]   Feelings of the mind, whether leading to ACTION or not, are the parts of our nature, <sub> sub
622o048 ciations.-- often feelings which do not lead to ACTION are repressed thus avarice. &c &c.-- [RHC] 8)
624o50v ur moral taste p. 152. Reason never can lead to ACTION.-- p. 164. Ld. Shatsbury under term of Reflex
626o052 .B. If feeling or emotion rises from heredetary ACTION on body.-- This feeling, when instinctive will
626o052 .-- This feeling, when instinctive will lead to ACTION.-- the passion rising from weariness leads  to
629o55v ses Mackintosh gives & try it.-- p. 241 (1) Any ACTION by habit may be thought wrong.-- & conscience
638j059 born. with tendency to make animal perform some ACTION.-- as well as gain it. by habit.-- New  theory
027r028 Chiloe In the endless cycle of revolutions. by ACTIONS of rivers currents. & sea beaches. All mineral
028r029 rated; is it washed from the solid rock by the ACTIONS of Springs or more probably by some unknown Vo
264c079 w.-- see its passion & rage, sulkiness, & very ACTIONS of despair; «let him look at savage,  roasting
292c171 e may be called instinctive.-- But why do some ACTIONS become heredetary & instinctive & not others.-
292c171 ained slowly.-- therefore it can only be those ACTIONS, which Many successive generations are impelle
301c199 es.-- Knowing that animals have some reason, & ACTIONS habitual. it surely is not worthy interpositio
312c232 wonderful   According to my view <ins> beccause ACTIONS are constant they are instincts, & not ':' insti
313c236 ects analogous.?-- «(Even plants have habitual ACTIONS.-- «this very important in considering how ch
313c236 considering how children come to suck or other ACTIONS in foetus of Mammalia, or chick eat») Generati
315c241 able analogies might be drawn between habitual ACTIONS of plants «when exciting cause is absent» & me
315c242 ion of fibre)-- it is most remarkable habitual ACTIONS in plants, it allows of any degree in lowest a
377d139 itch, for some feeling must urge them to these ACTIONS. «These facts may, be turned to ridicule, or m
522m013 ost.-- most delicate people do most indelicate ACTIONS,-- as if «these emotions» acquired.-- this may
525m023 eir dirt, running home.-- in these cases their ACTIONS do not look like fear, but shame.-- I cannot r
526m030 oves because its appetites urges it to certain ACTIONS, which are modified by circumstances, & thus t
533m059 what   they have been accustomed to, in certain ACTIONS-- the difficulty of getting on a horse on  the
536m072 rhaps, though from not having pain or pleasure ACTIONS unavoidable & only to be changed by habits). n
538m080 man.-- If one could remember all ones farthers ACTIONS, as one does those in second childhood, <they>
539m082 complicated trades can hardly be considered as ACTIONS otherwise than habitual.-- instances?? The pos
540m084 ng his mouth in romps, <so> he smiles. Many of ACTIONS as hiccough & yawn are probably merely coorgan
542m094 to animals.-- Curious to trace, which of these ACTIONS are habitual, & which now connected physical r
545m104 uire idea to order muscles to do <certain> the ACTIONS.¿ is it the <becom> impression becoming very o
545m108 which «now» excites the expression.-- Habitual ACTIONS are the reverse of intellectual, there is no c
550m127 ater-- innate September.1-- If one performs some ACTIONS, which are pleasant, every concomitant circums
552m132 scle Sept. 8th. I am tempted to say that those ACTIONS which have been found necessary for long gener
564n003 to it--!! Man moreover who reasons much on his ACTIONS, makes his conscience far more sensitive. ulit
564n003 cience far more sensitive. ulitmate effects of ACTIONS.→ till at last he face «instinct of» hunger, «
572n033 stinct to muscular movement.-- say instinctive ACTIONS. senses. notions &c Octob 30th-- Dreamt somebo
574n042 pendently of habits.-- the limits of these two ACTIONS either on form or brain very hard to define.--
577n050 nts, must be association.-- a certain round of ACTIONS, & closing of the leaves,
577n050 on   from want of stimulus, after certain other ACTIONS, & hence becomes associated with them.-- The e
577n051 ll help my theory of sensitive Plants Habitual ACTIONS, (independent of mind) in the intestinal funct
578n057 rfect over voluntary muscles, these convulsive ACTIONS-- (except in weak people & hysterical people i
578n057 ple & hysterical people inclined to convulsive ACTIONS).-- But, the Lachrymal gland is «not» under vo
581n066 fects on the mind, accompanying certain bodily ACTIONS|CD.¿ but what first caused this bodily action.
582n068 to   the performance of a number of prearranged ACTIONS, which will bring about a certain result, whil
582n068 in result, while the creature performing those ACTIONS neither knows nor intends the result they will
587n086 soc of ideas & emotions. rather ideas & bodily ACTIONS make the emotions.-- p. 272. Some remarks appl
588n090 -- Vol II p. 445. If we compare the judgments & ACTIONS of a young animal with an old.-- (dog horse, s
589n092 to action & yet our emotions being only bodily ACTIONS associated with ideas.-- A sigh, is an abortiv
591n099 ve «as a boy» wondered why all abnormal sexual ACTIONS or even impulses. (where sensations of individ
591n100 eld   in abhorrence.-- all this makes analogy of ACTIONS with <&> against benevolent & parental instinc
601o009 n.-- man moving leg when asleep-- «or habitual ACTIONS» perhaps polypi-- (so that lower animals are s
602o011 I Discourse (p 115) a very good passage. about ACTIONS & decisions bein the result of sagacity, or in
607o025 s, natural as usual (a) one well feels how many ACTIONS are not determined by what is called free will
615o038 n) or an association of pleasures with certain ACTIONS performed by your parents, conscience This «X»
619o042 n instinct gives great pleasure, & <an> «such» ACTIONS being prevented by <necessity> «some force» gi
619o043 y regard to his own interest. likewise if such ACTIONS were prevented by force he would feel pain. [.
621o047 cience rebukes malevolent feelings, as much as ACTIONS, therefore Sir J. M. talks too much about the
622o048 e the instinctive, ones, <which either lead to ACTIONS or not, as feeling of cowardice? «This is  not
624o051 , of parents strives* to same end.-- & general ACTIONS of community must frequently team same end.--
625o052 onscience.-- I believe that certain feelings & ACTIONS are implanted in us. & that doing them gives p
626o052 nctive feelings will doubtless lead to similar ACTIONS which in prior <races> generations led to thei
628o53v nalogous to education of child,-- causing many ACTIONS to be considered right & wrong,-- to be associ
112a089 here one alone has been formed--Look at the now ACTIVE volcanos & see what high they are «See Athenae
345d044 legs,   & face & tail, just like species.-- high ACTIVE breedin[g] [not located] half breed liable  to
558m151 are love.-- fear for others acting in unison.-- ACTIVE assistance. &c &c. it comes to Miss Martineaus
570n025 inine.-- both accompanied by depending head., & ACTIVE vessels of skin.-- What difference is there be
619o042 n, we see in other animals they consist in such ACTIVE sympathy that the individual forgets itself, &
023r011 -- «Has this fault determined side of volcanic ACTIVITY.» That axis was produced, from a fissure in a
032r040 tant rising with successive periods of greater ACTIVITY & rest.--Such changes could be shown (as repr
043r070 spaces   of the globes & "periods" of increased ACTIVITY.-- such as that of 1835.-- State the three «o
603o11b ied to instinct.-- p. 134. a painted must not a ACTORS, or a scene in garden.-- yet both beautiful! p
289c162 any insects may be told by their larvae) but the ACTS of condensing must alter method of generation--
415o065 n as Mastodon to Man. Man ACTS on. & is acted on by «the» organic and inorgani
538o080 double   individuality implied by habit, when one ACTS unconsciously with respect to most energetic se
607o025 (as generally used) is not there present, but he ACTS from motives, nearly as usual (a) one well feel
619o042 he individual forgets itself, & aids & defends & ACTS for others at its own expense.-- Moreover <the>
635j55r we know nothing of the will of the Deity. how it ACTS & whether constant or inconstant like that of M
109a082 sea. off coast of England.-- Sea must always on ACTUAL beach act same way.-- a little further from be
531m052 t <her> inherited & therefore remains, when the ACTUAL movement does not take place.-- A start is HAB
565n006 n, is like the grin of the Hyaena from fear, no ACTUAL intention to bite at moment, but mere symbol o
090a022 many, (one even 3000) This ninety includes all ACTUALLY counted.-- The weight «or size» is given of 2
109a083 . independent of currents.-- mud going out can ACTUALLY be seen.-- ¿ The preservation of dikes & ledg
153g056 ng.» neighboring rock plains & [...] sandstone ACTUALLY resting on them on summit of hill rounded, si
244c021 hern regions.-- «it would now represent. what <ACTUALLY is» «has» taken place with quadrupeds» p. 118
254c048 & likewise those much higher in scale. So Owen ACTUALLY believes in this view!!! p. 392.-- except gen
```

Page ***(Key Word)***

268c096 count of Tit mouse feeding young of redstart & ACTUALLY driving away parent birds.-- showing how blin
382d157 . whether a Heautandrous animal is <evidently> ACTUALLY split in two-- keeping sexes separate. Owen s
385d163 ertile» without she know & LIKES HIM & then is ACTUALLY obliged to be held.-- like she wolf of Hunter
388d173 ng present is-- it also shows that semen. must ACTUALLY reach the ovum.-- [Why in making a bud, which
470t176 nglish, planted within few yards of each other ACTUALLY produced hybrids-- My Father remembered when
511q018 ed-- not caring whether good or bad.-- are any ACTUALLY rejected?? (8) Get Sir. R. Heron to give me P
617o40v nt of the <force exhibited in every> phenomena ACTUALLY apprehensible by sense. 5) There is nothing a
434e129 ll sometimes produce a few seeds,-- -- Ruscus ACULEATUS. a dioecious plant, in which the Male plant s
065r138 eater care. vegetation & climate of Tristan D. ACUNEHA. Kerguelen Land. Prince Edwards Isld. Marion &
102a059 ast is driven on to it.-- rollers at Tristan d'.ACUNHA.-- silting up. channels on coast of England--
218b192 der are formed by nature. Carmichael. Tristan D'ACUNHA, a list of its Flora. is given Mr Don remarked
219b195 rted.-- (Glaciers might have acted at Tristan d'ACUNHA-- Carmichael Linn. Transacts. Vol XII.-- The A
225b218 African form of plant being found in Tristan D'ACUNHA. may be said to deceive man. as likely as foss
225b220 ieve none.-- Canary islds.? Madeira? «Tristan d'ACUNHA?» «Iceland?--» The Connection between Mauritiu
226b222 ndo Po.). with plants of St. Helena & Tristan D'ACUNHA, resolves itself into question of proportion o
269c100 xamine Abrolhos Flora with this view) Tristan D'ACUNHA, St Helena &c &c. Juan Fernandez A communicati
358d074 pect to ancient geography of Atlantic Tristan D'ACUNHA ditto Juan Fernandez do Mitchell. Australia Vo
480z008 n. Transacts Vol XII. p 496. Birds at Tristan d'ACUNHA.-- (Turdus Guayanensis?)) Emberiza Brasiliensi
036r051 ng & prolonged). NB, Is it generally known. the ACUTE chirping sound produced in walking over the san
622o048 m of society broken..-- & how far more <feelin> ACUTE the feeling really is.-- All these associated «
271c106 heredetary whilst species have changed Argumentum AD absurdum. The creative American halo has extende
546m108 dest memory-- also low faculty of understanding. ADAM Smith (.D. Stewart life of. p. 27), says <sympa
559m155 rth reading Copied <Smith> «D. Stewart» lives of ADAM Smith Reid, &c worth reading. as giving abstrac
595n184 rds being expressive, (Vol. 4 of Works) [blank] "ADAM Smith Moral Sentiments" much on life & characte
609o29v croaching on views in second volume of Malthus). ADAM Smith also talks of the necessity of these pass
171b004 thus being modified,-- therefore generation to ADAPT & alter the race to changing world.-- On other
376d135 edgings, must be to sort out proper structure & ADAPT it to change.-- to do that for form which Malth
473s07r ogs--cattle? As we see the frame of animals can ADAPT itself to course of life, «as in trades» there
583n070 make cell of certain form. (& especially as it ADAPT its cell to circumstances), it must have impuls
063r133 howing non Creation does not bear upon solely ADAPTATION of animals.--extinction in same manner may n
171b003 see generation here seems a means to vary. or ADAPTATION.-- Again we <believe> «know» in course of ge
176b021 nges not result of will of animal, but law of ADAPTATION as much as acid & alkali organized beings re
180b038 nces are continued & produce according to the ADAPTATION of such circumstances & therefore that death
180b039 ary to what would appear from America) of non ADAPTATION of circumstances.-- Vide two. pages back. Di
182b046 ition of every animal is partly due to direct ADAPTATION & partly to heredetary taint;. hence the res
185b055 eagles.-- This is but carrying on. attempt at ADAPTATION of each element.-- May this not be explained
187b064 perpetuate certain peculiarities, (therefore ADAPTATION), & to obliterate accidental varieties, & to
189b069 cies.-- In proof that structure is not simple ADAPTATION, armadilloes «&» & Megatherium. each with sa
190b078 N. Africa An originality is given (& power of ADAPTATION) is given by true generation, throughe means
219b194 rmediate position.-- We cannot consider it as ADAPTATION because volcanic isld. whilst <neig> Africa,
222b205 e of Mammalia for Reptiles, which can only be ADAPTATION to changing world:-- I cannot for a moment d
223b210 of species less altered prevents the complete ADAPTATION which would ensue A. B. C. D. -- (A) crossin
227b225 typical structure.-- Every species is due to ADAPTATION + heredetary structure. latter far chief ele
227b227 to discover causes of change.-- the manner of ADAPTATION (wish of parents??) instinct & structure bec
230b236 . All variation of animal is either effect or ADAPTATION,∴ animal best fitted to that country when ch
240c004 ching to nature of Monster, heredetary. other ADAPTATION.-- Mr Yarrell says, that after breeding in p
258c060 y.-- Animals having wide range, by preventing ADAPTATION owing to crossing, with unseasoned people. w
259c065 certain very unfavourable conditions.-- as an ADAPTATION, but adaptation during earliest existence; i
259c065 avourable conditions.-- as an adaptation, but ADAPTATION during earliest existence; if whole life the
259c065 g earliest existence; if whole life then real ADAPTATION The case of heredetary disease, is on the sa
259c065 ents have healthy children the other case is <ADAPTATION> «change» during life of parent, & therefore
259c065 herefore being always necessary may be called ADAPTATION With respect to my theory of generation, fac
259c066 ht into cold country, «& there acquired» then ADAPTATION.-- No Carrion Vultures in Australia!! Wilson
261c072 an exception) can only be explained by direct ADAPTATION to animals wants & not as change in typical
265c084 heredetary‖ With respect to question what is ADAPTATION.-- Ermine, ptarmigan hare becoming white in
265c085 aws, as sulphuric acid disorganizes wood, but ADAPTATION.-- albino however is monster. yet albino may
265c085 nster. yet albino may so far be considered an ADAPTATION as best attempt of nature, colouring matter
265c085 alpine district & dwarf plants from seed, one ADAPTATION, other monster.-- The only way of judging wh
265c086 clearly attribute to heredetary origin & not ADAPTATION. to its habits.-- Few will dispute that it i
282c140 ter,-- we here suppose these changes of <use> ADAPTATION greater than those heredetary ones.-- which
293c175 rving its relations.-- the wonderful power of ADAPTATION given to organization.-- This really perhaps
294c175 t New York «instance of the fine relations of ADAPTATION of animals & the country they inhabit.--» &
297c186 ge of night jar. like owls. analogy in habits ADAPTATION to nocturnal habits-- to cats &c.-- must be
314c236 ds.-- I can scarcely doubt final cause is the ADAPTATION of species to circumstances by principles, w
314c238 . & not the organ itself How except by direct ADAPTATION has such a change been effected.-- the consc
318c252 ,---- Bachman has seen webbed Shrews. case of ADAPTATION.-- (case of Squirrel from extreme north turn
318c252 te like Hares??--) I never saw more beautiful ADAPTATION for snow-- like snow shoes. than feet & hind
338d024 to) prevents his taking any form of Malaria-- ADAPTATION & species-like,-- -- Says Negro-- thick skin
341d030 ven much smaller than in other Struthios. was ADAPTATION to little Movement.-- nocturnal crawling bir
343d036 duce changes of form in the organic world, as ADAPTATION. & these changing affect each other, & their
365d107 of effects of country, (& no monstrosity, or ADAPTATION to unhealthy state of womb).-- One can perce
365d107 es or species., all the structure of which is ADAPTATION to habits (& habit second nature) may be mor
397e004 mall-- because change in forms is <al> solely ADAPTATION of whole of one race to some change of circu
399e009 Animal No structure will last. without it is ADAPTATION to whole life of animal, & not if it be sole
404e026 luence on parent affecting offspring.--‖ & as ADAPTATION,-- however mysterious such is case‖. therefo
423e097 large part of the complexity of structure is ADAPTATION. though perhaps difference between jaguar &
432e124 must be made with great caution; owing to its ADAPTATION to the surrounding circumstances According t
507q15v ood case as showing how simple, but beautiful ADAPTATION might be arrived at.= Any book with drawing
636j56v ign agency-- as insects, as wonderful case of ADAPTATION.! There would not have been any Dioecious pl
636j57r 32. gives Woodpecker as instance of beautiful ADAPTATION.-- & then Chamelion, which feeding on same f
637j57v simply statement of productiveness, & laws of ADAPTATION‖ p. 234. The non-absorbing Camel's stomach i
637j58r ong necks.-- p. 236. Marsupial bones especial ADAPTATION, to «young».-- good God & yet Mails have the
638j58v t instinct is-- consider this I look at every ADAPTATION, as the surviving one of ten, thousand trial
638j58v ammae.» to the then existing conditions.-- An ADAPTATION made by intellect this process is shortened,
195b099 vertebrae in neck same cause, such beautiful ADAPTATIONS yet other animals live so well.-- This view
225b219 mind produce any change in offspring? if so ADAPTATIONS of species by generation explained? NB. Look
265c083 le are sometimes heredetary,-- yet these not ADAPTATIONS «they are counteracted by nature by crossing
293c174 t insects lodging there. Now these exquisite ADAPTATIONS can hardly be accounted for by My method of
632j53r o deny laws.-- The whole universe is full of ADAPTATIONS.-- but these are, I believe, only direct con
633j53v But are we certain that these are necessary ADAPTATIONS.-- May they not be accidental? We have good
636j57r , that animals could not exist without these ADAPTATIONS.--fossil forms show such losses.-- Consider
637j57v ld & yet talk of perfection Get instances of ADAPTATIONS in varieties.-- greyhound to hare.-- waterdo
637j57v anks-- cowslip to <banks> fields-- these are ADAPTATIONS just as much as Woodpecker. --only we here s
637j58r the long spinous processes in Giraffe &c, as ADAPTATIONS to long necks-- why they may as well say, «l
637j58r them. What trash p. 237. Gives as Summary of ADAPTATIONS Horny point to chickens beak, to break egg.
640j28v f old age, after breeding season, or gaining ADAPTATIONS, but for youth most necessary: the fertility
056r111 lic ores.--which mingling & separating is well ADAPTED to use of mankind.--<Hutton show> Earthquakes
176b024 rrangement.-- if each Main stem of the tree is ADAPTED for these three elements, there will be certai
180b037 hat variety of ostrich, Petise may not be well ADAPTED, & thus perish out, or on other hand like Orph
181b045 ages.-- The Creator has made tribes of animals ADAPTED preeminently for each element, but it seems la
181b045 tible with such structure are in minor degrees ADAPTED for. other elements. every part would probably
193b090 tances) & those alone preserved which are well ADAPTED, This would account for each tribe <being> «ac
202b130 Good argument for species not being so closely ADAPTED. Near the Caspian «Province of Ghilan» wooded
223b211 s offspring of A. becoming a good species well ADAPTED to locality A. but it is instead a stunted & d
223b211 is instead a stunted & diseased form a plant, ADAPTED to A. B. C. D.-- Destroy plants B. C. D. & A w
258c060 th people from cold, children would not become ADAPTED to climate.-- Descent. or true relationship, t
262c073 ahiti. thistle. Pampas. show how nicely things ADAPTED--.-- These «aberrant» varieties will be formed
267c095 Cryptogamic plants.--Owing to plants not being ADAPTED to Air!! p. 11--&c. valuable paper on quadrupe

```
289c160 tries-- nightingale do.-- all shows how nicely ADAPTED species to localities⫫-- p. 390,. young <grebe
298c191 central  parts become occupied by a third best ADAPTED kind.-- lower species would then revert to pri
302c203 g to such interm: father-species, being little ADAPTED to some physical change.-- If Patagonia became
347d049 ll as profound. because he says length of days ADAPTED to duration of sleep of man.!!! whole universe
347d049 duration of sleep of man.!!! whole universe so ADAPTED!!! & not man to Planets.-- instance of arrogan
357d074 ate that. the <p> mechanism by which seeds are ADAPTED for long transportation, seems «?» to imply kn
365d106 an hardly be thought that the cross would have ADAPTED it to changing circumstances.-- More  probably
365d107 re conformable to the structure which has been ADAPTED to former changes. than a mere monstrosity pro
375d135 usand wedges trying force <into> every kind of ADAPTED structure into the gaps <of> in the oeconomy o
379d151 a] History of the Caspian. Fresh Water Fish!! ¿ADAPTED to salt water?-- peculiar species, crabs & mol
386d167 changes  of nature are slow. if animals became ADAPTED to every minute change, they would not be fitt
412e057 variations, as long as each shall be perfectly ADAPTED to circumstances of times. & from  persistency
416e071 slower.--  No domesticated animal is perfectly ADAPTED to external conditions.-- (hence great variati
416e072 ance & in species are only ancient & perfectly ADAPTED races L'Institut 1838. p. 394. Rhinoceros «tic
422e092 eminently  adaptive.-- does it not mean lately ADAPTED or transformed. & hence not indicative of true
431e122 hange, because its number show it is perfectly ADAPTED; it where few stray ones. are, that change may
444e158 egg, or larva. or foetus to perfect animal are ADAPTED by foreknowledge, so must the mutations of spe
461t041 itions, some species will undergo & yet remain ADAPTED.-- it does away with difficulty of rabbits  of
574n039 lines.  of poetry.-- <signs> sounds singularly ADAPTED to subject see <A> ⫫ I think this argument mig
587n087 nfitness of the objects then viewed. to organs ADAPTED to other objects. (as that senna is necessaril
587n087 at senna is necessarily disagreeable to organs ADAPTED to like sugar, acid, &c, which may be  doubted
633j54r w my theory makes all organic beings perfectly ADAPTED to all situations, where in accordance to cert
637j58r cks-- why they may as well say, «long» neck is ADAPTED to long necks.-- p. 236. Marsupial bones espec
639j28v f the Grallae <are> «have been made» long «(as ADAPTED to)» because their food lies deep.-- I say  it
294c177 the allowing at same time true species.» & its ADAPTION to classification & affinities, its extension
341d029 th asked whether structure of pelvis & was not ADAPTIVE structure, like little wings of Auks which do
341d029 red that all characters might be considered as ADAPTIVE & that he did not see where the line could be
421e091 ifications of these organs to mistake a merely ADAPTIVE to an essential character--" How little clear
422e092 difference between an essential character & an ADAPTIVE. one.-- are not the essential ones  eminently
422e092 e. one.-- are not the essential ones eminently ADAPTIVE.-- does it not mean lately adapted or transfo
025r108 e whole history of Europe, with America; I might ADD I have drawn all my illustrations from  America,
031r038 again  & covered no sign of upheaval To Cleavage ADD other instances in old world of symetrical struc
040r062 use such are found in perfection on that side.-- ADD from M. Lesson. character of Flora to New Zealan
043r071 ions & durability of similar causes go together. ADD. <">" <from> "in the same line" to "from the epo
232b246 one»,   oh we will take a day from the equator to ADD to the mean of the other.-- If the the world had
239c002 ring would be chesnut.-- On this principle I may ADD, that fact of half cross with parents, going bac
276c123 Mention persecution of early Astronomers.-- then ADD chief good of individual scientific men is to pu
283c146 of formation of species & genera, is probably to ADD to quantum of life possible with certain preexis
350d054 . It has been argued Man first civilized. the <ADD> this in note. ¿mere conjecture?-- Australians.--
382d158 mb, or ovarium.-- or testicles in female.-- the <ADD> presence of both testes & ovaria in Hermaphrodi
410e052 this,  having myself aided in such sins» (do not ADD name, without reference to description),  except
549m118 but the» «&» the sensual enjoyment of the minute ADD to the happiness.-- but as they are not recollec
549m122 lieving it. to be true, & then acting on it, will ADD to happiness.-- Men having some instincts as rev
118a103 grand  idea of amount denudation.-- This may be ADDED to any place where dikes described-- {P} Cordil
132a139 he cold «bottom of» ocean. (with fresh sediment ADDED to bottom) be caused, by absence of circulating
290c162 er in those animals, where much change has been ADDED. «as it speaks to amount of change only & not k
310c222 ,-- If I be asked by what power the creator has ADDED thought to <an> so many animals of different ty
313c235 ang outang of (in June 1838 when young male was ADDED good instance of instinct showing itself, not f
318c253 ecoming common-- likewise of the Hirundo fulva (ADDED by Audubon in Appendix) showing WHAT CHANGES ar
351d056 ch move imperfectly has eye-point, but Broderip ADDED it has been stated that stationary Spondylus ha
372d129 Planaria,  the whole grown to that part.-- claw ADDED to crab, tail to lizard,-- healing of wound.--
372d129 ng body is present-- in generation something is ADDED from one part of the body «(or of other  <like>
378d147 own.-- Sexual Selection If masculine character. ADDED to species,. we can see why young & female alik
387d168 lation impresses offspring more & more with the ADDED «like Lord Moretons case & Dr. Andrew Smith,» d
387d169 me force as first pair, but to this tendency is ADDED <that> the 3d tendency from first pair.-- Now i
390d178 ange in form. ideosyncrasy or dispositions were ADDED or substracted at each, or in several generatio
390d179 es of forms to acquire differences: if none are ADDED object failed, & then by that corelation of str
391d174 sity that there should be something «each time» ADDED to that kind of generation, which typifies  the
391d175 ther & sister are somewhat different) should be ADDED to each individual before he can procreate. the
392d175 differences.  shows that difference need not be ADDED EACH TIME. but after some time]CD What kind  of
418e083 over [not located] it utterly untold,-- what is ADDED to the composition of the atom, to make it aliv
418e084 the  succesive modifications of structure being ADDED to the germ, at a time, (as even in  childhood)
418e084 such modifications, become as much fixed, as if ADDED to old individuals,, during thousands of centur
576n047 the dog, or chiefly habitual as in man), for it ADDED much to the happiness of his life, & the chance
601o019 ) Consciousness is sensation No. 2. with memory ADDED to it, man in sleep not conscious, nor  child--
606o023 assures of harmonious colours &c &c surely to be ADDED. Lessings Laocoon p. 125-- says new subjects ar
406e037 n strayed hang their tails.-- November lst.-- ADDENDA to Journal. I show erratic blocks  transported
115a096 ch unite the homogenious crystals, . must aid in ADDING effects to common heat.-- Where there are clif
343d039 9th. With respect to the Deluge it may be worth ADDING in note than amongst the Mammalia of Europe th
400e013 ke.--, showing effects of cultivation gradually ADDING up. & four more generations before they  began
422e095 became complicated, they opened fresh, means of ADDING to their complexity.-- but yet there is no «NE
612o035 one  common one always satisfactory, though not ADDING to positive knowledge. lessening amount of ign
384d161 (1) From praeternatural situation of parts (2) ADDITION of parts, (3) deficiency of parts (4) combine
384d161 of parts, (3) deficiency of parts (4) combined ADDITION & deficiency of parts, as in  Hermaphrodites,
392d175 es having fertile offspring without coition or ADDITION of differences. shows that difference need no
418e080 developement  of one sex on one side, than the ADDITION of other organ, in which case the hermaphrodi
618o41v should imply «X» the existence of something in ADDITION to matter is because our knowledge of  matter
078r174 t with care to 3 species. I think I have much ADDITIONAL information ⫫ Guanaxuato, which has  yielded
201b126 II  talking of annelidae.-- <">The fact is an ADDITIONAL illustration of that axiom in Natural Histor
444e157 of  the snout of the mole & Pig in having two ADDITIONAL bones to give strength to it.-- p. 139. Doub
445e159 e means of the medium in which they live do. "ADDITIONS". p. 454.-- does really attribute metamorphos
081rIBC usen in 1819 Kotzebue 1816 Constant log always ADDITIVE to convert French Toise into English ft. 0.80
415e070 can  be made out, that varieties are generally ADDITIVE, & not abortive: with reference to the non-ne
434e128 y handed to offspring.-- Whewell's anniversary ADDRESS 1839, p. 9.,-- talks about fossil Infusoria be
063r132 red» by different methods, associated life only ADDS one other method where the division is not perf
605o18v rror & wonder so often concomitant with sublime. ADDS not a little to the effect: as when we look  at
303c205 is true,» there are no genera. if Mammalia are ADDUCED. say oh look to your fossils, now if extinctio
428e111 says  Laurels have not been so. (which is case ADDUCED by Herbert) because not reared by seedlings.--
131a135 ions on Mountains of the Moon. by Dr. Nichol-- ADDUCES the case to show Sir. J. Herschel's theory wro
632j53r fer to the author if I use these facts p. 280. ADDUCES provision of seeds for transportation  through
051r095 circumstance  determines whether an animal will ADHERE to a certain part. Apropos to question does an
051r095 a certain part. Apropos to question does animal ADHERE to rock because it does not decompose. or vice
634j54v ins it «"By» saying "It is the determination to ADHERE to a plan once adopted; & it is from these ver
632j53r rough the air.-- cocoa nut by water «fucus for ADHESION».-- as examples of design.-- perhaps they are
402e020 s have spread from the N. end of Borneo to the ADJACENT island-- In Sooloo we find the elephant--  in
304c206 s of habit affect particular organs.-- of two ADJOINING families & not all organs blending away.-- ++
451e177 een Borneo & Java) Lat 5 degree. 50 minute S. ADJOINING it are several small islands. abounding  with
500q10a efore last» beans & peas were planted in rows ADJOINING & seeds gathered there were planted «last yea
640j167 joined, nature & climate very different, from ADJOINING coast. Admirable explanation is thus offered.
243c019 n eût été bien éloigné, il y a peu d'années, d'ADMETTRE que ces oiseaux eussent leurs représentants d
033r043 boldt. S. American Geolog. in Daubeny. P. 349 ADMIRABLE little table showing long PERIODS of great vi
210b158 it  to languages But how do plants cross?-- = ADMIRABLE discussion Von Buch says from Humboldt,, in La
252c042 Desert of Korto & Steppes of Kordofan p. 401. ADMIRABLE letter from Macleay to Bicheno much excellent
280c134 her parent.-- Shows instinct (Sir J. Sebright ADMIRABLE essay) heredetary Young wild ducks.-- lose as
317c250 e lower region of the Canary islands-- p. 250 ADMIRABLE table of plants of St. Jago showing many comm
344d040 y <older> geological epochs.-- There are some ADMIRABLE tables on Geograph distribution of reptiles i
344d041 foetus.-- August 23d The Rev R. Jones gave an ADMIRABLE harrier from Ireland to Brighton Park--first
424e100 imestone &c &c.-- L'Institut 1838.-- p. 290-- ADMIRABLE paper on geographical distribution of Crustac
```

Page **(Key Word)**
481z009 on with Falkland. good also for Journal.-- 18 ADMIRABLE engravings in Meyen Zoology on animal of Camp
640j167 climate very different, from adjoining coast. ADMIRABLE explanation is thus offered.-- From these vie
256c054 in distinguishing which were species, (theory ADMIRABLY) yet a glance would tell from which country,-
134a141 irgin p. 59. dip of Clay slate in T del Fuego ADMIRALTY Sound. SE dip. much p. 136. Rocks on Western
291c166 operty of matter? It is our arrogance, it our ADMIRATION of ourselves.-- The idea of foetus being of
578n054 d or bad. either giving a beggar, & expecting ADMIRATION or an act of cowardice, or cheating.-- one d
546m108 re. August 26th. I cannot help. thinking horses ADMIRE a wide prospect.-- The very superiority of man
559m154 der each object an act of separate creation. we ADMIRE it more. because we can compare it to the stan
581n064 music, colours we must suppose <we> «Pea-hens» ADMIRE peacock's tail, as much as we do.-- touch appa
600o08v difference between moral sense & conscience? we ADMIRE what is right by one & are ordered to do it by
620o045 conscience. & palliates the offence; one always ADMIRE the habit formed by «obediance to instinct» <c
570n027 negress beautiful,-- [male glow worm doubtless ADMIRES female. showing. no connection with male figur
573n036 when he tells us satellites &c &c «The Savage ADMIRES not a steam engine, but a piece» of coloured g
573n036 steam engine, but a piece» of coloured glass <&ADMIRES> is lost in astonishment at the artificer.--»
529m041 is much imagination in every view. if one were ADMIRING one in India. & a tiger stalked across the pl
632j53r e to successive developement I admit, but the ADMISSION is probably from ignorance]CD Who would ever
230b239 difficulty permanently transmitted.-- <It will ADMIT> a plant will admit of a certain quantity of ch
230b239 tly transmitted.-- <It will admit> a plant will ADMIT of a certain quantity of change at once. but af
259c064 lakes are American form, that one is brought to ADMIT the possibility (any great change in species is
294c176 ds, (forgets authority).-- Lonsdale is ready to ADMIT, permanent small alterations in wild animals, &
442e151 may take place by gemmation «My theory will not ADMIT this, now that tulips break by cultivation can
599o007 and chorus are utterly inexplicable-- I cannot <ADMIT> think reason sufficient to give up my theory--
623o050 om our reason, that all which <has> (as we must ADMIT) has been acquired, does possess the beneficial
625o052 cation & hence are forgotten-- only so far do I ADMIT its supremacy p. 37. Whewells gives Mackintosh'
626o052 &c» Butler's view given on conscience: I cannot ADMIT it.-- see notes to it by me..' 'p. 333 «& p. 37
632j53r ted with reference to successive developement I ADMIT, but the admission is probably from ignorance]C
046r079 ux of melted matter.--Volcanos perhaps may be ADMITTANCE of water, through the rent strata: «Mr Lyell
369d115 understand their oeconomy, is now universally ADMITTED.""-- p. 483. Owen thinks from climate of Aust
459tf02 of he-goat & sheep, it seems male gives form. ADMITTED by Linnaeus.-- seems to doubt its applicabili
632j53r d as the Alpine pinnacles.-- One thing must be ADMITTED there would not be these plants, if there was
608o026 motive power not in proper state.-- When the ADMONITION succeeds who does not recognize an accidenta
198b115 a point of great interest to prove animals not ADOPTED to each country.-- Provision for transportal o
634j54v is the determination to adhere to a plan once ADOPTED; & it is from these very circumstances, that w
361d095 milar.-- one species retained this character in ADULT stage, other alters entirely]CD In common sparr
614o037 mparison of ideas.-- As man has so very few (in ADULT life) instincts.-- this loss is compensated by
276c123 ic men is to push their science a few years in ADVANCE only of their age. (differently from literary
276c123 whose opinion they believe have endeavoured to ADVANCE cause of truth It is of the utmost importance
432e124 e hairy character of his forefathers only when ADVANCED in age, & therefore the children do not, (& i
533m061 ng read or thought of some such remarke as now ADVANCED; for I caught it like a flash.--. strange if
543m098 probably some genera in different orders more ADVANCED than others just as dog & Elephant most intel
588n090 ity.-- I suspect very strong argument might be ADVANCED, that animals have reasons, because they have
052r098 matter.-- Give various cases. [Fig. 6] {P} A ADVANCING coast to Seaward. Retreating case in excess a
423e095 done so from the new relations caused by the ADVANCING complexity of others.-- It may be said, why s
248c029 New Zealand Rat belong to There is this great ADVANTAGE in studying. Geograph. range of quadrupeds.:
355d067 alled. there can be no prediction.-- The only ADVANTAGE of discovering laws is to foretell what will
431e123 o male parent p. 8. his whole doctrine of the ADVANTAGE of crossing consists in the idea of the male
436e137 et if a seed were produced with infinitesimal ADVANTAGE it would have better chance of being propagat
483z013 ge of shells might perhaps be worked out with ADVANTAGE. with Cumms collections & my own: & Capt. Kin
530m045 herefore that their children have some little ADVANTAGE in these trades.-- Delirium seems to rest the
614o037 wonderful as at first appears, & no too great ADVANTAGE.; for superiority of memory does not depend o
527m034 nvention) such castles in the air are highly ADVANTAGEOUS, before real train of inventive thoughts ar
623o050 ving love of mother; the having received some ADVANTAGES from man. during many generations giving the
633j53v 284. it is hard on my theory of gain of small ADVANTAGES thus to explain the curling of the valves of
133a141 resting on film of molten rock.-- Voyages of ADVENTURE & Beagle vol I. p. 2 & 3. Porphyry at St. Ele
448e166 ways the case according to Brown.-- Voyage of ADVENTURE & Beagle Vol I. p. 306 Shells, as well as pla
419e085 f infinite numbers of individuals from one, is ADVERSE.-- Decemb. 25th.-- Lyell says the elevated she
327c265 arh. 20t. 1839. Philosophy of Blushing lately ADVERTISED. /6s Mrs Necker on Education preeminently wo
502q11v eeds raised.-- His Book.-- 32. Would wheat from AEGYPT ripen in Scotland?-- to show acclimatisation.-
172b006 is instinct for opposites to like each other AEGYPTIAN cats & dogs ibis same as formerly but separat
285c153 ally rapid, with respect to their effects The AEGYPTIAN animals domesticated «??», & therefore Most e
042r068 out, after rise from sea: <As did> as did those AERIAL Volcanos in Germany» In the Valle del Yeso it
268c100 each great class of animals having its aquatic, AERIAL &c type?-- This of consequence, because applic
273c111 st <m> strongly marked, there is a preeminently AERIAL,-- formed for flight & great movement in the a
273c112 is one most remarkable connection between these AERIAL representatives of the different families.-- t
273c112 hich is most wonderful of all. ¿whether in most AERIAL of swallows.» Milvulus, & still more wonderful
274c115 not discover any other clear relations besides AERIAL, & terrestial,-- How is it in water birds, the
283c143 e groups. where you have representations.-- The AERIAL type in each family is relation to elements &
058r115 lustre» ask Erasmus. whether electricity would AFFECT this. -- State the circumstances of appearance
234b261 iar to any one country do not species generally AFFECT different stations;-- this would be strong arg
248c033 hybrids.-- As we see external influences first AFFECT external [for]m, so will the internal parts be
299c194 ormity Primrose & Cowslip, quite wild, but they AFFECT different localities,-- latter on banks & in d
304c206 e generally perfect organs. do changes of habit AFFECT particular organs.-- of two adjoining families
343d036 organic world, as adaptation. & these changing AFFECT each other, & their bodies, by certain laws of
417e079 same species.--). some races of men. D'Orbigny. AFFECT the common progeny more than others.-- does th
426e106 s, & the fact that old varieties do not so much AFFECT first race, as it does indelibly the many subs
446e163 milarity of conditions-- & that no change would AFFECT them in short period & hence no change would e
491q01v -- does one branch of Cabbage being mongrelized AFFECT other branches-- The French Apple tree «with a
536m071 king udders of mare strictly analogous to men's AFFECT for womens breasts.∴ Dr Darwin's theory probab
544m104 uce children. or after he has useless. does not AFFECT race. argument for early education.-- fear of
574n042 y intelligible An habitual action must some way AFFECT the brain in a manner which can be transmitted
617o39v iderable degree of attention. How do the senses AFFECT us, except by internal consciousness 3) We mus
624o051 dance to beneficial tendency» will most readily AFFECT. the instincts, for they are in accordance wit
043r070 <At> Lyell. Vol I. P. 316. Earthquake of 1812 AFFECTED valley of Missisippi & New Madrid & Caraccas.
061r127 then for any spot of land. = Yet new creation AFFECTED by Halo of neighbouring continent: ≠ as if an
102a059 rtow would draw it outwards.-- form of breaker AFFECTED some way out to sea.--¿ effects on bottom a t
104a066 lways produce it) but where great thickness is AFFECTED, they would be far off In Discussion on dikes
248c030 inct not interfered with, or generative organs AFFECTED as with plants) no animals VERY different wil
248c033 mportant structures. <which are less obviously AFFECTED by external circumstances> these therefore wi
259c064 ange of species to species «although we see it AFFECTED» tempts one to bring one back to distinct cre
441e149 e curious law, Certainly Australian Dog is not AFFECTED by domestication, yet offspring are,-- if Aus
466t096 ants-- also goodness of flavour in fruit-- all AFFECTED by cultivation during life of individual. Jun
491q01v dise EVERY flower on melon & see whether fruit AFFECTED. Mr. B. seemed to say impregnation <caused> o
491q01v explains apples on side near other tree being AFFECTED.-- does one branch of Cabbage being mongreliz
523m018 others from drinking cold drink.-- then brain AFFECTED like getting suddenly into passion.-- There s
525m025 uthority of Mr Wynne that bitch's offspring is AFFECTED by previous marriages with impure breed.-- A
526m027 ers or teaching of others.-- (NB man much more AFFECTED by other fellow-animals, than any other anima
526m027 than any other animal & probably the only one AFFECTED by various knowledge which is not heredetary
530m044 cidents:-- After journey, a fit of = gout, has AFFECTED his memory of everything in <he a [...] Mr B>
532m054 ry fear) In these cases probably the system is AFFECTED, & by habit the mind tries to fix upon some o
536m073 organization. that organization may have been AFFECTED by circumstances & education, & by choice whi
557m149 shame frowning, & anguish,-- shyness not so.-- AFFECTED laughter.-- A dog who goes home from shooting
575n043 brain.?-- is this more wonderful than memory. AFFECTED by diseases. &c &c, double consciousness? Wha
584n073 inful thought crosses mind, before it can have AFFECTED respiration V E. p. 125 Wrong Entry Madagasca
590n093 <grief> sighing comes on before circulation is AFFECTED. p. 44.-- Jealousy. causes spasm in bile duct
596nIBC convulsions. are the muscles of the face first AFFECTED?-- Can shivering & trembling be considered co
058r118 omme une ceinture d'immenses remparts; toutes AFFECTENT une pente plus ou moins inclinée vers le riva
137a149 the effects of veins of slag in iron furnaces AFFECTING to some distance & blending with sandstone «s
191b082 r life, is in accordance with The Male animal AFFECTING all the Progeny of female insures often mixin

(Key Word)

213b171 ounds?-- Yarrell's remark about old varieties AFFECTING the cross most well worthy of observation.--
239cIFC y close species ever yet failed. About trades AFFECTING form of man. Could you get racehorse from Car
398e004 of small physical changes & oscillations, not AFFECTING organic forms, that the whole value of the ge
404e026 , or some one mentions of influence on parent AFFECTING offspring.--\ & as adaptation,-- however myst
410e051 parts, from the laws of variation of one part AFFECTING another.-- (I from looking at all facts as in
419e087 geology rests upon amount of physical change <AFFECTING whole bodies of species>, & only secondarily,
446e163 & hence no change would effect them, without AFFECTING all the individuals-- «-- hence there would b
567n014 clasping «& rubbed» his arm. & show signs of AFFECTING something like man. Has an oyster necessary n
641j29v Articulata, & Vertebrata, & Planaria, & light AFFECTING plants. in insects the end is gained by some
264c079 as if it understood every word said-- see its AFFECTION.-- to those it knew.-- see its passion & rage
522m013 or delirium.-- In Mania all idea of decency & AFFECTION are lost.-- most delicate people do most inde
532m056 to her old ways & became fat! What remarkable AFFECTION to a place.-- How like strong feelings of Man
549m119 e free from offence» -- pleasure of intellect AFFECTION excited, pleasure of imagination-- therefore
557n150 emble, like fear.-- Why does any great mental AFFECTION make body tremble. Why much laughter tears.--
565n009 kiness all negative expression? Expression of AFFECTION is accompanied by slight protrusion of lips,
577n052 apt to blush.-- -- The power of vivid mental AFFECTION, on separate organs most curiously shown in t
594n115 th another large dog his companion. Descent --AFFECTION & [...] Monkeys «Ogleby» seen Zool. Soc-- 183
614o037 , as love of virtue, of association, parental AFFECTION-- The very existence of mankind requires thes
228b231 make the black man other kind?» Animals with AFFECTIONS, imitation, fear <of death>. pain. sorrow fo
298c190 comes habitually passionate.-- the Key to the AFFECTIONS might perhaps thus be found-- a person who i
359d087 which is often the case, & why should organic AFFECTIONS always influence the sexual organs alone.--
524m020 closely analogous to Epilepsy & convulsion.-- AFFECTIONS of the thinking organs \ -- the action of br
525m026 edetary in any one family.-- In Aunt-- B. the AFFECTIONS «& N B affections very soon go in Maniacs» s
525m026 family.-- In Aunt-- B. the affections «& N B AFFECTIONS very soon go in Maniacs» seem to have failed
525m026 failed even more than the memory.-- therefore AFFECTIONS effect of organization which can hardly be d
593n109 her men. & hence idea of beauty.-- the social AFFECTIONS of animal taking man in place of other anima
604o017 ot quite perceive drift of Book.-- Sympathy & AFFECTIONS chiefly fail.-- Notices. struggle <between>
627o52v Butler & arguments of beneficial tendency of AFFECTIONS.-- If ever I write on these subjects consult
627o053 to point-- clearly shows this is true. p. 13. AFFECTIONS cannot be analysed into "power" &c &c &c-- &
628o054 e, which gives the instinct.--]CD p. 22. says AFFECTIONS, desires, & moral sense all different.-- P.
629o54v ess will always follow it.-- Would not the maternal AFFECTIONS (in a dog. & therefore not <instinct> «consc
332d001 did not see other cases.-- He thinks apoplexy AFFECTS people all over England at same periods When h
389d176 ion may impregnate one or many offspring.-- it AFFECTS the subsequent offspring, <when> though other
443e153 eties as we know that the kind of stock greatly AFFECTS the Graft.-- \(Plants circumstanced as the Gorz
080r178 the character of a Araucarian tribe, with point AFFIN of yew & intermediate Puncture one animal with
074r165 mboldt has urged phenomena in veins, chemical AFFINITIES like in composed rock. granites syenite» «st
181b043 ater between animals & Plants But yet besides AFFINITIES from three elements, from the «infinite» var
227b228 look to first germ-- --led to comprehend true AFFINITIES. My theory would give zest to recent & Fossi
281c139 d linear arrangement the central twigs dying, AFFINITIES would be <circular>, in broken circles.-- wh
291c165 or it may be heredetary & strictly point out AFFINITIES. conducct of Gould, remark of D'orbigny poin
292c170 cial knowledge «like myself» of <classi> real AFFINITIES. ie structure of the whole animal let him re
294c177 species. & & its adaption to classification & AFFINITIES, its extension.-- Von Buch. Travels, p. 306.
338d026 eing distributed after flood (!) according to AFFINITIES\ confounds, like Whewell affinity with anal
348d050 il groups natural (p 6) as expressing natural AFFINITIES\ Macleays plan of arrangement depends on the
348d051 in point." Now what is natural arrangement,-- AFFINITIES, what is that, amount of resemblance,-- how
421e091 affording very clear indications of its true AFFINITIES. We are least likely in the modifications of
553m138 part in structure» The monkeys understand the AFFINITIES of man, better than the boasted philosopher
176b024 ee elements, there will be certainly points of AFFINITY in each branch A species as soon as once form
204b139 mb or Octob 1837 Westwood has written paper on AFFINITY and analogy in Linnaean Transactions Mr Wynne
211b162 emarked Mere length of bill does not <indicate AFFINITY with snipes> indicate affinity, because simil
211b162 s not <indicate affinity with snipes> indicate AFFINITY, because similar habits produce similar struc
211b162 Rhyncus Would not relationship express, a real AFFINITY & affinity-- whales & fish.-- Progeny of Mank
211b162 ld not relationship express, a real affinity & AFFINITY-- whales & fish.-- Progeny of Manks cats with
258c062 n of Analogy. All the discussion <after> about AFFINITY & how one order first becomes developed & the
264c081 hout corresponding habits clearly showing true AFFINITY, for instance tail of ground woodpecker-- --
269c103 "Carus on the Kingdoms of Nature, their life & AFFINITY" in Scientific Memoirs I can see that perfect
274c115 observed even in Lessonia &c & In relations of AFFINITY all organs change together, in analogy certai
282c140 ach other, but to some external contingency.-- AFFINITY is the sum of all the relations, analogy is t
282c142 to rest of its family as in ground cuckoos, AFFINITY with respect to species of each other, becaus
282c142 lation of all the ground cuckoos. would not be AFFINITY, but the truth would never be discovered When
285c151 be first those of analogy, but will grow into AFFINITY.--but whether ever arrive at true affinity do
285c151 nto affinity.--but whether ever arrive at true AFFINITY doubtful A species is only fixed thing with r
289c162 ower developed forms.=» (NB waterhouse says of AFFINITY of many insects may be told by their larvae)
302c201 animal at present time having an intermediate AFFINITY between two classes.-- there may be some desc
302c201 gy, which when sufficiently Multipliied become AFFINITY yet often retaining a family likeness, & this
340d026 ording to affinities!. confounds, like Whewell AFFINITY with analogy-- Good table at end of distrib:
348d051 f difference is or can be settled,-- I believe AFFINITY may be taken literally,, though how far we ca
422e092 or transformed. & hence not indicative of true AFFINITY.-- -- Owen says Dugong connected with Pach
435e133 odgkins» p. 54. The axolotl, siren, & Proteus, AFFINITY to tadpoles. p. 210. Shows. that the action o
556m146 ing.-- -- good case of expression showing real AFFINITY in face of monkeys, horse & zebra. when going
038r059 have been elevated considerably. which shows an AFFLUX of inferior melted rocks to those parts. Are n
045r077 rea. we may believe the fluid matter instead of AFFLUX (always slightly oscillating as that of a spri
046r079 o ground, that the first phenomem. is an inward AFFLUX of melted matter.--Volcanos perhaps may be adm
046r079 olcanic. I consider latter as accidental on the AFFLUX of the former. -- Ascension. Vegetation? Rats
254c047 n remarks on Entozoa, the organs of generation, AFFORD the least certain indication of the perfection
287c158 ls.-- + whether variations in eye of vertebrate AFFORD better character, than variations in eye of mo
431e121 -- L. doubts.-- Lonsdale thinks Ammonites would AFFORD instance of such facts.-- Ask Phillips.-- The
635j56r . 420 thinks the great fecundity of germs is to AFFORD support to other beings.-- true, (& the doctri
640j167 e long legged race would prevail, even if have AFFORDED only 10th part before & now formed eighth par
284c149 cies, it is manifestly of less importance, as AFFORDING natural characters than among those groups, w
421e091 food of an animal, I have always regarded as AFFORDING very clear indications of its true affinities
068r148 California. Beagle Channel.--One need never be AFRAID of speculating on the sea The 24 ft. elevation
376d137 them try to pull up petticoats, & if woman not AFRAID clasp them round waist & look in their faces &
532m054 the night. being slightly unwell & felt so much AFRAID though my reason was laughing & told me there
557n149 who goes home from shooting. runs away. is not AFRAID the whole way. but ashamed of himself.-- Jealo
565n007 air «& raises its head, & pricks its ears» when AFRAID, though not every time really wishing to smell
041r064x ect of Salt water of the Salado.--Mem. in Owens AFRICA it is mentioned that the Elephant came towns d
044r072 d geology of N. America. India.--remembering S. AFRICA. Australia.. Oceanic Isles. Geology of whole w
044r073 1 distribution of Volcanic action, Australia S. AFRICA-- on one side. S. America on the other: The ex
048r085 the country in which the Rhinoceros lives in S. AFRICA. the same caution is applicable to the Siberia
048r086 individuals as in the half desert country of S. AFRICA. It would be well to quote Burchell. V. where
050r091 Is there not a sudden deepening on E. coast of AFRICA. as at Brazil [blank] What is nature of strip
051r097 cifick, as compared to whole E. America. <East> AFRICA. Australia. profoundly deep: a great fault or
057r113 use too cold:--With discussion of camel urge S. AFRICA productions.-- I think in Patagonia white beds
091a027 al said so. «Campbell the Poet» Accra. Coast of AFRICA. Clay Slate & Quartz. strike SSW & NNE dip 30
187b062 S. America.-- explains how Zebras reached South AFRICA-- It is a wonderful fact Horse, Elephant & Mas
189b072 se all die. == The fossil horse, generated in S. AFRICA Zebra.-- & continued.-- perished in America Al
190b077 to Southern hemisphere, does not look as if S. AFRICA peopled from N. Africa An originality is given
190b077 , does not look as if S. africa peopled from N. AFRICA An originality is given (& power of adaptation
191b083 ary event. & succession the extraordinary South AFRICA. proof of subsidence. & recent elevation: Pray
194b094 be made to approach the Colobes which in south AFRICA, appear to represent the semnopitheque of Indi
200b121 ility of common branching isld.? Accra, Coast of AFRICA. Clay slate. strike. SSW. & NNE. dip 30 degree
202b133 horse in S. America. or like living Edentata in AFRICA &c &c.-- Now if suppose world more perfectly c
209b157 ,3: Mem. Lyell on shells.-- {T} Genera In North AFRICA. I: 4,2 Iles Canaries I: 1,46 St. Helena I: I,
212b167 ng to periods.-- NB. Was Europe desert (like S. AFRICA) after Coal Period.-- ¿In those divisions of m
218b190 em. Capt. Owen's story of cats on West coast of AFRICA.-- changing hair-- The Edinburgh. Journal of N
219b194 adaptation because volcanic isld. whilst <neig> AFRICA, sandstone, & granite, (that is genera near Ca
226b220 Madagascar very good.-- Fernando Po & Coast of AFRICA. equally good.-- Small isld off New Guinea sam

```
*****************************************(Key Word)**************************************************
226b223 guar been in S. America.-- <East.> «W» Coast of AFRICA & <West.> «E» of America, ought to present gre
232b249 n neighbouring islets & a sub-genus in Southern AFRICA In same manner. Cuscus, (a sub genus of Phalan
242c018 from  S. America & identical with those from S. AFRICA. «M. Bibron doubts fact.--» My toad is same sp
244c021 e hairless kind, «said» to come originally from AFRICA p. 122. Mus decumanus, at Caroline Isld, & a R
250c037 continent.  Try amongst Europaean quadrupeds if AFRICA destroyed would not then some forms be peculia
250c038 in  Europe & America Some portion of the world «AFRICA» being left more equable. yet America preemine
250c038 h species to have been formed & spread to other AFRICA & East India Arch.-- but where these great ani
250c038 Marsupial  & Edentata increased most. Certainly AFRICA approaches Nearest to what is supposed to have
251c040 calities of certain parrots habitations India & AFRICA.-- NB. Any monograph like Gould on Trogons wor
253c046 separate isld very long» America & India deer.= AFRICA not.-- Africa Camels?? Africa Bears??-- Planti
253c046 very long» America & India deer.= Africa not.-- AFRICA Camels?? Africa Bears??-- Plantigrade Carnivor
253c046 ca & India deer.= Africa not.-- Africa Camels?? AFRICA Bears??-- Plantigrade Carnivora??-- «compare r
274c116 es certain <imper> parts changed Have <not>. S. AFRICA, Australia, & S. America very few forms in com
278c132 fitting country,-- if Toxodon had been found in AFRICA, the wonder have been same for S. America & Eu
278c132 mals not preserved, in central S. America & yet AFRICA & India???-- & Indian Islds.-- Sir J. Sebright
310c224 pecies Himalayas, 13,000 & Melville Isd.-- West AFRICA & India some plants same. --America.-- See Bro
311c227 's Voyage to Surat, floating isld. off coast of AFRICA .p 69. with tall grass. & p 72 hairy sheep-- E
313c233 & Europaean. different.-- thorax & head differ AFRICA Australia Parasites die, when brought over  on
316c246 s there are some shells common to West coast of AFRICA & E. S. America.-- get instances.-- very good
317c249 nkey of St Jago C. de Verd; same as on coast of AFRICA.-- Macleay tells me same thing p. 55. 40 leagu
322c269 velyns Sylva. skimmed, stupid Brownes travel in AFRICA; «well skimmed.» 1839 Jan 10t.-- All life of W
338d025 of  Hyaena? Hippotamus.? Indio-African, or pure AFRICA?-- ‖Fossil Elephant of Africa Most important u
338d025 -African, or pure Africa?-- ‖Fossil Elephant of AFRICA Most important under this view, & Hippotamus o
343d038 re becoming extinct. others though the negro of AFRICA is not loosing ground. Yet, as the tribes of t
343d038 ch member from slave trade, & colonization of S. AFRICA, so must the tribes become blended & prevent t
347d050 st-- 29th.-- Macleay in A. Smith's Zoolog.-- of AFRICA.-- p. 4. sticks to genus or group of any  kind
360d089 «twice made» between terrier & hairless dogs of AFRICA,-- some puppies hairless. some in patches, & s
407e039 ine change..-- Therefore I argue from this that AFRICA «& East Indian Archipelago»-- formerly were no
407e041 state  of vegetation, & conchology,-- shells of AFRICA ought most to resemble fossil ones of  Europe,
407e041 connects with Asia» between two polar lands,--; AFRICA not so equatorial..-- The fact of No. Mam: Pla
413e060 truly  bisexual. Buckland's Reliqu: Diluv. says AFRICA only place, where, Elephant, Rhinoceros, Hippo
414e064 in animals chiefly organization: though Cont of AFRICA & West Indies shows organization in Black Race
420e088 p. 414; M. Guyon thinks Monsters more common in AFRICA than in Europe especially with Europaeans sett
421e090 y Isld approaches more to neighbouring coast of AFRICA, than to other parts of that continent. in lik
421e091 like manner as Madagascar does to other side of AFRICA.-- (& Juan Fernandez to Chile??) Falklands  to
448e167 t there are Baboons in St Thomas on W. coast of AFRICA Owen Linn. Soc. April 2d. 1839 The Lepidosiren
459tflr tions, Boteler's Narrative Voyage East coast of AFRICA-- Vol II. p. 256-- wild cattle at Madagascar--
465t079 s country-- same argument to India & Europe-- & AFRICA!.-- any negative argument against-- monkey-man
471tfl1 rises Dr Andrew Smith says in the larks from S. AFRICA he can almost make series from end to end-- so
501g10a hard  in Europe, Walnut in America.-- Heaths in AFRICA; Hooker? are these genera less difficult, in o
540m086 ns of the Amazons & Brazil-- with the negros of AFRICA, (or again the black man of <B> Van Diemens la
029c032 says.  that Boulders do not occur in the South AFRICAN plains.-- Sydney no I believe the secondary? f
212b167 e Water Rat.-- ¿Consult Dr Smith History of S. AFRICAN Cattle Phillips Geology «p 81» in Lardens Ency
218b192 is given Mr Don remarked to me. that some good AFRICAN & some good S. American forms. (& daresays som
218b192 S.  American forms. (& daresays some of these <AFRICAN forms> forms would have some peculiarity.-- No
219b194 he great ocean, have made plants of American & AFRICAN form, merely because intermediate  position.--
225b218 an with gun. & Bustards &c &c!!! An American & AFRICAN form of plant being found in Tristan D'Acunha.
231b242 w by any means.-- Ostriches.-- Hippotamus only AFRICAN.-- American & African forms mingle in India  &
231b242 iches.-- Hippotamus only African.-- American & AFRICAN forms mingle in India & East Indian islds-- Mo
236b278 ordering continents same type collect cases.-- AFRICAN isld.-- «How is Juan Fernandez-- Humming Birds
295c183 ver old mare by being shown, young one.-- Many AFRICAN monkeys in Fernando Po-- no new forms only spe
296c184 Canary islands-- Endeavour to find out whether AFRICAN forms. (anyhow not Australian) on Peaks. Did C
312c233 fferent species to different,-- inguinal louse AFRICAN & Europaean. different.-- thorax & head differ
325c267 v. (good to trace Európean forms compared with AFRICAN <A> Annals Histoire Generalle et  particuliere
338d025 ely the fossil Mamalogy of Britain & Europe is AFRICAN. & the only difference is by the extinction of
338d025 -- what is range of Hyaena? Hippotamus.? Indio-AFRICAN, or pure Africa?-- ‖Fossil Elephant of  Africa
356d072 g crossed in wild state-- & the English & Some AFRICAN dove.-- The extinction of the S. American quad
416e072 tichorhinus» in Paris basin.-- its relation to AFRICAN Species <good observations.-- >, larger than a
484z014 bigny, that Serpent Eater-- or Secretary is S. AFRICAN representative of Caracaras of Americas.-- man
509q017 a Leone. cow. taking bulls. is it Domesticated AFRICAN Animal= Knows nothing [<...>] It is very impor
514q21. so differ in different countries-- on flora of AFRICAN Islds-- names of Plants found on mountains  of
603o012 rious. Chinese, S. American. Polynesians Jews, AFRICAN all sacrifices. How completely men must have p
032r042 especting the brecciated white stone of Chiloe, AFTER having examined the changes of pumice at Ascens
042r068 oreover, the Volcanos from sea there burst out, AFTER rise from sea: <As did> as did those aerial Vol
044r074 me geologically to vertical movements. In Cord: AFTER seeing small Bombs. without a vesicle. we may c
047r083 fall must vary proportionally Partial shrinking AFTER elevation in perfect conformity with «Mr Lyell'
058r116 istance of more than 500 leagues. A little time AFTER a bad earthquake in Chili; Areguipa in 82 was o
068r146 ing injections.--Old vents would keep open long AFTER emersion, but improbably long, that to be su
074r165 n.--metallic veins follow mountain chain. there AFTER NW <W>.-- «same chemical laws as in concretions
092a030 6 1837 High up the Essequibo, granite & quartz, AFTER passing sandstone Vol II. p. 69.-- Geograp Jour
102a062 mountain chains. accompanied by do.-- Give this AFTER supposition p. 461 «of Proceedings» List of col
106a076 hquake unaccompanied by Volcanos must be sought AFTER proofs of sinking.-- No Sweden!! swelling of ro
108a080 izontal oscillation. or so many shocks directly AFTER great shock -- It appears to me unphilosophical
151g042 plain of other, which must have been waterwork AFTER 3d lake.-- 4th shelf runs up some way on  great
181b041 eat against, any two of the 12. having progeny. AFTER that distant period.-- Hence if this is true, t
209b155 ies made by isolation; then their distribution (AFTER physical changes) would be in rays-- from certa
212b167 riods.-- NB. Was Europe desert (like S. Africa) AFTER Coal Period.-- ¿In those divisions of molluscs.
216b181 ton, where mare was influenced in this cross to AFTER births, like aphides.-- Case of boy with foetus
217b184 ring quite intermediate sometimes take strongly AFTER either parent. about <half & half tim> as often
217b184 -- Where two dogs have lined bitch directly one AFTER the other, puppies differ, & like both parents.
223b209 s,, as physical changes are gradual, is this if AFTER isolation (seed blown into desert) or separatio
240c004 tary. other adaptation.-- Mr Yarrell says, that AFTER breeding in pidgeons with very much care  that,
244c023 - Waigiou Speaking of Lepus Magellanicus says; <AFTER> "après un examen attentif, et forts surtout de
251c040 of latter from <Tarton> Barton.-- swifts return AFTER years to nest Vol II. p. 49. on the localities
258c062 for  definition of Analogy. All the discussion <AFTER> about affinity & how one order first becomes d
258c062 ording as parent types are present) must follow AFTER there is proof of the non creation of animals.-
265c083 ith other varieties»-- but <accidental> changes AFTER birth do not effect progeny-- many dogs in Engl
265c086 is  possible to have structure without habits-- AFTER seeing beetle with wings beneath soldered wing-
266c092 831. p 160. account of Bulbous root from Mummy: AFTER 2000 years, germinating.--!! Henslow doubts? GE
267c095 5. roe of Asterias in stomach.-- of Sammon remain AFTER rest of animal digested.-- Important do p.  98.
269c103 nder-- spontaneous generation not improbable.-- AFTER reading "Carus on the Kingdoms of Nature, their
285c151 e which one branch of the tree if live occupied AFTER its decay, will be occupied by the vigorous sho
286c156 carcely any traces of passage a difficulty, but AFTER all a slight one It will be necessary from mann
293c172 might  yet be transmitted.= Memory springing up AFTER long intervals of forgetfulness.-- after sleep,
293c172 ing up after long intervals of forgetfulness.-- AFTER sleep, «strong» analogies with memory in offspr
293c173 an by effort of Memory can remember how to swim AFTER having once learnt, & if that was a regular con
313c235 pregnation, therefore sexual passion must arise AFTER long interval very good case.-- habit is awaken
316c244 tions.-- +York's Minster story of storm of snow AFTER his brothers murder.--» good anecdote.-- Sowerb
332d003 ry thinks the half bred Alderney Cows take more AFTER Alderney that the Durham,, with which they have
333d007 s crossed with tame, offspring always take most AFTER wild.-- i.e that <alw> «no» domesticated ones h
333d008 dogs & Fox thinks they decidedly take much most AFTER Esquimaux.-- this agrees perfectly with Yarrell
339d026 any  improbability to animals being distributed AFTER flood (!) according to affinities!.  confounds,
350d054 mere conjecture?-- Australians.-- Americans. &c AFTER Decandolles idea Septemb. 1st. Macleay & Broder
359d087 e very wild & take <very little> in disposition AFTER their «pheasant» parents.-- (There are some 3/4
362d099 ious the rapidity of the change in 5 or 6 weeks AFTER castration, fresh horns begin to grow.-- Mr Yar
363d102 two  redpoles can hardly be told apart, so that AFTER di‾ferences were pointed out Selby confounded t
371d128 they  produced not be transplanted', & yet year AFTER year, successive roses & bud are produced, like
380d155 -- (as is seen in <fe> plumage of hybrid birds) AFTER animal has copulated., though no offspring, Mil
```

Page **(Key Word)**
392d175 hat difference need not be added EACH TIME. but AFTER some time]CD What kind of plants are Monooeciou
392d180 chs & live? In Muscovy ducks do young take most AFTER father or Mother according as they are crossed?
406e035 most grave source of doubt. in distinguishing <AFTER> which parent impresses offspring most is wheth
410e051 effected by sexes: All the above should follow AFTER discussion of crossing of <species> individuals
432e123 inese boars &c &c &c» p. 10 offspring take more AFTER father than mother; illustrated by the crossing
437e140 ought to her young ones the cub of a fox, which AFTER it had fought well & desperately bitten the you
449e169 en the male Chinense & female common goose took AFTER the common goose thus contradicting (probably)
450e175 f y Burramposter & S of Tropic-- By J. H. Moore AFTER quitting Bengal this fact is noticed in Cassay
459t013 of pecu> «drake» with the penguin duck. it took AFTER the Penguin in the form of its body & in the ma
460t013 e half of the cross, as above, take «generally» AFTER the swan-gander. one of these half-bred ganders
461t037 hence, the application of structure to purpose AFTER purpose would tend to render complex the series
466t096 king only hereditary characters, wh. come on in AFTER life of Plants-- also goodness of flavour in fr
468t112 ctary is same case as Azalea or Rhododendron xx AFTER several gloomy days. hot one, Bees almost P eve
468t112 at all were brushed by Bees & especially stigma AFTER bee had brushed over the anthers of long stamen
470t176 ated was introduced about '65 years ago--& soon AFTER mules abounded--so that palmated has now nearly
472s02r y similar-- June 2. 42 Maer <Thursday> Thursday AFTER watching 14 days. many times every day. many cl
472s02v fy, as if sucked?! opens & shuts end of sucker, AFTER having withdrawn it.-- Saw 4 more Bees at work-
493q003 important. {In crosses does male offspring take AFTER male parent & vice versâ = History of Tortoise-
496q05a s flaccid, as Koelreuter describes Kill Sparrow AFTER feeding on oats, give body to Hawk & sow pellet
504q013 5) Examine the Parnassia whose stamens move one AFTER other to flower & Menyanthes whose pollen burst
508q016 swered (5) About cross-bred races of men taking AFTER sex. A Smith. About species of Rhinoceros. beco
515q022 These Hybrids differ in colour of beak, taking AFTER male & female parent.-- Will they grow up in ot
516q24v isons & see in what order plants would reappear AFTER <th> being killed Experiments not connected wit
520m001 her has seen innumerable cases of people taking AFTER their parents, when the latter died so long bef
521m009 itto.-- Case of Mr Corbet of the <Hall> «Park», AFTER paralytic stroke. intellect impaired. <after pa
521m009 », after paralytic stroke. intellect impaired. <AFTER paralytic stroke> : . could converse well on an
523m015 ocket, & studying mathematics.-- My Father says AFTER insanity is over people often think no more abo
523m016 High Ercall, the thoughts of it, for some years AFTER, was far more painful than the thing itself. As
524m021 od, (but appear most capricious) as in delirium AFTER epilepsy, but in the failing from old age, they
530m044 y in sleep giving directions,-- & forgetfulness AFTER bad accidents:-- After journey, a fit of = gout
530m044 tions,-- & forgetfulness after bad accidents:-- AFTER journey, a fit of = gout, has affected his memo
532m056 ould not go out of house except with Caroline-- AFTER fortnight. continued to grow thin & did not see
539m081 was very much struck with an intense headache «AFTER good days work» which came on from reading «rev
539m082 ideas of FitzWilliam Musm. I was amused at this AFTER seven years interval. Augt. 15th. As child gain
539m083 instance of heredetary mind. I a Darwin & take AFTER my Father in heraldic principle. & Eras a Wedgw
542m094 tions.-- CD[like sighing to relieve circulation AFTER stillness.-- Now I conceive if organization wer
544m104 appens to man who does not produce children. or AFTER he has useless. does not affect race. argument
556m144 n account of blood. but all these idea came one AFTER other, without ever comparing them, I neither d
559m156 time every day.-- naturally close at that time AFTER long period.-- My Father about double conscious
563n001 owl. on the ground, at roost, in all seasons, & AFTER? he has done <g> crowing.-- instances of expres
570n025 xed with triumphant feeling so similar to shame AFTER asinine.-- both accompanied by depending head.,
570n025 skin.-- What difference is there between Squib AFTER having eaten meat on table, & criminal,-- who h
577n050 of the leaves, comes on from want of stimulus, AFTER certain other actions, & hence becomes associat
584n072 instinctive: the facts of memory of roads long AFTER once visited by horse & dogs. (even blind horse
625o50v evolence (but not cool malevolence). it is only AFTER reason comes into play that anger can be said t
640j28v is no way of eliminating the evils of old age, AFTER breeding season, or gaining adaptations, but fo
060r124 t retain same level) to a greater distance".--AFTERWARDS speaks of this phenomena in connection with
078r175 nte 85 degree to S. // Tasco 40 degree to NW (AFTERWARDS said to be «all with some exception» directe
096a040 ew Britain-- &c &c In Ascension for centuries AFTERWARDS it might be percieved on which side craters
152g047 ation from Axis, then rivers might deposit, & AFTERWARDS with greater cut through, not applicable to
199b120 tile> <,but producing> «before domestication, AFTERWARDS none or little with <fertile offspring> » fe
217b183 d her & produced a very large litter.-- never AFTERWARDS went in heat.-- This is good instance of sam
230b239 of a certain quantity of change at once. but AFTERWARDS will not alter. This need not apply to very
293c172 was given a Great coat & this he put on & we AFTERWARDS could understand «(language better instance)
294c175 d one was diseased in its loins & all were so AFTERWARDS, (forgets authority).-- Lonsdale is ready to
298c191 by crossing) with alpine form) lower species AFTERWARDS would probably often be destroyed.-- or regr
332d001 ication, were seized with it, & for ten years AFTERWARDS, he then did not see other cases.-- He think
334d012 cow walked in, then disappeared, & three days AFTERWARDS came again, bringing with her the other & yo
364d104 breed of cattle with white heads; which years AFTERWARDS occasionally went back-- (Effect of imaginat
371d118 e simplest transmission is direct instinct. & AFTERWARDS enlarged powers to meet with contingency.--
384d162 -yet may be presumed from hybridity of ferns) AFTERWARDS they can be seen distinct. (in dioecious pla
411e055 ns the laws probably would be. generalized, & AFTERWARDS by the examination of the special cases, und
451e178 by Heber) «from» which «all» mankind «(& yet AFTERWARDS says native tribes can live there)» flee dur
491q01v y pollen of different genus & then some hours AFTERWARDS of nearly related plant & see if first polle
522m012 ot take any effect at the time, but some time AFTERWARDS it calls up pain. or pleasure. & is often re
525m025 t «they all died»:-- she had kittens before & AFTERWARDS with tails. My father says, perfect deformit
530m045 sleep--; some doctors care it, by stimulus & AFTERWARDS patient sinks.-- When a muscle is moved very
546m111 per cutter, hunted in vain for it-- ten years AFTERWARDS whilst at a meal, she suddenly like a flash
586n079 efore experience or habit) could be formed or AFTERWARDS.-- child sucking whole wonder instinctive.--
620o044 in its nature is only temporary, & we do not AFTERWARDS think of it.-- Whatever the cause of this ma
620o044 truck in passion fades away, so that when man AFTERWARDS thinks why was such an instinct not followed
031r037 lts in Patagonia[,] enormous extent; if lowered AGAIN & covered no sign of upheaval To Cleavage add o
038r057 n when whole summit of mountain is blown off; & AGAIN when in great crater. different little craters
170bIFC in All useful pages cut out Dec. 7th. /1856/ (& AGAIN looked through April 21 1873) p. 26 30 41 46 50
171b003 on here seems a means to vary. or adaptation.-- AGAIN we <believe> «know» in course of generations ev
177b025 s dead; so that passages cannot be seen.-- this AGAIN offers contradiction to constant succession of
185b056 abidae, Crysomela, Scarabadae, & longicornes.-- AGAIN taking a subdivision of Heteromera same. thing
203b135 o Zoological provinces-- united-- & now divided AGAIN-- Weakest part of theory death of species witho
217b187 ies of Australian genus being found in Sumatra; AGAIN another of other Genus in Sandwich islands-- A
228b230 Ask Henslow for some plant, whose seeds go back AGAIN, not a monstrous plant, but any marked. variety
234b256 from. Devonshire, from another from Swansea.-- AGAIN Waterhouse finds certain varieties of a Harpalu
234b261 be strong argument for propagation of species-- AGAIN is there not similarity even in quite distinct
249c034 of[fspri]ng or such as not capable of producing AGAIN The Varieties of Cardoon are cases <sp> like th
259c065 have no tail (example probably not true).-- or AGAIN healthy parents have healthy children the other
265c085 mpt of nature, colouring matter being absent.-- AGAIN dwarf plant in alpine district & dwarf plants f
299c196 re instincts in rodents than in other animals & AGAIN in Mans mind, in different races, being unequal
306c212 long in a passion & looked out for him to come AGAIN very differently from dog. perhaps being in pas
333d007 resting as Esquimaux dog approaches to species. AGAIN he has seen several crosses between Esquimaux d
334d012 then disappeared, & three days afterwards came AGAIN, bringing with her the other & younger cow.-- {
340d026 trib: of <birds>. Anatidae.-- Consult this book AGAIN.-- Mine is a bold theory. which attempts to exp
349d052 clearly so.-- NB. This paper worth referring to AGAIN.-- According to my theory, every species in any
387d169 e same peculiarity in lesser degree C. & theirs AGAIN in lesser degree.--now if the <tw> second race
448e166 ighbouring & Senegal as sea.-- is remarkable.-- AGAIN the resemblance between the Superga & Paris, nu
465t081 ountry.-- in a descending series of strata This AGAIN shows how much forms depend on other forms Lyel
467t100 with ordinary divisions, & a few with one lobe AGAIN divided «Have dried some».-- some with no divis
468t111 alighted on base of filaments & reached nectar =AGAIN= between them, hence quite below stigma. & so a
472s02r roboscis, under sigma & draw it out over & over AGAIN & wipe off pollen. (as a needle becomes covered
528m035 eving a vivid castle in the air, or dreams real AGAIN explains insanity.-- Analysis of pleasures of s
528m038 waving perfectly parallel lines are elegant.-- AGAIN there is beauty in rhythm & symmetry, of forms-
532m054 o seize hold of objects to be frightened at.-- (AGAIN diseases of the heart are accompanied by much i
540m086 azons & Brazil-- with the negros of Africa, (or AGAIN the black man of <B> Van Diemens land & the ene
549m121 , main source of the intense happiness.-- it is AGAIN another question, whether this happiness is the
570n024 h a person on the ribs & how he gulps in air.-- AGAIN a master says I will see you damned first." the
578n056 ?)-- squeeze out tears. replaced & squeezed out AGAIN-- as power of mind by habit gets more perfect o
583n070 - p. 19. animals capable of education; (this is AGAIN assumed as more allied to reason than instinct.
587n087 ight be acquired. as the Turks have of Rhubarb: AGAIN on other hand, it is said people, who like swee
086a004 hen Mammoth & narrow toothed Mastodon.-- argue AGAINST the prejudice of not believing recent elevatio
103a064 ing formed by crust being too large & pitching AGAINST each other, is, I suspect much weakened by <vi

Page		(Key Word)	
103a065	Absence of Caverns, in Plutonic rocks argument	AGAINST	great bodies of vapour. according to Hopkins t
118a104	on of dikes have reached the surface Arguments	AGAINST	Herschel's view of cause of continental elevat
124a120	s p.119 on such strata {P} do p. 171. argument	AGAINST	lateral injection. from probability of fissure
126a122	e fact of a dumplin being bad conductor is {P}	AGAINST	my views-- if we had rod thus & judged by incr
126a124	an take place.-- ‖ the depth of frozen soil is	AGAINST	this view.-- however it is said in some of the
132a138	s. it ought, (according to M. Parrots argument	AGAINST	central heat to warm the ocean).-- and M. Parr
145g001	Journal of Agricl Dec 1837 Yet instances given	AGAINST	it-- Mere fact of many races of Animals in Bri
152g050	within 200 ft of level of 4th shelf= argument	AGAINST	river--composition &-- stratification argument
181b041	te which does not increase» it will be chances	AGAINST	any one «of them» having progeny living ten th
181b041	n> therefore the chances are excessively great	AGAINST	, any two of the 12. having progeny. after that
196b110	r passage to animals appears greatest argument	AGAINST	theory of analogies. States there is but one a
211b161	ions of beauty, therefore instinctive feelings	AGAINST	other «for sexual ends» species, whereas Man h
225b216	ng" doctrine absurd. (as equally are arguments	AGAINST	it-- namely how did otter live before being ma
225b218	l be said there are latent insects.-- as crows	AGAINST	man with gun. & Bustards &c &c!!! An American
230b235	«p. 154»-- 1838. Hybrid Ferns It may be argued	AGAINST	theory of changes that if so in approaching de
258c061	us (all order; cocks all warlike)» «thiis wars	AGAINST	in any class:, those points which are differen
280c136	et to produce whole generation unlike would go	AGAINST	the tendency.-- it tries to go back to grandfa
292c170	not staggered <">= I confess. no dissertation	AGAINST	these views, could possibly have had <to so co
301c199	ious of web. feet.-- p. 7. Mr Blyths arguments	AGAINST	squirrel using reason in hiding its food is ap
310c222	petrel-grebe. external» appears to be a puzzle	AGAINST	my theory,-- If I be asked by what power the c
312c228	nal.»-- consult on this point-- pigs always go	AGAINST	this, without «number of vertebrae» new acquis
340d026	ther-- & hybrids fertile inter se--No directly	AGAINST	Eyton's rule. ¿Are the hybrids similar inter s
343d037	cramped imagination that God created. (warring	AGAINST	those very laws he established in all <nature>
346d048	ng., bold.-- a Mr W: Hall remarked that it was	AGAINST	all rules their preserving character & breedin
364d103	l sing longest, & they in evident rivalry sing	AGAINST	each other, till it has been known one has kil
366d111	the uncles & aunts) & therefore does not tell	AGAINST	transmutation of species-- Will it against gen
366d111	ell against transmutation of species-- Will it	AGAINST	genera.-- How long will the wretched inhabitan
386d166	ich is not used is absorbed.-- this law acting	AGAINST	heredetary tendency causes abortive organs.--
406e035	heep & Moufflon of Corsica. <would not>, sadly	AGAINST	Yarrell's law.-- not so much against my modifi
406e035	t>, sadly against Yarrell's law.-- not so much	AGAINST	my modification of it-- Goat & Moufflon will n
411e055	s head» 27th November When summing up argument	AGAINST	my theory, doubtless, the presence of animals
415e066	other animal.-- Would anyone raise an argument	AGAINST	, my theory, should no fossil «very distinct sp
426e106	acts of grouse, & pheasant, & hooded crow goes	AGAINST	this. & wild hybrid plants. If many wild anima
432e124	we see same fact) go back, & this is argument	AGAINST	Blyth's doctrine of young birds retrogressing-
436e136	certain limits, but beyond these not.-- argue	AGAINST	this-- -- analogy will certainly allow variati
443e155	sperm.--]CD I utterly deny the right to argue	AGAINST	my theory, because it makes the world far olde
465t079	& Europe-- & Africa!,-- any negative argument	AGAINST	-- monkey-man, valueless.-- May not several gen
522m013	this may be doubted, whether rather not going	AGAINST	natural instincts.-- My Grand F. thought the f
523m014	om habit the feeling of anger must be directed	AGAINST	somebody.-- Have insane people any misgivings
523m015	ousness of insanity coming on.-- his struggles	AGAINST	it, his knowledge of the untruth of the idea,
535m070	xed laws of organization.-- M. le Comte argues	AGAINST	all contrivance-- it is what my views tend to.
552m131	rell's story of wheel horse in drays, scraping	AGAINST	cornice stone to cause friction Athenaeum 1838
554m138	y, when Keeper was away, take her chair & bang	AGAINST	the door to force it open, when she could not
564n003	sense, if it were proved instead of militating	AGAINST	the existence of such an attribute would be ra
568n019	ingdom of nature. If I want some good passages	AGAINST	, opposition of divines to progress of knowledg
577n049	itself.]CD & hence there is great probability	AGAINST	free action.-- on my view of free will, no one
591n100	n eating cause disgust, because he does not go	AGAINST	instinctive feeling, only does not fullfil, li
591n100	.-- all this makes analogy of actions with <&>	AGAINST	benevolent & parental instincts very clear.--
594n112	were not frightened at a dog.-- » The instinct	AGAINST	man is perhaps, as strong as against hawk, but
594n112	instinct against man is perhaps, as strong as	AGAINST	hawk, but the birds at Maer have learned that
627o053	s contemplation.--" Now Eugenius would contend	AGAINST	this-- but the pleasure a dog has in obeying i
635j56v	rs should not be exposed to weather.-- this is	AGAINST	my theory of frequent intermarriage.-- A plant
270c104	at. Hist) spiral structure in Echinodermata.--	AGASSIZ	says Infusoria «are» insecta.-- G. R. Treviran
414e060	sca (& plants???) been so little progressive «!	AGASSIZ	makes it wonderfully changed, since Cretaceous
640j167	riking anomaly to this. Have they wide ranges?	AGASSIZ	has shewn that they most widely differ» 3 A ve
100a053	the Key to the story.-- consider stalactites.--	AGATE	rings, crystallization transverse.-- or rather
098a046	. 86. et p 95.-- It is easy to prove. (pyrites,	AGATES	, calcareous balls) that concretions are connec
270c103	have position of organs of generation!!!. Mem.	AGAZIZ	. (INo Annals of Nat. Hist) spiral structure in
048r086	s was killed. -- In Patagonia, are all beds same	AGE	? is white substance triturated Porphyritic rock.
055r107	raters).--worn into mud & dust.--connection with	AGE	, & agreement with number of craters. No cliffs a
055r108	ght, of the solid lavas.--proportionally high to	AGE	. (we do not wonder to see tertiary plains consum
056r109	rk to the ocean, who would ever suppose that its	AGE	was limited? Who could suppose such trifling mea
057r113	ploughs [,] Volcanos as Marl-pits: Consider well	AGE	of Bones. = slowness of elevation proved at St J
136a147	important letters-- When I come to treat of the	AGE	of the Pampas Deposit, I may properly remark on
213b170	r back, fish approaching to reptiles at Silurian	AGE	-- How long back have insects been known? As Goul
241c016	. not extinct species. good Resumé do/p. .62 ???	AGE	of Deinotherium. p. 23.. Bull: Soc. Geolog. 1837
247c028	y generally distributed: Mem of great geological	AGE	-- Gastrobranchus «only» 2 species one in Norther
276c123	eir science a few years in advance only of their	AGE	. (differently from literary men.--) must remembe
278c131	ith teeth of seals and dugong, therefore immense	AGE	since breccia accumulated-- surely ask Owen to s
324c268	t has published his laws about sexes relative to	AGE	of Marriages Brown at end of Flinders & at end o
351d057	subject.-- Monstrosities, kind of determined by	AGE	of foetus.-- As Larvae may be more perfect (as w
354d062	the nerves, but in different parts according to	AGE	of individuals-- (see Mammae of Women) in differ
354d062	-- (see Mammae of Women) in different parts when	AGE	changes caterpillar to Butterfly.-- When two Var
355d067	<also> Hunters law of monstrosity with regard to	AGE	of foetus. distinct consideration) Now in differ
358d076	se facts are effects of castration on males & of	AGE	or castration on females.-- [not located] hen fr
379d151	. Wildness Reversion Q [not located] The present	AGE	is the one for large Cetacea, as the past for ot
388d172	is supported by change which takes place in old	AGE	of female assuming plumage of cock, & beards gro
432e124	aracter of his forefathers only when advanced in	AGE	, & therefore the children do not, (& in hairless
435e130	observed to change their sex,-- this effect from	AGE	, what Mr Knight [not located] the stigma retains
454e184	What annuals can be budded «& rendered of great	AGE	» as must be inferred from what Mr Knight says. H
466t095	» of characters being inherited at corresponing	AGE	& sex, opposed by cantering horses having colts
492q002	nnuals be budded. with reference to extension of	AGE	of individuals-- 9. Do plants in becoming double
524m021	rium after epilepsy, but in the failing from old	AGE	, they constantly do.-- In Mrs P. <...> of B. <ta
524m022	- My father's test of sincerity.-- People in old	AGE	. exceedingly sharp in some things, though so con
553m135	f the God (<thus> & hence arises the theological	AGE	of science in every nation according to M. le Co
580n062	on lost; yet many of the habits acquired in that	AGE	are retained through life" p. 200.-- "The desire
620o045	to some <subsequent> power (reason) obtained by	AGE	, which should show the child, which of its insti
623o051	ily attain instinct of pointing as a dog.-- also	AGE	has much influence.)-- & only that which is bene
640j28v	«There is no way of eliminating the evils of old	AGE	, after breeding season, or gaining adaptations,
640j28v	outh most necessary: the fertility of Man in old	AGE	keeps woman alive: for Man & woman are same: fer
049r089	te is much more perfect. than in believing mere	AGENCY	of dikes: & indeed when do these dikes lead to
397e003	can fall, or plant rise, without the immediate	AGENCY	of the deity. But we know from experience! tha
440e146	can say, how much structure is due to external	AGENCY	, without final cause. either in present, or pa
497q007	f countries during last century or two.-- where	AGENCY	of man not known.-- (3) How is Iris impregnate
534m062	something wrong in comparing these cases, when	AGENCY	is unknown, with simple exertion of intellectu
636j56v	the impregnation of Dioecious Plants by foreign	AGENCY	-- as insects, as wonderful case of adaptation.
106a074	idening valley.-- it is essentially a deepening	AGENT	{P} Therefore when we have valleys of this stru
616o039	dinary matter is that which action bears to the	AGENT	. Matter is by a metaphor said to attract; & hen
617o40v	nimate matter up to the action of some animated	AGENT	Now the phenomena of gravity are manifestly the
627o053	ation of one's offspring) is not the aim of the	AGENT	, for it does not enter into his contemplation.-
103a063	ogical Society-- Dikes have not been the moving	AGENTS	, because not wedge-formed.-- Hence fill up fis
415e065	n. & is acted on by «the» organic and inorganic	AGENTS	of this earth. like every other animal.-- Woul
435e133	Edwards on the Influence of <external> Physical	AGENTS	". «translated by Dr. Hodgkins» p. 54. The axol
435e134	e effects produced on organization. by physical	AGENTS	." p. 466. Many facts given of high temperature
057r114	ved.-- curious similarity of rocks of very diff.	AGES	. at Port Desire on plain. & interstratified.--
063r132	divided either at one moment or through lapse of	AGES	.--Therefore we are not so much surprised at see
174b015	greatest differences-- if separated from immens	AGES	possibly two distinct type, but each having its
175b016	he world.-- This view supposes that in course of	AGES	. & therefore changes. every animal has tendency

181b044 certainly is the case at least during subsequent AGES.-- The Creator has made tribes of animals adapt
192b088 n Stonefied slate, the father of all Mammalia in AGES long gone past.. & still more so «known» with f
193b090 . ¿-- whether every animal produces in course of AGES ten thousand varieties, (influenced itself perh
221b201 changes been so slow., that all have existed for AGES as metamorphic; & therefore according to Lyells
248c029 arter, where we know quadrupeds have existed for AGES.-- ∴ The most hypoth: part of my theory, that «
248c030 part of my theory, that «two» varieties of many AGES standing, will not readily breed together: The
285c152 is utterly rotten & obliterated in the course of AGES.-- As species is <certain> real thing with rega
576n047 to progressive. developement) on account of dark AGES.-- «effects of external circumstances» Look at
615o038 king pleasure in virtue because acquired in past AGES; seems to indicate that when we <return> turn i
623o050 ich has beneficial tendency through <all> «many» AGES. could be acquired, & we are certain from our r
638j059 a hundred schemes of structure, in the course of AGES «step by step».-- in Man, the nervous system, g
164g119 dular strata coral upside down strata coarse AGGLOMERATE [...] shells from [...] Wenlock Edge [blank]
098a047 crystallization, & therefore as a consequence AGGREGATED (I assume the same force which draws togethe
099a049 thematician to when two particles «would» be AGGREGATED, would they not attract strong. a third.-- &
098a047 rce crystallizes minerals in layer. therefore AGGREGATES them in layer.-- So that layer of feldspar i
350d054 s in two formations? by no way.?-- "Natura nihil AGIT frustra", as Sir Thomas Browne says "is the onl
206b146 ated to the (200dth year) degree. Then 200 years AGO, there were 200 people living who now have succe
470t176 sh,--the Palmated was introduced about '65 years AGO--& soon after mules abounded--so that palmated h
472s01v See separate note-- Elizabeth says several years AGO seeds were procured with the P. orientale in gar
472s02r ot case, on several flowers I examined some days AGO-- This Bee flew from yellow to yellow & purple h
477z001 ording to Charpentier de Cossigny. only 10 years AGO <no> snail was introduced to Mauritius. 18 Azara
542m094 & yawning. (common to other animals) scream of AGONY, sigh of discomfort & weariness. & meditative t
588n030 306 "the eyes are rolled upwards during mental AGONY, & whilst strong emotions of reverence & piety
191b080 t Indian archipelago.-- West Indies = Opossum & AGOUTI same as on continent-- 3 Paradupasi in common
058r117 ere patches as in Italy proved by Coral hypoth. AGREE with great continents). Voyage aux terres Austr
197b112 r has said each animal made for itself does not AGREE with old & modern types being constant. Cuvier'
233b252 ticulate than <M> Vertebrate. But how does this AGREE with longevity of species in Molluscs!!! When w
254c047 the perfection of the Species--! How does this AGREE with grand fact of Marsupial, low Cerebral stru
271c105 ame species. yet that it should so strictly <f> AGREE in habits with the Turdus Musicus «not found in
274c116 s out of the rest of the world?-- Will this not AGREE with Waterhouse & <birds> Mammalia.-- We have c
306c213 e. What productions Sandal. Wood Isd.? ought to AGREE with Java?? Terrestrial Planariae assuming brig
414e065 hat of animals transported by floating ice.-- I AGREE with Mr Lyell., man is not an intruder.-- : the
424e100 Northern forms-- & American ones & Europaean-- AGREE very much closer, than the present ones., which
429e115 19. do says lofty Alpine plants of & Pyrennees AGREE with those of Norway. Lapland & Greenland, but
480z008 e «species of» smaller petrels, are night birds AGREE. with <pe> nocturnal habits of Crustaceae Mr Br
571n028 ome children acquire it soon. & why do all men. AGREE ultimately?-- We acquire many notions unconscio
629o55v on what it is desirable to be taught,-- all are AGREED general utility (3) It is other question wheth
221b202 bout Miocene fossils some species being recent AGREEING with Senegal. whilst Crag <agrees with> accor
312c231 ther weight & size & they would produce number AGREEING almost to the point in question.-- «--merely
054r102 imestone with recent shells 200 ft, how exact AGREEMENT with Coquimbo; [not located] [not located] Is
055r107 worn into mud & dust.--connection with age, & AGREEMENT with number of craters. No cliffs at Ascensio
212b165 1837: p. 404. account of instinct of dogs.-- AGREEMENT & reason Some animals common to Mauritius & M
464tf6r -- Lund's Antilope in Brazil another point of AGREEMENT with. N. America & S., (¿ is the peculiar.
030r035 ntains, probably chiefly leaves.--This position AGREES with character of.. «in Basins from rivers. &
040r062 esson. character of Flora to New Zealand, which AGREES with St Helena in being unique, yet no quadrup
041r064 hey could have lived in so deep a sea.--Perhaps AGREES with formation of pebbles & vertical trees Gra
054r102 ribed by Pernetty: account of streams of stones AGREES with mine.--At Conception, cleavage E & W! at
055s107 ds of eruptions how immense the time!! How well AGREES with number of Craters!--At S. Cruz. there is
085a102 nds. from number of craters very ancient. which AGREES. with peculiar character of Vegetation. -- So
181b044 ation will be sudden-- Heaven know whether this AGREES with Nature: Cuidado The above speculations ar
186b057 ents) How far Does Waterhouse's representatives AGREES with breeding.. in irregular trees. & extincti
187b062 g animal die out in S. America; with no change, AGREES with belief. that Siberian animals lived in co
208b153 species few in Arctic. in proportion to genera. AGREES with late production of those regions, & conse
209b155 nges) would be in rays-- from certain sports.-- AGREES with old Linnaean doctrine & Lyells. to certai
222b202 eing recent agreeing with Senegal. whilst Crag <AGREES with> according to Beck has none recent, yet g
227b225 lieve the world older than geologists think. it AGREES. with excessive inequality of numbers of specie
229b234 rmer inhabitants of Mauritius Freycinet Voyage, AGREES. with several mammalia being peculiar (?) If.
231b244 s, far more than by non-embedment of remains-- ¿AGREES with non-blending of languages?-- Till man acq
283c145 such not having been long in blood?-- My theory AGREES with unequal distances between species. some f
333d008 ecidedly take much most after Esquimaux.-- this AGREES perfectly with Yarrell & no leading question w
346d048 in & in-- Nonsense a flock of more than 100.-- AGREES, «nearly» with. the account. given by Boethius
391d175 hence every individual is different). (All this AGREES well with my view of those «forms» slightly fa
484z014 arasitical on Vellellae in Atlantic Ocean Gould AGREES with D'Orbigny, that Serpent Eater-- or Secret
526m029 f early memory consisting of things seen, quite AGREES with my Fathers case of Mr Corbet of the Hall
527m032 s heredetary knowledge like that of man, & this AGREES with the stated fact, that «birds from» certai
538m080 se one forgets. what one performs habitually.-- AGREES with insanity, as in Dr Ash's case, when he st
549m121 y their fault.-- Whether this rule of happiness AGREES with that of New Testament is other question.-
608o026 uct. as being better & making him happier.-- he AGREES & yet does not.-- because motive power not in
613o036 him, & then jump over the gate & bring it. ---- AGREES with ONE animal [RHC] Kirby extends instinct t
629o55v ught instinctively; I say yes, & my explanation AGREES with last head.-- (4) It is other question, h
145g001 ndelibly than female p 367 Quarterly Journal of AGRICL Dec 1837 Yet instances given against it-- Mere
345d043 [g] oldest breed?.-- -- Quarterly Journal of AGRICULTURE p. 367. Dec. 1837. Generally-- received opin
525m040 ghts vary so much in different people., an AGRICULTURIST. in whose mind supply of food was evasive &
506g015 to distribute some of my questions amongst AGRICULTURISTS. whom he know.-- Col. le Couteur on Wheat.
556m144 when they hear ready, but if they see anything AHEAD. which cad cannot see. they do not move muscle.
599o008 a les facultes, q'il possede seul sur la terre. J'AI trouve son âme" &c-- -- Confesses these facultie
040r063 ery rare) to cause floods in valleys, which must AID in preserving the terraces <...> Molina's Case A
115a096 e, & which unite the homogenious crystals., must AID in adding effects to common heat.-- Where there
277c127 hem as (a). (b) until data be given.-- This will AID in preventing the chaos.-- will point out what t
277c127 chaos.-- will point out what to observe.-- will AID us in physiology, tell traveller what to observe
341d030 ent.-- clavicle scapula &c strongly developed to AID in breathing.-- Animals from Hobart Town mention
345d044 any breeds of animals in Britain shows, with the AID of seclusion in breeding. how easy races or vari
554m139 or on board the ship should not puzzle her-- with AID of teeth & hands.-- Descent 1838 It was very cur
608o026 & educate by putting contingencies in the way to AID motive power.--if incorrigably bad nothing will
123a116 The theory of veins will, I suspect be greatly AIDED by considering space formed-- great vacuum-- by
410e052 e is no cure «I may say all this, having myself AIDED in such sins» (do not add name, without referen
564n005 state of surface, or other means by which eyes, AIDED by experience is supposed in man to guide to kn
619o042 e sympathy that the individual forgets itself, & AIDS & defends & acts for others at its own expense.
243c019 îles Macquarie et Campbell (52 degree S) qui n'AIENT egalement leur espèces; et certainment on eût é
477z003 Azara Las Vinchuca or Benchuca. "Les individus AILES peuvent avoir <quatre> cinq lignes de long et v
233b252 forests dare to say that intellectuality is only AIM in this world [not located] T. Carlyle, saw with
583n069 t with in the lower animals.-- hence the general AIM of fable, & expression as cunningness of fox, in
627o053 the gratification of one's offspring) is not the AIM of the agent, for it does not enter into his own
599o008 in the polity of Nature than any other animal-- AIMÉ Martin de l'Education des Mères Vol. I. p. 198.
601o08a Master takes Hat de l'education des Mères par L AIMÉ Martin Leroy Lettres. Philosophique sur l'intel
032r041 accumulation is brought into play As in Ocean & AIR; there are «likewise» differences of temperature
147g019 iday morning ¼ past seven o'clock 29.642 Temp 55 AIR 50 degree? Friday. Inverorum about 20 ft above L
147g019 out 20 ft above Loch Tulla 29.804 Temp 62 degree AIR 60 degree Below Loch Tulla whole wide valley sca
147g21 erorum & King's House 28.935/82 degree A Temp of AIR 65 degree? Glenoe, 6 ft above high water mark 30
152g049 evel of «bed of» River 30.221/65 degree/ Temp of AIR 65 degree? There are two terraces on the East si
152g050 40 ft beneath general plain. 30.127 A 72 degree AIR 65 degree? at level of upper terrace The buttress
154g058 ith granite block «& Ben Erin» 29.287. 72 degree AIR 65? 70? (P) Where a buttress projects from side
159g086 upper (rather above)? shelf 29.290 A. 69 degree AIR 68 degree? Barom 29.008 A. 75 degree Air 70 degr
159g086 degree Air 68 degree? Barom 29.008 A. 75 degree AIR 70 degree? This station a little way down slope
160g091 Isthmus broad flat Loch {P} XX Barom 28.92 A 75 AIR 70 degree? Isthmus broad flat peat mass-- (gener
161g094 ass Divortium aquarium) Barom. 29.200 A.77 degree AIR 70 degree? Barom. 066 lower than last. A. 77
162g100 Loch Lochy near Letter Finlay Barom 30.267, A 68 AIR 65 degree? <.194 372 about 267 28.75 .105 I reac
162g103 ld Macphee> Saturday Morning 29.958 A 64 degree, AIR 60 «Evening do» The extreme right arm of River T

176b023 hing in the tree of life owing to three elements AIR, land & water, & the endeavour of each <one> typ
178b030 passed over, & where other species have <come> «"AIR" of that place. Will it be said, those have been
181b045 be not complete, if birds were fitted solely for AIR & fishes for water. If my idea of origin of Quin
267c095 ic plants.--Owing to plants not being adapted to AIR!! p. 11--&c. valuable paper on quadrupeds of Van
273c111 ial,-- formed for flight & great movement in the AIR, & likewise rasorial species, & likewise perchin
308c218 birds are the only two tribes fitted for water, AIR, & land, (Macleay has this remark) Mem. number 5
367d113 he power of inflating its body like balloon-- by AIR cells connected with cheek pouches.-- Hunter's A
433e126 ween two great distinct formations.-- particular AIR given. p. 246.-- 248 & p. 258 A beautiful case,
444a158 nct, for how could experience teach distances in AIR, in which it never touches objects.-- far better
502q012 roperly called Canadense-- would it grow in open AIR in Sweden. Linnaeus found 2 flower. which had an
527m033).-- Granny says she never builds castles in the AIR-- Catherine often, but not of an inventive class
527m034 ess of my stomach I observe a long castle in the AIR, is as hard work (abstracting it being done in o
527m034 as hard work (abstracting it being done in open AIR, with exercise &c no organs of sense being requi
527m034 proving powers of invention) such castles in the AIR are highly advantageous, before real train of in
527m035 to play & then perhaps the sooner castles in the AIR are banished the better.-- The facility with whi
527m035 etter.-- The facility with which a castle in the AIR is interrupted & utterly forgotten--, so as to f
528m035 the approach to believing a vivid castle in the AIR, or dreams real again explains insanity.--Analy
540m088 is.-- "The fledge-dove knows the prowlers of the AIR" &c &c &c so is conscience &c &c Coleridge,-- Za
544m103 s-- & which tells one of reality-- castle in the AIR, is more prolonged than dream. never fatiguing,--
547m111 9th. Went to Bed. & built «common" Castle in the AIR, of being compelled, from some quite imaginary c
548m115 trains of thought are in progress-- In castle of AIR the trouble «I well recollect" is in making thin
558m152 mouth wide open, each [lip] drawn back & driving AIR out of mouth «hairs erect on back» «wide open" w
558m152 Thus <sudden> «forcible prolonged» expulsion of AIR «dogs snarl much the same way» generic manifesta
565n007 rs down to horse.-- Horse snuffs «& snorts», the AIR «& raises its head, & pricks its ears" when afra
570n023 ed. touch a person on the ribs & how he gulps in AIR.-- Again a master says I will see you damned fir
584n072 ious the means of guiding themselves through the AIR,-- waterbirds, the bee to its nest,-- cats when
594n112 gs. hernes are common. not unlike in size in the AIR at a distance.-- How can such an instinct arise?
632j53r rovision of seeds for transportation through the AIR.-- cocoa nut by water «fucus for adhesion».-- as
447e165 to the Canada onion mentioned in Hort. Transact. AIRA caespitosa becomes vivaparous on mountains & ye
209b156 enres qui, sous ce climat, se divisent le plus AISEMENT en espèces distinctes et permanentes." p. 145
571n031 h struck in great avenue, resemblance to gloomy AISLE of Churche.-- these are Mayo's ideas.-- In lang
522m014 s almost involuntarily when a person is tired is AKIN to insanity.-- «I know the feeling also of depr
397e004 imes has been small-- because change in forms is <AL> solely adaptation of whole of one race to some
551m128 e door evidently did, for it did with far more ALACRITY <than> when something good was shown him, tha
609o029 feel between conscience & impulse) but shame «we ALAS know» is far easier conquered than the deeper &
290c163 on generation might appear an analogy NB Pyrrho-ALAUDA (bird of St Jago) of brown colour; lives on gr
346d047 n plumage of the Pelican.-- Mem pink spots on ALBATROSS, on some Gulls. Flamingo-- (Spoonbill Wader.
199b119 se ancestors hairy with one hairy child, and of ALBINO DISEASE being banished, & given to Portuguese.
265c085 huric acid disorganizes wood, but adaptation.-- ALBINO however is monster. yet albino may so far be c
265c085 ut adaptation.-- albino however is monster. yet ALBINO may so far be considered an adaptation as best
451e176 565. in a Paper by Lieut. Newbold.--» A Malayan ALBINO described "To this day the tomb of his grandfa
451e176 ay the tomb of his grandfather, who was also an ALBINO is held sacred by the credulous natives, & vow
451e176 ents were of the usual colour. His sister is an ALBINO like himself said not to be common-- probably,
104a066 of pressure of liquid rock. Andes discussion-- ALBITE certainly contains 6 per cent more silica than
104a066 rdillera, in Andite - containing 80 per cent of ALBITE 80/100 X 6/100 = 480 In Falkland islands. & ge
108a081 imborazo., & Pichincha. Melaphyre. = Andesite-- ALBITE & amphibole= Cook found Granite at Christmas S
244c023 urus does in North. Hemisphere.-- p. 158 Cuscus ALBUS. New Ireland ---- maculatus -- Waigiou Speaking
246c026 ot species.-- Vol :694. King-fisher of Europe, (ALCEDO ispida) from Molluccas. scarcely differs at al
481z010 ings in Meyen Zoology on animal of Campanularia ALCEDO stellata. Meyen p. 92.-- great Kingfisher of T
500q10a Gould-- go over the Pigeons, Philotis, Dacelo. ALCYONE, where there are very close species & see whet
266c092 tra. Cattle generally marked like those of the ALDERNEY breed, but size not larger than those of Blac
332d003 day.-- Mark at Shrewsbury thinks the half bred ALDERNEY Cows take more after Alderney that the Durham
332d003 ks the half bred Alderney Cows take more after ALDERNEY that the Durham,, with which they have been c
332d003 Durham,, with which they have been crossed--is ALDERNEY oldest breed-- He believes all pretty much al
031r038 symetrical structure. East India Archipelago. «ALEUTIAN Arch.--» V. Fitton. Australia: cases in Europ
582n068 a kind of mental pain & pleasure.-- The Revd. ALGERNON Wells Lecture on animal instinct. 1834: p. 15
472s02r w Fly 21 «this Heartease withered on Monday.--» ALIGHT on upper petals & insert proboscis, under sigm
468t111 ce» P on Fraxinella <Heartease> «small. Humble ALIGHTED on base of filaments & reached nectar =again=
466t094 ways base {a} of upper petal from facility of ALIGHTING? which is not differently coloured & to which
472s02v wer & stigma dusted.-- <I think> When It first ALIGHTS, it cleaned sucker & <I think> pollen was scra
266c091 nmad, the last to extend, their dominion, armed ALIKE with the Koran and the sword" ﴾quote Whewells B
280c136 han to produce that capable of producing itself ALIKE.-- in one case it changes one, in other it chan
332d003 rney oldest breed-- He believes all pretty much ALIKE.-- My Father Water-in the hair a century since
333d001 mon cat, exact variety unknown., three kittens, ALIKE each other, partaking <more» «very closely» of
333d008 are exactly intermediate in character & Kittens ALIKE each other.-- Even in children of parents <some
336d017 return to parent stock. but if both parents are ALIKE, offspring must be like Hence mutilations not h
342d033 ies.-- The hybrids do not vary (ie the hens all ALIKE & Cocks all alike) More than parent species-- M
342d033 do not vary (ie the hens all alike & Cocks all ALIKE) More than parent species-- Mr Blyth remarked o
368d114 - «(NB. most strange cocks & hens. being either ALIKE or very different in recently altered genera. G
378d147 Young birds, one may be sure cock & hen will be ALIKE-- I presume converse is not true for he says He
378d147 rse is not true for he says Hen & cock Starling ALIKE, yet young ones brown.-- Sexual Selection If ma
378d147 dded to species,. we can see why young & female ALIKE Good Ch 6 Keep Is it Male that assumes change,
391d174 parent. <but these buds do not procreate> & all ALIKE in one parent or tree, (but not in other trees.
449e169 - Eyton says that the young of two hatches «all ALIKE» between the male Chinense & female common goos
449e169 hey interbred. & the young kept constant. & all ALIKE Waterhouse says some of the Galapagos Heteromer
492q003 ing of Animals If two half bred animals exactly ALIKE be interbred will offspring be uniform.-- Mr Fo
493q004 Cross two half-bred animals. which are exactly ALIKE & see result.-- (3) Cross the Esquimaux dog. wi
088a014 se» has frequently heard that Herons bring eels ALIVE to their nests; & then they may picked up benea
191b082 limits of large animals-- Owls. transport mice ALIVE? Species formed by subsidence. Java & Sumatra.
244c022 rom Java.-- Hairs, & deer.-- Procured two makis ALIVE from there.-- Mem Waterhouse knows of some spec
264c079 not have lived when certain other animals were ALIVE, which have perished.-- Let man visit Ourang-ou
296c185 56. Peregrine Falcon holds birds for some time ALIVE ¿therefore other species mice & only kills them
418e083 dded to the composition of the atom, to make it ALIVE, & how the laws of generation were impressed on
437e139 Morne Mountains, it then dropped it & was found ALIVE.-- Stanleys Familiar History of Birds several c
439e142 obably would do-- or be with difficulty be kept ALIVE.-- Nevertheless much probably depends on circum
532m055 ng means from the senses to the mind being more ALIVE.-- How is it. with people nervous from illness.
640j28v ry: the fertility of Man in old age keeps woman ALIVE: for Man & woman are same: fertility of either
200b121 hich unite very different structure as. petrel & ALK. do not show the possibility of common branching
176b021 animal, but law of adaptation as much as acid & ALKALI organized beings represent a tree. irregularly
138a176 ng Carbonic Acid unite with «piece of cabbage» ALKLALI & precipitate silica / or charcoal charged wit
025r018 f. Europe, with America; I might add I have drawn ALL my illustrations from America, purposely to show
028r028 s. by actions of rivers currents. & sea beaches. ALL mineral masses must have a tendency. to mingle;
028r030 unknown Volcanic process? How does it come that ALL Lime is not accumulated in the Tropical oceans d
028r030 Organic powers. We know the waters of the ocean ALL are mingled. These reflections might be introduc
028r030 me sandstones, as in Australia.--Have Limestones ALL been dissolved. if so sea would separate them fr
029r033 o I believe the secondary? formations of Brazil, ALL originate from the decomposition of Granitic roc
030r036 e Tertiary strata there is Coal-- ¿ No shells in ALL cases. «.Mytilus.--» «at Guacho» «on N. Chile? W
033r043 mb up many parts, in James Isd.--Mem St Helena-- ALL Trachytic.--Daubeny P. 171. Vol I. Humboldt Ther
035r045 ling &c &c My results go to believe that much of ALL old strata of England. formed near surface: Mem
035r047 should have been kept; it shows that throughout ALL England, whole surface oscillated equably.-- The
038r057 of coatings; hence it will be necessary to state ALL arguments for believing that there must be a cen
038r057 en in great crater. different little craters are ALL burning, surely there must be «somewhere» below
038r059 lence crossing lines of crater, <arg² state that ALL the great Volcanoes. have been elevated consider
039r059 ot been points of eruption. Nobody supposes that ALL the dikes in Cornwall or in the coal measures ha
040r064 ations in the upheaval of Andes.--but as long as ALL below water no evidence--The depth of shells (wh
041r065 owns driving by the want of water.--I believe in ALL flat countries. years of drought are common.--Mr
041r066 excess.-- If veins {P} are secretionary, so are ALL those plates in Australia. New Red Sandstone. at
042r069 have destroyed regularity of slope of valleys.--ALL my observations of period «& manner" of elevatio

```
044r073  rials being consolidated; one inclines to belief ALL strata of Europe formed near coast. Humboldts qu
048r086  the  Rhinoceros was killed. -- In Patagonia, are ALL beds same age? is white substance triturated Por
049r087  alled Volcanic or Plutonic The cellular state of ALL the Porphyry specimens, must be well examined At
054r102  New Zeeland rich in particular genera of plants: ALL St. Catherine & coast Granite: P. 199;  Falkland
054r105  eagues North of Callao, called Marques, where in ALL appearances not many years since, the sea covere
056r110  vated country of granite, not <so> great«er» for ALL Europe, than from the Plata to Caraccas, which i
056r110  urope, than from the Plata to Caraccas, which is ALL of granite: In discussing circulation of fluid n
056r111  lime).  <Also Volcanos separate.> Volcanos blend ALL substances together; & products being similar ov
057r113  y.-- Sir J. Herschel. says. precip. of Sulph. B. ALL the infinitesimal cryst. arrange themselves in p
058r115  elevation.  Effects of great waves to obliterate ALL land marks.--At the first it would though be eas
059r120  ers in centre:-- Bailly talks of much granite on ALL East side of Van Diemen Land. All the Calcareous
059r120  uch granite on all East side of Van Diemen Land.  ALL the Calcareous rocks which harden by  themselves
059r121  Climate.!? or small Proportion of Alum: matter.--ALL pale cream colour.-- The Brecciated structure of
059r121  pale cream colour.-- The Brecciated structure of ALL the Pitchstone (which I have seen). is a kind of
059r121  f diff cont: & even in one case contained lime.--ALL bear close analogy to Obsidian, & all show chemi
059r121  ned lime.--All bear close analogy to Obsidian, & ALL show chemical action as well as effects of cooli
062r128  bterranean lakes, near Volcanoes. lakes of brine ALL inhabited: Go steadily through all the limits of
062r128  akes of brine all inhabited: Go steadily through ALL the limits of birds & animals in S. America. Zor
063r132  & eggs which become quite separate.--Considering ALL individuals of all species. as «each» one indivi
063r132  quite  separate.--Considering all individuals of ALL species. as «each» one individual «divided» by d
063r133  -- Dogs. Cats. Horses. Cattle. Goat. Asses. have ALL run wild & bred. no doubt with perfect success.-
064r137  on Buch has urged that Java volcanos differ from ALL others in quantity of Sulph. acid emitted: mem:
066r142  e Earthquakes in Cordoba. one of which dried up <ALL> a lake in neigbourhood of town Mr Murchison ins
068r147  propelling.  mass. If one inch can be raised then ALL can, for fresh layers of igneous rock replace st
073r161  d mixed with lead & copper is infinitely rare in ALL parts of the globe". p. 113 How utterly incompre
076r169  ks of pure silver not common in <S.> America: In ALL climates distribution of silver «in veins»  very
078r175  // Tasco 40 degree to NW (afterwards said to be «ALL with some exception» directed NW & SE). «Vol III
085a001  88. Salt deposited on windows of houses. & trees ALL injured on Eastern side, far inland.-- even 70 m
089a018  ral limestone, with interstices yet emptty.-- In ALL the mountains of Saint Marc et des Gonaïves it i
090a022  te come out before.-- What must be the effect of ALL the meteoric stone which must have fallen on the
090a022  ed as many, (one even 3000) This ninety includes ALL actually counted.-- The weight «or size» is give
096a040  Mackenzie  talks of gravel on basalt of Heckla-- ALL the Azores Isld. Von Buch p 359 stretched out NE
098a046  - now cleavage as suggested by Sir J. Hershel as ALL crystals obeying one law of crystallization. the
099a049  on evidently small scale of concretionary action ALL fluid at once, the films vertical. Ascertain law
103a064  ys thin out above which explains a difficulty.-- ALL De la Beche's reasoning of mountains being forme
106a075  valleys of this structure. as the inclination in ALL probability would be greater when flowing over (
106a075  ower channel instead of wider.-- This applies to ALL vallies (except mere talus «over cliffs edge» of
106a076  Fox  on increase of temperature at great depths. ALL Earthquake unaccompanied by Volcanoes must be sou
115a098  ard: if matter too coarse, then {P} that form.-- ALL this depending not on absolute <force» «size of»
120a107  (¿ metamorphic action at the bottom of the sea?) ALL this profoundly considered. study Hopkins. theor
122a113  ogous to my Valparaiso case. Consider profoundly ALL consequences of EXTREME FLUIDITY of earth.-- stu
133a140  M.  Parrot ends his paper like a fool.-- Feb 25' ALL facts show how slowly heat travels; &  therefore
147g021  30.380. 68 degree 65 degree? For comparison with ALL the measure before There some of the half rounde
148g022  ent but nature I am quite doubtful as I am of     ALL the Alluvium. At Mouth of Caledonian Canal oppos
148g023  per one 100 ft & other one 40-- ‖ traces of them ALL along <Glencoe>.-- towards Fort William yet in G
148g027  t terraces one above much inclined towards river ALL these composed-- where side ravine entered terra
148g028  ened if the side-streamlet had cut them out-- In ALL cases «I urge» deposition marine-: because if no
154g059  ine suppose continued across to {P} side removed ALL well & good, but how came river to do this vast
155g066  allings» of no particular hardness no wonder that ALL «three» lines «should be» EQUALLY preserved 2d o
156g067  , & some «other parts» Boulders of same granite, ALL on these three shelves soil is <the> usually sla
156g070  te & some other rocks at head of shelf 3d almost ALL granite pebbles Level of plain of 4th shelf at h
160g090  d gneiss «narrow sharp ridge with peak» I walked ALL round hill. Boulder about 20 ft. below summit <I
170bIFC  ervous» brain makes thought Glen Roy B C. Darwin ALL useful pages cut out Dec. 7th. /1856/ (& again I
170b001  oonomia Two kinds of generation the coeval kind, ALL individuals absolutely similar; for instance fru
171b002  world  subject to cycle of change, temperature & ALL circumstances which influence living beings.-- W
173b010  &c &c non fertility of hybridity &c &c <assuming ALL> if species (<a>) «(I)». <fr» may be derived fro
173b011  he type would be of the continent though species ALL different. In cases as Galapagos & Juan Fernande
173b012  es we can see why a form peculiar to continents; ALL bred in from one parent why. Myothera several sp
175b020  e may look at Megatheria, armadillos & sloths as ALL offsprings of some still older type some of  the
177b029  points of organization commenced branching.-- As ALL the species of some genera have died; have  they
177b029  the  species of some genera have died; have they ALL one determinate life dependent on genus,. that g
178b031  those  have been there created there;--» Are not ALL our «British» Shrews diff: species from the cont
179b034  ies (for species they certainly are according to ALL common language) will keep to their type: in ani
179b035  ne branch, & the monucule has definite life, then ALL die at one. period, which is not case,:. MONUCULE
180b037  the contemporary must have left no offspring at ALL, so as to keep number of species constant.-- Wit
181b043  hree elements, from the «infinite» variations, & ALL coming from one stock & obeying one law, they ma
184b054  ow according to my view. in S. America parent of ALL armadilloes might be brother to Megatherium.-- u
185b055  w of chance would cause this to have happened in ALL. but less in water birds-- carrion eagles.-- Thi
186b057  each  other, (at least in one point, in truth in ALL excepting specific character); and in passing fr
186b058  es to genera, each retains some one character of ALL its family; but why so? I can see no reason  for
188b068  For instance ever so many seeds of white flower. ALL would  come up white, though planted in same soil
189b072  n pippen. if produced by seed go on.-- otherwise ALL die.== The fossil horse, generated in S.  Africa
189b073  rica Zebra.-- & continued.-- perished in America ALL animals <are> of same species are bound together
191b082  is  in accordance with The Male animal affecting ALL the Progeny of female insures often mixing of in
192b083  te that most clearly». Fox tells me, that beyond ALL doubt seeds of Ribston Pippin, produce  Ribstone
192b085  than  we know varieties can produce.-- Therefore ALL genera MAY have had intermediate steps.--  Quote
192b086  it is other question, whether there have existed ALL those intermediate steps especially in those cla
192b089  rsupial animal in Stonefied slate, the father of ALL Mammalia in ages long gone past.. & still more s
193b093  I  p. 174. says from Swan river long South coast ALL that can be expected-- This answers Cuvier-- Per
194b096  other plants because this may be applied to show ALL the remarkable Australia genera, collected toget
198b114  at does the expression mean used by Cuvier, that ALL plants do receive intermixture.-- But how with «
198b114  says grand idea god giving laws & & then leaving ALL animals (though some may be) have not been creat
198b115  erent animals»'large Mammalia not being found on ALL to follow consequences.-- I cannot make out  his
199b116  pagation that out of the thousand of new insects ALL isld, (if act of fresh creation why not produced
199b117  . why out of the thousands of forms, should they ALL belong to same types already established. why ou
201b126  lustration of that axiom in Natural History that ALL be classified.-- Propagation explains  this.----
203b135  without apparent physical cause:-- Mem: Mastodon ALL over S. America. Hilaire does not seem(?) to con
204b141  season  bred readily with common ducks.--» Kirby ALL through Bridgewater errs greatly in thinking eve
205b145  do  "If hornless ram be put to horned ewe almost ALL the lambs will be hornless.--" does this apply t
206b147  though long distant, when of the present men (of ALL races) not more than a few will have successors.
206b148  ct in the two case,» May this not be extended to ALL animals first consider species of cats-- <& oth
213b169  > species. in domesticated <species> races.-- If ALL men were dead then monkeys make men.-- Men makes
217b188  Iceland no species to itself, a remark common to ALL northern islds.-- This is interesting, because I
217b188  is  interesting, because Iceland, must have been ALL ice in time of ice transported.-- This gives roo
218b189  to  impossibility, holds good in plants» between ALL different forms; therefore when from being put o
219b196  as  gret a difficulty to account for movement of ALL, by one law. as to account for each separate one
219b196  to account for each separate one, so to say that ALL Mammalia, were born from one stock, & since dist
220b199  No. 537. Feb. 1838. p. 107. Mr Blyth states that ALL «genera of» birds in «N.» America & Europe, whic
221b201  arkable.--¿ Have the changes been so slow,. that ALL have existed for ages as metamorphic; & therefor
222b204  ation almost infers, what we call improvement.-- ALL Mammalia from one stock, & now that one stock ca
222b206  & not include Mammalian remains.-- The Father of ALL insects gives same argument as father of Mammali
224b215  t not probable owing to mixture of races.-- When ALL mixed & physical changes (¿intellectual being ac
225b219  quadruped since days of Didelphis in Stonefield'. ALL lands united (Falkland Fox. ice). . Mauritius wh
228b232  im over whole world.-- «--the soul by consent of ALL is superadded, animals not got it, not look forw
229b232  origin  in <there> one common ancestor we may be ALL netted together.-- Hermaphrodite animals couple-
230b236  erences producing more fertile offspring.-- 1st. ALL variation of animal is either effect or adaptati
231b241  ontinent. Indian Rhinoceros. Java & Sumatra ones ALL different.-- Join Sumatra & Java <to>  together,
```

```
231b242 existing in East Indian Seas. Marsupials animals ALL show greater connexion in quadrupeds, <bu> plant
232b245 s in proportion to genera than in present seas, «ALL» The <one> species which survives any change may
232b247 he plants from extreme north, which according to ALL analogy would have been very unlike Southern Eur
233b249 Waterhous  remark Australia Fauna so far. Indian ALL the rest. Timor according to Mountain chain ough
233b250 bell.-- -- It was most curious to observe, that ALL the species of mice in S. America. which were ha
234b261 Dr.  Smith, he says, is deeply [not located] of <ALL> genera, «in all classes» are not a few only cos
234b261 ys, is deeply [not located] of <all> genera, «in ALL classes» are not a few only cosmopolites, & in g
235b262 f dogs in different countries a case in point.-- ALL cases like Irish & English Hare bear upon this.-
235b273 lants-- Would it not be possible to work through ALL genera, & see how many confined to certain count
239cIFC en between («beginning of» February & July 1838) ALL good References selected Dec. 13 1856 Also looke
240c004 h ease, is analogous to what occurs in plants.-- ALL these facts clearly point out two kinds of varie
240c004 other  birds & brought up by hand.-- These facts ALL account for [not located] Falkner Patagonia no d
240c013 an» & Celebes.-- ▐ Amboina; Viverra Zibetha. ▐-- ALL the Moluccas, Waggious New Guinea. New  Ireland,
241c015 er rongeur in Australia.-- p 67 ¿American forms? ALL Infusoria. not extinct species. good Resumé do/p
242c018 w S. Wales. <V.> p. 123 Crocodile at New Guinea. ALL the isles of Oceania have the Scincus with golde
243c019 ts dans de si hautes latitudes"., --¿translate?) ALL Australian forms have representatives (& instanc
245c024 as written Flora of Falkland Islds, where is it? ALL the Society isles have the same productions p 29
245c024 s much.-- The (p. 296) Columba Kurukuru found in ALL Malasia-- & oceania, offers many varieties in ea
246c026 cedo ispida) from Molluccas. scarcely differs at ALL from those of Europe, but beak rather  sharper.,
247c028 & New S. Wales Scincus Cyanurus «p 8 &». p 49 on ALL the Moluccas «New Guinea, New Ireland» & ««even»
248c033 bridity, namely the [not located] animals unite, ALL the change that has been accumulated cannot be t
251c041 o Denham Clapperton &c on Mammalia no doubt will ALL be included in Smiths work «do» Vol. IV p 273. M
252c045 immense importance of local faunas foundation of ALL our knowledge especially great continents»  Give
252c045 own good species Indian species so distinct that ALL analogy from each other I do not know how differ
254c048 rsupial, low Cerebral structure??-- «do» p. 390. ALL classes of Acrite exhibit lowest stages of anima
254c049 except  generation & digestion in Acrite Kingdom ALL organs blended together, & same organ «where eli
255c051 e.-- shape of wings have altered many times, but ALL have had feathers.-- if wing totally obliterated
255c052 scription of class is description of ancestor of ALL birds, & so for birds. We thus obtain an abstrac
255c053 incessant change in her offspring. ,has invented ALL kinds of plan to insure stability; but isolate y
256c055 nt in tracing history of Man.-- granted.--but if ALL other animals have been so formed, then man  may
257c056 ly 12,000 ft above sea in Bolivia; he examined-- ALL species & found "beaucoup des mêmes oiseaux. que
257c057 did  the otter live before it had its web-feet-- ALL Nature answers to the possibility.-- My views wi
258c060 country has some analogy in not gaudy colours so ALL changes may be considered in this light.--XX Zoo
258c061 victorious deer, hence males armed & pugnacious (ALL order; cocks all warlike)» «thiis wars against i
258c061 hence males armed & pugnacious (all order; cocks ALL warlike)» «thiis wars against in any class:, tho
258c062 e Abercrombie p. 172. for definition of Analogy. ALL the discussion <after> about affinity & how  one
259c063 at species must take to that particular habit.-- ALL structures either direct effect of habit, or her
259c064 of species-- Epidemic amongst trees. Plane trees ALL died certain year. Extreme difficulty of TRACING
260c069 ine of valuable facts,. regarding habits range & ALL kinds of information-- instinct Swainson's remar
261c069 be  studied. There is capital table of extent of ALL species Accumulate instances of one family sendi
264c080 into shrikes & at the other into Crows. yet ALL forming, according to Gould, good genus Gould se
267c093 introduced  from Norfolk Isld-- "& it now claims ALL the moist & fertile land of Tahiti, in spite of
272c109 n families.-- thus the banded tarsi is common to ALL the Laniadae & Muscicapidae of new World, but no
273c111 Gould), but the latter is obscure because nearly ALL are so.-- Thus in Hawks, there is a swallow, bot
273c112 he case in swallow??? which is most wonderful of ALL. ¿whether in most aerial of swallows.» Milvulus,
274c114 acters vary in degree in last instance hardly at ALL developed. Not confined to one species, but gene
274c115 d even in Lessonia &c & In relations of affinity ALL organs change together, in analogy certain parts
274c119 ave clear indication [not located] alone, but on ALL the general arguments-- Lamarck was the Hutton o
275c121 ieties is reduced to scarcely anything.-- almost ALL imagination-- He says he recollects all half Bre
275c121 - almost all imagination-- He says he recollects ALL half Breed cattle of L'Darnleys were most like pa
276c124 t number of vertebrae are produced, where, (& in ALL such structure) there cannot be gradation. See w
280c135 in  nature. <t>---- Many animals not breeding at ALL in domestication. throws great difficulty in way
281c138 o does it in real classification The relation of ALL cock birds in Gallinaceous having tendency to lo
281c138 ow propagation of forms.--just same way as <we> «ALL men» not all equally related to each other I can
281c138 n of forms.--just same way as <we> «all men» not ALL equally related to each other I cannot help thin
281c138 analogy  might be traced between relationship of ALL men now living & the classification of animals--
282c140 e external contingency.-- affinity is the sum of ALL the relations, analogy is the close relationship
282c142 preexisting  race, thus the analogy would not in ALL cases be produced, but would depend upon exclusi
282c142 ect to species of each other, because we suppose ALL descended from same.-- but if two original speci
282c142 species, each became ground then the relation of ALL the ground cuckoos. would not be affinity, but t
282c143 ined to such points by the vital laws.-- so that ALL character originally may «must» have had the cha
283c145 n compelled to breed hybrids produced--}» & then ALL that I want is granted.-- For. at Galapagos. mak
283c145 enera.-- let short billed one, be exaggerated, & ALL rest destroyed far remote genera. will be produc
284c149 if  general.-- Where any structure is general in ALL species in group we may suppose it is oldest, &
285c151 ary to show hybridity from few forms, parents of ALL species not possible in some detail,-- the relat
285c152 l were put together, they would not according to ALL analogy breed together.-- The bottom of the tree
286c155 y in arranging animals in paper as drying plant, ALL brought in one plane Fleming Quarterly review sa
286c156 ly any traces of passage a difficulty, but after ALL a slight one It will be necessary from manner Fl
287c158 in eye of mollusca. [ + ] These questions may be ALL disputable, but the one end of classification to
289c160 en so rare in some countries-- nightingale do.-- ALL shows how nicely adapted species to localities}-
291c167 female  it is a wonderful relation going through ALL Nature.-- Makes hermaphroditics. one step in ¿s
291c167 explain  genera & classes. if extinct forms were ALL fathers of present, then there would be  perfect
294c175 st one that bred one was diseased in its loins & ALL were so afterwards, (forgets authority).-- Lonsd
296c184 nyhow not Australian) on Peaks. Did Creator make ALL new yet forms like neighbouring Continent.  This
296c184 will  be plants & land animals. & land shells.-- ALL in short Extreme North = = to peak of Teyde in r
298c191 ng as island increases.-- upper parts attracting ALL the moisture.-- Henslow thinks if leaf of plant
298c192 w thinks if leaf of plant varies, «whole cross» ALL organs» vary in plant. The variation in characte
301c199 cling from short time.-- My theory must encounter ALL these difficulties.-- Knowing that animals  have
301c200 is  not applicable. The degree of development of ALL animals of same class being about equal.-- organ
302c202 «great  evil» from vast opposition in opinion on ALL subjects of classification, I must work out hypo
302c203 due to accident as submersion of land containing ALL of intermediate Father-species, & not, therefore
302c203 e physical change.-- If Patagonia became fertile ALL intermediate species living there would be destr
303c205 creation this would have been fair, but to place ALL that ever lived <on> into one list is unfair, [m
304c206 icular organs.-- of two adjoining families & not ALL organs blending away.-- +++ Hopeless work to sys
304c206 Hopeless  work to systematist. who believed that ALL his divisions merely marked his own ignorance.--
304c207 common with other birds reveal the secret.-- Now ALL the different forms of Synallaxis. trifling char
304c208 al structure have either been more rapid than in ALL other birds, or that it sprung from a branch hig
306c214 g-haired cats are supposed to come from there.-- ALL the sheep are thick-tailed The dogs called Persi
308c216 l. Soc Vol I It is capable of demonstration that ALL animals have never at any one time formed chain,
308c217 ivilized Man, May exclaim with Christian «we are ALL» Brothers in spirit-- all children of one father
308c217 with Christian «we are all» Brothers in spirit-- ALL children of one father.-- yet differences carrie
308c218 life of Mackintosh Vol II. p. 495)-- in fact, in ALL reasonings, of which human nature is the object,
309c220 f animals-- argue «opening» case. «thus» Educate ALL classes-- avoid the contamination of <cl> castes
313c236 ect the process.-- but why two sexes scission in ALL cases probably gemmation (.Ehrenberg) --not nece
315c243 nd this argument with his having canine teeth at ALL.-- This way of viewing the subject  important.--
315c243 zzler-- Under this point of view. expression «of ALL animals» becomes very curious.-- a dog  snarling
317c251 les N.?) replaced by three other species.-- Says ALL the hares West of <all> Rocky Mountains have pec
317c251 ree other species.-- Says all the hares West of <ALL> Rocky Mountains have peculiar character in extr
317c251 mbs, so that he first thougt only one species. & ALL hares on East side have other peculiar appearanc
319c255 Bachman  says he thinks the Mocking thrush beats ALL English birds in song.-- one of their thrushes e
322c270 OEconomy.  edited by Owen. read several papers-- ALL, that bear on any of my subjects Elie De Beaumon
323c269 ravel in Africa; «well skimmed.» 1839 Jan 10t.-- ALL life of W. Scott.--, except the V Volume.-- -- 1
325c267 by  Isid. Geoffroy. St. Hilaires. 1832. contains ALL his fathers views.-- Quoted by Owen.-- Hunter ha
331dIFC tes on Monkeys recognising Sexes of animals:]CD [ALL Selected Dec. 14-- 1856]CD Towards closer I first
332d001 other cases.-- He thinks apoplexy affects people ALL over England at same periods When he began pract
332d003 crossed--is  Alderney oldest breed-- He believes ALL pretty much alike.-- My Father Water-in the hair
334d011 her leader mare,-- this stallion though eager to ALL other mare had been entirely broken from their m
334d011 to  brothers & sisters in Mankind.-- The case of ALL blue eyed cats (Fox has seen repeated cases) bei
```

```
334d012 ich form a marked wild variety. doubtful whether ALL are white. Fox says the Half Muscovy Fox says  a
335d015 - If mules did breed, the offspring would «as in ALL other animals» be like either parent, or interme
335d016 ts offspring, which have been gained slowly, now ALL the mules have their whole <body> form of body g
337d021 speculations. Must not go back to first stock of ALL animals, but merely to classes where types exist
341d029 elation in one point or many) Owen answered that ALL characters might be considered as adaptive & tha
341d031 tells me line of Rocky Mountains separate almost ALL Mammals of N. America & many birds.-- which howe
342d033 species.--  The hybrids do not vary (ie the hens ALL alike & Cocks all alike) More than parent specie
342d033 brids do not vary (ie the hens all alike & Cocks ALL alike) More than parent species-- Mr Blyth remar
342d034 ell remarks he has somewhere met conjecture that ALL salt-water fish were once salt water (as they al
342d034 continents)  but Ogleby well answers that nearly ALL F. W. Fish are Abdominals. ∴that order first con
343d037 arring against those very laws he established in ALL <nature> organic nature) the Rhinoceros of  Java
345d042 hen case of avitism.++» Three gentlemen of party ALL thought with pigs &c, that hybrids were uncertai
345d043 ell fixed» breed,: Jones says Sussex cattle were ALL white headed, but this was bred out & now all ar
345d043 re all white headed, but this was bred out & now ALL are pure red, yet calf every now & then born wit
346d048 839.-- are bad breeders & subject to the rush as ALL animals which breed, in & in are-- colour white,
346d048 old.-- a Mr W: Hall remarked that it was against ALL rules their preserving character & breeding in &
348d050 genus expresses as now used almost any group.-- ALL groups natural (p 6) as expressing natural affin
349d052 tamus, solely owing to number of lost links. if ALL species know they would be innumerable does not
349d052 cial, as interlopement of Marsupials will change ALL.-- & so on no one will settle number of  primary
349d053 other subgenera will come from. common stock.-- ALL genera, common stock.-- so that value can only b
350d053 mount  upwards.-- «judged by analogy»-- Consider ALL this NB. How can local species as at Galapagos.,
350d055 female  «(I)» fixed & blind: -- Macleay observed ALL these facts prove that perfection of range has
352d058 ead-- hence there is no central radiating point, ALL united . (links in circle must be granted unequa
353d061 38 vol II p. 402. Mr Gould on Australian birds-- ALL Eagles. of Australia characterized by wedge tail
358d076 or less developed state.-- the female & young of ALL birds resemble each other in plumage «(that is w
361d095 th, at Zoolog. Meeting stated, that Green-finch, ALL linnets red-pole, goldfinch, hawfinch-- in nursl
361d095 t of Cross-Beak-- In lark if I understand right, ALL species have same character which is mottled  &
361d096 row, (if I understand rightly) young cock & hen, ALL nearly similar.-- in blackbird group young  like
362d099 pposes this a consequence of the female breeding ALL the year round. ask Colonel Sykes.-- Even our do
362d100 se on Domestic Pidgeon, in which it appears that ALL the <bird> varieties «,now know» were then  <pr>
363d101 n exaggeration of cooing.--» & compare them with ALL the varieties.-- Habits of rock pidgeon. (I susp
365d107 can perceive that Natural varieties or species., ALL the structure of which is adaptation to habits (
366d108 hanged, & if their characteristic qualities were ALL deeply imbued in them from long permanence, so t
366d108 ply imbued in them from long permanence, so that ALL their peculiarities must be transmitted if their
367d113 Saw  mule. apparently fathered by a donkey. with ALL four legs ringed with brown.-- animal like large
368d114 vidently the male which recedes from the species ALL females being most like offspring, Q (how is thi
370d116 (p. 451.)-- Wasps breed many females, but almost ALL die.-- bees breed but few, because they are kept
370d117 Collect cases of difficulty of growing plants in ALL parts of world, thus tea tree in Brazil must hav
371d127 arden having coloured offspring.-- but surely in ALL these cases an unseen change is produced in pare
372d129 in one stem.-- a bud may be transplanted & carry ALL these peculiarities not so a seed.-- Bud probabl
375d135 decreasing in number must effect instantaneously ALL the rest.-- One may say there is a force like  a
375d135 y thrusting out weaker ones. «The final cause of ALL this wedgings, must be to sort out proper struct
378d147 e metallic tints, such as Magpie, Jay, & perhaps ALL the rollers-- «He says» whenever metallic brilli
378d148 origin  of any identical bird-- for they were of ALL colours.-- they were "half wild-half tame,  they
379d152 he ewe must never be put to any other breed else ALL the lambs will deteriorate.-- Lord Moreton's Cas
381d156 ca, with pectinibranchiate order-- the Annelida. ALL others, <animals,> «are Dioeecous as» Cephalopod
381d156 branchiate molluscs.-- insects. spider crabs.-- (ALL these however do not require coition every gener
381d157 ose still more that they must in effect be so in ALL.-- 2 NB. In Pectinibr Mollusca.-- «or Cephalopod
382d158 n.-- Mine is much simpler.-- Hunter shows almost ALL animals subject to Hermaphroditism,-- those orga
385d163 as known many cases of bitch going to mongrel, & ALL subsequent litters having a throw of this mongre
387d169 e peculiarity for first time & if their <D & E> «ALL their offspring» inherit the same peculiarity in
388d173 - [Why in making a bud, which is to pass through ALL transformations, should there need two organs; w
390d178 t to differ from those of other.-- The upshot of ALL this is that effect of Male is to impress some d
390d178 onnected with the physical differences in almost ALL Male animals?]CD If the male «in the course of s
390d179 t breed, but the female at least (¿male?) looses ALL appetite.-- It is the comparison of each  animal
391d174 its  parent. <but these buds do not procreate> & ALL alike in one parent or tree, (but not in other t
391d174 ates those differences which are in harmony with ALL its previous changes, which mutilations are not)
391d174 tend that way «& from frequency of this tendency ALL mammalia must long have so existed.» with double
391d175 ferences (hence every individual is different). (ALL this agrees well with my view of those «forms» s
392d175 very  curious how this was superinduced? (Surely ALL are really dioecious..) only simple form of life
400d011 Thus  Hattica is great genus.-- because found in ALL quarters: his ideas not clear. In Australia from
400d013 -- says Decandoelle, distributed seeds of Dahlia ALL over Europe same year.-- he sowed them for  four
400d013 t suppose one aboriginal variety.-- for they are ALL made by fertilizing one plant with another-- Unc
401d014 y cabbage, where a great many will not return to ALL sorts of varieties, which he attributes to cross
401d014 er stocks. Says if any variety of apple be sown, ALL sorts come up from it. lately saw a nonpareil so
401d015 ver tried of separating apple tree entirely from ALL others-- so my experiment of strawberry not so a
401d016 any varieties, & probably would take long before ALL the stain would be got out of it.-- Now this  is
402d018 ls. which are met with near Canton" "Here, as in ALL Malay countries, I noticed a peculiarity in  the
402d019 ked, as if they had been broken". are born so in ALL Malay Countries W. Earl Eastern Seas. p 233 Octo
404d025 tempted to think that it must have been invented ALL at once.-- but naturalists if they had series pe
406d038 n of pustular disease following handling sheep-- ALL cases: d degree p. 354-- The most vicious dog. w
407d038 change,  than any part, (except Europe. in which ALL Tropical forms have been obliterated) of the wor
408d044 ands, having been purely result of elevation,-- «ALL» modern & wholly volcanic-- Azores might be prop
408d045 r part of coast of Somersetshire the Cockles are ALL apt to be diseased., & some of them symmetricall
409d048 s, as individuals, & though we may not trace out ALL the ill effects. -- we see it is not the order i
409d049 nt, or many anterior epochs.-- but we can see if ALL species, there would not be social animals. «thi
409d049 s I hope to show is «probably» the foundation of ALL that is most beautiful in the moral sentiments o
410d051 changes. = this could only be effected by sexes: ALL the above should follow after discussion of cros
410d052 one part affecting another.-- (I from looking at ALL facts as inducing towards law of  transmutation,
410d052 ons of country, most important & will be done to ALL countries,-- but naming mere «single specimens i
410d052 than  useless.-- yet there is no cure «I may say ALL this, having myself aided in such sins» (do  not
411d054 ow whether the individuals «forms» are permanet, ALL steps in the series, their relation to the exter
411d055 have  been fixed, to study the physical causes. «ALL Cuviers generalization. of teeth to kind of extr
412d058 Poland. &c.-- Three principles, will account for ALL (1) Grandchildren. like. grandfathers (2) Tenden
413d060 ble from their alliance to Articulata, which are ALL truly bisexual. Buckland's Reliqu: Diluv. says A
414d063 wise [not located] Are the feet of water-dogs at ALL more webbed than those of other dogs.-- if natur
416d071 terrestrial  animals.-- in shells?-- insects?.-- ALL!?!?!-- Worms? [Barnacles, aquatic., <yet> Crusta
417d076 shells of the common species: from one locality, ALL left whorled.-- He kept two to see if they would
417d078 ither: & c alike progenitors.-- in some families ALL the children like mother & in some like father «
418d082 ation were impressed on it.-- Seeing that <Man> «ALL vertebrates. [Müller's Physiolog. p.24.]CD» can
418d083 in  each species,-- & knowing from analogy, that ALL these very animals are descended from some one s
419d087 must serve to confound our chronology» «CONSIDER ALL THIS» Extinction & transmutation, two foundation
420d089 being six metamorphosed vertebrae, the parent of ALL vertebrate animals.-- must have been like some m
422d095 he simple animals to become complicated although ALL perhaps will have done so from the new relations
423e096 vegetable in one quarter of the world would kill ALL of the one herbivorous. & its one carnivorous de
423e097 plify the organization of the different beings, (ALL fishes to the state of the Ammocoetus) Crustacea
427e108 rms. once formed. would remain stationary, hence ALL present types are ancient. According to my views
427e109 ancient.  According to my views of <Dioecious p> ALL plants, being occasionally dioecious; & really d
428e111 's monpe-- my principle being the destruction of ALL the less hardy ones. & the <accidental> preserva
430e117 als.-- grateful & intelligent.-- The theory that ALL animals have sprung from few stocks. does not be
430e117 extinction  is denied.-- it will not account for ALL species. even if it will for all.-- Varieties ar
430e117 not account for all species. even if it will for ALL.-- Varieties are made in two ways-- local variet
430e119 s.-- the beak of this one has concentric striae, ALL the lower part rayed longitudinally «give woodcu
431e123 sses, render my view of the crossing of mosses & ALL others by action of wind difficult.-- Cline on t
433e127 es.-- The fact of tumbling pidgeons; flying high ALL together & then tumbling, far more wonderful tha
434e130 . which <to> seems to have taken place.-- Almost ALL Dioecious & Monooecious plants have  rudimentary
437e140 ll & desperately bitten the young ones, would in ALL probability have escaped".-- if it had not  been
437e141 ddenly to type when brought back to home. (& yet ALL the varieties of Brassica certainly not becoming
```

Page
(Key Word)
```
437e141 ds on character of antecedent races.-- «& yet in ALL probability the Brussels Sprout was slowly forme
438e142 [NB  TIME is element in change, as in Dahlias]CD ALL much varied breeds both plants & animals have lo
440e145 which changes the acorn into the oak.-- In short ALL which «Nutrition, growth & reproduction» is comm
440e145 «Nutrition,  growth & reproduction» is common to ALL living beings. vide Lamarck Vol II. p. 115. 4 fo
440e147 se» corresponding change in Birds of Paradise.-- ALL that we can say in such cases, is that the pluma
441e150 r of flowers, Hydrangea -- black bullfinches-- & ALL varieties must be presumed to be result of  such
441e151 ry organ is become fixed. & cannot vary.-- which ALL facts show to be absurd.-- As there are  plants,
442e153 ated by gemmation should be absolutely similar; [ALL the gorze in Norway ought to be thus characteriz
444e157 e Icthyosaurus 60 or 70 bones in the paddle, yet ALL in the arm are perfect.-- p. 144.-- Alludes to t
446e163 e no change would effect them, without affecting ALL the individuals-- «-- hence there would be  real
447e164 en this-- (Poa alpina vivaparous sometimes seeds ALL species of Lemna sometimes though very rarely fl
448e166 of animals on ice p. 643-- very curious table of ALL the castes from Stephenson at Lima The same nume
448e168 1839  The Lepidosiren-- Amblyrhyncus & Toxodon, <ALL> equally aberrant-- the two former connecting cl
449e169 mal?-- Eyton says that the young of two hatches  ALL alike» between the male Chinense & female common
449e169 m-- they interbred. & the young kept constant. & ALL alike Waterhouse says some of the Galapagos Hete
450e175 nce it appears there are shades of difference in ALL the isld, like in wild animals).-- There are pre
451e178 ferous region (mentioned by Heber) «from» which «ALL» mankind «(& yet afterwards says native tribes c
453e182 is found in Borneo.-- «p. 233» There, as well in ALL Malay countries «the» cats are born with the joi
454e184 es one, sex & sometimes. other, so as to become <ALL> monooecious.-- Are there not wild plants,  some
458t001 & Tortoises gigantic-- hyaena-- bear & ruminants ALL of larger size.-- the  law of large size establis
459tf1v been occupied [...] species, which has undergone ALL the changes. [im]portant view, copie[d] Gleaning
462f205 is incidentally said that a mongrel man may lose ALL traces of his parentage in «about» seven «7» gen
466t094 oscis within nectary «they do» they must disturb ALL anthers, wh otherwise lie protected by the hairy
466t094 visiting  it.» In Columbine nectaries are placed ALL round flower as they are in Crown-Imperial Lily
466t096 of  Plants-- also goodness of flavour in fruit-- ALL affected by cultivation during life of individua
467t103 ectly ripe & pollen abundant filaments & stamens ALL protrude «there is a brush at end of stigma, whi
467t104 wings  seem beautifully to protect sheath {a} In ALL these nectar seems to be at base of super  petal
467t105 swarmed with meligethes & small Staphylinidae on ALL their bodies pollen-- on a sulphur Broccoli  not
468t112 plant  to plant.»-- to my grt surprise-- I found ALL, stamens straightened pollen profusely shed; len
468t112 gthened & turned up «more than stamens», so that ALL were brushed by Bees & especially stigma after b
469t151 in  rows «close to each other» & seeds gathered «ALL» came up in 1840 true. Shrewsbury.--  Abberley--
469t151 seeds gathered same year came up true «in 1840»: ALL in together blossomed together The seeds of thes
470t178 with  violence in shower On many Papilionaceous; ALL wh. are in flower «I saw Bees;»-- on Monk's Hood
471tf09 ps (Lardner's E. vol. II p. 18.) capital list of ALL the fossil Mamm. of Europe-- Large Lizards in Na
472s01r foliage not «being» like the true P. bracteatum; ALL supposed to have been hybridised == Has tried se
472s01v were  procured with the P. orientale in garden & ALL came up hybridised. It is possible to raise them
472s01v ge & no black spot & a third considerably paler, ALL rest very similar-- June 2. 42 Maer «Thursday» T
473s03v n-- Fish one step lower in America-- How curious ALL negative laws of America of depth of organisms h
473s004 has  been «as» much subsidence as elevation then ALL continents of cretaceous periods, together  with
473s07v iarities may be early impressed & others later-- ALL poultry with same down-feathers. Zoology 1856 Sk
477z002 tatous (.4 pichye, pelud, mulita et mataco.) are ALL found south of 26 degree 30 minute. Lat -- -- do
483z013 irimus. of Geof.-- reference from Rüppel travels ALL Owens papers on Intestinal worms must be studied
484z016 of  same plumage.-- general red mark on wings of ALL-- Spix has described Philedon. allied to some of
492q002 Are the wild Bananas of Otaheite seedless;-- are ALL varieties seedless-- if so. how have varieties b
495q05a nsifolium «an annual» «sleep» «closes flower» on ALL gloomy days.-- The «garden» Coronella also sleep
495q05a ed.-- Examine scum of pond for seeds.-- 11. Soak ALL kinds of seeds for week in Salt. artificial wate
498q008 eas, beans, as raised, do the Seedsmen select at ALL from the plants? If not, I am surprised «plan» s
499q09v Lychnis. Butchers Broom-- «also, Vinca,» Examine ALL these, are they much frequented by Bees or Butte
499q09v ny plants which are known easily to be crossed & ALL monooecious plants.-- Hooker says Rafflesia is d
500q010 Does  the yellow white Butterfly deposit eggs in ALL varieties of Cabbage. (26) Do deer Keepers cross
504q13v ly the Peas to cross ought to be placed far from ALL other Peas, from Wiegman Shrewsbury (1)  Peas.--
506q015 nts grow near each other.-- ? Cannot remember at ALL. (37) Any cases of plants. which will not produc
509q017 Rhinoceros or Whale, than in old?? Falconer says ALL in cases. Owen. Have talked partially with him A
513q020 characters  are then intermediate or «sometimes» ALL on one side, as in crossing varieties Amongst va
522m010 f such a man.-- (My Father explained why he had ALL about him, but still maintained he had never hea
522m011 name of Child «of Kinlet» & married Miss A. B.-- ALL the same names as a few minutes before he mainta
522m013 , often ends in insanity or delirium.-- In Mania ALL idea of decency & affection are lost.-- most del
523m017 during  these two years that she had been insane ALL the time.-- There are numberless people insane o
523m018 trances on him» which are never generally, if at ALL discovered.-- <Sup> Sometimes comes on  suddenly
525m025 t Shrewsbury & its kittens <h> (in number 3) had ALL short tails; but one a little longer than rest «
525m025 t tails; but one a little longer than rest «they ALL died»:-- she had kittens before & afterwards wit
526m029 enes which I first recollect, «at Zoos» they are ALL things, which are brought to mind, by memory  of
526m029 mber the part of page.-- one is tempted to think ALL memory consists in a set of sketches. some real-
527m031 ts, one will rise according to law. How strange <ALL> «so many» birds singing in England, in Tierra d
531m050 (How  often one forgets where put one key. where ALL keys are placed) Memory cannot solely be  number
535m064 be  normal. with insects, but habit forgotten in ALL older species. The earwig & a doubtful one of Ac
536m070 ws of organization.-- M. le Comte argues against ALL contrivance-- it is what my views tend to.-- Whe
536m072 ing cannot doubt that they have free will, if so ALL animals., then an oyster has & a polype (& a pla
538m080 cond & unreasonable man.-- If one could remember ALL ones farthers actions, as one does those in seco
539m081 ther, & read so intently as to be unconscious of ALL around, yet there was no strain on the intellect
540m085 dog & horse & man yawn, makes me feel how <much> ALL animals <are> built on one structure.-- He who d
541m091 & think of qualities as flowers, cloth &c & with ALL this difficult EXPERIMENTIZE upon this effort.--
542m097 ften said» of language in man is very great from ALL animals-- but do not overrate-- animals communic
543m098 h other.-- Waterhouse says far more instincts in ALL of the Hymenoptera;. <therefore> than in other o
543m098 llectual.-- Hymenoptera typical insects. ie have ALL parts. Waterhouse Study well the greater  number
543m100 t mathematicians not being profound reasoners.-- ALL same fact-- for, as Jones observed, in playing c
543m100 ny places, & contingency a man has keep in mind. ALL is certain.-- there is judgment of probabilities
544m102 cters of dreams no surprise, at the violation of ALL <rules> relations of time, <identity,> place & p
544m103 ughts are broken-- Sir J. Franklin when starved, ALL party dreamt of <goo> feasts of good food-- The
544m103 ally moves.-- the willing therefore is ideal, as ALL other perceptions.-- The mind thinks with ex
545m106 -- I see monkeys grin with passion, that is show ALL the teeth: «& make noise not like pish, but like
547m113 tances were such one naturally would so so!) Now ALL these parallel trains of thought necessary heirs
547m114 , effect of not reasoning. effect of not having <ALL> other trains of thought, or memory from innumer
549m123 ry & no doubt were preservative, & are now, like ALL other structures slowly vanishing-- the mind  of
550m123 is no more perfect, than instincts of animals to ALL & changing contingencies, or bodies of either.--
550m125 back  to happy days, are they not those of which ALL our recollections are pleasant.-- Browne Religio
552m131 l images-- Mem Chiloe <pi> Sow, who carried from ALL parts straw to make its nest. Figs & Elephants,
552m132 e that on long run will do good.-- alter will in ALL cases to have & origin as well as rule will be g
555m142 rity & found everywhere (is it not present with ALL associated animals?) I doubted it in Fuegians, t
555m144 kind of wit, showing he had honourable wounds.-- ALL this was kind of wit.-- I changed I believe from
556m144 recovering from hanging on account of blood. but ALL these idea came one after other, without ever co
557m148 arched. & tail stiff.-- is shame, jealousy, envy ALL primitive feelings, no more to be analysed  than
558m153 ng arches wings-- as does black Swan.-- Goose do ALL species put their necks straight out & hiss.-- [
558m153 side part of nostril, when passion commences.-- <ALL> Nearly all will exclaim, your arguments are goo
558m153 nostril,  when passion commences.-- <All> Nearly ALL will exclaim, your arguments are good but look a
563nIFC or Species Theory» Dec. 16 1856 Looked through A ALL other Books May 1873-- October 2d.. 1838 Essays
563n001 ood pidgeons building near houses. yet so shy at ALL other times.-- Birth Hill shows it is evergreens
563n001 ground.-- Cock fowl. on the ground, at roost, in ALL seasons, & after? he has done <g> crowing.-- ins
565n009 compression of mouth showing action,-- sulkiness ALL negative expression? Expression of affection  is
567n014 ock in passion & sends blood to its breast &c &c ALL Science is reason acting «systematizing» on prin
567n016 is obliged carefully to separate its memory from ALL ordinary lines of association.-- is totally dist
568n019 pretty & what ugly than a cow--" «so it is with  ALL uneducated.--»-- Old man at Cambridge observed t
569n022 nied by dignity-- "no mon dieu," with a shrug-- "ALL I can say, I am very sorry so it is"-- does  not
571n027 mpressed on us.-- Surely we have taste naturally ALL has not been acquired by education. else why  do
571n028 e why do some children acquire it soon. & why do ALL men. agree ultimately?-- We acquire many notions
571n030 ctrine «of instinct» has been carried too far In ALL the foregoing cases most difficult to distinguis
572n032 overing body» &c &c is matter of custom."-- this ALL applies to bodily weakness & inferiority, but no
```

574n041 ays.= No doubt man has great tendency. to exert ALL senses, when thus stimulated, smell, as Sir. Ch.
578n056 does an injury of the spine-- that it paralyzes ALL muscular action -- «in man & animals» Blubbering
580n061 when in playing by memory. she does not think at ALL, whether she can or can not play the piece, she
581n063 en the phrase "heredetary habits." very clearly, ALL I must do is to generalize it, & see whether app
581n063 is to generalize it, & see whether applicable to ALL cases.-- & analogize it with ordinary habits tha
583n069 their parents offer best cases of instincts].CD ALL this may be true,, but relation of imitation & r
585n076 e pouted & whined, when, man went out of room.-- ALL theories of magnetic powe in birds, seeing the s
585n077 w, stonge henge raised, yet not instinct, but if ALL men placed stones in same position, it would be
586n078 n.-- «but does not apply to dogs.--» they may do ALL this instinctively «yes because power varies in
591n099 disobeyed-- I often have «as a boy» wondered why ALL abnormal sexual actions or even impulses. (where
591n100 ot fit, as cannabalism, is held in abhorrence.-- ALL this makes analogy of actions with «&» against b
592n103 uth in fright because nature intends to lay open ALL senses: <do> Horse prick his ears «& snort clear
593n111 water, suddenly, as if by word of command, they ALL took flight & flappered across pool to bed of fl
594n112 that probably few had ever before seen one, yet ALL-- flew to bed of flags. hernes are common. not u
599o005 s originally musical!!!??-- At least it appears ALL speculations of the origin of language.-- must p
600o008 ry Beau ideal, Mem. Negro, beau,--Jeffrey denies ALL Beau-- How does Hen determine which most beautif
602o011 partial results, & the impressions on them are <ALL> remembered, when the meaning or reasons are for
602o11b Upon the whole it seems."-- "that the object of «ALL» art is the realizing and embodying, what never
603o012 Chinese, S. American. Polynesians Jews, African ALL sacrifices. How completely men must have personi
608o026) difference is from imperfect condition of mind ALL motives do not. come into play.-- †It may be urg
609o030 t good «or rather what was necessary for good at ALL» is the «instinctive» moral senses: (& this alon
610o032 wonderful manner.-- as the hour of the day &c-- ALL habits must conduce to their health & comforts.--
613o035 directs <to> other parts of body. to do such.-- ALL this can take place & man not conscious as in sl
613o036 our intestines have? [RHC] 5) Kirby thinks that <ALL> there is one one instinct to all animals modifi
613o036 y thinks that <all> there is one one instinct to ALL animals modified according to species. This I su
613o036 aying that the thinking principle is the same in ALL animals. [LHC] «3)» Eyton told me that his retri
615o038 t corresponding change in «external» man; and as ALL men nearly same species, so general instincts ne
618o041 of sulphuric acid for metal of another person at ALL, we can only infer it from his its behaviour. Th
618o41v e to another, & <the> (or conceive it) & that is ALL we know of attraction, but we cannot see an atom
618o41v think: they are as incongruous as blue & weight; ALL that can be said that thought & organization run
622o048 child sees uniformly performed by the teachers & ALL around him, will be paramount,-- hence the law o
622o048 far more «feelin» acute the feeling really is.-- ALL these associated «habitual» feelings. become lik
623o050 social feelings.-- [LHC] According to my theory, ALL instincts demand some explanation [RHC] Although
623o050 but that which has beneficial tendency through <ALL> «many» ages. could be acquired, & we are certai
623o050 acquired, & we are certain from our reason, that ALL which <has> (as we must admit) has been acquired
624o051 C] *for it strives to give conduct beneficial to ALL the children, « <then> each himself» & parents,
624o051 then> each himself» & parents, & hence to nearly ALL the world.-- As conditions change, from civiliza
628o53v to commence with scarcely possible to teach it-- ALL dogs might be taught, but not cat, that is not a
628o53v the problem, of ethics-- [my answer would be to ALL such cases-- either, that from the necessities «
628o054 D p. 22. says affections, desires, & moral sense ALL different.-- P. 22. Butler & Mackintosh characte
628o54v re.-- at least point of «false» honour will stop ALL wish to gratify <it>-- anything contrary to it]C
629o54v <instinct> «conscience») equally <prefe> destroy ALL wish of outward gratification,-- see what cases
629o55v er question what it is desirable to be taught,-- ALL are agreed general utility (3) It is other quest
632j53r «then» believe the pappus of <th> any one seed. (ALL have not it) was DIRECTLY created. for transport
633j54r o arrest mud &c at deltas.-- Now my theory makes ALL organic beings perfectly adapted to all situatio
633j54r ry makes all organic beings perfectly adapted to ALL situations, where in accordance to certain laws
633j54r tor to the standard of one his weak creations.-- ALL such facts are merely relations of one general l
633j54r eaved by volcanic force, for these Marsh plants. ALL flow from some grand & simple laws.-- 4 «Study C
634j54v Many other good cases -- p. do]CD <Mac. remarks ALL Mammifers originally land--animals. as> 5 p. 314
634j55r inally land--animals. as> 5 p. 314. Mac. remarks ALL <land> Mammiferous animals originally terrestria
635j56r to accidents & any other cause.-- (& my theory [ALL PARTS OF ONE GREAT SYSTEM. C. D]CD [All this doe
635j56r theory [ALL PARTS OF ONE GREAT SYSTEM. C. D]CD ALL this does not explain death, but reproduction]CD
636j57r ick movements. (sliminess instead of barbs)-- In ALL these cases it should be remembered, that animal
637j57v -only we here see means-- but not in the other. ALL Bridgewater Treaties. are reduced simply statem
638j28r il, foot, sack. power of endurance &c & Camels? ALL good cases of correlations.-- [There must have be
639j28r nd in mole, & Mole cricket & rodents (?) p. 251. ALL animals run by hind legs-- Kangaroo. only a cari
318c255 ountains & on one lofty isolated spot on the ALLEGHANIES to which it migrats every year,; and probabl
472s01r here is such a thing as a species-- Jun 1. 1842 ALLEN W. sowed some years since gathered the seeds of
538m079 rgotten.» Such facts bear on such characters as ALLEN W. & Babington, both half ideotic in some respe
538m079 officer, horticulture & religious sects.-- yet ALLEN. W. remark about his slippers bad for fires, wh
147g016 hese upper patches if prolonged would intersect ALLEY above the 300 ft Alluvium <abo> by Loch Dochart
041r066 on a concreationary contraction: the fact is in ALLIANCE with those balls at Chiloe, full of sand.--th
240c014 Casoars, perroquets, establishes its «zoolog» ALLIANCE with New Holland. The Barbaroussa, (when youn
379d151 lluscs few.-- ¿are not some same-- what is the ALLIANCE with the Black sea.-- it would be ocean, what
413e060 Cirrhipedes are the more remarkable from their ALLIANCE to Articulata, which are all truly bisexual.
030r037 in the coraliferous mountain Limestone are they ALLIED to the jaws of the Cocos fish Rio Shells argum
070r153 mmon parent? why should two of the most closely ALLIED species occur in same country? In botany insta
187b065 in N. America & Asia, but many species closely ALLIED but different, because country separated since
195b100 ly to wander. Did it create two species closely ALLIED to Mus. coronata, but not coronata.-- We know
208b153 mals} See R. N. p. 130 Speculations on range of ALLIED species. p. 127. p. 132 There is no more wonde
216b181 spring black others white which is more closely ALLIED to case of cross of dogs.-- See paper in Philo
226b223 pecies. (¿are carnivorous Mamm: in Paris basin «ALLIED to present») more like present carnivora than
236b280 to some South American kinds.-- Are the closest ALLIED species always from distant countries, as Deca
246c027 s, Friendly Isles «& Hebrides») is very closely ALLIED to C. muscadivora., which lives in the Eastern
305c210 ing « <& Creat> sensible» principle (intimately ALLIED to one kind of organic matter.-- brain. & whic
310c224 rious case is Saxifrage, almost <same> <same> ALLIED» species Himalayas, 13,000 & Melville Isd.-- W
311c225 ist of birds in Europe & N. America, on closely ALLIED species. «replacing each other». good to consu
353d060 ing at animal, if there be many others somewhat ALLIED whether «like» parent stock, or not. Now wings
447e164 species periwinkle wants insects to impregnate ALLIED to Asclepias Turpin cell is individual May 29t
448e165 ht to require insects to impregnate it.-- it is ALLIED to Asclepias, where this is always the case ac
449e170 - that the «animals of» islands N. of Timor are ALLIED to the «type of genera in» islas de Sonda as w
454e184 p. 252. Is there any very sleepy Mimosa, nearly ALLIED to the Sensitive Plant.-- p. 290. Dr. Edwards
464t065 all wings, & surely the Apteryx is more closely ALLIED to the Struthonidae than any other forms-- Lun
465t089 62.-- some Mammals of Norfolk Crag. mentioned-- ALLIED Beaver to present forms.-- -- How many «tertia
482z012 the Mulita from 41 degree to 26 degree CLOSELY ALLIED species, therefore interlock.-- Testudo INDICU
484z016 on wings of all-- Spix has described Philedon. ALLIED to some of my birds-- These groups strictly Am
494q004 rd to former (8) Is form of globule of blood in ALLIED species similar.-- if not how is it in <allied
494q004 allied species similar.-- if not how is it in <ALLIED> varieties (9) Cross largest Malay with Bantam
581n064 be generalize.-- The tastes of man, same as in ALLIED Kingdoms-- "food, sm<e>ll.. (ourang-outang), mu
583n070 le of education; (this is again assumed as more ALLIED to reason than instinct.) Mr Wells I can see m
600o08a e wonderful, if, mind of animal was not closely ALLIED to that of men, when the five senses were the
602o11b tly true, but you do not convince me.--" Belief ALLIED to instinct.-- p. 134. a painted must not a ac
381d156 ipeds rotifers, trematode & cestoid Entozoa ALLOTRIANDROUS <or M> Mollusca, with pectinibranchiate or
222b205 ermediate in character, the same reasoning will ALLOW of decrease in character. (which perhaps is) Ca
300c198 ems to supply instincts,-- & those powers which ALLOW of, acquirement of language. heredetary & acqui
310c223 redetary & such definite thoughts, I will never ALLOW that because there is a chasm between Man (--<a
314c237 egment of the corolla being (probably) small to ALLOW it to lie on one side.-- but in other species,
360d093 ad mottled breasts, <when> of a sort that would ALLOW the offspring to have some different kind of mo
436e136 argue against this-- -- analogy will certainly ALLOW variation as much as «the» difference between <
440e147 plumage has not been so injurious to bird as to ALLOW any other kind of animal to usurp its place.--
447e164 redicament, as one, in which structure does not ALLOW of crossing with other individuals, «with facil
463t055 of Snake wonderful!! distinct!!-- He would not ALLOW such series showed passages-- yet in talking, c
491q01v at Spallanzani» Raise only single Plants & only ALLOW <few> one flower (5) Dr Fleming. Philosop. of Z
573n036 about our origin of a notion of a Deity We can ALLOW «satellites», planets, suns. universe, nay whol
591n101 ogue on Natural Religion.» however, he seems to ALLOW it is an instinct.» I suspect the endless round
606o023 ination through the medium of the eye"; he will ALLOW the secondary pleasures of harmonious colours &
627o053 ation of intellectual faculties-- Will Eugenius ALLOW this moral obligation? [2] [The improvement of
620o0045 4) as starvation, or fear of death, one makes ALLOWANCE & either excuses the «non-» following of ones

Page ***(Key Word)***
113a091 e may confidently infer that time has not been ALLOWED for lower beds to cool down. & then in 50000 y
198b113 thing) p. III G. St Hilaire Insects & Molluscs ALLOWED to be wide hiatus: states in one the sanguineo
250c038 . yet America preeminently equable. might have ALLOWED fresh species to have been formed & spread to
261c070 animals of N. America Hence it is universally ALLOWED that the discrimination of species is empirica
307c215 for if these abortive wings in the female are ALLOWED to the fully organized wings of the male rende
430e118 nts effectually the offspring are picked & not ALLOWED to cross.-- » Has nature any process analogous
595n117 heir power of imagination-- for it will not be ALLOWED they can dream, & not have day-dreams-- think
627o053 be subclassed as "disinterested" p. 14. It is ALLOWED, that we have conception of moral obligation «
294c177 t of the Work is its proof, «its limiting, the ALLOWING at same time true species.» & its adaption to
416e075 ces & variety? «Man picks the Male, instead of ALLOWING strength to get the day» The fertility of Ind
469t135 late in evening-- On rough calc. 280 flowers-- ALLOWING each Bee visits 10 flowers in «minute» each f
541m089 by decreasing headache) & found best plan was ALLOWING my mind to skip from subject to subject as qu
037r053 protect a rock, or that the rock not weathering ALLOWS such Compare the elevated estuary of the Plata
315c242 most remarkable habitual actions in plants, it ALLOWS of any degree in lowest animals --habitual act
445e158 s mentioned by Sir. J. Banks. p. 212.-- p. 282. ALLOWS this instinctive power in chicken, yet says it
537m076 , as in other races of mankind..-- p. 27. Mart. ALLOWS some universal feelings of right & wrong «(& t
112a089 nucleuse's of old volcanos within Cordillera-- ALLUDE to Lyell's view of not discovering dike one en
383d159 of becoming either sex.-- ‖ In my theory I must ALLUDE to separtion of sexes as very great difficulty
258c061 al-- Parrots in Macquarrie isld. vol III p 430 ALLUDED to by Capt. King do. p. 434. Table of birds fr
351d057 s., heart altered & umbilical cord,-- Broderip ALLUDED to Hunter's views on this subject.-- Monstrosi
442e152 hat little change is produced.-- The fact just ALLUDED to of Northern flowers, throws enormous diffic
490q001 uestion 1 Where has Duchesne described Atavism ALLUDED to by Dr. Holland-- <Jordan> Smith of Jordan H
101a058 he Delaware. is it Edentate? Phillips p 289.-- ALLUDES to big bones in interior at Falkland Isd.-- Pe
261c072 ck Owen talking of Plesiossaurus Plesiosaurus. ALLUDES to some structure in head, which he says (evid
267c094 they always like to cross their breed p. 333-- ALLUDES to the Macúsie breed no description given-- Ch
311c225 extending over 90 degree of Long. & Col. Sykes ALLUDES to some other case of 180 degree & great diff
353d061 «do» p. 403. & 404 vol II. do (p. <69> «71»). ALLUDES to Eyton's discovery of different number of ve
371d127 ices" &c &c Owen illustrates case of Dingo (he ALLUDES to the dholes or wild dogs of India) in Zoolog
422e092 of writings of Goethe.-- who maintains, that «ALLUDES to difference between fossil & recent Bull; li
444e157 Camel has not escaped Naturalists." Before he ALLUDES to the resemblances of the snout of the mole &
444e157 e, yet all in the arm are perfect.-- p. 144.-- ALLUDES to two theories;-- that species are the result
452e182 ngles of Borabhum & Dholbum.-- Vol do. p. 634, ALLUDES to fact stated by M. Tournal that skulls found
458t001 account of fossil frog, 40 inches in length--! ALLUDES to ancient gigantic salamanders-- Every order
633j54v dation of skeleton in Vertebrates & constantly ALLUDES «(& at p. 312)» to the abortive bones. He expl
096a041 t of Galapagos Isld. steep side to windward in ALLUSION to St. Helena discussion. Mr Brayley says he
297c186 ditto.-- Mag of Zoolog. & Bot. Vol. II p. 125 ALLUSION to abortive spiracles in Hemiptera do. p. 160
105a071 -. Will they introduce other causes to explain «ALLUVI» in valleys Lowe in his paper says land shells
048r086 plicable to the Siberia case We must not think ALLUVIAL plains «always» most favourable; In what part
087a010 shes Efflorescence nothing -- Study account.-- ALLUVIAL plains of Mississippi -- No Vol. I. p212. Cuv
150g036 4 not visible 3a 3a 3 Bouthoner 3 2 Terrace 3 ALLUVIALS 3a 3a 2 Mass 3 2 rather longer than 3a Sunday
028r031 fresh water at Iquiqui. not from rain, because ALLUVIUM saline; Mem: on coast of Northern Chili as sp
105a071 ity. stratification, If chain of lake. <a> the ALLUVIUM would form a succession of flights of steps;
116a100 pebbles-- it is clear gold occurs in submarine ALLUVIUM, or sublittoral formations. p. 150. at Portez
146g013 uth of upper end of Loch Dochart buttresses of ALLUVIUM or rather mass of well rounded pebbles in yel
147g015 kly studded with ridges & flat topped hill/ do ALLUVIUM. NB In one part pure sand in current cleavage
147g016 olonged would intersect alley above the 300 ft ALLUVIUM <abo> by Loch Dochart-- Rivers could not have
147g018 very considerable mass of waterworn pebbles in ALLUVIUM which without lake or sea could not be placed
147g020 tered with few very small & irregular hills of ALLUVIUM-- nothing very striking yet possibly sea more
148g022 ture I am quite doubtful of as I am of all the ALLUVIUM. At Mouth of Caledonian Canal opposite Loch L
151g041 ttle lines of Hill (judging from external form ALLUVIUM) descend from shelf 3d & almost meet, but are
151g043 elf runs up some way on great sloping plain of ALLUVIUM (much corroded by rivers) & not to head of pl
152g050 ee? at level of upper terrace The butresses of ALLUVIUM rise nearly up to Glen Collarig up within 200
155g063 much about «50 or 60 ft» «no doubt, a mound of ALLUVIUM nearly parallel--» Inclination of river must
159g087 about 60 higher-- There are however fringes of ALLUVIUM (?) still higher Slope of valley much more ge
160g088 , it is to the west of Glen Tarf What I called ALLUVIUM shows the ascending fringes {P} which makes m
160g093 which runs E & W) broad terrace «of pebbles? & ALLUVIUM» which appear perfectly level, <on op> dies a
154g057 bscure NB In Glen Collarig tidal channel, sides <ALM> 15 ft above bank or terrace, from terrace of 2d
108a080 reefs.-- Where vegetation luxuriant it might be ALMOST as well said probably much Carbonic Acid gaz h
147g017 arefully for Marine remains-- Some of the hills ALMOST appeared as if they belonged to double series
151g041 external form alluvium) descend from shelf 3d & ALMOST meet, but are separated by flat bottomed strai
155g062 hollowed out most Wednesday Shelf 3d dies away ALMOST imperceptibly on Glen Turrit side 2nd shelf ve
156g070 granite & some other rocks at head of shelf 3d ALMOST all granite pebbles Level of plain of 4th shel
157g075 ch of granite pebbles, & around which shelf 2d «ALMOST» forms it into island-- whole hill composed of
158g082 river doctrine <Little Hill with granite blocks ALMOST encircled> <fre> Gneiss cut smooth on sides of
174b014 odern animals same type as extinct which is law ALMOST proved.-- We can see «why» structure is common
205b145 p. 20. do "If hornless ram be put to horned ewe ALMOST all the lambs will be hornless.--" does this a
222b204 I believe very curious-- My idea of propagation ALMOST infers, what we call improvement, --All Mammal
252c045 f time-- Analogy from three first will give one ALMOST certain guide ∵ because time required too sepa
273c110 , then he says from long experience, you may be ALMOST sure, that there exist intermediate species,--
275c121 ld varieties is reduced to scarcely anything.-- ALMOST all imagination-- He says he recollects all ha
288c159 ogs howl most dismally, very rarely bark.-- are ALMOST useless not the least notion of hunting, or ke
288c160 n enclosing a common seems in part of-- to have ALMOST banished the Grasshopper Warbler-- --Yellow Wa
291c167 of some forms & succession of others, (which is ALMOST proved. Elephant has left no descendant in Eur
308c217 t of putting tea in pot made me go to tea chest ALMOST unconsciously.-- why do absent «Dr. Black. tea
310c224 most close The most curious case is Saxifrage, ALMOST <same> «closely allied» species Himalayas, 13,
312c231 ght & size & they would produce number agreeing ALMOST to the point in question.-- «--merely picking
322c270 Vol. of Memoirs on Geology of France.= on Etna. ALMOST reread the previous volume. & C. Prevost on L'
341d031 chman tells me line of Rocky Mountains separate ALMOST all Mammals of N. America & many birds.-- whic
342d034 l salt-water fish were once salt water (as they ALMOST must have been on elevation of continents) but
347d050 clearly shows that genus expresses as now used ALMOST any group.-- ‖all groups natural (p 6) as expr
357d073 e of Edentata-- Edentata & Marsupials have been ALMOST destroyed wherever other animals existed.-- At
364d103 e has killed itself.-- Q Sir. J. Sebright-- has ALMOST lost his Owl-Pigeons from infertility,-- Yarr
370d116 -- CD[(p. 451.)-- Wasps breed many females, but ALMOST all die.-- bees breed but few, because they ar
376d136 the cross between the Guaranis & Spaniards are ALMOST White from first generation., that with Quichu
382d158 arckian.-- Mine is much simpler.-- Hunter shows ALMOST all animals subject to Hermaphroditism,-- thos
390d178 this connected with the physical differences in ALMOST all Male animals?]CD If the male «in the cours
397e003 ons of what we call nature, have been conducted ALMOST! invariably according to fixed laws: And since
399e009 long ears-- & longer hind legs??? so that I was ALMOST doubtful which it was.-- do hind legs increase
405e032 of S. America & as it is <falle> embedded with ALMOST recent shells.-- shows that progression of cha
409e047 Having proved mens & brutes bodies on one type: ALMOST superfluous to consider minds.-- as difference
415e069 ht, with such chances be made intellectual, but ALMOST certainly not made into man.-- It is one thing
416e075 living [not located] A Greyhound might be made «ALMOST» without any relation to running hares.-- as i
430e120 ly different, in proportional dimensions, must <ALMOST> be considered merely varieties. & even Mr Sow
434e130 change, which <to> seems to have taken place.-- ALMOST all Dioecious & Monooecious plants have rudime
448e167 & Paris, numerically the same with recent & yet ALMOST wholly different, is same, as if Isthmus of Pa
451e178 this region between April & October & like man ALMOST (this looks inaccurate C. D) they will catch t
468t105 take flight-- Yet we have crosses-- I see Bees ALMOST every flower-- Blue-bells-- wild-raspberry--le
468t112 ron xx after several gloomy days. hot one, Bees ALMOST P every minute to Fraxinella «& from <flower>
468t112 n upwards & bend over stigma:-- but stigma «is» ALMOST roofed by united filaments.-- This flower host
471tf11 w Smith says in the larks from S. Africa he can ALMOST make series from end to end-- so that he is al
471tf11 ost make series from end to end-- so that he is ALMOST led to doubt. whether there is such a thing as
522m013 nd F. thought the feeling of anger, which rises ALMOST involuntarily when a person is tired is akin t
524m020 & suspicion are qualities, which my F says are ALMOST constantly present in people, likely to become
543m099 Players-- A man at Cambridge, during his time, ALMOST an absolute fool used to play regularly with D
546m110 compares it with Somnambulism.-- the young lady ALMOST equally in her senses in either state.-- does
546m111 Lady in perfect «mental» health.-- «Erasmus had ALMOST same thing happen to him about a knife. which
548m116 iousness, one would pity suffering in one state ALMOST as much as in the other,-- though she when wel
549m121 we obey literally New Testament future life is ALMOST the sole object--. -- I doubt whether the last

Page ***(Key Word)***
567n015 ces-- Vol I p. 334 Does a negress blush.-- I am ALMOST sure Fuegia Basket did. & Jemmy, when Chico pl
569n019 ed taste.-- Whilst music extremely primitive.-- ALMOST like tastes of mouth & smell. Descent of Man U
571n031 ntle things in gentle language, & vice versa.-- ALMOST proves that at earliest times there must have
574n041 sgusting lewd old man. ones tendency to kiss, & ALMOST bite, that which one sexually loves is probabl
589n091 ing heredetary form the instincts of animals.-- ALMOST identical with my theory-- no facts, & mingled
594n113 when stomach a little disordered) at thought of ALMOST anything ugly. baby-- association-- pouting ch
599o007 e effects of heredetary knowledge, has produced ALMOST → greater changes in the polity of Nature than
605o020 ow it originates, & by what means it becomes an ALMOST instantaneous perception.-- Taste has been sup
609o030 my view <says> unites both «& shows them to be ALMOST identical» + What has produced the greatest go
610o033 knowledge except our experience".-- is this not ALMOST a question whether we have any instincts, or r
614o037] As sexual instinct comes on late in life, man ALMOST alone in this case can perceive instinct. boy
620o045 the rule, that the passions & appetites should «ALMOST» always be sacrificed to the instincts.-- -- O
621o047 Now we know it is easy by association to give «ALMOST» any taste to a young person. or it is acciden
621o047 umstance.-- Thus a child may be taught to think ALMOST anything nasty. (<& accidentally> «by odd asso
621o047 from seeing conduct of others, the feeling that ALMOST (rarely if opposed to natural instincts) any a
622o049 se to the feeling of right & wrong.-- on which «ALMOST» any other might be grafted.-- Origin of the i
033r042 ery curious a structure: Have shells ever casts ALONE in Calcareous. rocks??--if so case precisely an
036r051 the prevailing movement being one of elevation ALONE.--In England much subsidence: hence difference;
042r068 delta of Indus», my belief in submarine tilting ALONE, must be modified. «Moreover, the Volcanos from
045r076 s which attend severe Earthquakes «1822¿ 1835?» ALONE, (& the general belief in N. Chili, where rains
063r130 : Furnarius. <Caracara> Calandria: inosculation ALONE shows not gradation;-- An argument for the Crus
097a044 .-- Mr Bollaert tells me, that the upper strata ALONE at Guantajaya contains salt see Geolog. proceed
112a089 how few isolated volcanos there are. where one ALONE has been formed--Look at the now active volcano
117a101 views of insensible oscillations of level will ALONE explain the immense amount of change which must
156g072 tly bending up each main valley.-- & that river ALONE had modified it-- perhaps however sea also,-- B
193b090 uenced itself perhaps by circumstances) & those ALONE preserved which are well adapted, This would ac
222b207 species most near in form to ancient; in shells ALONE can this comparison be instituted. People often
274c119 malia.-- We have clear indication [not located] ALONE, but on all the general arguments-- Lamarck was
359d087 c affections always influence the sexual organs ALONE.-- It is singular pheasant & fowl being so tota
365d105 rom common.-- Under this predicament, probably, ALONE would species cross in wild state.-- Is English
391d174 neration fails.-- How completely circumstances «ALONE» make changes or species!! CD[The view of <In>
435e134 dark caverns of Carniola p. 112. Man. "standing ALONE in the gift of intellect, he resembles, other m
440e145 imal owes its form, to that form being «the one ALONE» out of innumerable other ones, <alone> «which
440e145 «the one alone» out of innumerable other ones, <ALONE> «which has been» preserved.-- but be it rememb
441e151 northern latitudes, which are generated by buds ALONE or roots, & never flower, so there may be anima
466t091 June/41/, observed 3 plants of Caltha Palustris ALONE together. one had seed-pods turning brown, whil
504q014 from Wiegman Shrewsbury (1) Peas.-- Beans seeds ALONE remain to be compared-- Cabbages.-- kept true T
609o030 all» is the «instinctive» moral senses: (& this ALONE explains why our moral sense points <is> to rev
614o037 xual instinct comes on late in life, man almost ALONE in this case can perceive instinct. boy takes d
032r040 --Such changes could be shown (as represented), ALONG line of coast.--[Fig. 2] Mem San. Lorenzo; Vall
047r083 ikely to be coincidental than single elevations ALONG whole line of coast Darby mentions beds of mari
078r175 he other E & W.--veins richest not in ravins or ALONG gentle slopes. but on the most elevated summits
148g023 ne 100 ft & other one 40-- ‖ traces of them all ALONG <Glencoe>.-- towards Fort William yet in Glenco
162g099 hem-- but it may be said that a mound stretches ALONG, parallel to Shelf on opposite side & dies away
366d112 would doubt any law.-- Yet seeing the feathers ALONG one toe of the Pouter one thinks there is a law
467t104 stigma covered with pollen was pressed & rubbed ALONG whole breast-- {b} pressing either one or both
633j53v ed-pod of a desert plant (Anastatica) is rolled ALONG, & splits when it comes to a damp. place.-- Kol
464tf6v stence of Molina's Pudu or goat There is ibex of ALP Pyrenees &c-- (see Blyth's work on Ruminants,--
522m011 if at the other nothing.-- He could repeat the ALPHABET straight, but did not know [Z]CD when heard i
569n020 voice, as roaring for lion &c &c. (in same way ALPHABET. arose from letters, symbol of word beginning
447e164 vivapara F ovina-- propagated like oni[on] Poa ALPINA because vivaparous. Henslow has seen this-- (P
447e164 ecause vivaparous. Henslow has seen this-- (Poa ALPINA vivaparous sometimes seeds All species of Lemn
447e165 mountains & yet can be raised in gardens.-- Poa ALPINA, thougt generally vivaparous sometimes seeds.-
075r166 er hand, mine of Gualgayoc or Chota & Pasco in "ALPINE limestone" = "The wealth of the veins in most
219b195 ha-- Carmichael Linn. Transacts. Vol XII.-- The ALPINE plants of the Alps. must be <Alpine> «new form
219b195 XII.-- The Alpine plants of the Alps. must be <ALPINE> «new formations» because snow formerly descen
265c085 ing matter being absent.-- again dwarf plant in ALPINE district & dwarf plants from seed, one adaptat
291c168 and Isd they would change & make new species.-- ALPINE species being destroyed at Falkland Isds++.--
298c191 eling.-- There is great difficulty in Making an ALPINE species from one in lower country during gradu
298c191 (which must have been altered by crossing) with ALPINE form) lower species afterwards would probably
429e115 Transact Vol I. M. Ramond. p. 19. do says lofty ALPINE plants of & Pyrennees agree with those of Norw
464tf6v ntains, when the cold was intense just like the ALPINE plants-- In S. America. it appears from Lund m
497q007 crossing by Bees.-- Henslow.-- (1) Character of ALPINE Flora of Tierra del Fuego and Entomology of.--
510q017 ect facts for me-- what? What does Blume say on ALPINE Flora of Java? Has Schow written on double cre
514q21. s on Wellington Mountain described in Flinders= ALPINE Australia Flora= Banana's seedless-- 20 variet
632j53r e coral rock might have been uninhabited as the ALPINE pinnacles.-- One thing must be admitted there
633j54r lants require earth, why not created to live on ALPINE pinnacle? if we once to presume that God «crea
053r100 that contact of Granite & sedimentary rocks, in ALPS becomes metalliferous. Vol III Latter Part Are
089a020 Flori[da]> Excellent paper on Erratic blocks in ALPS. Memoires de la Soc. «de Geneva» Vol 3' P. II.
114a093 ound & penetrating their own range in Australian ALPS.-- Taylors Scientific Memoir, Part IV. p. 403 E
123a116 Buch Lyell. (under head of Delta) describes near ALPS great beds of rivers which must be like the Chi
129a131 l. Journal Vol XXI. p. 213. Beyond the limits of ALPS size of boulders sorted: ditto Murchisons case.
219b195 . Transacts. Vol XII.-- The Alpine plants of the ALPS. must be <Alpine> «new formations» because snow
295c178 t cat, rodents.» (Pachyderm in Portland stone of ALPS!!!!? No) p. 15 (Lyell's Pamphlet) Is man more ha
425e105 have peculiar forms.-- on the southern flanks of ALPS.-- many peculiar plants on single mountains, th
514q21. & Species theory-- peculiar Fauna?. {Australian ALPS--; are any Europaean forms found there-- Lindle
199b117 ousand of new insects all belong to same types ALREADY established. why out of the thousands of forms
224b215 monkeys & man may produce other species., man ALREADY has produced marked varieties & may someday pr
270c104 inorganic life".-- animals only live on matter ALREADY organized.-- This paper might be worth consult
411e056 ng some change in habits before form.-- I have ALREADY given various examples The Pipe-fish is instan
424e100 tion of Crustaceae.-- (I forget whether I have ALREADY referred to it.-- also on spermatic animalcule
512q019 ll young wild ones breed as well as,, as those ALREADY bred in cages. Get direction write to-- (2) Do
155g063 arallel--» Inclination of river must constantly ALTER with falling sea & so corrode plain into terrac
155g064 o corrode plain into terrace as regressed What <ALTER> a balance there must be in power of rivers eit
171b004 ing modified,-- therefore generation to adapt & ALTER the race to changing world.-- On other hand, ge
230b239 tity of change at once. but afterwards will not ALTER. This need not apply to very slow changes. with
276c124 us if compelled solely to fish. structure would ALTER.-- It is a difficulty how a different number of
289c162 y their larvae) but the acts of condensing must ALTER method of generation-- Heaven knows how.-- This
314c237 imals, which only propagate by scission can not ALTER much.?? Mr Brown showed me Bauer's drawings of
343d036 y keep perfect in these themselves.-- instincts ALTER, reason is formed, & the world peopled «with My
366d112 athers to grow there «That Mutilations will not ALTER form may be inferred from Australians knocking
473s07v , & then may just as well be born a tendency to ALTER or assume some form late in youth,-- only facts
552m132 rule is those that on long run will do good.-- ALTER will in all cases to have & origin as well as r
304c207 OLATION of range « <far more prob> » tends to ALTERATION views.-- ostriches do-- but then there may h
294c176 Lonsdale is ready to admit, permanent small ALTERATIONS in wild animals, & thinks Lyell has overlook
335d015 s «as supernumerary fingers» (that is slight ALTERATIONS of primitive stocks «relative to changes whi
335d015 ds between very near species (that is slight ALTERATIONS of primitive stock) are heredetary: «Hybrids
030r036 of Chonos. interesting from great quantity of ALTERED Carbonaceous shales Examine chart of Patagonia
046r078 averse granites, are granitic materials simply ALTERED by circumstances; & not in chemical nature, or
046r079 essure might have its «proportional» particles ALTERED.-- With respect to Volcanic theory. I want to
061r127 essif<e>: produced at one blow. if one species ALTERED: <altered> Mem: my idea of Volc: islands. elev
061r127 produced at one blow. if one species altered: <ALTERED> Mem: my idea of Volc: islands. elevated. then
121a111 r angles sharp-- yet with character completely ALTERED, & a crystalline structure superinduced Lyell
173b011 kind had in interval arrived) might have grown ALTERED Hence the type would be of the continent thoug
186b058 of atavism, where real structure obliged to be ALTERED, I can conceive colouring retained; therefore
219b195 ended lower, therefore species of lower genera ALTERED. or northern plants «No» CD[Mem. the antarctic
223b209 ntain chains &c the species have not been much ALTERED they will cross (perhaps more fertility & so m
223b210 desert?-- because the crossing of species less ALTERED prevents the complete adaptation which would e

Page		(Key Word)	
225b217	ine on glands of throat, (or colour of plumage	ALTERED	during passage of birds (where is this stateme
255c051	ngest part of structure.-- shape of wings have	ALTERED	many times, but all have had feathers.-- if wi
288c159	w whether breeds of oxen have deteriorated, or	ALTERED,	but it is certain that rams & bulls from Engl
294c178	trees ceasing to grow far N. becomes stunted,	ALTERED,	& lose (mere sickness)? fertility ¿because of
296c184	can migrate remaining constant in form, others	ALTERED	much.-- these others will be plants & land ani
298c191	revert to pristine form (which must have been	ALTERED	by crossing) with alpine form) lower species a
351d057	Foetus of man undergoes metamorphosis., heart	ALTERED	& umbilical cord,-- Broderip alluded to Hunter
368d114	ing either alike or very different in recently	ALTERED	genera. Guinea Fowl & Peaccocks.!!» other bird
371d127	ll beget young different in colour, form, & so	ALTERED	in disposition, as to be more easily trained u
434e127	ambling horses. Whether the body of parent be	ALTERED,	that is the Nisus formativus. (what does Mull
571n030	perhaps even latter may be vitiated. or rather	ALTERED.	The Reason why New Buildings look ugly is bec
271c106	<an> a statement in Mr Wynne's book, about not	ALTERING	breed of animals in certain countries.-- Fras
298c191	rafted with fresh arrivals..-- &c &c --Climate	ALTERING	as island increases.-- upper parts attracting
434e127	ativus. (what does Muller call it) succeeds in	ALTERING	<or> form of body, or whether it merely has t
442e153	Knights statements about fruit trees. grafted.	ALTERING	is hostile to this: but on other hand, fruit
526m030	are right about habitual exercise of the mind,	ALTERING	form of head, & thus these qualities become h
599o05v	ressive.-- We cannot doubt that language is an	ALTERING	element, we see words invented-- we see their
609o29v	ruder state of Society.-- Civilization is now	ALTERING	these instinctive passions--, which being unn
049r089	-- Daubeny. P 95. Glassy & Stony Pearlstones	ALTERNATE	together in contorted layers: Mem: Phillips M
162g104	nite (boulder), sloping buttresses, an[d] one	ALTERNATE	curved layer of fine sand & small angular-- r
035r045	nae <...> changes in rocks. connected with &	ALTERNATING	with obsidian must clearly be chemical diffe
040r062	er of such pebbles extending to seaward, the	ALTERNATING	with such matter at St Julians looks like su
114a094	«transversely fibrous» quartz. & iron stone	ALTERNATING.	bear on subject of cleavage Clay slate. a d
163g106	where Locks now are (32 ft rise) they found	ALTERNATING	layers of coarse & fine & many Sea shells. M
423e099	has been equal even in one bed, much less in	ALTERNATING	strata of sand & limestone &c &c.-- L'Instit
118a104	w of cause of continental elevations (I) the	ALTERNATION	of linear bands of movement in Indian & Paci
286c156	from manner Fleming treats subject to put in	ALTERNATIVE	of Man created by distinct miracle. Macleay
172b006	d that marrying in deteriorates a race, that is	ALTERS	it from some end which is good for Man.-- Let
224b215	physical changes (¿intellectual being acquired	ALTERS	case) other species or angels. produced «&» Ha
361d095	s retained this character in adult stage, other	ALTERS	entirely]CD In common sparrow young & female s
533m057	rent stock.-- (& phrenologists state that brain	ALTERS)	It is known that birds learn to sing & do not
029r033	quake shook the ship in a most violent manner.	ALTHOUGH	it lasted about a minute, there was no uncomm
033r042	been removed?--As shell out of its cast which,	ALTHOUGH	not very intelligble is a familiar case: If r
060r126	dstone of Andes fusible? no. mad dogs. Azores.	ALTHOUGH	kept in numbers. p. 124. Webster Consult W. P
065r140	n T. del Fuego, and connection of quadrupeds.--	ALTHOUGH	recent elevation, there may have been great s
192b083	hence possibility of reproducing any variety,	ALTHOUGH	many of the seeds will go back.-- Get instanc
220b197	uar & Tiger & Europe, as to produce same one.	ALTHOUGH	in plants, you cannot say that instinct perve
223b207	Nothing compared to the first thinking being.	ALTHOUGH	hard to draw line.-- -- not so great as betwe
226b222	ius? <In plants where do most species occur.)>	ALTHOUGH	the Horse has perished from S. America, the j
227b224	o genera.-- <we then cease to know the steps.>	ALTHOUGH	D E F. follow close to A. B. C. we cannot be
240c003	ety.-- -- He says of two varieties of pidgeon,	ALTHOUGH	having skulls so different, that they would b
259c064	culty of TRACING change of species to species	«ALTHOUGH	we see it affected» tempts one to bring one b
263c075	most laborious, & painful effort of the mind	(ALTHOUGH	this may appear an absurd saying) & will neve
303c205	r so creates animals, «, it will be said» that	ALTHOUGH	at any one there. are gaps. yet <what> altoge
355d067	vertebrae vary? «See Cuvier Ossemens Fossiles»	ALTHOUGH	no new fact be elicited by these speculation
384d162	ciple) one individual secretes two substances,	ALTHOUGH	organs for the double purpose are not disting
407e039	have survived a change, (& become transmuted),	ALTHOUGH	other parallel species in other continents mi
409e049	ent a social animal» a fact few will dispute,	[ALTHOUGH,	that it was the sole object, I will dispute,
422e095	cy in the simple animals to become complicated	ALTHOUGH	all perhaps will have done so from the new re
436e134	ollusca in cold parts of sea, like Cetaceae,--	ALTHOUGH	the Cephalopods, seem to have decreased since
467t099	in Kitchen Garden.-- As we see in Hybrids that	ALTHOUGH	anther «nor filaments» shrivel, yet stigma do
506q015	Any unproductive, where germen does not swell,	ALTHOUGH	there be pollen.-- or FEW. or bad seeds forme
520m001	ribing to his father & old Mrs Harrison, said,	ALTHOUGH	constantly seeing him, she was often struck w
534m063	rgoes transformation in the stem of Hollyhock,	ALTHOUGH	ordinary Habitat is Malva sylvestris. do. p.
541m090	rom subject to subject as quick as it chose.--	ALTHOUGH	thinking «& talking» for the moments with int
553m132	"Dogs learn sooner to take kangaroos than emu,	ALTHOUGH	young dogs get sadly torn in conflicts with t
556m146	rning round to kick» kicking they do the same.	ALTHOUGH	it is then quite useless-- Cats kneeding when
595n121	ow put a guaze over her head. & came near him,	ALTHOUGH	knowing it was Snow.-- Is this part of same f
606o20v	is metaphorically applied to this mental power	ALTHOUGH	taste must necessarily be acquired by a long
620o045	nature, & its effects lasting, whilst passions	ALTHOUGH	equally natural leave effects not lasting. By
620o046	man ought to follow certain lines of conduct,	<ALTHOUGH>	even when tempted not to do so, by other nat
623o050	[RHC] 9) We can thus explain love of place.--	ALTHOUGH	here we have not received pleasure from the p
623o050	y, all instincts demand some explanation [RHC]	ALTHOUGH	I cannot pretend to say how far & minutely ou
624o051	been so in some past time, hence passions]CD	«ALTHOUGH	perhaps useful at present to some extent.» He
638j58v	being perfect «or nearly so (except no in isd)	ALTHOUGH	having heredetary superfluities Man could exi
282c141	possessed by the different races of man, yet	ALTOGETHER	different.-- To make this case perfect, we m
290c165	I saw> .-- The attachment of dogs to man. not	ALTOGETHER	explained by F. Cuvier, «-- .Mem. Hensleighs
296c184	can arrive quick enough» Vegetation of peak--	ALTOGETHER	original. owing to being oldest. & having un
303c205	though at any one there. are gaps. yet <what>	ALTOGETHER	«he» has created a perfect chain. <Icthyo> .
444e157	ones to give strength to it.-- p. 139. Doubts	ALTOGETHER	the law of balancing of organs.-- In the Bat
459t009	d hog & of several species of deer, which are	ALTOGETHER	immaculate when grown up". Saw at Mr Bell's
059r121	would harden.--Climate.!? or small Proportion of	ALUM:	matter.--all pale cream colour.-- The Brecciat
537m075	cured of sucking his finger by rubbing them with	ALUM,	so more slowly does animal leave off <t> insti
028r028	adation of Feldspar & other minerals containing	ALUMEN.	--This matter accumulating in deep seas forms
038r058	; & from action of water probably not so much	ALUMINATED.	As argument in favor of lines of anticlinal
333d007	fspring always take most after wild.-- i.e that	<ALW>	«no» domesticated ones have been so long as wil
073r163	II. p. 130 Metals in Mexico rarely in secondary	ALWAY	in primitive & transition; the latter rarely ap
384d161	45) & that strength> In speaking of generation	ALWAY	put female first Will not even a fruit tree or
045r077	may believe the fluid matter instead of afflux	(ALWAYS	slightly oscillating as that of a spring) move
048r086	Siberia case We must not think alluvial plains	«ALWAYS»	most favourable; In what part of the globe ar
073r161	entary jasper.--do undulations (as Hutton says)	ALWAYS	come from without.-- "True native iron that to
076r169	carbonate of lime-- native silver in Mexico is	ALWAYS	accompanied by Sulp. silver sometimes by selen
081rIBC	ellinghausen in 1819 Kotzebue 1816 Constant log	ALWAYS	additive to convert French Toise into English
103a063	s from above unite with those from below. would	ALWAYS	thin out above which explains a difficulty.--
104a066	anticlinal lines near. (lateral pressure would	ALWAYS	produce it) but where great thickness is affec
109a082	bottom of sea. off coast of England.-- Sea must	ALWAYS	on actual beach act same way.-- a little furth
117a102	o see effects of degradation, «no» tides, water	ALWAYS	falling or at least not rising are there cliff
118a103	Dick says (.p 52) fringe of sublittoral deposit	ALWAYS	equal width --subject of fine paper this would
145g004	- {P} Veins, amygdaloidal-- as well as base not	ALWAYS	parallel to strata 3 or 4 seams/ 3 or 4 inches
179b033	st Indies «--:Humboldt. New Spain:--» Dr. Smith	ALWAYS	urges the distinct locality or Metropolis of e
233b252	scs!!! When we talk of higher orders, we should	ALWAYS	say, intellectually higher.-- But who with the
236b280	merican kinds.-- Are the closest allied species	ALWAYS	from distant countries, as Decandoelle says, n
259c065	hange» during life of parent, & therefore being	ALWAYS	necessary may be called adaptation With respec
267c094	t to sea Vol VII. p. 325-- Wild dogs of Guayana	ALWAYS	hunt in packs «30 or 40 together» colour reddi
267c094	like bull terrier-- Indian secured one, as they	ALWAYS	like to cross their breed p. 333-- alludes to
268c099	is.-- ¿Pelagic forms similar--birds??-- We must	ALWAYS	bear in mind proofs of most equable climate bo
271c106	tinents are not stationary «unerring proofs not	ALWAYS	continents».-- it is a plastic virtue.-- it is
273c112	t observe that this preeminent structure is not	ALWAYS	applicable to same habits, though swallow hawk
308c217	Carboniferous some perished before, then there	ALWAYS	have been gaps, & there now must be, ∴extincti
308c217	t was finished kept it in-- <right> left, but I	ALWAYS	for a week took of cover of right side, though
312c228	rdines Journal.»-- consult on this point-- pigs	ALWAYS	go against this, without «number of vertebrae»
316c244	e some restricted genera, but then they appears	ALWAYS	very small ones as Trigonia in Australia or Co
333d007	n a wild animal is crossed with tame, offspring	ALWAYS	take most after wild.-- i.e that <alw> «no» do
333d007	everal cases of foxes & dogs crossed, offspring	ALWAYS	more resembled foxes than dogs (Mem Jackall in
334d011	se chains, by being at first beaten from her, &	ALWAYS	accustomed to her.-- case parallel to brothers
358d075	a Vol I. p. 306 "The crows were amazingly bold,	ALWAYS	accompanying us from camp to camp; it was abso
358d076	ere one of the sexes is little developed, it is	ALWAYS	female which approaches in character to the la
359d087	often the case, & why should organic affections	ALWAYS	influence the sexual organs alone.-- It is sin

363d102 ose species, of birds, habits when well watched ALWAYS very different.-- the two redpoles can hardly
363d102 eir own songs, though imperfectly.-- Male birds ALWAYS second their songs, the ++ Cervus Campestris s
367d113 most common in vegetable feeders. because males ALWAYS armed in carnivora. Where females, are peacabl
370d115 uinea.--!! S. America.-- Such difficulties will ALWAYS occur if animals are thought to have been crea
372d129 mplest forms of budding. Why does Gecko produce ALWAYS different tail? An Individual bud may be thus
380d155 tch is in heat.-- Yarrell believes Gestation is ALWAYS some multiple of seven-- if woman does not men
391d175 hich induce «which must be external» change are ALWAYS of one nature species is formed if not.-- the
421e091 lated to the habits & food of an animal, I have ALWAYS regarded as affording very clear indications o
423e095 cies tending to dis-developement (some probably ALWAYS have done so, as the simplest fish &), my answ
437e140 e.-- «In Shiant Isld. it is said, that an Eagle ALWAYS procured its prey from another island.-- » p.
441e148 of concourse of two «individuals» & the action ALWAYS of two organs-- instead of one part «as» in pr
448e165 it.-- it is allied to Asclepias, where this is ALWAYS the case according to Brown.-- Voyage of Adven
455eIBC t. if so irritate them, «as by an insect coming ALWAYS at same time» see if by so doing can be made s
466t094 tle In a wild purple Geranium, I see Bees visit ALWAYS base {a} of upper petal from facility of aligh
472s01r but the pods have (except this one year (1827), ALWAYS been empty.-- See separate note-- Elizabeth sa
491q01v that it is in these that male organs (not being ALWAYS useful). fail-- Really good subject for experi
495q005 at a proportion springs up true.-- This in fact ALWAYS takes place in natural Hybrids of Cabbages (7)
496q006 e common Fig Dioecious-- are its female flowers ALWAYS barren-- if not how does impregnation take pla
514q21. ed Does Mormodes (one of the Catasetums) really ALWAYS hit stigma by projecting pollen-masses?-- = an
534m063 f wings of flies from intellect. but it does it ALWAYS instinctively or habitually.-- good Heavens is
538m079 me disposition recurs, such as -- -- of Trinity ALWAYS thinking people were calling him a bastard.--
538m079 -- when drunk.-- having really been so.-- some ALWAYS sentimental, some quarrelsome as B.e on board
544m101 ions-- such as child sucking, gives pleasure, & ALWAYS has done therefore sight of own child. (when f
547m113 s of thought necessary heirs of every action, & ALWAYS running on in mind, being absent. one could no
564n005 ll on back.-- To study Metaphysic, as they have ALWAYS been studied appears to me to be like puzzling
574n041 women. are described as biting: so do stallions ALWAYS.,= No doubt man has great tendency. to exert a
578n055 ne of having done either good or bad action, it ALWAYS bear some references to thoughts of other pers
580n061 it.-- «Weak people say I know it because I was ALWAYS told so in childhood.-- hence the belief in th
584n071 measure the cell; p. 22. instincts & structure ALWAYS go together: thus woodpecker: but this is not
586n080 dietary part of it,-- & faculty (faculty «being» ALWAYS heredetary helps this confusion.--) Hensleigh
612o035 .) Joining two difficulties into one common one ALWAYS satisfactory, though not adding to positive kn
615o038 HC] NB. Two dogs having very different instinct ALWAYS obtain peculiarities of external configuration
618o41v run in a parallel series: if blueness & weight ALWAYS went together. & as a thing grew blue it «uniq
620o044 he knows that the instinct. (or conscience) is ALWAYS present (which is indeed, often felt at very t
620o045 f ones conscience. & palliates the offence; one ALWAYS admire the habit formed by «obediance to insti
620o045 , that the passions & appetites should «almost» ALWAYS be sacrificed to the instincts.-- -- One does
621o046 hat the instinctive feeling in its nature being ALWAYS present. & his passion shortlived, it is to hi
632j53r och. Attribs of Deity. Vol: I it will be better ALWAYS to refer to the author if I use these facts p.
218b191 a distinct work on Hybridity under title of AMARYLLIDAE & Narcissus. Mr Donn considers Mr H. rather
358d075 hell. Australia Vol I. p. 306 "The crows were AMAZINGLY bold, always accompanying us from camp to cam
077r170 a is the same with that on surface of plains of AMAZON, no relation--there is more modern breccia, ch
037r056 Miers saw then near? Mem. La Condamaine on the AMAZONS. Consult Insist on the frequency of dikes in G
053r100 glish escarpment The great conglomerate of the AMAZONS & Orinoco mentioned by Humboldt under name of
540m086 vated table land of Peru the hot plains of the AMAZONS & Brazil-- with the negros of Africa, (or agai
486z018 insects in T. del Fuego.-- Hence it is odd that AMBER insects of Europe have Tropical Forms See p. 25
282c143 fect.-- the spirit of life must be every where AMBIENT. & <in> merely determined to such points by th
526m030 -- absolute free will, motive may be anything AMBITION, avarice, &c &c An animal improves because it
629o055 [In similar manner our desires become fixed to AMBITION. money, books &c &c.-- <]> the "secondary pas
290c163 good forests for beautiful birds.-- heredetary AMBLING horses, (if not looked at as instinctive) then
433e127 n tumbling, far more wonderful than heredetary AMBLING horses. Whether the body of parent be altered,
448e168 Linn. Soc. April 2d. 1839 The Lepidosiren-- AMBLYRHYNCUS & Toxodon, <all> equally aberrant-- the two
486z019 ade by dung beetles, like those from Chiloe. AMBLYRHYNCUS de marlin James_Isd-- Lutke Voyage Vol III
240c013 es from the Moluccas «Matchian» & Celebes.-- ‖ AMBOINA; Viverra Zibetha. ‖-- All the Moluccas, Waggio
241c014 fferent from that of the Marianna islands & at AMBOINA» I fancy there is marked wild breed of oxen at
242c018 treaks-- the lacerta vittata extends <to> from AMBOINA to New Ireland p. 23 Voyage of Coquille Lesson
245c024 rvus near Marianus new, & some rats & mice. In AMBOINA only Cuscus & Barbyroussa «NB» [islds. Springi
599o008 J'ai possede seul sur la terre. J'ai trouve son AME" &c-- -- Confesses these faculties of soul, trea
023r010 ions of those mountains on the continent of S. AMERICA is inadmissible «may have happened from incipi
024r015 Fathoms Vide facts in Beechey. on NW coast of AMERICA off Cape of Good Hope 70 fathoms 20 miles from
025r018 gists to compare whole history of Europe, with AMERICA; I might add I have drawn all my illustrations
025r018 ght add I have drawn all my illustrations from AMERICA, purposely to show what facts can be supported
025r018 when we see conclusions substantiated over S. AMERICA & Europe. we may believe them applicable to th
026r022 ar to have taken place in the Cordillera of S. AMERICA. Study Geolog: Map of Europe Conybeare. Introd
034r045 in soundings. on approaching the coast of NW. AMERICA P. 209--13 P & 444 «(Yanky Edit)» <I think>° At
039r060 he has not fully considered the subject.-- S. AMERICA in the form of the land decidedly bears the st
042r069 rces have acted with far more regularity in S. AMERICA: in France we have freshwater lakes unequally
043r071 territory vibrates from any one shock-- In S. AMERICA--continuity of space in formations & durabilit
044r072 s.-- «Humboldts. fragments.» Read geology of N. AMERICA. India.--remembering S. Africa. Australia.. Oc
044r073 action, Australia S. Africa-- on one side. S. AMERICA on the other: The extreme frequency of soft ma
045r077 .--When discussing connection of Pacifick & S. AMERICA.-- Volcanos must be considered as chemical re
048r085 er a lump of hard clay. -- In the History of S AMERICA we cannot dive into the causes of the losses o
051r097 nk] Beechey.--changes in bottom in NW coast of AMERICA. from shingle to sand &c &c. <Vol II> P. 209.
051r097 n» Shores of Pacifick, as compared to whole E. AMERICA. <East> Africa. Australia. profoundly deep: a
054r105 [not located] [not located] Isld near coast of AMERICA not reached. Juan. Galapagos. Cocos-- Ulloas v
060r125 e mineralogical difference of formations of S. AMERICA & Europe.-- If great chain of Volc. had been i
062r128 hrough all the limits of birds & animals in S. AMERICA. Zorilla: wide limits of Waders: Ascension. Ke
068r145 s in Polynesian Islds--. Volcanic plenty in S. AMERICA!! Metamorphic Volcanos only burst out where st
076r169 great blocks of pure silver not common in <S.> AMERICA: In all climates distribution of silver «in ve
076r169 rated: wonderful quantity of pure silver in S. AMERICA. Geology of Guanuaxuato.--Clay slate. passing
086a003 should not be surprise at Horse being found in N. AMERICA, when Mammoth & narrow toothed Mastodon.-- ag
086a005 odon Toxodon Is the general saline tendency of AMERICA connected with its elevation. vapour from belo
087a008 d be required to produce climate resembling S. AMERICA in Europaean latitudes.-- Will it be supposed
093a032 s. Mem sublimation of sulphur to form salts of AMERICA. -- The number of minute turbos in red earth w
101a057 e Rogers for Southern limits of Boulders in N. AMERICA do/p. 280. the gravel beds in England differen
106a076 p. 137. Three inosculating rivers in Southern AMERICA ¿ effect of subsidence-- <Is there same.> Inst
115a097 rough lake that near-- Obstruction Sound in S. AMERICA The very general absence of fragments «& pebbl
123a115 p 270.-- SPLENDID PAPER on fossil shells of S. AMERICA. Von Buch Lyell. (under head of Delta) describ
124a118 . p. 62. Dr. Daubeny on mountain Chains in N. AMERICA Erasmus suggested to me that Herschel's theory
128a130 ta on coast of do-- I believe?? coast of North AMERICA., like the Mexican Gulf. is fouled by bars of
129a131 When writing on Valleys. «Tertiary strata of S AMERICA» read parts of this work, though it is but poo
131a136 of water flowing from beneath frozen crust in AMERICA Richardson.-- From strata being not only verti
172b008 nd representative species; this we do in South AMERICA closely approaching.-- but as they inosculate,
174b013 one parent why. Myothera several species in S. AMERICA why, 2 of ostriches in. S. America-- This is a
174b013 ecies in S. America why, 2 of ostriches in. S. AMERICA-- This is answer to Decandoelle. (his argument
179b033 the existence of whiter tribes in centre of S. AMERICA shows this.-- <If> Is there a tendency in plan
179b035 olute end of certain form from considering. S. AMERICA (independent of external causes) does appear v
180b038 onsequence (contrary to what would appear from AMERICA) of non adaptation of circumstances.-- Vide tw
182b047 ifferences for instance of finches of Europe & AMERICA. &c &c The new system of Natural History wi
184b054 now living.-- Now according to my view. in S. AMERICA parent of all armadilloes might be brother to
187b062 erinduce a change? Seeing animal die out in S. AMERICA; with no change, agrees with belief. that Sibe
187b062 ries.-- Seeing how horse & Elephant reached S. AMERICA.-- explains how Zebras reached South Africa--
187b063 stance killed it over a tract from Spain to S. AMERICA.-- Never They die; without they change; like G
187b065 should expect that Bear & Foxes &c same in N. AMERICA & Asia, but many species closely allied but di
188b066 Mice.-- --- Animals common to South and North AMERICA.-- ¿are there any? Rhinoceros peculiar to Java
189b072 S. Africa Zebra.-- & continued.-- perished in S. AMERICA All animals <are> of same species are bound to
196b106 in some? Why did not fossil horse breed in S. AMERICA-- it will not do to say period unfavourable to
200b125 h.-- ‖.do p. 326. 2 Fossil species of ox in N. AMERICA: as well as 2 recent See Geolog. Proc. p. 569.
202b133 is exception to law of type. like horse in S. AMERICA. or like living Edentata in Africa &c &c.-- No

Page **(Key Word)**
202b134 al. we might have wanderers. (as Peccari in N. AMERICA) then if it is doomed that only one species of
203b135 nt physical cause:-- Mem: Mastodon all over S. AMERICA. Hilaire does not seem(?) to consider the monk
203b137 ow I think we may conclude from Australia & S. AMERICA. that only some mundine cause has destroyed an
219b196 produce «for the creator» two quadrupeds at S. AMERICA Jaguar & Tiger & Europe, as to produce same on
220b199 lyth states that all «genera of» birds in «N.» AMERICA & Europe, which have not their representative
226b221 y moderately large ones.-- Is the flora of <S. AMERICA> Tierra del Fuego like that of North Europe, m
226b222 cur.)> Although the Horse has perished from S. AMERICA, the jaguar has been left & Fox, & bear.-- If
226b223 roamed to Europe & Pachydermata from Europe to AMERICA., How strange would presence of Jaguar been in
226b223 ow strange would presence of Jaguar been in S. AMERICA.-- <East.> «W» Coast of Africa & <West.> «E» o
226b223 - <East.> «W» Coast of Africa & <West.> «E» of AMERICA, ought to present great contrast in forms.-- I
233b250 to observe, that all the species of mice in S. AMERICA. which were hard to distinguish came from clos
233b251 oductions. of great Fresh water lakes of North AMERICA If Parasite different, whilst man & his domest
242c018 said to have brought a tortoise & toad from S. AMERICA & identical with those from S. Africa. «M. Bib
246c025 far as Oualan.-- Wide space of sea, to East of AMERICA. would account for this.-- Coquille Voyage Say
249c036 ith, Mr Blyth's statement of birds of Europe & AMERICA, which are of different forms being migratory;
250c037 destroyed great <quadrupeds>; Pachyderm in S. AMERICA destroyed great Edentata or American form.-- I
250c037 mate having grown more extreme both in, N & S. AMERICA, is only common cause I can conceive of destru
250c037 ve of destruction of Great Animals in Europe and AMERICA Some portion of the world «Africa» being left
250c038 he world «Africa» being left more equable. yet AMERICA preeminently equable. might have allowed fresh
250c038 to have been condition of former whole world. AMERICA might have been string of islands.-- ¿Europe h
252c045 ertainly Get Closer species-- FOX IS & Mice of AMERICA «good case on account of varieties in N Americ
252c045 merica «good case on account of varieties in N AMERICA» «some doubt, from want of knowledge of time--
253c046 use time required too separate isld very long» AMERICA & India deer.= Africa not.-- Africa Camels?? A
255c050 'orbigny. Birds of prey, are distributed in S. AMERICA like other forms, but those inhabiting 3d zone
260c067 D'orbigny, Spix, &c might compare birds of N. AMERICA & South-- Any how temperate regions-- crows in
260c067 outh-- Any how temperate regions-- crows in N. AMERICA‖ Study Bonapartes list In the Zoological Journ
261c069 .-- like Synallaxis or Marsupial animals of N. AMERICA‖Hence it is universally allowed that the discr
271c105 abits with the Turdus Musicus «not found in N. AMERICA» whose Southern range is? <One> The black & wh
274c116 changed Have <not>. S. Africa, Australia, & S. AMERICA very few forms in common.,-- but each several
274c116 several with Europe & northern Asia & Northern AMERICA.-- may we not look to these Northern regions a
278c132 ter than Toxodon, Macrauchenia, &c compared to AMERICA.-- the wonder is that the Europaean forms, wer
278c132 nd in Africa, the wonder have been same for S. AMERICA & Europe.-- The difficulty is how came it anim
278c132 w came it animals not preserved, in central S. AMERICA & yet Africa & India???-- & Indian Islds.-- Si
291c167 left no descendant in Europe<)> Toxodon in S. AMERICA) is absolutely necessary to explain genera & c
291c168 ies or gradation.-- It is easy to see if South AMERICA grew very much hotter, then Brazilian species
295c183 o salamanders (D'orbigny Rapport. p. 11) in S. AMERICA so highly developed in North.-- Icthiology of
295c183 highly developed in North.-- Icthiology of S. AMERICA. more peculiar than its ornithology X p. 12 do
309c221 Isd.-- West Africa & India some plants same. --AMERICA.-- See Brown Congo Expedition: 400 Australian
310c224 Queen, more wonderful case Dwights' Travels in AMERICA, speaks of short legged sheep. heredetary proc
311c225 ks on Bonaparte's list of birds in Europe & N. AMERICA, on closely allied species. «replacing each ot
313c234 than molluscs, argue case both in Europe & S. AMERICA. «very difficult case» Does this law of durati
316c244 nes as Trigonia in Australia or Concholepas in AMERICA-- yet many countries have far more species th
316c245 Cyclostoma in Phillippines & Anphidesma in S. AMERICA-- yet there are a few Cyclostomes & a few Anph
316c245 iads (study De Ferrussac) are confined to <S.> AMERICA-- Mr Sowerby says there are some shells commo
316c246 shells common to West coast of Africa & E. S. AMERICA.-- get instances.-- very good anomaly in range
316c246 rding to Sowerby fine species) on coasts of N. AMERICA & England.-- but the fossils are not like, exc
316c247 e said American fossils more resemble those of AMERICA than of Europe, because the recent ones are so
316c247 s are so close. Was there continent between N. AMERICA & Europe?-- ‖.Norton has written on fossils of
316c247 urope?-- ‖.Norton has written on fossils of N. AMERICA.-- At the end of "White's Selbourne." many ref
317c251 he various hares «some since discovered» of N. AMERICA, & of the shrews.-- Dr Bachman told me. that n
318c252 this is precisely the case with the mice of S. AMERICA, with respect.. to the Cordillera,---- Bachman
318c254 birds (& this seems common «kind» migration of AMERICA) migrate singly flying few miles every day «ge
319c256 inks <many> more birds sing in England than in AMERICA, but the few of N. America are quite as beauti
319c256 in England than in America, but the few of N. AMERICA are quite as beautiful. The thrushes of N. Ame
319c256 ica are quite as beautiful. The thrushes of N. AMERICA. singing so well. & the mocking thrush being s
319c256 ing so very beautiful gret contrast with South AMERICA.-- In Home's History of Man at Maer, it is sai
319c257 n the Anteater of C of Good Hope & those of S. AMERICA,-- Are not some of the Australian fossils inte
323c269 cal causes: well skimmed Bartrams Travels in N AMERICA May 18th Stanley familiar History of Birds ---
326c266 Physical & Medical Researches. on Horse in N. AMERICA.-- Owen has it.-- Ld. Brougham. Dissertations
327c265 Views of the Cultivation of Fruit trees in N. AMERICA" in Lib. of Hort. Soc Mr Neil. has written goo
341d031 ky Mountains separate almost all Mammals of N. AMERICA & many birds.-- which however are most closely
343d038 ne «Guanaco» of the characteristic forms of S. AMERICA. With respect to future destinies of mankind,
343d039 lia of Europe the shells of do-- shells of S. AMERICA.-- shells of S. America.-- there is no appeara
343d039 of do-- shells of. N. America.-- shells of S. AMERICA.-- there is no appearance of sudden terminatio
357d072 als over whole world shows there is rule.-- S. AMERICA & Australia appear to have suffered most with
357d073 ., & the less developement of Marsupials in S. AMERICA. from presence of Edentata-- Edentata & Marsup
361d096 ty close repesentative species in England & N. AMERICA.-- the teal which some authors [not located] S
362d099 when horns not perfect-- (is not this so in S. AMERICA with C. Campestris<)> refer to my notes) & Mr
363d103 ommon duck were often caught wild off coast of AMERICA.-- showing hybrids can fare for themselves.‖ +
366d111 of structure.-- When will the musquitoes of S. AMERICA take an effect.-- would perfect impunity from
370d115 quench their thirst"-- But New Guinea.--!! S. AMERICA.-- Such difficulties will always occur if anim
405e032 M. angustidens be found to be inhabitant of S. AMERICA & as it is <falle» embedded with almost recent
405e032 r. Megatherium & Mastodon are coembedded in N. AMERICA. see my Journal for references In such cases a
406e035 scension most species like & identical with S. AMERICA. & many very close: see full paper. L'Institut
406e037 rs destroyed about same time in North & South. AMERICA.-- Whole wor[l]d, formerly possessed a climate
407e037 d, formerly possessed a climate compared to S. AMERICA at present days,, which S. America now does to
407e037 pared to S. America at present days,, which S. AMERICA now does to North. America & Europe.-- S. Amer
407e037 ent days,, which S. America now does to North. AMERICA & Europe.-- S. America favourable to Tropical
407e037 rica now does to North. America & Europe.-- S. AMERICA favourable to Tropical productions. The world
407e038 re so. yet climate of same order as that of S. AMERICA.-- (Explained by profound views of Lyell) Now
407e038 d by profound views of Lyell) Now «Equatorial» AMERICA from the «low» limits of blocks both North & S
407e040 inently equable & temperate climate, <that> of AMERICA, that the Mammalia of S. America are as difere
407e040 te, <that> of America, that the Mammalia of S. AMERICA are as diferent from the existing orders, as t
407e041 Europe, Consider probable form of land,-- -- S AMERICA, an island, «connects with Asia» between two p
407e041 t of No. Mam: Placent: insectivore being in S. AMERICA & Australia. reason, why: Marsupiata, when fir
408e042 rom N. to S. American forms. The climate of N. AMERICA. must have been equable & low-- more so than a
411e055 ls in <own> «the present» orders (not so in S. AMERICA, however) is very remarkable & none discovered
424e100 rch 5th. Lyell says «fossil» shells from North AMERICA, Scotland, Uddevalla. Many species same. & Nor
429e116 hatka, Siberia, or even of polar regions of N. AMERICA.-- if true curious on my view-- because these
458t001 are of large size established--«Australia,. S. AMERICA-- These strange forms,,. camels, giraffes. Siv
464tf6r in Brazil another point of agreement with. N. AMERICA & S. (¿ is the peculiar. N. American form)--
465t079 as intense just like the alpine plants-- In S. AMERICA. it appears from Lund more Mammals, than at pr
465t079 , <there are now> have been found fossil in S. AMERICA, there are now-- -- species in S. America. --
465t079 n S. America, there are now-- -- species in S. AMERICA. -- so see what a «mere» vestige, is preserved
465t081 mem. dogs «& pigs» in Polynesia; & dogs in S. AMERICA «Rengger.»-- now it is this very immigration
473s03v lephants-- Lyell says New Red Sandstone of. N. AMERICA is Red Sandstone. & Birds true! Plants in Devo
473s03v lants in our Devonian-- Fish one step lower in AMERICA-- How curious all negative laws of America of
473s03v in America-- How curious all negative laws of AMERICA-- of depth of organisms holding in America as in
473s03v ws of America of depth of organisms holding in AMERICA as in Britain. If there has been «as» much sub
480z006 es p. 81 & p 113 of 1834 On the passeres of S. AMERICA. D'.Orbigny. L'.Institut. No.-- 221 Good accou
480z007 837. No 212 Observations on the Raptores of S. AMERICA translated from D Orbigny no IV Mag. of Zoolog
480z007 s of prey have longer tails than old ones-- in AMERICA & sexes not of different size-- How does this
481z009 - Waterhouse remarks that no insectivore in S. AMERICA or Australia-- very curious.-- replaced by did
484z014 duction without Tropics in Northern & Southern AMERICA-- valuable & practicable deed Caricaridae wand
484z014 cticable deed Caricaridae wanderers.--?? in N. AMERICA?? Wilson N. American Ornithology must be studie
485z017 to move much further North on West coast of S. AMERICA. than on East.-- not being replaced by Brazili
486z018 rpid.-- On this principle tropical forms in N. AMERICA extend much further N. in N. America than in E
486z018 rms in N. America extend much further N. in N. AMERICA than in Europe-- Coleoptera especially require

486z019 or comparison of singing powers of birds of N. AMERICA & Europe. Entomolog. Transact. Vol I. p. 130.
501q10a ot in other: Rosa is hard in Europe, Walnut in AMERICA.-- Heaths in Africa; Hooker? are these genera
502q11v nipe Migratory-- probably united by Land to S. AMERICA (33) Ornithologum commonly but improperly call
510q017 they from West, like as between Australia & S. AMERICA? Sabine says North of Siberia, no sea-current,
514q21. lds-- names of Plants found on mountains of N. AMERICA similar to Lapland Plants --will get answer= I
552m131 beny on the direction of mountain chains in N. AMERICA Fear probably is connected with habitual stopp
571n029 poisonous?-- How did animals in «Australia» & AMERICA manage;-- This shows doctrine «of instinct» ha
582n067 n be cut off.-- <Horses> Colts cantering in S. AMERICA capital instance of heredetary habit:-- there
033r043 There is rather good abstract of Humboldt. S. AMERICAN Geolog. in Daubeny. P. 349 Admirable little t
067r142 man Journal. year 1835 excellent account of N. AMERICAN geology. Conybeare Lava in Cordillera & on Ea
087a010 -- The Guanaco the Camel.? Make note about N. AMERICAN bone not probably in salt marshes Efflorescen
088a015 Study Humboldt. Fragmens Asiatiques account of AMERICAN Volcanic action. -- Fragments of slate conver
165g126 / Why is the Tetrao scoticus & Tetrao-- not an AMERICAN form The union of two instincts crossing most
183b053 sed with it.-- ¿Whether extinction of great S. AMERICAN quadrupeds. part of some great system acting
185b056 (Gray) new <liza> species, belonging to true. AMERICAN genus Waterhouse says he is certain, that in
186b060 .-- extinct species of that country parents of AMERICAN.-- Now Genera of those two countries ought to
194b094 - Tooth of <Spi> of Sapajou-- NB Sapajou is S. AMERICAN form. therefore it is like case of great <rod
194b095 ones.-- Mem: Silurian fossils: ¿How are. South AMERICAN shells? ¿Do not plants, which have male & fem
195b098 bi[r]ds with plumage «&» tone of voice partly AMERICAN North & South.-- (& geographical <distri> div
202b133 n isles.-- Compares it to fossil Didelphis (S. AMERICAN genus) in plaster of Paris.-- Now this is exc
214b172 eme poles.-- Oh. Wealden.-- Wealden. Do the N. AMERICAN. Tertiary deposits present analogies to shell
214b174 lips. Lardner p. 289 It is certain, that North AMERICAN fossils bear the closest relation to those no
214b174 .-- See Rogers report to Brit Assoc <to> on N. AMERICAN Zoology-- A breed of Blood Hounds from Aston
217b187 Isl.-- Isle of Pines-- Australia.-- A <South> AMERICAN «form of» Lathyrus has one species in Europe
217b187 s one species in Europe Madagascar has several AMERICAN forms-- The above facts evidently show that M
218b192 d to me. that some good African & some good S. AMERICAN forms. (& daresays some of these <African for
219b194 peared in the great ocean, have made plants of AMERICAN & African form, merely because intermediate p
225b218 s against man with gun. & Bustards &c &c!!! An AMERICAN & African form of plant being found in Trista
231b242 ans.-- Ostriches.-- Hippotamus only African.-- AMERICAN & African forms mingle in India & East Indian
235b275 r-- Indian Bull?-- Do species of any genus. as AMERICAN or Indian genus inhabit different kind of loc
236b279 s found), decidedly next species to some South AMERICAN kinds.-- Are the closest allied species alway
241c015 in E. Indi: Arch: In New Zealand. a sturnus of AMERICAN form-- a Synallaxis. ¿American?). p. 159. & 1
241c015 d. a sturnus of American form-- a Synallaxis. ¿AMERICAN?). p. 159. & 160 «162» list of some birds of
241c015 A Dipus. & other rongeur in Australia.-- p 67 ¿AMERICAN forms? All Infusoria. not extinct. species. go
242c017 «des» couroucous et rupicole vert instances of AMERICAN forms in East. Ind: Archipelago. ▌Raffles. Ho
250c037 derm in S. America destroyed great Edentata or AMERICAN form.-- Is the Australian Dipus an American f
250c037 or American form.-- Is the Australian Dipus an AMERICAN form? The climate having grown more extreme b
251c042 . 4 species probably in Cuba (p 271 Viedo says AMERICAN dogs silent. Mem contrary assertion of Molina
256c054 I often disputed for a moment,-- Galapagos, S. AMERICAN-- -- genus.-- The circumstance of having two
259c064 &c.--, fresh water animals of great lakes are AMERICAN form, that one is brought to admit the possib
260c069 in district. which they frequent!!?-- Wilson's AMERICAN ornithology a mine of valuable facts,. regard
268c096 owing how blind a storge From what I see of S. AMERICAN birds. genera blend into each other in very s
271c106 e changed Argumentum ad absurdum. The creative AMERICAN halo has extended to Juan Fernandez in birds.
288c159 me country. may explain greater migrations, if AMERICAN & intersected wider & wider at Rio Plata. bir
300c197 eath, only of pain. of death acquired?. The S. AMERICAN dung beetles will each become the fathers of
305c209 shes of Galapagos having tone of voice like S. AMERICAN.-- Have not Ruffs & Reeves a remarkably varyi
310c223 ut not nearly so confined as now thought.-- N. AMERICAN, European & Chinese Genera & some species on
316c247 se however come from Siberia it cannot be said AMERICAN fossils more resemble those of America than o
318c253 curious history of first appearance of the S. AMERICAN Pipra Flycatcher, which is now becoming commo
325c267 on birds (facts about close species) Wilson's AMERICAN Ornithology Read Aristotle to see whether any
345d043 my Note Book.-- Why is not Tetrao Scoticus. an AMERICAN form (if so)?.-- A Sphepherd of Glen Turret.
353d061 are.-- good case these hares compared to North AMERICAN hares. Many species, separated by Mountains.
356d072 Some African dove.-- The extinction of the S. AMERICAN quadrupeds is difficulty on any theory-- with
357d073 continent since grown.-- This will explain. S. AMERICAN case & Didelphis being Mundine form., & the l
363d102 over Europe)-- The habits of some «same» North AMERICAN & Europaean birds «slight» different-- Barn O
376d136 from first generation., that with Quichuas the AMERICAN character is more tenacious. & does not disap
399e011 positively he has seen. a Calosoma. (very like AMERICAN form) in Stonesfield slate., & a Melolonittha
400e013 off Mexico with small Hares & raccoons.-- «S. AMERICAN form.--» off province of Guadalaxura-- Octobe
408e042 and well the relation of passage from N. to S. AMERICAN forms. The climate of N. America, must have b
419e086 more like those of Scandinavia, than of the N. AMERICAN species--Glacial period Dr. Beck says the me
423e098 eb. 24th. Monoceros, which Sowerby says, is an AMERICAN form.-- has several species in my fossils-- C
424e100 e known the Composites of Galapagos were South AMERICAN.-- several cases of species peculiar to separ
424e100 valla. Many species same. & Northern forms-- & AMERICAN ones & Europaean-- agree very much closer, th
425e104 at general forms.-- are the Labiata nearest to AMERICAN, or Indian groups?-- = Believes some Mediter
425e105 Owen. Fossil Mammalia. p. 55. talks of Tapirus AMERICAN form. found in Eocene beds of Paris Lyell has
427e109 nly: as stated by Capt. Graah») & break up. N. AMERICAN Conchology from Europaean., & the climate bei
448e166 well as plants «of Juan Fernandez» differ from AMERICAN Coast Vol II.-- <Reference> p. 251. about the
464tf6r with. N. America & S., (¿ is the peculiar. N. AMERICAN form)-- ¿Hunting leopard, how strange, anyone
483z013 scribed "Arion" of Ascension. p. do.-- some S. AMERICAN Reptiles are described Shells from Tahiti and
484z014 ridae wanderers.--?? in N. America?? Wilson N. AMERICAN Ornitholog must be studied before writing my
484z016 ed to some of my birds-- These groups strictly AMERICAN. Colouring on under side of wings It would be
486z019 ith T del Fuego Compare birds of do with <T> N AMERICAN & T. del. Fuego & Iceland Spix & Martius talk
511q018 Sir. R. H. supposes is now extinct-- (9) About. AMERICAN & Europaean common species, having somewhat o
526m029 t to mind, by memory of the scenes, (indeed my AMERICAN recollections are a collection of pictures).-
540m085 s about national character let him compare the AMERICAN whether in the cold regions of the North,-- t
540m086 copper coloured natives of New Zealand)-- the AMERICAN in Brazil is under same conditions as Negro o
545m106 this Baboon. knew women.-- Another little old AMERICAN monkey «(Mycelis)» I gave nut, but held it be
553m138 er) remarked that he had never seen any of the AMERICAN Monkey show any desire for women-- «very curi
555m142 ns of monkeys-- I could only perceive that the AMERICAN ones, often put on a peevish expression, but
555m143 espect resembles some of the old ones.- -- S. AMERICAN group sneer.-- Sept 21st Was witty in a dream
568n018 aste for singing with Mammalian structure. «-- AMERICAN monkeys utter pleasant plaintive cry-- The t
596n184 us, Niger. Cercopithecus make labial st st. S. AMERICAN monkeys. pull back skin from head very little
603o012 re.-- Origin is certainly curious. Chinese, S. AMERICAN. Polynesians Jews, African all sacrifices. Ho
473s03r .-- Sillimans Journal <vo> 1842. p. 142-- Sus AMERICANA & Hippotamus «with Megatherium & Mylodon» in
350d054 s in note. ¿mere conjecture?-- Australians.-- AMERICANS. &c After Decandolles idea Septemb. 1st. Macl
230b240 animals not soon being subjected to change in AMERICAS. perhaps merely gone back previous to fresh c
405e031 next 20 years none of his remain found in the AMERICAS probably did not.-- Octob. 25th. I observed i
408e042 e World.-- Europe perhaps less so, that either AMERICAS.-- If species change, we see external conditi
484z014 y is S. American representative of Caracaras of AMERICAS.-- manner of walking-- foot bill crest feathe
482z012 es come from Galapagos!!! Azara. Voyage dans l'AMERIQUE Merid. Tatu noir. abundant from Paraguay to 2
423e097 erent beings, (all fishes to the state of the AMMOCOETUS) Crustacea to---- ? &c) without reducing the
496q006 a-- Salt. Gypsum. Magnesium Iron Rust Carb. of AMMONIA.-- Horse Urine &c on associated plants. whe
043r071 from» "in the same line" to "from the epoch of AMMONITE to the present day. at Mauritius. (consult Bo
431e121 ast of France.-- L. doubts.-- Lonsdale thinks AMMONITES would afford instance of such facts.-- Ask Ph
182b047 ble the number «of steps» known). <for instance AMONG the Carabidae.-- instance in birds> Examine «go
284c149 mportance, as affording natural characters than AMONG those groups, where it remains less subject to
341d030 the young of this animal, which is so anomalous AMONG true deer, yet is spotted like so many deer.--
355d066 only in those character which are seen to <vary AMONG> be different in species of same genus." Law of
022r006 ne action of heat. it is so found in Anglesea, AMONGST the varying & dubious granites.--Wide limits o
023r012 ving been seen & from the contents of its maw, AMONGST which were things pitched over board early in
120a110 Notebook-- & scraps on Salsisbury Craigs. Kept AMONGST <old> papers read before societies.-- Sir. J H
177b027 it would only appear like circles;-- & insects AMONGST articulata.-- but in lower classes, perhaps a
204b141 has crossed Ducks & Widgeon & offspring either AMONGST themselves or with parent birds.-- «W. Fox. kn
250c037 re fully characterized, of each continent. Try AMONGST Europaean quadrupeds if Africa destroyed would
256c055 rickland & Hamilton-- found tertiary formation AMONGST Graecian isles, ¿See if type continued?-- See
259c064 ul case of extermination of species-- Epidemic AMONGST trees. Plane trees all died certain year. Extr
264c082 on inhabited by them-- Timor. Australian forms AMONGST birds Java. not so much-- Peculiarities of str
272c109 arkable how small detail in structure prevails AMONGST the same species & subgenera in families.-- th

Page **(Key Word)**
299c196 d.-- ¿is not Elephant intellectually developed AMONGST Pachydermata. like Man amongst Monkeys-- or do
299c196 ually developed amongst Pachydermata. like Man AMONGST Monkeys-- or dogs in Carnivora.-- Man in his a
343d039 the Deluge it may be worth adding in note than AMONGST the Mammalia of Europe the shells of do-- shel
375d135 lthus no one clearly perceived the great check AMONGST men.-- «Even a few years plenty, makes populat
378d148 k's eggs under common ducks, the young crossed AMONGST themselves, & I presume with common ducks. so
391d174 er bud produced by union of two common buds??? AMONGST buds each one exactly like its parent. <but th
477z002 Lat -- -- do. p. 207. La punaise was not known AMONGST Indian. introduced in Paraguay in 1769 introdu
480z008 tten in German.-- Stuttgart ranks these bodies AMONGST Vegetables in Linn. Soc.-- Mr. Donn Carmichael
490q001 ones not embedded?-- Do the Tame Parrots breed AMONGST the Indians Do the Savages select their dogs S
506q015 Ask Henslow to distribute some of my questions AMONGST agriculturists. whom he know.-- Col. le Couteu
513q020 mes» all on one side, as in crossing varieties AMONGST varieties cross one with abortive tail or horn
542m095 hats put up their hands, & as attention would AMONGST lowest savages clearly be directed chiefly by
570n027 o must idea of beauty.-- [Old Graecians living AMONGST naked figures, & observing powers common to sa
610o30v e, is strictly analogous to change of instinct AMONGST animals.-- Jan 13th. 1839 My father received a
610o033 . March 1840 p. 267--- says the great division AMONGST metaphysicians-- the school of Locke, Bentham,
095a039 n rocks «at Cobija.» At Iquique of elevation to AMOUNT of 30 ft.-- Mr Bollaert (at Roy. Institut) tal
117a101 lations of level will alone explain the immense AMOUNT of change which must have taken place, otherwi
118a103 ut off by denudation it gives one grand idea of AMOUNT denudation.-- This may be added to any place w
176b021 forms equable: this being due to subdivisions & AMOUNT of differences, so forms would be about equall
223b211 e Must confess. that we canot tell, what is the AMOUNT of difference, which improves. & checks it.--
227b225 sion from what we have seen. in old world, & on AMOUNT changes which may happen-- ‖ It leads you to b
231b241 in wild state to show that we do not know what AMOUNT of difference prevents breeding;--or as others
231b241 events breeding;--or as others we would exjudge it AMOUNT of varying in wild state.-- When breaking up «
239cIFC Also looked through April 23. 1873 Books About AMOUNT of difference: where hybrids produced have any
290c162 re much change has been added. «as it speaks to AMOUNT of change only & not kind» insects-- & vertebr
336d017 ng is in the blood, the more persistent.-- «any AMOUNT of change» shorter time less [s]o.-- the resul
348d051 atural arrangement,-- affinities, what is that, AMOUNT of resemblance,-- how can we estimate this amo
348d051 ount of resemblance,-- how can we estimate this AMOUNT, when <value> no scale of value of difference
355d069 nction of species, the capability of only small AMOUNT of change at any one time Seeing what Von Buch
356d069 t Medicines on organs, leads one to suspect any AMOUNT of change from eating different kinds of food:
364d104 ainly appears in domesticated animals, that the AMOUNT of variation is soon reached-- as in pidgeons
397e004 h. It cannot be objected to my theory, that the AMOUNT of change within historical times has been sma
418e080 maphroditism would not be perfect as in Ox. the AMOUNT of double sexual developement is spread over [
419e087 ooked that the chronology of geology rests upon AMOUNT of physical change <affecting whole bodies of
438e141 n crossing between varieties & species, yet the AMOUNT of may depend on many circumstances, time of d
442e151 generate once in a thousand generations.-- any AMOUNT of generation may take place by gemmation «My
461t037 jurious «(or requiring nutrition)» to a certain AMOUNT it will be so handed down«(».. as mammae of me
498q007 to cross. (5) It is most important to ascertain AMOUNT of variation in plants raised by Scions, as El
507q15v t down turning into Rye.-- 46). Book describing AMOUNT of Horticultural Variation? Henslow knows only
610o033 on whether we have any instincts, or rather the AMOUNT of our instincts-- surely in animals according
612o035 ugh not adding to positive knowledge. lessening AMOUNT of ignorance [RHC] The radicle of plants absor
218b189 p. 450 There is in nature a «real» repulsion «AMOUNTING to impossibility, holds good in plants» betwe
398e005 history of the world most closely, & know the AMOUNTS of change now in progress, will be the last to
073r163 between 18 degree & 22 degree N. = formations of AMPH: porphyry. greenstone[,] amygdaloid. basalt & o
037r055 tains <shells few corals Tortoise> «remains of AMPHIBIA, exclusively.» & Turtle bones. & the bones of
073r164 above porphyries characterized by no quartz & AMPHIBOLE frequently only vitreous felspar: = gold vein
076r170 ormable greenstone porphyrys & phonolites do. AMPHIBOLE quartz & mica very rare.-- ancient freestone
108a081 & Pichincha. Melaphyre. = Andesite-- Albite & AMPHIBOLE= Cook found Granite at Christmas Sound Vol XI
599o006 of Living. p. 264. "Architecture is a fine AMPLIFICATION of two ideas in nature; a developement of t
335d014 first hybrids may be compared to animal with AMPUTATED limb.-- Heredetary <thr> Six fingered people,
539m082 old indistinct ideas of FitzWilliam Musm. I was AMUSED at this after seven years interval. Augt. 15th
542m093 erect & stiff like that of turkey.-- he may be AMUSED, he need not express it, he may most earnestly
228b232 the most laborious work, our companion in our AMUSEMENTS. they may partake, from our origin in <there
437e139 ds got off from the young ones while they were AMUSING themselves with them, and one day a rabbit esc
567n013 y lead to something.-- October. 8th. Jenny was AMUSING herself--, by getting out ears of corn with he
073r163 = formations of amph: porphyry. greenstone[,] AMYGDALOID. basalt & other trap cover it to great thick
102a060 - Any one. who has studied rocks in detail as AMYGDALOID. calcareous rocks of Ascension, each particl
145g004] Salisbury Craigs V. Specimens-- {P} Veins, AMYGDALOIDAL-- as well as base not always parallel to st
071r158 shore of Pacific.--Isabelle's volcano, many AMYGDALOIDS.--Boussingualt «(Lyell)» cracks mountains fa
427e111 fect on offspring-- Mr. Herbert says «p 347. AMYYRALIDAE» Plants do not become acclimatised by crossi
299c194 ese cases,-- but absolute species formed. The ANAGALLIS perhaps, offers another case of permanent var
327cIBC ies de plantarum proesentum crypt. transitu et ANALOGIA commentatia Library of Useful Knowledge on Ho
278c129 another then those sections & subgenera; are ANALOGICAL, because we do not know, whether nearest spe
282c142 on exclusion.-- The same characters which are ANALOGICAL in a genus with respect to rest of its famil
282c143 iginally may «must» have had the character of ANALOGICAL.-- Gould says it is only in large groups. wh
460t017 , partly of the same external conditions (ie. ANALOGICAL structure) & partly the laws of organization
315c241 nstitut «1838» p. 184 Botany of Bonin. "grande ANALOGIE avec le flore du Japon", some Europaean & San
186b058 y; but why so? I can see no reason for these. ANALOGIES; CD[from the principle of atavism, where real
196b110 s appears greatest argument against theory of ANALOGIES. States there is but one animal: one set of o
214b172 Do the N. American. Tertiary deposits present ANALOGIES to shells of living seas.?-- Roxburgh. list o
282c141 ain degree in habits, yet we might have these ANALOGIES.-- We must two races of such men living in sa
293c172 als of forgetfulness.-- after sleep, «strong» ANALOGIES with memory in offspring.-- Some association
302c201 giving off of these two families, but we see ANALOGIES between fish.-- Birds same remarks. Character
315c241 ead in 1837. semestre I suspect some valuable ANALOGIES might be drawn between habitual actions of pl
352d059 es deep-- with respect to Macleay's theory of ANALOGIES-- <be> when it is considered the tree of life
354d062 may believe anything in such rigmaroles about ANALOGIES & number L'Institut p. 275. (1838) M. Blainvi
434e128 Didelphys says. "If we cannot reason from the ANALOGIES of the existing to the events of the past wor
641j29v f bee, sting of nettle.-- Are there any other ANALOGIES-- prickly plants or animals-- Exudation of fe
581n063 t, & see whether applicable to all cases.-- & ANALOGIZE it with ordinary habits that is my new part o
033r042 in Calcareous. rocks??--if so case precisely ANALOGOUS: fragments instead Peak of Teneriffe. also Co
069r150 rious fact felspar melted gneiss/// QUARTZ!!! ANALOGOUS to Von Buch. Basalt where Basalt. trachyte wh
072r160 llowness of <sep> Chiloe concretions somewhat ANALOGOUS to septa.-- would particle attracted towards
097a043 a dull & poor varnish, which I conceive to be ANALOGOUS to the black glazing observed by Humboldt on
098a047 in septaria. a kind of concretionary process (ANALOGOUS to layers of quartz & feldspar) within other
109a083 ing up.-- very good this will show effects.-- ANALOGOUS to broad flat sand beach. {P} -- De la Beches
119a104 mountains in the moon, which though not very ANALOGOUS (see Edinburgh. Phil. Journal <]CD>, no great
119a104 malayas, but great circular mountains, yet so ANALOGOUS, that as we see mountains formed (& mountains
121a113 . p. 12. proofs of small rise at Stockholm.-- ANALOGOUS to my Valparaiso case. Consider profoundly al
137a149 stance & blending with sandstone «said to be» ANALOGOUS to granite infiltering some of its constituen
209b156)» 25. plants. 36 St Helena, without ferns.-- ANALOGOUS to nearest continent: poorness in exact propo
220b198 The fact of plants going back hybrid plants; ANALOGOUS to Men. & dogs. Now if we take structure as c
221b202 Coniferous Woods before Dicotyledenous a fact ANALOGOUS to reptiles before Mammalia Think about Mioce
240c003 s of wings like the wild rock pidgeon.-- fact ANALOGOUS to Owen's Phil: remark of Apteryx having feat
240c004 being produced, & handed down, with ease, is ANALOGOUS to what occurs in plants.-- All these facts c
254c048 lowest stages of animal organization, "& are ANALOGOUS to the earliest conditions of the higher clas
263c078 is is a replacements in mental machinery-- so ANALOGOUS to what we see in bodily. that <I> it does no
281c139 egree; the tailors-- «in each branch would be ANALOGOUS to each other-- &c &c.-- V. p. 140» I should
282c141 be supposed to change. & make genera of bird. ANALOGOUS. animals would be possessed by the different
298c192 acter of leaf of plants is remarkable what is ANALOGOUS to it in animals?-- Babington says in most pl
300c198 The greater individuality of mind in man, is ANALOGOUS to greater individuality of bodies of some an
302c202 not easily told, for any small family. having ANALOGOUS characters, might be multiplied.-- we must ar
305c209 tatement nothing is explained.-- this is fact ANALOGOUS to mocking thrushes of Galapagos having tone
310c222 class» & full of second--this class of facts «ANALOGOUS to petrel-grebe. external» appears to be a pu
313c235 t, which had run wild in India. in Heber?) is ANALOGOUS to dormant instinct.-- (How wonderful a case
313c236 ympathetic nerves & nervous system of insects ANALOGOUS.?-- («Even plants have habitual actions.--» «
344d041 one impregnation sufficing to several births ANALOGOUS to superfoetation, & to successive fertile of
354d065 resemblance as if the "variation in one, was ANALOGOUS to specific character of other species in gen
356d070 ccasionally (as Fox says) same fruit trees is ANALOGOUS to some hybrids breedings-- there is tendency

Page ***(Key Word)**
362d099 ag. who rubs the skin off horns to fight-- is ANALOGOUS to the love of woman (as Mitchell remarks see
373d132 eon (yes) surely) secrete milk? from stomach. ANALOGOUS to other males feeding young, & to abortive <
387d169 of which receives <young> «eggs» in belly.-- ANALOGOUS to men having mammae.-- There is an analogy b
400e012 ery good case, might be made out of variation ANALOGOUS to specific variations.-- Kerr's Collect of V
409e046 ellectual than fox, wolf &c &c-- is precisely ANALOGOUS case to man, exceeding monkeys;-- Having prov
430e118 allowed to cross.-- » Has nature any process ANALOGOUS-- -- if so she can produce great ends-- But h
437e141 a certainly not becoming Brussels Sprouts) is ANALOGOUS to Primrose & Cowslip suddenly changing into
446e162 ct of fresh bud-- here then is case of change ANALOGOUS to change in grafted trees «:so is not effect
447e165 opagated in a garden, which is case precisely ANALOGOUS to the Canada onion mentioned in Hort. Transa
490q001 acter of the extinct land-shells of Madeira-- ANALOGOUS or quite distinct from recent ones-- I presum
524m019 brings on suddenly in head, a frame of mind, ANALOGOUS to those feelings. which may be considered as
524m020 hen a disagreeable thought occurs, is closely ANALOGOUS to Epilepsy & convulsion.-- affections of the
528m035 cause train cannot be discovered-- is closely ANALOGOUS to my Fathers positive statement that insanit
530m045 -- Delirium seems to rest the sensorium.-- -- ANALOGOUS to sleep--; some doctors care it, by stimulus
533m058 what frowning means) if so this is precisely ANALOGOUS or identical, with bird knowing a cat, the fi
536m071 ove? Stallion licking udders of mare strictly ANALOGOUS to men's affect for womens breasts.∴ Dr Darwi
538m078 eir insanity» there seem other cases somewhat ANALOGOUS, & which I think will lead to fact of old peo
538m080 from the ordinary state of mind, is probably ANALOGOUS to the double individuality implied by habit,
541m091 EXPERIMENTIZE upon this effort.-- it looks so ANALOGOUS to muscle in one position great fatigue.-- ma
564n006 ed between eyes & upper lip., is most clearly ANALOGOUS to a panther I saw in garden uncovering its t
570n026 we have been most generally accustomed to:-- ANALOGOUS case to my idea of conscience.-- deduction fr
574n042 a manner which can be transmitted.-- this is ANALOGOUS to a blacksmith having children with strong a
575n046 , when new ideas will not enter. is something ANALOGOUS. to instinct, to the permanence of old herede
581n063 n's Christian name, writing for the surname,, ANALOGOUS to instinctive memory, & consequently instinc
582n066 Mozart. fine music is evidently considered as ANALOGOUS to glowing conversation of several people.--
582n066 rubbery. when other people are about: this is ANALOGOUS to young pigs hiding themselves; & heredetary
584n072 ven blind horses & dogs) shows it is somewhat ANALOGOUS to memory. Shrugging shoulders seems sign of
600o08v ssion. like avarice.-- Is there not something ANALOGOUS to imperiousness of Conscience: in Maternal i
610o30v The change <of> our moral sense, is strictly ANALOGOUS to change of instinct amongst animals.-- Jan
614o037 amarck's theory true [RHC] Acquired instincts ANALOGOUS «(& replace)» to experience gained by man in
615o36v Heredetary effect of former tropical climate ANALOGOUS to inflorescence of Tropical plants when impo
618o041 y apprehensible by sense. 5) There is nothing ANALOGOUS to this in the relation of thought, perceptio
628o53v f the instinct of a shepherd dog, is strictly ANALOGOUS to education of child,-- causing many actions
638j58v intellect this process is shortened, but yet ANALOGOUS, no savage ever made a perfect hinge.-- reaso
116a099 ers to species non decrite de petites corbules ANALOGUE living in mouth of Plate. p. 26. Geology of A
215b178 iless cat of Isle of Man mentioned in Loudons (ANALOGUE of Blood hound-- Bull. Soc. Geolog. 1834. p.
216b179 4. p. 217. Java Fossils 10 out of twenty have ANALOGUES uses this word «for» similar in the Indian se
255c050 ore commonly are the same species, instead of ANALOGUES.-- <as> in other classes this evidently relat
257c056 és en Patagonie. ou au moins des espèces tres-ANALOGUES,-- quand ce nétaient pas tout à fait les môme
353d061 by wedge tails.-- many of the hawks <to> are ANALOGUES to <Bustard> Europaean birds. also «do» p. 40
641j29v of Chamelion. crabs Crabs & Mollusca we have ANALOGUES The stillness p. 276) of flight of Owl remark
039r060 he famous eruption of Rialeja, & the more true ANALOGY from the Galapagos-- Mr Lyell. P. 111 & 113. «
039r061 formations: Mem the envelopes at Coquimbo. the ANALOGY is now perfect <The grand propulsion of fluid
039r061 evates a continent> We are more abound to take ANALOGY of movements of W coast in explaining plains b
059r121 en in one case contained lime.--All bear close ANALOGY to Obsidian, & all show chemical action as wel
069r152 - Oh the vast power of the ocean! Make a grand ANALOGY between Wealden & Bolivia Transportal of congl
070r155 crust thin.--Concepcion earthquake Draw close ANALOGY Lake of Cordill: of Copiápò & Desaguadero.--th
073r164 ic porphyry. = several parts of N. Spain great ANALOGY to Hungary. = Veins of Zimapan offer zeolite.
088a013 ce turned over axis or hinge no doubt fluid.-- ANALOGY as continental elevations slow. so would line
100a055 , would Babbage.-- Webster Phillips insists of ANALOGY between Australia & Oolitic period.-- comparis
146g012 egs & tail like species in colouring Strike an ANALOGY between pleasures of association, & passions,
196b110 that there must be very great gaps.-- yet some ANALOGY The existence of plants, & their passage to an
201b129 e produced fe[w] [not located] The relation of ANALOGY of Maclay &c. appears to me the same, as the i
202b130 from external influence.--...... Hence name of ANALOGY, the structures in the two animals bearing rel
204b139 837 Westwood has written paper on affinity and ANALOGY in Linnaean Transactions Mr Wynne distinctly s
210b158 t why they should be manu> Does it not present ANALOGY to what takes place from time? Von Buch distin
210b159 ic but I . 2,53.-- In know varieties. there is ANALOGY to species & genera.-- for instance three kind
210b161 certain laws according to species. present an ANALOGY to production of species.-- Animals have no no
214b173 land.-- Phillips. Lardner Encyclop. insists on ANALOGY between Australia and «fossils of» Oolitic Ser
232b247 nts from extreme north, which according to all ANALOGY would have been very unlike Southern Europaean
252c045 od species Indian species so distinct that all ANALOGY from each other I do not know how different Su
252c045 «some doubt, from want of knowledge of time-- ANALOGY from three first will give one almost certain
258c060 e form, (but is modified), the relationship of ANALOGY is a divellent power & tends to make forms rem
258c060 owers.-- Every animal in cold country has some ANALOGY in not gaudy colours so all changes may be con
258c061 nothing of much interest XX. Hence relation of ANALOGY may chiefly be looked for in the aberrant grou
258c061 t from each other, & resemble some other class ANALOGY» The resemblances relationship, the dissenblan
258c061 e resemblances relationship, the dissenblances ANALOGY,-- See Abercrombie p. 172. for definition of A
258c061 y,-- See Abercrombie p. 172. for definition of ANALOGY. All the discussion <after> about affinity & h
274c115 ons of affinity all organs change together, in ANALOGY certain parts perfect of typical structures ce
277c126 should be mark to every species. only known by ANALOGY genera of course distinct. analogy from every
277c126 ly known by analogy genera of course distinct. ANALOGY from every country & class tells us that ¿O. M
277c127 t two ostriches good species because interlock ANALOGY to be guide. in islands. species.-- each descr
277c128 many ideas of causes of change.-- The mark of ANALOGY would be empirical because as soon as two spec
281c138 ated to each other I cannot help thinking good ANALOGY might be traced between relationship of all me
281c139 in each group is quite fatal.-- Relations of ANALOGY being those last obtained.-- less firmly fixed
282c140 ome genera, which are most fixed in others. In ANALOGY it is not the relation to bear to each other,
282c140 y.-- affinity is the sum of all the relations, ANALOGY is the close relationship in some one.-- imagi
282c142 but might hire the preexisting race, thus the ANALOGY would not in all cases be produced, but would
285c151 nt ones) the characters will be first those of ANALOGY, but will grow into affinity.--but whether eve
285c152 put together, they would not according to all ANALOGY breed together.-- The bottom of the tree of li
288c158 &c &c-- Macleay rests his whole groundwork of ANALOGY on its concurrence in parallel parts of his se
290c162 t classification on generation might appear an ANALOGY NB Pyrrho-alauda (bird of St Jago) of brown co
291c166 t anything but structure of brain heredetary,. ANALOGY points out to this.-- love of the deity effect
293c173 s its loss ¿ if it explains its acquirement.-- ANALOGY. a bird can swim without being web footed yet
294c176 show species-- I fear argument must rest upon ANALOGY & absence of varieties in a wild state-- it ma
294c177 y descent. or two of the willow wrens &c &c. & ANALOGY will necessarily explain the rest.-- Lonsdale
295c178 ds attract females by song. do they by beauty, ANALOGY of man if so war not [not located] Erasmus say
297c186 p. 160. soft plumage of night jar. like owls. ANALOGY in habits adaptation to nocturnal habits-- to
302c201 tion between two such classes will be those of ANALOGY, which when sufficiently Multipliied become af
302c202 ween fish.-- Birds same remarks. Characters of ANALOGY.-- last acquired,-- or aberrant, therefore mor
302c202 ersely. WHERE CHARACTER VARIABLE it is (one of ANALOGY or) LATELY ACQUIRED. In pigs number of vertebr
302c202 «of equal value» must be so from characters of ANALOGY.-- see my notes on p. 37. of Macleay. wonderfu
340d026 nities!. confounds, like Whewell affinity with ANALOGY-- Good table at end of distrib: of <birds>. An
350d053 e for genera, & so mount upwards.-- «judged by ANALOGY»-- Consider all this NB. How can local species
382d158 ery animals, but unequally developed.-- surely ANALOGY of molluscs. & neuter bee would shew this-- (D
387d170 analogous to men having mammae.-- There is an ANALOGY between caterpillars with respect to moths, &
411e056 transportation will be answered) one looks to ANALOGY for cause in plants. where innumberable indivi
416e070 called» progressive tendency law.-- In animals ANALOGY leads one to suppose that seminal fluid fluid,
417e077 n the Systema Naturae.-- Mr. Knight makes this ANALOGY between grafting & sexual union-- Looking at s
418e083 s, different in each species,-- & knowing from ANALOGY, that all these very animals are descended fro
427e110 gions? Arctic forms have travelled S. From the ANALOGY of the animal kingdom I should suppose, that t
434e128 no foundation for our science".-- <it is only ANALOGY.> but experience has shown we can & that analo
434e128 alogy.> but experience has shown we can & that ANALOGY is sure guide & my theory explains why it is s
436e136 ut beyond these not.-- argue against this-- -- ANALOGY will certainly allow variation as much as «the
441e149 offspring are,-- if Australian Dog, could bud, ANALOGY tells us, <be> «offspring» would be similar to
446e162 stocks in this case».-- & strong case showing ANALOGY of production by gemmation & by seed-- which H
447e164 es. &-- very heavy seeds.-- as Cocos do mer.-- ANALOGY shows some most important end.-- Festuca vivap
462tf4v e Mataco-armadillo & the woodlouse-- -- a good ANALOGY-- sea-Crustacea-- Tullus. Athenaeum 1839 p. 77

Page
(Key Word)
543m101 s. as other parts of structure. C. D.27 Can an ANALOGY be drawn between «heredetary» associated pleas
545m106 ut such movements in skin of eyebrow important ANALOGY with man.-- I see monkeys grin with passion, t
559m153 ifference so great... "Ay Sir there is much in ANALOGY, we never find out." This unwillingness to con
566n012 state of science, grand idea: as before having ANALOGY to guide one to conclusion that any one fact w
567n013 nd before they can talk, so do many animals.-- ANALOGY probably false, may lead to something.-- Octob
575n044 this is most important: can there be stronger ANALOGY that the tendency to hybrid greyhound to hunt
575n045 27th. November.-- Think, whether there is any ANALOGY between grief & pain-- certain ideas hurting b
579n060 nt. The Blushing of Camelion & Octopus; strong ANALOGY with my view of blushing-- in former irritatio
591n100 alism, is held in abhorrence.-- all this makes ANALOGY of actions with <&> against benevolent & paren
604o017 n> when insanity is coming on «Thinks clearest ANALOGY between dreams & insanity.» D. Stewart on the
613o036 nization of brain; «[LHC] not used by Kirby» (:ANALOGY:-- as races are formed or modification of exte
625o50v if Mackintosh.-- p. 262. Some good remarks, on ANALOGY of pleasure of imagination «the utility part b
638j58v forms.-- the exuviae of the dead & extinct The ANALOGY between the works of art «or intellect» such a
639j28r have been deserts in the old world!]CD p. 252 ANALOGY of hand in mole, & Mole cricket & rodents   (?)
639j28r , in mouth of swine & in stomach of lobsters-- ANALOGY in Flamingo & Duck, Ornithorhyncus «externally
641j29v these views will bear on geology-- There is an ANALOGY between fang of snake, (jaw of spider?) sting
278c129 must be a true cleft-- putting out of case the ANALOGYS.-- If genus does not Mean this it means nothi
311c226 e (including brain & other organs difficult to ANALYSE will not this separate facts about abortive o
407e040 cene of Paris! (Great Edentata at that period) ANALYSE this,-- consider state of vegetation, & concho
558m150 as yourself". "love thy neighbour as thyself". ANALYSE this out.-- bearing in mind many new relations
565n007 - Why does suspicion look obliquely.-- who can ANALYSE suspicion-- yet who does not recognise look of
567n015 <ins> like instinct, preeminently so-- who can ANALYSE the sensation, when meeting a stranger. who on
570n025 th both have shame-- Animals have not modesty. ANALYSE this.-- «Excellent-- my theory of blushing sol
573n039 she said nothing but shrugged her shoulders.-- ANALYSE this.-- Miss C. quite aware & indignant with M
578n054 would marry Miss. O. B.-- Mrs. B. A. blushed. ANALYSE this:-- Let a person have committed any «conce
606o025 r is the effect of a motive.-- [-- must be so, ANALYSE (a) ones feelings when wagging one's finger--
608o026 cause man has power of action, & he can seldom ANALYSE his motives (originally mostly INSTINCTIVE, &
615o36v was there any cause, & if surprise was felt.-- ANALYSE feelings. Mr Wynne says, that beyond doubt cou
075r167 not been vein «of iron» discovered?-- Klaproth ANALYSED silver ores from Peru consisted of native sil
557m148 sy, envy all primitive feelings, no more to be ANALYSED than fear or anger? I should think shame woul
557m148 ger? I should think shame would be more easily ANALYSED than jealousy, because less discoverable in a
614o037 s even the most complicated instinct. might be ANALYSED into steps, as species change.-- Must be so i
627o053 hows this is true. p. 13. Affections cannot be ANALYSED into "power" &c &c &c-- & if termed "selfish"
071r156 ken, by earthquake. but no serious injury. -- <ANALYSIS of Atacama. Iron in Edinburgh. Phisoph. Trans
095a037 son's Bay. 2. species Vol VI. Geograph. Journ. ANALYSIS of Poenig Voyage Valparaiso Dr. Gillies in MS
111a087 328 VI. p. 365. Meyen on Chile must be studied ANALYSIS of Voyage: many observations on heights of va
516q23v l substances-- (Athenaeum (40) p. 823 chemical ANALYSIS of Peat (2) Athenaeum 1840 p. 777. Decaying w
528m036 air, or dreams real again explains insanity.-- ANALYSIS of pleasures of scenery.-- There is absolute
556m146 essing cool irony, not biting? What is Emotion ANALYSIS of expression of desire-- is there not protru
206b148 ries, the other will become extinct.-- Who can ANALYZE causes, dislike to marriage, heredetary diseas
468t106 rb blossom swarming with small Staphylinidae-- ANASPIS, Melegethes, Leptuse-- Diptera & small Hymenop
633j53v tal.-- p. 285 the seed-pod of a desert plant (ANASTATICA) is rolled along, & splits when it comes to
326c266 d art.. on <Etn> Entozoa by Owen in Encyclop. of ANAT & Physiology.-- Dampier. probably worth readin
367d113 n Mare inland Owen says that Bell in Encyclop of ANAT & Phys. describes, a high-flying bat, which has
386d166 hole.-- if cut off nerves in snail. (Encyclop of ANAT & Phys) can make a head: the other part may sur
459tf02 . III p. 83. Paper translated from Meckel. Comp. ANAT.-- From Buffon cross of he-goat & sheep, it see
338d025 use. contemporaries. In introduction to Eytons ANATIDAE.-- recurs to idea of only animals from distan
340d026 gy-- Good table at end of distrib: of «birds». ANATIDAE.-- Consult this book again.-- Mine is a bold
479z005 Falklands birds Discussion of Firola.-- Salpa ANATIFS without shells.! p 442.-- Planariae p 451.-- m
309c222 arrangement.-- Thinks passages very rare., in ANATOMICAL structure.-- «the passages between-- owls &
369d115 rvation of the living habits of animals, with ANATOMICAL & Zoological research, in order to establish
633j54v m some grand & simple laws.-- 4 «Study Cuviers ANATOMIE Comparé» p 308. Traces the gradation of skele
198b114 ideas about propagation His work. Philosophie ANATOMIQUE. 2d Vol about monsters worth reading NB well
509q017 ith him Ask him to introduce me to some Human ANATOMIST. Has he dissected any animal often, which has
223b208 eleton of a Negro-- had been found what would ANATOMISTS have said.-- ¿where is Pentland's account of
227b228 would give zest to recent & Fossil Comparative ANATOMY, & it would lead to study of instincts, herede
481z010 og. of Voyage of Astrolabe must be studied for ANATOMY. of. corals.-- nevertheless the details appear
515q022 ter se, & with parents & of Chinese geese. (2) ANATOMY of muscles of stumps of tailess dogs & cats.--
186b058 uring of crysomela may be going back to common ANCESTOR of Crysom. & Heterom, but I cannot understand
229b232 partake, from our origin in <there> one common ANCESTOR we may be all netted together.-- Hermaphrodit
255c052 to this description of class is description of ANCESTOR of all birds, & so for birds. We thus obtain
286c154 o raise on the same shelf-- to (look at common ANCESTOR, (scarcely[)] conceivable in savages) Has not
199b119 Mission to Ava. account of HAIRY man. because ANCESTORS hairy with one hairy child, and of albino DIS
295c178 mphlet) Is man more hairy than woman. because ANCESTORS so, or has he assumed that character -- -- fe
390d179 the comparison of each animal with its<elf> «ANCESTORS», & not its comparison «of difference» with o
571n027 e clear evidence, of the general ideas of our ANCESTORS being impressed on us.-- Surely we have taste
574n041 ion of saliva, is probably due to our distant ANCESTORS having been like dogs to bitches.-- How comes
222b206 osed that form, which has wandered least from ANCESTRAL form. If so are present typical species most
566n010 do they indicate mind & body retrograding to ANCESTRAL type of consciousness &c &c.-- Lavater. (Holc
468t111 of Cistus Speedwell to Rhododendron-- <Loasa» «ANCHUSA»-- speedwell Iris-- Azalea. Rhododendron. Fraxin
468t111 dwell Iris-- Azalea. Rhododendron. Fraxinella s ANCHUSA <never» «once» P on Fraxinella «Heartease» «sm
502q11v s-- does not elm. does it «in» melon-- «Loasa» ANCHUSA «Campanula» &c & dead-nettle.-- Lithospernum.
023r011 he Pacifick is 60 miles distant from the grand ANCIENT volcanic axis «of the Andes».-- «Has this faul
063r130 istinct species inosculate, so must we believe ANCIENT ones: «.·» not gradual change or degeneration.
076r170 lites do. amphibole quartz & mica very rare.-- ANCIENT freestone & breccia is the same with that. on s
077r170 to destruction of porphyries. whereas other to ANCIENT rock.--this N degree 2. superimposed on N degr
717r171 en No. 2. might be mistaken for Porphyry above ANCIENT freestone, limestone & «many» «other secondary
085a002 Volcanic islands. from number of craters very ANCIENT. which agrees. with peculiar character of Vege
098a044 Bulletin de la Soc. Geolog: 1833-34. p. 35.-- ANCIENT Lake Lemagne in Auvergne Proofs from Phryganea
180b037 genera would be formed.-- bearing relation to ANCIENT types.-- with several extinct forms, for if ea
180b037 several extinct forms, for if each species «an ANCIENT (1)» is capable of making, 13 recent forms.--
199b117 e classified.-- Propagation explains this.---- ANCIENT Flora thought to more uniform than existing. E
222b207 e present typical species most near in form to ANCIENT; in shells alone can this comparison be instit
227b224 s intervened.-- (2d) By character of any «two» ANCIENT geography species tell of Physical relations i
275c119 > I use (new step in induction) as keystone of ANCIENT geography species to see whether any my views very
325c267 ead Aristotle to see whether any my views very ANCIENT? Study with profound care, abortive organs pro
346d048 early» with. the. account. given by Boethius of ANCIENT Caledonian Cattle. Ch 3. Instinct I'Institut.
358d074 ic.--» to nearest continent.-- With respect to ANCIENT geography of Atlantic Tristan D'Acunha ditto J
407e040 to their productions.-- Hence it is, from the ANCIENT preeminently equable & temperate climate, <tha
416e072 depend upon inheritance & in species are only ANCIENT & perfectly adapted races L'Institut 1838.   p.
427e108 remain stationary, hence all present types are ANCIENT. According to my views of <Dioecious p> all pl
430e117 g from few stocks. does not bear, the least on ANCIENT generic forms.-- the animals in Eocene period
439e145 uraged-- » Wilkinsons Manners & Customs of the ANCIENT Egyptians Vol III. p. 33-- They had several br
458t001 fossil frog, 40 inches in length--! alludes to ANCIENT gigantic salamanders-- Every order (except wha
541m093 ling real bitter sarcasm.-- <These> Seeing how ANCIENT these expressions are, it is no wonder that th
553m136 st in different degrees in races.-- whether in ANCIENT Greeks, with their mystical but sublime views,
579n058 they laugh.-- Has frowning anything to do with ANCIENT movement of ears A man shivers, from fear, has
634j55r -- How came Bats also.? before birds? They are ANCIENT.-- Are Cetaceae found in Paris Basin?.-- NB) T
357d073 produced commonly in Nature. both in Sweden & ANCIENTLY in Britain) between hen Caperailkie & cock Bl
571n028 ting them & reasoning on them (as justice?? as ANCIENTS did high forehead sign of exalted character??
451e177 ct for my theory Cocos Isld & Preparis between ANDAMAN & Pegu. <have> abound with monkeys & squirrels
515q022 ts different?--. Important.-- Oct. 44 Tell J. ANDERSON'S statement of English Horses having fewer ver
022r007 ch Chlorite in some of the dikes.--P 432. as in ANDES. In Dampier's voyage there is a mine of metereo
023r011 nt from the grand ancient volcanic axis «of the ANDES».-- «Has this fault determined side of volcanic
034r044 ulphuric vapours in East Indian Volcanos Gypsum ANDES Mem. Beechey. account of regular change in soun
036r052 specially rocky parts of central Patagonia Does ANDES in Chili. separate geographical ranges of plant
040r064 e may have been oscillations in the upheaval of ANDES.--but as long as all below water no evidence--T
042r069 lly elevated, which movements if present in the ANDES, would have destroyed regularity of slope of va

33

Page **(Key Word)***
```
060r125 would the rocks have been. The red Sandstone of ANDES fusible? no. mad dogs. Azores. although kept in
064r136 rge enormous quantity of matter from CREVICE of ANDES--therefore flowed towards it. a mass on each si
088a016 ry important V. Malte brun -- Main character of ANDES Metamorphic action -- Mem: red sand of Europe n
103a065 s. argues in favour of pressure of liquid rock. ANDES discussion-- Albite certainly contains 6 per ce
107a079 tz Roy's Case of S. Maria & Tubul applicable to ANDES & Patagonia-- On Lyells idea of whole centre of
108a079 hen crust of solid earth would be thicker.-- PP ANDES mark the line between sinking & rising areas.--
119a104 urgh. Phil. Journal <JCD>, no great chains like ANDES or Himalayas, but great circular mountains, yet
120a107 all pieces-- mem coal-field.-- the structure of ANDES. where we believe we can trace the outlines of
122a115 rates on shore of Loch Lochy very like those of ANDES Speculate under head of Beagle Channel. on orig
149g030 ach side of the two valleys corresponding as in ANDES, composed of sand & perfectly rounded stones--
408e045 h Snow on Himmalaya-- Humboldt bones at 7800 in ANDES-- parallel & curious facts.-- The Himmalaya. ca
436e137 Sandstone-- Van Diemen's land.-- Porphyries of ANDES. A familiar History of Birds by the Rev. E. Sta
091a026 another on Mexican Trachyte <roc> lava called ANDESITE. Red Coral in the Mediterranean 700 feet deep
108a081 ocks of Chimborazo., & Pichincha. Melaphyre. = ANDESITE-- Albite & amphibole= Cook found Granite at C
065r138 Isld.> South Shetland Cape Possession. Syenite¿ ANDITE?-- Degrading of inland bays. like St. Julian &
104a066 mon felspar therefore on axis of Cordillera, in ANDITE - containing 80 per cent of Albite 80/100 X 6/
278c129 ies banished by this test.-- Excepting where an ANDREW Smith, «Richardson» a Vaillant, a D'orbigny ha
376d136 t disappear for Many generations Sept. 29th Dr. ANDREW. Smith «Remarks on extraordinary curiosity of
387d168 e with the added <like Lord Moretons case & Dr. ANDREW Smith,» difference.-- If A. B. C. D. E be <off
471t211 illiams. Narrative of Missionary enterprises Dr ANDREW Smith says in the larks from S. Africa he can
507q016 uller's plants ,Teazle Dr. Holland ; My Father. ANDREW Smith (1) Are cross-births, or other accidents
508q016 -- How are livers obscure organ. no answer?-- 3 ANDREW Smith, about tamed wild animals breeding at th
508q016 ond limits of the metropolis of each-- Cause?-- ANDREW. Smith. (6) What size book Gallesio storia del
592n101 st. of Religion" on its origin in Human mind.-- ANDREW Smith says hen doves & the female chamaeleon c
316c244 orm of snow after his brothers murder.--» good ANECDOTE --.Sowerby.--. Geographical range of shells l
376d136 ry curiosity of Monkeys». The Baboon of which ANECDOTES have been told is Cyanocephalus Porcarius.--
507q15v be arrived at.= Any book with drawing of Seed. ANEMONE with, tuft-- Bull Rush-- Dandelion-- Sycamore.
067r143 1: Buenos Ayres 1836: W. Parish?? «by Pedro de ANGELIS.» This work is reviewed in present Edinburgh M
123a117 segregation in Salisbury Craigs Letter from M ANGELIS. B. Ayres. 3d. May. states remains found in ma
213b169 en were dead then monkeys make men.-- Men makes ANGELS-- Those species which have long remained are t
224b215 al being acquired alters case) other species or ANGELS. produced «&» Has the Creator since the Cambri
616o038 ems to indicate that when we <return» turn into ANGELS. this imperfect memory may become perfect & we
522m012 1» intemperance.-- often accompanied by extreme ANGER, at not being understood.-- My F. says there is
522m013 instincts.-- My Grand F. thought the feeling of ANGER, which rises almost involuntarily when a person
523m014 ciousness not just.-- From habit the feeling of ANGER must be directed against somebody.-- Have insan
531m051 ot pa.> In reflecting over an insane feeling of ANGER which came over me, when listening one evening
531m051 when tired --« how true the heart the scene of ANGER.--» to the pianoforte, it seemed solely to be f
535m069 t savage attribute thunder & lightening to Gods ANGER.-- (∴ more poetry in that state of mind: the Ch
536m070 anting not to feel angry-- such efforts prevent ANGER, but observing eyes thus unconsciously discover
539m084 - My handwriting same as Grandfather. Aug. 16th ANGER «Rage» in worst form is described by Spenser (F
542m094 is crying-- peculiar not common?--» no bark of ANGER nor have monkeys & many other animals,-- but ye
549m122 ness.-- Men having some instincts as revenge «& ANGER», which experience shows it must for his happin
550m124 nstinct, would probably feel but little that of ANGER or revenge.-- they are incompatible & the forme
557m147 sed pricks his ears?--).-- How is expression of ANGER in species of swans, in parrots & &c ---- pea
557m148 e feelings, no more to be analysed than fear or ANGER? I should think shame would be more easily anal
565m008 le one attacks him Contempt, when there is some ANGER «& respect to opponent» is showed by same movem
594n113 gly. baby- association-- pouting child same as ANGER, lips not compressed sullen, protruded. determi
609o029 er animals & therefore to progenitor far back. (ANGER <to> at the very beginning, & therefore most de
619o043 n. he would feel pain, which would generally be ANGER, as he would be tempted to interfere, but with
625o50v in child or animal it is equally proper to obey ANGER as benevolence (but not cool malevolence). it i
625o50v ). it is only after reason comes into play that ANGER can be said to be wrong.--. for then only is it
106a073 not great.-- Is there more degradation at first ANGLE owing to momentum. which the water has obtained
121a111 angular fragments of rock, which retained their ANGLES sharp-- yet with character completely altered,
021r005 calls it a Fucus. P «Vol 1 287» P 379. Henslow ANGLESEA, nodules in Clay Slate. major axis 2.½ ft.--
022r006 ve undergone action of heat. it is so found in ANGLESEA, amongst the varying & dubious granites.--Wid
022r007 Plas Newydd dike.--Mem tres Montes. ((Henslow ANGLESEA)) great variety in nature of a dike.--Mem. at
091a025 e structure (& veins appearing): mem. Henslows ANGLESEA solution of silex also shewn. No 3d of Ed. N.
115a097 e deep.-- Henslow has deposited specimens from ANGLESEA in Geolog. Soc. if numbered compare them with
290c163 round, the minute gets into the road at right ANGLLES, how pleased it is, just like man. emotions ve
306c214 vels in Asia Minor of the domestic animals-- At ANGORA «centre of Asia Minor» are the fine-haired goa
536m070 eiving myself skipping when wanting not to feel ANGRY-- such efforts prevent anger, but observing eye
542m094 ve monkeys & many other animals,-- but yet when ANGRY it is hard not to growl out some sound even if
545m107 ost curious <like» «remember» the expostulatory ANGRY look of black spider monkey when touched, also
552m131 father remark, a tired man. involuntarily feels ANGRY, when brain is pumping force to legs & body, &
555m143 fferent disposition from others, slow cautious, ANGRY cross look, followed by protrusion of lips, in
557m147 t, to when in a passion.-- dog tail curled when ANGRY & very stiff. back arched. just contrary. when
558m152 arch their backs-- Bengal tiger. when slightly ANGRY. curls tip of tail.-- do two cats arch their ba
565n008 p are depressed & opposite muscles used to when ANGRY sneering is in progress.<--> the hypothesis of
582n066 rigidity of muscles.-- man cannot be said to be ANGRY.--» «He may have pain or pleasure these are sen
435e134 at tadpole increased in size now the Proteus ANGUIFORMIS. he remarks lives in dark caverns of Carniol
557m149 amping. grinding teeth.-- in shame frowning, & ANGUISH,-- shyness not so.-- affected laughter.-- A do
026r020 blocks" not sufficient distinction is given to ANGULAR & rounded.-- Fox Philosoph. Transactions on me
035r048 nce; as fractures, consequent on grand rise, & ANGULAR displacement, consequent of injection of fluid
121a111 se) shows power of segregation.-- & has heated ANGULAR fragments of rock, which retained their angles
156g068 > «make out» composition of shelves: generally ANGULAR except near head of valley fragments which had
157g074 ony parts; appearance chiefly cause by fall of ANGULAR masses from above on soft shelf-- 29.330 A 84
162g104 ne alternate curved layer of fine sand & small ANGULAR-- rounded pebbles-- dip sideward, & inwards--
528m038 ms or to the repetition of similar forms as in ANGULAR leaves.-- (this Rhythmical beauty is shown by
253c046 rimigenius over so wide a range, & Mastodon ANGUSTIDENS.-- Ogleby has facts to show that Australian
405e032 & gold fish-- Octob. 26th. If. hereafter. M. ANGUSTIDENS be found to be inhabitant of S. America & as
424e102 th of the Elbe.-- yet they meet in one wood in ANHAULT. & there every year produce hybrids-- now this
337d023 om one stock.-- Theory of Geograph. Distrib: of <ANI> organic beings.-- Animals «of same classes» dif
454e184 malcule. has described instrument for galvanize <ANI> them-- Cross Irish & Common Hare Decandoelle ha
526m028 «Catherine thinks that children like looking at <ANI> pictures, an early taste, of animals. they know
040r062 at St Julians looks like such?--destructive to ANIMAL life.--Patagonia In the Chonos Islds we must i
051r095 e a trifling circumstance determines whether an ANIMAL will adhere to a certain part. Apropos to ques
051r095 ere to a certain part. Apropos to question does ANIMAL adhere to rock because it does not decompose.
063r132 ether ordinary. hermaphrodite. or by cutting an ANIMAL in two. (gemmiparous. by nature or accident).
078r174 name of Sagitta Triptera D'Orbigny has figured ANIMAL with setae like my undescribed[.] p. 140. Flèc
080r178 point affin of yew & intermediate Puncture one ANIMAL with recent dead body of other. & see if same
175b016 t in course of ages. & therefore changes. every ANIMAL has tendency to change.-- this difficult to pr
175b019 epoch-- How is this Ehrenbarg? every successive ANIMAL is branching upwards different types of organi
176b021 equally numerous. changes not result of will of ANIMAL, but law of adaptation as much as acid & alkal
182b046 wman the (7) Man studied The condition of every ANIMAL is partly due to direct adaptation & partly to
183b052 remières generations" go back to type of either ANIMAL when crossed with it.-- ¿Whether extinction of
185b055 t.-- May this not be explained on principle, of ANIMAL having come to island. where it could live-- b
187b062 hing in country to superinduce a change? Seeing ANIMAL die out in S. America; with no change, agrees
189b073 though produced either sooner or later.-- Prove ANIMAL like plants:-- trace gradation between associa
189b074 will be complete.-- It is absurd to talk of one ANIMAL being higher than another.-- We consider those
191b079 on a porcupine on Echidna-- Good to study Regne ANIMAL for Geography.-- The motion of the earth must
191b082 gether for life, is in accordance with The Male ANIMAL affecting all the Progeny of female insures of
192b087 chidna & Hedgehog]CD-- As we have one Marsupial ANIMAL in Stonefied slate, the father of all Mammalia
192b089 ia as Heterodox as ornithorhyncus. If this last ANIMAL bred-- might not new classes be brought into p
193b090 atized, climatizes the child. ¿-- whether every ANIMAL produces in course of ages ten thousand variet
195b101 cular destiny.-- In same manner God orders each ANIMAL created with certain form in certain country¿
195b101 certain laws such as inevitable consequen let ANIMAL be created, then by the fixed laws of generati
197b111 st theory of analogies. States there is but one ANIMAL: one set of organ:-- the others «animals» crea
197b111 eneration as a short process, by which man «one ANIMAL» passes from worm to man; «highest» as typical
```

197b112 ; Is Vol of Fish p. 59.]CD Cuvier has said each ANIMAL made for itself does not agree with old & mode
200b125 . J. Morse found in Virginia p. 325. July 1828. ANIMAL now confined to extreme North.-- \.do p. 326.
204b141 ough Bridgewater errs greatly in thinking every ANIMAL born to consume this or that thing.-- There is
205b143 climbing fish. p. 122 A Terrestrial annelidous ANIMAL p. 347. Vol I.-- compare with my planariae Lea
205b145 l be hornless.--" does this apply to where same ANIMAL breeds often with same female p. 28 "It <is> w
221b200 applicable to insects & &c?-- (p. 23 do.-- On ANIMAL - Confervae-- p. 23 p. 267. Dela Beche. Geolog
223b209 ccount of Bolivian human species?-- Small «new» ANIMAL mentioned from Fernando Po Zoolog. Proceedings
224b214 thought (plants) & living thing with thoughts (ANIMAL). «∴ my theory very distinct from Lamarcks» Wi
228b229 ucing crab. is the offspring of a male & female ANIMAL of one variety going back ¿whether this going
230b235 essation of female offspring: applicable to any ANIMAL-- Athenaeum. <Jan> «p. 154»-- 1838. Hybrid Fer
230b236 more fertile offspring.-- 1st. All variation of ANIMAL is either effect or adaptation,∴ animal best f
230b236 tion of animal is either effect or adaptation,∴ ANIMAL best fitted to that country when change has ta
231b244 - Till man acquired reason, he would be limited ANIMAL in range-- & hence probability of starting fro
254c048 All classes of Acrite exhibit lowest stages of ANIMAL organization, "& are analogous to the earliest
255c052 We thus obtain an abstract idea of a bird-- An ANIMAL with skeleton of such general forms.-- The hyb
256c055 Speech. Feb 1838 thinks gradation between Man & ANIMAL, small point in tracing history of Man.-- gran
257c058 less to speculate «not only» about beginning of ANIMAL life.: generally, but even about great divisio
258c060 to make forms remote antagonist powers.-- Every ANIMAL in cold country has some analogy in not gaudy
264c079 y would.-- the world now being fit, for such an ANIMAL.--man, (rude, uncivilized man) might not have
267c095 rias in stomach. of Sammon remain after rest of ANIMAL digested.-- Important do p. 98. on a quaternar
270c103 rest of]CD universe «which Carus considers big ANIMAL» becomes more developed in higher animals than
270c104 wn by Carus how intermediate plants are between ANIMAL life & "inorganic life".-- animals only live o
282c143 ion-- millions in few days-- one doubt that one ANIMAL can really produce so great an effect.-- the s
285c153 tain physical changes at last become unfit, the ANIMAL cannot change quick enough & perishes.-- Lyell
286c154 ow black, often wished to consider him as other ANIMAL-- it is the way of mankind. & I believe those
292c169 better than others.-- The more complicated the ANIMAL the more subject to variation. therefore sexes
292c170 ssi> real affinities. ie structure of the whole ANIMAL let him read Mr Swainson's on the Classificati
293c174 corelations are not, however, perfect, else one ANIMAL would not cause misery to other.-- else smell
294c176 those changes permanent so would the change in ANIMAL be permanent.-- It will be easy to prove persi
302c201 My theory drives me to say that there can be no ANIMAL at present time having an intermediate affinit
302c201 mily likeness, & this I believe the case. = any ANIMAL really connecting the fish & Mammalia, must be
303c204 umenbach & Prichard -- Now we might expect that ANIMAL halfway between man & monkey, would have diffe
305c211 tween races, & recurrent habits in animals.-- --ANIMAL Magnetism-- principles of irritations sleep wa
322c270 Abercrombie on the Intellectual powers. Hunters ANIMAL OEconomy. edited by Owen. read several papers-
323c289 Breeding of Animals -- Spallanzani's Essays on ANIMAL Reproduction -- Treatise on Domestic pidgeons
326c268 connected with Natural Theology.-- on instinct & ANIMAL, intelligence.-- very good. Endlicher has publ
333d006 fect is the same.-- Fox thinks that when a wild ANIMAL is crossed with tame, offspring always take mo
335d014 w breed.-- the first hybrids may be compared to ANIMAL with amputated limb.-- Heredetary <thr> Six fi
335d016 y fail.-- In last page. I should have said, "an ANIMAL <acquires <th> any new> is <only> able to tran
336d017 ve organ.-- Same Prop. better enunciated. "An ANIMAL Either parent cannot transmit to its offspring
336d017 er time less [s]o.-- the result of this is that ANIMAL would endeavour to return to parent stock. but
336d018 if change be in blood long, it becomes part of ANIMAL &» by a succession of <such changes> generatio
341d030 by Ogleby, who observed that the young of this ANIMAL, which is so anomalous among true deer, yet is
345d042 r thought that a <pure blooded> «"first blood"» ANIMAL must have gone on for many years, before deser
347d049 must be some law, that whatever organization an ANIMAL has, it tends to multiply & IMPROVE on it.-- A
352d088 g of Cochineal?? Is there some law in nature an ANIMAL may acquire organs, but lose them with more di
353d060 merely a few pages.-- Hence (p. 59) looking at ANIMAL, if there be many others somewhat allied wheth
359d088 ng in & in like our pidgeons) The male of every ANIMAL certainly seems chiefly to impress the young m
360d091 lice & fleas. sticking on them, but never in an ANIMAL, that had long been in confinement-- is this e
367d113 donkey. with all four legs ringed with brown.-- ANIMAL like large, heavily made cream coloured ass.--
367d113 cells connected with cheek pouches.-- Hunter's ANIMAL OEconomy p. 45 "One of the most general marks
369d115 racter.-- Chapt I. Also Latent Character Hunter ANIMAL Economy p. 482 (Same book) Owen says "the nece
372d130 icial division.-- On this view each particle of ANIMAL must have structure of whole comprehended in i
377d140 rapidity of change of form, & instincts in the ANIMAL kingdom.-- It is the unit of our calendar.-- e
380d153 t proportion of sexes, at birth & causes. If an ANIMAL breeds young her growth is immediately checked
380d154 latent instincts even in brain of male.-- Every ANIMAL surely is hermaphrodite-- (as is seen in <fe>
380d155 is seen in <fe> plumage of hybrid birds) After ANIMAL has copulated., though no offspring, Milk some
381d156 pon is cut, it increases in size prodigiously-- ANIMAL OEconomy by, Hunter. (edited by Owen) p. 34.--
382d157 thought by Owen to ask. whether a Heautandrous ANIMAL is <evidently> actually split in two-- keeping
384d161 ery different from either parent bird-- Hunters ANIMAL OEconomy. (by Owen) p. 44. Classification of M
386d165 ng probably not education.-- Cannot I find some ANIMAL with definite life & split it, & see whether i
389d176 ite & each have offspring by same mother.-- one ANIMAL will fecundate female for several births, & ev
390d179 es all appetite.-- It is the comparison of each ANIMAL with its<elf> «ancestors», & not its compariso
398e005 changes of species «are» not as «without every ANIMAL preserved.».-- the latter pages in the history
399e009 ot located] Study introduction to Cuviers Regne ANIMAL No structure will last. without it is adaptati
399e009 last. without it is adaptation to whole life of ANIMAL, & not if it be solely to womb, as in monster.
402e018 s or badgers, deer, apes, baboons, monkeys & an ANIMAL probably a tapir p. 233. dogs in Borneo-- <bro
403e023 says it is nonsense to say take a tooth of any ANIMAL (as Toxodon) & say its relations.-- if we know
406e036 354-- The most vicious dog. will not attack any ANIMAL except, dog when absent from its master.-- dog
409e049 «whether he was or not. He is present a social ANIMAL-- a fact few will dispute, [although, that it w
415e065 norganic agents of this earth. like every other ANIMAL.-- Would anyone raise an argument against, my
416e071 t this rather effect of liquid semen: therefore ANIMAL life commenced in the Water!» It is a beautifu
416e071 erfectly & infinitely slower.-- No domesticated ANIMAL is perfectly adapted to external conditions.--
417e076 descendant of <Mammferus> «Mammiferous» <vert> ANIMAL, which would find its place in the Systema Nat
418e080 an be impregnated externally-- My view of every ANIMAL being Hermaphrodite-- probably will recieve il
418e084 s,-- each of us, then <is as old, as the oldest ANIMAL>, have passed through as many changes, as has
420e089 must have been like some molluscous «bisexual» ANIMAL with a vertebra only & no head-- !! Handwritin
420e090 y'-- Athenaeum .1839. p. <8>36.-- A crustaceous ANIMAL is mentioned which inhabits the Pinna of Rio J
421e091 ost remotely related to the habits & food of an ANIMAL, I have always regarded as affording very clea
424e103 Australian dog an instance of a half reclaimed ANIMAL.-- The dogs, which have run wild have, have do
427e110 forms have travelled S. From the analogy of the ANIMAL kingdom I should suppose, that the pollen of c
427e110 would be different.-- same way one variety of <ANIMAL> dog does not prefer other. but produces great
432e124 ng circumstances According to my theory no land ANIMAL with fluid seeds can be true hermaphrodite Man
439e144 mine his preservation-- if killed by some other ANIMAL, then that quality which saved him, would be t
440e145 budding» tree, & every buzzing insect & grazing ANIMAL owes its form, to that form being «the one alo
440e147 injurious to bird as to allow any other kind of ANIMAL to usurp its place.-- & therefore the degree o
441e150 theory is, the absolute necessity, that every <ANIMAL> «organic being» should cross with another.--
444e157 result of circumstances,;-- or the will of the ANIMAL. p. 145. Seems to argue, that as the transform
444e158 ns from the egg, or larva. or foetus to perfect ANIMAL are adapted by foreknowledge, so must the muta
449e169 remains of the Hamster.-- is not this Siberian ANIMAL?-- Eyton says that the young of two hatches «a
458t002 with pony mare & produced a very pretty little ANIMAL, showing something of Mule In its ears-- ((thi
463t055 ut it is not possible to imagine what habits an ANIMAL could have had with such structure.-- perhaps
478z004 ire 133 1803. on Pennatula showing it to be one ANIMAL In Australia I was assured wild dog copulates
479z005 N. p 24 Bougainville Voyage round world no land ANIMAL besides Wolf at Falkland ∴ black rabbits not i
479z005 Sagitta Triptera M. D'Orbigny has described my ANIMAL with teeth {P} p. 140. Fléche of Quoy et Gaima
481z009 .-- 18 Admirable engravings in Meyen Zoology on ANIMAL of Campanularia Alcedo stellata. Meyen p. 92.-
481z010 anariae) Sagittella, or Fleche «p. 8» my little ANIMAL with horns. Madrepores p. 26 Nullipora p. 29--
493q003 s.-- Good observation-- examine semen of Hybrid ANIMAL. in comparison with Weigh skeleton of Tame Duc
496q05a beauty of flowers-- contrasted by Kirby-- with ANIMAL reproductive system.-- -- cover flower-- put a
509q017 . cow. taking bulls. is it Domesticated African ANIMAL= Knows nothing [<...>] It is very important to
509q017 e to some Human Anatomist. Has he dissected any ANIMAL often, which has abortive bone. (ask more abou
511q018 vrne, &c Could by selection a different looking ANIMAL be formed-- not caring whether good or bad.--
516q23v Will, an extract of peat do to preserve fungi or ANIMAL substances-- (Athenaeum (40) p. 823 chemical a
526m027 ffected by other fellow-animals, than any other ANIMAL & probably the only one affected by various kn
526m030 ive may be anything ambition, avarice, &c &c An ANIMAL improves because its appetites urges it to cer
531m053 hear or see. which frightens. them.-- Now every ANIMAL moves quickly away from any sudden sound or no
534m061 say germ within egg, cannot think-- as well as ANIMAL born with instinctive knowledge.-- but if so,
536m071 beating: one ceasing, so does other.-- What an ANIMAL like taste of likes smell of,∴ Hyaena likes sm

537m075 by rubbing them with alum, so more slowly does ANIMAL leave off <t> instinct, when attended with bad
537m076 consequence of man, like deer &c, being social ANIMAL, & this conscience or instinct may be most fir
539m082 «or trick» so much more easily than man, so may ANIMAL obtain it far more easily, in proportion to va
542m096 nctive-- child does not sneer. because no young ANIMAL has canine teeth.-- A dog when he barks puts h
564n003 nscience.-- Therefore I say grant reason to any ANIMAL with social & sexual instinct «& yet with pass
564n003 nt.--Mem. Bee how different instinct a solitary ANIMAL still different.-- ¶ Different nations having
574n043 ake the case of Jenner's <Hyaena> Jackall.-- an ANIMAL not destined by nature to exist. & carrying «l
577n049 tid to be a cause of itself.-- [by my theory no ANIMAL. as now existing can be cause of itself.]CD &
579n060 an excitement of surface under the will? of the ANIMAL.(-- Jan 21. 1839. Herchel's Discourse p. 35. O
582n067 wide difference, between the means by which an ANIMAL performs an instinct, & its impulse to do it.-
582n067 ust be present on any hypothesis whatever]CD an ANIMAL may so far be said to will to perform an insti
582n068 pleasure.-- The Revd. Algernon Wells Lecture on ANIMAL instinct. 1834: p. 15. "To act from instinct i
586n081 experience» directed to certain quarter"-- "An ANIMAL has faculty of walking. which in man is learnt
588n090 f we compare the judgments & actions of a young ANIMAL with an old.-- (dog horse, sow) we perceive gr
593n107 usic now excite our feelings.-- How does Social ANIMAL recognize «& take pleasure in» other animal, (
593n107 ial animal recognize «& take pleasure in» other ANIMAL, (especiall as in some <instinct> «insects» wh
593n107 doubt it may be attempted to be said that young ANIMAL learns parent smell & look so by association r
593n109 ence idea of beauty.-- the social affections of ANIMAL taking man in place of other animals is hostil
599o07v changes in the polity of Nature than any other ANIMAL-- Aimé Martin de l'Education des Mères Vol. I.
600o08a us.-- It would indeed be wonderful, if, mind of ANIMAL was not closely allied to that of men, when th
604o016 or the article. A Planaria must be looked at as ANIMAL, with consciousness,, it choosing food-- crawl
611o034 on given in full.-- [LHC] ¿Has any vegetable or ANIMAL matter been formed by the union of simple non-
612o035 of willing. These +willings are common to every ANIMAL instinctive and unavoidable.-- +Can the word w
612o035 rallina are not two kinds of life vegetable and ANIMAL strictly united? [RHC] It is easy to conceive
613o036 over the gate & bring it. ---- Agrees with ONE ANIMAL, [RHC] Kirby extends instinct to plants, but su
619o042 as a Naturalist would at any other mammiferous ANIMAL. It may be concluded that he has parental, con
619o042 if we judge him by his habits, as <if> another ANIMAL. These instincts consist of a feeling of love
620o045 ung child to be in passion, any more than in an ANIMAL.-- which shows that. it is owing to some <subs
620o049 , as we see in dogs & pidgeons.-- But as man is ANIMAL at head of series in which «special» instincts
625o50v short lived Passions' State broadly in child or ANIMAL it is equally proper to obey anger as benevole
638j059 brain should not be born. with tendency to make ANIMAL perform some action.-- as well as gain it. by
373d132 individual a spontaneous generation: what is ANIMALCULAR semen-- but this-- -- the <nerve> living ner
386d167 internal organs in animals. One «invisible» ANIMALCULE in four days could form 2. cubic stone. like
431e123 is true?-- most strange.-- Does not spermatic ANIMALCULE in Mosses, render my view of the crossing of
454e184 p. 290. Dr. Edwards in his essay on Spermatic ANIMALCULE. has described instrument for galvanize <ani
424e100 already referred to it.-- also on spermatic ANIMALCULES in Musci frondosi, et hepatici,-- in Chara,
048r086 the globe are there such vast numbers of wild ANIMALS. both species & individuals as in the half des
057r115 en bottom or beach of sea to explain preserved ANIMALS.--Mem: stream of water in the country.-- Sir J
062r128 Go steadily through all the limits of birds & ANIMALS in S. America. Zorilla: wide limits of Waders:
062r129 nowing it to be a desert.-- Tempted to believe ANIMALS created for a definite time:--not extinguished
063r132 urprised at seeing Zoophite producing distinct ANIMALS. still partly united. & eggs which become quit
063r133 eation does not bear upon solely adaptation of ANIMALS.--extinction in same manner may not depend.--T
064r134 quite in deserts.--Much struck with number of ANIMAL[S] at Cape of Good Hope Says at Santos «M Birch
072r159 sh nor on Norway, or Spitzbergen.--Spitzbergen ANIMALS (?). ≠ The Hollowness of <sep> Chiloe concreti
145g001 given against it-- Mere fact of many races of ANIMALS in Britain shows that either races soon made o
149g031 delling power of sea N of Valparaiso are those ANIMALS subject to much variation which have lately ac
170b002 st office in organization (especially in lower ANIMALS, where mind, & therefore relations to other li
171b004 e effect of accidental injuries, <on> which if ANIMALS lived for ever would be endless (that is with
172b007 ll dare say what result According to this view ANIMALS, on separate islands, ought to become differen
174b014 ent countries. Propagation explains why modern ANIMALS same type as extinct which is law almost prove
178b031 rabbit may perhaps be instance of domesticated ANIMALS having effected; a change whi[ch] the Fr. natu
179b034 l common language) will keep to their type: in ANIMALS so far removed, with instinct in lieu of reaso
179b034 ke marriage.-- as Dr Smith remarked Man & wild ANIMALS in this respect are differently circumstanced.
179b035 &c &c-- If we <suppose» «grant» similarity of ANIMALS in one country owing to springing from one bra
181b043 rtebrate and Articulata. still greater between ANIMALS & Plants But yet besides affinities from three
181b043 , they may approach,-- some birds may approach ANIMALS, & some of the vertebrata invertebrates-- Such
181b045 sequent ages.-- The Creator has made tribes of ANIMALS adapted preeminently for each element, but it
183b051 . point in view.--;whether highly domesticated ANIMALS like races of man.-- M. Flourens. .Journal des
186b059 .-- It would be curious to know in plants, (or ANIMALS) whether, <in> races have tendency to keep to
186b061 hange, superinduced, or new species; therefore ANIMALS would perish, if there was nothing in country
187b062 h no change, agrees with belief. that Siberian ANIMALS lived in cold countries & therefore not killed
188b066 nt applies to England.-- Mem. Shew Mice.-- --- ANIMALS common to South and North America.-- ¿are ther
189b073 ebra.-- & continued.-- perished in America All ANIMALS <are> of same species are bound together just
189b073 gradation between associated & non associated ANIMALS.-- & the story will be complete.-- It is absur
191b080 n formations-- England & Europe Ireland common ANIMALS-- + + + for instance tertiary deposits between
191b081 ported-- Mem plants on Coral islets.-- Next to ANIMALS land plants.-- & life shorter or change greater
191b081 would be interesting to trace limits of large ANIMALS-- Owls. transport mice alive? Species formed b
193b092 ning of Volume on Geographical distribution of ANIMALS Brown Geograph Journal. Vol I p. 174. says fro
195b099 seeds.-- The same remarks applicable to fossil ANIMALS same type, armadillo like covering created.--
195b099 me cause, such beautiful adaptations yet other ANIMALS live so well.-- This view of propagation gives
195b100 ata, but not coronata.-- We know that domestic ANIMALS vary in countries, without any assignable reas
196b108 n created; but on other hand creation of small ANIMALS must have gone on since from parasitical natur
196b108 egetables & mass of insects could live without ANIMALS but not vice versâ. ¿could plants live without
196b109 thout carbonic acid gaz.)-- Yet unquestionably ANIMALS most dependent on vegetables. of the two great
196b110 gy The existence of plants, & their passage to ANIMALS appears greatest argument against theory of an
197b111 ut one animal: one set of organ:-- the others «ANIMALS» created with endless differences:-- does not
197b111 which can be traced in same organ in different ANIMALS in scale.-- In monsters «also» organs of lower
197b112 in scale.-- In monsters «also» organs of lower ANIMALS appear.-- yet nothing about propagation= I see
198b113 ituations as possible. for instance take birds ANIMALS, reptiles fish-- Conditions will not explain s
198b114 s the expression mean used by Cuvier, that all ANIMALS (though some may be) have not been created on
198b115 orth reading NB well to insist upon <different ANIMALS> large Mammalia not being found on all isld, (
198b115 en.-- It is a point of great interest to prove ANIMALS not adopted to each country.-- Provision for t
200b122 t is daily happening; that naturalist describe ANIMALS as species, for instance Australian dog: <yet
202b130 nce name of analogy, the structures in the two ANIMALS bearing relations to a third body., or common
202b130 or common end of structure A Race of domestic ANIMALS made from influences in one country is permane
203b137 ca. that only some mundine cause has destroyed ANIMALS over the whole world-- For instance gradual re
205b145 e. does this rule apply? A Treatise on Form of ANIMALS by Mr Cline "The character of both parents are
205b145 28 "It <is> wrong to regard a Native breed of ANIMALS. for in <the> proportion to their increase of
205b148 the two-case,» May this not be extended to all ANIMALS first consider species of cats.-- <& other tri
207b152 rt Town-- (from difference of races of men and ANIMALS] See R. N. p. 130 Speculations on range of all
210b160 t: compare Sicily & Galapagos!!-- Some of the ANIMALS peculiar. to Mauritius are not found at Bourbo
211b161 present an analogy to production of species.-- ANIMALS have no notions of beauty, therefore instincti
211b163 itut. Curious paper by M. Serres on Molluscous ANIMALS representing foetuses of Vertebrata, &c 1837 p
211b164 p. 370 Owen says Nonsense The distribut of big ANIMALS in East Indian Archipelago--, very good in con
212b166 of instinct of dogs.-- agreement & reason Some ANIMALS common to Mauritius & Madagascar.? Proceedings
213b168 e would have been many genera of monotrematous ANIMALS.-- p. 82 «There are many tables in Phillips of
218b190 ce appeared good with facts about changes when ANIMALS transported.) Mr Herbert's papers are in the H
223b208 e is that there is wide gap between Man & next ANIMALS in mind, more than in structures.-- If the ske
223b210 nt submits to more individual change, (as some ANIMALS do more than others, & cut off limbs & new one
223b210 ropagates varieties according to same law with ANIMALS?? Why are species not formed. during ascent of
224b213 with constant characters, together with other <ANIMALS> beings of very near structure.-- Hence specie
224b214 t species.-- The difference intellect of Man & ANIMALS not so great as between living thing without t
224b216 ince the Cambrian formations gone on creating ANIMALS with same general structure.-- miserable limit
225b219 drupeds for some facts.-- about dogs &c &c NB. ANIMALS very remote. ass & Horse. produce offspring ex
225b220 is & Williams. zoology of South Sea islands. any ANIMALS?-- I believe none.-- Canary islds.? Madeira? «
227b226 Nature of physical change between one group of ANIMALS & a successive one.-- It leads to knowledge wh
228b231 .-- hybridity showing connexion of two plants. ANIMALS-- whom we have made our slaves we do not like
228b231 olders wish to make the black man other kind?» ANIMALS with affections, imitation, fear <of death>. p

Page ***(Key Word)***

228b232 - «--the soul by consent of all is superadded, ANIMALS not got it, not look forward» if we choose to
228b232 e choose to let conjecture run wild then <our> ANIMALS our fellow brethren in pain, disease death & s
229b232 we may be all netted together.-- Hermaphrodite ANIMALS couple: argument for true molluscs coupling.--
230b240 ent in circle.-- Falkland Isd case good one of ANIMALS not soon being subjected to change in Americas
230b241 us to fresh change Get a good many examples of ANIMALS «or plants» very close (take Europaean birds.
231b242 Tapir existing in East Indian Seas. Marsupials ANIMALS all show greater connexion in quadrupeds, <bu>
231b244 form good species, so cannot the domesticated ANIMALS with him.!-- Modern origin shown by only one s
232b248 me Get Hope to give me an account of parasitic ANIMALS of beasts varying in different climates Those
232b248 age, they perceive the superiority of man over ANIMALS, without such resorts Mr Waterhouse has most c
234b256 arieties «of colour & size, but not for[m]» of ANIMALS.-- He says Stephens say he can at once tell by
235b262 her. Are there any cases, where «domesticated» ANIMALS separated. & long interbred <p> having great t
235b263 Diemens land people require so many, imported ANIMALS?-- At what. part of tree of life, can orders l
235b263 part of tree of life, can orders like birds & ANIMALS separate &c &c Work out Quinary system accordi
235b274 nduced-- why is every one so anxious to cross. ANIMALS from different quarters to prevent them taking
240c013 ated] Falkner Patagonia no description of wild ANIMALS, nor in Dobrizhoffer Abipones.-- Voyage. de L'
248c030 generative organs affected as with plants) no ANIMALS VERY different will breed together, so when we
248c033 the law of hybridity, namely the [not located] ANIMALS unite, all the change that has been accumulate
249c035 tions is this not connected with wide range of ANIMALS. Follow this out, where species of same genera
249c036 d,-- where shut up by themselves without other ANIMALS? but they were not shut up!! Extreme southern
250c037 n cause I can conceive of destruction of Great ANIMALS in Europe & America Some portion of the world
250c038 ica & East India Arch.-- but where these great ANIMALS had not spread then such tribes as Marsupial &
250c039 of large quadrupeds.-- common to two types of ANIMALS What reptiles coexisted with Palaeotherium in
256c053 s vary more,. than what makes species in other ANIMALS.--? Forster on South Sea, will probably contai
256c054 probably contain descriptions of domesticated ANIMALS in those regions Species so far are not natura
256c055 sexes is the check to distribution of birds & ANIMALS Mr Strickland & Hamilton-- found tertiary form
257c055 g history of Man.-- granted.--but if all other ANIMALS have been so formed, then man may be a miracle
258c060 al circumstances effecting the area equably.-- ANIMALS having wide range, by preventing adaptation ow
258c062 ow after there is proof of the non creation of ANIMALS.-- then argumen May be.-- subterranean lakes,
259c064 that the ground woodpecker &c.--, fresh water ANIMALS of great lakes are American form, that one is
260c068 wrong,-- Mr Yarrell says-- that some birds or ANIMALS are placed in white rooms to give tinge to off
261c069 to many genera.-- like Synallaxis or Marsupial ANIMALS of N. America Hence it is universally allowed
261c071 uring, when form has changed. Can be said that ANIMALS no notion of beauty, when does prefer most pow
261c072 can only be explained by direct adaptation to ANIMALS wants & not as change in typical structure?!!
263c077 general instincts, <as> & <moral> feelings as ANIMALS.-- they on other hand can reason-- but Man has
263c078 de man! Seclusion want &c & perhaps a train of ANIMALS of hundred generations of species to produce c
264c079 d man) might not have lived when certain other ANIMALS were alive, which have perished.-- Let man vis
266c092 ose of Arabia & Egypt.-- CIVETS CATS only wild ANIMALS on isld.-- Niether Hyenas; jackals monkeys-- c
268c096 Procyon.-- by Wiegman Classified catalogue of ANIMALS of Nepal read before Linnaean Soc. Feb. 1838.-
268c099 r to present. ¿cause of destruction of «great» ANIMALS?-- Show independency of shells to external fea
268c100 mere chance?-- or it like each great class of ANIMALS having its aquatic, aerial &c type?-- This of
270c103 s big animal» becomes more developed in higher ANIMALS than in vegetables. p. 243 radiate animals <tu
270c103 her animals than in vegetables. p. 243 radiate ANIMALS <tu> plants turned inside out. have position o
270c104 are between animal life & "inorganic life".-- ANIMALS only live on matter already organized.-- This
271c106 n Mr Wynne's book, about not altering breed of ANIMALS in certain countries.-- Fraser remarked to me
273c110 is not true, with shells?.?.?» It looks as if ANIMALS perished by errors.-- It is most wonderful how
275c119 of lofty genius Using geograph distribution of ANIMALS, <as> I use (new step in induction) as keyston
278c132 rica & Europe.-- The difficulty is how came it ANIMALS not preserved, in central S. America & yet Afr
280c134 y applicable to formation of instincts in wild ANIMALS, many species in one genus-- external circumst
280c135 grand apparent anomaly in nature. <t>---- Many ANIMALS not breeding at all in domestication. throws g
281c137 ld be best).-- Argue the case theoretically if ANIMALS did change excessively slowly. whether geologi
281c138 of all men now living & the classification of ANIMALS-- talking of men as related in the third & fou
282c141 ed.-- let these families take <dogs> <domestic ANIMALS> with them. they might be supposed to change.
282c141 d to change. & make genera of bird. analogous. ANIMALS would be possessed by the different races of m
285c153 d, with respect to their effects The AEgyptian ANIMALS domesticated «??». & therefore Most especially
286c154 Man. & external circumstances not variable.-- ANIMALS have voice, so has man. Not saltus. but hiatus
286c154 have voice, so has man. Not saltus. but hiatus ANIMALS expression of countenance. «[s]hare of sicknes
286c154 ey may convey much thus, Man has expression.-- ANIMALS signals. (rabbit stamping ground) Man signals.
286c154 gnals. (rabbit stamping ground) Man signals.-- ANIMALS understand the language, they know the crys of
286c155 is false There is same difficulty in arranging ANIMALS in paper as drying plant, all brought in one p
286c156 rder, or in next family? In considering fossil ANIMALS, what relation in classification in books, oug
287c157 e (of reason), she has also placed a series of ANIMALS on the steps that lead up to it" p. 20 ╫ +++hi
287c158 ance of the organs in same state. in different ANIMALS, & the value of those organs, when changed in
287c158 lue of those organs, when changed in different ANIMALS.-- + whether variations in eye of vertebrate a
289c161 hich no doubt be overcome, but until it is the ANIMALS are distinct species If any one is staggered a
290c162 tion is very good «general» character in those ANIMALS, where much change has been added. «as it spea
291c167 etween <mon> monoeecious & dioecious plants in ANIMALS it may be difficult to imagine how sexes were
292c169 e subject to variation. therefore sexes or two ANIMALS:-- «When» sexes <being in one normal-- when so
292c170 im read Mr Swainson's on the Classification of ANIMALS. & observe the character of the demonstrations
293c174 ws--]CD. Many diseases in common between man & ANIMALS-- Hydrophobia & cowpox, proof of common origi
294c175 nstance of the fine relations of adaptation of ANIMALS & the country they inhabit.--» & the first one
294c176 to admit, permanent small alterations in wild ANIMALS, & thinks Lyell has overlooked argument, that
294c176 ell has overlooked argument, that domesticated ANIMALS change a little with external influence-- & if
294c176 be easy to prove persistent Varieties in wild ANIMALS-- but how to show species-- I fear argument mu
296c184 red much.-- these others will be plants & land ANIMALS. & land shells.-- all in short Extreme North =
298c190 try to trace from simplest reasoning in lower ANIMALS many times produced, a general tendency produc
298c192 lants is remarkable what is analogous to it in ANIMALS?-- Babington says in most plants, even those o
299c196 dom in more instincts in rodents than in other ANIMALS & again in Mans mind, in different races, bein
300c197 & I believe true to consider him created from ANIMALS.-- Insects shamming death, most difficult case
300c197 ctive dread it is exceedingly doubtful whether ANIMALS have any fear of death, only of pain. of death
300c198 ous to greater individuality of bodies of some ANIMALS over those of others.-- the mind of different
300c198 over those of others.-- the mind of different ANIMALS less divided.-- But as man has heredetary tend
301c198 t of brutes p. Hard to say what is instinct in ANIMALS. « & what reason in precisely same way not po
301c199 counter all these difficulties.-- Knowing that ANIMALS have some reason, & actions habitual. it surel
301c200 t applicable. The degree of development of all ANIMALS of same class being about equal.-- organs of g
303c204 considering over the vicissitudes of present ANIMALS.-- He-will be bold. I will venture to say unph
303c205 t is now perfect.--]CD. the creator so creates ANIMALS, «, it will be said» that although at any one
305c211 emotions between races, & recurrent habits in ANIMALS.-- --Animal Magnetism-- principles of irritati
306c212 ypus & germ plant & seed.-- instincts in young ANIMALS, well developed, just like, habits easily gain
306c214 t in his Travels in Asia Minor of the domestic ANIMALS-- At Angora «centre of Asia Minor» are the fin
307c215 are» have «These abortive organs in some Males ANIMALS, Mammae in Men, capable of giving milk)» rudim
307c216 e any abortive organs produced in domesticated ANIMALS, in plants I presume there are.? get examples.
308c216 Vol I It is capable of demonstration that all ANIMALS have never at any one time formed chain, since
308c219 gyptian drawings & Old Testament» Domesticated ANIMALS having same idiosyncrasy, cause of fertility.-
308c219 imates &c. Do I mean that ideosyncracy of wild ANIMALS is generally different, because their differen
309c219 relationship, children of one parent, races of ANIMALS-- argue «opening» case. «thus» Educate all cla
309c220 mprove-- The areas of subsidence marked out by ANIMALS of same <spe> genera. is not equal to areas of
310c222 the creator has added thought to <an> so many ANIMALS of different types. I will confess my profound
310c223 «consequence of» for the scheme of nature) and ANIMALS that man has different origin. «Royal Institut
312c232 no other means whatever.--» Individual Men «& ANIMALS» could only exist by habit.-- therefore same p
312c233 m Hope says that genus of parasite to genus of ANIMALS different «+++ p 234», different species to di
313c233 a Parasites die, when brought over on tropical ANIMALS which accounts for the species changing which
314c237 nces by principles, which I have given ∴ Those ANIMALS, which only propagate by scission can not alte
315c241 ts «when exciting cause is absent» & memory of ANIMALS.-- (surely in plants movements effects of irri
315c242 s in plants, it allows of any degree in lowest ANIMALS --habitual action, in intestines subject to sy
315c243 iling modified laughing. Barking to tell other ANIMALS in associated kinds of good news. discovery of
315c243 - Under this point of view. expression «of all ANIMALS» becomes very curious.-- a dog snarling in pla
316c244 ty», only difference between the mind of man & ANIMALS.-- yet how faint in a Fuegian or Australian! w
316c247 .-- (connect these facts with identity of land ANIMALS.-- these however come from Siberia it cannot b

```
323c269 ecture on instinct -- Cline on the Breeding of ANIMALS -- Spallanzani's Essays on Animal Reproduction
324c268 ange Study Buffon on Varieties of Domesticated ANIMALS see if law's cannot be made out Find out  from
327cIBC uites Cline on the improvement of domesticated ANIMALS Fries de plantarum proesentum crypt.  transitu
331dIFC number of nipples in domesticated very fertile ANIMALS increased?-- Where offspring, heterogenous, in
331dIFC curious  notes on Monkeys recognising Sexes of ANIMALS:]CD [All Selected Dec. 14-- 1856]CD Towards cl
334d009 irst cross were equally fertile with pure bred ANIMALS.-- Mem. number of Mules.-- «He recollects  one
335d015 id breed, the offspring would «as in all other ANIMALS» be like either parent, or intermediate within
335d015 t long in blood.= The case of union of perfect ANIMALS is distinct case,-- gradation from physical im
335d016 to (perhaps increased). fertility.-- (but many ANIMALS are fertile, when offspring infertile,-- two c
336d018 ry,, but size of particular Muscles-- When two ANIMALS cross. each sends his own likeness, & the unio
336d019 his own law. So far is there any appearance of ANIMALS being created. it is probable if created at on
337d020 ld perish.-- The Varieties of the domesticated ANIMALS must be most complicated, because they are par
337d021 ations. Must not go back to first stock of all ANIMALS, but merely to classes where types exist for i
337d023 Geograph.  Distrib: of <ani> organic beings.-- ANIMALS «of same classes» differ in different countrie
338d025 n to Eytons Anatidae.-- recurs to idea of only ANIMALS from distant countries breeding!. «Mem 3 speci
339d025 urely wild Duck & «pintail» Widgeon!-- Divides ANIMALS «world into Zoological Provinces» according to
339d026 -- does not seem to think any improbability to ANIMALS being distributed after flood (!) according to
340d026 or  asserts to be explicable every instinct in ANIMALS. Heard at Zoolog Soc their Pintail & Common Du
341d030 a &c strongly developed to aid in breathing.-- ANIMALS from Hobart Town mentioned, it seems most of s
343d037 has  made a long succession of vile Molluscous ANIMALS-- How beneath the dignity of him, who «is supp
343d037 when Conditions are unfavourable to numbers of ANIMALS. as in changing from <hot>. Warm to cold, damp
343d039 cies from Egyptian Mummies & from the existing ANIMALS found fossil when Europe must have worn a quit
345d044 eir wars & rivalry.--» The very many breeds of ANIMALS in Britain shows, with the aid of seclusion in
346d048 are  bad breeders & subject to the rush as all ANIMALS which breed, in & in are-- colour white, unifo
347d049 ends to multiply & IMPROVE on it.-- Articulate ANIMALS must articulate. <i> in vertebrates tendency t
347d049 if generation is condensation of changes. then ANIMALS must tend to improve.-- yet fish same as, or l
348d051 nt or not,). p. 7. "The Natural arrangement of ANIMALS themselves is the question in point." Now what
350d056 does nothing in vain, therfore organ fitted to ANIMALS place in creation.-- thus senses, especially s
352d057 ade, but these facts are rare.-- 2d Sept Those ANIMALS which have many ABORTIVE organs, might be expe
352d059 ts.-- The present geographical distribution of ANIMALS countenances the belief of their extreme antiq
353d060 «philosopher», who has trace the structure of ANIMALS & plants.-- he get merely a few pages.-- Hence
356d069 e from eating different kinds of food: grazing ANIMALS who eat every species new.-- Sept. 8'. A Golde
356d071 many facts as mine‖ The facts about half breed ANIMALS being wilder than parents is very curious as p
357d073 ials have been almost destroyed wherever other ANIMALS existed.-- Athenaeum 1838. p. 654. Reason give
359d086 - it may be so, but this assumption as long as ANIMALS are healthy which is often the case, & why sho
359d087 sant & fowl being so totally infertile whereas ANIMALS further apart have bred inter se.-- These hybr
364d104 thers).-- It certainly appears in domesticated ANIMALS, that the amount of variation is soon reached-
365d106 k Cock, & other hybrids-- The fact of Egyptian ANIMALS not having changed is good-- I scarcely hesita
366d108 ture.-- According to my view, the domesticated ANIMALS would cease being fertile inter se., or at lea
369d115 combining  observation of the living habits of ANIMALS, with anatomical & Zoological research, in ord
369d115 rsupial. («but» also mice) & these being water ANIMALS <that this> structure <connected with  animals
369d115 animals  <that this> structure <connected with ANIMALS being compelled to travers) "May have referenc
370d115 rica.-- Such difficulties will always occur if ANIMALS are thought to have been created.-- it might a
372d130 .-- Ehrenberg considers artificial division of ANIMALS, as gemmation, I consider gemmation as artific
374d134 he  thinks deeply has assumed that increase of ANIMALS exactly proportional[1] to the number that  can
377d139 ontinent, when elephants lived. & when present ANIMALS-- lived.-- we know the great time, necessary t
378d147 changing  forms, <& loosing do> if so domestic ANIMALS ought to show them.-- Anyhow not connected wit
381d156 ibranchiate order-- the Annelida. All others, <ANIMALS,> «are Dioeecous as» Cephalopods, Pectinibranc
381d157 ious & Dioecious plants.-- NB. in Heautandrous ANIMALS <are> is there gradation of structure  leading
382d158 ine is much simpler.-- Hunter shows almost all ANIMALS subject to Hermaphroditism,-- those organs whi
382d158 to my mind.-- , that both are present in every ANIMALS, but unequally developed.-- surely analogy of
382d158 s. & neuter bee would shew this-- (Do any male ANIMALS give suck)--But this not distinctly stated  by
386d166 ably part of same general law, which makes two ANIMALS out of one & heals piece of skin.-- if the tai
386d167 cause the great changes of nature are slow. if ANIMALS became adapted to every minute change, they wo
386d167 y. 1838. p. 123. Ehrenberg. makes gemmation in ANIMALS very different from that of plants. (though la
386d167 plants. (though latter does sometimes occur in ANIMALS). latter the division taking place from outsid
386d167 vision taking place from outside inwards. & in ANIMALS, from inside to the outside.-- is this not own
386d167 imply to more importance of internal organs in ANIMALS‖. One «invisible» animalcule in four days coul
387d168 difference.-- If A. B. C. D. E be <offspring> «ANIMALS»: if «x» male impresses ovum <of> in A,  «with
387d169 escend to several generations» If A & B be two ANIMALS which have some peculiarity for first time & i
388d172 ence harelips heredetary. disease. extinction. ANIMALS in domestication (even Elephant) not breeding-
389d173 y generation applies only the more complicated ANIMALS.]CD p. 310 She wolf took dog. but had such ave
389d176 n> though other male may have copulated.-- two ANIMALS may unite & each have offspring by same mother
390d178 th the physical differences in almost all Male ANIMALS?]CD If the male «in the course of some generat
390d179 of breeding in, it is not merely the too close ANIMALS, which will not breed, but the female at least
390d179 really breeding in & in, but « <on> » breeding ANIMALS that have neither varied from their stock, for
398e005 -- on the contrary islands separated with some ANIMALS, &c.-- «if the change could be shown to be mor
402e018 r defunct companion".-- p. 229. Borneo.-- only ANIMALS he heard of pigs, small bears or badgers, deer
402e018 g of the latter being the same as the fox-like ANIMALS. which are met with near Canton" "Here, as in
402e020 culiar to them" do-- p. 368. "Several kinds of ANIMALS have spread from the N. end of Borneo to the a
403e023 let us look at facts. considering few domestic ANIMALS few. that have not <which> not, cows hornless,
404e024 their only escape is that rule applies to wild ANIMALS only. from which plain inference might be draw
406e036 cows, pigs & sheep.-- diseases common to men & ANIMALS cow pox.-- case in Spain of pustular disease f
406e037 ere.-- likewise far North in Southern.-- Great ANIMALS. of same two great orders destroyed about same
407e042 Molucca species.-- L'Institut 1837. p. 253, on ANIMALS of Antilles.-- (see Macleay in Zoolog. Journal
409e048 was  so changed. When discussing extinction of ANIMALS in Europe. :the forms themselves have been bas
409e048 exists  for one cause» of sexes «in separate» «ANIMALS»: for otherwise, there would be as many specie
409e049 see  if all species, there would not be social ANIMALS. «this is stated too strongly. for there would
409e049 few only social there could not be one body of ANIMALS, living with certainty on other» hence not soc
410e050 d not be improvement. «& hence not «be» higher ANIMALS» -- it was absolutely necessary that  Physical
411e055 against  my theory, doubtless, the presence of ANIMALS in <own» «the present» orders (not so in S. Am
414e063 n meet, they act precisely like two species of ANIMALS.-- they fight, eat each other, bring diseases
414e064 o gain the day.-- In man chiefly intellect, in ANIMALS chiefly organization: though Cont of Africa  &
414e065 erse.-- The range of man is not unlike that of ANIMALS transported by floating ice.-- I agree with Mr
415e067 es in mankind, «the more valuable domesticated ANIMALS» no doubt is owing to the rearing up of  every
416e070 the «so called» progressive tendency law.-- In ANIMALS analogy leads one to suppose that seminal flui
416e070 Planaria, they couple-- CD [lowest terrestrial ANIMALS-- in shells?-- insects?.-- all!??!?--  Worms?
416e071 cted by Plants). & as there are no fixed. land ANIMALS. so there are true hermaphrodites.-- I suspect
418e083 -- & knowing from analogy, that all these very ANIMALS are descended from some one single stock,--one
419e085 r kinds. (yet living in the salt?.)-- very few ANIMALS of any kind-- Fauna, must be very curious.-- W
420e089 phosed vertebrate, the parent of all vertebrate ANIMALS -- must have been like some molluscous «disexu
422e095 beaches» [not located] The enormous number of ANIMALS in the world depends, of their varied structur
422e095 there is no «NECESSARY» tendency in the simple ANIMALS to become complicated although all perhaps wil
423e096 s far from true.-- I doubt not if the simplest ANIMALS could be destroyed, the more highly  organized
426e107 ainst this. & wild hybrid plants. If many wild ANIMALS were crossed, there would probably be  perfect
426e107 s is false, [give instance of series from wild ANIMALS & plants]CD.-- Mr Marsh has some nephews,  who
427e108 rly stages of transmutations. the relations of ANIMALS & plants to each other would rapidly increase,
429e113 first  appearance of varieties of domesticated ANIMALS, yet as we know how many plants have been prod
429e113 duced (look at the Dahlias. we may infer it in ANIMALS.)-- Azara gives account of production of hornl
430e117 showing  small variations in offspring of wild ANIMALS.-- grateful & intelligent.-- The theory that a
430e117 grateful & intelligent.-- The theory that all ANIMALS have sprung from few stocks. does not bear, th
430e117 ear, the least on ancient generic forms.-- the ANIMALS in Eocene period could not have been direct pa
431e122 rates a plant he wishes to vary-- domesticated ANIMALS tend to vary. March 20th. Phillips in Lecture
431e123 of  wind difficult.-- Cline on the breeding of ANIMALS, p. 8. size of foetus in proportion to male pa
432e12 p.  12. Attempts to improve the native <breed> ANIMALS of any country must be made with great caution
436e136 of  Old World <If> It «may» be said, that wild ANIMALS will vary, according to my Malthusian views, w
438e142 ahlias]CD all much varied breeds both plants & ANIMALS have long been subjected to domestication.-- t
439e142 e of characters, formerly possessed-- <that is ANIMALS> «or rather the parents» having passed through
```

Page ***(Key Word)***
439e143 plants proves how much depends on instincts in ANIMALS.-- yet the existence of wild close species of
439e144 here is tendency to prevent the crossing.-- in ANIMALS where there is much facility in crossing there
441e151 lone or roots, & never flower, so there may be ANIMALS as Coralline, or others. which only generate o
445e159 es really attribute metamorphoses to habits of ANIMALS & takes series of flying mammifers-- says lemu
447e164 ng is much less,-- now certainly in the higher ANIMALS; changes seem to have been more rapid, & the f
448e166 I.-- <Reference> p. 251. about the drifting of ANIMALS on ice p. 643-- very curious table of all the
449e170 minary discourse to Fauna of Japan-- that the «ANIMALS of» islands N. of Timor are allied to the «typ
450e175 es of difference in all the isld, like in wild ANIMALS).-- There are prevailing colours in the differ
451e178 the Asiatic Soc. vol. I. p. 335. Catalogue of ANIMALS of Nepal by. B. Hodgson. p. 336 In the most pe
451e178 rld consistently reside & are bred. "take tame ANIMALS into this region between April & October & lik
451e180 ff. voyage through the Moluccas 1825--"No wild ANIMALS in Moa.--" Chapt.-- V.-- : do. Chat XXI. Wild
452e181 "-- Forrest Voyages. p. 39-- deer but no wild ANIMALS in Gilolo.-- p. 134: Birds of Paradise were fi
455eIBC or some one. Father-- diseases common to men & ANIMALS.--:likenesses of children CD[Does any annual g
460t017 ion «& sporting» in flowers & domestication of ANIMALS Aug. 26th.-- When it is said that there is evi
463t055 &c Bats are a great difficulty not only are no ANIMALS known with an intermediate structure, but it i
463t057 two bones of tibia into one.--) because if the ANIMALS were taken from which these series were drawn
463t059 is is not required..-- Waterhouse says perhaps ANIMALS of Fernando Noronha are found unknown coast in
473s07r pigeons--dogs--cattle? As we see the frame of ANIMALS can adapt itself to course of life, «as in tra
477z001 II.-- 35. Phil Trans Burrowing & boring marine ANIMALS-- CXVI. P 111 do Observations on Planarias by
478z003 oyage Vol p. 597 Many descriptions about lower ANIMALS of Falklands &c &c Bennett on Chinchillidae Zo
478z004 rm. luminous property-- Curious arrangement of ANIMALS in rays Par un officier du Roi Rapid growth of
483z013 rand paper. p. 387. "on Classification of such ANIMALS.."-- Voyage. Coquille's Voyage p 302 Vol II p.
492q003 n in Norway No Questions regarding Breeding of ANIMALS If two half bred animals exactly alike be inte
492q003 regarding Breeding of Animals If two half bred ANIMALS exactly alike be interbred will offspring be u
493q004 s effects of fatness.-- Experiment in crossing ANIMALS.-- &c (1) To cross some artificial male with <
493q004 reed or permanently.-- (2) Cross two half-bred ANIMALS. which are exactly alike & see result.-- (3) C
493q004 ux dog. with the hairless Brazilian or Persian ANIMALS of different heredetary constitution, to see w
494q004 oolog. Gardens-- with respect to conditions of ANIMALS & their general healthiness-- Fox's, Bears Bad
494q004 thiness-- Fox's, Bears Badgers,-- How few wild ANIMALS are propagated,, though valuable as show, & cu
502q11v characters of races of different vegetables & ANIMALS come on.-- Compare calves.: Compare young. bea
503q012 ack 37 Col. Sykes fertility of men & Europaean ANIMALS in India?-- about Chetah & other tame animals
503q012 animals in India?-- about Chetah & other tame ANIMALS not breeding when tame in India?-- does not kn
508q016 no answer?-- 3 Andrew Smith, about tamed wild ANIMALS breeding at the Cape.-- [About two vars: of Li
510q018 lancement Wm Yarrell (1) About non-breeding of ANIMALS in confinement, curious.-- foxes-- English ani
510q018 als in confinement, curious.-- foxes-- English ANIMALS. [Made no import. remark](2) Secondary male
511q018 seed-raisers (5) List of qualities in birds & ANIMALS for prizes.= Pidgeons. Canary birds-- Bantams.
512q020 hey fly in few hours Zoological Soc (1) Do the ANIMALS there, sometimes couple but not conceive :Bear
512q020 not conceive :Bears /Yes/ (2) Foxes & English ANIMALS & birds breed (3) In cases where Lions have br
526m027 .-- (NB man much more affected by other fellow-ANIMALS, than any other animal & probably the only one
526m028 looking at <ani> pictures, an early taste, of ANIMALS. they know.-- pleasure of imitation (common to
529m040 ined notion of land covered with ocean, former ANIMALS, slow force cracking surface &c truly poetical
529m041 looking at trees at [i.e., as] great compound ANIMALS united by wonderful & mysterious manner.-- The
531m053 alpitation, voiding urine because done by some ANIMALS in defence, &c Starting must be habitual «invo
532m057 collapse may be imitation of death, which many ANIMALS put on.-- The flush which accompanies passion
533m059 wn men can experience-- Instinctive walking of ANIMALS. that is the ready movement & co-relation of t
536m071 posterior at another.-- Why do bulls & horses, ANIMALS of different orders turn up their nostrils whe
536m072 nnot doubt that they have free will, if so all ANIMALS,, then an oyster has & a polype (& a plant in
538m081 , renders much less wondefful the instincts of ANIMALS-- Aug. 12th. 38. At the Athenaeum Club. was ve
540m085 b.-- We need not feel so much surprise at male ANIMALS smelling vaginae of females.-- when it is reco
540m085 horse & man yawn, makes me feel how <much> all ANIMALS <are> built on one structure.-- He who doubts
542m094 burst forth, this & yawning. (common to other ANIMALS) scream of agony, sigh of discomfort & wearine
542m094 no bark of anger nor have monkeys & many other ANIMALS,-- but yet when angry it is hard not to growl
542m094 se sounds which are involuntary, are common to ANIMALS.-- Curious to trace, which of these actions ar
542m097 aid" of language in man is very great from all ANIMALS-- but do not overrate-- animals communicate
542m097 great from all animals-- but do not overrate-- ANIMALS communicate to each other.-- Lonsdale's story
546m109 h birds. & ¿ howling monkeys-- smell with many ANIMALS-- see how a dog likes smell of Partridge--, ma
546m110 - Vide page 103, supra (by mistake) have lower ANIMALS these vivid thoughts In same book (p. 143) won
549m122 instincts-- like what is happening with other ANIMALS-- is far from odd nor is it odd he should have
550m123 d of man is no more perfect, than instincts of ANIMALS to all & changing contingencies, or bodies of
550m126 o free will. Mayo. Philosop. of Living p. 293. ANIMALS "have notion of property" -- their own propert
552m132 for long generation, (as friendship to fellow ANIMALS in social animals) are those which are good &
552m132 on, (as friendship to fellow animals in social ANIMALS) are those which are good & consequently give
555m142 rywhere (is it not present with all associated ANIMALS?) I doubted it in Fuegians, till I remembered
557m148 ed than jealousy, because less discoverable in ANIMALS.-- In the drawings of Voltaire why is under li
559m155 se.-- Read. Paper on consciousness in Brutes & ANIMALS. in Blackwood's Magazine June. 1838. Copied Mr
565n007 o growl of hounds».. <when> as fear to <man as> ANIMALS. comes at distance, mouth is placed open.-- He
566n011 inct in the chicken, just bursting from egg.-- ANIMALS have necessary notions. which of them? & curio
566n011 ecessary to explain origin of idea of deity.-- ANIMALS do not know they have 'these necessary notions
567n013 nected with <our> the willing of the simplest ANIMALS, as hydra towards light. being direct effect o
567n013 en understand before they can talk, so do many ANIMALS-- analogy probably false, may lead to somethi
567n014 ting «systematizing» on principles, which even ANIMALS practically know «art precedes science-- art i
567n015 Basket did. & Jemmy, when Chico plagued him-- «ANIMALS I should think would not have any emotion like
568n017 g it, (either if prevented, or overtempted.-- «ANIMALS have shyness with strangers» «as in case of te
570n025 y be said to have fear, both both have shame-- ANIMALS have not modesty. analyse this.-- «Excellent--
571n029 nt.-- Mental & Bodily Consider case of grazing ANIMALS knowing poisonous «herbs:» & man not.-- ¿no ve
571n029 ble good «for man» to eat poisonous?-- How did ANIMALS in «Australia» & America manage;-- This shows
576n049 instinct is transmitted.-- Arguing from man in ANIMALS is philosophical. viz. man is not a cause like
577n052 him, than of any one of his own sex.-- Hence, ANIMALS. not being such thinking people. do not blush.
578n056 it paralyzes all muscular action -- «in man & ANIMALS» Blubbering of a child (different in different
580n062 f science connected with Nat. Theology-- says ANIMALS have abstraction because they understand signs
580n062 concludes that difference of intellect between ANIMALS & men only in Kind.-- probably very important
582n069 ely a faculty in man not met with in the lower ANIMALS.-- hence the general aim of fable, & expressio
583n069 -- false instinctive pointing varies.-- p. 18. ANIMALS possess strong imitative faculty: pure instinc
583n070 mitation & reason must be thought of.-- p. 19. ANIMALS capable of education; (this is again assumed a
583n070 gained by instruction, or imitation.-- p. 20. ANIMALS may be called "creatures of instinct" with som
588n090 t very strong argument might be advanced, that ANIMALS have reasons, because they have memory.-- what
589n091 bits becoming heredetary form the instincts of ANIMALS.-- almost identical with my theory-- no facts,
591n098 en-- hence variation in character in different ANIMALS of same species.-- The general «(as I believe)
591n101 ent man Hume has section (IX) on the Reason of ANIMALS Essays Vol 2.-- «also on origin of religion or
593n107 instinctive feeling which is pleased by other ANIMALS smell & looks.-- no doubt it may be attempted
593n109 ections of animal taking man in place of other ANIMALS is hostile «is subversive of» to this view, &
593n109 e great distinguishing character between man & ANIMALS.-- [blank] Double consciousness. only extreme
601o009 itual actions" perhaps polypi-- (so that lower ANIMALS are sleeping higher animals & not plants as su
601o009 i-- (so that lower animals are sleeping higher ANIMALS & not plants as supposed by Buffon) Consciousn
604o016 light.-- Yet we can split Planaria into three ANIMALS, & this consciousness becomes multiplied with
608o026 ect of Education & mental capabilities.-- '(P) ANIMALS do attack the weak & sickly as we do the wicke
609o029 hich are strongest in man, are common to other ANIMALS & therefore to progenitor far back (anger <to
609o30v d? The origin of the social instinct «in man & ANIMALS» must be separately considered.-- The differen
610o30v rictly analogous to change of instinct amongst ANIMALS.-- Jan 13th. 1839 My father received a letter
610o033 rather the amount of our instincts-- surely in ANIMALS according to usual definition, there is much k
612o035 be able to exist as individual.-- [RHC] 2) In ANIMALS, growth of body precisely same as in plants, b
612o035 th of body precisely same as in plants, but as ANIMALS bear relation to less simple bodies, and to mo
612o035 out consciousness, for it is not evident, what ANIMALS have consciousness. These willings have relati
612o035 C] I here omit the case (if such there are) of ANIMALS enjoying only movements such as sensitive plan
612o035 ain stimulants without conscience in the lower ANIMALS, as in stomach, intestines & heart of man. [LH
612o035 in structure is the ganglionic system of lower ANIMALS & sympathetic of man [RHC] ¿How does conscious
613o036 ks that <all> there is one one instinct to all ANIMALS modified according to species. This I suppose

(Key Word)

```
613o036 that the thinking principle is the same in all ANIMALS. [LHC] «3)» Eyton told me that his retriever S
613o036 xternal form. so modifications of brain) As in ANIMALS no prejudices about souls, we see particular t
614o037 towards some final end.-- production of higher ANIMALS-- perhaps, say attribute of such higher animal
614o037 nimals-- perhaps, say attribute of such higher ANIMALS may be looking back, ∴ therefore consciousness
614o037 of memory does not depend on its length.: Many ANIMALS (as horses) very long & good memories-- but no
615o038 nt probably applies to particular instincts of ANIMALS, even in wild state; certainly to the domestic
616o038 conscious selves.-- Such memory may go back to ANIMALS which were changed into man .: they meet their
616o038 etion being some time governed by will in some ANIMALS, involuntary in others. [1]) Why may it not be
619o042 ithout regarding their origin, we see in other ANIMALS they consist in such active sympathy that the
623o050 y in races of man. & certainly in «species of» ANIMALS, in which case it undoubtedly is instinctive.
623o050 cts extend, yet as they are acquired by social ANIMALS, living under certain conditions, in this worl
623o050 n only be such, as are consistent» with social ANIMALS, that in which have a beneficial tendency, (no
623o051 ed> a beneficial tendency to them, as <social> ANIMALS of peculiar <kinds> «social feelings», & livin
629o055 iation from the instinctive. right & wrong.-- (ANIMALS excepting domesticated ones have no right & wr
634j54v D <Mac. remarks all Mammifers originally land--ANIMALS. as> 5 p. 314. Mac. remarks all <land> Mammife
634j55r 5 p. 314. Mac. remarks all <land> Mammiferous ANIMALS originally terrestrial.-- for we find even in
634j55r esulting from the will of the deity, to create ANIMALS on certain plans.-- is no explanation-- it has
635j56r rld? & the physical changes it was to undergo «ANIMALS feeding on each other &c &c».-- «(Causing deat
636j56v in the same predicament as a group of bisexual ANIMALS living on the borders of a country favourable
636j57r all these cases it should be remembered, that ANIMALS could not exist without these adaptations.--fo
639j28r mole, & Mole cricket & rodents (?) p. 251. all ANIMALS run by hind legs-- Kangaroo. only a caricature
640j167 e views quite exclude the idea of domesticated ANIMALS changing.-- From these views we can deduce why
641j29r e understand of the Physiological relations of ANIMALS. equatorial countries are supposed favourable
641j29v there any other analogies-- prickly plants or ANIMALS-- Exudation of fetid «& acrid» secretion in Mo
409e049 most beautiful in the moral sentiments of the ANIMATED beings.-- &c-- If man is one great object, fo
617o40v e in inanimate matter up to the action of some ANIMATED agent Now the phenomena of gravity are manife
617o40v he same as if every particle of matter were an ANIMATED being pulling every other particle by invisib
215b178 tes Lamarck priority refers to introduction to ANIMAUX Sans Vertèbres as latest authority. The case o
324c268 ney Bee Dutrochet Memoires sur les Vegetaux et ANIMAUX.-- on sleep & movements of Plant/ I£:4s Voyage
325c267 Anomalies de l'organization des Hommes. & les ANIMAUX.-- by Isid. Geoffroy. St. Hilaires. 1832. cont
463t063 t of it.-- Cuvier has grand sentence about the ANIMAUX fossiles-- being a mere fragment of the discov
601o08b Lettres. Philosophique sur l'intelligence des ANIMAUX-- & Le Parfait Chasseur, par Desgraviers, un V
241c016 uth Sea (Indio Polynes: <)> vegetation far East) ANN: des Sciences. Semptemb. 1825 Get Henslow to rea
109a084 No one can doubt. A-B once formed low coast.-- ANNALES des Mines. a translation of paper by rose on G
220b198 slow says. (Feb 1838) that few months since in ANNALES des Sciences. paper on Botany of Tahiti In Cha
326c266 od. Endlicher has published in first volume of ANNALES of Vienna. sketch of south Sea. Botany R. Brow
508q016 eding at the Cape.-- ‖About two vars: of Lion: ANNALES des Sciences‖ (4) Prolifickness of female, rel
267c095 Abstract of Infusoria. <p> do p. 62. Ehrenberg ANNALS of Nat. Hist. precursor of magazine??? p. 75.
268c096 of Nepal read before Linnaean Soc. Feb. 1838.-- ANNALS of Natural History. «vol I» p. 159 curious acc
270c103 n of organs of generation!!!. Mem. Agaziz. (INo ANNALS of Nat. Hist) spiral structure in Echinodermat
287c157 letter to Dr. Fleming. Philosophical Magazine & ANNALS. 1830 (?)." if she has put man on the throne (
298c189 gland & Ireland.-- curious in so wild a bird.-- ANNALS of Natural History Vol. I. p. 185 case of tit
310c225 eme difficulty in mundine geological chronology ANNALS of Natural History «Vol I??» p. 318. some rema
320c275 r Magazine of. Zoology & Botany & Continuation «ANNALS of Natural History» Skimmed Von Buch Travels W
324c268 oc.» F.. Cuvier on instincts L. Jenyns paper in ANNALS of Nat. History Prichard.-- Lawrence Des St.
325c267 trace Europëan forms compared with African <A> ANNALS Histoire Generalle et particuliere des Anomali
346d048 fell.-- gore to death the old & wounded,-- see ANNALS. vol. 2. 1839.-- are bad breeders & subject to
379d151 s father so that case ceases to be true avitism ANNALS of. Natural. History. p. 135. Natural History
386d167 to the slow great changes really in progress.-- ANNALS of Natural History. 1838. p. 123. Ehrenberg. m
387d169 pair, + the influence they themselves inherit./ ANNALS of Natural History .p. 96. Vol I. Notice the S
421e090 he spermatic fluid fertilized spawn of frogs.-- ANNALS of Natural History. (p 225. 1838.) account of
421e091 ) Falklands to southern portion.-- ‖.do p. 269. ANNALS of Nat. Hist 1838 on «a» freshwater fish pecul
461t025 haps rendered more complex & some simplified.-- ANNALS of Natural History. <no. XII.> Vol. 2. p. 96 &
243c019 rtainment on eût été bien éloigné, il y a peu d'ANNÉES, d'admettre que ces oiseaux eussent leurs repr
381d156 > Mollusca, with pectinibranchiate order-- the ANNELIDA. All others, <animals,> «are Dioeecous as» Ce
201b126 ys Mag of Zooly & Bot. p 65 Vol II talking of ANNELIDAE.-- <">The fact is an additional illustration
370d117 heir separation.-- thus Vertebrate blend with ANNELIDAE by some fish.-- But birds quite distinct.-- C
205b143 p. 123 A climbing fish. p. 122 A Terrestrial ANNELIDOUS animal p. 347. Vol I.-- compare with my plan
256c055 o Boblaye & Virlet.-- Whewell thinks (p 642) ANNIVERSARY Speech. Feb 1838 thinks gradation between Ma
434e128 are equally handed to offspring.-- Whewell's ANNIVERSARY address 1839, p. 9.,-- talks about fossil In
521m009 , not through hearing.-- Thus when dinner was ANNOUNCED he could not understand it, but the watch was
380d153 the offspring-- this is clearly the converse of ANNUAL being rendered biennial-- the hardness of life
386d165 en trees! How is this with buds of plants. does ANNUAL give buds.-- life may be thus prolonged bud be
389d176 throw itself off.--» as might be inferred from ANNUAL plant being prolonged till it has bred.-- Offs
455eIBC & animals.--:likenesses of children CD[Does any ANNUAL give buds, or tubers. Yes-- but these are same
495q05a rm found wild a The Leptosiphon densifolium «an ANNUAL» <sleep> «closes flower» on all gloomy days.--
501q011 er failing.-- answered by Gaertner (28) Can any ANNUAL or Biennial be grafted or cuttings taken or tu
500q010 es. Spallanzani Essay-- Figs 2 kinds of flower ANNUALLY.-- Periwinkle. (not asclepiadae. «in» Lindley
171b002 rfect; mothes apparently only born to breed.-- ANNUALS rendered perennial. &c &c.-- Yet Eunuchs nor «
390d177 lower) exotics brought from foreign country. (<ANNUALS> & so must those forms which are produced by b
454e184 gland, which do not perfect their seed?-- What ANNUALS can be budded «& rendered of great age» as mus
492q002 . how have varieties been formed?-- 8. Can any ANNUALS be budded. with reference to extension of age
495q005 ings vary much more than cuttings &c (4) Raise ANNUALS or common English plants in Hothouse & see wha
501q011 on this subject-- (31) Ask Henslow for list of ANNUALS to place in Hot house to see effect on general
350d055 (are not these differences in sex confined to ANNULOSA?) Remarked that young of Cirrhipedes can move
203b135 n «V. L. Institut p 245. 1837» we should have ANOMALIES. as Cape Anteater,.-- This supposes world div
325c267 Annals Histoire Generalle et particuliere des ANOMALIES de l'organization des Hommes. & les animaux.-
051r093 member many Corals?? Breccia--Stratification? ANOMALOUS action of ocean.--at Ascension. (where occass
051r094 worn into bare cliffs evident); the action is ANOMALOUS; It is wonderful to see Coral reef--or confer
206b149 ng ones.-- NB As Illustrations are there many ANOMALOUS lizards living; or of the tribe fish extinct.
341d030 ed that the young of this animal, which is so ANOMALOUS among true deer, yet is spotted like so many
349d052 ill ornithorhynchus come in circle?!!! p. 8-- ANOMALOUS structures, as in Hippotamus, solely owing to
640j167 slds are explained. On distinct Creation, how ANOMALOUS, that the smallest newest, & most wretched is
181b043 Such on form each side will yet present some ANOMALY & <the g> bearing stamp of <some> great main t
280c135 id.-- My theory thus explains a grand apparent ANOMALY in nature. <t>---- Many animals not breeding a
293c172 ing up image which had been past-- so great an ANOMALY in structure of brain not probable) put note.
316c246 & E. S. America.-- get instances.-- very good ANOMALY in range + + What circumstances have led to fo
359d088 not most like Penguin duck.-- which is strange ANOMALY in Yarrells law.-- it probably is explained by
401e016 uce cowslip, one is tempted to think here some ANOMALY-- I can fancy cowslip producing primrose retur
584n071 tructure ones; & hence we get over an apparent ANOMALY, for if anyone has taken the Woodpecker as an
637j58r merable eggs is explained by Malthus.-- [is it ANOMALY in me to talk of Final causes: consider this!-
640j167 at true, but do not fish offer a most striking ANOMALY to this. Have they wide ranges? Agassiz has sh
233b250 k «India» Monkeys of old World. Crocodiles. ANOPLOTHERIUM.-- <M. Jerrod> & Dumeril great work on Rept
458t001 nge forms., camels, giraffes. Sivatherium & ANOPLOTHERIUM, with existing, or nearly existing forms of
316c245 s «++» p. 246» as Cyclostoma in Phillippines & ANPHIDESMA in S. America-- yet there are a few Cyclosto
316c245 car-- yet there are a few Cyclostomes & a few ANPHIDESMAS.-- this is remarkable.-- Fish & drift sea we
523m017 hed from whims passion &c by coming on suddenly. ANS no.-- because often, if not generally, does not
521m007 t meals, no [not located] There is a case of Mr ANSON. who told a story of hunting «-- habitual fits.
522m011 nswered To be sure I do.-- What became of him.-- ANSW Had large fortune left. him. took name of Child
148g025 sked this question in many ways & received same ANSWER Thought lambs most like MOTHER!-- the cross no
174b013 ca why, 2 of ostriches in. S. America-- This is ANSWER to Decandoelle. (his argument applies only to
175b016 - this difficult to prove cats &c from Egypt no ANSWER because time short & no great change has happe
198b115 produced on New Zealand; if generated <No> «an» ANSWER <could> «can» be given.-- It is a point of gre
225b217 orms.-- Opponent will say. show them me, I will ANSWER yes, if you will show me every step between bu
277c128 t must take place-- such a classification would ANSWER every purpose, & would present many ideas of c
298c189 e «most» like Australian.-- Curious this +ready ANSWER, without any leading question.-- [+] This migh
346d047 y. I asked this in many ways, but received same ANSWER.-- Thought lambs were more like father than Mo
423e095 lways have done so, as the simplest fish &), my ANSWER is because, if we begin with the simplest form
```

Page ***(Key Word)***

446e162 h by the latter very rarely») here is a case in ANSWER to Mr Knights doctrine.-- Case like Corallina-
463t057 tion will probably reject this theory-- (I must ANSWER it by rooting out curious cases of intermediat
495q05a ravel walk» will drift many seeds= Necessary to ANSWER Wiessenborns doctrine of Equivocal Generation
508q016 ace of gout.-- How are livers obscure organ. no ANSWER?-- 3 Andrew Smith, about tamed wild animals br
514q21. N. America similar to Lapland Plants --will get ANSWER= Is pollen of cultivated Orchis & Asclepias &-
547m112 w what was difference between Castle & dream No ANSWER shows our profound ignorance in so simple case
616o039 out these faculties & indeed until we know what ANSWER they would give in support of their view it is
617o39v ishes to fully understand this subject, but the ANSWER to it would require a considerable degree of a
628o53v p my word"-- gives the problem, of ethics-- [my ANSWER would be to all such cases-- either, that from
109a083 s argument of low coast gaining & high loosing ANSWERED by this -- No one can doubt. A-B once formed
250c039 rld??? Lyells Principles, must be abstracted & ANSWERED Much might be argued what is not cause of des
341d029 e. whether relation in one point or many) Owen ANSWERED that all characters might be considered as ad
351d056 ionary Spondylus has eye-points-- Macleay then ANSWERED, because nature leaves vestiges of what she d
354d062 r to show Stonesfield Didelphis not Didelphis «ANSWERED satisfactorily by. Valenciennes.» The change
411e056 s for individuals (Mem: transportation will be ANSWERED) one looks to analogy for cause in plants. wh
493q003 imary colours occur in relation with species-- ANSWERED «by Henslow» see notes In varieties is there
501q011 about stigma in similar manner ever failing.-- ANSWERED by Gaertner (28) Can any annual or Biennial b
508q016 s of female, relation to healthiness? & father ANSWERED (5) About cross-bred races of men taking afte
514q21. better. or "constructive action on germ." '=?? ANSWERED Does Mormodes (one of the Catasetums) really
514q21. ys hit stigma by projecting pollen-masses?-- = ANSWERED = Has Ophrys nectary?= Bunbury says no «hollo
521m009 ted out the Gardener & said, who is tha? Mr C. ANSWERED why do you not know, that is A. B my gardener
522m010 know whom Mr Child «of Kinlett» had married.-- ANSWERED never heard of such a man.-- (My Father expla
522m011 en said you remember Jack Baldwin at school.-- ANSWERED To be sure I do.-- What became of him.-- Answ
523m017 two years before, to prove she was not insane, ANSWERED she had known it at time & had bought arsenic
537m077 cotchman will his country or Swis.-- it may be ANSWERED effects of education, may be opposed undoubte
192b089 fore & that is all that can be expected-- This ANSWERS Cuvier-- Perhaps the father of Mammalia as Het
257c057 r live before it had its web-feet-- all Nature ANSWERS to the possibility.-- My views will explain no
342d034 en on elevation of continents) but Ogleby well ANSWERS that nearly all F. W. Fish are Abdominals. ∴th
491q01v The French Apple tree «with abortive stamens» ANSWERS first question in negative.-- Questions Regard
505q014 Nitrate of Soda under Beech.-- Lychnis dioica ANSWERS this question= (5) Open more Horned oranges.=
093a034 ms to consider that elevation & eruptions are ANTAGONIST forces. but they are parts of one force, one
258c060 divellent power & tends to make forms remote ANTAGONIST powers.-- Every animal in cold country has s
577n049 that the life & will of a conferva is not an ANTAGONIST quality to life & mind of man.-- & we do not
060r125 June 1780, Sept. 21st. 1817.--p 371. Webster ANTARCTIC veg:-- Study Ulloa to see if Indian habitatio
219b195 altered. or northern plants «No» CD[Mem. the ANTARCTIC flora must formerly have been separated by sh
510q017 creations & where? How are current & winds in ANTARCTIC ocean: are they from West, like as between Au
641j29r «Narwhal» Polar bear. Walrus, great Seals of ANTARCTIC seas. (on other hand Spermaceti Whale & Manat
247c028 n Hemisphere 2d in southern --p. 71 Chimera-- ANTARCTICA «also Taeniatole austral» «caught» Chile, Va
203b135 p 245. 1837» we should have anomalies. as Cape ANTEATER,.-- This supposes world divided into Zoologic
319c257 says the Macrotherium of Europe is between the ANTEATER of C of Good Hope & those of S. America,-- Ar
261c070 seed-- not solely effects of climate on some «ANTECEDENT races, perhaps not on now existing» Mr Gould
437e141 ng into each other, & depends on character of ANTECEDENT races.-- «& yet in all probability the Bruss
240c014 bits Celebes & few of the larger islands.-- -- ANTELOPE in Celebes, Bourou new species of Axis.-- «Ce
174b014 ather, the result, would be as it is.-- Hence ANTELOPES at C. of Good Hope-- Marsupials. at Australia
266c092 ommon to either coast not found here not even ANTELOPES, though common on coast of Arabia not even an
266c093 es, though common on coast of Arabia not even ANTELOPES though common on islets off Arabian Coast.--
182b049 lopment gives final cause for enormous periods ANTERIOR to Man.. difficult for man to be unprejudiced
268c099 quable climate both in S. & N. Hemisphere just ANTERIOR to present. ¿cause of destruction of «great»
302c201 sh & Mammalia, must be sprung from some source ANTERIOR to giving off of these two families, but we s
409e049 ct <uni> world, either at the present, or many ANTERIOR epochs.-- but we can see if all species, ther
601o009 - Evidence of consciousness, <t> movements «¿» ANTERIOR to any direct sensation, in order to avaoid i
603o013 so that much of EVERY language shows traces of ANTERIOR state?? Edinburgh Review Vol 18. (1st Article
304c207 ame parent with other birds,, or branched off ANTERIORLY think what principles are there to guide in
466t093 to it-- The Humbles in crawling out brush over ANTHER & pistil & one I SAW IMPREGNATE by pollen with
467t099 en Garden.-- As we see in Hybrids that although ANTHER «nor filaments» shrivel, yet stigma does not,
466t093 . {P} Stamens & pistils curve upwards, so that ANTHERS & stigma lie in fairway to nectary.-- Is not t
466t094 d «necessary» cross directly over the bunch of ANTHERS & pistils, but these <do> do not bend up-- In
466t094 within nectary «they do» they must disturb all ANTHERS, wh otherwise lie protected by the hairy black
466t099 : filaments united in whole length to corolla--ANTHERS minute, distinctly doubled, brown, but with no
467t103 rom extremity pollen, or pollen comes out with ANTHERS & stigma in slit» -- As I think they do in Bro
468t112 pecially stigma after bee had brushed over the ANTHERS of long stamens {P} as stamens grow old «& she
502q012 in Sweden. Linnaeus found 2 flower. which had ANTHERS removed, did not become impregnated. (34) Any
635j56v er, where such care seems to be taken that the ANTHERS should not be exposed to weather.-- this is ag
280c135 his true?? My views, which would even lead to ANTICIPATE mules is very important for L.yell said to m
431e122 here few stray ones. are, that change may be ANTICIPATED, & this would look like fresh Creation. the
038r059 aluminated. As argument in favor of lines of ANTICLINAL violence crossing lines of crater, <arg> sta
078r175 evated summits, where mountains most torn.--(¿ANTICLINAL line?). -- Mines of Catorce «(Principal vein
104a066 kness of <strata> not great, one can conceive ANTICLINAL lines near. (lateral pressure would always p
120a107 of earthquakes.-- by the narrowness which the ANTICLINAL lines are apart-- the curvatures of the stra
127a125 st-clinal. West clinal. S.-clinal. N-clinal & ANTICLINAL «synclinal--» line.-- <ditto of synclinal> s
407e042 cies.-- L'Institut 1837. p. 253, on animals of ANTILLES.-- (see Macleay in Zoolog. Journal. for those
464tf6r the Struthonidae than any other forms-- Lund's ANTILOPE in Brazil another point of agreement with. N.
075r167 ulphuretted silver, arsenical grey copper, and ANTIMONY, horn silver, black silver & red silver, do n
062r129 rats offering in the history of rats, in the ANTIPODES a parallel case.-- Should urge that extinct L
521m008 to play by morbid action.-- Old Elspeth's «in ANTIQUARY» power of repeating poetry in her dotage is f
575n046 etches two birds out at once.-- Old People-- (ANTIQUARY Vol II. p. 77) remembering things of youth, w
576n046 he acquirement of new ideas.-- Walter Scott «(ANTIQUARY)» Vol II p. 126 says seals knit their brows w
530m044 tion of idea.-- Mr Blakeway has mentioned in ANTIQUITIES of Shrewsbury something about big noses & na
352d059 mals countenances the belief of their extreme ANTIQUITY (ie much intervening physical change).-- dist
405e035 r having never [not located] ARGUMENT REAL of ANTIQUITY of reasonable cosmopolite man.-- L'Institut.
438e141 ow that <time> the fixity of characters «from ANTIQUITY» prevents their variation, which is not impro
641j29v separated Arctic genera, there is evidence of ANTIQUITY & extinction of such forms-- these views will
498q008 are not artificially impregnated. Abberley says ANTS-- Enquire (13) Do any of same species of Willow
520mIFC q 36 Grt. Marlborough Str.-- (p. 64. On <insect> ANTS getting on Table. Col. Sykes) Private Finished.
534m063 with simple exertion of intellectual faculty) if ANTS had at once made this leap it would have been i
067r143 eare Lava in Cordillera & on Eastern plains «by ANTUCO». Athenaeum April 1836 (p302) Coleccion de obr
235b274 t change be superinduced-- why is every one so ANXIOUS to cross. animals from different quarters to p
296c184 - Endeavour to find out whether African forms. (ANYHOW not Australian) on Peaks. Did Creator make all
378d147 o> if so domestic animals ought to show them.-- ANYHOW not connected with habits According as child i
453e183 this good question-- single, or half double.-- ANYHOW fertile because they «are» raised by seed.-- W
547m112 are closed probably by sleep & not vica versa. ANYHOW I might have been quite still, & not attending
553m135 Metaphysicks shows that such a view cannot be, ANYHOW, easily overturned.-- so ready is change from.
292c170 alia, than in shells ¿ univalves or bivalves.-- ANYMAN No VI. Magazine of Zoology & Botany p. 566 wan
556m145 th two eyes.." I think this cannot be disputed ANYMORE in men. than in animals.-- In the drawings of
572n034 separately, neglecting time, & general sense, ANYMORE than connected with general tendency of the dr
618o41v t be said that the blueness caused the weight, ANYMORE that weight, the blueness, still less between
263c075 an absurd saying) & will never be conquered by ANYONE (if has any kind of prejudices) <without> who
283c145 haps I feel the impossibility of this more than ANYONE.-- no turn the Zebra into the Quagga.-- «let t
292c170 at abstract of Swainson on Classification "Let ANYONE even with a very superficial knowledge «like m
415e066 of this earth. like every other animal.-- Would ANYONE raise an argument against, my theory, should n
463t055 d with such structure.-- perhaps greatest Could ANYONE. have foreseen, sailing, climbing & mud-walkin
464tf6r American form)-- ¿Hunting leopard, how strange, ANYONE, would have thought isolated species Mr Blyth,
578n054 ies on, by association, the question, "one will ANYONE, especially a women think of my face,"? to one
578n055 let the possibility of this being discovered by ANYONE, especiall if it be a person. whose opinion he
584n071 hence we get over an apparent anomaly,, for if ANYONE has taken the Woodpecker as an example fitted
617o39v stion which ought to be clearly comprehended by ANYONE who wishes to fully understand this subject, b
622o048 t explanation of law of honour from Paley [RHC] ANYONE, who will reflect must feel, how like to injur
275c121 nce about old varieties is reduced to scarcely ANYTHING.-- almost all imagination-- He says he recoll

291c166 heredetary<)>.-- it is difficult to imagine it ANYTHING but structure of brain heredetary,. analogy p
350d054 s p. 20 There are no grotesques in nature; not ANYTHING framed to fill up empty cantons, & unnecessar
354d062 ement & not the Quinary.-- any one may believe ANYTHING in such rigmaroles about analogies & number L
440e147 of explaining, much structure, than attempting ANYTHING about habits-- no one can be shocked at absen
523m015 omewhere on horse being insane at the sight of ANYTHING scarlet.-- dogs ideotic.-- dotage.--» Doctor
525m023 a friendly dog).-- they feel shame, when doing ANYTHING which is wrong.-- as eating meat., doing thei
526m030 urges him,-- absolute free will, motive may be ANYTHING ambition, avarice, &c &c An animal improves b
530m046 y thoughts.-- An intentionally recollection of ANYTHING is solely by association, & association is pr
531m053 ause Nancy tells me very young babies start at ANYTHING they hear or see. which frightens. them.-- No
548m116 - though she when well did not recollect <it> «ANYTHING».-- if one was subject to this disease onesel
554m139 ermint.--» Perfect understand voice.-- will do ANYTHING.-- will take & give food to Tommy, or anythin
554m139 anything.-- will take & give food to Tommy, or ANYTHING of any sort.-- I saw Tommy picking his nose w
556m144 ly start when they hear ready, but if they see ANYTHING ahead. which cad cannot see, they do not move
560mIBC muscle) very early in life Do they wink, when ANYTHING placed before their eyes, very young. before
566n010 have them because not beautiful-- is there -- ANYTHING in these absurd ideas.-- do they indicate min
570n024 he wishes to show, he is determined not to say ANYTHING. he presses his lips together & shrugs his sh
579n058 ildren smile before they laugh.-- Has frowning ANYTHING to do with ancient movement of ears A man shi
591n097 - wagged its tail «a little» when attending to ANYTHING or excited.-- so do young dingos, as I saw wa
591n097 young dingos, as I saw wag tail when watching ANYTHING-- Keeper does not think they drop their ears.
593n105 ome secondary one-- blood being disagreeable & ANYTHING disagreeable being pursued.-- A dog turning r
594n113 mach a little disordered) at thought of almost ANYTHING ugly. baby-- association-- pouting child same
595n121 this part of same feeling which make us think ANYTHING ugly-- a beau-ideal feeling. Same effect as a
608o027 profound humility, one deserves no credit for ANYTHING. (yet one takes it for beauty & good temper),
610o033 sources of knowledge.-- whether <we th there> "ANYTHING can be <any> «the» object of «our» knowledge
621o047 -- Thus a child may be taught to think almost ANYTHING nasty. (<& accidentally> «by odd association»
626o052 ears to me rather rigmarole.-- He does not say ANYTHING about any principles born in us.-- Great diff
628o54v e» honour will stop all wish to gratify <it>-- ANYTHING contrary to it]CD NB. the very end of conscie
387d168 with only difference of time) is the above law ANYWAYS connected with the case of successive copulati
446e162 trine.-- Case like Corallina-- «Does it flower ANYWHERE?-- Yes on the continent is there more variati
199b117 xisting. Ed. N. Philos J. p. <191» «p 191» No. 5. AP 1827 F. Cuvier says. "But we could only produce
199b118 into an instinct." Ed. N. Phi. J. p 297, No 8 Jan-AP. 1828 -- I take higher grounds & say life is sho
120a107 y the narrowness which the anticlinal lines are APART-- the curvatures of the strata.¿ the enormous f
172b007 ght to become different if kept long enough.-- «APART, with slightly differen circumstances.--» Now G
255c050 food, at the Mission of Mojos (even 20 leagues APART from each other.-- this bird was well known for
284c148 digiously heavy (where trees of such Nature far APART. Must have travelled by <dead> «each» trees dyi
359d087 ng so totally infertile whereas animals further APART have bred inter se.-- These hybrids are very wi
363d102 ifferent.-- the two redpoles can hardly be told APART, so that after differences were pointed out Sel
311c227 rthy of investigation.-- Institut 1838. p. 174. APERCÚ very good on insectiferous <insects> quadruped
023r011 lcanos have been said to be merely accidental APERTURES still open.--The fault like appearance «arisi
402e018 he heard of pigs, small bears or badgers, deer, APES, baboons, monkeys & an animal probably a tapir
216b181 influenced in this cross to after births, like APHIDES.-- Case of boy with foetus developed in the br
344d040 , which have fertile offspring. Entomostraca & APHIDES. The extreme difference of sexes. is probably
344d041 successive fertile offspring in Entomostraca & APHIDES Developement of sexes in Caterpillar. very val
392d175 ng the upper hand. <]CD> & forming species)-- [APHIDES having fertile offspring without coition or ad
491q01v 4 May we no suppose, that certain plants, like APHIDES produce impregnated young ones; & that it is i
639j28r ion?) C. D. Spines in Hedge Hog & Echidna.. & APHRODITES C. D. Endless cases.-- Macculloch p. 260 int
639j28r s. C D woodcuts stones swallowed by birds & by APHYSIA. C D p. 258. «grinding» teeth in <stomach of>
534m063 f this country (thus Dahlias by snails)-- <The> APION radiolum undergoes transformation in the stem o
566n010 I Octav. Edit)-- certainly neither a Minerva or APOLLO would have them because not beautiful-- is the
546m111 membered she had put it in branch of tree, & APOLOGISING to party, went out & found it there!!! Lady
332d001 , he then did not see other cases.-- He thinks APOPLEXY affects people all over England at same perio
450e178 d: shows they will propagate get dimensions-- do APP. p 73 State of Muar in Malacca.-- speaks of Rhin
411e056 ble individuals can be produced. & yet sexual APPARATUS.-- My account of Cirrus cinereus of the Falkl
056r109 ire island so encircled, the one slow cause is APPARENT. I confess I never see such islands whose in
203b135 eakest part of theory death of species without APPARENT physical cause:-- Mem: Mastodon all over S. A
261c071 the latter most important in obviating a great APPARENT difficulty-- preservation of colouring, when
280c135 ght be said.-- My theory thus explains a grand APPARENT anomaly in nature. <t>---- Many animals not b
430e118 ends-- But how.-- «-- .Make the difficulty APPARENT by cross-questioning.-- » even if placed on I
506q015 produce seed in this country-- where cause not APPARENT-- Any where pollen is not produced or small i
547m112 e mind is wholly absorbed with one idea (hence APPARENT vividness) & there being no other parallel tr
553m135 of causation, to give a cause (& no one being APPARENT, one fixes on imaginary beings, many vicariou
553m136 es who thus argue, make the same mistake, more APPARENT however to us, as does that philosopher who s
584n071 ore the structure does; & hence we get over an APPARENT anomaly,, for if anyone has taken the Woodpec
134a144 do {T} Journal of Asiatic Soc Vol V. p. p 96. APPARENTLY good geological paper. by Malcolmson-- worth
149g032 ed «base» pebbles of quartz & other rocks not APPARENTLY in situ <& in> hill being gneiss <& also> al
158g079 like head of Glen Guoy nor is horizontal line APPARENTLY continuation of upper terrace -- on right ha
171b002 ids where every thing else is perfect; mothes APPARENTLY only born to breed.-- annuals rendered peren
307c215 ale rendered abortive in the womb.-- if these APPARENTLY useless organs do indicate such origin, then
343d037 ther by developement of new forms in one., or APPARENTLY so. by the extinction of prominent ones in <
367d113 will not be produced.-- Sept. 17th. Saw mule. APPARENTLY fathered by a donkey. with all four legs rin
465t079 ormous periods may elapse, even in situations APPARENTLY favourable for the preservation of shells; w
581n064 ire peacock's tail, as much as we do.-- touch APPARENTLY. ourang outang very fond of soft, silk-handk
026r022 the nature of strata & Organic remains does not APPEAR to have taken place in the Cordillera of S. Am
030r037 rise In Cordillera, the dikes do not generally APPEAR to have fallen into lines of faults I do not t
039r060 ark as many distinct elevations; hence it would APPEAR he has not fully considered the subject.-- S.
053r101 must be quoted at length. The Lines of Mountain APPEAR to me to be effect of expansions acting at gre
073r163 ay in primitive & transition; the latter rarely APPEAR in central Cordillera. particularly between 18
077r171 ins into which the valley of Marfil is divided, APPEAR to have a decided influence on the richness of
150g038 elow the Houses The Hills in this neighbourhood APPEAR very round-topped with much drainage & far mor
157g073 mica slate, only sand blow away]CD where lines APPEAR to cross stony parts; appearance chiefly cause
160g093 W) broad terrace «of pebbles? & Alluvium» which APPEAR perfectly level, <on op> dies away on gradual
177b027 en.-- so that in Mammalia «birds» it would only APPEAR like circles;-- & insects amongst articulata.-
177b029 Monad has not definite existence.-- There does APPEAR some connection shortness of existence, «in» p
179b035 . America (independent of external causes) does APPEAR very probable:-- Mem: Horse, Llama. &c &c-- If
180b038 pecies is a consequence (contrary to what would APPEAR from America) of non adaptation of circumstanc
182b049 ending range, reason & futurity. it does as yet APPEAR clim In Mr Gould Australian work some most cur
194b094 to approach the Colobes which in south Africa, APPEAR to represent the semmoipthegue of India.-- Too
197b112 e.-- In monsters «also» organs of lower animals APPEAR.-- yet nothing about propagation= I see nothin
207b150 Flora) and Dicotyledenous, which «nearly» first APPEAR «(p 321)» at Tertiary epock p. 330. Fossil Inf
214b173 tralia and «fossils of» Oolitic Series does not APPEAR to me very strong What is Osteopora platycepha
263c075 & painful effort of the mind (although this may APPEAR an absurd saying) & will never be conquered by
268c095 paper on quadrupeds of Van Diemen's land, which APPEAR diff. from Australia (Waterhouse <disputes thi
290c162 ts. At first classification on generation might APPEAR an analogy NB Pyrrho-alauda (bird of St Jago)
305c209 e of the ostriches were to die, then they would APPEAR isolated.| In my birds from S. Hemisphere ther
357d072 d shows there is rule.-- S. America & Australia APPEAR to have suffered most with respect to extincti
383d159 y Hunter.-- Do testes, & ovaria when they first APPEAR occupy their proper positions,-- this would be
406e036 he similarity of offspring to Parents same laws APPEAR to hold good. with regard to marriage of indiv
451e178 e there & do not pine visibly. p. 337. it would APPEAR as if p. 345. The Ceylonese Elephant [...] sau
473s07v ucture can be concepcional, as limbs &c &c only APPEAR late in pregnancy, & then may just as well be
478z003 inq lignes de long et volent. p. 208 Fleas only APPEAR in winter in Paraguay p 207 Slight notice on h
481z010 anatomy. of. corals.-- nevertheless the details APPEAR very trifling Also Berre «p. 8» (I think Plana
496q006 on associated plants. when proportional number APPEAR equal-- & see whether proportions will vary, w
497q007 ; what part of stigma?-- (4) As Papil. flowers APPEAR difficult to cross, are there unusually many s
511q018 oy secondary character believe No or did result APPEAR without his wish Has since recrossed this bree
524m019 cure for madness is forgetfulness.-- which does APPEAR a real difference, between oddity & madness.--
524m021 ty, the ideas do not go back to childhood, (but APPEAR most capricious) as in delirium after epilepsy
549m120 sanity &c unhappy-- perhaps not so much as they APPEAR & perhaps partly their fault.-- Whether this r
594n112 e.-- How can such an instinct arise?? «it would APPEAR that an instinct long remains, if no steps are

Page ***(Key Word)***
613o036 e same, & the means very similar.-- It does not APPEAR more than saying that the thinking principle i
614o037 s, therefore reward in good life [RHC] Instinct APPEAR like heredetary memory; but first memory in ma
615o038 iguration. [RHC] General-- Instincts, certainly APPEAR a sort of acquired memory. a permanent secreti
023r011 idental apertures still open.--The fault like APPEARANCE «arising from the manner of horizontal uphea
038r058 uite different from the Porphyries: certainly APPEARANCE leads me to believe mere fissures filled up.
038r058 s me to believe mere fissures filled up.--the APPEARANCE will here be the strongest argument:--¿ Cons
041r067 - In Patagonia. the blending of pebbles & the APPEARANCE of travelling may be owing to successive tra
058r115 ld affect this. -- State the circumstances of APPEARANCE at Concepcion [.] no sign of elevation. Effe
157g074 y]CD where lines appear to cross stony parts; APPEARANCE chiefly cause by fall of angular masses from
223b207 l event of intellectual Man appearing..-- the APPEARANCE of insects with other senses is more wonderf
245c023 puensis «partly domesticated» like in general APPEARANCE to Siamese kind.-- but considered good speci
266c091 Hindu's, in the same tracts;. & that in their APPEARANCE & manners they are as opposite as day & nigh
272c107 s, having spiny legs & running quick & generl APPEARANCE of blattae other Hemiptera stikingly resembl
288c159 glish cross readily-- think about half way in APPEARANCE.-- bark about half way «in tone»-- the nativ
317c249 nches differed considerably in their colour & APPEARANCE Every now & then a short-tailed cat.¿cut? ha
318c253 udubon there is most curious history of first APPEARANCE of the S. American Pipra Flycatcher, which i
336d019 s contradict his own law. So far is there any APPEARANCE of animals being created. it is probable if
343d039 merica.-- shells of S. America.-- there is no APPEARANCE of sudden termination of existence.-- nor is
345d044 loured like Magellanic Fox.-- peculiar hair & APPEARANCE-- good case of Provincial Breed-- Highland S
355d068 . shells (see Paper in Geolog Transacts) same APPEARANCE with Secondary Species distinct-- but close.
358d085 gs. & the men cannot «hardly» tell any sex by APPEARANCE.-- The silver & common pheasant crossed, has
360d090 om <bitch dog do> father dog. & hence general APPEARANCE of face & tail somewhat like dog-- though it
413e059 Babbage 2d Edit, p. 226.-- Herschel calls the APPEARANCE of new species. the mystery of mysteries. &
426e107 of domesticated ie new varieties destroys the APPEARANCE of this series & makes one think that one la
429e113 Whether we can or not trace history of first APPEARANCE of varieties of domesticated animals, yet as
460t013 offspring with» so much of the swan-goose in APPEARANCE Bell at Hornsey (though only ⅓ of blood). th
511q018 common species, having somewhat of different APPEARANCE.-- {will introduce it in work} Whether <Yar>
538m079 knowledge or other & unusual line-- both odd APPEARANCE about eyes.-- one botanist & great knowledge
577n051 is intimately concerned with thinking of ones APPEARANCE,--does the thought drive blood to surface ex
044r074 ll Bombs. without a vesicle. we may consider APPEARANCES of eruption at bottom.--solution under high
054r105 orth of Callao, called Marques, where in all APPEARANCES not many years since, the sea covered above
147g020 probably than river-- No exact terraces but APPEARANCES, as if valley had been filled with sloping b
318c252 & all hares on East side have other peculiar APPEARANCES-- Now this is precisely the case with the mi
122a115 Gregory Bay). Shropshire case where lamination APPEARED.-- Lyells Denmark -- L'Institut (1838) p. 26
147g017 for Marine remains-- Some of the hills almost APPEARED as if they belonged to double series Whole ve
218b190 dinburgh. Journal of Natural History-- Preface APPEARED good with facts about changes when animals tr
219b193 plants <grow closely> When this volcanic point APPEARED in the great ocean, have made plants of Ameri
280c135 ell said to me, the fact of existence of mules APPEARED to him most strange.-- This even might be sai
290c164 nnot doubt laws of change, Will be known.-- It APPEARED to me that half between fowls & pheasants, is
333d005 nown common house cat.-- had four Kittens. two APPEARED «so» very like common cat, that they were kil
333d007 <likewise more resembled the wolf than dog.--> APPEARED to be intermediate between two Parents.-- thi
453e183 k Dr Holland cases where peculiarity has first APPEARED.-- "Storia della Riproduzione Vegetale". by G
472s02v cker & <I think> pollen was scraped off, which APPEARED like Heartease pollen.-- the pollen appeared
472s02v h appeared like Heartease pollen.-- the pollen APPEARED chaffy, as if sucked?! opens & shuts end of s
511q018 d this breed.-- Have secondary male characters APPEARED.= (4) Does he know any seed-raisers (5) List
535m064 thus acting 4 to 6 weeks. The deserted broods APPEARED healthy-- This remarkable case may be normal.
594n112 wild-ducks would have fled equally if man had APPEARED-- though instinct so firmly implanted, birds
026r021 line of veins & cleavage is importants; veins APPEARING a galvanic phenomenon, so probably will the C
090a025 granite showing schistose structure (& veins APPEARING): mem. Henslows Anglesea solution of silex al
222b207 lk of the wonderful event of intellectual Man APPEARING..-- the appearance of insects with other sens
318c254 ry near coast & others the mountains, & these APPEARING to remain about a fortnight, «See Silliman's
035r046 ted in different basins; little or no relation APPEARS to <exist> be made out, but in those belonging
068r147 C, now grown solid.) Red Sea near Kosir, land APPEARS elevated. Geograph. Journal p 202 Vol IV When
074r164 opal:-- Veins in Limestone & Grauwacke: Silver APPEARS far more abundant in the upper limestone, whic
095a038 ect crater <--> near summit,-- much pumice --. APPEARS to be outside of the Cordillera-- Near the Pla
108a080 o many shocks directly after great shock -- It APPEARS to me unphilosophical to think calcareous spri
132a139 . Parrots own hypothesis some such explanation APPEARS to me necessary) as M. Parrots shows from vari
156g071 y 29.581 A 82 75 degree? From this point plain APPEARS like one uniform slope slightly bending up eac
160g093 ut narrow shelves just like road of Glen Roy-- APPEARS to lip with moss On this terrace «station perh
161g095 agments, then clear. this bit to eye certainly APPEARS level with road, & with piece of excised rock
161g097 above it-- (NB the buttress or pass at Isthmus APPEARS above level of shelf certainly) I took another
170b002 f what the original molecule has done).-- This APPEARS highest office in organization (especially in
185b055 w Species. Cuvier examined it. There certainly APPEARS attempt in each dominant structure to accomoda
190b077 die away, & partake of Indian character There APPEARS in Australia great abundance of species of few
196b110 xistence of plants, & their passage to animals APPEARS greatest argument against theory of analogies.
198b113 ments p 66]CD With unknown limits, every tribe APPEARS fitted for as many situations as possible. for
201b129 located] The relation of Analogy of Maclay &c. APPEARS to me the same, as the irregularities in the d
202b131 then forest 120 ft Micaceous rocks. subsidence APPEARS indicated.--- p. 36.-- Geograp. Journ. Vol IV.
246c027 snakes on isles of central Pacific, yet there APPEARS to be one at Botouma from account of natives,
276c122 varieties laws of change-- whether beak (as it APPEARS to me). colour of plumage & laws, which might
283c144 nts our ignorance is indeed profound & such it APPEARS.-- Is there not some statement about diversity
306c213 tralia» by a bank of soundings of which «there APPEARS to be one line, in which» greatest depth <appe
306c213 ears to be one line, in which» greatest depth <APPEARS to be> is not more than 60F. & «in» the whole
310c222 of facts «analogous to petrel-grebe. external» APPEARS to be a puzzle against my theory,-- If I be as
316c244 here are some restricted genera, but then they APPEARS always very small ones as Trigonia in Australi
326c266 nt diseases. 4to 1801.-- quoted by do.-- There APPEARS to be good art.. on <Etn> Entozoa by Owen in E
342d033 e offspring produce heterogenous offspring. It APPEARS certain that hybrid Muscovy & Common duck have
362d100 765? Treatise on Domestic Pidgeon, in which it APPEARS that all the <bird> varieties «,now know» were
364d104 pheasants have white feathers).-- It certainly APPEARS in domesticated animals, that the amount of va
388d172 -- stuff.!-- How much opposed. the Quagga case APPEARS to that of «2» dog begetting different puppies
417e077 es.--]CD CD[The similarity of child to parent APPEARS to follow same law in two of «the» same <speci
436e136 - : then comes question of genera It certainly APPEARS that swallows have decreased in numbers, what
450e175 the Sambawa, Java & Sumatra breeds, (.Hence it APPEARS there are shades of difference in all the isld
452e181 ated-- spins thread of cotton.-- do p. 583, It APPEARS probable,«?» that the Hippopotamus occurs in I
460t015 l at Hornsey (though only ⅓ of blood). that it APPEARS about half way between swan-goose & common goo
465t079 ust like the alpine plants-- In S. America. it APPEARS from Lund more Mammals, than at present «in Eu
515q021 n» which some organ is absent by abortion, but APPEARS in abortive state either in the species, or in
544m103 , & senses tell us it is not real.= = dreaming APPEARS clearly rest of the mind, with all other facul
547m112 Castle would not have turned into dream.-- It APPEARS to me, that the mind is wholly absorbed with o
551m127 red spectral illusion & insanity the connexion APPEARS to me vague-- Delirium of every degree of inte
564n005 y Metaphysic, as they have always been studied APPEARS to me to be like puzzling at Astronomy without
572n033 & <running> «running» over imaginary words: it APPEARS as if the mind had dwealt on each word separat
588n090 ng emotions of reverence & piety are felt." it APPEARS to me mere consequence of stooping, as sign of
599o005 e signs originally muscical!!!??-- At least it APPEARS all speculations of the origin of language.--
605o19v y to such emotions. this same term.-- Hence it APPEARS, that when certain causes, as great height, et
605o19v ations before mentioned. we call sublime.-- It APPEARS to me, that we may often trace the source of t
614o037 Heredetary memory not so wonderful as at first APPEARS, & no too great advantage.; for superiority of
625o052 y habit-- [LHC] 11) Whewells preface. [RHC] It APPEARS that Sir. J. & others think there is distinct
626o052 gulates feelings, as of cowardice.-- the whole APPEARS to me rather rigmarole.-- He does not say anyt
251c041 ura-- new genus of Mam: found in Sumatra p. 452 APPEND to Denham Clapperton &c on Mammalia no doubt w
429e114 om authors (Ramond. Hort. Transact Vol I. p. 17 APPEND) that in the Pyrenees, that the Rhododendron f
022r006 limits of this mineral in Australia. Fitton's APPENDIX Would Slate. & unstratified rocks show any di
190b076 go back-- there is an end to species.-- «Brown APPENDIX» A most remarkable observation of Mr Brown, a
269c102 tch of plants of New Holland, supplementary to APPENDIX to Flinders Voyage by Brown.-- great space se
317c248 ritish Aviary" or Bird Keepers Companion Study APPENDIX (& only appendix) of Congo Expedition, NB. I
317c248 Bird Keepers Companion Study Appendix (& only APPENDIX) of Congo Expedition, NB. I met an old man--,
317c251 worth studying, with respect to forms.-- Study APPENDIX to Tuckey's Expedition Journal of the Academy
318c253 wise of the Hirundo fulva (added by Audubon in APPENDIX) showing WHAT CHANGES are taking place & how

Page **(Key Word)**
321c275 o--- & Lisle's Husbandry Tuckeys voyage reread APPENDIX Ovington Voyage to Surinam. Voyage Congo expe
321c275 Voyage Congo expedition: Zaire except Brown's APPENDIX & excellent table of Canary isld: Plants Home
450e176 , & deer South part of Mildanao.-- Q Horse do. APPENDIX. p. 43. «& 45» the Breed of elephants <of> in
390d179 d, but the female at least (¿male?) looses all APPETITE.-- It is the comparison of each animal with i
619o043 .CD But should he prevented by some passion or APPETITE, what would be the result? In a dog we see a
619o043 result? In a dog we see a struggle between its APPETITE, or love of exercise & its love of its puppie
620o044 he will be forced to reflect on his choice: an APPETITE gratified gives only short pleasure. passion
526m030 avarice, &c &c An animal improves because its APPETITES urges it to certain actions, which are modifi
526m031 ich are modified by circumstances, & thus the APPETITES themselves become changed.-- appetites urge t
526m031 us the appetites themselves become changed.-- APPETITES urge the man, but indefinitely, he chooses (b
620o045 ation one gains the rule, that the passions & APPETITES should «almost» always be sacrificed to the i
620o046 n when tempted not to do so, by other natural APPETITES.-- he is monster, or unnatural if malevolent,
625o052 egulates our feelings steadily & not like our APPETITES & passion, which receive enjoyment from grati
390d178 ut it fails.-- therefore «each» seedling of one APPLE ought to differ from those of other.-- The upsh
401e014 turnips or other stocks. Says if any variety of APPLE be sown, all sorts come up from it. lately saw
401e015 varieties may be due to impregnation from other APPLE trees.-- now seeds of crab produce crab, so tha
401e015 of crab produce crab, so that some effect from APPLE trees is produced.-- Thinks probably experiment
401e015 obably experiment was never tried of separating APPLE tree entirely from all others-- so my experimen
427e110 en of any kind would fertilize it» fertilize an APPLE somewhat more readily, «than other apples» but
427e110 elibly stain offspring-- it would not reach one APPLE sooner than <other:> «that of another» apple. o
427e110 ne apple sooner than <other:> «that of another» APPLE. only effect produced would be different.-- sam
491q01v mongrelized affect other branches-- The French APPLE tree «with abortive stamens» answers first ques
504q013 rsts before flower is open-- -- No (6) There is APPLE with branch in middle of tree with flowers near
504q013 der-nut ‖. Laburnum ‖. Dodecatheon ‖ . Castrate APPLE & pear to see if pollen naturally carried, on a
401e017 ers. by chance <often> sometimes graft pears on APPLES. they will live but not flourish-- a medlar ma
427e110 ize an apple somewhat more readily, «than other APPLES» but probably would more indelibly stain offsp
491q01v d symmetry in cone-- The «above Exper» explains APPLES on side near other tree being affected.-- does
515q021 same question with regard to Primroses. (4) Do APPLES "sport" in fruit, or time of leafing (5) Do th
459tf02 admitted by Linnaeus.-- seems to doubt its APPLICABILITY to common mule & hinnus-- in one case basta
025r018 over S. America & Europe. we may believe them APPLICABLE to the world.-- My general opinion from the
048r085 ceros lives in S. Africa: the same caution is APPLICABLE to the Siberia case We must not think alluvi
052r099 ia «Lat degree ( )», he has rocks on surface. APPLICABLE to Patagonia. During a period of subsidence
065r139 of inland bays. like St. Julian & Port Desire APPLICABLE to Craters of Elevation.--The longer diamete
096a040 percieved on which side craters were low --¿ APPLICABLE to Auvergne??? The fact of Galapagos Isld. s
100a052 .> {P} on the diagonal of BK.-- ‖ This is not APPLICABLE. it does not explain CLEAVAGE of rock-- nor
105a070 og «1837» p. 320. paper on shrinking of Clay. APPLICABLE to Cleavage. C. Prevost.-- In Cordillera. a
107a079 onductor) Fitz Roy's Case of S. Maria & Tubul APPLICABLE to Andes & Patagonia-- On Lyells idea of who
136a147 Ladner Vol. II p. 80-- some remarks on dikes: APPLICABLE to Cordillera Phillips in Lardner Vol II. p.
152g048 t, & afterwards with greater cut through, not APPLICABLE to Glen Roy Lake, must have remained very lo
181b044 th Nature: Cuidado The above speculations are APPLICABLE to non progressive development which certain
195b099 w birds do arrive & seeds.-- The same remarks APPLICABLE to fossil animals same type, armadillo like
221b200 undergone most metamorphosis Islds X Is this APPLICABLE to insects &c &c?-- (p. 23 do.-- On animal -
230b235 in California cessation of female offspring: APPLICABLE to any animal-- Athenaeum. <Jan> «p. 154»--
257c059 mption? not solely producing like itself, not APPLICABLE to monsters:-- Are monstrosity heredetary??.
259c063 ace of plants--Lamark's willing absurd, ∵ not APPLICABLE to plant, Epidemics of South Sea, wonderful
268c100 erial &c type?-- This of consequence, because APPLICABLE to N. Hemisphere (NB.-- Examine Abrolhos Flo
269c101 h, (& Erman's surprise that it is not 700) is APPLICABLE to metamorphs theory suppose when rhinoceros
273c112 xes <are> «have» same plumage.-- <no> this is APPLICABLE to swallow-hawk, «this not the case in swall
273c112 that this preeminent structure is not always APPLICABLE to same habits, though swallow hawk, ,milvul
280c134 - instincts of many kinds in dogs, is clearly APPLICABLE to formation of instincts in wild animals, m
297c186 be acquired by my theory-- else my theory not APPLICABLE [not located] p. 428. Ouzel sometimes builds
301c199 t squirrel using reason in hiding its food is APPLICABLE to any habitual action. even which Man perfo
301c199 xternal contingencies, where the habit is not APPLICABLE. The degree of development of all animals of
304c208 ng from a branch high up.-- this argument not APPLICABLE to apterix.-- but source of error for if som
332d004 ere he had been accustomed to turn down.-- -- APPLICABLE to birds migrations & Australian Savages.--
352d057 xpected to have larvae more perfect-- this is APPLICABLE to young of Cochineal?? Is there some law in
414e064 ave been exterminated on principles. strictly APPLICABLE to the universe.-- The range of man is not u
417e078 fathers to children of mankind, no doubt are APPLICABLE to likenesses, when species & races are cros
550m126 chance & will of Deity are confounded.-- well APPLICABLE to free will. Mayo. Philosop. of Living p. 2
574n042 trong arms, outliving the weaker ones, may be APPLICABLE to the formation of instincts, independently
581n063 I must do is to generalize it, & see whether APPLICABLE to all cases.-- & analogize it with ordinary
588n089 ons make the emotions.-- p. 272. Some remarks APPLICABLE to my theory of happiness.-- Bell on the Han
602o11v re my ideas of a general notion of everything APPLICABLE to the high idea «p. 131.» in Tragic acting-
639j28v 260 intimates canines no special use to Man. APPLICABLE to Bell's sneering-theory.-- p. 263. This ki
461t037 es than it would be eliminated, & hence, the APPLICATION of structure to purpose after purpose would
616o038 : ¿ False,-- secretion in both involuntary, <APPLICATION in> «ejection only has» will: there must be
044r073 on of instability of ground at present. day.-- APPLIED by me geologically to vertical movements. In C
194b096 f pollen from other plants because this may be APPLIED to show all plants do receive intermixture.--
265c087 ly not of the use to which wings are generally APPLIED.-- Therefore argument not destroyed even if th
585n075 ers, that we only do know that it is one, when APPLIED in peculiar manner.--]CD April 3d. 1839 The G
585n076 seeing the sun &c are absolutely useless when APPLIED to birds, which have been carried in hampers.
606o20v result, that the term taste is metaphorically APPLIED to this mental power Although taste must neces
102a060 do with arrangement of particles in rock. This APPLIES to cleavage & concretions.-- Septaria in concr
106a075 ut a narrower channel instead of wider.-- This APPLIES to all vallies (except mere talus «over cliffs
174b013 - This is answer to Decandoelle. (his argument APPLIES only to hybridity.-- genera being usually pecu
187b064 varieties is accomodation). Now this argument APPLIES to species.-- If individual cannot procreate,
187b065 ce time of extinct quadrupeds:-- same argument APPLIES to England.-- Mem. Shew Mice.-- --- Animals co
218b189 akes variety these are vitiated.-- This barely APPLIES to plants Female pig apt to produce monsters i
220b197 ne distinction of species would fail. But this APPLIES only to coition & not production. But who can
280c134 t authority because written on dog-Breaking.-- APPLIES it to national character.-- N.B. If two specie
389d173 necessity for change. in process by generation APPLIES only the more complicated animals.]CD p. 310 S
397e003 f nature with which we are acquainted."-- this APPLIES to one species-- I would apply it not only to
403e024 ch is absurd.-- their only escape is that rule APPLIES to wild animals only. from which plain inferen
514q21. osa?-- good-- Norfolk Isd-- geology. volcanic? APPLIES to my geology & Species theory-- peculiar Faun
528m038 ian to be single cause) this symmetry & rhythm APPLIES to the view as a whole.-- Colour «& light» has
572n032 g body» &c &c is matter of custom."-- this all APPLIES to bodily weakness & inferiority, but now we c
584n072 so called instincts to which my theory no way APPLIES.-- it is the acquirement of a new sense,-- bat
615o038 ncts nearly same; which same argument probably APPLIES to particular instincts of animals. even in wi
067r144 ossils decidedly belong to old Silurian system. APPLY degradation of landlocked harbors to Craters of
125a121 iew of transmission of heat by gases-- does not APPLY it to thickness of crust.-- {P} if crust were m
174b015 ood Hope-- Marsupials. at Australia-- Will this APPLY to whole organic kingdom, when our planet first
205b144 s no awk in Southern hemisphere. does this rule APPLY? A Treatise on Form of Animals by Mr Cline "The
205b145 st all the lambs will be hornless.--" does this APPLY to where same animal breeds often with same fem
230b239 e. but afterwards will not alter. This need not APPLY to very slow changes. without crossing.-- Now a
313c226 «very difficult case» Does this law of duration APPLY to utter extinction or rapidity of specific cha
335d013 o go back, or have none-- the argument does not APPLY to first parents, because they are not new bree
370d117 ish a point as a probability by induction, & to APPLY it as hypothesis to other points. & see whether
388d170 food.-- is real. difference-- but this does not APPLY to potato.-- With respect to offspring being de
397e003 inted."-- this applies to one species-- I would APPLY it not only to population & depopulation, but e
428e111 eared by seedlings.-- Now my principle does not APPLY to any plant reared artificially, & only very p
480z007 a & sexes not of different size-- How does this APPLY to pale brown Caracara Krauss on Corallinae fro
491q01v in rich soil & propagate from their seed 3. To APPLY pollen of different genus & then some hours aft
586n078 which way they go; & so return.-- «but does not APPLY to dogs.--» they may do all this instinctively
586n080 rence, which clearly ought to be separated-- We APPLY instinct to one part. or another-- but (an inst
588n090 faculty if not reason.-- or does this reasoning APPLY chiefly to recollection. yet a dog hunting for
604o018 reat power-- -- 2 From these & other reasons we APPLY to God the notion of living in lofty regions. 3
604o018 ssociation of power &c &c with height, we often APPLY the term sublime, where there is no real sublim
605o18v the original cause of these feelings & thus we APPLY to them the metaphorical term sublime 7 So that

Page
(Key Word)*
605o019 gh a complicated series of associations that we APPLY to such emotions. this same term.-- Hence it ap
622o049 RHC] But the love is instinctive, & how does it APPLY to mother loving child, from whom, she has neve
623o050 s instinctive. But does not Hartley explanation APPLY perfectly to origin of these instincts.-- the h
452e182 d by M. Tournal that skulls found near Vienna APPOXIMAT to Negro form; those from Rhine to the Caribs
617o39v r external what senses in the way in which we APPREHEND the force of inanimate bodies. How we identif
617o040 n the course of its DIRECTION, & thus when we APPREHEND force in inanimate matter we feel dissatisfie
617o040 force (be it remembered) being a phenomenon APPREHENDED by the same faculty with matter & being nece
617o40v orce exhibited in every> phenomena actually APPREHENSIBLE by sense. 5) There is nothing analogous to
119a106 c judging from what we see when trap in dike & APPROACH other rocks. & trap at least as hot as lava--
181b043 ing from one stock & obeying one law, they may APPROACH.-- some birds may approach animals, & some of
181b043 g one law, they may approach,-- some birds may APPROACH animals, & some of the vertebrata invertebrat
194b094 ection of fossil "singe", it cannot be made to APPROACH the Colobes which in south Africa, appear to
198b113 us developed. (Owen's idea) states these class APPROACH on the confines? Balanidae?-- --- I cannot un
223b210 ecies not formed. during ascent of mountain or APPROACH of desert?-- because the crossing of species
354d065 any varieties of sheep «evidently artificial» APPROACH in character to goats.-- or dogs to foxes. (y
400e011 arters: his ideas not clear. In Australia from APPROACH to Asiatic [...]t in part near Timor, & to Eu
421e090 rocco «Mr Forbes says the Fauna»-- (near Oran) APPROACH in character to Canary Isld.-- ie Canary Isld
528m035 sanity is only cured by forgetfulness.-- & the APPROACH to believing a vivid castle in the air, or dr
533m059 cious of recollecting it-- this may be nearest APPROACH to <the> such instincts which full grown men
250c038 l & Edentata increased most. Certainly Africa APPROACHES Nearest to what is supposed to have been con
333d007 .-- this is very interesting as Esquimaux dog APPROACHES to species. Again he has seen several crosse
337d020 is picked out, & few even of local varieties APPROACHES quite to wild local variety.-- our Europaean
355d066 ts variety of horse, dun-coloured with stripe APPROACHES to ass.» or fowls to the several aboriginal
358d076 s little developed, it is always female which APPROACHES in character to the larva, or less developed
421e090 in character to Canary Isld.-- ie Canary Isld APPROACHES more to neighbouring coast of Africa, than t
602o011 al reason".-- This power of the mind, faintly APPROACHES to instinct How strange it, that Nature shou
034r045 . account of regular change in soundings. on APPROACHING the coast of NW. America P. 209--13 P & 444
041r064 and Seco at B. Ayres; mention about the deer APPROACHING the wells.-- the effect of Salt water of the
172b008 species; this we do in South America closely APPROACHING.-- but as they inosculate, we must suppose t
177b028 well characterized parts belonging to each)» APPROACHING another. <Petrels have divided themselves in
213b170 e most organized fishes lived far back, fish APPROACHING to reptiles at Silurian age-- How long back
230b235 gued against theory of changes that if so in APPROACHING desert country or ascending mountain you oug
240c004 arly point out two kinds of varieties.-- One APPROACHING to nature of Monster, heredetary. other adap
619o043 ill be presently shown.-- This then is moral APPROBATION, as far as it goes.].CD But should he preven
621o046 & likewise <that the> then receive the moral APPROBATION of his fellow men.-- [RHC] 6) Hence man must
583n070 mean are called "creatures of reason", more APPROPRIATELY they would be "creatures of habit."-- CD[as
537m075 l manal sense.-- «from difference of action of APPROVED» Yet as, I think, the opposite side has been
606o023 he beauties developed in a work of art are not APPROVED by the eye itself, but by the imagination thr
628o53v ed right & wrong,-- to be associated with the APPROVING or disapproving instinct-- which were not ori
411e054 life,-- the end of Natural History, will be APPROXIMATED to.-- Treating of the formal laws of corela
089a018 jours precedes et suivis, queque temps avant et APRES, par de petites secousses."-- Tom 54. p. 106 do
244c023 u Speaking of Lepus Magellanicus says; <after> "APRES un examen attentif, et forts surtout de l'opini
323c269 n Domestic pidgeons 30th Lives of Hayd & Mozart «APRI 25th Lockarts life of Napoleon.» April 5d Dr. E
516q023 . Experiments in Garden Sow stones of Standard APRICOT grafted on what, & see what comes up.-- [Unnum
067r143 lera & on Eastern plains «by Antuco». Athenaeum APRIL 1836 (p302) Coleccion de obras. 2 Vols fol: Bue
112a090 aph. Journal.--» A meeting of the Geograph Soc, APRIL 9 1838. Letter from M. Erhman stating that the
170bIFC out Dec. 7th. /1856/ (& again looked through APRIL 21 1873) p. 26 30 41 46 50 54 56 67 69 76 79 91
183b052 of man.-- M. Flourens. .Journal des Savants.-- APRIL 1837. p. 243 it is said as well known fact that
184b054 atherium.-- uncle now dead. Bulletin Geologique APRIL 1837. p. 216 Deshayes on change in shells from
239cIFC ences selected Dec. 13 1856 Also looked through APRIL 23. 1873 Books About amount of difference: wher
323c269 & Mozart «Apri 25th Lockarts life of Napoleon.» APRIL 5d Dr. Edwards of <tær> influence of Physical c
327c265 <Edin> British & Foreign Medical Review No XIV. APRIL 1839.-- Review on "Walker on intermarriage" pri
432e126 Pamplet. (published in Philosop. Journal <Mar> APRIL 1st 1839) by Sedgwick & Murchison; which is a b
434e129 theory explains why it is sure guide.-- Lychnis APRIL 3d.-- Henslow tells me following facts: believe
435e133 s, when pollen even most remote is put to it.-- APRIL 6th "Dr. Edwards on the Influence of <external>
436e136 but not so much these, because circumstances» <APRIL 12th..> Cestracion, Port Jackson Shark-- Owen t
448e168 St Thomas on W. coast of Africa Owen Linn. Soc. APRIL 2d. 1839 The Lepidosiren-- Amblyrhyncus & Toxod
451e178 ed. "take tame animals into this region between APRIL & October & like man almost (this looks inaccur
585n075 is one, when applied in peculiar manner.--]CD APRIL 3d. 1839 The Giraffe kicks with front legs & kn
595n121 ts acting on us-- <The Baby> «Effie Wedgwood» APRIL 28th 1840 was frightened at wild beasts in Zool
051r095 ether an animal will adhere to a certain part. APROPOS to question does animal adhere to rock because
196b104 dine genera, Bats .Foxes. Mus are birds that are APT to wander & of easy transportal.-- Waders & Wate
199b119 est.--In first settling a country.-- people very APT to be split up into many isolated races. ¿are th
218b190 ated.-- This barely applies to plants Female pig APT to produce monsters in Isle of France-- -- Madag
347d049 variety will be harder to vary, & therefore more APT to be extinguished.--???» Mayo (.Philosop of Liv
408e045 rt of coast of Somersetshire the Cockles are all APT to be diseased., & some of them symmetrically.--
515q021 flower are reduced from normal number, they are APT to vary in number in individuals of same species
532m054 olt has once been frightened & started much more APT, this partly owing to heart? readily taking same
557m149 ce shy people (shame of ridicule) are singularly APT to catch tricks.-- so are people in passion my F
570n024 e respiration when immensely immersed-- mechanic APT to sigh.-- & hence carried on as trick) «Shruggi
577n052 hinking people. do not blush.-- sensitive people APT to blush.-- -- The power of vivid mental affecti
255c051 reat difficulty in propagation.-- Feathers on, APTERIX because we may suppose longest part of structu
263c074 narians to deceive himself.-- Give the case of APTERIX-- split, depress & elevate & enlarge New Zeala
263c074 e enlarge New Zealand; a division of nature of APTERIX, many genera & species-- The believing that mo
304c207 ? A Question of immense difficulty is, whether APTERIX descends from same parent with other birds,, o
304c207 but then there may have existed series between APTERIX & other birds.-- will having many trifling cha
304c208 nch high up.-- this argument not applicable to APTERIX.-- but source of error for if some of the ostr
307c215 an produce in sex, what she does in species of APTERIX This is important for if these abortive wings
352d059 law.-- Penguins wing perhaps not abortive???. APTERIX certainly.-- Lyell's excellent view of geology
436e135 seem to have decreased since earliest times-- APTERIX has a most perfect Struthio head pulled out. y
509q017 ccygis-- Turbinated bones? False ribs Wings of APTERIX clavicle in--? Combs in combless Poultry-- Te
189b070 as in space. (Mem: Galapagos). Little wings of APTERYX Dacelo & Kingfisher same colours Strong colour
211b162 s of pigs. difference in skeletons: VERY GOOD. APTERYX. a good instance probably of rudimentary bones
233b251 ingless birds S. Continents-- Ostriches. Dodo. APTERYX Penguin-- Logger-headed Duck-- Large proportio
240c003 on.-- fact analogous to Owen's Phil: remark of APTERYX having feathers.-- It is possible, time being
341d029 drawn-- thus the most remarkable character in APTERYX, small respiratory system; even much smaller t
464t065 uthomidous Bird from New Zealand-- <so> not an APTERYX, yet it shows the Apteryx is not «quite» isola
464t065 ealand-- <so> not an Apteryx, yet it shows the APTERYX is not «quite» isolated in its present localit
464t065 st other birds, with small wings, & surely the APTERYX is more closely allied to the Struthonidae tha
509q017 bula: in Man any abortive bones??? do. Wing in APTERYX ‖‖ no as Os Coccygis-- Turbinated bones? False
154g057 } Shelf of Glen Guoy flat peat plain divortium AQUARIUM-- tidal channel-- 12ft obscure obscure NB In
157g078 wards Loch Spey 29.297 A 79.½ 29.316 divortium AQUARIUM «about 12 ft higher than last station» 29.316
157g078 ation» 29.316 true terrace «2d» near divortium AQUARIUM is a lip with it-- Dick right-- Mac mistook te
158g079 -- Granite such as boulder on <thes> Divortium AQUARIUM Peaty Mass of this point very nearly like head
160g089 fect old Loch, making <several> two divortiums AQUARIUM, viz two branches of River Bought & between on
161g094 too low» (to test last on Peat-Mass Divortium AQUARIUM) Barom. 29.200 A.77 degree Air 70 degree? Baro
162g103 River Tarf <it> Has a very long, flat divatium AQUARIUM with, left of Bright.-- like bed of lake with
268c100 it like each great class of animals having its AQUATIC, aerial &c type?-- This of consequence, becaus
282c142 now if one of these races had become eminently AQUATIC--; «NB, aquatic, i.e relation to elements & no
282c142 ese races had become eminently aquatic--; «NB, AQUATIC, i.e relation to elements & not minding partic
360d093 rtaking of character of other.-- <so> the most AQUATIC & most terrestrial species, might be harder to
416e071 s?-- insects?.-- all!??!?-- Worms? [Barnacles. AQUATIC.. <yet> Crustacean, & true hermaphrodites] CD
422e092 t was a Pachyderm. which was the origin of the AQUATIC Mammifers» p. 306, the Dugongs cannot be unite
422e092 united with true Cetacea or whales.-- but are AQUATIC Pachyderms. & Walrus-- aquatic seal.-- (Consul
422e092 ales.-- but are aquatic Pachyderms. & Walrus-- AQUATIC seal.-- (Consult this passage, when considerin
458t001 um, with existing, or nearly existing forms of AQUATIC reptiles most strange, & shows as in shells so
486z019 ays only one Reptile in Kamtchatka (Salamandra AQUATICA). Compare with T del Fuego Compare birds of d
100a054 ner Encyclop.--» absolutely considers gneiss an AQUEO deposit resulting from disintegrated granite!!!

```
****************************************(Key Word)****************************************
490q001 m of other species Hooker says the species of AQUILEGIA vary much in their spurs & Ranunculus in  the
266c092 ack cattle. Not have hump like those of India & ARABIA p. 202-- sheep have not the enormous tails, wh
266c092 ot the enormous tails, which disfigure those of ARABIA & Egypt.-- CIVETS CATS only wild animals on is
266c092 e not even Antelopes, though common on coast of ARABIA not even antelopes though common on islets off
266c093 - Vol VI. p. 89.-- Lieut Wellsted "on coast of ARABIA between Ras Mohammed & Jeddah". sheep numerous
182b048 89.-- Lieut. Wellsted obtained many sheep from ARABIAN coast. "These were of two kinds one <with> whi
266c093 not even antelopes though common on islets off ARABIAN Coast.-- Vol VI. p. 89.-- Lieut Wellsted "on
309c221 uotes Burkhardt to show black colour of certain ARABS.-- NB avoid quoting these hackneyed cases Mr Ed
071r156 oduct.=> <Did Peruvian Indians use arrows or ARAUCANIANS?--> If wood now preserved over world Dicotyl
217b187 in  Van Diemen's land and Tierra del Fuego.-- ARAUCARIA, species. Brazil {P} Chile, Norfolk Isl.-- Is
080r178 st to Seaward partaking of the character of a ARAUCARIAN tribe, with point affin of yew & intermediat
416e071 ence great variation in each birth) from man ARBITRARILY destroying certain forms & not others.-- Ter
195b098 outh.-- (& geographical <distri> division are ARBITRARY, & not permanent. this might be made very str
262c074 re power (2.typical 3.subtypical) where power ARBITRARY. leaves door open for Quinarians to deceive h
281c137 up his species & genera very finely" show how ARBITRARY & optional operation it is.-- show how finely
640j167 , fortunately, to take their ideas, which are ARBITRARY & empirical, from their own Faunas, which  in
543m099 an  absolute fool used to play regularly with D'ARBLAY of Christ of great genius, & yet invariably us
061r128 ontrast low limit of Palms, evergreen trees, ARBORESCENT grasses, parasitic plants, Cacti: & with lim
122a113 th.-- study different forms of earth as shown by ARC.-- read Herschels astronomy with oscillations of
266c092 for some time to flower at their own periods.-- ARCANA of Science & Art. 1831. p 160. account of Bulb
028r031 h <I did not see one dike in the whole Galapagos ARCH; because no sections> same cause as no colour S
031r038 cal structure. East India Archipelago. «Aleutian ARCH.--» V. Fitton. Australia: cases in Europe.-- Au
128a128 l first fall-- the problem will be falling of an ARCH weighted in its centre.-- Will not abrasion  of
241c015 common  to New Guinea & rest of isle in E. Indi: ARCH: In New Zealand. a sturnus of American form-- a
242c017 lt, Reinwardt «Forrest» authors on E. I«ndian». A«RCH.» Borneo & Sumatra both seem to have  elephant
243c019 representatives (& instances given) in East Ind. ARCH:-- Birds of New Zealand absolutely different.--
250c038 een formed & spread to other Africa & East India ARCH.-- but where these great animals had not spread
450e175 .-- Ox & hog natives of Borneo Notices of Indian ARCH. Singapore 1837. By J. H. ? do. p. 189. «190» N
500q10a aces or doubtful species; how is this at Canarys ARCH-- it is so at Galapagos.-- Ireland, doubtful sp
558m152 estation of great passion.-- I do not think they ARCH their backs-- Bengal tiger. when slightly angry
558m152 slightly angry. curls tip of tail.-- do two cats ARCH their back when fighting, & not with dog. when
558m152 when  fear might enter?-- I believe common Swan, ARCH raises neck & depresses chin-- strikes with win
486z020 il H. Wedgwood says in <14th» «13th.» Vol of ARCHAEOLOGIA arrow=heads described in Suffolk as lying u
557m147 - dog tail curled when angry & very stiff. back ARCHED. just contrary. when pleased tail loose & wagg
557m147 accid, when furious «with fright» back absurdly ARCHED. & tail stiff.-- is shame, jealousy, envy  all
599o006 he thoughts expressed in Fingals cave, & in the ARCHED & leafy forests" Very good!. I grant that  the
558m153 aises neck & depresses chin-- strikes with wing ARCHES wings-- as does black Swan.-- Goose do all spe
021r005 e confervae, is very common within E. Indian ARCHIPELAGO, no minute description, calls it a Fucus.  P
023r009 s distant! Where are Hippotami found in that ARCHIPELAGO? Such have never been observed in  Australia
031r038 ld world of symetrical structure. East India ARCHIPELAGO. «Aleutian Arch.--» V. Fitton. Australia: ca
043r070 account of extensive areas. -- P. 322 In any ARCHIPELAGO. & neigbouring Volcanos. eruption from «more
188b067 to Swainson to certain islets in East Indian ARCHIPELAGO. Dr Smith considers probable two Northern spe
191b080 & down.-- Elephants in Ceylon-- East Indian ARCHIPELAGO.-- West Indies = Opossum & Agouti same as on
191b081 orter or change greater-- In the East Indian ARCHIPELAGO it would be interesting to trace limits of l
211b164 The  distribut of big Animals in East Indian ARCHIPELAGO--, very good in connection with Von Buch Vol
225b219 us fact. of absence of quadrupeds East India ARCHIPELAGO very good on opposite tendency.-- Study Elli
240c013 The New Holland species are not found in the ARCHIPELAGO-- Former statements to such effects false In
242c017 rt instances of American forms in East. Ind: ARCHIPELAGO. Raffles. Horsfield. Diard. Duvaucel. Lesch
264c080 &c-- but of course they might be blended, if ARCHIPELAGO turned into continent &c &.-- There is beaut
327cIBC aean plants transported.-- Crawford. Eastern ARCHIPELAGO.. probably some account Raffles. Sir. S do. d
402e019 d be worth skimming over with regard to this ARCHIPELAGO Octob. 13th.-- Kotzebues first Voyage. Vol I
407e039 I argue from this that Africa «& East Indian ARCHIPELAGO»-- formerly were not so «very» EQUABLE, or s
408e045 - Mem. elevation & subsidence of East Indian ARCHIPELAGO. now rising. On a particular part of coast o
450e174 t by domestication.-- "Notices of the Indian ARCHIPELAGO" Published at Singapore in 1837. by Mr. J. H
452e181 singular  thing that throughout the Moluccas ARCHIPELAGO they are only to be found on the isld of Bat
569n022 y Waitz (In Theil. V) in describing Caroline ARCHIPELAGO. Dn 75 cf., p 268 without, however, very sin
191b080 & Australia From the consideration of these ARCHIPELAGOS ups & downs in full conformity with Europae
599o006 rigin.-- Mayo Philosophy of Living. p. 264. "ARCHITECTURE is a fine amplification of two ideas in nat
603o11b n garden.-- yet both beautiful! p. 136. Says ARCHITECTURE does not come under imitative art [my  view
379d151 to  continent-- Original Paper, worth studying. ARCHIV. fur. Naturgeschichte. September 11' Generatio
087a008 opaedia-- Lately elevated When Siberia went up. ARCTIC land went down.-- Probably more Arctic land wo
087a008 went up. Arctic land went down.-- Probably more ARCTIC land would be required to produce climate rese
208b153 ck. & Lyell. most curious law of species few in ARCTIC. in proportion to genera. agrees with late pro
226b221 w species. The number of genera on islands & on ARCTIC shores evidently due to «the» chance of some o
265c084 ine, ptarmigan hare becoming white in winter of ARCTIC countries few will say it is direct effect, <o
373d133 . Dr. Beck on numerical proportion in shells in ARCTIC Ocean. p. 350 Grallae in Wealden. oldest birds
427e109 the climate being now less extreme, than before ARCTIC forms would retreat: effect on snow of  arctic
427e109 e arctic forms would retreat: effect on snow of ARCTIC climate in far north regions? Arctic forms hav
427e109 on snow of arctic climate in far north regions? ARCTIC forms have travelled S. From the analogy of th
448e166 ag & Touraine beds, the one with neighbouring & ARCTIC sea, & the other with neighbouring & Senegal a
641j29r ones «of large size» <to> are best nourished by ARCTIC regions-- Whales. «Narwhal» Polar bear. Walrus
641j29v mains in the Wealden? In the strongly separated ARCTIC genera, there is evidence of antiquity & extin
579n059 ars A man shivers, from fear, sublimity, sexual ARDOUR.-- a man cries from grief, joy. & sublimity. J
045r077 pes explanation of low Barometer? In a subsiding AREA. we may believe the fluid matter instead of aff
061r128 ≠ as if any creation «taking place» over certain AREA must have peculiar character: Contrast low limi
099a048 ncretions more vertically than laterally. -- <In AREA of this> {P} If surface covered with oil should
258c059 lation, from general circumstances effecting the AREA equably.-- Animals having wide range, by preven
306c213 rs to be> is not more than 60F. & «in» the whole AREA,. 120 is greatest (about 200 miles  distant).--
409e048 basis  of argument of change.-- now take greater AREA of water & snow line descent. My theory gives g
499q010 tioned by Humboldt in his account of extensive AREAS. -- P. 322 In any archipelago. & neigbouring Vo
043r070 ological table the precise periods over immense AREAS. (& the counterbalancing variations) of rain. =
055r108 ns.-- We must suppose everywhere--, in granitic AREAS &c &c volcanos {P} fissure dike.-- thus dikes t
104a061 PP Andes mark the line between sinking & rising AREAS.-- In Earthquake: if Subsidence we should not ex
108a079 (double influence) & mankind must improve-- The AREAS of subsidence marked out by animals of same <sp
309c220 y animals of same <spe> genera. is not equal to AREAS of elevation: Marked out by existence of elevat
309c220 A little time after a bad earthquake in Chili; AREQUIPA in 82 was overthrown, & 86. Lima. next year Q
058r116 f anticlinal violence crossing lines of crater, <ARG> state that all the great Volcanoes. have been e
038r059 mon there accompanied by molybdated lead & «ARGENTIFEROUS lead»; sulfated Barytes very «un»common  in
075r167 lenite.-- in New Spain, contrary to Europe. ARGENTIFEROUS lead not abundant. = considerable  quantity
076r169 er mass of well rounded pebbles in yellowish ARGILLACEOUS or sandy soil-- These Buttresses formed ves
146g014 rica, when Mammoth & narrow toothed Mastodon.-- ARGUE against the prejudice of not believing recent e
086a004 d, they would be far off In Discussion on dikes ARGUE impossibility on fissure going right through su
104a067 y are varieties, (though that would be best).-- ARGUE the case theoretically if animals did change ex
281c137 e if creator had so created them.-- People will ARGUE & fortify their minds with such sentences as "o
283c145 gous characters, might be multiplied.-- we must ARGUE reversely. WHERE CHARACTER VARIABLE it is  (one
302c202 ess fertile. according to Mr Herbert's views.-- ARGUE <argue> case of abortive organs to mules in the
309c219 tile. according to Mr Herbert's views.-- Argue <ARGUE> case of abortive organs to mules in their geni
309c219 hip, children of one parent, races of animals-- ARGUE «opening» case. «thus» Educate all classes-- av
313c234 law of mammals shorter duration, than molluscs, ARGUE case both in Europe & S. America. «very difficu
333d009 ect» spaniels & setters are produced. one would ARGUE the whole effect of race was determined by male
354d065 pensis. Philippines Man have varies the range-- ARGUE the case of Probability. has Creator made rat f
355d066 the several aboriginal species «or ducks» (here ARGUE if it be said domestic fowls are descended from
407e039 ve survived this mundine change..-- Therefore I ARGUE from this that Africa «& East Indian Archipelag
436e136 within  certain limits, but beyond these not.-- ARGUE against this-- -- analogy will certainly  allow
443e155 fluid  sperm.-- ]CD I utterly deny the right to ARGUE against my theory, because it makes the world f
444e158 ;-- or the will of the Animal. p. 145. Seems to ARGUE, that as the transformations from the egg, or l
465t080 h a casualty as, bones of Mammalia in caves:-- :ARGUE first case of bones (New Red Sandstone) &  then
```

```
553m136 rding to M. le Comte).-- Those savages who thus ARGUE, make the same mistake, more apparent however t
554m141 take instance of Dray Horse going down hill.-- (ARGUE sophism of association. Kenyon, & then go on to
564m005 body.-- we must bring some stable foundation to ARGUE from.-- Octob. 4th. Seeing some drawings «in La
195b103 mbers & distribution of the species!! It may be ARGUED representative species chiefly found where bar
230b235 . <Jan> «p. 154»-- 1838. Hybrid Ferns It may be ARGUED against theory of changes that if so in approa
250c039 es, must be abstracted & answered Much might be ARGUED what is not cause of destruction of large quad
350d054 ture is the art of God" Septemb I,. It has been ARGUED Man first civilized. <note> add this in  note.
419e085 in the World»,-- the production of vitality, as ARGUED by Müller from propagation of infinite numbers
443e156 r lives-- Being myself a geologist, I have thus ARGUED to myself, till I can honestly reject such fal
554m141 enyon, & then go on to show, that if Cart horse ARGUED from this into a theory of friction & gravity.
619o042 s.-- [LHC] ----- p. 113. Mackintosh Grotius has ARGUED nearly so [RHC] The history of every race of m
171b002 er life has not come into play)--See Zoonomia ARGUEMENTS, fails in hybrids where every thing else is
103a065 to Hopkins theory.-- general presence of dikes. ARGUES in favour of pressure of liquid rock. Andes di
216b180 h says he is sure of the case at Cape.-- McClay ARGUES from it Black & White species.-- For, says  he
535m070 m» as fixed laws of organization.-- M. le Comte ARGUES against all contrivance-- it is what my  views
537m075 ad effects Martineau. How to observe, p. 21-26. ARGUES «with examples» very justly there is no univer
634j54v & abandoned. Such is Man's philosophy. when he ARGUES about his Creator! p. 309. says the ribs in Dr
576n049 e means by which an instinct is transmitted.-- An ARGUING from man to animals is philosophical. viz. man
258c062 s proof of the non creation of animals.-- then ARGUMEN May be.-- subterranean lakes, hot spring &c &c
030r037 llied to the jaws of the Cocos fish Rio Shells ARGUMENT for rise In Cordillera, the dikes do not gene
038r058 up.--the appearance will here be the strongest ARGUMENT:--¿ Consider causes for subaqueous crater bei
038r059 n of water probably not so much aluminated. As ARGUMENT in favor of lines of anticlinal violence cros
056r111 larity of Volcanic products «over whole world» ARGUMENT, as well as separating causes by water.--Or r
062r129 death not to change of circumstances; reversed ARGUMENT. knowing it to be a desert.-- Tempted to beli
063r131 : inosculation alone shows not gradation;-- An ARGUMENT for the Crust of globe being thin, may be dra
087a012 paean risings. Pacific great land. -- Will use ARGUMENT of proof of slow corrosion of valley of <Pata
103a065 mass.-- Absence of Caverns, in Plutonic rocks ARGUMENT against great bodies of vapour. according  to
106a075 reat over) with very gently sloping sides This ARGUMENT is partly taken from Delabechs Theoretical Re
109a083 ervation certainly is lessened.-- Coral flats. ARGUMENT for Heaping up.-- very good this will show ef
109a083 to  broad flat sand beach. {P} -- De la Beches ARGUMENT of low coast gaining & high loosing  answered
124a120 . Elements p.119 on such strata {P} do p. 171. ARGUMENT against lateral injection. from probability o
131a136 but  turned over in many parts of the world.-- ARGUMENT strong in favour of thin crust theory.-- What
132a138 resumè  p. 536)-- «NB. I cannot understand the ARGUMENT, that cold «oceans» «lakes» bottom. if not co
132a138 continents. it ought, (according to M. Parrots ARGUMENT against central heat to warm the ocean).-- an
152g050 llarig up within 200 ft of level of 4th shelf= ARGUMENT against river--composition &-- stratification
152g051 against  river--composition &-- stratification ARGUMENT detritus-- {P} where buttresses on 4th shelf:
153g051 cause of upper edge of cliff» Others below it-- ARGUMENT for lake «or sea» at successive levels--  {P}
174b013 America--  This is answer to Decandoelle. (his ARGUMENT applies only to hybridity.-- genera being usu
175b016 has happened I look at two ostriches as strong ARGUMENT of possibility of such change,-- as we see th
187b064 e even in varieties is accomodation). Now this ARGUMENT applies to species.-- If individual cannot pr
187b065 rated since time of extinct quadrupeds;-- same ARGUMENT applies to England.-- Mem. Shew Mice.-- --- A
194b096 from  other plants.-- Does not Lyell give some ARGUMENT about varieties being difficult to keep on ac
196b110 s, & their passage to animals appears greatest ARGUMENT against theory of analogies. States there  is
202b130 in one country is permanent in another.-- Good ARGUMENT for species not being so closely adapted. Nea
222b206 emains.-- The Father of all insects gives same ARGUMENT as father of Mammalia; but have improvement i
229b232 tted together.-- Hermaphrodite animals couple: ARGUMENT for true molluscs coupling.-- Dr. Smith's Inf
230b240 ill be overturned.-- Hence extreme difficulty, ARGUMENT in circle.-- Falkland Isd case good one of an
234b261 ect different stations;-- this would be strong ARGUMENT for propagation of species-- Again is there n
248c030 standing, will not readily breed together: The ARGUMENT must thus be taken, as «in» wild state (where
265c087 which wings are generally applied.-- Therefore ARGUMENT not destroyed even if these shrivelled  wings
276c124 its sometimes go before structures.-- the only ARGUMENT can be, a bird practising imperfectly some ha
294c176 in wild animals, & thinks Lyell has overlooked ARGUMENT, that domesticated animals change a little wi
294c176 ild animals-- but how to show species-- I fear ARGUMENT must rest upon analogy & absence of varieties
294c176 of  varieties in a wild state-- it may be said ARGUMENT will explain very close Species in islds. nea
296c184 brids between pheasants & Black fowl.-- use as ARGUMENT possibly some few hybrids in nature.-- «p. 47
304c208 r that it sprung from a branch high up.-- this ARGUMENT not applicable to apterix.-- but source of er
306c212 teach  it taste, but that is much more general ARGUMENT.-- & therefore down the stream followed ebb tid
307c215 e tendency in species, this is capital & novel ARGUMENT.-- Are there any abortive organs in neuter be
307c216 e converted «into female», it will be splendid ARGUMENT old female, turning into cock, abortive spurs
307c216 rgans even in Haustellata & mandibulata.--!! -- ARGUMENT, when general argument is extended from speci
307c216 ta & mandibulata.--!! --Argument, when general ARGUMENT is extended from species to genera & classes.
313c234 new genus have been made, & only species, good ARGUMENT for origin of man one.-- Is the extinction o
315c243 reat canine teeth.-- (This may be made capital ARGUMENT if man does move muscles for uncovering canin
315c243 muscles for uncovering canines) --. Blend this ARGUMENT with his having canine teeth at all.-- This w
335d013 pring will tend to go back, or have none-- the ARGUMENT does not apply to first parents, because they
352d058 n every thing.-- PURE HYPOTHESIS be careful.-- ARGUMENT for circularity of groups. When <species of>
370d117 degenerated.  as must spices &c &c The line of ARGUMENT «often» pursued throughout my theory is to es
373d132 ans» «mammae» in male Mammalia:-- ¿is not this ARGUMENT, for Mammalia recent creation.-- why. what te
380d153 e vis formativa had completed them-- (but this ARGUMENT is VERY WEAK without knowing whether if kept
383d159 occupy their proper positions,-- this would be ARGUMENT for developement of either.-- (Mammae or shea
405e035 long  separated, or having never [not located] ARGUMENT REAL of antiquity of reasonable cosmopolite m
409e048 rope. :the forms themselves have been basis of ARGUMENT of change.-- now take greater area of water &
411e055 under this head» 27th November When summing up ARGUMENT against my theory, doubtless, the presence of
415e066 ke every other animal.-- Would anyone raise an ARGUMENT against, my theory, should no fossil «very di
424e102 em varieties & what says Jenyns to it?-- -- In ARGUMENT of origin of Wolf, difference of mind is most
426e105 ies never reappear when once extinct-- Lyell's ARGUMENT about <Tertiary> Isld «neighbouring» formed i
432e124 s kittens we see same fact) go back, & this is ARGUMENT against Blyth's doctrine of young birds retro
433e126 y than during the deposition of the beds-- The ARGUMENT must be thus put, shall we give up whole syst
448e167 ant» with Touraine «which as L. says is strong ARGUMENT for their contemporaneous»-- how is this with
463t057 a parallel case) Waterhouse remarked, that any ARGUMENT for transmut, from one organ graduating  into
465t079 » vestige, is preserved in this country-- same ARGUMENT to India & Europe-- & Africa!,-- any negative
465t079 to  India & Europe-- & Africa!,-- any negative ARGUMENT against-- monkey-man, valueless.-- May not se
524m019 which  comes on from bodily causes.-- It is an ARGUMENT for materialism. that cold water brings on su
536m073 o may free will make change in man.-- the real ARGUMENT fixes on heredetary disposition & instincts--
536m073 incts--.-- Put it so.-- Probably some error in ARGUMENT, should be grateful if it were pointed out.--
543m101 hows that new instinct can originate.-- strong ARGUMENT for brain bringing thought, & not merely inst
544m104 or after he has useless. does not affect race. ARGUMENT for early education.-- fear of death!!! as Mo
558m151 kintosh on Moral sense & emotions.-- The whole ARGUMENT of expression more than any other point of st
574n039 arly adapted to subject see <A> I think this ARGUMENT might be used to show language had a beginnin
581n065 c, imitative of the things.-- CD I may put the ARGUMENT,, that many learned men seem to consider ther
583n070 lities of imitation & education may be used as ARGUMENT.-- for instinctive knowledge is not gained by
588n090 , as sign of humility.-- I suspect very strong ARGUMENT might be advanced, that animals have reasons,
589n091 uch hypothesis.-- see M.S. notes, where strong ARGUMENT in favour of brain forming the instincts,-- c
593n111 e consciousness. only extreme step of an ideal ARGUMENT held in one's own mind, & Dr. Hollands  story
599o005 hese speculations are utterly valueless-- then ARGUMENT fails-- if they have, then language was progr
599o05v origin  in names of People.-- Sound of words-- ARGUMENT of original formation.-- declension &c  often
613o036 instincts  being acquired, is a most important ARGUMENT, to show that they result from organization o
614o037 f so high a mind without further end just same ARGUMENT. without indeed we are step towards some fina
615o038 , so general instincts nearly same; which same ARGUMENT probably applies to particular instincts of a
618o41v , because both due to some common cause:-- The ARGUMENT reduces itself to what is cause & effect:  it
624o051 hings easiest become instinctive, this part of ARGUMENT fails, or rather is weak.-- [RHC] Better simp
035r047 rtiary. being less than secondary:-- consider ARGUMENTS for oscillation of level independent of miner
038r057 ings; hence it will be necessary to state all ARGUMENTS for believing that there must be a central co
055r108 sediment Look at St Helena!!-- There are some ARGUMENTS which strike the mind with force.--the  exact
118a104 proportion  of dikes have reached the surface ARGUMENTS against Herschel's view of cause of continent
173b010 form  (2). &c.-- <{> Then (remembering Lyells ARGUMENTS of transportal) <continents>' island near cont
225b216 ks "willing" doctrine absurd. (as equally are ARGUMENTS against it-- namely how did otter live before
274c119 n [not located] alone, but on all the general ARGUMENTS-- Lamarck was the Hutton of Geology. he had f
```

301c199 is conscious of web. feet.-- p. 7. Mr Blyths ARGUMENTS against squirrel using reason in hiding its f
558m153 mences.-- <All> Nearly all will exclaim, your ARGUMENTS are good but look at the immense difference.
584n071 pecker as an example fitted for climbing, his ARGUMENTS partly fall, when a species is found which do
627o52v -- Hume's Inquiry-- good abstract of Butler & ARGUMENTS of beneficial tendency of affections.-- If ev
271c106 habits heredetary whilst species have changed ARGUMENTUM ad absurdum. The creative American halo has
116a099 gue living in mouth of Plate. p. 26. Geology of ARICA <Schit> Schmidtmeyer travels into Chile p 29. g
483z013 302 Vol II p. 302. Vaginulus of Lima described "ARION" of Ascension. p. do.-- some S. American Reptil
060r124 arthquake at Demerara. The earthquakes "seem to ARISE from some efforts in the land to lift itself hi
308c219 generally different, because their difference. ARISE a good deal from climate & habits, & therefore
313c235 out impregnation, therefore sexual passion must ARISE after long interval very good case.-- habit is
535m070 sees this, one suspects that our will may <be> «ARISE from» as fixed laws of organization.-- M. le Co
536m073 t.-- My wish to improve my temper, what does it ARISE from but organization. that organization may ha
538m081 howing what a train of though[t] action &c will ARISE from physical action on the brain, renders much
551m128 smus» says in Phaedo that our "necessary ideas" ARISE from the preexistence of the soul, are not deri
558m150 t, as being less so will.-- May not moral sense ARISE from our enlarged capacity <acting> «yet being
558m151 e principle of charity.-- ¿ May not idea of God ARISE from our confused idea of "ought." joined with
594n112 e air at a distance.-- How can such an instinct ARISE?? «it would appear that an instinct long remain
308c219 abits which would. have formed them, would have ARISEN under different climates &c. Do I mean that id
404o225 d become obscure & therefore it might thus have ARISEN, & M. Edwards p. 330 distinctly states that th
541m089 ted with death!-- How has this instinctive fear ARISEN? 19th. When I went down to Woollich I was tryi
637j57v is puzzler p. do says inconvenience would have ARISEN had « <not> some» some insects <not> not been
528m036 agination, (as in hearing music), this probably ARISES from (1) harmony of colours, <whi> & their abs
530m046 nvoluntary memory, as in sleep.-- a new thought ARISES?? compounded of the involuntary thoughts.-- An
553m135 ning the direct will of the God (<thus> & hence ARISES the theological age of science in every nation
608o026 the way of Contingencies.--but his desire to do ARISES from motives.--& his knowledge that it is good
617o40v nly movement can.-- 4) the source from which it ARISES. But coming round to the <subjective> aspect o
023r010 he Volcanos of the chain of the Cordilleras as ARISING from «the expulsion of fluid nucleus through»
023r011 rtures still open.--The fault like appearance «ARISING from the manner of horizontal upheaval» of the
315c243 iated kinds of good news. discovery of prey.-- ARISING no doubt from want of assistance.-- crying is
622o049 r J) explains our love of another, as pleasure ARISING from association from having received benefits
505q014 ng= Put pot of boiled earth on top of House =ARISTOLOCHIA, plant wh require insects to impregnate it
325c267 e species) Wilson's American Ornithology Read ARISTOTLE to see whether any my views very ancient? Stu
160g092 aracter in these mountains & not ridges) between ARM of Glen Bright flowing into E. end of L. Oich, &
162g103 64 degree, air 60 «Evening do» The extreme right ARM of River Tarf <it> Has a very long, flat divatiu
372d130 ration.-- Why crab can produce claw. but man not ARM. hard to say-- if it were possible to support th
372d131 hard to say-- if it were possible to support the ARM of Man, when cut off , it would produce another
372d131 es likeness of twin bear on this subject? A mans ARM would produce arm if supported., <so> & in makin
372d131 n bear on this subject? A mans arm would produce ARM if supported., <so> & in making «true» bud some
403e022 Deformations are particularly common.-- without ARM, <skin> hands thumb,-- one leg, hare lip &c &c.
444e157 rus 60 or 70 bones in the paddle, yet all in the ARM are perfect.-- p. 144.-- Alludes to two theories
567n014 en fondling the keeper., clasping «& rubbed» his ARM. & show signs of affecting something like man. H
195b099 marks applicable to fossil animals same type, ARMADILLO like covering created.-- passage for vertebra
462tf4v . We see the same object gained by the Mataco-ARMADILLO & the woodlouse-- -- a good analogy-- sea-Cru
087a009 an latitudes.-- Will it be supposed that the ARMADILLOES have eaten out the Megatherium. -- The Guana
184b054 ding to my view. in S. America parent of all ARMADILLOES might be brother to Megatherium.-- uncle now
189b069 oof that structure is not simple adaptation, ARMADILLOES «&» & Megatherium. each with same kind of co
175b020 gatherium nothing. We may look at Megatheria, ARMADILLOS & sloths as all offsprings of some still old
256c061 seals, hence deer victorious deer, hence males ARMED & pugnacious (all order; cocks all warlike)» «t
266c091 f Muhammad, the last to extend, their dominion, ARMED alike with the Koran and the sword" |quote Whew
367d113 mmon in vegetable feeders. because males always ARMED in carnivora. Where females, are peacable-- (Me
259c066 th respect to my theory of generation, fact of ARMLESS parent not having armless child, shows than th
259c066 generation, fact of armless parent not having ARMLESS child, shows than there is reference to more t
366d112 on. -- «<as Hunter supposes with Monsters)» if ARMLESS cat can propagate, ie with the chance of two b
574n042 gous to a blacksmith having children with strong ARMS.-- The other principle of those children. which
574n042 ose children. which chance? produced with strong ARMS, outliving the weaker ones, may be applicable t
588n089 d up & down-- in latter case they struggle their ARMS-- do. p. 306 "the eyes are rolled upwards duri
595n115 comforted until the Keeper took it <her> in his ARMS & carried to see.-- [blank] A Dog «whilst» drea
612o035 b by physical laws of endosmic & exosmic juices. ARMS of polypus, show either local or general will,
538m079 f Irish Politics, «both bad jokers.--» the other ARMY officer, horticulture & religious sects.-- yet
559m155 British Plants A Volume published by Colonel in ARMY in Jersey.-- very curious facts abo
240c013 o such effects false In New Guinea. a Kangaroo d'AROE (Didelphis Brunii) which as yet had only been f
240c013 nii) which as yet had only been found in isle of AROE & Solor), «Vol I» likewise new species of Param
569n020 roaring for lion &c &c. (in same way alphabet. AROSE from letters, symbol of word beginning with the
157g075 which shelf 3d form beach of granite pebbles, & AROUND which shelf 2d «almost» forms it into island--
539m081 & read so intently as to be unconscious of all AROUND, yet there was no strain on the intellectual p
622o048 sees uniformly performed by the teachers & all AROUND him, will be paramount,-- hence the law of hon
570n024 igh.-- & hence carried on as trick) «Shrugging AROUSED acting» Octob 25. Why is modesty, mixed with t
026r021 cing that cryst of glassy felspar in Phonolite ARRANGE themselves in determinate planes ∴ such action
057r115 cip. of Sulph. B. all the infinitesimal cryst. ARRANGE themselves in planes. «Mem silky lustre» ask E
101a056 curious exception in Wealden.-- Would crystals ARRANGE themselves in that direction, in which most su
285c150 ion must be given It would not be difficult to ARRANGE children of same parents in a circle,-- & «her
048r085 To the Horse. = One might fancy that it was so ARRANGED from the forseight of the works of man Feelin
049r088 equent. as in dikes. In Granite great crystals ARRANGED on sides. V. Lyell P 355 Vol III. constitutio
102a060 eavage & concretions.-- Septaria in concretion ARRANGED in planes, case of separation.-- the branchin
278c130 series. Could I not give Catalogue of Mammalia ARRANGED according to my own methods. Dasyurus being f
370d116 e bred first, «then males--» How has this been ARRANGEMENT-- Neuters are true females, but with parts li
100a052 GE of rock-- nor the Falkland case, nor. the ARRANGEMENT of particles of granite in Henslow's Grit, y
102a060 are how little common Gravity has to do with ARRANGEMENT of particles in rock. This applies to cleava
176b023 ains. & subdivision <six> three more, double ARRANGEMENT.-- if each Main stem of the tree is adapted
177b027 but in lower classes, perhaps a more linear ARRANGEMENT.-- ¿How is it that there come aberant specie
252c045 specially great continents" Give Specimen of ARRANGEMENT, 3 <5> Species Rhinoceros Cape town good spe
267c095 ested.-- Important do p. 98. on a quaternary ARRANGEMENT of Cryptogamic plants.--Owing to plants not
281c139 . p. 140» I should think meaning of circular ARRANGEMENT was only so far true as avoided linear arran
281c139 ement was only so far true as avoided linear ARRANGEMENT the central twigs dying, affinities would be
286c155 o as. yet.-- We now know what is the natural ARRANGEMENT, it is the classification of <arrangement> r
286c155 al arrangement, it is the classification of <ARRANGEMENT> relationship; latter word meaning descent.-
286c155 is taken by Fleming as emblem of dichotomous ARRANGEMENT which is false There is same difficulty in a
292c170 any p. 566 wants to see absurdity of Quinary ARRANGEMENT let him look at abstract of Swainson on Clas
309c222 Blyth does not believe in circular or linear ARRANGEMENT.-- Thinks passages very rare., in anatomical
348d050 ressing natural affinities| Macleays plan of ARRANGEMENT depends on the organs judged to be of import
348d051 ether important or not,). p. 7. "The Natural ARRANGEMENT of animals themselves is the question in poi
348d051 the question in point." Now what is natural ARRANGEMENT,-- affinities, what is that, amount of resem
349d051 ns.≠ p. 7. «In» Some <of> cases the circular ARRANGEMENT from fewness of forms-- Cannot be discovered
354d062 . 69. A Dr Macdonald believes the Quaternary ARRANGEMENT & not the Quinary.-- any one may believe any
370d116 ale, according to importance of divisions in ARRANGEMENT, of the perfection of their separation.-- th
478z004 oc on glow worm. luminous property-- Curious ARRANGEMENT of animals in rays Par un officier du Roi Ra
513q21. Hunter use expression of "male principle of ARRANGEMENT."-- would not male or female "constructive p
286c155 nt which is false There is same difficulty in ARRANGING animals in paper as drying plant, all brought
633j54r arises as long rigmarole about plants being created to ARREST mud &c at deltas.-- Now my theory makes all or
633j54r m. If we once venture to say plants created to <ARREST> "prevent" the valuable soil in its seaward co
633j54r general law. the plants were no more created to ARREST the earth, than the earth revolves to form rai
633j54r e once to presume that God «created plants to» ARRESTS earth, (like a Dutchman plants them to stop th
245c025 different species than those sinking, because ARRIVAL of any one plant might make conditions in any
173b010 species same as nearest land, which were late ARRIVALS others old ones, (of which none of same kind
298c191 often be destroyed.-- or regrafted with fresh ARRIVALS..-- &c &c --Climate altering as island increa
195b098 ence Must exist in such spots. We know birds do ARRIVE & seeds.-- The same remarks applicable to foss
195b100 creative force know that «these» species could, ARRIVE-- did it only create those kinds not so likely
285c151 but will grow into affinity.--but whether ever ARRIVE at true affinity doubtful A species is only fi

Page **(Key Word)**
```
296c184 ling Isd «shows where proper dampness seeds can ARRIVE quick enough» Vegetation of peak--  altogether
398e005 ese subjects. should be absolutely necessary to ARRIVE at right conclusion about species Changes of 1
443e154 mation do not now undergo metamorphoses, but to ARRIVE at their present structure they must have <don
173b011 s, (of which none of same kind had in interval ARRIVED) might have grown altered Hence the type would
219b193 os. Many trees Compositae, because seeds first ARRIVED «Ferns ditto.--» & hence formed trees]CD & wou
299c193 ermanent plants are. & this conclusion must be ARRIVED at, when one sees a plant like Paris quadrifol
344d041 . The extreme difference of sexes. is probably ARRIVED at in case of insects as glowworm The case of
356d069 pretend  to no originality of idea-- (though I ARRIVED at them quite independently & have used them s
507q15v how  simple, but beautiful adaptation might be ARRIVED at.= Any book with drawing of Seed. Anemone wi
629o055 incts explains the feeling of right & wrong.-- ARRIVED at first <rationally> by feeling-- reasoned on
640j167 in progress or is, present with respect to new ARRIVERS, the small body of species would far more eas
062r129 monly seen. at long distances; generally first ARRIVES:-- New Zealand rats offering in the history of
023r010 s floating on the surface of the ocean, before ARRIVING at the Abrolhos shoals ▌ -- N.B. The view of
122a144 e degrading action must prevent internal fluid ARRIVING at equilibrium so soon from; crust being  cut
446e163 during  hundreds of years, without fresh seeds ARRIVING.»-- throws a very great difficulty in my theo
286c154 the  crys of pain, as well as we.-- It is our ARROGANCE, to raise on the same shelf-- to (look at com
291c166 than  gravity a property of matter? It is our ARROGANCE, it our admiration of ourselves.-- The idea o
300c196 Monkeys--  or dogs in Carnivora.-- Man in his ARROGANCE thinks himself a great work. worthy the inter
565n009 s decision lies in wrinkles about the nose, & ARROGANCE in upper lip. <The> Children having  peculiar
106a073 s obtained.-- If {P} inclination be great where ARROW stands the force immediately deflected from (B)
486z020 wood says in <14th>·<13th.» Vol of Archaeologia ARROW=heads described in Suffolk as lying under strat
570n026 ilarity of the earliest arts.-- Mem.-- Stokes-- ARROW heads &c &c October 27th Consult the VII discou
071r156 . Volcanic product.=> <Did Peruvian Indians use ARROWS or Araucanians?--> If wood now preserved  over
085a002 inland.--  even 70 miles from salt water. Mr. ARROWSMITH tells me, that Himalayas penetrated like Bol
523m017 answered she had known it at time & had bought ARSENIC for that purpose.-- this found to be true.-- H
075r167 oxide of Iron in Mexico. sulphuretted silver, ARSENICAL grey copper, and antimony, horn silver, black
477z002 le negro, et le pajero) l'yaguaré «the zorilla-ARSKINK» Le quiyá (Coipu) viscacha.-- A. Patagonicus l
179b034 of  reason, there would probably be repugnance & ART required to make marriage.-- as Dr Smith remarke
266c092 ower at their own periods.-- Arcana of Science & ART. 1831. p 160. account of Bulbous root from Mummy
279c133 off--  so that not propagated by nature.-- Whole ART of making varieties may be inferred from  <this>
300c197 mming death, most difficult case to iimagine how ART acquired.-- They reason however on this to a deg
321c270 s M«artineaus» How to observe Mayo Philosophy of ART.. on <Etn> Entozoa by Owen in Encyclop. of Anat.
326c266 1801.-- quoted by do.-- There appears to be good ART.. on <Etn> Entozoa by Owen in Encyclop. of Anat.
350d054 , & unnecessary spaces" p 23. "for Nature is the ART of God" Septemb I,. It has been argued Man first
365d107 se <restr> restricted in their range by men & by ART.-- the former only giving average of effects  of
365d107 r changes. than a mere monstrosity propagated by ART. Yarrell told me of a cat & of a dog, born witho
525m023 icking up a stone, though only acquired rules by ART.-- like the law of honour.-- they feel  pleasure
567n014 principles, which even animals practically know «ART precedes science-- art is experience & observati
567n014 animals practically know «art precedes science-- ART is experience & observation.--» in balancing a b
602o11v it, that Nature should have so little to do with ART (p 128) R. compares a view taken by a camera obs
602o11b the whole it seems."-- "that the object of «all» ART is the realizing and embodying, what never exist
603o11b Says  Architecture does not come under imitative ART [my view says yes. <old> mass of rock--]CD or po
605o020 ir receiving pleasures from beauties of nature & ART." But as we often see people who are susceptible
606o022 ult.-- Lessings Laocoon. 2d Lect-- The object of ART., sculpture & painting, is beauty.-- which he th
606o023 ocoon p. 75 "The beauties developed in a work of ART are not approved by the eye itself, but by the i
609o030 a  hive of Bees without their instincts.-- Gives ART to when I say How social instincts generated? Th
638j58v dead & extinct The analogy between the works of ART «or intellect» such as hinge, & hinge of  shell.
209B157 Fernandez Galapagos =Radack Islds = ∴ Islands & ARTIC are in same relation. We find species few in pr
622r020 guration have resembled Chiloe In De La Beche, ARTICLE "Erratic blocks" not sufficient distinction is
202b133 to associate together" There is long rigmarole ARTICLE by S Hilaire on wonder of finding Monkey in Fr
327c265 in lib. of Hort. Soc Mr Neil. has written good ARTICLE on Horticulture in Edinburgh. Encyclop.-- The
539m081 manner  in which my head got well when reading ARTICLE by Boz.-- now in this I was interested as  was
603o014 anterior state?? Edinburgh Review Vol 18. (1st ARTICLE) on Taste «EXCELLENT». Deficient in not explai
603o014 sion--  statues not painted-- <music> very good ARTICLE-- why flower beautiful? ¿even to children S. J
604o015 1756--  Ceased in 1758-- Read the Review or the ARTICLE. A Planaria must be looked at as animal,  with
177b027 only  appear like circles;-- & insects amongst ARTICULATA.-- but in lower classes, perhaps a more line
181b043 ammalia, Still greater between Vertebrate and ARTICULATA. still greater between animals & Plants  But
227b225 y of numbers of species in divisions. look at ARTICULATA!!!--! It leads to Nature of physical  change
413e060 is the more remarkable from their alliance to ARTICULATA, which are all truly bisexual. Buckland's Re
641j29v & Mammalia. The eye being formed in Mollusca, ARTICULATA, & Vertebrata, & Planaria, & light affecting
223B208 perfect  insect & former hard to tell whether ARTICULATE or intestinal, or even a mite.-- a bee «comp
233b252 e not so. greater facilities of change in the ARTICULATE than <M> Vertebrate. But how does this agree
347d049 has,  it tends to multiply & IMPROVE on it.-- ARTICULATE animals must articulate. <i> in vertebrates
347d049 ly & IMPROVE on it.-- Articulate animals must ARTICULATE. <i> in vertebrates tendency to improve in i
222b206 Mammalia;  but have improvement in system of ARTICULATION. ¿whether type of each order may not be sup
629o55v tando,]CD & if passions makes one break these ARTIFICAL rules, get remorse-- ((hence desires do not i
573n036 ass <&admires> is lost in astonishment at the ARTIFICER.--» Our faculties are more fitted to recogniz
566n011 ? & curiosity «strongly shewn in the numerous ARTIFICES to take birds & beasts».-- very necessary  to
107a077 of rock from Heat. Specific gravities of many ARTIFICIAL limestones produced by Sir J. Hal. End of pa
349d052 ps from a head, as subkingdom.-- -- evidently ARTIFICIAL, as interlopement of Marsupials will  change
354d065 of this. Do any varieties of sheep «evidently ARTIFICIAL» approach in character to goats.-- or dogs t
359d087 .-- Former strange mishaped bird-- looks very ARTIFICIAL breed-- but Mr Miller says that breeds large
360d091 mistake about Yarrell's law, as it is local (not ARTIFICIAL variation) which impresses offspring most.--
372d130 t method in generation.-- Ehrenberg considers ARTIFICIAL division of animals, as gemmation, I conside
372d130 nimals, as gemmation, I consider gemmation as ARTIFICIAL division.-- On this view each particle of an
427e109 Dioecia & Monooecia. ought to be preeminently ARTIFICIAL.-- Would not subsidence of Greenland render
435e130 Polygamia: Monooecia & Dioecia, preeminently ARTIFICIAL, so that even some species only in genera <a
493q004 t in crossing animals.-- &c (1) To cross some ARTIFICIAL male with <old> female of old breed & see re
495q05a 11. Soak all kinds of seeds for week in Salt. ARTIFICIAL water.-- 12. Plant two races of Cabbages nea
496q05a reproductive system.-- -- cover flower-- put ARTIFICIAL flowers-- also do with honey-- What is use o
528m036 t is a beautiful object one knows from seeing ARTIFICIAL lights in the night.-- from the mere exercis
628o054 arts of the emotive part of man, may be quite ARTIFICIAL, as avarice love of gold.-- love of fame-- Y
392d180 ds with moths, where fecundation can be made ARTIFICIALLY.-- Are hybrids pintail & common ducks. simi
428e111 principle does not apply to any plant reared ARTIFICIALLY, & only very partially to the Zizanias in i
498q008 -- (12) At Maer Cowcumbers in frames are not ARTIFICIALLY impregnated. Abberley says Ants-- Enquire (
505q014 Flower.  (as it is required to impregnate it ARTIFICIALLY.)-- Asclepias-- Flowers not seeding= Put po
264c079 im look at savage, roasting his parent, naked, ARTLESS, not improving yet improvable» & then let  him
228b231 nt state. with what he is as former species. His ARTS around not let him have taken him over whole world.
570n026 n's reasons: shewn by similarity of the earliest ARTS.-- Mem.-- Stokes-- arrow heads &c &c October 27
505q014 essary to fruit--; become well shaped by care 13 ARUM before pollen is shed can you find flys  dusted
073r164 . pyenite. native sulphur.. fluor spar. bayte. ASBESTOS garnets.--carb & chrom. of lead. orpiment. ch
311c227 sheep-- Edinburgh. Transact. Vol IX p. 107. an ASCARIS inhabits the eyes of horses in India in  which
025r017 aordinary freshness of the streams of Lava in ASCENCION known to be inactive 300 years? No Volcanic E
044r072 h have been seen to form in atmosphere.--Mem. ASCENCION. concretions & Galapagos.-- «Humboldts. fragm
036r050 a strange story; I believe it was necessary to ASCEND the hill,--but my recollection is imperfect &
036r051 the sand: I am nearly sure, it is necessary to ASCEND the hill.-- The absence of Second form, except
349d051 forms--  Cannot be discovered <un>till <in> «we ASCEND to» subgenera & families, <even in Cetionidae>
534m062 .-- table being removed a little further, they ASCENDED about a foot & then leapt across. (Col Sykes
160g088 of Glen Tarf What I called Alluvium shows the ASCENDING fringes {P} which makes me think it submarine
230b235 s that if so in approaching desert country or ASCENDING mountain you ought to have a gradation of spe
178b031 i[ch] the Fr. naturalists thought was species <ASCENSI> Study Lesson Voyage of Coquille.-- Dr.  Smith
032r042 fter having examined the changes of pumice at ASCENSION In Calc: sandstone at Ascension, each particl
033r042 of  pumice at Ascension In Calc: sandstone at ASCENSION, each particles coated by pellucid envelope o
035r045 . 209--13 P & 444 «(Yanky Edit)» <I think>· At ASCENSION, the laminae <...> changes in rocks. connecte
039r060 e most remarkable feature in the structure of ASCENSION» give as an example the great subsidence at t
046r079 as accidental on the afflux of the former. -- ASCENSION. Vegetation? Rats & Mices. At St Helena there
049r089 gy some such fact stated to exist in Peru. -- ASCENSION At Ischia there is a pumiceous conglomerate w
```

(Key Word)
050r090 ture of which is doubtful. P. 180. I think my ASCENSION case very doubtful. -- In Iceland Bladders of
051r093 tratification? Anomalous action of ocean.--at ASCENSION. (where occassionally most tremendous surf &
055r107 p. 252 Urge cliff form of land, in St Helena. ASCENSION. Azores. («sandstone first gives» half demoli
055r107 greement with number of craters. No cliffs at ASCENSION (or modern streams of St Jgo) yet no historic
062r129 n S. America. Zorilla: wide limits of Waders: ASCENSION. Keeling: at sea so commonly seen. at long di
096a040 lcanos Salomon Isld,-- New Britain-- &c &c In ASCENSION for centuries afterwards it might be percieve
100a054 at gneiss of Rio Concretions in Pumice bed at ASCENSION instance of hollow concretions & concretion f
102a060 in detail as amygdaloid. calcareous rocks of ASCENSION, each particle coated. &c will be aware how l
354d065 case of Probability. has Creator made rat for ASCENSION.-- The Galapagos mouse probably transported l
406e035 d.-- p. do.-- Fish of Teneriffe. St. Helena & ASCENSION most species like & identical with S. America
483z013 . 302. Vaginulus of Lima described "Arion" of ASCENSION. p. do.-- some S. American Reptiles are descr
604o018 ng of Sublimity is height. & with the idea of ASCENSION we associate something extraordinary & of gre
223b210 th animals?? Why are species not formed. during ASCENT of mountain or approach of desert?-- because t
099a049 action all fluid at once, the films vertical. ASCERTAIN law of attraction of particles of same nature
498q007 fficult to cross. (5) It is most important to ASCERTAIN amount of variation in plants raised by Scion
085a001 Patagonia seaward, at mouth of S. Cruz. from ASCERTAINED inclination. of plains: Lias in Shropshire.
126a122 ed.» how much matter separates them, this is ASCERTAINED by conducting powers-- we judge from the sur
227b224 ny curious points of speculation; for having ASCERTAINED means of transport, we should then know whet
277c126 ulator & O. Patagonicus. till neutral ground ASCERTAINED, call them varieties. but two ostriches good
277c127 er to this country, without range, or habits ASCERTAINED-- put them as (a). (b) until data be given.-
280c135 stication. throws great difficulty in way of ASCERTAINING about hybrids.-- & is a very remarkable fac
410e051 t see the deductions which are possible.)-- ASCERTAINMENT of closest species (& naming them) with rel
254c049 Polypi, Surely not correct view of Flustra or ASCIDIA spicule in sponge. stomachs in infusoria, gene
500q010 kinds of flower annually.-- Periwinkle. (not ASCLEPIADAE. «in» Lindley) (24) Do Bees distinguish spec
435e133 R. Brown found the <poll> masses of pollen of ASCLEPIAS placed on Orchis (so very different) that the
447e164 iwinkle wants insects to impregnate allied to ASCLEPIAS Turpin cell is individual May 29th.-- -- Hens
448e165 e insects to impregnate it.-- it is allied to ASCLEPIAS, where this is always the case according to B
505q014 is required to impregnate it artificially.)-- ASCLEPIAS-- Flowers not seeding= Put pot of boiled eart
514q21. get answer= Is pollen of cultivated Orchis & ASCLEPIAS &-- carnosa?-- good-- Norfolk Isd-- geology.
290c164 common gull in garden at Zoology Soc. it's pale ASH grey back, like a black bird washed, Whilst tips
401e017 rish-- a medlar may be Grafted on pear. Mountain-ASH & white Thorn! Species not being observed to cha
525m024 a dog doing what he ought not to do, & looking ASHAMED of himself.-- Squib at Maer, used to betray hi
525m024 uib at Maer, used to betray himself by looking ASHAMED before it was known he had been on the table,-
557m148 ink one can remonstrate with a dog, & make him ASHAMED of himself, in manner quite different from fea
557m149 g. runs away. is not afraid the whole way. but ASHAMED of himself.-- Jealousy probably originally ent
620o046 e says for shame (& the «old» dog really feels ASHAMED?) not so puppy, we <do> try to teach him & str
524m019 hey are insane & that their idea is wrong.-- (Dr ASHE, the Birmingham Doctor), in this precisely like
042r067 black scoriaceous rocks of R Chupat. & fall of ASHES of Falkner, ¿how far is the distance?-- Fossil
044r072 owed from centre-- Pisolitic balls occur in the ASHES which fill up theatre of Pompeei (?). -- Such h
127a126 137. Lord Tullamore found Sulph of Soda in peat ASHES in Ireland dikes in mountains. «(not on contine
548m117 ones expression of double self, though as in Dr ASH'S case, one here was conscious of the two states
538m080 ms habitually.-- Agrees with insanity, as in Dr ASH'S case, when he struggled as it were with a secon
187b065 expect that Bear & Foxes &c same in N. America & ASIA, but many species closely allied but different,
203b136 n Germany & thinks even now in central & Eastern ASIA beyond the Ganges & perhaps even in India p. 26
274c116 mmon.,-- but each several with Europe & northern ASIA & Northern America.-- may we not look to these
306c214 Hamilton will give an account in his Travels in ASIA Minor of the domestic animals-- At Angora «cent
306c214 t of the domestic animals-- At Angora «centre of ASIA Minor» are the fine-haired goats. which it is s
306c214 ersian «greyhounds» are Kurdish & come also from ASIA Minor.-- tail like setters. long ears-- colours
402e020 same families and genera, that are natives of S. ASIA, but many of the species are peculiar to them"
407e041 land.-- -- S America, an island, «connects with ASIA» between two polar lands,--; Africa not so equa
134a143 n large lake-- Berghaus Chart of do Journal of ASIATIC Society Vol I. {T} p. 145. on salt mines of Pu
134a144 . 798 do Vol 7. p. <52> 363. do {T} Journal of ASIATIC Soc Vol V. p. p 96. apparently good geological
205b142 definite information) West of Rocky Mountains ASIATIC types discoverable.-- Bridgewater Treatise p 8
320c276 ished March 1838 Whole of Geographical Journal ASIATIC Journal to end of 1837. re«a»d-- contains very
400e011 ideas not clear. In Australia from approach to ASIATIC [...]t in part near Timor, & to Europaean in V
451e176 s as well as Tapir.-- <do do p 75» «Journal of ASIATIC Soc.. Vol V. p. 565. in a Paper by Lieut. Newb
451e177 Vol II. p. 527.-- <Scientific Soci> Journal of ASIATIC Society Vol I. p. 261. <J> Catalogue of Birds
451e178 re 5,283 attached to its body-- Journal of the ASIATIC Soc. vol. I. p. 335. Catalogue of animals of N
452e181 keys <at> near shore of Magindanao Journal of [ASIATIC Soc] [...] p [...]-- most wonderful instinct,
499q010 Vomica is eaten by a Buceros in East Indies-- ASIATIC Researches} (23) Talk about Thyme. Horned Oran
558m152 le in different trades &c &c &c I observed the ASIATIC Leopard. quarrelling. mouth wide open, each [l
088a015 hamisso in Kotzebue. Study Humboldt. Fragmens ASIATIQUES account of American Volcanic action. -- Frag
105a073 little the force required to move <it> «stream» ASIDE is not great.-- Is there more degradation at fi
570n025 h triumphant feeling so similar to shame after ASININE.-- both accompanied by depending head., & acti
042r067 gle travels on the Chesil bank. V. De la Beche). ASK Capt. F.: R: how the swell, generally & during g
058r115 arrange themselves in planes. «Mem silky lustre» ASK Erasmus. whether electricity would affect this.
100a055 refers to salt as being produced by local heat, ASK Capt. Beaufort, whether, water flashing into ste
104a068 mountain chain case parallel to Banda Oriental. ASK Lyell for sentence.-- Origin of Breccia, introdu
132a137 f Chile, with that of the passage of the moon.-- ASK Hopkins. M. Parrot, Mem. Acad. Imp. des Sciences
191b083 a. proof of subsidence. & recent elevation: Pray ASK Dr. Smith.-- «to state that most clearly». Fox t
192b088 ne rocks. we can only expect some steps.-- I may ASK whether the series is not more perfect by the di
225b218 dian Cow with bump & pigs foot with cloven hoof) ASK Entomologists whether they know of any case of i
228b230 bells & see what offspring would come from them. ASK Henslow for some plant, whose seeds go back agai
232b248 ise on land shells in salt water & lizards do.-- ASK Eyton to procure me some Get Hope to give me
235b273 ed to certain countries-- so on with families.-- ASK Royle about Indian Cattle with humps.-- ¿To be s
278c131 e immense age since breccia accumulated-- surely ASK Owen to see whether species same, excessive impr
362d099 uence of the female breeding all the year round. ASK Colonel Sykes.-- Even our domesticated cattle ha
382d157 ther), Hunter <asks> p. 36 is thought by Owen to ASK. whether a Heautandrous animal is «evidently» ac
385d163 ters having a throw of this mongrel.-- I did not ASK the question.-- His bitch will not take «& if sh
392d180 squimaux (& Australian) dogs with common dogs--> ASK my father to look out for instances of Avitism E
409e047 1 Fuego is to be converted into civilized man.-- ASK the missionaries about Australians yet slow prog
409e047 w progress has done so.-- Show a savage a dog, & ASK him, how wolf was so changed. When discussing ex
430e120 nsdale evidently inclines to think it Hybrid.!!! ASK Woodward Mr Lonsdale says Trigonia costata & elo
431e121 Ammonites would afford instance of such facts.-- ASK Phillips.-- The more I think, the more convinced
453e183 seed.-- Where has Duchesne described Atavism.-- ASK Dr Holland cases where peculiarity has first app
501q011 what Mr. Herbert observe on this subject-- (31) ASK Henslow for list of annuals to place in Hot hous
501q011 generative organs of great Heat (32) Can Henslow ASK question of Col. Le. Couteur about Wheat-- Chang
502q11v or for sterility Land Birds Madeira Migratory-- ASK Gould about N. Zealand, as Cuculus lucidus is.--
502q11v Gould about N. Zealand, as Cuculus lucidus is.-- ASK Sulivan about Falklands Isds.-- Snipe Migratory-
503q012 f domestication-- said to require Selection (36) ASK Mr Gowen he asks Mr Herbert, how many generations
503q012 - said to require Selection (36) Ask Mr Gowen to ASK Mr Herbert, how many generations any hybrid has
503q012 tions any hybrid has <been> reproduced itself.-- ASK Gray to ask Mr Riley to experimentise on hybridi
503q012 brid has <been> reproduced itself.-- Ask Gray to ASK Mr Riley to experimentise on hybridising ferns,
506q015 being propagated many years, by cuttings.-- (40) ASK Henslow to distribute some of my questions among
507q016 accidents of delivery-- inheritable.?-- Bell cd ASK Accouchers Is any peculiarity in milk teeth inhe
509q017 1 in cases. Owen. Have talked partially with him ASK him to introduce me to some Human Anatomist. Has
509q017 cted any animal often, which has abortive bone. (ASK more about the lowest cervical vertebrae process
514q21. phrys nectary?= Bunbury says no «hollow» spur.-- ASK about Pinks & Solanum impregnation before flower
586n082 more surprise at it & feels no more inclined to ASK [not located] If dislike, distaste. & disapprova
616o039 opriety of the expression. They would do well to ASK themselves the converse of the <expr> question a
148g025 , but they thought the breed liable to vary-- I ASKED this question in many ways & received same answ
165g125 ow the lambs because most like Mother in face-- ASKED stated this generally the case Wednesday 12/ &
257c057 t might be doubted were possible,-- it has been ASKED how did the otter live before it had its web-fe
310c222 ars to be a puzzle against my theory,-- If I be ASKED by what power the creator has added thought to
341d029 osest.-- (& two Rheas still closer).-- Mr Blyth ASKED whether structure of pelvis & was not adaptive
346d047 n[g] [not located] half breed liable to vary. I ASKED this in many ways, but received same answer.--
460t019 long since present forms existed, but if it be ASKED how this complexity from a few types originated
521m009 exclaimed, why it is dinner time.-- » My father ASKED him whether he had gardener of name A. B., &c &

521m009 d of such a man & had no gardener.-- My F. then ASKED Mr C. to come to the window & pointed out the G
522m010 in if early association were called up.-- My F. ASKED him, did he know whom Mr Child «of Kinlett» had
523m017 er, was far more painful than the thing itself. ASKED my F. whether insanity is not distinguished fro
530m043 a bond, yet when he paid the Attorneys bill, he ASKED what bond he could have had. yet during whole i
382d157 ans barren in one plant & not in other), Hunter <ASKS> p. 36 is thought by Owen to ask. whether a Hea
601o009 to the primary sensation.-- man moving leg when ASLEEP-- «or habitual actions» perhaps polypi-- (so t
470t177 n Thistles many (curious because a Composite) ASPARAGUS very small flowers & as much shut up, frequen
505q014 ked Pea.-- intends to breed from it and large ASPARAGUS: result? = failed to germinate 16 Will plant
218b191 ties & each one will have a peculiar «constant» ASPECT. That is varieties, though of trifling order a
616o39v ely what are called it's subjective & objective ASPECT. The subjective aspect of bodily action is rev
617o39v s subjective & objective aspect. The subjective ASPECT of bodily action is revealed to us by the effo
617o40o do without it as well as we can. <The objective ASPECT of> bodily action as recognised by our externa
617o40v it arises. But coming round to the <subjective> ASPECT of action as known by the exertion of our own
618o041 by internal consciousness, & have no objective ASPECT. If thought bore the same relation to the brai
617o39v e of inanimate bodies. How we identify the two ASPECTS as different phases of the same object of thou
225b219 cts.-- about dogs &c &c NB. Animals very remote. ASS & Horse. produce offspring exactly intermediate.
311c225 placing each other». good to consult p. 326 wild ASS extending over 90 degree of Long. & Col. Sykes a
355d066 of horse, dun-coloured with stripe approaches to ASS.» or fowls to the several aboriginal species «or
366d112 inking to keep out flies might be used» The wild ASS has no cross. how comes it that the tame donkey
367d113 - animal like large, heavily made cream coloured ASS.-- stripe on back also.-- legs reminded me stron
417e075 owing to the domestication of both.-- Now in the ASS-- there is little tendency to vary. & hence offs
424e103 ught not to be able to hybridise the Camel» like ASS «same way some plants vary more than others» & h
458t002 ogical Gardens. informs me that a hybrid between ASS & Zebra, crossed with pony mare & produced a ver
459tf02 tail limbs-- in the mules, these parts resemble ASS. (& part of body mare)-- -- this may be, perhaps
493q003 into B. from B. into A. as takes place in mules ASS & horse-- important. {In crosses does male offsp
512q020 iller says Wombwalls were (4) About fertility of ASS-zebra-horse= (4) About fertility of ass-zebra-ho
513q020 ility of ass-zebra-horse= (4) About fertility of ASS-zebra-horse= (5) About callosities on Camels-hor
558m153 correctly than this for C. Sphynx.-- In the wild ASS there is a curious drawing out of the side part
567n014 ience & observation.--» in balancing a body & an ASS knows one side of triangle shorter than two. V.
568n016 le shorter than third. is this necessary notion, ASS has it.-- When one is «simply» habituated «in li
092a028 Syria Geolog. Proc. p. 541. year 1837 In Upper ASSAM. Geolog Proc p. 566 1837.-- Tertiary <bea> form
251c041 n of Cats Vol III. <p.> p 233, stated that the "ASSEEL Gazal. (Bos Gazoeus) does not mix with the Gob
568n016 llow as consequences on habitual or instinctive ASSENT to propositions, which are the result of our s
640j167 terion.-- Hence it is highly unphilosophical to ASSERT, that they are not species, until their breedi
036r050 &c &c. There is a Hill. near Copiapò which is ASSERTED to make a noise,--My impression. is not very
251c042 Viedo says American dogs silent. Mem contrary ASSERTION of Molina) (p. 277.). probably another in Jam
310c224 rld Athenaeum June 3d 1838. quotes M. Turpins ASSERTION that globules of milk produce a plant capable
573n037 dropped his head when he meant to eat, hence ASSERTION.-- but nodding is less strongly marked than n
340d026 s a bold theory. which attempts to explain, or ASSERTS to be explicable every instinct in animals. He
063r133 ot perfect.-- Dogs. Cats. Horses. Cattle. Goat. ASSES. have all run wild & bred. no doubt with perfec
585n075 uts down its ear. good to contrast with horses, ASSES, <mi> Zebras &c &c.-- Here there is kicker but
195b100 mestic animals vary in countries, without any ASSIGNABLE reason.-- Astronomers might formerly have sa
546m111 a meal, she suddenly like a flash without any ASSIGNABLE cause, remembered she had put it in branch o
277c126 es it is reverting to old plan, but reason now ASSIGNED for doing so There should be mark to every sp
570n023 e implies negation, without violence, without ASSIGNING or understanding reason.-- surprise with nega
260o067 many birds of different kind have been know to ASSIST in feeding young cuckoo; as if there was storg
446e160 sing equally well. and <in> these reciprocally ASSIST in domestic cares, as building nest, sitting o
540m089 sure at pain of others, with rational desire to ASSIST them,-- otherwise as he remarks sympathy could
608o028 sickly as we do the wicked.--we ought to pity & ASSIST & encourage.-- crying is a puzzler-- Under this
315c243 very of prey.-- arising no doubt from want of ASSISTANCE.-- crying is a puzzler-- Under this point of
558m151 -- fear for others acting in unison.-- active ASSISTANCE. &c &c. it comes to Miss Martineaus one prin
594n115 tellect Lyell has seen a little dog go to the ASSISTANCE & bite a big dog. which was fast struggling
269c101 velled that thickness in that period & no ways ASSISTED by fluid currents which, may take place in Me
092a029 s Paris. 1500 ft high Mr Bird in paper to Brit. ASSOC: has shewn how electrical currents tend to depo
214b174 living in the sea.-- See Rogers report to Brit ASSOC <to> on N. American Zoology-- A breed of Blood
326c266 4to. contains much on dogs.-- Reports of Brit. ASSOC.-- some important Papers. Dr. Mayo. Pathology o
587n089 f Brain.-- Mackintosh first clearly insisted on ASSOC of ideas & emotions. rather ideas & bodily acti
202b132 the English cattle, nor could we get them to ASSOCIATE together" There is long rigmarole article by
551m129 .-- different from sneer-- How easily. horses ASSOCIATE sounds may be seen by omnibus Horses startin
604o018 ty is height. & with the idea of ascension we ASSOCIATE something extraordinary & of great power-- --
063r132 ne individual «divided» by different methods, ASSOCIATED life only adds one other method where the di
189b073 animal like plants:-- trace gradation between ASSOCIATED & non associated animals.-- & the story will
189b073 s:-- trace gradation between associated & non ASSOCIATED animals.-- & the story will be complete.-- I
278c131 n 1000 ft. & many hundred miles from the sea, ASSOCIATED with teeth of seals and dugong, therefore im
315c243 ed laughing. Barking to tell other animals in ASSOCIATED with the ducks.-- most strange voice often i
341d032 s bird, did not seem to know itself,. at last ASSOCIATED with some other, as pointing with smell.= Th
466t095 d.» Probably every such «new» quality becomes ASSOCIATED plants. when proportional number appear equa
496q006 Rust Carb. of Ammonia.-- Horse Urine &c &c on ASSOCIATED pleasures & pains & emotions-- such as child
543m101 Can an analogy be drawn between «heredetary» ASSOCIATED is kept up with waking thought.-- Ld Brougha
544m102 when intense, or when so near waking. that an ASSOCIATED animals?) I doubted it in Fuegians, till I r
555m142 found everywhere (is it not present with all ASSOCIATED pleasure &c accompanying such memory.-- A Me
569n021 ess & multiplicity of things remembered & the ASSOCIATED with them.-- The establishment of this princ
577n050 after certain other actions, & hence becomes ASSOCIATED with reason: [N B. insects which have never
583n069 is not imitative: imitations seems invariably ASSOCIATED with ideas.-- A sigh, is an abortive groan.-
589n092 & yet our emotions being only bodily actions ASSOCIATED with God. these phenomena we (feel & ?) call
604o018 . 3 Infinity eternity. darkness, power. being ASSOCIATED sensations so often gives, when excited by o
605o18v lorrying, which height. by its accompanying & ASSOCIATED with it. as the idea of Deity. with vastness
605o19v of an object itself or to the ideas excited & ASSOCIATED «habitual» feelings. become like the instinc
622o048 lin> acute the feeling really is.-- All these ASSOCIATED during lifetime). so is our moral taste p. 1
624o50v due to <habit> heredetary habit (& modified & ASSOCIATED with the approving or disapproving instinct-
628o53v tions to be considered right & wrong.-- to be ASSOCIATED with the approving or disapproving instinct-
628o054 or that I have been taught or habituated to ASSOCIATICAL, the emotions of this instinct, with that l
073r162 products of Solfatarasi.} some general laws. ASSOCIATION of lead & silver. Sulp. of Barytes: Fluoric.
146g012 uring Strike an analogy between pleasures of ASSOCIATION. & passions, such as love-- dislike & <f> pa
234b256 -- CD [There was notice in report of British ASSOCIATION of 1838. (Newcastle) about somebody who had
255c050 ence. «This excellent case of memory without ASSOCIATION..» Instinct goes before structure (habits of
293c172 » analogies with memory in offspring.-- Some ASSOCIATION in such cases recall the idea. «or simple st
293c172 ou have heard conversation before. is strong ASSOCIATION recalling up image which had been past-- so
313c235 erval very good case.-- habit is awakened by ASSOCIATION (case of Elephant, which had run wild in Ind
366d111 influence propagation of species.-- Case of ASSOCIATION very disagreeable hearing maed servant clean
455eIBC of sleeping someway useful.-- it is only the ASSOCIATION which is useless. Granfather's Handwriting,
494q005 sensitive & sleeping species, & see whether ASSOCIATION can be given (2) do the stamina of C. Specio
522m010 ersation could catch up a new train if early ASSOCIATION were called up.-- My F. asked him, did he kn
526m029 t of the Hall understanding. (on hearing old ASSOCIATION brought up) by sight & not by hearing One is
528m039 known by autumn, on clear day.-- 3d pleasure ASSOCIATION warmth, exercise, birds singings.-- 4th. Ple
530m046 onally recollection of anything is solely by ASSOCIATION, & association is probably a physical effect
530m046 tion of anything is solely by association, & ASSOCIATION is probably a physical effect of brain the «
534m062 es, or ears (language mere means of exciting ASSOCIATION.)-- or of memory of such sensations, & memor
538m078 sness, explained by Dr Dewar on principle of ASSOCIATION.--«fully bears out my fathers doctrine about
539m082 o one & smelt the peculiar smell of Picture. ASSOCIATION with much pleasure immediately thrilled acro
547m111 shed, at the time. & could trace no chain of ASSOCIATION» Mayo Philos. seems certain that muscular, m
550m127 ce calls up pleasure. or pleasure or pain of ASSOCIATION.-- now if one has these feelings, without be
550m127 these feelings, without being aware of their ASSOCIATION «ie heredetary», does one not call them inst
554m139 this was permitted & eat it.-- good case of ASSOCIATION.-- «Listened with great attention to Harmoni
554m141 y Horse going down hill.-- (argue sophism of ASSOCIATION. Kenyon, & then go on to show, that if Cart
555m142 abit cannot be traced) V. D. p. 111, case of ASSOCIATION. Sept. 16th Zoological Gardens-- Endeavoured
567n016 parate its memory from all ordinary lines of ASSOCIATION.-- is totally distinct from learning it by h
569n020 nderstanding languages seem simplits case of ASSOCIATION.-- Elephant often given food & word open you

(Key Word)

```
574n041 es saliva to flow «yes, certainly»-- curious ASSOCIATION: I have seen Nina licking her chops.-- someo
574n041 aring music. to certain degree sexual.-- The ASSOCIATION of saliva, is probably due to our distant an
574n041 en like dogs to bitches.-- How comes such an ASSOCIATION in man.-- it is bare fact, on my theory inte
577n050 had  not it.-- The memory of Plants, must be ASSOCIATION,-- a certain round of actions take place eve
577n050 em.-- The establishment of this principle of ASSOCIATION will help my theory of sensitive Plants Habi
578n054 ally a most modest person one carries on, by ASSOCIATION, the question, "one will anyone, especially
578n057 y power, (or only very little so) & hence by ASSOCIATION, there pour out tears, & there is slight con
587n089 does not lead to action.-- p. 248. Theory of ASSOCIATION. owing to time when entered brain, try conti
593n107 oung animal learns parent smell & look so by ASSOCIATION receives pleasure. This [blank] will not  do
594n113 ) at thought of almost anything ugly. baby-- ASSOCIATION-- pouting child same as anger, lips not comp
595n117 & sweat, when hearing merely hunting horn-- ASSOCIATION or imagination [blank] [not located] Ernest
596n184 hink well worth studying-- "Thomas Brown" on ASSOCIATION worthy of close study.-- full of practical o
604o018 ena we (feel & ?) call sublime.-- 4 From the ASSOCIATION of power &c &c with height, we often apply t
614o037 any general instincts, as love of virtue, of ASSOCIATION, parental affection-- The very existence of
615o038 ence national character, love of country, of ASSOCIATION &c stronger in some than others-- Hence supe
615o038 lants of certain kinds such secretion) or an ASSOCIATION of pleasures with certain actions performed
619o043 ould feel pain. [.By a very slight change in ASSOCIATION if others injured these objects, without his
619o043 d know that many experienced pleasure,> & by ASSOCIATION he would feel part of that pleasure, which t
620o045 qually natural leave effects not lasting. By ASSOCIATION one gains the rule, that the passions & appe
621o047 ong in his mind.-- Now we know it is easy by ASSOCIATION to give «almost» any taste to a young person
621o047 st anything nasty. (<& accidentally> «by odd ASSOCIATION» comes to this conclusion, not owing to pecu
622o094 ur love of another, as pleasure arising from ASSOCIATION from having received benefits from this pers
624o051 ong, though, that part, which is acquired by ASSOCIATION from education & imitation, has often been p
540m087 ideas  are habitual, nor recalled by obvious ASSOCIATIONS. as by reading a book.-- Consider this.-- "
543m099 lculators, from the confined nature of their ASSOCIATIONS (it is not so in punning) are people of ver
605o019 t that it is through a complicated series of ASSOCIATIONS that we apply to such emotions. this same t
605o19v wonderment) <which> «this» emotion, from the ASSOCIATIONS before mentioned. we call sublime.-- It app
622o048 r nature, <sub> subject to their instincts & ASSOCIATIONS.-- often feelings which do not lead to acti
098a047 ion, & therefore as a consequence aggregated (I ASSUME the same force which draws together two partic
119a105 tendency to irregularity,--. Why does Sir John ASSUME it to be constant.-- It is to be profoundly co
232b246 does  not measure time but physical changes (we ASSUME like weather on long average tolerably> unifor
361d096 that female of some water birds, (as Phalarope) ASSUME for breeding a more brilliant plumage than mal
378d147 bright blue.-- «thus young of» Many of the pies ASSUME the metallic tints, such as Magpie, Jay, & per
387d168 degree»  -- Then when (C) unites with Male (X) «ASSUME that every peculiarity has a tendency to desce
473s07v may just as well be born a tendency to alter or ASSUME some form late in youth,-- only facts can deci
595n125 ns of communication) in man, than sucking.-- [I ASSUME a child pouts who has never seen others pout]C
295c178 ry than woman. because ancestors so, or has he ASSUMED that character -- -- female & young seem  most
295c178 oung seem most like mean characters the others ASSUMED--+++ ‖Daines Barrington says cock birds attrac
305c211 prin> thinking principle. seems to be given or ASSUMED according to a more extended relations of  the
308c216 e time formed chain, since if cretceous period ASSUMED, then some perished before, Carboniferous some
374d134 not doubt, every one till he thinks deeply has ASSUMED that increase of animals exactly proportiona[1
583n070 . animals capable of education; (this is again ASSUMED as more allied to reason than instinct.) Mr We
305c210 ncies of organic matter & chiefly heat), which ASSUMES a multitude of forms «each having acting princ
378d147 & female alike Good Ch 6 Keep Is it Male that ASSUMES a change, & is the offspring brought back to ear
432e124 d seeds can be true hermaphrodite Man probably ASSUMES the hairy character of his forefathers only wh
125a121 es includes opposite directions. Mem. S. Cruz. ASSUMING from Sir. W. Herschel's views earth originall
173b010 nct &c &c &c non fertility of hybridity &c &c <ASSUMING all» if species (<a>) «(I)». <fr> may be deri
199b116 1 otherwise not so numerous: quote from Lyell: ASSUMING truth of quadrupeds being created on small sp
306c214 ght to agree with Java?? Terrestrial Planariae ASSUMING bright colours., good instance of colours dep
384d162 :-- ∴ developed instincts of Capon. & power of ASSUMING male plumage in females., & female plumage in
388d172 change  which takes place in old age of female ASSUMING plumage of cock, & beards growing on old wome
195b104 them-- two hypotheses fresh creations is mere ASSUMPTION, it explains nothing further, points gained
257c059 genera of fish &c &c at present day.-- It is ASSUMPTION to say generation produces young ones capabl
257c059 ng young ones like itself, but ¿whether great ASSUMPTION? not solely producing like itself, not appli
359d086 organic  difference.-- it may be so, but this ASSUMPTION as long as animals are healthy which is ofte
419e087 e bodies of species>, & only secondarily,, by ASSUMPTION well grounded, on time;-- therefore the mere
478z004 showing it to be one animal In Australia I was ASSURED wild dog copulates <.> freely with tame: comes
385d163 at Willis «Grt. Marlborough Str, Hair dresser, ASSURES me he has known many cases of bitch going to m
025c022 in. Mem Concepcion Says Echinites. Encrinites. ASTERIAE, usually petrified into a peculiar cream-colo
267c095 . Hist. precursor of magazine??? p. 75. roe of ASTERIAS in stomach. of Sammon remain after rest of an
508q016 d through females, like Hydrocele Dr. H. thinks ASTHMA in females takes place of gout.-- How are live
214b175 American Zoology-- A breed of Blood Hounds from ASTON Hall close to Birmingham, and supposed to be de
104a068 rface Lyell remarked to me that Kylow (?) was ASTONISHED with him that <th> gneiss, mica-slate of who
546m111 h he had hid some years before.-- was greatly ASTONISHED, at the time. & could trace no chain of asso
593n111 & flappered across pool to bed of flags I was ASTONISHED & having looked round saw at considerable di
610o031 e had written Wilson & pointed it out; he was ASTONISHED, & said how very odd.-- --could not think wh
035r047 preserved  at Patagonia. The English fact is ASTONISHING consult book itself. P. xx: same fact is ind
426e108 ts]CD.-- Mr Marsh has some nephews, who are ASTONISHINGLY like to some distant cousins, the nearest b
056r109 ands whose inclination natural [...] deepest ASTONISHMENT.» Perhaps scarcely a pebble might remain to
573n036 ece» of coloured glass <&admires> is lost in ASTONISHMENT at the artificer.--~ Our faculties are more
240c013 nor  in Dobrizhoffer Abipones.-- Voyage. de L'ASTROLABE Zoologie. p. 60. Vol I. Cynocephalus.  niger.
320c276 n read Voyage a l'isle de Frances Voyage de l'ASTROLABE Partie Zoologique Pernety. voyage a 1 isle Ma
481z010 n Lyell's possession» of Zoolog. of Voyage of ASTROLABE must be studied for anatomy. of. corals.-- ne
287c157 bears  to Cuvier that relation of theoretical ASTRONOMER to plain observer‖ +++. between Mammalia & f
409e049 ar from the geologist the history, & from the ASTRONOMER that the moon probably is uninhabited]CD & i
573n036 r result of laws. yet we placidly believe the ASTRONOMER, when he tells us satellites &c &c «The Sava
195n101 countries,  without any assignable reason.-- ASTRONOMERS might formerly have said that God ordered,
276c123 o with forms.-- Mention persecution of early ASTRONOMERS.-- then added chief good of individual scienti
342d036 a magnificent view one can take of the world ASTRONOMICAL <& unknown> causes, modified by unknown one
122a113 ms of earth as shown by arc.--read Herschels ASTRONOMY with oscillations of level.-- {P} will  point
564n005 studied  appears to me to be like puzzling at ASTRONOMY without Mechanics.-- Experience shows the pro
071r156 hquake. but no serious injury. -- <Analysis of ATACAMA. Iron in Edinburgh. Phisoph. Transactions. = M
186b058 for these. analogies; CD[from the principle of ATAVISM, where real structure obliged to be altered, I
186b059 ach> either parent, (this is what French call (ATAVISM) Probably this is first step in dislike to uni
257c059 ers:-- Are monstrosity heredetary??.? Does not ATAVISM relate to this law.?-- Local varieties  formed
259c065 ity (any great change in species is reduced by ATAVISM) Even a deformity may be looked at as the feed
259c066 here is reference to more than offspring (like ATAVISM) & shows my «view of» generation right?-- If p
453e183 raised by seed.-- Where has Duchesne described ATAVISM.-- ask Dr Holland cases where peculiarity  has
490q001 porary Question 1 Where has Duchesne described ATAVISM alluded to by Dr. Holland-- <Jordan> Smith  of
591n097 er cows--> Mr. Hamilton on vital laws (in the ATHAENAEUM Library) describes effects of emotions-- fea
614o037 ery.-- [LHC] This Materialism does not tend to ATHEISM. intulity of so high a mind without further e
536m074 of  a new kind, because he would tend to be an ATHEIST. Man thus believing, <yet> would more earnestl
067r143 n Cordillera & on Eastern plains «by Antuco». ATHENAEUM April 1836 (p302) Coleccion de obras. 2 Vols
092a029 inct to Volcanic will explain their solution. ATHENAEUM M. 516 1837 High up the Essequibo, granite &
106a076 aken from Dalbechs Theoretical Researches.-- ATHENAEUM. 1838-- p. 137. Three inosculating rivers in
112a090 active volcanos & see what high they are <See ATHENAEUM. 1838. p 274. probably will be published in t
124a118 iberia (with refer to Metamor) wrong entrance ATHENAEUM. 1838. p. 652. Dr. Daubeny on mountain Chains
128a130 mma. Transactions of the Maryland Academy (at ATHENAEUM.) I. Part. 1 Vol.-- some notices on modern Te
129a132 ad parts of this work, though it is but poor. ATHENAEUM. 1838 p. 791 -- Most curious account of great
131a136 .-- ground ice-- subterranean isothermal line ATHENAEUM. 1839. p. 52. On Frozen soil of Siberia.-- fa
220b199 ooks with guns) of the Bustards in Germany.-- ATHENAEUM. No. 537. Feb. 1838. p. 107. Mr Blyth states
230b235 female  offspring: applicable to any animal-- ATHENAEUM. <Jan> «p. 154»-- 1838. Hybrid Ferns It may b
310c224 stralian plants found in other parts of world ATHENAEUM June 3d 1838. quotes M. Turpins assertion tha
338d024 tand the vast number of domesticated races.-- ATHENAEUM. p. 505. some (very poor account) of plants o
346d048 age might possibly be made out.-- August 25th ATHENAEUM (1838) p. 611. Ld. Tankerville account of wil
357d073 t destroyed wherever other animals existed.-- ATHENAEUM 1838. p. 654. Reason given for supposing Tetr
379d152 ee Marten. <Owen.> See Hunter's Owen-- In the ATHENAEUM Numbers 406, 407, 409, Quetelet papers are gi
388d172 ion (even Elephant) not breeding-- remarkable ATHENAEUM 1838. p 653. Ehrenberg<h> thinks multiplicati
```

```
420e090 hat minute details of structure heredetary'-- ATHENAEUM .1839. p. <8>36.-- A crustaceous animal is me
446e161 ird of more splendid plumage than the male.-- ATHENAEUM May 18. 1839. p. 377.-- Statement that the cl
449e173 hinks he has seen specimen at Paris Museum.-- ATHENAEUM: 1839. p. 451. Sheep Merinos from Cape of Goo
461tf03 ding much further geographically than others. ATHENAEUM p. 605 Mr. Macgillivray says "<A Thrush &> Bl
462tf4r «its» natural state to mate with a thrush"-- ATHENAEUM 1839. p. 708.-- Shrew, found by M. Lartet sam
462tf05 - -- a good analogy-- sea-Crustacea-- Tullus. ATHENAEUM 1839 p. 772-- A curious theoretical French bo
516g23v do  to preserve fungi or animal substances-- (ATHENAEUM (40) p. 823 chemical analysis of Peat (2) Ath
516g23v eum (40) p. 823 chemical analysis of Peat (2) ATHENAEUM 1840 p. 777. Decaying wood absorbs oxygen & f
539m081 instincts  of animals-- Aug. 12th. 38. At the ATHENAEUM Club. was very much struck with an intense he
552m131 aping against cornice stone to cause friction ATHENAEUM 1838. p. 652. Dr Daubeny on the direction  of
599o005 ritten about the year 1837 & Earlier-- ]CD in ATHENAEUM "Smart-- Beginning of a new School of Metaphy
358d074 tinent.-- With respect to ancient geography of ATLANTIC Tristan D'Acunha ditto Juan Fernandez do Mitc
483z012 oolog. Journ Vol I. p. 125, owls seen crossing ATLANTIC. fact taken from Jenner (1825) Phils: Transac
483z013 ae velellae (Less) parasitical on Vellellae in ATLANTIC Ocean Gould agrees with D'Orbigny, that Serpe
540m086 e conditions as Negro on the other side of the ATLANTIC. Why then is he so different-- in organizatio
044r072 ompeei (?). -- Such have been seen to form in ATMOSPHERE.--Mem. Ascencion. concretions & Galapagos.--
371d128 t Loddiges 1279 roses kept in same soil. same ATMOSPHERE?-- may they produced not be transplanted?, &
055r108 ought down by every torrent proves the decay ATMOSPHERE of the most solid rocks.--The grand cliffs o
086a007 ow much is temperature of world regulated by ATMOSPHERIC currents?-- chiefly clearly by sun's positio
182b046 - if in any in the Cryptogamic flora but not ATMOSPHERIC type. Hence probably only four, is not  this
115a097 unted for by great molecular attraction of every ATOM in rock On a coast, the shallower the water, th
418e083 ntold,-- what is added to the composition of the ATOM, to make it alive, & how the laws of generation
618o41v all  we know of attraction, but we cannot see an ATOM think: they are as incongruous as blue & weight
021rIFC y Thursday 29th gale Lyell's Geology The living ATOMS having definite existence, those that have unde
115a096 ed. The chemical action which gives polarity to ATOMS in slates that cleave, & which unite the homoge
224b212 ood species, & hence the importance Naturalists ATTACH to Geographical range of species.-- Definition
354d065 - It should be observed with what facility mice ATTACH themselves to man. Sept 7th. --- I was struck l
225b218 roduced plant, which any insects hav[e] become ATTACHED to.-- that insect <not> being called <Phitoph
451e177 of  a tick «in India» & found there were 5,283 ATTACHED to its body-- Journal of the Asiatic Soc. vol
051r095 e. or vice versâ. Clay slates unfavourable to ATTACHMENT of many bodies [blank] Beechey.--changes  in
290c165 n.pheasants. have crossed.-- </I saw>..-- The ATTACHMENT of dogs to man. not altogether explained  by
557m149 first  try «to» attract female, (or object of ATTACHMENT) & then failing to drive away rival.--  Fear
406e036 degree  p. 354-- The most vicious dog. will not ATTACK any animal except, dog when absent from its ma
531m052 nt muscular exertion» which accompanies violent ATTACK,-- Even the worm when trod upon turneth,, here
563n002 ing when I struck the Keeper) may be tempted to ATTACK him from jealousy. (Pincher & Nina)-- or to la
608o026 cation & mental capabilities.-- '(P) Animals do ATTACK the weak & sickly as we do the wicked.--we oug
492q002 nsact.-- 4.. Are any varieties of Cabbages not ATTACKED in bad years from Caterpillars. 5. Whether Ro
534m063 Westwood remarks that some imported plants are ATTACKED by insects & snails of this country (thus Dah
430e117 ion.-- Poet Cowper, describes his tame Hares, ATTACKING a sick one like Chillingham bulls are describ
564n005 s the problem of the mind cannot be solved by ATTACKING the citadel itself.-- the mind is function of
613o036 n,-- pointers method of standing,-- method of ATTACKING peccari-- --retriever-- produced as soon as b
565n007 look  obliquely so does dog. when a little one ATTACKS him Contempt, when there is some anger «& resp
623o051 nknown conditions) (for pig will not so readily ATTAINED instinct of pointing as a dog.-- also age  has
126a122 II but an equilibrium is supposed to have been ATTAINED.» how much matter separates them, this is asc
126a123 if  bad conductor-- III But equilibrium is not ATTAINED, & if cold water did not percolate surface, w
627o053 "The pleasure which results when the object is ATTAINED (the gratification of one's offspring) is not
185b055 s. Cuvier examined it. There certainly appears ATTEMPT in each dominant structure to accomodate itsel
185b055 s-- carrion eagles.-- This is but carrying on. ATTEMPT at adaptation of each element.-- May this  not
222b203 orth young having very different characters is ATTEMPT at returning to parent stock.-- I think we may
259c065 Even  a deformity may be looked at as the best ATTEMPT of nature under certain very unfavourable cond
265c085 may so far be considered an adaptation as best ATTEMPT of nature, colouring matter being absent.-- ag
267c094 st & fertile land of Tahiti, in spite of every ATTEMPT to check its increase. The <bush> woodlands fo
312c228 grown  up.-- Are mules homogenious owing to no ATTEMPT to keep up offspring, are not half lion & tige
535m069 lastly  the tracing facts to laws. without any ATTEMPT to know their nature.-- Reviewer considers thi
605o019 e 7 So that in this Essay. D. Stewart does not ATTEMPT «by one common principle» to explain the vario
098a047 concretion.--  state last page thus. point of ATTEMPTED crystallization, & therefore as a consequence
234b255 then with snout lift up latch & back.-- Frogs ATTEMPTED to be introduced I isle of France p. 170. «Fi
370d115 t to have been created.-- it might as well be ATTEMPTED to be shown from peculiarities of climate cau
593n107 r animals smell & looks.-- no doubt it may be ATTEMPTED to be said that young animal learns parent sm
440e147 bable way of explaining, much structure, than ATTEMPTING anything about habits-- no one can be shocke
088a013 f veins prove, that there are at least several ATTEMPTS at elevation From the lost & turned about pos
340d026 his book again.-- Mine is a bold theory. which ATTEMPTS to explain, or asserts to be explicable every
432e124 is with what highland shepherds said.-- p. 12. ATTEMPTS to improve the native <breed> animals of  any
638j58v e.-- reason, & not death rejects the imperfect ATTEMPTS. In the «Bee» Mollusca the nervous system  is
045r076 er on the great rise). -- The great rains which ATTEND severe Earthquakes «1822¿ 1835?» alone, (& the
045r077 much more curious & perplexing. than those that ATTEND Eruptions: Mr F. Scopes explanation of low Bar
515q021 or  does fruit merely not ripen.-- The point to ATTEND to is whether good & plenty of pollen is produ
415e068 ous.-- What a chance it, has been, (with what ATTENDANT organization, Hand & throat) that has made a
255c050 l part of Voyage of??? A Urubu, (with one leg) ATTENDED the distribution of food, at the Mission of M
332d003 such disorders accompanied with some fever, be ATTENDED by the transmission of large number of  worms
537m075 lowly does animal leave off <t> instinct, when ATTENDED with bad effects Martineau. How to observe, p
520m001 obably that they should have imitated.-- when ATTENDING Mr Dryden Corbet, he could not help thinking,
547m112 . anyhow I might have been quite still, & not ATTENDING to bodily sensation & yet the Castle would no
591n097 rs like dog-- wagged its tail «a little» when ATTENDING to anything or excited.-- so do young dingos,
620o044 sure guide.-- Hence conscience is improved by ATTENDING & reasoning on its action, & on the results f
590n093 on."-- p. 37. The increase of Bilary secretion ATTENDS passion p. 39. The sweat that accompanies fear
590n093 at accompanies fear is the same, as that which ATTENDS great weakness.-- <Diarrhaea> & syncope p. 42.
244c023 us Magellanicus say; <after> "après un examen ATTENTIF, et forts surtout de l'opinion du baron Cuvie
120a110 Greenstone of Salisbury Craigs well worthy of ATTENTION-- rear Glen Roy Notebook-- & scraps on Salsis
524m022 ntly perceived when my Father to distract her ATTENTION took her «left» hand to pretend to feel her p
542m095 savages without hats put up their hands, & as ATTENTION would amongst lowest savages clearly be direc
552m131 ng of breath to hear any sound.-- attitude of ATTENTION «So intimately connected is passion with send
554m139 d case of association.-- «Listened with great ATTENTION to Harmonicon. & readily put it. when  guided
592n103 s.» following pages contain remarks worthy of ATTENTION p. 15. 20. 40. 61. CD[a person is here said t
608o027 arm.-- Believer in these views will pay great ATTENTION to Education.-- 4) These views are directly o
617o39v to  it would require a considerable degree of ATTENTION. How do the senses affect us, except by inter
183b053 nce declined.-- Read his theory of the Earth ATTENTIVELY Cuvier objects to-<tran> propagation of spec
263c077 -- -- wonderful Man. "divino ore versus coelum ATTENTUS" is an exception.-- He is Mammalian.-- his <h
055r106 not occur on the beaches. Perhaps these facts ATTEST a <more> decided elevation of sea's bottom. be
101a058 erior at Falkland Isd.-- Peron does as if well ATTESTED.-- There is no difference between dike & moun
453e182 rom Rhine to the Caribs.-- Vol II p. 650. Long ATTESTED account of fall of fish in India.-- Windsor E
552m131 bitual stopping of breath to hear any sound.-- ATTITUDE of attention «So intimately connected is pass
530m043 My father signed a bond, yet when he paid the ATTORNEYS bill, he asked what bond he could have had. y
099a049 rticles <would> are aggregated, would they not ATTRACT strong. a third.-- & this would make layers.--
099a051 icle would tend to meet at <B. but if particls ATTRACT each other in some increasing ratio in proport
219b196 ourse. a parallel fact to Blood-Hounds. Before ATTRACT of Gravity discovered. it might have been said
295c178 ssumed--+++ ‖Daines Barrington says cock birds ATTRACT females by song. do they by beauty, analogy of
557m149 bly originally entirely sexual; first try «to» ATTRACT female, (or object of attachment) & then faili
616o039 to  the agent. Matter is by a metaphor said to ATTRACT; & hence if thought &c bore the same  relation
072r160 somewhat analogous to septa.-- would particle ATTRACTED towards space tend to form ring. [Fig. 10] {P
368d114 emales then fight for male) & are merely most ATTRACTED). -- singing best sign of most vigorous males
298c191 e altering as island increases.-- upper parts ATTRACTING all the moisture.-- Henslow thinks if leaf o
616o39v on» by which we come to conceive of matter as ATTRACTING & shew that the groundwork <of this> is enti
048r084 --The motion is most wonderful, from chemical ATTRACTION, as a blade of grass penetrating by action o
099a049 at once, the films vertical. Ascertain law of ATTRACTION of particles of same nature: then get mathem
099a050 fore < Ŝ of inclination «varies with chemical ATTRACTION &c.» becomes measure of force. < ∴ where lit
104a068 being lost by evaporation therefore capillary ATTRACTION would bring water with salt to surface Lyell
115a097 ra» p 94) be accounted for by great molecular ATTRACTION of every atom in rock On a coast, the shallo
```

```
195b101 but how much more simple, & sublime power let ATTRACTION act according to certain laws such are inevi
611b034 principle for connect of electricity chemical ATTRACTION, heat & gravity is probable.-- And the Organ
614o037 nction of organ, as bile of liver.-- ¿ is the ATTRACTION of carbon. hydrogen <&c> in certain definite
616o039 (especially the cerebral portions of it) that ATTRACTION has to ordinary matter. The relation of attr
616o039 ction has to ordinary matter. The relation of ATTRACTION to ordinary matter is that which action bear
616o039 t &c bore the same relation to the brain that ATTRACTION does to matter, it might with equal propriet
618o041 of the brain. We cannot perceive the thought ATTRACTION of sulphuric acid for metal of another perso
618o041 ? the brain only objectively. We do not know ATTRACTION objectively 6) The reason why thought &c. sh
618o41v he> (or conceive it) & that is all we know of ATTRACTION, but we cannot see an atom think: they are a
468L105 minute--not excessively abundant flowers not ATTRACTIVE, very small--stigma rather large & rough-- f
635j56v the doctrine of checks & my theory) Macculloch. ATTRIB. Vol I. p. 330. Mentions the many cases, as in
636j57r ce of beginning in organic world.-- Macculloch. ATTRIB. of Deity. Vol I p. 232. gives Woodpecker as i
632j53r strations of the Attributes of God Macculloch. ATTRIBS of Deity. Vol: I it will be better always to r
073r161 .-- "True native iron that to which we cannot ATTRIBUTE a meteoric origin & which is constantly found
265c086 nostril of Puffinuria I think we may clearly ATTRIBUTE to heredetary origin & not adaptation. to its
445e159 y live do. "Additions". p. 454.-- does really ATTRIBUTE metamorphoses to habits of animals & takes se
535m069 stic virtue, «&c» (Very true, no doubt savage ATTRIBUTE thunder & lightening to Gods anger.-- ∴ more
564n003 f militating against the existence of such an ATTRIBUTE would be rather favourable to it--!! Man more
614o037 - production of higher animals-- perhaps, say ATTRIBUTE of such higher animals may be looking back, ∴
616o39v ch thought or memory. might be in like manner ATTRIBUTED to the brain. There are two modes of percept
617o040 body of the individual to whom the action is ATTRIBUTED; force (be it remembered) being a phenomenon
401e014 ot return to all sorts of varieties, which he ATTRIBUTES to crossing.-- Cape Broccolli can hardly be
428e113 rses, which Eclipse? has begotten <?> «Walker ATTRIBUTES this to effect of male sex on locomotive sys
602o11b the imagination".-- Macculloch Vol I. p. 115. ATTRIBUTES of Deity. on Belief.-- you belief things you
632j53r cculloch 1837 Proofs and Illustrations of the ATTRIBUTES of God Macculloch. Attribs of Deity. Vol: I
634j55r rst occupy the Poles? Is this origin of Polar ATTRIBUTES of the Cetaceae.-- How came Bats also.? befo
638j28r nct, returning to Kirby's view.-- Macculloch. ATTRIBUTES of Deity Vol I. p. 251-- stomach hump, kinds
058r118 moins inclinée vers le rivage de la mer, tandis, AU contraire, que vers le centre de' l ile, elles p
209b156 of Flora of islands; "ou bien encore on pourrait AU plus en conclure quels sont les genres qui, sous
257c056 x. que nous avions déjà observés en Patagonie. ou AU moins des espèces tres-analogues.-- quand ce nêt
065r138 Land. Prince Edwards Isld. Marion & Crozet.. L. AUCKLAND. Macqueries.--Sandwich Isd-- Specimens of roc
590n094 owper, whole sentences spoken & believed to be AUDIBLE. one has good ground to call imagination a fac
318c253 snow.-- Acclimatisation.-- Bachman tells me in AUDUBON there is most curious history of first appeara
318c253 mmon-- likewise of the Hirundo fulva (added by AUDUBON in Appendix) showing WHAT CHANGES are taking p
575n045 he shudder of pleasure. from pleasure of music AUDUBON IV Vol of Ornith. Biog. case of Newfoundland d
326c266 Papers. Dr. Mayo. Pathology of Human. Mind.-- AUDUBONS. Ornithological Biography. 4. Volumes well wo
342d036 sing with females not thus characterized.-- 16th AUG.-- What a magnificent view one can take of the w
346d047 man confirmed my account of the Shepherd dogs.-- AUG. 24th. Was struck with pink shade on plumage of
460t017 ed showed no sexual inclination for each other-- AUG. 20th The Echnida & Hedgehog Tenrec both having
460t019 sporting" in flowers & domestication of animals AUG. 26th.-- When it is said that there is evidence
510q017 iberia, no sea-current, icebergs travel by wind. AUG. St. Hilaire Bot. p. 787. position of embryo in
534m062 place in brain. when sensation is perceived. = = AUG. 7th--38. Transactions of the Entomological Soci
539n081 s much less wondefful the instincts of animals-- AUG. 12th. 38. At the Athenaeum Club. was very much
539n082 g with his toe to perform some difficult task.-- AUG. 12th. When in National Institution & not feelin
539n083 hout direct consciousness?) change its habits.-- AUG. 16th. As instance of heredetary mind. I a Darwi
539n084 tion same.-- My handwriting same as Grandfather. AUG. 16th Anger «Rage» in worst form is described by
541m092 t have this high faculty, yet not clever people. AUG. 21st. 38 When a dog in play has his mouth open
423e096 ico-geographico changes must tend sometimes to AUGMENT & sometimes to simplify structures:= Without e
539m082 I was amused at this after seven years interval. AUGT. 15th. As child gains habit «or trick» so much
092a030 Journal Earthquake at Melville Isld New Holland AUGUS 1d to 3d & 19 1827 Geograp Journ There are some
121a112 racks?--? How came there ever to be cracks 11th AUGUST. 1838 Near Woolwich there are plains & valleys
122a115 became thicker, then when fluid moved [...] {P} AUGUST 25. I saw metamorphic conglomerates on shore o
343d037 said let there be light & there was light".-- » AUGUST 17th Two regions may be Zoolo-geographically d
344d041 tuses, as young of Marsup. is sucking foetus.-- AUGUST 23d The Rev R. Jones gave an admirable harrier
346d048 - laws of plumage might possibly be made out.-- AUGUST 25th Athenaeum (1838) p. 611. Ld. Tankerville
347d049 ard so as to resist external influence.-- 27th. AUGUST. There must be some law, that whatever organiz
347d050 & not man to Planets.-- instance of arrogance!! AUGUST-- 29th.-- Macleay in A. Smith's Zoolog.-- of A
540m089 n naughty.-- Why does person cry for joy? 17th. AUGUST Montaigne (Vol. I) has well observed, one does
543m099 wers of communicating knowledge to each other-- AUGUST 23d. Jones said the great calculators, from th
545m01k in those races, where it is customary to die-- AUGUST 24th. As some impressions «Hume» become uncons
546m108 y because does not like Burke explain pleasure. AUGUST 26th. I cannot help. thinking horses admire a
547m111 digestive nervous influence replace each other AUGUST 29th. Went to Bed. & built <common Castle in
548m117 se, one here was conscious of the two states.-- AUGUST 30th.-- It is singular when looking at a table
159g085 hypothesis Thursday. from Glen Turrit to Fort AUGUSTUS Barom on upper (rather above)? shelf 29.290 A
163g107 them himself-- Sand with tide ripple Near Fort AUGUSTUS hill & fringe as if it has been filled up «at
484z015 ral account-- ¿ Do not the Penguins replace the <AUK> Guillemout of the northern Hemisphere, & the Pu
341d029 was not adaptive structure, like little wings of AUKS which does not make that bird a Penguin.-- (i.e
484z015 he Puffinuria, the Awks.-- What structure do the AUKS bear traces of.-- like Puffinuria does of Petre
521m008 ating poetry in her dotage is fact of same sort. AUNT. B. ditto.-- Case of Mr Corbet of the <Hall> «P
522m012 ow [Z]CD when heard isolately.-- In old people. (AUNT. B.) when they hear a thing it often does not t
525m026 insanity are heredetary in any one family.-- In AUNT-- B. the affections «& N B affections very soon
539m083 le. & Eras a Wedgwood in many respects & some of AUNT Sarahs. cranks, & so is Catherine in some respec
366d111 species; & therefore to genera (& the uncles & AUNTS) & therefore does not tell against transmutatio
482z011 oliosoma -- novozelandiae -- histrinicus Vultus AURA Excessively inaccurate Saw a Chouette a huppe c
203b136 as produced by climate?-- M. Baer (thinks) the AUROCK, was found in Germany & thinks even now in cen
182b050 ecies between Australia & Van Diemen's land. & AUSTRAL & New Zealand Mr Gould says in sub-genera, the
247c028 --p. 71 Chimera-- Antarctica «also Taeniatole AUSTRAL» «caught»-- Chile, Van Diemen's land & Cape of G
243c021 different points.-- Consult Voyage aux terres AUSTRALES Chap XXXIX tom IV p. 273 2d Edit Consult Latr
324c268 & movements of Plant/ I£:4s Voyage aux terres AUSTRALES. Chapt. XIX. tom IV. p 273 Latreille Geograph
022r006 ous granites.-- Wide limits of this mineral in AUSTRALIA. Fitton's appendix Would Slate. & unstratifie
023r009 Archipelago? Such have never been observed in AUSTRALIA Dampier also repeatedly talks about the immen
028r030 hypothetical origin of some sandstones, as in AUSTRALIA.--Have Limestones all been dissolved. if so s
031r038 ia Archipelago. «Aleutian Arch.--» V. Fitton. AUSTRALIA: cases in Europe.-- Auvergne. very little Pum
041r066 are secretionary, so are all those plates in AUSTRALIA. New Red Sandstone. at Bahia in modern sandst
044r072 of N. America. India.--remembering S. Africa. AUSTRALIA. Oceanic Isles. Geology of whole world will
044r073 ns.--Unequal distribution of Volcanic action, AUSTRALIA S. Africa-- on one side. S. America on the ot
051r097 compared to whole E. America. <East> Africa. AUSTRALIA. profoundly deep: a great fault or rather man
053r101 Ralix Islds? In my Cleavage paper Dr Fittons AUSTRALIA case must be quoted at length. The Lines of M
079r177 apagos. no Hydrophobia at Quito. P 281. do do AUSTRALIA, C. of Good Hope.--Azores Isds «nor at St Hel
100a055 - Webster Phillips insists of analogy between AUSTRALIA & Oolitic period.-- comparison rather loose.
113a092 ferred subsidence; Mem my remarks on coast of AUSTRALIA.-- Great NW. dip in SE part of Australia.-- P
113a092 t of Australia.-- Great NW. dip in SE part of AUSTRALIA.-- Probably a case of rivers turning round &
114a093 nella Examine Iron stone of C. of Good Hope & AUSTRALIA/ and mud of salt-lakes of Rio Negro--Mr Bower
174b015 Antelopes at C. of Good Hope-- Marsupials. at AUSTRALIA-- Will this apply to whole organic kingdom, w
174b015 but each having its representatives-- «as in AUSTRALIA» This presupposes time when no Mammalia exist
182b050 close but certainly distinct species between AUSTRALIA & Van Diemen's land. & Austral & New Zealand
190b077 partake of Indian character There appears in AUSTRALIA great abundance of species of few genera or f
191b080 3 Paradupasi in common to Van Diemen's Land & AUSTRALIA From the consideration of these archipelagos
193b093 wan river long South coast all the remarkable AUSTRALIA genera, collected together Man has no heredet
203b137 ice of a swallow I think we may conclude from AUSTRALIA & S. America. that only some mundine cause ha
214b173 Lardner Encyclop. insists on analogy between AUSTRALIA and «fossils of» Oolitic Series does not appe
217b187 zil {P} Chile, Norfolk Isl.-- Isle of Pines-- AUSTRALIA.-- A <South> American «form of» Lathyrus has
226b220 by canoes Ceylon & India.-- Van Diemen's land AUSTRALIA. England & Europe.-- It will be well worth wh
233b249 land. see Coquilles Voyage), Waterhous remark AUSTRALIA Fauna so far. Indian all the rest. Timor acco
235b264 e elements How is Fauna of Van Diemens Land & AUSTRALIA [blank] Falconer's remarks on influence of cl
241c015 t with stiff tail like woodpecker.-- Birds of AUSTRALIA. Many in common ¿species? with New Guinea.--
241c015 L'Institut. 1838. A Dipus. & other rongeur in AUSTRALIA.-- p 67 ¿American forms? All Infusoria. not e
243c019 ferent.-- --Philedon circinnatus not found in AUSTRALIA only New Zealand-- Norfolk. Isd. & New Caledo
```

Page   ***(Key Word)***
246c027 cely differs more from, <Van Diemen's land.> AUSTRALIA more than Van Diemen's land.-- Vol II p. 8 no
253c046 hat Australian dog introduced by savages into AUSTRALIA.-- What are they? Colonel Montagu probably co
260c067 ed» then adaptation.-- No Carrion Vultures in AUSTRALIA!! Wilsons Ornithology, Vol III. p. 226 Wilson
264c080 some genus of yellow & brown-breasted bird in AUSTRALIA &c &c-- but of course they might be, blended,
264c080 .-- There is beautiful gradations of forms in AUSTRALIA leading on one side into shrikes & at the oth
268c095 of Van Diemen's land, which appear diff. from AUSTRALIA (Waterhouse <disputes this>) says differently
269c102 th coasts of New Holland.-- «Compare birds of AUSTRALIA with plants, with this object in view» The in
272c107 in Chalcididous insect, which I brought from AUSTRALIA, probably live in flowers & has Elytra. forme
274c113 g legged cuckoos with claw like lark, (one in AUSTRALIA is called swamp pheasant) Goatsucker--, parro
274c116 <imper> parts changed Have <not>. S. Africa, AUSTRALIA, & S. America very few forms in common.,-- bu
278c130 y own methods. Dasyurus being found fossil in AUSTRALIA, & only one tree species (Mitchell's authorit
278c130 ly one tree species (Mitchell's authority) in AUSTRALIA, & several in Van Diemen's land is most impor
306c213 ammalia & reptiles &c Timor is connected with AUSTRALIA «map to King's Australia» by a bank of soundi
306c213 or is connected with Australia «map to King's AUSTRALIA» by a bank of soundings of which «there appea
311c225 of Latd. p 355 Echidna of Van Diemen's land & AUSTRALIA different Temminck Fauna Japonica (?!) 82 mam
311c225 Japonica (?!) 82 mammalia 293 Phalangista of AUSTRALIA & Van Diemen's land diff.-- Habits can only b
313c233 ean. different.-- thorax & head differ Africa AUSTRALIA Parasites die, when brought over on tropical
314c239 e (Palm & Phormium tenax) as in New Zealand & AUSTRALIA, some SPECIES of Australian GENERA.. good case
314c239 ut Goulds case of birds of Van Diemens land & AUSTRALIA.-- The wombat (Brown) is found in Isd of Bass
315c240 Mush room & other cryptogamic plants same in AUSTRALIA & Europe.-- if creation be absolute thing, th
316c244 appears always very small ones as Trigonia in AUSTRALIA or Concholepas in America,-- yet many countri
319c257 rmediate between those of Van Diemen's land & AUSTRALIA proper.-- Irish Elk case of fossil geographic
321c275 story of Selbourne References at end Dr. Lang AUSTRALIA «trash» skimmed Macleay's Horae Entomologica
322c270 e many marginal notes <Rengger &c> Mitchell's AUSTRALIA Walter Scotts life I & 2d & 3rd Volumes Aberc
339d025 rovinces» according to varieties of Man.? «In AUSTRALIA. plants E & W very different.-- Man not so, b
341d030 seems most of species from there now found in AUSTRALIA New species of Moschus, characterized by Ogle
353d061 of Paradoxurus common to Van Diemen's land & AUSTRALIA well developed <tits> «Mammae» in male ourang
353d061 Mr Gould on Australian birds-- all Eagles. of AUSTRALIA characterized by wedge tails.-- many of the h
357d072 ole world shows there is rule.-- S. America & AUSTRALIA appear to have suffered most with respect to
357d072 forms.-- From observing way the Marsupials of AUSTRALIA have branched out into orders one is strongly
358d075 an D'Acunha ditto Juan Fernandez do Mitchell. AUSTRALIA Vol I. p. 306 "The crows were amazingly bold.
365d107 ave taken to separate. Van Diemen's land from AUSTRALIA &c &c Sept. 14th. When Macleay says their is
366d111 How long will the wretched inhabitants of NW. AUSTRALIA, go on blinking their eyes. without extermina
369d115 tted.""-- p. 483. Owen thinks from climate of AUSTRALIA, & from Ornithorhyncus & Hydromys not being M
400e011 ound in all quarters: his ideas not clear. In AUSTRALIA from approach to Asiatic [...]t in part near
407e041 m: Placent: insectivore being in S. America & AUSTRALIA. reason, why: Marsupiata, when first introduc
414e064 there gives them preponderance. intellect in AUSTRALIA to the white.-- The peculiar skulls of the me
436e136 Cestracion, Port Jackson Shark-- Owen thinks AUSTRALIA part of Old World <If> It «may» be said, that
458t001 size.-- the law of large size established-- «AUSTRALIA., S. America--» These strange forms., camels,
478z004 . on Pennatula showing it to be one animal In AUSTRALIA I was assured wild dog copulates <.> freely w
481z009 remarks that no insectivore in S. America or AUSTRALIA-- very curious.-- replaced by didelphidae Sku
510q017 ic ocean: are they from West, like as between AUSTRALIA & S. America? Sabine says North of Siberia, in
514q21. ington Mountain described in Flinders- Alpine AUSTRALIA Flora= Banana's seedless-- 20 varieties in mo
552m132 given.-- Descent of Man Moral Sense Mitchell AUSTRALIA Vol I, p 292 "Dogs learn sooner to take kanga
571m029 man» to eat poisonous?-- How did animals in «AUSTRALIA» & America manage;-- This shows doctrine «of
114a093 urning round & penetrating their own range in AUSTRALIAN Alps.-- Taylors Scientific Memoir, Part IV.
174b015 is presuppose time when no Mammalia existed; AUSTRALIAN; Mamm were produced from propagation from di
182b050 urity. it does as yet appear clim In Mr Gould AUSTRALIAN work some most curious cases. of close but c
200b122 ist describe animals as species, for instance AUSTRALIAN dog: <yet when that> or Falkland rabbit.-- T
212b165 former-- Mr Martens of Zoolog Soc told me an AUSTRALIAN dog he had, used to burrow like fox.-- a sor
217b187 d] Mr Don gave me instances of one species of AUSTRALIAN genus being found in Sumatra; again another
233b249 Timor according to Mountain chain ought to be AUSTRALIAN.?-- Mr Gould has been struck with similar ex
233b250 s offspring of cats sometimes heterogenous.-- AUSTRALIAN dog jumped into tub leaving only nose above
239c001 as no doubt that same thing would happen with AUSTRALIAN dog & any of our common varieties. He has no
243c019 de si hautes latitudes"., --¿translate?) All AUSTRALIAN forms have representatives (& instances give
243c020 ia peculiar species of cassicans: (¿cassicans AUSTRALIAN form? p. 27. many fish of Taiti found at <Ne
250c037 yed great Edentata or American form.-- Is the AUSTRALIAN Dipus an American form? The climate having g
253c046 angustidens.-- Ogleby has facts to show that AUSTRALIAN dog introduced by savages into Australia.--
264c082 variety of station inhabited by them-- Timor. AUSTRALIAN forms amongst birds Java. not so much-- Pecu
267c095 given-- Ch. 2. dogs L'Institut. 1838. p. 67. AUSTRALIAN rodents Abstract of Infusoria. <p> do p. 82.
296c184 o find out whether African forms. (anyhow not AUSTRALIAN) on Peaks. Did Creator make all new yet form
298c189 st like fox.-- He felt sure the half breed of AUSTRALIAN dogs, would be «most» like Australian.-- Cur
298c189 reed of Australian dogs, would be «most» like AUSTRALIAN.-- Curious this +ready answer, without any l
306c213 f difference Major Mitchell is not aware that AUSTRALIAN dogs ever hunt in company -- marked differen
310c224 --America.-- See Brown Congo Expedition: 400 AUSTRALIAN plants found in other parts of world Athenae
314c238 w Holland. As in N. Zealand-- Some species of AUSTRALIAN Genera Some species same (Palm & Phormium te
314c239 s in New Zealand & Australia, some SPECIES of AUSTRALIAN GENERA. good case. rather large flora. (150?
314c239 (150?) Mr Brown did not observe scarcely any AUSTRALIAN character in Timor plants, yet it seems ther
316c244 an & animals.-- yet how faint in a Fuegian or AUSTRALIAN! why not gradation.-- no greater difficulty
319c257 & those of S. America,-- Are not some of the AUSTRALIAN fossils intermediate between those of Van Di
332d004 n down.-- -- applicable to birds migrations & AUSTRALIAN Savages.-- W. D. Fox has a cat. which he bou
353d061 ural History. 1838 vol II p. 402. Mr Gould on AUSTRALIAN birds-- all Eagles. of Australia characteriz
355d066 character to goats.-- or dogs to foxes. (yes AUSTRALIAN dog) or donkeys to Zebras.-- «Mr Herberts va
392d180 dens; <Buffalo & common cattle-- Esquimaux (& AUSTRALIAN dogs with common dogs--> Ask my father to l
404e026 arent may be become favourable to offspring:\ AUSTRALIAN dogs have mottled coloured puppies case of t
424e103 ts puppies to be fondled.-- and we see in the AUSTRALIAN dog an instance of a half reclaimed animal.-
441e149 this.-- This would be curious law, Certainly AUSTRALIAN Dog is not affected by domestication, yet of
441e149 ted by domestication, yet offspring are,-- if AUSTRALIAN Dog, could bud, analogy tells us, <be> «offs
513q21. mature at same time on same plant --Flora of AUSTRALIAN Mountains.-- Is setting of fruit. cross Conc
514q21. geology & Species theory-- peculiar Fauna?. {AUSTRALIAN Alps--; are any Europaean forms found there-
553m137 wretched fears & strange superstitions of an AUSTRALIAN savage or one of Tierra del Fuego.-- Mr Mill
584n072 ating to one spot, this is indeed instinct.-- AUSTRALIAN man, may be called instinctive: the facts of
590n094 sight of instrument.-- Bennett's Wanderings, AUSTRALIAN Dog does not Bark-- quotes Gardner's Music o
350d054 <note> add this in note. ¿mere conjecture?-- AUSTRALIANS.-- Americans. &c After Decandolles idea Sept
366d112 ons will not alter form may be inferred from AUSTRALIANS knocking out teeth.-- the account of the peo
409e047 civilized man.-- ask the missionaries about AUSTRALIANS yet slow progress has done so.-- Show a sava
061r127 y less surprised at new creation for large.--AUSTRALIA'S = if for volc. isld. then for any spot of la
479z006 nsiders Dasyprus villosus is true Peludo Cavia AUSTRALIS. Dorbigny Vol II, p 24 Proceedings of Zoolog
058r118 gree with great continents). Voyage aux terres AUSTRALS Vol. I. p. 54. M. Bailly says."en effet toute
342d035 d as whole dynasties have been featured by the AUSTRIAN lip & the Bourbon nose". if this be not imagi
533m060 -- In Review (Edinburgh) of Froude's life. that AUTHOR remarks, that writing down his confessions of
632j53r Vol: I it will be better always to refer to the AUTHOR if I use these facts p. 280. adduces provision
215b178 roduction to Animaux Sans Vertèbres as latest AUTHORITY. The case of the tailess cat of Isle of Man m
278c130 ustralia, & only one tree species (Mitchell's AUTHORITY) in Australia, & several in Van Diemen's land
280c134 ases effect. it.-- Sir J. Sebright excellent AUTHORITY because written on dog-Breaking.-- applies it
294c175 its loins & all were so afterwards, (forgets AUTHORITY).-- Lonsdale is ready to admit, permanent sma
298c192 oast of Ireland, are absolutely (& who better AUTHORITY) similar with those over whole of country.--
424e101 re not to be trusted.-- -- Lyell tells me, on AUTHORITY of Beck, that Hooded crow & Carrion crow. hav
525m025 e S.) is very heredetary.-- My father says on AUTHORITY of Mr Wynne that bitch's offspring is affecte
242c017 henault Kuhl. Van-Hasselt, Reinwardt «Forrest» AUTHORS on E. I«ndian». A«rch.»\ Borneo & Sumatra both
324c268 es 8vo p. 181.-- See (p. 17) for references to AUTHORS about E. Indian Islands. consult Dr Horsfield
361d096 in England & N. America.-- the teal which some AUTHORS [not located] September 13th The passion of th
429e114 rmly each holds its place.-- When we hear from AUTHORS (Ramond. Hort. Transact Vol I. p. 17 Append) t
058r118 outes les montagnes de cette île se developper AUTOUR d'elle comme une ceinture d'immenses remparts;
569n022 ot in Library no good There is a Lutké's Voyage AUTOUR du Monde (1826-9) Paris. 1835 Quoted repeatedl
466t099 he Lemon-thyme.-- equally abortive as it was in AUTUMN: filaments united in whole length to corolla--
528m039 light» has very much to do, as may be known by AUTUMN, on clear day.-- 3d pleasure association warmt
031r038 h.--» V. Fitton. Australia: cases in Europe.-- AUVERGNE. very little Pumice, though Trachyte. same fa

(Key Word)
```
096a040 which  side craters were low --¿ applicable to AUVERGNE??? The fact of Galapagos Isld. steep side  to
098a044 log: 1833-34. p. 35.-- Ancient Lake Lemagne in AUVERGNE Proofs from Phryganea NB. Sedgwick talks of C
058r118 ral hypoth. agree with great continents). Voyage AUX terres Australs Vol. I. p. 54. M. Bailly says."e
243c021 culiar to the different points.-- Consult Voyage AUX terres Australes Chap XXXIX tom IV p. 273 2d Edi
324c268 x.-- on sleep & movements of Plant/ I£:4s Voyage AUX terres australes. Chapt. XIX. tom IV. p 273 Latr
199b119 Philos. Journ.-- Paper by Crawford on Mission to AVA. account of HAIRY man. because ancestors hairy w
450e175 r quitting Bengal this fact is noticed in Cassay AVA Pegue-- seldom equals 13 hands-- those of Lao  &
040r063 ndergone the same process.-- Neither lakes or AVALANCHES (Glaciers very rare) to cause floods in vall
089a018 esque toujours precedes et suivis, queque temps AVANT et apres, par de petites secousses."-- Tom  54.
601o009 » anterior to any direct sensation, in order to AVAOID it-- beetles feigning death upon seeing an obj
526m030 te free will, motive may be anything ambition, AVARICE, &c &c An animal improves because its appetite
600o008 eautiful cock, which best singer-- Remember.-- AVARICE a compounded passion gained in life time]CD 3.
600o08v nscience, an heredatary compound passion. like AVARICE.-- Is there not something analogous to imperio
622o048 which do not lead to action are repressed thus AVARICE. &c &c.-- [RHC] 8) in the beginning I mentione
628o054 otive part of man, may be quite artificial, as AVARICE love of gold.-- love of fame-- Yes Hartley exp
183b052 p. 243 it is said as well known fact that "serin AVEC le chardonneret, avec la linotte, avec le verdi
183b052 ell known fact that "serin avec le chardonneret, AVEC la linotte, avec le verdier" &, <fr> silver gol
183b052 at "serin avec le chardonneret, avec la linotte, AVEC le verdier" &, <fr> silver gold & common pheasa
315c241 «1838» p. 184 Botany of Bonin. "grande analogie AVEC le flore du Japon", some Europaean & Sandwich s
571m031 natural rise-- I was also much struck in great AVENUE, resemblance to gloomy aisle of Churche.-- the
090a023 l weight recorded is 473. pounds (taking about AVERAGE when several are given), this will give nearly
090a023 al are given), this will give nearly 19 pounds AVERAGE for each stone. that fell, that was weighed,;
206b146 and  at present day, every ten living souls on AVERAGE are related to the (200dth year) degree.  Then
232b246 ysical changes (we assume like weather on long AVERAGE tolerably> uniform).-- Comparing fossils  with
284c147 f desert, open ocean, &c this probably on long AVERAGE, equal quantity, 2d on relations of heat & col
284c147 stations & diversity--.-- this perhaps on long AVERAGE equal.-- The Cocos do Mar on the Mahé island,.
365d107 ange by men & by art.-- the former only giving AVERAGE of effects of country, (& no monstrosity, or a
375d135 wheat  for making brandy.--» take Europe on an AVERAGE, every species must have same number killed, y
431e123 e male being smaller, & the female larger than AVERAGE size: (surely this is very limited view, thoug
582n067 According  to my view marrying late, will make AVERAGE of life longer.-- for short-lived constitution
194b093 proportion instinct more. reason less. so will AVERSION be L. Institut «1837. No 246» a section of fo
255c051 cklings and chickens) Young water ouzels hence AVERSION to generation, before great difficulty in pro
389d173 als.]CD p. 310 She wolf took dog. but had such AVERSION to it, that she was held Hunters Eoeconomy So
070r153 columnar  & orbicular in basalt.-- When we see AVESTRUZ two species. certainly different. not insensi
317c248 e.-- Also some few facts at end of "The British AVIARY" or Bird Keepers Companion Study Appendix (& o
321c275 s Wisdom of God references at end-- The British AVIARY.--do-- & Lisle's Husbandry Tuckeys voyage rer
025r016 lo [13 degree 22 minute S.] 9 120 {T} Garcia de AVILA [lighthouse] [12 degree 35 minute S.] 9 124 Ita
257c056 s & found "beaucoup des mêmes oiseaux. que nous AVIONS déjà observés en Patagonie. ou au moins des es
191b083 ures often mixing of individuals. Here we have AVITISM the ordinary event. & succession the extraordi
291c166 Materialist!-- Read Barclay on organization!! AVITISM in mental structure or disposition. & avitism
291c166 Avitism  in mental structure or disposition. & AVITISM in corporeal structure are facts full of meani
344d042 ve been acquired, & hence this is then case of AVITISM.++» Three gentlemen of party all thought  with
367d112 within it, is forming «& this must be so, else AVITISM could hardly ever occur.--».-- & if that canno
379d151 imself is not like-- now this is clear case of AVITISM. but then ¿ was «not» the expression of <fathe
379d151 from his father so that case ceases to be true AVITISM Annals of. Natural. History. p. 135. Natural H
392d180 --> Ask my father to look out for instances of AVITISM Examine English weeds in Hot. Houses will they
371d118 eveloped tails, & one with beak turned up like AVOCETTE. here is what [not located] that it shall beg
309c220 ue «opening» case. «thus» Educate all classes-- AVOID the contamination of <cl> castes. improve the w
309c221 rdt to show black colour of certain Arabs.-- NB AVOID quoting these hackneyed cases Mr Ed Blyth  does
531m052 t take place.-- A start is HABITUAL movement to AVOID any danger-- Fear, shamming death, or running a
531m053 be habitual «involuntary» movement from wish to AVOID some danger-- but it is instinctive because Nan
532m057 te» effect of short -- but violent action.-- To AVOID stating how far, I believe, in Materialism, say
536m070 puts  himself still, & walks hard.-- «He cannot AVOID sending will of action to muscles. any more tha
560mIBC oung, before experience can have taught. them to AVOID danger Do they frown, when they first see it? C
573n038 o walk to door without touching table.-- cannot AVOID it.-- curious mixture of voluntary & involuntar
281c139 f circular arrangement was only so far true as AVOIDED linear arrangement the central twigs dying, af
468t111 = between them, hence quite below stigma. & so AVOIDED it.» On certain days Humble seem to frequent c
584n072 - it is the acquirement of a new sense,-- bats AVOIDING strings «in the dark» as well might be called
493q003 heir wing bones & see if relation is same good, AVOIDS effects of fatness.-- Experiment in crossing a
477z003 chuca or Benchuca. "Les individus ailes peuvent AVOIR <quatre> cinq lignes de long et volent. p.  208
276c123 st remember that if they believe & do not openly AVOW their belief. they do as much to retard, as tho
544m102 ther in manner <they> quite different from when AWAKE.-- peculiar sensation as flying. (No memory  of
548m114 ing by senses.-- As sleep <is> only one idea is AWAKE, when one is awake many necessarily are.,  when
548m115 sleep  <is> only one idea is awake, when one is AWAKE many necessarily are., when one is deeply reaso
572n033 d not gather general sense of this page.-- Now <AWAKE> «when awake» I could not picture to myself rea
572n033 general sense of this page.-- Now <awake> «when AWAKE» I could not picture to myself reading French b
313c235 after long interval very good case.-- habit is AWAKENED by association (case of Elephant, which had r
528m039 of imagination, which correspond to those <he> AWAKENED during music.-- connection with poetry, abund
532m053 ear must be simple instinctive feeling: I have AWAKENED in the night. being slightly unwell & felt so
539m083 the  consciousness of double individual is not AWAKENED.-- The habitual individual remembers things d
542m095 awaking  from sleep see how a dog yawns when he AWAKES. & streching & yawning can be explained from t
601o08a 177  good passage in French on what dog dreams, AWAKES-- does when Master takes Hat de l'education de
542m095 like sneering does.-- is yawning habitual from AWAKING from sleep see how a dog yawns when he awakes.
102a060 of  Ascension, each particle coated. &c will be AWARE how little common Gravity has to do with arrang
306c213 passion  chief difference Major Mitchell is not AWARE that Australian dogs ever hunt in company -- ma
366d111 ften as she touched handle, though really fully AWARE she was not coming in, could not help being per
524m019 e done in passion.-- People are constantly well AWARE that they are insane & that their idea is wrong
550m127 -- now if one has these feelings, without being AWARE of their association «ie heredetary», does  one
573n039 her  shoulders.-- analyse this.-- Miss C. quite AWARE & indignant with Mrs C. but had no influence ov
032r039 the band of greatest action not having been worn AWAY.--If the level of the sea was to sink by very s
045r077 slightly  oscillating as that of a spring) moves AWAY.--Will geology ever succeed in showing a direct
053r100 nterstratified with sediment.--& escarpment worn AWAY like english escarpment The great  conglomerate
11a090 per four hundred feet of strata having conducted AWAY the heat of surface. & if conducting powers had
150g038 & far more earthy than what is usual-- Lines die AWAY where slope less., best developed on steep east
151g039 hy slope, two circumstances rarely united.-- die AWAY also, without any cause, must be tides. &c. {P}
155g062 st has hollowed out most Wednesday Shelf 3d dies AWAY almost imperceptibly on Glen Turrit side 2nd sh
157g073 f Beagle Channel when mica slate, only sand blow AWAY]CD where lines appear to cross stony parts; app
160g093 vium» which appear perfectly level, <on op> dies AWAY on gradual slope-- : on N side.. dies away on r
160g093 dies  away on gradual slope-- : on N side.. dies AWAY on rocky place, but narrow shelves just like ro
162g100 along, parallel to Shelf on opposite side & dies AWAY on the steep & rocky gully of last stream Frida
163g107 lled up «at» 30 ft. higher with pebbles now worn AWAY-- The above shells must have been about 60 ft a
190b076 forms. reduce towards Northern Eastern end & die AWAY, & partake of Indian character There appears in
268c096 use feeding young of redstart & actually driving AWAY parent birds.-- showing how blind a storge From
304c206 two adjoining families & not all organs blending AWAY.-- +++ Hopeless work to systematist, who believ
461t041 ies will undergo & yet remain adapted.-- it does AWAY with difficulty of rabbits of England remaining
531m053 id any danger-- Fear, shamming death, or running AWAY. accompanied with want of muscular exertion, pa
531m053 rightens. them.-- Now every animal moves quickly AWAY from any sudden sound or noise, & therefore bra
532m057 uscles, are not these effects of violent running AWAY, & must not <this> «running away» have been usu
532m057 violent running away, & must not <this> «running AWAY» have been usual effects of fear.-- the state o
535m064 ng them from «the» sun & enemies-- would not fly AWAY, but bit pencil when touched with it-- do not k
554m138 us.-- Mr Yarrell has seen Jenny, when Keeper was AWAY, take her chair & bang against the door to forc
555m143 fe, & then made many jokes. about not having run AWAY &c having faced death like a hero, & then I had
557m148 erent from fear; there is no inclination to jump AWAY.-- it is, ill-defined fear.-- Yet one knows one
557m149 ghter.-- A dog who goes home from shooting. runs AWAY. is not afraid the whole way. but ashamed of hi
557m149 or object of attachment) & then failing to drive AWAY rival.-- Fear is open mouthed to hear. though i
563n002 him from jealousy. (Pincher & Nina)-- or to take AWAY food &c &c-- Now if dogs mind were so framed th
565n010 must  be very cautious. Remember how Lavater ran AWAY with new Lavaters,-- Ye Gods!:-- says fleshy li
569n023 made no reply, but shrugged his shoulders & went AWAY."-- he implies negation, without violence, with
```

Page ***(Key Word)***
620o044 d dinner, or from a blow struck in passion fades AWAY, so that when man afterwards thinks why was suc
637j58r harder. so must those with weak beaks be sifted AWAY.-- 4 & the species, like 10,000 others, perish.
204b139 id about half way. Eyton says Hybrid about half AWAYS & result the same Indian cattle & common produc
057r112 --Yet neglecting these final causes.--What more AWFUL scourges to mankind than the Volcano & Earthqua
205b144 er Does the odd Petrel of F. del F. take form of AWK, because there is no awk in Southern hemisphere.
205b144 F. del F. take form of awk, because there is no AWK in Southern hemisphere. does this rule apply? A
177b028 ivided themselves into many species, so have the AWKS, there is particular circumstances, to which> i
484z015 f the northern Hemisphere, & the Puffinuria, the AWKS.-- What structure do the auks bear traces of.--
564n005 Here then, that faculty, whether for position of AXE of eyes, state of surface, or other means by whi
201b126 >The fact is an additional illustration of that AXIOM in Natural History that all aberrant & osculant
350d054 ir Thomas Browne says "is the only indisputable AXIOM in Philosophy<"> Religio Medici. Vol II. Sir T
021r005 . Henslow Anglesea, nodules in Clay Slate. major AXIS 2.½ ft.-- singular structure of nodule, constit
021r005 e, constitution «same as» of slate same.--longer AXIS in line of Cleavage. laminae fold round them; Q
023r011 60 miles distant from the grand ancient volcanic AXIS «of the Andes».-- «Has this fault determined si
023r011 ault determined side of volcanic activity.» That AXIS was produced, from a fissure in a deep & theref
032r041 ward.--If matter proceeds from great depth. from AXIS to surface must gain a Westerly current:--If gr
088a013 hickness not very great; where piece turned over AXIS or hinge no doubt fluid.-- analogy as continent
099a048 with oil should shrink. film parallel to longer AXIS. But if great depth NB. Prof <Henslow> Sedgwick
101a058 - There is no difference between dike & mountain AXIS. except in relative <strata> size with superinc
101a058 ncumbent strata. where they have yielded conical AXIS of mountain.-- only when dikes reach near the s
103a065 apital discussion might be made between dikes & «AXIS of» mountain-chain in proportion to weight of s
104a066 ent more silica than common felspar therefore on AXIS of Cordillera, in Andite - containing 80 per ce
129a132 es long I in with.» which must have been from an AXIS, «20 ft at least in depth» near mouth of Columb
152g047 d shells Important contingency if elevation from AXIS, then rivers might deposit, & afterwards with g
241c014 -- -- Antelope in Celebes, Bourou new species of AXIS.-- «Cervus moluccensis is different from that o
362d099 orns begin to grow.-- Mr Yarrell says the «male» AXIS of India, breeds at times when horns not perfec
435e133 ents". «translated by Dr. Hodgkins» p. 54. The AXOLOTL, siren, & Proteus, affinity to tadpoles. p. 21
559m153 g & outang, & dare to say difference so great... "AY Sir there is much in analogy, we never find out.
041r064 on of pebbles & vertical trees Grand Seco at B. AYRES; mention about the deer approaching the wells.-
067r143 6 (p302) Coleccion de obras. 2 Vols fol: Buenos AYRES 1836: W. Parish?? «by Pedro de Angelis.» This w
095a037 a certainly is found with the Mactra. at Buenos AYRES at the Zoolog: Soc: Terebratula from Hudson's B
123a117 n in Salisbury Craigs Letter from M Angelis. B. AYRES. 3d. May. states remains found in many part.--
123a117 -- large quadruped bigger than ox.-- at Buenos AYRES 20½ quadras from river; 20 varas from surface i
244c022 p. 139. Vespertilio bonar«i»ensis (from Buenos AYRES) replaces <Vesp.> holds same relation with equa
465t080 tagonia-- Beds of La Plata. (except close to B. AYRES).-- If we may take this as guide, the shells pr
466t093 is so in Kidney Bean. How is it generally.-- In AZALEA <do> «it is so» <Though I saw no Bees «several
468t111 odendron-- <Loasa> «Anchusa»-- speedwell Iris-- AZALEA. Rhodendron. Fraxinella to Anchusa <never> «on
468t112 inella. with respect to nectary is same case as AZALEA or Rhododendron xx after several gloomy days.
496q05a rs as do not seed or seed rarely-- Magnolias. «AZALEAS» & plants grown under unfavourable circumstanc
060r126 n numbers. p. 124. Webster Consult W. Parish. & AZARA about dry season[.] 1791. seen commonly bad ove
271c105 ern range is? <One> The black & white thrush of AZARA builds its nest in <same country> something sam
429e113 at the Dahlias. we may infer it in animals.)-- AZARA gives account of production of hornless cattle-
477z001 rvations on Planariae by Johnson CXII. & CXV do AZARA Voyage Vol I p. 196. According to Charpentier d
477z002 ago <no> snail was introduced to Mauritius. 18 AZARA Voyage Vol. I. p. 279 Thinks the Moruffetes of
477z003 Paraguay in 1769 introduce in Governor's tran?? AZARA Las Vinchuca or Benchuca. "Les individus ailes
482z012 two species of Tortoises come from Galapagos!!! AZARA. Voyage dans l'Amerique Merid. Tatu noir. abund
251c041 pecies domesticated, strangely contradictory to AZARAS fact of conduct of wild & tame horses.-- p. 24
320c276 nzie's Iceland Molinas Chile Falkners Patagonia AZARAS Voyages & Quadrupeds of Paraguay Dobrizhoffer.
055r107 ge cliff form of land, in St Helena. Ascension. AZORES. («sandstone first gives» half demolished crat
060r126 e red Sandstone of Andes fusible? no. mad dogs. AZORES. although kept in numbers. p. 124. Webster Con
074r165 mans account of steam acting on trachytes. also AZORES. We here have case of such vapours washing a r
079r177 uito. P 281. do do Australia, C. of Good Hope.--AZORES Isds «nor at St Helena.--» Humboldt. New Spain
096a040 e talks of gravel on basalt of Heckla-- All the AZORES Isld. Von Buch p 359 stretched out NE & SW.--
408e044 f elevation,-- «all» modern & wholly volcanic-- AZORES might be prophecied to have this character.--
511q018 ologists, Bhem & Glöger Consul Hunt, birds from AZORES or Madeira Mr. Blyth (1) Mentions some breeder
100a055 ort, whether, water flashing into steam, would BABBAGE.-- Webster Phillips insists of analogy between
413e059 er this (Cetaceae) with reference to my theory BABBAGE 2d Edit, p. 226.-- Herschel calls the appearan
530m044 omewhere heard (Hunter?) that pulse of new born BABIES of labouring classes are slower than those of
531m053 s instinctive because Nancy tells me very young BABIES start at anything they hear or see. which frig
532m053 oing it.-- Fanny Hensleigh doubts whether young BABIES start.-- ‖ If children wink. it is instinct Fe
533m058 f acquiring language.-- Hensleigh. W. says that BABIES know a frown very early in life, <before they>
560mIBC erns in the hothouse at home Natural History of BABIES-- Do babies start, (ie useless sudden movement
560mIBC hothouse at home Natural History of Babies-- Do BABIES start, (ie useless sudden movement of muscle)
588n089 piness.-- Bell on the Hand p. 191 Says <childr> BABIES have an instinctive fear of falling.-- &p. 193
298c192 arkable what is analogous to it in animals?-- BABINGTON says in most plants, even those on Guernsey &
299c195 ants many of the Female flowers unimpregnated BABINGTON We see gradation to mans mind in Vertebrate K
489qIFC p. 14.-- Father. And. Smith Dr. Holland p. 16 BABINGTON-- Gould ---- 10.(a) J. Gray ---- 17 Yarrell--
500q10a ecies from Van Diemen's Land? or New Zealand? BABINGTON about differences of Irish & British Species
538m079 h facts bear on such characters as Allen W. & BABINGTON, both half ideotic in some respects & with st
538m079 bad for fires, what is wrong in his head. & BABINGTON'S silly joking The possibility of the brain ha
306c212 t were the stream 1000 miles long.-- a monkey. (BABOON) at Z. Gardens upon being beaten behaved very
315c243 th. no doubt a habit gained by formerly being a BABOON with great canine teeth.-- (This may be made c
376d136 rks on extraordinary curiosity of Monkeys». The BABOON of which anecdotes have been told is Cyanoceph
376d137 r it, & made it meet in front.-- Dr Smith every BABOON & monkey, big & little that ever he saw knew w
539m084 Metaphysic must flourish.-- He who understands BABOON <will> would do more towards metaphysics than
545m105 -- do/ I was much struck with observing how the BABOON (<Macaco> «Cyanocephalus Sphynx Linnaeus») con
545m106 they cannot remain still.-- I do not doubt this BABOON. knew women.-- Another little old American mon
550m123 f our evil passions!!-- The Devil under form of BABOON is our grandfather!-- A man, who perfectly obe
553m137 es of monkeys go in groups. thus the pig-tailed BABOON, shoved out its lip, looking absurdly sulky «a
402e018 d of pigs, small bears or badgers, deer, apes, BABOONS, monkeys & an animal probably a tapir p. 233.
448e167 . p 155. By inference I imagine that there are BABOONS in St Thomas on W. coast of Africa Owen Linn.
445e158 t says it is evidently acquired by experience in BABY Lamarck. Vol II p. 152.-- Philosophie Zoologie.
542m096 scurely with the wish to make it out?-- Seeing a BABY (like Hensleigh's) smile & frown, who can doubt
594n113 disordered) at thought of almost anything ugly. BABY-- association-- pouting child same as anger, li
595n121 deal feeling. Same effect as acting on us-- <The BABY> «Effie Wedgwood» April 28th 1840 was frightene
481z009 agonia. Mem:-- S. Cruz. Molina Vol. I. p. 244. BACCALAO. migratory fish.-- See Kings drawings.-- for
317c251 scovered» of N. America, & of the shrews.-- Dr BACHMAN told me. that near Charlestown ?three species,
318c252 America, with respect.. to the Cordillera,---- BACHMAN has seen webbed Shrews. case of adaptation.--
318c253 tted for regions of snow.-- Acclimatisation.-- BACHMAN tells me in Audubon there is most curious hist
318c254 tnight, «See Silliman's Journal 1837. Paper by BACHMAN.» that is succession of birds.-- «in» some spe
319c255 rer like the first pair of Pipra flycatcher.-- BACHMAN says he thinks the Mocking thrush beats all En
328cIBC ces. Silliman's Journal. during 1837. paper by BACHMAN on migration of birds Temminck has written "Co
341d031 us like some facts of Mr Blyth on birds.-- Dr. BACHMAN tells me line of Rocky Mountains separate almo
341d031 g by some permanent white streaks.-- &c &c Dr. BACHMAN has crossed cock Guinea Fowl with Pea <cock> H
342d032 n duck a variety of Muscovy) with goose!!) Dr. BACHMAN regularly breeds «in Carolina» for his table M
342d034 - Are Pheasant & Grouse homogeneous? I observe Dr BACHMAN calls these Hybrids new species. Yarrell says
363d103 er. & Moschus &c & -- like young blackbirds Dr BACHMAN told me that 1/2 Muscovy & common duck were of
031r038 rachyte. same fact in Galapagos. Daubeny P 24 V. BACK of page 1 of New Zealand Geological Notes. at S
157g077 wo branches unite in upper Glen Roy» very little BACK from line 2d; little action since «that shelf»
179b033 <If> Is there a tendency in plants hybrids to go BACK?-- If so Men & plants together would establish
180b039 n adaptation of circumstances.-- Vide two. pages BACK. Diagram The largeness of present genera render
181b040 not rapidly increasing.-- If we thus go very far BACK to look to the source of the Mammalian type of
181b041 esent day many are relatives, so that by tracing BACK. the <descen> fathers would be reduced to small
183b052 être feconds. dès les premières generations" go BACK to type of either animal when crossed with it.-
186b058 <heteromera> colouring of crysomela may be going BACK to common ancestor of Crysom. & Heterom, but I
190b075 such marriages or offspring show tendency to go BACK-- there is an end to species.-- «Brown Appendix
192b083 any variety, although many of the seeds will go BACK.-- Get instances of a variety of fruit tree or
213b170 90. it seems the most organized fishes lived far BACK, fish approaching to reptiles at Silurian age--

(Key Word)
213b171 proaching to reptiles at Silurian age-- How long BACK have insects been known? As Gould remarked to m
220b197 em Lord Moreton's Mare. The fact of plants going BACK hybrid plants; analogous to Men. & dogs. Now if
228b229 e laws of life".-- Where we have near genera far BACK, as well as at present time, we might expect co
228b229 g of a male & female animal of one variety going BACK ¿whether this going back may not be owing to cr
228b229 al of one variety going back ¿whether this going BACK may not be owing to cross from other trees.????
228b230 them. Ask Henslow for some plant, whose seeds go BACK again, not a monstrous plant, but any marked. v
230b240 ected to change in Americas. perhaps merely gone BACK previous to fresh change Get a good many exampl
234b255 put legs over, & then with snout lift up latch & BACK.-- Frogs attempted to be introduced I isle of F
239c002 has no doubt that Chesnut, for many generations BACK, were crossed with Bay mare, only bay a few gen
239c002 add, that fact of half cross with parents, going BACK to either parent is lucidly explained.-- Mr Yar
248c030 even in domesticated varieties a tendency to go BACK to oldest race, which evidently is tending to s
255c051 trifling marks.-- such as the bands on pidgeons BACK.-- According to this description of class is de
257c056 rly, we cannot guess causes of change.-- hump on BACK of cow!!-- &c &c D'orbigny (p 108) says having
259c064 ough we see it affected» tempts one to bring one BACK to distinct creations.-- it is only be recollec
280c133 d, yet seems to grant, that difficult & other go BACK to either parent.-- Shows instinct (Sir J. Sebr
280c136 would go against the tendency.-- it tries to go BACK to grandfather, but if too unlike its own paren
280c136 les bred or two certain varieties, they would go BACK to grandfather, which is true) & infertility is
290c164 ull in garden at Zoology Soc. it's pale ash grey BACK, like a black bird washed, Whilst tips of prima
293c172 reflection or consciousness of reasoning to tell BACK from front. &c or use of button holes it would
304c207 then be told whether not descended from long way BACK.-- aberrant forms produced where many species <
335d013 would sometimes vibrate-- «seeing no tea brought BACK memory» old habit of putting tea in pot made me
335d013 e parents.-- therefore offspring will tend to go BACK, or have none-- the argument does not apply to
336d019 a bud could not be taken, without it either went BACK, or not being perfect would perish.-- The Varie
337d021 intermarriage.-- In my speculations. Must not go BACK to first stock of all animals, but merely to cl
344d042 ed Beagle Staghound. «†† ».:the grandchildren went BACK to either paret, & breed not fixed. though she
358d086 h> the breast of which is like common pheasant & BACK like silver.-- But the hen hybrid of this bird,
364d104 heads; which rears afterwards occasionally went BACK-- (Effect of imagination on mother. white peele
367d113 ge, heavily made cream coloured ass.-- stripe on BACK also.-- legs reminded me strongly of Zebra.-- M
377d147 e <p> Kingfisher (.p. 169) has the colour on its BACK bright blue.-- «thus young of» Many of the pies
378d147 that assumes change, & is the offspring brought BACK to earlier type by Mother?-- do these differenc
379d151 as the past for other Mammalia. & still further BACK reptiles & Cephalopoda: Old Jones remarked to m
383d160 -- Head like silver except in not having tuft,-- BACK like do.-- but the black lines on each feather
383d160 en? & not purple?-- legs pale coloured.-- In the BACK feathers, we have character very different from
398e006 es which the government is subject to.-- further BACK we obtain here & there in order a scattered pag
432e124 not, (& in hairless kittens we see same fact) go BACK, & this is argument against Blyth's doctrine of
437e141 s Sprout returning suddenly to type when brought BACK to home. (& yet all the varieties of Brassica c
460t015 een swan-goose & common goose.-- the stripe down BACK pretty plain in in these <half> <3/4> bred ones
470t178 x Down, 1854, Sept.) In Spanish Broom by pulling BACK Wings, pollen is ejected with violence in showe
484z014 ts-- Does the Secretary, make noise & throw head BACK M Edwards,--on polypi of Tubulipores L'Institut
493q003 «wh was female» with tinge of tortoise-shell «on BACK.--» = Length of intestine in Persian Cat-- , in
503q012 o experimentise on hybridising ferns, tying them BACK to back 37 Col. Sykes fertility of men & Europa
503q012 mentise on hybridising ferns, tying them back to BACK 37 Col. Sykes fertility of men & Europaean anim
505q014 inches. From sole of foot to shoulder on line of BACK, height 17½'. The Greyhound. was in length (mea
508q016 more common in man than in female-- (8) In Hump-BACK ever heredetary (9) Are the works of Berhave (t
524m021 ell understood. In insanity, the ideas do not go BACK to childhood, (but appear most capricious) as i
524m021 nd childhood full of meaning:-- Dreams do not go BACK to childhood-- People, my Father says, do not d
525m023 der their duty.-- as carrying a basket, bringing BACK game, or picking up a stone, though only acquir
532m056 py. in five weeks was so thin, that she was sent BACK to Shrewsbury,, then immediately fell into her
545m107 der same circumstances, threw itself down on its BACK & kicked & cryed like naughty child.-- Do monke
548m109 ses.; than the <small> fact that no one, looking BACK to his life, would say how many good dinners or
547m111 vaguely thought of packing up.-- was lying on my BACK fell to sleep for second & wakened.-- had very
548m116 use in this state, the consciousness does not go BACK to former periods so «as» to <make> <give> one
550m125 on-- by saying what is Happiness?-- When we look BACK to happy days, are they not those of which all
557m147 sion.-- dog tail curled when angry & very stiff. BACK arched. just contrary. when pleased tail loose
557m147 pleased, erect its tail & make it very stiff «& BACK» when savage «no» & ready to dash at prey strec
557m147 reched out & flaccid, when furious «with fright» BACK absurdly arched. & tail stiff.-- is shame, jeal
558m152 . quarrelling. mouth wide open, each [lip] drawn BACK & driving air out of mouth «hairs erect on back
558m152 back & driving air out of mouth «hairs erect on BACK» «wide open» with prodigious force.-- making gr
558m152 gry. curls tip of tail.-- do two cats arch their BACK when fighting, & not with dog. when fear might
564n005 - young partridge can run even with its shell on BACK.-- To study Metaphysic, as they have always bee
570n024 s nothing. if he did go to reply. he would throw BACK his shoulders. he wishes to show, he is determi
585n075 The Giraffe kicks with front legs & knocks with BACK of Head, yet never puts down its ear. good to c
585n077 & then taken other way-- would not find its way BACK.-- ?? this is not instinct, but a faculty, or s
586n079 umber of house though one cannot remember it.]CD BACK, without consciousness & by habit, such habit o
596n184 gradation towards man.» Macacus especially pulls BACK skin of whole forehead & 2 ears.-- emotions of
596n184 cus make labial st st. S. American monkeys. pull BACK skin from head very little Does blood go in <bo
609o029 n to other animals & therefore to progenitor far BACK, (anger <to> at the very beginning, & therefore
609o030 of what has generally been best for our good far BACK.-- (much further than we can look forward: henc
614o037 attribute of such higher animals may be looking BACK, ∴ therefore consciousness, therefore reward in
616o038 mperfect memory may become perfect & we may look BACK to definite action or to our conscious selves.-
616o038 or to our conscious selves.-- Such memory may go BACK to animals which were changed into man ∴ they m
639j28v l get the upper hand. though continually dragged BACK to old type by intermarrying with ordinary race
234b255 I isle of France p. 170. «Fish introduced» Hump BACKED race of cows from Madagascar-- p 173. Vol I. V
400e012 8 found cattle in Table Bay with Humps on their BACKS & big tailed sheep do Vol 10. p. 373, «& 374» S
558m152 great passion.-- I do not think they arch their BACKS-- Bengal tiger. when slightly angry. curls tip
206b146 ople <might be> being related within 200 years BACKWARD might be calculated & this number elimanated
222b202 ion of species by travelling of climates & the BACKWARD & forward introduction of species.-- When spe
206b146 were progenitors of present people, and so on BACKWARDS to one progenitor, who might have continued b
206b146 might have continued breeding from eternity «BACKWARDS.--» If population was increasing between each
391d175 ies is formed if not.-- the changes oscillate BACKWARDS & forwards & are individual differences (henc
058r116 of more than 500 leagues. A little time after a BAD earthquake in Chili; Arequipa in 82 was overthro
060r126 & Azara about dry season[.] 1791. seen commonly BAD over whole world. «(Was it so in Sydney, consult
077r172 veta madre of [misnumbered page] Dr D. remarks. BAD conductor of Heat do of Electricity Does not iro
126a122 may be very wrong,-- The fact of a dumplin being BAD conductor is {P} against my views-- if we had ro
126a123 surface does not become hot?-- this looks as if BAD conductor-- III But equilibrium is not attained,
132a139 rots shows from variation in strata earth a very BAD conductor.-- shows p. 516 that subterranean spri
279c133 lood.-- +† thinks difficulty in crossing race.-- BAD effects of incestuous intercourse...-- excellent
316c244 uture state, that when good enough for Heaven or BAD enough for Hell.-- «†glimpses bursting on mind &
343d037 e» said let there be light & there was light.-- «BAD taste (whom it has been declared "he said let th
346d048 old & wounded,-- see Annals. vol. 2. 1839.-- are BAD breeders & subject to the rush as all animals wh
399e010 trongly suspect, that breeding in & in, produces BAD effects solely, because of similarity, because i
443e154 worn-out kinds, & quotes from Pliny, that it is BAD to graft from top shoots-- If prolongation of l
492q002 .. Are any varieties of Cabbages not attacked in BAD years from Caterpillars. 5. Whether Roses impreg
506q015 ot swell, although there be pollen.-- or FEW. or BAD seeds formed; badness may be merely not ripening
511q018 ng animal be formed-- not caring whether good or BAD.-- are any actually rejected?? (8) Get Sir. R. H
520m002 elow par in intellect frequently <are> have very BAD memories for things which happened in early infa
530m044 sleep giving directions,-- & forgetfulness after BAD accidents:-- After journey, a fit of = gout, has
537m075 nged may be inferred from expression. "relict of BAD habit." as child is cured of sucking his finger
537m075 nimal leave off <t> instinct, when attended with BAD effects Martineau. How to observe, p. 21-26. arg
538m079 anist & great knowledge of Irish Politics, «both BAD jokers.--» the other army officer, horticulture
538m079 sects.-- yet Allen. W. remark about his slippers BAD for fires, what is wrong in his head. & Babingto
548m115 y.-- one cannot bring it to one self.-- nor of a BAD dream, when that is not recollected, nor of the
565n007 g, but laughing involuntary.-- When one fear any BAD news, «though in a letter» why is person painted
572n034 if he has been cowardly, or has injured another BAD, vindictive.-- or lied &c &c Are the facts (abou
573n039 - flow it will.-- My father told Miss. C. of the BAD conduct of Mrs C. (her brother's wife). & she sa
578n054 y face,"? to one moral conduct.-- either good or BAD. either giving a beggar, & expecting admiration
578n055 y one suspects one of having done either good or BAD action, it always bear some references to though
589n091 ndividual played a little, & something destroyed BAD brain. see p. 90.-- The relation of reason to or

605o20v e power of discriminating & respecting good from BAD. And it is manifestly from this fact & the insta
608o026 in the way to aid motive power.--if incorrigibly BAD nothing will cure him' 3) disgusted. with them.
609o029 nquered than the deeper & worser feelings. These BAD feelings no doubt orginally necessary revenge wa
620o045 lence,, when not urged to it by passion, shows a BAD child.-- Hence there are certain instincts point
494q004 w. & curiosities!! What is price of fox. otter. BADGER &c &c &c.-- (11) Keep. Tumbling pigeons. cross
402e018 only animals he heard of pigs, small bears or BADGER, deer, apes, baboons, monkeys & an animal prob
494q004 als & their general healthiness-- Fox's, Bears BADGERS,-- How few wild animals are propagated,, thoug
335d015 herefore a mule can have no offspring.= but as «BADLY» deformed people & as mutilations «(produced ve
573n038 tten with mice.-- A person with St Vitus' dance BADLY, told should have shilling to walk to door with
575n045 ll not enter water, till he sees. whether birds BADLY wounded, or only winged.-- fetches two birds ou
506q015 here be pollen.-- or FEW. or bad seeds formed; BADNESS may be merely not ripening= (38) Have Dioeciou
203b136 as a wanderer, but as produced by climate?-- M. BAER (thinks) the Aurock, was found in Germany & thi
338d024 count) of plants of Nova Zenbla -- in review of BAERS work Edinburgh. Royal. Transact.-- p. 297. Vol
398e006 yet with <gov> symmetry «& regular laws» that BAFFLES idea of revolution.-- My very theory requires
497q006 roportions not effect of Chance Maer.= (12) Take BAG of soil from centre of woods «especially if date
027f027 lk New Providence more hilly than others of the BAHAMA consists of rock & sand mixed with sea shells-
080r180 oaling of coasts 93 action of sea on coast. 27. BAHAMA Isd De Lucs travels Beauforts Karamania Capt.
024r016 . of C. Frio Except at Abrolhos. [18 degree S.] BAHIA [12 degree 57 minute S.] 8 200 Morro S Paulo [1
037r056 there, hills of Basalt & other Volcanic rocks. BAHIA, Rio de Jan: B. Oriental? level surface not dis
041r066 hose plates in Australia. New Red Sandstone. at BAHIA in modern sandstone. a circle.,{P}, had in its
042r067 liff marking a pause» When mentioning pumice of BAHIA Blanca, mention black scoriaceous rocks of R Ch
042r068 ous bones so likewise «of miocene period».--Mem BAHIA blanca P. 204 Vol III. Lyell Owing to «open» fa
051r093 beach) deposits «calcareous» encrustations; At BAHIA ferruginous.--At Pernambuco (great swell & turb
051r094 t protect surface; On «hard» exposed rocks near BAHIA, whole surface to where highest spray (there pa
093a033 d earth with volutas. prove regular mud bank at BAHIA Blanca. <fl> Flustra identical. recent & bone b
134a142 bones in Falklands Some of the Tosca nodules at BAHIA Blanca Mr. Malcolmson says are like Kankaer Sou
135a146 mation of shore of Coromandel. just same as. at BAHIA Blanca-- letter in drawer with important letter
481z008 uta found in not less than 7 fathoms water. Mem BAHIA Blanca. De la Beche theoretical researches Comp
484z015 warblers of Europe-- Study profoundly shells of BAHIA Blanca & Southern Hemisphere It is most interes
058r118). Voyage aux terres Australs Vol. I. p. 54. M. BAILLY says."en effet toutes les montagnes de cette i
059r120 t fell in.--Says posterior craters in centre:-- BAILLY talks of much granite on all East side of Van
155g064 plain into terrace as regressed What «alter» a BALANCE there must be in power of rivers either bringi
429e115 here is a contest. & a grain of sand turns the BALANCE.-- Hort. Transact Vol I. M. Ramond. p. 19. do
510q017 riation in varieties. G. St. Hilaires law of BALANCEMENT Wm Yarrell (1) About non-breeding of animals
641j29r <bri> abounding with species-- there will be a BALANCE, continents have been split up.-- who can deci
444e157 to it.-- p. 139. Doubts altogether the law of BALANCING of organs.-- In the Batracian Order the «32»
567n014 ence-- art is experience & observation.--» in BALANCING a body & an ass knows one side of triangle sh
617o40v riginate in any point an opposition of forces BALANCING each other & moving in opposite directions. W
244c023 surtout de l'opinion du baron Cuvier, nous ne BALANÇONS pas a la regarder comme une espèce distincte!
027r024 ilosoph Transact: at R. de Janeiro. Coquimbo. BALANIDAE. at Concepcion. Humb: Pers. N. vii P. 56 Serp
198b113 states these class approach on the confines? BALANIDAE?-- --- I cannot understand whether. G. H. thi
275c121 but same marks on wings are Blue Pouters & small BALD Heads Mr Yarrell will mention in his work I am
522m011 d of him).-- My F. then said you remember Jack BALDWIN at school.-- Answered To be sure I do.-- What
497q06v rmed. (15). What is History of Viburnum. or snow-BALL-tree. what would result from seeds being sown=
499q009 with pollen-- in what state (whole or broken) is BALL of pollen on Bees thighs (18) Place pin's heads
152g047 at very head of valley indicates new terrace BALLIVARD 2 miles North of Grant town to Forrest road c
367d113 which has the power of inflating its body like BALLOON-- by air cells connected with cheek pouches.--
041r066 contraction: the fact is in alliance with those BALLS at Chiloe, full of sand.--the <scale> «quantity
044r072 dy of lavas have flowed from centre-- Pisolitic BALLS occur in the Ashes which fill up theatre of Pom
070r153 columns. show that granite when weathering into BALLS. must exhibit orbicular structure.--When we rec
079r176 Tasco vein in Mica Slate & overlying Limestone BALLS of Silver ore occur in do veins. At Huantajaia.
098a046 is easy to prove. (pyrites, agates, calcareous BALLS) that concretions are connected with a crystall
486z019 ntomolog. Transact. Vol I. p. 130. Col Sykes on BALLS made by dung beetles, like those from Chiloe. A
512q019 ord cow similar to reverse cross.-- Sow cast-up=BALLS of Hawks or even owls.-- How long do seeds rema
503q012 atia Royle & Horsfield (35) Talk about races of BANANA & yet seedless-- no light Henslow or Royle, la
492q002 ot some one written on them? 7... Are the wild BANANAS of Otaheite seedless;-- are all varieties seed
514q21. described in Flinders: Alpine Australia Flora-- BANANA'S seedless-- 20 varieties in mountains of Tahit
032r039 extension of the waves. by the part beneath the BAND of greatest action not having been worn away.--
032r039 line (2). The part (o) which was before beneath BAND. of greatest action. would now by degrees be ex
032r040 st of Chile.-- Must first explain «top of» tidal BAND of action. This case differs. I think. from Pat
051r094 by mere force of waves: & action on upper tidal BAND, I do not see how to account for oceans power.--
096a039 ks of quantities of shells at Iquique. ‹Ceylon›. BAND of Volcanic action in Iceland parellel to Green
151g040 t «doubtless worn into coincidence» has beach or BAND of pebbles on line of 4th shelf.-- Even on Laud
153g055 of Granite 28.362 68 degree 60 degree Granite--«BAND» 4 X 3 X 2 «feet» & 2 deep Another rather small
273c110 rump feathers.--; <then> & one species has small BAND & others large, then he says from long experien
304c207 forms of Synallaxis. trifling characters as red BAND on wing show to be from one parent.-- same form
363d101 he Pidgeons, trace the washing out of the forked BAND, like in plumage of ducks.-- Mr Yarrell says in
104a068 ontorted yet no mountain chain case parallel to BANDA Orientel. ask Lyell for sentence.-- Origin of B
233b250 main divisions of cats. Tortoise shell-& grey-BANDED. ¿species?-- thinks offspring of cats sometime
272c109 ame species & subgenera in families.-- thus the BANDED tarsi is common to all the Laniadae & Muscicap
035r047 xx: same fact is indeed shewn? by the parallel BANDS of formations on any Geolog Map: Quoted from Da
118a104 nental elevations (I) the alternation of linear BANDS of movement in Indian & Pacific Oceans.-- (2d--
255c051 ermanence in many trifling marks.-- such as the BANDS on pidgeons back.-- According to this descripti
318c253 rious case-- -- the birds seem to follow narrow BANDS, certain kinds as gallinules taking the low cou
318c254 e of Selbournes Rock Ouzels.-- ..If the line or BANDS of country (These facts show the Normal conditi
554m138 en Jenny, when Keeper was away, take her chair & BANG against the door to force it open, when she cou
292c171 t of reason implies diversity & therefore would BANISH individual, but general ones might yet be tran
114a095 as plainly as Temple of Serapis. (now we have BANISHED diluvial waves). & likewise <tells,> «offers
199b119 h one hairy child, and of albino DISEASE being BANISHED, & given to Portuguese. priest.--In first set
199b119 es. ¿are there any instance of peculiar people BANISHED by rest?-- ∴ most monstrous form has tendency
278c128 owever be much <shu> surer, when false species BANISHED by this test.-- Excepting where an Andrew Smi
288c160 ing a common seems in part of-- to have almost BANISHED the Grasshopper Warbler-- --Yellow Wagtail ne
527m035 then perhaps the sooner castles in the air are BANISHED the better.-- The facility with which a castl
042r067 vailing swell, (as Shingle travels on the Chesil BANK. V. De la Beche). Ask Capt. F.: R: how the swel
052r098 and soundings introduce this discussion.--Brazil BANK. (& I believe SE coast of Madagascar. where a -
093a033 bos in red earth with volutas. prove regular mud BANK at Bahia Blanca. <fl> Flustra identical. recent
154q057 Collaria tidal channel, sides <alm> 15 ft above BANK or terrace, from terrace of 2d shelf Level of s
163q106 ove sea-- Loch Ness 40 ft above do. When cutting BANK where Locks now are (32 ft rise) they found alt
306c213 ed with Australia «map to King's Australia» by a BANK of soundings of which «there appears to be one
593n111 . I saw many coots & waterhens feeding on grassy BANK some way from water, suddenly, as if by word of
027r027 d mixed with sea shells--about 500 Isd. & great BANKS. effect of Elevation. United service Journal In
047r082 before the higher part. -- Does the sea fall on BANKS as a Bore wave rushes up? (NB. Earthquake wave
048r084 f coast Darby mentions beds of marine shells on BANKS of Red River Louisiana. V. Lyell. Vol I. P. 191
050r091 more common. the shoaler the water & nearer the BANKS Is there not a sudden deepening on E. coast of
116a100 in beds of river, but in shelving «successive» BANKS <above> 30 ft or so above bed of river. formed
130a134 onia {P} B subsidence; <as in> be cautious. mud BANKS & sand. dunes.-- in these littoral deposits the
299c194 t they affect different localities.-- latter on BANKS & in damp parts.-- both propagated by seeds.--
427e111 edling with hardier constitution.-- Now Sir. J. BANKS. says Zizania in 16 generations did become, acc
445e158 hell stuck to its tail" as mentioned by Sir. J. BANKS. p. 212.-- p. 282. Allows this instinctive powe
637j57v bull dog to bulls.-- primrose to <open fields> BANKS-- cowslip to <banks> fields-- these are adaptat
637j57v - primrose to <open fields> banks-- cowslip to <BANKS> fields-- these are adaptations just as much as
275c120 nued by picking chickens of each brood.-- These BANTAM feathers at last got ducky, then took white Ch
275c121 hers at last got ducky, then took white Chinese BANTAM crossed & got some yellow & others yellowish &
312c231 were not permanent, in the new cross.-- In the BANTAM clubs, they used to fix on the kind wanted, co
363d101 long fur.-- feathers on legs of Ptarmigan & in BANTAM.-- {CD[In the Pidgeons, trace the washing o
494q004 ‹allied› varieties (9) Cross largest Malay with BANTAM-- will egg kill Hen Bantam.-- Cross common Fow
494q004 s largest Malay with Bantam-- will egg kill Hen BANTAM.-- Cross common Fowl with Dorking (10) Statist
275c120 J. Sebright first got {P} point on hackles on BANTAMS by crossing with common Polish cock «is not th

275c121 by picking the yellow one & crossing with duck BANTAMS procured old variety.-- The pidgeons which hav
511q018 his features-- in negro & white (3) About the BANTAMS at Zoolog Soc.-- did Sir. J. Sebright select t
511q018 animals for prizes-- = Pidgeons. Canary birds-- BANTAMS.-- (6) <Mad> Porto Santo Rabbit. Descript. of
513q020 arlyroussa with tame.-- (7) About fertility of BANTAMS from different countries-- Do the Peacocks cros
555m144 ging to head cut off. «there was the feeling of BANTER & joking» because the whole train of Dr Monro
113a092 d me a river <near> W. of Port Philip. which had BAR at mouth excavated in solid rock.-- 4 & 5 fathom
613o036 iever Sailor he has seen push a hare through the BAR of a gate before him, & then jump over the gate
240c014 its «zoolog» alliance with New Holland. The BARBAROUSSA, (when young very like the Siam race with lo
636j57r ept «in» quick movements. (sliminess instead of BARBS)-- In all these cases it should be remembered,
245c024 to puzzle naturalists.-- p. 372. Bourous. the BARBYROUSA; a Cervus near Marianus new, & some rats & m
245c024 & some rats & mice. In Amboina only Cuscus & BARBYROUSSA «NB» [islds. Springing up more likely to <M>
291c166 ct of organization. oh you Materialist!-- Read BARCLAY on organization!! Avitism in mental structure
051r094 signs of degradation; (soft substances worn into BARE cliffs evident); the action is anomalous; It is
114a095 phic action which I give at C. of Good Hope.-- A BARE hill of greenstone, if we know origin of greens
332d005 outh, said to come from coast of Guinea, -- ears BARE. skin black & wrinkled-- fur short. (tail cut o
560m156 ry.-- <fe> Shame, independent of fear: colour of BARE nails--, & of eyes.-- Do female monkeys care fo
574n041 -- How comes such an association in man.-- it is BARE fact, on my theory intelligible An habitual act
218b189 en Man makes variety these are vitiated.-- This BARELY applies to plants Female pig apt to produce mo
212b165 d, used to burrow like fox.-- a sort of internal BARK. would remain for long time together in tub of
246c025 g, black & white, ears short & straight-- do not BARK p. 433. birds & bats have certainly travelled f
288c159 readily-- think about half way in appearance.-- BARK about half way «in tone»-- the native dogs howl
288c159 the native dogs howl most dismally, very rarely BARK.-- are almost useless not the least notion of h
540m084 so does man.-- dogs laughs for joy, so does dog BARK. (not shout) when opening his mouth in romps, <
541m092 8 When a dog in play has his mouth open ready to BARK, & lip twisted up, in that peculiar manner they
542m094 en.-- How is crying-- peculiar not common?--» no BARK of anger nor have monkeys & many other animals,
590n094 -- Bennett's Wanderings, Australian Dog does not BARK-- quotes Gardner's Music of nature to show bark
315c243 ing the subject important.-- Laughing modified BARKING., smiling modified laughing. Barking to tell o
315c243 modified barking., smiling modified laughing. BARKING to tell other animals in associated kinds of g
542m096 is way, when opening mouth between interval of BARKING, now this is smile. With respect to sneering
590n094 ark-- quotes Gardner's Music of nature to show BARKING not natural. (Vol I. p. 234) Vol. II p 153. «d
542m096 young animal has canine teeth.-- A dog when he BARKS puts his lips in peculiar position, & he holds
513q020 els-horses. &c &c Rhinoceros= (6) Cross. Sus BARLYROUSSA with tame.-- (7) About fertility of Bantams
363d102 American & Europaean birds «slight» different-- BARN Owl <the> in the former place breeds in <flags>
416e071 - in shells?-- insects?.-- all!??!?-- Worms? [BARNACLES, aquatic., <yet> Crustacean, & true hermaphro
363d102 flags» «thick vegetation» in swamps-- (owing to BARNS, perhaps, not being left open to them,-- . In
159g086 sis Thursday, from Glen Turrit to Fort Augustus BAROM upper (rather above)? shelf 29.290 A. 69 deg
159g086 bove)? shelf 29.290 A. 69 degree Air 68 degree? BAROM 29.008 A. 75 degree Air 70 degree? This station
160g090 these & Glen Tarf Hill «Cairn <taw> leer peak» BAROM 28.700. A.75 degree 75 degree? Boulder, much co
160g091 more Haberclador Isthmus broad flat Loch {P} XX BAROM 28.92 A 75 Air 70 degree? Isthmus broad flat pe
160g092 into west end with obscure terraces on one side BAROM 29.200 A 80 70 degree? for about 3/4 of mile on
161g094 » (to test last on Peat-Mass Divortium aquarum) BAROM. 29.200 A.77 degree Air 70 degree? Barom. 066 l
161g094 uarum) Barom. 29.200 A.77 degree Air 70 degree? BAROM. 066 lower than last. but A 77 station was <a f
161g096 n dead level «by eye» to moss-- on this terrace BAROM. 29.264 A 82 75 degree? This last measurement t
161g098 inus .008 ---- .192 Loch Lochy 4 ft above water BAROM: 30.372 A 76 degree 75 degree? The River <the>
162g100 ast stream Friday Loch Lochy near Letter Finlay BAROM 30.267, A 68 Air 65 degree? <.194 372 about 267
045r077 nd Eruptions: Mr P. Scopes explanation of low BAROMETER? In a subsiding area. we may believe the flui
156g072 had modified it-- perhaps however sea also,-- BAROMETER on shelf 3d. 29.455 A 83 degree ∴ plain of 4
026r022 of sea near T. del Fuego.-- Is there account of BARON Roussin's voyage.-- In Europe proofs of many os
244c023 amen attentif, et forts surtout de l'opinion du BARON Cuvier, nous ne balançons pas a la regarder com
381d157 s-- (& cultivation might make one set of organs BARREN in one plant & not in other), Hunter <asks> p.
496q006 n Fig Dioecious-- are its female flowers always BARREN-- if not how does impregnation take place male
540m089 em,-- otherwise as he remarks sympathy could be BARREN. & lead people from scenes of distress.-- see
637j58r inal causes: consider this!--]CD consider these BARREN Virgins p. 235. talks of the long spinous proc
105a071 hts of steps; if one lake then we must suppose BARRIER in the very part, where barrier least probable
105a071 e must suppose barrier in the very part, where BARRIER least probable.-- The sea harmonizes well with
147g016 Dochart-- Rivers could not have deposited it. BARRIER of lake very lofty, & no trace of it; to the S
148g028 se if not chain of lake & if so there would be BARRIER-- recollect the case of loch <in> <below> <by>
150g035 cases) but then if gradually drained, where is BARRIER {P} great waterworn frame terrace 4 & not visi
244c021 demonstrate.; not distance, makes species but BARRIER.-- --it would make strong contrast with southe
269c102 by Brown.-- great pause seems to act per se as BARRIER-- Mem. Tartary & China.--, both coasts of New
328cIBC tt on Cattle-- (Waterhouse has it) shells from BARRIER isld many relations with a living Matica & man
195b103 ued representative species chiefly found where BARRIER «& what are barriers but» interruption of com
195b103 ecies chiefly found where barriers «& what are BARRIERS but» interruption of communication. or when c
275c119 ime «& forms» distribution tells of horizontal BARRIERS-- Mr Yarrell.-- says my view of varieties is
295c178 an characters the others assumed--+++ ‖Daines BARRINGTON says cock birds attract females by song. do
072r159 ld Greenland, close to W of Jan Meyen Isld.--Mr BARROW thinks N & S. line connects western isles of S
128a130 th America, like the Mexican Gulf. is fouled by BARS of sand & shallow lagoon.-- when describing Coa
273c110 marked colouring of plumage (as «black & white» BARS on wings of trogons are lengthened rump feather
251c040 l of maize. (get limits of latter from <Tarton> BARTON.-- swifts return after years to nest Vol II. p
323c269 er» influence of Physical causes: well skimmed BARTRAMS Travels in N America May 18th Stanley familia
446e160 of spontaneous generation.-- Introduction to BARTRAM'S Travels p. XXIII. <Some birds> Both sexes of
073r162 l laws. association of lead & silver. Sulp. of BARYTES:-- Fluoric. Barytes:-- Humboldt. New Spain. Vol
073r162 n of lead & silver. Sulp. of Barytes: Fluoric. BARYTES:-- Humboldt. New Spain. Vol III. p. 130 Metals
075r167 lybdated lead & «argentiferous lead»; sulfated BARYTES very «un»common in Mexico. Fluor spar only in
049r087 ries of faults. [Fig. 4] {P} In Cordill: should BASAL lavas be called Volcanic or Plutonic The cellul
398e005 ess of conviction, absolutely necessary as the «BASAL» foundation stone of further inductive reasonin
037r056 -- M. Video exception, but even there, hills of BASALT & other Volcanic rocks. Bahia, Rio de Jan: B.
055r107 is no occasion to wonder what has become of the BASALT. Gone into fine sediment Look at St Helena!!--
069r150 lted gneiss/// QUARTZ!!! Analogous to Von Buch. BASALT where Basalt. trachyte where trachyte. There m
069r150 / QUARTZ!!! Analogous to Von Buch. Basalt where BASALT. trachyte where trachyte. There must have been
070r153 recollect connection of columnar & orbicular in BASALT.-- When we see Avestruz two species. certainly
073r163 ns of amph: porphyry. greenstone[,] amygdaloid. BASALT & other trap cover it to great thickness. = Co
093a031 7. 1837.-- The most infusible first injected.-- BASALT: last because it could reach the surface. befo
093a032 ivine a preexisting mineral.-- Mem. Galapagos ∴ BASALT deepest?? Marcel Serres L'Institut. 1837. p 33
096a039 Isd. p. 351. NB. Mackenzie talks of gravel on BASALT of Heckla-- All the Azores Isld. Von Buch p 35
110a086 se in part. block not crystallized Salband like BASALT. full of circular cryst of glassy felspar diff
031r038 at St. Helena. This structure was very clear at BASE of great lava cliffs [Fig. I] line of high tida
032r039 o it, & the result would [be] a uniform slope to BASE of cliff (Z). to which point the waves would ba
033r043 al crater: at Teneriffe Wall of Porph. Lava with BASE of Pitchstone; Mem Galapagos. chiefly red glass
033r043 s. chiefly red glassy scoriae.--could walk roun BASE:--not universal: could not climb up many stairs,
110a086 leavage, veins of pyrites, few curious fissures; BASE in part. block not crystallized Salband like ba
110a086 rent from either fragment or dike, blackish grey BASE. crystals from fragment disseminated on that si
110a086 ion DISTINCT from dike junction mechanical: DIKE BASE reddish feldspathes with grenish. black specks
110a086 mica.-- large cryst of Hornblende blending into BASE-- Salband might have oozed out of cleavage plat
145g004 Specimens-- {P} Veins, amygdaloidal-- as well as BASE not always parallel to strata 3 or 4 seams/ 3 o
149g032 e» of Meal-- Derry there were perfectly rounded «BASE» pebbles of quartz & other rocks not apparently
177b025 life should perhaps be called the coral of life. BASE of branches dead; so that passages cannot be se
177b029 on», «species from many» <therefore> changes and BASE of branches being dead from which they bifurcat
214b175 Galton, have one of the vertebra, about 2/3 from BASE of tail, enlarged two very considerably, so tha
303c204 o or father of negro probably was first black at BASE of nails & over white of eyes,-- + + +, Will he
466t093 siding it».-- In yellow day lily, the Bees visit BASE of upper petal, though not differently coloured
466t094 a wild purple Geranium, I see Bees visit always BASE {a} of upper petal from facility of alighting?
467t104 ct sheath {a} In all these nectar seems to be at BASE of upper petal & the curvature of <an> pistil,
468t111 raxinella <Heartease> «small. Humble alighted on BASE of filaments & reached nectar =again= between t
472s01v 0 one brighter with mere traces of black spot at BASE, one paler with less riged foliage & no black s
531m051 ess of brothers children shows that sympathy is BASED as Burke maintains on pleasure in beholding the
568n017 r in case of any fantastic custom» «Probably BASHFULNESS is connected with some disturbed habit» [Thu
030r034 en coast. Perhaps as at Concepcion. favoured by BASIN formed by outlying rocks; (such as between Moch

```
226b222  -- Hence this must have been condition of Paris BASIN land.-- (How is this with Fernando Po.). with p
226b223  on of species. (¿are carnivorous Mamm: in Paris BASIN «allied to present») more like present carnivor
416e072  1838. p. 394. Rhinoceros «tichorhinus» in Paris BASIN.-- its relation to African Species <good observ
634j55r  They are ancient.-- Are Cetaceae found in Paris BASIN?.-- NB) The explanation of types of structure i
030r035  .--This position agrees with character of.. «in BASINS from rivers. & natural position» position at N
035r046  the height of same beds, deposited in different BASINS; little or no relation appears to <exist> be m
409e048  mals in Europe. :the forms themselves have been BASIS of argument of change.-- now take greater  area
525m023  what  they consider their duty.-- as carrying a BASKET, bringing back game, or picking up a stone, th
567n015  Does a negress blush.-- I am almost sure Fuegia BASKET did. & Jemmy, when Chico plagued him-- Animals
314c239  stralia.-- The wombat (Brown) is found in Isd of BASS' Straits The common Mush room & other cryptogam
377d139  the  great time, necessary to form channel & (& BASSES St) yet no change in English species-- time no
459tf02  cability to common mule & hinnus-- in one case BASTARD of wolf & dog had more form of male, & another
538m079  nity always thinking people were calling him a BASTARD.-- when drunk.-- having really been so.-- some
201b129  rnal influences.-- For instance he says wings of BAT, are from external influence.--...... Hence name
367d113  ncyclop of Anat & Phys. describes, a high-flying BAT, which has the power of inflating its body  like
399e010  . Narrative of Miss. Enterprise, p. 497. Vampire BAT abound in the Navigators & at Manguia, but are u
452e181  elago they are only to be found on the isld of BATCHIAN near SE. end of Gilolo.-- "-- Forrest Voyages
293c175  e theory.-- There is breed of tailless cats near BATH. Lonsdale do. says Sheep could not live for som
076r168  the  surface of the earth."--p. 156. Mines of BATOPILAS in New Biscay, "Nature, exhibits the same min
242c018  nd p. 23 Voyage of Coquille Lesson No (p. 24) BATRACHIAN in isles of Great ocean says in conformity w
444e157  ther the law of balancing of organs.-- In the BATRACHIAN Order the «32» ribs are wanting. p. 144 in th
196b104  connected.--  No doubt in birds, mundine genera, BATS .Foxes. Mus are birds that are apt to wander  &
221b200  fined species in Sicicly. Jan: 1838 L'.Institut. BATS, in Eocene beds, very like present species. p
246c025  s short & straight-- do not bark p. 433. birds & BATS have certainly travelled from East Indies, isld
463t055  nd-- instead of saying as brain is created &c &c BATS are a great difficulty not only are no  animals
584m072  plies.-- it is the acquirement of a new sense,-- BATS avoiding strings «in the dark» as well might be
634j55r  of  Polar attributes of the Cetaceae.-- How came BATS also.? before birds? They are ancient.-- Are Ce
641j29v  markable, [gained by very different process from BATS. CD]CD. «Macculloch says no other bird could ca
428e112  be  no weeding or encouragement, but a vigorous BATTLE between strong & weak March 11th. Yarrell's la
314c237  ssion can not alter much.?? Mr Brown showed me BAUER'S drawings of a curious plant where a tube consi
513q21.  in unopened flower-- <doubt> disbelieve this in BAUERS case of orchidiae Where does J. Hunter use exp
023r009  nk extreamly."--This shark was caught in Shark's BAY. Lat 25 degree. The nearest of the E. Indian Isl
023r012  ard early in the passage!!-- M. Labillardiere in BAY of Legrand, (SW part). describes a Small granite
037r054  ompare the elevated estuary of the Plata. to the BAY of Bengal. dimensions? Strong currents off the G
054r105  the beach--"This is particularly observable in a BAY about five leagues North of Callao, called Marqu
054r105  t. <"> The rocks in the most inland part of this BAY are perforated & smoothed like those washed by t
057r114  with ice thrown.-- Capt Ross found in Possession BAY in 73 degree 39 N. living worms in the mud which
095a037  es at the Zoolog: Soc: Terebratula from Hudson's BAY. 2. species Vol VI. Geograph. Journ. Analysis of
105a069  ed with width. for besides more surface exposed. BAY more open to turbulence. Bull. Soc. Geolog «1837
122a115  tones scattered irregularly.-- (Mem near Gregory BAY). Shropshire case where lamination appeared.-- L
207b151  St. Helena. as. Pineaster & Mimosa called Botany BAY Willow V. Dr Royle introductory remarks to Himal
211b163  efore like dogs.-- Ogleby says, Wolves at Hudson BAY breed with dogs.-- the bitches never being kille
239c002  ut, for many generations back, were crossed with BAY mare, only bay a few generations, that offspring
239c002  nerations back, were crossed with Bay mare, only BAY a few generations, that offspring would be chesn
400e012  «p. 46» Capt Davis in 1598 found cattle in Table BAY with Humps on their backs & big tailed sheep  do
419e086  mb. 25th.-- Lyell says the elevated shells in BAYFIELDS district are much more like those of Scandina
065r139  ssession. Syenite¿ Andite?-- Degrading of inland BAYS. like St. Julian & Port Desire applicable to Cr
148g027  e side ravine entered terraces formed successive BAYS but plains sloped centre-wards which would  not
073r164  ammalite. pyenite. native sulphur.. fluor spar. BAYTE. asbestos garnets.--carb & chrom. of lead. orpi
092a028  pper Assam. Geolog Proc p. 566 1837.-- Tertiary <BEA> formation twenty species same as Paris. 1500 ft
047r082  t very evident movement.--The swelling first on BEACH I cannot understand, without (cs <[...]> raised
048r084  motion in the «loose» bed of pebbles. (On a sea BEACH under a cascade, one can understand pebbles thu
051r093  ccassionally most tremendous surf & loose sandy BEACH) deposits «calcareous» encrustations; At  Bahia
051r095  t manifest example of degradation I ever saw on BEACH near Callao.--From Sir. H Davy experiment on th
054r105  sea judge from the pebbles such as those on the BEACH-- "This is particularly observable in a bay abou
057r115  zing point!!! Remember idea of frozen bottom of BEACH of sea to explain preserved animals.--Mem: stre
058r116  -At the first it would though be easy to see on BEACH successive lines of sea weed-- Histoire Naturel
105a071  ards by successive torrent spread out. by sea-- BEACH action -- no one will dispute. sea. once came t
109a082  f coast of England.-- Sea must always on actual BEACH act same way.-- a little further from beach act
109a082  ual beach act same way.-- a little further from BEACH action probably modified by form of waves & cur
109a083  ll show effects.-- analogous to broad flat sand BEACH. {P} -- De la Beches argument of low coast gain
114a095  esumption» it has been excessively slow because BEACH line chief cause of denudation, but does not te
116a100  ova project on plain, like <re> a reef on a sea BEACH-- «p. 151» first discovered «very small» bits o
149g032  4th  Shelf a little lower down the hillock with BEACH & channel precisely as with Isld-- {P} do  they
151g040  he summit «doubtless worn into coincidence» has BEACH or band of pebbles on line of 4th shelf.-- Even
157g075  of «that» hill, in front of which shelf 3d form BEACH of granite pebbles, & around which shelf 2d «al
200b124  ge enough for land birds, seeds picked from the BEACH by the birds; most seeds germinating.--- It wou
028r028  olutions. by actions of rivers currents. & sea BEACHES. All mineral masses must have a tendency. to m
055r106  earth.--Moreover that such do not occur on the BEACHES. Perhaps these facts attest a <more> decided e
377d140  the matter removed by the waves of the sea, on BEACHES-- we really, measure the rapidity of change of
422e092  ike fossil & recent shells of the <new> raised BEACHES-- [not located] The enormous number of  animals
024r015  f Good Hope 70 fathoms 20 miles from the shore? BEAGLE Coast of Brazil? where not rivers in my  Coral
068r148  02 Vol IV When recollecting Gulf of California. BEAGLE Channel.--One need never be afraid of speculat
122a115  ery like those of Andes Speculate under head of BEAGLE Channel. on origin of mud with stones scattere
133a141  n film of molten rock.-- Voyages of Adventure & BEAGLE vol I. p. 2 & 3. Porphyry at St. Elena. p.  6.
157g073  d by its being lower,-- [no pebbles in parts of BEAGLE Channel when mica slate, only sand blow away]C
163g109  ed in little spots Speculate on «under head of» BEAGLE Channel. Forchammers (Lyells Denmark) Shrewsbu
344d042  happened from her looks thougt she was halfbred BEAGLE Staghound. «++»:the grandchildren went back to
429e115  near London.-- what makes the line, as trees in BEAGLE Channel.-- it is not elements.-- we cannot bel
448e166  ase according to Brown.-- Voyage of Adventure & BEAGLE Vol I. p. 306 Shells, as well as plants «of Ju
538m079  s sentimental, some quarrelsome as B.e on board BEAGLE, some merry goodhumoured as self.-- «When Miss
246c026  carcely differs at all from those of Europe, but BEAK rather sharper., & rather longer in proportion,
276c122  tudy of local varieties laws of change-- whether BEAK (as it appears to me). colour of plumage & laws
283c145  s very short legs & long tail «short much curved BEAK.-», other very long beak, with short., let the
283c145  ail «short much curved beak.--», other very long BEAK, with short., let these only have progeny  with
304c207  wing show to be from one parent.-- same forms of BEAK &c without these trifles. it would not then  be
361d095  ch-- in nursling plumage resembled that of Cross-BEAK-- In lark if I understand right, all species ha
363d101  , & see whether feathered legs.-- «Carruncles on BEAK & in Muscovy duck» crested feather, pouters, la
371d118  s, saw several fully developed tails, & one with BEAK turned up like Avocette. here is what [not loca
430e119  ate between I. concentricus & I. sulcatus.-- the BEAK of this one has concentric striae, all the lowe
515q022  ent.-- May. 44 These Hybrids differ in colour of BEAK, taking after male & female parent.-- Will they
637j58r  s Summary of adaptations Horny point to chickens BEAK, to break egg. shells-- why chicken could not h
207b151  p of trees carry duckling to the water in their BEAKS, & the young one <inland> directly by instinct.
637j58r  egg shells grow harder. so must those with weak BEAKS be sifted away.-- 4 & the species, like  10,000
466t093  n fairway to nectary.-- Is not this so in Kidney BEAN. How is it generally.-- In Azalea <do> «it is s
467t103  requent» & seem to act, something like on Kidney BEAN-- they go to nectar at foot of upper petal stand
469t151  rly Magazine-- &c. double-blossomed «& dwarf-fan BEAN» bean, were planted in rows, & seeds gathered s
469t151  gazine-- &c. double-blossomed «& dwarf-fan Bean» BEAN, were planted in rows, & seeds gathered same ye
501q011  - (30) Get Abberley to plant SINGLE Peas, Kidney BEAN & Bean, intertwined, «without sticks»-- in refe
501q011  Get Abberley to plant SINGLE Peas, Kidney Bean & BEAN, intertwined, «without sticks»-- in reference t
467t103  to clover & once this happened.-- And in common BEANS it is wonderful {a} how the Humbles force down
467t103  Humbles force down the wings most violently: in BEANS the wings seem beautifully to protect sheath {a
467t104  Lupine-- Seen Bees on Potato & several times on BEANS Rough.--green-cabbage «in flower»-- swarmed wit
498q008  when «seed of» the varieties of Cabbages, peas, BEANS, as raised, do the Seedsmen select at all  from
500q10a  .-- Q.30) March 1842. <Last> Year «before last» BEANS & peas were planted in rows adjoining & seeds g
502q11v  mals come on.-- Compare calves.: Compare young. BEANS. cabbages.-- History of Pheasant-fowl. Hen colo
504q014  other Peas, from Wiegman Shrewsbury (1) Peas.-- BEANS seeds alone remain to be compared-- Cabbages.--
059r121  ff cont: & even in one case contained lime.--All BEAR close analogy to Obsidian, & all show  chemical
```

063r133 perfect success.--showing non Creation does not BEAR upon solely adaptation of animals.--extinction
090a021 ompared to those of falling stones.--¿ does this BEAR upon the sorting of matter. in making trachyte
102a060 ase of separation.-- the branching cracks-- only BEAR relations to VEINS in primitive rocks-- Are sub
107a078 up.----My view of Volcanos &c &c This view will BEAR much reflection on method of cooling--Very diff
114a094 rsely fibrous» quartz. & iron stone alternating. BEAR on subject of cleavage Clay slate. a distinct f
117a101 ns formed near coast, limestone deep water. will BEAR on formations. during elevation & depression. C
127a126 change of form as the result of heat.-- will it BEAR on central fluidity.-- do p. 137. Lord Tullamor
129a131 ulders sorted: ditto Murchisons case.--¿ does it BEAR on Patagonia? «Facts about subsided forests.--
187b065 o issue, so with species.-- I should expect that BEAR & Foxes &c same in N. America & Asia, but many
208b154 uently not many yet multiplied: NB How does this BEAR with law referred to by Richardson in Report ab
212b164 Java.-- at Leyden series from several islands.-- BEAR peculiar to Sumatra & not found on Java-- Monke
214b174 . 289 It is certain, that North American fossils BEAR the closest relation to those now living in the
224b212 ence, which improves. & checks it.-- It does not BEAR any precise relation to structures Mem. Eyton's
226b222 om S. America, the jaguar has been left & Fox, & BEAR.-- If I had not discovered channel of communica
235b262 in point.-- All cases like Irish & English Hare BEAR upon this.-- Why do Van Diemens land people req
255c053 ustrated or rather a new principle is brought to BEAR. If man created as now. languages. would surely
268c099 ¿Pelagic forms similar--birds??-- We must always BEAR in mind proofs of most equable climate both in
282c140 in others. In analogy it is not the relation to BEAR to each other, but to some external contingency
284c148 lants peculiar to these isld.-- ¿Brown can «not» BEAR the least salt water.-- Nuts prodigiously heavy
294c177 st,-- Lonsdale says he has seen in old Book last BEAR in England killed in year 1000. reference to su
322c270 edited by Owen. read several papers-- all. that BEAR on any of my subjects Elie De Beaumonts. 10 Vol
372d131 skin grows over a wound.-- Does likeness of twin BEAR on this subject? A mans arm would produce arm i
408e044 d. Reliquiae Diluvianae. p. 222. Bones of Horse. BEAR & Deer at 16000 ft. with Snow on Himmalaya-- Hu
410e050 ividuals.-- so that the changes should be slow & BEAR relation to the whole changes of country, & not
430e117 ll animals have sprung from few stocks. does not BEAR, the least on ancient generic forms.-- the anim
436e134 an live.-- Lyell says that naked cuttle fish now BEAR a very large proportion to other mollusca in co
458t001 ils-- Ruminants. & Tortoises gigantic-- hyaena-- BEAR & ruminants all of larger size.-- the law of la
465t079 founded in the caves? It is highly important, to BEAR in mind that enormous periods may elapse, even
484z015 ffinuria, the Awks.-- What structure do the auks BEAR traces of.-- like Puffinuria does of Petrel?--
485z018 nd further north, because during winter they can BEAR the cold when torpid.-- On this principle tropi
504q013 given (3) in Heartease (4) Does the Thyme BEAR abortive stamens every year & Spring. & within
516q23v absorbs oxygen & forms Carbonic Acid. will this BEAR on Petrifaction?-- [blank] Questions & Experime
538m079 e heard, so very probably forgotten.» Such facts BEAR on such characters as Allen W. & Babington, bot
578n055 having done either good or bad action, it always BEAR some references to thoughts of other person Dec
586n082 & so in some senses, is sight--CD [The faculties BEAR so close a relation to the senses, that one fee
593n107 e saucer-shaped depression.-- [blank] Does music BEAR any relation to the period when men. communicat
611o034 xternal accidents, & in such cases modifications BEAR fixed relation to such accidents. But such tiss
612o035 body precisely same as in plants, but as animals BEAR relation to less simple bodies, and to more ext
641j29r shed by arctic regions-- Whales.-- «Narwhal» Polar BEAR. Walrus, great Seals of Antarctic seas. (on oth
641j29r numerous in cold Oceans I think not.-- Does this BEAR on, the absence of their remains in the Wealden
641j29v ty & extinction of such forms-- these views will BEAR on geology-- There is an analogy between fang o
388d172 n old age of female assuming plumage of cock, & BEARDS growing on old women = Stags horns & testes cu
029r034 ion northwards of the Coal in Chili as clearly BEARING a relation to present position of <Coal> Fores
180b036 ter distinction Thus genera would be formed.-- BEARING relation to ancient types-- with several exti
181b043 h side will yet present some anomaly & <the g> BEARING stamp of <some> great main type, & the gradati
188b068 anted in same soil with blue. Now this is same BEARING with Dr. Smith's fact of races of men tendency
202b130 of analogy, the structures in the two animals BEARING relations to a third body., or common end of s
305c211 sary.--) which is modified into endless forms, BEARING a close relation in degree & kind to the endle
355d067 laws is to foretell what will happen & to see BEARING of scattered facts.-- What takes place in the
558m150 thy neighbour as thyself". Analyse this out.-- BEARING in mind many new relations from language.-- th
039r061 -- S. America in the form of the land decidedly BEARS the stamp of recent elevation. which is differe
062r130 : The same kind of relation that common ostrich BEARS to (Petisse. & diff kinds of Fourmillier): exti
251c041 » Zoolog Journal Vol 2 p 221. Horsfield on two BEARS very close species, inhabiting Borneo & Sumatra
253c046 dia deer.-- Africa not.-- Africa Camels?? Africa BEARS??-- Plantigrade Carnivora??-- «compare rodents
255c052 of such general forms.-- The hybridity of ferns BEARS on my doctrine of cross-generation. The inferti
287c157 n.-- Linn: Transact Vol XIV.--., p. 24. Lamarck BEARS to Cuvier that relation of theoretical astronom
296c184 anges «no near lofty country»?? p. 475 NB. This BEARS on fossils of Europe., those species which can
308c217 ps, & there now must be, .extinction of species BEARS relation to existence of genera &c &c Two savag
353d060 e ostrich not. The peculiar «Malacca» «Malacca» BEARS, <are> belong to same section with with those o
388d172 ible.-- it should be observed that transmission BEARS no relation to utility of change-- hence hareli
402e018 Borneo.-- only animals he heard of pigs, small BEARS or badgers, deer, apes, baboons, monkeys & an a
408e045 arallel & curious facts.-- The Himmalaya. case, BEARS on the vast changes even in that quarter of the
425e103 of, «easily» making tolerably fertile hybrids, BEARS relation to capability of variation?? my theory
434e130 ecious plant, in which the Male plant sometimes BEARS female flowers, the organs are most clearly abo
494q004 of animals & their general healthiness-- Fox's, BEARS Badgers,-- How few wild animals are propagated,
512q020 imals there, sometimes couple but not conceive :BEARS /Yes/ (2) Foxes & English animals & birds breed
538m078 y Dr Dewar on principle of association.--«fully BEARS out my fathers doctrine about people forgetting
577n051 f mind) in the intestinal functions &c &c.-- BEARS. the same relation to true memory, that the for
601o009 ect.-- are Planariae conscious.-- Consciousness BEARS some relation to time & memory Reynolds X disco
611o034 ation to such accidents. But such tissue <must> BEARS relation to whole, that is enough must be prese
616o039 raction to ordinary matter is that which action BEARS to the agent. Matter is by a metaphor said to a
629o55v -- seeing how shame is accompanied by blushing, BEARS some relation to others 5) if so, it is perhaps
512q019 An ugly calf <turns> sometimes turns into fine BEAST. would its offspring have ugly calves. also tur
232b248 e to give me an account of parasitic animals of BEASTS varying in different climates Those will not o
453e182 ampier. Vol I. p. 320. says no wild (carnivora) BEASTS on Phillipines. Forrest somewhere says same.--
459tf1r . 256-- wild cattle at Madagascar-- «p. 121» No BEASTS of Prey. any country should during [...] condi
512q019 spring have ugly calves. also turning into fine BEASTS.-- For comparison with hybrids, is offspring o
566n011 shewn in the numerous artifices to take birds & BEASTS».-- necessary to explain origin of idea o
595n121 edgwood» April 28th 1840 was frightened at wild BEASTS in Zoolog. Garden [blank] [not located] A chil
600o08v separation between soul of man. & intellect of BEASTS, not clear.-- ¿does not Mackintosh make great
536m070 action to muscles, any more than «prevent» heart BEAT» remember how Pincher does just the same; I not
543m099 Christ of great genius, & yet invariably used to BEAT him-- The son of a Fruiterer in Bond St. was so
567n015 it connected with surprise.-- heart beginning to BEAT-- children inherit it <ins> like instinct, pree
306c212 .-- a monkey. (Baboon) at Z. Gardens upon being BEATEN behaved very differently from a dog.-- more li
334d011 he same cart in loose chains, by being at first BEATEN from her, & always accustomed to her.-- case p
532m057 sensation of fear is accompanied by «troubled» BEATING of heart, sweat, trembling of muscles, are not
536m070 walk then lightly as to endeavur to stop heart BEATING: one ceasing, so does other.-- What an animal
586n081 nsiders breathing instinctive, certainly heart BEATING may be considered also such.-- heredetary habi
319c255 er.-- Bachman says he thinks the Mocking thrush BEATS all English birds in song.-- one of their thrus
207b151 can dive & conceal themselves in the grass.-- BEATSON St. Helena says no trees succeed well at St
595n121 me feeling which make us think anything ugly-- a BEAU-ideal feeling. Same effect as acting on us-- <T
599o008 n des Mères Vol. I. p. 198.-- "Moralité, raison, BEAU ideal, infini conscience; voila l'homme separe
600o008 a struggle in man.-- two souls in one body-- (2) BEAU ideal, refers chiefly to moral, beau desires co
600o008 body-- (2) Beau ideal, refers chiefly to moral, BEAU desires conscience & love.-- [With regard to or
600o008 es conscience & love.-- [With regard to ordinary BEAU ideal, Mem. Negro, beau,--Jeffrey denies all Be
600o008 [With regard to ordinary Beau ideal, Mem. Negro, BEAU,--Jeffrey denies all Beau-- How does Hen determ
600o008 eau ideal, Mem. Negro, beau,--Jeffrey denies all BEAU-- How does Hen determine which most beautiful c
606o024 floating idea.-- as statue of beauty, is of the "BEAU ideal", my instinctive impression 1) September
257c056 in Bolivia; he examined-- all species & found "BEAUCOUP des mêmes oiseaux. que nous avions déjà obser
100a055 alt as being produced by local heat, Ask Capt. BEAUFORT, whether, water flashing into steam, would Ba
080r181 sea on coast. 27. Bahama Isd De Lucs travels BEAUFORTS Karamania Capt. Ross. & Scoresby deep soundin
127a127 in chains. may be effects of subsidence Elie de BEAUM. Memoires of French Geolog. Cantal Vol III I? p
322c270 all, that bear on any of my subjects Elie De BEAUMONTS. 10 Vol. of Memoirs on Geology of France.= on
605o020 eptibility from Blair receiving pleasures from BEAUTIES of nature & art." But as we often see people
606o023 soms.-- how comes it there? Laocoon p. 75 "The BEAUTIES developed in a work of art are not approved b
172b005 are. species are constant over whole country; BEAUTIFUL law of intermarriages <separating> partaking
195b099 assage for vertebrae in neck same cause, such BEAUTIFUL adaptations yet other animals live so well.--
197b112 aw of crustacea-- with respect to mouth those BEAUTIFUL passages from one to other organ.-- Cuvier on

233b252 h the face of the earth covered with the most BEAUTIFUL savannahs & forests dare to say that intellec
260c068 y. is as effectual as a cold one. in checking BEAUTIFUL colours of species-- Mem. St. Jago--solitary
264c080 pelago turned into continent &c &.-- There is BEAUTIFUL gradations of forms in Australia leading on o
290c163 some effect.-- Maldonado as good forests for BEAUTIFUL birds.-- heredetary ambling horses, (if not l
293c174 me corelation. but. the whole Mechanism is so BEAUTIFUL.-- The corelations are not, however, perfect,
318c252 urning white like Hares??--) I never saw more BEAUTIFUL adaptation for snow-- like snow shoes. than f
319c256 erica, but the few of N. America are quite as BEAUTIFUL. The thrushes of N. America. singing so well.
319c256 g so well. & the mocking thrush being so very BEAUTIFUL gret contrast with South America.-- In Home's
358d085 s.-- [not located] hen freely.-- here we have BEAUTIFUL proof of the breeding in & in (like «courage
409e049 «probably» the foundation of all that is most BEAUTIFUL in the moral sentiments of the animated being
416e071 animal life commenced in the Water!» It is a BEAUTIFUL part of my theory, that «domesticated» races.
432e126 1st 1839) by Sedgwick & Murchison; which is a BEAUTIFUL instance of forms, intercalated between two g
433e126 articular air given. p. 246.-- 248 & p. 258 A BEAUTIFUL case, showing the gradation from one grand sy
436e137 decreased in numbers, what cause?? Seeing the BEAUTIFUL seed of a Bull Rush I thought, surely no "for
507q15v ucture.= good case as showing how simple, but BEAUTIFUL adaptation might be arrived at.= Any book wit
528m036 , especially when coloured.-- that light is a BEAUTIFUL object one knows from seeing artificial light
528m037 k.-- (2d) form. some forms seem instinctively BEAUTIFUL «as round, ovals»;-- then there the pleasure
536m071 ly wrong, otherwise horses would have idea of BEAUTIFUL forms.-- With respect to free will, seeing a
566n010 Minerva or Apollo would have them because not BEAUTIFUL-- is there -- anything in these absurd ideas.
569n019 s works of imitation.-- Hence pleasure in the BEAUTIFUL. (distinct from sexual beauty) is acquired ta
570n027 negro, similarly treated would think negress BEAUTIFUL,-- [male glow worm doubtless admires female.
578n057 ion? At end of Burke's essay on the sublime & BEAUTIFUL there are some notes. & likewise on Wordswort
600o008 all Beau-- How does Hen determine which most BEAUTIFUL cock, which best singer-- Remember.-- avarice
603o11b ot a actors, or a scene in garden.-- yet both BEAUTIFUL! p. 136. Says Architecture does not come unde
603o014 nted-- <music> very good article-- why flower BEAUTIFUL? ¿even to children S. Jenyn's Inquiry into th
636j57r Vol I p. 232. gives Woodpecker as instance of BEAUTIFUL adaptation.-- & then Chamelion, which feeding
467t103 ings most violently: in Beans the wings seem BEAUTIFULLY to protect sheath (a) In all these nectar se
211b161 uction of species.-- Animals have no notions of BEAUTY, therefore instinctive feelings against other
213b171 sects been known? As Gould remarked to me, the "BEAUTY of species is their exactness,' but do not kno
261c071 changed. Can be said that animals no notion of BEAUTY, when does prefer most powerful buck Owen talk
295c178 cock birds attract females by song. do they by BEAUTY, analogy of man if so war not [not located] Er
368d114 Guinea Fowl & Peacocks.!!» other birds display BEAUTY of plumage.-- (The females (as Owen observes)
496q05a theory of insect-like Orchis-- & final cause of BEAUTY of flowers-- contrasted by Kirby-- with animal
527m032 rds return to same quarter for many years]CD.-- BEAUTY is instinctive feeling, & thus cuts the Knot:-
527m032 for our acquiring «the instinct» our notion of BEAUTY & negroes another; but it does not explain the
528m038 (1) harmony of colours, <whi> & their absolute BEAUTY. (which is as real a cause as in music) from t
528m038 ly parallel lines are elegant.-- Again there is BEAUTY in rhythm & symmetry, of forms-- the beauty of
528m038 is beauty in rhythm & symmetry, of forms-- the BEAUTY of some as Norfolk Isd fir shows this, or sea
528m038 fir shows this, or sea weed, &c &c-- this gives BEAUTY to a single tree,-- & the leaves of the foregr
528m038 & the leaves of the foreground either owe their BEAUTY to absolute forms or to the repetition of simi
528m038 forms as in angular leaves,-- (this Rhythmical BEAUTY is shown by Humboldt from occurrence in Mexica
569n019 leasure in the beautiful. (distinct from sexual BEAUTY) is acquired taste.-- Whilst music extremely p
570n026 discourse by Sir J. Reynolds.-- Is our idea of BEAUTY, that which we have been most generally accust
570n027 le figure]CD-- As forms change, so must idea of BEAUTY.-- [Old Graecians living amongst naked figures
593n109 e obscure picture of other men. & hence idea of BEAUTY.-- the social affections of animal taking man
593n109 atching stones. in some degree is so.-- idea of BEAUTY of music are great distinguishing character be
606o022 -- The object of art., sculpture & painting, is BEAUTY.-- which he thinks is a better definition than
606o024 sion of pain cannot be represented. But what is BEAUTY?-- it is an ideal standard, by which real obje
606o024 he embodying of a floating idea.-- as statue of BEAUTY, is of the "beau ideal", my instinctive impres
608o027 s no credit for anything. (yet one takes it for BEAUTY & good temper), nor ought one to blame others.
465t089 ome Mammals of Norfolk Crag. mentioned-- allied BEAVER to present forms.-- -- How many «tertiary» est
623o051 endency [RHC] 10) that the instincts of bees & BEAVERS «& deer» have <been formed> a beneficial tende
318c255 er which way to fly.-- There is a kind of Wren (BEBYK??) which seems common in Rocky Mountains & on o
122a114 g cut of-- if part of «cold» crust under ocean, BECAME thicker, then when fluid moved [...] {P} Augus
190b079 > Mr Don remarked to me, that he though species BECAME obscurer as knowledge increased, but geneta st
282c142 from same.-- but if two original species, each BECAME ground then the relation of all the ground cur
302c203 adapted to some physical change.-- If Patagonia BECAME fertile all intermediate species living there
365d106 s.-- More probably during known changes climate BECAME unfit for. subalpina, or some Northern species
373d132 lant.-- The Marsupial structure shows that they BECAME Mammalia. through a different series of change
386d167 he great. changes of nature are slow. if animals BECAME adapted to every minute change, they would not
415e066 f the Ornithorhyncus be found.;-- yet until man BECAME cosmopolite, he would probably be confined in
422e095 ed structure & complexity.-- hence as the forms BECAME complicated, they opened fresh, means of addin
522m011 n at school.-- Answered To be sure I do.-- What BECAME of him.-- Answ Had large fortune left. him, to
532m056 ury, then immediately fell into her old ways & BECAME fat! What remarkable affection to a place.-- H
640j167 x puppies, if a hare was introduced, or <a spe> BECAME more numerous. (from death of its destroyer),
027r023 re found. -- The above corelations remarkable BECAUSE the formations are now seen in regular descend
028r031 rity of fresh water at Iquiqui. not from rain, BECAUSE alluvium saline; Mem: on coast of Northern Chi
028r031 not see one dike in the whole Galapagos Arch; BECAUSE no sections> same cause as no colour Sir J. He
031r037 lts in Cordillera, as in English Coal field -- BECAUSE lowered & raised--so on--but gradually & simpl
032r041 case differs. I think. from Patagonian steps, BECAUSE the deposition & accumulation is brought into
039r061 y of movements of W coast in explaining plains BECAUSE such are found in perfection on that side.-- A
041r065 . having lived over whole bottom is important; BECAUSE in this latter case. we cannot judge whether s
051r095 Apropos to question does animal adhere to rock BECAUSE it does not decompose. or vice versâ. Clay sla
057r113 cate well that Horse at least has not perished BECAUSE too cold:--With discussion of camel urge S. Af
075r167 silver & red silver, do not name native silver BECAUSE not very abundant.--\ muriated silver. which i
091a025 4. Fact of dust blown far out to sea valuable; BECAUSE transportal of Minute seeds-- L. Institut. p.
093a031 most infusible first injected.-- Basalt: last BECAUSE it could reach the surface. before being coole
103a063 ciety-- Dikes have not been the moving agents, BECAUSE not wedge-formed.-- Hence fill up fissures-- I
105a072 hells should be preserved in it-- some error? (BECAUSE more recent) ------ Coquimbo on. other hand?--
107a077 st much thinner beneath ocean than above it no BECAUSE heat proceeds from great body of mass.-- The l
114a095 rs a presumption» it has been excessively slow BECAUSE beach line chief cause of denudation, but does
114a095 ently metamorphosed than other deposits.-- NB. BECAUSE lowest. first accumulated in bed of ocean With
122a114 cannot be equilibrium of fluid, but of solid. BECAUSE if of fluid, the waters of the ocean would obe
148g028 ut-- In all cases «I urge» deposition marine-- BECAUSE if not chain of lake & if so there would be ba
153g051 esses on 4th shelf: others «lines not so level BECAUSE of upper edge of cliff» Others below it--argum
165g125 m Glen Turret said he learnt to know the lambs BECAUSE most like Mother in face-- asked stated this g
175b016 ifficult to prove cats &c from Egypt no answer BECAUSE time short & no great change has happened I lo
181b041 aving progeny living ten thousand years hence; BECAUSE at present day many are relatives, so that by
183b051 me country. How is propagation of wolf & Dog. (BECAUSE being believed same species) if they do not br
187b065 but many species closely allied but different. BECAUSE country separated since time of extinct quadru
191b081 tion of Mammalia more valuable than any other. BECAUSE less easily transported-- Mem plants on Coral
192b087 bably the new species would have been more perfect. BECAUSE in each there is possibility of such organizat
194b096 to keep on account of pollen from other plants BECAUSE this may be applied to show all plants do rece
199b119 wford on Mission to Ava. account of HAIRY man. BECAUSE ancestors hairy with one hairy child, and of a
200b123 But what a character is this? Race permanent, BECAUSE every trifle heredetary, without some cause of
205b144 the odd Petrel of F. del F. take form of awk, BECAUSE there is no awk in Southern hemisphere. does t
211b162 icate affinity with snipes? indicate affinity, BECAUSE similar habits produce similar structure.-- Me
217b188 to all northern islds.-- This is interesting, BECAUSE Iceland, must have been all ice in time of ice
219b193 . Fernandez. Galapagos. Many trees Compositae, BECAUSE seeds first arrived «Ferns ditto.--» & hence f
219b194 made plants of American & African form, merely BECAUSE intermediate position.-- We cannot consider it
219b194 osition.-- We cannot consider it as adaptation BECAUSE volcanic isld. whilst <neig> Africa, sandstone
219b195 of the Alps. must be <Alpine> «new formations» BECAUSE snow formerly descended lower, therefore speci
223b210 ng ascent of mountain or approach of desert?-- BECAUSE the crossing of species less altered prevents
239c002 th common pidgeon, offspring most like latter, BECAUSE oldest variety.-- -- He says of two varieties
245c025 <M> have different species than those sinking, BECAUSE arrival of any one plant might make conditions
252c045 ree first will give one almost certain guide ∴ BECAUSE time required too separate isld very long» Ame
255c051 ficulty in propagation.-- Feathers on, Apterix BECAUSE we may suppose longest part of structure.-- sh
268c100 quatic, aerial &c type?-- This of consequence, BECAUSE applicable to N. Hemisphere (NB.-- Examine Abr

273c111 se perching (Gould), but the latter is obscure BECAUSE nearly all are so.-- Thus in Hawks, there is a
275c120 in England lost courage-- (Bull-dogs are used BECAUSE they have no scent!) Mr Wynne) at end of chase
277c126 them varieties. but two ostriches good species BECAUSE interlock analogy to be guide. in islands. spe
277c128 ange.-- The mark of analogy would be empirical BECAUSE as soon as two species were placed in differen
278c129 en those sections & subgenera; are analogical, BECAUSE we do not know, whether nearest species of eac
280c134 ct. it.-- Sir J.. Sebright excellent authority BECAUSE written on dog-Breaking.-- applies it to natio
282c142 ffinity with respect to species of each other, BECAUSE we suppose all descended from same.-- but if t
285c151 ed by the vigorous shoots from each branch No: BECAUSE decay in that species <shows> is effects of un
290c163 ined by habit Talent &c in man not heredetary, BECAUSE crossed with women with pretty faces When hors
290c164 pheasants, is most like pheasant., I think so BECAUSE very 3/4 bred.-- (hence hybrids in this case h
291c165 ts become important element in classification, BECAUSE structure has tendency to follow it, or it may
294c178 d, altered, & lose (mere sickness)? fertility ¿BECAUSE offspring too unlike.--?? Memoire by Charles D
295c178 yell's Pamphlet) Is man more hairy than woman. BECAUSE ancestors so, or has he assumed that character
307c216 Zoolog Transactions» & Hunter on this subject) BECAUSE if so as she can be converted «into female», i
308c218 ct, there is really no natural starting place, BECAUSE there is nothing more elementary than that com
308c219 ncracy of wild animals is generally different, BECAUSE their difference. arise a good deal from clima
310c223 uch definite thoughts, I will never allow that BECAUSE there is a chasm between Man (--<a> «&» chasm«
313c233 ng which accounts for the species changing « ∴ BECAUSE mammalia can subsist where parasites cannot» R
316c247 more resemble those of America than of Europe, BECAUSE the recent ones are so close. Was there contin
335d013 the argument does not apply to first parents, BECAUSE they are not new breed.-- the first hybrids ma
335d015 eredetary: «Hybrids of» Varieties is different BECAUSE not long in blood.= The case of union of perfe
336d019 grandfather's theory of Mules not heredetary, BECAUSE generation -- highest point of organization] C
337d020 domesticated animals must be most complicated, BECAUSE they are partly local & then the local ones ar
338d025 t under this view, & Hippotamus of Madagascar: BECAUSE. contemporaries. In introduction to Eytons Ana
345d043 of Glen Turret. said he learnt to know lambs, BECAUSE in their faces they were most like their mothe
347d049 Philosop of Living) quote Whewell as profound. BECAUSE he says length of days adapted to duration of
348d050 malia Edentata¦ We do (p 6) say such is group. BECAUSE it has such characters of importance, "but we
351d056 ndylus has eye¬points-- Macleay then answered, BECAUSE nature leaves vestiges of what she does-- does
352d058 ed . (links in circle must be granted unequal, BECAUSE fossil) Now what is group without centre but c
367d113 f Bull.-- is most common in vegetable feeders. BECAUSE males always armed in carnivora. Where females
370d116 les, but almost all die.-- bees breed but few, BECAUSE they are kept in security.-- Hunter doubts abo
377d140 Heaverns.-- Is not puma, same colour as Lion. BECAUSE inhabitant of plain & Jaguar of woods &c like
385d163 Last litters are considered the most valuable. BECAUSE smallest sized dogs.-- one litter big & then s
385d165 ather in the daughters! This last remark good. BECAUSE showing probably not education.-- Cannot I fin
386d167 to obliterate differences. final cause of this BECAUSE the great changes of nature are slow. if anima
391d174 whole course of change from simplest form.-- (BECAUSE by this process it separates those differences
397e004 hange within historical times has been small-- BECAUSE change in forms is <a1> solely adaptation of w
399e010 breeding in & in, produces bad effects solely, BECAUSE of similarity, because in every country, where
399e010 ces bad effects solely, because of similarity, BECAUSE in every country, where only pair has been int
400e011 distribution-- Thus Hattica is great genus.-- BECAUSE found in all quarters: his ideas not clear. In
403e022 good. (ie invariable) in some classes-- it is BECAUSE every part is under change, now one part now a
403e023 change in number. (even species do not this). BECAUSE it has been so pronounced ex cathedrâ. let us
404e024 irectly incorrect The case of my mice is good, BECAUSE it is an involuntary variation made by man, co
423e095 done so, as the simplest fish &), my answer is BECAUSE, if we begin with the simplest forms & suppose
428e111 ot been so. (which is case adduced by Herbert) BECAUSE not reared by seedlings.-- Now my principle do
429e116 of N. America.-- if true curious on my view-- BECAUSE these points were last connected with those no
431e122 s is most common, we need not look for change, BECAUSE its number show it is perfectly adapted; it wh
436e135 ing reduced in numbers, but not so much these, BECAUSE circumstances» <April 12th..> Cestracion, Port
442e151 k by cultivation can a form become permanent?» BECAUSE its very essence is that little change is prod
443e155 rly deny the right to argue against my theory, BECAUSE it makes the world far older than what Geologi
447e164 a F ovina-- propagated like oni[on] Poa alpina BECAUSE vivaparous.-- Henslow has seen this-- (Poa alpin
448e165 reely.-- The periwinkle seldom produces seeds, BECAUSE it is thought to require insects to impregnate
453e183 ion-- single, or half double.-- anyhow fertile BECAUSE they «are» raised by seed.-- Where has Duchesn
461t041 rs) in old beds & existing species is valuable BECAUSE it shows no innate power of change & it also s
463t057 e into skull., two bones of tibia into one.--) BECAUSE if the animals were taken from which these ser
470t177 many Bees & Humbles--on Thistles many (curious BECAUSE a Composite) Asparagus very small flowers & as
485z018 heat, the tropical forms extend further north, BECAUSE during winter they can bear the cold when torp
501q011 s can be obtained-- I name these three plants. BECAUSE they cannot be crossed, I think, I expect, exc
521m008 nds.-- Miss C. memory cannot be called memory, BECAUSE she did not remembered, it was an habitual act
523m017 ms passion &c by coming on suddenly. Ans no.-- BECAUSE often, if not generally, does not really come
526m030 ng ambition, avarice, &c An animal improves BECAUSE its appetites urges it to certain actions, whi
527m035 in real train of thought this does not happen. BECAUSE papers, &c &c round one. one recalls the castl
528m035 ls the castle by going to beginning of castle» BECAUSE train cannot be discovered-- is closely analog
530m044 a [...] Mr B> journey. short time previous,-- BECAUSE, pain prevents repetition of idea.-- Mr Blakew
531m050 asily forgotten, (if unconnected with fear &c) BECAUSE people think that the importance of the event
531m050 ory cannot solely be number of times repeated, BECAUSE some people can remember poetry when once read
531m053 muscular exertion, palpitation, voiding urine BECAUSE done by some animals in defence, &c Starting m
531m053 h to avoid some danger-- but it is instinctive BECAUSE Nancy tells me very young babies start at anyt
533m057 degrees of talent, which are heredetary are so BECAUSE brain of child resemble, parent stock.-- (& ph
533m059 n a horse on the left side (not good example,) BECAUSE leg is right handed.-- In Review (Edinburgh) o
536m074 uld make a man a predestinarian of a new kind, BECAUSE he would tend to be- an atheist. Man thus belie
539m083 embers things done in the other habitual state BECAUSE it will (without direct consciousness?) change
542m096 these are instinctive-- child does not sneer. BECAUSE no young animal has canine teeth.-- A dog when
543m099 in insecta-- not connected with transformation BECAUSE Spiders have many,-- great powers of communica
546m108 ike them--. hence sympathy very unsatisfactory BECAUSE does not like Burke explain pleasure. August 2
548m115 Somnambulist. (if he had been unhappy)-- it is BECAUSE in this state, the consciousness does not go b
552m130 p-- (one can dream of intense scarlet??) is it BECAUSE one then has no immediate comparison with perc
555m144 ff. «there was the feeling of banter & joking» BECAUSE the whole train of Dr Monro experiment about h
557m148 e would be more easily analysed than jealousy, BECAUSE less discoverable in animals than latter.-- Ye
559m154 n act of separate creation. we admire it more. BECAUSE we can compare it to the standard of our own m
564n003 conscience would not have been same with mans BECAUSE original instincts different.--Mem. Bee how di
566n010 ly neither a Minerva or Apollo would have them BECAUSE not beautiful-- is there -- anything in these
571n030 red. The Reason why New Buildings look ugly as BECAUSE there is some connection between them, & great
575n044 es of dogs on their instincts, most important, BECAUSE they obey the same laws, as the crossing of ja
576n049 s not a cause like a deity, as M. Cousin says. BECAUSE if so ourang outang.-- oyster & zoophyte: it i
577n052 neself.-- «blushing» is connected with sexual, BECAUSE each sex thinks more of what another thinks of
580n061 t ones modify it.-- «Weak people say I know it BECAUSE I was always told so in childhood.-- hence the
580n062 Nat. Theology.-- says animals have abstraction BECAUSE they understand signs.-- very profound.-- conc
586n078 gs.--» they may do all this instinctively «yes BECAUSE power varies in breeds,» something of kind one
586n079 ration, <only> «only» more wonderful in young, BECAUSE can not have been taught, where to go-- the ac
588n090 might be advanced, that animals have reasons, BECAUSE they have memory.-- what use this faculty if n
591n099 in normal cases) are held in abhorrence it is BECAUSE instincts to woman is not followed; good case
591n100 ence.-- Why does not man eating cause disgust, BECAUSE he does not go against instinctive feeling, on
592n103 [a person is here said to open mouth in fright BECAUSE nature intends to lay open all senses: <do> Ho
600o08v and sport &c &c -- The Bitch does not so act, BECAUSE maternal instinct gives most pleasure. but bec
600o08v use maternal instinct gives most pleasure. but BECAUSE most imperious.-- It would indeed be wonderful
602o011 ve reason, though he feels he is right-- it is BECAUSE each decision &c is made up of many partial re
608o026 ing him happier.-- he agrees & yet does not.-- BECAUSE motive power not in proper state.-- When the a
608o026 om contingencies a mans character may change-- BECAUSE motive power changes with organization The gen
608o026 he general delusion about free will obvious.-- BECAUSE man has power of action, & he can seldom analy
608o027 to blame others.-- This view will not do harm, BECAUSE no one can be really fully convinced of its tr
609o030 forward «& to the general action» -- certainly BECAUSE it is the result of what has generally been be
615o038 y «the» general kind taking pleasure in virtue BECAUSE acquired in past ages; seems to indicate that
616o039 converse of the <expr> question above stated, BECAUSE there are living bodies without these facultie
618o41v xistence of something in addition to matter is BECAUSE our knowledge of matter is quite insufficient
618o41v t it now be said, that blueness caused weight, BECAUSE both due to some common cause:-- The argument
618o41v and effect has relation to forces & mentality BECAUSE effort is felt [LHC] 1) May 5th. 1839.-- Maer
625o052 ar it has independent existence. & is supreme. BECAUSE it is «a» part of our nature, <not> which regu

635j55r refore utterly useless-- it foretells nothing» BECAUSE we know nothing of the will of the Deity. how
639j28v <are> «have been made» long «(as adapted to)» BECAUSE their food lies deep.-- I say it is «as» simpl
641j29r same principles that islands are favourable,) BECAUSE it must take so long to change species-- yet t
312c232 ble., not wonderful According to my view <ins> BECCAUSE actions are constant they are instincts, & no
025r019 king over the lists of organic remains in De La BECHE, for the older formations I must believe they «
026r020 ts configuration have resembled Chiloe In De La BECHE, article "Erratic blocks" not sufficient distin
042r067 as Shingle travels on the Chesil bank. V. De la BECHE). Ask Capt. F.: R: how the swell, generally & d
221b201 o.-- On animal - Confervae-- p. 23 p. 267. Dela BECHE. Geolog. Researches. facts of salt-water shells
481z008 s than 7 fathoms water. Mem Bahia Blanca. De la BECHE theoretical researches Compare land shells of G
103a064 above which explains a difficulty.-- All De la BECHE'S reasoning of mountains being formed by crust b
109a083 nalogous to broad flat sand beach. {P} -- De la BECHES argument of low coast gaining & high loosing a
110a085 of Soda mixed.-- Turner's Chemistry p. 206 Both BECK & Deshayes saw fossil shells from West Indies &
208b153 cies Paris Tertiary Shells in India!? A p. 28 Dr BECK. & Lyell. most curious law of species few in Ar
222b202 Senegal. whilst Crag <agrees with> according to BECK has none recent, yet genera same.-- Speculate o
265c084 produce like children. Lyell has story from.-- BECK about six fingered children heredetary With re
281c137 - show how finely the series is graeduated.-- Dr BECK doubt if local varieties should be remembered,
373d133 nursed in Mould.-- Lyells Elements. p. 290. Dr. BECK on numerical proportion in shells in Arctic Oce
413e059 to support of parents December 2d Lyell tells me BECK considers the characteristics of the Tropical F
419e086 n of the N. American species--Glacial period Dr. BECK says the shells in Scandinavia from height of 2
424e101 oser, than the present ones., which according to BECK are different.-- Subsidence of Greenland-- case
424e101 be trusted.-- -- Lyell tells me, on authority of BECK, that Hooded crow & Carrion crow. have in Europ
545m105 uscles to do <certain> the actions.¿ is it the <BECOM> impression becoming very often unconscious, wh
028r031 ine: Mem: on coast of Northern Chili as springs BECOME rarer, so does the rain, therefore such «rain»
035r048 whole surface oscillated equably.-- These facts BECOME easy if we look at the action as a deep & exte
049r088 n nature, -- Does not granite at C. Tres Montes BECOME more siliceous in close contact? -- «Cordiller
051r093 in the most violent surfs: in both latter cases BECOME petrified, & increase. -- In Southern regions
053r100 d of subsidence the shingle of Patagonia would BECOME more or less interstratified with sediment.--&
055r107 . Cruz. there is no occasion to wonder what has BECOME of the Basalt. Gone into fine sediment Look at
063r132 inct animals. still partly united. & eggs which BECOME quite separate.--Considering all individuals o
106a074 o tendency to widen course until inclination is BECOME comparatively small, & when that is case force
118a103 t every dike. which has not formed volcanos. or BECOME scoriform. has thinned upwards & is now cut of
126a123 n for thousands of years, that surface does not BECOME hot?-- this looks as if bad conductor-- III Bu
126a123 if cold water did not percolate surface, would BECOME hotter.-- hence temperature ought to increase
171b003 ee <living beings>. the young of living beings, BECOME permanently changed or subject to variety, acc
172b007 his view animals, on separate islands, ought to BECOME different if kept long enough.-- «apart, with
188b069 uet slightly different localities, so that they BECOME useful to know what is species.-- In proof tha
205b145 <the> proportion to their increase of size they BECOME worse in form, less hardy, & more liable to di
206b148 with successors «for» centuries, the other will BECOME extinct.-- Who can analyze causes, dislike to
210b158 uch distinctly states that permanent varieties. BECOME species. p.147 «p. 150»-- not being crossed wi
225b218 e of introduced plant, which any insects hav[e] BECOME attached to.-- that insect «not» being called
227b227 overed, for speculating on future. !.fish never BECOME a man.-- Does not require fresh creation!-- If
248c034 ties «produced by slow causes, without picking» BECOME more & more impressed in blood with time, then
258c060 ossed with people from cold, children would not BECOME adapted to climate.-- Descent. or true relatio
266c088 ding the ground.-- why do beetles & birds & <f> BECOME dull coloured in sterile countries.-- Gould in
282c141 ead of mere colour & trifling form & head &c to BECOME greatly changed. in structure & even to certai
282c142 ry but separated, now if one of these races had BECOME eminently aquatic--: «NB, aquatic, i.e relatio
285c153 ture being «necessarily» excessively slow, they BECOME firmly embedded in the constitution, which oth
285c153 etary & fixed, certain physical changes at last BECOME unfit, the animal cannot change quick enough &
291c165 an gained & heredetary. «problem solved» habits BECOME important element in classification, because s
292c171 e called instinctive.-- But why do some actions BECOME heredetary & instinctive & not others.-- We ev
293c173 that was a regular contingency the brain would BECOME webfooted & there would be no act of memory.--
298c191 on -- mountain side» of which the central parts BECOME occupied by a third best adapted kind.-- lower
300c197 quired?. The S. American dung beetles will each BECOME the fathers of many species.-- a few eggs tran
302c201 of analogy, which when sufficiently Multipliied BECOME affinity yet often retaining a family likeness
303c205 n> into one list is unfair, [moreover what will BECOME of the future creations, if the list is now pe
312c232 gs, I am told, go to heat, take dog. but do not BECOME impregnated & puppies delicate-- they cross si
336d018 <such changes> generations, these small changes BECOME multiplied, & great change be effected, but in
343d038 & colonization of S. Africa, so must the tribes BECOME blended & prevent that strong separation which
348d050 tant characters break down in certain species & BECOME worthless-- Mammalia Edentata We do (p 6) say
352d058 with abortive organ of any kind few.-- » hence BECOME EXTINCT, & hence the IMPROVEMENTS of every typ
384d161 tter only being developed, when the first <are> BECOME of use <Great characteristic of male greater
384d162 s are there abortive sexual organs?): they then BECOME so related to each other, as never to be able
404e025 ries perfect, would expect this structure would BECOME obscure & therefore it might thus have arisen,
404e026 ance & unfavourable conditions to parent may be BECOME favourable to offspring: Australian dogs have
407e039 ot have been able to have survived a change, (& BECOME transmuted), although other parallel species i
418e084 he organization is pliable, such modifications, BECOME as much fixed, as if added to old individuals,
422e095 o «NECESSARY» tendency in the simple animals to BECOME complicated although all perhaps will have don
427e111 Herbert says «p 347. Amyyralidae» Plants do not BECOME acclimatised by crossing, or by accidental pro
427e111 r. J. Banks. says Zizania in 16 generations did BECOME, acclimatized. & says Laurels have not been so
431e122 ips in Lecture in Royal Institution says shells BECOME less in number. (¿ species, or individuals) th
434e130 organs are most clearly abortive, so that they BECOME so by suppression of one organ. (here language
439e144 ssing as I think,, or that these varieties have BECOME as fixed as species, & prefer their own pollen
441e151 uch a monstrous conclusion, that every organ is BECOME fixed. & cannot vary.-- which all facts show t
442e151 now that tulips break by cultivation can a form BECOME permanent?» because its very essence is that l
454e184 sometimes one, sex & sometimes. other, so as to BECOME <all> monooecious.-- Are there not wild plants
461t037 & therefore probably any structure would rather BECOME accomodated to new circumstances than it would
467t099 prised at Henslow's remark that pistil does not BECOME abortive. Examined in microscope--some of the
492q002 ividuals-- 9. Do plants in becoming double ever BECOME monooecious-- loosing one sex & not other: whi
496q006 ver-fuge Groundsil.-- gilly flower will break & BECOME double.-- There is a double Crows-foot. or Ran
498q007 poor flower, but not fruit-- -- Do not orchards BECOME unproductive from poorness of soil.-- yet crab
502q012 nd 2 flower. which had anthers removed, did not BECOME impregnated. (34) Any recent information about
505q014 ir horns» is impregnation necessary to fruit--; BECOME well shaped by care 13 Arum before pollen is s
524m020 almost constantly present in people, likely to BECOME insane.-- now this is well worth considering,
526m030 , altering form of head, & thus these qualities BECOME heredetary.-- When a man says I will improve m
526m031 circumstances, & thus the appetites themselves BECOME changed.-- appetites urge the man, but indefin
529m040 th about science being sufficiently habitual to BECOME poetical) the botanist might so view plants &
545m104 o die-- August 24th. As some impressions «Hume» BECOME unconscious. so may some ideas.-- ie habits, w
545m105 , if so (think of this). study what impressions BECOME unconscious those which are viewed with little
571n029 ired, by certain foods being habitual-- & hence BECOME heredetary; on same principle we know many tas
571n029 redetary; on same principle we know many tastes BECOME acquired during life time:-- the latter corres
574n040 form an idea of a mind so limited as Birgos to BECOME absorbed by one end of Cocoa nut.-- November 2
576n047 ce to each man is small. Man's intellect is not BECOME superior to that of the Greeks.-- (which seems
593n107 especiall as in some <instinct> «insects» which BECOME in imago state social) by smell or looks. but
603o013 is promptness «of consequence» hence languages BECOME corrupt, & whole classes of words «are abbrevi
616o038 rn» turn into angels. this imperfect memory may BECOME perfect & we may look back to definite action
622o048 is.-- All these associated «habitual» feelings. BECOME like the instinctive, ones, <which either lead
623o051 d a beneficial tendency during past races could BECOME instinctive.-- [LHC] x It is probably That bec
624o051 NB. Until, it can be shewn, what things easiest BECOME instinctive, this part of argument fails, or r
629o055 y education.-- CD[In similar manner our desires BECOME fixed to ambition. money, books &c &c.-- <]> t
633j53v & domesticated variations show us accidents may BECOME heredetary [produce some peculiarity in seed v
634j54v & it is from these very circumstances, that we BECOME satisfied respecting an original thought, or d
636j56v other varieties was prevented Do races of peas BECOME intermixed & gardner have hybrid seedlings] p.
639j28v eep.-- I say it is «as» simple consequence they BECOME long. not at once, but by steps. of which we h
050r091 hat generally in North part of Brazil. <gravel BECOMES> sand less & gravel more common. the shoaler t
053r100 ontact of Granite & sedimentary rocks, in Alps BECOMES metalliferous. Vol III Latter Part Are there E
099a050 lination «varies with chemical attraction &c.» BECOMES measure of force. < ∴ where little inclination
105a069 wer of river & sea.; the former as its channel BECOMES wider looses its cutting power. (as does it wh
105a069 utting power. (as does it when the inclination BECOMES less & ∴ tends to finite power) whereas sea. o

```
*********************************************(Key Word)**********************************************
107a078 rom great body of mass.-- The last speculation BECOMES important with respect to thickness of crust b
113a092 l equilibrium, the height of lava (habitually) BECOMES measure of force in that part.-- Important as
171b003 in  course of generations even mind & instinct BECOMES influenced.-- child of savage not civilized ma
227b227 ation (wish of parents??) instinct & structure BECOMES full of speculation & line of observation.-- V
258c062 n <after> about affinity & how one order first BECOMES developed & then another-- (according as paren
270c103 CD universe «which Carus considers big animal» BECOMES more developed in higher animals than in veget
292c171 thing.  see men walking in sleep».-- an action BECOMES habitual is probably first stage, & an habitua
293c173 with  much practice & led on by circumstanc it BECOMES web footed, now Man by effort of Memory can re
294c178 . 306. account of trees ceasing to grow far N. BECOMES stunted, altered, & lose (mere sickness)? fert
298c190 . Such as man getting habitually into passion, BECOMES habitually passionate.-- the Key to the affect
302c203 would  be destroyed, & N & S. existing species BECOMES father of genera-- whatever the cause is. <the
313c236 foetus  of Mammalia, or chick eat») Generation BECOMES necessary, when organs of parent are concentra
315c243 his point of view. expression «of all animals» BECOMES very curious.-- a dog snarling in play.-- Hens
336d018 herited.-- «but if change be in blood long, it BECOMES part of animal &» by a succession of <such cha
337d021 w how the first eye is formed.-- how one nerve BECOMES sensitive to light.-- (Mem whole plant may  be
357d073 - (Curious the readiness with which this genus BECOMES crossed. ¿is red game an hybrid?-- When I show
384d162 roditism takes place.-- thus one organ in each BECOMES obliterated, & sexes as in Vertebrate tak plac
387d168 ck having been repeated several times, that it BECOMES fixed in blood.-- Looking at ovum of mother &
408e043 reat effect on them, & therefore extermination BECOMES part of same law.-- When we know what a  great
408e043 e whole system may. produce-- ? When a species BECOMES rarer, as it progresses towards extermination.
447e165 n mentioned in Hort. Transact. Aira caespitosa BECOMES vivaparous on mountains & yet can be raised in
466t095 y improved.» Probably every such «new» quality BECOMES associated with some other, as pointing with s
472s02r r & over again & wipe off pollen. (as a needle BECOMES covered) so whole sides of flower & stigma dus
530m046 When  a muscle is moved very often, the motion BECOMES habitual & involuntary.-- when a thought is th
530m046 ary.-- when a thought is thought very often it BECOMES habitual & involuntary.-- that is  involuntary
565n007 mes at distance, mouth is placed open.-- Hence BECOMES instinctive to fear., as ears down to horse.--
577n050 stimulus, after certain other actions, & hence BECOMES associated with them.-- The establishment of t
604o016 naria into three animals, & this consciousness BECOMES multiplied with the organisms structure, it lo
605o020 s gained how it originates, & by what means it BECOMES an almost instantaneous perception.-- Taste ha
606o20v ents & observations. & yet, like in vision, it BECOMES so instantaneous. that we cannot ever perceive
623o051 ome instinctive.-- [LHC] x It is probably That BECOMES instinctive, which is repeated under many gene
624o051 y must frequently teach same end.-- Hence this BECOMES the law of right & wrong, though, that part, w
110a086 isseminated on that side of salband. gradually BECOMING finer grained & more compact on that side-- s
175b018 Man  gains ideas. the simplest cannot help..-- BECOMING more complicated,¿ & if we look to first orig
223b211 ving crossed with (C) prevents offspring of A. BECOMING a good species well adapted to locality A. bu
232b245 change of those forms, which have succeeded in BECOMING habituated to colder climate whilst others di
265c084 n what is adaptation.-- Ermine, ptarmigan hare BECOMING white in winter of Arctic countries few  will
318c253 the S. American Pipra Flycatcher, which is now BECOMING common-- likewise of the Hirundo fulva (added
343d038 s of mankind, some of species or varieties are BECOMING extinct. others though the negro of Africa is
380d153 Yarrell does not know of any case of old Male. BECOMING like female, though many of old female becomi
380d154 ecoming like female, though many of old female BECOMING like cocks.-- It is very singular. so many Ga
383d159 to  show whole body imbued with possibility of BECOMING either sex.-- ▌ In my theory I must allude to
434e128 ss 1839, p. 9.-- talks about fossil Infusoria BECOMING extinct not so soon as other forms.-- p. 36..
437e141 et all the varieties of Brassica certainly not BECOMING Brussels Sprouts) is analogous to Primrose  &
454e184 Dr. Holland. Are there instances of plants, in BECOMING double loosing fertility if, sometimes one, s
485z016 birds--  rather indefinite letter Mem Orpheus--BECOMING tyrant-- flycatcher-- shown by habits & pluma
492g002 ension of age of individuals-- 9. Do plants in BECOMING double ever become monooecious-- loosing  one
508g016 ter sex. A Smith. About species of Rhinoceros. BECOMING rare beyond limits of the metropolis of each-
545m105 in> the actions.¿ is it the <becom> impression BECOMING very often unconscious, which makes the  idea
589n091 intelligence  less.-- p. 325 «to 29».-- Habits BECOMING heredetary form the instincts of animals.-- a
595n117 similarity  in mind.-- think of Eyton's horses BECOMING <white> with <lather> <foame> & sweat, when h
048r084 with Tosca. which implies motion in the «loose» BED of pebbles. (On a sea beach under a cascade, one
077r171 ato to SW. with respect to latter doubts whether BED or vein (very like that of Spital of Schemnitz i
089a020 Memoires de la Soc. «de Geneva» Vol 3' P. II. -- BED. of elevated shells on the Senegal. L Institot p
093a033 ia Blanca. <fl> Flustra identical. recent & bone BED.-- November 8th 1877 (Memoranda so far distribut
100a054 e!!! Look at gneiss of Rio Concretions in Pumice BED at Ascension instance of hollow concretions & co
114a095 sits.-- NB. because lowest. first accumulated in BED of ocean With the exception of sandstone rare to
116a100 ing «successive» banks <above> 30 ft or so above BED of river. formed of rounded pebbles-- it is clea
147g020 ances, as if valley had been filled with sloping BED of rubbish Friday Highest part of road between I
152g049 ear Loch Tring-- Tuesday Bridge of Roy Level of «BED of» River 30.221/65 degree/ Temp of air 65 degre
152g049 ere are two terraces on the East side of river & BED of river about 40 ft beneath general plain. 30.1
162g103 at divatium aquarium with, left of Bright.-- like BED of lake with trace of terraces on each side High
399e006 o have lasted for its time: but we ought in same BED if very thick to find some change in upper & low
399e006 layers.--  good objection to my theory: a modern BED at present might be very thick & yet have same f
423e098 ingle turnip in a garden is sufficent to spoil a BED of Cauliflower.-- (How curious it would be to ma
423e099 ls-- CD[If cases of one variety in upper part of BED-- & another in lower is very rare, the conclusio
423e099 ot rate of deposition has been equal even in one BED, much less in alternating strata of sand & limes
434e129 ioicous. yet parts only very slightly abortive & BED of female flowers will sometimes produce a few s
532m056 d of her & of servant of Richard & of Mary & her BED brought from Shrewsbury) yet for a fortnight con
532m056 any  room, would not sleep at night even when in BED room-- grew very thin, would not go out of house
547m111 nfluence replace each other August 29th. Went to BED. & built «common» Castle in the air, of being co
573n038 esire to swallow.-- tells himself not to turn in BED. will turn in bed.-- in case spittle, effect of
573n038 - tells himself not to turn in bed. will turn in BED.-- in case spittle, effect of thought is to make
593n111 they  all took flight & flappered across pool to BED of flags I was astonished & having looked  round
594n112 few  had ever before seen one, yet all-- flew to BED of flags. hernes are common. not unlike in  size
162g102 66 degree 30.095 .0458 or 6 difference between BEDROCK & Loch Ness <30.100> <Donald Macphee> Saturday
029r034 present  position of <Coal> Forests. These thick BEDS of Lignite stratified with substances so like t
035r045 and. formed near surface: Mem Patagonian pebbles BEDS, most unfavourable to preservation of bones  &c
035r046 introduct to Geolog--"Between the height of same BEDS, deposited in different basins; little or no re
035r047 & dependent: & then how wonderful level «of same BEDS» should have been kept; it shows that throughou
041r064 nce--The depth of shells (which being packed. in BEDS) lived there, makes it very doubtful whether th
048r084 vations along whole line of coast Darby mentions BEDS of marine shells on banks of Red River Louisian
048r086 Rhinoceros  was killed. -- In Patagonia, are all BEDS same age? is whole substance triturated Porphyr
055r106 test a <more> decided elevation of sea's bottom. BEDS of shells. 2 - 3 toises thick.--Vol II. p.  252
057r114 Africa productions.-- I think in Patagonia white BEDS having proceeded from gravel proved.--  curious
067r143 nt Edinburgh March 1835 Sir W. Parish says. that BEDS of shells are found on whole coast from P. Indi
075r166 st part totally independent of the nature of the BEDS they intersect". = In the Guatemala part. (& Ch
076r170 te. passing into talcose & chloritic slate. with BEDS of syenite & <sep> serpentine dipping to SW  at
077r171 says fragments from roof & penetrating overlying BEDS tells the secret.-- p. 189. "The small ravins i
101a057 of  Boulders in N. America do/p. 280. the gravel BEDS in England different from Boulder beds-- What i
101a057 he gravel beds in England different from Boulder BEDS-- What is Osteopora platycephalus (Harlan) foun
110a087 of the true rocks, most probably from the gneiss BEDS in the mica slate.-- Geograph. Journal. Vol  IV
113a091 y infer that time has not been allowed for lower BEDS to cool down. & then in 50000 years the depth w
115a096 ndstone rare to have any horizontal non cleaving BEDS. metamorphosed. The chemical action which gives
116a099 on D'Orbigny's Voyage. good section of Rio Negro BEDS.-- -- refers to species non decrite de  petites
116a100 o Chile p 29. gold is not sought for in Chile in BEDS of river, but in shelving «successive» banks <a
121a112 lls on surface of Patagonia, yet none in shingle BEDS. Lyell on Sweden. p. 12. proofs of small rise a
123a116 (under  head of Delta) describes near Alps great BEDS of rivers which must be like the Chilian ones.-
134a141 «living» shells. on coast of do p 8.-- soft Clay BEDS near C. Virgin p. 59. dip of Clay slate in T de
138a153 league inshore both N & S of Lima.--judges from «BEDS of» sand & gravel & shells. p. 47. do has table
221b200 Scicily.  Jan: 1838 L'.Institut. Bats, in Eocene BEDS, very like present species., p 8. ¿Are  mundine
425e105 talks  of Tapirus American form. found in Eocene BEDS of Paris Lyell has remarked species never reapp
433e126 more  probably than during the deposition of the BEDS-- The argument must be thus put, shall we  give
448e166 ecies and subgenera) between the Crag & Touraine BEDS, the one with neighbouring & Arctic sea, &  the
448e167 unting for them.-- Think over this-- The Superga BEDS have many shells in common «& are not far dista
448e167 r contemporaneous»-- how is this with the Eocene BEDS.-- see Lyells tables Bennetts Wandering Vol II.
461t041 some  species-- (especially of mammifers) in old BEDS & existing species is valuable because it shows
465t080 sts of Chile, excepting Concepcion-- Patagonia-- BEDS of La Plata. (except close to B. Ayres).-- If w
```

Page
(Key Word)
189b074 ctual faculties} most developed, as highest.-- A BEE doubtless would when the instincts were.-- relat
223b208 er articulate or intestinal, or even a mite.-- a BEE «compared with cheese mite» with its wonderful i
307c216 ument.-- Are there any abortive organs in neuter BEE, (There is paper by Yarrell «in Zoolog Transacti
324c268 ntroduction to the Natural system Bevan on Honey BEE Dutrochet Memoires sur les Vegetaux et animaux.-
382d158 eveloped.-- surely analogy of molluscs. & neuter BEE would shew this-- (Do any male animals give suck
388d172 us instances of corelation in structure = Neuter BEE having both sexes abortive that = same tendency
466t093 I SAW IMPREGNATE by pollen with which <bees> «a BEE» was dusted over. {P} Stamens & pistils curve up
466t094 en & pistils have no relation. In Monk's Hood, a BEE entering long nectary, would «necessary» cross d
466t094 wer division of nectary: «wh. itself resembles a BEE, but does not prevent bees visiting it.» In Colu
467t103 nectar at foot of upper petal standing on «I saw BEE go to two species of Lupine,» two wings. & when
467t104 llow pollen protrudes at sheath.-- At last I saw BEE collecting pollen from <sheath> Keel of Lupine--
467t105 ot very abundant. not very small-- Saw one small BEE; saw another on Cabbage--white Butterflies suck
468t112 l were brushed by Bees & especially stigma after BEE had brushed over the anthers of long stamens {P}
468t135 ing-- On rough calc. 280 flowers-- allowing each BEE visits 10 flowers in «minute» each flower will b
470t178 of Butterfly Orchis & Listera? Bryony saw common BEE on: Linn. Trans 18. p. 133 Westwood on the Fulgo
472s02r y day. many clumps of heartseases, never saw any BEE go to them. Yesterday remarked that many flowers
472s02r several flowers I examined some days ago-- This BEE flew from yellow to yellow & purple heartease wi
472s02r ow to yellow & purple heartease without doubt.-- BEE, not large, very dusky & broad never saw such a
496q05a al flowers-- also do with honey-- What is use of BEE Larkspur= =Toad Orchis= How many flowers in minu
515q022 in Sand-walk, on which I think I have never seen BEE visit. Experiments in Garden Sow stones of Stand
564n003 mans because original instincts different.--Mem. BEE how different instinct a solitary animal still d
583n069 & expression as cunningness of fox, industry of BEE &c &c-- p. 15."instincts act with unerring preci
583n070 they would be "creatures of habit."-- CD[as the BEE makes its cells, by means of ordinary senses & m
584n072 ng themselves through the air,-- waterbirds, the BEE to its nest,-- cats when carried in confinement,
586n081 ired instinctively" So with <sight> sight-- so a BEE has the faculty of building «regular» cells-- [b
594n115 eness. [blank] Circumstances having given to the BEE its instinct is not <more> less wonderful than m
637j57v ts <not> not been provided. «with proboscis» «as BEE & butterfly» inconvenience.! extinction, utter e
638j58v ess superadded This is similar idea, to cells of BEE, corresponding to <every> «one or any»-- brain m
638j059 ot death rejects the imperfect attempts. In the «BEE» Mollusca the nervous system is endowed with the
641j29v between fang of snake, (jaw of spider?) sting of BEE, sting of nettle.-- Are there any other analogie
505q014 's. fact.= (4) Effects of Nitrate of Soda under BEECH.-- Lychnis dioica answers this question= (5) Op
024r015 h of Mocha; 19 miles. 65 Fathoms Vide facts in BEECHEY. on NW coast of America off Cape of Good Hope
034r045 ours in East Indian Volcanos Gypsum Andes Mem. BEECHEY. account of regular change in soundings. on ap
051r097 avourable to attachment of many bodies [blank] BEECHEY.--changes in bottom in NW coast of America. fr
458t002 substances-- useful perversion of instincts-- BEECHEY'S Voyage Vol I. p. 499. «4to. Edit»-- Horses in
129a132 hot heads &c heat beneath the sea.-- CD[did not BEECHY have some such case]CD what would be the chanc
291c167 «es». is strongly supported by wonderful fact of BEES changing the sex by feeding.-- no it is develop
309c221 glow-worm knowing female good case of instinct. BEES turning neuter into Queen, more wonderful case
313c235 ous to dormant instinct.-- (How wonderful a case BEES developing sex of neuters) species may have had
326c266 hy. 4. Volumes well worth reading Bevans work on BEES, new Edit 1838 Harlaam. Physical & Medical Rese
370d116 rganization. no more can be said.... In paper on BEES in same work. (it is said that some kind lay <p
370d116 Wasps breed many females, but almost all die.-- BEES breed but few, because they are kept in securit
401e014 t with another-- Uncle John says he has no doubt BEES fertilize enormous number of plants-- it is sca
466t093 ry marked by orange freckles on {a} upper petal; BEES & flies seen directed to it-- The Humbles in cr
466t093 til & one I SAW IMPREGNATE by pollen with which <BEES> «a bee» was dusted over. {P} Stamens & pistils
466t093 ly.-- In Azalea <do> «it is so» <Though I saw no BEES «several» visiting it>.-- In yellow day lily, t
466t093 severaly visiting it>.-- In yellow day lily, the BEES visit base of upper petal, though not different
466t094 end up a little In a wild purple Geranium, I see BEES visit always base {a} of upper petal from facil
466t094 but these <do> do not bend up-- In Lark-spur, if BEES put proboscis within nectary «they do» they mus
466t094 wh. itself resembles a Bee, but does not prevent BEES visiting it.» In Columbine nectaries are placed
467t103 ortive stamen are of useful height.-- In Lupine, BEES «frequent» & seem to act, something like on Kid
467t103 certainly when over-ripe & half withered-- I saw BEES going to clover & once this happened.-- And in
467t104 cting pollen from <sheath> Keel of Lupine-- Seen BEES on Potato & several times on Beans Rough.--gree
468t105 legs & take flight-- Yet we have crosses-- I see BEES almost every flower-- Blue-bells-- wild-raspber
468t112 dodendron xx after several gloomy days. hot one, BEES almost P every minute to Fraxinella «& from <fl
468t112 «more than stamens», so that all were brushed by BEES & especially stigma after bee had brushed over
469t135 wer stalks for ten minutes. it was visited by 13 BEES-- & each examined very many flowers.= 22d.-- /d
469t135 g several succeeding days <many> «most numerous» BEES visited this same bunch & on this day in five m
469t135 n Humbles came & each visited many flowers-- Saw BEES frequent these flowers till late in evening-- O
470t177 es <in our garden> show no trace of palmation!!? BEES at Wild St Johns Wort--Scabies, Cyanoglossum--R
470t177 rt--Scabies, Cyanoglossum--Reseda wild very many BEES & Humbles--on Thistles many (curious because a
470t177 flowers & as much shut up, frequented by «many» BEES & Humbles-- «Humbles & common» On silene, many
470t177 Many Humbles on hedge Linaria= (Plenty of Humble BEES on Phlox Down, 1854, Sept.) In Spanish Broom by
470t178 any Papilionaceous; all wh. are in flower «I saw BEES;»-- on Monk's Hood, brushing over stamen «Egg T
472s02v sucker, after having withdrawn it.-- Saw 4 more BEES at work-- another odd genus-- & a small common
495q05a such dioecious individ--small orifice (8) Carry BEES, powdered with starch & Carmine & experimentise
496q05a come up true-- whilst others are crossed.-- Are BEES, guided by smell-- or sight.-- --. touching Mr B
497q06v ich have abortive parts, whether such vary.-- Do BEES go to Sweet Peas, IMPORTANT, for if so, as thes
497q06v hese can be raised true, there is no crossing by BEES.-- Henslow.-- (1) Character of alpine Flora of
498q008 grt degree from charcoal & good treatment (8) Do BEES frequent Cabbages «& Cowcumber's out of doors.»
498q008 d <plan> such plants do not degenerate,-- as the BEES will mingle the infinitesimal varieties which m
499q009 of grains of pollen in any one flower (17) Catch BEES, Butterflies-- Syrphus-- Meligethes & see wheth
499q009 hat state (whole or broken) is ball of pollen on BEES thighs (18) Place pin's heads with Bird lime ne
499q09 » Examine all these, are they much frequented by BEES or Butterflies or little insect?= or is pollen
500q010 iwinkle. (not asclepiadae. «in» Lindley) (24) Do BEES distinguish species, they do not varieties.-- (
505q014 rough Penny Cyclopaedia) Abberley says that some BEES are smaller & more vicious. Will try to get me
505q014 ok at:-- Was once offered a hive. of these small BEES-- at Sundorne has large Bees July/42/ Mark has
505q014 ive. of these small Bees-- at Sundorne has large BEES July/42/ Mark has six day's puppy of small true
585n077 which might be «possibly» acquired by habit. so BEES in building cells, must have some means of meas
609o030 ept for the moral sense, any more than a hive of BEES without their instincts.-- Gives art to when I
623o051 eficial tendency [RHC] 10) that the instincts of BEES & beavers «& deer» have <been formed> a benefic
344d040 ome larvae of insects-- (¿glowworms) breeding-- <BEET> imago state fertile at once.-- Consider this w
265c086 to have structure without habits-- after seeing BEETLE with wings beneath soldered wing-cases-- Yet t
573n036 itted to recognize the wonderful structure of a BEETLE than a Universe.-- November 20th Saw the young
192b084 mined.-- so with useless wings under elytra of BEETLES.-- born from beetles with wings.& modified.--
192b084 ess wings under elytra of beetles.-- born from BEETLES with wings.& modified.-- if simple creation, s
265c088 to suppose from beholding the ground.-- why do BEETLES & birds & <f> become dull coloured in sterile
300c197 pain. of death acquired?. The S. American dung BEETLES will each become the fathers of many species.--
310c223 era & some species on Himalaya.-- some English BEETLES, birds & a fox most close The most curious cas
486z019 Vol I. p. 130. Col Sykes on balls made by dung BEETLES, like those from Chiloe. Amblyrhyncus de marli
601o009 any direct sensation, in order to avaoid it-- BEETLES feigning death upon seeing an object.-- are Pl
214b173 of living seas.?-- Roxburgh. list of plants in BEETSONS St. Helena. -- Galapagos--Juan Fernandez Falk
023r010 ish bones floating on the surface of the ocean, BEFORE arriving at the Abrolhos shoals \ -- N.B. The
028r032 cted by Cordillera, where that action commenced BEFORE any great accumulation of such matter.-- Dr A.
032r039 l movements to line (2). The part (o) which was BEFORE beneath band. of greatest action. would now by
047r081 {P} form present, i e a part below «mean» level BEFORE the higher part. -- Does the sea fall on banks
086a006 t Lakes» Siberia must be read as well as Pallas BEFORE Geology is written Cuvier. Europe possessed a
089a017 f Heat bubbles volatilized at bottom, condensed BEFORE rising?-- Mem. granite heated.-- Metamorphic a
090a021 sorting of matter. in making trachyte come out BEFORE -- What must be the effect of all the meteoric
093a031 asalt: last because it could reach the surface. BEFORE being cooled.-- Berzelius. L'Institut. [1837 p
117a102 aily be scene of ruin in late Natical Magazine (BEFORE June 1838) that 70. F were obtained 100 miles
120a110 lsisbury Craigs. Kept amongst <old> papers read BEFORE societies.-- Sir. J Hall Vol VI. p 173. (Ed. T
147g021 65 degree? For comparison with all the measure BEFORE There some of the half rounded gravel nearly a
148g026 account of the «YOUNG» Shepherd dogs Saturday. BEFORE coming to Bridge of Spean, hills of «sea», gra
156g068 near head of valley fragments which had fallen BEFORE lake drained could be told from «some of» thos
157g076 ranite veins & quartz, & garnets.-- Boulders as BEFORE certainly must have <come> «been drifted» here
175b017 them in space, so might they in time As I have BEFORE said isolate species <& give even less change>
192b088 erfect by the discovery of fossil Mammalia than BEFORE & that is all that can be expected-- This answ

194b095 k a very strong case might be made out of world BEFORE zoological divisions.-- Mem. species doubtful
195b098 ds so permanent a breath cannot reside in space BEFORE island existed.-- Such an influence Must exist
199b120 ge <if offspring not fertile> <,but producing> «BEFORE domestication, afterwards none or little with
219b196 s intercourse. a parallel fact to Blood-Hounds. BEFORE Attract of Gravity discovered. it might have b
221b202 removed?? Is the prevalence of Coniferous Woods BEFORE Dicotyledenous a fact analogous to reptiles be
221b202 ore Dicotyledenous a fact analogous to reptiles BEFORE Mammalia Think about Miocene fossils some spec
225b216 rguments against it-- namely how did otter live BEFORE being made otter-- why to be sure there were a
228b229 Question, which every naturalist ought to have BEFORE him, when dissecting a whale, or classifyng a
255c051 of memory without association..» Instinct goes BEFORE structure (habits of ducklings and chickens) Y
255c051 oung water ouzels hence aversion to generation, BEFORE great difficulty in propagation.-- Feathers on
257c057 ble,-- it has been asked how did the otter live BEFORE it had its web-feet-- all Nature answers to th
268c096 n Classified catalogue of animals of Nepal read BEFORE Linnaean Soc. Feb. 1838.-- Annals of Natural H
274c113 s not this structure., instance of habits going BEFORE structure).-- even one kingfisher-- Gould has
276c124 ost importance to show that habits sometimes go BEFORE structures.-- the only argument can be, a bird
293c172 when you feel sure you have heard conversation BEFORE. is strong association recalling up image whic
301c190 cts precede structure.--duckling runs to water. BEFORE it is concious of web. feet.-- p. 7. Mr Blyth
308c216 if cretceous period assumed, then some perished BEFORE, Carboniferous some perished before, then ther
308c217 me perished before, Carboniferous some perished BEFORE, then there always have been gaps, & there now
332d004 umber of worms the child not having passed them BEFORE.-- Hence disordered intestines are not healthy
344d042 spring came out one big & one small. Now Jones, BEFORE this happened from her looks thougt she was ha
345d042 lood'» animal must have gone on for many years, BEFORE deserves <name> «to be so called»,-- the short
364d104 e peeled rods mentioned in old Testament placed BEFORE sheep-- it has been thought that silver Pheasa
364d105 w races.-- In Scandinavia besides the Rakhekna, BEFORE mentioned between Capercailzie & Black Cock.--
380d153 as a <female which have> larvae which have bred BEFORE the vis formativa had completed them-- (but th
391d175 t different) should be added to each individual BEFORE he can procreate. these changes may be effect
400e013 same year.-- he sowed them for four generation BEFORE they broke.--, showing effects of cultivation
400e013 on gradually adding up. & four more generations BEFORE they began to double.-- at present time Uncle
401e016 from many varieties, & probably would take long BEFORE all the stain would be got out of it.-- Now th
406e035 ng most is whether mother has had any offspring BEFORE.-- -- now this is never stated.-- Regarding th
411e055 , however) is very remarkable & none discovered BEFORE them in any part of World.-- Wealden to boot.--
411e056 is interesting as showing some change in habits BEFORE. form.-- I have already given various examples
427e109 an., & the climate being now less extreme, than BEFORE arctic forms would retreat: effect on snow of
443e157 that of the Camel has not escaped Naturalists." BEFORE he alludes to the resemblances of the snout of
463t063 cription of my fossils makes same such remark & BEFORE the conclusion of his work-- Lund makes his wo
466t095 These qualities have been given to foetus <fr> BEFORE sex developed-- Double flowers & colours break
472s02r large, very dusky & broad never saw such a one BEFORE-- Saw Fly 21 «this Heartease withered on Monda
483z013 rms must be studied in Vol I, Zoolog: Transact. BEFORE writing on Planariae or Polypi & is especially
484z014 ? Wilson N. American Ornitholog must be studied BEFORE writing my general account-- ¿ Do not the Peng
500q10a ests of sexes.-- Q.30) March 1842. <Last> Year «BEFORE last» beans & peas were planted in rows adjoin
504q013 ther to flower & Menyanthes whose pollen bursts BEFORE flower is open-- -- No (6) There is apple with
505q014 to fruit--; become well shaped by care 13 Arum BEFORE pollen is shed can you find flys dusted with p
513q21. rossed R. BROWN-- will pollen act on any flower BEFORE stigmas expanded-- in reference to Lobelia & C
514q21. spur.-- Ask about Pinks & Solanum impregnation BEFORE flower open. (An. des Sci Where is Boerhaave's
520m001 ter their parents, when the latter died so long BEFORE, that it is extremely improbably that they sho
522m011 iss A. B.-- all the same names as a few minutes BEFORE he maintained he had never heard of.-- Thus in
523m017 of husband connection with housemaid two years BEFORE, to prove she was not insane, answered she had
524m022 what they think of most. intently.-- criminals BEFORE execution.-- Widows not of their husbands-- My
525m024 Maer, used to betray himself by looking ashamed BEFORE it was known he had been on the table,-- guilt
525m025 er than rest «they all died»:-- she had kittens BEFORE & afterwards with tails. My father says, perfe
527m034 uch castles in the air are highly advantageous, BEFORE real train of inventive thoughts are brought i
533m058 s that babies know a frown very early in life, <BEFORE they> (I think I have seen same thing before t
533m058 , <before they> (I think I have seen same thing BEFORE they could understand. what frowning means) if
534m061 - then thinking consists of sensation of images BEFORE your eyes, or ears (language mere means of exc
538m079 repeated things, which none about her had EVER BEFORE heard, so very probably forgotten.» Such facts
541m090 that whole effort consists in keeping one idea BEFORE your mind steadily., & not merely thinking int
541m091 eavur to keep any simple idea as scarlet steady BEFORE mind for period, «if the scarlet was before on
541m091 ady before mind for period, «if the scarlet was BEFORE one effort less» one is obliged to repeat the
546m111 him about a knife. which he had hid some years BEFORE.-- was greatly astonished, at the time. & coul
554m139 curious to see her take bread from a visitor, & BEFORE eating «everytime», look up to «keeper» see wh
555m144 train of Dr Monro experiment about hanging came BEFORE me showing impossibility of person recovering
560mIBC arly in life Do they wink, when anything placed BEFORE their eyes, very young, before experience can
560mIBC anything placed before their eyes, very young, BEFORE experience can have taught them to avoid dange
563n001 rgreens they seek Cock Pheasant claps his wings BEFORE? crowing & only in breeding season & on the gr
566n011 ng really useful to them: this must be studied. BEFORE my view of origin of evil passions.-- Man gett
566n012 of theological state of science, grand idea: as BEFORE having analogy to guide one to conclusion that
567n013 oes such notion commence?-- Children understand BEFORE they can talk, so do many animals.-- analogy p
573n037 atest surprise at emotions in her countenance-- BEFORE they can have learnt by experience, that movem
577n052 riously shown in the sudden cures of tooth ache BEFORE being drawn,-- My father «even» believes that
578n054 of cowardice, or cheating.-- one does not blush BEFORE utter stranger,-- or habitual friends.-- but h
579n058 to my theory of smile. remember children smile BEFORE they laugh.-- Has frowning anything to do with
579n059 ys he loves a person-- do not the features pass BEFORE him marked, with the habitual expressemotions,
580n062 lats: "the recollections of the infant likewise BEFORE two years are soon lost; yet many of the habit
581n064 expressing their want, pleasure, or pains long BEFORE they can speak-- or understand-- thinks so it
584n071 er: but this is not so,, the instincts may vary BEFORE the structure does; & hence we get over an app
584n073 mences as soon as painful thought crosses mind, BEFORE it can have affected respiration V E. p. 125 W
586n079 g performed or emotion felt in early childhood (BEFORE experience or habit) could be formed or afterw
589n092 muscles of voice than respiration.-- like sigh BEFORE false sneeze.-- "A Dissertation on the Influen
590n093 circulation-- no, for «grief» sighing comes on BEFORE circulation is affected. p. 44.-- Jealousy. ca
593n107 y relation to the period when men. communicated BEFORE language was invented,-- were musical notes th
594n112 rare « <s.> » here,, that probably few had ever BEFORE seen one, yet all-- flew to bed of flags. hern
605o19v) <which> «this» emotion, from the associations BEFORE mentioned. we call sublime.-- It appears to me
613o036 has seen push a hare through the bar of a gate BEFORE him, & then jump over the gate & bring it. ---
614o037 perceive instinct. boy takes delight in mammae BEFORE any reason had told him this distinctive mark,
634j25r ributes of the Cetaceae.-- How came Bats also.? BEFORE birds? They are ancient.-- Are Cetaceae found
638j059 Man, the nervous system, gains that knowledge, BEFORE hand. & can in idea (with consciousness.) <th>
640j167 d prevail, even if have afforded only 10th part BEFORE & now formed eighth part.-- or if other prey d
065r139 elonged to an Englishman.--On 8th of March cove BEGAN to freeze. correspond to September ¿Did I make
332d002 people all over England at same periods When he BEGAN practice, he remember during a year or two he s
397e003 ly according to fixed laws: And since the world BEGAN, the causes of population & depopulation have b
400e013 adding up. & four more generations before they BEGAN to double.-- at present time Uncle J. does not
522m011 had never heard of.-- Thus in many things if he BEGAN at one end, he knew the whole subject.-- if at
336d018 , & the union makes hybrid, in fact the parents BEGET child like themselves. expression of countenanc
371d127 cette. here is what [not located] that it shall BEGET young different in colour, form, & so altered t
388d172 d. the Quagga case appears to that of <2» dog BEGETTING different puppies out of same mother.-- The v
578n054 conduct.-- either good or bad. either giving a BEGGAR, & expecting admiration or an act of cowardice
056r111 well as separating causes by water.--Or rather BEGIN & explain how water separates.--(intertropics a
308c218 with which our speculations must end as well as BEGIN" &c &c then centre is every where & then circum
362d099 e in 5 or 6 weeks after castration, fresh horns BEGIN to grow.-- Mr Yarrell says the «male» Axis of I
423e095 e simplest fish &), my answer is because, if we BEGIN with the simplest forms & suppose them to have
550m125 ater proportion of <each» «every» man's time.-- BEGIN discussion-- by saying what is Happiness?-- Whe
615o36v fowl» totally different habits from Europaean. BEGIN to prowl about in the evening «seldom leave the
170bIFC written in January 183[8]: probably ended in BEGINNING of February Zoonomia Two kinds of generation
193b092 - I have abstracted Mr Swainson's trash.-- at BEGINNING of Volume on Geographical distribution of ani
218b192 ed by water & studded with others.-- we see a BEGINNING to isld. Graham isld.-- we know many seeds, m
239cIFC the world.-- Charles Darwin written between («BEGINNING of» February & July 1838) All good References
257c058 . it is useless to speculate «not only» about BEGINNING of animal life.: generally, but even about gr
503q013 ouse «ANY male branch.» --¿number of seeds in BEGINNING of November 1841.-- Trees above male? (2) Res
528m035 round one. one recalls the castle by going to BEGINNING of castle» because train cannot be discovered

567n015 ody.-- is it connected with surprise.-- heart BEGINNING to beat-- children inherit it <ins> like inst
568n019 ew. 1827? In Water Scotts life.. Tom Purdie, (BEGINNING of Vol V) «finally» says "he knew no more wha
569n020 alphabet. arose from letters, symbol of word BEGINNING with the sound of letter)-- crying yawning la
574n039 argument might be used to show language had a BEGINNING, which my theory requires. There probably is
599o005 ar 1837 & Earlier-- JCD in Athenaeum "Smart-- BEGINNING of a new School of Metaphysic,"-- give my doc
609o029 progenitor far back, (anger <to> at the very BEGINNING, & therefore most deeply impressed). shame pe
622o049 ressed thus avarice. &c &c.-- [RHC] 8) in the BEGINNING I mentioned only three instincts.-- I am far
636j56v e equally simple series, & therefore trace of BEGINNING in organic world.-- Macculloch. Attrib. of De
159g087 he slope is continued some hundred feet lower & BEGINS about 60 higher-- There are however fringes of
388d170 arries with stock of food.-- the generalization BEGINS low.-- it goes through transformation. nearly
429e115 he Pyrenees, that the Rhododendron ferrugineum. BEGINS at 1600 metres precisely & stops at 2600. & ye
548m117 ue idea something is not there, & then when one BEGINS eating one perceives butter or salt is not the
428e113 number of good race-horses, which Eclipse? has BEGOTTEN <?> «Walker attributes this to effect of male
306c212 nkey. (Baboon) at Z. Gardens upon being beaten BEHAVED very differently from a dog.-- more like man.
618o041 son at all, we can only infer it from his its BEHAVIOUR. Thought is only known subjectively? -- ? the
555m143 & then I had some confused idea of showing scar BEHIND (.instead of front) (having changed hanging in
556m145 t> inspiration & quickly retracting tongue from BEHIND upper & little between incisors.-- like <W[...
570n023 with body distended.-- intolerable to be poked BEHIND, without ones chest, being distended. touch a
599o007 thrill, which runs throug every fibre, when one BEHOLD the last rays of & & or grand chorus are utter
265c088 e to habits.-- one is tempted to suppose from BEHOLDING the ground.-- why do beetles & birds & <f> be
531m051 hy is based as Burke maintains on pleasure in BEHOLDING the misfortunes of others.-- In young childre
602o011) a very good passage. about actions & decisions BEIN the result of sagacity, or intuition. when indi
034r044 Daubeny Von Buch is very strong about Trachyte BEING the most inferior rocks--The stream at Portillo
035r047 ubeny P 402: likewise, mean height of tertiary. BEING less than secondary:-- consider arguments for o
036r051 e Volc: in harmony with the prevailing movement BEING one of elevation alone.--In England much subsid
037r052 ted beneath low water in the Southern ocean not BEING buoyed with Kelp.-- With respect to degradation
038r058 ument:--¿ Consider causes for subaqueous crater BEING of diff: form subaerial one?--In former not so
040r062 to New Zealand, which agrees with St Helena in BEING unique, yet no quadrupeds.-- Is the white matt
040r064 w water no evidence--The depth of shells (which BEING packed, in beds) lived there, makes it very dou
041r065 shell's concretions, living only in that spot & BEING cause of concretion; or being only preserved in
041r065 ly in that spot & being cause of concretion; or BEING only preserved in that part. having lived over
041r066 , full of sand.--the «scale» «quantity of iron» BEING there in excess.-- If veins {P} are secretionar
044r073 other: The extreme frequency of soft materials BEING consolidated; one inclines to belief all strata
047r081 ber of waves & in wind, instead of sea's bottom BEING in motion what difference? In watching heavy sw
048r086 P. Desire). = Where talking of such substances BEING worn into channels. mention submarine channels.
055r106 pecion 50 toises above the sea. = talks of them BEING packed clean. & without earth.--Moreover that s
056r111 canos blend all substances together; & products BEING similar over whole world, general circulation.
063r130 n latter time. (or changes consequent on lapse) BEING the relation.--As in first cases distinct speci
063r131 gradation;-- An argument for the Crust of globe BEING thin, may be drawn. from. Cordillera. rocks.--W
086a003 nmixed is very pleasing; owing to the movements BEING of one order. -- There should not be surprise a
086a003 order. -- There should not be surprise at Horse BEING found in America, when Mammoth & narrow toothed
088a016 L. Institut 1837.-- Helms remark on common salt BEING found on low hills East of Cordillera very impo
092a031 ere are some ideas about order of injected rock BEING determined by fusibility in. L Institut p 247.
093a031 last because it could reach the surface. before BEING cooled.-- Berzelius. L'Institut. [1837 p. 297]C
096a041 y says he can give me facts respecting lime <n> BEING heated without parting with Carb. Acid.-- Mr Ma
100a055 er-- Phillips Lardner p. 197. refers to salt as BEING produced by local heat, Ask Capt. Beaufort, whe
102a061 rst the more fusible substance, & then the next BEING sucked out. In Cleavage discussion, state broad
103a064 lty.-- All De la Beche's reasoning of mountains BEING formed by crust being too large & pitching agai
103a064 's reasoning of mountains being formed by crust BEING too large & pitching against each other, is, I
103a065 the crust might be considered a level.-- Dikes BEING last action. (effect of horizontal movement) he
103a065 tersect metallic dikes: It is an important view BEING subsequent to dislocation of strata. A capital
104a068 Salt on surface of plains due to whole moisture BEING lost by evaporation therefore capillary attract
110a086 ted of hornblende (?) & felspar, (some crystals BEING red) «with» cleavage, veins of pyrites, few cur
112a090 stating that the mean temp at Yakous in Siberia BEING -8 Reaumur.-- there ought to be 32 degree Fah.
113a090 e Fah. at a greater depth than 400. & the limit BEING 400 ft. shows that the strata have very unusual
113a091 ither better or worse & the depth of 32 degree. BEING little we may confidently infer that time has n
115a097 94. Von Buch's Travels account of Norway chain BEING broken through like that near-- Obstruction Sou
122a114 uid arriving at equilibrium so soon from; crust BEING cut of-- if part of «cold» crust under ocean, b
124a120 lateral injection. from probability of fissures BEING prolonged to surface. see p. 181 on do subject
126a122 this may be very wrong.-- The fact of a dumplin BEING bad conductor is {P} against my views-- if we h
131a136 ozen crust in America Richardson.-- From strata BEING not only vertical, but turned over in many part
132a139 cold of ocean accounted for, by the circulation BEING greater, than the transmission from ocean's bot
148g024 to be impressed Case of Birch Wood by Inverorum BEING determined by sheep & not deer When Black faced
149g032 other rocks not apparently in situ <& in> hill BEING gneiss <& also> also near summit on Hill on sid
157g073 es, but I believe this is chiefly caused by its BEING lower,-- [no pebbles in parts of Beagle Channel
171b004 grown individual «with fixed organization» thus BEING modified,-- therefore generation to adapt & alt
174b013 is argument applies only to hybridity.-- genera BEING usually peculiar to same country, different gen
176b021 son for supposing number of forms equable: this BEING due to subdivisions & amount of differences, so
176b022 , as we may suppose is the case. their creation BEING dependent on definite laws, then those which ha
177b029 many» <therefore> changes and base of branches BEING dead from which they bifurcated.-- Type of Eoce
180b037 thus perish out, or on other hand like Orpheus. BEING favourable many they might be produced.-- This requi
183b051 try. How is propagation of wolf & Dog. (because BEING believed same species) if they do not breed rea
189b072 ction between great groups.-- Speculate on land BEING grouped towards centres near Equator in former
189b074 complete.-- It is absurd to talk of one animal BEING higher than another.-- We consider those, where
190b077 (long separated.-- Proteaceae & other forms (?) BEING common to Southern hemisphere, does not look as
190b078 ry step of progressive increase of organization BEING imitated in the womb, which has been passed thr
191b082 dence continually forming species.-- Man & wife BEING constant together for life, is in accordance wi
193b090 new classes be brought into play.-- The father BEING climatized, climatizes the child. ¿-- whether e
193b090 ell adapted, This would account for each tribe <BEING> «acting as» in vacuum to each other p. 306.--.
194b096 es not Lyell give some argument about varieties BEING difficult to keep on account of pollen from oth
197b112 r itself does not agree with old & modern types BEING constant. Cuvier's theory of Conditions of exis
198b115 ist upon <different animals> large Mammalia not BEING found on all isld, (if act of fresh creation wh
199b116 quote from Lyell: assuming truth of quadrupeds BEING created on small spots of land, of the same typ
199b119 iry with one hairy child, and of albino DISEASE BEING banished, & given to Portuguese. priest.--In fi
202b130 ent in another.-- Good argument for species not BEING so closely adapted. Near the Caspian «Province
206b146 ors.-- Then the chance of 200 people <might be> BEING related within 200 years backward might be calc
210b158 varieties. become species. p.147 «p. 150»-- not BEING crossed with others.-- Compares it to languages
211b163 Hudson bay breed with dogs.-- the bitches never BEING killed by them, whilst they eat up the dogs.--
217b187 me instances of one species of Australian genus BEING found in Sumatra; again another of other Genus
217b188 dently show that Mr D. wonders at these species BEING wanderers.-- Iceland no species to itself. a re
218b189 etween all different forms; therefore when from BEING put on island. & fresh species made. parents do
219b196 ognize, may be thought to explain nothing.-- it BEING as easy to produce «for the creator» two quadru
221b202 mmalia Think about Miocene fossils some species BEING recent. agreeing with Senegal. whilst Crag <agre
223b207 of Man. Nothing compared to the first thinking BEING. although hard to draw line.-- -- not so great
224b215 hen all mixed & physical changes (¿intellectual BEING acquired alters case) other species or angels.
225b216 s against it-- namely how did otter live before BEING made otter-- why to be sure there were a thousa
225b218 hav[e] become attached to.-- that insect «not» BEING called <Phitophagous> omniphitophagous. But it
225b218 ds &c &c!!! An American & African form of plant BEING found in Tristan D'Acunha. may be said to decei
226b221 the» chance of some one of the different orders BEING able to survive or chance having transported th
227b225 We may foretell species. limits of good species BEING known.-- It explains the blending of two genera
227b227 on & line of observation.-- View of generation. BEING condensation, test of highest organization inte
229b234 Freycinet Voyage, agrees. with several mammalia BEING peculiar (?) If. Henslow discusses possibility
230b240 Falkland Isd case good one of animals not soon BEING subjected to change in Americas. perhaps merely
240c003 Apteryx having feathers.-- It is possible, time BEING an element in the transmission of form, may exp
240c003 he transmission of form, may explain mule & pig BEING half way. Yet dogs sometimes like father, somet
240c004 es like mother. The fact of great monstrosities BEING produced, & handed down, with ease, is analogou
249c036 Europe & America, which are of different forms BEING migratory; also with Temminks fact of forms bei
249c036 ing migratory; also with Temminks fact of forms BEING within Tropics.-- Europaean birds at Japan. con

250c038 pe & America Some portion of the world «Africa» BEING left more equable. yet America preeminently equ
258c061 species may not be made by a little more vigour BEING given to the chance offspring who have any slig
259c062 en how is this effected by-- for instance, fish BEING excessively abundant & tempting the Jaguar to u
259c065 on> «change» during life of parent, & therefore BEING always necessary may be called adaptation With
262c073 of reasoning &c &c.-- Study the wars of organic BEING.-- the fact of guavas having overrun-- Tahiti.
263c074 breed (if mankind destroyed) some intellectual BEING though not MAN.-- is as difficult to understand
264c079 ight not,-- but probably would.-- the world now BEING fit, for such an animal.--man, (rude, unciviliz
265c085 ion as best attempt of nature, colouring matter BEING absent.-- again dwarf plant in alpine district
265c087 heredetary. The circumstance of aberrant groups BEING small it is truism. for it not so not aberrant.
267c095 ent of Cryptogamic plants.--Owing to plants not BEING adapted to Air!! p. 11--&c. valuable paper on g
268c100 ing seemed to consider it owing to one of each, BEING fitted for transport ¿may it not be explained b
269c101 c. in February or March 1838 on soil in Siberia BEING frozen to 400 ft in depth, (& Erman's surprise
278c130 arranged according to my own methods. Dasyurus BEING found fossil in Australia, & only one tree spec
279c133 ost important, showing effects of peculiarities BEING long in blood.-- ++ thinks difficulty in crossi
279c133 ..-- excellent observations of sickly offspring BEING cut off-- so that not propagated by nature.-- W
281c139 h group is quite fatal.-- ¶Relations of analogy BEING those last obtained.-- less firmly fixed & ther
281c139 to change.-- may account for certain organs not BEING fixed, <whi> in some genera, which are most fix
284c149 st, & therefore lest subject to variation.-- + <BEING> good for generic divisions [+] ought genus to
285c152 only fixed thing with reference to other living BEING.-- one species May have passed through a thousa
285c153 ertility must settle it.-- Changes in structure BEING «necessarily» excessively slow, they become fir
285c153 s «made by» of Nature & Man.-- The constitution BEING heredetary & fixed, certain physical changes at
291c166 ification.-- Thought (or desires more properly) BEING heredetary>.-- it is difficult to imagine it
291c166 re are facts full of meaning.-- Why is thought. BEING a secretion of brain, more wonderful than gravi
291c167 r admiration of ourselves.-- The idea of foetus BEING of one «botho sex«es». is strongly supported by
291c168 then Brazilian species would migrate south ward BEING ready made.-- & so destroy individuals. whereas
291c168 uld change & make new species.-- alpine species BEING destroyed at Falkland Isds++.-- Mem Lyell hypot
292c169 therefore sexes or two animals:-- «When» sexes <BEING in one normal-- when so the> are united (which
292c171 eflect much over my view of particular instinct BEING memory transmitted without consciousness «a mos
293c173 acquirement.-- Analogy. a bird can swim without BEING web footed yet with much practice & led on by c
295c183 seen old Stallion tempted to cover old mare by BEING shown, young one.-- Many African monkeys in Fer
296c184 etation of peak-- altogether original. owing to BEING oldest. & having undergone changes «no near lof
299c196 imals & again in Mans mind, in different races, BEING unequally developed.-- ¿is not Elephant intelle
300c197 where ground thick.-- shamming death it is but BEING motionless. How is instinctive dread it is exce
300c198 eason & instinct very just, but these faculties BEING viewed as replacing each other it is hiatus & n
301c200 ree of development of all animals of same class BEING about equal.-- organs of generation about equal
302c203 e, solely owing to such interm: father-species, BEING little adapted to some physical change.-- If Pa
306c212 s long.-- a monkey. (Baboon) at Z. Gardens upon BEING beaten behaved very differently from a dog.-- m
306c212 o come again very differently from dog. perhaps BEING in passion chief difference Major Mitchell is n
311c226 ntly worthy of study on the idea of those parts BEING most easily mostrified, which last produced --i
312c227 tail.-- in raptorial birds, & tigers & sharks, BEING spotted, & colours of little value Dr Smith if
312c232 ort time in some extent counterpart, mutilation BEING «variation» produced in shortest possible time.
314c237 tils & stamens united into long organ, moved on BEING touched, so as to protect itself, one segment o
314c237 s to protect itself, one segment of the corolla BEING (probably) small to allow it to lie on one side
315c243 nine teeth. no doubt a habit gained by formerly BEING a baboon with great canine teeth.-- (This may b
319c256 America. singing so well. & the mocking thrush BEING so very beautiful gret contrast with South Amer
325c266 -it is not effect, as Lyell suggested, of organ BEING worn out as. otherwise old whores would not hav
334d010 old, explained by Stallions, (according to Fox) BEING guided entirely by their smell.-- Fox says he k
334d011) & worked in the same cart in loose chains, by BEING at first beaten from her, & always accustomed t
334d011 ll blue eyed cats (Fox has seen repeated cases) BEING deaf curious case of corelation of imperfect st
335d014 mportant, as showing above facts as first cross BEING new species, ¶ -- Are not dreadful monsters, ab
335d015 are long in blood so will they remain, a mule «BEING new species» will have no tendency to have offs
335d016 have seen mules could have no offspring, & this BEING case, owing to the corelations of system, the o
336d019 law. So far is there any appearance of animals BEING created. it is probable if created at once. <wd
336d019 not have offspring-- On the idea of generation BEING a <slip> «bud» from parent. if whole parent not
336d019 t be taken, without it either went back, or not BEING perfect would perish.-- The Varieties of the do
338d024 «that strength of» hair goes with colour. black BEING strongest.-- V. p. 63. Note Book M'. for case o
339d026 not seem to think any improbability to animals BEING distributed after flood (!) according to affini
344d040 Transactions of birds of Java Caterpillars not BEING fertile is same as children not being so.-- con
344d040 llars not being fertile is same as children not BEING so.-- consider this with reference to "new spec
345d043 this resemblance gained. ¿depends upon mother BEING[G] oldest breed?.-- -- Quarterly Journal of Agri
347d050 p. 4. sticks to genus or group of any kind not BEING perfect till circular. p. 5 Most clearly shows
349d051 ers VARY most easily:-- those which do not vary BEING foundation for chief divisions.≠ p. 7. «In» Som
349d053 line of descent.-- <& here limits of varieties BEING constant. it would be exceedingl wrong to call,
350d055 n seems common, but that perfection consists in BEING able to reproduce Here there is some error-- Ob
352d060 's excellent view of geology, of each formation BEING merely a page torn out of a history, & the geol
352d060 y a page torn out of a history, & the geologist BEING obliged to fill up the gaps.-- is possibly the
356d070 .-- ¶Say my Grandfathers expression of generat. BEING highest end of organization good expression but
356d071 cts as mine¶ The facts about half breed animals BEING wilder than parents is very curious as pointing
357d073 This will explain. S. American case & Didelphis BEING Mundine form., & the less developemente of Marsu
357d074 land would have no plants were it not for seeds BEING floated about.-- I must state that. the <p> mec
359d087 organs alone.-- It is singular pheasant & fowl BEING so totally infertile whereas animals further ap
362d099 which never «dry up &» peel off their skin (not BEING wanted for war) & hence never fall off.¶ Curiou
363d102 tion» in swamps-- (owing to barns, perhaps, not BEING left open to them,-- ¶. In singing birds, part
365d106 fit for. subalpina, or some Northern species, & BEING restricted species has been Made.-- In the hybr
366d108 o my view, the domesticated animals would cease BEING fertile inter se., or at least show repugnance
366d111 lly aware she was not coming in, could not help BEING perfectly distracted «Referred to <other> Book
366d112 ss cat can propagate, ie with the chance of two BEING born at same time, & make breed, one would doub
367d113 umstance, perhaps, equally so, is this strength BEING directed to one part more than another, which p
368d114 orous males.-- «(NB. most strange cocks & hens. BEING either alike or very different in recently alte
368d114 male which recedes from the species all females BEING most like offspring, Q (how is this with those
369d115 Australia, & from Ornithorhyncus & Hydromys not BEING Marsupial. («but» also mice) & these being wate
369d115 not being Marsupial. («but» also mice) & these BEING water animals <that this> structure <connected
369d115 s <that this> structure <connected with animals BEING compelled to travers> "May have reference to th
371d127 there of such changes, not acquired by parent, BEING handed down? Are not Loddiges 1279 roses kept i
372d129 ot doubt, the> Do plants loose any qualities by BEING buds-- , more than if whole branch transplanted
380d153 spring-- this is clearly the converse of annual BEING rendered biennial-- the hardness of life in fem
383d159 egree of abortion).-- Insecta.-- hermaphrodite, BEING not only dimidiate, but quarter-grown seems to
384d114 marks into primary & secondary, the latter only BEING developed, when the first <are> become of use¶
385d163 between hybrids & inter se offspring in latter BEING unhealthy.--» males «bred in & in» never lose p
386d115 ual give buds.-- life may be thus prolonged bud BEING formed & one part dying for great length of tim
388d171 ot apply to potato.-- With respect to offspring BEING determined by imagination of Mother.-- We see i
388d171 e see in a litter every possible variation from BEING very near mother, & some very near father.-- no
388d173 how simply instinctive the feeling of other sex BEING present is-- it also shows that semen. must act
389d176 off.--» as might be inferred from annual plant BEING prolonged till it has bred.-- Offspring like bo
389d177 s far as parturition is concerned.-- Generation BEING means to propagate & perpetuate differences (of
390d179 m parent stock.-- The very theory of generation BEING the passing through whole series of forms to ac
391d174 n.-- At present I can only say the whole object BEING to acquire differences «indifferently of what k
391d174 cies!! CD[The view of <In> each Man or mammalia BEING abortive hermaphrodite simplifys case much; & o
391d174 ase much; & originally <her> each hermaphrodite BEING simple (Are not Coniferous trees generally dioe
402e017 n pear. Mountain-ash & white Thorn! Species not BEING observed to change «is very great difficulty» i
402e017 k strata. can only be explained, by such strata BEING merely leaf. if one river did pour sediment in
402e018 neo-- <brought probably by Chinese>, "the breed BEING of the latter being the same as the fox-like an
402e018 bly by Chinese>, "the breed being of the latter BEING the same as the fox-like animals. which are met
404e026 hole class. Case of Mexican greyhounds.-- young BEING habituated. instance such as Hunter, or some on
407e041 l..-- The fact of No. Mam: Placent: insectivore BEING in S. America & Australia. reason, why: Marsupi
409e046 genera come from New Holland, ¿Sydney? The dog BEING so much more intellectual than fox, wolf &c &c-
410e053 genera &c &c can never be told, without species BEING described.-- but then permanent varieties in sa
412e057 instance of part of the hermaphrodite structure BEING retained in the male.-- <like» «far» more than

415e066 onfined in locality like Ornithorhyncus,: since BEING cosmopolite, we do find his remains.-- Lima.--
415e066 e do find his remains.-- Lima.-- caves.-- There BEING no fossils, the only way, that I can see to dis
415e067 wards fatal diseases, & such constitutions only BEING cleared off by fatal diseases.-- The Value of a
417e077 sexual union-- Looking at simple generation as BEING the action of two organs in one body.-- or in t
418e080 mpregnated externally-- My view of every animal BEING Hermaphrodite-- probably will recieve illustrat
418e084 th the the succesive modifications of structure BEING added to the germ, at a time, (as even in child
420e089 ped.-- ». Hairy.-- could move his ears The head BEING six metamorphosed vertebrae, the parent of all
421e091 his paper on the Dugong, "The generative organs BEING those which are most remotely related to the ha
426e108 like to some distant cousins, the nearest blood BEING a great great-grandfather.-- -- Little Miss Hib
427e109 ording to my views of <Dioecious p> all plants, BEING occasionally dioecious; & really dioecious plan
427e109 casionally dioecious; & really dioecious plants BEING effect of abortion of one sex.-- Linnaean class
427e109 rican Conchology from Europaean., & the climate BEING now less extreme, than before arctic forms woul
428e111 he Zizanias in Sir. J's ponds-- my principle BEING the destruction of all the less hardy ones. & t
430e119 w grounds are those, which are common & nearest BEING common to other parts of the world-- March 16th
431e123 ge of crossing consists in the idea of the male BEING smaller, & the female larger than average size:
436e135 st efficient in producing new species; also one BEING reduced in numbers, but not so much these, beca
436e137 esimal advantage it would have better chance of BEING propagated & so &c. The greatest difficulty to
437e139 tory of Birds several cases on record of stoats BEING carried (p. 121) & dropped having wounded the b
437e139 "Sometimes it seems hares, rabbits, rats & not BEING sufficiently weakened by wounds got off from th
440e145 ct & grazing animal owes its form, to that form BEING «the one alone» out of innumerable other ones,
441e150 a result of such laws.-- The effect of one part BEING greatly developed on another, must not be overl
441e150 bsolute necessity, that every <animal> «organic BEING» should cross with another.-- to escape it «in
443e154 - If prolongation of life by gemmation <can be> BEING impossible. can be overturned, then the conclus
443e155 the inference from some plants & some mollusca BEING hermaphrodite is, that intercourse every time i
443e156 relation in duration of a planet to our lives-- BEING myself a geologist, I have thus argued to mysel
453e183 <Cocks> The possibility of different varieties BEING raised by seed is highly odd-- as it is not so
461t037 ong as any structure can be handed down without BEING absolutely injurious «(or requiring nutrition)»
463t063 has grand sentence about the Animaux fossiles-- BEING a mere fragment of the discoveries to come-- Ow
466t095 variety acquired> « <character> » of characters BEING inherited at corresponding age & sex, opposed by
472s01r came up true.-- colour of flower & foliage not «BEING» like the true P. bracteatum; all supposed to h
485z017 n West coast of S. America. than on East.-- not BEING replaced by Brazilian Species.-- Mem Turdus Mag
491q01v es; & that it is in these that male organs (not BEING always useful). fail-- Really good subject for
491q01v Exper» explains apples on side near other tree BEING affected.-- does one branch of Cabbage being mo
491q01v ee being affected.-- does one branch of Cabbage BEING mongrelized affect other branches-- The French
494q004 my Father know any case of quick or slow pulse BEING heredetary. (6) In the last 1000 years how many
497q06v or snow-ball-tree. what would result from seeds BEING sown= See in Cultivated Plants, as Pentstemon,
497q007 lks of the several great & natural Families, as BEING difficult to cross. (5) It is most important to
506q015 Flemings statement of Sweet Williams & Stocks, BEING propagated many years, by cuttings.-- (40) Ask
508q016 e cases of diseases, generally occurring in man BEING transmitted through females, like Hydrocele Dr.
513q21. Knights notion of pollen & stigma generally not BEING mature at same time on same plant --Flora of Au
516q24v in what order plants would reappear after <th> BEING killed Experiments not connected with Species T
522m012 e.-- often accompanied by extreme anger, at not BEING understood.-- My F. says there is perfect grada
523m015 This «N B. I have read paper somewhere on horse BEING insane at the sight of anything scarlet.-- dogs
527m032 n in a state of nature.-- Singing of birds, not BEING instinctive, is heredetary knowledge like that
527m033 in same manner vivid & grand. the frame of mind BEING just kept up by the music of the poetry.-- (the
527m034 tle in the air, is as hard work (abstracting it BEING done in open air, with exercise &c no organs of
527m034 n open air, with exercise &c no organs of sense BEING required) as the closest train of geological th
529m040 &c truly poetical. (V. Wordsworth about science BEING sufficiently habitual to become poetical) the b
530m046 effect of brain the «similar remark» thoughts, BEING functions of same part of brain, or the tendenc
532m054 tinctive feeling: I have awakened in the night. BEING slightly unwell & felt so much afraid though my
532m054 to heart? readily taking same movements, senses BEING on the look out, & the conveying means from the
532m055 the conveying means from the senses to the mind BEING more alive.-- How is it. with people nervous fr
534m062 tretched themselves from wall to table.-- table BEING removed a little further, they ascended about a
535m069 iversally» thought to be the will of a superior BEING; whose natures can only be rudely traced out. W
537m076 y is natural. consequence of man, like deer &c, BEING social animal, & this conscience or instinct ma
537m077 ost firmly fixed, but it will not prevent other BEING engrafted.-- No one doubts patriotism & family
540m088 g of sympathy.-- Mem: Burke's idea of Sympathy. BEING real pleasure at pain of others, with rational
543m100 yer.-- Peacocks remark about mathematicians not BEING profound reasoners.-- all same fact-- for, as J
546m109 n's taste for smell of flowers, owing to parent BEING fruit eater.-- origin of colours?-- Nothing sho
547m111 to Bed. & built «common» Castle in the air, of BEING compelled, from some quite imaginary cause to s
547m112 ith one idea (hence apparent vividness) & there BEING no other parallel trains of ideas connected wit
547m113 s of every action, & always running on in mind, BEING absent. one could not compare the castle with t
550m124 ne degrees of happiness-- Entire happiness. not BEING so desirable as <broken> intense happiness even
550m127 ation.-- now if one has these feelings, without BEING aware of their association «ie heredetary», doe
552m131 legs & body, & especially, when to whole body, BEING failed, & not to any particular muscle Sept. 8t
553m135 s long repeated, without the powers of the mind BEING EQUAL to the smallest casuistical doubts.-- The
553m135 ur idea of causation, to give a cause (& no one BEING apparent, one fixes on imaginary beings, many v
558m150 at sympathetic nerve. most subject to habit, as BEING less so will.-- May not moral sense arise from
558m150 arise from our enlarged capacity <acting> «yet BEING obscurely guided» or strong instinctive sexual,
566m010 .-- Book IV, Chapt I on passions of mankind, as BEING really useful to them: this must be studied. be
567m012 purely theological.-- Origin of cause & effect BEING a necessary notion is it connected with <our> t
567m013 the simplelst animals, as hydra towards light. BEING direct effect of some law.-- have plants any no
569m020 the sound of letter)-- crying yawning laughing BEING necessary sounds... not produced by will <by> b
570m023 lerable to be poked behind, without ones chest, BEING distended. touch a person on the ribs & how he
571m027 evidence, of the general ideas of our ancestors BEING impressed on us.-- Surely we have taste natural
571m029 to my theory must be acquired, by certain foods BEING habitual-- & hence become heredetary; on same p
575m044 ourage & staunchness of greyhounds.-- bull-dogs BEING preferred from not having any smell.-- 27th. No
575m046 t, to the permanence of old heredetary ideas.-- BEING lower faculty than the acquirement of new ideas
577m052 f any one of his own sex.-- Hence, animals. not BEING such thinking people. do not blush.-- sensitive
577m052 shown in the sudden cures of tooth ache before BEING drawn,-- My father «even» believes that the gen
578m055 r it with sorrow,-- let the possibility of this BEING discovered by anyone, especiall if it be a pers
579m059 feeling, something like sexual feelings-- love BEING an emotion does it regard «is it influenced by»
586m079 w how much of the wonder consists in the action BEING performed or emotion felt in early childhood (b
586m080 he heredetary part of it,-- & faculty (faculty «BEING» always heredetary helps this confusion.--) Hen
588m089 - &p. 193. that they perceive the difference on BEING carried up or downstairs, or dangled up & down-
589m092 reason not leading to action & yet our emotions BEING only bodily actions associated with ideas.-- A
592m105 end gained «& therefore the» cause, and origin BEING so is not odd.; for instance wild cattle & deer
593m105 tto-- it is probably some secondary one-- blood BEING disagreeable & anything disagreeable being purs
593m105 lood being disagreeable & anything disagreeable BEING pursued.-- A dog turning round & round is some
595m127 k] Goldsmiths Essays No XV,, on sounds of words BEING expressive, (Vol. 4 of Works) [blank] "Adam Smi
602o011 asons are forgotten. Our happiness &c, our well-BEING depends upon the "habitual reason".-- This powe
604o018 regions. 3 Infinity eternity. darkness. power. BEING associated with God. these phenomena we (feel &
607o025 an hearing Bible for first time, & great effect BEING produced.-- the wax was soft,-- the condition o
607o025 ,-- the condition of mind which leads to motion BEING inclined that way]CD one sees this law in man i
608o026 o persuade person to change line of conduct. as BEING better & making him happier.-- he agrees & yet
608o027 he will know his happiness lays in doing good & BEING perfect, & therefore will not be tempted, from
609o29v heck.-- to licentiousness jealousy, & every one BEING married to keep up population. with the existen
609o29v ow altering these instinctive passions--, which BEING unnecessary we call vicious.-- (jealousy in a d
610o034 that govern in the inorganic world; life itself BEING, the capability of such matter obeying a certai
613o036 I suppose he deduces from the ends in each case BEING the same, & the means very similar.-- It does n
613o036 ut Reason Will Consciousness Definite instincts BEING acquired, is a most important argument, to show
616o038 nly has» will: there must be cases of secretion BEING some time governed by will in some animals, inv
617o040 anifestation of force i.e. movement? capable of BEING traced to the body of the individual to whom th
617o040 action is attributed; force (be it remembered) BEING a phenomenon apprehended by the same faculty wi
617o040 n apprehended by the same faculty with matter & BEING necessarily exhibited in & by matter. The pheno
617o40v as if every particle of matter were an animated BEING pulling every other particle by invisible strin
619o042 nct gives great pleasure, & <an> «such» actions BEING prevented by <necessity> «some force» give pain
619o043 on if others injured these objects, without his BEING able to prevent it, he would likewise feel pain

620o045 t ought to be followed is a consequence of that BEING part of our nature, & its effects lasting, whil
621o046 him, that the instinctive feeling in its nature BEING always present. & his passion shortlived, it is
625o50v gy of pleasure of imagination «the utility part BEING blended & lost» & moral sense.-- My theory expl
625o052 anted in us. & that doing them gives pleasure & BEING prevented uneasiness, & that this is the feelin
633j54r ff p. 292. Mac. has long rigmarole about plants BEING created to arrest mud &c at deltas.-- Now my th
638j58v viving one of ten, thousand trials.-- each step BEING perfect «or nearly so (except no in isd) althou
641j29v cts «Carabids & Staphylini» & Mammalia. The eye BEING formed in Mollusca, Articulata, & Vertebrata, &
051r094 (there pale green confervae) coated with living BEINGS; In smooth seas (& even turbulent as at St Hel
171b003 ture & all circumstances which influence living BEINGS.-- We see <living beings>. the young of living
171b003 which influence living beings.-- We see <living BEINGS>. the young of living beings, become permanent
171b003 .-- We see <living beings>. the young of living BEINGS, become permanently changed or subject to vari
176b021 f adaptation as much as acid & alkali organized BEINGS represent a tree. irregularly branched some br
224b213 stant characters, together with other <animals> BEINGS of very near structure.-- Hence species may be
305c211 egree & kind to the endless forms of the living BEINGS.-- We see thus Unity in thinking and acting pr
337d023 - Theory of Geograph. Distrib: of <ani> organic BEINGS.-- Animals «of same classes» differ in differe
409e049 autiful in the moral sentiments of the animated BEINGS.-- &c-- If man is one great object, for which
423e097 e to simplify the organization of the different BEINGS, (all fishes to the state of the Ammocoetus) C
423e097 --- ? &c) without reducing the number of living BEINGS-- but there is the strongest possible to incre
429e114 ieve in the dreadful «but quiet» war of organic BEINGS. going on the peaceful woods. & smiling fields
440e142 growth & reproduction» is common to all living BEINGS. vide Lamarck Vol II. p. 115. 4 four laws Who
443e154 d on us.-- My theory only requires that organic BEINGS propagated by gemmation do not now undergo met
443e155 t is singular there is no true hermaphrodite in BEINGS with which have fluid sperm.--]CD I utterly d
460t019 irst origin of the world.-- our present organic BEINGS are the descendants, <slightly> «a good deal»
553m135 & no one being apparent, one fixes on imaginary BEINGS, many vicarious, like ourselves) that savages
611o034 s-- According to the individual forms of living BEINGS, matter is united in different modification, p
611o034 to them-- [RHC] In the simplest forms of living BEINGS namely «one individual» vegetables, the vital
633j54r &c at deltas.-- Now my theory makes all organic BEINGS perfectly adapted to all situations, where in
635j56r ecundity of germs is to afford support to other BEINGS.-- true, (& the doctrine of checks & my theory
640j167 ines life:.» «With respect to whether Galapagos BEINGS are species. it should be remembered that Natu
032r040 uld be made, & so on.-- This is grounded on the BELIEF of constant rising with successive periods of
042r068 of Earthquakes. «on Chili & delta of Indus», my BELIEF in submarine tilting alone, must be modified.
044r073 t materials being consolidated; one inclines to BELIEF all strata of Europe formed near coast. Humbol
045r076 Earthquakes «1822¿ 1835?» alone, (& the general BELIEF in N. Chili, where rains are so infrequent; so
187b062 out in S. America; with no change, agrees with BELIEF. that Siberian animals lived in cold countries
227b227 of things might have easily been formed.-- With BELIEF of <change.> transmutation & geographical grou
276c123 that if they believe & do not openly avow their BELIEF. they do as much to retard, as those, whose op
352d059 phical distribution of animals countenances the BELIEF of their extreme antiquity (ie much intervenin
401e017 primrose producing cowslip Uncle J. says common BELIEF. that female plant impresses main features on
580n061 I was always told so in childhood.-- hence the BELIEF in the many strange religions.» Emma W. says t
602o11b cculloch Vol I. p. 115. Attributes of Deity. on BELIEF.-- you belief things you can give no proof for
602o11b p. 115. Attributes of Deity. on Belief.-- you BELIEF things you can give no proof for, & one often
602o11b perfectly true, but you do not convince me.--" BELIEF allied to instinct.-- p. 134. a painted must n
183b052 loup et du chien, que celui de la chevre et du BELIEF, cessent d être feconds. dès les premières gen
025r018 substantiated over S. America & Europe. we may BELIEVE them applicable to the world.-- My general opi
026r019 n De la Beche, for the older formations I must BELIEVE they «the limestones» have been formed in shal
029r033 cur in the South African plains.-- Sydney no I BELIEVE the secondary? formations of Brazil, all origi
035r045 those of rapid cooling &c &c My results go to BELIEVE that much of all old strata of England. formed
036r050 lled "Bramidor"(?).--it was a strange story; I BELIEVE it was necessary to ascend the hill,--but my r
038r058 e Porphyries: certainly appearance leads me to BELIEVE mere fissures filled up.--the appearance will
041r065 nt came towns driving by the want of water.--I BELIEVE in all flat countries. years of drought are co
044r071 (consult Bory «dip of strata on East») cannot BELIEVE in a great explosion, nor would sea remove mor
045r075 ution of silex & many other phenomena I do not BELIEVE that the extraordinary fissures of the ground
045r077 of low Barometer? In a subsiding area. we may BELIEVE the fluid matter instead of efflux (always sli
051r094 greatest action; I now having seen Pernambuco BELIEVE much is owing to protection of Organic product
052r098 introduce this discussion.--Brazil bank: (& I BELIEVE SE coast of Madagascar. where a -40 line «sho
062r129 ument. knowing it to be a desert.-- Tempted to BELIEVE animals created for a definite time:--not exti
063r130 cases distinct species inosculate, so must we BELIEVE ancient ones: «.» not gradual change or degene
098a048 th layer of flint on calc.: sandstone. (& as I BELIEVE most strata) (Hence endless passages from gnei
120a107 coal-field.-- the structure of Andes. where we BELIEVE we can trace the outlines of what were fluid u
128a130 s on modern Tertiary strata on coast of do-- I BELIEVE?? coast of North America., like the Mexican Gu
145g006 eams/ 3 or 4 inches thick-- {P} 35 degree is a BELIEVE about greatest dip of sandstone in upper part
155g063 t faintly on East side of Glen Turrit, where I BELIEVE they end in upwards inclined plains, as in Cor
155g063 wards inclined plains, as in Corry. & as «as I BELIEVE in side ravine above houses of Roy» Macculloch
157g073 ear> only usually contains many pebbles, but I BELIEVE this is chiefly caused by its being lower,-- [
171b003 ms a means to vary. or adaptation.-- Again we <BELIEVE> «know» in course of generations even mind & i
172b008 loe, fox,-- Inglish & Irish Hare.-- As we thus BELIEVE species vary, <in> changing climate we ought t
174b014 common in certain countries when we can hardly BELIEVE necessary, but if it was necessary to one fore
195b098 rmanent. this might be made very strong. if we BELIEVE the Creator creates by any laws. which, I thin
196b105 ad existed in different Continents In plants I BELIEVE not..-- It is a very great puzzle why Marsupia
222b204 tween land shells of Porto Santo & Madeira-- I BELIEVE very curious-- My idea of propagation almost i
225b220 . zoology of South Sea islds. any animals?-- I BELIEVE none.-- Canary islds.? Madeira? «Tristan d'Acu
227b225 t changes which may happen-- ‖ It leads you to BELIEVE the world older than geologists think. it agre
263c075 y of destruction; then he will choose & firmly BELIEVE in his new faith of the lesser of the difficul
276c123 om literary men.--) must remember that if they BELIEVE & do not openly avow their belief. they do as
276c123 s much to retard, as those, whose opinion they BELIEVE have endeavoured to advance cause of truth It
286c154 s other animal-- it is the way of mankind. & I BELIEVE those who soar above Such prejudices, yet have
300c196 the interposition of a deity, more humble & I BELIEVE true to consider him created from animals.-- I
302c201 et often retaining a family likeness, & this I BELIEVE the case. = any animal really connecting the f
309c222 ing these hackneyed cases Mr Ed Blyth does not BELIEVE in circular or linear arrangement.-- Thinks pa
348d051 value of difference is or can be settled,-- I BELIEVE affinity may be taken literally,, though how f
534d062 y arrangement & not the Quinary.-- any one may BELIEVE anything in such rigmaroles about analogies &
357d072 hed out into orders one is strongly tempted to BELIEVE, one or two were landed as at present in New I
384d180 ed. {P} & much broader., & <more ro> «three, I BELIEVE instead of» two lines.-- «faintly edged with r
385d164 hair-- showing foetus has gone on growing-- I BELIEVE same has happened in boys bodies.-- Lavaters.
397e003 with the most liberal! spirit of philosophy to BELIEVE that no stone can fall, or plant rise, without
429e113 onditions» sudden loosing of horns.-- I do not BELIEVE this Nature's plan.-- Whether we can or not tr
429e114 - ¿& others?-- March 12th-- It is difficult to BELIEVE in the dreadful «but quiet» war of organic bei
429e115 gle Channel.-- it is not elements.-- we cannot BELIEVE in such a line., it is other plants. a broad
433e127 hall we give up whole system, of transmut., or BELIEVE that time has been much greater, & that system
511q018 Sebright select to destroy secondary character BELIEVE No or did result appear without his wish Has s
512q019 in cages. Get direction write to-- (2) Does he BELIEVE. Stanley's fact of Hawks distributing live Mam
523m018 ess.-- ira furor brevis est.-- My father quite BELIEVE my grand F doctrine is true, that the only cur
526m030 p) by sight & not by hearing One is tempted to BELIEVE phrenologists are right about habitual exercis
527m031 hat is frame of mind owing to.--<)>-- I verily BELIEVE free-will & chance are synonymous.-- Shake ten
529m042 ing white lead. who was most violently purged «BELIEVE worms were passed off.» & vomited, but who whe
530m049 ducing a train of thought.-- [not located] Fox BELIEVE. cats discover birds nests & watch them till th
532m057 violent action.-- To avoid stating how far, I BELIEVE., in Materialism, say only that emotions, insti
547m113 castle with them, therefore could not doubt or BELIEVE.-- When I say trains, it may be instantaneous
551m127 the impression on its senses.-- insane people BELIEVE they hear as well see things which have no exi
555m144 nds.-- all this was kind of wit.-- I changed I BELIEVE from hanging to head cut off. «there was the f
557m147 . when pleased tail loose & wagging-- if as (I BELIEVE) Hunter says. neither fox. nor wolf wag their
558m152 ng, & not with dog. when fear might enter?-- I BELIEVE common Swan, arch raises neck & depresses chin
573m036 pecial act, or result of laws. yet we placidly BELIEVE the Astronomer, when he tells us satellites &c
576n048 epting from favourable circumstances!» We must BELIEVE, that it require a far higher & far more compl
591n099 animals of same species.-- The general «(as I BELIEVE)» contempt at suicide. (even when no relatives
609n29v the necessity of these passions, but refers (I BELIEVE) to present day & not to ruder state of Societ
625o052 there is distinct faculty, of conscience.-- I BELIEVE that certain feelings & actions are implanted
632j53r rse is full of adaptations.-- but these are, I BELIEVE, only direct consequences of still higher laws

Page **(Key Word)**

632j53r uences of still higher laws.-- I do not «then» BELIEVE the pappus of <th> any one seed. (all have not
183b051 w is propagation of wolf & Dog. (because being BELIEVED same species) if they do not breed readily. p
304c206 away.-- +++ Hopeless work to systematist, who BELIEVED that all his divisions merely marked his own
304c206 gh physiology would profit. if the series were BELIEVED to past into each other-- Different classes K
311c227 may be seen swimming about. A Smith is firmly BELIEVED in representation. certain birds in many fami
436e135 time (as cause of change) which can hardly be BELIEVED, then, uniformity in «geological» formation.
556m144 ever comparing them, I neither doubted them or BELIEVED them.-- Believing consists in the comparison
590m094 n one sees in Cowper, whole sentences spoken & BELIEVED to be audible. one has good ground to call im
608o027 e does is independent of himself to do harm.-- BELIEVER in these views will pay great attention to Ed
138a153 re El Clima del Lima par Dr. H. Unanùe says he BELIEVES the sea has formerly stood three hundred feet
179b033 tinct locality or Metropolis of every species: BELIEVES in repugnance in crossing of species in wild
229b233 Abyssinia to extreme South coast. Elephant he BELIEVES is mentioned by old writers on extreme Northe
254c048 e those much higher in scale. So Owen actually BELIEVES in this view!!! p. 392.-- except generation &
274c113 e Humming bird? the woodpeckers Gould says, he BELIEVES does. but also on fruit.-- The Rasorial type
298c189 ng whether instinct, or reason?? Gould says he BELIEVES that he has seen half fox & dog. & that it wa
332d003 ve been crossed--is Alderney oldest breed-- He BELIEVES all pretty much alike.-- My Father Water-in t
337d022 say there is distinct Creation required if he BELIEVES «hyaena & squirrel» seal & mouse, elephant, c
345d043 their faces they were most like their mothers BELIEVES this resemblance general. ¿depends upon mothe
354d062 Mountains. & & &c.-- do. p. 69. A Dr Macdonald BELIEVES the Quaternary arrangement & not the Quinary.
380d155 Mammae & even when bitch is in heat.-- Yarrell BELIEVES Gestation is always some multiple of seven--
423e098 fields, some for cauliflower &c.-- Uncle John BELIEVES one single turnip in a garden is sufficient to
425e104 rch 6th. Mr Bentham says in Sandwich Islds. he BELIEVES, there are, many cases of genera peculiar to
425e104 ata nearest to American, or Indian groups?-- = BELIEVES some Mediterranean. but chiefly mountainous «
434e129 April 3d.-- Henslow tells me following facts: BELIEVES that «only» red Lychnis grows in <south> Wale
453e182 n India.-- Windsor Earl-- Eastern Seas p. 229. BELIEVES the <Rhinoceros> «Tapir» is found in Borneo.-
464tf6r ve thought isolated species Mr Blyth, however, BELIEVES in the existence of Molina's Pudu or goat The
529m042 o.-- My Father says there is case on record he BELIEVES in Philosoph. Transactions, of ideot 18 years
553m137 t he thinks not sulkiness-- this expression he BELIEVES is common to that group.-- this is very impor
577n053 th ache before being drawn,-- My father «even» BELIEVES that the general talking about any disease te
579n058 nted, prove of the difference, which my theory BELIEVES in.-- From the manner short-sighted people fr
038r057 will be necessary to state all arguments for BELIEVING that there must be a central core of melted r
049r089 lava to Granite is much more perfect. than in BELIEVING mere agency of dikes: & indeed when do these
086a004 astodon.-- argue against the prejudice of not BELIEVING recent elevation, yet sea shells at tops of m
090a021 to Cordillera I see Brewster speculates from BELIEVING meteorolite but old Planet, that inside our g
263c074 ature of Apteryx, many genera & species-- The BELIEVING that monkey would breed (if mankind destroyed
528m035 y cured by forgetfulness.-- & the approach to BELIEVING a vivid castle in the air, or dreams real aga
536m074 ause he would tend to be an atheist. Man thus BELIEVING, <yet> would more earnestly pray "deliver us
547m114 babilities in a dream, effect of doubting nor BELIEVING, effect of not reasoning. effect of not havin
549m122 mortify yourself do not tend to this-- though BELIEVING it to be true, & then acting on it, will add
556m144 m, I neither doubted them or believed them.-- BELIEVING consists in the comparison of ideas, connecte
069r150 h <gneiss>.--(Mica Slate) [Fig. 9] {P} ((3) like BELL of Quillota.) (A) in this strata may be older t
178b031 rews diff: species from the continent look over. BELL, & L. Jenyns. Falkland rabbit may perhaps be in
212b165 th only nose projecting.-- would pull the garden BELL, & then run into Kennel to watch who would come
216b183 as blood hounds from other parts of England. Mr BELL of Oxford St'-. had a very fine blood hound bit
217b184 spaniel «produce litter like both parents» & Mr BELL has half bloodhound & greyhound.-- Where two do
225b219 f species by generation explained? NB. Look over BELL.-- On Quadrupeds for some facts.-- about dogs &c &
233b250 ped into tub leaving only nose above it-- pulled BELL.-- -- It was most curious to observe, that all
315c243 ccidental-- My first thought of sea side-- Study BELL on Expression & the Zoonomia, for if the former
367d113 . Quagga & Ld Moreton Mare ringed Owen says that BELL in Encyclop of Anat & Phys. describes, a high-f
424e103 Wolf, difference of mind is most relied on, but BELL has some account of wolf in Zoolog. Gardens, wh
443e157 till I can honestly reject such false reasoning BELL Bridgewater's Treatise on the Hand.-- p. 94.--
460t013 ng with» so much of the swan-goose in appearance BELL at Hornsey (though only ¼ of blood). that it ap
495q05a or other accidents of delivery-- inheritable.?-- BELL-glass-- sow these seeds & see if they will come
507q018 w often one cannot tell whether one has rung the BELL, when one recollects circumstances were such o
547m113 y senses fail first, as whether I had pulled the BELL??)-- It may be deception to say the mind <think
574n041 senses, when thus stimulated, smell, as Sir. Ch. BELL says, & hearing music. to certain degree sexual
588n089 remarks applicable to my theory of happiness.-- BELL on the Hand p. 191 Says <childr> babies have an
592n103 ansactions Vol 44. 1746-47. Paper. like. Sir Ch. BELL on Expression «First Croonian Lectures by Parso
081r181 razil.-- Did Melaspena publish his travels? BELLINGHAUSEN in 1819 Kotzebue 1816 Constant log always a
542m094 n if it be inarticulate.-- the maniac shouts & BELLOWS with passion.-- It is not a little remarkable
228b230 orth trying to isolate some plants, under glass BELLS & see what offspring would come from them. Ask
236b279 haracter of Miocene Mammalia--of Europe Mem. Mr BELL'S case of Sub Himalayan land emys, decidedly an
323c269 of Birds -- Mackintoshs' Ethical Philospohy -- BELL'S Bridgewater Treatise -- Wilkinsons Egyptian re
390d179 e.-- Horse & Cattle Library of Useful Knowledge BELL'S Quadrupeds the effects of breeding in, it is n
421e091 e wild Chillingham Cattle, with reference to Mr BELL'S statement of the tame ones.-- An instance of
459t013 altogether immaculate when grown up". Saw at Mr BELL'S at Hornsey the offspring of a Black & white <d
468t116 crosses-- I see Bees almost every flower-- Blue- BELL'S-- wild-raspberry--leeks-- Flowers which thought
639j28v es canines no special use to Man. Applicable to BELL'S sneering-theory.-- p. 263. This kind of doctri
386d167 knows how to make a head. & head & tail, & the BELLY both head & tail,--no wonder there should be sy
387d169 sh the male of which receives <young> «eggs» in BELLY.-- analogous to men having mammae.-- There is a
067r144 . younger. says that Falkland fossils decidedly BELONG to old Silurian system. Apply degradation of l
162g099 f it-- the terraces of which, last measurements BELONG are so complicated, that nothing can be made o
199b116 ion that out of the thousand of new insects all BELONG to same types already established. why out of
244c021 tatous!!! p. 120.-- Most of the dogs of Payta-- BELONG to the hairless kind, «said» to come originall
248c029 sported. ¿What section does the New Zealand Rat BELONG to There is this great advantage in studying.
353d060 . The peculiar <Malacca> «Malacca» bears, <are> BELONG to same section with with those of India-- Wat
065r139 ell preserved. that it was thought not to have BELONGED to an Englishman.--On 8th of March cove began
147g107 - Some of the hills almost appeared as if they BELONGED to double series Whole very obscure but it is
035r046 appears to <exist> be made out, but in those BELONGING to the same district there seems. I think, li
177b028 in each genus «(with well characterized parts BELONGING to each)» approaching another. <Petrels have
185b056 phyressa bilineata (Gray) new <liza> species, BELONGING to true. American genus Waterhouse says he is
226b221 .-- shrews, & when big continent many species BELONGING to its own genera Therefore if in small tract
265c088 - Tenioptera rufiventris, is instance of bird BELONGING to family with peculiar coloured plumage, whe
274c113 as very different habits, though preeminently BELONGING to this type, ¿the Humming bird? the woodpeck
266c088 d insist much upon knowing to what type a bird BELONGS.-- I conceive without knowing from which count
153g055 to Glen Fintec a kind of landing place is formed BEN Erin summit. 27. 813. 65 55 degree? Boulder of Gra
153g056 on them on summit of hill rounded, site N W W of BEN Erin {P} Shelf of Glen Guoy flat peat plain divo
154g058 Glen Guoy form comparison with granite block «& BEN Erin» 29.287. 72 degree Air 65? 70? {P} Where a
160g091 ghest point joining this hill to others 3000? if BEN Erin is 3500 boulder Cairn leet more Habercelador
477z003 uce in Governor's tran?? Azara Las Vinchuca or BENCHUCA. "Les individus ailes peuvent avoir <quatre>
153g052 ive levels-- {P} Shelf opposite Glen collarig at BEND & here most accumulations At gentler bends road
156g068 ally slaty Point of rounded not scooped rock on <BEND> of 3(a) Cannot <see> «make out» composition of
466t093 tal, though not differently coloured-- & stamens BEND up a little In a wild purple Geranium, I see Be
466t094 unch of anthers & pistils, but these <do> do not BEND up-- In Lark-spur, if Bees put proboscis within
468t112 ow old «& shed some pollen». they turn upwards & BEND over stigma:-- but stigma «is» almost roofed by
156g071 plain appears like one uniform slope slightly BENDING up each main valley.-- & that river alone had
153g052 ig at bend & here most accumulations At gentler BENDS roads disappear The normal condition of 4th she
032r039 er lateral extension of the waves. by the part BENEATH the band of greatest action not having been wo
032r039 nts to line (2). The part (o) which was before BENEATH band. of greatest action. would now by degrees
037r052 . Urge the entire absence of any rock situated BENEATH low water in the Southern ocean not being buoy
040r062 que, yet no quadrupeds. -- Is the white matter BENEATH pebbles. the degraded matter of such pebbles e
063r131 , may be drawn. from. Cordillera. rocks.--When BENEATH water.--together with hypothetical case of Bra
068r146 of rock ∴ «in earliest stage» when covered up BENEATH ocean).--The first dislocations & eruptions ca
068r146 y happen during first movements, and therefore BENEATH ocean, for subsequently there is a coating of
088a014 live to their nests; & then they may picked up BENEATH the trees---- Are any Fish seed-eaters. This i
107a077 effects of Elevation if not crust much thinner BENEATH ocean than above it no because heat proceeds f
107a078 I think from dislocation taking place chiefly BENEATH water & volcanos. crust must be thinner «under

126a123 -- hence temperature ought to increase rapidly BENEATH level of sea.-- deep seated springs «spring re
126a123 ter show, that water does percolate, & springs BENEATH sea-- → According to this latter view the rod
127a127 antal Vol III I? p. 246. on formation of cones BENEATH sea.-- with reference to old submarine orifice
129a132 1838.   several case given of hot heads &c heat BENEATH the sea.-- CD[did not Beechy have some such ca
131a136 soil of Siberia.-- facts of water flowing from BENEATH frozen crust in America Richardson.-- From st
132a138 does M. Parrot suppose there is no volcanicity BENEATH lakes)?» Suppose ocean represents proper <stat
132a138 > now as we know volcanic action prevails more BENEATH the sea, <than» <&» on coast lines, than on co
133a139 water.-- & therefore that temperature of earth BENEATH <of Sahara de» a dry desert, would be very hig
137a151 & p. 142 / p. 155. the increase of temperature BENEATH the sea, is probably much more rapid than bene
137a151 eath the sea, is probably much more rapid than BENEATH continents In Berlin Transactions (1832. or 3?
152g049 East side of river & bed of river about 40 ft BENEATH general plain. 30.127 A 72 degree Air 65 degre
158g081 nite in Upper Glen Roy great plain about 60 ft BENEATH shelf peat on pebbles tidal plain as sea gradu
202b131 e.-- Demerara. In note. Demerara. 10 «12» feet BENEATH surface forest trees fallen «kind well known,
265c086 ithout habits-- after seeing beetle with wings BENEATH soldered wing-cases-- Yet these wings may be o
343d037 ng succession of vile Molluscous animals-- How BENEATH the dignity of him, who «is supposed to have»
623o050 nt» with social animals, that in which have a BENEFICIAL tendency, (not to any one individual but to
623o050 the Law of Utility Nothing but that which has BENEFICIAL tendency through <all» «many» ages. could be
623o050 st admit) has been acquired, does possess the BENEFICIAL tendency [RHC} 10) that the instincts of bee
623o051 bees & beavers «& deer» have <been formed> a BENEFICIAL tendency to them, as <social» animals of pec
623o051 , & hence <must have» «only that which» had a BENEFICIAL tendency during past races could become inst
623o051 e has much influence.)-- & only that which is BENEFICIAL to race, will have reoccurred'. NB. Until, i
624o051 rather is weak.-- [RHC] Better simply put it, BENEFICIAL tendency in every instinct to the species in
624o051 ight.-- [LHC] *for it strives to give conduct BENEFICIAL to all the children, « <then> each himself»
624o051 ects both.-- <such» changes «in accordance to BENEFICIAL tendency» will most readily affect. the inst
627o52v uiry-- good abstract of Butler & arguments of BENEFICIAL tendency of affections.-- If ever I write on
622o049 g child, from whom, she has never received any BENEFIT.-- Yet I think there is much truth in doctrine
622o049 arising from association from having received BENEFITS from this person.-- [LHC} p. 254. &c &c [RHC]
591n100 al instincts very clear.-- even to the cold or BENEVOLO- continent man Hume has section (IX) on the R
619o042 sist of a feeling of love <and sympathy> «or BENEVOLENCE» to the object in question. Without regardin
619o042 erefore in man we should expect that acts of BENEVOLENCE towards fellow <living> creatures, or of kin
625o50v animal it is equally proper to obey anger as BENEVOLENCE (but most cool malevolence). it is only after
591n100 his makes analogy of actions with <&> against BENEVOLENT & parental instincts very clear.-- even to t
621o048 s & others, (as the parents are instinctively BENEVOLENT) they will teach to be wrong or right; this
037o154 he elevated estuary of the Plata. to the Bay of BENGAL. dimensions? Strong currents off the Galapagos
135a145 es from coast-- quote passage to show abundance BENGAL Journal. Vol 4. 1835. p. 437. Tours by Benza N
135a145 s-- Much inform. on. decomposition of granite-- BENGAL. J. vol 7. p. 522. Mountain c near Caubul. par
450e175 r & S of Tropic-- By J. H. Moore after quitting BENGAL this fact is noticed in Cassay Ava Pegue-- sel
458t001 ther's Handwriting, to compare with my own.-- E BENGAL Journal Vol 7. p. 658-- Falconer on Sub. Him.
458t002 ave great prototype!!.-- Copied Vol II p. 502. «BENGAL Journal» The Taylor Bird uses pieces of thread
558m152 ssion.-- I do not think they arch their backs-- BENGAL tiger. when slightly angry. curls tip of tail.
267c093 of goats" Geograp. Journ. Vol VII. p. 216. Mr BENNETT Voyage round world, 20 years have scarcely ela
478z004 iptions about lower animals of Falklands &c &c BENNETT on Chinchillidae Zoolog Transacts. worth readi
506q15v oes he know Botanist who does-- What is Ruppia BENNETT says in same state. of flower 45. Charlsworth.
448e167 this with the Eocene beds.-- see Lyells tables BENNETTS Wandering Vol II. p 155. By inference I imagi
590n094 carious tooth cured by sight of instrument.-- BENNETT'S Wanderings, Australian Dog does not Bark-- qu
472s01v ised. It is possible to raise them pure for Miss BENT three years since gave her some She means to tr
425e104 ariation?? my theory says so.--» March 6th. Mr BENT says in Sandwich Islds. he believes, there are
610o033 amongst metaphysicians-- the school of Locke, BENTHAM, & Hartley, &. the school of Kant. to Colerid
509q017 ribs.) & does its abortion vary, according to BENTHAM'S Remark. Horse or cow.-- degree of soldering o
515q021 vidual by chance & under domestication.-- N.B. BENTHAMS remarks, where parts of flower are reduced fr
135a145 e Bengal Journal. Vol 4. 1835. p. 437. Tours by BENZA Neilgherries-- Much inform. on. decomposition o
494q005 . Speciosissimus collapse during sleep & do of BERBERIS-- (latter I think certainly not) (3) Sow seed
111a087 ulders of GRANITE" "direction of strata on the BERBICE N. 35 degree. E. dip to NW to 80 degree faults
485z018 eproduced.-- Milne Edwards p. 138 on Polypi.-- BERENICA &c &c L'Institut, 1838 p. 46 Macleay Horae En
134a142 pines there is volcano on isld in large lake-- BERGHAUS chart of do Journal of Asiatic Society Vol I.
508q016 Rump-back ever heredetary (9) Are the works of BEHAVE (treating of heredetary diseases) translated.
289c161 en he will cease to doubt :Scales into Teeth in BERING Pike (Waterhouse) Magazine of Zooly & Bot-- Vo
138a151 ably much more rapid than beneath continents In BERLIN Transactions (1832. or 3?) there is an account
480z007 no IV Mag. of Zoolog & Botany p. 356 Lesson on BERRE. do-- Magazine of Zoolog & Botany. Vol I p. 358
481z010 ertheless the details appear very trifling Also BERRE «p. 8» (I think Planariæ) Sagittella, or Flech
504q014 .-- kept true Try experiment (30/p.11) (2) Yew BERRIES germinate?-- Yew trees sexes-- (3) Get Holyhoa
271c105 n there have not a tree in which it builds, a BERRY on which it feeds. or insects it devours is sam
296c184 ome few hybrids in nature.-- «p. 473» Webb &. BERTHELOT. must be studied on Canary islands-- Endeavou
335d014 detary <thr> Six fingered people, <Hill> <Lord BERWICK' family with defective palates. heredetary & t
525m025 «stammering in my Father family» (as in Lord BERWICK'S family) are heredetary.-- other deformities a
093a032 uld reach the surface. before being cooled.-- BERZELIUS. L'Institut. [1837 p. 297]CD thinks Olivine a
130a133 crust yield easily. & if easily must be thin: <BESIDE mere fracture? A Elevation as in Patagonia {P}
037r055 g which like the secondary strata of England, «BESIDES ordinary marine remains» may contains <shells
090a024 the perturbation be serious? if so other cause BESIDES thin vapour bringing planets to an end? Fragme
105a069 f sea, cutting power increased with width. for BESIDES more surface exposed. bay more open to turbule
105a070 cks, that there has been no tumultuous rush.-- BESIDES general improbability. stratification, If chai
181b043 still greater between animals & Plants But yet BESIDES affinities from three elements, from the «infi
222b204 like mother Has Lowe written any other papers BESIDES one in Latin one <of> «on» Madeira-- any gener
274c115 I could not discover any other clear relations BESIDES aerial, & terrestial,-- How is it in water bir
325c267 Hunter has written quarto. work on Physiology BESIDES the papers collected by Owen. (at Shrewsbury)
364d105 as in pidgeons no new races.-- In Scandinavia BESIDES the Rakhekna, before mentioned between Caperca
400e012 Spaniards says no Tortoises in other «places» BESIDES Galapagos do. p. 376. Isle Tres Marias off Mex
479z005 Bougainville Voyage round world no land animal BESIDES Wolf at Falkland ∴ black rabbits not indigenou
506q15v flowers of Keeling Dioecious, or Monooecious, BESIDES the Nettle. at Galapagos-- Dioecious.-- Carex.
548m115 necessarily are,. when one is deeply reasoning BESIDES these (which must be present, though one is no
045r074 es might go on. & not a bubbles on the surface BESPEAK the changes. -- metallic veins solution of sil
538m079 often been repeated: Now it is remarked that A. BESSY repeated things, which none about her had EVER
100a053 e grand centre.-- A Stalactite of Gypsum, is the BEST case of cleavage.-- Phillips (113) «Lardner Enc
151g039 hat is usual-- Lines die away where slope less., BEST developed on steep earthy slope, two circumstan
173b009 ruer even than in Lamarck's time. Gray's remark, BEST known species. (as some common land shells) Mos
230b236 f animal is either effect or adaptation,∴ animal BEST fitted to that country when change has taken pl
259c065 tavism) Even a deformity may be looked at as the BEST attempt of nature under certain very unfavourab
263c075 y down the subject without long meditation-- His BEST chance is to have profoundly over the enormous
265c085 albino may so far be considered an adaptation as BEST attempt of nature, colouring matter being absen
281c187 d that they are varieties. (though that would be BEST).-- Argue these theoretically if animals did
286c154 an, who has debased his Nature «& violates every BEST instinctive feeling» by making slave of his fel
298c191 ich the central parts become occupied by a third BEST adapted kind.-- lower species would then revert
312c231 n, we must [not located] Henry Thompson tells me BEST way to improve cattle is to cross between a goo
337d020 en to fresh country & breed confined. to certain BEST individuals.-- scarcely any breed but what some
350d053 e group genus & other subgenus,,--> Propagation, BEST rule for genera, & so mount upwards.-- «judged
368d114 r male) & are merely best attracted). -- singing BEST sign of most vigorous males.-- «(NB. most stran
405e031 darker,, so that whole colour is changed, these BEST marked characters are partly retained, therefor
414e064 the more deadly struggle,, namely which have the BEST fitted organization, or instincts (ie intellect
443e153 has observed that to graft from the roots to the BEST way to get young trees, from worn-out kinds, &
490q001 spurs & Ranunculus in the nectaries. The former BEST for my experiment on Selection. Experiments in
527m032 ct, that «birds from» certain districts have the BEST song. [Migratory birds return to same quarter f
540m088 with sensations of sorrowful delight, very like BEST feeling of sympathy.-- Mem: Burke's idea of Sym
541m089 (testing success by decreasing headache) & found BEST plan was allowing my mind to skip from subject
583n069 nsects which have never seen their parents offer BEST cases of instincts].CD all this may be true,, b
600o008 s Hen determine which most beautiful cock, which BEST singer-- Remember.-- avarice a compounded passi
609o030 ause it is the result of what has generally been BEST for our good far back.-- (much further than we
620o045 hould show the child, which of its instincts are BEST to be followed.-- Yet even at this time, malevo
641j29r Mammifers-- Marine ones «of large size» <to> are BEST nourished by arctic regions-- Whales. «Narwhal»

473s05r er, tell them from L. Groznerat, «on road to BETHGELLERT» wh flows by Tremadoc. but can tell them fro
525m024 ng ashamed of himself.-- Squib at Maer, used to BETRAY himself by looking ashamed before it was known
038r057 ld of fluid rock.--In the discussion it will be BETTER not to refer to Lyell. but merely to state the
055r108 the exact yearly rise of the great rivers prove BETTER than any meterological table the precise perio
107a078 thinner «under water» but cause most difficult (BETTER conductor) Fitz Roy's Case of S. Maria & Tubul
113a090 eat of surface. & if conducting powers had been BETTER then 32 degree would have been found lower.--
113a091 right to consider the conducting powers either BETTER or worse & the depth of 32 degree. being littl
125a121 rust.-- {P} if crust were metal then thinner if BETTER conductor, then still thinner ⁒ The problem is
126a122 dgement be-- Does condensed metal, conduct heat BETTER than plain?-- Mem 1000 {P} how easily water pe
261c070 water, is a good instance of connate instinct, BETTER than child sucking or even ducklings & fowls--
287c158 whether variations in eye of vertebrate afford BETTER character, than variations in eye of mollusca.
292c169 n at Falklands some probably would stand change BETTER than others.-- The more complicated the animal
293c172 on & we afterwards could understand «(language BETTER instance)» he had done this without reflection
298c192 on West coast of Ireland, are absolutely (& who BETTER authority) similar with those over whole of co
312c231 breed, & that first offspring thus produced are BETTER, than those bred in & in.-- which looks as if
336d017 g & therefore no generative organ.-- Same Prop. BETTER enunciated.-- "An animal Either parent cannot
380d154 A capon will sit upon eggs, as well as, & often BETTER than a female.-- this is full of interest; for
419e086 of present seas.-- now in this country we have BETTER means of judging of slowness of physical chang
436e137 uced with infinitesimal advantage it would have BETTER chance of being propagated & so &c. The greate
439e143 d in letter to Henslow) fertilizing each other, BETTER than the pollen of same flower,-- as it tends
441e150 .-- The existence of "laws of organization" had BETTER be shown-- soil on colour of flowers, Hydrange
445e158 n air, in which it never touches objects.-- far BETTER case than chicken pecking fly.-- "whilst the s
449e173 on from those of Europe-- for they stand India. BETTER than the latter-- Forrest Voyage p. 323. Soolo
458t002 s of thread, picked «up» instead of spinning-- BETTER case than English birds, using cotton &c inste
514q21. not male or female "constructive principle" be BETTER. or "constructive action on æarm." '=?? answer
527m035 the sooner castles in the air are banished the BETTER.-- The facility with which a castle in the air
553m138 » The monkeys understand the affinities of man, BETTER than the boasted philosopher himself it is chi
570n023 one inspire, when surprise, can one resist blow BETTER with body distended.-- intolerable to be poked
580n061 he can or can not play the piece, she plays <f> BETTER than when she tries is not this precisely the
580n062 & progressive nature of intellect indication of BETTER life p. 207 March 16th.-- Is not that kind of
586n080 er-- but (an instinctus means stained in?). had BETTER refer to to the heredetary part of it,-- & fac
595n125 r young crocodile snapping.-- these I think are BETTER instances of instincts (highly useful as only
606o022 e & painting, is beauty.-- which he thinks is a BETTER definition than Winkleman's. who says it is si
608o026 uade person to change line of conduct. as being BETTER & making him happier.-- he agrees & yet does n
624o051 t of argument fails, or rather is weak.-- [RHC] BETTER simply put it, beneficial tendency in every in
632j53r Macculloch. Attribs of Deity. Vol: I it will be BETTER always to refer to the author if I use these f
473s05v ake S. of Moel Siabod. wh. flows into Conway by BETTWS there joins streams from Capel-Curig-- Mr Bu
324c268 Man. Lindlys introduction to the Natural system BEVAN on Honey Bee Dutrochet Memoires sur les Vegetau
326c266 ogical Biography. 4. Volumes well worth reading BEVANS work on Bees, new Edit 1838 Harlaam. Physical
363d101 e a group of white speckles on elbow joint-- in BEWICK drawing the the rock Pidgeon has not: now how
025r016 s not seem to consider this a very shoal coast. BEYOND the 10 or 12 leagues sea deepens suddenly. coa
047r081 wave «with same general character» reaches far BEYOND coast, which has been raised.--It must be cons
094a036 lakes.-- Mr Murchison. M.S. Chapter on drift.-- BEYOND region of great boulders, pebbles of granite c
129a131 case.-- Ed. New. Phil. Journal Vol XXI. p. 213. BEYOND the limits of Alps size of boulders sorted: di
192b083 to state that most clearly». Fox tells me, that BEYOND all doubt seeds of Ribston Pippin, produce Rib
203b136 any & thinks even now in central & Eastern Asia BEYOND the Ganges & perhaps even in India p. 261. L.
306c213 greatest (about 200 miles distant).-- directly BEYOND produced line of Timor 215 degree. What produc
436e136 my Malthusian views, within certain limits, but BEYOND these not.-- argue against this-- -- analogy w
508q016 ith. About species of Rhinoceros. becoming rare BEYOND limits of the metropolis of each-- Cause?-- An
615o36v s felt.-- analyse feelings. Mr Wynne says, that BEYOND doubt courage is heredetary in fowls & not eff
511q018 hether homogeneous (About German ornithologists, BHEM & Glöger Consul Hunt, birds from Azores or Made
607o025 lled like chance) circumstances. As man hearing BIBLE for first time, & great effect being produced.--
242c018 rica & identical with those from S. Africa. «M. BIBRON doubts fact.--» My toad is same species Coquil
256c054 C. D E., or A C D E H. Very striking to see M. BIBRON looking over reptiles he often had difficulty
482z012 NDICUS not fossil at Isle of France: «Jerrold?» BIBRON Zoolog. Journ Vol I. p. 125, owls seen crossin
482z012 f Mammalidae &c &c &c-- The French «Jerrold?» BIBRON coworker of Dumeril» who is writing with Dumer
252c042 dofan p. 401. Admirable Letter from Macleay to BICHENO much excellent detail & fine, views about Spec
129a132 f Columbia river-- Read Mr Parker's Book.-- M. BICHOFFS Papers, in Edinburgh New Phil. Journ 1838. se
209b156 ernandez]CD. From study of Flora of islands; "ou BIEN encore on pourrait au plus en conclure quels so
243c019 galement leur espèces; et certainment on eût été BIEN éloigné, il y a peu d'années, d'admettre que ce
299c195 ce «offspring» variety. wild carrot. made into BIENNIAL domesticated kind with large root by sowing i
380d153 clearly the converse of annual being rendered BIENNIAL-- the hardness of life in female Moth &c Mr Y
491q01v ing. Philosop. of Zoolog. vol 1. p. 427-- says BIENNIAL-wall-flowers & scarlet Lychnis can be propaga
501q011 -- answered by Gaertner (28) Can any annual or BIENNIAL be grafted or cuttings taken or tuber-- talk
177b029 d base of branches being dead from which they BIFURCATED.-- Type of Eocene with respect to Miocene of
023r009 great many teeth, 2 of them, 8 inches long, & as BIG as a mans thumb, the rest not above half so long
101a058 re. is it Edentate? Phillips p 289.-- Alludes to BIG bones in interior at Falkland Isd.-- Peron does
211b164 1837 p. 370 Owen says Nonsense The distribut of BIG Animals in East Indian Archipelago--, very good
226b221 lish & Irish Hare.-- Galapagos.-- shrews, & when BIG continent many species belonging to its own gene
270c103 from rest of]CD universe «which Carus considers BIG animal» becomes more developed in higher animals
344d042 o breed from her, but her offspring came out one BIG & one small. Now Jones, before this happened fro
376d137 meet in front.-- Dr Smith every baboon & monkey, BIG & little that ever he saw knew women.-- he has r
385d163 uable. because smallest sized dogs.-- one litter BIG & then second small & so.-- Says, there is breed
400e012 cattle in Table Bay with Humps on their backs & BIG tailed sheep do Vol 10. p. 373, «& 374» Spaniard
529m042 a fall of ideotcy.-- The story of the Corbets & BIG noses, quite conjectural, in Blakeways book of S
530m044 ned in Antiquities of Shrewsbury something about BIG noses & name Corbet, perhaps nonsense.-- look to
530m049 over birds nests & watch them till the young are BIG enough to eat.-- There was blackbirds nest, near
594n115 seen a little dog go to the assistance & bite a BIG dog. which was fast struggling with another larg
061r127 w) Speculate on neutral ground of 2. ostriches; BIGGER one encroaches on smaller.--change not progres
066r141 he way it stands gales = very strong. Stones as BIGGER than a man's head.-- Kerguelen 40 by 20 league
123a117 great Dasypus near Canelones -- large quadruped BIGGER than ox.-- at Buenos Ayres 20½ quadras from ri
150g035 Spean most clear & upper line running up great BIGHT just as Dick shows NB. Lake gradually draining
590n093 uence of the Passion."-- p. 37. The increase of BILARY secretion attends passion p. 39. The sweat tha
590n093 is affected. p. 44.-- Jealousy. causes spasm in BILE duct, & throws bile in circulation p. 75. Halle
590n093 -- Jealousy. causes spasm in bile duct, & throws BILE in circulation p. 75. Haller says tooth ache, e
614o037 e it may be, seems as much function of organ, as BILE of liver.-- ¿ is the attraction of carbon. hydr
185b056 at the Galapagos. Fernando Noronha Ophyresna BILINEATA (Gray) new <liza> species, belonging to true.
211b162 y bones.-- As Waterhouse remarked Mere length of BILL does not <indicate affinity with snipes> indica
265c086 ily, having it with very different habits-- Thus BILL & nostril of Puffinuria I think we may clearly
484z014 aracaras of Americas.-- manner of walking-- foot BILL crest feathering on legs-- habits-- Does the Se
530m043 er signed a bond, yet when he paid the Attorneys BILL, he asked what bond he could have had. yet duri
021r005 e commencement of world.-- [not located] La. BILLARDIERE mentions the floating marine confervae, is v
283c145 species & there will be two genera.-- let short BILLED one, be exaggerated, & all rest destroyed far
319c256 ey have one with very sweet notes.-- Their soft-BILLED birds are inferior to ours, & our lark ranks v
386d167 ur days could form 2. cubic stone. like that of BILLIN.-- <Generation--> V. p. 152 It is very singula
639j28v a kind of doctrine runs through Macculloch, the BILLS of the Grallae <are> «have been made» long «(as
415e066 ver whether the parent of man was quadruped or BIMANOUS,, is to see, what parts of structure abortive
415e068 monkeys have equal chance that progenitor was BIMANOUS, or quadrumanous.-- What a chance it, has bee
250c039 sted with Palaeotherium in Paris quarries & at BINSTEAD. Mem. recent Crocodiles with Palaeotherium in
575n045 from pleasure of music Audubon IV Vol of Ornith. BIOG. case of Newfoundland dogs. who will not enter
326c266 gy of Human. Mind.-- Audubons. Ornithological BIOGRAPHY. 4. Volumes well worth reading Bevans work on
270c104 s Infusoria «are» insecta.-- G. R. Treviranus, BIOLOGIE referred to,. as compilation of action of org
148g024 Slate too hard & uneven to be impressed Case of BIRCH Wood by Inverorum being determined by sheep & n
064r134 extinction of species than of individual.-- Mr BIRCHELL says Elephant lives on very wretched cou[n]tr
064r134 imal[s] at Cape of Good Hope Says at Santos «M BIRCHELS» at foot of range some miles from shore. rock
092a029 on twenty species same as Paris. 1500 ft high Mr BIRD in paper to Brit. Assoc: has shewn how electric
255c050 s (even 20 leagues apart from each other.-- this BIRD was well known for its impudence. «This excelle
255c052 for birds. We thus obtain an abstract idea of a BIRD-- An animal with skeleton of such general forms

```
260c067 resisted,  when hearing crys of hunger of little BIRD, in same way Wilson (p. 5). describes many kind
260c068 same  way Wilson (p. 5). describes many kinds of BIRD uniting together in pursuit of Blue-Jay, when <
260c068 urs of species-- Mem. St. Jago--solitary Halcyon BIRD of passage.-- M. Coronata of Latham, wrong.-- M
260c069 ne green of vegetation,-- ¿account for colour of BIRD in district. which they frequent!!?-- Wilson's
264c080 urnarii.-- some genus of yellow & brown-breasted BIRD in Australia &c &c-- but of course they might b
265c088 errant.-- Tenioptera rufiventris, is instance of BIRD belonging to family with peculiar coloured plum
266c088 -- Gould insist much upon knowing to what type a BIRD belongs.-- I conceive without knowing from whic
273c111 s.-- It is most wonderful how in every family of BIRD,  even the most <m> strongly marked, there is a
273c111 c would instantly fill up their place.-- Humming BIRD there is strongly marked variety,:.in the Tyran
273c112 ilvulus, & still more wonderfully to the Humming BIRD, which is one instance of its whole family wher
273c113 cole (¿connecte[d] with Chionis), yet the Tropic BIRD, has very different habits, though preeminently
274c113 reeminently belonging to this type, ¿the Humming BIRD? the woodpeckers Gould says, he believes  does.
274c114 with long tarsi.-- «Ground woodpecker» Secretary BIRD.-- & Millisuga. Kingii very rasorial for type.-
276c124 before structures.-- the only argument can be, a BIRD practising imperfectly some habit, which the wh
282c141 ey might be supposed to change. & make genera of BIRD. analogous. animals would be possessed by the d
283c143 ion to elements & not habits as shown by frigate BIRD & flying eagle.-- Hawk Gould seemed to think, t
283c144 eagle.--  Hawk Gould seemed to think, that widow BIRD. replaced Birds of Paradise-- if such fantastic
288c160 zed> ceased their migrations lost?? I conceive a BIRD Migrating from Falkland Isd regularly to main l
289c160 h the water.-- capital instances of typical land BIRD, having habits of a Grebe, structures might fol
289c161 ctures might follow.-- examine structure of this BIRD & get account of habits My definition <in wild>
290c163 ration might appear an analogy NB Pyrrho-alauda (BIRD of St Jago) of brown colour; lives on ground, c
290c164 ology Soc. it's pale ash grey back, like a black BIRD washed, Whilst tips of primaries black, by exam
293c173 s ¿ if it explains its acquirement.-- Analogy. a BIRD can swim without being web footed yet with much
298c189 ole of England & Ireland.-- curious in so wild a BIRD.-- Annals of Natural History Vol. I. p. 185 cas
303c205 titut 1838. p. 128. Extraordinary genus. Mesites BIRD from Madagascar uniting pidgeons & gallinaceous
317c248 some few facts at end of "The British Aviary" or BIRD Keepers Companion Study Appendix (& only append
341d029 ke little wings of Auks which does not make that BIRD a Penguin.-- (i.e. whether relation in one poin
341d030 ptation to little Movement.-- nocturnal crawling BIRD.-- Wings reduced to rudiment.-- clavicle scapul
341d032 ven years produced even an egg.-- a most curious BIRD, did not seem to know itself,. at last associat
342d034 alls these Hybrids new species. Yarrell says the BIRD fanciers say the throw of any two species cross
358d075 ot, & carried off by one of these birds" Case of BIRD of different family. having very same habits in
358d086 & back like silver.-- But the hen hybrid of this BIRD, has long tail figure, & some degree of whitene
359d087 ombay» & Canada Goose.-- Former strange mishaped BIRD-- looks very artificial breed-- but Mr Miller s
360d092 more difficult to propagate-- <snow> «as» if one BIRD had very bright red breast & other very  bright
362d100 estic Pidgeon, in which it appears that all the <BIRD> varieties «,now know> were then <pr> existing.
364d103 can  fare for themselves.‖ + + first year.-- The BIRD fanciers match their birds to see which will si
378d148 ossible to say which was origin of any identical BIRD-- for they were of all colours.-- they were "ha
383d160 ." Hybrid between Silver & Common Pheasant. Male BIRD, said to be infertile.-- spurs rather smaller t
383d160 have character very different from either parent BIRD-- Hunters Animal OEconomy. (by Owen) p. 44. Cla
385d164 pheasant, half fowls.-- eggs fertile, but parent BIRD will never sit on them.-- May be just worth rem
393dIBC es. which cross & are fertile heterogenous? When BIRD fanciers say the throw of two varieties is once
437e139 ng carried (p. 121) & dropped having wounded the BIRD. p. 124-- Mr Willoughby found a dead lamb «& ha
437e140 e the old Eagle could not find it..-- The parent BIRD another day brought to her young ones the cub o
440e147 is that the plumage has not been so injurious to BIRD as to allow any other kind of animal to usurp i
446e160 of this, & the female of the icterus minor is a BIRD of more splendid plumage than the male.-- Athen
458t002 opied Vol II p. 502. «Bengal Journal» The Taylor BIRD uses pieces of thread, picked «up» instead  of
464t065 nkeys= Owen has described a greatt Struthonidous BIRD from New Zealand-- <so> not an Apteryx, yet  it
499q009 ollen on Bees thighs (18) Place pin's heads with BIRD lime near male yew tree & see whether they catc
533m058 o this is precisely analogous or identical, with BIRD knowing a cat, the first it sees it.-- it is fr
554m141 liged to go in with a stick, if he drops it, the BIRD will fly at him-- Knowledge.-- Sept. 13th It wi
582n068 t,.--" this not wholly true, for we must grant a BIRD knows what is about when building its nest;  it
586n078 e does it by reason & experience, or habit.-- so BIRD migrating to certain quarter is instinct, but h
586n079 tive.-- carrier pidgeon just as wonderful in old BIRD as new.-- migration, <only> «only» more wonderf
586n081 to  volition.-- like plants going to sleep.-- "A BIRD has the faculty of finding its way, which in ce
620o046 <duty> their nature.-- When a pointer spring his BIRD. one says for shame (& the «old» dog really fee
626o052 g that instinct cannot be said to guide will. as BIRD building nest, but supplies it-- instinctive fe
641j29v cess from Bats. CD]CD. «Macculloch says no other BIRD could catch mouse by night» Sailing lizards. sq
062t128 nhabited: Go steadily through all the limits of BIRDS & animals in S. America. Zorilla: wide limits o
111a088 to Geograph Journal Vol VII p. 279. Carcases of BIRDS drifting out to sea do p. 358. changed sounding
171b004 fluenced.-- child of savage not civilized man.--BIRDS rendered wild <through> generations, acquire id
172b007 cumstances.--» Now Galapagos Tortoises, Mocking BIRDS; Falkland Fox-- Chiloe, fox,-- Inglish & Irish
177b026 be  traced right down to simple organization.-- BIRDS-- not. {P} We may fancy, according to shortness
177b027 ttom of branches deaden.-- so that in Mammalia «BIRDS» it would only appear like circles;-- & insects
181b042 - for instance there would be great gap between BIRDS & mammalia, Still greater between Vertebrate an
181b043 ck & obeying one law, they may approach,-- some BIRDS may approach animals, & some of the  vertebrata
181b045 . every part would probably be not complete, if BIRDS were fitted solely for air & fishes for  water.
182b047 for instance among the Carabidae.-- instance in BIRDS? Examine «good» collection of insects with this
185b055 this to have happened in all. but less in water BIRDS-- carrion eagles.-- This is but carrying on. at
191b081 plants  on Coral islets.-- Next to animals land BIRDS.-- & life shorter or change greater-- In the Ea
195b098 tive power acted at Galapagos it so acted that BI[R]DS with plumage «&» tone of voice partly American
195b098 an  influence Must exist in such spots. We know BIRDS do arrive & seeds.-- The same remarks applicabl
195b100 on  mens' breasts.-- How does it come wandering BIRDS. such sandpipers. not new at Galapagos.-- did t
196b104 ained if any facts are connected.-- No doubt in BIRDS, mundine genera, Bats .Foxes. Mus are birds tha
196b104 in  birds, mundine genera, Bats .Foxes. Mus are BIRDS that are apt to wander & of easy transportal.--
196b105 & draw up tables-- Instinct may confine certain BIRDS which have wide powers of flight; but are there
197b112 n= I see nothing like grandfather of Mammalia & BIRDS &c ‖ p. 32. reference to M Edwards. law of crus
198b113 many  situations as possible. for instance take BIRDS animals, reptiles fish-- Conditions will not ex
200b124 acts.-- As soon as island large enough for land BIRDS, seeds picked from the beach by the birds; most
200b124 land  birds, seeds picked from the beach by the BIRDS; most seeds germinating.--- It would be curious
204b141 spring either amongst themselves or with parent BIRDS.-- «W. Fox. know of case of male widgeon, winge
212b164 .-- Dr. Horsfield At India House, collection of BIRDS from Java.-- at Leyden series from several isla
220b199 8. p. 107. Mr Blyth states that all «genera of» BIRDS in «N.» America & Europe, which have not  their
225b217 (or colour of plumage altered during passage of BIRDS (where is this statement I remember. L. Jenyns.
230b241 animals  «or plants» very close (take European BIRDS. Mr Goulds' case of Willow wren) & others varyi
231b242 s?? Royles case of Himalayan, plants ¿migratory BIRDS, he told me some story of crane from Holland!!!
233b249 s been struck with similar extension of form in BIRDS.-- | Waterhouse thinks two main divisions of ca
233b251 -- His book Probably worth studying.-- Wingless BIRDS S. Continents-- Ostriches. Dodo. Apteryx Pengui
234b256 times introduced by ice <no> «only few» pigs.-- BIRDS mentioned. but few.-- CD [There was notice in r
234b256 about somebody who had made great collection of BIRDS of Iceland. --M. Gaimard, however, will  settle
235b263 At  what. part of tree of life, can orders like BIRDS & animals separate &c &c Work out Quinary syste
236b278 frican isld.-- «How is Juan Fernandez-- Humming BIRDS» types of former dogs. character of Miocene Mam
240c004 fficulty to rear them, eggs hatched under other BIRDS» brought up by hand.-- These facts all account
241c015 .-- Parroket with stiff tail like woodpecker.-- BIRDS of Australia. Many in common. ¿species? with New
241c015 ¿American?). p. 159. & 160 «162» list of some BIRDS of Tongatabou. & New Ireland.-- Gould will here
241c015 New  Ireland.-- Gould will hereafter know about BIRDS of N. Zealand L'Institut. 1838. A Dipus. & othe
243c019 atives (& instances given) in East Ind. Arch:-- BIRDS of New Zealand absolutely different.-- --Philed
246c025 te, ears short & straight-- do not bark p. 433. BIRDS & bats have certainly travelled from East Indie
246c025 no  reptiles. p 460 & very doubtful whether any BIRDS Except. Dodo!!-- in Mauritius Lesson &c p. 620.
249c036 may  be connected with, Mr Blyth's statement of BIRDS of Europe & America, which are of different for
249c036 fact of forms being within Tropics.-- Europaean BIRDS at Japan. connected with Europaean forms on Him
251c040 sacts. (read November 20th) Paper by Jenner, on BIRDS seen far at sea, migrations of species, geese k
253c046 ably contains some facts about close species of BIRDS. Zoolog. Transact. Vol I p. 165.-- <a> "an acco
255c050 Must be studied.-- <Three p. 7. Am.> D'orbigny. BIRDS of prey, are distributed in S. America like oth
255c052 tion of class is description of ancestor of all BIRDS, & so for birds. We thus obtain an abstract ide
255c052 description of ancestor of all birds, & so for BIRDS. We thus obtain an abstract idea of a bird-- An
256c055 aving two sexes is the check to distribution of BIRDS & animals Mr Strickland & Hamilton-- found tert
257c057 tudy> The circumstance of ground woodpeckers.-- BIRDS that cannot fly &c &c. seem clearly to indicate
257c058 epochs, & developement of lizards.-- As we have BIRDS impressions in Red Sandstone. great lizards  in
```

Page **(Key Word)**
258c061 0 alluded to by Capt. King do. p. 434. Table of BIRDS from Cuba Vigors.-- nothing of much interest XX
260c067 Ornithology, D'orbigny, Spix, &c might compare BIRDS of N. America & South-- Any how temperate regio
260c067 I read a curious account to show that very many BIRDS of different kind have been know to assist in f
260c068 ing together in pursuit of Blue-Jay, when <one> BIRDS hears <dis> crys of distress of other parents.-
260c068 of Latham, wrong,-- Mr Yarrell says-- that some BIRDS or animals are placed in white rooms to give ti
261c071 Furnarius, Synallaxis &c &c. sure to unite the BIRDS into group.-- it is same as Yarrell's remark ab
262c073 no passages; nature is full off them.-- wading BIRDS partially webbed. &c &c.--)-- & in round of cha
264c082 cognizes habits in making out classification of BIRDS Birds vary much (more than shells) owing to var
264c082 es habits in making out classification of birds BIRDS vary much (more than shells) owing to variety o
264c082 bited by them-- Timor. Australian forms amongst BIRDS Java. not so much-- Peculiarities of structure.
266c088 e from beholding the ground.-- why do beetles & BIRDS & <f> become dull coloured in sterile countries
266c088 onceive without knowing from which country many BIRDS come it would be impossible to classify them.--
267c094 ed solely of this shrub".-- p. 229. carcases of BIRDS drifting out to sea Vol VII. p. 325-- Wild dogs
268c096 oung of redstart & actually driving away parent BIRDS.-- showing how blind a storge From what I see o
268c096 w blind a storge From what I see of S. American BIRDS. genera blend into each other in very same dist
268c099 ical distribution is.-- ¿Pelagic forms similar--BIRDS??-- We must always bear in mind proofs of most
269c102 hina.--, both coasts of New Holland.-- «Compare BIRDS of Australia with plants, with this object in v
271c106 American halo has extended to Juan Fernandez in BIRDS. but ¿whether to same island in plants?-- What
271c107 y,, that you never find two «similar» groups of BIRDS in two countries, without intermediate ones occ
272c109 dant,-- the tail in cock peacock,. widowbird.-- BIRDS of Paradise. Trogons.-- the one feather in wing
274c115 ides aerial, & terrestial,-- How is it in water BIRDS, there are walking forms in water birds,-- but
274c115 n water birds, there are walking forms in water BIRDS,-- but no web forms in <water> land birds,,-- G
274c115 water birds,-- but no web forms in <water> land BIRDS,,-- Grups of very different value have their re
274c116 world?-- Will this not agree with Waterhouse & <BIRDS> Mammalia.-- We have clear indication [not loca
281c138 in real classification The relation of all cock BIRDS in Gallinaceous having tendency to lon[g] or pe
283c144 ould seemed to think, that widow bird. replaced BIRDS of Paradise-- if such fantastic «sexual» orname
284c149 h remark that a resemblanc between some form in BIRDS is visible, when young, but not when old.-- thu
286c156 classification in books, ought they to hold,-- BIRDS having web-feet, where we see scarcely any trac
288c159 instinct» modified.-- The partial migrations of BIRDS in same country. may explain greater migrations
288c159 rican & intersected wider & wider at Rio Plata. BIRDS which had originally crossed would continue to
290c163 fect.-- Maldonado as good forests for beautiful BIRDS.-- heredetary ambling horses, (if not looked at
290c163 tions very similar.-- Geology. Transact. Vol V. BIRDS bones-- in strata of Tilgate forest Seeing comm
295c178 thers assumed--+++ ‖Daines Barrington says cock BIRDS attract females by song. do they by beauty, ana
296c185 gazine of Zoology p. 56. Peregrine Falcon holds BIRDS for some time alive ¿therefore other species mi
302c201 families, but we see analogies between fish.-- BIRDS same remarks. Characters of analogy.-- last acq
303c205 from Madagascar uniting pidgeons & gallinaceous BIRDS & parrots.-- legs of pidgeons perfect.-- &c &c.
304c206 fferent degrees of closeness. -- look how close BIRDS! look at Mammals: how wide.-- therefore birds y
304c206 e birds! look at Mammals: how wide.-- therefore BIRDS younger???? or «have» not «been» exposed to so
304c207 er Apterix descends from same parent with other BIRDS,, or branched off anteriorly think what princip
304c207 may have existed series between apterix & other BIRDS.-- will having many trifling characters, in com
304c207 many trifling characters, in common with other BIRDS reveal the secret.-- Now all the different form
304c208 e have either been more rapid than in all other BIRDS, or that it sprung from a branch high up.-- thi
305c209 to die, then they would appear isolated.‖ In my BIRDS from S. Hemisphere there are some godwits which
305c209 & Reeves a remarkably varying plumage for wild BIRDS-- At Zoolog Gardens there is half Jackal & Scot
308c218 . tea & sugar» people. reverse habits Insects & BIRDS are the only two tribes fitted for water, air,
310c223 me species on Himalaya.-- some English beetles, BIRDS & a fox most close The most curious case is Sax
311c225 ??» p. 318. some remarks on Bonaparte's list of BIRDS in Europe & N. America, on closely allied speci
311c227 h is firmly believed in representation. certain BIRDS in many families, «+very often in number 5» wil
312c227 n number 5» will have long tail.-- in raptorial BIRDS, & tigers & sharks, being spotted, & colours of
314c239 (Hostile fact) Be cautious about Goulds case of BIRDS of Van Diemens land & Australia.-- The wombat (
317c249 an--, who told me that the mules between canary BIRDS & goldfinches differed considerably in their co
318c253 ix) showing WHAT CHANGES are taking place & how BIRDS are extending their ranges. «even migratory bir
318c253 rds are extending their ranges. «even migratory BIRDS, lik swallows» -- degree/ migrations of birds h
318c253 y birds, lik swallows» -- degree/ migrations of BIRDS he mentioned many most curious case-- -- the bi
318c253 ds he mentioned many most curious case-- -- the BIRDS seem to follow narrow bands, certain kinds as g
318c254 1837. Paper by Bachman.» that is succession of BIRDS.-- «in» some species «a Tanagra» Males come fir
318c254 n flocks. «as in English Nightingales» -- other BIRDS (& this seems common «kind» migration of Americ
318c254 w miles every day «generally by night» -- other BIRDS which is strictly diurnal, migrates singly by n
318c254 rates singly by night.-- others in flock, these BIRDS seem clearly directed by kind of country; «kind
318c255 condition of Migration) gradually separated the BIRDS might yet remember which way to fly.-- There is
319c255 he thinks the Mocking thrush beats all English BIRDS in song.-- one of their thrushes exceeds our bl
319c256 one with very sweet notes.-- Their soft-billed BIRDS are inferior to ours, & our lark ranks very hig
319c256 very high.-- Upon the whole thinks <many> more BIRDS sing in England than in America, but the few of
323c269 N America May 18th Stanley familiar History of BIRDS -- Mackintoshs' Ethical Philospohy -- Bell's Br
325c267 f Paraguay. account of wild cattle & Montagu on BIRDS (facts about close species) Wilson's American O
328cIBC . during 1837. paper by Bachman on migration of BIRDS Temminck has written "Coup d'oeil sur la Faune
332d004 een accustomed to turn down.-- -- applicable to BIRDS migrations & Australian Savages.-- W. D. Fox ha
340d026 ith analogy-- Good table at end of distrib: of <BIRDS>. Anatidae.-- Consult this book again.-- Mine i
340d029 ater difference in than in many large orders of BIRDS. The Emu & Cassowary closest.-- Ostrich & Rhea
341d030 .-- very curious like some facts of Mr Blyth on BIRDS.-- Dr. Bachman tells me line of Rocky Mountains
341d031 eparate almost all Mammals of N. America & many BIRDS.-- which however are most closely represented.-
342d035 eologically? Owen says relation of Osteology of BIRDS to Reptiles shown in osteology of young Ostrich
344d040 gors has given list in Linnaean Transactions of BIRDS of Java Caterpillars not being fertile is same
353d061 ory. 1838 vol II p. 402. Mr Gould on Australian BIRDS-- all Eagles. of Australia characterized by wed
353d061 hawks <to> are analogues to <Bustard> Europaean BIRDS. also «do» p. 403. & 404 vol II. do (p. <69> «7
358d075 om a boiling pot, & carried off by one of these BIRDS" Case of bird of different family. having very
358d076 ss developed state.-- the female & young of all BIRDS resemble each other in plumage «(that is where
359d087 their «pheasant» parents.-- (There are some 3/4 BIRDS «of», which I think there must be some mistake
361d096 ?)-- Yarrell observed that female of some water BIRDS, (as Phalarope) assume for breeding a more bril
363d102 ks.-- Mr Yarrell says in very close species, of BIRDS, habits when well watched always very different
363d102 abits of some «same» North American & Europaean BIRDS «slight» different-- Barn Owl <the> in the form
363d102 s, not being left open to them,-- ‖. In singing BIRDS, part instinctive & part acquired,-- thus Yarre
363d102 ing their own songs, though imperfectly.-- Male BIRDS always second their songs, the ++ Cervus Campes
364d103 + + first year.-- The bird fanciers match their BIRDS to see which will sing longest, & they in evide
364d103 lity,-- Yarrell says in such case they exchange BIRDS with some other fancier, thus getting fresh blo
365d105 o be trod, & in many parts of Scandinavia these BIRDS are very far from common.-- Under this predicam
367d113 are peacable-- (Mem Lucanus & Copris &c).-- In BIRDS singing of cocks settle point.-- (do the female
368d114 tered genera. Guinea Fowl & Peaccocks.!!» other BIRDS display beauty of plumage.-- (The females (as O
368d114 -- (The females (as Owen observes) in Raptorial BIRDS largest.-- p. 47. («"> is evidently the male wh
370d117 ebrate blend with Annelidae by some fish.-- But BIRDS quite distinct.-- Collect cases of difficulty o
371d118 of variation.-- Saw his collection of «Humming» BIRDS, saw several fully developed tails, & one with
373d133 Arctic Ocean. p. 350 Grallae in Wealden. oldest BIRDS. p. 411 -- Decapod Crust in Muschelkalk, & 5 ge
377d140 itant of plain & Jaguar of woods &c like ground BIRDS [not located] :Hence, also structure not really
378d147 henever metallic brilliancy is present in Young BIRDS, one may be sure cock & hen will be alike-- I p
380d154 ks.-- It is very singular. so many Gallinaceous BIRDS have cock & hen plumage so different, yet the C
380d154 hrodite-- (as is seen in <fe> plumage of hybrid BIRDS) After animal has copulated., though no offspri
392d180 ova of fishes & Mollusca «& Frogs» pass through BIRDS stomachs & live? In Muscovy ducks do young take
426e108 of voice by Man. we should remember, that even BIRDS can imitate the sounds surprisingly well-- In e
429e114 ur gardens (opportunities of escape for foreign BIRDS & insects) which are propagated with very littl
432e125 s is argument against Blyth's doctrine of young BIRDS retrogressing-- Uncovering the canine teeth, or
436e138 d.-- Porphyries of Andes. A familiar History of BIRDS by the Rev. E. Stanley Vol I. p. 72.-- Goldfinc
437e139 was found alive.-- Stanleys Familiar History of BIRDS several cases on record of stoats being carried
440e147 f habits, so no <cause> corresponding change in BIRDS of Paradise.-- All that we can say in such case
446e160 troduction to Bartram's Travels p. XXIII. <Some BIRDS> Both sexes of some birds sing equally well. an
446e160 avels p. XXIII. <Some birds> Both sexes of some BIRDS sing equally well. and <in> these reciprocally
448e168 »-- Fish & reptiles in former case-- Reptiles & BIRDS & Mamm. in ornityhyrhycus-- is not this right?-
450e174 s-- geese polygamous (¿when wild) but only some BIRDS are so when wild-- wild ducks monogamous; tame
451e177 Asiatic Society Vol I. p. 261. <J> Catalogue of BIRDS of India. -- p. 555. Lieut. Hutton counted, the

Page **(Key Word)***

452e181 - deer but no wild animals in Gilolo.-- p. 134: BIRDS of Paradise were first procured from Gilolo p.
458t002 instead of spinning-- better case than English BIRDS, using cotton &c instead of natural substances--
464t065 esent locality-- there have been at least other BIRDS, with small wings, & surely the Apteryx is more
473s03v ed Sandstone of. N. America is Red Sandstone. & BIRDS true! Plants in Devonian-- How strange no plant
479z005 imits of Nullipora Discussion good on Falklands BIRDS Discussion of Firola,-- Salpa Anatifs without s
480z007 . 358. D'.Orbigny <considers> states that young BIRDS of prey have longer tails than old ones-- in Am
480z008 Donn Carmichael Linn. Transacts Vol XII. p 496. BIRDS at Tristan d'Acunha.-- (Turdus Guayanensis?)) E
480z008 of the «species of» smaller petrels, are night BIRDS agree. with <pe> nocturnal habits of Crustaceae
481z009 ory fish.-- See Kings drawings.-- for real name BIRDS of Iceland. Mackenzie. p 345 for comparison wit
482z011 species mentioned) p. 205. only 9. Terrestrial BIRDS at Falklands Isd 8 waders. 22 palmipedes: out o
482z011 9. passeres! Says the thrush & another species! BIRDS of passage!! sylvia macloviana, 2d like sylvia
483z012 ner (1825) Phils: Transact.-- "on Migrations of BIRDS".-- 18 do. Vol III p. 422. letter from Capt Kin
483z012 18 do. Vol III p. 422. letter from Capt King on BIRDS of St of Magellan. Very inaccurate & Vol IV p.
484z015 s of.-- like Puffinuria does of Petrel?-- Study BIRDS of Europe for other representatives of this cla
484z016 ix has described Philedon. allied to some of my BIRDS-- These groups strictly American. Colouring on
485z016 nia compared with those of England.-- or ground BIRDS-- rather indefinite letter Mem Orpheus--becomin
485z017 o very similar to some of the Fluvicolae?-- The BIRDS seem to move much further North on West coast o
486z019 te Book (C) for comparison of singing powers of BIRDS of N. America & Europe. Entomolog. Transact. Vo
486z019 dra aquatica). Compare with T del Fuego Compare BIRDS living in the forests of Brazil H. Wedgwood sa
486z020 T. del. Fuego & Iceland Spix & Martius talk of BIRDS 13. Mr. Herbert says Crocuses are very difficul
497q006 .-- but seeds continually dropping in woods. by BIRDS in New Zealand, plants so few-- Range of mundan
500q10a this year copied Gould.-- Number of species of BIRDS in New Zealand, plants so few-- Range of mundan
500q10a d, plants so few-- Range of mundane genera, «in BIRDS» in accordance with range of species?-- Are the
502q11v short turf. for abortion. or for sterility Land BIRDS Madeira Migratory-- ask Gould about N. Zealand,
511q018 know any seed-raisers (5) List of qualities in BIRDS & animals for prizes.= Pidgeons. Canary birds--
511q018 n birds & animals for prizes.= Pidgeons. Canary BIRDS-- Bantams.-- (6) <Mad> Porto Santo Rabbit. Desc
511q018 rman ornithologists, Bhem & Glöger Consul Hunt, BIRDS from Azores or Madeira Mr. Blyth (1) Mentions s
512q019) Mentions some breeder who raises many English BIRDS-- will young wild ones breed as well as,, as th
512q019 --. How long do seeds remain in stomach of BIRDS-- Mem: how many miles they fly in few hours Zoo
512q020 eive :Bears /Yes/ (2) Foxes & English animals & BIRDS breed (3) In cases where Lions have bred,, have
521m008 ogan's memory of the tune, might be compared to BIRDS singing, or some instinctive <or> sounds.-- Mis
527m031 e according to law. How strange <all> «so many» BIRDS singing in England, in Tierra del Fuego not one
527m031 , in Tierra del Fuego not one.-- now as we know BIRDS learn from each other «though different species
527m032 y they learn in a state of nature.-- Singing of BIRDS, not being instinctive, is heredetary knowledge
527m032 man, & this agrees with the stated fact, that «BIRDS from» certain districts have the best song. [Mi
527m032 ertain districts have the best song. [Migratory BIRDS return to same quarter for many years]CD.-- Bea
528m039 ay.-- 3d pleasure association warmth, exercise, BIRDS singings.-- 4th. Pleasure of imagination, which
530m049 ught.-- [not located] Fox believe cats discover BIRDS nests & watch them till the young are big enoug
531m049 st the cat could If cats will «ever» eat little BIRDS, this most curious instance of reason & abstine
533m058 gists state that brain alters) It is known that BIRDS learn to sing & do not acquire it instinctively
535m064 her, so that two mothers to one group.-- (as in BIRDS blind storge-- They continue till death, thus a
543m097 nguage is considerable, thus carthorse & dog.-- BIRDS many cries. monkeys communicate much to each ot
546m109 stes, he partakes, taste for musical sound with BIRDS. & ¿ howling monkeys-- smell with many animals-
550m126 " -- their own property. (--regarding food & in BIRDS & beasts».-- very necessary to explain origin o
566n011 trongly shewn in the numerous artifices to take BIRDS voice & taste for singing with Mammalian struct
568n018 howl in harmony-- frogs chirp in do-- union of BIRDS badly wounded, or only winged.-- fetches two bi
575n045 who will not enter water, till he sees. whether BIRDS out at once.-- Old People-- (Antiquary Vol II.
575n045 ds badly wounded, or only winged.-- fetches two BIRDS, seeing the sun &c are absolutely useless when
585n076 out of room.-- all theories of magnetic powe in BIRDS, which have been carried in hampers. if they ha
585n076 e sun &c are absolutely useless when applied to BIRDS at Maer have learned that he is not dangerous--
594n112 is perhaps, as strong as against hawk, but the BIRDS soon <dis> learn to disobey it-- I have seen ha
594n112 appeared-- though instinct so firmly implanted, BIRDS? They are ancient.-- Are Cetaceae found in Pari
634j55r of the Cetaceae.-- How came Bats also.? before BIRGOS opening a Cocoa nut shell at one end.-- Childr
639j28r lla. & Mantis. C D woodcuts stones swallowed by BIRGOS to become absorbed by one end of Cocoa nut.--
574n040 wers & the fixing of habits,-- for instance the BIRMINGHAM, and supposed to be descended from a breed k
574n040 can hardly form an idea of a mind so limited as BIRMINGHAM Doctor praising his sister who confined him.
214b175 reed of Blood Hounds from Aston Hall close to BIRTH place.--Some general reflections might be intro
523n014 - It must be so from the curious story of the BIRTH do not effect progeny-- many dogs in England mu
524m019 e & that their idea is wrong.-- (Dr Ashe, the BIRTH & causes. If an animal breeds young her growth
044r073 Fortunate for this science. that Europe was its BIRTH or rather]CD §It should be observed that the co
265c083 her varieties»-- but <accidental> changes after BIRTH from man arbitrarily destroying certain forms
379d152 s there mentioned about proportion of sexes, at BIRTH the the succesive modifications of structure be
389d173 umber of globules: generally sufficient for one BIRTH Hill shows it is evergreens they seek Cock Phea
416e071 al conditions.-- (hence great variation in each BIRTHS, like aphides.-- Case of boy with foetus devel
418e083 e single stock,--one is led to suspect that the BIRTHS analogous to superfoetation, & to successive f
418e083 n their present forms, are closely related-- By BIRTHS, & even produce fertile offspring-- DESIRE LOS
563n001 g near houses. yet so shy at all other times.-- BIRTHS.-- it is the latter only that one refers to in
216b181 here mare was influenced in this cross to after BISCAY, "Nature, exhibits the same minerals <as> ther
344d041 e case of one impregnation sufficing to several BISCHOFF. On the effects of meteoric waters on the tem
389d176 -- one animal will fecundate female for several BISCHOFF. On the effects of meteoric waters on the tem
417e079 blance is permanent, or the similarity at first BISCHOFF. On the effects of meteoric waters on the tem
507q016 Holland ; My Father. Andrew Smith (1) Are cross-BISEXUAL» animal with a vertebra only & no head-- !! H
076r168 the earth."--p. 156. Mines of Batopilas in New BISEXUAL animals living on the borders of a country fa
137a151 [blank] Ed: New. Phil J. 1838. p. 132. «& 134» BISHOOFS Paper.-- Weelsted told me of some large fresh
170b001 es, probably polypi, gemmiparous propagation. BISON, at some resemblance as if the "variation in on
413e060 ir alliance to Articulata, which are all truly BIT to eye certainly appears level with road, & with
420e089 nimals.-- must have been like some molluscous BIT of chopped horse hair with legs & take flight--
636j56v plant is in the same predicament as a group of BIT, Valerian-- Urtica Dioica Sorrell. Lychnis. Butc
130a133 ely we here have proofs of hot bottom.-- Study BITCH which would never take the dog. But at last a r
354d065 at the Indian cattle with Bump. together with BITCH directly one after the other, puppies differ, &
161g095 e, then concealed by fragments, then clear. this BITCH & «perfect» spaniels & setters are produced. on
468t105 covered with pollen-- «Thrips» about as large as BITCH-- tried to breed from her, but her offspring ca
499q08v of plants necessary &c &c (a) Mercurialis-- Frog BITCH do do> father dog. & hence general appearance
535m084 om «the» sun & enemies-- would not fly away, but BITCH than do do: Monkey thus examine each other sexes
185b183 ell of Oxford St'-. had a very fine blood hound BITCH, for some feeling must urge them to these actio
217b184 dhound & greyhound.-- Where two dogs have lined BITCH is in heat.-- Yarrell believes Gestation is alw
333d009 ly intermediate.-- Where two dogs line the same BITCH going to mongrel, & all subsequent litters havi
344d041 rrier from Ireland to Brighton Park--first rate BITCH will not take «& if she did take, probably woul
360d090 ange» is effected)-- the one in garden is from BITCH: or dog defending companion. (mem Cyanocephalu
376d138 . but did not seen to evince more lewdness for BITCH does not so act, because maternal instinct give
377d139 esire may be said to be more definite than with BITCHES puppies are less purely bred owing to <first>
380d155 ing, Milk sometimes comes in Mammae & even when BITCHES never being killed by them, whilst they eat up
385d163 dresser, assures me he has known many cases of BITCH'S offspring is affected by previous marriages wi
385d163 is mongrel.-- I did not ask the question.-- His BITE influence propagation of species.-- Case of Ass
563n001 stopped by fleas. also by greater temptation as BITE.-- the senseless grin of passion, is like the g
600o08v ring over love of Master and sport &c &c -- The BITE at moment, but mere symbol of readiness, & ther
179b032 each copulation producing its effect; as when BITE, that which one sexually loves is probably conn
211b163 s, Wolves at Hudson bay breed with dogs.-- the BITE.-- Here there is kicker but not BITE.-- Henslow remarks that Chimpanze pouted & whin
574n041 our distant ancestors having been like dogs to BITE a big dog. which was fast struggling with anoth
525m025 - My father says on authority of Mr Wynne that BITING? What is Emotion analysis of expression of des
366d111 .-- would perfect impunity from muskitoes BITING: so do stallions always..= No doubt man has gr
565n006 panther I saw in garden uncovering its teeth to
565n006 of the Hyaena from fear, no actual intention to
574n041 ng lewd old man. ones tendency to kiss, & almost
585n075 m> Zebras &c &c.-- Here there is kicker but not
594n115 ell has seen a little dog go to the assistance &
556m145 per with mouth shut. expressing cool irony, not
574n041 th & jaws. Lascivious women. are described as

Page
(Key Word)
094a036 te clearly effect of remodelling same manner. as BITS of Patagonian boulders might be transported.--
116a100 a beach-- «p. 151» first discovered «very small» BITS of red granite between 40 & 50 from Portezuelo.
437e140 x, which after it had fought well & desperately BITTEN the young ones, would in all probability have
293c174 ce of difference in races of men.-- Wax of Ear. BITTER perhaps to prevent insects lodging there. Now
541m092 joying a satirical. laugh.-- when snarling real BITTER sarcasm.-- <These> Seeing how ancient these ex
040m063 Patagonia In the Chonos Islds we must imagine BITUMINOUS shales have been metamorphised, as in Brazil
577n051 ue memory, that the formation of a hinge «in a BIVALVE shell» does to reason.-- an inflamed membrane
292c169 ter in Mammalia, than in shells ¿ univalves or BIVALVES.-- Anyman No VI. Magazine of Zoology & Botany
100a052 d they not unite in B. K.> {P} on the diagonal of BK.-- ‖ This is not applicable. it does not explain
042r067 When mentioning pumice of Bahia Blanca, mention BLACK scoriaceous rocks of R Chupat. & fall of Ashes
042r068 lkner. ¿how far is the distance?-- Fossil bones BLACK as if from peat.--yet cetaceous bones so likewi
075r167 senical grey copper, and antimony, horn silver, BLACK silver & red silver, do not name native silver
080r178 same effects, as with man Does Indian rubber & BLACK lead unite chemically like grease & mercury [bl
097a043 arnish, which I conceive to be analogous to the BLACK glazing observed by Humboldt on the granitic ro
110a086 al: DIKE base reddish feldspathes with grenish. BLACK specks of hornblende, large irregular cryst of
146g011 ss thunder storm, many <hundred> thousand tuns. BLACK faced sheep, sometimes mottled with white black
146g011 Black faced sheep, sometimes mottled with white BLACK legs & tail like species in colouring Strike an
148g025 rorum being determined by sheep & not deer When BLACK faced sheep are crossed with English my informa
179b032 having once borne Mongrels he has thus seen the BLACK blood come out from the grandfather, (when the
179b034 es are blended «by» by intermarriages, then the BLACK & white is so far gone, that the species (for s
182b048 These were of two kinds one <with> white with a BLACK face, & similar to those brought from Abyssinia
216b179 ermarriages with people. either a little nearer BLACK or white as it may happen.-- Dr Smith says he i
216b180 re of the case at Cape.-- McClay argues from it BLACK & White species.-- For, says he Seeds of hybrid
216b181 of pidgeons. fowls. rabbits cats &c &c.-- When BLACK & white men cross some offspring black others w
216b181 .-- When black & white men cross some offspring BLACK others white which is more closely allied to ca
225b217 ignorant. as why millet seed turns a Bullfinch BLACK, or iodine on glands of throat, (or colour of p
228b231 quals.-- «Do not slave holders wish to make the BLACK man other kind?» Animals with affections, imita
234b255 celand. are seldom seen with horns" --- p. 341. BLACK Fox sometimes introduced by ice <no> «only few»
246c025 and of large size, resemble, chien-loup.--long, BLACK & white, ears short & straight-- do not bark p.
247c029 e as variously coloured as a herd in England"-- BLACK & Grey varieties of rabbits thus handed down fo
261c070 lings & fowls-- When talking of races of Man.-- BLACK men, black bull finches from linseed-- not sole
261c070 ls-- When talking of races of Man.-- black men, BLACK bull finches from linseed-- not solely effects
266c092 derney breed, but size not larger than those of BLACK cattle. Not have hump like those of India & Ara
267c093 sheep numerous "of two kinds one white with «a» BLACK face, & similar to those brought from Abyssinia
271c105 N. America» whose Southern range is? <One> The BLACK & white thrush of Azara builds its nest in <sam
273c110) there is any marked colouring of plumage (as «BLACK & white» bars on wings of trogons are lengthene
286c154 tinctive feeling» by making slave of his fellow BLACK, often wished to consider him as other animal--
290c184 at Zoology Soc. it's pale ash grey back, like a BLACK bird washed, Whilst tips of primaries black, by
290c184 e a black bird washed, Whilst tips of primaries BLACK, by examining series I cannot doubt laws of cha
296c184 450. 4 instances of hybrids between pheasants & BLACK fowl.-- use as argument possibly some few hybri
303c204 .= (Negro or father of negro probably was first BLACK at base of nails & over white of eyes,-- + + +,
308c217 hest almost unconsciously.-- why do absent «Dr. BLACK. tea & sugar» people. reverse habits Insects &
309c221 d) p. 4.-- do. p. 186. quotes Burkhardt to show BLACK colour of certain Arabs.-- NB avoid quoting the
312c228 spotted, & colours of little value Dr Smith if BLACK & white Man crosses; children heterogenous, he
312c228 rs ditto. (see Griffith) & half Muscovy ducks, «BLACK cock & pheasant see Jardines Journal.»-- consul
319c257 women (¿ north end of the Oural mountains) have BLACK nipples to their breasts.-- L' Institut, 1838,
332d005 o come from coast of Guinea, -- ears bare. skin BLACK & wrinkled-- fur short. (tail cut off in progen
338d024 o-- thick skinned My hairdresser (Willis) says <BLACK> «that strength of» hair goes with colour. blac
338d024 lack> «that strength of» hair goes with colour. BLACK being strongest.-- V. p. 63. Note Book M'. for
341d031 h is represented by one not differing except by BLACK line,-- A Bunting by one only differing by some
345d043 n born with white head (,or «short-horned with» BLACK lip) & then calf «in both cases» is killed. Not
345d044 d case of Provincial Breed-- Highland Sheep jet BLACK legs, & face & tail, just like species.-- high
346d047 father than Mother.-- The cross not so hardy as BLACK faced, but more tendency to fatten-- This man c
356d072 > 8th Sept Yarrell told me he had just heard of BLACK game & Ptarmigan having crossed in wild state--
357d073 ntly in Britain) between hen Caperailkie & cock BLACK-cock.-- (Curious the readiness with which this
362d100 but very strange races» of them have the forked BLACK mark of the Rock Pidgeon,-- several have a grou
364d104 ib. of Useful. Knowledge that sheep originally. BLACK. & Yarrell thinks the occasional production of
364d104 . & Yarrell thinks the occasional production of BLACK lambs is owing to old <story> return.-- The Rev
364d105 khekna, before mentioned between Capercailzie & BLACK Cock.-- The latter has crossed with the Ptarmig
365d105 n England no doubt the cross between Pheasant & BLACK game is owing to their rarity., a single female
365d105 state.-- Is English red Grouse. a cross between BLACK Game. &, the subalpina of Sweden, (which in sum
365d106 s has been Made.-- In the hybrid grouse between BLACK Cock & Ptarmigan (probably subalpina.) former h
365d106 - be careful, See to hybrids between Pheasant & BLACK Cock, & other hybrids-- The fact of Egyptian an
379d151 e not some same-- what is the alliance with the BLACK sea.-- it would be ocean, what is land to conti
383d160 t in not having tuft,-- back like do.-- but the BLACK lines on each feather instead of coming to poin
383d160 two lines.-- clearly edged with reddish brown» BLACK marks on tail much <blacker> «broader.-- » Brea
414e064 t of Africa & West Indies shows organization in BLACK Race there gives them preponderance. intellect
441e150 shown-- soil on colour of flowers, Hydrangea -- BLACK bullfinches-- & all varieties must be presumed
450e174 & small green parrots. June 26th-- Yarrell.;-- BLACK Swan «in domestication & nature» strictly monog
459t013 Saw at Mr Bell's at Hornsey the offspring of a BLACK & white «duck of pecu» «drake» with the penguin
466t094 nthers, who otherwise lie protected by the hairy BLACK lip of lower division of nectary: «wh. itself r
472s01v tion in the 60 one brighter with mere traces of BLACK spot at base, one paler with less riged foliage
472s01v at base, one paler with less riged foliage & no BLACK spot & a third considerably paler, all rest ver
479z005 world no land animal besides Wolf at Falkland ∴ BLACK rabbits not indigenous p 112 M Lesson--Voyage o
506g014 30 inches Examine Keel of Common & Wild Duck-- BLACK Duck & Penguin Henslow &c (36) Has not H. raise
540m086 azil-- with the negros of Africa, (or again the BLACK man of <B> Van Diemens land & the energetic cop
545m107 ike» «remember» the expostulatory angry look of BLACK spider monkey when touched, also another monkey
554m141 most important step in progression.-- The male BLACK Swan is very fierce when female is sitting the
555m143 some men have this power abortive muscles) The BLACK Spider Monkey, very different disposition from
558m153 chin-- strikes with wing arches wings-- as does BLACK Swan.-- Goose do all species put their necks st
595n115 n had S[...] been dead about two months. saw a «BLACK» spider monkey brought it at opposite end of ho
284c149 t not when old.-- thus speckled form of young BLACKBIRD. good remark if general.-- Where any structur
319c255 in song.-- one of their thrushes exceeds our BLACKBIRD, but our blackbird exceeds their other thrush
319c255 their thrushes exceeds our blackbird, but our BLACKBIRD exceeds their other thrushes-- yet they have
361d096 y) young cock & hen, all nearly similar.-- in BLACKBIRD group young like some of the species-- (¿do t
461tf03 im.-- p. 605 Mr. Macgillivray says "<A Thrush &> BLACKBIRD have been known in <their> <its> natural stat
363d103 th fallow? deer. & Moschus &c & -- like young BLACKBIRDS Dr Bachman told me that 1/2 Muscovy & common
530m049 the young are big enough to eat.-- There was BLACKBIRDS nest, near hot-house at Shrewsbury, which th
383d160 with reddish brown» black marks on tail much <BLACKER> «broader.-- » Breast red like Common pheasant
097a043 "superficially coated by a thin pellicle of a BLACKISH colour like a dull & poor varnish, which I co
110a086 elspar different from either fragment or dike, BLACKISH grey base. crystals from fragment disseminate
405e031 » Deer. which were of a nearly uniform <dusky> BLACKISH brown.-- yet retained a trace of horizontal m
574m042 can be transmitted.-- this is analogous to a BLACKSMITH having children with strong arms.-- The othe
530m045 arities of form in trades (,as sailor tailor BLACKSMITHS?) are likewise hereditary, & therefore that
530d056 sight connected with locomotion.-- «Mem. Dr. BLACKWELL (Abercrmbies) comparison of sight to threads.
325c267 consult. Paper on Consciousness in Brutes in BLACKWOOD. June 1838 H. C. Watson on Geograph. Distrib:
559m155 per on consciousness in Brutes & Animals. in BLACKWOOD'S Magazine June. 1838. Copied Mr H. C. Watson
415e057 ortive.-- Remember my fathers remark about the BLADDER.-- The numbers of fatal diseases in mankind, «
504q013 f accident?? (7) Which. Rhododendrum seeds?-- BLADDER-nut ‖. Laburnum ‖. Dodecatheon ‖. Castrate ap
050r090 my Ascension case very doubtful. -- In Iceland BLADDERS of Lava are described, & many minute craters
048r084 most wonderful, from chemical attraction, as a BLADE of grass penetrating by action of Organic power
311c227 s> quadrupeds geographical range very good.-- BLAINVILLE Ovington's Voyage to Surat, floating isld. o
354d062 alogies & number L'Institut p. 275. (1838) M. BLAINVILLE has written paper to show Stonesfield Didelp
605o020 to consist of "an exquisite susceptibility from BLAIR receiving pleasures from beauties of nature & a
530m044 ousae, pain prevents repetition of idea.-- Mr BLAKEWAY has mentioned in Antiquities of Shrewsbury so
529m042 he Corbets & big noses, quite conjectural, in BLAKEWAYS book of Sheriffs.-- July 22d. 1838 No Deliriu
608o027 it for beauty & good temper), nor ought one to BLAME others.-- This view will not do harm, because n
042r067 arking a pause» When mentioning pumice of Bahia BLANCA, mention black scoriaceous rocks of R Chupat.

(Key Word)*********************************

```
042r068 nes so likewise «of miocene period».--Mem Bahia BLANCA P. 204 Vol III. Lyell Owing to «open» faults i
057r113 rgeon's? I really should think probably that B. BLANCA & M. Hermoso contemp:.--Inculcate well that Ho
093a033 h with volutas. prove regular mud bank at Bahia BLANCA. <fl> Flustra identical. recent & bone bed.--
134a142 in Falklands Some of the Tosca nodules at Bahia BLANCA Mr. Malcolmson says are like Kankaer South of
135a146 of  shore of Coromandel. just same as. at Bahia BLANCA-- letter in drawer with important letters-- Wh
481z008 und in not less than 7 fathoms water. Mem Bahia BLANCA. De la Beche theoretical researches Compare la
484z015 rs of Europe-- Study profoundly shells of Bahia BLANCA & Southern Hemisphere It is most interesting t
050r092 deepening  on E. coast of Africa. as at Brazil [BLANK] What is nature of strip of Mountain  Limestone
051r096 ates unfavourable to attachment of many bodies [BLANK] Beechey.--changes in bottom in NW coast of Ame
080r179 ck lead unite chemically like grease & mercury [BLANK] NB. P. 73. General reflections on the  geology
137a148 ineral veins p. 125 to 129 & p. 135--160 & 162 [BLANK] Ed. New. Phil J. 1838. p. 72. on metallic vapo
137a150 filtering some of its constituents into chert. [BLANK] Ed: New. Phil J. 1838. p. 132. «& 134» Bischof
138a152 f Sellow Geolog. Observat. in Southern Brazil. [BLANK] «p. 4. (Lyells Book)» Observacones sobre El C
138a154 r comparison with the moon at some future time [BLANK] Sir. J. Halls Paper on the consolidation of st
138a156 sand  red hot & brine was boiling on the top-- [BLANK] Would rotting wood by yielding Carbonic Acid u
139a177 ilica / or charcoal charged with carbonic acid [BLANK] Many interesting experiments might be tried by
146g008 plained by my idea-- highest part must project [BLANK] {P} Path East End near Holyrood Palace In same
163g110 ebbles brought by different cause: from mud.-- [BLANK] {P} muddy nodular strata coral upside down str
164g121 glomerate [...] shells from [...] Wenlock Edge [BLANK] L. Lochy 12 ft 96 L. Oich 12 84 {P} 29.958 - 1
235b265 s How is Fauna of Van Diemens Land & Australia [BLANK] Falconer's remarks on influence of climates, s
236b277 ed by seed.-- Lychnis.-- Flax.-- Read Swainson [BLANK] In production of varieties is it not per saltu
319c258 Irish Elk case of fossil geographical range.-- [BLANK] Books examined: with ref: to Species Most of t
420e088 ndu.-- I suspect good case of fossil filling up BLANK.-- CD[not between existing series of species of
516q024 ted on what, & see what comes up.-- [Unnumbered [BLANK] Experiment Cover patch of ground, with differe
516qIBC rbonic Acid. will this bear on Petrifaction?--  [BLANK] Questions & Experiments Expression M Charles D
593r106 object  is to make saucer-shaped depression.-- [BLANK] Does music bear any relation to the period whe
593n108 look so by association receives pleasure. This [BLANK] will not do for insects. if this view holds go
593n110 tinguishing character between man & animals.--  [BLANK] Double consciousness. only extreme step of  an
594n114 ned to do nothing. & so manifesting sulleness.  [BLANK] Circumstances having given to the Bee its inst
595n116 took  it <her> in his arms & carried to see.-- [BLANK] A Dog «whilst» dreaming, growling. & yelpings.
595n118 rely hunting horn-- association or imagination [BLANK] [not located] Ernest W. playing with Snow. whe
595n122 as frightened at wild beasts in Zoolog. Garden [BLANK] [not located] A child crying. frowning, poutin
595n126 child  pouts who has never seen others pout]CD [BLANK] Goldsmiths Essays No XV,, on sounds of words b
595n128 s of words being expressive, (Vol. 4 of Works) [BLANK] "Adam Smith Moral Sentiments" much on life & c
635j55v Man.--  the cause given we know not the effect [BLANK] 6 p. 412. Macculloch explains the shortness of
554m140 pped. will cover herself with straw, or with a BLANKET.-- these cases of commonly using, foreign bodi
272c107 ny legs & running quick & generl appearance of BLATTAE other Hemiptera stikingly resemble Coleoptera.
056r111 t fix lime). <Also Volcanos separate.> Volcanos BLEND all substances together; & products being simil
268c096 ge From what I see of S. American birds. genera BLEND into each other in very same district.-- <The s
315c243 n does move muscles for uncovering canines) --. BLEND this argument with his having canine teeth at a
370d117 rfection of their separation.-- thus Vertebrate BLEND with Annelidae by some fish.-- But birds  quite
408e044 - worth going there for.-- «Gales of wind would BLEND species» Buckland. Reliquiae Diluvianae. p. 222
179b033 can doubt that lesser trifling differences are BLENDED «by» by intermarriages, then the black & white
199b120 eases. In intermarriages; smallest differences BLENDED, rather stronger tendency to imitate one of th
254c049 ation & digestion in Acrite Kingdom all organs BLENDED together, & same organ «where eliminated is» o
264c080 Australia  &c &c-- but of course they might be BLENDED, if archipelago turned into continent &c  &.--
276c125 le.-- if habits & structure similar would have BLENDED together Mem Mr Herberts law; habits determini
343d038 zation of S. Africa, so must the tribes become BLENDED & prevent that strong separation which otherwi
494q004 cross  them with other breed.-- (12) About the BLENDED instincts Remote Experiments-- Plants Raise se
515q022 ogs & cats.-- (3) Hounds-- varying-- (4) About BLENDED instincts of the geese which he crossed; espec
625o50v leasure of imagination «the utility part being BLENDED & lost». & moral sense.-- My theory explains bo
041r067 ns which project outwards.-- In Patagonia. the BLENDING of pebbles & the appearance of travelling may
110a086 & scales.-- large cryst of Hornblende BLENDING into base-- Salband might have oozed out of c
137a149 in  iron furnaces affecting to some distance & BLENDING with sandstone «said to be» analogous to gran
227b225 of good species being known.-- It explains the BLENDING of two genera-- It explains typical structure
231b244 by non-embedment of remains-- ¿agrees with non-BLENDING of languages?-- Till man acquired reason,  he
304c206 .-- of two adjoining families & not all organs BLENDING away.-- +++ Hopeless work to systematist, who
268c096 tually driving away parent birds.-- showing how BLIND a storge From what I see of S. American  birds.
332d004 e this climate-- .) -- July 23d. Eyton, a stone BLIND horse, seemed to perceive turn on road where No
350d055 Crustacean,  like Trilobite. (Polirus??) female BLIND & of quite different form from male with eyes!-
350d055 about & see, parent «(2)», female «(I)» fixed & BLIND: -- Macleay observed all these facts prove that
535m064 o that two mothers to one group.-- (as in birds BLIND storge-- They continue till death, thus  acting
579n059 otions, which make us love him, or her.-- it is BLIND feeling, something like sexual feelings-- love
584n072 long  after once visited by horse & dogs. (even BLIND horses & dogs) shows it is somewhat analogous t
587n088 iple of liking, as simply heredetary habit.-- A BLIND man might be born with idea of scarlet, as well
545m108 comparison  of ideas-- one follows other as in BLINDEST memory-- also low faculty of understanding, he
366d111 e wretched inhabitants of NW. Australia, go on BLINKING their eyes. without extermination, & change o
366d112 .-- the account of the people on the NW. Coast BLINKING to keep out flies might be used» The wild ass
110a086 of pyrites, few curious fissures; base in part. BLOCK not crystallized Salband like basalt. full of c
112a088 z in connection with Fitz Roys fact of elevated BLOCK of stone.-- & Caldcleughs collection of facts S
129a131 hssets. p. 133 The most wonderful case of great BLOCK of rock moved by gale-- When writing on Valleys
153g056 X  3 X 2 «feet» & 2 deep Another rather smaller BLOCK 30 ft «above» & other 50 ft lower & other small
154g058 shelf of Glen Guoy form comparison with granite BLOCK «& Ben Erin» 29.287. 72 degree Air 65? 70?  {P}
157g074 with last measurement of shelf of 3d:-- granite BLOCK a yard across. On side of «that» hill, in front
159g084 & 3d shelf Mountain <Mica> «composed of» Gneiss BLOCK on 2d shelf & below it some way; several  large
026r020 sembled Chiloe In De La Beche, article "Erratic BLOCKS" not sufficient distinction is given to angula
076r169 of  silver procured from martial pyrites; great BLOCKS of pure silver not common in <S.> America:  In
089a020 point  of Flori[da]» Excellent paper on Erratic BLOCKS in Alps. Memoires de la Soc. «de Geneva» Vol 3
105a070 rosion. but we have evidence in distribution of BLOCKS, that there has been no tumultuous rush.-- bes
105a071 character  of mouth of valleys &c; Pampas.-- If BLOCKS above their parent rocks. would be prove of su
120a110 t down Mention absolute elevation of Patagonian BLOCKS (1200 ft??). Scotland at least 2200. Jura 4000
158g082 ain on river doctrine «Little Hill with granite BLOCKS almost encircled» <fre> Gneiss cut smooth on s
159g084 s {P} River Gorge 4th Sh side of valley Granite BLOCKS on this side (return) between 2d & 3d shelf Mo
159g087 e than in Glen Roy, & partly shut in No Granite BLOCKS in higher parts?? Bought Glen name of Glen  by
406e037 mber 1st..-- Addenda to Journal. I show erratic BLOCKS transported far S. in Northern.  Hemisphere.--
407e038 w «Equatorial» America from the «low» limits of BLOCKS both North & South, has probably undergone a g
179b032 once  borne Mongrels he has thus seen the black BLOOD come out from the grandfather, (when the mother
214b175 Assoc  <to> on N. American Zoology-- A breed of BLOOD Hounds from Aston Hall close to Birmingham, and
215b178 f Isle of Man mentioned in Loudons (analogue of BLOOD hound-- Bull. Soc. Geolog. 1834. p. 217. Java F
216b182 view of races not mingling?-- In Foxes case of BLOOD Hounds, a little mingling would probably have b
216b182 g would probably have been good, namely such as BLOOD hounds from other parts of England. Mr Bell  of
216b183 ngland. Mr Bell of Oxford St'-. had a very fine BLOOD hound bitch which would never take the dog. But
217b184 rents.-- Fox told me of case of mare covered by BLOOD horse & Carthorse two folds [not located] Mr Do
219b196 from Incestuous intercourse. a parallel fact to BLOOD-Hounds. Before Attract of Gravity discovered. i
248c034 ithout picking» become more & more impressed in BLOOD with time, then generation will «only»  produce
249c034 ot varieties. which are not deeply impressed on BLOOD., will cross & produce fertile offspring in the
275c120 & crossed & recrossed, till there was a dash of BLOOD, with whole form of grey hound.-- picking out f
279c133 showing  effects of peculiarities being long in BLOOD.-- ++ thinks difficulty in crossing race.-- bad
280c136 right way of viewing it.-- Variety when long in BLOOD, gets stronger & stronger, so that though by gr
283c144 es.-- explained by such not having been long in BLOOD?-- My theory agrees with unequal distances betw
335d013 nearer to common goose.-- What has long been in BLOOD, will remain in blood.-- --converse, what has n
335d013 .-- What has long been in blood, will remain in BLOOD.-- --converse, what has not been, will not rema
335d015 g law, that in proportion as things are long in BLOOD so will they remain, a mule «being new species»
335d015 of»  Varieties is different because not long in BLOOD.= The case of union of perfect animals is disti
336d017 lowly obtained NB. The longer a thing is in the BLOOD, the more persistent.-- «any amount of  change»
336d018 s more easily inherited.-- «but if change be in BLOOD long, it becomes part of animal &» by a success
345d042 rinkwater thought that a <pure blooded> «"first BLOOD"» animal must have gone on for many years, befo
364d103 rds with some other fancier, thus getting fresh BLOOD, without fresh feather, & consequent trouble in
```

Page ***(Key Word)***

387d168 epeated several times, that it becomes fixed in BLOOD.-- Looking at ovum of mother & ovum in offsprin
390d179 difference» with other sex. = The highest bred BLOOD-hound. would be infertile with highest bred of
426e108 ingly like to some distant cousins, the nearest BLOOD being a great great-grandfather.-- -- Little Mi
460t015 in appearance Bell at Hornsey (though only ¼ of BLOOD). that it appears about half way between swan-g
494q004 with regard to former (8) Is form of globule of BLOOD in allied species similar.-- if not how is it i
527m033 efore singing intermediate, who has not had his BLOOD run cold by singing).-- Granny says she never b
555m144 of person recovering from hanging on account of BLOOD. but all these idea came one after other, witho
557m148 fferent from that.-- like «slight» passion from BLOOD rushing in face, with less action of the heart.
567n014 tless has not.-- Turkey cock in passion & sends BLOOD to its breast &c &c All Science is reason actin
567n015 e blush.-- when extreme sensation of heat shows BLOOD is pumped over whole body.-- is it connected wi
577n051 ing of ones appearance,--does the thought drive BLOOD to surface exposed, face of man, face, neck-- «
578n055 se opinion he regards, <& see how> feel how the BLOOD gushed into his face,-- "as <she> «the» thought
578n055 his knowing «it», suddenly came across her, the BLOOD rushed to her face,"-- One blush if one thinks
593n105 s a ditto-- it is probably some secondary one-- BLOOD being disagreeable & anything disagreeable bein
596n184 keys. pull back skin from head very little Does BLOOD go in <body> face in pashion.?-- cry? Do people
345d042 uncertain. Mr Drinkwater thought that a <pure BLOODED> «"first blood"» animal must have gone on for
217b184 litter like both parents» & Mr Bell has half BLOODHOUND & greyhound.-- Where two dogs have lined bit
468t106 hich thought very unattractive-- Found Rhubarb BLOSSOM swarming with small Staphylinidae-- Anapsis, M
469t151 ury.-- Abberley-- Early Magazine-- &c. double-BLOSSOMED «& dwarf-fan Bean» bean, were planted in rows
469t151 year came up true «in 1840»: All in together BLOSSOMED together The seeds of these plants will be co
446e162 eatment will suddenly send forth quantities of BLOSSOMS--» The case of the Lemna, «and the vivaparous
061r127 ller.--change not progressif<e>: produced at one BLOW. if one species altered: <altered> Mem: my idea
157g073 rts of Beagle Channel when mica slate, only sand BLOW away]CD where lines appear to cross stony parts
570n023 does one inspire, when surprise, can one resist BLOW better with body distended.-- intolerable to be
620o044 ow soon the pleasure from good dinner, or from a BLOW struck in passion fades away, so that when man
038r057 grand eruption when whole summit of mountain is BLOWN off; & again when in great crater. different li
091a025 n. No 3d of Ed. N. Phil. J. p 194. Fact of dust BLOWN far out to sea valuable; because transportal of
218b192 - we know many seeds, might be transported some BLOWN--floating trees Thrushes & bunting & coots-- «(
223b209 s are gradual, is this if after isolation (seed BLOWN into desert) or separation by mountain chains &
626o052 passion rising from weariness leads to striking BLOWS.--' p. 224.-- Hume's Inquiry-- good abstract of
565n009 liar expression is remarkable. the pouting, & BLUBBERING-- sulkiness is same as pouting, <but> lesser
578n056 zes all muscular action -- «in man & animals» BLUBBERING of a child (different in different ones?) in
596nIBC Get facts about instincts of mongrel dogs Do BLUBBERING children, if of. convulsive tendency easily
188b068 come up white, though planted in same soil with BLUE. Now this is same bearing with Dr. Smith's fact
260c068 any kinds of bird uniting together in pursuit of BLUE-Jay, when <one> birds hears <dis> crys of distr
275c121 ch different skulls, but same marks on wings are BLUE Pouters & small Bald Heads Mr Yarrell will ment
334d011 brothers & sisters in Mankind.-- The case of all BLUE eyed cats (Fox has seen repeated cases) being d
360d092 d had very bright red breast & other very bright BLUE, it might be harder <to tr> for both parents to
365d106 ock & Ptarmigan (probably subalpina.) former has BLUE breast, latter reddish, hybrid purple-- be care
378d147 sher (.p. 169) has the colour on its back bright BLUE.-- «thus young of» Many of the pies assume the
468t106 have crosses-- I see Bees almost every flower-- BLUE-bells-- wild-raspberry--leeks-- Flowers which t
470t178 n Phlox though they examine it.--«Little Dusty & BLUE» Butterflies at Clover,--Veronica--, Ranunculus
502q11v sa «Campanula» &c & dead-nettle.-- Lithospernum. BLUE Gloss. it is not possible to see orifice of poi
506q015 nslow &c (36) Has not H. raised races of white & BLUE Linum-- did parent plants grow near each other.
594n115 838 remember with distress their companions-- a «BLUE» Gibbon. whose companion had S[...] been dead a
618o41v ot see an atom think: they are as incongruous as BLUE & weight; all that can be said that thought & o
618o41v & weight always went together. & as a thing grew BLUE it «uniquely» grew heavier yet it could not be
618o41v ht & organization run in a parallel series: if BLUENESS & weight always went together. & as a thing g
618o41v grew heavier yet it could not be said that the BLUENESS caused the weight, anymore that weight, the b
618o41v ss caused the weight, anymore that weight, the BLUENESS, still less between <action> «things» so diff
618o41v ation: But if the weight never came untill the BLUENESS had a certain intensity (& the experiment was
618o41v nt was varied) then might it now be said, that BLUENESS caused weight, because both due to some commo
510q017 iguel to collect facts for me-- what? What does BLUME say on alpine Flora of Java? Has Schow written
303c204 make abstract on this subject from Lawrence. BLUMENBACH & Prichard -- Now we might expect that anima
323c269 le's French Revolution 3? vols. oct: -- 26th BLUMENBACH'S Essay on Generation. Englis Transla -- The
567n015 Induct. Sciences-- Vol I p. 334 Does a negress BLUSH.-- I am almost sure Fuegia Basket did. & Jemmy,
567n015 I should think would not have any emotion like BLUSH.-- when extreme sensation of heat shows blood i
577n052 animals. not being such thinking people. do not BLUSH.-- sensitive people apt to blush.-- -- The powe
577n052 people. do not blush.-- sensitive people apt to BLUSH.-- -- The power of vivid mental affection, on s
578n053 ly to have one» you wont. ==)== No surer way to BLUSH, than particularly to wish not to do so.= = How
578n053 How directly personal remark will make any one BLUSH.-- Is there not some saying about a person even
578n054 n act of cowardice, or cheating.-- one does not BLUSH before utter stranger,-- or habitual friends.--
578n055 cross her, the blood rushed to her face,"-- One BLUSH if one thinks that any one suspects one of havi
579n060 irritation on a piece of skin cut off made the BLUSH come.-- it is an excitement of surface under th
578n054 your son would marry Miss. O. B.-- Mrs. B. A. BLUSHED. analyse this:-- Let a person have committed a
578n053 the dark-- «so modest a person.» A person who BLUSHES in the dark is proverbially a most modest pers
327c265 age" price 14s. Marh. 20t. 1839. Philosophy of BLUSHING lately advertised. /6s Mrs Necker on Educatio
556m144 h judgment. [What is the Philosophy of Shame & BLUSHING]CD «Does Elephant know shame-- dog knows triu
570n025 sty. analyse this.-- «Excellent-- my theory of BLUSHING solves this.--» The similarity of men's reaso
577n051 d membrane from local irritation to passion.-- BLUSHING is intimately concerned with thinking of ones
577n052 ry much connected with thinking of oneself.-- «BLUSHING» is connected with sexual, because each sex t
578n053 - Is there not some saying about a person even BLUSHING in the dark-- «so modest a person.» A person
579n060 onometer, is seen to be muscular movement. The BLUSHING of Camelion & Octopus; strong analogy with my
579n060 lion & Octopus; strong analogy with my view of BLUSHING-- in former irritation on a piece of skin cut
629o55v irst case-- seeing how shame is accompanied by BLUSHING, bears some relation to others 5) if so, it i
220b199 ny.-- Athenaeum. No. 537. Feb. 1838. p. 107. Mr BLYTH states that all «genera of» birds in «N.» Ameri
284c149 ?!)-- looks like subsidence.-- on the islets Mr BLYTH remark that a resemblanc between some form in b
309c222 -- NB avoid quoting these hackneyed cases Mr Ed BLYTH does not believe in circular or linear arrangem
340d029 r se.-- [not located] the «4» Struthionidae, Mr BLYTH remarked that greater difference in than in man
341d029 hea closest.-- (& two Rheas still closer).-- Mr BLYTH asked whether structure of pelvis & was not ada
341d030 many deer.-- very curious like some facts of Mr BLYTH on birds.-- Dr. Bachman tells me line of Rocky
342d033 Cocks all alike) More than parent species-- Mr BLYTH remarked only near species or varieties produce
361d095 ght cross easier than two last. Sept. 11. N Mr. BLYTH, at Zoolog. Meeting stated, that Green-finch, a
361d096 le.-- «My case of Caracara. N. Zelandiae.--» Mr. BLYTH stated «that there are» two ducks, which have p
464tf6r anyone, would have thought isolated species Mr BLYTH, however, believes in the existence of Molina's
489qIFC ould ---- 10.(a) J. Gray ---- 17 Yarrell---- 18 BLYTH---- 19-- Mr. Tollett {T} Zoolog Soc «Gardens» -
512q019 r Consul Hunt, birds from Azores or Madeira Mr. BLYTH (1) Mentions some breeder who raises many Engli
249c036 rth of 30 degree.--, may be connected with, Mr BLYTH'S statement of birds of Europe & America, which
300c198 hange of habits in Van Diemen's land. Study Mr BLYTH'S papers on Instinct.-- His distinction between
301c199 before it is conscious of web. feet.-- p. 7. Mr BLYTHS arguments against squirrel using reason in hid
432e124 same fact) go back, & this is argument against BLYTH'S doctrine of young birds retrogressing-- Uncove
464tf6v r goat There is ibex of Alp Pyrenees &c-- (see BLYTH'S work on Ruminants,-- these species must have m
022r008 could not cut it: in which we found the Head & BOANS of a Hippotomus; the hairy lips of which were s
023r012 its maw, amongst which were things pitched over BOARD early in the passage!!-- M. Labillardiere in Ba
538m079 always sentimental, some quarrelsome as B.e on BOARD Beagle, some merry goodhumoured as self.-- «Whe
554m139 y untying a very difficult knot-- the sailor on BOARD the ship could not puzzle her-- with aid of tee
203b136 . Institut. 1837 Mem. Sir F. Darwin cross breed BOARS were wilder than parents. which is same as Indi
431e123 ement) «give examples, pigs, with small chinese BOARS &c &c &c» p. 10 offspring take more after fathe
264c079 mproving yet improvable» & then let him dare to BOAST of his proud preeminence.-- «not understanding
553m138 erstand the affinities of man, better than the BOASTED philosopher himself it is chiefly shown in old
503q013 s not reverse possible?? Maer (1) Yew Trees near BOAT House «ANY male branch.» --¿number of seeds in
256c055 aecian isles, ¿See if type continued?-- See to BOBLAYE & Virlet.-- Whewell thinks (p 642) anniversary
051r093 Pernambuco (great swell & turbid water) organic BODIES protect like peat reef of sandstone.--Corals,
051r095 Clay slates unfavourable to attachment of many BODIES [blank] Beechey.--changes in bottom in NW coas
103a065 verns, in Plutonic rocks argument against great BODIES of vapour. according to Hopkins theory.-- gene
300c198 n man, is analogous to greater individuality of BODIES of some animals over those of others.-- the mi
343d036 on. & these changing affect each other, & their BODIES, by certain laws of harmony keep perfect in th
385d164 n growing-- I believe same has happened in boys BODIES.-- Lavaters. Essays on Phy. transl by Holcroft

385d164 p. 195. says children resemble parents in their BODIES "It is a fact equally well known, that we obse
409e047 xceeding monkeys;-- Having proved mens & brutes BODIES on one type: almost superfluous to consider mi
417e077 e action of two organs in one body.-- or in two BODIES, <th> we can as well understand the necessity
419e087 upon amount of physical change <affecting whole BODIES of species>, & only secondarily,, by assumptio
467t105 h meligethes & small Staphylinidae on all their BODIES pollen-- on a sulphur Broccoli not many do-- p
480z008 Seas written in German.-- Stuttgart ranks these BODIES amongst Vegetables in Linn. Soc.-- Mr. Donn Ca
550m123 of animals to all & changing contingencies, or BODIES of either.-- Our descent, then, is the origin
554m140 anket.-- these cases of commonly using, foreign BODIES, for end. most important step in progression.-
585n075 t thing is same, which touches two parts of our BODIES, «or touches one part. very quickly successive
612o035 ts, but as animals bear relation to less simple BODIES, and to more extended space, such powers of re
614o037 rtions, (different from what takes place out of BODIES) really less wonderful than thoughts-- One org
616o039 question about stated, because there are living BODIES without these faculties & indeed until we know
617o39v ay in which we apprehend the force of inanimate BODIES. How we identify the two aspects as different
263c078 ntal machinery-- so analogous to what we see in BODILY. that <I> it does not stagger me.-- What circu
520m001 16 1856 July 15th 1838 My father says he thinks BODILY complaints «& mental disposition» oftener go w
520m001 resembled his father in body, but his mother in BODILY & mental disposition.-- My father has seen in
524m019 n, ill-humour & depression, which comes on from BODILY causes.-- It is an argument for materialism. t
536m073 -- the free will (if so called) makes change in BODILY organization of oyster. so may free will make
536m073 -- Verily the faults of the fathers, corporeal & BODILY are visited upon the children.-- The above vie
540m088 en the mind is rendered ductile by grief, or by BODILY weakness, melts into tears, with sensations of
547m112 might have been quite still, & not attending to BODILY sensation & yet the Castle would not have turn
571n029 abitual, if heredetary, is pleasant.-- Mental & BODILY Consider case of grazing animals knowing poiso
572n032 &c is master of custom."-- this all applies to BODILY weakness & inferiority, but now we carry it on
572n032 mental inferiority-- when we do not expect any BODILY harm-- case of habitual action.-- L'Institut.
576n048 as an instinct.-- Instinct is a modification of BODILY structure «(connected with locomotion.)» «no,
579n058 ll be curious if it is so.-- frown with grief,¿ BODILY pain? frown shows the mind is intent on one ob
581n066 etary effects on the mind, accompanying certain BODILY actions]CD.¿ but what first caused this bodily
581n066 bodily actions]CD.¿ but what first caused this BODILY action. if the emotion was not first felt?-- «
587n089 ed on assoc of ideas & emotions. rather ideas & BODILY actions make the emotions.-- p. 272. Some rema
589n092 leading to action & yet our emotions being only BODILY actions associated with ideas.-- A sigh, is an
606o025 it in passion, love-- jealousy-- «as» effect of BODILY organisms-- one knows it, when one wishes to d
616o39v ere are two modes of perceptive action by which BODILY action is made known to us, revealing respecti
617o39v ve & objective aspect. The subjective aspect of BODILY action is revealed to us by the effort it cost
617o040 it as well as we can. <The objective aspect of> BODILY action as recognised by our external senses co
618o041 thought, perceptions, memory &c. either to our BODILY frame or the cerebral portion of it Thoughts,
618o041 me relation to the brain that force does to the BODILY frame, they could be perceived by the faculty
044r072 s.--wide valleys.--central peak small; yet great BODY of lavas have flowed from centre-- Pisolitic ba
047r082 shes up? (NB. Earthquake wave is an oscillation, BODY of water manifestly does not travel up.--) If t
065r139 55 minute. <only> one lichen. only production. a BODY which had long been buried, <see> from rotten s
065r139 ied in a mound» long consigned to the earth. yet BODY had scarcely undergone any decomposition: count
080r178 ntermediate Puncture one animal with recent dead BODY of other. & see if same effects, as with man Do
107a077 han above it no because heat proceeds from great BODY of mass.-- The last speculation becomes importa
171b005 d be considered that (that is with our present system of BODY & universe therefore final cause of life With t
202b130 in the two animals bearing relations to a third BODY., or common end of structure A Race of domestic
272c107 . formed from developement of some other part of BODY.-- there are hemipterous insects, having spiny
300c198 ary tendencies. his mind is still only a divided BODY. ¶p 3. language seems to supply instincts,-- &
336d017 ined slowly, now all the mules have their whole <BODY> form of body gained in one generation, so it i
336d017 ow all the mules have their whole <body> form of BODY gained in one generation, so it is impossible t
354d062 ar to butterfly-- is not more wonderful than the BODY of a man undergoing a constant round,--each par
367d113 g, though it has full share of Jackall shape of BODY.-- disposition wild, & fearful. though not so m
367d113 flying bat, which has the power of inflating its BODY like balloon-- by air cells connected with chee
371d118 l that childs nervous system should build up its BODY, like its parent, than that it should be provid
372d129 n the separated part every element of the living BODY is present-- in generation something is added f
372d129 neration something is added from one part of the BODY «(or of other <like» «similar» body)» to anothe
372d129 part of the body «(or of other <like» «similar» BODY)» to another part of body.-- [in plants does no
372d129 other <like» «similar» body)» to another part of BODY.-- [in plants does not whole individual change
383d159 dimidiate, but quarter-grown seems to show whole BODY imbued with possibility of becoming either sex.
389d177 means to propagate & perpetuate differences (of BODY, mind & constitution) is the end frustrated, wh
409e049 .& hence few only social there could not be one BODY of animals, living with certainty on other» hen
417e077 eration as being the action of two organs in one BODY.-- or in two bodies, <th> we can as well unders
426e107 of this series & makes one think that one large BODY of varieties are fertile & make mongrel, & othe
434e127 rful than heredetary ambling horses. Whether the BODY of parent be altered, that is the Nisus formati
434e127 uller call it) succeeds in altering <or> form of BODY, or whether it merely has tendency (as effects
451e177 India» & found there were 5,283 attached to its BODY-- Journal of the Asiatic Soc. vol. I. p. 335. C
459tf02 the mules, these parts resemble ass. (& part of BODY mare)-- -- this may be, perhaps. squeezed into
459t013 ck. it took after the Penguin in the form of its BODY & in the manner of walking but not waddling; it
496q05a scribes Kill Sparrow after feeding on oats, give BODY to Hawk & sow pellet. ejected. done Examine pol
520m001 ition» oftener go with colour, than with form of BODY.-- thus the late Colonel Leigton resembled his
520m001 the late Colonel Leigton resembled his father in BODY, but his mother in bodily & mental disposition.
522m014 ion, & both these give strength & comfort to the BODY» I know the feeling, thinking over somebody who
552m131 els angry, when brain is pumping force to legs & BODY, & especially, when to whole body, being failed
552m131 orce to legs & body, & especially, when to whole BODY, being failed, & not to any particular muscle S
557m150 fear.-- Why does any great mental affection make BODY tremble. Why much laughter tears.-- & shaking b
557m150 dy tremble. Why much laughter tears.-- & shaking BODY.-- Are those parts of body, as heart, & chest (
557m150 er tears.-- & shaking body.-- Are those parts of BODY, as heart, & chest (sobbing) which are most und
564n005 ng the citadel itself.-- the mind is function of BODY.-- we must bring some stable foundation to argu
566n010 in these absurd ideas.-- do they indicate mind & BODY retrograding to ancestral type of consciousness
567n014 t is experience & observation.--» in balancing a BODY & an ass knows one side of triangle shorter tha
567n015 nsation of heat shows blood is pumped over whole BODY.-- is it connected with surprise.-- heart begin
570n023 , when surprise, can one resist blow better with BODY distended.-- intolerable to be poked behind, wi
572n032 self less, but the manner, whether by bowing the BODY, kneeling, prostration «uncovering body» &c &c
572n032 wing the body, kneeling, prostration «uncovering BODY» &c &c is matter of custom."-- this all applies
575n043 ny that brain would be intermediate like rest of BODY? Can we deny relation of mind & brain. «Do we d
575n045 certain ideas hurting brain, like a wound hurts BODY.== (if you <think> «fear» you shall not have e-
577n053 , showing, effect of mind on individual parts of BODY-- tears flow from both, as when one burns end o
596n184 ack skin from head very little Does blood go in <BODY> face in pashion.?-- cry? Do people of weak int
600o008 which makes struggle in man.-- two souls in one BODY-- (2) Beau ideal, refers chiefly to moral, beau
601o009 rder. where the sensation is conveyed over whole BODY (which it may be in first case. as when the exc
612o035 as individual.-- [RHC] 2) In animals, growth of BODY precisely same as in plants, but as animals bea
613o035 resent, & where will directs <to> other parts of BODY. to do such.-- All this can take place & man no
614o037 eally less wonderful than thoughts-- One organic BODY likes one <m> kind more than another-- What is
616o039 ss memory &c. have the same relation to a living BODY (especially the cerebral portions of it) that a
617o040 ce i.e. movement? capable of being traced to the BODY of the individual to whom the action is attribu
626o052 eling or emotion rises from heredetary action on BODY.-- This feeling, when instinctive will lead to
638j58v y»-- brain making structure, instead of parts of BODY.-- Now we know what instinct is-- consider this
640j167 present with respect to new arrivers, the small BODY of species would far more easily be changed.--
514q21. on before flower open. (An. des Sci Where is BOERHAAVE'S paper on impregnation of violets.= Zostera=
346d048 - Agrees, «nearly» with. the account. given by BOETHIUS of ancient Caledonian Cattle. Ch 3. Instinct
123a115 L'Institut (1838) p. 268. Paper by Humboldt on BOGOTA. Cordillera,-- nothing.-- salt & coal near Bog
123a115 gota. Cordillera,-- nothing.-- salt & coal near BOGOTA; p 270.-- SPLENDID PAPER on fossil shells of S
150g037 n centre In Glen Collarig, on side of Hill of BOHUNTHINE upper road (2) extends as far nearly as hous
089a017 ertainly Volcanic-- CD[Might not bottom of ocean BOIL; yet heat never reach surface.-- Journal de Phy
505q014)-- Asclepias-- Flowers not seeding= Put pot of BOILED earth on top of House =Aristolochia, plant wh
138a155 of strata-- he heated sand red hot & brine was BOILING on the top-- [blank] Would rotting wood by yie
358d075 igilance a piece of pork 3 lb was taken from a BOILING pot, & carried off by one of these birds" Case
275c119 utton of Geology. he had few clear facts, but so BOLD in many such profound judgment, that he forseei
303c204 e viccissitudes of present animals.-- He-will be BOLD. I will venture to say unphilosophical. L'Insti
340d026 Anatidae.-- Consult this book again.-- Mine is a BOLD theory. which attempts to explain, or asserts t

Page
(Key Word)
346d048 uniform.--crafty, go in file, hide their young., BOLD.-- a Mr W: Hall remarked that it was against al
358d075 ustralia Vol I. p. 306 "The crows were amazingly BOLD, always accompanying us from camp to camp; it w
451e179 vated & higher forequarters: is said to be of a BOLDER & more generous temper-- Hodgson Koloff. voyag
069r152 ocean! Make a grand analogy between Wealden & BOLIVIA Transportal of conglomerate between two ranges
257c056 then in Chile & lastly 12,000 ft above sea in BOLIVIA; he examined-- all species & found "beaucoup d
414e064 he peculiar skulls of the men on the plains of BOLIVIA-- strictly fossil «& in Van Diemen's land»-- t
085a002 smith tells me, that Himalayas penetrated like BOLIVIAN Chain. Volcanic islands. from number of crate
223b209 s have said.-- ¿where is Pentland's account of BOLIVIAN human species?-- Small «new» animal mentioned
096a039 Iquique of elevation to amount of 30 ft.-- Mr BOLLAERT (at Roy. Institut) talks of quantities of she
097a044 on is given.-- Vol II. 2d Series. p. 221.-- Mr BOLLAERT tells me, that the upper strata alone at Guan
060r124 the <ground> land in the W Indies.--p. 200. BOLLINGBROKE voyage to the Demerary Earthquakes at St He
359d087 ir origin) Saw cross between Penguin Duck «from BOMBAY» & Canada Goose.-- Former strange mishaped bir
044r074 vertical movements. In Cord: after seeing small BOMBS. without a vesicle. we may consider appearances
044r074 ter in solution must be great: & in the fact of BOMBS in tufa there is proof of such gaz) steam conde
131a135 t. 1838 p. 360. on orbicular trap thought to be BOMBS submarine L'Institut 1838 p. 400. Observations
260c067 emperate regions-- crows in N. America Study BONAPARTES list In the Zoological Journal I read a curi
310c225 al History «Vol I??» p. 318. some remarks on BONAPARTE'S list of birds in Europe & N. America, on clo
244c022 which escaped there.-- p. 139. Vespertilio BONAR«I»ENSIS (from Buenos Ayres) replaces <Vesp.> holds
530m043 are absolutely forgotten.-- My father signed a BOND, yet when he paid the Attorneys bill, he asked
530m043 t when he paid the Attorneys bill, he asked what BOND he could have had. yet during whole illness, he
543m100 bly used to beat him-- The son of a Fruiterer in BOND St. was so great a fool that his Father only le
087a010 Guanaco the Camel.? Make note about N. American BONE not probably in salt marshes Efflorescence noth
093a033 t Bahia Blanca. <fl> Flustra identical. recent & BONE bed.-- November 8th 1877 (Memoranda so far dist
509q017 e dissected any animal often, which has abortive BONE. (ask more about the lowest cervical vertebrae
588n090 chiefly to recollection. yet a dog hunting for a BONE shows he has recollection.-- Lamarck. Phil. Zoo
594n112 & sparrow in Shrewsbury garden picking from same BONE A child born on the 1st March was frightened on
023r010 lks about the immense quantities of Cuttle fish BONES floating on the surface of the ocean, before ar
026r022 f Europe Conybeare. Introduct XII P. silicified BONES not common in Britain. Mem Concepcion Says Echi
035r045 bles beds, most unfavourable to preservation of BONES &c &c--Yet <silicified> turn over silicified wo
037r055 e> «remains of Amphibia, exclusively.» & Turtle BONES. & the bones of <two graniniverous> a herbivoro
037r055 f Amphibia, exclusively.» & Turtle bones. & the BONES of <two graniniverous> a herbivorous lizard.--
042r068 of Falkner. ¿how far is the distance?-- Fossil BONES black as if from peat.--yet cetaceous bones so
042r068 sil bones black as if from peat.--yet cetaceous BONES so likewise «of miocene period».--Mem Bahia bla
057r113 [,] Volcanos as Marl-pits: Consider well age of BONES. = slowness of elevation proved at St Julian. =
057r113 t elevation proved at St Julian. = do not these BONES differ as much nearly as the Eocene. = Should M
057r113 nearly as the Eocene. = Should Mr Owen consider BONES washed about much at Coll. of. Surgeon's? I rea
101a058 s it Edentate? Phillips p 289.-- Alludes to big BONES in interior at Falkland Isd.-- Peron does as if
109a082 ..] Subaqueous. removal, shown by the number of BONES lying at the bottom of sea. off coast of Englan
134a141 cks of S. Western Coast Vol II p. 277. on whale BONES in Falklands Some of the Tosca nodules at Bahia
173b009 o separate Every character continues to vanish, BONES instinct &c &c &c non fertility of hybridity &c
194b095 ons.-- Mem. species doubtful when known only by BONES.-- Mem: Silurian fossils: ¿How are. South Ameri
211b162 pteryx. a good instance probably of rudimentary BONES.-- As Waterhouse remarked Mere length of bill d
213b172 in strata could not now accumulate without seal-BONES & cetaceans.-- both found in every sea, from Eq
290c163 very similar.-- Geology. Transact. Vol V. Birds BONES-- in strata of Tilgate forest Seeing common gul
403e023 on in external form of varieties, do we suppose BONES will not change in number. (even species do not
408e044 pecies» Buckland. Reliquiae Diluvianae. p. 222. BONES of Horse. Bear & Deer at 16000 ft. with Snow on
408e044 at 16000 ft. with Snow on Himmalaya-- Humboldt BONES at 7800 in Andes-- parallel & curious facts.--
412e057 in the male.-- <like> «far» more than marsupial BONES, . & even more than Mammae, which have given mil
444e157 nout of the mole & Pig in having two additional BONES to give strength to it.-- p. 139. Doubts altoge
444e157 re wanting. p. 144 in the Icthyosaurus 60 or 70 BONES in the paddle, yet all in the arm are perfect.-
463c055 as the <brain> spinal marrow expands, so do the BONES <are created> expand-- instead of saying as bra
463t057 er is lost, <be> (as vertebrae into skull., two BONES of tibia into one.--) because if the animals we
465t080 shells preserved must be as much a casualty as. BONES of Mammalia in caves:-- :argue first case of bo
465t080 es of Mammalia in caves:-- :argue first case of BONES (New Red Sandstone) & then go on to shells-- A
493q003 Tame Duck & Wild Duck, & then weigh their wing BONES & see if relation is same good, avoids effects
509q017 oldering of tibia & fibula: in Man any abortive BONES??? do. Wing in Apteryx ‖ no as Os Coccygis-- T
509q017 ng in Apteryx ‖ no as Os Coccygis-- Turbinated BONES? False ribs Wings of Apterix: clavicle in--? Co
543m101 hepherd dogs-- Inherited Habits: Have Effect in BONES is valuable it shows that new instinct can orig
634j54v stantly alludes «(& at p. 312)» to the abortive BONES He explains it <"By> saying "It is the determi
637j58r k is adapted to long necks.-- p. 236. Marsupial BONES especial adaptation, to «young».-- good God & y
315c241 instincts L.' Institut «1838» p. 184 Botany of BONIN. "grande analogie avec le flore du Japon", some
095a039 s of very much of Gypsum.-- The officers of the BONITE. French discovery ship, found clear proofs of
480z006 as written account of Guanaca. In transaction of BONN Society M. Edwards on Corallines L'Institut 183
376d137 t. st noise.-- The Cercopithecus chinensis: (or BONNET faced monkey he has seen do this.-- These Monk
029r033 ambuco. EARTHQUAKE AT SEA.--Extract from the log-BOOK of the James Cruikshank, Captain John Young, on
035r047 tagonia. The English fact is astonishing consult BOOK itself. P. xx: same fact is indeed shewn? by th
112a088 leughs collection of facts See page 101. in Note BOOK (C) for some speculats on conducting powers of
124a118 bsidence in Demarara p. 131 (B.) Wrong Entrance. BOOK C. p. 101. On Frozen Soil of Siberia (with refe
129a132 near mouth of Columbia river-- Read Mr Parker's BOOK.-- M. Bichoffs Papers, in Edinburgh New Phil. J
138a153 rvat. in Southern Brazil. [blank] «p. 4. (Lyells BOOK)» Observaciones sobre El Clima del Lima par Dr.
139aIBC of superadded vital influence?-- See End of Note BOOK. called R. N.-- Massac[h]usset would be well wo
170bIFC 54 56 67 69 76 79 91 93 107 Ireland 113 117 This BOOK was commenced about July. 1837 p. 235. was writ
233b251 om Maurice & Madagascar & C. of Good Hope.-- His BOOK Probably worth studying.-- Wingless birds S. Co
247c028 s land & Cape of Good Hope V. p. 44 of this Note BOOK Rabbits introduced in 64, of very many colours,
271c106 ng bred in-- Mem, <an> a statement in Mr Wynne's BOOK, about not altering breed of animals in certain
294c177 ain the rest,-- Lonsdale says he has seen in old BOOK last Bear in England killed in year 1000. refer
322c270 - very poor Sir T. Browne's Religio Medici Lyell BOOK III There are many marginal notes <Rengger &c>
331dIFC f breed of pigs with solid feet.-- 1838 [In this BOOK some curious notes on Monkeys recognising Sexes
338d025 colour. black being strongest.-- V. p. 63. Note BOOK M'. for case of change in food in insects enter
340d026 of distrib: of <birds>. Anatidae.-- Consult this BOOK again.-- Mine is a bold theory. which attempts
345d043 both cases» is killed. Notes from Glen Roy Note BOOK.-- Why is not Tetrao Scoticus. an american form
362d100 to breed at particular times. Mr Yarrell has old BOOK 1765? Treatise on Domestic Pidgeon, in which it
366d111 being perfectly distracted «Referred to <other> BOOK M.» Is there any law of variation. -- «(as Hunt
369d115 ent Character Hunter Animal Economy p. 482 (Same BOOK) Owen says "the necessity of combining observat
462tf05 naeum 1839 p. 72-- A curious theoretical French BOOK review on politics in relation to the different
486z019 of Europe have Tropical Forms See p. 256 of Note BOOK (C) for comparison of singing powers of birds o
500q010 esirable as in Cattle in Chillingham Park-- What BOOK on varieties &c of deer. Contests of sexes.-- Q
501q011 ge of Soil-- crossing-- when seeds raised.-- His BOOK.-- 32. Would wheat from AEgypt ripen in Scotlan
503q012 says does» Royle In Royle's productive Resources BOOK no information & Hope about Silk worms. Varieti
507q15v p. 670-- oats cut down turning into Rye.-- 46). BOOK describing amount of Horticultural Variation? H
507q15v t beautiful adaptation might be arrived at.= Any BOOK with drawing of Seed. Anemone with, tuft-- Bull
508q016 s of each-- Cause?-- Andrew Smith. (6) What size BOOK Gallesio storia del Reproduzione.-- D. Holland
520mIFC e. Col. Sykes) Private Finished. Octob. 2d. This BOOK full of Metaphysics on Morals & Speculations on
526m029 of pictures).-- when one remembers a thing in a BOOK, one remember the part of page.-- one is tempte
529m042 ets & big noses, quite conjectural, in Blakeways BOOK of Sheriffs.-- July 22d. 1838 No Deliriums, yet
540m087 ecalled by obvious associations. as by reading a BOOK.-- Consider this.-- "The fledge-dove knows the
546m110 have lower animals these vivid thoughts In same BOOK (p. 143) wonderful case of perfect double consc
546m110 ct to Mr Mayo himself. she was one day reading a BOOK, with ivory paper cutter, which she valued, & s
559m155 varieties.-- Rev R. Jones has it.-- very curious BOOK-- Hume's essay on the Human Understanding well
566n010 on the wish to support a wife a ruling motive.-- BOOK IV, Chapt I on passions of mankind, as being re
572n033 otions &c Octob 30th-- Dreamt somebody gave me a BOOK in French I read the first page & pronounced ea
572n033 ke» I could not picture to myself reading French BOOK quickly, & <running> «running» over imaginary w
604o017 ssive emotions.-- Cannot quite perceive drift of BOOK.-- Sympathy & affections chiefly fail.-- Notice
610o031 luded it could not be so.--Looked at a direction BOOK, but could not find out-- Directed his letter,
610o031 -- remembered, that he had. looked in direction BOOK under head of Wilson, referred to Robert & foun
239cIFC Dec. 13 1856 Also looked through April 23. 1873 BOOKS About amount of difference: where hybrids produ
286c156 sil animals, what relation in classification in BOOKS, ought they to hold,-- Birds having web-feet, w
319c276 lk case of fossil geographical range.-- [blank] BOOKS examined: with ref: to Species Most of those wh

(Key Word)

326c265 maps. by Copenhagen Botanist of range of plants BOOKS quoted by Herbert. p. 338 Schiede in 1825. & La
390d179 om himself, for it should be observed that from BOOKS to read Buffon Suites de.-- Horse & Cattle Libr
563nIFC Theory» Dec. 16 1856 Looked through & all other BOOKS May 1873-- October 2d.. 1838 Essays on Natural
601o08b ns morales et physiologiques The first of these BOOKS I daresay good. 1. Sensation is the <conse> ord
629o055 er our desires become fixed to ambition. money, BOOKS &c &c.-- <]> the "secondary passion" of Hutches
411e055 d before them in any part of World.-- Wealden to BOOKS-- When one sees in Coralline powers of multipl
489qIFC Zoolog Soc «Gardens» ---- . 20 & Breeders Dr. BOOTT: R. Brown p. 21 Horticulturists p. 21--23 Eyton
514q21. less-- 20 varieties in mountains of Tahiti. Dr. BOOTT-- says caricas from every isld differs-- do the
452e181 ippopotamus occurs in India. in the Jungles of BORABHUM & Dholbum.-- Vol do. p. 634, alludes to fact
429e115 in such a line., it is other plants.-- a broad BORDER of Killed trees would form fringe.-- but there
507q15v ush-- Dandelion-- Sycamore. & seeds with «mere» BORDER-- & Humboldts spinning seed.-- (50) Any cases
236b278 ion of varieties is it not per saltum.-- Isld BORDERING continents same type collect cases.-- African
636j56v t as a group of bisexual animals living on the BORDERS of a country favourable to change.-- It might
047r082 higher part. -- Does the sea fall on banks as a BORE wave rushes up? (NB. Earthquake wave is an osci
616o039 metaphor said to attract; & hence if thought &c BORE the same relation to the brain that attraction
618o041 iousness, & have no objective aspect. If thought BORE the same relation to the brain that force does
193b091 paper on Geographical range Richardson-- Fauna BOREALIS. It is important the possibility of some isld
261c069 rmation-- instinct Swainson's remarks in Fauna BOREALIS must be studied. There is capital table of ex
324c268 pain-- Much about Castes & Richardson's Faun. BOREALIS Entomological Magazine (paper on Geograp. ran
226b223 ia; intermediate, see how that is.-- ¿are shell-BORING Molluscs, like Carnivorous Mammalia in their w
265c083 rd of any effect.-- New Hollanders have gone on BORING their noses. &c & This congenital changes show
477z001 ompsons-- Part II.-- 35. Phil Trans Burrowing & BORING marine animals-- CXVI. P 111 do Observations o
171b002 ry thing else is perfect; mothes apparently only BORN to breed.-- annuals rendered perennial. &c &c.-
192b084 so with useless wings under elytra of beetles.-- BORN from beetles with wings.& modified.-- if simple
192b084 ed.-- if simple creation, surely would have been BORN without them.= In some of the lower orders a pe
204b141 ridgewater errs greatly in thinking every animal BORN to consume this or that thing.-- There is some
219b196 separate one, so to say that all Mammalia, were BORN from one stock, & since distributed by such mea
259c063 t.-- perhaps in process of change.-- Are any men BORN with any peculiarity. or any race of plants--La
259c066 shows my «view of» generation right?-- If puppy BORN with thick coat monstrosity, if brought into co
285c150 hers' economy Dr. S. showed that savages are not BORN with any capacity for observation of tracks &c
289c162 ariety could be transmitted more easily in those BORN without coitus, than with.» ▌Might be given as
317c249 led cat.¿cut? has its offspring short tails /one BORN at Maer. Tuckeys voyage-- p. 36 "Cercopithecus
345d043 now all are pure red, yet calf every now & then BORN with white head (,or «short-horned with» black
366d108 ted by art. Yarrell told me of a cat & of a dog, BORN without front legs-- -- the former of which had
366d112 t can propagate, ie with the chance of two being BORN at same time, & make breed, one would doubt any
372d131 such process is effected.-- a child might be so BORN. but it would be very different from true gener
393dIBC peculiarity or variation common to any zoophyte «BORN in succession» which is not transmitted by gene
402e019 erally crooked, as if they had been broken". are «BORN so in all Malay Countries W. Earl Eastern Seas.
453e182 e, as well in all Malay countries «the» cats are BORN with the joints near the tip crooked.-- is the
473s07r there is no reason, why the peculiarities shd be BORN,-- may come in corresponding time of life of of
473s07v ar late in pregnancy, & then may just as well be BORN a tendency to alter or assume some form late in
524m021 herself near Drayton & Ternhill, (where she was BORN) though she never naturally talked of these pla
530mo44 has somewhere heard (Hunter?) that pulse of new BORN babies of labouring classes are slower than tho
534m061 erm within egg, cannot think-- as well as animal BORN with instinctive knowledge.-- but if so, yet th
556m145 We ought never to forget-- --; that every man is BORN with a portion of phsiognominical sensation, as
556m145 s certainly as every man who is not deformed. is BORN with two eyes.." I think this cannot be dispute
570n026 on from this would be that a mountaineer <takes> BORN out of country yet would love mountains, & a ne
587n088 simply heredetary habit.-- A blind man might be BORN with idea of scarlet, as well as remember it.--
594n113 Shrewsbury garden picking from same bone A child BORN on the 1st March was frightened on the 24th of
595n125 st as much instinctive as a bull <tr> calf, just BORN butting, or young crocodile snapping.-- these I
626o052 -- He does not say anything about any principles BORN in us.-- Great difference with my theory.-- see
638j059 no reason, why structure of brain should not be BORN. with tendency to make animal perform some acti
639j28w everal genera of Grallae Suppose six puppies are BORN «& it so chances, that one out of every hundred
639j28v hances, that one out of every hundred litters is BORN with long legs» & in the Malthusian rush for li
179b032 e less purely bred owing to <first> having once BORNE Mongrels he has thus seen the black blood come
242c017 ardt «Forrest» authors on E. I«ndian». A«rch.»▌ BORNEO & Sumatra both seem to have elephant & has ora
242c017 as orangs, ▌ Tapir common to Sumatra & Malacca▌ BORNEO & Malacca «& Cochin China» are said to have or
251c041 eld on two bears very close species, inhabiting BORNEO & Sumatra. differ only in form of white mark o
253c046 » & «Monkeys.» Fact of Elephant same species in BORNEO. Sumatra. India Ceylon-- perhaps shows great p
402e018 to examine their defunct companion".-- p. 229. BORNEO.-- only animals here heard of pigs, small bears
402e018 ys & an animal probably a tapir p. 233. dogs in BORNEO-- <brought probably by Chinese>, "the breed be
402e020 kinds of animals have spread from the N. end of BORNEO to the adjacent island-- In Sooloo we find the
450e174 but only in one part the northern peninsula of BORNEO.-- Ox & hog natives of Borneo Notices of India
450e174 hern peninsula of Borneo.-- Ox & hog natives of BORNEO Notices of Indian Arch. Singapore 1837. By J.
451e177 Directory. Vol II. p. 46 Carimon Java. (between BORNEO & Java) Lat 5 degree. 50 minute S. adjoining i
453e182 . Believes the <Rhinoceros> «Tapir» is found in BORNEO.-- «p. 233» There, as well in all Malay countr
044r071 onite to the present day. at Mauritius. (consult BORY «dip of strata on East») cannot believe in a gr
207b152 yle introductory remarks to Himalaya Mountains-- BORY St. Vincent Vol. III. p. 164. Lile de la Reunio
324c268 r in Annals of Nat. History Prichard.-- Lawrence BORY St. Vincent Vol III p. 164. on unfixed forms.
242c018 in isles of Great ocean says in conformity with BORY'S Views.-- <Says> D'Orbigny is said to have brou
229b234 oceros (?)-- Some paper in Institute on range of BOS in India.-- Range of Zebra?-- The Crocodile & To
241c014 d wild breed of oxen at Java.--.p. 140, calls it BOS. leucoprymnus. does not say whether wild or not
251c041 III. <p.> p 233, stated that the "Asseel Gazal. (BOS Gazoeus) does not mix with the Gobbah or village
634j54v till it must be abandoned for another".-- What BOSCH!! Put it to case of man. <&> The <design> deter
072r159 e connects western isles of Scotland & Iceland.--BOSH nor on Norway, or Spitzbergen.--Spitzbergen ani
628o54v lst the passions have no relation I think this <BOSHES> «nonsense»-- My theory of durableness will ex
577n051 face exposed, face of man, face, neck-- «upper» BOSOM in woman: like erection shyness is certainly ve
606o022 are judged; & how obtained.-- implanted in our BOSOMS.-- how comes it there? Laocoon p. 75 "The beau
323c269 Observations on morals by Eugenius Feb 14th. BO«S»WELL'S life of Johnson. 4. Vols 25th Phillips. Geol
201b129 5 «great monkey» Mr Johnston says Mag of Zooly & BOT. p 65 Vol II talking of annelidae.-- <">The fact
289c162 n Bering Pike (Waterhouse) Magazine of Zooly & BOT-- Vol II p. Dr Johnston <on> Entomostraca Daphni
296c184 us. p. 112. & paper on genus Magazine of Zool. & BOT.-- Vol I p. 450. 4 instances of hybrids between
297c186 emmants.-- Cephalopoda ditto.-- Mag of Zoolog. & BOT. Vol. II p 125 Allusion to abortive spiracles i
510q017 rrent, icebergs travel by wind. Aug. St. Hilaire BOT. p. 787. position of embryo in close species of
310c223 . «Royal Institution» Dr Royle seems to think BOTANICAL Provinces will turn out not nearly so confine
328cIBC Jardin du Roi Java fossils at same time Study BOTANICAL work on Buds & Gemmae. C D Charles Darwin 36
462t051 cattle in every part of England. &c &c NB. In BOTANICAL geography, there can be no sharp division of
548m114 ly reason -- <In a> Abercrombie's case of «in BOTANICAL Student» somnabulism, did reason about himsel
548m115 eam, when that is not recollected, nor of the BOTANICAL Somnambulist. (if he had been unhappy)-- it i
062r128 llera, where climate similar.-- I do not know BOTANICALLY = but picturesquely = Both N & S. great cont
236bIBC Introduct Dict. Science. Naturelle Geographie BOTANIQUE De Candoelle. Geol. Soc Horae Entomolgicae Li
326c266 rown. has curious coloured maps. by Copenhagen BOTANIST of range of plants Books quoted by Herbert. p.
506q15v ostera. Has he seen it in flower? does he know BOTANIST who does-- What is Ruppia Bennett-says in sam
538m041 sufficiently habitual to become poetical) the BOTANIST might so view plants & trees.-- I am sure I r
538m079 l line-- both odd appearance about eyes.-- one BOTANIST & great knowledge of Irish Politics, «both ba
070r153 losely allied species occur in same country? In BOTANY Instances diametrically opposite have been ins
207b151 ll at St. Helena. as. Pineaster & Mimosa called BOTANY Bay Willow V. Dr Royle introductory remarks to
217b189 between grouse & pheasant-- Magazine. Zoology & BOTANY Vol I p. 450 There is in nature a «real» repul
220b198 months since in Annales des Sciences, paper on BOTANY of Tahiti In Charlesworth Magazine Jan: 1830.
288c160 ormerly nearer.-- «Selby» Magazine of Zoology & BOTANY No XI p. 390. a slight change in enclosing a c
292c170 bivalves.-- Anyman No VI. Magazine of Zoology & BOTANY p. 566 wants to see absurdity of Quinary arran
313c235 on-- In the Entomostraca (Magazine of Zoology & BOTANY) where several generations are produced in suc
314c238 . 1833 Steph. Endlicker (He will give sketch of BOTANY of islands of south seas says so in preface.--
315c241 papers on instincts L.' Institut «1838» p. 184 BOTANY of Bonin. "grande analogie avec le flore du Ja
320c275 irs. published by Taylor Magazine of. Zoology & BOTANY & Continuation «Annals of Natural History» Ski
322c269 troduction to Natural Philosophy R. W. Darwin's BOTANY.-- References at end Mayo Pathology of the Hum
326c266 lume of Annales of Vienna. sketch of south Sea. BOTANY R. Brown. has curious coloured maps. by Copenh
358d074 eite in relation «See Gaudichauds Volume on the BOTANY of the Pacific.--» to nearest continent.-- Wit

Page **(Key Word)***
477z001 s in L. Jenyn's introduct to Mag of Zoology and BOTANY. Philosoph. Transacts. 3. papers connected wit
480z007 ranslated from D Orbigny no IV Mag. of Zoolog & BOTANY p. 356 Lesson on Berre. do-- Magazine of Zoolo
480z007 356 Lesson on Berre. do-- Magazine of Zoolog & BOTANY. Vol I p. 358. D'.Orbigny <considers> states t
459tf1r -- ((this is good case as showing gradations, BOTELER'S Narrative Voyage East coast of Africa-- Vol I
246c027 entral Pacific, yet there appears to be one at BOTOUMA from account of natives, & probably on Oualan.
023r011 ient elevation.» The volcanos originated in the BOTTOM of the ocean. & the present Volcanos have been
023r011 in a deep & therefore weak part of the ocean's BOTTOM. With respect to Sharks distributing fossil re
025r019 upwards. that life is exceedingly rare, at the BOTTOM of the sea.--«certainly data insufficient, yet
026r021 e at Hobart town between the older strata & the BOTTOM of sea near T. del Fuego.-- Is there account o
030r035 ng these matters are not now collecting, in the BOTTOM of an open & not deep sea.--(Character of coas
041r065 preserved in that part. having lived over whole BOTTOM is important; because in this latter case. we
044r074 cle. we may consider appearances of eruption at BOTTOM.--solution under high pressure of gazes. espec
047r081 in number of waves & in wind, instead of sea's BOTTOM being in motion what difference? In watching h
051r095 lao.--From Sir. H Davy experiment on the copper BOTTOM. we see a trifling circumstance determines whe
051r097 ent of many bodies [blank] Beechey.--changes in BOTTOM in NW coast of America. from shingle to sand &
055r106 acts attest a <more> decided elevation of sea's BOTTOM. beds of shells. 2 - 3 toises thick.--Vol II.
057r115 below freezing point!!! Remember idea of frozen BOTTOM or beach of sea to explain preserved animals.-
089a017 hells --¿ action of Heat bubbles volatilized at BOTTOM, condensed before rising?-- Mem. granite heate
089a017 sandstone.-- Certainly Volcanic-- CD[Might not BOTTOM of ocean boil; yet heat never reach surface.--
102a059 ker affected some way out to sea.--¿ effects on BOTTOM a thing floating some way from coast is driven
109a082 oval, shown by the number of bones lying at the BOTTOM of sea. off coast of England.-- Sea must alway
120a107 of Glen Roy road. (¿ metamorphic action at the BOTTOM of the sea?) All this profoundly considered. s
126a124 would be much nearer the surface. especially at BOTTOM of great ocean, where the circulations from su
130a133 s a hot.--spring.-- Hot water would not lie. at BOTTOM.-- Surely we here have proofs of hot bottom.--
130a133 at bottom.-- Surely we here have proofs of hot BOTTOM.-- Study Bishoofs Paper.-- Weelsted told me of
132a138 rstand the argument, that cold <oceans> «lakes» BOTTOM. if not colder than mean of place, shows earth
132a139 ing greater, than the transmission from ocean's BOTTOM.-- (according to M.. Parrots own hypothesis so
132a139 be trusted than any others-- may not the cold «BOTTOM of» ocean. (with fresh sediment added to botto
132a139 bottom of» ocean. (with fresh sediment added to BOTTOM) be caused, by absence of circulating water.--
177b027 ness of life of species that in perfection, the BOTTOM of branches deaden.-- so that in Mammalia «bir
285c152 according to all analogy breed together.-- The BOTTOM of the tree of life is utterly rotten & oblite
536m071 mell of that fatty substance it scrapes off its BOTTOM.-- it is relic of same thing that makes one do
151g042 lf 3d & almost meet, but are separated by flat BOTTOMED strait. connecting flat on one side with irre
079r177 so at Gualgayoc. where many petrified shells BOUGAINVILLE says P 291.-- The Fuegians treat the "chefs
479z005 cier du Roi Rapid growth of Coral-- RN. p 24 BOUGAINVILLE Voyage round world no land animal besides W
159g087 tly shut in No Granite blocks in higher parts?? BOUGHT Glen name of Glen by which we descended, it is
160g089 raight isthmus connecting E & W connecting Glen BOUGHT & Glen Tarf a perfect old Loch, making <severa
160g089 o divortiums aquarum, viz two branches of River BOUGHT & between one of these & Glen Tarf Hill «Cairn
332d005 ralian Savages.-- W. D. Fox has a cat. which he BOUGHT in Portsmouth, said to come from coast of Guin
523m017 insane, answered she had known it at time & had BOUGHT arsenic for that purpose.-- this found to be t
151g040 iss of Moel Derry) on low hill between Inn & BOUHUNTHINE the summit «doubtless worn into coincidence»
101a057 280. the gravel beds in England different from BOULDER beds-- What is Osteopora platycephalus (Harlan
149g032 also> also near summit on Hill on side of Inn BOULDER of granite above 4th Shelf a little lower down
153g055 s formed Ben Erin summit 27.813. 65 55 degree? BOULDER of Granite 28.362 68 degree 60 degree Granite-
158g079 c mistook terrace also right-- Granite such as BOULDER on <thes> Divortium aquarum Peaty Mass of this
158g082 <fre> Gneiss cut smooth on sides of hill where BOULDER lies. buttresses «occur» high up on Shelf 2d «
159g085 emained, no peat supply.-- Consider profoundly BOULDER hypothesis Thursday, from Glen Turrit to Fort
160g090 eer peak» Barom 28.700. A.75 degree 75 degree? BOULDER, much covered by turf 2ft. 8- long of syenite
160g091 harp ridge with peak» I walked all round hill. BOULDER about 20 ft. below summit <Isthmus> {P} highes
160g091 this hill to others 3000? if Ben Erin is 3500 BOULDER Cairn leet more Haberclador Isthmus broad flat
162g104 aces on each side High up the Tarf (a Granite (BOULDER), sloping buttresses, an[d] one alternate curv
029r032 ation of such matter.-- Dr A. Smith says. that BOULDERS do not occur in the South African plains.-- S
057r114 re on plain. & interstratified.-- Urge fact of BOULDERS not in lower strata. only in upper. in accord
070r155 s that in Helm's travels accounts of travelled BOULDERS. from the Cordovise range. Signor Rozales tel
094a036 .S. Chapter on drift.-- Beyond region of great BOULDERS, pebbles of granite clearly effect of remodel
094a036 remodelling same manner. as bits of Patagonian BOULDERS might be transported.-- {T} On grooved rocks.
101a056 nd in Northern England influence dispersion of BOULDERS.-- See Rogers for Southern limits of Boulders
101a057 Boulders.-- See Rogers for Southern limits of BOULDERS in N. America do/p. 280. the gravel beds in E
110a087 E. Vol VI. p. 247. Mr. Schomburgk NW. numerous BOULDERS of GRANITE" "direction of strata on the Berbi
129a131 XXI. p. 213. Beyond the limits of Alps size of BOULDERS sorted: ditto Murchisons case.--¿ does it bea
151g040 <notch> roads very much this character.-- The BOULDERS (one of Gneiss remarkably water worn) are oft
153g056 other 50 ft lower & other smaller ones «these BOULDERS are decaying.» neighboring rock gneiss & [...
156g067 e lines in Glen Collarig, & some «other parts» BOULDERS of same granite, all on these three shelves s
157g076 with red granite veins & quartz, & garnets.-- BOULDERS as before certainly must have <come> «been dr
157g076 no granite-- (in valley «there are» granite) «BOULDERS» {P} cory stream hill with boulders river Rig
157g076 granite) «boulders» {P} cory stream hill with BOULDERS river Right Hand Cascade has <cut> «where two
189b073 n America All animals <are> of same species are BOUND together just like buds of plants, which die at
206b149 which stand between great groups, which we are BOUND to consider the increasing ones.-- NB As Illust
307c215 ess organs do indicate such origin, then we are BOUND to consider abortive organs of same tendency in
428e113 o effect of male sex on locomotive system» I am BOUND to insist honestly that the sudden, change from
055r108 nsumed) Where slope «plainly» indicates former BOUNDARY. (as in other unworn islands) we take in at o
342d035 s have been featured by the Austrian lip & the BOURBON nose". if this be not imagination.-- then old
210b160 nimals peculiar. to Mauritius are not found at BOURBOND Zoolog. Proceedings. <p> 1832.p. III Mr Owen
241c014 f the larger islands.-- -- Antelope in Celebes, BOUROU new species of Axis.-- «Cervus moluccensis is
246c027 -- Coquille Voyage The caswary, inhabits Ceram, BOUROU & especially New Guinea (replaces, Emeu) in No
245c024 in each place to puzzle naturalists.-- p. 372. BOUROUS. the Barbyrousa; a Cervus near Marianus new, &
071r158 fic.--Isabelle's volcano, many amygdaloids.--BOUSSINGUALT «(Lyell)» cracks mountains falling in.-- Ea
150g036 terworn frame terrace 4 4 not visible 3a 3a 3 BOUTHONER 3 2 Terrace 3 Alluvials 3a 3a 2 Mass 3 2 rath
032r041 changes of climate have happened. hurricane in BOWELS of earth cause:--<exp> does not explain cleava
119a105 n have been plunged so many miles deep into the BOWELS of the earth, as would be required by thermome
575n044 o deny identity of instinct.-- Habits import to BOWEN No one doubts that a cross of bull dogs-- incre
114a093 ralia/ and mud of salt-lakes of Rio Negro--Mr BOWERBANK-- Dr. A. Smith's curious specimens of «transv
572n032 aking yourself less, but the manner, whether by BOWING the body, kneeling, prostration «uncovering bo
499q009 ign-- Eyton has such a grove of Willows.-- (14) BOWMAN female branch At What distances from males, wi
547m112 s for moment, implied by «presence» my servant, «BOX» my own manner of ordering things to be done.--
547m112 rung for Covington. whether he had come & opened BOX, whether I had thought what clothes to take (how
216b181 s cross to after births, like aphides.-- Case of BOY with foetus developed in the breast.--looking as
572n034 he dream.-- It does not hurt the conscience of a BOY to swear, though reason may tell him not, but it
591n099 f-preservation is disobeyed-- I often have «as a BOY» wondered why all abnormal sexual actions or eve
614o037 almost alone in this case can perceive instinct. BOY takes delight in mammae before any reason had to
385d164 gone on growing-- I believe same has happened in BOYS bodies.-- Lavaters. Essays on Phy. transl by Ho
594n113 rightened on the 24th of May at Cresselly by the BOYS making faces at it, so much so that the nurse h
539m081 n which my head got well when reading article by BOZ.-- now in this I was interested as was I in the
061r127 n 1816 (?).-- Mr Owen's curious fact about Crust BRA in Brine Springs. (Henslow) Speculate on neutral
472s01r ome years since gathered the seeds of Papaver BRACTEATUM, & the Papaver oncitate was growing in same
472s01r flower & foliage not «being» like the true P. BRACTEATUM; all supposed to have been hybridised == Has
275c121 red cattle of L'Darnleys were most like parent BRAHMIN bulls-- Mr. Y. is inclined to think that the m
165g126 t remarkable ever obseved? Shows that <nervous> BRAIN makes thought Glen Roy B C. Darwin All useful p
227b226 ok for intermediate structures <between> say in BRAIN. between lowest Mammal & Reptile. (or between e
248c033 refore most permanent Owen [re]markable laws of BRAIN & manner of generation «& primary divisions of
263c076 to <per> force in one man the developement of a BRAIN capable of producing more glowing imagining or
291c166 fficult to imagine it anything but structure of BRAIN heredetary,. analogy points out to this.-- love
291c166 meaning.-- Why is thought. being a secretion of BRAIN, more wonderful than gravity a property of matt
293c172 cases recall the idea. «or simple structure in BRAIN people in fevers recollecting things utterly fo
293c172 been past-- so great an anomaly in structure of BRAIN not probable) put note. Sir W. Scott has writte
293c173 learnt, & if that was a regular contingency the BRAIN would become webfooted & there would be no act
305c211 imately allied to one kind of organic matter.-- BRAIN. & which <prin> thinking principle. seems to be
311c226 fication» as indication of structure (including BRAIN & other organs difficult to analyse) will not t

360d094 similar.-- this however is a sophism for their BRAIN or stomach would be different.-- Or if one spec
380d154 interest; for it shows latent instincts even in BRAIN of male.-- Every animal surely is hermaphrodite
463t055 sages-- yet in talking, constantly said as the <BRAIN> spinal marrow expands, so do the bones <are cr
463t055 nes <are created> expand-- instead of saying as BRAIN is created &c &c Bats are a great difficulty no
523m018 ing) in others from drinking cold drink.-- then BRAIN affected like getting suddenly into passion.--
524m020 tions of the thinking organs -- the action of BRAIN which gives sensation of pain, emits its power
530m046 & association is probably a physical effect of BRAIN the «similar remark» thoughts, being functions
530m046 mark» thoughts, being functions of same part of BRAIN, or the tendency to habit of producing a train
531m053 way from any sudden sound or noise, & therefore BRAIN has been accustomed to send a mandate to the mu
533m057 of talent, which are heredetary are so because BRAIN of child resemble, parent stock.-- (& phrenolog
533m057 le, parent stock.-- (& phrenologists state that BRAIN alters) It is known that birds learn to sing &
534m061 conscience, or instinct. Hensleigh says to say. BRAIN per se thinks is nonsense; yet who will venture
534m062 memory. is repetition of whatever takes place in BRAIN. when sensation is perceived.= = Aug. 7th--38.
538m080 Babington's silly joking The possibility of the BRAIN having whole train of thoughts, feeling & perce
538m081 ction. &c will arise from physical action on the BRAIN, renders much less wondetful the instincts of a
543m101 w instinct can originate.-- strong argument for BRAIN bringing thought, & not merely instinct; a sepa
552m131 k, a tired man. involuntarily feels angry, when BRAIN is pumping force to legs & body, & especially,
574m042 ble An habitual action must some way affect the BRAIN in a manner which can he transmitted.-- this is
574m043 e limits of these two actions either on form or BRAIN very hard to define.-- Consider the acquirement
575m043 it the provision for death.-- can we deny that BRAIN would be intermediate like rest of body? Can we
575m043 ke rest of body? Can we deny relation of mind & BRAIN. «Do we deny the mind of a greyhound & spaniel.
575m043 for mice from some peculiarity of structure of BRAIN.?-- is this more wonderful than memory. affecte
575m045 gy between grief & pain-- certain ideas hurting BRAIN, like a wound hurts body-- tears flow from both
587m089 eory of Association. owing to time when entered BRAIN, try contiguity of parts of Brain.-- Mackintosh
587m089 when entered brain, try contiguity of parts of BRAIN.-- Mackintosh first clearly insisted on assoc o
589m091 M.S. notes, where strong argument in favour of BRAIN forming the instincts,-- could brain make a tun
589m091 favour of brain forming the instincts,-- could BRAIN make a tune on the pianoforte, yes if every ind
589m091 dual played a little, & something destroyed bad BRAIN. see p. 90.-- The relation of reason to organs
613o036 , to show that they result from organization of BRAIN; «[LHC] not used by Kirby» (:analogy:-- as race
613o036 ification of external form. so modifications of BRAIN) As in animals no prejudices about souls, we to
613o036 ing peccari-- --retriever-- produced as soon as BRAIN developed, and as I have said, no soul superadd
615o038 Heathen race.-- But as no great modification in BRAIN would probably take place without corresponding
616o039 nce if thought &c bore the same relation to the BRAIN that attraction does to matter, it might with e
616o039 ht with equal propriety be said that the living BRAIN perceived, thought, remembered &c. Well the hea
616o39v mory. might be in like manner attributed to the BRAIN. There are two modes of perceptive action by wh
618o041 spect. If thought bore the same relation to the BRAIN that force does to the bodily frame, they could
618o041 could be perceived by the faculty by which the BRAIN is perceived but they are known by courses of a
618o041 ought & yet be ignorant of the existence of the BRAIN. We cannot perceive the thought attraction of s
618o041 r. Thought is only known subjectively? -- ? the BRAIN only objectively. We do not know attraction obj
638j58v of bee, corresponding to <every> «one or any»-- BRAIN making structure, instead of parts of body.-- N
638j059 se schemes.-- I see no reason, why structure of BRAIN should not be born. with tendency to make anima
572m033 ning powers than Europaean.-- Ideots. defective BRAINS.-- Erasmus does not liken term instinct to mus
575m043 nd of a greyhound & spaniel. differs from their BRAINS> then can we deny that the grand child dug for
036r050 onnected with movement of sand.--it is called "BRAMIDOR'(?).--it was a strange story; I believe it wa
176b024 re will be certainly points of affinity in each BRANCH A species as soon as once formed by separation
179b035 mals in one country owing to springing from one BRANCH, & the monucle has definite life, then all cle
281c139 parated from any degree; the tailors-- «in each BRANCH would be analogous to each other-- &c &c.-- V.
285c151 &c will probably upset it-- The space which one BRANCH of the tree if live occupied after its decay,
285c151 ll be occupied by the vigorous shoots from each BRANCH No: because decay in that species <shows> is e
304o208 ed much time elapsed & therefore descended from BRANCH high up.-- Such probabilities only guides.-- Y
304o208 an in all other birds, or that it sprung from a BRANCH high up.-- this argument not applicable to apt
372d129 qualities by being buds-- , more than if whole BRANCH transplanted? +.simplest forms of budding. Why
429e116 yrenees but are found no where else not even in BRANCH valleys-- M. Ramond offers no explanation.-- P
491q01v side near other tree being affected.-- does one BRANCH of Cabbage being mongrelized affect other bran
495q005 e will result hybrids-- (6) Dust flowers of one BRANCH of Cabbage with pollen of other, count seeds,
499q009 s such a grove of Willows.-- (14) Bowman female BRANCH At What distances from males, will female (a)
503q013 ?? Maer (1) Yew Trees near Boat House «ANY male BRANCH.» --¿number of seeds in beginning of November
504q013 flower is open-- -- No (6) There is apple with BRANCH in middle of tree with flowers near end of orc
504q013 flowers near end of orchard.= At Shrewsbury one BRANCH of Rhod. flowered later.-- effect of accident?
546m111 assignable cause, remembered she had put it in BRANCH of tree, & apologising to party, went out & fo
176b021 organized beings represent a tree. irregularly BRANCHED some branches far more branched.-- Hence Gene
176b021 e. irregularly branched some branches far more BRANCHED.-- Hence Genera.-- «as many terminal buds dyi
304c207 escends from same parent with other birds,, or BRANCHED off anteriorly think what principles are ther
357d072 observing way the Marsupials of Australia have BRANCHED out into orders one is strongly tempted to be
135a145 bul. parallel ranges. with here & there little BRANCHES at {P} from each side intercepting plain & di
157g077 river Right Hand Cascade has <cut> «where two BRANCHES unite in upper Glen Roy» very little back fro
160g089 king <several> two divortiums aquarum, viz two BRANCHES of River Bought & between one of these & Glen
175b020 ffsprings of some still older type some of the BRANCHES dying out.-- with this tendency to change, (&
176b021 gs represent a tree. irregularly branched some BRANCHES far more branched.-- Hence Genera.-- «as many
177b025 d perhaps be called the coral of life, base of BRANCHES dead; so that passages cannot be seen.-- this
177b027 e of species that in perfection, the bottom of BRANCHES deaden.-- so that in Mammalia «birds» it woul
177b029 ies from many» <therefore> changes and base of BRANCHES being dead from which they bifurcated.-- Type
317c249 of reed & trees p. 259 120 ft in length, some BRANCHES of Justicia still growing,) passed us. do. p.
337d021 ged together with some training in the earlier BRANCHES «as in common greyhound» & much intermarriage
491q01v anch of Cabbage being mongrelized affect other BRANCHES-- The French Apple tree «with abortive stamen
051c093 rock is buoyed by Kelp, now Kelp sends forth BRANCHING roots which must protect surface; On «hard» e
102a060 arranged in planes. case of separation.-- the BRANCHING cracks-- only bear relations to VEINS in prim
175b019 is this Ehrenberg? every successive animal is BRANCHING upwards different types of organization impro
176b023 fe of Mammalia.-- Would there not be a triple BRANCHING in the tree of life owing to three elements a
177b028 o favourable points of organization commenced BRANCHING.-- As all the species of some genera have die
198b133 thinks developent in quite straight line, or BRANCHING S. H What does the expression mean used by Cu
200b121 & alk. do not show the possibility of common BRANCHING off.? Accra, Coast of Africa. Clay slate. str
375d135 ood used for other purposes as wheat for making BRANDY.--» take Europe on an average, every species m
219b193 ng & coots-- «(Turdus Guyanensis?) (Emberiza BRASILIENSIS?) (Fulica Chloropus)--» might bring in stom
480z008 d'Acunha.-- (Turdus Guyanensis?) Emberiza BRASILIENSIS?)) Fulica Chloropus. says some of the «sp
124a119 tory force-- Erasmus says he has seen in making BRASS a piece of copper not melted absorb, zinc thrug
437e141 ught back to home. (& yet all the varieties of BRASSICA certainly not becoming Brussels Sprouts) is a
362d099 woman (as Mitchell remarks seen in savages) to BRAVE men.-- Effect of castration horns drop off., re
096a041 dward in allusion to St. Helena discussion. Mr BRAYLEY says he can give me facts respecting lime <n>
024r015 athoms 20 miles from the shore? Beagle Coast of BRAZIL? where not rivers in my Coral paper {T} league
025r016 10 or 12 leagues sea deepens suddenly. coast of BRAZIL generally.-- Mrs Power at Port Louis talked of
029r033 ydney no I believe the secondary? formations of BRAZIL, all originate from the decomposition of Grani
040r063 ituminous shales have been metamorphised, as in BRAZIL feruginous sandy ones have undergone the same
050r091 Roussin states that generally in North part of BRAZIL. <gravel becomes> sand less & gravel more comm
050r091 a sudden deepening on E. coast of Africa. as at BRAZIL [blank] What is nature of strip of Mountain Li
052r098 Falkland soundings introduce this discussion.--BRAZIL bank: (& I believe SE coast of Madagascar. whe
063r131 eath water.--together with hypothetical case of BRAZIL.-- Propagation. whether ordinary. hermaphrodit
081r181 eep soundings Gilbert Farquhar Mathison travels BRAZIL. Peru. Sandwich Isd Mawes travels down the Bra
081r181 azil. Peru. Sandwich Isd Mawes travels down the BRAZIL.-- Did Melaspena publish his travels? Bellingh
097a042 pose each other.-- on Direction of mountains in BRAZIL L.'Institut No degree 221 Lamellar dikes like
128a130 d & shallow lagoon.-- when describing Coast of. BRAZIL. Maldonado enter into this case.-- Ed. New. Ph
138a151 account of Sellow Geolog. Observat. in Southern BRAZIL. [blank] «p. 4. (Lyells Book)» Observaciones s
217b187 and and Tierra del Fuego.-- Araucaria, species. BRAZIL {P} Chile, Norfolk Isl.-- Isle of Pines-- Aust
370d117 plants in all parts of world, thus tea tree in BRAZIL must have degenerated. as must spices &c &c Th
464tf6r nidae than any other forms-- Lund's Antilope in BRAZIL another point of agreement with. N. America &
486z020 Martius talk of birds singing in the forests of BRAZIL H. Wedgwood says in <14th> «13th.» Vol of Arch
540m086 le land of Peru the hot plains of the Amazons & BRAZIL-- with the negros of Africa, (or again the bla
540m086 oured natives of New Zealand)-- the American in BRAZIL is under same conditions as Negro on the other

Page   **(Key Word)**
```
291c168 if  South America grew very much hotter, then BRAZILIAN species would migrate south ward being  ready
473s006 ays Miers has described in Linn: Transacts. a BRAZILIAN plant «in Lin. Transacts. it has three stamen
485z017 merica. than on East.-- not being replaced by BRAZILIAN Species.-- Mem Turdus Magellanicus.-- C, <Chi
493q003 » = Length of intestine in Persian Cat-- , in BRAZILIAN «toothless» dog-- I. St. Hilaire says  length
493q004 3) Cross the Esquimaux dog. with the hairless BRAZILIAN or Persian animals of different heredetary co
554m139 escent 1838 It was very curious to see her take BREAD from a visitor, & before eating «everytime», lo
047r083 ove as). -- In great Calabrian wave did not sea BREAD first? I can imagine from local form of coast (
056r141 in Obstruction Sound, in the narrow parts which BREAK through the N & South lines the tides form eddi
348d050 be  the oldest) ["The most important characters BREAK down in certain species & become worthless-- Ma
427e109 here & there only: as stated by Capt. Graahy) & BREAK up. N. American Conchology from Europaean. & t
442e151 «My theory will not admit. this, now that tulips BREAK by cultivation can a form become permanent?» be
446e162 are cultivated during several years & then they BREAK-- -- each tulip is the <of> product of fresh bu
496q006 aisy. Fever-fuge Groundsil.-- gilly flower will BREAK & become double.-- There is a double Crows-foot
629o55v prepuce, crepitando,]CD & if passions makes one BREAK these artifical rules, get remorse-- ((hence de
633j53v ons some hybrid, whose flower great tendency to BREAK off p. 292. Mac. has long rigmarole about plant
637j58r of adaptations Horny point to chickens beak, to BREAK egg. shells-- why chicken could not have  lived
102a059 lf, undertow would draw it outwards.-- form of BREAKER affected some way out to sea.--¿ effects on bo
051r094 nderful to see Coral reef--or confervae in the BREAKERS or in waterfall: Excepting by removal of larg
231b241 dge it amount of varying in wild state.-- When BREAKING up the primeval.» continent. Indian Rhinocer
280c134 ght excellent authority because written on dog-BREAKING.-- applies it to national character.-- N.B. I
466t095 efore sex developed-- Double flowers & colours BREAKING only hereditary characters, wh. come on in af
047r081 e? In watching heavy swell, sea retreats & then BREAKS: i e to form a wave in ocean. is not this [Fig
192b084 oreign country.-- When one sees nipple on man's BREAST. one does not say some use, but <no.> sex  not
216b181 des.-- Case of boy with foetus developed in the BREAST.--looking as if many ova-- impregnated at once
251c041 & Sumatra. differ only in form of white mark on BREAST: p. 234.-- good case.-- p. 526. (ref) To Temmi
358d085 sant crossed, has a cock (infertile) <with> the BREAST of which is like common pheasant & back like s
360d092 te-- <now> «as» if one bird had very bright red BREAST & other very bright blue, it might be harder <
365d106 Ptarmigan (probably subalpina.) former has blue BREAST, latter reddish, hybrid purple-- be careful, S
383d160 lack marks on tail much <blacker> «broader.-- » BREAST red like Common pheasant.-- lower part of brea
383d160 reast red like Common pheasant.-- lower part of BREAST, each feather is fine metallic green. <from> w
467t104 ed with pollen was pressed & rubbed along whole BREAST-- {b} pressing either one or both of Pea's win
556m146 s-- Cats kneeding when old, like kittens at the BREAST now if horns were to grow on horses, they must
567n014 .-- Turkey cock in passion & sends blood to its BREAST &c &c All Science is reason acting «systematiz
264c080 as my Furnarii.-- some genus of yellow & brown-BREASTED bird in Australia &c &c-- but of course  they
341d031 r are most closely represented.-- Thus the red BREASTED thrush is represented by one not differing ex
195b099 -- or it may be of use.-- like Mammae on mens' BREASTS.-- How does it come wandering birds. such sand
319c257 e Oural mountains) have black nipples to their BREASTS.-- L' Institut, 1838, p. 230 says the Macrothe
360d093 iarities; that if <one had a> both had mottled BREASTS, <when> of a sort that would allow the offsprd
536m071 strictly  analogous to men's affect for womens BREASTS.·. Dr Darwin's theory probably wrong, otherwise
195b098 gical character of these islands so permanent a BREATH cannot reside in space before island existed.-
552m131 probably is connected with habitual stopping of BREATH to hear any sound.-- attitude of attention «So
035r048 equable effects.--though so immense to short BREATHED traveller» Mountains, which in size are grain
341d030 vicle scapula &c strongly developed to aid in BREATHING.-- Animals from Hobart Town mentioned, it see
586n081 helps  this confusion.--} Hensleigh considers BREATHING instinctive, certainly heart beating may be c
050r093 ales. was it reef. -- I remember many Corals?? BRECCIA--Stratification? Anomalous action of ocean.--a
056r110 than outer parts.-- The common occurrence of a BRECCIA of primitive rocks between that formation  and
076r170 quartz & mica very rare.-- ancient freestone & BRECCIA is the same with that on surface of plains of
077r170 s of Amazon, no relation--there is more modern BRECCIA, chiefly owing to destruction of porphyries. w
104a069 Orientel.  ask Lyell for sentence.-- Origin of BRECCIA, introduce in Cordillera discussion, deep sea,
278c131 seals and dugong, therefore immense age since BRECCIA accumulated-- surely ask Owen to see whether s
032r042 y of world.-- I feel no doubt. respecting the BRECCIATED white stone of Chiloe, after having examined
059r121 f Alum: matter.--all pale cream colour.-- The BRECCIATED structure of all the Pitchstone (which I hav
063r133 Horses. Cattle. Goat. Asses. have all run wild & BRED. no doubt with perfect success.--showing non Cr
173b012 e can see why a form peculiar to continents; all BRED in from one parent why. Myothera several specie
179b032 effect;  as when bitches puppies are less purely BRED-- might not new classes be brought into play.--
192b089 Heterodox as ornithorhyncus. If this last animal BRED-- readily with common ducks.--» Kirby all through
204b141 e widgeon, winged & turned on pool, first season BRED into each other, the females loose desire,  and
215b176 ct, Very curious case = W. D. Fox. When dogs are BRED in & no new ones introduced would not change be
235b274 -- ¿To be solved if horses sent to India. & long BRED in-- Mem, <an> a statement in Mr Wynne's  book,
271c106 y Man.-- effect of external contingencies & long BRED <greyhound>. bull-dog. & crossed & recrossed, t
275c120 f chase would not run up hill-- he took thorough BRED cattle of L'Darnleys were most like parent Brah
275c121 all imagination-- he says he recollects all half BRED or two certain varieties, they would go back to
280c136 sible-- (Hence we might expect even if two mules BRED.-- (hence hybrids in this case have bred). Whit
290c164 most like pheasant., I think so because very 3/4 BRED). White & common pheasants. have crossed.-- </I
290c164 ery 3/4 bred.-- (hence hybrids in this case have BRED one was diseased in its loins & all were so aft
294c175 he country they inhabit.--» & the first one that BRED in & in.-- which looks as if qualities were not
312c231 t offspring thus produced are better, than those BRED Alderney Cows take more after Alderney that the
332d003 on one day.-- Mark at Shrewsbury thinks the half BRED with unknown common house cat.-- had four Kitte
333d005 grels Hybridisim Fox has half Persian cat. which BRED Persian.-- Here then we have clear case of hete
333d006 ly resembled in form of tail, fur &c to the half BRED animals.-- Mem. number of Mules.-- «He recollec
334d009 this  first cross were equally fertile with pure BRED Persians «cat» favour the Persian side.-- Theor
335d015 monsters,  abortive, just like mules. Fox's half BRED, & surely wild Duck & «pintail» Widgeon!-- Divi
339d025 ecies of grouse»! Has not Goldfinch & Greenfinch BRED (just like common mules) & lay many eggs but ne
342d032 hey are produced in full equal Numbers with pure BRED out & now all are pure red, yet calf every  now
345d043 ussex cattle were all white headed, but this was BRED inter se.-- These hybrids are very wild & take
359d087 lly infertile whereas animals further apart have BRED first, «then males--» how has this been arrange
370d116 doubts about production of Queens.-- Neuters are BRED before the vis formativa had completed them-- (
380d153 bug,  as a <female which have> larvae which have BRED in & in» never lose passion. (Mem: so it was sa
385d163 e offspring in latter being unhealthy.--» males «BRED into each other.-- This is somehow connected (T
389d176 ed from annual plant being prolonged till it has BRED.-- Offspring like both father & mother, or very
389d177 near relations, & therefore those very close are BRED into each other.-- This is somehow connected (T
389d177 ems case, for by careful observing cattle can be BRED in & in.)-- [The loss of passion in hybrids. pe
390d179 on «of difference» with other sex. = The highest BRED Blood-hound. would be infertile with highest br
390d179 red Blood-hound. would be infertile with highest BRED of other ¿ breed.= Therefore it is not really b
393dIBC smitted by generation?? Is it «chiefly» in high. BRED dogs ie. (bred in & in) that one copulation wit
393dIBC ration?? Is it «chiefly» in high. bred dogs ie. (BRED in & in) that one copulation with other dogs re
393dIBC Does male fail in passion.-- Disposition of half BRED Cattle at Cinbermere? How is Jackall & dog at Z
399e010 ere only pair has been introduced, & have freely BRED, they have not lost power of producing. William
451e178 mammifers in the world consistently reside & are BRED. "take tame animals into this region between Ap
460t013 h common goose produce full as many eggs as pure BRED common.-- the half of the cross, as above, take
460t013 erally» after the swan-gander. Has this been BRED ganders. crossed with common goose <to has> «pr
460t015 down  back pretty plain in in these <half> «3/4» BRED ones-- The brothers & sisters half-breed showed
492q003 stions regarding Breeding of Animals If two half BRED animals exactly alike be interbred will offspri
493q004 first breed or permanently.-- (2) Cross two half-BRED animals. which are exactly alike & see result.-
494q005 aterpillars (1) Shake a sleeping mimosa, or half BRED mimosa (a) between sensitive & sleeping species
508q016 o healthiness? & father answered (5) About cross-BRED races of men taking after sex. A Smith. About s
512q019 ng wild ones breed as well as,, as those already BRED in cages. Get direction write to-- (2) Does he
512q020 mals & birds breed (3) In cases where Lions have BRED, have they been raised from young ones, bred in
512q020 ave bred, have they been raised from young ones, BRED in captivity --Mr Miller says Wombwalls were (4
515q022 same  species Eyton (1) Number of eggs-- of half-BRED geese-- inter se, & with parents & of Chinese g
515q022 ntal horses. {About the leaping of Irish Horses, BRED in this country. {Chinese Dog's Head to send Co
148q025 arents» (& not like dogs), but they thought the BRED liable to vary-- I asked this question in  many
171b002 else is perfect; mothes apparently only born to BREED.-- annuals rendered perennial. &c &c.-- Yet Eun
183b051 use being believed same species) if they do not BREED readily. point in view.--¿whether highly domest
193b093 eredetary prejudices «or instinc» to conquer or BREED together:-- Man has no limits to desire, in pro
196b106 es & died off in some? Why did not fossil horse BREED in S. America-- it will not do to say period un
200b122 se most rare, or when placed together they will BREED.-- But what a character is this? Race permanent
203b136 261. L. Institut. 1837 Mem. Sir F. Darwin cross BREED boars were wilder than parents. which is same a
```

204b139 says that the mixture between Chinese & English BREED. decidedly exceedingly prolific & hybrid about
205b145 female p. 28 "It <is> wrong to enlarge a Native BREED of animals. for in <the> proportion to their in
211b163 like dogs.-- Ogleby says, Wolves at Hudson bay BREED with dogs.-- the bitches never being killed by
213b171 do not known varieties do the same, May you not BREED, ten thousand grey hounds & will they not be gr
214b175 t to Brit Assoc <to> on N. American Zoology-- A BREED of Blood Hounds from Aston Hall close to Birmin
214b175 Birmingham, and supposed to be descended from a BREED known to be there since the time of Charles.,--
222b203 ion of species.-- When species cross & «hybrid» BREED, their offspring show tendency to return to one
231b244 of small creations.-- Will Dromedaries & Camels BREED?-- As man has not had time to form good species
235b275 e races of plants run wild or nearly so, which <BREED> do not intermix,--any cultivated plants produc
241c014 ands & at Amboina» I fancy there is marked wild BREED of oxen at Java.--.p. 140, calls it Bos. leucop
248c030 rieties of many ages standing, will not readily BREED together: The argument must thus be taken, as «
248c030 as with plants) no animals VERY different will BREED together, so when we grant «(which can be shown
248c030 there will be presumption that they would not. BREED together.-- We see even in domesticated varieti
263c074 era & species-- The believing that monkey would BREED (if mankind destroyed) some intellectual being
266c092 tle generally marked like those of the Alderney BREED, but size not larger than those of Black cattle
267c094 secured one, as they always like to cross their BREED p. 333-- alludes to the Macúsie breed no descri
267c094 oss their breed p. 333-- alludes to the Macúsie BREED no description given-- Ch. 2. dogs L'Institut.
271c106 tatement in Mr Wynne's book, about not altering BREED of animals in certain countries.-- Fraser remar
275c120 arieties. unnatural circumstance Ld Orfords had BREED of greyhounds fleestest in England lost courage
278c129 know, whether nearest species of each might not BREED:-- Genus must be a true cleft-- putting out of
283c145 inct & (even though <fertile> when compelled to BREED hybrids produced--)» & then all that I want is
285c152 gether, they would not according to all analogy BREED together.-- The bottom of the tree of life is u
288c159 ucing, instead of breeding from original Durham BREED.-- Native dogs & English cross readily-- think
293c175 greatest difficulty to whole theory.-- There is BREED of tailless cats near Bath. Lonsdale do. says S
296c185 abstract true are wonderfully absurd.-- p. 565 <BREED> Scotch wild Cattle. breed freely with the tame
296c185 ly absurd.-- p. 565 <breed> Scotch wild Cattle. BREED freely with the tame Vol II. Magazine of Zoolog
298c189 t it was most like fox.-- He felt sure the half BREED of Australian dogs, would be «most» like Austra
312c231 cross between a good bull & <be> the provincial BREED, & that first offspring thus produced are bette
331dIFC r.?-- Mem. for Eyton.-- Sir. R. Heron's case of BREED of pigs with solid feet.-- 1838 [In this Book s
332d003 hich they have been crossed--is Alderney oldest BREED-- He believes all pretty much alike.-- My Fathe
334d009 says a cousin «one of Mr Strutt» of his used to BREED to Common & Muscovy Ducks.-- English. <Common>
334d012 Park «or in the Duke of Marlborough» there is a BREED of white-tailed squirrels, which form a marked
335d013 pply to first parents, because they are not new BREED.-- the first hybrids may be compared to animal
335d015 e.-- Theory of abortive hybrids.-- If mules did BREED, the offspring would «as in all other animals»
337d020 hen the local ones are taken to fresh country & BREED confined. to certain best individuals.-- scarce
337d020 ed. to certain best individuals.-- scarcely any BREED but what some individuals are picked out.-- in
337d020 dividuals are picked out.-- in a really natural BREED, not one is picked out, & few even of local var
340d026 ard at Zoolog Soc their Pintail & Common Ducks, BREED one with another-- & hybrids fertile inter se--
342d033 e varying element). Then do those SPECIES which BREED most freely. & produce somewhat fertile offspri
344d041 d to Brighton Park--first rate bitch-- tried to BREED from her, but her offspring came out one big &
344d042 ∴the grandchildren went back to either paret, & BREED not fixed. though she resembled a harrier & her
345d042 gone on for 50 or 70? years-- now «well fixed» BREED.; Jones says Sussex cattle were all white heade
345d043 ce general. ¿depends upon mother bein[g] oldest BREED?.---- Quarterly Journal of Agriculture p. 367.
345d044 iar hair & appearance-- good case of Provincial BREED-- Highland Sheep jet black legs, & face & tail,
346d047 es.-- high active breedin[g] [not located] half BREED liable to vary. I asked this in many ways, but
346d048 ders & subject to the rush all animals which BREED, in & in are-- colour white, uniform.--crafty,
356d071 ude so many facts as mine. The facts about half BREED animals being wilder than parents is very curio
359d087 r strange misshaped bird--"looks very artificial BREED-- but Mr Miller says that breeds larger numbers
362d100 - Even our domesticated cattle have tendency to BREED at particular times. Mr Yarrell has old book 17
364d104 ame story about some Southern «see p. 43 supra» BREED of cattle with white heads; which years afterwa
366d112 e chance of two being born at same time, & make BREED, one would doubt any law.-- Yet seeing the feat
370d116 rly are made for larvae.-- CD[(p. 451.)- Wasps BREED many females, but almost all die.-- bees breed
370d116 breed many females, but almost all die.-- bees BREED but few, because they are kept in security.-- H
379d152 own that the ewe must never be put to any other BREED else all the lambs will deteriorate.-- Lord Mor
385d163 big & then second small & so.-- Says, there is BREED of Fowls called everlasting layer--. or Polish
385d163 of Fowls called everlasting layer--. or Polish BREED. (he thinks half pheasant, half fowls.-- eggs f
387d169 -- Now if two of third pair of same peculiarity BREED they will have same influence as first pair + t
390d179 ot merely the too close animals, which will not BREED, but the female at least (¿male?) looses all ap
390d179 would be infertile with highest bred of other ¿ BREED.= Therefore it is not really breeding in & in,
390d179 at have neither varied from their stock, for to BREED (as Sir J. Sebright urges?) one with opposed ch
390d179 one with opposed characters is by impliance to BREED two which have each varied from parent stock.--
393dIBC hey cannot tell first result., or that «hybrid» BREED is uncertain Is there any peculiarity or variat
402e018 in Borneo-- <brought probably by Chinese», "the BREED being of the latter being the same as the fox-l
406e035 y modification of it-- Goat & Moufflon will not BREED.-- p. do.-- Fish of Teneriffe. St. Helena & Asc
414e063 much.-- (yet one cross, & the permanence of his BREED is destroyed)-- -- When two races of men meet,
417e076 eft whorled.-- He kept two to see if they would BREED, It is difficult to think of ¿¿Plato & Socrates
428e112 ns,, as in horses, would not care so much about BREED.-- what can «however» be more striking, about i
430e118 produced, but by training, & crossing & keeping BREED pure.-- «& so in plants effectually the offspri
432e124 said.-- p. 12. Attempts to improve the native <BREED> animals of any country must be made with great
450e176 anao.-- Q Horse do. Appendix. p. 43. «& 45» the BREED of elephants <of> in little isld of Sooloo.-- s
459t013 ing resembled the drake.-- another of same half BREED resembled the plumage of drake still more.-- So
460t015 > «3/4» bred ones-- The brothers & sisters half BREED showed no sexual inclination for each other-- A
482z017 pecies of Mephites 4 distinct Camelidae. do not BREED together Mag: of Zoolog & B. Vol. II. p. 127. L
490q001 recent ones not embedded?-- Do the Tame Parrots BREED amongst the Indians Do the Savages select their
493q004 s some artificial male with <old> female of old BREED & see result.-- According to Mr Walker the form
493q004 to Mr Yarrell the latter ought: either in first BREED or permanently.-- (2) Cross two half-bred anima
494q004) Keep. Tumbling pigeons. cross them with other BREED.-- (12) About the blended instincts Remote Expe
500q010 ties of Cabbage. (26) Do deer Keepers cross the BREED-- desirable as in Cattle in Chillingham Park--
505q014 y? (15) Abberley has a hooked Pea.-- intends to BREED from it and large Asparagus: result? = failed t
511q018 ppear without his wish Has since recrossed this BREED.-- Have secondary male characters appeared.= (4
511q018 ng them with common pigs=[it is a Lincolnshire BREED]CD-- Sir. R. H. supposes is now extinct= (9) Ab
512q019 aises many English birds-- will young wild ones BREED as well as,, as those already bred in cages. Ge
512q020 Bears /Yes/ (2) Foxes & English animals & birds BREED (3) In cases where Lions have bred, have they b
525m025 g is affected by previous marriages with impure BREED.-- A cat had its tail cut off at Shrewsbury & i
615o36v individual force in any individual.-- His Malay BREED «of fowl» totally different habits from Europae
639j28v thusian rush for life, only two of them live to BREED, if circumstances determine that, the long legg
512q019 Azores or Madeira Mr. Blyth (1) Mentions some BREEDER who raises many English birds-- will young wil
346d048 wounded,-- see Annals. vol. 2. 1839.-- are bad BREEDERS & subject to the rush as all animals which br
360d089 w. chief trust must be in general knowledge of BREEDERS, where their interest is concerned. Same man
428e112 rue, as enuntiated by him to me, for otherwise BREEDERS who only care for first generations,, as in h
489qIFC Tollett {T} Zoolog Soc «Gardens» -----. . 20 & BREEDERS Dr. Boott: R. Brown p. 21 Horticulturists p.
180b038 rmanent varieties produced by <inter> confined BREEDING & changing circumstances are continued & prod
186b057 Does Waterhouse's representatives agrees with BREEDING.. in irregular trees. & extinction of forms.?
206b146 ds to one progenitor, who might have continued BREEDING from eternity «backwards.--» If population wa
220b197 her is evidently an instinct-- & this prevents BREEDING. now domestication depends on perversion of i
230b239 uments imperfect) or horizontally & then cross BREEDING prevents perfect change.-- It is scarcely pos
231b241 do not known what amount of difference prevents BREEDING;--or as others would exjudge it amount of var
240c004 ther adaptation.-- Mr Yarrell says, that after BREEDING in pidgeons with very much care that, it requ
280c135 nt anomaly in nature. <t>---- Many animals not BREEDING at all in domestication. throws great difficu
288c159 is evident to be worth introducing, instead of BREEDING from original Durham breed.-- Native dogs & E
293c174 ns can hardly be accounted for by My method of BREEDING there must be some corelation. but. the whole
312c233 litters or of father & child are thought long BREEDING in. Must not trust him Hope says that genus o
323c269 A. Wells. Lecture on instinct -- Cline on the BREEDING of Animals -- Spallanzani's Essays on Animal
338d023 perhaps so great, as separation on <be> inter-BREEDING, for otherwise we could not understand the va
338d025 to idea of only animals from distant countries BREEDING!. <Mem 3 species of grouse»! Has not Goldfinc
344d010 this in some larvae of insects-- (¿glowworms) BREEDING-- <beet> imago state fertile at once.-- Consi
345d044 in Britain shows, with the aid of seclusion in BREEDING. how easy races or varieties are made.-- The
345d044 face & tail, just like species.-- high active BREEDIN[G] [not located] half breed liable to vary. I a

```
346d048 against all rules their preserving character & BREEDING in & in-- Nonsense a flock of more than 100.-
358d085 freely.-- here we have beautiful proof of the BREEDING in & in (like «courage in dogs» EFFEMINATE me
358d086 s like a Male.-- Thus castration, hybridity, & BREEDING in & in tend to produce same effects.-- CD[Ma
359d086 roduce same effects.-- CD[May it be said, that BREEDING in & in tends to produce unhealthiness,-- «or
359d088 ood species, or local variety, & not effect of BREEDING in & in like our pidgeons) The male of  every
361d096 of some water birds, (as Phalarope) assume for BREEDING a more brilliant plumage than male.-- «My cas
362d099 rell supposes this a consequence of the female BREEDING all the year round. ask Colonel Sykes.-- Even
366d108 tile inter se., or at least show repugnance to BREEDING if instincts unchanged, & if their characteri
379d152 ation Mr Yarrell says it is well known that in BREEDING very pure South Down that the ewe must  never
380d153 Seep p. 84. Hens «like»-- Cocks from effect of BREEDING in & in.-- Mr Yarrell does not know of any ca
388d172 . Animals in domestication (even Elephant) not BREEDING-- remarkable Athenaeum 1838. p 653. Ehrenberg
389d173 t she was held Hunters Eoeconomy So with inter-BREEDING as told by Willis Q Proved facts relating  to
389d177 happen with hybrids?]CD Plants must stand much BREEDING in & in (those which have solitary flower) ex
390d177 as cryptogamia & hydras,-- (this repugnance to BREEDING in & in seems connected with more developed f
390d178 rom what it received» (for it is probable that BREEDING in & in would not be deletereous if the relat
390d179 ful Knowledge Bell's Quadrupeds the effects of BREEDING in, it is not merely the too close animals, w
390d179 of  other ¿ breed.= Therefore it is not really BREEDING in & in, but « <on> » breeding animals that h
390d179 t is not really breeding in & in, but « <on> » BREEDING animals that have neither varied from their s
399e010 in  any rabbits One may strongly suspect, that BREEDING in & in, produces bad effects solely, because
431e123 rs by action of wind difficult.-- Cline on the BREEDING of Animals, p. 8. size of foetus in proportio
492q003 gorze  common in Norway No Questions regarding BREEDING of Animals If two half bred animals exactly a
494q004 ss common Fowl with Dorking (10) Statistics of BREEDING in Zoolog. Gardens-- with respect to conditio
503q012 India?-- about Chetah & other tame animals not BREEDING when tame in India?-- does not know About Yak
508q016 er?-- 3 Andrew Smith, about tamed wild animals BREEDING at the Cape.-- ⟨About two vars: of Lion: Anna
510q018 es law of Balancement Wm Yarrell (1) About non-BREEDING of animals in confinement, curious.-- foxes--
511q018 skin-- Van. Voorst often writes to Lowe (7) In BREEDING. pointers. Bull-Dogs. Spaniels-- Grey-hounds-
563n001 sant claps his wings before? crowing & only in BREEDING season & on the ground.-- Cock fowl. on the g
584n071 ,-- subsidiary to food & temperature molting & BREEDING instincts, sexual, social, «subordinate to,»
640j28v way of eliminating the evils of old age, after BREEDING season, or gaining adaptations, but for youth
640j167 assert, that they are not species, until their BREEDING together has been tried.-- With respect to th
356d070 same fruit trees is analogous to some hybrids BREEDINGS-- there is tendency to reproduce in each case
145g002 like Magellanic fox.. an instance of Provincial BREEDS. [3] Veins of Segregation in Salisbury  Craigs
181b039 ppose only each species in each generation only BREEDS; like individuals in a country not rapidly inc
205b145 rnless.--" does this apply to where same animal BREEDS often with same female p. 28 "It <is> wrong to
217b183 st probably have occurred to every one) of rare BREEDS of dogs, from owners great care of them. Fox s
217b183 are of them. Fox says when two dogs of opposite BREEDS are crossed, sometimes offspring quite interme
239cIFC roduced, by picking offspring? Instances of old BREEDS taking greatest effect Account of the [...] th
279c133 organ (.see marks on pages).-- Crosses of diff: BREEDS succeed, yet seems to grant, that difficult  &
288c159 completed Major Mitchell, does not know whether BREEDS of oxen have deteriorated, or altered, but  it
342d032 of Muscovy) with goose!!) Dr. Bachman regularly BREEDS «in Carolina» for his table Muscovy & common d
344d042 d was pure Harrier.-- «The peculiarities of our BREEDS must have been acquired, & hence this is  then
345d044 ony with their wars & rivalry.--» The very many BREEDS of animals in Britain shows, with the aid of s
358d087 very artificial breed-- but Mr Miller says that BREEDS larger numbers, & rears an unusual number  out
362d099 ow.-- Mr Yarrell says the «male» Axis of India, BREEDS at times when horns not perfect-- (is not this
363d102 different-- Barn Owl <the> in the former place BREEDS in <flags» «thick vegetation» in swamps-- (owi
380d153 rtion of sexes, at birth & causes. If an animal BREEDS young her growth is immediately checked-- the
438e142 ent in change, as in Dahlias]CD all much varied BREEDS both plants & animals have long been subjected
439e145 ent Egyptians Vol III. p. 33-- They had several BREEDS of dogs.-- like greyhound-- fox-dog-- turnspit
450e175 Siam inferior to those of Pegu-- in Sumatra two BREEDS both small -- Java pony occasionally reaches 1
450e175 omewhat larger than the Sambawa, Java & Sumatra BREEDS, (.Hence it appears there are shades of differ
451e178 he Malaria & die.-- On the other hand there are BREEDS of Men the Thârû & the Dhangar who can live th
465t081 igration of other races, so it is with domestic BREEDS. (though in this case crossing has had somewha
493q004 ration, period of gestation differ in different BREEDS of dogs. Cattle, (Indian & Common) &c:  length
512q019 any treadèe?-- Difference in lambs of different BREEDS Is there any difference in breeds of Cattle  &
512q019 of  different breeds Is there any difference in BREEDS of Cattle & sheep in the sprouting of the horn
512q019 of the horns. at different periods in different BREEDS--?? or in individual case: subject to  disease
586n078 this instinctively «yes because power varies in BREEDS,» something of kind oneself knows in walking [
449e168 es & colour.-- Eyton has observed same thing in BRENT Goose. Eyton says some of the pidgeons in commo
228b232 jecture run wild then <our» animals our fellow BRETHREN in pain, disease death & suffering «& famine»
523m018 tween enthusiasm passion & madness.-- ira furor BREVIS est.-- My father quite believe my grand F doct
090a021 n to Guyana said to extend to Cordillera I see BREWSTER speculates from believing meteorolite but old
641j29v pecies-- yet this is contradicted by continents <BRI> abounding with species-- there will be a balanc
244c021 118. wild pigs of Falklands, generally "red of BRICK" hair, very stiff, p. 120-- Coati roux  common.
148g026 YOUNG» Shepherd dogs Saturday. Before coming to BRIDGE of Span, hills of «sea», gravel, current clea
148g027 bbish at head of Loch Dochart <Nea> Above Spean BRIDGE many flat terraces one above much inclined tow
148g029 the erosion may often be due to rivers-- By Roy BRIDGE, a tongue of flat land, with terraces of  each
150g035 aining off would form plains such as those near BRIDGE Roy (& other cases) but then if gradually drai
152g049 dge of which they cut near Loch Tring-- Tuesday BRIDGE of Roy Level of «bed of» River 30.221/65 degre
606o025 hen one wishes to do some action (as jump off a BRIDGE to save another) & yet dare not -- one could d
204b141 dily with common ducks.-- Kirby all through BRIDGEWATER errs greatly in thinking every animal born t
205b142 ocky Mountains Asiatic types discoverable.-- Kirby BRIDGEWATER Treatise There are some good accounts of pas
205b143 o vary where intermixture precluded.-- Kirby BRIDGEWATER Treatise p 85. Parasite of Negroes different
266c091 ith the Koran and the sword" ⟨quote Whewells BRIDGEWATER treatise, (p..26). about plants from Cape of
323c269 -- Mackintoshs' Ethical Philosophy -- Bell's BRIDGEWATER Treatise -- Wilkinsons Egyptian remains skim
637j57v here  see means-- but not in the other. ⟨All BRIDGEWATER Treatises. are reduced simply statement of p
443e157 n honestly reject such false reasoning Bell BRIDGEWATER'S Treatise on the Hand.-- p. 94.-- "The resem
160g092 ese mountains & not ridges) between arm of Glen BRIGHT flowing into E. end of L. Oich, & waters flowi
162g103 very long, flat divatium aquarium with, left of BRIGHT.-- like bed of lake with trace of terraces  on
306c214 gree with Java?? Terrestrial Planarise assuming BRIGHT colours., good instance of colours dependent o
360d092 to  propagate-- <now> <as» if one bird had very BRIGHT red breast & other very bright blue, it might
360d092 ne bird had very bright red breast & other very BRIGHT blue, it might be harder <to tr> for both pare
368d114 ose females which put on (like some waders) the BRIGHT plumage.-- «thinks» Hence specific character m
377d147 Kingfisher (.p. 169) has the colour on its back BRIGHT blue.-- «thus young of» Many of the pies assum
459t013 g; its colour was darker than the penguin & the BRIGHT feathers on its wing resembled the drake.-- an
494q005 -- Plants Raise seedlings surrounded by various BRIGHT colours, any effect? and silk caterpillars (1)
565n009 the expectant eye. looking to distant object, BRIGHTENED & moistened by emotion,-- why does emotion m
472s01v try  this year. Little variation in the 60 one BRIGHTER with mere traces of black spot at base, one p
344d041 ones gave an admirable harrier from Ireland to BRIGHTON Park--first rate bitch-- tried to breed  from
378d147 all the rollers-- «He says» whenever metallic BRILLIANCY is present in Young birds, one may be sure c
401e017 offspring. & male the lesser peculiarities.-- BRILLIANCY of inflorescence Gardeners. by chance <often
361d096 ds, (as Phalarope) assume for breeding a more BRILLIANT plumage than male.-- «My case of Caracara. N.
061r127 ?).-- Mr Owen's curious fact about Crust Bra in BRINE Springs. (Henslow) Speculate on neutral  ground
062r128 .--subterranean lakes, near Volcanoes. lakes of BRINE all inhabited: Go steadily through all the limi
138a155 solidation of strata-- he heated sand red hot & BRINE was boiling on the top-- [blank] Would  rotting
030r035 rom shore. V. Chart) Every winter torrents must BRING much vegetable matter from thickly wooded mount
088a014 > «Waterhouse» has frequently heard that Herons BRING eels alive to their nests; & then they may pick
104a068 vaporation therefore capillary attraction would BRING water with salt to surface Lyell remarked to me
219b193 riza Brasiliensis?) (Fulica Chloropus)--» might BRING in stomach-- &c &c. (Mem discover what kinds of
259c064 ies «although we see it affected» tempts one to BRING one back to distinct creations.-- it is only be
414e063 ecies of animals.-- they fight, eat each other, BRING diseases to each other &c, but then comes the m
437e139 rrying great weight. p. 125 is said that Eagles BRING rabbits & hares to the young ones to exercise t
524m022 What  fails first?-- How is this?-- Does memory BRING in old ideas <I have elsewhere remarked do> Dog
548m115 are for the pains of ones infancy.-- one cannot BRING it to one self-- nor of a bad dream, when that
564n005 self.-- the mind is function of body.-- we must BRING some stable foundation to argue from.-- Octob.
582n068 of  a number of prearranged actions, which will BRING about a certain result, while the creature perf
613o036 a gate before him, & then jump over the gate & BRING it. ---- Agrees with ONE animal [RHC] Kirby ext
090a024 serious? if so other cause besides thin vapour BRINGING planets to an end? Fragmentary granite showin
117a102 0. F were obtained 100 miles E of Staten land. BRINGING up pebbles 2 inches long?-- L'Institut.  1838
```

```
155g064 alance there must be in power of rivers either BRINGING more «detritus» than they corrode or vice ver
222b203 en what are called varieties.-- NB. one mother BRINGING forth young having very different   characters
263c075 s &c &c. this multiplication of little means & BRINGING the mind to grapple with great effect produce
334d012 sappeared, & three days afterwards came again, BRINGING with her the other & younger cow.-- {P} Fox s
525m023 y consider their duty.-- as carrying a basket, BRINGING back game, or picking up a stone, though only
539m082 much pleasure immediately thrilled across me, BRINGING up old indistinct ideas of FitzWilliam Musm.
543m101 nct can originate.-- strong argument for brain BRINGING thought, & not merely instinct, a separate th
524m019 is an argument for materialism. that cold water BRINGS on suddenly in head, a frame of mind, analogou
605o18v n excited by other means, as moral excellences, BRINGS to our recollection the original cause of thes
636j56v dner have hybrid seedlings} p. 333. Macculloch. BRINGS forward. the impregnation of Dioecious Plants
092a029 same   as Paris. 1500 ft high Mr Bird in paper to BRIT. Assoc: has shewn how electrical currents tend
214b174 se now living in the sea.-- See Rogers report to BRIT Assoc <to> on N. American Zoology-- A breed of
326c266 sitory. 4to. contains much on dogs.-- Reports of BRIT. Assoc.-- some important Papers. Dr. Mayo. Path
026r022 ntroduct XII P. silicified bones not common in BRITAIN. Mem Concepcion Says Echinites. Encrinites. As
096a040 e p. 406. List of Volcanos Salomon Isld.-- New BRITAIN-- &c &c In Ascension for centuries afterwards
145g002 nst it-- Mere fact of many races of Animals in BRITAIN shows that either races soon made or crosses d
196b107 , & yew very different from any found in great BRITAIN, British varieties are also found in Ireland--
242c017 w Isds.-- p. 22. New Calidonia-- New Ireland & BRITAIN same kind of dog, with those of New S. Wales.
338d025 tered by mistake Surely the fossil Mamalogy of BRITAIN & Europe is African. & the only difference is
345d044 rivalry.--» The very many breeds of animals in BRITAIN shows, with the aid of seclusion in breeding.
357d073 monly in Nature. both in Sweden & anciently in BRITAIN) between hen Caperailkie & cock Black-cock.--
473s03v of depth of organisms holding in America as in BRITAIN. If there has been «as» much subsidence as ele
326c265 of Spontaneous Hybrids. where? Sweet. Hortus BRITANNICUS. has remarks on acclimatizing of Plants. Her
036r050 of pebbles.--Plains. off coast of Patagonia.-- BRITISH channel &c &c. There is a Hill. near Copiapö w
163g108 e Mr H. C. Watson Geographical distribution of BRITISH Plants Shropshire Quartz what substance is col
178b031 been   there created there;--» Are not all our «BRITISH» Shrews diff: species from the continent look
196b107 ery different from any found in great Britain, BRITISH varieties are also found in Ireland-- There mu
234b256 . but few.-- CD [There was notice in report of BRITISH Association of 1838. (Newcastle) about somebod
317c248 to these.-- Also some few facts at end of "The BRITISH Aviary" or Bird Keepers Companion Study Append
321c275 ca Ray's Wisdom of God references at end-- The BRITISH Aviary.--do-- & Lisle's Husbandry Tuckeys voy
325c267 une 1838 H. C. Watson on Geograph. Distrib: of BRITISH Plants. Humes Essay on H. Understanding (somet
327c265 ticulture in Edinburgh. Encyclop.-- The <Edin> BRITISH & Foreign Medical Review No XIV. April 1839.--
500q10a ealand? Babington about differences of Irish & BRITISH Species & British & distant parts of Europe.--
500q10a about differences of Irish & British Species & BRITISH & distant parts of Europe.-- Gould-- go over t
559m155 r H. C. Watson on Geographical distribution of BRITISH Plants A Volume published by Colonel in army o
064r136 rds it. a mass on each side 3000 ft thick & 150 BROAD. neglecting Cordillera itself now remaining-- L
109a083 ery good this will show effects.-- analogous to BROAD flat sand beach. {P} -- De la Beches argument o
155g062 mperceptibly on Glen Turrit side 2nd shelf very BROAD «& cut out. produced» from same «cause» as «gre
160g091 500 boulder Cairn leet more Haberclador Isthmus BROAD flat Loch {P} XX Barom 28.92 A 75 Air 70 degree
160g092 {P}   XX Barom 28.92 A 75 Air 70 degree? Isthmus BROAD flat peat mass-- (general character in these mo
160g093 ne» S side of «this» Isthmus (which runs E & W) BROAD terrace «of pebbles? & Alluvium» which appear p
429d115 elieve in such a line., it is other plants.-- a BROAD border of Killed trees would form fringe.-- but
472s02r e without doubt.-- Bee, not large, very dusky & BROAD never saw such a one before-- Saw Fly 21 «this
383d160 ming to point {P} are more rounded. {P} & much BROADER., & «more ro» «three, I believe instead of» tw
383d160 ish brown» black marks on tail much «blacker» «BROADER.-- » Breast red like Common pheasant.-- lower
102a062 eing sucked out. In Cleavage discussion, state BROADLY indication of new law acting in certain direct
294c177 se-- &c <Lonsdale says. that first shee» State BROADLY scarcely any novelty in my theory, only slight
625o50v ive feelings & our short lived Passions' State BROADLY in child or animal it is equally proper to obe
467t105 idae on all their bodies pollen-- on a sulphur BROCCOLI not many do-- pollen not very abundant. not v
401e014 ties, which he attributes to crossing.-- Cape BROCCOLLI can hardly be reared without greatest care be
350d055 After Decandolles idea Septemb. 1st. Macleay is BRODERIP were talking of some Crustacean, like Trilobi
351d056 ten, which move imperfectly has eye-point, but BRODERIP added it has been stated that stationary Spon
351d057 amorphosis., heart altered & umbilical cord,-- BRODERIP alluded to Hunter's views on this subject.--
481z008 e. with <pe> nocturnal habits of Crustaceae Mr BRODERIP says that Voluta found in not less than 7 fat
400e013 - he sowed them for four generation before they BROKE.--, showing effects of cultivation gradually ad
054r102 lade covered by quartzose sandstones: refers to BROKEN hill described by Pernetty: account of streams
075r167 y volcanic rocks. //St Helena has been slightly BROKEN up, & has there not been vein «of iron» discov
107a078 es important with respect to thickness of crust BROKEN up.----My view of Volcanos &c &c This view wil
115a097 on Buch's Travels account of Norway chain being BROKEN through like that near-- Obstruction Sound  in
214b176 ably, so that any person would say the tail was BROKEN-- This came so often «that» it was difficult t
281c139 twigs dying, affinities would be <circular>, in BROKEN circles.-- which in each group is quite fatal.
332d003 glands.--   My Father has seen case of pleurisy, BROKEN limb «in children» & other such disorders acco
334d011 hough eager to all other mare had been entirely BROKEN from their mares, (though horsing every month)
402e019 tail are generally crooked, as if they had been BROKEN". are born so in all Malay Countries W. Earl E
465t079 able for the preservation of shells; where land BROKEN, rivers entering.-- & yet no shells-- now look
499q009 re dusted with pollen-- in what state (whole or BROKEN) is ball of pollen on Bees thighs (18) Place p
544m103 ous connect past, present & future thoughts are BROKEN-- Sir J. Franklin when starved, all party drea
550m124 -- Entire happiness. not being so desirable as <BROKEN» intense happiness even with some pain,-- comp
622o048 cience, is the feeling of any custom of society BROKEN..-- & how far more «feelin» acute the feeling
636j56v nts were formed. Macculloch says, life, forms a BROKEN, recurrent series, whilst the habitation «or w
206b150 «Read Buckland» L'..Institut «1837.» p 319 -- BRONGNIART.-- no dicotyledenous plants & few Monocot in
275c120 feathers   continued by picking chickens of each BROOD.-- These bantam feathers at last got ducky, the
535m064 em «nor helping larva from egg» watching them, BROODING over them, preserving them from «the» sun & e
535m064 l death, thus acting 4 to 6 weeks. The deserted BROODS appeared healthy-- This remarkable case may be
162g099 flows   into canal between L. Lochy & Oich. is a BROOK on the Lochy side of it-- the terraces of which
162g101 did not grow at first-- relics destroyed.-- the BROOK <about> Head of which is so interesting. enters
196b107 that   Ireland possesses varieties of the furze, BROOM, & yew very different from any found in great B
467t103 hers & stigma in slit» -- As I think they do in BROOM & certainly when over-ripe & half withered-- I
470t178 ble Bees on Phlox Down, 1854, Sept.) In Spanish BROOM by pulling back Wings, pollen is ejected with v
499q09v rian-- Urtica Dioica Sorrell. Lychnis. Butchers BROOM-- «also, Vinca,» Examine all these, are they mu
633j53v hus to explain the curling of the valves of the BROOM.-- or the springing of other seeds.-- But are w
184b054 S. America parent of all armadilloes might be BROTHER to Megatherium.-- uncle now dead. Bulletin Geo
312c233 nated & puppies delicate-- they cross sister & BROTHER of same litter, those. of different litters ar
391d175 that some «but not great» difference (for even BROTHER & sister are somewhat different) should be add
181b040 f we take «a man from.» any large family of 12 BROTHER & sisters--in a state which does not increase
308c217 d Man, May exclaim with Christian «we are all» BROTHERS in spirit-- all children of one father.-- yet
316c244 ork's Minster story of storm of snow after his BROTHERS murder.--» good anecdote --.Sowerby.--. Geogr
334d011 & always accustomed to her.-- case parallel to BROTHERS & sisters in Mankind.-- The case of all blue
374d134 of hybrids inter se, the first cross generally BROTHERS & sisters, & therefore somewhat unfavourable-
460t015 plain in in these <half> «3/4» bred ones-- The BROTHERS & sisters half-breed showed no sexual inclina
531m051 e pleasure children show in the naughtiness of BROTHERS children shows that sympathy is based as Burk
573n039 ld Miss. C. of the bad conduct of Mrs C. (her BROTHER'S wife). & she said nothing but shrugged her sh
326c266 on Horse in. N. America.-- Owen has it.-- Ld. BROUGHAM. Dissertations on subject of Science connecte
544m102 ssociated is kept up with waking thought.-- Ld BROUGHAM thinks no dreams except at this time. how doe
580n062 e double-conscious kept playing so well.-- Lr. BROUGHAM «Dissert.» on subject of science connected wi
032r041 teps, because the deposition & accumulation is BROUGHT into play As in Ocean & Air; there are «likewi
051r095 for   oceans power.--excepting when pebbles are BROUGHT into play; most manifest example of degradatio
055r108 of rain. = The Bulk of sediment «daily» yearly BROUGHT down by every torrent proves the decay atmosph
065r138 eries.--Sandwich Isd-- Specimens of rocks were BROUGHT home in Capt. Forster expedition from <Decepti
120a109 ion about action of rivers.-- Excess of matter BROUGHT down Mention absolute elevation of Patagonian
163g109 ewsbury rubbish.-- Speculate on origin pebbles BROUGHT by different cause: from mud.-- [blank] {P} mu
182b048 h» white with a black face, & similar to those BROUGHT from Abyssinia, & others dark brown, with long
192b089 is last animal bred-- might not new classes be BROUGHT into play.-- The father being climatized, clim
226b220 t see Coquille's Voyage.-- Galapagos mouse (?) BROUGHT by canoes Ceylon & India.-- Van Diemen's land
240c004 to rear them, eggs hatched under other birds & BROUGHT up by hand.-- These facts all account for [not
242c018 ry's Views.-- «Says» D'Orbigny is said to have BROUGHT a tortoise & toad from S. America & identical
255c053 lan is frustrated or rather a new principle is BROUGHT to bear. If man created as now. languages. wou
259c064 of   great lakes are American form, that one is BROUGHT to admit the possibility (any great change in
```

Page **(Key Word)***
```
259c066 If  puppy born with thick coat monstrosity, if BROUGHT into cold country, «& there acquired» then ada
267c093 white  with «a» black face, & similar to those BROUGHT from Abyssinia; the others dark brown with lon
272c107 of other, thus in Chalcididous insect, which I BROUGHT from Australia, probably live in flowers & has
277c127 e has done least part.-- that he will not have BROUGHT home new species. until, he can show range & h
286c154 unequal  life,-- stimulated by same passions-- BROUGHT into the world same way» they may convey  much
286c155 rranging animals in paper as drying plant, all BROUGHT in one plane Fleming Quarterly review says nat
292c170 views,  could possibly have had <to so convin> BROUGHT so much conviction to my mind.-- Reflect  much
308c217 rate> would sometimes vibrate-- «seeing no tea BROUGHT back memory» old habit of putting tea in pot m
313c233 ad differ Africa Australia Parasites die, when BROUGHT over on tropical animals which accounts for th
378d147 t Male that assumes change, & is the offspring BROUGHT back to earlier type by Mother?-- do these dif
390d177 in  (those which have solitary flower) exotics BROUGHT from foreign country. (<annuals> & so must tho
402e018 mal probably a tapir p. 233. dogs in Borneo-- <BROUGHT probably by Chinese>, "the breed being of  the
409c049 n is one great object, for which the world was BROUGHT into present state.-- <whether he war or  not.
424e103 some account of wolf in Zoolog. Gardens, which BROUGHT its puppies to be fondled.-- and we see in the
437e140 ld not find it.-- The parent bird another day BROUGHT to her young ones the cub of a fox, which afte
437e141 ussels Sprout returning suddenly to type when BROUGHT back to home. (& yet all the varieties of Bras
521m008 n habitual action of thought-secreting organs, BROUGHT into play by morbid action.-- Old Elspeth's «i
526m029 lect, «at Zoos» they are all things, which are BROUGHT to mind, by memory of the scenes, (indeed my A
526m029 all understanding. (on hearing old association BROUGHT up) by sight & not by hearing One is tempted t.
527m035 s, before real train of inventive thoughts are BROUGHT into play & then perhaps the sooner castles in
532m056 were  sound. Caroline tells me that Nina, when BROUGHT from Shrewsbury to Clayton, (though so fond of
532m056 er & of servant of Richard & of Mary & her bed BROUGHT from Shrewsbury) yet for a fortnight continued
543m101 e can thus trace causation of thought.-- it is BROUGHT within <our own> limits of examination.-- obey
595n115 about  two months. saw a «black» spider monkey BROUGHT it at opposite end of house. & commenced a mos
557m147 s, cheerful face».-- Man when at ease has smooth BROW contrary to wrinkled: (a horse when winnowing &
070r155 opiapo, as well as in latter.-- According to Mr BROWN, a person (whom I met at S. W. P.) the Cordille
073r161 arated from them with steel instruments." In R. BROWN (Collect: «of F. W.») where the stalactiform ma
075r167 ver ores from Peru consisted of native silver & BROWN oxide of Iron in Mexico. sulphuretted silver, a
182b048 to  those brought from Abyssinia, & others dark BROWN, with long clotted hair resembling that of goat
190b076 ncy to go back-- there is an end to species.-- «BROWN Appendix» A most remarkable observation of Mr B
190b076 n Appendix» A most remarkable observation of Mr BROWN, about peculiarities of Flora. on East &  West.
193b093 Volume  on Geographical distribution of animals BROWN Geograph Journal. Vol I p. 174. says from  Swan
264c080 ities, as my Furnarii.-- some genus of yellow & BROWN-breasted bird in Australia &c &c-- but of cours
267c093 o those brought from Abyssinia; the others dark BROWN with long clotted hair resembling that of goats
267c094 nt in packs «30 or 40 together» colour reddish <BROWN>. ears long.-- like bull terrier-- Indian secur
269c102 supplementary to Appendix to Flinders Voyage by BROWN.-- great space seems to act per se as barrier--
284c148 d.-- no other plants peculiar to these isld.-- ¿BROWN can «not» bear the least salt water.-- Nuts pro
290c163 n analogy NB Pyrrho-alauda (bird of St Jago) of BROWN colour; lives on ground, colour of  habitation.
310c224 rica & India some plants same. --America.-- See BROWN Congo Expedition: 400 Australian plants found i
314c237 propagate  by scission can not alter much.?? Mr BROWN showed me Bauer's drawings of a curious plant w
314c238 islands  of south seas says so in preface.-- Mr BROWN says character of Flora, N. Zealand & N. Caledo
314c239 ENERA'. good case. rather large flora,. (150?) Mr BROWN did not observe scarcely any Australian charact
314c239 of  'Van Diemens land & Australia.-- The combat (BROWN) is found in Isd of Bass'' Straits The common Mu
324c268 s laws about sexes relative to age of Marriages BROWN at end of Flinders & at end of the Congo Voyage
326c266 nales of Vienna. sketch of south Sea. Botany R. BROWN. has curious coloured maps. by Copenhagen Botan
367d113 red by a donkey. with all four legs ringed with BROWN.-- animal like large, heavily made cream colour
378d147 says  Hen & cock Starling alike, yet young ones BROWN.-- Sexual Selection If masculine character. add
383d160 ad of» two lines.-- «faintly edged with reddish BROWN» black marks on tail much <blacker> «broader.--
405e031 which were of a nearly uniform <dusky> blackish BROWN.-- yet retained a trace of horizontal mark on f
435e133 not located] the stigma retains its power.-- R. BROWN found the <poll> masses of pollen of  Asclepias
448e165 ias, where this is always the case according to BROWN,-- Voyage of Adventure & Beagle Vol I. p. 306 S
466t091 stris alone together. one had seed-pods turning BROWN, whilst both others were in nearly full  flower
467t099 to corolla--anthers minute, distinctly doubled, BROWN, but with no pollen.-- Common Thyme growing clo
480z007 of different size-- How does this apply to pale BROWN Caracara Krauss on Corallinae from S. Seas writ
489qqIFC «Gardens»  ---- . . 20 & Breeders Dr. Boott: R. BROWN p. 21 Horticulturists p. 21--23 Eyton p. 22 Sch
496q05a s guided by smell-- or sight.-- --. touching Mr BROWN theory of insect-like Orchis-- & final cause of
513q21. l species, as dioecious plants, when crossed R. BROWN-- will pollen act on any flower before stigmas
596n184 y" I should think well worth studying-- "Thomas BROWN" on Association worthy of close study.-- full o
350d054 .?-- "Natura nihil agit frustra", as Sir Thomas BROWNE says "is the only indisputable axiom in Philos
550m126 of  which all our recollections are pleasant.-- BROWNE Religio Medici, p. 21-24. Curious passages sho
321c270 ons Imaginary Conversations-- very poor Sir T. BROWNE'S Religio Medici Lyell Book III There are  many
322c269 the Human. Mind Evelyns Sylva. skimmed, stupid BROWNES travel in Africa; «well skimmed.» 1839 Jan 10t
350d054 in Philosophy<"> Religio Medici. Vol II. Sir T. BROWNE'S Works p. 20 There are no grotesques in nature
405e031 state; thus mark on ear of cats, colour can be BROWNISH do Saw what was said to be hybrid between sil
321c275 Surinam. Voyage Congo expedition: Zaire except BROWN'S Appendix & excellent table of Canary isld: Pla
506q015 fecundated,  as mass of pollen is requisite.-- BROWN'S paper 43. Any flowers of Keeling Dioecious, or
576n046 Antiquary)» Vol II p. 126 says seals knit their BROWS when incensed.-- A Dog may hesitate to  jump  in
503q012 about  Sugar-Cane Edwards says does not seed-- «BRUCE says does» Royle In Royle's productive Resource
086a006 ed with its elevation. vapour from below-- Malte BRUN «Salt Lakes» Siberia must be read as well as Pa
088a018 hills East of Cordillera very important V. Malte BRUN -- Main character of Andes Metamorphic action -
402e019 ercopithecus., & skins of galiopithecus.-- Malte BRUN. Vol <I> II p.,133: at Samar SE of Luçon,  many
402e019 of  Luçon, many monkeys, buffaloes &c &c-- Malte BRUN. would be worth skimming over with regard to th
240c013 lse In New Guinea. a Kangaroo d'Aroe (Didelphis BRUNII) which as yet had only been found in isle of A
592n103 o, if I follow up this subject & a reference to BRUN'S work.-- Shutting eyes in contempt opposite act
466t093 en directed to it-- The Humbles in crawling out BRUSH over anther & pistil & one I SAW IMPREGNATE  by
467t103 nt filaments & stamens all protrude <there is a BRUSH at end of stigma, which forces out from extremi
468t112 urned up «more than stamens», so that all were BRUSHED by Bees & especially stigma after bee had brus
468t112 shed by Bees & especially stigma after bee had BRUSHED over the anthers of long stamens {P} as stamen
470t178 are  in flower «I saw Bees;»-- on Monk's Hood, BRUSHING over stamen «Egg Tree»--I think never on  the
437e141 here was great swarm of mice.-- May 4th.-- The BRUSSELS Sprout returning suddenly to type when brough
437e141 e varieties of Brassica certainly not becoming BRUSSELS Sprouts) is analogous to Primrose & Cowslip s
438e141 ecedent races.-- «& yet in all probability the BRUSSELS Sprout was slowly formed.-- » if it shall  be
300c198 erefore mans mind not so different from that of BRUTES p. Hard to say what is instinct in animals.  «
325c267 ght be worth consult. Paper on Consciousness in BRUTES In Blackwood. June 1838 H. C. Watson on Geogra
375d134 ecies as inference from Malthus.-- «increase of BRUTES, must be prevented soley by positive checks, e
409e047 man,  exceeding monkeys;-- Having proved mens & BRUTES bodies on one type: almost superfluous to cons
558m151 to show hiatus in mind not saltus between man & BRUTES) no one can doubt this connexion.-- look at fa
559m155 final  cause.-- Read. Paper on consciousness in BRUTES & Animals. in Blackwood's Magazine June. 1838.
470t178 g, curved nectar of Butterfly Orchis & Listera? BRYONY saw common Bee on: Linn. Trans 18. p. 133 West
231b242 nimals all show greater connexion in quadrupeds, <BU> plants do not follow by any means.-- Ostriches
045r074 haps these mighty changes might go on. & not a BUBBLES on the surface bespeak the changes. -- metalli
072r159 ¿what  does he mean?) Consult Dr Holland about BUBBLES. -- No Volcanic action on coast line of Old Gr
089a017 of  Europe no fossil shells --¿ action of Heat BUBBLES volatilized at bottom, condensed before rising
283c145 their  minds with such sentences as "oh turn a BUCCINUM into a Tiger."-- but perhaps I feel the impos
374d134 ian-- several existing genera. Nautilus turbo. BUCCINUM. turritela. terebratula, orbiculas, with man
499q010 ds good to eat. (even Nux Vomica is eaten by a BUCEROS in East Indies-- Asiatic Researches) (23) Talk
034r044 ic. from Humboldt: Comparison P 361. Daubeny Von BUCH is very strong about Trachyte being the most in
064r137 maining-- Lyell « <p 419> p 428» states that Von BUCH has urged that Java volcanos differ from all ot
069r150 spar melted gneiss/7/ QUARTZ!!! Analogous to Von BUCH. Basalt where Basalt. trachyte where  trachyte.
096a039 el to Greenland: Mem.¿ Greenland subsiding.) Von BUCH Canary Isd. p. 351.. NB. Mackenzie talks of gra
096a040 l on basalt of Heckla-- All the Azores Isld. Von BUCH p 359 stretched out NE & SW.-- Von Buch. Can. I
096a040 sld. Von Buch p 359 stretched out NE & SW.-- Von BUCH. Can. Ile p. 406. List of Volcanos Salomon Isld
097a042 o degree 221 Lamellar dikes like Mica Slate Von. BUCH. Canary Isd. p 170.-- Mem. Cordillera Can Green
123a115 LENDID PAPER on fossil shells of S. America. Von BUCH Lyell. (under head of Delta) describes near Alp
209b156 innaean doctrina & Lyells. to certain extent Von BUCH,-- Canary Isles: French Edit. Flora of Islds v
210b158 esent analogy to what takes place from time? Von BUCH distinctly states that permanent varieties. bec
210b159 ow do plants cross?-- = admirable discussion Von BUCH says from Humboldt, in Laponia. genera to speci
211b164 Archipelago--,  very good in connection with Von BUCH Volcanic chart & my idea of double line of inte
```

294c178 lassification & affinities, its extension.-- Von BUCH. Travels, p. 306. account of trees ceasing to g
320c275 inuation «Annals of Natural History» Skimmed Von BUCH Travels Whites Natural History of Selbourne Ref
355d068 condary Species distinct-- but close.-- Mem. Von BUCH on Cordillera fossils same remark. ¿was there f
356d069 amount of change at any one time Seeing what Von BUCH (Humboldt). G. St. Hilaire, & Lamarck have writ
442e153 Norway ought to be thus characterized study Von BUCH.]CD Now Mr Knights statements about fruit trees
115a097 ocks. when writing on Falkland Islds p. 94. Von BUCH'S Travels account of Norway chain being broken t
261c071 notion of beauty, when does prefer most powerful BUCK Owen talking of Plesiosaurus Plesiosaurus. all
094a036 35 Sir J Hall Trans. Phils Royal Ed. Vol 7 Dr BUCKLAND Reliquiae Diluvianae p. 201. & seq Murc Trans
206b149 rees; or in certain shell cephalopoda.-- «Read BUCKLAND» L'..Institut «1837.» p 319 -- Brongniart.--
408e044 ere for.-- «Gales of wind would blend species» BUCKLAND. Reliquiae Diluvianae. p. 222. Bones of Horse
413e060 to Articulata, which are all truly bisexual. BUCKLAND'S Reliqu: Diluv. says Africa only place, where
305c211 ther, as well as Man & child, polypus & polypus, BUD & bud, polypus & germ plant & seed.-- instincts
305c211 as well as Man & child, polypus & polypus, bud & BUD, polypus & germ plant & seed.-- instincts in you
336d019 ring-- On the idea of generation being a <slip> «BUD» from parent. if whole parent not entirely embue
336d019 le parent not entirely embued with the change, a BUD could not be taken, without it either went back.
355d068 facts.-- What takes place in the formation of a BUD-- the very same must take place in copulation--
355d068 same plant<s>)-- now in some Polypi we see young BUD changing into ovules.-- Captain Grants. Himalaya
371d128 nted?, & yet year after year, successive roses & BUD are produced, like parent stock, or if different
371d128 s though transmitting them with such facility to BUD.-- this must be owing to their unity in one stem
372d129 is must be owing to their unity in one stem.-- a BUD may be transplanted & carry all these peculiarit
372d129 & carry all these peculiarities not so a seed.-- BUD probably is like cutting off tail of Planaria, t
372d130 cko produce always different tail? An Individual BUD may be thus produced from the growth of one part
372d131 oduce am if supported., <so> & in making «true» BUD some such process is effected.-- a child might b
384d162 r some generations Theory of sexes (woman makes, BUD, man puts primordial vivifying principle) one in
386d165 s annual give buds.-- life may be thus prolonged BUD being formed & one part dying for great length o
388d170 tate.-- When it is said. that difference between BUD & seed, that latter carries with stock of food.-
388d173 must actually reach the ovum.-- [Why in making a BUD, which is to pass through all transformations, s
388d173 , should there need two organs; whilst in common BUD there is no such need.-- one would <one> suppose
388d173 ssed through transformation, & was received into BUD matured by female;<]CD> such view no ways explai
390d177 e many flowers same Spath, as they have only one BUD.-- Every individual foetus would reproduce its k
390d178 Male is to impress some difference: to make the BUD of the woman, not a bud in every respect.-- [Is
390d178 difference: to make the bud of the woman, not a BUD in every respect.-- [Is this connected with the
391d174 pra <v. infra> p 179, continued from Is a flower BUD produced by union of two common buds??? Amongst
441e148 o organs-- instead of one part «as» in producing BUD.-- Fewer of the lately acquired peculiarities ar
441e149 n, yet offspring are,-- if Australian Dog, could BUD, analogy tells us, <be> «offspring» would be sim
446e162 eak-- -- each tulip is the <of> product of fresh BUD-- here then is case of change analogous to chang
454e184 o not perfect their seed?-- What annuals can be BUDDED «& rendered of great age» as must be inferred
492q002 varieties been formed?-- 8. Can any annuals be BUDDED. with reference to extension of age of individ
372d128 whole branch transplanted? +.simplest forms of BUDDING. Why does Gecko produce always different tail?
389d176 as, a superabundance of life, like tendency to BUDDING, which wishes to throw itself off.--» as might
390d177 s» & so must those forms which are produced by BUDDING «only» as cryptogamia & hydras.-- (this repug
390d178 l generations, the process would be similar to BUDDING. which is not object of generation-- therefor
440e145 kinds It seems absurd proposition, that every «BUDDING» tree, & every buzzing insect & grazing animal
139a180 ause such monstrous growth as oak galls or rose <BUDS> galls.-- is it not effect of superadded vital
171b003 are produced, though new individuals produced by BUDS are constant, hence we see generation here seem
176b021 re branched.-- Hence Genera.-- «as many terminal BUDS dying, as new ones generated» There is nothing
189b073 re> of same species are bound together just like BUDS of plants, which die at one time, though produc
313c236 eration (lateral with no relation to time) as in BUDS.-- I can scarcely doubt final cause is the adap
328cIBC ava fossils at same time Study Botanical work on BUDS & Gemmae. C D Charles Darwin 36 Great Marlborou
372d129 ubt, the> Do plants loose any qualities by being BUDS-- , more than if whole branch transplanted? +.s
384d161 se degenerate during its life so that successive BUDS do differ-- any variety is not handed down. but
386d165 fe.-- like Golden Pippen trees! How is this with BUDS of plants, does annual give buds.-- life may be
386d165 ow is this with buds of plants, does annual give BUDS.-- life may be thus prolonged bud being formed
390d177 seems connected with more developed forms) Study BUDS-- gemmae-- & monocotyledenous, do those which a
391d174 Is a flower bud produced by union of two common BUDS??? Amongst buds each one exactly like its paren
391d174 produced by union of two common buds??? Amongst BUDS each one exactly like its parent. <but these bu
391d174 uds each one exactly like its parent. <but these BUDS do not procreate> & all alike in one parent or
440e148 stops»-- Spallanzani's facts in connection with BUDS.-- They differ from possibility of concourse of
441e151 s, in northern latitudes, which are generated by BUDS alone or roots, & never flower, so there may be
446e162 produces itself «in England, as yet observed» by BUDS-- (the other three by buds & seeds «though by t
446e162 , as yet observed» by buds-- (the other three by BUDS & seeds «though by the latter very rarely») her
446e163 g constant, without crossing.-- & propagation by BUDS does not insure constancy of form.-- is the con
447e164 rying. A plant <producing> propagating itself by BUDS is in same predicament, as one, in which struct
448e165 urious facts about every part of plant producing BUDS, so that Turpin says each cell of plant is indi
448e165 vidual.-- Most plants which propagate rapidly by BUDS, layers &c &-- do not seed freely.-- The peri
455eIBC -:likenesses of children CD[Does any annual give BUDS, or tubers. Yes-- but these are same as trees.-
067r143 ril 1836 (p302) Coleccion de obras. 2 Vols fol: BUENOS Ayres 1836: W. Parish?? «by Pedro de Angelis.»
095a037 labiata certainly is found with the Mactra. at BUENOS Ayres at the Zoolog: Soc: Terebratula from Hud
123a117 nelones -- large quadruped bigger than ox.-- at BUENOS Ayres 20½ quadras from river; 20 varas from su
244c022 have.-- p. 139. Vespertilio bonar«i»ensis (from BUENOS Ayres) replaces <Vesp.> holds same relation wi
392d180 ia «with our common ones» in Zoolog. Gardens; <BUFFALO & common cattle-- Esquimaux (& Australian) dog
403e021 94 (1st Edit) of Sumatra has given account of BUFFALO of the East which differs from that of S. Euro
202b132 n. Vol IV. P II. p. 160. Melville Isd.-- "The BUFFALOES, introduced from Timor, herded separate from
402e019 I p.,133: at Samar SE of Luçon, many monkeys, BUFFALOES &c &c-- Malte Brun. would be worth skimming o
399e009 r shops., of same colour as a Hare, but pale! & BUFFER.-- with long ears-- & longer hind legs??? so t
324c268 logical Magazine (paper con Geograp. range Study BUFFON on Varieties of Domesticated animals see if la
327cIBC probably some account Raffles. Sir. S do. do-- BUFFON Suites Cline on the improvement of domesticate
344d040 Geograph distribution of reptiles in Suites de BUFFON.-- Vigors has given list in Limnaean Transacti
366d112 . how comes it that the tame donkey has. CD[old BUFFON should be read on Mare My view, why hybrids ar
390d179 r it should be observed that from Books to read BUFFON Suites de.-- Horse & Cattle Library of Useful
459f02 aper translated from Meckel. Comp. Anat.-- From BUFFON cross of he-goat & sheep, it seems male gives
601o009 ping higher animals & not plants as supposed by BUFFON) Consciousness is sensation No. 2. with memory
380d153 h &c Mr Y. says that Macleay considers the house BUG, as a «female which have» larvae which have bred
207b151 un. p. 325. Vol. IV. Ducks on rivers in Guiana. BUILD top of trees carry duckling to the water in the
271c105 placed to the North by other species.--) should BUILD a nest lined with mud, in forest where not a tr
371d118 ess wonderful that childs nervous system should BUILD up its body, like its parent, than that it shou
446e160 hese reciprocally assist in domestic cares, as BUILDING nest, sitting on eggs. & feeding & defending
563n001 ks.-- My Father says pea-hens do Wood pidgeons BUILDING near houses. yet so shy at all other times.--
582n068 we must grant a bird knows what is about when BUILDING its nest; it knows its object but not result
582n069 knows its object but not result (first time of BUILDING?), but not the means of performing it.-- p. 1
585n077 ht be «possibly» acquired by habit. so bees in BUILDING cells, must have some means of measuring cell
586n082 th <sight> sight-- so a Bee has the faculty of BUILDING «regular» cells-- [but this faculty <may poss
586n082 instinctively exerts in concert with others in BUILDING comb-- My faculty often will turn out to be i
626o052 instinct cannot be said to guide will. as bird BUILDING nest, but supplies it-- instinctive feelings
528m037 ective. which cannot be doubted if we look at BUILDINGS, even ugly ones.-- the pleasure from perspect
571n030 tiated. or rather altered. The Reason why New BUILDINGS look ugly is because there is some connection
271c105 ith mud, in forest where not a tree in which it BUILDS, a berry on which it feeds. or insects it devo
271c105 nge is? <One> The black & white thrush of Azara BUILDS its nest in <same country> something same mann
297c189 pplicable [not located] p. 428. Ouzel sometimes BUILDS nest without doom. Vol 2. Mag of Z. & B. p. 43
527m033 d run cold by singing).-- Granny says she never BUILDS castles in the air-- Catherine often, but not
540m085 awn, makes me feel how «much» all animals <are> BUILT on one structure.-- He who doubts about nationa
547m111 replace each other August 29th. Went to Bed. & BUILT «common» Castle in the air, of being compelled,
266c092 cana of Science & Art. 1831. p 160. account of BULBOUS root from Mummy: after 2000 years, germinating
495q005 ertainly not) (3) Sow seeds & place cuttings or BULBS in several different soils & temperatures & see
055r108 the counterbalancing variations) of rain. = The BULK of sediment «daily» yearly brought down by ever
105a070 re surface exposed. bay more open to turbulence. BULL. Soc. Geolog «1837» p. 320. paper on shrinking
108a081 here.-- [top portion page excised, not located] BULL:. Soc: Geolog. Tome IX 1837-8. p. 24. rocks of
117a101 of red granite between 40 & 50 from Portezuelo. BULL: Soc: Geolog. 1837. December. p. 91. a classifi

216b179 mentioned in Loudons (analogue of Blood hound-- BULL. Soc. Geolog. 1834. p. 217. Java Fossils 10 out
225b217 swer yes, if you will show me every step between BULL Dog & Greyhound). I should say the changes were
229b233 (very) aboriginal at Cape: crossed with English BULL. offspring very like common English.-- Hottento
235b274 prevent them taking peculiar character-- Indian BULL.?-- Do species of any genus. as American or Indi
241c016 esumé do/p. .62 ??? Age of Deinotherium. p. 23.. BULL: Soc. Geolog. 1837-8. Tom: IX.-- M. D'.Urville
261c070 When talking of races of Man.-- black men, black BULL finches from linseed-- not solely effects of cl
267c094 ether» colour reddish <brown>. ears long.-- like BULL terrier-- Indian secured one, as they always li
275c120 greyhounds fleestest in England lost courage-- (BULL-dogs are used because they have no scent!) Mr W
275c120 run up hill-- he took thorough bred <greyhound>. BULL-dog. & crossed & recrossed, till there was a da
312c231 way to improve cattle is to cross between a good BULL & <be> the provincial breed, & that first offsp
367d113 in fighting" instances thighs of cock & Neck of BULL.-- is most common in vegetable feeders. because
380d155 enstruate in the month, she will in 5 weeks.-- A BULL is never taken from his own field to bull a cow
380d155 s.-- A Bull is never taken from his own field to BULL a cow.-- -- a dog if led in string will not.--
422e092 t «Alludes to difference between fossil & recent BULL; like fossil & recent shells of the <new> raise
436e137 ers, what cause?? Seeing the beautiful seed of a BULL Rush I thought, surely no "fortuitous" growth c
505q014 July/42/ Mark has six day's puppy of small true BULL-Dog-- length from nose over head to root of tai
507q15v book with drawing of Seed. Anemone with, tuft-- BULL Rush-- Dandelion- Sycamore. & seeds with where
511q018 often writes to Lowe (7) In breeding. pointers. BULL-Dogs. Spaniels-- Grey-hounds-- is there ever an
512q019 parison with hybrids, is offspring of short-horn BULL & hereford cow similar to reverse cross.-- Sow
575n044 ts import to Bowen No one doubts that a cross of BULL dogs-- increase the courage & staunchness of gr
575n044 rease the courage & staunchness of greyhounds.-- BULL-dogs being preferred from not having any smell.
595n125 outing, «smiling», just as much instinctive as a BULL <tr> calf, just born butting, or young crocodil
637j57v -- greyhound to hare.-- waterdog hair to water-- BULL dog to bulls.-- primrose to <open fields> banks
098a044 ery conclusive proofs, but certainly probable. BULLETIN de la Soc. Geolog: 1833-34. p. 35.-- Ancient
184b014 t be brother to Megatherium.-- uncle now dead. BULLETIN Geologique April 1837. p. 216 Deshayes on cha
225b217 e are as ignorant. as why millet seed turns a BULLFINCH black, or iodine on glands of throat, (or col
441e150 oil on colour of flowers, Hydrangea -- black BULLFINCHES-- & all varieties must be presumed to be res
450e175 se. Forrest--. (p. 270) says many wild horses, BULLOCKS, & deer South part of Mildanao.-- Q Horse do.
275c121 tle of L'Darnleys were most like parent Brahmin BULLS-- Mr. Y. is inclined to think that the male com
288c159 ated, or altered, but it is certain that rams & BULLS from England fetch very <go> large price. as is
430e117 me Hares, attacking a sick one like Chillingham BULLS are described.-- His three have had VERY differ
509q017 (1) Particulars about Sierra Leone. cow. taking BULLS. is it Domesticated African Animal= Knows nothi
536m071 es one dog smell posterior at another.-- Why do BULLS & horses, animals of different orders turn up t
556m146 desire-- is there not protrusion of chin, like BULLS & horses.-- 1838 good instance of useless muscu
637j57v to hare.-- waterdog hair to water-- bull dog to BULLS.-- primrose to <open fields> banks-- cowslip to
056r109 as been corroded. -- If man could raise such a BULWARK to the ocean, who would ever suppose that its
225b217 s. talking of it) or how to make Indian Cow with BUMP & pigs foot with cloven hoof) Ask Entomologists
354d065 - I was struck looking at the Indian cattle with BUMP. together with Bison, at some resemblance as if
473s05r ths of the sea-- Maer. June/42/ June/42/-- Mr. BUNBURY says has heard the Trout from different lakes
473s006 ws & there joins streams from Capel-Curig-- Mr BUNBURY says Miers has described in Linn: Transacts. a
514q21. en-masses?- = answered = Has Ophrys nectary?= BUNBURY says no «hollow» spur.-- Ask about Pinks & Sol
466t094 tary, would «necessary» cross directly over the BUNCH of anthers & pistils, but these <do> do not ben
469t135 s «many» «most numerous» bees visited this same BUNCH & on this day in five minutes eleven Humbles ca
219b193 nsported some blown--floating trees Thrushes & BUNTING & coots-- «Turdus Guyanensis?) (Emberiza Bras
341d031 by one not differing except by black line,-- A BUNTING by one only differing by some permanent white
452e181 e first procured from Gilolo p. 253 In isld of BUNWOOD (18 miles in circum) there are hogs & monkeys
037r052 neath low water in the Southern ocean not being BUOYED with Kelp.-- With respect to degradation of ro
051r093 increase. -- In Southern regions every rock is BUOYED by Kelp, now Kelp sends forth branching roots
048r086 ountry of S. Africa. It would be well to quote BURCHELL. V. where the Rhinoceros was killed. -- In Pa
065r139 en. only production. a body which had long been BURIED, <see> from rotten state of coffin «buried in
065r139 been buried, <see> from rotten state of coffin «BURIED in a mound» long consigned to the earth. yet b
473s004 ether with their littoral deposits are probably BURIED in the depths of the sea-- Maer. June/42/ June
531m051 others children shows that sympathy is based as BURKE maintains on pleasure in beholding the misfortu
546m108 pathy very unsatisfactory because does not like BURKE explain pleasure. August 26th. I cannot help. t
566n010 (Holcroft Translat) Vol III. p.37, quotes from BURKE, who says on mimicking expression of emotions,
540m088 ht, very like best feeling of sympathy.-- Mem: BURKE'S idea of Sympathy. being real pleasure at pain
578n057 wn up people cry.-- What is emotion? At end of BURKE'S essay on the sublime & Beautiful there are som
309c221 n nations (quoted) p. 4.-- do. p. 186. quotes BURKHARDT to show black colour of certain Arabs.-- NB a
135a144 logical paper. by Malcolmson-- worth reading-- BURNETTS. vol 4. p. 193 in Lat 26 degree S. Wafer look
038r057 great crater. different little craters are all BURNING, surely there must be «somewhere» below a fiel
059r120 ts incontestably formed the parts of one whole BURNING mountain, & that the central part fell in.--Sa
575n045 hurts body-- tears flow from both, as when one BURNS end of nose with a hot razor.-- joy <p> a menta
450e175 «190» No full sized horse is found East of y BURRAMPOSTER & S of Tropic-- By J. H. Moore after quitti
507q15v - for woodcut-- 1 double hook-- -- Geum. Galium BURRH ≡ single hook; curved spines-- simple spines--
212b165 g Soc told me an Australian dog he had, used to BURROW like fox.-- a sort of internal bark. would rem
477z001 twood & Thompsons-- Part II.-- 35. Phil Trans BURROWING & boring marine animals-- CXVI. P 111 do Obse
042r068 odified. «Moreover, the Volcanos from sea there BURST out, after rise from sea: <As did> as did those
068r146 lenty in S. America!! Metamorphic Volcanos only BURST out where strata in act of dislocation (NB. dis
152g045 help {P} In Glen Collarig, by Dicks theory lake BURST in most improbable part & not in Pass, where sh
542m093 h [not] to do it, but an involuntary laugh will BURST forth, this & yawning. (common to other animals
316c244 or Heaven or bad enough for Hell.-- «+glimpses BURSTING on mind & giving rise to the wildest imaginat
566n011 which is a real instinct in the chicken, just BURSTING from egg.-- Animals have necessary notions. w
504q013 after other to flower & Menyanthes whose pollen BURSTS before flower is open-- -- No (6) There is app
267c094 ite of every attempt to check its increase. The <BUSH> woodlands for miles in extent are composed sol
499q009 ee whether they catch pollen-- <Ne> In Oenothera BUSH.-- (19) Theory of mock flowers in Hydrangea (20
036r052 n on land different Volney, P 351. Vol I. woody BUSHES. «gazelles» hares, grasshoppers & Rats. charac
353d061 ls.-- many of the hawks <to> are analogues to <BUSTARD> European birds. also «do» p. 403. & 404 vol
220b199 hereddetary fear (like rooks with guns) of the BUSTARDS in Germany.-- Athenaeum. No. 537. Feb. 1838.
225b218 nt insects.-- as crows against man with gun. & BUSTARDS & &c!!! An American & African form of plant
499q09v it, Valerian-- Urtica Dioica Sorrell. Lychnis. BUTCHERS Broom-- «also, Vinca,» Examine all these, are
627o52v --' p. 224.-- Hume's Inquiry-- good abstract of BUTLER & arguments of beneficial tendency of affectio
628o054 desires, & moral sense all different.-- P. 22. BUTLER & Mackintosh characterize the moral sense, by
626o052 remark on this point.-- [LHC] p. 194. «&c &c» BUTLER'S view given on conscience: I cannot admit it.-
151g043 ses where rivulet enters two great projecting BUTTRESSES, upper slope of which corresponds to shelf th
152g050 Air 65 degree? at level of upper terrace The BUTTRESSES of Alluvium rise nearly up to Glen Collarig u
346d048 am,-- habits peculiar,-- young one 203 days old BUTTED violently. & fell.-- gore to death the old & w
548m117 re, & then when one begins eating one perceives BUTTER or salt is not there.-- the reality does not r
467t105 one small Bee; saw another on Cabbage--white BUTTERFLIES suck nectar: «Maer June 41» Rhubarb. pollen
470t178 ough they examine it.--«Little Dusty & Blue» BUTTERFLIES at Clover,--Veronica--, Ranunculus in number
499q009 of pollen in any one flower (17) Catch Bees, BUTTERFLIES-- Syrphus-- Meligethes & see whether they ar
499q09v l these, are they much frequented by Bees or BUTTERFLIES or little insect?= or is pollen excessively
354d062 Valenciennes.» The change from caterpillar to BUTTERFLY-- is not more wonderful than the body of a ma
354d063 fferent parts when age changes caterpillar to BUTTERFLY.-- When two Varieties of bugs cross, Erasmus
470t178 t can get honey out of long, curved nectar of BUTTERFLY Orchis & Listera? Bryony saw common Bee on: L
500q010 o not varieties.-- (25) Does the yellow white BUTTERFLY deposit eggs in all varieties of Cabbage. (26
637j57v not been provided. «with proboscis» «as bee & BUTTERFLY» inconvenience.! extinction, utter extinction
595n125 uch instinctive as a bull <tr> calf, just born BUTTING, or young crocodile snapping.-- these I think
293c172 reasoning to tell back from front. &c or use of BUTTON holes it would be instinctive.-- My view of in
154g059 rin» 29.287. 72 degree Air 65? 70? {P} Where a BUTTRESS projects from side of hill if line suppose co
161g097 not far continuous flights above it-- (NB the BUTTRESS or pass at Isthmus appears above level of she
161g097 certainly) I took another measurement on short BUTTRESS but not continuous & it was 29.200 minus .008
146g013 de of Hill South of upper end of Loch Dochart BUTTRESSES of Alluvium or rather mass of well rounded p
146g014 yellowish argillaceous or sandy soil-- These BUTTRESSES formed vestige of irregular terrace perhaps
152g048 remained very long at 4th shelf from size of BUTTRESSES, to upper edge of which they cut near Loch T
153g051 stratification argument detritus-- {P} after BUTTRESSES on 4th shelf: others «lines not so level bec
156g072 ounded pebbles of granite & forming «sloping» BUTTRESSES Yet certainly shelf 4th <near> only usually
158g079 errace -- on right hand {P} rounded waterworn BUTTRESSES of granite obscure terrace 15 ft divortium m
158g082 t smooth on sides of hill where Boulder lies. BUTTRESSES «occur» high up on Shelf 2d «in Upper Glen R

158g083 nd strewed with pebbles Shelf 3d runs up with BUTTRESSES on each side «very little way» in Upper Glen
162g104 igh up the Tarf (a Granite (boulder), sloping BUTTRESSES, an[d] one alternate curved layer of fine sa
185b055 e were causes to induce great change. like the BUZZARD which has changed into Cara cara at the Galapa
440e145 roposition, that every «budding» tree, & every BUZZING insect & grazing animal owes its form, to that
066r141 head.-- Kerguelen 40 by 20 leagues. dimensions: BYNOE informs me that in Obstruction Sound, in the na
555m142 s?) I doubted it in Fuegians, till I remembered BYNOES story of the women.-- The Chillingham cattle (
109a084 discussing concretions Carbonate soda. formed by CA. of L. & Mur. of Soda mixed.-- Turner's Chemistr
138a176 by yielding Carbonic Acid unite with «piece of CABBAGE» alklali & precipitate silica / or charcoal ch
401e014 is scarcely possible to purchase seeds of any CABBAGE, where a great many will not return to all sor
401e015 so absurd.-- Thinks-- that such variety as red CABBAGE produced from passage from many varieties, & p
467t105 Potato & several times on Beans Rough.--green-CABBAGE «in flower»-- swarmed with meligethes & small
467t105 very small-- Saw one small Bee; saw another on CABBAGE--white Butterflies suck nectar: «Maer June 41»
491q01v ther tree being affected.-- does one branch of CABBAGE being mongrelized affect other branches-- The
495q005 n organs of generation (5) Place pollen of Red CABBAGE «mixed with own pollen» on flowers of other ca
495q005 lt hybrids-- (6) Dust flowers of one branch of CABBAGE with pollen of other, count seeds, & see how g
495q005 rids of Cabbages (7) Sow <daisy> seeds of wild CABBAGE in VERY rich soil, will plants abort?, does it
500q010 ite Butterfly deposit eggs in all varieties of CABBAGE. (26) Do deer Keepers cross the breed-- desira
440e146 either in present, or past generation.-- thus CABBAGES growing like Nepenthes.-- cases of pidgeons w
492q002 IV. Hort. Transact.-- 4.. Are any varieties of CABBAGES not attacked in bad years from Caterpillars.-
492q003 Mr Ford Has M. Sageret WRITTEN on crossing of CABBAGES, quoted by (as if oral) Decandoelle in V. Vol
495q005 ge «mixed with own pollen» on flowers of other CABBAGES & see whether there will result hybrids-- (6)
495q005 fact always takes place in natural Hybrids of CABBAGES (7) Sow <daisy> seeds of wild cabbage in VERY
495q05a lt. artificial water.-- 12. Plant two races of CABBAGES near each other-- & enclose one twig of each
498q008 charcoal & good treatment (8) Do bees frequent CABBAGES «& Cowcumber's out of doors.» much-- or the m
498q008 the nurseries, when «seed of» the varieties of CABBAGES, peas, beans, as raised, do the Seedsmen sele
502q11v & Fennel. Verbena Compare flower of different CABBAGES most carefully to see if variation equal in f
502q11v e on.-- Compare calves.: Compare young. beans. CABBAGES.-- History of Pheasant-fowl. Hen coloured lik
503q013 bove male? (2) Result of Edwards experiment in CABBAGES given (3)   in Heartease (4) Does the Th
504q014 s.-- Beans seeds alone remain to be compared-- CABBAGES.-- kept true Try experiment (30/p.11) (2) Yew
062r128 n trees, arborescent grasses, parasitic plants, CACTI: & with limits of no vegetation at S. Shetland
216b180 whether know varieties in plants do so.-- As in CACTI & &c.-- as in dogs investigate case of pidgeon
551m129 n by omnibus Horses starting, when door shut or CAD cries out "right." or Drinkwater's horse jumping
556m144 ear ready, but if they saw anything ahead. which CAD cannot see, they do not move muscle.-- reason CD
029r033 tude 61 deg. 22 min. W. mid. calm and clear. CAERMARTHEN Journal I look at the cessation northwards o
447e165 anada onion mentioned in Hort. Transact. Aira CAESPITOSA becomes vivaparous on mountains & yet can be
460t019 olite.-- insects, of do orders-- cheiroptera & CAETACEA in Eocene-- dicot. plants in coal measures.--
348d050 in inverse ratio to their variability.-- (Now CAETERIS paribus these will be the oldest) |"The most
216b182 impregnated at once.-- Dr. Smith considers the CAFFERS (like Englishmen) men of many countenances, as
218b189 s-- we see it even in men); the possibility of CAFFERS & Hottentots coexisting. proves this-- but whe
508q016 ble each other very closely, more closely than CAFFRES.= 13 Where are there any medical Statisics, pr
512q019 nes breed as well as,, as those already bred in CAGES. Get direction write to-- (2) Does he believe.
483z012 Vol IV p. 388. Domestic mouse of Egypt is Mus CAHIRIMUS. of Geof.-- reference from Rüppel travels All
160g090 Bought & between one of these & Glen Tarf Hill «CAIRN <taw> leer peak» Barom 28.700. A.75 degree 75 d
160g091 ill to others 3000? if Ben Erin is 3500 boulder CAIRN leet more Haberclador Isthmus broad flat Loch (
045r075 at the extraordinary fissures of the granite at CALABRIA were present at the Concepcion earthquake.--e
045r075 here there are no country newspapers)--At the CALABRIAN earthquake things pitched off the ground. «Ul
047r083 out (cs <[...]> raised above as). -- In great CALABRIAN wave did not sea break first? I can imagine f
063r130 ented.--Chiloe creeper: Furnarius. <Caracara> CALANDRIA: inosculation alone shows not gradation;-- An
027r027 ed] [not located] The frequency of shells in the CALC. Sandstone Concret, is connected with frequency
033r042 g examined the changes of pumice at Ascension In CALC: sandstone at Ascension, each particles coated
098a048 ar in gneiss is identical with layer of flint on CALC.: sandstone. (& as I believe most strata) (Henc
469t135 nt these flowers till late in evening-- On rough CALC. 280 flowers-- allowing each Bee visits 10 flow
023r012 rt). describes a Small granite Isd. capped by CALCAREOUS rock; following Curvature of hill; states co
033r042 a structure: Have shells ever casts alone in CALCAREOUS. rocks??--if so case precisely analogous: fr
051r093 remendous surf & loose sandy beach) deposits «CALCAREOUS» encrustations; At Bahia ferruginous.--At Pe
039r120 on all East side of Van Diemen Land. All the CALCAREOUS rocks which harden by themselves cannot be p
098a046 95.-- It is easy to prove. (pyrites, agates, CALCAREOUS balls) that concretions are connected with a
102a060 ho has studied rocks in detail as amygdaloid. CALCAREOUS rocks of Ascension, each particle coated. &c
105a072 Lowe in his paper says land shells found with CALCAREOUS matter & concretions on coast of Madeira.? H
108a080 -- It appears to me unphilosophical to think CALCAREOUS springs near coral reefs.-- Where vegetation
425e103 egree,-- how different to dog!-- (Hybrids of CALCEOLARIA.)-- «:CD[Does the Power of, «easily» making
132a137 shows first that data wholly insufficient to CALCULATE rate of increase of hares in earth's crust.--
209b157 Iles Canaries I: I,46 St. Helena I: I,15 {T} CALCULATE my Keeling Case: Juan Fernandez Galapagos =Ra
206b146 ng related within 200 years backward might be CALCULATED & this number elimanated say 150 people four
494q004 of man have there been.-- on what principles CALCULATED.-- in order to guess how many generations in
621o047 of man.-- [RHC] By interest I do not mean any CALCULATED pleasure, but the satisfaction of the mind,
206b147 his number would vary at each lustrum, & the CALCULATION of chance of the relationship of the progeni
499q009 a Yew fruit without impregnation.-- (16) Any CALCULATION of number of grains of pollen in any one flo
543m099 ach other-- August 23d. Jones said the great CALCULATORS, from the confined nature of their associati
246c026 ava & Phillippines, has variety at Madagascar, CALCUTTA & Sumatra,. but I do not see how it is known
112a088 tz Roys fact of elevated block of stone.-- & CALDCLEUGHS collection of facts See page 101. in Note Bo
243c019 tralia only New Zealand-- Norfolk. Isd. & New CALEDONIA peculiar species of cassicans: (¿cassicans Au
314c238 rown says character of Flora, N. Zealand & N. CALEDONIA with a dash of New Holland. As in N. Zealand-
339d025 Man not so, but N. & S. New Zealand & New +++ CALEDONIA. two races of Men, but not plants» will it ho
148g022 l of as I am of all the Alluvium. At Mouth of CALEDONIAN Canal opposite Loch Leven two terraces perha
346d048 th. the account. given by Boethius of ancient CALEDONIAN Cattle. Ch 3. Instinct L'Institut. p. 249. (
377d140 in the animal kingdom.-- It is the unit of our CALENDAR.-- epochs & creations, reduce themselves to t
345d043 ut this was bred out & now all are pure red, yet CALF every now & then born with white head (,or «sho
345d043 head (,or «short-horned with» black lip) & then CALF «in both cases» is killed. Notes from Glen Roy
512q019 igration of coots-- variation in hounds= An ugly CALF <turns> sometimes turns into fine beast. would
595n125 miling», just as much instinctive as a bull <tr> CALF, just born butting, or young crocodile snapping
242c017 common to Moluccas & Pelew Isds.-- p. 22. New CALIDONIA-- New Ireland & Britain same kind of dog, wit
068r148 ournal p 202 Vol IV When recollecting Gulf of CALIFORNIA. Beagle Channel.--One need never be afraid o
230b235 67. Dr. Coulter on decrease of population in CALIFORNIA cessation of female offspring: applicable to
186b059 ep to <each> either parent, (this is what French CALL (atavism) Probably this is first step in dislik
222b204 -- My idea of propagation almost infers, what we CALL improvement, --All Mammalia from one stock, & n
277c126 O. Patagonicus. till neutral ground ascertained, CALL them varieties. but two ostriches good species
349d053 being constant. it would be exceedingl wrong to CALL,, one group genus & other subgenus,--> Propaga
397e003 tom experience! that these generations of what we CALL nature, have been conducted almost! invariably
434e127 that. is the Nisus formativus. (what does Muller CALL it) succeeds in altering <or> form of body, or
550m127 their association «ie heredetary», does one not CALL them instinctive emotions?-- Dr Holland remarke
590n094 & believed to be audible. one has good ground to CALL imagination a faculty, a power, quite distinct
604o018 sociated with God. these phenomena we (feel & ?) CALL sublime.-- 4 From the association of power &c &
605o019 the various causes of those sensations, which we CALL metaphorically sublime, but that it is through
605o19v tion, from the associations before mentioned. we CALL sublime.-- It appears to me, that we may often
609o29v stinctive passions--, which being unnecessary we CALL vicious.-- (jealousy in a dog no one calls vice
616o038 rhaps should hardly be called memory; you cannot CALL the frame of mind which makes music pleasant, a
051r095 example of degradation I ever saw on beach near CALLAO.--From Sir. H Davy experiment on the copper bo
054r105 Juan. Galapagos. Cocos-- Ulloas voyage North of CALLAO, the country, to the distance of 3 or 4 league
054r105 observable in a bay about five leagues North of CALLAO, called Marques, where in all appearances not
036r050 in was connected with movement of sand.--it is CALLED "Bramidor"(?).--it was a strange story; I beli
045r076 nts (if <[...]> part of a regular system can be CALLED accidental; the proportional force of crust of
049r087 [Fig. 4] {P} In Cordill: should basal lavas be CALLED Volcanic or Plutonic The cellular state of all
054r105 le in a bay about five leagues North of Callao, CALLED Marques, where in all appearances not many yea
091a026 canoes & another on Mexican Trachyte <roc> lava CALLED Andesite. Red Coral in the Mediterranean 700 f
139aIBC radded vital influence?-- See End of Note Book. CALLED R. N.-- Massac[h]usset would be well worth vis
160g088 escended, it is to the west of Glen Tarf What I CALLED Alluvium shows the ascending fringes {P} which
162g101 of which is so interesting. enters by old tower CALLED Glengarry (Nead Roy told me) it is impossible

Page ***(Key Word)***
177b025 ach other.-- The tree of life should perhaps be CALLED the coral of life, base of branches dead; so t
207b151 d so well at St. Helena. as. Pineaster & Mimosa CALLED Botany Bay Willow V. Dr Royle introductory rem
213b169 rieties> < «Races» > Man in savage state may be CALLED, <species> species. in domesticated <species>
222b203 cy (only less strongly marked) between what are CALLED varieties.-- NB. one mother bringing forth you
225b218] become attached to.-- that insect «not» being CALLED <Phitophagous> omnraphagous. But it will b
230b240), for if they are different then, they will be CALLED species & mere producing fertile hybrids will
240c003 having skulls so different, that they would be CALLED genera., yet retains markings of wings like th
259c065 rent, & therefore being always necessary may be CALLED adaptation With respect to my theory of genera
274c113 ckoos with claw like lark, (one in Australia is CALLED swamp pheasant) Goatsucker--, parrots with cla
275c119 ing. consequence., was endowed with what may be CALLED the prophetic spirit in science--. the highest
283c144 relation to two continents as to be <replaced> CALLED into existence in two continents our ignorance
292c171 want of consciousness & will & therefore may be CALLED instinctive.-- But why do some actions become
306c214 here.-- All the sheep are thick-tailed The dogs CALLED Persian «greyhounds» are Kurdish & come also f
313c234 y of specific change.? <One> the first would be CALLED. generic & other specific extinction-- In the
332d003 er Water-in the hair a century since used to be CALLED Worm Fever, as used much more latley diseased
345d042 or many years, before deserves <name> «to be so CALLED»,-- the short horned cattle have gone on for 5
355d067 namely prediction.-- till facts are grouped. & CALLED. there can be no prediction.-- The only advant
385d163 ond small & so.-- Says, there is breed of Fowls CALLED everlasting layer--. or Polish breed. (he thin
415e070 with reference to the non-necessity of the «so CALLED» progressive tendency law.-- In animals analog
502q012 erica (33) Ornithologum commonly but improperly CALLED Canadense-- would it grow in open air in Swede
521m008 inctive <or> sounds.-- Miss C. memory cannot be CALLED memory, because she did not remembered, it was
522m010 catch up a new train if early association were CALLED up.-- My F. asked him, did he know whom Mr Chi
536m072 matter «(M. Le Compte)»-- the free will (if so CALLED) makes change in bodily organization of oyster
546m110 er cutter, which she valued, & she was suddenly CALLED to go on the lawn to see something, on her ret
583m070 truction, or imitation.-- p. 20. Animals may be CALLED "creatures of instinct" with some slight dash
583n070 ct" with some slight dash of reason so mean are CALLED "creatures of reason", more appropriately they
584m072 dark "it is inspiration."-- this is class of so CALLED instincts to which my theory no way applies.--
584n072 avoiding strings «in the dark» as well might be CALLED instinct,-- migrating to one spot, this is ind
584u072 is is indeed instinct.-- Australian man, may be CALLED instinctive: the facts of memory of roads long
607o025 ility of receiving impressions-- accidental (so CALLED like chance) circumstances. As man hearing Bib
607o025 how many actions are not determined by what is CALLED free will, but by strong invariable passions--
616o038 y meet their reward! X Perhaps should hardly be CALLED memory; you cannot call the frame of mind whic
616o39v de known to us, revealing respectively what are CALLED it's subjective & objective aspect. The subjec
538m079 s -- -- of Trinity always thinking people were CALLING him a bastard.-- when drunk.-- having really b
547m113 y be instantaneous changes in order <to every> CALLING up ideas of every late impression.-- (do the i
547m113 ing new ones from senses. & <comparing their> «CALLING up» old ones, to be sure of ones consciousness
608o025 (Mem: M. Le Compte case of Philosophy, & savage CALLING laws of nature chance) 2) difference is from i
620o044 an, from his memory & <pow> mental capacity of CALLING up past sensations, he will be forced to refle
554m138 - A very green monkey (from Senegal he thinks CALLITRIX Sebe??) he has seen place its head downwards
461t037 ill be so handed down«(».. as mammae of men «CALLOSITIES on Camels & Horses--».--«)» & therefore prob
513q020 bout fertility of ass-zebra-horse= (5) About CALLOSITIES on Camels-horses. &c &c Rhinoceros= (6) Cros
021r005 n E. Indian Archipelago, no minute description, CALLS it a Fucus. P «Vol I 287» P 379. Henslow Angles
074r164 more abundant in the upper limestone, which H. CALLS by several secondary names «Study Hoffmans acco
241c014 is marked wild breed of oxen at Java.--.p. 140, CALLS it Bos. leucoprymnus. does not say whether wild
342d034 heasant & Grouse homogeneous? I observe Bachman CALLS these Hybrids new species. Yarrell says the bir
413e059 o my theory Babbage 2d Edit, p. 226.-- Herschel CALLS the appearance of new species. the mystery of m
522m012 effect at the time, but some time afterwards it CALLS up pain. or pleasure. & is often recurred to &
539m083 is-- In the habitual train of thought one idea. CALLS up other, & the consciousness of double individ
550m127 ch are pleasant, every concomitant circumstance CALLS up pleasure. or pleasure or pain of association
608o025 these passions, weak, opposed & complicated one CALLS them free will--the chance of mechanical phenom
609o29v ry we call vicious.-- (jealousy in a dog no one CALLS vice). on same principle that Malthus had shown
623o050 the place, but merely in the place. & yet place CALLS up pleasure.-- [LHC] the instinct of sociabilit
029r033 as no uncommon ripple on the water. It was quite CALM at the time. Latitude 8 deg. 47 min. N: longitu
029r033 eg. 47 min. N: longitude 61 deg. 22 min. W. mid. CALM and clear. Caermarthen Journal I look at the ce
399e011 ety isles. Hope says positively he has seen. a CALOSOMA. (very like American form) in Stonesfield sla
466t091 ies-- few-- Maer June/41/, observed 3 plants of CALTHA Palustris alone together. one had seed-pods tu
473s07r cattle & sheep for horns & yet no difference in CALVES--how is this in young pigeons--dogs--cattle? A
502q11v fferent vegetables & animals come on.-- Compare CALVES.: Compare young. beans. cabbages.-- History of
512q019 into fine beast. would its offspring have ugly CALVES. also turning into fine beasts.-- For comparis
090a022 which must have fallen on the globe since the CAMBRIAN system In Ures dictionary between 1768 & 1818
224b216 angels. produced «&» Has the Creator since the CAMBRIAN formations gone on creating animals with same
436e137 ame type of shells in oldest formations?-- The CAMBRIAN formations do not however, extend round world
460t019 -- dicot. plants in coal measures.-- Shells in CAMBRIAN & Crust show how long since present forms exi
434e129 <south> Wales & certainly <old> only white in CAMBRIDGE, in some counties sometimes one & sometimes o
543m099 in the same way as are chess Players-- A man at CAMBRIDGE, during his time, almost an absolute fool use
568n019 so it is with all uneducated.--»-- Old man at CAMBRIDGE observed the ignorant. merely looked at pictu
041r064 n Owens Africa it is mentioned that the Elephant CAME towns driving by the want of water.--I believe
105a071 - beach action -- no one will dispute. sea. once CAME to Mendoza--. Will they introduce other causes
105a072 s matter & concretions on coast of Madeira.? How CAME it if this powder results from «decomposed sea»
121a112 ll has seen same thing-- Consider profoundly How CAME it. that Glen Roy district could have been elev
121a112 t fissure & unequal.-- where were cracks?--? How CAME there ever to do this vast quantity when during repo
154g060 oss to {P} side removed all well & good, but how CAME river to do this vast quantity when during repo
214b176 any person would say the tail was broken-- This CAME so often «that» it was difficult to obtain a li
215b177 lk--Fox tells me that it is generally said.= How CAME first species to go on.-- There never were any
233b250 ce in S. America. which were hard to distinguish CAME from closely neighbouring localities Institute
278c132 for S. America & Europe.-- The difficulty is how CAME it animals not preserved, in central S. America
288c160 rections: mysterious. Were the woodcocks,. which CAME Madeira & <seized> ceased their migrations lost
334d012 ed in, then disappeared, & three days afterwards CAME again, bringing with her the other & younger co
344d042 tch-- tried to breed from her, but her offspring CAME out one big & one small. Now Jones, before this
360d090 t go to heat. but parts swelled, though no fluid CAME from them.-- showing how gradually every «thing
378d148 colours.-- they were "half wild-half tame, they CAME to the windows to be fed, but still they have a
382d158 both present in every shade of perfection --How CAME it nipples <are> «though» abortive, are so plai
388d171 - now if one of these staid in the womb, when it CAME out. it might partake of shade of fathers chara
415e069 at a thing has been so, & another to show how it CAME to be so.-- I speak only of the former proposit
469t135 nch & on this day in five minutes eleven Humbles CAME up in 1840 true. Shrewsbury.-- Abberley-- Early
469t151 ows «close to each other» & seeds gathered «all» CAME up in 1840 true. colour of flower & foliage not «bein
469t151 were planted in rows, & seeds gathered same year CAME up true «in 1840»: All in together blossomed to
472s01r ng in same garden. & out of 60 seedlings not one CAME up true.-- colour of flower & foliage not «bein
472s01v e procured with the P. orientale in garden & all CAME up hybridised. It is possible to raise them pur
531m051 reflecting over an insane feeling of anger which CAME over me, when listening one evening when tired
539m081 an intense headache «after good days work» which CAME on from reading «review of» M. Comte Phil. whic
555m143 onfused manner. thought that a person was hung & CAME to life, & then made many jokes. about not havi
555m144 whole train of Dr Monro experiment about hanging CAME before me showing impossibility of person recov
556m144 hanging on account of blood. but all these idea CAME one after other, without ever comparing them, I
567n013 ust like child not knowing what to do with them, CAME several times & opened my hand, & put them in--
578n055 she> «the» thought of his knowing «it», suddenly CAME across her, the blood rushed to her face,"-- On
595n121 rightened when Snow put a guaze over her head. & CAME near him, although knowing it was Snow.-- Is th
618o41v thought & organization: But if the weight never CAME untill the blueness had a certain intensity (&
634j55r rigin of Polar attributes of the Cetaceae.-- How CAME Bats also.? before birds? They are ancient.-- A
057r113 perished because too cold:--With discussion of CAMEL urge S. Africa productions--. I think in Patago
087a009 e eaten out the Megatherium. -- The Guanaco the CAMEL.? Make note about N. American bone not probably
424e103 wild have, have done so in hot countries.-- CD[CAMEL does not vary «one ought not to be able to hybr
424e103 vary «one ought not to be able to hybridise the CAMEL» like ass «same way some plants vary more than
443e157 emblance of the foot the Ostrich to that of the CAMEL has not escaped Naturalists." Before he alludes
485z017 n D'Orbigny on species of Mephites 4 distinct CAMELIDAE. do not breed together Mag: of Zoolog & B. Vo
579n060 seen to be muscular movement. The Blushing of CAMELION & Octopus; strong analogy with my view of blu
086a005 e water"-- It may be worth noticing edentates & CAMELS in deserts & rodentia In Plata Mastodon Toxodo
231b244 titude of small creations.-- Will Dromedaries & CAMELS breed?-- As man has not had time to form good

253c046 ng» America & India deer.= Africa not.-- Africa CAMELS?? Africa Bears??-- Plantigrade Carnivora??-- «
458t001 Australia,. S. America--» These strange forms., CAMELS, giraffes. Sivatherium & Anoplotherium, with e
461t037 nded down«(».. as mammae of men «callosities on CAMELS & Horses--».--«}» & therefore probably any str
513q020 ty of ass-zebra-horse= (5) About callosities on CAMELS-horses. &c &c Rhinoceros= (6) Cross. Sus Barly
637j57v laws of adaptation[» p. 234. The non-absorbing CAMEL'S stomach is puzzler p. do says inconvenience wo
638j28r g nostril, foot, sack. power of endurance &c &c CAMELS? all good cases of corelations.-- [There must
602o11v with art (p 128) R. compares a view taken by a CAMERA obscura &c a Poussin.-- How are my ideas of a
358d075 were amazingly bold, always accompanying us from CAMP to camp; it was absolutely necessary to watch o
358d075 zingly bold, always accompanying us from camp to CAMP; it was absolutely necessary to watch our meat,
470t153 flowers of small Linaria in do Domestic do б CAMPANULA (two species)-- in do-- do 3 of do in about <
502q11v ot elm. does it «in» melon-- «Loasa» Anchusa «CAMPANULA» &c & dead-nettle.-- Lithospernum. Blue Gloss
481z009 ble engravings in Meyen Zoology on animal of CAMPANULARIA Alcedo stellata. Meyen p. 92.-- great Kingf
091a026 in some of. the twopenny periodical said so. «CAMPBELL the Poet» Accra. Coast of Africa. Clay Slate
243c019 Mais il n'y a pas jusqu'aux iles Macquarie et CAMPBELL (52 degree S) qui n'aient egalement leur espè
469t119 !xx In Phil Transact. about year 1778. Paper by CAMPER on Ourang-outang, has examined 7 says one spec
362d099 rfect-- (is not this so in S. America with C. CAMPESTRIS<)> refer to my notes) & Mr Yarrell supposes
363d103 irds always second their songs, the ++ Cervus CAMPESTRIS spotted white when a fawn compare with fallo
334d009 & Muscovy Ducks.-- English. <Common> «China» & CANADA Geese, & that they this first cross were equal
359d087 Saw cross between Penguin Duck «from Bombay» & CANADA Goose.-- Former strange mishaped bird-- looks
447e165 arden, which is case precisely analogous to the CANADA onion mentioned in Hort. Transact. Aira caespi
502q012) Ornithologum commonly but improperly called CANADENSE-- would it grow in open air in Sweden. Linnae
359d088 ommon duck-- Male Penguin was crossed with hen CANADIAN offspring, I should say in every respect most
148g022 am of all the Alluvium. At Mouth of Caledonian CANAL opposite Loch Leven two terraces perhaps upper
161g098 e source is a lip with the new shelf flows into CANAL between L. Lochy & Oich. is a brook on the Loch
209b157 lls.-- {T} Genera In North Africa. I: 4,2 Iles CANARIES I: I,46 St. Helena I: I,15 {T} Calculate my K
096a039 Greenland: Mem.¿ Greenland subsiding.) Von Buch CAMPBELL Isd. p. 351.. NB. Mackenzie talks of gravel on
097a042 & 221 Lamellar dikes like Mica Slate Von. Buch. CANARY Isd. p 170.-- Mem. Cordillera Can Greenstone d
209b156 ctrine & Lyells. to certain extent Von Buch,.-- CANARY Isles: French Edit. Flora of Islds very poor «
225b220 uth Sea islds. any animals?-- I believe none.-- CANARY islds.? Madeira? «Tristan d'Acunha?» «Iceland?
296c184 «p. 473» Webb &. Berthelot. must be studied on CANARY islands-- Endeavour to find out whether Africa
299c193 to themselves. this remarkable compare it with CANARY Islds. Galapagos.-- Iceland has same uniformit
317c249 n old man--, who told me that the mules between CANARY birds & goldfinches differed considerably in t
317c250 intances. which grow on the lower region of the CANARY islands-- p. 250 admirable table of plants of
317c250 le of plants of St. Jago showing many common to CANARY. isld., Europe, & St Jago upper region, & some
321c275 re except Brown's Appendix & excellent table of CANARY isld: Plants Home's History of Man Transaction
421e090 e Fauna»-- (near Oran) approach in character to CANARY Isld.-- ie Canary Isld approaches more to neig
421e090 ran) approach in character to Canary Isld.-- ie CANARY Isld approaches more to neighbouring coast of
511q018 ities in birds & animals for prizes.= Pidgeons. CANARY birds-- Bantams.-- (6) <Mad> Porto Santo Rabbi
500q10a many races or doubtful species; how is this at CANARYS Arch-- it is so at Galapagos.-- Ireland, doubt
332d002 ing a year or two he saw many cases of virulent CANCER in women, & since that time it has been rare d
577n053 lking about any disease tends to give it, as in CANCER, showing, effect of mind on individual parts o
236b1BC t. Science. Naturelle Geographie Botanique De CANDOELLÉ. Geol. Soc Horae Entomolgicae Linn: Soc. Geof
533m060 has obscurely occurred to me that Capt. F. R. CANDOUR & ready confession of error made him less repe
503q012 r Royle, latter says seedless-- Also about Sugar-CANE Edwards says does not seed-- «Bruce says does»
123a117 ains found in many part.-- great Dasypus near CANELONES -- large quadruped bigger than ox.-- at Bueno
315c243 mer shows that a man grinning is to exposes his CANINE teeth. no doubt a habit gained by formerly bei
315c243 it gained by formerly being a baboon with great CANINE teeth.-- (This may be made capital argument if
315c243 anines) --. Blend this argument with his having CANINE teeth at all.-- This way of viewing the subjec
432e125 e of young birds retrogressing-- Uncovering the CANINE teeth, or sneering, has no more relation to ou
542m093 t hard to keep his lip from stiffening over his CANINE teeth.-- He may feel satisfied with himself, &
542m096 ild does not sneer. because no young animal has CANINE teeth.-- A dog when he barks puts his lips in
542m096 re it is here continued when the uncovering the CANINE useless.-- The distinction «as often said» of
315c243 argument if man does move muscles for uncovering CANINES) --. Blend this argument with his having canin
639j28v . Endless cases.-- Macculloch p. 260 intimates CANINES no special use to Man. Applicable to Bell's sn
591n100 rs by habit (not instinct) think not fit, as CANNABALISM, is held in abhorrence.-- all this makes ana
120a108 springs &c &c--then if so, thermometer show it CAMT be ordinary heat, then there is something super
226b220 lle's Voyage.-- Galapagos mouse (?) brought by CANOES Ceylon & India.-- Van Diemen's land Australia.
223b211 niformity of species & we Must confess. that we CANOT tell, what is the amount of difference, which i
539m084 ry Queeene. CD 25 (Descript of Queen) «O» of Hell CANT IV or V.) as pale & trembling. & not as flushin
127a127 dence Elie de Beaum. Memoires of French Geolog. CANTAL Vol III I? p. 246. on formation of cones benea
466t095 osed by cantering horses having colts which can CANTER-- & DOGS trained to pursuit having PUPPIES wit
466t095 herited at corresponding age & sex, opposed by CANTERING horses having colts which can canter-- & DOGS
466t095 ve & doubtless not confined to sex.-- «Is not CANTERING a congenital peculiarity improved.» Probably
582n067 utions will then be cut off.-- <Horses> Colts CANTERING in S. America capital instance of heredetary
215b176 loose desire, and it is required to give the CANTHAIRIDES and milk--Fox tells me that it is generally
402e018 s the fox-like animals. which are met with near CANTON" "Here, as in all Malay countries, I noticed a
350d054 n nature; not anything framed to fill up empty CANTONS, & unnecessary spaces" p 23. "for Nature is th
458t001 forms are long preserved.-- vol VI. p. 539. Dr CANTOR'S account of fossil frog, 40 inches in length--
198b113 in status (Perhaps consideration of range of CAPABILITIES past & present might tell something) p. III
608o026 is good for him effect of Education & mental CAPABILITIES.-- '(P) Animals do attack the weak & sickly
355d069 nt with respect to extinction of species, the CAPABILITY of only small amount of change at any one ti
371d118 s Garden. 1279 varieties of roses!!! proof of CAPABILITY of variation.-- Saw his collection of «Hummi
425e103 tolerably fertile hybrids, bears relation to CAPABILITY of variation?? my theory says so.--» March 6
527m034 he closest train of geological thought.-- the CAPABILITY of such trains of thought makes a discoverer
543m097 they communicate not easy to know,-- but this CAPABILITY of understanding language is considerable, t
607o025 cation under the influence of others-- varied CAPABILITY of receiving impressions-- accidental (so ca
611o034 n the inorganic world; life itself being, the CAPABILITY of such matter obeying a certain & peculiar
180b037 forms, for if each species «an ancient (I)» is CAPABLE of making, 13 recent forms.-- Twelve of the co
249c034 en generation will «only» produce an offspring CAPABLE of producing <two> such as itself.-- therefore
249c034 l either produce no of[fspri]ng or such as not CAPABLE of producing again The Varieties of Cardoon ar
257c059 SUMPTION to say generation produces young ones CAPABLE of producing young ones like itself, but ¿whet
263c076 > force in one man the developement of a brain CAPABLE of producing more glowing imagining or more pr
280c136 offspring unlike itself, than to produce that CAPABLE of producing itself alike.-- in one case it ch
289c162 ston <on> Entomostraca Daphnia, produce young, CAPABLE of producing young many times & lay two sorts
307c215 e organs in some Males animals, Mammae in Men, CAPABLE of giving milk)» rudimentary wings. so nature
308c216 car!!!!!! Proceedings of Geol. Soc Vol I It is CAPABLE of demonstration that all animals have never a
310c224 ssertion that globules of milk produce a plant CAPABLE of growing!! & propagating itself. In Tropical
379d152 n's Case.-- When cows have twins, <one> though CAPABLE of producing both <male> pair of male & female
389d176 e & female as foetus one sex; & therefore both CAPABLE of propagating, but one is rendered abortive a
412e057 re innumerable variations>. Every structure is CAPABLE of innumerable variations, as long as each sha
541m090 .-- this greatest mental effort, of which I am CAPABLE-- I suspect from these facts that whole effort
553m136 . of which we profane «degnen» in thinking not CAPABLE to <do> produce every effect. of every kind wh
583n070 & reason must be thought of. p. 19. animals CAPABLE of education; (this is again assumed as more a
617o040 s in the manifestation of force i.e. movement? CAPABLE of being traced to the body of the individual
536m072 y to be direct effect of organization, by the CAPACITIES its senses give it of pain or pleasure, if s
285c150 . S. showed that savages are not born with any CAPACITY for observation of tracks &c &c Dr. S. has so
558m150 -- May not moral sense arise from our enlarged CAPACITY <acting> «yet being obscurely guided» or stro
620o044 .-- Not so man, from his memory & <pow> mental CAPACITY of calling up past sensations, he will be for
022r007 P. 417 Veins of quartz exceedingly rare Mem C. [CAPE] Turn P. 434 & 419 As Limestone passes into sch
024r015 ide facts in Beechey. on NW coast of America off CAPE of Good Hope 70 fathoms 20 miles from the shore
064r134 eserts.--Much struck with number of animal[s] at CAPE of Good Hope Says at Santos «M Birchels» at foo
065r138 expedition from <Deception Isld.> South Shetland CAPE Possession. Syenite¿ Andite?-- Degrading of inl
089a020 enegal. L Institut p. 192.-- (1837. Peninsula of CAPE Verd. volcanic.-- Isle of Gory. rocks encrusted
203b135 stitut p 245. 1837" we should have anomalies. as CAPE Anteater,.-- This supposes world divided into Z
216b180 appen.-- Dr Smith says he is sure of the case at CAPE.-- McClay argues from it Black & White species.
219b194 rica, sandstone, & granite, (that is genera near CAPE) see if there are any species same as T. del Fu
229b233 h's Information Long Horned (very) aboriginal at CAPE: crossed with English Bull. offspring very like
229b233 Hottentots say great tailed sheep aboriginal at CAPE & a thinner-tailed kind further inland.-- NB. T

Page
(Key Word)
247c028 ole austral» «caught» Chile, Van Diemen's land & CAPE of Good Hope V. p. 44 of this Note Book Rabbits
252c042 . travels (what language?) Hyena «venatica» <of> CAPE found in Desert of Korto & Steppes of Kordofan
252c045 pecimen of arrangement, 3 <5> Species Rhinoceros CAPE town good species Indian species so distinct th
266c091 Bridgewater treatise, (p.,26). about plants from CAPE of Good Hope continuing for some time to flower
310c224 gating itself. In Tropical countries (as St Jago CAPE de Verds) the shells in equal periods with Euro
317c250 isld., Europe, & St Jago upper region, & some to CAPE.-- some proper well-worth studying, with respec
401e014 of varieties, which he attributes to crossing.-- CAPE Broccolli can hardly be reared without greatest
449e173 m.-- Athenaeum: 1839. p. 451. Sheep Merinos from CAPE of Good Hope,. has different constitution from
478z003 the genera-- Cyclops p. 134. and p. 115 In white CAPE Pidgeon's stomach small shells (patella) sea we
497q007 eguminosae.-- Herbert explains numerous spec. of CAPE Heath by facility. ¿Knight take opposite view.
508q016 Smith, about tamed wild animals breeding at the CAPE.-- ﬞAbout two vars: of Lion: Annales des Scienc
508q016 rs Scientific Memoirs (11) And. Smith Savages at CAPE any selection of Males in «cattle» or in Killin
473s05r N. Wales can be distinguished-- & Jackson here (CAPEL-Curig) says that he can certainly tell Trout fr
473s05r ys that he can certainly tell Trout from Ogwen, CAPEL Curig & some other lakes, (different waters) He
473s05v nto Conway by Bettws & there joins streams from CAPEL-Curig-- Mr Bunbury says Miers has described in
357d073 n Sweden & anciently in Britain) between hen CAPERAILKIE & cock Black-cock.-- (Curious the readiness
364d105 sides the Rakhekna, before mentioned between CAPERCAILZIE & Black Cock.-- The latter has crossed with
104d068 moisture being lost by evaporation therefore CAPILLARY attraction would bring water with salt to sur
103d065 w being subsequent to dislocation of strata. A CAPITAL discussion might be made between dikes & «axis
110d086 these «lines» Description of rocks in Lyells'. CAPITAL Norway case.-- The fragment. consisted of horn
249c035 replacing species. Dr. Smith will give me some CAPITAL information. ¿Carnivora of New & Old word. do n
261c069 ks in Fauna Borealis must be studied. There is CAPITAL table of extent of all species Accumulate inst
289c160 «ring ouzels» dive instant touch the water.-- CAPITAL instances of typical land bird, having habits
307c215 ve organs of same tendency in species, this is CAPITAL & novel argument.-- Are there any abortive org
315c243 n with great canine teeth.-- (This may be made CAPITAL argument if man does move muscles for uncoveri
337d022 r.-- <Transactions of the Entomological Soc> A CAPITAL passage might be made from comparison of Man,
425e104 biatae, some of these species are described.-- CAPITAL case,-- for Sandwich Isld are very similar to
471tf09 other". Phillips (Lardner's E. vol. II p. 18.) CAPITAL list of all the fossil Mamm. of Europe-- Large
495q05a uled in squares to facilitate investigation.-- CAPITAL in middle of ploughed field-- on hills.-- 10 S
564n003 ith passion» he must have conscience-- this is CAPITAL view.-- Dogs conscience would not have been sa
582n067 off.-- <Horses> Colts cantering in S. America CAPITAL instance of heredetary habit:-- there must, ho
380d154 wary & Guinea Fowl cannot be distinguished.-- A CAPON will sit upon eggs, as well as, & often better
381d156 nation in Man, has strange effect.-- Directly a CAPON is cut, it increases in size prodigiously-- Ani
384d162 an is hermaphrodite:-- ∴ developed instincts of CAPON. & power of assuming male plumage in females.
023r012 rand, (SW part). describes a Small granite Isd. CAPPED by Calcareous rock; following Curvature of hil
537m074 man intentionally can wag his finger from real CAPRICE. it is chance, which way it will be, but yet i
524m021 do not go back to childhood, (but appear most CAPRICIOUS) as in delirium after epilepsy, but in the f
250c040 titudes!? Zoological Journal.--- Vol I. p. 81. CAPROMYS, West Indian isld. p. 120. «ref.» Philosop. T
251c042 in Smiths work «do» Vol. IV p 273. Macleay on CAPROMYS. 4 species probably in Cuba (p 271 Viedo says
515q021 ost cultivated show Heartease produce as large CAPSULES of seed, as the commoner kinds-- Cattle are h
023r012 rks distributing fossil remains: Sharks followed CAPT. Henry's vessel from the Friendly Isles. to Syd
042z067 travels on the Chesil bank. V. De la Beche). Ask CAPT. F.: R: how the swell, generally & during gales
057r114 pper. in accordance in Europe with ice theory.-- CAPT Ross found in Possession Bay in 73 degree 39 N.
061r126 1826.27.28. grt. drought at Sydney. which caused CAPT. Sturt expedition. --¿ Another one in 1816 (?)
064r135 els. there no outer coast.--important effect.-? CAPT. FitzRoy. -- Limited Volcanic action & limited
065r138 ch Isd-- Specimens of rocks were brought home in CAPT. Forster expedition from <Deception Isld.> Sout
065r140 les of Porphyry.--Falklands.--off East Coast. -- CAPT. Cook found soundings. (end of 2d voyage outsid
080r181 . Bahama Isd De Lucs travels Beauforts Karamania CAPT. Ross. & Scoresby deep soundings Gilbert Farquh
100a055 ers to salt as being produced by local heat. Ask CAPT. Beaufort, whether, water flashing into steam
218b180 ump.-- p 173. Voyage par. un Officier du Roi Mem. CAPT. Owen's story of cats on West coast Africa.-
253c047 > ̄an account of the MANELESS lion of Guzerat by CAPT. W. Shee. considered merely variety.-- red form
258c061 in Macquarie isld. vol III p 430 alluded to by CAPT. King do. p. 434. Table of birds from Cuba Vigo
400e012 tions.-- Kerr's Collect of Voyages Vol 8 «p. 46» CAPT Davis in 1598 found cattle in Table Bay with Hu
427e109 w line there «& there & there only: as stated by CAPT. Graah») & break up. N. American Conchology fro
483z012 of Birds".-- 18 do. Vol III p. 422. letter from CAPT King on birds of St of Magellan. Very inaccurat
483z013 th advantage. with Cumms collections & my own: & CAPT. King's p 453-- Planariae velellae (Less) paras
533m060 e humble.-- it has obscurely occurred to me that CAPT. F. R. candour & ready confession of error made
029r033 act from the log-book of the James Cruikshank, CAPTAIN John Young, on her voyage from Demerara to Lon
355d068 olypi we see young bud changing into ovules.-- CAPTAIN Grants. Himalaya. shells (see Paper in Geolog
306c213 ill say. not species.-- organs of generation a CAPTIAL character. (Owen) not for first & grandest div
512q020 ave they been raised from young ones, bred in CAPTIVITY --Mr Miller says Wombwalls were (4) About fer
185b055 change. like the Buzzard which has changed into CARA cara at the Galapagos. Fernando Noronha Ophyres
185b055 ge. like the Buzzard which has changed into Cara CARA at the Galapagos. Fernando Noronha Ophyress bi
182b047 er «of steps» known). <for instance among the CARABIDAE.-- instance in birds> Examine «good» collecti
185b056 atives (which at first would be mistaken for) CARABIDAE, Crysomela, Scarabadae, & longicornes.-- Agai
641j29v enid «& acrid» secretion in Mollusca. insects <CARABIDS & Staphylini» & Mammalia. The eye being forme
063r130 it & represented.--Chiloe creeper: Furnarius. <CARACARA> Calandria: inosculation alone shows not grad
358d075 ving very same habits in some respects as this CARACARA.-- Sept. 9th. It is worthy of observation tha
361d096 ore brilliant plumage than male.-- «My case of CARACARA. N. Zelandiae.--» Mr Blyth stated «that there
480z007 erent size-- How does this apply to pale brown CARACARA Krauss on Corallinae from S. Seas written in
484z014 or Secretary is S. African representative of CARACARAS of Americas.-- manner of walking:-- foot bill
043r070 2 affected valley of Missisippi & New Madrid & CARACCAS.-- Is this mentioned by Humboldt in his accou
056r110 eat«er» for all Europe, than from the Plata to CARACCAS, which is all of granite: In discussing circu
066r142 Sandstone. he could not distinguish from stone CARADOC from tower of third Silurian division--Togethe
044r074 gazes. especially the most abundant. Sulp. Hyd: CARB: A. Mur: A. = (& this effect of water thus hold
074r164 sulphur.. fluor spar. bayte. asbestos garnets.--CARB & chrom. of lead. orpiment. chrysop[r]ase. opal
079r176 mur of Silv.[,] Sulph. of do.[,]galena[,]quartz, CARB. of Lime. accompany.-- Ulloa has said silver in
096a041 cting lime <n> being heated without parting with CARB. Acid.-- Mr Malcolmson in Paper on India gives
096a041 dia gives reason for knowing that Mur. Soda. and CARB of lime decompose each other.-- on Direction of
098a047 same force which draws together two particles of CARB. of Lime, tends to crystallize them as seen in
496q006 rate of Soda-- Salt. Gypsum. Magnesium Iron Rust CARB. of Ammonia.-- Horse Urine &c &c on associated
530m043 about his own health.-- his complaint was CARBBUNCL on <Head> Neck.-- He has seen other cases of
123a116 the Chilian ones.-- Septemb. 2d.-- Sulphur like CARBON must go round of dissemination & separation in
614o037 gan, as bile of liver.-- ¿ is the attraction of CARBON, hydrogen <&c> in certain definite proportions
030r036 . interesting from great quantity of altered CARBONACEOUS shales Examine chart of Patagonian coast to
022r007 les have a tendency to change their position? CARBONATE of Lime disseminated through the great Plas N
076r168 Norway.--namely dendritic silver intersecting CARBONATE of lime-- native silver in Mexico is always a
109a084 think good p. 411 When discussing concretions CARBONATE soda. formed by Ca. of L. & Mur. of Soda mixe
108a080 it might be almost as well said probably much CARBONIC Acid gaz here.-- Top portion page excised, &
138a176 e top-- [blank] Would rotting wood by yielding CARBONIC Acid unite with «piece of cabbage» alklali &
138a176 precipitate silica / or charcoal charged with CARBONIC acid [blank] Many interesting experiments mig
196b109 but not vice versã. ¿could plants live without CARBONIC acid gaz.)-- Yet unquestionably animals most
516q23v 0 p. 777. Decaying wood absorbs oxygen & forms CARBONIC Acid. will this bear on Petrifaction?-- [blan
308c216 period assumed, then some perished before, CARBONIFEROUS some perished before, then there always hav
316c247 n fossiles-- «(so much the more remarkable ∵ CARBONIFEROUS ones similar?)» Now this is very remarkable
447e164 &c)-- Is there greater resemblance between CARBONIFEROUS. «& recent» mollusca, than between the corr
202b131 surface forest trees fallen «kind well known, CARBONIZED»--; clay fifty feet, then forest 120 ft Mica
041r065 ommon.--Mr Lyell has mentioned the drifting of CARCASES putrid. In Rio paper. when discussing probabl
111a088 folk Isd into Geograph Journal Vol VII p. 279. CARCASES of birds drifting out to sea do p. 358. chang
267c094 are composed solely of this shrub".-- p. 229. CARCASES of birds drifting out to sea Vol VII. p. 325-
328cIBC g Matica & many shells of Genera Corlula Cham. CARDIUM. Porcellus Turbo. Cerithium Jardin du Roi Java
490q001 of shells of Sandwich group {Sowerby monstrous CARDIUM-- does it remind him of other species Hooker s
249c035 ot capable of producing again The Varieties of CARDOON are cases <sp> like those of Primrose & Cowsli
065r138 na Stromboli & Vesuvius Investigate with greater CARE. vegetation & climate of Tristan D. Acuneha. Ke
078r174 uoy et Gaimard.--D'Orbigny has described it with CARE to 3 species. I think I have much additional in
217b183 y one) of rare breeds of dogs, from owners great CARE of them. Fox says when two dogs of opposite bre
240c004 , that after breeding in pidgeons with very much CARE that, it requires the greatest difficulty to re
285c153 sticated «??», & therefore Most especially under CARE of Man. & external circumstances not variable.-

325c267 r any my views very ancient? Study with profound CARE, abortive organs produced in domesticated plant
401e014 Broccolli can hardly be reared without greatest CARE be taken to prevent fertilization from turnips
410e053 - The traces of changes in forms of organs, will CARE little for species, except so far as wanting na
410e053 of the corelation of parts in individuals, will CARE little, <in> whether the individual be species
428e112 ed by him to me, for otherwise breeders who only CARE for first generations,, as in horses, would not
428e112 for first generations,, as in horses, would not CARE so much about breed.-- what can «however» be mo
429e114 & insects) which are propagated with very little CARE.-- & which might spread themselves, as well as
505q014 tion necessary to fruit--; become well shaped by CARE 13 Arum before pollen is shed can you find flys
530m045 nsorium.-- -- analogous to sleep--; some doctors CARE it, by stimulus & afterwards patient sinks.-- W
548m115 Consciousness is curious problem., one does not CARE for the pains of ones infancy.-- one cannot bri
560m156 of bare nails--, & of eyes.-- Do female monkeys CARE for men.-- Have we any ferns in the hothouse at
633j53v some peculiarity in seed vessel]CD if man takes CARE they are not detrimental.-- NB. One limit to th
635j56v y cases, as in Papilionaceous flower, where such CARE seems to be taken that the anthers should not b
352d058 ould explain every thing.-- PURE HYPOTHESIS be CAREFUL.-- Argument for circularity of groups. When <s
365d106 lue breast, latter reddish, hybrid purple-- be CAREFUL, See to hybrids between Pheasant & Black Cock,
389d177 is somehow connected (This seems case, for by CAREFULLY observing cattle can be bred in & in.)-- [The
147g017 f it; to the Sea more probable I did not look CAREFULLY for Marine remains-- Some of the hills almost
155g065 evel during any oscillation must have been so CAREFULLY preserved as to have thrown water in same «dr
322c269 two voyages, skimmed well. do Lutke's Voyage. CAREFULLY read.-- Reynolds Discourses Lessing's Laocoon
502q11v ena Compare flower of different Cabbages most CAREFULLY to see if variation equal in flower with leav
567n016 learning facts for induction. one is obliged CAREFULLY to separate its memory from all ordinary line
030r034 uch as between Mocha & main land). <[...]> At CARELMAPU.--Within Chiloe:-- On open coast, near where
446e160 and <in> these reciprocally assist in domestic CARES, as building nest, sitting on eggs. & feeding &
506q15v besides the Nettle. at Galapagos-- Dioecious.-- CAREX.-- We may presume Nettle spreads by seeds- (44)
452e182 ppoximat to Negro form; those from Rhine to the CARIBS.-- Vol II p. 650. Long attested account of fal
507q15v iation? Henslow knows only on Citrons 47. Ficus CARICA Henslow presumes females produce. Polygam. tri
484z014 uthern America-- valuable & practicable deed CARICARIDAE wanderers.--?? in N. America?? Wilson N. Ame
514q21. eties in mountains of Tahiti. Dr. Boott-- says CARICAS from every isld differs-- do they also differ
639j28r l animals run by hind legs-- Kangaroo. only a CARICATURE; Penguin.-- Pincers in Scorpion & Crust in S
451e177 ls.-- Horsbrugh E. I. Directory. Vol II. p. 46 CARIMON Java. (between Borneo & Java) Lat 5 degree. 50
511q018 from a different looking animal be formed-- not CARING whether good or bad.-- are any actually reject
590n093 ation p. 75. Haller says tooth ache, even from CARIOUS tooth cured by sight of instrument.-- Bennett'
234b255 ity is only aim in this world [not located] T. CARLYLE, saw with his own eyes. new gate. opening towa
572n033 abitual action.-- L'Institut. 1838. p. 340. Mr CARLYLE says that negro certainly has less reasoning p
323c269 on Hybrid Mixtures: Marginal notes. -- 20th. CARLYLE'S French Revolution 3? vols. oct: -- 26th Blume
218b192 hough of trifling order are formed by nature. CARMICHAEL. Tristan D'Acunha, a list of its Flora. is g
219b195 aciers might have acted at Tristan d'Acunha-- CARMICHAEL Linn. Transacts. Vol XII.-- The Alpine plant
480z008 s amongst Vegetables in Linn. Soc.-- Mr. Donn CARMICHAEL Linn. Transacts Vol XII. p 496. Birds at Tri
495q05a orifice (8) Carry Bees, powdered with starch & CARMINE & experimentise on their returning powers-- th
502q11v ossible to see orifice of poison-tube-- so put CARMINE in spirits & then experimentise: for gradation
037r056 piapo.--Sydney. K. G. Sound. C. of Good Hope.--CARNATIC It has been common practice of geologist. Lye
026r020 Transactions on metallic veins. 1830 P. 399.--CARNE. Geolog. Trans: Cornwall «Vol II» It is a fact
435e134 guiformis. he remarks lives in dark caverns of CARNIOLA p. 112. Man. "standing alone in the gift of i
204b141 vision «do» we not see a splitting in orders, CARNIVORA, rodents &c, JUST COMMENCING. Kirby says (not
226b223 basin «allied to present») more like present CARNIVORA than Pachydermata If my theory true, we get (
249c035 Smith will give me some capital information ¿CARNIVORA of New & Old word. do not form two sections i
253c046 Africa Camels?? Africa Bears??-- Plantigrade CARNIVORA??-- «compare rodents of two countries» & «Mon
299c196 ermata. like Man amongst Monkeys-- or dogs in CARNIVORA.-- Man in his arrogance thinks himself a grea
367d113 etable feeders. because males always armed in CARNIVORA. Where females, are peacable-- (Mem Lucanus &
453e182 m [...] Dampier. Vol II. p. 320. says no wild (CARNIVORA) beasts on Phillipines. Forrest somewhere say
226b223 that is :--are shell-boring Molluscs, like CARNIVOROUS Mammalia in their wide range & in their dura
226b223 range & in their duration of species. (are CARNIVOROUS Mamm: in Paris basin «allied to present») mo
423e096 d kill all of the one herbivorous. & its one CARNIVOROUS devourer.;-- it is quite clear that a large
514q21. Is pollen of cultivated Orchis & Asclepias &-- CARNOSA?-- good-- Norfolk Isd-- geology. volcanic? App
342d032 with goose!!) Dr. Bachman regularly breeds «in CAROLINA» for his table Muscovy & common ducks-- they
342d033 Common duck have been shot wild (escaped from CAROLINA?) off New York. therefore instincts not imper
569n022 series or harmonious prose.-- Lutké Voyage in CAROLINAS Vol II p. 132. offered to take a savage, said
244c022 iginally from Africa p. 122. Mus decumanus, at CAROLINE Isld, & a Roussette p. 136. Isle of France.--
246c026 eties P. 708. Columba Oceanica (Less) inhabits CAROLINE < «NB. The» > isld. (.perhaps Phillippines &
532m056 effort of will whilst their minds were sound. CAROLINE tells me that Nina, when brought from Shrewsb
532m056 ry thin, would not go out of house except with CAROLINE-- After fortnight. continued to grow thin & d
569n022 epeatedly by Waitz (In Theil. V) in describing CAROLINE Archipelago. Dn 75 cf., p 268 without, howeve
060r125 ons of vegetation.--«I can find nothing.» Mem CAROLINES quotation from Temple Urge the mineralogical
403e022 , a territory in the small isld of Eap in the CAROLINES, are remarkably short.-- & Deformations are p
308c217 all children of one father.-- yet differences CARRIED a long way. --Case of Habit I kept my tea in r
358d075 e of pork 3 lb was taken from a boiling pot, & CARRIED off by one of these birds" Case of bird of dif
358d085 (like «courage in dogs» EFFEMINATE men),-- if CARRIED much further, if by the process this were poss
437e139 Birds several cases on record of stoats being CARRIED (p. 121) & dropped having wounded the bird. p.
451e179 aul forests by having a smaller, lighter head, CARRIED more elevated & higher forequarters: is said t
499q09v r says Rafflesia is dioecious & Pollen must be CARRIED by some insect-- (21) Are there many instances
504q013 strate apple & pear to see if pollen naturally CARRIED, on account of Van Mons views-- Also PEAS-- N.
552m131 in Spectral images-- Mem Chiloe <pi> Sow, who CARRIED from all parts straw to make its nest. Pigs &
570n024 ely immersed-- mechanic apt to sigh.-- & hence CARRIED on as trick) «Shrugging aroused acting» Octob
571n029 ;-- This shows doctrine «of instinct» has been CARRIED too far In all the foregoing cases most diffic
584n072 - waterbirds, the bee to its nest,-- cats when CARRIED in confinement,-- carrier pidgeons proverbiall
584n072 n confinement,-- carrier pidgeons proverbially CARRIED to long distance in dark "it is inspiration."-
585n076 useless when applied to birds, which have been CARRIED in hampers. if they have not known the directi
588n089 93. that they perceive the difference on being CARRIED up or downstairs, or dangled up & down-- in la
595n115 d until the Keeper took it <her> in his arms & CARRIED to see.-- [blank] A Dog «whilst» dreaming, gro
596nIBC easily fall into convulsions A carrier pidgeon CARRIED & turned round & round in fainting state would
603o012 ction towards <man> «a king» <changed into> is CARRIED on toward deity.-- & as king might like cruel
629o055 steps forgotten, habit formed,-- & such habits CARRIED on to other feelings, such as temperance, acqu
584n072 ts nest.-- cats when carried in confinement,-- CARRIER pidgeons proverbially carried to long distance
586n079 s.-- child sucking whole wonder instinctive.-- CARRIER pidgeon just as wonderful in old bird as new.-
596nIBC ulsive tendency easily fall into convulsions A CARRIER pidgeon carried & turned round & round in fain
388d170 hat difference between bud & seed, that latter CARRIES with stock of food.-- the generalization begin
578n054 dark is proverbially a most modest person one CARRION on, by association, the question, "one will an
185b055 ave happened in all. but less in water birds-- CARRION eagles.-- This is but carrying on. attempt at
260c067 ntry, «& there acquired» then adaptation.-- No CARRION Vultures in Australia!! Wilsons Ornithology, V
294c177 s rather further.-- once grant good species as CARRION crow & rook formed by descent. or two of the w
424e101 s me, on authority of Beck, that Hooded crow & CARRION crow. have in Europe different ranges-- latter
424e102 ese crows are mixed in England-- for I presume CARRION Crow is found in Edinburgh.-- Why does Fleming
299c195 & their seeds produce <offspring> variety. wild CARROT. made into biennial domesticated kind with lar
441e149 probably would-- at least the experiment of the CARROT seems to show this.-- This would be curious la
490q01v ing &c Plants 1 Repeat the French experiment of CARROT 2 {also try Primrose & Cowslip in rich soil &
502q11v ion in structure Compare flowers of wild & tame CARROT-- Parsley & Fennel. Verbena Compare flower of
363d101 Pidgeons--, & see whether feathered legs.-- «CARRUNCLES on beak & in Muscovy duck» crested feather,
120a109 n went on at greater rate, not only river would CARRY further its own matter. but would cut wide gorg
207b151 . Ducks on rivers in Guiana. build top of trees CARRY duckling to the water in their beaks, & the you
372d129 nity in one stem.-- a bud may be transplanted & CARRY all these peculiarities not so a seed.-- Bud pr
495q05a ect-- such dioecious individ--small orifice (8) CARRY Bees, powdered with starch & Carmine & experime
495q05a experimentise on their returning powers-- then CARRY them in Electrical machine, reversing the poles
572n032 es to bodily weakness & inferiority, but now we CARRY it on to mental inferiority-- when we do not ex
594n113 g faces at it, so much so that the nurse had to CARRY it out of the room. nearly 3 months old. What i
090a023 ach stone. that fell, that was weighed,; <but> CARRYING on this ratio I can count 90 stones which hav
185b055 in water birds-- carrion eagles.-- This is but CARRYING on. attempt at adaptation of each element.--
437e138 t.-- p. 120 An Eagle is said to have been seen CARRYING a lamb two miles towards the Morne Mountains,
437e139 the side of Eagles nest, which shows power of CARRYING great weight. p. 125 is said that Eagles brin

Page ***(Key Word)***
525m023 hen doing. what they consider their duty.-- as CARRYING a basket, bringing back game, or picking up a
575n043 - an animal not destined by nature to exist. & CARRYING «like other hybrids» with <the> it the provis
239cIFC ecting form of man. Could you get racehorse from CART horse by picking without change of habits Mr Ya
334d011 though horsing every month) & worked in the same CART in loose chains, by being at first beaten from
554m141 sociation. Kenyon, & then go on to show, that if CART horse argued from this into a theory of frictio
334d010 entirely by their smell.-- Fox says he knew «a» CARTER well, who placed his stallion as second horse
217b184 d me of case of mare covered by blood horse & CARTHORSE two folds [not located] Mr Don gave me instan
543m097 understanding language is considerable, thus CARTHORSE & dog.-- birds many cries. monkeys communicat
269c103 ous generation not improbable.-- After reading "CARUS on the Kingdoms of Nature, their life & affinit
270c103 nctness of laws from rest of]CD universe «which CARUS considers big animal» becomes more developed in
270c104 on inorganic It is very remarkable as shown by CARUS how intermediate plants are between animal life
270c104 l speculations are entered in upon life. Namely CARUS.-- How remarkable that Turdus Magellanicus. in
210b161 > 1832.p. III Mr Owen suggested to me, that the <CAS> production «of monsters» (which, Hunter says ow
048r084 loose» bed of pebbles. (On a sea beach under a CASCADE, one can understand pebbles thus coated.--The
157g077 ory stream hill with boulders river Right Hand CASCADE has <cut> «where two branches unite in upper G
032r040 . If now the ocean should suddenly fall, (3) the CASE would be as at first. & according to the greate
032r041 irst explain «top of» tidal band of action. This CASE differs. I think. from Patagonian steps, becaus
033r042 .--form resembles the husks at Coquimbo: in that CASE, may not central and rather differently constit
033r042 ich, although not very intelligbe is a familiar CASE: If refiltered with other matter how very curio
033r042 s ever casts alone in Calcareous. rocks??--if so CASE precisely analogous: fragments instead Peak of
039r059 not tilted strata.-- It will be well to urge the CASE of St Helena, where dikes certainly have not be
040r063 st aid in preserving the terraces <...> Molina's CASE At Vesuvius. Vol III P. 124. Lyell. dikes have
040r063 e Mem. St Helena; probably more abundant in this CASE from intersecting a mass probably cold & not wa
041r065 hole bottom is important; because in this latter CASE, we cannot judge whether such fossils. lived in
048r085 a: the same caution is applicable to the Siberia CASE We must not think alluvial plains «always» most
049r087 dence.--The sudden increased dip is not parallel CASE to Isle of White. but rather to one out of a se
050r090 which, is doubtful. P. 180. I think my Ascension CASE very doubtful. -- In Iceland Bladders of Lava a
052r098 6] {P} A advancing coast to Seaward. Retreating CASE in excess as first case. When discussing Falkla
052r098 t to Seaward. Retreating case in excess as first CASE. When discussing Falkland soundings introduce t
053r101 Islds? In my Cleavage paper Dr Fittons Australia CASE must be quoted at length, The Lines of Mountain
056r110 and the secondary (stated in Playfair to be the CASE p. 51). presupposes an elevated country of gran
059r121 terlineal spaces are of diff cont: & even in one CASE contained lime.--All bear close analogy to Obsi
062r129 the history of rats, in the antipodes a parallel CASE.-- Should urge that extinct Llama owed its deat
062r130 urmillier): extinct Guanaco to recent: in former CASE position, in latter time. (or changes consequen
063r131 -When beneath water.--together with hypothetical CASE of Brazil.-- Propagation. whether ordinary. her
071r158 lake was absorbed.--Earthquakes felt. different CASE from shore of Pacific.--Isabelle's volcano, man
074r165 m acting on trachytes. also Azores. We here have CASE of such vapours washing a rock» Veins concretio
074r165 determined by fissures as in septaria. (& Chiloe CASE, at least corelation)--Galapagos vein. vein of
074r165 y action, conjoined with other» «(state simplest CASE. concretions of clay iron stone; iron pyrite in
074r165 ted chemical law & steam of salts, quite curious CASE of oxided Iron by Mitterschlich. Vol. II Journa
098A046 f crystallization. therefore concretions in this CASE laminar. hence the thick wedges of feldspar in
099c051 inclined layer!!!.-- The separation in the Ponza CASE of Scrope parallel to walls of dykes-- Mem. lam
100a052 not explain CLEAVAGE of rock-- nor the Falkland CASE, nor. the arrangement of particles of granite i
100a053 nd centre.-- A Stalactite of Gypsum, is the best CASE of cleavage.-- Phillips (113) «Lardner Encyclop
102a060 ns.-- Septaria in concretion arranged in planes, CASE of separation.-- the branching cracks-- only be
104a068 om of Norway was contorted yet no mountain chain CASE parallel to Banda Orientel. ask Lyell for sente
106a073 ely deflected from (B) which would not have been CASE. if inclination small.-- The power of widening
106a074 on is become comparatively small, & when that is CASE force is lessened. therefore rivers very ineffe
107a079 use most difficult (better conductor) Fitz Roy's CASE of S. Maria & Tubul applicable to Andes & Patag
110a086 Description of rocks in Lyells'. Capital Norway CASE.-- The fragment. consisted of hornblende (?) &
113a092 at NW. dip in SE part of Australia.-- Probably a CASE of rivers turning round & penetrating their own
120a109 d Sandstone. look as if a surface deposit.-- The CASE of the shingle in the great Chilian valleys mus
121a111 ed crystals of ice were formed-- (like my gypsum CASE) shows power of segregation.-- & has heated ang
121a113 rise at Stockholm.-- analogous to my Valparaiso CASE. Consider profoundly all consequences of EXTREM
122a115 regularly.-- (Mem near Gregory Bay). Shropshire CASE where lamination appeared.-- Lyells Denmark --
128a129 ater to counterbalance How strongly the Glen Roy CASE shows that the figure of the world has just the
128a130 bing Coast of. Brazil. Maldonado enter into this CASE.-- Ed. New. Phil. Journal Vol XXI. p. 213. Beyo
129a131 f Alps size of boulders sorted: ditto Murchisons CASE.--¿ does it bear on Patagonia? «Facts about sub
129a131 port on Massacuhssets. p. 133 The most wonderful CASE of great block of rock moved by gale-- When wri
129a132 pers, in Edinburgh New Phil. Journ 1838. several CASE given of hot heads &c heat beneath the sea.-- C
130a133 eath the sea.-- CD[did not Beechy have some such CASE]CD what would be the chance in sounding over a
131a135 untains of the Moon. by Dr. Nichol-- adduces the CASE to show Sir. J. Herschel's theory wrong.-- Geog
132a138 and M. Parrot does conjecture that in Scoresby's CASE volcanicity has warmed it. Is not cold of ocean
148g024 not Mica Slate too hard & uneven to be impressed CASE of Birch Wood by Inverorum being determined by
148g029 e & if so there would be barrier-- recollect the CASE of loch <in> <below> «by» pass of Glencoe-- the
152g045 in Pass, where shallowest In Glen Collarig good CASE of shelves entering «on» one side ravine. Are t
165g125 Mother in face-- asked stated this generally the CASE Wednesday 12/ & 3/ Why is the Tetrao scoticus &
172b008 like a variety produced-- --[every grade in that CASE surely is not Produced?-- <Granting> Species ac
176b022 nad definite existence, as we may suppose is the CASE. their creation being dependent on definite law
179b035 life, then all die at one. period, which is not CASE. MONUCULE NOT DEFINITE LIFE I think {P} Case m
180b036 ot case,∴ MONUCULE NOT DEFINITE LIFE I think {P} CASE must be that one generation then should be as m
181b044 n progressive development which certainly is the CASE at least during subsequent ages.-- The Creator
186b057 rees. & extinction of forms.?? It is in simplest CASE saying every species in genus resembles each ot
194b094 apajou is S. American form. therefore it is like CASE of great <rodent> edentate [has been doubted?]C
194b095 rum for Mammalia.-- I really think a very strong CASE might be made out of world before zoological di
204b141 mselves or with parent birds.-- «W. Fox. knew of CASE of male widgeon, winged & turned on pool, first
206b148 bt a different set of causes must act in the two CASE,» May this not be extended to all animals first
209b157 I,46 St. Helena I: I,15 {T} Calculate my Keeling CASE: Juan Fernandez Galapagos =Radack Islds = ∴ Isl
214b176 btain a litter without this defect, Very curious CASE = W. D. Fox. When dogs are bred into each other
215b178 Animaux Sans Vertèbres as latest authority. The CASE of the tailess cat of Isle of Man mentioned in
216b180 it may happen.-- Dr Smith says he is sure of the CASE at Cape.-- McClay argues from it Black & White
216b180 o.-- As in cacti &c &c.-- as in dogs investigate CASE of pidgeons. fowls. rabbits cats &c &c.-- When
216b181 ack others white which is more closely allied to CASE of cross of dogs.-- See paper in Philosoph. Tra
216b181 d in this cross to after births, like aphides.-- CASE of boy with foetus developed in the breast.--lo
216b182 on to his view of races not mingling?-- In Foxes CASE of Blood Hounds, a little mingling would probab
217b183 This is good instance of same fact in Mr Galtons CASE.-- It explains the loss & expense. (must probab
217b184 f tim» as often one way as other.-- He has known CASE of good pointer & rough water spaniel «produce
217b184 es differ, & like both parents.-- Fox told me of CASE of mare covered by blood horse & Carthorse two
222b205 low of decrease in character. (which perhaps is) CASE with fish-- as some of the most perfect kinds t
224b215 cal changes (¿intellectual being acquired alters CASE) other species or angels. produced «&» Has the
225b218 hoof} Ask Entomologists whether they know of any CASE of introduced plant, which any insects hav[e] b
228b230 & Golden Pippin &c produce real crabs, & in each CASE similar or mere mongrels.-- It really would be
230b236 tion of species, now this notoriously is not the CASE, you have stunted species, but not such as woul
230b240 e difficulty, argument in circle.-- Falkland Isd CASE good one of animals not soon being subjected to
230b241 ts» very close (take Europaean birds. Mr Goulds' CASE of Willow wren) & others varying in wild state
231b242 slds-- Monkeys different not travellers?? Royles CASE of Himalayan, plants ¿migratory birds, he told
235b262 nced; varieties of dogs in different countries a CASE in point.-- All cases like Irish & English Hare
236b279 er of Miocene Mammalia--of Europe Mem. Mr Bell's CASE of Sub Himalayan land emys, decidedly an Indian
246c027 ch lives in the Eastern Moluccas, New Guinea.-- (CASE of replacement)-- Coquille Voyage The caswary,
249c034 l cross & produce fertile offspring in the first CASE it will either produce no of[fspri]ng or such a
249c035 & Cowslip run wild, The two species of Clenomys. CASE of replacing species. Dr. Smith will give me so
251c041 in form of white mark on breast: p. 234.-- good CASE.-- p. 526. (ref) To Temminck Monograph. Mammal;
252c045 Closer species-- FOX IS & Mice of America «good CASE on account of varieties in N America» «some dou
255c060 as well known for its impudence. «This excellent CASE of memory without association..» Instinct goes
257c057 e nétaient pas tout à fait les mêmes." This good CASE. of replacement under peculiar conditions-- of
259c064 able to plant, Epidemics of South Sea, wonderful CASE of extermination of species-- Epidemic amongst
259c065 xistence; if whole life then real adaptation The CASE of heredetary disease, is on the same principle
259c065 healthy parents have healthy children the other CASE is <adaptation> «change» during life of parent,

263c074 en for Quinarians to deceive himself.-- Give the CASE of Apterix-- split, depress & elevate & enlarge
273c112 his is applicable to swallow-hawk, «this not the CASE in swallow??? which is most wonderful of all. ¿
278c129 d:-- Genus must be a true cleft-- putting out of CASE the Analogys.-- If genus does not Mean this it
280c136 that capable of producing itself alike.-- in one CASE it changes one, in other it changes thousands i
281c137 ieties, (though that would be best).-- Argue the CASE theoretically if animals did change excessively
282c141 of man, yet altogether different.-- To make this CASE perfect, we must suppose men instead of mere co
285c150 uced & made young.--¶ father must be left out of CASE, that difference occurring.-- It will be necess
290c164 because very 3/4 bred.-- (hence hybrids in this CASE have bred). White & common pheasants. have cros
298c189 bird.-- Annals of Natural History Vol. I. p. 185 CASE of tit lark placing withered grass over nest, w
299c194 es formed. The Anagallis perhaps, offers another CASE of permanent varieties in wild state-- The two.
300c191 nimals.-- Insects shamming death, most difficult CASE to iimagine how art acquired.-- They reason how
302c201 etaining a family likeness, & this I believe the CASE. = any animal really connecting the fish & Mamm
307c216 these cases, have plain meaning & none in other CASE! Savigny has shown same fundamental organs even
308c217 father.-- yet differences carried a long way. --CASE of Habit I kept my tea in right hand side for--
308c218 number 5 here most evident!!? examine into this CASE D. Jeffrey (life of Mackintosh Vol II. p. 495)-
309c219 according to Mr Herbert's views.-- Argue <argue> CASE of abortive organs to mules in their genitals &
309c219 f one parent, races of animals-- argue «opening» CASE. «thus» Educate all classes-- avoid the contami
309c221 different.-- Male glow-worm knowing female good CASE of instinct. bees turning neuter into Queen, mo
309c221 . bees turning neuter into Queen, more wonderful CASE Dwights' Travels in America, speaks of short le
310c224 etles, birds & a fox most close The most curious CASE is Saxifrage, almost <same> «closely allied» sp
311c225 gree of Long. & Col. Sykes alludes to some other CASE of 180 degree & great diff of Latd. p 355 Echid
313c234 f mammals shorter duration, than molluscs, argue CASE both in Europe & S. America. «very difficult ca
313c234 ase both in Europe & S. America. «very difficult CASE» Does this law of duration apply to utter extin
313c235 passion must arise after long interval very good CASE.-- habit is awakened by association (case of El
313c235 y good case.-- habit is awakened by association CASE of Elephant, which had run wild in India. in He
313c235 nalogous to dormant instinct.-- (How wonderful a CASE bees developing sex of neuters) species may hav
314c239 stralia, some SPECIES of Australian GENERA. good CASE. rather large flora. (150?) Mr Brown did not ob
314c239 yptus!-- (Hostile fact) Be cautious about Goulds CASE of birds of Van Diemens land & Australia.-- The
318c252 peculiar appearances-- Now this is precisely the CASE with the mice of S. America, with respect.. to
318c252 Cordillera,---- Bachman has seen webbed Shrews. CASE of adaptation.-- (case of Squirrel from extreme
318c252 n has seen webbed Shrews. case of adaptation.-- (CASE of Squirrel from extreme north turning white li
318c253 grations of birds he mentioned many most curious CASE-- -- the birds seem to follow narrow bands, cer
319c257 an Diemen's land & Australia proper.-- Irish Elk CASE of fossil geographical range.-- [blank] Books e
331d1FC ds greater.?-- Mem. for Eyton.-- Sir. R. Heron's CASE of breed of pigs with solid feet.-- 1838 [In th
332d001 her mention, than for ten years he never saw one CASE of malignant erysipelas spreading over the head
332d002 disease.-- but now (July 1838) he has seen more CASE in a month, than in several previous years, two
332d006 diseased Mesenteric glands.-- My Father has seen CASE of pleurisy, broken limb «in children» & other
333d006 the half bred Persian.-- Here then we have clear CASE of heterogenous offspring from one impregnation
334d009 ined by male: & How completely is Lord Moreton's CASE opposed to this fact & views.-- Fox says a cous
334d010 t se> plentiful, second absolutely sterile.-- My CASE of Stallion, according to Erasmus preferring yo
334d011 t beaten from her, & always accustomed to her.-- CASE parallel to brothers & sisters in Mankind.-- Th
334d011 parallel to brothers & sisters in Mankind.-- The CASE of all blue eyed cats (Fox has seen repeated ca
334d011 (Fox has seen repeated cases) being deaf curious CASE of corelation of imperfect structure.-- Fox say
335d015 ies is different because not long in blood.= The CASE of union of perfect animals is distinct case,--
335d016 The case of union of perfect animals is distinct CASE,-- gradation from physical impossibility to (pe
335d016 seen mules could have no offspring, & this being CASE, owing to the corelations of system, the organs
338d025 k being strongest.-- V. p. 63. Note Book M'. for CASE of change in food in insects entered by mistake
344d041 e difference of sexes. is probably arrived at in CASE of insects as glowworm The case of one impregna
344d041 ly arrived at in case of insects as glowworm The CASE of one impregnation sufficing to several births
344d042 ds must be acquired, & hence this is then CASE of avitism.+↑» Three gentlemen of party all tho
345d044 ellanic Fox.-- peculiar hair & appearance-- good CASE of Provincial Breed-- Highland Sheep jet black
353d061 ber of vertebrae in Irish & English Hare.-- good CASE these hares compared to North American hares. M
354d065 hilippines Man have varies the range-- Argue the CASE of Probability. has Creator made rat for Ascens
356d070 eedings-- there is tendency to reproduce in each CASE, but something prevents the completion.-- ¶Say
357d073 nt since grown.-- This will explain. S. American CASE & Didelphis being Mundine form., & the less dev
358d075 oiling pot, & carried off by one of these birds' CASE of bird of different family. having very same h
359d087 s long as animals are healthy which is often the CASE, & why should organic affections always influen
360d090 fearful. though not so much as in Jackall.-- In CASE where Jackall was father resemblance much neare
361d096 eeding a more brilliant plumage than male.-- «My CASE of Caracara. N. Zelandiae.--» Mr Blyth stated «
364d103 idgeons from infertility,-- Yarrell says in such CASE they exchange birds with some other fancier, th
366d111 e transmitted if their [not located] <The> repre CASE common to many good species; & therefore to gen
366d111 skitoes bite influence propagation of species.-- CASE of Association very disagreeable hearing maed s
369d114 ve no secondary characters.--)p. 49. (wonderful CASE of Pea hen. taking feathers of Peacock & spurs-
370d116 <pu> up honey even for single rainy day-- & from CASE of wasps, is supposed cells properly are made f
371d127 o the (required) offices" &c &c Owen illustrates CASE of Dingo (he alludes to the dholes or wild dogs
377d139 But he thinks other monkeys make st.-- noise¶ In CASE of woman instinctive desire may be said to be m
379d151 t Sir J. himself is not like-- now this is clear CASE of avitism. but then ¿ was «not» the expression
379d151 > Sir W. itself received from his father so that CASE ceases to be true avitism Annals of. Natural. H
379d152 all the lambs will deteriorate.-- Lord Moreton's CASE.-- When cows have twins, <one> though capable o
380d153 eding in & in.-- Mr Yarrell does not know of any CASE of old Male. becoming like female, though many
387d168 ime) is the above law anyways connected with the CASE of successive copulation impresses offspring mo
387d168 g more & more with the added «like Lord Moretons CASE & Dr. Andrew Smith,» difference.-- If A. B. C.
388d172 father.-- stuff.!-- How much opposed. the Quagga CASE appears to that of «2» dog begetting different
389d173 le;<]CD> such view no ways explains Ld. Moretons CASE: without the nervous matter consists of infinit
389d177 h other.-- This is somehow connected (This seems CASE, for by careful observing cattle can be bred in
389d177 ion in hybrids. perhaps connected with this same CASE (& not merely as I have stated it) it is certai
389d177 produce same effect as too little.-- in (latter CASE female often takes males but does not produce)
391d174 mammalia being abortive hermaphrodite simplifys CASE much; & originally <her> each hermaphrodite bei
391d175 is there some law about sexes of twins in former CASE.-- (many monster are really twins.)-- It is abs
399e006 t have same fossils. does not Lonsdale know some CASE of change in vertical series: Look at whole Gla
400e012 ries, & character of fur-- I am sure a very good CASE, might be made out of variation analogous to sp
404e024 or instinct, now this is directly incorrect The CASE of my mice is good, because it is an involuntar
404e026 modification of an organ present in whole class. CASE of Mexican greyhounds.-- young being habituated
404e026 -¶ as adaptation,-- however mysterious such is CASE¶. therefore chance & unfavourable conditions to
404e026 :¶ Australian dogs have mottled coloured puppies CASE of this.-- tendency in «manner of» life to be m
406e036 p.-- diseases common to men & animals cow pox.-- CASE in Spain of pustular disease following handling
408e045 des-- parallel & curious facts.-- The Himmalaya. CASE, bears on the vast changes even in that quarter
409e046 al than fox, wolf &c &c-- is precisely analogous CASE to man, exceeding monkeys;-- Having proved mens
418e080 side, than the addition of other organ, in which CASE the hermaphroditism would not be perfect as in
420e088 ards Hyaena.-- see Comte Rendu.-- I suspect good CASE of fossil filling up blank.-- CD[not between ea
423e096 a & fish & reptiles.?-- supposing such to be the CASE, it proves the law of developement in partial c
424e101 Beck are different.-- Subsidence of Greenland-- CASE of splitting of two regions-- -- are there any
424e102 r produce hybrids-- now this is independent good CASE, but very odd since these crows are mixed in En
425e104 , some of these species are described.-- Capital CASE,-- for Sandwich Isld are very similar to Galapa
426e108 reat great-grandfather.-- -- Little Miss Hibbert CASE of Hindoism coming out more than in mother or i
428e111 ized. & says Laurels have not been so. (which is CASE adduced by Herbert) because not reared by seedl
431e121 tions Portland Stone &c &c.--if? so «it is» good CASE:-- in Sowerby Min. Conch. it is however, said t
433e126 ar air given. p. 246.-- 248 & p. 258 A beautiful CASE, showing the gradation from one grand system to
437e138 ing their way to the Caymans from Honduras. good CASE. of migrating.-- shows my theory insufficient.--
441e151 hould cross with another.-- to escape it «in any CASE» we must draw such a monstrous conclusion, that
443e153 al commerce «The fact of Corallina & Halimeda is CASE in point».-- The relation of these «sexual» fun
445e158 in which it never touches objects.-- far better CASE than chicken pecking fly.-- "whilst the shell s
446e162 ds «though by the latter very rarely») here is a CASE in answer to Mr Knights doctrine.-- Case like C
446e162 re is a case in answer to Mr Knights doctrine.-- CASE like Corallina--«Does it flower anywhere?-- Ye
446e162 is the <of> product of fresh bud-- here then is CASE of change analogous to change in grafted trees
446e162 s «:so is not effect of different stocks in this CASE».-- & strong case showing analogy of production
446e162 ct of different stocks in this case».-- & strong CASE showing analogy of production by gemmation & by
446e163 uddenly send forth quantities of blossoms--» The CASE of the Lemna, «and the vivaparous grasses, whic

Page **(Key Word)**
447e164 therefore my theory does require crossing.-- The CASE of Lemna shows dispersion of germs is not end o
447e165 he has seen it propagated in a garden, which is CASE precisely analogous to the Canada onion mention
448e165 is allied to Asclepias, where this is always the CASE according to Brown.-- Voyage of Adventure & Bea
448e168 Toxodon «In orders»-- Fish & reptiles in former CASE-- Reptiles & Birds & Mamm. in ornityhyrhycus--
449e170 ives genera.-- <it is not transportation> now in CASE of large [not located] Mr Greenough on his Map
458t002 hread, picked «up-» instead of spinning-- better CASE than English birds, using cotton &c instead of
458t002 g something of Mule in its ears-- ((this is good CASE as showing gradations, Boteler's Narrative Voya
459tf02 s applicability to common mule & hinnus-- in one CASE bastard of wolf & dog had more form of male, &
463t057 e,, & supposing much extinction. give a parallel CASE) Waterhouse remarked, that any argument for tra
465t080 y as, bones of Mammalia in caves:-- :argue first CASE of bones (New Red Sandstone) & then go on to sh
465t081 , so it is with domestic breeds. (though in this CASE crossing has had somewhat to do with it. mem. d
468t112 ally Fraxinella. with respect to nectary is same CASE as Azalea or Rhododendron xx after several gloo
472s02r , on one of which pollen was routed. wh. was not CASE, on several flowers I examined some days ago--
491q01v & see if first pollen produces any effect, as in CASE of woodpidgeon & Hen. mentioned by Mr Knight. V
494q004 &c: length of life. (5) Does my Father know any CASE of quick or slow pulse being heredetary. (6) In
495q005 se on Primrose seeds-- it really is an important CASE-- cross with cowslip pollen.-- as these are wil
498q007 suspect Elms.-- & Orchidacaeous plants no other CASE.-- (6) Will plant accustomed to rich soil, when
507q15v es-- or seed-cases with similar structure.= good CASE as showing how simple, but beautiful adaptation
512q019 periods in different breeds--?? or in individual CASE: subject to disease in youth.-- Mr Tollett-- ab
513q21. pened flower-- <doubt> disbelieve this in Bauers CASE of orchidiae Where does J. Hunter use expressio
521m007 s shown about meals, no [not located] There is a CASE of Mr Anson. who told a story of hunting «-- ha
521m009 r dotage is fact of same sort. Aunt. B. ditto.-- CASE of Mr Corbet of the <Hall> «Park», after paraly
523m014 f the injustness of their hatreds, as <if> in my CASE.-- It must be so from the curious story of the
523m017 t generally, does not really come on suddenly.-- CASE of Mrs. C. O. who threw herself out of the wind
523m018 e numberless people insane of particular ideas, «CASE of Shrewsbury gentleman, unnatural union with t
523m018 up> Sometimes comes on suddenly from <I> (in one CASE ipecacuhan-- not acting) in others from drinkin
525m024 ,-- guilty conscience.-- Not probable in Squib's CASE any direct fear.-- My father thinks that selfis
526m029 ing of things seen, quite agrees with my Fathers CASE of Mr Corbet of the Hall understanding. (on hea
529m042 . Deer in Parks ditto.-- My Father says there is CASE on record he believes in Philosoph. Transaction
529m042 ut quite sensible & no ways an ideot.-- «in this CASE must have been functional.--» He has some idea
535m064 hest, when no tea do. p. 233. Mr Lewis describes CASE of insects «a Perga» of Terebrantia, laying egg
535m064 serted broods appeared healthy-- This remarkable CASE may be normal. with insects, but habit forgotte
538m078 gle families. Edinburgh. Phil. Transact. p. 365. CASE of double consciousness, one only «little» less
538m080 bitually.-- Agrees with insanity, as in Dr Ash's CASE, when he struggled as it were with a second & u
546m110 e vivid thoughts In same book (p. 143) wonderful CASE of perfect double consciousness Mayo compares i
546m110 nary state is concerned.?-- Mr. Mayo told me the CASE of a lady, (whose name was told me, who told th
547m112 answer shows our profound ignorance in so simple CASE.-- There was memory, for it related to past ide
548m114 ay be imperfectly reason -- <In a> Abercrombie's CASE of «in Botanical Student» somnabulism, did reas
548m116 «as» to <make> «give» one individuality in this CASE.-- But now in Mayo's «p. 140» case of double co
548m116 ality in this case.-- But now in Mayo's «p. 140» CASE of double consciousness, one would pity sufferi
548m117 xpression of double self, though as in Dr Ashe's CASE, one here was conscious of the two states.-- Au
554m139 see whether, this was permitted & eat it.-- good CASE of association.-- «Listened with great attentio
555m142 (& Porpoises) have not charity-- is it in former CASE instinct to destroy contagious disease.-- (Usef
555m142 heredetary habit cannot be traced) V. D. p. 111, CASE of Association. Sept. 16th Zoological Gardens--
556m146 ntinue to put down ears, when kicking.-- -- good CASE of expression showing real affinity in face of
556m146 hem close on head, when going to fight, in which CASE expression resembles a fox-- I can conceive the
557m149 ar is open mouthed to hear. though in individual CASE. nothing can be heard.-- Shame would never make
559m154 tandard of our own minds. which ceases to be the CASE when we consider the formation of laws invoking
568n017 .-- «animals have shyness with strangers» «as in CASE of temperance, or real virtue, that is action w
568n017 experience shows will be for general good, or in CASE of any fantastic custom» «Probably bashfulness
569n020 ent of Man Understanding languages seem simplits CASE of Association.-- Elephant often given food & w
570n026 e been most generally accustomed to:-- analogous CASE to my idea of conscience.-- deduction from this
571n029 edetary, is pleasant.-- Mental & Bodily Consider CASE of grazing animals knowing poisonous «herbs:» &
572n032 iority-- when we do not expect any bodily harm-- CASE of habitual action.-- L'Institut. 1838. p. 340.
573n038 mself not to turn in bed. will turn in bed.-- in CASE spittle, effect of thought is to make saliva fl
574n043 f instinct by dogs, would show habit.-- Take the CASE of Jenner's <Hyaena> Jackall.-- an animal not d
575n045 leasure of music Audubon IV Vol of Ornith. Biog. CASE of Newfoundland dogs. who will not enter water,
588n089 or downstairs, or dangled up & down-- in latter CASE they struggle their arms.-- do. p. 306 "the eye
591n099 because instincts to woman is not followed; good CASE of instinctive conscience.-- Why does not man e
601o009 nveyed over whole body (which it may be in first CASE. as when the excised heart is pricked) and cert
608o025 nce of mechanical phenomena.-- (Mem: M. Le Comte CASE of Philosophy, & savage calling laws of nature
612o035 s much as inflorescence.-- [LHC] I here omit the CASE (if such there are) of animals enjoying only mo
613o036 This I suppose he deduces from the ends in each CASE being the same, & the means very similar.-- It
614o037 comes on late in life, man almost alone in this CASE can perceive instinct. boy takes delight in mam
623o050 n. & certainly in «species of» animals, in which CASE it undoubtedly is instinctive. But does not Har
629o55v t, shame. right & wrong comes into mind in first CASE-- seeing how shame is accompanied by blushing,
634j54v abandoned for another".-- What bosch!! Put it to CASE of man. <&> The <design> determination of a God
636j56v nts by foreign agency-- as insects, as wonderful CASE of adaptation.! There would not have been any D
640j167 empirical, from their own Faunas, which in this CASE is only true criterion.-- Hence it is highly un
640j167 uld possess species to themselves.-- Probably no CASE in world like Galapagos. no hurricanes.-- islds
030r036 tiary strata there is Coal-- ¿ No shells in all CASES. «.Mytilus.--» «at Guacho» «on N. Chile? Washin
031r038 elago. «Aleutian Arch.--» V. Fitton. Australia: CASES in Europe.-- Auvergne. very little Pumice, thou
037r056 uency of dikes in Granitic countries, enumerate CASES. -- M. Video exception, but even there, hills o
051r093 vive, in the most violent surfs: in both latter CASES become petrified, & increase. -- In Southern re
052r098 ard transportal of drift matter.-- Give various CASES. [Fig. 6] {P} A advancing coast to Seaward. Ret
052r098 --40 line <shows> runs at equal distance?) 1st CASES. -- The terraces in Valleys of Chili may be wit
063r130 uent on lapse) being the relation.--As in first CASES distinct species inosculate, so must we believe
090a023 s each year.-- but instead of 90 stones in many CASES there were flights of stones of large numbers (
090a023 e flights of stones of large numbers (& how few CASES recorded if we say «100» <5>0 lbs a year too li
098a048 Why not horizontal? Why have particles in such CASES moved more laterally than vertically, in concre
148g028 if the side-streamlet had cut them out-- In all CASES «I urge» deposition marine-- because if not cha
150g035 m plains such as those near Bridge Roy (& other CASES) but then if gradually drained, where is barrie
173b011 the continent though species all different. In CASES as Galapagos & Juan Fernandez. When continet of
182b050 m In Mr Gould Australian work some most curious CASES. of close but certainly distinct species betwee
235b262 e hemisphere. more than in other. Are there any CASES, where «domesticated» animals separated. & long
235b262 s in different countries a case in point.-- All CASES like Irish & English Hare bear upon this.-- Why
236b278 .-- Isld bordering continents same type collect CASES.-- African isld.-- «How is Juan Fernandez-- Hum
249c035 of producing again The Varieties of Cardoon are CASES <sp> like those of Primrose & Cowslip run wild,
265c086 seeing beetle with wings beneath soldered wing-CASES-- Yet these wings may be of some use.-- Nature
280c134 s in one genus-- external circumstances in both CASES effect. it.-- Sir J.. Sebright excellent author
282c142 xisting race, thus the analogy would not in all CASES be produced, but would depend upon exclusion.--
293c172 memory in offspring.-- Some association in such CASES recall the idea. «or simple structure in brain
293c172 next generation. [NB what are those Marvellous CASES, when you feel sure you have heard conversation
299c194 there are not Many intermediate shades in these CASES,-- but absolute species formed. The Anagallis p
307c216 ll abortive organs «of very same kind» in these CASES, have plain meaning & none in other case! Savig
309c221 rtain Arabs.-- NB avoid quoting these hackneyed CASES Mr Ed Blyth does not believe in circular or lin
313c236 he process.-- but why two sexes scission in all CASES probably gemmation (.Ehrenberg) --not necessary
316c246 ut the fossils are not like, except in very few CASES, those of Tertiary Europaean fossils-- «(so muc
332d001 ten years afterwards, he then did not see other CASES.-- He thinks apoplexy affects people all over E
332d002 e, he remember during a year or two he saw many CASES of virulent cancer in women, & since that time
333d007 one under present form.-- Fox has seen several CASES of foxes & dogs crossed, offspring always more
334d011 se of all blue eyed cats (Fox has seen repeated CASES) being deaf curious case of corelation of imper
345d043 rt-horned with» black lip) & then calf «in both CASES» is killed. Notes from Glen Roy Note Book.-- Wh
349d051 tion for chief divisions.≠ p. 7. «In» Some <of> CASES the circular arrangement from fewness of forms-
370d117 ome fish.-- But birds quite distinct.-- Collect CASES of difficulty of growing plants in all parts of
371d127 g coloured offspring.-- but surely in all these CASES an unseen change is produced in parents--colour
385d163 Str, Hair dresser, assures me he has known many CASES of bitch going to mongrel, & all subsequent lit
405e032 America. see my Journal for references In such CASES as at Galapagos. where different islets have di

406e036 pustular disease following handling sheep-- all CASES: d degree p. 354-- The most vicious dog. will n
411e055 & afterwards by the examination of the special CASES, under which the individual steps in the series
423e099 rm.-- has several species in my fossils-- CD[If CASES of one variety in upper part of bed-- & another
424e100 tes of Galapagos were South American.-- several CASES of species peculiar to separate islets.-- March
424e101 of splitting of two regions-- -- are there any CASES of union of two regions in modern times.-- this
425e104 in Sandwich Islds. he believes, there are, many CASES of genera peculiar to the group having species
437e139 e.-- Stanleys Familiar History of Birds several CASES on record of stoats being carried (p. 121) & dr
440e146 tion.-- thus cabbages growing like Nepenthes.-- CASES of pidgeons with tufts &c &c here there is no f
440e147 irds of Paradise.-- All that we can say in such CASES, is that the plumage has not been so injurious
448e167 , is same, as if Isthmus of Panama.-- These two CASES highly improbable.-- yet I can see no other way
453e183 as Duchesne described Atavism.-- ask Dr Holland CASES where peculiarity has first appeared.-- "Storia
462t051 rp division of partition as between Mammalia in CASES such as that of Java & Sumatra Nov 15th Waterho
463t057 eory-- (I must answer it by rooting out curious CASES of intermediate structure, & supposing much ex
506q015 ach other.-- ? Cannot remember at all. (37) Any CASES of plants. which will not produce seed in this
507q15v e hook; curved spines-- simple spines-- or seed-CASES with similar structure.= good case as showing h
507q15v border-- & Humboldts spinning seed.-- (50) Any CASES of wild varieties plants growing together. unde
508q016 ou know whether any trace in germ. (2) Any more CASES of diseases, generally occurring in man being t
508q016 Haemorragic tendency, independent of heredetary CASES, more common in man than in female-- (8) In Hum
509q017 os or Whale, than in old?? Falconer says all in CASES. Owen. Have talked partially with him Ask him t
512q020 2) Foxes & English animals & birds breed (3) In CASES where Lions have bred, have they been raised fr
520m001 l disposition.-- My father has seen innumerable CASES of people taking after their parents, when the
522m012 a of time had been disturbed.-- These foregoing CASES of «mental» failures very general effect of «ear
525m023 at., doing their dirt, running home.-- in these CASES their actions do not look like fear, but shame.
530m043 s carbbuncl on «Head» Neck.-- He has seen other CASES of similar nature.-- --like FitzRoy in sleep gi
532m054 accompanied by much involuntary fear) In these CASES probably the system is affected, & by habit the
533m060 ss grateful.-- How comes this tendency in these CASES? How did my mind feel it was wrong (& it was no
534m062 e-- there is something wrong in comparing these CASES, when agency is unknown, with simple exertion o
537m077 effects of education, may be opposed undoubted CASES of heredetary pride & in single families. Edinb
538m078 ple forgetting their insanity» there seem other CASES somewhat analogous, & which I think will lead t
552m132 t on long run will do good.-- alter will in all CASES to have & origin as well as rule will be given.
554m140 herself with straw, or with a blanket.-- these CASES of commonly using, foreign bodies, for end. mos
560m156 one. about heredetary tricks & gestures, other CASES like D. Corbet; «do» ideots form habits readily
571n030 » has been carried too far In all the foregoing CASES most difficult to distinguish. between prejudic
581n063 generalize it, & see whether applicable to all CASES.-- & analogize it with ordinary habits that is
583n069 which have never seen their parents offer best CASES of instincts].CD all this may be true,, but rel
591n099 sensations of individual are same as in normal CASES) are held in abhorrence it is because instincts
610o032 th. 1839.-- My father says he has heard of many CASES of ideots knowing things, which are often repea
611o034 s not modified by external accidents, & in such CASES modifications bear fixed relation to such accid
614o037 ike heredetary memory; but first memory in many CASES cannot be acquired by experience for child suck
616o038 ion in» «ejection only has» will: there must be CASES of secretion being some time governed by will i
620o048 ces of country, so will the conscience in these CASES.-- Those instructions, which the child sees uni
628o53v em, of ethics-- [my answer would be to all such CASES-- either, that from the necessities «& good» of
629o54v y all wish of outward gratification,-- see what CASES Mackintosh gives & try it.-- p. 241 (1) Any act
634j54v e muscles to his ears.-- p. 313 Many other good CASES -- p. do]CD <Mac. remarks all Mammifers origina
635j56v lloch. Attrib. Vol I. p. 330. Mentions the many CASES, as in Papilionaceous flower, where such care s
636j57r ts. (sliminess instead of barbs)-- In all these CASES it should be remembered, that animals could not
638j28r sack. power of endurance &c &c Camels? all good CASES of corelations.-- [There must have been deserts
639j28r edge Hog & Echidna. & Aphrodites C. D. Endless CASES.-- Macculloch p. 260 intimates canines no speci
240c014 wise new species of Parameles, which joined to CASOAAS, perroquets, establishes its «zoolog» alliance
202b131 species not being so closely adapted. Near the CASPIAN «Province of Ghilan» wooded district cattle wi
379d151 tural. History. p. 135. Natural History of the CASPIAN. Fresh Water Fish!! ¿adapted to salt water?--
419e085 38. p. 412. M. Eichwald has published Fauna of CASPIAN.-- fishes fresh water kinds. (yet living in th
450e175 e after quitting Bengal this fact is noticed in CASSAY Ava Pegue-- seldom equals 13 hands-- those of
243c020 olk. Isd. & New Caledonia peculiar species of CASSICANS: (¿cassicans Australian form? p. 27. many fis
243c020 ew Caledonia peculiar species of cassicans: (¿CASSICANS Australian form? p. 27. many fish of Taiti fo
340d029 than in many large orders of birds. The Emu & CASSOWARY closest.-- Ostrich & Rhea closest.-- (& two R
380d154 have cock & hen plumage so different, yet the CASSOWARY & Guinea Fowl cannot be distinguished.-- A ca
033r042 ted lime have been removed?--As shell out of its CAST which, although not very intelligble is a famil
282c142 ades.--» then the second race would not obtain a CAST of washing men-- but might hire the preexisting
496q006 & kill them in hour or two «My Father made hens CAST Holly-seed & they grew» (9) Place. Snap-Dragon.
512q019 l & hereford cow similar to reverse cross.-- Sow CAST-up-balls of Hawks or even owls.-- How long do s
309c220 e all classes-- avoid the contamination of <c1> CASTES. improve the women. (double influence) & manki
324c268 zRoy To be read Humbold. New Spain-- Much about CASTES &c Richardson's Faun. Borealis Entomological M
448e166 s on ice p. 643-- very curious table of all the CASTES from Stephenson at Lima The same numerical rel
527m034 t, from weakness of my stomach I observe a long CASTLE in the air, is as hard work (abstracting it be
527m035 anished the better.-- The facility with which a CASTLE in the air is interrupted & utterly forgotten-
528m035 ecause papers, &c &c round one. one recalls the CASTLE in the air by going to beginning of castle» because train
528m035 one recalls the castle by going to beginning of CASTLE» because train cannot be discovered-- is close
528m035 tfulness.-- & the approach to believing a vivid CASTLE in the air, or dreams real again explains insa
544m103 consciousness-- & which tells one of reality-- CASTLE in the air, is more prolonged than dream. neve
547m111 ther August 29th. Went to Bed. & built «common» CASTLE in the air, of being compelled, from some quit
547m111 iving orders.-- Now what was difference between CASTLE & dream No answer shows our profound ignorance
547m112 , & not attending to bodily sensation & yet the CASTLE would not have turned into dream.-- It appears
547m113 n mind, being absent. one could not compare the CASTLE with them, therefore could not doubt or believ
548m115 of other trains of thought are in progress-- In CASTLE of air the trouble «I well recollect» is in ma
571n030 was much struck with this, when viewing Windsor CASTLE which rises naturally & hence sublimely from n
527m034 and improving powers of invention) such CASTLES in the air-- Catherine often, but not of an in
527m035 re brought into play & then perhaps the sooner CASTLES in the air are highly advantageous, before rea
527m035 re brought into play & then perhaps the sooner CASTLES in the air are banished the better.-- The faci
504q013 ?-- Bladder-nut ♫. Laburnum ♫. Dodecatheon ♫ . CASTRATE apple & pear to see if pollen naturally carri
384d162 male plumage in females. & female pigeon in CASTRATED male.-- «Men giving milk--» Sept. 25th Young
358d076 ually"= Opposed to these facts are effects of CASTRATION on males & of age or castration on females.-
358d076 re effects of castration on males & of age or CASTRATION on females.-- [not located] hen freely.-- he
358d086 some degree of whiteness like a Male.-- Thus CASTRATION, hybridity, & breeding in & in tend to produ
362d099 ks seen in savages) to brave men.-- Effect of CASTRATION horns drop off., replaced by hairy ones. whi
362d099 rapidity of the change in 5 or 6 weeks after CASTRATION, fresh horns begin to grow.-- Mr Yarrell say
033r042 how very curious a structure: Have shells ever CASTS alone in Calcareous. rocks??--if so case precis
039r061 ays that in N. Pliocene formation of Limestone, CASTS of shells, as in some older formations: Mem th
465t080 guide, the shells preserved must be as much a CASUALTY as, bones of Mammalia in caves::-- :argue firs
553m135 wers of the mind being EQUAL to the smallest CASUISTICAL doubts.-- The history of Metaphysics shows
246c027 -- (Case of replacement)-- Coquille Voyage The CASWARY, inhabits Ceram, Bourou & especially New Guine
215b178 res as latest authority. The case of the tailess CAT of Isle of Man mentioned in Loudons (analogue of
295c178 Pachydermata &c & other Mammals.-- «otter; civet CAT, rodents.-- (Pachyderm in Portland stone of Alps!
317c249 our & appearance Every now & then a short-tailed CAT.¿cut? has its offspring short tails /one born at
332d005 grations & Australian Savages.-- W. D. Fox has a CAT. which he bought in Portsmouth, said to come fro
333d005 aps rather different».-- crossed with <un>common CAT, exact variety unknown., three kittens, alike ea
333d005 sely» of form of mother: more than of the Common CAT.-- Ch IX Mongrels Hybridisim Fox has half Persia
333d005 - Ch IX Mongrels Hybridisim Fox has half Persian CAT. which bred with unknown common house cat.-- had
333d005 ersian cat. which bred with unknown common house CAT.-- had four Kittens. two appeared «so» very like
333d006 four Kittens. two appeared «so» very like common CAT, that they were killed, & other two very closely
335d015 tive, just like mules. Fox's half bred Persians «CAT» favour the Persian side.-- Theory of abortive h
366d108 strosity propagated by art. Yarrell told me of a CAT & of a dog, born without front legs-- -- the for
366d112 «(as Hunter supposes with Monsters)» if armless CAT can propagate, ie with the chance of two being b
380d155 led in string will not.-- some of the tigers.-- CAT, though caterwhalling. & put into female, when m
428e113 bt if wild species ever formed like short-tailed CAT or dog has been without recurrent tendency in ex
446e160 & defending their young.-- The oriolus (icterus CAT.) is an instance of this, & the female of the ic
493q003 Cats. as only one sex so coloured = I have grey-CAT «wh was female» with tinge of tortoise-shell «on
493q003 ll «on back.--» = Length of intestine in Persian CAT-- , in Brazilian «toothless» dog-- I. St. Hilair

(Key Word)
```
525m025 ted by previous marriages with impure breed.-- A CAT had its tail cut off at Shrewsbury & its kittens
531m049 ds nest, near hot-house at Shrewsbury, which the CAT was seen by Hubberley to visit daily to see  how
531m049 daily to see how the young got on. this nest the CAT could If cats will «ever» eat little birds, this
533m058 sely analogous or identical, with bird knowing a CAT, the first it sees it.-- it is frightened withou
557m147 &c  &c -- -- peacock & turkey cock in passion.-- CAT when pleased, erect its tail & make it very stif
628o53v to  teach it-- all dogs might be taught, but not CAT, that is not act by gusto, though by fear it mig
268c096 on the genus Procyon.-- by Wiegman Classified CATALOGUE of animals of Nepal read before Linnaean Soc.
278c130 used,  when there is series. CATALOGUE of Mammalia arranged according to my own meth
451e177 Journal of Asiatic Society Vol I. p. 261. <J> CATALOGUE of Birds of India. -- p. 555. Lieut. Hutton c
451e178 - Journal of the Asiatic Soc. vol. I. p. 335. CATALOGUE of animals of Nepal by. B. Hodgson. p. 336 In
514q21. erm." '=?? answered Does Mormodes (one of the CATASETUMS) really always hit stigma by projecting poll
419e087 ow change + & Therefore precludes effects of CATASTROPHES, which must serve to confound our chronolog
273c112 me habits, though swallow hawk, ,milvulus,, may CATCH insects on the wing & pratencole (¿connecte[d]
451e178 n almost (this looks inaccurate C. D) they will CATCH the Malaria & die.-- On the other hand there ar
499q009 mber of grains of pollen in any one flower (17) CATCH Bees, Butterflies-- Syrphus-- Meligethes & see
499q009 Bird lime near male yew tree & see whether they CATCH pollen-- <Ne> In Oenothera bush.-- (19)  Theory
522m010 ng.-- Mr Corbet, however, in conversation could CATCH up a new train if early association were called
557m149 eople (shame of ridicule) are singularly apt to CATCH tricks.-- so are people in passion my F. rubbin
641j29v ts. CD]CD. «Macculloch says no other bird could CATCH mouse by night» Sailing lizards. squirrels & Op
258c061 he aberrant groups.-- It is having walking fly CATCHER, woodpecker &c & which causes the confusion in
293c173 on between individual objects as Ichneumon & CATERPILLAR, though our ignorance, may make us think so,
344d041 omostraca & Aphides Developement of sexes in CATERPILLAR. very valuable facts-- they are eating foetu
354d062 factorily by. Valenciennes.» The change from CATERPILLAR to butterfly-- is not more wonderful than th
354d063 f Women) in different parts when age changes CATERPILLAR to Butterfly.-- When two Varieties of dogs c
372d131 ifferent from true generation.-- there is no CATERPILLAR state; the vast difference of two kinds of g
387d170 monkey & men.-- each man passess through its CATERPILLAR state. the monkey represents this state.-- W
489qIFC ewlands Lymington Hants. Habits of different CATERPILLAR races. --Name of Italian who sold eggs.-- Te
344d040 st in Linnaean Transactions of birds of Java CATERPILLARS not being fertile is same as children not b
387d170 having mammae.-- There is an analogy between CATERPILLARS with respect to moths, & monkey & men.-- ea
492q002 s of Cabbages not attacked in bad years from CATERPILLARS. 5. Whether Roses impregnate each other. wh
494q005 various bright colours, any effect? and silk CATERPILLARS (1) Shake a sleeping mimosa, or half bred m
380d155 ll not.-- some of the tigers.-- cat, though CATERWHALLING. & put into female, when muzzled, he is dis
403e023 ot this). because it has been so pronounced ex CATHEDRA. let us look at facts. considering few domest
024r016 oral paper {T} leagues Fathoms Parallel of St CATHERINE [27 degree 30 minute S.] 18--70 Paranagua [25
054r102 rich in particular genera of plants: All St. CATHERINE & coast Granite: P. 199; Falkland account  of
526m028 ng it up.-- equall true the two statements.-- CATHERINE remarks that pleasure received from works  of
526m028 ow differences between memory & imagination. «CATHERINE thinks that children like looking at <ani> pi
527m033 ny says she never builds castles in the air-- CATHERINE often, but not of an inventive class.-- Now t
539m083 spects a some of Aunt Sarahs. cranks, & so is CATHERINE in some respects--. good instances.-- when ed
078r175 s most torn.--(¿anticlinal line?). -- Mines of CATORCE «(Principal veins)» 25 degree to 30 degree  to
063r133 ethod where the division is not perfect.-- Dogs. CATS. Horses. Cattle. Goat. Asses. have all run wild
172b006 tinct for opposites to like each other AEgyptian CATS & dogs ibis same as formerly but separate a pai
175b016 as tendency to change.-- this difficult to prove CATS &c from Egypt no answer because time short & ho
206b148 xtended to all animals first consider species of CATS.-- -<& other tribes>.-- &c &c Exclude mothers  &
211b163 y & affinity-- whales & fish.-- Progeny of Manks CATS without tails: some long & some short: therefor
216b181 ogs investigate case of pidgeons. fowls. rabbits CATS &c &c.-- When black & white men cross some offs
218b190 ar un Officier du Roi Mem. Capt. Owen's story of CATS on West coast of Africa.-- changing hair-- The
233b250 rds.-- | Waterhouse thinks two main divisions of CATS. Tortoise shell--& grey-banded. ¿species?-- thi
233b250 --& grey-banded. ¿species?-- thinks offspring of CATS sometimes heterogenous.-- Australian dog jumped
251c041 . Mammal; «&to» good facts about distribution of CATS Vol III. <p.> p 233, stated that the "Asseel Ga
266c092 hich disfigure those of Arabia & Egypt.-- CIVETS CATS only wild animals on isld.-- Niether Hyenas; ja
293c175 ty to whole theory.-- There is breed of tailless CATS near Bath. Lonsdale do. says Sheep could not li
297c186 gy in habits adaptation to nocturnal habits-- to CATS &c.-- must be acquired by my theory-- else my t
306c214 ransported from their country.-- the long-haired CATS are supposed to come from there.-- All the shee
334d011 sisters in Mankind.-- The case of all blue eyed CATS (Fox has seen repeated cases) being deaf curiou
402e019 causes are most obscure. without doubt:-- Vide CATS «p 10» the joints near the tip of the tail  are
405e031 ould vary, if in wild state; thus mark on ear of CATS, colour can be brownish do Saw what was said to
453e182 233» There, as well in all Malay countries «the» CATS are born with the joints near the tip crooked.--
477z002 m those of La Plata or Paraguay.-- do. p. 365. 3 CATS (mbara caya. le negro, et le pajero) l'yaguaré
493q003 parent  & vice versã = History of Tortoise-shell CATS. as only one sex so coloured = I have  grey-cat
493q003 I. St. Hilaire says length differs in different CATS.-- Good observation-- examine semen of Hybrid a
515q022 ) Anatomy of muscles of stumps of tailess dogs & CATS.-- (3) Hounds-- varying-- (4) About blended ins
530m049 a  train of thought.-- [not located] Fox believe CATS discover birds nests & watch them till the youn
531m049 how the young got on. this nest the cat could If CATS will «ever» eat little birds, this most curious
556m146 do the same. although it is then quite useless-- CATS kneeding when old, like kittens at the breast n
558m152 when slightly angry. curls tip of tail.-- do two CATS arch their back when fighting, & not with  dog.
581n064 ng outang very fond of soft, silk-handkerchief-- CATS & dogs fond of slight tickling sensation.--  in
584n072 gh the air,-- waterbirds, the bee to its nest,-- CATS when carried in confinement,-- carrier pidgeons
063r133 division is not perfect.-- Dogs. Cats. Horses. CATTLE. Goat. Asses. have all run wild & bred. no dou
200b123 causes are most obscure. without doubt:-- Vide CATTLE: The grand fact is to establish whether in cro
202b131 he Caspian «Province of Ghilan» wooded district CATTLE with humps as in India. Geograph J. Vol III. P
202b132 ed from Timor, herded separate from the English CATTLE, nor could we get them to associate together'
203b136 re wilder than parents. which is same as Indian CATTLE ∴ tameness not heredetary?, having been gained
204b139 about half always & result the same Indian CATTLE & common produced very fine Hybrid  offspring,
212b167 Rat.-- ¿Consult Dr Smith History of S. African CATTLE Phillips Geology «p 8l» in Lardens Encyclop. *
234b255 ficier du Roi.-- Mackenzie Travel. p. 280. says CATTLE in Iceland. «"are very» like «those of Ice Hig
235b273 - so on with families.-- Ask Royle about Indian CATTLE with humps.--¿To be solved if horses sent  to
247c029 ntroduced in 64, of very many colours, like the CATTLE, -->which I say "are as variously coloured  as
266c092 l V. p 201 Wellsted. Memoir on isld of Socotra. CATTLE generally marked like those of the Alderney br
266c092 breed,  but size not larger than those of Black CATTLE. Not have hump like those of India & Arabia p.
275c121 agination-- He says he recollects all half Bred CATTLE of L'Darnleys were most like parent Brahmin bu
296c185 onderfully absurd.-- p. 565 <breed> Scotch wild CATTLE. breed freely with the tame Vol II. Magazine o
312c231 ed] Henry Thompson tells me best way to improve CATTLE is to cross between a good bull & <be> the pro
320c275 rterly Sir J. Sebright's Pamphlets Wilkinson on CATTLE not abstracted Scientific Memoirs. published b
325c267 engger on Mammalia of Paraguay. account of wild CATTLE & Montagu on birds (facts about close species)
328cIBC s de la sonde et de L'empire du Japon Wowett on CATTLE-- (Waterhouse has it) shells from Barrier isld
345d042 es <name> «to be so called».-- the short horned CATTLE have gone on for 50 or 70? years-- now «well f
345d042 rs-- now «well fixed» breed,: Jones says Sussex CATTLE were all white headed, but this was bred out &
346d048 (1838)  p. 611. Ld. Tankerville account of wild CATTLE of Chillingham,-- habits peculiar,-- young one
346d048 ccount. given by Boethius of ancient Caledonian CATTLE. Ch 3. Instinct L'Institut. p. 249. (1838). Eg
354d065 Sept 7th. -- I was struck looking at the Indian CATTLE with Bump. together with Bison, at some resemb
362d100 und. ask Colonel Sykes.-- Even our domesticated CATTLE have tendency to breed at particular times. Mr
363d101 nant has described them)-- [Study horns of wild CATTLE.-- plumage of fowls-- long ears of rabbits.  &
364d104 about  some Southern «see p. 43 supra» breed of CATTLE with white heads; which years afterwards occas
389d177 cted (This seems case, for by careful observing CATTLE can be bred in & in.)-- [The loss of passion i
390d179 from  Books to read Buffon Suites de.-- Horse & CATTLE Library of Useful Knowledge Bell's Quadrupeds
392d180 mon ones» in Zoolog. Gardens; <Buffalo & common CATTLE-- Esquimaux (& Australian) dogs with common do
393dIBC ale fail in passion.-- Disposition of half bred CATTLE at Cinbermere? How is Jackall & dog at Z. Gard
400e012 Voyages  Vol 8 «p. 46» Capt Davis in 1598 found CATTLE in Table Bay with Humps on their backs & big t
421e091 . 283. on the dark ears of the wild Chillingham CATTLE, with reference to Mr Bell's statement of  the
429e113 - Azara gives account of production of hornless CATTLE-- ¿& others?-- March 12th-- It is difficult to
451e180 s in Moa.--" Chapt.-- V.-- : do. Chat XXI. Wild CATTLE & Hogs on Timor-land-- monkeys do not exist. t
459tflr ge East coast of Africa-- Vol II. p. 256-- wild CATTLE at Madagascar-- «p. 121» No beasts of Prey. an
462t051 bjection, when one thinks of different kinds of CATTLE in every part of England. &c &c NB. In botanic
473s07r an Plant.-- June /42/-- June/42/ You can select CATTLE & sheep for horns & yet no difference in calve
473s07r in  calves--how is this in young pigeons--dogs--CATTLE? As we see the frame of animals can adapt itse
493q004 f gestation differ in different breeds of dogs. CATTLE, (Indian & Common) &c: length of life. (5) Doe
500q010 deer  Keepers cross the breed-- desirable as in CATTLE in Chillingham Park-- What Book on varieties &
```

508q016 mith Savages at Cape any selection of Males in «CATTLE» or in Killing the worst =or in dogs= (12) Do
512q019 ent breeds Is there any difference in breeds of CATTLE & sheep in the sprouting of the horns. at diff
515q021 large capsules of seed, as the commoner kinds-- CATTLE are horned, Suffolk have <abortive> «no» horns
555m142 ed Bynoes story of the women.-- The Chillingham CATTLE (& Porpoises) have not charity-- is it in form
592m105 origin being so is not odd.; for instance wild CATTLE & deer pursuing a wounded one.-- porpoises a d
135a145 ite-- Bengal. J. vol 7. p. 522. Mountain c near CAUBUL. parallel ranges. with here & there little bra
022r008 o» Dampier's last voyage to New Holland P 127.--CAUGHT a shark 11 ft long. "Its maw was like a leathe
023r009 f jelly which stank extreamly."--This shark was CAUGHT in Shark's Bay. Lat 25 degree. The nearest of
247c028 Chimera-- Antarctica «also Taeniatole austral» «CAUGHT» Chile, Van Diemen's land & Cape of Good Hope
363d103 ld me that 1/2 Muscovy & common duck were often CAUGHT wild off coast of America.-- showing hybrids c
482z011 cisticola.-- Embriza melanodera-- a linnet not CAUGHT.-- Troglogdytis Furnarius.-- Sturnus Magellani
533m061 ght of some such remarke as now advanced; for I CAUGHT it like a flash.--. strange if judgment remain
423e098 outhern Counties have whole fields, some for CAULIFLOWER &c.-- Uncle John believes one single turnip
423e098 p in a garden is sufficent to spoil a bed of CAULIFLOWER.-- (How curious it would be to make enquirie
024r013 out K. Georges Sound The idea of the water at CAUQUENES. coming from the [...] Cordillera & flowing T
543m101 eparate thing superadded.-- we can thus trace CAUSATION of thought.-- it is brought within <our own>
553m135 urned.-- so ready is change from. our idea of CAUSATION, to give a cause (& no one being apparent, on
558m151 of "ought." joined with necessary notion of "CAUSATION" in reference to this "ought," «as well as th
564n004 eason acting on (<not social instinct>) but a CAUSATION. & «perhaps» an instinct of conscience, feeli
579n060 rchel's Discourse p. 35. On origin of idea of CAUSATION; «succession of night & day does not give not
028r031 rer, so does the rain, therefore such «rain» is CAUSE, hence at least my ideas water is absorbed into the e
028r031 whole Galapagos Arch; because no sections» same CAUSE as no colour Sir J. Herschels idea of escape of
032r041 not be a circulation «however slow & weak.»; «(CAUSE of not accumulation of Coral limestone in inter
032r041 ate have happened. hurricane in bowels of earth CAUSE:--<exp> does not explain cleavage lines./ possi
040r063 her lakes or Avalanches (Glaciers very rare) to CAUSE floods in valleys, which must aid in preserving
041r065 s concretions, living only in that spot & being CAUSE of concretion; or being only preserved in that
053r101 depths (mem: profound earthquakes), which would CAUSE parallel lines, but the rectangular intersectio
056r109 see an entire island so encircled, the one slow CAUSE is apparent. «I confess I never see such island
056r110 pebble might remain to tell of these losses.-- CAUSE of chimney. to crater. as at Galapagos. St. Hel
058r117 levations as constantly going on we shall see a CAUSE for Volcanos part of same phenomena lasting so
090a024 would the perturbation be serious? if so other CAUSE besides thin vapour bringing planets to an end?
107a078 lcanos. crust must be thinner «under water» but CAUSE most difficult (better conductor) Fitz Roy's Ca
110a085 connected with variation of compass & these may CAUSE «or be effect of» elevation & subsidence. exami
114a095 been excessively slow because beach line chief CAUSE of denudation, but does not tell period.-- I ca
118a104 he surface Arguments against Herschel's view of CAUSE of continental elevations (I) the alternation o
128a129 tical equilibrium This will be only a modifying CAUSE. {P} land protuberant water to counterbalance H
139a180 ngth of life &c &c Will any inorganic substance CAUSE such monstrous growth as oak galls or rose <bud
151g039 ces rarely united.-- die away also, without any CAUSE, must be tides. &c. {P} <notch> roads very much
152g046 on level with shelves effect of corrosion & not CAUSE. Monday a rapid descent of a terrace except at
153g053 4th [shelf] river 4th [shelf] Could earthquake CAUSE collection of sediment? Where ravines enter sid
155g062 elf very broad «& cut out, produced» from same «CAUSE» as «great» spit <is> or plain <now> formed on
157g074 appear to cross stony parts; appearance chiefly CAUSE by fall of angular masses from above on soft sh
163g109 peculate on origin pebbles brought by different CAUSE: from mud.-- [blank] {P} muddy nodular strata c
171b005 esent system of body & universe therefore final CAUSE of life With this tendency to vary by generatio
182b049 of goats." Progressive development gives final CAUSE for enormous periods anterior to Man.. difficul
185b055 rat with land structures; :-law of chance would CAUSE this to have happened in all. but less in water
195b099 g created.-- passage for vertebrae in neck same CAUSE, such beautiful adaptations yet other animals l
200b123 , because every trifle heredetary, without some CAUSE of change,, yet such causes are most obscure. w
203b135 eory death of species without apparent physical CAUSE:-- Mem: Mastodon all over S. America. Hilaire a
203b137 Australia & S. America. that only some mundine CAUSE has destroyed animals over the whole world-- Fo
250c037 extreme both in, N & S. America, is only common CAUSE I can conceive of destruction of Great Animals
250c039 ted & answered Much might be argued what is not CAUSE of destruction of large quadrupeds.-- common to
258c060 wing to crossing, with unseasoned people. would CAUSE destruction.-- simile Man living in hot countri
268c099 n S. & N. Hemisphere just anterior to present. ¿CAUSE of destruction of «great» animals?-- Show indep
276c123 pinion they believe have endeavoured to advance CAUSE of truth It is of the utmost importance to show
276c125 cy to divide, which often enough repeated would CAUSE an unequal number of vertebrae?-- ¿Where two ver
292c170 here offered, & he must be a zealous man in the CAUSE if his faith is not staggered <">= I confess. n
292c171 ften «to be habitual» or of great importance to CAUSE long memory.-- structure is only gained slowly.
293c174 ot, however, perfect, else one animal would not CAUSE misery to other.-- else smell of Man would be d
303c203 species becomes father of genera-- whatever the CAUSE is. <the> any osculant species. which survived
308c219 Domesticated animals having same idiosyncrasy, CAUSE of fertility.-- varieties not produced as by na
309c219 eir genitals & even to a limb not used The only CAUSE of similarity in individuals «we know of:», is
314c236 time) as in buds.-- I can scarcely doubt final CAUSE is the adaptation of species to circumstances b
315c241 tween habitual actions of plants «when exciting CAUSE is absent» & memory of animals.-- (surely in pl
342d036 l <& unknown> causes, modified by unknown ones. CAUSE changes in geography & changes of climate super
369d114 . taking feathers of Peacock & spurs-- no final CAUSE here.-- & therefore different from Hunter I sho
370d115 mpted to be shown from peculiarities of climate CAUSE of N. Zealand not having any Mammalia.-- Type o
375d135 g gaps by thrusting out weaker ones. «The final CAUSE of all this wedgings, must be to sort out prope
386d167 e sympathy in human frame.-- «one of» The final CAUSE of sexes to obliterate differences. final cause
386d167 cause of sexes to obliterate differences. final CAUSE of this because the great changes of nature are
409e048 snow line descent. My theory gives great final CAUSE «I do not wish to say only cause, but one great
409e048 es great final cause «I do not wish to say only CAUSE, but one great final cause,-- nothing probably
409e048 not wish to say only cause, but one great final CAUSE,-- nothing probably exists for one cause» of se
409e048 final cause,-- nothing probably exists for one CAUSE» of sexes «in separate» «animals»: for otherwis
411e056 tion will be answered) one looks to analogy for CAUSE in plants. where innumberable individuals can b
417e078 dren like mother & in some like father «What is CAUSE of this.-- » (Lord Moretons law holds with diff
436e135 more efficient in making species, than time (as CAUSE of change) which can hardly be believed, then,
436e136 s that swallows have decreased in numbers, what CAUSE?? Seeing the beautiful seed of a Bull Rush I th
440e146 ucture is due to external agency, without final CAUSE. either in present, or past generation.-- thus
440e146 idgeons with tufts &c &c here there is no final CAUSE yet it must be effect of some condition of exte
440e146 age in pidgeons yet no change of habits, so no «CAUSE» corresponding change in Birds of Paradise.-- A
440e147 its-- no one can be shocked at absence of final CAUSE of beauty in man & wings under united elytra The l
441e150 ther, must not be overlooked.-- it makes fourth CAUSE or law of change.-- The weakest part of my theo
496q05a Mr Brown theory of insect-like Orchis-- & final CAUSE of beauty of flowers-- contrasted by Kirby-- wi
506q015 h will not produce seed in this country-- where CAUSE not apparent-- Any where pollen is not produced
508q016 rare beyond limits of the metropolis of each-- CAUSE?-- Andrew Smith. (6) What size book Gallesio st
528m036 i> & their absolute beauty. (which is as real a CAUSE as in music) from the splendour of light, espec
528m038 m occurrence in Mexican & Graecian to be single CAUSE) this symmetry & rhythm applies to the view as
540m087 en is he so different-- in organization.-- Same CAUSE as colour & shape & ideosyncracy.-- Look at the
540m087 look at them both semi-civilized-- Perhaps one CAUSE of the intense labour of original inventive tho
542m095 rows to see things in dark. & hence is this the CAUSE of expression of surprise-- viz seeing somethin
544m103 vividness, rapidity, novelty of separate ideas CAUSE fatigue to the mind,-- it is solely the compari
545m107 itor did, when excited or disturbed by the same CAUSE, which «now» excites the expression.-- Habitual
546m111 he suddenly like a flash without any assignable CAUSE, remembered she had put it in branch of tree, &
547m111 , of being compelled, from some quite imaginary CAUSE to start at once to Shrewsbury, vaguely though
552m131 rse in drays, scraping against cornice stone to CAUSE friction Athenaeum 1838. p. 652. Dr Daubeny on
553m135 s change from. our idea of causation, to give a CAUSE (& no one being apparent, one fixes on imaginar
559m154 rise at last even to the perception of a final CAUSE.-- Read. Paper on consciousness in Brutes & Ani
560mIBC r has not had it. but where grandfather was the CAUSE by his intemperance. <No.> Cannot say.-- Privat
567n012 y itself is now purely theological.-- Origin of CAUSE & effect being a necessary notion is it connect
567n013 effect of some law.-- have plants any notion of CAUSE & effect, «they have habitual action. which dep
568n017 it would feel «subsequent» sorrow, whatever the CAUSE had been]CD-- «Also» When one is prevented perf
576n049 to animals is philosophical. viz. man is not a CAUSE like a deity, as M. Cousin says. because if so
577n049 f man.-- & we do not suppose an hydatid to be a CAUSE of itself.-- [by my theory no animal. as now ex
577n049 [by my theory no animal. as now existing can be CAUSE of itself.]CD & hence there is great probabilit
579n060 ccession of night & day does not give notion of CAUSE,» do p. 135.-- on the importance of a name, wit
591n100 stinctive conscience.-- Why does not man eating CAUSE disgust, because he does not go against instinc
592n105 ble, <both> in the end gained «& therefore the» CAUSE, and origin being so is not odd.; for instance

```
605o18v lences, brings to our recollection the original CAUSE of these feelings & thus we apply to them the m
615o36v different  way of showing it, nor was there any CAUSE, & if surprise was felt.-- analyse feelings. Mr
618o41v caused  weight, because both due to some common CAUSE:-- The argument reduces itself to what is cause
618o41v cause:-- The argument reduces itself to what is CAUSE & effect: it merely is «invariable» priority of
618o41v first,  we should not think night an effect.]CD CAUSE and effect has relation to forces & mentality b
620o044 e do not afterwards think of it.-- Whatever the CAUSE of this may be, everyone must know, how soon th
620o044 bey the conscience is extremely great [LHC] The CAUSE perhaps lies in its frequency & in its consisti
628o53v instinctive  in me (& as a consequence, but not CAUSE gives me [3] pleasure) or that I have been taug
628o054 will be for the general good, that is, the same CAUSE, which gives the instinct.--]CD p. 22. says aff
635j55r constant  or inconstant like that of Man.-- the CAUSE given we know not the effect [blank] 6 p., 412.
635j56r ons, just as liability to accidents & any other CAUSE.-- (& my theory [ALL PARTS OF ONE GREAT SYSTEM.
637j58r et min study Malthus & Decandoelle.-- The Final CAUSE of innumerable eggs is explained by Malthus.--
640j167 merous. (from death of its destroyer), or other CAUSE, the long legged race would prevail, even if ha
061r126 ips.» 1826.27.28. grt. drought at Sydney. which CAUSED Capt. Sturt expedition. -- ¿ Another one in 18
132a139 ocean. (with fresh sediment added to bottom) be CAUSED, by absence of circulating water.-- & therefor
157g073 ins many pebbles, but I believe this is chiefly CAUSED by its being lower,-- [no pebbles in parts of
332d001 lignant erysipelas spreading over the head, not CAUSED by a wound, when suddenly during one time he h
422e095 erhaps will have done so from the new relations CAUSED by the advancing complexity of others.-- It ma
441e148 hanges in fruit trees. mentioned by Mr K may be CAUSED by the diversity of stocks, on which they  are
491q01v it affected. Mr. B. seemed to say impregnation <CAUSED> of some seeds, caused symmetry in cone--  The
491q01v med to say impregnation <caused> of some seeds, CAUSED symmetry in cone-- The «above Exper»  explains
523m016 e would not have been so.-- In Mr Hardinge, was CAUSED by thinking over the misery of an illness at R
535m069 thosoma grisea described [not located] as first CAUSED by will of Gods. «or God» secondly that  these
581n066 ying certain bodily actions]CD.¿ but what first CAUSED this bodily action. if the emotion was not fir
618o41v vier yet it could not be said that the blueness CAUSED the weight, anymore that weight, the blueness,
618o41v aried) then might it now be said, that blueness CAUSED weight, because both due to some common cause:
038r058 ill here be the strongest argument:--¿ Consider CAUSES for subaqueous crater being of diff: form suba
043r071 of  space in formations & durability of similar CAUSES go together. add. <">" <from> "in the same lin
048r085 he History of S America we cannot dive into the CAUSES of the losses of the «species of» Mastodons. w
056r111 er whole world» argument, as well as separating CAUSES by water.--Or rather begin & explain how water
057r112 hemical instrument.--Yet neglecting these final CAUSES.--What more awful scourges to mankind than the
105a071 ce came to Mendoza--. Will they introduce other CAUSES to explain «alluvi» in valleys Lowe in his pap
119a105 egular figure to be that of equilibrium,-- What CAUSES that of tendency to irregularity,--. Why  does
175b017 e change probably <change> vary quicker Unknown CAUSES of change. Volcanic isld.-- Electricity Each s
179b035 onsidering. S. America (independent of external CAUSES) does appear very probable:-- Mem: Horse, Llam
185b055 to island. where it could live-- but there were CAUSES to induce great change. like the Buzzard which
200b123 detary, without some cause of change,, yet such CAUSES are most obscure. without doubt:-- Vide cattle
206b148 he other will become extinct.-- Who can analyze CAUSES, dislike to marriage, hereditary disease, effe
206b148 se, effects of contagions & accidents: yet some CAUSES are evident, as for instance one man killing a
206b148 two  fine families «no doubt a different set of CAUSES must act in the two case,» May this not be ext
525b217 should say the changes were effects of external CAUSES, of which we are as ignorant. as why millet se
227b227 al grouping we are led to endeavour to discover CAUSES of change.-- the manner of adaptation (wish of
227b228 ur crossing & what prevents it--» & generation, CAUSES of change «in order» to know what we have come
248c034 ly heredetary.-- If varieties «produced by slow CAUSES, without picking» become more & more impressed
257c056 we know uses of organs clearly, we cannot guess CAUSES of change.-- hump on back of cow!!-- &c &c D'o
258c061 ving walking fly catcher, woodpecker &c & which CAUSES the confusion in this system of nature-- Wheth
277c128 er every purpose, & would present many ideas of CAUSES of change.-- The mark of analogy would be empi
303c204 hat Tertiary geology has obeyed rules of modern CAUSES. & considering over the viccissitudes of prese
323c269 1 5d Dr. Edwards of <ter> influence of Physical CAUSES: well skimmed Bartrams Travels in N America Ma
332d001 ctober 2d As a proof. what <trifling> «unknown» CAUSES act upon people. My Father mention, than for t
342d036 can  take of the world Astronomical <& unknown> CAUSES, modified by unknown ones. cause changes in ge
343d036 e superadded to change of climate from physical CAUSES.-- these superinduce changes of form in the or
375d135 population in Men increase, & an ordinary crop. CAUSES. a dearth then in Spring, like food used for ot
379d152 mentioned about proportion of sexes, at birth & CAUSES. If an animal breeds young her growth is immed
386d166 .-- this law acting against heredetary tendency CAUSES abortive organs.-- the origin of this law is p
390d178 ative had come from different quarters) then it CAUSES <to> a secretion of something someways differe
397e003 g to fixed laws: And since the world began, the CAUSES of population & depopulation have been probabl
410e052 al interest,-- "the great end must be the law & CAUSES".-- A philosopher, would as soon tur
411e055 e series have been fixed, to study the physical CAUSES. «All Cuviers generalization. of teeth to kind
413e059 passage  upon problem.! Hurrah.-- "intermediate CAUSES" The Sexual system of the Cirrhipedes is the m
449e104 ame, as Galapagos facts &c &c.-- & it shows the CAUSES which give same species to different isld.  is
523m016 han of a dream.-- Insanity is produced by moral CAUSES (ideotcy by fear. Chile earth quakes). in peop
524m019 humour & depression, which comes on from bodily CAUSES.-- It is an argument for materialism. that col
527m033 nct the ears (rhythm & pleasant sound per se) & CAUSES the mind to create short vivid flashes of imag
528m037 ld one. every time one looks at it.-- these two CAUSES very weak.-- (2d) form. some forms seem instin
590n093 ore circulation is affected. p. 44.-- Jealousy. CAUSES spasm in bile duct, & throws bile in circulati
605o19v by one common principle» to explain the various CAUSES of those sensations, which we call metaphorica
605o19v ame term.-- Hence it appears, that when certain CAUSES, as great height, eternity, &c &c. produces an
605o020 ople who are susceptible of pleasure from these CAUSES who are not men of taste & the reverse of this
637j58r althus.-- [is it anomaly in me to talk of Final CAUSES: consider this!--]CD consider these barren Vir
297c185 o many species., same circumstances., which by CAUSING death, makes the group aberrant When species r
628o53v is strictly analogous to education of child,-- CAUSING many actions to be considered right & wrong,--
635j56r rgo «animals feeding on each other &c &c»-- -- «(CAUSING death to some, &c &c)» These are reasons, just
048r085 ch the Rhinoceros lives in S. Africa: the same CAUTION is applicable to the Siberia case We must  not
120a109 ing cliffs. on each side, such as now exist.-- CAUTION about action of rivers.-- Excess of matter bro
432e124 animals of any country must be made with great CAUTION: owing to its adaptation to the surrounding ci
130a134 n as in Patagonia (P] B subsidence; <as in> be CAUTIOUS. mud banks & sand. dunes.-- in these littoral
314c239 s there may be Eucalyptus!-- (Hostile fact) Be CAUTIOUS about Goulds case of birds of Van Diemens lan
555m143 , very different disposition from others, slow CAUTIOUS, angry cross look, followed by protrusion of
565n010 just  what smile is to laugh.-- I must be very CAUTIOUS. Remember how Lavater ran away with new Lavat
641j29r rmaceti Whale & Manatee.-- Naturalists must be CAUTIOUS.-- <some others>: study these facts read Lacé
599o006 evelopement of the thoughts expressed in Fingals CAVE, & in the arched & leafy forests" Very good!. I
325c267 losophie d'Histoire Naturelle Marcel de Serres CAVERNÉS d'Ossements 3d. Edit. Octav. (good to trace E
103a065 on to weight of super [...] mass.-- Absence of CAVERNS, in Plutonic rocks argument against great bodi
435e134 Proteus  anguiformis. he remarks lives in dark CAVERNS of Carniola p. 112. Man. "standing alone in th
278c130 n of two continents & death of form in one. The CAVES are at a height of more than 1000 ft. & many hu
415e066 cosmopolite,  we do find his remains.-- Lima.-- CAVES.-- There being no fossils, the only way, that I
465t079 several generations have been confounded in the CAVES? It is highly important, to bear in mind that e
465t080 be  as much a casualty as, bones of Mammalia in CAVES:-- :argue first case of bones (New Red Sandston
479z006 bigny considers Dasypus villosus is true Peludo CAVIA Australis. Dorbigny Vol II, p 24 Proceedings of
055r106 ficient proof, that the sea formed these large CAVITIES", &c &c &c Vol II. Chapt VIII. p. 97 at Potos
477z002 a Plata or Paraguay.-- do. p. 365. 3 cats (mbara CAYA. le negro, et le pajero) l'yaguaré «the zorilla
089a020 f Gory. rocks encrusted with serpula-- Isle of CAYENNE. Syenite & diorite, covered with iron clay com
437e138 ry.-- p. 103. Turtles finding their way to the CAYMANS from Honduras. good case of migrating.-- shows
507q016 ther accidents of delivery-- inheritable.?-- Bell CD ask Accouchers Is any peculiarity in milk  teeth
641j29r ble, [gained by very different process from Bats. CD]CD. «Macculloch says no other bird could catch n
209b156 plus  en conclure quels sont les genres qui, sous CE climat, se divisent le plus aisément en espèces
257c056 . ou au moins des espèces tres-analogues,-- quand CE nétaient pas tout à fait le mêmes." This good c
227b224 f <genus.> structures in two genera.-- <we then CEASE to know the steps.> although D E F. follow clos
289c161 ook at wings & orbits of penguin & then he will CEASE to doubt :Scales into Teeth in Bering Pike (Wat
366d108 ding to my view, the domesticated animals would CEASE being fertile inter se,. or at least show-repug
173b011 ic existed might have been Monsoons.. when they CEASED importation ceased & changes commenced.-- or I
173b011 ve been Monsoons.. when they ceased importation CEASED & changes commenced.-- or intermediate land ex
288c160 e the woodcocks,. which came Madeira & <seized> CEASED their migrations lost?? I conceive a bird Migr
325c267 uced in domesticated plants; where function has CEASED to be used as tendril into stump Library of us
402e018 W. Earl's Eastern seas. p. 206-- shot a monkey, CEASED their cries. "many of them descending to exami
537m076 den» changes rare,-- as when Polynesian mothers CEASED to destroy their offspring--¿ yet perhaps if t
604o015 wed by Johnson in the Literary Magazine. 1756-- CEASED in 1758-- Read the Review or the article. A Pl
```

379d151 W. itself received from his father so that case CEASES to be true avitism Annals of. Natural. History
559m154 pare it to the standard of our own minds. which CEASES to be the case when we consider the formation
565n006 e <&> «whilst» laughing in glass. & then as one CEASES, or stops the noise , the face clearly passes
294c178 -- Von Buch. Travels, p. 306. account of trees CEASING to grow far N. becomes stunted, altered, & los
536m070 htly as to endeavur to stop heart beating: one CEASING, so does other.-- What an animal like taste of
058r118 tte ile se developpent autour d'elle comme une CEINTURE d'immenses remparts; toutes affectent une pen
232b249 ista New Holland form) is found in many island CELEBES «Waggiou» &c &c. (See Lyell. Vo‖ III p. 30) di
240c013 s. niger. comes from the Moluccas «Matchian» & CELEBES.-- ‖ Amboina; Viverra Zibetha.-- ‖-- All the Mol
240c014 am race with long nozzle & few hairs) inhabits CELEBES & few of the larger islands.-- -- Antelope in
240c014 & few of the larger islands.-- -- Antelope in CELEBES, Bourou new species of Axis.-- «Cervus molucce
450e175 -- Phillipines Pony somewhat resembles that of CELEBES is somewhat larger than the Sambawa, Java & Su
450e175 The horse is only found wild in the plains of CELEBES. (but language shows that probably not origina
447e164 insects to impregnate allied to Asclepias Turpin CELL is individual May 29th.-- -- Henslow says he ha
448e165 f plant producing buds, so that Turpin says each CELL of plant is individual.-- Most plants which pro
583n070 scles, we cannot look at him, as machine to make CELL of certain form. (& especially as it adapt its
583n070 l of certain form. (& especially as it adapt its CELL to circumstances), it must have impulse to make
583n071 o circumstances), it must have impulse to make a CELL in certain way, which way its organs are suffic
583n071 or hence it must some way be able to measure the CELL; p. 22. instincts & structure always go togethe
367d113 wer of inflating its body like balloon-- by air CELLS connected with cheek pouches.-- Hunter's Animal
370d116 e rainy day-- & from case of wasps, is supposed CELLS properly are made for larvae.-- CD[(p. 451.)--
485z018 - <Rep> do p. 324. Polypi shorter duration than CELLS.-- reproduced.-- Milne Edwards p. 138 on Polypi
583n070 "creatures of habit."-- CD[as the bee makes its CELLS, by means of ordinary senses & muscles, we cann
585n077 ossibly» acquired by habit. so bees in building CELLS, must have some means of measuring cells, which
585n077 ilding cells, must have some means of measuring CELLS, which is faculty, they use this faculty instin
586n082 so a Bee has the faculty of building «regular» CELLS-- [but this faculty <may possibly be> «probably
638j58v nsciousness superadded This is similar idea, to CELLS of bee, corresponding to <every» «one or any»--
049r087 basal lavas be called Volcanic or Plutonic The CELLULAR state of all the Porphyry specimens, must be
183b052 On sait que le "métis" du loup et du chien, que CELUI de la chevre et du belier, cessent d être fecon
622o048 cts, or heredetary habits fully explains the CEMENTATION of habits into instincts. [RHC] Feelings of
104a066 des discussion-- Albite certainly contains 6 per CENT more silica than common felspar therefore on ax
104a066 xis of Cordillera, in Andite - containing 80 per CENT of Albite 80/100 X 6/100 = 480 In Falkland isla
081rIBC 1. 10 Metre 3. 0. 11 lig[nes] Decimetre 3. 8 CENTIMETRE 4.4 {T} C. Darwin R. N. Range of Sharks Noth
033r042 s the husks at Coquimbo: in that case, may not CENTRAL and rather differently constituted lime have b
036r052 ditto for Patagonia, especially rocky parts of CENTRAL Patagonia Does Andes in Chili. separate geogra
038r057 l arguments for believing that there must be a CENTRAL core of melted rock--I think the strongest is
038r058 ate these reasons, & saying that they refer to CENTRAL nucleus & that envelopes no doubt existed. The
042r067 & during gales would tend to travel on a <me> CENTRAL line of Patagonia. «NB. Mr Lyell P. 211 Vol II
044r072 number of dikes in the cliffs.--wide valleys.--CENTRAL peak small; yet great body of lavas have flowe
059r120 arts of one whole burning mountain, & that the CENTRAL part fell in.--Says posterior craters in centr
073r163 tive & transition; the latter rarely appear in CENTRAL Cordillera. particularly between 18 degree & 2
098a044 e Geolog. proceedings Lake let out by steps in CENTRAL France not very conclusive proofs, but certain
100a053 llization transverse.-- or rather radiating to CENTRAL point. can cleavage be radiation from some gra
110a087 Journal. Vol IV (p 321) Mr Hillhouse describes CENTRAL granitic ridge of Guayana as NW / SE. Vol VI.
127a126 form as the result of heat.-- will it bear on CENTRAL fluidity.-- do p. 137. Lord Tullamore found Su
132a138 older than mean of place, shows earth not with CENTRAL heat.--» «(does M. Parrot suppose there is no
132a138 ght, (according to M. Parrots argument against CENTRAL heat to warm the ocean).-- and M. Parrot does
203b136 ock, was found in Germany & thinks even now in CENTRAL & Eastern Asia beyond the Ganges & perhaps eve
203b137 reduction of temperature from geographical or CENTRAL heat.-- But then shells-- Mr Yarrell says that
246c027 en's land.-- Vol II p. 8 no snakes on isles of CENTRAL Pacific, yet there appears to be one at Botoum
278c132 culty is how came it animals not preserved, in CENTRAL S. America & yet Africa & India???-- & Indian
281c139 so far true as avoided linear arrangement the CENTRAL twigs dying, affinities would be <circular>, i
298c191 one species «on -- mountain side» of which the CENTRAL parts become occupied by a third best adapted
352d058 ther probably will be dead-- hence there is no CENTRAL radiating point, all united . (links in circle
032r041 erences of temperature «at equal distances from CENTRE of rotation» & a <circulation owing» rotation
044r072 small; yet great body of lavas have flowed from CENTRE-- Pisolitic balls occur in the Ashes which fil
058r118 ge de la mer, tandis, au contraire, que vers le CENTRE de' l île, elles presentent une coupe abrupte
059r119 t formées de couches parallelles et inclinées du CENTRE d'ile, vers la mer; ces couches ont entre ell
059r120 entral part fell in.--Says posterior craters in CENTRE:-- Bailly talks of much granite on all East si
100a053 oint. can cleavage be radiation from some grand CENTRE.-- A Stalactite of Gypsum, is the best case of
107a079 to Andes & Patagonia-- On Lyells idea of whole CENTRE of earth same heat, then change in form of flu
108a079 f earth same heat, then change in form of fluid CENTRE would lift with it isothermal line, but if hea
108a079 lift with it isothermal line, but if heat from CENTRE, then crust of solid earth would be thicker.--
113a090 have very unusual conducting power of heat from CENTRE.-- But is this not wrong? we know mean of surf
128a128 blem will be falling of an arch weighted in its CENTRE.-- Will not abrasion of land on one side. prod
148g028 rraces formed successive bays but plains sloped CENTRE-wards which would not have happened if the sid
150g037 hallow channel 50 ft wide & river get formed in CENTRE In Glen Collarig, on side of Hill of Bohunthin
179b033 T. del Fuego. the existence of whiter tribes in CENTRE of S. America shows this.-- <If> Is there a te
190b076 West. ends of New Holland. diminishing towards CENTRE (p. 586)-- Parallel 33 degree-35 degree, sourc
306c214 Asia Minor of the domestic animals-- At Angora «CENTRE of Asia Minor» are the fine-haired goats. whic
308c218 culations must end as well as begin" &c &c then CENTRE is every where & then circumference no where--
352d058 qual, because fossil) Now what is group without CENTRE but circle, two or three lines deep-- with res
497q006 ect of Chance Maer.= (12) Take Bag of soil from CENTRE of woods «especially if date of wood be known»
189b072 ups.-- Speculate on land being grouped towards CENTRES near Equator in former periods & then splittin
246c026 cept. Good!!-- in Mauritius Lesson &c p. 620. CENTROPUS (Coucal) of Java & Phillippines, has variety
194b095 e facts were established it would go to show a CENTRUM for Mammalia.-- I really think a very strong c
096a040 Isld,-- New Britain-- &c &c In Ascension for CENTURIES afterwards it might be percieved on which isld
206b148 t two fine families one with successors «for» CENTURIES, the other will become extinct.-- Who can ana
418e084 dded to old individuals,, during thousands of CENTURIES,-- each of us, then <is as old, as the oldest
433e126 . show some great change who can say how many CENTURIES elapsed between each of these gaps, far more
332d003 ty much alike.-- My Father Water-in the hair a CENTURY since used to be called Worm Fever, as used mu
497q007 nown changes in Flora of countries during last CENTURY or two.-- where agency of man not known.-- (3)
206b149 or of coniferous trees; or in certain shell CEPHALOPODA.-- «Read Buckland» L'..Institut «1837.» p 31
297c186 . & Horses few forms. & they are remnants.-- CEPHALOPODA ditto.-- Mag of Zoolog. & Bot. Vol. II p. 12
374d133 mestone (how different from plants!) But the CEPHALOPODA depart more widely from living forms.-- p 45
379d151 er Mammalia. & still further back reptiles & CEPHALOPODA: Old Jones remarked to me, that one of the c
381d157 in all.-- 2 NB. In Pectinibr Mollusca.-- «or CEPHALOPODA» are there abortive traces of other «sexual»
423e096 then has there been a retrograde movement in CEPHALOPODA & fish & reptiles.?-- supposing such to be t
212b168 luscs. where species now least in number (as CEPHALOPODS,) in last tertiary epochs most genera dead?
250c040 mportant paper by Dillwyn, on replacement of CEPHALOPODS & Trachilidous Molluscs. by each other in se
381d156 a. All others, <animals,> «are Dioeecous as» CEPHALOPODS, Pectinibranchiate molluscs.-- insects. spid
436e135 parts of sea, like Cetaceae,-- although the CEPHALOPODS, seem to have decreased since earliest times
246c027 cement)-- Coquille Voyage The caswary, inhabits CERAM, Bourou & especially New Guinea (replaces, Emeu
088a071 transport of Fish Let a Hawk fly at Heron.-- CERATOPHYTES common in Northern seas p. 312. Chamisso in
479z006 vellous stories Ulloa's Voyage Vol I, p. 168 CERATOPHYTES common in Northern sea. Chamisso in Kotzebu
317c249 /one born at Maer. Tuckeys voyage-- p. 36 "CERATOPHYTES saboeus" said to be monkey of St Jago C. de
376d137 n their faces & Mak the st. st noise.-- The CERCOPITHECUS chinensis: (or bonnet faced monkey he has s
402e019 ome peculiar <M> p. 359. At Manilla a small CERCOPITHECUS., & skins of galiopithecus.-- Malte Brun. V
596m184 arcely able St.-- » Cyanocephalus, macacus. CERCOPITHECUS? very much., «Keeper says some of the monke
596m184 handed??]CD» Cyanocephalus, Macacus, Niger. CERCOPITHECUS make labial st st. S. American monkeys. pul
189b074 than another.-- We consider those, where the {CEREBRAL structure intellectual faculties} most develo
254c047 s this agree with grand fact of Marsupial, low CEREBRAL structure??-- «do» p. 390. All classes of Acr
616o039 same relation to a living body (especially the CEREBRAL portions of it) that attraction has to ordina
618o041 , memory &c. either to our bodily frame or the CEREBRAL portion of it Thoughts, perception &c. are mo
328cIBC enera Corluia Cham. Cardium. Porcellus Turbo. CERITHIUM Jardin du Roi Java fossils at same time Study
078r175 s, and the slope of the mountains (flaqueza del CERRO) have been parallel to the direction & inclinat
095a038 in Sir. W. Parish Possession. talks of <hill» «CERRO» of Diamante near stream of same name. with imp
051r095 determines whether an animal will adhere to a CERTAIN part. Apropos to question does animal adhere t
061r128 nent: ≠ as if any creation «taking place» over CERTAIN area must have peculiar character: Contrast lo

```
068r147 ontinent.-- change of volcanic focus.-- <it is CERTAIN, if strata can be> Problem dislocate strata wi
075r167 very  «un»common in Mexico. Fluor spar only in CERTAIN mines. «Vol. III» "In general it is observed b
102a062 state  broadly indication of new law acting in CERTAIN directions predominantly, connection with magn
146g013 situ of even imperishable pebbles/ I am nearly CERTAIN there were none on surface of any hill Thursda
147g017 to  double series Whole very obscure but it is CERTAIN there must once have been very considerable ma
151g041 esis impossible to explain absence of lines in CERTAIN parts.-- At the Pass of Glen Collarig two litt
174b014 ved.-- We can see ««why» structure is common in CERTAIN countries when we can hardly believe necessary
178b032 son Voyage of Coquille.-- Dr. Smith says he is CERTAIN that when White Men & Hottentots or Negros cro
179b035 ed.-- ¿Is this shortness of life of species in CERTAIN orders connected with gaps in the series of co
179b035 from same epoch certainly» The absolute end of CERTAIN form from considering. S. America (independent
185b056 to  true. American genus Waterhouse says he is CERTAIN, that in insects, each family, however many th
187b064 uals.-- Why does individual die, to perpetuate CERTAIN peculiarities, (therefore adaptation), & to ob
188b067 -Mem Parrots peculiar according to Swainson to CERTAIN islets in East Indian archipelago Dr Smith com
195b101 ame manner God orders each animal created with CERTAIN form in certain country, but how much more sim
195b101 rders each animal created with certain form in CERTAIN country, but how much more simple, & sublime p
195b101 sublime  power let attraction act according to CERTAIN laws such are inevitable consequen let  animal
198b105 enera, & draw up tables-- Instinct may confine CERTAIN birds which have wide powers of flight; but ar
206b148 of Pachydermata, or of coniferous trees; or in CERTAIN shell cephalopoda.-- ««Read Buckland» L'..Insti
209b155 fter physical changes) would be in rays-- from CERTAIN sports.-- Agrees with old Linnaean doctrine &
209b155 Agrees with old Linnaean doctrine & Lyells. to CERTAIN extent Von Buch,.-- Canary Isles: French Edit.
210b161 ir origin to very early stage) & which, follow CERTAIN laws according to species. present an  analogy
213b172 st well worthy of observation.-- I think it is CERTAIN strata could not now accumulate without seal-b
214b174 is it Edentate? Phillips. Lardner p. 289 It is CERTAIN, that North American fossils bear the  closest
230b239 ted.-- «It will admit» a plant will admit of a CERTAIN quantity of change at once. but afterwards wil
234b256 ver, will settle this.-- Waterhouse says he is CERTAIN there are local varieties «of colour & size, b
234b256 another from Swansea.-- Again Waterhouse finds CERTAIN varieties of a Harpalus. common at South  end,
235b273 through all genera, & see how many confined to CERTAIN countries-- so on with families.-- Ask Royle a
251c040 rs to nest Vol II. p. 49. on the localities of CERTAIN parrots habitations India & Africa.-- NB.  Any
252c045 Analogy  from three first will give one almost CERTAIN guide ʼ. because time required too separate isl
254c047 oa, the organs of generation, afford the least CERTAIN indication of the perfection of the Species--!
259c064 - Epidemic amongst trees. Plane trees all died CERTAIN year. Extreme difficulty of TRACING change of
259c065 looked  at as the best attempt of nature under CERTAIN very unfavourable conditions.-- as an adaptati
264c079 de, uncivilized man) might not have lived when CERTAIN other animals were alive, which have perished.
271c106 s book, about not altering breed of animals in CERTAIN countries.-- Fraser remarked to me at Zoologic
274c115 ffinity all organs change together, in analogy CERTAIN parts perfect of typical structures certain <i
274c115 gy certain parts perfect of typical structures CERTAIN <imper> parts changed Have <not>. S. Africa, A
280c136 we  might expect even if two mules bred or two CERTAIN varieties, they would go back to  grandfather,
281c139 fore most subject to change.-- may account for CERTAIN organs not being fixed, <whi> in some  genera,
282c141 become greatly changed. in structure & even to CERTAIN degree in habits, yet we might have these anai
283c146 obably to add to quantum of life possible with CERTAIN preexisting laws.-- .If only one kind of plant
285c152 erated in the course of ages.-- As species is <CERTAIN> real thing with regard to contemporaries-- fe
285c153 .-- The constitution being heredetary & fixed, CERTAIN physical changes at last become unfit. the ani
288c159 oxen  have deteriorated, or altered, but it is CERTAIN that rams & bulls from England fetch very <go>
303c205 n the Chalk Those who say «philosphically to a CERTAIN extent.-- nothing but experience. will, tell u
305c210 spirit,  prevalent over this word, (subject to CERTAIN contingencies of organic matter & chiefly heat
309c221 186.  quotes Burkhardt to show black colour of CERTAIN Arabs.-- NB avoid quoting these hackneyed case
311c227 A  Smith is firmly believed in representation. CERTAIN birds in many families, «+very often in number
318c253 se--- -- the birds seem to follow narrow bands, CERTAIN kinds as gallinules taking the low country nea
335d015 be  like either parent, or intermediate within CERTAIN small limits (within which limits they might r
337d020 re taken to fresh country & breed confined. to CERTAIN best individuals.-- scarcely any breed but wha
338d025 & the only difference is by the extinction of CERTAIN forms from Northern part & not by fresh creati
342d033 ing produce heterogenous offspring. It appears CERTAIN that hybrid Muscovy & Common duck have been sh
343d036 changing affect each other, & their bodies, by CERTAIN laws of harmony keep perfect in these themselv
348d050 "The  most important characters break down in CERTAIN species & become worthless-- Mammalia Edentata
416e071 in each birth) from man arbitrarily destroying CERTAIN forms & not others.-- Term variety may be used
436e136 vary, according to my Malthusian views, within CERTAIN limits, but beyond these not.-- argue  against
446e162 clined to think very close.-- «A fruit tree by CERTAIN treatment will suddenly send forth quantities
461t037 tely injurious «(or requiring nutrition)» to a CERTAIN amount it will be so handed down«(«.. as mamma
468t111 hence quite below stigma. & so avoided it.» On CERTAIN days Humble seem to frequent certain  flowers,
468t111 d it.» On certain days Humble seem to frequent CERTAIN flowers, to day early, the great scarlet Poppy
491q01v IV Hort. Transact.-- 4 May we no suppose, that CERTAIN plants, like Aphides produce impregnated young
526m030 mal improves because its appetites urges it to CERTAIN actions, which are modified by  circumstances,
527m032 agrees with the stated fact, that «birds from» CERTAIN districts have the best song. [Migratory birds
533m059 way from what they have been accustomed to, in CERTAIN actions-- the difficulty of getting on a horse
543m100 , & contingency a man has keep in mind. all is CERTAIN.-- there is judgment of probabilities, therefo
545m104 hich must require idea to order muscles to do <CERTAIN> the actions.¿ is it the <becom> impression be
547m111 ce no chain of association» Mayo Philos. seems CERTAIN that muscular, mental, <&> digestive nervous i
571n029 h, according to my theory must be acquired, by CERTAIN foods being habitual-- & hence become heredeta
574n041 ll, as Sir. Ch. Bell says, & hearing music. to CERTAIN degree sexual.-- The association of saliva, is
575n045 er there is any analogy between grief & pain-- CERTAIN ideas hurting brain, like a wound hurts body--
576n048 or plants have instincts» «either» to obtain a CERTAIN en<s>«d»: & intellect is a modification of <in
577n050 The memory of Plants, must be association,-- a CERTAIN round of actions take place every day, & Losi
577n050 leaves,  comes on from want of stimulus, after CERTAIN other actions, & hence becomes associated with
581n066 e heredetary effects on the mind, accompanying CERTAIN bodily actions]CD.; but what first caused this
582n068 prearranged  actions, which will bring about a CERTAIN result, while the creature performing those ac
583n070 cannot look at him, as machine to make cell of CERTAIN form. (& especially as it adapt its cell to ci
583n071 ances), it must have impulse to make a cell in CERTAIN way, which way its organs are sufficient for h
586n078 & experience, or habit.-- so bird migrating to CERTAIN quarter is instinct, but his knowledge of that
586n081 d has the faculty of finding its way, which in CERTAIN species is instinctively «not least by experi
586n081 ctively «not least by experience» directed to CERTAIN quarter"-- "An animal has faculty of  walking.
601o009 ase.-- as when the excised heart is pricked) and CERTAIN action. (only evidence. when not consciousness
605o19v this  same term.-- Hence it appears, that when CERTAIN causes, as great height, eternity, &c &c. prod
610o30v test good.-- Therefore rule of happiness is to CERTAIN degree <of> right.-- The change <of> our moral
610o034 ts of Life in the abstract is matter united by CERTAIN laws different from those., that govern in the
611o034 being,  the capability of such matter obeying a CERTAIN & peculiar system of movements. different from
611o034 nitely (<like> «as» chemical laws,) as long as CERTAIN contingencies are present, (contingencies as h
611o034 wth <extres> tissue <[...]> unites matter into CERTAIN form; invariable, as long as not modified by e
612o035 nceive such movements & choice, & obedience to CERTAIN stimulants without conscience in the lower ani
614o037 is  the attraction of carbon. hydrogen <&c> in CERTAIN definite proportions, (different from what tak
615o038 ught, (or under contingencies of stimulants of CERTAIN kinds such secretion) or an association of ple
615o038 secretion) or an association of pleasures with CERTAIN actions performed by your parents,  conscience
618o41v he weight never came untill the blueness had a CERTAIN intensity (& the experiment was varied) then m
620o045 passion,  shows a bad child.-- Hence there are CERTAIN instincts pointing out lines of conduct to oth
620o046 ngthen his instincts.-- so man ought to follow CERTAIN lines of conduct, <although> even when tempted
621o047 n must have a feeling, that he ought to follow CERTAIN lines of conduct, & he must soon necessarily l
623o050 y are acquired by social animals, living under CERTAIN conditions, in this world, they <will  conform
623o050 <all> «many» ages. could be acquired, & we are CERTAIN from our reason, that all which <has> (as we m
623o051 liar <kinds> «social feelings», & living under CERTAIN conditions; by my theory they have been formed
625o052 tinct faculty, of conscience.-- I believe that CERTAIN feelings & actions are implanted in us. & that
633j53v or  the springing of other seeds.-- But are we CERTAIN that these are necessary adaptations.-- May th
633j54r pted to all situations, where in accordance to CERTAIN laws they can live.-- Hence the mistake they a
634j55r om the will of the deity, to create animals on CERTAIN plans.-- is no explanation-- it has not the ch
025r019 exceedingly rare, at the bottom of the sea.--«CERTAINLY data insufficient, yet good» «(I suspect frag
038r058 strata.  quite different from the Porphyries: CERTAINLY appearance leads me to believe mere  fissures
039r059 ll to urge the case of St Helena, where dikes CERTAINLY have not been points of eruption. Nobody supp
070r153 n basalt.-- When we see Avestruz two species. CERTAINLY different. not insensible change.-- Yet one i
089a017 ted.-- Metamorphic action in red sandstone.-- CERTAINLY Volcanic-- CD[Might not bottom of ocean boil,
095a037 Geolog Soc Vol 2. p 257 {T} The Pota: labiata CERTAINLY is found with the Mactra. at Buenos Ayres at
```

```
098a044 entral France not very conclusive proofs, but CERTAINLY probable. Bulletin de la Soc. Geolog: 1833-34
104a066 ure of liquid rock. Andes discussion-- Albite CERTAINLY contains 6 per cent more silica than common f
109a083 oval--??? the difficulty of such preservation CERTAINLY is lessened.-- Coral flats. argument for Heap
110a087 ot occur in either dike or fragment. junction CERTAINLY most distinct on dike side.-- oozed from one
157g073 of granite & forming «sloping» buttresses Yet CERTAINLY shelf 4th <near> only usually contains many p
157g076 ins & quartz, & garnets.-- Boulders as before CERTAINLY must have <come> «been drifted» here: on very
161g095 led by fragments, then clear. this bit to eye CERTAINLY appears level with road, & with piece of exci
161g097 pass   at Isthmus appears above level of shelf CERTAINLY) I took another measurement on short buttress
176b024 apted for these three elements, there will be CERTAINLY points of affinity in each branch A species a
179b034 far  gone, that the species (for species they CERTAINLY are according to all common language) will ke
179b035 es of connection? «if stating from same epoch CERTAINLY» The absolute end of certain form from consid
181b044 plicable to non progressive development which CERTAINLY is the case at least during subsequent ages.-
182b050 an work some most curious cases. of close but CERTAINLY distinct species between Australia & Van Diem
185b055 G.   as new Species. Cuvier examined it. There CERTAINLY appears attempt in each dominant structure to
220b197 zation «especially connected with generation CERTAINLY is.= The dislike of two species to each other
246c025 aight-- do not bark p. 433. birds & bats have CERTAINLY travelled from East Indies, isld, as far as O
250c038 ribes as Marsupial & Edentata increased most. CERTAINLY Africa approaches Nearest to what is supposed
252c045 ncrease of knowledge would probably tell more CERTAINLY Get Closer species-- FOX IS & Mice of America
305c210 dens there is half Jackal & Scotch Terrier.-- CERTAINLY more like Jackall in gait, size, fur.: manner
352d059 enguins wing perhaps not abortive???. Apteryx CERTAINLY.-- Lyell's excellent view of geology. of each
359d088 n like our pidgeons) The male of every animal CERTAINLY seems chiefly to impress the young most  with
364d104 de other pheasants have white feathers).-- It CERTAINLY appears in domesticated animals, that the amo
389d177 case (& not merely as I have stated it) it is CERTAINLY very remarkable that too much difference shou
403e023 ngeners then we can.-- now on my theory this «CERTAINLY» can be accounted for, on any other it is the
414e060 - Stonesfield????). Have Mammalia?? My theory CERTAINLY requires progression, otherwise [not located]
415e069 such chances be made intellectual, but almost CERTAINLY not made into man.-- It is one thing to prove
434e129 t «only» red Lychnis grows in <south> Wales & CERTAINLY <old> only white in Cambridge, in some counti
436e136 e not.-- argue against this-- -- analogy will CERTAINLY allow variation as much as «the» difference b
436e136 pidgeons-- : then comes question of genera It CERTAINLY appears that swallows have decreased in numbe
437e141 to home. (& yet all the varieties of Brassica CERTAINLY not becoming Brussels Sprouts) is analogous t
441e149 ms to show this.-- This would be curious law, CERTAINLY Australian Dog is not affected by domesticati
447e164 hen necessity of crossing is much less,-- now CERTAINLY in the higher animals; changes seem to have b
467t103 gma in slit»-- As I think they do in Broom & CERTAINLY when over-ripe & half withered-- I saw Bees g
473s05r & Jackson here (Capel-Curig) says that he can CERTAINLY tell Trout from Ogwen, Capel Curig & some oth
494q005 ring sleep & do of Berberis-- (latter I think CERTAINLY not) (3) Sow seeds & place cuttings or  bulbs
538m078 ld people singing songs of their childhood. & CERTAINLY of Miss Cogan, & fully corroborates the  fact
556m145 th a portion of phsiognominical sensation, as CERTAINLY as every man who is not deformed. is born wit
566n010 ote sensuality (p 192 Vol. III Octav. Edit)-- CERTAINLY neither a Minerva or Apollo would have them b
572n033 tut. 1838. p. 340. Mr Carlyle says that negro CERTAINLY has less reasoning powers than Europaean.-- I
574n041 h.-- Sexual desire makes saliva to flow «yes, CERTAINLY»-- curious association: I have seen Nina lick
577n052 per» bosom in woman: like erection shyness in CERTAINLY very much connected with thinking of oneself.
580n061 aps of space-- in latter respect he thinks he CERTAINLY has observed that some people of very weak in
586n081 -) Hensleigh considers breathing instinctive, CERTAINLY heart beating may be considered also  such.--
603o012 es cruel.-- Something wrong here.-- Origin is CERTAINLY curious. Chinese, S. American. Polynesians Je
609o030 look far forward «& to the general action» -- CERTAINLY because it is the result of what has generall
615o038 lar instincts of animals. even in wild state; CERTAINLY to the domesticated.-- [LHC] NB. Two dogs hav
615o038 nal configuration. [RHC] General-- Instincts, CERTAINLY appear a sort of acquired memory. a premature
616o039 Well the heart is said to feel Now this would CERTAINLY be a startling expression, & so foreign to th
623o050 This feeling seems to vary in races of man. & CERTAINLY in «species of» animals, in which case it und
243c019 ee S) qui n'aient egalement leur espèces; et CERTAINMENT on eût été bien éloigné, il y a peu d'années
263c075 mous difficulty of reproductions of species & CERTAINTY of destruction; then he will choose & firmly
409e049 could not be one body of animals, living with CERTAINTY on other» hence not social instincts, which a
509q017 has  abortive bone. (ask more about the lowest CERVICAL vertebrae process developed into ribs.) & doe
241c014 lope in Celebes, Bourou new species of Axis.-- «CERVUS moluccensis is different from that of the Mari
245c024 turalists.-- p. 372. Bourous. the Barbyrousa; a CERVUS near Marianus new, & some rats & mice. In Ambo
363d103 -- Male birds always second their songs, the ++ CERVUS Campestris spotted white when a fawn compare w
058r118 ne coupe abrupte et souvent tailée a pic. Toutes CES montagnes sont formées de couches parallèles  et
059r119 eles et inclinées du centre d'ile, vers la mer; CES couches ont entre elles une correspondance exact
243c019 ien éloigné, il y a peu d'années. d'admettre que CES oiseaux eussent leurs représentants dans de si h
029c034 and  clear. Caermarthen Journal I look at the CESSATION northwards of the Coal in Chili as clearly be
230b235 ulter on decrease of population in California CESSATION of female offspring: applicable to any animal
183b052 du chien, que celui de la chevre et du belier, CESSENT d être feconds. dès les premières generations"
381d156 as in plants) Cirrhipeds rotifers, trematode & CESTOID Entozoa Allotriandrous <or M> Mollusca, with p
436e136 these,   because circumstances» <April 12th..> CESTRACION, Port Jackson Shark-- Owen thinks Australia
379d151 located]   The present age is the one for large CETACEA, as the past for other Mammalia. & still furth
422e092 p. 306, the Dugongs cannot be united with true CETACEA or whales.-- but are aquatic Pachyderms. & Wal
641j29r me others>: study these facts read Lacépède on CETACEA & Geographical Distrib of larger Seals-- Are P
222b206 world-- I cannot for a moment doubt, but what CETACEAE & Phocae now replace Saurians of Secondary ep
413e059 als» «species» of large size,-- consider this (CETACEAE) with reference to my theory Babbage 2d Edit,
422e092 s passage, when considering origin of northern CETACEAE).-- -- \.do. p. 318 M. Pictet of writings of
436e134 n to other molluscs in cold parts of sea, like CETACEAE.-- although the Cephalopods, seem to have dec
634j55r originally  terrestrial.-- for we find even in CETACEAE traces of hind extremities.-- How are we to
634j55r les? Is this origin of Polar attributes of the CETACEAE.-- How came Bats also.? before birds? They ar
634j55r s also.? before birds? They are ancient.-- Are CETACEAE found in Paris Basin?-- NB) The explanation
213b172 could not now accumulate without seal-bones & CETACEANS.-- both found in every sea, from Equatorial t
042r068 e?-- Fossil bones black as if from peat.--yet CETACEOUS bones so likewise «of miocene period».--Mem B
349d051 «we ascend to» subgenera & families, «even in CETIONIDAE» «in the Cetionidae».-- when will ornithorhy
349d051 nera & families, «even in Cetionidae» «in the CETIONIDAE».-- when will ornithorhynous come in circle
058r118 . Bailly says."en effet toutes les montagnes de CETTE île se development autour d'elle comme une cein
096a039 tut) talks of quantities of shells at Iquique. <CEYLON>. Band of Volcanic action in Iceland  parellel
191b080 rth must be excessive up & down.-- Elephants in CEYLON-- East Indian archipelago.-- West Indies = Opo
226b220 Voyage.-- Galapagos mouse (?) brought by canoes CEYLON & India.-- Van Diemen's land Australia. Englan
253c046 Elephant same species in Borneo. Sumatra. India CEYLON-- perhaps shows great persistency of character
451e179 ly. p. 337. it would appear as if p. 345. The CEYLONESE Elephant [...] saul forests by having a small
291c165 ster.-- heredetary tameness as well as wildness-- CF Sir J. Sebright.-- love. of man gained & heredet
569n022 eil. V) in describing Caroline Archipelago. Dn 75 CF., p 268 without, however, very sincere grief-- "
267c094 ludes to the Macûsie breed no description given-- CH. 2. dogs L'Institut. 1838. p. 67. Australian cat
333d005 of form of mother: more than of the Common cat.-- CH IX Mongrels Hybridisim Fox has half Persian cat.
346d048 . given by Boethius of ancient Caledonian Cattle. CH 3. Instinct L'Institut. p. 249. (1838). Eggs dis
378d147 pecies,. we can see why young & female alike Good CH 6 Keep Is it Male that assumes change, & is the
461t037 ology p. 97. for Man Chapt see Yarrell Syngnathus CH 6 I presume, from my theory, as long as any stru
461t037 urpose would tend to render complex the series.-- CH 6 Upland geese would transplant seeds very far.--
574n041 all  senses, when thus stimulated, smell, as Sir. CH. Bell says, & hearing music. to certain degree s
592n103 h. Transactions Vol 44. 1746-47. Paper. like. Sir CH. Bell on Expression «First Croonian Lectures  by
059r119 duire a des hauteurs communes sur le revers de CHACUNE des montagnes qui forment les vallées ou les s
444e158 so  must the mutations of species.!!-- p. 203 CHAETODON squirting water at fly.-- instinct, for how c
437e138 ary, in southern stays only winter.-- Jays & CHAFFINCHES sometimes migratory.-- p. 103. Turtles findi
472s02v ed like Heartease pollen.-- the pollen appeared CHAFFY, as if sucked?! opens & shuts end of sucker, a
023o010 hoals  -- N.B. The view of the Volcanos of the CHAIN of the Cordilleras as arising from «the expulsi
060r125 f formations of S. America & Europe.-- If great CHAIN of Volc. had been in action during secondary pe
074r165 n of secretion.--metallic veins follow mountain CHAIN. there after NW <W>.-- «same chemical laws as i
085a002 lls me, that Himalayas penetrated like Bolivian CHAIN. Volcanic islands. from number of craters  very
088a013 ntal elevations slow. so would line of mountain CHAIN be Mr <Lyell> «Waterhouse» has frequently heard
103a065 ight be made between dikes «axis of» mountain-CHAIN in proportion to weight of super [...]  mass.--
104a068 kingdom of Norway was contorted yet no mountain CHAIN case parallel to Banda Orientel. ask Lyell  for
145a070 sides general improbability. stratification, If CHAIN of lake. <a> the alluvium would form a successi
115a097 lds p. 94. Von Buch's Travels account of Norway CHAIN being broken through like that near-- Obstructi
148g028 ses «I urge» deposition marine-- because if not CHAIN of lake & if so there would be barrier-- recoll
```

(Key Word)
```
230b236 species (except perhaps in some plants & then a CHAIN of steps is found in same mountain).-- How is t
233b249 ndian all the rest. Timor according to Mountain CHAIN ought to be Australian.?-- Mr Gould has been st
303c205 et <what> altogether «he» has created a perfect CHAIN. <Icthyo> .+ + +. supra & next page It is a fac
308c216 t all animals have never at any one time formed CHAIN, since if cretceous period assumed, then some p
547m111 eatly astonished, at the time. & could trace no CHAIN of association» Mayo Philos. seems certain that
036r049 ed cracks must be filled up by dikes & mountain CHAINS.-- Introduce part of the above in Patagonian p
102a062 zontal elevations, so are injection of mountain CHAINS. accompanied by do.-- Give this after supposit
119a104 s (see Edinburgh. Phil. Journal <]CD>, no great CHAINS like Andes or Himalayas, but great circular mo
124a118 thenaeum. 1838. p. 652. Dr. Daubeny on mountain CHAINS in N. America Erasmus suggested to me that Her
127a127 continents)» prove elevation.-- great mountain CHAINS. may be effects of subsidence Elie de Beaum. M
128a128 be sought for below the sea mark.-- If mountain CHAINS are matter piled up. over crevice from effect
223b209 ed blown into desert) or separation by mountain CHAINS &c the species have not been much altered they
334d011 every month) & worked in the same cart in loose CHAINS, by being at first beaten from her, & always a
377d140 orces in raising continents, & forming mountain-CHAINS, when we estimate the matter removed by the wa
552m131 p. 652. Dr Daubeny on the direction of mountain CHAINS in N. America Fear probably is connected with
554m138 has seen Jenny, when Keeper was away, take her CHAIR & bang against the door to force it open, when
272c107 bly accompanied by habits) of other, thus in CHALCIDIOUS insect, which I brought from Australia, pro
027r027 connected with frequency of shells in flints in CHALK New Providence more hilly than others of the Ba
028r030 ould separate them from indissoluble rocks? Has CHALK ever been dissolved? Singularity of fresh water
059r120 harden by themselves cannot be pure. for if so CHALK would harden.--Climate.!? or small Proportion o
109a084 emoirs Edited by Taylor Ehrenbergh on flints in CHALK must be studied-- though I do not think good p.
118a103 - P p217. Pentlands Fossils & Meyens --<Jura &> CHALK When we consider parallelism of dikes (Hopkins)
303c205 ect.-- &c &c.-- do p. 136. Ichthyosaurus in the CHALK Those who say «philosphically to a certain exte
030r035 .--Within Chiloe:-- On open coast, near where CHALLENGER was lost: I know no reason for supposing the
328cIBC a living Matica & many shells of Genera Corlula CHAM. Cardium. Porcellus Turbo. Cerithium Jardin du
592n101 d.-- Andrew Smith says hen doves & the female CHAMAELEON court the males by odd gestures. In one of t
639j28r il. petrel & Whale in some respects CHAMAELION like power in Octopus & Chamaelion.-- C. D.
639j28r e respects Chamaelion like power in Octopus & CHAMAELION.-- C. D. Sucking feet in Frog. Walrus. Fly.
639j28r il. in Monkeys & Marsupials. Harvest mouse & (CHAMAELION?) C. D. Spines in Hedge Hog & Echidna.. & Ap
636j57r as instance of beautiful adaptation.-- & then CHAMELION, which feeding on same food, differs in every
641j29v very different method. in pedunculated eye of CHAMELION. crabs Crabs & Mollusca we have analogues The
088a015 - Ceratophytes common in Northern seas p. 312. CHAMISSO in Kotzebue. Study Humboldt. Fragmens Asiatiq
193b091 «acting as» in vacuum to each other p. 306.--. CHAMISSO on Kamschatka quadrupeds Kotezebues first Voy
200b124 tto in Plants.<==> It will be well to refer to CHAMISSO Vol III p. 155. about quantities of seeds in
229b234 t him to discuss those mention[ed] by Lesson & CHAMISSO.-- Geograph Journal Vol V. P. I. p. 67. Dr. C
247c028 e at one of the Pelew Islds.-- killed a woman. CHAMISSO p. 189 Tome III: Kotzebue.-- p 22. a Gecko on
247c028 uoy & Gaimard in Sandwich isld. & according to CHAMISSO in Radack isld.-- p. 69. Sharks very generall
479c006 I, p. 168 Ceratophytes common in Northern sea. CHAMISSO in Kotzebue p. 312 Leaches on leaves in Sumat
130a133 Beechy have some such case]CD what would be the CHANCE in sounding over a continent to fall across a
185b055 hionis water rat with land structures; :+law of CHANCE would cause this to have happened in all. but
202b134 at only one species of family has offspring the CHANCE is that these wanderers would not, but where o
206b146 ople living who now have successors.-- Then the CHANCE of 200 people <might be> being related within
206b147 ould vary at each lustrum, & the calculation of CHANCE of the relationship of the progenitors would h
226b221 lands & on Arctic shores evidently due to «the» CHANCE of some one of the different orders being able
226b221 f the different orders being able to survive or CHANCE having transported them to new station.-- When
226b221 s are formed of those genera.-- & hence by same CHANCE few representative species. this must happen.
258c061 made by a little more vigour being given to the CHANCE offspring who have any slight peculiarity of s
263c075 the subject without long meditation-- His best CHANCE is to have profoundly over the enormous diffic
268c100 for transport ¿may it not be explained by mere CHANCE?-- or it like each great class of animals havi
318c255 o which it migrats every year,; and probably a «CHANCE» wanderer like the first pair of Pipra flycatc
348d051 s to insulate <them> it".-- .i.e what characters CHANCE to be heredetary whether important or not,). p
366d112 ers)» if armless cat can propagate, is with the CHANCE of two being born at same time, & make breed,
401e017 es.-- brilliancy of inflorescence Gardeners. by CHANCE «often» sometimes graft pears on apples. they
404e026 ,-- however mysterious such is case¶. therefore CHANCE & unfavourable conditions to parent may be bec
415e068 he species.: therefore Man & monkeys have equal CHANCE that progenitor was bimanous, or quadrumanous.
415e068 genitor was bimanous, or quadrumanous.-- What a CHANCE it, has been, (with what attendant organizatio
436e137 th infinitesimal advantage it would have better CHANCE of being propagated & so &c. The greatest diff
496q006 h will show that such proportions not effect of CHANCE Maer.= (12) Take Bag of soil from centre of wo
515q021 either in the species, or in the individual by CHANCE & under domestication.-- N.B. Benthams remarks
526m027 there is, as we fancy there is such a thing as CHANCE.-- chance governs the descent of a farthing, f
526m027 as we fancy there is such a thing as chance.-- CHANCE governs the descent of a farthing, free will d
527m031 d owing to.--<)>-- I verily believe free-will & CHANCE are synonymous.-- Shake ten thousand grains of
536m072 n or pleasure, if so free will is to mind, what CHANCE is to matter «(M. Le Compte)»-- the free will
537m074 lly can wag his finger from real caprice. it is CHANCE, which way it will be, but yet it is settled b
550m126 , p. 21-24. Curious passages showing how easily CHANCE & will of Deity are confounded.-- well applica
574m042 -- The other principle of those children. which CHANCE? produced with strong arms, outliving the weak
576m047 added much to the happiness of his life, & the CHANCE, of so dreadful a consequence to each man is s
607o025 eiving impressions-- accidental (so called like CHANCE) circumstances. As man hearing Bible for first
608o025 sed & complicated one calls them free will--the CHANCE of mechanical phenomena.-- (Mem: M. Le Comte c
608o025 of Philosophy, & savage calling laws of nature CHANCE) 2) difference is from imperfect condition of
181b041 in a state which does not increase» it will be CHANCES against any one «of them» having progeny livin
181b041 ed to small percentage.-- & <in» therefore the CHANCES are excessively great against, any two of the
228b231 ia yet he may be found:-- We must not compare «CHANCES of embedment in» man in present state. with wh
262c073 ds partially webbed. &c &c.--)-- & in round of CHANCES every family will have some aberrant groups.--
415e069 an.-- CD [any monkey probably might, with such CHANCES be made intellectual, but almost certainly not
447e163 rm exceedingly difficult to vary.-- the run of CHANCES, would prevent it varying. A plant «producing»
639j28v Grallae Suppose six puppies are born <& it so CHANCES, that one out of every hundred litters is born
022r006 city? Would minute particles have a tendency to CHANGE their position? Carbonate of Lime disseminated
034r045 s Gypsum Andes Mem. Beechey. account of regular CHANGE in soundings. on approaching the coast of NW.
061r127 , ostriches; bigger one encroaches on smaller.--CHANGE not progressif<e>: produced at one blow. if on
062r129 d urge that extinct Llama owed its death not to CHANGE of circumstances; reversed argument. knowing i
062r129 eated for a definite time::not extinguished by CHANGE of circumstances: The same kind of relation th
063r130 o must we believe ancient ones: «.» not gradual CHANGE or degeneration. from circumstances: if one sp
063r130 ration. from circumstances: if one species does CHANGE into another it must be per saltum--or species
068r146 so long, that to be surrounded by continent.-- CHANGE of volcanic focus.-- <it is certain, if strata
069r149 ation at Concepcion. from impossibility of such CHANGE having taken place unrecorded must be insensib
070r153 wo species. certainly different. not insensible CHANGE.-- Yet one is urged to look to common parent?
107a079 s idea of whole centre of earth same heat, then CHANGE in form of fluid centre would lift with it iso
117a101 level will alone explain the immense amount of CHANGE which must have taken place, otherwise the wor
127a126 ed to 0 degree!)-- shows effects of pressure in CHANGE of form as the result of heat.-- will it bear
171b002 generation.-- We know world subject to cycle of CHANGE, temperature & all circumstances which influen
172b008 .-- but as they inoculate, we must suppose the CHANGE is effected at once, -- something like a varie
175b016 rm exceedingly difficult. every animal has tendency to CHANGE.-- this difficult to prove cats &c from Egypt
175b016 m Egypt no answer because time short & no great CHANGE has happened I look at two ostriches as strong
175b017 iches as strong argument of possibility of such CHANGE,-- as we see them in space, so might they in t
175b017 e before said isolate species <& give even less CHANGE> especially with some change probably <change>
175b017 <& give even less change> especially with some CHANGE probably <change> vary quicker Unknown causes
175b017 s change> especially with some change probably <CHANGE> vary quicker Unknown causes of change. Volcan
175b017 robably <change> vary quicker Unknown causes of CHANGE. Volcanic isld.-- Electricity Each species cha
175b020 the branches dying out.-- with this tendency to CHANGE, (& to multiplications when isolated, requires
176b024 species as soon as once formed by separation or CHANGE in part of country. repugnance to intermarriag
178b031 ance of domesticated animals having effected; a CHANGE whi[ch] the Fr. naturalists thought was specie
184b054 letin Geologique April 1837. p. 216 Deshayes on CHANGE in shells from salt & F. Water-- on what is sp
185b055 ld live-- but there were causes to induce great CHANGE. like the Buzzard which has changed into Cara
186b061 to be similar ¿Law: existence definite without CHANGE, superinduced, or new species; therefore anima
186b061 f there was nothing in country to superinduce a CHANGE? Seeing animal die out in S. America; with no
187b062 e? Seeing animal die out in S. America; with no CHANGE agrees with belief. that Siberian animals liv
187b063 n to S. America.-- Never They die; without they CHANGE; like Golden Pippens, it is a generation of sp
```

```
*********************************************(Key Word)********************************************
187b064 accidental varieties, & to accomodate itself to CHANGE, (for of course change even in varieties is ac
187b064 to  accomodate itself to change, (for of course CHANGE even in varieties is accomodation). Now this a
191b081 Next to animals land birds.-- & life shorter or CHANGE greater-- In the East Indian Archipelago it wo
192b085 - by steps so insensible, that each is not more CHANGE than we know varieties can produce.-- Therefor
199b118 hort for this object & others, viz not too much CHANGE In Number 6'.? of E.d. N. Philos. Journ.-- Pap
200b123 every  trifle heredetary, without some cause of CHANGE,, yet such causes are most obscure. without do
222b205 whole scale of Zoology may not be perfecting by CHANGE of Mammalia for Reptiles, which can only be ad
223b209 he reason why there is not perfect gradation of CHANGE in species,, as physical changes are  gradual,
223b210 cies or not. A plant submits to more individual CHANGE, (as some animals do more than others, & cut o
224b213 ith one kind of herbage & one with other, might CHANGE organization of stomach & hence remain distinc
225b219 aracter on offspring? Does the mind produce any CHANGE in offspring? if so adaptations of species  by
227b226 Articulata!!!--! It leads to Nature of physical CHANGE between one group of animals & a successive on
227b227 ight have easily been formed.-- With belief of <CHANGE.> transmutation & geographical grouping we are
227b227 g we are led to endeavour to discover causes of CHANGE.-- the manner of adaptation (wish of parents??
227b228 g & what prevents it--» & generation, causes of CHANGE «in order» to know what we have come from & to
228b228 es> structure in species, might lead to laws of CHANGE, which would then be main object of study,  to
230b236 ation,∴ animal best fitted to that country when CHANGE has taken place, Nature [not located] Any chan
230b239 hange has taken place, Nature [not located] Any CHANGE suddenly acquired is with difficulty permanent
230b239 it> a plant will admit of a certain quantity of CHANGE at once. but afterwards will not alter. This n
230b239 slow changes. without crossing.-- Now a gradual CHANGE can only be traced geologically (& then monume
230b239 zontally & then cross breeding prevents perfect CHANGE.-- It is scarcely possible to get evidence  of
230b240 good one of animals not soon being subjected to CHANGE in Americas. perhaps merely gone back previous
230b241 cas. perhaps merely gone back previous to fresh CHANGE Get a good many examples of animals «or plants
231b243 Hence  far greater discordance in latter-- Have CHANGE in form.-- This probably explains crag & mioce
231b243 - The descendants left in cooling climate might CHANGE twice over, whereas those which migrated a lit
232b245 one  point.-- In the crag we see the process of CHANGE of those forms, which have succeeded in becomi
232b245 eas, «All» The <one> species which survives any CHANGE may undergo indefinite change., (marking in th
232b245 hich survives any change may undergo indefinite CHANGE., (marking in their history an eocene  miocene
232b246 cies must be compared to neighboring sea.-- For CHANGE of species does not measure time but  physical
232b247 secular  refrigeration in chief part instead of CHANGE from insular to extreme climate, <more norther
233b252 ed quadrupeds are not so. greater facilities of CHANGE in the articulate than <M> Vertebrate. But how
235b274 long breed in & no new ones introduced would not CHANGE be superinduced-- why is every one so  anxious
235b275 s inhabit different kind of localities.-- if so CHANGE The GRAND QUESTION Are there races of plants r
239cIFC et racehorse from Cart horse by picking without CHANGE of habits Mr Yarrell «Give it as his theory» t
242c017 in species must be very slow, owing to physical CHANGE, slow & offspring not picked.-- as men do. whe
248c033 namely the [not located] animals unite, all the CHANGE that has been accumulated cannot be transmitte
255c053 nature conscious of the principle of incessant CHANGE.-- hump on back of cow!!-- &c &c D'orbigny  (p
257c056 es of organs clearly, we cannot guess causes of CHANGE.-- Are any men born with any peculiarity. or a
259c063 bined» effect of habit.-- perhaps in process of CHANGE of species to species «although we see it affe
259c064 ied certain year. Extreme difficulty of TRACING CHANGE in species is reduced by atavism) Even a defor
259c065 is  brought to admit the possibility (any great CHANGE» during life of parent, & therefore being alwa
259c065 ealthy children the other case is <adaptation> «CHANGE» of organization! The little turtle, without it
261c070 untries., may give thread to conduct to laws of CHANGE in typical structure?!! Whewell «in Comment/ f
261c072 by  direct adaptation to animals wants & not as CHANGE have travelled that thickness in that period &
269c101 e + 6 degree??., therefore 34 degree degrees of CHANGE together, in analogy certain parts perfect  of
274c115 ssonia &c & In relations of affinity all organs CHANGE-- whether beak (as it appears to me). colour o
276c122 ?! By profound study of local varieties laws of CHANGE.-- The mark of analogy would be empirical beca
277c128 urpose, & would present many ideas of causes of CHANGE in England from Rhinoceros Elephants &c in the
278c132 o discover whether dog found at Swan river. The CHANGE excessively slowly. whether geologists would n
281c137 .-- Argue the case theoretically if animals did CHANGE -- may account for certain organs not being fi
281c139 - less firmly fixed & therefore most subject to CHANGE, yet, as external conditions over whole  world.
282c140 one.-- imagine the men to have greater power of CHANGE, would be same-- Yet each family might have it
282c140 onstitution of man originally similar limits of CHANGE, & make genera of bird. analogous., animals wou
282c141 c animals» with them. they might be supposed to CHANGE quick enough & perishes.-- Lyell has show such
285c153 changes at last become unfit, the animal cannot CHANGE in organizaton. But the classfication must chi
287c158 elationship. & by so doing discover the laws of CHANGE in enclosing a common seems in part of-- to ha
288c160 zine of Zoology & Botany No XI p. 390. a slight CHANGE has been added. «as it speaks to amount of cha
290c162 general» character in those animals, where much CHANGE only & not kind» insects-- & vertebrata & plan
290c162 ange has been added. «as it speaks to amount of CHANGE, Will be known.-- It appeared to me that  half
290c164 ack, by examining series I cannot doubt laws of CHANGE & make new species.-- alpine species being des
291c168 individuals,  wheras in Falkland Isd they would CHANGE in Scicily.-- Splendid Harmony these views-- d
291c168 d at Falkland Isds++.-- Mem Lyell hypothesis of CHANGE better than others.-- The more complicated the
292c169 ++. Even at Falklands some probably would stand CHANGE cannot be great, otherwise it would be unlimit
292c169 (which probably is first stage) the tendency to CHANGE in animal be permanent.-- It will be easy to p
294c176 overlooked  argument, that domesticated animals CHANGE of habits in Van Diemen's land. Study Mr Blyth
294c176 nce-- & if those changes permanent so would the CHANGE of climate from physical causes.-- these super
300c197 few  eggs transported to the Str of Magellan.-- CHANGE of climate especially of Mammalia As eve
302c203 -species, being little adapted to some physical CHANGE all.-- & so on no one will settle number of pr
313c234 nt for origin of man one.-- Is the extinction & CHANGE of species two very different  considerations,
313c234 ply to utter extinction or rapidity of specific CHANGE.? <One> the first would be called. generic & o
314c238 self How except by direct adaptation has such a CHANGE been effected.-- the consciousness of the plan
325c267 cted by Owen. (at Shrewsbury) Yarrells Paper on CHANGE of plumage in Hen Pheasants <Zoological> Philo
336d017 not transmit to its offspring any <peculiarity? CHANGE» from the form which it inherits from its paren
336d017 he blood, the more persistent.-- «any amount of CHANGE» shorter time less [s]o.-- the result of  this
336d018 e slowly obtained & hereditary; <but if> if the CHANGE be congenital (that is most slowly obtained wi
336d018 ividual) it is more easily inherited.-- «but if CHANGE be in blood long, it becomes part of animal &»
336d018 these  small changes become multiplied, & great CHANGE be effected, but in a mule these conditions ar
336d019 t. if whole parent not entirely embued with the CHANGE, a bud could not be taken, without it either w
338d025 trongest.-- V. p. 63. Note Book M'. for case of CHANGE in food in insects entered by mistake Surely t
342d036 in geography & changes of climate superadded to CHANGE of climate from physical causes.-- this super
349d052 artificial, as interlopement of Marsupials will CHANGE all.-- & so on no one will settle number of pr
352d059 extreme antiquity (ie much intervening physical CHANGE).-- distribution especially of Mammalia As eve
354d062 «Answered satisfactorily by. Valenciennes.» The CHANGE from caterpillar to butterfly-- is not more we
355d069 species, the capability of only small amount of CHANGE at any one time Seeing what Von Buch (Humboldt
356d069 s on organs, leads one to suspect any amount of CHANGE from eating different kinds of food: grazing a
360d090 om them.-- showing how gradually every <thing» «CHANGE» is effected)-- the one in garden is from <bit
361d096 he species-- (¿do these facts indicate that the CHANGE is effected through the male??)-- Yarrell obse
362d099 ce never fall off.║ Curious the rapidity of the CHANGE in 5 or 6 weeks after castration, fresh horns
362d100 e been kept perfect-- also to trace the laws of CHANGE in this time.-- the impossibility of discoveri
365d106 tate to say that if there had been considerable CHANGE, it would have been greater puzzle, than none,
366d111 n blinking their eyes. without extermination, & CHANGE of structure.-- When will the musquitoes of S.
371d127 ring.-- but surely in all these cases an unseen CHANGE is produced in parents--colour is a doubtful s
372d129 of body.-- [in plants does not whole individual CHANGE into generative organs?]CD it is of no consequ
376d135 t be to sort out proper structure & adapt it to CHANGE.-- to do that for form which Malthus shows, is
377d139 ecessary to form channel & (& Basses St) yet no CHANGE in English species-- time no element in making
377d139 in  English species-- time no element in making CHANGE, only in fixing it: only circumstances. a cont
377d140 on beaches-- we really, measure the rapidity of CHANGE of form, & instincts in the animal  kingdom.--
378d147 le alike Good Ch 6 Keep Is it Male that assumes CHANGE, & is the offspring brought back to earlier ty
386d167 slow. if animals became adapted to every minute CHANGE, they would not be fitted to the slow great ch
388d172 nt &c is abortive hermaphrodite is supported by CHANGE which takes place in old age of female assumin
388d172 at transmission bears no relation to utility of CHANGE-- hence harelips heredetary, disease. extincti
389d173 uld be observed that the constant necessity for CHANGE. in process by generation applies only the mor
390d178 e its kind was it not for the necessity of some CHANGE.-- ║ Without some small change in form. ideosy
390d178 ecessity of some change.-- ║ Without some small CHANGE in form. ideosyncrasy or dispositions were add
391d174 generation,  which typifies the whole course of CHANGE from simplest form.-- (Because by this process
391d174 are not). but why should it demand some further CHANGE?-- Man properly is hermaphrodite (hence monstr
391d175 umstances which induce «which must be external» CHANGE are always of one nature species is formed  if
397e004 ot be objected to my theory, that the amount of CHANGE within historical times has been small-- becau
```

397e004 ithin historical times has been small-- because CHANGE in forms is <al> solely adaptation of whole of
397e004 solely adaptation of whole of one race to some CHANGE of circumstances; now we know how slowly & ins
397e004 n progress.-- we feel interest in discovering a CHANGE of level of a few feet during last two thousan
398e004 years in Italy, but what «changes» would such a CHANGE produce in climate vegetation &c.-- It is the
398e005 f the world most closely, & know the amounts of CHANGE now in progress, will be the last to object to
398e005 to object to this theory on the score of small CHANGE.-- on the contrary islands separated with some
398e005 ands separated with some animals, &c.-- «if the CHANGE could be shown to be more rapid, I should   say
398e006 rder a scattered page; we find <great> sensible CHANGE in the institutions. & we suppose not only rev
399e006 we ought in same bed if very thick to find some CHANGE in upper & lower layers.-- good objection to m
399e006 me fossils. does not Lonsdale know some case of CHANGE in vertical series: Look at whole Glacial peri
402e017 sh & white Thorn! Species not being observed to CHANGE «is very great difficulty» in thick strata, ca
403e022 ome classes-- it is because every part is under CHANGE, now one part now another Macleay says it is n
403e023 form of varieties, do we suppose bones will not CHANGE in number. (even species do not this). because
405e032 most recent shells.-- shows that progression of CHANGE in Mollusca is somewhat similar in two hemisph
407e038 North & South, has probably undergone a greater CHANGE, than any part, (except Europe. in which all T
407e039 te. might not have been able to have survived a CHANGE, (& become transmuted), although other paralle
407e039 her continents might have survived this mundine CHANGE.-- Therefore I argue from this that Africa «&
408e043 aps less so, that either Americas.-- If species CHANGE, we see external conditions have great effect
409e048 forms themselves have been basis of argument of CHANGE.-- now take greater area of water & snow line
410e052 t,-- "the great end must be the law & causes of CHANGE".-- A philosopher, would as soon turn   tailor,
410e053 er to, to those forms. where the termination of CHANGE occurs.-- those discovering the formal laws of
411e054 be important in understanding laws of specific CHANGE\.-- When the laws of change are known.-- -- th
411e054 ng laws of specific change\.-- When the laws of CHANGE are known.-- -- then primary forms may be spec
411e056 Falklands Isld. is interesting as showing some CHANGE in habits before form.-- I have already given
413e058 ldren. like. grandfathers (2) Tendency to small CHANGE.. «especially with physical change» (3) Great
413e058 ncy to small change.. «especially with physical CHANGE» (3) Great fertility in proportion to support
419e086 n in any other, & yet 200-300 ft elevation & no CHANGE & even no loss of species.-- It must never be
419e087 nology of geology rests upon amount of physical CHANGE <affecting whole bodies of species>, & only se
419e087 much less, (though <the> it also the effect of CHANGE) than a slow gradation in form, «which must be
419e087 radation in form, «which must be effect of slow CHANGE + & Therefore precludes effects of catastrophe
428e113 I am bound to insist honestly that the sudden, CHANGE from Primrose to Cowslip is great difficulty.
431e122 ny species is most common, we need not look for CHANGE, because its number show it is perfectly adapt
431e122 tly adapted; it where few stray ones. are, that CHANGE may be anticipated, & this would look like fre
433e126 from limestone to sandstone &c. show some great CHANGE who can say how many centuries elapsed between
434e130 n of one organ. (here language forces on us the CHANGE, which <to> seems to have taken place.-- Almos
435e130 ture.-- Some willow trees have been observed to CHANGE their sex,-- this effect from age, what Mr Kni
436e135 cient in making species, than time (as cause of CHANGE) which can hardly be believed, then, uniformit
438e142 t & Cuvier on Mummies]CD [NB TIME is element in CHANGE, as in Dahlias]CD all much varied breeds both
440e146 we see these strange plumage in pidgeons yet no CHANGE of habits, so no <cause> corresponding change
440e147 o change of habits, so no <cause> corresponding CHANGE in Birds of Paradise.-- All that we can say in
441e150 e overlooked.-- it makes fourth cause or law of CHANGE.-- The weakest part of my theory is, the absol
442e152 anent?» because its very essence is that little CHANGE is produced.-- The fact just alluded to of Nor
446e162 of> product of fresh bud-- here then is case of CHANGE analogous to change in grafted trees «:so is n
446e162 bud-- here then is case of change analogous to CHANGE in grafted trees «:so is not effect of differe
446e163 y owing to similarity of conditions-- & that no CHANGE would affect them in short period & hence no c
446e163 ge would affect them in short period & hence no CHANGE would effect them, without affecting all the i
450e174 ducks monogamous; tame ones highly polygamous-- CHANGE of instinct by domestication.-- "Notices of th
461t041 is valuable because it shows no innate power of CHANGE & it also shows, what enormous changes of cond
501q011 ask   question of Col. Le. Couteur about Wheat-- CHANGE of Soil-- crossing-- when seeds raised,-- His
536m072 e Compte)»-- the free will (if so called) makes CHANGE in bodily organization of oyster. so may free
536m073 y organization of oyster. so may free will make CHANGE in man.-- the real argument fixes on heredetar
539m083 because it will (without direct consciousness?) CHANGE its habits.-- Aug. 16th. As instance of herede
549m122 nces are so conditioned as they are effecting a CHANGE in his instincts-- like what is happening with
553m135 ot be, anyhow, easily overturned.-- so ready is CHANGE from. our idea of causation, to give a cause (
570n027 g. no connection with male figure]CD-- As forms CHANGE, so must idea of beauty.-- [Old Graecians livi
608o026 e urged how often one try to persuade person to CHANGE line of conduct. as being better & making him
608o026 erials. From contingencies a mans character may CHANGE-- because motive power changes with organizati
609o30v an & savage,-- is that former is endeavoring to CHANGE that part of the moral sense which experience
610o30v appiness is to certain degree <of> right.-- The CHANGE <of> our moral sense, is strictly analogous to
610o30v <of> our moral sense, is strictly analogous to CHANGE of instinct amongst animals.-- Jan 13th. 1839
614o037 tinct. might be analysed into steps, as species CHANGE.-- Must be so if Lamarck's theory true [RHC] A
615o038 would probably take place without corresponding CHANGE in «external» man; and as all men nearly same
619o043 by force he would feel pain. [.By a very slight CHANGE in association if others injured these objects
624o051 hence to nearly all the world.-- As conditions CHANGE, from civilization, education changes, & proba
624o051 to hunt one pig sooner than other, rather than CHANGE hunting instinct. *Our tastes in mouth by my t
628o054 es Hartley explains this & Mackintosh shows the CHANGE produced.-- 4) p 38 Conscience checks the wish
636j56v iving on the borders of a country favourable to CHANGE.-- It might be concluded that Plants would be
640j167 many peculiar species. for as long as physical CHANGE is in progress or is, present with respect to
641j29r re favourable,) because it must take so long to CHANGE species-- yet this is contradicted by continen
046r078 ature, or has a subterranean fluid mass itself CHANGED.--No.-- Yet the fluid granitic mass under <[.
112a088 rcases of birds drifting out to sea do p. 358. CHANGED soundings in Mouth of S. Cruz in connection wi
171b003 the young of living beings, become permanently CHANGED or subject to variety, according «to» circumst
176b022 endent on definite laws, then those which have CHANGED most. «owing to the accident of positions» mus
185b055 nduce great change. like the Buzzard which has CHANGED into Cara cara at the Galapagos. Fernando Noro
261c071 lty-- preservation of colouring, when form has CHANGED. Can be said that animals no notion of beauty,
265c088 peculiar coloured plumage, where colours have CHANGED, in accordance to habits.-- one is tempted to s
271c105 ts show, habits heredetary whilst species have CHANGED Argumentum ad absurdum. The creative American
274c115 ct of typical structures certain <imper> parts CHANGED Have <not>. S. Africa, Australia, & S. America
282c141 ur & trifling form & head &c to become greatly CHANGED. in structure & even to certain degree in habi
287c158 ent animals, & the value of those organs, when CHANGED in different animals.-- + whether variations i
310c224 equal   periods with Europe would probably have CHANGED much less.-- Here is an element of extreme dif
334d012 r Swan river, lost his <on> two cows entirely, CHANGED his residence a great many miles.-- yet one da
337d020 y the effect of <sev> local variety many times CHANGED together with some training in the earlier bra
352d059 use,   every abortive organ must have been once CHANGED.-- what is abortive? when it does not perform
365d106 rids-- The fact of Egyptian animals not having CHANGED is good-- I scarcely hesitate to say that if t
405e031 on each side darker,. so that whole colour is CHANGED, these best marked characters are partly retai
409e047 how a savage a dog, & ask him, how wolf was so CHANGED. When discussing extinction of animals in Euro
414e060 tle progressive «!Agassiz makes it wonderfully CHANGED, since Cretaceous period, whether progressive
423e095 with the simplest forms & suppose them to have CHANGED, these very changes <len> tend to give rise to
526m031 tances, & thus the appetites themselves become CHANGED.-- appetites urge the man, but indefinitely, h
536m072 n or pleasure actions unavoidable & only to be CHANGED by habits). now free will of oyster, one can f
537m075 t is settled by reason.--\ How slow habits are CHANGED may be inferred from expression. "relict of ba
537m076 this view, where the moral sense seems to have CHANGED suddenly-- but are not such <suddem> changes r
542m094 illness.-- Now I conceive if organization were CHANGED, I conceive sighing might yet remain just like
555m143 howing scar behind <.instead of front) (having CHANGED hanging into his head cut off) as kind of wit,
555m144 urable wounds.-- all this was kind of wit.-- I CHANGED I believe from hanging to head cut off. «there
603o012 races»-- thinks action towards <man> «a king» <CHANGED into> is carried on toward deity.-- & as king
616o038 Such memory may go back to animals which were CHANGED into man .:. they meet their reward! X Perhaps s
640j167 small body of species would far more easily be CHANGED.-- Hence the Galapagos Islds are explained. On
640j167 ld offer few species, or rather be very slowly CHANGED & vertebrata much so.-- so far true, but do no
021rIFC ose that have undergone the greatest number of CHANGES towards perfection (namely mammalia) must have
032r040 sive periods of greater activity & rest.--Such CHANGES could be shown (as represented), along line of
032r041 eties of substances ejected from same point. & CHANGES. «(changes in variation?)» as in Cordillera.--
032r041 bstances ejected from same point. & changes. « (CHANGES in variation?» as in Cordillera.-- From poles
032r041 urface must gain a Westerly current:--If great CHANGES of climate have happened. hurricane in bowels
032r042 ite stone of Chiloe, after having examined the CHANGES of pumice at Ascension In Calc: sandstone at A
035r045 it)» <I think> At Ascension, the laminae <...> CHANGES in rocks. connected with & alternating with ob
045r074 ch gaz) steam condensed.--Perhaps these mighty CHANGES might go on. & not a bubbles on the surface be

Page **(Key Word)**
045r074 on. & not a bubbles on the surface bespeak the CHANGES. -- metallic veins solution of silex & many ot
051r097 to attachment of many bodies [blank] Beechey.--CHANGES in bottom in NW coast of America. from shingle
059r123 shows powers have acted from great depths, so CHANGES, acting in those lines. must now proceed from
063r130 : in former case position, in latter time. (or CHANGES consequent on lapse) being the relation.--As i
094a035 mportant question with respect to my theory of CHANGES. of granites into Trachytes.-- Mention Osorno
099a050 not vertical ∵ combined with gravity.-- hence CHANGES in dip of no sort of consequence.-- Therefore
110a085 m to be recent species-- Lyell-- Some internal CHANGES are in process. connected with variation of co
173b012 nsoons.. when they ceased importation ceased & CHANGES commenced.-- or intermediate land existed.-- o
175b016 w supposes that in course of ages. & therefore CHANGES. every animal has tendency to change.-- this d
175b018 nge. Volcanic isld.-- Electricity Each species CHANGES. does it progress. Man gains ideas. the simple
176b021 ces, so forms would be about equally numerous. CHANGES not result of will of animal, but law of adapt
177b029 perfect<ion>, «species from many» <therefore> CHANGES and base of branches being dead from which the
184b054 . very good Has not Macculloch written on same CHANGES in Fish Mem. Rabbit of Falklands described by
195b102 rm of one country to another.-- let geological CHANGES go at such a rate, so will be the numbers & di
195b103 interruption of communication. or when country CHANGES. Will it said that Volcanic soil of Galapagos
197b111 rom worm to man; «highest» as typical of <sa>. CHANGES, which can be traced in same organ in differen
209b155 ation; then their distribution (after physical CHANGES) would be in rays-- from certain sports.-- Agr
218b190 story-- Preface appeared good with facts about CHANGES when animals transported.) Mr Herbert's papers
221b201 & Volcanic rocks. most remarkable.--¿ Have the CHANGES been so slow., that all have existed for ages
223b209 t gradation of change in species,, as physical CHANGES are gradual, is this if after isolation (seed
224b214 stomach & hence remain distinct. Where country CHANGES rapidly, we should expect most species.-- The
224b215 mixture of races.-- When all mixed & physical CHANGES (¿intellectual being acquired alters case) oth
225b217 etween bull Dog & Greyhound). I should say the CHANGES were effects of external causes, of which we a
227b225 m what we have seen. in old world, & on amount CHANGES which may happen-- ‖ It leads you to believe t
227b226 great divisions) thus a knowledge of possible CHANGES is discovered, for speculating on future. !.fi
230b235 ybrid Ferns It may be argued against theory of CHANGES that if so in approaching desert country or as
230b239 ll not alter. This need not apply to very slow CHANGES. without crossing.-- Now a gradual change can
232b246 of species does not measure time but physical CHANGES (we assume like weather on long average tolera
242c017 sons' remarks on the Floras can be trusted The CHANGES in species must be very slow, owing to physica
254c048 itions of the higher classes, during which the CHANGES of the ovum or embryo succeed each other with
257c057 fly &c &c. seem clearly to indicate those very CHANGES which at first it might be doubted were possib
258c060 y has some analogy in not gaudy colours so all CHANGES may be considered in this light.--XX Zoolog. J
265c083 ssing with other varieties»-- but <accidental> CHANGES after birth do not effect progeny-- many dogs
265c083 ne on boring their noses. &c & This congenital CHANGES show that grandson is determined, when child i
280c136 le of producing itself alike.-- in one case it CHANGES one, in other it changes thousands in futurity
280c136 ike.-- in one case it changes one, in other it CHANGES thousands in futurity.-- This is right way of
282c140 ave its own character,-- we here suppose these CHANGES of <use> adaptation greater than those heredet
282c140 y ones.-- which would elapse; during time such CHANGES had elapsed.-- let these families take <dogs>
285c152 one species May have passed through a thousand CHANGES, keeping distinct from other & if a first & la
285c153 o contemporaries-- fertility must settle it.-- CHANGES in structure being «necessarily» excessively s
285c153 ion being heredetary & fixed, certain physical CHANGES at last become unfit, the animal cannot change
285c153 ugh & perishes.-- Lyell has show such Physical CHANGES will be unequally rapid, with respect to their
294c176 a little with external influence-- & if those CHANGES permanent so would the change in animal be per
296c184 nal. owing to being oldest. & having undergone CHANGES «no near lofty country»?? p. 475 NB. This bear
304c206 ate. species have generally perfect organs. do CHANGES of habit affect particular organs.-- of two ad
304c208 ich which is not isolated, we must suppose the CHANGES from typical structure have either been more r
318c253 va (added by Audubon in Appendix) showing WHAT CHANGES are taking place & how birds are extending the
335d015 t alterations of primitive stocks «relative to CHANGES which every species undergoes») & hybrids betw
336d018 mes part of animal &» by a succession of <such CHANGES> generations, these small changes become multi
336d018 ion of <such changes> generations, these small CHANGES become multiplied, & great change be effected,
342d036 known> causes, modified by unknown ones. cause CHANGES in geography & changes of climate superadded t
342d036 by unknown ones. cause changes in geography & CHANGES of climate superadded to change of climate fro
343d036 mate from physical causes.-- these superinduce CHANGES of form in the organic world, as adaptation. &
347d049 intellect,-- if generation is condensation of CHANGES. then animals must tend to improve.-- yet fish
354d063 e Mammae of Women) in different parts when age CHANGES caterpillar to Butterfly.-- When two Varieties
365d106 ng circumstances.-- More probably during known CHANGES climate became unfit for. subalpina, or some N
365d107 the structure which has been adapted to former CHANGES. than a mere monstrosity propagated by art. Ya
371d127 ct, but what other instances are there of such CHANGES, not acquired by parent, being handed down? Ar
372d130 may produced by having undergone, the endless CHANGES, which its parents have.-- -- Not this is effe
373d132 became Mammalia, through a different series of CHANGES from the placentates, Having Hair. like true M
374d134 live.--» We ought to be far from wondering of CHANGES in number of species, from small changes in na
375d134 ng of changes in number of species, from small CHANGES in nature of locality. Even the energetic lang
386d167 erences. final cause of this because the great CHANGES of nature are slow. if animals became adapted
386d167 ge, they would not be fitted to the slow great CHANGES really in progress.-- Annals of Natural Histor
391d174 ces which are in harmony with all its previous CHANGES, which mutilations are not). but why should it
391d174 s.-- How completely circumstances «alone» make CHANGES or species!! CD[The view of <In> each Man or m
391d175 each individual before he can procreate. these CHANGES may be effect of differences of parents, or ex
391d175 of one nature species is formed if not.-- the CHANGES oscillate backwards & forwards & are individua
397e004 nces; now we know how slowly & insensibly such CHANGES are in progress.-- we feel interest in discove
398e004 ng last two thousand years in Italy, but what «CHANGES» would such a change produce in climate vegeta
398e004 &c.-- It is the circumstance of small physical CHANGES & oscillations, not affecting organic forms, t
398e005 ry to arrive at right conclusion about species CHANGES of level &c are easily recorded, but changes o
398e005 s Changes of level &c are easily recorded, but CHANGES of species «are» not as «without every animal
398e006 y are perfect, we obtain a glimpse only of the CHANGES which the government is subject to.-- further
402e017 nt in one spot, for <whole> many epochs-- such CHANGES would be observed.-- G. W. Earl's Eastern seas
408e045 acts.-- The Himmalaya. case, bears on the vast CHANGES even in that quarter of the world-- -- Mem. el
410e050 ithout sexual crossing, there would be endless CHANGES, & hence no feature would be deeply impressed
410e050 » -- it was absolutely necessary that Physical CHANGES should act not on individuals, but on masses o
410e050 s, but on masses of individuals.-- so that the CHANGES should be slow & bear relation to the whole ch
410e050 es should be slow & bear relation to the whole CHANGES of country, & not to the local changes. = this
410e051 e whole changes of country, & not to the local CHANGES. = this could only be effected by sexes: All t
410e053 varieties not in same country.-- The traces of CHANGES in forms of organs, will care little for speci
410e053 discover physical laws of such corelations, & CHANGES of individual organs, must know whether the in
416e071 rs.-- Term variety may be used to gradation of CHANGES which gradation shows it to be the effect of a
418e084 he oldest animal>, have passed through as many CHANGES, as has any species.-- Decemb. 21th.-- L'Insti
419e086 etter means of judging of slowness of physical CHANGES, than in any other, & yet 200-300 ft elevation
423e095 rms & suppose them to have changed, these very CHANGES <len> tend to give rise to others.-- Why then
423e096 fill their places.-- The Geologico-geographico CHANGES must tend sometimes to augment & sometimes to
433e126 e grand system to another: in each system, the CHANGES from limestone to sandstone &c. show some grea
438e142 of some may resist the means Man can offer of CHANGES.-- as desert «or rock» plant probably would do
439e142 rather the parents» having passed through many CHANGES.-- It is very important Mr Herberts fact about
440e145 stery is this,-- the law of growth, that which CHANGES the acorn into the oak.-- In short all which «
441e148 doubtful whether any are transmitted, for the CHANGES in fruit trees. mentioned by Mr K may be cause
447e164 h less,-- now certainly in the higher animals; CHANGES seem to have been more rapid, & the facility f
459tf1v ied [...] species, which has undergone all the CHANGES. [im]portant view, copie[d] Gleanings of Scien
461t041 power of change & it also shows, what enormous CHANGES of conditions, some species will undergo & yet
497q007 ¿genera in intermediate country (2) Any known CHANGES in Flora of countries during last century or t
522m013 upon by Emetics.-- people recognized,-- sudden CHANGES of disposition, like people in violent intoxic
537m076 e changed suddenly-- but are not such «sudden» CHANGES rare,-- as when Polynesian mothers ceased to d
541m091 thought.-- Examine frame of mind in following CHANGES during fall of sea.-- Is the effort greater if
547m113 .-- When I say trains, it may be instantaneous CHANGES in order <to every> calling up ideas of every
599o07v etary knowledge, has produced almost → greater CHANGES in the polity of Nature than any other animal-
608o026 ns character may change-- because motive power CHANGES with organization The general delusion about f
624o051 onditions change, from civilization, education CHANGES, & probably likewise instincts, for the same l
624o051 incts, for the same law effects both.-- <such> CHANGES «in accordance to beneficial tendency» will mo
635j56r the growing size of the world? & the physical CHANGES it was to undergo «animals feeding on each oth
641j29r must be destructive to species, when physical CHANGES are in progress; (on the same principles that
171b004 refore generation to adapt & alter the race to CHANGING world.-- On other hand, generation destroys t

Page **(Key Word)**

172b008 Hare.-- As we thus believe species vary, <in> CHANGING climate we ought to find representative speci
180b038 ieties produced by <inter> confined breeding & CHANGING circumstances are continued & produce accordi
218b190 wen's story of cats on West coast of Africa.-- CHANGING hair-- The Edinburgh. Journal of Natural Hist
222b205 for Reptiles, which can only be adaptation to CHANGING world:-- I cannot for a moment doubt, but wha
264c081 ructure may be obliterating, whilst habits are CHANGING-- or structure may be obtaining, whilst habit
291c167 s strongly supported by wonderful fact of bees CHANGING the sex by feeding.-- no it is developing an
313c233 ropical animals which accounts for the species CHANGING which accounts for the species changing « ∴ b
313c233 pecies changing which accounts for the species CHANGING « ∴ because mammalia can subsist where parasi
343d036 m in the organic world, as adaptation. & these CHANGING affect each other, & their bodies, by certain
343d037 are unfavourable to numbers of animals. as in CHANGING from <hot>. Warm to cold, damp to dry.-- Thus
355d068 lant<s>)-- now in some Polypi we see young bud CHANGING into ovules.-- Captain Grants. Himalaya. shel
365d106 hought that the cross would have adapted it to CHANGING circumstances.-- More probably during known c
378d147 ther?-- do these differences indicate, species CHANGING forms, <& loosing do> if so domestic animals
430e118 o some influence, & this would take place from CHANGING country: but greyhound. & poutter Pidgeons «r
437e141 s) is analogous to Primrose & Cowslip suddenly CHANGING into each other, & depends on character of an
550m123 re perfect, than instincts of animals to all & CHANGING contingencies, or bodies of either.-- Our des
640j167 quite exclude the idea of domesticated animals CHANGING.-- From these views we can deduce why small i
036r050 es.--Plains. off coast of Patagonia.-- British CHANNEL &c &c. There is a Hill. near Copiapò which is
068r148 V When recollecting Gulf of California. Beagle CHANNEL.--One need never be afraid of speculating on t
105a069 osive power of river & sea.; the former as its CHANNEL becomes wider looses its cutting power. (as do
106a074 if inclination small.-- The power of widening CHANNEL depends on power of deflection with stream ret
106a075 its tendency would <cut> be to cut a narrower CHANNEL instead of wider.-- This applies to all vallie
122a115 those of Andes Speculate under head of Beagle CHANNEL. on origin of mud with stones scattered irregu
149g032 f a little lower down the hillock with beach & CHANNEL precisely as with Isld-- {P} do they extend ro
150g037 Collarig, when water up to shelf very shallow CHANNEL 50 ft wide & river get formed in centre In Gle
154g057 uoy flat peat plain divortium aquarium-- tidal CHANNEL-- 12ft obscure obscure NB In Glen Collarig tid
154g057 12ft obscure obscure NB In Glen Collarig tidal CHANNEL, sides <alm> 15 ft above bank or terrace, from
157g073 being lower,-- [no pebbles in parts of Beagle CHANNEL when mica slate, only sand blow away]CD where
163g109 ttle spots Speculate on «under head of» Beagle CHANNEL. Forchammers (Lyells Denmark) Shrewsbury rubbi
226b223 left & Fox, & bear.-- If I had not discovered CHANNEL of communication by which great Edentate might
299c193 wood far from any other plants of same species CHANNEL Islds (& probably Isle of Man) no plants pecul
377d139 d.-- we know the great time, necessary to form CHANNEL, (& Basses St) yet no change in English speci
429e115 don.-- what makes the line, as trees in Beagle CHANNEL.-- it is not elements.-- we cannot believe in
043r069 confined to that country. Read description of CHANNELS or grooves in rocks at Costorphine hills. to
048r086 ere talking of such substances being worn into CHANNELS. mention submarine channels. such as that in
048r087 es being worn into channels. mention submarine CHANNELS. such as that in front of Sts. of Magellan In
064r135 ze the fact glaciers most abundant in interior CHANNELS. there no outer coast.--important effect.--?
102a059 -- rollers at Tristan d'.Acunha.-- silting up. CHANNELS on coast of England-- Any one. who has studie
229b233 nakes. with hinder teeth perforated for poison CHANNELS, but not having them, instance of useless str
029r033 rom the decomposition of Granitic rocks Mem. CHANTICLEERS voyage at <[...] Maranh> Pernambuco. EARTHQ
277c127 ata be given.-- This will aid in preventing the CHAOS.-- will point out what to observe.-- will aid u
304c206 ng a series, which would render our knowledge a CHAOS: who will doubt this if series now existed from
036r052 eparate geographical ranges of plants. V. Lyell. CHAP XI Vol II. Urge the entire absence of any rock
243c021 nt points.-- Consult Voyage aux terres Australes CHAP XXXIX tom IV p. 273 2d Edit Consult Latreille.
055r106 formed these large cavities", &c &c &c Vol II. CHAPT VIII. p. 97 at Potosi the veins run from North
324c268 ts of Plant/ If:4s Voyage aux terres australes. CHAPT. XIX. tom IV. p 273 Latreille Geographie des in
369d114 cede in organization from specific character.-- CHAPT I. Also Latent Character Hunter Animal Economy
451e180 the Moluccas 1825--"No wild animals in Moa.--" CHAPT.-- V.-- : do. Chat XXI. Wild cattle & Hogs on T
461f026 order of the fish.-- Embryology p. 97. for Man CHAPT see Yarrell Syngnathus Ch 6 I presume, from my
566n010 h to support a wife a ruling motive.-- Book IV, CHAPT I on passions of mankind, as being really usefu
094a036 ew Volcanos now in lakes.-- Mr Murchison. M.S. CHAPTER on drift.-- Beyond region of great boulders, p
323c269 Vol. Philo. Zoology «references at end of each CHAPTER» Crabbes Life June Is. King & FitzRoy To be re
454e184 em-- Cross Irish & Common Hare Decandoelle has CHAPTER on sensitive plants; Physiology Get Habberley
572n035 to conceal one's thought.-- Macculloch in his CHAPTER on the Existence of a Deity has an expression
600o008 ating of infinites not definable.-- Has little CHAPTER on each faculty of Soul.-- (I) <Conscience> «M
296c184 bouring Continent. This fact speaks volumes. 2 CHAPTERS. translated by Hooker.-- my theory explains t
036r052 & Rats. characteristic of the deserts of Syria <CHARA> ditto for Patagonia, especially rocky parts of
424e100 nimalcules in Musci frondosi, et hepatici,-- in CHARA, in Marchantia & Hypnum «Prof:» Don would have
030r035 g, in the bottom of an open & not deep sea.--(CHARACTER of coast regular & <not very> rather deep sou
030r035 ly chiefly leaves.--This position agrees with CHARACTER of.. «in Basins from rivers. & natural positi
040r062 erfection on that side.-- Add from M. Lesson. CHARACTER of Flora to New Zealand, which agrees with St
047r081 te movement» Moreover wave «with same general CHARACTER» reaches far beyond coast, which has been rai
061r128 g place» over certain area must have peculiar CHARACTER: Contrast low limit of Palms, evergreen trees
066r142 Silurian division--Together with same general CHARACTER of fossils deception complete.= Silliman Jour
080r178 W. NW & S.--last to Seaward partaking of the CHARACTER of a Araucarian tribe, with point affin of ye
085a002 ers very ancient. which agrees. with peculiar CHARACTER of Vegetation. -- So accustomed to utter conf
088a016 rdillera very important V. Malte brun -- Main CHARACTER of Andes Metamorphic action -- Mem: red sand
100a053 ary dip & inclination.-- which last is strong CHARACTER.-- A discussion on concretions and cleavage c
105a071 east probable.-- The sea harmonizes well with CHARACTER of mouth of valleys &c; Pampas.-- If blocks a
121a111 which retained their angles sharp-- yet with CHARACTER completely altered, & a crystalline structure
151g039 e tides. &c. {P} <notch> roads very much this CHARACTER.-- The boulders (one of Gneiss remarkably wat
160g092 gree? Isthmus broad flat peat mass-- (general CHARACTER in these mountains & not ridges) between arm
173b009 land shells) Most difficult to separate Every CHARACTER continues to vanish, bones instinct &c &c &c
186b057 one point, in truth in all excepting specific CHARACTER); and in passing from species to genera, each
186b058 from species to genera, each retains some one CHARACTER of all its family; but why so? I can see no r
190b076 n Eastern end & die away, & partake of Indian CHARACTER There appears in Australia great abundance of
195b098 is shown by the very facts of the Zoological CHARACTER of these islands so permanent a breath cannot
200b122 placed together they will breed.-- But what a CHARACTER is this? Race permanent, because every trifle
205b145 Treatise on Form of Animals by Mr Cline "The CHARACTER of both parents are observed in their offspri
220b198 s different species, dogs not, but if we take CHARACTER of offspring. Hogs not different. some dogs d
222b203 endency to return to one parent, this is only CHARACTER., & yet we find this same tendency (only less
222b205 ideas<)> of perfection); but intermediate in CHARACTER, the same reasoning will allow of decrease in
222b205 the same reasoning will allow of decrease in CHARACTER. (which perhaps is) Case with fish-- as some
223b211 of slightly different species & intermediate CHARACTER:-- For instance two wrens forced to haunt two
224b213 e good ones & differ scarcely in any external CHARACTER:-- of offsprings accounts for uniformity of spec
225b219 .!! Can the wishing of the Parent produce any CHARACTER on offspring? Does the mind produce any chang
227b224 ow whether former lands intervened.-- (2d) By CHARACTER of any «two» ancient fauna, we may form some
235b274 rent quarters to prevent them taking peculiar CHARACTER-- Indian Bull?-- Do species of any genus. as
236b278 nandez-- Humming Birds» types of former dogs. CHARACTER of Miocene Mammalia--of Europe Mem. Mr Bell's
253c046 a Ceylon-- perhaps shows great persistency of CHARACTER. Hence Elephas primigenious over so wide a ra
280c134 ten on dog-Breaking.-- applies it to national CHARACTER.-- N.B. If two species were excessively old,
282c140 be same-- Yet each family might have its own CHARACTER, -- we here suppose these changes of <use> ada
282c143 such points by the vital laws.-- so that all CHARACTER originally may «must» have had the character
282c143 character originally may «must» have had the CHARACTER of analogical.-- Gould says it is only in lar
284c149 used by Naturalists in their test of value of CHARACTER-- Macleys rule is converse, <when> value of c
284c149 r-- Macleys rule is converse, <when> value of CHARACTER depends on non-variation, & not on extension
287c158 variations in eye of vertebrate afford better CHARACTER, than variations in eye of mollusca. [+] Th
290c162 «method» of generation is very good «general» CHARACTER in those animals, where much change has been
292c170 the Classification of animals. & observe the CHARACTER of the demonstrations offered of the singular
295c178 because ancestors so, or has he assumed that CHARACTER -- -- female & young seem most like mean char
297c185 Aberrant groups few in numbers & vary much in CHARACTER, divided into many small genera ∵circumstance
298c192 «all organs» vary in plant. The variation in CHARACTER of leaf of plants is remarkable what is analo
299c195 cation. here we have generative organs. first CHARACTER.-- In dioecious plants many of the Female flo
302c202 multiplied.-- we must argue reversely. WHERE CHARACTER VARIABLE it is (one of analogy or) LATELY ACQ
305c210 r.; manner in which ears droop like dog older CHARACTER & manner of wagging tail.-- habitual movement
306c213 not species.-- organs of generation a captial CHARACTER. (Owen) not for first & grandest divisions. b
314c238 outh seas says so in preface.-- Mr Brown says CHARACTER of Flora, N. Zealand & N. Caledonia with a da
314c239 Brown did not observe scarcely any Australian CHARACTER in Timor plants, yet it seems there may be Eu

```
317c251 s West of <all> Rocky Mountains have peculiar CHARACTER in extreme length of ears & length of  limbs,
333d006 gnation, or two impregnations one giving half CHARACTER & other more of English, but the effect is th
333d008 alf Lion & Tigers are exactly intermediate in CHARACTER & Kittens alike each other.-- Even in childre
336d019 at once. <wd> according to ordinary laws, the CHARACTER of offspring would vary, or rather they would
341d029 ine could be drawn-- thus the most remarkable CHARACTER in Apteryx, small respiratory system; even mu
346d048 hat it was against all rules their preserving CHARACTER & breeding in & in-- Nonsense a flock of more
348d050 mportance, "but we say such happens to be the CHARACTER, of no matter of what importance, which preva
354d065 "variation in one, was analogous to specific CHARACTER of other species in genus."-- Is there any la
355d066 s of sheep «evidently artificial» approach in CHARACTER to goats.-- or dogs to foxes. (yes Australian
355d066 whether variations are produced only in those CHARACTER which are seen to <vary among> be different i
358d074 art of» system of great harmony. The peculiar CHARACTER of St. Helena.-- contrast with otaheite in re
358d076 oped, it is always female which approaches in CHARACTER to the larva, or less developed state.--  the
358d085 up.-- Yet odd they should have so much sexual CHARACTER as they have This character of not having sex
358d085 ve so much sexual character as they have This CHARACTER of not having sexual plumage is very common b
360d093 ent kind of mottle, each feather partaking of CHARACTER of other.-- <so> the most aquatic & most terr
361d095 if I understand right, all species have same CHARACTER which is mottled, & not like any existing spe
361d095 g) quite similar.-- one species retained this CHARACTER in adult stage, other alters entirely]CD In c
368d114 the bright plumage.-- «thinks» Hence specific CHARACTER most perfect in <male> «hermaphrodite» (Fishe
369d114 females  recede in organization from specific CHARACTER.-- Chapt I. Also Latent Character Hunter Anim
369d114 om specific character.-- Chapt I. Also Latent CHARACTER Hunter Animal Economy p. 482 (Same book) Owen
376d136 generation,. that with Quichuas the American CHARACTER is more tenacious. & does not disappear for M
378d147 g ones brown.-- Sexual Selection If masculine CHARACTER. added fo species,. we can see why young & fe
383d160 ale coloured.-- In the back feathers, we have CHARACTER very different from either parent bird-- Hunt
385d165 law.-- "How often do we find in the son, the CHARACTER, constitution, & most of the moral  qualities
385d165 f the father!. In how many daughters does the CHARACTER of the mother revive! Or the character of the
385d165 es the character of the mother revive! Or the CHARACTER of the mother in the son, & of the father  in
388d171 ame out. it might partake of shade of fathers CHARACTER.-- according to this view more semen to one c
400e012 take  any colour?)-- length of tail varies, & CHARACTER of fur-- I am sure a very good case, might be
403e022 per Oct 14th Macleay says, that <every> «any» CHARACTER even colour is good. (ie invariable) in  some
408e044 nic-- Azores might be prophecied to have this CHARACTER.-- worth going there for.-- «Gales of wind wo
421e090 bes says the Fauna»-- (near Oran) approach in CHARACTER to Canary Isld.-- ie Canary Isld approaches m
421e091 to  mistake a merely adaptive to an essential CHARACTER--" How little clear meaning has this compared
422e092 - What is the difference between an essential CHARACTER & an adaptive. one.-- are not the essential o
432e124 hermaphrodite Man probably assumes the hairy CHARACTER of his forefathers only when advanced in age,
436e135 Struthio head pulled out. yet feathers retain CHARACTER? If separation in horizontal direction is far
437e141 ddenly changing into each other, & depends on CHARACTER of antecedent races.-- «& yet in all probabil
446e162 the  continent is there more variation in its CHARACTER.?» No--well characterized.-- Tulips are culti
466t095 er flowers-- My view of <variety acquired> « <CHARACTER>  » of characters being inherited at correspon
490q007 Dr. Holland-- <Jordan> Smith of Jordan Hill-- CHARACTER of the extinct land-shells of Madeira-- analo
490q001 e in one country or district than in another? CHARACTER of shells of Sandwich group {Sowerby monstrou
497q007 here is no crossing by Bees.-- Henslow.-- (1) CHARACTER of alpine Flora of Tierra del Fuego and Entom
498q007 Horned oranges so? --Yes, my Father lost this CHARACTER in grt degree from charcoal & good  treatment
501q10a t such variation is not a generic or specific CHARACTER,, but contingent on country.-- How is it in P
511q018 Sir.  J. Sebright select to destroy secondary CHARACTER= Cross Rumpless fowls & Dorking fowls,-- or t
513q020 &c &c--if so probably a variety, not specific CHARACTER I felt it)-- this is kind of conscience, is o
533m061 was  not merely morally wrong, but hurting my CHARACTER let him compare the American whether in the c
540m085 one structure.-- He who doubts about national CHARACTER??? Why may not our heredetary nature thus ac
571n028 as ancients did high forehead sign of exalted CHARACTER is invariability.-- if explained by habits, u
588n090 at differs-- not <reason> «instinct», for its CHARACTER in different animals of same species.-- The g
591n098 ne nothing will frighten-- hence variation in CHARACTER between man & animals.-- [blank] Double consc
593n109 a of beauty of music are great distinguishing CHARACTER Humes Dissertation on the Passions. "Hartle
595n184 "Adam  Smith Moral Sentiments" much on life & CHARACTER.-- Hence Lessings shows expression of pain ca
606o022 s. who says it is simplicity with grandeur of CHARACTER may change-- because motive power changes wit
608o026 prepared materials. From contingencies a mans CHARACTER, love of country, of association &c stronger
615o038 ly modified in many countries, hence national CHARACTER of a physical law, «& is therefore utterly us
635j55r n plans.-- is no explanation-- it has not the CHARACTERISED species periwinkle wants insects to impregn
447e164 [bu]t  the one does on the continent-- well CHARACTERISTIC of the deserts of Syria <chara> ditto for
036r052 hes, «gazelles» hares, grasshoppers & Rats. CHARACTERISTIC forms of S. America. With respect to futur
343d038 Fuego has <not> «only» one «Guanaco» of the CHARACTERISTIC qualities were all deeply imbued in them f
366d108 breeding if instincts unchanged, & if their CHARACTERISTIC of male greater strength, (p 45) & that st
384d161 when  the first <are> become of use <Great CHARACTERISTIC of one kind of intellect is that when an i
580n061 eas.-- Have children loose ideas of time?-- CHARACTERISTICS of the Tropical Forms in shells. are numer
413e059 ember 2d Lyell tells me Beck considers the CHARACTERISTICS;-- they are more truly sensations??. a kin
582n068 e not love & hate emotions; what are their CHARACTERIZE the moral sense, by its "supremacy",-- I ma
628o054 all  different.-- P. 22. Butler & Mackintosh CHARACTERIZED by no quartz & amphibole frequently only vi
073r164 ld & silver." «p. 131» The above porphyries CHARACTERIZED parts belonging to each)» approaching anoth
177b028 e aberant species in each genus «(with well CHARACTERIZED, of each continent. Try amongst Europaean q
250c037 reme southern points of S. Hemisphere fully CHARACTERIZED by Ogleby, who observed that the young of t
341d030 found  in Australia New species of Moschus, CHARACTERIZED.-- 16th Aug.-- What a magnificent view  one
342d035 verbears the crossing with females not thus CHARACTERIZED by wedge tails.-- many of the hawks <to> ar
353d061 Australian birds-- all Eagles. of Australia CHARACTERIZED study Von Buch.]CD Now Mr Knights statement
442e153 ; [and the gorze in Norway ought to be thus CHARACTERIZED.-- Tulips are cultivated during several yea
446e162 more variation in its character.?» No--well CHARACTERIZED» dream, in continuation of waking thought--
547m111 had  very clear & pretty vivid «& perfectly CHARACTERS of both parents, & these infinite in  Number
172b005 w of intermarriages <separating> partaking of CHARACTERS is attempt at returning to parent stock.-- I
222b203 er bringing forth young having very different CHARACTERS, together with other <animals> beings of ver
224b213 es: one that remains «at large» with constant CHARACTERS, wild pigs. said by Forrest to swim from one
245c023 nd.-- but considered good species from dental CHARACTERS vary in degree in last instance hardly at al
274c114 ial for type.-- Now here I must observe these CHARACTERS which are analogical in a genus with respect
282c142 , but would depend upon exclusion.-- The same CHARACTERS, as do not vary in the species of it.-- «whe
284c149 visions [+] ought genus to be founded on such CHARACTERS» rule are used by Naturalists in their  test
284c149 it.-- «where does such occur?» now some such CHARACTERS than among those groups, where it remains le
284c149 stly of less importance, as affording natural CHARACTERS will be first those of analogy, but will gro
285c151 group  & connection of even distant. ones) the CHARACTERS which unite these of older standing than con
286c156 contains  many Linnaean genera-- Now are the CHARACTERS of analogy.-- last acquired,-- or aberrant,
295c178 cter -- -- female & young seem most like mean CHARACTERS the others assumed--+++ ‖Daines Barrington s
302c202 analogies between fish.-- Birds same remarks. CHARACTERS of analogy.-- see my notes on p. 37. of Macl
302c202 told,  for any small family. having analogous CHARACTERS, might be multiplied.-- we must argue revers
302c202 two  circles «of equal value» must be so from CHARACTERS of analogy-- see my notes on p. 37. of Macl
304c207 ix & other birds.-- will having many trifling CHARACTERS, in common with other birds reveal the secre
304c207 l the different forms of Synallaxis. trifling CHARACTERS as red band on wing show to be from one pare
341d029 in  one point or many) Owen answered that all CHARACTERS might be considered as adaptive & that he di
348d050 hese will be the oldest) ‖"The most important CHARACTERS break down in certain species & become worth
348d050 (p  6) say such is group. because it has such CHARACTERS of importance, "but we say such happens to b
348d050 p & serves to insulate <them> it".-- i.e what CHARACTERS chance to be heredetary whether important or
349d051 cited.-- It will rest upon the discovery what CHARACTERS VARY most easily:-- those which do not  vary
349d052 nking any group good, though not circular, if CHARACTERS can be established-- clearly so.-- NB.  This
369d114 le» «hermaphrodite» (Fishes have no secondary CHARACTERS.-- )p. 49. (wonderful case of Pea hen. takin
371d128 yet  they cannot transmit through seeds these CHARACTERS though transmitting them with such  facility
390d179 (as  Sir J. Sebright urges?) one with opposed CHARACTERS is by impliance to breed two which have each
399e011 paean genera--.-- Hope has ideas about generic CHARACTERS. dominant. predominant &c having relation to
405e031 at whole colour is changed, these best marked CHARACTERS are partly retained, therefore colours  vary
438e141 e difficult to show that <time> the fixity of CHARACTERS «from antiquity» prevents their variation, w
439e142 n circumstances favouring the reappearance of CHARACTERS, formerly possessed-- <that is animals» «or
466t095 view of <variety acquired> « <character> » of CHARACTERS being inherited at corresponding age & sex, o
466t095 le flowers & colours breaking only hereditary CHARACTERS, wh. come on in after life of Plants-- also
501q011 enerally fails-- perhaps indexed by secondary CHARACTERS-- in double flower. do Henslow Speaking of T
502q11v leaves.--  strawberries How <soon> «early» do CHARACTERS of races of different vegetables & animals c
```

Page **************************************(Key Word)***

506q015) Have Dioecious plants any secondary, sexual CHARACTERS.-- Stature, position of flowers-- Their smel
511q018 [Made no import. remark]CD (2) Secondary male CHARACTERS.-- does male transmit to male more of his fe
511q018 e recrossed this breed.-- Have secondary male CHARACTERS appeared.= (4) Does he know any seed-raisers
513q020 -- or tailless dogs & fox, to see whether the CHARACTERS are then intermediate or «sometimes» all on
538m079 probably forgotten.» Such facts bear on such CHARACTERS as Allen W. & Babington, both half ideotic i
544m102 unt for dogs & men speaking in their sleep.-- CHARACTERS of dreams no surprise, at the violation of a
138a176 of cabbage» alklali & precipitate silica / or CHARCOAL charged with carbonic acid [blank] Many inter
498q007 Father lost this character in grt degree from CHARCOAL & good treatment (8) Do bees frequent Cabbage
183b052 said as well known fact that "serin avec le CHARDONNERET, avec la linotte, avec le verdier" &, <fr>
138a176 ge» alklali & precipitate silica / or charcoal CHARGED with carbonic acid [blank] Many interesting ex
555m142 -- Miss Martineau (How to Observe p. 213) says CHARITY is found everywhere (is it not present with al
555m142 The Chillingham cattle (& Porpoises) have not CHARITY-- is it in former case instinct to destroy con
558m151 . it comes to Miss Martineaus one principle of CHARITY.-- ¿ May not idea of God arise from our confus
569n020 ome distant power of the mind-- superstition & CHARITY & prayer, or eloquent request. Reason in simpl
214b175 om a breed known to be there since the time of CHARLES. .-- and now in the possession of Mr Howard Gal
239cIFC atest effect Account of the [...] the world.-- CHARLES Darwin written between («beginning of» Februar
295c178 ¿because offspring too unlike.--?? Memoire by CHARLES D'orbigny on Plastic Clay of Paris contains ma
331dIFC ime Study Botanical work on Buds & Gemmae. C D CHARLES Darwin 36 Great Marlborough St Did Eytons <int
520mIFC - [blank] Questions & Experiments Expression M CHARLES Darwin Esq 36 Grt. Marlborough Str.-- (p. 64.
560mIBC danger Do they frown, when they first see it? CHARLES Darwin 36 Great Marlborough St Has my Father e
563nIFC ion N What are they sexual difference in monkeys.-- CHARLES Darwin [Private.]CD (Metaphysics & Expression)
317c251 the shrews.-- Dr Bachman told me. that near CHARLESTOWN ?three species, near New York. (600 miles N.
220b198 s des Sciences, paper on Botany of Tahiti In CHARLESWORTH Magazine Jan: 1830. most curious paper on h
506q15v ia Bennett says in same state. of flower 45. CHARLSWORTH. vol II. p. 670-- oats cut down turning into
495q05a Wiessenborns doctrine of Equivocal Generation CHARLWORTH p. 377. Have paper ruled in squares to facil
477z001 V do Azara Voyage Vol I p. 196. According to CHARPENTIER de Cossigny. only 10 years ago <no> snail wa
030r035 ings, 60-100 fathoms 2 & 3 miles from shore. V. CHART) Every winter torrents must bring much vegetabl
030r036 quantity of altered Carbonaceous shales Examine CHART of Patagonian coast to see proportional cliff &
134a142 ere is volcano on isld in large lake-- Berghaus CHART of do Journal of Asiatic Society Vol I. {T} p.
211b164 very good in connection with Von Buch Volcanic CHART & my idea of double line of intersection.-- Dr.
036r050 onsult. reconsult Geolog. Map of Europe Consult CHARTS for distribution of pebbles.--Plains. off coas
275c120 ecause they have no scent!) Mr Wynne) at end of CHASE would not run up hill-- he took thorough bred c
310c223 hts, I will never allow that because there is a CHASM between Man (--<a> «&» chasm«s» <necessary to a
310c223 ecause there is a chasm between Man (--<a> «&» CHASM«S» <necessary to account> «consequence of» for t
601o08b sur l'intelligence des Animaux-- & Le Parfait CHASSEUR, par Desgraviers, un Vol 8vo Keratry-- Induct
566n010 ion of heart⸜ Malthus on Pop. p. 32, origin of CHASTITY in women.-- rationally explained.-- on the wi
195b103 ws would produce species so close as Patagonian <CHAT> & Galapagos orpheus.= Put this strong so many
451e180 -"No wild animals in Moa.--" Chapt.-- V.-- : do. CHAT XXI. Wild cattle & Hogs on Timor-land-- monkeys
578n054 xpecting admiration or an act of cowardice, or CHEATING.-- one does not blush before utter stranger,-
256c055 .-- The circumstance of having two sexes in the CHECK to distribution of birds & animals Mr Stricklan
267c094 le land of Tahiti, in spite of every attempt to CHECK its increase. The <bush> woodlands for miles in
375d134 ase, whilst no checks prevail, but the positive CHECK of famine & consequently death.. population in
375d135 e of Malthus no one clearly perceived the great CHECK amongst men.-- «Even a few years plenty, makes
549m122 h experience shows it must for his happiness to CHECK-- that is external circumstances are so conditi
609o29v he vice of intemperance, circumstances made the CHECK.-- to licentiousness jealousy, & every one bein
628o54v n, whilst <the> no desire of gratification will CHECK the consciences desire for virtue.-- [I expect
032r039 f highest tidal action. this will at length be CHECKED by increased vertical <height> thickness (DZ)
210b160 o. ⟦Yes « Fox» ‖ The creative power seems to be CHECKED when islands are near continent: compare Siici
380d153 animal breeds young her growth is immediately CHECKED-- the vis formativa goes entirely to the offsp
403e021 rickly Limonia trifoliata, which cannot now be CHECKED".-- Mar[s]den p. 94 (1st Edit) of Sumatra has
260c068 ert country. is as effectual as a cold one. in CHECKING beautiful colours of species-- Mem. St. Jago-
224b212 is the amount of difference, which improves. & CHECKS it.-- It does not bear any precise relation to
375d134 of brutes, must be prevented soley by positive CHECKS, excepting that famine may stop desire.--» in
375d134 Nature production does not increase, whilst no CHECKS prevail, but the positive check of famine & co
609o029 t orginally necessary revenge was justice.-- No CHECKS were necessary to the vice of intemperance. ci
609o29v lation. with the existences of so many positive CHECKS.-- (This is encroaching on views in second vol
628o54v shows the change produced.-- 4) p 38 Conscience CHECKS the wish to <other> outward gratification, whi
635s56r ort to other beings.-- true, (& the doctrine of CHECKS & my theory) Macculloch. Attrib. Vol I. p. 330
367d113 body like balloon-- by air cells connected with CHEEK pouches.-- Hunter's Animal OEconomy p. 45 "One
557m147 re so teaching expression «as constant smiles, CHEERFUL face».-- Man when at ease has smooth brow con
223b208 estinal. or even a mite.-- a bee «compared with CHEESE mite" with its wonderful intincts <might well
079r177 gainville says P 291.-- The Fuegians treat the "CHEFS d'oeuvre de l[']industrie humaine, comme ils tr
460t019 rsupial in Oolite.-- insects, of do orders-- CHEIROPTERA & caetacea in Eocene-- dicot. plants in coal
035r045 th & alternating with obsidian must clearly be CHEMICAL differences. & not those of rapid cooling &c
046r078 S. America. -- Volcanos must be considered as CHEMICAL retorts.--neglecting the first production of
046r078 ials simply altered by circumstances; & not in CHEMICAL nature, or has a subterranean fluid mass itse
048r084 us coated.--The motion is most wonderful, from CHEMICAL attraction, as a blade of grass penetrating b
056r112 errestrial renovation & so is Volcano a useful CHEMICAL instrument.--Yet neglecting these final cause
059r121 All bear close analogy to Obsidian, & all show CHEMICAL action as well as effects of cooling [misnum
074r165 ow mountain chain. there after NW <W>.-- «same CHEMICAL laws as in concretions perhaps makes intersec
074r165 chest-- Humboldt has urged phenomena in veins, CHEMICAL affinities like in composed rock. granites sy
074r165 w how metals may be transported by complicated CHEMICAL law & steam of salts, quite curious case of c
099a050 e.-- Therefore < S of inclination «varies with CHEMICAL attraction &c.» becomes measure of force. < ∴
115a096 rizontal non cleaving beds. metamorphosed. The CHEMICAL action which gives polarity to atoms in slate
269c102 ew" The intimate relation of Life with laws of CHEMICAL combination, & the universality of latter ren
516q23v or animal substances-- (Athenaeum (40) p. 823 CHEMICAL analysis of Peat (2) Athenaeum 1840 p. 777. D
611o034 bably one principle for connect of electricity CHEMICAL attraction, heat & gravity is probable.-- And
611o034 es, the vital laws act definitely (<like» «as» CHEMICAL laws,) as long as certain contingencies and p
080r178 ith man Does Indian rubber & black lead unite CHEMICALLY like grease & mercury [blank] NB. P. 73. Gen
109a084 by Ca. of L. & Mur. of Soda mixed.-- Turner's CHEMISTRY p. 206 Both Beck & Deshayes saw fossil shells
137a149 anite infiltering some of its constituents into CHERT. [blank] Ed: New. Phil J. 1838. p. 132. «& 134»
042r067 om prevailing swell, (as Shingle travels on the CHESIL bank. V. De la Beche). Ask Capt. F.: R: how th
239c002 of our common varieties. He has no doubt that CHESNUT, for many generations back, were crossed with
239c002 bay a few generations, that offspring would be CHESNUT.-- On this principle I may add, that fact of h
543m099 very limited intellects, & in the same way are CHESS Players-- A man at Cambridge, during his time,
543m100 eft him a guinea a week. yet. he was inimitable CHESS player.-- Peacocks remark about mathematicians
543m100 same fact-- for, as Jones observed, in playing CHESS however many places, & contingency a man has ke
308c217 d habit of putting tea in pot made me go to tea CHESS almost unconsciously.-- why do absent <Dr. Blac
535m063 have done when taking lid off <tea> side of tea CHEST, when no tea do. p. 233. Mr Lewis describes cas
557m150 ng body.-- Are those parts of body, as heart, & CHEST (sobbing) which are most under great sympatheti
570n023 -- intolerable to be poked behind, without ones CHEST, being distended. touch a person on the ribs &
503q012 ty of men & Europaean animals in India?-- about CHETAH & other tame animals not breeding when tame in
183b052 le "métis" du loup et du chien, que celui de la CHEVRE ou du belier, cessent d être feconds. dès les
313c236 suck or other actions in foetus of Mammalia, or CHICK eat») Generation becomes necessary, when organs
445e158 never touches objects.-- far better case than CHICKEN pecking fly.-- "whilst the shell stuck to its
445e158 12.-- p. 282. Allows this instinctive power in CHICKEN, yet says it is evidently acquired by experien
564n004 s told by W of Downing. Coll. that he had seen CHICKEN only hatched few hours placed on table & when
564n005 ide to knowledge, was transmitted perfectly to CHICKEN so as to seize small moving object like fly.--
566n011 true instinct, which is a real instinct in the CHICKEN, just bursting from egg.-- Animals have necess
585n074 present in youth" (Mem. Mr Worsley's story of CHICKEN) to know that which we touch & what [...] the
637j58r t to chickens beak, to break egg. shells-- why CHICKEN could not have lived had it not been so.-- let
255c051 goes before structure (habits of ducklings and CHICKENS) Young water ouzels hence aversion to generat
275c120 diminished, but feathers continued by picking CHICKENS of each brood.-- These bantam feathers at las
637j58r Gives as Summary of adaptations Horny point to CHICKENS beak, to break egg. shells-- why chicken coul
567n015 am almost sure Fuegia Basket did. & Jemmy, when CHICO plagued him-- Animals I should think would not
114a095 it has been excessively slow because beach line CHIEF cause of denudation, but does not tell period.-
227b225 o adaptation + heredetary structure. latter far CHIEF element.∴ little service habits in classificatio
232b247 he world had cooled by secular refrigeration in CHIEF part instead of change from insular to extreme

276c123 on persecution of early Astronomers.-- then add CHIEF good of individual scientific men is to push th
306c212 differently from dog. perhaps being in passion CHIEF difference Major Mitchell is not aware that Aus
349d051 -- those which do not vary being foundation for CHIEF divisions.≠ p. 7. «In» Some «of» cases the circ
360d089 sex> Individual instances trouble Yarrels law. CHIEF trust must be in general knowledge of breeders,
467t100 stigmas of {P} shape of ordinary Labiatae --the CHIEF part with ordinary divisions, & a few with one
603o013 personified the deity.-- H. Tooke has shown one CHIEF object of language is promptness «of consequenc
030r035 matter from thickly wooded mountains, probably CHIEFLY leaves.--This position agrees with character o
033r043 . Lava with base of Pitchstone; Mem Galapagos. CHIEFLY red glassy scoriae.--could walk round base:--n
077r170 on, no relation--there is more modern breccia, CHIEFLY owing to destruction of porphyries. whereas ot
086a007 of world regulated by atmospheric currents?-- CHIEFLY clearly by sun's position = If equatorial stre
107a078 ct. PP-- I think from dislocation taking place CHIEFLY beneath water & volcanos. crust must be thinne
157g073 y contains many pebbles, but I believe this is CHIEFLY caused by its being lower,-- [no pebbles in pa
157g074 lines appear to cross stony parts; appearance CHIEFLY cause by fall of angular masses from above on
161g095 piece of excised rock lost at point of valley CHIEFLY from rockiness When on other side {P} Shelf A
195b103 cies!! It may be argued representative species CHIEFLY found where barriers «& what are barriers but»
203b138 en mingled with newer, hybrid variety partakes CHIEFLY of the former Eyton's paper on Hybrids Loudon'
248c033 xternal circumstances> these therefore will be CHIEFLY heredetary.-- If varieties «produced by slow c
249c036 riod. Have the Edentata & Marsupial forms been CHIEFLY preserved,-- where shut up by themselves witho
258c061 uch interest XX. Hence relation of analogy may CHIEFLY be looked for in the aberrant groups.-- It is
287c158 nge in organization. But the classfication must CHIEFLY rest on these same organs,-- habits, range. &c
299c195 difference of <locality>. station & varieties CHIEFLY produced by cultivating parent in rich soils &
303c204 forms on that isld.-- The races of men differ CHIEFLY in <size> colour, form of head «& features» (h
305c210 t to certain contingencies of organic matter & CHIEFLY heat), which assumes a multitude of forms «eac
359d088 eons) The male of every animal certainly seems CHIEFLY to impress the young most with its form & disp
393dIBC hich is not transmitted by generation?? Is it «CHIEFLY» in high. bred dogs ie. (bred in & in) that on
414e064 ie intellect in man?) to gain the day.-- In man CHIEFLY intellect, in animals chiefly organization: th
414e064 he day.-- In man chiefly intellect, in animals CHIEFLY organization: though Cont of Africa & West Ind
425e104 n groups?-- = Believes some Mediterranean, but CHIEFLY mountainous «this is very important. (Sicily e
542m095 uld amongst lowest savages clearly be directed CHIEFLY by objects of vision.-- Does the contraction &
554m138 ter than the boasted philosopher himself it is CHIEFLY shown in old male.-- A very green monkey (from
576n047 & Whether wholly instinctive as in the dog, or CHIEFLY habitual as in man), for it added much to the
588n090 if not reason.-- or does this reasoning apply CHIEFLY to recollection. yet a dog hunting for a bone
600o008 two souls in one body-- (2) Beau ideal, refers CHIEFLY to moral, beau desires conscience & love.-- [W
604o017 y.-- Prevailing idea. owing to loss of will.-- CHIEFLY excited by passive emotions.-- Cannot quite pe
604o017 erceive drift of Book.-- Sympathy & affections CHIEFLY fail.-- Notices. struggle <between> when insan
183b052 fowls..-- "On sait que le "métis" du loup et du CHIEN, que celui de la chevre et du belier, cessent d
246c025 4. dogs of New Zealand of large size, resemble, CHIEN-loup.--long, black & white, ears short & straig
171b004 ions even mind & instinct becomes influenced.-- CHILD of savage not civilized man.--birds rendered wi
193b090 .-- The father being climatized, climatizes the CHILD. ¿-- whether every animal produces in course of
199b119 IRY man. because ancestors hairy with one hairy CHILD, and of albino DISEASE being banished, & given
222b203 with trifling differences of expression -- one CHILD like father another like mother Has Lowe writte
259c066 tion, fact of armless parent not having armless CHILD, shows than there is reference to more than off
261c070 good instance of connate instinct, better than CHILD sucking or even ducklings & fowls-- When talkin
265c083 changes show that grandson is determined, when CHILD is.-- shows that generation implies more than m
265c084 -- shows that generation implies more than mere CHILD, but that child should produce like children. ⫝
265c084 neration implies more than mere child, but that CHILD should produce like children. ⫝Lyell has story
293c172 ir W. Scott has written about it]CD If we saw a CHILD do some action-- which its father. had done hab
301c199 any habitual action. even which Man performs.-- CHILD striking a post in passion.-- Habit instinct ga
305c211 y have some relation together, as well as Man & CHILD, polypus & polypus, bud & bud, polypus & germ p
306c212 l developed, just like, habits easily gained in CHILD hood.-- Young salmons. first a species which li
312c233 ter, those. of different litters or of father & CHILD are thought long breeding in. Must not trust hi
332d004 y the transmission of large number of worms the CHILD not having passed them before.-- Hence disorder
336d018 e union makes hybrid, in fact the parents beget CHILD like themselves. expression of countenances, or
372d131 g «true» bud some such process is effected.-- a CHILD might be so born. but it would be very differen
378d148 - Anyhow not connected with habits According as CHILD is like parent, so is species old: Hence <young
388d171 ter.-- according to this view more semen to one CHILD. more like father.-- stuff.!-- How much opposed
417e077 e grafting of trees.--]CD CD[The similarity of CHILD to parent appears to follow same law in two of
417e078 & races are crossed.-- Now these laws are, that CHILD may be either like father or mother, independen
522m010 lled up.-- My F. asked him, did he know whom Mr CHILD «of Kinlett» had married.-- Answered never hear
522m011 Answ Had large fortune left. him, took name of CHILD «of Kinlett» & married Miss A. B.-- all the same
532m054 nd tries to fix upon some object:-- When a man, CHILD or colt has once been frightened & started much
533m057 t, which are heredetary are so because brain of CHILD resemble, parent stock.-- (& phrenologists stat
533m058 t.-- it is frightened without knowing why-- the CHILD dislikes the frown without knowing why-- a man
537m075 rred from expression. "relict of bad habit." as CHILD is cured of sucking his finger by rubbing them
539m082 this after seven years interval. Augt. 15th. As CHILD gains habit «or trick» so much more easily than
542m096 & frown, who can doubt these are instinctive-- CHILD does not sneer. because no young animal has can
544m101 sociated pleasures & pains & emotions-- such as CHILD sucking, gives pleasure, & always has done ther
544m101 asure, & always has done therefore sight of own CHILD. (when frame in condition to receive pleasure)
545m107 down on its back & kicked & cryed like naughty CHILD.-- Do monkeys cry?-- «they whine like children.
549m118 looks like free will.-- V. last page. A healthy CHILD is «more» entirely happy (contentmt is differen
550m124 r, the more pleasant.-- Simple happiness «as of CHILD» is large proportion of pleasant to unpleasant
564n004 s heart those rules, which he wills to give his CHILD.-- Octob 3d. Was told by W of Downing. Coll. th
565n007 who does not recognise look of suspicion, even CHILD will do so.-- Contempt look obliquely so does d
567n013 corn with her teeth from the straw, & just like CHILD not knowing what to do with them, came several
567n013 al times & opened my hand, & put them in-- like CHILD. Tommy's face, now ill, has expression of langu
573n037 an a Monkey.-- November 20th Saw the youngest CHILD of H. W. constantly. when refusing food, turn h
575n043 m their brains» then can we deny that the grand CHILD dug for mice from some peculiarity of structure
578n056 ar action -- «in man & animals» Blubbering of a CHILD (different in different ones?) in the most perf
586n079 ence or habit) could be formed or afterwards.-- CHILD sucking whole wonder instinctive.-- carrier pid
587n088 & chimpanze. pout.-- Former, whines just like a CHILD. Get a Dictionary & make a list of every word,
588n090 ence same, but less in degree, as between man & CHILD.--]» what differs-- not <reason> «instinct», fo
594n113 w in Shrewsbury garden picking from same bone A CHILD born on the 1st March was frightened on the 24t
594n113 ost anything ugly. baby-- association-- pouting CHILD same as anger, lips not compressed sullen, prot
595n125 easts in Zoolog. Garden [blank] [not located] A CHILD crying. frowning, pouting, «smiling», just as m
595n125 munication) in man, than sucking.-- [I assume a CHILD pouts who has never seen others pout]CD [blank]
601o009 ry added to it, man in sleep not conscious, nor CHILD-- Evidence of consciousness, <t> movements «¿»
614o037 many cases cannot be acquired by experience for CHILD sucking.-- And is it more wonderful that memory
620o045 .-- One does not feel it wrong in very young CHILD to be in passion, any more than in an animal.--
620o045 (reason) obtained by age, which should show the CHILD, which of its instincts are best to be followed
620o045 , when not urged to it by passion, shows a bad CHILD.-- Hence there are certain instincts pointing o
621o047 uited from some trifling circumstance.-- Thus a CHILD may be taught to think almost anything nasty. (
621o047 nasty in taste, & right & wrong in action, so a CHILD may be taught, or will acquire from seeing cond
621o048 Hence, what parents think will be good for the CHILD in the long run, & for themselves & others, (as
622o048 in these cases.-- Those instructions, which the CHILD sees uniformly performed by the teachers & all
622o049 stinctive, & how does it apply to mother loving CHILD, from whom, she has never received any benefit.
625o50v gs & our short lived Passions' State broadly in CHILD or animal it is equally proper to obey anger as
628o53v herd dog, is strictly analogous to education of CHILD,-- causing many actions to be considered right
399e009 e solely to womb, as in monster. or solely to CHILDHOOD, or solely to manhood,-- it will decrease & b
418e084 ing added to the germ, at a time, (as even in CHILDHOOD) when the organization is pliable, such modif
524m021 ood. In insanity, the ideas do not go back to CHILDHOOD, (but appear most capricious) as in delirium
524m021 petition is not necessary)-- the words second CHILDHOOD full of meaning:-- Dreams do not go back to c
524m021 d full of meaning:-- Dreams do not go back to CHILDHOOD-- People, my Father says, do not dream of wha
533m059 feels, pleasure. in seeing the scenes of his CHILDHOOD. without knowing why-- had not conscious of re
538m078 to fact of old people singing songs of their CHILDHOOD. & certainly of Miss Cogan, & fully corrobora
538m080 farthers actions, as one does those in second CHILDHOOD, <they> or when drunk they would not be more
580n061 say I know it because I was always told so in CHILDHOOD.-- hence the belief in the many strange relig
586n079 tion being performed or emotion felt in early CHILDHOOD (before experience or habit) could be formed
588n089 y of happiness.-- Bell on the Hand p. 191 Says <CHILDR> babies have an instinctive fear of falling.--

178b032 entots or Negros cross at C. of Good. Hope the CHILDREN cannot be made intermediate, the first childr
178b032 hildren cannot be made intermediate, the first CHILDREN partake more of the mother, the later ones of
179b032 other was nearly quite white) in the two first CHILDREN How is this in West Indies «--:Humboldt. New
258c060 if continually crossed with people from cold, CHILDREN would not become adapted to climate.-- Descen
259c065 true).-- or again healthy parents have healthy CHILDREN the other case is <adaptation> «change» durin
265c084 mere child, but that child should produce like CHILDREN. Lyell has story from.-- Beck about six fing
265c084 yell has story from.-- Beck about six fingered CHILDREN heredetary With respect to question what is
285c150 be given It would not be difficult to arrange CHILDREN of same parents in a circle,-- & «hermaphrodi
298c190 be found-- a person who is habitually kind to CHILDREN,--<get> «increases» general instinctive feeli
308c217 hristian «we are all» Brothers in spirit-- all CHILDREN of one father.-- yet differences carried a lo
309c219 in individuals «we know of:», is relationship, CHILDREN of one parent, races of animals-- argue «open
312c228 e value Dr Smith if black & white Man crosses; CHILDREN heterogenous, he feels sure of this, first of
313c236 ns.--» «this very important in considering how CHILDREN come to suck or other actions in foetus of Ma
315c242 The vividness of first <thoughts> «memory» in CHILDREN or rather their memory. very remarkably-- sce
325c266 is. with respect to licentiousness, destroying CHILDREN. --it is not effect, as Lyell suggested, of o
325c266 rn out as. otherwise old whores would not have CHILDREN Turners embassy to Thibet, perhaps worth read
332d003 her has seen case of pleurisy, broken limb «in CHILDREN» & other such disorders accompanied with some
333d008 haracter & Kittens alike each other.-- Even in CHILDREN of parents <some> one sometimes resembles one
344d040 Java Caterpillars not being fertile is same as CHILDREN not being so.-- consider this with reference
358d076 s where the female differs from the male?)».-- CHILDREN & women = "women recognized inferior intellec
379d151 oda: Old Jones remarked to me, that one of the CHILDREN of Sir J. H. was so very like Sir W. whilst S
385d164 on Phy. transl by Holcroft Vol I. p. 195. says CHILDREN resemble parents in their bodies "It is a fac
385d164 erve in the temper, especially of the youngest CHILDREN, a striking <resemblance> similarity to the t
417e078 e laws, therefore, of likenesses of fathers to CHILDREN of mankind, no doubt are applicable to likene
417e078 r like progenitors.-- in some families all the CHILDREN like mother & in some like father «What is ca
417e079 t one refers to in speaking of resemblances of CHILDREN to their parents.-- Lord Moreton's law cannot
432e124 ers only when advanced in age, & therefore the CHILDREN do not, (& in hairless kittens we see same fa
455eIBC eases common to men & animals.--:likenesses of CHILDREN CD[Does any annual give buds, or tubers. Yes-
509q017 rs, when men & women have long worked, whether CHILDREN, who have not worked have any peculiar config
525m025 of the foetus.-- some mothers. have first dead CHILDREN, then children which were short term, & lastl
525m025 - some mothers. have first dead children, then CHILDREN which were short term, & lastly healthy ones.
526m028 true.-- mem Erasmus & mine taste for music.-- CHILDREN like hearing a story told though they remembe
526m028 n memory & imagination. «Catherine thinks that CHILDREN like looking at <ani> pictures, an early tast
530m045 re likewise heredetary, & therefore that their CHILDREN have some little advantage in these trades.--
531m051 ry when once read over.-- The extreme pleasure CHILDREN show in the naughtiness of brothers children
531m051 e children show in the naughtiness of brothers CHILDREN shows that sympathy is based as Burke maintai
531m051 eholding the misfortunes of others.-- In young CHILDREN, the violent passions they go into, shows how
532m053 eigh doubts whether young babies start.-- If CHILDREN wink. it is instinct Fear must be simple inst
535m069 nsiders this profoundly true.-- How is it with CHILDREN.-- Now it is not a little remarkable that the
536m073 thers, corporeal & bodily are visited upon the CHILDREN.-- The above views would make a man a predest
537m076 ring--¿ yet perhaps if they had murdered their CHILDREN, this moral sense, would have been so much, a
540m089 .-- see how a crowd collects at an accident,-- CHILDREN with other children naughty.-- Why does perso
540m089 collects at an accident,-- children with other CHILDREN naughty.-- Why does person cry for joy? 17th.
542m094 weariness. & meditative tranquility. «whine of CHILDREN. puppies do so dogs nearly silent, so with me
544m104 hing which happens to man who does not produce CHILDREN. or after he has useless. does not affect rac
545m107 ty child.-- Do monkeys cry?-- «they whine like CHILDREN.--» Expression, is an heredetary habitual mov
549m118 to instinctive feelings.-- for man losing his CHILDREN, any more than to dog losing his puppies-- The
565n009 bout the nose, & arrogance in upper lip. <The> CHILDREN having peculiar expression is remarkable. the
567n013 confidence» when does such notion commence?-- CHILDREN understand before they can talk, so do many a
567n015 ted with surprise.-- heart beginning to beat-- CHILDREN inherit it <ins> like instinct, preeminently
571n028 t been acquired by education. else why do some CHILDREN acquire it soon. & why do all men. agree ulti
573n037 at she has constantly observed that very young CHILDREN. express the greatest surprise at emotions in
574n040 Birgos opening a Cocoa nut shell at one end.-- CHILDREN & old people get into habits.-- we probably c
574n042 ed.-- this is analogous to a blacksmith having CHILDREN with strong arms.-- The other principle of th
574n042 th strong arms.-- The other principle of those CHILDREN, which chance? produced with strong arms, out
579n058 - With respect to my theory of smile. remember CHILDREN smile before they laugh.-- Has frowning anyth
580n061) have only possessed very loose ideas.-- Have CHILDREN no difficulty in expressing their want,
581n064 th. Gardiner's Music of Nature. p. 31. remarks CHILDREN loose ideas of time?-- Characteristic of one
582n066 s to glowing conversation of several people.-- CHILDREN have an uncommon pleasure in hiding themselve
585n074 N. Necker has remarks on the means. by which CHILDREN learn (probably not only experience,, but als
587n088 a of scarlet, as well as remember it.-- Why do CHILDREN pout & not men-- orang-outang & chimpanze. po
592n103 losing both eyelids express contempt. p. 76.-- CHILDREN have been tickled into excessive laughter & s
596nIBC about instincts of mongrel dogs Do blubbering CHILDREN, if of. convulsive tendency easily fall into
603o014 good article-- why flower beautiful? ¿even to CHILDREN S. Jenyn's Inquiry into the Origin of Evil. R
615o36v tion of instincts But habits acquired even by <CHILDREN> «plants»! [RHC] 7) As definite instincts mod
619o043 creatures, or of kindness to wife [RHC] 2) and CHILDREN would give him pleasure, without any regard t
621o046 ster, or unnatural if malevolent, or hates his CHILDREN without some passion.-- If his passions stron
624o051 strives to give conduct beneficial to all the CHILDREN, <then> each himself» & parents, & hence to
536m074 o good> «to improve his organization» for his CHILDREN'S sake & for the effect of his example on othe
371d118 it will solve them.-- It is less wonderful that CHILDS nervous system should build up its body, like
030r036 in all cases. «.Mytilus.--» «at Guacho» «on N. CHILE? Washington.--» Mem: Micaceous formation of Cho
032r040 Lorenzo; Valley of Copiapò & parts of coast of CHILE.-- Must first explain «top of» tidal band of ac
111a087 dike. V. VII. p. 316 & 328 VI. p. 365. Meyen on CHILE Geograph. Journal Vol. VII p. 216.-- Guava tree
111a087 age: many observations on heights of valleys in CHILE must be studied Analysis of Voyage: many observ
116a100 logy of Arica <Schit> Schmidtmeyer travels into CHILE p 29. gold is not sought for in Chile in beds o
116a100 vels into Chile p 29. gold is not sought for in CHILE, excepting Concepcion-- Patagonia-- Beds of La
118a103 nstitut. (1838) p. 216 M. Gay on the Geology of CHILE.-- P p217. Pentlands Fossils & Meyens --<Jura &
132a137 ld be to compare, the time of the earthquake of CHILE, with that of the passage of the moon.-- Ask Ho
148g024 lencoe in parts no trace of them-- Mem Coast of CHILE--¿ is not Mica Slate too hard & uneven to be im
217b187 rra del Fuego.-- Araucaria, species. Brazil {P} CHILE, Norfolk Isl.-- Isle of Pines-- Australia.-- A
247c028 - Antarctica «also Taeniatole austral» «caught» CHILE, Van Diemen's land & Cape of Good Hope V. p. 44
257c056 ving observed B. Tricolor in Patagonia. then in CHILE & lastly 12,000 ft above sea in Bolivia; he exa
277c126 ds (b). F. del Fuego differ from (C) Chiloe (E) CHILE. rupestris-- good species it is reverting to
320c276 is so said to have Mackenzie's Iceland Molinas CHILE Falkners Patagonia Azaras Voyages & Quadrupeds
421e091 to other side of Africa.-- (& Juan Fernandez to CHILE??) Falklands to southern portion.-- do, 269
465t080 et no shells-- now look at Scotland-- coasts of CHILE in beds of river, but in shelving «successive»
477z002 Voyage Vol. I. p. 279 Thinks the Moruffetes of CHILE different from those of La Plata or Paraguay.--
481z010 great Kingfisher of Tierra del Fuego killed in CHILE. Dobrizhoffer._ Vol I p. 310. History of the Ab
483z013 n Reptiles are described Shells from Tahiti and CHILE The North & S. Range of shells might perhaps be
523m016 y is reproduced by moral causes (ideotcy by fear. CHILE earth quakes). in people, who, probably otherwi
535m069 r.-- ¿: more poetry in that state of mind: the CHILENO says the mountains are as God made them,-- nex
028r031 ause alluvium saline; Mem; on coast of Northern CHILI as springs become rarer, so does the rain, ther
029r034 look at the cessation northwards of the Coal in CHILI as clearly bearing a relation to present positi
036r052 rocky parts of central Patagonia Does Andes in CHILI. separate geographical ranges of plants. V. Lye
042r068 of France, & unequal action of Earthquakes. «on CHILI & delta of Indus», my belief in submarine tilti
043r070 State the three «or 4» fields of Earthquakes in CHILI:-- Chiloe. Concepcion. Valparaiso (Copiapò & Gu
045r075 uption at time of great Lima earthquake» In the CHILI earthquakes if rise was more <than> inland than
045r076 1822¿ 1835?» alone, (& the general belief in N. CHILI, where rains are so infrequent; so as to exclai
046r080 omond water oscillated between 2 & 3 ft. (as in CHILI lake). Therefore motion of sea ought to be cons
052r099 ance?) 1st cases. -- The terraces in Valleys of CHILI may be with much truth compared to the step = f
058r116 on states that the same earthquake has run from CHILI to Quito a distance of more than 500 leagues. A
058r116 eagues. A little time after a bad earthquake in CHILI; Arequipa in 82 was overthrown, & 86. Lima. nex
183b051 lata Partridge «or Orpheus» was introduced into CHILI. in present states. <they> it might continue &
120a109 eposit.-- The case of the shingle in the great CHILIAN valleys must be profoundly considered. if elev
123a116 ps great beds of rivers which must be like the CHILIAN ones.-- Septemb. 2d.-- Sulphur like carbon mus
149g030 red to deposit this Remember however the great CHILIAN valley Aconuga, must there have deposited much
346d048 1. Ld. Tankerville account of wild cattle of CHILLINGHAM,-- habits peculiar,-- young one 203 days old
421e091 nd. do p. 283. on the dark ears of the wild CHILLINGHAM Cattle, with reference to Mr Bell's statemen

430e117 es his tame Hares, attacking a sick one like CHILLINGHAM bulls are described.-- His three have had VE
500q010 cross the breed-- desirable as in Cattle in CHILLINGHAM Park-- What Book on varieties &c of deer. Co
555m142 remembered Bynoes story of the women.-- The CHILLINGHAM cattle (& Porpoises) have not charity-- is i
026r020 vation must in its configuration have resembled CHILOE In De La Beche, article "Erratic blocks" not s
027r028 ed into siliceous pyritous & coaly matter. Mem: CHILOE In the endless cycle of revolutions. by action
030r034 cha & main land). <[...]> At Carelmapu.--Within CHILOE:-- On open coast, near where Challenger was lo
032r042 doubt. respecting the brecciated white stone of CHILOE, after having examined the changes of pumice a
035r046 cified" turn over silicified wood. Cordilleras, CHILOE. &c seems the organic structure most easily pr
041r066 on: the fact is in alliance with those balls at CHILOE, full of sand.--the <scale> «quantity of iron»
043r069 Costorphine hills. to compare with Galapagos.--CHILOE. M. Hermoso. & Coral reefs (imperfect in latte
043r071 three «or 4» fields of Earthquakes in Chili:-- CHILOE. Concepcion. Valparaiso (Copiapò & Guasco). ye
048r087 s. such as that in front of Sts. of Magellan In CHILOE curvilinear strata subsidence.--The sudden inc
063r130 s important, each its own limit & represented.--CHILOE creeper: Furnarius. <Caracara> Calandria: inos
072r160 tzbergen animals (?). ≠ The Hollowness of <sep> CHILOE concretions somewhat analogous to septa.-- wou
074r165 <dt> determined by fissures as in septaria. (& CHILOE case, at least corelation)--Galapagos vein. ve
075r166 ds they intersect". = In the Guatemala part. (& CHILOE do) no veins discovered. Humboldt suggests cov
172b007 apagos Tortoises, Mocking birds; Falkland Fox-- CHILOE, fox.-- Inglish & Irish Hare.-- As we thus bel
277c126 (a) Falklands (b). F. del Fuego differ from (C) CHILOE (E) Chile.. rupestris -- good species it is re
486z019 on balls made by dung beetles, like those from CHILOE. Amblyrhyncus de marlin James Isd-- Lutke Voya
552m131 vivid, «a reality» as in Spectral images-- Mem CHILOE <pi> Sow, who carried from all parts straw to
485z017 es.-- Mem Turdus Magellanicus.-- C, <Chingolo> CHIMANGO-- Diuca?? See Report <by> on D'Orbigny on spe
108a081 Soc: Geolog. Tome IX 1837-8. p. 24. rocks of CHIMBORAZO., & Pichincha. Melaphyre. = Andesite-- Albit
247c028 in Northern Hemisphere 2d in southern --p. 71 CHIMERA-- Antarctica «also Taeniatole austral» «caught
056r110 ght remain to tell of these losses.-- Cause of CHIMNEY. to crater. as at Galapagos. St. Helena.-- [Fi
585n076 s kicker but not bite.-- Henslow remarks that CHIMPANZE pouted & whined, when, man went out of room.-
587n088 y do children pout & not men-- orang-outang & CHIMPANZE. pout.-- Former, whines just like a child. Ge
596n184 of the monkeys move <its> «the» ears but <not> CHIMPAZE. does not gradation towards man.» Macacus esp
556m146 xpression of desire-- is there not protrusion of CHIN, like bulls & horses.-- 1838 good instance of u
558m152 elieve common Swan, arch raises neck & depresses CHIN-- strikes with wing arches wings-- as does blac
242c017 o Sumatra & Malacca, Borneo & Malacca «& Cochin CHINA» are said to have orang-utang & Pongo in common
269c102 seems to act per se as barrier-- Mem. Tartary & CHINA.--, both coasts of New Holland.-- «Compare bird
322c269 eas. .Octob12th.-- Sir G. Staunton's Embassy to CHINA. Oct. 12t Kotzebue's two voyages, skimmed well.
334d009 to Common & Muscovy Ducks.-- English. <Common> «CHINA» & Canada Geese, & that they this first cross w
334d013 her & younger cow.-- {P} Fox says when common & CHINA goose are crossed the neck is not intermediate
392d180 ina & Common Geese «how are their instincts?» & CHINA & Common Geese «how are their instincts?» «Chin
478z004 lower animals of Falklands &c &c Bennett on CHINCHILLIDAE Zoolog Transacts. worth reading Cuvier's Me
449e169 ng of two hatches «all alike» between the male CHINENSE & female common goose took after the common g
376d137 s & Mak the st. st noise.-- The Cercopithecus CHINENSIS: (or bonnet faced monkey he has seen do this.
204b139 Wynne distinctly says that the mixture between CHINESE & English Breed. decidedly exceedingly prolifi
275c121 am feathers at last got ducky, then took white CHINESE Bantam crossed & got some yellow & others yell
310c223 ined as now thought.-- N. American, European & CHINESE Genera & some species on Himalaya.-- some Engl
312c228 like <parents> Mother.-- like dogs Smith knew CHINESE hairless dog & common spaniel crossed.-- 3 pup
312c228 on spaniel crossed.-- 3 puppies PERFECTLY like CHINESE & 3 perfectly like spaniel even when grown up.
342d033 hybrid pheasants & grouse different.--» (if so CHINESE pigs & common must be considered as distant sp
402e018 p. 233. dogs in Borneo-- <brought probably by CHINESE>, "the breed being of the latter being the sam
431e123 true element) «give examples, pigs, with small CHINESE boars &c &c &c» p. 10 offspring take more afte
513q020 erent countries= Do the Peacocks cross.= Young CHINESE or Penguin Duck in very young state for skelet
515q022 alf-bred geese-- inter se, & with parents & of CHINESE geese. (2) Anatomy of muscles of stumps of tai
515q022 eaping of Irish Horses, bred in this country. {CHINESE Dog's Head to send Cover common Pea (& Sweet. P
571n031 ntimate connection between sound & language.-- CHINESE. simplest Language. Much pantomimic gesture??
603o012 ing wrong here.-- Origin is certainly curious. CHINESE, S. American. Polynesians Jews, African all sa
392d180 ina & Common Geese «how are their instincts?» «CHINESES & Common Pigs.--» Experimentalize on crossing o
485z017 lian Species.-- Mem Turdus Magellanicus.-- C, <CHINGOLO> Chimango-- Diuca?? See Report <by> on D'Orbi
185b055 r.-- mountain tringas.-- Upland goose.-- water CHIONIS water rat with land structures; :»law of chanc
273c113 ts on the wing & pratencole (¿connecte[d] with CHIONIS), yet the Tropic bird, has very different habi
568n018 e in music-- do monkeys howl in harmony-- frogs CHIRP in do-- union of birds voice & taste for singin
036r051 olonged). NB, Is it generally known. the acute CHIRPING sound produced in walking over the sand: I am
545m106 the teeth: «& make noise not like pish, but like CHIT-chit-chit, quickly uncovering their teeth, this
545m106 eeth: «& make noise not like pish, but like chit-CHIT-chit, quickly uncovering their teeth, this the
545m106 «& make noise not like pish, but like chit-chit-CHIT, quickly uncovering their teeth, this the Keepe
077r172 great pressure. (? heat!) unknown to us. ▌M. CHLADNI.--on meteoric Mexican stone. Journal des Mines
022r007 scales of chlorites--Mem. Maldonado P 375 Much CHLORITE in some of the dikes.--P 432. as in Andes.--
022r007 419 As Limestone passes into schist scales of CHLORITES--Mem. Maldonado P 375 Much Chlorite in some o
076r170 nuaxuato.--Clay slate. passing into talcose & CHLORITIC slate. with beds of syenite & <sep> serpentin
219b193 Guyanensis?) (Emberiza Brasiliensis?) (Fulica CHLOROPUS)--» might bring in stomach-- &c &c. (Mem disc
480z008 yanensis?)) Emberiza Brasiliensis (?)) Fulica CHLOROPUS. says some of the «species of» smaller petrel
305c211 extended relations of the individuals, whereby CHOICE with memory. or reason? is necessary.--) which
536m073 am affected by circumstances & education, & by CHOICE which at that time organization gave me to wil
612o035 ? [RHC] It is easy to conceive such movements & CHOICE, & obedience to certain stimulants without con
620o044 sensations, he will be forced to reflect on his CHOICE: an appetite gratified gives only short pleasu
022r007 a)) great variety in nature of a dike.--Mem. at CHONOS & Concepcion. P. 417 Veins of quartz exceeding
030r036 ile? Washington.--» Mem: Micaceous formation of CHONOS. interesting from great quantity of altered Ca
037r056 level surface not disturbed.--Whole West coast. CHONOS to Copiapo.--Sydney. K. G. Sound. C. of Good H
040r063 --destructive to animal life.--Patagonia In the CHONOS Islds we must imagine bituminous shales have b
458t002 oyage Vol I. p. 499. «4to. Edit»-- Horses in Lao CHOO so small, that person with long legs can hardly
228b232 ed, animals not got it, not look forward» if we CHOOSE to let conjecture run wild then <our> animals
263c075 pecies & certainty of destruction; then he will CHOOSE & firmly believe in his new faith of the lesse
316c244 gradation.-- no greater difficulty for Deity to CHOOSE, when perfect enough for future state, that wh
302c200 ed on my theory, = otherwise mere fact creator CHOOSES so to create.-- It is very remarkable, with so
526m031 - appetites urge the man, but indefinitely, he CHOOSES (but what makes him fix!? <)>-- frame of mind,
526m031 m fix!? <)>-- frame of mind, though perhaps he CHOOSES wrongly,-- & what is frame of mind owing to.--
604o016 e looked at as animal, with consciousness,, it CHOOSING food-- crawling from light.-- Yet we can spli
468t105 ith pollen-- «Thrips» about as large as bit of CHOPPED horse hair with legs & take flight-- Yet we ha
574n041 rious association: I have seen Nina licking her CHOPS.-- someone has described slovering <gum> «teeth
599o007 , when one behold the last rays of & & or grand CHORUS are utterly inexplicable-- I cannot <admit> th
541m090 to skip from subject to subject as quick as it CHOSE.-- although thinking «& talking» for the moment
075r166 s. In Peru. on other hand, mine of Gualgayoc or CHOTA & Pasco in "alpine limestone" = "The wealth of
482z011 nicus Vultus aura Excessively inaccurate Saw a CHOUETTE a huppe courte talks of nine terrestrial Turd
543m099 te fool used to play regularly with D'Arblay of CHRIST of great genius, & yet invariably used to beat
308c217 pecies means» civilized Man, May exclaim with CHRISTIAN «we are all» Brothers in spirit-- all childre
581n063 it. as my father trying to remember the man's CHRISTIAN name, writing for the surname,, analogous to
610o031 any letters to»-- could not <remember> <read> CHRISTIAN name; fancied it looked like. W. but conclude
610o031 ead of Wilson, referred to Robert & found his CHRISTIAN name was Wilson!!-- How curious an inward. un
615o038 er in some than others-- Hence superiority of CHRISTIAN over Heathen race.-- But as no great modifica
065r140 2d voyage outside coast of T. del Fuego. off. CHRISTMAS sound.-- «(Think some 60 fathoms, none thick
108a081 e-- Albite & amphibole= Cook found Granite at CHRISTMAS Sound Vol XIV. (My Edition) p 500. Well descr
074r164 .. fluor spar. bayte. asbestos garnets.--carb & CHROM. of lead. opriment. chrysop[r]ase. opal:-- Vein
310c225 t of extreme difficulty in mundine geological CHRONOLOGY Annals of Natural History «Vol I??» p. 318.
398e004 forms, that the whole value of the geological CHRONOLOGY depends, that most sublime discovery of the
419e087 ecies.-- It must never be overlooked that the CHRONOLOGY of geology rests upon amount of physical cha
419e087 atastrophes, which must serve to confound our CHRONOLOGY» «CONSIDER ALL THIS» Extinction & transmutat
579n059 hen a man keeps perfect. time in walking, to CHRONOMETER, is seen to be muscular movement. The Blushi
074r164 garnets.--carb & chrom. of lead. opriment. CHRYSOP[R]ASE. opal:-- Veins in Limestone & Grauwacke: Si
042r067 ia Blanca, mention black scoriaceous rocks of R CHUPAT. & fall of Ashes of Falkner, ¿how far is the d
070r155 to near Salta. & not far from Tucama[n]. & at CHUQUISACA. half across the continent.--He states plain
571n031 n great avenue, resemblance to gloomy aisle of CHURCHE.-- these are Mayo's ideas.-- In language, the
393d1BC passion.-- Disposition of half bred Cattle at CINBERMERE? How is Jackall & dog at Z. Gardens D E Fini
411e056 & yet sexual apparatus.-- My account of Circus CINEREUS of the Falklands Isld. is interesting as show

```
477z003 uca. "Les individus ailes peuvent avoir <quatre> CINQ lignes de long et volent. p. 208 Fleas only app
243c019 w Zealand absolutely different.-- --Philedon CIRCINNATUS not found in Australia only New Zealand-- No
041r086 Red  Sandstone. at Bahia in modern sandstone. a CIRCLE,.{P}, had in its middle a short <fissure> «vei
230b240 turned.-- Hence extreme difficulty, argument in CIRCLE.-- Falkland Isd case good one of animals not s
285c150 ficult to arrange children of same parents in a CIRCLE.-- & «hermaphrodites» father « <mother> & gr
349d051 atomiadae».-- when wild ornithorhynchus come in CIRCLE?!!! p. 8-- Anomalous structures, as in Hippota
352d058 central radiating point, all united . (links in CIRCLE must be granted unequal, because fossil) Now w
352d058 se fossil) Now what is group without centre but CIRCLE, two or three lines deep-- with respect to Mac
177b027 in  Mammalia «birds» it would only appear like CIRCLES;-- & insects amongst articulata.-- but in lowe
281c139 ing, affinities would be <circular>, in broken CIRCLES.-- which in each group is quite fatal.-- «Rela
283c144 statement about diversity of forms in aberrant CIRCLES.-- explained by such not having been long in b
288c158 s of his series, ie, cannot be discovered till CIRCLES completed Major Mitchell, does not know whethe
302c202 of last page. <an> osculant groups between two CIRCLES «of equal value» must be so from characters of
110a086 not  crystallied Salband like basalt. full of CIRCULAR cryst of glassy felspar different from either
119a104 reat chains like Andes or Himalayas, but great CIRCULAR mountains, yet so analogous, that as we see m
281c139 &c  &c.-- V. p. 140» I should think meaning of CIRCULAR arrangement was only so far true as avoided l
281c139 the central twigs dying, affinities would be <CIRCULAR>, in broken circles.-- which in each group is
309c222 ackneyed cases Mr Ed Blyth does not believe in CIRCULAR or linear arrangement.-- Thinks passages very
347d050 us or group of any kind not being perfect till CIRCULAR. p. 5 Most clearly shows that genus expresses
349d051 ief divisions.# p. 7. «In» Some <of> cases the CIRCULAR arrangement from fewness of forms-- Cannot be
349d052 Edwards, thinking any group good, though not CIRCULAR, if characters can be established-- clearly s
352d058 - PURE HYPOTHESIS be careful.-- Argument for CIRCULARITY of groups. When <species of> a group of spec
132a139 nt added to bottom) be caused, by absence of CIRCULATING water.-- & therefore that temperature of ear
032r041 qual distances from centre of rotation» & a <CIRCULATION owing> rotation in fluid matter of globe. mu
032r041 n fluid matter of globe. must there not be a CIRCULATION «however slow & weak.»; «(cause of not accum
056r111 ccas, which is all of granite: In discussing CIRCULATION of fluid nucleus,--the similarity of Volcani
056r111 ucts being similar over whole world, general CIRCULATION. But Volcanic action separates some  sulphur
125a121 e temperature may be kept up far higher from CIRCULATION of heated fluid or gases under pressure.-- {
132a138 . Is not cold of ocean accounted for, by the CIRCULATION being greater, than the transmission from oc
542m094 ical relations.-- CD[like sighing to relieve CIRCULATION after stillness.-- Now I conceive if organiz
590p093 from  grief. is method of increasing languid CIRCULATION-- no, for <grief> sighing comes on before ci
590n093 on-- no, for <grief> sighing comes on before CIRCULATION is affected. p. 44.-- Jealousy. causes spasm
590n093 causes  spasm in bile duct, & throws bile in CIRCULATION p. 75. Haller says tooth ache, even from car
126a124 pecially at bottom of great ocean, where the CIRCULATIONS from surface can take place.-- \ the depth
452e181 m Gilolo p. 253 In isld of Bunwood (18 miles in CIRCUM] there are hogs & monkeys <at> near shore of M
308c218 in" &c &c then centre is every where & then CIRCUMFERENCE no where-- as long as this is so-- !!Metaph
293c173 eb footed yet with much practice & led on by CIRCUMSTANC it becomes web footed, now Man by effort  of
051r095 ment on the copper bottom. we see a trifling CIRCUMSTANCE determines whether an animal will adhere to
171b003 hanged or subject to variety, according <to» CIRCUMSTANCE,-- seeds of plants sown in rich soil,  many
187b063 rent quarters.-- Will Mr Lyell say that some CIRCUMSTANCE killed it over a tract from Spain to S. Ame
207b150 . Fossil Infusoria found of unknown forms, a CIRCUMSTANCE undiscovered by Ehrenbergh.-- «Marcel Serre
236b280 ary a little, but such should not be general CIRCUMSTANCE.--In. insects «in England» surely it is not
256c054 ,-- Galapagos, S. American-- -- genus.-- The CIRCUMSTANCE of having two sexes is the check to distrib
257c057 arly» same kind country distant. <Study> The CIRCUMSTANCE of ground woodpeckers.- birds that  cannot
265c087 cted with habits, & how much heredetary. The CIRCUMSTANCE of aberrant groups being small it is truism
275c120 t I state.-- &c picking varieties. unnatural CIRCUMSTANCE Ld Orfords had breed of greyhounds fleestes
367d113 ength <of> «of make in» the males; & another CIRCUMSTANCE, perhaps, equally so, is this strength bein
398e004 roduce in climate vegetation &c.-- It is the CIRCUMSTANCE of small physical changes & oscillations, n
411e054 external  world, & every possible contingent CIRCUMSTANCE.-- \the laws of variation of races, may be
550m127 tions, which are pleasant, every concomitant CIRCUMSTANCE calls up pleasure. or pleasure or pain of a
621o047 is  accidentally acquired from some trifling CIRCUMSTANCE.-- Thus a child may be taught to think almo
179b034 ild animals in this respect are differently CIRCUMSTANCED.-- ¿Is this shortness of life of species in
235b262 ing great tendency to vary? Is not man thus CIRCUMSTANCED; varieties of dogs in different countries a
443e153 stock  greatly affects the Graft.-- \Plants CIRCUMSTANCED as the Gorze must be propagated by its root
046r078 s, are granitic materials simply altered by CIRCUMSTANCES; & not in chemical nature, or has a subterr
058r115 electricity would affect this. -- State the CIRCUMSTANCES of appearance at Concepcion [.] no sign of
062r129 tinct Llama owed its death not to change of CIRCUMSTANCES; reversed argument. knowing it to be a dese
062r129 finite time:--not extinguished by change of CIRCUMSTANCES: The same kind of relation that common ostr
063r130 .'» not gradual change or degeneration. from CIRCUMSTANCES: if one species does change into another it
151g039 , best developed on steep earthy slope, two CIRCUMSTANCES rarely united.-- die away also, without any
171b002 bject to cycle of change, temperature & all CIRCUMSTANCES; which influence living beings.-- We see <li
172b007 ng enough.-- «apart, with slightly differen CIRCUMSTANCES.--» Now Galapagos Tortoises, Mocking birds;
177b028 cies, so have the awks, there is particular CIRCUMSTANCES, to which> is it an index of the point when
180b038 ced by <inter> confined breeding & changing CIRCUMSTANCES are continued & produce according to the ad
180b038 produce according to the adaptation of such CIRCUMSTANCES & therefore that death of species is a cons
180b039 d appear from America) of non adaptation of CIRCUMSTANCES.-- Vide two. pages back. Diagram The largen
193b090 nd varieties, (influenced itself perhaps by CIRCUMSTANCES) & those alone preserved which are well ada
227b228 o closest examination of hybridity «to what CIRCUMSTANCES> favour crossing & what prevents it--» & gen
248c033 ich are less obviously affected by external CIRCUMSTANCES> these therefore will be chiefly heredetary
252c045 tematic naturalists get clear indication of CIRCUMSTANCES in Geography to help in distinguishing empi
258c059 lowness, even where isolation, from general CIRCUMSTANCES effecting the area equably.-- Animals havin
263c078 ly. that <I> it does not stagger me.-- What CIRCUMSTANCES may have been necessary to have made man! S
280c134 imals, many species in one genus-- external CIRCUMSTANCES in both cases effect. it.-- Sir J.. Sebrigh
285c153 st especially under care of Man. & external CIRCUMSTANCES not variable.-- Animals have voice, so has
297c185 character,  divided into many small genera :CIRCUMSTANCES not favourable to many species., same circu
297c185 ances not favourable to many species., same CIRCUMSTANCES, which by causing death, makes the group a
304c208 only  guides.-- Yet trifles are produced by CIRCUMSTANCES. Spines on Echidna.-- when it can be traced
304c208 then  probably heredetary & not produced by CIRCUMSTANCES in ostrich which is not isolated, we must s
314c236 final cause is the adaptation of species to CIRCUMSTANCES by principles, which I have given ∴ Those a
316c246 nces.-- very good anomaly in range + + What CIRCUMSTANCES have led to formation of new species some f
338d023 sport in the species itself, & in the local CIRCUMSTANCES of the two countries in times present & pas
365d106 the cross would have adapted it to changing CIRCUMSTANCES.-- More probably during known changes clima
371d128 y.-- I presume most of these roses, without CIRCUMSTANCES very unfavourable, will <deteriorated> cont
377d139 t in making change, only in fixing it: only CIRCUMSTANCES «alone» make changes or species!! CD[The vi
391d174 failing, generation fails.-- How completely CIRCUMSTANCES during life.-- if the circumstances which i
391d175 fect of differences of parents, or external CIRCUMSTANCES which induce «which must be external» chang
391d175 xternal circumstances during life.-- if the CIRCUMSTANCES; now we know how slowly & insensibly such c
397e004 tion of whole of one race to some change of CIRCUMSTANCES of times. & from persistency «owing to thei
412e057 long  as each shall be perfectly adapted to CIRCUMSTANCES, as shown by difficulty in forging, yet han
420e089 ndwriting is determined by most complicated CIRCUMSTANCES According to my theory no land animal  with
432e124 owing  to its adaptation to the surrounding CIRCUMSTANCES» <April 12th..> Cestracion, Port Jackson Sh
436e135 in  numbers, but not so much these, because CIRCUMSTANCES, time of domestication [see Wikinson on dog
438e142 ecies, just the amount of may depend on many CIRCUMSTANCES favouring the reappearance of characters, f
439e142 ve.-- Nevertheless much probably depends on CIRCUMSTANCES. results of complicated laws of organizatio
440e146 ust be effect of some condition of external CIRCUMSTANCES,;-- or the will of the Animal. p. 145. Seem
444e157 theories;-- that species are the result of CIRCUMSTANCES than it would be eliminated, & hence, the a
461t037 ture would rather become accomodated to new CIRCUMSTANCES» as Hyacinths in glasses &c &c  Experiments
496q05a «Azaleas» & plants grown under unfavourable CIRCUMSTANCES, which are modified by «which that time
526m030 t to certain actions, which are modified by CIRCUMSTANCES, & thus the appetites themselves become cha
536m073 that organization may have been affected by CIRCUMSTANCES & education, & by choice which at that time
545m107 atch my face. The ourang outang, under same CIRCUMSTANCES, threw itself down on its back & kicked & c
547m112 arallel trains of ideas connected with past CIRCUMSTANCES.-- as whether I really was going to Shrewsb
547m113 one has rung the bell.. when one recollects CIRCUMSTANCES were such one naturally would so so!) Now a
549m122 r his happiness to check-- that is external CIRCUMSTANCES are so conditioned as they are effecting  a
576m047 ccount of dark ages.-- «effects of external CIRCUMSTANCES» Look at Spain now.-- man's intellect might
576m047 y to progression, excepting from favourable CIRCUMSTANCES!» We must believe, that it require a far hi
583p070 form. (& especially as it adapt its cell to CIRCUMSTANCES); it must have impulse to make a cell in ce
594n115 othing. & so manifesting sulleness. [blank] CIRCUMSTANCES having given to the Bee its instinct is not
```

607o025 ssions-- accidental (so called like chance) CIRCUMSTANCES. As man hearing Bible for first time, & gre
609o029 were necessary to the vice of intemperance, CIRCUMSTANCES made the check.-- to licentiousness jealous
622o048 this teaching may be curiously modified by CIRCUMSTANCES of country, so will the conscience in these
623o051 ; by my theory they have been formed by the CIRCUMSTANCES, which have led to the peculiarities, & hen
634j54v plan once adopted; & it is from these very CIRCUMSTANCES, that we become satisfied respecting an ori
639j28v or life, only two of them live to breed, if CIRCUMSTANCES determine that, the long legged one shall r
411e056 oduced. & yet sexual apparatus.-- My account of CIRCUS cinereus of the Falklands Isld. is interesting
350d055 onfined to annulosa?) Remarked that young of CIRRHIPEDES can move & see, parent fixed,-- young of spo
413e060 ntermediate causes" The Sexual system of the CIRRHIPEDES is the more remarkable from their alliance t
381d156 ans formed to fecundate female (as in plants) CIRRHIPEDS rotifers, trematode & cestoid Entozoa Allotr
337d022 mestica of North Europe is replaced by the F. CISALPINA in Italy, which is so like that difference wo
482z011 f passage!! sylvia macloviana, 2d like sylvia CISTICOLA.-- Embriza melanodera-- a linnet not caught.-
468t111 ine two species of Larkspur -- two varieties of CISTUS Speedwell to Rhododendron-- <Loasa> «Anchusa»-
564n005 of the mind cannot be solved by attacking the CITADEL itself.-- the mind is function of body.-- we m
507q15v Horticultural Variation? Henslow knows only on CITRONS 47. Ficus carica Henslow presumes females prod
295c178 a of Pachydermata &c & other Mammals.-- «otter; CIVET cat, rodents.» (Pachyderm in Portland stone of
266c092 ils, which disfigure those of Arabia & Egypt.-- CIVETS CATS only wild animals on isld.-- Niether Hyen
262c072 hewell «in Comment/ few will dispute--» says CIVILIZATION heredetary; ie instincts of wisdom virtue?
581n065 - thinks so it must have been in the dawn of CIVILIZATION-- thinks many words, roar, scrape, crack, &
609o29v esent day & not to ruder state of Society.-- CIVILIZATION is now altering these instinctive passions-
624o051 all the world.-- As conditions change, from CIVILIZATION, education changes, & probably likewise ins
171b004 nct becomes influenced.-- child of savage not CIVILIZED man.--birds rendered wild <through> generatio
308c217 l, unless it were fixed what a species means» CIVILIZED Man, May exclaim with Christian «we are all»
311c226 strified, which last produced --insane men in CIVILIZED countries-- this is well worthy of investigat
350d054 God" Septemb I,. It has been argued Man first CIVILIZED. <note> add this in note. ¿mere conjecture?--
409e047 t of Tierra del Fuego is to be converted into CIVILIZED man.-- ask the missionaries about Australians
540m087 at them both savage-- look at them both semi-CIVILIZED-- Perhaps one cause of the intense labour of
553m136 ult to prove that» this innate idea of God in CIVILIZED nations has not been improved by culture « <w
609o30v parately considered.-- The difference between CIVILIZED man & savage,-- is that former is endeavoring
309c220 Educate all classes-- avoid the contamination of <CL> castes. improve the women. (double influence) &
267c093 Guava introduced from Norfolk Isld-- "& it now CLAIMS all the moist & fertile land of Tahiti, in spi
251c041 Mam: found in Sumatra p. 452 Append to Denham CLAPPERTON &c on Mammalia no doubt will all be included
563n001 shows it is evergreens they seek Cock Pheasant CLAPS his wings before? crowing & only in breeding se
327cIBC ary of Useful Knowledge on Horse & Cow & Sheep CLARKE'S Travels.-- Temminck Hist. Nat. des Pigeons et
513q21. e stigmas expanded-- in reference to Lobelia & CLARKIA-- Peas time of impregnation.-- says many flowe
376d137 y to pull up petticoats., & if woman not afraid CLASP them round waist & look in their faces & Mak th
567n014 g The Cyanocephalus when fondling the keeper., CLASPING «& rubbed» his arm. & show signs of affecting
176b023 & water, & the endeavour of each <one> typical CLASS to extend his domain into the other domains. &
177b029 ndent on genus,. that genus upon another, whole CLASS would die out, therefore not.-- Monad has not d
198b113 r nervous developed. (Owen's idea) states these CLASS approach on the confines? Balanidae?-- ---I ca
254c048 he greatest rapidity"-- so we find species each CLASS successively present modifications, typical of
255c051 dgeons back.-- According to this description of CLASS is description of ancestor of all birds, & so f
258c061 cocks all warlike)» «thiis wars against in any CLASS:, those points which are different from each ot
258c061 ifferent from each other, & resemble some other CLASS analogy» The resemblances relationship, the dis
288c100 plained by mere chance?-- or it like each great CLASS of animals having its aquatic, aerial &c type?-
277c126 f course distinct. analogy from every country & CLASS tells us that ¿O. Modulator &. O. Patagonicus. t
301c200 the degree of development of all animals of same CLASS being about equal.-- organs of generation about
310c222 ediate groups often have full structure «of one CLASS» & full of second--this class of facts «analogo
310c222 structure «of one class» & full of second--this CLASS of facts «analogous to petrel-grebe. external>
372d131 is often thought wonderful. it is part of same CLASS of facts that the skin grows over a wound.-- Do
404e025 iration, or rather ventilation peculiar to <the CLASS> «some orders» of crustacea, one is tempted to
404e025 imple modification of an organ present in whole CLASS. Case of Mexican greyhounds.-- young being habi
427e109 being effect of abortion of one sex.-- Linnaean CLASS Dioecia & Monooecia. ought to be preeminently a
484z015 rds of Europe for other representatives of this CLASS--. Pyrocephalus & many Tyrannulae-- replaces wa
507q15v e) Smith says many trees in Tropics are of this CLASS.-- (48) .Where «published» list of spontaneous
527m033 air-- Catherine often, but not of an inventive CLASS.-- Now that I have a test of hardness of though
584n072 distance in dark "it is inspiration."-- this is CLASS of so called instincts to which my theory no wa
177b027 - & insects amongst articulata.-- but in lower CLASSES, perhaps a more linear arrangement.-- ¿How is
192b086 l those intermediate steps especially in those CLASSES where species not numerous. (NB in those class
192b086 asses where species not numerous. (NB in those CLASSES with few species greatest jumps strongest mark
192b089 ncus. If this last animal bred-- might not new CLASSES be brought into play.-- The father being clima
234b261 deeply [not located] of <all> genera, «in all CLASSES» are not a few only cosmopolites, & in genera
254c048 l, low Cerebral structure??-- «do» p. 390. All CLASSES of Acrite exhibit lowest stages of animal orga
254c048 ogous to the earliest conditions of the higher CLASSES, during which the changes of the ovum or embry
254c048 y present modifications, typical of succeeding CLASSES & likewise those much higher in scale. So Owen
255c050 species, instead of analogues.-- «as» in other CLASSES this evidently relates to greater range of suc
271c106 ue.-- it is expression for ignorance Two grand CLASSES of varieties.; one where offspring picked, one
291c167 a) is absolutely necessary to explain genera & CLASSES. if extinct forms were all fathers of present,
302c201 me having an intermediate affinity between two CLASSES.-- there may be some descendant of some interm
302c201 e line.-- the only connection between two such CLASSES will be those of analogy, which when sufficien
304c206 e believed to past into each other-- Different CLASSES Keep to their types. with different degrees of
305c210 ssion in the developement in instincts in the <CLASSES> «orders» of insects, so is there none of reas
307c216 argument is extended from species to genera & CLASSES. p. 479. fragment of tusk «& Molar tooth» of H
309c220 als-- argue «opening» case. «thus» Educate all CLASSES-- avoid the contamination of <cl> castes. impr
309c220 «extinct?» genera of shells.-- duration in two CLASSES however different.-- Male glow-worm knowing fe
337d021 k to first stock of all animals, but merely to CLASSES where types exist for if so. it will be necess
337d023 b: of <ani> organic beings.-- Animals «of same CLASSES» differ in different countries in exact propor
403e022 r even colour. is good. (ie invariable) in some CLASSES-- it is because every part is under change, no
423e096 , it proves the law of developement in partial CLASSES is far from true.-- I doubt not if the simples
448e168 > equally aberrant-- the two former connecting CLASSES like Toxodon <In orders>-- Fish & reptiles in
530m044 r?) that pulse of new born babies of labouring CLASSES are slower than those of gentlefolks. & that p
603o013 uence» hence languages become corrupt, & whole CLASSES of words «are abbreviations» he thus derives f
609o030 he female October d. 1838 perhaps insist?? Two CLASSES of moralists: one says our rule of life is wha
634j55r - NB) The explanation of types of structure in CLASSES-- as resulting from the will of the deity, to
287c158 the laws of change in organization. But the CLASSFICATION must chiefly rest on these same organs,-- h
292c170 a very superficial knowledge «like myself» of <CLASSI> real affinities. ie structure of the whole an
117a101 Bull: Soc. Geolog. 1837. December. p, 91. a CLASSIFICATION of Europaean strata according to compositi
136a147 operly remark on the superiority of Lyell's CLASSIFICATION to that of Phillips as given p. 13. Vol II
227b225 far chief element:. little service habits in CLASSIFICATION. or rather the fact that they are not far
264c082 I see quite recognizes habits in making out CLASSIFICATION of birds Birds vary much (more than shells
277c128 will be slow. but must take place-- such a CLASSIFICATION would answer every purpose, & would presen
281c138 ribable, yet holds good, so does it in real CLASSIFICATION The relation of all cock birds in Gallinac
281c138 en relationship of all men now living & the CLASSIFICATION of animals-- talking of men as related in
286c155 what is the natural arrangement, it is the CLASSIFICATION of <arrangement> relationship; latter word
286c156 onsidering fossil animals, what relation in CLASSIFICATION in books, ought they to hold,-- Birds havi
287c158 s may be all disputable, but the one end of CLASSIFICATION to express relationship. & by so doing dis
290c162 » insects-- & vertebrata & plants. At first CLASSIFICATION on generation might appear an analogy NB P
291c165 solved" habits become important element in CLASSIFICATION, because structure has tendency to follow
291c165 D'orbigny point out importance of habits in CLASSIFICATION.-- Thought (or desires more properly) bein
292c170 ent let him look at abstract of Swainson on CLASSIFICATION "Let anyone even with a very superficial k
292c170 le animal let him read Mr Swainson's on the CLASSIFICATION of animals. & observe the character of the
294c177 same time true species.» & its adaption to CLASSIFICATION & affinities, its extension.-- Von Buch. T
299c195 pigonous. Perigonous &c-- very important in CLASSIFICATION. here we have generative organs. first cha
302c202 st opposition in opinion on all subjects of CLASSIFICATION, I must work out hypothesis, & compare it
311c226 's land diff.-- Habits can only be used «in CLASSIFICATION» as indication of structure (including bra
384d161 - Hunters Animal OEconomy. (by Owen) p. 44. CLASSIFICATION of Monsters. (1) From praeternatural situa
483z013 pi & is especially grand paper. p. 387. "on CLASSIFICATION of such animals.."-- Voyage. Coquille's Vo
199b117 of the thousands of forms, should they all be CLASSIFIED.-- Propagation explains this.---- Ancient Fl

268c096 y) do. do. on the genus Procyon.-- by Wiegman CLASSIFIED catalogue of animals of Nepal read before Li
381d156 my by, Hunter. (edited by Owen) p. 34.-- Owen CLASSIFIES Hermaphrodites. Cryptandrous. (only female o
266c088 ntry many birds come it would be impossible to CLASSIFY them.-- I would [not located] Musalman's of t
555m142 Sept. 16th Zoological Gardens-- Endeavoured to CLASSIFY expressions of monkeys-- I could only perceiv
543m101 highest intellectual powers of perceiving & CLASSIFYING distinct resemblances.-- The facts of half i
228b229 have before him, when dissecting a whale, or CLASSIFYNG a mole, a fungus, or an infusorian. is "What
341d030 crawling bird.-- Wings reduced to rudiment.-- CLAVICLE scapula &c strongly developed to aid in breat
509q017 Turbinated bones? False ribs Wings of Apterix: CLAVICLE in--? Combs in combless Poultry-- Teeth in fo
274c113 is wonderfully shown in long legged cuckoos with CLAW like lark, (one in Australia is called swamp ph
274c113 alled swamp pheasant) Goatsucker--, parrots with CLAW like lark (NB The La jeune veuve parrot though
372d129 ail of Planaria, the whole grown to that part.-- CLAW added to crab, tail to lizard,-- healing of wou
372d130 ng to do with generation.-- Why crab can produce CLAW. but man not arm. hard to say-- if it were poss
021r005 «Vol I 287» P 379. Henslow Anglesea, nodules in CLAY Slate. major axis 2.½ ft.-- singular structure
048r084 rating by action of Organic power a lump of hard CLAY. -- In the History of S America we cannot dive
049r087 quartz veins, there contemp--yet similar ones in CLAY. Slates contemporaneous others subsequent. as i
051r095 ck because it does not decompose. or vice versâ. CLAY slates unfavourable to attachment of many bodie
074r165 ith other» «(state simplest case. concretions of CLAY iron stone; iron pyrite in a fossil» Insist str
076r170 e silver in S. America. Geology of Guanuaxuato.--CLAY slate. passing into talcose & chloritic slate.
077r171 ny» «other secondary» rocks. Vein traverses both CLAY slate, Porphyry North 52 W, & is nearly the sam
089a020 of Cayenne. Syenite & diorite, covered with iron CLAY common to Guyana said to extend to Cordillera I
091a027 so. «Campbell the Poet» Accra. Coast of Africa. CLAY Slate & Quartz. strike SSW & NNE dip 30 degree
091a027 tion Hills & strata SE. direction of transitions CLAY slate &c nearly vertical Linear earthquake 500
105a070 Soc. Geolog «1837» p. 320. paper on shrinking of CLAY. applicable to Cleavage. C. Prevost.-- In Cordi
114a094 n stone alternating. bear on subject of cleavage CLAY slate. a distinct formation deep «& therefore e
114a094 r in Secondary in Europe. gneiss-- metamorphosed CLAY slate.-- --shale in shall sea. Lyell confounds
114a095 not tell period.-- I cannot help suspecting that CLAY-slates have been more frequently metamorphosed
121a111 r. J Hall Vol VI. p 173. (Ed. Transact) has seen CLAY stiff enough <to form> for potters to use. in w
134a141 few «living» shells. on coast of do p 8.-- soft CLAY beds near C. Virgin p. 59. dip of Clay slate in
134a141 8.-- soft Clay beds near C. Virgin p. 59. dip of CLAY slate in T del Fuego Admiralty Sound. SE dip. m
200b121 f common branching off.? Accra, Coast of Africa. CLAY slate. strike. SSW. & NNE. dip 30 degree - 80 d
202b131 st trees fallen «kind well known, carbonized»--; CLAY fifty feet, then forest 120 ft Micaceous rocks.
295c178 ike.--?? Memoire by Charles D'orbigny on Plastic CLAY of Paris contains many genera of Pachydermata &
486z020 bed in Suffolk as lying under strata of gravel & CLAY about 10 feet in thickness.-- (March, 1842) Que
075r166 is contained in a primitive slate, covered by a CLAYEY porphyry, containing grenats. In Peru. on othe
048r086 phyritic rock. s (mem white tufas with purple CLAYSTONES of P. Desire). = Where talking of such subst
532m056 me that Nina, when brought from Shrewsbury to CLAYTON, (though so fond of her & of servant of Richar
055r106 ses above the sea. = talks of them being packed CLEAN. & without earth.--Moreover that such do not oc
472s02v dusted.-- <I think> When It first alights, it CLEANED sucker & <I think> pollen was scraped off, whi
366d111 ciation very disagreeable hearing maed servant CLEANING door outside, as often as she touched handle,
029r033 . N: longitude 61 deg. 22 min. W. mid. calm and CLEAR. Caermarthen Journal I look at the cessation no
031r038 l Notes. at St. Helena. This structure was very CLEAR at base of great lava cliffs [Fig. I] line of h
042r088 forphyry has been upheaved in a dry form It is CLEAR the forces have acted with far more regularity
077r172) resist oxidation?-- Mem Sir W. P. stone It is CLEAR to me, there are laws of solution & deposition
095a039 ers of the Bonite. French discovery ship, found CLEAR proofs of shells & waterworn rocks «at Cobija.»
116a100 bed of river. formed of rounded pebbles-- it is CLEAR gold occurs in submarine alluvium, or sublittor
150g035 n plain red talus line on N. side of Spean most CLEAR & upper line running up great bight just as Dic
161g095 ost by slope, then concealed by fragments, then CLEAR. this bit to eye certainly appears level with r
252c045 re [not located] The systematic naturalists not CLEAR indication of circumstances in Geography to hel
274c115 tion in Museum-- I could not discover any other CLEAR relations besides aerial, & terrestial,-- How i
274c116 e with Waterhouse & <birds> Mammalia.-- We have CLEAR indication [not located] alone, but on all the
275c119 - Lamarck was the Hutton of Geology. he had few CLEAR facts, but so bold in many such profound judgme
333d006 c to the half bred Persian.-- Here then we have CLEAR case of heterogeneous offspring from one impregn
338d024 nsact.-- p. 297. Vol 9. Dr. Ferguson seems most CLEAR that the ideosyncracy of the Negro (& partly Mu
379d151 whilst Sir J. himself is not like-- now this is CLEAR case of avitism. but then ¿ was «not» the expre
400e011 -- because found in all quarters: his ideas not CLEAR. In Australia from approach to Asiatic [...]t i
421e091 nimal, I have always regarded as affording very CLEAR indications of its true affinities. We are leas
421e091 daptive to an essential character--" How little CLEAR meaning has this compared to what it might have
423e097 & its one carnivorous devourer.;-- it is quite CLEAR that a large part of the complexity of structur
528m039 very much to do, as may be known by autumn, on CLEAR day.-- 3d pleasure association warmth, exercise
547m111 fell to sleep for second & wakened.-- had very CLEAR & pretty vivid «& perfectly characterized» drea
571n027 The existence of taste in human mind. is to me CLEAR evidence, of the general ideas of our ancestors
591n100 &> against benevolent & parental instincts very CLEAR.-- even to the cold or benevelo- continent man
600o08v some transcendental kind-- (5) Conscience, not CLEAR-- Then these last heads. of separation between
600o08v between soul of man. & intellect of beasts, not CLEAR.-- ¿does not Mackintosh make great difference b
415e067 atal diseases, & such constitutions only being CLEARED off by fatal diseases.-- The Value of a group
604o017 e <between> when insanity is coming on «Thinks CLEAREST analogy between dreams & insanity.» D. Stewar
029r034 e cessation northwards of the Coal in Chili as CLEARLY bearing a relation to present position of <Coa
035r045 onnected with & alternating with obsidian must CLEARLY be chemical differences. & not those of rapid
086a007 d regulated by atmospheric currents?-- chiefly CLEARLY by sun's position = If equatorial streams of w
094a036 d region of great boulders, pebbles of granite CLEARLY effect of remodelling same manner. as bits of
113a091 LOW but successive transmission of temperature CLEARLY prove possibility of metamorphic theory On the
191b083 ion: Pray ask Dr. Smith.-- «to state that most CLEARLY». Fox tells me, that beyond all doubt seeds of
240c004 us to what occurs in plants.-- All these facts CLEARLY point out two kinds of varieties.-- One approa
257c056 s to other view.-- Till we know uses of organs CLEARLY, we cannot guess causes of change.-- hump on b
257c057 odpeckers.-- birds that cannot fly &c &c. seem CLEARLY to indicate those very changes which at first
264c081 ribe,-- structure without corresponding habits CLEARLY showing true affinity, for instance tail of gr
265c086 us bill & nostril of Puffinuria I think we may CLEARLY attribute to heredetary origin & not adaptatio
265c087 some use.-- Nature is never extravagant though CLEARLY not of the use to which wings are generally ap
280c134 instance-- instincts of many kinds in dogs, is CLEARLY applicable to formation of instincts in wild a
318c254 by night.-- others in flock, these birds seem CLEARLY directed by kind of country; «kinds of migrati
347d050 ind not being perfect till circular. p. 5 Most CLEARLY shows that genus expresses as now used almost
349d052 t circular. if characters can be established-- CLEARLY so.-- NB. This paper worth referring to again.
375d135 - yet until the one sentence of Malthus no one CLEARLY perceived the great check amongst men.-- «Even
380d153 ativa goes entirely to the offspring-- this is CLEARLY the converse of annual being rendered biennial
407e042 ltiplied, specifically & individually.-- I see CLEARLY from F. R. it will be highly necessary to show
434e130 imes bears female flowers, the organs are most CLEARLY abortive, so that they become so by suppressio
486z018 her order is comparatively rare.-- These views CLEARLY explain rarity of insects in T. del Fuego.-- H
542m095 s, & as attention would amongst lowest savages CLEARLY be directed chiefly by objects of vision.-- Do
544m103 es tell us it is not real.= = dreaming appears CLEARLY rest of the mind, with all other faculties: «V
564n006 contracted between eyes & upper lip., is most CLEARLY analogous to a panther I saw in garden uncover
565n006 n as one ceases, or stops the noise , the face CLEARLY passes into smile-- laugh long prior to talkin
581n063 has given the phrase "heredetary habits." very CLEARLY, all I must do is to generalize it, & see whet
586n080 cts point out some essential difference, which CLEARLY ought to be separated-- We apply instinct to o
587n089 ntiguity of parts of Brain.-- Mackintosh first CLEARLY insisted on assoc of ideas & emotions. rather
589n092 our faculties have been given us to exist, is CLEARLY seen. in the absurdity of a tree having reason
617o39v ect of thought is a question which ought to be CLEARLY comprehended by anyone who wishes to fully und
627o053 ng its instinct,-- as young pointer to point-- CLEARLY shows this is true. p. 13. Affections cannot b
398e005 s & to <ponder> conceive the result with that CLEARNESS of conviction, absolutely necessary as the «b
592n103 all senses: <do> Horse prick his ears «& snort CLEARS nostrils» when frightened, does not hair & rab
021r005 ame as» of slate same.--longer axis in line of CLEAVAGE. laminae fold round them; Quote this. Valpara
026r021 ks The frequent coincidence of line of veins & CLEAVAGE is importants; veins appearing a galvanic phe
026r021 ng a galvanic phenomenon, so probably will the CLEAVAGE be There is a resemblance at Hobart town betw
031r038 lowered again & covered no sign of upheaval To CLEAVAGE add other instances in old world of symetrica
032r041 bowels of earth cause:--<exp> does not explain CLEAVAGE lines./ possibly general symetry of world.--
053r101 Earthquakes in the Radack & Ralix Islds? In my CLEAVAGE paper Dr Fittons Australia case must be quote
054r102 e & coast Granite: P. 199; Falkland account of CLEAVAGE differs wonderfully from mine: phyllade cover
054r102 ms of stones agrees with mine.--At Conception, CLEAVAGE E & W! at Payta. talcose slates, do at latter
098a045 ry.-- it may <of> come of use in discussion so faces CLEAVAGE &c Geolog Transacts. Vol III. p I. p. 86. et

098a046 re connected with a crystalline process.-- now CLEAVAGE as suggested by Sir J. Hershel is all crystal
099a049 e no effect, on particles of equal weight.--)¿ CLEAVAGE not vertical ∵ combined with gravity.-- hence
100a052 ¶ This is not applicable. it does not explain CLEAVAGE of rock-- nor the Falkland case, nor. the arr
100a053 g character.-- A discussion on concretions and CLEAVAGE conjoined very good.-- It is the Key to the s
100a053 e.-- or rather radiating to central point. can CLEAVAGE be radiation from some grand centre.-- A Stal
100a053 -- A Stalactite of Gypsum, is the best case of CLEAVAGE.-- Phillips (113) «Lardner Encyclop.--» absol
102a060 angement of particles in rock. This applies to CLEAVAGE & concretions.-- Septaria in concretion arran
102a062 ubstance, & then the next being sucked out. In CLEAVAGE discussion, state broadly indication of new l
105a070 320, paper on shrinking of Clay. applicable to CLEAVAGE. C. Prevost.-- In Cordillera. a rush of water
110a086 ?) & Felspar, (some crystals being red) «with» CLEAVAGE, veins of pyrites, few curious fissures; base
110a086 ng into base-- Salband might have oozed out of CLEAVAGE plates: the crystals must have recrystallized
114a094 . & iron stone alternating. bear on subject of CLEAVAGE Clay slate. a distinct formation deep «& ther
120a108 there is something superadded, that which give CLEAVAGE to rocks.-, but lava shows the rocks really
147g015 alluvium. NB In one part pure sand in current CLEAVAGE-- in other irregular horizontal strata I supp
148g026 idge of Spean, hills of «sea», gravel, current CLEAVAGE, & pretty well rounded stones, mixed with som
115a096 on which gives polarity to atoms in slates that CLEAVE, & which unite the homogenious crystals., must
115a096 n of sandstone rare to have any horizontal non CLEAVING beds. metamorphosed. The chemical action whic
278c129 of each might not breed:-- Genus must be a true CLEFT-- putting out of case the Analogys.-- If genus
249c035 rimrose & Cowslip run wild, The two species of CLENOMYS. case of replacing species. Dr. Smith will gi
541m092 their head must have this high faculty, yet not CLEVER people. Aug. 21st. 38 When a dog in play has h
030r036 e chart of Patagonian coast to see proportional CLIFF & low or sloping land What are the "palatal Tri
032r039 he result would [be] a uniform slope to base of CLIFF (Z). to which point the waves would not reach.
042r067 «NB. Mr Lyell P. 211 Vol III. talks of line of CLIFF marking a pause» When mentioning pumice of Bahi
055r107 hells. 2 - 3 toises thick.--Vol II. p. 252 Urge CLIFF form of land, in St Helena. Ascension. Azores.
153g051 rs «lines not so level because of upper edge of CLIFF» Others below it--argument for lake «or sea» at
031r038 structure was very clear at base of great lava CLIFFS [Fig. I] line of high tidal action {P} NB. pat
044r072 nally--I did not see any number of dikes in the CLIFFS.--wide valleys.--central peak small; yet great
051r094 of degradation; (soft substances worn into bare CLIFFS evident); the action is anomalous; It is wonde
055r107 ith age, & agreement with number of craters. No CLIFFS at Ascension (or modern streams of St Jgo) yet
055r108 atmospheric of the most solid rocks.--The grand CLIFFS of a thousand feet in height, of the solid lav
104a069 illera discussion, deep sea, fragments fall off CLIFFS. but then how spread abroad?-- There is thus w
106a075 applies to all vallies (except mere talus «over CLIFFS edge» of which limit cannot be great over) wit
115a096 dding effects to common heat.-- Where there are CLIFFS there ought to be creeks & mouths of rivers ou
117a102 always falling or at least not rising are there CLIFFS. Sir L. Dick says (.p 52) fringe of sublittora
120a109 s own matter. but would cut wide gorge. leaving CLIFFS, on each side, such as now exist.-- caution ab
278c131 r species same, excessive improbability. Mem in CLIFTS list a rat said to have been found!! rodents o
192b049 range, reason & futurity. it does as yet appear CLIM In Mr Gould Australian work some most curious c
138a153 k] «p. 4. (Lyells Book)» Observaciones sobre El CLIMA del Lima par Dr. H. Unanüe says he believes the
209b156 en conclure quels sont les genres qui, sous ce CLIMAT, se divisent le plus aisément en espèces disti
032r041 gain a Westerly current:--If great changes of CLIMATE have happened. hurricane in bowels of earth ca
059r121 cannot be pure. for if so Chalk would harden.--CLIMATE.!? or small Proportion of Alum: matter.--all p
062r128 eat contrast of two sides of Cordillera, where CLIMATE similar.--I do not know botanically = but pict
062r128 y = Both N & S. great contrast. from nature of CLIMATE. -- Perpetual snow.--subterranean lakes, near V
065r138 us Investigate with greater care. vegetation & CLIMATE of Tristan D. Acuneha. Kerguelen Land. Prince
087a008 more Arctic land would be required to produce CLIMATE resembling S. America in Europaean latitudes.-
127a125 -- But Siberia was once thawed. & hence. (when CLIMATE hotter) was cooled to greater depth.-- Now the
172b008 As we thus believe species vary, <in> changing CLIMATE we ought to find representative species; this
203b135 r the monkey as a wanderer, but as produced by CLIMATE?-- M. Baer (thinks) the Aurock, was found in G
206b148 aybe overlooked mere variations consequent on CLIMATE &c-- the whole races act towards each other, a
231b243 g & miocene.-- The descendants left in cooling CLIMATE might change twice over, whereas those which m
232b245 ave succeeded in becoming habituated to colder CLIMATE whilst others died out, or moved towards equat
232b247 part instead of change from insular to extreme CLIMATE, <more northern> Iceland would have possessed
250c037 Is the Australian Dipus an American form? The CLIMATE having grown more extreme both in, N & S. Amer
258c060 rom cold, children would not become adapted to CLIMATE.-- Descent. or true relationship, tends to kee
261c070 l finches from linseed-- not solely effects of CLIMATE on some «antecedent races, perhaps not on now
268c099 ust always bear in mind proofs of most equable CLIMATE both in S. & N. Hemisphere just anterior to pr
298c191 - or regrafted with fresh arrivals..-- &c &c --CLIMATE altering as island increases.-- upper parts a
309c219 cause their difference. arise a good deal from CLIMATE & habits, & therefore less fertile. according
324c268 g. ditto Falconers remarks on the influence of CLIMATE White's regular gradation in Man. Lindlys intr
332d004 sites of Tropical countries cannot endure this CLIMATE-- .) -- July 23d. Eyton, a stone blind horse,
342d036 ones. cause changes in geography & changes of CLIMATE superadded to change of climate from physical
342d036 y & changes of climate superadded to change of CLIMATE from physical causes.-- these superinduce chan
360d091 d long been in confinement-- is this effect of CLIMATE, or state in which they are kept?-- Is there a
365d106 mstances.-- More probably during known changes CLIMATE became unfit for. subalpina, or some Northern
369d115 ersally admitted."-- p. 483. Owen thinks from CLIMATE of Australia, & from Ornithorhyncus & Hydromys
370d115 be attempted to be shown from peculiarities of CLIMATE cause of N. Zealand not having any Mammalia.--
398e004 what «changes» would such a change produce in CLIMATE vegetation &c.-- It is the circumstance of sma
404e024 mmon to every individual & therefore effect of CLIMATE.-- Octob 19th. When reading. L'Institut: .1838
406e037 America.-- Whole wor[l]d, formerly possessed a CLIMATE compared to S. America at present days., which
407e038 ductions. The world formerly much more so. yet CLIMATE of same order as that of S. America.--`(Explai
407e039 the world. from the <Tropical> Equable kind of CLIMATE to the extreme.-- Therefore species, which wer
407e039 ch were fitted for such a preeminently equable CLIMATE. might not have been able to have survived a c
407e040 m the ancient preeminently equable & temperate CLIMATE, <that> of America, that the Mammalia of S. Am
408e042 n of passage from N. to S. American forms. The CLIMATE of N. America, must have been equable & low--
427e109 al.-- Would not subsidence of Greenland render CLIMATE less extreme. (& so account for descent of sno
427e109 N. American Conchology from Europaean., & the CLIMATE being now less extreme, than before arctic for
427e109 forms would retreat: effect on snow of arctic CLIMATE in far north regions? Arctic forms have travel
446e161 eum May 18. 1839. p. 377.-- Statement that the CLIMATE is on the decline, as far as vegetation is con
615o36v ferent.-- Heredetary effect of former tropical CLIMATE analogous to inflorescence of Tropical plants
640j167 no hurricanes.-- islds never joined, nature & CLIMATE very different, from adjoining coast. Admirabl
076r169 pure silver not common in <S.> America: In all CLIMATES distribution of silver «in veins» very unequa
175b019 pretty similar over whole world under similar CLIMATES & as far as world has been uniform, at former
205b142 orse & Ox have different parasite in different CLIMATES.-- Hunbt. Humboldt: Vol V. P II. p 565. Consu
222b202 on multiplication of species by travelling in different CLIMATES the backward & forward introduction of spec
232b248 rasitic animals of beasts varying in different CLIMATES Those will not object to my theory, those the
235b272 lia [blank] Falconer's remarks on influence of CLIMATES, situations &c on 242. Hook Smellie Philos of
308c219 formed them, would have arisen under different CLIMATES &c. Do I mean that ideosyncracy of wild anima
193b090 sses be brought into play.-- The father being CLIMATIZED, climatizes the child. ¿-- whether every ani
193b090 ght into play.-- The father being climatized, CLIMATIZES the child. ¿-- whether every animal produces
033r043 ould walk round base:--not universal: could not CLIMB up many parts, in James Isd.--Mem St Helena-- A
584n071 ly fall, when a species is found which does not CLIMB CD[.instinct may be divided into migration,-- s
205b143 ch emigrates over lands is a siluris, p. 123 A CLIMBING fish. p. 122 A Terrestrial annelidous animal
463t055 greatest Could anyone. have foreseen, sailing, CLIMBING & mud-walking fish? difficult-- yet suggested
584n071 taken the Woodpecker as an example fitted for CLIMBING, his arguments partly fall, when a species is
127a125 water {P} thawed at + in isothermal curve. East-CLINAL. West CLINAL. S.-clinal. N-clinal & anticlinal
127a125 wed at + isothermal curve. East-clinal. West CLINAL. S.-clinal. N-clinal & anticlinal «synclinal--
127a125 isothermal curve. East-clinal. West clinal. S.-CLINAL. N-clinal & anticlinal «synclinal--» line.-- <
127a125 l curve. East-clinal. West clinal. S.-clinal. N-CLINAL & anticlinal «synclinal--» line.-- «ditto of S
127a125 ynclinal--» line.-- <ditto of synclinal> simply CLINAL lines. dipping so & so or may be used East-cli
127a125 inal lines. dipping so & so or may be used East-CLINAL lines & c & .-- But Siberia was once thawed. &
205b145 rule apply? A Treatise on Form of Animals by Mr CLINE "The character of both parents are similar in a
323c269 a -- The Revd. A. Wells. Lecture on instinct -- CLINE on the Breeding of Animals -- Spallanzani's Ess
327cIBC account. Raffles. Sir. S do. do-- Buffon Suites CLINE on the improvement of domesticated animals Frie
431e123 ses & all others by action of wind difficult.-- CLINE on the breeding of Animals, p. 8. size of foetu
272c108 highs with same peculiar structure & habits of CLINGING to rushes similar.-- The question which I mor
303c204 points. female genital organs «in some monkeys CLITORIS wonderfully produced».-- make abstract on thi
382d158 ovaria in Hermaphrodite,-- but not of poenis & CLITORIS, shows to my mind.-- , that both are present
580n061 hat some people of very weak intellect (As Miss CLIVE) have only possessed very loose ideas.-- Have c

147g019 5> «636» Temp. 62 Friday morning ¼ past seven o'CLOCK 29.642 Temp 55 Air 50 degree? Friday. Inverorum
049r088 constitution of veins, is there said granite in CLOSE contact varies in nature, -- Does not granite a
049r088 nite at C. Tres Montes become more siliceous in CLOSE contact? -- «Cordillera???» Porphyry at Valpara
059r121 t: & even in one case contained lime.--All bear CLOSE analogy to Obsidian, & all show chemical action
070r155 lation'. crust thin.--Concepcion earthquake Draw CLOSE Analogy Lake of Cordill: of Copiápö & Desaguade
072r159 Volcanic action on coast line of Old Greenland, CLOSE to W of Jan Meyen Isld.--Mr Barrow thinks N & S
103a064 I suspect much weakened by <vi> considering how CLOSE the dislocations occur & therefore that the cru
117a102 stitut. 1838 p. 151. Formations of Payta extend CLOSE to Guayaquil.-- modern shells of Cobija doubtfu
182b050 uld Australian work some most curious cases. of CLOSE but certainly distinct species between Australi
188b069 endency to keep to one line Dr Smith says very. CLOSE species generally frequet slightly different lo
195b103 at external conditions would produce species so CLOSE as Patagonian <Chat> & Galapagos orpheus.= Put
214b175 ology-- A breed of Blood Hounds from Aston Hall CLOSE to Birmingham, and supposed to be descended fro
227b224 ease to know the steps.> although D E F. follow CLOSE to A. B. C. we cannot be sure that structure (C
230b241 good many examples of animals «or plants» very CLOSE (take Europaean birds. Mr Goulds' case of Willo
239cIFC of difference: where hybrids produced have any CLOSE species ever yet failed. About trades affecting
251c041 urnal Vol 2. p 221. Hursfield on two bears very CLOSE species, inhabiting Borneo & Sumatra. differ on
253c046 onel Montagu probably contains some facts about CLOSE species of Birds. Zoolog. Transact. Vol I p. 16
264c080 ys» Gould seems to think that many species when CLOSE come from different localities, as my Furnarii.
276c125 n unequal number of vertebrae-- ¿Where two very CLOSE species inhabit same country are not habits dif
276c125 different, (Mem: Gould's Willow Wren) but where CLOSE species inhabit different countries habits simi
282c140 is the sum of all the relations, analogy is the CLOSE relationship in some one.-- imagine the men to
285c151 ible in some detail,-- the relations to islands CLOSE species, «on these isld» &c will probably upset
294c176 ate-- it may be said argument will explain very CLOSE Species in islds. near continent, Must we resor
304c206 ith different degrees of closeness. -- look how CLOSE birds! look at Mammals: how wide.-- therefore b
305c209 S. Hemisphere there are some godwits which are CLOSE to European species, and the sexes of which var
305c211 which is modified into endless forms, bearing a CLOSE relation in degree & kind to the endless forms
310c223 aya.-- some English beetles, birds & a fox most CLOSE The most curious case is Saxifrage, almost <sam
316c247 than of Europe, because the recent ones are so CLOSE. Was there continent between N. America & Europ
325c267 of wild cattle & Montagu on birds (facts about CLOSE species) Wilson's American Ornithology Read Ari
331dIFC als:]CD [All Selected Dec. 14-- 1856]CD Towards CLOSE I first thought of selection owing to struggle
355d068 ppearance with Secondary Species distinct-- but CLOSE.-- Mem. Von Buch on Cordillera fossils same rem
361d096 d «that there are» two ducks, which have pretty CLOSE representative species in England & N. America.-
363d102 in plumage of ducks.-- Mr Yarrell says in very CLOSE species, of birds, habits when well watched alw
389d176 -- Offspring like both father & mother, or very CLOSE to either.-- Male & female as foetus one sex; &
389d177 ed, when near relations, & therefore those very CLOSE are bred into each other.-- This is somehow con
390d179 ffects of breeding in, it is not merely the too CLOSE animals, which will not breed, but the female a
400e011 o Europaean in Van Diemens land, where there is CLOSE species of elater.-- Where this collection is p
406e035 s like & identical with S. America. & many very CLOSE: see full paper. L'Institut 1838. p. 338 A most
407e042 must. Lesson I remember says Mariana Deer very CLOSE to a Molucca species. -- L'Institut 1837. p. 253
439e143 stincts in animals.-- yet the existence of wild CLOSE species of plants shows there is tendency to pr
446e162 seed-- which Henslow is inclined to think very CLOSE.-- «A fruit tree by certain treatment will sudd
465t080 cepcion-- Patagonia-- Beds of La Plata. (except CLOSE to B. Ayres).-- If we may take this as guide, t
467t099 own, but with no pollen.-- Common Thyme growing CLOSE by is equally abortive--and both growing within
469t151 t, Early frame, Groom's Dwarf. planted in rows «CLOSE to each other» & seeds gathered «all» came up i
482z011 - Sturnus Magellanicus.-- p. 210. Scolopax very CLOSE to ours Rengger's work of Mammali: of Paraguay
492q022 s. 5. Whether Roses impregnate each other. when CLOSE planted together: <do> Can Holyoak be raised di
493q003 e the> is the ground much manured In species of CLOSE genus do more than three primary colours occur
498q009 plant to sow & try to get other species <near> CLOSE to each other.-- As they are dioecious, if no h
500q10a Philotis, Dacelo. Alcyone, where there are very CLOSE species & see whether they come from islds. or
510q017 St. Hilaire Bot. p. 787. position of embryo in CLOSE species of Hilianthemum differs greatly-- how v
539m082 n & not feeling much enthusiasm, happened to go CLOSE to one & smelt the peculiar smell of Picture. a
556m146 ased.-- is it opposite movement to drawing them CLOSE on head, when going to fight, in which case exp
559m156 ry regularly at one time every day.-- naturally CLOSE at that time after long period.-- My Father abo
565n008 les will want much confirmation. A grave person CLOSE those muscles, which wrinkle when smile.-- Hope
586n082 ome senses, is sight--CD [The faculties bear so CLOSE a relation to the senses, that one feels no mor
596n184 dying-- "Thomas Brown" on Association worthy of CLOSE study.-- full of practical observations Ourang
127a126 118 water no--. oil will freeze if cooled in a CLOSED globule of glass. (oil may be cooled to 0 degr
545m107 also another monkey to dog. I showed nut & then CLOSED my mem. expression of fury, jump to scratch my
547m112 of ordering things to be done.-- The senses are CLOSED probably by sleep & not vica versa. anyhow I m
070r153 k to common parent? why should two of the most CLOSELY allied species occur in same country? In botan
172b008 sentative species; this we do in South America CLOSELY approaching.-- but as they inosculate, we must
187b065 &c same in N. America & Asia, but many species CLOSELY allied but different, because country separate
195b100 so likely to wander. Did it create two species CLOSELY allied to Mus. coronata, but not coronata.-- W
202b130 ther.-- Good argument for species not being so CLOSELY adapted. Near the Caspian «Province of Ghilan»
206b149 decreasing population at any one moment fewer CLOSELY related;'. (few species of genera) ultimately f
216b181 ome offspring black others white which is more CLOSELY allied to case of cross of dogs.-- See paper i
219b193 reator <on volcanic island.> make plants <grow CLOSELY> When this volcanic point appeared in the grea
233b250 rica. which were hard to distinguish came from CLOSELY neighbouring localities Institute 1838. p 38.
246c027 perhaps, Friendly Isles «& Hebrides») is very CLOSELY allied to C. muscadivora., which lives in the
310c224 most curious case is Saxifrage, almost <same> «CLOSELY allied» species Himalayas, 13,000 & Melville I
311c225 rte's list of birds in Europe & N. America, on CLOSELY allied species. «replacing each other». good t
333d005 tens, alike each other, partaking <more> «very CLOSELY» of form of mother: more than of the Common ca
333d006 n cat, that they were killed, & other two very CLOSELY resembled in form of tail, fur &c to the half
341d031 America & many birds.-- which however are most CLOSELY represented.-- Thus the red breasted thrush is
374d133 ediate between fish & reptiles-- yet osteology CLOSELY resembles reptiles.-- p. 432 some plants in co
389d176 offspring-- DESIRE LOST when male & female too CLOSELY related: this most important with regard to th
398e005 ose who have studied history of the world most CLOSELY, & know the amounts of change now in progress,
418e083 cies & individuals in their present forms, are CLOSELY related-- By birth the the succesive modificat
464t065 with small wings, & surely the Apteryx is more CLOSELY allied to the Struthonidae than any other form
482z012 e, then the Mulita from 41 degree to 26 degree CLOSELY allied species, therefore interlock.-- Testudo
508q016 Hottentots generally resemble each other very CLOSELY, more closely than Caffres.= 13 Where are ther
508q016 nerally resemble each other very closely, more CLOSELY than Caffres.= 13 Where are there any medical
524m020 itching when a disagreeable thought occurs, is CLOSELY analogous to Epilepsy & convulsion.-- affectio
528m035 astle» because train cannot be discovered-- is CLOSELY analogous to my Fathers positive statement tha
600o08a ndeed be wonderful, if, mind of animal was not CLOSELY allied to that of men, when the five senses we
304c206 eep to their types. with different degrees of CLOSENESS. -- look how close birds! look at Mammals: ho
461t041 ry far.-- Sept 31. The identity <of> (or only CLOSENESS) of some species-- (especially of mammifers)
252c045 nowledge would probably tell more certainly Get CLOSER species-- FOX IS & Mice of America «good case
341d029 -- Ostrich & Rhea closest.-- (& two Rheas still CLOSER).-- Mr Blyth asked whether structure of pelvis
424e101 - & American ones & Europaean-- agree very much CLOSER, than the present ones., which according to Be
495q05a he Leptosiphon densifolium «an annual» «sleep» «CLOSES flower» on all gloomy days.-- The «garden» Cor
214b174 certain, that North American fossils bear the CLOSEST relation to those now living in the sea.-- See
227b228 edetary, whole metaphysics.-- it would lead to CLOSEST examination of hybridity «to what circumstance
236b280 pecies to some South American kinds.-- Are the CLOSEST allied species always from distant countries,
340d029 any large orders of birds. The Emu & Cassoway CLOSEST.-- Ostrich & Rhea closest.-- (& two Rheas stil
341d029 The Emu & Cassowary closest.-- Ostrich & Rhea CLOSEST.-- (& two Rheas still closer).-- Mr Blyth aske
410e052 ctions which are possible.)-- Ascertainment of CLOSEST species (& naming them) with relation to habit
527m034 e &c no organs of sense being required) as the CLOSEST train of geological thought.-- the capability
577n050 rtain round of actions take place every day, & CLOSING of the leaves, comes on from want of stimulus,
592n103 & rabbit depress. them from squatting.-- p. 64 CLOSING both eyelids express contempt. p. 76.-- childr
638j28r p. 251-- stomach hump, kinds of foot. power of CLOSING nostril, foot, sack. power of endurance &c &c
541m091 peat the word, & think of qualities as flowers, CLOTH &c & with all this difficult EXPERIMENTIZE upon
547m112 come & opened box, whether I had thought what CLOTHES to take (how often one cannot tell whether one
182b048 from Abyssinia, & others dark brown, with long CLOTTED hair resembling that of goats." Progressive de
267c093 rom Abyssinia; the others dark brown with long CLOTTED hair resembling that of goats". Geograp. Journ.
529m043 inflammatory diseases, where there has been so CLOUD on the mind, every occurrence for a day or two
225b217 w to make Indian Cow with bump & pigs foot with CLOVER hoof) Ask Entomologists whether they know of a
467t103 over-ripe & half withered-- I saw Bees going to CLOVER & once this happened.-- And in common Beans it

470t178 amine it.--«Little Dusty & Blue» Butterflies at CLOVER,--Veronica--, Ranunculus in numbers =what inse
501q011 ays yes. (29) Are there RACES of Lupine, Stocks CLOVER, to experimentize on by sowing near each other
539m081 cts of animals-- Aug. 12th. 38. At the Athenaeum CLUB. was very much struck with an intense headache
312c231 ot permanent, in the new cross.-- In the Bantam CLUBS, they used to fix on the kind wanted, colouring
472s02r odd dusky humble (with pollen) on legs go from CLUMP to clump, & insect proboscis in many flowers, o
472s02r y humble (with pollen) on legs go from clump to CLUMP, & insect proboscis in many flowers, on one of
499q010 s of rare green Cotton Plant-- How large «area» CLUMP there? Distinguishable from <other> clumps from
472s02r er watching 14 days. many times every day. many CLUMPS of heartseases, never saw any Bee go to them.
499q010 nsect-- (21) Are there many instances of single CLUMPS of plants in counties, as of rare green Cotton
499q010 area» clump there? Distinguishable from <other> CLUMPS from other parts? Don says Irish, Scotch & Eng
367d112 if that cannot be formed, genetal organs by that CO-relation of parts, will not be produced.-- Sept.
410e051 f my theory, laws probably will be discovered. of CO relation of parts, from the laws of variation of
520m002 eneral manner of holding hands &c &c.-- Mr Dryden CO said he could not remember his father.-- My fath
533m059 walking of animals. that is the ready movement & CO-relation of the proper muscles. may be illustrat
029c034 ournal I look at the cessation northwards of the COAL in Chili are clearly bearing a relation to prese
029c034 early bearing a relation to present position of <COAL> Forests. These thick beds of Lignite stratifie
029c034 f Lignite stratified with substances so like the COAL measures in England (Excepting Conglomerates?)
030r036 cepcion where there are Tertiary strata there is COAL-- ¿ No shells in all cases. «Mytilus.--» «at G
031r037 hink so many faults in Cordillera, as in English COAL field -- because lowered & raised--so om-but g
039r059 upposes that all the dikes in Cornwall or in the COAL measures have been conduits to volcanoes.-- Tal
120a107 cking by vertical planes into small pieces-- mem COAL-field.-- the structure of Andes. where we belie
123a115 boldt on Bogota. Cordillera,-- nothing.-- salt & COAL near Bogota; p 270.-- SPLENDID PAPER on fossil
206b150 art.-- no dicotyledenous plants & few Monocot in COAL formation? p. 320 <Think> States Cryptogam. Flo
212b167 .-- NB. Was Europe desert (like S. Africa) after COAL Period.-- ¿In those divisions of molluscs. wher
257c058 in do.-- <Wood> <Dicot wood> Coniferous wood in COAL Measure.-- highest fish in Old Red Sandstone.--
374d133 reptiles have been found. p. 426 Sauroid fish in COAL, true fish, & not intermediate between fish & r
374d133 sely resembles reptiles.-- p. 432 some plants in COAL supposed to be intermediate between Coniferous
460t019 iroptera & caetacea in Eocene-- dicot. plants in COAL measures.-- Shells in Cambrian & Crust show how
027r028 would-- wood converted into siliceous pyritous & COALY matter. Mem: Chiloe In the endless cycle of rev
115a098 thin out, both inland & seaward: if matter too COARSE, then {P} that form.-- All this depending not
116a099 undance of matter would have same effect as too COARSE. Read Kylau on Granite Edinburgh Philosophical
163g106 e (32 ft rise) they found alternating layers of COARSE & fine & many Sea shells. My informant saw the
164g119 } muddy nodular strata coral upside down strata COARSE agglomerate [...] shells from [...] Wenlock Ed
024r015 ¿ Submarine currents Find instances; The whole COAST of New Holland shoals much: Dampier remarks on
024r015 much: Dampier remarks on great flats on the NW COAST:-- 8 leagues, from Sydney 90 fathoms La Peyrous
024r015 miles. 65 Fathoms Vide facts in Beechey. on NW COAST of America off Cape of Good Hope 70 fathoms 20
025r016 Hope 70 fathoms 20 miles from the shore? Beagle COAST of Brazil? where not rivers in my Coral paper {
025r016 Francisco [10 degree 32 minute S.] 10 50 Whole COAST to Olinda [8 degree S.] 9-10 = 30-40 {T} at twi
025r016 C. Rock. [5 degree 29 minute S.] still shoaler, COAST composed of sand dunes. 15--15 Does not seem to
025r016 -15 Does not seem to consider this a very shoal COAST. Beyond the 10 or 12 leagues sea deepens sudden
028r031 yond the 10 or 12 leagues sea deepens suddenly. COAST of Brazil generally.-- Mrs Power at Port Louis
030r034 not from rain, because alluvium saline; Mem: on COAST of Northern Chili as springs become rarer, so d
030r035 of limestone?» have been collected on the open COAST. Perhaps as at Concepcion. favoured by basin fo
030r035 <[...]> At Carelmapu.--Within Chiloe:-- On open COAST, near where Challenger was lost: I know no reas
030r035 ottom of an open & not deep sea.--(Character of COAST regular & <not very> rather deep soundings, 60-
030r036 ion at N. S. Wales & Van Diemen's land.-- Whole COAST S. of Concepcion where there are Tertiary strat
030r036 Carbonaceous shales Examine chart of Patagonian COAST to see proportional cliff & low or sloping land
032r040 could be shown (as represented), along line of COAST.--[Fig. 2] Mem San. Lorenzo; Valley of Copiapò
032r040 Mem San. Lorenzo; Valley of Copiapò & parts of COAST of Chile.-- Must first explain «top of» tidal b
034r045 regular change in soundings. on approaching the COAST of NW. America P. 209-13 P & 444 «(Yanky Edit)
036r050 harts for distribution of pebbles.--Plains. off COAST of Patagonia.-- British channel &c &c. There is
037r056 ental? level surface not disturbed.--Whole West COAST. Chonos to Copiapo.--Sydney. K. G. Sound. C. of
039r061 e more abound to take analogy of movements of W COAST. in explaining plains because such are found in
044r073 ines to belief all strata of Europe formed near COAST. Humboldts quotation of instability of ground a
045r075 thquakes if rise was more <than> inland than on COAST it would be invariably discovered; this may be
047r081 with same general character» reaches far beyond COAST, which has been raised.--It must be considered
047r083 a break first? I can imagine from local form of COAST (as seen in swell) the undertow & overfall must
047r083 ntal than single elevations along whole line of COAST Darby mentions beds of marine shells on banks o
050r091 the Banks Is there not a sudden deepening on E. COAST of Africa. as at Brazil [blank] What is nature
051r094 ion of Organic productions. = Yet everywhere on COAST (Il Defonsos «Kelp») rocks show signs of degrad
051r097 odies [blank] Beechey.--changes in bottom in NW COAST of America. from shingle to sand &c &c. <Vol II
052r097 ather many faults.-- Necessary form; as long as COAST line fixed.--[Fig. 5] {P} * Slope necessary for
052r098 -- Give various cases. [Fig. 6] {P} A advancing COAST to Seaward. Retreating case in excess as first
052r098 this discussion.--Brazil bank: & I believe SE COAST of Madagascar. where a --40 line <shows> runs a
054r102 articular genera of plants: All St. Catherine & COAST Granite: P. 199; Falkland account of cleavage d
054r105 Coquimbo; [not located] [not located] Isld near COAST of America not reached. Juan. Galapagos. Cocos-
054r105 ry, to the distance of 3 or 4 leagues «from the COAST» may be concluded to have been covered by the s
054r105 ma & the extent of a league & a half a long the COAST. <"> The rocks in the most inland part of this
064r135 t abundant in interior channels. there no outer COAST.--important effect.-- Capt. FitzRoy. -- Limite
065r140 Mem. pebbles of Porphyry.--Falklands.--off East COAST. -- Capt. Cook found soundings. (end of 2d voya
065r140 Cook found soundings.- (end of 2d voyage outside COAST of T. del Fuego. off. Christmas sound. -- «(Thi
067r143 sh says. that beds of shells are found on whole COAST from P. Indio to Quilmes. & at least seven mile
072r159 Holland about bubbles. -- No Volcanic action on COAST line of Old Greenland, close to W of Jan Meyen
073r163 alt & other trap cover it to great thickness. = COAST of Acapulco granitic rock.--in parts of table g
079r177 e, marevellous statements on, Vol I, P. 168. on COAST of Guayaquil, same as Galapagos. no Hydrophobia
080r180 gradual shoaling of coasts 93 action of sea on COAST. 27. Bahama Isd De Lucs travels Beauforts Karam
091a027 periodical said so. «Campbell the Poet» Accra. COAST of Africa. Clay Slate & Quartz. strike SSW & NN
100a055 p. 213) form of escarpment relation kept to sea COAST ∴ curious exception in Wealden.-- Would crystal
102a059 ocean. as pebbles would be lifted up & down. on COAST itself, undertow would draw it outwards.-- form
102a059 ffects on bottom a thing floating some way from COAST is driven on to it.-- rollers at Tristan d'.Acu
102a059 at Tristan d'.Acunha.-- silting up. channels on COAST of England-- Any one. who has studied rocks in
105a069 less & ∴ tends to finite power] whereas sea. on COAST, as long as exposed to waves of sea, cutting po
105a072 s found with calcareous matter & concretions on COAST of Madeira.? how came it if this powder results
109a082 number of bones lying at the bottom of sea. off COAST of England.-- Sea must always on actual beach a
109a083 sand beach. {P} -- De la Beches argument of low COAST gaining & high loosing answered by this -- No o
109a083 y this -- No one can doubt. A-B once formed low COAST.-- Annales des Mines. a translation of paper by
113a092 Mitchell inferred subsidence; Mem my remarks on COAST of Australia.-- Great NW. dip in SE part of Aus
115a098 molecular attraction of every atom in rock On a COAST, the shallower the water, the greater power of
117a101 thinks sand with vegetable remains formed near COAST, limestone deep water. will bear on formations.
120a109 be profoundly considered. if elevation near COAST? How can Herschel consider figure of earth stat
122a113 ain littoral mountains & volcanos.-- Why on one COAST of do-- I believe?? coast of North America., li
128a130 ol.-- some notices on modern Tertiary strata on COAST of do-- I believe?? coast of North America., li
128a130 rn Tertiary strata on coast of do-- I believe?? COAST of North America., like the Mexican Gulf. us to
128a130 ars of sand & shallow lagoon.-- when describing COAST of. Brazil. Maldonado enter into this case.-- E
130a133 d told me of some large fresh Water springs off COAST of Persia In Glen Roy paper I show crust yield
132a138 on prevails more beneath the sea, <than> «&» on COAST lines, than on continents. it ought, (according
133a141 yry at St. Elena. p. 6. few «living» shells. on COAST of do p 8.-- soft Clay beds near C. Virgin p. 5
134a141 ty Sound. SE dip. much p. 136. Rocks on Western COAST p. 204 do. do p. 210. Height on road from Valpa
134a141 to   & Rocks p. 375. on the soundings on outer COAST of T. del. Fuego.-- p 385 Rocks of S. Western C
134a141 t of T. del. Fuego.-- p 385 Rocks of S. Western COAST Vol II p. 277. on whale bones in Falklands Some
135a144 inland a great many sea shells some miles from COAST-- quote passage to show abundance Bengal Journa
148g024 yet in Glencoe in parts no trace of them-- Mem COAST of Chile--¿ is not Mica Slate too hard & uneven
182b048 ieut. Wellsted obtained coast many sheep from Arabian COAST. "These were of two kinds one <with> white with
193b093 . Vol I p. 174. says from Swan river long South COAST all the remarkable Australia genera, collected
200b121 he possibility of common branching of? Accra. COAST of Africa. Clay slate. strike. SSW. & NNE. dip
218b190 du Roi Mem. Capt. Owen's story of cats on West COAST of Africa.-- changing hair-- The Edinburgh Jou
226b220 uritius & Madagascar very good.-- Fernando Po & COAST of Africa. equally good.-- Small isld off New G
226b223 nce of Jaguar been in S. America.-- <East.> «W» COAST of Africa & <West.> «E» of America, ought to pr

229b233 hinoceros range from Abyssinia to extreme South COAST. Elephant he believes is mentioned by old write
229b233 is mentioned by old writers on extreme Northern COAST. Hippopotamus do.-- Giraffe do.-- Range of East
266c092 ther Hyenas; jackals monkeys-- common to either COAST not found here not even Antelopes, though commo
266c092 found here not even Antelopes, though common on COAST of Arabia not even antelopes though common on i
266c093 n antelopes, though common on islets off Arabian COAST.-- ∥Vol VI. p. 89.-- Lieut Wellstead "on coast o
266c093 n Coast.-- ∥Vol VI. p. 89.-- Lieut Wellstead "on COAST of Arabia between Ras Mohammed & Jeddah". sheep
268c099 --» very good. Study D'Orbigny. & range on West COAST «Guayaquil & Peru» Henslow in talking of so man
298c192 n most plants, even those on Guernsey & on West COAST of Ireland, are absolutely (& who better author
311c227 Ovington's Voyage to Surat, floating isld. off COAST of Africa .p 69. with tall grass. & p 72 hairy
316c246 werby says there are some shells common to West COAST of Africa & E. S. America.-- get instances.-- v
317c249 to be monkey of St Jago C. de Verd; same as on COAST of africa.-- Macleay tells me same thing p. 55.
318c253 kinds as gallinules taking the low country near COAST & others the mountains, & these appearing to re
332d005 hich he bought in Portsmouth, said to come from COAST of Guinea, -- ears bare. skin black & wrinkled-
363d103 uscovy & common duck were often caught wild off COAST of America.-- showing hybrids can fare for them
366d112 t teeth.-- the account of the people on the NW. COAST blinking to keep out flies might be used» The w
408e045 rchipelago. now rising. On a particular part of COAST of Somersetshire the Cockles are all apt to be
421e090 ie Canary Isld approaches more to neighbouring COAST of Africa, than to other parts of that continen
431e121 however, said they have been found together on COAST of France.-- L. doubts.-- Lonsdale thinks Ammon
448e166 plants «of Juan Fernandez» differ from American COAST Vol II.-- <Reference> p. 251. about the driftin
448e167 agine that there are Baboons in St Thomas on W. COAST of Africa Owen Linn. Soc. April 2d. 1839 The Le
459tf1r ing gradations, Boteler's Narrative Voyage East COAST of Africa-- Vol II. p. 256-- wild cattle at Mad
463t059 s animals of Fernando Noronha are found unknown COAST in front of it.-- Cuvier has grand sentence abo
485z017 e Birds seem to move much further North on West COAST of S. America. than on East.-- not being replac
640j167 nature & climate very different, from adjoining COAST. Admirable explanation is thus offered.-- From
080r180 logy of the world P. 14-91. gradual shoaling of COASTS 93 action of sea on coast. 27. Bahama Isd De L
269c102 r se as barrier-- Mem. Tartary & China.--, both COASTS of New Holland.-- «Compare birds of Australia
316c246 same (or according to Sowerby fine species) on COASTS of N. America & England.-- but the fossils are
465t080 ing.-- & yet no shells-- now look at Scotland-- COASTS of Chile, excepting Concepcion-- Patagonia-- B
189b070 illoes «&» & Megatherium. each with same kind of COAT.-- If we could tell, I do not doubt even colour
259c066 of» generation right?-- If puppy born with thick COAT monstrosity, if brought into cold country, «& t
293c172 nct.-- Even if savage takes. & was given a Great COAT & this he put on & we afterwards could understa
376d136 us Porcarius.-- this Monkey did not like a great COAT made for it at first, but in two or three days
033r042 In Calc: sandstone at Ascension, each particles COATED by pellucid envelope of Lime.--form resembles
048r084 I. P. 191 State at St Helena. pebbles entirely COATED with Tosca. which implies motion in the «loose
048r084 nder a cascade, one can understand pebbles thus COATED.--The motion is most wonderful, from chemical
051r094 here highest spray (there pale green confervae) COATED with living beings; In smooth seas (& even tur
097a043 ulett Scrope. talks of Trachyte, "superficially COATED by a thin pellicle of a blackish colour like a
102a060 d. calcareous rocks of Ascension, each particle COATED. &c will be aware how little common Gravity ha
244c021 rally "red of brick" hair, very stiff, p. 120-- COATI roux common. near Concepcion. some tatous!!! p.
482z011 luminous Sertularia Lesson Zoolog. Coq: p. 120 COATI Roux. Tatous & perhaps Yagourundi near Concepci
068r146 ore beneath ocean; for subsequently there is a COATING of solidifying igneous rocks which would be to
038r057 siders (P 84 Vol III.) whole of Etna series of COATINGS; hence it will be necessary to state all argu
077r172 ectricity Does not iron, combined with nickel & COBALT (meteoric) resist oxidation?-- Mem Sir W. P. s
095a039 nd clear proofs of shells & waterworn rocks «at COBIJA.» At Iquique of elevation to amount of 30 ft.-
117a102 a extend close to Guayaquil.-- modern shells of COBIJA doubtful. Examine well shores of lakes. to see
450e173 ts. wild hogs-- spotted deer, no loonies, but COCATORES & small green parrots. June 26th-- Yarrell.:-
509q017 rtive bones??? do. Wing in Apteryx ∥ no as Os COCCYGIS-- Turbinated bones? False ribs Wings of Apter
242c017 ommon to Sumatra & Malacca∥ Borneo & Malacca «& COCHIN China» are said to have orang-utang & Pongo in
350d055 nt fixed.-- young of sponges move.-- young of COCHINEAL insects move about & see, parent «(2)», femal
352d057 more perfect-- this is applicable to young of COCHINEAL?? Is there some law in nature an animal may a
272c109 pecies is concerned superabundant,-- the tail in COCK peacock,. widowbird.-- Birds of Paradise. Trogo
275c120 ackles on Bantams by crossing with common Polish COCK «is not that old variety» & then recrossing off
281c138 es it in real classification The relation of all COCK birds in Gallinaceous having tendency to lon[g]
295c178 the others assumed--+++ ∥Daines Barrington says COCK birds attract females by song. do they by beaut
307c216 ll be splendid argument old female, turning into COCK, abortive spurs. growing.-- Are there any abort
312c228 tto. (see Griffith) & half Muscovy ducks, «black COCK & pheasant see Jardines Journal.»-- consult or
341d031 t white streaks.-- &c &c Dr. Bachman has crossed COCK Guinea Fowl with Pea <cock> Hen.-- offspring fe
341d031 . Bachman has crossed cock Guinea Fowl with Pea <COCK> Hen.-- offspring female, yet so infertile heve
357d073 anciently in Britain) between hen Caperailkie & COCK Black-cock.-- (Curious the readiness with which
357d073 in Britain) between hen Caperailkie & cock Black-COCK.-- (Curious the readiness with which this genus
358d085 e.-- The silver & common pheasant crossed, has a COCK (infertile) <with> the breast of which is like
361d095 in tree sparrow, (if I understand rightly) young COCK & hen, all nearly similar.-- in blackbird group
364d105 a, before mentioned between Capercailzie & Black COCK.-- The latter has crossed with the Ptarmigan. s
365d106 been Made.-- In the hybrid grouse between Black COCK & Ptarmigan (probably subalpina.) former has bl
365d106 careful, See to hybrids between Pheasant & Black COCK, & other hybrids-- The fact of Egyptian animals
367d113 iately employed in fighting" instances thighs of COCK & Neck of Bull.-- is most common in vegetable f
378d147 iancy is present in Young birds, one may be sure COCK & hen will be alike-- I presume converse is not
378d147 I presume converse is not true for he says Hen & COCK Starling alike, yet young ones brown.-- Sexual
380d154 s very singular. so many Gallinaceous birds have COCK & hen plumage so different, yet the Cassowary &
383d159 brids between Common & Silver Pheasant, one like COCK & other like Hen.-- one doubts whether they are
385d163 never lose passion. (Mem: so it was said little COCK «yet very odd loosing visible powers» in Zoolog
388d172 es out of same mother.-- The view that man & «¢ COCK» pheasant &c is abortive hermaphrodite is suppo
388d172 s place in old age of female assuming plumage of COCK, & beards growing on old women = Stags horns &
502q11v s.-- History of Pheasant-fowl. Hen coloured like COCK-pheasant: said not to sit on own eggs Flowers i
523m018 hrewsbury gentleman, unnatural union with turkey COCK.-- was restrained by remonstrances on him» whic
557m147 t swans, in parrots &c &c ---- peacock & turkey COCK. in passion.-- Cat when pleased, erect its tail
563n001 s.-- Birth Hill shows it is evergreens they seek COCK Pheasant claps his wings before? crowing & only
563n001 ing & only in breeding season & on the ground.-- COCK fowl. on the ground, at roost, in all seasons,
567n104 lant though it moves doubtless has not.-- Turkey COCK in passion & sends blood to its breast &c &c Al
600o008 au-- How does Hen determine which most beautiful COCK, which best singer-- Remember.-- avarice a comp
564n004 w hours placed on table & when fly ran past it. COCKED its head, & picked it-- Here then, that facult
408e045 particular part of coast of Somersetshire the COCKLES are all apt to be diseased., & some of them sy
258c061 eer, hence males armed & pugnacious (all order; COCKS all warlike)» «thiis wars against in any class:
342d033 he hybrids do not vary (ie the hens all alike & COCKS all alike) More than parent species-- Mr Blyth
368d114 Mem Lucanus & Copris &c).-- In birds singing of COCKS settle point.-- (do the females then fight for
368d114 gn of most vigorous males.-- «(NB. most strange COCKS & hens. being either alike or very different in
377d139 ant with interest» Hyaena. thinks, when pleased COCKS his ears., when frighten depresses them.-- Engl
380d153 uld have wings.--).-- Seep p. 84. Hens «like»-- COCKS from effect of breeding in & in.-- Mr Yarrell d
380d154 female, though many of old female becoming like COCKS.-- It is very singular. so many Gallinaceous bi
453e182 orphous plants, abounding in volcanic islds.-- <COCKS> The possibility of different varieties being r
574n040 of habits,-- for instance the Birgos opening a COCOA nut shell at one end.-- Children & old people g
574n040 ited as Birgos to become absorbed by one end of COCOA nut.-- November 27th.-- Sexual desire makes sal
632j53r of seeds for transportation through the air.-- COCOA nut by water «fucus for adhesion».-- as example
030r037 in Limestone are they allied to the jaws of the COCOS fish Rio Shells argument for rise In Cordillera
054r105 coast of America not reached. Juan. Galapagos. COCOS-- Ulloas voyage North of Callao, the country, t
284c148 --.-- this perhaps on long average equal.-- The COCOS do Mar on the Mahé island,, one the higher part
447e164 -- likewise grasses. &-- very heavy seeds.-- as COCOS do mer.-- Analogy shows some most important end
451e177 ather first of race & if so, fact for my theory COCOS Isld & Preparis between Andaman & Pegu. <have>
263c077 But Man-- -- wonderful Man. "divino ore versus COELUM attentus" is an exception.-- He is Mammalian.-
405e032 vestigate whether. Megatherium & Mastodon are COEMBEDDED in N. America. see my Journal for references
170b001 f February Zoonomia Two kinds of generation the COEVAL kind, all individuals absolutely similar; for
250c039 common to two types of animals What reptiles COEXISTING with Palaeotherium in Paris quarries & at Bin
218b189 men); the possibility of Caffers & Hottentots COEXISTING. proves this-- but when Man makes variety th
065r139 ad long been buried, <see> from rotten state of COFFIN «buried in a mound» long consigned to the eart
524m021 early impressions most durable.-- (but Miss COGAN shows that repetition is not necessary)-- the w
538m078 g songs of their childhood. & certainly of Miss COGAN, & fully corroborates the fact of her not <reme
538m079 , some merry goodhumoured as self.-- «When Miss COGAN has remembered her song, then the song was to h
521m008 y an habitual disease of the muscles.???> Miss COGAN'S memory of the tune, might be compared to birds

```
026r021 can  take place in melted rocks The frequent COINCIDENCE of line of veins & cleavage is importants; v
151g040 Bouhunthine  the summit «doubtless worn into COINCIDENCE» has beach or band of pebbles on line of 4th
305c209 kable manner as Europaean species = singular COINCIDENCE if distinct creation.-- ie.-- a mere stateme
047r082 ot travel up.--) If these view are right the COINCIDENCE retreat at Portugal & Madeira (Lyell. vol I
047r083 riods of rest & vice versâ more likely to be COINCIDENTAL than single elevations along whole line of
477z002 jero) l'yaguaré «the zorilla-arskink» le quiyá (COIPU) viscacha.-- A. Patagonicus les tatous (.4 pich
220b197 f species would fail. But this applies only to COITION & not production. But who can say, whether off
381d156 der crabs.-- (all these however do not require COITION every generation)-- Epizoa & the nematoid Ento
392d175 s)-- [Aphides having fertile offspring without COITION or addition of differences. shows that differe
289c162 e transmitted more easily in those born without COITUS, than with.» ❙Might be given as a hopeless dif
311c225 326 wild ass extending over 90 degree of Long. & COL. Sykes alludes to some other case of 180 degree
486z019 ca & Europe. Entomolog. Transact. Vol I. p. 130. COL Sykes on balls made by dung beetles, like those
501q011 s of great Heat (32) Can Henslow ask question of COL. Le. Couteur about Wheat-- Change of Soil-- cros
503q012 on hybridising ferns, tying them back to back 37 COL. Sykes fertility of men & Europaean animals in I
506q015 uestions amongst agriculturists. whom he know.-- COL. le Couteur on Wheat.-- (41) Have any monooeciou
520mIFC tr.-- (p. 64. On <insect> Ants getting on Table. COL. Sykes) Private Finished. Octob. 2d. This Book f
534m062 Entomological  Society of London Vol. I. p. 106. COL. Sykes on Formica indefessa placed table in cups
534m062 based abased about a foot & then leapt across. (COL Sykes compares this with pidgeons finding their
040r063 t in this case from intersecting a mass probably COLD & not warm as sides of a crater as Vesuvius.--
057r113 that Horse at least has not perished because too COLD:--With discussion of camel urge S. Africa produ
086a007 f Heaven why are tops of Equatorial mountains so COLD.-- Siberia no plants to it, lately raised above
120a108 had  travelled some hundred miles through nearly COLD rock.-- in volcano the pool is not deep.  --Hot
120a109 ometer. Is not common salt more soluble in <hot> COLD than hot water with «-- especially if very  hot
122a114 m so soon from; crust being cut of-- if part of «COLD» crust under ocean, became thicker, then when f
126a123 ctor-- III But equilibrium is not attained, & if COLD water did not percolate surface, would become h
126a123 springs «spring requires connected column.--» of COLD water show. that water does percolate, & spring
132a138 6)-- «NB. I cannot understand the argument, that COLD <oceans> «lakes» bottom. if not colder than mea
132a138 coresby's case volcanicity has warmed it. Is not COLD of ocean accounted for, by the circulation bein
132a139 less to be trusted than any others-- may not the COLD «bottom of» ocean. (with fresh sediment added t
187b062 rees with belief. that Siberian animals lived in COLD countries & therefore not killed by <Siberia  a
187b062 tries & therefore not killed by <Siberia a more> COLD countries.-- Seeing how horse & Elephant reache
258c060 untries, if continually crossed with people from COLD, children would not become adapted to climate.-
258c060 orms remote antagonist powers.-- Every animal in COLD country has some analogy in not gaudy colours s
260c068 of language Desert country. is as effectual as a COLD one. in checking beautiful colours of species--
284c147 erage, equal quantity, 2d on relations of heat & COLD. therefore probably fewer now than  formerly.--
316c245 Some» genera confined to hot countries & many to COLD.-- Hence latitude is more important element tha
343d038 s of animals. as in changing from <hot>. Warm to COLD, damp to dry.-- Thus Tierra del Fuego has <not>
375d135 ame number killed, year with year, by hawks. by. COLD &c--. even one species of hawk decreasing in n
432e125 powers E. frowns prodigiously when drinking very COLD water «frowns connected with pain, as well as i
436e134 ear a very large proportion to other mollusca in COLD parts of sea, like Cetaceae,-- although the Cep
464tf6v must  have migrated to these mountains, when the COLD was intense just like the alpine plants-- In S.
485z018 r north, because during winter they can bear the COLD when torpid.-- On this principle tropical forms
523m018 ipecacuhan-- not acting) in others from drinking COLD drink.-- then brain affected like getting sudde
524m019 auses.-- It is an argument for materialism. that COLD water brings on suddenly in head, a frame of mi
527m033 ging intermediate, who has not had his blood run COLD by singing).-- Granny says she never builds cas
540m085 cter let him compare the American whether in the COLD regions of the North,-- the elevated table land
591n100 t & parental instincts very clear.-- even to the COLD or benevelo- continent man Hume has section (IX
634j55r as.-- Were they then killed out «by the increase COLD», & did mammifers then take their place?  Would
641j29r trib of larger Seals-- Are Porpoises numerous in COLD Oceans I think not.-- Does this bear on, the ab
132a138 ment, that cold <oceans> «lakes» bottom. if not COLDER than mean of place, shows earth not with centr
232b245 which  have succeeded in becoming habituated to COLDER climate whilst others died out, or moved towar
067r143 ains «by Antuco». Athenaeum April 1836 (p302) COLECCION de obras. 2 Vols fol: Buenos Ayres 1836: W. P
272c108 of blattae other Hemiptera stikingly resemble COLEOPTERA.-- Donacia.-- some orthopterous insects & so
486z018 uch further N. in N. America than in Europe-- COLEOPTERA especially require a greater duration of Hea
540m088 s of the air" &c &c &c so is conscience &c &c COLERIDGE,-- Zapoyla p. 117, Galignani Edition Fine poe
610o033 Bentham, & Hartley, &. the school of Kant. to COLERIDGE, is regarding the sources of knowledge.-- wh
057r113 ould Mr Owen consider bones washed about much at COLL. of. Surgeon's? I really should think probably
564n004 his child.-- Octob 3d. Was told by W of Downing. COLL. that he had seen chicken only hatched few hour
455eIBC sleeping mimosa-- do stamina of C. Speciosus. COLLAPSE at night. if so irritate them, «as by an inse
494q005 given  (2) do the stamina of C. Speciosissimus COLLAPSE during sleep & do of Berberis-- (latter I thi
532m057 ave been usual effects of fear.-- the state of COLLAPSE may be imitation of death, which many animals
150g037 Mass  3 2 rather longer than 3a Sunday In Glen COLLARIG, when water up to shelf very shallow  channel
150g037 0 ft wide & river get formed in centre In Glen COLLARIG, on side of Hill of Bohunthine upper road (2)
151g041 lines  in certain parts.-- At the Pass of Glen COLLARIG two little lines of Hill (judging from extern
152g045 with some line subsequent to shelf {P} In Glen COLLARIG, by Dicks theory lake burst in most improbabl
152g045 e part & not in Pass, where shallowest In Glen COLLARIG good case of shelves entering «on» one side r
152g050 e butresses of Alluvium rise nearly up to Glen COLLARIG up within 200 ft of level of 4th shelf= argum
153g052 at successive levels-- {P} Shelf opposite Glen COLLARIG at bend & here most accumulations At  gentler
154g057 ldal channel-- 12ft obscure obscure NB In Glen COLLARIG tidal channel, sides <alm> 15 ft above bank o
156g067 er but quite as perfect as those lines in Glen COLLARIG, & some «other parts» Boulders of same granit
073r161 rom them with steel instruments." In R. Brown (COLLECT: «of F. W.») where the stalactiform masses hav
236b278 saltum.-- Isld bordering continents same type COLLECT cases.-- African isld.-- «How is Juan Fernande
370d117 ae by some fish.-- But birds quite distinct.-- COLLECT cases of difficulty of growing plants in all p
400e012 ion analogous to specific variations.-- Kerr's COLLECT of Voyages Vol 8 «p. 46» Capt Davis in 1598 fo
510q017 ished? does he understand English.-- Miguel to COLLECT facts for me-- what? What does Blume say on al
029r034 merates?) «& absence of limestone?» have been COLLECTED on the open coast. Perhaps as at Concepcion.
163g109 sh Plants Shropshire Quartz what substance is COLLECTED in little spots Speculate on «under head of»
193b093 th coast all the remarkable Australia genera, COLLECTED together Man has no heredetray prejudices «or
325c267 quarto. work on Physiology besides the papers COLLECTED by Owen. (at Shrewsbury) Yarrells Paper on ch
469t151 ed together The seeds of these plants will be COLLECTED & resown.-- Humble 22 flowers of Egg Tree  in
030r035 eason for supposing these matters are not now COLLECTING, in the bottom of an open & not deep sea.--(
301c200 ted.-- An Entomologist going into a country & COLLECTING thousands & tens of thousands New insects, p
467t104 llen protrudes at sheath.-- At last I saw Bee COLLECTING pollen from <sheath> Keel of Lupine-- Seen B
112a088 t of elevated block of stone.-- & Caldcleughs COLLECTION of facts See page 101. in Note Book (C)  for
153g053 elf] river 4th [shelf] Could earthquake cause COLLECTION of sediment? Where ravines enter side by, op
173b009 ng> Species according to Lamarck disappear as COLLECTION made perfect.-- truer even than in Lamarck's
182b047 arabidae.-- instance in birds> Examine «good» COLLECTION of insects with this in view.-- Geogr. Journ
212b164 intersection.-- Dr. Horsfield At India House, COLLECTION of Birds from Java.-- at Leyden series  from
234b256 (Newcastle) about somebody who had made great COLLECTION of birds of Iceland. --M. Gaimard, however,
371d118 ! proof of capability of variation.-- Saw his COLLECTION of «Humming» birds, saw several fully develo
400e011 here is close species of elater.-- Where this COLLECTION is particularly rich. «as in Lucanidae» <no>
526m029 enes, (indeed my American recollections are a COLLECTION of pictures).-- when one remembers a thing i
103a063 supposition  p. 461 «of Proceedings» List of COLLECTIONS in Geological Society. Pumice at South Shetl
483z013 aps be worked out with advantage. with Cumms COLLECTIONS & my own: & Capt. King's p 453-- Planariae v
252c045 inguishing empirically what is species.-- The COLLECTOR is directed to study localities of isld.-«im
304c206 isions merely marked his own ignorance.-- the COLLECTOR who plodding at making a series, which  would
540m089 ple from scenes of distress.-- see how a crowd COLLECTS at an accident,-- children with other childre
509q017 state:  Mr. Horner. On Mr Tremenheres Scottish COLLIERS, when men & women have long worked, whether c
194b094 sil "singe", it cannot be made to approach the COLOBES which in south Africa, appear to represent the
094a036 cimen of rock from Costorphine at Geolog. Soc: COLONEL Imrie Transact Wern. Soc. Vol. 2. p. 35 Sir  J
253c046 ed by savages into Australia.-- What are they? COLONEL Montagu probably contains some facts about clo
362d099 of the female breeding all the year round. ask COLONEL Sykes.-- Even our domesticated cattle have ten
520m001 olour, than with form of body.-- thus the late COLONEL Leigton resembled his father in body, but  his
559m155 bution of British Plants A Volume published by COLONEL in army on "Wheat." in Jersey.-- very  curious
343d038 pushing  into each other from slave trade, & COLONIZATION of S. Africa, so must the tribes become ble
028r031 gos Arch; because no sections> same cause as no COLOUR Sir J. Herschels idea of escape of Heat preven
059r121 all Proportion of Alum: matter.--all pale cream COLOUR.-- The Brecciated structure of all the Pitchst
```

097a043 icially coated by a thin pellicle of a blackish COLOUR like a dull & poor varnish, which I conceive t
189b070 f coat.-- If we could tell, I do not doubt even COLOUR hereditary in time as in space. (Mem: Galapago
225b217 finch black, or iodine on glands of throat, (or COLOUR of plumage altered during passage of birds (wh
234b256 ays he is certain there are local varieties «of COLOUR & size, but not for[m]» of animals.-- He says
239cIFC two pigeons,-- «with specks» which cross & keep COLOUR on wing Effects of colour on parent, white roo
239cIFC s» which cross & keep colour on wing Effects of COLOUR from those of New Holland.-- The New Holland s
240c013 ve phalangista, which differ in «form & head &» COLOUR slightly different. Who can say whether specie
246c026 rather sharper., & rather longer·in proportion, COLOUR of bird in distinct. which they frequent!!?--
260c069 -- the fine green of vegetation.-- ¿account for COLOUR reddish <brown>. ears long.-- like bull terrie
267c094 uayana always hunt in packs «30 or 40 together» COLOUR of plumage & laws, which might probably be red
276c122 of change-- whether beak (as it appears to me). COLOUR & trifling form & head &c to become greatly ch
282c141 se perfect, we must suppose men instead of mere COLOUR of plumage in same remarkable manner as Europa
290c163 ogy NB Pyrrho-alauda (bird of St Jago) of brown COLOUR; lives on ground, colour of habitation. Must h
290c163 d of St Jago) of brown colour; lives on ground, COLOUR of habitation. Must have some effect.-- Maldon
303c204 ld.-- The races of men differ chiefly in <size> COLOUR, form of head «& features» (hence intellect?)
303c204 tween man & monkey, would have differed in hair COLOUR + + + form of head «& features»;. but likewise
305c209 uropean species, and the sexes of which vary in COLOUR of plumage in same remarkable manner as Europa
309c221 4.-- do. p. 186. quotes Burkhardt to show black COLOUR of certain Arabs.-- NB avoid quoting these hac
317c249 ds & goldfinches differed considerably in their COLOUR & appearance Every now & then a short-tailed c
338d024 says «that strength of» hair goes with   COLOUR. black being strongest.-- V. p. 63. Note Book
346d048 rush as all animals which breed, in & in are-- COLOUR white, uniform.--crafty, go in file, hide thei
371d127 located] that it shall beget young different in COLOUR, form, & so altered in disposition, as to be m
371d127 cases an unseen change is produced in parents--COLOUR is a doubtful subject, but what other instance
377d140 our system in the Heaverns.-- Is not puma, same COLOUR as Lion. because inhabitant of plain & Jaguar
377d147 e young of the <p> Kingfisher (.p. 169) has the COLOUR on its back bright blue.-- «thus young of» Man
399e009 ndoubtedly rabbits in poulterer shops., of same COLOUR as a Hare, but paler & buffer.-- with long ear
400e012 much,-- kind of fur-- (do tips of ears take any COLOUR?)-- length of tail varies, & character of fur-
403e022 Macleay says, that <every> «any» character even COLOUR is good. (ie invariable) in some classes-- it
405e031 ar {P} mark on each side darker,, so that whole COLOUR is changed, these best marked characters are p
405e031 ry, if in wild state; thus mark on ear of cats, COLOUR can be brownish do Saw what was said to be hyb
441e150 of organization" had better be shown-- soil on COLOUR of flowers, Hydrangea -- black bullfinches-- &
449e168 has laid in domestication eggs of two shapes & COLOUR.-- Eyton has observed same thing in Brent Goos
451e176 made at it. Both his parents were of the usual COLOUR. His sister is an albino like himself said not
459t013 in the manner of walking but not waddling; its COLOUR was darker than the penguin & the bright feath
472s01r . & out of 60 seedlings not one came up true.-- COLOUR of flower & foliage not «being» like the true
511q018 s.-- (6) <Mad> Porto Santo Rabbit. Descript. of COLOUR «& length of ears» & skeleton, & skin= Van. Vo
515q022 ither parent.-- May. 44 These Hybrids differ in COLOUR of beak, taking after male & female parent.--
520m001 mplaints «& mental disposition» oftener go with COLOUR, than with form of body.-- thus the late Colon
528m039 etry & rhythm applies to the view as a whole.-- COLOUR «& light» has very much to do, as may be known
540m087 so different-- in organization.-- Same cause as COLOUR & shape & ideosyncracy.-- Look at the Indian i
560m156 pit, or cry.-- <fe> Shame, independent of fear: COLOUR of bare nails--, & of eyes.-- Do female monkey
026r023 eriae, usually petrified into a peculiar cream-COLOURED Limestone: the strange substitution of matter
145g002 t Salisbury Craigs The Highland shepherds dogs COLOURED like Magellanic fox.. an instance of Provinci
247c029 e the cattle, <">which I say "are as variously COLOURED as a herd in England"-- Black & Grey varietie
265c088 ance of bird belonging to family with peculiar COLOURED plumage, where colours have changed in accord
266c088 nd.-- why do beetles & birds & <f> become dull COLOURED in sterile countries.-- Gould insist much upo
326c266 tch of south Sea. Botany R. Brown. has curious COLOURED maps. by Copenhagen Botanist of range of plan
345d044 ieties are made.-- The Highland Shepherd dogs, COLOURED like Magellanic Fox.-- peculiar hair & appear
355d066 o Zebras.-- «Mr Herberts variety of horse, dun-COLOURED with stripe approaches to ass.» or fowls to t
359d089 here were four, two like each other & two dark-COLOURED & different.-- -- the former were the parents
367d113 brown.-- animal like large, heavily made cream COLOURED ass.-- stripe on back also.-- legs reminded m
371d127 r wild dogs of India) in Zoolog. Garden having COLOURED offspring.-- but surely in all these cases an
383d160 sant. <wh wh> why green? & not purple?-- legs pale COLOURED.-- In the back feathers, we have character ve
404e026 le to offspring:‖ Australian dogs have mottled COLOURED puppies case of this.-- tendency in «manner o
466t093 it base of upper petal, though not differently COLOURED-- & stamens bend up a little In a wild purple
466t094 acility of alighting? which is not differently COLOURED & to which stamen & pistils have no relation.
493q003 ory of Tortoise-shell Cats. as only one sex so COLOURED = I have grey-cat «wh was female» with tinge
502q11v ans. cabbages.-- History of Pheasant-fowl. Hen COLOURED like cock-pheasant: said not to sit on own eg
528m036) from the splendour of light, especially when COLOURED.-- that light is a beautiful object one knows
540m086 of <B> Van Diemens land & the energetic copper COLOURED natives of New Zealand)-- the American in Bra
573n036 ge admires not a steam engine, but a piece» of COLOURED glass <&admires> is lost in astonishment at t
146q011 with white black legs & tail like species in COLOURING Strike an analogy between pleasures of associ
185b058 ructure obliged to be altered, I can conceive COLOURING retained; therefore probably in some <heterom
186b058 ined; therefore probably in some <heteromera> COLOURING of crysomela may be going back to common ance
234b256 s Stephens say he can at once tell by general COLOURING a group of Nebria complanata from. Devonshire
261c071 a great apparent difficulty-- preservation of COLOURING. when form has changed. But that anim
265c085 ered an adaptation as best attempt of nature, COLOURING matter being absent.-- again dwarf plant in a
273c100 oped» family (Gould says) there is any marked COLOURING of plumage (as «black & white» bars on wings
312c231 m clubs, they used to fix on the kind wanted, COLOURING of each feather weight & size & they would pr
408e043 hen we know what a great effect. light has in COLOURING plants,-- who can say. what <light> «colours»
484z016 of my birds-- These groups strictly American. COLOURING on under side of wings It would be interestin
189b070 ttle wings of Apteryx Dacelo & Kingfisher same COLOURS Strong odour of negroes, a point of real repug
247c029 te Book Rabbits introduced in 64, of very many COLOURS, like the cattle, <">which I say "are as vario
258c060 in cold country has some analogy in not gaudy COLOURS so all changes may be considered in this light
260c068 effectual as a cold one. in checking beautiful COLOURS of species-- Mem. St. Jago--solitary Halcyon b
265c088 o family with peculiar coloured plumage, where COLOURS have changed in accordance to habits.-- one is
306c214 h Java?? Terrestrial Planariae assuming bright COLOURS., good instance of colours dependent on locali
306c214 iae assuming bright colours., good instance of COLOURS dependent on localities.-- -- Hamilton will gi
306c214 m Asia Minor.-- tail like setters. long ears-- COLOURS vary, but form constant.-- The females of some
312c227 ial birds, & tigers & sharks, being spotted, & COLOURS of little value Dr Smith if black & white Man
363d101 d feather, pouters, fan tails are found in any COLOURS of plumage &c &c «Pouting pidgeon exaggeration
378d148 n of any identical bird-- for they were of all COLOURS.-- they were "half wild-half tame, they came t
405e031 rked characters are partly retained, therefore COLOURS vary in same Manner as they would vary, if in
408e043 colouring plants,-- who can say. what <light> «COLOURS». acting. by a most delicate organ, on the who
450e175 like in wild animals).-- There are prevailing COLOURS in the different islands.-- The horse is only
466t095 s <fr> before sex developed-- Double flowers & COLOURS breaking only hereditary characters, wh. come
493q003 cies of close genus do more than three primary COLOURS occur in relation with species-- answered «by
494q005 s Raise seedlings surrounded by various bright COLOURS, any effect? and silk caterpillars (1) Shake a
528m036 sic), this probably arises from (1) harmony of COLOURS, <whi> & their absolute beauty. (which is as r
546m109 owing to parent being fruit eater.-- origin of COLOURS?-- Nothing shows one how little happiness depe
581n064 doms-- "food, sm<e>ll. (ourang-outang), music, COLOURS we must suppose <we> «Pea-hens» admire peacock
606o023 ll allow the secondary pleasures of harmonious COLOURS &c &c surely to be added. Lessings Laocoon p.
154g060 ttle more {P} now that it has got to the rock of COLS if--. why should it deposits River terraces oft
532m054 to fix upon some object:-- When a man, child or COLT has once been frightened & started much more ap
466t095 g age & sex, opposed by cantering horses having COLTS which can canter-- & DOGS trained to pursuit ha
582n067 constitutions will then be cut off.-- <Horses> COLTS cantering in S. America capital instance of her
245c024 about this Lesson insists much.-- The (p. 296) COLUMBA Kurukuru found in all Malasia-- & oceania, off
246c026 ho can say whether species & varieties P. 708. COLUMBA Oceanica (Less) inhabits Caroline < «NB. The»
129a132 axis, «20 ft at least in depth» near mouth of COLUMBA river-- Read Mr Parker's Book.-- M. Bichoffs
466t094 e, but does not prevent bees visiting it.» In COLUMBINE nectaries are placed all round flower as they
126a123 deep seated springs «spring requires connected COLUMN.-- of cold water show, that water does petrol
070r153 ar structure.--When we recollect connection of COLUMNAR & orbicular in basalt.-- When we see Avestruz
070r153 except by trees The structure of ice in COLUMNS. show that granite when weathering into balls.
586n082 tively exerts in concert with others in building COMB-- My faculty often will turn out to be instinct
269c102 imate relation of Life with laws of Chemical COMBINATION, & the universality of latter member-- spont
077r172 uctor of Heat do of Electricity Does not iron, COMBINED with nickel & cobalt (meteoric) resist oxidat
099a049 s of equal weight.--); cleavage not vertical ‥ COMBINED with gravity.-- hence changes in dip of no so
259c063 ither direct effect of habit, or heredetary «& COMBINED» effect of habit.-- perhaps in process of cha

```
335d016 pring infertile,-- two considerations are here COMBINED). In last page, we have seen mules could have
384d161 addition of parts, (3) deficiency of parts (4) COMBINED addition & deficiency of parts, as in Hermaph
622o049 o ridiculous. [RHC] the social instinct may be COMBINED with feeling towards one as a leader,-- the c
369d115 . 482 (Same book) Owen says "the necessity of COMBINING observation of the living habits of animals,
509q017 ribs Wings of Apterix: clavicle in--? Combs in COMBLESS Poultry-- Teeth in foetal state: Mr. Horner.
509q017 es? False ribs Wings of Apterix: clavicle in--? COMBS in combless Poultry-- Teeth in foetal state: Mr
028r029 ly by some unknown Volcanic process? How does it COME that all Lime is not accumulated in the Tropica
073r161 jasper.--do undulations (as Hutton says) always COME from without.-- "True native iron that to which
090a021 r upon the sorting of matter. in making trachyte COME out before.-- What must be the effect of all th
098a045 ients) as uniting with cretionary.-- it may <of> COME of use in discussion on Cleavage &c Geolog Tran
136a147 letter in drawer with important letters-- When I COME to treat of the age of the Pampas Deposit, I ma
157g076 rnets.-- Boulders as before certainly must have <COME> «been drifted» here: on very summit no granite
171b002 ind, & therefore relations to other life has not COME into play)--See Zoonomia arguements, fails in h
177b028 more linear arrangement.-- ¿How is it that there COME aberant species in each genus «(with well chara
178b030 es have passed over, & where other species have <COME> «"air" of that place. Will it be said, those h
179b032 borne Mongrels he has thus seen the black blood COME out from the grandfather, (when the mother was
182b050 nd Mr Gould says in sub-genera, they undoubtedly COME from same countries.-- In mundine genera, the n
183b051 .-- In mundine genera, the nearest species often COME very remote quarters. (NB. of Plata Partridge «
185b055 not be explained on principle, of animal having COME to island. where it could live-- but there were
188b068 ce ever so many seeds of white flower. all would COME up white, though planted in same soil with blue
195b100 e.-- like Mammae on mens' breasts.-- How does it COME wandering birds. such sandpipers. not new at Ga
212b165 bell, & then run into Kennel to watch who would COME to the door-- would constantly do this, so was
228b228 causes of change «in order» to know what we have COME from & to what we tend.-- this & «direct» exami
228b230 ts, under glass bells & see what offspring would COME from them. Ask Henslow for some plant, whose se
244c021 f Payta-- belong to the hairless kind, «said» to COME originally from Africa p. 122. Mus decumanus, a
257c058 division, our «only» question is not, how there COME to be fishes & quadrupeds, but how there come t
257c058 re come to be fishes & quadrupeds, but how there COME to be, many genera of fish &c &c at present day
262c072 of wisdom virtue? «like senses of savages» (How COME its some countries patriotic?)-- but more espec
263c077 s not a deity, his end «under present form» will COME, (or how dredfully we are deceived) then he is
264c080 ould seems to think that many species when close COME from different localities, as my Furnarii.-- so
266c088 ve without knowing from which country many birds COME it would be impossible to classify them.-- I wo
277c127 them) which are found together.-- If two species COME over to this country, without range, or habits
299c194 sown in dry station it will for some generations COME up so.-- there are not Many intermediate shades
306c212 tinued long in a passion & looked out for him to COME again very differently from dog. perhaps being
306c214 country.-- the long-haired cats are supposed to COME from there.-- All the sheep are thick-tailed Th
306c214 e dogs called Persian «greyhounds» are Kurdish & COME also from Asia Minor.-- tail like setters. long
313c236 «this very important in considering how children COME to suck or other actions in foetus of Mammalia,
316c247 s with identity of land animals.-- these however COME from Siberia it cannot be said American fossils
318c254 of birds.-- «in» some species «a Tanagra» Males COME first & the females in flocks. «as in English N
332d005 as a cat. which he bought in Portsmouth, said to COME from coast of Guinea, -- ears bare. skin black
337d022 even «hyaena & squirrel» seal & mouse, elephant, COME from one stock.-- Theory of Geograph. Distrib:
349d051 «in the Cetoniadae»,-- when will ornithorhynchus COME in circle?!!! p. 8-- Anomalous structures, as i
349d053 ne stock, & that stock with other subgenera will COME from. common stock.-- all genera, common stock.
360d091 is Keeper has seen when sickly tigers have first COME over, insects somewhat like «between» lice & fl
390d178 in would not be deletereous if the relative had COME from different quarters) then it causes <to> a
401e015 Says if any variety of apple be sown, all sorts COME up from it. lately saw a nonpareil sowed by Mr
408e046 Lamark surprised to see how many Tropical genera COME from New Holland, ¡Sydney? The dog being so muc
411e055 generalization. of teeth to kind of extremities COME under this head» 27th November When summing up
449e170 says some of the Galapagos Heteromerous insects COME very near to Patagonian species-- p. 18. of Tem
463t063 es-- being a mere fragment of the discoveries to COME-- Owen in his description of my fossils makes s
466t095 colours breaking only hereditary characters, wh. COME on in after life of Plants-- also goodness of f
473s07r reason, why the peculiarities shd be born,-- may COME in corresponding time of life of offspring-- No
482z012 with Dumeril says that two species of Tortoises COME from Galapagos!!! Azara. Voyage dans l'Amerique
495q05a bell-glass-- sow these seeds & see if they will COME up true-- whilst others are crossed.-- Are Bees
500q10a there are very close species & see whether they COME up, not found in wood.-- but seeds continually
502q11v cters of races of different vegetables & animals COME on.-- Compare calves.: Compare young. beans. ca
521m009 n & had no gardener.-- My F. then asked Mr C. to COME to the window & pointed out the Gardener & said
523m017 because often, if not generally, does not really COME on suddenly.-- Case of Mrs. C. O. who threw her
529m040 ures & therefore imagining pleasure of imitation COME into play.-- the train of thoughts vary no doub
547m112 whether I had rung for Covington. whether he had COME & opened box, whether I had thought what clothe
549m121 I should better the last be right. The two rules COME very near each other.→ The rules to mortify you
551m129 ockarts life of W. Scott Vol VII p. 35 "as ideas COME. & the pulse rises, or as they flag & something
579n060 tation on a piece of skin cut off made the blush COME.-- it is an excitement of surface under the wil
603o11b th beautiful! p. 136. Says Architecture does not COME under imitative art [my view says yes. <old> ma
608o026 imperfect condition of mind all motives do not. COME into play.-- †It may be urged how often one try
613o035 does consciousness commence; where other senses COME into play, when relation is kept up with distan
616o39v out the mode «of perceptive action» by which we COME to conceive of matter as attracting & shew that
046r078 alt. lime, are spread over «whole» surface; how COMES it they do not flow out together? How are they
090a023 we say «100» <5>0 lbs a year too little.-- How COMES it none in fossil state? suppose «100» <5>0£ x
126a123 tter be driven upwards & so conduct heat?-- How COMES it in volcanos that have gone on for thousands
240c013 be Zoologie. p. 60. Vol I. Cynocephalus. niger. COMES from the Moluccas «Matchian» & Celebes.-- ◊ Amb
366d112 s might be used» The wild ass has no cross. how COMES it that the tame donkey has. CD[old Buffon shou
380d155 copulated., though no offspring, Milk sometimes COMES in Mammae & even when bitch is in heat.-- Yarre
414e063 ther, bring diseases to each other &c, but then COMES the more deadly struggle,, namely which have th
436e136 <pi> species,-- for instance pidgeons-- : then COMES question of genera It certainly appears that sw
439e144 where there is much facility in crossing there COMES the impediment of instinct-- the possibility of
467t103 ich forces out from extremity pollen, or pollen COMES out with anthers & stigma in slit»-- As I thin
478z004 ssured wild dog copulates <.> freely with tame: COMES to houses on purpose Mr J. Murray has given pap
482z017 to those of mine from T. del Fuego p. 141. How COMES it salt water so soon putrifies?? p. 319. on Hy
516q023 of Standard Apricot grafted on what, & see what COMES up.-- [Unnumbered blank] Experiment Cover patch
523m018 erally, if at all discovered.-- <Sup> Sometimes COMES on suddenly from <I> (in one case ipecacuhan--
524m019 ike the passion, ill-humour & depression, which COMES on from bodily causes.-- It is an argument for
531m053 send a mandate to the muscles & when the noise COMES it cannot help doing it.-- Fanny Hensleigh doub
533m060 ding to make myself in act less grateful.-- How COMES this tendency in these cases? How did my mind f
558m151 cting in unison.-- active assistance. &c &c. it COMES to Miss Martineaus one principle of charity.--
565m007 of hounds». <when> as fear to «man as» animals. COMES at distance, mouth is placed open.-- Hence beco
574n041 cestors having been like dogs to bitches.-- How COMES such an association in man.-- it is bare fact,
577n050 take place every day, & closing of the leaves, COMES on from want of stimulus, after certain other a
582n068 it does not do it.-- My theory explains how it COMES that the heart is the seat of the emotions.-- b
590n093 g languid circulation-- no, for «grief» sighing COMES on before circulation is affected. p. 44.-- Jea
606o022 how obtained.-- implanted in our bosoms.-- how COMES it there? Laocoon p. 75 "The beauties developed
612o035 and are subject to accident; the sexual willing COMES on period of year as much as inflorescence.-- [
614o037 instinctive feelings. [LHC] As sexual instinct COMES on late in life, man almost alone in this case
621o047 g nasty. (<& accidentally» «by odd association» COMES to this conclusion, not owing to peculiarity of
625o50v not cool malevolence). it is only after reason COMES into play that anger can be said to be wrong.--
629o55v how the feeling of ought, shame. right & wrong COMES into mind in first case-- seeing how shame is a
633j53v (Anastatica) is rolled along, & splits when it COMES to a damp. place.-- Kolreuter mentions some hyb
376d136 t at first, but in two or three days learn its COMFORT & though could not put it on, yet threw it ove
522m014 so of depression, & both these give strength & COMFORT to the body» I know the feeling, thinking over
595n115 commenced a most lamentable howls & & was not COMFORTED until the Keeper took it <her> in his arms &
610o032 &c-- All habits must conduce to their health & COMFORTS.-- Both ideots, old People & those of weak in
024r013 orges Sound The site of the water at Cauquenes. COMING from the [...] Cordillera & flowing The gradua
071r156 most abundant tree-- Metamorphic action: <most> COMING so near surface most important There is map of
079r177 . 58» At Acapulco earthquakes are recognized as COMING from three directions. from W. NW & S.--last t
148g026 t of the «YOUNG» Shepherd dogs Saturday. Before COMING to Bridge of Spean, hills of «sea», gravel, cu
175b019 of organization improving as Owen says simplest COMING in & most perfect «& others» occasionally dyin
181b043 elements, from the «infinite» variations, & all COMING from one stock & obeying one law, they may app
```

Page **(Key Word)**
366d111 d handle, though really fully aware she was not COMING in, could not help being perfectly distracted
383d160 but the black lines on each feather instead of COMING to point {P} are more rounded. {P} & much broa
426e108 ther.-- -- Little Miss Hibbert case of Hindoism COMING out more than in mother or indeed grandmother;
431e120 nsidered merely varieties. & even Mr Sowerby is COMING to this conclusion, from specimens in grades,
455eIBC at night. if so irritate them, «as by an insect COMING always at same time» see if by so doing can be
523m015 father his feeling of consciousness of insanity COMING on.-- his struggles against it, his knowledge
523m017 y is not distinguished from whims passion &c by COMING on suddenly. Ans no.-- because often, if not g
604o017 -- Notices. struggle <between> when insanity is COMING on «Thinks clearest analogy between dreams & i
617o40v can.-- 4) the source from which it arises. But COMING round to the <subjective> aspect of action as
593n111 ome way from water, suddenly, as if by word of COMMAND, they all took flight & flappered across pool
058r118 agnes de cette île se developpent autour d'elle COMME une ceinture d'immenses remparts; toutes affect
079r177 t the "chefs d'oeuvre de l['']industrie humaine, COMME ils traitent les loix de la nature & ses phenom
244c023 ron Cuvier, nous ne balançons pas a la regarder COMME une espèce distincte! p. 171. Sus papuensis «pa
567n013 ends on such confidence» when does such notion COMMENCE?-- Children understand before they can talk,
568n018 s part of reason Octob. 19th. Did our language COMMENCE with singing-- is this origin of our pleasure
613o035 mpathetic of man [RHC] ¿How does consciousness COMMENCE; where other senses come into play, when rela
628o53v inally, if the shepherd dog had no instinct to COMMENCE with scarcely possible to teach it-- all dogs
028r032 contradicted by Cordillera, where that action COMMENCED before any great accumulation of such matter.
170bIFC 76 79 91 93 107 Ireland 113 117 This Book was COMMENCED about July. 1837 p. 235. was written in Janua
173b012 when they ceased importation ceased & changes COMMENCED.-- or intermediate land existed.-- or they ma
177b028 whence, two favourable points of organization COMMENCED branching.-- As all the species of some gener
416e071 effect of liquid semen: therefore animal life COMMENCED in the Water!» It is a beautiful part of my t
566n012 connected with law.-- as soon as any enquiry COMMENCED, for instance probably such a thing as thunde
569n020 d as perfectly as a man.-- Probably, language COMMENCED in some necessary connexion between things &
595n115 monkey brought it at opposite end of house. & COMMENCED a most lamentable howls & & was not comforted
599o005 without it.-- quotes Ld Mondobbo.-- language COMMENCED in whole sentences.-- signs-- ¿ were signs or
021rIFC w supposes the simplest infusoria same since COMMENCEMENT of world.-- [not located] La. billardiere m
440e148 f growth «Lamark. Vol II. p. 120. observes it COMMENCES only, when growth stops».-- Spallanzani's fac
558m153 out of the side part of nostril, when passion COMMENCES.-- <All> Nearly all will exclaim, your argume
584n073 f helplessness E. says she can perceive sigh, COMMENCES as soon as painful thought crosses mind, befo
204b141 itting in orders, Carnivora, rodents &c, JUST COMMENCING. Kirby says (not definite information) West
262c072 as change in typical structure?!! Whewell «in COMMENT/ few will dispute--» says civilization heredet
327cIBC tarum proesentum crypt. transitu et analogia COMMENTATIA Library of Useful Knowledge on Horse & Cow &
443e155 hey must have <done> been propagated by sexual COMMERCE «The fact of Corallina & Halimeda is case in
152g047 d 2 miles North of Grant town to Forrest road COMMINUTED shells Important contingency if elevation fr
578n055 A. blushed. analyse this:-- Let a person have COMMITTED any «concealed» action he should not, & let h
021r005 mentions the floating marine confervae, is very COMMON within E. Indian Archipelago, no minute descri
026r019 lomerates: Yet this view is directly opposed to COMMON opinion The Tertiary formation South of the Ma
026r022 onybeare. Introduct XII P. silicified bones not COMMON in Britain. Mem Concepcion Says Echinites. Enc
038r057 . Sound. C. of Good Hope.--Carnatic It has been COMMON practice of geologist. Lyell considers (P 84 V
041r065 eve in all flat countries. years of drought are COMMON.--Mr Lyell has mentioned the drifting of carca
050r091 razil. <gravel becomes> sand less & gravel more COMMON. the shoaler the water & nearer the Banks Is t
056r110 ists degradation longer than outer parts.-- The COMMON occurrence of a breccia of primitive rocks bet
062r130 f circumstances: The same kind of relation that COMMON ostrich bears to (Petisse. & diff kinds of Fou
070r153 nsensible change.-- Yet one is urged to look to COMMON parent? why should two of the most closely all
075r167 -‖ muriated silver. which is so rare in Europe. COMMON there accompanied by molybdated lead & «argent
076r169 artial pyrites; great blocks of pure silver not COMMON in <S.> America: In all climates distribution
088a015 of Fish Let a Hawk fly at Heron.-- Ceratophytes COMMON in Northern seas p. 312. Chamisso in Kotzebue.
088a016 nde p. 248. L. Institut 1837.-- Helms remark on COMMON salt being found on low hills East of Cordille
089a020 enne. Syenite & diorite, covered with iron clay COMMON to Guyana said to extend to Cordillera I see B
102a060 ch particle coated. &c will be aware how little COMMON Gravity has to do with arrangement of particle
104a066 certainly contains 6 per cent more silica than COMMON felspar therefore on axis of Cordillera, in An
115a096 enious crystals., must aid in adding effects to COMMON heat.-- Where there are cliffs there ought to
120a109 ot. & therefore I doubt the thermometer. Is not COMMON salt more soluble in <hot> cold than hot water
145g006 rt «of Salisbury Craigs» 25 degree perhaps most COMMON-- Will not curved form of hill be explained by
173b009 me. Gray's remark, best known species. (as some COMMON land shells) Most difficult to separate Every
174b014 almost proved.-- We can see «why» structure is COMMON in certain countries when we can hardly believ
179b034 for species they certainly are according to all COMMON language) will keep to their type: in animals
183b052 linotte, avec le verdier" &, <fr> silver gold & COMMON pheasants & fowls..-- "On sait que le "métis"
186b058 ra> colouring of crysomela may be going back to COMMON ancestor of Crysom. & Heterom, but I cannot un
188b066 ies to England.-- Mem. Shew Mice.-- --- Animals COMMON to South and North America.-- ¿are there any?
190b077 separated.-- Proteaceae & other forms (?) being COMMON to Southern hemisphere, does not look as if S.
191b080 & Agouti same as on continent-- 3 Paradupasi in COMMON to Van Diemen's Land & Australia From the cons
191b080 Europaean formations-- England & Europe Ireland COMMON animals-- + + + for instance tertiary deposits
192b087 e been grade between pig & tapir, yet from some COMMON progenitor,-- Now if the intermediate ranks ha
194b097 not the slightest right to say there never was COMMON progenitor to Mammalia & fish. when there now
200b121 s. petrel & alk. do not show the possibility of COMMON branching off.? Accra, Coast of Africa. Clay s
202b130 animals bearing relations to a third body., or COMMON end of structure A Race of domestic animals ma
204b139 ut half aways & result the same Indian cattle & COMMON produced very fine Hybrid offspring, much larg
204b141 turned on pool, first season bred readily with COMMON ducks.--» Kirby all through Bridgewater errs g
206b150 p. 320 <Think> States Cryptogam. Flora formerly COMMON to New Holland?! p. 320. Says Coniferous struc
209b155 story of species between Indian cow with hump & COMMON;-- between Esquimaux & European dog? Yet man h
212b166 inct of dogs.-- agreement & reason Some animals COMMON to Mauritius & Madagascar.? Proceedings of Zoo
217b188 erers.-- Iceland no species to itself, a remark COMMON to all northern islds.-- This is interesting,
219b195 space from mountains low down, therefore plants COMMON take an example from T. del Fuego.-- Ellis (?)
224b214 rom Lamarcks» Without two species will generate COMMON kind, which is not probable; then monkeys will
229b232 hey may partake, from our origin in <there> one COMMON ancestor we may be all netted together.-- Herm
229b233 crossed with English Bull. offspring very like COMMON English.-- Hottentots say great tailed sheep a
234b256 terhouse finds certain varieties of a Harpalus. COMMON at South end, but <rare> «absent from» near Lo
239c001 g would happen with Australian dog & any of our COMMON varieties. He has no doubt that Chesnut, for m
239c002 ell states that if any odd pidgeon crossed with COMMON pidgeon, offspring most like latter, because o
241c015 like woodpecker.-- Birds of Australia. Many in COMMON ¿species? with New Guinea.-- Many <genera>. ki
241c015 pecies? with New Guinea.-- Many <genera>. kinds COMMON to New Guinea & rest of isle in E. Indi: Arch:
242c017 oth seem to have elephant & has orangs, ‖ Tapir COMMON to Sumatra & Malacca‖ Borneo & Malacca «& Coch
242c017 China» are said to have orang-utang & Pongo in COMMON‖-- Galiopithecus common to Moluccas & Pelew Is
242c017 orang-utang & Pongo in common‖-- Galiopithecus COMMON to Moluccas & Pelew Isds.-- p. 22. New Calidon
244c021 of brick" hair, very stiff, p. 120-- Coati roux COMMON. near Concepcion. some tatous!!! p. 120.-- Mos
247c028 «New Guinea, New Ireland» & «even» Java. & very COMMON on Otaheite-- according «stated in note to p 2
250c037 n more extreme both in, N & S. America, is only COMMON cause I can conceive of destruction of Great A
250c039 not cause of destruction of large quadrupeds.-- COMMON to two types of animals What reptiles coexiste
266c092 ls on isld.-- Niether Hyenas; jackals monkeys-- COMMON to either coast not found here not even Antelo
266c092 coast not found here not even Antelopes, though COMMON on coast of Arabia not even antelopes though c
266c093 on on coast of Arabia not even antelopes though COMMON on islets off Arabian Coast.-- ‖Vol VI. p. 89.
268c099 nal features of land by seeing how many species COMMON to Patagonia desert & Tierra del Fuego. & fore
272c109 bgenera in families.-- thus the banded tarsi is COMMON to all the Laniadae & Muscicapidae of new Worl
274c116 rica, Australia, & S. America very few forms in COMMON.,-- but each several with Europe & northern As
275c120 P} point on hackles on Bantams by crossing with COMMON Polish cock «is not that old variety» & then r
286c154 gance, to raise on the same shelf-- to (look at COMMON ancestor, (scarcely[)] conceivable in savages)
288c160 ny No XI p. 390. a slight change in enclosing a COMMON seems in part of-- to have almost banished the
289c160 llow Wagtail never seen in one district, though COMMON on another, (golden creted wren so rare in som
290c164 irds bones-- in strata of Tilgate forest Seeing COMMON gull in garden at Zoology Soc. it's pale ash g
290c164 (hence hybrids in this case have bred). White & COMMON pheasants. have crossed.-- </I saw> .-- The at
293c174 o, but only between laws--]CD. Many diseases in COMMON between man & animals-- Hydrophobia &c cowpox,
293c174 man & animals-- Hydrophobia &c cowpox, proof of COMMON origin of Man.-- different contagious diseases
304c207 rds.-- will having many trifling characters, in COMMON with other birds reveal the secret.-- Now all
312c228 .-- like dogs Smith knew chinese hairless dog & COMMON spaniel crossed.-- 3 puppies PERFECTLY like ch
315c240 at (Brown) is found in Isd of Bass' Straits The COMMON Mush room & other cryptogamic plants same in A
316c246 merica.-- Mr Sowerby says there are some shells COMMON to West coast of Africa & E. S. America.-- get

Page		(Key Word)	
317c250	irable table of plants of St. Jago showing many	COMMON	to Canary. isld., Europe, & St Jago upper regi
318c253	merican Pipra Flycatcher, which is now becoming	COMMON--	likewise of the Hirundo fulva (added by Audu
318c254	lish Nightingales» -- other birds (& this seems	COMMON	«kind» migration of America) migrate singly fl
318c255	- There is a kind of Wren (Bebyk??) which seems	COMMON	in Rocky Mountains & on one lofty isolated spo
331dIFC	etween Lion & Tiger at litter as numerous as in	COMMON	lion? Are the number of nipples in domesticate
333d005	ry closely» of form of mother: more than of the	COMMON	cat.-- Ch IX Mongrels Hybridisim Fox has half
333d005	x has half Persian cat. which bred with unknown	COMMON	house cat.-- had four Kittens. two appeared «s
333d006	- had four Kittens. two appeared «so» very like	COMMON	cat, that they were killed, & other two very c
333d007	as seen several crosses between Esquimaux dog &	COMMON	dogs & Fox thinks they decidedly take much mos
334d009	usin «one of Mr Strutt» of his used to breed to	COMMON	& Muscovy Ducks.-- English. <Common> «China» &
334d009	to breed to Common & Muscovy Ducks.-- English.	<COMMON>	«China» & Canada Geese, & that they this firs
334d013	er the other & younger cow.-- {P} Fox says when	COMMON	& China goose are crossed the neck is not inte
334d013	e in its peculiar long neck, but much nearer to	COMMON	goose.-- What has long been in blood, will rem
337d021	th some training in the earlier branches «as in	COMMON	greyhound» & much intermarriage.-- In my specu
340d026	in animals. Heard at Zoolog Soc their Pintail &	COMMON	Ducks, breed one with another-- & hybrids fert
342d032	ly breeds «in Carolina» for his table Muscovy &	COMMON	ducks-- they are produced in full equal Number
342d032	in full equal Numbers with pure bred (just like	COMMON	mules) & lay many eggs but never produce inter
342d033	ts & grouse different.--» (if so chinese pigs &	COMMON	must be considered as distant species?? or is
342d033	pring. It appears certain that hybrid Muscovy &	COMMON	duck have been shot wild (escaped from Carolin
349d053	that stock with other subgenera will come from.	COMMON	stock.-- all genera, common stock.-- so that v
349d053	era will come from. common stock.-- all genera,	COMMON	stock.-- so that value can only be judged of i
350d055	ction of individual, though such relation seems	COMMON,	but that perfection consists in being able to
353d061	- Waterhouse knows three species of Paradoxurus	COMMON	to Van Diemen's land & Australia well develope
358d085	character of not having sexual plumage is very	COMMON	by hybrids, that are infertile.-- thus the com
358d085	mmon by hybrids, that are infertile.-- thus the	COMMON.	pheasant & fowl when crossed never even lay e
358d085	dly» tell any sex by appearance.-- The silver &	COMMON	pheasant crossed, has a cock (infertile) <with
358d086	(infertile) <with> the breast of which is like	COMMON	pheasant & back like silver.-- But the hen hyb
359d088	sual number out of any one nest. even more than	COMMON	duck-- Male Penguin was crossed with hen Canad
361d095	ter in adult stage, other alters entirely]CD In	COMMON	sparrow young & female similar plumage.-- in t
363d103	lackbirds Dr Bachman told me that 1/2 Muscovy &	COMMON	duck were often caught wild off coast of Ameri
365d105	ts of Scandinavia these birds are very far from	COMMON.	-- Under this predicament, probably, alone wou
366d111	smitted if their [not located] <The> every case	COMMON	to many good species; & therefore to genera (&
367d113	tances thighs of cock & Neck of Bull.-- is most	COMMON	in vegetable feeders. because males always arm
378d148	te[r]ton «p. 197» put 12 wild duck's eggs under	COMMON	ducks, the young crossed amongst themselves, &
378d148	ng crossed amongst themselves, & I presume with	COMMON	ducks. so often, that it was impossible to say
383d159	helming.-- Seeing in Gardens of Hybrids between	COMMON	& Silver Pheasant, one like cock & other like
383d159	hrodites, like J. Hunters. Free Marten N.B. the	COMMON	mule must often have been dissected Zoolog. Ga
383d160	olog. Garden. Sept 16." Hybrid between Silver &	COMMON	Pheasant. Male bird, said to be infertile.-- s
383d160	il much <blacker> «broader.-- » Breast red like	COMMON	pheasant.-- lower part of breast, each feather
388d173	ations, should there need two organs; whilst in	COMMON	bud there is no such need.-- one would <one> s
391d174	d from Is a flower bud produced by union of two	COMMON	buds??? Amongst buds each one exactly like its
392d180	g as they are crossed? & How is it with China &	COMMON	Geese «how are their instincts?» «Chineses & C
392d180	on Geese «how are their instincts?» «Chineses &	COMMON	Pigs.--» Experimentize on crossing of the seve
392d180	species of wild fowl <in Z> of India «with our	COMMON	ones» in Zoolog. Gardens; <Buffalo & common ca
392d180	our common ones» in Zoolog. Gardens; <Buffalo &	COMMON	cattle-- Esquimaux (& Australian) dogs with co
392d180	mon cattle-- Esquimaux (& Australian) dogs with	COMMON	dogs--> Ask my father to look out for instance
393dIBC	n be made artificially.-- Are hybrids pintail &	COMMON	ducks. similar inter se? Zoolog. Gardens Are t
393dIBC	uncertain Is there any peculiarity or variation	COMMON	to any zoophyte «born in succession» which is
401e017	ut not primrose producing cowslip Uncle J. says	COMMON	belief. that female plant impresses main featu
403e022	rkably short.-- & Deformations are particularly	COMMON.	-- without arm, <skin> hands thumb,-- one leg,
404e024	use it is an involuntary variation made by man,	COMMON	to every individual & therefore effect of clim
406e036	ometimes ½ way. Ed. New-Phil. Transact. Rabies,	COMMON	to men, dogs, horses cows, pigs & sheep.-- dis
406e036	en, dogs, horses cows, pigs & sheep.-- diseases	COMMON	to men & animals cow pox.-- case in Spain of p
417e075	ength to get the day» The fertility of Indian &	COMMON	Oxen, which one must think deserve the name of
417e076	Sowerby <tel> showed me many land shells of the	COMMON	species: from one locality, all left whorled.-
417e079	s.--). some races of men. D'Orbigny. affect the	COMMON	progeny more than others.-- does this more ref
420e088	tut 1838. p. 414; M. Guyon thinks Monsters more	COMMON	in Africa than in Europe especially with Europ
420e088	ing series of species of dogs & Hyaena.-- but a	COMMON	point, whence both may have descended.-- Jan.
430e119	which grow in low grounds are those, which are	COMMON	& nearest being common to other parts of the w
430e119	nds are those, which are common & nearest being	COMMON	to other parts of the world-- March 16th. Mr L
431e122	out.?-- In the place where any species is most	COMMON,	we need not look for change, because its numb
440e145	all which «Nutrition, growth & reproduction» is	COMMON	to all living beings. vide Lamarck Vol II. p.
448e167	ver this-- The Superga beds have many shells in	COMMON	«& are not far distant» with Touraine «which a
449e169	Brent Goose. Eyton says some of the pidgeons in	COMMON	Dovecot are very like a Himalaya species -- le
449e169	«all alike» between the male Chinense & female	COMMON	goose took after the common goose thus contrad
449e169	e Chinense & female common goose took after the	COMMON	goose thus contradicting (probably) Yarrells l
451e176	sister is an albino like himself said not to be	COMMON--	probably, I should think grandfather first o
454e184	rument for galvanize <ani> them-- Cross Irish &	COMMON	Hare Decandoelle has chapter on sensitive plan
455eIBC	sses.-- Leighton or some one. Father-- diseases	COMMON	to men & animals.--:likenesses of children CD[
459tf02	Linnaeus.-- seems to doubt its applicability to	COMMON	mule & hinnus-- in one case bastard of wolf &
460t013	h on vars & species The <male> swan-gander with	COMMON	goose produce full as many eggs as pure bred c
460t013	on goose produce full as many eggs as pure bred	COMMON.	-- the half of the cross, as above, take «gene
460t013	r. one of these half-bred ganders. crossed with	COMMON	goose <to has> «produce offspring with» so muc
460t015	it appears about half way between swan-goose &	COMMON	goose.-- the stripe down back pretty plain in
467t099	istinctly doubled, brown, but with no pollen.--	COMMON	Thyme growing close by is equally abortive--an
467t103	going to clover & once this happened.-- And in	COMMON	Beans it is wonderful {a} how the Humbles forc
467t104	athyrus pratensis yellow saw stigma project» In	COMMON	Pea saw Humble so press down sheath, that stig
468t105	ry small--stigma rather large & rough-- flowers	COMMON--	many winged thrips, covered with pollen-- «T
470t177	requented by «many» Bees & Humbles-- «Humbles &	COMMON»	On silene, many plants of wh. have abortive s
470t178	»--I think never on the Galeum saxatile & other	COMMON	kind--I think not on Phlox though they examine
470t178	ectar of Butterfly Orchis & Listera? Bryony saw	COMMON	Bee on: Linn. Trans 18. p. 133 Westwood on the
472s02v	re Bees at work-- another odd genus-- & a small	COMMON	Humble-- & more of same fly Two more of the fl
479z006	ories Ulloa's Voyage Vol I, p. 168 Ceratophytes	COMMON	in Northern sea. Chamisso in Kotzebue p. 312 L
490q001	ssil (perhaps not embedded ¿ are there any very	COMMON	recent ones not embedded?-- Do the Tame Parrot
492q002	eds ever raised? 11. Is not non-flowering gorze	COMMON	in Norway No Questions regarding Breeding of A
493q004	in different breeds of dogs. Cattle, (Indian &	COMMON)	&c: length of life. (5) Does my Father know a
494q004	with Bantam-- will egg kill Hen Bantam.-- Cross	COMMON	Fowl with Dorking (10) Statistics of breeding
495q005	much more than cuttings &c (4) Raise annuals or	COMMON	English plants in Hothouse & see what effect o
496q006	Experiments Questions concerning Plants Is the	COMMON	Fig Dioecious-- are its female flowers always
506q014	e way) 47½-- in heigt 30 inches Examine Keel of	COMMON	& Wild Duck-- Black Duck & Penguin Henslow &c
508q016	tendency, independent of heredetary cases, more	COMMON	in man than in female-- (8) In Hump-back ever
511q018	iculars regarding effects of crossing them with	COMMON	pigs= [it is a Lincolnshire Breed]CD-- Sir. R.
511q018	is now extinct= (9) About. American & Europaean	COMMON	species, having somewhat of different appearan
515q022	this country. {Chinese Dog's Head to send Cover	COMMON	Pea (& Sweet Pea) for several generations unde
525m026	doubted, when seeing Nina with her puppy.-- The	COMMON	remark that fat men are goodnatured, & vice ve
526m028	of animals. they know.-- pleasure of imitation (	COMMON	to monkey), & not imagination.--» Thinking ove
528m037	e mere exercise of the organ of sight, which is	COMMON	to every kind of view-- as likewise is novelty
534m063	. 228 Newport says Dr Darwin mistaken in saying	COMMON	wasp cuts off wings of flies from intellect. b
542m093	untary laugh will burst forth, this & yawning. (	COMMON	to other animals) scream of agony, sigh of dis
542m094	nt, so with men.-- How is crying-- peculiar not	COMMON?	--» no bark of anger nor have monkeys & many o
542m094	le that those sounds which are involuntary, are	COMMON	to animals.-- Curious to trace, which of these
543m100	babilities, therefore this judgment gives a man	COMMON	sense, & the highest intellectual powers of pe
547m111	e each other August 29th. Went to Bed. & built	«COMMON»	Castle in the air, of being compelled, from s
553m137	not sulkiness-- this expression he believes is	COMMON	to that group.-- this is very important as sho
558m152	ot with dog. when fear might enter?-- I believe	COMMON	Swan, arch raises neck & depresses chin-- stri
568n018	cry--» The taste of recurring sounds in Harmony	COMMON	to t[he] whole kingdom of nature. If I want so
569n022	arroïlap quelque mal qu'on y fût."-- Expression	COMMON	to Savage & Frenchman, unaccompanied by dignit
570n027	iving amongst naked figures, & observing powers	COMMON	to savages???].CD-- The existence of taste in

Page ***(Key Word)***
580n062 0.-- "The desire of glory, immortal fame, &c so COMMON in the young are symptoms of the infinite & pr
594n112 one, yet all-- flew to bed of flags. hernes are COMMON. not unlike in size in the air at a distance.-
603o012 p. 405.-- Speculates on origin of sacrifices. «COMMON to many races»-- thinks action towards <man> «
605o019 this Essay. D. Stewart does not attempt «by one COMMON principle» to explain the various causes of th
609o029 Those emotions which are strongest in man, are COMMON to other animals & therefore to progenitor far
612o035 wondrous power of willing. These +willings are COMMON to every animal instinctive and unavoidable.--
612o035 ht or shade.) Joining two difficulties into one COMMON one always satisfactory, though not adding to
618o41v lueness caused weight, because both due to some COMMON cause:-- The argument reduces itself to what i
515q021 ease produce as large capsules of seed, as the COMMONER kinds-- Cattle are horned, Suffolk have <abor
022r006 aiso Granitic nodules in Gneiss. Epidote seems COMMONLY to occur where rocks have undergone action of
046r078 n.--Are we to consider that the dikes which so COMMONLY (state facts) traverse granites, are granitic
052r099 ine Mr Sorrell says that numerous icebergs are COMMONLY stranded on shores of Georgia «Lat degree ( )
060r126 Parish. & Azara about dry season[.] 1791. seen COMMONLY bad over whole world. «(Was it so in Sydney,
062r129 imits of Waders: Ascension. Keeling: at sea so COMMONLY seen. at long distances; generally first arri
255c050 d zone of <latit> height & 3d of latitude more COMMONLY are the same species, instead of analogues.--
357d073 Tetrao media or Rakkelhan is hybrid (produced COMMONLY in Nature. both in Sweden & anciently in Brit
502q012 united by Land to S. America (33) Ornithologum COMMONLY but improperly called Canadense-- would it gr
594n140 ith straw, or with a blanket.-- these cases of COMMONLY using, foreign bodies, for end. most importan
059r119 ndes, on les voit se reproduire a des hauteurs COMMUNES sur le revers de chacune des montagnes qui fo
553m136 hrough the goodness of God knowledge has been COMMUNICAT to us».-- & that it does exist in different
542m097 all animals-- but do not overrate-- animals COMMUNICATE to each other.-- Lonsdale's story of Snails,
543m097 ression of man's face.-- <That> How far they COMMUNICATE not easy to know,-- but this capability of u
543m097 carthorse & dog.-- birds many cries. monkeys COMMUNICATE much to each other.-- Waterhouse says far mo
046r080 a ought to be considered as a plain movement COMMUNICATED to it as well as by the vertical as lateral
523m015 scarlet.-- dogs ideotic.-- dotage.--» Doctor COMMUNICATED to my grandfather his feeling of consciousn
593n107 ic bear any relation to the period when men. COMMUNICATED before language was invented.-- were musica
275c121 -- Mr. Y. is inclined to think that the male COMMUNICATES the external resemblances, than the female.
543m099 ecause Spiders have many,-- great powers of COMMUNICATING knowledge to each other-- August 23d. Jones
190b075 countries direct relation to facilities of COMMUNICATION. Have races of Plants. ever been crossed rea
195b103 s «& what are barriers but» interruption of COMMUNICATION. or when country changes. Will it said that
226b223 bear.-- If I had not discovered channel of COMMUNICATION by which great Edentate might have roamed t
269c101 D'Acunha, St Helena &c &c. Juan Fernandez A COMMUNICATION to Geograph. Soc. in February or March 1838
332d001 rters of the county, who had had no sort of COMMUNICATION, were seized with it, & for ten years after
521m010 ardener.-- Thus was he in every respect, no COMMUNICATION could be held by means of hearing.-- Mr Cor
572m035 ctive.-- or lied &c &c Are the facts (about COMMUNICATION of ideas, &c) of expression lawless, whilst
595m125 f instincts (highly useful as only means of COMMUNICATION) in man, than sucking.-- [I assume a child
612o035 extended. Hence a sensorium, which receives COMMUNICATION from without, & gives wondrous power of wil
617o040 essence of whose existence consists in its COMMUNICATION to other matter in the course of its DIRECT
260c068 s> crys of distress of other parents.-- Shows COMMUNITY of language Desert country. is as effectual a
305c211 between those individuals thus endowed, & the COMMUNITY of mind, even in the tendency to delicate emo
624o051 strives* to same end.-- & general actions of COMMUNITY must frequently teach same end.-- Hence this
459tf02 s. Vol. III p. 83. Paper translated from Meckel. COMP. Anat.-- From Buffon cross of he-goat & sheep,
110a086 lband. gradually becoming finer grained & more COMPACT on that side-- separation DISTINCT from dike j
228b232 »; our slaves in the most laborious work, our COMPANION in our amusements. they may partake, from our
290c165 cuts the matter short by saying man cannot be COMPANION but master.-- heredetary tameness as well as
317c248 t end of "The British Aviary" or Bird Keepers COMPANION Study Appendix (& only appendix) of Congo Exp
402e018 y of them descending to examine their defunct COMPANION".-- p. 229. Borneo.-- only animals he heard o
563n002 reater temptation as bitch: or dogs defending COMPANION (mem Cyanocephalus. Sphynx howling when I st
594n115 as fast struggling with another large dog his COMPANION. Descent --Affection & [...] Monkeys «Ogleby»
594n115 ess their companions-- a «blue» Gibbon. whose COMPANION had S[...] been dead about two months. saw a
594n115 Zool. Soc-- 1838 remember with distress their COMPANIONS-- a «blue» Gibbon. whose companion had S[...
306c213 is not aware that Australian dogs ever hunt in COMPANY -- marked difference with dogs of La Plata & G
227b228 My cheory would give zest to recent & Fossil COMPARATIVE Anatomy, & it would lead to study of instinc
356d071 th any other. it should be observed not what COMPARATIVE difficulties (as long as not overwhelming) W
356d071 ficulties (as long as not overwhelming) What COMPARATIVE solutions & linking of facts-- Savages over
106a074 to widen course until inclination is become COMPARATIVELY small, & when that is case force is lessene
486z018 ort summer far N. where this other order is COMPARATIVELY rare.-- These views clearly explain rarity
025r018 eface, it might be well to urge, geologists to COMPARE whole history of Europe, with America; I might
037r054 k, or that the rock not weathering allows such COMPARE the elevated estuary of the Plata. to the Bay
043r069 s or grooves in rocks at Costorphine hills. to COMPARE with Galapagos.--Chiloe. M. Hermoso. & Coral r
115a097 mens from Anglesea in Geolog. Soc. if numbered COMPARE them with my rocks. when writing on Falkland I
132a137 -- What a curious investigation it would be to COMPARE, the time of the earthquake-- Chile with this
157g074 from above on soft shelf-- 29.330 A 84 degree COMPARE this with last measurement of shelf of 3d:-- g
205b143 Terrestrial annelidous animal p. 347. Vol I.-- COMPARE with my planariae Leaches out of water Does th
210b160 to be checked when islands are near continent: COMPARE Siicily & Galapagos!!-- Some of the animals pe
225b219 on Subsidence New is only hope.-- New Zealand «COMPARE to Van Diemen's land.» glorious fact. of absen
228b231 acraucheina yet he may be found:-- We must not COMPARE «chances of embedment in» man in present state
253c046 ?? Africa Bears??-- Plantigrade Carnivora?-- «COMPARE rodents of two countries» & «Monkeys.» Fact of
260c067 ilson's Ornithology, D'orbigny, Spix, &c might COMPARE birds of N. America & South-- Any how temperat
269c102 ary & China.--, both coasts of New Holland.-- «COMPARE birds of Australia with plants, with this obje
299c193 plants peculiar to themselves. this remarkable COMPARE it with Canary Islds. Galapagos.-- Iceland has
302c202 classification, I must work out hypothesis, & COMPARE it with resuts. if I acted otherwise, my premi
363d101 «Pouting pidgeon exaggeration of cooing.--» & COMPARE them with all the varieties.-- Habits of rock
363d103 ++ Cervus Campestris spotted white when a fawn COMPARE with fallow? deer. & Moschus &c & -- like youn
432e123 the crossing of hornless sheep with horned.-- COMPARE this with what highland shepherds said.-- p. 1
455eIBC which is useless. Granfather's Handwriting, to COMPARE with my own.-- E Bengal Journal Vol 7. p. 658-
481z008 hia Blanca. De la Beche theoretical researches COMPARE land shells of Galapagos different islds.-- Wa
486z019 e Reptile in Kamtchatka (Salamandra aquatica). COMPARE with T del Fuego Compare birds of do with <T>
486z019 Salamandra aquatica). Compare with T del Fuego COMPARE birds of do with <T> N American & T. del. Fueg
502q11v then experimentise: for gradation in structure COMPARE flowers of wild & tame carrot-- Parsley & Fenn
502q11v wild & tame carrot-- Parsley & Fennel. Verbena COMPARE flower of different Cabbages most carefully to
502q11v s of different vegetables & animals come on.-- COMPARE calves.: Compare young. beans. cabbages.-- His
502q11v getables & animals come on.-- Compare calves.: COMPARE young. beans. cabbages.-- History of Pheasant-
540m085 He who doubts about national character let him COMPARE the American whether in the cold regions of th
547m113 unning on in mind, being absent. one could not COMPARE the castle with them, therefore could not doub
559m153 use of language, & judge only by what you see. COMPARE, the Fuegian & Ourang & outang, & dare to say
559m154 te creation. we admire it more. because we can COMPARE it to the standard of our own minds. which mak
588m090 Lamarck. Phil. Zoolog.-- Vol II p. 445. If we COMPARE the judgments & actions of a young animal with
633j54v and & simple laws.-- 4 «Study Cuviers Anatomie COMPARE» p 308. Traces the gradation of skeleton in Ve
051r097 13. 444 «Yanky edition» Shores of Pacifick, as COMPARED to whole E. America. <East> Africa. Australia
052r099 ces in Valleys of Chili may be with much truth COMPARED to the step = formed streams of lava at St Ja
090a021 r globe melted magnetic metals. ∴ earthy crust COMPARED to those of falling stones.--¿ does this bear
223b207 ferent probably & introduction of Man. Nothing COMPARED to the first thinking being. although hard to
223b208 culate or intestinal, or even a mite.-- a bee «COMPARED with cheese mite» with its wonderful intincts
232b245 die out or move South ward.... species must be COMPARED to neighboring sea.-- For change of species d
278c132 oceros Elephants &c in the most modern period, COMPARED to Faunas of these countries, greter than Tox
278c132 untries, greter than Toxodon, Macrauchenia, &c COMPARED to America.-- the wonder is that the Europaea
281c139 the third & fourth degree.-- a species must be COMPARED to family entirely separated from any degree;
325c267 3d. Edit. Octav. (good to trace Europèan forms COMPARED with African <A> Annals Histoire Generalle et
335d014 are not new breed.-- the first hybrids may be COMPARED to animal with amputated limb.-- Heredetary <
353d061 Irish & English Hare.-- good case these hares COMPARED to North American hares. Many species, separa
406e037 -- Whole wor[l]d, formerly possessed a climate COMPARED to S. America at present days, which S. Amer
421e091 character--" How little clear meaning has this COMPARED to what it might have.-- What is the differen
485z016 he small finches walk at Maldonado & Patagonia COMPARED with those of England.-- or ground birds-- ra
496q05a it?? good=!! Examine pollen of double flowers. COMPARED with single & see whether grains flaccid, as
504q014 ury (1) Peas.-- Beans seeds alone remain to be COMPARED-- Cabbages.-- kept true Try experiment (30/p.
521m008 ???> Miss Cogan's memory of the tune, might be COMPARED to birds singing, or some instinctive <or> so
545m106 s the Keeper thinks is from pleasure, & may be COMPARED to laughing» they dance with passion, ie. ner

Page **(Key Word)**
549m118 appiness-- so is it <with an> when same man is COMPARED to peasant.-- To make greatest number of plea
550m124 easant mental sensations in any given time «-- COMPARED to what other people experience.--» But then
550m125 oken> intense happiness even with some pain,-- COMPARED to what others experience in same time.-- Ple
550m126 p. 127. Talks of difficulty of his own drawing COMPARED to a friend, whose who family can draw-- says
551m127 well see things which have no existence.-- He COMPARED spectral illusion & insanity the connexion ap
563n002 if dogs mind were so framed that he constantly COMPARED his impressions, & wished he had done so & so
624o50v ury under term of Reflex Senses seems to have <COMPARED> «perceived» the comparison between our insti
202b133 e.-- of genus peculiar to East Indian isles.-- COMPARES it to fossil Didelphis (S. American genus) in
210b158 47 «p. 150»-- not being crossed with others.-- COMPARES it to languages But how do plants cross?-- =
534m062 d about a foot & then leapt across. (Col Sykes COMPARES this with pidgeons finding their way home-- t
546m110 rful case of perfect double consciousness Mayo COMPARES it with Somnambulism.-- the young lady almost
602o11v hould have so little to do with art (p 128) R. COMPARES a view taken by a camera obscura &c a Poussin
139a180 any interesting experiments might be tried by COMPARING Zoophite to plants.-- grafting length of life
232b246 eather on long average tolerably> uniform).-- COMPARING fossils with whole world. would be like in a
356d071 between acquired & heredetary tameness.-- In COMPARING my theory with any other. it should be observ
534m062 their way home-- there is something wrong in COMPARING these cases, when agency is unknown, with sim
547m113 . & likewise gaining new ones from senses. & <COMPARING their> «calling up» old ones, to be sure of o
548m114 are invented as in imagination, & in rigidly COMPARING each step as in reasoning-- hence delirium &
548m115 ct» is in making things somewhat probable. in COMPARING every step, & inventing new means,-- therefor
556m144 these idea came one after other, without ever COMPARING them, I neither doubted them or believed them
027r024 rs. N. vii P. 56 Serpentine form: of Cuba for COMPARISON (?) with St Pauls [not located] [not located
033r043 DS of great violence volcanic. from Humboldt: COMPARISON P 361. Daubeny Von Buch is very strong about
100a055 analogy between Australia & Oolitic period.-- COMPARISON rather loose.-- perhaps worth Says from Lard
138a153 arthquake, during two years.-- will serve for COMPARISON with the moon at some future time [blank] Si
147g021 h water mark 30.380. 68 degree 65 degree? For COMPARISON with all the measure before There some of th
154g058 of 2d shelf Level of shelf of Glen Guoy form COMPARISON with granite block «& Ben Erin» 29.287. 72 d
222b207 in form to ancient; in shells alone can this COMPARISON be instituted. People often talk of the wond
232b246 ld. would be like in a Meteorologic table «in COMPARISON of temperature of two countries» finding a v
278c128 subgenera is empirical, & is judged solely by COMPARISON with other genera in other families.-- it wi
337d022 cal Soc> A capital passage might be made from COMPARISON of Man, with expression <of a> of Monkey, «w
350d056 comotion.-- «Mem. Dr. Blackwell (Abercrmbies) COMPARISON of sight to threads.--» Hence the Pecten, wh
390d179 ast (¿male?) looses all appetite.-- It is the COMPARISON of each animal with its<elf> «ancestors», &
390d179 h animal with its<elf> «ancestors», & not its COMPARISON «of difference» with other sex. = The highes
481z009 l name Birds of Iceland. Mackenzie. p 345 for COMPARISON with Falkland. good also for Journal.-- 18 A
484z014 ubulipores L'Institut-- 1838 p. 75 A detailed COMPARISON of production without Tropics in Northern &
484z016 n under side of wings It would be interesting COMPARISON to find how many of the small finches walk a
486z019 ropical Forms See p. 256 of Note Book (C) for COMPARISON of singing powers of birds of N. America & E
493q003 ervation-- examine semen of Hybrid animal. in COMPARISON with Weigh skeleton of Tame Duck & Wild Duck
512q019 calves. also turning into fine beasts.-- For COMPARISON with hybrids, is offspring of short-horn bul
513q020 tail or horn, with another & see result, for COMPARISON with natural species, as dioecious plants, w
544m103 cause fatigue to the mind,-- it is solely the COMPARISON, with past ideas. which makes consciousness-
545m108 are the reverse of intellectual, there is no COMPARISON of ideas-- one follows other as in blindest
552m130 et??) is it because one then has no immediate COMPARISON with perceptions, & that on[e] fancies the i
556m144 or believed them.-- Believing consists in the COMPARISON of ideas, connected with judgment. [What is
568n018 ved. direct effect of improving organization, COMPARISON of sensations would first take place, whethe
569n021 t. Reason in simplets form probably is single COMPARISON by senses of any two objects-- they by VIVID
614o037 good memories-- but on its multiplicity & the COMPARISON of ideas.-- As man has so very few (in adult
624o50v nses seems to have <compared> «perceived» the COMPARISON between our instinctive feelings & our short
110a085 es are in process. connected with variation of COMPASS & these may cause «or be effect of» elevation
586n078 t they do know from look of Heavens, points of COMPASS, & they do know which way they go; & so return
586n079 by habit, such habit of knowledge of points of COMPASS may be instinctive. it is a test to know how m
181b045 but it seems law that such tribes, as far as COMPATIBLE with such structure are in minor degrees ada
276c124 en Milvulus forficatus Tyrannus Sulphureus if COMPELLED solely to fish. structure would alter.-- It i
283c145 ir own instinct & (even though <fertile> when COMPELLED to breed hybrids produced--)» & then all that
369d115 this> structure <connected with animals being COMPELLED to travers> "May have reference to the Great
369d115 ich the Mammalia of N. S. Wales are generally COMPELLED to traverse in order to quench their thirst"-
547m111 & built «common» Castle in the air, of being COMPELLED, from some quite imaginary cause to start at
614o037 ew (in adult life) instincts.-- this loss is COMPENSATED by vast power of memory, reason &. & many ge
270c104 G. R. Treviranus, Biologie referred to,. as COMPILATION of action of organic nature on inorganic It
349d052 ne will settle number of primary divisions.-- COMPLAINS (p. 53) of M. Edwards, thinking any group goo
522m013 nct, different also from delirium, a peculiar COMPLAINT stomach not acted upon by Emetics.-- people r
530m043 en able to direct about his own health.-- his COMPLAINT was carbbuncl on <Head> Neck.-- He has seen o
520m001 uly 15th 1838 My father says he thinks bodily COMPLAINTS «& mental disposition» oftener go with colou
234b256 e tell by general colouring a group of Nebria COMPLANATA from. Devonshire, from another from Swansea.
066r142 th same general character of fossils deception COMPLETE.= Silliman Journal. year 1835 excellent accou
181b045 her elements. every part would probably be not COMPLETE, if birds were fitted solely for air & fishes
189b073 non associated animals.-- & the story will be COMPLETE.-- It is absurd to talk of one animal being h
223b210 crossing of species less altered prevents the COMPLETE adaptation which would ensue A. B. C. D. -- (
288c158 series, ie, cannot be discovered till circles COMPLETED Major Mitchell, does not know whether breeds
380d153 which have bred before the vis formativa had COMPLETED them-- (but this argument is VERY WEAK withou
121a111 ained their angles sharp-- yet with character COMPLETELY altered, & a crystalline structure superindu
288c159 east notion of hunting, or keeping watch. how COMPLETELY «nature & instinct» modified.-- The partial
334d009 effect of race was determined by male: & How COMPLETELY is Lord Moreton's case opposed to this fact
391d174 that object failing, generation fails.-- How COMPLETELY circumstances «alone» make changes or specie
471tf08 d it wd be difficult to point out a family so COMPLETELY natural & one whose groups pass so insensibl
603o012 Polynesians Jews, African all sacrifices. How COMPLETELY men must have personified the deity.-- H. To
356d070 duce in each case, but something prevents the COMPLETION.-- ‖Say my Grandfathers expression of genera
308c218 use there is nothing more elementary than that COMPLEX nature itself with which our speculations must
460t019 stock (which from action & reaction grew more COMPLEX)» some perhaps rendered more complex & some si
460t019 grew more complex)» some perhaps rendered more COMPLEX & some simplified.-- Annals of Natural History
461t037 to purpose after purpose would tend to render COMPLEX the series.-- Ch 6 Upland geese would transpla
422e095 he world depends, of their varied structure & COMPLEXITY.-- hence as the forms became complicated, th
422e095 , they opened fresh, means of adding to their COMPLEXITY.-- but yet there is no «NECESSARY» tendency
423e095 rom the new relations caused by the advancing COMPLEXITY of others.-- It may be said, why should ther
423e096 mes to simplify structures:= Without enormous COMPLEXITY, it is impossible to cover whole surface of
423e097 -- it is quite clear that a large part of the COMPLEXITY of structure is adaptation. though perhaps d
443e155 - The relation of these «sexual» functions to COMPLEXITY is evident, yet the inference from some plan
460t019 ce in the organic world of infinite & growing COMPLEXITY from a few types, it must not be supposed th
460t019 nt forms existed, but if it be asked how this COMPLEXITY from a few types originated, we must go to t
074r165 .B. To show how metals may be transported by COMPLICATED chemical law & steam of salts, quite curious
162g099 es of which, last measurements belong are so COMPLICATED, that nothing can be made out of them-- but
175b018 . the simplest cannot help..-- becoming more COMPLICATED,; & if we look to first origin there must be
177b026 in progress.-- «no only makes it excessively COMPLICATED.» {P} Is it thus fish can be traced right do
292c169 stand change better than others.-- The more COMPLICATED the animal the more subject to variation. th
301c200 equal.-- organs of generation about equally COMPLICATED.-- An Entomologist going into a country & co
337d020 ies of the domesticated animals must be most COMPLICATED, because they are partly local & then the lo
389d173 process by generation applies only the more COMPLICATED animals.]CD p. 310 She wolf took dog. but ha
420e089 head-- !! Handwriting is determined by most COMPLICATED circumstances, as shown by difficulty in for
422e095 re & complexity.-- hence as the forms became COMPLICATED, they opened fresh, means of adding to their
422e095 RY» tendency in the simple animals to become COMPLICATED although all perhaps will have done so from
440e146 dition of external circumstances. results of COMPLICATED laws of organization: as we see these strang
539m082 o variableness or power of intellect.-- Some COMPLICATED trades can hardly be considered as actions o
576n048 eve, that it require a far higher & far more COMPLICATED organization to learn Greek, that to have it
605o019 phorically sublime, but that it is through a COMPLICATED series of associations that we apply to such
608o025 sions-- when these passions, weak, opposed & COMPLICATED one calls them free will--the chance of mech
614o037 n:> individual-- [LHC] Perhaps even the most COMPLICATED instinct. might be analysed into steps, as s
463t055 a & Sumatra Nov 15th Waterhouse showed me the COMPONENT vertebrae of the head of Snake wonderful!! di
025r016 . [5 degree 29 minute S.] still shoaler, coast COMPOSED of sand dunes. 15--15 Does not seem to consid

Page **(Key Word)**
074r165 henomena in veins, chemical affinities like in COMPOSED rock. granites syenite» «strangling &c of vei
089a019 onaïves it is difficult to find stone not thus COMPOSED on the NE part more like marble requires poli
148g027 ne above much inclined towards river all these COMPOSED-- where side ravine entered terraces formed s
149g030 of the two valleys corresponding as in Andes, COMPOSED of sand & perfectly rounded stones-- lake req
157g075 2d «almost» forms it into island-- whole hill COMPOSED of remarkable gneiss with red granite veins &
159g084 return) between 2d & 3d shelf Mountain <Mica> «COMPOSED of» Gneiss Block on 2d shelf & below it some
267c094 . The <bush> woodlands for miles in extent are COMPOSED solely of this shrub".-- p. 229. carcases of
219b193 . Helena. J. Fernandez. Galapagos. Many trees COMPOSITAE, because seeds first arrived «Ferns ditto.--
470t177 Humbles--on Thistles many (curious because a COMPOSITE) Asparagus very small flowers & as much shut
221b200 st persistent?? do.-- The most perfect Plants COMPOSITES.--!!«good» those which have undergone most m
424e100 tia & Hypnum «Prof:» Don would have known the COMPOSITES of Galapagos were South American.-- several
114a094 discussion -- I see Lyell talks of different COMPOSITION using difference in metamorphic action which
117a101 ssification of Europaean strata according to COMPOSITION thinks sand with vegetable remains formed ne
152g050 level of 4th shelf= argument against river--COMPOSITION &-- stratification argument detritus-- {P} w
156g068 ck on <bend> of 3(a) Cannot <see> «make out» COMPOSITION of shelves: generally angular except near he
418e083 d] it utterly untold,-- what is added to the COMPOSITION of the atom, to make it alive, & how the law
529m041 xcited by looking at trees at [i.e., as] great COMPOUND animals united by wonderful & mysterious mann
600o08v y other.-- I suspect conscience, an heredetary COMPOUND passion. like avarice.-- Is there not somethi
530m046 memory, as in sleep.-- a new thought arises?? COMPOUNDED of the involuntary thoughts.-- An intentiona
600o008 ck, which best singer-- Remember.-- avarice a COMPOUNDED passion gained in life time]CD 3. The Infini
227b228 elligible.--may look to first germ-- --led to COMPREHEND true affinities. My theory would give zest t
432e125 » No one but a practised geologist can really COMPREHEND how old the world is, as the measurements re
372d130 ticle of animal must have structure of whole COMPREHENDED in itself.-- it must have the knowledge how
617o39v ught is a question which ought to be clearly COMPREHENDED by anyone who wishes to fully understand th
594n113 ation-- pouting child same as anger, lips not COMPRESSED sullen, protruded. determined to do nothing.
565n009 gree, no smile, no frown showing thought, no COMPRESSION of mouth showing action,-- sulkiness all neg
536m072 ll is to mind, what chance is to matter «(M. Le COMPTE)»-- the free will (if so called) makes change
420e088 n, a fossil dog-- leading towards Hyaena.-- see COMTE Rendu.-- I suspect good case of fossil filling
535m070 se from» as fixed laws of organization.-- M. le COMTE argues against all contrivance-- it is what my
539m081 work» which came on from reading «review of» M. COMTE Phil. which made me «endeavour to» remember, &
553m135 e of science in every nation according to M. le COMTE).-- Those savages who thus argue, make the same
608o025 e chance of mechanical phenomena.-- (Mem: M. Le COMTE case of Philosophy, & savage calling laws of na
376d136 populousness, on the energy of Man» D.'Orbigny. COMTES Rendus p. 569. 1838 says the cross between the
566n012 cessary notions any more than «a» Savage M. Le COMTE'S idea of theological state of science, grand id
207b151 one <inland> directly by instinct. can dive & CONCEAL themselves in the grass.-- Beatson St. Helena
541m093 it is no wonder that they are so difficult to CONCEAL.-- a man «insulted» may forgive his enemy & no
572n035 ed-- no one can say expression was invented to CONCEAL one's thought.-- Macculloch in his Chapter on
161g095 t moss most distinct then lost by slope, then CONCEALED by fragments, then clear. this bit to eye cer
578n055 lyse this:-- Let a person have committed any «CONCEALED» action he should not, & let him be thinking
159g085 p of spit between river & dry Corry Scarcely CONCEIVABLE. if Hill between Corry so much cut Granite c
286c154 -- to (look at common ancestor, (scarcely[)] CONCEIVABLE in savages) Has not the white Man, who has d
097a043 ish colour like a dull & poor varnish, which I CONCEIVE to be analogous to the black glazing observed
104a066 hic & thickness of <strata> not great, one can CONCEIVE anticlinal lines near. (lateral pressure woul
186b058 re real structure obliged to be altered, I can CONCEIVE colouring retained; therefore probably in som
250c037 in, N & S. America, is only common cause I can CONCEIVE of destruction of Great Animals in Europe & A
266c088 upon knowing to what type a bird belongs.-- I CONCEIVE without knowing from which country many birds
288c160 ra & <seized> ceased their migrations lost?? I CONCEIVE a bird Migrating from Falkland Isd regularly
398e005 ifficulty of multiplying effects & to <ponder> CONCEIVE the result with that clearness of conviction,
471tf07 of divine power"?.--"of their use difficult to CONCEIVE any idea" Linn. Trans. 18. p. 163. "D. Dod on
512q020 Do the animals there, sometimes couple but not CONCEIVE :Bears /Yes/ (2) Foxes & English animals & bi
542m094 o relieve circulation after stillness.-- Now I CONCEIVE if organization were changed, I conceive sigh
542m094 Now I conceive if organization were changed, I CONCEIVE sighing might yet remain just like sneering d
557m147 which case expression resembles a fox-- I can CONCEIVE the opposite muscles would act, to when in a
612o035 nd animal strictly united? [RHC] It is easy to CONCEIVE such movements & choice, & obedience to certa
616o39v ode «of perceptive action» by which we come to CONCEIVE of matter as attracting & shew that the groun
618o41v ee a particle move one to another, & <the> (or CONCEIVE it) & that is all we know of attraction, but
409e047 k difficulty-- I have felt some difficulty in CONCEIVING how inhabitant of Tierra del Fuego is to be
076r169 ual sometimes disseminated <[...]> sometimes CONCENTRATED: wonderful quantity of pure silver in S. Am
313c236 becomes necessary, when organs of parent are CONCENTRATED in different parts, & scission cannot effec
430e119 cus & I. sulcatus.-- the beak of this one has CONCENTRIC striae, all the lower part rayed longitudina
430e119 ne, which is exactly intermediate between I. CONCENTRICUS & I. sulcatus.-- the beak of this one has c
022r007 ariety in nature of a dike.--Mem. at Chonos & CONCEPCION. P. 417 Veins of quartz exceedingly rare Mem
026r022 . silicified bones not common in Britain. Mem CONCEPCION Says Echinites. Encrinites. Asteriae, usuall
027r024 ct: at R. de Janeiro. Coquimbo. Balanidae. at CONCEPCION. Humb: Pers. N. vii P. 56 Serpentine form: o
030r034 en collected on the open coast. Perhaps as at CONCEPCION. favoured by basin formed by outlying rocks;
030r036 ales & Van Diemen's land.-- Whole coast S. of CONCEPCION where there are Tertiary strata there is Coa
043r071 4» fields of Earthquakes in Chili:-- Chiloe. CONCEPCION. Valparaiso (Copiapò & Guasco). yet whole te
045r075 of the ground at Calabria were present at the CONCEPCION earthquake.--expatiate on difficulty of evid
047r082 . 471) is explained. also the similar fact at CONCEPCION? Read the various accounts & see if fall is
058r115 . -- State the circumstances of appearance at CONCEPCION [.] no sign of elevation. Effects of great w
069r149 peculating on the sea The 24 ft. elevation at CONCEPCION. from impossibility of such change having ta
070r154 alls & feeling shows undulation.: crust thin.--CONCEPCION earthquake Draw close Analogy Lake of Cordil
070r155 Rozales tells me at seven oclock Novem <5th> CONCEPCION most violently shaken, by earthquake. but no
244c021 very stiff, p. 120-- Coati roux common. near CONCEPCION. some tatous!!! p. 120.-- Most of the dogs o
465t080 look at Scotland-- coasts of Chile, excepting CONCEPCION-- Patagonia-- Beds of La Plata. (except clos
482z011 Coati Roux. Tatous & perhaps Yagourundi near CONCEPCION!!!.-- (no species mentioned) p. 205. only 9.
515q021 ty of pollen is produced. & 2d if so, whether CONCEPCION takes place,-- the mere fact of seeds ripeni
473s07v No peculiarity in external structure can be CONCEPCIONAL, as limbs &c &c only appear late in pregnan
055r106 iners to be the richest Vol II 147 Shells at CONCEPECION 50 toises above the sea. = talks of them bei
054r102 nt of streams of stones agrees with mine.--At CONCEPTION, cleavage E & W! at Payta. talcose slates, d
513q21. alian Mountains.-- Is setting of fruit. cross CONCEPTION--(‖ I could extract nothing from him)‖ Does
569n021 s of any two objects-- they by VIVID power of CONCEPTION between one or two absent things.-- reason p
627o053 nterested" p. 14. It is allowed, that we have CONCEPTION of moral obligation «when grown up???» & the
272c109 markable structure which as far so species is CONCERNED superabundant,-- the tail in cock peacock,. w
360d089 nowledge of breeders, where their interest is CONCERNED. Same man had crossed Jackal & dog-- (offspri
389d177 is rendered abortive as far as parturition is CONCERNED.-- Generation being means to propagate & perp
435e133 s. p. 210. Shows. that the action of light is CONCERNED with the developement of form; but that tadpo
446e161 te is on the decline, as far as vegetation is CONCERNED, in parts of the Northern «French» expedition
546m110 unconsciously as far as the ordinary state is CONCERNED.?-- Mr. Mayo told me the case of a lady, (who
577n051 ritation to passion.-- Blushing is intimately CONCERNED with thinking of ones appearance,--does the t
496q006 cinths in glasses &c &c Experiments Questions CONCERNING Plants Is the common Fig Dioecious-- are its
085aIFC th 1839 As far as p 140-- abstracted as far as CONCERNS "Geolog Observat on Volcanic islands & Coral
586n082 save wax.]CD which it instinctively exerts in CONCERT with others in building comb-- My faculty ofte
431e121 c.--if? so «it is» good case:-- in Sowerby Min. CONCH. it is however, said they have been found toget
316c245 ift sea weed-- may transport ova of shells.-- CONCHIFERA. hermaphrodites-- eggs in groups.. Have Dioe
316c244 very small ones as Trigonia in Australia or CONCHOLEPAS in America,-- yet many countries have far mo
407e041 alyse this,-- consider state of vegetation, & CONCHOLOGY,-- shells of Africa ought most to resemble f
427e109 ated by Capt. Graah») & break up. N. American CONCHOLOGY from Europaean., & the climate being now les
119a104 s are effect of continental elevations) we may CONCLUDE that elevation is independent of spreading ou
203b137 doing the service of a swallow I think we may CONCLUDE from Australia & S. America. that only some m
206b147 e different formula for each lustrum.-- We may CONCLUDE that there will be a period though long dista
544m103 nd thinks with extraordinary rapidity-- We may CONCLUDE that neither number, vividness, rapidity, nov
585n076 ch they STARTED, they cannot return.-- Hence I CONCLUDE. pidgeon taken little way, whirled, & then ta
054r105 nce of 3 or 4 leagues «from the coast» may be CONCLUDED to have been covered by the sea judge from th
197b111 ces:-- does not say propagated, but must have CONCLUDED so= Evidently «or hints» considers generation
610o031 hristian name; fancied it looked like. W. but CONCLUDED it could not be so.--Looked at a direction bo
619o042 ld at any other mammiferous animal, it may be CONCLUDED that he has parental, conjugal and social ins
636j56v a country favourable to change.-- It might be CONCLUDED that Plants would be subject to extreme varia

580n062 use they understand signs.-- very profound.-- CONCLUDES that difference of intellect between animals
209b156 slands; "ou bien encore on pourrait au plus en CONCLURE quels sont les genres qui, sous ce climat, se
299c193 ts show how very permanent plants are. & this CONCLUSION must be arrived at, when one sees a plant li
398e005 ld be absolutely necessary to arrive at right CONCLUSION about species Changes of level &c are easily
423e099 of bed-- & another in lower is very rare, the CONCLUSION will be that our greatest formations <are> h
431e120 being impossible. & even Mr Sowerby is coming to this CONCLUSION, from specimens in grades, now L. says the T
441e151 t «in any case» we must draw such a monstrous CONCLUSION, that every organ is become fixed. & cannot
443e154 being impossible. can be overturned, then the CONCLUSION that the two kinds of generation have some m
463t063 y fossils makes same such remark & before the CONCLUSION of his work-- Lund makes his wonderful disco
566n012 dea: as before having analogy to guide one to CONCLUSION that any one fact was connected with law.--
621o047 identally» «by odd association» comes to this CONCLUSION, not owing to peculiarity of organ of taste,
025r018 d from that part of the globe: & when we see CONCLUSIONS substantiated over S. America & Europe. we m
098a044 e let out by steps in Central France not very CONCLUSIVE proofs, but certainly probable. Bulletin de
606o025 er motives prevent the action see Abercrombie CONCLUSIVE remarks p. 205 & 206.)CD Motives are units i
417e079 gnated externally; nor can it be a necessary CONCOMITANT, with moths, which can be impregnated extern
550m127 orms some actions, which are pleasant, every CONCOMITANT circumstance calls up pleasure. or pleasure
605o18v y 5 The emotions of terror & wonder so often CONCOMITANT with sublime. adds not a little to the effec
441e148 with buds.-- They differ from possibility of CONCOURSE of two «individuals» & the action always of t
027r027 The frequency of shells in the Calc. Sandstone CONCRET, is connected with frequency of shells in flin
041r065 ns, living only in that spot & being cause of CONCRETION; or being only preserved in that part. havin
098a047 to layers of quartz & feldspar) within other CONCRETION.-- state last page thus. point of attempted
100a054 at Ascension instance of hollow concretions & CONCRETION filled with unconsolidated matter-- Phillips
102a060 lies to cleavage & concretions.-- Septaria in CONCRETION arranged in planes, case of separation.-- th
041r066 ure {P} in sandstone: evidently depend on a CONCRETIONARY contraction: the fact is in alliance with t
059r121 itchstone (which I have seen). is a kind of CONCRETIONARY structure, for the interlineal spaces are o
074r165 case of such vapours washing a rock» Veins CONCRETIONARY; concretions <dt> determined by fissures as
074r165 ng &c of veins can only be accounted for by CONCRETIONARY action, conjoined with other» «(state simpl
098a047 r in gneiss.-- Veins in septaria. a kind of CONCRETIONARY process (analogous to layers of quartz & fe
099a049 to stratification evidently small scale of CONCRETIONARY action all fluid at once, the films vertica
027r023 range substitution of matter in shells, like CONCRETIONS & laminae, show what movements take place in
041r065 hells. multitudes.-- The question of shell's CONCRETIONS, living only in that spot & being cause of c
044r072 seen to form in atmosphere.--Mem. Ascencion. CONCRETIONS & Galapagos.-- «Humboldts. fragmens.» Read g
072r160 nimals (?). ≠ The Hollowness of <sep> Chiloe CONCRETIONS somewhat analogous to septa.-- would particl
074r165 vapours washing a rock» Veins concretionary; CONCRETIONS <dt> determined by fissures as in septaria.
074r165 re after NW <W>.-- «same chemical laws as in CONCRETIONS perhaps makes intersections richest-- Humbol
074r165 conjoined with other» «(state simplest case. CONCRETIONS of clay iron stone; iron pyrite in a fossil»
098a046 ve. (pyrites, agates, calcareous balls) that CONCRETIONS are connected with a crystalline process.--
098a046 beying one law of crystallization. therefore CONCRETIONS in this case laminar. hence the thick wedges
099a048 ses moved more laterally than vertically, in CONCRETIONS more vertically than laterally. -- <In Area
100a053 last is strong character.-- A discussion on CONCRETIONS and cleavage conjoined very good.-- It is th
100a054 sintegrated granite!!! Look at gneiss of Rio CONCRETIONS in Pumice bed at Ascension instance of hollo
100a054 n Pumice bed at Ascension instance of hollow CONCRETIONS & concretion filled with unconsolidated matt
102a060 articles in rock. This applies to cleavage & CONCRETIONS.-- Septaria in concretion arranged in planes
105a072 s land shells found with calcareous matter & CONCRETIONS on coast of Madeira.? How came it if this pos
109a084 h I do not think good p. 411 When discussing CONCRETIONS Carbonate soda. formed by Ca. of L. & Mur. o
288c158 rests his whole groundwork of analogy on its CONCURRENCE in parallel parts of his series, ie, cannot
037r056 one was found.-- Miers saw then near? Mem. La CONDAMAINE on the Amazons. Consult Insist on the freque
227b227 of observation.-- View of generation. being CONDENSATION, test of highest organization intelligible.
347d049 to improve in intellect,-- if generation is CONDENSATION of changes. then animals must tend to impro
044r074 mbs in tufa there is proof of such gaz) steam CONDENSED.--Perhaps these mighty changes might go on. &
089a017 action of Heat bubbles volatilized at bottom, CONDENSED before rising?-- Mem. granite heated.-- Metam
119a104 tion of continents, if globe be considered as CONDENSED vapour.-- inequities are required to start a
126a122 at, how wrong, would our judgement be-- Does CONDENSED metal, conduct heat better than plain?-- Mem
137a149 New. Phil J. 1838. p. 72. on metallic vapours CONDENSED from furnaces do/p. 84 on the effects of vein
289c162 may be told by their larvae) but the acts of CONDENSING must alter method of generation-- Heaven kno
289c162 tinct creation.-- Generation may be viewed as CONDENSOR, «+++ must (on my theory) =supported by foeta
124a117 er proved that Siberia must have been in same CONDITION for long period Subsidence in Demarara p. 131
153g052 s At gentler bends roads disappear The normal CONDITION of 4th shelf, some way below House of Glen Ro
182b046 subject has Mr Newman the (7) Man studied The CONDITION of every animal is partly due to direct adapt
226b222 ny large islands.-- Hence this must have been CONDITION of Paris basin land.-- (How is this with Fern
250c038 ches Nearest to what is supposed to have been CONDITION of former whole world. America might have bee
296c184 ll.-- St. Helena (& flora of Galapagos?) same CONDITION. Keeling Isd «shows where proper dampness see
318c254 bands of country (These facts show the Normal CONDITION of Migration) gradually separated the birds m
440e146 no final cause yet it must be effect of some CONDITION of external circumstances. results of complic
462t051 smut. theory, on the grounds of similarity in CONDITION in Java & Sumatra & dissimilarity of forms--
544m101 therefore sight of own child. (when frame in CONDITION to receive pleasure) gives pleasure, ie. love
607o025 ect being produced.-- the wax was soft,-- in CONDITION of mind which leads to motion being inclined
608o026 ature chance) 2) difference is from imperfect CONDITION of mind all motives do not. come into play.--
549m122 heck-- that is external circumstances are so CONDITIONED as they are effecting a change in his instin
195b103 soil of Galapagos under equator that external CONDITIONS would produce species so close as Patagonian
197b112 dern types being constant. Cuvier's theory of CONDITIONS of existence is thought to account resemblan
198b113 instance take birds animals, reptiles fish-- CONDITIONS will not explain status (Perhaps considerati
201b126 hese form genera-- this is «from unfavourable CONDITIONS» there are many gaps. & those forms which «n
245c025 , because arrival of any one plant might make CONDITIONS in any one isld different]CD.-- p. 414. dogs
248c033 ivisions of insects» 2. Relation, of external CONDITIONS, & of succession: the <first> latter is most
254c048 rganization, "& are analogous to the earliest CONDITIONS of the higher classes, during which the chan
257c057 This good case. of replacement under peculiar CONDITIONS-- of «nearly» same kind country distant. <St
259c065 mpt of nature under certain very unfavourable CONDITIONS.-- as an adaptation, but adaptation during e
272c108 Entomologicae will tell this.-- What peculiar CONDITIONS the Staphylinidae on St. Pauls Rocks must be
282c140 have greater power of change yet, as external CONDITIONS over whole world. similar-- & constitution o
285c151 at species <shows> is effects of unfavourable CONDITIONS, (hence rise & depression of importance, in
302c200 as has gone on,, No greater gaps.-- external CONDITIONS, to be sure, have remained somewhat similar.
336d019 great change be effected, but in a mule these CONDITIONS are not fulfilled.-- «[My grandfather's the
338d023 times present & past. The effect of physical CONDITIONS of country is not perhaps so great, as separa
343d037 <latte> one: The latter will take place when CONDITIONS are unfavourable to numbers of animals. as i
404e026 uch is case. therefore chance & unfavourable CONDITIONS to parent may be become favourable to offspr
408e043 mericas.-- If species change, we see external CONDITIONS have great effect on them, & therefore exter
410e052) with relation to habits, ranges. & external CONDITIONS of country, most important & will be done to
416e071 cated animal is perfectly adapted to external CONDITIONS.-- (hence great variation in each birth) fro
416e072 fect of a gradation in difference in external CONDITIONS.-- -- as in plant up a mountain-- In races t
429e113 s been without recurrent tendency in external CONDITIONS» sudden loosing of horns.-- I do not believe
441e148 growth-- generation; & more of the effects of CONDITIONS on the «propagating» constitution. but not s
441e149 ilar to <f> first form.-- The great effect of CONDITIONS on offspring, but not on individuals is very
446e163 rm.-- is the constancy owing to similarity of CONDITIONS-- & that no change would affect them in shor
459tflv asts of Prey. any country should during [...] CONDITIONS-- every spot is occupied & has been occupied
460t017 s, is the effect, partly of the same external CONDITIONS (ie. analogical structure) & partly the laws
461t041 nge & it also shows, what enormous changes of CONDITIONS, some species will undergo & yet remain adap
494q004 breeding in Zoolog. Gardens-- with respect to CONDITIONS of animals & their general healthiness-- Fox
507q15v varieties plants growing together. under same CONDITIONS.-- like cowslip & primrose, but less strongl
540m086 aland)-- the American in Brazil is under same CONDITIONS as Negro on the other side of the Atlantic.
623o050 uired by social animals, living under certain CONDITIONS, in this world, they <will conform to the la
623o051 ds» «social feelings», & living under certain CONDITIONS; by my theory they have been formed by the c
623o051 arents, & hence to nearly all the world.-- As CONDITIONS) (for pig will not so readily attain instinc
624o051 d exist without Mammae.» to the then existing CONDITIONS change, from civilization, education changes
638j58v Orbigny. L'Institut. No.-- 221 Good account of CONDOR by Humboldt Zoologie Recuiel-- Meyen has writt
480o006 l I p. 310. History of the Abipones-- says «the CONDOR <it> is found in the Tucuman mountains The fo
481z010 -- do p. 393. <">The wild, small fowls at Pulo CONDORE "crow like ours, but much more small & shrill"
453e182

291c165 be heredetary & strictly point out affinities. CONDUCCT of Gould, remark of D'orbigny point out impor
610o032 -- as the hour of the day &c-- All habits must CONDUCE to their health & comforts.-- Both ideots, old
126a122 would our judgement be-- Does condensed metal, CONDUCT heat better than plain?-- Mem 1000 {P} how eas
126a123 ld not the fluid matter be driven upwards & so CONDUCT heat?-- How comes it in volcanos that have gon
251c041 ted, strangely contradictory to Azaras fact of CONDUCT of wild & tame horses.-- p. 246-- Gmmura-- new
261c070 es from distant countries., may give thread to CONDUCT to laws of change of organization! The little
544m102 No memory of past events?) or influence on our CONDUCT, the links which when conscious connect past,
573n039 it will.-- My father told Miss. C. of the bad CONDUCT of Mrs C. (her brother's wife). & she said not
578n054 ially a women think of my face,"? to one moral CONDUCT.-- either good or bad. either giving a beggar.
608o026 n one try to persuade person to change line of CONDUCT. as being better & making him happier.-- he ag
620o044 on its action, & on the results following our CONDUCT.-- If the temptation to disobey the conscience
620o045 re are certain instincts pointing out lines of CONDUCT to other men, [RHC] 5) which are natural (& wh
620o046 cts.-- so man ought to follow certain lines of CONDUCT, <although> even when tempted not to do so, by
621o047 ling, that he ought to follow certain lines of CONDUCT, & he must soon necessarily learn that it is h
621o047 ild may be taught, or will acquire from seeing CONDUCT of others, the feeling that almost (rarely if
624o051 rule of right.-- [LHC] *for it strives to give CONDUCT beneficial to all the children, « <then> each
628o53v from the necessities «& good» of society such CONDUCT is instinctive in me (& as a consequence, but
628o054 e emotions of this instinct, with that line of CONDUCT, & if taught rightly, it will be for the gener
113a090 the upper four hundred feet of strata having CONDUCTED away the heat of surface. & if conducting pow
397e003 operations of what we call nature, have been CONDUCTED almost! invariably according to fixed laws: A
022r006 fied rocks show any difference in facility of CONDUCTING Electricity? Would minute particles have a t
112a088 e 101. in Note Book (C) for some speculats on CONDUCTING powers of rock-- -- Geograph Journal Vol IV
113a090 0 ft. shows that the strata have very unusual CONDUCTING power of heat from centre.-- But is this not
113a090 ving conducted away the heat of surface. & if CONDUCTING powers had been better then 32 degree would
113a091 und lower.-- We have no right to consider the CONDUCTING powers either better or worse & the depth of
126a122 matter separates them, this is ascertained by CONDUCTING powers-- we judge from the surface, & say 60
077r172 adre of [misnumbered page] Dr D. remarks. bad CONDUCTOR of Heat do of Electricity Does not iron, comb
107a078 under water" but cause most difficult (better CONDUCTOR) Fitz Roy's Case of S. Maria & Tubul applicab
125a121 P} if crust were metal then thinner if better CONDUCTOR, then still thinner → The problem is, you hav
126a122 very wrong,-- The fact of a dumplin being bad CONDUCTOR is {P} against my views-- if we had rod thus
126a123 e does not become hot?-- this looks as if bad CONDUCTOR-- III But equilibrium is not attained, & if c
132a139 ows from variation in strata earth a very bad CONDUCTOR.-- shows p. 516 that subterranean springs giv
039r059 in Cornwall or in the coal measures have been CONDUITS to volcanoes.-- Talking of the cricket valley
491q01v ation <caused> of some seeds, caused symmetry in CONE-- The «above Exper» explains apples on side nea
049r088 ; Epidote -- Must we look at regular greenstone CONES at S. T. del Fuego as nucleus of a Volcano or a
127a127 olog. Cantal Vol III I? p. 246. on formation of CONES beneath sea.-- with reference to old submarine
577n049 for it can be shown that the life & will of a CONFERVA is not an antagonist quality to life & mind o
021r005 La. billardiere mentions the floating marine CONFERVAE, is very common within E. Indian Archipelago,
051r094 face to where highest spray (there pale green CONFERVAE) coated with living beings; In smooth seas (&
051r094 malous; It is wonderful to see Coral reef-or CONFERVAE-- in the breakers or in waterfall: Excepting by
221b200 to insects &c &c?-- (p. 23 do.-- On animal - CONFERVAE-- p. 23 p. 267. Dela Beche. Geolog. Researche
056r109 encircled, the one slow cause is apparent. «I CONFESS I never see such islands whose inclination nat
223b211 s accounts for uniformity of species & we Must CONFESS. that we canot tell, what is the amount of dif
292c170 the cause if his faith is not staggered <">= I CONFESS. no dissertation against these views, could po
310c222 an> so many animals of different types. I will CONFESS my profound ignorance.-- but seeing such passi
428e112 ion of accidental hardy seedlings: (which are CONFESSED to by Herbert) to sift out the weaker ones: t
600o008 ul sur la terre. J'ai trouve son âme" &c-- -- CONFESSES these faculties of soul, treating of infinite
533m060 curred to me that Capt. F. R. candour & ready CONFESSION of error made him less repentant.-- In makin
533m060 . that author remarks, that writing down his CONFESSIONS of sins. did not make him more humble.-- it
567n013 y have habitual action. which depends on such CONFIDENCE» when does such notion commence?-- Children
113a091 the depth of 32 degree. being little we may CONFIDENTLY infer that time has not been allowed for low
026r020 aypo at one period of elevation must in its CONFIGURATION have resembled Chiloe In De La Beche, artic
509q017 dren, who have not worked have any peculiar CONFIGURATION.-- Hooker <Meta> Metaphysics of Morphology.
615o038 nct always obtain peculiarities of external CONFIGURATION. [RHC] General-- Instincts, certainly appea
196b105 tinize genera, & draw up tables-- Instinct may CONFINE certain birds which have wide powers of flight
043r069 tion Volcanic action, must be more exclusively CONFINED to that country. Read description of channels
180b038 at the permanent varieties produced by <inter> CONFINED breeding & changing circumstances are continu
194b095 been doubted?]CD & opossum found in Europe now CONFINED to southern hemisphere.-- If these facts were
200b125 ound in Virginia p. 325. July 1828. Animal now CONFINED to extreme North.-- ▓.do p. 326. 2 Fossil spe
220b199 ous.-- Some general statements about mundine & CONFINED genera.-- Lyell has remarked about no confine
220b200 confined genera.-- Lyell has remarked about no CONFINED species in Sicily. Jan: 1838 L'.Institut. Ba
235b273 ble to work through all genera, & see how many CONFINED to certain countries-- so on with families.--
272c109 must be placed under Gould says most subgenera CONFINED to continent, though we have seen species «of
274c114 in last instance hardly at all developed. Not CONFINED to one species, but generally small genus. ¿a
278c129 liminated. where every species of a section is CONFINED to one continent & every species to another t
310c223 otanical Provinces will turn out not nearly so CONFINED as now thought.-- N. American, European & Chi
316c245 ranges than other plants.-- Many «Some» genera CONFINED to hot countries & many to cold.-- Hence lati
316c245 ment. thus the Naiads (study De Ferrussac) are CONFINED to <S.> American.-- Mr Sowerby says there are
337d020 local ones are taken to fresh country & breed CONFINED. to certain best individuals-- scarcely any
350d055 with eyes!-- (are not these differences in sex CONFINED to annulosa?) Remarked that young of Cirrhipe
415o066 l man became cosmopolite, he would probably be CONFINED in locality like Ornithorhyncus.; since being
466t095 th the same powers instinctive & doubtless not CONFINED to sex.-- «Is not cantering a congenital pecu
500q10a rict.-- About <endemic &> wandering species of CONFINED genera By my theory in volcanic or rising isl
523m014 the Birmingham Doctor praising his sister who CONFINED him. & yet disinheriting her.-- This «N B. I
543m099 3d. Jones said the great calculators, from the CONFINED nature of their associations-- This is not so in
316c245 de.-- But in land & F W shells there is more CONFINEMENT. thus the Naiads (study De Ferrussac) are co
360d091 ut never in an animal, that had long been in CONFINEMENT-- is this effect of climate, or state in whi
380d155 en muzzled, he is disabled.-- so Elephant in CONFINEMENT, & so imagination in Man, has strange effect
510q018 Yarrell (1) About non-breeding of animals in CONFINEMENT, curious.-- foxes-- English animals. [Made n
527m031 ach other «though different species» when in CONFINEMENT, so may they learn in a state of nature.-- S
584n072 the bee to its nest.-- cats when carried in CONFINEMENT,-- carrier pidgeons proverbially carried to
198b113 wen's idea) states these class approach on the CONFINES? Balanidae?-- --- I cannot understand whether
565n008 ypothesis of opposite muscles will want much CONFIRMATION. A grave person close those muscles, which
148q026 not so hardy but more easily fatten, This man CONFIRMED the account of the «YOUNG» Shepherd dogs Satu
346d047 faced, but more tendency to fatten-- This man CONFIRMED my account of the Shepherd dogs.-- Aug. 24th.
553m132 an emu, although young dogs get sadly torn in CONFLICTS with the former. But it is one thing for a sw
623o050 certain conditions, in this world, they <will CONFORM to the law,> «can only be such, as are consist
076r170 to SW at 45 degree to 50 degree-- covered by CONFORMABLE greenstone porphyrys & phonolites do. amphib
365d107 ture) may be more in constitutional.,-- more CONFORMABLE to the structure which has been adapted to f
047r083 Partial shrinking after elevation in perfect CONFORMITY with <Mr Lyell's> idea of an injected mass o
191b080 ion of these archipelagos ups & downs in full CONFORMITY with Europaean formations-- England & Europe
242c018 4) batrachian in isles of Great ocean says in CONFORMITY with Bory's Views.-- <Says> D'Orbigny is sai
419e087 s effects of catastrophes, which must serve to CONFOUND our chronology» «CONSIDER ALL THIS» Extinctio
541m090 people are fickle & full of levity (¿ do I not CONFOUND action & thought here?) The opposite extreme
363d102 that after differences were pointed out Selby CONFOUNDED them, yet can readily be told by incubation
419e087 on & transmutation, two foundations, hitherto CONFOUNDED,. of geology.-- L'Institut 1838. p. 414; M.
465t079 less.-- May not several generations have been CONFOUNDED in the caves? It is highly important, to bea
550m126 showing how easily chance & will of Deity are CONFOUNDED-- well applicable to free will. Mayo. Philo
114a094 sed clay slate.-- -shale in shall sea. Lyell CONFOUNDS these introduce discussion -- I see Lyell tal
339d026 ted after flood (!) according to affinities!. CONFOUNDS, like Whewell affinity with analogy-- Good ta
524m022 e. exceedingly sharp in some things, though so CONFUSED in others-- Mrs P. when in state as above de
555m143 y sneer.-- Sept 21st Was witty in a dream in a CONFUSED manner. thought that a person was hung & came
555m143 ing faced death like a hero, & then I had some CONFUSED idea of showing scar behind (.instead of fron
558m151 harity.-- ¿ May not idea of God arise from our CONFUSED idea of "ought". joined with necessary notion
086a003 cter of Vegetation. -- So accustomed to utter CONFUSION in Europe, that the simplicity of Ventana's «
228b229 , as well as at present time, we might expect CONFUSION of species.-- Important. For instance take Vo
258c061 fly catcher, woodpecker &c & which causes the CONFUSION in this system of nature-- Whether species ma
586n080 (faculty «being» always heredetary helps this CONFUSION.--) Hensleigh considers breathing instinctive

403e023 oxodon) & say its relations.-- if we know its CONGENERS then we can.-- now on my theory this «certain
265c083 rs have gone on boring their noses. &c & This CONGENITAL changes show that grandson is determined, wh
336d018 ained & hereditary; <but if> if the change be CONGENITAL (that is most slowly obtained with respect t
466t095 ss not confined to sex.-- «Is not cantering a CONGENITAL peculiarity improved.» Probably every such «
050r090 -- Ascension At Ischia there is a pumiceous CONGLOMERATE with small & large fragments, nature of whi
053r100 worn away like english escarpment The great CONGLOMERATE of the Amazons & Orinoco mentioned by Humbo
069r151 where trachyte. There must have been as much CONGLOMERATE on West of Peuquenes as on East. Where gone
069r151 Where gone to.?-- There must have been some CONGLOMERATE East of Portillo Where gone to? Intermediat
069r152 ogy between Wealden & Bolivia Transportal of CONGLOMERATE between two ranges mysterious!-- Mem. SUBSI
146g013 pleasure; mem. Shepherd dogs The Patches of CONGLOMERATE on S. Ventana, excellent instance, how acci
026r019 e been formed in shallow water: so have the CONGLOMERATES: Yet this view is directly opposed to commo
027r023 ted rocks P xv. mentions in what formations CONGLOMERATES are found.--- The above oscillations remark
029r034 ike the Coal measures in England (Excepting CONGLOMERATES?) -«& bsence of limestone?» have been colle
031r038 high tidal action {P} NB. patches of modern CONGLOMERATES [Fig. 2] {P} The action of sea A. B. will b
122a115 oved [...] {P} August 25. I saw metamorphic CONGLOMERATES on shore of Loch Lochy very like those of A
165g124 och Oich 92 each Loch 8 ft. The Metamorphic CONGLOMERATES near Loch Lochy would be well worth examini
208b154 ar to C. de Verd's.--? NO Macleay Name given in CONGO Expedition We need not expect to find <species>
310c224 India some plants same. --America.-- See Brown CONGO Expedition: 400 Australian plants found in othe
317c248 s Companion Study Appendix (& only appendix) of CONGO Expedition, NB. I met an old man--, who told me
321c275 ead Appendix Ovington Voyage to Surinam. Voyage CONGO expedition: Zaire except Brown's Appendix & exc
324c268 riages Brown at end of Flinders & at end of the CONGO Voyage Decandoelle. Philosophie. or Geographica
033r043 xi has a <[...]> cylinder placed on the rim of CONICAL crater: at Teneriffe Wall of Porph. Lava with
049r088 eus of a Volcano or as an injected mass.--From CONICAL form I incline to <latter> former; & thus occu
049r089 dikes: & indeed when do these dikes lead to a CONICAL mass. will this conical mass be granite? Why n
049r089 these dikes lead to a conical mass. will this CONICAL mass be granite? Why not more probably greenst
101a058 superincumbent strata. where they have yielded CONICAL axis of mountain.-- only when dikes reach near
159g086 a little way down slope of obscure terraces (& CONICAL hills on same) of «semi» waterworn & some part
471tf08 ans. 18. p. 163. "D. Dod on two new genera of CONIFERAE".-- referring to the 3 main divisions & speak
071r156 r world Dicotyledones far preponderant, if so CONIFEROUS must formerly have been most abundant tree--
206b149 tribe fish extinct. or of Pachydermata, or of CONIFEROUS trees; or in certain shell cephalopoda.-- «R
207b150 formerly common to New Holland?! p. 320. Says CONIFEROUS structure intermediate between vascular or c
221b202 yells doctrine removed?? Is the prevalence of CONIFEROUS Woods before Dicotyledenous a fact analogous
257c058 e. great lizards in do.-- <Wood> <Dicot wood> CONIFEROUS wood in Coal Measure.-- highest fish in Old
374d133 s in coal supposed to be intermediate between CONIFEROUS trees & Lycopodiums.-- p. 437. Many. existin
391d174 her» each hermaphrodite being simple (Are not CONIFEROUS trees generally dioecious oldest forms) why
529m042 The story of the Corbets & big noses, quite CONJECTURAL, in Blakeways book of Sheriffs.-- July 22d.
132a138 heat to warm the ocean).-- and M. Parrot does CONJECTURE that in Scoresby's case volcanicity has warm
228b232 got it, not look forward» if we choose to let CONJECTURE run wild then <our> animals our fellow breth
342d034 ertain-- Yarrell remarks he has somewhere met CONJECTURE that all salt-water fish were once salt wate
350d054 rst civilised. <note> add this in note. ¿mere CONJECTURE?-- Australians.-- Americans. &c After Decand
074r165 nly be accounted for by concretionary action, CONJOINED with other» <state simplest case. concretion
100a053 r.-- A discussion on concretions and cleavage CONJOINED very good.-- It is the Key to the story.--co
619o042 mal, it may be concluded that he has parental, CONJUGAL and social instincts, and perhaps others.-- [
622o049 ed with feeling towards one as a leader,-- the CONJUGAL feeling may be directed towards one or more.-
261c070 nt running to the water, is a good instance of CONNATE instinct, better than child sucking or even du
292c168 .-- Splendid Harmony these views-- did Lamarck CONNECT extermination of some forms with his views.--
316c247 nes similar?)» Now this is very remarkable.-- (CONNECT these facts with identity of land animals.-- t
544m102 on our conduct, the links which when conscious CONNECT past, present & future thoughts are broken-- S
611o034 - The inorganic are probably one principle for CONNECT of electricity chemical attraction, heat & gra
027r027 of shells in the Calc. Sandstone Concret, is CONNECTED with frequency of shells in flints in Chalk N
035r045 scension, the laminae <...> changes in rocks. CONNECTED with & alternating with obsidian must clearly
036r050 stinct, from some of the lower orders; it was CONNECTED with movement of sand.--it is called "Bramido
064r137 it must be said, that lines of elevation have CONNECTED <lines> «points» of eruption [.] give instanc
068r146 strata in act of dislocation (NB. dislocation CONNECTED with fluidity of rock ∴ «in earliest stage» w
069r150 ofoundly the sandstone of the Portillo line.--CONNECTED with <gneiss>.--(Mica Slate) [Fig. 9] {P} ((3
086a005 don Is the general saline tendency of America CONNECTED with its elevation. vapour from below-- Malte
098a046 gates, calcareous balls) that concretions are CONNECTED with a crystalline process.-- now cleavage as
110a085 Lyell-- Some internal changes are in process. CONNECTED with variation of compass & these may cause «
126a123 f sea.-- deep seated springs «spring requires CONNECTED column.--» of cold water show, that water doe
179b035 hortness of life of species in certain orders CONNECTED with gaps in the series of connection? «if st
195b104 thing further, points gained if any facts are CONNECTED.-- No doubt in birds, mundine genera, Bats .F
220b197 inct perverted, yet organization «especially» CONNECTED with generation certainly is.= The dislike of
248c033 ession: the <first> latter is most intimately CONNECTED with important structures. <which are less ob
249c033 ld word. do not form two sections is this not CONNECTED with wide range of animals. Follow this out,
249c036 l districts???? north of 30 degree.--, may be CONNECTED with, Mr Blyth's statement of birds of Europe
249c036 g within Tropics.-- Europaean birds at Japan. CONNECTED with Europaean forms on Himalaya??-- This is
265c087 n we should never know how much structure was CONNECTED with habits, & how much heredetary. The circu
273c112 may catch insects on the wing & pratencole (¿CONNECTE[D] with Chionis), yet the Tropic bird, has ver
305c210 manner of wagging tail.-- habitual movements CONNECTED with mind There is no progression in the deve
306c213 rtebrata, but mammalia & reptiles &c Timor is CONNECTED with Australia «map to King's Australia» by a
326c268 Brougham. Dissertations on subject of Science CONNECTED with Natural Theology.-- on instinct & animal
350d056 in creation.-- thus senses, especially sight CONNECTED with locomotion.-- «Mem. Dr. Blackwell (Aberc
367d113 nflating its body like balloon-- by air cells CONNECTED with cheek pouches.-- Hunter's Animal OEconom
369d115 se being water animals <that this> structure <CONNECTED with animals being compelled to travers> «May
378d147 stic animals ought to show them.-- Anyhow not CONNECTED with habits According as child is like parent
387d168 difference of time) is the above law anyways CONNECTED with the case of successive copulation impres
389d176 ant with regard to theory, showing generation CONNECTED with whole system, «as if there was, a supera
389d177 e are bred into each other.-- This is somehow CONNECTED (This seems case, for by careful observing ca
389d177 .)-- [The loss of passion in hybrids. perhaps CONNECTED with this same case (& not merely as I have s
390d177 -- (this repugnance to breeding in & in seems CONNECTED with more developed forms) Study buds-- gemma
390d178 woman, not a bud in every respect.-- [Is this CONNECTED with the physical differences in almost all M
422e092 ve of true affinity.-- -- -- Owen says Dugong CONNECTED with Pachydermata.-- <it was a Pachyderm. whi
425e105 plants on single mountains, though these are CONNECTED with other mountains laterally.-- Owen. Fossi
429e116 s on my view-- because these points were last CONNECTED with those northern regions-- do p. 21 says.
432e125 giously when drinking very cold water «frowns CONNECTED with pain, as well as intense thought.--» No
477z001 y and Botany. Philosoph. Transacts. 3. papers CONNECTED with transform of Crust-- Westwood & Thompson
498q008 own?-- Linnaeus has shown that each pistil is CONNECTED with separate division of germen <?>-- (11) M
516q23v ppear after <th> being killed Experiments not CONNECTED with Species Theory (1) Will an extract of pe
533m058 not acquire it instinctively. may not this be CONNECTED with their power of acquiring language.-- Hen
541m089 pain, but one only fears that pain, which is CONNECTED with death!-- How has this instinctive fear a
542m094 ch of these actions are habitual, & which now CONNECTED physical relations.-- CD[like sighing to reli
543m099 he greater number of insects in insecta-- not CONNECTED with transformation because Spiders have many
547m112 there being no other parallel trains of ideas CONNECTED with past circumstances.-- as whether I reall
552m131 ountain chains in N. America Fear probably is CONNECTED with habitual stopping of breath to hear any
552m131 sound.-- attitude of attention «So intimately CONNECTED is passion with sending force to muscles, tha
556m144 elieving consists in the comparison of ideas, CONNECTED with judgment. [What is the Philosophy of Sha
566n012 guide one to conclusion that any one fact was CONNECTED with law.-- as soon as any enquiry commenced,
567n012 cause & effect being a necessary notion is it CONNECTED with <our> the willing of the simplest anima
567n015 hows blood is pumped over whole body.-- is it CONNECTED with surprise.-- heart beginning to beat-- ch
568n017 ny fantastic custom» «Probably bashfulness is CONNECTED with some disturbed habit» [Thus shepherd dog
570n024 is shoulders & walks off.-- I think shrugging CONNECTED with many emotions.-- (Explanation of sighing
572n034 eglecting time, & general sense, anymore than CONNECTED with general tendency of the dream.-- It does
574n041 te, that which one sexually loves is probably CONNECTED with flow of saliva, & hence with action of m
576n048 tinct is a modification of bodily structure «(CONNECTED with locomotion.)» «no, for plants have insti
577n052 like erection shyness is certainly very much CONNECTED with thinking of oneself.-- «blushing» is con
577n052 ted with thinking of oneself.-- «blushing» is CONNECTED with sexual, because each sex thinks more of
580n062 Lr. Brougham «Dissert.» on subject of science CONNECTED with Nat. Theology.-- says animals have abstr
622o048 or not, as feeling of cowardice> «This is not CONNECTED with sense» instantaneous so declaring it is

Page **(Key Word)***
151g042 t, but are separated by flat bottomed strait. CONNECTING flat on one side with irregular gravel plain
160g088 above station! There is long straight isthmus CONNECTING E & W connecting Glen Bought & Glen Tarf a p
160g089 ere is long straight isthmus connecting E & W CONNECTING Glen Bought & Glen Tarf a perfect old Loch,
196b110 duce from extreme difficulty of hypothesis of CONNECTING Mollusca & vertebrata, that there must be ve
302c201 this I believe the case. = any animal really CONNECTING the fish & Mammalia, must be sprung from som
448e168 odon, <all> equally aberrant-- the two former CONNECTING classes like Toxodon «In orders»-- Fish & re
045r077 rising, when another falls.--When discussing CONNECTION of Pacifick & S. America. -- Volcanos must b
050r090 worth reading Some earthquakes of Sumatra no CONNECTION with a neighbouring Volcano of Priamang.--Ma
055r107 demolished craters).--worn into mud & dust.--CONNECTION with age, & agreement with number of craters
060r124 nce".--Afterwards speaks of this phenomena in CONNECTION with "the shooting upwards" of the <ground>
065r140 of land shows subsidence in T. del Fuego, and CONNECTION of quadrupeds.--although recent elevation, t
070r153 hibit orbicular structure.--When we recollect CONNECTION of columnar & orbicular in basalt.-- When we
102a062 w acting in certain directions predominantly, CONNECTION with magnetism &c counteracting gravity.-- A
112a088 358. changed soundings in Mouth of S. Cruz in CONNECTION with Fitz Roys fact of elevated block of sto
177b029 definite existence.-- There does appear some CONNECTION shortness of existence, «in» perfect<ion>, «
179b035 n orders connected with gaps in the series of CONNECTION? «if stating from same epoch certainly» The
189b071 epugnance.-- Waterhouse says there is no TRUE CONNECTION between great groups.-- Speculate on land be
211b164 ls in East Indian Archipelago--, very good in CONNECTION with Von Buch Volcanic chart & my idea of do
226b220 Madeira? «Tristan d'Acunha?» «Iceland?--» The CONNECTION between Mauritius & Madagascar very good.--
227b224 auna, we may form some idea of <origin under> CONNECTION of those two countries Hence India, Mexico &
231b243 ate regions & tropics are only related by one CONNECTION.-- viz descent.-- Hence far greater discorda
250c039 ent Crocodiles with Palaeotherium in India--: CONNECTION with Latitudes!? Zoological Journal.--- Vol
273c112 e exaggerated.-- There is one most remarkable CONNECTION between these aerial representatives of the
278c130 en's land is most important as showing former CONNECTION of two continents & death of form in one. Th
285c151 e & depression of importance, in each group & CONNECTION of even distant ones) the characters will be
302c201 cendant of some intermediate link.-- the only CONNECTION between two such classes will be those of an
308c219 this is so-- !!Metaphysics!!! Mrs Somerville, CONNECTION of Physical Sciences p 276 May be worth glan
440e148 when growth stops».-- Spallanzani's facts in CONNECTION with buds.-- They differ from possibility of
523m017 ndow to kill herself from jealousy of husband CONNECTION with housemaid two years before, to prove sh
525m026 ious an illtempered fat man looks, shows same CONNECTION between organization & mind.-- thinking over
528m039 espond to those <he> awakened during music.-- CONNECTION with poetry, abundance, fertility, rustic li
553m137 t group.-- this is very important as showing <CONNECTION> that expression mean SOMETHING.-- Hunt (the
570n027 ow worm doubtless admires female. showing. no CONNECTION with male figure]CD-- As forms change, so mu
571n030 Buildings look ugly is because there is some CONNECTION between them, & great masses of rock.-- I wa
571n031 earliest times there must have been intimate CONNECTION between sound & language.-- Chinese. simples
574n040 ch my theory requires. There probably is some CONNECTION between very limited reasoning powers & the
544m102 ations of time, <identity,> place & personal CONNECTIONS-- ideas are strung together in manner <they>
072r159 Jan Meyen Isld.--Mr Barrow thinks N & S. line CONNECTS western isles of Scotland & Iceland.--Bosh no
407e041 able form of land,-- -- S America, an island, «CONNECTS with Asia» between two polar lands,--; Africa
228b230 wn by hybridity of ferns.-- hybridity showing CONNEXION of two plants. Animals-- whom we have made ou
231b242 ian Seas. Marsupials animals all show greater CONNEXION in quadrupeds, <bu> plants do not follow by a
540m085 cough & yawn are probably merely coorganic as CONNEXION of mammae & womb.-- We need not feel so much
551m127 He compared spectral illusion & insanity the CONNEXION appears to me vague-- Delirium of every degre
558m151 point of structure takes its value. from its CONNEXION with mind, (to show hiatus in mind not saltus
558m151 s between man & Brutes) no one can doubt this CONNEXION.-- look at faces of people in different trade
569n020 robably, language commenced in some necessary CONNEXION between things & voice, as roaring for lion &
193b093 n has no heredetray prejudices «or instinc» to CONQUER or breed together:-- Man has no limits to desi
263c075 may appear an absurd saying) & will never be CONQUERED by anyone (if has any kind of prejudices) <wi
609o029 pulse) but shame «we alas know» is far easier CONQUERED than the deeper & worser feelings. These bad
619o043 love of its puppies: the latter generally soon CONQUERS, & the dog [RHC] 3) probably thinks no more o
621o047 ity of organ of taste, for when grown up often CONQUERS it). It will be only rarely that it thinks th
525m024 was known he had been on the table,-- guilty CONSCIENCE.-- Not probable in Squib's case any direct f
533m061 ing my character I felt it)-- this is kind of CONSCIENCE, is obscure memory of having read or thought
534m061 ent remains, where reason is forgotten. it is CONSCIENCE, or instinct. Hensleigh says to say. Brain p
537m075 hown-- see Mackintosh.-- Must grant, that the CONSCIENCE varies in different races.-- no more wonderf
537m076 an, like deer &c, being social animal, & this CONSCIENCE or instinct may be most firmly fixed, but it
537m077 not act up to them, no doubt disobeys & hurts CONSCIENCE more than other.-- A Scotchman will his coun
540m088 knows the prowlers of the air" &c &c &c so is CONSCIENCE &c &c Coleridge,-- Zapoyla p. 117, Galignani
549m119 -- These thoughts are most pleasant. when the CONSCIENCE tells our [mind], good has been done-- «& co
549m119 nce tells our [mind], good has been done-- «& CONSCIENCE free from offence» -- pleasure of intellect
550m124 grandfather!-- A man, who perfectly obeys his CONSCIENCE or instinct, would probably feel but little
564n002 elieved, he would be sorry or have a troubled CONSCIENCE.-- Therefore I say grant reason to any anima
564n003 al instinct «& yet with passion» he must have CONSCIENCE-- this is capital view.-- Dogs conscience wo
564n003 ave conscience-- this is capital view.-- Dogs CONSCIENCE would not have been same with mans because o
564n003 er who reasons much on his actions, makes his CONSCIENCE far more sensitive. ulitmate effects of acti
564n004 of» death & for the satisfaction of following CONSCIENCE, obeying habits, & dread of misery of future
564n004) but a causation. & «perhaps» an instinct of CONSCIENCE, feeling in his heart those rules, which he
570n026 accustomed to:-- analogous case to my idea of CONSCIENCE.-- deduction from this would be that a mount
572n034 tendency of the dream.-- It does not hurt the CONSCIENCE of a Boy to swear, though reason may tell hi
572n034 reason may tell him not, but it does hurt his CONSCIENCE, if he has been cowardly, or has injured ano
576n046 life,-- if he meditated on this, it would be CONSCIENCE. A man, might not <t> do so even to save a f
591n100 man is not followed; good case of instinctive CONSCIENCE.-- Why does not man eating cause disgust, be
599o008 198.-- "Moralité, raison, beau ideal, infini CONSCIENCE; voila l'homme separe de la matiere et du te
600o008 ittle Chapter on each faculty of Soul.-- (I) <CONSCIENCE> «Moral Sentiments» imperative sense of duty
600o008 ideal, refers chiefly to moral, beau desires CONSCIENCE & love.-- [With regard to ordinary Beau idea
600o08v y. (4) Reason, some transcendental kind-- (5) CONSCIENCE, not clear-- Then these last heads. of separ
600o08v h make great difference between moral sense & CONSCIENCE? we admire what is right by one & are ordere
600o08v & are ordered to do it by other.-- I suspect CONSCIENCE, an heredetary compound passion. like avaric
600o08v e not something analogous to imperiousness of CONSCIENCE: in Maternal instinct domineering over love
609o029 (does it originate in a doubting feel between CONSCIENCE & impulse) but shame «we alas know» is far e
612o035 ce, & obedience to certain stimulants without CONSCIENCE in the lower animals, as in stomach, intesti
614o037 weak so as to be overcome easily by reason.-- CONSCIENCE is one of these instinctive feelings. [LHC]
615o038 th certain actions performed by your parents, CONSCIENCE This «X» memory especially «the» general kin
620o044 her time, for he knows that the instinct. (or CONSCIENCE) is always present (which is indeed, often f
620o044 ime it is disobeyed) & is sure guide.-- Hence CONSCIENCE is improved by attending & reasoning on its
620o044 r conduct.-- If the temptation to disobey the CONSCIENCE is extremely great [LHC] The cause perhaps l
620o045 & either excuses the «non-» following of ones CONSCIENCE. & palliates the offence; one always admire
620o045 the habit formed by «obediance to instinct» <CONSCIENCE>., or rather the strengthened instinct, even
621o047 osed by some natural passion.-- (a) [LHC] The CONSCIENCE rebukes malevolent feelings, as much as acti
622o048 fied by circumstances of country, so will the CONSCIENCE in these cases.-- Those instructions, which
622o048 o will reflect must feel, how like to injured CONSCIENCE, is the feeling of any custom of society bro
625o052 & others think there is distinct faculty, of CONSCIENCE.-- I believe that certain feelings & actions
626o052 contact with will" is unintelligible to me.-- CONSCIENCE regulates feelings, as of cowardice.-- the w
626o052 [LHC] p. 194. «&c &c» Butler's view given on CONSCIENCE: I cannot admit it.-- see notes to it by me.
628o54v ckintosh shows the change produced.-- 4) p 38 CONSCIENCE checks the wish to <other> outward gratifica
628o54v nything contrary to it]CD NB. the very end of CONSCIENCE is stop to wishes of passion &c. whilst the
629o54v ctions (in a dog. & therefore not <instinct> «CONSCIENCE») equally <prefe> destroy all wish of outwar
629o55v Any action by habit may be thought wrong.-- & CONSCIENCE will imperiously say so, & produce shame & r
629o55v desires do not intervene between this kind of CONSCIENCE & the will, though «this» conscience does be
629o55v kind of conscience & the will, though «this» CONSCIENCE does between the desires & will?)) (2) It is
628o54v e> no desire of gratification will check the CONSCIENCES desire for virtue.-- [I expect there is some
255c053 does.-- One is tempted to exclaim that nature CONSCIOUS of the principle of incessant change in her o
301c199 ucture.--duckling runs to water. before it is CONSCIOUS of web. feet.-- p. 7. Mr Blyths arguments aga
533m059 f his childhood without knowing why-- had not CONSCIOUS of recollecting it-- this may be nearest appr
544m102 nfluence on our conduct, the links which when CONSCIOUS connect past, present & future thoughts are b
548m115 ese (which must be present, though one is not CONSCIOUS of them, else one would not stand) a crowd of
548m117 lf, though as in Dr Ashe's case, one here was CONSCIOUS of the two states.-- August 30th.-- It is sin
580n061 is not this precisely the same, as the double-CONSCIOUS kept playing so well.-- Lr. Brougham «Dissert

601o009 2. with memory added to it, man in sleep not CONSCIOUS, nor child-- Evidence of consciousness, <t> m
601o009 death upon seeing an object.-- are Planariae CONSCIOUS.-- Consciousness bears some relation to time
611o034 least shows a local will, though perhaps not CONSCIOUS sensation. [RHC] During growth <extres> tissu
613o035 do such.-- All this can take place & man not CONSCIOUS as in sleep; or in sleep is man momentarily c
613o035 s as in sleep; or in sleep is man momentarily CONSCIOUS, but is memory gone?-- Where pain & pleasure
616o038 we may look back to definite action or to our CONSCIOUS selves.-- Such memory may go back to animals
617o40v of our own power & consciousness of it we are CONSCIOUS that we ourselves can originate in any point
292c171 r instinct being memory transmitted without CONSCIOUSNESS «a most possible thing. see men walking in
292c171 stage, & an habitual action implies want of CONSCIOUSNESS & will & therefore may be called instinctiv
293c172 ce)» he had done this without reflection or CONSCIOUSNESS of reasoning to tell back from front. &c or
314c238 tion has such a change been effected.-- the CONSCIOUSNESS of the plant that this part must be protect
325c267 e on Roses might be worth consult. Paper on CONSCIOUSNESS in Brutes in Blackwood. June 1838 H. C. Wat
521m007 rom one generation to another, also without CONSCIOUSNESS, as instincts are, is not so very wonderful
522m014 d me, plotting speeches, yet with a sort of CONSCIOUSNESS not just.-- From habit the feeling of anger
523m015 mmunicated to my grandfather his feeling of CONSCIOUSNESS of insanity coming on.-- his struggles agai
533m060 tary feelings of gratitude, I had a sort of CONSCIOUSNESS I was not right; though I never realized th
538m078 rgh. Phil. Transact. p. 365. Case of double CONSCIOUSNESS, one only «little» less perfect than other,
538m078 onsider this profoundly, may throw light on CONSCIOUSNESS, explained by Dr Dewar on principle of asso
539m083 te trains going on in the mind as in double CONSCIOUSNESS may really explain what habit is-- In the h
539m083 of thought one idea. calls up other, & the CONSCIOUSNESS of double individual is not awakened.-- The
539m083 itual state because it will (without direct CONSCIOUSNESS?) change its habits.-- Aug. 16th. As instan
544m103 he comparison, with past ideas. which makes CONSCIOUSNESS-- & which tells one of reality-- castle in
544m103 eam. never fatiguing,-- else it is only our CONSCIOUSNESS, & senses tell us it is not real.= = dreami
546m110 k (p. 143) wonderful case of perfect double CONSCIOUSNESS Mayo compares it with Somnambulism.-- the y
547m112 d to past idea.-- there was a kind of ideal CONSCIOUSNESS for moment, implied by «presence» my servan
547m113 > «calling up» old ones, to be sure of ones CONSCIOUSNESS.-- Mayo observe no improbabilities in a dre
548m115 work-- though dreams do that One Reflective CONSCIOUSNESS is curious problem., one does not care for
548m116 unhappy)-- it is because in this state, the CONSCIOUSNESS does not go back to former periods so «as»
548m116 - But now in Mayo's «p. 140» case of double CONSCIOUSNESS, one would pity suffering in one state almo
548m116 ity is <much> «somewhat» the same as double CONSCIOUSNESS, as shown in the tendency to forget the ins
559m155 rception of a final cause.-- Read. Paper on CONSCIOUSNESS in Brutes & Animals. in Blackwood's Magazin
559m156 after long period.-- My Father about double CONSCIOUSNESS.-- & somnambulism. Do people when inhaling
566n010 nd & body retrograding to ancestral type o̶ CONSCIOUSNESS &c &c.-- Lavater. (Holcroft Translat) Vol I
575n043 memory. affected by diseases. &c &c, double CONSCIOUSNESS? What other explanation-- can we suppose so
586n079 gh one cannot remember it.]CD back, without CONSCIOUSNESS & by habit, such habit of knowledge of poin
593n111 ter between man & animals.-- [blank] Double CONSCIOUSNESS. only extreme step of an ideal argument hel
593n111 emens hearing other man speaks. shows, that CONSCIOUSNESS of personnal identity is by no means a nece
601o009 ontraction (that is the only evidence. when CONSCIOUSNESS is absent) in fibres united with nervous fi
601o009 nd certain action. (only evidence. when not CONSCIOUSNESS) are produced in consequence having some re
601o009 animals & not plants as supposed by Buffon) CONSCIOUSNESS is sensation No. 2. with memory added to it
601o009 leep not conscious, nor child-- Evidence of CONSCIOUSNESS, <t> movements «¡» anterior to any direct s
601o009 ing an object.-- are Planariae conscious.-- CONSCIOUSNESS bears some relation to time & memory Reynol
604o016 Planaria must be looked at as animal, with CONSCIOUSNESS,, it choosing food-- crawling from light.--
604o016 n split Planaria into three animals, & this CONSCIOUSNESS becomes multiplied with the organisms struc
604o016 ith the organisms structure, it looks as if CONSCIOUSNESS an effects of sufficient perfection of orga
604o016 sufficient perfection of organization & if CONSCIOUSNESS, individuality.-- Quotes D. Stewarts System
612o035 le.-- +Can the word willing be used without CONSCIOUSNESS, for it is not evident, what animals have c
613o035 nimals & sympathetic of man [RHC] ¿How does CONSCIOUSNESS. These willings have relation to external c
613o035 Where pain & pleasure is felt there must be CONSCIOUSNESS commence; where other senses come into play
613o035 s??? ? [LHC] ¿Can insects live with no more CONSCIOUSNESS??? ? [LHC] ¿Can insects live with no more c
613o036 ning of Words; must be made out Reason Will CONSCIOUSNESS than our intestines have? [RHC] 5) Kirby th
614o037 er animals may be looking back, .˙. therefore CONSCIOUSNESS Definite instincts being acquired, is a mos
616o039 not be said that thought perceptions will, CONSCIOUSNESS therefore reward in good life [RHC] Instin
617o39v t it costs us to exert force or by internal CONSCIOUSNESS memory &c. have the same relation to a livi
617o39v do the senses affect us, except by internal CONSCIOUSNESS; the objective, by our external what senses
617o40v as known by the exertion of our own power & CONSCIOUSNESS 3) We must endeavour to do without it as we
618o041 ve action-- they are known only by internal CONSCIOUSNESS of it we are conscious that we ourselves ca
638j58v ntellect, but organization, with mysterious CONSCIOUSNESS, & have no objective aspect. If thought bor
638j059 knowledge, before hand. & can in idea (with CONSCIOUSNESS superadded This is similar idea, to cells o
601o009 hese books I daresay good. 1. Sensation is the <CONSE> form these schemes.-- I see no reaso
601o009 hese books I daresay good. 1. Sensation is the <CONSE> ordering contraction (that is the only evidenc
228b232 e taken him over whole world.-- «--the soul by CONSENT of all is superaded, animals not got it, not
195b101 according to certain laws such as are inevitable CONSEQUEN let animal be created, then by the fixed laws
098a047 attempted crystallization, & therefore as a CONSEQUENCE aggregated (I assume the same force which dr
099a050 ravity.-- hence changes in dip of no sort of CONSEQUENCE.-- Therefore < S of inclination «varies with
105a072 urse.-- the latter (it is generally said) is CONSEQUENCE of <rapid> slow course, & with slow course s
180b038 ances & therefore that death of species is a CONSEQUENCE (contrary to what would appear from America)
268c100 aving its aquatic, aerial &c type?-- This of CONSEQUENCE, because applicable to N. Hemisphere (NB.--
275c119 y such profound judgment, that he forseeing. CONSEQUENCE., was endowed with what may be called the pr
281c137 grandfather, which is true) & infertility is CONSEQUENCE.-- The simple expression of such a naturalis
310c223 (--<a> «&» chasm«s» <necessary to account> «CONSEQUENCE of» for the scheme of nature) and animals th
362d099 er to my notes) & Mr Yarrell supposes this a CONSEQUENCE of the female breeding all the year round. a
372d129 hange into generative organs?]CD it is of no CONSEQUENCE if it does= <I do not doubt, the> Do plants
416e071 t may be said that true hermaphroditism is a CONSEQUENCE of non-locomotion-- (contradicted by Plants)
443e155 ite is, that intercourse every time is of no CONSEQUENCE in that degree of developement.-- [It is sin
537m076 hout temptation.-- This probably is natural. CONSEQUENCE of man, like deer &c, being social animal, &
569n021 or two absent things.-- reason probably mere CONSEQUENCE of vividness & multiplicity of things rememb
576n047 of his life, & the chance, of so dreadful a CONSEQUENCE to each man is small. Man's intellect is not
588n090 nce & piety are felt." it appears to me mere CONSEQUENCE of stooping, as sign of humility.-- I suspec
601o009 nce. when not consciousness) are produced in CONSEQUENCE having some relation to the primary sensatio
603o013 e chief object of language is promptness «of CONSEQUENCE» hence languages become corrupt, & whole cla
608o028 fe is a reward or retribution.-- it may be a CONSEQUENCE but nothing further.-- October <8> 2d. 1838
620o045 that the instinct ought to be followed is a CONSEQUENCE of that being part of our nature, & its effe
628o53v ty such conduct is instinctive in me (& as a CONSEQUENCE, but not cause gives me [3] pleasure) or tha
639j28v ir food lies deep.-- I say it is «as» simple CONSEQUENCE they become long. not at once, but by steps.
122a113 my Valparaiso case. Consider profoundly all CONSEQUENCES of EXTREME FLUIDITY of earth.-- study fifte
198b114 d giving laws & & then leaving all to follow CONSEQUENCES.-- I cannot make out his ideas about propag
568n016 eart. Do not our necessary notions follow as CONSEQUENCES on habitual or instinctive assent to propos
632j53r ons.-- but these are, I believe, only direct CONSEQUENCES of still higher laws.-- I do not «then» bel
035r048 to their proper insignificance; as fractures, CONSEQUENT on grand rise, & angular displacement, conse
035r048 equent on grand rise, & angular displacement, CONSEQUENT of injection of fluid rock.-- Try on globe.
063r130 er case position, in latter time. (or changes CONSEQUENT on lapse) being the relation.--As in first c
206b148 these races may be overlooked mere variations CONSEQUENT on climate &c-- the whole races act towards
364d103 getting fresh blood, without fresh feather, & CONSEQUENT trouble in obliterating the fresh feather, b
404e024 ility «of hybrids receive no explanation» was CONSEQUENT on mind or instinct, now this is directly in
531m052 ent movement; may not passion be the feeling «CONSEQUENT on the violent muscular exertion» which acco
531m052 no feeling of passion, but muscular exertion CONSEQUENT on the injury & consequently excited action
545m107 xpression, is an heredetary habitual movement CONSEQUENT on some action, which the progenitor did, wh
208b153 ees with late production of those regions, & CONSEQUENTLY not many yet multiplied: NB How does this b
375d134 prevail, but the positive check of famine & CONSEQUENTLY death.. population in increase at geometric
531m052 muscular exertion consequent on the injury & CONSEQUENTLY excited action of heart.-- now this is the
552m132 n social animals) are those which are good & CONSEQUENTLY give pleasure, & not as Paleys rule is thos
581n063 surname,, analogous to instinctive memory, & CONSEQUENTLY instinctive action.-- Sir. J. Sebright. has
025r016 omposed of sand dunes. 15--15 Does not seem to CONSIDER this a very shoal coast. Beyond the 10 or 12
035r047 ight of tertiary. being less than secondary:-- CONSIDER arguments for oscillation of level independen
038r058 arance will here be the strongest argument:--¿ CONSIDER causes for subaqueous crater being of diff: f
044r074 seeing small Bombs. without a vesicle. we may CONSIDER appearances of eruption at bottom.--solution
046r078 f we look to a constant revolution.--Are we to CONSIDER that the dikes which so commonly (state facts

```
046r079 ic rocks are generated as often as Volcanic. I CONSIDER latter as accidental on the afflux of the for
057r113 akes act as ploughs [,] Volcanos as Marl-pits: CONSIDER well age of Bones. = slowness of elevation pr
057r113 as much nearly as the Eocene. = Should Mr Owen CONSIDER bones washed about much at Coll. of. Surgeon'
069r150 or instance) expansion of solid matter by Heat CONSIDER profoundly the sandstone of the Portillo line
093a034 uanaco of P. St. Julian. -- Mr Scrope seems to CONSIDER that elevation & eruptions are antagonist for
100a053 ned very good.-- It is the Key to the story.-- CONSIDER stalactites.-- agate rings, crystallization t
113a091 d have been found lower.-- We have no right to CONSIDER the conducting powers either better or worse
118a103 ands Fossils & Meyens --<Jura &> Chalk When we CONSIDER parallelism of dikes (Hopkins) & that every d
121a112 es.. Smith of Jordanhill has seen same thing-- CONSIDER profoundly How came it. that Glen Roy distric
122a113 Stockholm.--  analogous to my Valparaiso case. CONSIDER profoundly all consequences of EXTREME FLUIDI
122a114 volcanos.-- Why on one coast? How can Herschel CONSIDER figure of earth statical.-- if platform of me
124a117 rman about great depths of frozen soil. p. 211 CONSIDER proved that Siberia must have been in same co
159b085 Granite could have remained, no peat supply.-- CONSIDER profoundly Boulder hypothesis Thursday, from
189b074 of  one animal being higher than another.-- We CONSIDER those, where the {cerebral structure intellec
203b135 l over S. America. Hilaire does not seem(?) to CONSIDER the monkey as a wanderer, but as produced by
206b148 May  this not be extended to all animals first CONSIDER species of cats.-- <& other tribes>.-- &c  &c
206b149 nd between great groups, which we are bound to CONSIDER the increasing ones.-- NB As Illustrations ar
219b194 ely because intermediate position.-- We cannot CONSIDER it as adaptation because volcanic isld. whils
228b231 whom we have made our slaves we do not like to CONSIDER our equals.-- «Do not slave holders wish to m
249c036 Himalaya??--  This is very remarkable, when we CONSIDER number of quadrupeds in Eocene period. Have t
268c100 lking of so many families on Keeling seemed to CONSIDER it owing to one of each, being fitted for tra
281c137 rieties should be remembered, therefore do not CONSIDER it as proved that, they are varieties, (though
286c154 ing slave of his fellow black, often wished to CONSIDER him as other animal-- it is the way of mankin
300c197 on of a deity, more humble & I believe true to CONSIDER him created from animals.-- Insects  shamming
307c215 do  indicate such origin, then we are bound to CONSIDER abortive organs of same tendency in  species,
344d040 ng fertile is same as children not being so.-- CONSIDER this with reference to "new species &  hybrid
344d040 eeding-- <beet> imago state fertile at once.-- CONSIDER this with reference to those insects, which h
350d053 a, & so mount upwards.-- «judged by analogy»-- CONSIDER all this NB. How can local species as at Gala
372d130 rtificial division of animals, as gemmation, I CONSIDER gemmation as artificial division.-- On this v
407e040 Great Edentata at that period) Analyse this,-- CONSIDER state of vegetation, & conchology,-- shells o
407e041 ought  most to resemble fossil ones of Europe, CONSIDER probable form of land,-- -- S America, an isl
409e047 utes bodies on one type: almost superfluous to CONSIDER minds.-- as difference between mind of a  dog
413e059 ls, & <individuals> «species» of large size,-- CONSIDER this (Cetacea) with reference to my theory B
419e087 which  must serve to confound our chronology» «CONSIDER ALL THIS» Extinction & transmutation, two fou
424e102 Crow is found in Edinburgh.-- Why does Fleming CONSIDER them varieties & what says Jenyns to it?-- --
445e159 ans, has skin between its legs.-- -- strangely CONSIDER existing «long-organized» forms as parent for
525m023 do>  Dogs take pleasure, when doing. what they CONSIDER their duty.-- as carrying a basket,  bringing
538m078 ess perfect than other, absolutely two people. CONSIDER this profoundly, may throw light on conscious
540m087 obvious  associations. as by reading a book.-- CONSIDER this.-- "The fledge-dove knows the prowlers o
553m135 ike ourselves) that savages (mem York Minster) CONSIDER the thunder & lightning the direct will of th
559m154 ogy, we never find out." This unwillingness to CONSIDER Creator as governing by laws is probably that
559m154 verning by laws is probably that as long as we CONSIDER each object an act of separate creation. we a
559m154 own minds. which ceases to be the case when we CONSIDER the formation of laws invoking laws. & giving
571n029 if heredetary, is pleasant.-- Mental & Bodily CONSIDER case of grazing animals knowing poisonous «he
574n043 either on form or brain very hard to define.-- CONSIDER the acquirement of instinct by dogs, would sh
579n058 o not short sighted people squinny-- when they CONSIDER profoundly,-- this will be curious if it is s
581n065 t the argument,, that many learned men seem to CONSIDER there is good evidence in the structure of la
636j57r adaptations.--fossil forms show such losses.-- CONSIDER ground Woodpecker stiff tailed cormorant: pal
637j58r [is it anomaly in me to talk of Final causes:  CONSIDER this!--]CD consider these barren Virgin p. 2
637j58r e to talk of Final causes: consider this!--]CD CONSIDER these barren Virgins p. 235. talks of the lon
638j58v arts of body.-- Now we know what instinct is-- CONSIDER this I look at every adaptation, as the survi
076r169 o Europe. argentiferous lead not abundant. = CONSIDERABLE quantity of silver procured from martial py
147g018 it is certain there must once have been very CONSIDERABLE mass of waterworn pebbles in Alluvium which
298t191 adual elevation of isld.-- We must imagine a CONSIDERABLE range of one species «on -- mountain  side»
365d106 rcely hesitate to say that if there had been CONSIDERABLE change, it would have been greater  puzzle,
543m097 this capability of understanding language is CONSIDERABLE, thus carthorse & dog.-- birds many  cries.
593n111 was  astonished & having looked round saw at CONSIDERABLE distance a very large hawk, which are  «so»
617o39v ubject, but the answer to it would require a CONSIDERABLE degree of attention. How do the senses affe
038f059 all  the great Volcanoes. have been elevated CONSIDERABLY. which shows an afflux of inferior melted r
214b176 out 2/3 from base of tail, enlarged two very CONSIDERABLY, so that any person would say the tail  was
317c249 between  canary birds & goldfinches differed CONSIDERABLY in their colour & appearance Every now & th
430e120 dale says Trigonia costata & elongata thougt CONSIDERABLY different, in proportional dimensions, must
469t135 hen each flower is visited 30 times a day is CONSIDERABLY under mark, & this has now gone on 14 days.
472s01v less riged foliage & no black spot & a third CONSIDERABLY paler, all rest very similar-- June 2. 42 M
595n117 tching paws" which they only do when <great> CONSIDERABLY excited, shows their power of imagination--
038r057 f melted rock--I think the strongest is the CONSIDERATION of the state at a grand eruption when whole
100a052 granite  in Henslow's Grit, yet it is worth CONSIDERATION. especially effect of gravity, versus  some
112a089 subsidence of the land in Guiana, worthy of CONSIDERATION. When discussing nucleuse's of old volcanos
191b080 n to Van Diemen's Land & Australia From the CONSIDERATION of these archipelagos ups & downs in full c
198b113 Conditions will not explain status (Perhaps CONSIDERATION of range of capabilities past & present mig
213b168 genera in fossil & recent state, well worth CONSIDERATION--» Tabulate Mammalia on this principle. <Va
355d067 sity with regard to age of foetus. distinct CONSIDERATION) Now in different SPECIES of genus Sus.  do
465t081 dstone) & then go on to shells-- A profound CONSIDERATION of method by which races of men have been e
313c234 tion & change of species two very different CONSIDERATIONS, with respect to law of mammals shorter du
335d016 re fertile, when offspring infertile.-- two CONSIDERATIONS are here combined). In last page, we  have
039r060 tions; hence it would appear he has not fully CONSIDERED the subject.-- S. America in the form of the
045r076 (as  wells as in the Cordillera), they may be CONSIDERED as accidents (if <[...]> part of a regular s
046r078 of Pacifick & S. America. -- Volcanos must be CONSIDERED as chemical retorts.--neglecting the first p
046r080 li lake). Before motion of sea ought to be CONSIDERED as a plain movement communicated to it as we
047r081 coand coast, which has been raised.--It must be CONSIDERED as an oscillation, from violence. Is it  not
073r163 veins  visible:--"Porphyries of Mexico may be CONSIDERED for most parts as rock eminently rich in min
103a064 ons occur & therefore that the crust might be CONSIDERED a level.-- Dikes being last action. (effect
119a104 in first formation of continents, if globe be CONSIDERED as condensed vapour.-- inequlities are requi
119a105 e it to be constant.-- It is to be physically CONSIDERED), metamorphic rocks at surface. & great heigt
120a107 t the bottom of the sea?) All this profoundly CONSIDERED. study Hopkins. theory of dikes may throw so
120a109 the  great Chilian valleys must be profoundly CONSIDERED. if elevation near coast more than at interi
245c023 in  general appearance to Siamese kind.-- but CONSIDERED good species from dental characters, wild pi
253c047 he MANELESS lion of Guzerat by Capt. W. Shee. CONSIDERED merely variety.-- red form of skull very sli
258c060 gy in not gaudy colours so all changes may be CONSIDERED in this light.--XX Zoolog. Journal-- Parrots
264c081 & habits go together. This must be profoundly CONSIDERED.-- Structure may be obliterating, whilst hab
265c085 however  is monster. yet albino may so far be CONSIDERED an adaptation as best attempt of nature, col
337d021 sensitive to light.-- (Mem whole plant may be CONSIDERED as one large eye-- have they smell, do plant
341d029 y) Owen answered that all characters might be CONSIDERED, as adaptive & that he did not see where  the
342d033 rent.--» (if so chinese pigs & common must be CONSIDERED as distant species?? or is time the  varying
352d059 cleay's theory of analogies-- <be> when it is CONSIDERED the tree of life must be erect not pressed o
381d157 Would not Ferns according to this doctrine be CONSIDERED as really cryptandrous, & they have hybrids-
385d163 & Kings at Otaheite) <Think> Last litters are CONSIDERED the most valuable. because smallest sized do
430e120 in  proportional dimensions, must <almost> be CONSIDERED merely varieties. & even Mr Sowerby is comin
524m019 nd, analogous to those feelings. which may be CONSIDERED as truly spiritual.-- a person twitching when
539m082 lect.-- Some complicated trades can hardly be CONSIDERED as actions otherwise than habitual.-- instan
565n008 t mingled with disgust, when ones opponent is CONSIDERED. as quite insignificant, & when pride makes p
581n063 let the proof of heredetariness in habits. be CONSIDERED. as grand sept if it can be generalize.-- Th
582n066 ife of Hayd & Mozart. fine music is evidently CONSIDERED as analogous to glowing conversation of seve
586n081 g instinctive, certainly heart beating may be CONSIDERED also such.-- heredetary habit, is a part nev
596nIBC irst affected?-- Can shivering & trembling be CONSIDERED convulsive.-- is convulsion. are involuntary
609o30v nstinct «in man & animals» must be separately CONSIDERED.-- The difference between civilized man & sa
617o040 ited in & by matter. The phenomena of gravity CONSIDERED in themselves consist in a force  manifested
617o040 very other particle; but FORCE, <objectively> CONSIDERED, by our external senses is a phenomenon  the
```

628o53v cation of child,-- causing many actions to be CONSIDERED right & wrong,-- to be associated with the a
063r132 united. & eggs which become quite separate.--CONSIDERING all individuals of all species. as «each» on
103a064 h other, is, I suspect much weakened by <vi> CONSIDERING how close the dislocations occur & therefore
123a116 of veins will, I suspect be greatly aided by CONSIDERING space formed-- great vacuum-- by dike.-- Mem
179b035 ainly» The absolute end of certain form from CONSIDERING S. America (independent of external causes)
182b049 t for man to be unprejudiced about self, but CONSIDERING power, extending range, reason & futurity. i
264c081 w habits must form most important element in CONSIDERING to which tribe,-- structure without correspo
286c156 f stamens.-- in order, or in next family? In CONSIDERING fossil animals, what relation in classificat
303c204 geology has obeyed rules of modern causes. & CONSIDERING over the viccissitudes of present animals.--
313c236 habitual actions.--» «this very important in CONSIDERING how children come to suck or other actions i
374d134 any extinct forms & Trilobites Sept 25th. In CONSIDERING infertility of hybrids inter se, the first c
403e023 ronounced ex cathedrâ. let us look at facts. CONSIDERING few domestic animals few. that have not <whi
422e092 aquatic   seal.-- (Consult this passage, when CONSIDERING origin of northern Cetaceae).-- -- ℕ.do. p.
423e097 ence between jaguar & tiger may not be so.-- CONSIDERING the Kingdom of nature as it now is, it would
524m020 y to become insane.-- now this is well worth CONSIDERING, if pride & suspicion can be well understood
592n101 d of doubts & scepticisms might be solved by CONSIDERING the origin of reason. as gradually developed
038r057 has   been common practice of geologist. Lyell CONSIDERS (P 84 Vol III.) whole of Etna series of coati
039r060 alapagos-- Mr Lyell. P. 111 & 113. «seems to» CONSIDERS that successive terraces mark as many distinc
046r079 of   water, through the rent strata: «Mr Lyell CONSIDERS that Plutonic rocks are generated as often as
046r080 ound). Michell (Philos: Transacts) «seems to» CONSIDERS that fall first movement (as in Peru 1746).--
053r101 ngular intersections are singular-- M. Lesson CONSIDERS the Sandstone & Granite districts to be separ
058r116 was   overthrown, & 86. Lima. next year Quito. CONSIDERS these earthquakes travel in order.-- If we lo
093d032 epest?? Marcel Serres L'Institut. 1837. p 331 CONSIDERS that Mercury & Sulpuret of Iron has been subl
100a054 illips (113) <Gardner Encyclop.--» absolutely CONSIDERS gneiss an aqueo deposit resulting from disint
119a105 sea.-- as no sea exists there.-- But Sir John CONSIDERS an irregular figure to be that of equilibrium
188b067 in islets in East Indian archipelago Dr Smith CONSIDERS probable two Northern species replace <No> So
197b111 must   have concluded so= Evidently «or hints» CONSIDERS generation as a short process, by which man «
207b151 nbergh.-- <Marcel Serres p. 331. L'Institut-- CONSIDERS that mercu» Geo. Joun. p. 325. Vol. IV. Ducks
207b152 que toute la terre ferme de lancien monde".-- CONSIDERS forms in recent volcanic islets not well fixe
216b182 f many ova-- impregnated at once.-- Dr. Smith CONSIDERS the Caffers (like Englishmen) men of many cou
218b191 der title of Amaryllidae & Narcissus. Mr Donn CONSIDERS Mr H. rather wild Mr Donn remarks to me. that
270c103 of laws from rest of]CD universe «which Carus CONSIDERS big animal» becomes more developed in higher
372d130 ed by short method in generation.-- Ehrenberg CONSIDERS artificial division of animals, as gemmation,
380d153 ife in female Moth &c Mr Y. says that Macleay CONSIDERS the house bug, as a <female which have> larva
413e059 rt of parents December 2d Lyell tells me Beck CONSIDERS the characteristics of the Tropical Forms in
479z006 leaves in Sumatra Marsden. p. 311 D'.Orbigny CONSIDERS Dasypus villosus is true Peludo Cavia Austral
480z007 of Zoolog & Botany. Vol I p. 358. D'.Orbigny <CONSIDERS> states that young birds of prey have longer
535m069 any   attempt to know their nature.-- Reviewer CONSIDERS this profoundly true.-- How is it with childr
586n081 heredetary helps this confusion.--) Hensleigh CONSIDERS breathing instinctive, certainly heart beatin
065r139 tten state of coffin «buried in a mound» long CONSIGNED to the earth. yet body had scarcely undergone
119a105 cks at surface. & great heigth on mountains.-- CONSIST of rocks with fossils,, therefore formed near
605o020 rception.-- Taste has been supposed by some to CONSIST of "an exquisite susceptibility from Blair rec
605o20v the   reverse of this. taste evidently does not CONSIST of this. but rather in the power of discrimina
617o040 phenomena   of gravity considered in themselves CONSIST in a force manifested in every particle of mat
619o042 abits, as <if> another animal. These instincts CONSIST of a feeling of love <and sympathy> «or benevo
619o042 ing their origin, we see in other animals they CONSIST in such active sympathy that the individual fo
075r167 ed?-- Klaproth analysed silver ores from Peru CONSISTED of native silver & brown oxide of Iron in Mex
110a086 Lyells'. Capital Norway case.-- The fragment. CONSISTED of hornblende (?) & felspar, (some crystals b
623o050 onform to the law,> «can only be such, as are CONSISTENT» with social animals, that in which have a b
451e178 of   12.-- the largest mammifers in the world CONSISTENTLY reside & are bred. "take tame animals into
314c237 er's drawings of a curious plant where a tube CONSISTING of pistils & stamens united into long organ,
526m029 l-- some fancied.-- this fact of early memory CONSISTING of things seen, quite agrees with my Fathers
620o044 cause   perhaps lies in its frequency & in its CONSISTING in desire gratified & therefore as soon as d
027z027 rovidence more hilly than others of the Bahama CONSISTS of rock & sand mixed with sea shells--about 5
350d055 uch relation seems common, but that perfection CONSISTS in being able to reproduce Here there is some
389d173 Ld. Moretons case: without the nervous matter CONSISTS of infinite number of globules: generally suf
431e123 is whole doctrine of the advantage of crossing CONSISTS in the idea of the male being smaller, & the
526m029 of   page.-- one is tempted to think all memory CONSISTS in a set of sketches. some real-- some fancie
534m061 knowledge   acquired by senses,-- then thinking CONSISTS of sensation of images before your eyes, or e
541m090 - I suspect from these facts that whole effort CONSISTS in keeping one idea before your mind steadily
556m144 her doubted them or believed them.-- Believing CONSISTS in the comparison of ideas, connected with ju
586n079 e. it is a test to know how much of the wonder CONSISTS in the action being performed or emotion felt
617o040 ly action as recognised by our external senses CONSISTS in the manifestation of force i.e. movement?
617o040 is a phenomenon the essence of whose existence CONSISTS in its communication to other matter in the c
323c269 t. Mungo Park-- travels Feb 12. Sir. H. Davy CONSOLATIONS in Travels -- Observations on morals by Eug
580n062 ortant work.-- Feb. 12. 1839. Sir. H. Davy -- CONSOLATS: "the recollections of the infant likewise be
044r073 he extreme frequency of soft materials being CONSOLIDATED; one inclines to belief all strata of Europ
138a155 ure time [blank] Sir. J. Halls Paper on the CONSOLIDATION of strata-- he heated sand red hot & brine
399e011 late., & a Melolonittha-- In marl from «Lake» CONSTANCE species of Europaean genera=.-- Hope has idea
446e163 sing.-- & propagation by buds does not insure CONSTANCY of form.-- is the constancy owing to similari
446e163 s does not insure constancy of form.-- is the CONSTANCY owing to similarity of conditions-- & that no
021rIFC a) must have a shorter duration, than the more CONSTANT: This view supposes the simplest infusoria to
032r040 , & so on.-- This is grounded on the belief of CONSTANT rising with successive periods of greater act
046r078 wise must separate ingredients if we look to a CONSTANT revolution.--Are we to consider that the dike
081rIBC s travels? Bellinghausen in 1819 Kotzebue 1816 CONSTANT log always additive to convert French Toise i
119a105 gularity,--. Why does Sir John assume it to be CONSTANT.-- It is to be profoundly considered, metamor
171b003 d, though new individuals produced by buds are CONSTANT, hence we see generation here seems a means t
171b005 cy to vary by generation, why are. species are CONSTANT over whole country; beautiful law of intermar
172b006 it   is very doubtful whether they would remain CONSTANT; is it not said that marrying in deteriorates
177b026 be   seen.-- this again offers contradiction to CONSTANT succession of germs in progress.-- «no only m
180b037 spring at all, so as to keep number of species CONSTANT.-- With respect to extinction we can easily s
191b082 ontinually forming species.-- Man & wife being CONSTANT together for life, is in accordance with The
195b097 rms as ornithorhyncus The type of organization CONSTANT in the shells.-- The question if creative pow
197b112 f does not agree with old & modern types being CONSTANT. Cuvier's theory of Conditions of existence i
206b146 e liable to disease" If population of place be CONSTANT «say 2000» and at present day, every ten livi
215b177 first   species to go on.-- There never were any CONSTANT species Both males & females. lose desire. Na
218b191 er localities & each one will have a peculiar «CONSTANT» aspect. That is varieties, though of triflin
224b213 n of Species: one that remains «at large» with CONSTANT characters, together with other <animals> bei
248c033]m, so will the internal parts be of longest [CONSTA]NT & therefore most permanent Owen [re]markable
269c103 tly develops multiplicity="> [(his definition "CONSTANT manifestation of unity through multiplicity"
286c156 cters which unite these of older standing than CONSTANT number of stamens.-- in order, or in next fam
296c184 pe., those species which can migrate remaining CONSTANT in form, others altered much.-- these others
306c214 ke setters. long ears-- colours vary, but form CONSTANT.-- The females of some moths, like glowworm <
312c232 ccording to my view <ins> beccause actions are CONSTANT they are instincts, & not ∵ instincts, consta
312c232 onstant they are instincts, & not ∵ instincts, CONSTANT. ¿ whether mutilations non-heredetary & varia
349d053 f descent.-- <& here limits of varieties being CONSTANT. it would be exceeding wrong to call,, one g
354d062 wonderful   that the body of a man undergoing a CONSTANT round,--each particle is placed in place of l
381d156 Entozoa-- Therefore highness in scale has no «CONSTANT» relation to separtion of sexes, as may be se
389d173 h or rather]CD ⫿It should be observed that the CONSTANT necessity for change. in process by generatio
397e003 opulation & depopulation have been probably as CONSTANT as any of the laws of nature with which we ar
446e163 y in my theory, here we have a plant remaining CONSTANT, without crossing.-- & propagation by buds do
449e169 giving form-- they interbred. & the young kept CONSTANT. & all alike Waterhouse says some of the Gala
501q10a are either numerous or even where few are they CONSTANT: this very important for it wd show that such
557m147 urrence of pleasure so teaching expression «as CONSTANT smiles, cheerful face».-- Man when at ease ha
563n002 sobeyed a wish which was part of his system, & CONSTANT, for a wish which was only short & might othe
635j55r f the will of the Deity. how it acts & whether CONSTANT or inconstant like that of Man.-- the cause g
635j56r duction]CD though such a scheme. would require CONSTANT miracles.-- p. 420 thinks the great fecundity
058r117 ravel in order.-- If we look at Elevations as CONSTANTLY going on we shall see a cause for Volcanos p

060r124 elf higher & to grow upwards; for the land is CONSTANTLY pushing the sea (which of course must retain
073r161 cannot attribute a meteoric origin & which is CONSTANTLY found mixed with lead & copper is infinitely
155g063 nearly parallel--» Inclination of river must CONSTANTLY alter with falling sea & so corrode plain in
175b018 e must be progress. if we suppose monads are «CONSTANTLY» formed ¿would they not be pretty similar ov
212b165 l to watch who would come to the door-- would CONSTANTLY do this, so was obliged to be removed.-- In
269c103 h respect to life generally.-- where <">unity CONSTANTLY develops multiplicity<"> [(his definition "c
463t055 such series showed passages-- yet in talking, CONSTANTLY said as the <brain> spinal marrow expands, s
520m001 his father & old Mrs Harrison, said, although CONSTANTLY seeing him, she was often struck with this f
524m019 what they have done in passion.-- People are CONSTANTLY well aware that they are insane & that their
524m020 ion are qualities, which my F says are almost CONSTANTLY present in people, likely to become insane.-
524m021 ilepsy, but in the failing from old age, they CONSTANTLY do.-- In Mrs P. <...> of B. <talked of.> tho
532m056 for a fortnight continued wretchedly unhappy, CONSTANTLY whined, would not remain quiet in any room,
545m105 on (<Macaco> «Cyanocephalus Sphynx Linnaeus») CONSTANTLY moved the skin of forehead over eyes, at eve
563n002 &c-- Now if dogs mind were so framed that he CONSTANTLY compared his impressions, & wished he had do
573m037 November 20th Saw the youngest child of H. W. CONSTANTLY. when refusing food, turn his head first to
573n037 ed than negation Marianne. says. that she has CONSTANTLY observed that very young children. express t
633j54v es the gradation of skeleton in Vertebrates & CONSTANTLY alludes «(& at p. 312)» to the abortive bone
137a149 analogous to granite infiltering some of its CONSTITUENTS into chert. [blank] Ed: New. Phil J. 1838.
033r042 case, may not central and rather differently CONSTITUTED lime have been removed?--As shell out of its
463t057 s tooth also a difficult), the whole mind is CONSTITUTED that a difficulty makes greater impression,
021r005 axis 2.½ ft.-- singular structure of nodule, CONSTITUTION «same as» of slate same.--longer axis in li
049r088 s arranged on sides. V. Lyell P 355 Vol III. CONSTITUTION of veins, is there said granite in close co
282c140 nal conditions over whole world. similar-- & CONSTITUTION of man originally similar limits of change,
285c153 ely slow, they become firmly embedded in the CONSTITUTION, which other marked difference in the varie
285c153 e varieties «made by» of Nature & Man.-- The CONSTITUTION being heredetary & fixed, certain physical
385d165 often do we find in the son, the character, CONSTITUTION, & most of the moral qualities of the fathe
389d177 te & perpetuate differences (of body, mind & CONSTITUTION) is the end frustrated, when near relations
427e111 cidental production of seedling with hardier CONSTITUTION.-- Now Sir. J. Banks. says Zizania in 16 ge
438e142 long been subjected to domestication.-- the CONSTITUTION of some may resist the means Man can offer
441e148 e effects of conditions on the «propagating» CONSTITUTION. but not structure of the parents.-- Thus w
449e173 rinos from Cape of Good Hope,. has different CONSTITUTION from those of Europe-- for they stand India
493q004 n or Persian animals of different heredetary CONSTITUTION, to see whether offspring infertile.-- (4)
526m027 e will every action determined by heredetary CONSTITUTION, example of others or teaching of others.--
607o025 units in the universe. [Effect of heredetary CONSTITUTION.-- education under the influence of others-
365d107 bits (& habit second nature) may be more in CONSTITUTIONAL.,-- more conformable to the structure whic
415e067 ary tendency towards fatal diseases, & such CONSTITUTIONS only being cleared off by fatal diseases.--
582n067 e average of life longer.-- for short-lived CONSTITUTIONS will then be cut off.-- <Horses> Colts cant
514q21. of arrangement."-- would not male or female "CONSTRUCTIVE principle" be better. or "constructive acti
514q21. male "constructive principle" be better. or "CONSTRUCTIVE action on germ." '=?? answered Does Mormode
511q018 ous {About German ornithologists, Bhem & Glöger CONSUL Hunt, birds from Azores or Madeira Mr. Blyth (
035r047 at Patagonia. The English fact is astonishing CONSULT book itself. P. xx: same fact is indeed shewn?
036r049 n Patagonian paper; & part in grand discussion CONSULT. reconsult Geolog. Map of Europe Consult chart
036r050 ssion Consult. reconsult Geolog. Map of Europe CONSULT charts for distribution of pebbles.--Plains. o
037r056 then near? Mem. La Condamaine on the Amazons. CONSULT Insist on the frequency of dikes in Granitic c
044r071 of Ammonite to the present day. at Mauritius. (CONSULT Bory «dip of strata on East») cannot believe i
060r126 res. although kept in numbers. p. 124. Webster CONSULT W. Parish. & Azara about dry season[.] 1791. s
061r126 y bad over whole world. «(Was it so in Sydney, CONSULT history? Phillips.» 1826.27.28. grt. drought a
072r159 of lava flowing up Hill; ¿what does he mean?) CONSULT Dr Holland about bubbles. -- No Volcanic actio
205b142 imates.-- Hunbt. Humboldt! Vol V. P II. p 565. CONSULT-- Says types most subject to vary where interm
212b167 e has information respecting the Water Rat.-- ¿CONSULT Dr Smith History of S. African Cattle Phillips
243c021 continents peculiar to the different points.-- CONSULT Voyage aux terres Australes Chap XXXIX tom IV
244c021 res Australes Chap XXXIX tom IV p. 273 2d Edit CONSULT Latreille. Geographie des insectes, in 8 degre
311c225 llied species. «replacing each other». good to CONSULT p. 326 wild ass extending over 90 degree of Lo
312c228 black cock & pheasant see Jardines Journal.»-- CONSULT on this point-- pigs always go against this, w
324c268 references to authors about E. Indian Islands. CONSULT Dr Horsfield Silliman's Journal Rengger on Mam
325c267 e of Dahlias Mrs. Gore on Roses might be worth CONSULT. Paper on Consciousness in Brutes in Blackwood
340d026 ble at end of distrib: of <birds>. Anatidae.-- CONSULT this book again.-- Mine is a bold theory. whic
422e092 quatic Pachyderms. & Walrus-- aquatic seal.-- (CONSULT this passage, when considering origin of north
570n026 Mem.-- Stokes-- arrow heads &c &c October 27th CONSULT the VII discourse by Sir J. Reynolds.-- Is our
627o52v ffections.-- If ever I write on these subjects CONSULT <following> pages. <p. 231> marked in my Macki
332d002 h, than in several previous years, two having CONSULTED him on one day.-- Mark at Shrewsbury thinks t
270c104 lready organized.-- This paper might be worth CONSULTING, if any Metaphysical speculations are entere
204b141 errs greatly in thinking every animal born to CONSUME this or that thing.-- There is some much highe
055r108 age. (we do not wonder to see tertiary plains CONSUMED) Where slope «plainly» indicates former bound
059r121 tructure, for the interlineal spaces are of diff CONT: & even in one case contained lime.--All bear c
414e064 tellect, in animals chiefly organization: though CONT of Africa & West Indies shows organization in B
049r089 e rocks are seen to graduate into granites the <CONTA> passage from lava to Granite is much more perf
049r088 ution of varies, is there said granite in close CONTACT varies in nature, -- Does not granite at C. Tr
049r088 C. Tres Montes become more siliceous in close CONTACT? -- «Cordillera???» Porphyry at Valparaiso; Ep
053r100 rs about shells at Quillota Lyell, states that CONTACT of Granite & sedimentary rocks, in Alps become
625o052 gives Mackintosh's theory: the remarks about "CONTACT with will" is unintelligible to me.-- conscien
206b148 e to marriage, heredetary disease, effects of CONTAGIONS & accidents: yet some causes are evident, as
293c174 x, proof of common origin of Man.-- different CONTAGIOUS diseases, where habits of people nearly simi
555m142 ty-- is it in former case instinct to destroy CONTAGIOUS disease.-- (Useful to use term instinct, whe
076r168 eru, that those oxidated masses of iron. which CONTAIN silver are peculiar to that part of the veins,
256c054 nimals.--? Forster on South Sea, will probably CONTAIN descriptions of domesticated animals in those
465t089 any «tertiary» estuary & Lacrustine formations CONTAIN fossils,-- mammals-- a few only -- & how many
592n103 Croonian Lectures by Parsons.» following pages CONTAIN remarks worthy of attention p. 15, 25. 40. 61.
059r121 l spaces are of diff cont: & even in one case CONTAINED lime.--All bear close analogy to Obsidian, &
075r166 Sciences? -- H. says in Potosi the silver is CONTAINED in a primitive slate, covered by a clayey por
385d164 (Paper in Vol I of Irish Royal Academy) have CONTAINED perfect teeth & hair-- showing foetus has gon
028r028 from degradation of Feldspar & other minerals CONTAINED Alumen.--This matter accumulating in deep se
075r166 rimitive slate, covered by a clayey porphyry, CONTAINING grenats. In Peru. on other hand, mine of Gua
104a066 therefore on axis of Cordillera, in Andite - CONTAINING 80 per cent of Albite 80/100 X 6/100 = 480 I
124a119 exactly parallel to limestone & volcanic rock CONTAINING magnesia Lyell. Elements p.119 on such strat
302c203 imes be due to accident as submersion of land CONTAINING all of intermediate Father-species, & not, t
037c055 England, «besides ordinary marine remains» may CONTAINS <shells few corals Tortoise> «remains of Amph
097a044 me, that the upper strata alone at Guantajaya CONTAINS salt see Geolog. proceedings Lake let out by
104a066 quid rock. Andes discussion-- Albite certainly CONTAINS 6 per cent more silica than common felspar th
157g073 es Yet certainly shelf 4th <near> only usually CONTAINS many pebbles, but I believe this is chiefly c
253o046 lia.-- What are they? Colonel Montagu probably CONTAINS some facts about close species of Birds. Zool
286c156 ing Quarterly review says nat: fam: of Willows CONTAINS many Linnaean genera.-- Now are the character
295c178 by Charles D'orbigny on Plastic Clay of Paris CONTAINS many genera of Pachydermata &c & other Mammal
320c276 urnal Asiatic Journal to end of 1837. re«a»d-- CONTAINS very little Macleay's letter to Dr. Fleming.
325c267 maux.-- by Isid. Geoffroy. St. Hilaires. 1832. CONTAINS all his fathers views.-- Quoted by Owen.-- Hu
326c266 hould lik it. The Sportsman's Repository. 4to. CONTAINS much on dogs.-- Reports of Brit. Assoc.-- som
415e069 of each volume of Whewells Inductive History. CONTAINS many most valuable references See if any law
309c220 ase. «thus» Educate all classes-- avoid the CONTAMINATION of <c> castes. improve the women. (double
049r087 ges marked by do.» discuss quartz veins, there CONTEMP--yet similar ones in Clay. Slates contemporane
057r113 uld think probably that B. Blanca & M. Hermoso CONTEMP:.--Inculcate well that Horse at least has not
627o053 f the agent, for it does not enter into his CONTEMPLATION.--" Now Eugenius would contend against this
043r070 occur at same time: this is contrasted to CONTEMPORANEOUS action over larger spaces of the globes &
049r087 contemp--yet similar ones in Clay. Slates CONTEMPORANEOUS others subsequent. as in dikes. In Granite
448e167 ch as L. says is strong argument for their CONTEMPORANEOUS»-- how is this with the Eocene beds.-- see
285c152 cies is <certain> real thing with regard to CONTEMPORARIES-- fertility must settle it.-- Changes in s
338d025 view, & Hippotamus of Madagascar: because. CONTEMPORARIES. In introduction to Eytons Anatidae.-- rec
181b039 genera renders it probable that <the> «many» CONTEMPORARY, would have left scarcely any type of their
273c110 uppose, that species in same group generally CONTEMPORARY «++». «This would lead one to expect that f

180b037 of making, 13 recent forms.-- Twelve of the CONTEMPORARYS must have left no offspring at all, so as t
551m129 ough he does not pout. pushes out both lips in CONTEMPT <&> disgust & defiance.-- different from snee
565n007 se look of suspicion, even child will do so.-- CONTEMPT look obliquely so does dog. when a little one
565n008 ely so does dog. when a little one attacks him CONTEMPT, when there is some anger «& respect to oppon
565n008 de & is therefore of the snarling order.-- But CONTEMPT mingled with disgust, when ones opponent is c
591n099 f same species.-- The general «(as I believe)» CONTEMPT at suicide. (even when no relatives left to l
592n103 quatting.-- p. 64 closing both eyelids express CONTEMPT. p. 76.-- children have been tickled into exc
592n103 a reference to Brun's work.-- Shutting eyes in CONTEMPT opposite action to opening eyes in fear The e
633j54r l in its seaward course,-- we sink into such CONTEMPTIBLE queries, as why should the earth have drift
6270053 into his contemplation.--" Now Eugenius would CONTEND against this-- but the pleasure a dog has in o
549m118 ge. A healthy child is «more» entirely happy (CONTENTMT is different it refers to wishes for future)
023r012 to Sydney; know by having been seen & from the CONTENTS of its maw, amongst which were things pitched
429e115 lled trees would form fringe.-- but there is a CONTEST. & a grain of sand turns the balance.-- Hort.
500q010 gham Park-- What Book on varieties &c of deer. CONTESTS of sexes.-- Q.30) March 1842. <Last> Year «be
587n089 iation. owing to time when entered brain, try CONTIGUITY of parts of Brain.-- Mackintosh first clearl
023r010 d by the elevations of those mountains on the CONTINENT of S. America is inadmissible «may have happe
039r061 nd propulsion of fluid rock, which elevates a CONTINENT> We are more abound to take analogy of moveme
061r127 new creation affected by Halo of neighbouring CONTINENT: ≠ as if any creation «taking place» over cer
068r146 improbably so long, that to be surrounded by CONTINENT.-- change of volcanic focus.-- <it is certain
070r155 m Tucama[n]. & at Chuquisaca. half across the CONTINENT.--He states plains of Mendoza smooth. Sir W.
087a011 erefore to the South sinks.---- Mediterranean CONTINENT corresponding to Europaean risings. Pacific g
130a133 D what would be the chance in sounding over a CONTINENT to fall across a hot.--spring.-- Hot water wo
173b011 grown altered Hence the type would be of the CONTINENT though species all different. In cases as Gal
178b030 brids propagating freely In Isld neighbouring CONTINENT where some species have passed over, & where
178b031 l our «British» Shrews diff: species from the CONTINENT look over. Bell, & L. Jenyns. Falkland rabbit
191b080 .-- West Indies = Opossum & Agouti same as on CONTINENT-- 3 Paradupasi in common to Van Diemen's Land
203b138 ne. Gould on Motacilla,.. «species peculiar to CONTINENT & England» Loudon Mag: Septemb or Octob 1837
209b156 Helena, without ferns.-- analogous to nearest CONTINENT: poorness in exact proportion to distance (?)
210b160 wer seems to be checked when islands are near CONTINENT: compare Siicily & Galapagos!!-- Some of the
226b221 Irish Hare.-- Galapagos.-- shrews, & when big CONTINENT many species belonging to its own genera Ther
227b227 man.-- Does not require fresh creation!-- If CONTINENT had sprung up round Galapagos on Pacific side
231b241 ild state.-- When breaking up «the primeval.» CONTINENT. Indian Rhinoceros. Java & Sumatra ones all d
250c037 of S. Hemisphere fully characterized, of each CONTINENT. Try amongst Europaean quadrupeds if Africa d
264c080 might be blended, if archipelago turned into CONTINENT &c &.-- There is beautiful gradations of form
272c109 d under Gould says most subgenera confined to CONTINENT, though we have seen species «of subgenera» s
278c129 every species of a section is confined to one CONTINENT & every species to another then those section
294c176 ill explain very close Species in islds. near CONTINENT, Must we resort to quite different origin whe
296c184 ator make all new yet forms like neighbouring CONTINENT. This fact speaks volumes. 2 Chapters. transl
316c247 cause the recent ones are so close. Was there CONTINENT between N. America & Europe?-- ̈.Norton has w
357d073 wo were landed as at present in New Ireland & CONTINENT since grown.-- This will explain. S. American
358d074 e on the Botany of the Pacific.--» to nearest CONTINENT.-- With respect to ancient geography of Atlan
377d139 ghten depresses them.-- England was united to CONTINENT, when elephants lived. & when present animals
379d151 ack sea.-- it would be ocean, what is land to CONTINENT-- Original Paper, worth studying. Archiv. fur
421e091 coast of Africa, than to other parts of that CONTINENT. in like manner as Madagascar does to other s
446e162 lina-- «Does it flower anywhere?-- Yes on the CONTINENT is there more variation in its character.?» N
447e164 very rarely flower [bu]t the one does on the CONTINENT-- well characterised species periwinkle wants
591n100 tinctive feeling, only does not fullfil, like CONTINENT man.-- a man eating what others by habit (not
591n100 s very clear.-- even to the cold or benevelo- CONTINENT man Hume has section (IX) on the Reason of an
088a013 r axis or hinge no doubt fluid.-- analogy as CONTINENTAL elevations slow. so would line of mountain c
118a104 rguments against Herschel's view of cause of CONTINENTAL elevations (I) the alternation of linear ban
119a104 mountains formed (& mountains are effect of CONTINENTAL elevations) we may conclude that elevation i
202b133 &c &c.-- Now if suppose world more perfectly CONTINENTAL. we might have wanderers. (as Peccari in N.
226b222 act we have many species, we may insure mass CONTINENTAL or many large islands.-- Hence this must hav
515q022 Horses having fewer vertebrae in tail, than CONTINENTAL horses. {About the leaping of Irish Horses,
058r117 taly proved by Coral hypoth. agree with great CONTINENTS.) Voyage aux terres Australs Vol. I. p. 54.
119a104 -- (2d--) does not explain first formation of CONTINENTS. if globe be considered as condensed vapour.
127a127 ashes in Ireland dikes in mountains. «(not on CONTINENTS)» prove elevation.-- great mountain chains.
132a138 h the sea, <than> «&» on coast lines, than on CONTINENTS it ought, (according to M. Parrots argument
137a151 sea, is probably much more rapid than beneath CONTINENTS In Berlin Transactions (1832. or 3?) there i
173b010 remembering Lyells arguments of transportal) <CONTINENTS> island near continents might have some spec
173b010 ents of transportal) <continents> island near CONTINENTS might have some species same as nearest land
173b012 of species we can see why a form peculiar to CONTINENTS; all bred in from one parent why. Myothera s
196b105 ful if the two Rheas had existed in different CONTINENTS In plants I believe not..-- It is a very gre
199b116 pots of land, of the same type with the great CONTINENTS we get a means of knowing movements.-- How c
233b251 Probably worth studying.-- Wingless birds S. CONTINENTS-- Ostriches. Dodo. Apteryx Penguin-- Logger-
236b278 ieties is it not per saltum.-- Isld bordering CONTINENTS same type collect cases.-- African isld.-- «
243c020 But he says shells towards extremities of the CONTINENTS peculiar to the different points.-- Consult
252c045 ndation of all our knowledge especially great CONTINENTS» Give Specimen of arrangement, 3 <5> Species
271c106 same island in plants?-- What is this halo.-- CONTINENTS are not stationary «unerring proofs not alwa
271c106 re not stationary «unerring proofs not always CONTINENTS».-- it is a plastic virtue.-- it is expressi
278c130 important as showing former connection of two CONTINENTS & death of form in one. The caves are at a h
283c144 ornaments. have so intimate a relation to two CONTINENTS as to be <replaced> called into existence in
283c144 to be <replaced> called into existence in two CONTINENTS our ignorance is indeed profound & such it a
342d034 as they almost must have been on elevation of CONTINENTS) but Ogleby well answers that nearly all F.
355d068 was there formerly one great sea, & two Polar CONTINENTS Marsupial. Edentata.-- Pachydermata &c &c--
377d140 of «earthquakes», elevating forces in raising CONTINENTS, & forming mountain-chains, when we estimate
407e039 ed), although other parallel species in other CONTINENTS might have survived this mundine change..--
473s004 en «as» much subsidence as elevation then all CONTINENTS of cretaceous periods, together with their l
641j29r change species-- yet this is contradicted by CONTINENTS <bri> abounding with species-- there will be
641j29r nding with species-- there will be a balance, CONTINENTS have been split up.-- who can decide their l
173b011 . In cases as Galapagos & Juan Fernandez. When CONTINET of Pacific existed might have been Monsoons..
304c206 ???? or «have» not «been» exposed to so many CONTINGENCIES??? A Question of immense difficulty is, whe
271c106 e counteracted by Man.-- effect of external CONTINGENCIES & long bred in-- Mem, <an> a statement in M
301c199 s.? p. 8. mistakes of instinct are external CONTINGENCIES, where the habit is not applicable. The deg
305c210 evalent over this word, (subject to certain CONTINGENCIES of organic matter & chiefly heat), which as
371d118 , than that it should be provided with many CONTINGENCIES how to act-- so with the mind. the simplest
550m123 than instincts of animals to all & changing CONTINGENCIES, or bodies of either.-- Our descent, then,
608o026 l spark falling on prepared materials. From CONTINGENCIES a mans character may change-- because motiv
608o026 e & be †A man may put himself in the way of CONTINGENCIES.--but his desire to do arises from motives.
608o026 ought to pity & assist & educate by putting CONTINGENCIES in the way to aid motive power.--if incorri
611o034 ke» «as» chemical laws,) as long as certain CONTINGENCIES are present, (contingencies as heat light &
611o034 long as certain contingencies are present, (CONTINGENCIES as heat light &c). [LHC] This is true as lo
612o035 s. These willings have relation to external CONTINGENCIES, as much as growth of tissue and are subjec
615o038 a permanent secretion of thought, (or under CONTINGENCIES of stimulants of certain kinds such secreti
152g047 to Forrest road comminuted shells Important CONTINGENCY if elevation from Axis, then rivers might de
282c140 to bear to each other, but to some external CONTINGENCY.-- affinity is the sum of all the relations,
293c173 having once learnt, & if that was a regular CONTINGENCY the brain would become webfooted & there wou
371d118 t. & afterwards enlarged powers to meet with CONTINGENCY.-- Sept. 23rd. Saw in Loddiges Garden. 1279
377d139 ge, only in fixing it: only circumstances, & CONTINGENCY of time. When we multiply the effects of «ea
543m100 ved, in playing chess however many places, & CONTINGENCY a man has keep in mind. all is certain.-- th
549m118 st number of pleasant thoughts, he must have CONTINGENCY of good food, no pain,-- <but the» «&» the s
411e054 ation to the external world, & every possible CONTINGENT circumstance.-- ̈the laws of variation of ra
501q10a is not a generic or specific character,, but CONTINGENT on country.-- How is it in Patella or Oyster
263c078 of hundred generations of species to produce CONTINGENTS proper.-- Present monkeys might not,-- but p
621o047 therefore Sir J. M. talks too much about the CONTINUITY to will. (a) The origin of passions too stro
191b082 nct. two species made elevation & subsidence CONTINUALLY forming species.-- Man & wife being constant
258c060 on.-- simile Man living in hot countries, if CONTINUALLY crossed with people from cold, children woul
497q006 inds come up, not found in wood.-- but seeds CONTINUALLY dropping in woods. by birds 13. Mr. Herbert

639j28v legged race will get the upper hand. though CONTINUALLY dragged back to old type by intermarrying wi
158g079 Glen Guoy nor is horizontal line apparently CONTINUATION of upper terrace -- on right hand {P} round
320c275 ed by Taylor Magazine of. Zoology & Botany & CONTINUATION «Annals of Natural History» Skimmed Von Buc
547m111 vivid «& perfectly characterized» dream, in CONTINUATION of waking thought-- my servant was in the r
183b051 into Chili. in present states. <they> it might CONTINUE & thus two species be created) & live in same
288c159 lata. birds which had originally crossed would CONTINUE to cross, means of knowing directions: myster
371d128 stances very unfavourable, will <deteriorated> CONTINUE of same variety as long as life lasts, yet th
535m064 one group.-- (as in birds blind storge-- They CONTINUE till death, thus acting 4 to 6 weeks. The des
556m146 if horns were to grow on horses, they must yet CONTINUE to put down ears, when kicking.-- -- good cas
109a082 form of waves & currents.-- but this must be CONTINUED. no currents & elevation have same effect, a
154g059 ss projects from side of hill if line suppose CONTINUED across to {P} side removed all well & good, b
159g086 river could not have deposited» the slope is CONTINUED some hundred feet lower & begins about 60 hig
180b038 onfined breeding & changing circumstances are CONTINUED & produce according to the adaptation of such
189b072 ssil horse, generated in S. Africa Zebra.-- & CONTINUED.-- perished in America All animals <are> of s
206b146 n backwards to one progenitor, who might have CONTINUED breeding from eternity «backwards.--» If popu
256c055 ormation amongst Graecian isles, ¿See if type CONTINUED?-- See to Boblaye & Virlet.-- Whewell thinks
275c120 ff spring. till size diminished, but feathers CONTINUED by picking chickens of each brood.-- These ba
306c212 very differently from a dog.-- more like man. CONTINUED long in a passion & looked out for him to com
391d174 some differences. V. Supra <v. infra> p 179, CONTINUED from Is a flower bud produced by union of two
532m056 brought from Shrewsbury) yet for a fortnight CONTINUED wretchedly unhappy, constantly whined, would
532m056 house except with Caroline-- After fortnight. CONTINUED to grow thin & did not seem quite happy. in f
542m096 g it when useless.-- <&> therefore it is here CONTINUED when the uncovering the canine useless.-- The
173b009 s) Most difficult to separate Every character CONTINUES to vanish, bones instinct &c &c non fertil
032r040 ngular mass removed vary.--The gradual rising CONTINUING. a another sloping platform would be made, &
266c091 (p..26). about plants from Cape of Good Hope CONTINUING for some time to flower at their own periods
542m096 g the very essence of an habitual movement is CONTINUING it when useless.-- <&> therefore it is here
043r071 vibrates from any one shock-- In S. America--CONTINUITY of space in formations & durability of simil
161g097 rrect) there were several obscure but not far CONTINUOUS flights above it-- (NB the buttress or pass
161g097 another measurement on short buttress but not CONTINUOUS & it was 29.200 minus .008 ----- .192 Loch Lo
181b042 groups the greater the gaps (or solutions of CONTINUOUS structure) <between them.»-- for instance t
049r089 ssy & Stony Pearlstones alternate together in CONTORTED layers: Mem: Phillips Mineralogy some such fa
104a068 ss, mica-slate of whole kingdom of Norway was CONTORTED yet no mountain chain case parallel to Banda
111a087 dip to NW to 80 degree faults with red wacke CONTORTED evidently like. V. VII. p. 316 & 328 VI. p. 3
160g090 ar;-- whole hill dark grey fine grained. Much CONTORTED gneiss «narrow sharp ridge with peak» I walke
542m095 - Does the contraction & wrinkling of the skin CONTRACT iris?-- same way as one lifts up eyebrows to
564n006 the manner in which whole skin or muscles are CONTRACTED between eyes & upper lip., is most clearly a
041r066 ndstone: evidently depend on a concretionary CONTRACTION: the fact is in alliance with those balls at
315c242 ough means injection of fluid different from CONTRACTION of fibre)-- it is most remarkable habitual a
542m095 ted chiefly by objects of vision.-- Does the CONTRACTION & wrinkling of the skin contract iris?-- sam
601o009 y good. 1. Sensation is the <conse> ordering CONTRACTION (that is the only evidence. when consciousne
551m128 of it as wonderful that Elephants understand CONTRACTS.-- but W. Fox's dog that shut the door eviden
336d019 nization] CD» false.-- The creator would thus CONTRADICT his own law. So far is there any appearance
028r032 sedimentary rocks, & hence Volcanic action, CONTRADICTED by Cordillera, where that action commenced
352d058 organs, but lose them with more difficulty, «CONTRADICTED by abortive organs, but number of species w
416e071 ditism is a consequence of non-locomotion-- (CONTRADICTED by Plants). & as there are no fixed. land a
641j29r take so long to change species-- yet this is CONTRADICTED by continents <bri> abounding with species-
449e169 mmon goose took after the common goose thus CONTRADICTING (probably) Yarrells law & Walkers of the ma
177b026 assages cannot be seen.-- this again offers CONTRADICTION to constant succession of germs in progress
216b182 y countenances, as hybrid once. Is not this CONTRADICTION to his view of races not mingling?-- In Fox
251c041 latter same species domesticated, strangely CONTRADICTORY to Azaras fact of conduct of wild & tame ho
092a029 , if in solution. My view of metamorphic in CONTRADISTINCT to Volcanic will explain their solution. A
050r091 dations for Coral reefs.--does he mean in CONTRADISTINCTION to sand?? B. Roussin states that generall
058r118 inclinêe vers le rivage de la mer, tandis, au CONTRAIRE, que vers le centre de' l ile, elles presente
076r169 silver sometimes by selenite.-- in New Spain, CONTRARY to Europe. argentiferous lead not abundant. =
180b038 refore that death of species is a consequence (CONTRARY to what would appear from America) of non ada
251c042 ba (p 271 Viedo says American dogs silent. Mem CONTRARY assertion of Molina) (p. 277.). probably anot
398e005 theory on the score of small change.-- on the CONTRARY islands separated with some animals, &c.-- «i
557m147 led when angry & very stiff. back arched. just CONTRARY. when pleased tail loose & wagging-- if as (I
557m147 rful face».-- Man when at ease has smooth brow CONTRARY to wrinkled: (a horse when winnowing & please
628o54v will stop all wish to gratify <it>-- anything CONTRARY to it]CD NB. the very end of conscience is st
061r128 ver certain area must have peculiar character: CONTRAST low limit of Palms, evergreen trees, arboresc
062r128 limits of no vegetation at S. Shetland = Great CONTRAST of two sides of Cordillera, where climate sim
062r128 ically = but picturesquely = Both N & S. great CONTRAST. from nature of climate. = Perpetual snow.--s
223b209 rnando Po Zoolog. Proceedings October (?) 1837 CONTRAST New Zealand with Tasmania The reason why ther
226b223 <West.> «E» of America, ought to present great CONTRAST in forms.-- India; intermediate, see how that
244c021 species but barrier.-- --it would make strong CONTRAST with southern regions.-- «it would now repres
319c256 he mocking thrush being so very beautiful gret CONTRAST with South America.-- In Home's History of Ma
358d074 rmony. The peculiar character of St. Helena.-- CONTRAST with otaheite in relation «See Gaudichauds Vo
583n069 ncts act with unerring precision".-- no p. 17. CONTRAST the invariability of instinctive powers in in
585n075 of Head, yet never puts down its ear. good to CONTRAST with horses, asses, <mi> Zebras &c &c.-- Here
043r070 ce <...> does not occur at same time: this is CONTRASTED to contemporaneous action over larger spaces
496q05a Orchis-- & final cause of beauty of flowers-- CONTRASTED by Kirby-- with animal reproductive system.-
499q010 naeus says so great percentage of seeds have CONTRIVANCE for transportal, does he include seeds good
535m070 ganization.-- M. le Comte argues against all CONTRIVANCE-- it is what my views tend to.-- When a man
608o028 suppose that the sins of a man, are under his CONTROL, & that a future life is a reward or retributi
434e128 as other forms.-- p. 36.. speaking about the CONTROVERSY on Didelphys says. "If we cannot reason from
228b229 pecies.-- Important. For instance take Voluta & CONUS (??) which now run together, were not both gene
206b149 w genera (for otherwise the relationship would CONVERGE sooner) & lastly perhaps some one single one.
274c115 for insects & structure for vegetation.-- In CONVERSATION in Museum-- I could not discover any other
293c172 ous cases, when you feel sure you have heard CONVERSATION before. is strong association recalling up
294c177 differences. «the opinion of many people in CONVERSATION.» the whole object of the Work is its proof
522m010 y means of hearing.-- Mr Corbet, however, in CONVERSATION could catch up a new train if early associa
544m104 he pain.-- How was this instinct gained.? by CONVERSATION-- ∴ modified in those races, where it is cu
582n066 evidently considered as analogous to glowing CONVERSATION of several people.-- Children have an uncom
321c270 g Several of Water Savage Landons Imaginary CONVERSATIONS-- very poor Sir T. Browne's Religio Medici
284c149 r test of value of character-- Macleys rule is CONVERSE, <when> value of character depends on non-var
335d013 long been in blood, will remain in blood.--- CONVERSE, what has not been, will not remain,-- yet of
378d147 y be sure cock & hen will be alike-- I presume CONVERSE is not true for he says Hen & cock Starling a
380d153 ntirely to the offspring-- this is clearly the CONVERSE of annual being rendered biennial-- the hardn
521m009 t impaired. <after paralytic stroke> : could CONVERSE well on any subject when once started,-- coul
616o039 sion. They would do well to ask themselves the CONVERSE of the <expr> question above stated, because
081rIBC Kotzebue 1816 Constant log always additive to CONVERT French Toise into English ft. 0.8058372 French
119a105 ical scale.-- (for the temp must be immense to CONVERT rock into gneiss &c judging from what we see w
027r028 urnal In the Iron sand formation <would> wood CONVERTED into siliceous pyritous & coaly matter. Mem:
088a015 erican Volcanic action. -- Fragments of slate CONVERTED into crystals of Hornblende p. 248. L. Instit
307c216 on this subject) because if so as she can be CONVERTED «into female», it will be splendid argument o
314c237 side.-- but in other species, this segment is CONVERTED into hood which possesses power of movement.
342d034 F. W. Fish are Abdominals. ∴that order first CONVERTED-- is it an old order Geologically? Owen says
409e047 g how inhabitant of Tierra del Fuego is to be CONVERTED into civilized man.-- ask the missionaries ab
286c154 ons-- brought into the world same way» they may CONVEY much thus, Man has expression.-- animals signa
375d134 ic language of «Malthus» «Decandoelle» does not CONVEY the warring of the species as inference from M
601o009 sation of higher order. where the sensation is CONVEYED over whole body (which it may be in first cas
126a124 r view the rod is reversed, upper part metal «CONVEYING heat in one direction only, like water below
532m054 ovements, senses being on the look out, & the CONVEYING means from the senses to the mind being more
292c170 sibly have had <to so convin> brought so much CONVICTION to my mind.-- Reflect much over my view of p
398e005 r> conceive the result with that clearness of CONVICTION, absolutely necessary as the «basal» foundat
292c170 nst these views, could possibly have had <to so CONVIN> brought so much conviction to my mind.-- Refl
602o11b what you say is perfectly true, but you do not CONVINCE me.--" Belief allied to instinct.-- p. 134. a

(Key Word)
431e122 -- Ask Phillips.-- The more I think, the more CONVINCED I am, that extinction plays greater part then
608o027 t do harm, because no one can be really fully CONVINCED of its truth. except man who has thought very
524m020 ht occurs, is closely analogous to Epilepsy & CONVULSION.-- affections of the thinking organs ̧ -- th
578n056 n to remove disagreeable impression like true CONVULSION. (Hence pass into convulsions?)-- squeeze ou
596n1BC ng & trembling be considered convulsive.-- is CONVULSION. are involuntary movement of voluntary muscl
578n056 ssion like true convulsion. (Hence pass into CONVULSIONS?)-- squeeze out tears. replaced & squeezed o
592n103 en tickled into excessive laughter & so into CONVULSIONS.-- «Paper» must be referred to, if I follow
596n1BC if of. convulsive tendency easily fall into CONVULSIONS A carrier pidgeon carried & turned round & r
596n1BC ould it then know its direction.-- In slight CONVULSIONS. are the muscles of the face first affected?
578n056 perfect fainting, sphincters are loosed is a CONVULSIVE action to remove disagreeable impression lik
578n057 ts more perfect over voluntary muscles, these CONVULSIVE actions-- (except in weak people & hysterica
578n057 n weak people & hysterical people inclined to CONVULSIVE actions).-- But, the lachyrmal gland is «not
578n057 tion, there pour out tears, & there is slight CONVULSIVE wrinkling of some of the muscles «or twitchi
596n1BC f mongrel dogs Do blubbering children, if of. CONVULSIVE tendency easily fall into convulsions A carr
596n1BC ed?-- Can shivering & trembling be considered CONVULSIVE.-- is convulsion. are involuntary movement o
473s05v hem from lake S. of Moel Siabod. wh. flows into CONWAY by Bettws & there joins streams from Capel-Cur
026r022 ra of S. America. Study Geolog: Map of Europe CONYBEARE. Introduct XII P. silicified bones not common
035r046 organic structure most easily preserved.-- Mr CONYBEARE introduct to Geolog--"Between the height of s
067r142 835 excellent account of N. American geology. CONYBEARE Lava in Cordillera & on Eastern plains «by An
363d101 plumage &c &c «Pouting pidgeon exaggeration of COOING.-» & compare them with all the varieties.-- H
065r140 Porphyry.--Falklands.--off East Coast. -- Capt. COOK found soundings. (end of 2d voyage outside coas
108a081 cha. Melaphyre. = Andesite-- Albite & amphibole-- COOK found Granite at Christmas Sound Vol XIV. (My E
113a091 that time has not been allowed for lower beds to COOL down. & then in 50000 years the depth will be g
556m145 ip curled over upper with mouth shut. expressing COOL irony, not biting? What is Emotion analysis of
625o50v lly proper to obey anger as benevolence (but not COOL malevolence). it is only after reason comes int
059r123 ave been melted with little pressure. & perhaps COOLED suddenly.-- As the rude symmetry of the globe
093a031 ecause it could reach the surface. before being COOLED.-- Berzelius. L'Institut. [1837 p. 297]CD thin
127a125 once thawed. & hence. (when climate hotter) was COOLED to greater depth.-- Now the <inf> subterranean
127a126 my Vol 8. p. 118 water no--. oil will freeze if COOLED in a closed globule of glass. (oil may be cool
127a126 ooled in a closed globule of glass. (oil may be COOLED to 0 degree!)-- shows effects of pressure in c
174b015 to whole organic kingdom, when our planet first COOLED.-- Countries longest separated greatest differ
232b247 o the mean of the other.-- If the world had COOLED by secular refrigeration in chief part instead
035r045 be chemical differences. & not those of rapid COOLING &c &c My results go to believe that much of al
059r121 all show chemical action as well as effects of COOLING [misnumbering, no page 122] In Igneous rocks.-
107a078 is view will bear much reflection on method of COOLING--Very difficult subject. PP-- I think from dis
120a108 f dikes may throw some light.-- thin dikes not COOLING if they had travelled some hundred miles throu
125a121 Herschel's views earth originally fluid, then COOLING process must go from surface towards the inter
231b243 ains crag & miocene.-- The descendants left in COOLING climate might change twice over, whereas those
540m084 ctions as hiccough & yawn are probably merely COORGANIC as connexion of mammae & womb.-- We need not
219b193 some blown--floating trees Thrushes & bunting & COOTS--«(Turdus Guyanensis?) (Emberiza Brasiliensis?
495q05a .-- 10 Shoot tame duck on pond with Duck-weed-- COOTS-- waterhens-- examine dog, which has swum-- on
512q019 early habits in Dorsetshire sheep migration of COOTS-- variation in hounds= An ugly calf <turns> som
593n111 part of man's mind.-- At Maer. Pool. I saw many COOTS & waterhens feeding on grassy bank some way fro
326c266 otany R. Brown. has curious coloured maps. by COPENHAGEN Botanist of range of plants Books quoted by
032r040 f coast.--[Fig. 2] Mem San. Lorenzo; Valley of COPIAPO & parts of coast of Chile.-- Must first explai
036r050 - British channel &c &c. There is a Hill. near COPIAPO which is asserted to make a noise,--My impress
037r056 ce not disturbed.--Whole West coast. Chonos to COPIAPO.--Sydney. K. G. Sound. C. of Good Hope.--Carna
043r071 es in Chili:-- Chiloe. Concepcion. Valparaiso (COPIAPO & Guasco). yet whole territory vibrates from a
070r155 thquake Draw close Analogy Lake of Cordill: of COPIAPO & Desaguadero.--three ridges in Copiapo, as we
070r155 ll: of Copiapo & Desaguadero.--three ridges in COPIAPO, as well as in latter.-- According to Mr Brown
135a144 . p. 193 in Lat 26 degree S. Wafer looking for COPIAPO. found inland a great many sea shells some mil
193b091 n Kamschatka quadrupeds Kotzebues first Voyage COPIED into list Entomological Magazine paper on Geog
458t002 order (except whales) have great prototype!!.-- COPIED Vol II p. 502. «Bengal Journal» The Taylor Bir
459f1v s undergone all the changes. [im]portant view, COPIE[D] Gleanings of Sciences. Vol. III p. 83. Paper
500q10a athered & these are now to be planted this year COPIED Gould.-- Number of species of Birds in New Zea
559m155 & Animals. in Blackwood's Magazine June. 1838. COPIED Mr H. C. Watson on Geographical distribution o
559m155 y on the Human Understanding well worth reading COPIED <Smith> <O Stewart> lives of Adam Smith Reid,
051r095 ear Callao.--From Sir. H Davy experiment on the COPPER bottom. we see a trifling circumstance determi
073r161 n & which is constantly found mixed with lead & COPPER is infinitely rare in all parts of the globe".
075r167 in Mexico. sulphuretted silver, arsenical grey COPPER, and antimony, horn silver, black silver & red
124a119 mus says he has seen in making brass a piece of COPPER not melted absorb, zinc thrugout its thickness
540m086 ack man of <B> Van Diemens land & the energetic COPPER coloured natives of New Zealand)-- the America
367d113 a. Where females, are peacable-- (Mem Lucanus & COPRIS &c).-- In birds singing of cocks settle point.
388d173 lopement of gemma.-- The manner in which Frogs COPULATE & fish shows how simply instinctive the feeli
380d155 fe» plumage of hybrid birds) After animal has COPULATED., though no offspring, Milk sometimes comes i
389d170 offspring, --when> though other male may have COPULATED.-- two animals may unite & each have offsprin
478z004 ne animal In Australia I was assured wild dog COPULATES <.> freely with tame: comes to houses on purp
179b032 ones of the father; is not this owing to each COPULATION producing its effect; as when bitches puppie
355d068 n of a bud-- the very same must take place in COPULATION-- (Man & woman separate parts of same plant<
387d168 anyways connected with the case of successive COPULATION impresses offspring more & more with the add
389d176 lis Q Proved facts relating to Generation One COPULATION may impregnate one or many offspring.-- it a
393dIBC n high. bred dogs ie. (bred in & in) that one COPULATION with other dogs renders subsequent progeny f
482z011 eference to a luminous Sertularia Lesson Zoolog. COQ: p. 120 Coati Roux. Tatous & perhaps Yagourundi
178b031 t was species <Ascensi> Study Lesson Voyage of COQUILLE.-- Dr. Smith says he is certain that when Whi
242c017 as men do. when making varieties.-- Voyage of COQUILLE Zoolog. p 19.. Tapir, «des» couroucous et ru
242c018 o> from Amboina to New Ireland p. 23 Voyage of COQUILLE Lesson No (p. 24) batrachian in isles of Grea
243c019 Bibron doubts fact.--» My toad is same species COQUILLE Voyage p. 25 Mais il n'y a pas jusqu'aux îles
245c023 good species, with different number of teats» COQUILLE Voyage Durville has written Flora of Falkland
246c025 to East of America. would account for this.-- COQUILLE Voyage Says no reptiles. p 460 & very doubtfu
246c027 oluccas, New Guinea.-- (Case of replacement)-- COQUILLE Voyage The caswary, inhabits Ceram, Bourou &
320c276 louines Zoological Journal 5 Vols Voyage de la COQUILLE Zoological Transactions. <done> up to parts p
479z005 bbits not indigenous p 112 M Lesson--Voyage of COQUILLE wide limits of Nullipora Discussion good on F
226b220 od.-- Small isld off New Guinea same fact see COQUILLE'S Voyage.-- Galapagos mouse (?) brought by can
239b249 ferent isld. (as far East as New Ireland. see COQUILLES Voyage), Waterhous remark Australia Fauna so
241c018 t the pages from about 8 to 20 of Zoologie of COQUILLE'S Voyage to see if Lessons' remarks on the Flo
433d013 n Classification of such animals.."-- Voyage. COQUILLE'S Voyage p 302 Vol II p. 302. Vaginulus of Lim
027r024 s paper. Philosoph Transact: at R. de Janeiro. COQUIMBO. Balanidae. at Concepcion. Humb: Pers. N. vii
033r042 envelope of Lime.--form resembles the husks at COQUIMBO: in that case, may not central and rather dif
039r061 in some older formations: Mem the envelopes at COQUIMBO: the analogy is now perfect <The grand propul
054r102 recent shells 200 ft, how exact agreement with COQUIMBO; [not located] [not located] Isld near coast
105a072 it-- some error? (because more recent) ------ COQUIMBO on. other hand?-- The widening a valley depen
024r015 Beagle Coast of Brazil? where not rivers in my CORAL paper {T} leagues Fathoms Parallel of St Cather
028r030 flections might be introduced either in note in CORAL Paper or hypothetical origin of some sandstones
032r041 r slow & weak.»; «(cause of not accumulation of CORAL limestone in intertropical)» hence varieties of
043r069 compare with Galapagos.--Chiloe. M. Hermoso. & CORAL reefs (imperfect in latter). <At> Lyell. Vol I.
050r091 nk that Volcanic eruptions form foundations for CORAL reefs.--does he mean in contradistinction to sa
051r094 the action is anomalous; It is wonderful to see CORAL reef--or confervae in the breakers or in waterf
058r117 vements (not mere patches as in Italy proved by CORAL hypoth. agree with great continents). Voyage au
085aIFC concerns "Geolog Observat on Volcanic islands & CORAL Formation Lyell's Salband p. 86 Shells near Woo
089a018 10. Mountains on west side of Domingo formed of CORAL limestone, with interstices yet emptty.-- In al
091a026 exican Trachyte <roc> lava called Andesite. Red CORAL in the Mediterranean 700 feet deep in some of.
108a080 nphilosophical to think calcareous springs near CORAL reefs.-- Where vegetation luxuriant it might be
109a083 y of such preservation certainly is lessened.-- CORAL flats. argument for Heaping up.-- very good thi
164g119 e: from mud.-- [blank] {P} muddy nodular strata CORAL upside down strata coarse agglomerate [...] she
177b025 - The tree of life should perhaps be called the CORAL of life, base of branches dead; so that passage
191b081 because less easily transported-- Mem plants on CORAL islets.-- Next to animals land birds.-- & life
227b224 ies Hence India, Mexico & Europe. one gret sea (CORAL reefs.: shallow water at Melville Isd. (3d) We k
403e022 p 363 account of Flora of pacific, given in my CORAL paper Oct 14th Macleay says, that <every> «any»

478z004 in rays Par un officier du Roi Rapid growth of CORAL-- RN. p 24 Bougainville Voyage round world no 1
632j53r les of design.-- perhaps they are so.-- but the CORAL rock might have been uninhabited as the Alpine
030r036 What are the "palatal Tritores" found in the CORALIFEROUS mountain Limestone are they allied to the j
051r093 otect like peat reef of sandstone.--Corals, & CORALLINA survive, in the most violent surfs: in both 1
443e155 en propagated by sexual commerce «The fact of CORALLINA & Halimeda is case in point».-- The relation
446e162 in answer to Mr Knights doctrine.-- Case like CORALLINA-- «Does it flower anywhere?-- Yes on the cont
612o035 l will, & stomach likewise «does». [LHC] ¿ in CORALLINA are not two kinds of life vegetable and anima
480z008 s this apply to pale brown Caracara Krauss on CORALLINAE from S. Seas written in German.-- Stuttgart
411e056 World.-- Wealden to boot.-- When one sees in CORALLINE powers of multiplication of individuals, & ye
441e151 s, & never flower, so there may be animals as CORALLINE, or others. which only generate once in a tho
480z007 In transaction of Bonn Society M. Edwards on CORALLINES L'Institut 1837. No 212 Observations on the
027r024 ng steps Mem.; rapidity of germination in young CORALS.--vide L. Jackson's paper. Philosoph Transact:
037r055 dinary marine remains» may contains <shells few CORALS Tortoise» «remains of Amphibia, exclusively.»
050r093 ne in N. Wales. was it newf. -- I remember many CORALS?? Breccia--Stratification? Anomalous action of
051r093 ic bodies protect like peat reef of sandstone.--CORALS, & Corallina survive, in the most violent surf
447e164 lants & true hermaphrodite Mollusca, & probably CORALS.-- these forms that ought to be very persisten
481z010 e of Astrolabe must be studied for anatomy. of. CORALS.-- nevertheless the details appear very trifli
521m009 d it, but the watch was <seen> shown him.-- «<Mr CORB» the servant showed him watch & said dinner is
520m001 hould have imitated.-- when attending Mr Drydem CORBET, he could not help thinking, he was prescribin
521m009 fact of same sort. Aunt. B. ditto.-- Case like Mr CORBET of the <Hall» «Park», after paralytic stroke.
522m010 ication could be held by means of hearing.-- Mr CORBET, however, in conversation could catch up a new
526m029 s seen, quite agrees with my Fathers case of Mr CORBET of the Hall understanding. (on hearing old ass
530m044 of Shrewsbury something about big noses & name CORBET, perhaps nonsense.-- look to it My father has
560m156 redetary tricks & gestures, other cases like D. CORBET; «do» ideots form habits readily?? Do the Oura
529m042 red from a fall of ideotcy.-- The story of the CORBETS & big noses, quite conjectural, in Blakeways b
116a099 -- -- refers to species non decrite de petites CORBULES analogue living in mouth of Plate. p. 26. Geo
044r074 ied by me geologically to vertical movements. In CORD: after seeing small Bombs. without a vesicle. we
351d057 ergoes metamorphosis., heart altered & umbilical CORD,-- Broderip alluded to Hunter's views on this s
049r087 one out of a series of faults. [Fig. 4] {P} In CORDILL: should basal lavas be called Volcanic or Plut
070r155 ncepcion earthquake Draw close Analogy Lake of CORDILL: of Copiápò & Desaguadero.--three ridges in Co
024r013 the water at Cauquenes. coming from the [...] CORDILLERA & flowing The gradual shoaling of the water
026r022 ns does not appear to have taken place in the CORDILLERA of S. America. Study Geolog: Map of Europe C
028r032 cks, & hence Volcanic action, contradicted by In CORDILLERA, where that action commenced before any grea
030r037 he Cocos fish Rio Shells argument for rise In CORDILLERA, the dikes do not generally appear to have f
031r037 es of faults I do not think so many faults in CORDILLERA, as in English Coal field -- because lowered
032r041 t. & changes. «(changes in variation?)» as in CORDILLERA.-- From poles to Equator current downwards &
045r076 t eruptions <both> at sea (as wells as in the CORDILLERA), they may be considered as accidents (if <[
049r088 s become more siliceous in close contact? -- «CORDILLERA???» Porphyry at Valparaiso; Epidote -- Must
062r128 S. Shetland = Great contrast of two sides of CORDILLERA, where climate similar.--I do not know botan
063r131 rust of globe being thin, may be drawn. from. CORDILLERA. rocks.--When beneath water.--together with
064r136 ch side 3000 ft thick & 150 broad. neglecting CORDILLERA itself now remaining-- Lyell « <p 419» p 428
064r137 cid emitted: mem: Grand gypseous formation of CORDILLERA In describing structure of Cordillera it mus
064r137 tion of Cordillera In describing structure of CORDILLERA it must be said, that lines of elevation hav
067r143 unt of N. American geology. Conybeare Lava in CORDILLERA & on Eastern plains «by Antuco». Athenaeum A
067r144 henomenon. showing line of disturbance inside CORDILLERA: It is not therefore so wonderful that volca
069r149 d must be insensible. Quantity of matter from CORDILLERA HORIZONTAL movement of fluid matter not. (Co
070r155 Brown, a person (whom I met at S. W. P.) the CORDILLERA extend to near Salta. & not far from Tucama(
071r157 o near surface most important There is map of CORDILLERA by Humboldt in Geolog. Society Sir Woodbine
073r163 ansition; the latter rarely appear in central CORDILLERA. particularly between 18 degree & 22 degree
078r176 ception» directed NW & SE). «Vol III» Mexican CORDILLERA "immense variety of Porphyries which are des
079r176 at some of the richest gold mines on ridge of CORDILLERA near Pataz, also at Gualgayoc. where many pe
088a016 common salt being found on low hills East of CORDILLERA very important V. Malte brun -- Main charact
089a020 iron clay common to Guyana said to extend to CORDILLERA I see Brewster speculates from believing met
095a038 much pumice --. appears to be outside of the CORDILLERA-- Near the Planchon talks of very much of Gy
097a042 ca Slate Von. Buch. Canary Isd. p 170.-- Mem. CORDILLERA Can Greenstone dikes. be residue of quartzos
099a051 l to walls of dykes-- Mem. laminated dikes in CORDILLERA.!!!-- In stratum OP. let force drag particls
104a066 lica than common felspar therefore on axis of CORDILLERA, in Andite - containing 80 per cent of Albit
104a069 r sentence.-- Origin of Breccia, introduce in CORDILLERA discussion, deep sea, fragments fall off cli
105a070 lay. applicable to Cleavage. C. Prevost.-- In CORDILLERA. a rush of water will account for filling up
112a089 discussing nucleuse's of old volcanos within CORDILLERA-- allude to Lyell's view of not discovering
118a103 ribed-- {P} Cordillera. St Helena &c &c.-- in CORDILLERA. St Helena &c &c.-- in Cordillera, it is at
118a103 dded to any place where dikes described-- {P} CORDILLERA. it is at once evident only small proportion
123a115 (1838) p. 268. Paper by Humboldt on Bogota. CORDILLERA,-- nothing.-- salt & coal near Bogota; p 270
127a127 - with reference to old submarine orifices in CORDILLERA Geograph. Journal vol II. p 89. at Madras. s
136a147 p. 80-- some remarks on dikes: applicable to CORDILLERA Phillips in Lardner Vol II. p. 81. «&83» Som
318c252 the mice of S. America, with respect.. to the CORDILLERA,---- Bachman has seen webbed Shrews. case of
355d068 cies distinct-- but close.-- Mem. Von Buch on CORDILLERA fossils same remark. ¿was there formerly one
023r010 The view of the Volcanos of the chain of the CORDILLERAS as arising from «the expulsion of fluid nucl
035r046 -Yet <silicified> turn over silicified wood. CORDILLERAS, Chiloe. &c seems the organic structure most
052r099 t St Jago. C de Verds Quartz pebbles in the CORDILLERAS look as if some peaks elevated.-- Greywacke.
066r142 ce Sir W. Parish says they have Earthquakes in CORDOBA. one of which dried up <all> a lake in neigbou
067r144 to Quilmes. & at least seven miles inland. The CORDOBA earthquake a very remarkable phenomenon. showi
071r157 uman life.--Temple mentions some earthquake at CORDOVA.-- There the Cordova earthquake in which lake
071r157 ntions some earthquake at Cordova.-- There the CORDOVA earthquake in which lake was absorbed.--Earthq
116a100 150. at Portezuelo, extremity of mountains of CORDOVA project on plain, like <re> a reef on a sea be
070r155 vels accounts of travelled boulders. from the CORDOVISE range. Signor Rozales tells me at seven ocloc
038r057 ments for believing that there must be a central CORE of melted rock--I think the strongest is the co
074r165 ures as in septaria. (& Chiloe case, at least CORELATION is no act of memory.-- [There is no CORELATION between individual objects as Ichneumon & ca
293c173 ere would be no act of memory.-- [There is no CORELATION between individual objects as Ichneumon & ca
293c174 r by My method of breeding there must be some CORELATION. but. the whole Mechanism is so beautiful.--
334d011 en repeated cases] being deaf curious case of CORELATION of imperfect structure.-- Fox says in «Lord»
388d172 n = Stags horns & testes curious instances of CORELATION in structure = Neuter bee having both sexes
391d179 none are added object failed, & then by that CORELATION of structure desire fails. Every individual
410e053 s.-- those discovering the formal laws of the CORELATION of parts in individuals, will care little, <
411e054 oximated to.-- Treating of the formal laws of CORELATION of parts & organs it may serve perfectly to
293c174 . the whole Mechanism is so beautiful.-- The CORELATIONS are not, however, perfect, else one animal w
335d016 o offspring, & this being case, owing to the CORELATIONS of system, the organs of generation would ne
397e003 & production of new forms.-- their number & CORELATIONS Octob. 4th. It cannot be objected to my theo
410e053 riety, but to discover physical laws of such CORELATIONS, & changes of individual organs, must know w
638j28r of endurance &c &c Camels? all good cases of CORELATIONS.-- [There must have been deserts in the old
328cIBC s with a living Matica & many shells of Genera CORLULA Cham. Cardium. Porcellus Turbo. Cerithium Jard
636j57r es.-- Consider ground Woodpecker stiff tailed CORMORANT: pain & disease in world & yet talk of perfec
301c199 ition of deity to teach squirrel to kill ears of CORN according to my views, habits give structure,..
567n013 ny was amusing herself--, by getting out ears of CORN with her teeth from the straw, & just like chil
565n008 e makes person extremely self-sufficient,-- the CORNER of lower lip are depressed & opposite muscles
552m131 tory of wheel horse in drays, scraping against CORNICE stone to cause friction Athenaeum 1838. p. 652
467t104 f <an> pistil, etc lies in gangway» In Lotus CORNICULATUS saw Humble press down wings which ejects po
026r020 ic veins. 1830 P. 399.-- Carne. Geolog. Trans: CORNWALL «Vol II» It is a fact worth noticing that cry
039r059 ruption. Nobody supposes that all the dikes in CORNWALL or in the coal measures have been conduits to
314c237 d, so as to protect itself, one segment of the COROLLA being (probably) small to allow it to lie on o
466t099 in autumn: filaments united in whole length to COROLLA--anthers minute, distinctly doubled, brown, bu
135a146 alcolmson has described formation of shore of COROMANDEL. just same as. at Bahia Blanca-- letter in d
195b100 d it create two species closely allied to Mus. CORONATA, but not coronata.-- We know that domestic an
195b100 ecies closely allied to Mus. coronata, but not CORONATA.-- We know that domestic animals vary in coun
260c068 . Jago--solitary Halcyon bird of passage.-- M. CORONATA of Latham, wrong.-- Mr Yarrell says that so
495q05a es flower» on all gloomy days.-- The «garden» CORONELLA also sleeps on ditto-- Cover them up periodic
291c166 mental structure or disposition. & avitism in CORPOREAL structure are facts full of meaning.-- Why is
536m073 e to will-- Verily the faults of the fathers, CORPOREAL & bodily are visited upon the children.-- The.

569n020 ry sounds... not produced by will <by> but by CORPOREAL structure.-- Devotional feelings, probably so
161g097 rns out too low, (NB .260 would have been more CORRECT) there were several obscure but not far contin
254c049 ften repeated, as mouths in Polypi, Surely not CORRECT view of Flustra or Ascidia spicule in sponge.
526m028 though they remember it so well that they can CORRECT every detail, yet they have not imagination en
570n024 motions.-- (Explanation of sighing is probably CORRECT, to relieve respiration when immensely immerse
558m153 mps with passion. can expression be used more CORRECTLY than this for C. Sphynx.-- In the wild ass th
624o051 to the species in which it occurs. [or, more CORRECTLY «in which it» has been so in some past time,
065r139 shman.--On 8th of March cove began to freeze. CORRESPOND to September ¿Did I make any observations on
151g044 o shelf the truncation & the upper shores may CORRESPOND with some line subsequent to shelf {P} In Gl
321c270 ology Gibbons life on himself Hume's do, with CORRESPOND. with Rousseau Miss M«artineaus» How to obse
528m039 ngings.-- 4th. Pleasure of imagination, which CORRESPOND to those <he> awakened during music.-- conne
571n029 ecome acquired during life time:-- the latter CORRESPOND to fashions in ideal taste & the former to t
059r119 ers la mer; ces couches ont entre elles une CORRESPONDANCE exacte, et lorsquelles se trouvent interro
087a011 he South sinks.---- Mediteranean continent CORRESPONDING to Europaean risings. Pacific great land. -
149g030 th terraces of each side of the two valleys CORRESPONDING as in Andes, composed of sand & perfectly r
264c081 idering to which tribe,-- structure without CORRESPONDING habits clearly showing true affinity, for i
352d059 be erect not pressed on paper, to study the CORRESPONDING points.-- The present geographical distribu
446d147 eons yet no change of habits, so no <cause> CORRESPONDING change in Birds of Paradise.-- All that we
447e164 rous. -«& recent» mollusca, than between the CORRESPONDING acalepha?--- But if Acalepha do not cross th
473s07r he peculiarities shd be born,-- may come in CORRESPONDING time of life of offspring-- No peculiarity
615o038 in brain would probably take place without CORRESPONDING change in «external» man; and as all men ne
638j58v dded This is similar idea, to cells of bee, CORRESPONDING to <every> «one or any»- brain making stru
151g043 t projecting butresses, upper slope of which CORRESPONDS to shelf the truncation & the upper shores m
466t095 haracter? » of characters being inherited at CORRESPONDING age & sex, opposed by cantering horses havi
538m078 hildhood. & certainly of Miss Cogan, & fully CORROBORATES the fact of her not «remembering which» «re
155g063 er must constantly alter with falling sea & so CORRODE plain into terrace as regressed What <alter> a
155g064 vers either bringing more «detritus» than they CORRODE or vice versa Same inclination when serpentine
056r109 in at once the stupendous mass which has been CORRODED. -- If man could raise such a bulwark to the
151g043 e way on great sloping plain of alluvium (much CORRODED by rivers) & not to head of plain.-- but belo
153g053 of Glen Roy, seems to be which higher up on is CORRODED {P} 4th [shelf] river 4th [shelf] Could earth
087a012 t land. -- Will use argument of proof of slow CORROSION of valley of <Patagonia.> S Cruz -- from terr
152g046 necks of land on level with shelves effect of CORROSION & not cause. Monday a rapid descent of a terr
603o013 ptness «of consequence» hence languages become CORRUPT, & whole classes of words «are abbreviations»
155g063 ieve they end in upwards inclined plains, as in CORRY. & as «as I believe in side ravine above houses
159g084 6 ft across) on top of spit between river & dry CORRY Scarcely conceivable. if Hill between Corry so
159g085 dry Corry Scarcely conceivable. if Hill between CORRY so much cut Granite could have remained, no pea
406e035 ortant account of cross of sheep & Moufflon of CORSICA. <would not>, sadly against Yarrell's law.-- n
157g076 - (in valley «there are» granite) «boulders» {P} CORY stream hill with boulders river Right Hand Casc
405e035 ed] ARGUMENT REAL of antiquity of reasonable COSMOPOLITE man.-- L'Institut. 1838. p. 338.«V[ide]» Imp
415e066 thorhyncus be found.;-- yet until man became COSMOPOLITE, he would probably be confined in locality 1
415e066 n locality like Ornithorhyncus.: since being COSMOPOLITE, we do find his remains.-- Lima.-- caves.--
234b261 genera, «in all classes» are not a few only COSMOPOLITES, & in genera peculiar to any one country do
477z001 yage Vol I p. 196. According to Charpentier de COSSIGNY. only 10 years ago <no> snail was introduced
430e120 rid.!!! Ask Woodward Mr Lonsdale says Trigonia COSTATA & elongata thougt considerably different, in p
431e120 , from specimens in grades, now L. says the T. COSTATUS is in England found in the Inferior Oolite, &
043f069 scription of channels or grooves in rocks at COSTORPHINE hills. to compare with Galapagos.--Chiloe. M
094o396 {T} On grooved rocks. Specimen of rock from COSTORPHINE at Geolog. Soc: Colonel Imrie Transact Wern.
617o39v odily action is revealed to us by the effort it COSTS us to exert force or by internal consciousness;
033r043 ous: fragments instead Peak of Teneriffe. also COTOPAXI has a <[...]> cylinder placed on the rim of c
452e181 how could it have originated-- spins thread of COTTON.-- do p. 583; It APPEARS probable,«?» that the
458t002 pinning-- better case than English birds, using COTTON &c instead of natural substances-- useful perv
499q010 clumps of plants in counties, as of rare green COTTON Plant-- How large «area» clump there? Distingu
246c026 o!!-- in Mauritius Lesson &c p. 620. Centupus (COUCAL) of Java & Phillippines, has variety at Madaga
059r119 ée a pic. Toutes ces montagnes sont formées de COUCHES paralleles et inclinées du centre d'ile, vers
059r119 t inclinées du centre d'ile, vers la mer; ces COUCHES ont entre elles une correspondance exacte, et
022r008 sack, very thick & so tough that a sharp knife COULD not cut it: in which we found the Head & Boans
023r013 reous rock; following Curvature of hill; states COULD discover no shells: nothing said about K. Georg
032r040 riods of greater activity & rest.--Such changes COULD be shown (as represented), along line of coast.
033r043 ne; Mem Galapagos. chiefly red glassy scoriae.--COULD walk round base:--not universal: could not clim
033r043 coriae.--could walk round base:--not universal: COULD not climb up many parts, in James Isd.--Mem St
041r064 ived there, makes it very doubtful whether they COULD have lived in so deep a sea.--Perhaps agrees wi
056r109 pendous mass which has been corroded. -- If man COULD raise such a bulwark to the ocean, who would ev
056r109 ould ever suppose that its age was limited? Who COULD suppose such trifling means could efface & obli
056r109 limited? Who could suppose such trifling means COULD efface & obliterate so grand a work?--In valley
066r142 hat taking up a piece of Falkland Sandstone. he COULD not distinguish from stone Caradoc from lower o
070r154 et it not be overlooked that except by trees, I COULD not see trace of Subsidence at Uspallata.--¿If
093a031 sible first injected.-- Basalt: last because it COULD reach the surface. before being cooled.-- Berze
106a074 retain its force if inclination be great. There COULD «not» be great deflection in a "rapid".-- is a
121a112 profoundly How came it. that Glen Roy district COULD have been elevated without fissure & unequal.--
146g010 ear Holyrood Palace In same way at top the trap COULD be traced Grey in front on wall perhaps wall ob
146g014 rhaps near 300 ft above Loch.-- From this point COULD be followed up to neighbourhood of Tyndrum wher
147e106 300 ft Alluvium <abo> by Loch Dochart-- Rivers COULD not have deposited it. Barrier of lake very lof
147e018 n pebbles in Alluvium which without lake or sea COULD not be placed in present position Thursday Even
153g053 n is corroded {P} 4th [shelf] river 4th [shelf] COULD earthquake cause collection of sediment? Where
156g068 fragments which had fallen before lake drained COULD be told from «some of» those since fallen. «on
159g085 able. if Hill between Corry so much cut Granite COULD have remained, no peat supply.-- Consider profo
159g086 & some partly well worn pebbles-- «which river COULD not have deposited» the slope is continued some
161g096 ntle mossy slope, which from a distance hid it, COULD be followed for at least 2 miles on dead level
185l055 iple, of animal having come to island. where it COULD live-- but there were causes to induce great ch
189b070 gatherium. each with same kind of coat.-- If we COULD tell, I do not doubt even colour hereditary in
195b100 id the creative force which that «these» species COULD, arrive-- did it only create those kinds not so
196b108 ance none--?, of the <vebtetrata» «vertebrates» COULD exist without plants & insects had been created
196b108 ct we may say that vegetables & mass of insects COULD live without animals but not vice versâ. ¿could
196b109 could live without animals but not vice versâ. ¿COULD plants live without carbonic acid gaz.)-- Yet u
198b115 on New Zealand; if generated <No> can» answer <COULD> «can» be given.-- It is a point of great inter
199b118 «p 191» No. 5. Ap 1827 F. Cuvier says. "But we COULD only produce domestic individuals & not races,
202b132 r, herded separate from the English cattle, nor COULD we get them to associate together" There is lon
213b172 of observation.-- I think it is certain strata COULD not now accumulate without seal-bones & cetacea
227b225 o A. B. C. we cannot be sure that structure (C) COULD pass into (D).-- We may foretell species. limit
239c1FC yet failed. About trades affecting form of man. COULD you get racehorse from Cart horse by picking wi
248c029 ection, with house mice. It is wonderful how it COULD have been transported. ¿What section does the N
260c067 ing young cuckoo; as if there was storge, which COULD not be resisted, when hearing crys of hunger of
265c087 nt not destroyed even if these shrivelled wings COULD be shown to be of some use. If we only had Puff
274c115 for vegetation.-- In conversation in Museum-- I COULD not discover any other clear relations besides
278c130 should be some term used, when there is series. COULD I not give Catalogue of Mammalia arranged accor
289c162 «It would be curious to know whether a variety COULD be transmitted more easily in those born withou
292c170 I confess. no dissertation against these views, COULD possibly have had <to so convin> brought so muc
293c172 n a Great coat & this he put on & we afterwards COULD understand «(language better instance)» he had
294c175 ailless cats near Bath. Lonsdale do. says Sheep COULD not live for some time at New York «instance of
309c221 from an accident. New England farmer,-- useful COULD not leap fences:-- Dr Lang on Polynesian nation
312c232 r means whatever.-- » Individual Men «& animals» COULD only exist by habit.-- therefore same principle
335d016 ere combined). In last page, we have seen mules COULD have no offspring, & this being case, owing to
336d019 rent not entirely embued with the change, a bud COULD not be taken, without it either went back, or n
338d023 ration on <be> inter-breeding, for otherwise we COULD not understand the vast number of domesticated
341d029 s adaptive & that he did not see where the line COULD be drawn-- thus the most remarkable character i
366d108 ir J. Sebright. thought if he had had a pair he COULD have produced from these.-- this instance of mo
366d108 se.-- this instance of monstrous variety. which COULD not have been persistent in nature.-- According
366d111 hough really fully aware she was not coming in, COULD not help being perfectly distracted «Referred t

367d112 it, is forming «& this must be so, else avitism COULD hardly ever occur.--».-- & if that cannot be fo
376d136 in two or three days learn its comfort & though COULD not put it on, yet threw it over it, & made it
386d167 imals|. One «invisible» animalcule in four days COULD form 2. cubic stone. like that of Billin.-- <Ge
398e005 parated with some animals, &c.-- «if the change COULD be shown to be more rapid, I should say there w
409e049 merable species. .& hence few only social there COULD not be one body of animals, living with certain
410e050 would be deeply impressed on it, & hence there COULD not be improvement. «& hence not «be» higher an
410e051 of country, & not to the local changes. = this COULD only be effected by sexes: All the above should
420e089 s originally quadru<manous> «ped.-- ». Hairy.-- COULD move his ears The head being six metamorphosed
423e096 rom true.-- I doubt not if the simplest animals COULD be destroyed, the more highly organized ones. w
430e117 t generic forms.-- the animals in Eocene period COULD not have been direct parents of any of ours,--
436e137 l Rush I thought, surely no "fortuitous" growth COULD have produced these innumerable seeds-- yet if
437e140 rabbit escaped into a hole, where the old Eagle COULD not find it..-- The parent bird another day bro
441e149 cation, yet offspring are,-- if Australian Dog, COULD bud, analogy tells us, <be> «offspring» would b
444e158 don squirting water at fly.-- instinct, for how COULD experience teach distances in air, in which it
452e181] [...] p [...] -- most wonderful instinct, how COULD it have originated-- spins thread of cotton.--
463t055 s not possible to imagine what habits an animal COULD have had with such structure.-- perhaps greates
463t055 ave had with such structure.-- perhaps greatest COULD anyone. have foreseen, sailing, climbing & mud-
511q018 r any degeneration?? HOUNDS. Eyton Mr Wynne, &c COULD by selection a different looking animal be form
513q21. .-- Is setting of fruit. cross Conception-- (| I COULD extract nothing from him)| Does impregnation ev
520m001 imitated.-- when attending Mr Dryden Corbet, he COULD not help thinking, he was prescribing to his fa
520m002 of holding hands &c &c.-- Mr Dryden Co said he COULD not remember his father.-- My father thinks. pe
521m009 ntellect impaired. <after paralytic stroke> : . COULD converse well on any subject when once started,
521m009 nverse well on any subject when once started,-- COULD receive a new train through eyesight, though, n
521m009 gh hearing,-- Thus when dinner was announced he COULD not understand it, but the watch were <seen> sho
521m010 Thus was he in every respect, no communication COULD be held by means of hearing.-- Mr Corbet, howev
522m010 hearing.-- Mr Corbet, however, in conversation COULD catch up a new train if early association were
522m011 whole subject.-- if at the other nothing.-- He COULD repeat the alphabet straight, but did not know
530m043 paid the Attorneys bill, he asked what bond he COULD have had. yet during whole illness, he had been
531m049 to see how the young got on. this nest the cat COULD If cats will «ever» eat little birds, this most
532m055 s» of infancy, it is very doubtful whether they COULD recollect these same things from any effort of
533m058 ey> (I think I have seen same thing before they COULD understand. what frowning means) if so this is
538m080 were with a second & unreasonable man.-- If one COULD remember all ones farthers actions, as one does
540m089 assist them,-- otherwise as he remarks sympathy COULD be barren. & lead people from scenes of distres
545m106 at every emotion & <look> «turn» of the head. I COULD not perceive «any» distinct wrinkle, but such m
546m110 go on the lawn to see something, on her return COULD not find paper cutter, hunted in vain for it--
547m111 efore.-- was greatly astonished, at the time. & COULD trace no chain of association» Mayo Philos. see
547m113 & always running on in mind, being absent. one COULD not compare the castle with them, therefore I
547m113 uld not compare the castle with them, therefore COULD not doubt or believe.-- When I say trains, it m
554m138 ang against the door to force it open, when she COULD not succeed of herself.-- <The male> «I saw» Je
554m139 y difficult knot-- the sailor on board the ship COULD not puzzle her-- with aid of teeth & hands.-- D
555m142 eavoured to classify expressions of monkeys-- I COULD only perceive that the American ones, often put
572m033 ounced each word distinctly. woke instantly but COULD not gather general sense of this page.-- Now <a
572m033 ense of this page.-- Now <awake> «when awake» I COULD not picture to myself reading French book quick
577m049 free action.-- on my view of free will, no one COULD discover he had not it.-- The memory of Plants,
586m079 in early childhood (before experience or habit) COULD be formed or afterwards.-- child sucking whole
589m091 ent in favour of brain forming the instincts,-- COULD brain make a tune on the pianoforte, yes if eve
599o005 origin of language-- & effect of reason. reason COULD not have existed without it.-- quotes Ld Mondob
606o025 a bridge to save another) &yet dare not -- one COULD do it, but other motives prevent the action see
609o030 rule may sometimes be hard to tell) + + Society COULD not go on except for the moral sense, any more
610o031 he had long known & directed many letters to»-- COULD not «remember» «read» Christian name; fancied i
610o031 me; fancied it looked like. W. but concluded it COULD not be so.--Looked at a direction book, but cou
610o031 uld not be so.--Looked at a direction book, but COULD not find out-- Directed his letter, & I observe
610o031 ut; he was astonished, & said how very odd.-- -- COULD not think what had put Wilson into his head.--
618o041 brain that force does to the bodily frame, they COULD be perceived by the faculty by which the brain
618o41v ing grew blue it «uniquely» grew heavier yet it COULD not be said that the blueness caused the weight
623o050 beneficial   tendency through <all> «many» ages. COULD be acquired, & we are certain from our reason,
623o051 ch» had a beneficial tendency during past races COULD become instinctive.-- [LHC] x It is probably Th
636j57r ese cases it should be remembered, that animals COULD not exist without these adaptations.--fossil fo
637j58r ickens beak, to break egg. shells-- why chicken COULD not have lived had it not been so.-- let egg sh
638j58v d) although having heredetary superfluities Man COULD exist without Mammae.» to the then existing con
641j29v rom Bats. CD]CD. «Macculloch says no other bird COULD catch mouse by night» Sailing lizards. squirrel
230b235 so.-- Geograph Journal Vol V. P. I. p. 67. Dr. COULTER on decrease of population in California cessat
090a023 as weighed,; <but> carrying on this ratio I can COUNT 90 stones which have fallen in the 50 years. ∴
495q005 of one branch of Cabbage with pollen of other, COUNT seeds, & see how great a proportion springs up
090a022 e even 3000) This ninety includes all actually COUNTED.-- The weight «or size» is given of 25 stones.
437e140 sland.-- » p. 175., 28 sho[r]t eared owls were COUNTED in a field, where there was great swarm of mic
451e177 ue of Birds of India.-- p. 555. Lieut. Hutton COUNTED, the ova of a tick «in India» & found there we
065r139 dy had scarcely undergone any decomposition: COUNTENANCE so well preserved. that it was thought not t
286c154 Not saltus. but hiatus animals expression of COUNTENANCE. «[s]hare of sickness,-- death, unequal life
453e182 oldt. Vol I. p. 275. says Teneriffe does not COUNTENANCE the theory of polymorphous plants, abounding
573m037 ess the greatest surprise at emotions in her COUNTENANCE-- before they can have learnt by experience,
216b182 rs the Caffers (like Englishmen) men of many COUNTENANCES, as hybrid once. Is not this contradiction
336d018 s beget child like themselves. expression of COUNTENANCES, organic diseases, mental disposition, stat
352d059 present geographical distribution of animals COUNTENANCES the belief of their extreme antiquity (ie m
265c083 etary.-- yet these not adaptations «they are COUNTERACTED by nature by crossing with other varieties»
271c106 e latter made by man & Nature; but cannot be COUNTERACTED by Man.-- effect of external contingencies
102a062 predominantly, connection with magnetism &c COUNTERACTING gravity.-- As volcanic eruptions are accomp
128a129 ifying cause. {P} land protuberant water to COUNTERBALANCE How strongly the Glen Roy case shows that
498q008 b- not these infinitesimal varieties, which COUNTERBALANCE each other? (10) Is number of pollen-grain
055c108 precise periods over immense areas. (& the COUNTERBALANCING variations) of rain. = The Bulk of sedime
312c232 ations produced in short time in some extent COUNTERPART, mutilation being «variation» produced in sh
423e098 seeds to London is that people in the southern COUNTIES have whole fields, some for cauliflower &c.--
434e129 rtainly <old> only white in Cambridge, in some COUNTIES sometimes one & sometimes other.-- there is s
499q010 e many instances of single clumps of plants in COUNTIES, as of rare green Cotton Plant-- How large «a
037r056 Insist on the frequency of dikes in Granitic COUNTRIES, enumerate cases. -- M. Video exception, for
041r065 by the want of water.--I believe in all flat COUNTRIES. years of drought are common.--Mr Lyell has m
064r134 irchell says Elephant lives on very wretched COU[N]TRIES thinly covered by vegetation. Rhinoceros cha
174b013 r to same country, different genera different COUNTRIES. Propagation explains why modern animals same
174b014 can see «why» structure is common in certain COUNTRIES when we can hardly believe necessary, but if
174b015 anic kingdom, when our planet first cooled.-- COUNTRIES longest separated greatest differences-- if s
182b050 n sub-genera, they undoubtedly come from same COUNTRIES.-- In mundine genera, the nearest species oft
186b060 arents of American.-- Now Genera of those two COUNTRIES ought to be similar ¿Law: existence definite
187b062 h belief. that Siberian animals lived in cold COUNTRIES & therefore not killed by <Siberia a more> co
187b062 therefore not killed by <Siberia a more> cold COUNTRIES.-- Seeing how horse & Elephant reached S. Ame
190b075 the instincts were.-- relation of type in two COUNTRIES direct relation to facilities of communicatio
191b080 + + +. Ireland longer separated., Hare of two COUNTRIES different.-- Ireland & Isle of Man possessed
195b100 nata.-- We know that domestic animals vary in COUNTRIES, without any assignable reason.-- Astronomers
220b199 each other, are migratory species from warmer COUNTRIES. When will this paper be published it will be
224b212 hat between Species from «moderately» distant COUNTRIES. there is no test but generation, «(but exper
227b224 dea of <origin under> connection of those two COUNTRIES Hence India, Mexico & Europe. one gret sea (C
232b246 ic table «in comparison of temperature of two COUNTRIES» finding a very hot day, «in one», oh we will
234b261 s there not similarity even in quite distinct COUNTRIES in same hemisphere. more than in other. Are t
235b262 circumstanced; varieties of dogs in different COUNTRIES a case in point.-- All cases like Irish & Eng
235b273 ll genera, & see how many confined to certain COUNTRIES-- so on with families.-- Ask Royle about Indi
236b280 he closest allied species always from distant COUNTRIES, as Decandoelle says, no he only says sometim
253c046 tigrade Carnivora??-- «compare rodents of two COUNTRIES» & «Monkeys.» Fact of Elephant same species i
258c060 cause destruction.-- simile Man living in hot COUNTRIES, if continually crossed with people from cold
261c070 ry, & the examination of species from distant COUNTRIES., may give thread to conduct to laws of chang

262c072 ? «like senses of savages» (How come its some COUNTRIES patriotic?)-- but more especially the powers
265c084 migan hare becoming white in winter of Arctic COUNTRIES few will say it is direct effect, <of> accord
266c088 & birds & <f> become dull coloured in sterile COUNTRIES.-- Gould insist much upon knowing to what typ
271c106 bout not altering breed of animals in certain COUNTRIES.-- Fraser remarked to me at Zoological Societ
271c107 ver find two «similar» groups of birds in two COUNTRIES, without intermediate ones occurring in inter
276c125 en) but where close species inhabit different COUNTRIES habits similar ¿law?-- probable.-- if habits
278c132 st modern period, compared to Faunas of these COUNTRIES, greter than Toxodon, Macrauchenia, &c compar
289c160 another, (golden creted wren so rare in some COUNTRIES-- nightingale do.-- all shows how nicely adap
296c184 = to peak of Teyde in relation to surrounding COUNTRIES & present tropical countries. p. 564. an abst
296c184 n to surrounding countries & present tropical COUNTRIES. p. 564. an abstract of Mr Swainsons views. w
298c192 ry.-- some species are larger &c in different COUNTRIES. These facts show how very permanent plants a
310c224 f growing!! & propagating itself. In Tropical COUNTRIES (as St Jago Cape de Verds) the shells in equa
311c226 which last produced --insane men in civilized COUNTRIES-- this is well worthy of investigation.-- Ins
316c244 stralia or Concholepas in America,-- that many COUNTRIES have far more species than other countries «+
316c244 ny countries have far more species than other COUNTRIES «++ p. 246» as Cyclostoma in Phillippines & A
316c245 plants.-- Many «Some» genera confined to hot COUNTRIES & many to cold.-- Hence latitude is more impo
332d004 healthy to worms, (like parasites of Tropical COUNTRIES cannot endure this climate-- .) -- July 23d.
337d023 Animals «of same classes» differ in different COUNTRIES in exact proportion to the time they have bee
338d023 self, & in the local circumstances of the two COUNTRIES in times present & past. The effect of physic
338d025 - recurs to idea of only animals from distant COUNTRIES breeding!. «Mem 3 species of grouse»! Has not
402e018 met with near Canton" "Here, as in all Malay COUNTRIES, I noticed a peculiarity in the cats «p 10» t
402e019 ey had been broken". are born so in all Malay COUNTRIES. W. Earl Eastern Seas. p 233 Octob 12th Kotzeb
410e052 country, most important & will be done to all COUNTRIES.-- but naming mere «single specimens in» skin
424e103 which have run wild have, have done so in hot COUNTRIES.-- CD[Camel does not vary «one ought not to b
453e182 orner.-- «p. 233» There, as well in all Malay COUNTRIES «the» cats are born with the joints near the
497q007 ate country (2) Any known changes in Flora of COUNTRIES during last century or two.-- where agency of
501q10a er? are these genera less difficult, in other COUNTRIES, where species are either numerous or even wh
508q016 proportion of diseases (heredetary?) in diff. COUNTRIES in same races Mr. Gray General Questions (1)
513q020 (7) About fertility of Bantams from different COUNTRIES-- Do the Peacocks cross.= Young Chinese or Pen
514q21. ld differs-- do they also differ in different COUNTRIES-- on flora of African Islds-- names of Plants
615o038 indoo population.-- Slightly modified in many COUNTRIES, hence national character, love of country, o
641j29r hysiological relations of animals. equatorial COUNTRIES are supposed favourable to terrestrial Mammif
035r048 e movement of viscid nucleus, which in any one COUNTRY would produce equable effects.--«though so imm
043r069 ion, must be more exclusively confined to that COUNTRY. Read description of channels or grooves in ro
045r075 out eruptions of Volcanos. (where there are no COUNTRY newspapers)--At the Calabrian earthquake thing
045r075 his may be mentioned with general slope of the COUNTRY; (perhaps generally over whole world) Yet erup
048r085 is removed by reflecting on the nature of the COUNTRY in which the Rhinoceros lives in S. Africa: th
048r086 th species & individuals as in the half desert COUNTRY of S. Africa. It would be well to quote Burche
054r105 os. Cocos-- Ulloas voyage North of Callao, the COUNTRY, to the distance of 3 or 4 leagues «from the c
056r110 to be the case p. 51). presupposes an elevated COUNTRY of granite, not <so> great«er» for all Europe,
057r115 reserved animals.--Mem: stream of water in the COUNTRY.-- Sir J. Herschel. says. precip. of Sulph. B.
070r153 the most closely allied species occur in same COUNTRY? In botany instances diametrically opposite ha
172b005 tion, why are. species are constant over whole COUNTRY; beautiful law of intermarriages <separating>
173b012 nd existed.-- or they may represent some large COUNTRY long separated.-- On this idea of propagation
174b013 idity.-- genera being usually peculiar to same COUNTRY, different genera different countries. Propaga
176b024 once formed by separation or change in part of COUNTRY. repugnance to intermarriage <increases it-->
179b035 <suppose» «grant» similarity of animals in one COUNTRY owing to springing from one branch, & the monu
181b039 generation only breeds; like individuals in a COUNTRY not rapidly increasing.-- If we thus go very f
183b051 & thus two species be created) & live in same COUNTRY. How is propagation of wolf & Dog. (because be
185b055 ions as possible.-- Why should we have in open COUNTRY a ground «do. <w> parrot.--» woodpecker-- a de
186b060 ype of organization.-- extinct species of that COUNTRY parents of American.-- Now Genera of those two
186b061 animals would perish, if there was nothing in COUNTRY to superinduce a change? Seeing animal die out
187b065 species closely allied but different, because COUNTRY separated since time of extinct quadrupeds:--
192b083 ety of fruit tree or plant run wild in foreign COUNTRY.-- When one sees nipple on man's breast. one d
195b101 ch animal created with certain form in certain COUNTRY, but how much more simple, & sublime power let
195b102 ansportal be such & so will be the form of one COUNTRY to another.-- let geological changes go at suc
195b103 rs but» interruption of communication. or when COUNTRY changes. Will it said that Volcanic soil of Ga
198b115 interest to prove animals not adopted to each COUNTRY.-- Provision for transportal otherwise not so
199b119 en to Portuguese. priest.--In first settling a COUNTRY.-- people very apt to be split up into many is
202b130 f domestic animals made from influences in one COUNTRY is permanent in another.-- Good argument for s
224b214 tion of stomach & hence remain distinct. Where COUNTRY changes rapidly, we should expect most species
230b235 ry of changes that if so in approaching desert COUNTRY or ascending mountain you ought to have a grad
230b236 ect or adaptation,∴ animal best fitted to that COUNTRY when change has taken place. Nature [not locat
234b261 cosmopolites, & in genera peculiar to any one COUNTRY do not species generally affect different stat
239c001 to be oldest which have long been known in any COUNTRY; he states that Esquimaux dog when crossed wit
256c054 admirably) yet a glance would tell from which COUNTRY,-- I often disputed for a moment,-- Galapagos.
257c057 er peculiar conditions-- of «nearly» same kind COUNTRY distant. <Study> The circumstance of ground wo
258c060 mote antagonist powers.-- Every animal in cold COUNTRY has some analogy in not gaudy colours so all c
259c066 h thick coat monstrosity, if brought into cold COUNTRY, «& there acquired» then adaptation.-- No Carr
260c068 parents.-- Shows community of language Desert COUNTRY. is as effectual as a cold one. in checking be
266c088 longs.-- I conceive without knowing from which COUNTRY many birds come it would be impossible to clas
271c105 white thrush of Azara builds its nest in <same COUNTRY> something same manner, much mud.-- These fact
271c107 ut intermediate ones occurring in intermediate COUNTRY-- ie. mundine groups.-- Waterhouse tells me in
276c125 e-- ¿Where two very close species inhabit same COUNTRY are not habits different, (Mem: Gould's Willow
277c126 genera of course distinct. analogy from every COUNTRY & class tells us that ¿O. Modulator & O. Patag
277c127 d together.-- If two species come over to this COUNTRY, without range, or habits ascertained-- put th
278c132 orms, were able to escape to some more fitting COUNTRY,-- if Toxodon had been found in Africa, the wo
282c141 - We must two races of same men living in same COUNTRY but separated, now if one of these races had b
283c145 a into the Quagga.-- «let them be wild in same COUNTRY with their own instinct & (even though <fertil
288c159 ied.-- The partial migrations of birds in same COUNTRY. may explain greater migrations, if American &
294c175 fine relations of adaptation of animals & the COUNTRY they inhabit.--» & the first one that bred one
296c184 est. & having undergone changes «no near lofty COUNTRY»?? p. 475 NB. This bears on fossils of Europe.
298c191 in Making an alpine species from one in lower COUNTRY during gradual elevation of isld.-- We must im
298c192 er authority) similar with those over whole of COUNTRY.-- some species are larger &c in different cou
301c200 ly complicated.-- An Entomologist going into a COUNTRY. & collecting thousands & tens of thousands New
306c214 ch it is said cannot be transported from their COUNTRY.-- the long-haired cats are supposed to come f
318c253 ds, certain kinds as gallinules taking the low COUNTRY near coast & others the mountains, & these app
318c254 , these birds seem clearly directed by kind of COUNTRY; «kinds of migration quite different in specie
318c254 ay about a fortnight in one particular part of COUNTRY, like White of Selbournes Rock Ouzels.-- .If
318c254 urnes Rock Ouzels.-- .If the line or bands of COUNTRY (These facts show the Normal condition of Migr
337d020 local & then the local ones are taken to fresh COUNTRY & breed confined. to certain best individuals.
338d023 parated; together with physical differences of COUNTRY: the time of separation depends on facility of
338d023 t & past. The effect of physical conditions of COUNTRY is not perhaps so great, as separation on <be>
365d107 - the former only giving average of effects of COUNTRY (& no monstrosity, or adaptation to unhealthy
390d177 solitary flower) exotics brought from foreign COUNTRY, (<annuals> & so must those forms which are pr
399e010 olely, because of similarity, because in every COUNTRY, where only pair has been introduced, & have f
410e050 e slow & bear relation to the whole changes of COUNTRY, & not to the local changes. = this could only
410e052 on to habits, ranges. & external conditions of COUNTRY, most important & will be done to all countrie
410e053 cribed.-- but then permanent varieties in same COUNTRY, must be distinguished, from permanent varieti
410e053 nguished, from permanent varieties not in same COUNTRY.-- The traces of changes in forms of organs, w
419e086 y same as those of present seas.-- now in this COUNTRY we have better means of judging of slowness of
430e118 fluence, & this would take place from changing COUNTRY: but greyhound. & poutter Pidgeons «race-horse
432e124 s to improve the native <breed> animals of any COUNTRY must be made with great caution; owing to its
459tflv t Madagascar-- «p. 121» No beasts of Prey. any COUNTRY should during [...] conditions-- every spot is
465t079 ee what a «mere» vestige, is preserved in this COUNTRY-- same argument to India & Europe-- & Africa!,
465t081 o perfect gradation can be expected in any one COUNTRY.-- in a descending series of strata This Afric
490q001 dual Shell or insect or group vary more in one COUNTRY or district than in another? Character of shel
497q007 oved possible point.-- ¿genera in intermediate COUNTRY (2) Any known changes in Flora of countries du

```
501q10a s. vary & hard to separate specifically in one COUNTRY & not in other: Rosa is hard in Europe, Walnut
501q10a eric or specific character,, but contingent on COUNTRY.-- How is it in Patella or Oysters or Helix. O
506q015 of plants. which will not produce seed in this COUNTRY-- where cause not apparent-- Any where pollen
515q022 bout the leaping of Irish Horses, bred in this COUNTRY. {Chinese Dog's Head to send Cover common  Pea
534m063 lants are attacked by insects & snails of this COUNTRY (thus Dahlias by snails)-- <The> Apion radiolu
537m077 cience more than other.-- A Scotchman will his COUNTRY or Swis.-- it may be answered effects of educa
570n026 ould be that a mountaineer <takes> born out of COUNTRY yet would love mountains, & a negro, similarly
615o038 y countries, hence national character, love of COUNTRY, of association & stronger in some than other
622o048 may  be curiously modified by circumstances of COUNTRY, so will the conscience in these cases.-- Thos
636j56v of bisexual animals living on the borders of a COUNTRY favourable to change.-- It might be  concluded
332d001 three  patients at very distant quarters of the COUNTY, who had had no sort of communication, were se
599o007 y-- Viewing from eminence. the wide expanse, of COUNTY, netted with edges & crowded with towns & thor
328cIBC hman on migration of birds Temminck has written "COUP d'oeil sur la Faune des iles de la sonde et  de
058r118 vers  le centre de' l ile, elles presentent une COUPE abrupte et souvent tailée a pic. Toutes ces mon
229b232 be all netted together.-- Hermaphrodite animals COUPLE: argument for true molluscs coupling.-- Dr. Sm
416e070 theory]  CD without the hermaphrodites mutually COUPLE,-- now how is it-- in Planaria, they  couple--
416e070 lly couple,-- now how is it-- in Planaria, they COUPLE-- CD [lowest terrestrial animals.- in shells?
512q020 logical Soc (1) Do the animals there, sometimes COUPLE but not conceive :Bears /Yes/ (2) Foxes & Engl
229b232 ite animals couple: argument for true molluscs COUPLING.-- Dr. Smith's Information Long Horned (very)
204b140 bly-- Esquimaux dog & Pointer «Game-fowls have COURAGE independently of individual force» Mr Wynne ha
275c120 breed of greyhounds fleestest in England lost COURAGE-- (Bull-dogs are used because they have no sce
358d085 beautiful proof of the breeding in & in (like «COURAGE in dogs» EFFEMINATE men),-- if carried much fu
575n044 oubts that a cross of bull dogs-- increase the COURAGE & staunchness of greyhounds.-- bull-dogs being
615o36v yse feelings. Mr Wynne says, that beyond doubt COURAGE is heredetary in fowls & not effect of feeling
242o017 yage of Coquille. Zoolog, p 19 . Tapir, «des» COUROUCOUS et rupicole vert instances of American forms
089a018 urnal de Physique, et D Histoire Naturelle, C«O»URREJOLLES. 11th Observ.-- Les grands tremblemens  de
060r124 he land is constantly pushing the sea (which of COURSE must retain same level) to a greater distance"
105a072 ?-- The widening a valley depends on serpentine COURSE.-- the latter (it is generally said) is conseq
105a072 generally  said) is consequence of <rapid> slow COURSE, & with slow course small erosive power. there
105a073 consequence of <rapid> slow course, & with slow COURSE small erosive power. therefore tendency of run
105a073 deepen not to widen valley.-- Why is serpentine COURSE result of little inclination??----It is simply
106a074 on.-- Therefore stream has no tendency to widen COURSE until inclination is become comparatively smal
115a098 fineness.  then most regular slope-- {P} if not COURSE enough flat top. ended by abrupt slope {P} eac
171b003 . or adaptation.-- Again we <believe> <know» in COURSE of generations even mind & instinct becomes in
175b016 rest of the world.-- This view supposes that in COURSE of ages.-- therefore changes. every animal has
187b064 ties, & to accomodate itself to change, (for of COURSE change even in varieties is accomodation). Now
193b090 the child. ;-- whether every animal produces in COURSE of ages ten thousand varieties, (influenced it
200b122 ; one when they can proved descendant, which of COURSE most rare, or when placed together they will b
264c080 brown-breasted bird in Australia &c &c-- but of COURSE they might be blended, if archipelago turned i
277c126 every  species. only known by analogy genera of COURSE distinct. analogy from every country & class t
285c152 of  life is utterly rotten & obliterated in the COURSE of ages.-- As species is <certain> real  thing
390d178 almost all Male animals?]CD If the male «in the COURSE of some generations» has gained some differenc
391d174 at kind of generation, which typifies the whole COURSE of change from simplest form.-- (Because by th
417e078 ffspring of true <pare> hermaphrodite, would of COURSE be like either, that is both parents, for they
473s07r we see the frame of animals can adapt itself to COURSE of life, «as in trades» there is no reason, wh
617o040 sts in its communication to other matter in the COURSE of its DIRECTION, & thus when we apprehend for
633j54r est> «prevent» the valuable soil in its seaward COURSE,-- we sink into such contemptible queries,  as
638j059 f trying a hundred schemes of structure, in the COURSE of ages «step by step».-- in Man, the  nervous
618o041 h the brain is perceived but they are known by COURSES of action quite independent of each other. A p
592n101 ew Smith says hen doves & the female chamaeleon COURT the males by odd gestures. In one of the six (?
482z011 a Excessively inaccurate Saw a Chouette a huppe COURTE talks of nine terestrial Turdus falklandii  s
334d009 case opposed to this fact & views.-- Fox says a COUSIN «one of Mr Strutt» of his used to breed to Com
576n049 al. viz. man is not a cause like a deity, as M. COUSIN says. because if so ourang outang.-- oyster  &
426e108 ws, who are astonishingly like to some distant COUSINS, the nearest blood being a great great-grandfa
501q011 Heat (32) Can Henslow ask question of Col. Le. COUTEUR about Wheat.-- Change of Soil-- crossing-- when
506q015 mongst agriculturists. whom he know.-- Col. le COUTEUR on Wheat.-- (41) Have any monooecious or dioec
065r139 have belonged to an Englishman.--On 8th of March COVE began to freeze. correspond to September ¿Did I
073r163 . greenstone[,] amygdaloid. basalt & other trap COVE it to great thickness. = Coast of Acapulco gran
295c183 rasmus says he has seen old Stallion tempted to COVER old mare by being shown, young one.-- Many Afri
308c217 - <right> left, but. I always for a week took of COVER of right side, though my hand <vibrate> would s
423e096 ithout enormous complexity, it is impossible to COVER whole surface of world with life.-- for otherwi
495q05a The «garden» Coronella also sleeps on ditto-- COVER them up periodically & see effect-- such dioeci
496q05a y Kirby-- with animal reproductive system.-- -- COVER flower-- put artificial flowers-- also do  with
515q022 ed in this country. {Chinese Dog's Head to send COVER common Pea (& Sweet Pea) for several generation
515q022 al generations under net & see if get sterile-- COVER that little Ervum in Sand-walk, on which I thin
516q24v what  comes up.-- [Unnumbered blank] Experiment COVER patch of ground, with different salts & poisons
554m140 hen she thinks she is going to be whipped. will COVER herself with straw, or with a blanket.--  these
031r037 tagonia[,] enormous extent; if lowered again & COVERED no sign of upheaval To Cleavage add other inst
054r102 eavage differs wonderfully from mine: phyllade COVERED by quartzose sandstones: refers to broken hill
054r102 at latter place. sandy. sandstone with gypsum, COVERED by limestone with recent shells 200 ft, how ex
054r105 «from the coast» may be concluded to have been COVERED by the sea judge from the pebbles such as thos
054r105 all  appearances not many years since, the land COVERED above half a league of what is now Terra Firma
064r134 hant lives on very wretched cou[n]tries thinly COVERED by vegetation. Rhinoceros quite in deserts.--M
068r146 th fluidity of rock ∴ «in earliest stage» when COVERED up beneath ocean).--The first dislocations & e
075r166 the  silver is contained in a primitive slate, COVERED by a clayey porphyry, containing granites. In P
075r166 loe do) no veins discovered. Humboldt suggests COVERED up by volcanic rocks. //St Helena has been sli
076r170 tine dipping to SW at 45 degree to 50 degree-- COVERED by conformable greenstone porphyrys & phonolit
089a020 serpula-- Isle of Cayenne. dentite & diorite, COVERED with iron clay common to Guyana said to extend
099a048 laterally. -- <In Area of this> {P} If surface COVERED with oil should shrink. film parallel to longe
160g090 m 28.700. A.75 degree 75 degree? Boulder, much COVERED by turf 2ft. 8- long of syenite with pinkish f
217b184 ke both parents.-- Fox told me of case of mare COVERED by blood horse & Carthorse two folds [not loca
233b252 y higher.-- But who with the face of the earth COVERED with the most beautiful savannahs & forests da
467t104 a saw Humble so press down sheath, that stigma COVERED with pollen was pressed & rubbed along whole b
468t105 & rough-- flowers common-- many winged thrips, COVERED with pollen-- «Thrips» about as large as bit o
472s02v again & wipe off pollen. (as a needle becomes COVERED) so whole sides of flower & stigma dusted.-- <
495q05a hich way they fly.-- (9) I have noticed leaves COVERED with Honey-dew dusted with pollen of neighbour
495q05a of neighbouring grass-- Spread sheets of Paper. COVERED with some sticky stuff in flat places & see wh
529mo04 - I a geologist have illdefined notion of land COVERED with ocean, former animals, slow force crackin
146g010 ell-- denuded.-- «of hard metamorph» path only COVERING Great Slip, 10 years since three hundred feet
195b099 le to fossil animals same type, armadillo like COVERING created.-- passage for vertebrae in neck same
551m129 , or as they flag & something like a snow-haze. COVERS my whole imagination." Septembe. 3d Why when o
541m090 following out such an idea, as effect of sea on COVES when waters had fallen, as in my Glen Roy paper
547m112 s going to Shrewsbury, whether I had rung for COVINGTON. whether he had come & opened box, whether I
209b155 find  trace or history of species between Indian COW with hump & Common;-- between Esquimaux & Europe
225b217 L.  Jenyns. talking of it) or how to make Indian COW with bump & pigs foot with cloven hoof) Ask Ento
257c056 cannot guess causes of change.-- hump on back of COW!!-- &c &c D'orbigny (p 108) says haying observed
287c157 but not saltus-- when Linnaeus put whale between COW & hawk a frolicsome saltus. «p. 19»| Macleay see
325c267 l into stump Library of useful knowledge. Horse, COW, Sheep.-- Verey. Philosophie d'Histoire Naturell
327cIBC mmentatia Library of Useful Knowledge on Horse & COW & Sheep Clarke's Travels.-- Temminck Hist.  Nat.
334d012 sidence a great many miles.-- yet one day <th> a COW walked in, then disappeared, & three days afterw
334d012 ame again, bringing with her the other & younger COW.-- {P} Fox says when common & China goose are cr
380d155 Bull is never taken from his own field to bull a COW.-- -- a dog if led in string will not.-- some of
406e036 pigs & sheep.-- diseases common to men & animals COW pox.-- case in Spain of pustular disease followi
503q012 ds-- Dogs &c &c 38 Does only male yak cross with COW: is not reverse possible?? Maer (1) Yew Trees ne
509q017 al Questions (1) Particulars about Sierra Leone. COW. taking bulls. is it Domesticated African Animal
509q017 on vary, according to Bentham's Remark. Horse or COW.-- degree of soldering of tibia & fibula: in Man
512q019 rids, is offspring of short-horn bull & hereford COW similar to reverse cross.-- Sow cast-up-balls of
568n019 knew  no more what was pretty & what ugly than a COW--" «so it is with all uneducated.--»-- Old man a
```

578n054 a beggar, & expecting admiration or an act of COWARDICE, or cheating.-- one does not blush before utt
622o048 either lead to actions or not, as feeling of COWARDICE> «This is not connected with sense» instantan
626o052 to me.-- conscience regulates feelings, as of COWARDICE.-- the whole appears to me rather rigmarole.-
572n034 ut it does hurt his conscience, if he has been COWARDLY, or has injured another bad, vindictive.-- or
591n098 ears.-- -- George the lion is extraordinarily COWARDLY.-- the other one nothing will frighten-- henc
498q008 d treatment (8) Do bees frequent Cabbages «& COWCUMBER'S out of doors.» much-- or the minute Orthopt.
498q008 esume only stigma impregnable.-- (12) At Maer COWCUMBERS in frames are not artificially impregnated.
505q014 plenty of seed & these Seeds of unimpregnated COWCUMBERS will they seed.?-- (11) Abberley has planted
482g012 idae &c &c &c-- The French <Jerrold?> «Bibrons COWORKER of Dumeril» who is writing with Dumeril says
430e117 lleys-- M. Ramond offers no explanation.-- Poet COWPER, describes his tame Hares, attacking a sick on
590n094 n Ourang.-- see his Travels.-- When one sees in COWPER, whole sentences spoken & believed to be audib
293c174 n common between man & animals-- Hydrophobia &c COWPOX, proof of common origin of Man.-- different co
234b255 ce p. 170. «Fish introduced» Hump backed race of COWS from Madagascar-- p 173. Vol I. Voyage à France
332d003 Mark at Shrewsbury thinks the half bred Alderney COWS take more after Alderney that the Durham, with
334d012 ays a settler near Swan river, lost his <on> two COWS entirely, changed his residence a great many mi
379d152 will deteriorate.-- Lord Moreton's Case.-- When COWS have twins, <one> though capable of producing b
403e023 domestic animals few. that have not <which> not, COWS hornless, (horses not) If they give up infertil
406e036 l. Transact. Rabies, common to men, dogs, horses COWS, pigs & sheep.-- diseases common to men & anima
542m097 each other.-- Lonsdale's story of Snails, Fox of COWS, & many of insects-- they likewise must underst
591n097 inct from self. «or will» [not located] <& other COWS--> Mr. Hamilton on vital laws (in the Athaenaeu
249c035 ardoon are cases <sp> like those of Primrose & COWSLIP run wild, The two species of Clenomys. case of
299c194 agos.-- Iceland has same uniformity Primrose & COWSLIP, quite wild, but they affect different localit
401e016 ously different from primrose suddenly produce COWSLIP, one is tempted to think here some anomaly-- I
401e016 mpted to think here some anomaly-- I can fancy COWSLIP producing primrose return to old stock, but no
401e016 eturn to old stock, but not primrose producing COWSLIP Uncle J. says common belief. that female plant
428e113 estly that the sudden, change from Primrose to COWSLIP is great difficulty. «I should doubt if wild s
437e141 g Brussels Sprouts) is analogous to Primrose & COWSLIP suddenly changing into each other, & depends o
491q01v ch experiment of Carrot 2 {also try Primrose & COWSLIP in rich soil & propagate from their seed 3. To
495q005 -- it really is an important case-- cross with COWSLIP pollen.-- as these are wild varieties. Is any
507q15v rowing together. under same conditions.-- like COWSLIP & primrose, but less strongly marked.-- 31. Pl
637j57v to bulls.-- primrose to <open fields> banks-- COWSLIP to <banks> fields-- these are adaptations just
327c265 rossing Oats, &c «Horticultural Transacts.--» Mr COXE "Views of the Cultivation of Fruit trees in. N.
228b229 dant. Seed of Ribston Pippin tree <go> producing CRAB. is the offspring of a male & female animal of
372d129 a, the whole grown to that part.-- claw added to CRAB, tail to lizard,-- healing of wound.-- reproduc
372d130 ut this has nothing to do with generation.-- Why CRAB can produce claw. but man not arm. hard to say-
401e015 regnation from other apple trees.-- now seeds of CRAB produce crab, so that some effect from apple tr
401e015 m other apple trees.-- now seeds of crab produce CRAB, so that some effect from apple trees is produc
427e110 mal kingdom I should suppose, that the pollen of CRAB, would POSSIBLY <No, for pollen of any kind wou
441e148 but not structure of the parents.-- Thus would a CRAB tree vary if planted in rich soil, I presume no
323c269 o. Zoology «references at end of each Chapter» CRABBES Life June 1s. King & FitzRoy To be read Humbol
228b230 Ribston Pippin & Golden Pippin &c produce real CRABS, & in each case similar or mere mongrels.-- It
379d151 h!! ¿adapted to salt water?-- peculiar species, CRABS & molluscs few.-- ¿are not some same-- what is
381d156 , Pectinibranchiate molluscs.-- insects. spider CRABS.-- (all these however do not require coition ev
498q007 ecome unproductive from poorness of soil.-- yet CRABS probably would grow there (7) Where parts of fr
641j29v erent method. in pedunculated eye of Chamelion. CRABS Crabs & Mollusca we have analogues The stillnes
641j29v method. in pedunculated eye of Chamelion. crabs CRABS & Mollusca we have analogues The stillness p. 2
581n065 civilization-- thinks many words, roar, scrape, CRACK, &c, imitative of the things.-- CD[I may put th
120a107 mous faults & facility with which the earth is CRACKING by vertical planes into small pieces-- mem co
529m040 covered with ocean, former animals, slow force CRACKING surface &c truly poetical. (V. Wordsworth abo
036r049 ch will represent the dilatation, which dilated CRACKS must be filled up by dikes & mountain chains.-
071r158 cano, many amygdaloids.--Boussingualt «(Lyell)» CRACKS mountains falling in.-- Earthquakes at. Quito.
102a060 in planes, case of separation.-- the branching CRACKS-- only bear relations to VEINS in primitive ro
121a112 levated without fissure & unequal.-- where were CRACKS?--? How came there ever to be cracks 11th Augu
121a112 where were cracks?--? How came there ever to be CRACKS 11th August. 1838 Near Woolwich there are cla
346d048 ch breed, in & in are-- colour white, uniform.--CRAFTY, go in file, hide their young,. bold.-- a Mr W
222b202 ecies being recent agreeing with Senegal. whilst CRAG <agrees with> according to Beck has none recent
231b243 -- Have change in form.-- This probably explains CRAG & miocene.-- The descendants left in cooling cl
232b245 probability of starting from one point.-- In the CRAG we see the process of change of those forms, wh
448e166 tion (both in species and subgenera) between the CRAG & Touraine beds, the one with neighbouring & Ar
465t089 Mag. May. 1840 p. 362.-- some Mammals of Norfolk CRAG. mentioned-- allied Beaver to present forms.--
120a110 veins of segregation in Greenstone of Salisbury CRAIGS well worthy of attention-- rear Glen Roy Noteb
120a110 rear Glen Roy Notebook-- & scraps on Salsisbury CRAIGS. Kept amongst <old> papers read before societi
123a116 Mem. however. veins of segregation in Salisbury CRAIGS Letter from M Angelis. B. Ayres. 3d. May. slat
145g002 races soon made or crosses difficult Salisbury CRAIGS The Highland shepherds dogs coloured like Mage
145g003 l breeds. [3] Veins of Segregation in Salisbury CRAIGS [4] Salisbury Craigs V. Specimens-- {P} Veins,
145g004 f Segregation in Salisbury Craigs [4] Salisbury CRAIGS V. Specimens-- {P} Veins, amygdaloidal-- as we
145g006 st dip of sandstone in upper part «of Salisbury CRAIGS» 25 degree perhaps most common-- Will not curv
343d036 to the future-- How far grander than idea from CRAMPED imagination that God created.-- (warring against
231b242 ants ¿migratory birds, he told me some story of CRANE from Holland!!! in stomach --or in feathers--se
539m083 edgwood in many respects & some of Aunt Sarahs. CRANKS, & so is Catherine in some respects--. good in
033r043 a <[...]> cylinder placed on the rim of conical CRATER: at Teneriffe Wall of Porph. Lava with base of
038r057 est argument:--¿ Consider causes for subaqueous CRATER. different little craters are all burning, sur
038r058 est argument:--¿ Consider causes for subaqueous CRATER being of diff: form subaerial one?--In former
038r059 lines of anticlinal violence crossing lines of CRATER, <arg> state that all the great Volcanoes. hav
040r063 g a mass probably cold & not warm as sides of a CRATER as Vesuvius.-- There may have been oscillation
056r110 to tell of these losses.-- Cause of chimney. to CRATER. as at Galapagos. St. Helena.-- [Fig. 7] {P} e
095a038 amante near stream of same name. with imperfect CRATER <--> near summit,-- much pumice --. appears to
038r057 & again when in great crater. different little CRATERS are all burning, surely there must be «somewhe
050r090 Bladders of Lava are described, & many minute CRATERS as at Galapagos. <|> Sir George Mackenzie must
055r107 ores. («sandstone first gives» half demolished CRATERS).--worn into mud & dust.--connection with age,
055r107 onnection with age, & agreement with number of CRATERS!--At S. Cruz. there is no occasion to wonder w
055r107 ense the time!! How well agrees with number of CRATERS!-At S. Cruz. there is no occasion to wonder w
059r120 that the central part fell in.--Says posterior CRATERS in centre:-- Bailly talks of much granite on a
065r139 s. like St. Julian & Port Desire applicable to CRATERS of Elevation.--The longer diameter of Deceptio
067r144 em. Apply degradation of landlocked harbors to CRATERS of elevation.-- Lyell suggested to me that no
085a002 livian Chain. Volcanic islands. from number of CRATERS very ancient. which agrees. with peculiar char
096a040 afterwards it might be percieved on which side CRATERS were low --¿ applicable to Auvergne??? The fac
199b119 mber 6'.? of E.d. N. Philos. Journ.-- Paper by CRAWFORD on Mission to Ava. account of HAIRY man. beca
327cIBC 71. account of Europaean plants transported.-- CRAWFORD. Eastern Archipelago. probably some account R
284c148 each» trees dying & mountain torrents.-- but to CRAWL up an hill, then by deaths?!)-- looks like subs
341d030 was adaptation to little Movement.-- nocturnal CRAWLING bird.-- Wings reduced to rudiment.-- clavicle
466t093 s & flies seen directed to it-- The Humbles in CRAWLING out brush over anther & pistil & one I SAW IM
604o016 nimal, with consciousness,, it choosing food-- CRAWLING from light.-- Yet we can split Planaria into
026r023 es. Asteriae, usually petrified into a peculiar CREAM-coloured Limestone: the strange substitution of
059r121 or small Proportion of Alum: matter.--all pale CREAM colour.-- The Brecciated structure of all the P
367d113 d with brown.-- animal like large, heavily made CREAM coloured ass.-- stripe on back also.-- legs rem
305c210 subordinate laws.-- There is one thinking « <& CREAT> sensible» principle (intimately allied to one
195b100 hat «these» species could, arrive-- did it only CREATE those kinds not so likely to wander. Did it cr
195b100 ate those kinds not so likely to wander. Did it CREATE two species closely allied to Mus. coronata, b
302c200 ry, = otherwise mere fact creator chooses so to CREATE.-- It is very remarkable, so much death,
356d072 ulty on any theory-- without God is supposed to CREATE & destroy without rule-- But what does he in t
527m033 m & pleasant sound per se) & causes the mind to CREATE short vivid flashes of images & thoughts.-- Po
634j55r s-- as resulting from the will of the deity, to CREATE animals on certain plans.-- is no explanation-
061r127 Volc: islands. elevated. then peculiar plants CREATED. if for such mere points; then any mountain, o
062r129 t to be a desert.-- Tempted to believe animals CREATED for a definite time:--not extinguished by chan
178b030 place. Will it be said, those have been there CREATED there;--» Are not all our «British» Shrews dif
183b051 <they> it might continue & thus two species be CREATED) & live in same country. How is propagation of
195b099 sil animals same type, armadillo like covering CREATED.-- passage for vertebrae in neck same cause, s

Page **(Key Word)**
195b101 stiny.-- In same manner God orders each animal CREATED with certain form in certain country, but how
195b101 ws such are inevitable consequen let animal be CREATED, then by the fixed laws of generation, such wi
196b108 could exist without plants & insects had been CREATED; but on other hand creation of small animals m
197b111 imal: one set of organ:-- the others «animals» CREATED with endless differences:-- does not say propa
198b114 all animals (though some may be) have not been CREATED on the same plan. [Second resumé well worth st
199b116 from Lyell: assuming truth of quadrupeds being CREATED on small spots of land, of the same type with
248c029 g. Geograph. range of quadrupeds.: that either CREATED in each point, or migrated from those quarter,
256c053 her a new principle is brought to bear. If man CREATED as now. languages. would surely have been more
283c145 some wide. which is strange if creator had so CREATED them.-- People will argue & fortify their mind
286c156 ng treats subject to put in alternative of Man CREATED by distinct miracle. Macleay letter to Dr. Fle
300c197 , more humble & I believe true to consider him CREATED from animals.-- Insects shamming death, most d
303c205 here. are gaps. yet <what> altogether «he» has CREATED a perfect chain. <Icthyo> .+ + +. supra & next
313c234 » Read Entomological Transactions Why if louse CREATED should not new genus have been made, & only sp
315c240 port & creation exists..-- pooh. May have been CREATED at many spots & since disseminated See. Habits
336d019 o far is there any appearance of animals being CREATED. it is probable if created at once. <wd> accor
336d019 ce of animals being created. it is probable if CREATED at once. <wd> according to ordinary laws, the
343d037 er than idea from cramped imagination that God CREATED. (warring against those very laws he establish
370d115 ways occur if animals are thought to have been CREATED.-- it might as well be attempted to be shown f
398e006 lutions, but entire obliterations & fresh laws CREATED., & yet with <gov> symmetry «& regular laws» t
463t055 n> spinal marrow expands, so do the bones <are CREATED> expand-- instead of saying as brain is create
463t055 reated> expand-- instead of saying as brain is CREATED &c &c Bats are a great difficulty not only are
573n036 laws,, but the smallest insect, we wish to be CREATED at once by special act, provided with its inst
632j53r > any one seed. (all have not it) was DIRECTLY CREATED. for transportation. it follows from some more
632j53r ral law.-- [that the laws of propagation, were CREATED with reference to successive developement I ad
633j54r 92. Mac. has long rigmarole about plants being CREATED to arrest mud &c at deltas.-- Now my theory ma
633j54r ws they can live.-- Hence the mistake they are CREATED for them. If we once venture to say plants cre
633j54r ted for them. If we once venture to say plants CREATED to <arrest> «prevent» the valuable soil in its
633j54r fted; why should plants require earth, why not CREATED to live on alpine pinnacle? if we once to pres
633j54r pine pinnacle? if we once to presume that God «CREATED plants to» arrests earth, (like a Dutchman pla
633j54r ns of one general law. the plants were no more CREATED to arrest the earth, than the earth revolves t
195b098 be made very strong. if we believe the Creator CREATES by any laws. which, I think is shown by the ve
303c205 the list is now perfect.--]CD. the creator so CREATES animals, «, it will be said» that although at
224b216 Creator since the Cambrian formations gone on CREATING animals with same general structure.-- misera
061r127 mountain, one is falsely less surprised at new CREATION for large.--Australia's = if for volc. isld.
061r127 lc. isld. then for any spot of land. = Yet new CREATION affected by Halo of neighbouring continent: ≠
061r128 by Halo of neighbouring continent: ≠ as if any CREATION «taking place» over certain area must have pe
063r133 d. no doubt with perfect success.--showing non CREATION does not bear upon solely adaptation of anima
176b022 xistence, as we may suppose is the case. their CREATION being dependent on definite laws, then those
192b084 rom beetles with wings.& modified.-- if simple CREATION, surely would have been born without them.= I
196b108 & insects had been created; but on other hand CREATION of small animals must have gone on since from
198b115 not being found on all isld, (if act of fresh CREATION why not produced on New Zealand; if generated
225b219 to Pig & Dogs. My theory will make me deny the CREATION of any new quadruped since days of Didelphis
227b227 h never become a man.-- Does not require fresh CREATION!-- If continent had sprung up round Galapagos
231b243 t if so.-- Now this is difficult to explain by CREATION-- or we must suppose a multitude of small cre
258c062 t) must follow after there is proof of the non CREATION of animals.-- then argumen May be.-- subterra
283c146 forms with recent, we have nothing to do with CREATION.-- <On> The end of formation of species & gen
289c162 n as a hopeless difficulty, except as distinct CREATION.-- Generation may be viewed as condensor, «++
303c204 ils & over white of eyes,-- + + +, Will he say CREATION is at end, seeing that Tertiary geology has o
303c205 r fossils, now if extinction had gone, without CREATION this would have been fair, but to place all t
305c209 ean species = singular coincidence if distinct CREATION.-- ie.-- a mere statement nothing is explaine
315c240 ogamic plants same in Australia & Europe.-- if CREATION be absolute thing, the creation must take pla
315c240 & Europe.-- if creation be absolute thing, the CREATION must take place on[ly] when creator sees. the
315c240 rwise no relation between means of Transport & CREATION exists..-- pooh. May have been Created at man
337d021 o plants emit odour solely for others parts of CREATION) & another nerve to finest vibration of sound
337d022 who imitates.-- who will say there is distinct CREATION required if he believes «hyaena & squirrel» s
338d025 ertain forms from Northern part & not by fresh CREATION of new forms.-- what is range of Hyaena? Hipp
350d056 ain, therfore organ fitted to animals place in CREATION.-- thus senses, especially sight connected wi
373d132 :-- ¿is not this argument, for Mammalia recent CREATION.-- why. what tendency can there be for aborti
431e122 y be anticipated, & this would look like fresh CREATION. the gardener separates a plant he wishes to
559m154 as we consider each object an act of separate CREATION. we admire it more. because we can compare it
640j167 the Galapagos Islds are explained. On distinct CREATION, how anomalous, that the smallest newest, & m
195b104 &. others replace them-- two hypotheses fresh CREATIONS is mere assumption, it explains nothing furth
231b243 ion-- or we must suppose a multitude of small CREATIONS.-- Will Dromedaries & Camels breed?-- As man
259c064 ted» tempts one to bring one back to distinct CREATIONS.-- it is only be recollecting that the ground
303c205 air, [moreover what will become of the future CREATIONS, if the list is now perfect.--]CD. the creato
377d140 -- It is the unit of our calendar.-- epochs & CREATIONS, reduce themselves to the revolutions of our
510q017 ne Flora of Java? Has Schow written on double CREATIONS & where? How are current & winds in Antarctic
633j54r r the creator to the standard of one his weak CREATIONS.-- All such facts are merely relations of one
195b098 tion constant in the shells.-- The question if CREATIVE power acted at Galapagos it so acted that bi[
195b100 ch sandpipers. not new at Galapagos.-- did the CREATIVE force know that «these» species could, arrive
210b160 y will be this relation also.‖Yes « Fox» ‖ The CREATIVE power seems to be checked when islands are ne
271c106 ecies have changed Argumentum ad absurdum. The CREATIVE American halo has extended to Juan Fernandez
181b045 he case at least during subsequent ages.-- The CREATOR has made tribes of animals adapted preeminentl
195b098 s might be made very strong. if we believe the CREATOR creates by any laws. which, I think is shown b
219b193 erns ditto.--» & hence formed trees]CD & would CREATOR <on volcanic island.> make plants <grow closel
219b196 othing.-- it being as easy to produce «for the CREATOR» two quadrupeds at S. America Jaguar & Tiger &
224b216 other species or angels. produced «&» Has the CREATOR since the Cambrian formations gone on creating
283c145 es. some fine & some wide. which is strange if CREATOR had so created them.-- People will argue & for
296c184 n forms. (anyhow not Australian) on Peaks. Did CREATOR make all new yet forms like neighbouring Conti
302c200 explained on my theory, = otherwise mere fact CREATOR chooses so to create.-- It is very remarkable,
303c205 eations, if the list is now perfect.--]CD. the CREATOR so creates animals, «, it will be said» that a
310c222 t my theory,-- If I be asked by what power the CREATOR has added thought to <an> so many animals of d
315c240 hing, the creation must take place on[ly] when CREATOR sees. the means of transport fail.-- otherwise
336d019 ighest point of organization] CD» false.-- The CREATOR would thus contradict his own law. So far is t
354d065 the range-- Argue the case of Probability. has CREATOR made rat for Ascension.-- The Galapagos mouse
553m136 t philosopher who says the innate knowledge of CREATOR <is> «has been» implanted in us (<by> ¿ indivi
559m154 ever find out." This unwillingness to consider CREATOR as governing by laws is probably that as long
633j54r hem to stop the moving sand) we <do> lower the CREATOR to the standard of one his weak creations.-- A
634j54v of a God-head.-- the designs of an omnipotent CREATOR, exhausted & abandoned. Such is Man's philosop
634j54v is Man's philosophy. when he argues about his CREATOR! p. 309. says the ribs in Draco support the fl
582n068 h will bring about a certain result, while the CREATURE performing those actions neither knows nor in
583n070 or imitation.-- p. 20. Animals may be called "CREATURES of instinct" with some slight dash of reason
583n070 ome slight dash of reason so mean are called "CREATURES of reason", more appropriately they would be
583n070 of reason", more appropriately they would be "CREATURES of habit."-- CD[as the bee makes its cells, b
619o042 t acts of benevolence towards fellow <living> CREATURES, or of kindness to wife [RHC] 2) and children
549m118 of <happi> pleasure of such thoughts We give no CREDIT to instinctive feelings.-- for man losing his
608o027 ld teach one profound humility, one deserves no CREDIT for anything. (yet one takes it for beauty & g
451e176 who was also an albino is held sacred by the CREDULOUS natives, & vow made at it. Both his parents w
115a096 eat.-- Where there are cliffs there ought to be CREEKS & mouths of rivers ought to be deep.-- Henslow
063r130 ant, each its own limit & represented.--Chiloe CREEPER: Furnarius. <Caracara> Calandria: inosculation
484z016 gonian Furnarius.-- into Oxyurus, by Maldonado CREEPER of same plumage.-- general red mark on wings o
127a125 the <inf> subterranean isothermal line must be CREEPING «pushing» up to «the» line of ice.-- Hence fu
629o55v sly so, for Marquesans think only of prepuce, CREPITANDO,]CD & if passions makes one break these arti
594n113 st March was frightened on the 24th of May at CRESSELLY by the boys making faces at it, so much so th
484z014 as of Americas.-- manner of walking-- foot bill CREST feathering on legs-- habits-- Does the Secretar
363d101 legs.-- «Carruncles on beak & in Muscovy duck» CRESTED feather, pouters, fan tails are found in any c
414e060 «!Agassiz makes it wonderfully changed, since CRETACEOUS period, whether progressive I know not.» (&
473s004 ubsidence as elevation then all continents of CRETACEOUS periods, together with their littoral deposi

308c216 never at any one time formed chain, since if CRETCEOUS period assumed, then some perished before, Ca
289c160 one district, though common on another, (golden CRETED wren so rare in some countries-- nightingale d
098a045 (∴ separation of ingredients) as uniting with CRETIONARY.-- it may <of> come of use in discussion on
064r136 ferous-- Urge enormous quantity of matter from CREVICE of Andes--therefore flowed towards it. a mass
128a128 - If mountain chains are matter piled up. over CREVICE from effect of general elevation,-- when subsi
039r060 e been conduits to volcanoes.-- Talking of the CRICKET valley «the most remarkable feature in the str
639j28r rld!]CD p. 252 analogy of hand in mole, & Mole CRICKET & rodents (?) p. 251. all animals run by hind
402e018 tern seas. p. 206-- shot a monkey, ceased their CRIES. "many of them descending to examine their defu
543m097 onsiderable, thus carthorse & dog.-- birds many CRIES. monkeys communicate much to each other.-- Wate
551m129 omnibuss Horses starting, when door shut or cad CRIES out "right." or Drinkwater's horse jumping when
579n059 s, from fear, sublimity, sexual ardour.-- a man CRIES from grief, joy. & sublimity. January 6th.-- Wh
570n025 ween Squib after having eaten meat on table, & CRIMINAL,-- who has stolen. neither, or both may be ea
524m022 dream of what they think of most. intently.-- CRIMINALS before execution.-- Widows not of their husba
608o027 sgusted. with them. Yet it is right to punish CRIMINALS; but solely to deter others.-- It is not more
220b188 s to Men. & dogs. Now if we take structure as CRITERION of species Hogs different species, dogs not.
640j167 r own Faunas, which in this case is only true CRITERION.-- Hence it is highly unphilosophical to asse
285c150 ion of tracks &c &c Dr. S. has some remarkable CROCHETS about instincts. whenever instinct is mention
229b234 ange of Bos in India.-- Range of Zebra?-- The CROCODILE & Tortise former inhabitants of Mauritius Fre
242c018 dog, with those of New S. Wales. <V.> p. 123 CROCODILE at New Guinea. All the isles of Oceania have
247c028 udy Silliman.--» Vol II. p. 10. it seems that CROCODILE was washed on shore at one of the Pelew Islds
403e021 m that of S. Europe p. 189. The gaut, kind of CROCODILE, sometimes wanders from Pellew to Eap.-- Ther
584n071 ion ducks & turtles running to water,-- young CROCODILE snapping-- p. 28. how curious the means of gu
595n125 a bull <tr> calf, just born butting, or young CROCODILE snapping.-- these I think are better instance
233b250 ils of Sewalick «India» Monkeys of old World. CROCODILES. Anoplotherium.-- <M. Jerrod> & Dumeril grea
250c039 in Paris quarries & at Binstead. Mem. recent CROCODILES with Palaeotherium in India--: connection wi
497q06v opping in woods. by birds 13. Mr. Herbert says CROCUSES are very difficult to cross--: are there race
402e019 joints near the tip of the tail are generally CROOKED, as if they had been broken". are born so in a
453e182 he» cats are born with the joints near the tip CROOKED.-- is the form [...] Dampier. Vol I. p. 320. s
592n103 Paper. like. Sir Ch. Bell on Expression «First CROONIAN Lectures by Parsons.» following pages contain
375d135 makes population in Men increase, & an ordinary CROP. causes a dearth then in Spring, like food used
251c040 of species, geese killed in Newfoundland, with CROPS full of maize. (get limits of latter from <Tart
148g026 me answer Thought lambs most like MOTHER!-- the CROSS not so hardy but more easily fatten, This man c
157g074 e, only sand blow away]CD where lines appear to CROSS stony parts; appearance chiefly cause by fall o
178b032 tain that when White Men & Hottentots or Negros CROSS at C. of Good. Hope the children cannot be made
179b033 wild state.-- No doubt «C. D.» wild men do not CROSS readily, distinctness of tribes in T. del Fuego
203b136 ia p. 261. L. Institut. 1837 Mem. Sir F. Darwin CROSS breed boars were wilder than parents. which is
204b140 e between parents.-- How easily does Wolf & Dog CROSS? Mr Yarrel thinks oldest variety impresses the
210b158 s.-- Compares it to languages But how do plants CROSS?-- = admirable discussion Von Buch says from Hu
213b170 se ¿Lyell?, which have wide range and therefore CROSS & keep similar. But this is difficulty; This im
213b171 rell's remark about old varieties affecting the CROSS most well worthy of observation.-- I think it i
216b179 Negro & white will return to native stock (the CROSS often whiter<>> than white parent) the mulattos
216b181 s. rabbits cats &c &c.-- When black & white men CROSS some offspring black others white which is more
216b181 s white which is more closely allied to case of CROSS of dogs.-- See paper in Philosoph. Transaction
216b181 Ld. Moreton, where mare was influenced in this CROSS to after births, like aphides.-- Case of boy wi
218b189 on island. & fresh species mate. parents do not CROSS-- we see it even in men); the possibility of Ca
222b203 forward introduction of species.-- When species CROSS & «hybrid» breed, their offspring show tendency
223b209 he species have not been much altered they will CROSS (perhaps more fertility & so make that sudden s
228b229 ck ¿whether this going back may not be owing to CROSS from other trees.???? Do the seeds of Ribston P
230b239 hen monuments imperfect) or horizontally & then CROSS breeding prevents perfect change.-- It is scarc
235b274 e superinduced-- why is every one so anxious to CROSS. animals from different quarters to prevent the
239cIFC C N[a]me of two pigeons,-- «with specks» which CROSS & keep colour on wing Effects of colour on pare
239c002 On this principle I may add, that fact of half CROSS with parents, going back to either parent is lu
249c034 which are not deeply impressed on blood., will CROSS & produce fertile offspring in the first case i
255c052 The hybridity of ferns on my doctrine of CROSS-generation. The infertility of crosse & cross,
255c052 f cross-generation. The infertility of crosse & CROSS, is method of nature to prevent the picking of
267c094 er-- Indian secured one, as they always like to CROSS their breed p. 333-- alludes to the Macusie bre
288c159 original Durham breed.-- Native dogs & English CROSS readily-- think about half way in appearance.--
288c159 which had originally crossed would continue to CROSS, means of knowing directions: mysterious. When
298c192 Henslow thinks if leaf of plant varies, <whole CROSS> «all organs» vary in plant. The variation in c
312c231 mpson tells me best way to improve cattle is to CROSS between a good bull & <be> the provincial breed
312c231 as if qualities were not permanent, in the new CROSS.-- In the Bantam clubs, they used to fix on the
312c233 ot become impregnated & puppies delicate-- they CROSS sister & brother of same litter, those. of diff
331dIFC dency to return to either parent.? Is the first CROSS, which makes hybrids. productive like geese?--
334d009 «China» & Canada Geese, & that they this first CROSS were equally fertile with pure bred animals.--
334d010 ts? This indicates a remarkable law, that first CROSS <not se> plentiful, second absolutely sterile.-
335d014 ey «abortive» twins.-- ¶ The fertility of first CROSS, as stated by Fox, is very important, as showin
335d014 very important, as showing above facts as first CROSS being new species, ▌ -- Are not dreadful monste
336d018 t size of particular Muscles-- When two animals CROSS. each sends his own likeness, & the union makes
346d047 lambs were more like father than Mother.-- The CROSS not so hardy as Black faced, but more tendency
354d063 llar to Butterfly.-- When two Varieties of dogs CROSS, Erasmus says it took a hal[e] Institut. 1837. p.
359d087 there must be some mistake in their origin) Saw CROSS between Penguin Duck «from Bombay» & Canada Goo
360d093 & most terrestrial species, might be harder to CROSS than two less opposed in habits, though externa
360d094 tail.-- I can readily see that two first might CROSS easier than two last. Sept. 11. N Mr. Blyth, at
361d095 awfinch-- in nursling plumage resembled that of CROSS-Beak-- In lark if I understand right, all speci
365d105 s given figure of it.-- In England no doubt how CROSS between Pheasant & Black game is owing to their
365d105 this predicament, probably, alone would species CROSS in wild state.-- Is English red Grouse. a cross
365d105 cross in wild state.-- Is English red Grouse. a CROSS between Black Game. &, the subalpina of Sweden,
365d106 probably, for it can hardly be thought that the CROSS would have adapted it to changing circumstances
366d112 ep out flies might be used» The wild ass has no CROSS. how comes it that the tame donkey has. CD[old
374d134 ring infertility of hybrids inter se, the first CROSS generally brothers & sisters, & therefore somew
376d136 D.'Orbigny. Comtes Rendus p. 569. 1838 says the CROSS between the Guaramis & Spaniards are almost Whi
393dIBC Gardens Are the hybrids of those species. which CROSS & are fertile heterogenous? When bird fanciers
405e035 tut. 1838. p. 338.«V[ide]» Important account of CROSS of sheep & Moufflon of Corsica. <would not>, sa
414e063 ment. even without time can do much.-- (yet one CROSS, & the permanence of his breed is destroyed)--
426e106 think that a <do> variety of one species would CROSS easier with 2d species, than two perfect specie
430e118 ually the offspring are picked & not allowed to CROSS.-- » Has nature any process analogous-- -- if s
430e118 how.-- -- «--.Make the difficulty apparent by CROSS-questioning.-- » even if placed on Isld-- if &c
441e150 ity, that every <animal> «organic being» should CROSS with another.-- to escape it «in any case» we m
447e164 orresponding acalepha?-- But if Acalepha do not CROSS there would by my theory gradation of form from
454e184 described instrument for galvanize <ani> them-- CROSS Irish & Common Hare Decandoelle has chapter on
459f02 anslated from Meckel. Comp. Anat.-- From Buffon CROSS of he-goat & sheep, it seems male gives form. a
460t013 any eggs as pure bred common.-- the half of the CROSS, as above, take «generally» after the swan-gand
466t094 a bee entering long nectary, would «necessary» CROSS directly over the bunch of anthers & pistils, b
493q004 .-- Experiment in crossing animals.-- &c (1) To CROSS some artificial male with <old> female of old b
493q004 ht: either in first breed or permanently.-- (2) CROSS two half-bred animals. which are exactly alike
493q004 ls. which are exactly alike & see result.-- (3) CROSS the Esquimaux dog. with the hairless Brazilian
494q004 r.-- if not how is it in <allied> varieties (9) CROSS largest Malay with Bantam-- will egg kill Hen B
494q004 Malay with Bantam-- will egg kill Hen Bantam.-- CROSS common Fowl with Dorking (10) Statistics of bre
494q004 Badger &c &c &c.-- (11) Keep. Tumbling pigeons. CROSS them with other breed.-- (12) About the blended
495q005 imrose seeds-- it really is an important case-- CROSS with cowslip pollen.-- as these are wild variet
497q06v Mr. Herbert says Crocuses are very difficult to CROSS.-- are there races-- if so plant them together.
497q007 ma?-- (4) As Papil. flowers appear difficult to CROSS, are there unusually many species in genera of
497q007 great & natural Families, as being difficult to CROSS. (5) It is most important to ascertain amount o
498q008 thopt.-- important, as we know how readily they CROSS.-- (9) In the nurseries, when «seed of» the var
500q010 all varieties of Cabbage. (26) Do deer Keepers CROSS the breed-- desirable as in Cattle in Chillingh
501q011 tize on by sowing near each other & see whether CROSS can be obtained-- I name these three plants. be
503q012 ther Hybrids-- Dogs &c &c 38 Does only male yak CROSS with cow: is not reverse possible?? Maer (1) Ye
504q13v lso PEAS-- N.B. I think very likely the Peas to CROSS ought to be placed far from all other Peas, fro

507q15v & any with peculiarities of structure rendering CROSS impregnation difficult or reverse (.49) List of
507q016 e Dr. Holland ; My Father. Andrew Smith (1) Are CROSS-births, or other accidents of delivery-- inheri
508q016 ion to healthiness? & father answered (5) About CROSS-bred races of men taking after sex. A Smith. Ab
512q019 ort-horn bull & hereford cow similar to reverse CROSS.-- Sow cast-up-balls of Hawks or even owls.-- H
513q020 osities on Camels-horses. &c &c Rhinoceros= (6) CROSS. Sus Barlyroussa with tame.-- (7) About fertili
513q020 ntams from different countries= Do the Peacock's CROSS.= Young Chinese or Penguin Duck in very young s
513q020 so probably a variety, not specific character= CROSS Rumpless fowls & Dorking fowls,-- or tailless d
513q020 ide, as in crossing varieties Amongst varieties CROSS one with abortive tail or horn, with another &
513q21. of Australian Mountains.-- Is setting of fruit. CROSS Conception--《Ʌ I could extract nothing from him
555m143 t disposition from others, slow cautious, angry CROSS look, followed by protrusion of lips, in which
564n006 ngs «in Lavater, P. cii Vol III» of excessively CROSS-half furious faces «which may be described as a
575n044 .-- Habits import to Bowen No one doubts that a CROSS of bull dogs-- increase the courage & staunchne
255c052 octrine of cross-generation. The infertility of CROSS & cross, is method of nature to prevent the pi
148g025 by sheep & not deer When Black faced sheep are CROSSED with English my informant said the lambs were
161g095 e further on, where three [...] abutted Having CROSSED the mouth, (deep) of above valley this road le
183b052 rations" go back to type of either animal when CROSSED with it.-- ¿Whether extinction of great S. Ame
190b075 communication Have races of Plants. ever been CROSSED really, if there is any difficulty in such mar
204b141 ndependently of individual force» Mr Wynne has CROSSED Ducks & Widgeon & offspring either amongst the
210b158 es. become species. p.147 «p. 150»-- not being CROSSED with others.-- Compares it to languages But ho
217b183 Fox says when two dogs of opposite breeds are CROSSED, sometimes offspring quite intermediate someti
223b211 B. C. D. -- (A) crossing with (B) (& B having CROSSED with (C) prevents offspring of A. becoming a g
229b233 rmation Long Horned (very) aboriginal at Cape: CROSSED with English Bull. offspring very like common
239c001 any country; he states that Esquimaux dog when CROSSED with pointer produces offspring much nearer Es
239c002 that Chesnut, for many generations back, were CROSSED with Bay mare, only bay a few generations, tha
239c002 d.-- Mr Yarrell states that if any odd pidgeon CROSSED with common pidgeon, offspring most like latte
258c060 le Man living in hot countries, if continually CROSSED with people from cold, children would not beco
275c120 he took thorough bred <greyhound>. bull-dog. & CROSSED & recrossed, till there was a dash of blood, w
275c121 last got ducky, then took white Chinese Bantam CROSSED & got some yellow & others yellowish & white v
288c159 wider at Rio Plata. birds which had originally CROSSED would continue to cross, means of knowing dire
290c163 habit Talent &c in man not heredetary, because CROSSED with women with pretty faces When horse goes a
290c164 ase have bred). White & common pheasants. have CROSSED.-- </ I saw> .-- The attachment of dogs to man.
312c228 ith knew chinese hairless dog & common spaniel CROSSED.-- 3 puppies PERFECTLY like chinese & 3 perfec
332d003 ey that the Durham,, with which they have been CROSSED--is Alderney oldest breed-- He believes all pr
333d005 t; tone of voice. perhaps rather different».-- CROSSED with <un>common cat, exact variety unknown., t
333d006 same.-- Fox thinks that when a wild animal is CROSSED with tame, offspring always take most after wi
333d007 .-- Fox has seen several cases of foxes & dogs CROSSED, offspring always more resembled foxes than do
334d013 .-- {P} Fox says when common & China goose are CROSSED the neck is not intermediate in its peculiar l
341d031 rmanent white streaks.-- &c &c Dr. Bachman has CROSSED cock Guinea Fowl with Pea <cock> Hen.-- offspr
342d034 bird fanciers say the throw of any two species CROSSED is uncertain-- Yarrell remarks he has somewher
356d072 ad just heard of Black game & Ptarmigan having CROSSED in wild state-- & the English & Some African d
357d073 us the readiness with which this genus becomes CROSSED ¿is red game an hybrid?-- When I show that is
358d085 rtile.-- thus the common. pheasant & fowl when CROSSED never even lay eggs. & the men cannot «hardly»
358d085 by appearance.ᴸ- The silver & common pheasant CROSSED, has a cock (infertile) <with> the breast of w
359d088 even more than common duck-- Male Penguin was CROSSED with hen Canadian offspring, I should say in e
360d090 here their interest is concerned. Same man had CROSSED Jackal & dog-- (offspring did not go to. heat.
364d105 en Capercailzie & Black Cock.-- The latter has CROSSED with the Ptarmigan. subalpina in wild state.--
378d148 wild duck's eggs under common ducks, the young CROSSED? & How is it with China & Common Geese «how ar
392d180 t after father or Mother according as they are CROSSED.-- Now these laws are, that child may be eithe
417e078 icable to likenesses, when species & races are CROSSED.-- Make hybrid plants. If many wild animals were
426e107 wild hybrid plants. If many wild animals were CROSSED, there would probably be perfect series, from
455eIBC ents. about raising plants. where they cannot «CROSSED» etc.-- Make Hybrid mosses.-- Leighton or some
458t002 informs me that a hybrid between ass & Zebra, CROSSED with pony mare & produced a very pretty little
460t013 e swan-gander. one of these half-bred ganders. CROSSED with common goose <to has> «produce offspring
495q05a if they will come up true-- whilst others are CROSSED.-- Are Bees guided by smell-- or sight.-- --.
499q09v - Also any plants which are known easily to be CROSSED & all monoeecious plants.-- Hooker says Raffle
501q011 ame these three plants. because they cannot be CROSSED, I think, I expect, except by very minute inse
513q020 ith natural species, as dioecious plants, when CROSSED R. BROWN-- will pollen act on any flower befor
515q022 About blended instincts of the geese which he CROSSED; especially if the hybrids were recrossed with
543m101 acts of half instincts. when two varieties are CROSSED as in Shepherd dogs-- Inherited Habits: Have E
145g002 n Britain shows that either races soon made or CROSSES difficult Salisbury Craigs The Highland shephe
279c133 en of important organ (.see marks on pages).-- CROSSES of diff: breeds succeed, yet seems to grant, t
312c228 of little value Dr Smith if black & white Man CROSSES; children heterogenous, he feels sure of this,
333d007 proaches to species. Again he has seen several CROSSES between Esquimaux dog & common dogs & Fox thin
360d089 the three little ones.-- Keeper said in <two> CROSSES <twice made» between terrier & hairless dogs o
468t105 rse hair with legs & take flight-- Yet we have CROSSES-- I see Bees almost every flower-- Blue-bells-
493q003 es place in mules ass & horse-- important. {In CROSSES does male offspring take after male parent & v
584n073 ot an American form The union of two instincts CROSSES mind, before it can have affected respiration
038r059 ument in favor of lines of anticlinal violence CROSSING lines of crater, <arg> state that all the gre
165g126 ot an American form The union of two instincts CROSSING most remarkable ever observed? Shows that <ner
179b033 is of every species: believes in repugnance in CROSSING of species in wild state.-- No doubt «C. D.»
200b123 tle: The grand fact is to establish whether in CROSSING very opposite races, whether you would expect
216b181 r in Philosoph. Transaction on a quagga & mare CROSSING by Ld. Moreton, where mare was influenced in
223b210 mountain or approach of desert?-- because the CROSSING of species less altered prevents the complete
223b211 daptation which would ensue A. B. C. D. -- (A) CROSSING with (B) (& B having crossed with (C) prevent
227b228 ion of hybridity «to what circumstances favour CROSSING & what prevents it--» & generation, causes of
230b239 s need not apply to very slow changes. without CROSSING.-- Now a gradual change can only be traced ge
258c060 wide range, by preventing adaptation owing to CROSSING, with unseasoned people. would cause destruct
265c083 daptations «they are counteracted by nature by CROSSING with other varieties»-- but <accidental> chan
275c120 y hound.-- picking out finest of each litter & CROSSING them with finest greyhounds.-- Sir. J. Sebrig
275c120 t first got {P} point on hackles on Bantams by CROSSING with common Polish cock «is not that old vari
275c121 & white varieties by picking the yellow one & CROSSING with duck bantams procured old variety.-- The
279c133 being long in blood.-- ++ thinks difficulty in CROSSING race.-- bad effects of incestuous intercourse
298c191 gination.-- then old peculiarity overbears the CROSSING) with alpine form) lower species afterwards w
326c265 apers Wiegman has published German Pamphlet on CROSSING Oats, &c «Horticultural Transacts.--» Mr Coxe
342d035 trouble in obliterating the fresh feather, by CROSSING with females not thus characterized.-- 16th A
364d103 » «Chinese & Common Pigs.--» Experimentize on CROSSING-- It seems from Lib. of Useful. Knowledge tha
392d180 all sorts of varieties, which he attributes to CROSSING of the several species of wild fowl <in Z> of
401e014 n of sexes rigidly necessary.-- Without sexual CROSSING.-- Cape Broccolli can hardly be reared withou
410e050 ll the above should follow after discussion of CROSSING, there would be endless changes, & hence no f
410e051 y as in plants) therefore, great difficulty in CROSSING of <species> individuals. with respect to rep
416e070 yralidae» Plants do not become acclimatised by CROSSING [& this most important obstacle to my theory]
427e111 ave not been thus produced, but by training, & CROSSING, or by accidental production of seedling with
430e118 t p. 8. his whole doctrine of the advantage of CROSSING & keeping breed pure.-- «& so in plants effec
431e123 ic animalcule in Mosses, render my view of the CROSSING of mosses & all others by action of wind diff
431e123 e atter father than mother; illustrated by the CROSSING consists in the idea of the male being smalle
432e123 t does not seem to recognize any difference in CROSSING of hornless sheep with horned-- compare this
438e141 ype Mr Herbert showing the extreme facility of CROSSING, between varieties & species, yet the amount o
439e143 plants shows there is tendency to prevent the CROSSING, in plants proves how much depends on instinc
439e143 is this so) shows either there is not so much CROSSING.-- in animals where there is much facility in
439e144 - in plants where there is much facility in CROSSING there comes the impediment of instinct-- the
446e163 re we have a plant remaining constant, without CROSSING.-- & propagation by buds does not insure cons
447e164 , as one, in which structure does not allow of CROSSING with other individuals, «with facility»-- suc
447e164 ht to be very persistent, & then necessity of CROSSING is much less,-- now certainly in the higher a
447e164 ies to other: therefore my theory does require CROSSING.-- The case of Lemna shows dispersion of germ
465t081 is with domestic breeds. (though in this case CROSSING has had somewhat to do with it. mem. dogs «&
483z012 Bibron Zoolog. Journ Vol I. p. 125, owls seen CROSSING Atlantic. fact taken from Jenner (1825) Phils
490q01v for my experiment on Selection. Experiments in CROSSING &c Plants 1 Repeat the French experiment of C

492q003 e uniform.-- Mr Ford Has M. Sageret WRITTEN on CROSSING of Cabbages, quoted by (as if oral) Decandoel
493q004 od, avoids effects of fatness.-- Experiment in CROSSING animals.-- &c (1) To cross some artificial ma
494q004 ny generations in Mammalia. in group effect of CROSSING.-- (7) Are the Eggs of the Penguin Duck quite
497q06v f so, as these can be raised true, there is no CROSSING by Bees.-- Henslow.-- (1) Character of alpine
501q011 ol. Le. Couteur about Wheat-- Change of Soil-- CROSSING-- when seeds raised.-- His Book.-- 32. Would
505q014 te-flowered var. with abortive stamens.-- show CROSSING & ¿heredetary? (15) Abberley has a hooked Pea
511q018 vided, & more particulars regarding effects of CROSSING them with common pigs= [it is a Lincolnshire
513q020 rmediate or «sometimes» all on one side, as in CROSSING varieties Amongst varieties cross one with ab
514q021 hybrids healthy: number of generations: about CROSSING of plants; especially Papilionaceous order (2
515q021 rines (does reverse happen?) what is effect of CROSSING peaches & nectarines: same question with rega
575n044 - can we suppose some essence. The facts about CROSSING races of dogs on their instincts, most import
575n044 rtant, because they obey the same laws, as the CROSSING of jackall & Fox & wolf & dog.-- the only tes
585n075 successively.--» [& we know from experiment of CROSSING fingers, that we only do know that it is one,
586n079 not have been taught, where to go-- the act of CROSSING the sea in dark night & not loosing its direc
636j56v uld be subject to extreme variation as long as CROSSING with other varieties was prevented Do races o
294c177 er further.-- once grant good species as carrion CROW & rook formed by descent. or two of the willow
424e101 yell tells me, on authority of Beck, that Hooded CROW & Carrion crow. have in Europe different ranges
424e101 on authority of Beck, that Hooded crow & Carrion CROW. have in Europe different ranges-- latter not g
424e102 ows are mixed in England-- for I presume Carrion CROW is found in Edinburgh.-- Why does Fleming consi
426e106 ecies; but facts of grouse, & pheasant, & hooded CROW goes against this. & wild hybrid plants. If man
453e182 . 393. <">The wild, small fowls at Pulo Condore "CROW like ours, but much more small & shrill".-- Hum
615o36v evening "seldom leave their perch till evening» CROW different.-- Heredetary effect of former tropic
540m089 ead people from scenes of distress.-- see how a CROWD collects at an accident,-- children with other
548m115 conscious of them, else one would not stand) a CROWD of other trains of thought are in progress-- In
599o007 e wide expense, of county, netted with edges & CROWDED with towns & thoroughfares, I grant that man,
563n001 hey seek Cock Pheasant claps his wings before? CROWING & only in breeding season & on the ground.-- C
563n001 oost, in all seasons, & after? he has done <g> CROWING.-- instances of expression.-- Octob. 3d. Dog o
466t094 ries are placed all round flower as they are in CROWN-Imperial Lily & many other flowers-- My view of
225b218 it will be said there are latent insects.-- as CROWS against man with gun. & Bustards &c &c!!! An Am
260c067 . America & South-- Any how temperate regions-- CROWS in N. America‖ Study Bonapartes list In the Zoo
264c080 one side into shrikes & at the other into into CROWS. yet all forming, according to Gould, good genu
358d075 andez do Mitchell. Australia Vol I. p. 306 "The CROWS were amazingly bold, always accompanying us fro
424e102 independent good case, but very odd since these CROWS are mixed in England-- for I presume Carrion Cr
496q006 will break & become double.-- There is a double CROWS-foot. or Ranunculus.= (1) Try.. Nitrate of Sod
613o036 particular trains of thoughts as fear of man,-- CROWS fear gun,-- pointers method of standing,-- meth
065r138 . Kerguelen Land. Prince Edwards Isld. Marion & CROZET. L. Auckland. Macqueries.--Sandwich Isd-- Spec
603o012 carried on toward deity.-- & as king might like CRUEL pleasure, so sacrifices cruel.-- Something wron
603o012 s king might like cruel pleasure, so sacrifices CRUEL.-- Something wrong here.-- Origin is certainly
029r033 SEA.--Extract from the log-book of the James CRUIKSHANK, Captain John Young, on her voyage from Deme
399e009 will decrease & be driven outwards in the grand CRUSH of population.-- Octob 10th. Saw. two undoubted
045r076 be called accidental; the proportional force of CRUST of globe & injecting matter on the great rise).
061r127 one in 1816 (?).-- Mr Owen's curious fact about CRUST Bra in Brine Springs. (Henslow) Speculate on ne
063r131 lone shows not gradation;-- An argument for the CRUST of globe being thin, may be drawn. from. Cordil
070r154 not see trace of Subsidence at Uspallata.-- ¿If CRUST very thick would there be undulation? would it
070r154 ibration? but walls & feeling shows undulation∴. CRUST thin.--Concepcion earthquake Draw close Analogy
090a021 side our globe melted magnetic metals. ∴ earthy CRUST compared to those of falling stones.--¿ does th
103a064 Beche's reasoning of mountains being formed by CRUST being too large & pitching against each other,
103a064 ose the dislocations occur & therefore that the CRUST might be considered a level.-- Dikes being last
107a077 e moves upward from effects of Elevation if not CRUST much thinner beneath ocean than above it no bec
107a078 becomes important with respect to thickness of CRUST broken up.----My view of Volcanos &c &c This vi
107a078 taking place chiefly beneath water & volcanos. CRUST must be thinner «under water» but cause most di
108a079 isothermal line, but if heat from centre, then CRUST of solid earth would be thicker.-- PP Andes mar
122a114 nal fluid arriving at equilibrium so soon from; CRUST being cut of-- if part of «cold» crust under oc
122a114 on from; crust being cut of-- if part of «cold» CRUST under ocean, became thicker, then when fluid mo
125a121 at by gases-- does not apply it to thickness of CRUST.- {P} if crust were metal then thinner if bett
125a121 es not apply it to thickness of crust.-- {P} if CRUST were metal then thinner if better conductor, th
130a133 gs off coast of Persia In Glen Roy paper I show CRUST yield easily. & if easily must be thin: <beside
131a136 a.-- facts of water flowing from beneath frozen CRUST in America Richardson.-- From strata being not
131a136 the world.-- argument strong in favour of thin CRUST theory.-- What a curious investigation it would
132a137 calculate rate of increase of heats in earth's CRUST.-- yet heat does increase,-- but in Ocean does
136a147 ol II. p. 81. «&83» Some remarks on thinness of CRUST as implied by meeting with granite every-where.
373d133 lae in Wealden. oldest birds. p. 411 -- Decapod CRUST in Muschelkalk, & 5 genera of reptiles.-- <M> p
460t019 plants in coal measures.-- Shells in Cambrian & CRUST show how long since present forms existed, but
477z001 ransacts. 3. papers connected with transform of CRUST-- Westwood & Thompsons-- Part II.-- 35. Phil Tr
478z003 een Lat 56 degree and 57 degree only inhabitant CRUST Entomost of the genera-- Cyclops p. 134. and p.
639j28r a caricature; Penguin.-- Pincers in Scorpion & CRUST in Squilla. & Mantis. C D woodcuts stones swall
197b112 ds &c ‖ p. 32. reference to M Edwards. law of CRUSTACEA-- with respect to mouth those beautiful passa
205b143 unts of passages of legs into mouth-pieces of CRUSTACEA. Vol II. p 75 a Fish which emigrates over lan
404e025 tion peculiar to <the class» «some orders» of CRUSTACEA, one is tempted to think that it must have be
423e097 , (all fishes to the state of the Ammocoetus) CRUSTACEA to--- ? &c) without reducing the number of li
462tf4v llo & the woodlouse-- -- a good analogy-- sea-CRUSTACEA-- Tullus. Athenaeum 1839 p. 772-- A curious
424e100 mirable paper on geographical distribution of CRUSTACEAE.-- (I forget whether I have already referred
481z008 ht birds agree. with <pe> nocturnal habits of CRUSTACEAN Mr Broderip says that Voluta found in not le
350d055 1st. Macleay & Broderip were talking of some CRUSTACEAN, like Trilobite. (Polirus??) female blind &
416e071 all!??!?-- Worms? [Barnacles, aquatic. <yet> CRUSTACEAN, & true hermaphrodites] CD «It may be said t
420e090 heredetary'-- Athenaeum .1839. p. <8>36.-- A CRUSTACEOUS animal is mentioned which inhabits the Pinna
055r107 ! How well agrees with number of Craters!--At S. CRUZ. There is no occasion to wonder what has become
065r140 er ¿Did I make any observations on springs at S. CRUZ.???-- Form of land shows subsidence in T. del F
085a001 e extension of Patagonia seaward. at mouth of S. CRUZ. from ascertained inclination. of plains: Lias
087a012 of of slow corrosion of valley of <Patagonia.> S CRUZ.-- from terrace like structure-- Intersection o
112a088 sea do p. 358. changed soundings in Mouth of S. CRUZ. in connection with Fitz Roys fact of elevated b
125a120 e 90 miles includes opposite directions. Mem. S. CRUZ. Assuming from Sir. W. Herschel's views earth o
481z009 lphidae Skunk inhabitant of Patagonia. Mem:-- S. CRUZ. Molina Vol. I. p. 244. Baccalao. migratory fis
540m089 n with other children naughty.-- Why does person CRY for joy? 17th. August Montaigne (Vol. I) has wel
545m107 kicked & cryed like naughty child.-- Do monkeys CRY?-- «they whine like children.--» Expression, is
560m156 ce when woman present? Do they pout, or spit, or CRY.-- <fe> Shame, independent of fear: colour of ba
568m018 e. <-- American monkeys utter pleasant plaintive CRY--» The taste of recurring sounds in Harmony comm
578n057 hy does joy & OTHER EMOTION make grown up people CRY.-- What is emotion? At end of Burke's essay on t
596n184 ttle Does blood go in <body> face in pashion.?-- CRY? Do people of weak intellects easily fall into h
545m107 ances, threw itself down on its back & kicked & CRYED like naughty child.-- Do monkeys cry?-- «they w
315c243 -- arising no doubt from want of assistance.-- CRYING is a puzzler-- Under this point of view. expre
542m094 do so dogs nearly silent, so with men.-- How is CRYING-- peculiar not common?--» no bark of anger nor
569n020 l of word beginning with the sound of letter)-- CRYING yawning laughing being necessary sounds... not
595n125 in Zoolog. Garden [blank] [not located] A child CRYING. frowning, pouting, «smiling», just as much in
327c1BC esticated animals Fries de plantarum proesentum CRYPT. transitu et analogia commentata Library of Us
381d156 en) p. 34.-- Owen classifies Hermaphrodites. CRYPTANDROUS. (only female organs visible). Oyster. cryst
381d157 f structure leading to supposition, that the CRYPTANDROUS are really, Heautandrous.-- How is fecundat
381d157 ing to this doctrine be considered as really CRYPTANDROUS, & they have hybrids-- this is most importa
206b150 ocot in Coal formation? p. 320 <Think> States CRYPTOGAM. Flora formerly common to New Holland?! p. 32
207b150 us structure intermediate between vascular or CRYPTOGAM. (original Flora) and Dicotyledenous, which «
390d177 orms which are produced by budding «only» as CRYPTOGAMIA & hydras,-- (this repugnance to breeding in
182b046 r Proportion terrestrial,-- if in any in the CRYPTOGAMIC flora but not atmospheric type. Hence probab
210b159 there 144 genera & 365 species of plants not CRYPTOGAMIC but I . 2,53.-- In know varieties. there is
267c095 ant do p. 98. on a quaternary arrangement of CRYPTOGAMIC plants.--Owing to plants not being adapted t
315c240 f Bass' Straits The common Mush room & other CRYPTOGAMIC plants same in Australia & Europe.-- if crea
316c244 owerby.--. Geographical range of shells like CRYPTOGAMIC plants. of Marine kinds, there are some rest
447e164 other individuals, «with facility»-- such as CRYPTOGAMIC plants & true hermaphrodite Mollusca, & prob
260c067 torge, which could not be resisted, when hearing CRYS of hunger of little bird, in same way Wilson (p

260c068 ursuit of Blue-Jay, when <one> birds hears <dis> CRYS of distress of other parents.-- Shows community
286c154 - animals understand the language, they know the CRYS of pain, as well as we.-- It is our arrogance,
186b058 ysomela may be going back to common ancestor of CRYSOM. & Heterom, but I cannot understand the univer
185b056 ch at first would be mistaken for) Carabidae, CRYSOMELA, Scarabadae, & longicornes.-- Again taking a
185b058 re probably in some <heteromera> colouring of CRYSOMELA may be going back to common ancestor of Cryso
272c108 tructure like Curculionidae.-- Are there any CRYSOMELIDA, with similar habits. But the Horae Entomol
026r021 nwall «Vol II» It is a fact worth noticing that CRYST of glassy felspar in Phonolite arrange themselv
057r115 ays. precip. of Sulph. B. all the infinitesimal CRYST. arrange themselves in planes. «Mem silky lustr
059r123 no page 122] In Igneous rocks.--which have the CRYST of glassy F. fractured. have been melted with l
110a086 stallized Salband like basalt. full of circular CRYST of glassy felspar different from either fragmen
110a086 sh. black specks of hornblende, large irregular CRYST of reddish felspar. & scales. of mica.-- large
110a086 of reddish felspar. & scales. of mica.-- large CRYST of Hornblende blending into base-- Salband migh
098a046 balls) that concretions are connected with a CRYSTALLINE process.-- now cleavage as suggested by Sir
121a111 - yet with character completely altered, & a CRYSTALLINE structure superinduced Lyell on Sweden p. 5.
098a046 Hershel is all crystals obeying one law of CRYSTALLIZATION. therefore concretions in this case lamina
098a047 - state last page thus. point of attempted CRYSTALLIZATION, & therefore as a consequence aggregated (
100a053 ry.-- consider stalactites.-- agate rings, CRYSTALLIZATION transverse.-- or rather radiating to centr
098a047 her two particles of Carb. of Lime, tends to CRYSTALLIZE them as seen in stalactite).-- some force cr
110a086 ew curious fissures; base in part. block not CRYSTALLIZED Salband like basalt. full of circular cryst
098a047 ze them as seen in stalactite).-- some force CRYSTALLIZES minerals in layer. therefore aggregates the
102a061 per quartzose ones & felspar.?? Are the great CRYSTALLS, the layers first of felspar & then quartz
049r088 hers subsequent. as in dikes. In Granite great CRYSTALS arranged on sides. V. Lyell P 355 Vol III. co
088a015 c action. -- Fragments of slate converted into CRYSTALS of Hornblende p. 248. L. Institut 1837.-- Hel
098a046 cleavage as suggested by Sir J. Hershel is all CRYSTALS obeying one law of crystallization. therefore
101a056 coast ∴ curious exception in Wealden.-- Would CRYSTALS arrange themselves in that direction, in whic
110a086 . consisted of hornblende (?) & felspar, (some CRYSTALS being red) «with» cleavage, veins of pyrites,
110a086 m either fragment or dike, blackish grey base. CRYSTALS from fragment disseminated on that side of sa
110a086 d might have oozed out of cleavage plates: the CRYSTALS must have recrystallized, as such do not occu
115a096 tes that cleave, & which unite the homogenious CRYSTALS., must aid in adding effects to common heat.-
121a111 or potters to use. in which great Knife formed CRYSTALS of ice were formed-- (like my gypsum case) sh
047r082 ling first on beach I cannot understand, without (CS <[...]> raised above as).-- In great Calabrian
437e140 t bird another day brought to her young ones the CUB of a fox, which after it had fought well & despe
027r024 on. Humb: Pers. N. vii P. 56 Serpentine form: of CUBA for comparison (?) with St Pauls [not located]
251c042 273. Macleay on Capromys. 4 species probably in CUBA (p 271 Viedo says American dogs silent. Mem con
258c061 to by Capt. King do. p. 434. Table of birds from CUBA Vigors.-- nothing of much interest XX. Hence re
408e042 -- (see Macleay in Zoolog. Journal. for those of CUBA.-- It is important to understand well the relat
386d157 nvisible» animalcule in four days could form 2. CUBIC stone. like that of Billin.-- <Generation--> V.
260c067 kind have been know to assist in feeding young CUCKOO; as if there was storge, which could not be re
274c113 orial type is wonderfully shown in long legged CUCKOOS with claw like lark, (one in Australia is call
282c142 ith respect to rest of its family as in ground CUCKOOS, is affinity with respect to species of each o
282c142 ame ground then the relation of all the ground CUCKOOS. would not be affinity, but the truth would ne
502q11v ira Migratory-- ask Gould about N. Zealand, as CUCULUS lucidus is.-- Ask Sulivan about Falklands Isds
181b044 - Heaven know whether this agrees with Nature: CUIDADO The above speculations are applicable to non p
235b275 early so, which <breed> do not intermix,--any CULTIVATED plants produced by seed.-- Lychnis.-- Flax.-
403e021 Vol III p. 77. Many foreign plants have been CULTIVATED in Guahon. (Mariannes), "for example the pri
429e115 & stops at 2600. & yet know that plant can be CULTIVATED with ease near London.-- what makes the line
446e162 acter.?» No--well characterized.-- Tulips are CULTIVATED during several years & then they break-- --
497q06v at would result from seeds being sown= See in CULTIVATED Plants, as Pentstemon, which have abortive p
514q21. apland Plants --will get answer= Is pollen of CULTIVATED Orchis & Asclepias &-- carnosa?-- good-- Nor
515q021 in fruit, or time of leafing (5) Do the most CULTIVATED show Heartease produce as large capsules of
299c195 ty». station & varieties chiefly produced by CULTIVATING parent in rich soils & their seeds produce <
327c265 cultural Transacts.--» Mr Coxe "Views of the CULTIVATION of Fruit trees in. N. America" in Lib. of Ho
381d157 genus some dioecious & some monooecious-- (& CULTIVATION might make one set of organs barren in one p
400e013 ion before they broke.--, showing effects of CULTIVATION gradually adding up. & four more generations
434e127 hether it merely has tendency (as effects of CULTIVATION on successive generations of plants) to do s
442e151 ill not admit this, now that tulips break by CULTIVATION can a form become permanent?» because its ve
466t096 odness of flavour in fruit-- all affected by CULTIVATION during life of individual. June 1st 1841. Ma
505q014 er has seeds produced good pollen? Yes «From CULTIVATION lost their horns» is impregnation necessary
549m121 there said of intellectual <ple [...] hope» CULTIVATION, main source of the intense happiness.-- it
325c267 al> Philosop. Transactions. 1827 Paxton on the CULTURE of Dahlias Mrs. Gore on Roses might be worth c
553m136 in civilized nations has not been improved by CULTURE « <was> who feel the most implicit faith that
489qIFC 22 Schomburgk.---- 1 Jordan Smith. p 1. Sowerby CUMING. -- p 1 Owen p 17 Hooker p. 17 (T) Mrs. Whitby
483z013 ight perhaps be worked out with advantage. with CUMMS collections & my own: & Capt. King's p 453-- Pl
583n069 ce the general aim of fable, & expressed as CUNNINGNESS of fox, industry of bee &c &c-- p. 15."insti
534m062 Col. Sykes on Formica indefessa placed table in CUPS of water which they waded. or swam across.-- th
342d035 in osteology of young Ostrich. 16th. D Israeli (CUR of Literat. Vol II p 11) accidentally says "--is
272c108 ra, which have habits & part structure like CURCULIONIDAE.-- Are there any Crysomelidae, with similar
410e052 in» skins worse than useless.-- yet there is no CURE «I may say all this, having myself aided in suc
523m018 lieve my grand F doctrine is true, that the only CURE for madness is forgetfulness.-- which does appe
608o026 motive power.--if incorrigably bad nothing will CURE him' 3) disgusted. with them. Yet it is right t
528m035 athers positive statement that insanity is only CURED by forgetfulness.-- & the approach to believing
529m042 e has some idea of a son of Dr. Prietly who was CURED from a fall of ideotcy.-- The story of the Corb
537m075 expression. "relict of bad habit." as child is CURED of sucking his finger by rubbing them with alum
590n093 Haller says tooth ache, even from carious tooth CURED by sight of instrument.-- Bennett's Wanderings,
577n052 arate organs most curiously shown in the sudden CURES of tooth ache before being drawn,-- My father «
473s05r es can be distinguished-- & Jackson here (Capel-CURIG) says that he can certainly tell Trout from Ogw
473s05r t he can certainly tell Trout from Ogwen, Capel CURIG & some other lakes, (different waters) He canno
473s05v nway by Bettws & there joins streams from Capel-CURIG-- Mr Bunbury says Miers has described in Linn:
523m015 f the idea, namely his poverty.-- his manner of CURING it. by keeping the sum-total of his accounts i
494q004 are propagated,, though valuable as show. & CURIOSITIES!! What is price of fox. otter. Badger &c &c
313c235 sed.-- we see in the Entomostraca. The sexual CURIOSITY of the orang outang of (in June 1838 when you
376d136 h Dr. Andrew. Smith «Remarks on extraordinary CURIOSITY of Monkeys». The Baboon of which anecdotes ha
376d137 y he has seen do this.-- These Monkeys had no CURIOSITY to pull up trousers of men. Evidently knew <m
566n011 mals have necessary notions. which of them? & CURIOSITY «strongly shewn in the numerous artifices to
033r042 case: If refiltered with other matter how very CURIOUS a structure: Have shells ever casts alone in C
045r076 e having a pretty severe shock). are much more CURIOUS & perplexing. than those that attend Eruptions
057r114 te beds having proceeded from gravel proved.-- CURIOUS similarity of rocks of very diff. ages. at Por
061r127 ion. -- ¿ Another one in 1816 (?).-- Mr Owen's CURIOUS fact about Crust Bra in Brine Springs. (Henslo
069r150 be older than (B).-- Most important view Urge CURIOUS fact felspar melted gneiss/// QUARTZ!!! Analog
074r165 mplicated chemical law & steam of salts, quite CURIOUS case of oxided Iron by Mitterschlich. Vol. II
100a055 orm of escarpment relation kept to sea coast ∴ CURIOUS exception in Wealden.-- Would crystals arrange
110a086 ng red) «with» cleavage, veins of pyrites, few CURIOUS fissures; base in part. block not crystallized
114a094 es of Rio Negro--Mr Bowerbank-- Dr. A. Smith's CURIOUS specimens of «transversely fibrous» quartz. &
124a119 sorb, zinc thrugout its thickness.-- this most CURIOUS with respect to epigmous action.-- if the zinc
124a119 ith 90 percent of lead. it would be still more CURIOUS to know whether it would be absorbed.-- if so
129a132 it is but poor. Athenaeum. 1838 p. 791 -- Most CURIOUS account of great subsidence «20 miles long I i
132a137 trong in favour of thin crust theory.-- What a CURIOUS investigation it would be to compare, the time
182b050 ear clim In Mr Gould Australian work some most CURIOUS cases. of close but certainly distinct species
186b059 md the universality of such law.-- It would be CURIOUS to know in plants, (or animals) whether, <in>
200b125 birds; most seeds germinating.---- It would be CURIOUS experiment to know whether soaking seeds in sa
208b153 ells in India!? A p. 28 Dr Beck. & Lyell. most CURIOUS law of species few in Arctic. in proportion to
211b163 . whilst they eat up the dogs.-- L.' Institut. CURIOUS paper by M. Serres on Molluscous animals repre
214b176 t to obtain a litter without this defect, Very CURIOUS case = W. D. Fox. When dogs are bred into each
220b188 ahiti In Charlesworth Magazine Jan: 1830. most CURIOUS paper on heredetary fear (like rooks with guns
220b199 . When will this paper be published it will be CURIOUS.-- Some general statements about mundine & con
222b204 ells of Porto Santo & Madeira-- I believe very CURIOUS-- My idea of propagation almost infers, what w
227b224 istory of earth «within recent times.». & many CURIOUS points of speculation; for having ascertained
232b249 s, without such resorts Mr Waterhouse has most CURIOUS facts about the distribution of Lemurs in Mada

233b250 nose above it-- pulled bell.-- -- It was most CURIOUS to observe, that all the species of mice in S.
260c067 partes list In the Zoological Journal I read a CURIOUS account to show that very many birds of differ
268c096 8.-- Annals of Natural History. «vol I» p. 159 CURIOUS account of Tit mouse feeding young of redstart
272c109 radise. Trogons.-- the one feather in wing the CURIOUS feathers in tail of Edolius.-- Remarkable how
289c162 s-- one remaining through winter. «It would be CURIOUS to know whether a variety could be transmitted
293c174 seases, where habits of people nearly similar. CURIOUS instance of difference in races of men.-- Wax
298c189 in numbers over whole of England & Ireland.-- CURIOUS in so wild a bird.-- Annals of Natural History
298c189 alian dogs, would be «most» like Australian.-- CURIOUS this +ready answer, without any leading questi
310c224 ish beetles, birds & a fox most close The most CURIOUS case is Saxifrage, almost <same> «closely alli
314c237 ch.?? Mr Brown showed me Bauer's drawings of a CURIOUS plant where a tube consisting of pistils & sta
315c243 view. expression «of all animals» becomes very CURIOUS.-- a dog snarling in play.-- Hensleigh says th
318c253 n.-- Bachman tells me in Audubon there is most CURIOUS history of first appearance of the S. American
318c253 ee/ migrations of birds he mentioned many most CURIOUS case-- -- the birds seem to follow narrow band
326c266 nna. sketch of south Sea. Botany R. Brown. has CURIOUS coloured maps. by Copenhagen Botanist of range
331dIFC igs with solid feet.-- 1838 [In this Book some CURIOUS notes on Monkeys recognising Sexes of animals:
334d011 cats (Fox has seen repeated cases) being deaf CURIOUS case of corelation of imperfect structure.-- F
337d022 , «when offended» who loves, who fears, who is CURIOUS &c &c &c who imitates.-- who will say there is
341d030 deer, yet is spotted like so many deer.-- very CURIOUS like some facts of Mr Blyth on birds.-- Dr. Ba
341d032 in seven years produced even an egg.-- a most CURIOUS bird, did not seem to know itself,. at last as
356d071 reed animals being wilder than parents is very CURIOUS as pointing out difference between acquired &
357d073 between hen Caperailkie & cock Black-cock.-- (CURIOUS the readiness with which this genus becomes cr
362d099 being wanted for war) & hence never fall off.▌ CURIOUS the rapidity of the change in 5 or 6 weeks aft
362d100 ize pidgeons» in 1834-- now this would be most CURIOUS to show that in sixty years-- (how many genera
388d172 ds growing on old women = Stags horns & testes CURIOUS instances of corelation in structure = Neuter
392d175 of plants are Monooecious or dioecious.-- very CURIOUS how this was superinduced? (Surely all are rea
398c005 further inductive reasoning is immense. It is CURIOUS that geology. by giving proper ideas of these
404e025 : .1838 p. 329-- Milne Edwards, description of CURIOUS mechanism of respiration, or rather ventilatio
408e045 - Humboldt bones at 7800 in Andes-- parallel & CURIOUS facts.-- The Himmalaya. case, bears on the vas
419e085 few animals of any kind-- Fauna, must be very CURIOUS.-- With respect to the non-development of Moll
421e090 of metamorphosis in the young of Syngnathus.= CURIOUS as showing generality of law. even in fish: ▌.
423e098 ufficent to spoil a bed of Cauliflower.-- (How CURIOUS it would be to make enquiries of some of these
429e116 even of polar regions of N. America.-- if true CURIOUS on my view-- because these points were last co
441e149 the carrot seems to show this.-- This would be CURIOUS law, Certainly Australian Dog is not affected
441e149 s on offspring, but not on individuals is very CURIOUS & important.-- The existence of "laws of organ
443e153 rze must be propagated by its roots: now it is CURIOUS Mr K. has observed that to graft from the root
448e165 vaporous sometimes seeds.-- ▌There are endless CURIOUS facts about every part of plant producing buds
448e166 t the drifting of animals on ice p. 643-- very CURIOUS table of all the castes from Stephenson at Lim
461t025 & p. 451. 1839-- Translation of P. Fries most CURIOUS paper on the Pipe-fish-- which he divides into
461t025 up. Mammalia, which <do> have not sack,-- Most CURIOUS facts & this paper deserves fresh study & whol
462tfC5 -Crustacea-- Tullus. Athenaeum 1839 p. 772-- A CURIOUS theoretical French book review on politics in
463t057 this theory-- (I must answer it by rooting out CURIOUS cases of intermediate structure,, & supposing
470t177 ld very many Bees & Humbles--on Thistles many (CURIOUS because a Composite) Asparagus very small flow
473s03v evonian-- Fish one step lower in America-- How CURIOUS all negative laws of America of depth of organ
478z004 to Royal Soc on glow worm. luminous property-- CURIOUS arrangement of animals in rays Par un officier
481z009 insectivore in S. America or Australia-- very CURIOUS.-- replaced by didelphidae Skunk inhabitant of
510q018 About non-breeding of animals in confinement, CURIOUS.-- foxes-- English animals. [Made no import. r
523m014 , as <if> in my case.-- It must be so from the CURIOUS story of the Birmingham Doctor praising his si
531m049 f cats will «ever» eat little birds, this most CURIOUS instance of reason & abstinence.-- My Father r
542m094 hich are involuntary, are common to animals.-- CURIOUS to trace, which of these actions are habitual,
545m107 tween fingers, the peevish expression was most CURIOUS <like> «remember» the expostulatory angry look
548m115 dreams do that One Reflective Consciousness is CURIOUS problem., one does not care for the pains of o
550m126 e pleasant.-- Browne Religio Medici, p. 21-24. CURIOUS passages showing how easily chance & will of D
553m138 rican Monkey show any desire for women-- «very CURIOUS. as they depart in structure» The monkeys unde
554m138 ts-- just like Jenny with Tommy ourang.-- Very CURIOUS.-- Mr Yarrell has seen Jenny, when Keeper was
554m139 d of teeth & hands.-- Descent 1838 It was very CURIOUS to see her take bread from a visitor, & before
557m147 fox. nor wolf wag their tails, &c. it is very CURIOUS recurrence of pleasure so teaching expression
558m153 is for C. Sphynx.-- In the wild ass there is a CURIOUS drawing out of the side part of nostril, when
559m155 Colonel in army on "Wheat." in Jersey.-- very CURIOUS facts about early production of foreign seeds.
559m155 many varieties.-- Rev R. Jones has it.-- very CURIOUS book.-- Hume's essay on the Human Understandin
573n038 r without touching table.-- cannot avoid it.-- CURIOUS mixture of voluntary & involuntary movements.-
574n041 desire makes saliva to flow «yes, certainly»-- CURIOUS association: I have seen Nina licking her chop
579n058 when they consider profoundly,-- this will be CURIOUS if it is so.-- frown with grief,¿ bodily pain?
584n072 water,-- young crocodile snapping-- p. 28. how CURIOUS the means of guiding themselves through the ai
602o010 ion to time & memory Reynolds X discourse very CURIOUS as showing "the perfection of this science of
603o012 -- Something wrong here.-- Origin is certainly CURIOUS. Chinese, S. American. Polynesians Jews, Afric
610o031 & found his Christian name was Wilson!!-- How CURIOUS an inward. unconscious memory.-- Jan 14th. 183
401e016 e state would be got out of it.-- Now this is CURIOUSLY different from primrose suddenly produce cows
577n052 vid mental affection, on separate organs most CURIOUSLY shown in the sudden cures of tooth ache befor
621o048 ch to be wrong or right; this teaching may be CURIOUSLY modified by circumstances of country, so will
556m145 -- In the drawings of Voltaire why is under lip CURLED over upper with mouth shut. expressing cool ir
557m147 les would act, to when in a passion.-- dog tail CURLED when angry & very stiff. back arched. just con
633j53v f gain of small advantages thus to explain the CURLING of the valves of the broom.-- or the springing
558m152 heir backs-- Bengal tiger. when slightly angry. CURLS tip of tail.-- do two cats arch their back when
032r041 n?)» as in Cordillera.-- From poles to Equator CURRENT downwards & to West.--From Equator to poles. n
032r041 pth. from axis to surface must gain a Westerly CURRENT:--If great changes of climate have happened. h
147g015 hill/ do alluvium. NB In one part pure sand in CURRENT cleavage-- in other irregular horizontal strat
148g026 up to Bridge of Spean, hills of «sea», gravel, CURRENT cleavage, & pretty well rounded stones, mixed
510q017 w written on double creations & where? How are CURRENT & winds in Antarctic ocean: are they from West
510q017 America? Sabine says North of Siberia, no sea-- CURRENT, icebergs travel by wind. Aug. St. Hilaire Bot
024r014 tence of some moving <point> power ¿ Submarine CURRENTS Find instances; The whole coast of New Hollan
027r028 ess cycle of revolutions. by actions of rivers CURRENTS. & sea beaches. All mineral masses must have
037r055 lata. to the Bay of Bengal. dimensions? Strong CURRENTS off the Galapagos.--strata must be accumulati
086a007 temperature of world regulated by atmospheric CURRENTS?-- chiefly clearly by sun's position = If equ
092a029 paper to Brit. Assoc: has shewn how electrical CURRENTS tend to deposit metals, if in solution. My vi
109a082 ch action probably modified by form of waves & CURRENTS.-- but this must be continued. no currents &
109a082 s & currents.-- but this must be continued. no CURRENTS & elevation have same effect, a tendency dire
109a083 lique) outwards may be granted. independent of CURRENTS.-- mud going out can actually be seen.-- ¿ Th
115a098 the water, the greater power of oscillations & CURRENTS.-- if matter was «successively» given of ever
115a098 epending not on absolute <force> «size of» of <CURRENTS> «fragments» but relative to currents. Small
115a098 of» of <currents> «fragments» but relative to CURRENTS. Small lakes have power of levelling their sh
116a099 kes have power of levelling their shores where CURRENTS very weak??-- too great an abundance of matte
269c101 ess in that period & no ways assisted by fluid CURRENTS which, may take place in Metamorphic action.-
023c013 ite Isd. capped by Calcareous rock; following CURVATURE of hill; strata could discover no shells: not
119a106 luid pool.-- (this is shown by the softness & CURVATURE of quartz rock?) also by my phenomena of eart
467t104 ctar seems to be at base of upper petal & the CURVATURE of <an> pistil, etc lies in gangway= In Lotus
120a107 ss which the anticlinal lines are apart-- the CURVATURES of the strata.¿ the enormous faults & facili
126a124 ria.-- from water {P} thawed at + in isothermal CURVE. East-clinal. West clinal. S-clinal. N-clinal
466t093 «a bee» was dusted over. {P} Stamens & pistils CURVE upwards, so that anthers & stigma lie in fairwa
035r048 ck.-- Try on globe. with slip paper a gradually CURVED enlargement. see its increased length. which wi
146g007 raigs» 25 degree perhaps most common-- Will not CURVED form of hill be explained by my idea-- highest
162g104 ulder), sloping buttresses, an[d] one alternate CURVED layer of fine sand & small angular-- rounded p
283c145 ich has very short legs & long tail «short much CURVED beak.-», other very long beak, with short. . |
470t178 numbers =what insect can get honey out of long, CURVED nectar of Butterfly Orchis & Listera? Bryony s
507q15v he hook-- -- Geum. Galium Burrh ≡ single hook; CURVED spines-- simple spines-- or seed-cases with si
048r087 that in front of Sts. of Magellan In Chiloe CURVILINEAR strata subsidence.--The sudden increased dip
232b249 a sub-genus in Southern Africa In same manner. CUSCUS, (a sub genus of Phalangista New Holland form)
244c023 p. lasiurus does in North. Hemisphere.-- p. 158 CUSCUS albus. New Ireland ---- maculatus -- Waigiou S
245c024 rianus new, & some rats & mice. In Amboina only CUSCUS & Barbyroussa «NB» [islds. Springing up more l

568n017 e for general good, or in case of any fantastic CUSTOM» «Probably bashfulness is connected with some
572n032 rostration «uncovering body» &c &c is matter of CUSTOM."-- this all applies to bodily weakness & infe
622o048 ke to injured conscience, is the feeling of any CUSTOM of society broken..-- & how far more <feelin>
544m104 tion-- ∴ modified in those races, where it is CUSTOMARY to die-- August 24th. As some impressions «Hu
439e145 be the one encouraged-- » Wilkinsons Manners & CUSTOMS of the Ancient Egyptians Vol III. p. 33-- They
022r008 ry thick & so tough that a sharp knife could not CUT it: in which we found the Head & Boans of a Hipp
106a075 ng over (B) than when at (C) its tendency would CUT: be to cut a narrower channel instead of wider.--
106a075 than when at (C) its tendency would <cut> be to CUT a narrower channel instead of wider.-- This appl
118a103 r become scoriform. has thinned upwards & is now CUT off by denudation it gives one grand idea of amo
120a109 er would carry further its own matter. but would CUT wide gorge. leaving cliffs, on each side, such a
122a114 rriving at equilibrium so soon from; crust being CUT of-- if part of «cold» crust under ocean, became
148g028 ould not have happened if the side-streamlet had CUT them out-- In all cases «I urge» deposition mari
152g048 rivers might deposit, & afterwards with greater CUT through, not applicable to Glen Roy Lake, must h
152g048 size of buttresses, to upper edge of which they CUT near Loch Tring-- Tuesday Bridge of Roy Level of
155g062 ibly on Glen Turrit side 2nd shelf very broad «& CUT out, produced» from same «cause» as «great» side
155g064 ne might remove, what above straight line «only» CUT deep gorge on sea hypothesis, if gullies not now
157g077 hill with boulders river Right Hand Cascade has <CUT> «where two branches unite in upper Glen Roy» ve
158g082 th granite blocks almost encircled» <fre> Gneiss CUT smooth on sides of hill where Boulder lies. butt
159g085 rcely conceivable. if Hill between Corry so much CUT Granite could have remained, no peat supply.-- C
170ßIFC es thought Glen Roy B C. Darwin All useful pages CUT out Dec. 7th. /1856/ (& again looked through Apr
171b002 ls rendered perennial. &c &c.-- Yet Eunuchs nor «CUT» Stallions nor nuns are longer lived Why is life
189b072 erate «other species», their race is not utterly CUT off:-- like golden pippen. if produced by seed g
223b210 change, (as some animals do more than others, & CUT off limbs & new ones are formed) but yet propaga
259c065 redetary disease, is on the same principle that, CUT a sheeps tail off plenty of times & you will hav
265c083 England must have been lopped off & sheeps tails CUT yet there is no record of any effect.-- New Holl
279c133 excellent observations of sickly offspring being CUT off-- so that not propagated by nature.-- Whole
317c249 appearance Every now & then a short-tailed cat.¿CUT? has its offspring short tails /one born at Maer
332d005 s bare. skin black & wrinkled-- fur short. (tail CUT off in progeny peculiar) limbs very long, eyes v
372d131 it were possible to support the arm of Man, when CUT off , it would produce another man.-- That the e
381d156 n Man, has strange effect.-- Directly a Capon is CUT, it increases in size prodigiously-- Animal OEco
386d166 of the part, of what is good for the whole.-- if CUT off nerves in snail. (Encyclop of Anat & Phys) <
506q15v of flower 45. Charlsworth. vol II. p. 670-- oats CUT down turning into Rye.-- 46). Book describing am
525m025 arriages with impure breed.-- A cat had its tail CUT off at Shrewsbury & its kittens <h> (in number 3
555m143 of front) (having changed hanging into his head CUT off) as kind of wit, showing he had honourable w
555m144 wit.-- I changed I believe from hanging to head CUT off. «there was the feeling of banter & joking»
579n060 ushing-- in former irritation on a piece of skin CUT off made the blush come.-- it is an excitement o
582n067 er.-- /Mem. Hensleighs objection.--» it is more, he CUT off.-- <Horses> Colts cantering in S. America ca
290c165 --.Mem. Hensleighs objection.--» it is more, he CUTS the matter short by saying man cannot be compar
527m032 ears]CD.-- Beauty is instinctive feeling, & thus CUTS off wings of flies from intellect. but it does
534m063 rt says Dr Darwin mistaken in saying common wasp CUTS off the Knot:-- Sir J. Reynolds explanation may per
546m110 he was one day reading a book, with ivory paper CUTTER, which she valued, & she was suddenly called t
546m110 e something, on her return could not find paper CUTTER, hunted in vain for it-- ten years afterwards
063r132 gation. whether ordinary. hermaphrodite. or by CUTTING an animal in two. (gemmiparous. by nature or a
105a069 former as its channel becomes wider looses its CUTTING power. (as does it when the inclination become
105a069 on coast, as long as exposed to waves of sea, CUTTING power increased with width. for besides more s
163g106 ft above sea-- Loch Ness 40 ft above do. When CUTTING bank where Locks now are (32 ft rise) they fou
372d129 iarities not so a seed.-- Bud probably is like CUTTING off tail of Planaria, the whole grown to that
491q01v flowers & scarlet Lychnis can be propagated by CUTTINGS.-- Try.-- Important as discovering function o
495q005 r I think certainly not) (3) Sow seeds & place CUTTINGS or bulbs in several different soils & tempera
495q005 will be.-- will seedlings vary much more than CUTTINGS &c (4) Raise annuals or common English plants
501q011 (28) Can any annual or Biennial be grafted or CUTTINGS taken or tuber-- talk about Mr Knights theory
506q015 iams & Stocks, being propagated many years, by CUTTINGS.-- (40) Ask Henslow to distribute some of my
023r010 epeatedly talks about the immense quantities of CUTTLE fish bones floating on the surface of the ocea
436e134 which fish &c can live.-- Lyell says that naked CUTTLE fish now bear a very large proportion to other
086a006 ead as well as Pallas before Geology is written CUVIER. Europe possessed a great edentata. -- How muc
087a010 uvial plains of Mississippi -- No Vol. I. p212. CUVIER Oss Foss Wide range of Mammalia really very im
183b053 ned.-- Read his theory of the Earth attentively CUVIER objects to <tran> propagation of species, by s
184b054 Falklands described by Q. & G. as new Species. CUVIER examined it. There certainly appears attempt i
192b089 that is all that can be expected-- This answers CUVIER-- Perhaps the father of Mammalia as Heterodox
197b112 e beautiful passages from one to other organ-- CUVIER on opposite side; Is Vol of Fish p. 59.]CD Cuv
197b112 vier on opposite side; Is Vol of Fish p. 59.]CD CUVIER has said each animal made for itself does not
198b114 hing S. H What does the expression mean used by CUVIER, that all animals (though some may be) have no
199b118 N. Philos J. p. <191» «p 191» No. 5. Ap 1827 F. CUVIER says. "But we could only produce domestic indi
244c023 ttentif, et forts surtout de l'opinion du baron CUVIER, nous ne balançons pas a la regarder comme une
287c157 : Transact Vol XIV.--¾. p. 24. Lamarck bears to CUVIER that relation of theoretical astronomer to pla
290c165 of dogs to man. not altogether explained by F.. CUVIER, «--.Mem. Hensleighs objection.--» it is more
324c268 b. «in Dict. Sciences. Nat. in Geolog Soc.» F.. CUVIER on instincts L. Jenyns paper in Annals of Nat.
355d067 t SPECIES of genus Sus. do vertebrae vary? «See CUVIER Ossemens Fossiles» Although no new fact be eli
438e142 domestication [see Wikinson on dogs of Egypt & CUVIER on Mummies]CD [NB TIME is element in change, a
463t063 ronha are found unknown coast in front of it.-- CUVIER has grand sentence about the Animaux fossiles-
473s03r stodon longirostris in miocene like in Europe-- CUVIER never found remains of Sus with Elephants-- Ly
197b112 agree with old & modern types being constant. CUVIER'S theory of Conditions of existence is thought
399e009 al period? [not located] Study introduction to CUVIERS Regne Animal No structure will last. without i
411e055 been fixed, to study the physical causes. «All CUVIERS generalization. of teeth to kind of extremitie
478z004 Chinchillidae Zoolog Transacts. worth reading CUVIER'S Memoire 133 1803. on Pennatula showing it to
633j54v flow from some grand & simple laws.-- 4 «Study CUVIERS Anatomie Comparé» p 308. Traces the gradation
376d136 Baboon of which anecdotes have been told is CYANOCEPHALUS Porcarius.-- this Monkey did not like a gre
545m105 ck with observing how the Baboon (<Macaco» «CYANOCEPHALUS Sphynx Linnaeus») constantly moved the skin
563n002 as bitch: or dogs defending companion. (mem CYANOCEPHALUS. Sphynx howling when I struck the Keeper) m
567n014 , has expression of languor & suffering The CYANOCEPHALUS when fondling the keeper., clasping «& rubb
596n184 -- or skin of head.-- «scarcely able St.-- » CYANOCEPHALUS, macacus. Cercopithecus? very much., «Keepe
596n184 d.-- «[Are monkeys <are> right-handed??]CD» CYANOCEPHALUS, Macacus, Niger. Cercopithecus make labial
470t177 tion!!? Bees at Wild St Johns Wort--Scabies, CYANOGLOSSUM--Reseda wild very many Bees & Humbles--on T
247c028 ineatus (p 45) Moluccas & New S. Wales Scincus CYANURUS «p 8 &». p 49 on all the Moluccas «New Guinea
027r028 tous & coaly matter. Mem: Chiloe In the endless CYCLE of revolutions. by actions of rivers currents.
171b002 h object generation.-- We know world subject to CYCLE of change, temperature & all circumstances whic
505q014 t-house. will it seed?-- (Skim through Penny CYCLOPAEDIA) Abberley says that some Bees are smaller &
478z003 only inhabitant crust Entomost of the genera-- CYCLOPS p. 134. and p. 115 In white Cape Pidgeon's sto
316c245 e species than other countries «++ p. 246» as CYCLOSTOMA in Phillippines & Anphidesma in S. America--
316c245 phidesma in S. America-- yet there are a few CYCLOSTOMES & a few Anphidesmas.-- this is remarkable.--
033r043 Peak of Teneriffe. also Cotopaxi has a <[...]> CYLINDER placed on the rim of conical crater: at Tener
240c013 yage. de L'Astrolabe Zoologie. p. 60. Vol I. CYNOCEPHALUS. niger. comes from the Moluccas «Matchiam»
505q014 11) Abberley has planted seeds of pale green CYNOGLOSSUM. never germinated 12 Does the horned orange.
277c127 Take instances of most disputed shells, such as CYRENA This is reform which probably will be slow. bu
381d156 tandrous. (only female organs visible). Oyster. CYSTIC Entozoa. Echinoderms. Acalephes. Polyps. Spong
189b070 pace. (Mem: Galapagos). Little wings of Apteryx DACELO & Kingfisher same colours Strong odour of negr
500q10a urope.-- Gould-- go over the Pigeons, Philotis, DACELO. Alcyone, where there are very close species &
400e013 e John-- says Decandoelle, distributed seeds of DAHLIA all over Europe same year.-- he sowed them for
325c267 p. Transactions. 1827 Paxton on the culture of DAHLIAS Mrs. Gore on Roses might be worth consult. Pap
429e113 ow many plants have been produced (look at the DAHLIAS]CD all much varied breeds both plants & animal
438e142 ummies]CD [NB TIME is element in change, as in DAHLIAS by snails)-- <The> Apion radiolum undergoes tr
534m063 cked by insects & snails of this country (thus DAHLIAS by snails).-- <The> Apion radiolum undergoes tr
055r108 ng variations) of rain. = The Bulk of sediment «DAILY» yearly brought down by every torrent proves th
117a101 ust have taken place, otherwise the world would DAILY be scene of ruin in late Natical Magazine (befo
200b122 ?).-- Ed. Phi.l.. N. J. p. 410 <Nov» 1828 It is DAILY happening; that naturalist describe animals as
531m049 y, which the cat was seen by Hubberley to visit DAILY to see how the young got on. this nest the cat
549m120 nts of the minute make large <parts> portion of DAILY <happiness» «pleasure». A wise man will try to
295c178 t like mean characters the others assumed--+++ ‖DAINES Barrington says cock birds attract females by

495q005 s place in natural Hybrids of Cabbages (7) Sow <DAISY> seeds of wild cabbage in VERY rich soil, will
496q006 d be effect-- (10) Try in how many generations. DAISY. Fever-fuge Groundsil.-- gilly flower will brea
508q016 f heredetary diseases) translated. (10) About DALTONISM in the MALE Troughtons.-- paper in Taylors Sc
204b140 very fine Hybrid offspring, much larger than the DAM, from those imported by Ld. Powis Hybrid dogs of
570n024 ps in air.-- Again a master says I will see you DAMNED first." the man shrugs his shoulders & replies
299c194 ect different localities,-- latter on banks & in DAMP parts.-- both propagated by seeds.-- There are
343d038 nimals. as in changing from <hot>. Warm to cold, DAMP to dry.-- Thus Tierra del Fuego has <not> «only
633j53v ca) is rolled along, & splits when it comes to a DAMP. place.-- Kolreuter mentions some hybrid, whose
023r010 go? Such have never been observed in Australia DAMPIER also repeatedly talks about the immense quanti
024r015 s; The whole coast of New Holland shoals much: DAMPIER remarks on great flats on the NW coast:-- 8 le
326c266 by Owen in Encyclop. of Anat. & Physiology.-- DAMPIER. probably worth reading Lessings. Laoccaon.--
453e182 ints near the tip crooked.-- is the form [...] DAMPIER. Vol I. p. 320. says no wild (carnivora) beast
022r008 in some of the dikes.--P 432. as in Andes. In DAMPIER'S voyage there is a mine of metereology with re
022r008 f winds & storms:--«in Volney's travels also» DAMPIER'S last voyage to New Holland P 127.--Caught a s
296c184 ame condition. Keeling Isd «shows where proper DAMPNESS seeds can arrive quick enough» Vegetation of
545m106 m pleasure, & may be compared to laughing» they DANCE with passion, ie. nervous impulse to action is
573n038 ike Kitten with mice.-- A person with St Vitus' DANCE badly, told should have shilling to walk to doo
507q15v ing of Seed. Anemone with, tuft-- Bull Rush-- DANDELION-- Sycamore. & seeds with «mere» border-- & Hu
299c194 .-- both propagated by seeds.-- There are two DANDELIONS, which just lately have been shewn to be sam
531m052 ce.-- A start is HABITUAL movement to avoid any DANGER-- Fear, shamming death, or running away. accom
531m053 «involuntary» movement from wish to avoid some DANGER-- but it is instinctive because Nancy tells me
560mIBC before experience can have taught them to avoid DANGER Do they frown, when they first see it? Charles
594n112 the birds at Maer have learned that he is not DANGEROUS-- wild-ducks would have fled equally if man h
588n089 fference on being carried up or downstairs, or DANGLED up & down-- in latter case they struggle their
515q021 ve» «no» horns by abortion, but sometimes have DANGLING ones.-- Is there any genus of plants, «in» wh
243c019 tre que ces oiseaux eussent leurs représentants DANS de si hautes latitudes"., --¿translate?} All Au
482z012 f Tortoises come from Galapagos!!! Azara. Voyage DANS l'Amerique Merid. Tatu noir. abundant from Para
289c162 Bot-- Vol II p. Dr Johnston <on> Entomostraca DAPHNIA, produce young, capable of producing young man
048r084 han single elevations along whole line of coast DARBY mentions beds of marine shells on banks of Red
172b007 many enemies. so as often to intermarry who will DARE say what result According to this view animals,
233b252 ered with the most beautiful savannahs & forests DARE to say that intellectuality is only aim in this
264c079 ss, not improving yet improvable» & then let him DARE to boast of his proud preeminence.-- «not under
559m153 u see. compare, the Fuegian & Ourang & outang, & DARE to say difference so great... "Ay Sir there is
606o025 ion (as jump off a bridge to save another) & yet DARE not -- one could do it, but other motives preve
638j58v species, like 10,000 others, perish. & who will DARE to say that this is an infringement on the wisd
601o08b s et physiologiques The first of these books I DARESAY good. 1. Sensation is the <conse> ordering con
218b192 good African & some good S. American forms. (& DARESAYS some of these <African forms> forms would hav
160g090 ng of syenite with pinkish felspar;-- whole hill DARK grey fine grained. Much contorted gneiss «narro
182b048 imilar to those brought from Abyssinia, & others DARK brown, with long clotted hair resembling that o
267c093 ilar to those brought from Abyssinia; the others DARK brown with long clotted hair resembling that of
359d089 these there were four, two like each other & two DARK-coloured & different.-- -- the former were the
421e091 ter fish peculiar to Ireland. [do p. 283. on the DARK ears of the wild Chillingham Cattle, with refer
435e134 now the Proteus anguiformis. he remarks lives in DARK caverns of Carniola p. 112. Man. "standing alon
542m095 me way as one lifts up eyebrows to see things in DARK. & hence is this the cause of expression of sur
576n047 osed to progressive. developement) on account of DARK ages.-- «effects of external circumstances» Loo
578n053 some saying about a person even blushing in the DARK-- «so modest a person.» A person who blushes in
578n053 so modest a person.» A person who blushes in the DARK is proverbially a most modest person one carrie
584n072 idgeons proverbially carried to long distance in DARK "it is inspiration."-- this is class of so call
584n072 of a new sense,-- bats avoiding strings «in the DARK» as well might be called instinct,-- migrating
586n080 ht, where to go-- the act of crossing the sea in DARK night & not loosing its direction, equally wond
405e031 tail. & kind of semilunar {P} mark on each side DARKER,, so that whole colour is changed, these best
459t013 ner of walking but not waddling; its colour is DARKER than the penguin & the bright feathers on its
260c068 d in white rooms to give tinge to offspring.-- DARKNESS effect on human offspring.-- white, snow.-- t
604o018 living in lofty regions. 3 Infinity eternity. DARKNESS, power. being associated with God. these phen
275c121 e says he recollects all half Bred cattle of L'DARNLEYS were most like parent Brahmin bulls-- Mr. Y.
091a027 degree. <strike> «direction<?>»ESE-- CD [In the DARWAR. transition Hills & strata SE. direction of tr
081rIBC 1 lig[nes] Decimetre 3. 8 Centimetre 4.4 {T} C. DARWIN R. N. Range of Sharks Nothing For any Purpose
139aIBC worth visiting really good account of ice.-- C. DARWIN A. Glen Roy Generally received opinion that ma
170bIFC hat <nervous> brain makes thought Glen Roy B C. DARWIN All useful pages cut out Dec. 7th. /1856/ (& a
203b136 in India p. 261. L. Institut. 1837 Mem. Sir F. DARWIN cross breed boars were wilder than parents. wh
239cIFC ffect Account of the [...] the world.-- Charles DARWIN written between («beginning of» February & Jul
331dIFC dy Botanical work on Buds & Gemmae. C D Charles DARWIN 36 Great Marlborough St Did Eytons <intermedia
520mIFC k] Questions & Experiments Expression M Charles DARWIN Esq 36 Grt. Marlborough Str.-- (p. 64. On <ins
534m063 is Malva sylvestris. do. p. 228 Newport says Dr DARWIN mistaken in saying common wasp cuts off wings
539m083 Aug. 16th. As instance of heredetary mind. I a DARWIN & take after my Father in heraldic principle.
560mIBC Do they frown, when they first see it? Charles DARWIN 36 Great Marlborough St Has my Father ever kno
563nIFC hat are sexual difference in monkeys.-- Charles DARWIN [Private.]CD (Metaphysics & Expression) Select
322c269 hels. introduction to Natural Philosophy R. W. DARWIN'S Botany.-- References at end Mayo Pathology of
536m071 logous to men's affect for womens breasts.∴ Dr DARWIN'S theory probably wrong, otherwise horses would
632j53r assion" of Hutcheson unfolded by D. Hartley.-- DARWIN'S Abstract of John Macculloch 1837 Proofs and I
275c120 ull-dog. & crossed & recrossed, till there was a DASH of blood, with whole form of grey hound.-- pick
314c238 acter of Flora, N. Zealand & N. Caledonia with a DASH of New Holland. As in N. Zealand-- Some species
557m147 very stiff «& back» when savage «no» & ready to DASH at prey streched out & flaccid, when furious «w
583n070 called "creatures of instinct" with some slight DASH of reason so mean are called "creatures of reas
123a117 ay. states remains found in many part.-- great DASYPUS near Canelones -- large quadruped bigger than
479z006 n Sumatra Marsden. p. 311 D'.Orbigny considers DASYPUS villosus is true Peludo Cavia Australis. Dorbi
278c130 Mammalia arranged according to my own methods. DASYURUS being found fossil in Australia, & only one t
025r019 ngly rare, at the bottom of the sea.--«certainly DATA insufficient, yet good» «(I suspect fragments o
132a137 et Naturelles. Tom I. p 501.-- shows first that DATA wholly insufficient to calculate rate of increa
277c127 habits ascertained-- put them as (a). (b) until DATA be given.-- This will aid in preventing the cha
497q006 Bag of soil from centre of woods «especially if DATE of wood be known» & other odd places & see what
031r038 mice, though Trachyte. same fact in Galapagos. DAUBENY P 24 V. back of page 1 of New Zealand Geologic
033r043 in James Isd.--Mem St Helena-- All Trachytic.--DAUBENY P. 171. Vol I. Humboldt There is long discussi
033r043 d abstract of Humboldt. S. American Geolog. in DAUBENY. P. 349 Admirable little table showing long PE
033r043 nce volcanic. from Humboldt: Comparison P 361. DAUBENY Von Buch is very strong about Trachyte being t
034r044 stream at Portillo Pass example of do? <Poor> DAUBENY good account of ejected granitic fragments P.
035r047 s of formations on any Geolog Map: Quoted from DAUBENY P 402: likewise, mean height of tertiary. bein
036r051 lection is imperfect & was recalled by note in DAUBENY. P. 438., of similar fact near the Red Sea.--w
049r089 given to the numerous hills of greenstone? -- DAUBENY. P 95. Glassy & Stony Pearlstones alternate to
124a118 r) wrong entrance Athenaeum. 1838. p. 652. Dr. DAUBENY on mountain Chains in N. America Erasmus sugge
552m131 e to cause friction Athenaeum 1838. p. 652. Dr DAUBENY on the direction of mountain chains in N. Amer
021rIFC e / July 1835. the excess of harbour = 180 See DAUBISSON both Volumes, and Molina 1st Vol & Lyell Sail
385d165 e moral qualities of the father! In how many DAUGHTERS does the character of the mother revive! Or t
385d165 the mother in the son, & of the father in the DAUGHTERS! This last remark good. because showing proba
400e012 -- Kerr's Collect of Voyages Vol 8 «p. 46» Capt DAVIS in 1598 found cattle in Table Bay with Humps on
051r095 on I ever saw on beach near Callao.--From Sir. H DAVY experiment on the copper bottom. we see a trifl
323c269 .-- -- 19t. Mungo Park-- travels Feb 12. Sir. H. DAVY Consolations in Travels -- Observations on mora
580n062 ly very important work.-- Feb. 12. 1839. Sir. H. DAVY -- Consolats: "the recollections of the infant
581n065 astonishment-- thinks so it must have been in the DAWN of civilization-- thinks many words, roar, scra
043r071 e" to "from the epoch of Ammonite to the present DAY. at Mauritius. (consult Bory «dip of strata on E
044r073 s quotation of instability of ground at present. DAY.-- applied by me geologically to vertical moveme
181b041 ing ten thousand years hence; because at present DAY many are relatives, so that by tracing back. the
206b146 n of place be constant «say 2000» and at present DAY, every ten living souls on average are related t
206b147 more than a few will have successors. at present DAY. in looking at two fine families one with succes
222b206 sible to suppose such an accumulation at present DAY & not include Mammalian remains.-- The Father of
232b246 temperature of two countries" finding a very hot DAY, «in one», oh we will take a day from the equato
232b246 ding a very hot day, «in one», oh we will take a DAY from the equator to add to the mean of the other
257c058 come to be, many genera of fish &c &c at present DAY.-- It is ASSUMPTION to say generation produces y

```
266c091 eir appearance & manners they are as opposite as DAY & night: yet we know how remote the periods at w
278c131 habitants most important!! like Dipus of present DAY??! Major Mitchell does not think that dog was fo
316c246 cattered over whole world Many shells at present DAY same (or according to Sowerby fine species) on c
318c254 f America] migrate singly flying few miles every DAY «generally by night» -- other birds which is str
332d002 previous  years, two having consulted him on one DAY -- Mark at Shrewsbury thinks the half bred Alder
334d012 nged his residence a great many miles.-- yet one DAY <th> a cow walked in, then disappeared, & three
370d116 ome kind lay <pu> up honey even for single rainy DAY-- & from case of wasps, is supposed cells proper
414e064 , or instincts (ie intellect in man) to gain the DAY.-- In man chiefly intellect, in animals  chiefly
416e075 he Male, instead of allowing strength to get the DAY» The fertility of Indian & Common Oxen, which on
437e139 they  were amusing themselves with them, and one DAY a rabbit escaped into a hole, where the old Eagl
437e140 le could not find it..-- The parent bird another DAY brought to her young ones the cub of a fox, whic
451e176 Newbold.--» A Malayan albino described "To this  DAY the tomb of his grandfather, who was also an alb
466t093 saw  no Bees «several» visiting it».-- In yellow DAY lily, the Bees visit base of upper petal, though
468t111 days Humble seem to frequent certain flowers, to DAY early, the great scarlet Poppy-- So that, finall
469t135 numerous» bees visited this same bunch & on this DAY in five minutes eleven Humbles came & each visit
469t135 tes-- say then each flower is visited 30 times a DAY is considerably under mark, & this has now gone
472s02r hursday After watching 14 days. many times every DAY. many clumps of heartseases, never saw any Bee g
472s02r ed that many flowers had suddenly withered, & to DAY saw very odd dusky humble (with pollen) on  legs
495q05a n flat places & see whether wind, on «dry» windy DAY, «flower garden on gravel walk» will drift  many
528m039 much  to do, as may be known by autumn, on clear DAY.-- 3d pleasure association warmth, exercise, bir
529m043 een no cloud on the mind, every occurrence for a DAY or two are absoluteley forgotten.-- My father si
546m110 ho told the fact to Mr Mayo himself. she was one DAY reading a book, with ivory paper cutter, which s
559m156 nt if irritated very regularly at one time every DAY.-- naturally close at that time after long perio
577n050 n,-- a certain round of actions take place every DAY, & closing of the leaves, comes on from want  of
579n060 gin of idea of causation; «succession of night & DAY does not give notion of cause,» do p. 135.-- on
595n117 t will not be allowed they can dream, & not have DAY-dreams-- think well over this;-- it shows simila
609o29v hese passions, but refers (I believe) to present DAY & not to ruder state of Society.-- Civilization
610o032 ated in a wonderful manner.-- as the hour of the DAY &c-- All habits must conduce to their health & c
615o36v I think Pincher shows surprise, walking home one DAY met him, with Mark riding instantly followed, me
618o41v iority of one to other: no not only thus, for if DAY was first, we should not think might an effect.]
225b219 me  deny the creation of any new quadruped since DAYS of Didelphis in Stonefield;. all lands united (F
282c143 on the the enormous production-- millions in few DAYS-- one doubt that one animal can really  produce
334d012 <th>  a cow walked in, then disappeared, & three DAYS afterwards came again, bringing with her the ot
346d048 Chillingham,--  habits peculiar.-- young one 203 DAYS old butted violently. & fell.-- gore to death t
347d049 mprove.-- yet fish same as, or lower than in old DAYS: «for a very old variety will be harder to vary
347d049 e Whewell as profound. because he says length of DAYS adapted to duration of sleep of man.!!! whole u
376d136 t coat made. for it at first, but in two or three DAYS learn its comfort & though could not put it on,
386d167 in  animals‖. One «invisible» animalcule in four DAYS could form 2. cubic stone. like that of Billin.
407e037 ssed a climate compared to S. America at present DAYS,, which S. America now does to North. America &
468t111 quite below stigma. & so avoided it.» On certain DAYS Humble seem to frequent certain flowers, to day
468t112 s Azalea or Rhododendron xx after several gloomy DAYS. hot one, Bees almost P every minute to Fraxine
469t135 many flowers.= 22d.-- /during several succeeding DAYS. <many> «most numerous» bees visited this same b
469t135 nsiderably under mark, & this has now gone on 14 DAYS. (except some wet ones/ & wd go on longer-- Woo
472s02r 2. 42 Maer <Thursday> Thursday After watching 14 DAYS. many times every day. many clumps of heartseas
472s02r was not case, on several flowers I examined some DAYS ago-- This Bee flew from yellow to yellow & pur
495q05a an annual» <sleep> «closes flower» on all gloomy DAYS.-- The «garden» Coronella also sleeps on ditto-
505q014 t Sundorne has large Bees July/42/ Mark has six DAY'S puppy of small true Bull-Dog-- length from nose
535m064 aying eggs on leaves of Eucalyptus, watching few DAYS till larva excluded, then though not feeding th
539m081 much struck with an intense headache «after good DAYS work» which came on from reading «review of» M.
546m109 ........ he had had, he would say how many happy DAYS. he spent in such a place.-- Vide page 103, supr
550m125 what  is Happiness?-- When we look back to happy DAYS, are they not those of which all our recollecti
025r019 ts of shells will generally be found to be old & DEAD)» «(I have not kept a record)» In looking  over
080r178 w & intermediate Puncture one animal with recent DEAD body of other. & see if same effects, as with o
134a141 ight on road from Valparaiso to Santiago p. 328. DEAD trees on Isthmus of Pen. Tres Montes.-- as by s
161g096 id it, could be followed for at least 2 miles on DEAD level «by eye» to moss-- on this terrace Barom.
177b025 ps be called the coral of life, base of branches DEAD; so that passages cannot be seen.-- this  again
177b029 » <therefore> changes and base of branches being DEAD from which they bifurcated.-- Type of Eocene wi
184b054 oes might be brother to Megatherium.-- uncle now DEAD. Bulletin Geologique April 1837. p. 216 Deshaye
212b168 ephalopods,) in last tertiary epochs most genera DEAD? --Examine into this «in Phillips».-- According
213b169 domesticated  <species> races.-- If all men were DEAD then monkeys make men.-- Men makes angels-- Tho
228b231 imitation, fear <of death>. pain. sorrow for the DEAD.-- respect We have no more reason to expect the
284c148 f such Nature far apart. Must have travelled by <DEAD> «each» trees dying & mountain torrents.--  but
352d058 roup of species is made. father probably will be DEAD-- hence there is no central radiating point, al
437e139 wounded the bird. p. 124-- Mr Willoughby found a DEAD lamb «& hare» by the side of Eagles nest, which
502q11v it «in» melon» «Loasa» Anchusa «Campanula» &c & DEAD-nettle.-- Lithospernum. Blue Gloss. it is not p
505q014 amens will be produced in individual plants 17 A DEAD-nettle in Hot-house. will it seed?-- (Skim thro
524m022 bove described, (forgetting that her husband was DEAD) yet instantly perceived when my Father to dist
525m025 nesses of the foetus.-- some mothers. have first DEAD children, then children which were short  term,
594m115 a «blue» Gibbon. whose companion had S[...] been DEAD about two months. saw a «black» spider monkey b
638j58v hole rocks nay very mountains are formed of such DEAD & extinct forms.-- the exuviae of the dead & ex
638j58v such  dead & extinct forms.-- the exuviae of the DEAD & extinct The analogy between the works of art
177b027 cies that in perfection, the bottom of branches DEADEN.-- so that in Mammalia «birds» it would only a
414e063 eases to each other &c, but then comes the more DEADLY struggle, namely which have the best fitted o
334d011 ue eyed cats (Fox has seen repeated cases) being DEAF curious case of corelation of imperfect structu
592n102 ournal paper showing that the signs invented for DEAF & dumb school & used between Indian tribes  are
308c219 ifferent, because their difference. arise a good DEAL from climate & habits, & therefore less fertile
460t019 c beings are the descendants, <slightly> «a good DEAL» modified <& Many Forms lost; if> «of this  old
539m084 s flushing & with muscles rigid.-- How is this? DEALT with p. 241 Origin of man now proved.-- Metaphy
548m116 el sympathy. as for for the heard suffering of a DEAR friend-- this gives one strong idea of what ind
565n009 light protrusion of lips, as if going to say "my DEAR," just what smile is to laugh.-- I must be very
375d135 n in Men increase, & an ordinary crop. causes a DEARTH then in Spring, like food used for other purpo
062r129 case.-- Should urge that extinct Llama owed its DEATH not to change of circumstances; reversed argume
176b022 ew ones generated» There is nothing stranger in DEATH of species, than individuals If we suppose mona
180b038 aptation of such circumstances & therefore that DEATH of species is a consequence (contrary to what w
203b135 -- & now divided again-- Weakest part of theory DEATH of species without apparent physical cause:-- M
228b231 ?» Animals with affections, imitation, fear <of DEATH>. pain. sorrow for the dead.-- respect We have
228b232 r> animals our fellow brethren in pain, disease DEATH & suffering «& famine»; our slaves in the most
278c130 s showing former connection of two continents & DEATH of form in one. The caves are at a height of mo
281c138 culiar tails strange ¿¿/Genus only natural from DEATH or slow propagation of forms.--just same way as
286c154 ression of countenance. «[s]hare of sickness,-- DEATH, unequal life,-- stimulated by same  passions--
291c167 me flowers hermaphrodites & others not??? ‖ The DEATH of some forms & succession of others, (which is
297c185 species., same circumstances., which by causing DEATH, makes the group aberrant When species rare  we
300c197 er him created from animals.-- Insects shamming DEATH, most difficult case to iimagine how art  acquir
300c197 er only dropping where ground thick.-- shamming DEATH it is but being motionless. How is  instinctive
300c197 ingly doubtful whether animals have any fear of DEATH, only of pain. of death acquired?. The S. Ameri
300c197 nimals have any fear of death, only of pain. of DEATH acquired?. The S. American dung beetles will ea
302c200 o create.-- It is very remarkable, with so much DEATH, as has gone on,, No greater gaps.-- external c
304c208 d where many species <osculant> but. where much DEATH, may be inferred much time elapsed & therefore
346d048 03 days old butted violently. & fell.-- gore to DEATH the old & wounded,-- see Annals. vol. 2. 1839.--
375d134 but the positive check of famine & consequently DEATH.. population iin increase at geometrical ratio i
531m052 L movement to avoid any danger-- Fear, shamming DEATH, or running away. accompanied with want of musc
532m057 ar.-- the state of collapse may be imitation of DEATH, which many animals put on.-- The flush which a
535m064 (as  in birds blind storge-- They continue till DEATH, thus acting 4 to 6 weeks. The deserted broods
540m089 e (Vol. I) has well observed, one does not fear DEATH from its pain, but one only fears that pain, wh
541m089 e only fears that pain, which is connected with DEATH!-- How has this instinctive fear arisen?  19th.
544m101 s pleasure, ie. love.-- & so pain gives fear of DEATH. Mayo Philosophy of Living. p. 140-- Dreams goo
544m104 t race. argument for early education.-- fear of DEATH!!! as Montaigne observes. distinct from pain, f
544m104 pain, for one hates pain from this fear-- & not DEATH for the pain.-- How was this instinct  gained.?
```

555m143 okes. about not having run away &c having faced DEATH like a hero, & then I had some confused idea of
564n004 till at last he face «instinct of» hunger, «of» DEATH & for the satisfaction of following conscience,
575n043 other hybrids» with <the> it the provision for DEATH.-- can we deny that brain would be intermediate
601o009 ation, in order to avaoid it-- beetles feigning DEATH upon seeing an object.-- are Planariae consciou
615o038 Parental feelings weakened in Otahiati; fear of DEATH in Hindoo population.-- Slightly modified in ma
620o045 e forgotten. [RHC] 4) as starvation, or fear of DEATH, one makes allowance & either excuses the «non-
622o049 ple as I have said.-- [LHC] instinctive fear of DEATH: of hoarding.. Ld. Kames, which Sir. J. says is
635j56r imals feeding on each other &c &c».-- «(Causing DEATH to some, &c &c)» These are reasons, just as lia
635j56r REAT SYSTEM. C. D]CD [All this does not explain DEATH, but reproduction]CD though such a scheme. woul
638j58v vage ever made a perfect hinge.-- reason, & not DEATH rejects the imperfect attempts. In the «Bee» Mo
640j167 roduced, or <a spe» became more numerous. (from DEATH of its destroyer), or other cause, the long leg
175b020 , (& to multiplications when isolated, requires DEATHS of species to keep numbers of forms equable:--
284c148 in torrents.-- but to crawl up an hill, then by DEATHS?!)-- looks like subsidence.-- on the islets Mr
286c154 ble in savages) Has not the white Man, who has DEBASED his Nature «& violates every best instinctive
106a076 there same.> Institute. 1838 p. 40 or Phil Mag. DEC 1837. p. 520 Mr Fox on increase of temperature a
145g001 ly than female p 367 Quarterly Journal of Agricl DEC 1837 Yet instances given against it-- Mere fact
170bIFC ht Glen Roy B C. Darwin All useful pages cut out DEC. 7th. /1856/ (& again looked through April 21 18
239cIFC bruary & July 1838) All good References selected DEC. 13 1856 Also looked through April 23. 1873 Book
331dIFC s recognising Sexes of animals:]CD [All Selected DEC. 14-- 1856]CD Towards close I first thought of s
345d043 ?.-- -- Quarterly Journal of Agriculture p. 367. DEC. 1837. Generally-- received opinion that male im
397eIFC Gardens D E Finished July 10th 1839.-- Selected. DEC 15 1856 [not located] Epidemics-- seem intimatel
520mIFC ls & Speculations on Expression -- 1838 Selected DEC 16 1856 July 15th 1838 My father says he thinks
563nIFC sics & Expression) Selected «for Species Theory» DEC. 16 1856 Looked through & all other Books May 18
174b013 ostriches in. S. America-- This is answer to DECANDOELLE. (his argument applies only to hybridity.--
236b280 ed species always from distant countries, as DECANDOELLE says, no he only says sometimes we might exp
324c268 end of Flinders & at end of the Congo Voyage DECANDOELLE. Philosopher. or Geographical distrib. «in D
375d134 y. Even the energetic language of <Malthus» DECANDOELLE» does not convey the warring of the species
400e013 dalaxura-- October 11th.-- Uncle John-- says DECANDOELLE, distributed seeds of Dahlia all over Europe
454e184 anize <ani> them-- Cross Irish & Common Hare DECANDOELLE has chapter on sensitive plants; Physiology
492g003 crossing of Cabbages, quoted by (as if oral) DECANDOELLE in V. Vol of Hort. Transacts & M. Sageret is
637j57v n, utter extinction! let him study Malthus & DECANDOELLE.-- The Final cause of innumerable eggs is ex
350d054 ecture?-- Australians.-- Americans. &c After DECANDOLLES idea Septemb. 1st. Macleay & Broderip were t
373d133 50 Grallae in Wealden. oldest birds. p. 411 -- DECAPOD Crust in Muschelkalk, & 5 genera of reptiles.-
055r108 yearly brought down by every torrent proves the DECAY atmospheric of the most solid rocks.--The grand
155g066 so, these lines & even water-scooped rock «only DECAY from fragment falling» of no particular hardnes
285c151 e branch of the tree if live occupied after its DECAY, will be occupied by the vigorous shoots from e
285c151 he vigorous shoots from each branch No: because DECAY in that species <shows> is effects of unfavoura
163g108 ls must have been about 60 ft above sea-- soon DECAYED on exposure Mr H. C. Watson Geographical distr
153g056 lower & other smaller ones «these boulders are DECAYING.» neighboring rock gneiss & [...] sandstone a
516q23v al analysis of Peat (2) Athenaeum 1840 p. 777. DECAYING wood absorbs oxygen & forms Carbonic Acid. wi
414e060 Haena &c are found together.-- Read this Work-- DECB. 4th.-- Why has the organization of fishes & Mo
225b218 eing found in Tristan D'Acunha. may be said to DECEIVE man. as likely as fossils in old rocks for sam
262c074 arbitrary. leaves door open for Quinarians to DECEIVE himself.-- Give the case of Apterix-- split, d
263c077 sent form» will come, (or how dredfully we are DECEIVED) then he is no exception.-- he possesses some
060r124 st now proceed from great depths.--important.-- DECEMB 10. 1802. Earthquake at Demerara. The earthqua
418e085 through as many changes, as has any species.-- DECEMB. 21th.-- L'Institut 1838. p. 412. M. Eichwald
419e086 numbers of individuals from one, is adverse.-- DECEMB. 25th.-- Lyell says the elevated shells in Bay
578n056 ear some references to thoughts of other person DECEMB. 27th.-- Fear loose the sphincter muscles, onl
117a101 & 50 from Portezuelo. Bull: Soc. Geolog. 1837. DECEMBER. p. 91. a classification of Europaean strata
413e059 fertility in proportion to support of parents DECEMBER 2d Lyell tells me Beck considers the characte
415e069 as in «races of» Dogs, so in species, & in Man DECEMBER 16th. The end of each volume of Whewells Indu
522m013 n insanity or delirium.-- In Mania all idea of DECENCY & affection are lost.-- most delicate people d
065r138 rought home at Capt. Forster expedition from <DECEPTION Isld.> South Shetland Cape Possession. Syenit
065r139 Craters of Elevation.--The longer diameter of DECEPTION Isl is six Geographical miles and width 2 & ½
066r142 gether with same general character of fossils DECEPTION complete.= Silliman Journal. year 1835 excell
547m113 whether I had pulled the bell??)-- It may be DECEPTION to say the mind <thinks> quicker in sleep, it
473s07v ssume some form late in youth,-- only facts can DECIDE-- some peculiarities may be early impressed &
641j29r lance, continents have been split up.-- who can DECIDE their limits.-- To show how little we understa
055r106 e beaches. Perhaps these facts attest a <more> DECIDED elevation of sea's bottom. beds of shells. 2-
077r171 valley of Marfil is divided, appear to have a DECIDED influence on the richness of the veta madre of
039r060 subject.-- S. America in the form of the land DECIDEDLY bears the stamp of recent elevation. which is
067r144 Sowerby. younger. says that Falkland fossils DECIDEDLY belong to old Silurian system. Apply degradat
204b139 the mixture between Chinese & English Breed. DECIDEDLY exceedingly prolific & hybrid about half way.
236b279 m. Mr Bell's case of Sub Himalayan land emys, DECIDEDLY an Indian form of Tortoise.-- On other hand.
236b279 Germany. (where Mr Murchison fox was found), DECIDEDLY next species to some South American kinds.--
333d007 Esquimaux dog & common dogs & Fox thinks they DECIDEDLY take much most after Esquimaux.-- this agrees
081rIBC Hectometre 51. 1. 10 Metre 3. 0. 11 lig[nes] DECIMETRE 3. 8 Centimetre 4.4 {T} C. Darwin R. N. Range
602o011 ough he feels he is right-- it is because each DECISION &c is made up of many partial results, & the
602o011 (p 115) a very good passage. about actions & DECISIONS bein the result of sagacity, or intuition. wh
110a085 Deshayes saw fossil shells from West Indies & DECLARE them to be recent species-- Lyell-- Some inter
343d037 here was light.-- «bad taste {whom it has been DECLARED "he said let there be light & there was light
622o048 is not connected with sense» instantaneous so DECLARING it is right or wrong.-- «[just as in taste o
599o05v of words-- argument of original formation.-- DECLENSION &c often show traces of origin.-- Mayo Philo
446e161 p. 377.-- Statement that the climate is on the DECLINE, as far as vegetation is concerned, in parts o
183b053 eds declining as great reptiles must have once DECLINED.-- Read his theory of the Earth attentively C
183b053 r whole world, the period of great quadrupeds DECLINING as great reptiles must have once declined.--
051r095 oes animal adhere to rock because it does not DECOMPOSE. or vice versâ. Clay slates unfavourable to a
096a041 for knowing that Mur. Soda. and Carb of Lime DECOMPOSE each other.-- on Direction of mountains in Br
105a072 ra.? How came it if this powder results from «DECOMPOSED sea» shells, that land shells should be pres
121a111 ell on Sweden p. 5. «& 7.» violet strata from DECOMPOSED muscles.. Smith of Jordanhill has seen same
029r033 ormations of Brazil, all originate from the DECOMPOSITION of Granitic rocks Mem. Chanticleers voyage
065r139 earth. yet body had scarcely undergone any DECOMPOSITION: countenance so well preserved. that it was
135a145 rs by Benza Neilgherries-- Much inform. on. DECOMPOSITION of granite-- Bengal. J. vol 7. p. 522. Moun
222b205 in character, the same reasoning will allow of DECREASE in character. (which perhaps is) Case with fi
230b235 aph Journal Vol V. P. I. p. 67. Dr. Coulter on DECREASE of population in California cessation of fema
326c266 y Malthus.-- Heberdens Observat. on increase & DECREASE of different diseases. 4to 1801.-- quoted by
399e009 to childhood, or solely to manhood,-- it will DECREASE & be driven outwards in the grand crush of po
622o049 at head of series in which «special» instincts DECREASE, I should think they were very few & general
436e135 eee.-- although the Cephalopods, seem to have DECREASED since earliest times-- Apterix has a most per
436e136 enera It certainly appears that swallows have DECREASED in numbers, what cause?? Seeing the beautiful
206b149 xclude mothers & then try this as simile In a DECREASING population. at any one moment fewer closely r
375d135 wks. by. cold &c--.. even one species of hawk DECREASING in number must effect instantaneously all th
541m089 mind as much as possible (testing success by DECREASING headache) & found best plan was allowing my
116a099 of Rio Negro beds.-- -- refers to species non DECRITE de petites corbules analogue living in mouth o
244c022 d» to come originally from Africa p. 122. Mus DECUMANUS, at Caroline Isld, & a Roussette p. 136. Isle
196b110 at Kingdoms.-- Principes de Zool: Philosop:-- I DEDUCE from extreme difficulty of hypothesis of conne
640j167 ted animals changing.-- From these views we can DEDUCE why small islands. should possess many peculia
613o036 dified according to species. This I suppose he DEDUCES from the ends in each case being the same, & t
570n026 -- analogous case to my idea of conscience.-- DEDUCTION from this would be that a mountaineer <takes>
410e051 towards law of transmutation, cannot see the DEDUCTIONS which are possible.)-- Ascertainment of clos
484z014 hern & Southern America-- valuable & practicable DEED Caricaridae wanderers.--?? in N. America?? Wils
023r011 ty.» That axis was produced, from a fissure in a DEEP & therefore weak part of the ocean's bottom. Wi
028r029 containing Alumen.--This matter accumulating in DEEP seas forms slates: How is the Lime separated; i
030r035 t now collecting, in the bottom of an open & not DEEP sea.--(Character of coast regular & <not very>
030r035 -(Character of coast regular & <not very> rather DEEP soundings, 60-100 fathoms 2 & 3 miles from shor
035r048 facts become easy if we look at the action as a DEEP & extensive movement of viscid nucleus, which i
041r064 ery doubtful whether they could have lived in so DEEP a sea.--Perhaps agrees with formation of pebble
051r097 E. America. <East> Africa. Australia. profoundly DEEP: a great fault or rather many faults.-- Necessa

```
066r140 grow  on shoals like Fucus giganteus! 24 fathoms DEEP 24» under 50. Kerguelen Land, = the way it stan
080r181 avels Beauforts Karamania Capt. Ross. & Scoresby DEEP soundings Gilbert Farquhar Mathison travels Bra
091a026 ndesite. Red Coral in the Mediterranean 700 feet DEEP in some of. the twopenny periodical said so. «C
104a069 of  Breccia, introduce in Cordillera discussion, DEEP sea, fragments fall off cliffs. but then how sp
113a092 t mouth excavated in solid rock.-- 4 & 5 fathoms DEEP. perfectly still water. Major Mitchell inferred
114a094 ect of cleavage Clay slate. a distinct formation DEEP «& therefore extensive» water ∴ not formed in m
115a096 ught to be creeks & mouths of rivers ought to be DEEP.-- Henslow has deposited specimens from Anglese
117a101 h vegetable remains formed near coast, limestone DEEP water. will bear on formations. during elevatio
119a105 whether they can have been plunged so many miles DEEP into the bowels of the earth, as would be requi
120a108 h nearly cold rock.-- in volcano the pool is not DEEP. --Hot springs &c &c--then if so, thermometer s
126a123 ught to increase rapidly beneath level of sea.-- DEEP seated springs «spring requires connected colum
153g055 e 60 degree Granite--«band» 4 X 3 X 2 «feet» & 2 DEEP Another rather smaller block 30 ft «above» & ot
155g064 ight remove, what above straight line «only» cut DEEP gorge on sea hypothesis, if gullies not now for
161g095 e three [...] abutted Having crossed the mouth, (DEEP) of above valley this road level with Peat moss
352d059 up without centre but circle, two or three lines DEEP-- with respect to Macleay's theory of analogies
639j28v » long «(as adapted to)» because their food lies DEEP.-- I say it is «as» simple consequence they bec
105a073 e power. therefore tendency of running water to DEEPEN not to widen valley.-- Why is serpentine cours
050r091 ater & nearer the Banks Is there not a sudden DEEPENING on E. coast of Africa. as at Brazil [blank] N
106a034 ual in widening valley.-- it is essentially a DEEPENING agent {P} Therefore when we have valleys of t
025r016 y shoal coast. Beyond the 10 or 12 leagues sea DEEPENS suddenly. coast of Brazil generally.-- Mrs Pow
431e122 less in number. (¿ species, or individuals) the DEEPER one goes-- surely is this true?-- most strange
609o029 «we alas know» is far easier conquered than the DEEPER & worser feelings. These bad feelings no doubt
056r109 e such islands whose inclination natural [...] DEEPEST astonishment.» Perhaps scarcely a pebble might
093a032 preexisting mineral.-- Mem. Galapagos ∴ Basalt DEEPEST?? Marcel Serres L'Institut. 1837. p 331 Consid
234b256 ent from» near London. = Dr. Smith, he says, is DEEPLY [not located] of «all» genera, «in all classes
249c034 roduce hybrids but not varieties. which are not DEEPLY impressed on blood., will cross & produce fert
366d108 d, & if their characteristic qualities were all DEEPLY imbued in them from long permanence, so that a
374d134 28th. «I do not doubt, every one till he thinks DEEPLY has assumed that increase of animals exactly p
410e050 be endless changes, & hence no feature would be DEEPLY impressed on it, & hence there could not be im
539m081 ich made me «endeavour to» remember, & to think DEEPLY, & the immediate manner in which my head got w
548m115 one is awake many necessarily are., when one is DEEPLY reasoning besides these (which must be present
609o029 er <to> at the very beginning, & therefore most DEEPLY impressed). shame perhaps an exception. (does
041r064 trees  Grand Seco at B. Ayres; mention about the DEER approaching the wells.-- the effect of Salt wan
148g024 ood by Inverorum being determined by sheep & not DEER When Black faced sheep are crossed with English
244c022 cs from Madagascar. Monkey from Java.-- Hairs, & DEER.-- Procured two makis alive from there.-- Mem W
253c046 red too separate isld very long» America & India & DEER.= Africa not.-- Africa Camels?? Africa Bears??-
258c061 cture. «hence seals take victorious seals, hence DEER victorious deer, hence males armed & pugnacious
258c061 als take victorious seals, hence deer victorious DEER, hence males armed & pugnacious (all order; coc
341d030 of this animal, which is so anomalous among true DEER, yet is spotted like so many deer.-- very curio
341d030 ous among true deer, yet is spotted like so many DEER.-- very curious like some facts of Mr Blyth on
363d103 s spotted white when a fawn compare with fallow? DEER. & Moschus &c & -- like young blackbirds Dr Bac
402e018 nimals he heard of with pigs, small bears or badgers, DEER. apes, baboons, monkeys & an animal probably a
405e031 25th. I observed in Windsor Park.-- the «Fallow» DEER. which were of a nearly uniform <dusky> blackis
407e042 all, genera must. Lesson I remember says Mariana & DEER very close to a Molucca species.-- L'Institut 1
408e044 quiae Diluvianae. p. 222. Bones of Horse. Bear & DEER at 16000 ft. with Snow on Himmalaya-- Humboldt
450e173 Sooloo. imported elephants. wild hogs-- spotted DEER, no loonies, but cocatores & small green parrot
450e175 t--. (p. 270) says many wild horses, bullocks, & DEER South part of Mildanao.-- Q Horse do. Appendix.
451e177 ing it are several small islands. abounding with DEER-- Horsburgs. Vol II. p. 527.-- <Scientific Soci
452e181 E. end of Gilolo.-- "-- Forrest Voyages. p. 39-- DEER but no wild animals in Gilolo.-- p. 134: Birds
459t009 gs of Science Vol III. p 320. Mr Hodgson on Musk DEER-- young spotted <like in> "prettty much as we s
459t009 he young of the wild hog & of several species of DEER, which are altogether immaculate when grown up"
500q010 eposit eggs in all varieties of Cabbage. (26) Do DEER Keepers cross the breed-- desirable as in Cattl
500q010 Chillingham  Park-- What Book on varieties &c of DEER. Contests of sexes.-- Q.30) March 1842. <Last>
529m041 would  be excited, & how the scenery would rise. DEER in Parks ditto.-- My Father says there is case
537m076 is probably is natural. consequence of man. like DEER &c, being social animal, & this conscience or i
592m105 being so is not odd.; for instance wild cattle & DEER pursuing a wounded one.-- porpoises a ditto-- i
623o051 RHC] 10) that the instincts of bees & beavers «& DEER» have <been formed> a beneficial tendency to th
214b176 t was difficult to obtain a litter without this DEFECT, Very curious case = W. D. Fox. When dogs are
335d014 red people, <Hill» «Lord Berwick» family with DEFECTIVE palates. heredatary & therefore exceptions. t
572n033 ss reasoning powers than Europaean.-- Ideots. DEFECTIVE brains.-- Erasmus does not liken term instinc
531m053 voiding  urine because done by some animals in DEFENCE, &c Starting must be habitual «involuntary» mo
446e160 s building nest. sitting on eggs. & feeding & DEFENDING their young.-- The oriolus (icterus Cat.) is
525m023 g their instincts naturally.-- (generosity in DEFENDING a friendly dog).-- they feel shame, when doin
563n022 also  by greater temptation as bitch: or dogs DEFENDING companion. (mem Cyanocephalus. Sphynx howling
619o042 y that the individual forgets itself, & aids & DEFENDS & acts for others at its own expense.-- Moreov
551m129 pushes out both lips in contempt <&> disgust & DEFIANCE.-- different from sneer-- How easily. horses
384d161 situation of parts (2) addition of parts, (3) DEFICIENCY of parts (4) combined addition & deficiency
384d161 ) deficiency of parts (4) combined addition & DEFICIENCY of parts, as in Hermaphrodites, (shows my do
520m002 t Mr Dryden C. is good instance as he is very DEFICIENT, he was nearly 9 years old. when his father d
603o014 w Vol 18. (1st Article) on Taste «EXCELLENT». DEFICIENT in not explaining the possibility of <handsom
600o008 faculties  of soul, treating of infinites not DEFINABLE.-- Has little Chapter on each faculty of Soul
574n043 wo actions either on form or brain very hard to DEFINE.-- Consider the acquirement of instinct by dog
529m040 in whose mind supply of food was evasive & ill DEFINED thought would receive pleasure from thinking o
557m148 e is no inclination to jump away,-- it is, ill-DEFINED fear.-- Yet one knows oneself it is quite diff
021rIFC h gale Lyell's Geology The living atoms having DEFINITE existence, those that have undergone the grea
062r129 rt.-- Tempted to believe animals created for a DEFINITE time:--not extinguished by change of circumst
176b022 species,  than individuals If we suppose monad DEFINITE existence, as we may suppose is the case. the
176b022 is the case. their creation being dependent on DEFINITE laws, then those which have changed most. «ow
177b029 would  die out, therefore not.-- Monad has not DEFINITE existence.-- There does appear some connectio
179b035 o springing from one branch, & the monucle has DEFINITE life, then all die at one. period, which is n
179b035 one. period, which is not case,∴ MONUCULE NOT DEFINITE LIFE I think {P} Case must be that one genera
186b061 countries  ought to be similar ¡Law: existence DEFINITE without change, superinduced, or new species;
205b142 , rodents &c, JUST COMMENCING. Kirby says (not DEFINITE information) West of Rocky Mountains  Asiatic
263c078 Man has reasoning powers in excess. instead of DEFINITE instincts.-- this is a replacements in mental
310c223 ing such passions acquired & heredatary & such DEFINITE thoughts, I will never allow that because the
377d139 oman instinctive desire may be said to be more DEFINITE than with bitch, for some feeling must urge t
386d165 ot education.-- Cannot I find some animal with DEFINITE life & split it, & see whether it retains san
541m092 s scarlet?-- How can people dwell on pain ¿ no DEFINITE idea. nor is an emotion.-- People who can mul
613o036 ds, must be made out Reason Will Consciousness DEFINITE instincts being acquired, is a most important
614o037 attraction of carbon. hydrogen <&c> in certain DEFINITE proportions, (different from what takes place
615o038 uired even by <children> «plants»!. [RHC] 7) As DEFINITE instincts modified by heredatary;-- so succes
616o038 emory may become perfect & we may look back to DEFINITE action or to our conscious selves.-- Such mem
611o034 ne individual» vegetables, the vital laws act DEFINITELY (<like> «as» chemical laws,) as long as cert
224b213 ts attach to Geographical range of species.-- DEFINITION of Species: one that remains «at large» with
258c061 lances analogy,-- See Abercrombie p. 172. for DEFINITION of Analogy. All the discussion «after» about
269c103 ity constantly develops multiplicity<"> [(his DEFINITION "constant manifestation of unity through mul
285c150 nstincts. whenever instinct is mentioned some DEFINITION must be given It would not be difficult to a
287c158 ltus. «p. 19»¶ Macleay seems to limit Lamarck DEFINITION of relations to settling the relative import
289c161 cture of this bird & get account of habits My DEFINITION <in wild> of species. has nothing to do with
548m117 o reason what there should be & discover loss DEFINITION of happiness the number of pleasant ideas pa
554m141 t. 13th It will be good to give Abercrombie's DEFINITION of "reason" & "reasoning," & take instance o
606o022 ing, is beauty.-- which he thinks is a better DEFINITION than Winkleman's. who says it is simplicity
610o033 tincts-- surely in animals according to usual DEFINITION, there is much knowledge without experience.
611o034 m inorganic movements.-- See Lamarck for this DEFINITION given in full.-- [LHC] ¿Has any vegetable or
106a073 reat where arrow stands the force immediately DEFLECTED from (B) which would not have been case. if i
106a073 aining its force, now it will be evident that DEFLECTED stream cannot retain its force if inclination
106a074 power of widening channel depends on power of DEFLECTION with stream retaining its force, now it will
106a074 lination be great. There could «not» be great DEFLECTION in a "rapid".-- is a familiar illustration.-
```

051r094 nic productions. = Yet everywhere on coast (IL DEFONSOS «Kelp») rocks show signs of degradation; (sof
403e022 in the Carolines, are remarkably short.-- & DEFORMATIONS are particularly common.-- without arm, «sk
335d015 a mule can have no offspring.= but as «badly» DEFORMED people & as mutilations «(produced very quick
556m145 ensation, as certainly as every man who is not DEFORMED. is born with two eyes.." I think this cannot
335d015 ules (as real mule) have offspring,-- slight DEFORMITIES «as supernumerary fingers» (that is slight a
525m025 rd Berwick's family) are heredetary.-- other DEFORMITIES are illnesses of the foetus.-- some mothers.
259c065 ange in species is reduced by atavism) Even a DEFORMITY may be looked at as the best attempt of natur
389d177 takes males but does not produce) tendency to DEFORMITY ¿this does not happen with hybrids?]CD Plants
525m025 fterwards with tails. My father says, perfect DEFORMITY, as an extra number of fingers.-- hare lip or
402e018 ies. "many of them descending to examine their DEFUNCT companion".-- p. 229. Borneo.-- only animals h
029r033 water. It was quite calm at the time. Latitude 8 DEG. 47 min. N: longitude 61 deg. 22 min. W. mid. ca
029r033 he time. Latitude 8 deg. 47 min. N: longitude 61 DEG. 22 min. W. mid. calm and clear. Caermarthen Jou
384d161 male first Will not even a fruit tree or rose DEGENERATE during its life so that successive buds do d
498q008 not, I am surprised <plan> such plants do not DEGENERATE,-- as the Bees will mingle the infinitesimal
370d117 of world, thus tea tree in Brazil must have DEGENERATED. as must spices &c &c The line of argument «
063r130 ieve ancient ones: «.'.» not gradual change or DEGENERATION. from circumstances: if one species does ch
511q018 . Spaniels-- Grey-hounds-- is there ever any DEGENERATION?? HOUNDS. Eyton Mr Wynne, &c Could by selec
553m136 his most magnificent laws. of which we profane «DEGNEM» in thinking not capable to <do> produce every
328r028 se sand from the finer matter resulting from DEGRADATION of Feldspar & other minerals containing Alum
037r053 ot being buoyed with Kelp.-- With respect to DEGRADATION of rocks--It may be a question. whether orga
051r094 ast (IL DEFONSOS «Kelp») rocks show signs of DEGRADATION; (soft substances worn into bare cliffs evid
051r095 brought into play; most manifest example of DEGRADATION I ever saw on beach near Callao.--From Sir.
056r110 effect of heat on inner wall, hence resists DEGRADATION longer than outer parts.-- The common occurr
067r144 cidedly belong to old Silurian system. Apply DEGRADATION of landlocked harbors to Craters of elevatio
105a073 «stream» aside is not great.-- Is there more DEGRADATION at first angle owing to momentum. which the
117a102 mine well shores of lakes. to see effects of DEGRADATION, «no» tides, water always falling or at leas
201b129 to me the same, as the irregularities in the DEGRADATION of structure of Lamarck, which he says depen
040r062 s. -- Is the white matter beneath pebbles. the DEGRADED matter of such pebbles extending to seaward,
065r139 Shetland Cape Possession. Syenite¿ Andite?-- DEGRADING of inland bays. like St. Julian & Port Desire
122a114 aw. & lie over the platform:-- On my view the DEGRADING action must prevent internal fluid arriving a
021rIFC   R. N up to 1 DEGREE / July 1835. the excess of harbor = 180 See Da
023r009 "--This shark was caught in Shark's Bay. Lat 25 DEGREE. The nearest of the E. Indian Islands. namely
024r016 T] leagues Fathoms Parallel of St Catherine [27 DEGREE 30 minute S.] 18--70 Paranagua [25 degree 42 m
024r016 e [27 degree 30 minute S.] 18--70 Paranagua [25 DEGREE 42 minute S.] 12--40 St Sebastian [23 degree 5
024r016 25 degree 42 minute S.] 12--40 St Sebastian [23 DEGREE 52 minute S.] 12 50 {T} Joatingua SE [23 degre
024r016 degree 52 minute S.] 12 50 {T} Joatingua SE [23 DEGREE 52 minute S.] 5 35 R. de Janeiro SE [23 degree
024r016 degree 52 minute S.] 5 35 R. de Janeiro SE [23 DEGREE 58 minute S.] 18 77 C. Frio [23 degree S.] 7 6
024r016 o SE [23 degree 58 minute S.] 18 77 C. Frio [23 DEGREE S.] 7 60 {T} Soundings about same as last to N
024r016 s last to N. of C. Frio Except at Abrolhos. [18 DEGREE S.] Bahia [12 degree 57 minute S.] 8 200 Morro
024r016 io Except at Abrolhos. [18 degree S.] Bahia [12 DEGREE 57 minute S.] 8 200 Morro S Paulo [13 degree 2
025r016 12 degree 57 minute S.] 8 200 Morro S Paulo [13 DEGREE 22 minute S.] 9 120 {T} Garcia de Avila [light
025r016 S.] 9 120 {T} Garcia de Avila [lighthouse] [12 DEGREE 35 minute S.] 9 124 Itapicuru [R.] [11 degree
025r016 2 degree 35 minute S.] 9 124 Itapicuru [R.] [11 DEGREE 46 minute S.] 9 200 R. Real [11 degree 31 minu
025r016 [R.] [11 degree 46 minute S.] 9 200 R. Real [11 DEGREE 31 minute S.] & [R.] Sergipe [11 degree 10 min
025r016 eal [11 degree 31 minute S.] & [R.] Sergipe [11 DEGREE 10 minute S.] 20 190 R. San Francisco [10 degr
025r016 egree 10 minute S.] 20 190 R. San Francisco [10 DEGREE 32 minute S.] 10 50 Whole coast to Olinda [8 d
025r016 ee 32 minute S.] 10 50 Whole coast to Olinda [8 DEGREE S.] 9-10 = 30-40 {T} at twice or 18-20 <60>--8
025r016 aler N. of Olinda.--a little WNW of C. Rock. [5 DEGREE 29 minute S.] still shoaler, coast composed of
052r099 are commonly stranded on shores of Georgia «Lat DEGREE ( )», he has rocks on surface. applicable to P
057r114 eory.-- Capt Ross found in Possession Bay in 73 DEGREE 39 N. living worms in the mud which he drew up
065r139 iles and width 2 & ¼ miles S. Shetland. Lat. 62 DEGREE 55 minute. «only» one lichen. only production.
073r163 in central Cordillera. particularly between 18 DEGREE & 22 degree N. = formations of amph: porphyry.
073r163 Cordillera. particularly between 18 degree & 22 DEGREE N. = formations of amph: porphyry. greenstone[
076r170 syenite & <sep> serpentine dipping to SW at 45 DEGREE to 50 degree-- covered by conformable greensto
076r170 ep> serpentine dipping to SW at 45 degree to 50 DEGREE-- covered by conformable greenstone porphyrys
077r170 phyries. whereas other to ancient rock.--this N DEGREE 2. superimposed on N degree I. even No. 2. mig
077r170 cient rock.--this N degree 2. superimposed on N DEGREE I. even No. 2. might be mistaken for Porphyry
078r175 e?). -- Mines of Catorce «(Principal veins)» 25 DEGREE to 30 degree to NE. vein of Moran 84 degree NE
078r175 of Catorce «(Principal veins)» 25 degree to 30 DEGREE to NE. vein of Moran 84 degree NE. of Real del
078r175 25 degree to 30 degree to NE. vein of Moran 84 DEGREE NE. of Real del Monte 85 degree to S. // Tasco
078r175 ein of Moran 84 degree NE. of Real del Monte 85 DEGREE to S. // Tasco 40 degree to NW (afterwards sai
078r175 . of Real del Monte 85 degree to S. // Tasco 40 DEGREE to NW (afterwards said to be «all with some ex
091a027 a. Clay Slate & Quartz. strike SSW & NNE dip 30 DEGREE - 80 degree Ed. N. Phil Journ. p. 410. 1828 Ed
091a027 e & Quartz. strike SSW & NNE dip 30 degree - 80 DEGREE Ed. N. Phil Journ. p. 410. 1828 Ed. N. P. J. p
091a027 1828. gneiss in India (falls of Garsipa) dip 30 DEGREE. <strike> «direction<?>»ESE-- CD [In the Darwa
097a042 Direction of mountains in Brazil L.'Institut No DEGREE 221 Lamellar dikes like Mica Slate Von. Buch.
111a087 NITE" "direction of strata on the Berbice N. 35 DEGREE. E. dip to NW to 80 degree faults with red wac
111a087 on the Berbice N. 35 degree. E. dip to NW to 80 DEGREE faults with red wacke contorted evidently dike
113a090 iberia being -8 Reaumur.-- there ought to be 32 DEGREE Fah. at a greater depth than 400. & the limit
113a091 & if conducting powers had been better then 32 DEGREE would have been found lower.-- We have no righ
113a091 powers either better or worse & the depth of 32 DEGREE. being little we may confidently infer that ti
115a098 --.-- if matter was «successively» given of every DEGREE of fineness. then most regular slope-- {P} if
126a122 ers-- we judge from the surface, & say 60 ft to DEGREE.-- but this may be very wrong,-- The fact of a
126a124 heat in one direction only, like water below 39 DEGREE» & lower part glass.-- then the high temperatu
127a126 losed globule of glass. (oil may be cooled to 0 DEGREE!)-- shows effects of pressure in change of for
135a144 rth reading-- Burnetts. vol 4. p. 193 in Lat 26 DEGREE S. Wafer looking for Copiapo. found inland a g
145g006 rata 3 or 4 seams/ 3 or 4 inches thick-- {P} 35 DEGREE as I believe about greatest dip of sandstone i
145g006 andstone in upper part «of Salisbury Craigs» 25 DEGREE perhaps most common-- Will not curved form of
147g019 ning ½ past seven o'clock 29.642 Temp 55 Air 50 DEGREE? Friday. Inverorum about 20 ft above Loch Tull
147g019 rum about 20 ft above Loch Tulla 29.804 Temp 62 DEGREE Air 60 degree Below Loch Tulla whole wide vall
147g019 t above Loch Tulla 29.804 Temp 62 degree Air 60 DEGREE Below Loch Tulla whole wide valley scattered w
147g021 road between Inverorum & King's House 28.935/82 DEGREE A Temp of Air 65 degree? Glenoe, 6 ft above hi
147g021 King's House 28.935/82 degree A Temp of Air 65 DEGREE? Glenoe, 6 ft above high water mark 30.380. 68
147g021 ? Glenoe, 6 ft above high water mark 30.380. 68 DEGREE 65 degree? For comparison with all the measure
147g021 6 ft above high water mark 30.380. 68 degree 65 DEGREE? For comparison with all the measure before Th
152g049 Bridge of Roy Level of «bed of» River 30.221/65 DEGREE/ Temp of air 65 degree? There are two terraces
152g049 «bed of» River 30.221/65 degree/ Temp of air 65 DEGREE? There are two terraces on the East side of ri
152g050 about 40 ft beneath general plain. 30.127 A 72 DEGREE Air 65 degree? at level of upper terrace The b
152g050 eneath general plain. 30.127 A 72 degree Air 65 DEGREE? at level of upper terrace The butresses of Al
153g055 g place is formed Ben Erin summit 27.813. 65 55 DEGREE? Boulder of Granite 28.362 68 degree 60 degree
153g055 813. 65 55 degree? Boulder of Granite 28.362 68 DEGREE 60 degree Granite--«band» 4 X 3 X 2 «feet» & 2
153g055 degree? Boulder of Granite 28.362 68 degree 60 DEGREE Granite--«band» 4 X 3 X 2 «feet» & 2 deep Anot
154g058 ison with granite block «& Ben Erin» 29.287. 72 DEGREE Air 65? 70? {P} Where a buttress projects from
156g071 h shelf at head of Lower Glenroy 29.581 A 82 75 DEGREE? From this point plain appears like one unifor
156g072 sea also.-- Barometer on shelf 3d. 29.455 A 83 DEGREE ∴ plain of 4 minute shelf slope, above «line o
157g074 r masses from above on soft shelf-- 29.330 A 84 DEGREE compare this with last measurement of shelf of
157g077 terrace near Loch Spey <29.35161> 29.360? A 79 DEGREE 75 degree? A little below Divortium on slope t
157g077 ear Loch Spey <29.35161> 29.360? A 79 degree 75 DEGREE? A little below Divortium on slope towards Loc
159g086 rom on upper (rather above)? shelf 29.290 A. 69 DEGREE Air 68 degree? Barom 29.008 A. 75 degree Air 7
159g086 rather above)? shelf 29.290 A. 69 degree Air 68 DEGREE? Barom 29.008 A. 75 degree Air 70 degree? This
159g086 A. 69 degree Air 68 degree? Barom 29.008 A. 75 DEGREE Air 70 degree? This station a little way down
159g086 Air 68 degree? Barom 29.008 A. 75 degree Air 70 DEGREE? This station a little way down slope of obscu
160g090 Hill «Cairn <taw> leer peak» Barom 28.700. A.75 DEGREE 75 degree? Boulder, much covered by turf 2ft.
160g090 n <taw> leer peak» Barom 28.700. A.75 degree 75 DEGREE? Boulder, much covered by turf 2ft. 8- long of
160g091 broad flat Loch {P} XX Barom 29.2 A 75 Air 70 DEGREE? Isthmus broad flat peat mass-- (general chara
160g092 scure terraces on one side Barom 29.200 A 80 70 DEGREE? for about 3/4 of mile on <one> S side of «thi
161g094 Peat-Mass Divortium aquarum) Barom. 29.200 A.77 DEGREE Air 70 degree? Barom. 066 lower than last. but
161g094 rtium aquarum) Barom. 29.200 A.77 degree Air 70 DEGREE? Barom. 066 lower than last. but A 77 station

Page
(Key Word)

161g096 to moss-- on this terrace Barom. 29.264 A 82 75 DEGREE? This last measurement turns out too low, (NB
161g098 Loch Lochy 4 ft above water Barom: 30.372 A 76 DEGREE? 75 degree? The River <the> of which the source
161g098 y 4 ft above water Barom: 30.372 A 76 degree 75 DEGREE? The River <the> of which the source is a lip
162g100 hy near Letter Finlay Barom 30.267, A 68 Air 65 DEGREE? <.194 372 about 267 28.75 .105 I reached> 29.
162g102 my new shelf, from road: Loch Ness 30.140. A 66 DEGREE 30.095 .0458 or 6 difference between bedrock &
162g103 > <Donald Macphee> Saturday Morning 29.958 A 64 DEGREE, air 60 «Evening do» The extreme right arm of
190b076 minishing towards centre (p. 586)-- Parallel 33 DEGREE-35 degree. source of forms. reduce towards Nor
190b076 towards centre (p. 586)-- Parallel 33 degree-35 DEGREE, source of forms. reduce towards Northern East
193b092 Essai sur la Geographie des Plants. I Vol in 4 DEGREE.-- I have abstracted Mr Swainson's trash.-- at
200b121 Africa. Clay slate. strike. SSW. & NNE. dip 30 DEGREE - 80 degree (?).-- Ed. Phi.l.. N. J. p. 410 <N
200b121 y slate. strike. SSW. & NNE. dip 30 degree - 80 DEGREE (?).-- Ed. Phi.l.. N. J. p. 410 <Nov» 1828 It
202b131 India. Geograph J. Vol III. P. I. p. 17 (Lat 37 DEGREE about) Vol IV P. I. Geograp. Journal. Voyage u
206b146 uls on average are related to the (200dth year) DEGREE. Then 200 years ago, there were 200 people liv
243c019 a pas jusqu'aux iles Macquarie et Campbell (52 DEGREE S) qui n'aient egalement leur espèces; et cert
244c021 onsult Latreille. Geographie des insectes, in 8 DEGREE. p. 181 «who says insects Indian, like. Plants
249c036 ar to word to special districts???? north of 30 DEGREE.--, may be connected with, Mr Blyth's statemen
269c101 ry suppose when rhinocerose lived. mean temp 60 DEGREE mean, then temp at depth of four hundred feet
269c101 temp at depth of four hundred feet would be 60 DEGREE + 6 degree??., therefore 34 degree degrees of
269c101 pth of four hundred feet would be 60 degree + 6 DEGREE??., therefore 34 degree degrees of change have
269c101 would be 60 degree + 6 degree??., therefore 34 DEGREE degrees of change have travelled that thicknes
274c114 ow here I must observe thes characters vary in DEGREE in last instance hardly at all developed. Not
281c138 talking of men as related in the third & fourth DEGREE.-- a species must be compared to family entire
281c139 compared to family entirely separated from any DEGREE; the tailors-- «in each branch would be analog
282c141 greatly changed. in structure & even to certain DEGREE in habits, yet we might have these analogies.-
300c197 rt acquired.-- They reason however on this to a DEGREE. Mem Spider only dropping where ground thick.-
301c200 gencies, where the habit is not applicable. The DEGREE of development of all animals of same class be
305c211 into endless forms, bearing a close relation in DEGREE & kind to the endless forms of the living bein
306c213).-- directly beyond produced line of Timor 215 DEGREE. What productions Sandal. Wood Isd.? ought to
311c225 od to consult p. 326 wild ass extending over 90 DEGREE of Long. & Col. Sykes alludes to some other ca
311c225 & Col. Sykes alludes to some other case of 180 DEGREE & great diff of Latd. p 355 Echidna of Van Die
315c242 le habitual actions in plants, it allows of any DEGREE in lowest animals --habitual action, in intest
318c253 ranges. «even migratory birds, lik swallows» -- DEGREE/ migrations of birds he mentioned many most cu
358d086 brid of this bird, has long tail figure, & some DEGREE of whiteness like a Male.-- Thus castration, h
383d159 e or sheath of Horses poenis reduced to extreme DEGREE of abortion).-- Insecta.-- hermaphrodite, bein
387d168 e peculiarity» that in (B) to <a slight> «some» DEGREE, & likewise ovum in (B) <an C> that in (C) «in
387d168 ewise ovum in (B) <an C> that in (C) «in lesser DEGREE» -- Then when (C) unites with Male (X) «assume
387d169 fspring» inherit the same peculiarity in lesser DEGREE C. & theirs again in lesser degree.--now if th
387d169 ty in lesser degree C. & theirs again in lesser DEGREE.--now if the <tw> second race both have this p
402e019 of insects of St. Peter & St. Pauls in Lat' 53 DEGREE yet fauna like that 60 degree & 70 degree of E
402e019 Pauls in Lat' 53 degree yet fauna like that 60 DEGREE & 70 degree of Europe.-- Many Europaean insect
402e019 t' 53 degree yet fauna like that 60 degree & 70 DEGREE of Europe.-- Many Europaean insects-- list giv
406e036 disease following handling sheep-- all cases: d DEGREE p. 354-- The most vicious dog. will not attack
423e097 strongest possible to increase them, hence the DEGREE of developement is either stationary or more p
424e103 plants vary more than others» & horse in lesser DEGREE,-- how different to dog!-- (Hybrids of Calceol
440e147 of animal to usurp its place.-- & therefore the DEGREE of injuriousness must have been exceedingly sm
443e155 rcourse every time is of no consequence in that DEGREE of developement.-- [It is singular there is no
451e177 46 Carimon Java. (between Borneo & Java) Lat 5 DEGREE. 50 minute S. adjoining it are several small i
477z002 d, mulita et mataco.) are all found south of 26 DEGREE 30 minute. Lat -- -- do. p. 207. La punaise wa
478z003 28 degree North p. 239 In ocean between Lat 56 DEGREE North p. 239 In ocean between Lat 56 degree an
478z003 28 degree North p. 239 In ocean between Lat 56 DEGREE and 57 degree only inhabitant crust Entomost o
478z003 th p. 239 In ocean between Lat 56 degree and 57 DEGREE only inhabitant crust Entomost of the genera--
482z012 Merid. Tatu noir. abundant from Paraguay to 27 DEGREE, then the Mulita from 41 degree to 26 degree C
482z012 Paraguay to 27 degree. then the Mulita from 41 DEGREE to 26 degree CLOSELY allied species, therefore
482z012 27 degree, then the Mulita from 41 degree to 26 DEGREE CLOSELY allied species, therefore interlock.--
498z007 so? --Yes, my Father lost this character in grt DEGREE from charcoal & good treatment (8) Do bees fre
509z017 according to Bentham's Remark. Horse or cow.-- DEGREE of soldering of tibia & fibula: in Man any abo
548m117 sing through mind in given time.-- intensity to DEGREE of <happi> pleasure of such thoughts We give n
551m127 nnexion appears to me vague-- Delirium of every DEGREE of intensity-- «in old man, he had just seen m
565m009 - sulkiness is same as pouting, <but> lesser in DEGREE, no smile, no frown showing thought, no compre
574m041 Sir. Ch. Bell says, & hearing music. to certain DEGREE sexual.-- The association of saliva, is probab
588m090 -- «(& is not this difference same, but less in DEGREE, as between man & child.--)» what differs-- no
593m109 to this view, & fowls hatching stones. in some DEGREE is so.-- idea of beauty of music are great dis
610o30v od.-- Therefore rule of happiness is to certain DEGREE <of> right.-- The change <of> our moral sense,
617o39v t the answer to it would require a considerable DEGREE of attention. How do the senses affect us. ar
032r039 beneath band. of greatest action. would now by DEGREES be exposed to it, & the result would [be] a un
181b045 as compatible with such structure are in minor DEGREES adapted for. other elements. every part would
269c101 e 60 degree + 6 degree??., therefore 34 degree DEGREES of change have travelled that thickness in tha
304c206 nt classes Keep to their types. with different DEGREES of closeness. -- look how close birds! look at
533m057 Materialism, say only that emotions, instincts DEGREES of talent, which are heredetary are so because
550m124 easant & unpleasant, in same time,-- therefore DEGREES of happiness-- Entire happiness. not being so
553m136 at to us».-- & that it does exist in different DEGREES in races.-- whether in Ancient Greeks, with th
241c016 ct species. good Resumé do/p. .62 ??? Age of DEINOTHERIUM. p. 23.. Bull: Soc. Geolog. 1837-8. Tom: IX
263c077 s> origin has not been indefinite-- he is not a DEITY, his end «under present form» will come, (or ho
291c166 ary,. analogy points out to this.-- love of the DEITY effect of organization. oh you Materialist!-- R
300c196 elf a great work. worthy the interposition of a DEITY, more humble & I believe true to consider him c
301c199 itual. it surely is not worthy interposition of DEITY to teach squirrel to kill ears of corn accordin
316c244 rling in play.-- Hensleigh says the love of the DEITY & thought of him «or eternity», only difference
316c244 no greater difficulty for DEITY to choose, when perfect enough for future state
397e003 plant rise, without the immediate agency of the DEITY. But we know from experience! that these operat
550m126 us passages showing how easily chance & will of DEITY are confounded.-- well applicable to free will.
564m004 re thinking of injured moral sense.-- Notion of DEITY effect of reason acting on <Not social instinc
566m011 .-- very necessary to explain origin of idea of DEITY.-- Animals do not know they have 'these necessa
572m035 Macculloch in his Chapter on the Existence of a DEITY has an expression the very same as mine about o
572m035 same as mine about our origin of a notion of a DEITY We can allow «satellites», planets, suns. unive
576m049 s philosophical. viz. man is not a cause like a DEITY, as M. Cousin says. because if so ourang outang
602o11b ion".-- Macculloch Vol I. p. 115. Attributes of DEITY. on Belief.-- you belief things you can give no
603o012 n> «a king» <changed into> is carried on toward DEITY.-- & as king might like cruel pleasure, so sacr
603o012 s. How completely men must have personified the DEITY.-- H. Tooke has shown one chief object of langu
605o19v as excited & associated with it. as the idea of DEITY. with vastness of Eternity. which superiority w
632j53r of the Attributes of God Macculloch. Attribs of DEITY Vol: I it will be better always to refer to th
634j55r in classes-- as resulting from the will of the DEITY, to create animals on certain plans.-- is no ex
635j55r ing» because we know nothing of the will of the DEITY. how it acts & whether constant or inconstant l
636j57r ning in organic world.-- Macculloch. Attrib. of DEITY Vol I p. 232. gives Woodpecker as instance of
638j28r ng to Kirby's view.-- Macculloch. Attributes of DEITY Vol I. p. 251-- stomach hump, kinds of foot. po
257c056 und "beaucoup des mêmes oiseaux. que nous avions DÉJA observés en Patagonie. ou au moins des espèces
025r017 rs? No Volcanic Earthquakes or Hot Springs in T. DEL Fuego = The Wager's Earthquake the most Southern
026r021 een the older strata & the bottom of sea near T. DEL Fuego.-- Is there account of Baron Roussin's voy
042r068 d those aerial Volcanos in Germany» In the Valle DEL Yeso it is probable that point of Porphyry has b
049r088 ust we look at regular greenstone cones at S. T. DEL Fuego as nucleus of a Volcano or as an injected
052r099 ted.-- Greywacke. as a general fact absent in T. DEL Fuego, excepting in Port Famine Mr Sorrell says
065r140 . Cruz.???-- Form of land shows subsidence in T. DEL Fuego, and connection of quadrupeds.--although r
065r140 soundings. (end of 2d voyage outside coast of T. DEL Fuego. off. Christmas sound. -- «(Think some 60
078r175 ravins, and the slope of the mountains (flaqueza DEL cerro) have been parallel to the direction & inc
078r175 egree to NE. vein of Moran 84 degree NE. of Real DEL Monte 85 degree to S. // Tasco 40 degree to NW (
134a141 es near C. Virgin p. 59. dip of Clay slate in T DEL Fuego Admiralty Sound. SE dip. much p. 136. Rock
134a141 ks p. 375. on the soundings on outer coast of T. DEL. Fuego.-- p 385 Rocks of S. Western Coast Vol II
138a153 . 4. (Lyells Book)» Observaciones sobre El Clima DEL Lima par Dr. H. Unanúe says he believes the sea
179b033 not cross readily, distinctness of tribes in T. DEL Fuego. the existence of whiter tribes in centre

205b144 e Leaches out of water Does the odd Petrel of F. DEL F. take form of awk, because there is no awk in
217b187 nus with species in Van Diemen's land and Tierra DEL Fuego.-- Araucaria, species. Brazil {P} Chile, N
219b194 ar Cape) see if there are any species same as T. DEL Fuego & C. of Good Hope show possibility of tran
219b195 therefore plants common take an example from T. DEL Fuego.-- Ellis (?) says Tahitian kings. would ha
226b221 arge ones.-- Is the flora of <S. America> Tierra DEL Fuego like that of North Europe, many genera & f
268c099 many species common to Patagonia desert & Tierra DEL Fuego. & forest «Parrots in Macquarrie Isd.--» v
277c126 Opetiorhyncus. fulginosus. (a) Falklands (b). F. DEL Fuego differ from (C) Chiloe (E) Chile. rupestr
343d038 <hot>. Warm to cold, damp to dry.-- Thus Tierra DEL Fuego has <not> «only» one «Guanaco» of the char
409e047 ifficulty in conceiving how inhabitant of Tierra DEL Fuego is to be converted into civilized man.-- a
481z010 llata. Meyen. p. 92.-- great Kingfisher of Tierra DEL Fuego killed in Chile. Dobrizhoffer., Vol I p. 3
485z017 e &c &c. with reference to those of mine from T. DEL Fuego p. 141. How comes it salt water so soon pu
486z018 se views clearly explain rarity of insects in T. DEL Fuego.-- Hence it is odd that Amber insects of E
486z019 Kamtchatka (Salamandra aquatica). Compare with T DEL Fuego Compare birds of do with <T> N American &
486z019 ego Compare birds of do with <T> N American & T. DEL. Fuego & Iceland Spix & Martius talk of birds si
497q007 nslow.-- (1) Character of alpine Flora of Tierra DEL Fuego and Entomology of.-- most important, as fo
508q016 Andrew Smith. (6) What size book Gallesio storia DEL Reproduzione.-- D. Holland (7) Is Haemorragic te
527m031 1> «so many» birds singing in England, in Tierra DEL Fuego not one.-- now as we know birds learn from
553m137 titions of an Australian savage or one of Tierra DEL Fuego.-- Mr Miller (superintendent of the Zoolog
221b201 . 23 do.-- On animal - Confervae-- p. 23 p. 267. DELA Beche. Geolog. Researches. facts of salt-water
106a075 ping sides This argument is partly taken from DELABECHS Theoretical Researches.-- Athenaeum. 1838-- p
101a057 Osteopora platycephalus (Harlan) found on the DELAWARE. is it Edentate? Phillips p 289.-- Alludes to
214b173 is Osteopora platycephalus. (Harlan) found on DELAWARE is it Edentate? Phillips. Lardner p. 289 It i
523m016 of an illness at Rome, when by accidental <was> DELAY of money, he was «only» NEARLY thrown into a ho
390d178 probable that breeding in & in would not be DELETEREOUS if the relative had come from different quar
305c211 the community of mind, even in the tendency to DELICATE emotions between races, & recurrent habits in
312c233 e dog. but do not become impregnated & puppies DELICATE-- they cross sister & brother of same litter,
408e043 say. what <light> «colours». acting. by a most DELICATE organ, on the whole system may. produce-- ? W
522m013 l idea of decency & affection are lost.-- most DELICATE people do most indelicate actions,-- as if «t
540m088 melts into tears, with sensations of sorrowful DELIGHT, very like best feeling of sympathy.-- Mem: Bu
614o037 in this case can perceive instinct. boy takes DELIGHT in mammae before any reason had told him this
522m013 . Mania is quite distinct, different also from DELIRIUM, a peculiar complaint stomach not acted upon
522m013 iolent intoxication, often ends in insanity or DELIRIUM.-- In Mania all idea of decency & affection a
524m021 childhood, (but appear most capricious) as in DELIRIUM after epilepsy, but in the failing from old a
530m045 have some little advantage in these trades.-- DELIRIUM seems to rest the sensorium.-- -- analogous t
541m090 est on each.-- ∴ my father. is right in saying DELIRIUM rest-- therefore dreams thus act.-- ∴ weak mi
548m114 ly comparing each step as in reasoning-- hence DELIRIUM & sleep mental rest. though. most vivid & rap
551m127 & insanity the connexion appears to me vague-- DELIRIUM of every degree of intensity-- «in old man, h
593n111 one's own mind, & Dr. Hollands story of man in DELIRIUM tremens hearing other man speaks. shows, that
529m043 akeways book of Sheriffs.-- July 22d. 1838 No DELIRIUMS, yet in some inflammatory diseases, where the
536m074 us believing, <yet> would more earnestly pray "DELIVER us from temptation,' he would be most humble,
507q016 th (1) Are cross-births, or other accidents of DELIVERY-- inheritable.?-- Bell cd ask Accouchers Is a
453e183 where peculiarity has first appeared.-- "Storia DELLA Riproduzione Vegetale". by Gallesio. Pisa 1816
042r068 e, & unequal action of Earthquakes. on Chili & DELTA of Indus», my belief in submarine tilting alone
123a116 s of S. America. Von Buch Lyell. (under head of DELTA) describes near Alps great beds of rivers which
633j54r about plants being created to arrest mud &c at DELTAS.-- Now my theory makes all organic beings perf
343d039 uite different figure 19th. With respect to the DELUGE it may be worth adding in note than amongst th
608o026 ve power changes with organization The general DELUSION about free will obvious.-- because man has po
391d174 , which mutilations are not). but why should it DEMAND some further change?-- Man properly is hermaph
623o050 .-- [LHC] According to my theory, all instincts DEMAND some explanation [RHC] Although I cannot prete
124a118 n same condition for long period Subsidence in DEMARARA p. 131 (B.) Wrong Entrance. Book C. p. 101. O
029r033 kshank, Captain John Young, on her voyage from DEMARARA to London:-- "Feb. 12, 1835. At 10h. 15m. a s
060r124 .--important.-- Decemb 10. 1802. Earthquake at DEMERARA. The earthquakes "seem to arise from some eff
202b131 l. Voyage up the Massaroony by W. Hillhouse.-- DEMERARA. In note. Demerara. 10 «12» feet beneath surf
202b131 ssaroony by W. Hillhouse.-- Demerara. In note. DEMERARA. 10 «12» feet beneath surface forest trees fa
060r124 W Indies.--p. 200. Bollingbroke voyage to the DEMERARY Earthquakes at St Helena. 1756. June 1780, Se
055r107 ension. Azores. («sandstone first gives» half DEMOLISHED craters).--worn into mud & dust.--connection
244c021 of fish & shells in tropical sea, it. would DEMONSTRATE.; not distance, makes species but barrier.--
308c216 eedings of Geol. Soc Vol I It is capable of DEMONSTRATION that all animals have never at any one time
292c170 of animals. & observe the character of the DEMONSTRATIONS offered of the singular views there offere
076r168 in the veins of Kongsberg in Norway.--namely DENDRITIC silver intersecting carbonate of lime-- nativ
251c041 parents of Mam: found in Sumatra p. 452 Append to DENHAM Clapperton &c on Mammalia no doubt will all be
430e117 e, & unequal of any of ours,-- even if extinction is DENIED.-- it will not account for all species. even i
600o008 ordinary Beau ideal, Mem. Negro, beau,--Jeffrey DENIES all Beau-- How does Hen determine which most b
122a116 shire case where lamination appeared.-- Lyells DENMARK -- L'Institut (1838) p. 268. Paper by Humbold
163g109 r head of» Beagle Channel. Forchammers (Lyells DENMARK) Shrewsbury rubbish.-- Speculate on origin peb
565n010 th new Lavaters,-- Ye Gods!:-- says fleshy lips DENOTE sensuality (p 192 Vol. III Octav. Edit)-- cert
495q05a termediate form found wild a The Leptosiphon DENSIFOLIUM «an annual» <sleep> «closes flower» on all g
245c023 iamese kind.-- but considered apogee species from DENTAL characters, wild pigs. said by Forrest to swim
114a095 sively slow because beach line chief cause of DENUDATION, but does not tell period.-- I cannot help s
118a103 form. has thinned upwards & is now cut off by DENUDATION it gives one grand idea of amount denudation
118a103 denudation it gives one grand idea of amount DENUDATION.-- This may be added to any place where dike
146g010 perhaps wall oblique The hill has been well-- DENUDED.-- «of hard metamorph» path only covering Grea
225b219 Reference to Pig & Dogs. My theory will make me DENY the creation of any new quadruped since days of
403e024 st sense. <es> «as» test of species.-- they must DENY species which is absurd.-- their only escape is
443e155 ngs with which have fluid sperm.--]CD I utterly DENY the right to argue against my theory, because i
575n043 with <the> it the provision for death.-- can we DENY that brain would be intermediate like rest of b
575n043 would be intermediate like rest of body? Can we DENY relation of mind & brain. «Do we deny the mind
575n043 dy? Can we deny relation of mind & brain. «Do we DENY the mind of a greyhound & spaniel. differs from
575n043 spaniel. differs from their brains» then can we DENY that the grand child dug for mice from some pec
575n044 mice-- Jenners Jackall Have we somewhat right to DENY identity of instinct.-- Habits import to Bowen
632j53r ision for transportation:-- But I do not want to DENY laws.-- The whole universe is full of adaptatio
374d133 how different from plants!) But the Cephalopoda DEPART more widely from living forms.-- p 458 Upper S
553m138 w any desire for women-- «very curious. as they DEPART in structure» The monkeys understand the affin
041r066 eins of this figure {P} in sandstone: evidently DEPEND on a concretionary contraction: the fact is in
063r133 of animals.--extinction in same manner may not DEPEND.--There is no more wonder in extinction of spe
220b197 on. But who can say, whether offspring does not DEPEND on mind or instinct of parent. Mem Lord Moreto
282c142 y would not in all cases be produced, but would DEPEND upon exclusion.-- The same characters which ar
415e068 fatal diseases.-- The Value of a group does not DEPEND on the number of the species.: therefore Man &
416e072 plant up a mountain-- In races the differences DEPEND upon inheritance & in species are only ancient
424e101 on of two regions in modern times.-- this would DEPEND on negative evidence of fossil remains, & ther
438e142 ween varieties & species, yet the amount of may DEPEND on many circumstances, time of domestication [
465t081 eries of strata This again shows how much forms DEPEND on other forms Lyell's Paper, in Taylor's Jour
538m078 ngs, & drunkedness, show what trains of thought DEPEND on state of turn In drunkedness same dispositi
614o037 advantage.; for superiority of memory does not DEPEND on its length.: Many animals (as horses) very
439e144 he presence of the Leopard & Tiger together DEPENDED on some nice qualifications each possess., &
035r047 f level independent of mineralogical nature & DEPENDENT: & then how wonderful level «of same beds» sh
176b022 may suppose is the case. their creation being DEPENDENT on definite laws, then those which have chang
177b029 have died; have they all one determinate life DEPENDENT on genus,. that genus upon another, whole cla
196b109 acid gaz.)-- Yet unquestionably animals most DEPENDENT on vegetables. of the two great Kingdoms.-- P
306c214 ing bright colours., good instance of colours DEPENDENT on localities.-- -- Hamilton will give an acc
115a098 er too coarse, then {P} that form.-- All this DEPENDING not on absolute <force> «size of» of <current
570n025 to shame after asinine.-- both accompanied by DEPENDING head., & active vessels of skin.-- What diffe
105a072 quimbo on. other hand?-- The widening a valley DEPENDS on serpentine course.-- the latter (it is gene
106a074 ination small.-- The power of widening channel DEPENDS on power of deflection with stream retaining i
201b129 adation of structure of Lamarck, which he says DEPENDS on external influences.-- For instance he says
220b197 -- & this prevents breeding. now domestication DEPENDS on perversion of instinct (in plants domestica
284c147 antity of life on planet at different periods, DEPENDS,-- on relations of desert, open ocean, &c this
284c147 fewer now than formerly.-- The number of forms DEPENDS on the external relations (a fixed quantity) &

284c149 ys rule is converse, <when> value of character DEPENDS on non-variation, & not on extension ¿these go
338d023 differences of country: the time of separation DEPENDS on facility of transport in the species itself
345d043 ir mothers believes this resemblance general. ¿DEPENDS upon mother bein[g] oldest breed?.-- -- Quarte
348d050 tural affinities\ Macleays plan of arrangement DEPENDS on the organs judged to be of importance in in
398e004 t the whole value of the geological chronology DEPENDS, that most sublime discovery of the genius of
422e095 d] The enormous number of animals in the world DEPENDS, of their varied structure & complexity.-- hen
437e141 & Cowslip suddenly changing into each other, & DEPENDS on character of antecedent races.-- «& yet in
439e142 ty be kept alive.-- Nevertheless much probably DEPENDS on circumstances favouring the reappearance of
439e143 acility of crossing, in plants proves how much DEPENDS on instincts in animals.-- yet the existence o
546m108 rospect.-- The very superiority of man perhaps DEPENDS on the number of sources of pleasure & innate
546m109 ours?-- Nothing shows one how little happiness DEPENDS on the senses.; than the <small> fact that no
567n013 se & effect, «they have habitual action. which DEPENDS on such confidence» when does such notion comm
602o011 re forgotten. Our happiness &c, our well-being DEPENDS upon the "habitual reason".-- This power of th
397e003 the world began, the causes of population & DEPOPULATION have been probably as constant as any of th
397e003 -- I would apply it not only to population & DEPOPULATION, but extermination & production of new form
092a029 soc: has shewn how electrical currents tend to DEPOSIT metals, if in solution. My view of metamorphic
100a054 yclop.--» absolutely considers gneiss an aqueo DEPOSIT resulting from disintegrated granite!!! Look a
118a103 Sir L. Dick says (.p 52) fringe of sublittoral DEPOSIT always equal width --subject of fine paper thi
120a109 eps in New Red Sandstone. look as if a surface DEPOSIT.-- The case of the shingle in the great Chilia
136a147 When I come to treat of the age of the Pampas DEPOSIT, I may properly remark on the superiority of L
149g030 & perfectly rounded stones-- lake required to DEPOSIT this Remember however the great Chilian valley
152g047 ency if elevation from Axis, then rivers might DEPOSIT, & afterwards with greater cut through, not ap
500q010 ieties.-- (25) Does the yellow white Butterfly DEPOSIT eggs in all varieties of Cabbage. (26) Do deer
035r046 to Geolog--"Between the height of same beds, DEPOSITED in different basins; little or no relation ap
085a001 tlyer.-- Linn: Transact. Vol. 8. p. 288. Salt DEPOSITED on windows of houses. & trees all injured on
115a097 ths of rivers ought to be deep.-- Henslow has DEPOSITED specimens from Anglesea in Geolog. Soc. if nu
147g016 <abo> by Loch Dochart-- Rivers could not have DEPOSITED it. Barrier of lake very lofty, & no trace of
149g030 great Chilian valley Acongua, must there have DEPOSITED much-- On other hand remember modelling power
159g086 ll worn pebbles-- «which river could not have DEPOSITED» the slope is continued some hundred feet low
162g104 - rounded pebbles-- dip sideward, & inwards-- DEPOSITED» when water stood at higher Loch Keeper tells
423e099 that our greatest formations <are> have been DEPOSITED in a period (say 10,000 years) which is suffi
032r041 . I think. from Patagonian steps, because the DEPOSITION & accumulation is brought into play As in Oc
077r172 is clear to me, there are laws of solution & DEPOSITION under great pressure. (? heat!) unknown to u
148g028 mlet had cut them out-- In all cases «I urge» DEPOSITION marine-- because if not chain of lake & if s
149g032 of Spean» difficult to explain on <formation> DEPOSITION in lake On the summit «& on Spean side» of M
423e099 modified organic forms.-- we know not rate of DEPOSITION has been equal even in one bed, much less in
433e126 these gaps, far more probably than during the DEPOSITION of the beds-- The argument must be thus put,
051r093 ally most tremendous surf & loose sandy beach) DEPOSITS «calcareous» encrustations; At Bahia ferrugin
114a095 been more frequently metamorphosed than other DEPOSITS.-- NB. because lowest. first accumulated in b
130a134 . mud banks & sand. dunes.-- in these littoral DEPOSITS there probably would be marked line of separa
134a143 of Punjab p. 149. on the <salt mines> «saline DEPOSITS» of India p. 503. On Indian Saline Deposits.
134a143 ne deposits» of India p. 503. On Indian Saline DEPOSITS. Vol II. p. 23. p. 77 do Vol III p. 36. do --
154g060 as got to the rock of cols if--. why should it DEPOSITS River terraces often descend by flights the t
191b080 d common animals-- + + + for instance tertiary DEPOSITS between East Indian islets-- (+ + +. Ireland
214b172 alden.-- Wealden. Do the N. American. Tertiary DEPOSITS present analogies to shells of living seas.?-
473s004 etaceous periods, together with their littoral DEPOSITS are probably buried in the depths of the sea-
263c074 e himself.-- Give the case of Apterix-- split, DEPRESS & elevate & enlarge New Zealand; a division of
592n103 trils» when frightened, does not hair & rabbit DEPRESS. them from squatting.-- p. 64 closing both eye
565n008 elf-sufficient,-- the corner of lower lip are DEPRESSED & opposite muscles used to when angry sneerin
377d139 , when pleased cocks his ears., when frighten DEPRESSES them.-- England was united to Continent, when
558m152 ?-- I believe common Swan, arch raises neck & DEPRESSES chin-- strikes with wing arches wings-- as do
117a101 . will bear on formations. during elevation & DEPRESSION. C. Prevost.-- My views of insensible oscill
285c151 cts of unfavourable conditions. (hence rise & DEPRESSION of importance, in each group & connection of
522m014 in to insanity.-- «I know the feeling also of DEPRESSION, & both these give strength & comfort to the
524m019 this precisely like the passion, ill-humour & DEPRESSION, which comes on from bodily causes.-- It is
593n105 erself.-- the object is to make saucer-shaped DEPRESSION-- [blank] Does music bear any relation to t
032r041 asden to the Eastward.--If matter proceeds from great DEPTH. from axis to surface must gain a Westerly curr
040r064 but as long as all below water no evidence--The DEPTH of shells (which being packed, in beds) lived t
098a048 ink. film parallel to longer axis. But if great DEPTH than 400. & the limit being 400 ft. shows that
113a090 - there ought to be 32 degree Fah. at a greater DEPTH than 400. & the limit being 400 ft. shows that
113a091 conducting powers either better or worse & the DEPTH of 32 degree. being little we may confidently i
113a091 er beds to cool down. & then in 50000 years the DEPTH will be greater than <5000.> 400.-- These facts
126a124 rculations from surface can take place.-- \ the DEPTH of frozen soil is against this view.-- however
127a125 ce. (when climate hotter) was cooled to greater DEPTH.-- Now the <inf> subterranean isothermal line m
127a125 N. when soil frozen for greater length of time DEPTH of ice ought to be less.-- Memoir of the Irish
129a132 must have been from an axis, «20 ft at least in DEPTH» near mouth of Columbia river-- Read Mr Parker'
269c101 38 on soil in Siberia being frozen to 400 ft in DEPTH, (& Erman's surprise that it is not 700) is app
269c101 e lived. mean temp 60 degree mean, then temp at DEPTH of four hundred feet would be 60 degree + 6 deg
306c213 here appears to be one line, in which» greatest DEPTH <appears to be> is not more than 60F. & «in» th
473s03v a-- How curious all negative laws of America of DEPTH of organisms holding in America as in Britain.
053r101 o me to be effect of expansions acting at great DEPTHS (mem: profound earthquakes), which would cause
059r123 of the globe shows powers have acted from great DEPTHS, so changes, acting in those lines. must now p
059r123 ing in those lines. must now proceed from great DEPTHS.--important.-- Decemb 10. 1802. Earthquake at
106a076 520 Mr Fox on increase of temperature at great DEPTHS. All Earthquake unaccompanied by Volcanos must
124a117 II. (1838) p 212. Facts from Erman about great DEPTHS of frozen soil. p. 211 Consider proved that Si
473s004 ir littoral deposits are probably buried in the DEPTHS of the sea-- Maer. June/42/ June/42/-- Mr. Bun
565n009 hy does emotion make tears fall?? Lavater says DERISION lies in wrinkles about the nose, & arrogance
551m128 se from the preexistence of the soul, are not DERIVABLE from experience.-- read monkeys for preexiste
173b010 uming all> if species (<a>) «(I)». <fr> may be DERIVED from form (2). &c.-- <(> Then (remembering Lye
190b078 passed through to form that species.-- <Man is DERIVED from Monad, each fresh--> Mr Don remarked to m
528m037 ugly ones.-- the pleasure from perspective is DERIVED in a river from seeing how the serpentine line
603o013 e classes of words «are abbreviations» he thus DERIVES from nouns & verbs-- so that much of EVERY lan
149g032 lake On the summit «& on Spean side» of Meal-- DERRY there were perfectly rounded «base» pebbles of
151g040 te neighbourhood, (as granite or gneiss of Moel DERRY) on low hill between Inn & Bouhunthine the summ
070r155 close Analogy Lake of Cordill: of Copiápo & DESAGUADERO.--three ridges in Copiapo, as well as in lat
181b041 ny are relatives, so that by tracing back. the <DESCEN> fathers would be reduced to small percentage.
151g041 of Hill (judging from external form alluvium) DESCEN from shelf 3d & almost meet, but are separated
154g061 -. why should it deposits River terraces often DESCEND by flights the terraces if the largest has hol
359d089 il duck) from which they were descended-- they DESCEND from 1/2 pintail <into> «by» duck, into pintai
367d168 ssume that every peculiarity has a tendency to DESCEND to several generations» If A & B be two races
200b122 g to them it is not; one when they can proved DESCENDANT, which of course most rare, or when placed t
291c157 (which is almost proved. Elephant has left no DESCENDANT in Europe<)> Toxodon in S. America] is absol
302c201 nity between two classes.-- there may be some DESCENDANT of some intermediate link.-- the only connec
417e076 ing the Immortality of the Soul as the linear DESCENDANT of <Mammferus» «Mammiferous» <vert> animal,
231b243 This probably explains crag & miocene.-- The DESCENDANTS left in cooling climate might change twice o
292c169 species the greater number probably have no DESCENDANTS on earth.-- +++. Even at Falklands some prob
460t019 world.-- our present organic beings are the DESCENDANTS, <slightly> «a good deal» modified <& Many F
160g088 parts?? Bought Glen name of Glen by which we DESCENDED, it is to the west of Glen Tarf What I called
214b175 Hall close to Birmingham, and supposed to be DESCENDED from a breed known to be there since the time
219b195 lpine» «new formations» because snow formerly DESCENDED lower, therefore species of lower genera alte
282c142 species of each other, because we suppose all DESCENDED from same.-- but if two original species, eac
304c207 rifles. it would not then be told whether not DESCENDED from long way back.-- aberrant forms produced
304c208 may be inferred much time elapsed & therefore DESCENDED from branch high up.-- Such probabilities onl
349d052 eory, every species in any sub-genus will be. DESCENDED from one stock, & that stock with other subge
355d066 (here argue if it be said domestic fowls are DESCENDED from several stock, then species are fertile\
359d089 er wild or Pintail duck) from which they were DESCENDED-- they descend from 1/2 pintail <into> «by» d
418e083 from analogy, that all these very animals are DESCENDED from some one single stock,--one is led to su
420e088 a.-- but a common point, whence both may have DESCENDED.-- Jan. 6th The rudiment of a tail, shows man

027r023 ecause the formations are now seen in regular DESCENDING steps Mem.; rapidity of germination in young
402e018 t a monkey, ceased their cries. "many of them DESCENDING to examine their defunct companion".-- p. 22
465t081 on can be expected in any one country.-- in a DESCENDING series of strata This again shows how much f
304c207 tion of immense difficulty is, whether Apterix DESCENDS from same parent with other birds,, or branch
152g046 ffect of corrosion & not cause. Monday a rapid DESCENT of a terrace except at very head of valley ind
231b243 pics are only related by one connection.-- viz DESCENT.-- Hence far greater discordance in latter-- H
258c060 hildren would not become adapted to climate.-- DESCENT. or true relationship, tends to keep to specie
286c155 arrangement> relationship; latter word meaning DESCENT.-- A tree is taken by Fleming as emblem of dic
294c177 good species as carrion crow & rook formed by DESCENT. or two of the willow wrens &c &c. & analogy w
349d053 n only be judged of in each «separate» line of DESCENT.-- <& here limits of varieties being constant.
409e048 .-- now take greater area of water & snow line DESCENT. My theory gives great final cause «I do not w
427e109 render climate less extreme. (& so account for DESCENT of snow line there «& there & there only: as s
526m027 s such a thing as chance.-- chance governs the DESCENT of a farthing, free will determines our throwi
550m123 ging contingencies, or bodies of either.-- Our DESCENT, then, is the origin of our evil passions!!--
552m132 have & origin as well as rule will be given.-- DESCENT of Man Moral Sense Mitchell Australia Vol I, p
554m139 not puzzle her-- with aid of teeth & hands.-- DESCENT 1838 It was very curious to see her take bread
569n019 mitive.-- almost like tastes of mouth & smell. DESCENT of Man Understanding languages seem simplits c
594n115 ruggling with another large dog his companion. DESCENT --Affection & [...] Monkeys «Ogleby» seen Zool
182b047 c The new system of Natural History will be to DESCRIBE limits of form. (& where possible the number
200b122 v> 1828 It is daily happening; that naturalist DESCRIBE animals as species, for instance Australian d
040r063 dikes have a parting of pitchstone; which is DESCRIBED as very rare Mem. St Helena; probably more ab
050r090 doubtful. -- In Iceland Bladders of Lava are DESCRIBED, & many minute craters as at Galapogos. <|> S
054r102 y quartzose sandstones: refers to broken hill DESCRIBED by Pernetty: account of streams of stones agr
078r174 40. Flèche of Quoy et Gaimard.--D'Orbigny has DESCRIBED it with care to 3 species. I think I have muc
090a022 lling; many of these were not single, but are DESCRIBED as many, (one even 3000) This ninety includes
108a081 stmas Sound Vol XIV. (My Edition) p 500. Well DESCRIBED [top portion page excised, not located] -- do
118a103 -- This may be added to any place where dikes DESCRIBED-- {P} Cordillera. St Helena &c &c.-- in Cordi
135a146 pkins fissure at {P} .-- G. J. Malcolmson has DESCRIBED formation of shore of Coromandel. just same a
184b054 same changes in Fish Mem. Rabbit of Falklands DESCRIBED by Q. & G. as new Species. Cuvier examined it
363d101 abits of rock pidgeon. (I suspect Pennant has DESCRIBED them)-- [Study horns of wild cattle.-- plumag
410e053 c &c can never be told, without species being DESCRIBED.-- but then permanent varieties in same count
425e104 rk on the Labiatae, some of these species are DESCRIBED.-- Capital case,-- for Sandwich Isld are very
430e117 tacking a sick one like Chillingham bulls are DESCRIBED.-- His three have had VERY different disposit
451e176 Paper by Lieut. Newbold.--» A Malayan albino DESCRIBED "To this day the tomb of his grandfather, who
453e183 ey «are» raised by seed.-- Where has Duchesne DESCRIBED Atavism.-- ask Dr Holland cases where peculia
454e184 rds in his essay on Spermatic animalcule. has DESCRIBED instrument for galvanize <ani> them-- Cross I
464t065 gative facts are valueless= monkeys= Owen has DESCRIBED a greatt Struthonidous Bird from New Zealand-
473s006 from Capel-Curig-- Mr Bunbury says Miers has DESCRIBED in Linn: Transacts. a Brazilian plant «in Lin
473s006 between Orchis & other plants-- & Wallich has DESCRIBED Indian Plant.-- June /42/-- June/42/ You can
479z005 the name of Sagitta Triptera M. D'Orbigny has DESCRIBED my animal with teeth {P} p. 140. Flèche of Qu
483z013 Voyage p 302 Vol II p. 302. Vaginulus of Lima DESCRIBED "Arion" of Ascension. p. do.-- some S. Americ
483z013 nsion. p. do.-- some S. American Reptiles are DESCRIBED Shells from Tahiti and Chile The North & S. R
484z016 - general red mark on wings of all-- Spix has DESCRIBED Philedon. allied to some of my birds-- These
486z020 14th> «13th.» Vol of Archaeologia arrow=heads DESCRIBED in Suffolk as lying under strata of gravel &
490q001 gs.-- Temporary Question 1 Where has Duchesne DESCRIBED Atavism alluded to by Dr. Holland-- <Jordan>
514q21. tera= Are dwarf plants on Wellington Mountain DESCRIBED in Flinders-- Alpine Australia Flora= Banana's
524m022 ed in others.-- Mrs P. when in state as above DESCRIBED, (forgetting that her husband was dead) yet i
535m064 earwig & a doubtful one of Acanthosoma grisea DESCRIBED [not located] as first caused by will of Gods
539m084 ther. Aug. 16th Anger «Rage» in worst form is DESCRIBED by Spenser (Faery Queene. CD 25 (Descript of
556m145 ise which C. Sphynx made at Z. Gardens may be DESCRIBED as partaking of <st.> made by <ret> inspirati
564n006 sively cross-half furious faces «which may be DESCRIBED as an exaggerated habitual sneer» the manner
574n041 ve seen Nina licking her chops.-- someone has DESCRIBED slovering <gum> <teeth>less-jaws. as picture
574n041 tion of mouth & jaws.-- Lascivious women. are DESCRIBED as biting: so do stallions always.-- = No doubt
277c127 logy to be guide. in islands. species.-- each DESCRIBER giving his test namely differ as much as thos
410e052 losopher, would as soon turn tailor, as, mere DESCRIBER of species, from its garments, without some e
410e052 me, without reference to description), except DESCRIBERS having some high theoretical interest,-- "th
410e052 ts garments, without some end.-- Respect good DESCRIBERS like Richardson.-- The relations of numbers
023r012 . Labillardiere in Bay of Legrand, (SW part). DESCRIBES a Small granite Isd. capped by Calcareous roc
110a087 eograph. Journal. Vol IV (p 321) Mr Hillhouse DESCRIBES central granitic ridge of Guayana as NW / SE.
123a116 merica. Von Buch Lyell. (under head of Delta) DESCRIBES near Alps great beds of rivers which must be
260c068 er of little bird, in same way Wilson (p. 5). DESCRIBES many kinds of bird uniting together in pursui
367d113 en says that Bell in Encyclop of Anat & Phys. DESCRIBES, a high-flying bat, which has the power of in
430e117 Ramond offers no explanation.-- Poet Cowper. DESCRIBES his tame Hares, attacking a sick one like Chi
496q05a e & see whether grains flaccid, as Koelreuter DESCRIBES Kill Sparrow after feeding on oats, give body
535m064 f tea chest, when no tea do. p. 233. Mr Lewis DESCRIBES case of insects «a Perga» of Terebrantia, lay
563n001 2d.. 1838 Essays on Natural History Waterton DESCRIBES. pheasant springing from nest & leaving no tr
591n097 ton on vital laws (in the Athaenaeum Library) DESCRIBES effects of emotions-- fear giving goose skin-
064r137 em: Grand gypseous formation of Cordillera In DESCRIBING structure of Cordillera it must be said, tha
128a130 uled by bars of sand & shallow lagoon.-- when DESCRIBING Coast of. Brazil. Maldonado enter into this
507q15v - oats cut down turning into Rye.-- 46). Book DESCRIBING amount of Horticultural Variation? Henslow k
569n022 5 Quoted repeatedly by Waitz (In Theil. V) in DESCRIBING Caroline Archipelago. Dn 75 cf., p 268 witho
571n031 deas.-- In language. the possibility of poets DESCRIBING gentle things in gentle language, & vice ver
511q018 rds-- Bantams.-- (6) <Mad> Porto Santo Rabbit. DESCRIPT. of colour «& length of ears» & skeleton, & s
539m084 is described by Spenser (Faery Queene. CD 25 (DESCRIPT of Queen) «O» of Hell Cant IV or V.) as pale
021r005 mmon within E. Indian Archipelago, no minute DESCRIPTION, calls it a Fucus. P «Vol I 287» P 379. Hens
043r069 e exclusively confined to that country. Read DESCRIPTION of channels or grooves in rocks at Costorphi
097a043 ut> on one of the Ponza isles. but no minute DESCRIPTION is given.-- Vol II. 2d Series. p. 221.-- Mr
110a086 levation & subsidence. examine these «lines» DESCRIPTION of rocks in Lyells'. Capital Norway case.--
240c013 count for [not located] Falkner Patagonia no DESCRIPTION of wild animals, nor in Dobrizhoffer Abipone
255c051 bands on pidgeons back.-- According to this DESCRIPTION of class is description of ancestor of all b
255c051 -- According to this description of class is DESCRIPTION of ancestor of all birds, & so for birds. We
267c094 eed p. 333-- alludes to the Macúsie breed no DESCRIPTION given-- Ch. 2. dogs L'Institut. 1838. p. 67.
404e025 g. L'Institut: .1838 p. 329-- Milne Edwards, DESCRIPTION of curious mechanism of respiration, or rath
410e052 sins» (do not add name, without reference to DESCRIPTION), except describers having some high theoret
463t063 ent of the discoveries to come-- Owen in his DESCRIPTION of my fossils makes same such remark & befor
256c054 Forster on South Sea, will probably contain DESCRIPTIONS of domesticated animals in those regions Sp
478z003 ora. Quoy. Freycinets Voyage Vol p. 597 Many DESCRIPTIONS about lower animals of Falklands &c Benn
048r086 mals. both species & individuals as in the half DESCRIPT country of S. Africa. It would be well to quot
062r129 mstances; reversed argument. knowing it to be a DESERT.-- Tempted to believe animals created for a de
133a139 mperature of earth beneath <of Sahara de> a dry DESERT, would be very high.-- M. Parrot ends his pape
185b055 try a ground «do. <w> parrot.--» woodpecker-- a DESERT. Kingfisher.-- mountain tringas.-- Upland goos
212b167 Mollusca according to periods.-- NB. Was Europe DESERT (like S. Africa) after Coal Period.-- ¿In thos
223b209 al. is this if after isolation (seed blown into DESERT) or separation by mountain chains &c the speci
223b210 ormed. during ascent of mountain or approach of DESERT?-- because the crossing of species less altere
230b235 nst theory of changes that if so in approaching DESERT country or ascending mountain you ought to hav
252c042 language?) Hyena «venatica» <of> Cape found in DESERT of Korto & Steppes of Kordofan p. 401. Admirab
260c068 of other parents.-- Shows community of language DESERT country. is as effectual as a cold one. in che
268c099 by seeing how many species common to Patagonia DESERT & Tierra del Fuego. & forest «Parrots in Macqu
284c147 t different periods, depends,-- on relations of DESERT, open ocean, &c this probably on long average,
438e142 resist the means Man can offer of changes.-- as DESERT «or rock» plant probably would do-- or be with
535m064 d to other larvae, when two groups near. mother DESERT one sometimes & go to other, so that two mothe
633j53v are not detrimental.-- p. 285 the seed-pod of a DESERT plant (Anastatica) is rolled along, & splits w
535m064 inue till death, thus acting 4 to 6 weeks. The DESERTED broods appeared healthy-- This remarkable cas
036r052 es, grasshoppers & Rats. characteristic of the DESERTS of Syria <chara> ditto for Patagonia, especial
064r134 nly covered by vegetation. Rhinoceros quite in DESERTS.--Much struck with number of animal[s] at Cape
086a005 It may be worth noticing edentates & camels in DESERTS & rodentia In Plata Mastodon Toxodon Is the ge
639j28r cases of corelations.-- [There must have been DESERTS in the old world!]CD p. 252 analogy of hand in
417e075 of Indian & Common Oxen, which one must think DESERVE the name of species, may be owing to the littl

Page **(Key Word)**
345d042 nimal must have gone on for many years, before DESERVES <name> «to be so called»,-- the short horned
461t025 ve not sack,-- Most curious facts & this paper DESERVES fresh study & whole order of the fish.-- Embr
608o027 s view should teach one profound humility, one DESERVES no credit for anything. (yet one takes it for
601o08b nce des Animaux-- & Le Parfait Chasseur, par DESGRAVIERS, un Vol 8vo Keratry-- Inductions morales et
110a085 mixed.-- Turner's Chemistry p. 206 Both Beck & DESHAYES saw fossil shells from West Indies & declare
184b054 w dead. Bulletin Geologique April 1837. p. 216 DESHAYES on change in shells from salt & F. Water-- on
215b178 e" developing his ideas on passage of forms.-- DESHAYES states Lamarck priority refers to introductio
216b179 s this word «for» similar in the Indian sea.-- DESHAYES.-- Mr McClay is inclined to think that offspr
632j53r by water «fucus for adhesion».-- as examples of DESIGN.-- perhaps they are so.-- but the coral rock m
634j54v me satisfied respecting an original thought, or DESIGN, pursued to its utmost exhaustion, & till it m
634j54v -- What bosch!! Put it to case of man. <&> The <DESIGN> determination of a God-head.-- the designs of
634j54v he <design> determination of a God-head.-- the DESIGNS of an omnipotent creator, exhausted & abandone
500q010 bbage. (26) Do deer Keepers cross the breed-- DESIRABLE as in Cattle in Chillingham Park-- What Book
550m124 of happiness-- Entire happiness. not being so DESIRABLE as <broken> intense happiness even with some
629o55v & will?)) (2) It is other question what it is DESIRABLE to be taught,-- all are agreed general utilit
048r086 s (mem white tufas with purple Claystones of P. DESIRE). = Where talking of such substances being wor
057r114 similarity of rocks of very diff. ages. at Port DESIRE on plain. & interstratified.-- Urge fact of Bo
065r139 egrading of inland bays. like St. Julian & Port DESIRE applicable to Craters of Elevation.--The longe
194b093 nquer or breed together:-- Man has no limits to DESIRE, in proportion instinct more. reason less. so
215b176 ogs are bred into each other, the females loose DESIRE, and it is required to give the canthairides a
215b177 any constant species Both males & females. lose DESIRE. Native dog not found in V. Diemen's land J. d
375d134 positive checks, excepting that famine may stop DESIRE.--» in Nature production does not increase, wh
377d139 make st.-- noise| In case of woman instinctive DESIRE may be said to be more definite than with bitc
389d176 eral births, & even produce fertile offspring-- DESIRE LOST when male & female too closely related: t
391d179 failed, & then by that corelation of structure DESIRE fails. Every individual except by incestuous m
540m089 real pleasure at pain of others, with rational DESIRE to assist them,-- otherwise as he remarks symp
553m138 never seen any of the American Monkey show any DESIRE for women-- «very curious. as they depart in s
556m146 ting? What is Emotion analysis of expression of DESIRE-- is there not protrusion of chin, like bulls
573n038 t swallow spittle. will have involuntary flow & DESIRE to swallow.-- tells himself not to turn in bed
574n041 one end of Cocoa nut.-- November 27th.-- Sexual DESIRE makes saliva to flow «yes, certainly»-- curiou
580n062 t age are retained through life" p. 200.-- "The DESIRE of glory, immortal fame, &c so common in the y
587n088 make a list of every word, expressing a mental <DESIRE> «quality» &c &c Mackintosh Ethics p. 97. on D
608o026 t himself in the way of Contingencies.--but his DESIRE to do arises from motives.--& his knowledge th
620o044 ps lies in its frequency & in its consisting in DESIRE gratified & therefore as soon as desire is ful
620o044 ting in desire gratified & therefore as soon as DESIRE is fullfilled, pleasure forgotten. [RHC] 4) as
628o54v <other> outward gratification, whilst <the> no DESIRE of gratification will check the consciences de
628o54v ire of gratification will check the consciences DESIRE for virtue.-- [I expect there is some fallacy
291c166 nce of habits in classification.-- Thought (or DESIRES more properly) being heredetary<)>.-- it is di
600o008 (2) Beau ideal, refers chiefly to moral, beau DESIRES conscience & love.-- [With regard to ordinary
628o054 ves the instinct.--]CD p. 22. says affections, DESIRES, & moral sense all different.-- P. 22. Butler
629o55v k these artifical rules, get remorse-- ((hence DESIRES do not intervene between this kind of conscien
629o55v ill, though «this» conscience does between the DESIRES & will?)) (2) It is other question what it is
629o055 uired by education.-- CD[In similar manner our DESIRES become fixed to ambition. money, books &c &c.-
264c079 s passion & rage, sulkiness, & very actions of DESPAIR; «let him look at savage, roasting his parent,
437e140 b of a fox, which after it had fought well & DESPERATELY bitten the young ones, would in all probabil
620o045 fluous, as one man trying to save another in DESPERATION.-- This shows, that our feeling, that the in
541m093 r more difficult to to look tranquil.-- He may DESPISE a man & say nothing, but without a most distin
574n043 of Jenner's <Hyaena> Jackall.-- an animal not DESTINED by nature to exist. & carrying «like other hy
343d038 c forms of S. America. With respect to future DESTINIES of mankind, some of species or varieties are
195b101 ordered, each planet to move in its particular DESTINY.-- In same manner God orders each animal creat
078r176 lera "immense variety of Porphyries which are DESTITUTE of quartz, & wh abound both in hornblend & vi
223b211 iseased form a plant, adapted to A. B. C. D.-- DESTROY plants B. C. D. & A will soon form good specie
230b240 cies & mere producing fertile hybrids will not DESTROY that evidence, as so many plants produce hybri
291c168 ld migrate south ward being ready made.-- & so DESTROY individuals, wheras in Falkland Isd they would
356d072 y theory-- without God is supposed to create & DESTROY without rule-- But what does he in this world
511q018 t Zoolog Soc.-- did Sir. J. Sebright select to DESTROY secondary character believe No or did result a
537m076 s rare,-- as when Polynesian mothers ceased to DESTROY their offspring--¿ yet perhaps if they had mur
555m142 not charity-- is it in former case instinct to DESTROY contagious disease.-- (Useful to use term inst
629o54v e not <instinct> «conscience») equally <prefe> DESTROY all wish of outward gratification,-- see what
042r069 movements if present in the Andes, would have DESTROYED regularity of slope of valleys.--All my obser
162g101 eather When it did not grow at first-- relics DESTROYED.-- the Brook <about> Head of which is so inte
203b157 S. America. that only some mundine cause has DESTROYED animals over the whole world-- For instance g
250c037 t. Try amongst Europaean quadrupeds if Africa DESTROYED would not then some forms be peculiar to it,
250c037 be peculiar to it, so on, & so on.-- Whatever DESTROYED great <quadrupeds>; Pachyderm in S. America d
250c037 d great <quadrupeds>; Pachyderm in S. America DESTROYED great Edentata or American form.-- Is the Aus
263c074 believing that monkey would breed (if mankind DESTROYED) some intellectual being though not MAN.-- is
265c087 e generally applied.-- Therefore argument not DESTROYED even if these shrivelled wings could be shown
283c145 short billed one, be exaggerated, & all rest DESTROYED far remote genera. will be produced. As we kn
291c168 ge & make new species.-- alpine species being DESTROYED at Falkland Isds++.-- Mem Lyell hypothesis of
298c191 er species afterwards would probably often be DESTROYED.-- or regrafted with fresh arrivals..-- &c &c
302c203 ll intermediate species living there would be DESTROYED, & N & S. existing species becomes father of
357d073 tata-- Edentata & Marsupials have been almost DESTROYED wherever other animals existed.-- Athenaeum 1
406e037 rn.-- Great animals. of same two great orders DESTROYED about same time in North & South. America.--
414e063 t one cross, & the permanence of his breed is DESTROYED)-- -- When two races of men meet, they act pr
423e096 I doubt not if the simplest animals could be DESTROYED, the more highly organized ones. would soon b
589n091 every individual played a little, & something DESTROYED bad brain. see p. 90.-- The relation of reaso
640j167 spe> became more numerous. (from death of its DESTROYER), or other cause, the long legged race would
465t081 this very immigration which tends to make the DESTROYERS vary; so that we here see reasons-- why no p
325c266 ion of Paris. with respect to licentiousness, DESTROYING children. --it is not effect, as Lyell sugge
416e071 variation in each birth) from man arbitrarily DESTROYING certain forms & not others.-- Term variety m
171b004 to changing world.-- On other hand, generation DESTROYS the effect of accidental injuries, <on> which
426e107 intervention of domesticated ie new varieties DESTROYS the appearance of this series & makes one thi
071r157 utterly overthrown by earthquake with great DESTRUCTION of human life.--Temple mentions some earthqu
077r170 ere is more modern breccia, chiefly owing to DESTRUCTION of porphyries. whereas other to ancient rock
250c037 rica, is only common cause I can conceive of DESTRUCTION of Great Animals in Europe & America Some po
250c039 ed Much might be argued what is not cause of DESTRUCTION of large quadrupeds.-- common to two types o
258c060 rossing, with unseasoned people. would cause DESTRUCTION.-- simile Man living in hot countries, if co
263c075 y of reproductions of species & certainty of DESTRUCTION; then he will choose & firmly believe in his
268c099 misphere just anterior to present. ¿cause of DESTRUCTION of «great» animals?-- Show independency of s
357d072 what does he in this world without rule? The DESTRUCTION of the great Mammals over whole world shows
428e111 n in Sir. J's ponds-- my principle being the DESTRUCTION of all the less hardy ones. & the <accidenta
040r062 such matter at St Julians looks like such?--DESTRUCTIVE to animal life.--Patagonia In the Chonos Isl
641j29r t widely differ» 3 A very wide range must be DESTRUCTIVE to species, when physical changes are in pro
541m090 & thought here?) The opposite extreme of this DESULTORY thought is following out such an idea, as eff
102a060 of England-- Any one. who has studied rocks in DETAIL as amygdaloid. calcareous rocks of Ascension,
192b085 era MAY have had intermediate steps.-- Quote in DETAIL some good instances. But it is other question,
252c042 e letter from Macleay to Bicheno much excellent DETAIL & fine, views about Species-- MUST BE STUDIED:
272c109 hers in tail of Edolius.-- Remarkable how small DETAIL in structure prevails amongst the same species
285c151 ms, parents of all species not possible in some DETAIL,-- the relations to islands close species, «on
526m028 remember it so well that they can correct every DETAIL, yet they have not imagination enough to <up>
484z014 olypi of Tubulipores L'Institut-- 1838 p. 75 A DETAILED comparison of production without Tropics in N
420e089 said to be heredetary. shows well what minute DETAILS of structure heredetary'-- Athenaeum .1839. p.
481z010 ed for anatomy. of. corals.-- nevertheless the DETAILS appear very trifling Also Berre «p. 8» (I thin
028r029 Lime is not accumulated in the Tropical oceans DETAINED by Organic powers. We know the waters of the
608o027 it is right to punish criminals; but solely to DETER others.-- It is not more strange that there sho
379d152 t to any other breed else all the lambs will DETERIORATE.-- Lord Moreton's Case.-- When cows have twi
391d174 what kind, either progressive improvement or DETERIORATE» that object failing, generation fails.-- Ho
576n047 k at Spain now.-- man's intellect might well DETERIORATE. «CD[in my theory there is no absolute tende

Page
(Key Word)
288c159 l, does not know whether breeds of oxen have DETERIORATED, or altered, but it is certain that rams &
371d128 thout circumstances very unfavourable, will <DETERIORATED> continue of same variety as long as life l
172b006 in constant; is it not said that marrying in DETERIORATES a race, that is alters it from some end whi
026r021 y felspar in Phonolite arrange themselves in DETERMINATE planes ∴ such action can take place in melte
177b029 of some genera have died; have they all one DETERMINATE life dependent on genus,. that genus upon an
634j54v nes. He explains it <"By> saying "It is the DETERMINATION to adhere to a plan once adopted; & it is f
634j54v h!! Put it to case of man. <&> The <design> DETERMINATION of a God-head.-- the designs of an omnipote
439e144 & that tiger springing an inch further would DETERMINE his preservation-- if killed by some other an
460t017 ation in every possible way.-- the laws which DETERMINE the kinds of monstrosity, & determine the kin
460t017 s which determine the kinds of monstrosity, & DETERMINE the kind of variation «& sporting» in flowers
600o008 beau,--Jeffrey denies all Beau-- How does Hen DETERMINE which most beautiful cock, which best singer-
639j28v y two of them live to breed, if circumstances DETERMINE that, the long legged one shall rather oftene
023r011 lcanic axis «of the Andes».-- "Has this fault DETERMINED side of volcanic activity.» That axis was pr
074r165 a rock» Veins concretionary; concretions <dt> DETERMINED by fissures as in septaria. (& Chiloe case,
092a031 some ideas about order of injected rock being DETERMINED by fusibility in. L Institut p 247. 1837.--
148g024 pressed Case of Birch Wood by Inverorum being DETERMINED by sheep & not deer When Black faced sheep a
192b084 I say some use, but <no.> sex not having been DETERMINED.-- so with useless wings under elytra of bee
265c083 This congenital changes show that grandson is DETERMINED, when child is.-- shows that generation impl
282c143 fe must be every where ambient. & <in> merely DETERMINED to such points by the vital laws.-- so that
333d009 one would argue the whole effect of race was DETERMINED by male: & How completely is Lord Moreton's
351d057 ews on this subject.-- Monstrosities, kind of DETERMINED by age of foetus.-- As Larvae may be more pe
360d089 the former preponderated <which seems owing «DETERMINED» by the sex> Individual instances trouble Ya
388d171 to potato.-- With respect to offspring being DETERMINED by imagination of Mother.-- We see in a litt
420e089 a vertebra only & no head-- !! Handwriting is DETERMINED by most complicated circumstances, as shown
439e143 s «with parent species» false, which makes it DETERMINED by a facility in returning to old type Mr He
526m027 ne doubts existence of free will every action DETERMINED by heredetary constitution, example of other
570n024 back his shoulders. he wishes to show, he is DETERMINED not to say anything. he presses his lips tog
594n113 anger, lips not compressed sullen, protruded. DETERMINED to do nothing. & so manifesting sulleness. [
607o025 l (a) one well feels how many actions are not DETERMINED by what is called free will, but by strong i
051r095 copper bottom. we see a trifling circumstance DETERMINES whether an animal will adhere to a certain p
404e026 of» life to be mottled + heredetary tendency DETERMINES the puppies to be so.-- [not located] Did ma
526m027 governs the descent of a farthing, free will DETERMINES our throwing it up.-- equall true the two st
620o044 ifling he feels remorse.-- He reasons on it & DETERMINES to act more wisely other time, for he knows
640j28v Man & woman are same: fertility of either sex DETERMINES life:.» «With respect to whether Galapagos b
277c125 blended together Mem Mr Herberts law; habits DETERMINING fertility Scheme for abolishing specific nam
633j53v e good reason to know that they would not be DETRIMENTAL accidents, & domesticated variations show us
633j53v eed vessel]CD if man takes care they are not DETRIMENTAL.-- NB. One limit to the transmission of abor
633j53v rtive organs will be as long as they are not DETRIMENTAL.-- p. 285 the seed-pod of a desert plant (An
152g051 river--composition &-- stratification argument DETRITUS-- {P} where buttresses on 4th shelf: others «
155g064 st be in power of rivers either bringing more «DETRITUS» than they corrode or vice versa Same inclina
151g039 sual-- Lines die away where slope less., best DEVELOPED on steep earthy slope, two circumstances rare
189b074 rebral structure intellectual faculties} most DEVELOPED, as highest.-- A bee doubtless would when the
198b113 one the sanguineous system, in other nervous DEVELOPED. (Owen's idea) states these class approach on
216b181 rths, like aphides.-- Case of boy with foetus DEVELOPED in the breast.--looking as if many ova-- impr
254c049 tive energies easily expended & no one system DEVELOPED <is> not surprising to find many forms in Acr
258c062 about. affinity & how one order first becomes DEVELOPED & then another-- (according as parent types a
270c103 hich Carus considers big animal» becomes more DEVELOPED in higher animals than in vegetables. p. 243
273c110 not found in Old-World--. + + If in any «well DEVELOPED» family (Gould says) there is any marked colo
274c114 vary in degree in last instance hardly at all DEVELOPED. Not confined to one species, but generally s
289c162 ust (on my theory) =supported by foetal lower DEVELOPED forms.=» (NB waterhouse says of affinity of m
295c183 bigny Rapport. p. 11) in S. America so highly DEVELOPED in North.-- Icthiology of S. America. more pe
299c196 ans mind, in different races, being unequally DEVELOPED.-- ¿is not Elephant intellectually developed
299c196 developed.-- ¿is not Elephant intellectually DEVELOPED amongst Pachydermata. like Man amongst Monkey
306c212 nt & seed.-- instincts in young animals, well DEVELOPED, just like, habits easily gained in child hoo
341d030 d to rudiment.-- clavicle scapula &c strongly DEVELOPED to aid in breathing.-- Animals from Hobart To
353d061 common to Van Diemen's land & Australia well DEVELOPED <tits> «Mammae» in male ourang-outang. other
358d076 t in insects where one of the sexes is little DEVELOPED, it is always female which approaches in char
358d076 approaches in character to the larva, or less DEVELOPED state.-- the female & young of all birds rese
370d116 uters are true females, but with parts little DEVELOPED.-- Sept. 19th <Are> There is no scale, accord
371d118 lection of «Humming» birds, saw several fully DEVELOPED tails, & one with beak turned up like Avocett
382d158 h are present in every animals, but unequally DEVELOPED-- surely analogy of molluscs. & neuter bee w
384d161 to primary & secondary, the latter only being DEVELOPED, when the first <are> become of use¶ <Great c
384d162 .∴ Every man & woman is hermaphrodite:-- ∴ DEVELOPED instincts of Capon. & power of assuming male
390d177 to breeding in & in seems connected with more DEVELOPED forms) Study buds-- gemmae-- & monocotyledeno
441e150 laws.-- The effect of one being greatly DEVELOPED on another, must not be overlooked.-- it make
466t095 ies have been given to foetus <fr> before sex DEVELOPED-- Double flowers & colours breaking only here
509q017 e about the lowest cervical vertebrae process DEVELOPED into ribs.) & does its abortion vary, accordi
592n101 onsidering the origin of reason. as gradually DEVELOPED. see Hume on Sceptical Philosophy. Hume has w
606o023 w comes it there? Laocoon p. 75 "The beauties DEVELOPED in a work of art are not approved by the eye
613o036 ari-- --retriever-- produced as soon as brain DEVELOPED, and as I have said, no soul superadded, so [
279c133 orted by Mr Wilkinson. = Milking heredetary, DEVELOPEMEN of important organ (.see marks on pages).--
257c057 I explain no Mammalia in secondary-epocks, & DEVELOPEMENT of lizards.-- As we have birds impressions
259c063 ar to use its feet much in swimming, & every DEVELOPEMENT giving greater vigour to the parent so tend
263c076 rganization as to <per> force in one man the DEVELOPEMENT of a brain capable of producing more glowin
272c107 ly live in flowers & has Elytra. formed from DEVELOPEMENT of some other part of body.-- there are hem
305c210 ted with mind There is no progression in the DEVELOPEMENT in instincts in the <classes> «orders» of i
343d037 ay be Zoolo-geographically divided either by DEVELOPEMENT of new forms in one,. or apparently so. by
344d041 fertile offspring in Entomostraca & Aphides DEVELOPEMENT of sexes in Caterpillar. very valuable fact
357d073 & Didelphis being Mundine form., & the less DEVELOPEMENT of Marsupials in S. America. from presence
383d159 oper positions,-- this would be argument for DEVELOPEMENT of either.-- (Mammae or sheath of Horses po
388d172 thinks multiplication by division «only> is DEVELOPEMENT of gemma.-- The manner in which Frogs copul
418e080 ts is it not easier to understand ¿perfect?? DEVELOPEMENT of one sex on one side, than the addition o
418e080 erfect as in Ox. the amount of double sexual DEVELOPEMENT is spread over [not located] it utterly unt
423e095 e at any time as many species tending to dis-DEVELOPEMENT (some probably always have done so, as the
423e096 ng such to be the case, it proves the law of DEVELOPEMENT in partial classes is far from true.-- I do
423e097 ssible to increase them, hence the degree of DEVELOPEMENT is either stationary or more probably incre
435e133 at the action of light is concerned with the DEVELOPEMENT of form; but that tadpole increased in size
443e155 time is of no consequence in that degree of DEVELOPEMENT.-- [It is singular there is no true hermaph
543m098 th this view) therefore there is Instinctual DEVELOPEMENT) in one order, as there is Intellectual in h
576n047 eeks.-- (which seems opposed to progressive. DEVELOPEMENT) on account of dark ages.-- «effects of ext
599o006 time amplification of two ideas in nature; a DEVELOPEMENT of the thoughts expressed in Fingals cave,.
632j53r n, were created with reference to successive DEVELOPEMENT I admit, but the admission is probably from
198b113 --- I cannot understand whether. G. H. thinks DEVELOPENT in quite straight line. or branching S. H Wh
215b178 written "opuscule" entitled "Paleontographie" DEVELOPING his ideas on passage of forms.-- Deshayes st
291c187 bees changing the sex by feeding.-- no it is DEVELOPING an hybrid female it is a wonderful relation
313c237 ormant instinct.-- (How wonderful a case bees DEVELOPING sex of neuters) species may have had their i
445e159 ecessity of supposing some inward progressive DEVELOPING power.-- My theory leaves quite untouched th
181b044 eculations are applicable to non progressive DEVELOPMENT which certainly is the case at least during
182b049 hair resembling that of goats." Progressive DEVELOPMENT gives final cause for enormous periods anter
196b108 found in Ireland-- There must be progressive DEVELOPMENT; for instance none--?, of the <vebtetrata> »
301c200 e the habit is not applicable. The degree of DEVELOPMENT of all animals of same class being about equ
419e085 t be very curious.-- With respect to the non-DEVELOPMENT of Mollusca, which I have sometimes speculat
058r118 n effet toutes les montagnes de cette ile se DEVELOPPENT autour d'elle comme une ceinture d'immenses
269c103 to life generally.-- where <">unity constantly DEVELOPS multiplicity<"> [(his definition "constant ma
384d161 I observes "every species has a disposition to DEVIATE from Nature in a manner peculiar to itself" <I
629o055 me relation to others 5) if so, it is perhaps DEVIATION from the instinctive. right & wrong.-- (anima
550m123 hen, is the origin of our evil passions!!-- The DEVIL under form of Baboon is our grandfather!-- A ma
473s03v rica is Red Sandstone. & Birds true! Plants in DEVONIAN-- How strange no plants in our Devonian-- Fis

473s03v nts in Devonian-- How strange no plants in our DEVONIAN-- Fish one step lower in America-- How curiou
234b256 colouring a group of Nebria complanata from. DEVONSHIRE, from another from Swansea.-- Again Waterhou
569n020 ed by will <by> but by corporeal structure.-- DEVOTIONAL feelings, probably some distant power of the
587n089 > «quality» &c &c Mackintosh Ethics p. 97. on DEVOTIONAL feeling p. 103-- Abstraction p. 152. Percept
423e096 of the one herbivorous. & its one carnivorous DEVOURER.;-- it is quite clear that a large part of th
271c105 ilds, a berry on which it feeds. or insects it DEVOURS is same species. yet that it should so strictl
495q05a .-- (9) I have noticed leaves covered with Honey-DEW dusted with pollen of neighbouring grass= Spread
538m078 y throw light on consciousness, explained by Dr DEWAR on principle of association.--«fully bears out
451e178 r hand there are breeds of Men the Thârû & the DHANGAR who can live there & do not pine visibly. p. 3
452e181 occurs in India. in the Jungles of Borabhum & DHOLBUM.-- Vol do. p. 634, alludes to fact stated by M
371d127 en illustrates case of Dingo (he alludes to the DHOLES or wild dogs of India) in Zoolog. Garden havin
302c202 ith resuts. if I acted otherwise, my premises <in DI> would be disputed.-- according to Principles of
100a052 mity would they not unite in B. K.> {P} on the DIAGONAL of BK.-- ⅄ This is not applicable. it does no
180b039 tion of circumstances.-- Vide two. pages back. DIAGRAM The largeness of present genera renders it pro
591n101 ion or polytheism, at p. 424 Vol. II «Sect XV. DIALOGUE on Natural Religion.» however, he seems to al
085a038 Parish Possession. talks of <hill> «cerro» of DIAMANTE near stream of same name. with imperfect crat
085r139 pplicable to Craters of Elevation.--The longer DIAMETER of Deception Isl is six Geographical miles an
070r153 occur in same country? In botany instances DIAMETRICALLY opposite have been instanced: it is Let it
242c017 in East. Ind: Archipelago. ⫲Raffles. Horsfield. DIARD. Duvaucel. Leschenault Kuhl. Van=Hasselt, Reinw
590n093 ame, as that which attends great weakness.-- <DIARRHAEA> & syncope p. 42. Sighing from grief. is meth
513q21. ime of impregnation.-- says many flowers are DICHOGAMOUS Zostera-- Knights notion of pollen & stigma
286c155 t.-- A tree is taken by Fleming as emblem of DICHOTOMOUS arrangement which is false There is same dif
117a102 or at least not rising are there cliffs. Sir L. DICK says (.p 52) fringe of sublittoral deposit alwa
150g035 lear & upper line running up great bight just as DICK shows NB. Lake gradually draining off would for
157g078 e «2d» near divortium aquarum is a lip with it-- DICK right-- Mac mistook terrace also right-- Granit
151g041 pebbles on line of 4th shelf.-- Even on Lauder DICKS Hypothesis impossible to explain absence of lin
152g045 ne subsequent to shelf {P} In Glen Collarig, by DICKS theory lake burst in most improbable part & not
257c058 n Red Sandstone. great lizards in do.-- <Wood> <DICOT wood> Coniferous wood in Coal Measure.-- highes
460t019 do orders-- cheiroptera & caetacea in Eocene-- DICOT. plants in coal measures.-- Shells in Cambrian
206b150 .Institut «1837.» p 319 -- Brongniart.-- no DICOTYLEDENOUS plants & few Monocot in Coal formation? p.
207b150 vascular or cryptogam. (original Flora) and DICOTYLEDENOUS, which «nearly» first appear «(p 321)» at
221b202 s the prevalence of Coniferous Woods before DICOTYLEDENOUS a fact analogous to reptiles before Mammal
071r156 anians?--> If wood now preserved over world DICOTYLEDONES far preponderant, if so coniferous must for
236bIBC iate genera we might expect.-- Lindley Introduct DICT. Science. Naturelle Geographie Botanique De Can
324c268 oelle. Philosophie. or Geographical distrib. «in DICT. Sciences. Nat. in Geolog Soc.» F.. Cuvier on i
090a022 n the globe since the Cambrian system In Ures DICTIONARY between 1768 & 1818. that is fifty years-- 9
587n088 ut.-- Former, whines just like a child. Get a DICTIONARY & make a list of every word, expressing a me
481z009 a or Australia-- very curious.-- replaced by DIDELPHIDAE Skunk inhabitant of Patagonia. Mem:-- S. Cru
202b133 to East Indian isles.-- Compares it to fossil DIDELPHIS (S. American genus) in plaster of Paris.-- No
225b219 e creation of any new quadruped since days of DIDELPHIS in Stonefield. all lands united (Falkland Fox
240c013 fects false In New Guinea. a Kangaroo d'Aroe (DIDELPHIS Brunii) which as yet had only been found in i
354d062 inville has written paper to show Stonesfield DIDELPHIS not Didelphis «Answered satisfactorily by. Va
354d062 itten paper to show Stonesfield Didelphis not DIDELPHIS «Answered satisfactorily by. Valenciennes.» T
357d073 rown.-- This will explain. S. American case & DIDELPHIS being Mundine form., & the less developement
434e128 .-- p. 36.. speaking about the controversy on DIDELPHYS says. "If we cannot reason from the analogies
079r177 phenomenes."-- Ulloa's Voyage, Shell fish purple DIE, marevellous statements on, Vol I, P. 168. on co
150g038 age & far more earthy than what is usual-- Lines DIE away where slope less., best developed on steep
151g039 earthy slope, two circumstances rarely united.-- DIE away also, without any cause, must be tides. &c.
177b029 nus., that genus upon another, whole class would DIE out, therefore not.-- Monad has not definite exi
179b035 ranch, & the monucle has definite life, then all DIE at one. period, which is not case,∴ MONUCULE NOT
187b062 n country to superinduce a change? Seeing animal DIE out in S. America; with no change, agrees with b
187b063 r a tract from Spain to S. America.-- Never They DIE; without they change; like Golden Pippens, it is
187b064 generation of individuals.-- Why does individual DIE, to perpetuate certain peculiarities, (therefore
189b072 ppen. if produced by seed go on.-- otherwise all DIE.== The fossil horse, generated in S. Africa Zebr
189b073 e bound together just like buds of plants, which DIE at one time, though produced either sooner or la
190b076 of forms. reduce towards Northern Eastern end & DIE away, & partake of Indian character There appear
195b104 miles distant.-- Absolute knowledge that species DIE &. others replace them-- two hypotheses fresh cr
232b245 ene miocene & pliocene epoch), whilst others may DIE out or move South ward.... species must be compa
305c209 ce of error for if some of the ostriches are to DIE, then they would appear isolated.⫲ In my birds f
313c233 thorax & head differ Africa Australia Parasites DIE, when brought over on tropical animals which acc
370d116 451.)-- Wasps breed many females, but almost all DIE.-- bees breed but few, because they are kept in
431e122 part then transmutation.-- Do species migrate or DIE out.?-- In the place where any species is most c
451e178 s inaccurate C. D) they will catch the Malaria & DIE.-- On the other hand there are breeds of Men the
479z005 léche of Quoy et Gaimard Ulloa shell fish Purple DIE Marvellous stories Ulloa's Voyage Vol I, p. 168
544m104 odified in those races, where it is customary to DIE-- August 24th. As some impressions «Hume» become
177b029 nching.-- As all the species of some genera have DIED; have they all one determinate life dependent o
196b106 misphere. Were they produced in several places & DIED off in some? Why did not fossil horse breed in
232b245 oming habituated to colder climate whilst others DIED out, or moved towards equator.-- «or some speci
259c064 pecies-- Epidemic amongst trees. Plane trees all DIED certain year. Extreme difficulty of TRACING cha
520m001 ople taking after their parents, when the latter DIED so long before, that it is extremely improbably
520m002 ient, he was nearly 9 years old. when his father DIED.-- The omnipotence of habit is shown about meal
525m025 ils; but one a little longer than rest «they all DIED»:-- she had kittens before & afterwards with ta
059r120 y talks of much granite on all East side of Van DIEMEN Land. All the Calcareous rocks which harden by
030r036 atural position» position at N. S. Wales & Van DIEMEN'S land.-- Whole coast S. of Concepcion where th
182b050 ainly distinct species between Australia & Van DIEMEN'S land. & Austral & New Zealand Mr Gould says i
191b080 s on continent-- 3 Paradupasi in common to Van DIEMEN'S Land & Australia From the consideration of th
207b152 anic islets not well fixed.-- Peron thinks Van DIEMEN'S land long separated from Hobart Town-- {from
215b177 males. lose desire. Native dog not found in V. DIEMEN'S land J. de Physique. Tom 59. p 467. Peron G.
217b187 Sandwich islands-- A genus with species in Van DIEMEN'S land and Tierra del Fuego.-- Araucaria, speci
225b219 ew is only hope.-- New Zealand «compare to Van DIEMEN'S land.» glorious fact. of absence of quadruped
226b220 se (?) brought by canoes Ceylon & India.-- Van DIEMEN'S land Australia. England & Europe.-- It will b
235b263 sh & English Hare bear upon this.-- Why do Van DIEMENS land people require so many, imported animals?
235b264 cording to three elements How is Fauna of Van DIEMENS Land & Australia [blank] Falconer's remarks on
246c027 -- New Guinea scarcely differs more from, <Van DIEMENS land.⫲> Australia more than Van Diemen's land
246c027 <Van Diemen's land.⫲> Australia more than Van DIEMEN'S land.-- Vol II p. 8 no snakes on isles of cen
247c028 «also Taeniatole austral» «caught» Chile, Van DIEMEN'S land & Cape of Good Hope V. p. 44 of this Not
268c095 p. 11--&c. valuable paper on quadrupeds of Van DIEMEN'S land, which appear diff. from Australia (Wate
278c130 ll's authority) in Australia, & several in Van DIEMEN'S land is most important as showing former conn
278c131 chell does not think that dog was found in Van DIEMEN'S Land.-- V. ls. Number of Geographical Journal
300c197 the Str of Magellan.-- Change of habits in Van DIEMEN'S land. Study Mr Blyth's papers on Instinct.--
311c225 ree & great diff of Latd. p 355 Echidna of Van DIEMEN'S land & Australia different Temminck Fauna Jap
311c225 82 mammalia 293 Phalangista of Australia & Van DIEMEN'S land diff.-- Habits can only be used «in clas
314c239 Be cautious about Goulds case of birds of Van DIEMEN'S land & Australia.-- The wombat (Brown) is foun
319c257 lian fossils intermediate between those of Van DIEMEN'S land & Australia proper.-- Irish Elk case of
353d061 ows three species of Paradoxurus common to Van DIEMEN'S land & Australia well developed <tits> «Mamma
365d107 time which it must have taken to separate. Van DIEMEN'S land from Australia &c &c Sept. 14th. When Ma
400e011 ..]t in part near Timor, & to Europaean in Van DIEMENS land, where there is close species of elater.-
414e064 plains of Bolivia-- strictly fossil «& in Van DIEMEN'S land»-- they have been exterminated on princi
436e137 Quartz of Falkland.-- Old Red Sandstone-- Van DIEMEN'S land.-- Porphyries of Andes. A familiar Histo
500q10a - Are there any fine doubtful species from Van DIEMEN'S Land? or New Zealand? Babington about differe
540m086 of Africa, (or again the black man of <B> Van DIEMENS land & the energetic copper coloured natives o
155g062 largest has hollowed out most Wednesday Shelf 3d DIES away almost imperceptibly on Glen Turrit side 2
160g093 Alluvium» which appear perfectly level, <on op> DIES away on gradual slope-- : on N side.. dies away
160g093 n op> dies away on gradual slope-- : on N side.. DIES away on rocky place, but narrow shelves just li
162g100 ches along, parallel to Shelf on opposite side & DIES away on the steep & rocky gully of last stream
371d128 roduced, like parent stock, or if different DIETIORATING very slowly.-- I presume most of these ros
569n022 & Frenchman, unaccompanied by dignity-- "no mon DIEU," with a shrug-- "all I can say, I am very sorr
305c211 g and acting principle in the various shades of <DIF> separation between those individuals thus endow

```
407e040 merica, that the Mammalia of S. America are as DIFFERENT from the existing orders, as the Eocene of Pa
038r058 ¿ Consider causes for subaqueous crater being of DIFF: form subaerial one?--In former not so much; or
057r114 el proved.-- curious similarity of rocks of very DIFF. ages. at Port Desire on plain. & interstratifi
059r121 ary structure, for the interlineal spaces are of DIFF cont: & even in one case contained lime.--All b
060r125 . had been in action during secondary period how DIFF. would the rocks have been. The red Sandstone o
062r130 elation that common ostrich bears to (Petisse. & DIFF kinds of Fourmillier): extinct Guanaco to recen
178b031 eated there;--» Are not all our «British» Shrews DIFF: species from the continent look over. Bell, &
268c059 on quadrupeds of Van Diemen's land, which appear DIFF. from Australia (Waterhouse <disputes this)> sa
279c133 ortant organ (.see marks on pages).-- Crosses of DIFF: breeds succeed, yet seems to grant, that diffi
311c225 alludes to some other case of 180 degree & great DIFF of Latd. p 355 Echidna of Van Diemen's land & A
311c225 293 Phalangista of Australia & Van Diemen's land DIFF.-- Habits can only be used «in classification»
508q016 atisics, proportion of diseases (heredetary?) in DIFF. countries in same races Mr. Gray General Quest
057r113 ation proved at St Julian. = do not these bones DIFFER as much nearly as the Eocene. = Should Mr Owen
064r137 ates that Von Buch has urged that Java volcanos DIFFER from all others in quantity of Sulph. acid emi
217b184 ned bitch directly one after the other, puppies DIFFER, & like both parents.-- Fox told me of case of
224b213 r structure.-- Hence species may be good ones & DIFFER scarcely in any external character:-- for inst
240c013 ew Guinea. New Ireland, have phalangista, which DIFFER in «form & head &» colour from those of New Ho
251c041 ery close species, inhabiting Borneo & Sumatra. DIFFER only in form of white mark on breast: p. 234.-
277c126 us. fulginosus. (a) Falklands (b). F. del Fuego DIFFER from (C) Chiloe (E) Chile.. rupestris -- good
277c127 pecies.-- each describer giving his test namely DIFFER as much as those (naming them) which are found
303c204 uropaean forms on that isld.-- The races of men DIFFER chiefly in <size> colour, form of head «& feat
313c233 African & Europaean. different.-- thorax & head DIFFER Africa Australia Parasites die, when brought o
337d023 ni» organic beings.-- Animals «of same classes» DIFFER in different countries in exact proportion to
384d161 rate during its life so that successive buds do DIFFER-- any variety is not handed down. but is hande
390d178 therefore «each» seedling of one apple ought to DIFFER from those of other.-- The upshot of all this
440e148 lanzani's facts in connection with buds.-- They DIFFER from possibility of concourse of two «individu
448e160 6 Shells, as well as plants «of Juan Fernandez» DIFFER from American Coast Vol II.-- <Reference> p. 2
493q004 mber of pulse, Respiration, period of gestation DIFFER in different breeds of dogs. Cattle, (Indian &
514q21. caricas from every isld differs-- do they also DIFFER in different countries-- on flora of African I
515q022 sed with either parent.-- May. 44 These Hybrids DIFFER in colour of beak, taking after male & female
640j167 ranges? Agassiz has shewn that they most widely DIFFER» 3 A very wide range must be destructive to sp
303c204 nimal halfway between man & monkey, would have DIFFERED in hair colour + + + form of head «& features
317c249 t the mules between canary birds & goldfinches DIFFERED considerably in their colour & appearance Eve
172b007 t if kept long enough.-- «apart, with slightly DIFFEREN circumstances.--» Now Galapagos Tortoises, Mo
022r006 ix Would Slate. & unstratified rocks show any DIFFERENCE in facility of conducting Electricity? Would
036r051 ion alone.--In England much subsidence: hence DIFFERENCE; action on land different Volney, P 351. Vol
047r081 instead of sea's bottom being in motion what DIFFERENCE? In watching heavy swell, sea retreats & the
060r125 quotation from Temple Urge the mineralogical DIFFERENCE of formations of S. America & Europe.-- If g
064r136 local elevations of the land in Europe-- Urge DIFFERENCE of plutonic rocks & Volcanic metalliferous--
101a058 Peron does as if well attested.-- There is no DIFFERENCE between dike & mountain axis. except in rela
105a069 then how spread abroad?-- There is thus wide DIFFERENCE between erosive power of river & sea.; the f
114a094 ee Lyell talks of different composition using DIFFERENCE in metamorphic action which I give at C. of
162g102 ch Ness 30.140. A 66 degree 30.095 .0458 or 6 DIFFERENCE between bedrock & Loch Ness <30.100> <Donald
207b152 land long separated from Hobart Town-- {from DIFFERENCE of races of men and animals} See R. N. p. 13
210b159 . do the seeds of marked varieties produce no DIFFERENCE. if they do.--there probably will be this re
211b162 . «by Eyton» Account of three, kinds of pigs. DIFFERENCE in skeletons: VERY GOOD. Apteryx. a good ins
222b204 of» «on» Madeira-- any general observations-- DIFFERENCE of species between land shells of Porto Sant
223b208 s wonderful intincts <might well say how> The DIFFERENCE is that there is wide gap between Man & next
224b212 ss. that we canot tell, what is the amount of DIFFERENCE, which improves. & checks it.-- It does not
224b214 rapidly, we should expect most species.-- The DIFFERENCE intellect of Man & animals not so great as b
231b241 te to show that we do not know what amount of DIFFERENCE prevents breeding;--or as others would exjud
239cIFC through April 23. 1873 Books About amount of DIFFERENCE: where hybrids produced have any close speci
285c150 ung.--¶ father must be left out of case, that DIFFERENCE occurring.-- It will be necessary to show hy
285c153 edded in the constitution, what other marked DIFFERENCE in the varieties «made by» of Nature & Man.-
293c174 of people nearly similar. Curious instance of DIFFERENCE in races of men.-- Wax of Ear, bitter perhap
299c195 s in wild state-- The two. former produced by DIFFERENCE of <locality>. station & varieties chiefly p
306c212 ntly from dog. perhaps being in passion chief DIFFERENCE Major Mitchell is not aware that Australian
306c213 ustralian dogs ever hunt in company -- marked DIFFERENCE with dogs of La Plata & Guyana¶ people will
308c219 animals is generally different, because their DIFFERENCE. arise a good deal from climate & habits, &
316c244 he deity & thought of him «or eternity», only DIFFERENCE between the mind of man & animals.-- yet how
337d022 F. cisalpina in Italy, which is so like that DIFFERENCE would not be discovered by an unscientific o
338d025 gy of Britain & Europe is African. & the only DIFFERENCE is by the extinction of certain forms from N
340d029 Struthionidae, Mr Blyth remarked that greater DIFFERENCE in than in many large orders of birds. The E
344d041 ffspring. Entomostraca & Aphides. The extreme DIFFERENCE of sexes. is probably arrived at in case of
348d051 his amount, when <value> no scale of value of DIFFERENCE is or can be settled.-- I believe affinity m
349d052 they would be innumerable¶ does not know any DIFFERENCE between permanent variety & species!! (given
356d071 than parents is very curious as pointing out DIFFERENCE between acquired & heredetary tameness.-- In
359d086 ealthiness.-- «or» to perpetuate some organic DIFFERENCE.-- it may be so, but this assumption as long
360d092 syncrasy I have hitherto thought that a small DIFFERENCE <of any kind>, if very firmly fixed from lon
360d092 if very firmly fixed from long time, made no DIFFERENCE what its kind was.-- but if it were opposed
360d092 its kind was.-- but if it were opposed to the DIFFERENCE in other sex, it would be much more difficul
365d107 &c Sept. 14th. When Macleay says their is no DIFFERENCE between <t> "permanent varieties" & species.
372d131 on.-- there is no caterpillar state; the vast DIFFERENCE of two kinds of generation shown by their ha
385d163 dily & are subject to fits.-- «there is great DIFFERENCE between hybrids & inter se offspring in latt
387d168 ion--> V. p. 152 It is very singular the same DIFFERENCE from parental stock having been repeated sev
387d168 ilar to the several ova in mother. (with only DIFFERENCE of time) is the above law anyways connected
387d168 «like Lord Moretons case & Dr. Andrew Smith,» DIFFERENCE.-- If A. B. C. D. E be <offspring> «animals»
388d170 epresents this state.-- When it is said. that DIFFERENCE between bud & seed, that latter carries with
388d170 wants independent supply of food.-- is real. DIFFERENCE-- but this does not apply to potato.-- With
389d177 it is certainly very remarkable that too much DIFFERENCE should produce same effect as too little.--
390d178 oubt there is tendency to propagate the whole DIFFERENCE: to make the bud of the woman, not a bud in
390d178 his is that effect of Male is to impress some DIFFERENCE.-- «from what it received» (for it is probable
390d178 e course of some generations» has gained some DIFFERENCE.-- «from what it received» (for it is probable
390d179 ts<elf> «ancestors», & not its comparison «of DIFFERENCE» with other sex. = The highest bred Blood-ho
391d175 bsolutely necessary that some «but not great» DIFFERENCE (for even brother & sister are somewhat diff
392d175 oition or addition of differences. shows that DIFFERENCE need not be added EACH TIME. but after some
409e040 e: almost superfluous to consider minds.-- as DIFFERENCE between mind of a dog & a porpoise was not t
416e072 n shows it to be the effect of a gradation in DIFFERENCE in external conditions.-- = as in plant up
416e075 tructure is fully practised & perfected Hence DIFFERENCE between races & variety? «Man picks the Male
417e077 two varieties, & this we might expect, as the DIFFERENCE between man & woman is «indeed» (independent
422e092 compared to what it might have.-- What is the DIFFERENCE between an essential character & an adaptive
422e092 of Goethe.-- who maintains, that «Alludes to DIFFERENCE between fossil & recent Bull; like fossil &
423e097 ty of structure is adaptation. though perhaps DIFFERENCE between jaguar & tiger may not be so.-- Cons
424e102 ns to it?-- -- In argument of origin of Wolf, DIFFERENCE of mind is most relied on. but Bell has some
434e129 etimes one & sometimes other.-- there is some DIFFERENCE of habit between these varieties, so that th
436e136 ll certainly allow variation as much as «the» DIFFERENCE between <pi> species.-- for instance pidgeon
438e141 as Mr Herbert does not seem to recognize any DIFFERENCE in crossing between varieties & species, yet
443e154 kinds of generation have some most important DIFFERENCE is forced on us.-- My theory only requires t
450e175 reeds, (.Hence it appears there are shades of DIFFERENCE in all the isld, like in wild animals).-- Th
473s07r can select cattle & sheep for horns & yet no DIFFERENCE in calves--how is this in young pigeons--dog
493q003 Henslow»» see notes In varieties is there any DIFFERENCE in off spring from A. into B. from B. into A
512q019 ut hybrid pheasants treading-- any treadè?-- DIFFERENCE in lambs of different breeds Is there any di
512q019 nce in lambs of different breeds is there any DIFFERENCE in breeds of Cattle & sheep in the sprouting
524m019 is forgetfulness.-- which does appear a real DIFFERENCE, between oddity & madness.-- but then people
537m075 ly there is no universal moral sense.-- «from DIFFERENCE of action of approved» Yet as, I think, the
539m081 as no strain on the intellectual powers-- the DIFFERENCE is of a man wagging his foot & working with
547m111 ed in hurriedly giving orders.-- Now what was DIFFERENCE between Castle & dream No answer shows our p
558m153 ur arguments are good but look at the immense DIFFERENCE. between man,-- forget the use of language,
```

Page **(Key Word)**
559m153 the Fuegian & Ourang & outang, & dare to say DIFFERENCE so great... "Ay Sir there is much in analogy
563nIFC te. Expression M Expression N What are sexual DIFFERENCE in monkeys.-- Charles Darwin [Private.]CD (M
570n025 nding head., & active vessels of skin.-- What DIFFERENCE is there between Squib after having eaten me
579n058 r shyness, having been invented, prove of the DIFFERENCE, which my theory believes in.-- From the man
580n062 tand signs.-- very profound.-- concludes that DIFFERENCE of intellect between animals & men only in K
582n067 nconsciously of any end.-- N B. There is wide DIFFERENCE, between the means by which an animal perfor
586n080 & old.-- These facts point out some essential DIFFERENCE, which clearly ought to be separated-- We ap
588n089 of falling.-- &p. 193. that they perceive the DIFFERENCE on being carried up or downstairs, or dangle
588n090 an old.-- (dog horse, sow) we perceive great DIFFERENCE.-- «(& is not this difference same, but less
588n090 perceive great difference.-- «(& is not this DIFFERENCE same, but less in degree, as between man & c
600o08v not clear.-- ¿does not Mackintosh make great DIFFERENCE between moral sense & conscience? we admire
608o026 y, & savage calling laws of nature chance) 2) DIFFERENCE is from imperfect condition of mind all moti
609o30v animals» must be separately considered.-- The DIFFERENCE between civilized man & savage,-- is that fo
616o038 nced by this heredetary kind of memory.-- The DIFFERENCE between heredetary memory & individual secre
616o038 ividual secretion of thought, may be no more «DIFFERENCE» than sexual intercourse in plants is involu
626o052 hing about any principles born in us.-- Great DIFFERENCE with my theory.-- see p. 349.-- remark on th
032r041 play As in Ocean & Air; there are «likewise» DIFFERENCES of temperature «at equal distances from cent
035r045 ating with obsidian must clearly be chemical DIFFERENCES. & not those of rapid cooling &c &c My resul
174b015 oled.-- Countries longest separated greatest DIFFERENCES-- if separated from immens ages possibly two
176b021 : this being due to subdivisions & amount of DIFFERENCES, so forms would be about equally numerous. c
179b033 ted:-- no one can doubt that lesser trifling DIFFERENCES are blended «by» by intermarriages, then the
182b047 heredetary taint;. hence the resemblances & DIFFERENCES for instance of finches of Europe & America.
197b111 -- the others «animals» created with endless DIFFERENCES:-- does not say propagated, but must have co
199b120 ell as diseases. In intermarriages; smallest DIFFERENCES blended, rather stronger tendency to imitate
222b203 t it so--?? It holds good even with trifling DIFFERENCES of expression -- one child like father anoth
230b236 in).-- How is this explained by law of small DIFFERENCES producing more fertile offspring.-- 1st. All
294c177 arcely any novelty in my theory, only slight DIFFERENCES. «the opinion of many people in conversation
308c217 n spirit-- all children of one father.-- yet DIFFERENCES carried a long way. --Case of Habit I kept m
337d023 have been separated; together with physical DIFFERENCES of country: the time of separation depends o
350d055 t form from male with eyes!-- (are not these DIFFERENCES in sex confined to annulosa?) Remarked that
363d102 oles can hardly be told apart, so that after DIFFERENCES were pointed out Selby confounded them, yet
378d147 t back to earlier type by Mother?-- do these DIFFERENCES indicate, species changing forms, <& loosing
386d167 e of» The final cause of sexes to obliterate DIFFERENCES. final cause of this because the great chang
389d177 ration being means to propagate & perpetuate DIFFERENCES (of body, mind & constitution) is the end fr
390d178 pect.-- [Is this connected with the physical DIFFERENCES in almost all Male animals?]CD If the male «
390d179 ing through whole series of forms to acquire DIFFERENCES: if none are added object failed, & then by
391d179 tuous marriage has acquired from father some DIFFERENCES. V. Supra <v. infra> p 179, continued from I
391d174 (Because by this process it separates those DIFFERENCES which are in harmony with all its previous c
391d174 n only say the whole object being to acquire DIFFERENCES «indifferently of what kind, either progress
391d175 an procreate. these changes may be effect of DIFFERENCES of parents, or external circumstances during
391d175 illate backwards & forwards & are individual DIFFERENCES (hence every individual is different). (All
392d175 ile offspring without coition or addition of DIFFERENCES. shows that difference need not be added EAC
416e072 -- as in plant up a mountain-- In races the DIFFERENCES depend upon inheritance & in species are onl
417e077 n & woman is «indeed» (independent of sexual DIFFERENCES) a variety. The offspring of true <pare> her
500q10a emen's Land? or New Zealand? Babington about DIFFERENCES of Irish & British Species & British & dista
526m028 worth thinking over.-- it. will perhaps show DIFFERENCES between memory & imagination. «Catherine thi
035r046 Between the height of same beds, deposited in DIFFERENT basins; little or no relation appears to <exi
036r051 subsidence: hence difference; action on land DIFFERENT Volney, P 351. Vol I. woody bushes, «gazelles
038r057 n is blown off; & again when in great crater. DIFFERENT little craters are all burning, surely there
038r058 .--? Are not the dikes in upper strata. quite DIFFERENT from the Porphyries: certainly appearance lea
039r061 bears the stamp of recent elevation. which is DIFFERENT from what Mr Lyell supposes. Lyell P 116 Vol
063r132 pecies. as «each» one individual «divided» by DIFFERENT methods, associated life only adds one other
070r153 - When we see Avestruz two species. certainly DIFFERENT. not insensible change.-- Yet one is urged to
071r158 n which lake was absorbed.--Earthquakes felt. DIFFERENT case from shore of Pacific.--Isabelle's volca
101a057 America do/p. 280. the gravel beds in England DIFFERENT from Boulder beds-- What is Osteopora platyce
110a086 alt. full of circular cryst of glassy felspar DIFFERENT from either fragment or dike, blackish grey b
114a094 introduce discussion -- I see Lyell talks of DIFFERENT composition using difference in metamorphic a
122a113 quences of EXTREME FLUIDITY of earth.-- study DIFFERENT forms of earth as shown by arc.-- read Hersch
163g109 ish.-- Speculate on origin pebbles brought by DIFFERENT cause: from mud.-- [blank] {P} muddy nodular
172b007 animals, on separate islands, ought to become DIFFERENT if kept long enough.-- «apart, with slightly
173b011 would be of the continent though species all DIFFERENT. In cases as Galapagos & Juan Fernandez. When
174b013 enera being usually peculiar to same country, DIFFERENT genera different countries. Propagation expla
174b013 ly peculiar to same country, different genera DIFFERENT countries. Propagation explains why modern an
174b015 ian; Mamm were produced from propagation from DIFFERENT set, as the rest of the world.-- This view su
175b019 every successive animal is branching upwards DIFFERENT types of organization improving as Owen says
187b063 & Mastodon dying out about same time in such DIFFERENT quarters.-- Will Mr Lyell say that some circu
187b065 a & Asia, but many species closely allied but DIFFERENT, because country separated since time of exti
188b069 ery. close species generally frequet slightly DIFFERENT localities, so that they become useful to kno
191b080 land longer separated., Hare of two countries DIFFERENT.-- Ireland & Isle of Man possessed Elk not En
196b105 een wonderful if the two Rheas had existed in DIFFERENT Continents In plants I believe not..-- It is
196b107 ses varieties of the furze, broom, & yew very DIFFERENT from any found in great Britain, British vari
197b111 changes, which can be traced in same organ in DIFFERENT animals in scale.-- In monsters «also» organs
198b115 onsters worth reading NB well to insist upon <DIFFERENT animals> large Mammalia not being found on al
200b121 age.-- ¿whether those genera which unite very DIFFERENT structure as. petrel & alk. do not show the p
205b142 ridgewater Treatise p 85. Parasite of Negroes DIFFERENT from European.-- Horse & Ox have different pa
205b142 es different from European.-- Horse & Ox have DIFFERENT parasite in different climates.-- Hunbt. Humb
205b142 pean.-- Horse & Ox have different parasite in DIFFERENT climates.-- Hunbt. Humboldt? Vol V. P II. p 5
206b147 he relationship of the progenitors would have DIFFERENT formula for each lustrum.-- We may conclude t
206b148 , just like the two fine families «no doubt a DIFFERENT set of causes must act in the two case,» May
212b166 Zoolog. Soc June 1837 p. 53. an Irish Rat.-- DIFFERENT from English. Waterhouse has information resp
218b189 ossibility, holds good in plants» between all DIFFERENT forms; therefore when from being put on islan
220b198 e take structure as criterion of species Hogs DIFFERENT species, dogs not, but if we take character o
220b198 t if we take character of offspring. Hogs not DIFFERENT. some dogs different.-- Henslow says. (Feb 18
220b198 r of offspring. Hogs not different. some dogs DIFFERENT.-- Henslow says. (Feb 1838) that few months s
222b203 . one mother bringing forth young having very DIFFERENT characters is attempt at returning to parent
223b207 other senses is more wonderful. its mind more DIFFERENT probably & introduction of Man. Nothing compa
223b211 species! The increased fertility of slightly DIFFERENT species & intermediate character of offspring
226b221 dently due to «the» chance of some one of the DIFFERENT orders being able to survive or chance having
230b240 out impregnating each other), for if they are DIFFERENT then, they will be called species & mere prod
231b241 t. Indian Rhinoceros. Java & Sumatra ones all DIFFERENT.-- Join Sumatra & Java <to> together, by elev
231b242 mingle in India & East Indian islds-- Monkeys DIFFERENT not travellers?? Royles case of Himalayan, pl
231b243 to the southward would merely be specifically DIFFERENT if so.-- Now this is difficult to explain by
232b248 unt of parasitic animals of beasts varying in DIFFERENT climates Those will not object to my theory,
233b249 s «Waggiou» &c &c. (See Lyell. Vol III p. 30) DIFFERENT species in different isld. (as far East as Ne
233b249 ee Lyell. Vol III p. 30) different species in DIFFERENT isld. (as far East as New Ireland. see Coquil
233b252 resh water lakes of North America If Parasite DIFFERENT, whilst man & his domesticated quadrupeds are
234b261 y one country do not species generally affect DIFFERENT stations;-- this would be strong argument for
235b262 man thus circumstanced; varieties of dogs in DIFFERENT countries a case in point.-- All cases like I
235b274 s every one so anxious to cross. animals from DIFFERENT quarters to prevent them taking peculiar char
235b275 ny genus. as American or Indian genus inhabit DIFFERENT kind of localities.-- if so change The GRAND
240c003 rieties of pidgeon, although having skulls so DIFFERENT, that they would be called genera., yet retai
241c014 new species of Axis.-- «Cervus moluccensis is DIFFERENT from that of the Marianna islands & at Amboin
243c019 Ind. Arch:-- Birds of New Zealand absolutely DIFFERENT.-- --Philedon circinnatus not found in Austra
243c020 extremities of the continents peculiar to the DIFFERENT points.-- Consult Voyage aux terres Australes
245c023 sld to another--] «It is a good species, with DIFFERENT number of teats» Coquille Voyage Durville has
245c025 [islds. Springing up more likely to <M> have DIFFERENT species than those sinking, because arrival o
245c025 e plant might make conditions in any one isld DIFFERENT]CD.-- p. 414. dogs of New Zealand of large si
246c026 rather longer in proportion, colour slightly DIFFERENT. Who can say whether species & varieties P. 7

```
248c030 gans affected as with plants) no animals VERY DIFFERENT will breed together, so when we grant «(which
249c034 oducing <two> such as itself.-- therefore two DIFFERENT varieties will produce hybrids but not variet
249c036 nt of birds of Europe & America, which are of DIFFERENT forms being migratory; also with Temminks fac
252c045 all analogy from each other I do not know how DIFFERENT Sumatra Java ---------- do from Indian increa
253c047 ly variety.-- red form of skull very slightly DIFFERENT.-- Q Zoolog. T. V. I. p 389. Owen remarks on
258c061 against in any class:, those points which are DIFFERENT from each other, & resemble some other class
260c067 rious account to show that very many birds of DIFFERENT kind have been know to assist in feeding youn
264c080 think that many species when close come from DIFFERENT localities, as my Furnarii.-- some genus of y
265c086 ne member of that family, having it with very DIFFERENT habits-- Thus bill & nostril of Puffinuria I
273c112 n between these aerial representatives of the DIFFERENT families.-- that sexes <are> «have» same plum
273c113 with Chionis), yet the Tropic bird, has very DIFFERENT habits, though preeminently belonging to this
274c115 forms in <water> land birds,,-- Grups of very DIFFERENT value have their represetatives;, the rasoria
275c121 d old variety.-- The pidgeons which have such DIFFERENT skulls, but same marks on wings are Blue Pout
276c124 cture would alter.-- It is a difficulty how a DIFFERENT number of vertebrae are produced, where, (& i
276c125 e species inhabit same country are not habits DIFFERENT, (Mem: Gould's Willow Wren) but where close s
276c125 Willow Wren) but where close species inhabit DIFFERENT countries habits similar ¿law?-- probable.--
278c128 because as soon as two species were placed in DIFFERENT subgenera, then it would be useless, but the
280c135 hybrids, whereas two newer ones, even if more DIFFERENT might do so.-- <whi> is this true?? My views,
282c141 analogous. animals would be possessed by the DIFFERENT races of man, yet altogether different.-- To
282c141 by the different races of man, yet altogether DIFFERENT.-- To make this case perfect, we must suppose
284c147 so many.-- The quantity of life on planet at DIFFERENT periods, depends,-- on relations of desert, o
284c149 o together? Therefore value of organs vary in DIFFERENT group. & Not known in single ones--. viz. Mac
287c158 ve importance of the organs in same state. in DIFFERENT animals, & the value of those organs, when ch
287c158 & the value of those organs, when changed in DIFFERENT animals.-- + whether variations in eye of ver
293c174 a &c cowpox, proof of common origin of Man.-- DIFFERENT contagious diseases, where habits of people n
294c176 slds. near continent, Must we resort to quite DIFFERENT origin when species rather further.-- once gr
294c177 year 1000. reference to succession of types ¿DIFFERENT species;-- Horse-- &c <Lonsdale says. that fi
298c192 e of country.-- some species are larger &c in DIFFERENT countries. These facts show how very permanen
299c194 imrose & Cowslip, quite wild, but they affect DIFFERENT localities,-- latter on banks & in damp parts
299c196 han in other animals & again in Mans mind, in DIFFERENT races, being unequally developed.-- ¿is not E
300c198 e animals over those of others.-- the mind of DIFFERENT animals less divided.-- But as man has herede
300c198 ry & acquirable.-- therefore mans mind not so DIFFERENT from that of brutes p. Hard to say what is in
304c206 eries were believed to past into each other-- DIFFERENT classes Keep to their types. with different d
304c206 - Different classes Keep to their types. with DIFFERENT degrees of closeness. -- look how close birds
304c207 other birds reveal the secret.-- Now all the DIFFERENT forms of Synallaxis. trifling characters as r
308c219 ld. have formed them, would have arisen under DIFFERENT climates &c. Do I mean that ideosyncracy of w
308c219 hat ideosyncracy of wild animals is generally DIFFERENT, because their difference. arise a good deal
309c220 of shells.-- duration in two classes however DIFFERENT.-- Male glow-worm knowing female good case of
310c222 has added thought to <an> so many animals of DIFFERENT types. I will confess my profound ignorance.--
310c223 he scheme of nature) and animals that man has DIFFERENT origin. «Royal Institution» Dr Royle seems to
311c235 355 Echidna of Van Diemen's land & Australia DIFFERENT Temminck Fauna Japonica (?!) 82 mammalia 293
312c233 ss sister & brother of same litter, those. of DIFFERENT litters or of father & child are thought long
312c233 ys that genus of parasite to genus of animals DIFFERENT «+++ p 234», different species to different,-
312c233 te to genus of animals different «+++ p 234», DIFFERENT species to different,-- inguinal louse Africa
312c233 s different «+++ p 234», different species to DIFFERENT,-- inguinal louse African & Europaean. differ
313c233 ferent,-- inguinal louse African & Europaean. DIFFERENT.-- thorax & head differ Africa Australia Para
313c234 s the extinction & change of species two very DIFFERENT considerations, with respect to law of mammal
313c236 ry, when organs of parent are concentrated in DIFFERENT parts, & scission cannot effect the process.-
315c242 irritability, though means injection of fluid DIFFERENT from contraction of fibre)-- it is most remar
316c254 by kind of country; «kinds of migration quite DIFFERENT in species of same genus.» The Muscicapa soli
326c266 Heberdens Observat. on increase & decrease of DIFFERENT diseases. 4to 1801.-- quoted by do.-- There a
333d005 ge, very fierce to dogs «otherwise habits not DIFFERENT; tone of voice. perhaps rather different».--
333d005 not different; tone of voice. perhaps rather DIFFERENT».-- crossed with <un>common cat, exact variet
335d015 ck) are heredetary: «Hybrids of» Varieties is DIFFERENT because not long in blood.= The case of union
337d023 beings.-- Animals «of same classes» differ in DIFFERENT countries in exact proportion to the time the
339d025 ies of Man.? «In Australia. plants E & W very DIFFERENT.-- Man not so, but N. & S. New Zealand & New
342d033 ngs.-- «are not the hybrid pheasants & grouse DIFFERENT.--» (if so chinese pigs & common must be cons
343d039 und fossil when Europe must have worn a quite DIFFERENT figure 19th. With respect to the Deluge it ma
350d055 rilobite. (Polirus??) female blind & of quite DIFFERENT form from male with eyes!-- (are not these di
353d061 . <69> «71»). alludes to Eyton's discovery of DIFFERENT number of vertebrae in Irish & English Hare.--
354d062 of last by the ordering of the nerves, but in DIFFERENT parts according to age of individuals-- (see
354d062 age of individuals-- (see Mammae of Women) in DIFFERENT parts when age changes caterpillar to Butterf
355d066 e character which are seen to <vary among> be DIFFERENT in species of same genus." Law of monstrosity
355d067 age of foetus. distinct consideration) Now in DIFFERENT SPECIES of genus Sus. do vertebrae vary? «See
356d069 ory of Medicine, the extraordinary effects of DIFFERENT Medicines on organs, leads one to suspect any
356d069 e to suspect any amount of change from eating DIFFERENT kinds of food: grazing animals who eat every
358d075 ed off by one of these birds" Case of bird of DIFFERENT family. having very same habits in some respe
359d089 ur, two like each other & two dark-coloured & DIFFERENT.-- -- the former were the parents of the thre
360d093 t that would allow the offspring to have some DIFFERENT kind of mottle, each feather partaking of cha
360d094 a sophism for their brain or stomach would be DIFFERENT.-- Or if one species left its type in having
363d102 f birds, habits when well watched always very DIFFERENT.-- the two redpoles can hardly be told apart,
363d102 me» North American & Europaean birds «slight» DIFFERENT-- Barn Owl <the> in the former place breeds i
368d114 ange cocks & hens. being either alike or very DIFFERENT in recently altered genera. Guinea Fowl & Pea
369d114 & spurs-- no final cause here.-- & therefore DIFFERENT from Hunter I should say females recede in or
371d127 what [not located] that it shall beget young DIFFERENT in colour, form, & so altered in disposition,
371d128 & bud are produced, like parent stock, or if DIFFERENT dieteriorating very slowly.-- I presume most
372d129 rms of budding. Why does Gecko produce always DIFFERENT tail? An Individual bud may be thus produced
372d131 child might be so born. but it would be very DIFFERENT from true generation.-- there is no caterpil
373d132 re shows that they became Mammalia, through a DIFFERENT series of changes from the placentates, Havin
374d133 nera of shells in the mountain limestone (how DIFFERENT from plants!) But the Cephalopoda depart more
380d154 Gallinaceous birds have cock & hen plumage so DIFFERENT, yet the Cassowary & Guinea Fowl cannot be di
382d158 uble, only modified. those which perform very DIFFERENT, are both present in every shade of perfectio
383d160 In the back feathers, we have character very DIFFERENT from either parent bird-- Hunters Animal OEco
386d167 3. Ehrenberg. makes gemmation in animals very DIFFERENT from that of plants. (though latter does some
388d172 dga case appears to that of «2» dog begetting DIFFERENT puppies out of same mother.-- The view that m
390d178 be deleterious if the relative had come from DIFFERENT quarters) then it causes <to> a secretion of
390d178 causes <to> a secretion of something someways DIFFERENT from himself, for it should be observed that
391d175 rence (for even brother & sister are somewhat DIFFERENT) should be added to each individual before he
391d175 vidual differences (hence every individual is DIFFERENT). (All this agrees well with my view of those
401e016 uld be got out of it.-- Now this is curiously DIFFERENT from primrose suddenly produce cowslip, one i
405e032 ferences In such cases as at Galapagos. where DIFFERENT islets have different forms it is either effe
405e032 as at Galapagos. where different islets have DIFFERENT forms it is either effects of having been lon
406e036 individuals, & varieties of same species & so DIFFERENT species-- sometimes like one parent & sometim
407e039 & therefore present state of world is not so DIFFERENT, with <d> regard to their productions.-- Henc
417e078 y of its sex, or half way between, or someway DIFFERENT from either: & or like progenitors.-- in some
417e079 se of this.-- » (Lord Moretons law holds with DIFFERENT species, & individuals of same species.--). s
418e083 ciple, which gives rise to the sexual organs, DIFFERENT in each species,-- & knowing from analogy, th
423e097 possible to simplify the organization of the DIFFERENT beings, (all fishes to the state of the Ammoc
424e101 he present ones., which according to Beck are DIFFERENT.-- Subsidence of Greenland-- case of splittin
424e101 at Hooded crow & Carrion crow. have in Europe DIFFERENT ranges-- latter not going North of the Elbe.-
424e103 than others» & horse in lesser degree,-- how DIFFERENT to day!-- (Hybrids of Calceolaria.)-- «:CD[Do
426e106 - Is there any relation between the fact that DIFFERENT species produce abundantly infertile hybrids,
427e110 another».-- apple. only effect produced would be DIFFERENT.-- same way one variety of <animal> dog does
430e117 ulls are described.-- His three have had VERY DIFFERENT dispositions: this is important as showing sm
430e120 igonia costata & elongata thougt considerably DIFFERENT, in proportional dimensions, must <almost> be
434e129 rieties, so that they have been thought to be DIFFERENT species. Lychnis dioica, generally dioicous.
435e133 pollen of Asclepias placed on Orchis (so very DIFFERENT) that the granules exserted their tubes: now
446e162 change in grafted trees «:so is not effect of DIFFERENT stocks in this case».-- & strong case showing
```

Page **(Key Word)**
448e167 ally the same with recent & yet almost wholly DIFFERENT, is same, as if Isthmus of Panama.-- These tw
449e170 those which are identical, as those which are DIFFERENT.-- now this is same, as Galapagos facts &c &c
449e170 t shows the causes which give same species to DIFFERENT isld. is the same as that which gives genera.
449e173 1. Sheep Merinos from Cape of Good Hope,. has DIFFERENT constitution from those of Europe-- for they
450e175 imals).-- There are prevailing colours in the DIFFERENT islands.-- The horse is only found wild in th
453e183 volcanic islds.-- <Cocks> The possibility of DIFFERENT varieties being raised by seed is highly odd-
462tf05 ch book review on politics in relation to the DIFFERENT races of men, some more intellectual than oth
462t051 valueless this objection, when one thinks of DIFFERENT kinds of cattle in every part of England. &c
473s05r /-- Mr. Bunbury says has heard the Trout from DIFFERENT lakes of N. Wales can be distinguished-- & Ja
473s05r from Ogwen, Capel Curig & some other lakes, (DIFFERENT waters) He cannot, however, tell them from L.
477z002 Vol. I. p. 279 Thinks the Moruffetes of Chile DIFFERENT from those of La Plata or Paraguay.-- do. p.
480z007 ils than old ones-- in America & sexes not of DIFFERENT size-- How does this apply to pale brown Cara
481z008 l researches Compare land shells of Galapagos DIFFERENT islds.-- Waterhouse remarks that no insectivo
489qIFC . Whitby. Newlands Lymington Hants. Habits of DIFFERENT caterpillar races. --Name of Italian who sold
491q01v opagate from their seed 3. To apply pollen of DIFFERENT genus & then some hours afterwards of nearly
493q003 » dog-- I. St. Hilaire says length differs in DIFFERENT cats.-- Good observation-- examine semen of H
493q004 the hairless Brazilian or Persian animals of DIFFERENT heredetary constitution, to see whether offsp
493q004 e, Respiration, period of gestation differ in DIFFERENT breeds of dogs. Cattle, (Indian & Common) &c:
495q005 ow seeds & place cuttings or bulbs in several DIFFERENT soils & temperatures & see what the effect wi
498q008 n & flower at same time. Has H. seen group of DIFFERENT species growing White Mullein good plant to s
500q10a pecies & see whether they come from islds. or DIFFERENT parts or same district.-- About <endemic &> w
502q11v - Parsley & Fennel. Verbena Compare flower of DIFFERENT Cabbages most carefully to see if variation e
502q11v How <soon> «early» do characters of races of DIFFERENT vegetables & animals come on.-- Compare calve
511q018 UNDS. Eyton Mr Wynne, &c Could by selection a DIFFERENT looking animal be formed-- not caring whether
511q018 Europaean common species, having somewhat of DIFFERENT appearance.-- {will introduce it in work} Whe
512q019 ading-- any treadèe?-- Difference in lambs of DIFFERENT breeds Is there any difference in breeds of C
512q019 tle & sheep in the sprouting of the horns. at DIFFERENT periods in different breeds--?? or in individ
512q019 routing of the horns. at different periods in DIFFERENT breeds--?? or in individual case: subject to
513q020 h tame.-- (7) About fertility of Bantams from DIFFERENT countries= Do the Peacocks cross.= Young Chin
514q21. m every isld differs-- do they also differ in DIFFERENT countries-- on flora of African Islds-- names
515q022 parent.-- Will they grow up in other respects DIFFERENT?--. Important.-- Oct. 44 Tell J. Anderson's s
516q24v blank] Experiment Cover patch of ground, with DIFFERENT salts & poisons & see in what order plants wo
522m013 nsane. at some time. Mania is quite distinct, DIFFERENT also from delirium, a peculiar complaint stom
526m028 asure received from works of imagination very DIFFERENT from the inventive power,-- this, though very
527m031 s we know birds learn from each other «though DIFFERENT species» when in confinement, so may they lea
529m040 lay.-- the train of thoughts vary no doubt in DIFFERENT people., an agriculturist, in whose mind supp
533m059 y the extreme difficulty of moving muscles in DIFFERENT way from what they have been accustomed to, i
536m071 another.-- Why do bulls & horses, animals of DIFFERENT orders turn up their nostrils when excited by
537m075 .-- Must grant, that the conscience varies in DIFFERENT races.-- no more wonderful than dogs should h
537m075 es.-- no more wonderful than dogs should have DIFFERENT instincts.-- Fact most opposed to this view,
538m080 , <they> or when drunk they would not be more DIFFERENT, & yet they would make one's father & self on
540m086 other side of the Atlantic. Why then is he so DIFFERENT-- in organization.-- Same cause as colour & s
543m098 tellectual in human-- probably some genera in DIFFERENT orders more advanced than others just as dog
544m102 as are strung together in manner <they> quite DIFFERENT from when awake.-- peculiar sensation as flyi
549m118 child is «more» entirely happy (contentmt is DIFFERENT it refers to wishes for future) than perhaps
551m129 th lips in contempt <&> disgust & defiance.-- DIFFERENT from sneer-- How easily. horses associate sou
553m136 communicat to us».-- & that it does exist in DIFFERENT degrees in races.-- whether in Ancient Greeks
555m143 ortive muscles) The black Spider Monkey, very DIFFERENT disposition from others, slow cautious, angry
557m148 make him ashamed of himself, in manner quite DIFFERENT from fear; there is no inclination to jump aw
557m148 ned fear.-- Yet one knows oneself it is quite DIFFERENT from that.-- like «slight» passion from blood
558m151 this connexion.-- look at faces of people in DIFFERENT trades &c &c &c I observed the Asiatic Leopar
564n003 een same with mans because original instincts DIFFERENT.--Mem. Bee how different instinct a solitary
564n003 e original instincts different.--Mem. Bee how DIFFERENT instinct a solitary animal still different.--
564n003 ow different instinct a solitary animal still DIFFERENT.-- ‖ Different nations having different moral
564n003 stinct a solitary animal still different.-- ‖ DIFFERENT nations having different moral sense, if it w
564n003 still different.-- ‖ Different nations having DIFFERENT moral sense, if it were proved instead of mil
578n056 -- «in man & animals» Blubbering of a child (DIFFERENT in different ones?) in the most perfect faint
578n056 animals» Blubbering of a child (different in DIFFERENT ones?) in the most perfect fainting, sphincte
587n089 p. 103-- Abstraction p. 152. Perception very DIFFERENT from emotion.-- The former is used with regar
591n098 ll frighten-- hence variation in character in DIFFERENT animals of same species.-- The general «(as I
610o034 the abstract is matter united by certain laws DIFFERENT from those., that govern in the inorganic wor
611o034 ing a certain & peculiar system of movements. DIFFERENT from inorganic movements.-- See Lamarck for t
611o034 l forms of living beings, matter is united in DIFFERENT modification, peculiarities of external form
611o034 , peculiarities of external form impressed, & DIFFERENT laws of movements. [LHC] Hence there are two
614o037 drogen <&c> in certain definite proportions, (DIFFERENT from what takes place out of bodies) really l
615o36v I don't think this only pleasure; for it was DIFFERENT way of showing it, nor was there any cause, &
615o36v dividual.-- His Malay breed «of fowl» totally DIFFERENT habits from Europaean. begin to prowl about i
615o36v «seldom leave their perch till evening» crow DIFFERENT.-- Heredetary effect of former tropical clima
615o038 omesticated.-- [LHC] NB. Two dogs having very DIFFERENT instinct always obtain peculiarities of exter
617o39v te bodies. How we identify the two aspects as DIFFERENT phases of the same object of thought is a que
618o41v ness, still less between <action> «things» so DIFFERENT as action thought & organization: But if the
622o048 tiquettes of Society.-- [LHC] Sir J. M. gives DIFFERENT explanation of law of honour from Paley [RHC]
622o049 .-- It will be hard to discover this, for the DIFFERENT races of man may have different instincts, as
622o049 this, for the different races of man may have DIFFERENT instincts, as we see in dogs & pidgeons.-- Bu
628o054 . says affections, desires, & moral sense all DIFFERENT.-- P. 22. Butler & Mackintosh characterize th
640j167 .-- islds never joined, nature & climate very DIFFERENT, from adjoining coast. Admirable explanation
641j29v ts. in insects the end is gained by some very DIFFERENT method. in pedunculated eye of Chamelion. cra
641j29v of flight of Owl remarkable, [gained by very DIFFERENT process from Bats. CD]CD. «Macculloch says no
033r042 bo: in that case, may not central and rather DIFFERENTLY constituted lime have been removed?--As shel
179b034 arked Man & wild animals in this respect are DIFFERENTLY circumstanced.-- ¿Is this shortness of life
268c095 Australia (Waterhouse <disputes this>) says DIFFERENTLY) do. do. on the genus Procyon.-- by Wiegman
276c123 e a few years in advance only of their age. (DIFFERENTLY from literary men.--) must remember that if
306c212 at Z. Gardens upon being beaten behaved very DIFFERENTLY from a dog.-- more like man. continued long
306c212 sion & looked out for him to come again very DIFFERENTLY from dog. perhaps being in passion chief dif
466t093 e Bees visit base of upper petal, though not DIFFERENTLY coloured-- & stamens bend up a little In a w
466t094 tal from facility of alighting? which is not DIFFERENTLY coloured & to which stamen & pistils have no
341d031 red breasted thrush is represented by one not DIFFERING except by black line,-- A Bunting by one only
341d031 except by black line,-- A Bunting by one only DIFFERING by some permanent white streaks.-- &c &c Dr.
032r041 plain «top of» tidal band of action. This case DIFFERS. I think. from Patagonian steps, because the d
054r102 Granite: P. 199; Falkland account of cleavage DIFFERS wonderfully from mine: phyllade covered by qua
246c026 rope, (Alcedo ispida) from Moluccas. scarcely DIFFERS at all from those of Europe, but beak rather s
246c027 in North of New Holland.-- New Guinea scarcely DIFFERS more from, <Van Diemen's land.‖> Australia mor
303c204 r, which I do not know whether it <would have> DIFFERS in present races, & form of feet.= (Negro or f
358d076 ch other in plumage «(that is where the female DIFFERS from the male?)».-- children & women = "women
384d161 phrodites, (shows my doctrine of Hermaphrodite DIFFERS from Hunter)-- Hunter (p. 45) observes "every
403e021 has given account of Buffalo of the East which DIFFERS from that of S. Europe p. 189. The gaut, kind
493q003 n «toothless» dog-- I. St. Hilaire says length DIFFERS in different cats.-- Good observation-- examin
510q017 ion of embryo in close species of Hilianthemum DIFFERS greatly-- how very interesting to see if any v
514q21. hiti. Dr. Boott-- says caricas from every isld DIFFERS-- do they also differ in different countries--
575n043 «Do we deny the mind of a greyhound & spaniel. DIFFERS from their brains» then can we deny that the g
588n090 ss in degree, as between man & child.--)» what DIFFERS-- not <reason> «instinct», for its character i
636j57r & then Chamelion, which feeding on same food, DIFFERS in every respect, except «in» quick movements.
089a018 mountains of Saint Marc et des Gonaïves it is DIFFICULT to find stone not thus composed on the NE par
107a078 ar much reflection on method of cooling--Very DIFFICULT subject. PP-- I think from dislocation taking
107a078 must be thinner «under water» but cause most DIFFICULT (better conductor) Fitz Roy's Case of S. Mari
145g002 shows that either races soon made or crosses DIFFICULT Salisbury Craigs The Highland shepherds dogs
149g032 Roy double terrace river «& to West of Spean» DIFFICULT to explain on <formation> deposition in lake
173b009 wn species. (as some common land shells) Most DIFFICULT to separate Every character continues to vani

175b016 . every animal has tendency to change.-- this DIFFICULT to prove cats &c from Egypt no answer because
182b049 cause for enormous periods anterior to Man.. DIFFICULT for man to be unprejudiced about self, but co
188b068 slow says. that when race once established so DIFFICULT to root out.-- For instance ever so many seed
194b096 yell give some argument about varieties being DIFFICULT to keep on account of pollen from other plant
214b176 was broken-- This came so often «that» it was DIFFICULT to obtain a litter without this defect, Very
231b243 e specifically different if so.-- Now this is DIFFICULT to explain by creation-- or we must suppose a
263c074 me intellectual being though not MAN.-- is as DIFFICULT to understand as Lyells doctrine of slow move
278c129 , a D'orbigny has travelled this will be most DIFFICULT. Sub-genera so far may be eliminated. where e
280c133 iff: breeds succeed, yet seems to grant, that DIFFICULT & other go back to either parent.-- Shows ins
280c136 rkable fact. show influence of mind It is not DIFFICULT to see that it is less repugnant to nature to
285c150 some definition must be given It would not be DIFFICULT to arrange children of same parents in a circ
291c166 s more properly) being heredetary<)>.-- it is DIFFICULT to imagine it anything but structure of brain
291c167 cious & dioecious plants in animals it may be DIFFICULT to imagine how sexes were separated.-- in pla
300c197 from animals.-- Insects shamming death, most DIFFICULT case to iimagine how art acquired.-- They rea
311c226 of structure (including brain & other organs DIFFICULT to analyse) will not this separate facts abou
313c234 argue case both in Europe & S. America. «very DIFFICULT case» Does this law of duration apply to utte
360d092 ifference in other sex, it would be much more DIFFICULT to propagate-- <now> «as» if one bird had ver
417e076 He kept two to see if they would breed, It is DIFFICULT to think of ¿¿Plato & Socrates, when discussi
429e114 less cattle-- ¿& others?-- March 12th-- It is DIFFICULT to believe in the dreadful «but quiet» war of
431e123 sing of mosses & all others by action of wind DIFFICULT.-- Cline on the breeding of Animals, p. 8. si
438e141 Sprout was slowly formed.-- » if it shall be DIFFICULT to show that <time> the fixity of characters
447e163 & hence most persistent-- if form exceedingly DIFFICULT to vary.-- the run of chances, would prevent
463t057 reseen, sailing, climbing & mud-walking fish? DIFFICULT-- yet suggested. (vipers tooth also a difficu
463t057 fficult-- yet suggested. (vipers tooth also a DIFFICULT), the whole mind is constituted that a diffic
471tf07 nifestation of divine power"?.--"of their use DIFFICULT to conceive any idea" Linn. Trans. 18. p. 163
471tf08 arity «in structure» he says "indeed it wd be DIFFICULT to point out a family so completely natural &
497q06v birds 13. Mr. Herbert says Crocuses are very DIFFICULT to cross.-- are there races-- if so plant the
497q007 art of stigma?-- (4) As Papil. flowers appear DIFFICULT to cross, are there unusually many species in
497q007 he several great & natural Families, as being DIFFICULT to cross. (5) It is most important to ascerta
501q10a aths in Africa; Hooker? are these genera less DIFFICULT, in other countries, where species are either
507q15v ies of structure rendering cross impregnation DIFFICULT or reverse (.49) List of seeds Gaertner de fr
539m081 s foot & working with his toe to perform some DIFFICULT task.-- Aug. 12th. When in National Instituti
541m091 ualities as flowers, cloth &c & with all this DIFFICULT EXPERIMENTIZE upon this effort.-- it looks so
541m093 essions are, it is no wonder that they are so DIFFICULT to conceal.-- a man «insulted» may forgive hi
541m093 h to strike him, but he will find it far more DIFFICULT to to look tranquil.-- He may despise a man &
553m136 ind which surrounds us. Moreover «it would be DIFFICULT to prove that» this innate idea of God in civ
554m139 lf.-- <The male> «I saw» Jenny untying a very DIFFICULT knot-- the sailor on board the ship could not
571n030 rried too far In all the foregoing cases most DIFFICULT to distinguish. between prejudices of youth f
263c075 elieve in his new faith of the lesser of the DIFFICULTIES Once grant that «species» one genus may pas
301c199 t time.-- My theory must encounter all these DIFFICULTIES.-- Knowing that animals have some reason, &
356d071 . it should be observed not what comparative DIFFICULTIES (as long as not overwhelming) What comparat
370d115 st"-- But New Guinea.--!! S. America.-- Such DIFFICULTIES will always occur if animals are thought to
612o035 effect of heat, light or shade.) Joining two DIFFICULTIES into one common one always satisfactory, th
045r075 t at the Concepcion earthquake.--expatiate on DIFFICULTY of evidence about eruptions of Volcanos. (wh
072r160 eable & of so great tenacity, that it is with DIFFICULTY that a few fragments can be separated from t
103a063 would always thin out above which explains a DIFFICULTY.-- All De la Beche's reasoning of mountains
109a083 ce in showing not subaqueous removal--??? the DIFFICULTY of such preservation certainly is lessened.-
171b00a as ditto. V. Zoonomia.-- There may be unknown DIFFICULTY with full grown individual «with fixed organ
190b075 ts. ever been crossed really, if there is any DIFFICULTY in such marriages or offspring show tendency
196b110 es de Zool: Philosop:-- I deduce from extreme DIFFICULTY of hypothesis of connecting Mollusca & verte
210b158 elation. We find species few in proportion to DIFFICULTY of transport. For instance the temperate par
213b170 d therefore cross & keep similar. But this is DIFFICULTY; This immutability of some species. In Phill
219b196 red. it might have been said it was as gret a DIFFICULTY to account for movement of all, by one law.
225b219 nited (Falkland Fox. ice).. Mauritius what a DIFFICULTY-- where elevation Subsidence New is only hop
230b239 located) Any change suddenly acquired is with DIFFICULTY permanently transmitted.-- <It will admit> a
230b240 le fabric will be overturned.-- Hence extreme DIFFICULTY, argument in circle.-- Falkland Isd case goo
240c004 very much care that, it requires the greatest DIFFICULTY to rear them, eggs hatched under other birds
255c051 ls have aversion to generation, before great DIFFICULTY in propagation.-- Feathers on, Apterix becau
256c054 M. Bibron looking over reptiles he often had DIFFICULTY in distinguishing which were species, (theor
259c064 s. Plane trees all died certain year. Extreme DIFFICULTY of TRACING change of species to species «alt
261c071 most important in obviating a great apparent DIFFICULTY-- preservation of colouring, when form has c
263c075 hance is to have profoundly over the enormous DIFFICULTY of reproductions of species & certainty of d
276c124 ely to fish. structure would alter.-- It is a DIFFICULTY how a different number of vertebrae are prod
278c132 have been same for S. America & Europe.-- The DIFFICULTY is how came it animals not preserved, in cen
279c133 eculiarities being long in blood.-- ++ thinks DIFFICULTY in crossing race.-- bad effects of incestuou
280c135 reeding at all in domestication. throws great DIFFICULTY in way of ascertaining about hybrids.-- & is
286c155 mous arrangement which is false There is same DIFFICULTY in arranging animals in paper as drying plan
286c156 where we see scarcely any traces of passage a DIFFICULTY, but after all a slight one It will be neces
289c162 us, than with.» Might be given as a hopeless DIFFICULTY, except as distinct creation.-- Generation m
293c175 organization.-- This really perhaps greatest DIFFICULTY to whole theory.-- There is breed of tailles
298c191 general instinctive feeling.-- There is great DIFFICULTY in Making an alpine species from one in lowe
304c207 so many contingences??? A Question of immense DIFFICULTY is, whether Apterix descends from same paren
310c225 ed much less.-- Here is an element of extreme DIFFICULTY in mundine geological chronology Annals of N
316c244 r Australian! why not graduation.-- no greater DIFFICULTY for Deity to choose, when perfect enough for
352d058 l may acquire organs, but lose them with more DIFFICULTY, «contradicted by abortive organs, but numbe
356d072 e extinction of the S. American quadrupeds is DIFFICULTY on any theory-- without God is supposed to c
370d117 But birds quite distinct.-- Collect cases of DIFFICULTY of growing plants in all parts of world, thu
383b159 st allude to separtion of sexes as very great DIFFICULTY, then give speculation to show that it is no
398e005 f geological reasoning, extremely faulty» The DIFFICULTY of multiplying effects & to <ponder> conceiv
400e011 articularly rich. «as in Lucanidae» <no> less DIFFICULTY in establishing good groups.-- ears varying
402e017 s not being observed to change «is very great DIFFICULTY» in thick strata, can only be explained, by
409e047 t thougt overwhelming.-- yet I will not shirk DIFFICULTY-- I have felt some difficulty in conceiving
409e047 will not shirk difficulty-- I have felt some DIFFICULTY in conceiving how inhabitant of Tierra del F
416e070 id, (& not dry as in plants) therefore, great DIFFICULTY in crossing [& this most important obstacle
420e089 y most complicated circumstances, as shown by DIFFICULTY in forging, yet handwriting said to be hered
428e113 den, change from Primrose to Cowslip is great DIFFICULTY. «I should doubt if wild species ever formed
430e118 duce great ends-- But how.-- «-- .Make the DIFFICULTY apparent by cross-questioning.-- » even if p
436e137 nce of being propagated & so &c. The greatest DIFFICULTY to my theory, is same type of shells in olde
439e142 or rock» plant probably would do-- or be with DIFFICULTY be kept alive.-- Nevertheless much probably
442e152 luded to of Northern flowers, throws enormous DIFFICULTY in the way of Mr Knights. theory «without se
446e163 fresh seeds arriving.»-- throws a very great DIFFICULTY in my theory, here we have a plant remaining
461t041 rgo & yet remain adapted.-- it does away with DIFFICULTY of rabbits of England remaining same (if so)
463t055 ng as brain is created &c &c Bats are a great DIFFICULTY not only are no animals known with an interm
463t057 ficult), the whole mind is constituted that a DIFFICULTY makes greater impression, than the grouping
533m059 er muscles. may be illustrated by the extreme DIFFICULTY of moving muscles in different way from what
533m059 been accustomed to, in certain actions-- the DIFFICULTY of getting on a horse on the left side (not
550m126 ship.-- Scott's Life. Vol I, p. 127. Talks of DIFFICULTY of his own drawing compared to a friend, who
581n064 ic of Nature. p. 31. remarks children have no DIFFICULTY in expressing their want, pleasure, or pains
267c095 stomach. of Sammon remain after rest of animal DIGESTED.-- Important do p. 98. on a quaternary arrang
254c048 in this view!!! p. 392.-- except generation & DIGESTION in Acrite Kingdom all organs blended together
547m111 los. seem certain that muscular, mental, <&> DIGESTIVE nervous influence replace each other August 2
343d037 n of vile Molluscous animals-- How beneath the DIGNITY of him, who «is supposed to have» said let the
569n022 common to Savage & Frenchman, unaccompanied by DIGNITY-- "no mon dieu," with a shrug-- "all I can say
022r007 Lime disseminated through the great Plas Newydd DIKE.--Mem tres Montes. ((Henslow Anglesea)) great v
022r007 (Henslow Anglesea)) great variety in nature of a DIKE.--Mem. at Chonos & Concepcion. P. 417 Veins of
028r031 er is absorbed into the earth <I did not see one DIKE in the whole Galapagos Arch; because no section
101a058 well attested.-- There is no difference between DIKE & mountain axis. except in relative <strata> si
102a061 impact, «it» would look like it. Are greenstone DIKE in Granite residual matter of upper quartzose o

```
104a067 --, in granitic areas &c &c volcanos {P} fissure DIKE.-- thus dikes terminated Solubility of fluids v
110a086 glassy felspar different from either fragment or DIKE, blackish grey base. crystals from fragment dis
110a086 compact  on that side-- separation DISTINCT from DIKE junction mechanical: DIKE base reddish feldspat
110a086 paration DISTINCT from dike junction mechanical: DIKE base reddish feldspathes with grenish. black sp
110a087 e recrystallized, as such do not occur in either DIKE or fragment. junction certainly most distinct o
110a087 or fragment. junction certainly most distinct on DIKE side.-- oozed from one of the true rocks, most
111a087 degree faults with red wacke contorted evidently DIKE. V. VII. p. 316 & 328 VI. p. 365. Meyen on Chil
112a089 lera-- allude to Lyell's view of not discovering DIKE one end granite & other trap.-- It is in the mo
118a103 ider parallelism of dikes (Hopkins) & that every DIKE. which has not formed volcanos. or become scori
119a106 gneiss  &c judging from what we see when trap in DIKE & approach other rocks. & trap at least as hot
123a116 by  considering space formed-- great vacuum-- by DIKE.-- Mem. however. veins of segregation in Salisb
022r007 m. Maldonado P 375 Much Chlorite in some of the DIKES.--P 432. as in Andes. In Dampier's voyage there
030r037 Rio Shells argument for rise In Cordillera, the DIKES do not generally appear to have fallen into lin
036r049 tion, which dilated cracks must be filled up by DIKES & mountain chains.-- Introduce part of the abov
037r056 the Amazons. Consult Insist on the frequency of DIKES in Granitic countries, enumerate cases. -- M. V
038r058 d pebbles & on which trees grew.--? Are not the DIKES in upper strata. quite different from the Porph
038r059 ferior melted rocks to those parts. Are not the DIKES generally vertical? if so posterior to elevatio
039r059 ll be well to urge the case of St Helena, where DIKES certainly have not been points of eruption. The
039r059 oints of eruption. Nobody supposes that all the DIKES in Cornwall or in the coal measures have been c
040r063 lina's Case At Vesuvius. Vol III P. 124. Lyell. DIKES have a parting of pitchstone; which is describe
044r071 ly than externally--I did not see any number of DIKES in the cliffs.--wide valleys.--central peak sma
046r078 istant revolution.--Are we to consider that the DIKES which so commonly (state facts) traverse granit
049r087 Slates contemporaneous others subsequent. as in DIKES. In Granite great crystals arranged on sides. V
049r089 more  perfect. than in believing mere agency of DIKES: & indeed when do these dikes lead to a conical
049r089 ng mere agency of dikes: & indeed when do these DIKES lead to a conical mass. will this conical  mass
094a035 e other. -- Is the felspar glassy in greenstone DIKES which rise through granite.-- a most  important
097a042 ns in Brazil L.'Institut No degree 221 Lamellar DIKES like Mica Slate Von. Buch. Canary Isd. p 170.--
097a042 ry Isd. p 170.-- Mem. Cordillera Can Greenstone DIKES. be residue of quartzose vein in higher  parts?
099a051 ope parallel to walls of dykes-- Mem. laminated DIKES in Cordillera.!!!-- In stratum OP. let force dr
102a059 e yielded conical axis of mountain.-- only when DIKES reach near the surface. that strata yield.-- In
103a063 Pumice  at South Shetland. Geological Society-- DIKES have not been the moving agents, because not we
103a063 not wedge-formed.-- Hence fill up fissures-- If DIKES effect of horizontal elevation excepting fissur
103a065 e that the crust might be considered a level.-- DIKES being last action. (effect of horizontal moveme
103a065 al movement) hence generally intersect metallic DIKES: It is an important view being subsequent to di
103a065 ata. A capital discussion might be made between DIKES & «axis of» mountain-chain in proportion to wei
103a065 ording to Hopkins theory.-- general presence of DIKES. argues in favour of pressure of liquid rock. A
104a067 ffected, they would be far off In Discussion on DIKES argue impossibility on fissure going right thro
104a067 rying hardness,-- takes time to trace) from few DIKES which have given rise to eruptions.-- We must s
104a067 c areas &c &c volcanos {P} fissure dike.-- thus DIKES terminated Solubility of fluids varies with tem
109a083 t can actually be seen.-- ¿ The preservation of DIKES & ledges of first-rate importance in showing no
118a103 -<Jura &> Chalk When we consider parallelism of DIKES (Hopkins) & that every dike. which has not form
118a103 udation.-- This may be added to any place where DIKES described-- {P} Cordillera. St Helena &c  &c.--
118a103 it  is at once evident only small proportion of DIKES have reached the surface Arguments against Hers
120a107 profoundly considered. study Hopkins. theory of DIKES may throw some light.-- thin dikes not  cooling
120a108 s. theory of dikes may throw some light.-- thin DIKES not cooling if they had travelled some hundred
127a127 re found Sulph of Soda in peat ashes in Ireland DIKES in mountains. «(not on continents)» prove eleva
136a147 llips in Ladner Vol. II p. 80-- some remarks on DIKES: applicable to Cordillera Phillips in Lardner V
036r049 ts increased length. which will represent the DILATATION, which dilated cracks must be filled up by d
036r049 th. which will represent the dilatation, which DILATED cracks must be filled up by dikes & mountain c
128a129 e of the world has just that form which forces DILEMMA. Transactions of the Maryland Academy (at Athe
250c040 nsacts 1823. Read June 5th) important paper by DILLWYN, on replacement of Cephalopods & Trachilidous
413e060 hich are all truly bisexual. Buckland's Reliqu: DLLUV. says Africa only place, where, Elephant, Rhino
056r109 re whether fissures may not have helped it. or DILUVIAL waves. but when we see an entire island so en
114a095 ly as Temple of Serapis. (now we have banished DILUVIAL waves). & likewise stells,> «offers a presump
094a036 . Phils Royal Ed. Vol 7 Dr Buckland Reliquae DILUVIANAE p. 201. & seq Murc Trans Geolog Soc Vol 2. p
408e044 wind would blend species» Buckland. Reliquae DILUVIANAE. p. 222. Bones of Horse. Bear & Deer at 1600
037r054 d estuary of the Plata. to the Bay of Bengal. DIMENSIONS? Strong currents off the Galapagos.--strata
066r141 n a man's head.-- Kerguelen 40 by 20 leagues. DIMENSIONS: Bynoe informs me that in Obstruction Sound,
430e120 hougt considerably different, in proportional DIMENSIONS, must <almost> be considered merely varietie
450e176 been  imported: shows they will propagate get DIMENSIONS-- do App. p 73 State of Muar in Malacca.-- s
480z006 ccount of habits of Tubularia. p 52. May 1836 DIMENSIONS of immense Tortoises p. 81 & p 113 of 1834 O
383d159 ).-- Insecta.-- hermaphrodite, being not only DIMIDIATE, but quarter-grown seems to show whole body i
640j167 r prey diminished, total number of dogs. would DIMINISH, whilst the long legged variety would prevail
275c120 iety» & then recrossing off spring. till size DIMINISHED, but feathers continued by picking  chickens
640j167 & now formed eighth part.-- or if other prey DIMINISHED, total number of dogs. would diminish, whils
190b076 Flora.  on East & West. ends of New Holland. DIMINISHING towards centre (p. 586)-- Parallel 33 degree
371d127 quired) offices" &c &c Owen illustrates case of DINGO (he alludes to the dholes or wild dogs of India
591n097 attending to anything or excited.-- so do young DINGOS, as I saw wag tail when watching anything-- Ke
521m009 sight, though, not through hearing,-- Thus when DINNER was announced he could not understand it,  but
521m009 «<Mr  Corb» the servant showed him watch & said DINNER is ready, what, what.-- then showed the  watch
521m009 ed the watch upon which he exclaimed, why it is DINNER time.-- » My father asked him whether he had g
620o044 yone must know, how soon the pleasure from good DINNER, or from a blow struck in passion fades  away,
546m109 king back to his life, would say how many good DINNERS or......... he had had, he would say how  many
427e109 ffect of abortion of one sex.-- Linnaean class DIOECIA & Monooecia. ought to be preeminently artifici
435e130 gans, even more so than Polygamia: Monooecia & DIOECIA, preeminently artificial, so that even some s
291c167 ts we have a step between <mon> monoeecious & DIOECIOUS plants in animals it may be difficult to imag
291c167 ants we have some flowers monoeecious & others DIOECIOUS. some flowers hermaphrodites & others  not???
299c195 have generative organs. first character.-- In DIOECIOUS plants more of the Female flowers unimpregnat
316c245 ifera. hermaphrodites-- eggs in groups.. Have DIOECIOUS plants more restricted ranges than other plan
381d157 ion of sexes, as may be seen in Monoeecious & DIOECIOUS plants.-- NB. in Heautandrous animals <are> i
381d157 imple.-- as in plants even in same genus some DIOECIOUS & some monoeecious-- (& cultivation might mak
384d162 ns) afterwards they can be seen distinct. (in DIOECIOUS plants are there abortive sexual organs?): th
381d174 ng simple (Are not Coniferous trees generally DIOECIOUS oldest forms) why are twin in man more like «
392d175 ime]CD What kind of plants are Monoeecious or DIOECIOUS.-- very curious how this was superinduced? (S
392d175 this was superinduced? (Surely all are really DIOECIOUS..) only simple form of life are  monooeecious.
427e109 types  are ancient. According to my views of <DIOECIOUS p> all plants, being occasionally  dioecious;
427e109 <Dioecious  p> all plants, being occasionally DIOECIOUS; & really dioecious plants being effect of ab
427e109 lants, being occasionally dioecious; & really DIOECIOUS plants being effect of abortion of one sex.--
434e129 produce a few seeds,-- -- Ruscus aculeatus. a DIOECIOUS plant, in which the Male plant sometimes bear
434e130 <to>  seems to have taken place.-- Almost all DIOECIOUS & Monooecious plants have rudimentary abortiv
447e164 facility for inter marriage is greater (Hence DIOECIOUS plants highest.-- Palms &c &c)-- Is there gre
454e184 ous.-- Are there not wild plants, some partly DIOECIOUS? Mushroom Hybrids? Any «wild» plants in Engla
495q05a over them up periodically & see effect-- such DIOECIOUS individ--small orifice (8) Carry Bees, powder
496q006 Questions concerning Plants Is the common Fig DIOECIOUS-- are its female flowers always barren-- if n
498q009 ies <near> close to each other.-- As they are DIOECIOUS, if no hybrids were produced by seed, we migh
499q009 y of mock flowers in Hydrangea (20) As Hop is DIOECIOUS-- seedsmen who raise Hop-seed-- may know some
499q09v onooecious plants.-- Hooker says Rafflesia is DIOECIOUS & Pollen must be carried by some insect-- St
506q015 badness may be merely not ripening-- (38) Have DIOECIOUS plants any secondary, sexual characters.-- St
506q015 teur on Wheat.-- (41) Have any monooecious or DIOECIOUS plants the Papilionaceous structure of flower
506q15v e.-- Brown's paper 43. Any flowers of Keeling DIOECIOUS, or Monooecious, besides the Nettle. at Galap
506q15v nooecious, besides the Nettle. at Galapagos-- DIOECIOUS.-- Carex.-- We may presume Nettle spreads  by
513q020 sult, for comparison with natural species, as DIOECIOUS Plants, when crossed R. BROWN-- will pollen a
636j56v cculloch. brings forward. the impregnation of DIOECIOUS Plants by foreign agency-- as insects, as won
636j56v of adaptation.! There would not have been any DIOECIOUS plants, had there been no insects. The  right
636j56v insects «¿when were Palms formed?» as soon as DIOECIOUS Plants were formed. Macculloch says, life, fo
381d156 r-- the Annelida. All others, <animals,> «are DIOEECOUS as» Cephalopods, Pectinibranchiate molluscs.-
507q15v Papilionaceous plants,-- whether many mono or DIOEIOUS plants, & any with peculiarities of structure
434e129 e been thought to be different species. Lychnis DIOICA, generally dioicous. yet parts only very sligh
```

499q09v c (a) Mercurialis-- Frog Bit, Valerian-- Urtica DIOICA Sorrell. Lychnis. Butchers Broom-- «also, Vinc
505q014 fects of Nitrate of Soda under Beech.-- Lychnis DIOICA answers this question= (5) Open more Horned or
434e129 e different species. Lychnis dioica, generally DIOICOUS. yet parts only very slightly abortive & bed
089a020 sted with serpula-- Isle of Cayenne. Syenite & DIORITE, covered with iron clay common to Guyana said
109a084 a translation of paper by rose on Greenstone, DIORITE, &c most important.:-- must be studied.-- Scie
044r071 to the present day. at Mauritius. (consult Bory «DIP of strata on East») cannot believe in a great ex
048r087 ilinear strata subsidence.--The sudden increased DIP is not parallel case to Isle of White. but rathe
091a027 of Africa. Clay Slate & Quartz. strike SSW & NNE DIP 30 degree - 80 degree Ed. N. Phil Journ. p. 410.
091a027 5. Oct. 1828. gneiss in India (falls of Garsipa) DIP 30 degree. <strike> «direction<?»ESE-- CD [In t
099a050 ical ∵ combined with gravity.-- hence changes in DIP of no sort of consequence.-- Therefore < S of in
100a052 ct of gravity, versus some fault explaining vary DIP & inclination.-- which last is strong character.
111a087 ection of strata on the Berbice N. 35 degree. E. DIP to NW to 80 degree faults with red wacke contort
113a092 em my remarks on coast of Australia.-- Great NW. DIP in SE part of Australia.-- Probably a case of ri
134a141 f do p 8.-- soft Clay beds near C. Virgin p. 59. DIP of Clay slate in T del Fuego Admiralty Sound. SE
134a141 of Clay slate in T del Fuego Admiralty Sound. SE DIP. much p. 136. Rocks on Western Coast p. 204 do.
145g006 hick-- {P} 35 degree is I believe about greatest DIP of sandstone in upper part «of Salisbury Craigs»
162a104 of fine sand & small angular-- rounded pebbles-- DIP sideward, & inwards-- deposited when water stood
200b121 Coast of Africa. Clay slate. strike. SSW. & NNE. DIP 30 degree - 80 degree (?).-- Ed. Phi.l.. N. J. p
076r170 slate. with beds of syenite & <sep> serpentine DIPPING to SW at 45 degree to 50 degree-- covered by c
127a125 e.-- <ditto of synclinal> simply clinal lines. DIPPING so & so or may be used Ran-Clinal lines & c &
468t106 Staphylinidae-- Anapsis, Melegethes, Leptuse-- DIPTERA & small Hymenoptera Saw Humble go from great S
241c015 w about birds of N. Zealand L'Institut. 1838. A DIPUS. & other rongeur in Australia.-- p 67 ¡American
250c037 Edentate or American form.-- Is the Australian DIPUS an American form? The climate having grown more
278c131 ! rodents cold inhabitants most important!! like DIPUS of present day??! Major Mitchell does not think
045r077 s away.--Will geology ever succeed in showing a DIRECT relation of a part of globe rising, when anoth
109a083 rrents & elevation have same effect, a tendency DIRECT (or oblique) outwards may be granted. independ
182b046 The condition of every animal is partly due to DIRECT adaptation & partly to heredetary taint;. henc
190b075 incts were.-- relation of type in two countries DIRECT relation to facilities of communication Have r
228b228 we have come from & to what we tend.-- this & «DIRECT» examination of direct passages of <species> s
228b228 what we tend.-- this & «direct» examination of DIRECT passages of <species> structure in species, mi
259c063 that particular habit.-- All structures either DIRECT effect of habit, or heredetary «& combined» ef
261c072 ently is an exception) can only be explained by DIRECT adaptation to animals wants & not as change in
265c084 n winter of Arctic countries few will say it is DIRECT effect, <of> according to Physical laws, as su
314c238 movement. & not the organ itself How except by DIRECT adaptation has such a change been effected.--
371d118 so with the mind. the simplest transmission is DIRECT instinct. & afterwards enlarged powers to meet
430e117 he animals in Eocene period could not have been DIRECT parents of any of ours,-- even if extinction i
525m024 conscience.-- Not probable in Squib's case any DIRECT fear.-- My father thinks that selfishness, pri
530m043 . yet during whole illness, he had been able to DIRECT about his own health.-- his complaint was carb
536m072). now free will of oyster, one can fancy to be DIRECT effect of organization, by the capacities its
539m083 e other habitual state because it will (without DIRECT consciousness?) change its habits.-- Aug. 16th
547m113 deas of every late impression.-- (do the ideas, DIRECT effect of perception by senses fail first, as
553m135 k Minster) consider the thunder & lightning the DIRECT will of the God (<thus> & hence arises the the
567n013 implelst animals, as hydra towards light. being DIRECT effect of some law.-- have plants any notion o
568n018 in performing it.-- As soon as memory improved. DIRECT effect of improving organization, comparison o
601o009 onsciousness, <t> movements «¿» anterior to any DIRECT sensation, in order to avaoid it-- beetles fei
611o034 movement of sensitive plant can be shewn to be DIRECT physical effect of touch & not irritability, w
612o035 r sunflower to sun? ∵ I should think there. was DIRECT «physical» effects of more or less turgid vess
632j53r f adaptations.-- but these are, I believe, only DIRECT consequences of still higher laws.-- I do not
078r175 fterwards said to be «all with some exception» DIRECTED NW & SE). «Vol III» Mexican Cordillera "immen
252c045 mpirically what is species.-- The Collector is DIRECTED to study localities of isld.--immense import
318c254 t.-- others in flock, these birds seem clearly DIRECTED by kind of country; <kinds of migration quite
367d113 e, perhaps, equally so, is this strength being DIRECTED to one part more than another, which part is
466t093 freckles on {a} upper petal; bees & flies seen DIRECTED to it-- The Humbles in crawling out brush ove
523m014 ust.-- From habit the feeling of anger must be DIRECTED against somebody.-- Have insane people any mi
542m095 ention would amongst lowest savages clearly be DIRECTED chiefly by objects of vision.-- Does the cont
586n081 es is instinctively «not least by experience» DIRECTED to certain quarter"-- "An animal has faculty
610o031 rom Mr Roberts-- «a person he had long known & DIRECTED many letters to»-- could not <remember> «read
610o031 at a direction book, but could not find out-- DIRECTED his letter, & I observed he had written Wilso
617o040 a force manifested in every particle of matter DIRECTED towards every other particle; but FORCE, <obj
622o049 one as a leader,-- the conjugal feeling may be DIRECTED towards one or more.-- It will be hard to dis
055r106 ttle to the West: the veins which follow this DIRECTION are thought by the <oldest> «most intelligent
078r175 , which has yielded the most metal, where the DIRECTION of ravins, and the slope of the mountains (fl
078r175 flaqueza del cerro) have been parallel to the DIRECTION & inclination of the vein".-- at Zacatecas th
078r175 ein".-- at Zacatecas the veta grande has same DIRECTION as Guanax.--the other E & W.--veins richest n
091a027 a (falls of Garsipa) dip 30 degree. <strike> «DIRECTION<?»ESE-- CD [In the Darwar. transition Hills
091a027 [In the Darwar. transition Hills & strata SE. DIRECTION of transitions clay slate &c nearly vertical
097a042 . and Carb of lime decompose each other.-- on DIRECTION of mountains in Brazil L.'Institut No degree
099a050 re little inclination, little force & varying DIRECTION.--> Therefore in PILE of mud from Trapiches.
101a056 .-- Would crystals arrange themselves in that DIRECTION, in which most substance lies <.-->? Phillips
111a087 Schomburgk NW. numerous boulders of GRANITE" "DIRECTION of strata on the Berbice N. 35 degree. E. dip
126a124 rsed, upper part metal «conveying heat in one DIRECTION only, like water below 39 degree» & lower par
436e135 retain character? If separation in horizontal DIRECTION is far more efficient in making species, than
512q019 well as, as those already bred in cages. Get DIRECTION write to-- (2) Does he believe. Stanley's fac
552m131 ion Athenaeum 1838. p. 652. Dr Daubeny on the DIRECTION of mountain chains in N. America Fear probabl
585n076 arried in hampers. if they have not known the DIRECTION in which they STARTED, they cannot return.--
586n080 ssing the sea in dark night & not loosing its DIRECTION, equally wonderful in young & old.-- These fa
596nIBC ound in fainting state would it then know its DIRECTION.-- In slight convulsions. are the muscles of
610o031 ut concluded it could not be so.--Looked at a DIRECTION book, but could not find out-- Directed his l
610o031 is head.-- remembered, that he had. looked in DIRECTION book under head of Wilson, referred to Robert
617o040 nication to other matter in the course of its DIRECTION, & thus when we apprehend force in inanimate
079r177 rthquakes are recognized as coming from three DIRECTIONS. from W. NW & S.--last to Seaward partaking
102a062 oadly indication of new law acting in certain DIRECTIONS predominantly, connection with magnetism &c
123a120 celand stream. the 90 miles includes opposite DIRECTIONS. Mem. S. Cruz. Assuming from Sir. W. Hersche
288c159 sed would continue to cross, means of knowing DIRECTIONS: mysterious. Were the woodcocks.. which came
530m043 ilar nature.--like FitzRoy in sleep giving DIRECTIONS.-- & forgetfulness after bad accidents:-- Af
617o40v ces balancing each other & moving in opposite DIRECTIONS. We are satisfied therefore, if we can trace
026r019 r: so have the Conglomerates: Yet this view is DIRECTLY opposed to common opinion The Tertiary format
108a080 much horizontal oscillation. or so many shocks DIRECTLY after great shock -- It appears to me unphilo
207b151 water in their beaks, & the young one <inland> DIRECTLY by instinct. can dive & conceal themselves in
217b184 & greyhound.-- Where two dogs have lined bitch DIRECTLY one after the other, puppies differ, & like b
306c213 . 120 is greatest (about 200 miles distant).-- DIRECTLY beyond produced line of Timor 215 degree. Wha
340d026 with another-- & hybrids fertile inter se--No DIRECTLY against Eyton's rule. ¿Are the hybrids simila
381d156 & so imagination in Man, has strange effect.-- DIRECTLY a Capon is cut, it increases in size prodigio
404e024 as consequent on mind or instinct, now this is DIRECTLY incorrect The case of my mice is good, becaus
466t094 entering long nectary, would «necessary» cross DIRECTLY over the bunch of anthers & pistils, but thes
578n053 than particularly to wish not to do so.= = How DIRECTLY personal remark will make any one blush.-- Is
608o028 t attention to Education.-- 4) These views are DIRECTLY opposed & inexplicable if we suppose that the
632j53r us of <th> any one seed. (all have not it) was DIRECTLY created. for transportation. it follows from
451e177 with monkeys & squirrels.-- Horsbrugh E. I. DIRECTORY. Vol II. p. 46 Carimon Java. (between Borneo
613o035 re many such objects are present, & where will DIRECTS <to> other parts of body. to do such.-- All th
525m023 g which is wrong.-- as eating meat., doing their DIRT, running home.-- in these cases their actions d
260c068 in pursuit of Blue-Jay, when <one> birds hears <DIS> crys of distress of other parents.-- Shows comm
423e095 re not be at any time as many species tending to DIS-developement (some probably always have done so,
594n112 though instinct so firmly implanted, birds soon <DIS> learn to disobey it-- I have seen hawk & sparro
380d155 alling. & put into female, when muzzled, he is DISABLED.-- so Elephant in confinement, & so imaginati
540m085 is recollected that smell of ones own pud. not DISAGREE.-- Ourang outang at Zoolog Gardens touched pu
293c174 isery to other.-- else smell of Man would be DISAGREEABLE to Musquitoes We never may be able to trace
366d111 ation of species.-- Case of Association very DISAGREEABLE hearing maed servant cleaning door outside,

524m020 truly spritual.-- a person twitching when a DISAGREEABLE thought occurs, is closely analogous to Epi
578n056 are loosed is a convulsive action to remove DISAGREEABLE impression like true convulsion. (Hence pas
587n087 other objects. (as that senna is necessarily DISAGREEABLE to organs adapted to like sugar, acid, &c
593n105 is probably some secondary one-- blood being DISAGREEABLE & anything disagreeable being pursued.-- A
593n105 ry one-- blood being disagreeable & anything DISAGREEABLE being pursued.-- A dog turning round & roun
153g052 ere most accumulations At gentler bends roads DISAPPEAR The normal condition of 4th shelf, some way b
173b009 ed?-- <Granting> Species according to Lamarck DISAPPEAR as collection made perfect.-- truer even than
376d136 rican character is more tenacious. & does not DISAPPEAR for Many generations Sept. 29th Dr. Andrew. S
334d012 es.-- yet one day <th> a cow walked in, then DISAPPEARED, & three days afterwards came again, bringin
470t176 es abounded--so that palmated has now nearly DISAPPEARED. <& old English> But these mules <in our gar
373d132 endency can there be for abortive organ ever DISAPPEARING??-- Have Marsupiata abortive Mammae?.-- My
527m035 utterly forgotten--, so as to feel a severe DISAPPOINTMENT «in real train of thought this does not ha
587n087 to ask [not located] If dislike, distaste. & DISAPPROVAL. were not something more than the unfitness
628o53v ng.-- to be associated with the approving or DISAPPROVING instinct-- which were not originally, if th
513q21. larly take place in unopened flower-- <doubt> DISBELIEVE this in Bauers case of orchidiae Where does
531m051 ianoforte, it seemed solely to be feelings of DISCOMFORT, especially about heart as of excited action
542m094 on to other animals) scream of agony, sigh of DISCOMFORT & weariness. & meditative tranquility. «whin
231b243 nnection.-- viz descent.-- Hence far greater DISCORDANCE in latter-- Have change in form.-- This prob
302c202 cordant. with fact there stated, only in most DISCORDANT groups. The formation of genera may sometime
449e170 an species-- p. 18. of Temmincks. Preliminary DISCOURSE to Fauna of Japan-- that the «animals of» isl
570n026 rrow heads &c &c October 27th Consult the VII DISCOURSE by Sir J. Reynolds.-- Is our idea of beauty,
579n060 ll? of the animal.(-- Jan 21. 1839. Herchel's DISCOURSE p. 35. On origin of idea of causation; «succe
602o010 ars some relation to time & memory Reynolds X DISCOURSE very curious as showing "the perfection of th
602o011 ut sleep-- Nerves.-- Volition &c Reynold XIII DISCOURSE (p 115) a very good passage. about actions &
322c269 do Lutke's Voyage. carefully read.-- Reynolds DISCOURSES Lessing's Laocoon Whewells-- inductive Histo
023r013 ock; following Curvature of hill; states could DISCOVER no shells: nothing said about K. Georges Soun
219b193 ropus)--» might bring in stomach-- &c &c. (Mem DISCOVER what kinds of seeds. these plants) [Mem Fact
227b227 ographical grouping we are led to endeavour to DISCOVER causes of change.-- the manner of adaptation
274c115 ion.-- In conversation in Museum-- I could not DISCOVER any other clear relations besides aerial, & t
278c131 nd.-- V. 1s. Number of Geographical Journal to DISCOVER whether dog found at Swan river. The change i
287c158 ication to express relationship. & by so doing DISCOVER the laws of change in organizaton. But the cl
348d051 e taken literally,, though how far we can ever DISCOVER the real relationship is doubtful.-- not till
410e053 r the individual be species or variety, but to DISCOVER physical laws of such corelations, & changes
415e066 ng no fossils, the only way, that I can see to DISCOVER whether the parent of man was quadruped or bi
530m049 n of thought.-- [not located] Fox believe cats DISCOVER birds nests & watch them till the young are b
536m070 t anger, but observing eyes thus unconsciously DISCOVER struggle of feeling.-- It is as much effort t
548m117 does not stop to reason what there should be & DISCOVER loss Definition of happiness the number of pl
577n049 ction.-- on my view of free will, no one could DISCOVER he had not it.-- The memory of Plants, must b
608o026 IVE, & therefore now great effort of reason to DISCOVER then: this is important explanation) he think
622o049 cted towards one or more.-- It will be hard to DISCOVER this, for the different races of man may have
205b142 ation) West of Rocky Mountains Asiatic types DISCOVERABLE.-- Bridgewater Treatise p 85. Parasite of N
557m148 easily analysed than jealousy, because less DISCOVERABLE in animals than latter.-- Yet I think one c
045r075 > inland than on coast it would be invariably DISCOVERED; this may be mentioned with general slope of
075r166 In the Guatemala part. (& Chiloe do) no veins DISCOVERED. Humboldt suggests covered up by volcanic ro
075r167 roken up, & has there not been vein «of iron» DISCOVERED?-- Klaproth analysed silver ores from Peru c
116a100 e <re> a reef on a sea beach-- «p. 151» first DISCOVERED «very small» bits of red granite between 40
183b053 ng, why not have some intermediate forms been DISCOVERED. between palaeotherium, megalonyx mastodon,
219b196 ct to Blood-Hounds. Before Attract of Gravity DISCOVERED. it might have been said it was as gret a di
226b222 r has been left & Fox, & bear.-- If I had not DISCOVERED channel of communication by which great Eden
227b226 ions) thus a knowledge of possible changes is DISCOVERED, for speculating on future. !.fish never bec
282c142 not be affinity, but the truth would never be DISCOVERED When one reads in Ehrenbergs Paper on Infuso
288c158 n parallel parts of his series, ie, cannot be DISCOVERED till circles completed Major Mitchell, does
317c251 1837 account of the various hares «some since DISCOVERED» of N. America, & of the shrews.-- Dr Bachma
337d022 which is so like that difference would not be DISCOVERED by an unscientific observer.-- <Transactions
347d049 3. Instinct L'Institut. p. 249. (1838). Eggs DISCOVERED to Taenia.-- hard so as to resist external i
349d051 arrangement from fewness of forms-- Cannot be DISCOVERED «un»till <in> «we ascend to» subgenera & fam
410e051 f effects of my theory, laws probably will be DISCOVERED. of co relation of parts, from the laws of v
411e055 . America, however) is very remarkable & none DISCOVERED before them in any part of World.-- Wealden
414e065 same relation as Mastodon to Man. were to be DISCOVERED. Man acts on. & is acted on by «the» organic
523m018 on him» which are never generally, if at all DISCOVERED.-- <Sup> Sometimes comes on suddenly from <I
528m035 beginning of castle» because train cannot be DISCOVERED-- is closely analogous to my Fathers positiv
578n055 h sorrow,-- let the possibility of this being DISCOVERED by anyone, especiall if it be a person. whos
527m034 capability of such trains of thought makes a DISCOVERER, & therefore (independent of improving power
554m141 o a theory of friction & gravity. it would be DISCOVERER "reasoning" or "reasoning"-- only rather mor
463c063 maux fossiles-- being a mere fragment of the DISCOVERIES to come-- Owen in his description of my foss
464t063 usion of his work-- Lund makes his wonderful DISCOVERIES= negative facts are valueless= monkeys= Owen
112a089 n Cordillera-- allude to Lyell's view of not DISCOVERING dike one end granite & other trap.-- It is i
355d067 an be no prediction.-- The only advantage of DISCOVERING laws is to foretell what will happen & to se
362d100 change in this time.-- the impossibility of DISCOVERING their origin.-- I see only some «but very st
397e004 anges are in progress.-- we feel interest in DISCOVERING a change of level of a few feet during last
410e053 re the termination of change occurs.-- those DISCOVERING the formal laws of the corelation of parts i
491g01v ropagated by cuttings.-- Try.-- Important as DISCOVERING function of seeds-- (6) To hybridise EVERY f
095a039 Gypsum.--   The officers of the Bonite. French DISCOVERY ship, found clear proofs of shells & waterwor
192b088 whether the series is not more perfect by the DISCOVERY of fossil Mammalia than before & that is all
315c243 her animals in associated kinds of good news. DISCOVERY of prey.-- arising no doubt from want of assi
349d051 nowledge is elicited.-- it will rest upon the DISCOVERY what characters VARY most easily:-- those whi
353d061 vol II. do (p. <69> «71»). alludes to Eyton's DISCOVERY of different number of vertebrae in Irish & E
398e004 logical chronology depends, that most sublime DISCOVERY of the genius of man Those who have studied h
605o20v consist of this. but rather in the power of DISCRIMINATION & respecting good from bad. And it is mani
261c070 ca Hence it is universally allowed that the DISCRIMINATION of species is empirical. show this by inst
049r087 At M. Video «facts of Passages marked by do.» DISCUSS quartz veins, cements hitherto-- yet similar ones
229b234 ling standing transport.-- <tr but> Get him to DISCUSS those mention[ed] by Lesson & Chamisso.-- Geog
229b234 veral mammalia being peculiar (?) If. Henslow DISCUSSES possibility of seeds of Keeling standing tran
041r065 ifting of carcases putrid. In Rio paper. when DISCUSSING probable rise of land: Mention M. Gay's fact
045r077 rt of globe rising, when another falls.--When DISCUSSING connection of Pacifick & S. America. -- Volc
052r098 Retreating case in excess as first case. When DISCUSSING Falkland soundings introduce this discussion
056r111 lata to Caraccas, which is all of granite: In DISCUSSING circulation of fluid nucleus,--the similarit
109a084 died-- though I do not think good p. 411 When DISCUSSING concretions Carbonate soda. formed by Ca. of
112a089 land in Andes, worthy of consideration. When DISCUSSING nucleuse's of old volcanos within Cordillera
409e048 dog, & ask him, how wolf was so changed. When DISCUSSING extinction of animals in Europe. :the forms
417e076 ifficult to think of ¿¿Plato & Socrates, when DISCUSSING the Immortality of the Soul as the linear de
022r008 is a mine of metereology with respect to the DISCUSSION of winds & storms;--«in Volney's travels als
033r043 Daubeny P. 171. Vol I. Humboldt There is long DISCUSSION on Pumice «& Obsidian:» in the I Vol. Humb:
036r049 he above in Patagonian paper; & part in grand DISCUSSION Consult. reconsult Geolog. Map of Europe Con
038r057 mewhere» below a field of fluid rock.--In the DISCUSSION it will be better not to refer to Lyell. but
052r098 discussing Falkland soundings introduce this DISCUSSION.--Brazil bank: (& I believe SE coast of Mada
057r113 east has not perished because too cold:--With DISCUSSION of camel urge S. Africa productions.-- I thi
096a041 ep side to windward in allusion to St. Helena DISCUSSION. Mr Brayley says he can give me facts respec
098a045 with cretionary.-- it may <of> come of use in DISCUSSION on Cleavage &c Geolog Transacts. Vol III. A
100a053 nation.-- which last is strong character.-- A DISCUSSION on concretions and cleavage conjoined very g
101a056 ies <.-->? Phillips. Lardner's p. 270-4, good DISCUSSION showing present form of land in Northern Eng
102a062 & then the next being sucked out. In Cleavage DISCUSSION, state broadly indication of new law acting
103a065 ubsequent to dislocation of strata. A capital DISCUSSION might be made between dikes & «axis of» moun
103a065 s in favour of pressure of liquid rock. Andes DISCUSSION-- Albite certainly contains 6 per cent more
104a067 ickness is affected, they would be far off In DISCUSSION on dikes argue impossibility on fissure open
104a069 -- Origin of Breccia, introduce in Cordillera DISCUSSION, deep sea, fragments fall off cliffs. but th
114a094 in shall sea. Lyell confounds these introduce DISCUSSION -- I see Lyell talks of different compositio
136a147 where. Phillips in Ladner Vol II p. 125. Good DISCUSSION on mineral veins p. 125 to 129 & p. 135--160

210b158 guages But how do plants cross?-- = admirable DISCUSSION Von Buch says from Humboldt, in Laponia. gen
258c062 ie p. 172. for definition of Analogy. All the DISCUSSION <after> about affinity & how one order first
308c217 of genera &c &c Two savages, two species.-- --DISCUSSION ustil, unless it were fixed what a species m
410e051 d by sexes: All the above should follow after DISCUSSION of crossing of <species> individuals. with r
479z005 --Voyage of Coquille wide limits of Nullipora DISCUSSION good on Falklands birds Discussion of Firola
479z005 Nullipora Discussion good on Falklands birds DISCUSSION of Firola,-- Salpa Anatifs without shells.!
482z011 Mammal1: of Paraguay must be most important a DISCUSSION of geographical distribution of Mammalidae &
550m125 portion of <each> «every» man's time.-- Begin DISCUSSION-- by saying what is Happiness?-- When we loo
199b119 tors hairy with one hairy child, and of albino DISEASE being banished, & given to Portuguese. priest.
205b145 me worse in form, less hardy, & more liable to DISEASE" If population of place be constant «say 2000»
206b148 nalyze causes, dislike to marriage, heredetary DISEASE, effects of contagions & accidents: yet some c
228b232 hen <our> animals our fellow brethren in pain, DISEASE death & suffering «& famine»; our slaves in th
259c065 fe then real adaptation The case of heredetary DISEASE., on the same principle that, cut a sheeps t
332d002 r in women, & since that time it has been rare DISEASE.-- but now (July 1838) he has seen more case i
386d166 sorb a useless member.-- in fact they do it in DISEASE & injury.-- The sympathy of part is probably p
388d172 utility of change-- hence harelips heredetary DISEASE. extinction. Animals in domestication (even El
406e036 & animals cow pox.-- case in Spain of pustular DISEASE following handling sheep-- all cases: d degree
512q019 t breeds--?? or in individual case: subject to DISEASE in youth.-- Mr Tollett-- about selection for m
521m007 wonderful.-- <Now is not epilepsy an habitual DISEASE of the muscles.???> Miss Cogan's memory of the
548m116 <it> «anything».-- if one was subject to this DISEASE oneself, one would only feel sympathy. as for
555m142 in former case instinct to destroy contagious DISEASE.-- (Useful to use term instinct, when origin o
560mIBC gh St Has my Father ever known <intemperance> «DISEASE» in grandchild, when father has not had it. bu
577n053 n» believes that the general talking about any DISEASE tends to give it, as in cancer, showing, effec
608o027 hat there should be necessary. wickedness than DISEASE. This view should teach one profound humility,
636j57r ound Woodpecker stiff tailed cormorant: pain & DISEASE in world & yet talk of perfection Get instance
223b211 d to locality A. but it is instead a stunted & DISEASED form a plant, adapted to A. B. C. D.-- Destro
294c175 inhabit.--» & the first one that bred one was DISEASED in its loins & all were so afterwards, (forge
332d003 be called Worm Fever, as used much more latley DISEASED Mesenteric glands.-- My Father has seen case
408o045 of Somersetshire the Cockles are all apt to be DISEASED., & some of them symmetrically.-- it is easy
608o026 ke a sickly one(P)-- We cannot help loathing a DISEASED offensive object, so we view wickedness.-- it
198b119 ous form has tendency to propagate, as well as DISEASES.. In intermarriages; smallest differences blen
293c174 us think so, but only between laws--]CD. Many DISEASES in common between man & animals-- Hydrophobia
293c174 f common origin of Man.-- different contagious DISEASES, where habits of people nearly similar. Curio
326c266 Observat. on increase & decrease of different DISEASES. 4to 1801.-- quoted by do.-- There appears to
336d018 hemselves. expression of countenances, organic DISEASES, mental disposition, stature, are slowly obta
406e036 mon to men, dogs, horses cows, pigs & sheep.-- DISEASES common to men & animals cow pox.-- case in Sp
414e063 f animals.-- they fight, eat each other, bring DISEASES to each other &c, but then comes the more dea
415e067 mark about the Bladder.-- The numbers of fatal DISEASES in mankind, «the more valuable domesticated a
415e067 up of every heredetary tendency towards fatal DISEASES, & such constitutions only being cleared off
415e067 constitutions only being cleared off by fatal DISEASES.-- The Value of a group does not depend on th
455eIBC ybrid mosses.-- Leighton or some one. Father-- DISEASES common to men & animals.--:likenesses of chil
508q016 ether any trace in germ. (2) Any more cases of DISEASES, generally occurring in man being transmitted
508q016 e the works of Berhave (treating of heredetary DISEASES) translated. (10) About Daltonism in the MALE
508q016 are there any medical Statisics, proportion of DISEASES (heredetary?) in diff. countries in same race
529m043 d. 1838 No Deliriums, yet in some inflammatory DISEASES, where there has been no cloud on the mind, e
532m054 hold of objects to be frightened at.-- (again DISEASES of the heart are accompanied by much involunt
575n043 s this more wonderful than memory. affected by DISEASES. &c &c, double consciousness? What other expl
266c092 02-- sheep have not the enormous tails, which DISFIGURE those of Arabia & Egypt.-- CIVETS CATS only w
551m129 not pout. pushes out both lips in contempt <&> DISGUST & defiance.-- different from sneer-- How easil
565n008 he snarling order.-- But contempt mingled with DISGUST, when ones opponent is considered as quite ins
591n100 ve conscience.-- Why does not man eating cause DISGUST, because he does not go against instinctive fe
594n113 rdity, why does one laugh at it-- sensation of DISGUST with nausea, (when stomach a little disordered
608o027 if incorrigably bad nothing will cure him' 3) DISGUSTED. with them. Yet it is right to punish crimina
377d139 may, be turned to ridicule, or may be thought DISGUSTING, but to philosophic naturalist pregnant with
574n041 overing <gum> «teeth»less-jaws. as picture of DISGUSTING lewd old man. ones tendency to kiss, & almos
523m014 praising his sister who confined him. & yet DISINHERITING her.-- This «N B. I have read paper somewhe
100a054 ders gneiss an aqueo deposit resulting from DISINTEGRATED granite!!! Look at gneiss of Rio Concretion
627o053 if termed "selfish", must be subclassed as "DISINTERESTED" p. 14. It is allowed, that we have concept
146g012 res of association, & passions, such as love-- DISLIKE & <f> passion of hatred To fulfil an instinct
186b059 call (atavism) Probably this is first step in DISLIKE to union, offspring not well intermediate Lyel
206b148 will become extinct.-- Who can analyze causes, DISLIKE to marriage, heredetary disease, effects of co
220b197 connected with generation certainly is.= The DISLIKE of two species to each other is evidently an i
567n015 on, when meeting a stranger. who one may like. DISLIKE, or be indifferent about, yet feel shy.-- not
587n087 feels no more inclined to ask [not located] If DISLIKE, distaste. & disapproval. were not something m
587n087 hand, it is said people, who like sweet things DISLIKE others.-- dogs dislike perfume] I should think
587n087 , who like sweet things dislike others.-- dogs DISLIKE perfume I should think, great principle of li
533m058 is frightened without knowing why-- the child DISLIKES the frown without knowing why-- a man as in G
068r147 .-- <it is certain, if strata can be> Problem DISLOCATE strata without ejection of the fluid propelli
068r146 lcanos only burst out where strata in act of DISLOCATION (NB. dislocation connected with fluidity of
068r146 out where strata in act of dislocation (NB. DISLOCATION connected with fluidity of rock ∴ «in earlie
103a065 It is an important view being subsequent to DISLOCATION of strata. A capital discussion might be mad
107a078 g--Very difficult subject. PP-- I think from DISLOCATION taking place chiefly beneath water & volcano
068r146 » when covered up beneath ocean).--The first DISLOCATIONS & eruptions can only happen during first mo
103a064 h weakened by <vi> considering how close the DISLOCATIONS occur & therefore that the crust might be c
288c159 half way «in tone»-- the native dogs howl most DISMALLY, very rarely bark.-- are almost useless not t
594n112 so firmly implanted, birds soon <dis> learn to DISOBEY it-- I have seen hawk & sparrow in Shrewsbury
620o044 following our conduct.-- If the temptation to DISOBEY the conscience is extremely great [LHC] The ca
563n002 had done so & so for his interest, & found he DISOBEYED a wish which was part of his system, & consta
591m099 ing that the instinct of self-preservation is DISOBEYED-- I often have «as a boy» wondered why all ab
620o044 hich is indeed, often felt at very time it is DISOBEYED) & is sure guide.-- Hence conscience is impro
537m077 se strong, & does not act up to them, no doubt DISOBEYS & hurts conscience more than other.-- A Scotc
332d004 child not having passed them before.-- Hence DISORDERED intestines are not healthy to worms, (like p
594n113 f disgust with nausea, (when stomach a little DISORDERED) at thought of almost anything ugly. baby--
332d003 urisy, broken limb «in children» & other such DISORDERS accompanied with some fever, be attended by t
423e096 he more highly organized ones. would soon be DISORGANIZED to fill their places.-- The Geologico-geogr
265c084 ccording to Physical laws, as sulphuric acid DISORGANIZES wood, but adaptation.-- albino however is m
101a056 nt form of land in Northern England influence DISPERSION of Boulders.-- See Rogers for Southern limit
447e164 s require crossing.-- The case of Lemna shows DISPERSION of germs is not end of seminal reproduction.
035r048 actures, consequent on grand rise, & angular DISPLACEMENT, consequent of injection of fluid rock.-- T
368d114 enera. Guinea Fowl & Peacocks.!!» other birds DISPLAY beauty of plumage.-- (The females (as Owen obs
471tf07 ates the strange forms which the thorax & head DISPLAYS.-- most fantastic & use unknown-- "«when we
551m129 Ourang in <Zoolog» Gardens pouts. partly out DISPLEASURE (& partly out of I do not know what when it
291c166 rganization!! Avitism in mental structure or DISPOSITION, & avitism in corporeal structure are facts
336d018 on of countenances, organic diseases, mental DISPOSITION, stature, are slowly obtained & hereditary;
359d087 ybrids are very wild & take <very little» in DISPOSITION after their «pheasant» parents.-- (There are
359d088 ly to impress the young most with its form & DISPOSITION Saw three young ducks, like each other,-- (&
360d090 t has full share of Jackall shape of body.-- DISPOSITION wild, & fearful. though not so much as in Ja
371d127 g different in colour, form, & so altered in DISPOSITION, as to be more easily trained up to the (req
384d161 Hunter (p. 45) observes "every species has a DISPOSITION to deviate from Nature in a manner peculiar
393dIBC progeny faulty. Does male fail in passion.-- DISPOSITION of half bred Cattle at Cinbermere? How is Ja
520m001 r says he thinks bodily complaints «& mental DISPOSITION» oftener go with colour, than with form of b
520m001 r in body, but his mother in bodily & mental DISPOSITION.-- My father has seen innumerable cases of p
522m013 cs.-- people recognized,-- sudden changes of DISPOSITION, like people in violent intoxication, often
536m073 man.-- the real argument fixes on heredetary DISPOSITION & instincts--.-- Put it so.-- Probably some
538m079 depend on state of turn In drunkedness same DISPOSITION recurs, such as -- -- of Trinity always thin
555m143 les) The black Spider Monkey, very different DISPOSITION from others, slow cautious, angry cross look
390d178 t some small change in form. ideosyncrasy or DISPOSITIONS were added or substracted at each, or in se
430e117 scribed.-- His three have had VERY different DISPOSITIONS: this is important as showing small variati

Page **(Key Word)**
```
287c158 of mollusca. [ + ] These questions may be all DISPUTABLE, but the one end of classification to expres
105a071 read out. by sea-- beach action -- no one will DISPUTE. sea. once came to Mendoza--. Will they introd
262c072 cal structure?!! Whewell «in Comment/ few will DISPUTE--» says civilization heredetary; ie  instincts
265c086 in & not adaptation. to its habits.-- Few will DISPUTE that it is possible to have structure  without
409e049 He is present a social animal» a fact few will DISPUTE, [although, that it was the sole object, I wil
409e049 [although, that it was the sole object, I will DISPUTE, when I hear from the geologist the history, &
555m141 ng" or "reasoning"-- only rather more steps.-- DISPUTE about words.-- Miss Martineau (How to  Observe
256c054 lance would tell from which country,-- I often DISPUTED for a moment,-- Galapagos, S. American-- -- g
277c127 n show range & habits-- Take instances of most DISPUTED shells, such as Cyrena This is reform which p
302c202 acted otherwise, my premises <in di> would be DISPUTED.-- according to Principles of last page. <an>
535m063 tinctively or habitually.-- good Heavens is it DISPUTED that a wasp has this much intellect. yet habi
556m145 s born with two eyes." I think this cannot be DISPUTED anymore in men. than in animals.-- In the dra
268c095 which appear diff. from Australia (Waterhouse «DISPUTES this)> says differently) do. do. on the genus
617o040 apprehend  force in inanimate matter we feel DISSATISFIED until we can point out How can force be rec
383d159 ten N.B. the common mule must often have been DISSECTED Zoolog. Garden. Sept 16." Hybrid between Silv
509q017 introduce  me to some Human Anatomist. Has he DISSECTED any animal often, which has abortive bone. (a
228b229 ery naturalist ought to have before him, when DISSECTING a whale, or classifyng a mite, a fungus,  or
022r007 to  change their position? Carbonate of Lime DISSEMINATED through the great Plas Newydd dike.--Mem tr
076r169 of  silver «in veins» very unequal sometimes DISSEMINATED <[...]> sometimes concentrated: wonderful q
110a086 , blackish grey base. crystals from fragment DISSEMINATED on that side of salband. gradually becoming
236b280 s, no he only says sometimes we might expect DISSEMINATED species to vary a little, but such should n
315c240 May  have been Created at many spots & since DISSEMINATED See. Habits of Malay fowls p 5. (note) in s
123a116 2d.--  Sulphur like carbon must go round of DISSEMINATION & separation in volcanos.-- if so why not m
258c061 analogy» The resemblances relationship, the DISSENBLANCES analogy,-- See Abercrombie p. 172. for defi
580n062 onscious kept playing so well.-- Lr. Brougham «DISSERT.» on subject of science connected with Nat. T
292c170 is faith is not staggered <">= I confess. no DISSERTATION against these views, could possibly have ha
579n057 e are some notes. & likewise on Wordsworth's DISSERTATION on Poetry.-- The expression of shame-facedn
590n093 ation.-- like sigh before false sneeze.-- "A DISSERTATION on the Influence of the Passion."-- p.  37.
596h184 Sentiments" much on life & character "Humes DISSERTATION on the Passions." "Hartley" I should  think
326c266 N.  America.-- Owen has it.-- Ld. Brougham. DISSERTATIONS on subject of Science connected with Natura
462t051 similarity in condition in Java & Sumatra & DISSIMILARITY of forms-- yet how valueless this objection
028r030 s, as in Australia.--Have Limestones all been DISSOLVED. if so sea would separate them from indissolu
028r031 from  indissoluble rocks? Has Chalk ever been DISSOLVED? Singularity of fresh water at Iquiqui. not f
042r067 t. & fall of Ashes of Falkner, ¿how far is the DISTANCE?-- Fossil bones black as if from peat.--yet c
052r098 ascar. where a --40 line <shows> runs at equal DISTANCE?) 1st cases. -- The terraces in Valleys of Ch
054r105 as voyage North of Callao, the country, to the DISTANCE of 3 or 4 leagues «from the coast» may be con
058r116 same  earthquake has run from Chili to Quito a DISTANCE of more than 500 leagues. A little time after
060r124 of course must retain same level) to a greater DISTANCE".--Afterwards speaks of this phenomena in con
137a149 ins of slag in iron furnaces affecting to some DISTANCE & blending with sandstone «said to be» analog
161g096 A  at head of Gentle mossy slope, which from a DISTANCE hid it, could be followed for at least 2 mile
209b156 est continent: poorness in exact proportion to DISTANCE (?). & similarity of type (?).-- [«Mem» Juan
244c021 s in tropical sea, it. would demonstrate.; not DISTANCE, makes species but barrier.-- --it would make
528m037 seeing  how the serpentine lines narrow in the DISTANCE.-- & even on paper two waving perfectly paral
565n007 . <when> as fear to «man as» animals. comes at DISTANCE, mouth is placed open.-- Hence becomes instin
584n072 carrier  pidgeons proverbially carried to long DISTANCE in dark "it is inspiration."-- this is  class
593n111 shed & having looked round saw at considerable DISTANCE a very large hawk, which are «so» rare « <s.>
594n112 are common. not unlike in size in the air at a DISTANCE.-- How can such an instinct arise?? «it would
032r041 ikewise» differences of temperature «at equal DISTANCES from centre of rotation» & a <circulation owi
062r129 on. Keeling: at sea so commonly seen. at long DISTANCES; generally first arrives:-- New Zealand  rats
283c145 ong in blood?-- My theory agrees with unequal DISTANCES between species. some fine & some wide. which
369d115 to  travers» "May have reference to the Great DISTANCES which the Mammalia of N. S. Wales are general
444e158 y.-- instinct, for how could experience teach DISTANCES in air, in which it never touches  objects.--
499q009 Willows.--  (14) Bowman female branch At What DISTANCES from males, will female (a) Willows or Yews s
566n011 grown  years, thinking he instinctively knows DISTANCES:. is good instance of obtaining <that> «a» fa
023r009 e E. Indian Islands. namely Java is 1000 miles DISTANT! Where are Hippotami found in that Archipelago
023r011 aval» of the shore of the Pacifick is 60 miles DISTANT from the grand ancient volcanic axis «of the A
181b041 any  two of the 12. having progeny. after that DISTANT period.-- Hence if this is true, that the grea
195b103 pheus.= Put this strong so many thousand miles DISTANT.-- Absolute knowledge that species die &. othe
206b147 nclude that there will be a period though long DISTANT, when of the present men (of all races) not mo
224b212 xplains that between Species from «moderately» DISTANT countries. there is no test but generation, «(
236b280 .-- Are the closest allied species always from DISTANT countries, as Decandoelle says, no he only say
257c057 iar conditions-- of «nearly» same kind country DISTANT. <Study> The circumstance of ground woodpecker
261c070 t my theory, & the examination of species from DISTANT countries., may give thread to conduct to laws
285c151 importance, in each group & connection of even DISTANT ones) the characters will be first those of an
306c213 whole  area,. 120 is greatest (about 200 miles DISTANT).-- directly beyond produced line of Timor 215
332d001 during  one time he had three patients at very DISTANT quarters of the county, who had had no sort of
338d025 natidae.-- recurs to idea of only animals from DISTANT countries breeding!. «Mem 3 species of grouse»
342d033 so chinese pigs & common must be considered as DISTANT species?? or is time the varying element). The
377d139 to say give me-- the other when Dr. Smith more DISTANT.-- But he thinks other monkeys make st.-- nois
426e108 me nephews, who are astonishingly like to some DISTANT cousins, the nearest blood being a great great
448e167 beds have many shells in common «& are not far DISTANT» with Touraine «which as L. says is strong arg
500q10a erences of Irish & British Species & British & DISTANT parts of Europe.-- Gould-- go over the Pigeons
565n009 smile.-- Hope is the expectant eye. looking to DISTANT object, brightened & moistened by emotion,-- w
569n020 tructure.-- Devotional feelings, probably some DISTANT power of the mind-- superstition & charity & p
574n041 association of saliva, is probably due to our DISTANT ancestors having been like dogs to bitches.--
613o035 come  into play, when relation is kept up with DISTANT object. where many such objects are present, &
587n087 more inclined to ask [not located] If dislike, DISTASTE. & disapproval. were not something more  than
385d163 held.-- like she wolf of Hunter.-- young take DISTEMPER very readily & are subject to fits.-- «there
570n023 urprise, can one resist blow better with body DISTENDED.-- intolerable to be poked behind, without on
570n023 to be poked behind, without ones chest, being DISTENDED. touch a person on the ribs & how he gulps in
036r050 d to make a noise,--My impression. is not very DISTINCT, from some of the lower orders; it was connec
039r060 onsiders that successive terraces mark as many DISTINCT elevations; hence it would appear he has  not
063r130 lapse)  being the relation.--As in first cases DISTINCT species inosculate, so must we believe ancien
063r132 so much surprised at seeing Zoophyte producing DISTINCT animals. still partly united. & eggs which be
110a086 ained & more compact on that side-- separation DISTINCT from dike junction mechanical: DIKE base redd
110a087 ther dike or fragment. junction certainly most DISTINCT on dike side.-- oozed from one of the true ro
114a094 ing. bear on subject of cleavage Clay slate. a DISTINCT formation deep «& therefore extensive»  water
161g095 ove valley this road level with Peat moss most DISTINCT then lost by slope, then concealed by fragmen
174b015 s-- if separated from immens ages possibly two DISTINCT type, but each having its representatives-- «
179b033 oldt. New Spain:--» Dr. Smith always urges the DISTINCT locality or Metropolis of every species: beli
182b050 ome most curious cases. of close but certainly DISTINCT species between Australia & Van Diemen's land
191b082 ava & Sumatra. Rhinoceros. Elevate & join keep DISTINCT. two species made elevation & subsidence cont
218b191 rs are in the Horticultural Transactions and a DISTINCT work on Hybridity under title of  Amaryllidae
224b213 change  organization of stomach & hence remain DISTINCT. Where country changes rapidly, we should exp
224b214 hing with thoughts (animal). «∴ my theory very DISTINCT from Lamarcks» Without two species will gener
234b261 -- Again is there not similarity even in quite DISTINCT countries in same hemisphere. more than in ot
250c039 nds.-- ¿Europe has many species but not genera DISTINCT from rest of world??? Lyells Principles, must
252c045 ceros Cape town good species Indian species so DISTINCT that all analogy from each other I do not kno
259c064 e it affected» tempts one to bring one back to DISTINCT creations-- it is only be recollecting  that
277c126 pecies. only known by analogy genera of course DISTINCT. analogy from every country & class tells us
285c152 ave passed through a thousand changes, keeping DISTINCT from other & if a first & last individual wer
286c156 ubject to put in alternative of Man created by DISTINCT miracle. Macleay letter to Dr. Fleming. Philo
289c161 t be overcome, but until it is the animals are DISTINCT species If any one is staggered at feathers &
289c162 t be given as a hopeless difficulty, except as DISTINCT creation.-- Generation may be viewed as conde
305c209 as Europaean species = singular coincidence if DISTINCT creation.-- le.-- a mere statement nothing is
335d016 lood.= The case of union of perfect animals is DISTINCT case,-- gradation from physical impossibility
337d022 &c &c &c who imitates.-- who will say there is DISTINCT Creation required if he believes «hyaena & sq
343d036 s formed, & the world peopled «with Myriads of DISTINCT forms» from a period short of eternity to the
```

343d039 ce. otherwise in 10,000 years Negro probably a DISTINCT species-- We know how long a Mammal may go on
355d067 w of monstrosity with regard to age of foetus. DISTINCT consideration) Now in different SPECIES of ge
355d068 nsacts) same appearance with Secondary Species DISTINCT-- but close.-- Mem. Von Buch on Cordillera fo
370d117 with Annelidae by some fish.-- But birds quite DISTINCT.-- Collect cases of difficulty of growing pla
384d162 ybridity of ferns) afterwards they can be seen DISTINCT. (in dioecious plants are there abortive sexu
415e066 ent against, my theory, should no fossil «very DISTINCT species» of the Ornithorhyncus be found.;-- y
433e126 tance of forms, intercalated between two great DISTINCT formations.-- particular air given. p. 246.--
463t055 ent vertebrae of the head of Snake wonderful!! DISTINCT!!-- He would not allow such series showed pas
485z017 ort <by> on D'Orbigny on species of Mephites 4 DISTINCT Camelidae. do not breed together Mag: of Zool
490q001 ct land-shells of Madeira-- analogous or quite DISTINCT from recent ones-- I presume some recent not
492q002 e planted together: <do> Can Holyoak be raised DISTINCT by seed-- Heartease. 6. -- Do not species of
522m013 rybody is insane. at some time. Mania is quite DISTINCT, different also from delirium, a peculiar com
541m093 espise a man & say nothing, but without a most DISTINCT will, he will find it hard to keep his lip fr
543m101 ntellectual powers of perceiving & classifying DISTINCT resemblances.-- The facts of half instincts.
544m104 ion.-- fear of death!!! as Montaigne observes. DISTINCT from pain, for one hates pain from this fear-
545m106 «turn» of the head. I could not perceive «any» DISTINCT wrinkle, but such movements in skin of eyebro
567n016 ll ordinary lines of association.-- is totally DISTINCT from learning it by heart. Do not our necessa
569n019 imitation.-- Hence pleasure in the beautiful. (DISTINCT from sexual beauty) is acquired taste.-- Whil
590n094 to call imagination a faculty, a power, quite DISTINCT from self. «or will» [not located] <& other c
601o009 ited with nervous filaments.-- ¿plants? yes by DISTINCT mechanism 2. Sensation of higher order. where
625o052 t appears that Sir. J. & others think there is DISTINCT faculty, of conscience.-- I believe that cert
640j167 -- Hence the Galapagos Islds are explained. On DISTINCT Creation, how anomalous, that the smallest ne
244c023 balançons pas a la regarder comme une espèce DISTINCTE! p. 171. Sus papuensis «partly domesticated»
209b156 imat, se divisent le plus aisément en espèces DISTINCTES et permanentes." p. 145. In Humboldt great w
026r020 che, article "Erratic blocks" not sufficient DISTINCTION is given to angular & rounded.-- Fox Philoso
180b036 . the finest gradation, B & D rather greater DISTINCTION Thus genera would be formed.-- bearing relat
220b197 lly reproductive organs) & therefore the one DISTINCTION of species would fail. But this applies only
300c198 . Study Mr Blyth's papers on Instinct.-- His DISTINCTION between reason & instinct very just, but the
523m018 ting suddenly into passion.-- There seems no DISTINCTION between enthusiasm passion & madness.-- ira
542m096 hen the uncovering the canine useless.-- The DISTINCTION «as often said» of language in man is very g
615o037 n mammae before any reason had told him this DISTINCTIVE mark, it is downright instinct, leading to t
204b139 and analogy in Linnaean Transactions Mr Wynne DISTINCTLY says that the mixture between Chinese & Engl
210b158 alogy to what takes place from time? Von Buch DISTINCTLY states that permanent varieties. become spec
342d035 Literat. Vol II p 11) accidentally says "--is DISTINCTLY marked as whole dynasties have been featured
382d158 (Do any male animals give suck)--But this not DISTINCTLY stated by Hunter.-- Do testes, & ovaria when
404e025 t might thus have arisen, & M. Edwards p. 330 DISTINCTLY states that the flipper is a mere simple mod
453e183 oaks, flaxes &c &c? Mr Herbert in letter says DISTINCTLY, that Hollyoak reproduce each other. & yet I
467t099 d in whole length to corolla--anthers minute, DISTINCTLY doubled, brown, but with no pollen.-- Common
572n033 I read the first page & pronounced each word DISTINCTLY. woke instantly but could not gather general
179b033 doubt «C. D.» wild men do not cross readily, DISTINCTNESS of tribes in T. del Fuego. the existence of
270c103 ity through multiplicity" this unity,-- this DISTINCTNESS of laws from rest of]CD universe «which Car
066r142 a piece of Falkland Sandstone. he could not DISTINGUISH from stone Caradoc from lower of third Silur
233b250 es of mice in S. America. which were hard to DISTINGUISH came from closely neighbouring localities In
500q010 (not asclepiadae. «in» Lindley) (24) Do Bees DISTINGUISH species, they do not varieties.-- (25) Does
571n030 In all the foregoing cases most difficult to DISTINGUISH. between prejudices of youth from <here> hab
499q010 s Irish, Scotch & English plants generally DISTINGUISHABLE.= What structure of seeds.-- (Paris) (22)
317c250 horbia so near Piscataria as scarcely to be DISTINGUISHED from it.-- & several old acquaintances. whi
350d053 How can local species as at Galapagos., be DISTINGUISHED from temporal species as in two formations?
380d154 , yet the Cassowary & Guinea Fowl cannot be DISTINGUISHED. -- A capon will sit upon eggs, as well as,
384d162 hough organs for the double purpose are not DISTINGUISHED. --yet may be presumed from hybridity of fe
410e053 ermanent varieties in same country, must be DISTINGUISHED, from permanent varieties not in same count
473s05r out from different lakes of N. Wales can be DISTINGUISHED-- & Jackson here (Capel-Curig) says that he
523m017 itself. Asked my F. whether insanity is not DISTINGUISHED from whims passion &c by coming on suddenly
252c045 on of circumstances in Geography to help in DISTINGUISHING empirically what is species.-- The Collect
256c054 ng over reptiles he often had difficulty in DISTINGUISHING which were species, (theory admirably) yet
406e035 38. p. 338 A most grave source of doubt. in DISTINGUISHING <after> which parent impresses offspring m
593n109 is so.-- idea of beauty of music are great DISTINGUISHING character between man & animals.-- [blank]
524m022 ead) yet instantly perceived when my Father to DISTRACT her attention took her «left» hand to pretend
366d111 not coming in, could not help being perfectly DISTRACTED «Referred to <other> Book M.» Is there any l
260c068 Blue-Jay, when <one> birds hears <dis> crys of DISTRESS of other parents.-- Shows community of langua
540m089 could be barren. & lead people from scenes of DISTRESS.-- see how a crowd collects at an accident,--
594n115 s «Ogleby» seen Zool. Soc-- 1838 remember with DISTRESS their companions-- a «blue» Gibbon. whose com
195b098 rtly American North & South.-- (& geographical <DISTRI> division are arbitrary, & not permanent. this
241c016 eolog. 1837-8. Tom: IX.-- M. D'.Urville on the DISTRIB of Ferns in South Sea (Indio Polynes: <}> vege
324c268 yage Decandoelle. Philosophie. or Geographical DISTRIB. «in Dict. Sciences. Nat. in Geolog Soc.» F..
325c267 Blackwood. June 1838 H. C. Watson on Geograph. DISTRIB: of British Plants. Humes Essay on H. Understa
337d023 nt, come from one stock.-- Theory of Geograph. DISTRIB: of <ani> organic beings.-- Animals «of same c
340d026 l affinity with analogy-- Good table at end of DISTRIB: of <birds>. Anatidae.-- Consult this book aga
641j29r facts read Lacépède on Cetacea & Geographical DISTRIB of larger Seals-- Are Porpoises numerous in co
209b157 rmanentes". p. 145. In Humboldt great work de DISTRIBUT. Plantarum. relation of genera to species in
211b164 ebrata, &c 1837 p. 370 Owen says Nonsense The DISTRIBUT of big Animals in East Indian Archipelago--,
506q015 any years, by cuttings.-- (40) Ask Henslow to DISTRIBUTE some of my questions amongst agriculturists.
085aIFC wich Nothing on any Subject As far as p. 33. DISTRIBUTED to several subjects. Feb 24th 1839 As far as
093a033 e bed.-- November 8th 1877 (Memoranda so far DISTRIBUTED to various subjects) Dr. A. Smith informs me
219b196 Mammalia, were born from one stock, & since DISTRIBUTED by such means as we can recognize, may be th
247c028 Radack isld.-- p. 69. Sharks very generally DISTRIBUTED: Mem of great geological age-- Gastrobranchu
255c050 ree p. 7. Am.> D'orbigny. Birds of prey, are DISTRIBUTED in S. America like other forms, but those in
339d026 to think any improbability to animals being DISTRIBUTED after flood (!) according to affinities!. co
400e013 tober 11th.-- Uncle John-- says Decandoelle, DISTRIBUTED seeds of Dahlia all over Europe same year.--
495q05a pools & rivers-- every kind of seed must be DISTRIBUTED.-- Examine scum of pond for seeds.-- 11. Soa
023r012 f the ocean's bottom. With respect to Sharks DISTRIBUTING fossil remains: Sharks followed Capt. Henry
512q019 (2) Does he believe. Stanley's fact of Hawks DISTRIBUTING live Mamals (3) Do most Hawks eat stomach.
036r050 ult Geolog. Map of Europe Consult charts for DISTRIBUTION of pebbles.--Plains. off coast of Patagonia
044r073 arth;--Scarcity of Organic remains.--Unequal DISTRIBUTION of Volcanic action, Australia S. Africa-- o
076r169 not common in <S.> America: In all climates DISTRIBUTION of silver «in veins» very unequal sometimes
105a070 rge by slow erosion. but we have evidence in DISTRIBUTION of blocks, that there has been no tumultuou
163g108 yed on exposure Mr H. C. Watson Geographical DISTRIBUTION of British Plants Shropshire Quartz what su
191b081 ? Negative facts tell for little) Geographic DISTRIBUTION of Mammalia more valuable than any other, b
193b092 sh.-- at beginning of Volume on Geographical DISTRIBUTION of animals Brown Geograph Journal. Vol I p.
195b102 go at such a rate, so will be the numbers & DISTRIBUTION of the species!! It may be argued represent
209b155 es. If species made by isolation; then their DISTRIBUTION (after physical changes) would be in rays--
232b249 Waterhouse has most curious facts about the DISTRIBUTION of Lemurs in Madagascar, on neighbouring is
251c041 ck Monograph. Mammal. «4to» good facts about DISTRIBUTION of Cats Vol III. <p.> p 233, stated that th
255c050 e of??? A Urubu, (with one leg) attended the DISTRIBUTION of food, at the Mission of Mojos (even 20 l
256c055 umstance of having two sexes is the check to DISTRIBUTION of birds & animals Mr Strickland & Hamilton
263c076 on. study unit of type-- Study geographical DISTRIBUTION study relation of fossil with recent. the f
268c099 Milvulus [not located] element geographical DISTRIBUTION of animals, <as> I use (new step in inducti
275c119 est endowment of lofty genius Using geograph DISTRIBUTION of horizontal barriers-- Mr Yarrell.-
275c119 tell of Physical relations in time «& forms» DISTRIBUTION tells of horizontal barriers-- Mr Yarrell.-
344d040 There are some admirable tables on Geograph DISTRIBUTION of reptiles in Suites de Buffon.-- Vigors h
352d059 esponding points.-- The present geographical DISTRIBUTION of animals countenances the belief of their
352d059 ity (ie much intervening physical change).-- DISTRIBUTION especially of Mammalia As every organ is mo
400e011 edominant &c having relation to geographical DISTRIBUTION-- Thus Hattica is great genus.-- because fo
424e100 .-- p. 290-- admirable paper on geographical DISTRIBUTION of Crustaceae.-- (I forget whether I have a
482z011 most important a discussion of geographical DISTRIBUTION of Mammalidae &c &c &c-- The French <Jerrol
559m155 1838. Copied Mr H. C. Watson on Geographical DISTRIBUTION of British Plants A Volume published by Col
035r046 e made out, but in those belonging to the same DISTRICT there seems. I think, little ground for skept

Page **(Key Word)***
121a112 Consider profoundly How came it. that Glen Roy DISTRICT could have been elevated without fissure & un
202b131 . Near the Caspian «Province of Ghilan» wooded DISTRICT cattle with humps as in India. Geograph J. Vo
260c069 f vegetation,-- ¿account for colour of bird in DISTRICT. which they frequent!!?-- Wilson's American o
265c085 er being absent.-- again dwarf plant in alpine DISTRICT & dwarf plants from seed, one adaptation, oth
268c096 rds. genera blend into each other in very same DISTRICT.-- <The same> <Mem> Tennioptera & Tyrannula (
288c160 r Warbler-- Yellow Wagtail never seen in one DISTRICT, though common on another, (golden creted wre
419e086 -- Lyell says the elevated shells in Bayfields DISTRICT are much more like those of Scandinavia, than
490q001 or insect or group vary more in one country or DISTRICT than in another? Character of shells of Sandw
500q10a ey come from islds. or different parts or same DISTRICT.-- About <endemic &> wandering species of con
053r101 - M. Lesson considers the Sandstone & Granite DISTRICTS to be separated by profound valley [.] Sydney
249c036 fact of no forms peculiar to word to special DISTRICTS???? north of 30 degree.--, may be connected w
527m032 th the stated fact, that «birds from» certain DISTRICTS have the best song. [Migratory birds return t
466t094 t proboscis within nectary «they do» they must DISTURB all anthers, wh otherwise lie protected by the
067r144 very remarkable phenomenon. showing line of DISTURBANCE inside Cordillera: It is not therefore so wo
037r056 a, Rio de Jan: B. Oriental? level surface not DISTURBED.--Whole West coast. Chonos to Copiapo.--Sydne
522m012 aken place.-- as if the idea of time had been DISTURBED.-- These foregoing cases of «mental» failure
545m107 on, which the progenitor did, when excited or DISTURBED by the same cause, which «now» excites the ex
568n017 «Probably bashfulness is connected with some DISTURBED habit» [Thus shepherd dog. has pleasure in fo
569n022 wife would be grieved-- "il leva les epaules et DIT qu'il valait mieux rester a Farroilap quelque ma
036r052 characteristic of the deserts of Syria <chara> DITTO for Patagonia, especially rocky parts of centra
127a125 l. N-clinal & anticlinal «synclinal--» line.-- <DITTO of synclinal> simply clinal lines. dipping so d
129a131 ond the limits of Alps size of boulders sorted: DITTO Murchisons case.--¿ does it bear on Patagonia?
171b004 dered wild <through> generations, acquire ideas DITTO. V. Zoonomia.-- There may be unknown difficulty
200b123 ces, whether you would expect equal fertility-- DITTO in Plants.<===> It will be well to refer to Cham
219b193 Compositae, because seeds first arrived «Ferns DITTO.--» & hence formed trees]CD & would creator <on
297c186 s few forms. & they are remnants-- Cephalopoda DITTO.-- Mag of Zoolog. & Bot. Vol. II p. 125 Allusio
312c228 o keep up offspring, are not half lion & tigers DITTO. (see Griffith) & half Muscovy ducks, «black co
324c268 types. Smellie. Philosophy of Zoology Flemming. DITTO Falconers remarks on the influence of climate W
358d074 ancient geography of Atlantic Tristan D'Acunha DITTO Juan Fernandez do Mitchell. Australia Vol I. p.
495q05a y days.-- The «garden» Coronella also sleeps in DITTO-- Cover them up periodically & see effect-- suc
521m008 ry in her dotage is fact of same sort. Aunt. B. DITTO.-- Case of Mr Corbet of the «Hall» «Park», afte
529m041 ed, & how the scenery would rise. Deer in Parks DITTO.-- My Father says there is case on record he be
592n105 le & deer pursuing a wounded one.-- porpoises a DITTO--. it is probably some secondary one-- blood bei
485z017 Turdus Magellanicus.-- C, <Chingolo> Chimango-- DIUCA?? See Report <by> on D'Orbigny on species of Me
318c254 lly by night -- other birds which is strictly DIURNAL, migrates singly by night.-- others in flock,
162g103 t arm of River Tarf <it> Has a very long, flat DIVATIUM aquarum with, left of Bright.-- like bed of l
048r085 d clay. -- In the History of S America we cannot DIVE into the causes of the losses of the «species o
207b151 the young one <inland> directly by instinct. can DIVE & conceal themselves in the grass.-- Beatson St
289c160 alities‖-- p. 390,. young «grebes» «ring ouzels» DIVE instant touch the water.-- Capital instances of
258c060 s modified). the relationship of Analogy is a DIVELLENT power & tends to make forms remote antagonist
283c144 appears.-- Is there not some statement about DIVERSITY of forms in aberrant circles.-- explained by
284c147 ixed quantity) & on subdivision of stations & DIVERSITY--.-- this perhaps on long average equal.-- Th
292c171 same way-- The improvement of reason implies DIVERSITY & therefore would banish individual, but gene
441e148 trees. mentioned by Mr K may be caused by the DIVERSITY of stocks, on which they are grafted.-- No th
581n065 ow simple an explanation it offers of radical DIVERSITY of tongues.-- [Emotions are the heredetary ef
276c124 rae much lengthened &c there may be tendency to DIVIDE, which often enough repeated would cause an un
063r132 . by nature or accident). we see an individual DIVIDED either at one moment or through lapse of ages.
063r132 uals of all species. as «each» one individual «DIVIDED» by different methods, associated life only ad
077r171 mall ravins into which the valley of Marfil is DIVIDED, appear to have a decided influence on the ric
177b028 g to each)» approaching another. <Petrels have DIVIDED themselves into many species, so have the awks
203b135 lies. as Cape Anteater,.-- This supposes world DIVIDED into Zoological provinces-- united-- & now div
203b135 ded into Zoological provinces-- united-- & now DIVIDED again-- Weakest part of theory death of specie
297c185 roups few in numbers & vary much in character, DIVIDED into many small genera ʹ:circumstances not favo
300c198 f others.-- the mind of different animals less DIVIDED.-- But, as man has heredetary tendencies. his m
300c198 eredetary tendencies. his mind is still only a DIVIDED body. ℕp 3. language seems to supply instincts
343d037 t 17th Two regions may be Zoolo-geographically DIVIDED either by developement of new forms in one., o
467t100 rdinary divisions, & a few with one lobe again DIVIDED «Have dried some».-- some with no division in
584n071 found which does not climb CD[.instinct may be DIVIDED into migration,-- subsidiary to food & tempera
339d025 red, & surely wild Duck & «pintail» Widgeon!-- DIVIDES animals «world into Zoological Provinces» acco
384d161 <not> of each species not similarly subject-- ‖DIVIDES sexual marks into primary & secondary, the lat
461t025 most curious paper on the Pipe-fish-- which he DIVIDES into two divisions, one of which are marsupial
135a145 hes at {P} from each side intercepting plain & DIVIDING it-- Hopkins fissure at {P} .-- G. J. Malcolm
471t07 ty of form in the same> organ "manifestation of DIVINE power"?.--"of their use difficult to conceive
568n019 want some good passages against, opposition of DIVINES to progress of knowledge. see Lyell on Scrope,
263c077 the fabric falls! But Man-- -- wonderful Man. "DIVINO ore versus coelum attentus" is an exception.--
209b156 quels   sont les genres qui, sous ce climat, se DIVISENT le plus aisément en espèces distinctes et per
063r132 ated life only adds one other method where the DIVISION is not perfect.-- Dogs. Cats. Horses. Cattle.
066r142 rom stone Carado: from lower of third Silurian DIVISION--Together with same general character of foss
195b098 ican North & South.-- (& geographical <distri> DIVISION are arbitrary, & not permanent. this might be
204b141 ch higher generalization in view. In Marsupial DIVISION «do» we not see a splitting in orders, Carniv
229b233 ner-tailed kind further inland.-- NB. There is DIVISION of snakes, with hinder teeth perforated for p
257c058 animal life.: generally, but even about great DIVISION, our <only> question is, how there come the l
262c073 five in each group absurd.-- the mere fact of DIVISION of lesser & more power (2.typical 3.subtypica
263c074 it, depress & elevate & enlarge New Zealand; a DIVISION of nature of Apterix, many genera & species--
274c114 e there not many ground woodpeckers?-- In each DIVISION Gould thinks he can trace structure for insec
372d130 n generation.-- Ehrenberg considers artificial DIVISION of animals, as gemmation, I consider gemmatio
372d130 gemmation, I consider gemmation as artificial DIVISION.-- On this view each particle of animal must
386d167 r does sometimes occur in animals). latter the DIVISION taking place from outside inwards. & in anima
388d172 . p 653. Ehrenberg<h> thinks multiplication by DIVISION <only> is developement of gemma.-- The manner
462t051 In botanical geography, there can be no sharp DIVISION of partition as between Mammalia in cases suc
466t094 lie protected by the hairy black lip of lower DIVISION of nectary: «wh. itself resembles a Bee, but
467t100 gain divided «Have dried some».-- some with no DIVISION in young flowers. The abortive stamen are of
498q008 wn that each pistil is connected with separate DIVISION of germen <?>-- (11) Must pollen grain be who
610o033 er Review. March 1840 p. 267--- says the great DIVISION amongst metaphysicians-- the school of Locke,
194b095 might be made out of world before zoological DIVISIONS.-- Mem. species doubtful when known only by b
212b168 ike S. Africa) after Coal Period.-- ¿In those DIVISIONS of molluscs. where species now least in numbe
227b225 excessive inequality of numbers of species in DIVISIONS. look at Articulata!!!--! It leads to Nature
227b226 Reptile. (or between extremities of any great DIVISIONS) thus a knowledge of possible changes is disc
233b250 form in birds.-- ǀ Waterhouse thinks two main DIVISIONS of cats. Tortoise shell--& grey-banded. ¿spec
248c033 ws of Brain & manner of generation «& primary DIVISIONS of insects» 2. Relation, of external conditio
284c149 ct to variation.-- + <being> good for generic DIVISIONS [+] ought genus to be founded on such charact
304c206 ork to systematist, who believed that all his DIVISIONS merely marked his own ignorance.-- the collec
306c213 al character. (Owen) not for first & grandest DIVISIONS. but for ones of very high order. not for ver
339d026 says No genera.-- thinks there are some small DIVISIONS.# p. 7. «In» Some <of> cases the circular arr
349d051 which do not vary being foundation for chief DIVISIONS.# p 7. «In» Some <of> cases the circular arr
349d052 & so on no one will settle number of primary DIVISIONS.--Complains (p. 53) of M. Edwards, thinking
370d116 There is no scale, according to importance of DIVISIONS in arrangement, of the perfection of their se
461t025 on the Pipe-fish-- which he divides into two DIVISIONS, one of which are marsupial & the other have
467t100 inary Labiatae --the chief part with ordinary DIVISIONS, & a few with one lobe again divided «Have dr
471t068 nera of coniferae".-- referring to the 3 main DIVISIONS & speaking of their similarity «in structure»
154g057 n Erin {P} Shelf of Glen Guoy flat peat plain DIVORTIUM aquarum-- tidal channel-- 12ft obscure obscu
157g078 29.360? A 79 degree 75 degree? A little below DIVORTIUM on slope towards Loch Spey 29.297 A 79.¾ 29.3
157g078 slope towards Loch Spey 29.297 A 79.¾ 29.316 DIVORTIUM aquarum «about 12 ft higher than last station
157g078 n last station» 29.316 true terrace «2d» near DIVORTIUM aquarum is a lip with it-- Dick right-- Mac m
158g079 lso right-- Granite such as boulder on <thes> DIVORTIUM aquarum Peaty Mass of this point very nearly
158g079 n buttresses of granite obscure terrace 15 ft DIVORTIUM my measurements simply case of Loch Spey Forms
161g094 haps 6 ft too low» (to test last on Peat-Mass DIVORTIUM aquarum) Barom. 29.200 A.77 degree Air 70 deg
160g089 Tarf a perfect old Loch, making <several> two DIVORTIUMS aquarum, viz two branches of River Bought &

```
569n022 (In Theil. V) in describing Caroline Archipelago. DN 75 cf., p 268 without, however, very sincere gri
240c013 gonia no description of wild animals, nor in DOBRIZHOFFER Abipones.-- Voyage. de L'Astrolabe Zoologie
320c276 onia Azaras Voyages & Quadrupeds of Paraguay DOBRIZHOFFER. Abipom<e>nes. Edinburgh New. Phil Journal.
481z010 gfisher of Tierra del Fuego killed in Chile. DOBRIZHOFFER.. Vol I p. 310. History of the Abipones-- s
146g013 day On side of Hill South of upper end of Loch DOCHART buttresses of Alluvium or rather mass of well
147g016 alley  above the 300 ft Alluvium <abo> by Loch DOCHART-- Rivers could not have deposited it.  Barrier
148g027 te irregular very like rubbish at head of Loch DOCHART <Nea> Above Spean Bridge many flat terraces on
523m014 be  so from the curious story of the Birmingham DOCTOR praising his sister who confined him. & yet di
523m015 anything  scarlet.-- dogs ideotic.-- dotage.--» DOCTOR communicated to my grandfather his feeling of
524m019 their idea is wrong.-- (Dr Ashe, the Birmingham DOCTOR), in this precisely like the passion, ill-humo
530m045 the sensorium.-- -- analogous to sleep--; some DOCTORS care it, by stimulus & afterwards patient sink
158g081 ea gradually retired, hard to explain on river DOCTRINE <Little Hill with granite blocks almost encir
209b155 rom certain sports.-- Agrees with old Linnaean DOCTRINE & Lyells. to certain extent Von Buch,.-- Cana
221b201 s metamorphic; & therefore according to Lyells DOCTRINE removed?? Is the prevalence of Coniferous Woo
225b216 h respect to how species are. Lamaks "willing" DOCTRINE absurd. (as equally are arguments against it-
255c052 al forms.-- The hybridity of ferns bears on my DOCTRINE of cross-generation. The infertility of cross
263c074 MAN.-- is as difficult to understand as Lyells DOCTRINE of slow movements &c &c. this multiplication
311c226 is separate facts about abortive organs &c The DOCTRINE of monsters is preeminently worthy of study o
344d040 r this with reference to "new species & hybrid DOCTRINE"-- I have read there are exceptions to this i
381d157 en to water? Would not Ferns according to this DOCTRINE be considered as really cryptandrous, & they
384d161 ency of parts, as in Hermaphrodites, (shows my DOCTRINE of Hermaphrodite differs from Hunter)-- Hunte
431e123 s in proportion to male parent p. 8. his whole DOCTRINE of the advantage of crossing consists in the
432e125 t) go back, & this is argument against Blyth's DOCTRINE of young birds retrogressing-- Uncovering the
446e162 arely») here is a case in answer to Mr Knights DOCTRINE.-- Case like Corallina-- «Does it flower anyw
495q05a t many seeds= Necessary to answer Wiessenborns DOCTRINE of Equivocal Generation Charlworth p. 377. Ha
523m018 evis est.-- My father quite believe my grand F DOCTRINE is true, that the only cure for madness is fo
538m078 e of association.--«fully bears out my fathers DOCTRINE about people forgetting their insanity» there
571n029 in  «Australia» & America manage;-- This shows DOCTRINE «of instinct» has been carried too far In all
622o049 benefit.-- Yet I think there is much truth in DOCTRINE, for [RHC] 9) We can thus explain love of pla
635j56r afford support to other beings.-- true, (& the DOCTRINE of checks & my theory) Macculloch. Attrib. Vo
639j28v Bell's sneering-theory.-- p. 263. This kind of DOCTRINE runs through Macculloch, the bills of the Gra
266c091 their  forefathers;-- the first to escape the DOCTRINES of Muhammad, the last to extend, their domini
599o005 ing of a new School of Metaphysic,"-- give my DOCTRINES about origin of language-- & effect of reason
471tf08 conceive  any idea" Linn. Trans. 18. p. 163. "D. DOD on two new genera of coniferae".-- referring to
504q013 dendrum seeds??-- Bladder-nut ║. Laburnum ║. DODECATHEON ║. Castrate apple & pear to see if pollen n
233b251 ing.-- Wingless birds S. Continents-- Ostriches. DODO. Apteryx Penguin-- Logger-headed Duck-- Large p
246c025 p  460 & very doubtful whether any birds Except. DODO!!-- in Mauritius Lesson &c p. 620. Centropus (C
362d099 [not  located] September 13th The passion of the DOE to the victorious stag. who rubs the skin off ho
183b051 ve in same country. How is propagation of wolf & DOG. (because being believed same species) if they d
200b122 ribe animals as species, for instance Australian DOG: <yet when that> or Falkland rabbit.-- There is
204b140 ediate between parents.-- How easily does Wolf & DOG cross? Mr Yarrel thinks oldest variety impresses
204b140 mpresses the offspring most forcibly-- Esquimaux DOG & Pointer «Game-fowls have courage independently
209b155 th hump & Common;-- between Esquimaux & European DOG? Yet man has had no interest in perpetuating the
212b165 - Mr Martens of Zoolog. Soc told me an Australian DOG he had, used to burrow like fox.-- a sort of int
215b177 pecies Both males & females. lose desire. Native DOG not found in V. Diemen's land J. de Physique. To
217b183 ine blood hound bitch which would never take the DOG. But at last a rough-haired shepherd dog lined h
217b183 ake the dog. But at last a rough-haired shepherd DOG lined her & produced a very large litter.-- neve
225b217 yes, if you will show me every step between bull DOG & Greyhound). I should say the changes were effe
233b250 ing of cats sometimes heterogenous.-- Australian DOG jumped into tub leaving only nose above it-- pul
239c001 n known in any country; he states that Esquimaux DOG when crossed with pointer produces offspring muc
239c001 ubt that same thing would happen with Australian DOG & any of our common varieties. He has no doubt t
242c017 w Caledonia-- New Ireland & Britain same kind of DOG, with those of New S. Wales. <V.> p. 123 Crocodi
253c046 dens.-- Ogleby has facts to show that Australian DOG introduced by savages into Australia.-- What are
275c120 p hill-- he took thorough bred <greyhound>. bull-DOG. & crossed & recrossed, till there was a dash of
278c131 resent day??! Major Mitchell does not think that DOG was found in Van Diemen's Land.-- V. 1s. Number
278c131 mber of Geographical Journal to discover whether DOG found at Swan river. The change in England from
280c134 Sebright  excellent authority because written on DOG-Breaking.-- applies it to national character.--
298c189 uld says he believes that he has seen half fox & DOG. & that it was most like fox.-- He felt sure the
305c210 ait, size, fur.; manner in which ears droop like DOG. older character & manner of wagging tail.-- habi
305c210 son in order of <ver> Mammals.-- Mem Elephants & DOG.-- There is one living spirit, prevalent over th
306c212 pon being beaten behaved very differently from a DOG.-- more like man. continued long in a passion &
306c212 out  for him to come again very differently from DOG. perhaps being in passion chief difference Major
312c228 Mother.--  like dogs Smith knew chinese hairless DOG & common spaniel crossed.-- 3 puppies PERFECTLY
312c232 g eared little dogs, I am told, go to heat, take DOG. but do not become impregnated & puppies delicat
315c243 ssion «of all animals» becomes very curious.-- a DOG snarling in play.-- Hensleigh says the love of t
333d007 has  seen in a show half Wolf & «half Esquimaux» DOG which <likewise more resembled the wolf than dog
333d007 dog which <likewise more resembled the wolf than DOG.--> appeared to be intermediate between two Pare
333d007 Parents.-- this is very interesting as Esquimaux DOG approaches to species. Again he has seen several
333d007 in he has seen several crosses between Esquimaux DOG & common dogs & Fox thinks they decidedly take m
355d066 er to goats.-- or dogs to foxes. (yes Australian DOG) or donkeys to Zebras.-- «Mr Herberts variety of
360d090 rest is concerned. Same man had crossed Jackal & DOG-- (offspring did not go to heat. but parts swell
360d090 is  effected)-- the one in garden is from <bitch DOG do> father dog. & hence general appearance of fa
360d090 the  one in garden is from <bitch dog do> father DOG. & hence general appearance of face & tail somew
360d090 general  appearance of face & tail somewhat like DOG-- though it has full share of Jackall shape of b
366d108 opagated by art. Yarrell told me of a cat & of a DOG, born without front legs-- the former of whic
376d138 not  seen to evince more lewdness for bitch than DOG: Monkey thus examine each other sexes.-- <by tak
380d155 r taken from his own field to bull a cow.-- -- a DOG if led in string will not.-- some of the tigers.
388d172 opposed.  the Quagga case appears to that of «2» DOG begetting different puppies out of same mother.-
389d173 ore complicated animals.]CD p. 310 She wolf took DOG. but had such aversion to it, that she was held
393dIBC half bred Cattle at Cinbermere? How is Jackall & DOG at Z. Gardens D E Finished July 10th 1839.-- Sel
406e036 -- all cases: d degree p. 354-- The most vicious DOG. will not attack any animal except, dog when abs
406e036 vicious  dog. will not attack any animal except, DOG when absent from its master.-- dogs when strayed
409e046 pical genera come from New Holland, ¿Sydney? The DOG being so much more intellectual than fox, wolf &
409e047 onsider minds.-- as difference between mind of a DOG & a porpoise was not thougt overwhelming.-- yet
409e047 yet slow progress has done so.-- Show a savage a DOG, & ask him, how wolf was so changed. When discus
420e088 o. p. 419, «long» account of Hyaenodon, a fossil DOG-- leading towards Hyaena.-- see Comte Rendu.-- I
424e103 es to be fondled.-- and we see in the Australian DOG an instance of a half reclaimed animal.-- The do
424e103 rs» & horse in lesser degree,-- how different to DOG!-- (Hybrids of Calceolaria.)-- «:CD[Does the Pow
427e110 be different.-- same way one variety of <animal> DOG does not prefer other. but produces greater effe
428e113 ild species ever formed like short-tailed cat or DOG has been without recurrent tendency in external
439e145 d several breeds of dogs.-- like greyhound-- fox-DOG-- turnspit & two other kinds It seems absurd pro
441e149 This  would be curious law, Certainly Australian DOG is not affected by domestication, yet offspring
441e149 omestication, yet offspring are,-- if Australian DOG, could bud, analogy tells us, <be> «offspring» w
459tf02 on mule & hinnus-- in one case bastard of wolf & DOG had more form of male, & another of both progeni
478z004 to be one animal In Australia I was assured wild DOG copulates <.> freely with tame: comes to houses
493q003 tine in Persian Cat-- , in Brazilian «toothless» DOG-- I. St. Hilaire says length differs in differen
493q004 ly alike & see result.-- (3) Cross the Esquimaux DOG. with the hairless Brazilian or Persian animals
495q05a ond with Duck-weed-- coots-- waterhens-- examine DOG, which has swum-- on pools & rivers-- every kind
505q014 /42/ Mark has six day's puppy of small true Bull-DOG-- length from nose over head to root of tail 28½
525m023 naturally.-- (generosity in defending a friendly DOG).-- they feel shame, when doing anything which i
525m024 emember instances, but I feel sure I have seen a DOG doing what he ought not to do, & looking ashamed
536m071 ttom.-- it is relic of same thing that makes one DOG smell posterior at another.-- Why do bulls & hor
540m084 > would do more towards metaphysics than Locke A DOG whines, & so does man.-- dogs laughs for joy, so
540m084 s, & so does man.-- dogs laughs for joy, so does DOG bark. (not shout) when opening his mouth in romp
540m085 pud. of young male & smelt its fingers. Seeing a DOG & horse & man yawn, makes me feel how <much> all
541m092 lty, yet not clever people. Aug. 21st. 38 When a DOG in play has his mouth open ready to bark, & lip
542m095 wning habitual from awaking from sleep see how a DOG yawns when he awakes. & streching & yawning can
542m096 r. because no young animal has canine teeth.-- A DOG when he barks puts his lips in peculiar position
```

(Key Word)

543m097 nding language is considerable, thus carthorse & DOG.-- birds many cries. monkeys communicate much to
543m098 fferent orders more advanced than others just as DOG & Elephant most intellectual.-- Hymenoptera typi
545m107 ider monkey when touched, also another monkey to DOG. I showed nut & then closed my mem. expression o
546m109 ng monkeys-- smell with many animals-- see how a DOG likes smell of Partridge--, man's taste for smel
549m118 -- for man losing his children, any more than to DOG losing his puppies-- This looks like free will.
550m126 an draw-- says friend viewed him as Newfoundland DOG would Greyhound about dread of water-- innate Se
551m128 t Elephants understand contracts.-- but W. Fox's DOG that shut the door evidently did, for it did wit
553m132 with the former. But it is one thing for a swift DOG to overtake an emu, & [not located] notion, are
556m144 Shame & Blushing]CD «Does Elephant know shame-- DOG knows triumph.--» Sept. 23rd. Horses in Omnibus
556m146 y, horse & zebra. when going to kick.-- Why does DOG put down ears, when pleased.-- is it opposite mo
557m147 osite muscles would act, to when in a passion.-- DOG tail curled when angry & very stiff. back arched
557m148 latter.-- Yet I think one can remonstrate with a DOG, & make him ashamed of himself, in manner quite
557m149 uish,-- shyness not so.-- affected laughter.-- A DOG who goes home from shooting. runs away. is not a
558m152 o cats arch their back when fighting, & not with DOG. when fear might enter?-- I believe common Swan,
558m153 .-- [Hyaena pisses from fear so does man.-- & so DOG]CD Man grins & stamps with passion. can express
563n001 crowing.-- instances of expression.-- Octob. 3d. DOG obeying instinct of running hare is stopped by f
565n007 ld will do so.-- Contempt look obliquely so does DOG. when a little one attacks him Contempt, when th
568n017 nected with some disturbed habit» [Thus shepherd DOG. has pleasure in following its instinct & pain i
575n044 laws, as the crossing of jackall & Fox & wolf & DOG.-- the only test this is most important: can the
576n046 6 says seals knit their brows when incensed.-- A DOG may hesitate to jump in to save his masters life
576n047 of right (& Whether wholly instinctive as in the DOG, or chiefly habitual as in man), for it added mu
588n090 s reasoning apply chiefly to recollection. yet a DOG hunting for a bone shows he has recollection.--
588n090 ents & actions of a young animal with an old.-- (DOG horse, sow) we perceive great difference.-- «(&
589n092 en. in the absurdity of a tree having reason: or DOG, having high powers without hand or voice.-- the
590n094 f instrument.-- Bennett's Wanderings, Australian DOG does not Bark-- quotes Gardner's Music of nature
591n097 tame young wolf & it never dropped its ears like DOG-- wagged its tail «a little» when attending to a
593n105 eable & anything disagreeable being pursued.-- A DOG turning round & round is some old instinct «perv
594n112 ays, «her» tame rabbits were not frightened at a DOG.-- » The instinct against man is perhaps, as str
594n113 l than man his intellect Lyell has seen a little DOG go to the assistance & bite a big DOG. which was
594n115 n a little dog go to the assistance & bite a big DOG. which was fast struggling with another large do
594n115 og. which was fast struggling with another large DOG his companion. Descent --Affection & [...] Monke
595n117 <her> in his arms & carried to see.-- [blank] A DOG «whilst» dreaming, growling. & yelpings. «& twit
601o08a ons-- p 176 & 177 good passage in French on what DOG dreams, awakes-- does when Master takes Hat de l
609o29v ng unnecessary we call vicious.-- (jealousy in a DOG no one calls vice). on same principle that Malth
619o043 sion or appetite, what would be the result? In a DOG we see a struggle between its appetite, or love
619o043 ppies: the latter especially soon conquers, & the DOG [RHC] 3) probably thinks no more of it.-- Not so
620o046 spring his bird. one says for shame (& the «old» DOG really feels ashamed?) not so puppy, we <do> try
623o051 would contend against this-- but the pleasure a DOG.-- also age has much influence.)-- & only that w
624o051 ncts, for they are in accordance with it. thus a DOG may be trained to hunt one pig sooner than other
627o053 nct-- which were not originally, if the shepherd DOG has in obeying its instinct,-- as young pointer
628o53v] [The improvement of the instinct of a shepherd DOG, is strictly analogous to education of child,--
628o53v nct-- which were not originally, if the shepherd DOG'had no instinct to commence with scarcely possib
629o54v in it.-- Would not the maternal affections (in a DOG. & therefore not <instinct> «conscience») equall
637j57v eyhound to hare.-- waterdog hair to water-- bull DOG to bulls.-- primrose to <open fields> banks-- co
060r126 een. The red Sandstone of Andes fusible? no. mad DOGS. Azores. although kept in numbers. p. 124. Webs
063r133 ther method where the division is not perfect.-- DOGS. Cats. Horses. Cattle. Goat. Asses. have all ru
145g002 ifficult Salisbury Craigs The Highland shepherds DOGS coloured like Magellanic fox.. an instance of P
146g012 To fulfil an instinct a pleasure; mem. Shepherd DOGS The Patches of Conglomerate on S. Ventana, exce
148g025 each other «& half between parents» (& not like DOGS), but they thought the breed liable to vary-- I
148g026 an confirmed the account of the «YOUNG» Shepherd DOGS Saturday. Before coming to Bridge of Spean, hil
172b006 or opposites to like each other AEgyptian cats & DOGS ibis same as formerly but separate a pair & pla
204b140 the dam, from those imported by Ld. Powis Hybrid DOGS offspring seldom intermediate between parents.-
211b163 ut tails: some long & some short: therefore like DOGS.-- Ogleby says, Wolves at Hudson bay breed with
211b163 .-- Ogleby says, Wolves at Hudson bay breed with DOGS.-- the bitches never being killed by them, whil
211b163 ver being killed by them, whilst they eat up the DOGS.-- L.' Institut. Curious paper by M. Serres on
212b165 ' Institut. 1837: p. 404. account of instinct of DOGS.-- agreement & reason Some animals common to Ma
215b176 this defect, Very curious case = W. D. Fox. When DOGS are bred into each other, the females loose des
216b180 es in plants do so.-- As in cacti &c &c.-- as in DOGS investigate case of pidgeons. fowls. rabbits ca
216b181 which is more closely allied to case of cross of DOGS.-- See paper in Philosoph. Transaction on a qua
217b183 ly have occurred to every one) of rare breeds of DOGS, from owners great care of them. Fox says when
217b183 rom owners great care of them. Fox says when two DOGS of opposite breeds are crossed, sometimes offsp
217b184 ell has half bloodhound & greyhound.-- Where two DOGS. have lined bitch directly one after the other,
220b198 ts going back hybrid plants; analogous to Men. & DOGS. Now if we take structure as criterion of speci
220b198 as criterion of species Hogs different species, DOGS not, but if we take character of offspring. Hog
220b198 character of offspring. Hogs not different. some DOGS different.-- Henslow says. (Feb 1838) that few
224b212 ecise relation to structures Mem. Eyton's Hogs & DOGS.-- The passage in last page explains that betwe
225b219 over Bell on Quadrupeds for some facts.-- about DOGS &c &c NB. Animals very remote. ass & Horse. pro
225b219 pring exactly intermediate.-- Reference to Pig & DOGS. My theory will make me deny the creation of an
235b262 ary? Is not man thus circumstanced; varieties of DOGS in different countries a case in point.-- All c
236b278 Juan Fernandez-- Humming Birds» types of former DOGS. character of Miocene Mammalia--of Europe Mem.
240c003 form, may explain mule & pig being half way. Yet DOGS sometimes like father, sometimes like mother. T
244c021 Concepcion. some tatous!!! p. 120.-- Most of the DOGS of Payta-- belong to the hairless kind, «said»
246c025 nditions in any one isld different]CD.-- p. 414. DOGS of New Zealand of large size, resemble, chien-l
251c042 cies probably in Cuba (p 271 Viedo says American DOGS silent. Mem contrary assertion of Molina) (p. 2
265c083 changes after birth do not effect progeny-- many DOGS in England must have been lopped off & sheeps t
267c094 birds drifting out to sea Vol VII. p. 325-- Wild DOGS of Guayana always hunt in packs «30 or 40 toget
267c094 the Macúsie breed no description given-- Ch. 2. DOGS L'Institut. 1838. p. 67. Australian rodents Abs
275c120 hounds fleetest in England lost courage-- (Bull- DOGS are used because they have no scent!) Mr Wynne»
280c134 bbits good instance-- instinct of many kinds in DOGS, is clearly applicable to formation of instinct
282c141 changes had elapsed.-- let these families take <DOGS> «domestic animals» with them. they might be su
288c159 of breeding from original Durham breed.-- Native DOGS & English cross readily-- think about half way
288c159 ce.-- bark about half way «in tone»-- the native DOGS howl most dismally, very rarely bark.-- are alm
290c165 . have crossed.-- </I saw> .-- The attachment of DOGS to man. not altogether explained by F. Cuvier.
298c189 fox.-- He felt sure the half breed of Australian DOGS, would be «most» like Australian.-- Curious thi
299c196 ngst Pachydermata. like Man amongst Monkeys-- or DOGS in Carnivora.-- Man in his arrogance thinks him
306c213 ence Major Mitchell is not aware that Australian DOGS ever hunt in company.-- marked difference with
306c213 s ever hunt in company -- marked difference with DOGS of La Plata & Guayana] people will say. not spec
306c214 from there.-- All the sheep are thick-tailed The DOGS called Persian «greyhounds» are Kurdish & come
312c228 rst offspring most like <parents> Mother.-- like DOGS Smith knew chinese hairless dog & common spanie
312c232 rtest possible time. Mr Willis Long eared little DOGS, I am told, go to heat, take dog. but do not be
326c266 he Sportsman's Repository. 4to. contains much on DOGS.-- Reports of Brit. Assoc.-- some important Pap
333d005 limbs very long, eyes very large, very fierce to DOGS «otherwise habits not different; tone of voice.
333d007 nt form.-- Fox has seen several cases of foxes & DOGS crossed, offspring always more resembled foxes
333d007 ssed, offspring always more resembled foxes than DOGS (Mem Jackall in Zoolog Garden) He has seen in a
333d007 n several crosses between Esquimaux dog & common DOGS & Fox thinks they decidedly take much most afte
333d009 ther & are not exactly intermediate.-- Where two DOGS line the same bitch & «perfect» spaniels & sett
345d044 s or varieties are made.-- The Highland Shepherd DOGS, coloured like Magellanic Fox.-- peculiar hair
346d047 -- This man confirmed my account of the Shepherd DOGS.-- Aug. 24th. Was struck with pink shade on blu
354d063 aterpillar to Butterfly.-- When two Varieties of DOGS cross, Erasmus says it look lik[e] Institut. 18
355d066 artificial» approach in character to goats.-- or DOGS to foxes. (yes Australian dog) or donkeys to Ze
358d085 proof of the breeding in & in (like «courage in DOGS» EFFEMINATE men),-- if carried much further, if
360d089 crosses «twice made» between terrier & hairless DOGS of Africa,-- some puppies hairless. some in pat
371d127 case of Dingo (he alludes to the dholes or wild DOGS of India) in Zoolog. Garden having coloured off
376d138 tail of Hen; which lived with it.-- also of <a> DOG«S». but did not seen to evince more lewdness for
385d163 idered the most valuable. because smallest sized DOGS.-- one litter big & then second small & so.-- S
392d180 ffalo & common cattle-- Esquimaux (& Australian) DOGS with common dogs--> Ask my father to look out f
392d180 ttle-- Esquimaux (& Australian) dogs with common DOGS--> Ask my father to look out for instances of A
393dIBC ed by generation?? Is it «chiefly» in high. bred DOGS ie. (bred in & in) that one copulation with oth

Page **(Key Word)**
393dIBC e. (bred in & in) that one copulation with other DOGS renders subsequent progeny faulty. Does male fa
402e018 ns, monkeys & an animal probably a tapir p. 233. DOGS in Borneo-- <brought probably by Chinese>, "the
404e026 y be become favourable to offspring:| Australian DOGS have mottled coloured puppies case of this.-- t
406e036 . Ed. New-Phil. Transact. Rabies, common to men, DOGS, horses cows, pigs & sheep.-- diseases common t
406e037 nimal except, dog when absent from its master.-- DOGS when strayed hang their tails.-- November 1st..
414e063 n, otherwise [not located] Are the feet of water-DOGS at all more webbed than those of other dogs.--
414e063 ater-dogs at all more webbed than those of other DOGS.-- if nature had had the picking she would make
415e069 ly of the former proposition.-- as in «races of» DOGS, so in species, & in Man December 16th. The end
420e088 .-- CD[not between existing series of species of DOGS & Hyaena.-- but a common point, whence both may
424e103 og an instance of a half reclaimed animal.-- The DOGS, which have run wild have, have done so in hot
438e142 mstances, time of domestication [see Wikinson on DOGS of Egypt & Cuvier on Mummies]CD [NB TIME is ele
439e145 ians Vol III. p. 33-- They had several breeds of DOGS.-- like greyhound-- fox-dog-- turnspit & two ot
465t081 se crossing has had somewhat to do with it. mem. DOGS «& pigs» in Polynesia; & dogs in S. America «Re
465t081 o do with it. mem. dogs «& pigs» in Polynesia; & DOGS in S. America «Rengger.» -- now it is this very
466t095 ntering horses having colts which can canter-- & DOGS trained to pursuit having PUPPIES with the same
473s07r ference in calves--how is this in young pigeons--DOGS--cattle? As we see the frame of animals can ada
490q001 amongst the Indians Do the Savages select their DOGS Sowerby Entomologist Does individual Shell or i
493q004 eriod of gestation differ in different breeds of DOGS. Cattle, (Indian & Common) &c: length of life.
503q012 a?-- does not know About Yaks. & other Hybrids-- DOGS &c &c 38 Does only male yak cross with cow: is
508q016 Males in «cattle» or in Killing the worst =or in DOGS= (12) Do Hottentots generally resemble each oth
511q018 n writes to Lowe (7) In breeding. pointers. Bull-DOGS. Spaniels-- Grey-hounds-- is there ever any deg
513q020 ss Rumpless fowls & Dorking fowls,-- or tailless DOGS & fox, to see whether the characters are then i
515q022 ese. (2) Anatomy of muscles of stumps of tailess DOGS & cats.-- (3) Hounds-- varying-- (4) About blen
515q022 of Irish Horses, bred in this country. {Chinese DOG'S Head to send Cover common Pea (& Sweet Pea) for
523m015 being insane at the sight of anything scarlet.-- DOGS ideotic.-- dotage.--» Doctor communicated to my
525m023 ring in old ideas <I have elsewhere remarked do> DOGS take pleasure, when doing. what they consider t
537m075 ies in different races.-- no more wonderful than DOGS should have different instincts.-- Fact most op
540m084 hysics than Locke A dog whines, & so does man.-- DOGS laughs for joy, so does dog bark. (not shout) w
542m094 e tranquility. «whine of children. puppies do so DOGS nearly silent, so with men.-- How is crying-- p
542m097 ressions, sounds, & signal movements.-- some say DOGS understand expression of man's face.-- <That> H
543m101 s. when two varieties are crossed as in Shepherd DOGS-- Inherited Habits: Have Effect in Bones is val
544m102 ams except at this time. how does he account for DOGS & men speaking in their sleep.-- Characters of
550m126 rding food & in birds of place for nest.)-- with DOGS "have notion of masters property"-- is not this
552m132 Man Moral Sense Mitchell Australia Vol I, p 292 "DOGS learn sooner to take kangaroos than emu, althou
553m132 ooner to take kangaroos than emu, although young DOGS get sadly torn in conflicts with the former. Bu
558m152 <sudden> «forcible prolonged» expulsion of air «DOGS snarl much the same way» generic manifestation
563n001 y fleas, also by greater temptation as bitch: or DOGS defending companion. (mem Cyanocephalus. Sphynx
563n002 er & Nina)-- or to take away food &c &c-- Now if DOGS mind were so framed that he constantly compared
564n003 e must have conscience-- this is capital view.-- DOGS conscience would not have been same with mans b
565n007 time really wishing to smell its enemy.-- Man & DOGS show triumph (& pride) same way walk erect & st
574n041 ly due to our distant ancestors having been like DOGS to bitches.-- How comes such an association in
574n043 efine.-- Consider the acquirement of instinct by DOGS, would show habit.-- Take the case of Jenner's
575n044 some essence. The facts about crossing races of DOGS on their instincts, most important, because the
575n044 port to Bowen No one doubts that a cross of bull DOGS-- increase the courage & staunchness of greyhou
575n044 the courage & staunchness of greyhounds.-- bull-DOGS being preferred from not having any smell.-- 27
575n045 bon IV Vol of Ornith. Biog. case of Newfoundland DOGS. who will not enter water, till he sees. whethe
581n064 ng very fond of soft, silk-handkerchief-- cats & DOGS fond of slight tickling sensation.-- in savages
584n072 mory of roads long after once visited by horse & DOGS. (even blind horses & dogs) shows it is somewha
584n072 ce visited by horse & dogs. (even blind horses & DOGS) shows it is somewhat analogous to memory. Shru
586n078 y they go; & so return.-- «but does not apply to DOGS.--» they may do all this instinctively «yes bec
587n087 people, who like sweet things dislike others.-- DOGS dislike perfume) I should think, great principl
596nIBC into habits Get facts about instincts of mongrel DOGS Do blubbering children, if of. convulsive tende
615o038 ; certainly to the domesticated.-- [LHC] NB. Two DOGS having very different instinct always obtain pe
622o049 f man may have different instincts, as we see in DOGS & pidgeons.-- But as man is animal at head of s
628o53v ommence with scarcely possible to teach it-- all DOGS might be taught, but not cat, that is not act b
640j167 .-- or if other prey diminished, total number of DOGS. would diminish, whilst the long legged variety
176b023 avour of each <one> typical class to extend his DOMAIN into the other domains. & subdivision <six> th
176b023 ical class to extend his domain into the other DOMAINS. & subdivision <six> three more, double arrang
195b100 us. coronata, but not coronata.-- We know that DOMESTIC animals vary in countries, without any assign
199b118 827 F. Cuvier says. "But we could only produce DOMESTIC individuals & not races, without the occurenc
202b130 rd body., or common end of structure A Race of DOMESTIC animals made from influences in one country i
282c141 had elapsed.-- let these families take <dogs> «DOMESTIC animals» with them. they might be supposed to
306c214 an account in his Travels in Asia Minor of the DOMESTIC animals-- At Angora «centre of Asia Minor» ar
323c269 s Essays on Animal Reproduction -- Treatise on DOMESTIC pidgeons 30th Lives of Hayd & Mozart «Apri 25
355d066 l species «or ducks» (here argue if it be said DOMESTIC fowls are descended from several stock, then
362d100 mes. Mr Yarrell has old book 1765? Treatise on DOMESTIC Pidgeon, in which it appears that all the <bi
378d147 , species changing forms, <& loosing do> if so DOMESTIC animals ought to show them.-- Anyhow not conn
403e023 athedrâ. let us look at facts. considering few DOMESTIC animals few. that have not <which> not, cows
446e160 ly well. and <in> these reciprocally assist in DOMESTIC cares, as building nest, sitting on eggs. & f
465t081 g to immigration of other races, so it is with DOMESTIC breeds. (though in this case crossing has had
470t153 in do Humble 24 flowers of small Linaria in do DOMESTIC do 6 Campanula (two species)-- in do-- do 3 o
483z012 ry inaccurate & Vol IV p. 91.-- Vol IV p. 388. DOMESTIC mouse of Egypt is Mus Cahirimus. of Geof.-- r
337d022 sible.-- Mr Spence remarks that the Fringilla DOMESTICA of North Europe is replaced by the F. cisalpi
178b031 . Falkland rabbit may perhaps be instance of DOMESTICATED animals having effected; a change whi[ch] t
183b051 eed readily. point in view.--¿whether highly DOMESTICATED animals like races of man.-- M. Flourens. .
213b169 e state may be called, <species> species. in DOMESTICATED <species> races.-- If all men were dead the
231b244 had time to form good species, so cannot the DOMESTICATED animals with him.!-- Modern origin shown by
233b252 rica If Parasite different, whilst man & his DOMESTICATED quadrupeds are not so. greater facilities o
235b262 e than in other. Are there any cases, where «DOMESTICATED» animals separated. & long interbred <p> ha
245c023 èce distincte! p. 171. Sus papuensis «partly DOMESTICATED» like in general appearance to Siamese kind
248c030 would not. breed together.-- We see even in DOMESTICATED varieties a tendency to go back to oldest r
251c041 or village Gazal.-- ¿is latter same species DOMESTICATED, strangely contradictory to Azaras fact of
256c054 h Sea, will probably contain descriptions of DOMESTICATED animals in those regions Species so far are
285c153 spect to their effects The AEgyptian animals DOMESTICATED «??», & therefore Most especially under car
294c176 & thinks Lyell has overlooked argument, that DOMESTICATED animals change a little with external influ
299c195 ng> variety. wild carrot. made into biennial DOMESTICATED kind with large root by sowing it at wrong
307c216 -- Are there any abortive organs produced in DOMESTICATED animals, in plants I presume there are.? ge
308c219 earliest Egyptian drawings & Old Testament» DOMESTICATED animals having same idiosyncrasy, cause of
324c268 Geograp. range Study Buffon on Varieties of DOMESTICATED animals see if law's cannot be made out Fin
325c267 h profound care, abortive organs produced in DOMESTICATED plants; where function has ceased to be use
327cIBC -- Buffon Suites Cline on the improvement of DOMESTICATED animals Fries de plantarum proesentum crypt
331dIFC in common lion? Are the number of nipples in DOMESTICATED very fertile animals increased?-- Where off
333d007 take most after wild.-- i.e that <alw> «no» DOMESTICATED ones have been so long as wild one under pr
337d020 perfect would perish.-- The Varieties of the DOMESTICATED animals must be most complicated, because t
338d024 e we could not understand the vast number of DOMESTICATED races.-- Athenaeum. p. 505. some (very poor
362d100 he year round. ask Colonel Sykes.-- Even our DOMESTICATED cattle have tendency to breed at particular
364d104 e white feathers).-- It certainly appears in DOMESTICATED animals, that the amount of variation is so
366d108 stent in nature.-- According to my view, the DOMESTICATED animals would cease being fertile inter se.
415e067 atal diseases in mankind, «the more valuable DOMESTICATED animals» no doubt is owing to the rearing u
416e071 » It is a beautiful part of my theory, that «DOMESTICATED» races. of <a> organics. are made by percis
416e071 far more perfectly & infinitely slower.-- No DOMESTICATED animal is perfectly adapted to external con
426e107 might be obtained>: but the intervention of DOMESTICATED ie new varieties destroys the appearance of
429e113 history of first appearance of varieties of DOMESTICATED animals, yet as we know how many plants hav
431e122 rdener separates a plant he wishes to vary-- DOMESTICATED animals tend to vary. March 20th. Phillips
509q017 about Sierra Leone. cow. taking bulls. is it DOMESTICATED African Animal= Knows nothing [<...>] It is
615o038 nimals. even in wild state; certainly to the DOMESTICATED.-- [LHC] NB. Two dogs having very different
629o055 inctive. right & wrong.-- (animals excepting DOMESTICATED ones have no right & wrong except instincti
633j53v t they would not be detrimental accidents, & DOMESTICATED variations show us accidents may become her

Page
**************************************(Key Word)*************************************
```
640j167 y: NB. These views quite exclude the idea of DOMESTICATED animals changing.-- From these views we can
199b120 pring not fertile> <,but producing> «before DOMESTICATION, afterwards none or little with <fertile of
199b120 iage never probably excepting from «strict» DOMESTICATION offspring not fertile. or at least most rar
220b197 an instinct-- & this prevents breeding. now DOMESTICATION depends on perversion of instinct (in plant
220b197 epends on perversion of instinct (in plants DOMESTICATION on perversion of structures especially repr
264c079 perished.-- Let man visit Ourang-outang in DOMESTICATION, hear expressive whine, see its intelligenc
280c135 <t>---- Many animals not breeding at all in DOMESTICATION. throws great difficulty in way of ascertai
388d172 heredetary, disease. extinction. Animals in DOMESTICATION (even Elephant) not breeding-- remarkable A
417e075 ganization, in the two races,. owing to the DOMESTICATION of both.-- Now in the ass-- there is little
418e080 e-- probably will recieve illustration from DOMESTICATION of Monooecious plants, & abortion of others
438e142 f may depend on many circumstances, time of DOMESTICATION [see Wikinson on dogs of Egypt & Cuvier  on
438e142 lants & animals have long been subjected to DOMESTICATION.-- the constitution of some may resist  the
441e149 Certainly Australian Dog is not affected by DOMESTICATION, yet offspring are,-- if Australian Dog, co
449e168 Yarrell  knows of a Gull, which has laid in DOMESTICATION eggs of two shapes & colour.-- Eyton has ob
450e174 ots. June 26th-- Yarrell.:-- Black Swan «in DOMESTICATION & nature» strictly monogamous-- geese polyg
450e174 s highly polygamous-- change of instinct by DOMESTICATION.-- "Notices of the Indian Archipelago" Publ
460t017 kind of variation «& sporting» in flowers & DOMESTICATION of animals Aug. 26th.-- When it is said tha
503q012 Hope about Silk worms. Varieties effects of DOMESTICATION-- said to require Selection (36) Ask Mr Gow
515q021 ies, or in the individual by chance & under DOMESTICATION.-- N.B. Benthams remarks, where parts of fl
185b055 ed it. There certainly appears attempt in each DOMINANT structure to accomodate itself to as many sit
400e011 a=.-- Hope has ideas about generic characters. DOMINANT. predominant &c having relation to geographic
600o08v iousness of Conscience: in Maternal instinct DOMINEERING over love of Master and sport &c &c -- The B
089a018 p.  106 do-- p. 110. Mountains on west side of DOMINGO formed of coral limestone, with interstices ye
266c091 ctrines of Muhanmad, the last to extend, their DOMINION, armed alike with the Koran and the sword" Kq
190b079 .-- <Man is derived from Monad, each fresh--> Mr DON remarked to me, that he though species became ob
217b187 ood horse & Carthorse two folds [not located] Mr DON gave me instances of one species of Australian g
218b192 istan D'Acunha, a list of its Flora. is given Mr DON remarked to me. that some good African & some go
219b193 s of seeds. these plants) [Mem Fact stated by Mr DON in island, Teneriffe, St. Helena. J.  Fernandez.
424e100 tici,-- in Chara, in Marchantia & Hypnum «Prof:» DON would have known the Composites of Galapagos wer
499q010 nguishable from <other> clumps from other parts? DON says Irish, Scotch & English plants generally di
272c108 her Hemiptera stikingly resemble Coleoptera.-- DONACIA.-- some orthopterous insects & some third, hav
162g102 ifference between bedrock & Loch Ness <30.100> <DONALD Macphee> Saturday Morning 29.958 A 64  degree,
366d112 ld ass has no cross. how comes it that the tame DONKEY has. CD[old Buffon should be read on Mare My v
367d113 Sept.  17th. Saw mule. apparently fathered by a DONKEY. with all four legs ringed with brown.-- anima
556m146 of  expression showing real affinity in face of DONKEY, horse & zebra. when going to kick.-- Why does
355d066 s.-- or dogs to foxes. (yes Australian dog) or DONKEYS to Zebras.-- «Mr Herberts variety of horse, du
218b191 idity under title of Amaryllidae & Narcissus. Mr DONN considers Mr H. rather wild Mr Donn remarks  to
218b191 arcissus. Mr Donn considers Mr H. rather wild Mr DONN remarks to me. that give him a species from Ire
480z008 se bodies amongst Vegetables in Linn. Soc.-- Mr. DONN Carmichael Linn. Transacts Vol XII. p 496. Bird
615c36v five minutes every now and then howled.-- Now I DON'T think this only pleasure; for it was  different
297c189 ted] p. 428. Ouzel sometimes builds nest without DOOM. Vol 2. Mag of Z. & B. p. 431. Missel thrush la
202b134 erers. (as Peccari in N. America) then if it is DOOMED that only one species of family has  offspring
212b165 n run into Kennel to watch who would come to the DOOR-- would constantly do this, so was obliged to b
262c074 ical 3.subtypical) where power arbitrary. leaves DOOR open for Quinarians to deceive himself.--  Give
366d111 very  disagreeable hearing maed servant cleaning DOOR outside, as often as she touched handle, though
551m128 tand contracts.-- but W. Fox's dog that shut the DOOR evidently did, for it did with far more alacrit
551m129 ds may be seen by omnibus Horses starting, when DOOR shut or cad cries out "right." or  Drinkwater's
554m138 eper was away, take her chair & bang against the DOOR to force it open, when she could not succeed of
573n038 ance badly, told should have shilling to walk to DOOR without touching table.-- cannot avoid it.--  cu
498q008 Do bees frequent Cabbages «& Cowcumber's out of DOORS.» much-- or the minute Orthopt.-- important, as
479z006 sypus villosus is true Peludo Cavia Australis. DORBIGNY Vol II, p 24 Proceedings of Zoolog Soc. Impor
494q004 egg  kill Hen Bantam.-- Cross common Fowl with DORKING (10) Statistics of breeding in Zoolog. Gardens
513q020 not specific character= Cross Rumpless fowls & DORKING fowls,-- or tailless dogs & fox, to see whethe
165g125 ich hill is round & not merely thoughts laying DORMANT-- Man from Glen Turret said he learnt to  know
313c235 run  wild in India. in Heber?) is analogous to DORMANT instinct.-- (How wonderful a case bees develop
497q006 seeds are transported, or how long they remain DORMANT. if kinds come up, not found in wood.-- but se
521m007 Now if memory «of a tune & words» can thus lie DORMANT, during a whole life time, quite unconsciously
512q019 ection for milking-- loss of early habits in DORSETSHIRE sheep migration of coots-- variation in houn
521m008 «in Antiquary» power of repeating poetry in her DOTAGE is fact of same sort. Aunt. B. ditto.-- Case o
523m015 he sight of anything scarlet.-- dogs ideotic.-- DOTAGE.--» Doctor communicated to my grandfather his
532m055 ys he should think that in old people, in their DOTAGE, who sing the songs «& tales» of infancy, it i
492q002 lly fails first?-- Mal[e] 10. Henslow says semi-DOUBL flowers are those whose stamens are  monstrous,
147g017 he hills almost appeared as if they belonged to DOUBLE series Whole very obscure but it is certain th
149g031 e of A & B regular & even towards {P} Spean Roy DOUBLE terrace river «& to West of Spean» difficult t
176b023 other  domains. & subdivision <six> three more, DOUBLE arrangement.-- if each Main stem of the tree i
211b164 ction with Von Buch Volcanic chart & my idea of DOUBLE line of intersection.-- Dr. Horsfield At India
309c220 ntamination of <cl> castes. improve the women. (DOUBLE influence) & mankind must improve-- The  areas
382d158 nearly same function in both sexes.-- are never DOUBLE, only modified. those which perform very diffe
384d162 ecretes two substances, although organs for the DOUBLE purpose are not distinguished. --yet may be pr
391d174 y all mammalia must long have so existed.» with DOUBLE union-- At present I can only say the whole o
400e013 p. & four more generations before they began to DOUBLE.-- at present time Uncle J. does not suppose o
418e080 sm would not be perfect as in Ox. the amount of DOUBLE sexual developement is spread over [not locate
453e183 den.-- now this pood question-- single, or half DOUBLE.-- anyhow fertile because they «are» raised by
454e184 and. Are there instances of plants, in becoming DOUBLE loosing fertility if, sometimes one, sex & som
466t095 een given to foetus <fr> before sex developed-- DOUBLE flowers & colours breaking only hereditary cha
469t151 . Shrewsbury.-- Abberley-- Early Magazine-- &c. DOUBLE-blossomed «& dwarf-fan Bean» bean, were plante
492q002 f age of individuals-- 9. Do plants in becoming DOUBLE ever become monooecious-- loosing one sex & no
496q05a inute do they visit?? good=!! Examine pollen of DOUBLE flowers. compared with single & see whether gr
496q006 ery rich soil-- As they have little tendency to DOUBLE; what would be effect-- (10) Try in how many g
496q006 e Groundsil.-- gilly flower will break & become DOUBLE.-- There is a double Crows-foot. or Ranunculus
496q006 flower will break & become double.-- There is a DOUBLE Crows-foot. or Ranunculus.= (11) Try.. Nitrate
501q011 -- perhaps indexed by secondary characters-- in DOUBLE flower. do Henslow Speaking of Thyme doubts ab
507q15v t of seeds Gaertner de fruct:-- for woodcut-- 1 DOUBLE hook-- -- Geum. Galium Burrh = single hook; cu
510q017 y on alpine Flora of Java? Has Schow written on DOUBLE creations & where? How are current & winds  in
538m078 ies. Edinburgh. Phil. Transact. p. 365. Case of DOUBLE consciousness, one only «little» less  perfect
538m080 ary state of mind, is probably analogous to the DOUBLE individuality implied by habit, when one  acts
539m083 uite separate trains going on in the mind as in DOUBLE consciousness may really explain what habit is
539m083 ne idea. calls up other, & the consciousness of DOUBLE individual is not awakened.-- The habitual ind
546m110 In same book (p. 143) wonderful case of perfect DOUBLE consciousness Mayo compares it with Somnambuli
548m116 this case.-- But now in Mayo's «p. 140» case of DOUBLE consciousness, one would pity suffering in one
548m116 is.-- Insanity is <much> «somewhat» the same as DOUBLE consciousness, as shown in the tendency to for
548m117 to forget the insane idea; & ones expression of DOUBLE self, though as in Dr Ashe's case, one here wa
559m156 that  time after long period.-- My Father about DOUBLE consciousness.-- & somnambulism. Do people whe
575n043 erful than memory. affected by diseases. &c &c, DOUBLE consciousness? What other explanation-- can we
580n061 he tries is not this precisely the same, as the DOUBLE-conscious kept playing so well.-- Lr. Brougham
593n111 hing character between man & animals.-- [blank] DOUBLE consciousness. only extreme step of an ideal a
467k099 length  to corolla--anthers minute, distinctly DOUBLED, brown, but with no pollen.-- Common Thyme gro
032r042 possibly  general symetry of world.-- I feel no DOUBT. respecting the brecciated white stone of Chilo
038r058 ey refer to CENTRAL nucleus & that envelopes no DOUBT existed. These higher portions probably  formed
063r133 ttle. Goat. Asses. have all run wild & bred. no DOUBT with perfect success.--showing non Creation doe
088a013 great; where piece turned over axis or hinge no DOUBT fluid.-- analogy as continental elevations slow
108a083 g & high loosing answered by this-- No one can DOUBT. A-B once formed low coast.-- Annales des Mines
120a108 lava  shows the rocks really hot. & therefore I DOUBT the thermometer. Is not common salt more solubl
155g063 ouses of Roy» Maccullochs supernumerary shelf I DOUBT, much about «50 or 60 ft» «no doubt, a mound of
155g063 ary shelf I doubt, much about «50 or 60 ft» «no DOUBT, a mound of Alluvium nearly parallel--» Inclina
179b033 ance in crossing of species in wild state.-- No DOUBT «C. D.» wild men do not cross readily, distinct
179b033 d establish law. ≠as above stated:-- no one can DOUBT that lesser trifling differences are blended «b
189b070 same kind of coat.-- If we could tell, I do not DOUBT even colour hereditary in time as in space. (Me
```

192b083 at most clearly». Fox tells me, that beyond all DOUBT seeds of Ribston Pippin, produce Ribstone Pippi
196b104 points gained if any facts are connected.-- No DOUBT in birds, mundine genera, Bats .Foxes. Mus are
200b123 nge,, yet such causes are most obscure. without DOUBT:-- Vide cattle: The grand fact is to establish
206b148 e acted on, just like the two fine families «no DOUBT a different set of causes must act in the two c
222b206 tion to changing world:-- I cannot for a moment DOUBT, but what cetaceae & Phocae now replace Saurian
239c001 ell «Give it as his theory» tells me. he has no DOUBT that oldest variety, takes greatest effect on o
239c001 much nearer Esquimaux than Pointer.-- He has no DOUBT that same thing would happen with Australian do
239c001 an dog & any of our common varieties. He has no DOUBT that Chesnut, for many generations back, were c
251c041 2 Append to Denham Clapperton &c on Mammalia no DOUBT will all be included in Smiths work «do» Vol. I
252c045 ase on account of varieties in N America» «some DOUBT, from want of knowledge of time-- Analogy from
264c081 , according to Gould, good genus Gould seems to DOUBT how far structure & habits go together. This mu
281c137 how finely the series is graeduated.-- Dr Beck DOUBT if local varieties should be remembered, theref
282c143 normous production-- millions in few days-- one DOUBT that one animal can really produce so great an
289c161 instinctive impulse to keep separate, which no DOUBT be overcome, but until it is the animals are di
289c161 ngs & orbits of penguin & then he will cease to DOUBT :Scales into Teeth in Bering Pike (Waterhouse)
290c164 f primaries black, by examining series I cannot DOUBT Laws of change, Will be known.-- It appeared to
304c206 ch would render our knowledge a chaos: who will DOUBT this if series now existed from Man to Monads--
314c236 relation to time) as in buds.-- I can scarcely DOUBT final cause is the adaptation of species to cir
315c243 man grinning is to exposes his canine teeth. no DOUBT a habit gained by formerly being a baboon with
315c243 s of good news. discovery of prey.-- arising no DOUBT from want of assistance.-- crying is a puzzler-
365d105 Neilson has given figure of it.-- In England no DOUBT the cross between Pheasant & Black game is owin
366d112 eing born at same time, & make breed, one would DOUBT any law.-- Yet seeing the feathers along one to
372d129 D it is of no consequence if it does= <I do not DOUBT, the> Do plants loose any qualities by being bu
374d134 erefore somewhat unfavourable-- 28th. «I do not DOUBT, every one till he thinks deeply has assumed th
390d178 -- therefore passions fail.-- In fruit trees no DOUBT there is tendency to propagate the whole differ
401e014 plant with another-- Uncle John says he has no DOUBT bees fertilize enormous number of plants-- it i
406e035 L'Institut 1838. p. 338 A most grave source of DOUBT. in distinguishing <after> which parent impress
415e067 nd, «the more valuable domesticated animals» no DOUBT is owing to the rearing up of every heredetary
417e078 ikenesses of fathers to children of mankind, no DOUBT are applicable to likenesses, when species & ra
423e096 pement in partial classes is far from true.-- I DOUBT not if the simplest animals could be destroyed,
428e113 mrose to Cowslip is great difficulty. «I should DOUBT if wild species ever formed like short-tailed c
446e163 he Lemma, «and the vivaparous grasses, which no DOUBT are propagated during hundreds of years, withou
447e165 9th.-- -- Henslow says he has not the slightest DOUBT that Festuca vivapara is the same species with
459tf02 le gives form. admitted by Linnaeus.-- seems to DOUBT its applicability to common mule & hinnus-- in
471tf11 s from end to end-- so that he is almost led to DOUBT. whether there is such a thing as a species-- J
472s02r rom yellow to yellow & purple heartease without DOUBT.-- Bee, not large, very dusky & broad never saw
492q003 ort. Transacts & M. Sageret is referred to with DOUBT by Herbert Do forest-trees sport much in nurser
513q21. ever regularly take place in unopened flower-- <DOUBT> disbelieve this in Bauers case of orchidiae Wh
529m040 come into play.-- the train of thoughts vary no DOUBT in different people., an agriculturist, in whos
535m069 ns, such as plastic virtue, «&c» (Very true, no DOUBT savage attribute thunder & lightening to Gods a
536m072 ect to free will, seeing a puppy playing cannot DOUBT that they have free will, if so all animals., t
537m077 has these strong, & does not act up to them, no DOUBT disobeys & hurts conscience more than other.--
542m096 Baby (like Hensleigh's) smile & frown, who can DOUBT these are instinctive-- child does not sneer. b
545m106 limbs that they cannot remain still.-- I do not DOUBT this Baboon. knew women.-- Another little old A
547m113 mpare the castle with them, therefore could not DOUBT or believe.-- When I say trains, it may be inst
549m121 t future life is almost the sole object-- -- I DOUBT whether the last be right. The two rules come v
549m123 h lesser intellect they might be necessary & no DOUBT were preservative, & are now, like all other st
550m127 land remarked that insanity like sleep does not DOUBT the reality of the impression on its senses.--
558m151 ind not saltus between man & Brutes) no one can DOUBT this connexion.-- look at faces of people in di
574n041 scribed as biting: so do stallions always.,= No DOUBT man has great tendency. to exert all senses, wh
593n107 is pleased by other animals smell & looks.-- no DOUBT it may be attempted to be said that young anima
599o05v ave, then language was progressive.-- We cannot DOUBT that language is an altering element, we see wo
609o029 deeper & worser feelings. These bad feelings no DOUBT orginally necessary revenge was justice.-- No c
610o033 ere may be in men-- which the reviewer seems to DOUBT. [RHC] 1) Effects of Life in the abstract is ma
615o36v -- analyse feelings. Mr Wynne says, that beyond DOUBT courage is heredetary in fowls & not effect of
194b094 like case of great <rodent> edentate [has been DOUBTED?]CD & opossum found in Europe now confined to
257c057 those very changes which at first it might be DOUBTED were possible,-- it has been asked how did the
273c111 llow, both in structure & habits (it cannot be DOUBTED that if swallow perished) hawks & Milvulus &c
522m013 as if «these emotions» acquired.-- this may be DOUBTED, whether rather not going against natural inst
525m026 ons effect of organization which can hardly be DOUBTED, when seeing Nina with her puppy.-- The common
528m037 e the pleasure of perspective. which cannot be DOUBTED if we look at buildings, even ugly ones.-- the
537m074 he effect of his example on others. It may be DOUBTED whether a man intentionally can wag his finger
555m142 it not present with all associated animals?) I DOUBTED it in Fuegians, till I remembered Bynoes story
556m144 other, without ever comparing them, I neither DOUBTED them or believed them.-- Believing consists in
587n087 adapted to like sugar, acid, &c, which may be DOUBTED for possibly even taste of senna. might be acq
086a004 of mountains we ought to sympathize with. old DOUBTERS of what are fossil shells.-- accustomed to su
041r064 ng packed. in beds) lived there, makes it very DOUBTFUL whether they could have lived in so deep a se
050r090 th small & large fragments, nature of which is DOUBTFUL. P. 180. I think my Ascension case very doubt
050r090 ubtful. P. 180. I think my Ascension case very DOUBTFUL. -- In Iceland Bladders of Lava are described
117a102 close to Guayaquil.-- modern shells of Cobija DOUBTFUL. Examine well shores of lakes. to see effects
148g022 h as highest measurement but nature I am quite DOUBTFUL of as I am of all the Alluvium. At Mouth of C
172b006 a pair & place them on fresh isld. it is very DOUBTFUL whether they would remain constant; is it not
194b095 ld before zoological divisions.-- Mem. species DOUBTFUL when known only by bones.-- Mem: Silurian fos
246c025 Coquille Voyage Says no reptiles. p 460 & very DOUBTFUL whether any birds Except. Dodo!!-- in Mauriti
285c151 ity.--but whether ever arrive at true affinity DOUBTFUL A species is only fixed thing with reference
300c197 ss. How is instinctive dread it is exceedingly DOUBTFUL whether animals have any fear of death, only
334d012 d squirrels, which form a marked wild variety. DOUBTFUL whether all are white. Fox says the Half Musc
339d025 ot plants» will it hold good.-- Thinks Temmink DOUBTFUL when he says No genera.-- thinks there are so
348d051 we can ever discover the real relationship is DOUBTFUL.-- not till much knowledge is elicited.-- It
371d127 een change is produced in parents--colour is a DOUBTFUL subject, but what other instances are there o
399e009 s-- & longer hind legs??? so that I was almost DOUBTFUL which it was.-- do hind legs increase in any
441e148 y acquired peculiarities are transmitted it is DOUBTFUL whether any are transmitted, for the changes
500q10a ce with range of species?-- Are there any fine DOUBTFUL species from Van Diemen's Land? or New Zealan
500q10a g isld, there ought to be a good many races or DOUBTFUL species; how is this at Canarys Arch-- it is
500q10a narys Arch-- it is so at Galapagos.-- Ireland, DOUBTFUL species-- Does any genus of Plants. vary & ha
532m055 ing the songs «& tales» of infancy, it is very DOUBTFUL whether they could recollect these same thing
535m064 forgotten in all older species. The earwig & a DOUBTFUL one of Acanthosoma grisea described [not loca
547m114 serve no improbabilities in a dream, effect of DOUBTING nor believing, effect of not reasoning. effec
609o029 perhaps an exception. (does it originate in a DOUBTING feel between conscience & impulse) but shame
151g040 ow hill between Inn & Bouhunthine the summit «DOUBTLESS worn into coincidence» has beach or band of p
189b074 aculties} most developed, as highest.-- A bee DOUBTLESS would when the instincts were.-- relation of
357d074 «?» to imply knowledge of whole world-- if so DOUBTLESS «part of» system of great harmony. The peculi
358d085 by the process this were possible, the organs DOUBTLESS would shrivel up.-- Yet odd they should have
383d160 & part of shaft metallic green.-- This green DOUBTLESS, is effect of Metallic hue of silver pheasant.
411e055 r When summing up argument against my theory, DOUBTLESS, the presence of animals in <own> «the presen
466t095 ng PUPPIES with the same powers instinctive & DOUBTLESS not confined to sex.-- «Is not cantering a co
567n014 ssary notion of space-- plant though it moves DOUBTLESS has not.-- Turkey cock in passion & sends blo
570n027 ld think negress beautiful,-- [male glow worm DOUBTLESS admires female. showing. no connection with m
623o050 C] the instinct of sociability & sociability, DOUBTLESS grow together [RHC] This feeling seems to var
626o052 , but supplios it-- instinctive feelings will DOUBTLESS lead to similar actions which in prior <races
077r171 an--of Guanaxuato to SW. with respect to latter DOUBTS whether bed or vein (very like that of Spital
242c018 identical with those from S. Africa. «M. Bibron DOUBTS fact.--» My toad is same species Coquille Voya
266c092 mmy: after 2000 years, germinating.--!! Henslow DOUBTS? GEOGRAPHICAL JOURNAL. Vol V. p 201 Wellsted.
370d116 ew, because they are kept in security.-- Hunter DOUBTS about production of Queens.-- Neuters are bred
383d159 Pheasant, one like cock & other like Hen.-- one DOUBTS whether they are not Hermaphrodites, like J. H
431e121 ve been found together on coast of France.-- L. DOUBTS-- Lonsdale thinks Ammonites would afford inst
444e157 itional bones to give strength to it.-- p. 139. DOUBTS altogether the law of balancing of organs.-- I
501q011 in double flower. do Henslow Speaking of Thyme DOUBTS about stigma in similar manner ever failing.--

```
526m027 ation & mind.-- thinking over these things, one DOUBTS existence of free will every action determined
532m053 omes it cannot help doing it.-- Fanny Hensleigh DOUBTS whether young babies start.-- ¶ If children wi
537m077 ill not prevent other being engrafted.-- No one DOUBTS patriotism & family pride are heredetary., & t
540m085 animals <are> built on one structure.-- He who DOUBTS about national character let him compare the A
553m135 he mind being EQUAL to the smallest casuistical DOUBTS.-- The history of Metaphysicks shows that such
575n044 ty of instinct.-- Habits import to Bowen No one DOUBTS that a cross of bull dogs-- increase the coura
592n101 is an instinct.» I suspect the endless round of DOUBTS & scepticisms might be solved by considering t
623o050 ividual but to the whole past race).-- <no one> DOUBTS» « <I cannot> » [LHC] On the Law of Utility No
574n039 t had no influence over her.-- Hensleigh says. DOUGLAS. «& Spencer», an old Scotch Poet, has numerous
356d072 sed in wild state-- & the English & Some African DOVE.-- The extinction of the S. American quadrupeds
540m088 y reading a book.-- Consider this.-- "The fledge-DOVE knows the prowlers of the air" &c &c &c so is c
449e169 ose. Eyton says some of the pidgeons in common DOVECOT are very like a Himalaya species -- leuconotes
592n101 s origin in Human mind.-- Andrew Smith says hen DOVES & the female chamaeleon court the males by odd
046r080 to water keeping its level whilst land rose up & DOWN.--But from above reasons, do not think so  also
055r108 n. -- The Bulk of sediment «daily» yearly brought DOWN by every torrent proves the decay atmospheric o
081r181 travels Brazil. Peru. Sandwich Isd Mawes travels DOWN the Brazil.-- Did Melaspena publish his travels
087a008 elevated  When Siberia went up. Arctic land went DOWN.-- Probably more Arctic land would be  required
102a059 n in open ocean. as pebbles would be lifted up & DOWN. on coast itself, undertow would draw it outwar
113a091 time has not been allowed for lower beds to cool DOWN. & then in 50000 years the depth will be greate
120a109 out action of rivers.-- Excess of matter brought DOWN Mention absolute elevation of Patagonian blocks
125a121 knows  how far that may have penetrated,-- lower DOWN the temperature may be kept up far higher  from
149g032 OULDER of granite above 4th Shelf a little lower DOWN the hillock with beach & channel precisely as w
159g086 degree  Air 70 degree? This station a little way DOWN slope of obscure terraces (& conical hills on s
164g119 -- [blank] {P} muddy nodular strata coral upside DOWN strata coarse agglomerate [...] shells from [..
177b026 icated.-- {P} Is it thus fish can be traced right DOWN to simple organization.-- birds-- not. {P} We m
191b080 - The motion of the earth must be excessive up & DOWN.-- Elephants in Ceylon-- East Indian archipela
219b195 been separated by short space from mountains low DOWN, therefore plants common take an example from T
240c004 of  great monstrosities being produced, & handed DOWN, with ease, is analogous to what occurs in plan
247c029 -- Black & Grey varieties of rabbits thus handed DOWN for nearly 70 year. Galapagos Mouse not the sam
263c075 of prejudices) <without> who just takes up & lay DOWN the subject without long meditation-- His  best
306c212 that  is much more general argument) & therefore DOWN the stream followed ebb tide, therefore got int
306c212 owed ebb tide, therefore got into habit of going DOWN stream which would last were the stream 1000 mi
332d004 on Mascott,  where he had been accustomed to turn DOWN.-- -- applicable to birds migrations & Australi
348d050 he oldest) ¶"The most important characters break DOWN in certain species & become worthless-- Mammali
355d066 pponents will «are» not «able to» tie themselves DOWN, they can find loopholes) "It is well worthy of
371d127 ch changes, not acquired by parent, being handed DOWN? Are not Loddiges 1279 roses kept in same soil.
379d152 t is well known that in breeding very pure South DOWN that the ewe must never be put to any other bre
384d161 ssive buds do differ-- any variety is not handed DOWN. but is handed down for some generations Theory
384d161 -- any variety is not handed down. but is handed DOWN for some generations Theory of sexes (woman mak
460t015 between  swan-goose & common goose.-- the stripe DOWN back pretty plain in in these <half> «3/4» bred
461t037 y theory, as long as any structure can be handed DOWN without being absolutely injurious «(or requiri
461t037 ition)» to a certain amount it will be so handed DOWN«(»..as mammae of men «callosities on Camels  &
467t103 Beans  it is wonderful (a) how the Humbles force DOWN the wings most violently: in Beans the wings se
467t104 gangway=  In Lotus corniculatus saw Humble press DOWN wings which ejects pollen from tip of sheath.--
467t104 tigma project= In common Pea saw Humble so press DOWN sheath. that stigma covered with pollen was pre
470t177 n hedge Linaria= (Plenty of Humble Bees on Phlox DOWN, 1854, Sept.) In Spanish Broom by pulling  back
473s07v impressed & others later-- All poultry with same DOWN-feathers. Zoology 1856 Skimmed through & abstra
506q15v lower 45. Charlsworth. vol II. p. 670-- oats cut DOWN turning into Rye.-- 46). Book describing amount
533m060 Froude's life. that author remarks, that writing DOWN his confessions of sins. did not make him  more
541m089 this  instinctive fear arisen? 19th. When I went DOWN to Woollich I was trying to unbend my mind as m
545m107 g outang, under same circumstances, threw itself DOWN on its back & kicked & cryed like naughty child
554m141 "reasoning," & take instance of Dray Horse going DOWN hill.-- (argue sophism of association. Kenyon,
556m146 anying emotion.-- when horses fighting, they put DOWN ears, when «turning round to kick» kicking they
556m146 to grow on horses, they must yet continue to put DOWN ears, when kicking.-- -- good case of expressio
556m146 & zebra. when going to kick.-- Why does dog put DOWN ears, when pleased.-- is it opposite movement t
565n007 n.-- Hence becomes instinctive to fear., as ears DOWN to horse.-- Horse snuffs «& snorts», the air «&
576n048 anization to learn Greek, that to have it handed DOWN as an instinct.-- Instinct is a modification of
582n067 as in the young salmon to go towards the sea. or DOWN the stream; which it does unconsciously of  any
585n075 legs  & knocks with back of Head, yet never puts DOWN its ear. good to contrast with horses, asses, <
588n089 being  carried up or downstairs, or dangled up & DOWN-- in latter case they struggle their arms.-- do
593n105 & round is some old instinct «perverted» handed DOWN & down.-- mem. Nina used to get into hay & make
593n105 d is some old instinct «perverted» handed down & DOWN.-- mem. Nina used to get into hay & make a nest
633j54r th, than the earth revolves to form rain to wash DOWN earth, from the mountains upheaved by  volcanic
564n004 o give his child.-- Octob 3d. Was told by W of DOWNING. Coll. that he had seen chicken only hatched f
615o037 son had told him this distinctive mark, it is DOWNRIGHT instinct, leading to touch a particular organ
191b080 m the consideration of these archipelagos ups & DOWNS in full conformity with Europaean  formations--
588n089 erceive the difference on being carried up or DOWNSTAIRS, or dangled up & down-- in latter case  they
032r041 n Cordillera.-- From poles to Equator current DOWNWARDS & to West.--From Equator to poles. nearer the
105a071 ocks. would be prove of subsidence.-- removal DOWNWARDS by successive torrent spread out. by sea-- be
554m138 Callitrix  Sebe??) he has seen place its head DOWNWARDS to look up womens petticoats-- just like Jenn
634j54v ues about his Creator! p. 309. says the ribs in DRACO support the flying membrane?!!-- that the phala
099a051 kes in Cordillera.!!!-- In stratum OP. let force DRAG particls to line {P} AB, & likewise gravity MN.
639j28v ce will get the upper hand. though continually DRAGGED back to old type by intermarrying with ordinar
496q006 ns cast Holly-seed & they grew» (9) Place. Snap-DRAGON. (I have seen one monstrous) Fox Glove &  such
150g038 ighbourhood appear very round-topped with much DRAINAGE & far more earthy than what is usual-- Lines
155g065 lly preserved as to have thrown water in same «DRAINAGE» lines Mound of Gneiss though wonderful--<th
150g035 idge Roy (& other cases) but then if gradually DRAINED, where is barrier {P} great waterworn frame te
156g068 valley  fragments which had fallen before lake DRAINED could be told from «some of» those since falle
150g035 at bight just as Dick shows NB. Lake gradually DRAINING off would form plains such as those near Brid
459t013 he offspring of a black & white «duck of pecu» «DRAKE» with the penguin duck. it took after the Pengu
459t013 & the bright feathers on its wing resembled the DRAKE.-- another of same half breed resembled the plu
459t013 her of same half breed resembled the plumage of DRAKE still more.-- So Penguin impresses its form bot
070r155 s undulation.'. crust thin.--Concepcion earthquake DRAW close Analogy Lake of Cordill: of Copiapó & Des
102a059 ifted up & down. on coast itself, undertow would DRAW it outwards.-- form of breaker affected some wa
196b105 al.-- Waders & Waterfowl.-- scrutinize genera, & DRAW up tables-- Instinct may confine certain birds
223b207 ed to the first thinking being. although hard to DRAW line.-- -- not so great as between perfect inse
441e151 th another.-- to escape it «in any case» we must DRAW such a monstrous conclusion, that every organ i
472s02r n upper petals & insert proboscis, under sigma & DRAW it out over & over again & wipe off pollen. (as
550m126 awing compared to a friend, whose who family can DRAW-- says friend viewed him as Newfoundland dog wo
135a146 ndel. just same as. at Bahia Blanca-- letter in DRAWER with important letters-- When I come to  treat
362d100 existing.-- he has also some very fine recent DRAWING «of prize pidgeons» in 1834-- now this would b
363d101 p of white speckles on elbow joint-- in Bewick DRAWING the the rock Pidgeon has not: now how many wil
507q15v adaptation might be arrived at.= Any book with DRAWING of Seed. Anemone with, tuft-- Bull Rush-- Dand
550m126 Vol  I, p. 127. Talks of difficulty of his own DRAWING compared to a friend, whose who family can dra
556m146 rs, when pleased.-- is it opposite movement to DRAWING them close on head, when going to fight, in wh
558m153 . Sphynx.-- In the wild ass there is a curious DRAWING out of the side part of nostril, when  passion
308c219 egroes existed since time of earliest Egyptian DRAWINGS & Old Testament» Domesticated animals  having
314c237 n not alter much.?? Mr Brown showed me Bauer's DRAWINGS of a curious plant where a tube consisting of
481z009 p.  244. Baccalao. migratory fish.-- See Kings DRAWINGS.-- for real name Birds of Iceland. Mackenzie.
556m145 uted anymore in men. than in animals.-- In the DRAWINGS of Voltaire why is under lip curled over uppe
564n006 ation to argue from.-- Octob. 4th. Seeing some DRAWINGS «in Lavater, P. cii Vol III» of excessively c
025r018 ory of Europe, with America; I might add I have DRAWN all my illustrations from America, purposely to
063r131 ument for the Crust of globe being thin, may be DRAWN. from. Cordillera. rocks.--When beneath water.-
149g033 ld-- {P} do they extend round hill too low line DRAWN plain red talus line on N. side of Spean most c
315c241 stre I suspect some valuable analogies might be DRAWN between habitual actions of plants «when exciti
341d029 e & that he did not see where the line could be DRAWN-- thus the most remarkable character in Apteryx
404e024 imals only. from which plain inference might be DRAWN that whole infertility «of hybrids receive no e
463t057 animals were taken from which these series were DRAWN they would not be intermediate, but this is not
```

(Key Word)
543m101 r parts of structure. C. D.27 Can an analogy be DRAWN between «heredetary» associated pleasures & pai
558m152 opard. quarrelling. mouth wide open, each [lip] DRAWN back & driving air out of mouth «hairs erect on
577n052 in the sudden cures of tooth ache before being DRAWN,-- My father «even» believes that the general t
098a047 uence aggregated (I assume the same force which DRAWS together two particles of Carb. of Lime, tends
554m141 on of "reason" & "reasoning," & take instance of DRAY Horse going down hill.-- (argue sophism of asso
552m131 llect.-- mem: Yarrell's story of wheel horse in DRAYS, scraping against cornice stone to cause fricti
524m021 <...> of B. <talked of.> thought herself near DRAYTON & Ternhill, (where she was born) though she ne
300c197 it is but being motionless. How is instinctive DREAD it is exceedingly doubtful whether animals have
550m126 d him as Newfoundland dog would Greyhound about DREAD of water-- innate Septemb 1-- If one performs s
564m004 tion of following conscience, obeying habits, & DREAD of misery of future thinking of injured moral s
335d014 as first cross being new species,|-- Are not DREADFUL monsters, abortive, just like mules. Fox's ha
429e114 March 12th-- It is difficult to believe in the DREADFUL «but quiet» war of organic beings. going on t
576n047 the happiness of his life, & the chance, of so DREADFUL a consequence to each man is small. Man's int
542m093 - He may feel satisfied with himself, & though DREADING to say so, his step will grow erect & stiff l
523m015 r people often think no more about it than of a DREAM.-- Insanity is produced by moral causes (ideotc
524m021 k to childhood-- People, my Father says, do not DREAM of what they think of most. intently.-- crimina
544m103 ity-- castle in the air, is more prolonged than DREAM. never fatiguing,-- else it is only our conscio
547m111 lear & pretty vivid «& perfectly characterized» DREAM, in continuation of waking thought-- my servant
547m111 ers.-- Now what was difference between Castle & DREAM No answer shows our profound ignorance in so si
547m112 ion & yet the Castle would not have turned into DREAM.-- It appears to me, that the mind is wholly ab
547m114 ousness.-- Mayo observe no improbabilities in a DREAM, effect of doubting nor believing, effect of no
548m115 one cannot bring it to one self.-- nor of a bad DREAM, when that is not recollected, nor of the Botan
552m130) is the image not vivid as in sleep-- (one can DREAM of intense scarlet??) is it because one then ha
552m130 ies the image more vivid? Surely the image in a DREAM cannot truly be <more> «as» vivid, «a reality»
555m143 merican group sneer.-- Sept 21st Was witty in a DREAM in a confused manner. thought that a person was
572m034 ore than connected with general tendency of the DREAM.-- It does not hurt the conscience of a Boy to
595n117 agination-- for it will not be allowed they can DREAM, & not have day-dreams-- think well over this;-
544m103 ciousness, & senses tell us it is not real.= = DREAMING appears clearly rest of the mind, with all ot
595n117 rms & carried to see.-- [blank] A Dog «whilst» DREAMING, growling. & yelpings. «& twitching paws» whi
524m021 - the words second childhood full of meaning:-- DREAMS do not go back to childhood-- People, my Fathe
528m035 oach to believing a vivid castle in the air, or DREAMS real again explains insanity.-- Analysis of pl
541m090 r. is right in saying delirium rest-- therefore DREAMS thus act.-- ∴ weak minded people are fickle &
544m102 r of death. Mayo Philosophy of Living. p. 140-- DREAMS good account of «thinks» are recollected when
544m102 up with waking thought.-- Ld Brougham thinks no DREAMS except at this time. how does he account for d
544m102 & men speaking in their sleep.-- Characters of DREAMS no surprise, at the violation of all <rules> r
548m115 ne idea present is, perhaps, hard work-- though DREAMS do that One Reflective Consciousness is curiou
595n117 l not be allowed they can dream, & not have day-DREAMS, think well over this;-- it shows similarity
601o08a p 176 & 177 good passage in French on what dog DREAMS, «awakes-- does when Master takes Hat do l'educ
604o017 y is coming on «Thinks clearest analogy between DREAMS & insanity.» D. Stewart on the Sublime The lit
544m103 roken-- Sir J. Franklin when starved, all party DREAMT of «goo» feasts of good food-- The mind wills
572m033 nctive actions. senses. notions &c Octob 30th-- DREAMT somebody gave me a book in French I read the f
263c077 s end «under present form» will come, (or how DREDFULLY we are deceived) then he is no exception.-- h
365d105 e. &, the subalpina of Sweden, (which in summer DRESS somewhat resembles Red Grouse) it may be so-- b
385d163 oung man at Willis «Grt. Marlborough Str, Hair DRESSER, assures me he has known many cases of bitch g
057r114 73 degree 39 N. living worms in the mud which he DREW up from 1,000 f[athoms], & the temp of which wa
066r142 they have Earthquakes in Cordoba. one of which DRIED up «all> a lake in neigbourhood of town Mr Murc
467t100 ions, & a few with one lobe again divided «Have DRIED some».-- some with no division in young flowers
052r097 P} * Slope necessary for seaward transportal of DRIFT matter.-- Give various cases. [Fig. 6] {P} A ad
094a036 s now in lakes.-- Mr Murchison. M.S. Chapter on DRIFT.-- Beyond region of great boulders, pebbles of
316c245 few Anphidesmas.-- this is remarkable.-- Fish & DRIFT sea weed-- may transport ova of shells.-- Conch
495q05a windy day, «flower garden on gravel walk» will DRIFT many seeds= Necessary to answer Wiessenborns do
604o017 ed by passive emotions.-- Cannot quite perceive DRIFT of Book.-- Sympathy & affections chiefly fail.-
157g076 ers as before certainly must have <come> «been DRIFTED» here: on very summit no granite-- (in valley
633j54r emptible queries, as why should the earth have DRIFTED; why should plants require earth, why not crea
041r065 rought are common.--Mr Lyell has mentioned the DRIFTING of carcases putrid. In Rio paper. when discus
111a088 raph Journal Vol VII p. 279. Carcases of birds DRIFTING out to sea do p. 358. changed soundings in Mo
267c094 ly of this shrub".-- p. 229. carcases of birds DRIFTING out to sea Vol VII. p. 325-- Wild dogs of Gua
448e166 Coast Vol II.-- <Reference> p. 251. about the DRIFTING of animals on ice p. 643-- very curious table
523m018 uhan-- not acting) in others from drinking cold DRINK.-- then brain affected like getting suddenly in
432e125 our hearing powers E. frowns prodigiously when DRINKING very cold water «frowns connected with pain,
523m018 e case ipecacuhan-- not acting) in others from DRINKING cold drink.-- then brain affected like gettin
345d042 with pigs &c, that hybrids were uncertain. Mr DRINKWATER thought that a <pure blooded> «"first blood"
551m129 when door shut or cad cries out "right." or DRINKWATER'S horse jumping when word Jump said-- I saw t
557m149 le, (or object of attachment) & then failing to DRIVE away rival.-- Fear is open mouthed to hear. tho
577n051 thinking of ones appearance,--does the thought DRIVE blood to surface exposed, face of man, face, ne
102a059 bottom a thing floating some way from coast is DRIVEN on to it.-- rollers at Tristan d'.Acunha.-- si
126a123 or under surface, would not the fluid matter be DRIVEN upwards & so conduct heat?-- How comes it in v
399e009 , or solely to manhood,-- it will decrease & be DRIVEN outwards in the grand crush of population.-- O
302c201 have remained somewhat similar.--!!! My theory DRIVES me to say that there can be no animal at prese
041r065 it is mentioned that the Redstart came towns DRIVING by the want of water.--I believe in all flat c
268c096 Tit mouse feeding young of redstart & actually DRIVING away parent birds.-- showing how blind a storg
558m152 ling. mouth wide open, each [lip] drawn back & DRIVING air out of mouth «hairs erect on back» «wide o
231b244 ppose a multitude of small creations.-- Will DROMEDARIES & Camels breed?-- As man has not had time to
305c210 ckall in gait, size, fur.; manner in which ears DROOP like dog older character & manner of wagging ta
362d099 ages) to brave men.-- Effect of castration horns DROP off.. replaced by hairy ones. which never «dry
436e138 t not in sight of each other will sing till they DROP off their perch.-- p. 101-- Kingfisher in north
502q11v 1842> When nettle leaf. put into spirits, poison-DROP exudes-- does not elm. does it «in» melon-- «Lo
591n098 n watching anything-- Keeper does not think they DROP their ears.-- -- George the lion is extraordina
437e139 two miles towards the Morne Mountains, it then DROPPED it & was found alive.-- Stanleys Familiar Hist
437e139 s on record of stoats being carried (p. 121) & DROPPED having wounded the bird. p. 124-- Mr Willough
573n037 ther. & hence rotatory movement negation.-- he DROPPED his head when he meant to eat, hence assertion
591n097 he has often watche tame young wolf & it never DROPPED its ears like dog-- wagged its tail «a little»
300c197 n however on this to a degree. Mem Spider only DROPPING where ground thick.-- shamming death it is bu
497q006 up, not found in wood.-- but seeds continually DROPPING in woods. by birds 13. Mr. Herbert says Crocu
554m141 Keeper is obliged to go in with a stick, if he DROPS it, the bird will fly at him-- Knowledge.-- Sep
041r065 er.--I believe in all flat countries. years of DROUGHT are common.--Mr Lyell has mentioned the drifti
061r126 , consult history? Phillips.-- 1826.27.28. grt. DROUGHT at Sydney. which caused Capt. Sturt expedition
523m016 thrown into a hospital.-- My father was nearly DROWNED at High Ercall, the thoughts of it, for some y
538m079 nking people were calling him a bastard.-- when DRUNK.-- having really been so.-- some always sentime
538m080 does those in second childhood, <they> or when DRUNK they would not be more different, & yet they wo
538m078 cted it in perfect senses.-- These things, & DRUNKEDNESS, show what trains of thought depend on state
538m079 trains of thought depend on state of turn In DRUNKEDNESS same disposition recurs, such as -- -- of Tr
042r068 le that point of Porphyry has been upheaved in a DRY form It is clear the forces have acted with far
060r126 p. 124. Webster Consult W. Parish. & Azara about DRY season[.] 1791. seen commonly bad over whole wor
133a139 at temperature of earth beneath <of Sahara de> a DRY desert, would be very high.-- M. Parrot ends his
159g084 (one 6 ft across) on top of spit between river & DRY Corry Scarcely conceivable. if Hill between Corr
299c194 n shewn to be same.-- one grows in marsh & other DRY; yet if T. palustris be sown in dry station it w
299c194 arsh & other dry; yet if T. palustris be sown in DRY station it will for some generations come up so.
343d038 as in changing from <hot>. Warm to cold, damp to DRY.-- Thus Tierra del Fuego has <not> «only» one «G
362d099 drop off., replaced by hairy ones. which never «DRY up &» peel off their skin (not being wanted for
416e070 one to suppose that seminal fluid fluid, (& not DRY as in plants) therefore, great difficulty in cro
495q05a cky stuff in flat places & see whether wind, on «DRY» windy day, «flower garden on gravel walk» will
520m001 they should have imitated.-- when attending Mr DRYDEN Corbet, he could not help thinking, he was pre
520m002 , & general manner of holding hands &c &c.-- Mr DRYDEN Co said he could not remember his father.-- My
520m002 ich happened in early infancy-- of this fact Mr DRYDEN C. is good instance as he is very deficient, h
286c155 ame difficulty in arranging animals in paper as DRYING plant, all brought in one plane Fleming Quarte
074r165 washing a rock» Veins concretionary; concretions <DT> determined by fissures as in septaria. (& Chilo
206b146 n living souls on average are related to the (200DTH year) degree. Then 200 years ago, there were 200

022r006 is so found in Anglesea, amongst the varying & DUBIOUS granites.--Wide limits of this mineral in Aust
453e183 because they «are» raised by seed.-- Where has DUCHESNE described Atavism.-- ask Dr Holland cases whe
490q001 ho sold eggs.-- Temporary Question 1 Where has DUCHESNE described Atavism alluded to by Dr. Holland--
233b251 Ostriches. Dodo. Apteryx Penguin-- Logger-headed DUCK-- Large proportion of Water & small of land or
275c121 ieties by picking the yellow one & crossing with DUCK bantams procured old variety.-- The pidgeons wh
339d025 s not Goldfinch & Greenfinch bred, & surely wild DUCK & «pintail» Widgeon!-- Divides animals «world i
341d032 -- (In Zoolog Gardens there is hybrid of Penguin DUCK a variety of Muscovy) with goose!!) Dr. Bachman
342d033 It appears certain that hybrid Muscovy & Common DUCK have been shot wild (escaped from Carolina?) of
359d087 stake in their origin) Saw cross between Penguin DUCK «from Bombay» & Canada Goose.-- Former strange
359d088 umber out of any one nest. even more than common DUCK-- Male Penguin was crossed with hen Canadian of
359d088 I should say in every respect most like Penguin DUCK.-- which is strange anomaly in Yarrells law.--
359d089 - (& not very like either either wild or Pintail DUCK) from which they were descended-- they descend
359d089 nded-- they descend from 1/2 pintail <into> «by» DUCK, into pintail.-- Of these there were four, two
360d091 ing most.-- «& not time» thinking of the Penguin DUCK & Herberts law of ideosyncrasy I have hitherto
363d101 athered legs.-- «Carruncles on beak & in Muscovy DUCK» crested feather, pouters, fan tails are found
363d103 rds Dr Bachman told me that 1/2 Muscovy & common DUCK were often caught wild off coast of America.--
459t013 l's at Hornsey the offspring of a black & white <DUCK of pecu» «drake» with the penguin duck. it took
459t013 & white «duck of pecu» «drake» with the penguin DUCK. it took after the Penguin in the form of its b
493q003 nimal. in comparison with Weigh skeleton of Tame DUCK & Wild Duck, & then weigh their wing bones & se
493q003 mparison with Weigh skeleton of Tame Duck & Wild DUCK, & then weigh their wing bones & see if relatio
494q004 t of crossing.-- (7) Are the Eggs of the Penguin DUCK quite similar to those of another Duck. ¿in Pid
494q004 e Penguin Duck quite similar to those of another DUCK. ¿in Pidgeon?-- Mr. Miller said yes with regard
495q05a le of ploughed field-- on hills.-- 10 Shoot tame DUCK on pond with Duck-weed-- coots-- waterhens-- ex
495q05a ld-- on hills.-- 10 Shoot tame duck on pond with DUCK-weed-- coots-- waterhens-- examine dog, which h
496q006 male & female flower in same receptacle (8) Make DUCK eat Spawn, eggs of snail, row of fish & kill th
506q014 in heigt 30 inches Examine Keel of Common & Wild DUCK-- Black Duck & Penguin Henslow &c (36) Has not
506q014 nches Examine Keel of Common & Wild Duck-- Black DUCK & Penguin Henslow &c (36) Has not H. raised rac
513q020 Do the Peacocks cross.= Young Chinese or Penguin DUCK in very young state for skeleton== Does the tum
639j28r & in stomach of lobsters-- analogy in Flamingo & DUCK, Ornithorhyncus «externally». petrel & Whale in
207b151 on rivers in Guiana. build top of trees carry DUCKLING to the water in their beaks, & the young one
301c199 ture,.. habitual instincts precede structure.--DUCKLING runs to water. before it is conscious of web.
255c051 ..» Instinct goes before structure (habits of DUCKLINGS and chickens) Young water ouzels hence aversi
261c070 e instinct, better than child sucking or even DUCKLINGS & fowls-- When talking of races of Man.-- bla
204b141 ently of individual force» Mr Wynne has crossed DUCKS & Widgeon & offspring either amongst themselves
204b141 on pool, first season bred readily with common DUCKS.--» Kirby all through Bridgewater errs greatly
207b151 nsiders that mercu» Geo. Joun. p. 325. Vol. IV. DUCKS on rivers in Guiana. build top of trees carry d
264c082 tail of ground woodpecker-- -- but tail of some DUCKS aberrant from-- habits.-- Gould I see quite rec
280c134 Sebright admirable essay) heredetary Young wild DUCKS.-- lose as well as gain instincts. Wild & tame
312c228 n & tigers ditto. (see Griffith) & half Muscovy DUCKS, «black cock & pheasant see Jardines Journal.»-
334d009 trutt» of his used to breed to Common & Muscovy DUCKS-- English. <Common» «China» & Canada Geese, &
340d026 als. Heard at Zoolog Soc their Pintail & Common DUCKS, breed one with another-- & hybrids fertile int
341d032 em to know itself,. at last associated with the DUCKS-- most strange voice often in the night, like
342d032 ds «in Carolina» for his table Muscovy & common DUCKS-- they are produced in full equal Numbers with
355d066 or fowls to the several aboriginal species «or DUCKS» (here argue if it be said domestic fowls are d
359d089 ost with its form & disposition Saw three young DUCKS, like each other.-- (& not very like either eit
361d096 andiae.--» Mr Blyth stated «that there are» two DUCKS, which have pretty close representative species
363d101 hing out of the forked band, like in plumage of DUCKS.-- Mr Yarrell says in very close species, of bi
378d148 s & fowls.--??? Wate[r]ton «p. 197» put 12 wild DUCK'S eggs under common ducks, the young crossed amo
378d148 n «p. 197» put 12 wild duck's eggs under common DUCKS, the young crossed amongst themselves, & I pres
378d148 sed amongst themselves, & I presume with common DUCKS. so often, that it was impossible to say which
392d180 pass through birds stomachs & live? In Muscovy DUCKS do young take most after father or Mother accor
393dIBC de artificially.-- Are hybrids pintail & common DUCKS. similar inter se? Zoolog. Gardens Are the hybr
450e174 ld) but only some birds are so when wild-- wild DUCKS monogamous; tame ones highly polygamous-- chang
584n071 ssion.-- use of senses.-- knowledge of location DUCKS & turtles running to water,-- young crocodile s
594n112 er have learned that he is not dangerous-- wild-DUCKS would have fled equally if man had appeared-- t
275c121 each brood.-- These bantam feathers at last got DUCKY, then took white Chinese Bantam crossed & got s
590n093 ffected. p. 44.-- Jealousy. causes spasm in bile DUCT, & throws bile in circulation p. 75. Haller say
540m088 r a strain of music, when the mind is rendered DUCTILE by grief, or by bodily weakness, melts into te
104a068 ature ¿ with pressure? Salt on surface of plains DUE to whole moisture being lost by evaporation ther
148g029 «by» pass of Glencoe-- the erosion may often be DUE to rivers-- By Roy Bridge, a tongue of flat land
162g101 d> 29.090 Preservation of form of land very much DUE to Peat & Heather When it did not grow at first-
176b021 or supposing number of forms equable: this being DUE to subdivisions & amount of differences, so form
182b046 studied The condition of every animal is partly DUE to direct adaptation & partly to heredetary tain
226b221 f genera on islands & on Arctic shores evidently DUE to «the» chance of some one of the different ord
227b225 t explains typical structure.-- Every species is DUE to adaptation + heredetary structure. latter far
302c203 groups. The formation of genera may sometimes be DUE to accident as submersion of land containing all
401e015 obable that great part of those varieties may be DUE to impregnation from other apple trees.-- now se
440e146 . 4 four laws Who can say, how much structure is DUE to external agency, without final cause. either
574n041 sexual.-- The association of saliva, is probably DUE to our distant ancestors having been like dogs t
618o41v said, that blueness caused weight, because both DUE to some common cause:-- The argument reduces its
624o50v instinct. *Our tastes in mouth by my theory are DUE to «habit» heredetary habit (& modified & associ
628o054 its "supremacy",-- I make its supremacy, solely DUE to greater duration of impression of social inst
575n043 ir brains» then can we deny that the grand child DUG for mice from some peculiarity of structure of b
278c131 rom the sea, associated with teeth of seals and DUGONG, therefore immense age since breccia accumulat
421e091 en says <tha» «in abstract» in his paper on the DUGONG, "The generative organs being those which are
422e092 indicative of true affinity.-- -- Owen says DUGONG connected with Pachydermata.-- <it was a Pachy
422e092 e origin of the aquatic Mammifers» p. 306, the DUGONGS cannot be united with true Cetacea or whales.-
334d012 e.-- Fox says in «Lord» Exeter's Park «or in the DUKE of Marlborough» there is a breed of white-taile
097a043 d by a thin pellicle of a blackish colour like a DULL & poor varnish, which I conceive to be analogou
266c088 he ground.-- why do beetles & birds & <f» become DULL coloured in sterile countries.-- Gould insist m
273c112 instance of its whole family where female is not DULL.-- I must observe that this preeminent structur
592n102 paper showing that the signs invented for Deaf & DUMB school & used between Indian tribes are Many th
233b251 ld. Crocodiles. Anoplotherium.-- «M. Jerrod» & DUMERIL great work on Reptiles. M. J. says some reptil
482z012 c-- The French <Jerrold?> «Bibrons coworker of DUMERIL» who is writing with Dumeril says that two spe
482z012 brons coworker of Dumeril» who is writing with DUMERIL says that two species of Tortoises come from G
126a122 -- but this may be very wrong,-- The fact of a DUMPLIN being bad conductor is {P} against my views--
355d066 keys to Zebras.-- «Mr Herberts variety of horse, DUN-coloured with stripe approaches to ass.» or fowl
025r016 inute S.] still shoaler, coast composed of sand DUNES. 15--15 Does not seem to consider this a very s
130a134 sidence; <as in> be cautious. mud banks & sand. DUNES.-- in these littoral deposits there probably wo
300c197 nly of pain. of death acquired?. The S. American DUNG beetles will each become the fathers of many sp
486z019 nsact. Vol. I. p. 130. Col Sykes on balls made by DUNG beetles, like those from Chiloe. Amblyrhyncus d
043r071 America--continuity of space in formations & DURABILITY of similar causes go together. add. <">» <fr
227b225 far the most serviceable. We may speculate of DURABILITY of succession from what we have seen. in old
199b118 nsmission of a fortuitous modification, into a DURABLE form, of a fugitive want into a fundamental pr
524m021 F. says, shows that early impressions are most DURABLE.-- (but Miss Cogan shows that repetition is no
628o54v hink this <boshes» «nonsense»-- My theory of DURABLENESS will explain it.-- Would not the maternal af
021rIFC rfection (namely mammalia) must have a shorter DURATION, than the more constant: This view supposes t
226b223 vorous Mammalia in their wide range & in their DURATION of species. (¿are carnivorous Mamm: in Paris
309c220 nce of elevated <extinct?» genera of shells.-- DURATION in two classes however different.-- Male glow
313c234 ations, with respect to law of mammals shorter DURATION, than molluscs, argue case both in Europe & S
313c234 merica. «very difficult case» Does this law of DURATION apply to utter extinction or rapidity of spec
347d049 und. because he says length of days adapted to DURATION of a planet to our lives-- Being myself a geo
443e156 ologists.-- & what is older-- what relation in DURATION of a planet to our lives-- Being myself a geo
461t041 ith those of Spain & such facts-- This unequal DURATION is exactly same as some species extending muc
485z018 dra-- polypi-- <Rep> do p. 324. Polypi shorter DURATION than cells.-- reproduced.-- Milne Edwards p.
486z018 rope-- Coleoptera especially require a greater DURATION of Heat. hence musquitoes & knats abound duri
628o054 -- I make its supremacy, solely due to greater DURATION of impression of social instincts, than other
288c159 introducing, instead of breeding from original DURHAM breed.-- Native dogs & English cross readily--

332d003 Alderney Cows take more after Alderney that the DURHAM,, with which they have been crossed--is Aldern
245c024 ith different number of teats» Coquille Voyage DURVILLE has written Flora of Falkland Islds, where is
405e031 «Fallow» Deer. which were of a nearly uniform <DUSKY> blackish brown.-- yet retained a trace of hori
472s02r rs had suddenly withered, & to day saw very odd DUSKY humble (with pollen) on legs go from clump to c
472s02r heartease without doubt.-- Bee, not large, very DUSKY & broad never saw such a one before-- Saw Fly 2
055r107 ives» half demolished craters).--worn into mud & DUST.--connection with age, & agreement with number
091a025 o shewn. No 3d of Ed. N. Phil. J. p 194. Fact of DUST blown far out to sea valuable; because transpor
358d071 (Major <I> Mitchell p. 244. vol I) spit & throw DUST <according to my theory of generation (p. 175)
495q005 es & see whether there will result hybrids-- (6) DUST flowers of one branch of Cabbage with pollen of
466t093 REGNATE by pollen with which <bees> «a been was DUSTED over. [P} Stamens & pistils curve upwards, so
472s02v omes covered) so whole sides of flower & stigma DUSTED.-- <I think> When It first alights, it cleaned
495q05a 9) I have noticed leaves covered with Honey-dew DUSTED with pollen of neighbouring grass» Spread shee
499q009 s-- Syrphus-- Meligethes & see whether they are DUSTED with pollen-- in what state (whole or broken)
505q014 13 Arum before pollen is shed can you find flys DUSTED with pollen from other flowers? Can flys' esca
470t178 k not on Phlox though they examine it.--«Little DUSTY & Blue» Butterflies at Clover,--Veronica--. Ran
633j54r God «created plants to» arrests earth, (like a DUTCHMAN plants them to stop the moving sand) we <do>
324c268 tion to the Natural system Bevan on Honey Bee DUTROCHET Memoires sur les Vegetaux et animaux.-- on sl
525m023 e pleasure, when doing. what they consider their DUTY.-- as carrying a basket, bringing back game, or
600o008 nscience» «Moral Sentiments» imperative sense of DUTY-- which makes struggle in man.-- two souls in o
620o046 easure.) & which man ought to follow-- it is his DUTY to do so.-- So we say a pointer ought to stand
620o046 pointer ought to stand a <spaniels> «housedog's» DUTY is to watch the house.-- it is part of <duty> t
620o046 's» duty is to watch the house.-- it is part of <DUTY> their nature.-- When a pointer spring his bird
242c017 Ind: Archipelago. [Raffles. Horsfield. Diard. DUVAUCEL. Leschenault Kuhl. Van-Hasselt, Reinwardt <Fo
265c085 nature, colouring matter being absent.-- again DWARF plant in alpine district & dwarf plants from se
265c085 bsent.-- again dwarf plant in alpine district & DWARF plants from seed, one adaptation, other monster
469t151 er-- Woodfords Marrow fat, Early frame, Groom's DWARF. planted in rows «close to each other» & seeds
469t151 rley-- Early Magazine-- &c. double-blossomed «& DWARF-fan Bean» bean, were planted in rows, & seeds g
514q21. paper on impregnation of violets.= Zostera= Are DWARF plants on Wellington Mountain described in Flin
572n034 imaginary words: it appears as if the mind had DWEALT on each word separately, neglecting time, & ge
541m092 han if simple idea as scarlet?-- How can people DWELL on pain ¿ no definite idea. nor is an emotion.-
309c221 turning neuter into Queen, more wonderful case DWIGHTS' Travels in America, speaks of short legged sh
175b019 oming in & most perfect «& others» occasionally DYING out;--for instance secondary terebratula may ha
175b020 s of some still older type some of the branches DYING out.-- with this tendency to change, (& to mult
176b021 nched.-- Hence Genera.-- «as many terminal buds DYING, as new ones generated» There is nothing strang
187b062 is a wonderful fact Horse, Elephant & Mastodon DYING out about same time in such different quarters.
281c139 as avoided linear arrangement the central twigs DYING, affinities would be <circular>, in broken circ
284c148 art. Must have travelled by <dead> «each» trees DYING & mountain torrents.-- but to crawl up an hill.
386d155 y be thus prolonged bud being formed & one part DYING for great length of time.-- There is probably l
099a051 n the Ponza case of Scrope parallel to walls of DYKES-- Mem. laminated dikes in Cordillera.!!!-- In s
342d035 entally says "--is distinctly marked as whole DYNASTIES have been featured by the Austrian lip & the
032c039 checked by increased vertical <height> thickness (DZ} of mass to be removed & from the resistance off
334d011 e & another leader mare.-- this stallion though EAGER to all other mare had been entirely broken from
283c143 & not habits as shown by frigate Bird & flying EAGLE.-- Hawk Gould seemed to have think, that widow bird.
437e138 ng.-- shows my theory insufficient.-- p. 120 An EAGLE is said to have been seen carrying a lamb two m
437e140 day a rabbit escaped into a hole, where the old EAGLE could not find it.-- The parent bird another 2
437e140 e scene.-- «In Shiant Isld. it is said, that an EAGLE always procured its prey from another island.--
185b055 pened in all. but less in water birds-- carrion EAGLES.-- This is but carrying on. attempt at adaptat
353d061 l II p. 402. Mr Gould on Australian birds-- all EAGLES. of Australia characterized by wedge tails.--
437e139 ughby found a dead lamb «& hare» by the side of EAGLES nest, which shows power of carrying great weig
437e139 r of carrying great weight. p. 125 is said that EAGLES bring rabbits & hares to the young ones to exe
403e021 d of crocodile, sometimes wanders from Pellew to EAP.-- There is another great lizard. Kaluz. which i
403e021 great lizard. Kaluz. which is found at Pellew & EAP, but not at Feis (near island) do p. 190. The la
403e022 nts of Summagi, a territory in the small isld of EAP in the Carolines, are remarkably short.-- & Defo
293c174 instance of difference in races of men.-- Wax of EAR, bitter perhaps to prevent insects lodging there
405e031 they would vary, if in wild state; thus mark on EAR of cats, colour can be brownish do Saw what was
585n075 nocks with back of Head, yet never puts down its EAR. good to contrast with horses, asses, <mi> Zebra
312c232 duced in shortest possible time. Mr Willis Long EARED little dogs, I am told, go to heat, take dog. b
437e140 ey from another island.-- » p. 175., 28 sho[r]t EARED owls were counted in a field, where there was g
402e019 n broken!. are born so in all Malay Countries W. EARL Eastern Seas. p 233 Octob 12th Kotzebue's secon
453e182 sted account of fall of fish in India.-- Windsor EARL-- Eastern Seas p. 229. Believes the <Rhinoceros
337d021 mes changed together with some training in the EARLIER branches «as in common greyhound» & much inter
378d147 mes change, & is the offspring brought back to EARLIER type by Mother?-- do these differences indicat
599o004 aphysical points written about the year 1837 & EARLIER--]CD in Athenaeum "Smart-- Beginning of a new
068r146 location connected with fluidity of rock ∴ «in EARLIEST stage» when covered up beneath ocean).--The f
254c048 f animal organization, "& are analogous to the EARLIEST conditions of the higher classes, during whic
259c065 ons.-- as an adaptation, but adaptation during EARLIEST existence; if whole life then real adaptation
308c219 te of knowledge «Negroes existed since time of EARLIEST Egyptian drawings & Old Testament» Domesticat
436e135 the Cephalopods, seem to have decreased since EARLIEST times-- Apterix has a most perfect Struthio h
570n026 y of men's reasons: shewn by similarity of the EARLIEST arts.-- Mem.-- Stokes-- arrow heads &c &c Oct
571n031 anguage, & vice versa.-- almost proves that at EARLIEST times there must have been intimate connectio
322c270 siognomy ---- Octob 3d Malthus on Population W. EARLS'. Eastern Seas. .Octob12th.-- Sir G. Staunton's
012e018 pochs-- such changes would be observed.-- G. W. EARL'S Eastern seas. p. 206-- shot a monkey, ceased t
023r022 w, amongst whose were things pitched over board EARLY in the passage!!-- M. Labillardiere in Bay of L
210b161 s» (which, Hunter says owe their origin to very EARLY stage) & which, follow certain laws according t
276c123 I will do with forms.-- Mention persecution of EARLY Astronomers.-- then add chief good of individua
405e031 o.-- [not located] Did man spread over world as EARLY as Elephants &c &.-- if in next 20 years none o
427e108 s can imitate the sounds surprisingly well-- In EARLY stages of transmutations, the relations of anim
468t111 Humble seem to frequent certain flowers, to day EARLY, the great scarlet Poppy-- So that, finally Fra
469t151 ones/ & wd go on longer-- Woodfords Marrow fat, EARLY frame, Groom's Dwarf. planted in rows «close te
469t151 came up in 1840 true. Shrewsbury.-- Abberley- EARLY Magazine-- &c. double-blossomed «& dwarf-fan Be
473s07v ly facts can decide-- some peculiarities may be EARLY impressed & others later-- All poultry with sam
502q11v flower with leaves.-- strawberries How <soon> «EARLY» do characters of races of different vegetables
512q019 Tollett-- about selection for milking-- loss of EARLY habits in Dorsetshire sheep migration of coats
520m002 very bad memories for things which happened in EARLY infancy-- of this fact Mr Dryden C. is good ins
522m010 , in conversation could catch up a new train if EARLY association were called up.-- My F. asked him,
522m012 ses of «mental» failure very general effect of <EARLY> «slight habitual» intemperance.-- often accomp
524m021 alked of these places.-- My F. says, shows that EARLY impressions are most durable.-- (but Miss Cogan
526m028 hat children like looking at <ani> pictures, an EARLY taste, of animals. they know.-- pleasure of imi
526m029 tches. some real-- some fancied.-- this fact of EARLY memory consisting of things seen, quite agrees
533m058 ensleigh. W. says that babies know a frown very EARLY in life, <before they> (I think I have seen sam
544m104 has useless. does not affect race. argument for EARLY education.-- fear of death!!! as Montaigne obse
559m155 "Wheat." in Jersey.-- very curious facts about EARLY production of foreign seeds.-- many varieties.-
560mIBC rt, (ie useless sudden movement of muscle) very EARLY in life Do they wink, when anything placed befo
586n079 n the action being performed or emotion felt in EARLY childhood (before experience or habit) could be
623o050 ing received pleasure from some one «person» in EARLY infancy, during many generations giving love of
536m074 atheist. Man thus believing, <yet> would more EARNESTLY pray "deliver us from temptation,' he would b
542m093 e amused, he need not express it, he may most EARNESTLY wish [not] to do it, but an involuntary laugh
246c025 ize, resemble, chien-loup.--long, black & white, EARS short & straight-- do not bark p. 433. birds &
267c094 acks «30 or 40 together» colour reddish <brown>. EARS long.-- like bull terrier-- Indian secured one,
301c199 interposition of deity to teach squirrel to kill EARS of corn according to my views, habits give stru
305c210 ike Jackall in gait, size, fur.; manner in which EARS droop like dog older character & manner of wagg
306c214 also from Asia Minor.-- tail like setters. long EARS-- colours vary, but form constant.-- The female
317c251 ins have peculiar character in extreme length of EARS & length of limbs, so that he first thougt only
332d005 ortsmouth, said to come from coast of Guinea, -- EARS bare. skin black & wrinkled-- fur short. (tail
363d101 horns of wild cattle.-- plumage of fowls-- long EARS of rabbits. & long fur.-- feathers on legs of P
377d139 interest» Hyaena. thinks, when pleased cocks his EARS.. when frighten depresses them.-- England was u
399e009 olour as a Hare, but paler & buffer.-- with long EARS-- & longer hind legs??? so that I was almost do
400e012 > less difficulty in establishing good groups.-- EARS varying so much,-- kind of fur-- (do tips of ea

Page ***(Key Word)***
400e012 ars varying so much,-- kind of fur-- (do tips of EARS take any colour?)-- length of tail varies, & ch
420e089 uadru<manous> «ped.-- ». Hairy.-- could move his EARS The head being six metamorphosed vertebrae, the
421e091 ish peculiar to Ireland. \do p. 283. on the dark EARS of the wild Chillingham Cattle, with reference
432e125 ent wants. or structure, than the muscles of the EARS to our hearing powers E. frowns prodigiously wh
458t002 little animal, showing something of Mule in its EARS-- ((this is good case as showing gradations, Bo
459tf02 nitors-- the hinnus, resembles horse in its head EARS, tail limbs-- in the mules, these parts resembl
511q018 o Santo Rabbit. Descript. of colour «& length of EARS» & skeleton, & skin= Van. Voorst often writes t
527m033 of one scale.-- former pleases from instinct the EARS (rhythm & pleasant sound per se) & causes the m
534m061 ists of sensation of images before your eyes, or EARS (language mere means of exciting association.)-
555m143 e some of the old world ones move skin of head & EARS,-- ∴ some men have this power abortive muscles)
556m146 g emotion.-- when horses fighting, they put down EARS, when «turning round to kick» kicking they do t
556m146 ow on horses, they must yet continue to put down EARS, when kicking.-- -- good case of expression sho
556m146 bra. when going to kick.-- Why does dog put down EARS, when pleased.-- is it opposite movement to dra
557m147 ed: (a horse when winnowing & pleased pricks his EARS?--).-- How is expression of anger in species of
565n007 d open.-- Hence becomes instinctive to fear., as EARS down to horse.-- Horse snuffs «& snorts», the a
565n007 norts», the air «& raises its head, & pricks its EARS» when afraid, though not every time really wish
567n013 8th. Jenny was amusing herself--, by getting out EARS of corn with her teeth from the straw, & just l
579n058 frowning anything to do with ancient movement of EARS A man shivers, from fear, sublimity, sexual ard
591n097 en watche tame young wolf & it never dropped its EARS like dog-- wagged its tail «a little» when atte
591n098 anything-- Keeper does not think they drop their EARS.-- -- George the lion is extraordinarily coward
592n103 nds to lay open all senses: <do> Horse prick his EARS «& snort clears nostrils» when frightened, does
596n184 Keeper says some of the monkeys move <its> «the» EARS but <not> Chimpaze. does not gradation towards
596n184 especially pulls back skin of whole forehead & 2 EARS.-- emotions of every kind.-- «[Are monkeys <are
634j54v s ruber (a fish) Man has abortive muscles to his EARS.-- p. 313 Many other good cases -- p. do]CD <Ma
028r031 e, hence at least no water is absorbed into the EARTH <I did not see one dike in the whole Galapagos
032r041 f climate have happened. hurricane in bowels of EARTH cause:--<exp> does not explain cleavage lines./
044r073 insignificant islets--general movements of the EARTH;--Scarcity of Organic remains.--Unequal distrib
055r106 . = talks of them being packed clean. & without EARTH.--Moreover that such do not occur on the beache
065r139 offin «buried in a mound» long consigned to the EARTH. yet body had scarcely undergone any decomposit
076r168 art of the veins, nearest to the surface of the EARTH."--p. 156. Mines of Batopilas in New Biscay, "N
093a033 America. -- The number of minute turbos in red EARTH with volutas. prove regular mud bank at Bahia B
107a079 & Patagonia-- On Lyells idea of whole centre of EARTH same heat, then change in form of fluid centre
108a079 e, but if heat from centre, then crust of solid EARTH would be thicker.-- PP Andes mark the line betw
119a105 unged so many miles deep into the bowels of the EARTH, as would be required by thermometrical scale.-
120a107 ¿ the enormous faults & facility with which the EARTH is cracking by vertical planes into small piece
122a113 foundly all consequences of EXTREME FLUIDITY of EARTH.-- study different forms of earth as shown by a
122a113 E FLUIDITY of earth.-- study different forms of EARTH as shown by arc.-- read Herschels astronomy wit
122a114 one coast? How can Herschel consider figure of EARTH statical.-- if platform of mexico owes its elev
125a121 S. Cruz. Assuming from Sir. W. Herschel's views EARTH originally fluid, then cooling process must go
132a138 bottom. if not colder than mean of place, shows EARTH not with central heat.--» «(does M. Parrot supp
132a138 ocean represents proper <state> temperature of EARTH. at the freezing point.-- accounts for increase
132a138 the freezing point.-- accounts for increase on EARTH by volcanic action.-- <Why> now as we know volc
132a139 y) as M. Parrots shows from variation in strata EARTH a very bad conductor.-- shows p. 516 that subte
133a139 lating water.-- & therefore that temperature of EARTH beneath <of Sahara de> a dry desert, would be v
183b053 st have once declined.-- Read his theory of the EARTH attentively Cuvier objects to <tran> propagatio
191b080 Regne Animal for Geography.-- The motion of the EARTH must be excessive up & down.-- Elephants in Cey
227b224 theory true, we get (I) a horizontal history of EARTH «within recent times.». & many curious points o
233b252 ectually higher.-- But who with the face of the EARTH covered with the most beautiful savannahs & for
292c169 greater number probably have no descendants on EARTH.-- +++. Even at Falklands some probably would s
415e065 n by «the» organic and inorganic agents of this EARTH. like every other animal.-- Would anyone raise
505q014 lepias-- Flowers not seeding= Put pot of boiled EARTH on top of House =Aristolochia, plant wh require
523m016 roduced by moral causes (ideotcy by fear. Chile EARTH quakes). in people, who, probably otherwise wou
551m129 - I saw the ourang. take up a stone & pound the EARTH. Lockarts life of W. Scott Vol VII p. 35 "as id
633j54r to such contemptible queries, as why should the EARTH have drifted; why should plants require earth,
633j54r e earth have drifted; why should plants require EARTH, why not created to live on alpine pinnacle? if
633j54r to presume that God «created plants to» arrests EARTH, (like a Dutchman plants them to stop the movin
633j54r . the plants were no more created to arrest the EARTH, than the earth revolves to form rain to wash d
633j54r e no more created to arrest the earth, than the EARTH revolves to form rain to wash down earth, from
633j54r an the earth revolves to form rain to wash down EARTH, from the mountains upheaved by volcanic force,
025r017 or Hot Springs in T. del Fuego = The Wager's EARTHQUAKE the most Southern one I have heard of In a p
029r033 ticleers voyage at <[...] Maranh> Pernambuco. EARTHQUAKE AT SEA.--Extract from the log-book of the Ja
029r033 Feb. 12, 1835. At 10h. 15m. a severe shock of EARTHQUAKE shook the ship in a most violent manner. Alt
043r070 erfect in latter). <At> Lyell. Vol I. P. 316. EARTHQUAKE of 1812 affected valley of Missisippi & New
045r075 nd at Calabria were present at the Concepcion EARTHQUAKE.--expatiate on difficulty of evidence about
045r075 are no country newspapers)--At the Calabrian EARTHQUAKE things pitched off the ground. «Ulloa states
045r075 anos!! were in eruption at time of great Lima EARTHQUAKE» In the Chili earthquakes if rise was more <
046r080 movement (as in Peru 1746).--At great Lisbon EARTHQUAKE Loch Lomond water oscillated between 2 & 3 f
047r081 above reasons, do not think so also elevating EARTHQUAKE of Valparaiso. (1822) no great wave on recor
047r082 fall on banks as a Bore wave rushes up? (NB. EARTHQUAKE wave is an oscillation, body of water manife
057r112 awful scourges to mankind than the Volcano & EARTHQUAKE.--Earthquakes act as ploughs [,] Volcanos as
058r116 5. of French «?» Edition states that the same EARTHQUAKE has run from Chili to Quito a distance of mo
058r116 e than 500 leagues. A little time after a bad EARTHQUAKE in Chili; Arequipa in 82 was overthrown, & 8
060r124 great depths.--important.-- Decemb 10. 1802. EARTHQUAKE at Demerara. The earthquakes "seem to arise
067r144 s. & at least seven miles inland. The Cordoba EARTHQUAKE a very remarkable phenomenon. showing line o
067r144 c rocks at M. Video «Volcano in Pampas» Pasto EARTHQUAKE. Happened on January 20th. 1834 Mr Sowerby.
070r154 ing shows undulation∴ crust thin.--Concepcion EARTHQUAKE Draw close Analogy Lake of Cordill: of Copiá
070r155 em <5th> Concepcion most violently shaken, by EARTHQUAKE. but no serious injury. -- <Analysis of Atac
071r157 towards the Vermejo was utterly overthrown by EARTHQUAKE with great destruction of human life.--Templ
071r157 truction of human life.--Temple mentions some EARTHQUAKE at Cordova.-- There the Cordova earthquake i
071r157 me earthquake at Cordova.-- There the Cordova EARTHQUAKE in which lake was absorbed.--Earthquakes fel
092a028 nsitions clay slate &c nearly vertical Linear EARTHQUAKE 500 by 90.-- in Syria Geolog. Proc. p. 541.
092a030 ng sandstone Vol II. p. 69.-- Geograp Journal EARTHQUAKE at Melville Isld New Holland Augus 1d to 3d
106a076 increase of temperature at great depths. All EARTHQUAKE unaccompanied by Volcanos must be sought aft
108a080 the line between sinking & rising areas.-- In EARTHQUAKE if Subsidence we should not expect volcanos.
132a137 ation it would be to compare, the time of the EARTHQUAKE of Chile, with that of the passage of the mo
138a153 gravel & shells. p. 47. do has table of every EARTHQUAKE, during two years.-- will serve for comparis
153g053 roded {P} 4th [shelf] river 4th [shelf] Could EARTHQUAKE cause collection of sediment? Where ravines
025r017 known to be inactive 300 years? No Volcanic EARTHQUAKES or Hot Springs in T. del Fuego = The Wager's
042r068 eocene lakes of France, & unequal action of EARTHQUAKES. «on Chili & delta of Indus», my belief in s
043r070 of 1835.-- State the three «or 4» fields of EARTHQUAKES in Chili:-- Chiloe. Concepcion. Valparaiso (
045r075 time of great Lima earthquake» In the Chili EARTHQUAKES if rise was more <than> inland than on coast
045r076 ise). -- The great rains which attend severe EARTHQUAKES «1822¿ 1835?» alone, (& the general belief i
050r090 George Mackenzie must be worth reading Some EARTHQUAKES of Sumatra no connection with a neighbouring
053r101 metalliferous. Vol III Latter Part Are there EARTHQUAKES in the Radack & Ralix Islds? In my Cleavage
053r101 nsions acting at great depths (mem: profound EARTHQUAKES), which would cause parallel lines, but the
056r112 ll adapted to use of mankind.--<Hutton show> EARTHQUAKES part of necessary process of terrestrial ren
057r112 s to mankind than the Volcano & Earthquake.--EARTHQUAKES act as ploughs [,] Volcanos as Marl-pits: Co
058r116 & 86. Lima. next year Quito. considers these EARTHQUAKES travel in order.-- If we look at Elevations
060r124 Decemb 10. 1802. Earthquake at Demerara. The EARTHQUAKES "seem to arise from some efforts in the land
060r125 -p. 200. Bollingbroke voyage to the Demerary EARTHQUAKES at St Helena. 1756. June 1780, Sept. 21st. 1
064r135 itzRoy. -- Limited Volcanic action & limited EARTHQUAKES & great but local elevations of the land in
066r142 rse valleys Ice Sir W. Parish says they have EARTHQUAKES in Cordoba. one of which dried up <all> a la
071r158 dova earthquake in which lake was absorbed.--EARTHQUAKES felt. different case from shore of Pacific.-
071r158 alt «(Lyell)» cracks mountains falling in.-- EARTHQUAKES at Quito. tranquillity «at Mendoza» exceptio
079r177 oldt. New Spain Vol. IV. «p. 58» At Acapulco EARTHQUAKES are recognized as coming from three directio
119a106 ure of quartz rock?) also by my phenomena of EARTHQUAKES.-- by the narrowness which the anticlinal li
377d140 cy of time. When we multiply the effects of <EARTHQUAKES>, elevating forces in raising continents, &
132a137 ient to calculate rate of increase of heats in EARTH'S crust.-- yet heat does increase,-- but in Ocea

090a021 that inside our globe melted magnetic metals. ∴ EARTHY crust compared to those of falling stones.--¿
150g038 very round-topped with much drainage & far more EARTHY than what is usual-- Lines die away where slop
151g039 away where slope less., best developed on steep EARTHY slope, two circumstances rarely united.-- die
535m064 , but habit forgotten in all older species. The EARWIG & a doubtful one of Acanthosoma grisea describ
240c004 onstrosities being produced, & handed down, with EASE, is analogous to what occurs in plants.-- All t
429e115 00. & yet know that plant can be cultivated with EASE near London.-- what makes the line, as trees in
557m147 s constant smiles, cheerful face».-- Man when at EASE has smooth brow contrary to wrinkled: (a horse
360d094 -- I can readily see that two first might cross EASIER than two last. Sept. 11. N Mr. Blyth, at Zoolo
418e080 rs.-- ¿ in hemi-hermaphrodite insects is it not EASIER to understand ¿perfect?? developement of one s
426e106 that a <do> variety of one species would cross EASIER with 2d species, than two perfect species; but
609o029 ence & impulse) but shame «we alas know» is far EASIER conquered than the deeper & worser feelings. T
624o051 rred'. NB. Until, it can be shewn, what things EASIEST become instinctive, this part of argument fail
035r046 as, Chiloe. &c seems the organic structure most EASILY preserved.-- Mr Conybeare introduct to Geolog-
126a123 duct heat better than plain?-- Mem 1000 {P} how EASILY water percolates rocks,-- when pressure increa
130a133 of Persia In Glen Roy paper I show crust yield EASILY. & if easily must be thin: <beside mere fractu
130a133 Glen Roy paper I show crust yield easily. & if EASILY must be thin: <beside mere fracture> A Elevati
148g026 like MOTHER!-- the cross not so hardy but more EASILY fatten, This man confirmed the account of the
180b037 s constant.-- With respect to extinction we can EASILY see that variety of ostrich, Petise may not be
191b081 are more valuable than any other, because less EASILY transported-- Mem plants on Coral islets.-- Ne
196b105 ere any genera, mundine, which cannot transport EASILY.-- it would have been wonderful if the two Rhe
204b140 ring seldom intermediate between parents.-- How EASILY does Wolf & Dog cross? Mr Yarrel thinks oldest
227b227 fic side. the Oolite order of things might have EASILY been formed.-- With belief of <change.> transm
254c049 each joint of taenia worm.-- formative energies EASILY expended & no one system developed <is> not su
289c162 now whether a variety could be transmitted more EASILY in those born without coitus, than with.» ⟨Mig
301c199 bit instinct gained during life.-- do Elephants EASILY acquire habits is this the Key to their mental
302c202 -- last acquired,-- or aberrant, therefore more EASILY modified = = this is not easily told, for any
302c202 therefore more easily modified = = this is not EASILY told, for any small family. having analogous c
306c212 oung animals, well developed, just like, habits EASILY gained in child hood.-- Young salmons. first a
311c226 st study on the idea of those parts being most EASILY mostrified, which last produced --insane men i
336d018 ned with respect to that individual) it is more EASILY inherited.-- «but if change be in blood long,
349d051 st upon the discovery what characters VARY most EASILY:-- those which do not vary being foundation fo
371d127 orm, & so altered in disposition, as to be more EASILY trained up to the (required) offices" &c &c Ow
398e005 onclusion about species Changes of level &c are EASILY recorded, but changes of species «are» not as
414e063 g she would make <them> such a variety far more EASILY than man,-- though man's practiced judgment. e
425e103 ids of Calceolaria.)-- «:CD[Does the Power of, «EASILY» making tolerably fertile hybrids, bears relat
499g09v ated females.-- Also any plants which are known EASILY to be crossed & all monooecious plants.-- Hook
526m027 ans & therefore properly no free will.-- we may EASILY fancy there is, as we fancy there is such a th
531m050 her remarks that things of great importance are EASILY forgotten, (if unconnected with fear &c) becau
539m082 h. As child gains habit «or trick» so much more EASILY than man, so may animal obtain it far more eas
539m082 sily than man, so may animal obtain it far more EASILY, in proportion to variableness or power of int
550m126 Medici, p. 21-24. Curious passages showing how EASILY chance & will of Deity are confounded.-- well
551m129 isgust & defiance.-- different from sneer-- How EASILY. horses associate sounds may be seen by omnibu
553m135 sicks shows that such a view cannot be, anyhow, EASILY overturned.-- so ready is change from. our ide
557m148 ar or anger? I should think shame would be more EASILY analysed than jealousy, because less discovera
596nIBC n pashion.?-- cry? Do people of weak intellects EASILY fall into habits Get facts about instincts of
596nIBC blubbering children, if of. convulsive tendency EASILY fall into convulsions A carrier pidgeon carrie
614o037 stincts,: though very weak so as to be overcome EASILY by reason.-- Conscience is one of these instin
640j167 ivers, the small body of species would far more EASILY be changed.-- Hence the Galapagos Islds are ex
031r038 instances in old world of symetrical structure. EAST India Archipelago. «Aleutian Arch.--» V. Fitton
034r044 386 Mem. Lyell's fact about sulphuric vapours in EAST Indian Volcanos Gypsum Andes Mem. Beechey. acco
044r071 y. at Mauritius. (consult Bory «dip of strata on EAST») cannot believe in a great explosion, nor woul
051r097 s of Pacifick, as compared to whole E. America. <EAST> Africa. profoundly deep: a great fa
059r120 in centre:-- Bailly talks of much granite on all EAST side of Van Diemen Land. All the Calcareous roc
065r140 usly. Mem. pebbles of Porphyry.--Falklands.--off EAST Coast. -- Capt. Cook found soundings. (end of 2
069r151 as much conglomerate on West of Peuquenes as on EAST. Where gone to.?-- There must have been some co
069r151 ne to.?-- There must have been some conglomerate EAST of Portillo Where gone to? Intermediate space p
088a016 s remark on common salt being found on low hills EAST of Cordillera very important V. Malte brun -- M
127a125 from water {P} thawed at + in isothermal curve. EAST-clinal. West clinal. S.-clinal. N-clinal & anti
127a125 ply clinal lines. dipping to & so or may be used EAST-clinal lines & c & .-- But Siberia was once tha
146g009 dea-- highest part must project [Blank] {P} Path EAST End near Holyrood Palace In same way at top the
152g049 of air 65 degree? There are two terraces on the EAST side of river & bed of river about 40 ft beneat
155g062 d can be traced some way up, but most faintly on EAST side of Glen Turrit, where I believe they end i
188b067 uliar according to Swainson to certain islets in EAST Indian archipelago Dr Smith considers probable
190b076 on of Mr Brown, about peculiarities of Flora. on EAST & West. ends of New Holland. diminishing toward
191b080 be excessive up & down.-- Elephants in Ceylon-- EAST Indian archipelago.-- West Indies = Opossum & A
191b080 s-- + + + for instance tertiary deposits between EAST Indian islets-- (+ + +. Ireland longer separate
191b081 rds.-- & life shorter or change greater-- In the EAST Indian Archipelago it would be interesting to t
202b133 finding Monkey in France.-- of genus peculiar to EAST Indian isles.-- Compares it to fossil Didelphis
211b164 en says Nonsense The distribut of big Animals in EAST Indian Archipelago--, very good in connection w
225b219 s land. glorious fact. of absence of quadrupeds EAST India Archipelago very good on opposite tendenc
226b223 would presence of Jaguar been in S. America.-- <EAST.> «W» Coast of Africa & <West.> «E» of America,
229b234 Coast. Hippopotamus do.-- Giraffe do.-- Range of EAST Indian Rhinoceros (?)-- Some paper in Institute
231b242 Progress, & you will have two. Tapir existing in EAST Indian Seas. Marsupials animals all show greate
231b242 an.-- American & African forms mingle in India & EAST Indian islds-- Monkeys different not travellers
233b249 30) different species in different isld. (as far EAST as New Ireland. see Coquilles Voyage), Waterhou
241c016 in South Sea (Indio Polynes: <)> vegetation far EAST) Ann: des Sciences. Semptemb. 1825 Get Henslow
242c017 et rupicole vert instances of American forms in EAST. Ind: Archipelago. ⟨Raffles. Horsfield. Diard.
243c019 orms have representatives (& instances given) in EAST. Ind. Arch:-- Birds of New Zealand absolutely di
246c025 433. birds & bats have certainly travelled from EAST Indies, isld, as far as Oualan.-- Wide space of
246c025 isld, as far as Oualan.-- Wide space of sea, to EAST of America. would account for this.-- Coquille
250c038 s to have been formed & spread to other Africa & EAST India Arch.-- but where these great animals had
317c251 he first thougt only one species. & all hares on EAST side have other peculiar appearances-- Now this
403e021) of Sumatra has given account of Buffalo of the EAST which differs from that of S. Europe p. 189. Th
407e039 e..-- Therefore I argue from this that Africa «& EAST Indian Archipelago»-- formerly were not so «ver
408e045 of the world--.-- Mem. elevation & subsidence of EAST Indian Archipelago. now rising. On a particular
450e175 ? do. p. 189. «190» No full sized horse is found EAST of y Burramposter & S of Tropic-- By J. H. Moor
459f1r s showing gradations, Boteler's Narrative Voyage EAST coast of Africa-- Vol II. p. 256-- wild cattle
485z017 rther North on West coast of S. America. than on EAST.-- not being replaced by Brazilian Species.-- M
499q010 o eat. (even Nux Vomica is eaten by a Buceros in EAST Indies-- Asiatic Researches) (23) Talk about Th
067r143 can geology. Conybeare Lava in Cordillera & on EASTERN plains «by Antuco». Athenaeum April 1836 (p302
085a001 d on windows of houses. & trees all injured on EASTERN side, far inland.-- even 70 miles from salt wa
190b076 gree, source of forms. reduce towards Northern EASTERN end & die away, & partake of Indian character
203b136 ound in Germany & thinks even now in central & EASTERN Asia beyond the Ganges & perhaps even in India
246c027 allied to C. muscadivora., which lives in the EASTERN Moluccas, New Guinea.-- (Case of replacement)-
322c270 ---- Octob 3d Malthus on Population W. Earls'. EASTERN Seas. .Octob12th.-- Sir G. Staunton's Embassy
327cIBC t of Europaean plants transported.-- Crawford. EASTERN Archipelago. probably some account Raffles. Si
402e018 such changes would be observed.-- G. W. Earl's EASTERN Seas. p. 206-- shot a monkey, ceased their cri
402e019 n". are born so in all Malay Countries W. Earl EASTERN Seas. p 233 Octob 12th Kotzebue's second «1st»
453e182 unt of fall of fish in India.-- Windsor Earl-- EASTERN Seas p 229. Believes the <Rhinoceros» «Tapir»
032r041 Equator to poles. nearer the surface & to the EASTWARD.--If matter proceeds from great depth. from a
399e010 n the Natural History & at Manguia, but are unknown EASTWARD of the Navigators. Snakes occur there., but a
035r048 nature oscillated equally.-- These facts become EASY if we look at the action as a deep & extensive
058r116 all land marks.--At the first it would though be EASY to see on beach successive lines of sea weed--
098a046 Transacts. Vol. III. p I. p. 86. et p 95.-- It is EASY to prove. (pyrites, agates, calcareous balls) t
196b104 Foxes. Mus are birds that are apt to wander & of EASY transportal.-- Waders & Waterfowl.-- scrutinize
219b196 may be thought to explain nothing-- it being as EASY to produce «for the creator» two quadrupeds at
291c168 re would be perfect series or gradation.-- It is EASY to see if South America grew very much hotter,
294c176 the change in animal be permanent.-- It will be EASY to prove persistent Varieties in wild animals--

Page
(Key Word)

```
345d044 hows, with the aid of seclusion in breeding. how EASY races or varieties are made.-- The Highland She
408e045 diseased., & some of them symmetrically.-- it is EASY to get 50 of same kind of monstrosities.-- G. B
543m097 an's face.-- <That> How far they communicate not EASY to know,-- but this capability of understanding
612o035 egetable and animal strictly united? [RHC] It is EASY to conceive such movements & choice, & obedienc
621o047 e right & wrong in his mind.-- Now we know it is EASY by association to give «almost» any taste to  a
032r039 [Fig.  2] {P} The action of sea A. B. will be to EAT in the land in line of highest tidal action. thi
211b163 bitches  never being killed by them, whilst they EAT up the dogs.-- L.' Institut. Curious paper by M.
313c236 or other actions in foetus of Mammalia, or chick EAT») Generation becomes necessary, when organs of p
356d069 ing different kinds of food: grazing animals who EAT every species new.-- Sept. 8'' A Golden Pippen o
414o063 isely like two species of animals.-- they fight, EAT each other, bring diseases to each other &c, but
496o006 & female flower in same receptacle (8) Make Duck EAT Spawn, eggs of snail, row of fish & kill them in
499q010 e for transportal, does he include seeds good to EAT. (even Nux Vomica is eaten by a Buceros in  East
512q019 Hawks distributing live Mamals (3) Do most Hawks EAT stomach. of finches-- do they throw up pellets--
530m049 ts & watch them till the young are big enough to EAT.-- There was blackbirds nest, near hot-house  at
531m049 on. this nest the cat could if cats will «ever» EAT little birds, this most curious instance of reas
554m139 up to «keeper» see whether, this was permitted & EAT it.-- good case of association.-- «Listened with
571n029 s:» & man not.-- ¿no vegetable good «for man» to EAT poisonous?-- How did animals in «Australia» & Am
573n037 negation.-- he dropped his head when he meant to EAT, hence assertion.-- but nodding is less strongly
087a009 - Will it be supposed that the armadilloes have EATEN out the Megatherium. -- The Guanaco the Camel.?
499q010 include  seeds good to eat. (even Nux Vomica is EATEN by a Buceros in East Indies-- Asiatic Researche
570n025 difference  is there between Squib after having EATEN meat on table, & criminal,-- who has stolen. ne
484z014 Ocean Gould agrees with D'Orbigny, that Serpent EATER-- or Secretary is S. African representative of
546m109 r smell of flowers, owing to parent being fruit EATER.-- origin of colours?-- Nothing shows one how l
088a014 cked up beneath the trees---- Are any Fish seed-EATERS. This important in transport of Fish Let a Haw
344d041 in  Caterpillar. very valuable facts-- they are EATING foetuses, as young of Marsup. is sucking foetu
356d069 leads  one to suspect any amount of change from EATING different kinds of food: grazing animals who e
525m023 shame, when doing anything which is wrong.-- as EATING meat., doing their dirt, running home.-- in th
529m042 Philosoph.  Transactions, of ideot 18 years old EATING white lead. who was most violently purged «bel
548m117 something  is not there, & then when one begins EATING one perceives butter or salt is not there.-- t
554m139 to  see her take bread from a visitor, & before EATING «everytime», look up to «keeper» see whether,
591n100 e of instinctive conscience.-- Why does not man EATING cause disgust, because he does not go  against
591n100 y does not fullfil, like continent man.-- a man EATING what others by habit (not instinct) think  not
332d004 med to perceive turn on road where No houses to EATON Mascott, where he had been accustomed to turn d
306c212 l argument) & therefore down the stream followed EBB tide, therefore got into habit of going down str
191b079 How  remarkable spines, like on a porcupine on ECHIDNA-- Good to study Regne Animal for  Geography.-
192b087 s possibility of such organization. [Spines in ECHIDNA & Hedgehog]CD-- As we have one Marsupial  anima
304c208 ifles are produced by circumstances. Spines on ECHIDNA.-- when it can be traced through series then p
311c225 case of 180 degree & great diff of Latd. p 355 ECHIDNA of Van Diemen's land & Australia different Tem
373d132 r. like true Mammalia, no more wonderful. than ECHIDNA. & Hedgehog having spines.-- Does not male Pid
639j28r se & (Chamaelion?) C. D. Spines in Hedge Hog & ECHIDNA.. & Aphrodites C. D. Endless cases.-- Maccullo
026r022 es not common in Britain. Mem Concepcion Says ECHINITES. Encrinites. Asteriae, usually petrified into
270c103 No Analsis of Nat. Hist) spiral structure in ECHINODERMATA.-- Agassiz says Infusoria «are»  insecta.--
381d156 ale organs visible). Oyster. cystic Entozoa. ECHINODERMS. Acalephes. Polyps. Sponges Heautandrous, ma
460t017 ual inclination for each other-- Aug. 20th The ECHNIDA & Hedgehog Tenrec both having spines, is the e
428e113 ss, than the number of good race-horses, which ECLIPSE? has begotten <?> «Walker attributes this to e
284c150 ecies by another. supply place in each others' ECONOMY Dr. S. showed that savages are not born with a
369d115 - Chapt I. Also Latent Character Hunter Animal ECONOMY p. 482 (Same book) Owen says "the necessity of
091a025 s Anglesea solution of silex also shewn. No 3d of ED. N. Phil. J. p 194. Fact of dust blown far out t
091a027 uartz. strike SSW & NNE dip 30 degree - 80 degree ED. N. Phil Journ. p. 410. 1828 Ed. N. P. J. p. 105
091a027 egree - 80 degree Ed. N. Phil Journ. p. 410. 1828 ED. N. Phil J. p. 105. Oct. 1828. gneiss in India (fa
094a036 Soc.  Vol. 2. p. 35 Sir J Hall Trans. Phils Royal ED. Vol 7 Dr Buckland Reliquiae Diluvianae p.  201.
121a111 d before societies.-- Sir. J Hall Vol VI. p 173. (ED. Transact) has seen clay stiff enough <to  form>
129a131 ast of. Brazil. Maldonado enter into this case.-- ED. New. Phil. Journal Vol XXI. p. 213. Beyond  the
137a149 l veins p. 125 to 129 & p. 135--160 & 162 [blank] ED. New. Phil J. 1838. p. 72. on metallic vapours c
137a151 ring some of its constituents into chert. [blank] ED: New. Phil J. 1838. p. 132. «& 134» Bischoff. On
196b107 favourable to large quadrupeds--horse not large-- ED. New. Philosop J. No 3. p. 207 "It is not genera
199b117 ient Flora thought to more uniform than existing. ED. N. Philos J. p. <191> «p 191» No. 5. Ap 1827 F.
199b118 penity, of an accidental habit into an instinct." ED. N. Phi. J. p 297, No 8 Jan-Ap. 1828 -- I thank t
199b119 thers, viz not too much change In Number 6'.? of E.D. N. Philos. Journ.-- Paper by Crawford on Missio
200b121 ike. SSW. & NNE. dip 30 degree - 80 degree (?).-- ED. Phi.1.. N. J. p. 410 <Nov> 1828 It is daily hap
200b125 salt water &c has any tendency to form varieties? ED. N. Phil. J. Morse found in Virginia p. 325. Jul
309c222 rabs.-- NB avoid quoting these hackneyed cases Mr ED Blyth does not believe in circular or linear arr
406e036 e one parent & sometimes other & sometimes ½ way. ED. New-Phil. Transact. Rabies, common to men, dogs
465t081 en have been exterminated (see Pritchards paper) (ED. Phil. Journ. end of 1839) very important. it se
066r141 reak through the N & South lines the tides form EDDIES with its extreme force. Yet, no outlet at head
086a006 gy is written Cuvier. Europe possessed a great EDENTATA. -- How much is temperature of world regulate
196b106 .-- It is a very great puzzle why Marsupials & EDENTATA should only have left off springs <ne> in  or
202b133 type. like horse in S. America. or like living EDENTATA in Africa &c &c.-- Now if suppose world  more
249c036 umber of quadrupeds in Eocene period. Have the EDENTATA & Marsupial forms been chiefly preserved,-- w
250c037 peds»; Pachyderm in S. America destroyed great EDENTATA or American form.-- Is the Australian Dipus a
250c038 had not spread then such tribes as Marsupial & EDENTATA increased most. Certainly Africa approaches N
34d050 certain  species & become worthless-- Mammalia EDENTATA̷ We do (p 6) say such is group. because it ha
355d068 e great sea, & two Polar Continents Marsupial. EDENTATA.-- Pachydermata &c &c-- It is important  with
357d073 of  Marsupials in S. America. from presence of EDENTATA-- Edentata & Marsupials have been almost dest
357d073 als in S. America. from presence of Edentata-- EDENTATA & Marsupials have been almost destroyed where
407e040 xisting orders, as the Eocene of Paris! (Great EDENTATA at that period) Analyse this,-- consider stat
101a057 cephalus (Harlan) found on the Delaware. is it EDENTATE? Phillips p 289.-- Alludes to big bones in in
194b094 m. therefore it is like case of great <rodent> EDENTATE [has been doubted?]CD & opossum found in Euro
214b173 latycephalus. (Harlan) found on Delaware is it EDENTATE? Phillips. Lardner p. 289 It is certain, that
226b223 overed channel of communication by which great EDENTATE might have roamed to Europe & Pachydermata fr
086a005 able as the water"-- It may be worth noticing EDENTATES & camels in deserts & rodentia In Plata Masto
106a075 s to all vallies (except mere talus «over cliffs EDGE» of which limit cannot be great over) with very
152g048 g at 4th shelf from size of buttresses, to upper EDGE of which they cut near Loch Tring-- Tuesday Bri
153g051 elf: others «lines not so level because of upper EDGE of cliff» Others below it--argument for lake «o
164g119 arse agglomerate [...] shells from [...] Wenlock EDGE [blank] L. Lochy 12 ft 96 L. Oich 12 84 {P} 29.
383d160 ee, I believe instead of» two lines.-- «faintly EDGED with reddish brown» black marks on tail much <b
599o007 nence. the wide expanse, of county, netted with EDGES & crowded with towns & thoroughfares, I grant t
327c265 e on Horticulture in Edinburgh. Encyclop.-- The <EDIN> British & Foreign Medical Review No XIV. April
067r143 de Angelis.» This work is reviewed in present EDINBURGH March 1835 Sir W. Parish says. that beds of s
071r156 ious injury. -- <Analysis of Atacama. Iron in EDINBURGH. Phisoph. Transactions. = Mem: Olivine. Volca
107a077 d by Sir J. Hal. End of pages. p. 157. Vol VI EDINBURGH. Phil: Transacts.-- Does the isothermal {P} s
116a099 e effect as too coarse. Read Kylau on Granite EDINBURGH Philosophical Journal Rapport on  D'Orbigny's
119a104 he moon, which though not very analogous (see EDINBURGH. Phil. Journal <]CD>, no great chains like An
129a132 ad Mr Parker's Book.-- M. Bichoffs Papers, in EDINBURGH New Phil. Journ 1838. several case given of h
218b190 n West coast of Africa.-- changing hair-- The EDINBURGH. Journal of Natural History-- Preface appeare
311c227 .p 69. with tall grass. & p 72 hairy sheep-- EDINBURGH. Transact. Vol IX p. 107. an Ascaris inhabits
320c276 upeds of Paraguay Dobrizhoffer. Aipom<e>nes. EDINBURGH New. Phil Journal. about 13 numbers have been
327c265 . has written good article on Horticulture in EDINBURGH. Encyclop.-- The <Edin> British & Foreign Med
338d024 nts of Nova Zenbla -- in review of Baers work EDINBURGH. Royal. Transact.-- p. 297. Vol 9. Dr. Fergus
424e102 land-- for I presume Carrion Crow is found in EDINBURGH.-- Why does Fleming consider them varieties &
533m060 e,) because leg is right handed.-- In Review (EDINBURGH) of Froude's life. that author remarks,  that
538m078 ses of heredetary pride & in single families. EDINBURGH. Phil. Transact. p. 365. Case of double consc
603o014 ERY language shows traces of anterior state?? EDINBURGH Review Vol 18. (1st Article) on Taste «EXCELL
109a084 rtant.::-- must be studied.-- Scientific Memoirs EDITED by Taylor Ehrenbergh on flints in chalk must b
322c270 e Intellectual powers. Hunters Animal OEconomy. EDITED by Owen. read several papers-- all, that  bear
381d156 ize prodigiously-- Animal OEconomy by, Hunter. (EDITED by Owen) p. 34.-- Owen classifies Hermaphrodit
051r097 d &c &c. <Vol IV> P. 209. 211. 213. 444 «Yanky EDITION» Shores of Pacifick, as compared to whole E. A
058r116 urelle des Indes Acosta. p. 125. of French «?» EDITION states that the same earthquake has run from C
```

108a081 found Granite at Christmas Sound Vol XIV. (My EDITION) p 500. Well described [top portion page excis
540m088 e &c &c Coleridge,-- Zapoyla p. 117, Galignani EDITION Fine poetry, or a strain of music, when the mi
272c109 eather in wing the curious feathers in tail of EDOLIUS.-- Remarkable how small detail in structure pr
309c220 aces of animals-- argue «opening» case. «thus» EDUCATE all classes-- avoid the contamination of <cl>
608o026 we do the wicked.--we ought to pity & assist & EDUCATE by putting contingencies in the way to aid mot
327c265 Blushing lately advertised. /6s Mrs Necker on EDUCATION preeminently worthy of studying in Metaphysic
385d165 ast remark good. because showing probably not & EDUCATION.-- Cannot I find some animal with definite li
536m073 ion may have been affected by circumstances & EDUCATION, & by choice which at that time organization
537m077 ntry or Swis.-- it may be answered effects of EDUCATION, may be opposed undoubted cases of heredetary
539m083 ne in some respects--. good instances.-- when EDUCATION same.-- My handwriting same as Grandfather. A
544m104 ess. does not affect race. argument for early EDUCATION.-- fear of death!!! as Montaigne observes. di
571n028 taste naturally all has not been acquired by EDUCATION. else why do some children acquire it soon. &
583n070 st be thought of.-- p. 19. animals capable of EDUCATION; (this is again assumed as more allied to rea
583n070 reason: & then these qualities of imitation & EDUCATION may be used as argument.-- for instinctive kn
599o008 ture than any other animal-- Aimé Martin de l'EDUCATION des Mères Vol. I. p. 198.-- "Moralité, raison
601o08a ams, awakes-- does when Master takes Hat de l'EDUCATION des Mères par L Aimé Martin Leroy Lettres. Ph
607o025 iverse. [Effect of heredetary constitution,-- EDUCATION under the influence of others-- varied capabi
608o026 s knowledge that it is good for him effect of EDUCATION & mental capabilities.-- '(P) Animals do atta
608o027 er in these views will pay great attention to EDUCATION.-- 4) These views are directly opposed & inex
609o30v hat part of the moral sense which experience (EDUCATION is the experience of others) shows does not t
6240o51 our instinctive feelings of right & wrong,-- EDUCATION, of parents strives* to same end.-- & general
6240o51 t part, which is acquired by association from EDUCATION & imitation, has often been perverted from wa
6240o51 d.-- As conditions change, from civilization, EDUCATION changes, & probably likewise instincts, for t
628o53v t of a shepherd dog, is strictly analogous to EDUCATION of child,-- causing many actions to be consid
629o055 her feelings, such as temperance, acquired by EDUCATION.-- CD[In similar manner our desires become fi
065r138 of Tristan D. Acuneha. Kerguelen Land. Prince EDWARDS Isld. Marion & Crozet. L. Auckland. Macqueries
197b112 of Mammalia & birds &c \ p. 32. reference to M EDWARDS. law of crustacea-- with respect to mouth thos
323c269 25th Lockarts life of Napoleon.» April 5d Dr. EDWARDS of <ter> influence of Physical causes: well sk
349d052 f primary divisions.-- Complains (p. 53) of M. EDWARDS, thinking any group good, though not circular,
404e025 When reading. L'Institut: .1838 p. 329-- Milne EDWARDS, description of curious mechanism of respirati
404e025 re & therefore it might thus have arisen, & M. EDWARDS p. 330 distinctly states that the flipper is a
435e133 ven most remote is put to it.-- April 6th "Dr. EDWARDS on the Influence of <external> Physical agents
454e184 y allied to the Sensitive Plant.-- p. 290. Dr. EDWARDS in his essay on Spermatic animalcule. has desc
480z007 of Guanaca. In transaction of Bonn Society M. EDWARDS on Corallines L'Institut 1837. No 212 Observat
484z014 the Secretary, make noise & throw head back M EDWARDS,--on polypi of Tubulipores L'Institut-- 1838 p
485z018 ter duration than cells.-- reproduced.-- Milne EDWARDS p. 138 on Polypi.-- Berenica &c &c L'Institut,
503q012 , latter says seedless-- Also about Sugar-Cane EDWARDS says does not seed-- «Bruce says does» Royle I
503q013 vember 1841.-- Trees above male? (2) Result of EDWARDS experiment in Cabbages given (3) _____ in Hea
088a014 terhouse» has frequently heard that Herons bring EELS alive to their nests; & then they may picked up
056r109 ed? Who could suppose such trifling means could EFFACE & obliterate so grand a work?--In valleys one
027r027 with sea shells--about 500 Isd. & great banks. EFFECT of Elevation. United service Journal In the Ir
041r064 ion about the deer approaching the wells.-- the EFFECT of Salt water of the Salado.--Mem. in Owens Af
044r074 abundant. Sulp. Hyd: Carb: A. Mur: A. = (& this EFFECT of water thus holding matter in solution must
053r101 ength. The Lines of Mountain appear to me to be EFFECT of expansions acting at great depths (mem: pro
056r110 er. as at Galapagos. St. Helena.-- [Fig. 7] {P} EFFECT of heat on inner wall, hence resists degradati
064r135 rior channels. there no outer coast.--important EFFECT.--? Capt. FitzRoy.-- Limited Volcanic action
090a022 ng trachyte come out before.-- What must be the EFFECT of all the meteoric stone which must have fall
094a036 n of great boulders, pebbles of granite clearly EFFECT of remodelling same manner. as bits of Patagon
099a049 this would make layers.-- (Gravity can have no EFFECT, on particles of equal weight.--)¿ cleavage no
100a052 Grit, yet it is worth consideration. especially EFFECT of gravity, versus some fault explaining vary
103a063 dge-formed.-- Hence fill up fissures-- If dikes EFFECT of horizontal elevation excepting fissures fro
103a065 considered a level.-- Dikes being last action. (EFFECT of horizontal movement) hence generally inters
106a076 Three inosculating rivers in Southern America ¿ EFFECT of subsidence-- <Is there same.> Institute. 18
109a083 be continued. no currents & elevation have same EFFECT, a tendency direct (or oblique) outwards may b
110a085 h variation of compass & these may cause «or be EFFECT of» elevation & subsidence. examine these «lin
116a099 oo great an abundance of matter would have same EFFECT as too coarse. Read Kylau on Granite Edinburgh
119a104 hat as we see mountains formed (& mountains are EFFECT of continental elevations) we may conclude tha
120a109 . if elevation near coast more than at interior EFFECT would be such as present. to spread sheet of m
128a128 n chains are matter piled up. over crevice from EFFECT of general elevation,-- when subsidence takes
139a180 as oak galls or rose «buds» galls-- is it not EFFECT of superadded vital influence?-- See End of No
152m046 the lip, or necks of land on level with shelves EFFECT of corrosion & not cause. Monday a rapid desce
171b004 world.-- On other hand, generation destroys the EFFECT of accidental injuries, <on> which if animals
179b032 not this owing to each copulation producing its EFFECT; as when bitches puppis are less purely bred
230b236 pring.-- 1st. All variation of animal is either EFFECT or adaptation,. animal best fitted to that cou
231b243 t) are related by real relationship. as well as EFFECT of similar temperature.-- now those of tempera
239cIFC spring? Instances of old Breeds taking greatest EFFECT Account of the [...] the world.-- Charles Darw
239c001 as no doubt that oldest variety, takes greatest EFFECT on offspring. Thus presuming those varieties t
259c063 ater vigour to the parent so tending to produce EFFECT on offspring-- but WHOLE race of that species
259c063 articular habit.-- All structures either direct EFFECT of habit, or heredetary «& combined» effect of
259c063 ect effect of habit, or heredetary «& combined» EFFECT of habit.-- perhaps in process of change.-- Ar
260c068 te rooms to give tinge to offspring.-- Darkness EFFECT on human offspring.-- white, snow.-- the fine
263c075 means & bringing the mind to grapple with great EFFECT produced, is a most laborious, & painful effor
265c083 --- but «accidental» changes after birth do not EFFECT progeny-- many dogs in England must have been
265c083 sheeps tails cut yet there is no record of any EFFECT.-- New Hollanders have gone on boring their no
265c084 r of Arctic countries few will say it is direct EFFECT, <of> according to Physical laws, as sulphuric
271c106 & Nature; but cannot be counteracted by Man.-- EFFECT of external contingencies & long bred in-- Mem
280c134 ne genus-- external circumstances in both cases EFFECT. it.-- Sir J.. Sebright excellent authority be
282c143 that one animal can really produce so great an EFFECT.-- the spirit of life must be every where ambi
290c163 on ground, colour of habitation. Must have some EFFECT.-- Maldonado as good forests for beautiful bir
291c166 analogy points out to this.-- love of the deity EFFECT of organization. oh you Materialist!-- Read Ba
313c236 centrated in different parts, & scission cannot EFFECT the process.-- but why two sexes scission in a
325c266 icentiousness, destroying children. --it is not EFFECT, as Lyell suggested, of organ being worn out a
333d006 half character & other more of English, but the EFFECT is the same.-- Fox thinks that when a wild ani
333d009 setters are produced. one would argue the whole EFFECT of race was determined by male: & How complete
337d020 y unnatural-- Italian Greyhound is probably the EFFECT of <sev> local variety many times changed toge
338d023 the two countries in times present & past. The EFFECT of physical conditions of country is not perha
359d088 ey were a good species, or local variety, & not EFFECT of breeding in & in like our pidgeons) The mal
360d091 al, that had long been in confinement-- is this EFFECT of climate, or state in which they are kept?--
362d099 tchell remarks seen in savages) to brave men.-- EFFECT of castration horns drop off., replaced by hai
364d104 hich years afterwards occasionally went back-- (EFFECT of imagination on mother. white peeled rods me
366d111 When will the musquitoes of S. America take an EFFECT.-- would perfect impunity from muskitoes bite
375d135 n one species of hawk decreasing in number must EFFECT instantaneously all the rest.-- One may say th
376d135 that for form which Malthus shows, is the final EFFECT, (by means however of volition) of this populo
380d153 ngs.--).-- Seep p. 84. Hens «like»-- Cocks from EFFECT of breeding in & in.-- Mr Yarrell does not kno
380d155 nfinement, & so imagination in Man, has strange EFFECT.-- Directly a Capon is cut, it increases in si
381d157 ads one to suppose still more that they must in EFFECT be so in all.-- 2 NB. In Pectinibr Mollusca.--
383d160 shaft metallic green.-- This green doubtless is EFFECT of Metallic hue of silver pheasant. yet why gr
389d177 le that too much difference should produce same EFFECT as too little --- in (latter case female offen
390d178 bose of other.-- The upshot of all this is that EFFECT of Male is to impress some difference: to make
391d175 l before he can procreate. these changes may be EFFECT of differences of parents, or external circums
401e015 -- now seeds of crab produce crab, so that some EFFECT from apple trees is produced.-- Thinks probabl
404e024 by man, common to every individual & therefore EFFECT of climate.-- Octob 19th. When reading. L'Inst
408e043 s change, we see external conditions have great EFFECT on them, & therefore extermination becomes par
408e043 s part of same law.-- When we know what a great EFFECT. light has in colouring plants,-- who can say.
416e071 re true hermaphrodites.-- I suspect this rather EFFECT of liquid semen: therefore animal life commenc
416e072 n of changes which gradation shows it to be the EFFECT of a gradation in difference in external condi
419e087 ller tells much less, (though <the> it also the EFFECT of change) than a slow gradation in form, «whi
419e087) than a slow gradation in form, «which must be EFFECT of slow change + & Therefore precludes effects

```
427e109 ally dioecious; & really dioecious plants being EFFECT of abortion of one sex.-- Linnaean class Dioec
427e109 xtreme, than before arctic forms would retreat: EFFECT on snow of arctic climate in far north regions
427e110 ner than <other:> «that of another» apple. only EFFECT produced would be different.-- same way one va
427e110 dog does not prefer other. but produces greater EFFECT on offspring-- Mr. Herbert says «p 347. Amyyra
428e113 se? has begotten <?> «Walker attributes this to EFFECT of male sex on locomotive system» I am bound t
435e130 have  been observed to change their sex,-- this  EFFECT from age, what Mr Knight [not located] the sti
440e146 &c  here there is no final cause yet it must be  EFFECT of some condition of external circumstances. r
441e149 would be similar to <f> first form.-- The great  EFFECT of conditions on offspring, but not on individ
441e150 st be presumed to be result of such laws.-- The  EFFECT of one part being greatly developed on another
446e162 nalogous to change in grafted trees «:so is not  EFFECT of different stocks in this case».-- &  strong
446e163 ct them in short period & hence no change would  EFFECT them, without affecting all the  individuals--
460t017 da & Hedgehog Tenrec both having spines, is the  EFFECT, partly of the same external conditions (ie. a
491q01v elated plant & see if first pollen produces any  EFFECT, as in case of woodpidgeon & Hen. mentioned by
494q005 uess how many generations in Mammalia. in group  EFFECT of crossing.-- (7) Are the Eggs of the Penguin
494q005 lings surrounded by various bright colours, any  EFFECT? and silk caterpillars (1) Shake a sleeping mi
495q005 l different soils & temperatures & see what the  EFFECT will be.-- will seedlings vary much more  than
495q005 or common English plants in Hothouse & see what  EFFECT on organs of generation (5) Place pollen of Re
495q05a eps on ditto-- Cover them up periodically & see  EFFECT-- such dioecious individ--small orifice (8) Ca
496q006 y have little tendency to double; what would be  EFFECT-- (10) Try in how many generations. daisy. Fev
496q006 vary, which will show that such proportions not  EFFECT of Chance Maer.= (12) Take Bag of soil from ce
498q007 ion <lat> retrograde into leaves-- is this ever  EFFECT of want of nutrition.-- Horned oranges so? --Y
501q011 or list of annuals to place in Hot house to see  EFFECT on generative organs of great Heat (32) Can He
504q013 hrewsbury one branch of Rhod. flowered later.--  EFFECT of accident?? (7) Which. Rhododendrum seeds??-
515q021 into  Nectarines (does reverse happen?) what is  EFFECT of crossing peaches & nectarines: same questio
522m012 en they hear a thing it often does not take any  EFFECT at the time, but some time afterwards it calls
522m012 oregoing cases of «mental» failure very general  EFFECT of <early> «slight habitual» intemperance.-- o
525m026 en more than the memory.-- therefore affections  EFFECT of organization which can hardly be doubted, w
530m046 sociation, & association is probably a physical  EFFECT of brain the «similar remark» thoughts,  being
532m057 accompanies  passion & not sweat is the <state>  EFFECT of short -- but violent action.-- To avoid sta
536m072 free will of oyster, one can fancy to be direct  EFFECT of organization, by the capacities its  senses
537m074 organization» for his children's sake & for the  EFFECT of his example on others. It may be doubted w
541m090 ltory thought is following out such an idea, as  EFFECT of sea on coves when waters had fallen,  as in
543m101 ed as in Shepherd dogs-- Inherited Habits: Have  EFFECT in Bones is valuable it shows that new instinc
547m113 every  late impression.-- (do the ideas, direct  EFFECT of perception by senses fail first, as whether
547m114 .-- Mayo observe no improbabilities in a dream,  EFFECT of doubting nor believing, effect of not reaso
547m114 s in a dream, effect of doubting nor believing,  EFFECT of not reasoning. effect of not having <all> o
547m114 oubting nor believing, effect of not reasoning.  EFFECT of not having <all> other trains of thought, o
553m136 » in thinking not capable to <do> produce every  EFFECT, of every kind which surrounds us. Moreover «i
564m004 nking of injured moral sense.-- Notion of deity  EFFECT of reason acting on (<not social instinct>) bu
567m012 is  now purely theological.-- Origin of cause &  EFFECT being a necessary notion is it connected  with
567m013 t animals, as hydra towards light. being direct  EFFECT of some law.-- have plants any notion of cause
567m013 f some law.-- have plants any notion of cause &  EFFECT, «they have habitual action. which depends  on
568m018 orming it.-- As soon as memory improved. direct  EFFECT of improving organization, comparison of sensa
573m038 rn in bed. will turn in bed.-- in case spittle,  EFFECT of thought is to make saliva flow, & therefore
577m053 isease tends to give it, as in cancer, showing,  EFFECT of mind on individual parts of body.== (if you
582m068 neither  knows nor intends the result they will  EFFECT,.--" this not wholly true, for we must grant a
592m104 mpt opposite action to opening eyes in fear The  EFFECT of habitual movements in muscles of face, is w
595m121 hink anything ugly-- a beau-ideal feeling. Same  EFFECT as acting on us-- <The Baby» «Effie  Wedgwood»
599o005 give  my doctrines about origin of language--  & EFFECT of reason. reason could not have existed witho
605o18v comitant with sublime. adds not a little to the  EFFECT: as when we look at the vast ocean from any he
606o025 eptember 6th. 1838 Every action whatever is the  EFFECT of a motive.-- [-- must be so, analyse (a) one
606o025 one feels it in passion, love-- jealousy-- «as»  EFFECT of bodily organisms-- one knows it, when one w
607o025 5 & 206.]CD Motives are units in the universe.  [EFFECT of heredetary constitution,-- education  under
607o025 s. As man hearing Bible for first time, & great  EFFECT being produced.-- the wax was soft,-- the cond
608o026 tives.--& his knowledge that it is good for him  EFFECT of Education & mental capabilities.-- '(P) Ani
611o034 sitive plant can be shewn to be direct physical  EFFECT of touch & not irritability, which at least sh
612o035 ysical» effects of more or less turgid vessels;  EFFECT of heat, light or shade.) Joining two difficul
615o36v yond doubt courage is heredetary in fowls & not  EFFECT of feeling of individual force in any individu
615o36v erch till evening» crow different.-- Heredetary  EFFECT of former tropical climate analogous to inflor
618o41v The  argument reduces itself to what is cause &  EFFECT: it merely is «invariable» priority of one  to
618o41v if  day was first, we should not think night an  EFFECT.]CD Cause and effect has relation to forces  &
618o41v should  not think night an effect.]CD Cause and  EFFECT has relation to forces & mentality because eff
635j55r that  of Man.-- the cause given we know not the  EFFECT [blank] 6 p. 412. Macculloch explains the shor
172b008 they inosculate, we must suppose the change is  EFFECTED at once, -- something like a variety produced
178b031 aps be instance of domesticated animals having  EFFECTED; a change whi[ch] the Fr. naturalists thought
259c062 refore mud wood be inhabited, then how is this  EFFECTED by-- for instance, fish being excessively abu
314c238 pt by direct adaptation has such a change been  EFFECTED.-- the consciousness of the plant that this p
314c238 this  part must be protected however it may be  EFFECTED.-- Prodromus Florae Norfolkicas. 1833  Steph.
336d018 l changes become multiplied, & great change be  EFFECTED, but in a mule these conditions are not fullf
360d090 howing how gradually every <thing> «change» is  EFFECTED-- the one in garden is from <bitch dog do> f
361d096 - (¿do these facts indicate that the change is  EFFECTED through the male??)-- Yarrell observed that f
372d130 nges, which its parents have.-- -- Not this is  EFFECTED by short method in generation.-- Ehrenberg co
372d131 o> & in making «true» bud some such process is  EFFECTED.-- a child might be so born. but it would  be
381d157 are really, Heautandrous.-- How is fecundation  EFFECTED in latter; are <it> «organs» open to water? W
410e051 not to the local changes. = this could only be  EFFECTED by sexes: All the above should follow after d
258c059 n where isolation, from general circumstances  EFFECTING the area equably.-- Animals having wide range
549m122 circumstances  are so conditioned as they are  EFFECTING a change in his instincts-- like what is happ
499q009 el sure, that pollen of own kind is much more  EFFECTIVE than of foreign-- Eyton has such a grove of W
035r048 which in any one country would produce equable  EFFECTS.--«though so immense to short breathed travell
058r115 arance at Concepcion [.] no sign of elevation.  EFFECTS. EFFECTS of great waves to obliterate all land marks.--
059r121 bsidian, & all show chemical action as well a  EFFECTS, of cooling [misnumbering, no page 122] In Igne
080r178 with  recent dead body of other. & see if same  EFFECTS, as with man Does Indian rubber & black lead u
102a059 rm of breaker affected some way out to sea.--¿  EFFECTS on bottom a thing floating some way from coast
107a077 hermal {P} subterranean line moves upward from  EFFECTS of Elevation if not crust much thinner beneath
109a083 ent for Heaping up.-- very good this will show  EFFECTS -- analogous to broad flat sand beach. {P}  --
115a096 the  homogenious crystals., must aid in adding  EFFECTS to common heat.-- Where there are cliffs there
117a102 doubtful. Examine well shores of lakes. to see  EFFECTS of degradation, «no» tides, water always falli
127a126 lass. (oil may be cooled to 0 degree!)-- shows  EFFECTS of pressure in change of form as the result of
127a127 ove elevation.-- great mountain chains. may be  EFFECTS of subsidence Elie de Beaum. Memoires of Frenc
137a149 apours condensed from furnaces do/p. 84 on the  EFFECTS of veins of slag in iron furnaces affecting to
137a151 Phil J. 1838. p. 132. «& 134» Bischoff. On the  EFFECTS of meteoric waters on the temperature of the i
206b148 uses, dislike to marriage, heredetary disease,  EFFECTS of contagions & accidents: yet some causes are
225b217 og & (Greyhound). I should say the changes were  EFFECTS of external causes, of which we are as ignoran
239cIFC with specks» which cross & keep colour on wing  EFFECTS of colour on parent, white room How are variet
240c013 in the Archipelago-- Former statements to such  EFFECTS false In New Guinea. a Kangaroo d'Aroe (Didelp
261c070 , black bull finches from linseed-- not solely  EFFECTS of climate on some «antecedent races,  perhaps
279c133 Sebright-- pamphlet-- most important, showing  EFFECTS of peculiarities being long in blood.-- ++ thi
279c133 ++ thinks difficulty in crossing race.-- bad  EFFECTS of incestuous intercourse..-- excellent observ
285c151 h No: because decay in that species <shows> is  EFFECTS of unfavourable conditions. [hence rise & depr
285c153 will be unequally rapid, with respect to their  EFFECTS The AEgyptian animals domesticated <?>, & the
315c242 mory of animals.-- (surely in plants movements  EFFECTS of irritability, though means injection of flu
356d069 .-- The history of Medicine, the extraordinary  EFFECTS of different Medicines on organs, leads one to
358d076 or intellectually"= Opposed to these facts are  EFFECTS of castration on males & of age or  castration
358d086 idity, & breeding in & in tend to produce same  EFFECTS.-- CD[May it be said, that breeding in & in te
365d107 & by art.-- the former only giving average of  EFFECTS of country, (& no monstrosity, or adaptation t
377d140 s. a contingency of time. When we multiply the  EFFECTS of <earthquakes>, elevating forces in  raising
390d179 rary of Useful Knowledge Bell's Quadrupeds the  EFFECTS of breeding in, it is not merely the too close
398e005 xtremely faulty» The difficulty of multiplying  EFFECTS & to <ponder> conceive the result with that cl
```

399e010 y suspect, that breeding in & in, produces bad EFFECTS solely, because of similarity, because in ever
400e013 four generation before they broke.--, showing EFFECTS of cultivation gradually adding up. & four mor
405e032 erent islets have different forms it is either EFFECTS of having been long separated, or having never
409e048 als, & though we may not trace out all the ill EFFECTS. -- we see it is not the order in this perfect
410e051 pecies., when going N.orth & South Thinking of EFFECTS of my theory, laws probably will be discovered
419e087 effect of slow change + & Therefore precludes EFFECTS of catastrophes, which must serve to confound
434e127 of body, or whether it merely has tendency (as EFFECTS of cultivation on successive generations of pl
434e127 uccessive generations of plants) to do so, then EFFECTS are equally handed to offspring.-- Whewell's a
435e134 intellect, he resembles, other mammalia in the EFFECTS produced on organization. by physical agents."
441e148 No than by growth-- generation; & more of the EFFECTS of conditions on the «propagating» constitutio
493q003 g bones & see if relation is same good, avoids EFFECTS of fatness.-- Experiment in crossing animals.-
503q012 information & Hope about Silk worms. Varieties EFFECTS of domestication-- said to require Selection (
505q014 anted & Linum Perenne.-- Herbert's. fact.= (4) EFFECTS of Nitrate of Soda under Beech.-- Lychnis dioi
511q018 s foot undivided, & more particulars regarding EFFECTS of crossing them with common pigs= [it is a Li
532m057 rt, sweat, trembling of muscles, are not these EFFECTS of violent running away, & must not <this> «ru
532m057 must not <this> «running away» have been usual EFFECTS of fear.-- the state of collapse may be imitat
537m075 leave off <t> instinct, when attended with bad EFFECTS Martineau. How to observe, p. 21-26. argues «w
537m077 will his country or Swis.-- it may be answered EFFECTS of education, may be opposed undoubted cases o
553m135 ertake an emu, & [not located] notion, are not EFFECTS of impressions long repeated, without the powe
564n003 as his conscience far more sensitive. ultimate EFFECTS of actions.+ till at last the face «instinct of
576n047 ive. developement) on account of dark ages.-- «EFFECTS of external circumstances» Look at Spain now.--
581n066 ity of tongues.-- [Emotions are the heredetary EFFECTS on the mind, accompanying certain bodily actio
591n097 tal laws (in the Athaenaeum Library) describes EFFECTS of emotions-- fear giving goose skin-- & hair
599o007 ns & thoroughfares, I grant that man, from the EFFECTS of heredetary knowledge, has produced almost →
604o016 sms structure, it looks as if consciousness an EFFECTS of sufficient perfection of organization & if
608o026 rtant explanation] he thinks they have none.-- EFFECTS.-- One must view a wrecked man, like a sickly
610o034 -- which the reviewer seems to doubt. [RHC] 1) EFFECTS of Life in the abstract is matter united by ce
612o035 ∵ I should think there. was direct «physical» EFFECTS of more or less turgid vessels; effect of heat
620o045 quence of that being part of our nature, & its EFFECTS lasting, whilst passions although equally natu
620o045 whilst passions although equally natural leave EFFECTS not lasting. By association one gains the rule
624o051 probably likewise instincts, for the same law EFFECTS both.-- <such> changes «in accordance to benef
260c068 s community of language Desert country. is as EFFECTUAL as a cold one. in checking beautiful colours
430e118 sing & keeping breed pure.-- «& so in plants EFFECTUALLY the offspring are picked & not allowed to cr
358d085 the breeding in & in (like «courage in dogs» EFFEMINATE men),-- if carried much further, if by the p
058r118 rres Australs Vol. I. p. 54. M. Bailly says."en EFFET toutes les montagnes de cette ile se developpen
436e135 eparation in horizontal direction is far more EFFICIENT in making species, than time (as cause of cha
436e135 & separation of a few, probably would be most EFFICIENT in producing new species; also one being redu
595n121 ling. Same effect as acting on us-- <The Baby» EFFIE Wedgwood» April 28th 1840 was frightened at wil
087a010 American bone not probably in salt marshes EFFLORESCENCE nothing -- Study account.-- Alluvial plains
263c075 effect produced, is a most laborious, & painful EFFORT of the mind (although this may appear an absur
280c136 ts stronger & stronger, so that though by great EFFORT <pr> one unlike can be produced, yet to produc
293c173 y circumstanc it becomes web footed, now Man by EFFORT of Memory can remember how to swim after havin
532m055 they could recollect these same things from any EFFORT of will whilst their minds were sound. Carolin
536m070 y discover struggle of feeling.-- It is as much EFFORT to walk then lightly as to endeavur to stop he
541m090 as in my Glen Roy paper.-- this greatest mental EFFORT, of which I am capable-- I suspect from these
541m090 capable-- I suspect from these facts that whole EFFORT consists in keeping one idea before your mind
541m090 mind for period, «if the scarlet was before one EFFORT less» one is obliged to repeat the word, & thi
541m091 with all this difficult EXPERIMENTIZE upon this EFFORT.-- it looks so analogous to muscle in one posi
541m091 following changes during fall of sea.-- Is the EFFORT greater if the idea is abstract as love, (or a
608o026 nally mostly INSTINCTIVE, & therefore now great EFFORT of reason to discover them: this is important
617o39v spect of bodily action is revealed to us by the EFFORT it costs us to exert force or by internal cons
618o41v fect has relation to forces & mentality because EFFORT is felt [LHC] 1) May 5th. 1839.-- Maer Mackint
602r124 rara. The earthquakes "seem to arise from some EFFORTS in the land to lift itself higher & to grow up
536m070 skipping when wanting not to feel angry-- such EFFORTS prevent anger, but observing eyes thus unconsc
243c019 cquarie et Campbell (52 degree S) qui n'aient EGALEMENT leur espèces; et certainment on eût été bien
341d031 rtile never even in seven years produced even an EGG.-- a most curious bird, did not seem to know its
444e158 s to argue, that as the transformations from the EGG, or larva. or foetus to perfect animal are adapt
470t153 ll be collected & resown.-- Humble 22 flowers of EGG Tree in one minute Great Humble 17 flowers of La
470t178 w Bees;»-- on Monk's Hood, brushing over stamen «EGG Tree»--I think never on the Galeum saxatile & ot
494q004 eties (9) Cross largest Malay with Bantam-- will EGG kill Hen Bantam.-- Cross common Fowl with Dorkin
534m061 onsense; yet who will venture to say germ within EGG, cannot think-- as well as animal born with inst
535m064 though not feeding them «nor helping larva from EGG» watching them, brooding over them, preserving t
566n011 real instinct in the chicken, just bursting from EGG.-- Animals have necessary notions. which of them
637j58r aptations Horny point to chickens beak, to break EGG. shells-- why chicken could not have lived had i
637j58r n could not have lived had it not been so.-- let EGG shells grow harder. so must those with weak beak
063r132 oducing distinct animals. still partly united. & EGGS which become quite separate.--Considering all i
240c004 t requires the greatest difficulty to rear them, EGGS hatched under other birds & brought up by hand.
289c162 of producing young many times & lay two sorts of EGGS-- one remaining through winter. «It would be cu
300c197 each become the fathers of many species.-- a few EGGS transported to the Str of Magellan.-- Change of
316c245 rt ova of shells.-- Conchifera. hermaphrodites-- EGGS in groups.. Have Dioecious plants more restrict
342d032 th pure bred (just like common mules) & lay many EGGS but never produce inter se or with parent speci
347d049 ttle. Ch 3. Instinct L'Institut. p. 249. (1838). EGGS discovered to Taenia.-- hard so as to resist se
358d085 mon. pheasant & fowl when crossed never even lay EGGS. & the men cannot «hardly» tell any sex by appe
378d148 wls.--??? Wate[r]ton «p. 197» put 12 wild duck's EGGS under common ducks, the young crossed amongst t
380d154 cannot be distinguished.-- A capon will sit upon EGGS, as well as, & often better than a female.-- th
385d164 h breed. (he thinks half pheasant, half fowls.-- EGGS fertile, but parent bird will never sit on them
387d169 or Pipe fish the male of which receives <young> «EGGS» in belly.-- analogous to men having mammae.--
417e079 hes» & reptiles & those which <lay> «have» their EGGS, <inter>, impregnated externally; nor can it be
446e160 in domestic cares, as building nest, sitting on EGGS. & feeding & defending their young.-- The orio
449e168 knows of a Gull, which has laid in domestication EGGS of two shapes & colour.-- Eyton has observed sa
460t013 an-gander with common goose produce full as many EGGS as pure bred common.-- the half of the cross, a
489qIFC nt caterpillar races. --Name of Italian who sold EGGS.-- Temporary Question 1 Where has Duchesne desc
494q004 alia. in group effect of crossing.-- (7) Are the EGGS of the Penguin Duck quite similar to those of a
496q008 ower in same receptacle (8) Make Duck eat Spawn, EGGS of snail, row of fish & kill them in hour or tw
500q010 -- (25) Does the yellow white Butterfly deposit EGGS in all varieties of Cabbage. (26) Do deer Keepe
502q11v oured like cock-pheasant: said not to sit on own EGGS Flowers in short turf. for abortion. or for ste
515q022 individuals of same species Eyton (1) Number of EGGS-- of half-bred geese-- inter se, & with parents
535m064 case of insects «a Perga» of Terebrantia, laying EGGS on leaves of Eucalyptus, watching few days till
637j58r & Decandoelle.-- The Final cause of innumerable EGGS is explained by Malthus.-- [is it anomaly in me
175b016 change.-- this difficult to prove cats &c from EGYPT no answer because time short & no great change
266c092 ormous tails, which disfigure those of Arabia & EGYPT.-- CIVETS CATS only wild animals on isld.-- Nie
438e142 time of domestication [see Wikinson on dogs of EGYPT & Cuvier on Mummies]CD [NB TIME is element in c
483z012 ol IV p. 91.-- Vol IV p. 388. Domestic mouse of EGYPT is Mus Cahirimus. of Geof.-- reference from Rüp
308c219 wledge «Negroes existed since time of earliest EGYPTIAN drawings & Old Testament» Domesticated animal
323c269 y -- Bell's Bridgewater Treatise -- Wilkinsons EGYPTIAN remains skimmed -- Pliny Nat. Hist of World d
343d039 ow long a Mammal may go on as one species from EGYPTIAN Mummies & from the existing animals found fos
365d106 nt & Black Cock, & other hybrids-- The fact of EGYPTIAN animals not having changed is good-- I scarce
439e145 » Wilkinsons Manners & Customs of the Ancient EGYPTIANS Vol III. p. 33-- They had several breeds of d
114a093 -- Taylors Scientific Memoir, Part IV. p. 403 EHRENBERG on ferrugineous Gallionella Examine Iron ston
175b019 s been uniform, at former epoch-- How is this EHRENBERG? every successive animal is branching upwards
267c095 rodents Abstract of Infusoria. <p> do p. 62. EHRENBERG Annals of Nat. Hist. precursor of magazine???
313c236 es scission in all cases probably gemmation (.EHRENBERG) --not necessary to generation (lateral with
372d130 is effected by short method in generation.-- EHRENBERG considers artificial division of animals, as
386d167 s.-- Annals of Natural History. 1838. p. 123. EHRENBERG. makes gemmation in animals very different fr
109a084 tudied.-- Scientific Memoirs Edited by Taylor EHRENBERGH on flints in chalk must be studied-- though
207b150 unknown forms, a circumstance undiscovered by EHRENBERGH.-- <Marcel Serres p. 331. L'Institut-- consi
283c146 ote genera. will be produced. As we know from EHRENBERGH, there are fossil (see Scientific Memoirs &
388d172 breeding-- remarkable Athenaeum 1838. p 653. EHRENBERG<H> thinks multiplication by division <only> is

282c143 h would never be discovered When one reads in EHRENBERGS Paper on Infusoria on the the enormous produ
418e085 -- Decemb. 21th.-- L'Institut 1838. p. 412. M. EICHWALD has published Fauna of Caspian.-- fishes fres
640j167 ave afforded only 10th part before & now formed EIGHTH part.-- or if other prey diminished, total num
032r041 intertropical)» hence varieties of substances EJECTED from same point. & changes. «(changes in varia
034r044 example of do? <Poor> Daubeny good account of EJECTED granitic fragments P. 386 Mem. Lyell's fact ab
133a140 herefore the abysses where fluid rock has been EJECTED must remain fluid for an enormous period: now
470t178 Spanish Broom by pulling back Wings, pollen is EJECTED with violence in shower On many Papilionaceous
496q05a eding on oats, give body to Hawk & sow pellet. EJECTED. done Examine pollen of such flowers as do not
068r147 trata can be? Problem dislocate strata without EJECTION of the fluid propelling mass. If one inch can
616o038 cretion in both involuntary, <application in> «EJECTION only has» will: there must be cases of secret
467t104 corniculatus saw Humble press down wings which EJECTS pollen from tip of sheath.-- «Also in Lathyrus
138a153 [blank] «p. 4. (Lyells Book)» Observaciones sobre EL Clima del Lima par Dr. H. Unanùe says he believe
282c160 eater than those heredetary ones.-- which would ELAPSE; during time such changes had elapsed.-- let t
465t079 tant, to bear in mind that enormous periods may ELAPSE, even in situations apparently favourable for
267c093 ett Voyage round world, 20 years have scarcely ELAPSED since the Guava introduced from Norfolk Isld--
282c141 ich would elapse; during time such changes had ELAPSED.-- let these families take <dogs> «domestic an
304c208 t, where much death, may be inferred much time ELAPSED & therefore descended from branch high up.-- S
433e126 me great change who can say how many centuries ELAPSED between each of these gaps, far more probably
400e011 n Diemens land, where there is close species of ELATER.-- Where this collection is particularly rich
424e102 different ranges-- latter not going North of the ELBE.-- yet they meet in one wood in Anhault. & ther
363d101 on,-- several have a group of white speckles on ELBOW joint-- in Bewick drawing the the rock Pidgeon
092a029 r Bird in paper to Brit. Assoc: has shewn how ELECTRICAL currents tend to deposit metals, if in solut
495q05a n their returning powers-- then carry them in ELECTRICAL machine, reversing the poles test by suspend
022r006 how any difference in facility of conducting ELECTRICITY? Would minute particles have a tendency to c
058r115 nes. «Mem silky lustre» ask Erasmus. whether ELECTRICITY would affect this. -- State the circumstance
077r172] Dr D. remarks. bad conductor of Heat do of ELECTRICITY Does not iron, combined with nickel & cobalt
175b017 r Unknown causes of change. Volcanic isld.-- ELECTRICITY Each species changes. does it progress. Man
611o034 ic are probably one principle for connect of ELECTRICITY chemical attraction, heat & gravity is proba
528m037 paper two waving perfectly parallel lines are ELEGANT.-- Again there is beauty in rhythm & symmetry,
181b045 ribes of animals adapted preeminently for each ELEMENT, but it seems law that such tribes, as far as
185b055 but carrying on. attempt at adaptation of each ELEMENT.-- May this not be explained on principle, of
227b225 ation + heredetary structure. latter far chief ELEMENT.'. little service habits in classification. or r
240c003 ving feathers.-- It is possible, time being an ELEMENT in the transmission of form, may explain mule
264c081 From this view habits must form most important ELEMENT in considering to which tribe,-- structure wit
268c099 can we worked out into. Milvulus [not located] ELEMENT geographical distribution is.-- ¿Pelagic forms
291c165 tary. «problem solved» habits become important ELEMENT in classification, because structure has tende
310c225 probably have changed much less.-- Here is an ELEMENT of extreme difficulty in mundine geological ch
316c245 ny to cold.-- Hence latitude is more important ELEMENT than longitude.-- But in land & F W shells the
342d033 ed as distant species?? or is time the varying ELEMENT). Then do those SPECIES which breed most freel
372d129 oductive faculty + in the separated part every ELEMENT of the living body is present-- in generation
377d139 St) yet no change in English species-- time no ELEMENT in making change, only in fixing it: only circ
431e123 is is very limited view, though perhaps a true ELEMENT) «give examples, pigs, with small chinese boar
438e142 gs of Egypt & Cuvier on Mummies]CD [NB TIME is ELEMENT in change, as in Dahlias]CD all much varied br
599o05v - We cannot doubt that language is an altering ELEMENT, we see words invented-- we see their origin i
308c218 starting place, because there is nothing more ELEMENTARY than that complex nature itself with which o
124a120 one & volcanic rock containing magnesia Lyell. ELEMENTS p.119 on such strata {P} do p. 171. argument
176b023 e branching in the tree of life owing to three ELEMENTS air, land & water, & the endeavour of each <o
176b024 in stem of the tree is adapted for these three ELEMENTS, there will be certainly points of affinity i
181b043 & Plants But yet besides affinities from three ELEMENTS, from the «infinite» variations, & all coming
181b045 ucture are in minor degrees adapted for. other ELEMENTS. every part would probably be not complete, i
185b057 (NB I see Waterhouse thinks Quinary only three ELEMENTS) How far Does Waterhouse's representatives ag
197b112 count resemblances &'. quinary system, or three ELEMENTS p 66]CD With unknown limits, every tribe appe
235b263 &c Work out Quinary system according to three ELEMENTS How is Fauna of Van Diemens Land & Australia
282c142 ently aquatic--; «NB, aquatic, i.e relation to ELEMENTS & not minding particular trades.--» then the
283c143 The aerial type in each family is relation to ELEMENTS & not habits as shown by frigate Bird & flyin
321c270 ky's Voyage round world. 1803-6 Nothing Lyells ELEMENTS. of Geology Gibbons life on himself Hume's do,
373d133 <nerve> living nerve nursed in Mould.-- Lyells ELEMENTS. p. 290. Dr. Beck on numerical proportion in
429e115 line, as trees in Beagle Channel.-- it is not ELEMENTS.-- we cannot believe in such a line., it is o
133a141 nture & Beagle vol I. p. 2 & 3. Porphyry at St. ELENA. p. 6. few «living» shells. on coast of do p 8.
041r064 -Mem. in Owens Africa it is mentioned that the ELEPHANT came towns driving by the want of water.--I b
064r134 species than of individual.-- Mr Birchell says ELEPHANT lives on very wretched cou[n]tries thinly cov
187b062 a a more> cold countries.-- Seeing how horse & ELEPHANT reached S. America.-- explains how Zebras rea
187b062 d South Africa-- It is a wonderful fact Horse, ELEPHANT & Mastodon dying out about same time in such
229b233 s range from Abyssinia to extreme South coast. ELEPHANT he believes is mentioned by old writers on ex
242c017 ». Aerch.»|| Borneo & Sumatra both seem to have ELEPHANT & has orangs, || Tapir common to Sumatra & Mal
253c046 rodents of two countries» & «Monkeys.» Fact of ELEPHANT same species in Borneo. Sumatra. India Ceylon
291c167 succession of others, (which is almost proved. ELEPHANT has left no descendant in Europe<)> Toxodon i
299c196 nt races, being unequally developed.-- ¿is not ELEPHANT intellectually developed amongst Pachydermata
313c235 e.-- habit is awakened by association (case of ELEPHANT, which had run wild in India. in Heber?) is a
337d022 he believes «hyaena & squirrel» seal &_mouse ELEPHANT, come from one stock.-- Theory of Geograph. D
338d025 mus.? Indio-African, or pure Africa-- |Fossil ELEPHANT of Africa Most important under this view, & H
380d155 nto female, when muzzled, he is disabled.-- so ELEPHANT in confinement, & so imagination in Man, has
388d172 se. extinction. Animals in domestication (even ELEPHANT) not breeding-- remarkable Athenaeum 1838. p
402e020 to the adjacent island-- In Sooloo we find the ELEPHANT-- in Magindanao several kinds of the large mo
413e060 Reliqu: Diluv. says Africa only place, where, ELEPHANT, Rhinoceros, Hippot, Haena &c are found toget
414e065 geological history of man is as perfect as the ELEPHANT, if some genus. holding same relation as Mast
450e174 ingapore in 1837. by Mr. J. H. Moor-- -- p. 1. ELEPHANT. Rhinceros Leopard (but not Royal Tiger), &c
451e179 7. it would appear as if p. 345. The Ceylonese ELEPHANT [...] saul forests by having a smaller, light
543m098 orders more advanced than others just as dog & ELEPHANT most intellectual.-- Hymenoptera typical inse
556m144 is the Philosophy of Shame & Blushing]CD «Does ELEPHANT know shame-- dog knows triumph.--» Sept. 23rd
569n020 languages seem simplits case of Association.-- ELEPHANT often given food & word open your mouth said,
191b080 n of the earth must be excessive up & down.-- ELEPHANTS in Ceylon-- East Indian archipelago.-- West I
278c132 river. The change in England from Rhinoceros ELEPHANTS &c in the most modern period, compared to Fau
301c199 on.-- Habit instinct gained during life.-- do ELEPHANTS easily acquire habits is this the Key to thei
305c210 ne of reason in order of <ver> Mammals.-- Mem ELEPHANTS & dog.-- There is one living spirit, prevalen
377d139 them.-- England was united to Continent, when ELEPHANTS lived. & when present animals-- lived.-- we e
405e031 ocated] Did man spread over world as early as ELEPHANTS &c &.-- if in next 20 years none of his remai
450e173 ter-- Forrest Voyage p. 323. Sooloo. imported ELEPHANTS. wild hogs-- spotted deer, no loonies, but co
450e176 orse do. Appendix. p. 43. «& 45» the Breed of ELEPHANTS <of> in little isld of Sooloo.-- said to have
473s03r rope-- Cuvier never found remains of Sus with ELEPHANTS-- Lyell says New Red Sandstone of. N. America
551m128 his Principles talks of it as wonderful that ELEPHANTS understand contracts.-- but W. Fox's dog that
552m131 from all parts straw to make its nest. Pigs & ELEPHANTS, (both Pachyderms) much intellect.-- mem: Yar
253c046 ps shows great persistency of character. Hence ELEPHAS primigenious over so wide a range, & Mastodon
191b082 med by subsidence. Java & Sumatra. Rhinoceros. ELEVATE & join keep distinct. two species made elevati
263c074 -- Give the case of Apterix-- split, depress & ELEVATE & enlarge New Zealand; a division of nature of
037r054 he rock not weathering allows such Compare the ELEVATED estuary of the Plata. to the Bay of Bengal. d
038r059 state that all the great Volcanoes. have been ELEVATED considerably. which shows an afflux of inferi
042r068 Lyell Owing to «open» faults in mountains: to ELEVATED strata in eocene lakes of France, & unequal a
042r069 : in France we have freshwater lakes unequally ELEVATED, which movements if present in the Andes, wou
052r099 bbles in the Cordilleras look as if some peaks ELEVATED.-- Greywacke. as a general fact absent in T.
056r110 Playfair to be the case p. 51). presupposes an ELEVATED country of granite, not <so> great«er» for al
061r127 ered: <altered> Mem: my idea of Volc: islands. ELEVATED. then peculiar plants created. if for such me
068r147 grown solid.) Red Sea near Kosir, land appears ELEVATED. Geograph. Journal p 202 Vol IV When recollec
078r175 rvins or along gentle slopes. but on the most ELEVATED summits, where mountains most torn.--(¿anticl
087a007 evel of the Sea. Lyells Encyclopaedia-- Lately ELEVATED When Siberia went up. Arctic land went down.-
089a020 e la Soc. «de Geneva» Vol 3' P. II. -- Bed, of ELEVATED shells on the Senegal. L Institut p. 192.-- (
112a088 f S. Cruz in connection with Fitz Roys fact of ELEVATED block of stone.-- & Caldcleughs collection of
121a112 ame it. that Glen Roy district could have been ELEVATED without fissure & unequal.-- where were crack

309c220 areas of elevation: Marked out by existence of ELEVATED «extinct?» genera of shells.-- duration in tw
419e086 , is adverse.-- Decemb. 25th.-- Lyell says the ELEVATED shells in Bayfields district are much more li
451e179 y having a smaller, lighter head, carried more ELEVATED & higher forequarters: is said to be of a bol
540m085 hether in the cold regions of the North,-- the ELEVATED table land of Peru the hot plains of the Amaz
039r061 ect <The grand propulsion of fluid rock, which ELEVATES a continent> We are more abound to take analo
047r181 -But from above reasons, do not think so also ELEVATING Earthquake of Valparaiso. (1822) no great wav
377d140 hen we multiply the effects of <earthquakes>, ELEVATING forces in raising continents, & forming mount
023r010 nadmissible «may have happened from incipient ELEVATION.» The volcanos originated in the bottom of th
026r020 formation South of the Maypo at one period of ELEVATION must in its configuration have resembled Chil
027r027 ells--about 500 Isd. & great banks. effect of ELEVATION. United service Journal In the Iron sand form
035r047 oubled England; the more minute equalities of ELEVATION, may well be preserved at Patagonia. The Engl
036r051 ony with the prevailing movement being one of ELEVATION alone.--In England much subsidence: hence dif
039r061 the land decidedly bears the stamp of recent ELEVATION. which is different from what Mr Lyell suppos
042r069 --All my observations of period «& manner» of ELEVATION Volcanic action, must be more exclusively con
047r083 t vary proportionally Partial shrinking after ELEVATION in perfect conformity with <Mr Lyell's> idea
055r106 . Perhaps these facts attest a <more> decided ELEVATION of sea's bottom. beds of shells. 2 - 3 toises
057r113 ts: Consider well age of Bones. = slowness of ELEVATION proved at St Julian. = do not these bones dif
058r115 es of appearance at Concepcion [.] no sign of ELEVATION. Effects of great waves to obliterate all lan
064r137 of Cordillera it must be said, that lines of ELEVATION have connected <lines> «points» of eruption [
065r139 Julian & Port Desire applicable to Craters of ELEVATION.--The longer diameter of Deception Isl is six
065r140 nd connection of quadrupeds.--although recent ELEVATION, there may have been great subsidence previou
067r144 gradation of landlocked harbors to Craters of ELEVATION.-- Lyell suggested to me that no metals in Po
069r149 e afraid of speculating on the sea The 24 ft. ELEVATION at Concepcion. from impossibility of such cha
086a004 against the prejudice of not believing recent ELEVATION, yet sea shells at tops of mountains we ought
086a005 saline tendency of America connected with its ELEVATION. vapour from below-- Malte Brun «Salt Lakes»
088a013 , that there are at least several attempts at ELEVATION From the lost & turned about position of stra
093a034 . Julian. -- Mr Scrope seems to consider that ELEVATION & eruptions are antagonist forces. but they a
095a039 & waterworn rocks «at Cobija.» At Iquique of ELEVATION to amount of 30 ft.-- Mr Bollaert (at Roy. In
103a063 l up fissures-- If dikes effect of horizontal ELEVATION excepting fissures from above unite with thos
107a077 ubterranean line moves upward from effects of ELEVATION if not crust much thinner beneath ocean than
109a082 .-- but this must be continued. no currents & ELEVATION have same effect, a tendency direct (or obliq
110a085 f compass & these may cause «or be effect of» ELEVATION & subsidence. examine these «lines» Descripti
117a101 e deep water. will bear on formations. during ELEVATION & depression. C. Prevost.-- My views of insen
119a104 continental elevations) we may conclude that ELEVATION is independent of spreading out matter by act
120a109 ian valleys must be profoundly considered. if ELEVATION near coast more than at interior effect would
120a109 to spread sheet of matter over surface.-- if ELEVATION then went on at greater rate, not only river
120a110 xcess of matter brought down Mention absolute ELEVATION of Patagonian blocks (1200 ft??). Scotland at
122a114 th statical.-- if platform of mexico owes its ELEVATION to equilibrium.-- it cannot be equilibrium of
127a127 kes in mountains. «(not on continents)» prove ELEVATION.-- great mountain chains. may be effects of s
128a128 piled up. over crevice from effect of general ELEVATION.-- when subsidence takes place.-- Mountain wi
130a134 easily must be thin: <beside mere fracture> A ELEVATION as in Patagonia {P} B subsidence; <as in> be
152g047 ad comminuted shells Important contingency if ELEVATION from Axis, then rivers might deposit, & after
191b082 levate & join keep distinct. two species made ELEVATION & subsidence continually forming species.-- M
191b083 y South Africa. proof of subsidence. & recent ELEVATION: Pray ask Dr. Smith.-- «to state that most cl
225b219 . ice}. Mauritius what a difficulty-- where ELEVATION Subsidence New is only hope.-- New Zealand «c
298c130 cies from one in lower country during gradual ELEVATION of isld.-- We must imagine a considerable ran
309c220 f same <spe> genera. is not equal to areas of ELEVATION: Marked out by existence of elevated «extinct
342d034 salt water (as they almost must have been on ELEVATION of continents) but Ogleby well answers that n
408e044 o these islands, having been purely result of ELEVATION,-- «all» modern & wholly volcanic-- Azores mi
408e045 s even in that quarter of the world-- -- Mem. ELEVATION & subsidence of East Indian Archipelago. now
419e086 changes, than in any other, & yet 200-300 ft ELEVATION & no change & even no loss of species.-- It m
473s004 in. If there has been «as» much subsidence as ELEVATION then all continents of cretaceous periods, to
023r010 through» faults or fissures, produced by the ELEVATIONS of those mountains on the continent of S. Am
038r059 dikes generally vertical? if so posterior to ELEVATIONS? & not sources of lava streams.--Urge not ti
039r060 hat successive terraces mark as many distinct ELEVATIONS; hence it would appear he has not fully cons
047r083 sâ more likely to be coincidental than single ELEVATIONS along whole line of coast Darby mentions bed
058r117 earthquakes travel in order.-- If we look at ELEVATIONS as constantly going on we shall see a cause
064r135 ction & limited earthquakes & great but local ELEVATIONS of the land in Europe-- Urge difference of p
088a013 inge no doubt fluid.-- analogy as continental ELEVATIONS slow. so would line of mountain chain be Mr
102a062 canic eruptions are accompanied by horizontal ELEVATIONS, so are injection of mountain chains. accomp
118a104 ainst Herschel's view of cause of continental ELEVATIONS (I) the alternation of linear bands of movem
119a104 formed (& mountains are effect of continental ELEVATIONS) we may conclude that elevation is independe
231b241 rent.-- Join Sumatra & Java <to> together, by ELEVATIONS now in Progress, & you will have two. Tapir
124a118 fers no explanation of intermittent action of ELEVATORY force-- Erasmus says he has seen in making br
469t135 d this same bunch & on this day in five minutes ELEVEN Humbles came & each visited many flowers-- Saw
349d051 ship is doubtful.-- not till much knowledge is ELICITED.-- It will rest upon the discovery what chara
355d067 ier Ossemens Fossiles» Although no new fact be ELICITED by these speculation even if partly true they
127a127 at mountain chains. may be effects of subsidence ELIE De Beaum. Memoires of French Geolog. Cantal Vol
322c270 al papers-- all, that bear on any of my subjects ELIE De Beaumonts. 10 Vol. of Memoirs on Geology of
206b146 rs backward might be calculated & this number ELIMANATED say 150 people four hundred years since were
046r078 t they do not flow out together? How are they ELIMINATED.--«Sulphur last.--» Metallic veins likewise
254c049 organs blended together, & same organ «where ELIMINATED is» often repeated, as mouths in Polypi, Sur
278c129 l be most difficult. Sub-genera so far may be ELIMINATED. where every species of a section is confine
461t037 modated to new circumstances than it would be ELIMINATED, & hence, the application of structure to pu
640j28v ng with ordinary race.-- «There is no way of ELIMINATING the evils of old age, after breeding season,
459e144 their own pollen to that of other variety.-- «ELIZABETH & Hensleigh. seemed to think it absurd. that
472s01v 27), always been empty.-- See separate note-- ELIZABETH says several years ago seeds were procured wi
191b080 ies different.-- Ireland & Isle of Man possessed ELK not England. Did Ireland possesse Mastodons?? Ne
319c257 of Van Diemen's land & Australia proper.-- Irish ELK case of fossil geographical range.-- [blank] Boo
058r118 s montagnes de cette ile se development autour d'ELLE comme une ceinture d'immenses remparts; toutes
207b152 nt Vol. III. p. 164. Lile de la Reunion presente ELLE seule plus d'especes polymorphes que toute la t
058r118 is, au contraire, que vers le centre de' l ile, ELLES presentent une coupe abrupte et souvent taillée
059r119 ntre d'ile, vers la mer; ces couches ont entre ELLES une correspondance exacte, et lorsquelles se tr
602o1ov he highest enjoyment in mutilated statues In ELLIOTSON'S Physiology much about sleep-- Nerves.-- Voli
219b196 nts common take an example from T. del Fuego.-- ELLIS (?) says Tahitian kings. would hardly produce f
225b220 ipelago very good on opposite tendency.-- Study ELLIS & Williams. zoology of South Sea islds. any ani
502q11v put into spirits, poison-drop exudes-- does not ELM. does it «in» melon-- «Loasa» Anchusa «Campanula
498d007 ount of variation in plants raised by Scions, as ELMS. &c &c-- I have some reason to suspect Elms.--
498q007 , as Elms. &c &c-- I have some reason to suspect ELMS.-- & Orchidaceous plants no other case.-- (6)
243c019 t leur espèces; et certainment on eût été bien ELOIGNÉ, il y a peu d'années, d'admettre que ces oisea
430e120 k Woodward Mr Lonsdale says Trigonia costata & ELONGATA thougt considerably different, in proportiona
431e121 England found in the Inferior Oolite, & the T. ELONGATA in the uper formations Portland Stone &c &c.-
569n020 the mind-- superstition & charity & prayer, or ELOQUENT request. Reason in simplets form probably is
171b002 a arguements, fails in hybrids where every thing ELSE is perfect; mothes apparently only born to bree
224b215 marked varieties & may someday produce something ELSE., but not probable owing to mixture of races.--
230b240 evidence, as so many plants produce hybrids, or ELSE whole fabric will be overturned.-- Hence extrem
293c174 ul.-- The corelations are not, however, perfect, ELSE one animal would not cause misery to other.-- e
293c174 se one animal would not cause misery to other.-- ELSE smell of Man would be disagreeable to Musquitoe
297c186 --to cats &c.-- must be acquired by my theory-- ELSE my theory not applicable [not located] p. 428.
367d112 e ovum within it, is forming «& this must be so, ELSE avitism could hardly ever occur.--»--.-- & if tha
379d152 hat the ewe must never be put to any other breed ELSE all the lambs will deteriorate.-- Lord Moreton'
429e116 which penetrate Pyrenees but are found no where ELSE not even in branch valleys-- M. Ramond offers n
544m103 is more prolonged than dream. never fatiguing,-- ELSE it is only our consciousness, & senses tell us
548m115 be present, though one is not conscious of them, ELSE one would not stand) a crowd of other trains of
571n028 aturally all has not been acquired by education. ELSE why do some children acquire it soon. & why do
525m023 his?-- Does memory bring in old ideas <I have ELSEWHERE remarked do> Dogs take pleasure, when doing.
521m008 ns, brought into play by morbid action.-- Old ELSPETH'S «in Antiquary» power of repeating poetry in h
192b084 been determined.-- so with useless wings under ELYTRA of beetles.-- born from beetles with wings.& m

272c107 from Australia, probably live in flowers & has ELYTRA. formed from developement of some other part o
440e147 final cause mammae in man & wings under united ELYTRA The law of <growth> «generation» is only modif
093a034 orms me that in the year a Rhinoceros was found <EMB> in the mud, of the Salt river.-- in reference t
322c269 Eastern Seas. .Octob12th.-- Sir G. Staunton's EMBASSY to China. Oct. 12t Kotzebue's two voyages, ski
326c266 ise old whores would not have children Turners EMBASSY to Thibet, perhaps worth reading quoted by Mal
603o012 o primary sources, sight, & hearing-- Staunton EMBASSY Vol II p. 405.-- Speculates on origin of sacri
121a112 any shells in parts on surface, but I saw none EMBEDDED this point would be worth examining. to suppo
285c153 essarily» excessively slow, they become firmly EMBEDDED in the constitution, which other marked diffe
405e032 be inhabitant of S. America & as it is <falle> EMBEDDED with almost recent shells.-- shows that progr
490q001 sume some recent not found fossil (perhaps not EMBEDDED ¿ are there any very common recent ones not e
490q001 ed ¿ are there any very common recent ones not EMBEDDED?-- Do the Tame Parrots breed amongst the Indi
228b231 y be found:-- We must not compare «chances of EMBEDMENT in» man in present state. with what he is as
231b244 own by only one species, far more than by non-EMBEDMENT of remains-- ¿agrees with non-blending of lan
219b193 hes & bunting & coots-- «(Turdus Guyanensis?) (EMBERIZA Brasiliensis?) (Fulica Chloropus)--» might br
480z008 at Tristan d'Acunha.-- (Turdus Guyanensis?)) EMBERIZA Brasiliensis (?)) Fulica Chloropus. says some
286c155 eaning descent.-- A tree is taken by Fleming as EMBLEM of dichotomous arrangement which is false Ther
565n008 y same movement as sneering.-- it is then more «EMBLEM» manner of hurting opponent by insulting his p
602o11b the object of «all» art is the realizing and EMBODYING, what never existed but in the imagination".-
606o024 r subjects which we know, it is therefore the EMBODYING of a floating idea.-- as statue of beauty, is
482z011 sylvia macloviana, 2d like sylvia cisticola.-- EMBRIZA melanodera-- a linnet not caught.-- Troglogdyt
254c048 lasses, during which the changes of the ovum or EMBRYO succeed each other with the greatest rapidity"
372d131 off , it would produce another man.-- That the EMBRYO the thousandth of inch should produce a Newton
510q017 wind. Aug. St. Hilaire Bot. p. 787. position of EMBRYO in close species of Hilianthemum differs great
461t025 rves fresh study & whole order of the fish.-- EMBRYOLOGY p. 97. for Man Chapt see Yarrell Syngnathus
336d019 «bud» from parent. if whole parent not entirely EMBUED with the change, a bud could not be taken, wit
068r146 ections.--Old vents would keep open long after EMERSION, but improbably so long, that to be surrounde
522m013 a peculiar complaint stomach not acted upon by EMETICS.-- people recognized,-- sudden changes of disp
246c027 Ceram, Bourou & especially New Guinea (replaces, EMEU) in North of New Holland.-- New Guinea scarcely
205b143 ieces of Crustacea. Vol II. p 75 a Fish which EMIGRATES over lands is a siluris, p. 123 A climbing fi
599o007 sufficient to give up my theory-- Viewing from EMINENCE. the wide expanse, of county, netted with edg
073r163 xico may be considered for most parts as rock EMINENTLY rich in mines of gold & silver." «p. 131» The
282c142 parated, now if one of these races had become EMINENTLY aquatic--; «NB, aquatic, i.e relation to elem
422e092 n adaptive. one.-- are not the essential ones EMINENTLY adaptive.-- does it not mean lately adapted o
337d021 ed as one large eye-- have they smell, do plants EMIT odour solely for others parts of creation) & an
524m020 action of brain which gives sensation of pain, EMITS its power on the muscles in the twitching. Fric
064r137 fer from all others in quantity of Sulph. acid EMITTED: mem: Grand gypseous formation of Cordillera I
580n061 hence the belief in the many strange religions.» EMMA W. says that when in playing by memory. she doe
594n112 ains, if no steps are taken to eradicate it.--» «EMMA says, «her» tame rabbits were not frightened at
541m092 reater if the idea is abstract as love, (or an EMOTION not so) than if simple idea as scarlet?-- How
541m092 le dwell on pain ¿ no definite idea. not is an EMOTION.-- People who can multiply large numbers in th
545m105 moved the skin of forehead over eyes, at every EMOTION & <look> «turn» of the head. I could not perce
556m146 ut. expressing cool irony, not biting? What is EMOTION analysis of expression of desire-- is there no
556m146 stance of useless muscular tricks accompanying EMOTION.-- when horses fighting, they put down ears, w
565n009 g to distant object, brightened & moistened by EMOTION,-- why does emotion make tears fall?? Lavater
565n009 brightened & moistened by emotion,-- why does EMOTION make tears fall?? Lavater says derision lies i
567n015 im-- Animals I should think would not have any EMOTION like blush.-- when extreme sensation of heat s
578n057 les «or twitching».-- But why does joy & OTHER EMOTION make grown up people cry.-- What is emotion? A
578n057 ER EMOTION make grown up people cry.-- What is EMOTION? At end of Burke's essay on the sublime & Beau
579n059 something like sexual feelings-- love being an EMOTION does it regard «is it influenced by» other emo
581n066 t what first caused this bodily action. if the EMOTION was not first felt?-- «without «slight» flush,
586n079 nder consists in the action being performed or EMOTION felt in early childhood (before experience or
587n089 raction p. 152. Perception very different from EMOTION.-- The former is used with regard to the sense
605o19v anied with terror & wonderment} <which> «this» EMOTION, from the associations before mentioned. we ca
626o052 led to their formation.-- 'N.B. If feeling or EMOTION rises from heredetary action on body.-- This f
290c163 ght angles, how pleased it is, just like man. EMOTIONS very similar.-- Geology. Transact. Vol V. Bir
305c211 nity of mind, even in the tendency to delicate EMOTIONS between races, & recurrent habits in animals.
522m013 ple do most indelicate actions,-- as if «these EMOTIONS» acquired.-- this may be doubted, whether rat
532m057 far, I believe, in Materialism, say only that EMOTIONS, instincts degrees of talent, which are hered
544m101 en «heredetary» associated pleasures & pains & EMOTIONS-- such as child sucking, gives pleasure, & al
550m127 eredetary», does one not call them instinctive EMOTIONS?-- Dr Holland remarked that insanity like sle
558m151 hole world.-- Read Mackintosh on Moral sense & EMOTIONS.-- The whole argument of expression more than
566n010 rom Burke, who says on mimicking expression of EMOTIONS, he has felt the passions of a face «& mind»
568n017 t, by greater temptation, if memory of its own EMOTIONS. (which must be intimately united with reason
570n024 s off,-- I think shrugging connected with many EMOTIONS.-- (Explanation of sighing is probably correc
573n037 ung children. express the greatest surprise at EMOTIONS in her countenance-- before they can have lea
579n059 ion does it regard «is it influenced by» other EMOTIONS? When a man keeps perfect. time in walking, t
581n066 it offers of radical diversity of tongues.-- [EMOTIONS are the heredetary effects on the mind, accom
582n068 how it comes that the heart is the seat of the EMOTIONS.-- but are not love & hate emotions; what are
582n068 eat of the emotions.-- but are not love & hate EMOTIONS; what are their characteristics;-- they are m
587n089 osh first clearly insisted on assoc of ideas & EMOTIONS. rather ideas & bodily actions make the emoti
587n089 otions. rather ideas & bodily actions make the EMOTIONS.-- p. 272. Some remarks applicable to my theo
588o090 d upwards during mental agony, & whilst strong EMOTIONS of reverence & piety are felt." it appears to
589o092 of pure reason not leading to action & yet our EMOTIONS being only bodily actions associated with ide
591o097 n the Athaenaeum Library) describes effects of EMOTIONS-- fear giving goose skin-- & hair standing on
596n164 pulls back skin of whole forehead & 2 ears.-- EMOTIONS of every kind.-- «[Are monkeys <are> right-ha
600o08a he five senses were the same-- In its action-- EMOTIONS-- p 176 & 177 good passage in French on what
604o017 individuality.-- Quotes D. Stewarts System of EMOTIONS.-- T. Mayo-- Pathology of the Human Mind. Poo
604o017 to loss of will.-- chiefly excited by passive EMOTIONS.-- Cannot quite perceive drift of Book.-- Sym
605o18v ublime, where there is no real sublimity 5 The EMOTIONS of terror & wonder so often concomitant with
605o019 d series of associations that we apply to such EMOTIONS. this same term.-- Hence it appears, that whe
609o029 nothing further.-- October <8> 2d. 1838 Those EMOTIONS which are strongest in man, are common to oth
628o054 been taught or habituated to associatical, the EMOTIONS of this instinct, with that line of conduct,
628o054 this good?-- I should think some parts of the EMOTIONS of reflective part of man, may be quite artificial, as avari
328cIBC 'oeil sur la Faune des iles de la sonde et de L'EMPIRE du Japon Wowett on Cattle-- (Waterhouse has it
261c070 allowed that the discrimination of species is EMPIRICAL. show this by instances Once grant my theory,
277c128 ses of change.-- The mark of analogy would be EMPIRICAL because as soon as two species were placed in
278c128 be useless, but the formation of subgenera is EMPIRICAL, & is judged solely by comparison with other
640j167 y, to take their ideas, which are arbitrary & EMPIRICAL, from their own Faunas, which in this case is
252c045 enus in Geography to help in distinguishing EMPIRICALLY what is species.-- The Collector is directed
367d113 n another, which part is that most immediately EMPLOYED in fighting" instances thighs of cock & Neck
089a018 formed of coral limestone, with interstices yet EMPTY.-- In all the mountains of Saint Marc et des G
350d054 sques in nature; not anything framed to fill up EMPTY cantons, & unnecessary spaces" p 23. "for Natur
472s01r have (except this one year (1827), always been EMPTY.-- See separate note-- Elizabeth says several y
340d029 rence in than in many large orders of birds. The EMU & Cassowary closest.-- Ostrich & Rhea closest.--
553m132 p 292 "Dogs learn sooner to take kangaroos than EMU, although young dogs get sadly torn in conflicts
553m132 t it is one thing for a swift dog to overtake an EMU, & [not located] notion, are not effects of impr
236b279 Europe Mem. Mr Bell's case of Sub Himalayan land EMYS, decidedly an Indian form of Tortoise.-- On oth
058r118 ux terres Australs Vol. I. p. 54. M. Bailly says."EN effet toutes les montagnes de cette ile se devel
209b156 a of islands; "ou bien encore on pourrait au plus EN conclure quels sont les genres qui, sous ce
209b156 qui, sous ce climat, se divisent le plus aisément EN espèces distinctes et permanentes." p. 145. In H
257c056 des mêmes oiseaux. que nous avions déjà observés EN Patagonie. ou au moins des espèces très-analogue
056r109 al waves. but when we see an entire island so ENCIRCLED, the one slow cause is apparent. «I confess I
158g082 trine <Little Hill with granite blocks almost ENCIRCLED> <fre> Gneiss cut smooth on sides of hill whe
495q05a lant two races of Cabbages near each other-- & ENCLOSE one twig of each in bell-glass-- sow these see
288c160 ogy & Botany No XI p. 390. a slight change in ENCLOSING a common seems in part of-- to have almost ba
209b156 ez]CD. From study of Flora of islands; "ou bien ENCORE on pourrait au plus en conclure quels sont les
301c199 gh hoarding from short time.-- My theory must ENCOUNTER all these difficulties.-- Knowing that animal
439e144 hat quality which saved him, would be the one ENCOURAGED-- » Wilkinsons Manners & Customs of the Anci

428e112 eaker ones: there ought to be no weeding or ENCOURAGEMENT, but a vigorous battle between strong & wea
026r022 on in Britain. Mem Concepcion Says Echinites. ENCRINITES. Asteriae, usually petrified into a peculiar
061r127 on neutral ground of 2. ostriches; bigger one ENCROACHES on smaller.--change not progressif<e>: produ
609o29v ences of so many positive checks.-- (This is ENCROACHING on views in second volume of Malthus). Adam
051r093 & loose sandy beach) deposits «calcareous» ENCRUSTATIONS; At Bahia ferruginous.--At Pernambuco (grea
089a020 of Cape Verd. volcanic.-- Isle of Gory. rocks ENCRUSTED with serpula-- Isle of Cayenne. Syenite & dio
100a054 st case of cleavage.-- Phillips (113) «Lardner ENCYCLOP.--» absolutely considers gneiss an aqueo depo
212b167 ican Cattle Phillips Geology «p 81» in Lardens ENCYCLOP. Proportions between fossils & recent shells
214b173 nd Islds-- Kerguelen land.-- Phillips. Lardner ENCYCLOP. insists on analogy between Australia and «fo
326c266 s to be good art.. on <Etn> Entozoa by Owen in ENCYCLOP. of Anat. & Physiology.-- Dampier. probably w
327c265 ten good article on Horticulture in Edinburgh. ENCYCLOP.-- The <Edin> British & Foreign Medical Revie
367d113 Ld Moreton Mare ringed Says that Bell in ENCYCLOP of Anat & Phys. describes, a high-flying bat,
386d166 for the whole.-- if cut off nerves in snail. (ENCYCLOP of Anat & Phys) can make a head; the other pa
087a007 ately raised above level of the Sea. Lyells ENCYCLOPAEDIA-- Lately elevated When Siberia went up. Arc
065r140 -off East Coast. -- Capt. Cook found soundings. (END of 2d voyage outside coast of T. del Fuego. off.
090a024 cause besides thin vapour bringing planets to an END? Fragmentary granite showing schistose structure
107a077 ny artificial limestones produced by Sir J. Hal. END of pages. p. 157. Vol VI Edinburgh. Phil: Transa
112a089 lude to Lyell's view of not discovering dike one END granite & other trap.-- It is in the mountain ma
139a1BC not effect of superadded vital influence?-- See END of Note Book. called R. N.-- Massac[h]usset woul
146g009 highest part must project [Blank] {P} Path East END near Holyrood Palace In same way at top the trap
146g013 any hill Thursday On side of Hill South of upper END of Loch Dochart buttresses of Alluvium or rather
155g063 n East side of Glen Turrit, where I believe they END in upwards inclined plains, as in Corry. & as «a
160g092 dges) between arm of Glen Bright flowing into E. END of L. Oich, & waters flowing into west end with
160g092 to E. end of L. Oich, & waters flowing into west END with obscure terraces on one side Barom 29.200 A
172b006 deteriorates a race, that is alters it from some END which is good for Man.-- Let a pair be introduce
179b035 stating from same epoch certainly» The absolute END of certain form from considering. S. America (in
190b075 offspring show tendency to go back-- there is an END to species.-- «Brown Appendix» A most remarkable
190b076 source of forms. reduce towards Northern Eastern END & die away, & partake of Indian character There
202b130 ls bearing relations to a third body., or common END of structure A Race of domestic animals made fro
234b256 certain varieties of a Harpalus. common at South END, but <rare> «absent from» near London. = Dr. Smi
248c030 oldest race, which evidently is tending to same END, as the law of hybridity, namely the [not locate
263c077 has not been indefinite-- he is not a deity, his END «under present form» will come, (or how dredfull
275c120 e used because they have no scent!) Mr Wynne) at END of chase would not run up hill-- he took thoroug
283c146 we have nothing to do with CREATION.-- <On> The END of formation of species & genera, is probably to
287c158 ese questions may be all disputable, but the one END of classification to express relationship. & by
303c204 ite of eyes,-- + + +, Will he say creation is at END, seeing that Tertiary geology has obeyed rules o
308c218 x nature itself with which our speculations must END as well as begin" &c &c then centre is every whe
317c248 n has written on fossils of N. America.-- At the END of "White's Selbourne." many references very goo
317c248 " Often refer to these.-- Also some few facts at END of "The British Aviary" or Bird Keepers Companio
319c257 n at Maer, it is said the Samoyed women (¿ north END of the Oural mountains) have black nipples to th
319c276 o Species Most of those which have references at END; is so said to have Mackenzie's Iceland Molinas
320c276 Whole of Geographical Journal Asiatic Journal to END of 1837. re«a»d-- contains very little Macleay's
321c275 hites Natural History of Selbourne References at END Dr. Lang Australia «trash» skimmed Macleay's Hor
321c275 e Entomological Ray's Wisdom of God references at END-- The British Aviary.--do--- & Lisle's Husbandry
322c269 on Whewells-- inductive History.-- References at END of each Vol Herschels. introduction to Natural P
322c269 hilosophy R. W. Darwin's Botany.-- References at END Mayo Pathology of the Human. Mind Evelyns Sylva.
323c269 - Lamarck. II Vol. Philo. Zoology «references at END of each Chapter» Crabbes Life June 1s. King & Fi
324c268 bout sexes relative to age of Marriages Brown at END of Flinders & at end of the Congo Voyage Decando
324c268 o age of Marriages Brown at end of Flinders & at END of the Congo Voyage Decandoelle. Philosophie. or
340d026 ke Whewell affinity with analogy-- Good table at END of distrib: of <birds>. Anatidae.-- Consult this
355d067 ue they are of the greatest service, towards the END of science. namely prediction.-- till facts are
356d070 randfathers expression of generat. being highest END of organization good expression but does not inc
389d177 ifferences (of body, mind & constitution) is the END frustrated, when near relations, & therefore tho
402e020 Several kinds of animals have spread from the N. END of Borneo to the adjacent island-- In Sooloo we
410e052 ing some high theoretical interest,-- "the great END must be the law & causes of change".-- A philoso
410e052 iber of species, from its garments, without some END.-- Respect good describers like Richardson.-- Th
411e054 orms may be speculated on, & laws of life,-- the END of Natural History, will be approximated to." Th
415e069 Dogs, so in species, & in Man December 16th. The END of each volume of Whewells Inductive History. Co
447e164 e case of Lemna shows dispersion of germs is not END of seminal reproduction.-- likewise grasses. &--
447e164 ocos do mer.-- Analogy shows some most important END.-- Festuca vivapara F ovina-- propagated like on
452e181 nly to be found on the isld of Batchian near SE. END of Gilolo.-- "-- Forrest Voyages. p. 39-- deer b
465t081 minated (see Pritchards paper) (Ed. Phil. Journ. END of 1839) very important. it seems owing to immig
467t103 ents & stamens all protrude «there is a brush at END of stigma, which forces out from extremity polle
471tfl11 ks from S. Africa he can almost make series from END to end-- so that he is almost led to doubt. whet
471tfl11 S. Africa he can almost make series from end to END-- so that he is almost led to doubt. whether the
472s02v en appeared chaffy, as if sucked?! opens & shuts END of sucker, after having withdrawn it.-- Saw 4 mo
504q013 with branch in middle of tree with flowers near END of orchard.-- At Shrewsbury one branch of Rhod. f
522m011 ard of.-- Thus in many things if he began at one END, he knew the whole subject.-- if at the other no
554m140 ese cases of commonly using, foreign bodies, for END. most important step in progression.-- The male
574m040 ance the Birgos opening a Cocoa nut shell at one END.-- Children & old people get into hands.-- we p
574m040 d so limited as Birgos to become absorbed by one END of Cocoa nut.-- November 27th.-- Sexual desire m
575m045 s body-- tears flow from both, as when one burns END of nose with a hot razor.-- joy <p> a mental ple
578m057 make grown up people cry.-- What is emotion? At END of Burke's essay on the sublime & Beautiful ther
582m067 n the stream; which it does unconsciously of any END.-- N B. There is wide difference, between the me
591m097 ns-- fear giving goose skin-- & hair standing on END.-- July 20th Intelligent Keeper... Zoolog. Garde
592m105 are some instincts unintelligible, <both> in the END gained «& therefore the» cause, and origin being
595m115 w a «black» spider monkey brought it at opposite END of house. & commenced a most lamentable howls &
614o037 ism. inutility of so high a mind without further END just same argument. without indeed we are step t
614o037 t. without indeed we are step towards some final END.-- production of higher animals-- perhaps, say a
624o051 wrong,-- education, of parents strives* to same END.-- & general actions of community must frequentl
624o051 actions of community must frequently teach same END.-- Hence this becomes the law of right & wrong,
628o54v y <it>--- anything contrary to it]CD NB. the very END of conscience is stop to wishes of passion &c. w
641j29v anaria, & light affecting plants. in insects the END is gained by some very different method. in pedu
609o30v civilized man & savage,-- is that former is ENDEAVORING to change that part of the moral sense which
176b023 ng to three elements air, land & water, & the ENDEAVOUR of each <one> typical class to extend his dom
227b227 utation & geographical grouping we are led to ENDEAVOUR to discover causes of change.-- the manner of
296c184 erthelot. must be studied on Canary islands-- ENDEAVOUR to find out whether African forms. (anyhow no
336d017]o.-- the result of this is that animal would ENDEAVOUR to return to parent stock. but if both parent
539m081 ing «review of» M. Comte Phil. which made me «ENDEAVOUR to» remember, & to think deeply, & the immedi
617o040 , except by internal consciousness 3) We must ENDEAVOUR to do without it as well as we can.-- The obje
276c123 d, as those, whose opinion they believe have ENDEAVOURED to advance cause of truth It is of the utmos
555m142 Association. Sept. 18th Zoological Gardens-- ENDEAVOURED to classify expressions of monkeys-- I could
534m063 been instinctive, seeing that time is lost & ENDEAVOURS made must be experience & intellect.-- do. p
536m070 t is as much effort to walk then lightly as to ENDEAVUR to stop heart beating: one ceasing, so does o
541m091 with novel for a length of time.-- Then if one ENDEAVUR to keep any simple idea as scarlet. steady bef
115a098 ular slope-- {P} if not course enough flat top. ENDED by abrupt slope {P} each stratum would thin out
170bIFC p. 235. was written in January 183[8]: probably ENDED in beginning of February Zoonomia Two kinds of
500q10a or different parts or same district.-- About <ENDEMIC &> wandering species of confined genera By my
027r028 us pyritous & coaly matter. Mem: Chiloe In the ENDLESS cycle of revolutions. by actions of rivers cur
098a048 sandstone. (& as I believe most strata) (Hence ENDLESS passages from gneiss to granite): Why not hori
171b004 <on> which if animals lived for ever would be ENDLESS (that is with our present system of body & uni
197b111 of organ:-- the others «animals» created with ENDLESS differences:-- does not say propagated, but mu
305c211 eason? is necessary.--) which is modified into ENDLESS forms, bearing a close relation in degree & ki
305c211 aring a close relation in degree & kind to the ENDLESS forms of the living beings.-- We see thus Unit
372d130), or he may produced by having undergone, the ENDLESS changes, which its parents have.-- -- Not this
410e050 ary.-- Without sexual crossing, there would be ENDLESS changes, & hence no feature would be deeply im
448e165 rally vivaparous sometimes seeds.-- ¶There are ENDLESS curious facts about every part of plant produc
471tf07 tastic & use unknown.-- "<when we find such an ENDLESS variety of form in the same> organ "manifestat

```
592n101 ems to allow it is an instinct.» I suspect the ENDLESS round of doubts & scepticisms might be  solved
639j28r es in Hedge Hog & Echidna.. & Aphrodites C. D. ENDLESS cases.-- Macculloch p. 260 intimates canines n
326c266 instinct & animal, intelligence.-- very good. ENDLICHER has published in first volume of Annales of V
314c238 .-- Prodromus Florae Norfolkicae. 1833 Steph. ENDLICKER (He will give sketch of botany of islands  of
612o035 e radicle of plants absorb by physical laws of ENDOSMIC & exosmic juices. arms of polypus, show eithe
275c119 judgment, that he forseeing. consequence., was ENDOWED with what may be called the prophetic spirit i
305c211 dif> separation between those individuals thus ENDOWED, & the community of mind, even in the tendency
418e083 Physiolog.  p.24.]CD» can be traced to a germ, ENDOWED with the vital principle, which gives rise to
638j059 s. In the «Bee» Mollusca the nervous system is ENDOWED with the knowledge of trying a hundred schemes
275c119 he prophetic spirit in science--. the highest ENDOWMENT of lofty genius Using geograph distribution o
133a140 e> a dry desert, would be very high.-- M. Parrot ENDS his paper like a fool.-- Feb 25' All facts show
190b076 n, about peculiarities of Flora. on East & West. ENDS of New Holland. diminishing towards centre  (p.
211b161 e instinctive feelings against other «for sexual ENDS» species, whereas Man has such instincts very l
430e118 ocess analogous-- -- if so she can produce great ENDS-- But how.-- -- -- «-- .Make the difficulty appare
522m013 tion, like people in violent intoxication, often ENDS in insanity or delirium.-- In Mania all idea of
527m033 eeling in any one man.-- Music & poetry opposite ENDS of one scale.-- former pleases from instinct th
569n021 emory.-- A Melody on flute & Epic poem, opposite ENDS of series or harmonious prose.-- Lutké Voyage i
613o036 g to species. This I suppose he deduces from the ENDS in each case being the same, & the means very s
638j28r ower of closing nostril, foot, sack. power of ENDURANCE &c &c Camels? all good cases of corelations.-
332d004 s, (like parasites of Tropical countries cannot ENDURE this climate-- . ) -- July 23d. Eyton, a stone
172b007 air be introduced & increase slowly, from many ENEMIES. so as often to intermarry who will dare say w
535m064 ng over them, preserving them from «the» sun & ENEMIES-- would not fly away, but bit pencil when touc
584n071 ordinate to,» self preservation, (knowledge of ENEMIES), use of muscles, progression.-- use of senses
541m093 to conceal.-- a man «insulted» may forgive his ENEMY & not wish to strike him, but he will find it f
565n007 ough not every time really wishing to smell its ENEMY.-- Man & dogs show triumph (& pride) same way w
375d134 small changes in nature of locality. Even the ENERGETIC language of <Malthus> «Decandoelle» does  not
538m080 n one acts unconsciously with respect to more ENERGETIC self, & likewise one forgets. what one perfor
540m086 n the black man of <B> Van Diemens land & the ENERGETIC copper coloured natives of New Zealand)-- the
254c049 tion in each joint of taenia worm.-- formative ENERGIES easily expended & no one system developed <is
376d135 wever of volition) of this populousness, on the ENERGY of Man» D.'Orbigny. Comtes Rendus p. 569. 1838
547m111 ant was in the room. with my trunk out & I was ENGAGED in hurriedly giving orders.-- Now what was dif
573n036 atellites &c &c «The Savage admires not a steam ENGINE, but a piece» of coloured glass <&admires>  is
029r034 d with substances so like the Coal measures in ENGLAND (Excepting Conglomerates?) «& absence of limes
035r045 s go to believe that much of all old strata of ENGLAND. formed near surface: Mem Patagonian pebbles b
035r046 ent.-- Ireland & Isle of Man possessed Elk not ENGLAND. Did Ireland possesse Mastodons?? Negative fac
035r047 d have been kept; it shows that throughout all ENGLAND, whole surface oscillated equably.-- These fac
036r051 ing movement being one of elevation alone.--In ENGLAND much subsidence: hence difference; action on l
037r055 ccumulating which like the secondary strata of ENGLAND, «besides ordinary marine remains» may contain
101a056 ssion showing present form of land in Northern ENGLAND influence dispersion of Boulders.-- See Rogers
101a057 rs in N. America do/p. 280. the gravel beds in ENGLAND different from Boulder beds-- What is Osteopor
102a059 d'.Acunha.-- silting up. channels on coast of ENGLAND-- Any one. who has studied rocks in detail  as
109a082 bones lying at the bottom of sea. off coast of ENGLAND.-- Sea must always on actual beach act same wa
187b065 extinct quadrupeds:-- same argument applies to ENGLAND.-- Mem. Shew Mice.-- --- Animals common to Sou
191b080 in full conformity with Europaean formations-- ENGLAND & Europe Ireland common animals-- + + + for in
191b080 ent.-- Ireland & Isle of Man possessed Elk not ENGLAND. Did Ireland possesse Mastodons?? Negative fac
203b138 n Motacilla.. «species peculiar to Continent & ENGLAND» Loudon Mag: Septemb or Octob 1837 Westwood ha
218b182 amely such as blood hounds from other parts of ENGLAND. Mr Bell of Oxford St'-. had a very fine blood
218b191 s to me. that give him a species from Ireland, ENGLAND, Scotland & other localities & each one will h
226b220 Ceylon & India.-- Van Diemen's land Australia. ENGLAND & Europe.-- It will be well worth while to stu
236b280 not  be general circumstance.--In. insects «in ENGLAND» surely it is not-- intermediate genera we mig
247c029 I  say "are as variously coloured as a herd in ENGLAND"-- Black & Grey varieties of rabbits thus hand
265c083 ter birth do not effect progeny-- many dogs in ENGLAND must have been lopped off & sheeps tails cut y
275c120 d Orfords had breed of greyhounds fleestest in ENGLAND lost courage-- (Bull-dogs are used because the
276c123 What  the Frenchman. did for SPECIES-- between ENGLAND & France. I will do with forms.-- Mention pers
278c132 whether dog found at Swan river. The change in ENGLAND from Rhinoceros Elephants &c in the most moder
288c159 ered, but it is certain that rams & bulls from ENGLAND fetch very <go> large price. as is evident  to
294c177 dale says he has seen in old Book last Bear in ENGLAND killed in year 1000. reference to succession o
297c189 rush lately increased in numbers over whole of ENGLAND & Ireland.-- curious in so wild a bird.-- Anna
309c221 p. heredetary proceeding from an accident. New ENGLAND farmer,-- useful could not leap fences:-- Dr L
316c246 owerby fine species) on coasts of N. America & ENGLAND.-- but the fossils are not like, except in ver
319c256 pon the whole thinks <many> more birds sing in ENGLAND than in America, but the few of N. America are
332d001 .-- He thinks apoplexy affects people all over ENGLAND at same periods When he began practice, he rem
361d096 ich have pretty close representative species in ENGLAND & N. America.-- the teal which some authors [n
364d105 state.-- Neilson has given figure of it.-- In ENGLAND no doubt the cross between Pheasant & Black ga
377d139 cks his ears., when frighten depresses them.-- ENGLAND was united to Continent, when elephants lived.
424e102 e, but very odd since these crows are mixed in ENGLAND-- for I presume Carrion Crow is found in Edinb
431e121 s in grades, now L. says the T. costatus is in ENGLAND found in the Inferior Oolite, & the T. elongat
437e138 rch.-- p. 101-- Kingfisher in northern part of ENGLAND stationary, in southern stays only winter.-- J
446e162 4  species of Lemna only reproduces itself «in ENGLAND, as yet observed» by buds-- (the other three b
454e184 ecious? Mushroom Hybrids? Any «wild» plants in ENGLAND, which do not perfect their seed?-- What annua
461t041 .-- it does away with difficulty of rabbits of ENGLAND remaining same (if so) with those of Spain & s
462t051 of  different kinds of cattle in every part of ENGLAND. &c &c NB. In botanical geography, there can b
485z016 t Maldonado & Patagonia compared with those of ENGLAND.-- or ground birds-- rather indefinite letter
527m031 . How strange <all> «so many» birds singing in ENGLAND, in Tierra del Fuego none none.-- now as we know
323c269 oct:  -- 26th Blumenbach's Essay on Generation. ENGLIS Transla -- The Revd. A. Wells. Lecture on inst
031r037 not  think so many faults in Cordillera, as in ENGLISH Coal field-- because lowered & raised--so on
035r047 ation, may well be preserved at Patagonia. The ENGLISH fact is astonishing consult book itself. P. xx
081rIBC ed with sediment.--& escarpment worn away like ENGLISH escarpment The great conglomerate of the Amazo
081rIBC g always additive to convert French Toise into ENGLISH ft. 0.8058372 French metre into English ft. 0.
081rIBC e into English ft. 0.8058372 French metre into ENGLISH ft. 0.5159929 {T} Tolses Pieds Myriametre = 51
148q025 t deer When Black faced sheep are crossed with ENGLISH my informant said the lambs were nearly like e
202b132 ntroduced from Timor, herded separate from the ENGLISH cattle, nor could we get them to associate tog
204b139 inctly says that the mixture between Chinese & ENGLISH Breed. decidedly exceedingly prolific & hybrid
212b166 une 1837 p. 53. an Irish Rat.-- different from ENGLISH. Waterhouse has information respecting the Wat
226b221 epresentative system Of this we see example in ENGLISH & Irish Hare-- Galapagos.-- shrews, & when bi
229b233 Horned (very) aboriginal at Cape: crossed with ENGLISH Bull. offspring very like common English.-- Ho
229b233 with English Bull. offspring very like common ENGLISH.-- Hottentots say great tailed sheep aborigina
235b262 ries a case in point.-- All cases like Irish & ENGLISH Hare bear upon this.-- Why do Van Diemens land
288c159 ng from original Durham breed.-- Native dogs & ENGLISH cross readily-- think about half way in appear
310c223 nese Genera & some species on Himalaya.-- some ENGLISH beetles, birds & a fox most close The most cur
318c254 les come first & the females in flocks. «as in ENGLISH Nightingales»-- other birds (& this seems com
319c255 an says he thinks the Mocking thrush beats all ENGLISH birds in song.-- one of their thrushes exceeds
333d006 ions one giving half character & other more of ENGLISH, but the effect is the same.-- Fox thinks that
334d009 his used to breed to Common & Muscovy Ducks.-- ENGLISH. <Common> «China» & Canada Geese, & that  they
353d061 ry of different number of vertebrae in Irish & ENGLISH Hare.-- good case these hares compared to Nort
356d072 Ptarmigan having crossed in wild state-- & the ENGLISH & Some African dove.-- The extinction of the S
365d105 alone  would species cross in wild state.-- Is ENGLISH red Grouse. a cross between Black Game. &, the
377d139 form  channel & (& Basses St) yet no change in ENGLISH species-- time no element in making change, on
392d180 r to look out for instances of Avitism Examine ENGLISH weeds in Hot. Houses will they flower Make Hyb
458t002 d «up-» instead of spinning-- better case than ENGLISH birds, using cotton &c instead of natural subs
470t176 ther formerly planted «Turkey or» Palmated and ENGLISH, planted within few yards of each other actual
470t176 hen in the gardens, he knew there was none but ENGLISH,--the Palmated was introduced about '65 years
470t176 at palmated has now nearly disappeared. <& old ENGLISH" But these mules <in our garden> show no trace
495q005 e than cuttings &c (4) Raise annuals or common ENGLISH plants in Hothouse & see what effect on organs
499q010 mps from other parts? Don says Irish, Scotch & ENGLISH plants generally distinguishable.= What struct
510q017 of spaces: has he published? does he understand ENGLISH.-- Miguel to collect facts for me-- what? What
510q018 of  animals in confinement, curious.-- foxes-- ENGLISH animals. [Made no import. remark]CD (2) Second
512q019 lyth (1) Mentions some breeder who raises many ENGLISH birds-- will young wild ones breed as well as,
```

512q020 uple but not conceive :Bears /Yes/ (2) Foxes & ENGLISH animals & birds breed (3) In cases where Lions
515q022 ant.-- Oct. 44 Tell J. Anderson's statement of ENGLISH Horses having fewer vertebrae in tail, than Co
065r139 hat it was thought not to have belonged to an ENGLISHMAN.--On 8th of March cove began to freeze. corr
216b182 once.-- Dr. Smith considers the Caffers (like ENGLISHMEN) men of many countenances, as hybrid once. I
537m077 ly fixed, but it will not prevent other being ENGRAFTED.-- No one doubts patriotism & family pride ar
481z009 lkland. good also for Journal.-- 18 Admirable ENGRAVINGS in Meyen Zoology on animal of Campanularia A
616o038 es music pleasant, a memory; yet that frame is ENHANCED by memory of what has been heard; so love of
616o038 mory of what has been heard; so love of virtue ENHANCED by this heredetary kind of memory.-- The diff
541m092 y do, even more than in a real snarl, they are ENJOYING a satical. laugh.-- when snarling real bitt
512o035 o omit the case (if such there are) of animals ENJOYING only movements such as sensitive plants. (But
549m119 od food, no pain.-- <but the> «&» the sensual ENJOYMENT of the minute add to the happiness.-- but as
602o010 ct form" is the source of part of the highest ENJOYMENT in mutilated statues In Elliotson's Physiolog
625o052 t like our appetites & passion, which receive ENJOYMENT from gratification & hence are forgotten-- on
549m120 the happiness of a peasant, with whom sensual ENJOYMENTS of the minute make large <parts> portion of
205b145 often with same female p. 28 "It <is> wrong to ENLARGE a Native breed of animals. for in <the> propor
263c074 e case of Apterix-- split, depress & elevate & ENLARGE New Zealand; a division of nature of Apterix,
214b175 of the vertebra, about 2/3 from base of tail, ENLARGED two very considerably, so that any person wou
371d118 transmission is direct instinct. & afterwards ENLARGED powers to meet with contingency.-- Sept. 23rd
558m150 so will.-- May not moral sense arise from our ENLARGED capacity <acting> «yet being obscurely guided
035r048 on globe. with slip paper a gradually curved ENLARGEMENT see its increased length. which will represe
031r037 ally & simply raised No Faults in Patagonia[,] ENORMOUS extent; if lowered again & covered no sign of
064r136 plutonic rocks & Volcanic metalliferous-- Urge ENORMOUS quantity of matter from CREVICE of Andes--the
072r160 l VII p. 113 "Nature exhibited to the Mexicans ENORMOUS masses of Iron and Nickel, & these masses whi
120a107 are apart-- the curvatures of the strata.¿ the ENORMOUS faults & facility with which the earth is cra
133a140 rock has been ejected must remain fluid for an ENORMOUS period: now when we see how many points have
146g011 since three hundred feet in vertical height-- ENORMOUS mass thunder storm, many <hundred> thousand t
182b049 Progressive development gives final cause for ENORMOUS periods anterior to Man.. difficult for man t
263c075 His best chance is to have profoundly over the ENORMOUS difficulty of reproductions of species & cert
266c092 of India & Arabia p. 202-- sheep have not the ENORMOUS tails, which disfigure those of Arabia & Egyp
282c143 ds in Ehrenbergs Paper on Infusoria on the the ENORMOUS production-- millions in few days-- one doubt
365d106 have been greater puzzle, than none, for the «E»NORMOUS time which it must have taken to separate. Va
401e014 Uncle John says he has no doubt bees fertilize ENORMOUS number of plants-- it is scarcely possible to
422e095 of the <new> raised beaches» [not located] The ENORMOUS number of animals in the world depends, of th
423e096 t & sometimes to simplify structures:= Without ENORMOUS complexity, it is impossible to cover whole s
442e152 ct just alluded to of Northern flowers, throws ENORMOUS difficulty in the way of Mr Knights. theory «
461t041 o innate power of change & it also shows, what ENORMOUS changes of conditions, some species will unde
465t079 ? It is highly important, to bear in mind that ENORMOUS periods may elapse, even in situations appare
115a098 ss. then most regular slope-- {P} if not course ENOUGH flat top. ended by abrupt. slope {P} each strat
121a111 l VI. p 173. (Ed. Transact) has seen clay stiff ENOUGH <to form> for potters to use. in which great K
172b007 islands, ought to become different if kept long ENOUGH.-- «apart, with slightly differen circumstance
200b124 most important facts.-- As soon as island large ENOUGH for land birds, seeds picked from the beach by
276c124 &c there may be tendency to divide, which often ENOUGH repeated would cause an unequal number of vert
285c153 st become unfit, the animal cannot change quick ENOUGH» Vegetation of peak-- altogether original. owi
296c184 ws where proper dampness seeds can arrive quick ENOUGH» for future state, that when good enough for He
316c244 er difficulty for Deity to choose, when perfect ENOUGH for future state, that when good enough for He
316c244 perfect enough for future state, that when good ENOUGH for Heaven or bad enough for Hell.-- «+glimpse
316c244 state, that when good enough for Heaven or bad ENOUGH for Hell.-- «+glimpses bursting on mind & givi
526m028 ect every detail, yet they have not imagination ENOUGH to <up> recall up the image in their own mind,
530m049 birds nests & watch them till the young are big ENOUGH to eat.-- There was blackbirds nest, near hot-
611o034 tissue <must> bears relation to whole, that is ENOUGH must be present to be able to exist as individ
498o008 artificially impregnated. Abberley says Ants-- ENQUIRE (13) Do any of same species of Willows grow in
423e098 uliflower.-- (How curious it would be to make ENQUIRIES of some of these great seed-growers--).-- Fe
566n012 fact was connected with law.-- as soon as any ENQUIRY commenced, for instance probably such a thing
576n048 s have instincts» «either» to obtain a certain EN<S>«D»: & intellect is a modification of <intellect>
223b210 ed prevents the complete adaptation which would ENSUE A. B. C. D. -- (A) crossing with (B) (& B havin
632j53r intestines of a thrush were means sufficient to ENSURE propagation of Misseltoe?-- do p. 284. it is h
128a130 .-- when describing Coast of. Brazil. Maldonado ENTER into this case.-- Ed. New. Phil. Journal Vol XX
153g054 ake cause collection of sediment? Where ravines ENTER side by, opposite entrance into Glen Fintec a k
408e043 umber where then is the gap, for the new one to ENTER?-- The wonderful species of Galapagos, must be
558m152 when fighting, & not with dog. when fear might ENTER?-- I believe common Swan, arch raises neck & de
575m045 . Biog. case of Newfoundland dogs. who will not ENTER water, till he sees. whether birds badly wounde
575m046 bering things of youth, when new ideas will not ENTER. is something analogous. to instinct, to the pe
627o053 g) is not the aim of the agent, for it does not ENTER into his contemplation.--" Now Eugenius would c
148g027 s river all these composed-- where side ravine ENTERED terraces formed successive bays but plains slo
156g069 h's supposition;-- the old ravine, where water ENTERED are not proportionately large to those now for
270c104 nsulting, if any Metaphysical speculations are ENTERED in upon life. Namely Carus.-- How remarkable t
338d025 Book M'. for case of change in food in insects ENTERED by mistake Surely the fossil Mamalogy of Brita
587n089 248. Theory of Association. owing to time when ENTERED brain, try contiguity of parts of Brain.-- Mac
152g045 allowest In Glen Collarig good case of shelves ENTERING «on» one side ravine. Are the lip, or necks o
465t079 servation of shells; where land broken, rivers ENTERING «at» no shells-- now look at Scotland-- c
466t094 istils have no relation. In Monk's Hood, a bee ENTERING long nectary, would «necessary» cross directl
399e010 er of producing. Williams. Narrative of Miss. ENTERPRISE, p. 497. Vampire bat abound in the Navigator
471tf10 vigatores. Williams. Narrative of Missionary ENTERPRISES Dr Andrew Smith says in the larks from S. Af
150g038 the 3d below them opposite to where side ravine ENTERS On opposite side of valley both extend below t
151g043 head of plain.-- but below houses where rivulet ENTERS two great projecting butresses, upper slope of
162g101 Brook <about> Head of which is so interesting. ENTERS by old tower called Glengarry (Nead Roy told m
165g124 ters of the Tarf-- Kilfinnan Tower where stream ENTERS at head of which hill is round & not merely th
523m018 passion.-- There seems no distinction between ENTHUSIASM passion & madness.-- ira furor brevis est.--
539m082 en in National Institution & much philosophy much ENTHUSIASM, happened to go close to one & smelt the pec
037r052 s of plants. V. Lyell. Chap XI Vol II. Urge the ENTIRE absence of any rock situated beneath low water
056r109 elped it, or diluvial waves. but when we see an ENTIRE island so encircled, the one slow cause is app
398e006 tutions. & we suppose not only revolutions, but ENTIRE obliterations & fresh laws created., & yet wit
550m124 n same time,-- therefore degrees of happiness-- ENTIRE happiness. not being so desirable as <broken>
048t084 ell. Vol I. P. 191 State at St Helena. pebbles ENTIRELY coated with Tosca. which implies motion in th
054r102 ary formation of Payta: N. part of New Zeeland ENTIRELY volcanic!! New Zeeland rich in particular gen
281c139 degree.-- a species must be compared to family ENTIRELY separated from any degree; the tailors-- «in
334d010 by Stallions, (according to Fox) being guided ENTIRELY by their smell.-- Fox says he knew «a» carter
334d011 allion though eager to all other mare had been ENTIRELY broken from their mares, (though horsing ever
334d012 ettler near Swan river, lost his <on> two cows ENTIRELY, changed his residence a great many miles.--
336d019 <slip> «bud» from parent. if whole parent not ENTIRELY embued with the change, a bud could not be ta
361d095 ed this character in adult stage, other alters ENTIRELY]CD In common sparrow young & female similar p
369d115 l & Zoological research, in order to establish ENTIRELY their place in nature, as well as fully to un
380d153 s immediately checked-- the vis formativa goes ENTIRELY to the offspring-- this is clearly the conver
401e015 iment was never tried of separating apple tree ENTIRELY from all others-- so my experiment of strawbe
549m118 ill.-- V. last page. A healthy child is «more» ENTIRELY happy (contentmt is different it refers to wi
557m149 med of himself.-- Jealousy probably originally ENTIRELY sexual; first try «to» attract female, (or ob
616o39v acting & shew that the groundwork <of this> is ENTIRELY wanting by which thought or memory. might be
215b178 7. Peron G. St. Hilaire has written "opuscule" ENTITLED "Paleontographie" developing his ideas on pas
236bIBC phie Botanique De Candoelle. Geol. Soc Horae ENTOMOLOGICAE Linn: Soc. Geoffry-St. Hilaire Philosophy o
485z018 ca &c &c L'Institut, 1838 p. 46 Macleay Horae ENTOMOLOG. insects swarm in Lapland & Stizbergen wherev
486z019 nging powers of birds of N. America & Europe. ENTOMOLOG. Transact. Vol I. p. 130. Col Sykes on balls
321c275 ng Australia «trash» skimmed Macleay's Horae ENTOMOLOGICA Ray's Wisdom of God references at end-- The
272c108 melidae, with similar habits. But the Horae ENTOMOLOGICAE will tell this.-- What peculiar conditions
313b091 ds Kotezebues first Voyage Copied into list ENTOMOLOGICAL Magazine paper on Geographical range Richar
313c233 ia can subsist where parasites cannot» Read ENTOMOLOGICAL Transactions Why if louse created should no
321c275 s Home's History of Man Transactions of the ENTOMOLOGICAL Society Vol I. & 1s No of Vol II. (read mem
324c268 about Castes &c Richardson's Faun. Borealis ENTOMOLOGICAL Magazine (paper on Geograp. range Study Buf
337d022 scientific observer.-- <Transactions of the ENTOMOLOGICAL Soc> A capital passage might be made from c

534m062 eived.= = Aug. 7th--38. Transactions of the ENTOMOLOGICAL Society of London Vol. I. p. 106. Col. Syke
301c200 f generation about equally complicated.-- An ENTOMOLOGIST going into a country & collecting thousands
490q001 ans Do the Savages select their dogs Sowerby ENTOMOLOGIST Does individual Shell or insect or group va
225b218 with bump & pigs foot with cloven hoof) Ask ENTOMOLOGISTS whether they know of any case of introduced
497g007 acter of alpine Flora of Tierra del Fuego and ENTOMOLOGY of.-- most important, as furthest removed po
478z003 56 degree and 57 degree only inhabitant crust ENTOMOST of the genera-- Cyclops p. 134. and p. 115 In
289c162 of Zooly & Bot-- Vol II p. Dr Johnston <on> ENTOMOSTRACA Daphnia, produce young, capable of producin
313c235 generic & other specific extinction-- In the ENTOMOSTRACA (Magazine of Zoology & Botany) where severa
313c235 much more firmly impressed.-- we see in the ENTOMOSTRACA. The sexual curiosity of the orang outang o
344d040 those insects, which have fertile offspring. ENTOMOSTRACA & Aphides. The extreme difference of sexes.
344d041 tation, & to successive fertile offspring in ENTOMOSTRACA & Aphides Developement of sexes in Caterpil
254c047 t.-- Q Zoolog. T. V. I. p 389. Owen remarks on ENTOZOA, the organs of generation, afford the least ce
326c266 do.-- There appears to be good art... on <Etn> ENTOZOA by Owen in Encyclop. of Anat. & Physiology.--
381d156 . (only female organs visible). Oyster. cystic ENTOZOA. Echinoderms. Acalephes. Polyps. Sponges Heaut
381d156 ants) Cirrhipeds rotifers, trematode & cestoid ENTOZOA Allotriandrous <or M> Mollusca, with pectinibr
381d156 tion every generation)-- Epizoa & the nematoid ENTOZOA-- Therefore highness in scale has no «constant
124a118 eriod Subsidence in Demarara p. 131 (B.) Wrong ENTRANCE. Book C. p. 101. On Frozen Soil of Siberia (w
124a118 Soil of Siberia (with refer to Metamor) wrong ENTRANCE Athenaeum. 1838. p. 652. Dr. Daubeny on mount
153g054 ediment? Where ravines enter side by, opposite ENTRANCE into Glen Fintec a kind of landing place is f
059r119 du centre d'lile, vers la mer; ces couches ont ENTRE elles une correspondance exacte, et lorsquelles
584n073 can have affected respiration V E. p. 125 Wrong ENTRY Madagascar Lemur seemed to like Lavendar Water
037r056 the frequency of dikes in Granitic countries, ENUMERATE cases. -- M. Video exception, but even there,
471tf07 . Trans 18. p. 133 Westwood on the Fulgoridae ENUMERATES the strange forms which the thorax & head di
336d017 fore no generative organ.-- Same Prop. better ENUNCIATED.-- "An animal Either parent cannot transmit
428e112 h 11th. Yarrell's law must be partly true, as ENUNTIATED by him to me, for otherwise breeders who onl
033r042 t Ascension, each particles coated by pellucid ENVELOPE of Lime.--form resembles the husks at Coquimb
038r058 ing that they refer to CENTRAL nucleus & that ENVELOPES no doubt existed. These higher portions proba
039r061 shells, as in some older formations: Mem the ENVELOPES at Coquimbo. the analogy is now perfect <The
557m148 rdly arched. & tail stiff.-- is shame, jealousy, ENVY all primitive feelings, no more to be analysed
042c068 pen» faults in mountains: to elevated strata in EOCENE lakes of France, & unequal action of Earthquak
057r113 do not these bones differ as much nearly as the EOCENE. = Should Mr Owen consider bones washed about
178b030 eing dead from which they bifurcated.-- Type of EOCENE with respect to Miocene of Europe? Loudon. Jou
221b200 ies in Sicily. Jan: 1838 L'.Institut. Bats, in EOCENE beds, very like present species., p 8. ¿Are mu
232b245 ndefinite change., (marking in their history an EOCENE miocene & pliocene epoch), whilst others may d
249c036 kable, when we consider number of quadrupeds in EOCENE period. Have the Edentata & Marsupial forms be
407e040 re as diferent from the existing orders, as the EOCENE of Paris! (Great Edentata at that period) Anal
425e105 p. 55. talks of Tapirus American form. found in EOCENE beds of Paris Lyell has remarked species never
430e117 east on ancient generic forms.-- the animals in EOCENE period could not have been direct parents of a
448e167 r their contemporaneous»-- how is this with the EOCENE beds.-- see Lyells tables Bennetts Wandering V
460t019 sects, of do orders-- cheiroptera & caetacea in EOCENE-- dicot. plants in coal measures.-- Shells in
192b088 re so «known» with fishes & reptiles.-- In mere EOCINE rocks. we can only expect some steps.-- I may
389d173 uch aversion to it, that she was held Hunters EOECONOMY So with inter-breeding as told by Willis Q Pr
569n022 said his wife would be grieved-- "il leva les EPAULES et dit qu'il valait mieux rester a Farroïlap q
569n021 accompanying such memory.-- A Melody on flute & EPIC poem, opposite ends of series or harmonious pro
259c064 , wonderful case of extermination of species-- EPIDEMIC amongst trees. Plane trees all died certain y
259c064 's willing absurd, ∵ not applicable to plant, EPIDEMICS of South Sea, wonderful case of extermination
397e003 h 1839.-- Selected. Dec 15 1856 [not located] EPIDEMICS-- seem intimately related to famines., yet ve
022r006 e this. Valparaiso Granitic nodules in Gneiss. EPIDOTE seems commonly to occur where rocks have under
049r088 ct? -- «Cordillera???» Porphyry at Valparaiso; EPIDOTE -- Must we look at regular greenstone cones at
124a119 thickness.-- this most curious with respect to EPIGMOUS action.-- if the zinc were mixed with 90 perc
299c195 ing it at wrong time of year & manuring it.-- EPIGONOUS. Perigonous &c-- very important in classifica
521m007 s are, is not so very wonderful.-- <Now is not EPILEPSY an habitual disease of the muscles.???> Miss
524m020 eeable thought occurs, is closely analogous to EPILEPSY & convulsion.-- affections of the thinking or
524m021 t appear most capricious) as in delirium after EPILEPSY, but in the failing from old age, they consta
525m026 hort term, & lastly healthy ones.-- Insanity & EPILEPSY remain many generations in families.-- My fat
381d156 ever do not require coition every generation)-- EPIZOA & the nematoid Entozoa-- Therefore highness in
043r071 dd. <">" <from> "in the same line" to "from the EPOCH of Ammonite to the present day. at Mauritius. (
175b019 s & as far as world has been uniform, at former EPOCH-- How is this Ehrenberg? every successive anima
179b035 the series of connection? «if stating from same EPOCH certainly» The absolute end of certain form fro
222b206 ceae & Phocae now replace Saurians of Secondary EPOCH: it is impossible to suppose such an accumulati
232b245 g in their history an eocene miocene & pliocene EPOCH), whilst others may die out or move South ward.
426e105 ry> Isld «neighbouring» formed in the Tertiary «EPOCH» like Sicily not having species, if true import
212b168 st in number (as cephalopods,) in last tertiary EPOCHS most perfect kinds the shark. lived in remotest
222b205 most perfect kinds the shark. lived in remotest EPOCHS.-- ¿ lizards of secondary period in same predi
343d039 nor is there in the Tertiary «older> geological EPOCHS.-- There are some admirable tables on Geograph
377d140 al kingdom.-- It is the unit of our calendar.-- EPOCHS & creations, reduce themselves to the revoluti
402e017 did pour sediment in one spot, for <whole> many EPOCHS-- such changes would be observed.-- G. W. Earl
408e049 world, either at the present, or many anterior EPOCHS.-- but we can see if all species, there would
207b150 ich «nearly» first appear «(p 321)» at Tertiary EPOCH p. 330. Fossil Infusoria found of unknown forms
257c057 My views will explain no secondary-EPOCKS, & development of lizards.-- As we have birds
035r048 ucleus, which in any one country would produce EQUABLE effects.-«though so immense to short breathed
176b021 res deaths of species to keep numbers of forms EQUABLE:-- but is there any reason for supposing numbe
176b021 there any reason for supposing number of forms EQUABLE: this being due to subdivisions & amount of di
250c038 portion of the world «Africa» being left more EQUABLE. yet America preeminently equable. might have
250c038 ng left more equable. yet America preeminently EQUABLE. might have allowed fresh species to have been
268c099 ?-- We must always bear in mind proofs of most EQUABLE climate both in S. & N. Hemisphere just anteri
407e038 obliterated) of the world. from the <Tropical> EQUABLE kind of climate to the extreme.-- Therefore sp
407e039 ies, which were fitted for such a preeminently EQUABLE climate. might not have been able to have surv
407e039 ian Archipelago»-- formerly were not so «very» EQUABLE. or so tropical, & therefore present state of
407e040 .-- Hence it is, from the ancient preeminently EQUABLE & temperate climate, <that> of America, that t
408e042 rms. The climate of N. America, must have been EQUABLE & low-- more so than any other part of the Wor
035r047 roughout all England, whole surface oscillated EQUABLY.-- These facts become easy if we look at the a
258c059 from general circumstances effecting the area EQUABLY.-- Animals having wide range, by preventing ad
032r041 e are «likewise» differences of temperature «at EQUAL distances from centre of rotation» & a <circula
052r098 f Madagascar. where a --40 line <shows> runs at EQUAL distance?) 1st cases. -- The terraces in Valley
099a049 -- (Gravity can have no effect, on particles of EQUAL weight.--)¿ cleavage not vertical ∵ combined wi
118a103 ys (.p 52) fringe of sublittoral deposit always EQUAL width --subject of fine paper this would make.-
120a107 e outlines of what were fluid undulations-- the EQUAL movements of Glen Roy road. (¿ metamorphic acti
200b123 g very opposite races, whether you would expect EQUAL fertility-- ditto in Plants.<==> It will be wel
284c147 , open ocean, &c this probably on long average, EQUAL quantity, 2d on relations of heat & cold. there
284c147 s & diversity--.-- this perhaps on long average EQUAL.-- The Cocos do Mar on the Mahé island,, one th
301c200 opment of all animals of same class being about EQUAL.-- organs of generation about equally complicat
302c202 e. <an> osculant groups between two circles «of EQUAL value» must be so from characters of analogy.--
309c220 ked out by animals of same <spe> genera. is not EQUAL to areas of elevation: Marked out by existence
310c224 ntries (as St Jago Cape de Verds) the shells in EQUAL periods with Europe would probably have changed
342d032 covy & common ducks-- they are produced in full EQUAL Numbers with pure bred (just like common mules)
415e068 r of the species.: therefore Man & monkeys have EQUAL chance that progenitor was bimanous, or quadrum
423e099 orms.-- we know not rate of deposition has been EQUAL even in one bed, much less in alternating strat
496q006 ociated plants. when proportional number appear EQUAL-- & see whether proportions will vary, which wi
502q11v ent Cabbages most carefully to see if variation EQUAL in flower with leaves.-- strawberries How «soon
553m135 repeated, without the powers of the mind being EQUAL to the smallest casuistical doubts.-- The histo
616c039 n that attraction does to matter, it might with EQUAL propriety be said that the living brain perceiv
035r046 n happen in troubled England; the more minute EQUALITIES of elevation, may well be preserved at Patag
526m027 ing, free will determines our throwing it up.-- EQUALLY true the two statements.-- Catherine remarks t
102a061 - Are substances soluble under great pressure? EQUALLY with little pressure? An important question! I
155g066 s no wonder that all «three» lines «should be» EQUALLY preserved 2d or upper one more perfect in this
176b021 amount of differences, so forms would be about EQUALLY numerous. changes not result of will of animal
225b216 ies are. Lamaks "willing" doctrine absurd. (as EQUALLY are arguments against it-- namely how did otte

```
226b220  ar very good.-- Fernando Po & Coast of Africa. EQUALLY good.-- Small isld off New Guinea same fact se
281c138  orms.--just same way as <we> «all men» not all EQUALLY related to each other I cannot help thinking g
301c200  eing about equal.-- organs of generation about EQUALLY complicated.-- An Entomologist going into a co
334d009  anada Geese, & that they this first cross were EQUALLY fertile with pure bred animals.-- Mem.  number
367d113  n» the males; & another circumstance, perhaps, EQUALLY so, is this strength being directed to one par
385d164  resemble parents in their bodies "It is a fact EQUALLY well known, that we observe in the temper, esp
434e127  nerations of plants) to do so, the effects are EQUALLY handed to offspring.-- Whewell's anniversary a
446e160  II. <Some birds> Both sexes of some birds sing EQUALLY well. and <in> these reciprocally assist in do
448e168  he Lepidosiren-- Amblyrhyncus & Toxodon, <all> EQUALLY aberrant-- the two former connecting classes 1
466t099  une 1st 1841. Maer Examined the Lemon-thyme.-- EQUALLY abortive as it was in autumn: filaments united
467t099  no pollen-- Common Thyme growing close by is EQUALLY abortive--and both growing within Kitchen Gard
546m110  it  with Somnambulism.-- the young lady almost EQUALLY in her senses in either state.-- does this thr
586m080  sea in dark night & not loosing its direction, EQUALLY wonderful in young & old.-- These facts  point
594n112  is  not dangerous-- wild-ducks would have fled EQUALLY if man had appeared-- though instinct so firml
620o045  its  effects lasting, whilst passions although EQUALLY natural leave effects not lasting. By associat
625o50v  ssions' State broadly in child or animal it is EQUALLY proper to obey anger as benevolence (but not c
629o54v  dog.  & therefore not <instinct> «conscience») EQUALLY <prefe> destroy all wish of outward gratificat
636j56v  r world» simple series.-- My theory shows life EQUALLY simple series, & therefore trace of  beginning
228b231  made  our slaves we do not like to consider our EQUALS.-- «Do not slave holders wish to make the blac
450e175  is fact is noticed in Cassay Ava Pegue-- seldom EQUALS 13 hands-- those of Lao & Siam inferior to tho
032r041  variation?)» as in Cordillera.-- From poles to EQUATOR current downwards & to West.--From Equator to
032r041  to  Equator current downwards & to West.--From EQUATOR to poles. nearer the surface & to the Eastward
189b072  ate on land being grouped towards centres near EQUATOR in former periods & then splitting off.-- If s
195b103  it  said that Volcanic soil of Galapagos under EQUATOR that external conditions would produce species
232b245  imate whilst others died out, or moved towards EQUATOR.-- «or some species might then have been wande
232b246  day,  «in one», oh we will take a day from the EQUATOR to add to the mean of the other.-- If the  the
232b247  e»-- As Europaean forms have travelled towards EQUATOR, <th> so would the plants from extreme  north,
244c022  res) replaces <Vesp.> holds same relation with EQUATOR--that Vesp. lasiurus does in North. Hemisphere
048r085  the «species of» Mastodons. which ranged from EQUATORIAL plains to S. Patagonia. To the Megatherium.-
086a007  nts?-- chiefly clearly by sun's position = If EQUATORIAL streams of warm pole; in name of Heaven  why
086a007  warm  pole; in name of Heaven why are tops of EQUATORIAL mountains so cold.-- Siberia no plants to it
213b172  &  cetaceans.-- both found in every sea, from EQUATORIAL to extreme poles.-- Oh. Wealden.-- Wealden.
407e038  - (Explained by profound views of Lyell) Now «EQUATORIAL» America from the «low» limits of blocks bot
407e041  ia» between two polar lands,--; Africa not so EQUATORIAL..-- The fact of No. Mam: Placent: insectivor
641j29r  nd of the Physiological relations of animals. EQUATORIAL countries are supposed favourable to terrest
113a092  f metamorphic theory On the idea of statical EQUILIBRIUM, the height of lava (habitually) becomes mea
119a104  (& does not Hersche theory imply tendency to EQUILIBRIUM.) 3d. there are mountains in the moon, which
119a105  considers  an irregular figure to be that of EQUILIBRIUM.-- What causes that of tendency to irregular
122a114  if  platform of mexico owes its elevation to EQUILIBRIUM,-- it cannot be equilibrium of fluid, but of
122a114  its elevation to equilibrium.-- it cannot be EQUILIBRIUM of fluid, but of solid. because if of fluid,
122a114  tion must prevent internal fluid arriving at EQUILIBRIUM so soon from, crust being cut of-- if part o
126a122  wn far below surface, say 1000-- «III but an EQUILIBRIUM is supposed to have been attained.» how much
126a123  ?-- this looks as if bad conductor-- III But EQUILIBRIUM is not attained, & if cold water did not per
128a129  r on other. from tendency to regain statical EQUILIBRIUM This will be only a modifying cause. {P} lan
495q05a  Necessary  to answer Wiessenborns doctrine of EQUIVOCAL Generation Charlworth p. 377. Have paper rule
594n112  stinct long remains, if no steps are taken to ERADICATE it.--» «Emma says, «her» tame rabbits were no
421e091  instance  of a trifling peculiarity not to be ERADICATED.-- ¶.do. p. 305.-- Mr Owen says <tha> «in ab
539m083  &  take after my Father in heraldic principle. & ERAS a Wedgwood in many respects & some of Aunt Sara
058r115  e themselves in planes. «Mem silky lustre» ask ERASMUS. whether electricity would affect this. -- Sta
124a118  . Dr. Daubeny on mountain Chains in N. America ERASMUS suggested to me that Herschel's theory  offers
124a119  on of intermittent action of elevatory force-- ERASMUS says he has seen in making brass a piece of co
295c183  ty, analogy of man if so war not [not located] ERASMUS says he has seen old Stallion tempted to cover
326c266  ted in 1837) on limits of painting & poetry.-- ERASMUS thincks I should lik it. The Sportsman's Repos
334d010  y sterile.-- My case of Stallion, according to ERASMUS preferring young mare to old, explained by Sta
354d063  Butterfly.-- When two Varieties of dogs cross, ERASMUS says it look lik[e] Institut. 1837. p. 351. Pa
526m028  - this, though very odd is perhaps true.-- mem ERASMUS & mine taste for music.-- Children like hearin
546m111  t there!!! Lady in perfect «mental» health.-- «ERASMUS had almost same thing happen to him about a kn
551m128  m, than when merely ordered to do it.-- Plato «ERASMUS» says in Phaedo that our "necessary ideas" ari
572m033  than  Europaean.-- Ideots. defective brains.-- ERASMUS does not liken term instinct to muscular movem
523m016  ospital.-- My father was nearly drowned at High ERCALL, the thoughts of it, for some years after, was
352d059  when  it is considered the tree of life must be ERECT not pressed on paper, to study the correspondin
542m093  & though dreading to say so, his step will grow ERECT & stiff like that of turkey.-- he may be amused
557m147  k & turkey cock in passion.-- Cat when pleased, ERECT its tail & make it very stiff «& back» when sav
558m152  p] drawn back & driving air out of mouth «hairs ERECT on back» «wide open» with prodigious force.-- m
565n007  Man & dogs show triumph (& pride) same way walk ERECT & stiff, with head up.-- Why does suspicion loo
089a019  res polish to see structure.-- «He» Thought of ERECTING machine to see if water fell. -- «Keys off ex
577n051  man, face, neck-- «upper» bosom in woman: like ERECTION shyness is certainly very much connected with
112a090  the  Geograph Soc, April 9 1838. Letter from M. ERHMAN stating that the mean temp at Yakous in Siberi
153g055  len Fintec a kind of landing place is formed Ben ERIN summit 27.813. 65 55 degree? Boulder of Granite
153g056  hem on summit of hill rounded, site N N W of Ben ERIN (P} Shelf of Glen Guoy flat peat plain divortiu
154g058  n Guoy form comparison with granite block «& Ben ERIN» 29.287. 72 degree Air 65? 70? (P} Where a butt
160g091  t point joining this hill to others 3000? if Ben ERIN is 3500 boulder Cairn leet more Haberclador Ist
123a117  cal Journal Vol VIII. ( 1838) p 212. Facts from ERMAN about great depths of frozen soil. p. 211 Consi
269c101  in Siberia being frozen to 400° ft in depth, (& ERMAN'S surprise that it is not 700) is applicable to
265c084  With  respect to question what is adaptation.-- ERMINE, ptarmigan hare becoming white in winter of Ar
595n121  ssociation or imagination [blank] [not located] ERNEST W. playing with Snow. when 2½ years old. was f
105a070  ys-- subsequent opening a medial gorge by slow EROSION. but we have evidence in distribution of block
148g029  f loch <in> <below> «by» pass of Glencoe-- the EROSION may often be due to rivers-- By Roy Bridge,  a
105a069  broad?-- There is thus wide difference between EROSIVE power of river & sea.; the former as its chann
105a073  <rapid> slow course, & with slow course small EROSIVE power. therefore tendency of running water to
026r020  have resembled Chiloe In De La Beche, article "ERRATIC blocks" not sufficient distinction is given to
089a020  extreme point of Flori[da]> Excellent paper on ERRATIC blocks in Alps. Memoires de la Soc. «de Geneva
406e037  -- November 1st..-- Addenda to Journal. I show ERRATIC blocks transported far S. in Northern. Hemisph
616o039  t is impossible to shew satisfactorily it's ERRONEOUSNESS. it is a point of indifference 2) In the ab
105a072  at land shells should be preserved in it-- some ERROR? (because more recent) ------ Coquimbo on. othe
304c208  ment not applicable to apterix.-- but source of ERROR for if some of the ostriches were to die,  then
350d056  s in being able to reproduce Here there is some ERROR-- Observed, nature does nothing in vain, therfo
533m060  that  Capt. F. R. candour & ready confession of ERROR made him less repentant.-- In making too much p
536m073  ion & instincts--.-- Put it so.-- Probably some ERROR in argument, should be grateful if it were poin
273c110  shells?.?.?» It looks as if animals perished by ERRORS.-- It is most wonderful how in every family of
204b141  h common ducks.--» Kirby all through Bridgewater ERRS greatly in thinking every animal born to consum
038r057  t is the consideration of the state at a grand ERUPTION when whole summit of mountain is blown off; &
039r059  where  dikes certainly have not been points of ERUPTION. Nobody supposes that all the dikes in Cornwa
039r060  an  example the great subsidence at the famous ERUPTION of Rialeja, & the more true analogy from  the
043r070  22 In any archipelago. & neigbouring Volcanos. ERUPTION from «more than» one orifice <...> does not o
044r074  hout a vesicle. we may consider appearances of ERUPTION at bottom.--solution under high pressure of g
045r075  ground.  «Ulloa states that Volcanos!! were in ERUPTION at time of great Lima earthquake» In the Chil
064r137  f elevation have connected <lines> «points» of ERUPTION [.] give instance of Etna Stromboli & Vesuviu
045r075  e.--expatiate on difficulty of evidence about ERUPTIONS of Volcanos. (where there are no country news
045r076  try; (perhaps generally over whole world) Yet ERUPTIONS <both> at sea (as wells as in the Cordillera
045r077  curious  & perplexing. than those that attend ERUPTIONS: Mr P. Scopes explanation of low Barometer? I
050r001  a. M. De. Jonnes seems to think that Volcanic ERUPTIONS I form foundations for Coral reefs.--does he me
055r107  reams of St Jgo) yet no historical records of ERUPTIONS how immense the time!! Now well agrees with h
068r146  up  beneath ocean).--The first dislocations & ERUPTIONS can only happen during first movements, and t
093a034  Mr  Scrope seems to consider that elevation & ERUPTIONS are antagonist forces. but they are parts of
102a062  etism & counteracting gravity.-- As volcanic ERUPTIONS are accompanied by horizontal elevations,  so
104a020  race) from few dikes which have given rise to ERUPTIONS.-- We must suppose everywhere--, in granitic
515q022  er net & see if get sterile-- Cover that little ERVUM in Sand-walk, on which I think I have never see
```

332d001 ten years he never saw one case of malignant ERYSIPELAS spreading over the head, not caused by a wou
403e024 t) If they give up infertility in largest sense. <ES> «as» test of species.-- they must deny species
028r032 ame cause as no colour Sir J. Herschels idea of ESCAPE of Heat prevented by sedimentary rocks, & henc
266c091 the lands of their forefathers;-- the first to ESCAPE the doctrines of Muhammad, the last to extend,
278c132 onder is that the Europaean forms, were able to ESCAPE to some more fitting country,-- if Toxodon had
403e024 must deny species which is absurd.-- their only ESCAPE is that rule applies to wild animals only. fro
429e114 s introduced into our gardens (opportunities of ESCAPE for foreign birds & insects) which are propaga
441e151 «organic being» should cross with another.-- to ESCAPE it «in any case» we must draw such a monstrous
505q014 usted with pollen from other flowers? Can flys' ESCAPE from old flower= (14) Has planted seeds of Ger
244c022 .-- Mem Waterhouse knows of some species which ESCAPED there.-- p. 139. Vespertilio bonar«i»ensis (fr
342d033 rid Muscovy & Common duck have been shot wild (ESCAPED from Carolina?) off New York. therefore instin
437e139 ing themselves with them, and one day a rabbit ESCAPED into a hole, where the old Eagle could not fin
437e140 the young ones, would in all probability have ESCAPED".-- if it had not been shot by <some> «a» shep
443e157 foot the Ostrich to that of the Camel has not ESCAPED Naturalists." Before he alludes to the resembl
053r100 ore or less interstratified with sediment.--& ESCARPMENT worn away like english escarpment The great
053r100 ediment.--& escarpment worn away like english ESCARPMENT The great conglomerate of the Amazons & Orin
100a055 ps worth Says from Lardner's (p. 213) form of ESCARPMENT relation kept to sea coast ∴ curious excepti
453e183 seed is highly odd-- as it is not so with the ESCULENT vegetables-- how is it with hollyoaks, flaxes
091a027 f Garsipa) dip 30 degree. <strike> «direction<?>»ESE-- CD [In the Darwar. transition Hills & strata S
244c023 , nous ne balançons pas a la regarder comme une ESPECE distincte! p. 171. Sus papuensis «partly domes
207b152 Lile de la Reunion presente elle seule plus d'ESPECES polymorphes que toute la terre ferme de lancie
209b156 ous ce climat, se divisent le plus aisément en ESPECES distinctes et permanentes." p. 145. In Humbold
243c019 pbell (52 degree S) qui n'aient egalement leur ESPECES; et certainment on eût été bien éloigné, il y
257c056 ns déjà observés en Patagonie. ou au moins des ESPECES tres-analogues,-- quand ce nétaient pas tout à
637j58r apted to long necks.-- p. 236. Marsupial bones ESPECIAL adaptation, to «young».-- good God & yet Mail
578n055 ssibility of this being discovered by anyone, ESPECIALL if it be a person. whose opinion he regards,
593n107 recognize «& take pleasure in» other animal, (ESPECIALL as in some <instinct> «insects» which become
036r052 deserts of Syria <chara> ditto for Patagonia, ESPECIALLY rocky parts of central Patagonia Does Andes
044r073 s might be introduced on great size of ocean; ESPECIALLY Pacifick: insignificant islets--general move
044r074 ttom.--solution under high pressure of gazes. ESPECIALLY the most abundant. Sulp. Hyd: Carb: A. Mur:
100a052 enslow's Grit, yet it is worth consideration. ESPECIALLY effect of gravity, versus some fault explain
120a109 soluble in <hot> cold than hot water with «-- ESPECIALLY if very hot under high pressure.--» respect
126a124 temperature would be much nearer the surface. ESPECIALLY at bottom of great ocean, where the circulat
170b002 This appears highest office in organization (ESPECIALLY in lower animals, where mind, & therefore re
175b017 aid isolate species <& give even less change> ESPECIALLY with some change probably <change> vary quic
192b086 ere have existed all those intermediate steps ESPECIALLY in those classes where species not numerous.
220b197 ay that instinct perverted, yet organization «ESPECIALL» connected with generation certainly is.= Th
220b197 nts domestication on perversion of structures ESPECIALLY reproductive organs) & therefore the one dis
226b220 in & history of every terrestrial Mammalia.-- ESPECIALLY moderately large ones.-- Is the flora of <S.
246c027 Voyage The casswary, inhabits Ceram, Bourou & ESPECIALLY New Guinea (replaces, Emeu) in North of New
252c045 local faunas foundation of all our knowledge ESPECIALLY great continents» Give Specimen of arrangeme
262c072 ome its some countries patriotic?)-- but more ESPECIALLY the powers of reasoning &c &c.-- Study the w
285c153 n animals domesticated «??», & therefore Most ESPECIALLY under care of Man. & external circumstances
350d056 to animals place in creation.-- thus senses, ESPECIALLY sight connected with locomotion.-- «Mem. Dr.
352d059 intervening physical change).-- distribution ESPECIALLY of Mammalia As every organ is modified by us
385d164 ly well known, that we observe in the temper, ESPECIALLY of the youngest children, a striking <resemb
413e058 grandfathers (2) Tendency to small change.. «ESPECIALLY with physical change» (3) Great fertility in
420e088 Monsters more common in Africa than in Europe ESPECIALLY with Europaeans settled there L'Institut do.
461t041 y <of> (or only closeness) of some species-- (ESPECIALLY of mammifers) in old beds & existing species
468t112 stamens», so that all were brushed by Bees & ESPECIALLY stigma after bee had brushed over the anther
483z013 t. before writing on Planariae or Polypi & is ESPECIALLY grand paper. p. 387. "on Classification of s
486z018 M. in N. America than in Europe-- Coleoptera ESPECIALLY require a greater duration of Heat. hence mu
497q006 = (12) Take Bag of soil from centre of woods «ESPECIALLY if date of wood be known» & other odd places
514q021 ber of generations: about crossing of plants; ESPECIALLY Papilionaceous order (2) History of fruit tr
515q022 nded instincts of the geese which he crossed; ESPECIALLY if the hybrids were recrossed with either pa
528m036 use as in music) from the splendour of light, ESPECIALLY when coloured.-- that light is a beautiful o
531m051 t seemed solely to be feelings of discomfort, ESPECIALLY about heart as of excited action, accompanyi
547m114 gue of thinking is keeping up these trains,-- ESPECIALLY if they are invented as in imagination, & in
552m131 when brain is pumping force to legs & body, & ESPECIALLY, when to whole body, being failed, & not to
578n054 association, the question, "one will anyone, ESPECIALLY a women think of my face,"? to one moral con
583n070 , as machine to make cell of certain form. (& ESPECIALLY as it adapt its cell to circumstances), it m
596n184 aze. does not gradation towards man.» Macacus ESPECIALLY pulls back skin of whole forehead & 2 ears.-
609o29v Malthus had shown incontinence to be a vice & ESPECIALLY in the female October d. 1838 perhaps insist
615o038 d by your parents, conscience This «X» memory ESPECIALLY «the» general kind taking pleasure in virtue
616o039 &c. have the same relation to a living body (ESPECIALLY the cerebral portions of it) that attraction
520mIFC stions & Experiments Expression M Charles Darwin ESQ 36 Grt. Marlborough Str.-- (p. 64. On <insect> A
204b140 riety impresses the offspring most forcibly-- ESQUIMAUX dog & Pointer «Game-fowls have courage indepe
209b155 ween Indian cow with hump & Common;-- between ESQUIMAUX & European dog? Yet man has had no inteest i
239c001 ong been known in any country; he states that ESQUIMAUX dog when crossed with pointer produces offspr
239c001 d with pointer produces offspring much nearer ESQUIMAUX than Pointer.-- He has no doubt that same thi
333d007 rden) He has seen in a show half Wolf & «half ESQUIMAUX» dog which <likewise more resembled the wolf
333d007 en two Parents.-- this is very interesting as ESQUIMAUX dog approaches to species. Again he has seen
333d007 es. Again he has seen several crosses between ESQUIMAUX dog & common dogs & Fox thinks they decidedly
333d008 ox thinks they decidedly take much most after ESQUIMAUX.-- this agrees perfectly with Yarrell & no le
392d180 n Zoolog. Gardens; <Buffalo & common cattle-- ESQUIMAUX (& Australian) dogs with common dogs--> Ask m
493q004 e exactly alike & see result.-- (3) Cross the ESQUIMAUX dog. with the hairless Brazilian or Persian a
193b092 umboldt has written on the geography of plants. ESSAI sur la Geographie des Plants. I Vol in 4 degree
280c134 nt.-- Shows instinct (Sir J. Sebright admirable ESSAY) heredetary Young wild ducks.-- lose as well as
323c269 h Revolution 3? vols. oct: -- 26th Blumenbach's ESSAY on Generation. Englis Transla -- The Revd. A. W
325c267 on Geograph. Distrib: of British Plants. Humes ESSAY on H. Understanding (sometime) Du Stewart works
454e184 e Sensitive Plant.-- p. 290. Dr. Edwards in his ESSAY on Spermatic animalcule. has described instrume
500q010) Talk about Thyme. Horned Oranges. Spallanzani ESSAY-- Figs 2 kinds of flower annually.-- Periwinkle
559m155 R. Jones has it.-- very curious book.-- Hume's ESSAY on the Human Understanding well worth reading C
578n057 eople cry.-- What is emotion? At end of Burke's ESSAY on the sublime & Beautiful there are some notes
605o019 the metaphorical term sublime 7 So that in this ESSAY. D. Stewart does not attempt «by one common pri
605o020 ympathy. D. Stewart on taste The object of this ESSAY is to show how taste is gained how it originate
322c270 volume. & C. Prevost on L'Ile Julie Waterton's ESSAYS on Natural History. Octob 2d Transactions of R
323c269 ine on the Breeding of Animals -- Spallanzani's ESSAYS on Animal Reproduction -- Treatise on Domestic
385d164 e same has happened in boys bodies.-- Lavaters. ESSAYS on Phy. transl by Holcroft Vol I. p. 195. says
556m145 ..] what> person says "what a pity"-- Lavater's ESSAYS on Physiognomy translated by Holcroft «Vol I»
563n001 & all other Books May 1873-- October 2d.. 1838 ESSAYS on Natural History Waterton describes. pheasan
591n101 Hume has section (IX) on the Reason of animals ESSAYS Vol 2.-- «also on origin of religion or polyth
595n127 as never seen others pout]CD [blank] Goldsmiths ESSAYS No XV, on sounds of words being expressive, (
442e151 can a form become permanent?» because its very ESSENCE is that little change is produced.-- The fact
542m096 s is smile. With respect to sneering the very ESSENCE of an habitual movement is continuing it when
575n043 ? What other explanation-- can we suppose some ESSENCE. The facts about crossing races of dogs on the
617o040 ed, by our external senses is a phenomenon the ESSENCE of whose existence consists in its communicati
421e091 ese organs to mistake a merely adaptive to an ESSENTIAL character--" How little clear meaning has thi
422e092 ght have.-- What is the difference between an ESSENTIAL character & an adaptive. one.-- are not the a
422e092 l character & an adaptive. one.-- are not the ESSENTIAL ones eminently adaptive.-- does it not mean l
586n080 in young & old.-- These facts point out some ESSENTIAL difference, which clearly ought to be expect
106a074 very ineffectual in widening valley.-- it is ESSENTIALLY a deepening agent {P} Therefore when we have
092a030 r solution. Athenaeum M. 516 1837 High up the ESSEQUIBO, granite & quartz, after passing sandstone Vo
179b033 go back?-- If so Men & plants together would ESTABLISH law. #as above stated:-- no one can doubt tha
200b123 ut doubt:-- Vide cattle: The grand fact is to ESTABLISH whether in crossing very opposite races, whet
369d115 anatomical & Zoological research, in order to ESTABLISH entirely their place in nature, as well as fu
370d117 nt «often» pursued throughout my theory is to ESTABLISH a point as a probability by induction, & to a
188b068 r:-- Prof. Henslow says. that when race once ESTABLISHED so difficult to root out.-- For instance eve

194b095 o southern hemisphere.-- If these facts were ESTABLISHED it would go to show a centrum for Mammalia.-
199b117 new insects all belong to same types already ESTABLISHED. why out of the thousands of forms, should t
343d037 created. (warring against those very laws he ESTABLISHED in all <nature> organic nature) the Rhinocer
349d052 d, though not circular, if characters can be ESTABLISHED-- clearly so.-- NB. This paper worth referri
458t001 all of larger size.-- the law of large size ESTABLISHED-- «Australia,. S. America--» These strange f
240c014 ameles, which joined to Casoars, perroquets, ESTABLISHES its «zoolog» alliance with New Holland. The
400e011 h. «as in Lucanidae» <no> less difficulty in ESTABLISHING good groups.-- ears varying so much.-- kind
577n050 & hence becomes associated with them.-- The ESTABLISHMENT of this principle of Association will help
348d051 t is that, amount of resemblance,-- how can we ESTIMATE this amount, when <value> no scale of value o
377d140 continents, & forming mountain-chains, when we ESTIMATE the matter removed by the waves of the sea, o
306c212 Young salmons. first a species which lived in ESTUARIES its taste. taught it to go to <sea> salter wa
037r054 ot weathering allows such Compare the elevated ESTUARY of the Plata. to the Bay of Bengal. dimensions
465t089 ver to present forms.-- -- How many «tertiary» ESTUARY & Lacrustine formations contain fossils,-- mam
465t089 fossils,-- mammals-- a few only -- & how many ESTUARY formations are there in old Secondary Series--
243c019 nt egalement leur espèces; et certainment on eût. ETÉ bien éloigné, il y a peu d'années, d'admettre qu
538m080 d make one's father & self one person-- & thus ETERNAL punishment explained. These facts showing what
206b148 enitor, who might have continued breeding from ETERNITY «backwards.--» If population was increasing b
316c244 ays the love of the deity & thought of him «or ETERNITY», only difference between the mind of man & a
343d036 iads of distinct forms» from a period short of ETERNITY to the present time, to the future-- How far
600o08v D 3. The Infinite, -- lives by hopes, looks to ETERNITY. (4) Reason, some transcendental kind-- (5) C
604o018 notion of living in lofty regions. 3 Infinity ETERNITY. darkness, power. being associated with God.
605o19v rs, that when certain causes, as great height, ETERNITY, &c &c. produces an inward pride & glorying.
605o19v ith it. as the idea of Deity. with vastness of ETERNITY. which superiority we transfer to ourselves i
323c269 nley familiar History of Birds -- Mackintoshs' ETHICAL Philosophy -- Bell's Bridgewater Treatise -- W
325c267 th & giving abstract of their views Mackintosh ETHICAL Philos: Prostitution of Paris. with respect to
618o042 felt [LHC] 1) May 5th. 1839.-- Maer Mackintosh ETHICAL Philosophy [RHC] On the Moral Sense Looking at
587n089 ng a mental <desire> «quality» &c &c Mackintosh ETHICS p. 97. on Devotional feeling p. 103-- Abstract
628o53v ught I to keep my word»-- gives the problem, of ETHICS-- [my answer would be to all such cases-- eith
627o053 . 231» marked in my Mackintosh 1) Mackintosh's ETHICAL Philosophy p. 6-- "The pleasure which results
622o048 e paramount,-- hence the law of honour. & the ETIQUETTES of Society.-- [LHC] Sir J. M. gives differen
326c266 oted by do.-- There appears to be good art.. on <ETN> Entozoa by Owen in Encyclop. of Anat. & Physiol
038r057 logist. Lyell considers (P 84 Vol III.) whole of ETNA series of coatings; hence it will be necessary
064r137 lines> «points» of eruption [.] give instance of ETNA Stromboli & Vesuvius Investigate with greater c
322c270 ts. 10 Vol, of Memoirs on Geology of France.= on ETNA. Almost reread the previous volume. & C. Prevos
183b052 , que celui de la chevre et du belier, cessent d ETRE feconds. dès les premières generations" go back
314c239 er in Timor plants, yet it seems there may be EUCALYPTUS!-- (Hostile fact) Be cautious about Goulds c
535m064 rga» of Terebrantia, laying eggs on leaves of EUCALYPTUS, watching few days till larva excluded, then
323c269 ations in Travels -- Observations on morals by EUGENIUS Feb 14th. Bo«s»well's life of Johnson. 4. Vol
624o051 been perverted from want of reason.-- Hence as EUGENIUS says, slow growth of rule of right.-- [LHC] *
627o053 does not enter into his contemplation.--" Now EUGENIUS would contend against this-- but the pleasure
627o053 ome operation of intellectual faculties-- Will EUGENIUS allow this moral obligation? [2] [The improve
171b002 ed.-- annuals rendered perennial. &c &c.-- Yet EUNUCHS nor «cut» Stallions nor nuns are longer lived
317c250 's Journal) on the heights of St Jago found a EUPHORBIA so near Piscatoria as scarcely to be distingu
087a008 d to produce climate resembling S. America in EUROPAEAN latitudes.-- Will it be supposed that the arm
087a011 ---- Mediterranean continent corresponding to EUROPAEAN risings. Pacific great land. -- Will use argu
117a101 g. 1837. December. p. 91. a classification of EUROPAEAN strata according to composition thinks sand w
191b080 hipelagos ups & downs in full conformity with EUROPAEAN formations-- England & Europe Ireland common
230b241 mples of animals «or plants» very close (take EUROPAEAN birds. Mr Goulds' case of Willow wren) & othe
232b247 most peculiar Flora «& north of Europe»-- As EUROPAEAN forms have travelled towards Equator, <th> so
232b247 analogy would have been very unlike Southern EUROPAEAN ones.-- "a variation played on secular refrig
249c036 emminks fact of forms being within Tropics.-- EUROPAEAN birds at Japan. connected with Europaean form
249c036 s.-- Europaean birds at Japan. connected with EUROPAEAN forms on Himalaya??-- This is very remarkable
250c037 characterized, of each continent. Try amongst EUROPAEAN quadrupeds if Africa destroyed would not then
278c132 compared to America.-- the wonder is that the EUROPAEAN forms, were able to escape to some more fitti
303c203 number.-- Parallel of Japan near Himalaya, & EUROPAEAN forms on that isld.-- The races of men differ
305c209 olour of plumage in same remarkable manner as EUROPAEAN species = singular coincidence if distinct cr
312c233 cies to different,-- inguinal louse African & EUROPAEAN. different.-- thorax & head differ Africa Aus
315c241 grande analogie avec le flore du Japon", some EUROPAEAN & Sandwich species & some of Japan. I do not
316c247 , except in very few cases, those of Tertiary EUROPAEAN fossils-- «(so much the more remarkable ': car
327cIBC y's travels in Syria Vol I. p. 71. account of EUROPAEAN plants transported.-- Crawford. Eastern Archi
337d020 approaches quite to wild local variety.-- our EUROPAEAN varieties must be very unnatural-- Italian Gr
353d061 of the hawks <to> are analogues to <Bustard> EUROPAEAN birds. also «do» p. 403. & 404 vol II. do (p.
363d102 -- The habits of some «same» North American & EUROPAEAN birds «slight» different-- Barn Owl <the> in
399e011 ha-- In marl from «Lake» Constance species of EUROPAEAN genera»-- Hope has ideas about generic chara
400e011 ch to Asiatic [...]t in part near Timor, & to EUROPAEAN in Van Diemens land, where there is close spe
402e019 that 60 degree & 70 degree of Europe.-- Many EUROPAEAN insects-- list given,= some peculiar <M> p. 3
424e100 es same. & Northern forms-- & American ones & EUROPAEAN-- agree very much closer, than the present on
427e109 aah») & break up. N. American Conchology from EUROPAEAN., & the climate being now less extreme, than
503q012 back to back 37 Col. Sykes fertility of men & EUROPAEAN animals in India?-- about Chetah & other tame
511q018 upposes is now extinct=- (9) About. American & EUROPAEAN common species, having somewhat of different
514q21. peculiar Fauna?. {Australian Alps--; are any EUROPAEAN forms found there-- Lindley says that only on
572n033 egro certainly has less reasoning powers than EUROPAEAN.-- Ideots. defective brains.-- Erasmus does n
615o36v breed of fowl» totally different habits from EUROPAEAN. begin to prowl about in the evening «seldom
420e088 mmon in Africa than in Europe especially with EUROPAEANS settled there L'Institut do. p. 419, «long»
025r018 to urge, geologists to compare whole history of EUROPE, with America; I might add I have drawn all my
025r018 see conclusions substantiated over S. America & EUROPE. we may believe them applicable to the world.-
026r022 s there account of Baron Roussin's voyage.-- In EUROPE proofs of many oscillations of level, which in
026r022 Cordillera of S. America. Study Geolog: Map of EUROPE Conybeare. Introduct XII P. silicitied bones n
031r038 leutian Arch.--» V. Fitton. Australia: cases in EUROPE.-- Auvergne. very little Pumice, though Trachy
036r049 nd discussion Consult. reconsult Geolog. Map of EUROPE Consult charts for distribution of pebbles.--P
044r073 out simple.-- Fortunate for this science. that EUROPE was its birth place.--Some general reflections
044r073 solidated; one inclines to belief all strata of EUROPE formed near coast. Humboldts quotation of inst
056r110 country of granite, not <so> great«er» for all EUROPE, than from the Plata to Caraccas, which is all
057r114 n lower strata. only in upper. in accordance in EUROPE with ice theory.-- Capt Ross found in Possessi
060r125 ogical difference of formations of S. America & EUROPE.-- If great chain of Volc. had been in action
064r135 kes & great but local elevations of the land in EUROPE-- Urge difference of plutonic rocks & Volcanic
075r167 undant.--| muriated silver. which is so rare in EUROPE. common there accompanied by molybdated lead &
076r169 etimes by selenite.-- in New Spain, contrary to EUROPE. argentiferous lead not abundant. = considerab
086a003 etation. -- So accustomed to utter confusion in EUROPE, that the simplicity of Ventana's «Quartz.» un
086a006 ell as Pallas before Geology is written Cuvier. EUROPE possessed a great edentata. -- How much is tem
088a016 of Andes Metamorphic action -- Mem: red sand of EUROPE no fossil shells --¿ action of Heat bubbles vo
114a094 n modern formation & not ever in Secondary in EUROPE. gneiss-- metamorphosed clay slate.-- --shale
178b030 ed.-- Type of Eocene with respect to Miocene of EUROPE? Loudon. Journal. of Nat History.-- July. 1837
182b047 lances & differences for instance of finches in EUROPE & America. &c &c &c The new system of Natural
191b080 onformity with Europaean formations-- England & EUROPE Ireland common animals-- + + + for instance te
194b095 entate [has been doubted?]CD & opossum found in EUROPE now confined to southern hemisphere.-- If thes
212b167 hagous Mollusca according to periods.-- NB. Was EUROPE desert (like S. Africa) after Coal Period.-- ¿
217b187 American «form of» Lathyrus has one species in EUROPE Madagascar has several American forms-- The ab
220b197 » two quadrupeds at S. America Jaguar & Tiger & EUROPE, as to produce same one. «Although in plants,
220b199 es that all «genera of» birds in «N.» America & EUROPE, which have not their representative species i
226b220 India.-- Van Diemen's land Australia. England & EUROPE.-- It will be well worth while to study profou
226b223 S. America> Tierra del Fuego like that of North EUROPE, many genera & few species. The number of gene
226b223 on by which great Edentate might have roamed to EUROPE & Pachydermata from Europe to America., How st
226b223 might have roamed to Europe & Pachydermata from EUROPE to America., How strange would presence of Jag
227b224 on of those two countries Hence India, Mexico & EUROPE. one gret sea (Coral reefs:. shallow water at M
232b247 ave possessed a most peculiar Flora «& north of EUROPE»-- As Europaean forms have travelled towards E
236b278 former dogs. character of Miocene Mammalia--of EUROPE Mem. Mr Bell's case of Sub Himalayan land emys
246c026 eties & not species.-- Vol :694. King-fisher of EUROPE, (Alcedo ispida) from Molluccas. scarcely diff

```
246c026 olluccas. scarcely differs at all from those of EUROPE, but beak rather sharper., & rather longer  in
249c036 onnected with, Mr Blyth's statement of birds of EUROPE & America, which are of different forms  being
250c037 can conceive of destruction of Great Animals in EUROPE & America Some portion of the world «Africa» b
250c039 . America might have been string of islands.-- ¿EUROPE has many species but not genera distinct  from
274c116 y few forms in common.,-- but each several with EUROPE & northern Asia & Northern America.-- may we n
278c132 ica, the wonder have been same for S. America & EUROPE.-- The difficulty is how came it animals not p
291c167 most proved. Elephant has left no descendant in EUROPE<)> Toxodon in S. America) is absolutely necess
296c184 country»??  p. 475 NB. This bears on fossils of EUROPE., those species which can migrate remaining co
310c224 Cape de Verds} the shells in equal periods with EUROPE would probably have changed much less.--  Here
311c225 8. some remarks on Bonaparte's list of birds in EUROPE & N. America, on closely allied species. «repl
313c234 ter duration, than molluscs, argue case both in EUROPE & S. America. «very difficult case» Does  this
315c240 & other cryptogamic plants same in Australia & EUROPE.-- if creation be absolute thing, the creation
316c247 fossils  more resemble those of America than of EUROPE, because the recent ones are so close. Was the
316c247 close. Was there continent between N. America & EUROPE?-- ▌.Norton has written on fossils of N. Ameri
317c250 St.  Jago showing many common to Canary. isld., EUROPE, & St Jago upper region, & some to Cape.-- som
319c257 Institut, 1838, p. 230 says the Macrotherium of EUROPE is between the Anteater of C of Good Hope & th
337d022 e remarks that the Fringilla domestica of North EUROPE is replaced by the F. cisalpina in Italy, whic
338d025 mistake Surely the fossil Mamalogy of Britain & EUROPE is African. & the only difference is by the ex
343d039 s & from the existing animals found fossil when EUROPE must have worn a quite different figure  19th.
343d039 rth adding in note than amongst the Mammalia of EUROPE the shells of do-- shells of. N. America.-- sh
363d102 er-Wagtails mistake both species scattered over EUROPE)-- The habits of some «same» North American  &
375d135 er purposes as wheat for making brandy.--» take EUROPE on an average, every species must have same nu
400e013 candoelle, distributed seeds of Dahlia all over EUROPE same year.-- he sowed them for four generation
402e019 ee yet fauna like that 60 degree & 70 degree of EUROPE.-- Many Europaean insects-- list given,= some
403e021 ffalo of the East which differs from that of S. EUROPE. p. 189. The gaut, kind of crocodile, sometimes
407e037 , which S. America now does to North. America & EUROPE.-- S. America favourable to Tropical productio
407e038 ergone a greater change, than any part, (except EUROPE. in which all Tropical forms have been obliter
407e041 of Africa ought most to resemble fossil ones of EUROPE, Consider probable form of land,-- -- S Americ
408e042 w-- more so than any other part of the World.-- EUROPE perhaps less so, that either Americas.-- If sp
409e048 anged. When discussing extinction of animals in EUROPE. :the forms themselves have been basis of argu
420e088 n thinks Monsters more common in Africa than in EUROPE especially with Europaeans settled there L'Ins
424e101 Beck,  that Hooded crow & Carrion crow. have in EUROPE different ranges-- latter not going North of t
449e173 Hope., has different constitution from those of EUROPE-- for they stand India. better than the latter
465t079 ars from Lund more Mammals, than at present «in EUROPE we know there has been several successions  of
465t079 rved in this country-- same argument to India & EUROPE-- & Africa!,-- any negative argument against--
471tf09 p. 18.) capital list of all the fossil Mamm. of EUROPE-- Large Lizards in Navigatores. Williams. Narr
473s03r trata! Mastodon longirostris in miocene like in EUROPE-- Cuvier never found remains of Sus with Eleph
484z015 ike Puffinuria does of Petrel?-- Study Birds of EUROPE for other representatives of this class--. Pyr
484z015 phalus & many Tyrannulae-- replaces warblers of EUROPE-- Study profoundly shells of Bahia Blanca & So
486z018 ca extend much further N. in N. America than in EUROPE-- Coleoptera especially require a greater dura
486z018 Fuego.--  Hence it is odd that Amber insects of EUROPE have Tropical Forms See p. 256 of Note Book (C
486z019 ison of singing powers of birds of N. America & EUROPE. Entomolog. Transact. Vol I. p. 130. Col Sykes
500q10a & British Species & British & distant parts of EUROPE.-- Gould-- go over the Pigeons, Philotis, Dace
501q10a in  one country & not in other: Rosa is hard in EUROPE, Walnut in America.-- Heaths in Africa; Hooker
501q10a y «one» species of plant, vary in one region of EUROPE & less in another region-- (27) Which sex in M
205b142 atise p 85. Parasite of Negroes different from EUROPEAN.-- Horse & Ox have different parasite in diff
209b155 cow with hump & Common;-- between Esquimaux & EUROPEAN dog? Yet man has had no interest in perpetuat
305c209 here there are some godwits which are close to EUROPEAN species, and the sexes of which vary in colou
310c223 rly so confined as now thought.-- N. American, EUROPEAN & Chinese Genera & some species on Himalay.-
325c267 es d'Ossements 3d. Edit. Octav. (good to trace EUROPEAN forms compared with African <A> Annals Histoi
243c019 l y a peu d'années, d'admettre que ces oiseaux EUSSENT leurs représentants dans de si hautes latitude
243c019 'aient egalement leur espèces; et certainment on EUT été bien éloigné, il y a peu d'années, d'admettr
104a068 f plains due to whole moisture being lost by EVAPORATION therefore capillary attraction would bring w
529m040 griculturist, in whose mind supply of food was EVASIVE & ill defined thought would receive pleasure f
322c269 ences at end Mayo Pathology of the Human. Mind EVELYNS Sylva. skimmed, stupid Brownes travel in Afric
147g019 uld not be placed in present position Thursday EVENING ½ past 8 Tyndrum 29.<625> «636» Temp. 62 Frida
162g103 > Saturday Morning 29.958 A 64 degree, air 60 «EVENING do» The extreme right arm of River Tarf <it> H
469t135 - Saw Bees frequent these flowers till late in EVENING-- On rough calc. 280 flowers-- allowing each B
531m051 f anger which came over me, when listening one EVENING when tired «-- how true the heart the scene of
615o36v ts from Europaean. begin to prowl about in the EVENING «seldom leave their perch till evening» crow d
615o36v n  in the evening «seldom leave their perch till EVENING» crow different.-- Heredetary effect of former
191b083 individuals.  Here we have avitism the ordinary EVENT. & succession the extraordinary South Africa. p
222b207 instituted.   People often talk of the wonderful EVENT of intellectual Man appearing..-- the appearanc
531m050 because people think that the importance of the EVENT by itself will make it to be remembered. wherea
434e128 eason from the analogies of the existing to the EVENTS of the past world, we have no foundation for o
544m102 eculiar sensation as flying. (No memory of past EVENTS?) or influence on our conduct, the links which
547m114 ins of thought, or memory from innumerable late EVENTS.-- the fatigue of thinking is keeping up these
028r031 separate them from indissoluble rocks? Has Chalk EVER been dissolved? Singularity of fresh water at I
033r042 matter how very curious a structure: Have shells EVER casts alone in Calcareous. rocks??--if so  case
045r077 g as that of a spring) moves away.--Will geology EVER succeed in showing a direct relation of a  part
051r095 nto play; most manifest example of degradation I EVER saw on beach near Callao.--From Sir. H Davy exp
056r109 uld raise such a bulwark to the ocean, who would EVER suppose that its age was limited? Who could sup
114a094 ve» water ∴ not formed in modern formation & not EVER in Secondary in Europe. gneiss-- metamorphosed
121a112 unequal.--  where were cracks?--? How came there EVER to be cracks 11th August. 1838 Near Woolwich th
165g126 union  of two instincts crossing most remarkable EVER obseved? Shows that <nervous> brain makes thoug
171b004 dental injuries. <on> which if animals lived for EVER would be endless (that is with our present syst
188b068 blished so difficult to root out.-- For instance EVER so many seeds of white flower. all would come u
190b075 acilities of communication Have races of Plants. EVER been crossed really, if there is any difficulty
239c1FC e: where hybrids produced have any close species EVER yet failed. About trades affecting form of man.
285c151 alogy, but will grow into affinity.--but whether EVER arrive at true affinity doubtful A species is o
303c205 this would have been fair, but to place all that EVER lived <on> into one list is unfair, [moreover w
306c213 Major Mitchell is not aware that Australian dogs EVER hunt in company-- marked difference with dogs
348d051 y may be taken literally,, though how far we can EVER discover the real relationship is doubtful.- n
367d112 ng «& this must be so, else avitism could hardly EVER occur.-».-- & if that cannot be formed, geneta
373d132 y. what tendency can there be for abortive organ EVER disappearing??-- Have Marsupiata abortive Mamma
376d137 r Smith every baboon & monkey, big & little that EVER he saw knew women.-- he has repeatedly seen the
428e113 reat difficulty. «I should doubt if wild species EVER formed like short-tailed cat or dog has been wi
492q002 of individuals-- 9. Do plants in becoming double EVER become monooecious-- loosing one sex & not othe
492q002 whose  stamens are monstrous, how then are seeds EVER raised? 11. Is not non-flowering gorze common i
498q007 ification <lat> retrograde into leaves-- is this EVER effect of want of nutrition.-- Horned oranges s
501q011 g of Thyme doubts about stigma in similar manner EVER failing.-- answered by Gaertner (28) Can any an
507q15v produce.  Polygam. trioecia. (are female flowers EVER productive? Smith says many trees in Tropics ar
508q016 common  in man than in female-- (8) In Hump-back EVER heredetary (9) Are the works of Berhave (treati
511q018 rs. Bull-Dogs. Spaniels-- Grey-hounds-- is there EVER any degeneration?? HOUNDS. Eyton Mr Wynne, &c C
513q21. uld extract nothing from him)║ Does impregnation EVER regularly take place in unopened flower-- <doub
531m049 ng got on. this nest the cat could If cats will «EVER» eat little birds, this most curious instance o
538m079 Bessy  repeated things, which none about her had EVER before heard, so very probably forgotten.» Such
555m142 xpression, but not nearly so often <that> hardly EVER the expression of passion with open mouths like
556m144 but all these idea came one after other, without EVER comparing them, I neither doubted them or belie
560mIBC les Darwin 36 Great Marlborough St Has my Father EVER known <intemperance> «disease» in grandchild, w
576n047 o even to save a friend, or wife.-- yet he would EVER repent, & wished he had lost his life in  doing
594n112 «so»  rare « <s.> » here,, that probably few had EVER before seen one, yet all-- flew to bed of flags
606o021 ion, it becomes so instantaneous. that we cannot EVER perceive the various operations which the  mind
627o52v ments of beneficial tendency of affections.-- If EVER I write on these subjects consult <following> p
632j53r dmission is probably from ignorance]CD Who would EVER have thought that the intestines of a thrush we
638j58v ocess is shortened, but yet analogous, no savage EVER made a perfect hinge.-- reason, & not death rej
061r128 uliar character: Contrast low limit of Palms, EVERGREEN trees, arborescent grasses, parasitic plants,
563n001 at  all other times.-- Birth Hill shows it is EVERGREENS they seek Cock Pheasant claps his wings befo
```

(Key Word)*

385d163 & so.-- Says, there is breed of Fowls called EVERLASTING layer--. or Polish breed. (he thinks half ph
522m013 ation between sound people and insane.-- that EVERYBODY is insane. at some time. Mania is quite disti
620o044 nk of it.-- Whatever the cause of this may be, EVERYONE must know, how soon the pleasure from good di
530m044 , a fit of = gout, has affected his memory of EVERYTHING in <he a [...] Mr B> journey. short time pre
544m104 faculties: «Vide page 110, by mistake.» N B. EVERYTHING which happens to man who does not produce ch
571n029 in ideal taste & the former to true taste.-- EVERYTHING that is habitual, if heredetary, is pleasant
602o11v in.-- How are my ideas of a general notion of EVERYTHING applicable to the high idea «p. 131.» in Tra
554m139 r take bread from a visitor, & before eating «EVERYTIME», look up to «keeper» see whether, this was p
051r094 g to protection of Organic productions. = Yet EVERYWHERE on coast (Il Defonsos «Kelp») rocks show sig
104a067 ve given rise to eruptions.-- We must suppose EVERYWHERE--, in granitic areas &c &c volcanos {P} fiss
555m142 (How to Observe p. 213) says charity is found EVERYWHERE (is it not present with all associated anima
040r064 l of Andes.--but as long as all below water no EVIDENCE--The depth of shells (which being packed. in
045r075 epcion earthquake.--expatiate on difficulty of EVIDENCE about eruptions of Volcanos. (where there are
105a070 ng a medial gorge by slow erosion. but we have EVIDENCE in distribution of blocks, that there has bee
230b240 rfect change.-- It is scarcely possible to get EVIDENCE of two races of plants run wild.-- (for we kn
275c121 on in his work I am sorry to find Mr Yarrell's EVIDENCE, as so many plants produce hybrids, or else w
376d138 up tail» Mem: Ourang Jenny with Tommy.-- Good EVIDENCE about old varieties is reduced to scarcely an
424e101 modern times.-- this would depend on negative EVIDENCE of knowledge of Woman-- ¶ The noise st st. wh
460t019 als Aug. 26th.-- When it is said that there is EVIDENCE of fossil remains, & therefore not to be trus
571n027 istence of taste in human mind. is to me clear EVIDENCE in the organic world of infinite & growing co
581n065 any learned men seem to consider there is good EVIDENCE in the structure of language, that it was pro
601o009 <conse» ordering contraction (that is the only EVIDENCE. when consciousness is absent) in fibres unit
601o009 ed heart is pricked) and certain action. (only EVIDENCE. when not consciousness) are produced in cons
601o009 to it, man in sleep not conscious, nor child-- EVIDENCE of consciousness, <t> movements «¿» anterior
641j29v the strongly separated Arctic genera, there is EVIDENCE of antiquity & extinction of such forms-- the
047r082 s accounts & see if fall is not the first very EVIDENCE movement.--The swelling first on beach I canno
051r094 dation; (soft substances worn into bare cliffs EVIDENT); the action is anomalous; It is wonderful to
106a074 ith stream retaining its force, now it will be EVIDENT that deflected stream cannot retain its force
118a103 t Helena &c &c.-- in Cordillera, it is at once EVIDENT only small proportion of dikes have reached th
206b148 of contagions & accidents: yet some causes are EVIDENT, as for instance one man killing another.-- So
288c159 rom England fetch very <go» large price. as is EVIDENT to be worth introducing, instead of breeding f
308c218 cleay has this remark) Mem. number 5 here most EVIDENT!!? examine into this case D. Jeffrey (life of
364d103 irds to see which will sing longest, & they in EVIDENT rivalry sing against each other, till it has b
443e151 n of these «sexual» functions to complexity is EVIDENT, yet the inference from some plants & some mol
612o035 g be used without consciousness, for it is not EVIDENT, what animals have consciousness. These willin
041r066 ginous veins of this figure {P} in sandstone: EVIDENTLY depend on a concretionary contraction: the fa
099a049 dgwicks lamination parallel to stratification EVIDENTLY small scale of concretionary action all fluid
111a087 to 80 degree faults with red wacke contorted EVIDENTLY dike. V. VII. p. 316 & 328 VI. p. 365. Meyen
197b111 t say propagated, but must have concluded so= EVIDENTLY «or hints» considers generation as a short pr
217b187 has several American forms-- The above facts EVIDENTLY show that Mr D. wonders at these species bein
220b197 = The dislike of two species to each other is EVIDENTLY an instinct-- & this prevents breeding. now d
226b221 umber of genera on islands & on Arctic shores EVIDENTLY due to «the» chance of some one of the differ
248c030 s a tendency to go back to oldest race, which EVIDENTLY is tending to same end, as the law of hybridi
255c050 ad of analogues.-- <as> in other classes this EVIDENTLY relates to greater range of such forms.-- p.
261c072 des to some structure in head, which he says (EVIDENTLY) is an exception) can only be explained by dir
349d052 o many steps from a head, as subkingdom.-- -- EVIDENTLY artificial, as interloopment of Marsupials wi
354d065 e any law of this. Do any varieties of sheep «EVIDENTLY artificial» approach in character to goats.--
368d114 in Raptorial birds largest.-- p. 47. (<"> is EVIDENTLY the male which recedes from the species all f
376d137 had no curiosity to pull up trousers of men. EVIDENTLY knew «men» women, thinks perhaps by smell.--
382d157 wen to ask. whether a Heautandrous animal is <EVIDENTLY» actually split in two-- keeping sexes separa
430e120 unnamed this intermediate one.-- Mr Lonsdale EVIDENTLY inclines to think it Hybrid.!!! Ask Woodward
445e158 instinctive power in chicken, yet says it is EVIDENTLY acquired by experience in baby Lamarck. Vol I
542m095 be explained from too long rest of muscles.-- EVIDENTLY habitual when transferred, (also how often) t
551m128 tracts.-- but W. Fox's dog that shut the door EVIDENTLY did, for it did with far more alacrity <than>
582n066 > In the life of Hayd & Mozart. fine music is EVIDENTLY considered as analogous to glowing conversati
605o20v not men of taste & the reverse of this. taste EVIDENTLY does not consist of this. but rather in the p
302c202 tion. therefore lately acquired.-- I fear «great EVIL» from vast opposition in opinion on all subject
550m123 ither.-- Our descent, then, is the origin of our EVIL passions!!-- The Devil under form of Baboon is
566n011 his must be doubted. before my view of origin of EVIL passions.-- Man getting sight slowly,, but when
604o015 o children S. Jenyn's Inquiry into the Origin of EVIL. Reviewed by Johnson in the Literary Magazine.
640j28v ary race.-- «There is no way of eliminating the EVILS of old age, after breeding season, or gaining a
376d138 h it.-- also of <a> dog«s». but did not seen to EVINCE more lewdness for bitch than dog: Monkey thus
205b145 es," p. 20. do "If hornless ram be put to horned EWE almost all the lambs will be hornless.--" does t
379d152 n that in breeding very pure South Down that the EWE must never be put to any other breed else all th
403e023 s do not this). because it has been so pronounced EX cathedrâ. let us look at facts. considering few
054r102 red by limestone with recent shells 200 ft, how EXACT agreement with Coquimbo; [not located] [not loc
055r108 rguments which strike the mind with force.--the EXACT yearly rise of the great rivers prove better th
147g020 yet possibly sea more probably than river-- No EXACT terraces but appearances, as if valley had been
209b156 .-- analogous to nearest continent: poorness in EXACT proportion to distance (?). & similarity of typ
333d005 ther different».-- crossed with <un»common cat, EXACT variety unknown., three kittens, alike each oth
337d023 same classes» differ in different countries in EXACT proportion to the time they have been separated
059r119 ces couches ont entre elles une correspondance EXACTE, et lorsquelles se trouvent interrompues par q
124a119 to know whether it would be absorbed.-- if so EXACTLY parallel to limestone & volcanic rock containi
225b219 ls very remote. ass & Horse. produce offspring EXACTLY intermediate.-- Reference to Pig & Dogs. My th
275c120 s-- Mr Yarrell.-- says my view of varieties is EXACTLY what I state.-- &c picking varieties. unnatura
333d008 n was put.-- Fox thinks half Lion & Tigers are EXACTLY intermediate in character & Kittens alike each
333d008 s resembles one parent & one another & are not EXACTLY intermediate.-- Where two dogs line the same b
374d134 ks deeply has assumed that increase of animals EXACTLY proportiona[l] to the number that can live.--»
391d174 on of two common buds??? Amongst buds each one EXACTLY like its parent. <but these buds do not procre
430e119 oceranus from the Gault of Folkstone, which is EXACTLY intermediate between I. concentricus & I. sulc
461t041 Spain & such facts-- This unequal duration is EXACTLY same as some species extending much further ge
492q003 g Breeding of Animals If two half bred animals EXACTLY alike be interbred will offspring be uniform.--
493q004 .-- (2) Cross two half-bred animals. which are EXACTLY alike & see result.-- (3) Cross the Esquimaux
213b171 marked to me, the "beauty of species is their EXACTNESS,' but do not known varieties do the same. May
273c112 tail stiff.-- Swallow & goatsuckers likewise EXAGGERATED.-- There is one most remarkable connection b
283c145 ll be two genera.-- let short billed one, be EXAGGERATED, & all rest destroyed far remote genera. wil
564n006 furious faces «which may be described as an EXAGGERATED habitual sneer» the manner in which whole sk
363d101 ny colours of plumage &c &c «Pouting pidgeon EXAGGERATION of cooing.--» & compare them with all the v
571n028 ustice?? as ancients did high forehead sign of EXALTED character???) Why may not our heredetary natur
244c023 g of Lepus Magellanicus says: «after» "après un EXAMEN attentif, et forts surtout de l'opinion du bar
025r019 to the world.-- My general opinion from the EXAMINATION of soundings, from about 80 fathoms & upward
227b228 hole metaphysics.-- it would lead to closest EXAMINATION of hybridity «to what circumstances favour c
228b228 me from & to what we tend.-- this & <direct» EXAMINATION of direct passages of <species> structure in
261c070 his by instances Once grant my theory, & the EXAMINATION of species from distant countries. may give
355d066 ey can find loopholes) "It is well worthy of EXAMINATION whether variations are produced only in thos
411e055 y would be. generalized, & afterwards by the EXAMINATION of the special cases, under which the indivi
543m101 .-- it is brought within <our own> limits of EXAMINATION.-- obeys same laws. as other parts of struct
030r036 great quantity of altered Carbonaceous shales EXAMINE chart of Patagonian coast to see proportional
110a085 ause «or be effect of» elevation & subsidence. EXAMINE these «lines» Description of rocks in Lyells'.
114a093 . p. 403 Ehrenberg on ferruginous Gallionella EXAMINE Iron stone of C. of Good Hope & Australia/ and
117a102 Guayaquil.-- modern shells of Cobija doubtful. EXAMINE well shores of lakes. to see effects of degrad
182b047 ance among the Carabidae.-- instance in birds> EXAMINE «good» collection of insects with this in view
212b168 ,) in last tertiary epochs most genera dead? --EXAMINE into this «in Phillips».-- According to this,
269c100 ce, because applicable to N. Hemisphere (NB.-- EXAMINE Abrolhos Flora with this view) Tristan D'Acunh
289c161 habits of a Grebe, structures might follow.-- EXAMINE structure of this bird & get account of habits
308c218 his remark) Mem. number 5 here most evident!!! EXAMINE into this case D. Jeffrey (life of Mackintosh
376d137 women, thinks perhaps by smell.-- but monkeys EXAMINE sexes of every Has repeatedly seen one he kept

376d138 more   lewdness for bitch than dog: Monkey thus EXAMINE each other sexes,-- «by taking up tail» Mem: O
392d180 my father to look out for instances of Avitism EXAMINE English weeds in Hot. Houses will they flower
402e018 eased their cries. "many of them descending to EXAMINE their defunct companion".-- p. 229. Borneo.--
430e119 Then give my theory.-- excellently true theory EXAMINE list of St. Helena Plants & see whether those
470t178 common   kind--I think not on Phlox though they EXAMINE it.--«Little Dusty & Blue» Butterflies at Clov
493q003 iffers in different cats.-- Good observation-- EXAMINE semen of Hybrid animal. in comparison with Wei
495q05a k on pond with Duck-weed-- coots-- waterhens-- EXAMINE dog, which has swum-- on pools & rivers-- ever
495q05a rs-- every kind of seed must be distributed.-- EXAMINE scum of pond for seeds.-- 11. Soak all kinds o
496q05a many flowers in minute do they visit?? good=!! EXAMINE pollen of double flowers. compared with single
496q05a give body to Hawk & sow pellet. ejected. done EXAMINE pollen of such flowers as do not seed or seed
497q06v ter Mr Herbert says do about OEnothera.-- (14) EXAMINE pollen of those genera of which wild hybrids h
498q09v rell. Lychnis. Butchers Broom--«also, Vinca,» EXAMINE all these, are they much frequented by Bees or
504q013 ery year & Spring. & within garden //Yes// (5) EXAMINE the Parnassia whose stamens move one after oth
506q014 h (measured same way) 47¼-- in heigt 30 inches EXAMINE Keel of Common & Wild Duck-- Black Duck & Peng
541m091 plain excessive labour of inventive thought.-- EXAMINE frame of mind in following changes during fall
032r042 brecciated white stone of Chiloe, after having EXAMINED the changes of pumice at Ascension In Calc: s
049r087 te of all the Porphyry specimens, must be well EXAMINED At M. Video «facts of Passages marked by do.»
184b054 ds described by Q. & G. as new Species. Cuvier EXAMINED it. There certainly appears attempt in each d
257c056 le & lastly 12,000 ft above sea in Bolivia; he EXAMINED-- all species & found "beaucoup des mêmes ois
319c276 of fossil geographical range.-- [blank] Books EXAMINED: with ref: to Species Most of those which hav
466t099 during life of individual. June 1st 1841. Maer EXAMINED the Lemon-thyme.-- equally abortive as it was
467t100 s remark that pistil does not become abortive. EXAMINED in microscope--some of the stigmas of {P} sha
469t119 ar 1778. Paper by Camper on Ourang-outang, has EXAMINED 7 says one specimen had on one foot, a toe-na
469t135 en minutes. it was visited by 13 Bees-- & each EXAMINED very many flowers.= 22d.-- /during several su
472s02r routed. wh. was not case, on several flowers I EXAMINED some days ago-- This Bee flew from yellow to
121a112 I saw none embedded this point would be worth EXAMINING. to support. shells on surface of Patagonia,
165g124 lomerates near Loch Lochy would be well worth EXAMINING-- Inverness & waters of the Tarf-- Kilfinnan
290c164 rd washed, Whilst tips of primaries black, by EXAMINING series I cannot doubt laws of change, Will be
034r044 st inferior rocks--The stream at Portillo Pass EXAMPLE of do? <Poor> Daubeny good account of ejected
039r060 ture in the structure of Ascension» give as an EXAMPLE the great subsidence at the famous eruption of
051r095 n pebbles are brought into play; most manifest EXAMPLE of degradation I ever saw on beach near Callao
219b195 ains low down, therefore plants common take an EXAMPLE from T. del Fuego.-- Ellis (?) says Tahitian k
226b221 l explain representative system Of this we see EXAMPLE in English & Irish Hare.-- Galapagos.-- shrews
259c065 l off plenty of times & you will have no tail (EXAMPLE probably not true).-- or again healthy parents
403e021 e been cultivated in Guahon. (Mariannes), "for EXAMPLE the prickly Limonia trifoliata, which cannot n
526m027 action determined by heredetary constitution, EXAMPLE of others or teaching of others.-- (NB man muc
533m059 getting on a horse on the left side (not good EXAMPLE,) because leg is right handed.-- In Review (Ed
537m074 or his children's sake & for the effect of his EXAMPLE on others.‖ It may be doubted whether a man in
584n071 , for if anyone has taken the Woodpecker as an EXAMPLE fitted for climbing, his arguments partly fail
230b241 back previous to fresh change Get a good many EXAMPLES of animals «or plants» very close (take Europ
307c216 d animals, in plants I presume there are.? get EXAMPLES.-- for instance where a tendril passes into a
411e056 ts before form.-- I have already given various EXAMPLES The Pipe-fish is instance of part of the herm
431e123 ted view, though perhaps a true element) «give EXAMPLES, pigs, with small chinese boars &c &c &c» p.
537m075 tineau. How to observe, p. 21-26. argues «as EXAMPLES» very justly there is no universal moral sens
632j53r cocoa nut by water «fucus for adhesion».-- as EXAMPLES of design.-- perhaps they are so.-- but the c
286c155 ho soar above Such prejudices, yet have justly EXALTED nature of man. like to think his origin godli
113a092 ar> W. of Port Philip. which had bar at mouth EXCAVATED in solid rock.-- 4 & 5 fathoms deep. perfectl
409e046 f &c &c-- is precisely analogous case to man, EXCEEDING monkeys;-- Having proved mens & brutes bodies
349d053 mits of varieties being constant. it would be EXCEEDINGL wrong to call, one group genus & other subg
022r007 Chonos & Concepcion. P. 417 Veins of quartz EXCEEDINGLY rare Mem C. [Cape] Turn P. 434 & 419 As Lime
025r019 rom about 80 fathoms & upwards. that life is EXCEEDINGLY rare, at the bottom of the sea.--«certainly
204b139 e between Chinese & English Breed. decidedly EXCEEDINGLY prolific & hybrid about half way. Eyton says
300c197 g motionless. How is instinctive dread it is EXCEEDINGLY doubtful whether animals have any fear of de
440e147 e the degree of injuriousness must have been EXCEEDINGLY small.-- This is far more probable way of ex
447e163 t in world & hence most persistent-- if form EXCEEDINGLY difficult to vary.-- the run of chances, wou
524m022 er's test of sincerity.-- People in old age. EXCEEDINGLY sharp in some things, though so confused in
319c255 English birds in song.-- one of their thrushes EXCEEDS our blackbird, but our blackbird exceeds their
319c255 ushes exceeds our blackbird, but our blackbird EXCEEDS their other thrushes-- yet they have one with
605o18v gives, when excited by other means, as moral EXCELLENCES, brings to our recollection the original cau
067r142 eption complete.= Silliman Journal. year 1835 EXCELLENT paper on Erratic blocks in Alps. Memoires de
089a020 ell. -- <Keys off extreme point of Flori[da]> EXCELLENT instance, how accidental is the preservation
146g013 gs The Patches of Conglomerate on S. Ventana, EXCELLENT detail & fine, views about Species-- MUST BE
252c042 Admirable letter from Macleay to Bicheno much EXCELLENT case of memory without association..» Instinc
255c050 bird was well known for its impudence. «This EXCELLENT sketch of plants of New Holland, supplementar
269c102 c action.-- Geograph Journal. vol I. p. 17 &c EXCELLENT observations of sickly offspring being cut of
279c133 .-- bad effects of incestuous intercourse.-- EXCELLENT authority because written on dog-Breaking.--
280c134 in both cases effect. it.-- Sir J.. Sebright EXCELLENT PRINCIPLE OF ABORTION ISOLATION of range « <f
304c207 nciples are there to guide in this opinion?-- EXCELLENT table of Canary isld: Plants Home's History o
321c275 o expedition: Zaire except Brown's Appendix & EXCELLENT view of geology, of each formation being mine
352d060 not. abortive???. Apterix certainly.-- Lyell's EXCELLENT references in L. Jenyn's introduct to Mag of
477z001 856 Skimmed through & abstracted Zoology Some EXCELLENT account of blushing.-- this too confirms
570n025 -- Animals have not modesty. analyse this.-- «EXCELLENT-- my theory of blushing solves this.--» The s
603o014 nburgh Review Vol. 18. (1st Article) on Taste «EXCELLENT». Deficient in not explaining the possibility
430e118 on Isld-- if &c &c.-- Then give my theory.-- EXCELLENTLY true theory Examine list of St. Helena Plant
024r016 } Soundings about same as last to N. of C. Frio EXCEPT at Abrolhos. [18 degree S.] Bahia [12 degree 5
036r051 ascend the hill.-- The absence of Second form, EXCEPT near submarine Volc: in harmony with the preva
069r152 !-- Mem. SUBSIDENCE Uspallata of which no trace EXCEPT by trees The structure of ice in columns. show
070r154 instanced: it is Let it not be overlooked that EXCEPT by trees, I could not see trace of Subsidence
101a058 is no difference between dike & mountain axis. EXCEPT in relative «strata» size with superincumbent
106a075 nstead of wider.-- This applies to all vallies (EXCEPT mere talus «over cliffs edge» of which limit c
152g046 not cause. Monday a rapid descent of a terrace EXCEPT at very head of valley indicates new terrace B
156g068 out» composition of shelves: generally angular EXCEPT near head of valley fragments which had fallen
230b236 ed species, but not such as would make species (EXCEPT perhaps in some plants & then a chain of steps
234b255 ds of> the largest <sort> of our highland Sort, EXCEPT in one respect, that those of Iceland. are sel
246c025 ptiles. p 460 & very doubtful whether any birds EXCEPT. Dodo!!-- in Mauritius Lesson &c p. 620. Centr
254c048 wen actually believes in this view!!! p. 392.-- EXCEPT generation & digestion in Acrite Kingdom all o
289c162 ith.» ‖Might be given as a hopeless difficulty, EXCEPT as distinct creation.-- Generation may be view
314c238 s power of movement. & not the organ itself How EXCEPT by direct adaptation has such a change been ef
316c246 rica & England.-- but the fossils are not like, EXCEPT in very few cases, those of Tertiary Europaean
321c275 yage to Surinam. Voyage Congo expedition: Zaire EXCEPT Brown's Appendix & excellent table of Canary i
323c269 mmed.» 1839 Jan 10t.-- All life of W. Scott.--, EXCEPT the V Volume.-- 19t. Mungo Park-- travels F
341d031 sted thrush is represented by one not differing EXCEPT by black line,-- A Bunting by one only differi
383d160 ler than in <ma> silver male-- Head like silver EXCEPT in not having tuft,-- back like do.-- but the
391d179 ion of structure desire fails. Every individual EXCEPT by incestuous marriage has acquired from fathe
406e036 he most vicious dog. will not attack any animal EXCEPT, dog when absent from its master.-- dogs when
407e038 bly undergone a greater change, than any part, (EXCEPT Europe. in which all Tropical forms have been
410e052 ot add name, without reference to description), EXCEPT describers having some high theoretical intere
410e053 forms of organs, will care little for species, EXCEPT so far as wanting names to refer to, to those
458t001 to ancient gigantic salamanders-- Every order (EXCEPT whales) have great prototype!!.-- Copied Vol I
465t080 ing Concepcion-- Patagonia-- Beds of La Plata. (EXCEPT close to B. Ayres).-- If we may take this as g
469t135 ly under mark, & this has now gone on 14 days. (EXCEPT some wet ones/ & wd go on longer-- Woodfords M
472s01r several year to obtain seed, but the pods have (EXCEPT this one year (1827), always been empty.-- See
501q011 ause they cannot be crossed, I think, I expect, EXCEPT by very minute insects.-- (30) Get Abberley to
532m056 com-- grew very thin, would not go out of house EXCEPT with Caroline-- After fortnight. continued to
544m102 waking thought.-- Ld Brougham thinks no dreams EXCEPT at this time. how does he account for dogs & m
578n057 voluntary muscles, these convulsive actions-- (EXCEPT in weak people & hysterical people inclined to
608o027 one can be really fully convinced of its truth. EXCEPT man who has thought very much, & he will know
609o030 es be hard to tell) + + Society could not go on EXCEPT for the moral sense, any more than a hive of B

```
610o033 ng can be <any> «the» object of «our» knowledge EXCEPT our experience".-- is this not almost a questi
617o39v gree of attention. How do the senses affect us, EXCEPT by internal consciousness 3) We must endeavour
629o055 cepting domesticated ones have no right & wrong EXCEPT instinctive ones) Perhaps my theory of greater
636j57r feeding on same food, differs in every respect, EXCEPT «in» quick movements. (sliminess instead of ba
638j58v rials.-- each step being perfect «or nearly so (EXCEPT no in isd) although having heredetary superflu
029o034 stances so like the Coal measures in England (EXCEPTING Conglomerates?) «& absence of limestone?» hav
047r081 ot same as swell travelling across Pacifick.--EXCEPTING in number of waves & in wind, instead of sea
051r094 or confervae in the breakers or in waterfall: EXCEPTING by removal of large fragments by mere force o
051r095 do not see how to account for oceans power.--EXCEPTING when pebbles are brought into play; most mani
052o099 ke. as a general fact absent in T. del Fuego, EXCEPTING in Port Famine Mr Sorrell says that numerous
103a063 res-- If dikes effect of horizontal elevation EXCEPTING fissures from above unite with those from bel
186b057 ther, (at least in one point, in truth in all EXCEPTING specific character); and in passing from spec
199b116 of knowing movements.-- How can we understand EXCEPTING by propagation that out of the thousand of ne
199b120 » fertile offspring; marriage never probably EXCEPTING from «strict» domestication offspring not fer
278c129 , when false species banished by this test.-- EXCEPTING where an Andrew Smith, «Richardson» a Vaillan
295c183 ore peculiar than its ornithology X p. 12 do. EXCEPTING salmons L' Institut. Sorex from Mauritius. p.
375d134 , must be prevented soley by positive checks, EXCEPTING that famine may stop desire.--» in Nature pro
465t080 lls-- now look at Scotland-- coasts of Chile, EXCEPTING Concepcion-- Patagonia-- Beds of La Plata. (e
576n047 there is no absolute tendency to progression, EXCEPTING from favourable circumstances!» We must belie
629o055 om the instinctive. right & wrong.-- (animals EXCEPTING domesticated ones have no right & wrong excep
037r056 nitic countries, enumerate cases. -- M. Video EXCEPTION, but even there, hills of Basalt & other Volc
071r158 rthquakes at Quito. tranquillity «at Mendoza» EXCEPTION.--«formerly perhaps otherwise» Mendoza never
078r175 e to NW (afterwards said to be «all with some EXCEPTION» directed NW & SE). «Vol III» Mexican Cordill
100a055 carpment relation kept to sea coast ∴ curious EXCEPTION in Wealden.-- Would crystals arrange themselv
115a096 t. first accumulated in bed of ocean With the EXCEPTION of sandstone rare to have any horizontal non
202b133 can genus) in plaster of Paris.-- Now this is EXCEPTION to law of type. like horse in S. America. or
261c072 cture in head, which he says (evidently is an EXCEPTION) can only be explained by direct adaptation t
263c077 an. "divino ore versus coelum attentus" is an EXCEPTION.-- He is Mammalian.-- his <has> origin has no
263c077 how dredfully we are deceived) then he is no EXCEPTION.-- he possesses some of the same general inst
425e104 mountainous «this is very important. (Sicily EXCEPTION)-- see if this can be generalized.--» islds.,
609o029 fore most deeply impressed). shame perhaps an EXCEPTION. (does it originate in a doubting feel betwee
335d014 ith defective palates. heredetary & therefore EXCEPTIONS. to above law.-- Study what these monsters a
344d040 es & hybrid doctrine"-- I have read there are EXCEPTIONS to this in some larvae of insects-- (gslowwo
021rIFC             R. N up to 1 degree / July 1835. the EXCESS of harbor = 180 See Daubisson both Volumes, an
041r066 --the «scale» «quantity of iron» being there in EXCESS.-- If veins {P} are secretionary, so are all t
052r098 advancing coast to Seaward. Retreating case in EXCESS as first case. When discussing Falkland soundi
120a109 now exist.-- caution about action of rivers.-- EXCESS of matter brought down Mention absolute elevat.
263c077 nd can reason-- but Man has reasoning powers in EXCESS. instead of definite instincts.-- this is a re
191b080 Geography.-- The motion of the earth must be EXCESSIVE up & down.-- Elephants in Ceylon-- East India
227b225 d older than geologists think. it agrees with EXCESSIVE inequality of numbers of species in divisions
278c131 surely ask Owen to see whether species same, EXCESSIVE improbability. Mem in Clifts list a rat said
541m091 in one position great fatigue.-- may explain EXCESSIVE labour of inventive thought.-- Examine frame
592n103 mpt. p. 76.-- children have been tickled into EXCESSIVE laughter & so into convulsions.-- «Paper» mus
114a095 <tells,> «offers a presumption» it has been EXCESSIVELY slow because beach line chief cause of denud
177b026 on of germs in progress.-- «no only makes it EXCESSIVELY complicated.» {P} Is it thus fish can be tra
181b041 rcentage.-- & <in» therefore the chances are EXCESSIVELY great against, any two of the 12. having pro
259c062 this effected by-- for instance, fish being EXCESSIVELY abundant & tempting the Jaguar to use its fe
280c135 tional character.-- N.B. If two species were EXCESSIVELY old, they would not make hybrids, whereas tw
281c137 the case theoretically if animals did change EXCESSIVELY slowly. whether geologists would not find fo
285c153 .-- Changes in structure being «necessarily» EXCESSIVELY slow, they become firmly embedded in the con
468t105 er June 41» Rhubarb. pollen very minute--not EXCESSIVELY abundant flowers not attractive, very small-
482z011 -- novozelandiae -- histrionicus Vultus aura EXCESSIVELY inaccurate Saw a Chouette a huppe courte tal
499q09v Butterflies or little insect?= or is pollen EXCESSIVELY minute or abundant? do they seed plentifully
564n006 ome drawings «in Lavater, P. cii Vol III» of EXCESSIVELY cross-faced furious faces «which may be descr
364d103 infertility.-- Yarrell says in such case they EXCHANGE birds with some other fancier, thus getting f
108a081 ch Carbonic Acid gaz here.-- [top portion page EXCISED, not located] Bull:. Soc: Geolog. Tome IX 1837
108a082 ition) p 500. Well described [top portion page EXCISED, not located] -- do-- [...] Subaqueous. remova
161g095 ainly appears level with road, & with piece of EXCISED rock lost at point of valley chiefly from rock
601o009 dy (which it may be in first case. as when the EXCISED heart is pricked) and certain action. (only ev
593n107 the language of passion & hence does music now EXCITE our feelings.-- How does Social animal recogni
529m041 e in Kensington Gardens has often been greatly EXCITED by looking at trees at [i.e., as] great compou
529m041 across the plains, how ones feelings would be EXCITED, & how the scenery would rise. Deer in Parks d
531m052 gs of discomfort, especially about heart as of EXCITED action, accompanying violent movement; may not
531m052 ertion consequent on the injury & consequently EXCITED action of heart.-- now this is the oldest «her
532m055 ple nervous from illness., <the> it must be an EXCITED action in the involuntary mind which is startl
536m071 f different orders turn up their nostrils when EXCITED by love? Stallion licking udders of mare stril
545m107 on some action, which the progenitor did, when EXCITED or disturbed by the same cause, which «now» ex
549m119 om offence» -- pleasure of intellect affection EXCITED, pleasure of imagination-- therefore do these
550m125 ore usually refers to the sensations <it> when EXCITED by impressions, & not mental or ideal ones, <w
551m127 n, he had just seen mind went on RAMBLING till EXCITED by question.» Sept. 4th. Lyell in his Principl
591n097 tail «a little» when attending to anything or EXCITED.-- so do young dingos, as I saw wag tail when
595n117 » which they only do when <great» considerably EXCITED, shows their power of imagination-- for it wil
604o017 vailing idea. owing to loss of will.-- chiefly EXCITED by passive emotions.-- Cannot quite perceive d
605o18v g & associated sensations so often gives, when EXCITED by other means, as moral excellences, brings t
605o19v greatness of an object itself or to the ideas EXCITED & associated with it. as the idea of Deity.
579n060 skin cut off made the blush come.-- it is an EXCITEMENT of surface under the will? of the animal.(--
545m107 ed or disturbed by the same cause, which «now» EXCITES the expression.-- Habitual actions are the rev
315c241 drawn between habitual actions of plants when EXCITING cause is absent» & memory of animals.-- (sure
534m062 ore your eyes, or ears (language mere means of EXCITING association.)-- or of memory of such sensatio
045r076 Chili, where rains are so infrequent; so as to EXCLAIM, «as I have heard» how lucky! when they hear o
255c052 monstrosities as Man does.-- One is tempted to EXCLAIM that nature conscious of the principle of ince
293c172 hich its father. had done habitually we should EXCLAIM it was instinct.-- Even if savage takes. & was
308c217 fixed what a species means» civilized Man, May EXCLAIM with Christian «we are all» Brothers in spirit
558m153 hen passion commences.-- <All> Nearly all will EXCLAIM, your arguments are good but look at the immen
521m009 , what.-- then showed the watch upon which he EXCLAIMED, why it is dinner time.-- » My father asked h
206b148 r species of cats.-- <& other tribes>.-- &c &c EXCLUDE mothers & then try this as simile In a decreas
640j167 evall.-- Not separately: NB. These views quite EXCLUDED the idea of domesticated animals changing.-- F
535m064 es of Eucalyptus, watching few days till larva EXCLUDED, then though not feeding them «nor helping la
282c142 all cases be produced, but would depend upon EXCLUSION.-- The same characters which are analogical i
037r055 s few corals Tortoise> «remains of Amphibia, EXCLUSIVELY.» & Turtle bones. & the bones of <two granin
043r069 » of elevation Volcanic action, must be more EXCLUSIVELY confined to that country. Read description o
620o045 or fear of death, one makes allowance & either EXCUSES the «non-» following of ones conscience. & pal
524m022 y think of most. intently.-- criminals before EXECUTION.-- Widows not of their husbands-- My father's
437e139 les bring rabbits & hares to the young ones to EXERCISE them in killing them.-- "Sometimes it seems h
526m030 believe phrenologists are right about habitual EXERCISE of the mind, altering form of head, & thus th
527m034 k (abstracting it being done in open air, with EXERCISE &c no organs of sense being required) as the
528m036 rtificial lights in the night.-- from the mere EXERCISE of the organ of sight, which is common to eve
528m039 n clear day.-- 3d pleasure association warmth, EXERCISE, birds singings.-- 4th. Pleasure of imaginati
618o043 ee a struggle between its appetite, or love of EXERCISE & its love of its puppies: the latter general
574n041 s always.= No doubt man has great tendency. to EXERT all senses, when thus stimulated, smell, as Sir
617o39v is revealed to us by the effort it costs us to EXERT force or by internal consciousness; the objecti
531m052 he feeling «consequent on the violent muscular EXERTION» which accompanies violent attack,-- Even the
531m052 y there is no feeling of passion, but muscular EXERTION consequent on the injury & consequently excit
531m053 unning away. accompanied with want of muscular EXERTION, palpitation, voiding urine because done by s
534m062 ese cases, when agency is unknown, with simple EXERTION of intellectual faculty) if ants had at once
547m113 eep, it may do less work & yet do so, from the EXERTION of keeping up the memory of every late impres
617o40v <subjective> aspect of action as known by the EXERTION of our own power & consciousness of it we are
586n082 impulse to save wax.]CD which it instinctively EXERTS in concert with others in building comb-- My f
```

Page **(Key Word)**
334d012 n of imperfect structure.-- Fox says in «Lord» EXETER'S Park «or in the Duke of Marlborough» there is
634j54v head.-- the designs of an omnipotent creator, EXHAUSTED & abandoned. Such is Man's philosophy. when h
634j54v nal thought, or design, pursued to its utmost EXHAUSTION, & till it must be abandoned for another".--
070r153 that granite when weathering into balls. must EXHIBIT orbicular structure.--When we recollect connec
254c048 ructure??-- «do» p. 390. All classes of Acrite EXHIBIT lowest stages of animal organization, "& are a
072r160 out H. Kingdom N. Spa. Vol III p. 113 "Nature EXHIBITED to the Mexicans enormous masses of Iron and N
617o040 same faculty with matter & being necessarily EXHIBITED in & by matter. The phenomena of gravity cons
617o40v the fact to the mere statement of the <force EXHIBITED in every> phenomena actually apprehensible by
076r168 56. Mines of Batopilas in New Biscay, "Nature, EXHIBITS the same minerals <as> there, that are found
035r046 erent basins; little or no relation appears to <EXIST> be made out, but in those belonging to the sam
049r089 m: Phillips Mineralogy some such fact stated to EXIST in Peru. -- Ascension At Ischia there is a pumi
120a109 orge. leaving cliffs, on each side, such as now EXIST.-- caution about action of rivers.-- Excess of
181b040 wise> the successors of his relatives shall now EXIST.-- In same manner, if we take «a man from.» any
194b097 n progenitor to Mammalia & fish. when there now EXIST such strange forms as ornithorhyncus The type o
195b098 before island existed.-- Such an influence Must EXIST in such spots. We know birds do arrive & seeds.
196b108 one--?, of the <vebtetrata> «vertebrates» could EXIST without plants & insects had been created; but
273c110 experience, you may be almost sure, that there EXIST intermediate species.-- This is remarkable & wo
312c232 tever.--» Individual Men «& animals» could only EXIST by habit.-- therefore same principle transferab
337d021 all animals, but merely to classes where types EXIST for if so. it will be necessary to show how the
451e180 ld cattle & Hogs on Timor-land-- monkeys do not EXIST. there & it is a singular thing that throughout
553m136 ge has been communicat to us».-- & that it does EXIST in different degrees in races.-- whether in Anc
575n043 Jackall.-- an animal not destined by nature to EXIST. & carrying «like other hybrids» with <the> it
589n092 n-- or that our faculties have been given us to EXIST, is clearly seen. in the absurdity of a tree ha
611o034 e, that is enough must be present to be able to EXIST as individual.-- [RHC] 2) In animals, growth of
636j57r it should be remembered, that animals could not EXIST without these adaptations.--fossil forms show s
638j58v hough having heredetary superfluities Man could EXIST without Mammae.» to the then existing condition
038r058 r to CENTRAL nucleus & that envelopes no doubt EXISTED. These higher portions probably formed Islds f
173b011 gos & Juan Fernandez. When continet of Pacific EXISTED might have been Monsoons.. when they ceased im
173b012 ed & changes commenced.-- or intermediate land EXISTED.-- or they may represent some large country lo
174b015 tralia» This presupposes time when no Mammalia EXISTED; Australian; Mamm were produced from propagati
192b086 . But it is other question, whether there have EXISTED all those intermediate steps especially in tho
195b098 a breath cannot reside in space before island EXISTED.-- Such an influence Must exist in such spots.
196b105 would have been wonderful if the two Rheas had EXISTED in different Continents In plants I believe no
221b201 Have the changes been so slow., that all have EXISTED for ages as metamorphic; & therefore according
248c029 m those quarter, where we know quadrupeds have EXISTED for ages.-- ∴ The most hypoth: part of my theo
304c206 dge a chaos: who will doubt this if series now EXISTED from Man to Monads-- though physiology would p
304c207 iews.-- ostriches do-- but then there may have EXISTED series between apterix & other birds.-- will h
308c219 eas., it will show state of knowledge «Negroes EXISTED since time of earliest Egyptian drawings & Old
357d073 e been almost destroyed wherever other animals EXISTED.-- Athenaeum 1838. p. 654. Reason given for su
391d174 f this tendency all mammalia must long have so EXISTED. » with double union.-- At present I can only s
460t019 rian & Crust show how long since present forms EXISTED, but if it be asked how this complexity from a
599o005 ge-- & effect of reason. reason could not have EXISTED without it.-- quotes Ld Mondobbo.-- language c
602o11b art is the realizing and embodying, what never EXISTED but in the imagination".-- Macculloch Vol I. p
021rIFC ll's Geology The living atoms having definite EXISTENCE, those that have undergone the greatest numbe
024r014 he water to more than 100 fathoms. proves the EXISTENCE of some moving <point> power ¿ Submarine curr
176b022 than individuals If we suppose monad definite EXISTENCE. as we may suppose .is the case. their creatio
176b022 accident of positions» must in each state of EXISTENCE have shortest life.: Hence shortness of life
177b029 out, therefore not.-- Monad has not definite EXISTENCE.-- There does appear some connection shortnes
177b029 here does appear some connection shortness of EXISTENCE, «in» perfect<ion>, «species from many» <ther
179b033 , distinctness of tribes in T. del Fuego. the EXISTENCE of whiter tribes in centre of S. America show
181b039 y, would have left scarcely any type of their EXISTENCE in the present world.-- or we may suppose onl
186b061 those two countries ought to be similar ¿Law: EXISTENCE definite without change, superinduced, or new
196b110 st be very great gaps.-- yet some analogy The EXISTENCE of plants, & their passage to animals appears
197b112 ng constant. Cuvier's theory of Conditions of EXISTENCE is thought to account resemblances &. quinary
259c065 an adaptation, but adaptation during earliest EXISTENCE; if whole life then real adaptation The case
280c135 important for L.yell said to me, the fact of EXISTENCE of mules appeared to him most strange.-- This
283c144 wo continents as to be <replaced> called into EXISTENCE in two continents our ignorance is indeed pro
308c217 be, .extinction of species bears relation to EXISTENCE of genera &c &c Two savages, two species.-- «
309c220 ot equal to areas of elevation: Marked out by EXISTENCE of elevated «extinct?» genera of shells.-- du
343d039 ere is no appearance of sudden termination of EXISTENCE.-- nor is there in the Tertiary <older> geolo
439e143 ch depends on instincts in animals.-- yet the EXISTENCE of wild close species of plants shows there i
441e150 ndividuals is very curious & important.-- The EXISTENCE of "laws of organization" had better be shown
464tf6r ed species Mr Blyth, however, believes in the EXISTENCE of Molina's Pudu or goat There is ibex of Alp
526m027 ind.-- thinking over these things, one doubts EXISTENCE of free will every action determined by hered
551m127 ve they hear as well see things which have no EXISTENCE.-- He compared spectral illusion & insanity t
564n003 were proved instead of militating against the EXISTENCE of such an attribute would be rather favourab
571n027 serving powers common to savages???].CD-- The EXISTENCE of taste in human mind. is to me clear eviden
572n035 s thought.-- Macculloch in his Chapter on the EXISTENCE of a Deity has an expression the very same as
614o037 of association, parental affection-- The very EXISTENCE of mankind requires these instincts,: though
617o040 l senses is a phenomenon the essence of whose EXISTENCE consists in its communication to other matter
618o041 amiliar with thought & yet be ignorant of the EXISTENCE of the brain. We cannot perceive the thought
618o41v e reason why thought &c. should imply «X» the EXISTENCE of something in addition to matter is because
625o052 of right & wrong.-- so far it has independent EXISTENCE. & is supreme. because it is «a» part of our
609o29v being married to keep up population. with the EXISTENCES of so many positive checks.-- (This is encro
199b117 --- Ancient Flora thought to more uniform than EXISTING. Ed. N. Philos J. p. <191> «p 191» No. 5. Ap
231b242 ns now in Progress, & you will have two. Tapir EXISTING in East Indian Seas. Marsupials animals all s
261c070 on some «antecedent races, perhaps not on now EXISTING» Mr Gould says wherever any mark like red pat
302c203 cies living there would be destroyed, & N & S. EXISTING species becomes father of genera-- whatever t
343d039 s one species from Egyptian Mummies & from the EXISTING animals found fossil when Europe must have wo
361d095 ame character which is mottled, & not like any EXISTING species-- [In two herons, <both> plumage of b
362d100 he <bird> varieties «,now know» were then <pr> EXISTING.-- he has also some very fine recent drawing
374d133 oniferous trees & Lycopodiums.-- p. 437. Many. EXISTING genera of shells in the mountain limestone (h
374d134 ized.-- do. p. 461.-- Lower Silurian-- several EXISTING genera. Nautilus turbo. buccinum. turritella.
407e040 ammalia of S. America are as diferent from the EXISTING orders, as the Eocene of Paris! (Great Edenta
420e088 e of fossil filling up blank.-- CD[not between EXISTING series of species of dogs & Hyaena.-- but a c
434e128 "If we cannot reason from the analogies of the EXISTING to the events of the past world, we have no f
445e159 skin between its legs.-- -- strangely consider EXISTING «long-organized» forms as parent forms of exi
445e159 ting «long-organized» forms as parent forms of EXISTING highly organized forms-- this resulted from t
458t001 s, giraffes. Sivatherium & Anoplotherium, with EXISTING, or nearly existing forms of aquatic reptiles
458t001 rium & Anoplotherium, with existing, or nearly EXISTING forms of aquatic reptiles most strange, & sho
461t041 cies-- (especially of mammifers) in old beds & EXISTING species is valuable because it shows no innat
462tf4r 9. p. 708.-- Shrew, found by M. Lartet as now EXISTING species. We see the same object gained by the
577n049 e of itself.-- [by my theory no animal. as now EXISTING can be cause of itself.]CD & hence there is g
638j58v s Man could exist without Mammae.» to the then EXISTING conditions.-- An adaptation made by intellect
119a105 ng out matter by action of the sea.-- as no sea EXISTS there.-- But Sir John considers an irregular f
315c240 relation between means of Transport & creation EXISTS.-- pooh. May have been Created at many spots
409e048 , but one great final cause,-- nothing probably EXISTS for one cause» of sexes «in separate» «animals
231b241 ference prevents breeding;--or as others would EXJUDGE it amount of varying in wild state.-- When bre
612o035 f plants absorb by physical laws of endosmic & EXOSMIC juices. arms of polypus, show either local or
389d177 ing in & in (those which have solitary flower) EXOTICS brought from foreign country. (<annuals>& so
032r041 happened. hurricane in bowels of earth cause:--<EXP> does not explain cleavage lines./ possibly gene
553m137 endent of the Zoological Gardens) remarked that <EXP> the expression & noises of monkeys go in groups
463t055 l marrow expands, so do the bones <are created> EXPAND-- instead of saying as brain is created &c &c
513q21. - will pollen act on any flower before stigmas EXPANDED-- in reference to Lobelia & Clarkia-- Peas ti
463t055 , constantly said as the <brain> spinal marrow EXPANDS, so do the bones <are created> expand-- instea
599o007 up my theory-- Viewing from eminence. the wide EXPANSE, of county, netted with edges & crowded with t
069r149 L movement of fluid matter not (for instance) EXPANSION of solid matter by Heat Consider profoundly t
053r101 ines of Mountain appear to me to be effect of EXPANSIONS acting at great depths (mem: profound earthq

045r075 a were present at the Concepcion earthquake.--EXPATIATE on difficulty of evidence about eruptions of
108a080 as.-- In Earthquake if Subsidence we should not EXPECT volcanos.-- not so much horizontal oscillation
187b065 e, he has no issue, so with species.-- I should EXPECT that Bear & Foxes &c same in N. America & Asia
192b088 & reptiles.-- In mere eocine rocks. we can only EXPECT some steps.-- I may ask whether the series is
200b123 crossing very opposite races, whether you would EXPECT equal fertility-- ditto in Plants.<==> It will
208b154 leay Name given in Congo Expedition We need not EXPECT to find <species>, varieties, intermediate bet
224b214 tinct. Where country changes rapidly, we should EXPECT most species.-- The difference intellect of Ma
228b229 far back, as well as at present time, we might EXPECT confusion of species.-- Important. For instanc
228b231 r the dead.-- respect We have no more reason to EXPECT the father of man kind. than Macrauchenia yet
236b280 doelle says, no he only says sometimes we might EXPECT disseminated species to vary a little, but suc
236b280 surely it is not-- intermediate genera we might EXPECT.-- Lindley Introduct Dict. Science. Naturelle
273c110 ally contemporary «++.» «This would lead one to EXPECT that fossil forms would generally fill up gene
280c136 ts own parent this impossible-- (Hence we might EXPECT even if two mules bred or two certain varietie
303c204 Lawrence. Blumenbach & Prichard -- Now we might EXPECT that animal halfway between man & monkey, woul
404e025 t naturalists if they had series perfect, would EXPECT this structure would become obscure & therefor
417e077 , variety, as in two varieties, & this we might EXPECT, as the difference between man & woman is «ind
501q011 nts. because they cannot be crossed, I think, I EXPECT, except by very minute insects.-- (30) Get Abb
572n032 ny it on to mental inferiority-- when we do not EXPECT any bodily harm-- case of habitual action.-- L
619o042 p or hurting them.-- Therefore in man we should EXPECT that acts of benevolence towards fellow <livin
628o54v I check the consciences desire for virtue.-- [I EXPECT there is some fallacy here.-- at least point o
565n009 cles, which wrinkle when smile.-- Hope is the EXPECTANT eye. looking to distant object, brightened &
192b089 Mammalia than before & that is all that can be EXPECTED-- This answers Cuvier-- Perhaps the father of
352d057 mals which have many ABORTIVE organs, might be EXPECTED to have larvae more perfect-- this is applica
465t081 see reasons-- why no perfect gradation can be EXPECTED in any one country.-- in a descending series
578n054 either good or bad. either giving a beggar, & EXPECTING admiration or an act of cowardice, or cheatin
061r126 . drought at Sydney. which caused Capt. Sturt EXPEDITION-- ¿ Another one in 1816 (?).-- Mr Owen's c
065r138 s of rocks were brought home in Capt. Forster EXPEDITION from <Deception Isld.> South Shetland Cape P
208b154 de Verd's.--? NO Macleay Name given in Congo EXPEDITION We need not expect to find <species>, variet
310c224 ome plants same. --America.-- See Brown Congo EXPEDITION: 400 Australian plants found in other parts
317c248 ion Study Appendix (& only appendix) of Congo EXPEDITION, NB. I met an old man--, who told me that th
317c251 espect to forms.-- Study Appendix to Tuckey's EXPEDITION Journal of the Academy of Natural Sciences o
321c275 ndix Ovington Voyage to Surinam. Voyage Congo EXPEDITION: Zaire except Brown's Appendix & excellent t
446e161 concerned, in parts of the Northern «French» EXPEDITION,-- rather the reverse of facts stated by Smi
254c049 nt of taenia worm.-- formative energies easily EXPENDED & no one system developed <is> not surprising
217b183 t in Mr Galtons case.-- It explains the loss & EXPENSE, (must probably have occurred to every one) of
619o042 & aids & defends & acts for others at its own EXPENSE.-- Moreover <the> any action in accordance to
491q01v ome seeds, caused symmetry in cone-- The «above EXPER» explains apples on side near other tree being
224b212 tries. there is no test but generation, «(but EXPERIENCE according to each group)» whether good speci
273c110 l band & others large, then he says from long EXPERIENCE, you may be almost sure, that there exist in
303c205 osphically to a certain extent,-- nothing but EXPERIENCE. will, tell us. when group is true,» there a
352d059 when it does not perform that function which EXPERIENCE shows us it was for.-- Most important law.--
397e003 mediate agency of the deity. But we know from EXPERIENCE! that these operations of what we call natur
434e128 for our science".-- <it is only analogy.> but EXPERIENCE has shown we can & that analogy is sure guid
444e158 rting water at fly.-- instinct, for how could EXPERIENCE teach distances in air, in which it never to
445e158 chicken, yet says it is evidently acquired by EXPERIENCE in baby Lamarck. Vol II p. 152.-- Philosophi
533m059 <the> such instincts which full grown men can EXPERIENCE-- Instinctive walking of animals. that is th
534m063 g that time is lost & endeavours made must be EXPERIENCE & intellect.-- do. p. 157. Westwood remarks
549m122 ng some instincts as revenge «& anger», which EXPERIENCE shows it must for his happiness to check-- t
550m124 given time «-- compared to what other people EXPERIENCE.--» But then sensation may be more or less p
550m125 ven with some pain,-- compared to what others EXPERIENCE in same time.-- Pleasure more usually refers
551m128 existence of the soul, are not derivable from EXPERIENCE.-- read monkeys for preexistence <">-- They
560mIBC placed before their eyes, very young, before EXPERIENCE can have taught them to avoid danger Do they
564n005 face, or other means by which eyes, aided by EXPERIENCE is supposed in man to guide to knowledge, wa
564n005 ke puzzling, at Astronomy without Mechanics.-- EXPERIENCE shows the problem of the mind cannot be solv
567n014 actically know «art precedes science-- art is EXPERIENCE & observation.--» in balancing a body & an a
568n016 s, which are the result of our senses, or our EXPERIENCE.-- Two sides of a triangle shorter than thir
568n017 perance, or real virtue, that is action which EXPERIENCE shows will be for general good, or in case o
573n037 countenance-- before they can have learnt by EXPERIENCE, that movements of face are more expressive
585n074 s. by which children learn (probably not only EXPERIENCE,, but also «by an» instinct<ing> «which is o
586n078 stinct make watch, but he does it by reason & EXPERIENCE, or habit.-- so bird migrating to certain qu
586n079 ed or emotion felt in early childhood (before EXPERIENCE» or habit) could be formed or afterwards.-- c
586n081 main species is instinctively «not least by EXPERIENCE» directed to certain quarter"-- "An animal h
586n081 faculty of walking. which in man is learnt by EXPERIENCE is in other is acquired instinctively" So wi
602o11v s & sublime ideas independent of the senses & EXPERIENCE p. 142 "Upon the whole it seems."-- "that th
609o30v to change that part of the moral sense which EXPERIENCE (education is the experience of others) show
609o30v oral sense which experience (education is the EXPERIENCE of others) does not tend to greatest g
610o033 y» «the» object of «our» knowledge except our EXPERIENCE".-- is this not almost a question whether we
610o033 l definition, there is much knowledge without EXPERIENCE. so there may be in men-- which the reviewer
614o037 st memory in many cases cannot be acquired by EXPERIENCE for child sucking.-- And is it more wonderfu
614o037 Acquired instincts analogous «(& replace)» to EXPERIENCE gained by man in lifetime Heredetary memory
621o046 nstincts weak. he will have many struggles, & EXPERIENCE only will teach him, that the instinctive fe
529m039 ll scraps of poetry;-- former thoughts, & in EXPERIENCED people-- recall pictures & therefore imagini
619o043 e to his instincts, <he would know that many EXPERIENCED pleasure,> & by association he would feel pa
051r095 r saw on beach near Callao.--From Sir. H Davy EXPERIMENT on the copper bottom. we see a trifling circ
200b125 ost seeds germinating.--- It would be curious EXPERIMENT to know whether soaking seeds in salt water
401e015 om apple trees is produced.-- Thinks probably EXPERIMENT was never tried of separating apple tree ent
401e015 g apple tree entirely from all others-- so my EXPERIMENT of strawberry not so absurd.-- Thinks-- that
441e149 eds, I presume, probably would-- at least the EXPERIMENT of the carrot seems to show this.-- This wou
490q001 ulus in the nectaries. The former best for my EXPERIMENT on Selection. Experiments in crossing & Pla
490q01v nts in crossing & Plants 1 Repeat the French EXPERIMENT of Carrot 2 {also try Primrose & Cowslip in
491q01v lways useful). fail-- Really good subject for EXPERIMENT.--«to repeat Spallanzani» Raise only single
493q004 on is same good, avoids effects of fatness.-- EXPERIMENT in crossing animals.-- &c (1) To cross some
503q013 41.-- Trees above male? (2) Result of Edwards EXPERIMENT in Cabbages given (3) in Heartease (4
504q014 in to be compared-- Cabbages.-- kept true Try EXPERIMENT (30/p.11) (2) Yew Berries germinate?-- Yew t
516q24v at, & see what comes up.-- [Unnumbered blank] EXPERIMENT Cover patch of ground, with different salts
555m144 & joking» because the whole train of Dr Monro EXPERIMENT about hanging came before me showing impossi
585m075 very quickly successively.--» [& we know from EXPERIMENT of crossing fingers, that we only do know th
618o41v l the blueness had a certain intensity (& the EXPERIMENT was varied) then might it now be said, that
232b248 f the S. Hemisphere look as if heat gained? EXPERIMENTISE on land shells in salt water & lizards do.-
495q005 uch soil.-- Sow weeds in such soil.-- 7 (a) EXPERIMENTISE on Primrose seeds-- it really is an importa
495q05a arry Bees, powdered with starch & Carmine & EXPERIMENTISE on their returning powers-- then carry them
502q11v son-tube-- so put carmine in spirits & then EXPERIMENTISE: for gradation in structure Compare flowers
503q012 duced itself.-- Ask Gray to ask Mr Riley to EXPERIMENTISE on hybridising ferns, tying them back to ba
392d180 eir instincts?» «Chineses & Common Pigs.--» EXPERIMENTIZE on crossing of the several species of wild
501q011 re there RACES of Lupine, Stocks Clover, to EXPERIMENTIZE on by sowing near each other & see whether
541m091 flowers, cloth &c & with all this difficult EXPERIMENTIZE upon this effort.-- it looks so analogous t
139a180 with carbonic acid [blank] Many interesting EXPERIMENTS might be tried by comparing Zoophite to plan
420e090 as insects do flowers.-- Mem. Spallanzani's EXPERIMENTS showing how little of the spermatic fluid fe
455eIBC tive plants; Physiology Get Habberley to try EXPERIMENTS. about raising plants. where they cannot «cr
489q FC et in thickness.-- (March, 1842) Questions & EXPERIMENTS {T} Gowen, Royle, & Horsfield Sykes p. 12 Ma
490q01v former best for my experiment on Selection. EXPERIMENTS in crossing &c Plants 1 Repeat the French ex
494q005 d.-- (12) About the blended instincts Remote EXPERIMENTS-- Plants Raise seedlings surrounded by vario
496q006 circumstances, as Hyacinths in glasses &c &c EXPERIMENTS Questions concerning Plants Is the common Fi
516q023 n which I think I have never seen Bee visit. EXPERIMENTS in Garden Sow stones of Standard Apricot gra
516q23v lants would reappear after <th> being killed EXPERIMENTS not connected with Species Theory (1) Will a
516q BC bear on Petrifaction?-- [blank] Questions & EXPERIMENTS Expression M Charles Darwin Esq 36 Grt. Marl
606o20v necessarily be acquired by a long series of EXPERIMENTS & observations. & yet, like in vision, it be
032r040 opiapó & parts of coast of Chile.-- Must first EXPLAIN «top of» tidal band of action. This case diffe

Page **(Key Word)**
```
032r041 cane in bowels of earth cause:--<exp> does not EXPLAIN cleavage lines./ possibly general symetry of w
056r111 separating causes by water.--Or rather begin & EXPLAIN how water separates.--(intertropics at present
057r115 ember idea of frozen bottom or beach of sea to EXPLAIN preserved animals.--Mem: stream of water in th
092a029 metamorphic in contradistinct to Volcanic will EXPLAIN their solution. Athenaeum M. 516 1837 High  up
100a052 of BK.-- ┆ This is not applicable. it does not EXPLAIN CLEAVAGE of rock-- nor the Falkland case, nor.
105a071 Mendoza--  Will they introduce other causes to EXPLAIN «alluvi» in valleys Lowe in his paper says lan
117a101 of insensible oscillations of level will alone EXPLAIN the immense amount of change which must have t
119a104 in  Indian & Pacific Oceans.-- (2d--) does not EXPLAIN first formation of continents, if globe be con
122a113 he one which generally yields.-- Will this not EXPLAIN littoral mountains & volcanos.-- Why on one co
149g032 errace river «& to West of Spean» difficult to EXPLAIN on <formation> deposition in lake On the summi
151g041 Even  on Lauder Dicks Hypothesis impossible to EXPLAIN absence of lines in certain parts.-- At the Pa
158g081 tidal  plain as sea gradually retired, hard to EXPLAIN on river doctrine <Little Hill with granite bl
198b113 s animals, reptiles fish-- Conditions will not EXPLAIN status (Perhaps consideration of range of capa
216b179 <)> than white parent! the mulattos themselves EXPLAIN it by intermarriages with people. either a lit
219b196 h means as we can recognize, may be thought to EXPLAIN nothing.-- it being as easy to produce «for th
226b221 pecies. this must happen. & thus acquired will EXPLAIN representative system Of this we see example i
231b243 ly different if so.-- Now this is difficult to EXPLAIN by creation-- or we must suppose a multitude o
240c003 ng an element in the transmission of form, may EXPLAIN mule & pig being half way. Yet dogs  sometimes
257c057 re answers to the possibility.-- My views will EXPLAIN no Mammalia in secondary-epocks, & developemen
288c159 rtial migrations of birds in same country. may EXPLAIN greater migrations, if American & intersected
291c167 odon in S. America) is absolutely necessary to EXPLAIN genera & classes. if extinct forms were all fa
294c176 in a wild state-- it may be said argument will EXPLAIN very close Species in islds. near continent, M
294c177 willow wrens &c &c. & analogy will necessarily EXPLAIN the rest,-- Lonsdale says he has seen in old B
340d026 in.-- Mine is a bold theory. which attempts to EXPLAIN, or asserts to be explicable every instinct in
352d058 of  every type of organization. such law would EXPLAIN every thing.-- PURE HYPOTHESIS be careful.-- A
357d073 w Ireland & continent since grown.-- This will EXPLAIN. S. American case & Didelphis being Mundine fo
486z018 r is comparatively rare.-- These views clearly EXPLAIN rarity of insects in T. del Fuego.-- Hence it
527m032 n of beauty & negroes another; but it does not EXPLAIN the feeling in any one man.-- Music & poetry o
539m083 the mind as in double consciousness may really EXPLAIN what habit is-- In the habitual train of thoug
541m091 to muscle in one position great fatigue.-- may EXPLAIN excessive labour of inventive thought.-- Exami
546m108 ery unsatisfactory because does not like Burke EXPLAIN pleasure. August 26th. I cannot help. thinking
566n011 s to take birds & beasts».-- very necessary to EXPLAIN origin of idea of deity.-- Animals do not know
605o019 does  not attempt «by one common principle» to EXPLAIN the various causes of those sensations,  which
623o050 ch truth in doctrine, for [RHC] 9) We can thus EXPLAIN love of place.-- although here we have not rec
628o54v es» <nonsense>-- My theory of durableness will EXPLAIN it.-- Would not the maternal affections (in a
633j53v my  theory of gain of small advantages thus to EXPLAIN the curling of the valves of the broom.-- or t
634j55r ae traces of hind extremities.-- How are we to EXPLAIN this.-- Did reptiles first inhabit seas.-- Wer
635j56r F ONE GREAT SYSTEM. C. D]CD [All this does not EXPLAIN death, but reproduction]CD though such a them
047r082 Portugal  & Madeira (Lyell. vol I. P. 471) is EXPLAINED. also the similar fact at Concepcion? Read th
146g007 most common-- Will not curved form of hill be EXPLAINED by my idea-- highest part must project [Blank
185b055 adaptation of each element.-- May this not be EXPLAINED on principle, of animal having come to island
219b194 w possibility of transport. If some cannot be EXPLAINED more philosophical to state we do not know ho
225b219 g? if so adaptations of species by generation EXPLAINED? NB. Look over Bell on Quadrupeds for some fa
230b236 eps is found in same mountain).-- How is this EXPLAINED by law of small differences producing more fe
239c002 rents, going back to either parent is lucidly EXPLAINED.-- Mr Yarrell states that if any odd pidgeon
261c072 says  (evidently is an exception) can only be EXPLAINED by direct adaptation to animals wants & not a
268c100 ch, being fitted for transport ¿may it not be EXPLAINED by mere chance?-- or it like each great class
283c144 out diversity of forms in aberrant circles.-- EXPLAINED by such not having been long in blood?-- My t
290c165 The attachment of dogs to man. not altogether EXPLAINED by F. Cuvier, «--.Mem. Hensleighs objection.
302c200 w family & no new orders.-- Wonderful, partly EXPLAINED on my theory, = otherwise mere fact creator c
305c209 creation.-- ie.-- a mere statement nothing is EXPLAINED.-- this is fact analogous to mocking thrushes
334d010 ding to Erasmus preferring young mare to old, EXPLAINED by Stallions, (according to Fox) being guided
359d088 nge anomaly in Yarrells law.-- it probably is EXPLAINED by the vigour of their propagating powers. (a
402e017 reat difficulty» in thick strata, can only be EXPLAINED, by such strata being merely leaf, if one riv
407e038 imate of same order as that of S. America.-- (EXPLAINED by profound views of Lyell) Now «Equatorial»
522m010 wered never heard of such a man.-- (My Father EXPLAINED who he wa & all about him, but still maintain
538m078 profoundly, may throw light on consciousness, EXPLAINED by Dr Dewar on principle of association.--«fu
538m080 & self one person-- & thus eternal punishment EXPLAINED. These facts showing what a train of though[t
542m095 when  he awakes. & streching & yawning can be EXPLAINED from too long rest of muscles.-- evidently ha
566m010 32, origin of Chastity in women.-- rationally EXPLAINED.-- on the wish to support a wife a ruling mot
588m090 ct», for its character is invariability.-- if EXPLAINED by habits, useful to itself, how gained. reas
625o50v ie happiness-- yet this system not selfish.-- EXPLAINED by principles if Mackintosh.-- p. 262. Some g
637j58r lle.-- The Final cause of innumerable eggs is EXPLAINED by Malthus.-- [is it anomaly in me to talk of
640j161 y be chanced.-- Hence the Galapagos Islds are EXPLAINED. On distinct Creation, how anomalous, that th
039r061 nd to take analogy of movements of W coast in EXPLAINING plains because such are found in perfection
100a052 pecially effect of gravity, versus some fault EXPLAINING vary dip & inclination.-- which last is stro
113a092 measure of force in that part.-- Important as EXPLAINING want of levelness Major Mitchell showed me a
440e147 gly small.-- This is far more probable way of EXPLAINING, much structure, than attempting anything ab
603o014 ticle) on Taste «EXCELLENT». Deficient in not EXPLAINING the possibility of <handsome> «UGLY healthy»
103a063 from  below. would always thin out above which EXPLAINS a difficulty.-- All De la Beche's reasoning o
174b014 ferent genera different countries. Propagation EXPLAINS why modern animals same type as extinct which
187b062 ing how horse & Elephant reached S. America.-- EXPLAINS how Zebras reached South Africa-- It is a won
195b104 otheses fresh creations is mere assumption, it EXPLAINS nothing further, points gained if any facts a
199b117 , should they all be classified.-- Propagation EXPLAINS this.---- Ancient Flora thought to more unifo
217b183 instance of same fact in Mr Galtons case.-- It EXPLAINS the loss & expense, (must probably have occur
224b212 yton's Hogs & dogs.-- The passage in last page EXPLAINS that between Species from «moderately» distan
227b225 cies. limits of good species being known.-- It EXPLAINS the blending of two genera-- It explains typi
227b225 -- It explains the blending of two genera-- It EXPLAINS typical structure.-- Every species is due  to
231b243 latter-- Have change in form.-- This probably EXPLAINS crag & miocene.-- The descendants left in coo
280c135 e.-- This even might be said.-- My theory thus EXPLAINS a grand apparent anomaly in nature. <t>---- M
281c138 not  find fossils such as they are-- My theory EXPLAINS that family likeness, which as in absolute hu
293c173 it would be instinctive.-- My view of instinct EXPLAINS its loss ¿ if it explains its  acquirement.--
293c173 My  view of instinct explains its loss ¿ if it EXPLAINS its acquirement.-- Analogy. a bird can swim w
296c184 2  Chapters. translated by Hooker.-- my theory EXPLAINS this. but no other will.-- St. Helena (& flor
389d173 bud  matured by female;<]D> such view no ways EXPLAINS Ld. Moretons case: without the nervous matter
434e128 e can & that analogy is sure guide & my theory EXPLAINS why it is sure guide.-- Lychnis April 3d.-- H
491q01v s, caused symmetry in cone-- The «above Exper» EXPLAINS apples on side near other tree being affected
497q007 ny species in genera of Leguminosae.-- Herbert EXPLAINS numerous spec. of Cape Heath by facility. ¿Kn
528m035 vivid  castle in the air, or dreams real again EXPLAINS insanity.-- Analysis of pleasures of scenery.
582n068 ncomfortable if it does not do it.-- My theory EXPLAINS how it comes that the heart is the seat of th
609o030 the  «instinctive» moral senses: (& this alone EXPLAINS why our moral sense points <is> to  revenge).
622o048 heory of instincts, or heredetary habits fully EXPLAINS the cementation of habits into instincts. [RH
622o049 of the instincts Hartley, (according to Sir J) EXPLAINS our love of another, as pleasure arising from
625o50v ing blended & lost» & moral sense.-- My theory EXPLAINS both, perhaps, by habit-- [LHC] 11) Whewells
628o054 ice love of gold.-- love of fame-- Yes Hartley EXPLAINS this & Mackintosh shows the change produced.--
629o055 eory of greater permanence of social instincts EXPLAINS the feeling of right & wrong.-- arrived at fi
634j54v udes «(& at p. 312)» to the abortive bones. He EXPLAINS it <"By> saying "It is the determination to a
635j56r ow not the effect [blank] 6 p. 412. Macculloch EXPLAINS the shortness of life (peculiar to each speci
045r077 an those that attend Eruptions: Mr P. Scopes EXPLANATION of low Barometer? In a subsiding area. we ma
124a118 ested to me that Herschel's theory offers no EXPLANATION of intermittent action of elevatory force--
132a139 ding to M.. Parrots own hypothesis some such EXPLANATION appears to me necessary) as M. Parrots shows
404e024 hat whole infertility «of hybrids receive no EXPLANATION» was consequent on mind or instinct, now thi
429e116 even in branch valleys»-- M. Ramond offers no EXPLANATION.-- Poet Cowper, describes his tame Hares, at
463t057 e grouping of «many» facts with laws & their EXPLANATION will probably reject this theory-- (I must a
527m032 ing, & thus cuts the Knot:-- Sir J. Reynolds EXPLANATION may perhaps account for our acquiring «the i
570n024 k shrugging connected with many emotions.-- (EXPLANATION of sighing is probably correct, to relieve r
575n043 ses. &c &c, double consciousness? What other EXPLANATION-- can we suppose some essence. The facts abo
581n065 &c  -- <also g> if so & seeing how simple an EXPLANATION it offers of radical diversity of tongues.--
```

588n091 Philosop. Zoolog. «p. 284. Vol. II» -- gives EXPLANATION & instance of starting identical with mine,-
608o026 f reason to discover them: this is important EXPLANATION) he thinks they have none.-- Effects.-- One
622o048 f Society.-- [LHC] Sir J. M. gives different EXPLANATION of law of honour from Paley [RHC] Anyone, wh
623o050 ubtedly is instinctive. But does not Hartley EXPLANATION apply perfectly to origin of these instincts
623o050 ding to my theory, all instincts demand some EXPLANATION [RHC] Although I cannot pretend to say how f
629o55v ing is taught instinctively; I say yes, & my EXPLANATION agrees. with last head.-- (4) It is other qu
634j55r re Cetaceae found in Paris Basin?.-- NB) The EXPLANATION of types of structure in classes-- as result
634j55r to create animals on certain plans.-- is no EXPLANATION-- it has not the character of a physical law
640j167 y different, from adjoining coast. Admirable EXPLANATION is thus offered.-- From these views, one wou
621o047 rong for our present interest receive simple EXPLANATIONS from origin of man.-- [RHC] By interest I d
340d026 . which attempts to explain, or asserts to be EXPLICABLE every instinct in animals. Heard at Zoolog S
044r071 of strata on East») cannot believe in a great EXPLOSION, nor would sea remove more internally than ex
032r039 d. of greatest action. would now by degrees be EXPOSED to it, & the result would [be] a uniform slope
051r094 ng roots which must protect surface; On «hard» EXPOSED rocks near Bahia, whole surface to where highe
105a069 inite power) whereas sea. on coast, as long as EXPOSED to waves of sea, cutting power increased with
105a069 increased with width. for besides more surface EXPOSED. bay more open to turbulence. Bull. Soc. Geolo
304c206 erefore birds younger???? or «have» not «been» EXPOSED to so many contingences??? A Question of immen
577n051 ance,--does the thought drive blood to surface EXPOSED, face of man, face, neck-- «upper» bosom in wo
635j56v ems to be taken that the anthers should not be EXPOSED to weather.-- this is against my theory of fre
315c243 if the former shows that a man grinning is to EXPOSES his canine teeth. no doubt a habit gained by f
545m107 sion was most curious <like> «remember» the EXPOSTULATORY angry look of black spider monkey when touc
163g108 e been about 60 ft above sea-- soon decayed on EXPOSURE Mr H. C. Watson Geographical distribution of
616o039 d do well to ask themselves the converse of the <EXPR> question above stated, because there are livin
211b162 -- Mem. Ornitho Rhyncus Would not relationship EXPRESS, a real affinity & affinity-- whales & fish.--
287c158 sputable, but the one end of classification to EXPRESS relationship. & by so doing discover the laws
542m093 hat of turkey.-- he may be amused, he need not EXPRESS it, he may most earnestly wish [not] to do it,
573n037 constantly observed that very young children. EXPRESS the greatest surprise at emotions in her count
592n103 m from squatting.-- p. 64 closing both eyelids EXPRESS contempt. p. 76.-- children have been tickled
599o006 eas in nature; a developement of the thoughts EXPRESSED in Fingals cave, & in the arched & leafy fore
579n059 pass before him marked, with the habitual EXPRESSEMOTIONS, which make us love him, or her.-- it is b
347d050 circular. p. 5 Most clearly shows that genus EXPRESSES as now used almost any group.-- ∥all groups n
348d050 ost any group.-- ∥all groups natural (p 6) as EXPRESSING natural affinities∥ Macleays plan of arrange
533m060 too much profession, or rather in only fully EXPRESSING momentary feelings of gratitude, I had a sor
556m145 under lip curled over upper with mouth shut. EXPRESSING cool irony, not biting? What is Emotion anal
581n064 p. 31. remarks children have no difficulty in EXPRESSING their want, pleasure, or pains long before t
587n088 Get a Dictionary & make a list of every word, EXPRESSING a mental «desire» «quality» &c &c Mackintosh
198b114 traight line, or branching S. H What does the EXPRESSION mean used by Cuvier, that all animals (thoug
222b203 holds good even with trifling differences of EXPRESSION -- one child like father another like mother
271c106 ontinents».-- it is a plastic virtue.-- it is EXPRESSION for ignorance Two grand classes of varieties
276c122 e external resemblances, than the female. The EXPRESSION hybrid & fertile Hybrids, may be used to var
281c137 e) & infertility is consequence.-- The simple EXPRESSION of such a naturalist «splitting up his speci
286c154 e, so has man. Not saltus. but hiatus animals EXPRESSION of countenance. «[s]hare of sickness,-- deat
286c154 same way» they may convey much thus, Man has EXPRESSION.-- animals signals. (rabbit stamping ground)
315c243 My first thought of sea side-- Study Bell on EXPRESSION & the Zoonomia, for if the former shows that
315c243 ying is a puzzler-- Under this point of view. EXPRESSION «of all animals» becomes very curious.-- a d
336d018 fact the parents beget child like themselves. EXPRESSION of countenances, organic diseases, mental di
337d022 ge might be made from comparison of Man, with EXPRESSION <of a> of Monkey, «when offended» who loves,
356d070 events the completion.-- ∥Say my Grandfathers EXPRESSION of generat. being highest end of organizatio
356d070 nerat. being highest end of organization good EXPRESSION but does not include so many facts as mine∥
379d151 ear case of avitism. but then ¿ was «not» not EXPRESSION of <father> Sir W. itself received from his
513q2l. ts case of orchidiae Where does J. Hunter use EXPRESSION of «male principle of arrangement.»-- would
520m FC trifaction?-- [blank] Questions & Experiments EXPRESSION M Charles Darwin Esq 36 Grt. Marlborough Str
520mIFC ll of Metaphysics on Morals & Speculations on EXPRESSION-- 1838 Selected Dec 16 1856 July 15th 1838
537m075 slow habits are changed may be inferred from EXPRESSION. "relict of bad habit." as child is cured of
542m095 things in dark. & hence is this the cause of EXPRESSION of surprise-- viz seeing something obscurely
542m097 signal movements.-- some say dogs understand EXPRESSION of man's face.-- <That> How far they communi
545m106 nut, but held it between fingers, the peevish EXPRESSION was most curious <like> «remember» the expos
545m107 ey to dog. I showed nut & then closed my mem. EXPRESSION of fury, jump to scratch my face. The ourang
545m107 monkeys cry?-- «they whine like children.--» EXPRESSION, is an heredetary habitual movement conseque
545m107 ed by the same cause, which «now» excites the EXPRESSION.-- Habitual actions are the reverse of intel
548m116 he tendency to forget the insane idea; & ones EXPRESSION of double self, though as in Dr Ashe's case,
553m137 e Zoological Gardens) remarked that <exp> the EXPRESSION & noises of monkeys go in groups. thus the p
553m137 e to it,-- but he thinks not sulkiness-- this EXPRESSION he believes is common to that group.-- this
553m137 s very important as showing <connection> that EXPRESSION mean SOMETHING.-- Hunt (the intelligent Keep
555m142 hat the American ones, often put on a peevish EXPRESSION, but not nearly so often <that> hardly ever
555m142 ut not nearly so often <that> hardly ever the EXPRESSION of passion with open mouths like the old wor
556m146 rony, not biting? What is Emotion analysis of EXPRESSION of desire-- is there not protrusion of chin,
556m146 ut down ears, when kicking.-- -- good case of EXPRESSION showing real affinity in face of donkey, hor
556m146 e on head, when going to fight, in which case EXPRESSION resembles a fox-- I can conceive the opposit
557m147 y curious, recurrence of pleasure so teaching EXPRESSION «as constant smiles, cheerful face».-- Man w
557m147 owing & pleased pricks his ears?--).-- How is EXPRESSION of anger in species of swans, in parrots &c
558m151 ral sense & emotions.-- The whole argument of EXPRESSION more than any other point of structure takes
558m153 dog]:CD Man grins & stamps with passion. can EXPRESSION be used more correctly than this for C. Sphy
560m BC is intemperance. <No.> Cannot say.-- Private. EXPRESSION M Charles Darwin Esq 36 Grt. Marlborough Str [note: actually] -- see
563n FC ce. <No.> Cannot say.-- Private. Expression M EXPRESSION N What are sexual difference in monkeys.-- C
563nIFC -- Charles Darwin [Private.]CD (Metaphysics & EXPRESSION) Selected «for Species Theory» Dec. 16 1856
563n001 fter? he has done <g> crowing.-- instances of EXPRESSION.-- Octob. 3d. Dog obeying instinct of runnin
565n009 in upper lip. <The> Children having peculiar EXPRESSION is remarkable. the pouting, & blubbering-- s
565n009 outh showing action,-- sulkiness all negative EXPRESSION? Expression of affection is accompanied by s
565n009 action,-- sulkiness all negative expression? EXPRESSION of affection is accompanied by slight protru
566n010 .37, quotes from Burke, who says on mimicking EXPRESSION of emotions, he has felt the passions of a f
567n013 m in-- like child. Tommy's face, now ill, has EXPRESSION of languor & suffering The Cyanocephalus whe
569n022 ester a Farroïlap quelque mal qu'on y fût."-- EXPRESSION common to Savage & Frenchman, unaccompanied
569n022 accompany I will not. I am sorry I cannot.-- EXPRESSION leave «this» out not in Library no good Ther
572n035 e facts (about communication of ideas, &c) of EXPRESSION Lawless, whilst they are the only steady & u
572n035 universal. means. recognized-- no one can say EXPRESSION was invented to conceal one's thought.-- Mac
572n035 is Chapter on the Existence of a Deity has an EXPRESSION the very same as mine about our origin of a
579n058 on Wordsworth's dissertation on Poetry.-- The EXPRESSION of shame-facedness for shyness, having been
583n069 animals.-- hence the general aim of fable, & EXPRESSION as cunningness of fox, industry of bee &c &c
592n103 Vol 44. 1746-47. Paper. like. Sir Ch. Bell on EXPRESSION «First Croonian Lectures by Parsons» follow
592n104 een in shortsighted people.-- hence origin of EXPRESSION-- There are some instincts unintelligible, <
603o014 ndsome» «UGLY healthy» young woman, with good EXPRESSION-- statues not painted-- <music> very good ar
606o022 grandeur of character.-- Hence Lessings shows EXPRESSION of pain cannot be represented. But what is b
616o039 feel Now this would certainly be a startling EXPRESSION, & so foreign to the use of ordinary languag
616o039 those who would support the propriety of the EXPRESSION. They would do well to ask themselves the co
541m093 sarcasm.-- <These> Seeing how ancient these EXPRESSIONS are, it is no wonder that they are so diffic
542m097 s-- they likewise must understand each other EXPRESSIONS, sounds, & signal movements.-- some say dogs
555m142 Zoological Gardens-- Endeavoured to classify EXPRESSIONS of monkeys-- I could only perceive that the
596n BC ary muscles-- if so what is trembling palsy? EXPRESSIONS N [Old & USELESS notes about the moral sense
264c079 an visit Ourang-outang in domestication, hear EXPRESSIVE whine, see its intelligence when spoken; as
573n037 y experience, that movements of face are more EXPRESSIVE than movements of fingers.-- like Kitten wit
595n127 miths Essays No XV., on sounds of words being EXPRESSIVE, (Vol. 4 of Works) [blank] "Adam Smith Moral
023r010 chain of the Cordilleras as arising from «the EXPULSION of fluid nucleus through» faults or fissures,
558m152 e others-- Thus <sudden> «forcible prolonged» EXPULSION of air «dogs snarl much the same way» generic
293c174 s to prevent insects lodging there. Now these EXQUISITE adaptations can hardly be accounted for by My
605o020 e has been supposed by some to consist of "an EXQUISITE susceptibility from Blair receiving pleasures
435e133 n Orchis (so very different) that the granules EXSERTED their tubes: now Mr Herbert has shown that st
252c042 77.). probably another in Jamaica & perhaps one EXTANT at Leeward Isles. p. 388 Reference to Rüppel.

Page **(Key Word)**
```
070r155 person  (whom I met at S. W. P.) the Cordillera EXTEND to near Salta. & not far from Tucama[n]. & at
089a020 covered with iron clay common to Guyana said to EXTEND to Cordillera I see Brewster speculates from b
117a102 -- L'Institut. 1838 p. 151. Formations of Payta EXTEND close to Guayaquil.-- modern shells of  Cobija
149g033 & channel precisely as with Isld-- {P} do they EXTEND round hill too low line drawn plain red  talus
150g038 e ravine enters On opposite side of valley both EXTEND below the Houses The Hills in this neighbourho
176b023 & the endeavour of each <one> typical class to EXTEND his domain into the other domains. & subdivisi
266c091 o escape the doctrines of Muhanmad, the last to EXTEND, their dominion, armed alike with the Koran an
436e137 ions:-- The Cambrian formations do not however, EXTEND round world.-- Quartz of Falkland.-- Old Red S
485z018 rever there is extreme heat, the tropical forms EXTEND further north, because during winter they  can
486z018 On  this principle tropical forms in N. America EXTEND much further N. in N. America than in Europe--
623o050 pretend to say how far & minutely our instincts EXTEND, yet as they are acquired by social animals, l
138a153 t above its present level, & in many parts has EXTENDED a league inshore both N & S of  Lima.--judges
206b148 ses must act in the two case,» May this not be EXTENDED to all animals first consider species of cats
271c106 lata.--!! --Argument, when general argument is EXTENDED to Juan Fernandez in birds. but ¿whether to s
305c211 ems to be given or assumed according to a more EXTENDED relations of the individuals, whereby  choice
307c216 & channel precisely as with Isld-- {P} do they EXTENDED from species to genera & classes. p. 479. fra
612o035 ar relation to less simple bodies, and to more EXTENDED space, such powers of relation required to be
612o035 space, such powers of relation required to be EXTENDED. Hence a sensorium, which receives communicat
040r062 pebbles.  the degraded matter of such pebbles EXTENDING to seaward, the alternating with such  matter
182b049 prejudiced about self, but considering power, EXTENDING range, reason & futurity. it does as yet appe
311c225 each  other». good to consult p. 326 wild ass EXTENDING over 90 degree of Long. & Col. Sykes  alludes
318c253 WHAT CHANGES are taking place & how birds are EXTENDING their ranges. «even migratory birds, lik swal
461t041 qual duration is exactly same as some species EXTENDING much further geographically than others. Athe
150g037 , on side of Hill of Bohunthine upper road (2) EXTENDS as far nearly as house, the 3d below them  oppo
242c018 ncus with golden streaks-- the lacerta vittata EXTENDS <to> from Amboina to New Ireland p. 23  Voyage
613o036 ng it. ---- Agrees with ONE animal [RHC] Kirby EXTENDS instinct to plants, but surely instincts imply
032r039 the resistance offered to the greater lateral EXTENSION of the waves. by the part beneath the band of
085a001 Shells  near Woollich p. 112 Speculate on the EXTENSION of Patagonia seaward, at mouth of S. Cruz. fr
233b249 ian.?-- Mr Gould has been struck with similar EXTENSION of form in birds.-- | Waterhouse thinks two m
284c149 character  depends on non-variation, & not on EXTENSION ¿these go together? Therefore value of organs
294c177 adaption  to classification & affinities, its EXTENSION.-- Von Buch. Travels, p. 306. account of tree
492q002 Can  any annuals be budded. with reference to EXTENSION of age of individuals-- 9. Do plants in becom
035r048 ome easy if we look at the action as a deep & EXTENSIVE movement of viscid nucleus, which in any  one
043r070 this  mentioned by Humboldt in his account of EXTENSIVE areas. -- P. 322 In any archipelago. & neigbo
114a094 slate. a distinct formation deep «& therefore EXTENSIVE» water ∴ not formed in modern formation & not
031r037 imply raised No Faults in Patagonia[,] enormous EXTENT; if lowered again & covered no sign of upheava
054r105 half  a league of what is now Terra Firma & the EXTENT of a league & a half a long the coast. <"> The
209b155 with old Linnaean doctrine & Lyells. to certain EXTENT Von Buch,.-- Canary Isles: French Edit.  Flora
261c069 halk Those who say «philosphically to a certain EXTENT of all species Accumulate instances of one fam
267c094 its increase. The <bush> woodlands for miles in EXTENT are composed solely of this shrub".-- p.  229.
303c205 halk Those who say «philosphically to a certain EXTENT,-- nothing but experience. will, tell us. when
312c232 ary & variations produced in short time in some EXTENT counterpart, mutilation being «variation» prod
624o051 JCD «although perhaps useful at present to some EXTENT,.» Hence this is the law of our instinctive fee
414e064 il «& in Van Diemen's land»-- they have been EXTERMINATED on principles. strictly applicable to the u
465t081 on of method by which races of men have been EXTERMINATED (see Pritchards paper) (Ed. Phil. Journ. en
259o064 , Epidemics of South Sea, wonderful case of EXTERMINATION of species-- Epidemic amongst trees.  Plane
292c168 d Harmony these views-- did Lamarck connect EXTERMINATION of some forms with his views.-- as genera a
297c186 e group aberrant When species rare we infer EXTERMINATION, when group few in number of kind, extermin
297c186 mination. when group few in number of kind, EXTERMINATION.-- New forms made through probably an infin
366d111 stralia. go on blinking their eyes. without EXTERMINATION, & change of structure.-- When will the mus
397e003 not  only to population & depopulation, but EXTERMINATION & production of new forms.-- their number &
408e043 ions have great effect on them, & therefore EXTERMINATION becomes part of same law.-- When we know wh
408e043 ies becomes rarer, as it progresses towards EXTERMINATION. some other species must increase in number
151g041 ollarig two little lines of Hill (judging from EXTERNAL form alluvium) descend from shelf 3d & almost
179b035 m from considering. S. America (independent of EXTERNAL causes) does appear very probable:-- Mem: Hor
195b103 Volcanic  soil of Galapagos under equator that EXTERNAL conditions would produce species so close  as
201b129 structure of Lamarck, which he says depends on EXTERNAL influences.-- For instance he says wings of b
201b129 -- For instance he says wings of bat, are from EXTERNAL influence.--...... Hence name of analogy, the
224b213 cies may be good ones & differ scarcely in any EXTERNAL character:-- For instance two wrens forced to
225b217 und). I should say the changes were effects of EXTERNAL causes, of which we are as ignorant. as why m
248c033 t stock, than between two hybrids.-- As we see EXTERNAL influences first affect external [for]m, so w
248c033 .-- As we see external influences first affect EXTERNAL [for]m, so will the internal parts be of long
248c033 primary  divisions of insects» 2. Relation, of EXTERNAL conditions, & of succession: the <first> latt
248c033 uctures. <which are less obviously affected by EXTERNAL circumstances> these therefore will be chiefl
268c099 eat» animals?-- Show independency of shells to EXTERNAL features of land by seeing how many species c
271c106 but cannot be counteracted by Man.-- effect of EXTERNAL contingencies & long bred in-- Mem, <an> a st
275c121 clined to think that the male communicates the EXTERNAL resemblances, than the female. The expression
280c134 s in wild animals, many species in one genus-- EXTERNAL circumstances in both cases effect. it.-- Sir
282c140 he relation to bear to each other, but to some EXTERNAL contingency.-- affinity is the sum of all the
282c140 he men to have greater power of change yet, as EXTERNAL conditions over whole world. similar-- & cons
284c147 formerly.-- The number of forms depends on the EXTERNAL relations (a fixed quantity) & on subdivision
285c153 therefore Most especially under care of Man. & EXTERNAL circumstances not variable.-- Animals have vo
294c176 that domesticated animals change a little with EXTERNAL influence-- & if those changes permanent so w
301c199 mental powers.? p. 8. mistakes of instinct are EXTERNAL contingencies, where the habit is not applica
302c200 uch death, as has gone on., No greater gaps.-- EXTERNAL conditions, to be sure, have remained somewha
309c222 e.-- «the passages between-- owls & hawks only EXTERNAL» intermediate groups often have full structur
310c222 his class of facts «analogous to petrel-grebe. EXTERNAL» appears to be a puzzle against my  theory,--
347d049 s discovered to Taenia.-- hard so as to resist EXTERNAL influence.-- 27th. August. There must be some
391d175 es may be effect of differences of parents, or EXTERNAL circumstances during life.-- if the circumsta
391d175 the  circumstances which induce «which must be EXTERNAL» change are always of one nature species is f
403e023 might  be made-- why seeing great variation in EXTERNAL form of varieties, do we suppose bones will in
408e043 t either Americas.-- If species change, we see EXTERNAL conditions have great effect on them, & there
410e052 aming them) with relation to habits. ranges. & EXTERNAL conditions of country, most important & will
411e054 all steps in the series. their relation to the EXTERNAL world, & every possible contingent circumstan
416e071 No domesticated animal is perfectly adapted to EXTERNAL conditions.-- (hence great variation in each
416e072 be  the effect of a gradation in difference in EXTERNAL conditions.-- -- as in plant up a mountain--
428e113 or  dog has been without recurrent tendency in EXTERNAL conditions» sudden issuing of horns.-- I do n
435e133 -- April 6th "Dr. Edwards on the Influence of <EXTERNAL> Physical agents". «translated by Dr. Hodgkin
440e146 laws Who can say, how much structure is due to EXTERNAL agency, without final cause. either in presen
440e146 use yet it must be effect of some condition of EXTERNAL circumstances. results of complicated laws of
460t017 ving spines, is the effect, partly of the same EXTERNAL conditions (ie. analogical structure) & partl
473s07r time  of life of offspring-- No peculiarity in EXTERNAL structure can be concepcional, as limbs &c &c
549m122 s it must for his happiness to check-- that is EXTERNAL circumstances are so conditioned as they  are
576n047 pement) on account of dark ages.-- «effects of EXTERNAL circumstances» Look at Spain now.-- man's int
611o034 ed in different modification, peculiarities of EXTERNAL form impressed, & different laws of movements
611o034 n form; invariable, as long as not modified by EXTERNAL accidents, & in such cases modifications bear
612o035 consciousness. These willings have relation to EXTERNAL contingencies, as much as growth of tissue an
613o036 logy:-- as races are formed or modification of EXTERNAL form. so modifications of brain) As in animal
615o038 ly take place without corresponding change in «EXTERNAL» man; and as all men nearly same species,  so
615o038 ferent instinct always obtain peculiarities of EXTERNAL configuration. [RHC] General-- Instincts, cer
617o39v internal  consciousness; the objective, by our EXTERNAL what senses in the way in which we  apprehend
617o040 aspect  of> bodily action as recognised by our EXTERNAL senses consists in the manifestation of force
617o040 e; but FORCE, <objectively> considered, by our EXTERNAL senses is a phenomenon the essence of whose e
617o040 n point out How can force be recognized by our EXTERNAL senses--only movement can.-- 4) the source fr
044r071 on, nor would sea remove more internally than EXTERNALLY--I did not see any number of dikes in the cl
360d093 cross than two less opposed in habits, though EXTERNALLY similar.-- this however is a sophism for the
417e079 <lay> «have» their eggs, <inter>, impregnated EXTERNALLY; nor can it be a necessary concomitant, with
417e079 omitant, with moths, which can be impregnated EXTERNALLY-- My view of every animal being Hermaphrodit
```

639j28r - analogy in Flamingo & Duck, Ornithorhyncus «EXTERNALLY». petrel & Whale in some respects Chamaelion
062r129 antipodes a parallel case.-- Should urge that EXTINCT Llama owed its death not to change of circumst
062r130 ars to (Petisse. & diff kinds of Fourmillier): EXTINCT Guanaco to recent: in former case position, in
174b014 ation explains why modern animals same type as EXTINCT which is law almost proved.-- We can see «why»
180b037 ring relation to ancient types.-- with several EXTINCT forms, for if each species «an ancient (I)» is
186b060 beria; we must look to type of organization.-- EXTINCT species of that country parents of American.--
187b065 erent, because country separated since time of EXTINCT quadrupeds:-- same argument applies to England
206b148 cessors «for» centuries, the other will become EXTINCT.-- Who can analyze causes, dislike to marriage
206b149 anomalous lizards living; or of the tribe fish EXTINCT. or of Pachydermata, or of coniferous trees; o
241c015 ia.-- p 67 ¿American forms? All Infusoria. not EXTINCT species. good Resumé do/p. .62 ??? Age of Dein
291c167 tely necessary to explain genera & classes. if EXTINCT forms were all fathers of present, then there
292c168 ws.-- as genera are large probably only few of EXTINCT forms have generated species. & of 100 extinct
292c169 extinct forms have generated species. & of 100 EXTINCT species the greater number probably have no de
309c220 levation: Marked out by existence of elevated «EXTINCT?» genera of shells.-- duration in two classes
343d038 ind. some of species or varieties are becoming EXTINCT. others though the negro of Africa is not loos
352d058 ortive organ of any kind few.-- » hence become EXTINCT, & hence the IMPROVEMENTS of every type of org
374d134 turritella. terebratula, orbiculas, with many EXTINCT forms & Trilobites Sept 25th. In considering i
425e105 has remarked species never reappear when once EXTINCT-- Lyell's argument about «Tertiary» Isld «neig
434e128 p. 9.-- talks about fossil Infusoria becoming EXTINCT not so soon as other forms.-- p. 36.. speaking
490q001 ordan- Smith of Jordan Hill-- character of the EXTINCT land-shells of Madeira-- analogous or quite di
511q018 olmshire Breed[CD-- Sir. R. H. supposes is now EXTINCT= (9) About. American & Europaean common specie
638j58v s nay very mountains are formed of such dead & EXTINCT forms.-- the exuviae of the dead & extinct The
638j58v d & extinct forms.-- the exuviae of the dead & EXTINCT The analogy between the works of art «or intel
063r133 not bear upon solely adaptation of animals.--EXTINCTION in same manner may not depend.--There is no
063r133 r may not depend.--There is no more wonder in EXTINCTION of species than of individual.-- Mr Birchell
180b036 many species in same genus (as is). REQUIRES EXTINCTION. Thus between A. & B. immens gap of relation
180b037 number of species constant.-- With respect to EXTINCTION we can easily see that variety of ostrich, P
183b053 ither animal when crossed with it.-- ¿Whether EXTINCTION of great S. American quadrupeds. part of som
186b057 agrees with breeding.. in irregular trees. & EXTINCTION of forms.?? It is in simplest case saying ev
208b153 es. p. 127. p. 132 There is no more wonder in EXTINCTION of individuals than of species Paris Tertiar
303c205 adduced. say oh look to your fossils, now if EXTINCTION had gone, without creation this would have b
308c217 always have been gaps, & there now must be. ,:EXTINCTION of species bears relation to existence of ge
313c234 good argument for origin of man one.-- Is the EXTINCTION & change of species two very different consi
313c234 ase» Does this law of duration apply to utter EXTINCTION or rapidity of specific change.? <One» the f
313c234 rst would be called. generic & other specific EXTINCTION-- In the Entomostraca (Magazine of Zoology &
338d025 e is African. & the only difference is by the EXTINCTION of certain forms from Northern part & not by
343d037 f new forms in one., or apparently so. by the EXTINCTION of prominent ones in <latte» one: The latter
355d069 rmata &c &c-- It is important with respect to EXTINCTION of species, the capability of only small amo
356d072 te-- & the English & Some African dove.-- The EXTINCTION of the S. American quadrupeds is difficulty
357d072 appear to have suffered most with respect to EXTINCTION of larger forms.-- From observing way the Ma
388d172 change-- hence harelips heredetary, disease. EXTINCTION. Animals in domestication (even Elephant) no
409e048 him, how wolf was so changed. When discussing EXTINCTION of animals in Europe. :the forms themselves
419e087 confound our chronology» «CONSIDER ALL THIS» EXTINCTION & transmutation, two foundations, hitherto c
430e117 been direct parents of any of ours,-- even if EXTINCTION is denied.-- it will not account for all spe
431e122 e more I think, the more convinced I am, that EXTINCTION plays greater part then transmutation.-- Do
463t057 of intermediate structure, , & supposing much EXTINCTION. give a parallel case) Waterhouse remarked,
637j57v oboscis» «as bee & butterfly! inconvenience.! EXTINCTION. utter extinction! let him study Malthus & D
637j57v butterfly» inconvenience.! extinction, utter EXTINCTION! let him study Malthus & Decandoelle.-- The
641j29v ctic genera, there is evidence of antiquity & EXTINCTION of such forms-- these views will bear on geo
062r129 ve animals created for a definite time:--not EXTINGUISHED by change of circumstances: The same kind o
347d049 a harder to vary, & therefore more apt to be EXTINGUISHED.--???» Mayo (.Philosop of Living) quote Whe
525m025 tails. My father says, perfect deformity, as an EXTRA number of fingers.-- hare lip or imperfect roof
029r033 <[... Maranh» Fernambuco. EARTHQUAKE, AT SEA.--EXTRACT from the log-book of the James Cruikshank, Cap
513q21. setting of fruit. cross Conception--(↑ I could EXTRACT nothing from him)‖ Does impregnation ever regu
516q23v not connected with Species Theory (1) Will an EXTRACT of peat do to preserve fungi or animal substan
578n053 «fear» you shall not here--n, «or wish EXTRAORDINARILY to have one» you wont. ==)== No surer way
591n098 y drop their ears.-- -- George the lion is EXTRAORDINARILY cowardly.-- the other one nothing will fri
025r017 ly.-- Mrs Power at Port Louis talked of the EXTRAORDINARY freshness of the streams of Lava in Ascenci
045r075 y other phenomena I do not believe that the EXTRAORDINARY fissures of the ground at Calabria were pre
191b083 vitism the ordinary event. & succession the EXTRAORDINARY South Africa. proof of subsidence. & recent
303c205 y unphilosophical. L'Institut 1838. p. 128. EXTRAORDINARY genus. Mesites bird from Madagascar uniting
356d069 lowing work.-- The history of Medicine, the EXTRAORDINARY effects of different Medicines on organs, I
376d136 ns Sept. 29th Dr. Andrew. Smith «Remarks on EXTRAORDINARY curiosity of Monkeys». The Baboon of which
544m103 e other perceptions.-- The mind thinks with EXTRAORDINARY rapidity-- We may conclude that neither num
604o018 he idea of ascension we associate something EXTRAORDINARY & of great power-- -- 2 From these & other
265c087 wings may be of some use,-- Nature is never EXTRAVAGANT though clearly not of the use to which wings
023r009 o long; The maw was full of jelly which stank EXTREMELY."--This shark was caught in Shark's Bay. Lat
044r073 ca-- on one side. S. America on the other: The EXTREME frequency of soft materials being consolidated
066r141 N & South lines the tides form eddies with its EXTREME force. Yet, no outlet at head. Important in fo
089a019 ing machine to see if water fell. -- <Keys off EXTREME point of Flori[da]> Excellent paper on Erratic
122a113 case. Consider profoundly all consequences of EXTREME FLUIDITY of earth.-- study different forms of
162g103 ng 29.958 A 64 degree, air 60 «Evening do» The EXTREME right arm of River Tarf <it> Has a very long,
196b110 - Principes de Zool: Philosop:-- I deduce from EXTREME difficulty of hypothesis of connecting Mollusc
200b125 inia p. 325. July 1828. Animal now confined to EXTREME North.-- ↓.do p. 326. 2 Fossil species of ox i
213b172 -- both found in every sea, from Equatorial to EXTREME poles.-- Oh. Wealden-- Wealden. Do the N. Ame
229b233 species of Rhinoceros range from Abyssinia to EXTREME South coast. Elephant he believes is mentioned
229b233 ant he believes is mentioned by old writers on EXTREME Northern Coast. Hippopotamus do.-- Giraffe do.
230b240 else whole fabric will be overturned.-- Hence EXTREME difficulty, argument in circle.-- Falkland Isd
232b247 n chief part instead of change from insular to EXTREME climate, <more northern> Iceland would have po
232b247 towards Equator, <th> so would the plants from EXTREME north, which according to all analogy would ha
250c037 out other animals? but they were not shut up!! EXTREME southern points of S. Hemisphere fully charact
250c037 n American form? The climate having grown more EXTREME both in, N & S. America, is only common cause
258c059 te to this law.?-- Local varieties formed with EXTREME slowness, even where isolation, from general c
259c064 ngst trees. Plane trees all died certain year. EXTREME difficulty of TRACING change of species to spe
296c184 & land animals. & land shells.-- all in short EXTREME North = = to peak of Teyde in relation to surr
310c225 ave changed much less.-- Here is an element of EXTREME difficulty in mundine geological chronology An
317c251 ll> Rocky Mountains have peculiar character in EXTREME length of ears & length of limbs, so that me f
318c252 . case of adaptation.-- (case of Squirrel from EXTREME north turning white like Hares??--) I never sa
344d041 fertile offspring. Entomostraca & Aphides. The EXTREME difference of sexes. is probably arrived at in
352d059 on of animals countenances the belief of their EXTREME antiquity (ie much intervening physical change
383d159 (Mammae or sheath of Horses poenis reduced to EXTREME degree of abortion).-- Insecta.-- hermaphrodit
407e039 the <Tropical» Equable kind of climate to the EXTREME.-- Therefore species, which were fitted for su
427e109 ot subsidence of Greenland render climate less EXTREME. (& so account for descent of snow line there
427e109 from Europaean. , & the climate being now less EXTREME, than before arctic forms would retreat: effec
439e143 n returning to old type Mr Herbert showing the EXTREME facility of crossing, in plants proves how muc
485z018 warm in Lapland & Stizbergen wherever there is EXTREME heat, the tropical forms extend further north,
522m012 habitual» intemperance.-- often accompanied by EXTREME anger, at not being understood.-- My F. says t
531m051 can remember poetry when once read over.-- The EXTREME pleasure children show in the naughtiness of b
533m059 the proper muscles. may be illustrated by the EXTREME difficulty of moving muscles in different way
541m090 confound action & thought here?) The opposite EXTREME of this desultory thought is following out suc
565n006 mere symbol of readiness, & therefore done in EXTREME.-- Looking at ones face <&> «whilst» laughing
567n015 would not have any emotion like blush.-- when EXTREME sensation of heat shows blood is pumped over w
593n111 animals.-- [blank] Double consciousness. only EXTREME step of an ideal argument held in one's own mi
636j56v t be concluded that Plants would be subject to EXTREME variation as long as crossing with other varie
181b040 of the Mammalian type of organization; it is EXTREMELY improbable that any of <his relatives shall l
398e005 me link in our train of geological reasoning, EXTREMELY faulty» The difficulty of multiplying effects
520m001 en the latter died so long before, that it is EXTREMELY improbably that they should have imitated.--
565n008 uite insignificant, & when pride makes person EXTREMELY self-sufficient,-- the corner of lower lip ar

569n019 ual beauty) is acquired taste.-- Whilst music EXTREMELY primitive.-- almost like tastes of mouth & sm
620o044 f the temptation to disobey the conscience is EXTREMELY great [LHC] The cause perhaps lies in its fre
227b226 between lowest Mammal & Reptile. (or between EXTREMITIES of any great divisions) thus a knowledge of
243c020 gard to shells.-- But he says shells towards EXTREMITIES of the continents peculiar to the different
303c204 ead «& features»;. but likewise in length of EXTREMITIES, how are races in This respect upper & lower
411e055 Cuviers generalization. of teeth to kind of EXTREMITIES come under this head» 27th November When sum
634j55r for we find even in Cetacean traces of hind EXTREMITIES.-- How are we to explain this.-- Did reptile
116a100 ublittoral formations. p. 150. at Portezuelo, EXTREMITY of mountains of Cordova project on plain, lik
467t103 brush at end of stigma, which forces out from EXTREMITY pollen, or pollen comes out with anthers & st
611o034 s not conscious sensation. [RHC] During growth <EXTRES> tissue <[...]> unites matter into certain for
641j29v other analogies-- prickly plants or animals-- EXUDATION of fetid «& acrid» secretion in Mollusca. ins
502g11v When nettle leaf. put into spirits, poison-drop EXUDES-- does not elm. does it «in» melon-- «Loasa» A
638j58v are formed of such dead & extinct forms.-- the EXUVIAE of the dead & extinct The analogy between the
161g095 concealed by fragments, then clear. this bit to EYE certainly appears level with road, & with piece
161g096 followed for at least 2 miles on dead level, «by EYE» to moss-- on this terrace Barom. 29.264 A 82 75
287c158 in different animals.-- + whether variations in EYE of vertebrate afford better character, than vari
287c158 rate afford better character, than variations in EYE of mollusca. [+] These questions may be all di
293c175 trace the steps by which the organization of the EYE, passed from simpler stage to more perfect. pres
337d021 f so. it will be necessary to show how the first EYE is formed.-- how one nerve becomes sensitive to
337d021 (Mem whole plant may be considered as one large EYE-- have they smell, do plants emit odour solely f
351d056 --» Hence the Pecten, which move imperfectly has EYE-point, but Broderip added it has been stated tha
351d056 it has been stated that stationary Spondylus has EYE-points-- Macleay then answered, because nature l
565n009 hich wrinkle when smile.-- Hope is the expectant EYE. looking to distant object, brightened & moisten
586n082 namely the knowledge of size is merely judged by EYE, & use of limbs &c, or it result from mere impul
606o023 veloped in a work of art are not approved by the EYE itself, but by the imagination through the mediu
606o023 but by the imagination through the medium of the EYE"; he will allow the secondary pleasures of harmo
641j29v insects «Carabids & Staphylini» & Mammalia. The EYE being formed in Mollusca, Articulata, & Vertebra
641j29v d by some very different method. in pedunculated EYE of Chamelion. crabs Crabs & Mollusca we have ana
545m106 istinct wrinkle, but such movements in skin of EYEBROW important analogy with man.-- I see monkeys gr
542m095 skin contract iris?-- same way as one lifts up EYEBROWS to see things in dark. & hence is this the ca
555m143 he[y] move whole skin of head they do not move EYEBROWS.-- (I see some of the old world ones move ski
596n184 l of practical observations Ourang do not move EYEBROWS.-- or skin of head,-- "scarcely able St.-- »
334d011 ers & sisters in Mankind.-- The case of all blue EYED cats (Fox has seen repeated cases) being deaf c
592n103 ess. them from squatting.-- p. 64 closing both EYELIDS express contempt. p. 76.-- children have been
234b255 world [not located] T. Carlyle, saw with his own EYES. new gate. opening towards pig.-- latch on othe
303c204 was first black at base of nails & over white of EYES,-- + + +, Will he say creation is at end, seein
311c227 Transact. Vol IX p. 107. an Ascaris inhabits the EYES of horses in India in which it may be seen swim
333d005 il cut off in progeny peculiar) limbs very long, EYES very large, very fierce to dogs «otherwise habi
350d055 e blind & of quite different form from male with EYES!-- (are not these differences in sex confined t
366d111 habitants of NW. Australia, go on blinking their EYES. without extermination, & change of structure.-
534m061 king consists of sensation of images before your EYES, or ears (language mere means of exciting assoc
536m070 ngry-- such efforts prevent anger, but observing EYES thus unconsciously discover struggle of feeling
538m079 other & unusual line-- both odd appearance about EYES.-- one botanist & great knowledge of Irish Poli
545m105 eus») constantly moved the skin of forehead over EYES, at every emotion & <look> «turn» of the head.
542m130 or having looked at any object<)> one Shuts ones EYES) is the image not vivid as in sleep-- (one can
556m145 every man who is not deformed. is born with two EYES.-- " I think this cannot be disputed anymore in m
560m156 indenpendent of fear: colour of bare nails--, & of EYES.-- Do female monkeys care for men.-- Have we an
560mIBC Do they wink, when anything placed before their EYES, very young, before experience can have taught
564n005 en, that faculty, whether for position of axe of EYES, state of surface, or other means by which eyes
564n005 eyes, state of surface, or other means by which EYES, aided by experience is supposed in man to guid
564n006 ich whole skin or muscles are contracted between EYES & upper lip., is most clearly analogous to a pa
588n090 case they struggle their arms.-- do. p. 306 "the EYES are rolled upwards during mental agony, & whils
592n103 subject & a reference to Brun's work.-- Shutting EYES in contempt opposite action to opening eyes in
592n103 ting eyes in contempt opposite action to opening EYES in fear The effect of habitual movements in mus
521m009 e started,-- could receive a new train through EYESIGHT, though, not through hearing,-- Thus when din
178b030 Loudon. Journal. of Nat History.-- July. 1837. EYTON of Hybrids propagating freely In Isld neighbour
204b139 y exceedingly prolific & hybrid about half way. EYTON says Hybrid about half aways & result the same
211b162 y little. in Zoolog. Proceedings. Jan 1837, «by EYTON» Account of three, kinds of pigs. difference in
232b248 n land shells in salt water & lizards do.-- Ask EYTON to procure me some Get Hope to give me an accou
331dIFC ts are the number of seeds greater.?-- Mem. for EYTON.-- Sir. R. Heron's case of breed of pigs with s
332d004 es cannot endure this climate-- .) -- July 23d. EYTON, a stone blind horse, seemed to perceive turn o
449e168 rnityhyrhycus-- is not this right?-- June 18th. EYTON tells me, that Yarrell knows of a Gull, which h
449e168 in domestication eggs of two shapes & colour.-- EYTON has observed same thing in Brent Goose. Eyton s
449e169 - Eyton has observed same thing in Brent Goose. EYTON says some of the pidgeons in common Dovecot are
449e169 f the Hamster.-- is not this Siberian animal?-- EYTON says that the young of two hatches «all alike»
489qIFC Boott: R. Brown p. 21 Horticulturists p. 21--23 EYTON p. 22 Schomburgk.---- 1 Jordan Smith. p 1. Sowe
499q009 n kind is much more effective than of foreign-- EYTON has such a grove of Willows.-- (14) Bowman fema
511q018 unds-- is there ever any degeneration?? HOUNDS. EYTON Mr Wynne, &c Could by selection a different loo
515q022 o vary in number in individuals of same species EYTON (1) Number of eggs-- of half-bred geese-- inter
613o036 rinciple is the same in all animals. [LHC] «3)» EYTON told me that his retriever Sailor he has seen p
203b138 hybrid variety partakes chiefly of the former EYTON'S paper on Hybrids Loudon's Magazine. Gould on M
224b212 t bear any precise relation to structures Mem. EYTON'S Hogs & dogs.-- The passage in last page explai
276c124 structure) there cannot be gradation. See what EYTONS young pigs-- if vertebrae much lengthened &c t
331dIFC C D Charles Darwin 36 Great Marlborough St Did EYTONS <intermediate>. «hybrids, when» interbred. sho
338d025 ar: because. contemporaries. In introduction to EYTONS Anatidae.-- tecurs to idea of only animals fro
340d026 hybrids fertile inter se--No directly against EYTON'S rule. ¿Are the hybrids similar inter se-- [no
353d061 3. & 404 vol II. do (p. <69> «71»). alludes to EYTON'S discovery of different number of vertebrae in
595n117 his;-- it shows similarity in mind.-- think of EYTON'S horses becoming <white> with <lather> <foame>
583n069 n the lower animals.-- hence the general aim of FABLE, & expression as cunningness of fox, industry o
230b240 s so many plants produce hybrids, or else whole FABRIC will be overturned.-- Hence extreme difficulty
263c076 ning than other-- if this be granted!!) & whole FABRIC totters & falls.-- look abroad, study gradatio
263c077 ution study relation of fossil with recent. the FABRIC falls! But Man-- -- wonderful Man. "divino ore
165g125 nt to know the lambs because most like Mother in FACE-- asked stated this generally the case Wednesda
182b048 were of two kinds one <with> white with a black FACE, & similar to those brought from Abyssinia, & o
233b252 s say, intellectually higher.-- But who with the FACE of the earth covered with the most beautiful sa
267c093 numerous "of two kinds one white with «a» black FACE, & similar to those brought from Abyssinia; the
345d044 vincial Breed-- Highland Sheep jet black legs, & FACE & tail, just like species.-- high active breedi
360d090 og do» father dog. & hence general appearance of FACE & tail somewhat like dog-- though it has full s
542m097 .-- some say dogs understand expression of man's FACE.-- <That> How far they communicate not easy to
545m107 d my mem. expression of fury, jump to scratch my FACE. The ourang outang, under same circumstances, t
556m146 good case of expression showing real affinity in FACE of donkey, horse & zebra. when going to kick.--
557m147 eaching expression «as constant smiles, cheerful FACE».-- Man when at ease has smooth brow contrary t
557m148 t.-- like «slight» passion from blood rushing in FACE, with less action of the heart.-- tendency to m
564n004 e. ulitmate effects of actions.+ till at last he FACE «instinct of» hunger, «of» death & for the sati
565n006 , & therefore done in extreme.-- Looking at ones FACE <&> «whilst» laughing in glass. & then as one c
565n006 . & then as one ceases, or stops the noise , the FACE clearly passes into smile-- laugh long prior to
566n010 ssion of emotions, he has felt the passions of a FACE «& mind» sympathetics with internal organs, as
567n013 ned my hand, & put them in-- like child. Tommy's FACE, now ill, has expression of languor & suffering
573n037 can have learnt by experience, that movements of FACE are more expressive than movements of fingers.-
577n051 does the thought drive blood to surface exposed, FACE of man, face, neck-- «upper» bosom in woman: li
577n051 ght drive blood to surface exposed, face of man, FACE, neck-- «upper» bosom in woman: like erection s
578n054 "one will anyone, especially a women think of my FACE,"? to one moral conduct.-- either good or bad.
578n055 , & see how» feel how the blood gushed into his FACE,"-- «as <she» «the» thought of his knowing «it»,
578n055 uddenly came across her, the blood rushed to her FACE,"-- One blush if one thinks that any one suspec
592n104 r The effect of habitual movements in muscles of FACE, is well seen in shortsighted people.-- hence o
596n184 in from head very little Does blood go in <body> FACE in pashion.?-- cry? Do people of weak intellect
596nIBC -- In slight convulsions. are the muscles of the FACE first affected?-- Can shivering & trembling be
146g011 nder storm, many <hundred> thousand tuns. Black FACED sheep, sometimes mottled with white black legs

148g025 being determined by sheep & not deer When Black FACED sheep are crossed with English my informant sai
346d047 than Mother.-- The cross not so hardy as Black FACED, but more tendency to fatten-- This man confirm
376d137 oise.-- The Cercopithecus chinensis: (or bonnet FACED monkey he has seen do this.-- These Monkeys had
555m143 many jokes. about not having run away &c having FACED death like a hero, & then I had some confused i
579n058 ertation on Poetry.-- The expression of shame-FACEDNESS for shyness, having been invented, prove of t
290c163 edetary, because crossed with women with pretty FACES When horse goes a round, the minute gets into t
345d043 said he learnt to know lambs, because in their FACES they were most like their mothers believes this
376d137 t afraid clasp them round waist & look in their FACES & Mak the st. st noise.-- The Cercopithecus chi
558m151 tes) no one can doubt this connexion.-- look at FACES of people in different trades &c &c &c I observ
564n006 cii Vol III» of excessively cross-half furious FACES «which may be described as an exaggerated habit
594n113 the 24th of May at Cresselly by the boys making FACES at it, so much so that the nurse had to carry i
495q05a lworth p. 377. Have paper ruled in squares to FACILITATE investigation.-- Capital in middle of plough
190b075 n of type in two countries direct relation to FACILITIES of communication Have races of Plants. ever
233b252 s domesticated quadrupeds are not so. greater FACILITIES of change in the articulate than <M> Vertebr
022r006 e. & unstratified rocks show any difference in FACILITY of conducting Electricity? Would minute parti
120a107 rvatures of the strata.¿ the enormous faults & FACILITY with which the earth is cracking by vertical
338d023 of country: the time of separation depends on FACILITY of transport in the species itself, & in the
354d065 Zealand one-- It should be observed with what FACILITY mice attach themselves to man. Sept 7th. -- I
371d128 characters though transmitting them with such FACILITY to bud.-- this must be owing to their unity i
439e143 species» false, which makes it determined by a FACILITY in returning to old type Mr Herbert showing t
439e143 ing to old type Mr Herbert showing the extreme FACILITY of crossing, in plants proves how much depend
439e144 the crossing.-- in animals where there is much FACILITY in crossing there comes the impediment of ins
447e164 llow of crossing with other individuals, «with FACILITY»-- such as cryptogamic plants & true hermaphr
447e164 s; changes seem to have been more rapid, & the FACILITY for inter marriage is greater (Hence Dioeciou
466r094 Bees visit always base {a} of upper petal from FACILITY of alighting? which is not differently colour
497q007 rbert explains numerous spec. of Cape Heath by FACILITY. ¿Knight take opposite view. Gaertner talks o
527m035 tles in the air are banished the better.-- The FACILITY with which a castle in the air is interrupted
026r021 Carne. Geolog. Trans: Cornwall «Vol II» It is a FACT worth noticing that cryst of glassy felspar in
031r038 ergne. very little Pumice, though Trachyte. same FACT in Galapagos. Daubeny P 24 V. back of page 1 of
034r044 f ejected granitic fragments P. 386 Mem. Lyell's FACT about sulphuric vapours in East Indian Volcanos
035r047 may well be preserved at Patagonia. The English FACT is astonishing consult book itself. P. xx: same
035r047 is astonishing consult book itself. P. xx: same FACT is indeed shewn? by the parallel bands of forma
036r051 recalled by note in Daubeny. P. 438. of similar FACT near the Red Sea.--which occurred in a sandy pl
041r065 scussing probable rise of land: Mention M. Gay's FACT about shells: Hibernation of fresh water Shells
041r066 ently depend on a concretionary contraction: the FACT is in alliance with those balls at Chiloe, full
044r074 lding matter in solution must be great: & in the FACT of bombs in tufa there is proof of such gaz) st
047r082 l. vol I. P. 471) is explained. also the similar FACT at Concepcion? Read the various accounts & see
049r089 orted layers: Mem: Phillips Mineralogy some such FACT stated to exist in Peru. -- Ascension At Ischia
052r099 f some peaks elevated.-- Greywacke. as a general FACT absent in T. del Fuego, excepting in Port Famin
057r114 Port Desire on plain. & interstratified.-- Urge FACT of Boulders not in lower strata. only in upper.
061r127 - ¿ Another one in 1816 (?).-- Mr Owen's curious FACT about Crust Bra in Brine Springs. (Henslow) Spe
064r135 e as recent species. -- May I not generalize the FACT glaciers most abundant in interior channels. th
069r150 der than (B).-- Most important view Urge curious FACT felspar melted gneiss/// QUARTZ!!! Analogous to
074r165 pyrite in a fossil» Insist strongly on the grand FACT of Volcanic & non Volcanic. Then Solfataras. «M
091a025 ilex also shewn. No 3d of Ed. N. Phil. J. p 194. FACT of dust blown far out to sea valuable; because
096a041 aters were low --¿ applicable to Auvergne??? The FACT of Galapagos Isld. steep side to windward in al
112a088 in Mouth of S. Cruz in connection with Fitz Roys FACT of elevated block of stone.-- & Caldcleughs col
126a122 t to degree.-- but this may be very wrong,-- The FACT of a dumplin being bad conductor is {P} against
145g001 1 Dec 1837 Yet instances given against it-- Mere FACT of many races of Animals in Britain shows that
183b052 s.-- April 1837. p. 243 it is said as well known FACT that "serin avec le chardonneret, avec la linot
187b062 Zebras reached South Africa-- It is a wonderful FACT Horse, Elephant & Mastodon dying out about same
188b068 blue. Now this is same bearing with Dr. Smith's FACT of races of men tendency to keep to one line Dr
200b123 obscure. without doubt:-- Vide cattle:-- The grand FACT is to establish whether in crossing very opposi
201b128 Bot. p 65 Vol II talking of annelidae.-- <">The FACT is an additional illustration of that axiom in
217b183 ds weul in heat.-- This is good instance of same FACT in Mr Galtons case.-- It explains the loss & ex
219b193 discover what kinds of seeds. these plants) [Mem FACT stated by Mr Don in island, Teneriffe, St. Hele
219b196 produce from Incestuous intercourse. a parallel FACT to Blood-Hounds. Before Attract of Gravity disc
220b197 instinct of parent. Mem Lord Moreton's Mare. The FACT of plants going back hybrid plants; analogous t
221b202 ence of Coniferous Woods before Dicotyledenous a FACT analogous to reptiles before Mammalia Think abo
225b219 Zealand «compare to Van Diemen's land.» glorious FACT. of absence of quadrupeds East India Archipelag
226b220 . equally good.-- Small isld off New Guinea same FACT see Coquille's Voyage.-- Galapagos mouse (?) br
227b225 service habits in classification. or rather the FACT that they are not far the most serviceable. We
239c002 be chesnut.-- On this principle I may add, that FACT of half cross with parents, going back to eithe
240c003 markings of wings like the wild rock pidgeon:-- FACT analogous to Owen's Phil: remark of Apteryx hav
240c003 ometimes like father, sometimes like mother. The FACT of great monstrosities being produced, & handed
242c018 cal with those from S. Africa. Mr. Bibron doubts FACT.--" My toad is same species Coquille Voyage p.
249c036 ally wide range? Mice.-- Waterhouse's remarkable FACT of no forms peculiar to word to special distric
249c036 ferent forms being migratory; also with Temminks FACT of forms being within Tropics.-- Europaean bird
251c041 domesticated, strangely contradictory to Azaras FACT of conduct of wild & tame horses.-- p. 246-- Gm
253c046 «compare rodents of two countries» & «Monkeys.» FACT of Elephant same species in Borneo. Sumatra. In
254c047 of the Species--! How does this agree with grand FACT of Marsupial, low Cerebral structure??-- «do» p
259c066 ptation With respect to my theory of generation, FACT of armless parent not having armless child, sho
262c073 &c &c.-- Study the wars of organic being.-- the FACT of guavas having overrun-- Tahiti. thistle. Pam
262c073 for number five in each group absurd.-- the mere FACT of division of lesser & more power (2.typical 3
280c135 les is very important for L.yell said to me, the FACT of existence of mules appeared to him most stra
280c135 rtaining about hybrids.-- & is a very remarkable FACT. show influence of mind It is not difficult to
291c167 oth» sex«es». is strongly supported by wonderful FACT of bees changing the sex by feeding.-- no it is
296c184 new yet forms like neighbouring Continent. This FACT speaks volumes. 2 Chapters. translated by Hooke
302c200 partly explained on my theory = otherwise mere FACT creator chooses so to create.-- It is very rema
302c202 n p. 37. of Macleay. wonderfully accordant. with FACT there stated, only in most discordant groups. T
304c206 hain. <Icthyo> .+ + +. supra & next page It is a FACT pregnant with SOMETHING.? that intermediate. sp
305c209 a mere statement nothing is explained.-- this is FACT analogous to mocking thrushes of Galapagos havi
308c218 Jeffrey (life of Mackintosh Vol II. p. 495)-- in FACT, in all reasonings, of which human nature is th
314c239 yet it seems there may be Eucalyptus!-- (Hostile FACT) Be cautious about Goulds case of birds of Van
334d009 ompletely is Lord Moreton's case opposed to this FACT & views.-- Fox says a cousin «one of Mr Strutt»
336d018 s his own likeness, & the union makes hybrid, in FACT the parents beget child like themselves. expres
355d067 ? «See Cuvier Ossemens Fossiles» Although no new FACT be elicited by these speculation even if partly
356d069 ve used them since) the line of proof & reducing FACT to law only merit if merit there be in followin
365d106 een Pheasant & Black Cock, & other hybrids-- The FACT of Egyptian animals not having changed is good-
385d164 ildren resemble parents in their bodies "It is a FACT equally well known, that we observe in the temp
386d166 er part may surely absorb a useless member.-- in FACT they do it in disease & injury.-- The sympathy
388d172 tructure = Neuter bee having both sexes abortive FACT of same tendency. -- Mammae in man. having give
407e041 polar lands.--; Africa not so equatorial..-- The FACT of No. Mam: Placent: insectivore being in S. Am
409e049 he was or not. He is present a social animal» a FACT few will dispute, [although, that it was the so
426e106 -- March 9th-- Is there any relation between the FACT that different species produce abundantly infer
426e106 cies produce abundantly infertile hybrids, & the FACT that old varieties do not so much affect first
432e124 ldren do not, (& in hairless kittens we see same FACT) go back, & this is argument against Blyth's do
433e127 ems, are only leaves out of whole volumes.-- The FACT of tumbling pidgeons; flying high all together
439e143 many changes.-- It is very important Mr Herberts FACT about the hybrids (mentioned in letter to Hensl
442e152 essence is that little change is produced.-- The FACT just alluded to of Northern flowers, throws eno
443e155 e <done> been propagated by sexual commerce «The FACT of Corallina & Halimeda is case in point».-- Th
450e175 opic-- By J. H. Moore after quitting Bengal this FACT is noticed in Cassay Ava Pegue-- seldom equals
451e176 should think grandfather first of race & if so, FACT for my theory Cocos Isld & Preparis between And
452e182 Borabhum & Dholbum.-- Vol do. p. 634, alludes to FACT stated by M. Tournal that skulls found near Vie
483z012 ourn Vol I. p. 125, owls seen crossing Atlantic. FACT taken from Jenner (1825) Phils: Transact.-- "on
495q05a ow great a proportion springs up true.-- This in FACT always takes place in natural Hybrids of Cabbag
504q014 aks. races planted & Linum Perenne.-- Herbert's. FACT.= (4) Effects of Nitrate of Soda under Beech.--
512q019 ection write to-- (2) Does he believe. Stanley's FACT of Hawks distributing live Mamals (3) Do most H

```
515q021 f so, whether concepcion takes place,-- the mere FACT of seeds ripening has scarcely any no  relation
520m001 antly seeing him, she was often struck with this FACT.-- the resemblance was in odd twiching of muscl
520m002 things which happened in early infancy-- of this FACT Mr Dryden C. is good instance as he is very def
521m008 uary» power of repeating poetry in her dotage is FACT of same sort. Aunt. B. ditto.-- Case of Mr Corb
526m029 et of sketches. some real-- some fancied.-- this FACT of early memory consisting of things seen, quit
527m032 like  that of man, & this agrees with the stated FACT, that «birds from» certain districts have the b
537m075 ful than dogs should have different instincts.-- FACT most opposed to this view, where the moral sens
537m076 ersal feelings of right & wrong «(& therefore in FACT only limits moral sense)» which she seems to th
538m078 somewhat analogous, & which I think will lead to FACT of old people singing songs of their childhood.
538m078 ertainly of Miss Cogan, & fully corroborates the FACT of her not <remembering which> «repeating song»
543m100 ticians not being profound reasoners.-- all same FACT-- for, as Jones observed, in playing chess howe
546m109 ppiness depends on the senses.; than the <small> FACT that no one, looking back to his life, would sa
546m110 of a lady, (whose name was told me, who told the FACT to Mr Mayo himself. she was one day reading a b
566n012 analogy  to guide one to conclusion that any one FACT was connected with law.-- as soon as any enquir
574n041 w comes such an association in man.-- it is bare FACT, on my theory intelligible An habitual action m
605o20v ng good from bad. And it is manifestly from this FACT & the instantaneousness of the result, that the
617o40v we  prefer this metaphorical mode of stating the FACT to the mere statement of the <force exhibited i
024r015 ouse. South of Mocha; 19 miles. 65 Fathoms Vide FACTS in Beechey. on NW coast of America off Cape of
025r018 ustrations from America, purposely to show what FACTS can be supported from that part of the globe: &
035r048 land, whole surface oscillated equably.-- These FACTS become easy if we look at the action as a  deep
046r078 onsider that the dikes which so commonly (state FACTS) traverse granites, are granitic materials simp
049r087 y specimens, must be well examined At M. Video «FACTS of Passages marked by do.» discuss quartz veins
055r106 such do not occur on the beaches. Perhaps these FACTS attest a <more> decided elevation of sea's bott
096a041 lena discussion. Mr Brayley says he can give me FACTS respecting lime <n> being heated without partin
112a088 d block of stone.-- & Caldcleughs collection of FACTS See page 101. in Note Book (C) for some specula
113a091 depth will be greater than <5000.> 400.-- These FACTS of SLOW but successive transmission of temperat
123a117 > Geographical Journal Vol VIII. ( 1838) p 212. FACTS from Erman about great depths of frozen soil. p
129a131 Murchisons case.--¿ does it bear on Patagonia? «FACTS about subsided forests.-- Many repeated oscilla
131a136 aeum. 1839. p. 52. On Frozen soil of Siberia.-- FACTS of water flowing from beneath frozen crust in A
133a140 arrot ends his paper like a fool.-- Feb 25' All FACTS show how slowly heat travels; & therefore the a
191b080 land. Did Ireland possesse Mastodons?? Negative FACTS tell for little) Geographic distribution of Mam
194b095 now confined to southern hemisphere.-- If these FACTS were established it would go to show a  centrum
195b098 y any laws. which, I think is shown by the very FACTS of the Zoological character of these islands so
195b104 explains  nothing further, points gained if any FACTS are connected.-- No doubt in birds, mundine gen
200b124 o Holman: <at> Keeling these are most important FACTS.-- As soon as island large enough for land bird
217b187 dagascar has several American forms-- The above FACTS evidently show that Mr D. wonders at these spec
218b190 of Natural History-- Preface appeared good with FACTS about changes when animals transported.) Mr Her
221b201 - p. 23 p. 267. Dela Beche. Geolog. Researches. FACTS of salt-water shells living in absolutely fresh
225b219 ined? NB. Look over Bell on Quadrupeds for some FACTS.-- about dogs &c &c NB. Animals very remote. as
232b249 out such resorts Mr Waterhouse has most curious FACTS about the distribution of Lemurs in Madagascar,
240c004 analogous to what occurs in plants.-- All these FACTS clearly point out two kinds of varieties.-- One
240c004 under other birds & brought up by hand.-- These FACTS all account for [not located] Falkner Patagonia
251c041 (ref) To Temminck Monograph. Mammal; «4to» good FACTS about distribution of Cats Vol III. <p.> p 233,
253c046 e a range, & Mastodon angustidens.-- Ogleby has FACTS to show that Australian dog introduced by savag
253c046 re they? Colonel Montagu probably contains some FACTS about close species of Birds. Zoolog. Transact.
260c069 ilson's American ornithology a mine of valuable FACTS,. regarding habits range & all kinds of informa
271c105 untry> something same manner, much mud.-- These FACTS show, habits heredetary whilst species have cha
275c119 rck was the Hutton of Geology. he had few clear FACTS, but so bold in many such profound judgment, th
279c133 of making varieties may be inferred from <this> FACTS stated.-- + +. Fully supported by Mr Wilkinson.
291c166 sposition. & avitism in corporeal structure are FACTS full of meaning.-- Why is thought. being a secr
298c192 ies are larger &c in different countries. These FACTS show how very permanent plants are. & this conc
310c222 «of  one class» & full of second--this class of FACTS «analogous to petrel-grebe. external» appears t
311c226 ns difficult to analyse) will not this separate FACTS with identity of land animals.-- these however
316c247 » Now this is very remarkable.-- (connect these FACTS with abortive organs &c The doctrine of monste
317c248 of  God." Often refer to these.-- Also some few FACTS at end of "The British Aviary" or Bird  Keepers
318c254 els.--...If the line or bands of country (These FACTS show the Normal condition of Migration) gradual
325c267 uay. account of wild cattle & Montagu on birds (FACTS of close species) Wilson's American Ornithol
335d014 ted by Fox, is very important, as showing above FACTS as first cross being new species, ‖ -- Are not
341d030 ted like so many deer.-- very curious like some FACTS of Mr Blyth on birds.-- Dr. Bachman tells me li
344d041 lopement of sexes in Caterpillar. very valuable FACTS-- they are eating foetuses, as young of Marsup.
350d055 )» fixed & blind: -- Macleay observed all these FACTS prove that perfection of organs have nothing to
352d057 an parent, so may species retrograde, but these FACTS are rare.-- 2d Sept Those animals which have ma
355d067 s the end of science. namely prediction.-- till FACTS are grouped & called. there can be no predicti
355d067 what  will happen & to see bearing of scattered FACTS..-- What takes place in the formation of a bud-
356d070 on good expression but does not include so many FACTS as mine‖ The facts about half breed animals bei
356d071 but does not include so many facts as mine‖ The FACTS about half breed animals being wilder than par
356d071 elming) What comparative solutions & linking of FACTS-- Savages over whole world. (Major <I> Mitchell
358d076 ized inferior intellectually"= Opposed to these FACTS are effects of castration on males & of age  or
361d096 oup young like some of the species-- ¿do these FACTS indicate that the change is effected through th
372d131 thought  wonderful. it is part of same class of FACTS that the skin grows over a wound.-- Does likene
377d139 feeling must urge them to these actions. «These FACTS may, be turned to ridicule, or may be thought d
379d152 407,  409, Quetelet papers are given, & I think FACTS there mentioned about proportion of sexes, at b
389d176 with  inter-breeding as told by Willis Q Proved FACTS relating to Generation One copulation may impre
403e023 been  so pronounced ex cathedrâ. let us look at FACTS. considering few domestic animals few. that hav
408e045 ldt bones at 7800 in Andes-- parallel & curious FACTS.-- The Himmalaya. case, bears on the vast chang
410e051 art affecting another.-- (I from looking at all FACTS as inducing towards law of transmutation, canno
426e106 with  2d species, than two perfect species; but FACTS of grouse, & pheasant, & hooded crow goes again
431e121 thinks  Ammonites would afford instance of such FACTS.-- Ask Phillips.-- The more I think, the more c
434e129 Lychnis  April 3d.-- Henslow tells me following FACTS: believes that «only» red Lychnis grows in <sou
435e134 organization. by physical agents». p. 466. Many FACTS given of high temperature at which fish &c  can
440e148 ences only, when growth stops».-- Spallanzani's FACTS in connection with buds.-- They differ from pos
441e151 gan is become fixed. & cannot vary.-- which all FACTS show to be absurd.-- As there are plants, in no
446e161 rn «French» expedition.-- rather the reverse of FACTS stated by Smith of Jordan Hill.-- May 27th.-- H
448e165 s sometimes seeds.-- ‖There are endless curious FACTS about every part of plant producing buds, so th
449e170 are different.-- now this is same, as Galapagos FACTS &c &c.-- & it shows the causes which give  same
461t025 malia, which <do> have not sack,-- Most curious FACTS & this paper deserves fresh study & whole order
461t041 maining same (if so) with those of Spain & such FACTS-- This unequal duration is exactly same as some
463t057 alter or assume some form late in youth,-- only FACTS can decide-- some peculiarities may be early im
464t063 Lund makes his wonderful discoveries= negative FACTS are valueless= monkeys= Owen has described a gr
473t07v does he understand English.-- Miguel to collect FACTS for me-- what? What does Blume say on alpine Fl
510q017 . accounting for fossils). & lastly the tracing FACTS to laws. without any attempt to know their natu
535m069 before heard, so very probably forgotten.» Such FACTS bear on such characters as Allen W. & Babington
538m079 on-- & thus eternal punishment explained. These FACTS showing what a train of though[t] action &c wil
538m081 t, of which I am capable-- I suspect from these FACTS that whole effort consists in keeping one  idea
541m090 ving & classifying distinct resemblances.-- The FACTS of half instincts. when two varieties are cross
543m101 lism, did reason about himself-- but not about, FACTS gained or gaining by senses.-- As sleep <is> on
548m114 1 in army on "Wheat." in Jersey.-- very curious FACTS about early production of foreign seeds.-- many
559m155 f quite stranger.-- or less so.-- When learning FACTS for induction. one is obliged carefully to sepa
567n016 nother bad, vindictive.-- or lied &c &c Are the FACTS (about communication of ideas, &c) of expressio
572n035 explanation--  can we suppose some essence. The FACTS about crossing races of dogs on their instincts
575n044 Australian  man, may be called instinctive: the FACTS of memory of roads long after once visited by h
584n072 tion, equally wonderful in young & old.-- These FACTS point out some essential difference, which clea
586n080 animals.-- almost identical with my theory-- no FACTS, & mingled with much hypothesis.-- see M.S. not
589n091 of  weak intellects easily fall into habits Get FACTS about instincts of mongrel dogs Do blubbering c
596nIBC er always to refer to the author if I use these FACTS p. 280. adduces provision of seeds for transpor
632j53r standard  of one his weak creations.-- All such FACTS are merely relations of one general law. the pl
633j54r must  be cautious.-- <some others>: study these FACTS read Lacépède on Cetacea & Geographical Distrib
641j29r
```

599o008 me separe de la matiere et du temps! voila les FACULTES, q'il possede seul sur la terre. J'ai trouve
189b074 e, where the {cerebral structure intellectual FACULTIES} most developed, as highest.-- A bee doubtles
300c198 etween reason & instinct very just, but these FACULTIES being viewed as replacing each other it is hi
544m103 ears clearly rest of the mind, with all other FACULTIES: «Vide page 110, by mistake.» N B. Everything
573m036 lost in astonishment at the artificer.--» Our FACULTIES are more fitted to recognize the wonderful st
586n082 incts, & so in some senses, is sight--CD [The FACULTIES bear so close a relation to the senses, that
589n092 reason to organs of locomotion-- or that our FACULTIES have been given us to exist, is clearly seen.
600o008 J'ai trouve son âme" &c-- -- Confesses these FACULTIES of soul, treating of infinites not definable.
616o039 because there are living bodies without these FACULTIES & indeed until we know what answer they would
627o053 resolved into some operation of intellectual FACULTIES-- Will Eugenius allow this moral obligation?
372d129 to lizard,-- healing of wound.-- reproductive FACULTY + in the separated part every element of the l
526m030 so,-- is not this free will,-- he improves the FACULTY according to usual method, but what urges him,
534m063 unknown, with simple exertion of intellectual FACULTY) if ants had at once made this leap it would h
541m092 arge numbers in their head must have this high FACULTY, yet not clever people. Aug. 21st. 38 When a d
545m108 follows other as in blindest memory-- also low FACULTY of understanding. Adam Smith (.D. Stewart life
564n004 cocked its head, & picked it-- Here then, that FACULTY, whether for position of axe of eyes, state of
566n011 ces:, is good instance of obtaining <that> «a» FACULTY in the form of a true instinct, which is a rea
575n046 manence of old heredetary ideas.-- being lower FACULTY than the acquirement of new ideas.-- Walter Sc
582n069 of performing it.-- p. 14. There is scarcely a FACULTY in man not met with in the lower animals.-- he
583n069 ies.-- p. 18. Animals possess strong imitative FACULTY: pure instinct is not imitative: imitations se
585n077 its way back.-- ?? this is not instinct, but a FACULTY, or sense-- "We know not how, stonge henge rai
585n077 t have some means of measuring cells, which is FACULTY, they use this faculty instinctively; watchmak
585n077 asuring cells, which is faculty, they use this FACULTY instinctively; watchmaker has faculty by his i
585n077 use this faculty instinctively; watchmaker has FACULTY by his instruments to make toothed wheel. he m
586n078 stinct, but his knowledge of that quarter,, is FACULTY, whether by sun, & heavens, or magnetic virtue
586n080 ter refer to to the heredetary part of it,-- & FACULTY (faculty «being» always heredetary helps this
586n080 to to the heredetary part of it,-- & faculty (FACULTY «being» always heredetary helps this confusion
586n081 like plants going to sleep.-- "A bird has the FACULTY of finding its way, which in certain species i
586n081 directed to certain quarter"-- "An animal has FACULTY of walking. which in man is learnt by experien
586n081 vely" So with <sight> sight-- so a Bee has the FACULTY of building «regular» cells-- [but this facult
586n082 aculty of building «regular» cells-- [but this FACULTY «may possibly be» «probably is» instinctive, n
586n082 s in concert with others in building comb-- My FACULTY often will turn out to be instincts, & so in s
588n090 ons, because they have memory.-- what use this FACULTY if not reason.-- or does this reasoning apply
588n090 to itself, how gained. reason? or some unnamed FACULTY-- Lamarck. Philosop. Zoolog. «p. 284. Vol. II»
590n094 ble. one has good ground to call imagination a FACULTY, a power, quite distinct from self. «or will»
600o008 es not definable.-- Has little Chapter on each FACULTY of Soul.-- (I) <Conscience> «Moral Sentiments»
617o040 ed) being a phenomenon apprehended by the same FACULTY with matter & being necessarily exhibited in &
618o041 e bodily frame, they could be perceived by the FACULTY by which the brain is perceived but they are k
625o052 that Sir. J. & others think there is distinct FACULTY, of conscience.-- I believe that certain feeli
620o044 m good dinner, or from a blow struck in passion FADES away, so that when man afterwards thinks why wa
539m084 r «Rage» in worst form is described by Spenser (FAERY Queene. CD 25 (Descript of Queen) «O» of Hell C
113a090 being -8 Reaumur.-- there ought to be 32 degree FAH. at a greater depth than 400. & the limit being
220b197 & therefore the one distinction of species would FAIL. But this applies only to coition & not product
315c240 on[ly] when creator sees. the means of transport FAIL.-- otherwise no relation between means of Trans
335d016 stem, the organs of generation would necessarily FAIL.-- In last page. I should have said, "an animal
390d178 s not object of generation.-- therefore passions FAIL.-- In fruit trees no doubt there is tendency to
393dIBC ogs renders subsequent progeny faulty. Does male FAIL in passion.-- Disposition of half bred Cattle a
491q01v hese that male organs (not being always useful). FAIL-- Really good subject for experiment.--«to repe
547m113 the ideas, direct effect of perception by senses FAIL first, as whether I had pulled the bell??)-- It
580n060 igin of language My father says old people first FAIL in ideas of time, & perhaps of space-- in latte
604o017 e drift of Book.-- Sympathy & affections chiefly FAIL.-- Notices. struggle <between> when insanity is
239cIFC ybrids produced have any close species ever yet FAILED. About trades affecting form of man. Could you
390d179 o acquire differences: if none are added object FAILED, & then by that correlation of structure desire
505q014 to breed from it and large Asparagus: result? = FAILED to germinate 16 Will plant some of the Thyme w
525m028 ffections very soon go in Maniacs» seem to have FAILED even more than the memory.-- therefore affecti
552m131 & body, & especially, when to whole body, being FAILED, & not to any particular muscle Sept. 8th. 1 a
391d174 essive improvement or deteriorate» that object FAILING, generation fails.-- How completely circumstan
501q011 yme doubts about stigma in similar manner ever FAILING.-- answered by Gaertner (28) Can any annual or
524m021 ous) as in delirium after epilepsy, but in the FAILING from old age, they constantly do.-- In Mrs P.
557m149 tract female, (or object of attachment) & then FAILING to drive away rival.-- Fear is open mouthed to
171b002 s not come into play)--See Zoonomia arguements, FAILS in hybrids where every thing else is perfect; m
390d178 te the whole difference of parent, tree, but it FAILS.-- therefore «each» seedling of one apple ought
391d179 , & then by that corelation of structure desire FAILS. Every individual except by incestuous marriage
391d174 or deteriorate» that object failing, generation FAILS.-- How completely circumstances «alone» make ch
492q002 -- loosing one sex & not other: which generally FAILS first?-- Mal[e] 10. Henslow says semi-doubl flo
501q011 ther region-- (27) Which sex in Mules generally FAILS-- perhaps indexed by secondary characters-- in
524m022 left» hand to pretend to feel her pulse.-- What FAILS first?-- How is this?-- Does memory bring in ol
599o005 culations are utterly valueless-- then argument FAILS, or rather is weak.-- [RHC] Better simply put i
624o051 siest become instinctive, this part of argument FAILS, or rather is weak.-- [RHC] Better simply put i
522m012 disturbed.-- These foregoing cases of «mental» FAILURE very general effect of «early» «slight habitua
316c244 ce between the mind of man & animals.-- yet how FAINT in a Fuegian or Australian! why not gradation.-
578n056 ferent in different ones?) in the most perfect FAINTING, sphincters are loosed is a convulsive action
596nIBC rier pidgeon carried & turned round & round in FAINTING state would it then know its direction.-- In
155g062 th 2d & 3d can be traced some way up, but most FAINTLY on East side of Glen Turrit, where I believe t
383d160 o» «three, I believe instead of» two lines.-- «FAINTLY edged with reddish brown» black marks on tail
602o011 e "habitual reason".-- This power of the mind, FAINTLY approaches to instinct How strange it, that Na
303c205 had gone, without creation this would have been FAIR, but to place all that ever lived <on> into one
266c091 the Peninsula, are, generally speaking, a much FAIRER race than the Hindu's, in the same tracts;. &
616o039 ordinary language that the onus probandi might FAIRLY be laid with those who would support the propr
466t093 curve upwards, so that anthers & stigma lie in FAIRWAY to nectary.-- Is not this so in Kidney Bean. H
257c057 s tres-analogues,-- quand ce nétaient pas tout à FAIT les mêmes." This good case. of replacement unde
263c075 then he will choose & firmly believe in his new FAITH of the lesser of the difficulties Once grant th
292c170 & he must be a zealous man in the cause if his FAITH is not staggered <">= I confess. no dissertatio
553m136 d by culture « <was» who feel the most implicit FAITH that through the goodness of God knowledge has
482z011 22 palmipedes: out of the first 9.:4 raptores. FALCO poliosoma – novozelandiae -- histrionicus Vult
296c185 me Vol II. Magazine of Zoology p. 56. Peregrine FALCON holds birds for some time alive ¿therefore oth
458t001 ith my own.-- E Bengal Journal Vol 7. p. 658-- FALCONER on Sub. Him. fossils-- Ruminants. & Tortoises
508q017 in in young Rhinoceros or Whale, than in old?? FALCONER says all in cases. Owen. Have talked partiall
235b272 Fauna of Van Diemens Land & Australia [blank] FALCONER'S remarks on influence of climates, situations
324c268 mellie. Philosophy of Zoology Flemming. ditto FALCONERS remarks on the influence of climate White's r
052r098 case in excess as first case. When discussing FALKLAND soundings introduce this discussion.--Brazil
054r102 ts: All St. Catherine & coast Granite: P. 199; FALKLAND account of cleavage differs wonderfully from
066r142 n insisted strongly. that taking up a piece of FALKLAND Sandstone. he could not distinguish from ston
067r144 uary 20th. 1834 Mr Sowerby. younger. says that FALKLAND fossils decidedly belong to old Silurian syst
100a052 it does not explain CLEAVAGE of rock-- nor the FALKLAND case, nor. the arrangement of particles of gr
101a058 s p 289.-- Alludes to big bones in interior at FALKLAND Isd.-- Peron does as if well attested.-- Ther
104a066 80 per cent of Albite 80/100 X 6/100 = 480 In FALKLAND islands. & generally where rock metamorphic &
115a097 ed compare them with my rocks. when writing on FALKLAND Islds p. 94. Von Buch's Travels account of No
172b007 es.--» Now Galapagos Tortoises, Mocking birds; FALKLAND Fox-- Chiloe, fox,-- Inglish & Irish Hare.--
178b031 om the continent look over. Bell, & L. Jenyns. FALKLAND rabbit may perhaps be instance of domesticate
200b122 or instance Australian dog: <yet when that> or FALKLAND rabbit.-- There is only two ways of proving t
214b173 tsons St. Helena. -- Galapagos--Juan Fernandez FALKLAND Islds-- Kerguelen land.-- Phillips. Lardner E
225b219 of Didelphis in Stonefield.. all lands united-- FALKLAND Fox. ice). -- Mauritius what a difficulty-- wh
230b240 ence extreme difficulty, argument in circle.-- FALKLAND Isd case good one of animals not soon being s
245c024 Coquille Voyage Durville has written Flora of FALKLAND Islds, where is it? All the Society isles hav
288c160 ations lost?? I conceive a bird Migrating from FALKLAND Isd regularly to main land, proof of. land ha
291c168 dy made.-- & so destroy individuals, wheras in FALKLAND Isd they would change & make new species.-- a
291c168 w species.-- alpine species being destroyed at FALKLAND Isds++.-- Mem Lyell hypothesis of change in S

```
436e137  o not however, extend round world.-- Quartz of FALKLAND.-- Old Red Sandstone-- Van Diemen's land.-- P
479z005  age round world no land animal besides Wolf at FALKLAND ∴ black rabbits not indigenous p 112 M Lesson
481z009  Iceland. Mackenzie. p 345 for comparison with FALKLAND. good also for Journal.-- 18 Admirable engrav
482z011  huppe courte talks of nine terrestrial Turdus FALKLANDII & then 9. passeres! Says the thrush & anothe
065r140  idence previously. Mem. pebbles of Porphyry.--FALKLANDS.--off East Coast. -- Capt. Cook found soundin
134a141  estern Coast Vol II p. 277. on whale bones in FALKLANDS Some of the Tosca nodules at Bahia Blanca Mr.
184b054  ritten on same changes in Fish Mem. Rabbit of FALKLANDS described by Q. & G. as new Species. Cuvier e
244c021  n place with quadrupeds» p. 118. wild pigs of FALKLANDS, generally "red of brick" hair, very stiff, p
277c126  value.-- as in Opetiorhyncus. fulginosus. (a) FALKLANDS (b). F. del Fuego differ from (C) Chiloe  (E)
292c169  have  no descendants on earth.-- +++. Even at FALKLANDS some probably would stand change better  than
411e056  ratus.-- My account of Circus cinereus of the FALKLANDS Isld. is interesting as showing some change i
421e091  de of Africa.-- (& Juan Fernandez to Chile??) FALKLANDS to southern portion.-- ¶.do p. 269. Annals of
478z003  597  Many descriptions about lower animals of FALKLANDS &c &c Bennett on Chinchillidae Zoolog Transac
479z005  e wide limits of Nullipora Discussion good on FALKLANDS birds Discussion of Firola,-- Salpa Anatifs w
482z011  ntioned) p. 205. only 9. Terrestrial birds at FALKLANDS Isd 8 waders. 22 palmipedes: out of the first
502q11v  d, as Cuculus lucidus is.-- Ask Sulivan about FALKLANDS Isds.-- Snipe Migratory-- probably united by
042r067  riaceous rocks of R Chupat. & fall of Ashes of FALKNER, ¿how far is the distance?-- Fossil bones blac
240c013  d.-- These facts all account for [not located] FALKNER Patagonia no description of wild animals,  nor
320c276  said to have Mackenzie's Iceland Molinas Chile FALKNERS Patagonia Azaras Voyages & Quadrupeds of Para
032r040  ould not reach. If now the ocean should suddenly FALL, (3) the case would be as at first. & according
042r067  mention black scoriaceous rocks of R Chupat. & FALL of Ashes of Falkner, ¿how far is the distance?--
046r080  ll (Philos: Transacts) «seems to» considers that FALL first movement (as in Peru 1746).--At great Lis
047r082  on level before the higher part. -- Does the sea FALL on banks as a Bore wave rushes up? (NB. Earthqu
047r082  t Concepcion? Read the various accounts & see if FALL is not the first very evident movement.--The sw
104a069  ce in Cordillera discussion, deep sea, fragments FALL off cliffs. but then how spread abroad?-- There
128a128  en subsidence takes place.-- Mountain will first FALL-- the problem will be falling of an arch weight
130a133  ld be the chance in sounding over a continent to FALL across a hot.--spring.-- Hot water would not li
157g074  o cross stony parts; appearance chiefly cause by FALL of angular masses from above on soft shelf-- 29
362d099  ir skin (not being wanted for war) & hence never FALL off.¶ Curious the rapidity of the change in 5 o
397e003  pirit of philosophy to believe that no stone can FALL, or plant rise, without the immediate agency of
407e042  will be highly necessary to show that if species FALL, genera must. Lesson I remember says Mariana De
453e182  aribs.-- Vol II p. 650. Long attested account of FALL of fish in India.-- Windsor Earl-- Eastern Seas
529m042  dea of a son of Dr. Prietly who was cured from a FALL of ideotcy.-- The story of the Corbets & big no
541m091  xamine frame of mind in following changes during FALL of sea.-- Is the effort greater if the idea  is
565n009  stened by emotion,-- why does emotion make tears FALL?? Lavater says derision lies in wrinkles  about
584n071  xample fitted for climbing, his arguments partly FALL, when a species is found which does not climb C
596nIBC  ion.?-- cry? Do people of weak intellects easily FALL into habits Get facts about instincts of mongre
596nIBC  ring children, if of. convulsive tendency easily FALL into convulsions A carrier pidgeon carried & tu
628o54v  s desire for virtue.-- [I expect there is some FALLACY here.-- at least point of «false» honour  will
405e032  ound to be inhabitant of S. America & as it is <FALLE> embedded with almost recent shells.-- shows th
030r037  lera, the dikes do not generally appear to have FALLEN into lines of faults I do not think so many fa
090a022  ffect of all the meteoric stone which must have FALLEN on the globe since the Cambrian system In Ures
090a023  on  this ratio I can count 90 stones which have FALLEN in the 50 years. .: 90 x 19 = 1710 ÷ 50 = 34 po
156g068  except  near head of valley fragments which had FALLEN before lake drained could be told from «some o
156g068  rained could be told from «some of» those since FALLEN.  «on the 3 shelves» Solid rock is much notched
202b131  rara. 10 «12» feet beneath surface forest trees FALLEN «kind well known, carbonized»--; clay fifty fe
541m090  idea, as effect of sea on coves when waters had FALLEN, as in my Glen Roy paper.-- this greatest ment
071r158  oids.--Boussinguault «(Lyell)» cracks mountains FALLING in.-- Earthquakes at Quito. tranquillity «at M
090a021  ic metals. ∴ earthy crust compared to those of FALLING stones.--¿ does this bear upon the sorting of
090a022  years-- 90 «showers of» stones are recorded as FALLING; many of these were not single, but are descri
117a102  fects of degradation, «no» tides, water always FALLING or at least not rising are there cliffs. Sir L
128a128  Mountain will first fall-- the problem will be FALLING of an arch weighted in its centre.-- Will not
156g063  nclination of river must constantly alter with FALLING sea & so corrode plain into terrace as regress
156g068  n water-scooped rock «only decay from fragment FALLING» of no particular hardness no wonder that  all
588n089  ys <child?> babies have an instinctive fear of FALLING.-- &p. 193. that they perceive the  difference
608o026  eds who does not recognize an accidental spark FALLING on prepared materials. From contingencies a ma
363d103  mpestris spotted white when a fawn compare with FALLOW? & Moschus &c & -- like young blackbirds
405e031  Octob. 25th. I observed in Windsor Park,-- the «FALLOW» Deer. which were of a nearly uniform  <dusky>
045r077  elation of a part of globe rising, when another FALLS.--When discussing connection of Pacifick & S. A
091a027  d. N. P. J. p. 105. Oct. 1828. gneiss in India (FALLS of Garsipa) dip 30 degree. <strike> <direction<
263c076  if  this be granted!!) & whole fabric totters & FALLS.-- look abroad, study gradation. study unity of
263c077  tudy relation of fossil with recent. the fabric FALLS! But Man-- -- wonderful Man. "divino ore versus
240c013  Archipelago-- Former statements to such effects FALSE In New Guinea. a Kangaroo d'Aroe (Didelphis Bru
262c073  om of nature, where scheme not filled up, (most FALSE to say no passages; nature is full off  them.--
278c128  es.-- it will however be much <shu> surer, when FALSE species banished by this test.-- Excepting wher
286c155  g as emblem of dichotomous arrangement which is FALSE There is same difficulty in arranging animals i
336d019  eneration -- highest point of organization] CD» FALSE.-- The creator would thus contradict his own la
426e107  rwise & make on[ly] true hybrids.-- but this is FALSE, [give instance of series from wild animals & p
439e143  <i>nfertility of hybrids «with parent species» FALSE, which makes it determined by a facility in ret
443e156  gued to myself, till I can honestly reject such FALSE reasoning Bell Bridgewater's Treatise on the Ha
509q017  pteryx ‖‖ no as Os Coccygis-- Turbinated bones? FALSE ribs Wings of Apterix: clavicle in--? Combs  in
567n013  an talk, so do many animals.-- analogy probably FALSE, may lead to something.-- October. 8th. Jenny w
583n069  bility of reasoning power in one species man.-- FALSE instinctive pointing varies.-- p. 18. Animals p
589n092  s of voice than respiration.-- like sigh before FALSE sneeze.-- "A Dissertation on the Influence of t
616o038  e in plants is involuntary, in man voluntary: ¿ FALSE,-- secretion in both involuntary,  <application
628o54v  here is some fallacy here.-- at least point of «FALSE» honour will stop all wish to gratify <it>-- an
061r127  or such mere points; then any mountain, one is FALSELY less surprised at new creation for large.--Aus
286c156  in  one plane Fleming Quarterly review says nat: FAM: of Wolfius contains many Linnaean genera.-- Now
580n062  h life" p. 200.-- "The desire of glory, immortal FAME, &c so common in the young are symptoms of  the
628o054  e artificial, as avarice love of gold.-- love of FAME-- Yes Hartley explains this & Mackintosh shews
033r042  cast which, although not very intelligble is a FAMILIAR case: If refiltered with other matter how ver
106a074  «not» be great deflection in a "rapid".-- is a FAMILIAR illustration.-- Therefore stream has no tende
323c269  Bartrams Travels in N America May 18th Stanley FAMILIAR History of Birds -- Mackintoshs' Ethical Phil
436e138  -- Van Diemen's land.-- Porphyries of Andes. A FAMILIAR History of Birds by the Rev. E. Stanley Vol I
437e139  then  dropped it & was found alive.-- Stanleys FAMILIAR History of Birds several cases on record of s
618o041  pendent of each other. A person might be quite FAMILIAR with thought & yet be ignorant of the existen
190b077  ia great abundance of species of few genera or FAMILIES.-- (long separated.-- Proteaceae & other form
206b147  essors. at present day. in looking at two fine FAMILIES one with successors «for» centuries, the othe
206b148  ther, and are acted on, just like the two fine FAMILIES «no doubt a different set of causes must  act
235b273  any confined to certain countries-- so on with FAMILIES.-- Ask Royle about Indian Cattle with humps.-
268c100  uayaquil & Peru» Henslow in talking of so many FAMILIES on Keeling seemed to consider it owing to  one
272c109  evails amongst the same species & subgenera in FAMILIES.-- thus the banded tarsi is common to all the
273c112  these  aerial representatives of the different FAMILIES.-- that sexes <are> <have» same plumage.-- <n
278c128  olely by comparison with other genera in other FAMILIES.-- it will however be much <shu> surer,  when
282c141  ing time such changes had elapsed.-- let these FAMILIES take <dogs> «domestic animals» with them. the
302c201  ome source anterior to giving off of these two FAMILIES, but we see analogies between fish.-- Birds s
304c206  t affect particular organs.-- of two adjoining FAMILIES & not all organs blending away.-- +++ Hopeles
311c227  ieved in representation. certain birds in many FAMILIES, «+very often in number 5» will have long tai
349d051  veed «un»till <in> «we ascend to» subgenera & FAMILIES, «even in Cetionidae» «in the  Cetoniadae»,--
402e020  presents  us, for the most part, with the same FAMILIES and genera, that are natives of S. Asia,  but
417e078  from  either: & or like progenitors.-- in some FAMILIES all the children like mother & in some like f
497q007  Gaertner  talks of the several great & natural FAMILIES, as being difficult to cross. (5) It is  most
525m026  Insanity & Epilepsy remain many generations in FAMILIES.-- My fathers does not know whether trains of
537m077  ndoubted cases of heredetary pride & in single FAMILIES. Edinburgh. Phil. Transact. p. 365. Case of d
181b040  same manner, if we take «a man from.» any large FAMILY of 12 brothers & sisters «in a state which doe
185b056  house says he is certain, that in insects, each FAMILY, however many there may be, represent every ot
185b057  ccurs with regards to other tribes in that same FAMILY.-- (NB I see Waterhouse thinks Quinary only th
186b058  era, each retains some one character of all its FAMILY; but why so? I can see no reason for these. an
```

202b134) then if it is doomed that only one species of FAMILY has offspring the chance is that these wandere
261c069 tent of all species Accumulate instances of one FAMILY sending out structures into many genera.-- lik
262c073 webbed. &c &c.--)-- & in round of chances every FAMILY will have some aberrant groups.-- but as for n
265c085 abits, or heredetary is to see, whether a large FAMILY has it, & one member of that family, having it
265c086 her a large family has it, & one member of that FAMILY, having it with very different habits-- Thus b
265c088 a rufiventris, is instance of bird belonging to FAMILY with peculiar coloured plumage, where colours
273c110 in Old-World--. + + If in any «well developed» FAMILY (Gould says) there is any marked colouring of
273c111 by errors.-- It is most wonderful how in every FAMILY of bird,, even the most <m> strongly marked, t
273c112 umming bird, which is one instance of its whole FAMILY where female is not dull.-- I must observe tha
276c124 ectly some habit, which the whole rest of other FAMILY practise with a peculiar structure, then Milvu
281c138 sits such as they are-- My theory explains that FAMILY likeness, which as in absolute human family is
281c138 hat family likeness, which as in absolute human FAMILY is undescribable, yet holds good, so does it i
281c139 fourth degree.-- a species must be compared to FAMILY entirely separated from any degree; the tailor
282c140 ilar limits of change, would be same-- Yet each FAMILY might have its own character,-- we here suppos
282c142 alogical in a genus with respect to rest of its FAMILY as in ground cuckoos, is affinity with respect
283c143 have representations.-- The aerial type in each FAMILY is relation to elements & not habits as shown
286c156 stant number of stamens.-- in order, or in next FAMILY? In considering fossil animals, what relation
301c200 thousands New insects, perhaps scarcely one new FAMILY & no new orders,-- Wonderful, partly explained
302c201 ltipliied become affinity yet often retaining a FAMILY likeness, & this I believe the case. = any ani
302c202 fied = = this is not easily told, for any small FAMILY. having analogous characters, might be multipl
335d014 thr> Six fingered people, <Hill> «Lord Berwick» FAMILY with defective palates. heredetary & therefore
358d075 y one of these birds" Case of bird of different FAMILY. having very same habits in some respects as t
471tf08 says "indeed it wd be difficult to point out a FAMILY so completely natural & one whose groups pass
507q016 !!! very good Any peculiarity in the males of a FAMILY-- Where one tooth aborts, do you know whether
525m025 fect roof to the mouth «stammering in my Father FAMILY» (as in Lord Berwick's family) are heredetary.
525m025 ring in my Father family» (as in Lord Berwick's FAMILY) are heredetary.-- other deformities are illne
525m026 er trains of insanity as heredetary in any one FAMILY.-- In Aunt-- B. the affections «& N B affectio
537m077 r being engrafted.-- No one doubts patriotism & FAMILY pride are heredetary, & therefore he has thes
550m126 his own drawing compared to a friend, whose who FAMILY can draw-- says friend viewed him as Newfoundl
052r099 fact absent in T. del Fuego, excepting in Port FAMINE Mr Sorrell says that numerous icebergs are com
228b232 brethren in pain, disease death & suffering «& FAMINE»; our slaves in the most laborious work, our c
375d134 vented soley by positive checks, excepting that FAMINE may stop desire.--» in Nature production does
375d134 st no checks prevail, but the positive check of FAMINE & consequently death. population in increase
397e003 ocated} Epidemics-- seem intimately related to FAMINES., yet very inexplicable.-- do p. 529. "It acco
039r060 give as an example the great subsidence at the FAMOUS eruption of Rialeja, & the more true analogy f
363d101 eak & in Muscovy duck» crested feather, pouters, FAN tails are found in any colours of plumage &c &c
469t151 - Early Magazine-- &c. double-blossomed «& dwarf-FAN Bean» bean, were planted in rows, & seeds gather
526m029 onsists in a set of sketches. some real-- some FANCIED.-- this fact of early memory consisting of thi
610o031 -- could not <remember> «read» Christian name; FANCIED it looked like. W. but concluded it could not
364d103 such case they exchange birds with some other FANCIER, thus gettting fresh blood, without fresh feath
342d034 ese Hybrids new species. Yarrell says the bird FANCIERS say the throw of any two species crossed is u
364d103 re for themselves. | + + first year.-- The bird FANCIERS match their birds to see which will sing long
393d1BC ch cross & are fertile heterogenous? When bird FANCIERS say the throw of two varieties is uncertain d
552m130 iate comparison with perceptions, & that on[e] FANCIES the image more vivid? Surely the image in a dr
048r085 To the Megatherium.--To the Horse. = One might FANCY that it was so arranged from the forseight of t
177b027 simple organization.-- birds-- not. {P} We may FANCY, according to shortness of life of species that
241c014 om that of the Marianna islands & at Amboina» I FANCY there is marked wild breed of oxen at Java.--.p
401e016 e is tempted to think here some anomaly-- I can FANCY cowslip producing primrose return to old stock,
526m027 herefore properly no free will.-- we may easily FANCY there is, as we fancy there is such a thing as
526m027 ree will.-- we may easily fancy there is, as we FANCY there is such a thing as chance.-- chance gover
536m072 ed by habits). now free will of oyster, one can FANCY to be direct effect of organization, by the cap
641j29v ll bear on geology-- There is an analogy between FANG of snake, (jaw of spider?) sting of bee, sting
532m053 when the noise comes it cannot help doing it.-- FANNY Hensleigh doubts whether young babies start.--
283c144 ow bird. replaced Birds of Paradise-- if such FANTASTIC «sexual» ornaments. have so intimate a relati
471tf07 orms which the thorax & head displays.-- most FANTASTIC & use unknown.-- "<when we find such an endle
568n017 s will be for general good, or in case of any FANTASTIC custom» «Probably bashfulness is connected wi
363d103 wild off coast of America.-- showing hybrids can FARE for themselves. | + + first year.-- The bird fan
309c221 detary proceeding from an accident. New England FARMER,-- useful could not leap fences:-- Dr Lang on
081r181 Capt. Ross. & Scoresby deep soundings Gilbert FARQUHAR Mathison travels Brazil. Peru. Sandwich Isd M
569n022 es epaules et dit qu'il valait mieux rester a FARROILAP quelque mal qu'on y fût."-- Expression common
538m080 asonable man.-- If one could remember all ones FARTHERS actions, as one does those in second childhoo
526m027 g as chance.-- chance governs the descent of a FARTHING, free will determines our throwing it up.-- e
571n029 d during life time:-- the latter correspond to FASHIONS in ideal taste & the former to true taste.--
545m106 assion, ie. nervous impulse to action is sent so FAST to limbs that they cannot remain still.-- I do
594n115 go to the assistance & bite a big dog. which was FAST struggling with another large dog his companion
469t151 e wet ones/ & wd go on longer-- Woodfords Marrow FAT, Early frame, Groom's Dwarf. planted in rows «cl
525m026 ng Nina with her puppy.-- The common remark that FAT men are goodnatured, & vice versa Walter Scotts
525m026 a Walter Scotts remark how odious an illtempered FAT man looks, shows same connection between organiz
532m056 then immediately fell into her old ways & became FAT! What remarkable affection to a place.-- How lik
281c139 broken circles.-- which in each group is quite FATAL.-- {Relations of analogy being those last obtai
415c067 hers remark about the Bladder.-- The numbers of FATAL diseases in mankind, «the more valuable domesti
415c067 reating up of every heredetary tendency towards FATAL diseases, & such constitutions only being clear
415c067 & such constitutions only being cleared off by FATAL diseases.-- The Value of a group does not depen
178b032 rtake more of the mother, the later ones of the FATHER; is not this owing to each copulation producin
192b087 ve one Marsupial animal in Stonefied state, the FATHER of all Mammalia in ages long gone past.. & sti
192b089 be expected-- This answers Cuvier-- Perhaps the FATHER of Mammalia as Heterodox as ornithorhyncus. If
193b090 ght not new classes be brought into play.-- The FATHER being climatized, climatizes the child. ¿-- wh
222b203 ing differences of expression -- one child like FATHER another like mother Has Lowe written any other
222b206 sent day & not include Mammalian remains.-- The FATHER of all insects gives same argument as father o
222b206 he Father of all insects gives same argument as FATHER of Mammalia; but have improvement in system of
228b231 -- respect We have no more reason to expect the FATHER of man kind. than Macrauchenia yet he may be f
240c003 e & pig being half way. Yet dogs sometimes like FATHER, sometimes like mother. The fact of great mons
285c150 same parents in a circle,-- & «hermaphrodites» FATHER « <mother> » & grandfather <Might> Must be int
285c150 her <Might> Must be introduced & made young.--| FATHER must be left out of case, that difference occu
293c172 CD If we saw a child do some action-- which its FATHER. had done habitually we should exclaim it was
302c203 bmersion of land containing all of intermediate FATHER-species, & not, therefore, solely owing to suc
302c203 & not, therefore, solely owing to such interm: FATHER-species, being little adapted to some physical
302c203 be destroyed, & N & S. existing species becomes FATHER of genera-- whatever the cause is. <the> any o
303c204 antity & kind of hair forms of legs-- hence the FATHER of man kind probably possessed a structure in
303c204 rs in present races, & form of feet.= (Negro or FATHER of negro probably was first black at base of n
308c217 e all» Brothers in spirit-- all children of one FATHER.-- yet differences carried a long way. --Case
312c233 same litter, those. of different litters or of FATHER & child are thought long breeding in. Must not
332d001 <trifling> «unknown» causes act upon people. My FATHER mention, than for ten years he never saw one c
332d003 breed-- He believes all pretty much alike.-- My FATHER Water-in the hair a century since used to be c
332d003 ch more latley diseased Mesenteric glands.-- My FATHER has seen case of pleurisy, broken limb «in chi
346d047 ved same answer.-- Thought lambs were more like FATHER than Mother.-- The cross not so hardy as Black
352d058 . When <species of> a group of species is made. FATHER probably will be dead-- hence there is no cent
360d090 ted)-- the one in garden is from <bitch dog do> FATHER dog. & hence general appearance of face & tail
360d090 much as in Jackall.-- In case where Jackall was FATHER resemblance much nearer to Jackall.-- This Kee
379d151 vitism. but then ¿ was «not» the expression of <FATHER> Sir W. itself received from his father so tha
379d151 ion of <father> Sir W. itself received from his FATHER so that case ceases to be true avitism Annals
385d165 g <resemblance> similarity to the temper of the FATHER, or of the mother, or sometimes of both." If L
385d165 stitution, & most of the moral qualities of the FATHER! In how many daughters does the character of
385d165 he character of the mother in the son, & of the FATHER in the daughters! This last remark good. becau
388d171 n from being very near mother, & some very near FATHER.-- now if one of these staid in the womb, when
388d171 to this view more semen to one child. more like FATHER.-- stuff.!-- How much opposed. the Quagga case
389d176 olonged till it has bred.-- Offspring like both FATHER & mother, or very close to either.-- Male & fe
391d179 except by incestuous marriage has acquired from FATHER some differences. V. Supra <v. infra> p 179, c

```
392d180 live? In Muscovy ducks do young take most after FATHER or Mother according as they are crossed? & How
392d180 (& Australian) dogs with common dogs--> Ask my FATHER to look out for instances of Avitism Examine E
417e078 w these laws are, that child may be either like FATHER or mother, independently of its sex, or half w
417e078 ies all the children like mother & in some like FATHER «What is cause of this.-- » (Lord Moretons law
432e123 boars &c &c &c» p. 10 offspring take more after FATHER than mother; illustrated by the crossing of ho
455eIBC .-- Make Hybrid mosses.-- Leighton or some one. FATHER-- diseases common to men & animals.--:likeness
470t176 pollen gatherers & they seem slow= Maer 1840 My FATHER formerly planted «Turkey or» Palmated and Engl
470t176 ds of each other actually produced hybrids-- My FATHER remembered when in the gardens, he knew there
489qIFC .-- Shrewsbury p. 14 Henslow (2d time) p. 14.-- FATHER. And. Smith Dr. Holland p. 16 Babington-- Goul
494m004 ndian & Common) &c: length of life. (5) Does my FATHER know any case of quick or slow pulse being her
496q006 ail, row of fish & kill them in hour or two =My FATHER made hens cast Holly-seed & they grew» (9) Pla
498q007 nt of nutrition.-- Horned oranges so? --Yes, my FATHER lost this character in grt degree from charcoa
507q016 of the Fuller's plants ,Teazle Dr. Holland ; My FATHER. Andrew Smith (1) Are cross-births, or other a
508q016 ifickness of female, relation to healthiness? & FATHER answered (5) About cross-bred races of men tak
520m001 -- 1838 Selected Dec 16 1856 July 15th 1838 My FATHER says he thinks bodily complaints «& mental dis
520m001 .-- thus the late Colonel Leigton resembled his FATHER in body, but his mother in bodily & mental dis
520m001 his mother in bodily & mental disposition.-- My FATHER has seen innumerable cases of people taking af
520m001 ld not help thinking, he was prescribing to his FATHER & old Mrs Harrison, said, although constantly
520m002 .-- Mr Dryden Co said he could not remember his FATHER.-- My father thinks. people of weak minds, bel
520m002 Co said he could not remember his father.-- My FATHER thinks. people of weak minds, below par in int
520m002 deficient, he was nearly 9 years old. when his FATHER died.-- The omnipotence of habit is shown abou
521m007 story of hunting «-- habitual fits.--» which my FATHER thinks is mentioned in the Zoonomia.-- Now if
521m009 ich he exclaimed, why it is dinner time.-- » My FATHER asked him whether he had gardener of name A. B
522m010 ed.-- Answered never heard of such a man.-- (My FATHER explained who he wa & all about him, but still
523m015 nts in his pocket, & studying mathematics.-- My FATHER says after insanity is over people often think
523m016 was «only» NEARLY thrown into a hospital.-- My FATHER was nearly drowned at High Ercall, the thought
523m018 passion & madness.-- ira furor brevis est.-- My FATHER quite believe my grand F doctrine is true, tha
524m021 Dreams do not go back to childhood-- People, my FATHER says, do not dream of what they think of most.
524m022 sband was dead) yet instantly perceived when my FATHER to distract her attention took her «left» hand
525m024 probable in Squib's case any direct fear.-- My FATHER thinks that selfishness, pride & kind of folly
525m025 ly like (Mr George S.) is very heredetary.-- My FATHER says on authority of Mr Wynne that bitch's off
525m025 had kittens before & afterwards with tails. My FATHER says, perfect deformity, as an extra number of
525m025 r imperfect roof to the mouth «stammering in my FATHER family» (as in Lord Berwick's family) are here
529m042 e scenery would rise. Deer in Parks ditto.-- My FATHER says there is case on record he believes in Ph
530m043 or a day or two are absoluteley forgotten.-- My FATHER signed a bond, yet when he paid the Attorneys
530m044 name Corbet, perhaps nonsense.-- look to it My FATHER has somewhere heard (Hunter?) that pulse of ne
531m050 t curious instance of reason & abstinence.-- My FATHER remarks that things of great importance are ea
532m055 in the involuntary mind which is startled.-- My FATHER says he should think that in old people, in th
538m080 be more different, & yet they would make one's FATHER & self one person-- & thus eternal punishment
539m083 of heredetary mind. I a Darwin & take after my FATHER in heraldic principle. & Eras a Wedgwood in ma
541m090 » for the moments with interest on each.-- ∴ my FATHER. is right in saying delirium rest-- therefore
543m100 iterer in Bond St. was so great a fool that his FATHER only left him a guinea a week. yet. he was ini
559m156 ally close at that time after long period.-- My FATHER about double consciousness.-- & somnambulism.
560mIBC ? Charles Darwin 36 Great Marlborough St. Has my FATHER ever known <intemperance> «disease» in grandch
560mIBC wn <intemperance> «disease» in grandchild, when FATHER has not had it. but where grandfather was the
563m001 t springing from nest & leaving no tracks.-- My FATHER says pea-hens do Wood pidgeons building near h
573m039 n when wishing not to flow-- flow it will.-- My FATHER told Miss. C. of the bad conduct of Mrs C. (he
577m053 en cures of tooth ache before being drawn,-- My FATHER «even» believes that the general talking about
580m060 a name, with reference to origin of language My FATHER says old people first fail in ideas of time, &
581m063 operly, even when you cannot remember it. as my FATHER trying to remember the men's Christian name, w
610m031 f instinct amongst animals.-- Jan 13th. 1839 My FATHER received a letter from Mr Roberts-- «a person
610m032 ard. unconscious memory.-- Jan 14th. 1839.-- My FATHER says he has heard of many cases of ideots know
367d113 e produced.-- Sept. 17th. Saw mule. apparently FATHERED by a donkey. with all four legs ringed with b
181b041 latives, so that by tracing back. the «descen> FATHERS would be reduced to small percentage.-- & <in>
291c167 in genera & classes. if extinct forms were all FATHERS of present, then there would be perfect series
300c197 S. American dung beetles will each become the FATHERS of many species.-- a few eggs transported to t
325c267 Geoffroy. St. Hilaires. 1832. contains all his FATHERS views.-- Quoted by Owen-- Hunter has written
388d171 when it came out. it might partake of shade of FATHERS character.-- according to this view more semen
415e067 hat parts of structure abortive.-- Remember my FATHERS remark about the Bladder.-- The numbers of fat
417e078 e one.-- The laws, therefore, of likenesses of FATHERS to children of mankind, no doubt are applicabl
524m022 execution.-- Widows not of their husbands-- My FATHER'S test of sincerity.-- People in old age. excee
525m026 epsy remain many generations in families.-- My FATHERS does not know whether trains of insanity are h
526m029 onsisting of things seen, quite agrees with my FATHERS case of Mr Corbet of the Hall understanding. (
528m035 not be discovered-- is closely analogous to my FATHERS positive statement that insanity is only cured
536m073 ion gave me to will-- Verily the faults of the FATHERS, corporeal & bodily are visited upon the child
538m078 principle of association.--«fully bears out my FATHERS doctrine about people forgetting their insanit
024r014 gradual shoaling of the water to more than 100 FATHOMS. proves the existence of some moving <point> p
024r015 s on the NW coast:-- 8 leagues, from Sydney 90 FATHOMS La Peyrouse. South of Mocha; 19 miles. 65 Fath
024r015 homs La Peyrouse. South of Mocha; 19 miles. 65 FATHOMS Vide facts in Beechey. on NW coast of America
024r015 n NW coast of America off Cape of Good Hope 70 FATHOMS 20 miles from the shore? Beagle Coast of Brazi
024r015 where not rivers in my Coral paper {T} leagues FATHOMS Parallel of St Catherine [27 degree 30 minute
025r019 om the examination of soundings, from about 80 FATHOMS & upwards. that life is exceedingly rare, at t
030r035 lar & <not very> rather deep soundings, 60-100 FATHOMS 2 & 3 miles from shore. V. Chart) Every winter
057r114 g worms in the mud which he drew up from 1,000 F[ATHOMS], & the temp of which was below freezing poin
066r140 uego. off. Christmas sound. -- «(Think some 60 FATHOMS, none thicker than thumb» Sea weed said at Ker
066r140 sd. to grow on shoals like Fucus gigantous! 24 FATHOMS deep 24» under 50. Kerguelen Land, = the way i
113a092 bar at mouth excavated in solid rock.-- 4 & 5 FATHOMS deep. perfectly still water. Major Mitchell in
481z008 erip says that Voluta found in not less than 7 FATHOMS water. Mem Bahia Blanca. De la Beche theoretic
541m091 s so analogous to muscle in one position great FATIGUE-- may explain excessive labour of inventive t
544m103 ess, rapidity, novelty of separate ideas cause FATIGUE to the mind,-- it is solely the comparison, wi
547m114 or memory from innumerable late events.-- the FATIGUE of thinking is keeping up these trains,-- espe
544m103 the air, is more prolonged than dream. never FATIGUING,-- else it is only our consciousness, & sense
493q003 ee if relation is same good, avoids effects of FATNESS.-- Experiment in crossing animals.-- &c (1) To
148g026 OTHER!-- the cross not so hardy but more easily FATTEN, This man confirmed the account of the «YOUNG»
346d047 t so hardy as Black faced, but more tendency to FATTEN-- This man confirmed my account of the Shepher
536m071 of likes smell of,∴ Hyaena likes smell of that FATTY substance it scrapes off its bottom.-- it is re
023r011 be merely accidental apertures still open.--The FAULT like appearance «arising from the manner of hor
023r011 cient volcanic axis «of the Andes».-- «Has this FAULT determined side of volcanic activity.» That axi
051r097 st» Africa. Australia. profoundly deep: a great FAULT or rather many faults.-- Necessary form; as lon
100a052 tion. especially effect of gravity, versus some FAULT explaining vary dip & inclination.-- which last
549m120 t so much as they appear & perhaps partly their FAULT.-- Whether this rule of happiness agrees with t
023r010 g from «the expulsion of fluid nucleus through» FAULTS or fissures, produced by the elevations of tho
030r037 t generally appear to have fallen into lines of FAULTS I do not think so many faults in Cordillera, a
031r037 len into lines of faults I do not think so many FAULTS in Cordillera, as in English Coal field -- bec
031r037 raised--so on--but gradually & simply raised No FAULTS in Patagonia[,] enormous extent; if lowered ag
042r068 ia blanca P. 204 Vol III. Lyell Owing to «open» FAULTS in mountains: to elevated strata in eocene lak
049r087 of White. but rather to one out of a series of FAULTS. [Fig. 4] {P} In Cordill: should basal lavas b
051r097 . profoundly deep: a great fault or rather many FAULTS.-- Necessary form; as long as coast line fixed
111a087 Berbice N. 35 degree. E. dip to NW to 80 degree FAULTS with red wacke contorted evidently dike. V. VI
120a107 t-- the curvatures of the strata.¿ the enormous FAULTS & facility with which the earth is cracking by
536m073 time organization gave me to will-- Verily the FAULTS of the fathers, corporeal & bodily are visited
393dIBC tion with other dogs renders subsequent progeny FAULTY. Does male fail in passion.-- Disposition of h
398e005 in our train of geological reasoning, extremely FAULTY» The difficulty of multiplying effects & to <p
324c268 d. New Spain-- Much about Castes &c Richardson's FAUN. Borealis Entomological Magazine (paper on Geog
193b091 gazine paper on Geographical range Richardson-- FAUNA Borealis. It is important the possibility of so
227b224 vened.-- (2d) By character of any «two» ancient FAUNA, we may form some idea of <origin under> connec
233b249 e Coquilles Voyage), Waterhous remark Australia FAUNA so far. Indian all the rest. Timor according to
235b264 inary system according to three elements How is FAUNA of Van Diemens Land & Australia [blank] Falcone
```

261c069 of information-- instinct Swainson's remarks in FAUNA Borealis must be studied. There is capital tabl
311c225 an Diemen's land & Australia different Temminck FAUNA Japonica (?!) 82 mammalia 293 Phalangista of Au
402e019 of St. Peter & St. Pauls in Lat' 53 degree yet FAUNA like that 60 degree & 70 degree of Europe.-- Ma
402e020 .-- Kotzebues first Voyage. Vol II. p 367. "The FAUNA of the Sunda islands presents us, for the most
418e085 nstitut 1838. p. 412. M. Eichwald has published FAUNA of Caspian.-- fishes fresh water kinds. (yet li
419e085 in the salt?.)-- very few animals of any kind-- FAUNA, must be very curious.-- With respect to the mo
421e090 estrial mollusca of Morocco «Mr Forbes says the FAUNA»-- (near Oran) approach in character to Canary
449e170 - p. 18. of Temmincks. Preliminary discourse to FAUNA of Japan-- that the «animals of» islands N. of
514q21. plies to my geology & Species theory-- peculiar FAUNA?. (Australian Alps--; are any Europaean forms f
252c045 calities of isld.--«immense importance of local FAUNAS foundation of all our knowledge especially gre
278c132 hants &c in the most modern period, compared to FAUNAS of these countries, greter than Toxodon, Macra
640j167 which are arbitrary & empirical, from their own FAUNAS, which in this case is only true criterion.--
328cIBC birds Temminck has written "Coup d'oeil sur la FAUNE des iles de la sonde et de L'empire du Japon Wo
038r059 probably not so much aluminated. As argument in FAVOR of lines of anticlinal violence crossing lines
103a065 theory.-- general presence of dikes. argues in FAVOUR of pressure of liquid rock. Andes discussion--
131a136 n many parts of the world.-- argument strong in FAVOUR of thin crust theory.-- What a curious investi
227b228 examination of hybridity «to what circumstances FAVOUR crossing & what prevents it--» & generation, c
335d015 just like mules. Fox's half bred Persians «cat» FAVOUR the Persian side.-- Theory of abortive hybrids
589n091 sis.-- see M.S. notes, where strong argument in FAVOUR of brain forming the instincts,-- could brain
048r086 must not think alluvial plains «always» most FAVOURABLE; In what part of the globe are there such va
177b028 hich» is it an index of the point whence, two FAVOURABLE points of organization commenced branching.-
180b037 ish out, or on other hand like Orpheus. being FAVOURABLE many might be produced.-- This requires prin
297c185 ded into many small genera ·circumstances not FAVOURABLE to many species., same circumstances., which
404e026 favourable conditions to parent may be become FAVOURABLE to offspring; Australian dogs have mottled
407e037 does to North. America & Europe.-- S. America FAVOURABLE to Tropical productions. The world temporary
465t079 ods may elapse, even in situations apparently FAVOURABLE for the preservation of shells; where land b
564n003 xistence of such an attribute would be rather FAVOURABLE to it--!! Man moreover who reasons much on h
576n047 olute tendency to progression, excepting from FAVOURABLE circumstances.!» We must believe, that it req
636j56v al animals living on the borders of a country FAVOURABLE to change.-- It might be concluded that Plan
641j29r ess; (on the same principles that islands are FAVOURABLE,) because it must take so long to change spe
641j29r of animals. equatorial countries are supposed FAVOURABLE to terrestrial Mammifers-- Marine ones «of l
030r034 d on the open coast. Perhaps as at Concepcion. FAVOURED by basin formed by outlying rocks; (such as b
391d175 es well with my view of those «forms» slightly FAVOURED, getting the upper hand. <]CD> & forming spec
439e142 heless much probably depends on circumstances FAVOURING the reappearance of characters, formerly poss
363d103 s, the ++ Cervus Campestris spotted white when a FAWN compare with fallow? deer. & Moschus &c & -- li
380d154 animal surely is hermaphrodite-- (as is seen in <FE> plumage of hybrid birds) After animal has copul
560m156 woman present? Do they pout, or spit, or cry.-- <FE> Shame, independent of fear: colour of bare nail
220b198 zine Jan: 1830. most curious paper on heredetary FEAR (like rooks with guns) of the Bustards in Germa
228b231 other kind?» Animals with affections, imitation, FEAR <of death>. pain. sorrow for the dead.-- respec
294c176 es in wild animals-- but how to show species-- I FEAR argument must rest upon analogy & absence of va
300c197 is exceedingly doubtful whether animals have any FEAR of death, only of pain. of death acquired?. The
302c202 ect to variation. therefore lately acquired.-- I FEAR «great evil» from vast opposition in opinion on
523m016 Insanity is produced by moral causes (ideotcy by FEAR. Chile earth quakes). in people, who, probably
525m024 -- in these cases their actions do not look like FEAR, but shame.-- I cannot remember instances, but
525m024 ience.-- Not probable in Squib's case any direct FEAR.-- My father thinks that selfishness, pride & k
531m050 tance are easily forgotten, (if unconnected with FEAR &c) because people think that the importance of
531m052 start is HABITUAL movement to avoid any danger-- FEAR, shamming death, or running away. accompanied w
532m053 bies start.-- If children wink. it is instinct FEAR must be simple instinctive feeling: I have awak
532m054 of the heart are accompanied by much involuntary FEAR) In these cases probably the system is affected
532m057 like strong feelings of Man.-- The sensation of FEAR is accompanied by «troubled» beating of heart,
532m057 <this> «running away» have been usual effects of FEAR.-- the state of collapse may be imitation of de
540m089 ntaigne (Vol. I) has well observed, one does not FEAR death from its pain, but one only fears that pa
541m089 connected with death!-- How has this instinctive FEAR arisen? 19th. When I went down to Woollich I wa
544n101 ure) gives pleasure, ie. love.-- & so pain gives FEAR of death. Mayo Philosophy of Living. p. 140-- D
544n104 not affect race. argument for early education.-- FEAR of death!!! as Montaigne observes. distinct fro
544n104 distinct from pain, for one hates pain from this FEAR-- & not death for the pain.-- How was this inst
552n131 n the direction of mountain chains in N. America FEAR probably is connected with habitual stopping of
554n140 ong & will hide herself.-- I do not know whether FEAR or shame.-- When she thinks she is going to be
557n148 primitive feelings, no more to be analysed than FEAR or anger? I should think shame would be more ea
557n148 hamed of himself, in manner quite different from FEAR; there is no inclination to jump away,-- it is,
557n148 o inclination to jump away,-- it is, ill-defined FEAR.-- Yet one knows oneself it is quite different
557n149 ttachment) & then failing to drive away rival.-- FEAR is open mouthed to hear. though in individual c
557n150 d.-- Shame would never make person tremble, like FEAR.-- Why does any great mental affection make bod
558n151 ge.-- the social instinct more than mere love.-- FEAR for others acting in unison.-- active assistanc
558n152 h their back when fighting, & not with dog. when FEAR might enter?-- I believe common Swan, arch rais
558n153 necks straight out & hiss.-- [Hyaena pisses from FEAR so does man.-- & so dog]:CD Man grins & stamps
560m156 t, or spit, or cry.-- <fe> Shame, independent of FEAR: colour of bare nails--, & of eyes.-- Do female
565m006 of passion, is like the grin of the Hyaena from FEAR, no actual intention to bite at moment, but mer
565n007 p speaking, but laughing involuntary.-- When one FEAR any bad news, «though in a letter? why is perso
565n007 ht trys to listen to growl of hounds». <when> as FEAR to «man as» animals. comes at distance, mouth i
565n007 h is placed open.-- Hence becomes instinctive to FEAR, as ears down to horse.-- Horse snuffs «& snor
570n025 has stolen. neither. or both may be said to have FEAR, both both have shame-- Animals have not modest
578n053 on individual parts of body.== (if you <think> «FEAR» you shall not have e---n, «or wish extraordina
578n056 nces to thoughts of other person Decemb. 27th.-- FEAR loose the sphincter muscles, only on the princi
579n059 ith ancient movement of ears A man shivers, from FEAR. sublimity, sexual ardour.-- a man cries from g
588n089 p. 191 Says <childr> babies have an instinctive FEAR of falling.-- &p. 193. that they perceive the d
590n093 ttends question p. 39. The sweat that accompanies FEAR is the same, as that which attends great weakne
591n097 aenaeum Library) describes effects of emotions-- FEAR giving goose skin-- & hair standing on end.-- J
592n103 s in contempt opposite action to opening eyes in FEAR The effect of habitual movements in muscles of
613o036 t souls, we see particular trains of thoughts as FEAR of man,-- crows fear gun,-- pointers method of
613o036 cular trains of thoughts as fear of man,-- crows FEAR gun,-- pointers method of standing.-- method of
615o038 ones.-- Parental feelings weakened in Otahiati; FEAR of death in Hindoo population.-- Slightly modif
620o045 , pleasure forgotten. [RHC] 4) as starvation, or FEAR of death, one makes allowance & either excuses
622o049 re as simple as I have said.-- [LHC] instinctive FEAR of death: of hoarding.-- Ld. Kames, which Sir. J
628o53v but not cat, that is not act by gusto, though by FEAR it might be partly made.]CD p. 21. "Why ought I
360d090 f Jackall shape of body.-- disposition wild, & FEARFUL. though not so much as in Jackall.-- In case w
337d022 of a» of Monkey, «when offended» who loves, who FEARS, who is curious &c &c &c who imitates.-- who wi
541m089 does not fear death from its pain, but one only FEARS that pain, which is connected with death!-- How
553m137 eir mystical but sublime views, or the wretched FEARS & strange superstitions of an Australian savage
544m103 ranklin when starved, all party dreamt of <goo> FEASTS of good food-- The mind wills to do this & hea
272c109 wbird.-- Birds of Paradise. Trogons.-- the one FEATHER in wing the curious feathers in tail of Edoliu
312c231 d to fix on the kind wanted, colouring of each FEATHER weight & size & they would produce number agre
360d093 ng to have some different kind of mottle, each FEATHER partaking of character of other.-- <so> the mo
363d101 «Carruncles on beak & in Muscovy duck» crested FEATHER, pouters, fan tails are found in any colours o
364d103 ncier, thus getting fresh blood, without fresh FEATHER, & consequent trouble in obliterating the fres
364d103 & consequent trouble in obliterating the fresh FEATHER, by crossing-- It seems from Lib. of Useful. K
383d160 -- back like do.-- but the black lines on each FEATHER instead of coming to point {P} are more rounde
383d160 Common pheasant.-- lower part of breast, each FEATHER is fine metallic green. <from> with tip & part
363d101 dy Temmincks work on Pidgeons--, & see whether FEATHERED legs.--«Carruncles on beak & in Muscovy duck
484z014 ericas.-- manner of walking-- foot bill crest FEATHERING on legs-- habits-- Does the Secretary, make
231b242 ry of crane from Holland!!! in stomach --or in FEATHERS--seeds-- Two inhabitants of the Tropics, (wh
240c003 ogous to Owen's Phil: remark of Apteryx having FEATHERS.-- It is possible, time being an element in t
255c051 ion, before great difficulty in propagation.-- FEATHERS on, Apterix because we may suppose longest pa
255c051 ings have altered many times, but all have had FEATHERS.-- if wing totally obliterated. This may acco
272c109 Trogons.-- the one feather in wing the curious FEATHERS in tail of Edolius.-- Remarkable how small de
273c110 » bars on wings of trogons are lengthened rump FEATHERS.--; <then> & one species has small band & oth
275c120 crossing off spring. till size diminished, but FEATHERS continued by picking chickens of each brood.--
275c120 picking chickens of each brood.-- These bantam FEATHERS at last got ducky, then took white Chinese Ba

289c161 re distinct species If any one is staggered at FEATHERS & scales. passing into each other let him loo
363d101 of fowls-- long ears of rabbits. & long fur.-- FEATHERS on legs of Ptarmigan & in Bantam.--]CD CD[In
364d104 about. a house made other pheasants have white FEATHERS).-- It certainly appears in domesticated anim
366d112 eed. one would doubt any law.-- Yet seeing the FEATHERS along one toe of the Pouter one thinks there
366d112 w.,-- that there must have been a tendency for FEATHERS to grow there «That Mutilations will not alte
369d114 .--)p. 49. (wonderful case of Pea hen. taking FEATHERS of Peacock & spurs-- no final cause here.-- &
376d138 every Has repeatedly seen one he kept pull up FEATHERS of tail of Hen; which lived with it.-- also o
383d160 ot purple?-- legs pale coloured.-- In the back FEATHERS, we have character very different from either
436e135 s a most perfect Struthio head pulled out. yet FEATHERS retain character? If separation in horizontal
459t013 olour was darker than the penguin & the bright FEATHERS on its wing resembled the drake.-- another of
473s07v ed & others later-- All poultry with same down-FEATHERS. Zoology 1856 Skimmed through & abstracted Zo
039r060 ing of the cricket valley «the most remarkable FEATURE in the structure of Ascension» give as an exam
410e050 ng, there would be endless changes, & hence no FEATURE would be deeply impressed on it, & hence there
342d035 distinctly marked as whole dynasties have been FEATURED by the Austrian lip & the Bourbon nose". if t
268c099 als?-- Show independency of shells to external FEATURES of land by seeing how many species common to
303c204 ffer chiefly in <size> colour, form of head «& FEATURES» (hence intellect?) & what kinds of intellect
303c204 differed in hair colour + + + form of head «& FEATURES»;. but likewise in length of extremities, how
401e017 ommon belief. that female plant impresses main FEATURES on offspring. & male the lesser peculiarities
511q018 ters.-- does male transmit to male more of his FEATURES-- in negro & white (3) About the Bantams at Z
579n059 d. when he says he loves a person-- do not the FEATURES pass before him marked, with the habitual exp
029r033 Young, on her voyage from Demerara to London:-- "FEB. 12, 1835. At 10h. 15m. a severe shock of earthq
085aIFC s far as p. 33. distributed to several subjects. FEB 24th 1839 As far as p 140-- abstracted as far as
133a140 high.-- M. Parrot ends his paper like a fool.-- FEB 25' All facts show how slowly heat travels; & th
220b198 different. some dogs different.-- Henslow says. (FEB 1838) that few months since in Annales des Scien
220b199 f the Bustards in Germany.-- Athenaeum. No. 537. FEB. 1838. p. 107. Mr Blyth states that all «genera
256c055 et.-- Whewell thinks (p 642) anniversary Speech. FEB 1838 thinks gradation between Man & animal, smal
268c096 ue of animals of Nepal read before Linnaean Soc. FEB. 1838.-- Annals of Natural History. «vol I» p. 1
323c269 ept the V Volume.-- -- 19t. Mungo Park-- travels FEB 12. Sir. H. Davy Consolations in Travels -- Obse
323c269 in Travels -- Observations on morals by Eugenius FEB 14th. Bo«s»well's life of Johnson. 4. Vols 25th
423e098 iries of some of these great seed-growers--).-- FEB. 24th. Monoceros, which Sowerby says, is an Amer
580n062 only in Kind.-- probably very important work.-- FEB. 12. 1839. Sir. H. Davy-- Consolats: "the recol
170bIFC January 183[8]: probably ended in beginning of FEBRUARY Zoonomia Two kinds of generation the coeval k
239cIFC Charles Darwin written between («beginning O» FEBRUARY & July 1838) All good References selected Dec
269c101 Fernandez A communication to Geograph. Soc. in FEBRUARY or March 1838 on soil in Siberia being frozen
183b052 elui de la chevre et du belier, cessent d être FECONDS. dès les premières generations" go back to typ
381d156 . Sponges Heautandrous, male organs formed to FECUNDATE female (as in plants) Cirrhipeds rotifers, tr
389d176 e offspring by same mother.-- one animal will FECUNDATE female for several births, & even produce fer
445e159 lly hermaphrodite. & <thinks> even oyster may FECUNDATE each other, by the means of the medium in whi
506q015 f flower-- Ground nuts (42) How are Orchidiae FECUNDATED, as mass of pollen is requisite.-- Brown's p
381d157 ptandrous are really, Heautandrous.-- How is FECUNDATION effected in latter; are <it> «organs» open t
392d180 1 they flower Make Hybrids with moths, where FECUNDATION can be made artificially.-- Are hybrids pint
635j56r constant miracles.-- p. 420 thinks the great FECUNDITY of germs is to afford support to other beings
378d148 f wild-half tame, they came to the windows to be FED, but still they have a wariness about them quite
367d113 & Neck of Bull.-- is most common in vegetable FEEDS. because males always armed in carnivora. Wher
260c067 of different kind have been know to assist in FEEDING young cuckoo; as if there was storge, which co
268c099 y. «vol I» p. 159 curious account of Tit mouse FEEDING young of redstart & actually driving away pare
291c167 by wonderful fact of bees changing the sex by FEEDING.-- no it is developing an hybrid female it is
373d132 e milk? from stomach. analogous to other males FEEDING young, & to abortive <organs> «mammae» in male
446e160 ic cares, as building nest, sitting on eggs. & FEEDING & defending their young..-- The oriolus (icter
496q05a id, as Koelreuter describes Kill Sparrow after FEEDING on oats, give body to Hawk & sow pellet. eject
535m064 few days till larva excluded, then though not FEEDING them «nor helping larva from egg» watching the
593n111 -- At Maer. Pool. I saw many coots & waterhens FEEDING on grassy bank some way from water, suddenly,
635j56r he physical changes it was to undergo «animals FEEDING on each other &c &c».-- «(Causing death to som
636j57r eautiful adaptation.-- & then Chamelion, which FEEDING on same food, differs in every respect, except
271c105 a tree in which it builds, a berry on which it FEEDS. or insects it devours is same species. yet tha
032r042 e lines./ possibly general symetry of world.-- I FEEL no doubt. respecting the brecciated white stone
283c145 h turn a Buccinum into a Tiger."-- but perhaps I FEEL the impossibility of this more than anyone.-- n
293c172 n. [NB what are those Marvellous cases, when you FEEL sure you have heard conversation before. is str
397e004 & insensibly such changes are in progress.-- we FEEL interest in discovering a change of level of a
467t099 laments» shrivel, yet stigma does not, so we may FEEL somewhat «but little» less surprised at Henslow
498q009 s, if no hybrids were produced by seed, we might FEEL sure, that pollen of own kind is much more effe
524m022 her attention took her «left» hand to pretend to FEEL her pulse.-- What fails first?-- How is this?--
525m023 d rules by art.-- like the law of honour.-- they FEEL pleasure in obeying their instincts naturally.--
525m023 (generosity in defending a friendly dog).-- they FEEL shame, when doing anything which is wrong.-- as
525m024 but shame.-- I cannot remember instances, but I FEEL sure I have seen a dog doing what he ought not
527m035 r is interrupted & utterly forgotten--. so as to FEEL a severe disappointment «in real train of thoug
533m060 es this tendency in these cases? How did my mind FEEL it was wrong (& it was not merely morally wrong
536m070 y perceiving myself skipping when wanting not to FEEL angry-- such efforts prevent anger, but observi
540m085 nic as connexion of mammae & womb.-- We need not FEEL so much surprise at male animals smelling vagin
540m085 ngers. Seeing a dog & horse & man yawn, makes me FEEL how <much> all animals <are> built on one struc
542m093 from stiffening over his canine teeth.-- He may FEEL satisfied with himself, & though dreading to sa
546m108 putting ourselves in their situation, & then we FEEL like them--. hence sympathy very unsatisfactory
548m116 subject to this disease oneself, one would only FEEL sympathy. as for for the heard suffering of a d
550m124 obeys his conscience or instinct, would probably FEEL but little that of anger or revenge.-- they are
553m136 ons has not been improved by culture « <was> who FEEL the most implicit faith that through the goodne
567n015 may like. dislike, or be indifferent about, yet FEEL shy.-- not if quite stranger.-- or less so.-- W
568n017 must be intimately united with reason) it would FEEL «subsequent» sorrow, whatever the cause had bee
578n055 a person. whose opinion he regards, <& see how> FEEL how the blood gushed into his face,-- "as <she>
604o018 . being associated with God. these phenomena we (FEEL &?) call sublime.-- 4 From the association of
609o029 s an exception. (does it originate in a doubting FEEL between conscience & impulse) but shame «we ala
616o039 hought, remembered &c. Well the heart is said to FEEL Now this would certainly be a startling express
617o040 s when we apprehend force in inanimate matter we FEEL dissatisfied until we can point out How can for
619o043 if such actions were prevented by force he would FEEL pain.-- By a very slight change in association
619o043 his being able to prevent it, he would likewise FEEL pain.-- If he saw another man <say go> "acting
619o043 experienced pleasure,> & by association he would FEEL part of that pleasure, which the acter received
619o043 instincts from interference of passion. he would FEEL pain, which would generally be anger, as he wou
620o045 e sacrificed to the instincts.-- One does not FEEL it wrong in very young child to be in passion,
622o048 r from Paley [RHC] Anyone, who will reflect must FEEL, how like to injured conscience, is the feeling
622o048 ny custom of society broken..-- & how far more <FEELIN> acute the feeling really is.-- All these asso
048r085 rranged from the forseight of the works of man FEELING surprise at Mastodon inhabiting plains of Pata
070r154 n? would it not be mere vibration? but walls of FEELING shows undulation:. crust thin.--Concepcion eart
286c154 his Nature «& violates every best instinctive FEELING» by making slave of his fellow black, often wi
298c190 ildren,--<get> «increases» general instinctive FEELING.-- There is great difficulty in Making an alpi
377d139 to be more definite than with bitch, for some FEELING must urge them to these actions.-- «These facts
388d173 pulate & fish shows how simply instinctive the FEELING of other sex being present is-- it also shows
522m013 t natural instincts.-- My Grand F. thought the FEELING of anger, which rises almost involuntarily whe
522m014 on is tired is akin to insanity.-- «I know the FEELING also of depression, & both these give strength
522m014 ive strength & comfort to the body» I know the FEELING, thinking over somebody who has, perhaps, slig
523m014 rt of consciousness not just.-- From habit the FEELING of anger must be directed against somebody.--
523m015 .--» Doctor communicated to my grandfather his FEELING of consciousness of insanity coming on.-- his
527m032 ter for many years]CD.-- Beauty is instinctive FEELING & thus cuts the Knot:-- Sir J. Reynolds expla
527m032 & negroes another; but it does not explain the FEELING in any one man.-- Music & poetry opposite ends
531m051 s they go into, shows how truly an instinctive FEELING, <may not pa.> In reflecting over an insane fe
531m051 ng, <may not pa.> In reflecting over an insane FEELING of anger which came over me, when listening on
531m052 nying violent movement; may not passion be the FEELING «consequent on the violent muscular exertion»
531m052 trod upon turneth., here probably there is no FEELING of passion, but muscular exertion consequent o
532m053 it is instinct Fear must be simple instinctic FEELING: I have awakened in the night. being slightly
536m070 g eyes thus unconsciously discover struggle of FEELING.-- It is as much effort to walk then lightly a

538m080 y of the brain having whole train of thoughts, FEELING & perception separate, from the ordinary state
539m082 Aug. 12th. When in National Institution & not FEELING much enthusiasm, happened to go close to one &
549m088 ensations of sorrowful delight, very like best FEELING of sympathy.-- Mem: Burke's idea of Sympathy.
555m144 e from hanging to head cut off. «there was the FEELING of banter & joking» because the whole train of
564n004 sation. & «perhaps» an instinct of conscience, FEELING in his heart those rules, which he wills to pi
570n025 ctob 25. Why is modesty, mixed with triumphant FEELING so similar to shame after asinine.-- both acco
579n059 which make us love him, or her.-- it is blind FEELING, something like sexual feelings-- love being a
587n089 » &c &c Mackintosh Ethics p. 97. on Devotional FEELING p. 103-- Abstraction p. 152. Perception very d
591n099 n no relatives left to lament) is owing to the FEELING that the instinct of self-preservation is diso
591n100 st, because he does not go against instinctive FEELING, only does not fullfil, like continent man.--
593n107 ok, & therefore there must be some instinctive FEELING which is pleased by other animals smell & look
595n121 gh knowing it was Snow.-- Is this part of same FEELING which make us think anything ugly-- a beau-ide
595n121 ich make us think anything ugly-- a beau-ideal FEELING. Same effect as acting on us-- <The Baby> «Eff
615o36v courage is heredetary in fowls & not effect of FEELING of individual force in any individual.-- His M
619o042 > another animal. These instincts consist of a FEELING of love <and sympathy» «or benevolence» to the
620o045 another in desperation.-- This shows, that our FEELING, that the instinct ought to be followed is a c
621o046 ence only will teach him, that the instinctive FEELING in its nature being always present. & his pass
621o047 s fellow men.-- [RHC] 6) Hence man must have a FEELING, that he ought to follow certain lines of cond
621o047 ill acquire from seeing conduct of others, the FEELING that almost (rarely if opposed to natural inst
622o048 t feel, how like to injured conscience, is the FEELING of any custom of society broken..-- & how far
622o048 y broken..-- & how far more <feelin> acute the FEELING really is.-- All these associated «habitual» f
622o048 ones, <which either lead to actions or not, as FEELING of cowardice» «This is not connected with sens
622o049 [RHC] The social instinct may be combined with FEELING towards one as a leader,-- the conjugal feelin
622o049 eeling towards one as a leader,-- the conjugal FEELING may be directed towards one or more.-- It will
622o049 ave some, it is sufficient to give rise to the FEELING of right & wrong.-- on which «almost» any othe
625o052 being prevented uneasiness, & that this is the FEELING seems to vary in races of man. & certainly in
625o052 generations led to their formation.-- 'N.B. If FEELING of right & wrong.-- so far it has independent
626o052 n rises from heredetary action on body.-- This FEELING or emotion rises from heredetary action on bod
629o55v last head.-- (4) It is other question, how the FEELING, when instinctive will lead to action.-- the p
629o055 er permanence of social instincts explains the FEELING of ought, shame. right & wrong comes into mind
629o055 ht & wrong.-- arrived at first <rationally> by FEELING-- arrived at first <rational
211b161 ve no notions of beauty, therefore instinctive FEELING-- reasoned on, steps forgotten, habit formed,--
263o077 of the same general instincts, <as> & <moral> FEELINGS against other «for sexual ends» species, wher
524m019 y in head, a frame of mind, analogous to those FEELINGS as animals.-- they on other hand can reason--
529m041 a tiger stalked across the plains, how ones FEELINGS. which may be considered as truly spritual.--
531m051 ..-» to the pianoforte, it seemed solely to be FEELINGS would be excited, & how the scenery would ris
532m056 rkable affection to a place.-- How like strong FEELINGS of discomfort, especially about heart as of e
533m060 , or rather in only fully expressing momentary FEELINGS of Man.-- The sensation of fear is accompanie
537m076 mankind..-- p. 27. Mart. allows some universal FEELINGS of gratitude, I had a sort of consciousness I
549m118 such thoughts We give no credit to instinctive FEELINGS of right & wrong «(& therefore in fact only 1
550m127 or pain of association.-- now if one has these FEELINGS.-- for man losing his children, any more than
557m148 tiff.-- is shame, jealousy, envy all primitive FEELINGS, without being aware of their association «ie
569n020 <by> but by corporeal structure.-- Devotional FEELINGS, no more to be analysed than fear or anger? I
579n059 .-- it is blind feeling, something like sexual FEELINGS, probably some distant power of the mind-- su
593n107 e of passion & hence does music now excite our FEELINGS.-- love being an emotion does it regard «is it
605o18v o our recollection the original cause of these FEELINGS.-- How does Social animal recognize «& take p
606o025 f a motive.-- [-- must be so, analyse (a) ones FEELINGS & thus we apply to them the metaphorical term
609o029 far easier conquered than the deeper & worser FEELINGS when wagging one's finger-- one feels it in p
609o029 d than the deeper & worser feelings. These bad FEELINGS. These bad feelings no doubt orginally necess
614o037 ason.-- Conscience is one of these instinctive FEELINGS no doubt orginally necessary revenge was just
615o36v e any cause, & if surprise was felt.-- analyse FEELINGS. [LHC] As sexual instinct comes on late in li
615o038 succession so perhaps general ones.-- Parental FEELINGS. Mr Wynne says, that beyond doubt courage is
621o047 -- (a) [LHC] The conscience rebukes malevolent FEELINGS, as much as actions, therefore Sir J. M. talk
622o048 he cementation of habits into instincts. [RHC] FEELINGS. become like the instinctive, ones, <which ei
622o048 ect to their instincts & associations.-- often FEELINGS of the mind, whether leading to action or not
623o050 man. during many generations giving the social FEELINGS which do not lead to action are repressed thu
623o051 s <social> animals of peculiar <kinds> «social FEELINGS.-- [LHC] According to my theory, all instinct
624o051 ent.» Hence this is the law of our instinctive FEELINGS», & living under certain conditions; by my th
624o50v ceived» the comparison between our instinctive FEELINGS of right & wrong.-- education, of parents str
625o052 culty, of conscience.-- I believe that certain FEELINGS & our short lived Passions' State broadly in
625o052 part of our nature, <not> which regulates our FEELINGS & actions are implanted in us. & that doing t
626o052 s unintelligible to me.-- conscience regulates FEELINGS steadily & not like our appetites & passion,
626o052 d building nest, but supplies it-- instinctive FEELINGS, as of cowardice.-- the whole appears to me r
629o55v duce shame & remorse.-- [Thus pungency of one's FEELINGS will doubtless lead to similar actions which
629o055 it formed,-- & such habits carried on to other FEELINGS for indecency-- preposterously so, for Marque
312c228 & white Man crosses; children heterogenous, he FEELS sure of this, first offspring most like <parent
423e098 bably increases.-- Jan 29th. Uncle John says he FEELS sure, that the reason people send for their swe
533m059 thout knowing why-- a man as in Guy. Mannering. FEELS, pleasure. in seeing the scenes of his childhoo
552m131 grandfather remark, a tired man. involuntarily FEELS angry, when brain is pumping force to legs & bo
568n017 ife time» to any line of action, or thought one FEELS pain, at not performing it, (either if prevente
568n017 etary habit, (or moral sense, or instinct,) one FEELS pain, & vice versa pleasure in performing it.--
586n078 something of kind oneself knows in walking [one FEELS inclined to stop at right number of house thoug
586n082 ear so close a relation to the senses, that one FEELS no more surprise at it & feels no more inclined
586n082 senses, that one feels no more surprise at it & FEELS no more inclined to ask [not located] If dislik
602o011 . when individual cannot give reason, though he FEELS he is right-- it is because each decision &c is
606o025) ones feelings when wagging one's finger-- one FEELS it in passion, love-- jealousy-- «as» effect of
607o025 acts from motives, nearly as usual (a) one well FEELS how many actions are not determined by what is
620o044 llowed for a pleasure now though so trifling he FEELS remorse.-- He reasons on it & determines to act
620o046 ird. one says for shame (& the «old» dog really FEELS ashamed?) not so puppy, we <do> try to teach hi
055r108 ost solid rocks.--The grand cliffs of a thousand FEET in height, of the solid lavas.--proportionally
091a026 led Andesite. Red Coral in the Mediterranean 700 FEET deep in some of. the twopenny periodical said s
113a090 «so» that we must look at the upper four hundred FEET.-- The veins of segregation in Greenstone of Sa
120a110 s (1200 ft??). Scotland at least 2200. Jura 4000 FEET above its present level, & in many parts has ex
138a153 elieves the sea has formerly stood three hundred FEET in vertical height-- enormous mass thunder stor
146g011 overing Great Slip, 10 years since three hundred FEET «FEET» & 2 deep Another rather smaller block 30 ft «a
153g055 2 68 degree 60 degree Granite--«band» 4 X 3 X 2 «FEET» & 2 deep Another rather smaller block 30 ft «a
158g080 ts here side of Loch Spey Forms terrace about 60 FEET above Loch trace of this terrace «on «will> Gra
159g087 e deposited» the slope is continued some hundred FEET lower & begins about 60 higher-- There are howe
160g088 } which makes me think it submarine, 400 or more FEET above station! There is long straight isthmus c
161g094 han last. but A 77 station was <a few> «about 3» FEET <lower> too high about a quarter of a mile furt
202b131 illhouse.-- Demerara. In note. Demerara. 10 «12» FEET beneath surface forest trees fallen «kind well
202b131 ilem «kind well known, carbonized»--; clay fifty FEET, then forest 120 ft Micaceous rocks. subsidence
257c057 ked how did the otter live before it had its web-FEET-- all Nature answers to the possibility.-- My v
259c063 sively abundant & tempting the Jaguar to use its FEET much in swimming, & every developement giving g
269c101 degree mean, then temp at depth of four hundred FEET would be 60 degree + 6 degree??., therefore 34
286c156 in books, ought they to hold,-- Birds having web-FEET, where we see scarcely any traces of passage a
301c199 ng runs to water. before it is conscious of web. FEET.-- p. 7. Mr Blyths arguments against squirrel u
303c204 <would have> differs in present races, & form of FEET.= (Negro or father of negro probably was first
318c252 iful adaptation for snow-- like snow shoes. than FEET & hind legs of these white hares, fitted for re
331dIFC Sir. R. Heron's case of breed of pigs with solid FEET.-- 1838 [In this Book some curious notes on Mon
397e004 terest in discovering a change of level of a few FEET during last two thousand years in Italy, but wh
414e063 res progression, otherwise [not located] Are the FEET of water-dogs at all more webbed than those of
486z020 as lying under strata of gravel & clay about 10 FEET in thickness.-- (March, 1842) Questions & Exper
639j28r e power in Octopus & Chamaelion.-- C. D. Sucking FEET in Frog. Walrus. Fly. Gecko &c. Prehensile tail
601o009 ect sensation, in order to avaoid it-- beetles FEIGNING death upon seeing an object.-- are Planariae
403e021 aluz. which is found at Pellew & Eap, but not at FEIS (near island) do p. 190. The inhabitants of Sum

Page   **(Key Word)**
028r028 the finer matter resulting from degradation of FELDSPAR & other minerals containing Alumen.--This mat
098a046 n this case laminar. hence the thick wedges of FELDSPAR in gneiss.-- Veins in septaria. a kind of con
098a047 onary process (analogous to layers of quartz & FELDSPAR) within other concretion.-- state last page t
098a048 e aggregates them in layer.-- So that layer of FELDSPAR in gneiss is identical with layer of flint in
110a086 dike junction mechanical: DIKE base reddish FELDSPATHES with grenish. black specks of hornblende, la
059r120 whole burning mountain, & that the central part FELL in.--Says posterior craters in centre:-- Bailly
089a019 «He» Thought of erecting machine to see if water FELL. -- <Keys off extreme point of Flori[da]> Excel
090a023 ve nearly 19 pounds average for each stone. that FELL, that was weighed,; <but> carrying on this rati
346d048 ar,-- young one 203 days old butted violently. & FELL.-- gore to death the old & wounded,-- see Annal
532m056 e was sent back to Shrewsbury,, then immediately FELL into her old ways & became fat! What remarkable
547m111 ly thought of packing up.-- was lying on my back FELL to sleep for second & wakened.-- had very clear
228b232 let conjecture run wild then <our> animals our FELLOW brethren in pain, disease death & suffering «&
286c154 est instinctive feeling» by making slave of his FELLOW black, often wished to consider him as other a
526m027 f others.-- (NB man much more affected by other FELLOW-animals, than any other animal & probably the
552m132 ecessary for long generation, (as friendship to FELLOW animals in social animals) are those which are
619o042 should expect that acts of benevolence towards FELLOW <living> creatures, or of kindness to wife [RH
621o046 the> then receive the moral approbation of his FELLOW men.-- [RHC] 6) Hence man must have a feeling,
026r021 is a fact worth noticing that cryst of glassy FELSPAR in Phonolite arrange themselves in determinate
069r150 n (B).-- Most important view Urge curious fact FELSPAR melted gneiss/// QUARTZ!!! Analogous to Von Bu
073r164 no quartz & amphibole frequently only vitreous FELSPAR: = gold veins in a phonolitic porphyry. = seve
078r176 artz, & wh abound both in hornblend & vitreous FELSPAR".-- p. 215 Same metal in Tasco vein in Mica Sl
094a035 ce, one locally relieving the other. -- Is the FELSPAR glassy in greenstone dikes which rise through
102a061 nite residual matter of upper quartzose ones & FELSPAR.?? Are the great crystalls, & the layers first
102a061 Are the great crystalls, & the layers first of FELSPAR & then quartz &c, owing to separation having t
104a066 ly contains 6 per cent more silica than common FELSPAR therefore on axis of Cordillera, in Andite - c
110a086 -- The fragment. consisted of hornblende (?) & FELSPAR, (some crystals being red) «with» cleavage, ve
110a086 like basalt. full of circular cryst of glassy FELSPAR different. from either fragment or dike, black
110a086 f hornblende, large irregular cryst of reddish FELSPAR. & scales. of mica.-- large cryst of Hornblend
160g090 d by turf 2ft. 8- long of syenite with pinkish FELSPAR;-- whole hill dark grey fine grained. Much con
097a042 residue of quartzose vein in higher parts? & FELSPATHIC veins?-- Mr Poulett Scrope. talks of Trachyt
071r158 thquake in which lake was absorbed.--Earthquakes FELT. different case from shore of Pacific.--Isabell
298c189 alf fox & dog. & that it was most like fox.-- He FELT sure the half breed of Australian dogs, would b
409e047 ming.-- yet I will not shirk difficulty-- I have FELT some difficulty in conceiving how inhabitant of
532m054 e awakened in the night. being slightly unwell & FELT so much afraid though my reason was laughing &
533m061 merely morally wrong, but hurting my character I FELT it)-- this is kind of conscience, is obscure me
566n010 says on mimicking expression of emotions, he has FELT the passions of a face «& mind» sympathetics wi
581n066 this bodily action. if the emotion was not first FELT?-- «without «slight» flush, acceleration of pul
586n079 onsists in the action being performed or emotion FELT in early childhood (before experience or habit)
588n090 whilst strong emotions of reverence & piety are FELT." it appears to me mere consequence of stooping
613o035 , but is memory gone?-- Where pain & pleasure is FELT there must be consciousness??? ? [LHC] ¿Can ins
615o36v g it, nor was there any cause, & if surprise was FELT.-- analyse feelings. Mr Wynne says, that beyond
618o41v relation to forces & mentality because effort is FELT [LHC] 1) May 5th. 1839.-- Maer Mackintosh Ethic
620o044 ience) is always present (which is indeed, often FELT at very time it is disobeyed) & is sure guide.-
145ø001 at male impresses offspring more indelibly than FEMALE p 367 Quarterly Journal of Agricl Dec 1837 Yet
191b082 th The Male animal affecting all the Progeny of FEMALE insures often mixing of individuals. Here we h
194b096 ican shells? ¿Do not plants, which have male & FEMALE organs together, yet receive influence from ot
199b120 ertile. or at least most rarely & perhaps never FEMALE.--no offspring: physical impossibility to marr
205b145 ply to where same animal breeds often with same FEMALE p. 28 "It <is> wrong to enlarge a Native breed
218b190 e are vitiated.-- This barely applies to plants FEMALE pig apt to produce monsters in Isle of France-
228b229 o> producing crab. is the offspring of a male & FEMALE animal of one variety going back ¿whether this
230b235 crease of population in California cessation of FEMALE offspring: applicable to any animal-- Athenaeu
273c112 which is one instance of its whole family where FEMALE is not dull.-- I must observe that this preemi
275c121 ommunicates the external resemblances, than the FEMALE. The expression hybrid & fertile Hybrids, may
291c167 sex by feeding.-- no it is developing, an hybrid FEMALE it is a wonderful relation going through all N
295c178 tors so, or has he assumed that character -- -- FEMALE & young seem most like mean characters the oth
299c195 st character.-- In dioecious plants many of the FEMALE flowers unimpregnated Babington We see gradati
303c204 points for <t> a less time than other. points. FEMALE genital organs «in some monkeys clitoris wonde
307c215 is important for if these abortive wings in the FEMALE are allowed to the fully organized wings of th
307c216 ct) because if so as she can be converted «into FEMALE», it will be splendid argument old female, tur
307c216 «into female», it will be splendid argument old FEMALE, turning into cock, abortive spurs. growing.--
309c221 ses however different.-- Male glow-worm knowing FEMALE good case of instinct. bees turning neuter int
341d031 ck Guinea Fowl with Pea <cock> Hen.-- offspring FEMALE, yet so infertile never even in seven years pr
345d044 opinion that male impresses offspring more than FEMALE, yet instances given on opposite side,-- «The
350d055 of some Crustacean, like Trilobite. (Polirus?? FEMALE blind & of quite different form from male with
350d055 chineal insects move about & see, parent «(2)», FEMALE «(I)» fixed & blind: -- Macleay observed all t
358d076 of the sexes is little developed, it is always FEMALE which approaches in character to the larva, or
358d076 er to the larva, or less developed state.-- the FEMALE & young of all birds resemble each other in pl
358d076 emble each other in plumage «(that is where the FEMALE differs from the male?)».-- children & women »
361d095 er alters entirely]CD In common sparrow young & FEMALE similar plumage.-- in tree sparrow, (if I unde
361d096 ted through the male??)-- Yarrell observed that FEMALE of some water birds, (as Phalarope) assume for
362d099 & Mr Yarrell supposes this a consequence of the FEMALE breeding all the year round. ask Colonel Sykes
365d105 Black game is owing to their rarity., a single FEMALE in wood with Pheasants would sure to be trod,
378d147 cter. added to species,, we can see why young & FEMALE alike Good Ch 6 Keep Is it Male that assumes c
379d152 capable of producing both <male> pair of male & FEMALE.-- if there be one female, she will be free Ma
379d152 <male> pair of male & female.-- if there be one FEMALE, she will be free Marten. <Owen.> See Hunter's
380d153 ing rendered biennial-- the hardness of life in FEMALE Moth &c Mr Y. says that Macleay considers the
380d153 ays that Macleay considers the house bug, as a <FEMALE which have> larvae which have bred before the
380d153 not know of any case of old Male. becoming like FEMALE, though many of old female becoming like cocks
380d154 Male. becoming like female, though many of old FEMALE, becoming like cocks.-- It is very singular. so
380d154 it upon eggs, as well as, & often better than a FEMALE.-- this is full of interest; for it shows late
380d155 tigers.-- cat, though caterwhalling. & put into FEMALE, when muzzled, he is disabled.-- so Elephant i
381d156 classifies Hermaphrodites. Cryptandrous. (only FEMALE organs visible). Oyster. cystic Entozoa. Echin
381d156 s Heautandrous, male organs formed to fecundate FEMALE (as in plants) Cirrhipeds rotifers, trematode
382d158 of abortive womb, or ovarium-- or testicles in FEMALE.-- the <add> presence of both testes & ovaria
384d161 t strength> In speaking of generation alway put FEMALE first Will not even a fruit tree or rose degen
384d162 ordinate manner in the plants which have male & FEMALE flower on same stem.--» so that Molluscous her
384d162 & power of assuming male plumage in females.- & FEMALE plumage in castrated male.-- «Men giving milk-
388d172 orted by change which takes place in old age of FEMALE assuming plumage of cock, & beards growing on
389d173 nsformation, & was received into bud matured by FEMALE;<]CD> such view no ways explains Ld. Moretons
389d176 ing by same mother.-- one animal will fecundate FEMALE for several births, & even produce fertile off
389d176 uce fertile offspring-- DESIRE LOST when male & FEMALE too closely related: this most important. with
389d176 her & mother, or very close to either.-- Male & FEMALE as foetus one sex; & therefore both capable of
389d177 ce same effect as too little.-- in (latter case FEMALE often takes males but does not produce) tenden
390d179 oo close animals, which will not breed, but the FEMALE at least (¿male?) looses all appetite.-- It is
401e017 ucing cowslip Uncle J. says common belief. that FEMALE plant impresses main features on offspring. &
431e123 ts in the idea of the male being smaller, & the FEMALE larger than average size: (surely this is very
434e129 yet parts only very slightly abortive & bed of FEMALE flowers will sometimes produce a few seeds,--
434e130 plant, in which the Male plant sometimes bears FEMALE flowers, the organs are most clearly abortive,
446e160 us (icterus Cat.) is an instance of this, & the FEMALE of the icterus minor is a bird of more splendi
449e169 hatches «all alike» between the male Chinense & FEMALE common goose took after the common goose thus
493q003 y one sex so coloured = I have grey-cat «wh was FEMALE» with tinge of tortoise-shell «on back.--» = L
493q004 &c (1) To cross some artificial male with <old> FEMALE of old breed & see result.-- According to Mr W
496q006 ng Plants Is the common Fig Dioecious-- are its FEMALE flowers always barren-- if not how does impreg
496q006 if not how does impregnation take place male & FEMALE flower in same receptacle (8) Make Duck eat Sp
499q009 yton has such a grove of Willows.-- (14) Bowman FEMALE branch At What distances from males, will fema
499q009 emale branch At What distances from males, will FEMALE (a) Willows or Yews some poplar's produce.-- (
507q15v esumes females produce. Polygam. trioecia. (are FEMALE flowers ever productive) Smith says many trees
508q016 ion: Annales des Sciences‖ (4) Prolifickness of FEMALE, relation to healthiness? & father answered (5

508q016 of heredetary cases, more common in man than in FEMALE-- (8) In Hump-back ever heredetary (9) Are the
513q21. principle of arrangement."-- would not male or FEMALE "constructive principle" be better. or "constr
515q022 s differ in colour of beak, taking after male & FEMALE parent.-- Will they grow up in other respects
535m064 with it-- do not know their own larvae, but one FEMALE may be moved to other larvae, when two groups
554m141 sion.-- The male Black Swan is very fierce when FEMALE is sitting the Keeper is obliged to go in with
557m149 ginally entirely sexual; first try «to» attract FEMALE, (or object of attachment) & then failing to d
560m156 «& music».-- Have monkeys lice?-- picture.-- Do FEMALE monkeys not show signs of impatience when woma
560m156 f fear: colour of bare nails--, & of eyes.-- Do FEMALE monkeys care for men.-- Have we any ferns in t
570n027 beautiful,-- [male glow worm doubtless admires FEMALE. showing. no connection with male figure]CD--
592n101 Human mind.-- Andrew Smith says hen doves & the FEMALE chamaeleon court the males by odd gestures. In
609o29v n incontinence to be a vice & especially in the FEMALE October d. 1838 perhaps insist?? Two classes o
215b176 . Fox. When dogs are bred into each other, the FEMALES loose desire, and it is required to give the c
215b177 e never were any constant species Both males & FEMALES. lose desire. Native dog not found in V. Dieme
295c178 +++ ‖Daines Barrington says cock birds attract FEMALES by song. do they by beauty, analogy of man if
307q215 ears-- colours vary, but form constant.-- The FEMALES of some moths, like glowworm <are> have «These
318c254 ome species «a Tanagra» Males come first & the FEMALES in flocks. «as in English Nightingales»-- oth
342d035 en old peculiarity overbears the crossing with FEMALES not thus characterized.-- 16th Aug.-- What a m
358d076 castration on males & of age or castration on FEMALES-- [not located] hen freely.-- here we have be
367d113 because males always armed in carnivora. Where FEMALES, are peacable-- (Mem Lucanus & Copris &c).-- I
368d114 birds singing of cocks settle point.-- (do the FEMALES then fight for male) & are merely most attract
368d114 other birds display beauty of plumage.-- (The FEMALES (as Owen observes) in Raptorial birds largest.
368d114 ly the male which recedes from the species all FEMALES being most like offspring, Q (how is this with
368d114 most like offspring, Q (how is this with those FEMALES which put on (like some waders) the bright plu
369d114 & therefore different from Hunter I should say FEMALES recede in organization from specific character
370d116 for larvae.-- CD[(p. 451.)-- Wasps breed many FEMALES, but almost all die.-- bees breed but few, bec
370d116 how has this been arranged-- Neuters are true FEMALES, but with parts little developed.-- Sept. 19th
384d162 of Capon. & power of assuming male plumage in FEMALES., & female plumage in castrated male.-- «Men g
499q09v t? do they seed plentifully? Look for isolated FEMALES.-- Also any plants which are known easily to b
507q15v y on Citrons 47. Ficus carica Henslow presumes FEMALES produce. Polygam. trioecia. (are female flower
508q016 lly occurring in man being transmitted through FEMALES, like Hydrocele Dr. H. thinks asthma in female
508q016 emales, like Hydrocele Dr. H. thinks asthma in FEMALES takes place of gout.-- How are livers obscure
540m085 h surprise at male animals smelling vaginae of FEMALES.-- when it is recollected that smell of ones o
309c221 nt. New England farmer,-- useful could not leap FENCES:-- Dr Lang on Polynesian nations (quoted) p. 4
502q11v mpare flowers of wild & tame carrot-- Parsley & FENNEL. Verbena Compare flower of different Cabbages
338d024 inburgh. Royal. Transact.-- p. 297. Vol 9. Dr. FERGUSON seems most clear that the ideosyncracy of the
207b152 e plus d'especes polymorphes que toute la terre FERME de lancien monde".-- Considers forms in recent
046r080 a native mouse Did wave first retreat at Juan FERNANDEZ: the first great movement was one of rise (an
173b011 s all different. In cases as Galapagos & Juan FERNANDEZ. When continet of Pacific existed might have
209b156 (?). & similarity of type (?).-- [«Mem:» Juan FERNANDEZ]CD. From study of Flora of islands; "ou bien
209b157 a I: I,15 {T} Calculate my Keeling Case: Juan FERNANDEZ Galapagos =Radack Islds = ∴ Islands & Artic a
214b173 ts in Beetsons St. Helena. -- Galapagos--Juan FERNANDEZ Falkland Islds-- Kerguelen land.-- Phillips.
219b193 y Mr Don in island, Teneriffe, St. Helena. J. FERNANDEZ. Galapagos. Many trees Compositae, because se
236b278 collect cases.-- African isld.-- «How is Juan FERNANDEZ-- Humming Birds» types of former dogs. charac
269c100 view) Tristan D'Acunha, St Helena &c &c. Juan FERNANDEZ in birds. but ¿whether to same island in plan
271c106 e creative American halo has extended to Juan FERNANDEZ do Mitchell. Australia Vol I. p. 306 "The cro
356d074 raphy of Atlantic Tristan D'Acunha ditto Juan FERNANDEZ do Chile??) Falklands to southern portion.--
421c081 ascar does to other side of Africa.-- (& Juan FERNANDEZ to Chile??) Falklands to southern portion.--
448e166 I. p. 306 Shells, as well as plants «of Juan FERNANDEZ» differ from American Coast Vol II.-- «Refere
185b056 h has changed into Cara cara at the Galapagos. FERNANDO Noronha Ophyressa bilineata (Gray) new «liza»
223b209 n species?-- Small «new» animal mentioned from FERNANDO Po Zoolog. Proceedings October (?) 1837 Contr
226b220 on between Mauritius & Madagascar very good.-- FERNANDO Po & Coast of Africa. equally good.-- Small i
226b222 ition of Paris basin land.-- (How is this with FERNANDO Po.). with plants of St. Helena & Tristan D'A
295c183 ng shown, young one.-- Many African monkeys in FERNANDO Po-- no new forms only species!! No salamande
463t059 equired.-- Waterhouse says perhaps animals of FERNANDO Noronha are found unknown coast in front of i
209b156 or «(p. 145)» 25. plants. 36 St Helena, without «FERNS.-- analogous to nearest continent: poorness in
219b193 trees Compositae, because seeds first arrived «FERNS ditto.--» & hence formed trees]CD & would creat
228b230 ty of generation strongly shown by hybridity of FERNS.-- hybridity showing connexion of two plants. A
230b235 imal-- Athenaeum. <Jan> «p. 154»-- 1838. Hybrid FERNS It may be argued against theory of changes that
241c016 7-8. Tom: IX.-- M. D'.Urville on the Distrib of FERNS in South Sea (Indio Polynes: <>) vegetation far
255c052 leton of such general forms.-- The hybridity of FERNS bears on my doctrine of cross-generation. The i
381d157 ter; are <it> «organs» open to water? Would not FERNS according to this doctrine be considered as rea
384d162 uished. --yet may be presumed from hybridity of FERNS) afterwards they can be seen distinct. (in dioe
421e090 w. even in fish: ‖.do. p. 236-- on Hybridity in FERNS.-- ‖.do p. 250-- «speaking of» the terrestrial
503q012 to ask Mr Riley to experimentise on hybridising FERNS, tying them back to back 37 Col. Sykes fertilit
559m156 t parts of mosses & see if Hybrid can be made & FERNS.--» Would a sensitive plant if irritated very r
560m156 - Do female monkeys care for men.-- Have we any FERNS in the hothouse at home Natural History of Babi
114a093 entific Memoir, Part IV. p. 403 Ehrenberg on FERRUGINEOUS Gallionella Examine Iron stone of C. of Goo
429e115 that in the Pyrenees, that the Rhododendron FERRUGINEUM. begins at 1600 metres precisely & stops at
041r066 hether such fossils. lived in groups or not. FERRUGINOUS veins of this figure {P} in sandstone: evide
051r093 eposits «calcareous» encrustations; At Bahia FERRUGINOUS.--At Pernambuco (great swell & turbid water)
316c245 s more confinement. thus the Naiads (study De FERRUSSAC) are confined to <S.> America.-- Mr Sowerby s
199b120 ance «generally» to marriage <if offspring not FERTILE> <,but producing> «before domestication, after
199b120 domestication, afterwards none or little with <FERTILE offspring> » fertile offspring; marriage never
199b120 ards none or little with <fertile offspring> » FERTILE offspring; marriage never probably excepting f
199b120 ting from «strict» domestication offspring not FERTILE. or at least most rarely & perhaps never femal
230b234 ned by law of small differences producing more FERTILE offspring.-- 1st. All variation of animal is e
230b240 , they will be called species & mere producing FERTILE hybrids will not destroy that evidence, as so
249c034 eply impressed on blood., will cross & produce FERTILE offspring in the first case it will either pro
267c093 orfolk Isld-- "& it now claims all the moist & FERTILE land of Tahiti, in spite of every attempt to c
276c122 nces, then the female. The expression hybrid & FERTILE Hybrids, may be used to varieties, as well as
283c145 ountry with their own instinct & (even though <FERTILE> when compelled to breed hybrids produced--)»
302c203 to some physical change.-- If Patagonia became FERTILE all intermediate species living there would be
309c219 d deal from climate & habits, & therefore less FERTILE. according to Mr Herbert's views.-- Argue «arg
331dIFC Are the number of nipples in domesticated very FERTILE animals increased?-- Where offspring, heteroge
334d009 ese, & that they this first cross were equally FERTILE with pure bred animals.-- Mem. number of Mules
335d016 increased). fertility.-- (but many animals are FERTILE, when offspring infertile,-- two consideration
340d026 mmon Ducks, breed one with another-- & hybrids FERTILE inter se--No directly against Eyton's rule. ¿A
342d033 ES which breed most freely. & produce somewhat FERTILE offspring produce heterogenous offspring. It a
344d040 ctions of birds of Java Caterpillars not being FERTILE is same as children not being so.-- consider t
344d040 s-- (¿glowworms) breeding-- <beet> imago state FERTILE at once.-- Consider this with reference to tho
344d040 is with reference to those insects, which have FERTILE offspring. Entomostraca & Aphides. The extreme
344d041 s analogous to superfoetation, & to successive FERTILE offspring in Entomostraca & Aphides Developeme
355d066 descended from several stock, then species are FERTILE‖; as long as opponents will «are» not «able to
366d108 ew, the domesticated animals would cease being FERTILE‖ inter se., or at least show repugnance to bree
385d163 take «& if she did take, probably would not be FERTILE» without she know & LIKES HIM & then is actual
385d164 . (he thinks half pheasant, half fowls.-- eggs FERTILE, but parent bird will never sit on them.-- May
389d176 date female for several births, & even produce FERTILE offspring-- DESIRE LOST when male & female too
392d175 nd. <]CD> & forming species)-- [Aphides having FERTILE offspring without coition or addition of diffe
393dIBC he hybrids of those species. which cross & are FERTILE heterogenous? When bird fanciers say the throw
425e103 D[Does the Power of, «easily» making tolerably FERTILE hybrids, bears relation to capability of varia
426e107 one think that one large body of varieties are FERTILE & make mongrel, & other great series quite oth
453e183 of question-- single, or half double.-- anyhow FERTILE because they «are» raised by seed.-- Where has
511q018 ther Shaws hybrids between Trout & Salmon were FERTILE & whether homogeneous {About German ornitholog
420e090 ke some Mediterranean species).-- might these FERTILISE other shells, as insects do flowers.-- Mem. S
173b010 tinues to vanish, bones instinct &c &c &c non FERTILITY of hybridity &c &c «assuming all» if species
200b123 pposite races, whether you would expect equal FERTILITY-- ditto in Plants.<==> It will be well to ref
223b209 en much altered they will cross (perhaps more FERTILITY & so make that sudden step. species or not. A
223b211 A will soon form good species! The increased FERTILITY of slightly different species & intermediate

Page **(Key Word)**
```
246c033 tendency to revert to parent forms, & greater FERTILITY of hybrid & parent stock, than between two hy
277c125 ether Mem Mr Herberts law; habits determining FERTILITY Scheme for abolishing specific names & giving
285c152 n> real thing with regard to contemporaries-- FERTILITY must settle it.-- Changes in structure being
294c178 mes stunted, altered, & lose (mere sickness)? FERTILITY ¿because offspring too unlike.--?? Memoire by
308c219 ed animals having same idiosyncrasy, cause of FERTILITY.-- varieties not produced as by nature. if so
335d014 ers are:-- are they «abortive» twins.-- ∥ The FERTILITY of first cross, as stated by Fox, is very imp
335d016 hysical impossibility to (perhaps increased). FERTILITY.-- (but many animals are fertile, when offsp
413e058 . «especially with physical change» (3) Great FERTILITY in proportion to support of parents December
417e075 tead of allowing strength to get the day» The FERTILITY of Indian & Common Oxen, which one must think
454e184 stances of plants, in becoming double loosing FERTILITY if, sometimes one, sex & sometimes. other, so
503q012 ferns, tying them back to back 37 Col. Sykes FERTILITY of men & Europaean animals in India?-- about
512q020 ity --Mr Miller says Wombwalls were (4) About FERTILITY of ass-zebra-horse= (4) About fertility of as
513q020 About fertility of ass-zebra-horse= (4) About FERTILITY of ass-zebra-horse= (5) About callosities on
513q020 Cross. Sus Barlyroussa with tame.-- (7) About FERTILITY of Bantams from different countries= Do the P
528m039 a magic.-- connection with poetry, abundance, FERTILITY, rustic life, virtuous happiness.-- recall sc
529m040 t would receive pleasure from thinking of the FERTILITY.-- I a geologist have illdefined notion of la
640j28v daptations, but for youth most necessary: the FERTILITY of Man in old age keeps woman alive: for Man
640j28v keeps woman alive: for Man & woman are same: FERTILITY of either sex determines life:.» «With respec
401e014 d without greatest care be taken to prevent FERTILIZATION from turnips or other stocks. Says if any v
401e014 nother-- Uncle John says he has no doubt bees FERTILIZE enormous number of plants-- it is scarcely po
427e110 ld POSSIBLY «No, for pollen of any kind would FERTILIZE it» fertilize an apple somewhat more readily,
427e110 o, for pollen of any kind would fertilize it» FERTILIZE an apple somewhat more readily, «than other a
420e090 nts showing how little of the spermatic fluid FERTILIZED spawn of frogs.-- Annals of Natural History.
400e013 original variety.-- for they are all made by FERTILIZING one plant with another-- Uncle John says he
439e143 the hybrids (mentioned in letter to Henslow) FERTILIZING each other, better than the pollen of same f
040r063 shales have been metamorphised, as in Brazil FERUGINOUS sandy ones have undergone the same process.-
447e164 er.-- Analogy shows some most important end.-- FESTUCA vivapara F ovina-- propagated like oni[on] Poa
447e165 nslow says he has not the slightest doubt that FESTUCA vivapara is the same species with F. ovina, <&
288c159 ut it is certain that rams & bulls from England FETCH very <go> large price. as is evident to be wort
575n045 whether birds badly wounded, or only winged.-- FETCHES two birds out at once.-- Old People-- (Antiqua
641j29v gies-- prickly plants or animals-- Exudation of FETID «& acrid» secretion in Mollusca. insects «Carab
332d003 the hair a century since used to be called Worm FEVER, as used much more latley diseased Mesenteric g
332d003 n» & other such disorders accompanied with some FEVER, be attended by the transmission of large numbe
496q006 fect-- (10) Try in how many generations. daisy. FEVER-fuge Groundsil.-- gilly flower will break & bec
293c172 e idea. «or simple structure in brain people in FEVERS recollecting things utterly forgotten» --it is
206b149 le In a decreasing population at any one moment FEWER closely related;∴ (few species of genera) ultim
232b245 then have been wanderers.--» There ought to be FEWER species in proportion to genera than in present.
284c147 on relations of heat & cold. therefore probably FEWER now than formerly.-- The number of forms depend
402e020 agindanao several kinds of the large monkeys.-- FEWER <of th» «Mammalia» have passed to Paragua & in
441e148 -- instead of one part «as» in producing bud.-- FEWER of the lately acquired peculiarities are transm
515q022 . Anderson's statement of English Horses having FEWER vertebrae in tail, than Continental horses. {Ab
349d051 Some <of> cases the circular arrangement from FEWNESS of forms-- Cannot be discovered «un»till <in>
315c242 njection of fluid different from contraction of FIBRE)-- it is most remarkable habitual actions in pl
599o007 grant that the thrill, which runs throug every FIBRE, when one behold the last rays of & or grand
601o009 only evidence. when consciousness is absent) in FIBRES united with nervous filaments.-- ¿plants? yes
072r160 e scattered over the surface of the ground are FIBROUS. malleable & of so great tenacity, that it is
114a094 A. Smith's curious specimens of «transversely FIBROUS» quartz. & iron stone alternating. bear on sub
509q017 Horse or cow.-- degree of soldering of tibia & FIBULA: in Man any abortive bones??? do. Wing in Apte
541m090 ore dreams thus act.-- ∴ weak minded people are FICKLE & full of levity (¿ do I not confound action &
507q15v al Variation? Henslow knows only on Citrons 47. FICUS carica Henslow presumes females produce. Polyga
031r037 o many faults in Cordillera, as in English Coal FIELD -- because lowered & raised--so on--but gradual
038r057 rning, surely there must be «somewhere» below a FIELD of fluid rock.--In the discussion it will be be
120a107 by vertical planes into small pieces-- mem coal-FIELD.-- the structure of Andes. where we believe we
380d155 n 5 weeks.-- A Bull is never taken from his own FIELD to bull a cow.-- -- a dog if led in string will
437e140 . 175., 28 sho[r]t eared owls were counted in a FIELD, where there was great swarm of mice.-- May 4th
495q05a investigation.-- Capital in middle of ploughed FIELD-- on hills.-- 10 Shoot tame duck on pond with D
505q014 insects to impregnate it (7) History of Potato FIELD= (8) Abortive Thyme seeds weather wet--? Linum
043r090 such as that of 1835.-- State the three «or 4» FIELDS of Earthquakes in Chili:-- Chiloe. Concepcion.
423e098 that people in the southern Counties have whole FIELDS, some for cauliflower &c.-- Uncle John believe
429e114 beings. going on the peaceful woods. & smiling FIELDS.-- we must recollect the multitudes of plants
637j57v water-- bull dog to bulls.-- primrose to <open FIELDS> banks-- cowslip to <banks> fields-- these are
637j57v ose to <open fields> banks-- cowslip to <banks> FIELDS-- these are adaptations just as much as Woodpe
333d005 eculiar) limbs very long, eyes very large, very FIERCE to dogs «otherwise habits not different; tone
554m141 p in progression.-- The male Black Swan is very FIERCE when female is sitting the Keeper is obliged t
031r038 ure was very clear at base of great lava cliffs [FIG. I] line of high tidal action {P} NB. patches of
031r038 action {P} NB. patches of modern Conglomerates [FIG. 2] {P} The action of sea A. B. will be to eat l
032r040 e shown (as represented), along line of coast.--[FIG. 2] Mem San. Lorenzo; Valley of Copiapò & parts
047r081 reaks: i e to form a wave in ocean. is not this [FIG. 3] {P} form present, i e a part below «mean» le
049r087 e. but rather to one out of a series of faults. [FIG. 4] {P} In Cordill: should basal lavas be called
052r097 - Necessary form; as long as coast line fixed.--[FIG. 5] {P} * Slope necessary for seaward transporta
052r098 nsportal of drift matter.-- Give various cases. [FIG. 6] {P} A advancing coast to Seaward. Retreating
056r110 mmey. to crater. as at Galapagos. St. Helena.-- [FIG. 7] {P} effect of heat on inner wall, hence resi
068r147 g odd to find them injected by veins & mass[es] [FIG. 8] {P} (A. B, C, now grown solid.) Red Sea near
069r150 o line.--connected with <gneiss>.--(Mica Slate) [FIG. 9] {P} ((3) like Bell of Quillota.) (A) in this
072r160 icle attracted towards space tend to form ring. [FIG. 10] {P} motion from within and without H. Kingd
496q006 iments Questions concerning Plants Is the common FIG Dioecious-- are its female flowers always barren
362d099 victorious stag. who rubs the skin off horns to FIGHT--- is analogous to the love of woman (as Mitchel
368d114 g of cocks settle point.-- (do the females then FIGHT for male) & are merely most attracted)-- sing
414e063 t precisely like two species of animals.-- they FIGHT, eat each other, bring diseases to each other &
556m146 nt to drawing them close on head, when going to FIGHT, in which case expression resembles a fox-- I c
367d113 hich part is that most immediately employed in FIGHTING" instances thighs of cock & Bull.-- i
556m146 lar tricks accompanying emotion.-- when horses FIGHTING, they put down ears, when «turning round to k
558m152 ip of tail.-- do two cats arch their back when FIGHTING, & not with dog. when fear might enter?-- I b
500q010 about Thyme. Horned Oranges. Spallanzani Essay-- FIGS 2 kinds of flower annually.-- Periwinkle. (not
505q014 his question= (5) Open more Horned oranges.-- (6) FIGS, flower.--Passion Flower. (as it is required to
041r066 ved in groups or not. Ferruginous veins of this FIGURE {P} in sandstone: evidently depend on a concre
119a105 ts there.-- But Sir John considers an irregular FIGURE to be that of equilibrium,-- What causes that
122a114 .-- Why on one coast? How can Herschel consider FIGURE of earth statical.-- if platform of mexico owe
128a129 e How strongly the Glen Roy case shows that the FIGURE of the world has just that form which forces d
343d039 il when Europe must have worn a quite different FIGURE 19th. With respect to the Deluge it may be wor
358d086 But the hen hybrid of this bird, has long tail FIGURE, & some degree of whiteness like a Male.-- Thu
364d105 n. subalpina in wild state.-- Neilson has given FIGURE of it.-- In England no doubt the cross between
570n027 dmires female. showing. no connection with male FIGURE]CD-- As forms change, so must idea of beauty.-
078r174 ] Under name of Sagitta Triptera D'Orbigny has FIGURED animal with setae like my undescribed[, ] p. 14
570n027 beauty.-- [Old Graecians living amongst naked FIGURES, & observing powers common to savages??]. CD--
466t099 hyme.-- equally abortive as it was in autumn: FILAMENTS united in whole length to corolla--anthers mi
467t099 s we see in Hybrids that although anther «nor FILAMENTS» shrivel, yet stigma does not, so we may feel
467t103 ne flower is perfectly ripe & pollen abundant FILAMENTS & stamens all protrude «there is a brush at e
468t111 Heartease» «small. Humble alighted on base of FILAMENTS & reached nectar «again= between them, hence
468t112 ma:-- but stigma «is» almost roofed by united FILAMENTS.-- This flower hostile to intermarriage!!xx I
601o009 ness is absent) in fibres united with nervous FILAMENTS.-- ¿plants? yes by distinct mechanism 2. Sens
346d048 & in are-- colour white, uniform.--crafty, go in FILE, hide their young., bold.-- a Mr W: Hall remark
044r072 entre-- Pisolitic balls occur in the Ashes which FILL up theatre of Pompeei (?). -- Such have been se
103a063 moving agents, because not wedge-formed.-- Hence FILL up fissures-- If dikes effect of horizontal ele
273c110 one to expect that fossil forms would generally FILL up genera & not species, which is not true, wit
273c110 ow perished) hawks & Milvulus &c would instantly FILL up their place.-- Humming bird there is strongl
350d054 no grotesques in nature; not anything framed to FILL up empty cantons, & unnecessary spaces" p 23. "
352d060 t of a history, & the geologist being obliged to FILL up the gaps.-- is possibly the same with the <Z
```

423e096 ly organized ones. would soon be disorganized to FILL their places.-- The Geologico-geographico chang
036r049 nt the dilatation, which dilated cracks must be FILLED up by dikes & mountain chains.-- Introduce par
038r058 ly appearance leads me to believe mere fissures FILLED up.--the appearance will here be the strongest
100a054 ion instance of hollow concretions & concretion FILLED with unconsolidated matter-- Phillips Lardner
147g020 terraces but appearances, as if valley had been FILLED with sloping bed of rubbish Friday Highest par
163g107 r Fort Augustus hill & fringe as if it has been FILLED up «at» 30 ft. higher with pebbles now worn aw
262c073 rmed in any kingdom of nature, where scheme not FILLED up, (most false to say no passages; nature is
105a070 n Cordillera. a rush of water will account for FILLING up of valleys-- subsequent opening a medial go
420e088 e Comte Rendu.-- I suspect good case of fossil FILLING up blank.-- CD[not between existing series of
099a048 > {P} If surface covered with oil should shrink. FILM parallel to longer axis. But if great depth NB.
133a140 riod. one is led, to look at globe as resting on FILM of molten rock.-- Voyages of Adventure & Beagle
099a049 of concretionary action all fluid at once, the FILMS vertical. Ascertain law of attraction of partic
057r112 eful chemical instrument.--Yet neglecting these FINAL causes.--What more awful scourges to mankind th
171b005 our present system of body & universe therefore FINAL cause of life With this tendency to vary by gen
182b049 g that of goats." Progressive development gives FINAL cause for enormous periods anterior to Man.. di
314c236 ion to time) as in buds.-- I can scarcely doubt FINAL cause is the adaptation of species to circumsta
369d114 ea hen. taking feathers of Peacock & spurs-- no FINAL cause here.-- & therefore different from Hunter
375d135 forming gaps by thrusting out weaker ones. «The FINAL cause of all this wedgings, must be to sort out
376d135 to do that for form which Malthus shows, is the FINAL effect, (by means however of volition) of this
386d167 ould be sympathy in human frame.-- «one of» The FINAL cause of sexes to obliterate differences. final
386d167 final cause of sexes to obliterate differences. FINAL cause of this because the great changes of natu
409e048 ater & snow line descent. My theory gives great FINAL cause «I do not wish to say only cause, but one
409e048 «I do not wish to say only cause, but one great FINAL cause,-- nothing probably exists for one cause»
440e146 ch structure is due to external agency, without FINAL cause. either in present, or past generation.--
440e146 s of pidgeons with tufts &c &c here there is no FINAL cause yet it must be effect of some condition o
440e147 ut habits-- no one can be shocked at absence of FINAL cause mammae in man & wings under united elytra
496q05a ching Mr Brown theory of insect-like Orchis-- & FINAL cause of beauty of flowers-- contrasted by Kirb
559m154 giving rise at last even to the perception of a FINAL cause.-- Read. Paper on consciousness in Brutes
614o037 gument. without indeed we are step towards some FINAL end.-- production of higher animals-- perhaps,
637j58r ion! let him study Malthus & Decandoelle.-- The FINAL cause of innumerable eggs is explained by Malth
637j58r d by Malthus.-- [is it anomaly in me to talk of FINAL causes: consider this!--]CD consider these barr
468t112 day early, the great scarlet Poppy-- So that, FINALLY Fraxinella. with respect to nectary is same ca
568n019 cotts life. Tom Purdie, (beginning of Vol V) «FINALLY» says "he knew no more what was pretty & what
361d095 r. Blyth, at Zoolog. Meeting stated, that Green-FINCH, all linnets red-pole, goldfinch, hawfinch-- in
182b047 the resemblances & differences for instance of FINCHES of Europe & America. &c &c &c The new system o
261c070 lking of races of Man.-- black men, black bull FINCHES from linseed-- not solely effects of climate o
484z016 sting comparison to find how many of the small FINCHES walk at Maldonado & Patagonia compared with th
512q019 live Mamals (3) Do most Hawks eat stomach. of FINCHES-- do they throw up pellets-- (4) About hybrid
024r015 f some moving <point> power ¿ Submarine currents FIND instances; The whole coast of New Holland shoal
060r125 habitation above regions of vegetation.--«I can FIND nothing.» Mem Carolines quotation from Temple U
068r147 eous rock replace strata. & it is nothing odd to FIND them injected by veins & mass[es] [Fig. 8] {P}
089a019 of Saint Marc et des Gonaïves it is difficult to FIND stone not thus composed on the NE part more lik
172b008 species vary, <in> changing climate we ought to FIND representative species; this we do in South Ame
208b154 given in Congo Expedition We need not expect to FIND <species>, varieties, intermediate between ever
208b154 s, intermediate between every species.-- Who can FIND trace or history of species between Indian cow
209b157 lds = ∴ Islands & Artic are in same relation. We FIND species few in proportion to difficulty of tran
216b180 on hybrids) thus act.-- Now the point will be to FIND whether know varieties in plants do so.-- As in
222b203 to one parent, this is only character. & yet we FIND this same tendency (only less strongly marked)
254c048 d each other with the greatest rapidity"-- so we FIND species each class successively present modific
254c049 & no one system developed <is> not surprising to FIND many forms in Acrita,-- typical of other, (sure
271c107 ked to me at Zoological Society,, that you never FIND two «similar» groups of birds in two countries,
275c121 r Yarrell will mention in his work I am sorry to FIND Mr Yarrell's evidence about old varieties is re
281c137 excessively slowly. whether geologists would not FIND fossils such as they are-- My theory explains t
296c184 must be studied on Canary islands-- Endeavour to FIND out whether African forms. (anyhow not Australi
324c268 sticated animals see if law's cannot be made out FIND out from Statistical Society-- where M. Quetele
355d066 are» not «able to» tie themselves down, they can FIND loopholes) "It is well worthy of examination wh
385d165 , this is Lord. Moretons law.-- "How often do we FIND in the son, the character, constitution, & most
386d165 cause showing probably not education.-- Cannot I FIND some animal with definite life & split it, & se
398e006 btain here & there in order a scattered page; we FIND <great> sensible change in the institutions. &
399e006 time: but we ought in same bed if very thick to FIND some change in upper & lower layers-- good obj
402e020 of Borneo to the adjacent island-- In Sooloo we FIND the elephant-- in Magindanao several kinds of t
415e066 Ornithorhyncus,: since being cosmopolite, we do FIND his remains.-- Lima.-- caves.-- There being no
417e076 mferus» «Mammiferous» <vert> animal, which would FIND its place in the Systema Naturae.-- Mr. Knight
437e140 caped into a hole, where the old Eagle could not FIND it.-- The parent bird another day brought to h
471tf07 ays.-- most fantastic & use unknown.-- "<when we FIND such an endless variety of form in the same> or
484z016 e of wings It would be interesting comparison to FIND how many of the small finches walk at Maldonado
505q014 ed by care 13 Arum before pollen is shed can you FIND flys dusted with pollen from other flowers? Can
541m093 his enemy & not wish to strike him, but he will FIND it far more difficult to to look tranquil.-- He
542m093 thing, but without a most distinct will, he will FIND it hard to keep his lip from stiffening over hi
546m110 e lawn to see something, on her return could not FIND paper cutter, hunted in vain for it-- ten years
559m153 at... "Ay Sir there is much in analogy, we never FIND out." This unwillingness to consider Creator as
585n076 way, whirled, & then taken other way-- would not FIND its way back.-- ?? this is not instinct, but a
610o031 e so.--Looked at a direction book, but could not FIND out-- Directed his letter, & I observed he had
634j55r iferous animals originally terrestrial.-- for we FIND even in Cetacean traces of hind extremities.--
202b133 ng rigmarole article by S Hilaire on wonder of FINDING Monkey in France.-- of genus peculiar to East
232b246 in comparison of temperature of two countries» FINDING a very hot day, «in one», oh we will take a da
437e138 finches sometimes migratory.-- p. 103. Turtles FINDING their way to the Caymans from Honduras. good c
534m062 across. (Col Sykes compares this with pidgeons FINDING their way home-- there is something wrong in c
586n081 s going to sleep.-- "A bird has the faculty of FINDING its way, which in certain species is instincti
234b256 , from another from Swansea.-- Again Waterhouse FINDS certain varieties of a Harpalus. common at Sout
055r107 wonder what has become of the Basalt. Gone into FINE sediment Look at St Helena!!-- There are some a
118a103 littoral deposit always equal width --subject of FINE paper this would make.-- L'Institut. (1838) p.
160g090 ite with pinkish felspar;-- whole hill dark grey FINE grained. Much contorted gneiss «narrow sharp ri
162g104 buttresses, an[d] one alternate curved layer of FINE sand & small angular-- rounded pebbles-- dip si
163g106 rise) they found alternating layers of coarse & FINE & many Sea shells. My informant saw them himsel
204b139 lt the same Indian cattle & common produced very FINE Hybrid offspring, much larger than the dam, fro
206b147 ve successors. at present day. in looking at two FINE families one with successors «for» centuries, t
206b148 each other, and are acted on, just like the two FINE families «no doubt a different set of causes mu
216b183 s of England. Mr Bell of Oxford St'-. had a very FINE blood hound bitch which would never take the do
217b188 in time of ice transported.-- This gives room to FINE speculation.-- Are there many Northern genera p
252c042 from Macleay to Bicheno much excellent detail & FINE, views about Species-- MUST BE STUDIED: genera
260c069 effect on human offspring.-- white, snow.-- the FINE green of vegetation,-- ¿account for colour of b
283c145 ees with unequal distances between species. some FINE & some wide. which is strange if creator had so
294c175 live for some time at New York «instance of the FINE relations of adaptation of animals & the countr
306c214 imals-- At Angora «centre of Asia Minor» are the FINE-haired goats. which it is said cannot be transp
316c246 lls at present day same (or according to Sowerby FINE species) on coasts of N. America & England.-- b
334d009 -- «He recollects one hatch of hybrid geese very FINE.--"» How is it with plants? This indicates a rem
362d100 were then <pr> existing.-- he has also some very FINE recent drawing «of prize pidgeons» in 1834-- no
383d160 heasant.-- lower part of breast, each feather is FINE metallic green. <from» with tip & part of shaft
500q10a ccordance with range of species?-- Are there any FINE doubtful species from Van Diemen's Land? or New
512q019 ounds= An ugly calf <turns> sometimes turns into FINE beast. would its offspring have ugly calves. al
512q019 ts offspring have ugly calves. also turning into FINE beasts.-- For comparison with hybrids, is offsp
540m088 c Coleridge.-- Zapoyla p. 117, Galignani Edition FINE poetry, or a strain of music, when the mind is
582n066 rdner in his work» In the life of Hayd & Mozart. FINE music is evidently considered as analogous to g
599o006 Philosophy of Living. p. 264. "Architecture is a FINE amplification of two ideas in nature; a develop
281c137 uralist "splitting up his species & genera very FINELY" show how arbitrary & optional operation it is
281c137 rbitrary & optional operation it is.-- show how FINELY the series is graeduated.-- Dr Beck doubt if l
115a098 er was «successively» given of every degree of FINENESS. then most regular slope-- {P} if not course

Page ***(Key Word)***
028r028 The sea would separate quartzose sand from the FINER matter resulting from degradation of Feldspar &
110a086 ted on that side of salband. gradually becoming FINER grained & more compact on that side-- separatio
180b036 ween A. & B. immens gap of relation. C & B. the FINEST gradation, B & D rather greater distinction Th
275c120 d, with whole form of grey hound.-- picking out FINEST of each litter & crossing them with finest gre
275c120 out finest of each litter & crossing them with FINEST greyhounds.-- Sir. J. Sebright first got {P} p
337d021 or others parts of creation) & another nerve to FINEST vibration of sound.-- which is impossible.-- M
599o006 e; a developement of the thoughts expressed in FINGALS cave, & in the arched & leafy forests" Very go
537m074 doubted whether a man intentionally can wag his FINGER from real caprice. it is chance, which way it
537m075 of bad habit." as child is cured of sucking his FINGER by rubbing them with alum, so more slowly does
606o025 o, analyse (a) ones feelings when wagging one's FINGER-- one feels it in passion, love-- jealousy-- «
264c083 t so much-- Peculiarities of structure. as six FINGERED people are sometimes heredetary,-- yet these
265c084 ldren. ‖Lyell has story from.-- Beck about six FINGERED children heredetary‖ With respect to question
335d014 al with amputated limb.-- Heredetary <thr> Six FINGERED people, <Hill> «Lord Berwick» family with def
335d015 spring,-- slight deformities «as supernumerary FINGERS» (that is slight alterations of primitive stoc
525m025 says, perfect deformity, as an extra number of FINGERS.-- hare lip or imperfect roof to the mouth «st
540m085 Gardens touched pud. of young male & smelt its FINGERS. Seeing a dog & horse & man yawn, makes me fee
545m106 ey «(Mycelis)» I gave nut, but held it between FINGERS, the peevish expression was most curious <like
573n037 of face are more expressive than movements of FINGERS.-- like Kitten with mice.-- A person with St V
585n075 ely.--» [& we know from experiment of crossing FINGERS, that we only do know that it is one, when app
308c217 nd side for-- some month, & then when that was FINISHED kept it in-- <right> left, but I always for a
332d001 of selection owing to struggle July 15th. 1838 FINISHED. October 2d As a proof. what <trifling> «unkn
397eIFC ermere? How is Jackall & dog at Z. Gardens D E FINISHED July 10th 1839.-- Selected. Dec 15 1856 [not
520mIFC ct> Ants getting on Table. Col. Sykes) Private FINISHED. Octob. 2d. This Book full of Metaphysics on
105a069 when the inclination becomes less & ∴ tends to FINITE power) whereas sea. on coast, as long as expos
162g100 ly of last stream Friday Loch Lochy near Letter FINLAY Barom 30.267, A 68 Air 65 degree? <.194 372 ab
461t025 which undergo metamorphosis & are provided with FINS, & hence do not require sac.-- but the male in
153g054 ines enter side by, opposite entrance into Glen FINTEC a kind of landing place is formed Ben Erin sum
528m038 ry, of forms-- the beauty of some as Norfolk Isd FIR shows this, or sea weed, &c &c-- this gives beau
358d075 ssary to watch our meat, while in kettles on the FIRE, & on one occasion, not withstanding our vigila
538m079 yet Allen. W. remark about his slippers bad for FIRES, what is wrong in his head. & Babington's silly
023r009 ll sound and not petrified, and the jaw was also FIRM, out of which we pluckt a great many teeth, 2 o
054r105 overed above half a league of what is now Terra FIRMA & the extent of a league & a half a long the co
263c075 certainty of destruction; then he will choose & FIRMLY believe in his new faith of the lesser of the
281c139 ns of analogy being those last obtained.-- less FIRMLY fixed & therefore most subject to change.-- ma
285c153 ing «necessarily» excessively slow, they become FIRMLY embedded in the constitution, which other mark
311c227 which it may be seen swimming about. A Smith is FIRMLY believed in representation. certain birds in m
313c235 fancies, as well as men-- when habits much more FIRMLY impressed.-- we see in the Entomostraca. The s
360d092 that a small difference <of any kind>, if very FIRMLY fixed from long time, made no difference what
429e114 as our wild plants, we see how full nature. how FIRMLY each holds its place.-- When we hear from auth
537m077 imal, & this conscience or instinct may be most FIRMLY fixed, but it will not prevent other being eng
594n112 qually if man had appeared-- though instinct so FIRMLY implanted, birds soon <dis> learn to disobey i
479z005 iscussion good on Falklands birds Discussion of FIROLA,-- Salpa Anatifs without shells.! p 442.-- Pla
032r040 ould suddenly fall, (3) the case would be as at FIRST. & according to the greater or less time of res
032r040 ey of Copiapò & parts of coast of Chile.-- Must FIRST explain «top of» tidal band of action. This cas
046r078 considered as chemical retorts.--neglecting the FIRST production of Trachyte. look at Sulphur. salt.
046r079 to Volcanic theory. I want to ground, that the FIRST phenomem. is an inward afflux of melted matter.
046r080 . At St Helena there is a native mouse Did wave FIRST retreat at Juan Fernandez: the first great move
046r080 e Did wave first retreat at Juan Fernandez: the FIRST great movement was one of rise (any smaller pri
046r080 ilos: Transacts) «seems to» considers that fall FIRST movement (as in Peru 1746).--At great Lisbon Ea
046r080 ell as by the vertical as lateral movement.--At FIRST one would think movement. owing to water keepin
047r082 d the various accounts & see if fall is not the FIRST very evident movement.--The swelling first on b
047r082 the first very evident movement.--The swelling FIRST on beach I cannot understand, without (cs <[...
047r083). -- In great Calabrian wave did not sea break FIRST? I can imagine from local form of coast (as see
052r098 coast to Seaward. Retreating case in excess as FIRST case. When discussing Falkland soundings introd
055r107 d, in St Helena. Ascension. Azores. («sandstone FIRST gives» half demolished craters).--worn into mud
058r115 eat waves to obliterate all land marks.--At the FIRST it would though be easy to see on beach success
062r129 so commonly seen. at long distances; generally FIRST arrives:-- New Zealand rats offering in the his
063r130 consequent on lapse) being the relation.--As in FIRST cases distinct species inosculate, so must we b
068r146 est stage» when covered up beneath ocean).--The FIRST dislocations & eruptions can only happen during
068r146 dislocations & eruptions can only happen during FIRST movements, and therefore beneath ocean, for sub
093a031 n. L Institut p 247. 1837.-- The most infusible FIRST injected.-- Basalt: last because it could reach
102a061 elspar.?? Are the great crystalls, & the layers FIRST of felspar & then quartz &c, owing to separatio
102a061 o separation having taken place most gradually, FIRST the more fusible substance, & then the next bei
105a073 de is not great.-- Is there more degradation at FIRST angle owing to momentum. which the water has ob
109a083 seen.-- ¿ The preservation of dikes & ledges of FIRST-rate importance in showing not subaqueous remov
114a095 osed than other deposits.-- NB. because lowest. FIRST accumulated in bed of ocean With the exception
116a100 ain, like <re> a reef on a sea beach-- «p. 151» FIRST discovered «very small» bits of red granite bet
119a104 ian & Pacific Oceans.-- (2d--) does not explain FIRST formation of continents, if globe be considered
128a128 -- when subsidence takes place.-- Mountain will FIRST fall-- the problem will be falling of an arch w
132a137 ath. Phys. et Naturelles. Tom I. p 501.-- shows FIRST that data wholly insufficient to calculate rate
162g101 h due to Peat & Heather When it did not grow at FIRST-- relics destroyed.-- the Brook <about> Head of
174b015 apply to whole organic kingdom, when our planet FIRST cooled.-- Countries longest separated greatest
175b018 .-- becoming more complicated,; & if we look to FIRST origin there must be progress. if we suppose mo
178b032 e the children cannot be made intermediate, the FIRST children partake more of the mother, the later
179b032 bitches puppies are less purely bred owing to <FIRST> having once borne Mongrels he has thus seen th
179b032 n the mother was nearly quite white) in the two FIRST children How is this in West Indies --:Humbold
185b056 Heteromera, you have representatives (which at FIRST would be mistaken for) Carabidae, Crysomela, Sc
186b059 is what French call (atavism) Probably this is FIRST step in dislike to union, offspring not well in
193b091 -. Chamisso on Kamschatka quadrupeds Kotezebues FIRST Voyage Copied into list Entomological Magazine
199b119 ng banished, & given to Portuguese. priest.--In FIRST settling a country.-- people very apt to be spl
204b141 case of male widgeon, winged & turned on pool, FIRST season bred readily with common ducks.--» Kirby
206b147 between each lustrum, the number related at the FIRST start must be greater, & this number would vary
206b148 case,» May this not be extended to all animals FIRST consider species of cats.-- <& other tribes>.--
207b150 ginal Flora) and Dicotyledenous, which «nearly» FIRST appear «(p 321)» at Tertiary epock p. 330. Foss
215b177 x tells me that it is generally said.= How came FIRST species to go on.-- There never were any consta
219b193 Galapagos. Many trees Compositae, because seeds FIRST arrived «Ferns ditto.--» & hence formed trees]C
223b207 & introduction of Man. Nothing compared to the FIRST thinking being. although hard to draw line.-- -
227b227 highest organization intelligible.--may look to FIRST germ-- --led to comprehend true affinities. My
248c033 en two hybrids.-- As we see external influences FIRST affect external [for]m, so will the internal pa
248c033 , of external conditions, & of succession: the <FIRST> latter is most intimately connected with impor
249c034 , will cross & produce fertile offspring in the FIRST case it will either produce no of[fspri]ng or s
252c045 want of knowledge of time-- Analogy from three FIRST will give one almost certain guide ∵ because ti
257c057 clearly to indicate those very changes which at FIRST it might be doubted were possible,-- it has bee
258c062 scussion <after> about affinity & how one order FIRST becomes developed & then another-- (according a
266c091 both left the lands of their forefathers;-- the FIRST to escape the doctrines of Muhanmad, the last t
275c120 them with finest greyhounds.-- Sir. J. Sebright FIRST got {P} point on hackles on Bantams by crossing
285c151 on of even distant ones) the characters will be FIRST those of analogy, but will grow into affinity.-
285c152 and changes, keeping distinct from other & if a FIRST & last individual were put together, they would
290c162 & not kind» insects-- & vertebrata & plants. At FIRST classification on generation might appear an an
292c169 al-- when so the> are united (which probably is FIRST stage) the tendency to change cannot be great,
292c171 leep».-- an action becomes habitual is probably FIRST stage, & an habitual action implies want of con
294c175 of animals & the country they inhabit.--» & the FIRST one that bred one was diseased in its loins & a
294c177 rent species;-- Horse-- &c <Lonsdale says. that FIRST shee> State broadly scarcely any novelty in my
299c195 classification. here we have generative organs. FIRST character.-- In dioecious plants many of the Fe
303c204 f feet.= (Negro or father of negro probably was FIRST black at base of nails & over white of eyes,--
306c212 s easily gained in child hood.-- Young salmons. FIRST a species which lived in estuaries its taste. t
306c213 generation a captial character. (Owen) not for FIRST & grandest divisions. but for ones of very high
312c228 ; children heterogenous, he feels sure of this, FIRST offspring most like <parents> Mother.-- like do

```
312c231 a good bull & <be> the provincial breed, & that FIRST offspring thus produced are better, than those
313c234 tion or rapidity of specific change.? <One> the FIRST would be called. generic & other specific extin
315c242 ubject to sympathetic nerves-- The vividness of FIRST <thoughts> «memory» in children or rather their
315c242 markedly-- scenes in themselves accidental-- My FIRST thought of sea side-- Study Bell on Expression
317c251 me length of ears & length of limbs, so that he FIRST thougt only one species. & all hares on East si
318c253 me  in Audubon there is most curious history of FIRST appearance of the S. American Pipra Flycatcher,
318c254 rds.-- «in» some species «a Tanagra» Males come FIRST & the females in flocks. «as in English Nightin
318c255 ear.; and probably a «chance» wanderer like the FIRST pair of Pipra flycatcher.-- Bachman says he thi
326c266 igence.-- very good. Endlicher has published in FIRST volume of Annales of Vienna. sketch of south Se
331dIFC ny tendency to return to either parent.? Is the FIRST cross, which makes hybrids. productive like see
331dIFC [All Selected Dec. 14-- 1856]CD Towards close I FIRST cross owing to struggle July 15t
334d009 ommon> «China» & Canada Geese, & that they this FIRST cross were equally fertile with pure bred anima
334d010 h plants? This indicates a remarkable law, that FIRST cross <not se> plentiful, second absolutely ste
334d011 d in the same cart in loose chains, by being at FIRST beaten from her, & always accustomed to her.--
335d013 , or have none-- the argument does not apply to FIRST parents, because they are not new breed.-- the
335d013 parents,  because they are not new breed.-- the FIRST hybrids may be compared to animal with amputate
335d014 are they «abortive» twins.-- ‖ The fertility of FIRST cross, as stated by Fox, is very important,  as
335d014 x, is very important, as showing above facts as FIRST cross being new species, ‖ -- Are not  dreadful
337d021 iage.-- In my speculations. Must not go back to FIRST stock of all animals, but merely to classes whe
337d021 for if so. it will be necessary to show how the FIRST eye is formed.-- how one nerve becomes sensitiv
342d034 arly all F. W. Fish are Abdominals. ∴that order FIRST converted-- is it an old order Geologically? Ow
344d041 dmirable harrier from Ireland to Brighton Park--FIRST rate bitch-- tried to breed from her, but her o
345d042 . Mr Drinkwater thought that a <pure blooded> «"FIRST blood"» animal must have gone on for many years
350d054 art of God" Septemb 1,. It has been argued Man FIRST civilized. <note> add this in note. ¿mere conje
360d091 -- This Keeper has seen when sickly tigers have FIRST come over, insects somewhat like «between» lice
360d094 g very short tail.-- I can readily see that two FIRST might cross easier than two last. Sept. 11. N M
364d103 - showing hybrids can fare for themselves. ‖ + + FIRST year.-- The bird fanciers match their birds  to
370d116 about  production of Queens.-- Neuters are bred FIRST, «then males--» how has this been arranged-- Ne
374d134 onsidering infertility of hybrids inter se, the FIRST cross generally brothers & sisters, & therefore
376d136 the  Guaranis & Spaniards are almost White from FIRST generation., that with Quichuas the American ch
376d136 Monkey did not like a great coat made for it at FIRST, but in two or three days learn its comfort & t
383d159 ated by Hunter.-- Do testes, & ovaria when they FIRST appear occupy their proper positions,-- this wo
384d161 dary, the latter only being developed, when the FIRST <are> become of use‖ <Great characteristic of m
384d161 gth> In speaking of generation alway put female FIRST Will not even a fruit tree or rose degenerate d
387d169 be  two animals which have some peculiarity for FIRST time & if their <D & E> «all their offspring» i
387d169 rity strongly; they transmit with same force as FIRST pair, but to this tendency is added <that> the
387d169 s tendency is added <that> the 3d tendency from FIRST pair.-- Now if two of third pair of same peculi
387d169 uliarity breed they will have same influence as FIRST pair + tendency they inherited from second pair
393dIBC ties is uncertain do they mean  they cannot tell FIRST result., or that «hybrid» breed is uncertain Is
402e020 rd to this archipelago Octob. 13th.-- Kotzebues FIRST Voyage. Vol II. p 367. "The Fauna of the Sunda
407e041 rica & Australia. reason, why: Marsupiata, when FIRST introduced live & multiplied, specifically & in
417e079 resemblance  is permanent, or the similarity at FIRST births.-- it is the latter only that one refers
426e106 e fact that old varieties do not so much affect FIRST race, as it does indelibly the many  subsequent
428e112 to me, for otherwise breeders who only care for FIRST generations,, as in horses, would not care so m
429e113 plan.-- Whether we can or not trace history of FIRST appearance of varieties of domesticated animals
441e149 ls us, <be> «offspring» would be similar to <f> FIRST form.-- The great effect of conditions on offsp
451e176 e common-- probably, I should think grandfather FIRST of race & if so, fact for my theory Cocos  Isld
452e181 als in Gilolo.-- p. 134: Birds of Paradise were FIRST procured from Gilolo p. 253 In isld of  Bunwood
453e183 m.-- ask Dr Holland cases where peculiarity has FIRST appeared.-- "Storia della Riproduzione Vegetale
460t019 from  a few types originated, we must go to the FIRST origin of the world.-- our present organic bein
465t080 sualty as, bones of Mammalia in caves:-- :argue FIRST case of bones (New Red Sandstone) & then go  on
472s02v of  flower & stigma dusted.-- <I think> When It FIRST alights, it cleaned sucker & <I think> pollen w
482z011 lklands Isd 8 waders. 22 palmipedes: out of the FIRST 9.:4 raptores. Falco poliosoma -- novozelandiae
491q01v urs afterwards of nearly related plant & see if FIRST pollen produces any effect, as in case of woodp
491q01v ench Apple tree «with abortive stamens» answers FIRST question in negative.-- Questions Regarding Pla
492q002 sing one sex & not other: which generally fails FIRST?-- Mal[e] 10. Henslow says semi-doubl flowers a
493q004 rding to Mr Yarrell the latter ought: either in FIRST breed or permanently.-- (2) Cross two half-bred
524m022 hand to pretend to feel her pulse.-- What fails FIRST?-- How is this?-- Does memory bring in old idea
525m025 e illnesses of the foetus.-- some mothers. have FIRST dead children, then children which were short t
526m029 magination.--» Thinking over the scenes which I FIRST recollect, «at Zoos» they are all things, which
533m058 gous or identical, with bird knowing a cat, the FIRST it sees it.-- it is frightened without  knowing
535m069 f Acanthosoma grisea described [not located] as FIRST caused by will of Gods. «or God» secondly  that
547m113 eas, direct effect of perception by senses fail FIRST, as whether I had pulled the bell??)-- It may b
557m149 - Jealousy probably originally entirely sexual; FIRST try «to» attract female, (or object of attachme
560mIBC t them to avoid danger Do they frown, when they FIRST see it? Charles Darwin 36 Great Marlborough  St
568t018 ng organization, comparison of sensations would FIRST take place, whether to pursue immediate inclina
570c024 ir.-- Again a master says I will see you damned FIRST." the man shrugs his shoulders & replies nothin
572c033 mt somebody gave me a book in French I read the FIRST page & pronounced each word distinctly. woke in
573c037 . constantly. when refusing food, turn his head FIRST to one side & then to other. & hence rotatory m
580c060 to origin of language My father says old people FIRST fail in ideas of time, & perhaps of space-- in
581c066 companying certain bodily actions]CD.¿ but what FIRST caused this bodily action. if the emotion was n
581c066 used this bodily action. if the emotion was not FIRST felt?--«without «slight» flush, acceleration o
582c069 g its nest; it knows its object but not result (FIRST time of building?), but not the means of perfor
587c089 try  contiguity of parts of Brain.-- Mackintosh FIRST clearly insisted on assoc of ideas & emotions.
592c102 he males by odd gestures. In one of the six (?) FIRST Vol of Silliman's Journal paper showing that th
592c103 46-47. Paper. like. Sir Ch. Bell on Expression «FIRST Croonian Lectures by Parsons.» following  pages
596nIBC slight convulsions. are the muscles of the face FIRST affected?-- Can shivering & trembling be consid
601c08b atry-- Inductions morales et physiologiques The FIRST of these books I daresay good. 1. Sensation  is
601c009 is conveyed over whole body (which it may be in FIRST case. as when the excised heart is pricked) and
607c025 chance) circumstances. As man hearing Bible for FIRST time, & great effect being produced.-- the  wax
614c037 HC] Instinct appear like heredetary memory; but FIRST memory in many cases cannot be acquired by expe
614c037 fetime Heredetary memory not so wonderful as at FIRST appears, & no too great advantage.; for superio
618c041v one  to other: no not only thus, for if day was FIRST, we should not think night an effect.]CD Cause
629c55v ought,  shame. right & wrong comes into mind in FIRST case-- seeing how shame is accompanied by blush
629c055 ains the feeling of right & wrong.-- arrived at FIRST <rationally> by feeling-- reasoned on, steps fo
634j55r s.-- How are we to explain this.-- Did reptiles FIRST inhabit seas.-- Were they then killed out «by t
634j55r mammifers then take their place? Would they not FIRST occupy the Poles? Is this origin of Polar attri
023r010 dly talks about the immense quantities of Cuttle FIRST bones floating on the surface of the ocean, bef
030r037 mestone are they allied to the jaws of the Cocos FISH Rio Shells argument for rise In Cordillera, the
079r177 ature & ses phenomenes."-- Ulloa's Voyage, Shell FISH purple die, marevellous statements on, Vol I, P
088a014 they may picked up beneath the trees---- Are any FISH seed-eaters. This important in transport of Fis
088a014 Fish seed-eaters. This important in transport of FISH Let a Hawk fly at Heron.-- Ceratophytes  common
177b025 creases it--> settles it ¿We need not think that FISH & penguins really pass into each other.-- The t
177b026 akes it excessively complicated.» {P} Is it thus FISH can be traced right down to simple organization
184b054 od Has not Macculloch written on same changes in FISH Mem. Rabbit of Falklands described by Q. & G. a
194b097 there  never was common progenitor to Mammalia & FISH. when there now exist such strange forms as orn
197b112 ther organ.-- Cuvier on opposite side; Is Vol of FISH p. 59.]CD Cuvier has said each animal made  for
198b113 sible. for instance take birds animals, reptiles FISH-- Conditions will not explain status (Perhaps c
205b143 s into mouth-pieces of Crustacea. Vol II. p 75 a FISH which emigrates over lands is a siluris, p. 123
205b143 rates over lands is a siluris, p. 123 A climbing FISH. p. 122 A Terrestrial annelidous animal p. 347.
206b149 e many anomalous lizards living; or of the tribe FISH.extinct. or Pachydermata, or of coniferous t
211b162 p express, a real affinity & affinity-- whales & FISH.-- Progeny of Manks cats without tails: some lo
213b170 seems  the most organized fishes lived far back. FISH. approaching to reptiles in Silurian age-- How l
222b205 rease in character. (which perhaps is) Case with FISH-- as some of the most perfect kinds the  shark.
227b227 nges is discovered. for speculating on future. !.FISH never become a man.-- Does not require fresh cr
234b255 mpted to be introduced I isle of France p. 170. «FISH introduced» Hump backed race of cows from Madag
243c020 sicans: (¿cassicans Australian form? p. 27. many FISH of Taiti found at <New> Isle of France: xx inst
243c020 f wide range work this out-- L. Jenyns, about my FISH New Zealand & New Holland fish very  similar.--
```

243c020 Jenyns, about my fish New Zealand & New Holland FISH very similar.-- NB. Lesson method of generalizi
244c021 It would be very important to show wide range of FISH & shells in tropical sea, it. would demonstrate
257c058 wood> Coniferous wood in Coal Measure.-- highest FISH in Old Red Sandstone.-- Nautili in----. it is u
257c058 rupeds, but how there come to be, many genera of FISH &c &c at present day.-- It is ASSUMPTION to say
259c062 ed, then how is this effected by-- for instance, FISH being excessively abundant & tempting the Jagua
276c124 catus Tyrannus Sulphureus if compelled solely to FISH. structure would alter.-- It is a difficulty ho
302c201 eve the case. = any animal really connecting the FISH & Mammalia, must be sprung from some source ant
302c201 these two families, but we see analogies between FISH.-- Birds same remarks. Characters of analogy.--
316c245 mes & a few Anphidesmas.-- this is remarkable.-- FISH & drift sea weed-- may transport ova of shells.
342d034 has somewhere met conjecture that all salt-water FISH were once salt water (as they almost must have
342d034 s) but Ogleby well answers that nearly all F. W. FISH are Abdominals. ∴that order first converted-- i
347d049 hanges. then animals must tend to improve.-- yet FISH same as, or lower than in old days: «for a very
370d117 .-- thus Vertebrate blend with Annelidae by some FISH.-- But birds quite distinct.-- Collect cases of
374d133 n which reptiles have been found. p. 426 Sauroid FISH in Coal, true fish, & not intermediate between
374d133 ve been found. p. 426 Sauroid fish in Coal, true FISH, & not intermediate between fish & reptiles-- y
374d133 h in Coal, true fish, & not intermediate between FISH & reptiles-- yet osteology closely resembles re
379d151 135. Natural History of the Caspian. Fresh Water FISH!! ¿adapted to salt water?-- peculiar species, c
387d169 ry .p. 96. Vol I. Notice the Syngnathus, or Pipe FISH the male of which receives <young> «eggs» in be
388d173 of gemma.-- The manner in which Frogs copulate & FISH shows how simply instinctive the feeling of oth
405e032 what was said to be hybrid between silver & gold FISH-- Octob. 26th. If. hereafter. M. angustidens be
406e035 it-- Goat & Moufflon will not breed.-- p. do.-- FISH of Teneriffe. St. Helena & Ascension most speci
412e057 - I have already given various examples The Pipe-FISH is instance of part of the hermaphrodite struct
421e090 .= curious as showing generality of law. even in FISH: ⟨.do. p. 236-- on Hybridity in ferns.-- ⟨.do p
421e091 269. Annals of Nat. Hist 1838 on «a» freshwater FISH peculiar to Ireland. ⟨do p. 283. on the dark ea
423e095 me probably always have done so, as the simplest FISH &), my answer is because, if we begin with the
423e096 here been a retrograde movement in Cephalopoda & FISH &c can live.-- supposing, such to be the case, i
435e134 6. Many facts given of high temperature at which FISH now bear a very large proportion to other mollu
436e134 fish &c can live.-- Lyell says that naked cuttle FISH is really hermaphrodite. & <thinks> even oyster
445e159 ays it is not sufficiently proved that any shell FISH & reptiles in former case-- Reptiles & Birds &
448e168 er connecting classes like Toxodon «In orders»-- FISH in India.-- Windsor Earl-- Eastern Seas p. 229.
453e182 Vol II p. 650. Long attested account of fall of FISH-- which he divides into two divisions, one of w
461t025 ation of P. Fries most curious paper on the Pipe-FISH.-- Embryology p. 97. for Man Chapt see Yarrell
461t025 paper deserves fresh study & whole order of the FISH! difficult-- yet suggested. (vipers tooth also
463t055 . have foreseen, sailing, climbing & mud-walking FISH one step lower in America-- How curious all neg
473o3ov vonian-- How strange no plants in our Devonian-- FISH Purple die Marvellous stories Ulloa's Voyage Vo
479z005 P} p. 140. Fléche of Quoy et Gaimard Ulloa shell FISH.-- See Kings drawings.-- for real name Birds of
481z009 Cruz. Molina Vol. I. p. 244. Baccalao. migratory FISH & kill them in hour or two «My Father made hens
496q006 e (8) Make Duck eat Spawn, eggs of snail, row of FISH Man has abortive muscles to his ears.-- p. 313
634j54v e separate movements in the Holocentrus ruber (a FISH) is really hermaphrodite. & <thinks> even oyster
639j28r . D p. 258. «grinding» teeth in <stomach of> sun-FISH, in mouth of swine & in stomach of lobsters-- a
640j167 & vertebrata much so.-- so far true, but do not FISH offer a most striking anomaly to this. Have the
641j29v night» Sailing lizards. squirrels & Opossums «& FISH»: flying lizards.--Mammalia. C. D.--
246c026 y are varieties & not species.-- Vol :694. King-FISHER of Europe, (Alcedo ispida) from Molluccas. sca
181b045 complete, if birds were fitted solely for air & FISHES for water. If my idea of origin of Quinarian s
192b088 s long gone past.. & still more so «known» with FISHES & reptiles.-- In mere eocine rocks. we can onl
213b170 In Phillips. p. 90. it seems the most organized FISHES lived far back, fish approaching to reptiles a
257c058 ur <only> question is not, how there come to be FISHES & quadrupeds, but how there come to be, many g
287c157 omer to plain observer⟨ +++. between Mammalia & FISHES, one penguin, one tortoise shows hiatus-- but
369d114 aracter most perfect in <male> «hermaphrodite» (FISHES have no secondary characters.--)p. 49. (wonde
374d133 dely from living forms.-- p 458 Upper Silurian, FISHES oldest formation highly organized.-- do. p. 46
392d180 imple form of life are monooecious. Will ova of FISHES & Mollusca «& Frogs» pass through birds stomac
414e060 Work-- Decb. 4th.-- Why has the organization of FISHES & Mollusca (& plants???) been so little progre
417e079 parents.-- Lord Moreton's law cannot hold with FISHES, «& there are mule fishes» & reptiles & those
417e079 law cannot hold with fishes, «& there are mule FISHES» & reptiles & those which <lay> «have» their e
419e085 . M. Eichwald has published Fauna of Caspian.-- FISHES fresh water kinds. (yet living in the salt?.)-
423e097 the organization of the different beings, (all FISHES to the state of the Ammocoetus) Crustacea to--
023r011 anic activity.» That axis was produced, from a FISSURE in a deep & therefore weak part of the ocean's
041r066 one. a circle,.{P}, had in its middle a short <FISSURE> «vein» terminated each way, which little vein
104o067 In Discussion on dikes argue impossibility on FISSURE going right through superincumbent mass (varyi
104o067 ywhere--, in granitic areas &c &c volcanos {P} FISSURE dike.-- thus dikes terminated Solubility of fl
121a122 Roy district could have been elevated without FISSURE & unequal.-- where were cracks?--? How came th
135a145 ide intercepting plain & dividing it-- Hopkins FISSURE at {P} .-- G. J. Malcolmson has described form
023r010 expulsion of fluid nucleus through» faults or FISSURES, produced by the elevations of those mountain
038r058 certainly appearance leads me to believe mere FISSURES filled up.--the appearance will here be the s
045r075 nomena I do not believe that the extraordinary FISSURES of the ground at Calabria were present at the
056r109 nd a work?--In valleys one is not sure whether FISSURES may not have helped it, or diluvial waves. bu
074r165 concretionary; concretions <dt> determined by FISSURES as in septaria. (& Chiloe case, at least core
103a063 nts, because not wedge-formed.-- Hence fill up FISSURES-- If dikes effect of horizontal elevation exc
103a063 dikes effect of horizontal elevation excepting FISSURES from above unite with those from below. would
110a086 «with» cleavage, veins of pyrites, few curious FISSURES; base in part. block not crystallized Salband
124a120 against lateral injection. from probability of FISSURES being prolonged to surface. see p. 181 on do
264o079 not,-- but probably would.-- the world now being FIT, for such an animal.--man, (rude, uncivilized ma
450e175 bably not original there)-- shows these isld not FIT for horse. Forrest--. (p. 270) says many wild ho
530m044 tfulness after bad accidents:-- After journey, a FIT of = gout, has affected his memory of everything
591n100 ng what others by habit (not instinct) think not FIT, as cannabalism, is held in abhorrence.-- all th
606o024 sings Laocoon p. 125-- says new subjects are not FIT for painter or sculpture, but rather subjects wh
305c211 etism-- principles of irritations sleep walking. FITS, laught &c&c Man & Man may have some relation t
385d163 ung take distemper very readily & are subject to FITS.--» «there is great difference between hybrids &
521m007 Anson. who told a story of hunting «-- habitual FITS.--» which my Father thinks is mentioned in the
181b045 t would probably be not complete, if birds were FITTED solely for air & fishes for water. If my idea
198b113 66}CD With unknown limits, every tribe appears FITTED for as many situations as possible. for instan
230b236 al is either effect or adaptation,∴ animal best FITTED to that country when change has taken place, N
268c100 emed to consider it owing to one of each, being FITTED for transport ¿may it not be explained by mere
308c218 habits Insects & birds are the only two tribes FITTED for water, air, & land, (Macleay has this rema
318c252 es. than feet & hind legs of these white hares, FITTED for regions of snow.-- Acclimatisation.-- Bach
350d056 ed, nature does nothing in vain, therfore organ FITTED to animals place in creation.-- thus senses, &
377d147 [not located] :Hence, also structure not really FITTED for water, only habits & instincts-- The young
386d167 apted to every minute change, they would not be FITTED to the slow great changes really in progress.-
407e039 to the extreme.-- Therefore species, which were FITTED for such a preeminently equable climate. might
414e064 re deadly struggle,, namely which have the best FITTED organization, or instincts (ie intellect in ma
573m036 ent at the artificer.--» Our faculties are more FITTED to recognize the wonderful structure of a beet
584n071 f anyone has taken the Woodpecker as an example FITTED for climbing, his arguments partly fall, when
278c132 opaean forms, were able to escape to some more FITTING country.-- if Toxodon had been found in Africa
031r038 . East India Archipelago. «Aleutian Arch.--» V. FITTON. Australia: cases in Europe.-- Auvergne. very
022r006 es.--Wide limits of this mineral in Australia. FITTON'S appendix Would Slate. & unstratified rocks sh
053r101 Radack & Ralix Islds? In my Cleavage paper Dr FITTONS Australia case must be quoted at length. The L
107a079 ter» but cause most difficult (better conductor) FITZ Roy's Case of S. Maria & Tubul applicable to An
112a088 soundings in Mouth of S. Cruz in connection with FITZ Roys fact of elevated block of stone.-- & Caldc
134a141 sthmus of Pen. Tres Montes.-- as by subsidence ∥ FITZ Roy refers to ∥ & Rocks p. 375. on the sounding
064r135 re no outer coast.--important effect.--? Capt. FITZROY. -- Limited Volcanic action & limited earthqua
323c269 of each Chapter» Crabbes Life June Is. King & FITZROY To be read Humbold. New Spain-- Much about Cas
530m043 s seen other cases of similar nature.-- --like FITZROY in sleep giving directions,-- & forgetfulness
539m082 ross me, bringing up old indistinct ideas of FITZWILLIAM Musm. I was amused at this after seven years
056r111 n how water separates.--(intertropics at present FIX lime). <Also Volcanos separate.> Volcanos blend
312c231 e new cross.-- In the Bantam clubs, they used to FIX on the kind wanted, colouring of each feather we
526m031 but indefinitely, he chooses (but what makes him FIX!? <)>-- frame of mind, though perhaps he chooses
532m054 system is affected, & by habit the mind tries to FIX upon some object:-- When a man, child or colt ha
052r097 faults.-- Necessary form; as long as coast line FIXED.--[Fig. 5] {P} * Slope necessary for seaward tr

```
086a004 are  fossil shells.-- accustomed to such terms "FIXED as the land, stable as the water"-- It may be w
171b004 own difficulty with full grown individual «with FIXED organization» thus being modified,-- therefore
195b101 le consequen let animal be created, then by the FIXED laws of generation, such will be their successo
207b152 siders forms in recent volcanic islets not well FIXED.-- Peron thinks Van Diemen's land long separate
281c139 nalogy being those last obtained.-- less firmly FIXED & therefore most subject to change.-- may accou
281c139 nge.-- may account for certain organs not being FIXED, <whi> in some genera, which are most fixed  in
281c139 ing fixed, <whi> in some genera, which are most FIXED in others. In analogy it is not the relation to
284c147 r of forms depends on the external relations (a FIXED quantity) & on subdivision of stations & divers
285c152 ive at true affinity doubtful A species is only FIXED thing with reference to other living being.-- o
285c153 re & Man.-- The constitution being heredetary & FIXED, certain physical changes at last become unfit,
308c217 wo species.-- «discussion until, unless it were FIXED what a species means» civilized Man, May exclai
344d042 children went back to either paret, & breed not FIXED. though she resembled a harrier & her husband w
345d042 le have gone on for 50 or 70? years-- now «well FIXED» breed,: Jones says Sussex cattle were all whit
350d055 hat young of Cirrhipedes can move & see, parent FIXED,-- young of sponges move.-- young of Cochineal
350d055 ts move about & see, parent «(2)», female «(I)» FIXED & blind: -- Macleay observed all these facts pr
355d067 retrospective as showing what organs are little FIXED-- (<also> Hunters law of monstrosity with regar
360d092 small  difference <of any kind>, if very firmly FIXED from long time, made no difference what its kin
387d168 been repeated several times, that it becomes FIXED in blood.-- Looking at ovum of mother & ovum in
397e003 been  conducted almost! invariably according to FIXED laws: And since the world began, the causes of
411e055 ch the individual steps in the series have been FIXED, to study the physical causes. «All Cuviers gen
416e071 n- (contradicted by Plants). & as there are no FIXED, land animals. so there are true hermaphrodites
418e084 is  pliable, such modifications, become as much FIXED, as if added to old individuals,, during thousa
439e144 think,, or that these varieties have become as FIXED as species, & prefer their own pollen to that o
441e151 onstrous conclusion, that every organ is become FIXED. & cannot vary.-- which all facts show to be ab
535m069 n.-- Now it is not a little remarkable that the FIXED laws of nature should be «universally» thought
535m070 suspects that our will may <be> «arise from» as FIXED laws of organization.-- M. le Comte argues agai
537m077 this  conscience or instinct may be most firmly FIXED, but it will not prevent other being engrafted.
611o034 l accidents, & in such cases modifications bear FIXED relation to such accidents. But such tissue <mu
629o055 tion.-- CD[In similar manner our desires become FIXED to ambition. money, books &c &c.-- <]> the "sec
536m073 ee will make change in man.-- the real argument FIXES on heredetary disposition & instincts--.-- Put
553m135 , to give a cause (& no one being apparent, one FIXES on imaginary beings, many vicarious, like ourse
377d139 ies-- time no element in making change, only in FIXING it: only circumstances. a contingency of time.
574n040 ion between very limited reasoning powers & the FIXING of habits,-- for instance the Birgos opening a
417e075 the name of species, may be owing to the little FIXITY of organization, in the two races,. owing to t
438e141 f it shall be difficult to show that <time> the FIXITY of characters «from antiquity» prevents  their
093a033 volutas. prove regular mud bank at Bahia Blanca. <FL> Flustra identical. recent & bone bed.-- Novembe
496q05a ers. compared with single & see whether grains FLACCID, as Koelreuter describes Kill Sparrow after fe
557m147 ge «no» & ready to dash at prey streched out & FLACCID, when furious «with fright» back absurdly arch
551m129 35  "as ideas come & the pulse rises, or as they FLAG & something like a snow-haze. covers my whole i
363d102 - Barn Owl <the> in the former place breeds in <FLAGS> «thick vegetation» in swamps-- (owing to barns
593n111 1 took flight & flappered across pool to bed of FLAGS I was astonished & having looked round saw at c
594n112 ever  before seen one, yet all-- flew to bed of FLAGS. hernes are common. not unlike in size in the a
346d047 -- Mem pink spots on Albatross, on some Gulls. FLAMINGO-- (Spoonbill Wader. Ibis)-- laws of plumage m
639j28r of swine & in stomach of lobsters-- analogy in FLAMINGO & Duck, Ornithorhyncus «externally». petrel &
405e031 n.-- yet retained a trace of horizontal mark on FLANK.; & tail. & kind of semilunar {P} mark on each
425e105 islds.,  have peculiar forms.-- on the southern FLANKS of Alps.-- many peculiar plants on single moun
593n111 if by word of command, they all took flight & FLAPPERED across pool to bed of flags I was  astonished
073r175 ion of ravins, and the slope of the mountains (FLAQUEZA del cerro) have been parallel to the directio
533m061 remarke as now advanced; for I caught it like a FLASH.-- strange if judgement remains, where reason i
546m111 ftewards whilst at a meal, she suddenly like a FLASH without any assignable cause, remembered she ha
527m033 er se) & causes the mind to create short vivid FLASHES of images & thoughts.-- Poetry. the latter tho
100a055 local heat. Ask Capt. Beaufort, whether, water FLASHING into steam, would Babbage.-- Webster Phillips
041r065 driving  by the want of water.--I believe in all FLAT countries. years of drought are common.--Mr Lye
109a083 ood this will show effects.-- analogous to broad FLAT sand beach. {P} -- De la Beches argument of low
115a098 en most regular slope-- {P} if not course enough FLAT top. ended by abrupt slope {P} each stratum wou
147g015 f <plain> space is thickly studded with ridges & FLAT topped hill/ do alluvium. NB In one part pure s
148g027 ad of Loch Dochart <Nea> Above Spean Bridge many FLAT terraces one above much inclined towards  river
148g029 en be due to rivers-- By Roy Bridge, a tongue of FLAT land, with terraces of each side of the two val
151g042 rom shelf 3d & almost meet, but are separated by FLAT bottomed strait. connecting flat on one side wi
151g042 re separated by flat bottomed strait. connecting FLAT on one side with irregular gravel plain of othe
154g057 d, site N W of Ben Erin {P} Shelf of Glen Guoy FLAT peat plain divortium aquarium-- tidal channel--
160g091 oulder Cairn leet mere Haberclador Isthmus broad FLAT Loch {P} XX Barom 28.92 A 75 Air 70 degree? Ist
160g092 XX Barom 28.92 A 75 Air 70 degree? Isthmus broad FLAT peat mass-- (general character in these mountai
162g103 me right arm of River Tarf <it> Has a very long, FLAT divatium aquarium with, left of Bright.-- like b
495q05a eets of Paper. covered with some sticky stuff in FLAT places & see whether wind, on «dry» windy  day,
024r015 w Holland shoals much: Dampier remarks on great FLATS on the NW coast:-- 8 leagues, from Sydney 90 fa
109a083 uch preservation certainly is lessened.-- Coral FLATS. argument for Heaping up.-- very good this will
466t096 on  in after life of Plants-- also goodness of FLAVOUR in fruit-- all affected by cultivation  during
505g014 = (8) Abortive Thyme seeds weather wet--? Linum FLAVUM put in Spirits which plant seeds? (9) Melons f
235b275 cultivated plants produced by seed.-- Lychnis.-- FLAX.-- Read Swainson [blank] In production of varie
453e183 esculent vegetables-- how is it with hollyoaks, FLAXES &c &c? Mr Herbert in letter says distinctly, t
360d091 me over, insects somewhat like «between» lice & FLEAS. sticking on them, but never in an animal, that
478z003 <quatre> cinq lignes de long et volent. p. 208 FLEAS only appear in winter in Paraguay p 207  Slight
563n001 obeying  instinct of running hare is stopped by FLEAS, also by greater temptation as bitch: or dogs d
078r174 nimal with setae like my undescribed[.] p. 140. FLECHE of Quoy et Gaimard.--D'Orbigny has described i
479z005 has  described my animal with teeth {P} p. 140. FLECHE of Quoy et Gaimard Ulloa shell fish Purple die
481z010 Berre «p. 8» (I think Planariae) Sagittella, or FLECHE «p. 8» my little animal with horns. Madrepores
594n112 that he is not dangerous-- wild-ducks would have FLED equally if man had appeared-- though instinct s
540m088 . as by reading a book.-- Consider this.-- "The FLEDGE-dove knows the prowlers of the air" &c &c &c s
451e178 t afterwards says native tribes can live there)» FLEE during 8 months out of 12.-- the largest mammif
275c120 rcumstance Ld Orfords had breed of greyhounds FLEESTEST in England lost courage-- (Bull-dogs are used
284c149 known in single ones--. viz. Macleay letter to FLEMING p. 32 "where it (mode of generation) varies ac
286c155 ter word meaning descent.-- A tree is taken by FLEMING as emblem of dichotomous arrangement which  is
286c156 aper as drying plant, all brought in one plane FLEMING Quarterly review says nat: fam: of Willows con
286c156 a  slight one It will be necessary from manner FLEMING treats subject to put in alternative of Man cr
287c157 ted by distinct miracle. Macleay letter to Dr. FLEMING. Philosophical Magazine & Annals. 1830 (?)." i
320c276 - contains very little Macleay's letter to Dr. FLEMING. & Review of latter in Quarterly Sir J. Sebrig
424e102 Carrion Crow is found in Edinburgh.-- Why does FLEMING consider them varieties & what says Jenyns  to
491q01v le Plants & only allow <few> one flower (5) Dr FLEMING. Philosop. of Zoolog. vol 1. p. 427-- says bie
501q011 lk about Mr Knights theory with Henslow.-- Dr. FLEMING says yes. (29) Are there RACES of Lupine, Stoc
235b272 te regular gradat in Man poor trash Lyell 1024 FLEMINGS Philosophy of Zoolog Royle on Himalaya Plants
506g015 lar position.-- (39) What does he think of Dr. FLEMINGS statement of Sweet Williams & Stocks, being p
324c268 malayan. types. Smellie. Philosophy of Zoology FLEMING. ditto Falconers remarks on the influence  of
565n010 ran  away with new Lavaters,-- Ye Gods!:-- says FLESHY lips denote sensuality (p 192 Vol. III Octav.
472s02r eral flowers I examined some days ago-- This Bee FLEW from yellow to yellow & purple heartease withou
594n112 probably few had ever before seen one, yet all-- FLEW to bed of flags. hernes are common. not  unlike
073r161 ere soft, or redissolved soft.--/ is there any FLEXURE <fr> in the fragmentary jasper.--do undulation
366d112 he people on the NW. Coast blinking to keep out. FLIES might be used" The wild ass has no cross. how c
466t093 d by orange freckles on {a} upper petal; bees & FLIES seen directed to it-- The Humbles in crawling o
534m063 istaken in saying common wasp cuts off wings of FLIES from intellect. but it does it always instincti
196b105 confine certain birds which have wide powers of FLIGHT; but are there any genera, mundine, which cann
273c111 d, there is a preeminently aerial,-- formed for FLIGHT & great movement in the air, & likewise rasori
353d060 ther <like» parent stock, or not. Now wings for FLIGHT-- therefore ostrich not. The peculiar «Malacca
468t105 e as bit of chopped horse hair with legs & take FLIGHT-- Yet we have crosses-- I see Bees almost ever
593n111 ddenly, as if by word of command, they all took FLIGHT & flappered across pool to bed of flags I  was
641j29v usca we have analogues The stillness p. 276) of FLIGHT of Owl remarkable, [gained by very different p
090a023 instead  of 90 stones in many cases there were FLIGHTS of stones of large numbers (& how few cases re
105a071 e. <a> the alluvium would form a succession of FLIGHTS of steps; if one lake then we must suppose bar
```

154g061 ld it deposits River terraces often descend by FLIGHTS the terraces if the largest has hollowed out m
161g097 re were several obscure but not far continuous FLIGHTS above it-- (NB the buttress or pass at Isthmus
269c102 s of New Holland, supplementary to Appendix to FLINDERS Voyage by Brown.-- great space seems to act p
324c268 s relative to age of Marriages Brown at end of FLINDERS & at end of the Congo Voyage Decandoelle. Phi
514q21. arf plants on Wellington Mountain described in FLINDERS= Alpine Australia Flora= Banana's seedless--
098A048 f feldspar in gneiss is identical with layer of FLINT on calc.: sandstone. (& as I believe most strat
027r027 ncret, is connected with frequency of shells in FLINTS in Chalk New Providence more hilly than others
109a084 ientific Memoirs Edited by Taylor Ehrenbergh on FLINTS in chalk must be studied-- though I do not thi
404e025 & M. Edwards p. 330 distinctly states that the FLIPPER is a mere simple modification of an organ pres
119a106 ite shows that the metamorphic rocks have just FLOATED over the absolutely fluid pool.-- (this is sho
357d074 uld have no plants were it not for seeds being FLOATED about.-- I must state that. the <p> mechanism
021r005 .-- [not located] La. billardiere mentions the FLOATING marine confervae, is very common within E. In
023r010 ut the immense quantities of Cuttle fish bones FLOATING on the surface of the ocean, before arriving
102A059 e way out to sea.--¿ effects on bottom a thing FLOATING some way from coast is driven on to it.-- rol
218b192 w many seeds, might be transported some blown--FLOATING trees Thrushes & bunting & coots-- «(Turdus G
311c227 good.-- Blainville Ovington's Voyage to Surat, FLOATING isld. off coast of Africa p 69. with tall gr
414e065 n is not unlike that of animals transported by FLOATING ice.-- I agree with Mr Lyell., man is not an
606o024 ch we know, it is therefore the embodying of a FLOATING idea.-- as statue of beauty, is of the "beau
318c254 diurnal, migrates singly by night.-- others in FLOCK, these birds seem clearly directed by kind of c
346d048 rving character & breeding in & in-- Nonsense a FLOCK of more than 100.-- Agrees, «nearly» with. the
318c254 s «a Tanagra» Males come first & the females in FLOCKS, «as in English Nightingales» -- other birds (
339d026 mprobability to animals being distributed after FLOOD (!) according to affinities!. confounds, like W
040r063 kes or Avalanches (Glaciers very rare) to cause FLOODS in valleys, which must aid in preserving the t
040r062 n that side.-- Add from M. Lesson. character of FLORA to New Zealand, which agrees with St Helena in
182b046 ion terrestrial,-- if in any in the Cryptogamic FLORA but not atmospheric type. Hence probably only f
190b076 observation of Mr Brown, about peculiarities of FLORA. on East & West. ends of New Holland. diminishi
199b117 ified.-- Propagation explains this.---- Ancient FLORA thought to more uniform than existing. Ed. N. P
206b150 oal formation? p. 320 <Think> States Cryptogam. FLORA formerly common to New Holland?! p. 320. Says C
207b150 ediate between vascular or cryptogam. (original FLORA) and Dicotyledenous, which «nearly» first appea
209b156 extent Von Buch,.-- Canary Isles: French Edit. FLORA of Islds very poor «(p. 145)» 25. plants. 36 St
209b156 (?).-- [«Mem:» Juan Fernandez]CD. From study of FLORA of islands; "ou bien encore on pourrait au plus
218b192 re. Carmichael. Tristan D'Acunha, a list of its FLORA. is given Mr Don remarked to me. that some good
219b195 . or northern plants «No» CD[Mem. the antarctic FLORA must formerly have been separated by short spac
226b221 a.-- Especially moderately large ones.-- Is the FLORA of <S. America> Tierra del Fuego like that of N
232b247 n> Iceland would have possessed a most peculiar FLORA «& north of Europe»-- As Europaean forms have t
245c024 of teats» Coquille Voyage Durville has written FLORA of Falkland Islds, where is it? All the Society
269c100 icable to N. Hemisphere (NB.-- Examine Abrolhos FLORA with this view) Tristan D'Acunha, St Helena &c
296c184 plains this. but no other will.-- St. Helena (& FLORA of Galapagos?) same condition. Keeling Isd «sho
314c238 ays so in preface.-- Mr Brown says character of FLORA, N. Zealand & N. Caledonia with a dash of New H
314c239 S of Australian GENERA. good case. rather large FLORA. (150?) Mr Brown did not observe scarcely any A
403e022 leg, hare lip &c &c. in Vol II p 363 account of FLORA of pacific, given in my coral paper Oct 14th Ma
425e104 wich Isld are very similar to Galapagos-- study FLORA-- what general forms.-- are the Labiata nearest
497q007 g by Bees.-- Henslow.-- (1) Character of alpine FLORA of Tierra del Fuego and Entomology of.-- most i
497q007 n intermediate country (2) Any known changes in FLORA of countries during last century or two.-- wher
510q017 ts for me-- what? What does Blume say on alpine FLORA of Java? Has Schow written on double creations
513q21. y not being mature at same time on same plant --FLORA of Australian Mountains.-- Is setting of fruit.
514q21. ountain described in Flinders= Alpine Australia FLORA= Banana's seedless-- 20 varieties in mountains
514q21. do they also differ in different countries-- on FLORA of African Islds-- names of Plants found on mou
314c238 otected however it may be effected.-- Prodromus FLORAE Norfolkicae. 1833 Steph. Endlicker (He will gi
241c016 ille's Voyage to see if Lessons' remarks on the FLORAS can be trusted The changes in species must be
315c241 . 184 Botany of Bonin. "grande analogie avec le FLORE du Japon", some Europaean & Sandwich species &
089a019 e if water fell. -- <Keys off extreme point of FLORI[DA]> Excellent paper on Erratic blocks in Alps.
183b052 y domesticated animals like races of man.-- M. FLOURENS. .Journal des Savants.-- April 1837. p. 243 i
401e017 graft pears on apples. they will live but not FLOURISH-- a medlar may be Grafted on pear. Mountain-a
539m084 41 Origin of man now proved.-- Metaphysic must FLOURISH.-- He who understands baboon <will> would do
046r078 d over «whole» surface; how comes it they do not FLOW out together? How are they eliminated.--«Sulphu
573n038 told not swallow spittle. will have involuntary FLOW & desire to swallow.-- tells himself not to tur
573n038 ase spittle, effect of thought is to make saliva FLOW, & therefore thinking of subject, even when wis
573n039 re thinking of subject, even when wishing not to FLOW-- flow it will.-- My father told Miss. C. of th
573n039 king of subject, even when wishing not to flow-- FLOW it will.-- My father told Miss. C. of the bad c
574n041 - November 27th.-- Sexual desire makes saliva to FLOW «yes, certainly»-- curious association: I have
574n041 ch one sexually loves is probably connected with FLOW of saliva, & hence with action of mouth & jaws.
575n045 s hurting brain, like a wound hurts body-- tears FLOW from both, as when one burns end of nose with a
633j54r d by volcanic force, for these Marsh plants. All FLOW from some grand & simple laws.-- 4 «Study Cuvie
044r072 entral peak small; yet great body of lavas have FLOWED from centre-- Pisolitic balls occur in the Ash
064r136 tity of matter from CREVICE of Andes--therefore FLOWED towards it. a mass on each side 3000 ft thick
188b068 out.-- For instance ever so many seeds of white FLOWER. all would come up white, though planted in sa
266c091 m Cape of Good Hope continuing for some time to FLOWER at their own periods.-- Arcana of Science & Ar
384d162 e manner in the plants which have male & female FLOWER on same stem.--» so that Molluscous hermaphrod
389d177 uch breeding in & in (those which have solitary FLOWER) exotics brought from foreign country. («annua
391d174 V. Supra <v. infra> p 179, continued from Is a FLOWER bud produced by union of two common buds??? Am
392d180 Examine English weeds in Hot. Houses will they FLOWER Make Hybrids with moths, where fecundation can
439e143 zing each other, better than the pollen of same FLOWER,-- as it tends to show my view of <i>nfertilit
441e151 h are generated by buds alone or roots, & never FLOWER, so there may be animals as Coralline, or othe
446e162 ghts doctrine.-- Case like Corallina--«Does it FLOWER anywhere?-- Yes on the continent is there more
447e164 l species of Lemna sometimes though very rarely FLOWER [bu]t the one does on the continent-- well cha
466t091 g brown, whilst both others were in nearly full FLOWER Maer June/41/ Rhododendrum-- nectary marked by
466t094 t.» In Columbine nectaries are placed all round FLOWER as they are in Crown-Imperial Lily & many othe
467t103 pecies of Lupine,» two wings. & when the Lupine FLOWER is perfectly ripe & pollen abundant filaments
467t105 everal times on Beans Rough.--green-cabbage «in FLOWER»-- swarmed with meligethes & small Staphylinid
468t106 - Yet we have crosses-- I see Bees almost every FLOWER-- Blue-bells-- wild-raspberry--leeks-- Flowers
468t112 es almost P every minute to Fraxinella «& from <FLOWER> plant to plant.»-- to my grt surprise-- I fou
468t112 «is» almost roofed by united filaments.-- This FLOWER hostile to intermarriage!!xx In Phil Transact.
469t135 ./41/. Watched plants of Fraxinella, with seven FLOWER stalks for ten minutes. it was visited by 13 B
469t135 ing each Bee visits 10 flowers in «minute» each FLOWER will be visited in 28 minutes-- say then each
469t135 r will be visited in 28 minutes-- say then each FLOWER is visited 30 times a day is considerably unde
470t178 n shower On many Papilionaceous; all wh. are in FLOWER «I saw Bees;»-- on Monk's Hood, brushing over
472s01r 60 seedlings not one came up true.-- colour of FLOWER & foliage not «being» like the true P. bractea
472s02v (as a needle becomes covered) so whole sides of FLOWER & stigma dusted.-- <I think> When It first ali
491q01v Raise only single Plants & only allow <few> one FLOWER (5) Dr Fleming. Philosop. of Zoolog. vol 1. p.
491q01v ring function of seeds-- (6) To hybridise EVERY FLOWER on melon & see whether fruit affected. Mr. B.
495q05a osiphon densifolium «an annual» <sleep» «closes FLOWER» on all gloomy days.-- The «garden» Coronella
495q05a places & see whether wind, on «dry» windy day, «FLOWER garden on gravel walk» will drift many seeds=
496q05a y-- with animal reproductive system.-- cover FLOWER-- put artificial flowers-- also do with honey-
496q006 how does impregnation take place male & female FLOWER in same receptacle (8) Make Duck eat Spawn, eg
496q006 nerations. daisy. Fever-fuge Groundsil.-- gilly FLOWER will break & become double.-- There is a doubl
498q007 customed to rich soil, when placed in very poor FLOWER, but not fruit-- -- Do not orchards become unp
498q008 ame species of Willows grow in same situation & FLOWER at same time. Has H. seen group of different s
499q009 lation of number of grains of pollen in any one FLOWER (17) Catch Bees, Butterflies-- Syrphus-- Melig
500q010 ed Oranges. Spallanzani Essay-- Figs 2 kinds of FLOWER annually.-- Periwinkle. (not asclepiadae. «in»
501q011 aps indexed by secondary characters-- in double FLOWER. do Henslow Speaking of Thyme doubts about sti
502q11v tame carrot-- Parsley & Fennel. Verbena Compare FLOWER of different Cabbages most carefully to see if
502q11v ges most carefully to see if variation equal in FLOWER with leaves.-- strawberries How <soon> «early»
502q012 it grow in open air in Sweden. Linnaeus found 2 FLOWER. which had anthers removed, did not become imp
504q013 Parnassia whose stamens move one after other to FLOWER & Menyanthes whose pollen bursts before flower
504q013 flower. & Menyanthes whose pollen bursts before FLOWER is open-- -- No (6) There is apple with branch
505q014 stion= (5) Open more Horned oranges.= (6) Figs, FLOWER.--Passion Flower. (as it is required to impreg
505q014 ore Horned oranges.= (6) Figs, flower.--Passion FLOWER. (as it is required to impregnate it artificia

505q014 n from other flowers? Can flys' escape from old FLOWER= (14) Has planted seeds of Geranium pyrenaicum
506q015 of flowers-- Nectaries-- In Monooecious «order» FLOWER occupy particular position.-- (39) What does h
506q015 ioecious plants the Papilionaceous structure of FLOWER-- Ground nuts (42) How are Orchidiae fecundate
506q15v reads by seeds= (44) Zostera. Has he seen it in FLOWER? does he know Botanist who does-- What is Rupp
506q15v - What is Ruppia Bennett says in same state. of FLOWER 45. Charlsworth. vol II. p. 670-- oats cut dow
513q21. when   crossed R. BROWN-- will pollen act on any FLOWER before stigmas expanded-- in reference to Lobe
513q21. regnation ever regularly take place in unopened FLOWER-- <doubt> disbelieve this in Bauers case of or
514q21. - Ask about Pinks & Solanum impregnation before FLOWER open. (An. des Sci Where is Boerhaave's paper
515q021 of fruit trees far north in Scotland-- do they FLOWER-- do they live healthily, or does fruit merely
515q021 cation.-- N.B. Benthams remarks, where parts of FLOWER are reduced from normal number, they are apt t
603o014 s not painted-- <music> very good article-- why FLOWER beautiful? ¿even to children S. Jenyn's Inquir
633j53v place.-- Kolreuter mentions some hybrid, whose FLOWER great tendency to break off p. 292. Mac. has l
635j56v . Mentions the many cases, as in Papilionaceous FLOWER, where such care seems to be taken that the an
504q013 of orchard.= At Shrewsbury one branch of Rhod. FLOWERED later.-- effect of accident?? (7) Which. Rhod
505q014 nted seeds of Geranium pyrenaicum. small white-FLOWERING gorze common in Norway No Questions regarding
272c107 ich I brought from Australia, probably live in FLOWERS & has Elytra. formed from developement of some
291c167 sexes were separated.-- in plants we have some FLOWERS monoecious & others dioecious. some flowers he
291c167 me flowers monoecious & others dioecious. some FLOWERS hermaphrodites & others not??? ╓ The death of
299c195 cter.-- In dioecious plants many of the Female FLOWERS unimpregnated Babington We see gradation to ma
390d177 do those which are Monocotyledenous have many FLOWERS same Spath, as they have only one bud.-- Every
420e090 ht these fertilise other shells, as insects do FLOWERS.-- Mem. Spallanzani's experiments showing how
434e129 ts only very slightly abortive & bed of female FLOWERS will sometimes produce a few seeds,-- -- Ruscu
434e130 in which the Male plant sometimes bears female FLOWERS, the organs are most clearly abortive, so that
441e150 ation" had better be shown-- soil on colour of FLOWERS, Hydrangea -- black bullfinches-- & all variet
442e152 oduced.-- The fact just alluded to of Northern FLOWERS, throws enormous difficulty in the way of Mr K
442e152 of generations-- Gorze in Norway, which never FLOWERS!!-- <How did it get there? whether> According
460t017 etermine the kind of variation «& sporting» in FLOWERS & domestication of animals Aug. 26th.-- When i
466t094 s Asks about Pinks in Crown-Imperial Lily & many other FLOWERS-- My view of <variety acquired» « <character>
466t095 n to foetus <fr> before sex developed-- Double FLOWERS & colours breaking only hereditary characters,
467t100 dried some».-- some with no division in young FLOWERS. The abortive stamen are of useful height.-- I
468t105 . pollen very minute--not excessively abundant FLOWERS not attractive, very small--stigma rather larg
468t105 ive, very small--stigma rather large & rough-- FLOWERS common-- many winged thrips, covered with poll
468t106 flower-- Blue-bells-- wild-raspberry--leeks-- FLOWERS which thought very unattractive-- Found Rhubar
468t111 n certain days Humble seem to frequent certain FLOWERS, to day early, the great scarlet Poppy-- So th
469t135 visited by 13 Bees-- & each examined very many FLOWERS.= 22d.-- /during several succeeding days «many
469t135 inutes eleven Humbles came & each visited many FLOWERS-- Saw Bees frequent these flowers till late in
469t135 visited many flowers-- Saw Bees frequent these FLOWERS till late in evening-- On rough calc. 280 flow
469t135 wers till late in evening-- On rough calc. 280 FLOWERS-- allowing each Bee visits 10 flowers in «minu
469t135 alc. 280 flowers-- allowing each Bee visits 10 FLOWERS in «minute» each flower will be visited in 28
470t153 plants will be collected & resown.-- Humble 22 FLOWERS of Egg Tree in one minute Great Humble 17 flow
470t153 wers of Egg Tree in one minute Great Humble 17 FLOWERS of Larkspur on two plants in do Humble 24 flow
470t153 wers of Larkspur on two plants in do Humble 24 FLOWERS of small Linaria in do Domestic do 6 Campanula
470t177 ious because a Composite) Asparagus very small FLOWERS & as much shut up, frequented by «many» Bees &
472s02r y Bee go to them. Yesterday remarked that many FLOWERS had suddenly withered, & to day saw very odd d
472s02r rom clump to clump, & insect proboscis in many FLOWERS, on one of which pollen was routed. wh. was no
472s02r ollen was routed. wh. was not case, on several FLOWERS I examined some days ago-- This Bee flew from
472s02v on Humble-- & more pf same fly Two more of the FLOWERS withered.-- Sillimans Journal <vo> 1842. p. 14
491q01v of Zoolog. vol 1. p. 427-- says biennial-wall-FLOWERS & scarlet Lychnis can be propagated by cutting
491q002 hether the viviparous grasses & onion, produce FLOWERS, like the Oxalis from C. of Good Hope mentione
492q002 ls first?-- Mal[e] 10. Henslow says semi-doubl FLOWERS are those whose stamens are monstrous, how the
495q005 llen of Red Cabbage «mixed with own pollen»-- (6) Dust FLOWERS of other cabbages & see whether there will res
495q005 e whether there will result hybrids-- (6) Dust FLOWERS of one branch of Cabbage with pollen of other,
496q05a nsect-like Orchis-- & final cause of beauty of FLOWERS-- contrasted by Kirby-- with animal reproducti
496q05a is use of Bee Larkspur= =Toad Orchis= How many FLOWERS-- also do with honey-- What is use of Bee Lark
496q05a ive system.-- -- cover flower-- put artificial FLOWERS in minute do they visit?? good=!! Examine poll
496q05a they visit?? good=!! Examine pollen of double FLOWERS. compared with single & see whether grains fla
496q05a w pellet. ejected. done Examine pollen of such FLOWERS as do not seed or seed rarely-- Magnolias. «Az
496q006 s Is the common Fig Dioecious-- are its female FLOWERS always barren-- if not how does impregnation t
497q007 gnated.; which part of stigma?-- (4) As Papil. FLOWERS appear difficult to cross, are there unusually
499q009 <Ne> In Oenothera bush.-- (19) Theory of mock FLOWERS in Hydrangea (20) As Hop is Dioecious-- seedsm
502q11v erimentise: for gradation in structure Compare FLOWERS of wild & tame carrot-- Parsley & Fennel. Verb
502q11v ike cock-pheasant: said not to sit on own eggs FLOWERS in short turf. for abortion. or for sterility
504q013 re is apple with branch in middle of tree with FLOWERS near end of orchard.= At Shrewsbury one branch
505q014 an you find flys dusted with pollen from other FLOWERS? Can flys' escape from old flower= (14) Has pl
506q015 ary, sexual characters.-- Stature, position of FLOWERS-- Their smell-- form of flowers-- Nectaries--
506q015 e, position of flowers-- Their smell-- form of FLOWERS-- Nectaries-- In Monooecious «order» flower oc
507q15v f pollen is requisite.-- Brown's paper 43. Any FLOWERS of Keeling Dioecious, or Monooecious, besides
513q21. emales produce. Polygam. trioecia. (are female FLOWERS ever productive? Smith says many trees in Trop
541m091 ed to repeat the word, & think of qualities as FLOWERS are dichogamous Zostera-- Knights notion of po
546m109 smell of Partridge--, man's taste for smell of FLOWERS, owing to parent being fruit eater.-- origin o
024r013 Cauquenes. coming from the [...] Cordillera & FLOWING The gradual shoaling of the water to more than
072r159 wn.--no mountains Mackenzie has talked of lava FLOWING up Hill; ¿what does he mean?) Consult Dr Holla
106a075 ation in all probability would be greater when FLOWING over (B) than when at (C) its tendency would <
131a136 2. On Frozen soil of Siberia.-- facts of water FLOWING from beneath frozen crust in America Richardso
160g092 tains & not ridges) between arm of Glen Bright FLOWING into E. end of L. Oich, & waters flowing into
160g092 right flowing into E. end of L. Oich, & waters FLOWING into west end with obscure terraces on one sid
161g098 of which the source is a lip with the new shelf FLOWS into canal between L. Lochy & Oich. is a brook
473s05v from L. Groznerat, «on road to Bethgellert» wh FLOWS by Tremadoc. but can tell them from lake S. of
473s05v can tell them from lake S. of Moel Siabod. wh. FLOWS into Conway by Bettws & there joins streams fro
023r010 e Cordilleras as arising from «the expulsion of FLUID nucleus through» faults or fissures, produced b
032r041 f rotation» & a <circulation owing> rotation in FLUID matter of globe. must there not be a circulatio
035r048 ngular displacement, consequent of injection of FLUID rock-- Try on globe. with slip paper a gradual
038r057 rely there must be «somewhere» below a field of FLUID rock.--In the discussion it will be better not
039r061 analogy is now perfect <The grand propulsion of FLUID rock, which elevates a continent> We are more a
045r077 ometer? In a subsiding area. we may believe the FLUID matter instead of afflux (always slightly oscil
046r078 & not in chemical nature, or has a subterranean FLUID mass itself changed.--No. -- Yet the fluid gran
046r079 nean fluid mass itself changed.--No. -- Yet the FLUID granitic mass under <[...]> less pressure might
047r083 y with <Mr Lyell's> idea of an injected mass of FLUID rock In Patagonia plains. long periods of rest
056r111 is all of granite: In discussing circulation of FLUID nucleus,--the similarity of Volcanic products «
068r147 roblem dislocate strata without ejection of the FLUID propelling mass. If one inch can be raised then
069r149 matter from Cordillera. HORIZONTAL movement of FLUID matter not (for instance) expansion of solid ma
088a013 where piece turned over axis or hinge no doubt FLUID.-- analogy as continental elevations slow. so w
099a049 idently small scale of concretionary action all FLUID at once, the films vertical. Ascertain law of a
107a079 ntre of earth same heat, then change in form of FLUID centre would lift with it isothermal line, but
119a106 hic rocks have just floated over the absolutely FLUID pool.-- (this is shown by the softness & curvat
120a107 believe we can trace the outlines of what were FLUID undulations-- the equal movements of Glen Roy r
122a114 n to equilibrium.-- it cannot be equilibrium of FLUID, but of solid. because if of fluid, the waters
122a114 uilibrium of fluid, but of solid. because if of FLUID, the waters of the ocean would obey that Law. a
122a114 view the degrading action must prevent internal FLUID arriving at equilibrium so soon from; crust bei
122a114 d» crust under ocean, became thicker, then when FLUID moved [...] {P} August 25. I saw metamorphic co
125a121 from Sir. W. Herschel's views earth originally FLUID, then cooling process must go from surface towa
125a121 e kept up far higher from circulation of heated FLUID or gases under pressure.-- {P} Lyells view of t
126a123 ssure increased or under surface, would not the FLUID matter be driven upwards & so conduct heat?-- H
133a140 wly heat travels; & therefore the abysses where FLUID rock has been ejected must remain fluid for an
133a140 s where fluid rock has been ejected must remain FLUID for an enormous period: now when we see how man

269c101 thickness in that period & no ways assisted by FLUID currents which, may take place in Metamorphic a
315c242 ects of irritability, though means injection of FLUID different from contraction of fibre)-- it is mo
360d090 id not go to heat. but parts swelled, though no FLUID came from them.-- showing how gradually every <
416e070 imals analogy leads one to suppose that seminal FLUID fluid, (& not dry as in plants) therefore, grea
416e070 analogy leads one to suppose that seminal fluid FLUID, (& not dry as in plants) therefore, great diff
420e090 experiments showing how little of the spermatic FLUID fertilized spawn of frogs.-- Annals of Natural
432d124 nces According to my theory no land animal with FLUID seeds can be true hermaphrodite Man probably as
443e155 no true hermaphrodite in beings with which have FLUID sperm.--]CD I utterly deny the right to argue
068r146 of dislocation (NB. dislocation connected with FLUIDITY of rock ∴ «in earliest stage» when covered up
122a113 onsider profoundly all consequences of EXTREME FLUIDITY of earth.-- study different forms of earth as
127a126 the result of heat.-- will it bear on central FLUIDITY.-- do p. 137. Lord Tullamore found Sulph of S
104a068 ure dike.-- thus dikes terminated Solubility of FLUIDS varies with temperature ¿ with pressure? Salt
417e077 erstand the necessity of a relation between the FLUIDS of the two as in the grafting of trees.--]CD
073r164 stilbite. grammalite. pyenite. native sulphur.. FLUOR spar. bayte. asbestos garnets.--carb & chrom. o
075r167 d»; sulfated Barytes very «un»common in Mexico. FLUOR spar only in certain mines. «Vol. III» "In gene
073r162 ssociation of lead & silver. Sulp. of Barytes: FLUORIC. Barytes:-- Humboldt. New Spain. Vol III. p. 1
532m057 tion of death, which many animals put on.-- The FLUSH which accompanies passion & not sweat is the <s
581n066 emotion was not first felt?-- «without «slight» FLUSH, acceleration of pulse. or rigidity of muscles.
539m084 1 Cant IV or V.) as pale & trembling. & not as FLUSHING & with muscles rigid.-- How is this? dealt wi
093a033 . prove regular mud bank at Bahia Blanca. <fl> FLUSTRA identical. recent & bone bed.-- November 8th 1
254c049 s mouths in Polypi, Surely not correct view of FLUSTRA or Ascidia spicule in sponge. stomachs in infu
569n021 sure &c accompanying such memory.-- A Melody on FLUTE & Epic poem, opposite ends of series or harmoni
485z016 bits & plumage so very similar to some of the FLUVICOLAE?-- The Birds seem to move much further North
088a014 . This important in transport of Fish Let a Hawk FLY at Heron.-- Ceratophytes common in Northern seas
257c057 tance of ground woodpeckers.-- birds that cannot FLY &c &c. seem clearly to indicate those very chang
258c061 r in the aberrant groups.-- It is having walking FLY catcher, woodpecker &c & which causes the confus
318c255 arated the birds might yet remember which way to FLY.-- There is a kind of Wren (Bebyk??) which seems
444e158 species.!!-- p. 203 Chaetodon squirting water at FLY.-- instinct, for how could experience teach dist
445e158 objects.-- far better case than chicken pecking FLY.-- "whilst the shell stuck to its tail" as menti
472s02r dusky & broad never saw such a one before-- Saw FLY 21 «this Heartease withered on Monday.--» alight
472s02v genus-- & a small common Humble-- & more of same FLY Two more of the flowers withered.-- Sillimans Jo
495q05a by suspending magnet within & see which way they FLY.-- (9) I have noticed leaves covered with Honey-
512q019 n in stomach of birds-- Mem: how many miles they FLY in few hours Zoological Soc (1) Do the animals t
535m064 erving them from «the» sun & enemies-- would not FLY away, but bit pencil when touched with it-- do n
554m141 o in with a stick, if he drops it, the bird will FLY at him-- Knowledge.-- Sept. 13th It will be good
564n004 en only hatched few hours placed on table & when FLY ran past it. cocked its head, & picked it-- Here
564n005 chicken so as to seize small moving object like FLY.-- young partridge can run even with its shell o
639j28r hamaelion.-- C. D. Sucking feet in Frog. Walrus. FLY. Gecko &c. Prehensile tail. in Monkeys & Marsupi
203b137 lus forficetus <has a wide range> is a tyrant FLYCATCHER doing the service of a swallow I think we ma
318c253 of first appearance of the S. American Pipra FLYCATCHER, which is now becoming common-- likewise of
318c255 chance» wanderer like the first pair of Pipra FLYCATCHER.-- Bachman says he thinks the Mocking thrush
485z016 efinite letter Mem Orpheus--becoming tyrant-- FLYCATCHER-- shown by habits & plumage so very similar
273c111 variety,:.in the Tyrannidae.-- Milvulus-- Even FLYING woodpeckers, with powerful wings, but tail st
283c143 lements & not habits as shown by frigate Bird & FLYING eagle.-- Hawk Gould seemed to think, that wido
318c254 mon «kind» migration of America) migrate singly FLYING few miles every day «generally by night» -- ot
367d113 1 in Encyclop of Anat & Phys. describes, a high-FLYING bat, which has the power of inflating its body
433e127 whole volumes.-- The fact of tumbling pidgeons; FLYING high all together & then tumbling, far more we
445e159 orphoses to habits of animals & takes series of FLYING mammifers-- says lemur.-- volans, has skin bet
544m102 ferent from when awake.-- peculiar sensation as FLYING. (No memory of past events?) or influence on o
634j54v care p. 309. says the ribs in Draco support the FLYING membrane?!!-- that the phalanges have separate
641j29v Sailing lizards. squirrels & Opossums «& fish»: FLYING lizards.--Mammalia. C. D.--
505q014 care 13 Arum before pollen is shed can you find FLYS dusted with pollen from other flowers? Can flys
505q014 flys dusted with pollen from other flowers? Can FLYS' escape from old flower= (14) Has planted seeds
595n117 Eyton's horses becoming <white> with <lather> <FOAME> & sweat, when hearing merely hunting horn-- as
068r146 e surrounded by continent.-- change of volcanic FOCUS.-- <it is certain, if strata can be> Problem di
289c162 ndensor, «+++ must (on my theory) =supported by FOETAL lower developed forms.=» (NB waterhouse says o
509q017 icle in--? Combs in combless Poultry-- Teeth in FOETAL state: Mr. Horner. On Mr Tremenheres Scottish
216b181 after births, like aphides.-- Case of boy with FOETUS developed in the breast.--looking as if many o
291c167 , it our admiration of ourselves.-- The idea of FOETUS being of one «both» sex«es». is strongly suppo
313c236 g how children come to suck or other actions in FOETUS of Mammalia, or &c.) Generation becomes
344d041 eating foetuses, as young of Marsup. is sucking FOETUS.-- August 23d The Rev R. Jones gave an admirab
351d057 ot move per saltum-- yet does nothing in vain!! FOETUS of man undergoes metamorphosis., heart altered
351d057 .-- Monstrosities, kind of determined by age of FOETUS.-- As Larvae may be more perfect (as we use th
355d067 unters law of monstrosity with regard to age of FOETUS. distinct consideration) Now in different SPEC
367d112 , why hybrids are infertile. supposes that when FOETUS is forming the ovum within it, is forming «& t
385d164) have contained perfect teeth & hair-- showing FOETUS has gone on growing-- I believe same has happe
389d176 er, or very close to either.-- Male & female as FOETUS one sex; & therefore both capable of propagati
390d178 , as they have only one bud.-- Every individual FOETUS would reproduce its kind was it not for the ne
431e123 Cline on the breeding of Animals, p. 8. size of FOETUS to perfect animal are adapted by foreknowledge
444e158 the transformations from the egg, or larva. or FOETUS in proportion to male parent p. 8. his whole d
466t095 with smell.-- These qualities have been given to FOETUS <fr> before sex developed-- Double flowers & c
525m025 etary.-- other deformities are illnesses of the FOETUS.-- some mothers. have first dead children, the
211b163 y M. Serres on Molluscous animals representing FOETUSES of Vertebrata, &c 1837 p. 370 Owen says Nonse
344d041 rpillar. very valuable facts-- they are eating FOETUSES, as young of Marsup. is sucking foetus.-- Aug
067r143 eum April 1836 (p302) Coleccion de obras. 2 Vols FOL: Buenos Ayres 1836: W. Parish?? «by Pedro de Ang
021r005 same.--longer axis in line of Cleavage. laminae FOLD round them; Quote this. Valparaiso Granitic nod
217b184 of mare covered by blood horse & Carthorse two FOLDS [not located] Mr Don gave me instances of one s
472s01r ngs not one came up true.-- colour of flower & FOLIAGE not «being» like the true P. bracteatum; all s
472s01v black spot at base, one paler with less riged FOLIAGE & no black spot & a third considerably paler,
430e119 specimens of an Inoceranus from the Gault of FOLKSTONE, which is exactly intermediate between I. con
430e120 like I. sulcatus.-- Both species are found at FOLKSTONE.-- it is unnamed this intermediate one.-- r
055r106 inclining a little to the West: the veins which FOLLOW this direction are thought to be the <oldest> «mo
074r165 apagos vein. vein of secretion.--metallic veins FOLLOW mountain chain. there after NW <W>.-- «same ch
198b114 nd idea god giving laws & & then leaving all to FOLLOW consequences.-- I cannot make out his ideas ab
210b161 owe their origin to very early stage) & which, FOLLOW certain laws according to species. present an
227b224 then cease to know the steps.> although D E F. FOLLOW close to A. B. C. we cannot be sure that struc
231b242 ter connexion in quadrupeds, <bu> plants do not FOLLOW by any means.-- Ostriches.-- Hippotamus only A
249c035 this not connected with wide range of animals. FOLLOW this out, where species of same genera in two
258c062 -- (according as parent types are present) must FOLLOW after there is proof of the non creation of an
289c160 ird, having habits of a Grebe, structures might FOLLOW.-- examine structure of this bird & get accoun
291c165 assification, because structure has tendency to FOLLOW it, or it may be heredetary & strictly point o
318c253 d many most curious case-- -- the birds seem to FOLLOW narrow bands, certain kinds as gallinules taki
410e051 only be effected by sexes: All the above should FOLLOW after discussion of crossing of <species> indi
417e077 CD[The similarity of child to parent appears to FOLLOW same law in two of «the» same «species», varie
568n016 rning it by heart. Do not our necessary notions FOLLOW as consequences on habitual or instinctive ass
568n017 ts instinct & pain if held.-- if tempted not to FOLLOW it, by greater temptation, if memory of its ow
592n103 onvulsions.-- «Paper» must be referred to, if I FOLLOW up this subject & a reference to Brun's work.-
620o046 n present» give pleasure.) & which man ought to FOLLOW-- it is his duty to do so.-- So we say a point
620o046 m & strengthen his instincts.-- so man ought to FOLLOW certain lines of conduct, <although> even when
621o046 is passion shortlived, it is to his interest to FOLLOW the former; & likewise <that the> then receive
621o047 Hence man must have a feeling, that he ought to FOLLOW certain lines of conduct, & he must soon neces
621o047 on necessarily learn that it is his interest to FOLLOW it. even when opposed by some natural passion.
023r012 to Sharks distributing fossil remains: Sharks FOLLOWED Capt. Henry's vessel from the Friendly Isles.
146g014 300 ft above Loch.-- From this point could be FOLLOWED up to neighbourhood of Tyndrum where a large
161g096 slope, which from a distance hid it, could be FOLLOWED for at least 2 miles on dead level «by eye» t
306c212 general argument) & therefore down the stream FOLLOWED ebb tide, therefore got into habit of going d
555m143 from others, slow cautious, angry cross look, FOLLOWED by protrusion of lips, in which respect resem
591n099 rrence it is because instincts to woman is not FOLLOWED; good case of instinctive conscience.-- Why d

```
*************************************************(Key Word)*************************************************
615o36v me one day met him, with Mark riding instantly FOLLOWED, me and for five minutes every now and then h
620o044 afterwards thinks why was such an instinct not FOLLOWED for a pleasure now though so trifling he feel
620o045 hat our feeling, that the instinct ought to be FOLLOWED is a consequence of that being part of our na
620o045 e child, which of its instincts are best to be FOLLOWED.-- Yet even at this time, malevolence,, when
023r012 Small granite Isd. capped by Calcareous rock; FOLLOWING Curvature of hill; states could discover no s
356d069 g fact to law only merit if merit there be in FOLLOWING work.-- The history of Medicine, the extraord
388d172 an. having given milk. -- testis & ovaria The FOLLOWING views show that transmission of mutilation im
406e036 cow  pox.-- case in Spain of pustular disease FOLLOWING handling sheep-- all cases: d degree p. 354--
434e129 guide.-- Lychnis April 3d.-- Henslow tells me FOLLOWING facts: believes that «only» red Lychnis grows
541m090 opposite extreme of this desultory thought is FOLLOWING out such an idea, as effect of sea on coves w
541m091 inventive thought.-- Examine frame of mind in FOLLOWING changes during fall of sea.-- Is the effort g
564n004 hunger, «of» death & for the satisfaction of FOLLOWING conscience, obeying habits, & dread of misery
568n017 ed habit» [Thus shepherd dog. has pleasure in FOLLOWING its instinct & pain if held.-- if tempted not
592n103 ression «First Croonian Lectures by Parsons.» FOLLOWING pages contain remarks worthy of attention  p.
620o044 & reasoning on its action, & on the results FOLLOWING our conduct.-- If the temptation to disobey t
620o045 e makes allowance & either excuses the «non-» FOLLOWING of ones conscience. & palliates the offence;
627o52v -- If ever I write on these subjects consult <FOLLOWING> pages. <p. 231> marked in my Mackintosh 1) M
545m108 lectual, there is no comparison of ideas-- one FOLLOWS other as in blindest memory-- also low equally
632j53r ) was DIRECTLY created. for transportation. it FOLLOWS from some more general law.-- [that the taste
525m024 father thinks that selfishness, pride & kind of FOLLY like (Mr George S.) is very heredetary.-- My fa
532m056 n brought from Shrewsbury to Clayton, (though so FOND of her & of servant of Richard & of Mary & her
581n064 as we do.-- touch apparently. ourang outang very FOND of soft, silk-handkerchief-- cats & dogs fond o
581n064 ry fond of soft, silk-handkerchief-- cats & dogs FOND of slight tickling sensation.-- in savages othe
424e103 olog. Gardens, which brought its puppies to be FONDLED.-- and we see in the Australian dog an instanc
567n014 of  languor & suffering The Cyanocephalus when FONDLING the keeper., clasping «& rubbed» his arm. & s
255c050 ubu, (with one leg) attended the distribution of FOOD, at the Mission of Mojos (even 20 leagues apart
301c199 ents against squirrel using reason in hiding its FOOD is applicable to any habitual action. even whic
338d025 -- V. p. 63. Note Book M'. for case of change in FOOD in insects entered by mistake Surely the fossil
356d069 amount  of change from eating different kinds of FOOD: grazing animals who eat every species new.-- S
375d135 inary crop. causes a dearth then in Spring, like FOOD used for other purposes as wheat for making bra
377d138 mer signifies recognition with pleasure, as when FOOD is offered, as much as to say give me-- the oth
388d170 en bud & seed, that latter carries with stock of FOOD.-- the generalization begins low.-- it goes thr
388d170 s parent & therefore wants independent supply of FOOD.-- is real. difference-- but this does not appl
421e091 which  are most remotely related to the habits & FOOD of an animal, I have always regarded as affordi
529m040 ople., an agriculturist, in whose mind supply of FOOD was evasive & ill defined thought would receive
544m103 tarved, all party dreamt of <goo> feasts of good FOOD-- The mind wills to do this & hears that, but y
549m118 asant thoughts, he must have contingency of good FOOD, no pain.-- <but the» «&» the sensual enjoyment
550m126 of property" -- their own property. (--regarding FOOD & in birds of place for nest.)-- with dogs "hav
554m139 nd voice.-- will do anything.-- will take & give FOOD to Tommy, or anything of any sort.-- I saw Tomm
563n002 rom jealousy. (Pincher & Nina)-- or to take away FOOD &c &c-- Now if dogs mind were so framed that he
569n020 lits case of Association.-- Elephant often given FOOD & word open your mouth said, recognizes that so
573n037 oungest child of H. W. constantly. when refusing FOOD, turn his head first to one side & then to othe
581n064 The tastes of man, same as in Allied Kingdoms-- "FOOD, sm<e>ll. (ourang-outang), music, colours we mu
584n071 t may be divided into migration,-- subsidiary to FOOD & temperature molting & breeding instincts, sex
604o016 d at as animal, with consciousness,, it choosing FOOD-- crawling from light.-- Yet we can split Plana
636j57r ation.-- & then Chamelion, which feeding on same FOOD, differs in every respect, except «in» quick mo
639j28v been  made» long «(as adapted to)» because their FOOD lies deep.-- I say it is «as» simple consequenc
571n029 rding to my theory must be acquired, by certain FOODS being habitual-- & hence become heredetary;  on
133a140 be  very high.-- M. Parrot ends his paper like a FOOL.-- Feb 25' All facts show how slowly heat trave
543m099 t Cambridge, during his time, almost an absolute FOOL used to play regularly with D'Arblay of  Christ
543m100 he son of a Fruiterer in Bond St. was so great a FOOL that his Father only left him a guinea a  week.
064r134 Cape of Good Hope Says at Santos «M Birchels» «M FOOT of range some miles from shore. rock of oysters
225b217 f it) or how to make Indian Cow with bump & pigs FOOT with cloven hoof) Ask Entomologists whether the
443e157 on the Hand.-- p. 94.-- "The resemblance of the FOOT the Ostrich to that of the Camel has not escape
467t103 ething like on Kidney Bean, they go to nectar at FOOT of upper petal standing on «I saw Bee go to two
469t119 ang, has examined 7 says one specimen had on one FOOT, a toe-nail & two joints-- as it is on one foot
469t119 foot,  a toe-nail & two joints-- as it is on one FOOT probably monstruous & not a second species.-- <
484z014 of  Caracaras of Americas.-- manner of walking-- FOOT bill crest feathering on legs-- habits-- Does t
496q006 break & become double.-- There is a double Crows-FOOT. or Ranunculus.= (11) Try.. Nitrate of Soda-- S
505q014 r head to root of tail 28½. inches. From sole of FOOT to shoulder on line of back, height 17½/. The G
511q018 rejected?? (8) Get Sir. R. Heron to give me Pigs FOOT undivided, & more particulars regarding effects
534m062 removed  a little further, they ascended about a FOOT & then leapt across. (Col Sykes compares this w
539m081 powers--  the difference is of a man wagging his FOOT & working with his toe to perform some difficul
638j28r of  Deity Vol I. p. 251-- stomach hump, kinds of FOOT. power of closing nostril, foot, sack. power of
638j28r h hump, kinds of foot. power of closing nostril, FOOT, sack. power of endurance &c &c Camels? all goo
293c173 t.-- Analogy. a bird can swim without being web FOOTED yet with much practice & led on by circumstanc
293c173 practice & led on by circumstance it becomes web FOOTED, now Man by effort of Memory can remember how
120a109 sure.--» respect to formation of salt.?.--??? FOOTSTEPS in New Red Sandstone. look as if a surface de
421e090 ing of» the terrestrial mollusca of Morocco «Mr FORBES says the Fauna»-- (near Oran) approach in char
045r076 stem can be called accidental; the proportional FORCE of crust of globe & injecting matter on the gre
051r094 Excepting by removal of large fragments by mere FORCE of waves: & action on upper tidal band, I do no
055r108 e are some arguments which strike the mind with FORCE.--the exact yearly rise of the great rivers pro
066r141 th lines the tides form eddies with its extreme FORCE. Yet, no outlet at head. Important in forming t
093a034 re antagonist forces. but they are parts of one FORCE, one locally relieving the other. -- Is the fel
098a047 as  a consequence aggregated (I assume the same FORCE which draws together two particles of Carb. of
098a047 crystallize them as seen in stalactite).-- same FORCE crystallizes minerals in layer. therefore aggre
099a050 ith chemical attraction &c.» becomes measure of FORCE. < ∴ where little inclination, little force & v
099a050 of  force. < ∴ where little inclination, little FORCE & varying direction.--> Therefore in PILE of mu
099a051 ed dikes in Cordillera.!!!-- In stratum Of. let FORCE drag particls to line {P} AB, & likewise gravit
105a073 --It is simply as the inclination is little the FORCE required to move <it> «stream» aside is not gre
106a073 {P} inclination be great where arrow stands the FORCE immediately deflected from (B) which would not
106a074 n power of deflection with stream retaining its FORCE, now it will be evident that deflected stream c
106a074 evident that deflected stream cannot retain its FORCE if inclination be great. There could «not» be g
106a074 become comparatively small, & when that is case FORCE is lessened. therefore rivers very  ineffectua
113a092 height of lava (habitually) becomes measure of FORCE in that part.-- Important as explaining want of
115a098 hat form.-- All this depending not on absolute <FORCE> «size of» of «currents» «fragments» but relati
124a118 explanation of intermittent action of elevatory FORCE-- Erasmus says he has seen in making brass a pi
195b100 ipers. not new at Galapagos.-- did the creative FORCE know that «these» species could, arrive-- did i
204b140 -fowls have courage independently of individual FORCE» Mr Wynne has crossed Ducks & Widgeon & offspri
263c076 oint in ovum. has such organization as to <per» FORCE in one man the developement of a brain  capable
375d135 aneously all the rest.-- One may say there is a FORCE like a hundred thousand wedges trying force <in
375d135 s a force like a hundred thousand wedges trying FORCE <into» every kind of adapted structure into the
387d169 s peculiarity strongly; they transmit with same FORCE as first pair, but to this tendency is added <t
467t103 ommon Beans it is wonderful {a} how the Humbles FORCE down the wings most violently: in Beans the win
529m040 f land covered with ocean, former animals, slow FORCE cracking surface &c truly poetical. (V. Wordswo
552m131 So intimately connected is passion with sending FORCE to muscles, that in my grandfather remark, a ti
552m131 nvoluntarily feels angry, when brain is pumping FORCE to legs & body, & especially, when to whole bod
554m138 away, take her chair & bang against the door to FORCE it open, when she could not succeed of herself.
558m152 airs erect on back» «wide open» with prodigious FORCE.-- making growling, guggling noise. Puma did sa
615o36v in  fowls & not effect of feeling of individual FORCE in any individual.-- His Malay breed «of  fowl»
617o39v vealed to us by the effort it costs us to exert FORCE or by internal consciousness; the objective, by
617o39v hat senses in the way in which we apprehend the FORCE of inamimate bodies. How we identify the two as
617o040 xternal senses consists in the manifestation of FORCE i.e. movement? capable of being traced to the b
617o040 he individual to whom the action is attributed; FORCE (be it remembered) being a phenomenon apprehend
617o040 f gravity considered in themselves consist in a FORCE manifested in every particle of matter directed
617o040 tter directed towards every other particle; but FORCE, <objectively> considered, by our external sens
617o040 urse of its DIRECTION, & thus when we apprehend FORCE in inanimate matter we feel dissatisfied  until
617o040 eel dissatisfied until we can point out How can FORCE be recognized by our external senses--only move
```

617o40v We are satisfied therefore, if we can trace any FORCE in inanimate matter up to the action of some an
617o40v stating the fact to the mere statement of the <FORCE exhibited in every> phenomena actually apprehen
618o041 hought bore the same relation to the brain that FORCE does to the bodily frame, they could be perceiv
619o042 h» actions being prevented by «necessity» «some FORCE» give pain: for instance either protecting shee
619o043 est. likewise if such actions were prevented by FORCE he would feel pain. [.By a very slight change i
633j54r earth, from the mountains upheaved by volcanic FORCE, for these Marsh plants. All flow from some gra
224b213 ny external character:-- For instance two wrens FORCED to haunt two islands one with one kind of herb
443e154 neration have some most important difference is FORCED on us.-- My theory only requires that organic
620o044 acity of calling up past sensations, he will be FORCED to reflect on his choice: an appetite gratifie
042r068 has been upheaved in a dry form It is clear the FORCES have acted with far more regularity in S. Amer
093a034 sider that elevation & eruptions are antagonist FORCES. but they are parts of one force, one locally
128a129 he figure of the world has just that form which FORCES dilemma. Transactions of the Maryland Academy
377d140 ultiply the effects of <earthquakes>, elevating FORCES in raising continents, & forming mountain-chai
434e130 so by suppression of one organ. (here language FORCES on us the change, which <to> seems to have tak
467t103 trude «there is a brush at end of stigma, which FORCES out from extremity pollen, or pollen comes out
617o40v ves can originate in any point an opposition of FORCES balancing each other & moving in opposite dire
617o40v invisible strings & as on this supposition the FORCES manifested would be <sat> fundamentally accoun
618o41v an effect.JCD Cause and effect has relation to FORCES & mentality because effort is felt [LHC] 1) Ma
163g109 Speculate on «under head of» Beagle Channel. FORCHAMMERS (Lyells Denmark) Shrewsbury rubbish.-- Specu
558m152 Puma did same & & some others-- Thus <sudden> «FORCIBLE prolonged» expulsion of air «dogs snarl much
204b140 ks oldest variety impresses the offspring most FORCIBLY-- Esquimaux dog & Pointer «Game-fowls have co
492q003 ike be interbred will offspring be uniform.-- Mr FORD Has M. Sageret WRITTEN on crossing of Cabbages,
174b014 eve necessary, but if it was necessary to one FOREFATHER, the result, would be as it is.-- Hence Ante
266c091 eriods at which both left the lands of their FOREFATHERS;-- the first to escape the doctrines of Muha
432e124 probably assumes the hairy character of his FOREFATHERS only when advanced in age, & therefore the c
522m012 the idea of time had been disturbed.-- These FOREGOING cases of «mental» failure very general effect
571n030 instinct» has been carried too far In all the FOREGOING cases most difficult to distinguish. between
528m038 eauty to a single tree,-- & the leaves of the FOREGROUND either owe their beauty to absolute forms or
545m105 Sphynx Linnaeus») constantly moved the skin of FOREHEAD over eyes, at every emotion & <look> «turn» o
571n028 ing on them (as justice?? as ancients did high FOREHEAD sign of exalted character???) Why may not our
596n184 .» Macacus especially pulls back skin of whole FOREHEAD & 2 ears.-- emotions of every kind.-- «[Are m
192b083 f a variety of fruit tree or plant run wild in FOREIGN country.-- When one sees nipple on man's breas
327c265 in Edinburgh. Encyclop.-- The <Edin> British & FOREIGN Medical Review No XIV. April 1839.-- Review on
390d177 ich have solitary flower) exotics brought from FOREIGN country. (<annuals> & so must those forms whic
403e021 otzebue's Second Voyage do Vol III p. 77. Many FOREIGN plants have been cultivated in Guahon. (Marian
429e114 into our gardens (opportunities of escape for FOREIGN birds & insects) which are propagated with ver
499q009 len of own kind is much more effective than of FOREIGN-- Eyton has such a grove of Willows.-- (14) Bo
554m140 th a blanket.-- these cases of commonly using, FOREIGN bodies, for end. most important step in progre
559m155 - very curious facts about early production of FOREIGN seeds.-- many varieties.-- Rev R. Jones has it
616o039 ould certainly be a startling expression, & so FOREIGN to the use of ordinary language that the onus
636j56v rward. the impregnation of Dioecious Plants by FOREIGN agency-- as insects, as wonderful case of adap
444e158 or foetus to perfect animal are adapted by FOREKNOWLEDGE, so must the mutations of species.!!-- p. 2
451e179 lighter head, carried more elevated & higher FOREQUARTERS: is said to be of a bolder & more generous
463t055 ructure.-- perhaps greatest Could anyone. have FORESEEN, sailing, climbing & mud-walking fish? diffic
202b131 In note. Demerara. 10 «12» feet beneath surface FOREST trees fallen «kind well known, carbonized»--;
202b131 ell known, carbonized»--; clay fifty feet, then FOREST 120 ft Micaceous rocks. subsidence appears ind
268c099 ommon to Patagonia desert & Tierra del Fuego. & FOREST «Parrots in Macquarrie Isd.--» very good. Stud
271c105 cies.-- ») should build a nest lined with mud, in FOREST where not a tree in which it builds, a berry o
290c163 sact. Vol V. Birds bones-- in strata of Tilgate FOREST Seeing common gull in garden at Zoology Soc. i
492q003 Sageret is referred to with doubt by Herbert. Do FOREST-trees sport much in nursery gardens? <are the>
029r034 aring a relation to present position of <Coal> FORESTS. These thick beds of Lignite stratified with s
129a131 es it bear on Patagonia? «Facts about subsided FORESTS.-- Many repeated oscillations» Hitchcock Repor
233b252 th covered with the most beautiful savannahs & FORESTS dare to say that intellectuality is only aim i
290c163 on. Must have some effect.-- Maldonado as good FORESTS for beautiful birds.-- heredetary ambling hors
451e179 s if p. 345. The Ceylonese Elephant [...] saul FORESTS by having a smaller, lighter head, carried mor
486z020 nd Spix & Martius talk of birds singing in the FORESTS of Brazil H. Wedgwood says in <14th> «13th.» V
599o006 essed in Fingals cave, & in the arched & leafy FORESTS" Very good!. I grant that the thrill, which ru
227b225 at structure (C) could pass into (D).-- We may FORETELL species. limits of good species being known.-
355d067 - The only advantage of discovering laws is to FORETELL what will happen & to see bearing of scattere
635j55r cal law, «& is therefore utterly useless-- it FORETELLS nothing» because we know nothing of the will
276c124 tise with a peculiar structure, then Milvulus FORFICATUS Tyrannus Sulphureus if compelled solely to f
203b137 ?, having been gained in short time. Milvulus FORFICATUS <has a wide range> is a tyrant flycatcher do
424e100 n geographical distribution of Crustaceae.-- (I FORGET whether I have already referred to it.-- also
531m050 ereas it is the importance.-- people very often FORGET where money is placed.-- (How often one forget
548m116 uble consciousness, as shown in the tendency to FORGET the insane idea; & ones expression of double s
556m145 d by Holcroft «Vol I» .p. 86 "We ought never to FORGET-- --; that every man is born with a portion of
558m153 look at the immense difference. between man,-- FORGET the use of language, & judge only by what you
560m156 mbulism. Do people when inhaling Nitrous oxide, FORGET what they did when in this state, or remember
293c172 Memory springing up after long intervals of FORGETFULNESS.-- after sleep, «strong» analogies with mem
523m018 is true, that the only cure for madness is FORGETFULNESS.-- which does appear a real difference, bet
528m035 ve statement that insanity is only cured by FORGETFULNESS.-- & the approach to believing a vivid cast
530m043 ike FitzRoy in sleep giving directions,-- & FORGETFULNESS after bad accidents:-- After journey, a fit
294c175 seased in its loins & all were so afterwards, (FORGETS authority).-- Lonsdale is ready to admit, perm
531m050 forget where money is placed.-- (How often one FORGETS where put one key. where all keys are placed)
538m080 respect to more energetic self, & likewise one FORGETS. what one performs habitually.-- Agrees with i
619o042 st in such active sympathy that the individual FORGETS itself, & aids & defends & acts for others at
524m022 .-- Mrs P. when in state as above described, (FORGETTING that her husband was dead) yet instantly per
538m078 ly bears out my fathers doctrine about people FORGETTING their insanity» there seem other cases somew
420e089 cated circumstances, as shown by difficulty in FORGING, yet handwriting said to be heredetary. shows
541m099 o difficult to conceal.-- a man «insulted» may FORGIVE his enemy & not wish to strike him, but he wil
293c172 people in fevers recollecting things utterly FORGOTTEN»--it is scarcely more wonderful, that it sho
527m035 a castle in the air is interrupted & utterly FORGOTTEN--; so as to feel a severe disappointment «in
530m043 y occurrence for a day or two are absoluteley FORGOTTEN.-- My father signed a bond, yet when he paid
531m050 ks that things of great importance are easily FORGOTTEN. (if unconnected with fear &c) because people
534m061 strange if judgment remains, where reason is FORGOTTEN. it is conscience, or instinct. Hensleigh say
535m064 e case may be normal. with insects, but habit FORGOTTEN in all older species. The earwig & a doubtful
538m079 t her had EVER before heard, so very probably FORGOTTEN.» Such facts bear on such characters as Allen
602o011 > remembered, when the meaning or reasons are FORGOTTEN. Our happiness &c, our well-being depends upo
620o044 ore as soon as desire is fullfilled, pleasure FORGOTTEN. [RHC] 4) as starvation, or fear of death, on
625o052 eive enjoyment from gratification & hence are FORGOTTEN-- only so far do I admit its supremacy p. 37.
629o055 <rationally> by feeling-- reasoned on, steps FORGOTTEN, habit formed,-- & such habits carried on to
362d100 some «but very strange races» of them have the FORKED black mark of the Rock Pidgeon,-- several have
363d101 D[In the Pidgeons, trace the washing out of the FORKED band, like in plumage of ducks.--» Mr Yarrell s
027r024 Concepcion. Humb: Pers. N. vii P. 56 Serpentine FORM: of Cuba for comparison (?) with St Pauls [not
033r042 particles coated by pellucid envelope of Lime.--FORM resembles the husks at Coquimbo: in that case
036r051 sary to ascend the hill.-- The absence of Second FORM, except near submarine Volc: in harmony with th
038r058 ider causes for subaqueous crater being of diff: FORM subaerial one?--In former not so much; or no ra
039r060 ully considered the subject.-- S. America in the FORM of the land decidedly bears the stamp of recent
042r068 hat point of Porphyry has been upheaved in a dry FORM It is clear the forces have acted with far more
044r072 heatre of Pompeei (?). -- Such have been seen to FORM in atmosphere.--Mem. Ascencion. concretions & G
047r081 heavy swell, sea retreats & then breaks: i e to FORM a wave in ocean. is not this [Fig. 3] {P} form
047r081 o form a wave in ocean. is not this [Fig. 3] {P} FORM present, i e a part below «mean» level before t
047r083 id not sea break first? I can imagine from local FORM of coast (as seen in swell) the undertow & over
049r088 a Volcano or as an injected mass.--From conical FORM I incline to <latter> former; & thus occurring
050r091 e. Jonnes seems to think that Volcanic eruptions FORM foundations for Coral reefs.--does he mean in c
052r097 a great fault or rather many faults.-- Necessary FORM; as long as coast line fixed.--[Fig. 5] {P} * S
055r107 . 2 - 3 toises thick.--Vol II. p. 252 Urge cliff FORM of land, in St Helena. Ascension. Azores. («san
065r140 ake any observations on springs at S. Cruz.???-- FORM of land shows subsidence in T. del Fuego, and c

066r141 hich break through the N & South lines the tides FORM eddies with its extreme force. Yet, no outlet a
072r160 - would particle attracted towards space tend to FORM ring. [Fig. 10] {P} motion from within and with
093a032 nes of Vendarquas. Mem sublimation of sulphur to FORM salts of America. -- The number of minute turbo
100a055 se.-- perhaps worth Says from Lardner's (p. 213) FORM of escarpment relation kept to sea coast ∴ curi
101a056 dner's p. 270-4, good discussion showing present FORM of land in Northern England influence dispersio
102a059 coast itself, undertow would draw it outwards.-- FORM of breaker affected some way out to sea.--¿ eff
105a071 cation, If chain of lake. <a> the alluvium would FORM a succession of flights of steps; if one lake t
107a079 whole centre of earth same heat, then change in FORM of fluid centre would lift with it isothermal l
109a082 e further from beach action probably modified by FORM of waves & currents.-- but this must be continu
115a098 d & seaward: if matter too coarse, then {P} that FORM.-- All this depending not on absolute <force> «
121a111 3. (Ed. Transact) has seen clay stiff enough <to FORM> for potters to use. in which great Knife forme
127a126 egree!)-- shows effects of pressure in change of FORM as the result of heat.-- will it bear on centra
128a129 shows that the figure of the world has just that FORM which forces dilemma. Transactions of the Maryl
166g007 25 degree perhaps most common-- Will not curved FORM of hill be explained by my idea-- highest part
150g035 Dick shows NB. Lake gradually draining off would FORM plains such as those near Bridge Roy (& other c
151g041 two little lines of Hill (judging from external FORM alluvium) descend from shelf 3d & almost meet,
154g058 terrace of 2d shelf Level of shelf of Glen Guoy FORM comparison with granite block «& Ben Erin» 29.2
157g075 side of «that» hill, in front of which shelf 3d FORM beach of granite pebbles, & around which shelf
162g101 267 28.75 .105 I reached> 29.090 Preservation of FORM of land very much due to Peat & Heather When it
165g126 s the Tetrao scoticus & Tetrao-- not an American FORM The union of two instincts crossing most remark
173b010 if species (<a>) «(I)». <fr> may be derived from FORM (2). &c.-- <(> Then (remembering Lyells argumen
173b012 idea of propagation of species we can see why a FORM peculiar to continents; all bred in from one pa
179b035 ame epoch certainly» The absolute end of certain FORM from considering. S. America (independent of ex
182b047 of Natural History will be to describe limits of FORM. (& where possible the number «of steps» known)
190b078 ed in the womb, which has been passed through to FORM that species.-- <Man is derived from Monad, eac
194b094 of <Spi> of Sapajou-- NB Sapajou is S. American FORM. therefore it is like case of great <rodent> ed
195b101 nner God orders each animal created with certain FORM in certain country, but how much more simple, &
195b102 e powers of transportal be such & so will be the FORM of one country to another.-- let geological cha
199b118 ion of a fortuitous modification, into a durable FORM, of a fugitive want into a fundamental propenit
199b119 liar people banished by rest?-- ∴ most monstrous FORM has tendency to propagate, as well as diseases.
200b125 aking seeds in salt water &c has any tendency to FORM varieties? Ed. N. Phil. J. Morse found in Virgi
201b126 nly few in species, but every two or three these FORM genera-- this is «from unfavourable conditions»
205b144 t of water Does the odd Petrel of F. del F. take FORM of awk, because there is no awk in Southern hem
205b145 hemisphere. does this rule apply? A Treatise on FORM of Animals by Mr Cline "The character of both p
205b145 n to their increase of size they become worse in FORM, less hardy, & more liable to disease" If popul
217b187 Isle of Pines-- Australia.-- A <South> American «FORM of» Lathyrus has one species in Europe Madagasc
219b194 at ocean, have made plants of American & African FORM, merely because intermediate position.-- We can
222b206 ther type of each order may not be supposed that FORM, which has wandered least from ancestral form.
222b206 at form, which has wandered least from ancestral FORM. If so are present typical species most near in
222b207 . If so are present typical species most near in FORM to ancient; in shells alone can this comparison
223b211 cality A. but it is instead a stunted & diseased FORM a plant, adapted to A. B. C. D.-- Destroy plant
223b211 B. C. D.-- Destroy plants B. C. D. & A will soon FORM good species! The increased fertility of slight
225b218 h gun. & Bustards &c!!! An American & African FORM of plant being found in Tristan D'Acunha. may b
227b224 By character of any «two» ancient fauna, we may FORM some idea of <origin under> connection of those
231b243 r greater discordance in latter-- Have change in FORM.-- This probably explains crag & miocene.-- The
231b244 ies & Camels breed?-- As man has not had time to FORM good species, so cannot the domesticated animal
232b249 Cuscus, (a sub genus of Phalangista New Holland FORM) is found in many island Celebes «Waggiou» &c &
233b249 Gould has been struck with similar extension of FORM in birds.-- Waterhouse thinks two main divisi
234b256 are local varieties «of colour & size, but not FOR[M]» of animals.-- He says Stephens say he can at
236b279 of Sub Himalayan land emys, decidedly an Indian FORM of Tortoise.-- On other hand. fresh water torto
239cIFC species ever yet failed. About trades affecting FORM of man. Could you get racehorse from Cart horse
240c003 le, time being an element in the transmission of FORM, may explain mule & pig being half way. Yet dog
240c013 New Ireland, have phalangista, which differ in «FORM & head &» colour from those of New Holland.-- T
241c015 ndi: Arch: In New Zealand. a sturnus of American FORM-- a Synallaxis. ¿American?). p. 159. & 160 «162
243c020 iar species of cassicans: (¿cassicans Australian FORM? p. 27. many fish of Taiti found at <New> Isle
248c033 see external influences first affect external [FOR]M, so will the internal parts be of longest [cons
249c035 information ¿Carnivora of New & Old word. do not FORM two sections is this not connected with wide ra
250c037 S. America destroyed great Edentata or American FORM.-- Is the Australian Dipus an American form? Th
250c037 ican form.-- Is the Australian Dipus an American FORM? The climate having grown more extreme both in,
251c041 ies, inhabiting Borneo & Sumatra. differ only in FORM of white mark on breast: p. 234.-- good case.--
253c047 Capt. W. Shee. considered merely variety.-- red FORM of skull very slightly different.-- Q Zoolog. T
258c060 ue relationship, tends to keep to species to one FORM, (but is modified), the relationship of Analogy
259c064 fresh water animals of great lakes are American FORM, that one is brought to admit the possibility (
261c071 ent difficulty-- preservation of colouring, when FORM has changed. Can be said that animals no notion
263c077 nite-- he is not a deity, his end «under present FORM» will come, (or how dredfully we are deceived)
264c081 ightly preceed them-- From this view habits must FORM most important element in considering to which
275c120 ssed, till there was a dash of blood, with whole FORM of grey hound.-- picking out finest of each lit
278c130 g former connection of two continents & death of FORM in one. The caves are at a height of more than
282c141 st suppose men instead of mere colour & trifling FORM & head &c to become greatly changed. in structu
284c149 s Mr Blyth remark that a resemblanc between some FORM in birds is visible, when young, but not when o
284c149 e, when young, but not when old.-- thus speckled FORM of young blackbird. good remark if general.-- W
296c184 species which can migrate remaining constant in FORM, others altered much.-- these others will be pl
297c186 nfinite number of forms.-- therefore an isolated FORM probably a remnant.-- Pachydermata. & Horses fe
298c191 d.-- lower species would then revert to pristine FORM (which must have been altered by crossing) with
298c191 must have been altered by crossing) with alpine FORM) lower species afterwards would probably often
303c204 he races of men differ chiefly in <size> colour, FORM of head «& features»;. hence intellect?) & what
303c204 ther it <would have> differs in present races, & FORM of head «& features»;. but likewise in length o
303c204 monkey, would have differed in hair colour + + + FORM of feet.= (Negro or father of negro probably wa
306c214 tail like setters. long ears-- colours vary, but FORM constant.-- The females of some moths, like glo
333d005 e each other, partaking <more> «very closely» of FORM of mother: more than of the Common cat.-- Ch IX
333d006 re killed, & other two very closely resembled in FORM of tail, fur &c to the half bred Persian.-- Her
333d007 ones have been so long as wild one under present FORM.-- Fox has seen several cases of foxes & dogs c
334d012 here is a breed of white-tailed squirrels, which FORM a marked wild variety. doubtful whether all are
336d017 lowly, now all the mules have their whole <body> FORM of body gained in one generation, so it is impo
336d017 its offspring any «peculiarity» change from the FORM which it inherits from its parents «stock» with
338d024 Negro (& partly Mulatto) prevents his taking any FORM of Malaria-- adaptation & species-like,-- -- Sa
343d036 physical causes.-- these superinduce changes of FORM in the organic world, as adaptation. & these ch
345d043 Book.-- Why is not Tetrao Scoticus. an american FORM (if so)?.-- A Sphepherd of Glen Turret. said he
350d055 e. (Polirus??) female blind & of quite different FORM from male with eyes!-- (are not these differenc
357d073 lain. S. American case & Didelphis being Mundine FORM. & the less developement of Marsupials in S. A
359d088 seems chiefly to impress the young most with its FORM & disposition Saw three young ducks, like each
366d112 s to grow there «That Mutilations will not alter FORM, & so altered in disposition, as to be more eas
371d127] that it shall beget young different in colour, FORM, & which Malthus shows, is the final effect, (by m
376d135 structure & adapt it to change.-- to do that for FORM which Malthus shows, is the final effect, (by m
377d139 -- lived.-- we know the great time, necessary to FORM channel & (& Basses St) yet no change in Englis
377d140 s-- we really, measure the rapidity of change of FORM, & instincts in the animal kingdom.-- It is the
386d167 . One «invisible» animalcule in four days could FORM 2. cubic stone. like that of Billin.-- <Generat
390d178 of some change.-- Without some small change in FORM. ideosyncrasy or dispositions were added or sub
391d174 ypifies the whole course of change from simplest FORM.-- (Because by this process it separates those
392d175 (Surely all are really dioecious..) only simple FORM of life are monooecious. Will ova of fishes & M
398e006 ea of revolution.-- My very theory requires each FORM to have lasted for its time: but we ought in sa
399e011 ely he has seen. a Calosoma. (very like American FORM) in Stonesfield slate., & a Melolonittha-- In m
400e013 xico with small Hares & raccoons.-- «S. American FORM.--» off province of Guadalaxura-- October 11th.
403e023 be made-- why seeing great variation in external FORM of varieties, do we suppose bones will not chan
407e041 esemble fossil ones of Europe, Consider probable FORM of land,-- -- S America, an island, «connects w
411e056 eresting as showing some change in habits before FORM.-- I have already given various examples The Pi
419e087 o the effect of change) than a slow gradation in FORM, «which must be effect of slow change + & There
423e098 h. Monoceros, which Sowerby says, is an American FORM.-- has several species in my fossils-- CD[If ca

```
425e105 ossil Mammalia. p. 55. talks of Tapirus American FORM. found in Eocene beds of Paris Lyell has remark
429e115 r plants.-- a broad border of Killed trees would FORM fringe.-- but there is a contest. & a grain of
432e125 the sun & our lives, but to period necessary to FORM heaps of pebbles &c &c: the succession of organ
434e127 t does Muller call it) succeeds in altering <or> FORM of body, or whether it merely has tendency (as
435e133 n of light is concerned with the developement of FORM; but that tadpole increased in size now the Pro
440e145 & every buzzing insect & grazing animal owes its FORM, to that form being «the one alone» out of innu
440e145 g insect & grazing animal owes its form, to that FORM being «the one alone» out of innumerable other
441e149 , <be> «offspring» would be similar to <f> first FORM.-- The great effect of conditions on offspring,
442e153 this, now that tulips break by cultivation can a FORM become permanent?» because its very essence  is
446e163 propagation by buds does not insure constancy of FORM.-- is the constancy owing to similarity of cond
447e163 aps oldest in world & hence most persistent-- if FORM exceedingly difficult to vary.-- the run of cha
447e164 not. cross there would be my theory gradation of FORM from one species to other: therefore my  theory
449e169 bably) Yarrells law & Walkers of the male giving FORM-- they interbred. & the young kept constant.   &
452e182 that skulls found near Vienna appoximat to Negro FORM; those from Rhine to the Caribs.-- Vol II p. 65
453e182 n with the joints near the tip crooked.-- is the FORM [...] Dampier. Vol I. p. 320. says no wild (car
459tf02 on cross of he-goat & sheep, it seems male gives FORM. admitted by Linnaeus.-- seems to doubt its app
459tf02 nus-- in one case bazzard of wolf & dog had more FORM of male, & another of both progenitors-- the hi
459t013 e penguin duck. it took after the Penguin in the FORM of its body & in the manner of walking but  not
459t013 of  drake still more.-- So Penguin impresses its FORM both on vars & species The <male> swan-gander w
464tf6r N. America & S., (¿ is the peculiar. N. American FORM)-- ¿Hunting leopard, how strange, anyone, would
471tf07 own.-- "<when we find such an endless variety of FORM in the same> organ "manifestation of divine pow
473s07v well  be born a tendency to alter or assume some FORM late in youth,-- only facts can decide-- some p
493q004 breed & see result.-- According to Mr Walker the FORM of male ought to preponderate; according to  Mr
494q004 Mr. Miller said yes with regard to former (8) Is FORM of globule of blood in allied species similar.-
495q005 as these are wild varieties. Is any intermediate FORM found wild a The Leptosiphon densifolium «an an
506q015 .-- Stature, position of flowers-- Their smell-- FORM of flowers-- Nectaries-- In Monooecious «order»
520m001 l disposition» oftener go with colour, than with FORM of body.-- thus the late Colonel Leigton resemb
526m030 ht about habitual exercise of the mind, altering FORM of head, & thus these qualitics become heredeta
528m037 ooks at it.-- these two causes very weak.-- (2d) FORM. some forms seem instinctively beautiful «as ro
530m045 an those of gentlefolks. & that peculiarities of FORM in trades (,as sailor tailor blacksmiths?)  are
539m084 as  Grandfather. Aug. 16th Anger «Rage» in worst FORM is described by Spenser (Faery Queene. CD 25 (D
550m123 origin  of our evil passions!!-- The Devil under FORM of Baboon is our grandfather!-- A man, who perf
560m156 estures, other cases like D. Corbet; «do» ideots FORM habits readily?? Do the Ourang Outang like smel
566m011 instance  of obtaining <that> «a» faculty in the FORM of a true instinct, which is a real instinct in
569m021 prayer, or eloquent request. Reason in simplets  FORM probably is single comparison by senses of  an
574m040 people get into habits.-- we probably can hardly FORM an idea of a mind so limited as Birgos to becom
574m043 its.-- the limits of these two actions either on FORM or brain very hard to define.-- Consider the ac
583m070 look  at him, as machine to make cell of certain FORM. (& especially as it adapt its cell to circumst
588m091 .-- p. 325 «to 29».-- Habits becoming heredetary FORM the instincts of animals.- almost identical as
602o010 wing "the perfection of this science of abstract FORM" is the source of part of the highest enjoyment
611o034 ifferent modification, peculiarities of external FORM impressed, & different laws of movements. [LHC]
611o034 xtress> tissue <[...]> unites matter into certain FORM; invariable, as long as not modified by externa
613o036 as  races are formed or modification of external FORM. so modifications of brain) As in animals no pr
633j54r to  arrest the earth, than the earth revolves to FORM rain to wash down earth, from the mountains uph
638j059 e hand. & can in idea (with consciousness.) <th> FORM these schemes.-- I see no reason, why structure
410e053 ation of change occurs.-- those discovering the FORMAL laws of the corelation of parts in individuals
411e054 ory, will be approximated to.-- Treating of the FORMAL laws of corelation of parts & organs it may se
026r020 rectly opposed to common opinion The Tertiary FORMATION South of the Maypo at one period of elevation
027r028 tion. United service Journal In the Iron sand FORMATION <would> wood converted into siliceous pyritou
030r036 » «on N. Chile? Washington.--» Mem: Micaceous FORMATION of Chonos. interesting from great quantity of
039r061 Lyell P 116 Vol III, says that in N. Pliocene FORMATION of Limestone, casts of shells, as in some old
041r064 lived in so deep a sea.--Perhaps agrees with FORMATION of pebbles & vertical trees Grand Seco at  B.
054r102 [.] Sydney. -- Lesson Zoologie Grand tertiary FORMATION of Payta: N. part of New Zeeland entirely vol
056r110 of  a breccia of primitive rocks between that FORMATION and the secondary (stated in Playfair to be t
064r137 y of Sulph. acid emitted: mem: Grand gypseous FORMATION of Cordillera In describing structure of Cord
085aIFC "Geolog  Observat on Volcanic islands & Coral FORMATION Lyell's Salband p. 86 Shells near Woollich p.
092a028 am. Geolog Proc p. 566 1837.-- Tertiary <bea> FORMATION twenty species same as Paris. 1500 ft high Mr
114a094 on subject of cleavage Clay slate. a distinct FORMATION deep «& therefore extensive» water ∴ not form
114a094 efore extensive» water ∴ not formed in modern FORMATION & not ever in Secondary in Europe. gneiss-- m
119a104 cific Oceans.-- (2d--) does not explain first FORMATION of continents, if globe be considered as cond
120a109 f very hot under high pressure.--» respect to FORMATION of salt.?.--??? Footsteps in New Red Sandston
127a127 f French Geolog. Cantal Vol III I? p. 246. on FORMATION of cones beneath sea.-- with reference to old
135a146 ure at {P} .-- G. J. Malcolmson has described FORMATION of shore of Coromandel. just same as. at Bahi
149g032 «& to West of Spean» difficult to explain on <FORMATION> deposition in lake On the summit «& on Spean
206b150 o dicotyledenous plants & few Monocot in Coal FORMATION? p. 320 <Think> States Cryptogam. Flora forme
256c055 als Mr Strickland & Hamilton-- found tertiary FORMATION amongst Graecian isles, ¿See if type continue
276c122 be used to varieties, as well as species-- as FORMATION of species <of> gradual, so may we  suppose.,
278c128 subgenera,  then it would be useless, but the FORMATION of subgenera is empirical, & is judged solely
280c134 many  kinds in dogs, is clearly applicable to FORMATION of instincts in wild animals, many species in
283c146 othing to do with CREATION.-- <On> The end of FORMATION of species & genera, is probably to add to qu
302c203 e stated, only in most discordant groups. The FORMATION of genera may sometimes be due to accident as
316c246 y in range + + What circumstances have led to FORMATION of new species some few have been scattered o
352d060 -- Lyell's excellent view of geology, of each FORMATION being merely a page form out of a history,  &
355d068 f scattered facts..-- What takes place in the FORMATION of a bud-- the very same must take place in c
374d133 forms.-- p 458 Upper Silurian, fishes oldest FORMATION highly organized.-- do. p. 461.-- Lower Silur
409e049 ninhabited]CD & if my theory be true then the FORMATION of sexes rigidly necessary.-- Without sexual
412e057 imes. & from persistency «owing to their slow <FORMATION>» these variations tend to accumulate. «on any
436e135 be believed, then, uniformity in «geological» FORMATION. intelligible.-- «-- No. but the wandering  &
559m154 ch ceases to be the case when we consider the FORMATION of laws invoking laws. & giving rise at  last
574n042 ing the weaker ones, may be applicable to the FORMATION of instincts, independently of habits.--  the
577n051 s. the same relation to true memory, that the FORMATION of a hinge «in a bivalve shell» does to reaso
599o05v ople.-- Sound of words-- argument of original FORMATION.-- declension &c often show traces of origin.
626o052 ich in prior <races> generations led to their FORMATION.-- 'N.B. If feeling or emotion rises from her
026r019 organic remains in De la Beche, for the older FORMATIONS I must believe they «the limestones» have be
027r023 semiconsolidated rocks P xv. mentions in what FORMATIONS Conglomerates are found. -- The above oscill
027r023 The above oscillations remarkable because the FORMATIONS are now seen in regular descending steps Mem
029r033 plains.-- Sydney no I believe the secondary? FORMATIONS of Brazil, all originate from the decomposit
035r047 act is indeed shewn? by the parallel bands of FORMATIONS on any Geolog Map: Quoted from Daubeny P 402
039r061 Limestone, casts of shells, as in some older FORMATIONS: Mem the envelopes at Coquimbo. the  analogy
043r071 shock-- In S. America--continuity of space in FORMATIONS & durability of similar causes go  together.
060r125 m Temple Urge the mineralogical difference of FORMATIONS of S. America & Europe.-- If great chain of
073r163 rticularly between 18 degree & 22 degree N. = FORMATIONS of amph: porphyry. greenstone[,] amygdaloid.
116a100 occurs  in submarine alluvium, or sublittoral FORMATIONS. p. 150. at Portezuelo, extremity of mountai
117a101 ear coast, limestone deep water. will bear on FORMATIONS. during elevation & depression. C. Prevost.-
117a102 les 2 inches long?-- L'Institut. 1838 p. 151. FORMATIONS of Payta extend close to Guayaquil.-- modern
191b080 ups & downs in full conformity with Europaean FORMATIONS-- England & Europe Ireland common  animals--
219b195 ine plants of the Alps. must be <Alpine> «new FORMATIONS» because snow formerly descended lower, ther
224b216 oduced «&» Has the Creator since the Cambrian FORMATIONS gone on creating animals with same general s
350d053 distinguished from temporal species as in two FORMATIONS? by no way.?-- "Natura nihil agit  frustra".
423o099 are, the conclusion will be that our greatest FORMATIONS <are> have been deposited in a period (say 1
431e121 nferior Oolite, & the T. elongata in the uper FORMATIONS Portland Stone &c &c.--if? so «it is» good c
433e126 orms, intercalated between two great distinct FORMATIONS.-- particular air given. p. 246.- 248 & p.
436e137 o my theory, is same type of shells in oldest FORMATIONS:-- The Cambrian formations do not however, e
436e137 f shells in oldest formations:-- The Cambrian FORMATIONS do not however, extend round world.-- Quartz
465t089 - -- How many «tertiary» estuary & Lacrustine FORMATIONS contain fossils.-- mammals-- a few only -- &
465t089 -- mammals-- a few only -- & how many estuary FORMATIONS are there in old Secondary Series-- few-- Ma
380d153 g her growth is immediately checked-- the vis FORMATIVA goes entirely to the offspring-- this is clea
380d153 h have> larvae which have bred before the vis FORMATIVA had completed them-- (but this argument is VE
```

254c049 a, generation in each joint of taenia worm.-- FORMATIVE energies easily expended & no one system deve
434e127 body of parent be altered, that is the Nisus FORMATIVUS. (what does Muller call it) succeeds in alte
026r019 I must believe they «the limestones» have been FORMED in shallow water: so have the Conglomerates: Y
030r034 st. Perhaps as at Concepcion. favoured by basin FORMED by outlying rocks; (such as between Mocha & ma
035r045 believe that much of all old strata of England. FORMED near surface: Mem Patagonian pebbles beds, mos
038r058 o doubt existed. These higher portions probably FORMED Islds from which proceeded pebbles & on which
044r073 ed; one inclines to belief all strata of Europe FORMED near coast. Humboldts quotation of instability
052r099 i may be with much truth compared to the step = FORMED streams of lava at St Jago. C. de Verds Quartz
055r106 by the waves, a sufficient proof, that the sea FORMED these large cavities", &c &c &c Vol II. Chapt
059r119 ssures.--M. B. thinks these parts incontestably FORMED the parts of one whole burning mountain, & tha
089a018 do-- p. 110. Mountains on west side of Domingo FORMED of coral limestone, with intersticies yet emptt
103a063 e not been the moving agents, because not wedge-FORMED.-- Hence fill up fissures-- If dikes effect of
103a064 All De la Beche's reasoning of mountains being FORMED by crust being too large & pitching against ea
109a083 answered by this -- No one can doubt. A-B once FORMED low coast.-- Annales des Mines. a translation
109a084 411 When discussing concretions Carbonate soda. FORMED by Ca. of L. & Mur. of Soda mixed.-- Turner's
112a089 ed volcanos there are. where one alone has been FORMED--Look at the now active volcanos & see what hi
114a094 mation deep «& therefore extensive» water ∴ not FORMED in modern formation & not ever in Secondary in
116a100 » banks <above> 30 ft or so above bed of river. FORMED of rounded pebbles-- it is clear gold occurs i
117a101 composition thinks sand with vegetable remains FORMED near coast, limestone deep water. will bear on
118a103 ikes (Hopkins) & that every dike. which has not FORMED volcanos. or become scoriform. has thinned upw
119a104 ins, yet so analogous, that as we see mountains FORMED (& mountains are effect of continental elevati
119a105 ns.-- consist of rocks with fossils,, therefore FORMED near surface. whether they can have been plung
121a111 form» for potters to use. in which great Knife FORMED crystals of ice were formed-- (like my gypsum
121a111 n which great Knife formed crystals of ice were FORMED-- (like my gypsum case) shows power of segrega
123a116 I suspect be greatly aided by considering space FORMED-- great vacuum-- by dike.-- Mem. however. vein
146g014 h argillaceous or sandy soil-- These Buttresses FORMED vestige of irregular terrace perhaps near 300
148g027 e composed-- where side ravine entered terraces FORMED successive bays but plains sloped centre-wards
150g037 elf very shallow channel 50 ft wide & river get FORMED in centre In Glen Collarig, on side of Hill of
153g054 nce into Glen Fintec a kind of landing place is FORMED Ben Erin summit 27.813. 65 55 degree? Boulder
155g062 ame «cause» as «great» spit <is> or plain <now> FORMED on shelf 4th 2d & 3d can be traced some way up
155g065 eep gorge on sea hypothesis, if gullies not now FORMED «(Mac, hypoth,)» the level during any oscillat
156g069 ered are not proportionately large to those now FORMED in same spot by present torrents Maculloch wro
157g077 from line 2d; little action since «that shelf» FORMED Upper terrace near Loch Spey <29.35161> 29.360
175b018 progress. if we suppose monads are «constantly» FORMED ¿would they not be pretty similar over whole w
176b024 finity in each branch A species as soon as once FORMED by separation or change in part of country. re
180b036 rather greater distinction Thus genera would be FORMED.-- bearing relation to ancient types.-- with s
191b082 e animals-- Owls. transport mice alive? Species FORMED by subsidence. Java & Sumatra. Rhinoceros. Ele
218b191 That is varieties, though of trifling order are FORMED by nature. Carmichael. Tristan D'Acunha, a lis
219b193 se seeds first arrived «Ferns ditto.--» & hence FORMED trees]CD & would creator <on volcanic island.>
223b210 ore than others, & cut off limbs & new ones are FORMED) but yet propagates varieties according to sam
223b210 to same law with animals?? Why are species not FORMED. during ascent of mountain or approach of dese
226b221 he new island splits & grows larger species are FORMED of those genera.-- & hence by same chance few
227b227 e Oolite order of things might have easily been FORMED.-- With belief of <change.> transmutation & ge
250c038 . might have allowed fresh species to have been FORMED & spread to other Africa & East India Arch.--
257c055 granted.--but if all other animals have been so FORMED, then man may be a miracle, but induction lead
258c059 atavism relate to this law.?-- Local varieties FORMED with extreme slowness, even where isolation, f
262c073 adapted--.-- These «aberrant» varieties will be FORMED in any kingdom of nature, where scheme not fil
272c107 stralia, probably live in flowers & has Elytra. FORMED from developement of some other part of body.-
273c111 ongly marked, there is a preeminently aerial,-- FORMED for flight & great movement in the air, & like
294c177 once grant good species as carrion crow & rook FORMED by descent. or two of the willow wrens &c &c.
299c194 e shades in these cases,-- but absolute species FORMED. The Anagallis perhaps, offers another case of
308c216 ion that all animals have never at any one time FORMED chain, since if cretceous period assumed, then
308c219 by nature. if so. the habits which would. have FORMED them, would have arisen under different climat
337d021 will be necessary to show how the first eye is FORMED.-- how one nerve becomes sensitive to light.--
343d036 these themselves.-- instincts alter, reason is FORMED, & the world peopled «with Myriads of distinct
367d112 ld hardly ever occur.--».-- & if that cannot be FORMED, genetal organs by that co-relation of parts,
381d156 phes. Polyps. Sponges Heautandrous, male organs FORMED to fecundate female (as in plants) Cirrhipeds
386d165 ve buds.-- life may be thus prolonged bud being FORMED & one part dying for great length of time.-- T
391d175 nal» change are always of one nature species is FORMED if not.-- the changes oscillate backwards & fo
426e105 s argument about <Tertiary> Isld «neighbouring» FORMED in the Tertiary «epoch» like Sicily not having
427e108 rapidly increase, & hence number of forms. once FORMED. would remain stationary, hence all present ty
428e113 ifficulty. «I should doubt if wild species ever FORMED like short-tailed cat or dog has been without
438e141 all probability the Brussels Sprout was slowly FORMED.-- » if it shall be difficult to show that <ti
492q002 eties seedless-- if so. how have varieties been FORMED?-- 8. Can any annuals be budded. with referenc
497q06v of those genera of which wild hybrids have been FORMED. (15). What is History of Viburnum. or snow-ba
506q015 lthough there be pollen.-- or FEW. or bad seeds FORMED; badness may be merely not ripening= (38) Have
511q018 ould by selection a different looking animal be FORMED-- not caring whether good or bad.-- are any ac
581n065 tructure of language, that it was progressively FORMED. (--names like sounds)--. Horne Tookes tenses,
586n079 childhood (before experience or habit) could be FORMED by the union of simple non-organic matter, wit
611o034 [LHC] ¿Has any vegetable or animal matter been FORMED or afterwards.-- child sucking whole wonder in
613o036 C] not used by Kirby» (:analogy:-- as races are FORMED by obediance to instinct» <conscience>., or r
620o045 liates the offence; one always admire the habit FORMED by past recollections.-- Hence he has the righ
621o047 t the satisfaction of the mind, which is «much» FORMED a beneficial tendency to them, as <social> an
623o051 instincts of bees & beavers «& deer» have <been FORMED by circumstances, which have led to the pe
623o051 certain conditions; by my theory they have been FORMED, -- & such habits carried on to other feelings,
629o055 y feeling-- reasoned on, steps forgotten, habit FORMED, -- as soon as Dioecious Plants were formed. Mac
636j56v erence is, there were insects «¿when were Palms FORMED?» as soon as Dioecious Plants were formed. Mac
636j56v Palms formed?» as soon as Dioecious Plants were FORMED. Macculloch says, life, forms a broken, recurr
638j58v idence. when whole rocks nay very mountains are FORMED of such dead & extinct forms.-- the exuviae of
640j167 en if have afforded only 10th part before & now FORMED eighth part.-- or if other prey diminished, to
641j29v arabids & Staphylini» & Mammalia. The eye being FORMED in Mollusca, Articulata, & Vertebrata, & Plana
059r119 ouvent tailée a pic. Toutes ces montagnes sont FORMEES de couches parallèles et inclinées du centre d
059r119 nes sur le revers de chacune des montagnes qui FORMENT les vallées ou les scissures.--M. B. thinks th
038r058 s crater being of diff: form subaerial one?--In FORMER not so much; or no rapilli; & from action of w
046r079 sider latter as accidental on the afflux of the FORMER. -- Ascension. Vegetation? Rats & Mices. At St
049r088 mass.--From conical form I incline to <latter> FORMER; & thus occurring in groups.--As these greenst
055r108 lains consumed) Where slope «plainly» indicates FORMER boundary. (as in other unworn islands) we take
062r130 of Fourmillier): extinct Guanaco to recent: in FORMER case position, in latter time. (or changes con
105a069 ence between erosive power of river & sea.; the FORMER as its channel becomes wider looses its cuttin
175b019 climates & as far as world has been uniform, at FORMER epoch-- How is this Ehrenberg? every successiv
189b072 d being grouped towards centres near Equator in FORMER periods & then splitting off.-- If species gen
203b138 h newer, hybrid variety partakes chiefly of the FORMER Eyton's paper on Hybrids Loudon's Magazine. Go
212b164 und on Java-- Monkey peculiar to. latter not to FORMER-- Mr Martens of Zoolog Soc told me an Australi
223b208 .-- -- not so great as between perfect insect & FORMER hard to tell whether articulate or intestinal,
227b224 means of transport, we should then know whether FORMER lands intervened.-- (2d) By character of any «
228b231 nt in» man in present state. with what he is as FORMER species. His arts would not then have taken hi
229b234 ia.-- Range of Zebra?-- The Crocodile & Tortise FORMER inhabitants of Mauritius Freycinet Voyage, agr
236b278 How is Juan Fernandez-- Humming Birds» types of FORMER dogs. character of Miocene Mammalia--of Europe
240c013 land species are not found in the Archipelago-- FORMER statements to such effects false In New Guinea
250c038 t to what is supposed to have been condition of FORMER whole world. America might have been string of
278c130 Van Diemen's land is most important as showing FORMER connection of two continents & death of form i
299c195 of permanent varieties in wild state-- The two. FORMER produced by difference of <locality>. station
315c243 y Bell on Expression & the Zoonomia, for if the FORMER shows that a man grinning is to exposes his ca
359d087 en Penguin Duck «from Bombay» & Canada Goose.-- FORMER strange mishaped bird-- looks very artificial
359d089 other & two dark-coloured & different.-- -- the FORMER were the parents of the three little ones.-- K
360d089 s hairless. some in patches, & some hairy-- the FORMER preponderated <which seems owing <determined»
363d102 irds «slight» different-- Barn Owl <the> in the FORMER place breeds in <flags> «thick vegetation» in
365d106 en Black Cock & Ptarmigan (probably subalpina.) FORMER has blue breast, latter reddish, hybrid purple

Page ***(Key Word)***
```
365d107 estricted in their range by men & by art.-- the FORMER only giving average of effects of country, (&
365d107 able to the structure which has been adapted to FORMER changes. than a mere monstrosity propagated by
366d108 at & of a dog, born without front legs-- -- the FORMER of which had kittens with imperfect ones.-- no
377d138 orcarious., together with a grunting noise, the FORMER signifies recognition with pleasure, as when f
391d175 . Why is there some law about sexes of twins in FORMER case.-- (many monster are really twins.)-- It
415e069 how how it came to be so.-- I speak only of the FORMER proposition.-- as in «races of» Dogs, so in sp
448e168 cus & Toxodon, <all> equally aberrant-- the two FORMER connecting classes like Toxodon «In orders»--
448e168 s like Toxodon «In orders»-- Fish & reptiles in FORMER case-- Reptiles & Birds & Mamm. in ornityhyrhy
490q001 their spurs & Ranunculus in the nectaries. The FORMER best for my experiment on Selection. Experimen
494q004 n Pidgeon?-- Mr. Miller said yes with regard to FORMER (8) Is form of globule of blood in allied spec
527m033 -- Music & poetry opposite ends of one scale.-- FORMER pleases from instinct the ears (rhythm & pleas
529m039 irtuous happiness.-- recall scraps of poetry;-- FORMER thoughts, & in experienced people-- recall pic
529m040 e illdefined notion of land covered with ocean, FORMER animals, slow force cracking surface &c truly
545m105 nterest, & those which are viewed very often.-- FORMER do not give rise to ideas so much. as objects
548m116 is state, the consciousness does not go back to FORMER periods so «as» to <make> «give» one individua
550m124 anger or revenge.-- they are incompatible & the FORMER, the more pleasant.-- Simple happiness «as of
553m132 young dogs get sadly torn in conflicts with the FORMER. But it is one thing for a swift dog to overta
555m142 attle (& Porpoises) have not charity-- is it in FORMER case instinct to destroy contagious disease.--
560m156 hen in this state, or remember what they did in FORMER one. about heredetary tricks & gestures, other
571n029 ter correspond to fashions in ideal taste & the FORMER to true taste.-- Everything that is habitual,
579n060 s; strong analogy with my view of blushing-- in FORMER irritation on a piece of skin cut off made the
587n088 t & not men-- orang-outang & chimpanze. pout.-- FORMER, whines just like a child. Get a Dictionary &
587n089 . Perception very different from emotion.-- The FORMER is used with regard to the senses. Reason does
609o30v rence between civilized man & savage,-- is that FORMER is endeavoring to change that part of the mora
615o36v evening» crow different.-- Heredetary effect of FORMER tropical climate analogous to inflorescence of
621o046 shortlived, it is to his intrest to follow the FORMER; & likewise <that the> then receive the moral
071r156 edones far preponderant, if so coniferous must FORMERLY have been most abundant tree-- Metamorphic ac
071r158 Quito. tranquillity «at Mendoza» exception.--«FORMERLY perhaps otherwise» Mendoza never overthrown,
113a090 But is this not wrong? we know mean of surface FORMERLY much higher, «so» that we must look at the up
138a153 par Dr. H. Unanùe says he believes the sea has FORMERLY stood three hundred feet above its present le
172b006 each other AEgyptian cats & dogs ibis same as FORMERLY but separate a pair & place them on fresh isl
195b101 out any assignable reason.-- Astronomers might FORMERLY have said that God ordered, each planet to mo
206b150 mation? p. 320 <Think> States Cryptogam. Flora FORMERLY common to New Holland?! p. 320. Says Conifero
212b168 into this «in Phillips».-- According to this, FORMERLY there would have been many genera of monotreme
219b195 must be <Alpine> «new formations» because snow FORMERLY descended lower, therefore species of lower g
219b195 n plants «No» CD[Mem. the antarctic flora must FORMERLY have been separated by short space from mount
228b229 ) which now run together, were not both genera FORMERLY abundant. Seed of Ribston Pippin tree <go> pr
284c147 heat & cold. therefore probably fewer now than FORMERLY.-- The number of forms depends on the externa
288c160 larly to main land, proof of. land having been FORMERLY nearer.-- «Selby» Magazine of Zoology & Botan
315c243 s his canine teeth. no doubt a habit gained by FORMERLY being a baboon with great canine teeth.-- (Th
355d068 on Cordillera fossils same remark. ¿was there FORMERLY one great sea, & two Polar Continents Marsupi
406e037 me in North & South. America.-- Whole wor[l]d, FORMERLY possessed a climate compared to S. America at
407e038 favourable to Tropical productions. The world FORMERLY much more so. yet climate of same order as th
407e039 this that Africa «& East Indian Archipelago»-- FORMERLY were not so «very» EQUABLE. or so tropical, &
439e142 nces favouring the reappearance of characters, FORMERLY possessed-- <that is animals> «or rather the
470t176 atherers & they seem slow=- Maer 1840 My Father FORMERLY planted «Turkey or» Palmated and English, pla
534m062 ociety of London Vol. I. p. 106. Col. Sykes on FORMICA indefessa placed table in cups of water which
066r141 me force. Yet, no outlet at head. Important in FORMING transverse valleys Ice Sir W. Parish says they
156g072 ly with perfectly rounded pebbles of granite & FORMING «sloping» buttresses Yet certainly shelf 4th <
191b082 pecies made elevation & subsidence continually FORMING species.-- Man & wife being constant together
264c080 hrikes & at the other into into Crows. yet all FORMING, according to Gould, good genus Gould seems to
367d112 ds are infertile. supposes that when foetus is FORMING the ovum within it, is forming «& this must be
367d112 when foetus is forming the ovum within it, is FORMING «& this must be so, else avitism could hardly
375d135 gaps <of> in the oeconomy of Nature, or rather FORMING gaps by thrusting out weaker ones. «The final
377d140 es>, elevating forces in raising continents, & FORMING mountain-chains, when we estimate the matter r
392d175 htly favoured, getting the upper hand. <]CD> & FORMING species)-- [Aphides having fertile offspring w
589n091 ctes, where strong argument in favour of brain FORMING the instincts,-- could brain make a tune on th
028r029 Alumen.--This matter accumulating in deep seas FORMING slates: How is the Lime separated; is it washed
122a113 of EXTREME FLUIDITY of earth-- study different FORMS of earth as shown by arc.-- read Herschels astr
157g075 anite pebbles, & around which shelf 2d «almost» FORMS it into island-- whole hill composed of remarka
158g080 ivortium my measurements here side of Loch Spey FORMS terrace about 60 feet above Loch trace of this
176b021 , requires deaths of species to keep numbers of FORMS equable:-- but is there any reason for supposin
176b021 but is there any reason for supposing number of FORMS equable: this being due to subdivisions & amoun
176b021 due to subdivisions & amount of differences, so FORMS would be about equally numerous. changes not re
180b037 lation to ancient types.-- with several extinct FORMS, for if each species «an ancient (I)» is capabl
180b037 an ancient (I)» is capable of making, 13 recent FORMS.-- Twelve of the contemporarys must have left n
183b053 cies, by saying, why not have some intermediate FORMS been discovered. between palaeotherium, megalon
186b057 breeding.. in irregular trees. & extinction of FORMS.?? It is in simplest case saying every species
190b076 586)-- Parallel 33 degree-35 degree, source of FORMS. reduce towards Northern Eastern end & die away
190b077 milies.-- (long separated.-- Proteaceae & other FORMS (?) being common to Southern hemisphere, does n
192b085 er orders a perfect gradation can be found from FORMS marking good genera-- by steps so insensible, t
194b097 malia & fish. when there now exist such strange FORMS as ornithorhyncus The type of organization cons
199b117 lready established. why out of the thousands of FORMS, should they all be classified.-- Propagation e
201b126 urable conditions» there are many gaps. & those FORMS which «nevertheless» have produced species, hav
202b134 t these wanderers would not, but where original FORMS most numerous there would be wanderers.-- Some
207b150 epock p. 330. Fossil Infusoria found of unknown FORMS, a circumstance undiscovered by Ehrenbergh.-- <
207b152 e la terre ferme de lancien monde".-- Considers FORMS in recent volcanic islets not well fixed.-- Per
215b178 ontographie" developing his ideas on passage of FORMS.-- Deshayes states Lamarck priority refers to i
217b187 ecies in Europe Madagascar has several American FORMS-- The above facts evidently show that Mr D. won
218b189 ty, holds good in plants» between all different FORMS; therefore when from being put on island. & fre
218b192 that some good African & some good S. American FORMS. (& daresays some of these <African forms> form
218b192 rican forms. (& daresays some of these <African FORMS> forms would have some peculiarity.-- Now when
218b192 orms. (& daresays some of these <African forms> FORMS would have some peculiarity.-- Now when we hear
221b200 , very like present species., p 8. ¿Are mundine FORMS, longest persistent?? do.-- The most perfect Pl
225b217 y to be sure there were a thousand intermediate FORMS.-- Opponent will say. show them me, I will answ
226b223 of America, ought to present great contrast in FORMS.-- India; intermediate, see how that is.-- ¿are
231b242 - Hippotamus only African.-- American & African FORMS mingle in India & East Indian islds-- Monkeys a
232b245 the crag we see the process of change of those FORMS, which have succeeded in becoming habituated to
232b247 culiar Flora «& north of Europe»-- As Europaean FORMS have travelled towards Equator, <th> so would t
241c015 & other rongeur in Australia.-- p 67 ¿American FORMS? All Infusoria. not extinct. species. good Resum
242c017 uroucous et rupicole vert instances of American FORMS in East. Ind: Archipelago. [Raffles. Horsfield.
243c019 utes latitudes"., --;translate?) All Australian FORMS have representatives (& instances given) in Eas
248c033 tted;--- hence the tendency to revert to parent FORMS, & greater fertility of hybrid & parent. stock,
249c036 nge? Mice.-- Waterhouse's remarkable fact of no FORMS peculiar to word to special districts???? north
249c036 rds of Europe & America, which are of different FORMS being migratory; also with Temminks fact of for
249c036 rms being migratory; also with Temminks fact of FORMS being within Tropics.-- Europaean birds at Japa
249c036 opaean birds at Japan. connected with Europaean FORMS on Himalaya??-- This is very remarkable, when w
249c036 in Eocene period. Have the Edentata & Marsupial FORMS been chiefly preserved,-- where shut up by them
250c037 drupeds if Africa destroyed would not then some FORMS be peculiar to it, so on, & so on.-- Whatever d
254c049 stem developed <is> not surprising to find many FORMS in Acrita,-- typical of other, (surely rather p
255c050 prey, are distributed in S. America like other FORMS, but those inhabiting 3d zone of <latit> height
255c050 this evidently relates to greater range of such FORMS.-- p. 56: Ornithological part of Voyage of??? A
255c052 bird-- An animal with skeleton of such general FORMS.-- The hybridity of ferns bears on my doctrine
258c060 of Analogy is a divellent power & tends to make FORMS remote antagonist powers.-- Every animal in col
264c080 tinent &c &.-- There is beautiful gradations of FORMS in Australia leading on one side into shrikes &
264c082 f station inhabited by them-- Timor. Australian FORMS amongst birds Java. not so much-- Peculiarities
268c096 > Tennioptera & Tyrannula (NB work out how many FORMS Tyrannula can we worked out into. Milvulus [not
268c099 lement geographical distribution is.-- ¿Pelagic FORMS similar--birds??-- We must always bear in mind
```

273c110 ++.» «This would lead one to expect that fossil FORMS would generally fill up genera & not species, w
274c115 ,-- How is it in water birds, there are walking FORMS in water birds,-- but no web forms in <water> l
274c115 are walking forms in water birds,-- but no web FORMS in <water> land birds,.-- Grups of very differe
274c116 t>. S. Africa, Australia, & S. America very few FORMS in common.,-- but each several with Europe & no
275c119 y species tell of Physical relations in time «& FORMS» distribution tells of horizontal barriers-- Mr
276c123 CIES-- between England & France. I will do with FORMS.-- Mention persecution of early Astronomers.--
278c132 to America.-- the wonder is that the Europaean FORMS, were able to escape to some more fitting count
281c138 only natural from death or slow propagation of FORMS.--just same way as <we> «all men» not all equal
283c144 Is there not some statement about diversity of FORMS in aberrant circles.-- explained by such not ha
283c146 at there are Tertiary fossil Infusoria, of same FORMS with recent, we have nothing to do with CREATIO
284c147 obably fewer now than formerly.-- The number of FORMS depends on the external relations (a fixed quan
285c151 It will be necessary to show hybridity from few FORMS, parents of all species not possible in some de
289c162 my theory) =supported by foetal lower developed FORMS.=» (NB waterhouse says of affinity of many inse
291c167 maphrodites & others not??? ¶ The death of some FORMS & succession of others, (which is almost proved
291c167 cessary to explain genera & classes. if extinct FORMS were all fathers of present, then there would b
292c168 ews-- did Lamarck connect extermination of some FORMS with his views.-- as genera are large probably
292c169 s genera are large probably only few of extinct FORMS have generated species. & of 100 extinct specie
295c183 -- Many African monkeys in Fernando Po-- no new FORMS only species!! No salamanders (D'orbigny Rappor
296c184 islands-- Endeavour to find out whether African FORMS. (anyhow not Australian) on Peaks. Did Creator
296c184 tralian) on Peaks. Did Creator make all new yet FORMS like neighbouring Continent. This fact speaks v
297c186 oup few in number of kind, extermination.-- New FORMS made through probably an infinite number of for
297c186 rms made through probably an infinite number of FORMS.-- therefore an isolated form probably a remnan
297c186 robably a remnant.-- Pachydermata. & Horses few FORMS. & they are remnants.-- Cephalopoda ditto.-- Ma
303c203 -- Parallel of Japan near Himalaya, & Europaean FORMS on that isld.-- The races of men differ chiefly
303c204 hat kinds of intellect) quantity & kind of hair FORMS of legs-- hence the father of man kind probably
304c207 irds reveal the secret.-- Now all the different FORMS of Synallaxis. trifling characters as red band
304c207 band on wing show to be from one parent.-- same FORMS of beak &c without these trifles. it would not
304c207 er not descended from long way back.-- aberrant FORMS produced where many species <osculant> but, whe
305c211 r & chiefly heat), which assumes a multitude of FORMS, «each having acting principle» according to sub
305c211 is necessary.--) which is modified into endless FORMS, bearing a close relation in degree & kind to t
305c211 close relation in degree & kind to the endless FORMS of the living beings.-- We see thus Unity in th
317c250 ome proper well-worth studying, with respect to FORMS.-- Study Appendix to Tuckey's Expedition Journa
324c268 ce Bory St. Vincent. Vol III p. 164. on unfixed FORMS. Dr. Royle on Himalayan. types. Smellie. Philos
325c267 ements 3d. Edit. Octav. (good to trace Europèan FORMS compared with African <A> Annals Histoire Gener
338d025 only difference is by the extinction of certain FORMS from Northern part & not by fresh creation of n
338d025 om Northern part & not by fresh creation of new FORMS.-- what is range of Hyaena? Hippotamus.? Indio-
343d036 , & the world peopled «with Myriads of distinct FORMS» from a period short of eternity to the present
343d037 aphically divided either by developement of new FORMS in one., or apparently so. by the extinction of
343d038 not> «only» one «Guanaco» of the characteristic FORMS of S. America. With respect to future destinies
349d051 cases the circular arrangement from fewness of FORMS-- Cannot be discovered «un»till <in> «we ascend
357d072 fered most with respect to extinction of larger FORMS-- From observing way the Marsupials of Austral
372d129 e than if whole branch transplanted? +.simplest FORMS of budding. Why does Gecko produce always diffe
374d133 the Cephalopoda depart more widely from living FORMS.-- p 458 Upper Silurian, fishes oldest formatio
374d134 ella. terebratula, orbiculas, with many extinct FORMS & Trilobites Sept 25th. In considering infertil
378d147 do these differences indicate, species changing FORMS, <& loosing do> if so domestic animals ought to
390d177 rom foreign country. (<annuals> & so must those FORMS which are produced by budding «only» as cryptog
390d177 ing in & in seems connected with more developed FORMS) Study buds-- gemmae-- & monocotyledenous, do t
390d179 ation being the passing through whole series of FORMS to acquire differences: if none are added objec
391d174 not Coniferous trees generally dioecious oldest FORMS) why are twin in man more like «each other» tha
391d175). (All this agrees well with my view of those «FORMS» slightly favoured, getting the upper hand. <]C
397e003 pulation, but extermination & production of new FORMS.-- their number & corelations Octob. 4th. It ca
397e004 orical times have been small-- because change in FORMS is <al> solely adaptation of whole of one race
398e004 l changes & oscillations, not affecting organic FORMS, that the whole value of the geological chronol
405e032 alapagos. where different islets have different FORMS. it is either effects of having been long separa
407e038 any part, (except Europe. in which all Tropical FORMS have been obliterated) of the world. from the <
407e038 the relation of passage from N. to S. American FORMS. The climate of N. America, must have been equa
409e048 iscussing extinction of animals in Europe. :the FORMS themselves have been basis of argument of chang
410e053 not in same country.-- The traces of changes in FORMS of organs, will care little for species, except
410e053 t so far as wanting names to refer to, to those FORMS. where the termination of change occurs.-- thos
411e054 dual organs, must know whether the individuals «FORMS» are permanent, all steps in the series, their r
411e054 the laws of change are known.-- then primary FORMS may be speculated on, & laws of life,-- the end
413e059 k considers the characteristics of the Tropical FORMS in shells. are numerous species, numerous indiv
416e071 birth) from man arbitrarily destroying certain FORMS & not others.-- Term variety may be used to gra
418e083 h of the species & individuals in their present FORMS, are closely related-- By birth the the successi
422e095 r varied structure & complexity.-- hence as the FORMS became complicated, they opened fresh, means of
423e095 nswer is because, if we begin with the simplest FORMS & suppose them to have changed, these very chan
423e099 ent only to have most slightly modified organic FORMS.-- we know not rate of deposition has been equa
424e100 tland, Uddevalla. Many species same. & Northern FORMS-- & American ones & Europaean-- agree very much
425e104 milar to Galapagos-- study Flora-- what general FORMS-- are the Labiata nearest to American, or Indi
425e105 is can be generalized.--» islds., have peculiar FORMS.-- on the southern flanks of Alps.-- many pecul
427e108 other would rapidly increase, & hence number of FORMS. once formed. would remain stationary, hence al
427e109 mate being now less extreme, than before arctic FORMS would retreat: effect on snow of arctic climate
427e109 of arctic climate in far north regions? Arctic FORMS have travelled S. From the analogy of the anima
430e117 ks. does not bear, the least on ancient generic FORMS.-- the animals in Eocene period could not have
432e126 k & Murchison; which is a beautiful instance of FORMS, intercalated between two great distinct format
434e128 Infusoria becoming extinct not so soon as other FORMS.-- p. 36.. speaking about the controversy on Di
445e159 -- strangely consider existing «long-organized» FORMS as parent forms of existing highly organized fo
445e159 sider existing «long-organized» forms as parent FORMS of existing highly organized forms-- this resul
445e159 ms as parent forms of existing highly organized FORMS-- this resulted from the necessity of supposing
447e163 from one region to another.--» -- these simple FORMS perhaps oldest in world & hence most persistent
447e164 maphrodite Mollusca, & probably corals.-- these FORMS then ought to be very persistent,, & then neces
458t001 shed-- «Australia,. S. America--» These strange FORMS., camels, giraffes. Sivatherian & Anoplotherium
458t001 noplotherium, with existing, or nearly existing FORMS of aquatic reptiles most strange, & shows as in
458t001 eptiles most strange, & shows as in shells some FORMS are long preserved.-- vol VI. p. 539. Dr Cantor
460t019 in Cambrian & Crust show how long since present FORMS existed, but if it be asked how this complexity
460t019 ants, <slightly> «a good deal» modified <& Many FORMS lost; if> «of this old stock (which from action
462t051 condition in Java & Sumatra & dissimilarity of FORMS-- yet how valueless this objection, when one th
464t065 osely allied to the Struthonidae than any other FORMS-- Lund's Antilope in Brazil another point of ag
465t081 ding series of strata This again shows how much FORMS depend on other forms Lyell's Paper, in Taylor'
465t081 This again shows how much forms depend on other FORMS Lyell's Paper, in Taylor's Journ.-- Phil. Mag.
465t089 Tolk Crag. mentioned-- allied Beaver to present FORMS.-- -- How many «tertiary» estuary & Lacrustine
471t07 stwood on the Fulgoridae enumerates the strange FORMS which the thorax & head displays.-- most fantas
482t018 en wherever there is extreme heat, the tropical FORMS extend further north, because during winter the
482t018 cold when torpid.-- On this principle tropical FORMS in N. America extend much further N. in N. Amer
486z018 odd that Amber insects of Europe have Tropical FORMS See p. 256 of Note Book (C) for comparison of s
514q21. r Fauna?. {Australian Alps--; are any Europaean FORMS found there?-- Lindley says that only one pineap
516q23v eum 1840 p. 777. Decaying wood absorbs oxygen & FORMS Carbonic Acid. will this bear on Petrifaction?
528m037 - these two causes very weak.-- (2d) form. some FORMS seem instinctively beautiful «as round, ovals»;
528m038 Again there is beauty in rhythm & symmetry, of FORMS-- the beauty of some as Norfolk Isd fir shows t
528m038 foreground either owe their beauty to absolute FORMS or to the repetition of similar forms as in ang
528m038 absolute forms or to the repetition of similar FORMS as in angular leaves,-- this Rhythmical beauty
536m071 , otherwise horses would have idea of beautiful FORMS.-- With respect to free will, seeing a puppy pl
570n027 showing. no connection with male figure]CD-- As FORMS change, so must idea of beauty.-- [Old Graecian
610o034 ion of vital laws-- According to the individual FORMS of living beings, matter is united in different
610o034 nknown relation to them-- [RHC] In the simplest FORMS of living beings namely «one individual» vegeta
636j56v ious Plants were formed. Macculloch says, life, FORMS a broken, recurrent series, whilst the habitai
636j57r ld not exist without these adaptations.--fossil FORMS show such losses.-- Consider ground Woodpecker
638j58v ery mountains are formed of such dead & extinct FORMS.-- the exuviae of the dead & extinct The analog

641j29v e is evidence of antiquity & extinction of such FORMS-- these views will bear on geology-- There is a
206b147 onship of the progenitors would have different FORMULA for each lustrum.-- We may conclude that there
152g047 rrace Ballivard 2 miles North of Grant town to FORREST road comminuted shells Important contingency i
242c017 cel. Leschenault Kuhl. Van-Hasselt, Reinwardt «FORREST» authors on E. I«ndian». A«rch.»‖ Borneo & Sum
245c023 ies from dental characters, wild pigs. said by FORREST to swim from one isld to another--‖ «It is a g
450e173 for they stand India. better than the latter-- FORREST Voyage p. 323. Sooloo. imported elephants. wil
450e175 l there)-- shows these isld not fit for horse. FORREST-- (p. 270) says many wild horses, bullocks, &
452e181 isld of Batchian near SE. end of Gilolo.-- "-- FORREST Voyages. p. 39-- deer but no wild animals in G
453e182 ays no wild (carnivora) beasts on Phillipines. FORREST somewhere says same.-- do p. 393. <">The wild,
275c119 bold in many such profound judgment, that he FORSEEING. consequence., was endowed with what may be c
048r085 might fancy that it was so arranged from the FORSEIGHT of the works of man Feeling surprise at Masto
065r138 Specimens of rocks were brought home in Capt. FORSTER expedition from <Deception Isld.> South Shetla
256c054 than what makes species in other animals.--? FORSTER on South Sea, will probably contain descriptio
148g023 ‖ traces of them all along <Glencoe>.-- towards FORT William yet in Glencoe in parts no trace of the
159g085 Boulder hypothesis Thursday, from Glen Turrit to FORT Augustus Barom on upper (rather above)? shelf 2
163g107 nt saw them himself-- Sand with tide ripple Near FORT Augustus hill & fringe as if it has been filled
051r093 ns every rock is buoyed by Kelp, now Kelp sends FORTH branching roots which must protect surface; On
222b203 are called varieties.-- NB. one mother bringing FORTH young having very different characters is attem
446e162 it tree by certain treatment will suddenly send FORTH quantities of blossoms--» The case of the Lemna
542m093] to do it, but an involuntary laugh will burst FORTH, this & yawning. (common to other animals) scre
283c145 tor had so created them.-- People will argue & FORTIFY their minds with such sentences as "oh turn a
318c254 ountains, & these appearing to remain about a FORTNIGHT, «See Silliman's Journal 1837. Paper by Bachm
318c254 genus.» The Muscicapa solitaria stay about a FORTNIGHT in one particular part of country, like White
532m056 & her bed brought from Shrewsbury) yet for a FORTNIGHT continued wretchedly unhappy, constantly whin
532m056 go out of house except with Caroline-- After FORTNIGHT. continued to grow thin & did not seem quite
244c023 cus says; <after> "après un examen attentif. et FORTS surtout de l'opinion du baron Cuvier, nous ne b
199b118 t general laws of life, the transmission of a FORTUITOUS modification, into a durable form, of a fugi
436e137 ful seed of a Bull Rush I thought, surely no "FORTUITOUS" growth could have produced these innumerabl
044r073 ology of whole world will turn out simple.-- FORTUNATE for this science. that Europe was its birth p
640j167 ld be remembered that Naturalists are prone, FORTUNATELY, to take their ideas, which are arbitrary &
522m011 e I do.-- What became of him.-- Answ Had large FORTUNE left. him, took name of Child «of Kinlet» & ma
222b202 ies by travelling of climates & the backward & FORWARD introduction of species.-- When species cross
228b232 ll is superadded, animals not got it, not look FORWARD» if we choose to let conjecture run wild then
609o030 ha> of the rule of happiness we must look far FORWARD «& to the general action» -- certainly because
609o030 ood far back.-- (much further than we can look FORWARD: hence our <[...]> rule may sometimes be hard
636j56v e hybrid seedlings) p. 333. Macculloch. brings FORWARD. the impregnation of Dioecious Plants by forei
391d175 ed if not.-- the changes oscillate backwards & FORWARDS & are individual differences (hence every ind
087a010 ns of Mississippi -- No Vol. I. p212. Cuvier Oss FOSS Wide range of Mammalia really very important. h
023r012 n's bottom. With respect to Sharks distributing FOSSIL remains: Sharks followed Capt. Henry's vessel
042r068 f Ashes of Falkner, ¿how far is the distance?-- FOSSIL bones black as if from peat.--yet cetaceous bo
074r165 oncretions of clay iron stone; iron pyrite in a FOSSIL» Insist strongly on the grand fact of Volcanic
086a004 ht to sympathize with. old doubters of what are FOSSIL shells.-- accustomed to such terms "fixed as t
088a018 etamorphic action -- Mem: red sand of Europe no FOSSIL shells --¿ action of Heat bubbles volatilized
090a024 0 lbs a year too little.-- How comes it none in FOSSIL state? suppose «100» <5>0£ x 50,000 x <50 = 25
093a034 n the mud, of the Salt river.-- in reference to FOSSIL guanaco of P. St. Julian. -- Mr Scrope seems t
110a085 ner's Chemistry p. 206 Both Beck & Deshayes saw FOSSIL shells from West Indies & declare them to be r
123a115 & coal near Bogota; p 270.-- SPLENDID PAPER on FOSSIL shells of S. America. Von Buch Lyell. (under h
189b072 duced by seed go on.-- otherwise all die.== The FOSSIL horse, generated in S. Africa Zebra.-- & conti
192b088 series is not more perfect by the discovery of FOSSIL Mammalia than before & that is all that can be
194b094 sion be L. Institut «1837. No 246» a section of FOSSIL "singe", it cannot be made to approach the Col
195b099 rrive & seeds.-- The same remarks applicable to FOSSIL animals same type, armadillo like covering cre
196b106 several places & died off in some? Why did not FOSSIL horse breed in S. America-- it will not do to
200b125 now confined to extreme North.-- ‖.do p. 326. 2 FOSSIL species of ox in N. America: as well as 2 rece
202b133 peculiar to East Indian isles.-- Compares it to FOSSIL Didelphis (S. American genus) in plaster of Pa
207b150 irst appear «(p 321)» at Tertiary epoch p. 330. FOSSIL Infusoria found of unknown forms, a circumstan
213b168 e many tables in Phillips of numerous genera in FOSSIL & recent state, well worth consideration-- Ta
227b228 finities. My theory would give zest to recent & FOSSIL Comparative Anatomy, & it would lead to study
231b243 -- Two inhabitants of the Tropics, (whether one FOSSIL or not) are related by real relationship. as w
263c077 udy geographical distribution study relation of FOSSIL with recent. the fabric falls! But Man-- -- wo
273c110 orary «+», «This would lead one to expect that FOSSIL forms would generally fill up genera & not spe
278c130 cording to my own methods. Dasyurus being found FOSSIL in Australia, & only one tree species (Mitchel
283c146 produced. As we know from Ehrenbergh, there are FOSSIL (see Scientific Memoirs & L'Institut.) that th
283c146 Memoirs & L'Institut.) that there are Tertiary FOSSIL Infusoria, of same forms with recent, we have
286c156 .-- in order, or in next family? In considering FOSSIL animals, what relation in classification in bo
319c257 's land & Australia proper.-- Irish Elk case of FOSSIL geographical range.-- [blank] Books examined:
338d025 n food in insects entered by mistake Surely the FOSSIL Mamalogy of Britain & Europe is African. & the
338d025 Hippotamus.? Indio-African, or pure Africa?-- ‖FOSSIL Elephant of Africa Most important under this v
343d039 ptian Mummies & from the existing animals found ‖FOSSIL when Europe must have worn a quite different f
352d058 inks in circle must be granted unequal, because FOSSIL) Now what is group without centre but circle,
407e041 logy,-- shells of Africa ought most to resemble FOSSIL ones of Europe, Consider probable form of land
412e058 y structure.» L'Institut. 1838. p. 384. List of FOSSIL Mamm: from Poland. &c.-- Three principles, wil
414e064 of the men on the plains of Bolivia-- strictly FOSSIL «& in Van Diemen's land»-- they have been exte
415e066 raise an argument against, my theory, should no FOSSIL «very distinct species» of the Ornithorhyncus
420e088 itut do p. 419, «long» account of Hyaenodon, a FOSSIL dog-- leading towards Hyaena.-- see Comte Rend
420e088 na.-- see Comte Rendu.-- I suspect good case of FOSSIL filling up blank.-- CD[not between existing se
422e092 maintains, that «Alludes to difference between FOSSIL & recent Bull; like fossil & recent shells of
422e092 o difference between fossil & recent Bull; like FOSSIL & recent shells of the <new> raised beaches» [
424e100 ar to separate islets.-- March 5th. Lyell says «FOSSIL» shells from North America, Scotland, Uddevall
424e101 es.-- this would depend on negative evidence of FOSSIL remains, & therefore not to be trusted.-- -- L
425e105 nnected with other mountains laterally.-- Owen. FOSSIL Mammalia. p. 55. talks of Tapirus American for
434e128 anniversary address 1839, p. 9.,-- talks about FOSSIL Infusoria becoming extinct not so soon as othe
449e169 . History. 1839. p. 106.-- Waterhouse refers to FOSSIL remains of the Hamster.-- is not this Siberian
458t001 erved.-- vol VI. p. 539. Dr Cantor's account of FOSSIL frog, 40 inches in length--! alludes to ancien
465t079 ly two monkeys, <there are now> have been found FOSSIL in S. America, there are now-- -- species in S
471tf09 er's E. vol. II p. 18.) capital list of all the FOSSIL Mamm. of Europe-- Large Lizards in Navigatores
483z012 ies, therefore interlock.-- Testudo INDICUS not FOSSIL at Isle of France: <Jerrold?> Bibron Zoolog. J
490q001 m recent ones-- I presume some recent not found FOSSIL (perhaps not embedded ¿ are there any very com
636j57c als could not exist without these adaptations.-- FOSSIL forms show such losses.-- Consider ground Wood
355d067 s Sus. do vertebrae vary? «See Cuvier Ossemens FOSSILES» Although now new fact be elicited by these sp
463t063 -- Cuvier has grand sentence about the Animaux FOSSILES-- being a mere fragment of the discoveries to
041r065 this latter case. we cannot judge whether such FOSSILS. lived in groups or not. Ferruginous veins of
066r142 ision--Together with same general character of FOSSILS deception complete.= Silliman Journal. year 18
067r144 . 1834 Mr Sowerby. younger. says that Falkland FOSSILS decidedly belong to old Silurian system. Apply
118a103 y on the Geology of Chile.-- P p217. Pentlands FOSSILS & Meyens --<Jura &> Chalk When we consider par
119a105 t heigth on mountains.-- consist of rocks with FOSSILS., therefore formed near surface. whether they
194b095 tful when known only by bones.-- Mem: Silurian FOSSILS: ¿How are. South American shells? ¿Do not plan
201b126 olog. Proc. p. 569. 1837. Account of wonderful FOSSILS of India.-- & p. 545 «great monkey» Mr Johnsto
212b167 p 81» in Lardens Encyclop. Proportions between FOSSILS & recent shells between herbivorous & zoophago
214b173 lop. insists on analogy between Australia and «FOSSILS of» Oolitic Series does not appear to me very
214b174 dner p. 289 It is certain, that North American FOSSILS bear the closest relation to those now living
216b179 hound-- Bull. Soc. Geolog. 1834. p. 217. Java FOSSILS 10 out of twenty have ANALOGUES uses this word
221b202 o reptiles before Mammalia Think about Miocene FOSSILS some species being recent agreeing with Senega
225b218 unha. may be said to deceive man. as likely as FOSSILS in old rocks for same purpose.!! Can the whirl
232b246 long average tolerably> uniform).-- Comparing FOSSILS with whole world. would be like in a Meteorolo
233b250 ng localities Institute 1838. p 38. account of FOSSILS of Sewalick «India» Monkeys of old World. Croc
281c137 vely slowly. whether geologists would not find FOSSILS such as they are-- My theory explains that fam
296c184 near lofty country»?? p. 475 NB. This bears on FOSSILS of Europe., those species which can migrate re
303c205 . if Mammalia are adduced. say oh look to your FOSSILS, now if extinction had gone, without creation

316c246) on coasts of N. America & England.-- but the FOSSILS are not like, except in very few cases, those
316c247 in very few cases, those of Tertiary Europaean FOSSILS-- «(so much the more remarkable ∵ carboniferou
316c247 r come from Siberia it cannot be said American FOSSILS more resemble those of America than of Europe,
316c247 N. America & Europe?-- ‖.Norton has written on FOSSILS of N. America.-- At the end of "White's Selbou
319c257 f S. America,-- Are not some of the Australian FOSSILS intermediate between those of Van Diemen's lan
328cIBC Porcellus Turbo. Cerithium Jardin du Roi Java FOSSILS at same time Study Botanical work on Buds & Ge
355d068 nct-- but close.-- Mem. Von Buch on Cordillera FOSSILS same remark. ¿was there formerly one great sea
399e006 at present might be very thick & yet have same FOSSILS. does not Lonsdale know some case of change in
415e066 his remains.-- Lima.-- caves.-- There being no FOSSILS, the only way, that I can see to discover whet
423e099 an American form.-- has several species in my FOSSILS-- CD[If cases of one variety in upper part of
458t001 Journal Vol 7. p. 658-- Falconer on Sub. Him. FOSSILS-- Ruminants. & Tortoises gigantic-- hyaena-- b
463t063 veries to come-- Owen in his description of my FOSSILS makes same such remark & before the conclusion
465t089 tiary» estuary & Lacrustine formations contain FOSSILS. -- mammals-- a few only -- & how many estuary
535m069 step plastic <virtue> natures. accounting for FOSSILS). & lastly the tracing facts to laws. without
437e140 young ones the cub of a fox, which after it had FOUGHT well & desperately bitten the young ones, woul
128a130 st of North America., like the Mexican Gulf. is FOULED by bars of sand & shallow lagoon.-- when descr
022r006 s rocks have undergone action of heat. it is so FOUND in Anglesea, amongst the varying & dubious gran
022r008 hat a sharp knife could not cut it: in which we FOUND the Head & Boams of a Hippotomus; the hairy lip
023r009 Java is 1000 miles distant! Where are Hippotami FOUND in that Archipelago? Such have never been obser
025r019 I suspect fragments of shells will generally be FOUND to be old & dead)» «(I have not kept a record)»
027r023 . mentions in what formations Conglomerates are FOUND. -- The above oscillations remarkable because t
030r036 or sloping land What are the "palatal Tritores" FOUND in the coraliferous mountain Limestone are they
037r056 d in vain for a pebble of any sort; not one was FOUND.-- Miers saw then near? Mem. La Condamaine on t
039r061 f W coast in explaining plains because such are FOUND in perfection on that side.-- Add from M. Lesso
057r114 cordance in Europe with ice theory.-- Capt Ross FOUND in Possession Bay in 73 degree 39 N. living wor
065r140 yry.--Falklands.--off East Coast.-- Capt. Cook FOUND soundings. (end of 2d voyage outside coast of T
067r143 835 Sir W. Parish says. that beds of shells are FOUND on whole coast from P. Indio to Quilmes. & at l
073r161 tribute a meteoric origin & which is constantly FOUND mixed with lead & copper is infinitely rare in
076r168 exhibits the same minerals <as> there, that are FOUND in the veins of Kongsberg in Norway.--namely de
086a003 -- There should not be surprise at Horse being FOUND in America, when Mammoth & narrow toothed Masto
088a016 titut 1837.-- Helms remark on common salt being FOUND on low hills East of Cordillera very important
093a034 th informs me that in the year a Rhinoceros was FOUND <emb> in the mud, of the Salt river.-- in refer
095a037 Vol 2. p 257 {T} The Pota: labiata certainly is FOUND with the Mactra. at Buenos Ayres at the Zoolog:
095a039 officers of the Bonite. French discovery ship, FOUND clear proofs of shells & waterworn rocks «at Co
101a057 beds-- What is Osteopora platycephalus (Harlan) FOUND on the Delaware. is it Edentate? Phillips p 289
105a072 » in valleys Lowe in his paper says land shells FOUND with calcareous matter & concretions on coast o
108a081 elaphyne. = Andesite-- Albite & amphibole= Cook FOUND Granite at Christmas Sound Vol XIV. (My Edition
113a091 had been better then 32 degree would have been FOUND lower.-- We have no right to consider the condu
123a117 om M Angelis. B. Ayres. 3d. May. states remains FOUND in many part.-- great Dasypus near Canelones --
127a126 n central fluidity.-- do p. 137. Lord Tullamore FOUND Sulph of Soda in peat ashes in Ireland dikes in
135a144 in Lat 26 degree S. Wafer looking for Copiapo. FOUND inland a great many sea shells some miles from
163g106 ting bank where Locks now are (32 ft rise) they FOUND alternating layers of coarse & fine & many Sea
192b085 of the lower orders a perfect gradation can be FOUND from forms marking good genera-- by steps so in
194b095 odent> edentate [has been doubted?]CD & opossum FOUND in Europe now confined to southern hemisphere.-
195b103 It may be argued representative species chiefly FOUND where barriers «& what are barriers but» interr
196b107 the furze, broom, & yew very different from any FOUND in great Britain, British varieties are also fo
196b107 nd in great Britain, British varieties are also FOUND in Ireland-- There must be progressive developm
198b115 on <different animals> large Mammalia not being FOUND on all isld. (if act of fresh creation why not
200b125 ndency to form varieties? Ed. N. Phil. J. Morse FOUND in Virginia. p. 325. July 1828. Animal now confi
203b136 by climate?-- M. Baer (thinks) the Aurock, was FOUND in Germany & thinks even now in central & Easte
207b150 21)» at Tertiary epock p. 330. Fossil Infusoria FOUND of unknown forms, a circumstance undiscovered b
210b160 e of the animals peculiar. to Mauritius are not FOUND at Bourbond Zoolog. Proceedings‖. <p> 1832.p. I
212b164 veral islands.-- Bear peculiar to Sumatra & not FOUND on Java-- Monkey peculiar to. latter not to for
213b172 cumulate without seal-bones & cetaceans.-- both FOUND in every sea, from Equatorial to extreme poles.
214b173 trong What is Osteopora platycephalus. (Harlan) FOUND on Delaware is it Edentate? Phillips. Lardner p
215b177 th males & females. lose desire. Native dog not FOUND in V. Diemen's land J. de Physique. Tom 59. p 4
217b187 tances of one species of Australian genus being FOUND in Sumatra; again another of other Genus in San
223b208 ctures.-- If the skeleton of a Negro-- had been FOUND what would Anatomists have said.-- ¿where is Pe
225b218 &c!!! An American & African form of plant being FOUND in Tristan D'Acunha. may be said to deceive man
228b231 er of man kind. than Macrauchenia yet he may be FOUND:-- We must not compare «chances of embedment in
230b236 rhaps in some plants & then a chain of steps is FOUND in same mountain).-- How is this explained by l
232b249 a sub genus of Phalangista New Holland form) is FOUND in many island Celebes «Waggiow» &c &c. (See Ly
236b279 toise from Germany. (where Mr Murchison fox was FOUND), decidedly next species to some South American
240c013 New Holland.-- The New Holland species are not FOUND in the Archipelago-- Former statements to such
240c013 e (Didelphis Brunii) which as yet had only been FOUND in isle of Aroe & Solor), «Vol I» likewise new
243c019 olutely different.-- --Philedon circinnatus not FOUND in Australia only New Zealand-- Norfolk. Isd. &
243c020 cans Australian form? p. 27. many fish of Taiti FOUND at <New> Isle of France: xx instance of wide ra
245c024 n insists much.-- The (p. 296) Columba Kurukuru FOUND in all Malasia-- & oceania, offers many varieti
251c041 e horses.-- p. 246-- Gmnura-- new genus of Mam: FOUND in Sumatra. p. 452 Append to Denham Clapperton &
252c042 els (what language?) Hyena «venatica» <of> Cape FOUND in Desert of Korto & Steppes of Kordofan p. 401
256c055 n of birds & animals Mr Strickland & Hamilton-- FOUND tertiary formation amongst Graecian isles, ¿See
257c056 ove sea in Bolivia; he examined-- all species & FOUND "beaucoup des mêmes oiseaux. que nous avions dé
266c092 s; jackals monkeys-- common to either coast not FOUND here not even Antelopes, though common on coast
271c105 f> agree in habits with the Turdus Musicus «not FOUND in N. America» whose Southern range is? <One> T
272c109 e Laniadae & Muscicapidae of new World, but not FOUND together.-- If two species come over to this co
277c127 differ as much as those (naming them) which are FOUND fossil in Australia, & only one tree species (M
278c130 ged according to my own methods. Dasyurus being FOUND!! rodents old inhabitants most important!! like
278c131 ity. Mem in Clifts list a rat said to have been FOUND in Van Diemen's Land.-- V. 1s. Number of Geogra
278c131 y?! Major Mitchell does not think that dog was FOUND at Swan river. The change in England from Rhino
278c131 of Geographical Journal to discover whether dog FOUND in Africa, the wonder have been same for S. Ame
278c132 ome more fitting country.-- if Toxodon had been FOUND-- a person who is habitually kind to children.-
298c190 the Key to the affections might perhaps thus be FOUND in other parts of world Athenaeum June 3d 1838.
310c224 e Brown Congo Expedition: 400 Australian plants FOUND in Isd of Bass' Straits The common Mush room &
314c239 emens land & Australia.-- The wombat (Brown) is FOUND a Euphorbia so near Piscatoria as scarcely to b
317c250 ssor Smith's Journal) on the heights of St Jago FOUND. very good to see whether peculiar plants-- in
327c265 of Mauritius with locality in which each one is FOUND in Australia New species of Moschus, characteri
341d030 tioned, it seems most of species from there now FOUND fossil when Europe must have worn a quite diffe
343d039 om Egyptian Mummies & from the existing animals FOUND in any colours of plumage &c &c «Pouting pidgeo
363d101 y duck» crested feather, pouters, fan tails are FOUND. p. 426 Sauroid fish in Coal, true fish, & not
374d133 chstein oldest rock in which reptiles have been FOUND in all quarters: his ideas not clear. In Austra
400e011 bution-- Thus Hattica is great genus.-- because FOUND cattle in Table Bay with Humps on their backs &
400e012 ect of Voyages Vol 8 «p. 46» Capt Davis in 1598 FOUND at Pellew & Eap, but not at Feis (near island)
403e021 There is another great lizard. Kaluz. which is FOUND in the Americas probably did not.-- Octob. 25th
405e031 &c &.-- if in next 20 years none of his remain FOUND to be inhabitant of S. America & as it is <fall
405e032 - Octob. 26th. If. hereafter. M. angustidens be FOUND together.-- Read this Work-- Decb. 4th.-- Why h
413e060 ere, Elephant, Rhinoceros, Hippot, Haena &c are FOUND.;-- yet until man became cosmopolite, he would
415e066 very distinct species» of the Ornithorhyncus be FOUND in Edinburgh.-- Why does Fleming consider them
424e102 ixed in England-- for I presume Carrion Crow is FOUND in Eocene beds of Paris Lyell has remarked spec
425e105 ammalia. p. 55. talks of Tapirus American form. FOUND no where else not even in branch valleys-- M. R
429e116 & S. valleys, which penetrate Pyrenees but are FOUND at Folkstone.-- it is unnamed this intermediate
430e120 e woodcut) like I. sulcatus.-- Both species are FOUND in the Inferior Oolite, & the T. elongata in th
431e121 ades, now L. says the T. costatus is in England FOUND together on coast of France.-- L. doubts.-- Lon
431e121 Min. Conch. it is however, said they have been FOUND the <poll> masses of pollen of Asclepias placed
435e133 cated] the stigma retains its power.-- R. Brown FOUND alive.-- Stanleys Familiar History of Birds sev
437e139 s the Morne Mountains, it then dropped it & was FOUND a dead lamb «& hare» by the side of Eagles nest
437e139 having wounded the bird. p. 124-- Mr Willoughby FOUND at Timor.-- thinks he has seen specimen at Pari
449e173 on his Map of the World, has. written. Mastodon FOUND but only in one part the northern peninsula of
450e174 hinoceros Leopard (but not Royal Tiger), &c are FOUND but only in one part the northern peninsula of

Page ***(Key Word)***
450e175 . H. ? do. p. 189. «190» No full sized horse is FOUND East of y Burramposter & S of Tropic-- By J. H.
450e175 s in the different islands.-- The horse is only FOUND wild in the plains of Celebes. (but language sh
451e177 Hutton counted, the ova of a tick «in India» & FOUND there were 5,283 attached to its body-- Journal
452e181 ut the Moluccas Archipelago they are only to be FOUND on the isld of Batchian near SE. end of Gilolo.
452e182 lludes to fact stated by M. Tournal that skulls FOUND near Vienna appoximat to Negro form; those from
453e182 as p. 229. Believes the <Rhinoceros> «Tapir» is FOUND in Borneo.-- «p. 233» There, as well in all Mal
462tf4r th a thrush"-- Athenaeum 1839. p. 708.-- Shrew, FOUND by M. Lartet same as existing species. We see t
463t059 se says perhaps animals of Fernando Noronha are FOUND unknown coast in front of it.-- Cuvier has gran
465t079 yet only two monkeys, <there are now> have been FOUND fossil in S. America, there are now-- -- specie
468t106 eks-- Flowers which thought very unattractive-- FOUND Rhubarb blossom swarming with small Staphylinid
468t112 ower> plant to plant.»-- to my grt surprise-- I FOUND all, stamens straightened pollen profusely shed
473s03r ostris in miocene like in Europe-- Cuvier never FOUND remains of Sus with Elephants-- Lyell says New
477z002 s (.4 pichye, pelud, mulita et mataco.) are all FOUND south of 26 degree 30 minute. Lat -- -- do. p.
481z008 bits of Crustaceae Mr Broderip says that Voluta FOUND in not less than 7 fathoms water. Mem Bahia Bla
481z010 ory of the Abipones-- says «the Condor» <it> is FOUND in the Tucuman mountains The fourth Vol. «in Ly
490q001 ct from recent ones-- I presume some recent not FOUND fossil (perhaps not embedded ¿ are there any ve
495q005 se are wild varieties. Is any intermediate form FOUND wild a The Leptosiphon densifolium «an annual»
497q006 long they remain dormant. if kinds come up, not FOUND in wood.-- but seeds continually dropping in wo
502q012 - would it grow in open air in Sweden. Linnaeus FOUND 2 flower. which had anthers removed, did not be
514q21. s-- on flora of African Islds-- names of Plants FOUND on mountains of N. America similar to Lapland P
514q21. a?. {Australian Alps--; are any Europaean forms FOUND there-- Lindley says that only one pineaple Hor
523m017 e & had bought arsenic for that purpose.-- this FOUND to be true.-- Her Husband never suspected durin
529m042 off.» & vomited, but who when he recovered. was FOUND to be ignorant, but quite sensible & no ways an
541m089 ible (testing success by decreasing headache) & FOUND best plan was allowing my mind to skip from sub
546m111 nch of tree, & apologising to party, went out & FOUND it there!!! Lady in perfect «mental» health.--
552m132 mpted to say that those actions which have been FOUND necessary for long generation, (as friendship t
555m142 rtineau (How to Observe p. 213) says charity is FOUND everywhere (is it not present with all associat
563n002 wished he had done so & so for his interest, & FOUND he disobeyed a wish which was part of his syste
584n071 g, his arguments partly fall, when a species is FOUND which does not climb CD[.instinct may be divide
610o031 book under head of Wilson, referred to Robert & FOUND his Christian name was Wilson!!-- How curious a
634j55r before birds? They are ancient.-- Are Cetaceae FOUND in Paris Basin?.-- NB) The explanation of types
252c045 of isld.--«immense importance of local faunas FOUNDATION of all our knowledge especially great contin
349d051 most easily:-- those which do not vary being FOUNDATION for chief divisions.≠ p. 7. «In» Some <of> c
398e005 nviction, absolutely necessary as the «basal» FOUNDATION stone of further inductive reasoning is imme
409e049 ts, which as I hope to show is «probably» the FOUNDATION of all that is most beautiful in the moral s
434e128 g to the events of the past world, we have no FOUNDATION for our science".-- <it is only analogy.> bu
564n005 function of body.-- we must bring some stable FOUNDATION to argue from.-- Octob. 4th. Seeing some dra
050r091 seems to think that Volcanic eruptions form FOUNDATIONS for Coral reefs.--does he mean in contradist
419e087 ER ALL THIS» Extinction & transmutation, two FOUNDATIONS, hitherto confounded,. of geology.-- L'Insti
252c042 views about Species-- MUST BE STUDIED: genera FOUNDED in nature [not located] The systematic natural
284c149 od for generic divisions [+] ought genus to be FOUNDED on such characters, as do not vary in the spec
062r130 n ostrich bears to (Petisse. & diff kinds of FOURMILLIER): extinct Guanaco to recent: in former case
281c138 mals-- talking of men as related in the third & FOURTH degree.-- a species must be compared to family
441e150 on another, must not be overlooked.-- it makes FOURTH cause or law of change.-- The weakest part of
481z010 dor» <it> is found in the Tucuman mountains The FOURTH Vol. «in Lyell's possession» of Zoolog. of Voy
296c184 4 instances of hybrids between pheasants & Black FOWL.-- use as argument possibly some few hybrids in
341d031 aks.-- &c &c Dr. Bachman has crossed cock Guinea FOWL with Pea <cock> Hen.-- offspring female, yet so
358d085 hat are infertile.-- thus the common. pheasant & FOWL when crossed never even lay eggs. & the men can
359d087 sexual organs alone.-- It is singular pheasant & FOWL being so totally infertile whereas animals furt
368d114 ery different in recently altered genera. Guinea FOWL & Peaccocks.!!» other birds display beauty of p
380d154 plumage so different, yet the Cassowary & Guinea FOWL cannot be distinguished.-- A capon will sit upo
392d180 ntize on crossing of the several species of wild FOWL <in Z> of India «with our common ones» in Zoolo
494q004 antam-- will egg kill Hen Bantam.-- Cross common FOWL with Dorking (10) Statistics of breeding in Zoo
502q11v re young. beans. cabbages.-- History of Pheasant-FOWL. Hen coloured like cock-pheasant: said not to s
563n001 only in breeding season & on the ground.-- Cock FOWL. on the ground, at roost, in all seasons, & aft
615o36v l force in any individual.-- His Malay breed «of FOWL» totally different habits from Europaean. begin
183b052 rdier" &, <fr> silver gold & common pheasants & FOWLS..-- "On sait que le "métis" du loup et du chien
204b140 g most forcibly-- Esquimaux dog & Pointer «Game-FOWLS have courage independently of individual force»
216b180 &c.-- as in dogs investigate case of pidgeons. FOWLS. rabbits cats &c &c.-- When black & white men c
261c070 , better than child sucking or even ducklings & FOWLS-- When talking of races of Man.-- black men, bl
290c164 be known.-- It appeared to me that half between FOWLS & pheasants, is most like pheasant., I think so
315c240 spots & since disseminated See. Habits of Malay FOWLS p 5. (note) in some papers on instincts L.' Ins
355d066 dun-coloured with stripe approaches to ass.» or FOWLS to the several aboriginal species «or ducks» (h
355d066 s «or ducks» (here argue if it be said domestic FOWLS are descended from several stock, then species
363d101 em)-- [Study horns of wild cattle.-- plumage of FOWLS-- long ears of rabbits. & long fur.-- feathers
378d148 ir present plumage.-- How is this in Pidgeons & FOWLS.--??? Wate[r]ton «p. 197» put 12 wild duck's eg
385d163 en second small & so.-- Says, there is breed of FOWLS called everlasting layer--. or Polish breed. (h
385d164 or Polish breed. (he thinks half pheasant, half FOWLS.-- eggs fertile, but parent bird will never sit
453e182 here says same.-- do p. 393. <">The wild, small FOWLS at Pulo Condore "crow like ours, but much more
513q020 variety, not specific character= Cross Rumpless FOWLS & Dorking fowls,-- or tailless dogs & fox, to s
513q020 cific character= Cross Rumpless fowls & Dorking FOWLS,-- or tailless dogs & fox, to see whether the c
593n109 s is hostile «is subversive of» to this view, & FOWLS hatching stones. in some degree is so.-- idea o
615o36v ays, that beyond doubt courage is heredetary in FOWLS & not effect of feeling of individual force in
026r020 ent distinction is given to angular & rounded.-- FOX Philosoph. Transactions on metallic veins. 1830
106a076 ute. 1838 p. 40 or Phil Mag. Dec 1837. p. 520 Mr FOX on increase of temperature at great depths. All
145g002 Highland shepherds dogs coloured like Magellanic FOX.. an instance of Provincial breeds. [3] Veins of
172b007 Now Galapagos Tortoises, Mocking birds; Falkland FOX-- Chiloe, fox,-- Inglish & Irish Hare.-- As we t
172b007 Tortoises, Mocking birds; Falkland Fox-- Chiloe, FOX,-- Inglish & Irish Hare.-- As we thus believe sp
192b083 y ask Dr. Smith.-- «to state that most clearly». FOX tells me, that beyond all doubt seeds of Ribston
204b141 r amongst themselves or with parent birds.-- «W. FOX. knew of case of male widgeon, winged & turned o
210b159 there probably will be this relation also.‖Yes « FOX» ‖ The creative power seems to be checked when i
212b165 me an Australian dog he had, used to burrow like FOX.-- a sort of internal bark. would remain for lon
214b176 r without this defect, Very curious case = W. D. FOX. When dogs are bred into each other, the females
215b177 t is required to give the canthairides and milk--FOX tells me that it is generally said.= How came fi
217b183 breeds of dogs, from owners great care of them. FOX says when two dogs of opposite breeds are crosse
217b184 he other, puppies differ, & like both parents.-- FOX told me of case of mare covered by blood horse &
225b219 elphis in Stonefield∴ all lands united (Falkland FOX. ice). . Mauritius what a difficulty-- where ele
226b222 shed from S. America, the jaguar has been left & FOX, & bear.-- If I had not discovered channel of co
234b255 d. are seldom seen with horns" --- p. 341. Black FOX sometimes introduced by ice <no> «only few» pigs
236b279 water tortoise from Germany. (where Mr Murchison FOX was found), decidedly next species to some South
252c045 robably tell more certainly Get Closer species-- FOX IS & Mice of America «good case on account of va
298c189 n?? Gould says he believes that he has seen half FOX & dog. & that it was most like fox.-- He felt su
298c189 has seen half fox & dog. & that it was most like FOX.-- He felt sure the half breed of Australian dog
310c223 s on Himalaya.-- some English beetles, birds & a FOX most close The most curious case is Saxifrage, a
332d005 o birds migrations & Australian Savages.-- W. D. FOX has a cat. which he bought in Portsmouth, said t
333d005 n of the Common cat.-- Ch IX Mongrels Hybridisim FOX has half Persian cat. which bred with unknown co
333d006 r more of English, but the effect is the same.-- FOX thinks that when a wild animal is crossed with t
333d007 e been so long as wild one under present form.-- FOX has seen several cases of foxes & dogs crossed,
333d007 al crosses between Esquimaux dog & common dogs & FOX thinks they decidedly take much most after Esqui
333d008 ly with Yarrell & no leading question was put.-- FOX thinks half Lion & Tigers are exactly intermedia
334d009 d Moreton's case opposed to this fact & views.-- FOX says a cousin «one of Mr Strutt» of his used to
334d010 re to old, explained by Stallions, (according to FOX) being guided entirely by their smell.-- Fox say
334d010 to Fox) being guided entirely by their smell.-- FOX says he knew «a» carter well, who placed his sta
334d011 rs in Mankind.-- The case of all blue eyed cats (FOX has seen repeated cases) being deaf curious case
334d012 ous case of corelation of imperfect structure.-- FOX says in «Lord» Exeter's Park «or in the Duke of
334d012 ed wild variety. doubtful whether all are white. FOX says the Half Muscovy Fox says a settler near Sw
334d012 whether all are white. Fox says the Half Muscovy FOX says a settler near Swan river, lost his <on> tw
334d013 bringing with her the other & younger cow.-- {P} FOX says when common & China goose are crossed the n

335d014 .-- ¶ The fertility of first cross, as stated by FOX, is very important, as showing above facts as fi
345d044 Highland Shepherd dogs, coloured like Magellanic FOX.-- peculiar hair & appearance-- good case of Pro
356d070 Pippen or Ribston do producing occasionally (as FOX says) same fruit trees is analogous to some hybr
402e018 breed being of the latter being the same as the FOX-like animals. which are met with near Canton" "H
409e046 ey? The dog being so much more intellectual than FOX, wolf &c &c-- is precisely analogous case to man
437e140 other day brought to her young ones the cub of a FOX, which after it had fought well & desperately bi
439e145 y had several breeds of dogs.-- like greyhound-- FOX-dog-- turnspit & two other kinds It seems absurd
494q004 luable as show. & curiosities!! What is price of FOX. otter. Badger &c &c &c.-- (11) Keep. Tumbling p
496q006 Place. Snap-Dragon. (I have seen one monstrous) FOX Glove & such like in very rich soil-- As they ha
513q020 less fowls & Dorking fowls,-- or tailless dogs & FOX, to see whether the characters are then intermed
530m049 of producing a train of thought.-- [not located] FOX believe cats discover birds nests & watch them t
542m097 ate to each other.-- Lonsdale's story of Snails, FOX of cows, & many of insects-- they likewise must
557m147 g to fight, in which case expression resembles a FOX-- I can conceive the opposite muscles would act,
557m147 wagging-- if as (I believe) Hunter says. neither FOX. nor wolf wag their tails, &c. it is very curiou
575n044 obey the same laws, as the crossing of jackall & FOX & wolf & dog.-- the only test this is most impor
583n069 ral aim of fable, & expression as cunningness of FOX, industry of bee &c &c-- p. 15."instincts act wi
187b065 so with species.-- I should expect that Bear & FOXES &c same in N. America & Asia, but many species
196b104 ted.-- No doubt. in birds, mundine genera, Bats .FOXES. Mus are birds that are apt to wander & of easy
216b182 diction to his view of races not mingling?-- In FOXES case of Blood Hounds, a little mingling would p
333d007 r present form.-- Fox has seen several cases of FOXES & dogs crossed, offspring always more resembled
333d007 & dogs crossed, offspring always more resembled FOXES than dogs (Mem Jackall in Zoolog Garden) He has
355d066 l> approach in character to goats.-- or dogs to FOXES. (yes Australian dog) or donkeys to Zebras.-- «
510q018 -breeding of animals in confinement, curious.-- FOXES-- English animals. [Made no import. remark]CD (
512q020 etimes couple but not conceive :Bears /Yes/ (2) FOXES & English animals & birds breed (3) In cases wh
335d015 t dreadful monsters, abortive, just like mules. FOX'S half bred Persians «cat» favour the Persian sid
494q004 itions of animals & their general healthiness-- FOX'S, Bears Badgers,-- How few wild animals are prop
551m128 l that Elephants understand contracts.-- but W. FOX'S dog that shut the door evidently did, for it di
073r161 ft, or redissolved soft.--/ is there any flexure <FR> in the fragmentary jasper.--do undulations (as
173b010 ity &c &c «assuming all» if species (<a>) «(I)». <FR> may be derived from form (2). &c.-- <(> Then (r
178b031 ted animals having effected; a change whi[ch] the FR. naturalists thought was species <Ascensi> Study
183b052 ardonneret, avec la linotte, avec le verdier" &, <FR> silver gold & common pheasants & fowls..-- "On
466t095 mell.= These qualities have been given to foetus <FR> before sex developed-- Double flowers & colours
130a133 easily. & if easily must be thin: <beside mere FRACTURE} A Elevation as in Patagonia {P} B subsidence
059r123 ous rocks.--which have the cryst of glassy F. FRACTURED. have been melted with little pressure. & per
035r048 iew sink into their proper insignificance; as FRACTURES, consequent on grand rise, & angular displace
044r072 encion. concretions & Galapagos.-- «Humboldts. FRAGMENS.» Read geology of N. America. India.--remembe
088a015 p. 312. Chamisso in Kotzebue. Study Humboldt. FRAGMENS Asiatiques account of American Volcanic actio
110a086 f rocks in Lyells'. Capital Norway case.-- The FRAGMENT. consisted of hornblende (?) & felspar, (some
110a086 cryst of glassy felspar different from either FRAGMENT or dike, blackish grey base. crystals from fr
110a086 ent or dike, blackish grey base. crystals from FRAGMENT disseminated on that side of salband. gradual
110a087 llized, as such do not occur in either dike or FRAGMENT. junction certainly most distinct on dike sid
155g066 nes & even water-scooped rock «only decay from FRAGMENT falling» of no particular hardness no wonder
307c216 nded from species to genera & classes. p. 479. FRAGMENT of tusk «& Molar tooth» of Hippotamus from Ma
463t063 ence about the Animaux fossiles-- being a mere FRAGMENT of the discoveries to come-- Owen in his desc
073r161 ed soft.--/ is there any flexure <fr> in the FRAGMENTARY jasper.--do undulations (as Hutton says) alw
090a025 ides thin vapour bringing planets to an end?/ FRAGMENTARY granite showing schistose structure (& veins
025r019 inly data insufficient, yet good» «(I suspect FRAGMENTS of shells will generally be found to be old &
033r042 ous. rocks??--if so case precisely analogous: FRAGMENTS instead Peak of Teneriffe. also Cotopaxi has
034r044 oos? Daubeny good account of ejected granitic FRAGMENTS P. 386 Mem. Lyell's fact about sulphuric vapo
050r090 s a pumiceous conglomerate with small & large FRAGMENTS, nature of which is doubtful. P. 180. I think
051r094 r in waterfall: Excepting by removal of large FRAGMENTS by mere force of waves: & action on upper tid
072r160 nacity, that it is with difficulty that a few FRAGMENTS can be separated from them with steel instrum
077r171 pital of Schemnitz in Hungary.) Humboldt says FRAGMENTS from roof & penetrating overlying beds tells
088a015 iques account of American Volcanic action.-- FRAGMENTS of slate converted into crystals of Hornblend
104a069 introduce in Cordillera discussion, deep sea, FRAGMENTS fall off cliffs. but then how spread abroad?-
115a097 und in S. America The very general absence of FRAGMENTS «& pebbles» in mica slate & gneiss, can only
115a098 on absolute <force> «size of» of <currents» «FRAGMENTS» but relative to currents. Small lakes have p
121a111 power of segregation.-- & has heated angular FRAGMENTS of rock, which retained their angles sharp--
156g068 generally angular except near head of valley FRAGMENTS which had fallen before lake drained could be
161g095 istinct then lost by slope, then concealed by FRAGMENTS, then clear. this bit to eye certainly appear
150g036 y drained, where is barrier {P} great waterworn FRAME terrace 4 4 not visible 3a 3a 3 Bouthoner 3 2 T
386d167 l,--no wonder there should be sympathy in human FRAME.-- «one of» The final cause of sexes to obliter
469t151 & wd go on longer-- Woodfords Marrow fat, Early FRAME, Groom's Dwarf. planted in rows «close to each
473s07r s in young pigeons--dogs--cattle? As we see the FRAME of animals can adapt itself to course of life,
524m019 . that cold water brings on suddenly in head, a FRAME of mind, analogous to those feelings. which may
526m031 ely, he chooses (but what makes him fix!? <)>-- FRAME of mind, though perhaps he chooses wrongly,-- &
526m031 though perhaps he chooses wrongly,-- & what is FRAME of mind owing to.--<)>-- I verily believe free-
527m033 thoughts are in same manner vivid & grand. the FRAME of mind being just kept up by the music of the
541m091 xcessive labour of inventive thought.-- Examine FRAME of mind in following changes during fall of sea
544m101 ys has done therefore sight of own child. (when FRAME in condition to receive pleasure) gives pleasur
616o038 ld hardly be called memory; you cannot call the FRAME of mind which makes music pleasant, a memory; y
616o038 which makes music pleasant, a memory; yet that FRAME is enhanced by memory of what has been heard; s
618o041 t, perceptions, memory &c. either to our bodily FRAME or the cerebral portion of it Thoughts, percept
618o041 tion to the brain that force does to the bodily FRAME, they could be perceived by the faculty who whic
350d054 There are no grotesques in nature; not anything FRAMED to fill up empty cantons, & unnecessary spaces
563n022 take away food &c &c-- Now if dogs mind were so FRAMED that he constantly compared his impressions, &
498q008 tigma impregnable.-- (12) At Maer Cowcumbers in FRAMES are not artificially impregnated. Abberley say
042r068 ountains: to elevated strata in eocene lakes of FRANCE, & unequal action of Earthquakes. «on Chili &
042r069 cted with far more regularity in S. America: in FRANCE we have freshwater lakes unequally elevated, w
098a044 g. proceedings Lake let out by steps in Central FRANCE not very conclusive proofs. but certainly prob
202b133 cle by S Hilaire on wonder of finding Monkey in FRANCE.-- of genus peculiar to East Indian isles.-- C
209b157 ut. Plantarum. relation of genera to species in FRANCE is I: 5.7 in Laponia I: 2,3: Mem. Lyell on she
218b190 s Female pig apt to produce monsters in Isle of FRANCE-- Madagascar oxen with hump.-- p 173. Voyag
234b255 k.-- Frogs attempted to be introduced I isle of FRANCE p. 170. «Fish introduced» Hump backed race of
234b255 f cows from Madagascar-- p 173. Vol I. Voyage à FRANCE-- Par un Officier du Roi.-- Mackenzie Travel.
243c020 . 27. many fish of Taiti found at <New> Isle of FRANCE.-- the Tenrecs from Madagascar. Monkey from Ja
244c022 at Caroline Isld, & a Roussette p. 136. Isle of FRANCE.-- The Tenrecs from Madagascar. Monkey from Ja
247c028 813, a venemous snake was» one Gecko on Isle of FRANCE Scincus multilineatus (p 45) Moluccas & New S.
276c123 Frenchman. did for SPECIES-- between England & FRANCE. I will do with forms.-- Mention persecution o
322c270 De Beaumonts. 10 Vol. of Memoirs on Geology of FRANCE.= on Etna. Almost reread the previous volume.
431e121 said they have been found together on coast of FRANCE.-- L. doubts.-- Lonsdale thinks Ammonites woul
483z012 erlock.-- Testudo INDICUS not fossil at Isle of FRANCE: <Jerrold?> Bibron Zoolog. Journ Vol I. p. 125
320c276 t 13 numbers have been read Voyage à l'isle de FRANCES Voyage de l'Astrolabe Partie Zoologique Pernet
025r016 ergipe [11 degree 10 minute S.] 20 190 R. San FRANCISCO [10 degree 32 minute S.] 10 50 Whole coast to
544m103 present & future thoughts are broken-- Sir J. FRANKLIN when starved, all party dreamt of <goo> feast
271c107 tering breed of animals in certain countries.-- FRASER remarked to me at Zoological Society,. that yo
468t111 chusa» speedwell Iris-- Azalea. Rhodendron. FRAXINELLA to Anchusa <never> «once» P on Fraxinella <H
468t111 on. Fraxinella to Anchusa <never> «once» P on FRAXINELLA «Heartease» «small. Humble alighted on base
468t112 y, the great scarlet Poppy-- So that, finally FRAXINELLA «with respect to nectary is same case as Aza
468t112 days. hot one, Bees almost P every minute to FRAXINELLA «& from <flower> plant to plant.»-- to my gr
469t135 - <Saw> Maer. June 15./41/. Watched plants of FRAXINELLA, with seven flower stalks for ten minutes. i
158g082 ttle Hill with granite blocks almost encircled> <FRE> Gneiss cut smooth on sides of hill where Boulde
466t093 ne/41/ Rhododendrum-- nectary marked by orange FRECKLES on {a} upper petal; bees & flies seen directe
379d152 & female.-- if there be one female, she will be FREE Marten. <Owen.> See Hunter's Owen-- In the Athe
383d159 er they are not Hermaphrodites, like J. Hunters. FREE Marten N.B. the common mule must often have bee
526m027 nking over these things, one doubts existence of FREE will every action determined by heredetary cons
526m027 teach by the same means & therefore properly no FREE will.-- we may easily fancy there is, as we fan
526m027 nce.-- chance governs the descent of a farthing, FREE will determines our throwing it up.-- equall tr

526m030 powers of imagination, & does so,-- is not this FREE will,-- he improves the faculty according to us
526m030 to usual method, but what urges him,-- absolute FREE will, motive may be anything ambition, avarice,
527m031 frame of mind owing to.--<)>-- I verily believe FREE-will & chance are synonymous.-- Shake ten thous
536m072 have idea of beautiful forms.-- With respect to FREE will, seeing a puppy playing cannot doubt that
536m072 eing a puppy playing cannot doubt that they have FREE will, if so all animals., then an oyster has &
536m072 unavoidable & only to be changed by habits). now FREE will of oyster, one can fancy to be direct effe
536m072 es its senses give it of pain or pleasure, if so FREE will is to mind, what chance is to matter «(M.
536m072 what chance is to matter «(M. Le Compte)»-- the FREE will (if so called) makes change in bodily orga
536m073 change in bodily organization of oyster. so may FREE will make change in man.-- the real argument fi
549m118 than to dog losing his puppies-- This looks like FREE will.-- V. last page. A healthy child is «more»
549m119 s our [mind], good has been done-- «& conscience FREE from offence»-- --pleasure of intellect affectio
550m126 ll of Deity are confounded.-- well applicable to FREE will. Mayo. Philosop. of Living p. 293. Animals
577n049 f.]CD & hence there is great probability against FREE action.-- on my view of free will, no one could
577n049 probability against free action.-- on my view of FREE will, no one could discover he had not it.-- Th
607o025 ees this law in man in sommambulism or insanity. FREE will (as generally used) is not there present,
607o025 any actions are not determined by what is called FREE will, but by strong invariable passions-- when
608o025 ions, weak, opposed & complicated one calls them FREE will--the chance of mechanical phenomena.-- (Me
608o026 ges with organization The general delusion about FREE will obvious.-- because man has power of action
178b030 ory.-- July. 1837. Eyton of Hybrids propagating FREELY In Isld neighbouring continent where some spec
296c185 urd.-- p. 565 <breed> Scotch wild Cattle. breed FREELY with the tame Vol II. Magazine of Zoology p. 5
342d033 lement). Then do those SPECIES which breed most FREELY. & produce somewhat fertile offspring produce
358d085 e or castration on females.-- [not located] hen FREELY.-- here we have beautiful proof of the breedin
399e010 ry, where only pair has been introduced, & have FREELY bred, they have not lost power of producing. M
448e165 te rapidly by buds, layers &c & &-- do not seed FREELY.-- The periwinkle seldom produces seeds, becau
478z004 Australia I was assured wild dog copulates <.> FREELY with tame: comes to houses on purpose Mr J. Mu
076r170 amphibole quartz & mica very rare.-- ancient FREESTONE & breccia is the same with that on surface of
077r171 might be mistaken for Porphyry above ancient FREESTONE, limestone & <many> «other secondary» rocks.
065r139 o an Englishman.--On 8th of March cove began to FREEZE. correspond to September ¿Did I make any obser
127a126 rish Academy Vol 8. p. 118 water no--. oil will FREEZE if cooled in a closed globule of glass. (oil m
057r114 1,000 f[athoms], & the temp of which was below FREEZING point!!! Remember idea of frozen bottom or be
132a138 ts proper <state> temperature of earth. at the FREEZING point.-- accounts for increase on earth by vo
058r116 Histoire Naturelle des Indes Acosta. p. 125. of FRENCH «?» Edition states that the same earthquake ha
081rIBC ue 1816 Constant log always additive to convert FRENCH Toise into English ft. 0.8058372 French metre
081rIBC convert French Toise into English ft. 0.8058372 FRENCH metre into English ft. 0.5159929 {T} Toises Pi
095a039 y much of Gypsum.-- The officers of the Bonite. FRENCH discovery ship, found clear proofs of shells &
127a127 ffects of subsidence Elie de Beaum. Memoires of FRENCH Geolog. Cantal Vol III I? p. 246. on formation
186b059 to keep to <each> either parent, (this is what FRENCH call (atavism) Probably this is first step in
209b156 s. to certain extent Von Buch,-- Canary Isles-- FRENCH Edit. Flora of Islds very poor «(p. 145)» 25.
323c269 id Mixtures: Marginal notes. -- 20th. Carlyle's FRENCH Revolution 3? vols. oct: -- 26th Blumenbach's
446e161 etation is concerned, in parts of the Northern «FRENCH» expedition,-- rather the reverse of facts sta
462tf05 . Athenaeum 1839 p. 772-- A curious theoretical FRENCH book review on politics in relation to the dif
482z012 hical distribution of Mammalidae &c &c &c-- The FRENCH <Jerrold?> «Bibrons coworker of Dumeril» who i
490q01v Experiments in crossing &c Plants 1 Repeat the FRENCH experiment of Carrot 2 {also try Primrose & Co
491q01v e being mongrelized affect other branches-- The FRENCH Apple tree «with abortive stamens» answers fir
572n033 Octob 30th-- Dreamt somebody gave me a book in FRENCH I read the first page & pronounced each word d
572n033 en awake» I could not picture to myself reading FRENCH book quickly, & <running> «running» over imagi
601o08a action-- emotions-- p 176 & 177 good passage in FRENCH on what dog dreams, awakes-- does when Master
276c123 aws, which might probably be reduced What the FRENCHMAN. did for SPECIES-- between England & France.
569n022 qu'on y fût."-- Expression common to Savage & FRENCHMAN, unaccompanied by dignity-- "no mon dieu," wi
027r027 with St Pauls [not located] [not located] The FREQUENCY of shells in the Calc. Sandstone Concret, is
027r027 he Calc. Sandstone Concret, is connected with FREQUENCY of shells in flints in Chalk New Providence m
037r056 damaine on the Amazons. Consult Insist on the FREQUENCY of dikes in Granitic countries, enumerate cas
044r073 ne side. S. America on the other: The extreme FREQUENCY of soft materials being consolidated; one inc
391d174 te (hence monstrosities tend that way «& from FREQUENCY of this tendency all mammalia must long have
549m119 but. as they are not recollected whether from FREQUENCY, or inherent structure of mind. they make, ei
620o044 ely great [LHC] The cause perhaps lies in its FREQUENCY & in its consisting in desire gratified & the
026r021 such action can take place in melted rocks The FREQUENT coincidence of line of veins & cleavage is im
260c069 unt for colour of bird in district. which they FREQUENT!!?-- Wilson's American ornithology a mine of
467t103 tamen are of useful height.-- In Lupine, Bees «FREQUENT» & seem to act, something like on Kidney Bean
468t111 so avoided it.» On certain days Humble seem to FREQUENT certain flowers, to day early, the great scar
469t135 ce came & each visited many flowers-- Saw Bees FREQUENT these flowers till late in evening-- On rough
498q008 ree from charcoal & good treatment (8) Do bees FREQUENT Cabbages «& Cowcumber's out of doors.» much--
635j56v sed to weather.-- this is against my theory of FREQUENT intermarriage.-- A plant is in the same predi
470t177 paragus very small flowers & as much shut up, FREQUENTED by «many» Bees & Humbles-- «Humbles & common
499q09v lso, Vinca, <.> Examine all these, are they much FREQUENTED by Bees or Butterflies or little insect?= or
073r164 hyries characterized by no quartz & amphibole FREQUENTLY only vitreous felspar: = gold veins in a pho
088a014 mountain chain be Mr <Lyell> «Waterhouse» has FREQUENTLY heard that Herons bring eels alive to give
114a095 lp suspecting that clay-slates have been more FREQUENTLY metamorphosed than other deposits.-- NB. bec
205b145 in their offspring, but that of the male more FREQUENTLY predominates," p. 20. do "If hornless ram be
520m002 people of weak minds, below par in intellect FREQUENTLY <are> have very bad replacers for things whic
624o051 me end.-- & general actions of community must FREQUENTLY teach same end.-- Hence this becomes the law
188b069 ne Dr Smith says very. close species generally FREQUET slightly different localities, so that they be
028r031 ? Has Chalk ever been dissolved? Singularity of FRESH water at Iquiqui. not from rain, because alluvi
041r065 tion M. Gay's fact about shells: Hibernation of FRESH water Shells. multitudes.-- The question of she
068r147 ss. If one inch can be raised then all can, for FRESH layers of igneous rock replace strata. & it is
128a128 89. at Madras. surrounded by salt water. purest FRESH water must be sought for below the sea mark.--
130a133 ishoofs Paper.-- Weelsted told me of some large FRESH Water springs off coast of Persia In Glen Roy p
132a139 ers-- may not the cold «bottom of» ocean. (with FRESH sediment added to bottom) be caused, by absence
172b006 as formerly but separate a pair & place them on FRESH isld. it is very doubtful whether they would re
190b078 hat species.-- <Man is derived from Monad, each FRESH-> Mr Don remarked to me, that he though specie
195b104 ies die &. others replace them-- two hypotheses FRESH creations is mere assumption, it explains nothi
198b115 ammalia not being found on all isld, (if act of FRESH creation why not produced on New Zealand; if ge
218b189 rms; therefore when from being put on island. & FRESH species made. parents do not cross-- we see it
221b201 facts of salt-water shells living in absolutely FRESH water.-- origin of Fresh-water genera? The abse
221b201 s living in absolutely fresh water.-- origin of FRESH-water genera? The absence of lime in Plutonic &
227b227 . !.fish never become a man.-- Does not require FRESH creation!-- If continent had sprung up round Ga
230b241 Americas. perhaps merely gone back previous to FRESH change Get a good many examples of animals «or
233b251 or few quadrupeds-- Study Productions. of great FRESH water lakes of North America If Parasite differ
236b279 ly an Indian form of Tortoise.-- On other hand. FRESH water tortoise from Germany. (where Mr Murchiso
250c038 merica preeminently equable. might have allowed FRESH species to have been formed & spread to other A
259c064 recollecting that the ground woodpecker &c.--, FRESH water animals of great lakes are American form,
298c191 robably often be destroyed.-- or regrafted with FRESH arrivals..-- &c &c --Climate altering as island
337d020 partly local & then the local ones are taken to FRESH country & breed confined. to certain best indiv
338d025 on of certain forms from Northern part & not by FRESH creation of new forms.-- what is range of Hyaen
362d099 of the change in 5 or 6 weeks after castration, FRESH horns begin to grow.-- Mr Yarrell says the «mal
364d103 nge birds with some other fancier, thus getting FRESH blood, without fresh feather, & consequent trou
364d103 ther fancier, thus getting fresh blood, without FRESH feather, & consequent trouble in obliterating t
364d103 ather, & consequent trouble in obliterating the FRESH feather, by crossing-- It seems from Lib. of Us
379d151 istory. p. 135. Natural History of the Caspian. FRESH Water Fish!! ¿adapted to salt water?-- peculiar
398e006 ot only revolutions, but entire obliterations & FRESH laws created., & yet with <gov> symmetry «& reg
419e085 chwald has published Fauna of Caspian.-- fishes FRESH water kinds. (yet living in the salt?.)-- very
422e095 ce as the forms became complicated, they opened FRESH, means of adding to their complexity.-- but yet
431e122 ange may be anticipated, & this would look like FRESH Creation. the gardener separates a plant he wis
446e162 ey break-- -- each tulip is the <of> product of FRESH bud-- here then is case of change analogous to
446e163 re propagated during hundreds of years, without FRESH seeds arriving.-- throws a very great difficu
461t025 ack,-- Most curious facts & this paper deserves FRESH study & whole order of the fish.-- Embryology p
442e152 e way of Mr Knights. theory «without seeds are FRESHLY transported»-- throw over this theory, & the s
025r017 wer at Port Louis talked of the extraordinary FRESHNESS of the streams of Lava in Ascension known to

042r069 e regularity in S. America: in France we have FRESHWATER lakes unequally elevated, which movements if
421e091 \do p. 269. Annals of Nat. Hist 1838 on «a» FRESHWATER fish peculiar to Ireland. \do p. 283. on the
229b234 île & Tortise former inhabitants of Mauritius FREYCINET Voyage, agrees. with several mammalia being p
478z003 ny pebbles Mentions stinging Millepora. Quoy. FREYCINETS Voyage Vol p. 597 Many descriptions about lo
552m131 drays, scraping against cornice stone to cause FRICTION Athenaeum 1838. p. 652. Dr Daubeny on the dir
554m141 f Cart horse argued from this into a theory of FRICTION & gravity. it would be discoverer "reasoning"
021rIFC lumes, and Molina 1st Vol & Lyell Sailed, 27th <FRIDAY gale 29th> Friday Thursday 29th gale Lyell's G
021rIFC 1st Vol & Lyell Sailed, 27th <Friday gale 29th> FRIDAY Thursday 29th gale Lyell's Geology The living
147g019 vening ¼ past 8 Tyndrum 29.<625> «636» Temp. 62 FRIDAY morning ¼ past seven o'clock 29.642 Temp 55 Ai
147g019 ast seven o'clock 29.642 Temp 55 Air 50 degree? FRIDAY. Inverorum about 20 ft above Loch Tulla 29.804
147g021 ley had been filled with sloping bed of rubbish FRIDAY Highest part of road between Inverorum & King'
162g100 away   on the steep & rocky gully of last stream FRIDAY Loch Lochy near Letter Finlay Barom 30.267, A
548m116 pathy. as for for the heard suffering of a dear FRIEND-- this gives one strong idea of what individua
550m126 of difficulty of his own drawing compared to a FRIEND, whose who family can draw-- says friend viewe
550m126 d to a friend, whose who family can draw-- says FRIEND viewed him as Newfoundland dog would Greyhound
576n047 ence. A man, might not <t> do so even to save a FRIEND, or wife.-- yet he would ever repent, & wished
023r012 Sharks followed Capt. Henry's vessel from the FRIENDLY Isles. to Sydney; know by having been seen &
246c027 The» > isld. (.perhaps Phillippines & perhaps, FRIENDLY Isles «& Hebrides») is very closely allied to
247c027 probably on Oualan.-- «Mitchill says snakes on FRIENDLY isles. p 50. LX. Journal of Silliman» «Study
525m023 tincts naturally.-- (generosity in defending a FRIENDLY dog).-- they feel shame, when doing anything
578n054 not blush before utter stranger,-- or habitual FRIENDS.-- but half & half. Miss F. A. said to Mrs. B.
550m126 f masters property"-- is not this rather more FRIENDSHIP.-- Scott's Life. Vol I, p. 127. Talks of dif
552m132 been found necessary for long generation, (as FRIENDSHIP to fellow animals in social animals) are tho
182b046 ric type. Hence probably only four, is not this FRIES rule-- What subject has Mr Newman the (7) Man s
327cIBC line on the improvement of domesticated animals FRIES de plantarum proesentum crypt. transitu et anal
461t025 ol. 2. p. 96 & p. 451. 1839-- Translation of P. FRIES most curious paper on the Pipe-fish-- which he
283c143 relation to elements & not habits as shown by FRIGATE Bird & flying eagle.-- Hawk Gould seemed to th
557m147 prey streched out & flaccid, when furious «with FRIGHT» back absurdly arched. & tail stiff.-- is sham
592n103 . 61. CD[a person is here said to open mouth in FRIGHT because nature intends to lay open all senses:
377d139 na. thinks, when pleased cocks his ears., when FRIGHTEN depresses them.-- England was united to Conti
591n098 inarily cowardly.-- the other one nothing will FRIGHTEN-- hence variation in character in different a
532m054 thing, & tried to seize hold of objects to be FRIGHTENED at.-- (again diseases of the heart are accom
532m054 ct:-- When a man, child or colt has once been FRIGHTENED & started much more apt, this partly owing t
533m058 knowing a cat, the first it sees it.-- it is FRIGHTENED without knowing why-- the child dislikes the
592n103 prick his ears «& snort clears nostrils» when FRIGHTENED, does not hair & rabbit depress. them from s
594n112 t.--» «Emma says, «her» tame rabbits were not FRIGHTENED at a dog.-- » The instinct against man is pe
594n113 m same bone A child born on the 1st March was FRIGHTENED on the 24th of May at Cresselly by the boys
595n121 W. playing with Snow. when 2½ years old. was FRIGHTENED when Snow put a guaze over her head. & came
595n121 he Baby> «Effie Wedgwood» April 28th 1840 was FRIGHTENED at wild beasts in Zoolog. Garden [blank] [no
531m053 ies start at anything they hear or see. which FRIGHTENS. them.-- Now every animal moves quickly away
117a102 sing are there cliffs. Sir L. Dick says (.p 52) FRINGE of sublittoral deposit always equal width --su
163g107 Sand with tide ripple Near Fort Augustus hill & FRINGE as if it has been filled up «at» 30 ft. higher
429e115 ts.-- a broad border of Killed trees would form FRINGE.-- but there is a contest. & a grain of sand t
159g087 r & begins about 60 higher-- There are however FRINGES of alluvium (?) still higher Slope of valley m
160g088 arf What I called Alluvium shows the ascending FRINGES {P} which makes me think it submarine, 400 or
337d022 h is impossible.-- Mr Spence remarks that the FRINGILLA domestica of North Europe is replaced by the
024r016 de Janeiro SE [23 degree 58 minute S.] 18 77 C. FRIO [23 degree S.] 7 60 {T} Soundings about same as
024r016 60 {T} Soundings about same as last to N. of C. FRIO Except at Abrolhos. [18 degree S.] Bahia [12 de
458t001 -- vol VI. p. 539. Dr Cantor's account of fossil FROG, 40 inches in length--! alludes to ancient giga
499q09v tion of plants necessary &c (a) Mercurialis-- FROG Bit, Valerian-- Urtica Dioica Sorrell. Lychnis.
639j28r in Octopus & Chamaelion.-- C. D. Sucking feet in FROG. Walrus. Fly. Gecko &c. Prehensile tail. in Mon
234b255 over, & then with snout lift up latch & back.-- FROGS attempted to be introduced I isle of France p.
388d173 is developement of gemma.-- The manner in which FROGS copulate & fish shows how simply instinctive th
392d180 e monooecious. Will ova of fishes & Mollusca «& FROGS» pass through birds stomachs & live? In Muscovy
420e090 ttle of the spermatic fluid fertilized spawn of FROGS.-- Annals of Natural History. (p 225. 1838.) ac
568n018 leasure in music-- do monkeys howl in harmony-- FROGS chirp in do-- union of birds voice & taste for
287c157 when   Linnaeus put whale between cow & hawk a FROLICSOME saltus. «p. 19»\ Macleay seems to limit Lama
424e100 o it.-- also on spermatic animalcules in Musci FRONDOSI, et hepatici.-- in Chara, in Marchantia & Hyp
048r087 ls. mention submarine channels. such as that in FRONT on wall perhaps wall oblique The hill has been
146g010 ame way at top the trap could be traced Grey in FRONT on wall perhaps wall oblique The hill has been
157g075 block a yard across. On side of «that» hill, in FRONT of which shelf 3d form beach of granite pebbles
293c172 or consciousness of reasoning to tell back from FRONT. &c or use of button holes it would be instinct
366d108 rrell told me of a cat & of a dog, born without FRONT legs-- -- the former of which had kittens with
376d137 it on, yet threw it over it, & made it meet in FRONT.-- Dr Smith every baboon & monkey, big & little
463t059 of Fernando Noronha are found unknown coast in FRONT of it.-- Cuvier has grand sentence about the An
555m143 nfused idea of showing scar behind (.instead of FRONT) (having changed hanging into his head cut off)
585n075 er.--]CD April 3d. 1839 The Giraffe kicks with FRONT legs & knocks with back of Head, yet never puts
577n049 schel's Treatise) a "travelling instance" a-- "FRONTIER instance".-- for it can be shown that the lif
423e096 e surface of world with life.-- for otherwise a FROST if killing the vegetable in one quarter of the
533m060 eg is right handed.-- In Review (Edinburgh) of FROUDE'S life. that author remarks, that writing down
533m058 nguage.-- Hensleigh. W. says that babies know a FROWN very early in life, <before they> (I think I ha
533m058 ed without knowing why-- the child dislikes the FROWN without knowing why-- a man as in Guy. Mannerin
542m096 out?-- Seeing a Baby (like Hensleigh's) smile & FROWN, who can doubt these are instinctive-- child do
560mIBC o cean have taught them to avoid danger Do they FROWN, when they first see it? Charles Darwin 36 Grea
565n009 s pouting. <but> lesser in degree, no smile. no FROWN showing thought, no compression of mouth showin
579n058 eves in.-- From the manner short-sighted people FROWN, frowning must have some relation to short-sigh
579n058 ofoundly,-- this will be curious if it is so.-- FROWN with grief,¿ bodily pain? frown shows the mind
579n058 if it is so.-- frown with grief,¿ bodily pain? FROWN shows the mind is intent on one object.-- With
533m058 same thing before they could understand. what FROWNING means) if so this is precisely analogous or i
542m095 ow often) to the tale of a wearisome man.-- Is FROWNING, result of straining vision, as savages witho
557m149 hands.-- stamping. grinding teeth.-- in shame FROWNING, & anguish,-- shyness not so.-- affected laug
579n058 -- From the manner short-sighted people frown, FROWNING must have some relation to short-sightedness.
579n058 member children smile before they laugh.-- Has FROWNING anything to do with ancient movement of ears
595n125 . Garden [blank] [not located] A child crying. FROWNING, pouting, «smiling», just as much instinctive
432e125 he muscles of the ears to our hearing powers E. FROWNS prodigiously when drinking very cold water «fr
432e125 wns prodigiously when drinking very cold water «FROWNS connected with pain, as well as intense though
057r115 ch was below freezing point!!! Remember idea of FROZEN bottom or beach of sea to explain preserved an
124a117) p 212. Facts from Erman about great depths of FROZEN soil. p. 211 Consider proved that Siberia must
124a118 p. 131 (B.) Wrong Entrance. Book C. p. 101. On FROZEN Soil of Siberia (with refer to Metamor) wrong
126a124 s from surface can take place.-- \ the depth of FROZEN soil is against this view.-- however it is sai
127a125 «the» line of ice.-- Hence further N. when soil FROZEN for greater length of time depth of ice ought
131a136 nean isothermal line Athenaeum. 1839. p. 52. On FROZEN soil of Siberia.-- facts of water flowing from
131a136 Siberia.-- facts of water flowing from beneath FROZEN crust in America Richardson.-- From strata bei
269c101 February or March 1838 on soil in Siberia being FROZEN to 400 ft in depth, (& Erman's surprise that i
507q15v cult or reverse (.49) List of seeds Gaertner de FRUCT:-- for woodcut-- 1 double hook-- -- Geum. Galiu
498q007 robably would grow there (7) Where parts of FRUCTIFICATION <lat> retrograde into leaves-- is this eve
170b001 ll individuals absolutely similar; for instance FRUIT trees, probably polypi, gemmiparous propagation
192b083 s will go back.-- Get instances of a variety of FRUIT tree or plant run wild in foreign country.-- Wh
274c113 ckers Gould says, he believes does. but also on FRUIT.-- The Rasorial type is wonderfully shown in lo
327c265 nsacts.--» Mr Coxe "Views of the Cultivation of FRUIT trees. in N. America" in Lib. of Hort. Soc Mr N
356d070 on do producing occasionally (as Fox says) same FRUIT trees is analogous to some hybrids breedings--
384d181 neration alway put female first Will not even a FRUIT tree or rose degenerate during its life so that
390d178 of generation.-- therefore passions fail.-- In FRUIT trees no doubt there is tendency to propagate t
414e148 whether any are transmitted, for the changes in FRUIT trees. mentioned by Mr K may be caused by the d
442e153 dy Von Buch.]CD Now Mr Knights statements about FRUIT trees. grafted. altering is hostile to this: bu
442e153 altering is hostile to this: but on other hand, FRUIT trees are propagated by means, which wild plant
446e162 h Henslow is inclined to think very close.-- «A FRUIT tree by certain treatment will suddenly send fo
466t096 er life of Plants-- also goodness of flavour in FRUIT-- all affected by cultivation during life of in

491q01v o hybridise EVERY flower on melon & see whether FRUIT affected. Mr. B. seemed to say impregnation <ca
498q007 soil, when placed in very poor flower, but not FRUIT-- -- Do not orchards become unproductive from p
499q009 or Yews some poplar's produce.-- (15) Would Yew FRUIT without impregnation.-- (16) Any calculation of
505q014 um put in Spirits which plant seeds? (9) Melons FRUIT itself hybridised (10) one had no seeds, & two
505q014 lost their horns" is impregnation necessary to FRUIT--; become well shaped by care 13 Arum before po
513q21. -Flora of Australian Mountains.-- Is setting of FRUIT. cross Conception--(《 I could extract nothing f
515q021 especially Papilionaceous order (2) History of FRUIT trees far north in Scotland-- do they flower--
515q021 o they flower-- do they live healthily, or does FRUIT merely not ripen.-- The point to attend to is w
515q021 h regard to Primroses. (4) Do apples "sport" in FRUIT, or time of leafing (5) Do the most cultivated
546m109 ste for smell of flowers, owing to parent being FRUIT eater.-- origin of colours?-- Nothing shows one
543m100 et invariably used to beat him-- The son of a FRUITERER in Bond St. was so great a fool that his Fath
350d054 o formations? by no way.?-- "Natura nihil agit FRUSTRA", as Sir Thomas Browne says "is the only indis
255c053 ability; but isolate your species her plan is FRUSTRATED or rather a new principle is brought to bear
389d177 ces (of body, mind & constitution) is the end FRUSTRATED, when near relations, & therefore those very
021r005 w Anglesea, nodules in Clay Slate. major axis 2.½ FT.-- singular structure of nodule, constitution <s
022r008 t voyage to New Holland P 127.--Caught a shark 11 FT long. "Its maw was like a leathern sack, very th
046r080 hquake Loch Lomond water oscillated between 2 & 3 FT. (as in Chili lake). Therefore motion of sea oug
054r102 psum, covered by limestone with recent shells 200 FT, how exact agreement with Coquimbo; [not located
064r136 efore flowed towards it. a mass on each side 3000 FT thick & 150 broad. neglecting Cordillera itself
069r149 never be afraid of speculating on the sea The 24 FT. elevation at Concepcion. from impossibility of
081rIBC ays additive to convert French Toise into English FT. 0.8058372 French metre into English ft. 0.51599
081rIBC o English ft. 0.8058372 French metre into English FT. 0.5159929 {T} Toises Pieds Myriametre = 5130.,
092a028 bea> formation twenty species same as Paris. 1500 FT high Mr Bird in paper to Brit. Assoc: has shewn
095a039 Cobija.» At Iquique of elevation to amount of 30 FT.-- Mr Bollaert (at Roy. Institut) talks of quant
113a090 t a greater depth than 400. & the limit being 400 FT. shows that the strata have very unusual conduct
116a100 er, but in shelving «successive» banks <above> 30 FT or so above bed of river. formed of rounded pebb
120a110 ion absolute elevation of Patagonian blocks (1200 FT??). Scotland at least 2200. Jura 4000 feet.-- Th
126a122 ting powers-- we judge from the surface, & say 60 FT to degree.-- but this may be very wrong,-- The f
129a132 in with.» which must have been from an axis, «20 FT at least in depth» near mouth of Columbia river-
146q014 med vestige of irregular terrace perhaps near 300 FT above Loch.-- From this point could be followed
147q016 if prolonged would intersect alley above the 300 FT Alluvium <abo> by Loch Dochart-- Rivers could no
147q019 Temp 55 Air 50 degree? Friday. Inverorum about 20 FT above Loch Tulla 29.804 Temp 62 degree Air 60 de
147q021 .935/82 degree A Temp of Air 65 degree? Glenoe, 6 FT above high water mark 30.380. 68 degree 65 degre
148q023 ite Loch Leven two terraces perhaps upper one 100 FT & other one 40-- 《 traces of them all along <Gle
150q037 g, when water up to shelf very shallow channel 50 FT wide & river get formed in centre In Glen Collar
152q049 on the East side of river & bed of river about 40 FT beneath general plain. 30.127 A 72 degree Air 65
152q050 ium rise nearly up to Glen Collarig up within 200 FT of level of 4th shelf= argument against river--c
153q056 2 «feet» & 2 deep Another rather smaller block 30 FT «above» & other 50 ft lower & other smaller ones
153q056 her rather smaller block 30 ft «above» & other 50 FT lower & other smaller ones «these boulders are d
154q057 peat plain divortium aquarium-- tidal channel-- 12 FT obscure obscure NB In Glen Collarig tidal channe
154q057 NB In Glen Collarig tidal channel, sides <alm> 15 FT above bank or terrace, from terrace of 2d shelf
158q063 supernumerary shelf I doubt, much about «50 or 60 FT» «no doubt, a mound of Alluvium nearly parallel-
157q078 29.297 A 79.½ 29.316 divortium aquarum «about 12 FT higher than last station» 29.316 true terrace «2
158q079 aterworn buttresses of granite obscure terrace 15 FT divortium my measurements here side of Loch Spey
158q081 vers unite in Upper Glen Roy great plain about 60 FT beneath shelf peat on pebbles tidal plain as sea
159q084 lf & below it some way; several large ones (one 6 FT across) on top of spit between river & dry Corry
160q090 degree 75 degree? Boulder, much covered by turf 2 FT. 8- long of syenite with pinkish felspar;-- whol
160q091 h peak» I walked all round hill. Boulder about 20 FT. below summit <Isthmus> {P} highest point joinin
161q094 lip with moss On this terrace «station perhaps 6 FT too low» (to test last on Peat-Mass Divortium ag
161q098 & it was 29.200 minus .008 ---- .192 Loch Lochy 4 FT above water Barom: 30.372 A 76 degree 75 degree?
162q105 higher Loch Keeper tells me, that Loch Lochy is 8 FT below Loch Oich wh is 92 ft above sea-- Loch Nes
162q105 that Loch Lochy is 8 ft below Loch Oich wh is 92 FT above sea-- Loch Ness 40 ft above do. When cutti
162q105 ow Loch Oich wh is 92 ft above sea-- Loch Ness 40 FT above do. When cutting bank where Locks now are
163q106 ove do. When cutting bank where Locks now are (2 FT rise) they found alternating layers of coarse &
163q107 hill & fringe as if it has been filled up «at» 30 FT. higher with pebbles now worn away-- The above s
163q108 n away-- The above shells must have been about 60 FT above sea-- soon decayed on exposure Mr H. C. Wa
164q123 hells from [...] Wenlock Edge [blank] L. Lochy 12 FT 96 L. Oich 12 84 {P} 29.958 - 1.17 28.788 + 28.8
164q123 28.8 30.372 29.200 1.172 Loch Oich 92 each Loch 8 FT. The Metamorphic conglomerates near Loch Lochy w
202b131 , carbonized»--; clay fifty feet, then forest 120 FT Micaceous rocks. subsidence appears indicated.--
257c056 color in Patagonia. then in Chile & lastly 12,000 FT above sea in Bolivia; he examined-- all species
269c101 March 1838 on soil in Siberia being frozen to 400 FT in depth, (& Erman's surprise that it is not 700
278c130 one. The caves are at a height of more than 1000 FT. & many hundred miles from the sea, associated w
317c249 m land several patches of reed & trees p. 259 120 FT in length, some branches of Justicia still growi
408e044 nae. p. 222. Bones of Horse. Bear & Deer at 16000 FT. with Snow on Himmalaya-- Humboldt bones at 7800
419e086 he shells in Scandinavia from height of 200 & 300 FT are identically same as those of present seas.--
419e086 hysical changes, than in any other, & yet 200-300 FT elevation & no change & even no loss of species.
021r005 Archipelago, no minute description, calls it a FUCUS. P «Vol I 287» P 379. Henslow Anglesea, nodules
066r140 d said at Kerguelen Isd. to grow on shoals like FUCUS giganteus! 24 fathoms deep 24» under 50. Kergue
632j53r ortation through the air.-- cocoa nut by water «FUCUS for adhesion».-- as examples of design.-- perha
567n015 p. 334 Does a negress blush.-- I am almost sure FUEGIA Basket did. & Jemmy, when Chico plagued him--
264c079 preeminence.-- «not understanding language of FUEGIAN, puts on par with Monkeys» Gould seems to thin
316c244 he mind of man & animals.-- yet how faint in a FUEGIAN or Australian! why not gradation.-- no greater
559m153 ge, & judge only by what you see. compare, the FUEGIAN & Ourang & outang, & dare to say difference so
079r177 etrified shells Bougainville says P 291.-- The FUEGIANS treat the "chefs d'oeuvre de l[']industrie hu
555m142 with all associated animals?) I doubted it in FUEGIANS, till I remembered Bynoes story of the women.
025r017 o Volcanic Earthquakes or Hot Springs in T. del FUEGO = The Wager's Earthquake the most Southern one
026r021 he older strata & the bottom of sea near T. del FUEGO.-- Is there account of Baron Roussin's voyage.-
049r088 e look at regular greenstone cones at S. T. del FUEGO.-- as nucleus of a Volcano or as an injected mass.
052r099 - Greywacke. as a general fact absent in T. del FUEGO, excepting in Port Famine Mr Sorrell says that
065r140 z.???-- Form of land shows subsidence in T. del FUEGO, and connection of quadrupeds.--although recent
065r140 ings. (end of 2d voyage outside coast of T. del FUEGO. off. Christmas sound. -- «(Think some 60 fatho
134a141 ear C. Virgin p. 59. dip of Clay slate in T del FUEGO Admiralty Sound. SE dip. much p. 136. Rocks on
134a141 375. on the soundings on outer coast of T. del. FUEGO.-- p 385 Rocks of S. Western Coast Vol II p. 27
179b033 cross readily, distinctness of tribes in T. del FUEGO. the existence of whiter tribes in centre of S.
217b187 ith species in Van Diemen's land and Tierra del FUEGO.-- Araucaria, species. Brazil {P} Chile, Norfol
219b194 pe) see if there are any species same as T. del FUEGO & C. of Good Hope show possibility of transport
219b195 efore plants common take an example from T. del FUEGO.-- Ellis (?) says Tahitian kings. would hardly
226b221 ones.-- Is the flora of <S. America> Tierra del FUEGO like that of North Europe, many genera & few sp
268c099 species common to Patagonia desert & Tierra del FUEGO. & forest «Parrots in Macquarrie Isd.--» very g
277c126 orhyncus. fulginosus. (a) Falklands (b). F. del FUEGO differ from (C) Chiloe (E) Chile. rupestris --
343d038 >. Warm to cold, damp to dry.-- Thus Tierra del FUEGO has <not> «only» one «Guanaco» of the character
409e047 ulty in conceiving how inhabitant of Tierra del FUEGO is to be converted into civilized man.-- ask th
481z010 . Meyen p. 92.-- great Kingfisher of Tierra del FUEGO killed in Chile. Dobrizhoffer., Vol I p. 310. H
485z017 &c. with reference to those of mine from T. del FUEGO. p. 141. How comes it salt water so soon putrifi
486z018 ews clearly explain rarity of insects in T. del FUEGO.-- Hence it is odd that Amber insects of Europe
486z019 hatka (Salamandra aquatica). Compare with T del FUEGO Compare birds of do with <T> N American & T. de
486z019 mpare birds of do with <T> N American & T. del. FUEGO & Iceland Spix & Martius talk of birds singing
497q007 .-- (1) Character of alpine Flora of Tierra del FUEGO and Entomology of.-- most important, as furthes
527m031 o many» birds singing in England, in Tierra del FUEGO not one.-- now as we know birds learn from each
553m137 ns of an Australian savage or one of Tierra del FUEGO.-- Mr Miller (superintendent of the Zoological
496q006 - (10) Try in how many generations. daisy. Fever-FUGE Groundsil.-- gilly flower will break & become d
199b118 uitous modification, into a durable form, of a FUGITIVE want into a fundamental propenity, of an acci
146q012 ch as love-- dislike & <f> passion of hatred To FULFIL an instinct a pleasure; mem. Shepherd dogs The
277c126 subgenera. true value.-- as in Opetiorhyncus. FULGINOSUS. (a) Falklands (b). F. del Fuego differ from
471t07 ee on: Linn. Trans 18. p. 133 Westwood on the FULGORIDAE enumerates the strange forms which the thora
219b193 «(Turdus Guyanensis?) (Emberiza Brasiliensis?) (FULICA Chloropus)--» might bring in stomach-- &c &c.
480z008 urdus Guayanensis?)) Emberiza Brasiliensis (?)) FULICA Chloropus. says some of the «species of» small
023r009 mb, the rest not above half so long; The maw was FULL of jelly which stank extreamly."--This shark wa

041r066 fact is in alliance with those balls at Chiloe, FULL of sand.--the <scale> «quantity of iron» being
110a086 art. block not crystallized Salband like basalt. FULL of circular cryst of glassy felspar different f
171b004 Zoonomia.-- There may be unknown difficulty with FULL grown individual «with fixed organization» thus
191b080 nsideration of these archipelagos ups & downs in FULL conformity with Europaean formations-- England
227b227 (wish of parents??) instinct & structure becomes FULL of speculation & line of observation.-- View of
251c040 pecies, geese killed in Newfoundland, with crops FULL of maize. (get limits of latter from <Tarton> B
262c073 ed up, (most false to say no passages; nature is FULL off them.-- wading birds partially webbed. &c &
291c166 tion. & avitism in corporeal structure are facts FULL of meaning.-- Why is thought. being a secretion
309c222 ks only external» intermediate groups often have FULL structure «of one class» & full of second--this
310c222 roups often have full structure «of one class» & FULL of second--this class of facts «analogous to pe
342d032 le Muscovy & common ducks-- they are produced in FULL equal Numbers with pure bred (just like common
360d090 of face & tail somewhat like dog-- though it has FULL share of Jackall shape of body.-- disposition w
380d154 well as, & often better than a female.-- this is FULL of interest; for it shows latent instincts even
406e035 dentical with S. America. & many very close: see FULL paper. L'Institut 1838. p. 338 A most grave sou
429e114 emselves, as well as our wild plants, we see how FULL nature. how firmly each holds its place.-- When
450e175 Singapore 1837. By J. H. ? do. p. 189. «190» No FULL sized horse is found East of y Burramposter & S
460t013 The <male> swan-gander with common goose produce FULL as many eggs as pure bred common.-- the half of
466t091 turning brown, whilst both others were in nearly FULL flower Maer June/41/ Rhododendrum-- nectary mar
520mIFC l. Sykes) Private Finished. Octob. 2d. This Book FULL of Metaphysics on Morals & Speculations on Expr
524m021 n is not necessary)-- the words second childhood FULL of meaning.-- Dreams do not go back to childhoo
533m059 e nearest approach to <the> such instincts which FULL grown men can experience-- Instinctive walking
541m090 ms thus act.-- ∴ weak minded people are fickle & FULL of levity (¿ do I not confound action & thought
596n184 s Brown" on Association worthy of close study.-- FULL of practical observations Ourang do not move ey
611o034 ents.-- See Lamarck for this definition given in FULL.-- [LHC] ¿Has any vegetable or animal matter be
632j53r o not want to deny laws.-- The whole universe is FULL of adaptations.-- but these are, I believe, onl
507q15v less strongly marked.-- 31. Plant seeds of the FULLER'S plants ,Teazle Dr. Holland ; My Father. Andre
591n100 go against instinctive feeling, only does not FULLFIL, like continent man.-- a man eating what other
336d019 ected, but in a mule these conditions are not FULLFILLED.-- «[My grandfather's theory of Mules not he
620o044 re gratified & therefore as soon as desire is FULLFILLED, pleasure forgotten. [RHC] 4) as starvation,
039r060 ct elevations; hence it would appear he has not FULLY considered the subject.-- S. America in the for
250c037 t up!! Extreme southern points of S. Hemisphere FULLY characterized, of each continent. Try amongst E
279c133 ay be inferred from <this> facts stated.-- + +. FULLY supported by Mr Wilkinson. = Milking heredetary
307c215 abortive wings in the female are allowed to the FULLY organized wings of the male rendered abortive i
366d111 , as often as she touched handle, though really FULLY aware she was not coming in, could not help bei
369d115 lish entirely their place in nature, as well as FULLY to understand their oeconomy, is now universall
371d118 his collection of «Humming» birds, saw several FULLY developed tails, & one with beak turned up like
416e075 ecies every part of newly acquired structure is FULLY practised & perfected Hence difference between
533m060 n making too much profession, or rather in only FULLY expressing momentary feelings of gratitude, I h
538m078 ined by Dr Dewar on principle of association.--«FULLY bears out my fathers doctrine about people forg
538m078 f their childhood. & certainly of Miss Cogan, & FULLY corroborates the fact of her not <remembering w
608o027 will not do harm, because no one can be really FULLY convinced of its truth. except man who has thou
617o39v be clearly comprehended by anyone who wishes to FULLY understand this subject, but the answer to it w
622o048 C] My theory of instincts, or heredetary habits FULLY explains the cementation of habits into instinc
318c253 s now becoming common-- likewise of the Hirundo FULVA (added by Audubon in Appendix) showing WHAT CHA
325c267 organs produced in domesticated plants; where FUNCTION has ceased to be used as tendril into stump L
352d059 hat is abortive? when it does not perform that FUNCTION which experience shows us it was for.-- Most
352d158 tism,-- those organs which perform nearly same FUNCTION in both sexes-- are never double, only modif
455eIBC > see if by so doing can be made sensitive The FUNCTION of sleeping someway useful.-- it is only the
491q01v by cuttings.-- Try.-- Important as discovering FUNCTION of seeds-- (6) To hybridise EVERY flower on m
564n005 by attacking the citadel itself.-- the mind is FUNCTION of body.-- we must bring some stable foundati
614o037 owever unintelligible it may be, seems as much FUNCTION of organ, as bile of liver.-- ¿ is the attrac
529m042 ways an ideot.-- «in this case must have been FUNCTIONAL.--» He has some idea of a son of Dr. Prietly
443e155 e in point».-- The relation of these «sexual» FUNCTIONS to complexity is evident, yet the inference f
530m046 of brain the «similar remark» thoughts, being FUNCTIONS of same part of brain, or the tendency to hab
577n051 ions, (independent of mind) in the intestinal FUNCTIONS &c &c.-- bears. the same relation to true
199b118 to a durable form, of a fugitive want into a FUNDAMENTAL propenity, of an accidental habit into an in
307c216 & none in other case! Savigny has shown same FUNDAMENTAL organs even in Haustellata & mandibulata.--!
617o40v sition the forces manifested would be <sat> FUNDAMENTALLY accounted for, we prefer this metaphorical
516q23v eory (1) Will an extract of peat do to preserve FUNGI or animal substances-- (Athenaeum (40) p. 823 c
228b229 hen dissecting a whale, or classifyng a mite, a FUNGUS, or an infusorian. is "What are the laws of li
305c210 er.-- certainly more like Jackall in gait, size, FUR.; manner in which ears droop like dog older char
332d005 of Guinea, -- ears bare. skin black & wrinkled-- FUR short. (tail cut off in progeny peculiar) limbs
333d006 ther two very closely resembled in form of tail, FUR &c to the half bred Persian.-- Here then we have
363d101 plumage of fowls-- long ears of rabbits. & long FUR.-- feathers on legs of Ptarmigan & in Bantam.--
379d151 tinent-- Original Paper, worth studying. Archiv. FUR. Naturgeschichte. September 11' Generation Mr Ya
400e012 g good groups.-- ears varying so much,-- kind of FUR-- (do tips of ears take any colour?)-- length of
400e012 colour?)-- length of tail varies, & character of FUR-- I am sure a very good case, might be made out
557m147 y to dash at prey streched out & flaccid, when FURIOUS «with fright» back absurdly arched. & tail sti
564n006 ter, P. cii Vol III» of excessively cross-half FURIOUS faces «which may be described as an exaggerate
137a149 838. p. 72. on metallic vapours condensed from FURNACES do/p. 84 on the effects of veins of slag in i
137a149 /p. 84 on the effects of veins of slag in iron FURNACES affecting to some distance & blending with sa
264c080 en close come from different localities, as my FURNARII.-- some genus of yellow & brown-breasted bird
063r130 its own limit & represented.--Chiloe creeper: FURNARIUS. <Caracara> Calandria: inosculation alone sho
261c071 wherever any mark like red patch on wings of FURNARIUS, Synallaxis &c &c. sure to unite the birds &
482z011 anodera-- a linnet not caught.-- Troglogdytis FURNARIUS.-- Sturnus Magellanicus.-- p. 210. Scolopax v
484z016 ost interesting the way Synallaxis leads into FURNARIUS. by Patagonian Furnarius.-- into Oxyurus, by
484z016 ynallaxis leads into Furnarius. by Patagonian FURNARIUS.-- into Oxyurus, by Maldonado creeper of same
523m018 ion between enthusiasm passion & madness.-- ira FUROR brevis est.-- My father quite believe my grand
109a082 lways on actual beach act same way.-- a little FURTHER from beach action probably modified by form of
120a109 on at greater rate, not only river would carry FURTHER its own matter. but would cut wide gorge. leav
127a125 ing «pushing» up to «the» line of ice.-- Hence FURTHER N. when soil frozen for greater length of time
161g094 eet <lower> too high about a quarter of a mile FURTHER on, where three [...] abutted Having crossed t
188b067 es Orange river & says so far will I go and no FURTHER:-- Prof. Henslow says. that when race once est
195b104 ations is mere assumption, it explains nothing FURTHER, points gained if any facts are connected.-- N
229b233 eep aboriginal at Cape & a thinner-tailed kind FURTHER inland.-- NB. There is division of snakes. wit
294c176 to quite different origin when species rather FURTHER.-- once grant good species as carrion crow & r
358d085 ge in dogs» EFFEMINATE men),-- if carried much FURTHER, if by the process this were possible, the org
359d087 owl being so totally infertile whereas animals FURTHER apart have bred inter se.-- These hybrids are
379d151 tacea, as the past for other Mammalia. & still FURTHER back reptiles & Cephalopoda: Old Jones remarke
391d174 ations are not). but why should it demand some FURTHER change?-- Man properly is hermaphrodite (hence
398e005 y necessary as the «basal» foundation stone of FURTHER inductive reasoning is immense. It is curious
398e006 changes which the government is subject to.-- FURTHER back we obtain here & there in order a scatter
439e144 each possess., & that tiger springing an inch FURTHER would determine his preservation-- if killed b
461t041 is exactly same as some species extending much FURTHER geographically than others. Athenaeum p. 605 M
485z017 the Fluvicolae?-- The Birds seem to move much FURTHER North on West coast of S. America. than on Eas
485z018 ere is extreme heat, the tropical forms extend FURTHER north, because during winter they can bear the
486z018 ciple tropical forms in N. America extend much FURTHER N. in N. America than in Europe-- Coleoptera e
534m062 wall to table.-- table being removed a little FURTHER, they ascended about a foot & then leapt acros
608o028 ibution.-- it may be a consequence but nothing FURTHER.-- October <8> 2d. 1838 Those emotions which a
609o030 rally been best for our good far back.-- (much FURTHER than we can look forward: hence our <[...]> ru
614o037 o Atheism. inutility of so high a mind without FURTHER end just same argument. without indeed we are
497q007 Fuego and Entomology of.-- most important, as FURTHEST removed possible point.-- ¿genera in intermed
545m107 I showed nut & then closed my mem. expression of FURY, jump to scratch my face. The ourang outang, un
196b107 y known that Ireland possesses varieties of the FURZE, broom, & yew very different from any found in
093a031 ut order of injected rock being determined by FUSIBILITY in. L Institut p 247. 1837.-- The most infus
060r125 he rocks have been. The red Sandstone of Andes FUSIBLE? no. mad dogs. Azores. although kept in number
102a061 ing taken place most gradually, first the more FUSIBLE substance, & then the next being sucked out. I
569n022 ait mieux rester a Farroïlap quelque mal qu'on y FUT."-- Expression common to Savage & Frenchman, una

Page   ***(Key Word)***
138a153 will serve for comparison with the moon at some FUTURE time [blank] Sir. J. Halls Paper on the consol
227b226 sible changes is discovered, for speculating on FUTURE. !.fish never become a man.-- Does not require
228b229 our <past> speculations with respect to past & FUTURE. The Grand Question, which every naturalist ou
303c205 st is unfair, [moreover what will become of the FUTURE creations, if the list is now perfect.--]CD. t
316c244 ty for Deity to choose, when perfect enough for FUTURE state, that when good enough for Heaven or bad
343d036 d short of eternity to the present time, to the FUTURE-- How far grander than idea from cramped imagi
343d038 acteristic forms of S. America. With respect to FUTURE destinies of mankind, some of species or varie
544m102 ks which when conscious connect past, present & FUTURE thoughts are broken-- Sir J. Franklin when sta
549m118 (contentmt is different it refers to wishes for FUTURE) than perhaps well «regulated» philosopher-- y
549m121 - or whether if we obey literally New Testament FUTURE life is almost the sole object--. -- I doubt w
564n004 onscience, obeying habits, & dread of misery of FUTURE thinking of injured moral sense.-- Notion of d
568n018 whether to pursue immediate inclination or some FUTURE pleasure.-- hence judgment, which is part of r
608o028 sins of a man, are under his control, & that a FUTURE life is a reward or retribution.-- it may be a
182b049 t considering power, extending range, reason & FUTURITY. it does as yet appear clim In Mr Gould Austr
280c136 changes one, in other it changes thousands in FUTURITY.-- This is right way of viewing it.-- Variety
497q007 Heath by facility. ¿Knight take opposite view. GAERTNER talks of the several great & natural Families
501q011 in similar manner ever failing.-- answered by GAERTNER (28) Can any annual or Biennial be grafted or
507q15v ation difficult or reverse (.49) List of seeds GAERTNER de fruct:-- for woodcut-- 1 double hook-- --
078r174 ke my undescribed[.] p. 140. Flèche de Quoy et GAIMARD.--D'Orbigny has described it with care to 3 sp
234b256 ade great collection of birds of Iceland. --M. GAIMARD, however, will settle this.-- Waterhouse says
247c028 - according «stated in note to p 21» to Quoy & GAIMARD in Sandwich isld. & according to Chamisso in R
479z005 nimal with teeth {P} p. 140. Flèche de Quoy et GAIMARD Ulloa shell fish Purple die Marvellous stories
032r041 eeds from great depth. from axis to surface must GAIN a Westerly current:--If great changes of climat
280c134) heredetary Young wild ducks.-- lose as well as GAIN instincts. Wild & tame rabbits good instance--
414e064 anization, or instincts (ie intellect in man) to GAIN the day.-- In man chiefly intellect, in animals
625o50v ur passions are too strong for our instincts. to GAIN long-lived good, ie happiness-- yet this system
633j53v sseltoe?-- do p. 284. it is hard on my theory of GAIN of small advantages thus to explain the curling
638j059 to make animal perform some action.-- as well as GAIN it. by habit.-- New theory of instinct, returni
195b104 assumption, it explains nothing further, points GAINED if any facts are connected.-- No doubt in bird
203b136 Cattle.·. tameness not heredetary?, having been GAINED in short time. Milvulus forficetus <has a wide
232b247 Phenomena of the S. Hemisphere look as if heat GAINED> Experimentise on land shells in salt water &
290c163 etary power of Muscles.-- then we SEE structure GAINED by habit Talent &c in man not heredetary, beca
291c165 s wildness-- cf Sir J. Sebright.-- love. of man GAINED & heredetary. «problem solved» habits become i
292c171 tance to cause long memory.-- structure is only GAINED slowly.-- therefore it can only be those actio
301c199 ld striking a post in passion.-- Habit instinct GAINED during life.-- do Elephants easily acquire hab
306c212 imals, well developed, just like, habits easily GAINED in child hood.-- Young salmons. first a specie
315c243 s to exposes his canine teeth. no doubt a habit GAINED by formerly being a baboon with great canine t
335d016 eculiarities, to its offspring, which have been GAINED slowly, now all the mules have their whole <bo
336d017 the mules have their whole <body> form of body GAINED in one generation, so it is impossible to have
390d178 he male «in the course of some generations» has GAINED some difference «from what it received» (for i
462tf4v ame as existing species. We see the same object GAINED by the Mataco-armadillo & the woodlouse-- -- a
544m104 not death for the pain.-- How was this instinct GAINED? by conversation-- .·. modified in those races,
548m114 did reason about himself-- but not about, facts GAINED or gaining by senses.-- As sleep <is> only one
583n070 s I can see mentally refers by reason knowledge GAINED by reason: & then these qualities of imitation
583n070 as argument.-- for instinctive knowledge is not GAINED by instruction, or imitation.-- p. 20. Animals
588n090 - if explained by habits, useful to itself, how GAINED. reason? or some unnamed faculty-- Lamarck. Ph
592n105 ome instincts unintelligible, <both> in the end GAINED «& therefore the» cause, and origin being so i
600o008 nger-- Remember.-- avarice a compounded passion GAINED in life then]CD 3. The Infinite, -- lives by h
605o020 he object of this essay is to show how taste is GAINED how it originates, & by what means it becomes
614o037 instincts analogous «(& replace)» to experience GAINED by man in lifetime Heredetary memory not so wo
641j29v & light affecting plants. in insects the end is GAINED by some very different method. in pedunculated
641j29v stillness p. 276) of flight of Owl remarkable, [GAINED by very different process from Bats. CD]CD. wi
109a083 ach. {P} -- De la Beches argument of low coast GAINING & high loosing answered by this -- No one can
547m113 he memory of every late impression. & likewise GAINING new ones from senses. & <comparing their> «cal
548m114 about himself-- but not about, facts gained or GAINING by senses.-- As sleep <is> only one idea is aw
606o021 various operations which the mind undergoes in GAINING the result.-- Lessings Laocoon. 2d Lect-- The
640j28v he evils of old age, after breeding season, or GAINING adaptations, but for youth most necessary: the
175b018 ity Each species changes. does it progress. Man GAINS ideas. the simplest cannot help..-- becoming mo
539m082 fter seven years interval. Augt. 15th. As child GAINS habit «or trick» so much more easily than man,
620o045 l leave effects not lasting. By association one GAINS the rule, that the passions & appetites should
638j059 s «step by stop».-- in Man, the nervous system, GAINS that knowledge, before hand. & can in idea (wit
305c210 Scotch Terrier.-- certainly more like Jackall in GAIT, size, fur,, manner in which ears droop like do
028r031 he earth <I did not see one dike in the whole GALAPAGOS Arch; because no sections> same cause as no c
031r038 little Pumice, though Trachyte. same fact in GALAPAGOS. Daubeny P 24 V. back of page 1 of New Zealan
033r043 l of Porph. Lava with base of Pitchstone; Mem GALAPAGOS. chiefly red glassy scoriae.--could walk roun
037r055 f Bengal. dimensions? Strong currents off the GALAPAGOS.--strata must be accumulating which like the
039r060 of Rialeja, & the more true analogy from the GALAPAGOS-- Mr Lyell. P. 111 & 113. «seems to» consider
043r069 n rocks at Costorphine hills. to compare with GALAPAGOS.--Chiloe. M. Hermoso. & Coral reefs (imperfec
044r072 in atmosphere.--Mem. Ascencion. concretions & GALAPAGOS.-- «Humboldts. fragmens.» Read geology of N.
050r090 va are described, & many minute craters as at GALAPAGOS. <|> Sir George Mackenzie must be worth readi
054r105 Isld near coast of America not reached. Juan. GALAPAGOS. Cocos-- Ulloas voyage North of Callao, the c
056r110 losses.-- Cause of chimney. to crater. as at GALAPAGOS. St. Helena.-- [Fig. 7] {P} effect of heat on
074r165 ptaria. {& Chiloe case, at least corelation)--GALAPAGOS vein. vein of secretion.--metallic veins foll
079r177 Vol I, P. 168. on coast of Guayaquil, same as GALAPAGOS. no Hydrophobia at Quito. P 281. do do Austra
093a032 thinks Olivine a preexisting mineral.-- Mem. GALAPAGOS .·. Basalt deepest?? Marcel Serres L'Institut.
096a041 low --¿ applicable to Auvergne??? The fact of GALAPAGOS Isld. steep side to windward in allusion to S
172b007 with slightly differen circumstances.--» Now GALAPAGOS Tortoises, Mocking birds; Falkland Fox-- Chil
173b011 ent though species all different. In cases as GALAPAGOS & Juan Fernandez. When continet of Pacific ex
185b055 zzard which has changed into Cara cara at the GALAPAGOS. Fernando Noronha Ophyressa bilineata (Gray)
189b070 colour hereditary in time as in space. (Mem: GALAPAGOS). Little wings of Apteryx Dacelo & Kingfisher
195b098 ls.-- The question if creative power acted at GALAPAGOS it so acted that bi[r]ds with plumage «&» ton
195b100 wandering birds. such sandpipers. not new at GALAPAGOS.-- did the creative force know that «these» s
195b103 y changes. Will it said that Volcanic soil of GALAPAGOS under equator that external conditions would
195b103 oduce species so close as to Patagonian <Chat> GALAPAGOS orpheus.= Put this strong so many thousand mi
209b157 {T} Calculate my Keeling Case: Juan Fernandez GALAPAGOS =Radack Islds = .·. Islands & Artic are in sam
210b160 islands are near continent: compare Siicily & GALAPAGOS!!-- Some of the animals peculiar. to Mauritiu
214b173 gh. list of plants in Beetsons St. Helena. -- GALAPAGOS--Juan Fernandez Falkland Islds-- Kerguelen la
219b193 island, Teneriffe, St. Helena. J. Fernandez. GALAPAGOS. Many trees Compositae, because seeds first a
226b220 New Guinea same fact see Coquille's Voyage.-- GALAPAGOS mouse (?) brought by canoes Ceylon & India.--
226b221 his we see example in English & Irish Hare.-- GALAPAGOS -- shrews, & when big continent many species
227b227 creation!-- If continent had sprung up round GALAPAGOS on Pacific side. the Oolite order of things m
248c029 rabbits thus handed down for nearly 70 year. GALAPAGOS Mouse not the same section, with house mice.
256c054 h country.-- I often disputed for a moment,-- GALAPAGOS, S. American-- -- genus.-- The circumstance o
283c145 & then all that I want is granted.-- For. at GALAPAGOS. make ten species of Orpheus-- one of which h
296c184 . but no other will.-- St. Helena (& flora of GALAPAGOS?) same condition. Keeling Isd «shows where pr
299c193 this remarkable compare it with Canary Islds. GALAPAGOS.-- Iceland has same uniformity Primrose & Cow
305c209 this is fact analogous to mocking thrushes of GALAPAGOS. having tone of voice like S. American.-- Have
350d053 ider all this NB. How can local species as at GALAPAGOS. , be distinguished from temporal species as i
354d065 ty. has Creator made rat for Ascension.-- The GALAPAGOS mouse probably transported like the New Zeala
400e012 s says no Tortoises in other «places» besides GALAPAGOS do. p. 376. Isle Tres Marias off Mexico with
405e032 my Journal for references In such cases as at GALAPAGOS, where different islets have different forms
408e044 new one to enter?-- The wonderful species of GALAPAGOS, must be owing to these islands, having been
424e100 Prof:» Don would have known the Composites of GALAPAGOS were South American.-- several cases of speci
425e104 case,-- for Sandwich Isld are very similar to GALAPAGOS-- study Flora-- what general forms.-- are the
449e170 tant. & all alike Waterhouse says some of the GALAPAGOS Heteromerous insects come very near to Patago
449e170 e which are different.-- now this is same, as GALAPAGOS facts &c &c.-- & it shows the causes which gi
481z008 theoretical researches Compare land shells of GALAPAGOS different islds.-- Waterhouse remarks that no
482z012 says that two species of Tortoises come from GALAPAGOS!!! Azara. Voyage dans l'Amerique Merid. Tatu

500q10a es; how is this at Canarys Arch-- it is so at GALAPAGOS.-- Ireland, doubtful species-- Does any genus
506q15v cious, or Monooecious, besides the Nettle. at GALAPAGOS-- Dioecious.-- Carex.-- We may presume Nettle
640j167 x determines life:.» «With respect to whether GALAPAGOS beings are species. it should be remembered t
640j167 would far more easily be changed.-- Hence the GALAPAGOS Islds are explained. On distinct Creation, ho
640j167 themselves.-- Probably no case in world like GALAPAGOS. no hurricanes.-- islds never joined, nature
021rIFC and Molina 1st Vol & Lyell Sailed, 27th <Friday GALE 29th> Friday Thursday 29th gale Lyell's Geology
021rIFC ed, 27th <Friday gale 29th> Friday Thursday 29th GALE Lyell's Geology The living atoms having definit
129a131 t wonderful case of great block of rock moved by GALE-- When writing on Valleys. «Tertiary strata of
079r176 Humboldt says, mur of Silv.[,] Sulph. of do.[,]GALENA[,]guartz, Carb. of Lime. accompany.-- Ulloa ha
042r067 Capt. F.: R: how the swell, generally & during GALES would tend to travel on a <me> central line of
066r141 » under 50. Kerguelen Land, = the way it stands GALES = very strong. Stones as bigger than a man's he
408e044 ave this character.-- worth going there for.-- «GALES of wind would blend species» Buckland. Reliquia
470t178 ng over stamen «Egg Tree»--I think never on the GALEUM saxatile & other common kind--I think not on P
540m088 conscience &c &c Coleridge,-- Zapoyla p., 117, GALIGNANI Edition Fine poetry, or a strain of music, wh
242c017 id to have orang-utang & Pongo in common¶-- GALIOPITHECUS common to Moluccas & Pelew Isds.-- p. 22. N
402e019 Manilla a small Cercopithecus., & skins of GALIOPITHECUS.-- Malte Brun. Vol <I> II p.,133: at Samar
507q15v fruct:-- for woodcut-- 1 double hook-- -- Geum. GALIUM Burrh ≡ single hook; curved spines-- simple sp
453e183 ed.-- "Storia della Riproduzione Vegetale". by GALLESIO. Pisa 1816 p. 27. Dr. Holland. Are there inst
508q016 ch-- Cause?-- Andrew Smith. (6) What size book GALLESIO storia del Reproduzione.-- D. Holland (7) Is
281c138 sification The relation of all cock birds in GALLINACEOUS having tendency to lon[g] or peculiar tails
303c205 ites bird from Madagascar uniting pidgeons & GALLINACEOUS birds & parrots.-- legs of pidgeons perfect
380d154 g like cocks.-- It is very singular. so many GALLINACEOUS birds have cock & hen plumage so different,
327cIBC els.-- Temminck Hist. Nat. des Pigeons et des GALLINACES. Silliman's Journal. during 1837. paper by B
318c253 seem to follow narrow bands, certain kinds as GALLINULES taking the low country near coast & others t
114a093 r, Part IV. p. 403 Ehrenberg on ferrugineous GALLIONELLA Examine Iron stone of C. of Good Hope & Aust
139a180 ic substance cause such monstrous growth as oak GALLS or rose <buds> galls.-- is it not effect of sup
139a180 ch monstrous growth as oak galls or rose <buds> GALLS.-- is it not effect of superadded vital influen
214b175 rles.,-- and now in the possession of Mr Howard GALTON, have one of the vertebra, about 2/3 from base
217b183 at.-- This is good instance of same fact in Mr GALTONS case.-- It explains the loss & expense, (must
026r021 ns & cleavage is importants; veins appearing a GALVANIC phenomenon, so probably will the Cleavage be
454e184 atic animalcule. has described instrument for GALVANIZE <ani> them-- Cross Irish & Common Hare Decand
204b140 fspring most forcibly-- Esquimaux dog & Pointer «GAME-fowls have courage independently of individual
356d072 Sept Yarrell told me he had just heard of Black GAME & Ptarmigan having crossed in wild state-- & th
357d073 s with which this genus becomes crossed. ¿is red GAME an hybrid?-- When I show that island would have
365d105 land no doubt the cross between Pheasant & Black GAME is owing to their rarity., a single female in w
365d105 .-- Is English red Grouse. a cross between Black GAME. &, the subalpina of Sweden, (which in summer d
525m023 heir duty.-- as carrying a basket, bringing back GAME, or picking up a stone, though only acquired ru
460t013 its form both on vars & species The <male> swan-GANDER with common goose produce full as many eggs as
460t013 ross, as above, take «generally» after the swan-GANDER. one of these half-bred ganders, crossed with
460t013 after the swan-gander, one of these half-bred GANDERS. crossed with common goose <to has> «produce o
203b136 s even now in central & Eastern Asia beyond the GANGES & perhaps even in India p. 261. L. Institut. 1
612o035 t of man. [LHC] ¿How near in structure is the GANGLIONIC system of lower animals & sympathetic of man
467t104 al & the curvature of <an> pistil, etc lies in GANGWAY= In Lotus corniculatus saw Humble press down w
180b036 REQUIRES extinction. Thus between A. & B. immens GAP of relation. C & B. the finest gradation, B & D
181b042 ween them.».-- for instance there would be great GAP between birds & mammalia, Still greater between
223b208 II say how> The difference is that there is wide GAP between Man & next animals in mind, more than in
408e043 pecies must increase in number where then is the GAP, for the new one to enter?-- The wonderful speci
179b035 life of species in certain orders connected with GAPS in the series of connection? «if stating from s
181b042 rue, that the greater the groups the greater the GAPS (or solutions of continuous structure) «between
196b110 usca & vertebrata, that there must be very great GAPS.-- yet some analogy The existence of plants, &
201b126 is «from unfavourable conditions» there are many GAPS. & those forms which «nevertheless» have produc
302c200 with so much death, as has gone on,, No greater GAPS.-- external conditions, to be sure, have remain
303c205 ill be said» that although at any one there. are GAPS. yet <what> altogether «he» has created a perfe
308c217 ome perished before, then there always have been GAPS, & there now must be, .:extinction of species be
352d060 ry, & the geologist being obliged to fill up the GAPS.-- is possibly the same with the <Zoologist> «p
375d135 <into> every kind of adapted structure into the GAPS <of> in the oeconomy of Nature, or rather formi
375d135 of> in the oeconomy of Nature, or rather forming GAPS by thrusting out weaker ones. «The final cause
411e055 types, & limits of variation., & hence indicate GAPS.-- by this means the laws probably would be. ge
433e126 how many centuries elapsed between each of these GAPS, far more probably than during the deposition o
025r016 orro S Paulo [13 degree 22 minute S.] 9 120 {T} GARCIA de Avila [lighthouse] [12 degree 35 minute S.]
212b165 ter with only nose projecting.-- would pull the GARDEN bell, & then run into Kennel to watch who woul
290c164 strata of Tilgate forest Seeing common gull in GARDEN at Zoology Soc. it's pale ash grey back, like
333d007 esembled foxes than dogs (Mem Jackall in Zoolog GARDEN) He has seen in a show half Wolf & «half Esqui
360d090 very <thing> «change» is effected)-- the one in GARDEN is from <bitch dog do> father dog. & hence gen
371d118 with contingency.-- Sept. 23rd. Saw in Loddiges GARDEN. 1279 varieties of roses!!! proof of capabilit
371d127 to the dholes or wild dogs of India) in Zoolog. GARDEN having coloured offspring.-- but surely in all
383d160 mon mule must often have been dissected Zoolog. GARDEN. Sept 16." Hybrid between Silver & Common Phea
423e098 c.-- Uncle John believes one single turnip in a GARDEN is sufficent to spoil a bed of Cauliflower.--
447e165 n heights.-- yet he has seen it propagated in a GARDEN, which is case precisely analogous to the Cana
453e183 each other. & yet I presume seed raised in same GARDEN.-- now this good question-- single, or half do
467t099 ually abortive--and both growing within Kitchen GARDEN.-- As we see in Hybrids that although anther «
470t176 peared. <& old English> But these mules <in our GARDEN> show no trace of palmation!!? Bees at Wild St
472s01r tum, & the Papaver oncitate was growing in same GARDEN. & out of 60 seedlings not one came up true.--
472s01v go seeds were procured with the P. orientale in GARDEN & all came up hybridised. It is possible to na
495q05a eep» «closes flower» on all gloomy days.-- The «GARDEN» Coronella also sleeps on ditto-- Cover them u
495q05a & see whether wind, on «dry» windy day, «flower GARDEN on gravel walk» will drift many seeds= Necessa
504q013 abortive stamens every year & Spring. & within GARDEN //Yes// (5) Examine the Parnassia whose stamen
516q023 ink I have never seen Bee visit. Experiments in GARDEN Sow stones of Standard Apricot grafted on what
564n006 is most clearly analogous to a panther I saw in GARDEN uncovering its teeth to bite.-- the senseless
591n097 end.-- July 20th Intelligent Keeper... Zoolog. GARDEN told me. he has often watche tame young wolf &
594n112 y it-- I have seen hawk & sparrow in Shrewsbury GARDEN picking from same bone A child born on the 1st
595n121 h 1840 was frightened at wild beasts in Zoolog. GARDEN [blank] [not located] A child crying. frowning
603o11b 134. a painted must not a actors, or a scene in GARDEN.-- yet both beautiful! p. 136. Says Architectu
431e122 ed, & this would look like fresh Creation. the GARDENER separates a plant he wishes to vary-- domesti
521m009 r time.-- » My father asked him whether he had GARDENER of name A. B., &c &c. & he maintained he had
521m009 ined he had never heard of such a man & had no GARDENER.-- My F. then asked Mr C. to come to the wind
521m009 Mr C. to come to the window & pointed out the GARDENER & said, who is tha? Mr C. answered why do you
521m010 answered why do you not know, that is A. B my GARDENER.-- Thus was he in every respect, no communica
401e017 peculiarities.-- brilliancy of inflorescence GARDENERS. by chance <often> sometimes graft pears on a
305c210 bly varying plumage for wild birds-- At Zoolog GARDENS there is half Jackal & Scotch Terrier.-- certa
306c212 am 1000 miles long.-- a monkey. (Baboon) at Z. GARDENS upon being beaten behaved very differently fro
341d032 as Pea hen.-- about intermediate.-- (In Zoolog GARDENS there is hybrid of Penguin duck a variety of M
383d159 show that it is not overwhelming.-- Seeing in GARDENS of Hybrids between Common & Silver Pheasant, a
385d163 yet very odd loosing visible powers» in Zoolog GARDENS. & Kings at Otaheite) <Think> Last litters are
392d180 Z> of India «with our common ones» in Zoolog. GARDENS; <Buffalo & common cattle-- Esquimaux (& Austr
393dIBC tail & common ducks. similar inse? Zoolog. GARDENS Are the hybrids of those species. which cross
393dIBC ttle at Cinbermere? How is Jackall & dog at Z. GARDENS D E Finished July 10th 1839.-- Selected, Dec 1
424e103 , but Bell has some account of wolf in Zoolog. GARDENS, which brought its puppies to be fondled.-- an
429e114 t the multitudes of plants introduced into our GARDENS (opportunities of escape for foreign birds & i
447e165 vivaparous on mountains & yet can be raised in GARDENS.-- Poa alpina, thougt generally vivaparous som
458t002 hardly ride on them. Mr Miller-- in Zoological GARDENS. informs me that a hybrid between ass & Zebra,
470t176 ced hybrids-- My Father remembered when in the GARDENS, he knew there was none but English,--the Palm
489qIFC 18 Blyth-- 19-- Mr. Tollett {T} Zoolog Soc «GARDENS» ---- . . 20 & Breeders Dr. Boott: R. Brown p.
492q003 Herbert Do forest-trees sport much in nursery GARDENS? <are the> is the ground much manured In speci
494q004 Dorking (10) Statistics of breeding in Zoolog. GARDENS-- with respect to conditions of animals & thei
529m041 I am sure I remember my pleasure in Kensington GARDENS has often been greatly excited by looking at t
540m085 n pud. not disagree.-- Ourang outang at Zoolog GARDENS touched pud. of young male & smelt his fingers
551m129 reexistence <">-- The young Ourang in <Zoolog» GARDENS pouts. partly out displeasure (& partly out of

Page **(Key Word)***

553m137 -- Mr Miller (superintendent of the Zoological GARDENS) remarked that <exp> the expression & noises o
555m142 11, case of Association. Sept. 1826 Zoological GARDENS-- Endeavoured to classify expressions of monke
556m145 [The laughing noise which C. Sphynx made at Z. GARDENS may be described as partaking of <st.> made by
581n064 n.-- in savages other tastes few. March 16th. GARDNER'S Music of Nature. p. 31. remarks children hav
582n066 y have pain or pleasure these are sensations» <GARDNER in his work» In the life of Hayd & Mozart. fin
636j56v prevented Do races of peas become intermixed & GARDNER have hybrid seedlings} p. 333. Macculloch. bri
323c269 Phillips. Geology. Larder 2d vol.-- March 16. GARDNER'S Music of Nature--- -- Herbert on Hybrid Mixtur
590n094 erings, Australian Dog does not Bark-- quotes GARDNER'S Music of nature to show barking not natural.
410e052 ailor, as, mere describer of species, from its GARMENTS, without some end.-- Respect good describers
074r164 . native sulphur.. fluor spar. bayte. asbestos GARNETS.--carb & chrom. of lead. orpiment. chrysop[r]a
157g076 able gneiss with red granite veins & quartz, & GARNETS.-- Boulders as before certainly must have <com
265c087 to be of some use. If we only had Puffinuria GARROTTII & no other species-- as we have only ornithor
091a027 . p. 105. Oct. 1828. gneiss in India (falls of GARSIPA) dip 30 degree. <strike> «direction<?>»ESE-- C
125a121 far higher from circulation of heated fluid or GASES under pressure.-- {P} Lyells view of transmissi
125a121 e.-- {P} Lyells view of transmission of heat by GASES-- does not apply it to thickness of crust.-- {P
247c028 distributed: Mem of great geological age-- GASTROBRANCHUS «only» 2 species one in Northern Hemisher
234b255 located] T. Carlyle, saw with his own eyes. new GATE. opening towards pig.-- latch on other side.--
613o036 lor he has seen push a hare through the bar of a GATE before him, & then jump over the gate & bring i
613o036 e bar of a gate before him, & then jump over the GATE & bring it. ---- Agrees with ONE animal [RHC] K
572b033 h word distinctly. woke instantly but could not GATHER general sense of this page.-- Now <awake> when
469t151 planted in rows «close to each other» & seeds GATHERED «all» came up in 1840 true. Shrewsbury.-- Abb
469t151 -fan Bean» bean, were planted in rows, & seeds GATHERED same year came up true «in 1840»: All in toge
472s01r -- Jun 1. 1842 Allen W. sowed some years since GATHERED the seeds of Papaver bracteatum, & the Papave
500q10a & peas were planted in rows adjoining & seeds GATHERED there were planted «last year» pell mell, wit
500q10a «last year» pell mell, without sticks & seeds GATHERED & these are now to be planted this year copie
470t153 bout <3/4> of minute These latter were pollen GATHERERS & they seem slow= Maer 1840 My Father formerl
358d074 a.-- contrast with otaheite in relation «See GAUDICHAUDS Volume on the Botany of the Pacific.--» to n
258c060 animal in cold country has some analogy in not GAUDY colours so all changes may be considered in thi
430e119 owed me two specimens of an Inoceranus from the GAULT of Folkstone, which is exactly intermediate bet
403e021 which differs from that of S. Europe p. 189. The GAUT, kind of crocodile, sometimes wanders from Pell
118a103 this would make.-- L'Institut. (1838) p. 216 M. GAY on the Geology of Chile.-- P p217. Pentlands Fos
041r065 en discussing probable rise of land: Mention M. GAY'S fact about shells: Hibernation of fresh water S
044r074 the fact of bombs in tufa there is proof of such GAZ) steam condensed.--Perhaps these mighty changes
108a080 almost as well said probably much Carbonic Acid GAZ here.-- [top portion page excised, not located]
196b109 versâ. ¿could plants live without carbonic acid GAZ.)-- Yet unquestionably animals most dependent on
251c041 ts Vol III. <p.> p 233, stated that the "Asseel GAZAL. (Bos Gazoeus) does not mix with the Gobbah or
251c041 azoeus) does not mix with the Gobbah or village GAZAL.-- ¿is latter same species domesticated, strang
036r052 different Volney, P 351. Vol I. woody bushes, «GAZELLES» hares, grasshoppers & Rats. characteristic o
044r074 ion at bottom.--solution under high pressure of GASES. especially the most abundant. Sulp. Hyd: Carb:
251c041 p.> p 233, stated that the "Asseel Gazal. (Bos GAZOEUS) does not mix with the Gobbah or village Gazal
247c028 . Chamisso p. 189 Tome III: Kotzebue.-- p 22. a GECKO on St Helena.-- <in 1813, a venemous snake was>
247c028 t Helena.-- <in 1813, a venemous snake was> one GECKO on Isle of France Scincus multilineatus (p 45)
372d129 splanted? +.simplest forms of budding. Why does GECKO produce always different tail? An Individual bu
639j28r ion.-- C. D. Sucking feet in Frog. Walrus. Fly. GECKO &c. Prehensile tail. in Monkeys & Marsupials. H
251c040 n birds seen far at sea, migrations of species, GEESE killed in Newfoundland, with crops full of maiz
331d1FC rst cross, which makes hybrids. productive like GEESE?-- Are the number of kittens between Lion & Tig
334d009 ovy Ducks.-- English. <Common> «China» & Canada GEESE, & that they this first cross were equally fert
334d009 of Mules.-- «He recollects one hatch of hybrid GEESE very fine.--» How is it with plants? This indic
392d180 ey are crossed? & How is it with China & Common GEESE «how are their instincts?» «Chineses & Common P
450e174 n domestication & nature» strictly monogamous-- GEESE polygamous (¿when wild) but only some birds are
461t039 end to render complex the series.-- Ch 6 Upland GEESE would transplant seeds very far.-- Sept 31. The
515q022 species Eyton (1) Number of eggs-- of half-bred GEESE-- inter se, & with parents & of Chinese geese.
515q022 d geese-- inter se, & with parents & of Chinese GEESE. (2) Anatomy of muscles of stumps of tailess do
515q022 -- varying-- (4) About blended instincts of the GEESE which he crossed; especially if the hybrids wer
388d172 plication by division <only> is developement of GEMMA.-- The manner in which Frogs copulate & fish sh
328cIBC ils at same time Study Botanical work on Buds & GEMMAE. C D Charles Darwin 36 Great Marlborough St Di
390d177 nnected with more developed forms) Study buds-- GEMMAE-- & monocotyledenous, do those which are Monoc
313c236 why two sexes scission in all cases probably GEMMATION (.Ehrenberg) --not necessary to generation (1
372d130 considers artificial division of animals, as GEMMATION, I consider gemmation as artificial division.
372d130 division of animals, as gemmation, I consider GEMMATION as artificial division.-- On this view each p
386d167 tural History. 1838. p. 123. Ehrenberg. makes GEMMATION in animals very different from that of plants
442e151 -- any amount of generation may take place by GEMMATION «My theory will not admit this, now that tuli
442e153 e, that <species> «individuals> propagated by GEMMATION should be absolutely similar; [all the gorze
443e154 from top shoots.-- If prolongation of life by GEMMATION <can be> being impossible. can be overturned,
443e154 ly requires that organic beings propagated by GEMMATION do not now undergo metamorphoses, but to arri
446e162 strong case showing analogy of production by GEMMATION & by seed-- which Henslow is inclined to thin
063r132 maphrodite. or by cutting an animal in two. (GEMMIPAROUS. by nature or accident). we see an individua
170b001 ; for instance fruit trees, probably polypi, GEMMIPAROUS propagation. bisection of Planarias. &c &c.-
054r102 irely volcanic!! New Zeeland rich in particular GENERA of plants: All St. Catherine & coast Granite:
174b013 lle. (his argument applies only to hybridity.-- GENERA being usually peculiar to same country, differ
174b013 ing usually peculiar to same country, different GENERA different countries. Propagation explains why
176b021 anched some branches far more branched.-- Hence GENERA.-- «as many terminal buds dying, as new ones g
177b029 mmenced branching.-- As all the species of some GENERA have died; have they all one determinate life
180b036 radiation, B & D rather greater distinction Thus GENERA would be formed.-- bearing relation to ancient
181b039 o. pages back. Diagram The largeness of present GENERA renders it probable that <the> «many» contempo
182b050 d. & Austral & New Zealand Mr Gould says in sub-GENERA, they undoubtedly come from same countries.--
182b050 oubtedly come from same countries.-- In mundine GENERA, the nearest species often come very remote qu
186b057 ific character); and in passing from species to GENERA, each retains some one character of all its fa
186b060 cies of that country parents of American.-- Now GENERA of those two countries ought to be similar ¿La
190b077 in Australia great abundance of species of few GENERA or families.-- (long separated.-- Proteaceae &
190b079 ies became obscurer as knowledge increased, but GENERA stronger.-- Mr Waterhouse says no real <separ>
190b079 house says no real <separ> passage between good GENERA-- How remarkable spines, like on a porcupine o
192b085 gradation can be found from forms marking good GENERA-- by steps so insensible, that each is not mor
192b085 we know varieties can produce.-- Therefore all GENERA MAY have had innumerable steps.-- Quote in de
192b086 ith few species greatest jumps strongest marked GENERA? Reptiles?) For instance there never may have
193b093 r long South coast all the remarkable Australia GENERA, collected together Man has no heredetray prej
196b104 cts are connected.-- No doubt in birds, mundine GENERA, Bats .Foxes. Mus are birds that are apt to wa
196b105 transportal.-- Waders & Waterfowl.-- scrutinize GENERA, & draw up tables-- Instinct may confine certa
196b105 h have wide powers of flight; but are there any GENERA, mundine, which cannot transport easily.-- it
200b121 cal impossibility to marriage.-- ¿whether those GENERA which unite very different structure as. petre
201b126 w in species, but every two or three these form GENERA-- this is <from unfavourable conditions> there
206b149 moment fewer closely related;∴ (few species of GENERA) ultimately few genera (for otherwise the rela
206b149 elated;∴ (few species of genera) ultimately few GENERA (for otherwise the relationship would converge
206b149 single one.-- Will not this account for the odd GENERA with few species which stand between great gro
208b153 law of species few in Arctic. in proportion to GENERA. agrees with late production of those regions,
209b157 great work de distribut. Plantarum. relation of GENERA to species in France is I: 5.7 in Laponia I: 2
209b157 in Laponia I: 2,3: Mem. Lyell on shells.-- {T} GENERA In North Africa. I: 4,2 Iles Canaries I: 1,46
210b158 temperate parts of Teneriffe, the proportion of GENERA I: I. I can understand in «one» small island s
210b159 ussion Von Buch says from Humboldt, in Laponia. GENERA to species I. 2,3-- From Mackenzie Iceland the
210b159 ecies I. 2,3-- From Mackenzie Iceland there 144 GENERA & 365 species of plants not cryptogamic but I
210b159 n know varieties. there is analogy to species & GENERA.-- for instance three kinds of greyhound.-- In
212b168 (as cephalopods,) in last tertiary epochs most GENERA dead? --Examine into this «in Phillips».-- Acc
212b168 ng to this, formerly there would have been many GENERA of monotrematous animals.-- p. 82 «There are m
213b168 «There are many tables in Phillips of numerous GENERA in fossil & recent state, well worth considera
217b188 to fine speculation.-- Are there many Northern GENERA peculiar to itself-- on hybrids between grouse
219b194 t <neig> Africa, sandstone, & granite, (that is GENERA near Cape) see if there are any species same a
219b195 rly descended lower, therefore species of lower GENERA altered. or northern plants «No» CD[Mem. the a
220b199 7. Feb. 1838. p. 107. Mr Blyth states that all «GENERA of» birds in «N.» America & Europe, which have

220b199 ome general statements about mundine & confined GENERA.-- Lyell has remarked about no confined specie
221b201 absolutely fresh water.-- origin of Fresh-water GENERA? The absence of lime in Plutonic & Volcanic ro
222b202 es with> according to Beck has none recent, yet GENERA same.-- Speculate on multiplication of species
226b221 ierra del Fuego like that of North Europe, many GENERA & few species. The number of genera on islands
226b221 urope, many genera & few species. The number of GENERA on islands & on Arctic shores evidently due to
226b221 lits & grows larger species are formed of those GENERA.-- & hence by same chance few representative s
226b221 big continent many species belonging to its own GENERA Therefore if in small tract we have many speci
227b224 not be predicated of <genus.> structures in two GENERA.-- <we then cease to know the steps.> although
227b225 being   known.-- It explains the blending of two GENERA-- It explains typical structure.-- Every speci
228b229 hat are the laws of life".-- Where we have near GENERA far back, as well as at present time, we might
228b229 onus (??) which now run together, were not both GENERA formerly abundant. Seed of Ribston Pippin tree
232b245 here ought to be fewer species in proportion to GENERA than in present seas, «All» The <one> species
234b261 mith, he says, is deeply [not located] of <all> GENERA, «in all classes» are not a few only cosmopoli
234b261 classes» are not a few only cosmopolites, & in GENERA peculiar to any one country do not species gen
235b273 -- Would it not be possible to work through all GENERA, & see how many confined to certain countries-
236b280 ts «in England» surely it is not-- intermediate GENERA we might expect.-- Lindley Introduct Dict. Sci
240c003 skulls so different, that they would be called GENERA., yet retains markings of wings like the wild
241c015 ny in common ¿species? with New Guinea.-- Many <GENERA>. kinds common to New Guinea & rest of isle in
249c035 animals. Follow this out, where species of same GENERA in two words. have not species, generally wide
250c039 of islands.-- ¿Europe has many species but not GENERA distinct from rest of world??? Lyells Principl
252c042 & fine, views about Species-- MUST BE STUDIED: GENERA founded in nature [not located] The systematic
257c058 es & quadrupeds, but how there come to be, many GENERA of fish &c &c at present day.-- It is ASSUMPTI
261c069 of one family sending out structures into many GENERA.-- like Synallaxis or Marsupial animals of N.
263c074 Zealand, a division of nature of Apterix, many GENERA & species-- The believing that monkey would br
268c096 a storge From what I see of S. American birds. GENERA blend into each other in very same district.--
273c110 xcept that fossil forms would generally fill up GENERA & not species, which is not true, with shells?
277c126 be mark to every species. only known by analogy GENERA of course distinct. analogy from every country
278c128 al, & is judged solely by comparison with other GENERA in other families.-- it will however be much <
278c129 has travelled this will be most difficult. Sub-GENERA so far may be eliminated. where every species
281c137 f such a naturalist "splitting up his species & GENERA very finely" show how arbitrary & optional ope
281c139 r certain organs not being fixed, <whi> in some GENERA, which are most fixed in others. In analogy it
282c141 them. they might be supposed to change. & make GENERA.-- let short billed one, be exaggerated, & all
283c145 y have progeny with species & there will be two GENERA.-- will be produced. As we know from Ehrenbergh,
283c145 be exaggerated, & all rest destroyed far remote GENERA, is probably to add to quantum of life possibl
283c145 ATION.-- <On> The end of formation of species & GENERA.-- Now are the characters which unite these of
286c156 ays nat: fam: of Willows contains many Linnaean GENERA & classes. if extinct forms were all fathers o
291c167 S. America) is absolutely necessary to explain GENERA are large probably only few of extinct forms h
292c168 xtermination of some forms with his views.-- as GENERA of Pachydermata &c & other Mammals.-- «otter;
295c178 'orbigny on Plastic Clay of Paris contains many GENERA ·:circumstances not favourable to many species.
297c185 vary much in character, divided into many small GENERA may sometimes be due to accident as submersion
302c203 nly in most discordant groups. The formation of GENERA-- whatever the cause is. <the> any osculant sp
302c203 ed, & N & S. existing species becomes father of GENERA. if Mammalia are adduced. say oh look to your
303c205 ill, tell us. when group is true,» there are no GENERA & classes. p. 479. fragment of tusk «& Molar t
307c216 n general argument is extended from species to GENERA-- whatever the cause is. <the> any osculant sp
308c217 ction of species bears relation to existence of GENERA &c &c Two savages, two species.-- «discussion
309c220 subsidence marked out by animals of same <spe> GENERA. is not equal to areas of elevation: Marked ou
309c220 Marked out by existence of elevated «extinct?» GENERA of shells.-- duration in two classes however d
310c223 now thought.-- N. American, European & Chinese GENERA & some species on Himalaya.-- some English bee
314c238 . As in N. Zealand-- Some species of Australian GENERA Some species same (Palm & Phormium tenax) as i
314c239 Zealand & Australia, some SPECIES of Australian GENERA. good case. rather large flora. (150?) Mr Brow
316c244 nts. of Marine kinds, there are some restricted GENERA, but then they appears always very small ones
316c245 tricted ranges than other plants.-- Many «Some» GENERA confined to hot countries & many to cold.-- He
328cIBC relations with a living Matica & many shells of GENERA Corlula Cham. Cardium. Porcellus Turbo. Cerith
339d025 good.-- Thinks Temmink doubtful when he says No GENERA.-- thinks there are some small divisions.-- do
349d053 r subgenera will come from. common stock.-- all GENERA, common stock.-- so that value can only be jud
350d053 other subgenus,,--> Propagation, best rule for GENERA, & so mount upwards.-- «judged by analogy»-- C
366d111 as common to many good species; & therefore to GENERA (& the uncles & aunts) & therefore does not te
366d111 inst transmutation of species-- Will it against GENERA.-- How long will the wretched inhabitants of N
368d114 her alike or very different in recently altered GENERA. Guinea Fowl & Peacocks.!!» other birds displ
373d133 ds. p. 411 -- Decapod Crust in Muschelkalk, & 5 GENERA of reptiles.-- <Mr p. 417. Magnesian Limestone
374d133 s trees & Lycopodiums.-- p. 437. Many. existing GENERA of shells in the mountain limestone (how diffe
374d134 do. p. 461.-- Lower Silurian-- several existing GENERA. Nautilus turbo. buccinum. turritella. terebra
399e011 marl from «Lake» Constance species of Europaean GENERA-- Hope has ideas about generic characters. d
402e020 , for the most part, with the same families and GENERA, that are natives of S. Asia, but many of the
407e042 highly necessary to show that if species fall, GENERA must. Lesson I remember says Mariana Deer very
408e046 over Lamark surprised to see how many Tropical GENERA come from New Holland, ¿Sydney? The dog being
410e053 rdson.-- The relations of numbers of species to GENERA &c &c can never be told, without species being
425e104 ch Islds. he believes, there are, many cases of GENERA peculiar to the group having species peculiar
435e130 y artificial, so that even some species only in GENERA <are> have this structure.-- Some willow trees
436e136 or instance pidgeons-- : then comes question of GENERA It certainly appears that swallows have decrea
449e170 islands N. of Timor are allied to the «type of GENERA in» islas de Sonda as well by those which are
449e170 different isld. is the same as that which gives GENERA.-- <it is not transportation> now in case of l
471tf08 ea" Linn. Trans. 18. p. 163. "D. Dod on two new GENERA of coniferae".-- referring to the 3 main divis
478z003 57 degree only inhabitant crust Entomost of the GENERA-- Cyclops p. 134. and p. 115 In white Cape Pid
497q06v about OEnothera.-- (14) Examine pollen of those GENERA of which wild hybrids have been formed. (15).
497q007 portant, as furthest removed possible point.-- ¿GENERA in intermediate country (2) Any known changes
497q007 t to cross, are there unusually many species in GENERA of Leguminosae.-- Herbert explains numerous sp
500q10a n New Zealand, plants so few-- Range of mundane GENERA, «in Birds» in accordance with range of specie
500q10a About <endemic &> wandering species of confined GENERA By my theory in volcanic or rising isld, there
501q10a America.-- Heaths in Africa; Hooker? are these GENERA less difficult, in other countries, where spec
509q017 d's observation holds good, that in the mundane GENERA, the species <are> have wide range-- How is th
543m098 there is Intellectual in human-- probably some GENERA in different orders more advanced than others
639j28v of which we have manifold traces in the several GENERA of Grallae Suppose six puppies are born «& it
641j29v n the Wealden? In the strongly separated Arctic GENERA, there is evidence of antiquity & extinction o
025r019 may believe them applicable to the world.-- My GENERAL opinion from the examination of soundings, fro
032r041 xp> does not explain cleavage lines./ possibly GENERAL symetry of world.-- I feel no doubt. respectin
035r046 think, little ground for skepticism, as to the GENERAL truth of the proposition."-- If such can happe
044r073 cience. that Europe was its birth place.--Some GENERAL reflections might be introduced on great size
044r073 an; especially Pacifick: insignificant islets--GENERAL movements of the earth;--Scarcity of Organic r
045r075 ariably discovered; this may be mentioned with GENERAL slope of the country; (perhaps generally over
045r076 severe Earthquakes «1822¿ 1835?» alone, (& the GENERAL belief in N. Chili, ‹where rains are so infrequ
047r081 in absolute movement» Moreover wave «with same GENERAL character» reaches far beyond coast, which has
052r099 k as if some peaks elevated.-- Greywacke. as a GENERAL fact absent in T. del Fuego, excepting in Port
056r111 er: & products being similar over whole world, GENERAL circulation. But Volcanic action separates som
066r142 of third Silurian division--Together with same GENERAL character of fossils deception complete.= Sill
073r162 ies. Study well products of Solfataras[.] some GENERAL laws. association of lead & silver. Sulp. of B
076r168 uor spar only in certain mines. «Vol. III» "In GENERAL it is observed both in Mexico & Peru, that tho
080r180 cally like grease & mercury [blank] NB. P. 73. GENERAL reflections on the geology of the world P. 14-
086a005 ts & rodentia In Plata Mastodon Toxodon Is the GENERAL saline tendency of America connected with its
103a065 dies of vapour. according to Hopkins theory.-- GENERAL presence of dikes. argues in favour of pressur
105a070 t there has been no tumultuous rush.-- besides GENERAL improbability. stratification. If chain of lak
115a097 ear-- Obstruction Sound in S. America The very GENERAL absence of fragments «& pebbles» in mica slate
128a128 e matter piled up. over crevice from effect of GENERAL elevation.-- When subsidence takes place.-- Mo
152g049 de of river & bed of river about 40 ft beneath GENERAL plain. 30.127 A 72 degree Air 65 degree? at le
160g092 Air 70 degree? Isthmus broad flat peat mass-- (GENERAL character in these mountains & not ridges) bet
199b118 aces, without the occurence of one of the most GENERAL laws of life, the transmission of a fortuitous
220b199 paper be published it will be curious.-- Some GENERAL statements about mundine & confined genera.--
222b204 sides one in Latin one <of> «on» Madeira-- any GENERAL observations-- difference of species between l

224b216 formations gone on creating animals with same GENERAL structure.-- miserable limited view.-- With re
234b256 -- He says Stephens say he can at once tell by GENERAL colouring a group of Nebria complanata from. D
236b280 ecies to vary a little, but such should not be GENERAL circumstance.--In. insects «in England» surely
245c023 1. Sus papuensis «partly domesticated» like in GENERAL appearance to Siamese kind.-- but considered g
255c052 ea of a bird-- An animal with skeleton of such GENERAL forms.-- The hybridity of ferns bears on my do
258c059 h extreme slowness, even where isolation, from GENERAL circumstances effecting the area equably.-- An
263c077 no exception.-- he possesses some of the same GENERAL instincts, <as> & <moral> feelings as animals.
274c119 indication [not located] alone, but on all the GENERAL arguments-- Lamarck was the Hutton of Geology.
284c149 eckled form of young blackbird. good remark if GENERAL.-- Where any structure is general in all speci
284c149 od remark if general.-- Where any structure is GENERAL in all species in group we may suppose it is o
290c162 n∥ «Hence «method» of generation is very good «GENERAL» character in those animals, where much change
292c171 rsity & therefore would banish individual, but GENERAL ones might yet be transmitted.= Memory springi
298c190 soning in lower animals many times produced, a GENERAL tendency produced. Such as man getting habitua
298c190 abitually kind to children,--<get> «increases» GENERAL instinctive feeling.-- There is great difficul
306c212 essities teach it taste, but that is much more GENERAL argument) & therefore down the stream followed
307c216 austellata & mandibulata.--!! --Argument, when GENERAL argument is extended from species to genera &
345d043 t like their mothers believes this resemblance GENERAL. ¿depends upon mother bein[g] oldest breed?.--
360d089 is trouble Yarrels law. chief trust must be in GENERAL knowledge of breeders, where their interest is
360d090 den is from <bitch dog do> father dog. & hence GENERAL appearance of face & tail somewhat like dog--
367d113 unter's Animal OEconomy p. 45 "One of the most GENERAL marks is the superior strength <of> «of make i
386d166 The sympathy of part is probably part of same GENERAL law, which makes two animals out of one & heal
425e104 very similar to Galapagos-- study Flora-- what GENERAL forms.-- are the Labiata nearest to American,
484z014 n Ornitholog must be studied before writing my GENERAL account-- ¿ Do not the Penguins replace the <A
484z016 yurus, by Maldonado creeper of same plumage.-- GENERAL red mark on wings of all-- Spix has described
494q004 with respect to conditions of animals & their GENERAL healthiness-- Fox's, Bears Badgers,-- How few
509q017 ry?) in diff. countries in same races Mr. Gray GENERAL Questions (1) Particulars about Sierra Leone.
520m002 resemblance was in odd twiching of muscles, & GENERAL manner of holding hands &c &c.-- Mr Dryden Co
522m012 These foregoing cases of «mental» failure very GENERAL effect of <early> «slight habitual» intemperan
568n017 t is action which experience shows will be for GENERAL good, or in case of any fantastic custom» «Pro
571n027 in human mind. is to me clear evidence, of the GENERAL ideas of our ancestors being impressed on us.-
571n028 ay not our heredetary nature thus acquire some GENERAL notions, which are taste? Real taste in mouth,
572n032 happen.-- Reynolds Works. Vol I, p. 226-- "The GENERAL idea of showing respect is by making yourself
572n033 istinctly. woke instantly but could not gather GENERAL sense of this page.-- Now <awake> «when awake»
572n034 lt on each word separately, neglecting time, & GENERAL sense, anymore than connected with general ten
572n034 , & general sense, anymore than connected with GENERAL tendency of the dream.-- It does not hurt the
577n053 ng drawn,-- My father «even» believes that the GENERAL talking about any disease tends to give it, as
583n069 not met with in the lower animals.-- hence the GENERAL aim of fable, & expression as cunningness of f
591n099 er in different animals of same species.-- The GENERAL «(as I believe)» contempt at suicide. (even wh
602o11v obscura &c a Poussin.-- How are my ideas of a GENERAL notion of everything applicable to the high id
608o026 use motive power changes with organization The GENERAL delusion about free will obvious.-- because ma
609o030 f happiness we must look far forward «& to the GENERAL action» -- certainly because it is the result
612o035 juices. arms of polypus, show either local or GENERAL will, & stomach likewise «does». [LHC] ¿ in Co
614o037 ated by vast power of memory, reason &. & many GENERAL instincts, as love of virtue, of association,
615o038 fied by heredetary;-- so succession so perhaps GENERAL ones.-- Parental feelings weakened in Otahiati
615o038 l» man; and as all men nearly same species, so GENERAL instincts nearly same; which same argument pro
615o038 peculiarities of external configuration. [RHC] GENERAL-- Instincts, certainly appear a sort of acquir
615o038 s, conscience This «X» memory especially «the» GENERAL kind taking pleasure in virtue because acquire
622o049 decrease, I should think they were very few & GENERAL in their nature.-- So that we have some, it is
624o051 ducation, of parents strives* to same end.-- & GENERAL actions of community must frequently teach sam
628o054 nduct, & if taught rightly, it will be for the GENERAL good, that is, the same cause, which gives the
629o55v it is desirable to be taught,-- all are agreed GENERAL utility (3) It is other question whether any t
632j53r for transportation. it follows from some more GENERAL law.-- [that the laws of propagation, were cre
633j54r -- All such facts are merely relations of one GENERAL law. the plants were no more created to arrest
421e090 the young of Syngnathus.= curious as showing GENERALITY of law. even in fish: ⅄.do. p. 236-- on Hybr
204b141 or that thing.-- There is some much higher GENERALIZATION in view. In Marsupial division «do» we not
388d170 at latter carries with stock of food.-- the GENERALIZATION begins low.-- it goes through transformati
411e055 to study the physical causes. «All Cuviers GENERALIZATION, of teeth to kind of extremities come unde
064r135 nks them same as recent species. -- May I not GENERALIZE the fact glaciers most abundant in interior
581n063 ry habits." very clearly, all I must do is to GENERALIZE it, & see whether applicable to all cases.--
581n063 ts, be considered. as grand step if it can be GENERALIZE.-- The tastes of man, same as in Allied King
411e055 -- by this means the laws probably would be. GENERALIZED, & afterwards by the examination of the spec
425e104 ant. (Sicily exception)-- see if this can be GENERALIZED.--» islds., have peculiar forms.-- on the so
243c020 nd fish very similar.-- NB. Lesson method of GENERALIZING without tables or references highly unphilo
576n048 n of <intellect> «instinct»-- an unfolding & GENERALIZING of the means by which an instinct is transm
325c267 rms compared with African <A> Annals Histoire GENERALLE et particuliere des Anomalies de l'organizati
025r016 leagues sea deepens suddenly. coast of Brazil GENERALLY.-- Mrs Power at Port Louis talked of the extr
025r019 et good» «(I suspect fragments of shells will GENERALLY be found to be old & dead)» «(I have not kept
030r037 ment for rise In Cordillera, the dikes do not GENERALLY appear to have fallen into lines of faults I
036r051 --(the sound was long & prolonged). NB, Is it GENERALLY known. the acute chirping sound produced in w
038r059 elted rocks to those parts. Are not the dikes GENERALLY vertical? if so posterior to elevations? & ne
042r067 De la Beche). Ask Capt. F.: R: how the swell, GENERALLY & during gales would tend to travel on a <me>
045r075 d with general slope of the country; (perhaps GENERALLY over whole world) Yet eruptions <both> at sea
050r091 adistinction to sand?? B. Roussin states that GENERALLY in North part of Brazil. <gravel becomes> san
062r129 : at sea so commonly seen. at long distances; GENERALLY first arrives:-- New Zealand rats offering in
103a065 action. (effect of horizontal movement) hence GENERALLY intersect metallic dikes: It is an important
104a066 e 80/100 X 6/100 = 480 In Falkland islands. & GENERALLY where rock metamorphic & thickness of <strata
105a072 nds on serpentine course.-- the latter (it is GENERALLY said) is consequence of <rapid> slow course,
122a113 level.-- {P} will point {P} be the one which GENERALLY yields.-- Will this not explain littoral moun
145g001 good account of ice.-- C. Darwin A. Glen Roy GENERALLY received opinion that male impresses offsprin
156g068 nnot <see> «make out» composition of shelves: GENERALLY angular except near head of valley fragments
165g125 most like Mother in face-- asked stated this GENERALLY the case Wednesday 12/ & 3/ Why is the Tetrao
188b069 to one line Dr Smith says very. close species GENERALLY frequet slightly different localities, so tha
196b107 Ed. New. Philosop J. No 3. p. 207 "It is not GENERALLY known that Ireland possesses varieties of the
198b120 cy to imitate one of the parents; repugnance «GENERALLY» to marriage <if offspring not fertile> <,but
215b177 anthairides and milk--Fox tells me that it is GENERALLY said.= How came first species to go on.-- The
234b261 ra peculiar to any one country do not species GENERALLY affect different stations;-- this would be st
244c021 h quadrupeds» p. 118. wild pigs of Falklands, GENERALLY "red of brick" hair, very stiff, p. 120-- Coa
247c028 Chamisso in Radack isld.-- p. 69. Sharks very GENERALLY distributed: Mem of great geological age-- Ga
249c035 f same genera in two words. have not species, GENERALLY wide range? Mice.-- Waterhouse's remarkable f
257c058 e «not only» about beginning of animal life.: GENERALLY, but even about great division, our <only> Qu
265c087 ugh clearly not of the use to which wings are GENERALLY applied.-- Therefore argument not destroyed e
266c091 ot located] Musalman's of the Peninsula, are. GENERALLY speaking, a much fairer race than the Hindu's
266c092 1 Wellsted. Memoir on isld of Socotra. Cattle GENERALLY marked like those of the Alderney breed. but
269c103 fection may be talked of with respect to life GENERALLY.-- where <">unity constantly develops multipl
273c110 ad one to suppose, that species in same group GENERALLY contemporary «++.» «This would lead one to ex
273c110 ld lead one to expect that fossil forms would GENERALLY fill up genera & not species, which is not tr
274c114 l developed. not confined to one species, but GENERALLY small genus. ¿are there not many ground parro
304c206 h SOMETHING.? that intermediate. species have GENERALLY perfect organs. do changes of habit affect pa
308c219 o I mean that ideosyncracy of wild animals is GENERALLY different, because their difference. arise a
318c254 a) migrate singly flying few miles every day «GENERALLY by night» -- other birds which is strictly di
345d043 rly Journal of Agriculture p. 367. Dec. 1837. GENERALLY-- received opinion that male impresses offspr
369d115 stances which the Mammalia of N. S. Wales are GENERALLY compelled to traverse in order to quench thei
374d134 ertility of hybrids inter se, the first cross GENERALLY brothers & sisters, & therefore somewhat unfa
389d173 tter consists of infinite number of globules: GENERALLY sufficient for one birth or rather]CD ⅄It sho
391d174 rodite being simple (Are not Coniferous trees GENERALLY dioecious oldest forms) why are twin in man m
402e019 p 10» the joints near the tip of the tail are GENERALLY crooked, as if they had been broken". are bor
415e070 f any law can be made out, that varieties are GENERALLY additive, & not abortive: with reference to t
434e129 ught to be different species. Lychnis dioica, GENERALLY dioicous. yet parts only very slightly aborti

447e165 an be raised in gardens.-- Poa alpina, thougt GENERALLY vivaparous sometimes seeds.-- ‖There are endl
460t013 mon.-- the half of the cross, as above, take «GENERALLY» after the swan-gander. one of these half-bre
466t093 y.-- Is not this so in Kidney Bean. How is it GENERALLY.-- In Azalea <do> <it is so» «Though I saw no
492q002 ooecious-- loosing one sex & not other: which GENERALLY fails first?-- Mal[e] 10. Henslow says semi-d
499q010 arts? Don says Irish. Scotch & English plants GENERALLY distinguishable.-- What structure of seeds.--
501q011 s in another region-- (27) Which sex in Mules GENERALLY fails-- perhaps indexed by secondary characte
508q016 race in germ. (2) Any more cases of diseases, GENERALLY occurring in man being transmitted through fe
508q016 ing the worst =or in dogs= (12) Do Hottentots GENERALLY resemble each other very closely, more closel
513q21. s Zostera-- Knights notion of pollen & stigma GENERALLY not being mature at same time on same plant -
523m017 on suddenly. Ans no.-- because often, if not GENERALLY, does not really come on suddenly.-- Case of
523m018 ined by remonstrances on him» which are never GENERALLY, if at all discovered.-- <Sup> Sometimes come
570n026 idea of beauty, that which we have been most GENERALLY accustomed to:-- analogous case to my idea of
607o025 an in somnambulism or insanity. free will (as GENERALLY used) is not there present, but he acts from
609o030 ertainly because it is the result of what has GENERALLY been best for our good far back.-- (much furt
619o043 e of passion. he would feel pain, which would GENERALLY be anger, as he would be tempted to interfere
619o043 xercise & its love of its puppies: the latter GENERALLY soon conquers, & the dog [RHC] 3) probably th
356d070 mpletion.-- ‖Say my Grandfathers expression of GENERAT. being highest end of organization good expres
189b072 mer periods & then splitting off.-- If species GENERATE «other species», their race is not utterly cu
224b214 stinct from Lamarcks» Without two species will GENERATE common kind, which is not probable; then monk
441e151 be animals as Coralline, or others. which only GENERATE once in a thousand generations.-- any amount
046r079 : «Mr Lyell considers that Plutonic rocks are GENERATED as often as Volcanic. I consider latter as ac
176b021 .-- «as many terminal buds being, as new ones GENERATED» There is nothing stranger in death of specie
189b072 on.-- otherwise all die.== The fossil horse, GENERATED in S. Africa Zebra.-- & continued.-- perished
198b115 creation why not produced on New Zealand; if GENERATED <No> «an» answer <could> «can» be given.-- It
292c169 large probably only few of extinct forms have GENERATED species. & of 100 extinct species the greater
441e151 are plants, in northern latitudes, which are GENERATED by buds alone or roots, & never flower, so th
609o030 Gives art to whom I say How social instincts GENERATED? The origin of the social instinct «in man &
170b001 n beginning of February Zoonomia Two kinds of GENERATION the coeval kind, all individuals absolutely
171b002 lived Why is life short, Why such high object GENERATION.-- We know world subject to cycle of change,
171b003 s produced by buds are constant, hence we see GENERATION here seems a means to vary. or adaptation.--
171b004 rganization» thus being modified,-- therefore GENERATION to adapt & alter the race to changing world.
171b004 the race to changing world.-- On other hand, GENERATION destroys the effect of accidental injuries,
171b005 l cause of life With this tendency to vary by GENERATION, why are. species are constant over whole co
180b036 FINITE LIFE I think {P} Case must be that one GENERATION then should be as many living as now To do t
181b039 - or we may suppose only each species in each GENERATION only breeds; like individuals in a country n
187b063 out they change; like Golden Pippens, it is a GENERATION of species like generation of individuals.--
187b063 n Pippens, it is a generation of species like GENERATION of individuals.-- Why does individual die, t
190b078 iven (& power of adaptation) is given by true GENERATION, throughe means of every step of progressive
195b101 animal be created, then by the fixed laws of GENERATION, such will be their successors.-- let the po
197b111 concluded so= Evidently «or hints» considers GENERATION as a short process, by which man «one animal
220b197 yet organization «especially» connected with GENERATION certainly is.= The dislike of two species to
224b212 tely» distant countries. there is no test but GENERATION. «(but experience according to each group)»
225b219 in offspring? if so adaptations of species by GENERATION explained? NB. Look over Bell on Quadrupeds
227b227 speculation & line of observation.-- View of GENERATION being condensation, test of highest organiz
227b228 ances favour crossing & what prevents it--» & GENERATION, causes of change «in order» to know what we
228b230 berry produced by seed«s»??-- Universality of GENERATION strongly shown by hybridity of ferns.-- hybr
248c033 t Owen [re]markable laws of Brain & manner of GENERATION «& primary divisions of insects» 2. Relation
248c034 ore & more impressed in blood with time, then GENERATION will «only» produce an offspring capable of
254c047 p 389. Owen remarks on Entozoa, the organs of GENERATION, afford the least certain indication of the
254c048 lly believes in this view!!! p. 392.-- except GENERATION & digestion in Acrite Kingdom all organs ble
254c049 dia spicule in sponge. stomachs in infusoria, GENERATION in each joint of taenia worm.-- formative en
254c049 ts). (NB These views must lead to spontaneous GENERATION??) This whole Paper Must be studied.-- <Thre
255c051 hickens) Young water ouzels hence aversion to GENERATION, before great difficulty in propagation.-- F
255c052 ridity of ferns bears on my doctrine of cross-GENERATION. The infertility of crosse & cross, is metho
257c059 &c at present day.-- It is ASSUMPTION to say GENERATION produces young ones capable of producing you
259c066 alled adaptation With respect to my theory of GENERATION, fact of armless parent not having armless c
259c066 offspring (like atavism) & shows my «view of» GENERATION right?-- If puppy born with thick coat monst
265c084 on is determined, when child is.-- shows that GENERATION implies more than mere child, but that child
269c102 e universality of latter render-- spontaneous GENERATION not improbable.-- After reading "Carus on th
270c103 turned inside out. have position of organs of GENERATION!!!. Mem. Agaziz. (INo Annals of Nat. Hist) s
276c122 ion is infertile offspring. without organs of GENERATION?! By profound study of local varieties laws
280c136 unlike can be produced, yet to produce whole GENERATION unlike would go against the tendency.-- it t
284c149 ay letter to Fleming p. 32 "where it (mode of GENERATION) varies according to the species, it is mani
289c162 ss difficulty, except as distinct creation.-- GENERATION may be viewed as condensor. «+++ must (on my
289c162 t the acts of changing must alter method of GENERATION-- Heaven knows how.-- This reaction takes pl
290c162 akes place in every organ‖ «Hence «method» of GENERATION is very good «general» character in those an
290c162 rtebrata & plants. At first classification on GENERATION might appear an analogy NB Pyrrho-alauda (bi
293c172 nderful. that it should be remembered in next GENERATION. [NB what are those Marvellous cases, when y
301c200 of. same class being about equal.-- organs of GENERATION about equally complicated.-- An Entomologist
306c213 na‖ people will say. not species.-- organs of GENERATION a captial character. (Owen) not for first &
313c236 actions in foetus of Mammalia, or chick eat») GENERATION becomes necessary, when organs of parent are
313c236 bly gemmation (.Ehrenberg) --not necessary to GENERATION (lateral with no relation to time) as in bud
323c269 n 3? vols. oct: -- 26th Blumenbach's Essay on GENERATION. Englis Transla -- The Revd. A. Wells. Lectu
335d016 g to the corelations of system, the organs of GENERATION would necessarily fail.-- In last page. I sh
336d017 their whole <body> form of body gained in one GENERATION, so it is impossible to transmit them, & as
336d019 her's theory of Mules not hereditary, because GENERATION -- highest point of organization] CD» false.
336d019 hey would not have offspring-- On the idea of GENERATION being a <slip> «bud» from parent. if whole p
347d049 brates tendency to improve in intellect,-- if GENERATION is condensation of changes. then animals mus
356d071 spit & throw dust «according to my theory of GENERATION (p. 175) if> 8th Sept Yarrell told me he had
372d129 ry element of the living body is present-- in GENERATION something is added from one part of the body
372d130 -- -- Not this is effected by short method in GENERATION.-- Ehrenberg considers artificial division o
372d130 pair wounds-- but this has nothing to do with GENERATION.-- Why crab can produce claw. but man not ar
372d131 orn. but it would be very different from true GENERATION.-- there is no caterpillar state; the vast d
372d131 ar state; the vast difference of two kinds of GENERATION shown by their happening in same plant.-- Th
373d132 iew would make every individual a spontaneous GENERATION: what is animalcular semen-- but this-- -- t
376d136 ranis & Spaniards are almost White from first GENERATION., that with Quichuas the American character
379d152 . Archiv. fur. Naturgeschichte. September 11' GENERATION Mr Yarrell says it is well known that in bre
381d156 ll these however do not require coition every GENERATION)-- Epizoa & the nematoid Entozoa-- Therefore
384d161 ength, (p 45) & that strength> In speaking of GENERATION alway put female first Will not even a fruit
387d168 form 2. cubic stone. like that of Billin.-- <GENERATION--> V. p. 152 It is very singular the same di
389d173 constant necessity for change. in process by GENERATION applies only the more complicated animals.]C
389d176 as told by Willis Q Proved facts relating to GENERATION One copulation may impregnate one or many of
389d177 most important with regard to theory, showing GENERATION connected with whole system, «as if there wa
389d177 bortive as far as parturition is concerned.-- GENERATION being means to propagate & perpetuate differ
390d178 be similar to budding. which is not object of GENERATION.-- therefore passions fail.-- In fruit trees
390d179 aried from parent stock.-- The very theory of GENERATION being the passing through whole series of fo
391d174 e something exact time» added to that kind of GENERATION, which typifies the whole course of change f
391d174 rovement or deteriorate» that object failing. GENERATION fails.-- How completely circumstances «alone
393dIBC rn in succession» which is not transmitted by GENERATION?? Is it «chiefly» in high. bred dogs ie. (br
400e013 er Europe same year.-- he sowed them for four GENERATION before they broke.--, showing effects of cul
417e077 n grafting & sexual union-- Looking at simple GENERATION as being the action of two species in one bod
418e083 the atom, to make it alive, & how the laws of GENERATION were impressed on it.-- Seeing that <Man> «a
440e146 thout final cause. either in present, or past GENERATION.-- thus cabbages growing like Nepenthes.-- c
440e148 ings under united elytra The law of <growth» «GENERATION» is only modification, though important one,
441e148 which they are grafted.-- No than by growth-- GENERATION.; & more of the effects of conditions on the
442e151 ce in a thousand generations.-- any amount of GENERATION may take place by gemmation «My theory will
443e154 ed, then the conclusion that the two kinds of GENERATION have some most important difference is force
446e160 s quite untouched the question of spontaneous GENERATION.-- Introduction to Bartram's Travels p. XXII

```
495q005 ts in Hothouse & see what effect on organs of GENERATION (5) Place pollen of Red Cabbage «mixed  with
495q05a to  answer Wiessenborns doctrine of Equivocal GENERATION Charlworth p. 377. Have paper ruled in squar
521m007 e unconsciously of it, surely memory from one GENERATION to another, also without consciousness, as i
552m132 ions which have been found necessary for long GENERATION, (as friendship to fellow animals in  social
614o037 derful that memory should be transmitted from GENERATION.; than from hour to hour in <man:> individua
171b003 on.-- Again we <believe> «know» in course of GENERATIONS even mind & instinct becomes influenced.-- c
171b004 ivilized man.--birds rendered wild <through> GENERATIONS, acquire ideas ditto. V. Zoonomia.-- There m
183b022 r, cessent d être feconds. dès les premières GENERATIONS" go back to type of either animal when cross
239c002 ties. He has no doubt that Chesnut, for many GENERATIONS back, were crossed with Bay mare, only bay a
239c002 , were crossed with Bay mare, only bay a few GENERATIONS, that offspring would be chesnut.-- On  this
263c078 t &c & perhaps a train of animals of hundred GENERATIONS of species to produce contingents  proper.--
292c171 only be those actions, which Many successive GENERATIONS are impelled to do in same way-- The improve
299c194 tris be sown in dry station it will for some GENERATIONS come up so.-- there are not Many intermediat
313c235 (Magazine of Zoology & Botany) where several GENERATIONS are produced in succession (13?) without imp
336d018 animal  &» by a succession of <such changes> GENERATIONS, these small changes become multiplied, & gr
362d100 ious to show that in sixty years-- (how many GENERATIONS) the strangest peculiarities have been  kept
376d136 ore tenacious. & does not disappear for Many GENERATIONS Sept. 29th Dr. Andrew. Smith «Remarks on ext
384d161 not handed down. but is handed down for some GENERATIONS Theory of sexes (woman makes, bud, man  puts
387d168 liarity has a tendency to descend to several GENERATIONS» If A & B be two animals which have some pec
390d178 added  or substracted at each, or in several GENERATIONS, the process would be similar to budding. wh
390d178 imals?]CD If the male «in the course of some GENERATIONS» has gained some difference «from what it re
400e013 cultivation gradually adding up. & four more GENERATIONS before they began to double.-- at present ti
427e111 ion.-- Now Sir. J. Banks. says Zizania in 16 GENERATIONS did become, acclimatized. & says Laurels hav
428e112 r otherwise breeders who only care for first GENERATIONS,, in horses, would not care so much about
434e127 ncy (as effects of cultivation on successive GENERATIONS of plants) to do so, the effects are equally
442e151 hers. which only generate once in a thousand GENERATIONS.-- any amount of generation may take place b
442e152 uction of species may stop for any number of GENERATIONS-- Gorze in Norway, which never flowers!!-- <
462f205 traces of its parentage in «about» seven «7» GENERATIONS.-- so many!! Hensleigh objects to  transmut.
465t079 t-- monkey-man, valueless.-- May not several GENERATIONS have been confounded in the caves? It is hig
494q004 edetary. (6) In the last 1000 years how many GENERATIONS of man have there been.-- on what principles
494q004 les calculated.-- in order to guess how many GENERATIONS in Mammalia. in group effect of  crossing.--
495q005 ll plants abort?, does it require successive GENERATIONS to accustom them to such soil.-- Sow weeds i
496q006 what  would be effect-- (10) Try in how many GENERATIONS. daisy. Fever-fuge Groundsil.-- gilly flower
503q012 36) Ask Mr Gowen to ask Mr Herbert, how many GENERATIONS any hybrid has <been> reproduced itself.-- A
514q021 s (1) Are sterile hybrids healthy: number of GENERATIONS: about crossing of plants; especially Papili
515q022 d Cover common Pea (& Sweet Pea) for several GENERATIONS under net & see if get sterile-- Cover  that
525m026 lthy ones.-- Insanity & Epilepsy remain many GENERATIONS in families.-- My fathers does not know whet
623o050 e one «person» in early infancy, during many GENERATIONS giving love of mother; the having received s
623o050 ceived some advantages from man. during many GENERATIONS giving the social feelings.-- [LHC] Accordin
623o051 es instinctive, which is repeated under many GENERATIONS. (& under unknown conditions) (for pig  will
626o052 ad to similar actions which in prior <races> GENERATIONS led to their formation.-- 'N.B. If feeling o
248c030 state (where instinct not interfered with, or GENERATIVE organs affected as with plants) no animals V
299c195 ery important in classification. here we have GENERATIVE organs. first character.-- In dioecious plan
336d017 herefore mule has no offspring & therefore no GENERATIVE organ.-- Same Prop. better enunciated.-- "An
372d129 plants  does not whole individual change into GENERATIVE organs?]CD it is of no consequence if it doe
421e091 in abstract» in his paper on the Dugong, "The GENERATIVE organs being those which are most remotely r
501q011 nnuals to place in Hot house to see effect on GENERATIVE organs of great Heat (32) Can Henslow ask qu
284c149 est subject to variation.-- + <being> good for GENERIC divisions [+] ought genus to be founded on suc
313c234 ific change.? <One» the first would be called. GENERIC & other specific extinction-- In the Entomostr
399e011 s of Europaean genera=.-- Hope has ideas about GENERIC characters. dominant. predominant &c having re
430e117 ew stocks. does not bear, the least on ancient GENERIC forms.-- the animals in Eocene period could no
501q10a nt for it wd show that such variation is not a GENERIC or specific character,, but contingent on coun
558m152 xpulsion of air «dogs snarl much the same way» GENERIC manifestation of great passion.-- I do not thi
272c107 us insects, having spiny legs & running quick & GENERL appearance of blattae other Hemiptera stikingl
525m023 sure in obeying their instincts naturally.-- (GENEROSITY in defending a friendly dog).-- they feel sh
451e179 forequarters: is said to be of a bolder & more GENEROUS temper-- Hodgson Koloff. voyage through the M
367d112 ever  occur.--».-- & if that cannot be formed, GENETAL organs by that co-relation of parts, will  not
089a020 Erratic blocks in Alps. Memoires de la Soc. «de GENEVA» Vol 3' P. II. -- Bed, of elevated shells on t
303c204 for <t> a less time than other. points. female GENITAL organs «in some monkeys clitoris wonderfully p
308c219 gues- case of abortive organs to mules in their GENITALS & even to a limb not used The only cause of s
275c119 it in science--. the highest endowment of lofty GENIUS Using geograph distribution of animals, «as» I
398e004 ogy depends, that most sublime discovery of the GENIUS of man Those who have studied history of the w
543m099 play regularly with D'Arblay of Christ of great GENIUS, & yet invariably used to beat him-- The son o
209b106 s now To do this & to have many species in same GENRES qui, sous ce climat, se divisent le plus aisém
078r175 er E & W.--veins richest not in ravins or along GENTLE slopes. but on the most elevated summits, wher
159g087 vium (?)) still higher Slope of Valley much more GENTLE than in Glen Roy, & partly shut in No  Granite
161g096 en on other side {P} Shelf A Shelf A at head of GENTLE mossy slope, which from a distance hid it, cou
571n031 n language. the possibility of poets describing GENTLE things in gentle language, & vice versa.-- alm
571n031 ossibility of poets describing gentle things in GENTLE language, & vice versa.-- almost proves that a
530m044 f labouring classes are slower than those of GENTLEFOLKS. & that peculiarities of form in trades (,as
523m018 sane of particular ideas, «Case of Shrewsbury GENTLEMAN, unnatural union with turkey cock.-- was rest
345d042 hence  this is then case of avitism.++» Three GENTLEMEN of party all thought with pigs &c, that hybri
153c052 collarig  at bend & here most accumulations At GENTLER bends roads disappear The normal condition  of
106a075 of  which limit cannot be great over) with very GENTLY sloping sides This argument is partly taken fr
177b028 w is it that there come aberant species in each GENUS «(with well characterized parts belonging to ea
177b029 have they all one determinate life dependent on GENUS,. that genus upon another, whole class would di
177b029 one  determinate life dependent on genus,. that GENUS upon another, whole class would die out, theref
180b036 s now To do this & to have many species in same GENUS (as is). REQUIRES extinction. Thus between A. &
185b056 new <liza> species, belonging to true. American GENUS Waterhouse says he is certain, that in insects,
186b057 It  is in simplest case saying every species in GENUS resembles each other, (at least in one point, i
202b133 ire on wonder of finding Monkey in France.-- of GENUS peculiar to East Indian isles.-- Compares it to
202b133 -- Compares it to fossil Didelphis (S. American GENUS) in plaster of Paris.-- Now this is exception t
208b154 referred  to by Richardson in Report about each GENUS having its parent type in hotter parts of world
217b187 gave  me instances of one species of Australian GENUS being found in Sumatra; again another of  other
217b187 being  found in Sumatra; again another of other GENUS in Sandwich islands-- A genus with species in V
217b187 another  of other Genus in Sandwich islands-- A GENUS with species in Van Diemen's land and Tierra de
226b222 tself into question of proportion of species to GENUS If on one isld several species of same genus, s
226b222 to genus If on one isld several species of same GENUS, subsided land.-- Mauritius? <In plants where d
227b224 f every organ in A. B. C. three species «of one GENUS» can pass into each other «by steps we see»: bu
227b224 teps we see»: but this cannot be predicated of <GENUS.> structures in two genera.-- <we then cease to
232b249 s in Madagascar, on neighbouring islets & a sub-GENUS in Southern Africa In same manner. Cuscus, (a s
232b249 Southern  Africa In same manner. Cuscus, (a sub GENUS of Phalangista New Holland form) is found in ma
235b275 ar character-- Indian Bull?-- Do species of any GENUS. as American or Indian genus inhabit  different
235b275 Do  species of any genus. as American or Indian GENUS inhabit different kind of localities.-- if so c
251c041 of  wild & tame horses.-- p. 246-- Gmnira-- new GENUS of Mam: found in Sumatra p. 452 Append to Denha
256c054 ted for a moment,-- Galapagos, S. American-- -- GENUS.-- The circumstance of having two sexes is  the
263c076 the  difficulties Once grant that «species» one GENUS may pass into each other.-- grant that one inst
264c080 om  different localities, as my Furnarii.-- some GENUS of yellow & brown-breasted bird in Australia &c
264c080 rows. yet all forming, according to Gould, good GENUS Gould seems to doubt how far structure & habits
268c096 isputes this)> says differently) do. do. on the GENUS Procyon.-- by Wiegman Classified catalogue of a
274c114 ot confined to one species, but generally small GENUS. ¿are there not many ground parrots? are  there
278c129 ther nearest species of each might not breed:-- GENUS must be a true cleft-- putting out of case  the
278c129 cleft--  putting out of case the Analogys.-- If GENUS does not Mean this it means nothing.--There sho
280c134 instincts  in wild animals, many species in one GENUS-- external circumstances in both cases  effect.
281c138 tendency to lon[g] or peculiar tails strange ¿¿/GENUS only natural from death or slow propagation  of
282c142 -  The same characters which are analogical in a GENUS with respect to rest of its family as in ground
284c149 +  <being> good for generic divisions [+] ought GENUS to be founded on such characters, as do not var
295c183 titut. Sorex from Mauritius. p. 112. & paper on GENUS Magazine of Zool. & Bot.-- Vol I. p. 450. 4 ins
```

303c205 ophical. L'Institut 1838. p. 128. Extraordinary GENUS. Mesites bird from Madagascar uniting pidgeons
312c233 breeding in. Must not trust him Hope says that GENUS of parasite to genus of animals different «+++
312c233 t trust him Hope says that genus of parasite to GENUS of animals different «+++ p 234», different spe
313c234 ransactions Why if louse created should not new GENUS have been made, & only species, good argument f
318c254 of migration quite different in species of same GENUS.» The Muscicapa solitaria stay about a fortnigh
347d050 Smith's Zoolog.-- of Africa -- p. 4. sticks to GENUS or group of any kind not being perfect till cir
347d050 ect till circular. p. 5 Most clearly shows that GENUS expresses as now used almost any group.-- all
349d052 es!! (given in note.)-- Macleay <met> uses term GENUS when it is so many steps from a head, as subkin
349d052 ccording to my theory, every species in any sub-GENUS will be. descended from one stock, & that stock
350d053 t would be exceeding‚ wrong to call,, one group GENUS & other subgenus,--> Propagation, best rule fo
354d065 ogous to specific character of other species in GENUS."-- Is there any law of this. Do any varieties
355d066 to <vary among> be different in species of same GENUS." Law of monstrosity not prospective, but retro
355d067 inct consideration) Now in different SPECIES of GENUS Sus. do vertebrae vary? «See Cuvier Ossemens Fo
357d073 -cock.-- (Curious the readiness with which this GENUS becomes crossed. ¿is red game an hybrid?-- When
381d157 sexes very simple.-- as in plants even in same GENUS some dioecious & some monooecious-- (& cultivat
400e011 ographical distribution-- Thus Hattica is great GENUS.-- because found in all quarters: his ideas not
414e065 y of man is as perfect as the Elephant, if some GENUS. holding same relation as Mastodon to Man. were
472s02v awn it.-- Saw 4 more Bees at work-- another odd GENUS-- & a small common Humble-- & more of same fly
491q01v from their seed 3. To apply groups of different GENUS & then some hours afterwards of nearly related
493q003 is the ground much manured In species of close GENUS do more than three primary colours occur in rel
501q10a lapagos.-- Ireland, doubtful species-- Does any GENUS of Plants. vary & hard to separate specifically
515q021 ut sometimes have dangling ones.-- Is there any GENUS of, «in» which some organ is absent by a
207b151 erres p. 331. L'Institut-- considers that mercu> GEO. Joun. p. 325. Vol. IV. Ducks on rivers in Guian
483z012 88. Domestic mouse of Egypt is Mus Cahirimus. of GEOF. -- reference from Rüppel travels All Owens pape
325c267 nization des Hommes. & les animaux.-- by Isid. GEOFFROY. St. Hilaires. 1832. contains all his fathers
236bIBC oelle. Geol. Soc Horae Entomolgicae Linn: Soc. GEOFFRY-St. Hilaire Philosophy of Zoology Waterhouse B
182b048 ood» collection of insects with this in view.-- GEOGR. Journal Vol VI. P II. p 89.-- Lieut. Wellstec
092a030 artz, after passing sandstone Vol II. p. 69.-- GEOGRAP Journal Earthquake at Melville Isld New Hollan
092a030 ille Isld New Holland Augus 1d to 3d & 19 1827 GEOGRAP Journ There are some ideas about order of inje
202b131 P. I. p. 17 (Lat 37 degree about) Vol IV P. I. GEOGRAP. Journal. Voyage up the Massaroony by W. Hillh
202b132 cks. subsidence appears indicated.--- p. 36.-- GEOGRAP. Journ. Vol IV. P II. p. 160. Melville Isd.--
267c093 th long clotted hair resembling that of goats" GEOGRAP. Journ. Vol VII. p. 216. Mr Bennett Voyage rou
324c268 aun. Borealis Entomological Magazine (paper on GEOGRAP. range Study Buffon on Varieties of Domesticat
068r147 d.) Red Sea near Kosir, land appears elevated. GEOGRAPH. Journal p 202 Vol IV When recollecting Gulf
074r165 on by Mitterschlich. Vol. II Journal of Nat. & GEOGRAPH Sciences? -- H. says in Potosi the silver is
095a037 ebratula from Hudson's Bay. 2. species Vol VI. GEOGRAPH. Journ. Analysis of Poenig Voyage Valparaiso
110a087 ably from the gneiss beds in the mica slate.-- GEOGRAPH. Journal. Vol IV (p 321) Mr Hillhouse describ
111a087 ny observations on heights of valleys in Chile GEOGRAPH Journal Vol. VII p. 216.-- Guava trees, intr
111a088 wenty years since (1835) from Norfolk Isd into GEOGRAPH Journal Vol VII p. 279. Carcases of birds dri
112a089 me speculats on conducting powers of rock-- -- GEOGRAPH. Journal Vol IV p. 36. on subsidence of the la
112a090 1838. p 274. probably will be published in the GEOGRAPH. Journal.--» A meeting of the Geograph Soc, A
112a090 in the Geograph. Journal.--» A meeting of the GEOGRAPH Soc, April 9 1838. Letter from M. Erhman stat
128a128 erence to old submarine orifices in Cordillera GEOGRAPH. Journal vol II. p 89. at Madras. surrounded
131a135 ase to show Sir. J. Herschel's theory wrong.-- GEOGRAPH. Journal Vol. 8. p. 402.-- ground ice-- subte
193b093 on Geographical distribution of animals Brown GEOGRAPH. Journal, Vol I p. 174. says from Swan river l
202b131 wooded district cattle with humps as in India. GEOGRAPH J. Vol III. P. I. p. 17 (Lat 37 degree about)
230b235 cuss those mention[ed] by Lesson & Chamisso.-- GEOGRAPH Journal Vol V. P. I. p. 67. Dr. Coulter on de
248c029 to There is this great advantage in studying. GEOGRAPH. range of quadrupeds.: that either created in
269c101 elena &c &c. Juan Fernandez A communication to GEOGRAPH. Soc. in February or March 1838 on soil in Si
269c102 which, may take place in Metamorphic action.-- GEOGRAPH Journal. vol I. p. 17 &c excellent sketch of
275c119 -. the highest endowment of lofty genius Using GEOGRAPH distribution of animals, <as> I use (new step
325c267 Brutes in Blackwood. June 1838 H. C. Watson on GEOGRAPH. Distrib: of British Plants. Humes Essay on H
337d023 se, elephant, come from one stock.-- Theory of GEOGRAPH. Distrib: of <ani> organic beings.-- Animals
344d040 l epochs.-- There are some admirable tables on GEOGRAPH distribution of reptiles in Suites de Buffon.
191b081 e Mastodons?? Negative facts tell for little) GEOGRAPHIC distribution of Mammalia more valuable than
036r052 tral Patagonia Does Andes in Chili. separate GEOGRAPHICAL ranges of plants. V. Lyell. Chap XI Vol II.
065r139 -The longer diameter of Deception Isl is six GEOGRAPHICAL miles and width 2 & ½ miles S. Shetland. La
123a117 n interior..-- <The theory of [...] .> <The> GEOGRAPHICAL Journal Vol VIII. (1838) p 212. Facts from
163g108 a-- soon decayed on exposure Mr H. C. Watson GEOGRAPHICAL distribution of British Plants Shropshire Q
193b091 ed into list Entomological Magazine paper on GEOGRAPHICAL range Richardson-- Fauna Borealis. It is im
193b092 wainson's trash.-- at beginning of Volume on GEOGRAPHICAL distribution of animals Brown Geograph Jour
195b098 of voice partly American North & South.-- (& GEOGRAPHICAL <distri> division are arbitrary, & not perm
203b137 stance gradual reduction of temperature from GEOGRAPHICAL or central heat.-- But then shells-- Mr Yar
224b212 & hence the importance Naturalists attach to GEOGRAPHICAL range of species.-- Definition of Species:
227b227 .-- With belief of <change.> transmutation & GEOGRAPHICAL grouping we are led to endeavour to discove
263c076 study gradation. study unity of type-- Study GEOGRAPHICAL distribution study relation of fossil with
266c092 2000 years, germinating.--!! Henslow doubts? GEOGRAPHICAL JOURNAL. Vol V. p. 201 Wellsted. Memoir on i
268c099 ked out into. Milvulus [not located] element GEOGRAPHICAL distribution is.-- ¿Pelagic forms similar--
278c131 und in Van Diemen's Land.-- V. ls. Number of GEOGRAPHICAL Journal to discover whether dog found at Sw
311c227 y good on insectiferous <insects> quadrupeds GEOGRAPHICAL range very good.-- Blainville Ovington's Vo
316c244 hers murder.--» good anecdote --.Sowerby.--. GEOGRAPHICAL range of shells like Cryptogamic plants. of
319c257 Australia proper.-- Irish Elk case of fossil GEOGRAPHICAL range.-- [blank] Books examined: with ref:
320c276 e> up to parts published March 1838 Whole of GEOGRAPHICAL Journal Asiatic Journal to end of 1837. re«
324c268 he Congo Voyage Decandoelle. Philosophie. or GEOGRAPHICAL distrib. «in Dict. Sciences. Nat. in Geolog
352d059 tudy the corresponding points.-- The present GEOGRAPHICAL distribution of animals countenances the be
370d115 . Zealand not having any Mammalia.-- Type of GEOGRAPHICAL organization. no more can be said.... In pa
400e011 dominant. predominant & having relation to GEOGRAPHICAL distribution-- Thus Hattica is great genus.
424e100 Institut 1838.-- p. 290-- admirable paper on GEOGRAPHICAL distribution of Crustaceae.-- (I forget whe
482z011 aguay must be most important a discussion of GEOGRAPHICAL distribution of Mammalidae &c &c &c-- The F
559m155 gazine June. 1838. Copied Mr H. C. Watson on GEOGRAPHICAL distribution of British Plants A Volume pub
641j29r study these facts read Lacépède on Cetacea & GEOGRAPHICAL Distrib of larger Seals-- Are Porpoises num
343d037 ".-- » August 17th Two regions may be Zoolo-GEOGRAPHICALLY divided either by developement of new form
461t041 same as some species extending much further GEOGRAPHICALLY than others. Athenaeum p. 605 Mr. Macgilli
423e096 anized to fill their places.-- The Geologico-GEOGRAPHICO changes must tend sometimes to augment & som
193b092 tten on the geography of plants. Essai sur la GEOGRAPHIE des Plants. I Vol in 4 degree.-- I have abst
236bIBC -- Lindley Introduct Dict. Science. Naturelle GEOGRAPHIE Botanique De Candoelle. Geol. Soc Horae Ento
244c021 XXIX tom IV p. 273 2d Edit Consult Latreille GEOGRAPHIE des insectes, in 8 degree. p. 181 «who says
324c268 ustrales. Chapt. XIX. tom IV. p 273 Latreille GEOGRAPHIE des insectes. in 8vo p. 181.-- See (p. 17) for r
191b079 e on Echidna-- Good to study Regne Animal for GEOGRAPHY.-- The motion of the earth must be excessive
193b092 arge quadrupeds--" Humboldt has written on the GEOGRAPHY of plants. Essai sur la Geographie des Plants
252c045 ists get clear indication of circumstances in GEOGRAPHY to help in distinguishing empirically what is
275c119 new step (in induction) as keystone of ancient GEOGRAPHY species tell of Physical relations in time «&
342d036 s, modified by unknown ones. cause changes in GEOGRAPHY & changes of climate superadded to change of
358d074 nearest continent.-- With respect to ancient GEOGRAPHY of Atlantic Tristan D'Acunha ditto Juan Ferna
462t051 every part of England. &c &c NB. In botanical GEOGRAPHY, there can be no sharp division of partition
236bIBC ce. Naturelle Geographie Botanique De Candoelle. GEOL. Soc Horae Entomolgicae Linn: Soc. Geoffry-St.
307c216 Hippotamus from Madagascar!!!!!! Proceedings of GEOL. Soc Vol I It is capable of demonstration that
026r020 ctions on metallic veins. 1830 P. 399.-- Carne. GEOL. Trans: Cornwall «Vol II» It is a fact worth n
026r022 en place in the Cordillera of S. America. Study GEOL: Map of Europe Conybeare. Introduct XII P. sil
033r043 s rather good abstract of Humboldt. S. American GEOL. in Daubeny. P. 349 Admirable little table sho
035r046 t easily preserved.-- Mr Conybeare introduct to GEOLOG--"Between the height of same beds, deposited i
035r047 ewn? by the parallel bands of formations on any GEOLOG Map: Quoted from Daubeny P 402: likewise, mean
036r049 . & part in grand discussion Consult. reconsult GEOLOG. Map of Europe Consult charts for distribution
071r157 rtant There is map of Cordillera by Humboldt in GEOLOG. Society Sir Woodbine Parish informs me that t
085aIFC s far as p 140-- abstracted as far as concerns "GEOLOG Observat on Volcanic islands & Coral Formation
092a028 ertical Linear earthquake 500 by 90.-- in Syria GEOLOG. Proc. p. 541. year 1837 In Upper Assam. Geolo
092a028 Geolog. Proc. p. 541. year 1837 In Upper Assam. GEOLOG Proc p. 566 1837.-- Tertiary <bea> formation t
094a036 ved rocks. Specimen of rock from Costorphine at GEOLOG. Soc: Colonel Imrie Transact Wern. Soc. Vol. 2

Page   ***(Key Word)***
094a036 d Reliquiae Diluvianae p. 201. & seq Murc Trans GEOLOG Soc Vol 2. p 257 {T} The Pota: labiata certain
097a044 er strata alone at Guantajaya contains salt see GEOLOG. proceedings Lake let out by steps in Central
098a044 fs, but certainly probable. Bulletin de la Soc. GEOLOG: 1833-34. p. 35.-- Ancient Lake Lemagne in Auv
098a045 y <of> come of use in discussion on Cleavage &c GEOLOG Transacts. Vol III. p I. p. 86. et p 95.-- It
105a070 xposed. bay more open to turbulence. Bull. Soc. GEOLOG «1837» p. 320. paper on shrinking of Clay. app
108a081 portion page excised, not located] Bull:. Soc. GEOLOG. Tome IX 1837-8. p. 24. rocks of Chimborazo.,
115a097 enslow has deposited specimens from Anglesea in GEOLOG. Soc. if numbered compare them with my rocks.
117a101 ite between 40 & 50 from Portezuelo. Bull: Soc. GEOLOG. 1837. December. p. 91. a classification of Eu
127a127 of subsidence Elie de Beaum. Memoires of French GEOLOG. Cantal Vol III I? p. 246. on formation of con
138a151 ons (1832. or 3?) there is an account of Sellow GEOLOG. Observat. in Southern Brazil. [blank] «p. 4.
201b126 es of ox in N. America: as well as 2 recent See GEOLOG. Proc. p. 569. 1837. Account of wonderfull foss
216b179 n Loudons (analogue of Blood hound-- Bull. Soc. GEOLOG. 1834. p. 217. Java Fossils 10 out of twenty h
221b201 animal - Confervae-- p. 23 p. 267. Dela Beche. GEOLOG. Researches. facts of salt-water shells living
241c016 .62 ??? Age of Deinotherium.- p. 23. Bull: Soc. GEOLOG. 1837-8. Tom: IX.-- M. D'.Urville on the Distr
324c268 ographical distrib. «in Dict. Sciences. Nat. in GEOLOG Soc.» F.. Cuvier on instincts L. Jenyns paper
355d068 Captain Grants. Himalaya. shells (see Paper in GEOLOG Transacts) same appearance with Secondary Spec
031r038 Daubeny P 24 V. back of page 1 of New Zealand GEOLOGICAL Notes. at St. Helena. This structure was ver
103a063 . 461 «of Proceedings» List of collections in GEOLOGICAL Society. Pumice at South Shetland. Geologica
103a063 Geological Society. Pumice at South Shetland. GEOLOGICAL Society-- Dikes have not been the moving age
134a144 f Asiatic Soc Vol V. p. p 96. apparently good GEOLOGICAL paper. by Malcolmson-- worth reading-- Burne
195b102 be the form of one country to another.-- let GEOLOGICAL changes go at such a rate, so will be the nu
247c028 arks very generally distributed: Mem of great GEOLOGICAL age-- Gastrobranchus «only» 2 species one in
310c225 s an element of extreme difficulty in mundine GEOLOGICAL chronology Annals of Natural History «Vol I?
343d039 tence.-- nor is there in the Tertiary <older> GEOLOGICAL epochs.-- There are some admirable tables on
398e004 ng organic forms, that the whole value of the GEOLOGICAL chronology depends, that most sublime discov
398e005 hould say there was some link in our train of GEOLOGICAL reasoning, «xtremely faulty» The difficulty
414e065 th Mr Lyell., man is not an intruder.-- : the GEOLOGICAL history of man is as perfect as the Elephant
436e135 can hardly be believed, then, uniformity in «GEOLOGICAL» formation. intelligible.-- «-- No. but the
527m034 sense being required) as the closest train of GEOLOGICAL thought.-- the capability of such trains of
044r073 y of ground at present. day-- applied by me GEOLOGICALLY to vertical movements. In Cord: after seein
230b239 g.-- Now a gradual change can only be traced GEOLOGICALLY (& then monuments imperfect) or horizontall
342d034 t order first converted-- is it an old order GEOLOGICALLY? Owen says relation of Osteology of birds t
423e096 n be disorganized to fill their places.-- The GEOLOGICO-geographico changes must tend sometimes to au
184b054 er to Megatherium.-- uncle now dead. Bulletin GEOLOGIQUE April 1837. p. 216 Deshayes on change in she
038r057 ope.--Carnatic It has been common practice of GEOLOGIST. Lyell considers (P 84 Vol III.) whole of Etn
352d060 ng merely a page torn out of a history, & the GEOLOGIST being obliged to fill up the gaps.-- is possi
409e049 object, I will dispute, when I hear from the GEOLOGIST the history, & from the Astronomer that the m
432e125 as intense thought.--» No one but a practised GEOLOGIST can really comprehend how old the world is, a
443e156 ion of a planet to our lives-- Being myself a GEOLOGIST, I have thus argued to myself, till I can hon
529m040 leasure from thinking of the fertility.-- I a GEOLOGIST have illdefined notion of land covered with o
025r018 rd of In a preface, it might be well to urge, GEOLOGISTS to compare whole history of Europe, with Ame
227b225 It leads you to believe the world older than GEOLOGISTS think. it agrees with excessive inequality o
281c137 nimals did change excessively slowly. whether GEOLOGISTS would not find fossils such as they are-- My
443e155 ecause it makes the world far older than what GEOLOGISTS, think: it would be doing, what others but f
443e156 e doing, what others but fifty years since to GEOLOGISTS.-- & what is older-- what relation in durati
021rIFC y gale 29th> Friday Thursday 29th gale Lyell's GEOLOGY The living atoms having definite existence, th
044r072 ons & Galapagos.-- «Humboldts. fragmens.» Read GEOLOGY of N. America. India.--remembering S. Africa.
044r072 embering S. Africa. Australia.. Oceanic Isles. GEOLOGY of whole world will turn out simple.-- Fortuna
045r077 llating as that of a spring) moves away.--Will GEOLOGY ever succeed in showing a direct relation of a
067r142 al. year 1835 excellent account of N. American GEOLOGY. Conybeare Lava in Cordillera & on Eastern pla
076r170 nderful quantity of pure silver in S. America. GEOLOGY of Guanaxuato.--Clay slate. passing into talc
080r180 [blank] NB. P. 73. General reflections on the GEOLOGY of the world P. 14-91. gradual shoaling of coa
085a FC N. Range of Sharks Nothing For any Purpose A. GEOLOGY Note on Woolwich Nothing on any Subject As far
086a006 Siberia must be read as well as Pallas before GEOLOGY is written Cuvier. Europe possessed a great ed
116a099 ules analogue living in mouth of Plate. p. 26. GEOLOGY of Arica <Schit> Schmidtmeyer travels into Chi
118a103 ake.-- L'Institut. (1838) p. 216 M. Gay on the GEOLOGY of Chile.-- P p217. Pentlands Fossils & Meyens
212b167 Dr Smith History of S. African Cattle Phillips GEOLOGY «p 81» in Lardens Encyclop. Proportions betwee
275c119 general arguments-- Lamarck was the Hutton of GEOLOGY. he had few clear facts, but so bold in many s
290c163 it is, just like man. emotions very similar.-- GEOLOGY. Transact. Vol V. Birds bones-- in strata of T
303c204 e say creation is at end, seeing that Tertiary GEOLOGY has obeyed rules of modern causes. & consideri
321c270 round world. 1803-6 Nothing Lyells Elements of GEOLOGY Gibbons life on himself Hume's do, with corres
322c270 jects Elie De Beaumonts. 10 Vol. of Memoirs on GEOLOGY of France.= on Etna. Almost reread the previou
323c269 well's life of Johnson. 4. Vols 25th Phillips. GEOLOGY. Larder 2d vol.-- March 16. Gardner's Music of
352d060 Apterix certainly.-- Lyell's excellent view of GEOLOGY, of each formation being merely a page torn ou
398e005 ctive reasoning is immense. It is curious that GEOLOGY. by giving proper ideas of these subjects. sho
419e087 ust never be overlooked that the chronology of GEOLOGY rests upon amount of physical change <affectin
419e087 ion, two foundations, hitherto confounded,. of GEOLOGY.-- L'Institut 1838. p. 414; M. Guyon thinks Mo
514q21. Asclepias &-- carnosa?-- good-- Norfolk Isd-- GEOLOGY. volcanic? Applies to my geology & Species the
514q21. Norfolk Isd-- geology. volcanic? Applies to my GEOLOGY & Species theory-- peculiar Fauna?. {Australia
641j29v ction of such forms-- these views will bear on GEOLOGY-- There is an analogy between fang of snake, (
375d135 nsequently death.. population in increase at GEOMETRICAL ratio in FAR SHORTER time than 25 years-- ye
05O r090 & many minute craters as at Galapagos. <|> Sir GEORGE Mackenzie must be worth reading Some earthquak
525m024 hat selfishness, pride & kind of folly like (Mr GEORGE S.) is very heredetary.-- My father says on au
591n098 eeper does not think they drop their ears.-- -- GEORGE the lion is extraordinarily cowardly.-- the ot
023r013 ould discover no shells: nothing said about K. GEORGES Sound The idea of the water at Cauquenes. comi
052r099 us icebergs are commonly stranded on shores of GEORGIA «Lat degree ( )», he has rocks on surface. app
466b094 -- & stamens bend up a little In a wild purple GERANIUM, I see Bees visit always base {a} of upper pe
505q014 ape from old flower= (14) Has planted seeds of GERANIUM pyrenaicum. small white-flowered var. with ab
227b227 st organization intelligible.--may look to first GERM-- --led to comprehend true affinities. My theor
305c211 & child, polypus & polypus, bud & bud, polypus & GERM plant & seed.-- instincts in young animals, wel
418e083 Müller's Physiolog. p.24.]CD» can be traced to a GERM, endowed with the vital principle, which gives
418e084 ve modifications of structure being added to the GERM, at a time, (as even in childhood) when the org
507q016 e tooth aborts, do you know whether any trace in GERM. (2) Any more cases of diseases, generally occu
514q21. principle" be better. or "constructive action on GERM." '=?? answered Does Mormodes (one of the Catas
534m061 thinks is nonsense; yet who will venture to say GERM within egg, cannot think-- as well as animal bo
326c265 ce to Kolkreuter's Papers Wiegman has published GERMAN Pamphlet on crossing Oats, &c «Horticultural T
480z008 ra Krauss on Corallinae from S. Seas written in GERMAN.-- Stuttgart ranks these bodies amongst Vegeta
481z010 Nullipora p. 29-- In Meyen. Voyage round World GERMAN a reference to a luminous Sertularia Lesson Zo
511q018 almon were fertile & whether homogeneous {About GERMAN ornithologists, Bhem & Glöger Consul Hunt, bir
042r068 sea: <As did> as did those aerial Volcanos in GERMANY» In the Valle del Yeso it is probable that poi
203b136 e?-- M. Baer (thinks) the Aurock, was found in GERMANY & thinks even now in central & Eastern Asia we
220b199 fear (like rooks with guns) of the Bustards in GERMANY.-- Athenaeum. No. 537. Feb. 1838. p. 107. Mr B
236b279 se.-- On other hand. fresh water tortoise from GERMANY. (where Mr Murchison fox was found), decidedly
498q008 h pistil is connected with separate division of GERMEN <?>-- (11) Must pollen grain be whole, to impr
506q019 or small in quantity -- Any unproductive, where GERMEN does not swell, although there be pollen.-- or
504q014 true Try experiment (30/p.11) (2) Yew Berries GERMINATE?-- Yew trees sexes-- (3) Get Holyhoaks. races
505q014 m it and large Asparagus: result? = failed to GERMINATE 16 Will plant some of the Thyme with abortive
505q014 lanted seeds of pale green Cynoglossum. never GERMINATED 12 Does the horned orange. wh. never has see
200b124 cked from the beach by the birds; most seeds GERMINATING.--- It would be curious experiment to know w
266c092 f Bulbous root from Mummy: after 2000 years, GERMINATING.--!! Henslow doubts? GEOGRAPHICAL JOURNAL. V
027r024 n regular descending steps Mem.; rapidity of GERMINATION in young corals.--vide L. Jackson's paper. P
177b026 offers contradiction to constant succession of GERMS in progress.-- «no only makes it excessively co
447e164 ossing.-- The case of Lemna shows dispersion of GERMS is not end of seminal reproduction.-- likewise
635j56r iracles.-- p. 420 thinks the great fecundity of GERMS is to afford support to other beings.-- true, (
380d155 ven when bitch is in heat.-- Yarrell believes GESTATION is always some multiple of seven-- if woman d
493q004 s the number of pulse, Respiration, period of GESTATION differ in different breeds of dogs. Cattle, (
571n031 -- Chinese. simplest language. Much pantomimic GESTURE?? which would naturally happen.-- Reynolds Wor
560m156 y did in former one. about heredetary tricks & GESTURES, other cases like D. Corbet; «do» ideots form

GESTURES / GLEN

592n101 & the female chamaeleon court the males by odd GESTURES. In one of the six (?) first Vol of Silliman'
099a049 of attraction of particles of same nature: then GET mathematician to when two particles «would» are
150g037 to shelf very shallow channel 50 ft wide & river GET formed in centre In Glen Collarig, on side of Hi
192b083 iety, although many of the seeds will go back.-- GET instances of a variety of fruit tree or plant ru
199b116 d, of the same type with the great continents we GET a means of knowing movements.-- How can we under
202b132 d separate from the English cattle, nor could we GET them to associate together" There is long rigmar
227b224 arnivora than Pachydermata If my theory true, we GET (I) a horizontal history of earth «within recent
229b234 seeds of Keeling standing transport.-- <tr but> GET him to discuss those mention[ed] by Lesson & Cha
230b240 nts perfect change.-- It is scarcely possible to GET evidence of two races of plants run wild.-- (for
230b241 erhaps merely gone back previous to fresh change GET a good many examples of animals «or plants» very
232b240 ter & lizards do.-- Ask Eyton to procure me some GET Hope to give me an account of parasitic animals
239cIFC d. About trades affecting form of man. Could you GET racehorse from Cart horse by picking without cha
241c016 tion far East) Ann: des Sciences. Septemb. 1825 GET Henslow to read over the pages from about 8 to 2
251c040 lled in Newfoundland, with crops full of maize. (GET limits of latter from <Tarton> Barton.-- swifts
252c045 nature [not located] The systematic naturalists GET clear indication of circumstances in Geography t
252c045 of knowledge would probably tell more certainly GET Closer species-- FOX IS & Mice of America «good
289c161 might follow.-- examine structure of this bird & GET account of habits My definition <in wild> of spe
298c190 - a person who is habitually kind to children,--<GET> «increases» general instinctive feeling.-- Ther
307c216 ticated animals, in plants I presume there are.? GET examples.-- for instance where a tendril passes
316c246 ommon to West coast of Africa & E. S. America.-- GET instances.-- very good anomaly in range + + What
353d060 as trace the structure of animals & plants.-- he GET merely a few pages.-- Hence (p. 59) looking at a
408e045 , & some of them symmetrically.-- it is easy to GET 50 of same kind of monstrosities.-- G. B. Sowerb
416e075 picks the Male, instead of allowing strength to GET the day» The fertility of Indian & Common Oxen,
442e152 e in Norway, which never flowers!!-- <How did it GET there? whether? According to the above suggestio
443e153 that to graft from the roots is the best way to GET young trees, from worn-out kinds, & quotes from
450e176 to have been imported: shows they will propagate GET dimensions-- do App. p 73 State of Muar in Malac
455eIBC elle has chapter on sensitive plants; Physiology GET Habberley to try experiments. about raising plan
470t178 ronica--, Ranunculus in numbers =what insect can GET honey out of long, curved nectar of Butterfly Or
498q08v growing White Mullein good plant to sow & try to GET other species <near> close to each other.-- As t
501q011 I expect, except by very minute insects.-- (30) GET Abberley to plant SINGLE Peas, Kidney Bean & Bea
504q014) Yew Berries germinate?-- Yew trees sexes-- (3) GET Holyhoaks. races planted & Linum Perenne.-- Herb
505q014 ome Bees are smaller & more vicious. Will try to GET me some to look at:-- Was once offered a hive. o
511q018 r good or bad.-- are any actually rejected?? (8) GET Sir. R. Heron to give me Pigs foot undivided, &
512q019 eed as well as,, as those already bred in cages. GET direction write to-- (2) Does he believe. Stanle
514q21. s of N. America similar to Lapland Plants --will GET answer= Is pollen of cultivated Orchis & Asclepi
515q022 Pea) for several generations under net & see if GET sterile-- Cover that little Ervum in Sand-walk,
553m132 to take kangaroos than emu, although young dogs GET sadly torn in conflicts with the former. But it
574n040 oa nut shell at one end.-- Children & old people GET into habits.-- we probably can hardly form an id
584n071 s may vary before the structure does; & hence we GET over an apparent anomaly,, for if anyone has tak
587n088 panze. pout.-- Former, whines just like a child. GET a Dictionary & make a list of every word, expres
593n105 rverted> handed down & down.-- mem. Nina used to GET into hay & make a nest for herself.-- the object
596n1BC eople of weak intellects easily fall into habits GET facts about instincts of mongrel dogs Do blubber
629o55v passions makes one break these artificial rules, GET remorse-- (hence desires do not intervene betwe
637i57v pain & disease in world & yet talk of perfection GET instances of adaptations in varieties.-- greyhou
639j28v in ten thousand years the long legged race will GET the upper hand. though continually dragged back
280c136 way of viewing it.-- Variety when long in blood, GETS stronger & stronger, so that though by great ef
290c163 pretty faces When horse goes a round, the minute GETS into the road at right anglles, how pleased it
578n056 & squeezed out again-- as power of mind by habit GETS more perfect over voluntary muscles, these conv
298c190 uced, a general tendency produced. Such as man GETTING habitually into passion, becomes habitually pa
364d103 y exchange birds with some other fancier, thus GETTING fresh blood, without fresh feather, & consequen
391d175 th my view of those «forms» slightly favoured, GETTING the upper hand. <]CD> & forming species)-- [Ap
520mIFC t. Marlborough Str.-- (p. 64. On <insect> Ants GETTING on Table. Col. Sykes) Private Finished. Octob.
523m018 rinking cold drink.-- then brain affected like GETTING suddenly into passion.-- There seems no distin
533m059 med to, in certain actions-- the difficulty of GETTING on a horse on the left side (not good example,
566n011 fore my view of origin of evil passions.-- Man GETTING sight slowly,, but when in grown years, thinki
567n013 October. 8th. Jenny was amusing herself--, by GETTING out ears of corn with her teeth from the straw
507q15v ner de fruct:-- for woodcut-- 1 double hook-- -- GEUM. Galium Burrh = single hook; curved spines-- si
202b131 closely adapted. Near the Caspian «Province of GHILAN» wooded district cattle with humps as in India
594n115 ember with distress their companions-- a «blue» GIBBON. whose companion had S[...] been dead about tw
321c270 rld. 1803-6 Nothing Lyells Elements of Geology GIBBONS life on himself Hume's do, with correspond. wi
435e134 of Carniola p. 112. Man. "standing alone in the GIFT of intellect, he resembles, other mammalia in t
066r140 t Kerguelen Isd. to grow on shoals like Fucus GIGANTEUS! 24 fathoms deep 24» under 50. Kerguelen Land
458t001 on Sub. Him. fossils-- Ruminants. & Tortoises GIGANTIC-- hyaena-- bear & ruminants all of larger siz
458t001 rog, 40 inches in length--! alludes to ancient GIGANTIC salamanders-- Every order (except whales) hav
081r181 aramania Capt. Ross. & Scoresby deep soundings GILBERT Farquhar Mathison travels Brazil. Peru. Sandwi
095a038 ourn. Analysis of Poenig Voyage Valparaiso Dr. GILLIES in MS. letter in Sir. W. Parish Possession. ta
496q006 any generations. daisy. Fever-fuge Groundsil.-- GILLY flower will break & become double.-- There is a
452e181 e found on the isld of Batchian near SE. end of GILOLO.-- "-- Forrest Voyages. p. 39-- deer but no wi
452e181 st Voyages. p. 39-- deer but no wild animals in GILOLO.-- p. 134: Birds of Paradise were first procur
452e181 134: Birds of Paradise were first procured from GILOLO p. 253 In isld of Bunwood (18 miles in circum)
229b234 on extreme Northern Coast. Hippopotamus do.-- GIRAFFE do.-- Range of East Indian Rhinoceros (?)-- So
585n075 d in peculiar manner.--]CD April 3d. 1839 The GIRAFFE kicks with front legs & knocks with back of He
637j58r p. 235. talks of the long spinous processes in GIRAFFE &c, as adaptations to long necks-- why they ma
458t001 , S. America--» These strange forms., camels, GIRAFFES. Sivatherium & Anoplotherium, with existing,
399e006 se of change in vertical series: Look at whole GLACIAL period? [not located] Study introduction to Cu
419e086 Scandinavia, than of the N. American species--GLACIAL period Dr. Beck says the shells in Scandinavia
040r063 e same process.-- Neither lakes or Avalanches (GLACIERS very rare) to cause floods in valleys, which
064r135 cent species. -- May I not generalize the fact GLACIERS most abundant in interior channels. there no
219b195 al to state we do not know how transported.-- (GLACIERS might have acted at Tristan d'Acunha-- Carmic
256c054 ng which were species, (theory admirably) yet a GLANCE would tell from which country,-- I often dispu
308c219 ection of Physical Sciences p 276 May be worth GLANCING at, as she has no original ideas., it will sh
578n057 ed to convulsive actions).-- But, the lachyrmal GLAND is «not» under voluntary power, (or only very l
225b217 llet seed turns a Bullfinch black, or iodine on GLANDS of throat, (or colour of plumage altered durin
332d003 r, as used much more latley diseased Mesenteric GLANDS.-- My Father has seen case of pleurisy, broken
126a124 only, like water below 39 degree» & lower part GLASS.-- then the high temperature would be much near
127a126 il will freeze if cooled in a closed globule of GLASS. (oil may be cooled to 0 degree!)-- shows effec
228b230 d be worth trying to isolate some plants, under GLASS bells & see what offspring would come from them
495q05a each other-- & enclose one twig of each in bell-GLASS-- sow these seeds & see if they will come up tr
551m129 out of I do not know what when it looked at the GLASS) when pouting protrudes its lips into point-- m
565n006 - Looking at ones face <&> «whilst» laughing in GLASS. & then as one ceases, or stops the noise , the
573n036 es not a steam engine, but a piece» of coloured GLASS <&admires> is lost in astonishment at the artif
496q05a er unfavourable circumstances, as Hyacinths in GLASSES &c &c Experiments Questions concerning Plants
026r021 l II» It is a fact worth noticing that cryst of GLASSY felspar in Phonolite arrange themselves in det
033r043 base of Pitchstone; Mem Galapagos. chiefly red GLASSY scoriae.--could walk round base:--not universa
049r089 numerous hills of greenstone? -- Daubeny. P 95. GLASSY & Stony Pearlstones alternate together in cont
059r123 122] In Igneous rocks.--which have the cryst of GLASSY F. fractured. have been melted with little pre
094a035 locally relieving the other. -- Is the felspar GLASSY in greenstone dikes which rise through granite
110a086 Salband like basalt. full of circular cryst of GLASSY felspar different from either fragment or dike
097a043 which I conceive to be analogous to the black GLAZING observed by Humboldt on the granitic rocks of
459tf02 e all the changes. [im]portant view, copie[d] GLEANINGS of Sciences. Vol. III p. 83. Paper translated
459t009 ay be, perhaps. squeezed into Mr Walker's law GLEANINGS of Science Vol III. p 320. Mr Hodgson on Musk
120a107 were fluid undulations-- the equal movements of GLEN Roy road. (¿ metamorphic action at the bottom o
120a110 Salisbury Craigs well worthy of attention-- rear GLEN Roy Notebook-- & scraps on Salsisbury Craigs. K
121a112 me thing-- Consider profoundly How came it. that GLEN Roy district could have been elevated without f
128a129 uberant water to counterbalance How strongly the GLEN Roy case shows that the figure of the world has
130a133 large fresh Water springs off coast of Persia In GLEN Roy paper I show crust yield easily. & if easil
145g FC iting really good account of ice.-- C. Darwin A. GLEN Roy Generally received opinion that male impres
150g037 3a 3a 2 Mass 3 2 rather longer than 3a Sunday In GLEN Collarig, when water up to shelf very shallow c

```
150g037 annel 50 ft wide & river get formed in centre In GLEN Collarig, on side of Hill of Bohunthine upper r
151g041 ence of lines in certain parts.-- At the Pass of GLEN Collarig two little lines of Hill (judging from
152g045 espond with some line subsequent to shelf {P} In GLEN Collarig, by Dicks theory lake burst in most im
152g045 probable part & not in Pass, where shallowest In GLEN Collarig good case of shelves entering «on» one
152g048 ards with greater cut through, not applicable to GLEN Roy Lake, must have remained very long at 4th s
152g050 race The butresses of Alluvium rise nearly up to GLEN Collarig up within 200 ft of level of 4th shelf
153g052 r sea» at successive levels-- {P} Shelf opposite GLEN collarig at bend & here most accumulations At g
153g053 condition  of 4th shelf, some way below House of GLEN Roy, seems to be which higher up on is corroded
153g054 re ravines enter side by, opposite entrance into GLEN Fintec a kind of landing place is formed Ben Er
154g057 ill rounded, site N N W of Ben Erin {P} Shelf of GLEN Guoy flat peat plain divortium aquarium-- tidal
154g057 ium-- tidal channel-- 12ft obscure obscure NB In GLEN Collarig tidal channel, sides <alm> 15 ft above
154g058 race, from terrace of 2d shelf Level of shelf of GLEN Guoy form comparison with granite block «& Ben
155g062 esday Shelf 3d dies away almost imperceptibly on GLEN Turrit side 2nd shelf very broad «& cut out, pr
155g062 ed some way up, but most faintly on East side of GLEN Turrit, where I believe they end in upwards inc
156g067 ved 2d or upper one more perfect in this <part> «GLEN» than 3d. 3(a) less perfect than upper & lower
156g067 r & lower but quite as perfect as those lines in GLEN Collarig, & some «other parts» Boulders of same
157g077 ade has <cut> «where two branches unite in upper GLEN Roy» very little back from line 2d; little acti
158g079 eaty Mass of this point very nearly like head of GLEN Guoy nor is horizontal line apparently continua
158g080 te ridge or a modified Granite ridge» at head of GLEN Roy on same side where two rivers unite in Uppe
158g080 Roy on same side where two rivers unite in Upper GLEN Roy great plain about 60 ft beneath shelf  peat
158g082 butresses «occur» high up on Shelf 2d «in Upper GLEN Roy» In this upper part «about junction of Uppe
158g083 ttresses on each side «very little way» in Upper GLEN Roy at pass {P} River Gorge 4th Sh side of vall
158g085 der profoundly Boulder hypothesis Thursday, from GLEN Turrit to Fort Augustus Barom on upper  (rather
159g087 higher  Slope of valley much more gentle than in GLEN Roy, & partly shut in No Granite blocks in high
159g087 ut in No Granite blocks in higher parts?? Bought GLEN name of Glen by which we descended, it is to th
159g087 ite blocks in higher parts?? Bought Glen name of GLEN by which we descended, it is to the west of Gle
160g088 Glen by which we descended, it is to the west of GLEN Tarf What I called Alluvium shows the ascending
160g089 ong straight isthmus connecting E & W connecting GLEN Bought & Glen Tarf a perfect old Loch, making <
160g089 sthmus connecting E & W connecting Glen Bought & GLEN Tarf a perfect old Loch, making <several> two d
160g089 ranches of River Bought & between one of these & GLEN Tarf Hill «Cairn <taw> leer peak» Barom 28.700.
160g092 in  these mountains & not ridges) between arm of GLEN Bright flowing into E. end of L. Oich, & waters
160g093 ocky place, but narrow shelves just like road of GLEN Roy-- appears to lip with moss On this terrace
165g125 & not merely thoughts laying dormant-- Man from GLEN Turret said he learnt to know the lambs because
165g BC bseved? Shows that <nervous> brain makes thought GLEN Roy B C. Darwin All useful pages cut out Dec. 7
345d043 then  calf «in both cases» is killed. Notes from GLEN Roy Note Book.-- Why is not Tetrao Scoticus. an
345d043 cus. an american form (if so)?.-- A Sphepherd of GLEN Turret. said he learnt to know lambs, because i
541m090 of sea on coves when waters had fallen, as in my GLEN Roy paper.-- this greatest mental effort, of wh
148g023 t & other one 40-- ▌ traces of them all along <GLENCOE>.-- towards Fort William yet in Glencoe in par
148g023 along <Glencoe>.-- towards Fort William yet in GLENCOE in parts no trace of them-- Mem Coast of Chile
148g029 ect the case of loch <in> <below> «by» pass of GLENCOE-- the erosion may often be due to rivers-- By
162g101 is so interesting. enters by old tower called GLENGARRY (Nead Roy told me) it is impossible to see my
147g021 House 28.935/82 degree A Temp of Air 65 degree? GLENOE, 6 ft above high water mark 30.380. 68 degree
156g071 s Level of plain of 4th shelf at head of Lower GLENROY 29.581 A 82 75 degree? From this point plain a
158g083 his upper part «about junction of Upper & from GLENROY» near the upper shelfs ground strewed with peb
398e006 pages  in the history are perfect, we obtain a GLIMPSE only of the changes which the government is su
316c244 enough  for Heaven or bad enough for Hell.-- «+GLIMPSES bursting on mind & giving rise to the wildest
025r018 at facts can be supported from that part of the GLOBE.: & when we see conclusions substantiated over S
032r041 <circulation owing> rotation in fluid matter of GLOBE. must there not be a circulation «however  slow
035r048 consequent of injection of fluid rock.-- Try on GLOBE. with slip paper a gradually curved enlargement
045r076 accidental;  the proportional force of crust of GLOBE & injecting matter on the great rise.-- The g
045r077 cceed in showing a direct relation of a part of GLOBE rising, when another falls.--When discussing co
048r086 s «always» most favourable; In what part of the GLOBE are there such vast numbers of wild animals, bo
059r123 cooled  suddenly.-- As the rude symmetry of the GLOBE shows powers have acted from great depths, so c
063r131 s not gradation;-- An argument for the Crust of GLOBE being thin, may be drawn. from. Cordillera. roc
073r161 & copper is infinitely rare in all parts of the GLOBE". p. 113 How utterly incomprehensible that if m
090a021 ing meteorolite but old Planet, that inside our GLOBE melted magnetic metals. ∴ earthy crust compared
090a022 he meteoric stone which must have fallen on the GLOBE since the Cambrian system In Ures dictionary be
119a104 s not explain first formation of continents, if GLOBE be considered as condensed vapour.-- inequlitie
130a134 enc Math. Phy-- Nat. t. I, 1831. sur le temp du GLOBE on Volcanos &c worth reading. L'Institut.  1838
133a140 say the Tertiary period. one is led, to look at GLOBE as resting on film of molten rock.-- Voyages of
043r070 ontemporaneous action over larger spaces of the GLOBES & "periods" of increased activity.-- such as t
127a126 er no--. oil will freeze if cooled in a closed GLOBULE of glass. (oil may be cooled to 0 degree!)-- s
494q004 said  yes with regard to former (8) Is form of GLOBULE of blood in allied species similar.-- if not h
310c224 June 3d 1838. quotes M. Turpins assertion that GLOBULES of milk produce a plant capable of growing!!
389d173 nervous  matter consists of infinite number of GLOBULES: generally sufficient for one birth or rather
511q018 omogeneous {About German ornithologists, Bhem & GLOGER Consul Hunt, birds from Azores or Madeira  Mr.
468t112 case as Azalea or Rhododendron xx after several GLOOMY days. hot one, Bees almost P every minute to F
495q05a lium «an annual» «sleep» «closes flower» on all GLOOMY days.-- The «garden» Coronella also sleeps  on
571n031 lso much struck in great avenue, resemblance to GLOOMY aisle of Churche.-- these are Mayo's ideas.--
225b219 -- New Zealand «compare to Van Diemen's land.» GLORIOUS fact. of absence of quadrupeds East India Arc
605o18v ny height.-- 6 That the superiority & "inward GLORRYING, which height. by its accompanying & associat
580n062 retained through life" p. 200.-- "The desire of GLORY, immortal fame, &c so common in the young are s
605o19v t, eternity, &c &c. produces an inward pride & GLORYING. (often however accompanied with terror & won
605o19v we  may often trace the source of this "inward GLORYING" to the greatness of an object itself or to t
502q11v mpanula» &c & dead-nettle.-- Lithospernum. Blue GLOSS. it is not possible to see orifice of poison-tu
496q006 e. Snap-Dragon. (I have seen one monstrous) Fox GLOVE & such like in very rich soil-- As they have li
309c221 uration in two classes however different.-- Male GLOW-worm knowing female good case of instinct. bees
478z004 ose Mr J. Murray has given paper to Royal Soc on GLOW worm. luminous property-- Curious arrangement o
570n027 y treated would think negress beautiful,-- [male GLOW worm doubtless admires female. showing. no conn
263c076 elopement of a brain capable of producing more GLOWING imagining or more profound reasoning than othe
582n066 music  is evidently considered as analogous to GLOWING conversation of several people.-- Children hav
307c215 rm constant.-- The females of some moths, like GLOWWORM <are> have «These abortive organs in some Mal
344d041 . is probably arrived at in case of insects as GLOWWORM The case of one impregnation sufficing to sev
344d040 eptions to this in some larvae of insects-- (¿GLOWWORMS) breeding-- <beet> image state fertile at onc
251c041 ct of conduct of wild & tame horses.-- p. 246-- GMNURA-- new genus of Mam: found in Sumatra p. 452 Ap
021r005 hem; Quote this. Valparaiso Granitic nodules in GNEISS. Epidote seems commonly to occur where rocks h
069r150 andstone of the Portillo line.--connected with <GNEISS>.--(Mica Slate) [Fig. 9] {P} (13) like Bell of
069r150 important view Urge curious fact felspar melted GNEISS/// QUARTZ!!! Analogous to Von Buch. Basalt whe
073r163 lco granitic rock.--in parts of table granits & GNEISS with gold veins visible:--"Porphyries of Mexic
091a027 n. p. 410. 1828 Ed. N. P. J. p. 105. Oct. 1828. GNEISS in India (falls of Garsipa) dip 30 degree. <st
098a046 laminar.  hence the thick wedges of feldspar in GNEISS.-- Veins in septaria. a kind of concretionary
098a048 s them in layer.-- So that layer of feldspar in GNEISS is identical with layer of flint on calc.: san
098a048 lieve most shortly (Hence endless passages from GNEISS to granite): Why not horizontal? Why have part
100a054 113) «Lardner Encyclop.--» absolutely considers GNEISS an aqueo deposit resulting from  disintegrated
100a054 resulting from disintegrated granite!!! Look at GNEISS of Rio Concretions in Pumice bed at  Ascension
104a068 hat Kylow (?) was astonished with him that <th> GNEISS, mica-slate of whole kingdom of Norway was con
110a087 m one of the true rocks, most probably from the GNEISS beds in the mica slate.-- Geograph. Journal. V
114a094 rn formation & not ever in Secondary in Europe. GNEISS-- metamorphosed clay slate.-- --shale in shall
115a097 bsence of fragments «& pebbles» in mica slate & GNEISS, can only (see «supra» p 94) be accounted  for
119a105 r the temp must be immense to convert rock into GNEISS &c judging from what we see when trap in  dike
119a106 partly known-- <[...]> moreover gradation from GNEISS to granite shows that the metamorphic rocks ha
149g032 rocks  not apparently in situ <& in> hill being GNEISS <& also> also near summit on Hill on side of I
151g040 ery much this character.-- The boulders (one of GNEISS remarkably water worn) are often times of rock
151g040 not  in immediate neighbourhood, (as granite or GNEISS of Moel Derry) on low hill between Inn & Bouhu
153g056 «these boulders are decaying.» neighboring rock GNEISS & [...] sandstone actually resting on them  on
155g066 thrown  water in same «drainage» lines Mound of GNEISS though wonderful-- <that they are preserved> h
156g072 f This shelf at head where <granite &> «veined» GNEISS <unite> «occurs» abundantly with perfectly rou
157g075 into island-- whole hill composed of remarkable GNEISS with red granite veins & quartz, & garnets.--
```

```
*****************************************(Key Word)*****************************************
158g082 ill with granite blocks almost encircled> <fre> GNEISS cut smooth on sides of hill where Boulder lies
159g084 een 2d & 3d shelf Mountain <Mica> «composed of» GNEISS Block on 2d shelf & below it some way; several
160g090 ole hill dark grey fine grained. Much contorted GNEISS «narrow sharp ridge with peak» I walked all ro
188b067 rthern species replace <No> Southern kinds-- (I) GNU reeaches Orange river & says so far will I go an
035r045 es. & not those of rapid cooling &c &c My results GO to believe that much of all old strata of Englan
043r071 pace in formations & durability of similar causes GO together. add. <">" <from> "in the same line" to
045r074 am condensed.--Perhaps these mighty changes might GO on. & not a bubbles on the surface bespeak the c
062r128 es, near Volcanoes. lakes of brine all inhabited: GO steadily through all the limits of birds & anima
123a116 n ones.-- Septemb. 2d.-- Sulphur like carbon must GO round of dissemination & separation in volcanos.
125a121 earth originally fluid, then cooling process must GO from surface towards the interior -- who knows h
179b033 .-- <If> Is there a tendency in plants hybrids to GO back?-- If so Men & plants together would establ
181b040 in a country not rapidly increasing.-- If we thus GO very far back to look to the source of the Mamma
183b052 nt d être feconds. dès les premières generations" GO back to type of either animal when crossed with
188b067 I) Gnu reeaches Orange river & says so far will I GO and no further:-- Prof. Henslow says. that when
189b072 ut off:-- like golden pippen. if produced by seed GO on.-- otherwise all dif.== The fossil horse, gen
190b075 y in such marriages or offspring show tendency to GO back-- there is an end to species.-- «Brown Appe
192b083 cing any variety, although many of the seeds will GO back.-- Get instances of a variety of fruit tree
194b095 phere.-- If these facts were established it would GO to show a centrum for Mammalia.-- I really think
195b102 one country to another.-- let geological changes GO at such a rate, so will be the numbers & distrib
215b177 it is generally said.= How came first species to GO on.-- There never were any constant species Both
228b229 a formerly abundant. Seed of Ribston Pippin tree <GO> producing crab. is the offspring of a male & fe
228b230 rom them. Ask Henslow for some plant, whose seeds GO back again, not a monstrous plant, but any marke
248c030 see even in domesticated varieties a tendency to GO back to oldest race, which evidently is tending
264c081 s Gould seems to doubt how far structure & habits GO together. This must be profoundly considered.--
276c124 e utmost importance to show that habits sometimes GO before structures.-- the only argument can be, a
280c133 cceed, yet seems to grant, that difficult & other GO back to either parent.-- Shows instinct (Sir J.
280c136 ced, yet to produce whole generation unlike would GO against the tendency.-- it tries to go back to g
280c136 like would go against the tendency.-- it tries to GO back to grandfather, but if too unlike its own p
280c136 o mules bred or two certain varieties, they would GO back to grandfather, which is true) & infertilit
284c149 pends on non-variation, & not on extension ¿these GO together? Therefore value of organs vary in diff
288c159 ertain that rams & bulls from England fetch very <GO> large price. as is evident to be worth introduc
306c212 which lived in estuaries its taste. taught it to GO to <sea> salter water (& its necessities teach i
308c217 k memory» old habit of putting tea in pot made me GO to tea chest almost unconsciously.-- why do abse
312c228 s Journal.»-- consult on this point-- pigs always GO against this, without «number of vertebrae» new
312c232 ime. Mr Willis Long eared little dogs, I am told, GO to heat, take dog. but do not become impregnated
335d013 like parents,-- therefore offspring will tend to GO back, or have none-- the argument does not apply
337d021 uch intermarriage.-- In my speculations. Must not GO back to first stock of all animals, but merely t
343d039 distinct species-- We know how long a Mammal may GO on as one species from Egyptian Mummies & from t
346d048 ed, in & in are-- colour white, uniform.--crafty, GO in file, hide their young., bold.-- a Mr W: Hall
360d090 man had crossed Jackal & dog-- (offspring did not GO to heat. but parts swelled, though no fluid came
366d111 g will the wretched inhabitants of NW. Australia, GO on blinking their eyes. without extermination, &
432e124 do not, (& in hairless kittens we see same fact) GO back, & this is argument against Blyth's doctrin
460t019 s complexity from a few types originated, we must GO to the first origin of the world.-- our present
465t080 ue first case of bones (New Red Sandstone) & then GO on to shells-- A profound consideration of metho
467t103 seem to act, something like on Kidney Bean, they GO to nectar at foot of upper petal standing on «I
467t103 tar at foot of upper petal standing on «I saw Bee GO to two species of Lupine, two wings. & when the
468t111 Leptuse-- Diptera & small Hymenoptera Saw Humble GO from great Scarlet Poppy to Rhododendron-- from
469t135 now gone on 14 days, (except some wet ones/ & wd GO on longer-- Woodfords Marrow fat, Early frame, G
472s02r ay. many clumps of heartseases, never saw any Bee GO to them. Yesterday remarked that many flowers ha
472s02r y saw very odd dusky humble (with pollen) on legs GO from clump to clump, & insect proboscis in many
497q06v have abortive parts, whether such vary.-- Do Bees GO to Sweet Peas, IMPORTANT, for if so, as these ca
500q10a es & British & distant parts of Europe.-- Gould-- GO over the Pigeons, Philotis, Dacelo. Alcyone, whe
520m001 bodily complaints «& mental disposition» oftener GO with colour, than with form of body.-- thus the
524m021 be well understood. In insanity, the ideas do not GO back to childhood, (but appear most capricious)
524m021 second childhood full of meaning:-- Dreams do not GO back to childhood-- People, my Father says, do n
525m026 t-- B. the affections «& N B affections very soon GO in Maniacs» seem to have failed even more than t
531m051 s.-- In young children, the violent passions they GO into, shows how truly an instinctive feeling, <m
532m056 even when in bed room-- grew very thin, would not GO out of house except with Caroline-- After fortni
535m044 en two groups near. mother desert one sometimes & GO to other, so that two mothers to one group.-- (a
539m082 tution & not feeling much enthusiasm, happened to GO close to one & smelt the peculiar smell of Pictu
546m110 r, which she valued, & she was suddenly called to GO on the lawn to see something, on her return coul
548m116 because in this state, the consciousness does not GO back to former periods so «as» to <make> «give»
553m137 ked that <exp> the expression & noises of monkeys GO in groups. thus the pig-tailed baboon, shoved ou
554m141 e when female is sitting the Keeper is obliged to GO in with a stick, if he drops it, the bird will f
554m141 .-- (argue sophism of association. Kenyon, & then GO on to show, that if Cart horse argued from this
570n024 shrugs his shoulders & replies nothing. if he did GO to reply. he would throw back his shoulders. he
582n067 ve its legs so, as much as in the young salmon to GO towards the sea. or down the stream; which it do
584n071 ure the cell; p. 22. instincts & structure always GO together: thus woodpecker: but this is not so,,
586n078 points of compass, & they do know which way they GO; & so return.-- «but does not apply to dogs.--»
586n079 young, because can not have been taught, where to GO-- the act of crossing the sea in dark night & no
591n100 not man eating cause disgust, because he does not GO against instinctive feeling, only does not fullf
594n115 han man his intellect Lyell has seen a little dog GO to the assistance & bite a big dog. which was fa
596n184 . pull back skin from head very little Does blood GO in <body> face in pashion.?-- cry? Do people of
609o030 sometimes be hard to tell) + + Society could not GO on except for the moral sense, any more than a h
616o038 ion or to our conscious selves.-- Such memory may GO back to animals which were changed into man.: th
619o043 likewise feel pain.-- If he saw another man <say GO> «acting in» accordance to his instincts, <he wo
063r133 on is not perfect.-- Dogs. Cats. Horses. Cattle. GOAT. Asses. have all run wild & bred. no doubt with
406e035 w.-- not so much against my modification of it-- GOAT & Moufflon will not breed.-- p. do.-- Fish of T
459tf02 om Meckel. Comp. Anat.-- From Buffon cross of he-GOAT & sheep, it seems male gives form. admitted by
464tf6r r, believes in the existence of Molina's Pudu or GOAT There is ibex of Alp Pyrenees &c-- (see Blyth's
182b048 rown, with long clotted hair resembling that of GOATS." Progressive development gives final cause for
267c093 brown with long clotted hair resembling that of GOATS" Geograp. Journ. Vol VII. p. 216. Mr Bennett Vo
306c214 gora «centre of Asia Minor» are the fine-haired GOATS. which it is said cannot be transported from th
355d066 «evidently artificial» approach in character to GOATS.-- or dogs to foxes. (yes Australian dog) or do
274c113 , (one in Australia is called swamp pheasant) GOATSUCKER»-- parrots with claw like lark (NB The La je
273c112 powerful wings, but tail stiff.-- Swallow & GOATSUCKERS likewise exaggerated.-- There is one most re
251c001 seel Gazal. (Bos Gazoeus) does not mix with the GOBBAH or village Gazal.-- ¿is latter same species do
195b101 son.-- Astronomers might formerly have said that GOD ordered, each planet to move in its particular d
195b101 move in its particular destiny.-- In same manner GOD orders each animal created with certain form in
198b114 d resumé well worth studying] CD says grand idea GOD giving laws & & then leaving all to follow conse
317c248 many references very good. also "Rays Wisdom of GOD." Often refer to these.-- Also some few facts at
321c275 med Macleay's Horae Entomologica Ray's Wisdom of GOD references at end-- The British Aviary.--do-- &
343d037 grander than idea from cramped imagination that GOD created. (warring against those very laws he est
350d054 ecessary spaces" p 23. "for Nature is the art of GOD" Septemb I,. It has been argued Man first civili
356d072 quadrupeds is difficulty on any theory-- without GOD is supposed to create & destroy without rule-- B
403e023 be accounted for, on any other it is the will of GOD.-- Octob. 16th. A very strong passage might be m
535m069 ot located] as first caused by will of Gods. «or GOD» secondly that these are replaced by metaphysica
535m069 e of mind: the Chileno says the mountains are as GOD made them,-- next step plastic <virtue> natures.
553m135 r the thunder & lightning the direct will of the GOD (<thus> & hence arises the theological age of sc
553m136 ¿ individually or in race?) by a separate act of GOD, & not as a necessary integrant part of his most
553m136 be difficult to prove that» this innate idea of GOD in civilized nations has not been improved by cu
553m136 most implicit faith that through the goodness of GOD knowledge has been communicat to us».-- & that i
558m151 us one principle of charity.-- ¿ May not idea of GOD arise from our confused idea of "ought." joined
566m012 thing as thunder. would be placed to the will of GOD. Zoology itself is now purely theological.-- Ori
604o018 er-- -- 2 From these & other reasons we apply to GOD the notion of living in lofty regions. 3 Infinit
604o018 eternity. darkness, power. being associated with GOD. these phenomena we (feel & ?) call sublime.-- 4
632j53r 37 Proofs and Illustrations of the Attributes of GOD Macculloch. Attribs of Deity. Vol: I it will be
633j54r e on alpine pinnacle? if we once to presume that GOD «created plants to» arrests earth, (like a Dutch
```

```
634j54v case of man. <&> The <design> determination of a GOD-head.-- the designs of an omnipotent creator, ex
637j58r al bones especial adaptation, to <young».-- good GOD & yet Mails have them. What trash p. 237.  Gives
286c155 called nature of man. like to think his origin GODLIKE, at least every nation has. done so as. yet.--
535m069 scribed [not located] as first caused by will of GODS. «or God» secondly that these are replaced by m
535m069 o doubt savage attribute thunder & lightening to GODS anger.-- (∴ more poetry in that state of mind:
565n010 ber how Lavater: ran away with new Lavaters,-- Ye GODS!:-- says fleshy lips denote sensuality (p 192 V
305c209 In  my birds from S. Hemisphere there are some GODWITS which are close to European species, and the s
255c051 t case of memory without association.. Instinct GOES before structure (habits of ducklings and chick
290c163 crossed  with women with pretty faces When horse GOES a round, the minute gets into the road at right
338d024 er (Willis) says <black> «that strength of» hair GOES with colour. black being strongest.-- V. p. 63.
380d153 rowth is immediately checked-- the vis formativa GOES entirely to the offspring-- this is clearly the
388d170 k of food.-- the generalization begins low.-- it GOES through transformation, nearly independently of
426e106 ; but facts of grouse, & pheasant, & hooded crow GOES against this. & wild hybrid plants. If many wil
431e122 mber. (¿ species, or individuals) the deeper one GOES-- surely is this true?-- most strange.-- Does n
557m149 shyness not so.-- affected laughter.-- A dog who GOES home from shooting. runs away. is not afraid th
619o043 .-- This then is moral approbation, as far as it GOES.].CD But should he prevented by some passion or
422e092 ae).-- -- ∥.do. p. 318 M. Pictet of writings of GOETHE.-- who maintains, that «Alludes to difference
058r117 order.-- If we look at Elevations as constantly GOING on we shall see a cause for Volcanos part of sa
104a067 cussion on dikes argue impossibility on fissure GOING right through superincumbent mass (varying hard
109a083 may be granted. independent of currents.-- mud GOING out can actually be seen.-- ¿ The preservation
186b058 some <heteromera> colouring of crysomela may be GOING back to common ancestor of Crysom. & Heterom, b
220b197 nt. Mem Lord Moreton's Mare. The fact of plants GOING back hybrid plants; analogous to Men. & dogs. N
228b229 spring of a male & female animal of one variety GOING back ¿whether this going back may not be  owing
228b229 animal  of one variety going back ¿whether this GOING back may not be owing to cross from other trees
239c002 may  add, that fact of half cross with parents, GOING back to either parent is lucidly explained.-- M
274c113 und has not this structure., instance of habits GOING before structure).-- even one kingfisher-- Goul
291c167 ing an hybrid female it is a wonderful relation GOING through all Nature.-- Makes hermaphroditisms. o
301c200 on about equally complicated.-- An Entomologist GOING into a country & collecting thousands & tens of
306c212 followed  ebb tide, therefore got into habit of GOING down stream which would last were the stream 10
385d163 er, assures me he has known many cases of bitch GOING to mongrel, & all subsequent litters having a t
408e044 t be prophecied to have this character.-- worth GOING there for.-- «Gales of wind would blend species
410e051 . with respect to representative species., when GOING N.orth & South Thinking of effects of my theory
424e101 w. have in Europe different ranges-- latter not GOING North of the Elbe.-- yet they meet in one  wood
429e114 the dreadful «but quiet» war of organic beings. GOING on the peaceful woods. & smiling fields.-- we m
467t103 nly when over-ripe & half withered-- I saw Bees GOING to clover & once this happened.-- And in common
522m013 ired.-- this may be doubted, whether rather not GOING against natural instincts.-- My Grand F. though
528m035 ers, &c &c round one. one recalls the castle by GOING to beginning of castle» because train cannot be
539m083 ?? The possibility of two quite separate trains GOING on in the mind as in double consciousness may r
547m112 h past circumstances,-- as whether I really was GOING to Shrewsbury, whether I had rung for Covington
554m140 whether fear or shame.-- When she thinks she is GOING to be whipped. will cover herself with straw, o
554m141 n" & "reasoning," & take instance of Dray Horse GOING down hill.-- (argue sophism of association. Ken
556m146 affinity in face of donkey, horse & zebra. when GOING to kick.-- Why does dog put down ears, when pla
556m146 te movement to drawing them close on head, when GOING to fight, in which case expression resembles  a
565n009 accompanied by slight protrusion of lips, as if GOING to say "my dear," just what smile is to laugh.-
586n081 a part never subject to volition.-- like plants GOING to sleep.-- "A bird has the faculty of  finding
073r163 c rock.--in parts of table granits & gneiss with GOLD veins visible:--"Porphyries of Mexico may be co
073r163 or most parts as rock eminently rich in mines of GOLD & silver." «p. 131» The above porphyries charac
073r164 &  amphibole frequently only vitreous felspar: = GOLD veins in a phonolitic porphyry. = several parts
079r176 ompany.-- Ulloa has said silver in the highest & GOLD in the lowest. Humboldt states that some of the
079r176 lowest. Humboldt states that some of the richest GOLD mines on ridge of Cordillera near Pataz, also a
116a100 ca <Schit> Schmidtmeyer travels into Chile p 29. GOLD is not sought for in Chile in beds of river, bu
116a100 f river. formed of rounded pebbles-- it is clear GOLD occurs in submarine alluvium, or sublittoral fo
183b052 avec la linotte, avec le verdier" &, <fr> silver GOLD & common pheasants & fowls..-- "On sait que  le
405e032 Saw  what was said to be hybrid between silver & GOLD fish-- Octob. 26th. If. hereafter. M. angustide
628o054 man, may be quite artificial, as avarice love of GOLD.-- love of fame-- Yes Hartley explains this & M
187b063 ca.-- Never They die; without they change; like GOLDEN Pippens, it is a generation of species like ge
189b072 ies», their race is not utterly cut off:-- like GOLDEN pippen. if produced by seed go on.-- otherwise
192b083 of  Ribston Pippin, produce Ribstone Pippins, & GOLDEN Pippin, goldens-- hence-- sub-varieties & henc
228b230 her trees.???? Do the seeds of Ribston Pippin & GOLDEN Pippin &c produce real crabs, & in each case s
242c018 All  the isles of Oceania have the Scincus with GOLDEN streaks-- the lacerta vittata extends <to> fro
289c160 een in one district, though common on another, (GOLDEN creted wren so rare in some countries-- nighti
356d070 nimals who eat every species new.-- Sept. 8'. A GOLDEN Pippen or Ribston do producing occasionally (a
386d165 whether it retains same length of life.-- like GOLDEN Pippen trees! How is this with buds of plants,
192b083 in, produce Ribstone Pippins, & Golden Pippen, GOLDENS-- hence-- sub-varieties & hence possibility of
338d025 breeding!. «Mem 3 species of grouse»! Has not GOLDFINCH & Greenfinch bred, & surely wild Duck & «pint
361d095 ated, that Green-finch, all linnets red-pole, GOLDFINCH, hawfinch-- in nursling plumage resembled tha
317c249 old me that the mules between canary birds & GOLDFINCHES differed considerably in their colour & appe
436e138 Birds by the Rev. E. Stanley Vol I. p. 72.-- GOLDFINCHES placed near. but not in sight of each  other
595n127 uts who has never seen others onspit]CD [blank] GOLDSMITHS Essays No XV, on sounds of words being expr
089a018 y.-- In all the mountains of Saint Marc et des GONAIVES it is difficult to find stone not thus compos
055r107 ccasion to wonder what has become of the Basalt. GONE into fine sediment Look at St. Helena!'-- There
069r151 glomerate on West of Peuquenes as on East. Where GONE to.?-- There must have been some conglomerate E
069r151 ve been some conglomerate East of Portillo Where GONE to? Intermediate space protected.-- Oh the vast
126a123 nduct heat?-- How comes it in volcanos that have GONE, on for thousands of years, that surface does no
179b034 intermarriages, then the black & white is so far GONE, that the species (for species they certainly a
192b088 d slate, the father of all Mammalia in ages long GONE past.. & still more so «known» with fishes & re
196b108 n other hand creation of small animals must have GONE on since from parasitical nature of insects & w
224b216 &» Has the Creator since the Cambrian formations GONE on creating animals with same general structure
230b240 subjected  to change in Americas. perhaps merely GONE back previous to fresh change Get a good many e
265c083 s no record of any effect.-- New Hollanders have GONE on boring their noses. &c & This congenital cha
302c200 t is very remarkable, with so much death, as has GONE on,. No greater gaps.-- external conditions, to
303c205 y oh look to your fossils, now if extinction had GONE, without creation this would have been fair, bu
345d042 <pure  blooded> «"first blood"» animal must have GONE on for many years, before deserves <name> «to b
345d042 to be so called»,-- the short horned cattle have GONE on for 50 or 70? years-- now «well fixed» breed
385d164 tained perfect teeth & hair-- showing foetus has GONE on growing-- I believe same has happened in boy
469t135 a day is considerably under mark, & this has now GONE on 14 days. (except some wet ones/ & wd go on 1
613o035 leep is man momentarily conscious, but is memory GONE?-- Where pain & pleasure is felt there must be
544m103 r J. Franklin when starved, all party dreamt of <GOO> feasts of good food-- The mind wills to do this
024r015 s in Beechey. on NW coast of America off Cape of GOOD 70 fathoms 20 miles from the shore? Beagle
025r019 m of the sea.--«certainly data insufficient, yet GOOD» «(I suspect fragments of shells will generally
033r043 & Obsidian:» in the I Vol. Humb: There is rather GOOD abstract of Humboldt. S. American Geolog. in Da
034r044 m at Portillo Pass example of do? <Poor> Daubeny GOOD account of ejected granitic fragments P. 386 Me
037c056 . Chonos to Copiapo.--Sydney. K. G. Sound. C. of GOOD Hope.--Carnatic It has been common practice of
06r134 -Much struck with number of animal[s] at Cape of GOOD Hope Says at Santos «M Birchels» at foot of ran
079r177 rophobia at Quito. P 281. do do Australia, C. of GOOD Hope.--Azores Isds «nor at St Helena».--» Humbol
100a053 ssion on concretions and cleavage conjoined very GOOD.-- It is the Key to the story.-- consider stala
101a056 stance lies <.--»? Phillips. Lardner's p. 270-4, GOOD discussion showing present form of land in Nort
109a083 --Coral flats. argument for Heaping up.-- very GOOD this will show effects.-- analogous to broad fl
109a084 in chalk must be studied-- though I do not think GOOD p. 411 When discussing concretions Carbonate so
114a093 ugineous Gallionella Examine iron stone of C. of GOOD Hope & Australia/ and mud of salt-lakes of Rio
114a094 ence in metamorphic action which I give at C. of GOOD Hope.-- A bare hill of greenstone, if we know o
116a099 osophical Journal Rapport on D'Orbigny's Voyage. GOOD section of Rio Negro beds ---- refers to speci
134a144 ournal of Asiatic Soc Vol V. p. p 96. apparently GOOD geological paper. by Malcolmson-- worth reading
136a147 e every-where. Phillips in Ladner Vol II p. 125. GOOD discussion on mineral veins p. 125 to 129 & p.
139aIBC ssac[h]usset would be well worth visiting really GOOD account of ice.-- C. Darwin A. Glen Roy General
152g045 & not in Pass, where shallowest In Glen Collarig GOOD case of shelves entering «on» one side  ravine.
154g059 continued  across to {P} side removed all well & GOOD, but how came river to do this vast quantity wh
172b006 a race, that is alters it from some end which is GOOD for Man.-- Let a pair be introduced & increase
```

174b014 t, would be as it is.-- Hence Antelopes at C. of GOOD Hope-- Marsupials. at Australia-- Will this app
178b032 White Men & Hottentots or Negros cross at C. of GOOD. Hope the children cannot be made intermediate,
182b047 ong the Carabidae.-- instance in birds> Examine «GOOD» collection of insects with this in view.-- Geo
184b054 from salt & F. Water-- on what is species. very GOOD Has not Macculloch written on same changes in F
190b079 Waterhouse says no real <separ> passage between GOOD genera-- How remarkable spines, like on a porcu
191b079 arkable species, like on a porcupine on Echidna-- GOOD to study Regne Animal for Geography.-- The moti
192b085 erfect gradation can be found from forms marking GOOD genera-- by steps so insensible, that each is n
192b085 e had intermediate steps.-- Quote in detail some GOOD instances. But it is other question, whether th
202b130 uences in one country is permanent in another.-- GOOD argument for species not being so closely adapt
205b143 ded.-- Kirby Bridgewater Treatise There are some GOOD accounts of passages of legs into mouth-pieces
211b162 ee, kinds of pigs. difference in skeletons: VERY GOOD. Apteryx. a good instance probably of rudimenta
211b162 . difference in skeletons: VERY GOOD. Apteryx. a GOOD instance probably of rudimentary bones.-- As Wa
211b164 f big Animals in East Indian Archipelago--, very GOOD in connection with Von Buch Volcanic chart & my
216b182 unds, a little mingling would probably have been GOOD, namely such as blood hounds from other parts o
217b183 tter.- never afterwards went in heat.-- This is GOOD instance of same fact in Mr Galtons case.-- It
217b184 s often one way as other.-- He has known case of GOOD pointer & rough water spaniel «produce litter l
218b189 al» repulsion «amounting to impossibility, holds GOOD in plants» between all different forms; therefo
218b190 h. Journal of Natural History-- Preface appeared GOOD with facts about changes when animals transport
218b192 Flora. is given Mr Don remarked to me. that some GOOD African & some good S. American forms. (& dares
218b192 on remarked to me. that some good African & some GOOD S. American forms. (& daresays some of these «A
219b194 ere are any species same as T. del Fuego & C. of GOOD Hope show possibility of transport. If some can
221b200 ?? do.-- The most perfect Plants Composites.--!!«GOOD» those which have undergone most metamorphosis
222b203 ock.-- I think we may look at it so--?? It holds GOOD even with trifling differences of expression --
223b211 sed with (C) prevents offspring of A. becoming a GOOD species well adapted to locality A. but it is i
223b211 D.-- Destroy plants B. C. D. & A will soon form GOOD species! The increased fertility of slightly di
224b212 but experience according to each group)» whether GOOD species, & hence the importance Naturalists att
224b213 s of very near structure.-- Hence species may be GOOD ones & differ scarcely in any external characte
225b219 bsence of quadrupeds East India Archipelago very GOOD on opposite tendency.-- Study Ellis & Williams.
226b220 e Connection between Mauritius & Madagascar very GOOD.-- Fernando Po & Coast of Africa. equally good.
226b220 y good.-- Fernando Po & Coast of Africa. equally GOOD.-- Small isld off New Guinea same fact see Coqu
227b225 s into (D).-- We may foretell species. limits of GOOD species being known.-- It explains the blending
230b240 ficulty, argument in circle.-- Falkland Isd case GOOD one of animals not soon being subjected to chan
230b241 merely gone back previous to fresh change Get a GOOD many examples of animals «or plants» very close
231b244 Camels breed?-- As man has not had time to form GOOD species, so cannot the domesticated animals wit
233b251 reptiles   same from Maurice & Madagascar & C. of GOOD Hope.-- His book Probably worth studying.-- Win
239cIFC etween («beginning of» February & July 1838) All GOOD References selected Dec. 13 1856 Also looked th
241c015 rican forms? All Infusoria. not extinct species. GOOD Resumé do/p. .62 ??? Age of Deinotherium. p. 23
245c023 ral appearance to Siamese kind.-- but considered GOOD species from dental characters, wild pigs. said
245c023 est to swim from one isld to another--\ «It is a GOOD species, with different number of teats» Coquil
247c028 ral» «caught» Chile, Van Diemen's land & Cape of GOOD Hope V. p. 44 of this Note Book Rabbits introdu
251c041 only in form of white mark on breast: p. 234.-- GOOD case.-- p. 526. (ref) To Temminck Monograph. Ma
251c041 526. (ref) To Temminck Monograph. Mammal; «4to GOOD facts about distribution of Cats Vol III. <p.>
252c045 arrangement, 3 <5> Species Rhinoceros Cape town GOOD species Indian species so distinct that all ana
252c045 y Get Closer species-- FOX IS & Mice of America «GOOD case on account of varieties in N America» «som
257c057 and ce nétaient pas tout à fait les mêmes." This GOOD case. of replacement under peculiar conditions-
261c070 e, without its parent running to the water, is a GOOD instance of connate instinct, better than child
264c080 into Crows. yet all forming, according to Gould, GOOD genus Gould seems to doubt how far structure &
266c091 ter treatise, (p..26). about plants from Cape of GOOD Hope continuing for some time to flower at thei
268c099 go. & forest «Parrots in Macquarrie Isd.--» very GOOD. Study D'Orbigny. & range on West Coast «Guayaq
276c123 rsecution of early Astronomers.-- then add chief GOOD of individual sentific men is to push their s
277c126 differ from (C) Chiloe (E) Chile.. rupestris -- GOOD species it is reverting to old plan, but reason
277c126 ertained, call them varieties. but two ostriches GOOD species because interlock analogy to be guide.
280c134 e as well as gain instincts. Wild & tame rabbits GOOD instance-- instincts of many kinds in dogs, is
281c138 bsolute human family is undescribable, yet holds GOOD, so does it in real classification The relation
281c138 lly related to each other I cannot help thinking GOOD analogy might be traced between relationship of
284c149 en old.-- thus speckled form of young blackbird. GOOD remark if general.-- Where any structure is gen
284c149 therefore lest subject to variation.-- + <being> GOOD for generic divisions [+] ought genus to be fou
290c162 ery organ‖ «Hence «method» of generation is very GOOD «general» character in those animals, where muc
290c163 abitation. Must have some effect.-- Maldonado as GOOD forests for beautiful birds.-- heredetary ambli
294c176 origin when species rather further.-- once grant GOOD species as carrion crow & rook formed by descen
306c214 Terrestrial Planariae assuming bright colours., GOOD instance of colours dependent on localities.--
308c219 lly different, because their difference. arise a GOOD deal from climate & habits, & therefore less fe
309c221 wever different.-- Male glow-worm knowing female GOOD case of instinct. bees turning neuter into Quee
311c225 closely allied species. «replacing each other». GOOD to consult p. 326 wild ass extending over 90 de
311c227 estigation.-- Institut 1838. p. 174. Apercu very GOOD on insectiferous <insects> quadrupeds geographi
311c227 ous <insects> quadrupeds geographical range very GOOD.-- Blainville Ovington's Voyage to Surat, float
312c231 best way to improve cattle is to cross between a GOOD bull & <be> the provincial breed, & that first
313c234 ld not new genus have been made, & only species, GOOD argument for origin of man one.-- Is the extinc
313c234 xual passion must arise after long interval very GOOD case.-- habit is awakened by association (case
313c235 utang of (in June 1838 when young male was added GOOD instance of instinct showing itself, not from i
314c239 & Australia, some SPECIES of Australian GENERA'. GOOD case. rather large flora. (150?) Mr Brown did n
315c243 ing to tell other animals in associated kinds of GOOD news. discovery of prey.-- arising no doubt fro
316c244 when perfect enough for future state, that when GOOD enough for Heaven or bad enough for Hell.-- «tg
316c244 y of storm of snow after his brothers murder.--» GOOD anecdote --.Sowerby.--. Geographical range of s
316c246 Africa & E. S. America.-- get instances.-- very GOOD anomaly in range + + What circumstances have le
317c248 end of "White's Selbourne." many references very GOOD. also "Rays Wisdom of God." Often refer to thes
319c257 herium of Europe is between the Anteater of C of GOOD Hope & those of S. America,-- Are not some of t
325c267 de Serres Cavernes d'Ossements 3d. Edit. Octav. (GOOD to trace Europèan forms compared with African <
326c266 4to 1801.-- quoted by do.-- There appears to be GOOD art.. on <Etn> Entozoa by Owen in Encyclop. of
326c266 gy.-- on instinct & animal, intelligence.-- very GOOD. Endlicher has published in first volume of Ann
327c265 erica" in Lib. of Hort. Soc Mr Neil. has written GOOD article on Horticulture in Edinburgh. Encyclop.
327c265 s with locality in which each one is found. very GOOD to see whether peculiar plants-- in high points
339d025 . two races of Men, but not plants» will it hold GOOD.-- Thinks Temmink doubtful when he says No gene
340d026 confounds, like Whewell affinity with analogy-- GOOD table at end of distrib: of <birds>. Anatidae.-
345d044 e Magellanic Fox.-- peculiar hair & appearance-- GOOD case of Provincial Breed-- Highland Sheep jet b
349d052 plains (p. 53) of M. Edwards, thinking any group GOOD though not circular, if characters can be esta
353d061 t number of vertebrae in Irish & English Hare.-- GOOD case these hares compared to North American har
356d070 on of generat. being highest end of organization GOOD expression but does not include so many facts a
359d088 of their propagating powers. (as if they were a GOOD species, or local variety, & not effect of bree
365d106 e fact of Egyptian animals not having changed is GOOD-- I scarcely hesitate to say that if there had
366d111 ir [not located] <The> every case common to many GOOD species; & therefore to genera (&_the uncles &
376d138 taking up trail» Mem: Ourang Jenny with Tommy.-- GOOD evidence of knowledge of Woman-- ‖ The noise st
378d147 to species,. we can see why young & female alike GOOD Ch 6 Keep Is it Male that assumes change, & is
385d165 of the father in the daughters! This last remark GOOD. because showing probably not education.-- Cann
386d166 tem.-- of that knowledge of the part, of what is GOOD for the whole.-- if cut off nerves in snail. (E
399e006 k to find some change in upper & lower layers.-- GOOD objection to my theory: a modern bed at present
400e011 Lucanidae» <no> less difficulty in establishing GOOD groups.-- ears varying so much,-- kind of fur--
400e012 il varies, & character of fur-- I am sure a very GOOD case, might be made out of variation analogous
403e022 ays, that <every> «any» character even colour is GOOD. (ie invariable) in some classes-- it is becaus
404e024 his is directly incorrect The case of my mice is GOOD. because it is an involuntary variation made by
406e036 of offspring to Parents same laws appear to hold GOOD. with regard to marriage of individuals, & vari
410e052 , from its garments, without some end.-- Respect GOOD describers like Richardson.-- The relations of
416e072 Paris basin.-- its relation to African Species <GOOD observations.-- >, larger than any living [not
420e088 g towards Hyaena.-- see Comte Rendu.-- I suspect GOOD case of fossil filling up blank.-- CD[not betwe
424e102 y year produce hybrids-- now this is independent GOOD case, but very odd since these crows are mixed
428e113 triking, about indelibleness, than the number of GOOD race-horses, which Eclipse? has begotten <?> «W
431e121 formations Portland Stone &c &c.--if? so «it is» GOOD case:-- in Sowerby Min. Conch. it is however, s
437e138 finding   their way to the Caymans from Honduras. GOOD case of migrating.-- shows my theory insufficie

Page **(Key Word)**
449e173 enaeum: 1839. p. 451. Sheep Merinos from Cape of GOOD Hope,. has different constitution from those of
453e183 I presume seed raised in same garden.-- now this GOOD question-- single, or half double.-- anyhow fer
458t002 howing something of Mule in its ears-- ((this is GOOD case as showing gradations, Boteler's Narrative
460t019 rganic beings are the descendants, <slightly> «a GOOD deal» modified <& Many Forms lost; if» «of this
462tf4v d by the Mataco-armadillo & the woodlouse-- -- a GOOD analogy-- sea-Crustacea-- Tullus. Athenaeum 183
479z005 of Coquille wide limits of Nullipora Discussion GOOD on Falklands birds Discussion of Firola,-- Salp
480z006 f S. America. D'.Orbigny. L'.Institut. No.-- 221 GOOD account of Condor by Humboldt Zoologie Recuiel-
481z009 . Mackenzie. p 345 for comparison with Falkland. GOOD also for Journal.-- 18 Admirable engravings in
491q01v organs (not being always useful). fail-- Really GOOD subject for experiment.--«to repeat Spallanzani
491q002 ion, produce flowers, like the Oxalis from C. of GOOD Hope mentioned by Mr Herbert in vol IV. Hort. T
493q003 Hilaire says length differs in different cats.-- GOOD observation-- examine semen of Hybrid animal. i
493q003 weigh that of Merinos & see if relation is same GOOD, avoids effects of fatness.-- Experiment in cro
496q05a chis= How many flowers in minute do they visit?? GOOD=!! Examine pollen of double flowers. compared w
498q007 ost this character in grt degree from charcoal & GOOD treatment (8) Do bees frequent Cabbages «& Cowc
498q08v group of different species growing White Mullein GOOD plant to sow & try to get other species <near>
499q010 ntrivance for transportal, does he include seeds GOOD to eat. (even Nux Vomica is eaten by a Buceros
500q10a in volcanic or rising isld, there ought to be a GOOD many races or doubtful species; how is this   at
505q014 the horned orange. wh. never has seeds produced GOOD pollen? Yes <From cultivation lost their horns»
507q15v spines-- or seed-cases with similar structure.= GOOD case as showing how simple, but beautiful adapt
507q016 ny peculiarity in milk teeth inheritable!!! very GOOD Any peculiarity in the males of a family-- Wher
509q017 rtant to know, whether Gould's observation holds GOOD, that in the mundane genera, the species <are>
511q018 nt looking animal be formed-- not caring whether GOOD or bad.-- are any actually rejected?? (8) Get S
514q21. of cultivated Orchis & Asclepias &-- carnosa?-- GOOD-- Norfolk Isd-- geology. volcanic? Applies to m
515q021 y not ripen.-- The point to attend to is whether GOOD & plenty of pollen is produced. & 2d if so, whe
520m002 in early infancy-- of this fact Mr Dryden C. is GOOD instance as he is very deficient, he was nearly
533m059 ulty of getting on a horse on the left side (not GOOD example,) because leg is right handed.-- In Rev
535m063 it does it always instinctively or habitually.-- GOOD Heavens is it disputed that a wasp has this muc
536m074 he would be most humble, he would strive <to do GOOD> «to improve his organization» for his children
539m081 very much struck with an intense headache «after GOOD days work» which came on from reading «review o
539m083 s. cranks, & so is Catherine in some respects--. GOOD instances.-- when education same.-- My handwrit
544m102 eath. Mayo Philosophy of Living. p. 140-- Dreams GOOD account of «thinks» are recollected when intens
544m103 hen starved, all party dreamt of <goo> feasts of GOOD food-- The mind wills to do this & hears that,
546m109 ne, looking back to his life, would say how many GOOD dinners or......... he had had, he would say ho
549m118 f pleasant thoughts, he must have contingency of GOOD food, no pain,-- <but the> «&» the sensual enjo
549m119 pleasant. when the conscience tells our [mind], GOOD has been done-- «& conscience free from offence
549m120 is happiness. though he sees some «intellectual» GOOD men, from insanity &c unhappy-- perhaps not so
551m128 did with far more alacrity <than> when something GOOD was shown him, than when merely ordered to do i
552m132 w animals in social animals) are those which are GOOD & consequently give pleasure, & not as Paleys r
552m132 as Paleys rule is those that on long run will do GOOD.-- alter will in all cases to have & origin as
554m139 per» see whether, this was permitted & eat it.-- GOOD case of association.-- «Listened with great att
554m141 fly at him-- Knowledge.-- Sept. 13th It will be GOOD to give Abercrombie's definition of "reason" &
556m146 protrusion of chin, like bulls & horses.-- 1838 GOOD instance of useless muscular tricks accompanyin
556m146 et continue to put down ears, when kicking.-- -- GOOD case of expression showing real affinity in fac
558m153 All> Nearly all will exclaim, your arguments are GOOD but look at the immense difference. between man
566n011 , thinking he instinctively knows distances:, is GOOD instance of obtaining <that> «a» faculty in the
568n017 ction which experience shows will be for general GOOD, or in case of any fantastic custom» «Probably
568n019 to t[he] whole kingdom of nature. If I want some GOOD passages against, opposition of divines to prog
569n022 -- Expression leave «this» out post in Library no GOOD There is a Lutké's Voyage autour du Monde (1826
571n029 ng poisonous <herbs:» & man not.-- ¿no vegetable GOOD <for man» to eat poisonous?-- How did animals i
578n054 ink of my face,"? to one moral conduct.-- either GOOD or bad. either giving a beggar, & expecting adm
578n055 that any one suspects one of having done either GOOD or bad action, it always bear some references t
581n065 that many learned men seem to consider there is GOOD evidence in the structure of language, that it
585n075 with back of Head, yet never puts down its ear. GOOD to contrast with horses, asses, <mi> Zebras &c
590n094 ntences spoken & believed to be audible. one has GOOD ground to call imagination a faculty, a power,
591n099 t is because instincts to woman is not followed; GOOD case of instinctive conscience.-- Why does not
593n109 ank] will not do for insects. if this view holds GOOD-- then man, a socialist, does not know other me
599o006 gals cave, & in the arched & leafy forests» Very GOOD!. I grant that the thrill, which runs throug ev
601o08a he same-- In its action-- emotions-- p 176 & 177 GOOD passage in French on what dog dreams, awakes--
601o08b hysiologiques The first of these books I daresay GOOD. 1. Sensation is the <conse> ordering contracti
602o011 olition &c Reynold XIII Discourse (p 115) a very GOOD passage. about actions & decisions bein the res
603o014 y of <handsome> «UGLY healthy» young woman, with GOOD expression-- statues not painted-- <music» very
603o014 expression-- statues not painted-- <music» very GOOD article-- why flower beautiful? ¿even to childr
605o20v ther in the power of discriminating & respecting GOOD from bad. And it is manifestly from this fact &
608o026 arises from motives.--& his knowledge that it is GOOD for him effect of Education & mental capabiliti
608o027 dit for anything. (yet one takes it for beauty & GOOD temper), nor ought one to blame others.-- This
608o027 much, & he will know his happiness lays in doing GOOD & being perfect, & therefore will not be tempte
609o030 most identical» + What has produced the greatest GOOD «or rather what was necessary for good at all»
609o030 greatest good «or rather what was necessary for GOOD at all» is the «instinctive» moral senses: (& t
609o030 e result of what has generally been best for our GOOD far back.-- (much further than we can look forw
610o30v ience of others) shows does not tend to greatest GOOD.-- Therefore rule of happiness is to certain de
614o037 , ∴ therefore consciousness, therefore reward in GOOD life [RHC] Instinct appear like heredetary memo
614o037 ts length.: Many animals (as horses) very long & GOOD memories-- but on its multiplicity & the compar
615o36v Tropical plants when imported & plants sleeping GOOD show acquirement or obliteration of instincts B
620o044 , everyone must know, how soon the pleasure from GOOD dinner, or from a blow struck in passion fades
621o047 inks that nasty, which the natural tastes say is GOOD. yet horseflesh show that even this is possible
621o048 ng.-- [RHC] 7) Hence, what parents think will be GOOD for the child on the long run, & for themselves
625o50v too strong for our instincts. to gain long-lived GOOD, ie happiness-- yet this system not selfish.--
625o50v ined by principles if Mackintosh.-- p. 262. Some GOOD remarks, on analogy of pleasure of imagination
627o52v to striking blows.--' p. 224.-- Hume's Inquiry-- GOOD abstract of Butler & arguments of beneficial te
628o53v uch cases-- either, that from the necessities «& GOOD» of society such conduct is instinctive in me (
628o054 & if taught rightly, it will be for the general GOOD, that is, the same cause, which gives the insti
628o054 ts, than other passions, or instincts.-- is this GOOD?-- I should think some parts of the emotive par
633j53v aptations.-- May they not be accidental? We have GOOD reason to know that they would not be detriment
634j54v bortive muscles to his ears.-- p. 313 Many other GOOD cases -- p. do]CD <Mac. remarks all Mammifers o
637j58r rsupial bones especial adaptation, to «young»-.- GOOD God & yet Mails have them. What trash p. 237. G
638j28r foot, sack. power of endurance &c &c Camels? all GOOD cases of corelations.-- [There must have been d
538m079 arrelsome as B.e on board Beagle, some merry GOODHUMOURED as self.-- «When Miss Cogan has remembered
525m026 puppy.-- The common remark that fat men are GOODNATURED, & vice versa Walter Scotts remark how odiou
466t096 rs, wh. come on in after life of Plants-- also GOODNESS of flavour in fruit-- all affected by cultiva
553m136 feel the most implicit faith that through the GOODNESS of God knowledge has been communicat to us».-
185b055 esert. Kingfisher.-- mountain tringas.-- Upland GOOSE.-- water chionis water rat with land structures
334d013 younger cow.-- {P} Fox says when common & China GOOSE are crossed the neck is not intermediate in its
334d013 s peculiar long neck, but nearer to common GOOSE.-- What has long been in blood, will remain in
341d032 brid of Penguin duck a variety of Muscovy) with GOOSE!!) Dr. Bachman regularly breeds «in Carolina» f
359d087 oss between Penguin Duck «from Bombay» & Canada GOOSE.-- Former strange mishaped bird-- looks very ar
449e168 olour.-- Eyton has observed same thing in Brent GOOSE. Eyton says some of the pidgeons in common Dove
449e169 like» between the male Chinense & female common GOOSE took after the common goose thus contradicting
449e169 nse & female common goose took after the common GOOSE thus contradicting (probably) Yarrells law & Wa
460t013 rs & species The <male> swan-gander with common GOOSE produce full as many eggs as pure bred common.-
460t013 of these half-bred ganders. crossed with common GOOSE <to has» «produce offspring with» so much of th
460t013 s» «produce offspring with» so much of the swan-GOOSE in appearance Bell at Hornsey (though only ½ of
460t015 d). that it appears about half way between swan-GOOSE & common goose.-- the stripe down back pretty p
460t015 ears about half way between swan-goose & common-GOOSE.-- the stripe down back pretty plain in in thes
558m153 with wing arches wings-- as does black Swan.-- GOOSE do all species put their necks straight out & h
591n097 ry) describes effects of emotions-- fear giving GOOSE skin-- & hair standing on end.-- July 20th Inte
325c267 ions. 1827 Paxton on the culture of Dahlias Mrs. GORE on Roses might be worth consult. Paper on Consc
346d048 ung one 203 days old butted violently. & fell.-- GORE to death the old & wounded,-- see Annals. vol.
105a070 ing up of valleys-- subsequent opening a medial GORGE by slow erosion. but we have evidence in distri

120a109 arry further its own matter. but would cut wide GORGE. leaving cliffs, on each side, such as now exis
155g064 emove, what above straight line «only» cut deep GORGE on sea hypothesis, if gullies not now formed «(
158g083 little way» in Upper Glen Roy at pass {P} River GORGE 4th Sh side of valley Granite blocks on this si
089a020 837. Peninsula of Cape Verd. volcanic.-- Isle of GORGE. rocks encrusted with serpula-- Isle of Cayenne
442e152 pecies may stop for any number of generations-- GORZE in Norway, which never flowers!!-- <How did it
442e153 emmation should be absolutely similar; [all the GORZE in Norway ought to be thus characterized study
443e153 fects the Graft.-- ¶Plants circumstanced as the GORZE must be propagated by its roots: now it is curi
492q002 are seeds ever raised? 11. Is not non-flowering GORZE common in Norway No Questions regarding Breedin
154g060 lake it did but little more {P} now that it has GOT to the rock of cols if--. why should it deposits
228b232 oul by consent of all is superadded, animals not GOT it, not look forward» if we choose to let conjec
272c108 .-- some orthopterous insects & some third, have GOT thighs with same peculiar structure & habits of
275c120 with finest greyhounds.-- Sir. J. Sebright first GOT {P} point on hackles on Bantams by crossing with
275c121 s of each brood.-- These bantam feathers at last GOT ducky, then took white Chinese Bantam crossed &
275c121 ducky, then took white Chinese Bantam crossed & GOT some yellow & others yellowish & white varieties
306c212 ore down the stream followed ebb tide, therefore GOT into habit of going down stream which would last
401e016 ly would take long before all the stain would be GOT out of it.-- Now this is curiously different fro
437e139 rats & not being sufficiently weakened by wounds GOT off from the young ones while they were amusing
531m049 by Hubberley to visit daily to see how the young GOT on. nest the cat could If cats will «ever»
539m081 deeply, & the immediate manner in which my head GOT well when reading article by Boz.-- now in this
182b050 on & ftutirity. it does as yet appear clim In Mr GOULD Australian work some most curious cases. of clo
182b050 & Van Diemen's land. & Austral & New Zealand Mr GOULD says in sub-genera, they undoubtedly come from
203b138 mer Eyton's paper on Hybrids Loudon's Magazine. GOULD on Motacilla,. «species peculiar to Continent &
213b171 age-- How long back have insects been known? As GOULD remarked to me, the "beauty of species is their
233b249 to Mountain chain ought to be Australian.?-- Mr GOULD has been stuck with similar extension of form
241c015 t of some birds of Tongatabou. & New Ireland.-- GOULD will hereafter know about birds of N. Zealand L
251c040 ations India & Africa.-- NB. Any monograph like GOULD on Trogons worth studying.-- «do» Zoolog Journa
261c071 tecedent races, perhaps not on now existing» Mr GOULD says wherever any mark like red patch on wings
264c080 language of Fuegian, puts on par with Monkeys» GOULD seems to think that many species when close com
264c080 into into Crows. yet all forming, according to GOULD, good genus Gould seems to doubt how far struct
264c081 yet all forming, according to Gould, good genus GOULD seems to doubt how far structure & habits go to
264c082 ut tail of some ducks aberrant from-- habits.-- GOULD I see quite recognizes habits in making out cla
266c088 f> become dull coloured in sterile countries.-- GOULD insist much upon knowing to what type a bird be
272c109 linidae on St. Pauls Rocks must be placed under GOULD says most subgenera confined to continent, thou
273c110 World--. + + If in any «well developed» family (GOULD) says there is any marked colouring of plumage
273c111 likewise rasorial species, & likewise perching (GOULD), but the latter is obscure because nearly all
274c113 o this type, ¿the Humming bird? the woodpeckers (GOULD) says, he believes does. but also on fruit.-- Th
274c114 oing before structure).-- even one kingfisher-- GOULD has seen with long tarsi.-- «Ground woodpecker»
274c114 not many ground woodpeckers?-- In each division GOULD thinks he can trace structure for insects & str
283c143 «must» have had the character of analogical.-- GOULD says it is only in large groups. where you have
283c144 as shown by frigate Bird & flying eagle.-- Hawk GOULD seemed to think, that widow bird. replaced Bird
291c065 ry & strictly point out affinities. conducct of GOULD, remark of D'orbigny point out importance of ha
298c189 his most puzzling whether instinct, or reason?? GOULD says he believes that he has seen half fox & do
353d061 zine of Natural History. 1838 vol II p. 402. Mr GOULD on Australian birds-- all Eagles. of Australia
484z014 ess) parasitical on Vellellae in Atlantic Ocean GOULD agrees with D'Orbigny, that Serpent Eater-- or
489qIFC ather. And. Smith Dr. Holland p. 16 Babington-- GOULD ---- 10.(a) J. Gray ---- 17 Yarrell---- 18 Blyt
500q10a & these are now to be planted this year copied GOULD.-- Number of species of Birds in New Zealand, p
500q10a Species & British & distant parts of Europe.-- GOULD-- go over the Pigeons, Philotis, Dacelo. Alcyon
502q11v or sterility Land Birds Madeira Migratory-- ask GOULD about N. Zealand, as Cuculus lucidus is.-- Ask
230b241 or plants» very close (take Europaean birds. Mr GOULDS' case of Willow wren) & others varying in wild
276c125 t same country are not habits different, (Mem: GOULD'S Willow Wren) but where close species inhabit d
314c239 Eucalyptus!-- (Hostile fact) Be cautious about GOULDS case of birds of Van Diemens land & Australia.
363d102 by incubation & other peculiarities.-- (Mem.-- GOULDS Willow Wren.--) (Goulds story of Water-Wagtail
363d102 eoliarities.-- (Mem.-- Goulds Willow Wren.--) (GOULDS story of Water-Wagtails mistake both species s
509q017 [<...>] It is very important to know, whether GOULD'S observation holds good, that in the mundane ge
508q016 e Dr. H. thinks asthma in females takes place of GOUT.-- How are livers obscure organ. no answer?-- 3
530m044 after bad accidents:-- After journey, a fit of GOUT, has affected his memory of everything in <he a
398e006 obliterations & fresh laws created., & yet with <GOV> symmetry «& regular laws» that baffles idea of
610o034 ted by certain laws different from those., that GOVERN in the inorganic world; life itself being, the
573m036 , nay whole systems of universe <of man> to be GOVERNED by laws,, but the smallest insect, we wish to
616o038 ere must be cases of secretion being some time GOVERNED by will in some animals, involuntary in other
559m154 t." This unwillingness to consider Creator as GOVERNING by laws is probably that as long as we consid
398e006 btain a glimpse only of the changes which the GOVERNMENT is subject to.-- further back we obtain here
477z002 . introduced in Paraguay in 1769 introduce in GOVERNOR'S tran?? Azara Las Vinchuca or Benchuca. "Les
526m027 ancy there is such a thing as chance.-- chance GOVERS the descent of a farthing, free will determine
489qIFC ss.-- (March, 1842) Questions & Experiments {T} GOWEN, Royle, & Horsfield Sykes p. 12 Maer. p. 13 Que
503q012 ication-- said to require Selection (36) Ask Mr GOWEN to ask Mr Herbert, how many generations any hyb
427e109 there «& there & there only: as stated by Capt. GRAAH») & break up. N. American Conchology from Europ
235b272 ook Smellie Philos of Zoolog. 842 White regular GRADAT in Man poor trash Lyell 1024 Flemings Philosop
063r130 cara> Calandria: inosculation alone shows not GRADATION;-- An argument for the Crust of globe being t
119a106 emperature is partly known-- <[...]> moreover GRADATION from gneiss to granite shows that the metamor
180b036 B. immens gap of relation. C & B. the finest GRADATION, B & D rather greater distinction Thus genera
181b044 earing stamp of <some> great main type, & the GRADATION will be sudden-- Heaven know whether this agr
189b073 or later.-- Prove animal like plants:-- trace GRADATION between associated & non associated animals.-
192b085 them.= In some of the lower orders a perfect GRADATION can be found from forms marking good genera--
223b209 Tasmania The reason why there is not perfect GRADATION of change in species,, as physical changes ar
230b235 try or ascending mountain you ought to have a GRADATION of species, now this notoriously is not the c
256c055 s (p 642) anniversary Speech. Feb 1838 thinks GRADATION between Man & animal, small point in tracing
263c076 fabric totters & falls.-- look abroad, study GRADATION. study unity of type-- Study geographical dis
276c122 een no offspring & ordinary offspring:-- this GRADATION is infertile offspring. without organs of gen
276c124 re, (& in all such structure) there cannot be GRADATION. See what Eytons young pigs-- if vertebrae mu
291c168 Female flowers unimpregnated Babington We see GRADATION.-- It is easy to see if South America grew ve
299c196 resent, then there would be perfect series or GRADATION to mans mind in Vertebrate Kindgdom in more i
316c244 how faint in a Fuegian or Australian! why not GRADATION.-- no greater difficulty for Deity to choose,
324c268 s on the influence of climate White's regular GRADATION in Man. Lindlys introduction to the Natural s
335d016 union of perfect animals is distinct case,-- GRADATION from physical impossibility to (perhaps incre
381d157 -- NB. in Heautandrous animals <are> is there GRADATION of structure leading to supposition, that the
416e071 s & not others.-- Term variety may be used to GRADATION of changes which gradation shows it to be the
416e072 ety may be used to gradation of changes which GRADATION shows it to be the effect of a gradation in d
416e072 hich gradation shows it to be the effect of a GRADATION in difference in external conditions.-- -- as
419e087 he> it also the effect of change) than a slow GRADATION in form, «which must be effect of slow change
433e126 -- 248 & p. 258 A beautiful case, showing the GRADATION from one grand system to another: in each sys
447e164 calepha do not cross there would by my theory GRADATION of form from one species to other: therefore
465t081 so that we here see reasons-- why no perfect GRADATION can be expected in any one country.-- in a de
502q11v carmine in spirits & then experimentise: for GRADATION in structure Compare flowers of wild & tame c
522m013 ing understood.-- My F. says there is perfect GRADATION between sound people and insane.-- that every
596n184 <its> «the» ears but <not> Chimpaze. does not GRADATION towards man.» Macacus especially pulls back s
633j54v y Cuviers Anatomie Comparé» p 308. Traces the GRADATION of skeleton in Vertebrates & constantly allud
264c080 ned into continent &c &.-- There is beautiful GRADATIONS of forms in Australia leading on one side in
447e163 e individuals.-- «-- hence there would be real GRADATIONS in species from one region to another.--» --
458t002 in its ears-- ((this is good case as showing GRADATIONS, Boteler's Narrative Voyage East coast of Af
172b008 -- something like a variety produced-- --[every GRADE in that case surely is not Produced?-- <Grantin
192b086 ptiles?) For instance there never may have been GRADE between pig & tapir, yet from some common proge
431e120 is coming to this conclusion, from specimens in GRADES, now L. says the T. costatus is in England fou
024r014 coming from the [...] Cordillera & flowing The GRADUAL shoaling of the water to more than 100 fathoms
032r039 he level of the sea was to sink by very slow & GRADUAL movements to line (2). The part (o) which was
032r040 size of the triangular mass removed vary.--The GRADUAL rising continuing. a another sloping platform
063r130 late, so must we believe ancient ones: «.» not GRADUAL change or degeneration. from circumstances: if
080r180 lections on the geology of the world P. 14-91. GRADUAL shoaling of coasts 93 action of sea on coast.

```
160g093 h appear perfectly level, <on op> dies away on GRADUAL slope-- : on N side.. dies away on rocky place
203b137 ed animals over the whole world-- For instance GRADUAL reduction of tempereture from geographical  or
223b209 of change in species,, as physical changes are GRADUAL, is this if after isolation (seed blown into d
230b239 o very slow changes. without crossing.-- Now a GRADUAL change can only be traced geologically (& then
276c122 well as species-- as formation of species <of> GRADUAL, so may we suppose., that something intermedia
298c191 lpine species from one in lower country during  GRADUAL, elevation of isld.-- We must imagine a conside
031r037 field -- because lowered & raised--so on--but   GRADUALLY & simply raised No Faults in Patagonia[,] eno
035r048 fluid rock.-- Try on globe. with slip paper a   GRADUALLY curved enlargement see its increased  length.
102a061 , owing  to separation having taken place most  GRADUALLY, first the more fusible substance, & then the
110a086 ragment disseminated on that side of salband.   GRADUALLY becoming finer grained & more compact on that
150g035 ng up great bight just as Dick shows NB. Lake   GRADUALLY draining off would form plains such as  those
150g035 e near Bridge Roy (& other cases) but then if   GRADUALLY drained, where is barrier {P} great waterworn
158g081 eath shelf peat on pebbles tidal plain as sea   GRADUALLY retired, hard to explain on river doctrine <L
318c255 facts show the Normal condition of Migration)   GRADUALLY separated the birds might yet remember  which
360d090 though no fluid came from them.-- showing how   GRADUALLY every <thing> «change» is effected)-- the one
400e013 they broke.--, showing effects of cultivation   GRADUALLY adding up. & four more generations before the
592n101 olved by considering the origin of reason. as   GRADUALLY developed. see Hume on Sceptical  Philosophy.
04r088 groups.--As these greenstone rocks are seen to   GRADUATE into granites the <conta> passage from lava t
463t057 hat any argument for transmut. from one organ   GRADUATING into other is lost, <be> (as  vertebrae  into
256c055 & Hamilton-- found tertiary formation amongst   GRAECIAN isles, ¿See if type continued?-- See to Bobla
528m038 shown by Humboldt from occurrence in Mexican &   GRAECIAN to be single cause) this symmetry & rhythm  ap
570m027 forms  change, so must idea of beauty.-- [Old   GRAECIANS living amongst naked figures, & observing pow
281c137 ration it is.-- show how finely the series is   GRAEDUATED.-- Dr Beck doubt if local varieties should b
401e017 rescence Gardeners. by chance <often> sometimes GRAFT pears on apples. they will live but not flouris
443e153 know that the kind of stock greatly affects the GRAFT.-- ¶Plants circumstanced as the Gorze must be p
443e153 s: now it is curious Mr K. has observed that to GRAFT from the roots is the best way to get young tre
443e154 t kinds, & quotes from Pliny, that it is bad to GRAFT from top shoots.-- If prolongation of life by g
401e017 y will live but not flourish-- a medlar may be  GRAFTED on pear. Mountain-ash & white Thorn! Species n
441e148 then  is case of change analogous to change in  GRAFTED.-- No than by growth-- generation; & more of t
442e153 D Now Mr Knights statements about fruit trees.  GRAFTED. altering is hostile to this: but on other han
446e162 then  is case of change analogous to change in  GRAFTED trees «:so is not effect of different stocks i
501q011 by Gaertner (28) Can any annual or Biennial be  GRAFTED or cuttings taken or tuber-- talk about Mr Kni
516q023 ments in Garden Sow stones of Standard Apricot  GRAFTED on what, & see what comes up.-- [Unnumbered bl
622o049 wrong.-- on which «almost» any other might be   GRAFTED.-- Origin of the instincts Hartley, (according
139a180 ht be tried by comparing Zoophite to plants.--  GRAFTING length of life &c &c Will any inorganic subst
417e077 turae.-- Mr. Knight makes this analogy between  GRAFTING & sexual union-- Looking at simple generation
417e077 lation between the fluids of the two as in the  GRAFTING of trees.-- ]CD CD[The similarity of child to
218b192 dded with others.-- we see a beginning to isld. GRAHAM isld.-- we know many seeds, might be transport
429e115 ould form fringe.-- but there is a contest. &a  GRAIN of sand turns the balance.-- Hort. Transact Vol
498q008 arate division of germen <?>-- (11) Must pollen GRAIN be whole, to impregnate?-- I presume only stigm
110a086 that side of salband. gradually becoming finer  GRAINED & more compact on that side-- separation DISTI
160g090 h pinkish felspar;-- whole hill dark grey fine  GRAINED. Much contorted gneiss «narrow sharp ridge wit
035r048 reathed traveller» Mountains, which in size are GRAINS of sand, in this view sink into their proper i
301c199 idual p. 7. is not squirrel hoarding, & killing GRAINS. acquirable through hoarding from short time.-
496g05a ble flowers. compared with single & see whether GRAINS flaccid, as Koelreuter describes Kill  Sparrow
498q008 terbalance each other? (10) Is number of pollen GRAINS necessary to impregnate ordinary number of see
499q009 pregnation.-- (16) Any calculation of number of GRAINS of pollen in any one flower (17) Catch Bees, B
527m031 l & chance are synonymous.-- Shake ten thousand GRAINS of sand together & one will be uppermost:-- so
373d133 l proportion in shells in Arctic Ocean. p. 350  GRALLAE in Wealden. oldest birds. p. 411-- Decapod Cr
639j28v rine runs through Macculloch, the bills of the  GRALLAE <are> «have been made» long «(as adapted  to)»
639j28v have  manifold traces in, the several genera of GRALLAE Suppose six puppies are born «& it so chances,
073r164 . = Veins of Zimapan offer zeolite. stilbite.   GRAMMALITE. pyenite. native sulphur.. fluor spar. bayte
023r011 re of the Pacifick is 60 miles distant from the GRAND ancient volcanic axis «of the Andes».-- «Has t
035r048 per insignificance; as fractures, consequent on GRAND rise, & angular displacement, consequent of inj
036r049 art of the above in Patagonian paper; & part in GRAND discussion Consult. reconsult Geolog. Map of Eu
038r057 trongest is the consideration of the state at a GRAND eruption when whole summit of mountain is blown
039r061 es at Coquimbo. the analogy is now perfect <The GRAND propulsion of fluid rock, which elevates a cont
041r064 rees with formation of pebbles & vertical trees GRAND Seco at B. Ayres; mention about the deer approa
054r102 profound  valley [.] Sydney. -- Lesson Zoologie GRAND tertiary formation of Payta: N. part of New Zee
055r108 decay atmospheric of the most solid rocks.--The GRAND cliffs of a thousand feet in height, of the sol
056r109 uch trifling means could efface & obliterate so GRAND a work?--In valleys one is not sure whether fis
064r137 others in quantity of Sulph. acid emitted: mem: GRAND gypseous formation of Cordillera In  describing
069r152 ected.-- Oh the vast power of the ocean! Make a GRAND analogy between Wealden & Bolivia Transportal o
074r165 iron pyrite in a fossil» Insist strongly on the GRAND fact of Volcanic & non Volcanic. Then Solfatara
100a053 tral point. can cleavage be radiation from some GRAND centre.-- A Stalactite of Gypsum, is the best c
118a103 rds & is now cut off by denudation it gives one GRAND idea of amount denudation.-- This may be  added
198b114 an. [Second resumé well worth studying] CD says GRAND idea god giving laws & & then leaving all to fo
200b123 most obscure. without doubt:-- Vide cattle: The GRAND fact is to establish whether in crossing very o
228b229 speculations with respect to past & future. The GRAND Question, which every naturalist ought to  have
235b275 ifferent kind of localities.-- if so change The GRAND QUESTION Are there races of plants run wild  or
254c047 tion of the Species--! How does this agree with GRAND fact of Marsupial, low Cerebral structure??-- «
271c106 ic virtue.-- it is expression for ignorance Two GRAND classes of varieties.; one where offspring pick
280c135 even might be said.-- My theory thus explains a GRAND apparent anomaly in nature. <t>----- Many animal
399e009 -- it will decrease & be driven outwards in the GRAND crush of population.-- Octob 10th. Saw. two und
413e059 of new species. the mystery of mysteries. & has GRAND passage upon problem.! Hurrah.-- "intermediate
433e126 beautiful  case, showing the gradation from one GRAND system to another: in each system, the  changes
440e145 -- but be it remembered. how little part of the GRAND Mystery is this;-- the law of growth, that whic
463t063 ound unknown coast in front of it.-- Cuvier has GRAND sentence about the Animaux fossiles-- being a m
483z013 writing  on Planariae or Polypi & is especially GRAND paper. p. 387. "on Classification of such anima
522m013 ather not going against natural instincts.-- My GRAND F. thought the feeling of anger, which rises al
523m018 furor  brevis est.-- My father quite believe my GRAND F doctrine is true, that the only cure for madn
527m033 the  latter thoughts are in same manner vivid & GRAND. the frame of mind being just kept up by the mu
566m012 e Comte's idea of theological state of science, GRAND idea: as before having analogy to guide one  to
575m043 rs from their brains» then can we deny that the GRAND child dug for mice from some peculiarity of str
581m063 of  heredetariness in habits. be considered. as GRAND step if it can be generalize.-- The tastes of m
599o007 fibre,  when one behold the last rays of & & or GRAND chorus are utterly inexplicable-- I cannot <adm
633j54r rce, for these Marsh plants. All flow from some GRAND & simple laws.-- 4 «Study Cuviers Anatomie Comp
560mIBC Father ever known <intemperance> «disease» in  GRANDCHILD, when father has not had it. but where grand
344d042 she was halfbred Beagle Staghound. «++».;the    GRANDCHILDREN went back to either paret, & breed not fixe
412e058 Three  principles, will account for all (1)     GRANDCHILDREN. like. grandfathers (2) Tendency to small c
077r171 2 W, & is nearly the same with that of the veta GRANDE of Zacatecas, & veins of Tasco & Moran--of Gua
078r175 clination of the vein".-- at Zacatecas the veta GRANDE has same direction as Guanax.--the other E & W
315c241 ts L.' Institut «1838» p. 184 Botany of Bonin.  "GRANDE analogie avec le flore du Japon", some Europae
343d036 y to the present time, to the future-- How far  GRANDER than idea from cramped imagination that God cr
306c213 on a captial character. (Owen) not for first &  GRANDEST divisions. but for ones of very high order. n
606o022 an Winkleman's. who says it is simplicity with  GRANDEUR of character.-- Hence Lessings shows expressi
179b032 thus  seen the black blood come out from the    GRANDFATHER, (when the mother was nearly quite white) in
197b112 othing about propagation= I see nothing like    GRANDFATHER of Mammalia & birds &c ¶ p. 32. reference to
280c136 ainst the tendency.-- it tries to go back to    GRANDFATHER, but if too unlike its own parent this impos
280c136 two certain varieties, they would go back to    GRANDFATHER, which is true) & infertility is consequence
285c150 ,-- & «hermaphrodites» father « <mother> » &    GRANDFATHER <Might> Must be introduced & made  young.--¶
426e108 usins, the nearest blood being a great great-    GRANDFATHER.-- -- Little Miss Hibbert case of Hindoism c
451e176 lbino described "To this day the tomb of his    GRANDFATHER, who was also an albino is held sacred by th
451e176 not  to be common-- probably, I should think    GRANDFATHER first of race & if so, fact for my theory Co
523m015 otic.-- dotage.-- Doctor communicated to my     GRANDFATHER his feeling of consciousness of insanity com
539m083 hen education same.-- My handwriting same as    GRANDFATHER. Aug. 16th Anger «Rage» in worst form is des
550m123 ns!!-- The Devil under form of Baboon is our    GRANDFATHER!-- A man, who perfectly obeys his conscience
552m131 on with sending force to muscles, that in my    GRANDFATHER remark, a tired man. involuntarily feels ang
```

560mIBC child, when father has not had it. but where GRANDFATHER was the cause by his intemperance. <No.> Can
336d019 these conditions are not fullfilled.-- «[My GRANDFATHER'S theory of Mules not hereditary, because gen
356d070 something prevents the completion.-- ¶Say my GRANDFATHERS expression of generat. being highest end of
412e058 ill account for all (1) Grandchildren. like. GRANDFATHERS (2) Tendency to small change.. «especially
426e108 ism coming out more than in mother or indeed GRANDMOTHER; what is Mr S. S. parentage?-- Wonderful as
089a018 e Naturelle, C«o»urrejolles. 11th Observ.-- Les GRANDS tremblemens de terre sont presque toujours pre
265c083 noses. &c & This congenital changes show that GRANDSON is determined, when child is.-- shows that ge
455eIBC it is only the association which is useless. GRANFATHER'S Handwriting, to compare with my own.-- E Be
037r055 ively.» & Turtle bones. & the bones of <two GRANINIVEROUS» a herbivorous lizard.-- from the action of
023r012 n Bay of Legrand, (SW part). describes a Small GRANITE Isd. capped by Calcareous rock; following Curv
049r088 emporaneous others subsequent. as in dikes. In GRANITE great crystals arranged on sides. V. Lyell P 3
049r088 Vol. III. constitution of veins, is there said GRANITE in close contact varies in nature, -- Does not
049r088 in close contact varies in nature, -- Does not GRANITE at C. Tres Montes become more siliceous in clo
049r089 into granites the <conta» passage from lava to GRANITE is much more perfect. than in believing mere a
049r089 d to a conical mass. will this conical mass be GRANITE? Why not more probably greenstone? What probab
053r100 ells at Quillota Lyell, states that contact of GRANITE & sedimentary rocks, in Alps becomes metallife
053r101 singular-- M. Lesson considers the Sandstone & GRANITE districts to be separated by profound valley [
054r102 ar genera of plants: All St. Catherine & coast GRANITE: P. 199; Falkland account of cleavage differs
056r110 ase p. 51). presupposes an elevated country of GRANITE, not «so» great«er» for all Europe, than from
056r110 an from the Plata to Caraccas, which is all of GRANITE: In discussing circulation of fluid nucleus,--
059r120 rior craters in centre:-- Bailly talks of much GRANITE on all East side of Van Diemen Land. All the C
070r153 ees The structure of ice in columns. show that GRANITE when weathering into balls. must exhibit orbic
089a017 zed at bottom, condensed before rising?-- Mem. GRANITE heated.-- Metamorphic action in red sandstone.
090a025 vapour bringing planets to an end? Fragmentary GRANITE showing schistose structure (& veins appearing
092a030 . Athenaeum M. 516 1837 High up the Essequibo, GRANITE & quartz, after passing sandstone Vol II. p. 6
094a035 glassy in greenstone dikes which rise through GRANITE.-- a most important question with respect to m
094a036 -- Beyond region of great boulders, pebbles of GRANITE clearly effect of remodelling same manner. as
098a048 strata) (Hence endless passages from gneiss to GRANITE): Why not horizontal? Why have particles in su
100a052 and case, nor. the arrangement of particles of GRANITE in Henslow's Grit, yet it is worth considerati
100a054 an agueo deposit resulting from disintegrated GRANITE!!! Look at gneiss of Rio Concretions in Pumice
102a061 it» would look like it. Are greenstone dike in GRANITE residual matter of upper quartzose ones & fels
108a081 e. = Andesite-- Albite & amphibole= Cook found GRANITE at Christmas Sound Vol XIV. (My Edition) p 500
110a087 . 247. Mr. Schomburgk NW. numerous boulders of GRANITE" "direction of strata on the Berbice N. 35 deg
112a089 o Lyell's view of not discovering dike one end GRANITE & other trap.-- It is in the mountain masses w
116a099 have same effect as too coarse. Read Kylau on GRANITE Edinburgh Philosophical Journal Rapport on D'O
116a100 151» first discovered «very small» bits of red GRANITE between 40 & 50 from Portezuelo. Bull: Soc. Ge
119a106 wn-- <[...]> moreover gradation from gneiss to GRANITE shows that the metamorphic rocks have just flo
135a145 ilgherries-- Much inform. on. decomposition of GRANITE-- Bengal. J. vol 7. p. 522. Mountain c near Ca
136a147 n thinness of crust as implied by meeting with GRANITE every-where. Phillips in Ladner Vol II p. 125.
137a149 nding with sandstone «said to be» analogous to GRANITE infiltering some of its constituents into cher
149g032 near summit on Hill on side of Inn BOULDER of GRANITE above 4th Shelf a little lower down the hilloc
151g040 es of rock not in immediate neighbourhood, (as GRANITE or gneiss of Moel Derry) on low hill between I
153g055 n Erin summit 27.813. 65 55 degree? Boulder of GRANITE 28.362 68 degree Granite-«band» 4 X
153g055 Boulder of Granite 28.362 68 degree 60 degree GRANITE--«band» 4 X 3 X 2 «feet» & 2 deep Another rath
154g058 vel of shelf of Glen Guoy form comparison with GRANITE block «& Ben Erin» 29.287. 72 degree Air 65? 7
156g067 ollarig, & some «other parts» Boulders of same GRANITE, all on these three shelves suil is <the> usua
156g070 transported materials <into> on upper shelves GRANITE & some other rocks at head of shelf 3d almost
156g070 ome other rocks at head of shelf 3d almost all GRANITE pebbles Level of plain of 4th shelf at head of
156g072 «line of 4th» shelf This shelf at head where <GRANITE &> «veined» gneiss <unite> «occurs» abundantly
156g072 » abundantly with perfectly rounded pebbles of GRANITE & forming «sloping» buttresses Yet certainly s
157g074 e this with last measurement of shelf of 3d:-- GRANITE block a yard across. On side of «that» hill, i
157g075 hill, in front of which shelf 3d form beach of GRANITE pebbles, & around which shelf 2d «almost» form
157g075 le hill composed of remarkable gneiss with red GRANITE veins & quartz, & garnets.-- Boulders as befor
157g076 <come» «been drifted» here: on very summit no GRANITE-- (in valley «there are» granite) «boulders» (
157g076 ery summit no granite-- (in valley «there are» GRANITE) «boulders» {P} cory stream hill with boulders
158g079 Dick right-- Mac mistook terrace also right-- GRANITE such as boulder on <thes> Divortium aquarum Pe
158g079 right hand {P} rounded waterworn buttresses of GRANITE obscure terrace 15 ft divortium my measurement
158g080 et above Loch trace of this terrace «on <will> GRANITE ridge or a modified Granite ridge» at head of
158g080 terrace «on <will> Granite ridge or a modified GRANITE ridge» at head of Glen Roy on same side where
158g082 to explain on river doctrine <Little Hill with GRANITE blocks almost encircled» <fre» Gneiss cut smoo
159g084 at pass {P} River Gorge 4th Sh side of valley GRANITE blocks on this side (return) between 2d & 3d s
159g085 conceivable. if Hill between Corry so much cut GRANITE could have remained, no peat supply.-- Conside
159g087 e gentle than in Glen Roy, & partly shut in No GRANITE blocks in higher parts?? Bought Glen name of G
162g104 e of terraces on each side High up the Tarf (a GRANITE (boulder), sloping buttresses, an[d] one alter
219b194 canic isld. whilst <neig> Africa, sandstone, & GRANITE, (that is genera near Cape) see if there are a
022r006 und in Anglesea, amongst the varying & dubious GRANITES.--Wide limits of this mineral in Australia. F
046r078 dikes which so commonly (state facts) traverse GRANITES, are granitic materials simply altered by cir
049r088 ese greenstone rocks are seen to graduate into GRANITES the <conta> passage from lava to Granite is m
074r165 ns, chemical affinities like in composed rock. GRANITES syenite» «strangling &c of veins can only be
094a035 stion with respect to my theory of changes. of GRANITES into Trachytes.-- Mention Osorno in lake. few
021r005 aminae fold round them; Quote this. Valparaiso GRANITIC nodules in Gneiss. Epidote seems commonly to
029r033 razil, all originate from the decomposition of GRANITIC rocks Mem. Chanticleers voyage at <[...] Mara
034r044 of do? <Poor> Daubeny good account of ejected GRANITIC fragments P. 386 Mem. Lyell's fact about sulp
037r056 s. Consult Insist on the frequency of dikes in GRANITIC countries, enumerate cases. -- M. Video excep
046r078 commonly (state facts) traverse granites, are GRANITIC materials simply altered by circumstances; &
046r078 uid mass itself changed.--No. -- Yet the fluid GRANITIC mass under <[...]> less pressure might have i
073r163 ver it to great thickness. = Coast of Acapulco GRANITIC rock.--in parts of table granits & gneiss wit
097a043 the black glazing observed by Humboldt on the GRANITIC rocks of the Orinoco".-- <but> on one of the
104a067 eruptions.-- We must suppose everywhere--, in GRANITIC areas &c &c volcanos {P} fissure dike.-- thus
110a087 Vol IV (p 321) Mr Hillhouse describes central GRANITIC ridge of Guayana as NW / SE. Vol VI, p. 247.
073r163 of Acapulco granitic rock.--in parts of table GRANITS & gneiss with gold veins visible:--"Porphyries
527m033 o has not had his blood run cold by singing).-- GRANNY says she never builds castles in the air-- Cat
152g047 ndicates new terrace Ballivard 2 miles North of GRANT town to Forrest road comminuted shells Importan
179b035 :-- Mem: Horse, Llama. &c &c-- If we <suppose» «GRANT» similarity of animals in one country owing to
248c030 VERY different will breed together, so when we GRANT «(which can be shown probable,)» varieties may
261c070 ecies is empirical. show this by instances Once GRANT my theory, & the examination of species from di
263c076 ew faith of the lesser of the difficulties Once GRANT that «species» one genus may pass into each oth
263c076 «species» one genus may pass into each other.-- GRANT that one instinct to be acquired (if the medull
280c133 - Crosses of diff: breeds succeed, yet seems to GRANT, that difficult & other go back to either paren
294c176 rent origin when species rather further.-- once GRANT good species as carrion crow & rook formed by d
537m075 te side has been shown-- see Mackintosh.-- Must GRANT, that the conscience varies in different races.
564n002 r have a troubled conscience.-- Therefore I say GRANT reason to any animal with social & sexual insti
582n068 l effect,.--" this not wholly true, for we must GRANT a bird knows what is about when building its ne
599o007 & in the arched & leafy forests" Very good!. I GRANT that the thrill, which runs throug every fibre,
599o007 h edges & crowded with towns & thoroughfares, I GRANT that man, from the effects of heredetary knowle
109a083 a tendency direct (or oblique) outwards may be GRANTED. independent of currents.-- mud going out can
256c050 imal, small point in tracing history of Man.-- GRANTED.--but if all other animals have been so formed
263c076 ore profound reasoning than other-- if this be GRANTED!!) & whole fabric totters & falls.-- look abro
283c165 hybrids produced--)» & then all that I want is GRANTED.-- For. at Galapagos. make ten species of Orph
352d058 g point, all united . (links in circle must be GRANTED unequal, because fossil) Now what is group wit
173b009 grade in that case surely is not Produced?-- <GRANTING> Species according to Lamarck disappear as co
355d068 e see young bud changing into ovules.-- Captain GRANTS. Himalaya. shells (see Paper in Geolog Transac
435e133 placed on Orchis (so very different) that the GRANULES exserted their tubes: now Mr Herbert has show
263c075 ication of little means & bringing the mind to GRAPPLE with great effect produced, is a most laboriou
048r084 derful, from chemical attraction, as a blade of GRASS penetrating by action of Organic power a lump o
207b151 instinct. can dive & conceal themselves in the GRASS.-- Beatson St. Helena says no trees succeed so
298c189 ol. I. p. 185 case of tit lark placing withered GRASS over nest, when often looked at.-- this most pu
311c227 ting isld. off coast of Africa .p 69. with tall GRASS. & p 72 hairy sheep-- Edinburgh. Transact. Vol

495q05a th Honey-dew dusted with pollen of neighbouring GRASS= Spread sheets of Paper. covered with some stic
061r128 w limit of Palms, evergreen trees, arborescent GRASSES, parasitic plants, Cacti: & with limits of no
446e163 --» The case of the Lemna, «and the vivaparous GRASSES, which no doubt are propagated during hundreds
447e164 is not end of seminal reproduction.-- likewise GRASSES. &-- very heavy seeds.-- as Cocos do mer.-- An
491q002 s &c been obtained? 3.. Whether the viviparous GRASSES & onion, produce flowers, like the Oxalis from
288c160 ems in part of-- to have almost banished the GRASSHOPPER Warbler-- --Yellow Wagtail never seen in one
036r052 351.   Vol I. woody bushes, «gazelles» hares, GRASSHOPPERS & Rats. characteristic of the deserts of Sy
593n111 . Pool. I saw many coots & waterhens feeding on GRASSY bank some way from water, suddenly, as if by w
430e117 all variations in offspring of wild animals.-- GRATEFUL & intelligent.-- The theory that all animals
533m060 that I was tending to make myself in act less GRATEFUL.-- How comes this tendency in these cases? Ho
536m073 .-- Probably some error in argument, should be GRATEFUL if it were pointed out.-- My wish to improve
625o052 tes & passion, which receive enjoyment from GRATIFICATION & hence are forgotten-- only so far do I ad
627o053 ch results when the object is attained (the GRATIFICATION of one's offspring) is not the aim of the a
628o54v nscience checks the wish to <other> outward GRATIFICATION, whilst <the> no desire of gratification wi
628o54v rd gratification, whilst <the> no desire of GRATIFICATION will check the consciences desire for virtu
629o54v equally <prefe> destroy all wish of outward GRATIFICATION,-- see what cases Mackintosh gives & try it
620o044 forced to reflect on his choice: an appetite GRATIFIED gives only short pleasure. passion in its nat
620o044 n its frequency & in its consisting in desire GRATIFIED & therefore as soon as desire is fullfilled,
628o54v point of «false» honour will stop all wish to GRATIFY <it>-- anything contrary to it]CD NB. the very
533m060 n only fully expressing momentary feelings of GRATITUDE, I had a sort of consciousness I was not righ
074r164 . chrysop[r]ase. opal:-- Veins in Limestone & GRAUWACKE: Silver appears far more abundant in the uppe
406e035 see full paper. L'Institut 1838. p. 338 A most GRAVE source of doubt. in distinguishing <after> whic
565n008 opposite muscles will want much confirmation. A GRAVE person close those muscles, which wrinkle when
050r091 states that generally in North part of Brazil. <GRAVEL becomes> sand less & gravel more common. the s
050r091 th part of Brazil. <gravel becomes> sand less & GRAVEL more common. the shoaler the water & nearer th
057r114 k in Patagonia white beds having proceeded from GRAVEL proved.-- curious similarity of rocks of very
096a039 uch Canary Isd. p. 351.. NB. Mackenzie talks of GRAVEL on basalt of Heckla-- All the Azores Isld. Von
101a057 limits of Boulders in N. America do/p. 280. the GRAVEL beds in England different from Boulder beds--
138a153 th N & S of Lima.--judges from «beds of» sand & GRAVEL & shells. p. 47. do has table of every earthqu
148g022 e measure before There some of the half rounded GRAVEL nearly as high as highest measurement but natu
148g026 fore coming to Bridge of Spean, hills of «sea», GRAVEL, current cleavage, & pretty well rounded stone
151g042 ait. connecting flat on one side with irregular GRAVEL plain of other, which must have been waterworn
486z020 s described in Suffolk as lying under strata of GRAVEL & clay about 10 feet in thickness.-- (March, 1
495q05a her wind, on «dry» windy day, «flower garden on GRAVEL walk» will drift many seeds= Necessary to answ
107a077 Sweden!! swelling of rock from Heat. Specific GRAVITIES of many artificial limestones produced by Sir
099a049 trong. a third.-- & this would make layers.-- (GRAVITY can have no effect, on particles of equal weig
099a049 ght.--)¿ cleavage not vertical ∵ combined with GRAVITY.-- hence changes in dip of no sort of conseque
099a051 force drag particls to line {P} AB, & likewise GRAVITY MN. Then every particle would tend to meet at
100a052 t is worth consideration. especially effect of GRAVITY, versus some fault explaining vary dip & incli
102a060 cle coated. &c will be aware how little common GRAVITY has to do with arrangement of particles in roc
102a062 ly, connection with magnetism &c counteracting GRAVITY.-- As volcanic eruptions are accompanied by ho
219b196 rallel fact to Blood-Hounds. Before Attract of GRAVITY discovered. it might have been said it was as
291c166 eing a secretion of brain, more wonderful than GRAVITY a property of matter? It is our arrogance, it
554m141 e argued from this into a theory of friction & GRAVITY. it would be discoverer "reasoning" or "reason
611o034 ect of electricity chemical attraction, heat & GRAVITY is probable.-- And the Organic laws probably h
617o040 ily exhibited in & by matter. The phenomena of GRAVITY considered in themselves consist in a force ma
617o40v on of some animated agent Now the phenomena of GRAVITY are manifestly the same as if every particle o
185b056 Galapagos. Fernando Noronha Ophyressa bilineata (GRAY) new <liza> species, belonging to true. America
489qIFC . Holland p. 16 Babington-- Gould ---- 10.(a) J. GRAY ---- 17 Yarrell---- 18 Blyth---- 19-- Mr. Tolle
503q012 s any hybrid has <been> reproduced itself.-- Ask GRAY to ask Mr Riley to experimentise on hybridising
509q017 eredetary?) in diff. countries in same races Mr. GRAY General Questions (1) Particulars about Sierra
173b009 e perfect.-- truer even than in Lamarck's time. GRAY'S remark, best known species. (as some common la
356d069 of change from eating different kinds of food: GRAZING animals who eat every species new.-- Sept. 8'.
440e145 every «budding» tree, & every buzzing insect & GRAZING animal owes its form, to that form being «the
571n029 s pleasant.-- Mental & Bodily Consider case of GRAZING animals knowing poisonous «herbs:» & man not.-
080r178 ndian rubber & black lead unite chemically like GREASE & mercury [blank] NB. P. 73. General reflectio
022r007 ion? Carbonate of Lime disseminated through the GREAT Plas Newydd dike.--Mem tres Montes. ((Henslow A
022r007 dd dike.--Mem tres Montes. ((Henslow Anglesea)) GREAT variety in nature of a dike.--Mem. at Chonos &
023r009 the jaw was also firm, out of which we pluckt a GREAT many teeth, 2 of them, 8 inches long, & as big
024r015 of New Holland shoals much: Dampier remarks on GREAT flats on the NW coast:-- 8 leagues, from Sydney
027r027 & sand mixed with sea shells--about 500 Isd. & GREAT banks. effect of Elevation. United service Jour
028r032 dillera, where that action commenced before any GREAT accumulation of such matter.-- Dr A. Smith says
030r036 Micaceous formation of Chonos. interesting from GREAT quantity of altered Carbonaceous shales Examine
031r038 elena. This structure was very clear at base of GREAT lava cliffs [Fig. I] line of high tidal action
032r041 ace & to the Eastward.--If matter proceeds from GREAT depth. from axis to surface must gain a Westerl
032r041 is to surface must gain a Westerly current:--If GREAT changes of climate have happened. hurricane in
033r043 Admirable little table showing long PERIODS of GREAT violence volcanic. from Humboldt: Comparison P
038r057 ummit of mountain is blown off; & again when in GREAT crater. different little craters are all burnin
038r059 ssing lines of crater, <arg> state that all the GREAT Volcanoes. have been elevated considerably. whi
039r060 structure of Ascension> give as an example the GREAT subsidence at the famous eruption of Rialeja, &
044r071 ry «dip of strata on East») cannot believe in a GREAT explosion, nor would sea remove more internally
044r072 --wide valleys.--central peak small; yet GREAT body of lavas have flowed from centre-- Pisolit
044r073 Some general reflections might be introduced on GREAT size of ocean; especially Pacifick: insignifica
044r074 f water thus holding matter in solution based on GREAT: & in the fact of bombs in tufa there is proof
045r075 tes that Volcanos!! were in eruption at time of GREAT Lima earthquake» In the Chili earthquakes if ri
045r076 rce of crust of globe & injecting matter on the GREAT rise). -- The great rains which attend severe E
045r076 e & injecting matter on the great rise. -- The GREAT rains which attend severe Earthquakes «1822¿ 18
046r080 wave first retreat at Juan Fernandez: the first GREAT movement was one of rise (any smaller prior one
046r080 that fall first movement (as in Peru 1746).--At GREAT Lisbon Earthquake Loch Lomond water oscillated
047r081 o elevating Earthquake of Valparaiso. (1822) no GREAT wave on record. -- «also neighbouring sea must
047r083 nd, without (cs <[...]> raised above as). -- In GREAT Calabrian wave did not sea break first? I can i
049r088 eous others subsequent. as in dikes. In Granite GREAT crystals arranged on sides. V. Lyell P 355 Vol
051r093 stations; At Bahia ferruginous.--At Pernambuco (GREAT swell & turbid water) organic bodies protect li
051r097 a. <East> Africa. Australia. profoundly deep: a GREAT fault or rather many faults.-- Necessary form;
053r100 scarpment worn away like english escarpment The GREAT conglomerate of the Amazons & Orinoco mentioned
053r101 pear to me to be effect of expansions acting at GREAT depths (mem: profound earthquakes), which would
055r108 mind with force.--the exact yearly rise of the GREAT rivers prove better than any meterological tabl
058r115 Concepcion [.] no sign of elevation. Effects of GREAT waves to obliterate all land marks.--At the fir
058r117 s part of same phenomena lasting so long.-- The GREAT movements (not mere patches as in Italy proved
058r117 as in Italy proved by Coral hypoth. agree with GREAT continents). Voyage aux terres Australs Vol. I.
059r123 metry of the globe shows powers have acted from GREAT depths, so changes, acting in those lines. must
059r123 s, acting in those lines. must now proceed from GREAT depths.--important.-- Decemb 10. 1802. Earthqua
060r125 ence of formations of S. America & Europe.-- If GREAT chain of Volc. had been in action during second
062r128 & with limits of no vegetation at S. Shetland = GREAT contrast of two sides of Cordillera, where clim
062r128 w botanically = but picturesquely = Both N & S. GREAT contrast. from nature of climate. = Perpetual s
064r135 Limited Volcanic action & limited earthquakes & GREAT but local elevations of the land in Europe-- Ur
065r140 -although recent elevation, there may have been GREAT subsidence previously. Mem. pebbles of Porphyry
071r157 rmejo was utterly overthrown by earthquake with GREAT destruction of human life.--Temple mentions som
072r160 ce of the ground are fibrous. malleable & of so GREAT tenacity, that it is with difficulty that a few
073r163 [,] amygdaloid. basalt & other trap cover it to GREAT thickness. = Coast of Acapulco granitic rock.--
073r164 honolitic porphyry. = several parts of N. Spain GREAT analogy to Hungary. = Veins of Zimapan offer ze
076r169 antity of silver procured from martial pyrites; GREAT blocks of pure silver not common in <S.> Americ
077r172 , there are laws of solution & deposition under GREAT pressure. (? heat!) unknown to us. M. Chladni
086a006 e Geology is written Cuvier. Europe possessed a GREAT edentata. -- How much is temperature of world r
087a011 ent corresponding to Europaean risings. Pacific GREAT land. -- Will use argument of proof of slow cor
088a013 t position of strata, prooff thickness not very GREAT; where piece turned over axis or hinge no doubt
094a036 ison. M.S. Chapter on drift.-- Beyond region of GREAT boulders, pebbles of granite clearly effect of
099a048 ld shrink. film parallel to longer axis. But if GREAT depth NB. Prof <Henslow> Sedgwicks lamination p

102a061 primitive rocks-- Are substances soluble under GREAT pressure? equally with little pressure? An impo
102a061 er of upper quartzose ones & felspar.?? Are the GREAT crystals, & the layers first of felspar & then
103a065 of Caverns, in Plutonic rocks argument against GREAT bodies of vapour. according to Hopkins theory.-
104a066 re rock metamorphic & thickness of <strata> not GREAT, one can conceive anticlinal lines near. (later
104a066 ral pressure would always produce it) but where GREAT thickness is affected, they would be far off In
105a073 rce required to move <it> «stream» aside is not GREAT.-- Is there more degradation at first angle owi
106a073 the water has obtained.-- If {P} inclination be GREAT where arrow stands the force immediately deflec
106a074 tream cannot retain its force if inclination be GREAT. There could «not» be great deflection in a "ra
106a074 e if inclination be great. There could «not» be GREAT deflection in a "rapid".-- is a familiar illust
106a075 lus «over cliffs edge» of which limit cannot be GREAT over) with very gently sloping sides This argum
106a074 37. p. 520 Mr Fox on increase of temperature at GREAT depths. All Earthquake unaccompanied by Volcano
107a077 ean than above it no because heat proceeds from GREAT body of mass.-- The last speculation becomes im
108a080 l oscillation. or so many shocks directly after GREAT shock -- It appears to me unphilosophical to th
113a092 idence; Mem my remarks on coast of Australia.-- GREAT NW. dip in SE part of Australia.-- Probably a c
115a097 can only (see «supra» p 94) be accounted for by GREAT molecular attraction of every atom in rock On a
116a099 g their shores where currents vary weak??-- too GREAT an abundance of matter would have same effect a
119a104 alogous (see Edinburgh. Phil. Journal <]CD>, no GREAT chains like Andes or Himalayas, but great circu
119a104 >, no great chains like Andes or Himalayas, but GREAT circular mountains, yet so analogous, that as w
119a105 dly considered. metamorphic rocks at surface. & GREAT heigth on mountains.-- consist of rocks with fo
120a109 rface deposit.-- The case of the shingle in the GREAT Chilian valleys must be profoundly considered.
121a111 f enough <to form> for potters to use. in which GREAT Knife formed crystals of ice were formed-- (lik
123a116 yell. (under head of Delta) describes near Alps GREAT beds of rivers which must be like the Chilian o
123a116 be greatly aided by considering space formed-- GREAT vacuum-- by dike.-- Mem. however. veins of segr
123a117 3d. May. states remains found in many part.-- GREAT Dasypus near Camelones-- large quadruped bigg
123a117 Vol VIII. (1838) p 212. Facts from Erman about GREAT depths of frozen soil. p. 211 Consider proved t
125a120 . On Vertical trees. Uspallata.-- do p. 473. on GREAT Iceland stream. the 90 miles includes opposite
126a124 uch nearer the surface. especially at bottom of GREAT ocean, where the circulations from surface can
127a127 tains. «(not on continents)» prove elevation.-- GREAT mountain chains. may be effects of subsidence E
129a131 assacuhssets. p. 133 The most wonderful case of GREAT block of rock moved by gale-- When writing on V
129a132 henaeum. 1838 p. 791 -- Most curious account of GREAT subsidence «20 miles long I in with.» which mus
135a144 ee S. Wafer looking for Copiapo. found inland a GREAT many sea shells some miles from coast-- quote p
146g011 nuded.-- «of hard metamorph» path only covering GREAT Slip, 10 years since three hundred feet in vert
149g030 e required to deposit this Remember however the GREAT Chilian valley Aconguam, must there have deposit
150g035 ide of Spean most clear & upper line running up GREAT bight just as Dick shows NB. Lake gradually dra
150g036 then if gradually drained, where is barrier {P} GREAT waterworn frame terrace 4 4 not visible 3a 3a 3
151g042 after 3d lake.-- 4th shelf runs up some way on GREAT sloping plain of alluvium (much corroded by riv
151g043 in.-- but below houses where rivulet enters two GREAT projecting butresses, upper slope of which corr
155g062 oad «& cut out, produced» from same «cause» as «GREAT» spit <is> or plain <now> formed on shelf 4th 2
158g080 e side where two rivers unite in Upper Glen Roy GREAT plain about 60 ft beneath shelf peat on pebbles
175H016 &c from Egypt no answer because time short & no GREAT change has happened I look at two ostriches as
181b041 -- & <in> therefore the chances are excessively GREAT against, any two of the 12. having progeny. aft
181b042 «between them.».-- for instance there would be GREAT gap between birds & mammalia, Still greater bet
181b044 some anomaly & <the g> bearing stamp of <some> GREAT main type, & the gradation will be sudden-- Hea
183b053 when crossed with it.-- ¿Whether extinction of GREAT S. American quadrupeds. part of some great syst
183b053 n of great S. American quadrupeds. part of some GREAT system acting over whole world, the period of g
183b053 t system acting over whole world, the period of GREAT quadrupeds declining as great reptiles must hav
183b053 ld, the period of great quadrupeds declining as GREAT reptiles must have once declined.-- Read his th
185b055 it could live-- but there were causes to induce GREAT change. like the Buzzard which has changed into
186b060 n Miocen & so on.-- As I have traced the <type> GREAT quadrupeds to Siberia; we must look to type of
189b071 rhouse says there is no TRUE connection between GREAT groups.-- Speculate on land being grouped towar
190b077 of Indian character There appears in Australia GREAT abundance of species of few genera or families.
194b094 S. American form. therefore it is like case of GREAT <rodent> edentate [has been doubted?]CD & oposs
196b106 inents In plants I believe not.-- It is a very GREAT puzzle why Marsupials & Edentata should only ha
196b107 , broom, & yew very different from any found in GREAT Britain, British varieties are also found in Ir
196b109 nimals most dependent on vegetables. of the two GREAT Kingdoms.-- Principes de Zool: Philosop:-- I de
196b110 Mollusca & vertebrata, that there must be very GREAT gaps.-- yet some analogy The existence of plant
198b116 swer <could> «can» be given.-- It is a point of GREAT interest to prove animals not adopted to each c
199b116 small spots of land, of the same type with the GREAT continents we get a means of knowing movements.
201b126 ount of wonderful fossils of India.-- & p. 545 «GREAT monkey» Mr Johnston says Mag of Zooly & Bot. p
206b149 odd genera with few species which stand between GREAT groups, which we are bound to consider the incr
209b156 distinctes et permanentes." p. 145. In Humboldt GREAT work de distribut. Plantarum. relation of gener
217b183 every one) of rare breeds of dogs, from owners GREAT care of them. Fox says when two dogs of opposit
219b193 osely> When this volcanic point appeared in the GREAT ocean, have made plants of American & African f
223b208 being. although hard to draw line.-- -- not so GREAT as between perfect insect & former hard to tell
224b214 he difference intellect of Man & animals not so GREAT as between living thing without thought (plants
226b223 ot discovered channel of communication by which GREAT Edentate might have roamed to Europe & Pachyder
226b223 rica & <West.> «E» of America, ought to present GREAT contrast in forms.-- India; intermediate, see h
227b226 ammal & Reptile. (or between extremities of any GREAT divisions) thus a knowledge of possible changes
229b231 ring very like common English.-- Hottentots say GREAT tailed sheep aboriginal at Cape & a thinner-tai
233b251 codiles. Anoplotherium.-- <M. Jerrod> & Dumeril GREAT work on Reptiles. M. J. says some reptiles same
233b251 land or few quadrupeds-- Study Productions. of GREAT Fresh water lakes of North America II Parasite
234b256 f 1838. (Newcastle) about somebody who had made GREAT collection of birds of Iceland. --M. Gaimard, h
235b262 animals separated. & long interbred <p> having GREAT tendency to vary? Is not man thus circumstanced
240c004 like father, sometimes like mother. The fact of GREAT monstrosities being produced, & handed down, wi
242c018 quille Lesson No (p. 24) batrachian in isles of GREAT ocean says in conformity with Bory's Views.-- <
247c028 . 69. Sharks very generally distributed: Mem of GREAT geological age-- Gastrobranchus «only» 2 specie
248c029 oes the New Zealand Rat belong to There is this GREAT advantage in studying. Geograph. range of quadr
250c037 iar to it, so on, & so on.-- Whatever destroyed GREAT <quadrupeds>: Pachyderm in S. America destroyed
250c037 <quadrupeds>; Pachyderm in S. America destroyed GREAT Edentata or American form.-- Is the Australian
250c037 common cause I can conceive of destruction of GREAT Animals in Europe & America Some portion of the
250c038 her Africa & East India Arch.-- but where these GREAT animals had not spread then such tribes as Mars
252c045 unas foundation of all our knowledge especially GREAT continents> Give Specimen of arrangement, 3 <5>
253c046 n Borneo. Sumatra. India Ceylon-- perhaps shows GREAT persistency of character. Hence Elephas primige
255c051 ter ouzels hence aversion to generation, before GREAT difficulty in propagation.-- Feathers on, Apter
257c058 As we have birds impressions in Red Sandstone. GREAT lizards in do.-- <Wood> <Dicot wood> Coniferous
257c058 ning of animal life.: generally, but even about GREAT division, our <only> question is not, how there
257c059 producing young ones like itself, but ¿whether GREAT assumption? not solely producing like itself, n
259c064 ground woodpecker &c.--, fresh water animals of GREAT lakes are American form, that one is brought to
259c065 at one is brought to admit the possibility (any GREAT change in species is reduced by atavism) Even a
261c071 s.-- & the latter most important in obviating a GREAT apparent difficulty-- preservation of colouring
263c075 little means & bringing the mind to grapple with GREAT effect produced, is a most laborious, & painful
268c099 anterior to present. ¿cause of destruction of «GREAT» animals?-- Show independency of shells to exte
268c100 be explained by mere chance?-- or it like each GREAT class of animals having its aquatic, aerial &c
269c102 tary to Appendix to Flinders Voyage by Brown.-- GREAT space seems to act per se as barrier-- Mem. Tar
273c111 is a preeminently aerial,-- formed for flight & GREAT movement in the air, & likewise rasorial specie
280c135 ls not breeding at all in domestication. throws GREAT difficulty in way of ascertaining about hybrids
280c136 od, gets stronger & stronger, so that though by GREAT effort <pr> one unlike can be produced, yet to
282c143 one doubt that one animal can really produce so GREAT an effect.-- the spirit of life must be every w
292c169 s first stage) the tendency to change cannot be GREAT, otherwise it would be unlimited. We absolutely
292c171 they must be done often «to be habitual» or of GREAT importance to cause long memory.-- structure is
293c172 ion recalling up image which had been past-- so GREAT an anomaly in structure of brain not probable)
293c172 instinct.-- Even if savage takes. & was given a GREAT coat & this he put on & we afterwards could und
298c191 reases» general instinctive feeling.-- There is GREAT difficulty in Making an alpine species from one
300c196 nivora.-- Man in his arrogance thinks himself a GREAT work. worthy the interposition of a deity, more
302c202 variation. therefore lately acquired.-- I fear GREAT evil»-- from vast opposition in opinion on all su
311c225 ykes alludes to some other case of 180 degree & GREAT diff of Latd. p 355 Echidna of Van Diemen's lan
315c243 a habit gained by formerly being a baboon with GREAT canine teeth.-- (This may be made capital argum
331dIFC al work on Buds & Gemmae. C D Charles Darwin 36 GREAT Marlborough St Did Eytons <intermediate>. «hybr

Page
(Key Word)

```
334d012 <on> two cows entirely, changed his residence a GREAT many miles.-- yet one day <th> a cow walked in,
336d018 tions, these small changes become multiplied, & GREAT change be effected, but in a mule these conditi
338d023 hysical conditions of country is not perhaps so GREAT, as separation on <be> inter-breeding, for othe
355d068 ra fossils same remark. ¿was there formerly one GREAT sea, & two Polar Continents Marsupial. Edentata
357d072 this world without rule? The destruction of the GREAT Mammals over whole world shows there is rule.--
357d074 ole world-- if so doubtless «part of» system of GREAT harmony. The peculiar character of St. Helena.-
369d115 ompelled to travers> "May have reference to the GREAT distances which the Mammalia of N. S. Wales are
375d135 entence of Malthus no one clearly perceived the GREAT check amongst men.-- «Even a few years plenty,
376d136 ephalus Porcarius.-- this Monkey did not like a GREAT coat made for it at first, but in two or three
377d139 . & when present animals-- lived.-- we know the GREAT time, necessary to form channel & (& Basses St)
383d159 ory I must allude to separtion of sexes as very GREAT difficulty, then give speculation to show that
384d161 developed, when the first <are> become of use| <GREAT characteristic of male greater strength, (p 45)
385d163 very readily & are subject to fits.-- «there is GREAT difference between hybrids & inter se offspring
386d165 prolonged bud being formed & one part dying for GREAT length of time.-- There is probably law of natu
386d167 te differences. final cause of this because the GREAT changes of nature are slow. if animals became a
386d167 te change, they would not be fitted to the slow GREAT changes really in progress.-- Annals of Natural
391d175 - It is absolutely necessary that some «but not GREAT» difference (for even brother & sister are some
398e006 ere & there in order a scattered page; we find <GREAT> sensible change in the institutions. & we supp
400e011 to geographical distribution-- Thus Hattica is GREAT genus.-- because found in all quarters: his ide
401e014 sible to purchase seeds of any cabbage, where a GREAT many will not return to all sorts of varieties,
401e015 Mr Tollet so produce.-- thinks it probable that GREAT part of those varieties may be due to impregnat
402e017 ! Species not being observed to change «is very GREAT difficulty» in thick strata, can only be explai
403e021 wanders from Pellew to Eap.-- There is another GREAT lizard. Kaluz. which is found at Pellew & Eap,
403e023 very strong passage might be made-- why seeing GREAT variation in external form of varieties, do we
406e037 Hemisphere.-- likewise far North in Southern.-- GREAT animals. of same two great orders destroyed abo
406e037 North in Southern.-- Great animals. of same two GREAT orders destroyed about same time in North & Sou
407e040 m the existing orders, as the Eocene of Paris! (GREAT Edentata at that period) Analyse this,-- consid
408e043 species change, we see external conditions have GREAT effect on them, & therefore extermination becom
408e043 becomes part of same law.-- When we know what a GREAT effect. light has in colouring plants,-- who ca
409e048 a of water & snow line descent. My theory gives GREAT final cause «I do not wish to say only cause, b
409e048 cause «I do not wish to say only cause, but one GREAT final cause,-- nothing probably exists for one
409e049 ts of the animated beings.-- &c-- If man is one GREAT object, for which the world was brought into pr
410e052 s having some high theoretical interest,-- "the GREAT end must be the law & causes of change".-- A ph
413e058 change.. «especially with physical change» (3) GREAT fertility in proportion to support of parents D
416e070 luid fluid, (& not dry as in plants) therefore, GREAT difficulty in crossing [& this most important o
416e071 fectly adapted to external conditions.-- (hence GREAT variation in each birth) from man arbitrarily d
423e098 It would be to make enquiries of some of these GREAT seed-growers-- ).-- Feb. 24th. Monoceros, which
426e107 f varieties are fertile & make mongrel, & other GREAT series quite otherwise & make on[ly] true hybri
426e108 some distant cousins, the nearest blood being a GREAT great-grandfather.-- -- Little Miss Hibbert cas
426e108 istant cousins, the nearest blood being a great GREAT-grandfather.-- -- Little Miss Hibbert case of H
428e113 the sudden, change from Primrose to Cowslip is GREAT difficulty. «I should doubt if wild species eve
429e116 p. 21 says. many plants skirt each side of the GREAT N & S. valleys, which penetrate Pyrenees but ar
430e118 ny process analogous-- -- if so she can produce GREAT ends-- But how.-- -- «-- .Make the difficulty a
432e124 breed» animals of any country must be made with GREAT caution; owing to its adaptation to the surroun
433e126 ful instance of forms, intercalated between two GREAT distinct formations.-- particular air given. p.
433e126 anges from limestone to sandstone &c. show some GREAT change who can say how many centuries elapsed b
437e139 e of Eagles nest, which shows power of carrying GREAT weight. p. 125 is said that Eagles bring rabbit
437e140 d owls were counted in a field, where there was GREAT swarm of mice.-- May 4th.-- The Brussels Sprout
441e149 ring» would be similar to <f> first form.-- The GREAT effect of conditions on offspring, but not on i
446e163 without fresh seeds arriving.»-- throws a very GREAT difficulty in my theory, here we have a plant r
454e184 ed?-- What annuals can be budded «& rendered of GREAT age» as must be inferred from what Mr Knight sa
458t001 salamanders-- Every order (except whales) have GREAT prototype!!.-- Copied Vol II p. 502. «Bengal Jo
463t055 of saying as brain is created &c &c Bats are a GREAT difficulty not only are no animals known with a
468t111 Diptera & small Hymenoptera Saw Humble go from GREAT Scarlet Poppy to Rhododendron-- from Larkspur t
468t111 to frequent certain flowers, to day early, the GREAT scarlet Poppy-- So that, finally Fraxinella. wi
470t153 .-- Humble 22 flowers of Egg Tree in one minute GREAT Humble 17 flowers of Larkspur on two plants in
481z010 of Campanularia Alcedo stellata. Meyer p. 92.-- GREAT Kingfisher of Tierra del Fuego killed in Chile.
495q005 ge with pollen of other, count seeds. & see how GREAT a proportion springs up true.-- This in fact al
497q007 ke opposite view. Gaertner talks of the several GREAT & natural Families, as being difficult to cross
499q010 of seeds.-- (Paris) (22) When Linnaeus says so GREAT percentage of seeds have contrivance for transp
501q011 Hot house to see effect on generative organs of GREAT Heat (32) Can Henslow ask question of Col. Le.
529m041 eatly excited by trees at [i.e., as] compound GREAT compound animals united by wonderful & mysterio
531m050 abstinence.-- My Father remarks that things of GREAT importance are easily forgotten, (if unconnecte
538m079 oth odd appearance about eyes.-- one botanist & GREAT knowledge of Irish Politics, «both bad jokers.-
541m091 it looks so analogous to muscle in one position GREAT fatigue.-- may explain excessive labour of inve
542m096 tion «as often said» of language in man is very GREAT from all animals-- but do not overrate-- animal
543m099 ith transformation because Spiders have many,-- GREAT powers of communicating knowledge to each other
543m099 edge to each other-- August 23d. Jones said the GREAT calculators, from the confined nature of their
543m099 ed to play regularly with D'Arblay of Christ of GREAT genius, & yet invariably used to beat him-- The
543m100 him-- The son of a Fruiterer in Bond St. was so GREAT a fool that his Father only left him a guinea a
549m119 hese be happy-- & these pleasures are so very GREAT, that every one who has tasted them, will think
554m139 t.-- good case of association.-- "Listened with GREAT attention to Harmonicon. & readily put it. when
557m150 make person tremble, like fear.-- Why does any GREAT mental affection make body tremble. Why much la
558m150 s heart, & chest (sobbing) which are most under GREAT sympathetic nerve. most subject to habit, as be
558m152 arl much the same way» generic manifestation of GREAT passion.-- I do not think they arch their backs
559m153 & Ourang & outang, & dare to say difference so GREAT... "Ay Sir there is much in analogy, we never f
560mIBC rown, when they first see it? Charles Darwin 36 GREAT Marlborough St Has my Father ever known <intemp
571n030 ecause there is some connection between them, & GREAT masses of rock.-- I was much struck with this,
571n031 y from natural rise-- I was also much struck in GREAT avenue, resemblance to gloomy aisle of Churche.
574n041 ing: so do stallions always.= No doubt man has GREAT tendency. to exert all senses, when thus stimul
577n049 ing can be cause of itself.]CD & hence there is GREAT probability against free action.-- on my view o
587n087 others.-- dogs dislike perfume) I should think, GREAT principle of liking, as simply heredetary habit
588n090 mal with an old.-- (dog horse, sow) we perceive GREAT difference.-- «(& is not this difference same,
589n092 h powers without hand or voice.-- there is some GREAT puzzle in what Sir. J. M. says of pure reason n
590n093 mpanies fear is the same, as that which attends GREAT weakness.-- <Diarrhaea» & syncope p. 42. Sighin
593n109 ome degree is so.-- idea of beauty of music are GREAT distinguishing character between man & animals.
595n117 gs. «& twitching paws» which they only do when <GREAT> considerably excited, shows their power of ima
600o08W beasts, not clear.-- ¿does not Mackintosh make GREAT difference between moral sense & conscience? we
604o018 nsion we associate something extraordinary & of GREAT power-- 2 From these & other reasons we appl
605o19V Hence it appears, that when certain causes, as GREAT height, eternity, &c &c. produces an inward pri
607o025 stances. As man hearing Bible for first time, & GREAT effect being produced-- the wax was soft,-- th
608o026 (originally mostly INSTINCTIVE, & therefore now GREAT effort of reason to discover them: this is impo
608o027 to do harm.-- Believer in these views will pay GREAT attention to Education.-- 4) These views are di
610o033 stminster Review. March 1840 p. 267--- says the GREAT division amongst metaphysicians-- the school of
611o034 nt laws of movements. [LHC] Hence there are too GREAT <worlds, inor> systems of laws «in the world» t
614o037 not so wonderful as at first appears, & no too GREAT advantage.; for superiority of memory does not
615o038 ity of Christian over Heathen race.-- But as no GREAT modification in brain would probably take place
619o042 > any action in accordance to an instinct gives GREAT pleasure, & <an> «such» actions being prevented
620o044 mptation to disobey the conscience is extremely GREAT [LHC] The cause perhaps lies in its frequency &
626o052 say anything about any principles born in us.-- GREAT difference with my theory.-- see p. 349.-- rema
633j53v -- Kolreuter mentions some hybrid, whose flower GREAT tendency to break off p. 292. Mac. has long rig
635j56r y other cause.-- (& my theory [ALL PARTS OF ONE GREAT SYSTEM. C. D]CD [All this does not explain deat
635j56r require constant miracles.-- p. 420 thinks the GREAT fecundity of germs is to afford support to othe
641j29r regions-- Whales. «Narwhal» Polar bear. Walrus, GREAT Seals of Antarctic seas. (on other hand Spermac
032r039 e removed & from the resistance offered to the GREATER lateral extension of the waves. by the part be
032r040 case would be as at first. & according to the GREATER or less time of rest. so would the size of the
032r040 of constant rising with successive periods of GREATER activity & rest.--Such changes could be shown
056r110 poses an elevated country of granite, not <so> GREAT«ER» for all Europe, than from the Plata to Carac
```

060r124 (which of course must retain same level) to a GREATER distance".--Afterwards speaks of this phenomen
065r138 of Etna Stromboli & Vesuvius Investigate with GREATER care. vegetation & climate of Tristan D. Acune
106a075 as the inclination in all probability would be GREATER when flowing over (B) than when at (C) its ten
113a090 aumur.-- there ought to be 32 degree Fah. at a GREATER depth than 400. & the limit being 400 ft. show
113a091 down. & then in 50000 years the depth will be GREATER than <5000.> 400.-- These facts of SLOW but su
115a098 rock On a coast, the shallower the water, the GREATER power of oscillations & currents.-- if matter
120a109 r over surface.-- if elevation then went on at GREATER rate, not only river would carry further its o
127a125 . & hence. (when climate hotter) was cooled to GREATER depth.-- Now the <inf> subterranean isothermal
127a125 f ice.-- Hence further N. when soil frozen for GREATER length of time depth of ice ought to be less.-
132a139 ocean accounted for, by the circulation being GREATER, than the transmission from ocean's bottom.--
152g048 , then rivers might deposit, & afterwards with GREATER cut through, not applicable to Glen Roy Lake,
180b036 ion. C & B. the finest gradation, B & D rather GREATER distinction Thus genera would be formed.-- bea
181b042 tant period.-- Hence if this is true, that the GREATER the groups the greater the gaps (or solutions
181b042 this is true, that the greater the groups the GREATER the gaps (or solutions of continuous structure
181b042 d be great gap between birds & mammalia, Still GREATER between Vertebrate and Articulata. still great
181b043 eater between Vertebrate and Articulata. still GREATER between animals & Plants But yet besides affin
191b081 animals land birds.-- & life shorter or change GREATER-- In the East Indian Archipelago it would be i
206b147 the number related at the first start must be GREATER, & this number would vary at each lustrum, & t
231b242 East Indian Seas. Marsupials animals all show GREATER connexion in quadrupeds, <bu> plants do not fo
231b243 by one connection.-- viz descent.-- Hence far GREATER discordance in latter-- Have change in form.--
233b252 man & his domesticated quadrupeds are not so. GREATER facilities of change in the articulate than <M
248c033 ence the tendency to revert to parent forms, & GREATER fertility of hybrid & parent stock, than betwe
255c050 as> in other classes this evidently relates to GREATER range of such forms.-- p. 56: Ornithological p
259c063 much in swimming, & every developement giving GREATER vigour to the parent so tending to produce eff
282c140 ionship in some one.-- imagine the men to have GREATER power of change yet, as external conditions ov
282c140 here suppose these changes of <use> adaptation GREATER than those heredetary ones.-- which would elap
288c159 grations of birds in same country. may explain GREATER migrations, if American & intersected wider &
292c169 enerated species. & of 100 extinct species the GREATER number probably have no descendants on earth.-
292c169 unlimited. We absolutely know that tendency is GREATER in Mammalia, than in shells ¿ univalves or biv
300c198 ng each other it is hiatus & not saltus.-- The GREATER individuality of mind in man, is analogous to
300c198 individuality of mind in man, is analogous to GREATER individuality of bodies of some animals over t
302c200 kable, with so much death, as has gone on,, No GREATER gaps.-- external conditions, to be sure, have
316c244 Fuegian or Australian! why not gradation.-- no GREATER difficulty for Deity to choose, when perfect e
331d1FC eterogenous, in plants are the number of seeds GREATER.?-- Mem. for Eyton.-- Sir. R. Heron's case of
340d029 the "4» Struthionidae, Mr Blyth remarked that GREATER difference in than in many large orders of bir
365d106 d been considerable change, it would have been GREATER puzzle, than none, for the «»normous time whi
384d161 > become of use| <Great characteristic of male GREATER strength, (p 45) & that strength> In speaking
407e038 s both North & South, has probably undergone a GREATER change, than any part, (except Europe. in whic
409e048 e been basis of argument of change.-- now take GREATER area of water & snow line descent. My theory g
427e110 nimal> dog does not prefer other. but produces GREATER effect on offspring-- Mr. Herbert says «p 347,
431e122 the more convinced I am, that extinction plays GREATER part then transmutation.-- Do species migrate
433e127 transmut., or believe that time has been much GREATER, & that systems, are only leaves out of whole
447e164 re rapid, & the facility for inter marriage is GREATER (Hence Dioecious plants highest,-- Palms &c &c
447e164 ious plants highest,-- Palms &c &c)-- Is there GREATER resemblance between carboniferous. «& recent»
463t057 le mind is constituted that a difficulty makes GREATER impression, than the grouping of «many» facts
486z018 an in Europe-- Coleoptera especially require a GREATER duration of Heat. hence musquitoes & knats abo
541m092 ng changes during fall of sea.-- Is the effort GREATER if the idea is abstract as love, (or an emotio
543m099 . ie have all parts. Waterhouse Study well the GREATER number of insects in insecta-- not connected w
549m120 ed them, will think the sum total of happiness GREATER. even if mixed with some pain.-- than the happ
550m125 l or ideal ones, <which> «& these» must occupy GREATER proportion of <each> «every» man's time.-- Beg
563n001 t of running hare is stopped by fleas, also by GREATER temptation as bitch: or dogs defending compani
568n017 ain if held.-- if tempted not to follow it, by GREATER temptation, if memory of its own emotions. (wh
599o07v of heredetary knowledge, has produced almost → GREATER changes in the polity of Nature than any other
628o054 remacy",-- I make its supremacy, solely due to GREATER duration of impression of social instincts, th
629o055 except instinctive ones) Perhaps my theory of GREATER permanence of social instincts explains the fe
021rIFC inite existence, those that have undergone the GREATEST number of changes towards perfection (namely
032r039 of the waves. by the part beneath the band of GREATEST action not having been worn away.--If the lev
032r039 The part (o) which was before beneath band. of GREATEST action. would now by degrees be exposed to it
051r094 ent as at St Helena) I have mentioned point of GREATEST action; I now having seen Pernambuco believe
145g006 nches thick-- {P} 35 degree is I believe about GREATEST dip of sandstone in upper part «of Salisbury
174b015 et first cooled.-- Countries longest separated GREATEST differences-- if separated from immens ages p
192b086 umerous. (NB in those classes with few species GREATEST jumps strongest marked genera? Reptiles?) For
196b110 of plants, & their passage to animals appears GREATEST argument against theory of analogies. States
239cIFC king offspring? Instances of old Breeds taking GREATEST effect Account of the [...] the world.-- Char
239c001 me. he has no doubt that oldest variety, takes GREATEST effect on offspring. Thus presuming those var
240c004 eons with very much care that, it requires the GREATEST difficulty to rear them, eggs hatched under o
254c048 the ovum or embryo succeed each other with the GREATEST rapidity"-- so we find species each class suc
293c175 n given to organization.-- This really perhaps GREATEST difficulty to whole theory.-- There is breed
306c213 which «there appears to be one line, in which» GREATEST depth <appears to be> is not more than 60F. &
306c213 more than 60F. & «in» the whole area,. 120 is GREATEST (about 200 miles distant). directly beyond
355d067 peculation even if partly true they are of the GREATEST service, towards the end of science. namely p
401e014 -- Cape Broccolli can hardly be reared without GREATEST care be taken to prevent fertilization from t
423e099 is very rare, the conclusion will be that our GREATEST formations <are> have been deposited in a per
436e137 better chance of being propagated & so &c. The GREATEST difficulty to my theory, is same type of shel
463t055 could have had with such structure.-- perhaps GREATEST Could anyone. have foreseen, sailing, climbin
541m090 rs had fallen, as in my Glen Roy paper.-- this GREATEST mental effort, of which I am capable-- I susp
549m118 hen same man is compared to peasant.-- To make GREATEST number of pleasant thoughts, he must have con
573m037 observed that very young children. express the GREATEST surprise at emotions in her countenance-- bef
609o030 says our rule of life is what will produce the GREATEST happiness.-- The other says we have a moral s
609o030 o be almost identical» + What has produced the GREATEST good «or rather what was necessary for good a
610o30v e experience of others) shows does not tend to GREATEST good.-- Therefore rule of happiness is to cer
123a116 metals. The theory of veins will, I suspect be GREATLY aided by considering space formed-- great vacu
204b141 n ducks.--» Kirby all through Bridgewater errs GREATLY in thinking every animal born to consume this
282c141 ere colour & trifling form & head &c to become GREATLY changed. in structure & even to certain degree
441e150 t of such laws.-- The effect of one part being GREATLY developed on another,,must not be overlooked.-
443e153 her varieties & we know that the kind of stock GREATLY affects the Graft.-- ‖Plants circumstanced as
510q017 mbryo in close species of Hilianthemum differs GREATLY-- how very interesting to see if any variation
529m041 pleasure in Kensington Gardens has often been GREATLY excited by looking at trees at [i.e., as] grea
546m111 ife. which he had hid some years before.-- was GREATLY astonished, at the time.-- & could trace no chai
605o19v e the source of this "inward glorying" to the GREATNESS of an object itself or to the ideas excited &
464t065 ts are valueless= monkeys= Owen has described a GREATT Struthionidous Bird from New Zealand-- <so> not
289c160 tances of typical land bird, having habits of a GREBE, structures might follow.-- examine structure o
310c222 econd--this class of facts «analogous to petrel-GREBE. external» appears to be a puzzle against my th
289c160 dapted species to localities‖-- p. 390,. young <GREBES> «ring ouzels» dive instant touch the water.--
576n048 er & far more complicated organization to learn GREEK, that to have it handed down as an instinct.--
553m136 ifferent degrees in races.-- whether in Ancient GREEKS, with their mystical but sublime views, or the
576n047 intellect is not become superior to that of the GREEKS.-- (which seems opposed to progressive. develo
051r094 hole surface to where highest spray (there pale GREEN confervae) coated with living beings; In smooth
260c069 t on human offspring.-- white, snow.-- the fine GREEN of vegetation,-- ¿account for colour of bird in
361d095 1. N Mr. Blyth, at Zoolog. Meeting stated, that GREEN-finch, all linnets red-pole, goldfinch, hawfinc
383d160 r part of breast, each feather is fine metallic GREEN. <from> with tip & part of shaft metallic green
383d160 green. <from> with tip & part of shaft metallic GREEN.-- This green doubtless is effect of Metallic h
383d160 with tip & part of shaft metallic green.-- This GREEN doubtless is effect of Metallic hue of silver p
383d160 ect of Metallic hue of silver pheasant. yet why GREEN? & not purple?-- legs pale coloured.-- In the b
450e173 spotted deer, no loonies, but cocatores & small GREEN parrots. June 26th-- Yarrell.-- «& Black Swan «in
467t105 Bees on Potato & several times on Beans Rough.--GREEN-cabbage «in flower»-- swarmed with meligethes &
499q010 single clumps of plants in counties, as of rare GREEN Cotton Plant-- How large «area» clump there? Di
505q014 eed.?-- (11) Abberley has planted seeds of pale GREEN Cynoglossum. never germinated 12 Does the horne

554m138 mself it is chiefly shown in old male.-- A very GREEN monkey (from Senegal he thinks Callitrix Sebe??
339d025 Mem 3 species of grouse»! Has not Goldfinch & GREENFINCH bred, & surely wild Duck & «pintail» Widgeon
072r159 s. -- No Volcanic action on coast line of Old GREENLAND, close to W of Jan Meyen Isld.--Mr Barrow thi
096a039 and of Volcanic action in Iceland parellel to GREENLAND: Mem.¿ Greenland subsiding.) Von Buch Canary
096a039 ction in Iceland parellel to Greenland: Mem.¿ GREENLAND subsiding.) Von Buch Canary Isd. p. 351.. NB.
424e101 ording to Beck are different.-- Subsidence of GREENLAND-- case of splitting of two regions-- -- are t
427e109 inently artificial.-- Would not subsidence of GREENLAND render climate less extreme. (& so account fo
429e115 rennees agree with those of Norway. Lapland & GREENLAND, but not with those Kamtschatka, Siberia, or
449e173 tation» now in case of large [not located] Mr GREENOUGH on his Map of the World, has. written. Mastod
049r088 alparaiso; Epidote -- Must we look at regular GREENSTONE cones at S. T. del Fuego as nucleus of a Vol
049r088 former; & thus occurring in groups.--As these GREENSTONE rocks are seen to graduate into granites the
049r089 onical mass be granite? Why not more probably GREENSTONE? What probable origin can be given to the nu
049r089 origin can be given to the numerous hills of GREENSTONE? -- Daubeny. P 95. Glassy & Stony Pearlstone
073r163 22 degree N. = formations of amph: porphyry. GREENSTONE[,] amygdaloid. basalt & other trap cover it
076r170 degree to 50 degree-- covered by conformable GREENSTONE porphyrys & phonolites do. amphibole quartz
094a035 ieving the other. -- Is the felspar glassy in GREENSTONE dikes which rise through granite.-- a most i
097a042 uch. Canary Isd. p 170.-- Mem. Cordillera Can GREENSTONE dikes. he residue of quartzose vein in highe
102a061 ces from impact, «it» would look like it. Are GREENSTONE dike in Granite residual matter of upper gua
109a084 das Mines. a translation of paper by rose on GREENSTONE, diorite, &c most important.:-- must be stud
114a095 h I give at C. of Good Hope.-- A bare hill of GREENSTONE, if we know origin of greenstone tells subsi
114a095 bare hill of greenstone, if we know origin of GREENSTONE tells subsidence as plainly as Temple of Ser
120a110 Jura 4000 feet.-- The veins of segregation in GREENSTONE of Salisbury Craigs well worthy of attention
122a115 with stones scattered irregularly.-- (Mem near GREGORY Bay). Shropshire case where lamination appeare
075r166 late, covered by a clayey porphyry, containing GRENATS. In Peru. on other hand, mine of Gualgayoc or
110a086 mechanical: DIKE base reddish feldspathes with GRENISH. black specks of hornblende, large irregular c
219b196 ty discovered. it might have been said it was as GRET a difficulty to account for movement of all, by
227b224 two countries Hence India, Mexico & Europe. one GRET sea (Coral reefs:. shallow water at Melville Isd
319c256 ll. & the mocking thrush being so very beautiful GRET contrast with South America.-- In Home's Histor
278c132 period, compared to Faunas of these countries, GRETER than Toxodon, Macrauchenia, &c compared to Ame
038r058 ds from which proceeded pebbles & on which trees GREW.--? Are not the dikes in upper strata. quite di
291c168 gradation.-- It is easy to see if South America GREW very much hotter, then Brazilian species would
460t019 «of this old stock (which from action & reaction GREW more complex)» some perhaps rendered more compl
496q006 two «My Father made hens cast Holly-seed & they GREW» (9) Place. Snap-Dragon. (I have seen one monst
532m056 would not sleep at night even when in bed room-- GREW very thin, would not go out of house except wit
618o41v ness & weight always went together. & as a thing GREW blue it «uniquely» grew heavier yet it could no
618o41v t together. & as a thing grew blue it «uniquely» GREW heavier yet it could not be said that the bluen
075r167 f Iron in Mexico. sulphuretted silver, arsenical GREY copper, and antimony, horn silver, black silver
110a086 different from either fragment or dike, blackish GREY base. crystals from fragment disseminated on th
146g010 lace In same way at top the trap could be traced GREY in front on wall perhaps wall oblique The hill
160g090 syenite with pinkish felspar;-- whole hill dark GREY fine grained. Much contorted gneiss «narrow sha
213b171 ies do the same, May you not breed, ten thousand GREY hounds & will they not be greyhounds?-- Yarrell
233b250 ks two main divisions of cats. Tortoise shell--& GREY-banded. ¿species?-- thinks offspring of cats so
247c029 riously coloured as a herd in England"-- Black & GREY varieties of rabbits thus handed down for nearl
275c120 ll there was a dash of blood, with whole form of GREY hound.-- picking out finest of each litter & cr
290c164 mon gull in garden at Zoology Soc. it's pale ash GREY back, like a black bird washed, Whilst tips of
493q003 shell Cats. as only one sex so coloured = I have GREY-cat «wh was female» with tinge of tortoise-shel
511q018 (7) In breeding. pointers. Bull-Dogs. Spaniels-- GREY-hounds-- is there ever any degeneration?? HOUND
210b159 ecies & genera.-- for instance three kinds of GREYHOUND.-- In plants. do the seeds of marked varietie
217b184 both parents' & Mr Bell has half bloodhound & GREYHOUND.-- Where two dogs have lined bitch directly o
225b217 ou will show me every step between bull Dog & GREYHOUND). I should say the changes were effects of ex
275c120 ould not run up hill-- he took thorough bred <GREYHOUND>. bull-dog. & crossed & recrossed, till there
337d020 an varieties must be very unnatural-- Italian GREYHOUND is probably the effect of <sev> local variety
337d021 raining in the earlier branches «as in common GREYHOUND» & much intermarriage.-- In my speculations.
416e075 .-- >, larger than any living [not located] A GREYHOUND might be made «almost» without any relation t
416e075 ny relation to running hares.-- as an Italian GREYHOUND not so species every part of newly acquired s
430e118 s would take place from changing country: but GREYHOUND. & goutier Pidgeons «race-horse». have not be
439e145 33-- They had several breeds of dogs.-- like GREYHOUND-- fox-dog-- turnspit & two other kinds It see
506q014 to shoulder on line of back, height 17½/. The GREYHOUND. was in length (measured same way) 47¼-- in a
550m126 s friend viewed him as Newfoundland dog would GREYHOUND about dread of water-- innate Septemb 1-- If
575n043 on of mind & brain. «Do we deny the mind of a GREYHOUND & spaniel. differs from their brains» then ca
575n044 stronger analogy that the tendancy to hybrid GREYHOUND to hunt hares. «& leave the sheep» & jackall
637j57v Get instances of adaptations in varieties.-- GREYHOUND to hare.-- waterdog hair to water-- bull dog
213b171 , ten thousand grey hounds & will they not be GREYHOUNDS?-- Yarrell's remark about old varieties affe
275c120 nnatural circumstance Ld Orfords had breed of GREYHOUNDS fleestest in England lost courage-- (Bull-do
275c120 st of each litter & crossing them with finest GREYHOUNDS.-- Sir. J. Sebright first got {P} point on h
306c214 eep are thick-tailed The dogs called Persian «GREYHOUNDS» are Kurdish & come also from Asia Minor.--
404e026 organ present in whole class. Case of Mexican GREYHOUNDS.-- young being habituated. instance such as
575n044 dogs-- increase the courage & staunchness of GREYHOUNDS.-- bull-dogs being preferred from not having
052r099 Cordilleras look as if some peaks elevated.-- GREYWACKE. as a general fact absent in T. del Fuego, ex
540m088 of music, when the mind is rendered ductile by GRIEF, or by bodily weakness, melts into tears, with
569n023 Dn 75 cf., p 268 without, however, very sincere GRIEF-- "there is nothing more to be said."-- "made n
575n045 .-- Think, whether there is any analogy between GRIEF & pain-- certain ideas hurting brain, like a wo
579n058 this will be curious if it is so.-- frown with GRIEF,¿ bodily pain? frown shows the mind is intent o
579n059 r, sublimity, sexual ardour.-- a man cries from GRIEF, joy. & sublimity. January 6th.-- What passes i
590n093 ss.-- <Diarrhaea> & syncope p. 42. Sighing from GRIEF. is method of increasing languid circulation--
590n093 od of increasing languid circulation-- no, for <GRIEF> sighing comes on before circulation is affecte
569n022 fered to take a savage, said his wife would be GRIEVED-- "il leva las epaules et dit qu'il valait mie
312c228 spring, are not half lion & tigers ditto. (see GRIFFITH) & half Muscovy ducks, «black cock & pheasant
545m106 brow important analogy with man.-- I see monkeys GRIN with passion, that: is show all the teeth: «& ma
565n006 en uncovering its teeth to bite.-- the senseless GRIN of passion, is like the grin of the Hyaena from
565n006 n-- the senseless grin of passion, is like the GRIN of the Hyaena from fear, no actual intention to
557m149 le in passion my F. rubbing hands.-- stamping. GRINDING teeth.-- in shame frowning, & anguish,-- shyn
639j28r swallowed by birds & by Aphysia. C. D p. 258. «GRINDING» teeth in <stomach of> sun-fish, in mouth of
315c243 e Zoonomia, for if the former shows that a man GRINNING is to exposes his canine teeth. no doubt a ha
558m153 isses from fear so does man.-- & so dog]:CD Man GRINS & stamps with passion. can expression be used m
535m064 ies. The earwig & a doubtful one of Acanthosoma GRISEA described [not located] as first caused by wil
100a052 arrangement of particles of granite in Henslow's GRIT, yet it is worth consideration. especially effe
589n092 associated with ideas.-- A sigh, is an abortive GROAN.-- more power over muscles of voice than respir
469t151 on longer-- Woodfords Marrow fat, Early frame, GROOM'S Dwarf. planted in rows «close to each other» &
094a036 gonian boulders might be transported.-- {T} On GROOVED rocks. Specimen of rock from Costorphine at Ge
043r069 that country. Read description of channels or GROOVES in rocks at Costorphine hills. to compare with
350d054 l II. Sir T Browne's Works p. 20 There are no GROTESQUES in nature; not anything framed to fill up em
619o042 rhaps others.-- [LHC] ----- p. 113. Mackintosh GROTIUS has argued nearly so [RHC] The history of ever
035r046 the same district there seems. I think, little GROUND for skepticism, as to the general truth of the
044r073 ar coast. Humboldts quotation of instability of GROUND at present. day.-- applied by me geologically
045r075 believe that the extraordinary fissures of the GROUND at Calabria were present at the Concepcion ear
045r075 the Calabrian earthquake things pitched off the GROUND. «Ulloa states that Volcanos!! were in eruptio
046r079 d.-- With respect to Volcanic theory. I want to GROUND, that the first phenomem. is an inward afflux
046r080 s might have been owing to absolute movement of GROUND). Michell (Philos: Transacts) «seems to» consi
060r124 connection with "the shooting upwards" of the <GROUND> land in the W Indies.--p. 200. Bollingbroke v
061r127 n Brine Springs. (Henslow) Speculate on neutral GROUND of 2. ostriches; bigger one encroaches on smal
072r160 ses which are scattered over the surface of the GROUND are fibrous. malleable & of so great tenacity,
131a135 ry wrong.-- Geograph. Journal Vol. 8. p. 402.-- GROUND ice-- subterranean isothermal line Athenaeum.
158g083 of Upper & from Glenroy» near the upper shelfs GROUND strewed with pebbles Shelf 3d runs up with but
185b055 ossible.-- Why should we have in open country a GROUND «do. <w> parrot.--» woodpecker-- a desert. Kin
257c057 nd country distant. <Study> The circumstance of GROUND woodpeckers.-- birds that cannot fly &c &c. se
259c064 reations.-- it is only be recollecting that the GROUND woodpecker &c.--, fresh water animals of great
264c082 rly showing true affinity, for instance tail of GROUND woodpecker-- -- but tail of some ducks aberran

265c088 -- one is tempted to suppose from beholding the GROUND.-- why do beetles & birds & <f> become dull co
274c113 The La jeune veuve parrot though so much on the GROUND has not this structure., instance of habits go
274c114 kingfisher-- Gould has seen with long tarsi.-- «GROUND woodpecker" Secretary bird.-- & Millisuga. Kin
274c114 but generally small genus. ¿are there not many GROUND parrots? are there not many ground woodpeckers
274c114 ere not many ground parrots? are there not many GROUND woodpeckers?-- In each division Gould thinks h
277c126 at ¿O. Modulator & O. Patagonicus. till neutral GROUND ascertained, call them varieties. but two ostr
282c142 genus with respect to rest of its family as in GROUND cuckoos, is affinity with respect to species o
282c142 ame.-- but if two original species, each became GROUND then the relation of all the ground cuckoos. w
282c142 each became ground then the relation of all the GROUND cuckoos. would not be affinity, but the truth
286c154 expression.-- animals signals. (rabbit stamping GROUND) Man signals.-- animals understand the languag
290c163 uda (bird of St Jago) of brown colour; lives on GROUND, colour of habitation. Must have some effect.-
300c197 his to a degree. Mem Spider only dropping where GROUND thick.-- shamming death it is but being motion
343d038 thers though the negro of Africa is not loosing GROUND. Yet, as the tribes of the interior are pushin
377d140 e inhabitant of plain & Jaguar of woods &c like GROUND birds [not located] :Hence, also structure not
485z016 Patagonia compared with those of England.-- or GROUND birds-- rather indefinite letter Mem Orpheus--
492q003 sport much in nursery gardens? <are the> is the GROUND much manured In species of close genus do more
506q015 plants the Papilionaceous structure of flower-- GROUND nuts (42) How are Orchidiae fecundated, as mas
516q24v -- [Unnumbered blank] Experiment Cover patch of GROUND, with different salts & poisons & see in what
563n001 ore? crowing & only in breeding season & on the GROUND.-- Cock fowl. on the ground, at roost, in all
563n001 ing season & on the ground.-- Cock fowl. on the GROUND, at roost, in all seasons, & after? he has don
590n094 s spoken & believed to be audible. one has good GROUND to call imagination a faculty, a power, quite
636j57r ons.--fossil forms show such losses.-- Consider GROUND Woodpecker stiff tailed cormorant: pain & dise
032r040 ing platform would be made, & so on.-- This is GROUNDED on the belief of constant rising with success
419e087 cies>, & only secondarily,, by assumption well GROUNDED, on time;-- therefore the mere loss of specie
199b118 . J. p 297, No 8 Jan-Ap. 1828 -- I take higher GROUNDS & say life is short for this object & others,
430e119 a Plants & see whether those which grow in low GROUNDS are those, which are common & nearest being co
462t051 Hensleigh objects to transmut. theory, on the GROUNDS of similarity in condition in Java & Sumatra &
496q006 ry in how many generations. daisy. Fever-fuge GROUNDSIL.-- gilly flower will break & become double.--
288c158 abits, range. &c &c-- Maclay rests his whole GROUNDWORK of analogy on its concurrence in parallel pa
616o39v ceive of matter as attracting & shew that the GROUNDWORK <of this> is entirely wanting by which thoug
224b212 generation, «(but experience according to each GROUP)» whether good species, & hence the importance
227b226 leads to Nature of physical change between one GROUP of animals & a successive one.-- It leads to kn
234b256 say he can at once tell by general colouring a GROUP of Nebria complanata from. Devonshire, from ano
261c071 Synallaxis &c &c. sure to unite the birds into GROUP.-- it is same as Yarrell's remark about rock Pi
262c073 errant groups.-- but as for number five in each GROUP absurd.-- the mere fact of division of lesser &
273c110 would lead one to suppose, that species in same GROUP generally contemporary «++.» «This would lead o
281c139 <circular>, in broken circles.-- which in each GROUP is quite fatal.-- ¶Relations of analogy being t
284c149 here any structure is general in all species in GROUP we may suppose it is oldest, & therefore lest s
284c149 er? Therefore value of organs vary in different GROUP. & Not known in single ones--. viz. Maclay let
285c151 (hence rise & depression of importance, in each GROUP & connection of even distant ones) the characte
297c185 rcumstances., which by causing death, makes the GROUP aberrant When species rare we infer exterminati
297c186 When species rare we infer extermination, when GROUP few in number of kind, extermination.-- New for
303c205 ,-- nothing but experience. will, tell us. when GROUP is true,» there are no genera. if Mammalia are
347d050 Zoolog.-- of Africa -- p. 4. sticks to genus or GROUP of any kind not being perfect till circular. p.
348d050 ows that genus expresses as now used almost any GROUP.-- ¶all groups natural (p 6) as expressing natu
348d050 ss-- Mammalia Edentata¶ We do (p 6) say such is GROUP. because it has such characters of importance,
348d050 what importance, which prevails throughout the GROUP & serves to insulate <them> it".-- i.e what cha
349d052 - Complains (p. 53) of M. Edwards, thinking any GROUP good, though not circular, if characters can be
349d053 ant. it would be exceedingl wrong to call,, one GROUP genus & other subgenus,,--> Propagation, best r
352d058 for circularity of groups. When <species of> a GROUP of species is made. father probably will be dea
352d058 be granted unequal, because fossil) Now what is GROUP without centre but circle, two or three lines d
361d096 cock & hen, all nearly similar.-- in blackbird GROUP young like some of the species-- (¿do these fac
362d100 lack mark of the Rock Pidgeon,-- several have a GROUP of white speckles on elbow joint-- in Bewick dr
402e020 to Paragua & in Luçon the most northern of the GROUP the number is limited of the group, the number
403e021 thern of the group the number is limited of the GROUP, the number is very limited.-- Kotzebue's Secon
415e068 cleared off by fatal diseases.-- The Value of a GROUP does not depend on the number of the species.:
425e104 there are, many cases of genera peculiar to the GROUP having species peculiar to the separate islands
490q001 Entomologist Does individual Shell or insect or GROUP vary more in one country or district than in an
490q001 han in another? Character of shells of Sandwich GROUP [Sowerby monstrous Cardium-- does it remind him
494q004 r to guess how many generations in Mammalia. in GROUP effect of crossing.-- (7) Are the Eggs of the P
498q008 me situation & flower at same time. Has H. seen GROUP of different species growing White Mullein good
535m064 times & go to other, so that two mothers to one GROUP.-- (as in birds blind storge-- They continue ti
553m137 - this expression he believes is common to that GROUP.-- this is very important as showing connectio
555m143 esembles some of the old ones.-- -- S. American GROUP sneer.-- Sept 21st Was witty in a dream in a co
636j56v iage.-- A plant is in the same predicament as a GROUP of bisexual animals living on the borders of a
189b072 etween great groups.-- Speculate on land being GROUPED towards centres near Equator in former periods
355d067 f science. namely prediction.-- till facts are GROUPED. & called. there can be no prediction.-- The o
227b227 lief of <change.> transmutation & geographical GROUPING we are led to endeavour to discover causes of
463t057 difficulty makes greater impression, than the GROUPING of «many» facts with laws & their explanation
041r065 we cannot judge whether such fossils. lived in GROUPS or not. Ferruginous veins of this figure {P} i
049r088 incline to <latter> former; & thus occurring in GROUPS.--As these greenstone rocks are seen to gradua
181b042 .-- Hence if this is true, that the greater the GROUPS the greater the gaps (or solutions of continuo
189b071 says there is no TRUE connection between great GROUPS.-- Speculate on land being grouped towards cen
201b126 in Natural History that all aberrant & osculant GROUPS are not only few in species, but every two or
206b149 nera with few species which stand between great GROUPS, which we are bound to consider the increasing
258c061 alogy may chiefly be looked for in the aberrant GROUPS.-- It is having walking fly catcher, woodpecke
262c073 of chances every family will have some aberrant GROUPS.-- but as for number five in each group absurd
265c087 w much heredetary. The circumstance of aberrant GROUPS being small it is truism. for it not so not ab
271c107 cal Society,, that you never find two «similar» GROUPS of birds in two countries, without intermediat
271c107 occurring in intermediate country-- ie. mundine GROUPS.-- Waterhouse tells me in insects there are ma
283c143 of analogical.-- Gould says it is only in large GROUPS. where you have representations.-- The aerial
284c149 s affording natural characters than among those GROUPS, where it remains less subject to Variation" D
297c185 ls them when urged by hunger.-- p. 65. Aberrant GROUPS few in numbers & vary much in character, divid
302c202 rding to Principles of last page. <an> osculant GROUPS between two circles «of equal value» must be s
302c202 with fact there stated, only in most discordant GROUPS. The formation of genera may sometimes be due
309c222 ween-- owls & hawks only external» intermediate GROUPS often have full structure «of one class» & ful
316c245 shells.--\ Conchifera. hermaphrodites-- eggs in GROUPS.. Have Dioecious plants more restricted ranges
348d050 expresses as now used almost any group.-- ¶all GROUPS natural (p 6) as expressing natural affinities
352d058 HESIS be careful.-- Argument for circularity of GROUPS. When <species of> a group of species is made.
400e011 idae» <no> less difficulty in establishing good GROUPS.-- ears varying so much,-- kind of fur-- (do t
425e104 are the Labiata nearest to American, or Indian GROUPS?-- = Believes some Mediterranean, but chiefly
471f208 out a family so completely natural & one whose GROUPS pass so insensibly into each other". Phillips
484z016 ed Philedon. allied to some of my birds-- These GROUPS strictly American. Colouring on under side of
535m064 e female may be moved to other larvae, when two GROUPS near. mother desert one sometimes & go to othe
553m137 <exp> the expression & noises of monkeys go in GROUPS. thus the pig-tailed baboon, shoved out its li
217b189 genera peculiar to itself-- on hybrids between GROUSE & pheasant-- Magazine. Zoology & Botany Vol I
338d025 distant countries breeding!. «Mem 3 species of GROUSE»! Has not Goldfinch & Greenfinch bred, & surel
342d033 s offsprings.-- «are not the hybrid pheasants & GROUSE different.--» (if so chinese pigs & common mus
342d033 efore instincts not imperfect.-- Are Pheasant & GROUSE homogeneous? I observe Bachman calls these Hyb
365d105 d species cross in wild state.-- Is English red GROUSE. a cross between Black Game. &, the subalpina
365d105 , (which in summer dress somewhat resembles Red GROUSE) it may be so-- but very improbably, for it ca
365d106 stricted species has been Made.-- In the hybrid GROUSE between Black Cock & Ptarmigan (probably subal
426e106 species, than two perfect species; but facts of GROUSE, & pheasant, & hooded crow goes against this.
499q009 re effective than of foreign-- Eyton has such a GROVE of Willows.-- (14) Bowman female branch At What
060r124 e efforts in the land to lift itself higher & to GROW upwards; for the land is constantly pushing the
066r140 r than thumb» Sea weed said at Kerguelen Isd. to GROW on shoals like Fucus giganteus! 24 fathoms deep
162g101 very much due to Peat & Heather When it did not GROW at first-- relics destroyed.-- the Brook <about
219b193 would creator <on volcanic island.> make plants <GROW closely» When this volcanic point appeared in t

285c151 racters will be first those of analogy, but will GROW into affinity.--but whether ever arrive at true
294c178 ch. Travels, p. 306. account of trees ceasing to GROW far N. becomes stunted, altered, & lose (mere s
317c250 ed from it.-- & several old acquaintances. which GROW on the lower region of the Canary islands-- p.
362d099 r 6 weeks after castration, fresh horns begin to GROW.-- Mr Yarrell says the «male» Axis of India, br
366d112 there   must have been a tendency for feathers to GROW there «That Mutilations will not alter form may
372d130 d in itself.-- it must have the knowledge how to GROW, & therefore to repair wounds-- but this has no
430e119 t of St. Helena Plants & see whether those which GROW in low grounds are those, which are common & ne
468t112 over   the anthers of long stamens {P} as stamens GROW old «& shed some pollen». they turn upwards & b
498q007 rom poorness of soil.-- yet crabs probably would GROW there (7) Where parts of fructification <lat> r
498q008 - Enquire (13) Do any of same species of Willows GROW in same situation & flower at same time. Has H.
502q012 monly but improperly called Canadense-- would it GROW in open air in Sweden. Linnaeus found 2 flower.
506q015 races of white & Blue Linum-- did parent plants GROW near each other.-- ? Cannot remember at all. (3
515q022 , taking after male & female parent.-- Will they GROW up in other respects different?--. Important.--
532m056 pt with Caroline-- After fortnight. continued to GROW thin & did not seem quite happy. in five weeks
542m093 self, & though dreading to say so, his step will GROW erect & stiff like that of turkey.-- he may be
556m146 like kittens at the breast now if horns were to GROW on horses, they must yet continue to put down e
623o050 instinct of sociability & sociability, doubtless GROW together [RHC] This feeling seems to vary in ra
637j58r have lived had it not been so.-- let egg shells GROW harder. so must those with weak beaks be sifted
423e098 to make enquiries of some of these great seed-GROWERS--).-- Feb. 24th. Monoceros, which Sowerby say
299c193 when one sees a plant like Paris quadrifolium GROWING in one wood far from any other plants of same
307c216 old female, turning into cock, abortive spurs. GROWING.-- Are there any abortive organs produced in d
310c224 at globules of milk produce a plant capable of GROWING!! & propagating itself. In Tropical countries
317c249 ft in length, some branches of Justicia still GROWING,) passed us. do. p. 243 (, Professor Smith's J
370d117 ite distinct.-- Collect cases of difficulty of GROWING plants in all parts of world, thus tea tree in
385d164 fect teeth & hair-- showing foetus has gone on GROWING-- I believe same has happened in boys bodies.-
388d172 e of female assuming plumage of cock, & beards GROWING on old women = Stags horns & testes curious in
440e146 n present, or past generation.-- thus cabbages GROWING like Nepenthes.-- cases of pidgeons with tufts
447e165 with F. ovina, <& this> rendered vivaparous by GROWING on heights.-- yet he has seen it propagated in
460t019 is evidence in the organic world of infinite & GROWING complexity from a few types, it must not be su
467t099 led, brown, but with no pollen.-- Common Thyme GROWING close by is equally abortive--and both growing
467t099 growing close by is equally abortive--and both GROWING within Kitchen Garden.-- As we see in Hybrids
472s01r Papaver bracteatum, & the Papaver oncitate was GROWING in same garden. & out of 60 seedlings not one
498q008 e time. Has H. seen group of different species GROWING White Mullein good plant to sow & try to get o
507q15v eed.-- (50) Any cases of wild varieties plants GROWING together. under same conditions.-- like cowsli
635j56r f life (peculiar to each species) owing to the GROWING size of the world? & the physical changes it w
542m094 animals,-- but yet when angry it is hard not to GROWL out some sound even if it be inarticulate.-- th
565n007 one will perceive if in night trys to listen to GROWL of hounds». <when> as fear to «man as» animals.
558m152 k» «wide open» with prodigious force.-- making GROWLING, guggling noise. Puma did same & & some other
595n117 ied to see.-- [blank] A Dog «whilst» dreaming, GROWLING. & yelpings. «& twitching paws» which they on
068r147 by veins & mass[es] [Fig. 8] {P} (A. B. C, now GROWN solid.) Red Sea near Kosir, land appears elevat
171b004 ia.-- There may be unknown difficulty with full GROWN individual «with fixed organization» thus being
173b011 f same kind had in interval arrived) might have GROWN altered Hence the type would be of the continen
250c037 Lian Dipus an American form? The climate having GROWN more extreme both in, N & S. America, is only c
312c228 ke chinese & 3 perfectly like spaniel even when GROWN up.-- Are mules homogenious owing to no attempt
357d073 as at present in New Ireland & continent since GROWN.-- This will explain. S. American case & Didelp
372d129 is like cutting off tail of Planaria, the whole GROWN to that part.-- claw added to crab, tail to liz
383d159 phrodite, being not only dimidiate, but quarter-GROWN seems to show whole body imbued with possibilit
459t009 s of deer, which are altogether immaculate when GROWN up». Saw at Mr Bell's at Hornsey the offspring
496q05a or seed rarely-- Magnolias. «Azaleas» & plants GROWN under unfavourable circumstances, as Hyacinths
533m059 est approach to <the> such instincts which full GROWN men can experience-- Instinctive walking of ani
566n011 sions.-- Man getting sight slowly,, but when in GROWN years, thinking he instinctively knows distance
578n057 ching».-- But why does joy & OTHER EMOTION make GROWN up people cry.-- What is emotion? At end of Bur
621o047 wing to peculiarity of organ of taste, for when GROWN up often conquers it). It will be only rarely t
627o053 at we have conception of moral obligation «when GROWN up???» & the question is, whether this can be r
226b221 to new station.-- When the new island splits & GROWS larger species are formed of those genera.-- &
299c194 h just lately have been shewn to be same.-- one GROWS in marsh & other dry; yet if T. palustris be so
372d131 it is part of same class of facts that the skin GROWS over a wound.-- Does likeness of twin bear on t
434e129 llowing facts: believes that «only» red Lychnis GROWS in <south> Wales & certainly <old> only white i
139a180 ll any inorganic substance cause such monstrous GROWTH as oak galls or rose <buds> galls.-- is it not
372d130 An Individual bud may be thus produced from the GROWTH of one part, (not strictly new individual), or
380d153 t birth & causes. If an animal breeds young her GROWTH is immediately checked-- the vis formativa goe
436e137 f a Bull Rush I thought, surely no "fortuitous" GROWTH could have produced these innumerable seeds--
440e145 part of the Grand Mystery is this,-- the law of GROWTH, that which changes the acorn into the oak.--
440e145 into the oak.-- In short all «Nutrition, GROWTH & reproduction» is common to all living beings
440e148 in man & wings under united elytra The law of <GROWTH> «generation» is only modification, though imp
440e148 is only modification, though important one, of GROWTH «Lamark. Vol II. p. 120. observes it commences
440e148 ol II. p. 120. observes it commences only, when GROWTH stops».-- Spallanzani's facts in connection wi
441e148 stocks, on which they are grafted.-- No than by GROWTH-- generation; & more of the effects of conditi
478z004 of animals in rays Par un officier du Roi Rapid GROWTH of Coral-- RN. p 24 Bougainville Voyage round
611o034 h perhaps not conscious sensation. [RHC] During GROWTH <extres> tissue <[...]> unites matter into cer
612o035 to exist as individual.-- [RHC] 2) In animals, GROWTH of body precisely same as in plants, but as an
612o035 relation to external contingencies, as much as GROWTH of tissue and are subject to accident; the sex
624o051 want of reason.-- Hence as Eugenius says, slow GROWTH of rule of right.-- [LHC] *for it strives to g
473s05r waters) He cannot, however, tell them from L. GROZNERAT, «on road to Bethgellert» wh flows by Tremado
061r126 Sydney, consult history? Phillips.» 1826.27.28. GRT. drought at Sydney. which caused Capt. Sturt exp
385d163 n giving milk--» Sept. 25th Young man at Willis «GRT. Marlborough Str, Hair dresser, assures me he ha
468t112 inella «& from <flower> plant to plant.»-- to my GRT surprise-- I found all, stamens straightened pol
498q007 nges so? --Yes, my Father lost this character in GRT degree from charcoal & good treatment (8) Do bee
520mE2F & Experiments Expression M Charles Darwin Esq 36 GRT. Marlborough Str.-- (p. 64. On <insect> Ants get
377d138 so made by the C. porcarious., together with a GRUNTING noise, the former signifies recognition with
274c115 s,-- but no web forms in <water> land birds,,-- GRUPS of very different value have their representativ
030r036 l-- ¿ No shells in all cases. «.Mytilus.--» «at GUACHO» «on N. Chile? Washington.--» Mem: Micaceous f
400e013 ons.-- «S. American form.--» off province of GUADALAXURA-- October 11th.-- Uncle John-- says Decandoe
403e021 77. Many foreign plants have been cultivated in GUAHON. (Mariannes), "for example the prickly Limonia
075r166 ning grenats. In Peru. on other hand, mine of GUALGAYOC or Chota & Pasco in "alpine limestone" = "The
079r176 es on ridge of Cordillera near Pataz, also at GUALGAYOC. where many petrified shells Bougainville say
480z006 oologie Recuiel-- Meyen has written account of GUANACA. In transaction of Bonn Society M. Edwards on
062r130 Petisse. & diff kinds of Fourmillier): extinct GUANACO to recent: in former case position, in latter
087a009 dilloes have eaten out the Megatherium. -- The GUANACO the Camel.? Make note about N. American bone n
093a034 d, of the Salt river.-- in reference to fossil GUANACO of P. St. Julian. -- Mr Scrope seems to consid
343d038 -- Thus Tierra del Fuego has <not> «only» one «GUANAX.--» of the characteristic forms of S. America. Wi
078r175 Zacatecas the veta grande has same direction as GUANAX.--the other E & W.--veins richest not in ravin
077r171 de of Zacatecas, & veins of Tasco & Moran--of GUANAXUATO to SW. with respect to latter doubts whether
078r175 I think I have much additional information GUANAXUATO, which has yielded the most metal, where the
097a044 aert tells me, that the upper strata alone at GUANTAJAYA contains salt see Geolog. proceedings Lake 1
076r170 ity of pure silver in S. America. Geology of GUANUAXUATO.--Clay slate. passing into talcose & chlorit
376d136 Rendus p. 569. 1838 says the cross between the GUARANIS & Spaniards are almost White from first gener
043r071 li:-- Chiloe. Concepcion. Valparaiso (Copiapò & GUASCO). yet whole territory vibrates from any one sh
075r166 nature of the beds they intersect". = In the GUATEMALA part. (& Chiloe do) no veins discovered. Humb
111a088 s in Chile Geograph. Journal Vol. VII p. 216.-- GUAVA trees, introduced about twenty years since (183
267c093 world, 20 years have scarcely elapsed since the GUAVA introduced from Norfolk Isld-- "& it now claims
262c073 Study the wars of organic being.-- the fact of GUAVAS having overrun-- Tahiti. thistle. Pampas. show
110a087 Hillhouse describes central granitic ridge of GUAYANA as NW / SE. Vol VI. p. 247. Mr. Schomburgk NW.
267c094 ting out to sea Vol VII. p. 325-- Wild dogs of GUAYANA always hunt in packs «30 or 40 together» colou
480z008 p 496. Birds at Tristan d'Acunha.-- (Turdus GUAYANENSIS?)) Emberiza Brasiliensis (??) Fulica Chlorop
079r177 ous statements on, Vol I, P. 168. on coast of GUAYAQUIL, same as Galapagos. no Hydrophobia at Quito.
117a102 8 p. 151. Formations of Payta extend close to GUAYAQUIL.-- modern shells of Cobija doubtful. Examine
268c099 good. Study D'Orbigny. & range on West Coast «GUAYAQUIL & Peru» Henslow in talking of so many familie

(Key Word)

595n121 en 2½ years old. was frightened when Snow put a GUAZE over her head. & came near him, although knowin
298c192 - Babington says in most plants, even those on GUERNSEY & on West coast of Ireland, are absolutely (&
257c056 Till we know uses of organs clearly, we cannot GUESS causes of change.-- hump on back of cow!!-- &c
494q004 -- on what principles calculated.-- in order to GUESS how many generations in Mammalia. in group effe
558m152 pen" with prodigious force.-- making growling, GUGGLING noise. Puma did same & & some others-- Thus <
112a089 rnal Vol IV p. 36. on subsidence of the land in GUIANA, worthy of consideration. When discussing nucl
207b151 Geo. Joun. p. 325. Vol. IV. Ducks on rivers in GUIANA. build top of trees carry duckling to the wate
228b228 e, which would then be main object of study, to GUIDE our <past> speculations with respect to past &
252c045 y from three first will give one almost certain GUIDE ∵ because time required too separate isld very
277c127 es good species because interlock analogy to be GUIDE. in islands. species.-- each describer giving h
304c207 f anteriorly think what principles are there to GUIDE in this opinion?-- EXCELLENT PRINCIPLE OF ABORT
434e128 erience has shown we can & that analogy is sure GUIDE & my theory explains why it is sure guide.-- Ly
434e128 sure guide & my theory explains why it is sure GUIDE.-- Lychnis April 3d.-- Henslow tells me followi
465t080 pt close to B. Ayres).-- If we may take this as GUIDE, the shells preserved must be as much a casualt
564n005 eyes, aided by experience is supposed in man to GUIDE to knowledge, was transmitted perfectly to chic
566n012 cience, grand idea: as before having analogy to GUIDE one to conclusion that any one fact was connect
616o39v int of indifference 2) In the absence of such a GUIDE we can only <shew> point out the mode «of perce
620o044 en felt at very time it is disobeyed) & is sure GUIDE.-- Hence conscience is improved by attending &
626o052 emarks. showing that instinct cannot be said to GUIDE will. as bird building nest, but supplies it--
334d010 xplained by Stallions, (according to Fox) being GUIDED entirely by their smell.-- Fox says he knew «a
496q05a up true-- whilst others are crossed.-- Are Bees GUIDED by smell-- or sight.-- --. touching Mr Brown t
554m139 Attention to Harmonicon. & readily put it. when GUIDED by her own mouth.-- seemed to relish the smell
558m150 enlarged capacity ^acting» «yet being obscurely GUIDED» or strong instinctive sexual, parental & soci
582n068 ct. 1834: p. 15. "To act from instinct is to be GUIDED to the performance of a number of prearranged
304c208 from branch high up.-- Such probabilities only GUIDES.-- Yet trifles are produced by circumstances.
584n072 ile snapping-- p. 28. how curious the means of GUIDING themselves through the air,-- waterbirds, the
484z015 unt-- ¿ Do not the Penguins replace the <Auk> GUILLEMOST of the northern Hemisphere, & the Puffinuria
525m024 before it was known he had been on the table,-- GUILTY conscience.-- Not probable in Squib's case any
226b220 t of Africa. equally good.-- Small isld off New GUINEA same fact see Coquille's Voyage.-- Galapagos m
240c013 rra Zibetha. ‖-- All the Moluccas, Waggious New GUINEA. New Ireland, have phalangista, which differ i
240c013 Former statements to such effects false In New GUINEA. a Kangaroo d'Aroe (Didelphis Brunii) which as
241c015 of Australia. Many in common ¿species? with New GUINEA.-- Many <genera>. kinds common to New Guinea &
241c015 ew Guinea.-- Many <genera>. kinds common to New GUINEA & rest of isle in E. Indi: Arch: In New Zealan
242c018 e of New S. Wales. <V.> p. 123 Crocodile at New GUINEA. All the isles of Oceania have the Scincus wit
246c027 vora., which lives in the Eastern Moluccas, New GUINEA.-- (Case of replacement)-- Coquille Voyage The
246c027 aswary, inhabits Ceram, Bourou & especially New GUINEA (replaces, Emeu) in North of New Holland.-- Ne
246c027 (replaces, Emeu) in North of New Holland.-- New GUINEA scarcely differs more from, <Van Diemen's land
247c028 Cyanurus «p 8 &». p 49 on all the Moluccas «New GUINEA, New Ireland» & «even» Java. & very common on
332d005 ought in Portsmouth, said to come from coast of GUINEA, -- ears bare. skin black & wrinkled-- fur sho
341d031 e streaks.-- &c &c Dr. Bachman has crossed cock GUINEA Fowl with Pea <cock> Hen.-- offspring female,
368d114 e or very different in recently altered genera. GUINEA Fowl & Peacocks.!!» other birds display beaut
370d115 erse in order to quench their thirst"-- But New GUINEA.--!! S. America.-- Such difficulties will alwa
380d154 & hen plumage so different, yet the Cassowary & GUINEA Fowl cannot be distinguished.-- A capon will s
543m100 so great a fool that his Father only left him a GUINEA a week. yet. he was inimitable chess player.--
068r148 Geograph. Journal p 202 Vol IV When recollecting GULF of California. Beagle Channel.--One need never
128a130 ieve?? coast of North America., like the Mexican GULF. is fouled by bars of sand & shallow lagoon.--
290c164 ones-- in strata of Tilgate forest Seeing common GULL in garden at Zoology Soc. it's pale ash grey ba
449e168 ne 18th. Eyton tells me, that Yarrell knows of a GULL, which has laid in domestication eggs of two sh
155g065 ne «only» cut deep gorge on sea hypothesis, if GULLIES not now formed «(Mac, hypoth,)» the level duri
346d047 Pelican.-- Mem pink spots on Albatross, on some GULLS. Flamingo-- (Spoonbill Wader. Ibis)-- laws of p
162g100 opposite side & dies away on the steep & rocky GULLY of last stream Friday Loch Lochy near Letter Fi
570n023 distended. touch a person on the ribs & how he GULFS in air.-- Again a master says I will see you da
574n041 ng her chops-- someone has described slovering <GUN> «teethless-jaws. as picture of disgusting lewd
225b218 are latent insects.-- as crows against man with GUN. & Bustards &c &c!!! An American & African form
613o036 trains of thoughts as fear of man,-- crows fear GUN.-- pointers method of standing,-- method of atta
220b198 urious paper on heredetary fear (like rooks with GUNS) of the Bustards in Germany.-- Athenaeum. No. 5
154g057 ounded, site N N W of Ben Erin {P} Shelf of Glen GUOY flat peat plain divortium aquarium-- tidal chan
154g058 from terrace of 2d shelf Level of shelf of Glen GUOY form comparison with granite block «& Ben Erin»
158g079 Mass of this point very nearly like head of Glen GUOY nor is horizontal line apparently continuation
578n055 nion he regards, <& see how> feel how the blood GUSHED into his face,-- "as <she> «the» thought of hi
628o53v ight be taught, but not cat, that is not act by GUSTO, though by fear it might be partly made.]CD p.
533m059 ikes the frown without knowing why-- a man as in GUY. Mannering. feels, pleasure. in seeing the scene
089a020 ite & diorite, covered with iron clay common to GUYANA, said to extend to Cordillera I see Brewster sp
306c213 ny -- marked difference with dogs of La Plata & GUYANA‖ people will say. not species.-- organs of gen
219b193 g trees Thrushes & bunting & coots-- «(Turdus GUYANENSIS?) (Emberiza Brasiliensis?) (Fulica Chloropus
420e088 ded,. of geology.-- L'Institut 1838. p. 414; M. GUYON thinks Monsters more common in Africa than in E
253c047 165.-- <a> "an account of the MANELESS lion of GUZERAT by Capt. W. Shee. considered merely variety.--
064r137 in quantity of Sulph. acid emitted: mem: Grand GYPSEOUS formation of Cordillera In describing structu
034r044 about sulphuric vapours in East Indian Volcanos GYPSUM Andes Mem. Beechey. account of regular change
054r102 ates, do at latter place. sandy. sandstone with GYPSUM, covered by limestone with recent shells 200 f
095a038 llera-- Near the Planchon talks of very much of GYPSUM.-- The officers of the Bonite. French discover
100a053 ation from some grand centre.-- A Stalactite of GYPSUM, is the best case of cleavage.-- Phillips (113
121a111 e formed crystals of ice were formed-- (like my GYPSUM case) shows power of segregation.-- & has heat
496q006 Ranunculus.= (11) Try.. Nitrate of Soda-- Salt. GYPSUM. Magnesium Iron Rust Carb. of Ammonia.-- Horse
609o030 sense points <is> to revenge). In judging of <our HA> of the rule of happiness we must look far forwa
455eIBC s chapter on sensitive plants; Physiology Get HABBERLEY to try experiments. about raising plants. whe
160g091 if Ben Erin is 3500 boulder Cairn leet more HABERCLADOR Isthmus broad flat Loch {P} XX Barom 28.92 A
199B118 into a fundamental propenity, of an accidental HABIT into an instinct." Ed. N. Phi. J. p 297, No 8 J
259c063 ce of that species must take to that particular HABIT.-- All structures either direct effect of habit
259c063 habit.-- All structures either direct effect of HABIT, or heredetary «& combined» effect of habit.--
259c063 of habit, or heredetary «& combined» effect of HABIT.-- perhaps in process of change.-- Are any men
276c124 ment can be, a bird practising imperfectly some HABIT, which the whole rest of other family practise
290c163 r of Muscles.-- then we SEE structure gained by HABIT Talent &c in man not heredetary, because crosse
301c199 performs.-- child striking a post in passion.-- HABIT instinct gained during life.-- do Elephants eas
301c198 instinct are external contingencies, where the HABIT is not applicable. The degree of development of
304c208 es have generally perfect organs. do changes of HABIT affect particular organs.-- of two adjoining fa
306c212 he stream followed ebb tide, therefore got into HABIT of going down stream which would last were the
308c217 - yet differences carried a long way.-- «Case of HABIT I kept my tea in right hand side for-- some mon
308c217 brate-- «seeing no tea brought back memory» old HABIT of putting tea in pot made me go to tea chest a
312c232 Individual Men «& animals» could only exist by HABIT.-- therefore same principle transferable., not
313c235 ust arise after long interval very good case.-- HABIT is awakened by association (case of Elephant, w
315c243 ning is to exposes his canine teeth. no doubt a HABIT gained by formerly being a baboon with great ca
365d107 e structure of which is adaptation to habits (& HABIT second nature) may be more in constitutional.,-
434e129 sometimes other.-- there is some difference of HABIT between these varieties, so that they have been
478z003 ar in winter in Paraguay p 207 Slight notice on HABIT of Iguana. not pass Lat. 28 degree North p. 239
520m002 old. when his father died.-- The omnipotence of HABIT is shown about meals, no [not located] There is
522m014 t with a sort of consciousness not just.-- From HABIT the feeling of anger must be directed against s
530m046 tions of same part of brain, or the tendency to HABIT of producing a train of thought.-- [not located
532m054 ese cases probably the system is affected, & by HABIT the mind tries to fix upon some object:-- When
535m063 sputed that a wasp has this much intellect. yet HABIT may make it act wrong, as I have done when taki
535m064 emarkable case may be normal. with insects, but HABIT forgotten in all older species. The earwig & a
537m075 may be inferred from expression. "relict of bad HABIT." as child is cured of sucking his finger by ru
538m080 nalogous to the double individuality implied by HABIT, when one acts unconsciously with respect to mo
539m082 even years interval. Aug. 15th. As child gains HABIT «or trick» so much more easily than man, so may
539m083 in double consciousness may really explain what HABIT is-- In the habitual train of thought one idea.
555m142 to use term instinct, when origin of heredetary HABIT cannot be traced) V. D. p. 111, case of Associa
558m150 under great sympathetic nerve. most subject to HABIT, as being less so will.-- May not moral sense a
568n017 ly bashfulness is connected with some disturbed HABIT» [Thus shepherd dog. has pleasure in following

Page
(Key Word)
```
568n017 so» When one is prevented performing heredetary HABIT, (or moral sense, or instinct,) one feels pain,
574n043 the acquirement of instinct by dogs, would show HABIT.-- Take the case of Jenner's <Hyaena> Jackall.-
578n056 aced & squeezed out again-- as power of mind by HABIT gets more perfect over voluntary muscles, these
582n067 ng in S. America capital instance of heredetary HABIT:-- there must, however, be a mental impulse (th
583n070 more  appropriately they would be "creatures of HABIT."-- CD[as the bee makes its cells, by means  of
585n077 of things which might be «possibly» acquired by HABIT. so bees in building cells, must have some mean
586n078 atch, but he does it by reason & experience, or HABIT.-- so bird migrating to certain quarter is inst
586n079 emember it.]CD back, without consciousness & by HABIT, such habit of knowledge of points of compass m
586n079 CD back, without consciousness & by habit, such HABIT of knowledge of points of compass may be instin
586n079 n felt in early childhood (before experience or HABIT) could be formed or afterwards.-- child sucking
586n081 ating may be considered also such.-- heredetary HABIT, is a part never subject to volition.-- like pl
587n087 great principle of liking, as simply heredetary HABIT.-- A blind man might be born with idea of scarl
591n100 ke continent man.-- a man eating what others by HABIT (not instinct) think not fit, as cannabalism, i
620o045 &  palliates the offence; one always admire the HABIT formed by «obediance to instinct» <conscience>.
624o50v . *Our tastes in mouth by my theory are due to <HABIT> heredetary habit (& modified & associated duri
624o50v outh by my theory are due to <habit> heredetary HABIT (& modified & associated during lifetime). so i
625o50v l sense.-- My theory explains both, perhaps, by HABIT-- [LHC] 11) Whewells preface. [RHC] It  appears
629o55v tosh gives & try it.-- p. 241 (1) Any action by HABIT may be thought wrong.-- & conscience will imper
629o055 lly> by feeling-- reasoned on, steps forgotten, HABIT formed,-- & such habits carried on to other fee
638j059 I perform some action.-- as well as gain it. by HABIT.-- New Theory of instinct, returning to Kirby's
534m063 on in the stem of Hollyhook, although ordinary HABITAT is Malva sylvestris. do. p. 228 Newport says D
060r125 Antarctic veg;-- Study Ulloa to see if Indian HABITATION above regions of vegetation.--«I can find no
290c163 ) of brown colour; lives on ground, colour of HABITATION. Must have some effect.-- Maldonado as  good
636j56v forms  a broken, recurrent series, whilst the HABITATION «or world» simple series.-- My theory shows
251c040 p.  49. on the localities of certain parrots HABITATIONS India & Africa.-- NB. Any monograph like Gou
211b162 with snipes> indicate affinity, because similar HABITS produce similar structure.-- Mem. Ornitho Rhyn
227b225 cture. latter far chief element:. little service HABITS in classification. or rather the fact that the
239cIFC se from Cart horse by picking without change of HABITS Mr Yarrell «Give it as his theory» tells me. h
255c051 association..»  Instinct goes before structure (HABITS of ducklings and chickens) Young water  ouzels
260c069 rnithology a mine of valuable facts,. regarding HABITS range & all kinds of information-- instinct Sw
264c081 genus  Gould seems to doubt how far structure & HABITS go together. This must be profoundly considere
264c081 idered.-- Structure may be obliterating, whilst HABITS are changing-- or structure may be  obtaining,
264c081 hanging-- or structure may be obtaining, whilst HABITS slightly preceed them-- From this view  habits
264c081 t habits slightly preceed them-- From this view HABITS must form most important element in considerin
264c081 which  tribe,-- structure without corresponding HABITS clearly showing true affinity, for instance ta
264c082 ker-- -- but tail of some ducks aberrant from-- HABITS.-- Gould I see quite recognizes habits in maki
264c082 t from-- habits.-- Gould I see quite recognizes HABITS in making out classification of birds Birds va
265c085 ly way of judging whether structure is owing to HABITS, or heredetary is to see, whether a large fami
265c086 r of that family, having it with very different HABITS-- Thus bill & nostril of Puffinuria I think we
265c086 e to heredetary origin & not adaptation. to its HABITS.-- Few will dispute that it is possible to hav
265c086 e that it is possible to have structure without HABITS-- after seeing beetle with wings beneath solde
265c087 ever know how much structure was connected with HABITS, & how much heredetary. The circumstance of ab
265c088 ge, where colours have changed in accordance to HABITS.-- one is tempted to suppose from beholding th
271c105 es. yet that it should so strictly <f> agree in HABITS with the Turdus Musicus «not found in N. Ameri
271c105 hing same manner, much mud.-- These facts show, HABITS heredetary whilst species have changed Argumen
271c107 be taking on structure (probably accompanied by HABITS) of other, thus in Chalcididous insect,  which
272c108 have  got thighs with same peculiar structure & HABITS of clinging to rushes similar.-- The  question
272c108 mediately want are there Heteromera, which have HABITS & part structure like Curculionidae.-- Are the
272c108 dae.-- Are there any Crysomelidae, with similar HABITS. But the Horae Entomologicae will tell this.--
273c111 Hawks,  there is a swallow, both in structure & HABITS (it cannot be doubted that if swallow perished
273c112 ment structure is not always applicable to same HABITS, though swallow hawk, .milvulus,. may catch in
273c113 ionis), yet the Tropic bird, has very different HABITS, though preeminently belonging to this type, ¿
274c113 the ground has not this structure., instance of HABITS going before structure).-- even one kingfisher
276c124 uth It is of the utmost importance to show that HABITS sometimes go before structures.-- the only arg
276c125 very close species inhabit same country are not HABITS different, (Mem: Gould's Willow Wren) but wher
276c125 where close species inhabit different countries HABITS similar ¿law?-- probable.-- if habits & struct
276c125 countries habits similar ¿law?-- probable.-- if HABITS & structure similar would have blended togethe
277c125 ould have blended together Mem Mr Herberts law; HABITS determining fertility Scheme for abolishing sp
277c127 es come over to this country, without range, or HABITS ascertained-- put them as (a). (b) until  data
277c127 ht home new species. until, he can show range & HABITS-- Take instances of most disputed shells, such
282c141 anged. in structure & oven to certain degree in HABITS, yet we might have these analogies.-- We  must
283c143 pe in each family is relation to elements & not HABITS, as shown by frigate Bird & flying eagle.-- Haw
287c158 ation must chiefly rest on these same organs,-- HABITS, range. &c &c-- Macleay rests his whole ground
289c160 capital  instances of typical land bird, having HABITS of a Grebe, structures might follow.-- examine
289c161 examine structure of this bird & get account of HABITS My definition <in wild> of species. has nothin
291c165 e. of man gained & heredetary. «problem solved» HABITS become important element in classification, be
291c165 ld, remark of D'orbigny point out importance of HABITS in classification.-- Thought (or desires  more
293c174 of  Man.-- different contagious diseases, where HABITS of people nearly similar. Curious instance  of
297c186 oft plumage of night jar. like owls. analogy in HABITS adaptation to nocturnal habits-- to cats &c.--
297c186 owls. analogy in habits adaptation to nocturnal HABITS-- to cats &c.-- must be acquired by my theory-
300c197 transported to the Str of Magellan.-- Change of HABITS in Van Diemen's land. Study Mr Blyth's  papers
301c199 rel to kill ears of corn according to my views, HABITS give structure,.. habits precedes structure,..
301c199 according to my views, habits give structure,.. HABITS precedes structure,.. habitual instincts prece
301c199 ined during life.-- do Elephants easily acquire HABITS is this the Key to their mental powers.? p. 8.
305c211 to delicate emotions between races, & recurrent HABITS in animals.-- --Animal Magnetism-- principles
306c212 ts in young animals, well developed, just like, HABITS easily gained in child hood.-- Young  salmons.
308c217 absent «Dr. Black. tea & sugar» people. reverse HABITS Insects & birds are the only two tribes fitted
308c219 varieties not produced as by nature. if so. the HABITS which would. have formed them, would have aris
309c219 ir difference. arise a good deal from climate & HABITS, & therefore less fertile. according to Mr Her
311c226 ngista of Australia & Van Diemen's land diff.-- HABITS can only be used «in classification» as indica
313c235 have had their infancies, as well as men-- when HABITS much more firmly impressed.-- we see in the En
313c236 era. under sympathetic nerve may be instinct or HABITS. ¿are sympathetic nerves & nervous system of i
315c240 Created at many spots & since disseminated See. HABITS of Malay fowls p 5. (note) in some papers on i
333d005 eyes very large, very fierce to dogs «otherwise HABITS not different; tone of voice. perhaps rather d
346d048 rville account of wild cattle of Chillingham,-- HABITS peculiar,-- young one 203 days old butted viol
358d075 e of bird of different family. having very same HABITS in some respects as this Caracara.-- Sept. 9th
360d093 ght be harder to cross than two less opposed in HABITS, though externally similar.-- this however  is
363d101 ng.--» & compare them with all the varieties.-- HABITS of rock pidgeon. (I suspect Pennant has descri
363d102 r Yarrell says in very close species, of birds, HABITS when well watched always very different.-- The
363d102 stake both species scattered over Europe)-- The HABITS of some «same» North American & Europaean bird
365d107 s., all the structure of which is adaptation to HABITS (& habit second nature) may be more in constit
369d115 ecessity of combining observation of the living HABITS of animals, with anatomical & Zoological resea
377d147 lso structure not really fitted for water, only HABITS & instincts-- The young of the <p> Kingfisher
378d147 ought to show them.-- Anyhow not connected with HABITS According as child is like parent, so is speci
410e052 losest species (& naming them) with relation to HABITS, ranges. & external conditions of country, mos
411e056 Isld.  is interesting as showing some change in HABITS before form.-- I have already given various ex
421e091 ng those which are most remotely related to the HABITS & food of an animal, I have always regarded as
440e146 se strange plumage in pidgeons yet no change of HABITS, so no <cause> corresponding change in Birds o
440e147 much  structure, than attempting anything about HABITS-- no one can be shocked at absence of final ca
445e159 . 454.-- does really attribute metamorphoses to HABITS of animals & takes series of flying mammifers-
463t055 ructure, but it is not possible to imagine what HABITS an animal could have had with such structure.-
480z006 Proceedings of Zoolog Soc. Important account of HABITS of Tubularia. p 52. May 1836 dimensions of imm
481z008 els, are night birds agree. with <pe> nocturnal HABITS of Crustacean Mr Broderip says that Voluta fou
484z014 walking--  foot bill crest feathering on legs-- HABITS-- Does the Secretary, make noise & throw head
485z016 rpheus--becoming tyrant-- flycatcher-- shown by HABITS & plumage so very similar to some of the Fluvi
489qqIFC .  17 {T} Mrs. Whitby. Newlands Lymington Hants. HABITS of different caterpilar races. --Name of Ital
512q019 t-- about selection for milking-- loss of early HABITS in Dorsetshire sheep migration of coots-- vari
536m072 ure actions unavoidable & only to be changed by HABITS). now free will of oyster, one can fancy to be
```

537m075 e, but yet it is settled by reason.-- How slow HABITS are changed may be inferred from expression. "
539m083 will (without direct consciousness?) change its HABITS.-- Aug. 16th. As instance of heredetary mind.
543m101 ies are crossed as in Shepherd dogs-- Inherited HABITS: Have Effect in Bones is valuable it shows tha
545m104 me» become unconscious. so may some ideas.-- ie HABITS, which must require idea to order muscles to d
560m156 s, other cases like D. Corbet; «do» ideots form HABITS readily?? Do the Ourang Outang like smells «pe
564n004 e satisfaction of following conscience, obeying HABITS, & dread of misery of future thinking of injur
571n030 nguish. between prejudices of youth from <here> HABITS. & heredetary habits. & perhaps even latter ma
571n030 dices of youth from <here> habits. & heredetary HABITS. & perhaps even latter may be vitiated. or rat
574n040 n very limited reasoning powers & the fixing of HABITS,-- for instance the Birgos opening a Cocoa nut
574n040 ll at one end.-- Children & old people get into HABITS.-- we probably can hardly form an idea of a mi
574n042 to the formation of instincts, independently of HABITS.-- the limits of these two actions either on f
575n044 somewhat right to deny identity of instinct.-- HABITS import to Bowen No one doubts that a cross of
580n062 before two years are soon lost; yet many of the HABITS acquired in that age are retained through life
581n063 . J. Sebright. has given the phrase "heredetary HABITS." very clearly, all I must do is to generalize
581n063 le to all cases.-- & analogize it with ordinary HABITS that is my new part of the view.-- let the pro
581n063 the view.-- let the proof of heredetariness in HABITS. be considered. as grand step if it can be gen
588n090 s character is invariability.-- if explained by HABITS, useful to itself, how gained. reason? or some
589n091 dentical with mine,-- Lamarck. Vol II p. 319.-- HABITS more prevalent in proportion to intelligence l
589n091 rtion to intelligence less.-- p. 325 «to 29».-- HABITS becoming heredetary form the instincts of anim
596n1BC ? Do people of weak intellects easily fall into HABITS Get facts about instincts of mongrel dogs Do
610o032 erful manner.-- as the hour of the day &c-- All HABITS must conduce to their health & comforts.-- Bot
615o36v .-- His Malay breed «of fowl» totally different HABITS from European. begin to prowl about in the ev
615o36v ow acquirement or obliteration of instincts But HABITS acquired even by <children> «plants»! [RHC] 7)
619o042 race of man shows this, if we judge him by his HABITS, as <if> another animal. These instincts consi
622o048 CDs. [LHC] My theory of instincts. or heredetary HABITS fully explains the cementation of habits into
622o048 detary habits fully explains the cementation of HABITS into instincts. [RHC] Feelings of the mind, wh
629o055 ned on, steps forgotten, habit formed,-- & such HABITS carried on to other feelings, such as temperan
292c171 see men walking in sleep».-- an action becomes HABITUAL is probably first stage, & an habitual action
292c171 becomes habitual is probably first stage, & an HABITUAL action implies want of consciousness & will &
292c171 .-- We even see they must be done often «to be HABITUAL» or of great importance to cause long memory.
301c198 in precisely same way not possible to say what HABITUAL in men & what reasonable-- Same action may be
301c199 owing that animals have some reason, & actions HABITUAL. it surely is not worthy interposition of dei
301c199 give structure,.. habits precedes structure,.. HABITUAL instincts precede structure.--duckling runs t
301c199 reason in hiding its food is applicable to any HABITUAL action. even which Man performs.-- child stri
305c210 og older character & manner of wagging tail.-- HABITUAL movements connected with mind There is no pro
313c236 em of insects analogous.?-- («Even plants have HABITUAL actions.--» «this very important in consideri
315c241 some valuable analogies might be drawn between HABITUAL actions of plants «when exciting cause is abs
315c242 contraction of fibre)-- it is most remarkable HABITUAL actions in plants, it allows of any degree in
315c242 s, it allows of any degree in lowest animals --HABITUAL action, in intestines subject to sympathetic
521m007 e of Mr Anson. who told a story of hunting «-- HABITUAL fits.--» which my Father thinks is mentioned
521m007 t so very wonderful.-- <Now is not epilepsy an HABITUAL disease of the muscles.???> Miss Cogan's memo
521m008 ory, because she did not remembered, it was an HABITUAL action of thought-secreting organs, brought i
522m012 failure very general effect of <early> «slight HABITUAL» intemperance.-- often accompanied by extreme
526m030 mpted to believe phrenologists are right about HABITUAL exercise of the mind, altering form of head,
529m040 V. Wordsworth about science being sufficiently HABITUAL to become poetical) the botanist might so vie
530m046 muscle is moved very often, the motion becomes HABITUAL & involuntary.-- when a thought is thought ve
530m046 hen a thought is thought very often it becomes HABITUAL & involuntary,-- that is involuntary memory,
531m052 ual movement does not take place.-- A start is HABITUAL movement to avoid any danger-- Fear, shamming
531m053 y some animals in defence, &c Starting must be HABITUAL «involuntary» movement from wish to avoid som
539m082 hardly be considered as actions otherwise than HABITUAL.-- instances?? The possibility of two quite s
539m083 ness may really explain what habit is-- In the HABITUAL train of thought one idea. calls up other, &
539m083 ss of double individual is not awakened.-- The HABITUAL individual remembers things done in the other
539m083 individual remembers things done in the other HABITUAL state because it will (without direct conscio
540m087 nventive thought is that none of the idea are HABITUAL, nor recalled by obvious associations. as by
542m094 - Curious to trace, which of these actions are HABITUAL, & which now connected physical relations.--
542m095 t remain just like sneering does.-- is yawning HABITUAL from awaking from sleep see how a dog yawns w
542m095 ned from too long rest of muscles.-- evidently HABITUAL when transferred, (also how often) to the tal
542m096 ith respect to sneering the very essence of an HABITUAL movement is continuing it when useless.-- <&>
545m107 like children.--» Expression, is an heredetary HABITUAL movement consequent on some action, which the
545m108 e cause, which «now» excites the expression.-- HABITUAL actions are the reverse of intellectual, ther
552m131 in N. America Fear probably is connected with HABITUAL stopping of breath to hear any sound.-- attit
564n006 aces «which may be described as an exaggerated HABITUAL sneer" the manner in which whole skin or musc
567n013 lants any notion of cause & effect, «they have HABITUAL action. which depends on such confidence» whe
568n016 ur necessary notions follow as consequences on HABITUAL or instinctive assent to propositions, which
571n029 heory must be acquired, by certain foods being HABITUAL-- & hence become heredetary; on same principl
571n029 the former to true taste.-- Everything that is HABITUAL, if heredetary, is pleasant.-- Mental & Bodil
572n032 hen we do not expect any bodily harm-- case of HABITUAL action.-- L'Institut. 1838. p. 340. Mr Carlyl
574n042 it is bare fact, on my theory intelligible An HABITUAL action must some way affect the brain in a ma
576n047 r wholly instinctive as in the dog, or chiefly HABITUAL as in man), for it added much to the happines
577n051 iation will not help my theory of sensitive Plants HABITUAL actions, (independent of mind) in the intesti
578n054 one does not blush before utter stranger,-- or HABITUAL friends.-- but half & half. Miss F. A. said t
579n059 the features pass before him marked, with the HABITUAL expressemotions, which make us love him, or h
592n104 e action to opening eyes in fear The effect of HABITUAL movements in muscles of face, is well seen in
601o009 sensation.-- man moving leg when asleep-- «or HABITUAL actions» perhaps polypi-- (so that lower anim
602o011 happiness &c, our well-being depends upon the "HABITUAL reason".-- This power of the mind, faintly ap
622o048 the feeling really is.-- All these associated HABITUAL» feelings. become like the instinctive, ones,
113a092 of statical equilibrium, the height of lava (HABITUALLY) becomes measure of force in that part.-- Im
293c172 d do some action-- which its father. had done HABITUALLY we should exclaim it was instinct.-- Even if
298c190 eneral tendency produced. Such as man getting HABITUALLY into passion, becomes habitually passionate.
298c190 man getting habitually into passion, becomes HABITUALLY passionate.-- the Key to the affections migh
298c190 might perhaps thus be found-- a person who is HABITUALLY kind to children,--<get> «increases» general
534m063 llect. but it does it always instinctively or HABITUALLY.-- good Heavens is it disputed that a wasp h
538m080 lf, & likewise one forgets. what one performs HABITUALLY.-- Agrees with insanity, as in Dr Ash's case
232b245 those forms, which have succeeded in becoming HABITUATED to colder climate whilst others died out, or
404e026 ss. Case of Mexican greyhounds.-- young being HABITUATED. instance such as Hunter,. or some one mentio
568n017 ry notion, ass has it.-- When one is «simply» HABITUATED «in life time» to any line of action, or tho
628o054 e [3] pleasure) or that I have been taught or HABITUATED to associatical, the emotions of this instin
275c120 nds.-- Sir. J. Sebright first got {P} point on HACKLES on Bantams by crossing with common Polish cock
309c221 ur of certain Arabs.-- NB avoid quoting these HACKNEYED cases Mr Ed Blyth does not believe in circula
508g016 storia del Reproducione.-- D. Holland (7) Is HAEMORRAGIC tendency, independent of heredetary cases, m
413e060 nly place, where, Elephant, Rhinoceros, Hippot, HAENA &c are found together.-- Read this Work-- Decb.
182b048 byssinia, & others dark brown, with long clotted HAIR resembling that of goats." Progressive developm
218b190 tory of cats on West coast of Africa.-- changing HAIR-- The Edinburgh. Journal of Natural History-- P
244c021 wild pigs of Falklands, generally "red of brick" HAIR, very stiff, p. 120-- Coati roux common. near C
267c093 yssinia; the others dark brown with long clotted HAIR resembling that of goats" Geograp. Journ. Vol V
303c204 ?) & what kinds of intellect) quantity & kind of HAIR forms of legs-- hence the father of man kind pr
303c204 way between man & monkey, would have differed in HAIR colour + + + form of head «& features»;. but li
332d003 all pretty much alike.-- My Father Water-in the HAIR a century since used to be called Worm Fever, a
338d024 dresser (Willis) says <black> «that strength of» HAIR goes with colour. black being strongest.-- V. p
345d044 d dogs, coloured like Magellanic Fox.-- peculiar HAIR & appearance-- good case of Provincial Breed--
373d132 t series of changes from the placentares, Having HAIR. like true Mammalia, no more wonderful. than Ec
385d163 25th Young man at Willis «Grt. Marlborough Str, HAIR dresser, assures me he has known many cases of
385d164 sh Royal Academy) have contained perfect teeth & HAIR-- showing foetus has gone on growing-- I believ
468t105 «Thrips» about as large as bit of chopped horse HAIR with legs & take flight-- Yet we have crosses--
591m097 effects of emotions-- fear giving goose skin-- & HAIR standing on end.-- July 20th Intelligent Keeper
592n103 snort clears nostrils» when frightened, does not HAIR & rabbit depress. them from squatting.-- p. 64
637j57v ns in varieties.-- greyhound to hare.-- waterdog HAIR to water-- bull dog to bulls.-- primrose to <op
338d024 ies-like,-- -- Says Negro-- thick skinned My HAIRDRESSER (Willis) says <black> «that strength of» hai

Page **(Key Word)**
217b183 h would never take the dog. But at last a rough-HAIRED shepherd dog lined her & produced a very large
306c214 - At Angora «centre of Asia Minor» are the fine-HAIRED goats. which it is said cannot be transported
306c214 t be transported from their country.-- the long-HAIRED cats are supposed to come from there.-- All th
244c021 0.-- Most of the dogs of Payta-- belong to the HAIRLESS kind, «said» to come originally from Africa p
312c228 arents> Mother.-- like dogs Smith knew chinese HAIRLESS dog & common spaniel crossed.-- 3 puppies PER
360d089 n <two> crosses «twice made» between terrier & HAIRLESS dogs of Africa,-- some puppies hairless. some
360d089 rier & hairless dogs of Africa,-- some puppies HAIRLESS. some in patches, & some hairy-- the former p
432e124 in age, & therefore the children do not, (& in HAIRLESS kittens we see same fact) go back, & this is
493q004 esult.-- (3) Cross the Esquimaux dog. with the HAIRLESS Brazilian or Persian animals of different her
240c014 very like the Siam race with long nozzle & few HAIRS) inhabits Celebes & few of the larger islands.-
244c022 he Tenrecs from Madagascar. Monkey from Java.-- HAIRS, & deer.-- Procured two makis alive from there.
558m152 ch [lip] drawn back & driving air out of mouth «HAIRS erect on back» «wide open» with prodigious forc
022r008 we found the Head & Boans of a Hippotomus; the HAIRY lips of which were still sound and not petrifie
199b119 Paper by Crawford on Mission to Ava. account of HAIRY man. because ancestors hairy with one hairy chi
199b119 to Ava. account of HAIRY man. because ancestors HAIRY with one hairy child, and of albino DISEASE bei
199b119 of HAIRY man. because ancestors hairy with one HAIRY child, and of albino DISEASE being banished, &
295c178 ps!!!? No) p. 15 (Lyell's Pamphlet) Is man more HAIRY than woman. because ancestors so, or has he ass
311c227 coast of Africa .p 69. with tall grass. & p 72 HAIRY sheep-- Edinburgh. Transact. Vol IX p. 107. an
360d089 some puppies hairless. some in patches, & some HAIRY-- the former preponderated <which seems owing «
362d099 fect of castration horns drop off., replaced by HAIRY ones. which never «dry up &» peel off their ski
420e089 ws man was originally quadru<manous> «ped.-- ». HAIRY.-- could move his ears The head being six metam
432e124 be true hermaphrodite Man probably assumes the HAIRY character of his forefathers only when advanced
466t094 all anthers, wh otherwise lie protected by the HAIRY black lip of lower division of nectary: «wh. it
107a077 of many artificial limestones produced by Sir J. HAL. End of pages. p. 157. Vol VI Edinburgh. Phil: T
260c068 l colours of species-- Mem. St. Jago--solitary HALCYON bird of passage.-- M. Coronata of Latham, wron
023r009 ng, & as big as a mans thumb, the rest not above HALF so long; The maw was full of jelly which stank
048r086 ld animals. both species & individuals as in the HALF desert country of S. Africa. It would be well t
054r105 nces not many years since, the sea covered above HALF a league of what is now Terra Firma & the exten
054r105 is now Terra Firma & the extent of a league & a HALF a long the coast. <"> The rocks in the most inl
055r107 ena. Ascension. Azores. («sandstone first gives» HALF demolished craters).--worn into mud & dust.--co
070r155 alta. & not far from Tucama[n]. & at Chuquisaca. HALF across the continent.--He states plains of Mend
148g022 on with all the measure before There some of the HALF rounded gravel nearly as high as highest measur
148g025 nt said the lambs were nearly like each other «& HALF between parents» (& not like dogs), but they th
204b139 d. decidedly exceedingly prolific & hybrid about HALF way. Eyton says Hybrid about half aways & resul
204b139 & hybrid about half way. Eyton says Hybrid about HALF aways & result the same Indian cattle & common
217b184 etimes take strongly after either parent. about <HALF & half tim> as often one way as other.-- He has
217b184 take strongly after either parent. about <half & HALF tim> as often one way as other.-- He has known
217b184 «produce litter like both parents» & Mr Bell has HALF bloodhound & greyhound.-- Where two dogs have l
239c002 nut.-- On this principle I may add, that fact of HALF cross with parents, going back to either parent
240c003 ansmission of form, may explain mule & pig being HALF way. Yet dogs sometimes like father, sometimes
275c121 most all imagination-- He says he recollects all HALF Bred cattle of L'Darnleys were most like parent
288c159 ative dogs & English cross readily-- think about HALF way in appearance.-- bark about half way «in to
288c159 think about half way in appearance.-- bark about HALF way «in tone»-- the native dogs howl most disma
290c164 change, Will be known.-- It appeared to me that HALF between fowls & pheasants, is most like pheasan
298c189 reason?? Gould says he believes that he has seen HALF fox & dog. & that it was most like fox.-- He fe
298c189 & that it was most like fox.-- He felt sure the HALF breed of Australian dogs, would be «most» like
305c210 mage for wild birds-- At Zoolog Gardens there is HALF Jackal & Scotch Terrier.-- certainly more like
312c228 wing to no attempt to keep up offspring, are not HALF lion & tigers ditto. (see Griffith) & half Musc
312c228 e not half lion & tigers ditto. (see Griffith) & HALF Muscovy ducks, «black cock & pheasant see Jardi
332d003 him on one day.-- Mark at Shrewsbury thinks the HALF bred Alderney Cows take more after Alderney tha
333d005 Common cat.-- Ch IX Mongrels Hybridisim Fox has HALF Persian cat. which bred with unknown common hou
333d006 closely resembled in form of tail, fur &c to the HALF bred Persian.-- Here then we have clear case of
333d006 ne impregnation, or two impregnations one giving HALF character & other more of English, but the effe
333d007 Jackall in Zoolog Garden) He has seen in a show HALF Wolf & «half Esquimaux» dog which <likewise mor
333d007 oolog Garden) He has seen in a show half Wolf & «HALF Esquimaux» dog which <likewise more resembled t
333d008 rell & no leading question was put.-- Fox thinks HALF Lion & Tigers are exactly intermediate in chara
334d012 ty. doubtful whether all are white. Fox says the HALF Muscovy Fox says a settler near Swan river, los
335d015 adful monsters, abortive, just like mules. Fox's HALF bred Persians «cat» favour the Persian side.--
346d047 species.-- high active breedin[g] [not located] HALF breed liable to vary. I asked this in many ways
356d071 t include so many facts as mine The facts about HALF breed animals being wilder than parents is very
378d148 ird for they were of all colours.-- they were "HALF wild-half tame, they came to the windows to be
378d148 they were of all colours.-- they were "half wild-HALF tame, they came to the windows to be fed, but s
385d164 everlasting layer--. or Polish breed. (he thinks HALF pheasant, half fowls.-- eggs fertile, but paren
385d164 er--. or Polish breed. (he thinks half pheasant, HALF fowls.-- eggs fertile, but parent bird will nev
393dIBC lty. Does male fail in passion.-- Disposition of HALF bred Cattle at Cinbermere? How is Jackall & dog
417e078 e father or mother, independently of its sex, or HALF way between, or someway different from either:
424e103 nd we see in the Australian dog an instance of a HALF reclaimed animal.-- The dogs, which have run wi
453e183 me garden.-- now this good question-- single, or HALF double.-- anyhow fertile because they «are» rai
459t013 its wing resembled the drake.-- another of same HALF breed resembled the plumage of drake still more
460t013 uce full as many eggs as pure bred common.-- the HALF of the cross, as above, take «generally» after
460t013 «generally» after the swan-gander. one of these HALF-bred ganders. crossed with common goose <to has
460t015 (though only ¼ of blood). that it appears about HALF way between swan-goose & common goose.-- the st
460t015 - the stripe down back pretty plain in in these <HALF> «3/4» bred ones-- The brothers & sisters half-
460t015 <half> «3/4» bred ones-- The brothers & sisters HALF-breed showed no sexual inclination for each oth
467t103 nk they do in Broom & certainly when over-ripe & HALF withered-- I saw Bees going to clover & once th
492q003 o Questions regarding Breeding of Animals If two HALF bred animals exactly alike be interbred will of
493q004 r in first breed or permanently.-- (2) Cross two HALF-bred animals. which are exactly alike & see res
494q005 ilk caterpillars (1) Shake a sleeping mimosa, or HALF bred mimosa (a) between sensitive & sleeping sp
515q022 ls of same species Eyton (1) Number of eggs-- of HALF-bred geese-- inter se, & with parents & of Chin
538m079 on such characters as Allen W. & Babington, both HALF ideotic in some respects & with store of accura
543m101 lassifying distinct resemblances.-- The facts of HALF instincts. when two varieties are crossed as in
564n006 in Lavater, P. cii Vol III» of excessively cross-HALF furious faces «which may be described as an exa
578n054 ore utter stranger,-- or habitual friends.-- but HALF & half. Miss F. A. said to Mrs. B. A. how nice
578n054 er stranger,-- or habitual friends.-- but half & HALF. Miss F. A. said to Mrs. B. A. how nice it woul
344d042 re this happened from her looks thougt she was HALFBRED Beagle Staghound. «++»∴the grandchildren went
303c204 & Prichard -- Now we might expect that animal HALFWAY between man & monkey, would have differed in h
443e155 ed by sexual commerce «The fact of Corallina & HALIMEDA is case in point».-- The relation of these «s
094a036 el Imrie Transact Wern. Soc. Vol. 2. p. 35 Sir J HALL Trans. Phils Royal Ed. Vol 7 Dr Buckland Reliqu
121a111 gst <old> papers read before societies.-- Sir. J HALL Vol VI. p 173. (Ed. Transact) has seen clay sti
214b175 can Zoology-- A breed of Blood Hounds from Aston HALL close to Birmingham, and supposed to be descend
346d048 , go in file, hide their young., bold.-- a Mr W: HALL remarked that it was against all rules their pr
521m009 ort. Aunt. B. ditto.-- Case of Mr Corbet of the <HALL> «Park», after paralytic stroke. intellect impa
526m029 agrees with my Fathers case of Mr Corbet of the HALL understanding. (on hearing old association brou
590n093 bile duct, & throws bile in circulation p. 75. HALLER says tooth ache, even from carious tooth cured
138a155 th the moon at some future time [blank] Sir. J. HALLS Paper on the consolidation of strata-- he heate
061r127 any spot of land. = Yet new creation affected by HALO of neighbouring continent: ≠ as if any creation
271c106 ed Argumentum ad absurdum. The creative American HALO has extended to Juan Fernandez in birds. but ¿w
271c106 whether to same island in plants?-- What is this HALO.-- Continents are not stationary «unerring proo
256c055 istribution of birds & animals Mr Strickland & HAMILTON-- found tertiary formation amongst Graecian i
306c214 tance of colours dependent on localities.-- -- HAMILTON will give an account in his Travels in Asia M
591n097 . «or will» [not located] <& other cows--> Mr. HAMILTON on vital laws (in the Athaenaeum Library) des
585n076 n applied to birds, which have been carried in HAMPERS. if they have not known the direction in which
449e169 .-- Waterhouse refers to fossil remains of the HAMSTER.-- is not this Siberian animal?-- Eyton says t
075r166 porphyry, containing grenats. In Peru. on other HAND, mine of Gualgayoc or Chota & Pasco in "alpine
105a072 (because more recent) ------ Coquimbo on. other HAND?-- The widening a valley depends on serpentine
149g030 ongua, must there have deposited much-- On other HAND remember modelling power of sea N of Valparaiso
157g077 » {P} cory stream hill with boulders river Right HAND Cascade has <cut> «where two branches unite in
158g079 rently continuation of upper terrace -- on right HAND {P} rounded waterworn buttresses of granite obs

```
171b004 t & alter the race to changing world.-- On other HAND, generation destroys the effect of accidental i
180b037 be  well adapted, & thus perish out, or on other HAND like Orpheus. being favourable many might be pr
196b108 plants  & insects had been created; but on other HAND creation of small animals must have gone on sin
236b279 decidedly an Indian form of Tortoise.-- On other HAND. fresh water tortoise from Germany. (where Mr M
240c004  , eggs hatched under other birds & brought up by HAND.-- These facts all account for [not located] Fa
263c077 > & <moral> feelings as animals.-- they on other HAND can reason-- but Man has reasoning powers in ex
308c217 long way. --Case of Habit I kept my tea in right HAND side for-- some month, & then when that was fin
308c217 or a week took of cover of right side, though my HAND <vibrate> would sometimes vibrate-- «seeing  no
391d175 ose <forms» slightly favoured, getting the upper HAND. <]CD> & forming species)-- [Aphides having fer
415e068 it, has been, (with what attendant organization, HAND & throat) that has made a man.-- CD [any monkey
442e153 afted. altering is hostile to this: but on other HAND, fruit trees are propagated by means, which wil
443e157 lse reasoning Bell Bridgewater's Treatise on the HAND.-- p. 94.-- "The resemblance of the foot the Os
451e178 hey will catch the Malaria & die.-- On the other HAND there are breeds of Men the Thârû & the Dhangar
524m022 Father to distract her attention took her «left» HAND to pretend to feel her pulse.-- What fails firs
567n013 to  do with them, came several times & opened my HAND, & put them in-- like child. Tommy's face,  now
587n087 ed. as the Turks have of Rhubarb: again on other HAND, it is said people, who like sweet things disli
588n089 licable to my theory of happiness.-- Bell on the HAND p. 191 Says <childr> babies have an instinctive
589n092 aving reason: or dog, having high powers without HAND or voice.-- there is some great puzzle in  what
638i059 the nervous system, gains that knowledge, before HAND. & can in idea (with consciousness.) <th>  form
639j28r n deserts in the old world!]CD p. 252 analogy of HAND in mole, & Mole cricket & rodents (?) p. 251. a
639j28r nd years the long legged race will get the upper HAND. though continually dragged back to old type by
641j29r Walrus, great Seals of Antarctic seas. (on other HAND Spermaceti Whale & Manatee.-- Naturalists  must
240c004 e fact of great monstrosities being produced, & HANDED down, with ease, is analogous to what occurs i
247c029 gland"-- Black & Grey varieties of rabbits thus HANDED down for nearly 70 year. Galapagos Mouse not t
371d127 of  such changes, not acquired by parent, being HANDED down? Are not Loddiges 1279 roses kept in same
384d161 successive  buds do differ-- any variety is not HANDED down. but is handed down for some  generations
384d161 differ-- any variety is not handed down. but is HANDED down for some generations Theory of sexes (wom
434e127 ns of plants) to do so, the effects are equally HANDED to offspring.-- Whewell's anniversary  address
461t037 from my theory, as long as any structure can be HANDED down without being absolutely injurious «(or r
461t037 g nutrition)» to a certain amount it will be so HANDED down«(».. as mammae of men «callosities on Cam
533m059 t side (not good example,) because leg is right HANDED.-- In Review (Edinburgh) of Froude's life. tha
576n048 ed organization to learn Greek, that to have it HANDED down as an instinct.-- Instinct is a modificat
593n105 round  & round is some old instinct <perverted> HANDED down & down.-- mem. Nina used to get into  hay
596n184 ions of every kind.-- «[Are monkeys <are> right-HANDED??]CD» Cyanocephalus, Macacus, Niger. Cercopith
554m139 emed to relish the smell of Verbena & Pocket HANDKERCHIEF & liked the taste of Peppermint.--» Perfect
581n064 ently. ourang outang very fond of soft, silk-HANDKERCHIEF-- cats & dogs fond of slight tickling sensa
366d111 cleaning  door outside, as often as she touched HANDLE, though really fully aware she was not  coming
406e036 -- case in Spain of pustular disease following HANDLING sheep-- all cases: d degree p. 354-- The most
403e022 are  particularly common.-- without arm, <skin> HANDS thumb,-- one leg, hare lip &c &c. in Vol II p 3
450e175 noticed  in Cassay Ava Pegue-- seldom equals 13 HANDS-- those of Lao & Siam inferior to those of Pegu
450e175 both small -- Java pony occasionally reaches 13 HANDS.-- Phillipines Pony somewhat resembles that  of
520m002 wiching of muscles, & general manner of holding HANDS &c &c.-- Mr Dryden Co said he could not remembe
542m095 ng vision, as savages without hats put up their HANDS, & as attention would amongst lowest savages cl
554m139 ship could not puzzle her-- with aid of teeth & HANDS.-- Descent 1838 It was very curious to see  her
557n149 ricks.-- so are people in passion my F. rubbing HANDS.-- stamping. grinding teeth.-- in shame frownin
603o014 eficient in not explaining the possibility of <HANDSOME> «UGLY healthy» young woman, with good expres
420e089 » animal with a vertebra only & no head-- !! HANDWRITING is determined by most complicated circumstan
420e089 nces, as shown by difficulty in forging, yet HANDWRITING said to be heredetary. shows well what minut
455eIBC e association which is useless. Granfather's HANDWRITING, to compare with my own.-- E Bengal  Journal
539m083 good  instances.-- when education same.-- My HANDWRITING same as Grandfather. Aug. 16th Anger  «Rage»
406e037 when absent from its master.-- dogs when strayed HANG their tails.-- November 1st.-- Addenda to Jour
555m143 car behind (.instead of front) (having changed HANGING into his head cut off) as kind of wit, showing
555m144 is was kind of wit.-- I changed I believe from HANGING to head cut off. «there was the feeling of ban
555m144 e the whole train of Dr Monro experiment about HANGING came before me showing impossibility of person
555m144 howing impossibility of person recovering from HANGING on account of blood. but all these idea came o
489qIFC ooker p. 17 {T} Mrs. Whitby. Newlands Lymington HANTS. Habits of different caterpillar races.  --Name
035r046 eneral truth of the proposition."-- If such can HAPPEN in troubled England; the more minute equalitie
068r146 ).--The first dislocations & eruptions can only HAPPEN during first movements, and therefore  beneath
216b180 either a little nearer black or white as it may HAPPEN.-- Dr Smith says he is sure of the case at Cap
226b221 me chance few representative species. this must HAPPEN. &, thus acquired will explain representative s
227b225 en. in old world, & on amount changes which may HAPPEN--  It leads you to believe the world older th
239c001 ointer.-- He has no doubt that same thing would HAPPEN with Australian dog & any of our common variet
355d067 ge of discovering laws is to foretell what will HAPPEN & to see bearing of scattered facts..-- What t
389d177 t produce) tendency to deformity ¿this does not HAPPEN with hybrids?]CD Plants must stand much breedi
515q021 As  peaches sport into Nectarines (does reverse HAPPEN?) what is effect of crossing peaches & nectari
527m035 intment «in real train of thought this does not HAPPEN. because papers, &c &c round one. one  recalls
546m111 ental» health.-- «Erasmus had almost same thing HAPPEN to him about a knife. which he had hid some ye
571n031 Much pantomimic gesture?? which would naturally HAPPEN.-- Reynolds Works. Vol I, p. 226-- "The genera
023r010 tinent of S. America is inadmissible «may have HAPPENED from incipient elevation.» The volcanos origi
032r041 rly current:--If great changes of climate have HAPPENED. hurricane in bowels of earth cause:--<exp> d
067r144 M. Video «Volcano in Pampas» Pasto Earthquake. HAPPENED on January 20th. 1834 Mr Sowerby. younger. sa
148g028 lains sloped centre-wards which would not have HAPPENED if the side-streamlet had cut them out-- In a
175b016 nswer because time short & no great  change has HAPPENED I look at two ostriches as strong argument of
185b055 ures; :+law of chance would cause this to have HAPPENED in all. but less in water birds-- carrion eag
344d042 ut one big & one small. Now Jones, before this HAPPENED from her looks thougt she was halfbred Beagle
385d164 oetus has gone on growing-- I believe same has HAPPENED in boys bodies.-- Lavaters. Essays on Phy. tr
467t103 hered-- I saw Bees going to clover & once this HAPPENED.-- And in common Beans it is wonderful {a} ho
520m002 <are>  have very bad memories for things which HAPPENED in early infancy-- of this fact Mr Dryden  C.
539m082 nal Institution & not feeling much enthusiasm, HAPPENED to go close to one & smelt the peculiar smell
200b122 . Phi.l.. N. J. p. 410 <Nov> 1828 It is daily HAPPENING; that naturalist describe animals as species,
372d131 nce of two kinds of generation shown by their HAPPENING in same plant.-- The Marsupial structure show
549m122 ting a change in his instincts-- like what is HAPPENING with other animals-- is far from odd nor is i
348d050 uch characters of importance, "but we say such HAPPENS to be the character, of no matter of what impo
384d162 o be able to impregnate themselves (this never HAPPENS in plants «only in subordinate manner in the p
544m104 e page 110, by mistake.» N B. Everything which HAPPENS to man who does not produce children. or after
548m117 h mind in given time.-- intensity to degree of <HAPPI> pleasure of such thoughts We give no credit to
608o026 line of conduct. as being better & making him HAPPIER.-- he agrees & yet does not.-- because  motive
529m039 , abundance, fertility, rustic life, virtuous HAPPINESS.-- recall scraps of poetry;-- former thoughts
546m109 in of colours?-- Nothing shows one how little HAPPINESS depends on the senses.; than the <small> fact
548m117 there should be & discover loss Definition of HAPPINESS the number of pleasant ideas passing  through
549m118 - yet the philosopher has a much more intense HAPPINESS-- so is it <with an> when same man is compare
549m119 he sensual enjoyment of the minute add to the HAPPINESS.-- but as they are not recollected whether fr
549m120 has  tasted them, will think the sum total of HAPPINESS greater. even if mixed with some pain.-- than
549m120 ter. even if mixed with some pain.-- than the HAPPINESS of a peasant, with whom sensual enjoyments of
549m120 e minute make large <parts> portion of daily <HAPPINESS> «pleasure». A wise man will try to obtain th
549m120 pleasure». A wise man will try to obtain this HAPPINESS. though he sees some «intellectual» good men,
549m121 ps partly their fault.-- Whether this rule of HAPPINESS agrees with that of New Testament is other qu
549m121 hope» cultivation, main source of the intense HAPPINESS-- it is again another question, whether this
549m121 -- it is again another question, whether this HAPPINESS is the object of living.-- or whether if we o
549m122 to  be true, & then acting on it, will add to HAPPINESS.-- Men having some instincts as revenge «& an
549m122 nger», which experience shows it must for his HAPPINESS to check-- that is external circumstances are
550m124 ble & the former, the more pleasant.-- Simple HAPPINESS «as of child» is large proportion of pleasant
550m124 leasant, in same time,-- therefore degrees of HAPPINESS-- Entire happiness. not being so desirable as
550m124 me,-- therefore degrees of happiness-- Entire HAPPINESS. not being so desirable as <broken> intense h
550m125 s. not being so desirable as <broken> intense HAPPINESS even with some pain,-- compared to what other
550m125 time.-- Begin discussion-- by saying what is HAPPINESS?-- When we look back to happy days, are  they
576n047 habitual as in man), for it added much to the HAPPINESS of his life, & the chance, of so dreadful a c
588n089 272.  Some remarks applicable to my theory of HAPPINESS.-- Bell on the Hand p. 191 Says <childr> babi
```

```
602o011 hen the meaning or reasons are forgotten. Our HAPPINESS &c, our well-being depends upon the "habitual
608o027 who has thought very much, & he will know his HAPPINESS lays in doing good & being perfect, & therefo
609o030 ule of life is what will produce the greatest HAPPINESS.-- The other says we have a moral sense.-- Bu
609o030 venge). In judging of <our ha> of the rule of HAPPINESS we must look far forward «& to the general ac
610o30v ot tend to greatest good.-- Therefore rule of HAPPINESS is to certain degree <of> right.-- The change
625o50v or our instincts. to gain long-lived good, ie HAPPINESS-- yet this system not selfish.-- explained by
532m056 ht. continued to grow thin & did not seem quite HAPPY. in five weeks was so thin, that she was sent b
537m076 » which she seems to think «are» to make others HAPPY & wrong to injure them without temptation.-- Th
546m109 s or......... he had had, he would say how many HAPPY days he spent in such a place.-- Vide page 103,
549m118 . last page. A healthy child is «more» entirely HAPPY (contentmt is different it refers to wishes for
549m119 easure of imagination-- therefore do these & be HAPPY-- & these pleasures are so very great, that eve
550m125 aying what is Happiness?-- When we look back to HAPPY days, are they not those of which all our recol
021rIFC R. N up to 1 degree / July 1835. the excess of HARBOR = 180 See Daubisson both Volumes, and Molina 1
067r144 lurian system. Apply degradation of landlocked HARBORS to Craters of elevation.-- Lyell suggested to
048r084 penetrating by action of Organic power a lump of HARD clay. -- In the History of S America we  cannot
051r094 branching  roots which must protect surface; On «HARD» exposed rocks near Bahia, whole surface to whe
146g010 oblique  The hill must be well-- denuded.-- «of HARD metamorph» path only covering Great Slip, 10 ye
148g024 em-- Mem Coast of Chile--¿ is not Mica Slate too HARD & uneven to be impressed Case of Birch Wood  by
158g081 on pebbles tidal plain as sea gradually retired, HARD to explain on river doctrine <Little Hill  with
223b207 g compared to the first thinking being. although HARD to draw line.-- -- not so great as between perf
223b208 not  so great as between perfect insect & former HARD to tell whether articulate or intestinal, or ev
233b250 ll the species of mice in S. America. which were HARD to distinguish came from closely neighbouring l
301c198 ans mind not so different from that of brutes p. HARD to say what is instinct in animals. « & what e
347d049 ut. p. 249. (1838). Eggs discovered to Taenia.-- HARD so as to resist external influence.-- 27th. Aug
372d130 n.-- Why crab can produce claw. but man not arm. HARD to say-- if it were possible to support the arm
501q10a btful species-- Does any genus of Plants. vary & HARD to separate specifically in one country & not i
501q10a cifically in one country & not in other: Rosa is HARD in Europe, Walnut in America.-- Heaths in Afric
527m034 tomach I observe a long castle in the air, is as HARD work (abstracting it being done in open air, wi
536m070 n is in a passion he puts himself stiff, & walks HARD.-- «He cannot avoid sending will of action to m
542m093 ut without a most distinct will, he will find it HARD to keep his lip from stiffening over his canine
542m094 & many other animals,-- but yet when angry it is HARD not to growl out some sound even if it be inart
548m115 ting new means,-- therefore works of imagination HARD work,-- Keeping one idea present is, perhaps, h
548m115 rd work,-- Keeping one idea present is, perhaps, HARD work-- though dreams do that One Reflective Con
574m043 f these two actions either on form or brain very HARD to define.-- Consider the acquirement of instin
609o030 forward: hence our <[...]> rule may sometimes be HARD to tell) + + Society could not go on except for
622o049 ay be directed towards one or more.-- It will be HARD to discover this, for the different races of ma
633j53v ure propagation of Misseltoe?-- do p. 284. it is HARD on my theory of gain of small advantages thus t
059r120 Van Diemen Land. All the Calcareous rocks which HARDEN by themselves cannot be pure. for if so  Chalk
059r121 hemselves cannot be pure. for if so Chalk would HARDEN.--Climate.!? or small Proportion of Alum: matt
347d049 an in old days: «for a very old variety will be HARDER to vary, & therefore more apt to be extinguish
360d092 ed breast & other very bright blue, it might be HARDER <to tr> for both parents to transmit there pec
360d093 st aquatic & most terrestrial species, might be HARDER to cross than two less opposed in habits, thou
637j58r lived had it not been so.-- let egg shells grow HARDER. so must those with weak beaks be sifted away.
427e111 , or by accidental production of seedling with HARDIER constitution.-- Now Sir. J. Banks. says Zizani
523m016 ably otherwise would not have been so.-- In Mr HARDINGE, was caused by thinking over the misery of an
174b014 ture is common in certain countries when we can HARDLY believe necessary, but if it was necessary  to
219b196 l Fuego.-- Ellis (?) says Tahitian kings. would HARDLY produce from Incestuous intercourse. a paralle
274c114 hese characters vary in degree in last instance HARDLY at all developed. Not confined to one species,
293c174 ging there. Now these exquisite adaptations can HARDLY be accounted for by My method of breeding ther
358d085 crossed  never even lay eggs. & the men cannot «HARDLY» tell any sex by appearance.-- The silver & co
363d102 d always very different.-- the two redpoles can HARDLY be told apart, so that after differences  were
365d105 it may be so-- but very improbably. for it can HARDLY be thought that the cross would have adapted i
367d112 forming  «& this must be so, else avitism could HARDLY ever occur.--».-- & if that cannot be  formed,
401e014 he attributes to crossing.-- Cape Broccolli can HARDLY be reared without greatest care be taken to pr
436e135 ecies, than time (as cause of change) which can HARDLY be believed, then, uniformity in «geological»
458t002 o Choo so small, that person with long legs can HARDLY ride on them. Mr Miller-- in Zoological Garden
525m026 ore affections effect of organization which can HARDLY be doubted, when seeing Nina with her puppy.--
539m082 wer of intellect.-- Some complicated trades can HARDLY be considered as actions otherwise than habitu
555m142 vish expression, but not nearly so often <that> HARDLY ever the expression of passion with open mouth
574m040 & old people get into habits.-- we probably can HARDLY form an idea of a mind so limited as Birgos to
616o038 man  ∴ they meet their reward! X Perhaps should HARDLY be called memory; you cannot call the frame of
104a067 ing right through superincumbent mass (varying HARDNESS,-- takes time to trace) from few dikes  which
155g066 decay  from fragment falling» of no particular HARDNESS no wonder that all «three» lines «should be»
380d153 nverse of annual being rendered biennial-- the HARDNESS of life in female Moth &c Mr Y. says that Mac
527m034 n inventive class.-- Now that I have a test of HARDNESS of thought, from weakness of my stomach I obs
148g026 ught lambs most like MOTHER!-- the cross not so HARDY but more easily fatten, This man confirmed  the
205b145 ncrease of size they become worse in form, less HARDY, & more liable to disease" If population of pla
346d047 ore like father than Mother.-- The cross not so HARDY as Black faced, but more tendency to fatten-- T
428e111 principle being the destruction of all the less HARDY ones. & the <accidental> preservation of accide
428e112 . & the <accidental> preservation of accidental HARDY seedlings: (which are confessed to by  Herbert)
172b007 s; Falkland Fox-- Chiloe, fox,-- Inglish & Irish HARE.-- As we thus believe species vary, <in> changi
191b080 dian islets-- (+ + +. Ireland longer separated., HARE of two countries different.-- Ireland & Isle of
226b221 system Of this we see example in English & Irish HARE.-- Galapagos.-- shrews, & when big continent ma
235b262 case  in point.-- All cases like Irish & English HARE bear upon this.-- Why do Van Diemens land peopl
265c084 question what is adaptation.-- Ermine, ptarmigan HARE becoming white in winter of Arctic countries fe
353d061 different number of vertebrae in Irish & English HARE.-- good case these hares compared to North Amer
399e009 rabbits in poulterer shops., of same colour as a HARE, but paler & buffer.-- with long ears-- & longe
403e022 n.-- without arm, <skin> hands thumb,-- one leg, HARE lip &c &c. in Vol II p 363 account of Flora  of
437e139 ird. p. 124-- Mr Willoughby found a dead lamb «& HARE» by the side of Eagles nest, which shows  power
454e184 for  galvanize <ani> them-- Cross Irish & Common HARE Decandoelle has chapter on sensitive plants; Ph
525m025 fect deformity, as an extra number of fingers.-- HARE lip or imperfect roof to the mouth  «stammering
563n001 on.-- Octob. 3d. Dog obeying instinct of running HARE is stopped by fleas, also by greater temptation
613o036 me  that his retriever Sailor he has seen push a HARE through the bar of a gate before him, & then ju
637j57v nces of adaptations in varieties.-- greyhound to HARE.-- waterdog hair to water-- bull dog to bulls.-
640j167 n tried.-- With respect to the six puppies, if a HARE was introduced, or <a spe» became more numerous
388d172 bears no relation to utility of change-- hence HARELIPS heredetary, disease. extinction. Animals in d
036r052 Volney.  P 351. Vol 1. woody bushes, «gazelles» HARES, grasshoppers & Rats. characteristic of the des
317c251 Vol  VII. Part II/. 1837 account of the various HARES «some since discovered» of N. America, & of the
317c251 replaced by three other species.-- Says all the HARES West of <all> Rocky Mountains have peculiar cha
317c251 so that he first thougt only one species. & all HARES on East side have other peculiar  appearances--
318c252 Squirrel  from extreme north turning white like HARES??--) I never saw more beautiful adaptation  for
318c252 now shoes. than feet & hind legs of these white HARES, fitted for regions of snow.-- Acclimatisation.
353d061 brae in Irish & English Hare.-- good case these HARES compared to North American hares. Many species,
353d061 ood case these hares compared to North American HARES. Many species, separated by Mountains. & & &c.-
400e013 p.  376. Isle Tres Marias off Mexico with small HARES & raccoons.-- «S. American form.--» off provinc
416e075 e made «almost» without any relation to running HARES.-- as in Italian Greyhound not so species every
430e117 explanation.-- Poet Cowper, describes his tame HARES, attacking a sick one like Chillingham bulls ar
437e139 ght. p. 125 is said that Eagles bring rabbits & HARES to the young ones to exercise them in killing t
437e139 ise them in killing them.-- "Sometimes it seems HARES, rabbits, rats & not being sufficiently weakene
575n044 y that the tendency to hybrid greyhound to hunt HARES. «& leave the sheep» & jackall to skulk about &
326c266 rth reading Bevans work on Bees, new Edit 1838 HARLAAM. Physical & Medical Researches. on Horse in. N
101a057 Boulder beds-- What is Osteopora platycephalus (HARLAN) found on the Delaware. is it Edentate? Philli
214b173 e very strong What is Osteopora platycephalus. (HARLAN) found on Delaware is it Edentate? Phillips. L
572n032 l inferiority-- when we do not expect any bodily HARM-- case of habitual action.-- L'Institut.  1838.
608o027 ght one to blame others.-- This view will not do HARM, because no one can be really fully convinced o
608o027 ry thing he does is independent of himself to do HARM.-- Believer in these views will pay great atten
554m139 ociation.-- «Listened with great attention to HARMONICON. & readily put it. when guided to her own mo
569n021 flute & Epic poem, opposite ends of series or HARMONIOUS prose.-- Lutké Voyage in Carolinas Vol II p.
```

606o023 ye"; he will allow the secondary pleasures of HARMONIOUS colours &c &c surely to be added. Lessings L
087a011 Wide range of Mammalia really very important. HARMONIZES well with Lyells idea of intertropical land.
105a071 part, where barrier least probable.-- The sea HARMONIZES well with character of mouth of valleys &c;
036r051 of Second form, except near submarine Volc: in HARMONY with the prevailing movement being one of elev
291c168 ll hypothesis of change in Scicily.-- Splendid HARMONY these views-- did Lamarck connect exterminatio
343d036 each other, & their bodies, by certain laws of HARMONY keep perfect in these themselves.-- instincts
345d044 .-- «The theory of males impressing most is in HARMONY with their wars & rivalry.--» The very many br
357d074 ld-- if so doubtless «part of» system of great HARMONY. The peculiar character of St. Helena.-- contr
391d174 ss it separates those differences which are in HARMONY with all its previous changes, which mutilatio
528m036 hearing music), this probably arises from (1) HARMONY of colours, <whi> & their absolute beauty. (wh
568m018 of our pleasure in music-- do monkeys howl in HARMONY-- frogs chirp in do-- union of birds voice & t
568m018 intive cry--» The taste of recurring sounds in HARMONY common to t[he] whole kingdom of nature. If I
234b256 Again Waterhouse finds certain varieties of a HARPALUS. common at South end, but <rare> «absent from
344d041 August 23d The Rev R. Jones gave an admirable HARRIER from Ireland to Brighton Park--first rate bitc
344d042 ret, & breed not fixed. though she resembled a HARRIER & her husband was pure Harrier.-- «The peculia
344d042 she resembled a harrier & her husband was pure HARRIER.-- «The peculiarities of our breeds must have
520m001 ng, he was prescribing to his father & old Mrs HARRISON, said, although constantly seeing him, she wa
596m184 aracter "Humes Dissertation on the Passions." "HARTLEY" I should think well worth studying-- "Thomas
610o033 taphysicians-- the school of Locke, Bentham, & HARTLEY, &. the school of Kant. to Coleridge, is regar
622o049 er might be grafted.-- Origin of the instincts HARTLEY, (according to Sir J) explains our love of ano
623o050 se it undoubtedly is instinctive. But does not HARTLEY explanation apply perfectly to origin of these
628o054 as avarice love of gold.-- love of fame-- Yes HARTLEY explains this & Mackintosh shows the change pr
629o055 secondary passion" of Hutcheson unfolded by D. HARTLEY.-- Darwin's Abstract of John Macculloch 1837 P
639j28r &c.   Prehensile tail. in Monkeys & Marsupials. HARVEST mouse & (Chamaelion?) C. D. Spines in Hedge Ho
242c017 sfield. Diard. Duvaucel. Leschenault Kuhl. Van-HASSELT, Reinwardt «Forrest» authors on E. I«ndian». A
601o08a what dog dreams, awakes-- does when Master takes HAT de l'education des Mères par L Aimé Martin Leroy
334d009 s.-- Mem. number of Mules.-- «He recollects one HATCH of hybrid geese very fine.--» How is it with pl
461t025 nce do not require sac.-- but the male in these HATCH young-- are there not some. Marsup. Mammalia, w
240c004 res the greatest difficulty to rear them, eggs HATCHED under other birds & brought up by hand.-- Thes
564m004 f Downing. Coll. that he had seen chicken only HATCHED few hours placed on table & when fly ran past
449e169 ian animal?-- Eyton says that the young of two HATCHES «all alike» between the male Chinense & female
593n109 stile «is subversive of» to this view, & fowls HATCHING stones. in some degree is so.-- idea of beaut
582n068 s the seat of the emotions.-- but are not love & HATE emotions; what are their characteristics;-- the
608o026 it would however be more proper to pity than to HATE & be †A man may put himself in the way of Conti
544m104 Montaigne observes. distinct from pain, for one HATES pain from this fear-- & not death for the pain.
621o046 - he is monster, or unnatural if malevolent, or HATES his children without some passion.-- If his pas
146g012 ssions, such as love-- dislike & <f> passion of HATRED To fulfil an instinct a pleasure; mem. Shepher
523m014 ople any misgivings of the injustness of their HATREDS, as <if> in my case.-- It must be so from the
542m095 , result of straining vision, as savages without HATS put up their hands, & as attention would amongs
400e011 g relation to geographical distribution-- Thus HATTICA is great genus.-- because found in all quarter
224b213 l character:-- For instance two wrens forced to HAUNT two islands one with one kind of herbage & one
307c216 ny has shown same fundamental organs even in HAUSTELLATA & mandibulata.--!! --Argument, when general
243c019 oiseaux eussent leurs représentants dans de si HAUTES latitudes"., --¿translate?) All Australian for
059r119 res profondes, on les voit se reproduire a des HAUTEURS communes sur le revers de chacune des montagn
361d095 Green-finch, all linnets red-pole, goldfinch, HAWFINCH-- in nursling plumage resembled that of Cross
088a014 aters. This important in transport of Fish Let a HAWK fly at Heron.-- Ceratophytes common in Northern
273c112 me plumage.-- <no> this is applicable to swallow-HAWK, «this not the case in swallow??? which is most
273c112 always applicable to same habits, though swallow HAWK, ,milvulus,, may catch insects on the wing & pr
283c143 abits as shown by frigate Bird & flying eagle.-- HAWK Gould seemed to think, that widow bird. replace
287c157 t saltus-- when Linnaeus put whale between cow & HAWK a frolicsome saltus. «p. 19»| Macleay seems to
375d135 r, by hawks. by. cold &c--.. even one species of HAWK decreasing in number must effect instantaneousl
496q05a Kill Sparrow after feeding on oats, give body to HAWK & sow pellet. ejected. done Examine pollen of s
594n112 round saw at considerable distance a very large HAWK, which are <so> rare « <s.> » here, that proba
594n112 nct against man is perhaps. as strong as against HAWK, but the birds at Maer have learned that he is
594n112 rds soon <dis> learn to disobey it-- I have seen HAWK & sparrow in Shrewsbury garden picking from sam
273c111 is obscure because nearly all are so.-- Thus in HAWKS, there is a swallow, both in structure & habits
273c111 (it cannot be doubted that if swallow perished) HAWKS & Milvulus &c would instantly fill up their pla
309c222 cal structure.-- «the passages between-- owls & HAWKS only external» intermediate groups often have f
353d061 lia characterized by wedge tails.-- many of the HAWKS <to> are analogues to <Bustard> Europaean birds
375d135 ust have same number killed, year with year, by HAWKS. by. cold &c--.. even one species of hawk decre
512q019 ite to-- (2) Does he believe. Stanley's fact of HAWKS distributing live Mamals (3) Do most Hawks eat
512q019 t of Hawks distributing live Mamals (3) Do most HAWKS eat stomach. of finches-- do they throw up pell
512q019 imilar to reverse cross.-- Sow cast-up-balls of HAWKS or even owls.-- How long do seeds remain in sto
593n105 handed down & down.-- mem. Nina used to get into HAY & make a nest for herself.-- the object is to ma
323c269 n -- Treatise on Domestic pidgeons 30th Lives of HAYD & Mozart «Apri 25th Lockarts life of Napoleon.»
582n066 sensations» <Gardner in his work» In the life of HAYD & Mozart. fine music is evidently considered as
551m129 e rises, or as they flag & something like a snow-HAZE. covers my whole imagination." Septembe. 3d Why
022r008 rp knife could not cut it: in which we found the HEAD & Boans of a Hippotomus; the hairy lips of whic
066r141 les = very strong. Stones as bigger than a man's HEAD.-- Kerguelen 40 by 20 leagues. dimensions: Byno
066r141 eddies with its extreme force. Yet, no outlet at HEAD. Important in forming transverse valleys Ice Si
122a115 h Lochy very like those of Andes Speculate under HEAD of Beagle Channel. on origin of mud with stones
123a116 sil shells of S. America. Von Buch Lyell. (under HEAD of Delta) describes near Alps great beds of riv
148g027 d with some quite irregular very like rubbish at HEAD of Loch Dochart <Nea> Above Spean Bridge many f
151g043 n of alluvium (much corroded by rivers) & not to HEAD of plain.-- but below houses where rivulet ente
152g046 nday a rapid descent of a terrace except at very HEAD of valley indicates new terrace Ballivard 2 mil
156g068 sition of shelves: generally angular except near HEAD of valley fragments which had fallen before lak
156g070 > on upper shelves granite & some other rocks at HEAD of shelf 3d almost all granite pebbles Level of
156g071 l granite pebbles Level of plain of 4th shelf at HEAD of Lower Glenroy 29.581 A 82 75 degree? From th
156g072 f slope, above «line of 4th» shelf This shelf at HEAD where <granite &> «veined» gneiss <unite» «occu
158g079 quarum Peaty Mass of this point very nearly like HEAD of Glen Guoy nor is horizontal line apparently
158g080 l> Granite ridge or a modified Granite ridge» at HEAD of Glen Roy on same side where two rivers unite
161g096 kiness When on other side {P} Shelf A Shelf A at HEAD of Gentle mossy slope, which from a distance hi
162g101 at first-- relics destroyed.-- the Brook <about> HEAD of which is so interesting. enters by old tower
163g109 is collected in little spots Speculate on <under HEAD of» Beagle Channel. Forchammers (Lyells Denmark
165g124 he Tarf-- Kilfinnan Tower where stream enters at HEAD of which hill is round & not merely thoughts la
240c013 eland, have phalangista, which differ in «form & HEAD &» colour from those of New Holland.-- The New
261c072 aurus Plesiosaurus. alludes to some structure in HEAD, which he says (evidently is an exception) can
282c141 ose men instead of mere colour & trifling form & HEAD &c to become greatly changed. in structure & ev
303c204 of men differ chiefly in <size> colour, form of HEAD «& features» (hence intellect?) & what kinds of
303c204 would have differed in hair colour + + + form of HEAD «& features»;. but likewise in length of extrem
313c233 louse African & Europaean. different.-- thorax & HEAD differ Africa Australia Parasites die, when bro
332d001 case of malignant erysipelas spreading over the HEAD, not caused by a wound, when suddenly during on
345d043 e red, yet calf every now & then born with white HEAD (,or «short-horned with» black lip) & then calf
349d052 uses term genus when it is so many steps from a HEAD, as subkingdom.-- -- evidently artificial, as i
383d160 spurs rather smaller than in <ma> silver male-- HEAD like silver except in not having tuft,-- back l
386d166 s in snail. (Encyclop of Anat & Phys) can make a HEAD; the other part may surely absorb a useless mem
386d167 piece of skin.-- if the tail knows how to make a HEAD. & head & tail, & the belly both head & tail,--
386d167 skin.-- if the tail knows how to make a head. & HEAD & tail, & the belly both head & tail,--no wonde
386d167 to make a head. & head & tail, & the belly both HEAD & tail,--no wonder there should be sympathy in
411e055 of teeth to kind of extremities come under this HEAD» 27th November When summing up argument against
420e089 ous» «ped.-- ». Hairy.-- could move his ears The HEAD being six metamorphosed vertebrae, the parent o
420e089 cous «bisexual» animal with a vertebra only & no HEAD-- !! Handwriting is determined by most complica
436e135 iest times-- Apterix have a most perfect Struthio HEAD pulled out. yet feathers retain character? If s
451e179 [...] saul forests by having a smaller, lighter HEAD, carried more elevated & higher forequarters: i
459f02 progenitors-- the hinnus, resembles horse in its HEAD ears, tail limbs-- in the mules, these parts re
463t055 erhouse showed me the component vertebrae of the HEAD of Snake wonderful!! distinct!!-- He would not
471tf07 enumerates the strange forms which the thorax & HEAD displays.-- most fantastic & use unknown.-- "<w
484z014 habits-- Does the Secretary, make noise & throw HEAD back M Edwards,--on polypi of Tubulipores L'Ins

Page
(Key Word)
505g014 y of small true Bull-Dog-- length from nose over HEAD to root of tail 28½. inches. From sole of foot
515q022 ish Horses, bred in this country. {Chinese Dog's HEAD to send Cover common Pea (& Sweet Pea) for seve
524m019 terialism. that cold water brings on suddenly in HEAD, a frame of mind, analogous to those feelings.
526m030 habitual exercise of the mind, altering form of HEAD, & thus these qualities become heredetary.-- Wh
530m043 is own health.-- his complaint was carbbuncl on <HEAD> Neck.-- He has seen other cases of similar nat
538m079 his slippers bad for fires, what is wrong in his HEAD. & Babington's silly joking The possibility of
539m081 think deeply, & the immediate manner in which my HEAD got well when reading article by Boz.-- now in
541m092 - People who can multiply large numbers in their HEAD must have this high faculty, yet not clever peo
545m105 er eyes, at every emotion & <look> «turn» of the HEAD. I could not perceive «any» distinct wrinkle, b
554m138 e thinks Callitrix Sebe??) he has seen place its HEAD downwards to look up womens petticoats-- just l
555m143 d world ones.-- Though the[y] move whole skin of HEAD they do not move eyebrows.-- (I see some of the
555m143 - (I see some of the old world ones move skin of HEAD & ears,-- ∴ some men have this power abortive m
555m144 nd of wit.-- I changed I believe from hanging to HEAD cut off.-- «there was the feeling of banter & jok
556m146 is it opposite movement to drawing them close on HEAD, when going to fight, in which case expression
564n004 aced on table & when fly ran past it. cocked its HEAD, & picked it-- Here then, that faculty, whether
565n007 - Horse snuffs «& snorts», the air «& raises its HEAD, & pricks its ears» when afraid, though not eve
565n007 umph (& pride) same way walk erect & stiff, with HEAD up.-- Why does suspicion look obliquely.-- who
570n025 e after asinine.-- both accompanied by depending HEAD, & active vessels of skin.-- What difference i
573n037 f H. W. constantly. when refusing food, turn his HEAD first to one side & then to other. & hence rota
573n037 nce rotatory movement negation.-- he dropped his HEAD when he meant to eat, hence assertion.-- but no
585n075 affe kicks with front legs & knocks with back of HEAD, yet never puts down its ear. good to contrast
595n121 d. was frightened when Snow put a guaze over her HEAD. & came near him, although knowing it was Snow.
596n184 ations Ourang do not move eyebrows.-- or skin of HEAD,-- «scarcely able St.-- » Cyanocephalus, macacu
596n184 st st. S. American monkeys. pull back skin from HEAD very little Does blood go in <body> face in pas
610o031 - --could not think what had put Wilson into his HEAD.-- remembered, that he had. looked in direction
610o031 red, that he had. looked in direction book under HEAD of Wilson, referred to Robert & found his Chris
622o049 ee in dogs & pidgeons.-- But as man is animal at HEAD of series in which «special» instincts decrease
629o55v y; I say yes, & my explanation agrees. with last HEAD.-- (4) It is other question, how the feeling of
634j54v of man. <&> The <design> determination of a God-HEAD.-- the designs of an omnipotent creator, exhaus
539m081 eum Club. was very much struck with an intense HEADACHE «after good days work» which came on from rea
541m089 uch as possible (testing success by decreasing HEADACHE) & found best plan was allowing my mind to sk
233b251 nts-- Ostriches. Dodo. Apteryx Penguin-- Logger-HEADED Duck-- Large proportion of Water & small of la
345d043 breed,: Jones says Sussex cattle were all white HEADED, but this was bred out & now all are pure red,
129a132 New Phil. Journ 1838. several case given of hot HEADS &c heat beneath the sea.-- CD[did not Beechy ha
275c121 me marks on wings are Blue Pouters & small Bald HEADS Mr Yarrell will mention in his work I am sorry
364d104 rn «see p. 43 supra» breed of cattle with white HEADS; which years afterwards occasionally went back-
486z020 ays in <14th> «13th.» Vol of Archaeologia arrow=HEADS described in Suffolk as lying under strata of g
499q009 ball of pollen on Bees thighs (18) Place pin's HEADS with Bird lime near male yew tree & see whether
570n026 y of the earliest arts.-- Mem.-- Stokes-- arrow HEADS &c &c October 27th Consult the VII discourse by
600o08v d-- (5) Conscience, not clear-- Then these last HEADS. (5) separation between soul of man. & intellect
372d129 part.-- claw added to crab, tail to lizard,-- HEALING of wound.-- reproductive faculty + in the sepa
386d167 neral law, which makes two animals out of one & HEALS piece of skin.-- if the tail knows how to make
530m043 lness, he had been able to direct about his own HEALTH.-- his complaint was carbbuncl on <Head> Neck.
546m111 ut & found it there!!! Lady in perfect «mental» HEALTH.-- «Erasmus had almost same thing happen to hi
610o032 f the day &c-- All habits must conduce to their HEALTH & comforts.-- Both ideots, old People & those
515q021 h in Scotland-- do they flower-- do they live HEALTHILY, or does fruit merely not ripen.-- The point
494q004 ect to conditions of animals & their general HEALTHINESS-- Fox's, Bears Badgers,-- how few wild anima
508q016 es\ (4) Prolifickness of female, relation to HEALTHINESS? & father answered (5) About cross-bred race
259c065 o tail (example probably not true).-- or again HEALTHY parents have healthy children the other case i
259c065 bly not true).-- or again healthy parents have HEALTHY children the other case is <adaptation> «chang
332d004 before.-- Hence disordered intestines are not HEALTHY to worms, (like parasites of Tropical countrie
359d086 so, but this assumption as long as animals are HEALTHY which is often the case, & why should organic
514q021 neaple Horticulturists (1) Are sterile hybrids HEALTHY: number of generations: about crossing of plan
525m025 then children which were short term, & lastly HEALTHY ones.-- Insanity & Epilepsy remain many genera
535m064 ing 4 to 6 weeks. The deserted broods appeared HEALTHY-- This remarkable case may be normal. with ins
549m118 - This looks like free will.-- V. last page. A HEALTHY child is «more» entirely happy (contentmt is d
603o014 explaining the possibility of <handsome> «UGLY HEALTHY» young woman, with good expression-- statues n
109a083 ainly is lessened.-- Coral flats. argument for HEAPING up.-- very good this will show effects.-- anal
432e125 n & our lives,, but to period necessary to form HEAPS of pebbles &c &c: the succession of organisms t
045r076 exclaim, «as I have heard» how lucky! when they HEAR of a place having a pretty severe shock). are m
218b192 forms would have some peculiarity.-- Now when we HEAR that the whole island is volcanic surmounted by
264c079 -- Let man visit Ourang-outang in domestication, HEAR expressive whine, see its intelligence when spo
409e049 t it was the sole object, I will dispute, when I HEAR from the geologist the history, & from the Astr
429e114 ture. how firmly each holds its place.-- When we HEAR from authors (Ramond. Hort. Transact Vol I. p.
522m012 isolately.-- In old people. (Aunt. B.) when they HEAR a thing it often does not take any effect at th
531m053 ells me very young babies start at anything they HEAR or see. which frightens. them.-- Now every anim
551m127 sion on its senses.-- insane people believe they HEAR as well see things which have no existence.-- H
552m131 is connected with habitual stopping of breath to HEAR any sound.-- attitude of attention «So intimate
556m144 3rd. Horses in Omnibus instantly start when they HEAR ready, but if they see anything ahead. which ca
557m149 g to drive away rival.-- Fear is open mouthed to HEAR. though in individual case. nothing can be hear
565n007 -- why when person is listening is mouth open to HEAR well «as one will perceive if in night trys to
025r017 Wager's Earthquake the most Southern one I have HEARD of In a preface, it might be well to urge, geol
045r076 are so infrequent; so as to exclaim, «as I have HEARD» how lucky! when they hear of a place having a
088a014 chain be Mr <Lyell> «Waterhouse» has frequently HEARD that Herons bring eels alive to their nests; &
293c172 e Marvellous cases, when you feel sure you have HEARD conversation before. is strong association reca
340d026 rts to be explicable every instinct in animals. HEARD at Zoolog Soc their Pintail & Common Ducks, the
356d072 . 175) if> 6th Sept Yarrell told me he had just HEARD of Black game & Ptarmigan having crossed in wil
402e018 companion".-- p. 229. Borneo.-- only animals he HEARD of pigs, small bears or badgers, deer, apes, ba
473s05r Maer. June/42/ June/42/-- Mr. Bunbury says has HEARD the Trout from different lakes of N. Wales can
521m009 name A. B., &c &c. & he maintained he had never HEARD of such a man & had no gardener.-- My F. then a
522m010 hild «of Kinleth» had married.-- Answered never HEARD of such a man.-- (My Father explained who he wa
522m010 ll abuse him, but still maintained he had never HEARD of him).-- My F. then said you remember Jack Ba
522m011 a few minutes before he maintained he had never HEARD of.-- Thus in many things if he began at one en
522m011 alphabet straight, but did not know [Z]CD when HEARD isolately.-- In old people. (Aunt. B.) when the
530m044 nonsense.-- look to it My father has somewhere HEARD (Hunter?) that pulse of new born babies of labo
538m079 ed things, which none about her had EVER before HEARD, so very probably forgotten.» Such facts bear o
548m116 f, one would only feel sympathy. as for the the HEARD suffering of a dear friend-- this gives one str
557m149 hear. though in individual case. nothing can be HEARD.-- Shame would never make person tremble, like
610o032 mory.-- Jan 14th. 1839.-- My father says he has HEARD; of many cases of ideots knowing things, which a
616o038 at frame is enhanced by memory of what has been HEARD; so love of virtue enhanced by this heredetary
260c067 was storge, which could not be resisted, when HEARING crys of hunger of little bird, in same way Wil
366d111 ecies.-- Case of Association very disagreeable HEARING maed servant cleaning door outside, as often a
432e125 structure, than the muscles of the ears to our HEARING powers E. frowns prodigiously when drinking ve
521m009 ew train through eyesight, though, not through HEARING,-- Thus when dinner was announced he could not
521m010 ct, no communication could be held by means of HEARING.-- Mr Corbet, however, in conversation could c
526m028 rasmus & mine taste for music.-- Children like HEARING a story told though they remember it so well t
526m029 se of Mr Corbet of the Hall understanding. (on HEARING old association brought up) by sight & not by
526m029 old association brought up) by sight & not by HEARING One is tempted to believe phrenologists are ri
528m036 te pleasure independent of imagination, (as in HEARING music), this probably arises from (1) harmony
574n041 us stimulated, smell, as Sir. Ch. Bell says, & HEARING music. to certain degree sexual.-- The associa
593n111 Dr. Hollands story of man in Delirium tremens HEARING other man speaks. shows, that consciousness of
595n117 ng «white» with <lather> <foame> & sweat, when HEARING merely hunting horn-- association or imaginati
603o11b mitating song -- two primary sources, sight, & HEARING-- Staunton Embassy Vol II p. 405.-- Speculate
607o025 (so called like chance) circumstances. As man HEARING Bible for first time, & great effect being pro
260c068 gether in pursuit of Blue-Jay, when <one> birds HEARS <dis> crys of distress of other parents.-- Show
544m103 asts of good food-- The mind wills to do this & HEARS that, but yet scarcely really moves.-- the will
351d057 vain!! Foetus of man undergoes metamorphosis., HEART altered & umbilical cord,-- Broderip alluded to

(Key Word)

```
531m051 stening one evening when tired «-- how true the HEART the scene of anger.--» to the pianoforte, it se
531m051 to be feelings of discomfort, especially about HEART as of excited action, accompanying violent move
531m052 on  the injury & consequently excited action of HEART.-- now this is the oldest <her> inherited & the
532m054 s to be frightened at.-- (again diseases of the HEART are accompanied by much involuntary fear) In th
532m054 d & started much more apt, this partly owing to HEART? readily taking same movements, senses being on
532m057 of fear is accompanied by «troubled» beating of HEART, sweat, trembling of muscles, are not these eff
536m070 l of action to muscles, any more than «prevent» HEART beat» remember how Pincher does just the  same;
536m070 ort to walk then lightly as to endeavur to stop HEART beating: one ceasing, so does other.-- What  an
557m148 blood  rushing in face, with less action of the HEART.-- tendency to muscular movement, hence shy peo
557m150 - & shaking body.-- Are those parts of body, as HEART, & chest (sobbing) which are most under great s
564n004 haps» an instinct of conscience, feeling in his HEART, those rules, which he wills to give his child.-
566n010 sympathetics with internal organs, as action of HEART Malthus on Pop. p. 32, origin of Chastity in w
567n015 whole body.-- is it connected with surprise.-- HEART beginning to beat-- children inherit it <ins> l
568n016 tion.-- is totally distinct from learning it by HEART. Do not our necessary notions follow as consequ
582n068 it.-- My theory explains how it comes that the HEART is the seat of the emotions.-- but are not love
586n081 eigh considers breathing instinctive, certainly HEART beating may be considered also such.-- heredata
601o009 ch it may be in first case. as when the excised HEART is pricked) and certain action. (only evidence.
612o035 the  lower animals, as in stomach, intestines & HEART of man. [LHC] ¿How near in structure is the gan
616o039 ain perceived, thought, remembered &c. Well the HEART is said to feel Now this would certainly be a s
468t111 la to Anchusa <never> «once» P on Fraxinella <HEARTEASE> «small. Humble alighted on base of filaments
472s02r This Bee flew from yellow to yellow & purple <HEARTEASE without doubt.-- Bee, not large, very dusky &
472s02r ever saw such a one before-- Saw Fly 21 «this HEARTEASE withered on Monday.--» alight on upper petals
472s02v > pollen was scraped off, which appeared like HEARTEASE pollen.-- the pollen appeared chaffy, as if s
492q002 <do> Can Holyoak be raised distinct by seed-- HEARTEASE. 6.-- Do not species of wild Roses run  into
503q013 ds experiment in Cabbages given (3)       in HEARTEASE (4) Does the Thyme bear abortive stamens ever
515q021 me of leafing (5) Do the most cultivated show HEARTEASE produce as large capsules of seed, as the com
472s02r 4 days. many times every day. many clumps of HEARTSEASES, never saw any Bee go to them. Yesterday rem
022r006 ly to occur where rocks have undergone action of HEAT. it is so found in Anglesea, amongst the varyin
028r032 as  no colour Sir J. Herschels idea of escape of HEAT prevented by sedimentary rocks, & hence Volcani
056r110 Galapagos.  St.  Helena.-- [Fig. 7] {P} effect of HEAT on inner wall, hence resists degradation longer
069r149 not  (for instance) expansion of solid matter by HEAT Consider profoundly the sandstone of the Portil
077r172 isnumbered page] Dr D. remarks. bad conductor of HEAT do of Electricity. Does not iron, combined  with
077r172 f solution & deposition under great pressure. (? HEAT!) unknown to us. ▮ M. Chladni.--on meteoric Mex
088a016 ed sand of Europe no fossil shells --¿ action of HEAT bubbles volatilized at bottom, condensed before
089a017 olcanic-- CD[Might not bottom of ocean boil; yet HEAT never reach surface.-- Journal de Physique,  et
100a055 . 197. refers to salt as being produced by local HEAT, Ask Capt. Beaufort, whether, water flashing in
106a076 of  sinking.-- No Sweden!! swelling of rock from HEAT. Specific gravities of many artificial limeston
107a077 h thinner beneath ocean than above it no because HEAT proceeds from great body of mass.-- The last sp
107a079 a-- On Lyells idea of whole centre of earth same HEAT, then change in form of fluid centre would lift
108a079 entre would lift with it isothermal line, but if HEAT from centre, then crust of solid earth would be
113a090 the strata have very unusual conducting power of HEAT from centre.-- But is this not wrong? we know m
113a090 hundred feet of strata having conducted away the HEAT of surface. & if conducting powers had been bet
115a096 crystals., must aid in adding effects to common HEAT.-- Where there are cliffs there ought to be cre
120a108 hen if so, thermometer show it cannt be ordinary HEAT, then there is something superadded, that which
125a121 r pressure.-- {P} Lyells view of transmission of HEAT by gases-- does not apply it to thickness of cr
126a122 our judgement be-- Does condensed metal, conduct HEAT better than plain?-- Mem 1000 {P} how easily wa
126a123 the  fluid matter be driven upwards & so conduct HEAT?-- How comes it in volcanos that have gone on f
126a124 the rod is reversed, upper part metal «conveying HEAT in one direction only, like water below 39 degr
127a126 s of pressure in change of form as the result of HEAT.-- will it bear on central fluidity.-- do p. 13
129a132 . Journ 1838. several case given of hot heads &c HEAT beneath the sea.-- CD[did not Beechy have  some
132a137 ate of increase of heats in earth's crust.-- yet HEAT does increase,-- but in Ocean does not. (see re
132a138 than mean of place, shows earth not with central HEAT to warm the ocean).-- and M. Parrot does conjec
132a138 according to M. Parrots argument against central HEAT--» «(does M. Parrot suppose there is no volcan
133a140 like a fool.-- Feb 25' All facts show how slowly HEAT travels; & therefore the abysses where fluid ro
203b137 tion of tempereture from geographical or central HEAT.-- But then shells-- Mr Yarrell says that old r
217b183 a  very large litter.-- never afterwards went in HEAT.-- This is good instance of same fact in Mr Gal
232b24' - <The Phenomena of the S. Hemisphere look as if HEAT gained> Experimentise on land shells in salt wa
284c147 long average, equal quantity, 2d on relations of HEAT & cold. therefore probably fewer now than forme
305c210 ertain contingencies of organic matter & chiefly HEAT), which assumes a multitude of forms «each havi
312c232 Willis  Long eared little dogs, I am told, go to HEAT. take dog. but do not become impregnated & pupp
360d090 crossed  Jackal & dog-- (offspring did not go to HEAT. but parts swelled, though no fluid came from t
380d155 ometimes comes in Mammae & even when bitch is in HEAT.-- Yarrell believes Gestation is always some mu
485z018 n Lapland & Stizbergen wherever there is extreme HEAT, the tropical forms extend further north, becau
486z018 eoptera especially require a greater duration of HEAT. hence musquitoes & knats abound during short s
501q011 ouse to see effect on generative organs of great HEAT (32) Can Henslow ask question of Col. Le. Coute
567n015 emotion  like blush.-- when extreme sensation of HEAT shows blood is pumped over whole body.-- is  it
611o034 for   connect of electricity chemical attraction, HEAT & gravity is probable.-- And the Organic laws p
611o034 ain contingencies are present, (contingencies as HEAT light &c). [LHC] This is true as long as moveme
612o035 ffects of more or less turgid vessels; effect of HEAT, light or shade.) Joining two difficulties into
089a019 bottom, condensed before rising?-- Mem. granite HEATED.-- Metamorphic action in red sandstone.-- Cert
096a041 he  can give me facts respecting lime <n> being HEATED without parting with Carb. Acid.-- Mr Malcolms
121a111 ypsum case) shows power of segregation.-- & has HEATED angular fragments of rock, which retained thei
125a121 e may be kept up far higher from circulation of HEATED fluid or gases under pressure.-- {P} Lyells vi
138a155 Halls Paper on the consolidation of strata-- he HEATED sand red hot & brine was boiling on the top--
497q007 osae.-- Herbert explains numerous spec. of Cape HEATH by facility. ¿Knight take opposite view. Gaertn
615o038 n others-- Hence superiority of Christian over HEATHEN race.-- But as no great modification in  brain
162g101 vation of form of land very much due to Peat & HEATHER When it did not grow at first-- relics destroy
501q10a r: Rosa is hard in Europe, Walnut in America.-- HEATHS in Africa; Hooker? are these genera less diffi
132a137 y insufficient to calculate rate of increase of HEATS in earth's crust.-- yet heat does increase,-- b
381d156 zoa. Echinoderms. Acalephes. Polyps. Sponges HEAUTANDROUS, male organs formed to fecundate female (as
381d157 in  Monooecious & Dioecious plants.-- NB. in HEAUTANDROUS animals <are> is there gradation of structu
381d157 pposition, that the Cryptandrous are really, HEAUTANDROUS.-- How is fecundation effected in latter; a
382d157 > p. 36 is thought by Owen to ask. whether a HEAUTANDROUS animal is <evidently> actually split in two
086a007 If  equatorial streams of warm pole; in name of HEAVEN why are tops of Equatorial mountains so cold.-
181b044 eat main type, & the gradation will be sudden-- HEAVEN know whether this agrees with Nature:  Cuidado
289c162 of condensing must alter method of generation-- HEAVEN knows how.-- This reaction takes place in ever
316c244 ugh for future state, that when good enough for HEAVEN or bad enough for Hell.-- «†glimpses  bursting
535m063 it  always instinctively or habitually.-- and HEAVENS is it disputed that a wasp has this much intel
580n078 f that quarter,,, is faculty, whether by sun, & HEAVENS, or magnetic virtue,-- the most probably suppo
586n078 to pidgeons, is that they do know from look of HEAVENS, points of compass, & they do know which way t
377d140 selves on  the revolutions of our system in the HEAVENS.-- Is not puma, same colour as Lion.  because
618o41v her. & as a thing grew blue it «uniquely» grew HEAVIER yet it could not be said that the blueness cau
367d113 r legs ringed with brown.-- animal like large, HEAVILY made cream coloured ass.-- stripe on back also
047r081 om being in motion what difference? In watching HEAVY swell, sea retreats & then breaks: i e to  form
284c148 bear  the least salt water.-- Nuts prodigiously HEAVY (where trees of such Nature far apart. Must hav
447e164 inal reproduction.-- likewise grasses. &-- very HEAVY seeds-- as Cocos do mer.-- Analogy shows  some
313c235 se of Elephant, which had run wild in India. in HEBER?) is analogous to dormant instinct.-- (How wond
451e178 36 In the most pestiferous region (mentioned by HEBER) «from» which «all» mankind «(& yet  afterwards
326c266 t, perhaps worth reading quoted by Malthus.-- HEBERDENS Observat. on increase & decrease of different
246c027 haps Phillippines & perhaps, Friendly Isles «& HEBRIDES») is very closely allied to C. muscadivora.,
096a039 51. NB. Mackenzie talks of gravel on basalt of HECKLA-- All the Azores Isld. Von Buch p 359 stretche
081rIBC metre = 5130., 4. 5 inch Kilometre 513., 0. 5 HECTOMETRE 51. 1. 10 Metre 3. 0. 11 lig[nes] Decimetre
470t177 s of wh. have abortive stamens= Many Humbles on HEDGE Linaria= (Plenty of Humble Bees on Phlox  Down,
639j28r . Harvest mouse & (Chamaelion?) C. D. Spines in HEDGE Hog & Echidna. & Aphrodites C. D. Endless case
192b087 ity of such organization. [Spines in Echidna & HEDGEHOG]CD-- As we have one Marsupial animal in Stone
373d132 e Mammalia, no more wonderful. than Echidna. HEDGEHOG having spines.-- Does not male Pidgeon  (yes)
460t017 ation for each other-- Aug. 20th The Echnida & HEDGEHOG Tenrec both having spines, is the effect, par
032r039 ill at length be checked by increased vertical <HEIGHT> thickness (DZ) of mass to be removed & from t
```

Page ***(Key Word)***
035r046 Mr Conybeare introduce to Geolog--"Between the HEIGHT of same beds, deposited in different basins; 1
035r047 Map: Quoted from Daubeny P 402: likewise, mean HEIGHT of tertiary. being less than secondary:-- cons
055r108 rocks.--The grand cliffs of a thousand feet in HEIGHT, of the solid lavas.--proportionally high to a
113a092 theory On the idea of statical equilibrium, the HEIGHT of lava (habitually) becomes measure of force
134a141 6. Rocks on Western Coast p. 204 do. do p. 210. HEIGHT on road from Valparaiso to Santiago p. 328. de
146g011 , 10 years since three hundred feet in vertical HEIGHT-- enormous mass thunder storm, many <hundred>
255c050 forms, but those inhabiting 3d zone of <latit> HEIGHT & 3d of latitude more commonly are the same sp
278c130 ents & death of form in one. The caves are at a HEIGHT of more than 1000 ft. & many hundred miles fro
419e086 od Dr. Beck says the shells in Scandinavia from HEIGHT of 200 & 300 ft are identically same as those
467t100 oung flowers. The abortive stamen are of useful HEIGHT.-- In Lupine, Bees <frequent» & seem to act, s
505q014 From sole of foot to shoulder on line of back, HEIGHT 17½/. The Greyhound. was in length (measured s
604o018 the Sublime The literal meaning of Sublimity is HEIGHT. & with the idea of ascension we associate som
604o018 e.-- 4 From the association of power &c &c with HEIGHT, we often apply the term sublime, where there
605o18v ect: as when we look at the vast ocean from any HEIGHT.-- 6 That the superiority & "inward glorrying,
605o18v That the superiority & "inward glorrying, which HEIGHT. by its accompanying & associated sensations s
605o19v it appears, that when certain causes, as great HEIGHT, eternity, &c &c. produces an inward pride & g
111a087 udied Analysis of Voyage: many observations on HEIGHTS of valleys in Chile Geograph. Journal Vol. VII
317c250 o. p. 243 (, Professor Smith's Journal) on the HEIGHTS of St Jago found a Euphorbia so near Piscatori
447e165 na, <& this> rendered vivaparous by growing on HEIGHTS.-- yet he has seen it propagated in a garden,
506q014 und. was in length (measured same way) 47½-- in HEIGT 30 inches Examine Keel of Common & Wild Duck--
119a105 nsidered, metamorphic rocks at surface. & great HEIGTH on mountains.-- consist of rocks with fossils,
547m113 all these parallel trains of thought necessary HEIRS of every action, & always running on in mind, b
385d163 now & LIKES HIM & then is actually obliged to be HELD.-- like she wolf of Hunter.-- young take distem
389d173 k dog. but had such aversion to it, that she was HELD Hunters Eoeconomy So with inter-breeding as tol
451e176 mb of his grandfather, who was also an albino is HELD sacred by the credulous natives, & vow made at
521m010 s he in every respect, no communication could be HELD by means of hearing.-- Mr Corbet, however, in c
545m106 old American monkey «(Mycelis)» I gave not, but HELD it between fingers, the peevish expression was
568n017 has pleasure in following its instinct & pain if HELD.-- if tempted not to follow it, by greater temp
591n099 s of individual are same as in normal cases) are HELD in abhorrence it is because instincts to woman
591n100 (not instinct) think not fit, as cannabalism, is HELD in abhorrence.-- all this makes analogy of acti
593n111 iousness. only extreme step of an ideal argument HELD in one's own mind, & Dr. Hollands story of man
031r038 page 1 of New Zealand Geological Notes. at St. HELENA. This structure was very clear at base of grea
033r043 not climb up many parts, in James Isd.--Mem St HELENA-- All Trachytic.--Daubeny P. 171. Vol I. Humbo
039r059 trata.-- It will be well to urge the case of St HELENA, where dikes certainly have not been points of
040r062 r of Flora to New Zealand, which agrees with St HELENA in being unique, yet no quadrupeds. -- Is the
040r063 hstone; which is described as very rare Mem. St HELENA; probably more abundant in this case from inte
046r079 . -- Ascension. Vegetation? Rats & Mices. At St HELENA there is a native mouse Did wave first retreat
048r084 Louisiana. V. Lyell. Vol I. P. 191 State at St HELENA. pebbles entirely coated with Tosca. which imp
051r094 ings; In smooth seas (& even turbulent as at St HELENA) I have mentioned point of greatest action; l
055r107 --Vol II. p. 252 Urge cliff form of land, in St HELENA. Ascension. Azores. («sandstone first gives» h
055r107 the Basalt. Gone into fine sediment Look at St HELENA!!-- There are some arguments which strike the
056r110 use of chimney. to crater. as at Galapagos. St. HELENA.-- [Fig. 7] {P} effect of heat on inner wall,
060r125 gbroke voyage to the Demerary Earthquakes at St HELENA. 1756. June 1780, Sept. 21st. 1817.--p 371. We
075r167 ldt suggests covered up by volcanic rocks. //St HELENA has been slightly broken up, & has there not b
079r177 ralia, C. of Good Hope.--Azores Isds «nor at St HELENA.--» Humboldt. New Spain Vol. IV. «p. 58» At Ac
096a041 Isld. steep side to windward in allusion to St. HELENA discussion. Mr Brayley says he can give me fac
118a103 lace where dikes described-- {P} Cordillera. St HELENA &c &c.-- in Cordillera, it is at once evident
207b151 conceal themselves in the grass.-- Beatson St. HELENA, as. Pineaster & Mimosa called Botany Bay Will
207b151 St. Helena says no trees succeed so well at St. HELENA. as. Pineaster & Mimosa called Botany Bay Will
209b156 of Islds very poor «(p. 145)» 25. plants. 36 St HELENA, without ferns.-- analogous to nearest contine
209b157 North Africa. I: 4,2 Iles Canaries I: 1,46 St. HELENA I: 1,15 {T} Calculate my Keeling Case: Juan Fe
214b173 as.?-- Roxburgh. list of plants in Beetsons St. HELENA. -- Galapagos--Juan Fernandez Falkland Islds--
219b193 Fact stated by Mr Don in island, Teneriffe, St. HELENA. J. Fernandez. Galapagos. Many trees Composita
226b222 is this with Fernando Po.). with plants of St. HELENA & Tristan D'Acunha, resolves itself into quest
247c028 . 189 Tome III: Kotzebue.-- p 22. a Gecko on St HELENA.-- <in 1813, a venemous snake was> one Gecko o
269c100 lhos Flora with this view) Tristan D'Acunha, St HELENA &c &c. Juan Fernandez A communication to Geogr
296c184 theory explains this. but no other will.-- St. HELENA (& flora of Galapagos?) same condition. Keelin
358d074 of great harmony. The peculiar character of St. HELENA.-- contrast with otaheite in relation «See Gau
406e035 ll not breed.-- p. do.-- Fish of Teneriffe. St. HELENA & Ascension most species like & identical with
430e119 .-- excellently true theory Examine list of St. HELENA Plants & see whether those which grow in low g
501q10a n country.-- How is it in Patella or Oysters or HELIX. Or does any <one» species of plant, vary in on
316c244 at when good enough for Heaven or bad enough for HELL.-- «glimpses bursting on mind & giving rise to
539m084 (Faery Queene. CD 25 (Descript of Queen) «O» of HELL. Cant IV or V.) as pale & trembling. & not as fl
070r155 ins of Mendoza smooth. Sir W. P. states that in HELM'S travels accounts of travelled boulders. from t
088a016 stals of Hornblende p. 248. L. Institut 1837.-- HELMS remark on common salt being found on low hills
114a095 denudation, but does not tell period.-- I cannot HELP suspecting that clay-slates have been more freq
175b018 t progress. Man gains ideas. the simplest cannot HELP.-- becoming more complicated,; & if we look to
252c045 lear indication of circumstances in Geography to HELP in distinguishing empirically what is species.-
281c138 » not all equally related to each other I cannot HELP thinking good analogy might be traced between r
366d111 lly fully aware she was not coming in, could not HELP being perfectly distracted «Referred to <other>
520m001 -- when attending Mr Dryden Corbet, he could not HELP thinking, he was prescribing to his father & ol
532m053 to the muscles & when the noise comes it cannot HELP doing it.-- Fanny Hensleigh doubts whether youn
546m108 ke Burke explain pleasure. August 26th. I cannot HELP. thinking horses admire a wide prospect.-- The
565n006 ile-- laugh long prior to talking, hence one can HELP speaking, but laughing involuntary.-- When one
577n050 ablishment of this principle of Association will HELP my theory of sensitive Plants Habitual actions,
608o026 a wrecked man, like a sickly one(P)-- We cannot HELP loathing a diseased offensive object, so we vie
056r109 s one is not sure whether fissures may not have HELPED it, or diluvial waves. but when we see an enti
535m064 va excluded, then though not feeding them «nor HELPING larva from egg» watching them, brooding over t
584n073 to memory. Shrugging shoulders seems sign of HELPLESSNESS E. says she can perceive sigh, commences as
586n080 -- & faculty (faculty «being» always heredetary HELPS this confusion.--) Hensleigh considers breathin
418e080 Monooecious plants, & abortion of others.-- ¿ in HEMI-hermaphrodite insects is it not easier to under
272c108 ng quick & generl appearance of blattae other HEMIPTERA stikingly resemble Coleoptera.-- Donacia.-- s
297c186 . II p. 125 Allusion to abortive spiracles in HEMIPTERA do. p. 160. soft plumage of night jar. like o
272c107 ment of some other part of body.-- there are HEMIPTEROUS insects, having spiny legs & running quick &
190b077 ae & other forms (?) being common to Southern HEMISPHERE, does not look as if S. africa peopled from
194b095 ssum found in Europe now confined to Southern HEMISPHERE.-- If these facts were established it would
196b106 y have left off springs <ne> in or near South HEMISPHERE. Were they produced in several places & died
205b144 m of awk, because there is no awk in Southern HEMISPHERE. does this rule apply? A Treatise on Form of
232b247 ar refrigeration".-- <The Phenomena of the S. HEMISPHERE look as if heat gained> Experimentise on lan
234b261 rity even in quite distinct countries in same HEMISPHERE. more than in other. Are there any cases, wh
244c022 h equator--that Vesp. lasiurus does in North. HEMISPHERE.-- p. 158 Cuscus albus. New Ireland ----mac
247c028 strobranchus «only» 2 species one in Northern HEMISPHERE 2d in southern --p. 71 Chimera-- Antarctica
250c037 e not shut up!! Extreme southern points of S. HEMISPHERE fully characterized, of each continent. Try
268c099 roofs of most equable climate both in S. & N. HEMISPHERE just anterior to present. ¿cause of destruct
268c100 This of consequence, because applicable to N. HEMISPHERE (NB.-- Examine Abrolhos Flora with this view
271c105 markable that Turdus Magellanicus. in the. S. HEMISPHERE. (replaced to the North by other species.--)
305c209 y would appear isolated.| In my birds from S. HEMISPHERE there are some godwits which are close to Eu
406e037 ratic blocks transported far S. in Northern. HEMISPHERE.-- likewise far North in Southern.-- Great a
484z015 replace the «Auk» Guillemost of the northern HEMISPHERE, & the Puffinuria, the Awks.-- What structur
484z015 profoundly shells of Bahia Blanca & Southern HEMISPHERE It is most interesting the way Synallaxis le
405e032 hange in Mollusca is somewhat similar in two HEMISPHERES.-- It might be worth investigate whether. Me
325c267 rewsbury) Yarrells Paper on change of plumage in HEN Pheasants <Zoological> Philosop. Transactions. 1
341d031 man has crossed cock Guinea Fowl with Pea <cock> HEN.-- offspring female, yet so infertile never even
341d032 n the night, like peacock.-- tail as long as Pea HEN.-- about intermediate.-- (In Zoolog Gardens ther
357d073 . both in Sweden & anciently in Britain) between HEN Caperailkie & cock Black-cock.-- (Curious the re
358d085 of age or castration on females.-- [not located] HEN freely.-- here we have beautiful proof of the br
358d086 ke common pheasant & back like silver.-- But the HEN hybrid of this bird, has long tail figure, & som
359d088 than common duck-- Male Penguin was crossed with HEN Canadian offspring, I should say in every respec

```
361d096 sparrow,  (if I understand rightly) young cock & HEN, all nearly similar.-- in blackbird group  young
369d114 ary characters.-- )p. 49. (wonderful case of Pea HEN. taking feathers of Peacock & spurs-- no final c
376d138 dly seen one he kept pull up feathers of tail of HEN; which lived with it.-- also of <a> dog«s».  but
378d147 s present in Young birds, one may be sure cock & HEN will be alike-- I presume converse is not true f
378d147 ike-- I presume converse is not true for he says HEN & cock Starling alike, yet young ones brown.-- S
380d154 singular. so many Gallinaceous birds have cock & HEN plumage so different, yet the Cassoway & Guinea
383d159 on & Silver Pheasant, one like cock & other like HEN.-- one doubts whether they are not Hermaphrodite
491q01v produces any effect, as in case of woodpidgeon & HEN. mentioned by Mr Knight. Vol IV Hort. Transact.-
494q004 Cross  largest Malay with Bantam-- will egg kill HEN Bantam.-- Cross common Fowl with Dorking (10) St
502q11v ng. beans. cabbages.-- History of Pheasant-fowl. HEN coloured like cock-pheasant: said not to sit on
592n101 on its origin in Human mind.-- Andrew Smith says HEN doves & the female chamaeleon court the males by
600o008 Negro, beau,--Jeffrey denies all Beau-- How does HEN determine which most beautiful cock, which best
028r031 does the rain, therefore such «rain» is cause, HENCE at least no water is absorbed into the earth <I
028r032 scape of Heat prevented by sedimentary rocks, & HENCE Volcanic action, contradicted by Cordillera, wh
032r041 umulation of Coral limestone in intert{ropica)» HENCE varieties of substances ejected from same point
036r051 f elevation alone.--In England much subsidence: HENCE difference; action on land different Volney, P
038r057 84  Vol III.) whole of Etna series of coatings; HENCE it will be necessary to state all arguments for
039r060 sive terraces mark as many distinct elevations; HENCE it would appear he has not fully considered the
056r110 a.-- [Fig. 7] {P} effect of heat on inner wall, HENCE resists degradation longer than outer parts.--
098a046 on. therefore concretions in this case laminar. HENCE the thick wedges of feldspar in gneiss.-- Veins
098a048 calc.: sandstone. (& as I believe most strata) (HENCE endless passages from gneiss to granite): Why n
099a050 leavage not vertical ∵ combined with gravity.-- HENCE changes in dip of no sort of consequence.-- The
103a063 the  moving agents, because not wedge-formed.-- HENCE fill up fissures-- If dikes effect of horizonta
103a065 ng last action. (effect of horizontal movement) HENCE generally intersect metallic dikes: It is an im
126a123 d not percolate surface, would become hotter.-- HENCE temperature ought to increase rapidly beneath l
127a125 lines   & c & .-- But Siberia was once thawed. & HENCE. (when climate hotter) was cooled to greater de
127a125 e creeping «pushing» up to «the» line of ice.-- HENCE further N. when soil frozen for greater  length
171b003 new  individuals produced by buds are constant, HENCE we see generation here seems a means to vary. o
173b011 d in interval arrived) might have grown altered HENCE the type would be of the continent though speci
174b014 ne forefather, the result, would be as it is.-- HENCE Antelopes at C. of Good Hope-- Marsupials. at A
176b021 rly branched some branches far more branched.-- HENCE Genera.-- «as many terminal buds dying, as  new
176b023 in each state of existence have shortest life.; HENCE shortness of life of Mammalia.-- Would there no
181b041 them»  having progeny living ten thousand years HENCE; because at present day many are relatives,  so
181b042 2. having progeny. after that distant period.-- HENCE if this is true, that the greater the groups th
182b046 the Cryptogamic flora but not atmospheric type. HENCE probably only four, is not this Fries rule-- Wh
182b046 irect adaptation & partly to heredetary taint;. HENCE the resemblances & differences for instance of
192b083 ce Ribstone Pippins, & Golden Pippin, goldens-- HENCE-- sub-varieties & hence possibility of reproduc
192b083 olden Pippin, goldens-- hence-- sub-varieties & HENCE possibility of reproducing any variety, althoug
202b130 gs of bat,, are from external influence.--...... HENCE name of analogy, the structures in the two anim
219b193 because  seeds first arrived «Ferns ditto.--» & HENCE formed trees]CD & would creator <on volcanic is
224b212 cording to each group)» whether good species, & HENCE the importance Naturalists attach to Geographic
224b213 ther <animals> beings of very near structure.-- HENCE species may be good ones & differ scarcely in a
224b213 h other, might change organization of stomach & HENCE remain distinct. Where country changes rapidly,
226b221 larger  species are formed of those genera.-- & HENCE by same chance few representative species. this
226b222 nsure mass continental or many large islands.-- HENCE this must have been condition of Paris basin la
227b224 origin under> connection of those two countries HENCE India, Mexico & Europe. one gret sea (Coral ree
230b240 ids, or else whole fabric will be overturned.-- HENCE extreme difficulty, argument in circle.-- Falkl
231b243 nly related by one connection.-- viz descent.-- HENCE far greater discordance in latter-- Have change
231b244 reason, he would be limited animal in range-- & HENCE probability of starting from one point.-- In th
248c033 has  been accumulated cannot be transmitted;.-- HENCE the tendency to revert to parent forms, & great
253c046 - perhaps shows great persistency of character. HENCE Elephas primigenious over so wide a range, & Ma
255c051 s of ducklings and chickens) Young water ouzels HENCE aversion to generation, before great difficulty
258c061 rom Cuba Vigors.-- nothing of much interest XX. HENCE relation of analogy may chiefly be looked for i
258c061 who  have any slight peculiarity of structure. «HENCE seals take victorious seals, hence deer victori
258c061 structure.  «hence seals take victorious seals, HENCE seals take victorious seals, hence deer victori
258c061 e victorious seals, hence deer victorious deer, HENCE deer victorious deer, hence males armed & pugna
261c070 e Synallaxis or Marsupial animals of N. America HENCE males armed & pugnacious (all order; cocks  all
280c136 if too unlike its own parent this impossible-- (HENCE it is universally allowed that the discriminati
285c151 <shows> is effects of unfavourable conditions█ (HENCE we might expect even if two mules bred or two c
290c162 w.-- This reaction takes place in every organ█ (HENCE rise & depression of importance, in each  group
290c164 pheasant., I think so because very 3/4 bred.-- (HENCE «method» of generation is very good «general» c
303c204 ly in <size> colour, form of head «& features» (HENCE hybrids in this case have bred). White & common
303c204 ellect) quantity & kind of hair forms of legs-- HENCE intellect?) & what kinds of intellect) quantity
316c245 era confined to hot countries & many to cold.-- HENCE the father of man kind probably possessed a str
332d004 orms the child not having passed them before.-- HENCE latitude is more important element than longitu
336d018 both  parents are alike, offspring must be like HENCE disordered intestines are not healthy to worms,
344d042 rities of our breeds must have been acquired, & HENCE mutilations not heredetary,, but size of partic
350d056 Abercrmbies) comparison of sight to threads.--» HENCE this is then case of avitism.++» Three gentleme
352d058 pecies with abortive organ of any kind few.-- » HENCE the Pecten, which move imperfectly has eye-poin
352d058 an of any kind few.-- » hence become EXTINCT, & HENCE become EXTINCT, & hence the IMPROVEMENTS of eve
352d058 species is made. father probably will be dead-- HENCE the IMPROVEMENTS of every type of organization.
353d060 nimals & plants.-- he get merely a few pages.-- HENCE there is no central radiating point, all united
360d090 in  garden is from <bitch dog do> father dog. & HENCE (p. 59) looking at animal, if there be many oth
362d099 eel off their skin (not being wanted for war) & HENCE general appearance of face & tail somewhat like
368d114 ike some waders) the bright plumage.-- «thinks» HENCE never fall off.█ Curious the rapidity of the ch
377d147 ar of woods & like ground birds [not located] :HENCE, specific character most perfect in <male» «herm
378d148 ing as child is like parent, so is species old: HENCE also structure not really fitted for water, on
388d172 ission bears no relation to utility of change-- HENCE <young» Kingfisher & pigs, have long had  their
391d174 urther change?-- Man properly is hermaphrodite (HENCE harelips heredetary, disease, extinction. Anima
391d175 kwards & forwards & are individual differences (HENCE monstrosities tend that way «& from frequency o
407e040 ferent, with <d> regard to their productions.-- HENCE it is, from the ancient preeminently equable  &
409e049 gly. for there would be innumerable species. .& HENCE few only social there could not be one body  of
409e049 ody of animals, living with certainty on other» HENCE not social instincts, which as I hope to show i
410e050 ual crossing, there would be endless changes, & HENCE no feature would be deeply impressed on it, & h
410e050 e no feature would be deeply impressed on it, & HENCE there could not be improvement. «& hence not «b
410e050 it,  & hence there could not be improvement. «& HENCE not «be» higher animals» -- it was absolutely n
411e055 tly to specify types, & limits of variation., & HENCE indicate gaps.-- by this means the laws probabl
416e071 is perfectly adapted to external conditions.-- (HENCE great variation in each birth) from man arbitra
416e075 quired structure is fully practised & perfected HENCE difference between races & variety? «Man  picks
417e075 n the ass-- there is little tendency to vary. & HENCE offspring are hybrids,.-- Mr G. B. Sowerby <tel
422e092 es it not mean lately adapted or transformed. & HENCE not indicative of true affinity.-- -- -- Owen s
422e095 ends,  of their varied structure & complexity.-- HENCE as the forms became complicated, they opened fr
423e097 ere is the strongest possible to increase them, HENCE the degree of developemente is either stationary
427e108 plants  to each other would rapidly increase, & HENCE number of forms. once formed. would remain stat
427e108 of forms. once formed. would remain stationary, HENCE all present types are ancient. According to  my
446e163 t no change would affect them in short period & HENCE no change would effect them, without  affecting
446e163 em, without affecting all the individuals-- «-- HENCE there would be real gradations in species  from
447e163 -- these simple forms perhaps oldest in world & HENCE most persistent-- if form exceedingly difficult
447e164 , & the facility for inter marriage is greater (HENCE Dioecious plants highest,-- Palms &c &c)-- Is t
450e175 rger than the Sambawa, Java & Sumatra breeds, (.HENCE it appears there are shades of difference in al
461t025 dergo metamorphosis & are provided with fins, & HENCE, do not require sac.-- but the male in these hat
461t037 ew circumstances than it would be eliminated, & HENCE, the application of structure to purpose  after
468t111 ilaments & reached nectar =again= between them, HENCE quite below stigma. & so avoided it.» On certai
482z018 especially  require a greater duration of Heat. HENCE musquitoes & knats abound during short summer f
482z018 ly explain rarity of insects in T. del Fuego.-- HENCE it is odd that Amber insects of Europe have Tro
542m095 one  lifts up eyebrows to see things in dark. & HENCE is this the cause of expression of surprise-- v
546m108 in their situation, & then we feel like them--. HENCE sympathy very unsatisfactory because does not l
547m112 that the mind is wholly absorbed with one idea (HENCE apparent vividness) & there being no other para
```

548m114 n rigidly comparing each step as in reasoning-- HENCE delirium & sleep mental rest. though. most vivi
553m135 lightning the direct will of the God (<thus> & HENCE arises the theological age of science in every
557m149 of the heart.-- tendency to muscular movement, HENCE shy people (shame of ridicule) are singularly a
565n006 asses into smile-- laugh long prior to talking, HENCE one can help speaking, but laughing involuntary
565n007 als. comes at distance, mouth is placed open.-- HENCE becomes instinctive to fear., as ears down to h
568n018 mmediate inclination or some future pleasure.-- HENCE judgment, which is part of reason Octob. 19th.
569n019 rely looked at picture as works of imitation.-- HENCE pleasure in the beautiful. (distinct from sexua
570n024 immensely immersed-- mechanic apt to sigh.-- & HENCE carried on as trick) <Shrugging aroused acting>
571n029 e acquired, by certain foods being habitual-- & HENCE become heredetary; on same principle we know ma
571n030 viewing Windsor Castle which rises naturally & HENCE sublimely from natural rise-- I was also much s
573n037 n his head first to one side & then to other. & HENCE rotatory movement negation.-- he dropped his he
573n037 on.-- he dropped his head when he meant to eat, HENCE assertion.-- but nodding is less strongly marke
574n041 es is probably connected with flow of saliva, & HENCE with action of mouth & jaws.-- Lascivious women
577n049 l. as now existing can be cause of itself.]CD & HENCE there is great probability against free action.
577n050 ant of stimulus, after certain other actions, & HENCE becomes associated with them.-- The establishme
577n052 hinks of him, than of any one of his own sex.-- HENCE, animals. not being such thinking people. do no
578n056 disagreeable impression like true convulsion. (HENCE pass into convulsions?)-- squeeze out tears. re
578n057 der voluntary power, (or only very little so) & HENCE by association, there pour out tears, & there i
580n061 it because I was always told so in childhood.-- HENCE the belief in the many strange religions.» Emma
583n069 lty in man not met with in the lower animals.-- HENCE the general aim of fable, & expression as cunni
583n071 in way, which way its organs are sufficient for HENCE it must some way be able to measure the cell; p
584n071 instincts may vary before the structure does; & HENCE we get over an apparent anomaly,, for if anyone
585n076 on in which they STARTED, they cannot return.-- HENCE I conclude. pidgeon taken little way, whirled,
591n098 wardly.-- the other one nothing will frighten-- HENCE variation in character in different animals of
592n104 of face, is well seen in shortsighted people.-- HENCE origin of expression-- There are some instincts
593n107 -- were musical notes the language of passion & HENCE does music now excite our feelings.-- How does
593n109 does not know other men by smell, but by looks. HENCE. some obscure picture of other men. & hence ide
593n109 ks. hence. some obscure picture of other men.-- HENCE idea of beauty.-- the social affections of anim
603o013 ject of language is promptness «of consequence» HENCE languages become corrupt, & whole classes of wo
605o19v at we apply to such emotions. this same term.-- HENCE it appears, that when certain causes, as great
606o022 it is simplicity with grandeur of character.-- HENCE Lessings shows expression of pain cannot be rep
609o030 back.-- (much further than we can look forward: HENCE our <[...]> rule may sometimes be hard to tell)
611o034 impressed, & different laws of movements. [LHC] HENCE there are two great <worlds, inor> systems of l
612o035 uch powers of relation required to be extended. HENCE a sensorium, which receives communication from
615o038 ulation.-- Slightly modified in many countries, HENCE national character, love of country, of associa
615o038 f association & stronger in some than others-- HENCE superiority of Christian over Heathen race.-- B
616o039 ent. Matter is by a metaphor said to attract; & HENCE if thought &c bore the same relation to the bra
620o044 t very time it is disobeyed) & is sure guide.-- HENCE conscience is improved by attending & reasoning
620o045 ot urged to it by passion, shows a bad child.-- HENCE there are certain instincts pointing out lines
621o047 moral approbation of his fellow men.-- [RHC] 6) HENCE man must have a feeling, that he ought to follo
621o047 which is «much» formed by past recollections.-- HENCE he has the right & wrong in his mind.-- Now we
621o048 any action is either right or wrong.-- [RHC] 7) HENCE, what parents think will be good for the child
622o048 teachers & all around him, will be paramount,-- HENCE the law of honour. & the etiquettes of Society.
623o051 stances, which have led to the peculiarities, & HENCE <must have> «only that which» had a beneficial
624o051 ly «in which it» has been so in some past time, HENCE passions]CD «although perhaps useful at present
624o051 ough perhaps useful at present to some extent.» HENCE this is the law of our instinctive feelings of
624o051 of community must frequently teach same end.-- HENCE this becomes the law of right & wrong, though,
624o051 has often been perverted from want of reason.-- HENCE as Eugenius says, slow growth of rule of right.
624o051 e children, « <then> each himself» & parents, & HENCE to nearly all the world.-- As conditions change
625o052 n, which receive enjoyment from gratification & HENCE are forgotten-- only so far do I admit its supr
629o55v ne break these artifical rules, get remorse-- ((HENCE desires do not intervene between this kind of c
633j54r in accordance to certain laws they can live.-- HENCE the mistake they are created for them. If we on
640j167 s, which in this case is only true criterion.-- HENCE it is highly unphilosophical to assert, that th
640j167 of species would far more easily be changed.-- HENCE the Galapagos Islds are explained. On distinct
585n077 a faculty, or sense-- "We know not how, stonge HENGE raised, yet not instinct, but if all men placed
312c231 rtebrae» new acquisition, we must [not located] HENRY Thompson tells me best way to improve cattle is
023r012 ributing fossil remains: Sharks followed Capt. HENRY'S vessel from the Friendly Isles. to Sydney; kno
342d033 arent species.-- The hybrids do not vary (ie the HENS all alike & Cocks all alike) More than parent s
368d114 ost vigorous males.-- «(NB. most strange cocks & HENS. being either alike or very different in recent
380d153 if kept they would have wings.--).-- Seep p. 84. HENS «like»-- Cocks from effect of breeding in & in.
496q00G fish & kill them in hour or two «My Father made HENS cast Holly-seed & they grew» (9) Place. Snap-Dr
563n001 n nest & leaving no tracks.-- My Father says pea HENS do Wood pidgeons building near houses. yet so s
581n064 utang), music, colours we must suppose <we> «Pea-HENS» admire peacock's tail, as much as we do.-- tou
316c244 mes very curious.-- a dog snarling in play.-- HENSLOW says the love of the deity & thought of him «
439e144 llen to that of other variety.-- «Elizabeth & HENSLEIGH. seemed to think it absurd. that the presence
462c051 in «about» seven «7» generations.-- so many!! HENSLEIGH objects to transmut. theory, on the grounds o
532m053 noise comes it cannot help doing it.-- Fanny HENSLEIGH doubts whether young babies start.-- If chi
533m058 ted with their power of acquiring language.-- HENSLEIGH. W. says that babies know a frown. very early
534m061 is forgotten. it is conscience, or instinct. HENSLEIGH says to say. Brain per se thinks is nonsense;
574m039 with Mrs C. but had no influence over her.-- HENSLEIGH says. Douglas. «& Spencer», an old Scotch Poe
585n074 which we touch & what [...] the same.--(this HENSLEIGH therefore problem is how we know that thing i
586n081 g» always heredetary helps this confusion.--) HENSLEIGH considers breathing instinctive, certainly he
290c165 altogether explained by F. Cuvier, «-- .Mem. HENSLEIGHS objection--» it is more, he cuts the matter
542m096 e wish to make it out?-- Seeing a Baby (like HENSLEIGH'S) smile & frown, who can doubt these are inst
021r005 iption, calls it a Fucus. Z «Vol I 287» P 379. HENSLOW Anglesea, nodules in Clay Slate. major axis 2.
022r007 he great Plas Newydd dike.--Mem tres Montes. ((HENSLOW Anglesea)) great variety in nature of a dike.-
061r127 urious fact about Crust Bra in Brine Springs. ((HENSLOW) Speculate on neutral ground of 2. ostriches;
099a049 l to longer axis. But if great depth NB. Prof <HENSLOW> Sedgwicks lamination parallel to stratificati
115a097 creeks & multitude of rivers ought to be deep.-- HENSLOW has deposited specimens from Anglesea in Geolo
188b068 says so far will I go and no further:-- Prof. HENSLOW says. that when race once established so diffi
220b198 ng. Hogs not different. some dogs different.-- HENSLOW says. (Feb 1838) that few months since in Anna
228b230 & see what offspring would come from them. Ask HENSLOW for some plant, whose seeds go back again, not
229b234 . with several mammalia being peculiar (?) If. HENSLOW discusses possibility of seeds of Keeling stan
241c016 ar East) Ann: des Sciences. Semptemb. 1825 Get HENSLOW to read over the pages from about 8 to 20 of Z
266c092 from Mummy: after 2000 years, germinating.--!! HENSLOW doubts? GEOGRAPHICAL JOURNAL. Vol V. p 201 Wel
268c100 igny. & range on West Coast «Guayaquil & Peru» HENSLOW in talking of so many families on Keeling Isla
298c192 .-- upper parts attracting all the moisture.-- HENSLOW thinks if leaf of plant varies, «whole cross»
327c265 rthy of studying in Metaphysical point of view HENSLOW has list of plants of Mauritius with locality
434e129 ns why it is sure guide.-- Lychnis April 3d.-- HENSLOW tells me following facts: believes that «only»
439e143 fact about the hybrids (mentioned in letter to HENSLOW) fertilizing each other, better than the polle
446e162 stated by Smith of Jordan Hill.-- May 27th.-- HENSLOW One of the 4 species of Lemna only reproduces
446e162 y of production by gemmation & by seed-- which HENSLOW is inclined to think very close.-- «A fruit tr
447e164 ed like oni[on] Poa alpina because vivaparous. HENSLOW has seen this-- (Poa alpina vivaparous sometim
447e165 epias Turpin cell is individual May 29th.-- -- HENSLOW says he has not the slightest doubt that Festu
489qqIFC 13 Question &c. July. 1842.-- Shrewsbury p. 14 HENSLOW (2d time) p. 14.-- Father. And. Smith Dr. Holl
492q002 her: which generally fails first?-- Mal[e] 10. HENSLOW says semi-doubl flowers are those whose stamen
493q003 occur in relation with species-- answered «by HENSLOW» see notes In varieties is there any differenc
497q007 e raised true, there is no crossing by Bees.-- HENSLOW.-- (1) Character of alpine Flora of Tierra del
501q011 by secondary characters-- in double flower. do HENSLOW Speaking of Thyme doubts about stigma in simil
501q011 n or tuber-- talk about Mr Knights theory with HENSLOW.-- Dr. Fleming says yes. (29) Are there RACES
501q011 Mr. Herbert observe on this subject-- (31) Ask HENSLOW for list of annuals to place in Hot house to s
501q011 ct on generative organs of great Heat (32) Can HENSLOW ask question of Col. Le. Couteur about Wheat--
503q012 bout races of Banana & yet seedless-- no light HENSLOW or Royle, latter says seedless-- Also about Su
506q015 l of Common & Wild Duck-- Black Duck & Penguin HENSLOW &c (36) Has not H. raised races of white & Blu
507q15v describing amount of Horticultural Variation? HENSLOW to distribute some of my questions amongst agr
507q15v Henslow knows only on Citrons 47. Ficus carica HENSLOW knows only on Citrons 47. Ficus carica Henslow
585n074 emur seemed to like Lavender Water «very much» HENSLOW. N.. Necker has remarks on the means. by which

585n076 &c &c.-- Here there is kicker but not bite.-- HENSLOW remarks that Chimpanze pouted & whined, when,
091a025 schistose structure (& veins appearing): mem. HENSLOWS Anglesea solution of silex also shewn. No 3d
100a052 r. the arrangement of particles of granite in HENSLOW'S Grit, yet it is worth consideration. especial
467t099 feel somewhat «but little» less surprised at HENSLOW'S remark that pistil does not become abortive.
424e100 on spermatic animalcules in Musci frondosi, et HEPATICI,-- in Chara, in Marchantia & Hypnum «Prof:» D
539m083 ary mind. I a Darwin & take after my Father in HERALDIC principle. & Eras a Wedgwood in many respects
224b213 rced to haunt two islands one with one kind of HERBAGE & one with other, might change organization of
216b180 r, says the Seeds of hybrid lillies &c &c &, (V HERBERT on hybrids) thus act.-- Now the point will be
323c269 ol.-- March 16. Gardner's Music of Nature-- -- HERBERT on Hybrid Mixtures: Marginal notes. -- 20th. C
326c265 en Botanist of range of plants Books quoted by HERBERT p. 338 Schiede in 1825. & Lasch. Linn. in 182
326c265 nicus. has remarks on acclimatizing of Plants. HERBERT. p. 348. gives reference to Kolkreuter's Paper
427e111 but produces greater effect on offspring-- Mr. HERBERT says «p 347. Amyyralidæ» Plants do not become
428e111 ls have not been so. (which is case adduced by HERBERT) because not reared by seedlings.-- Now my pri
428e112 al hardy seedlings: (which are confessed to by HERBERT) to sift out the weaker ones: there ought to b
435e133 that the granules exserted their tubes: now Mr HERBERT has shown that stigma swells, when pollen even
438e141 their variation, which is not improbable as Mr HERBERT does not seem to recognize any difference in c
439e143 ined by a facility in returning to old type Mr HERBERT showing the extreme facility of crossing, in p
453e183 s-- how is it with hollyoaks, flaxes &c &c? Mr HERBERT in letter says distinctly, that Hollyoak repro
491q002 he Oxalis from C. of Good Hope mentioned by Mr HERBERT in vol IV. Hort. Transact.-- 4.. Are any varie
492q003 acts & M. Sageret is referred to with doubt by HERBERT Do forest-trees sport much in nursery gardens?
497q06v ontinually dropping in woods. by birds 13. Mr. HERBERT says Crocuses are very difficult to cross.-- a
497q06v t them together. & raise. seed.-- In letter Mr HERBERT says do about OEnothera.-- (14) Examine pollen
497q007 ually many species in genera of Leguminosae.-- HERBERT explains numerous spec. of Cape Heath by facil
501q011 d, «without sticks»-- in reference to what Mr. HERBERT observe on this subject.-- (31) Ask Henslow for
503q012 require Selection (36) Ask Mr Gowen to ask Mr HERBERT, how many generations any hybrid has <been> re
218b191 s about changes when animals transported.) Mr HERBERT'S papers are in the Horticultural Transactions
277c125 ure similar would have blended together Mem Mr HERBERT'S law; habits determining fertility Scheme for
309c219 ts, & therefore less fertile. according to Mr HERBERT'S views.-- Argue <argue> case of abortive organ
355d066 es Australian dog) or donkeys to Zebras.-- «Mr HERBERT variety of horse, dun-coloured with stripe ap
360d091 -- «& not time» thinking of the Penguin duck & HERBERTS law of ideosyncrasy I have hitherto thought t
439e143 hrough many changes.-- It is very important Mr HERBERTS fact about the hybrids (mentioned in letter t
504q014 t Holyhoaks. races planted & Linum Perenne.-- HERBERT'S. fact.= (4) Effects of Nitrate of Soda under
037r055 bones. & the bones of <two graniniverous> a HERBIVOROUS lizard.-- from the action of torrents. «mari
212b167 ions between fossils & recent shells between HERBIVOROUS & zoophagous Mollusca according to periods.-
423e096 arter of the world would kill all of the one HERBIVOROUS. & its one carnivorous devourer. ;-- it is qu
571n029 ider case of grazing animals knowing poisonous «HERBS:» & man not.-- ¿no vegetable good «for man» to
579n060 der the will? of the animal.(-- Jan 21. 1839. HERCHEL'S Discourse p. 35. On origin of idea of causati
247c029 , «"which I say "are as variously coloured as a HERD in England"-- Black & Grey varieties of rabbits
202b132 e Isd.-- "The buffaloes, introduced from Timor, HERDED separate from the English cattle, nor could we
241c015 ds of Tongatabou. & New Ireland.-- Gould will HEREAFTER know about birds of N. Zealand l'Institut. 18
405e032 between silver & gold fish-- Octob. 26th. If. HEREAFTER. M. angustidens be found to be inhabitant of
581n063 my new part of the view.-- let the proof of HEREDITARINESS in habits. be considered. as grand step if
182b046 s partly due to direct adaptation & partly to HEREDETARY taint:. hence the resemblances & differences
200b123 is this? Race permanent, because every trifle HEREDETARY, without some cause of change,, yet such cau
203b136 which is same as Indian Cattle.∴ tameness not HEREDETARY?, having been gained in short time. Milvulus
206b148 Who can analyze causes, dislike to marriage, HEREDETARY disease, effects of contagions & accidents:
220b198 rth Magazine Jan: 1830. most curious paper on HEREDETARY fear (like rooks with guns) of the Bustards
227b225 cture.-- Every species is due to adaptation + HEREDETARY structure. latter far chief element.: little
227b228 atomy, & it would lead to study of instincts, HEREDETARY. & mind heredetary, whole metaphysics.-- it
227b228 ead to study of instincts, heredetary. & mind HEREDETARY, whole metaphysics.-- it would lead to close
240c004 ties.-- One approaching to nature of Monster, HEREDETARY. other adaptation.-- Mr Yarrell says, that a
248c033 ircumstances> these therefore will be chiefly HEREDETARY.-- If varieties «produced by slow causes, wi
257c059 not applicable to monsters:-- Are monstrosity HEREDETARY??.? Does not atavism relate to this law.?--
259c063 structures either direct effect of habit, or HEREDETARY «& combined» effect of habit.-- perhaps in p
259c065 f whole life then real adaptation The case of HEREDETARY disease, is on the same principle that, cut
262c072 omment/ few will dispute--» says civilization HEREDETARY; ie instincts of wisdom virtue? «like senses
264c083 ructure. as six fingered people are sometimes HEREDETARY.-- yet these not adaptations «they are count
265c084 tory from.-- Beck about six fingered children HEREDETARY With respect to question what is adaptation
265c085 ging whether structure is owing to habits, or HEREDETARY is to see, whether a large family has it, &
265c086 uffinuria I think we may clearly attribute to HEREDETARY origin & not adaptation. to its habits.-- Fe
265c087 ructure was connected with habits, & how much HEREDETARY. The circumstance of aberrant groups being s
271c105 manner, much mud.-- These facts show, habits HEREDETARY whilst species have changed Argumentum ad ab
279c133 +. Fully supported by Mr Wilkinson. = Milking HEREDETARY, developemen of important organ (.see marks
280c134 ws instinct (Sir J. Sebright admirable essay) HEREDETARY Young wild ducks.-- lose as well as gain ins
282c140 hanges of <use> adaptation greater than those HEREDETARY ones.-- which would elapse; during time such
285c153 by» of Nature & Man.-- The constitution being HEREDETARY & fixed, certain physical changes at last be
290c163 donado as good forests for beautiful birds.-- HEREDETARY ambling horses, (if not looked at as instinc
290c163 oked at as instinctive) then must be owing to HEREDETARY power of Muscles.-- then we SEE structure ga
290c163 tructure gained by habit Talent &c in man not HEREDETARY, because crossed with women with pretty face
290c165 saying man cannot be companion but master.-- HEREDETARY tameness as well as wildness-- «Sir J. Seb
291c165 -of Sir J. Sebright.-- love. of man gained & HEREDETARY. «problem solved» habits become important el
291c165 cture has tendency to follow it, or it may be HEREDETARY & strictly point out affinities. conduct of
291c166 n.-- Thought (or desires more properly) being HEREDETARY<)>.-- it is difficult to imagine it anything
291c166 to imagine it anything but structure of brain HEREDETARY, analogy points out to this.-- love of the
292c171 instinctive.-- But why do some actions become HEREDETARY & instinctive & not others.-- We even see th
300c198 ferent animals less divided.-- But as man has HEREDETARY tendencies. his mind is still only a divided
300c198 wers which allow of, acquirement of language. HEREDETARY & acquirable.-- therefore mans mind not so d
304c208 it can be traced through series then probably HEREDETARY & not produced by circumstances In ostrich w
309c221 els in America, speaks of short legged sheep. HEREDETARY proceeding from an accident. New England far
310c223 orance.-- but seeing such passions acquired & HEREDETARY & such definite thoughts, I will never allow
312c232 nstincts, constant. ¿ whether mutilations non-HEREDETARY & variations produced in short time in some
335d014 be compared to animal with amputated limb.-- HEREDETARY <thr> Six fingered people, <Hill> «Lord Berw
335d014 «Lord Berwick» family with defective palates. HEREDETARY & therefore exceptions. to above law.-- Stud
335d015 is slight alterations of primitive stock) are HEREDETARY: «Hybrids of» Varieties is different because
336d018 offspring must be like Hence mutilations not HEREDETARY,, but size of particular Muscles-- When two
348d051 them> it".-- i.e what characters chance to be HEREDETARY whether important or not,). p. 7. "The Natur
356d071 as pointing out difference between acquired & HEREDETARY tameness.-- In comparing my theory with any
386d166 t used is absorbed.-- this law acting against HEREDETARY tendency causes abortive organs.-- the origi
388d172 elation to utility of change-- hence harelips HEREDETARY, disease. extinction. Animals in domesticati
404e026 tendency in «manner of» life to be mottled + HEREDETARY tendency determines the puppies to be so.--
415e067 no doubt is owing to the rearing up of every HEREDETARY tendency towards fatal diseases, & such cons
420e089 iculty in forging, yet handwriting said to be HEREDETARY. shows well what minute details of structure
420e089 . shows well what minute details of structure HEREDETARY'-- Athenaeum .1839. p. <8>36.-- A crustaceou
433e127 ther & then tumbling, far more wonderful than HEREDETARY ambling horses. Whether the body of parent b
493q004 ess Brazilian or Persian animals of different HEREDETARY constitution, to see whether offspring infer
494q004 er know any case of quick or slow pulse being HEREDETARY. (6) In the last 1000 years how many generat
505q014 ar. with abortive stamens.-- show crossing & ¿HEREDETARY? (15) Abberley has a hooked Pea.-- intends t
508q016 d (7) Is Haemorragic tendency, independent of HEREDETARY cases, more common in man than in female-- (
508q016 in man than in female-- (8) In Hump-back ever HEREDETARY (9) Are the works of Berhave (treating of he
508q016 ary (9) Are the works of Berhave (treating of HEREDETARY diseases) translated. (10) About Daltonism i
508q016 ny medical Statisics, proportion of diseases (HEREDETARY?) in diff. countries in same races Mr. Gray
525m024 e & kind of folly like (Mr George S.) is very HEREDETARY.-- My father says on authority of Mr Wynne t
525m024 her family» (as in Lord Berwick's family) are HEREDETARY.-- other deformities are illnesses of the fo
525m026 does not know whether trains of insanity are HEREDETARY in any one family.-- In Aunt-- B. the affect
526m027 tence of free will every action determined by HEREDETARY constitution, example of others or teaching
526m027 ne affected by various knowledge which is not HEREDETARY & instinctive) & the others <are> learnt. wh
526m030 g form of head, & thus these qualities become HEREDETARY.-- When a man says I will improve my powers
527m032 - Singing of birds, not being instinctive, is HEREDETARY knowledge like that of man, & this agrees wi

Page **(Key Word)**
530m045 (,as sailor tailor blacksmiths?) are likewise HEREDETARY, & therefore that their children have some l
533m057 tions, instincts degrees of talent, which are HEREDETARY are so because brain of child resemble, pare
536m073 e change in man.-- the real argument fixes on HEREDETARY disposition & instincts--.-- Put it so.-- Pr
537m077 - No one doubts patriotism & family pride are HEREDETARY., & therefore he has these strong, & does no
537m077 education, may be opposed undoubted cases of HEREDETARY pride & in single families. Edinburgh. Phil.
539m083 hange its habits.-- Aug. 16th. As instance of HEREDETARY mind. I a Darwin & take after my Father in h
543m101 ure. C. D.27 Can an analogy be drawn between «HEREDETARY» associated pleasures & pains & emotions-- s
545m107 hey whine like children.--» Expression, is an HEREDETARY habitual movement consequent on some action,
550m127 without being aware of their association «ie HEREDETARY», does one not call them instinctive emotion
555m142 (Useful to use term instinct, when origin of HEREDETARY habit cannot be traced) V. D. p. 111, case o
560m156 r remember what they did in former one. about HEREDETARY tricks & gestures, other cases like D. Corbe
568n017]CD-- «Also» When one is prevented performing HEREDETARY habit, (or moral sense, or instinct,) one fe
571n028 sign of exalted character???) Why may not our HEREDETARY nature thus acquire some general notions, wh
571n029 certain foods being habitual-- & hence become HEREDETARY; on same principle we know many tastes becom
571n029 true taste.-- Everything that is habitual, if HEREDETARY, is pleasant.-- Mental & Bodily Consider cas
571n030 een prejudices of youth from <here> habits. & HEREDETARY habits. & perhaps even latter may be vitiate
575n046 logous. to instinct, to the permanence of old HEREDETARY ideas.-- being lower faculty than the acquir
581n063 on.-- Sir. J. Sebright. has given the phrase "HEREDETARY habits." very clearly, all I must do is to g
581n066 cal diversity of tongues.-- [Emotions are the HEREDETARY effects on the mind, accompanying certain bo
582n066 analogous to young pigs hiding themselves; & HEREDETARY remains of savages state.-- N B. According t
582n067 s cantering in S. America capital instance of HEREDETARY habit:-- there must, however, be a mental im
585n077 position, it would be instinct-- instinct is HEREDETARY knowledge of things which might be «possibly
586n080 eans stained in?). had better refer to to the HEREDETARY part of it,-- & faculty (faculty «being» alw
586n080 rt of it,-- & faculty (faculty «being» always HEREDETARY helps this confusion.--) Hensleigh considers
586n081 heart beating may be considered also such.-- HEREDETARY habit, is a part never subject to volition.-
587n087 d think, great principle of liking, as simply HEREDETARY habit.-- A blind man might be born with idea
589n091 nce less.-- p. 325 «to 29».-- Habits becoming HEREDETARY form the instincts of animals.-- almost iden
599o007 hfares, I grant that man, from the effects of HEREDETARY knowledge, has produced almost → greater cha
600o08v to do it by other.-- I suspect conscience, an HEREDETARY compound passion. like avarice.-- Is there n
607o025 Motives are units in the universe. [Effect of HEREDETARY constitution,-- education under the influenc
614o037 eward in good life [RHC] Instinct appear like HEREDETARY memory; but first memory in many cases canno
614o037 ace)» to experience gained by man in lifetime HEREDETARY memory not so wonderful as at first appears,
615o36v . Mr Wynne says, that beyond doubt courage is HEREDETARY in fowls & not effect of feeling of individu
615o36v e their perch till evening» crow different.-- HEREDETARY effect of former tropical climate analogous
615o038 »! [RHC] 7) As definite instincts modified by HEREDETARY;-- so succession so perhaps general ones.--
616o038 een heard; so love of virtue enhanced by this HEREDETARY kind of memory.-- The difference between her
616o038 tary kind of memory.-- The difference between HEREDETARY memory & individual secretion of thought, ma
622o048 he mouth]CD» [LHC] My theory of instincts, or HEREDETARY habits fully explains the cementation of hab
624o50v stes in mouth by my theory are due to <habit> HEREDETARY habit (& modified & associated during lifeti
626o052 ion.-- 'N.B. If feeling or emotion rises from HEREDETARY action on body.-- This feeling, when instinc
633j53v cated variations show us accidents may become HEREDETARY [produce some peculiarity in seed vessel]CD
638j58v nearly so (except no in isd) although having HEREDETARY superfluities Man could exist without Mammae
193b093 stralia genera, collected together Man has no HEREDETRAY prejudices «or instinc» to conquer or breed
189b070 If we could tell, I do not doubt even colour HEREDITARY in time as in space. (Mem: Galapagos). Littl
336d018 l disposition, stature, are slowly obtained & HEREDITARY; <but if> if the change be congenital (that
336d019 led.-- «[My grandfather's theory of Mules not HEREDITARY, because generation -- highest point of orga
466t095 oped-- Double flowers & colours breaking only HEREDITARY characters, wh. come on in after life of Pla
512q019 ith hybrids, is offspring of short-horn bull & HEREFORD cow similar to reverse cross.-- Sow cast-up-b
063r132 of Brazil.-- Propagation. whether ordinary. HERMAPHRODITE. or by cutting an animal in two. (gemmiparo
194b096 ts do receive intermixture.-- But how with «HERMAPHRODITE» shells.!!!? We have not the slightest righ
229b232 n ancestor we may be all netted together.-- HERMAPHRODITE animals couple: argument for true molluscs
368d114 specific character most perfect in <male> «HERMAPHRODITE» (Fishes have no secondary characters.--)p
380d154 in brain of male.-- Every animal surely is HERMAPHRODITE-- (as is seen in <fe> plumage of hybrid bir
382d158 e <add> presence of both testes & ovaria in HERMAPHRODITE,-- but not of poenis & clitoris, shows to m
383d159 o extreme degree of abortion).-- Insecta.-- HERMAPHRODITE, being not only dimidiate, but quarter-grow
384d161 as in Hermaphrodites, (shows my doctrine of HERMAPHRODITE differs from Hunter)-- Hunter (p. 45) obser
384d162 tebrate tak place.-- ∴ Every man & woman is HERMAPHRODITE:-- ∴ developed instincts of Capon. & power
388d172 hat man & «or cock» pheasant &c is abortive HERMAPHRODITE is supported by change which takes place in
391d174 mand some further change?-- Man properly is HERMAPHRODITE (hence monstrosities tend that way «& from
391d174 of <In> each Man or mammalia being abortive HERMAPHRODITE simplifys case much; & originally <her> eac
391d174 implifys case much; & originally <her> each HERMAPHRODITE being simple (Are not Coniferous trees gene
412e057 es The Pipe-fish is instance of part of the HERMAPHRODITE structure being retained in the male.-- <li
417e078 es) a variety. The offspring of true <pare> HERMAPHRODITE, would of course be like either, that is bo
418e080 externally-- My view of every animal being HERMAPHRODITE-- probably will recieve illustration from d
418e080 s plants, & abortion of others.-- ¿ in hemi-HERMAPHRODITE insects is it not easier to understand ¿per
432e124 no land animal with fluid seeds can be true HERMAPHRODITE Man probably assumes the hairy character of
443e155 ence from some plants & some mollusca being HERMAPHRODITE is, that intercourse every time is of no co
443e155 opement.-- [It is singular there is no true HERMAPHRODITE in beings with which have fluid sperm.--]C
445e159 iently proved that any shell fish is really HERMAPHRODITE. & <thinks> even oyster may fecundate each
447e164 cility»-- such as cryptogamic plants & true HERMAPHRODITE Mollusca, & probably corals.-- these forms
285c150 children of same parents in a circle,-- & «HERMAPHRODITES» father « <mother> » & grandfather <Might>
291c167 monoecious & others dioecious. some flowers HERMAPHRODITES & others not??? ‖ The death of some forms
316c245 may transport ova of shells.-- Conchifera. HERMAPHRODITES-- eggs in groups.. Have Dioecious plants m
381d156 . (edited by Owen) p. 34.-- Owen classifies HERMAPHRODITES. Cryptandrous. (only female organs visible
383d159 like Hen.-- one doubts whether they are not HERMAPHRODITES, like J. Hunters. Free Marten N.B. the com
384d161 bined addition & deficiency of parts, as in HERMAPHRODITES, (shows my doctrine of Hermaphrodite diffe
416e070 rtant obstacle to my theory] CD without the HERMAPHRODITES mutually couple,-- now how is it-- in Plan
416e071 rnacles, aquatic., <yet> Crustacean, & true HERMAPHRODITES] CD «It may be said that true hermaphrodit
416e071 e no fixed. land animals. so there are true HERMAPHRODITES.-- I suspect this rather effect of liquid
382d158 Hunter shows almost all animals subject to HERMAPHRODITISM,-- those organs which perform nearly same
384d162 flower on same stem.--» so that Molluscous HERMAPHRODITISM takes place.-- thus one organ in each beco
416e071 rmaphrodites] CD «It may be said that true HERMAPHRODITISM is a consequence of non-locomotion-- (cont
418e080 addition of other organ, in which case the HERMAPHRODITISM would not be perfect as in Ox. the amount
291c167 relation going through all Nature.-- Makes HERMAPHRODITISMS. one step in <scale>. Series-- in plants
043r069 hills. to compare with Galapagos.--Chiloe. M. HERMOSO. & Coral reefs (imperfect in latter). <At> Lye
057r113 ally should think probably that B. Blanca & M. HERMOSO contemp:.--Inculcate well that Horse at least
594n112 efore seen one, yet all-- flew to bed of flags. HERNES are common. not unlike in size in the air at a
555m143 not having run away &c having faced death like a HERO, & then I had some confused idea of showing sca
088a014 mportant in transport of Fish Let a Hawk fly at HERON.-- Ceratophytes common in Northern seas p. 312.
511q018 .-- are any actually rejected?? (8) Get Sir. R. HERON to give me Pigs foot undivided, & more particul
088a014 <Lyell> «Waterhouse» has frequently heard that HERONS bring eels alive to their nests; & then they m
331dIFC of seeds greater.?-- Mem. for Eyton.-- Sir. R. HERON'S case of breed of pigs with solid feet.-- 1838
361d095 tled, & not like any existing species-- [In two HERONS, <both> plumage of both (nursling) quite simil
119a104 ulities are required to start with (& does not HERSCHE theory imply tendency to equilibrium.) 3d. the
057r115 -Mem: stream of water in the country.-- Sir J. HERSCHEL. says. precip. of Sulph. B. all the infinites
122a114 ntains & volcanos.-- Why on one coast? How can HERSCHEL consider figure of earth statical.-- if platf
413e059 erence to my theory Babbage 2d Edit, p. 226.-- HERSCHEL calls the appearance of new species. the myst
028r032 e no sections> same cause as no colour Sir J. HERSCHELS idea of escape of Heat prevented by sedimenta
118a104 es have reached the surface Arguments against HERSCHEL'S view of cause of continental elevations (I)
122a113 ferent forms of earth as shown by arc.-- read HERSCHELS astronomy with oscillations of level.-- {P} w
124a118 ns in N. America Erasmus suggested to me that HERSCHEL'S theory offers no explanation of intermittent
125a121 rections. Mem. S. Cruz. Assuming from Sir. W. HERSCHEL'S views earth originally fluid, then cooling p
131a135 Dr. Nichol-- adduces the case to show Sir. J. HERSCHEL'S theory wrong.-- Geograph. Journal Vol. 8. p.
322c269 tive History.-- References at end of each Vol HERSCHELS. introduction to Natural Philosophy R. W. Dar
577n049 & zoophyte: it is (I presume-- see p. 188 of HERSCHEL'S Treatise) a "travelling instance" a-- "front
523m017 me on suddenly.-- Case of Mrs. C. O. who threw HERSELF out of the window to kill herself from jealous
523m017 O. who threw herself out of the window to kill HERSELF from jealousy of husband connection with house
524m021 .-- In Mrs P. <...> of B. <talked of.> thought HERSELF near Drayton & Ternhill, (where she was born)

554m138 o force it open, when she could not succeed of HERSELF.-- <The male> «I saw» Jenny untying a very dif
554m140 is> then knows she has done wrong & will hide HERSELF.-- I do not know whether fear or shame.-- When
554m140 thinks she is going to be whipped. will cover HERSELF with straw, or with a blanket.-- these cases o
567n013 o something.-- October. 8th. Jenny was amusing HERSELF--, by getting out ears of corn with her teeth
593n105 m. Nina used to get into hay & make a nest for HERSELF.-- the object is to make saucer-shaped depress
098a046 process.-- now cleavage as suggested by Sir J. HERSELF is all crystals obeying one law of crystalliza
399e010 gators. Snakes occur there,, but are unknown in HERVEY or Society isles. Hope says positively he has
365d106 nimals not having changed is good-- I scarcely HESITATE to say that if there had been considerable ch
576n046 ls knit their brows when incensed.-- A Dog may HESITATE to jump in to save his masters life,-- if he
192b089 rs Cuvier-- Perhaps the father of Mammalia as HETERODOX as ornithorhyncus. If this last animal bred--
233b250 pecies?-- thinks offspring of cats sometimes HETEROGENOUS.-- Australian dog jumped into tub leaving o
312c228 Smith if black & white Man crosses; children HETEROGENOUS, he feels sure of this, first offspring mos
331dIFC ertile animals increased?-- Where offspring, HETEROGENOUS, in plants are the number of seeds greater.
333d006 d Persian.-- Here then we have clear case of HETEROGENOUS offspring from one impregnation ¿is this on
342d033 arked only near species or varieties produce HETEROGENOUS offsprings.-- «are not the hybrid pheasants
342d033 & produce somewhat fertile offspring produce HETEROGENOUS offspring. It appears certain that hybrid M
393dIBC of those species. which cross & are fertile HETEROGENOUS? When bird fanciers say the throw of two va
186b058 be going back to common ancestor of Crysom. & HETEROM, but I cannot understand the universality of s
185b056 ay be, represent every other; for instance in HETEROMERA, you have representatives (which at first wo
185b056 longicornes.-- Again taking a subdivision of HETEROMERA same. thing occurs with regards to other tri
186b058 louring retained; therefore probably in some <HETEROMERA> colouring of crysomela may be going back to
272c108 stion which I more immediately want are there HETEROMERA, which have habits & part structure like Cur
449e170 alike Waterhouse says some of the Galapagos HETEROMEROUS insects come very near to Patagonian specie
198b113 t Hilaire Insects & Molluscs allowed to be wide HIATUS: states in one the sanguineous system, in othe
286c154 Animals have voice, so has man. Not saltus.,but HIATUS animals expression of countenance. «[s]hare of
287c157 als on the steps that lead up to it" p. 20 ╲ +++HIATUS & saltus not syn.-- Linn: Transact Vol XIV.--╲
287c157 malia & fishes, one penguin, one tortoise shows HIATUS-- but not saltus-- when Linnaeus put whale bet
300c198 ties being viewed as replacing each other it is HIATUS & not saltus.-- The greater individuality of m
558m151 s value. from its connexion with mind, (to show HIATUS in mind not saltus between man & Brutes) no on
426e108 ng a great great-grandfather.-- -- Little Miss HIBBERT case of Hindoism coming out more than in mothe
041r065 of land: Mention M. Gay's fact about shells: HIBERNATION of fresh water Shells. multitudes.-- The que
540m084 n in romps, <so> he smiles. Many of actions as HICCOUGH a& they are probably merely coorganic as conne
161g096 ead of Gentle mossy slope, which from a distance HID it, could be followed for at least 2 miles on de
546m111 thing happen to him about a knife. which he had HID some years before.-- was greatly astonished, at
346d048 re--colour white, uniform.--crafty, go in file, HIDE their young., bold.-- a Mr W: Hall remarked tha
554m140 -- <but is> then knows she has done wrong & will HIDE herself.-- I do not know whether fear or shame.
195b099 so well.-- This view of propagation gives. <ro> HIDING place for many unintelligible structures. it m
301c199 yths arguments against squirrel using reason in HIDING its food is applicable to any habitual action.
582n066 people.-- Children have an uncommon pleasure in HIDING themselves & skulking about in shrubbery, when
582n066 ople are about: this is analogous to young pigs HIDING themselves; & heredatary remains of savages st
031r038 ar at base of great lava cliffs [Fig. I] line of HIGH tidal action {P} NB. patches of modern Conglome
044r074 pearances of eruption at bottom.--solution under HIGH pressure of gazes. especially the most abundant
055r108 t in height, of the solid lavas.--proportionally HIGH to age. (we do not wonder to see tertiary plain
092a028 formation twenty species same as Paris. 1500 ft HIGH Mr Bird in paper to Brit. Assoc: has shewn how
092a030 ll explain their solution. Athenaeum M. 516 1837 HIGH up the Essequibo, granite & quartz, after passi
109a083 -- De la Beches argument of low coast gaining & HIGH loosing answered by this -- No one can doubt. A
112a089 read--Look at the now active volcanos & see what HIGH they are «See Athenaeum. 1838. p 274. probably
120a109 hot water with «-- especially if very hot under HIGH pressure.--» respect to formation of salt.?.--?
126a124 below 39 degree» & lower part glass.-- then the HIGH temperature would be much nearer the surface. a
133a139 neath <of Sahara de> a dry desert, would be very HIGH.-- M. Parrot ends his paper like a fool.-- Feb
147g021 gree A Temp of Air 65 degree? Glenoe, 6 ft above HIGH water mark 30.380. 68 degree 65 degree? For com
148g022 There some of the half rounded gravel nearly as HIGH as highest measurement but nature I am quite do
158g082 s of hill where Boulder lies. buttresses «occur» HIGH up on Shelf 2d «in Upper Glen Roy" In this uppe
161g094 7 station was <a few> «about 3» feet <lower> too HIGH about a quarter of a mile further on, where thr
162g103 bed of lake with trace of terraces on each side HIGH up the Tarf (a Granite (boulder), sloping buttr
171b002 uns are longer lived Why is life short, Why such HIGH object generation.-- We know world subject to c
284c148 parts & only on those, & the islets separated at HIGH water.-- not other islands, nor any any other p
304c208 h time elapsed & therefore descended from branch HIGH up.-- Such probabilities only guides.-- Yet tri
304c208 all other birds, or that it sprung from a branch HIGH up.-- this argument not applicable to apterix.--
306c213 first & grandest divisions. but for ones of very HIGH order. not for vertebrata, but mammalia & repti
319c256 irds are inferior to ours, & our lark ranks very HIGH.-- Upon the whole thinks <many> more birds sing
327c265 d. very good to see whether peculiar plants-- in HIGH points Read Volney's travels in Syria Vol I. p.
345d044 black legs, & face & tail, just like species.-- HIGH active breedin[g] [not located] half breed liab
367d113 at Bell in Encyclop of Anat & Phys. describes, a HIGH-flying bat, which has the power of inflating it
393dIBC t transmitted by generation?? Is it «chiefly» in HIGH. bred dogs ie. (bred in & in) that one copulati
410e052 e to description), except describers having some HIGH theoretical interest,-- "the great end must be
433e127 volumes.-- The fact of tumbling pidgeons; flying HIGH all together & then tumbling, far more wonderfu
435e134 by physical agents." p. 466. Many facts given of HIGH temperature at which fish &c can live.-- Lyell
523m016 to a hospital.-- My father was nearly drowned at HIGH Ercall, the thoughts of it, some years afte
541m092 tiply large numbers in their head must have this HIGH faculty, yet not clever people. Aug. 21st. 38 W
571n028 reasoning on them (as justice? as ancients did HIGH forehead sign of exalted character???) Why may
589n092 bsurdity of a tree having reason: or dog, having HIGH powers without hand or voice.-- there is some g
602o11v a general notion of everything applicable to the HIGH idea «p. 131.» in Tragic acting-- CD [My idea.
614o037 ialism does not tend to Atheism. inutility of so HIGH a mind without further end just same argument.
038r058 ucleus & that envelopes no doubt existed. These HIGHER portions probably formed Islds from which proc
047r081 esent, i e a part below «mean» level before the HIGHER part.-- Does the sea fall on banks as a Bore
060r124 se from some efforts in the land to lift itself HIGHER & to grow upwards; for the land is constantly
097a042 eenstone dikes. be residue of quartzose vein in HIGHER parts? & felspathic veins?-- Mr Poulett Scrope
113a090 ot wrong? we know mean of surface formerly much HIGHER, «so» that we must look at the upper four hund
125a121 - lower down the temperature may be kept up far HIGHER from circulation of heated fluid or gases unde
153g053 way below House of Glen Roy, seems to be which HIGHER up on is corroded {P} 4th [shelf] river 4th [s
157g078 97 A 79.½ 29.316 divortium aquarium «about 12 ft HIGHER than last station» 29.316 true terrace «2d» ne
159g087 inued some hundred feet lower & begins about 60 HIGHER-- There are however fringes of alluvium (?) st
159g087 There are however fringes of alluvium (?) still HIGHER Slope of valley much more gentle than in Glen
159g087 Glen Roy, & partly shut in No Granite blocks in HIGHER parts?? Bought Glen name of Glen by which we d
162g104 ward, & inwards-- deposited when water stood at HIGHER Loch Keeper tells me, that Loch Lochy is 8 ft
163g107 fringe as if it has been filled up «at» 30 ft. HIGHER with pebbles now worn away-- The above shells
189b074 ete.-- It is absurd to talk of one animal being HIGHER than another.-- We consider those, where the {
199b118 . N. Phi. J. p 297, No 8 Jan-Ap. 1828 -- I take HIGHER grounds & say life is short for this object &
204b141 onsume this or that thing.-- There is some much HIGHER generalization in view. In Marsupial division
233b252 evity of species in Molluscs!!! When we talk of HIGHER orders, we should always say, intellectually h
233b252 er orders, we should always say, intellectually HIGHER.-- But who with the face of the earth covered
254c048 are analogous to the earliest conditions of the HIGHER classes, during which the changes of the ovum
254c048 cal of succeeding classes & likewise those much HIGHER in scale. So Owen actually believes in this vi
270c103 considers big animal» becomes more developed in HIGHER animals than in vegetables. p. 243 radiate ani
284c148 The Cocos do Mar on the Mahé island,, one the HIGHER parts & only on those, & the islets separated
410e050 ere could not be improvement. «& hence not «be» HIGHER animals» -- it was absolutely necessary that P
447e184 f crossing is much less,-- now certainly in the HIGHER animals; changes seem to have been more rapid,
451e179 smaller, lighter head, carried more elevated & HIGHER forequarters: is said to be of a bolder & more
576n048 tances!» We must believe, that it require a far HIGHER & far more complicated organization to learn G
601o009 ants? yes by distinct mechanism 2. Sensation of HIGHER order. where the sensation is conveyed over wh
601o009 ps polypii-- (so that lower animals are sleeping HIGHER animals & not plants as supposed by Buffon) Co
614o037 re step towards some final end.-- production of HIGHER animals-- perhaps, say attribute of such highe
614o037 higher animals-- perhaps, say attribute of such HIGHER animals may be looking back, ∴ therefore consc
632j53r e, I believe, only direct consequences of still HIGHER laws.-- I do not «then» believe the pappus of
032r039 ea A. B. will be to eat in the land in line of HIGHEST tidal action. this will at length be checked b
051r094 posed rocks near Bahia, whole surface to where HIGHEST spray (there pale green confervae) coated with
079r176 ime. accompany.-- Ulloa has said silver in the HIGHEST & gold in the lowest. Humboldt states that som

```
146g007 curved  form of hill be explained by my idea-- HIGHEST part must project [Blank] {P} Path East End ne
147g021 been filled with sloping bed of rubbish Friday HIGHEST part of road between Inverorum & King's House
148g022 e of the half rounded gravel nearly as high as HIGHEST measurement but nature I am quite doubtful of
160g091 oulder about 20 ft. below summit <Isthmus> {P} HIGHEST point joining this hill to others 3000? if Ben
170b002 he original molecule has done).-- This appears HIGHEST office in organization (especially in lower an
189b074 ure intellectual faculties} most developed, as HIGHEST.-- A bee doubtless would when the instincts we
197b111 ich man «one animal» passes from worm to man;  «HIGHEST» as typical of <sa>. changes, which can be tra
227b227 iew of generation. being condensation, test of HIGHEST organization intelligible.--may look to  first
257c058 Dicot wood> Coniferous wood in Coal Measure.-- HIGHEST fish in Old Red Sandstone.-- Nautili in----. i
275c119 called  the prophetic spirit in science--. the HIGHEST endowment of lofty genius Using geograph distr
336d019 of Mules not hereditary, because generation -- HIGHEST point of organization] CD» false.-- The creato
356d070 y my Grandfathers expression of generat. being HIGHEST end of organization good expression but does n
390d190 mparison «of difference» with other sex. = The HIGHEST bred Blood-hound. would be infertile with high
390d179 hest bred Blood-hound. would be infertile with HIGHEST bred of other ¿ breed.= Therefore it is not re
447e164 er marriage is greater (Hence Dioecious plants HIGHEST.-- Palms &c &c)-- Is there greater resemblance
543m100 this  judgment gives a man common sense, & the HIGHEST intellectual powers of perceiving & classifyin
602o010 of abstract form" is the source of part of the HIGHEST enjoyment in mutilated statues In  Elliotson's
145g002 made or crosses difficult Salisbury Craigs The HIGHLAND shepherds dogs coloured like Magellanic fox..
234b255 of Ice Highlands of> the largest <sort> of our HIGHLAND Sort, except in one respect, that those of Ic
345d044 g. how easy races or varieties are made.-- The HIGHLAND Shepherd dogs, coloured like Magellanic Fox.-
345d044 & appearance-- good case of Provincial Breed-- HIGHLAND Sheep jet black legs, & face & tail, just lik
432e123 ss sheep with horned.-- compare this with what HIGHLAND shepherds said.-- p. 12. Attempts to  improve
234b255 le in Iceland. «are very» like <those of Ice HIGHLANDS of> the largest <sort> of our highland  Sort,
183b051 do  not breed readily. point in view.--¿whether HIGHLY domesticated animals like races of man.-- M. F
243c020 od of generalizing without tables or references HIGHLY unphilosophical xx Says same remark with regar
295c183 ers (D'orbigny Rapport. p. 11) in S. America so HIGHLY developed in North.-- Icthiology of S. America
374d133 - p 458 Upper Silurian, fishes oldest formation HIGHLY organized.-- do. p. 461.-- Lower Silurian-- se
407e042 vidually.-- I see clearly from F. R. it will be HIGHLY necessary to show that if species fall, genera
423e096 e simplest animals could be destroyed, the more HIGHLY organized ones. would soon be disorganized  to
445e159 ng-organized» forms as parent forms of existing HIGHLY organized forms-- this resulted from the neces
448e167 ame, as if Isthmus of Panama.-- These two cases HIGHLY improbable.-- yet I can see no other way of ac
450e174 so when wild-- wild ducks monogamous; tame ones HIGHLY polygamous-- change of instinct by domesticati
453e183 of  different varieties being raised by seed is HIGHLY odd-- as it is not so with the esculent vegeta
465t079 ations have been confounded in the caves? It is HIGHLY important, to bear in mind that enormous perio
527m034 owers of invention) such castles in the air are HIGHLY advantageous, before real train of inventive t
595n125 hese I think are better instances of instincts (HIGHLY useful as only means of communication) in man,
640j167 this case is only true criterion.-- Hence it is HIGHLY unphilosophical to assert, that they are not s
381d156 n)-- Epizoa & the nematoid Entozoa-- Therefore HIGHNESS in scale has no «constant» relation to separt
198b113 t & present might tell something) p. III G. St HILAIRE Insects & Molluscs allowed to be wide  hiatus:
202b133 together" There is long rigmarole article by S HILAIRE on wonder of finding Monkey in France.-- of ge
203b135 al cause:-- Mem: Mastodon all over S. America.  HILAIRE does not seem(?) to consider the monkey as a w
215b178 nd J. de Physique. Tom 59. p 467. Peron G. St HILAIRE has written "opuscule" entitled "Paleontograph
236bIBC Soc  Horae Entomolgicae Linn: Soc. Geoffry-St. HILAIRE Philosophy of Zoology Waterhouse B C N[a]me of
356d069 e time Seeing what Von Buch (Humboldt). G. St. HILAIRE, & Lamarck have written I pretend to no origin
493q003 Cat--  , in Brazilian «toothless» dog-- I. St. HILAIRE says length differs in different cats.-- Good
510q017 sea-current, icebergs travel by wind. Aug. St. HILAIRE Bot. p. 787. position of embryo in close speci
325c267 ommes. & les animaux.-- by Isid. Geoffroy. St. HILAIRES. 1832. contains all his fathers views.-- Quot
510q017 g to see if any variation in varieties. G. St. HILAIRES law of Balancement Wm Yarrell (1) About non-b
510q017 787.  position of embryo in close species of HILIANTHEMUM differs greatly-- how very interesting to s
023r013 apped by Calcareous rock; following Curvature of HILL; states could discover no shells: nothing  said
036r050 f Patagonia.-- British channel &c &c. There is a HILL. near Copiapò which is asserted to make a noise
036r050 story;  I believe it was necessary to ascend the HILL,--but my recollection is imperfect & was recall
036r051 I  am nearly sure, it is necessary to ascend the HILL.-- The absence of Second form, except near subm
054r102 overed by quartzose sandstones: refers to broken HILL described by Permetty: account of streams of st
072r159 ountains Mackenzie has talked of lava flowing up HILL; ¿what does he mean?) Consult Dr Holland  about
095a038 . letter in Sir. W. Parish Possession. talks of <HILL> «cerro» of Diamante near stream of same  name.
114a095 action which I give at C. of Good Hope.-- A bare HILL of greenstone, if we know origin of  greenstone
146g007 ee perhaps most common-- Will not curved form of HILL be explained by my idea-- highest part must pro
146g010 d Grey in front on wall perhaps wall oblique The HILL has been well-- denuded.-- «of hard  metamorph»
146g013 nearly certain there were none on surface of any HILL Thursday On side of Hill South of upper end  of
146g013 none  on surface of any hill Thursday On side of HILL South of upper end of Loch Dochart buttresses o
147g015 ace is thickly studded with ridges & flat topped HILL/ do alluvium. NB In one part pure sand in curre
149g032 artz & other rocks not apparently in situ <& in> HILL being gneiss <& also> also near summit on  Hill
149g032 > hill being gneiss <& also> also near summit on HILL on side of Inn BOULDER of granite above 4th She
149g033 recisely as with Isld-- {P} do they extend round HILL too low line drawn plain red talus line on N. s
150g037 et formed in centre In Glen Collarig, on side of HILL of Bohunthine upper road (2) extends as far nea
151g040 ood, (as granite or gneiss of Moel Derry) on low HILL between Inn & Bouhunthine the summit «doubtless
151g041 At the Pass of Glen Collarig two little lines of HILL (judging from external form alluvium) descend f
153g056 sandstone  actually resting on them on summit of HILL rounded, site N N W of Ben Erin {P} Shelf of Gl
154g059 ? 70? {P} Where a buttress projects from side of HILL if line suppose continued across to {P} side re
157g075 - granite block a yard across. On side of «that» HILL, in front of which shelf 3d form beach of grani
157g075 h shelf 2d «almost» forms it into island-- whole HILL composed of remarkable gneiss with red  granite
157g076 «there  are» granite) «boulders» {P} cory stream HILL with boulders river Right Hand Cascade has <cut
158g082 tired, hard to explain on river doctrine <Little HILL with granite blocks almost encircled> <fre> Gne
158g082 t encircled> <fre> Gneiss cut smooth on sides of HILL where Boulder lies. buttresses «occur» high  up
159g085 tween river & dry Corry Scarcely conceivable. if HILL between Corry so much cut Granite could have re
160g090 River  Bought & between one of these & Glen Tarf HILL «Cairn «taw» leer peak» Barom 28.700. A.75 degr
160g090 8- long of syenite with pinkish felspar;-- whole HILL dark grey fine grained. Much contorted gneiss «
160g091 narrow sharp ridge with peak» I walked all round HILL. Boulder about 20 ft. below summit <Isthmus> {P
160g091 summit  <Isthmus> {P} highest point joining this HILL to others 3000? if Ben Erin is 3500 boulder Cai
163g107 mself-- Sand with tide ripple Near Fort Augustus HILL & fringe as if it has been filled up «at» 30 ft
165g124 innan Tower where stream enters at head of which HILL is round & not merely thoughts laying dormant--
275c120 ent!) Mr Wynne) at end of chase would not run up HILL-- he took thorough bred <greyhound>.  bull-dog.
284c148 dying  & mountain torrents.-- but to crawl up an HILL, then by deaths?!)-- looks like subsidence.-- o
335d014 d limb.-- Heredetary <thr> Six fingered people, <HILL> «Lord Berwick» family with defective  palates-
446e161 r the reverse of facts stated by Smith of Jordan HILL.-- May 27th.-- Henslow One of the 4 species  of
490q001 ded to by Dr. Holland-- <Jordan> Smith of Jordan HILL-- character of the extinct land-shells of Madei
554m141 oning," & take instance of Dray Horse going down HILL.-- <argue sophism of association. Kenyon, & the
563n001 r houses. yet so shy at all other times.-- Birth HILL shows it is evergreens they seek Cock  Pheasant
110a087 slate.-- Geograph. Journal. Vol IV (p 321) Mr HILLHOUSE describes central granitic ridge of Guayana a
202b131 grap. Journal.  Voyage up the Massaroony by W. HILLHOUSE.-- Demerara. In note. Demerara. 10 «12» feet
149g032 ranite above 4th Shelf a little lower down the HILLOCK with beach & channel precisely as with  Isld--
037r056 e cases. -- M. Video exception, but even there, HILLS of Basalt & other Volcanic rocks. Bahia, Rio de
043r069 of  channels or grooves in rocks at Costorphine HILLS. to compare with Galapagos.--Chiloe. M. Hermoso
049r089 at probable origin can be given to the numerous HILLS of greenstone? -- Daubeny. P 95. Glassy & Stony
088a016 Helms  remark on common salt being found on low HILLS East of Cordillera very important V. Malte brun
091a027 irection<?>»ESE-- CD [In the Darwar. transition HILLS & strata SE. direction of transitions clay slat
147g017 look carefully for Marine remains-- Some of the HILLS almost appeared as if they belonged to double s
147g020 alley scattered with few very small & irregular HILLS of alluvium-- nothing very striking yet possibl
148g026 ogs Saturday. Before coming to Bridge of Spean, HILLS of «sea», gravel, current cleavage, & pretty we
150g038 side of valley both extend below the Houses The HILLS in this neighbourhood appear very  round-topped
159g086 e way down slope of obscure terraces (& conical HILLS on same) of «semi» waterworn & some partly well
495q05a ion.-- Capital in middle of ploughed field-- on HILLS-- 10 Shoot tame duck on pond with  Duck-weed--
027r027 f shells in flints in Chalk New Providence more HILLY than others of the Bahama consists of rock & sa
207b151 Bay Willow V. Dr Royle introductory remarks to HIMALAYA Mountains-- Bory St. Vincent Vol. III. p. 164
235b272 ll 1024 Flemings Philosophy of Zoolog Royle on HIMALAYA Plants-- Would it not be possible to work thr
249c036 ds at Japan. connected with Europaean forms on HIMALAYA??-- This is very remarkable, when we consider
303c203 uld few in number.-- Parallel of Japan near HIMALAYA, & Europaean forms on that isld.-- The  races
```

310c223 n, European & Chinese Genera & some species on HIMALAYA.-- some English beetles, birds & a fox most c
355d068 ng bud changing into ovules.-- Captain Grants. HIMALAYA. shells (see Paper in Geolog Transacts) same
449e169 the pidgeons in common Dovecot are very like a HIMALAYA species -- leuconotes-- Magazine of Nat. Hist
231b242 eys different not travellers?? Royles case of HIMALAYAN, plants ¿migratory birds, he told me some sto
236b279 ammalia--of Europe Mem. Mr Bell's case of Sub HIMALAYAN land emys, decidedly an Indian form of Tortoi
324c268 ol III p. 164. on unfixed forms. Dr. Royle on HIMALAYAN. types. Smellie. Philosophy of Zoology Flemmi
085a002 rom salt water. Mr. Arrowsmith tells me, that HIMALAYAS penetrated like Bolivian Chain. Volcanic isla
119a104 Journal <JCD>, no great chains like Andes or HIMALAYAS, but great circular mountains, yet so analogo
310c224 frage, almost <same> «closely allied» species HIMALAYAS, 13,000 & Melville Isd.-- West Africa & India
408e044 Horse. Bear & Deer at 16000 ft. with Snow on HIMALAYA-- Humboldt bones at 7800 in Andes-- parallel
408e045 00 in Andes-- parallel & curious facts.-- The HIMALAYA. case, bears on the vast changes even in that
318c252 aptation for snow-- like snow shoes. than feet & HIND legs of these white hares, fitted for regions o
399e009 but paler & buffer.-- with long ears-- & longer HIND legs??? so that I was almost doubtful which it
399e009 so that I was almost doubtful which it was.-- do HIND legs increase in any rabbits One may strongly s
634j55r strial.-- for we find even in Cetaceae traces of HIND extremities.-- How are we to explain this.-- Di
639j28r cricket & rodents (?) p. 251. all animals run by HIND legs-- Kangaroo. only a caricature; Penguin.--
229b233 inland.-- NB. There is division of snakes. with HINDER teeth perforated for poison channels, but not
426e108 -grandfather.-- -- Little Miss Hibbert case of HINDOISM coming out more than in mother or indeed gran
615o038 feelings weakened in Otahiati; fear of death in HINDOO population.-- Slightly modified in many countr
266c091 enerally speaking, a much fairer race than the HINDU'S, in the same tracts; & that in their appearan
088a013 not very great; where piece turned over axis or HINGE no doubt fluid.-- analogy as continental elevat
577n051 elation to true memory, that the formation of a HINGE «in a bivalve shell» does to reason.-- an infla
638j58v between the works of art «or intellect» such as HINGE, & hinge of shell, works of laws of organizatio
638j58v he works of art «or intellect» such as hinge, & HINGE of shell, works of laws of organization is rema
638j58v ut yet analogous, no savage ever made a perfect HINGE.-- reason, & not death rejects the imperfect a
459tf02 ems to doubt its applicability to common mule & HINNUS-- in one case bastard of wolf & dog had more f
459tf02 rm of male, & another of both progenitors-- the HINNUS, resembles horse in its head ears, tail limbs-
197b111 ated, but must have concluded so= Evidently «or HINTS» considers generation as a short process, by wh
229b234 ilkinsons Egyptian remains skimmed -- Pliny Nat. HIPPOPOTAMUS do.-- Giraffe do.-- Range of East Indian Rh
452e181 do p. 583, it APPEARS probable,«?» that the HIPPOPOTAMUS occurs in India. in the Jungles of Borabhum
413e060 Africa only place, where, Elephant, Rhinoceros, HIPPOT, Haena &c are found together.-- Read this Work
023r009 namely Java is 1000 miles distant! Where are HIPPOTAMI found in that Archipelago? Such have never be
231b242 ts do not follow by any means.-- Ostriches.-- HIPPOPOTAMUS only African.-- American & African forms min
307c216 . p. 479. fragment of tusk «& Molar tooth» of HIPPOPOTAMUS from Madagascar!!!!!! Proceedings of Geol. S
338d025 tion of new forms.-- what is range of Hyaena? & HIPPOPOTAMUS.? Indio-African, or pure Africa?-- ‖Fossil E
338d025 t of Africa Most important under this view, & HIPPOPOTAMUS of Madagascar: because. contemporaries, In i
349d052 circle?!!! p. 8-- Anomalous structures, as in HIPPOPOTAMUS, solely owing to number of lost links.‖ if a
473s03r s Journal <vo> 1842. p. 142-- Sus americana & HIPPOPOTAMUS «with Megatherium & Mylodon» in post pliocen
022r008 t it: in which we found the Head & Boans of a HIPPOPOTOMUS; the hairy lips of which were still sound an
282c142 uld not obtain a cast of washing men-- but might HIRE the preexisting race, thus the analogy would no
318c253 which is now becoming common-- likewise of the HIRUNDO fulva (added by Audubon in Appendix) showing W
558m153 se do all species put their necks straight out & HISS.-- [Hyaena pisses from fear so does man.-- & so
267c095 nfusoria. <p> do p. 62. Ehrenberg Annals of Nat. HIST. precursor of magazine??? p. 75. roe of Asteria
270c103 generation!!!. Mem. Agaziz. (INo Annals of Nat. HIST) spiral structure in Echinodermata.-- Agassiz s
323c269 ilkinsons Egyptian remains skimmed -- Pliny Nat. HIST of World do -- Lamarck. II Vol. Philo. Zoology
327cIBC Horse & Cow & Sheep Clarke's Travels.-- Temminck HIST. Nat. des Pigeons et des Gallinaces. Silliman's
421e091 southern portion.-- ‖.do p. 269. Annals of Nat. HIST 1838 on «a» freshwater fish peculiar to Ireland
592n101 Sceptical Philosophy. Hume has written "Natural HIST. of Religion" on its origin in Human mind.-- An
058r116 to see on beach successive lines of sea weed-- HISTOIRE Naturelle des Indes Acosta. p. 125. of French
089a018 ver reach surface.-- Journal de Physique, et D HISTOIRE Naturelle, C«o»urzejolles. 11th Observ.-- Les
325c267 dge. Horse, Cow, Sheep.-- Verey. Philosophie d'HISTOIRE Naturelle Marcel de Serres Cavernes d'Ossemen
325c267 uropéan forms compared with African <A> Annals HISTOIRE Generale et particulière des Anomalies de l'
055r107 scension (or modern streams of St Jgo) yet no HISTORICAL records of eruptions how immense the time!!
397e004 o my theory, that the amount of change within HISTORICAL times has been small-- because change in for
025r018 t be well to urge, geologists to compare whole HISTORY of Europe, with America; I might add I have dr
048r085 f Organic power a lump of hard clay. -- In the HISTORY of S America we cannot dive into the causes of
061r126 er whole world. «(Was it so in Sydney, consult HISTORY? Phillips.» 1826.27.28. grt. drought at Sydney
062r129 st arrives:-- New Zealand rats offering in the HISTORY of rats, in the antipodes a parallel case.-- S
178b030 to Miocene of Europe? Loudon. Journal. of Nat HISTORY.-- July. 1837. Eyton of Hybrids propagating fr
182b047 & America. &c &c &c The new system of Natural HISTORY will be to describe limits of form. (& where p
201b126 ditional illustration of that axiom in Natural HISTORY that all aberrant & osculant groups are not on
208b154 between every species.-- Who can find trace or HISTORY of species between Indian cow with hump & Comm
212b167 respecting the Water Rat.-- ¿Consult Dr Smith HISTORY of S. African Cattle Phillips Geology «p 81» i
218b190 nging hair-- The Edinburgh. Journal of Natural HISTORY-- Preface appeared good with facts about chang
226b220 l worth while to study profoundly the origin & HISTORY of every terrestrial Mammalia.-- Especially mo
227b224 ata If my theory true, we get (I) a horizontal HISTORY of earth «within recent times.». & many curiou
232b245 undergo indefinite change.. (marking in their HISTORY an eocene miocene & pliocene epoch), whilst ot
256c055 n between Man & animal, small point in tracing HISTORY of Man.-- granted.--but if all other animals h
268c096 e Linnaean Soc. Feb. 1838.-- Annals of Natural HISTORY. «vol I» p. 159 curious account of Tit mouse f
298c189 curious in so wild a bird.-- Annals of Natural HISTORY. Vol. I. p. 185 case of tit lark placing wither
310c225 undine geological chronology Annals of Natural HISTORY «Vol I??» p. 318. some remarks on Bonaparte's
318c253 hman tells me in Audubon there is most curious HISTORY of first appearance of the S. American Pipra F
319c257 gret contrast with South America.-- In Home's HISTORY of Man at Maer, it is said the Samoyed women (
320c275 ogy & Botany & Continuation «Annals of Natural HISTORY» Skimmed Von Buch Travels Whites Natural Histo
321c275 story» Skimmed Von Buch Travels Whites Natural HISTORY of Selbourne References at end Dr. Lang Austra
321c275 excellent table of Canary isld: Plants Home's HISTORY of Man Transactions of the Entomological Socie
322c270 st on L'Ile Julie Waterton's Essays on Natural HISTORY. Octob 2d Transactions of Royal Irish Academy.
322c269 courses Lessing's Laocoon Whewells-- inductive HISTORY.-- References at end of each Vol Herschels. in
323c269 Travels in N America May 18th Stanley familiar HISTORY of Birds -- Mackintoshs' Ethical Philosophy --
324c268 on instincts L. Jenyns paper in Annals of Nat. HISTORY Prichard.-- Lawrence Bory St. Vincent. Vol III
352d060 ch formation being merely a page torn out of a HISTORY, & the geologist being obliged to fill up the
353d061 e with man.-- September 3d Magazine of Natural HISTORY. 1838 vol II p. 402. Mr Gould on Australian bi
356d069 rit if merit there be in following work.-- The HISTORY of Medicine, the extraordinary effects of diff
379d151 ceases to be true avitism Annals of. Natural. HISTORY. p. 135. Natural History of the Caspian. Fresh
379d151 m Annals of. Natural. History. p. 135. Natural HISTORY of the Caspian. Fresh Water Fish!! ¿adapted to
386d167 hanges really in progress.-- Annals of Natural HISTORY. 1838. p. 123. Ehrenberg. makes gemmation in a
387d169 ce they themselves inherit./ Annals of Natural HISTORY .p. 96. Vol I. Notice the Syngnathus, or Pipe
398e005 ry of the genius of man Those who have studied HISTORY of the world most closely, & know the amounts
398e005 animal preserved.».-- the latter pages in the HISTORY are perfect, we obtain a glimpse only of the c
409e049 ll dispute, when I hear from the geologist the HISTORY, & from the Astronomer that the moon probably
411e054 lated on, & laws of life,-- the end of Natural HISTORY, will be approximated to.-- Treating of the fo
414e065 l., man is not an intruder.-- : the geological HISTORY of man is as perfect as the Elephant, if some
415e069 . The end of each volume of Whewells Inductive HISTORY. Contains many most valuable references See if
421e090 fertilized spawn of frogs.-- Annals of Natural HISTORY. (p 225. 1838). account of metamorphosis in th
429e113 s Nature's plan.-- Whether we can or not trace HISTORY of first appearance of varieties of domesticat
436e138 emen's land.-- Porphyries of Andes. A familiar HISTORY of Birds by the Rev. E. Stanley Vol I. p. 72.-
437e139 pped it & was found alive.-- Stanleys Familiar HISTORY of Birds several cases on record of stoats bei
449e169 alaya species -- leuconotes-- Magazine of Nat. HISTORY. 1839. p. 106.-- Waterhouse refers to fossil r
461t025 complex & some simplified.-- Annals of Natural HISTORY. <no. XII.> Vol. 2. p. 96 & p. 451. 1839-- Tra
481z010 killed in Chile. Dobrizhoffer., Vol I p. 310. HISTORY of the Abipones-- says «the Condor» <it> is fo
493q003 ffspring take after male parent & vice versâ = HISTORY of Tortoise-shell Cats. as only one sex so col
497q06v h wild hybrids have been formed. (15). What is HISTORY of Viburnum. or snow-ball-tree. what would res
502q11v are calves.: Compare young. beans. cabbages.-- HISTORY of Pheasant-fowl. Hen coloured like cock-pheas
505q014 plant wh require insects to impregnate it (7) HISTORY of Potato field-- (8) Abortive Thyme seeds weat
515q021 of plants; especially Papilionaceous order (2) HISTORY of fruit trees far north in Scotland-- do they
553m135 QUAL to the smallest casuistical doubts.-- The HISTORY of Metaphysics shows that such a view cannot
560mIBC e we any ferns in the hothouse at home Natural HISTORY of Babies-- Do babies start, (ie useless sudde
563n001 May 1873-- October 2d.. 1838 Essays on Natural HISTORY Waterton describes. pheasant springing from ne

619o042 kintosh Grotius has argued nearly so [RHC] The HISTORY of every race of man shows this, if we judge h
482z011 aptores. Falco poliosoma -- novozelandiae -- HISTRIONICUS Vultus aura Excessively inaccurate Saw a Ch
514q21. s Mormodes (one of the Catasetums) really always HIT stigma by projecting pollen-masses?-- = answered
129a131 bsided forests.-- Many repeated oscillations» HITCHCOCK Report on Massacuhssets. p. 133 The most wond
360d092 uin duck & Herberts law of ideosyncrasy I have HITHERTO thought that a small difference <of any kind>
419e087 » Extinction & transmutation, two foundations, HITHERTO confounded,, of geology.-- L'Institut 1838. p
505q014 p to get me some to look at:-- Was once offered a HIVE. of these small Bees-- at Sundorne has large Be
609o030 o on except for the moral sense, any more than a HIVE of Bees without their instincts.-- Gives art to
301c199 ither in same individual p. 7. is not squirrel HOARDING, & killing grains. acquirable through hoardin
301c199 hoarding, & killing grains. acquirable through HOARDING from short time.-- My theory must encounter a
622o049 ve said.-- [LHC] instinctive fear of death: of HOARDING.. Ld. Kames, which Sir. J. says is so ridicul
026r021 will the Cleavage be There is a resemblance at HOBART town between the older strata & the bottom of
207b152 on thinks Van Diemen's land long separated from HOBART Town-- {from difference of races of men and an
341d030 y developed to aid in breathing.-- Animals from HOBART Town mentioned, it seems most of species from
435e133 external> Physical agents". «translated by Dr. HODGKINS» p. 54. The axolotl, siren, & Proteus, affini
451e178 . p. 335. Catalogue of animals of Nepal by. B. HODGSON. p. 336 In the most pestiferous region (mentio
451e179 aid to be of a bolder & more generous temper-- HODGSON Koloff. voyage through the Moluccas 1825--"No
459t009 's law Gleanings of Science Vol III. p 320. Mr HODGSON on Musk Deer-- young spotted <like in> "prettx
074r165 ich H. calls by several secondary names «Study HOFFMANS account of steam acting on trachytes. also Az
450e174 ne part the northern peninsula of Borneo.-- Ox & HOG natives of Borneo Notices of Indian Arch. Singap
459t009 "pretty much as we see in the young of the wild HOG & of several species of deer, which are altogeth
639j28r vest mouse & (Chamaelion?) C. D. Spines in Hedge HOG & Echidna.. & Aphrodites C. D. Endless cases.--
220b198 Now if we take structure as criterion of species HOGS different species, dogs not, but if we take cha
220b198 dogs not, but if we take character of offspring. HOGS not different. some dogs different.-- Henslow s
224b212 any precise relation to structures Mem. Eyton's HOGS & dogs.-- The passage in last page explains tha
450e173 Voyage p. 323. Sooloo. imported elephants. wild HOGS-- spotted deer, no loonies, but cocatores & sma
451e180 .--" Chapt.-- V.-- : do. Chat XXI. Wild cattle & HOGS on Timor-land-- monkeys do not exist. there & i
452e181 n isld of Bunwood (18 miles in circum) there are HOGS & monkeys <at> near shore of Magindanao Journal
385d164 s bodies.-- Lavaters. Essays on Phy. transl by HOLCROFT Vol I. p. 195. says children resemble parents
556m145 Lavater's Essays on Physiognomy translated by HOLCROFT «Vol I» .p. 86 "We ought never to forget-- --
566n010 stral type of consciousness &c &c.-- Lavater. (HOLCROFT Translat) Vol III. p.37, quotes from Burke, w
286c156 lation in classification in books, ought they to HOLD,-- Birds having web-feet, where we see scarcely
339d025 donia. two races of Men, but not plants» will it HOLD good.-- Thinks Temmink doubtful when he says No
406e036 rity of offspring to Parents same laws appear to HOLD good. with regard to marriage of individuals, &
417e079 en to their parents.-- Lord Moreton's law cannot HOLD with fishes, «& there are mule fishes» & reptil
532m054 ng & told me there was nothing, & tried to seize HOLD of objects to be frightened at.-- (again diseas
580n061 kind of intellect is that when an idea once take HOLD of the mind, no subsequent ones modify it.-- «W
228b231 t like to consider our equals.-- «Do not slave HOLDERS wish to make the black man other kind?» Animal
044r074 arb: A. Mur: A. = (& this effect of water thus HOLDING matter in solution must be great: & in the fac
414e065 is as perfect as the Elephant, if some genus. HOLDING same relation as Mastodon to Man. were to be d
473s03v negative laws of America of depth of organisms HOLDING in America as in Britain. If there has been «a
520m002 n odd twiching of muscles, & general manner of HOLDING hands &c &c.-- Mr Dryden Co said he could not
218b189 a «real» repulsion «amounting to impossibility, HOLDS good in plants» between all different forms; th
222b203 nt stock.-- I think we may look at it so--?? It HOLDS good even with trifling differences of expressi
244c022 ar«i»ensis (from Buenos Ayres) replaces <Vesp.> HOLDS same relation with equator--that Vesp. lasiurus
281c138 in absolute human family is undescribable, yet HOLDS good, so does it in real classification. The rel
296c185 II. Magazine of Zoology p. 56. Peregrine Falcon HOLDS birds for some time alive ¿therefore other spec
417e079 «What is cause of this.-- » (Lord Moretons law HOLDS with different species, & individuals of same s
429e114 plants, we see how full nature. how firmly each HOLDS its place.-- When we hear from authors (Ramond.
509q017 important to know, whether Gould's observation HOLDS good, that in the mundane genera, the species <
542m096 barks puts his lips in peculiar position, & he HOLDS them this way, when opening mouth between inter
593n109 s [blank] will not do for insects. if this view HOLDS good-- then man, a socialist, does not know oth
437e139 s with them, and one day a rabbit escaped into a HOLE, where the old Eagle could not find it..-- The
293c172 ng to tell back from front. &c or use of button HOLES it would be instinctive.-- My view of instinct
022r008 y's travels also» Dampier's last voyage to New HOLLAND P 127.--Caught a shark 11 ft long. "Its maw wa
024r015 urrents Find instances; The whole coast of New HOLLAND shoals much: Dampier remarks on great flats on
072r159 owing up Hill; ¿what does he mean?) Consult Dr HOLLAND about bubbles. -- No Volcanic action on coast
092a030 eograp Journal Earthquake at Melville Isld New HOLLAND Augus 1d to 3d & 19 1827 Geograp Journ There a
190b076 iarities of Flora. on East & West. ends of New HOLLAND. diminishing towards centre (p. 586)-- Paralle
206b150 States Cryptogam. Flora formerly common to New HOLLAND?! p. 320. Says Coniferous structure intermedia
231b242 ory birds, he told me some story of crane from New HOLLAND!!! in stomach --or in feathers--seeds.-- Two i
232b249 anner. Cuscus, (a sub genus of Phalangista New HOLLAND form) is found in many island Celebes «Waggiou
240c013 er in «form & head &» colour from those of New HOLLAND.-- The New Holland species are not found in th
240c013 &» colour from those of New Holland.-- The New HOLLAND species are not found in the Archipelago-- For
240c014 ts, establishes its «zoolog» alliance with New HOLLAND. The Barbaroussa, (when young very like the Si
243c020 t-- L. Jenyns, about my fish New Zealand & New HOLLAND fish very similar.-- NB. Lesson method of gene
246c027 ly New Guinea (replaces, Emeu) in North of New HOLLAND.-- New Guinea scarcely differs more from, <Van
269c102 I. p. 17 &c excellent sketch of plants of New HOLLAND, supplementary to Appendix to Flinders Voyage
269c102 -- Mem. Tartary & China.--, both coasts of New HOLLAND.-- «Compare birds of Australia with plants, wi
314c238 , N. Zealand & N. Caledonia with a dash of New HOLLAND. As in N. Zealand-- Some species of Australian
408e046 to see how many Tropical genera come from New HOLLAND, ¿Sydney? The dog being so much more intellect
453e183 Where has Duchesne described Atavism.-- ask Dr HOLLAND cases where peculiarity has first appeared.--
453e183 e Vegetale". by Gallesio. Pisa 1816 p. 27. Dr. HOLLAND. Are there instances of plants, in becoming do
489qIFC slow (2d time) p. 14.-- Father. And. Smith Dr. HOLLAND p. 16 Babington-- Gould ---- 10.(a) J. Gray --
490q001 s Duchesne described Atavism alluded to by Dr. HOLLAND-- <Jordan> Smith of Jordan Hill-- character of
507q016 Plant seeds of the Fuller's plants ,Teazle Dr. HOLLAND ; My Father. Andrew Smith (1) are cross-births
508q016 ze book Gallesio storia del Reproduzione.-- D. HOLLAND (7) Is Haemorragic tendency, independent of he
550m127 s one not call them instinctive emotions?-- Dr HOLLAND remarked that insanity like sleep does not dou
265c083 t yet there is no record of any effect.-- New HOLLANDERS have gone on boring their noses. &c & This c
593n111 n ideal argument held in one's own mind, & Dr. HOLLANDS story of man in Delirium tremens hearing othe
100a054 cretions in Pumice bed at Ascension instance of HOLLOW concretions & concretion filled with unconsoli
514q21. nswered = Has Ophrys nectary?= Bunbury says no «HOLLOW» spur.-- Ask about Pinks & Solanum impregnatio
154g061 end by flights the terraces if the largest has HOLLOWED out most Wednesday Shelf 3d dies away almost
072r160 Spitzbergen--Spitzbergen animals (?). ≠ The HOLLOWNESS of <sep> Chiloe concretions somewhat analogo
496c006 l them in hour or two «My Father made hens cast HOLLY-seed & they grew» (9) Place. Snap-Dragon. (I ha
534m063 iolum undergoes transformation in the stem of HOLLYHOCK, although ordinary Habitat is Malva sylvestri
453e183 &c? Mr Herbert in letter says distinctly, that HOLLYOAK reproduce each other. & yet I presume seed ia
453e183 with the esculent vegetables-- how is it with HOLLYOAKS, flaxes &c &c? Mr Herbert in letter says dist
200b124 p. 155. about quantities of seeds in sea; also HOLMAN: <at> Keeling these are most important facts.-
634j54v the phalanges have separate movements in the HOLOCENTRUS ruber (a fish) Man has abortive muscles to h
504q014 erries germinate?-- Yew trees sexes-- (3) Get HOLYHOAKS. races planted & Linum Perenne.-- Herbert's.
492q002 h other. when close planted together: <do> Can HOLYOAK be raised distinct by seed-- Heartease. 6. --
439e144 instinct-- the possibility of rearing by seeds HOLYOAKS-- (how far is this so) shows either there is
146g009 rt must project [Blank] {P} Path East End near HOLYROOD Palace In same way at top the trap could be t
065r138 --Sandwich Isd-- Specimens of rocks were brought HOME in Capt. Forster expedition from <Deception Isl
277c127 done least part.-- that he will not have brought HOME new species. until, he can show range & habits-
437e141 returning suddenly to type when brought back to HOME. (& yet all the varieties of Brassica certainly
525m023 ng.-- as eating meat., doing their dirt, running HOME.-- in these cases their actions do not look lik
534m062 es compares this with pidgeons finding their way HOME-- there is something wrong in comparing these c
557m149 ss not so.-- affected laughter.-- A dog who goes HOME from shooting. runs away. is not afraid the who
560m156 for men.-- Have we any ferns in the hothouse at HOME Natural History of Babies-- Do babies start, (i
615o36v organ.-- I think Pincher shows surprise, walking HOME one day met him, with Mark riding instantly fol
319c257 eautiful gret contrast with South America.-- In HOME'S History of Man at Maer, it is said the Samoyed
321c275 pendix & excellent table of Canary isld: Plants HOME'S History of Man Transactions of the Entomologic
599o008 raison, beau ideal, infini conscience; voila l'HOMME separe de la matiere et du temps! voila les fac
325c267 articuliere des Anomalies de l'organization des HOMMES. & les animaux.-- by Isid. Geoffroy. St. Hilai
342d033 incts not imperfect.-- Are Pheasant & Grouse HOMOGENEOUS? I observe Bachman calls these Hybrids new s
511q018 etween Trout & Salmon were fertile & whether HOMOGENEOUS {About German ornithologists, Bhem & Glöger

115a096 oms in slates that cleave, & which unite the HOMOGENIOUS crystals., must aid in adding effects to com
256c053 now. languages. would surely have been more HOMOGENIOUS.-- There must be some sophism. in Lyells sta
312c228 like spaniel even when grown up.-- Are mules HOMOGENIOUS owing to no attempt to keep up offspring, ar
437e138 Turtles finding their way to the Caymans from HONDURAS. good case of migrating.-- shows my theory in
428e113 sex on locomotive system» I am bound to insist HONESTLY that the sudden, change from Primrose to Cows
443e156 gist, I have thus argued to myself, till I can HONESTLY reject such false reasoning Bell Bridgewater'
324c268 lys introduction to the Natural system Bevan on HONEY Bee Dutrochet Memoires sur les Vegetaux et anim
370d116 me work. (it is said that some kind lay <pu> up HONEY even for single rainy day-- & from case of wasp
470t178 a--, Ranunculus in numbers =what insect can get HONEY out of long, curved nectar of Butterfly Orchis
495q05a y fly.-- (9) I have noticed leaves covered with HONEY-dew dusted with pollen of neighbouring grass= S
496q05a flower-- put artificial flowers-- also do with HONEY-- What is use of Bee Larkspur= =Toad Orchis= Ho
525m023 h only acquired rules by art.-- like the law of HONOUR.-- they feel pleasure in obeying their instinc
622o048 ound him, will be paramount,-- hence the law of HONOUR. & the etiquettes of Society.-- [LHC] Sir J. M
622o048 Sir J. M. gives different explanation of law of HONOUR from Paley [RHC] Anyone, who will reflect must
628o54v some fallacy here.-- at least point of «false» HONOUR will stop all wish to gratify <it>-- anything
555m144 head cut off) as kind of wit, showing he had HONOURABLE wounds.-- all this was kind of wit.-- I chan
306c212 eloped, just like, habits easily gained in child HOOD.-- Young salmons. first a species which lived i
314c237 in other species, this segment is converted into HOOD which possesses power of movement. & not the or
466t094 ich stamen & pistils have no relation. In Monk's HOOD, a bee entering long nectary, would «necessary»
470t178 all wh. are in flower «I saw Bees;»-- on Monk's HOOD, brushing over stamen «Egg Tree»--I think never
424e101 - -- Lyell tells me, on authority of Beck, that HOODED crow & Carrion crow. have in Europe different
426e105 ect species; but facts of grouse, & pheasant, & HOODED crow goes against this. & wild hybrid plants.
225b217 ake Indian Cow with bump & pigs foot with cloven HOOF) Ask Entomologists whether they know of any cas
235b272 on influence of climates, situations &c on 242. HOOK Smellie Philos of Zoolog. 842 White regular gra
507q15v eeds Gaertner de fruct.-- for woodcut-- 1 double HOOK-- -- Geum. Galium Burrh ≡ single hook; curved s
507q15v - 1 double hook---- Geum. Galium Burrh ≡ single HOOK; curved spines-- simple spines-- or seed-cases
505q014 how crossing & ¿heredetary? (15) Abberley has a HOOKED Pea.-- intends to breed from it and large Aspa
296c184 fact speaks volumes. 2 Chapters. translated by HOOKER.-- my theory explains this. but no other will.
489qIFC an Smith. p 1. Sowerby Cuming. -- p 1 Owen p 17 HOOKER p. 17 {T} Mrs. Whitby. Newlands Lymington Hant
490q001 s Cardium-- does it remind him of other species HOOKER says the species of Aquilegia vary much in the
499q09v asily to be crossed & all monooecious plants.-- HOOKER says Rafflesia is dioecious & Pollen must be c
501q10a Europe, Walnut in America.-- Heaths in Africa; HOOKER? are these genera less difficult, in other cou
510q017 e not worked have any peculiar configuration.-- HOOKER <Meta> Metaphysics of Morphology. ¶-- Schelgel
499q009 (19) Theory of mock flowers in Hydrangea (20) As HOP is Dioecious-- seedsmen who raise Hop-seed-- may
499q009 ea (20) As Hop is Dioecious-- seedsmen who raise HOP-seed-- may know something about proportion of pl
024r015 Beechey. on NW coast of America off Cape of Good HOPE 70 fathoms 20 miles from the shore? Beagle Coas
037r056 nos to Copiapo.--Sydney. K. G. Sound. C. of Good HOPE.--Carnatic It has been common practice of geolo
064r134 struck with number of animal[s] at Cape of Good HOPE Says at Santos «M Birchels» at foot of range so
079r177 bia at Quito. P 281. do do Australia, C. of Good HOPE.--Azores Isds «nor at St Helena.--» Humboldt. N
114a093 ous Gallionella Examine Iron stone of C. of Good HOPE & Australia/ and mud of salt-lakes of Rio Negro
114a094 in metamorphic action which I give at C. of Good HOPE.-- A bare hill of greenstone, if we know origin
174b014 uld be as it is.-- Hence Antelopes at C. of Good HOPE-- Marsupials. at Australia-- Will this apply to
178b032 Men & Hottentots or Negros cross at C. of Good. HOPE the children cannot be made intermediate, the f
219b194 re any species same as T. del Fuego & C. of Good HOPE show possibility of transport. If some cannot b
225b219 ficulty-- whence elevation Subsidence New is only HOPE.-- New Zealand «compare to Van Diemen's land.»
232b248 & lizards do.-- Ask Eyton to procure me some Get HOPE to give me an account of parasitic animals of b
233b251 iles same from Maurice & Madagascar & C. of Good HOPE.-- His book Probably worth studying.-- Wingless
247c028 «caught» Chile, Van Diemen's land & Cape of Good HOPE V. p. 44 of this Note Book Rabbits introduced i
266c091 reatise, (p..26). about plants from Cape of Good HOPE continuing for some time to flower at their own
312c233 are thought long breeding in. Must not trust him HOPE says that genus of parasite to genus of animals
319c257 m of Europe is between the Anteater of C of Good HOPE & those of S. America,-- Are not some of the Au
399e011 re,, but are unknown in Hervey or Society isles.-- HOPE says positively he has seen. a Calosoma. (very
399e011 «Lake» Constance species of Europaean genera=.-- HOPE has ideas about generic characters. dominant. p
409e049 on other» hence not social instincts, which as I HOPE to show is «probably» the foundation of all tha
449e173 in: 1839. p. 451. Sheep Merinos from Cape of Good HOPE., has different constitution from those of Euro
491q002 produce flowers, like the Oxalis from C. of Good HOPE mentioned by Mr Herbert in vol IV. Hort. Transa
503q012 yle's productive Resources Book no information & HOPE about Silk worms. Varieties effects of domestic
549m121 little is there said of intellectual <ple [...] HOPE> cultivation, main source of the intense happin
565n009 close those muscles, which wrinkle when smile.-- HOPE is the expectant eye. looking to distant object
289c162 thout coitus, than with.» ¶Might be given as a HOPELESS difficulty, except as distinct creation.-- Ge
304c206 families & not all organs blending away.-- +++ HOPELESS work to systematist, who believed that all hi
600o008 ed in life time]CD 3. The Infinite, -- lives by HOPES, looks to eternity. (4) Reason, some transcende
103a065 t against great bodies of vapour. according to HOPKINS theory.-- general presence of dikes. argues in
118a103 > Chalk When we consider parallelism of dikes (HOPKINS) & that every dike. which has not formed volca
120a107 he sea?) All this profoundly considered. study HOPKINS. theory of dikes may throw some light.-- thin
132a137 e, with that of the passage of the moon.-- Ask HOPKINS. M. Parrot, Mem. Acad. Imp. des Sciences. (Sc
135a145 m each side intercepting plain & dividing it-- HOPKINS fissure at {P} .-- G. J. Malcolmson has descri
236bIBC le Geographie Botanique De Candoelle. Geol. Soc HORAE Entomologicae Linn: Soc. Geoffry-St. Hilaire Phi
272c108 any Crysomelidae, with similar habits. But the HORAE Entomologicae will tell this.-- What peculiar c
321c275 nd Dr. Lang Australia «trash» skimmed Macleay's HORAE Entomologica Ray's Wisdom of God references at
485z018 - Berenica &c &c L'Institut, 1838 p. 46 Macleay HORAE Entomolo. insects swarm in Lapland & Stizberge
023r011 t like appearance «arising from the manner of HORIZONTAL upheaval» of the shore of the Pacifick is 60
069r149 sensible. Quantity of matter from Cordillera. HORIZONTAL movement of fluid matter not (for instance)
098a048 ess passages from gneiss to granite): Why not HORIZONTAL? Why have particles in such cases moved more
102a062 y.-- As volcanic eruptions are accompanied by HORIZONTAL elevations, so are injection of mountain cha
103a063 - Hence fill up fissures-- If dikes effect of HORIZONTAL elevation excepting fissures from above unit
103a065 level.-- Dikes being last action. (effect of HORIZONTAL movement) hence generally intersect metallic
108a080 we should not expect volcanos.-- not so much HORIZONTAL oscillation. or so many shocks directly afte
115a096 h the exception of sandstone rare to have any HORIZONTAL non cleaving beds. metamorphosed. The chemic
147g015 sand in current cleavage-- in HORIZONTAL strata I suppose these upper patches if prol
158g079 int very nearly like head of Glen Guoy nor is HORIZONTAL line apparently continuation of upper terrac
227b224 Pachydermata If my theory true, we get (I) a HORIZONTAL history of earth «within recent times.». & m
275c119 tions in time «& forms» distribution tells of HORIZONTAL barriers-- Mr Yarrell.-- says my view of var
405e031 ky> blackish brown.-- yet retained a trace of HORIZONTAL mark on flank.; & tail. & kind of semilunar
436e135 t feathers retain character? If separation in HORIZONTAL direction is far more efficient in making sp
230b239 geologically (& then monuments imperfect) or HORIZONTALLY & then cross breeding prevents perfect chan
075r167 ted silver, arsenical grey copper, and antimony, HORN silver, black silver & red silver, do not name
512q019 r compassion with hybrids, is offspring of short-HORN bull & hereford cow similar to reverse cross.--
513q020 mongst varieties cross one with abortive tail or HORN, with another & see result, for comparison with
595n117 er> <foame> & sweat, when hearing merely hunting HORN-- association or imagination [blank] [not locat
078r176 are destitute of quartz, & wh abound both in HORNBLEND & vitreous felspar".-- p. 215 Same metal in T
088a015 Fragments of slate converted into crystals of HORNBLENDE p. 248. L. Institut 1837.-- Helms remark on
110a086 tal Norway case.-- The fragment. consisted of HORNBLENDE (?) & felspar, (some crystals being red) «wi
110a086 ish feldspathes with grenish. black specks of HORNBLENDE, large irregular cryst of reddish felspar. &
110a086 felspar. & scales. of mica.-- large cryst of HORNBLENDE blending into base-- Salband might have ooze
581n065 progressively formed. (--names like sounds)--. HORNE Tookes tenses, &c &c -- <also g> if so & seeing
205b145 ominates," p. 20. do "If hornless ram be put to HORNED ewe almost all the lambs will be hornless.--"
229b233 lluscs coupling.-- Dr. Smith's Information Long HORNED (very) aboriginal at Cape: crossed with Englis
345d042 deserves <name> «to be so called».-- the short HORNED cattle have gone on for 50 or 70? years-- now
345d043 ery now & then born with white head (,or «short-HORNED with» black lip) & then calf «in both cases» i
432e123 ustrated by the crossing of hornless sheep with HORNED.-- compare this with what highland shepherds s
498q007 s-- is this ever effect of want of nutrition.-- HORNED oranges so? --Yes, my Father lost this charact
500q010 es-- Asiatic Researches) (23) Talk about Thyme. HORNED Oranges. Spallanzani Essay-- Figs 2 kinds of f
505q014 nis dioica answers this question= (5) Open more HORNED oranges.= (6) Figs, flower.--Passion Flower. (
505q014 green Cynoglossum. never germinated 12 Does the HORNED orange. wh. never has seeds produced good poll
515q021 les of seed, as the commoner kinds-- Cattle are HORNED, Suffolk have <abortive> «no» horns by abortio
509q017 n combless Poultry-- Teeth in foetal state: Mr. HORNER. On Mr Tremenheres Scottish Colliers, when men
205b145 e more frequently predominates," p. 20. do "If HORNLESS ram be put to horned ewe almost all the lambs

205b145 put to horned ewe almost all the lambs will be HORNLESS.--" does this apply to where same animal bree
403e023 c animals few. that have not <which> not, cows HORNLESS, (horses not) If they give up infertility in
429e113 imals.)-- Azara gives account of production of HORNLESS cattle-- ¿& others?-- March 12th-- It is diff
432e123 er than mother; illustrated by the crossing of HORNLESS sheep with horned.-- compare this with what h
234b255 ct, that those of Iceland. are seldom seen with HORNS" --- p. 341. Black Fox sometimes introduced by
362d099 e to the victorious stag. who rubs the skin off HORNS to fight-- is analogous to the love of woman (a
362d099 n savages) to brave men.-- Effect of castration HORNS drop off., replaced by hairy ones. which never
362d099 change in 5 or 6 weeks after castration, fresh HORNS begin to grow.-- Mr Yarrell says the «male» Axi
362d099 the «male» Axis of India, breeds at times when HORNS not perfect-- (is not this so in S. America wit
363d101 (I suspect Pennant has described them)-- [Study HORNS of wild cattle.-- plumage of fowls-- long ears
388d172 of cock, & beards growing on old women = Stags HORNS & testes curious instances of corelation in str
429e113 dency in external conditions» sudden loosing of HORNS.-- I do not believe this Nature's plan.-- Wheth
473s07r 2/-- June/42/ You can select cattle & sheep for HORNS & yet no difference in calves--how is this in y
481z010 ittella, or Fleche «p. 8» my little animal with HORNS. Madrepores p. 26 Nullipora p. 29-- In Meyen. V
505q014 d good pollen? Yes «From cultivation lost their HORNS» is impregnation necessary to fruit--; become w
512q019 reeds of Cattle & sheep in the sprouting of the HORNS. at different periods in different breeds--?? o
515q021 Cattle are horned, Suffolk have <abortive> «no» HORNS by abortion, but sometimes have dangling ones.-
556m146 ing when old, like kittens at the breast now if HORNS were to grow on horses, they must yet continue
459t013 immaculate when grown up". Saw at Mr Bell's at HORNSEY the offspring of a black & white «duck of pecu
460t013 o much of the swan-goose in appearance Bell at HORNSEY (though only ¼ of blood). that it appears abou
637j58r t trash p. 237. Gives as Summary of adaptations HORNY point to chickens beak, to break egg. shells--
451e177 gu. <have> abound with monkeys & squirrels.-- HORSBRUGH E. I. Directory. Vol II. p. 46 Carimon Java.
451e177 several small islands. abounding with deer-- HORSBURGS. Vol II. p. 527.-- <Scientific Soci> Journal
048r085 ns to S. Patagonia. To the Megatherium.--To the HORSE. = One might fancy that it was so arranged from
057r113 nca & M. Hermoso contemp:.--Inculcate well that HORSE at least has not perished because too cold:--Wi
086a003 f one order. -- There should not be surprise at HORSE being found in America, when Mammoth & narrow t
179b035 ernal causes) does appear very probable:-- Mem: HORSE, Llama. &c &c-- If we <suppose> «grant» similar
187b062 y «Siberia a more» cold countries.-- Seeing how HORSE & Elephant reached S. America.-- explains how Z
187b062 s reached South Africa-- It is a wonderful fact HORSE, Elephant & Mastodon dying out about same time
189b072 y seed go on.-- otherwise all die.== The fossil HORSE, generated in S. Africa Zebra.-- & continued.--
196b106 l places & died off in some? Why did not fossil HORSE breed in S. America-- it will not do to say per
196b106 to say period unfavourable to large quadrupeds--HORSE not large-- Ed. New. Philosop J. No 3. p. 207 "
202b133 s.-- Now this is exception to law of type. like HORSE in S. America. or like living Edentata in Afric
205b142 Parasite of Negroes different from European.-- HORSE & Ox have different parasite in different clima
217b184 -- Fox told me of case of mare covered by blood HORSE & Carthorse two folds [not located] Mr Don gave
225b219 about dogs &c &c NB. Animals very remote. ass & HORSE. produce offspring exactly intermediate.-- Refe
226b222 nts where do most species occur.)> Although the HORSE has perished from S. America, the jaguar has be
239cIFC form of man. Could you get racehorse from Cart HORSE by picking without change of habits Mr Yarrell
290c163 cause crossed with women with pretty faces When HORSE goes a round, the minute gets into the road at
294c177 ce to succession of types ¿different species;-- HORSE-- &c <Lonsdale says. that first shee» State bro
325c267 tendril into stump Library of useful knowledge. HORSE, Cow. Sheep.-- Verey. Philosophie d'Histoire Na
326c266 1838 Harlaam. Physical & Medical Researches. on HORSE in. N. America-- Owen has it.-- Ld. Brougham.
327cIBC ogia commentatia Library of Useful Knowledge on HORSE & Cow & Sheep Clarke's Travels.-- Temminck Hist
332d004 climate-- .) -- July 23d. Eyton, a stone blind HORSE, seemed to perceive turn on road where No house
334d010 carter well, who placed his stallion as second HORSE between <who> shaft mare & another leader mare,
335d066 or donkeys to Zebras.-- «Mr Herberts variety of HORSE, dun-coloured with stripe approaches to ass.» o
390d179 ved that from Books to read Buffon Suites de.-- HORSE & Cattle Library of Useful Knowledge Bell's Qua
408e044 uckland. Reliquiae Diluvianae. p. 222. Bones of HORSE. Bear & Deer at 16000 ft. with Snow on Himmalay
424e103 «same way some plants vary more than others» & HORSE in lesser degree,-- how different to dog!-- (Hy
430e118 ountry: but. greyhound. & poutter Pidgeons «race-HORSE». have not been thus produced, but by training,
450e175 837. By J. H. 7 do. p. 189. «190» No full sized HORSE is found East of y Burramposter & S of Tropic--
450e175 vailing colours in the different islands.-- The HORSE is only found wild in the plains of Celebes. (b
450e175 original there)-- shows these isld not fit for HORSE. Forrest--. (p. 270) says many wild horses, bul
450e175 s, bullocks, & deer South part of Mildanao.-- Q HORSE do. Appendix. p. 43. «& 45» the Breed of elepha
459tf02 her of both progenitors-- the hinnus, resembles HORSE in its head ears, tail limbs-- in the mules, th
468t105 len-- «Thrips» about as large as bit of chopped HORSE hair with legs & take flight-- Yet we have cros
493q003 . from B. into A. as takes place in mules ass & HORSE-- important. {In crosses does male offspring ta
496q006 Gypsum. Magnesian Iron Rust Carb. of Ammonia.-- HORSE Urine &c &c on associated plants. when proporti
509q017 s abortion vary, according to Bentham's Remark. HORSE or cow.-- degree of soldering of tibia & fibula
512q020 Wombwalls were (4) About fertility of ass-zebra-HORSE= (4) About fertility of ass-zebra-horse= (5) Ab
513q020 s-zebra-horse= (4) About fertility of ass-zebra-HORSE= (5) About callosities on Camels-horses.- &c &c
523m015 er.-- This «N B. I have read paper somewhere on HORSE being insane at the sight of anything scarlet.-
533m059 ertain actions-- the difficulty of getting on a HORSE on the left side (not good example,) because le
540m085 young male & smelt its fingers. Seeing a dog & HORSE & man yawn, makes me feel how «much» all animal
551m129 shut or cad cries out "right." or Drinkwater's HORSE jumping when word Jump said-- I saw the ourang.
552m131 much intellect.-- mem: Yarrell's story of wheel HORSE in drays, scraping against cornice stone to cau
554m141 "reason" & "reasoning," & take instance of Dray HORSE going down hill.-- (argue sophism of associatio
554m141 ion. Kenyon, & then go on to show, that if Cart HORSE argued from this into a theory of friction & gr
556m146 ession showing real affinity in face of donkey, HORSE & zebra. when going to kick.-- Why does dog put
557m147 t ease has smooth brow contrary to wrinkled: (a HORSE when winnowing & pleased pricks his ears?--).--
565n007 e becomes instinctive to fear., as ears down to HORSE.-- Horse snuffs «& snorts», the air «& raises i
565n007 instinctive to fear., as ears down to horse.-- HORSE snuffs «& snorts», the air «& raises its head,
584n072 s of memory of roads long after once visited by HORSE & dogs. (even blind horses & dogs) shows it is
588n090 & actions of a young animal with an old.-- (dog HORSE, sow) we perceive great difference.-- «(& is no
592n103 use nature intends to lay open all senses: <do> HORSE prick his ears «& snort clears nostrils» when f
621o047 ty, which the natural tastes say is good. yet HORSEFLESH show that even this is possible.-- So that a
063r133 here the division is not perfect.-- Dogs. Cats. HORSES. Cattle. Goat. Asses. have all run wild & bred
235b274 ut Indian Cattle with humps.-- ¿To be solved if HORSES sent to India. & long bred in & no new ones in
251c041 ictory to Azaras fact of conduct of wild & tame HORSES.-- p. 246-- Gmnura-- new genus of Mam: found i
290c163 rests for beautiful birds.-- heredetary ambling HORSES, (if not looked at as instinctive) then must b
297c186 ated form probably a remnant.-- Pachydermata. & HORSES few forms. & they are remnants.-- Cephalopoda
311c227 Vol IX p. 107. an Ascaris inhabits the eyes of HORSES in India in which it may be seen swimming abou
383d159 developement of either.-- (Mammae or sheath of HORSES poenis reduced to extreme degree of abortion).
403e023 few. that have not <which> not, cows hornless, (HORSES not) If they give up infertility in largest se
406e036 ew-Phil. Transact. Rabies, common to men, dogs, HORSES cows, pigs & sheep.-- diseases common to men &
428e112 ers who only care for first generations,, as in HORSES, would not care so much about breed.-- what ca
428e113 out indelibleness, than the number of good race-HORSES, which Eclipse? has begotten <?> «Walker attri
433e127 ing, far more wonderful than heredetary ambling HORSES. Whether the body of parent be altered, that i
450e175 t for horse. Forrest--. (p. 270) says many wild HORSES, bullocks, & deer South part of Mildanao.-- Q
458t002 - Beechey's Voyage Vol I. p. 499. «4to. Edit»-- HORSES in Lao Choo so small, that person with long le
461t037 «(».. as mammae of men «callosities on Camels & HORSES--».--«*)» & therefore probably any structure wo
466t095 at corresponding age & sex, opposed by cantering HORSES having colts which can canter-- & DOGS trained
513q020 ss-zebra-horse= (5) About callosities on Camels-HORSES. &c &c Rhinoceros= (6) Cross. Sus Barlyroussa
515q022 Oct. 44 Tell J. Anderson's statement of English HORSES having fewer vertebrae in tail, than Continent
515q022 aving fewer vertebrae in tail, than Continental HORSES. {About the leaping of Irish Horses, bred in t
515q022 Continental horses. {About the leaping of Irish HORSES, bred in this country. {Chinese Dog's Head to
536m071 og smell posterior at another.-- Why do bulls & HORSES, animals of different orders turn up their nos
536m071 .:. Dr Darwin's theory probably wrong, otherwise HORSES would have idea of beautiful forms.-- With res
546m108 pleasure. August 26th. I cannot help. thinking HORSES admire a wide prospect.-- The very superiority
551m129 defiance.-- different from sneer-- How easily. HORSES associate sounds may be seen by omnibuss Horse
551m129 horses associate sounds may be seen by omnibuss HORSES starting, when door shut or cad cries out "rig
556m144 know shame-- dog knows triumph.--» Sept. 23rd. HORSES in Omnibus instantly start when they hear read
556m146 - is there not protrusion of chin, like bulls & HORSES.-- 1838 good instance of useless muscular tric
556m146 tens at the breast now if horns were to grow on HORSES, they must yet continue to put down ears, when
582n067 rt-lived constitutions will then be cut off.-- <HORSES> Colts cantering in S. America capital instanc
584n072 after once visited by horse & dogs. (even blind HORSES & dogs) shows it is somewhat analogous to memo
585n075 never puts down its ear. good to contrast with HORSES, asses, <mi> Zebras &c &c.-- Here there is kic

595n117 it shows similarity in mind.-- think of Eyton's HORSES becoming <white> with <lather> <foame> & sweat
614o037 oes not depend on its length.: Many animals (as HORSES) very long & good memories-- but on its multip
211b164 my idea of double line of intersection.-- Dr. HORSFIELD At India House, collection of Birds from Java
242c017 an forms in East. Ind: Archipelago. ⟨Raffles. HORSFIELD. Diard. Duvaucel. Leschenault Kuhl. Van-Hasse
251c041 studying.-- «do» Zoolog Journal Vol 2. p 221. HORSFIELD on two bears very close species, inhabiting B
324c268 o authors about E. Indian Islands. consult Dr HORSFIELD Silliman's Journal Rengger on Mammalia of Par
489qIFC) Questions & Experiments {T} Gowen, Royle, & HORSFIELD Sykes p. 12 Maer. p. 13 Question &c. July. 18
503q012 information about pollen of Subularia Royle & HORSFIELD (35) Talk about races of Banana & yet seedles
334d011 been entirely broken from their mares, (though HORSING every month) & worked in the same cart in loos
327c265 vation of Fruit trees in. N. America" in Lib. of HORT. Soc Mr Neil. has written good article on Horti
429e114 its place.-- When we hear from authors (Ramond. HORT. Transact Vol I. p. 17 Append) that in the Pyre
429e115 contest. & a grain of sand turns the balance.-- HORT. Transact Vol I. M. Ramond. p. 19. do says loft
447e165 isely analogous to the Canada onion mentioned in HORT. Transact. Aira caespitosa becomes vivaparous o
454e184 n» as must be inferred from what Mr Knight says. HORT. Transat. V. II p. 252. Is there any very sleep
491q01v oodpidgeon & Hen. mentioned by Mr Knight. Vol IV HORT. Transact.-- 4 May we no suppose, that certain
491q002 of Good Hope mentioned by Mr Herbert in vol IV. HORT. Transact.-- 4.. Are any varieties of Cabbages
492q003 quoted by (as if oral) Decandoelle in V. Vol of HORT. Transacts & M. Sageret is referred to with dou
218b191 ransported.) Mr Herbert's papers are in the HORTICULTURAL Transactions and a distinct work on Hybridi
327c265 ished German Pamphlet on crossing Oats, &c «HORTICULTURAL Transacts.--» Mr Coxe "Views of the Cultiva
507q15v into Rye.-- 46). Book describing amount of HORTICULTURAL Variation? Henslow knows only on Citrons 47
327c265 rt. Soc Mr Neil. has written good article on HORTICULTURE in Edinburgh. Encyclop.-- The <Edin> Britis
538m079 «both bad jokers.--» the other army officer, HORTICULTURE & religious sects.-- yet Allen. W. remark a
489qIFC 20 & Breeders Dr. Boott: R. Brown p. 21 HORTICULTURISTS p. 21--23 Eyton p. 22 Schomburgk.---- 1 Jo
514q021 here-- Lindley says that only one pineaple HORTICULTURISTS (1) Are sterile hybrids healthy: number of
326c265 iven list of Spontaneous Hybrids. where? Sweet. HORTUS Britannicus. has remarks on acclimatizing of P
523m016 y of money, he was «only» NEARLY thrown into a HOSPITAL.-- My father was nearly drowned at High Ercal
314c239 ants, yet it seems there may be Eucalyptus!-- (HOSTILE fact) Be cautious about Goulds case of birds o
442e153 ements about fruit trees. grafted. altering is HOSTILE to this: but on other hand, fruit trees are pr
468t112 most roofed by united filaments.-- This flower HOSTILE to intermarriage!!xx In Phil Transact. about y
593n109 animal taking man in place of other animals is HOSTILE «is subversive of» to this view, & fowls hatch
025r017 e inactive 300 years? No Volcanic Earthquakes or HOT Springs in T. del Fuego = The Wager's Earthquake
119a106 dike & approach other rocks. & trap at least as HOT as lava-- of which temperature is partly known--
120a108 cold rock.-- in volcano the pool is not deep. --HOT springs &c &c--then if so, thermometer show it c
120a108 age to rocks.--, but lava shows the rocks really HOT. & therefore I doubt the thermometer. Is not com
120a109 thermometer. Is not common salt more soluble in <HOT> cold than hot water with «-- especially if very
120a108 not common salt more soluble in <hot> cold than HOT water with «-- especially if very hot under high
120a109 cold than hot water with «-- especially if very HOT under high pressure.--» respect to formation of
126a123 thousands of years, that surface does not become HOT?-- this looks as if bad conductor-- III But equi
129a132 urgh New Phil. Journ 1838. several case given of HOT heads &c heat beneath the sea.-- CD[did not Beec
130a133 ce in sounding over a continent to fall across a HOT.--spring.-- Hot water would not lie. at bottom.-
130a133 ver a continent to fall across a hot.--spring.-- HOT water would not lie. at bottom.-- Surely we here
130a133 lie. at bottom.-- Surely we here have proofs of HOT bottom.-- Study Bishoofs Paper.-- Weelsted told
138a155 the consolidation of strata-- he heated sand red HOT & brine was boiling on the top-- [blank] Would r
232b246 of temperature of two countries» finding a very HOT day, «in one», oh we will take a day from the eq
258c060 would cause destruction.-- simile Man living in HOT countries, if continually crossed with people fr
259c062 ls.-- then argumen May be.-- subterranean lakes, HOT spring &c &c inhabited therefore mud wood be inh
316c245 n other plants.-- Many «Some» genera confined to HOT countries & many to cold.-- Hence latitude is mo
343d037 able to numbers of animals. as in changing from <HOT>. Warm to cold, damp to dry.-- Thus Tierra del F
392d180 or instances of Avitism Examine English weeds in HOT. Houses will they flower Make Hybrids with moths
424e103 dogs, which have run wild have, have done so in HOT countries.-- CD[Camel does not vary «one ought n
468t112 ea or Rhododendron xx after several gloomy days. HOT one, Bees almost P every minute to Fraxinella «&
501q011 (31) Ask Henslow for list of annuals to place in HOT house to see effect on generative organs of grea
505q014 roduced in individual plants 17 A dead-nettle in HOT-house. will it seed?-- (Skim through Penny Cyclo
530m049 enough to eat.-- There was blackbirds nest, near HOT-house at Shrewsbury, which the cat was seen by H
540m086 the North,-- the elevated table land of Peru the HOT plains of the Amazons & Brazil-- with the negros
575n045 from both, as when one burns end of nose with a HOT razor.-- joy <p> a mental pleasure. with pleasur
495q005 (4) Raise annuals or common English plants in HOTHOUSE & see what effect on organs of generation (5)
560m156 nkeys care for men.-- Have we any ferns in the HOTHOUSE a more Natural History of Babies-- Do babies
178b032 mith says he is certain that when White Men & HOTTENTOTS or Negros cross at C. of Good. Hope the chil
218b189 it even in men); the possibility of Caffers & HOTTENTOTS coexisting. proves this-- but when Man makes
229b233 h Bull. offspring very like common English.-- HOTTENTOTS say great tailed sheep aboriginal at Cape &
508q016 or in Killing the worst =or in dogs= (12) Do HOTTENTOTS generally resemble each other very closely,
126a123 d water did not percolate surface, would become HOTTER.-- hence temperature ought to increase rapidly
127a125 Siberia was once thawed. & hence. (when climate HOTTER) was cooled to greater depth.-- Now the <inf>
208b154 port about each genus having its parent type in HOTTER parts of world is monkey. peculiar to C. de Ve
291c168 is easy to see if South America grew very much HOTTER, then Brazilian species would migrate south wa
215b178 of Man mentioned in Loudons (analogue of Blood HOUND-- Bull. Soc. Geolog. 1834. p. 217. Java Fossils
216b183 . Mr Bell of Oxford St'-. had a very fine blood HOUND bitch which would never take the dog. But at la
275c120 re was a dash of blood, with whole form of grey HOUND.-- picking out finest of each litter & crossing
390d179 rence» with other sex. = The highest bred Blood-HOUND. would be infertile with highest bred of other
213b171 the same, May you not breed, ten thousand grey HOUNDS & will they not be greyhounds?-- Yarrell's rem
214b175 <to» on N. American Zoology-- A breed of Blood HOUNDS from Aston Hall close to Birmingham, and suppo
216b182 of races not mingling?-- In Foxes case of Blood HOUNDS, a little mingling would probably have been go
216b182 d probably have been good, namely such as blood HOUNDS from other parts of England. Mr Bell of Oxford
219b196 ncestuous intercourse. a parallel fact to Blood-HOUNDS. Before Attract of Gravity discovered. it migh
511q018 breeding. pointers. Bull-Dogs. Spaniels-- Grey-HOUNDS-- is there ever any degeneration?? HOUNDS. Eyt
511q018 Grey-hounds-- is there ever any degeneration?? HOUNDS. Eyton Mr Wynne, &c Could by selection a diffe
512q019 etshire sheep migration of coots-- variation in HOUNDS= An ugly calf <turns> sometimes turns into fin
515q022 muscles of stumps of tailess dogs & cats.-- (3) HOUNDS-- varying-- (4) About blended instincts of the
565n007 perceive if in night trys to listen to growl of HOUNDS». <when> as fear to «man as» animals. comes at
496q006 Spawn, eggs of snail, row of fish & kill them in HOUR or two «My Father made hens cast Holly-seed & t
610o032 e often repeated in a wonderful manner.-- as the HOUR of the day &c-- All habits must conduce to thei
614o037 hould be transmitted from generation.; than from HOUR to hour in <man:> individual-- [LHC] Perhaps ev
614o037 transmitted from generation.; than from hour to HOUR in <man:> individual-- [LHC] Perhaps even the m
491q01v To apply pollen of different genus & then some HOURS afterwards of nearly related plant & see if fir
512q019 of birds-- Mem: how many miles they fly in few HOURS Zoological Soc (1) Do the animals there, someti
564n004 Coll. that he had seen chicken only hatched few HOURS placed on table & when fly ran past it. cocked
147g021 Highest part of road between Inverorum & King's HOUSE 28.935/82 degree A Temp of Air 65 degree? Gleno
150g037 unthine upper road (2) extends as far nearly as HOUSE, the 3d below them opposite to where side ravin
153g052 e normal condition of 4th shelf, some way below HOUSE of Glen Roy, seems to be which higher up on a
212b164 line of intersection.-- Dr. Horsfield At India HOUSE, collection of Birds from Java.-- at Leyden ser
248c029 ear. Galapagos Mouse not the same section, with HOUSE mice. It is wonderful how it could have been tr
333d005 alf Persian cat. which bred with unknown common HOUSE cat.-- had four Kittens. two appeared «so» very
364d104 has been thought that silver Pheasants about a HOUSE made other pheasants have white feathers).-- It
380d153 e Moth &c Mr Y. says that Macleay considers the HOUSE bug, as a <female which have> larvae which have
501q011 Ask Henslow for list of annuals to place in Hot HOUSE to see effect on generative organs of great Hea
503q013 reverse possible?? Maer (1) Yew Trees near Boat HOUSE «ANY male branch.» --¿number of seeds in beginn
505q014 not seeding= Put pot of boiled earth on top of HOUSE =Aristolochia, plant wh require insects to impr
505q014 ed in individual plants 17 A dead-nettle in Hot-HOUSE. will it seed?-- (Skim through Penny Cyclopaedi
530m049 h to eat.-- There was blackbirds nest, near Hot-HOUSE at Shrewsbury, which the cat was seen by Hubber
532m056 bed room-- grew very thin, would not go out of HOUSE except with Caroline-- After fortnight. continu
586n078 [one feels inclined to stop at right number of HOUSE though one cannot remember it.]CD back, without
595n115 ck» spider monkey brought it at opposite end of HOUSE. & commenced a most lamentable howls & & was no
620o046 a <spaniels> «housedog's» duty is to watch the HOUSE.-- it is part of <duty> their nature.-- When a
620o046 we say a pointer ought to stand a <spaniels> «HOUSEDOG'S» duty is to watch the house.-- it is part of
523m017 self from jealousy of husband connection with HOUSEMAID two years before, to prove she was not insane
085a001 t. Vol. 8. p. 288. Salt deposited on windows of HOUSES. & trees all injured on Eastern side, far inla
150g038 n opposite side of valley both extend below the HOUSES The Hills in this neighbourhood appear very ro

Page **************************************(Key Word)**************************************

```
151g043 by  rivers) & not to head of plain.-- but below HOUSES where rivulet enters two great projecting butr
155g063 Corry.  & as «as I believe in side ravine above HOUSES of Roy» Maccullochs supernumerary shelf I doub
332d004 horse, seemed to perceive turn on road where No HOUSES to Eaton Mascott, where he had been accustomed
392d180 tances of Avitism Examine English weeds in Hot. HOUSES will they flower Make Hybrids with moths, wher
478z004 ld dog copulates <.> freely with tame: comes to HOUSES on purpose Mr J. Murray has given paper to Roy
563n001 er says pea-hens do Wood pidgeons building near HOUSES. yet so shy at all other times.-- Birth Hill s
028r029 s matter accumulating in deep seas forms slates: HOW is the Lime separated; is it washed from the sol
028r029 more  probably by some unknown Volcanic process? HOW does it come that all Lime is not accumulated in
033r042 a familiar case: If refiltered with other matter HOW very curious a structure: Have shells ever casts
035r047 dent of mineralogical nature & dependent: & then HOW wonderful level «of same beds» should have been
042r067 e Chesil bank. V. De la Beche). Ask Capt. F.: R: HOW the swell, generally & during gales would tend t
042r067 rocks  of R Chupat. & fall of Ashes of Falkner, ¿HOW far is the distance?-- Fossil bones black as  if
045r076 infrequent;  so as to exclaim, «as I have heard» HOW lucky! when they hear of a place having a pretty
046r078 ur. salt. lime, are spread over «whole» surface; HOW comes it they do not flow out together? How  are
046r078 ace; how comes it they do not flow out together? HOW are they eliminated.--«Sulphur last.--» Metallic
051r095 aves: & action on upper tidal band, I do not see HOW to account for oceans power.--excepting when peb
054r102 covered  by limestone with recent shells 200 ft, HOW exact agreement with Coquimbo; [not located] [no
055r107 f St Jgo) yet no historical records of eruptions HOW immense the time!! How well agrees with number o
055r107 ical records of eruptions how immense the time!! HOW well agrees with number of Craters!--At S. Cruz.
056r111 ting causes by water.--Or rather begin & explain HOW water separates.--(intertropics at present fix l
060r125 Volc. had been in action during secondary period HOW diff. would the rocks have been. The red Sandsto
073r162 finitely rare in all parts of the globe". p. 113 HOW utterly incomprehensible that. if meteoric stones
074r165 fataras. «Mem: Micaceous iron ore.» N.B. To show HOW metals may be transported by complicated chemica
086a007 en Cuvier. Europe possessed a great edentata. -- HOW much is temperature of world regulated by atmosp
090a023 there were flights of stones of large numbers (& HOW few cases recorded if we say «100» <5>0 lbs a ye
090a023 ed if we say «100» <5>0 lbs a year too little.-- HOW comes it none in fossil state? suppose «100» <5>
092a029 is,  Mr Bird in paper to Brit. Assoc: has shewn HOW electrical currents tend to deposit metals, if i
102a060 scension, each particle coated. &c will be aware HOW little common Gravity has to do with arrangement
103a064 is,  I suspect much weakened by <vi> considering HOW close the dislocations occur & therefore that th
104a069 n, deep sea, fragments fall off cliffs. but then HOW spread abroad?-- There is thus wide difference b
105a072 reous matter & concretions on coast of Madeira.-- HOW came it if this powder results from «decomposed
112a089 in  the mountain masses we must look for that.-- HOW few isolated volcanos there are. where one alone
121a112 anhill has seen same thing-- Consider profoundly HOW came it. that Glen Roy district could have been
121a112 thout fissure & unequal.-- where were cracks?--? HOW came there ever to be cracks 11th August. 1838 N
122a114 ttoral mountains & volcanos.-- Why on one coast? HOW can Herschel consider figure of earth statical.-
125a121 o from surface towards the interior -- who knows HOW far that may have penetrated,-- lower down the t
126a122 equilibrium  is supposed to have been attained.» HOW much matter separates them, this is  ascertained
126a122 -- if we had rod thus & judged by increments at, HOW wrong, would our judgement be-- Does condensed m
126a123 , conduct heat better than plain?-- Mem 1000 {P} HOW easily water percolates rocks,-- when pressure i
126a123 id matter be driven upwards & so conduct heat?-- HOW comes it in volcanos that have gone on for thous
128a129 se. {P} land protuberant water to counterbalance HOW strongly the Glen Roy case shows that the figure
133a140 his  paper like a fool.-- Feb 25' All facts show HOW slowly heat travels; & therefore the abysses whe
133a140 in fluid for an enormous period: now when we see HOW many points have been penetrated by volcanic & t
146g013 Conglomerate  on S. Ventana, excellent instance, HOW accidental is the preservation in situ of even i
154g060 across  to {P} side removed all well & good, but HOW came river to do this vast quantity when  during
155g066 iss though wonderful-- <that they are preserved> HOW much more so, these lines & even water-scooped r
175b019 far as world has been uniform, at former epoch-- HOW is this Ehrenberg? every successive animal is br
177b028 r classes, perhaps a more linear arrangement.-- ¿HOW is it that there come aberant species in each ge
179b032 as nearly quite white) in the two first children HOW is this in West Indies «--:Humboldt. New Spain:-
183b051 two  species be created) & live in same country. HOW is propagation of wolf & Dog. (because being bel
186b057 e Waterhouse thinks Quinary only three elements) HOW far Does Waterhouse's representatives agrees wit
187b062 led by <Siberia a more> cold countries.-- Seeing HOW horse & Elephant reached S. America.-- explains
187b062 horse & Elephant reached S. America.-- explains HOW Zebras reached South Africa-- It is a wonderful
190b079 ys no real <separ> passage between good genera-- HOW remarkable spines, like on a porcupine on Echidn
194b095 n known only by bones.-- Mem: Silurian fossils: ¿HOW are. South American shells? ¿Do not plants, whic
194b096 o show all plants do receive intermixture.-- But HOW with «hermaphrodite» shells.!!!? We have not the
195b100 may be of use.-- like Mammae on mens' breasts.-- HOW does it come wandering birds. such sandpipers. n
195b101 reated with certain form in certain country, but HOW much more simple, & sublime power let attraction
199b116 ontinents we get a means of knowing movements.-- HOW can we understand excepting by propagation  that
204b140 offspring seldom intermediate between parents.-- HOW easily does Wolf & Dog cross? Mr Yarrel thinks o
208b154 ions, & consequently not many yet multiplied; NB HOW does this bear with law referred to by Richardso
210b158 ssed with others.-- Compares it to languages But HOW do plants cross?-- = admirable discussion Von Bu
213b171 , fish approaching to reptiles at Silurian age-- HOW long back have insects been known? As Gould rema
215b177 d milk--Fox tells me that it is generally said.= HOW came first species to go on.-- There never  were
219b194 ained more philosophical to state we do not know HOW transported.-- (Glaciers might have acted at Tri
223b208 ite» with its wonderful intincts <might well say HOW> The difference is that there is wide gap betwee
225b216 ure.-- miserable limited view.-- With respect to HOW species are. Lamaks "willing" doctrine absurd. (
225b216 d. (as equally are arguments against it-- namely HOW did otter live before being made otter-- why  to
225b217 atement I remember. L. Jenyns. talking of it) or HOW to make Indian Cow with bump & pigs foot with cl
226b222 must have been condition of Paris basin land.-- (HOW is this with Fernando Po.). with plants of St. H
226b223 o Europe & Pachydermata from Europe to America., HOW strange would presence of Jaguar been in S. Amer
226b223 at contrast in forms.-- India; intermediate, see HOW that is.-- ¿are shell-boring Molluscs, like Carn
230b236 n a chain of steps is found in same mountain).-- HOW is this explained by law of small differences pr
233b252 hange in the articulate than <M> Vertebrate. But HOW does this agree with longevity of species in Mol
235b264 k out Quinary system according to three elements HOW is Fauna of Van Diemens Land & Australia [blank]
235b273 ot be possible to work through all genera, & see HOW many confined to certain countries-- so on  with
236b278 ents same type collect cases.-- African isld.-- «HOW is Juan Fernandez-- Humming Birds» types of form
239cIFC on  wing Effects of colour on parent, white room HOW are varieties produced, by picking offspring? In
246c026 adagascar, Calcutta & Sumatra,. but I do not see HOW it is known that there are varieties & not specie
248c029 e same section, with house mice. It is wonderful HOW it could have been transported. ¿What section do
252c045 t that all analogy from each other I do not know HOW different Sumatra Java --------- do from Indian
254c047 n indication of the perfection of the Species--! HOW does this agree with grand fact of Marsupial, lo
257c057 ht be doubted were possible,-- it has been asked HOW did the otter live before it had its  web-feet--
257c058 bout great division, our <only> question is not, HOW come to be fishes & quadrupeds, but how th
257c058 t, how there come to be fishes & quadrupeds, but HOW there come to be, many genera of fish &c &c at p
258c062 ogy. All the discussion <after> about affinity & HOW one order first becomes developed & then another
259c062 inhabited  therefore mud wood be inhabited. then HOW is this effected by--- for instance, fish being e
260c067 might  compare birds of N. America & South-- Any HOW temperate regions-- crows in N. America Study B
262c072 ncts of wisdom virtue? «like senses of savages» (HOW come its some countries patriotic?)-- but more e
262c073 s having overrun-- Tahiti. thistle. Pampas. show HOW nicely things adapted--.-- These «aberrant» vari
263c077 ity, his end «under present form» will come, (or HOW dredfully we are deceived) then he is no excepti
264c081 ording to Gould, good genus Gould seems to doubt HOW far structure & habits go together. This must be
265c087 e only ornithorhyncus, then we should never know HOW much structure was connected with habits, & how
265c087 how  much structure was connected with habits, & HOW much heredetary. The circumstance of aberrant gr
266c091 they are as opposite as day & night: yet we know HOW remote the periods at which both left the  lands
268c096 & actually driving away parent birds.-- showing HOW blind a storge From what I see of S. American bi
268c096 same> <Mem> Tennioptera & Tyrannula (NB work out HOW many forms Tyrannula can we worked out into. Mil
268c099 of shells to external features of land by seeing HOW many species common to Patagonia desert & Tierra
270c104 norganic It is very remarkable as shown by Carus HOW intermediate plants are between animal life & "i
271c105 ations are entered in upon life. Namely Carus.-- HOW remarkable that Turdus Magellanicus. in the.  S.
272c109 urious feathers in tail of Edolius.-- Remarkable HOW small detail in structure prevails amongst the s
273c111 imals perished by errors.-- It is most wonderful HOW in every family of bird,, come the most <m> stro
274c115 clear  relations besides aerial, & terrestial,-- HOW is it in water birds, there are walking forms in
276c124 ish. structure would alter.-- It is a difficulty HOW a different number of vertebrae are produced, wh
278c132 ame for S. America & Europe.-- The difficulty is HOW came it animals not preserved, in central S. Ame
281c137 itting up his species & genera very finely" show HOW arbitrary & optional operation it is.-- show how
281c137 how arbitrary & optional operation it is.-- show HOW finely the series is graeduated.-- Dr Beck doubt
288c159 t the least notion of hunting, or keeping watch. HOW completely «nature & instinct» modified.-- The p
```

289c160 in some countries-- nightingale do.-- all shows HOW nicely adapted species to localities‖-- p. 390,.
289c162 g must alter method of generation-- Heaven knows HOW.-- This reaction takes place in every organ‖ «He
290c163 the minute gets into the road at right anglles, HOW pleased it is, just like man. emotions very simi
291c167 plants in animals it may be difficult to imagine HOW sexes were separated.-- in plants we have some f
293c173 footed, now Man by effort of Memory can remember HOW to swim after having once learnt, & if that was
294c176 prove persistent Varieties in wild animals-- but HOW to show species-- I fear argument must rest upon
298c192 rger &c in different countries. These facts show HOW very permanent plants are. & this conclusion mus
300c197 shamming death, most difficult case to iimagine HOW art acquired.-- They reason however on this to a
300c197 ck.-- shamming death it is but being motionless. HOW is instinctive dread it is exceedingly doubtful
303c204 atures»;. but likewise in length of extremities, HOW are races in This respect upper & lower, which I
304c206 es. with different degrees of closeness. -- look HOW close birds! look at Mammals: how wide.-- theref
304c206 eness. -- look how close birds! look at Mammals: HOW wide.-- therefore birds younger???? or «have» no
313c235 in Heber?) is analogous to dormant instinct.-- (HOW wonderful a case bees developing sex of neuters)
313c236 actions.--» «this very important in considering HOW children come to suck or other actions in foetus
314c238 sesses power of movement. & not the organ itself HOW except by direct adaptation has such a change be
316c244 ference between the mind of man & animals.-- yet HOW faint in a Fuegian or Australian! why not gradat
316c253 ppendix) showing WHAT CHANGES are taking place & HOW birds are extending their ranges. «even migrator
321c270 with correspond. with Rousseau Miss M«artineau» HOW to observe Mayo Philosophy of Art of Living Seve
334d009 e whole effect of race was determined by male: & HOW completely is Lord Moreton's case opposed to thi
334d009 collects one hatch of hybrid geese very fine.--» HOW is it with plants? This indicates a remarkable l
337d021 es exist for if so. it will be necessary to show HOW the first eye is formed.-- how one nerve becomes
337d021 necessary to show how the first eye is formed.-- HOW one nerve becomes sensitive to light.-- (Mem who
343d036 of eternity to the present time, to the future-- HOW far grander than idea from cramped imagination t
343d037 e a long succession of vile Molluscous animals-- HOW beneath the dignity of him, who «is supposed to
343d039 ears Negro probably a distinct species-- We know HOW long a Mammal may go on as one species from Egyp
345d044 in shows, with the aid of seclusion in breeding. HOW easy races or varieties are made.-- The Highland
348d051 finities, what is that, amount of resemblance,-- HOW can we estimate this amount, when <value> no sca
348d051 believe affinity may be taken literally,. though HOW far we can ever discover the real relationship i
350d053 s.-- «judged by analogy»-- Consider all this NB. HOW can local species as at Galapagos, be distingui
360d090 elled, though no fluid came from them.-- showing HOW gradually every <thing> «change» is effected)--
362d100 d be most curious to show that in sixty years-- (HOW many generations) the strangest peculiarities ha
363d101 Bewick drawing the the rock Pidgeon has not: now HOW many wild pidgeons have spangles on this part: t
366d111 smutation of species-- Will it against genera.-- HOW long will the wretched inhabitants of NW. Austra
366d112 flies might be used» The wild ass has no cross. HOW comes it that the tame donkey has. CD[old Buffon
368d114 pecies all females being most like offspring, Q (HOW is this with those females which put on (like so
370d116 Queens.-- Neuters are bred first, «then males--» HOW has this been arranged-- Neuters are true female
371d118 at it should be provided with many contingencies HOW to act-- so with the mind. the simplest transmis
372d130 rehended in itself.-- it must have the knowledge HOW to grow, & therefore to repair wounds-- but this
374d133 ting genera of shells in the mountain limestone HOW different from plants!) But the Cephalopoda depa
378d148 r & pies, have long had their present plumage.-- HOW is this in Pidgeons & fowls.--??? Wate[r]ton «p.
381d157 hat the Cryptandrous are really, Heautandrous.-- HOW is fecundation effected in latter; are <it> «org
382d158 are both present in every shade of perfection --HOW came it nipples <are> «though» abortive, are so
385d165 . can be trusted, this is Lord. Moretons law.-- "HOW often do we find in the son, the character, cons
385d165 & most of the moral qualities of the father! In HOW many daughters does the character of the mother
386d165 same length of life.-- like Golden Pippen trees! HOW is this with buds of plants, does annual give bu
386d167 f one & heals piece of skin.-- if the tail knows HOW to make a head. & head & tail, & the belly both
388d172 emen to one child. more like father.-- stuff.!-- HOW much opposed. the Quagga case appears to that of
388d173 The manner in which Frogs copulate & fish shows HOW simply instinctive the feeling of other sex bein
391d174 iorate» that object failing, generation fails.-- HOW completely circumstances «alone» make changes or
392d175 nts are Monooecious or dioecious.-- very curious HOW this was superinduced? (Surely all are really di
392d180 ather or Mother according as they are crossed? & HOW is it with China & Common Geese «how are their i
392d180 crossed? & How is it with China & Common Geese «HOW are their instincts?» «Chineses & Common Pigs.--
393d1BC - Disposition of half bred Cattle at Cinbermere? HOW is Jackall & dog at Z. Gardens D E Finished July
397e004 ace to some change of circumstances; now we know HOW slowly & insensibly such changes are in progress
408e046 Sowerby.-- Looking over Lamark surprised to see HOW many Tropical genera come from New Holland, ¿Syd
409e047 ulty-- I have felt some fluulty in conceiving HOW inhabitant of Tierra del Fuego is to be converte
409e047 s has done so.-- Show a savage a dog, & ask him, HOW wolf was so changed. When discussing extinction
415e069 rove that a thing has been so, & another to show HOW it came to be so.-- I speak only of the former p
416e070 ithout the hermaphrodites mutually couple.-- now HOW is it-- in Planaria, they couple-- CD [lowest te
418e083 the composition of the atom, to make it alive, & HOW the laws of generation were impressed on it.-- S
420e090 lowers.-- Mem. Spallanzani's experiments showing HOW little of the spermatic fluid fertilized spawn o
421e091 e a merely adaptive to an essential character--" HOW little clear meaning has this compared to what i
423e098 n is sufficent to spoil a bed of Cauliflower.-- (HOW curious it would be to make enquiries of some of
424e103 ry more than others» & horse in lesser degree,-- HOW different to dog!-- (Hybrids of Calceolaria.)--
429e113 arieties of domesticated animals, yet as we know HOW many plants have been produced (look at the Dahl
429e114 d themselves, as well as our wild plants, we see HOW full nature. how firmly each holds its place.--
429e114 well as our wild plants, we see how full nature. HOW firmly each holds its place.-- When we hear from
430e118 gous-- -- if so she can produce great ends-- But HOW-- -- «-- .Make the difficulty apparent by cross
432e125 but a practised geologist can really comprehend HOW old the world is, as the measurements refer not
433e126 sandstone &c. show some great change who can say HOW many centuries elapsed between each of these gap
439e143 e extreme facility of crossing, in plants proves HOW much depends on instincts in animals.-- yet how
439e144 the possibility of rearing by seeds Holyoaks-- HOW far is this so) shows either there is not so muc
440e145 ich has been» preserved.-- but be it remembered. HOW little part of the Grand Mystery is this,-- how
440e146 Lamarck Vol II. p. 115. 4 four laws Who can say, HOW much structure is due to external agency, withou
442e152 ions-- Gorze in Norway, which never flowers!!-- <HOW did it get there? whether> According to the abov
444e158 haetodon squirting water at fly.-- instinct, for HOW could experience teach distances in air, in whic
448e167 is strong argument for their contemporaneous»-- HOW is this with the Eocene beds.-- see Lyells table
452e181 c Soc] [...] p [...] -- most wonderful instinct, HOW could it have originated-- spins thread of cotto
453e183 - as it is not so with the esculent vegetables-- HOW is it with hollyoaks, flaxes &c &c? Mr Herbert i
460t019 coal measures.-- Shells in Cambrian & Crust show HOW long since present forms existed, but if it be a
460t019 in Java & Sumatra & dissimilarity of forms-- yet HOW this complexity from a few types originated, we
464t26r peculiar. N. American form)-- ¿Hunting leopard, HOW valueless this objection, when one thinks of dif
465t081 n a descending series of strata This again shows HOW strange, anyone, would have thought isolated spe
465t089 mentioned-- allied Beaver to present forms.-- -- HOW much forms depend on other forms Lyell's Paper,
465t089 ons contain fossils,-- mammals-- a few only -- & HOW many «tertiary» estuary & Lacrustine formations
466t093 way to nectary.-- Is not this so in Kidney Bean. HOW many estuary formations are there in old Seconda
467t103 pened.-- And in common Beans it is wonderful {a} HOW is it generally.-- In Azalea <do> «it is so» <Th
473s03v ed Sandstone. & Birds true! Plants in Devonian-- HOW the Humbles force down the wings most violently:
473s03v our Devonian-- Fish one step lower in America-- HOW strange no plants in our Devonian-- Fish one ste
473s07r & sheep for horns & yet no difference in calves--HOW curious all negative laws of America of depth of
480z007 nes-- in America & sexes not of different size-- HOW is this in young pigeons--dogs--cattle? As we se
484z016 wings It would be interesting comparison to find HOW does this apply to pale brown Caracara Krauss on
485z017 rence to those of mine from T. del Fuego p. 141. HOW many of the small finches walk at Maldonado & Pa
491q002 . 1. Uniformity of hybrid & Mongrel offspring 2. HOW comes it salt water so soon putrifies?? p. 319.
492q002 seedless;-- are all varieties seedless-- if so. HOW have late varieties of Peas &c been obtained? 3.
492q002 l flowers are those whose stamens are monstrous, HOW have varieties been formed?-- 8. Can any annuals
494q004 lse being heredetary. (6) In the last 1000 years HOW then are seeds ever raised? 11. Is not non-flowe
494q004 what principles calculated.-- in order to guess HOW many generations of man have there been.-- on wh
494q004 ule of blood in allied species similar.-- if not HOW many generations in Mammalia. in group effect of
494q004 ir general healthiness-- Fox's, Bears Badgers,-- HOW is it in <allied> varieties (9) Cross largest Ma
495q005 Cabbage with pollen of other, count seeds, & see HOW few wild animals are propagated,, though valuabl
496q05a ney-- What is use of Bee Larkspur= =Toad Orchis= HOW great a proportion springs up true.-- This in fa
496q006 -- are its female flowers always barren-- if not HOW many flowers in minute do they visit?? good=!! E
496q006 cy to double; what would be effect-- (10) Try in HOW does impregnation take place male & female flowe
497q006 see what plants will spring up which will show, HOW many generations. daisy. Fever-fuge Groundsil.--
497q006 p which will show, how seeds are transported, or HOW seeds are transported, or how long they remain d
497q007 y or two.-- where agency of man not known.-- (3) HOW long they remain dormant. if kinds come up, not
497q007 HOW is Iris impregnated.; which part of stigma?-- (4

Page **(Key Word)**
498q008 - or the minute Orthopt.-- important, as we know HOW readily they cross.-- (9) In the nurseries, when
499q010 nts in counties, as of rare green Cotton Plant-- HOW large «area» clump there? Distinguishable from <
500q10a ght to be a good many races or doubtful species; HOW is this at Canarys Arch-- it is so at Galapagos.
501q10a pecific character,, but contingent on country.-- HOW is it in Patella or Oysters or Helix. Or does an
502q11v tion equal in flower with leaves.-- strawberries HOW <soon> «early» do characters of races of differe
503q012 e Selection (36) Ask Mr Gowen to ask Mr Herbert, HOW many generations any hybrid has <been> reproduce
506q015 ionaceous structure of flower-- Ground nuts (42) HOW are Orchidiae fecundated, as mass of pollen is r
507q15v es with similar structure.= good case as showing HOW simple, but beautiful adaptation might be arrive
508q016 thinks asthma in females takes place of gout.-- HOW are livers obscure organ. no answer?-- 3 Andrew
509q017 dane genera, the species <are> have wide range-- HOW is this in «Plants??» Are abortive organs as <yo
510q017 ? Has Schow written on double creations & where? HOW are current & winds in Antarctic ocean: are they
510q017 close species of Hilianthemum differs greatly-- HOW very interesting to see if any variation in vari
512q019 ss.-- Sow cast-up-balls of Hawks or even owls.-- HOW long do seeds remain in stomach of birds-- Mem:
512q019 long do seeds remain in stomach of birds-- Mem: HOW many miles they fly in few hours Zoological Soc
524m022 pretend to feel her pulse.-- What fails first?-- HOW is this?-- Does memory bring in old ideas <I hav
525m026 e goodnatured, & vice versa Walter Scotts remark HOW odious an illtempered fat man looks, shows same
527m031 so in thoughts, one will rise according to law. HOW strange <all> «so many» birds singing in England
528m037 om perspective is derived in a river from seeing HOW the serpentine lines narrow in the distance.-- &
529m041 e in India. & a tiger stalked across the plains, HOW ones feelings would be excited, & how the scener
529m041 he plains, how ones feelings would be excited, & HOW the scenery would rise. Deer in Parks ditto.-- M
531m049 cat was seen by Hubberley to visit daily to see HOW the young got on. this nest the cat could If cat
531m050 ople very often forget where money is placed.-- (HOW often one forgets where put one key. where all k
531m051 ildren, the violent passions they go into, shows HOW truly an instinctive feeling, <may not pa.> In r
531m051 er me, when listening one evening when tired «-- HOW true the heart the scene of anger.--» to the pia
532m055 from the senses to the mind being more alive.-- HOW is it. with people nervous from illness., <the>
532m056 ame fat! What remarkable affection to a place.-- HOW like strong feelings of Man.-- The sensation of
532m057 short -- but violent action.-- To avoid stating HOW far, I believe, in Materialism, say only that em
533m060 s tending to make myself in act less grateful.-- HOW comes this tendency in these cases? How did my m
533m060 teful.-- How comes this tendency in these cases? HOW did my mind feel it was wrong (& it was not mere
535m069 re.-- Reviewer considers this profoundly true.-- HOW is it with children.-- Now it is not a little re
536m070 es, any more than «prevent» heart beat» remember HOW Pincher does just the same; I noticed this by pe
537m075 it will be, but yet it is settled by reason.-- HOW slow habits are changed may be inferred from exp
537m075 tinct, when attended with bad effects Martineau. HOW to observe, p. 21-26. argues «with examples» ver
539m084 bling. & not as flushing & with muscles rigid.-- HOW is this? dealt with p. 241 Origin of man now pro
540m085 . Seeing a dog & horse & man yawn, makes me feel HOW <much> all animals <are> built on one structure.
540m089 en. & lead people from scenes of distress.-- see HOW a crowd collects at an accident,-- children with
541m089 ears that pain, which is connected with death!-- HOW has this instinctive fear arisen? 19th. When I w
541m092 motion not so) than if simple idea as scarlet?-- HOW can people dwell on pain ¿ no definite idea. nor
541m093 n snarling real bitter sarcasm.-- <These> Seeing HOW ancient these expressions are, it is no wonder t
542m094 puppies do so dogs nearly silent, so with men.-- HOW is crying-- peculiar not common?--» no bark of a
542m095 is yawning habitual from awaking from sleep see HOW a dog yawns when he awakes. & streching & yawnin
542m095 es.-- evidently habitual when transferred, (also HOW often) to the tale of a wearisome man.-- Is frow
542m097 gs understand expression of man's face.-- <That> HOW far they communicate not easy to know,-- but thi
544m102 d Brougham thinks no dreams except at this time. HOW does he account for dogs & men speaking in their
544m104 ain from this fear-- & not death for the pain.-- HOW was this instinct gained.? by conversation-- ∴ m
545m105 interest.-- do/ I was much struck with observing HOW the Baboon (<Macaco> «Cyanocephalus Sphynx Linna
546m109 howling monkeys-- smell with many animals-- see HOW a dog likes smell of Partridge--, man's taste fo
546m109 eater.-- origin of colours?-- Nothing shows one HOW little happiness depends on the senses.; than th
546m109 that no one, looking back to his life, would say HOW many good dinners or......... he had had, he wou
546m109 ood dinners or......... he had had, he would say HOW many happy days he spent in such a place.-- Vide
547m112 box, whether I had thought what clothes to take (HOW often one cannot tell whether one has rung the b
550m126 ligio Medici, p. 21-24. Curious passages showing HOW easily chance & will of Deity are confounded.--
551m129 <&> disgust & defiance.-- different from sneer-- HOW easily. horses associate sounds may be seen by o
555m142 steps.-- dispute about words.-- Miss Martineau (HOW to Observe p. 213) says charity is found everywh
557m147 when winnowing & pleased pricks his ears?--).-- HOW is expression of anger in species of swans, in p
564n003 because original instincts different.--Mem. Bee HOW different instinct a solitary animal still diffe
565n010 is to laugh.-- I must be very cautious. Remember HOW Lavater ran away with new Lavaters,-- Ye Gods!:-
570n023 t, being distended. touch a person on the ribs & HOW he gulps in air.-- Again a master says I will se
571n029 ¿no vegetable good «for man» to eat poisonous?-- HOW did animals in «Australia» & America manage;-- T
574n041 nt ancestors having been like dogs to bitches.-- HOW comes such an association in man.-- it is bare f
578n053 lush, than particularly to wish not to do so.= = HOW directly personal remark will make any one blush
578n054 - but half & half. Miss F. A. said to Mrs. B. A. HOW nice it would be if your son would marry Miss. O
578n055 it be a person. whose opinion he regards, <& see HOW> feel how the blood gushed into his face,-- "as
578n055 rson. whose opinion he regards, <& see how> feel HOW the blood gushed into his face,-- "as <she> «the
581n065 Tookes tenses, &c &c -- <also g> if so & seeing HOW simple an explanation it offers of radical diver
582n068 table if it does not do it.-- My theory explains HOW it comes that the heart is the seat of the emoti
584n072 ng to water,-- young crocodile snapping-- p. 28. HOW curious the means of guiding themselves through
585n075 the same.--(this Hensleigh therefore problem is HOW we know that thing is same, which touches two pa
585n077 instinct, but a faculty, or sense-- "We know not HOW, stonge henge raised, yet not instinct, but if a
586n079 compass may be instinctive. it is a test to know HOW much of the wonder consists in the action being
588n090 ity.-- if explained by habits, useful to itself, HOW gained. reason? or some unnamed faculty-- Lamarc
593n107 on & hence does music now excite our feelings.-- HOW does Social animal recognize «& take pleasure in
594n112 . not unlike in size in the air at a distance.-- HOW can such an instinct arise?? «it would appear th
600o008 al, Mem. Negro, beau,--Jeffrey denies all Beau-- HOW does Hen determine which most beautiful cock, wh
602o11v ower of the mind, faintly approaches to instinct HOW strange it, that Nature should have so little to
602o11v a view taken by a camera obscura &c a Poussin.-- HOW are my ideas of a general notion of everything a
603o012 rican. Polynesians Jews, African all sacrifices. HOW completely men must have personified the deity.-
605o020 art on taste The object of this essay is to show HOW taste is gained how it originates, & by what mea
605o020 ect of this essay is to show how taste is gained HOW it originates, & by what means it becomes an alm
606o022 al standard, by which real objects are judged; & HOW obtained.-- implanted in our bosoms.-- how comes
606o022 ed; & how obtained.-- implanted in our bosoms.-- HOW comes it there? Laocoon p. 75 "The beauties deve
607o025 from motives, nearly as usual (a) one well feels HOW many actions are not determined by what is calle
608o026 tives do not. come into play.-- †It may be urged HOW often one try to persuade person to change line
609o030 thout their instincts.-- Gives art to when I say HOW social instincts generated? The origin of the so
610o031 lson & pointed it out; he was astonished, & said HOW very odd.-- --could not think what had put Wilso
610o031 Robert & found his Christian name was Wilson!!-- HOW curious an inward. unconscious memory.-- Jan 14t
612o035 as in stomach, intestines & heart of man. [LHC] ¿HOW near in structure is the ganglionic system of lo
613o035 tem of lower animals & sympathetic of man [RHC] ¿HOW does consciousness commence; where other senses
617o39v hich we apprehend the force of inamimate bodies. HOW we identify the two aspects as different phases
617o39v ould require a considerable degree of attention. HOW do the senses affect us, except by internal cons
617o040 tter we feel dissatisfied until we can point out HOW can force be recognized by our external senses--
620o044 er the cause of this may be, everyone must know, HOW soon the pleasure from good dinner, or from a bl
622o048 Paley [RHC] Anyone, who will reflect must feel, HOW like to injured conscience, is the feeling of an
622o048 he feeling of any custom of society broken..-- & HOW far more <feelin> acute the feeling really is.--
622o049 254. &c &c [RHC] But the love is instinctive, & HOW does it apply to mother loving child, from whom,
623o050 planation [RHC] Although I cannot pretend to say HOW far & minutely our instincts extend, yet as they
629o55v ees. with last head.-- (4) It is other question, HOW the feeling of ought, shame. right & wrong comes
629o55v t & wrong comes into mind in first case-- seeing HOW shame is accompanied by blushing, bears some rel
634j55r d even in Cetaceae traces of hind extremities.-- HOW are we to explain this.-- Did reptiles first inh
634j55r is origin of Polar attributes of the Cetaceae.-- HOW came Bats also.? before birds? They are ancient.
635j55r ecause we know nothing of the will of the Deity. HOW it acts & whether constant or inconstant like th
640j167 pagos Islds are explained. On distinct Creation, HOW anomalous, that the smallest newest, & most wret
641j29r lit up.-- who can decide their limits.-- To show HOW little we understand of the Physiological relati
214b175 of Charles.,-- and now in the possession of Mr HOWARD Galton, have one of the vertebra, about 2/3 fr
032r041 ter of globe. must there not be a circulation «HOWEVER slow & weak.»; «(cause of not accumulation of
123a116 space formed-- great vacuum-- by dike.-- Mem. HOWEVER. veins of segregation in Salisbury Craigs Lett
126a124 e depth of frozen soil is against this view.-- HOWEVER it is said in some of the papers that there ar
149g030 tones-- lake required to deposit this Remember HOWEVER the great Chilian valley Acongua, must there h

```
156g072 - & that river alone had modified it-- perhaps HOWEVER sea also,-- Barometer on shelf 3d. 29.455 A 83
159g087 eet lower & begins about 60 higher-- There are HOWEVER fringes of alluvium (?) still higher Slope of
185b056 s he is certain, that in insects, each family, HOWEVER many there may be, represent every other; for
202b134 most numerous there would be wanderers.-- Some HOWEVER might have offspring, & then «V. L. Institut p
234b256 collection of birds of Iceland. --M. Gaimard, HOWEVER, will settle this.-- Waterhouse says he is cer
265c085 id disorganizes wood, but adaptation.-- albino HOWEVER is monster. yet albino may so far be considere
278c128 with other genera in other families.-- it will HOWEVER be much <shu> surer, when false species banish
293c174 sm is so beautiful.-- The corelations are not, HOWEVER, perfect, else one animal would not cause mise
300c197 se to iimagine how art acquired.-- They reason HOWEVER on this to a degree. Mem Spider only dropping
309c220 ?» genera of shells.-- duration in two classes HOWEVER different.-- Male glow-worm knowing female goo
314c238 of the plant that this part must be protected HOWEVER it may be effected.-- Prodromus Florae Norfolk
316c247 e facts with identity of land animals.-- these HOWEVER come from Siberia it cannot be said American f
341d031 ll Mammals of N. America & many birds.-- which HOWEVER are most closely represented.-- Thus the red b
360d093 d in habits, though externally similar.-- this HOWEVER is a sophism for their brain or stomach would
376d135 Malthus shows, is the final effect, (by means HOWEVER of volition) of this populousness, on the ener
381d156 olluscs.-- insects. spider crabs.-- (all these HOWEVER do not require coition every generation)-- Epi
404e026 ent affecting offspring.--\ & as adaptation,-- HOWEVER mysterious such is case\. therefore chance & u
411e055 n> «the present» orders (not so in S. America, HOWEVER) is very remarkable & none discovered before t
428e112 ould not care so much about breed.-- what can «HOWEVER» be more striking, about indelibleness, than t
431e121 is» good case:-- in Sowerby Min. Conch. it is HOWEVER, said they have been found together on coast o
436e137 t formations:-- The Cambrian formations do not HOWEVER, extend round world.-- Quartz of Falkland.-- O
464t6r would have thought isolated species Mr Blyth, HOWEVER, believes in the existence of Molina's Pudu or
473s05r ome other lakes, (different waters) He cannot, HOWEVER, tell them from L. Groznerat, «on road to Beth
522m010 ould be held by means of hearing.-- Mr Corbet, HOWEVER, in conversation could catch up a new train if
543m100 act-- for, as Jones observed, in playing chess HOWEVER many places, & contingency a man has keep in m
553m136 us argue, make the same mistake, more apparent HOWEVER to us, as does that philosopher who says the i
569n023 aroline Archipelago. Dn 75 cf., p 268 without, HOWEVER, very sincere grief-- "there is nothing more t
582n067 al instance of heredetary habit:-- there must, HOWEVER, be a mental impulse (though unconscious of it
591n101 l. II «Sect XV. Dialogue on Natural Religion.» HOWEVER, he seems to allow it is an instinct.» I suspe
605o19v c. produces an inward pride & glorying. (often HOWEVER accompanied with terror & wonderment) <which>
608o026 sive object, so we view wickedness.-- it would HOWEVER be more proper to pity than to hate & be †A ma
614o037 said, no soul superadded, so [RHC] 6) thought, HOWEVER unintelligible it may be, seems as much functi
288c159 bark about half way «in tone»-- the native dogs HOWL most dismally, very rarely bark.-- are almost u
568n018 his origin of our pleasure in music-- do monkeys HOWL in harmony-- frogs chirp in do-- union of birds
615o36v wed, me and for five minutes every now and then HOWLED.-- Now I don't think this only pleasure; for i
546m109 takes, taste for musical sound with birds. & ¿ HOWLING monkeys-- smell with many animals-- see how a
563n002 efending companion. (mem Cyanocephalus. Sphynx HOWLING when I struck the Keeper) may be tempted to at
595n115 ite end of house. & commenced a most lamentable HOWLS & & was not comforted until the Keeper took it
079r176 one Balls of Silver core occur in do veins. At HUANTAJAIA. Humboldt says, mur of Silv.[,] Sulph. of do
531m049 ouse at Shrewsbury, which the cat was seen by HUBBERLEY to visit daily to see how the young got on. t
211b163 : therefore like dogs.-- Ogleby says, Wolves at HUDSON bay breed with dogs.-- the bitches never being
095a037 nos Ayres at the Zoolog: Soc: Terebratula from HUDSON'S Bay. 2. species Vol VI. Geograph. Journ. Anal
383d160 en.-- This green doubtless is effect of Metallic HUE of silver pheasant. yet why green? & not purple?
079r177 ans treat the "chefs d'oeuvre de l[']industrie HUMAINE, comme ils traitent les loix de la nature & se
071r157 rthrown by earthquake with great destruction of HUMAN life.--Temple mentions some earthquake at Cordo
223b209 aid.-- ¿where is Pentland's great account of Bolivian HUMAN species?-- Small «new» animal mentioned from Fe
260c068 o give tinge to offspring.-- Darkness effect on HUMAN offspring.-- white, snow.-- the fine green of v
281c138 ains that family likeness, which as in absolute HUMAN family is undescribable, yet holds good, so doe
308c218 p. 495)-- in fact, in all reasonings, of which HUMAN nature is the object, there is really no natura
322c269 tany.-- References at end Mayo Pathology of the HUMAN. Mind Evelyns Sylva. skimmed, stupid Brownes tr
326c266 - some important Papers. Dr. Mayo. Pathology of HUMAN. Mind.-- Audubons. Ornithological Biography. 4.
386d167 & tail,--no wonder there should be sympathy in HUMAN frame.-- «one of» The final cause of sexes to o
509g017 tially with him Ask him to introduce me to some HUMAN Anatomist. Has he dissected any animal often, w
543m098 ement in one order, as there is Intellectual in HUMAN-- probably some genera in different orders more
559m155 it.-- very curious book.-- Hume's essay on the HUMAN Understanding well worth reading Copied <Smith>
571n027 n to savages???].CD-- The existence of taste in HUMAN mind. is to me clear evidence, of the general i
592n101 en "Natural Hist. of Religion" on its origin in HUMAN mind.-- Andrew Smith says hen doves & the femal
604o017 ystem of Emotions.-- T. Mayo-- Pathology of the HUMAN Mind. Poor.-- on insanity.-- Prevailing idea. o
027r024 de Janeiro. Coquimbo. Balanidae. at Concepcion. HUMB: Pers. N. vii P. 56 Serpentine form: of Cuba fo
033r043 discussion on Pumice «& Obsidian:» in the I Vol. HUMB: There is rather good abstract of Humboldt. S.
300c196 work. worthy the interposition of a deity, more HUMBLE & I believe true to consider him created from
467t104 etc lies in gangway= In Lotus corniculatus saw HUMBLE press down wings which ejects pollen from tip
467t104 is yellow saw stigma project» In common Pea saw HUMBLE so press down sheath, that stigma covered with
468t111 thes, Leptuse-- Diptera & small Hymenoptera Saw HUMBLE go from great Scarlet Poppy to Rhododendron--
468t111 ver» «once» P on Fraxinella <Heartease> «small. HUMBLE alighted on base of filaments & reached nectar
468t111 below stigma. & so avoided it.» On certain days HUMBLE seem to frequent certain flowers, to day early
470t153 s of these plants will be collected & resown.-- HUMBLE 22 flowers of Egg Tree in one minute Great Hum
470t153 mble 22 flowers of Egg Tree in one minute Great HUMBLE 17 flowers of Larkspur on two plants in do Hum
470t153 mble 17 flowers of Larkspur on two plants in do HUMBLE 24 flowers of small Linaria in do Domestic do
470t177 mens= Many Humbles on hedge Linaria= (Plenty of HUMBLE Bees on Phlox Down, 1854, Sept.) In Spanish Br
472s02r suddenly withered, & to day saw very odd dusky HUMBLE (with pollen) on legs go from clump to clump,
472s02v at work-- another odd genus-- & a small common HUMBLE-- & more of same fly Two more of the flowers w
533m060 his confessions of sins. did not make him more HUMBLE.-- it has obscurely occurred to me that Capt.
536m074 "deliver us from temptation,' he would be most HUMBLE, he would strive <to do good> «to improve his
466t093 petal; bees & flies seen directed to it-- The HUMBLES in crawling out brush over anther & pistil & o
467t103 nd in common Beans it is wonderful {a} how the HUMBLES force down the wings most violently: in Beans
469t135 ame bunch & on this day in five minutes eleven HUMBLES came & each visited many flowers-- Saw Bees fr
470t177 es, Cyanoglossum--Reseda wild very many Bees & HUMBLES--on Thistles many (curious because a Composite
470t177 & as much shut up, frequented by «many» Bees & HUMBLES-- «Humbles & common» On silene, many plants of
470t177 hut up, frequented by «many» Bees & Humbles:-- «HUMBLES & common» On silene, many plants of wh. have a
470t177 many plants of wh. have abortive stamens= Many HUMBLES on hedge Linaria= (Plenty of Humble Bees on Ph
324c268 rabbes Life June Is. King & FitzRoy To be read HUMBOLD. New Spain-- Much about Castes &c Richardson's
033r043 elena-- All Trachytic.--Daubeny P. 171. Vol I. HUMBOLDT There is long discussion on Pumice «& Obsidia
033r043 I Vol. Humb: There is rather good abstract of HUMBOLDT. S. American Geolog. in Daubeny. P. 349 Admir
033r043 long PERIODS of great volcanic volcanic. from HUMBOLDT: Comparison F 361. Daubeny Von Buch is very s
043r070 New Madrid & Caraccas.-- Is this mentioned by HUMBOLDT is in his account of extensive areas. -- P. 322
053r100 lomerate of the Amazons & Orinoco mentioned by HUMBOLDT under name of Rothe-todte-liegende is perhaps
071r157 e most important There is map of Cordillera by HUMBOLDT in Geolog. Society Sir Woodbine Parish inform
073r163 silver. Sulp. of Barytes: Fluoric. Barytes:-- HUMBOLDT. New Spain. Vol III. p. 130 Metals in Mexico
074r165 cretions perhaps makes intersections richest-- HUMBOLDT has urged phenomena in veins, chemical affini
075r166 emala part. (& Chiloe do) no veins discovered. HUMBOLDT suggests covered up by volcanic rocks. //St H
077r171 like that of Spital of Schemnitz in Hungary.) HUMBOLDT says fragments from roof & penetrating overly
079r176 f Silver ore occur in do veins. At Huantajaia. HUMBOLDT says, mur of Silv.[,] Sulph. of do.[,]galena[
079r176 id silver in the highest & gold in the lowest. HUMBOLDT states that some of the richest gold mines on
079r177 Good Hope.--Azores Isds «nor at St Helena.--» HUMBOLDT. New Spain Vol. IV. «p. 58» At Acapulco earth
088a015 thern seas p. 312. Chamisso in Kotzebue. Study HUMBOLDT. Fragmens Asiatiques account of American Volc
091a026 eeds-- L. Institut. p. 209. May. 1837 Paper by HUMBOLDT on Quito Volcanoes & another on Mexican Trach
097a043 be analogous to the black glazing observed by HUMBOLDT on the granitic rocks of the Orinoco".-- <but
123a115 Denmark -- L'Institut ( 1838) p. 268. Paper by HUMBOLDT on Bogota. Cordillera,-- nothing.-- salt & co
179b032 first children How is this in West Indies «-: HUMBOLDT. New Spain:--» Dr. Smith always urges the dis
193b092 ity of some isld not having large quadrupeds-- HUMBOLDT has written on the geography of plants. Essai
205b142 erent parasite in different climates.-- Hunbt. HUMBOLDT? Vol V. P II. p 565. Consult-- Says types mos
209b156 espèces distinctes et permanentes." p. 145. In HUMBOLDT great work de distribut. Plantarum. relation
210b159 s?-- = admirable discussion Von Buch says from HUMBOLDT, in Laponia. genera to species I. 2,3-- From
356d069 f change at any one time Seeing what Von Buch (HUMBOLDT) G. St. Hilaire, & Lamarck have written I pr
408e044 r & Deer at 16000 ft. with Snow on Himmalaya-- HUMBOLDT bones at 7800 in Andes-- parallel & curious f
453e182 ow like ours, but much more small & shrill".-- HUMBOLDT. Vol I. p. 275. says Teneriffe does not count
480z006 .Institut. No.-- 221 Good account of Condor by HUMBOLDT Zoologie Recuiel-- Meyen has written account
```

(Key Word)

528m038 leaves,-- (this Rhythmical beauty is shown by HUMBOLDT from occurrence in Mexican & Graecian to be s
044r072 --Mem. Ascencion. concretions & Galapagos.-- «HUMBOLDTS. fragmens.» Read geology of N. America. India
044r073 elief all strata of Europe formed near coast. HUMBOLDTS quotation of instability of ground at present
507q15v on-- Sycamore. & seeds with «mere» border-- & HUMBOLDTS spinning seed.-- (50) Any cases of wild varie
545m104 omary to die-- August 24th. As some impressions «HUME» become unconscious. so may some ideas.-- ie ha
591n101 r.-- even to the cold or benevelo- continent man HUME has section (IX) on the Reason of animals Essay
592n101 he origin of reason. as gradually developed. see HUME on Sceptical Philosophy. Hume has written "Natu
592n101 lly developed. see Hume on Sceptical Philosophy. HUME has written "Natural Hist. of Religion" on its
321c270 lls Elements of Geology Gibbons life on himself HUME'S do, with correspond. with Rousseau Miss M«arti
325c267 Watson on Geograph. Distrib: of British Plants. HUMES Essay on H. Understanding (sometime) Du Stewart
559m155 .-- Rev R. Jones has it.-- very curious book.-- HUME'S essay on the Human Understanding well worth re
596n184 ith Moral Sentiments" much on life & character "HUMES Dissertation on the Passions." "Hartley" I shou
627o52v weariness leads to striking blows.--' p. 224.-- HUME'S Inquiry-- good abstract of Butler & arguments
588n090 to me mere consequence of stooping, as sign of HUMILITY.-- I suspect very strong argument might be ad
608o027 n disease. This view should teach one profound HUMILITY. one deserves no credit for anything. (yet on
236b278 es.-- African isld.-- «How is Juan Fernandez-- HUMMING Birds» types of former dogs. character of Mioc
273c111 ulus &c would instantly fill up their place.-- HUMMING bird there is strongly marked variety,:.in the
273c112 ws.» Milvulus, & still more wonderfully to the HUMMING bird, which is one instance of its whole famil
274c113 ough preeminently belonging to this type, ¿the HUMMING bird? the woodpeckers Gould says, he believes
371d118 ability of variation.-- Saw his collection of «HUMMING» birds, saw several fully developed tails, & o
524m019 octor), in this precisely like the passion, ill-HUMOUR & depression, which comes on from bodily cause
209b155 ce or history of species between Indian cow with HUMP & Common;-- between Esquimaux & European dog? Y
218b190 ters in Isle of France-- -- Madagascar oxen with HUMP.-- p 173. Voyage par un Officier du Roi Mem. Ca
234b255 duced I isle of France p. 170. «Fish introduced» HUMP backed race of cows from Madagascar-- p 173. Vo
257c056 ans clearly, we cannot guess causes of change.-- HUMP on back of cow!!-- &c &c D'orbigny (p 108) says
266c092 not larger than those of Black cattle. Not have HUMP like those of India & Arabia p. 202-- sheep hav
508q016 ases, more common in man than in female-- (8) In HUMP-back ever heredetary (9) Are the works of Berha
638j28r och. Attributes of Deity Vol I. p. 251-- stomach HUMP, kinds of foot. power of closing nostril, foot,
202b131 Province of Ghilan» wooded district cattle with HUMPS as in India. Geograph J. Vol III. P. I. p. 17 (
235b273 families.-- Ask Royle about Indian Cattle with HUMPS.-- ¿To be solved if horses sent to India. & lon
400e012 pt Davis in 1598 found cattle in Table Bay with HUMPS on their backs & big tailed sheep do Vol 10. p.
205b142 ave different parasite in different climates.-- HUNBT. Humboldt? Vol V. P II. p 565. Consult-- Says t
113a090 gher, «so» that we must look at the upper four HUNDRED feet of strata having conducted away the heat
120a108 n dikes not cooling if they had travelled some HUNDRED miles through nearly cold rock.-- in volcano t
138a153 s he believes the sea has formerly stood three HUNDRED feet above its present level, & in many parts
146g011 only covering Great Slip, 10 years since three HUNDRED feet in vertical height-- enormous mass thunde
146g011 al height-- enormous mass thunder storm, many <HUNDRED> thousand tuns. Black faced sheep, sometimes m
159g086 ot have deposited» the slope is continued some HUNDRED feet lower & begins about 60 higher-- There ar
206b146 d & this number elimanated say 150 people four HUNDRED years since were progenitors of present people
263c078 lusion want &c & perhaps a train of animals of HUNDRED generations of species to produce contingents
269c101 emp 60 degree mean, then temp at depth of four HUNDRED feet would be 60 degree + 6 degree??., therefo
278c130 s are at a height of more than 1000 ft. & many HUNDRED miles from the sea, associated with teeth of s
375d135 he rest.-- One may say there is a force like a HUNDRED thousand wedges trying force <into> every kind
638j059 stem is endowed with the knowledge of trying a HUNDRED schemes of structure, in the course of ages «s
639j28v e born «& it so chances, that one out of every HUNDRED litters is born with long legs» & in the Malth
446e163 grasses, which no doubt are propagated during HUNDREDS of years, without fresh seeds arriving.»-- th
555m143 in a confused manner. thought that a person was HUNG & came to life, & then made many jokes. about n
073r164 . = several parts of N. Spain great analogy to HUNGARY. = Veins of Zimapan offer zeolite. stilbite. g
077r171 vein (very like that of Spital of Schemnitz in HUNGARY.) Humboldt says fragments from roof & penetrat
260c067 ich could not be resisted, when hearing crys of HUNGER of little bird, in same way Wilson (p. 5). des
297c185 er species mice & only kills them when urged by HUNGER.-- p. 65. Aberrant groups few in numbers & var
564n004 of actions.→ till at last he face «instinct of» HUNGER, «of» death & for the satisfaction of followin
267c094 ea Vol VII. p. 325-- Wild dogs of Guayana always HUNT in packs «30 or 40 together» colour reddish <br
306c213 Mitchell is not aware the Australian dogs ever HUNT in company -- marked difference with dogs of La
511q018 bout German ornithologists, Bhem & Glöger Consul HUNT, birds from Azores or Madeira Mr. Blyth (1) Men
553m138 g <connection> that expression mean SOMETHING.-- HUNT (the intelligent Keeper) remarked that he had n
575n044 analogy that the tendency to hybrid greyhound to HUNT hares-- «& leave the sheep» & jackall to skulk a
575n044 . «& leave the sheep» & jackall to skulk about & HUNT mice-- Jenners Jackall Have we somewhat right t
624o051 accordance with it. thus a dog may be trained to HUNT one pig sooner than other, rather than change h
546m110 ing, on her return could not find paper cutter, HUNTED in vain for it-- ten years afterwards whilst a
210b161 that the <cas> production «of monsters» (which, HUNTER says owe their origin to very early stage) & w
307c216 is paper by Yarrell «in Zoolog Transactions» & HUNTER on this subject) because if so as she can be c
325c267 ains all his fathers views.-- Quoted by Owen.-- HUNTER has written quarto. work on Physiology besides
366d112 Book M.» Is there any law of variation. -- «(as HUNTER supposes with Monsters)» if armless cat can pr
369d114 final cause here.-- & therefore different from HUNTER I should say females recede in organization fr
369d115 fic character.-- Chapt I. Also Latent Character HUNTER Animal Economy p. 482 (Same book) Owen says "t
370d116 d but few, because they are kept in security.-- HUNTER doubts about production of Queens.-- Neuters a
381d156 ases in size prodigiously-- Animal OEconomy by, HUNTER. (edited by Owen) p. 34.-- Owen classifies Her
382d157 of organs barren in one plant & not in other), HUNTER <asks> p. 36 is thought by Owen to ask. whethe
382d158 rthy of a Lamarckian.-- Mine is much simpler.-- HUNTER shows almost all animals subject to Hermaphrod
382d158 s give suck)--But this not distinctly stated by HUNTER.-- Do testes, & ovaria when they first appear
384d161 shows my doctrine of Hermaphrodite differs from HUNTER)-- Hunter (p. 45) observes "every species has
384d161 octrine of Hermaphrodite differs from Hunter)-- HUNTER (p. 45) observes "every species has a disposit
385d163 actually obliged to be held.-- like she wolf of HUNTER.-- young take distemper very readily & are sub
404e026 nds.-- young being habituated. instance such as HUNTER, or some one mentions of influence on parent a
513q21. this in Bauers case of orchidiae Where does J. HUNTER use expression of "male principle of arrangeme
530m044 se.-- look to it My father has somewhere heard (HUNTER?) that pulse of new born babies of labouring c
557m147 leased tail loose & wagging-- if as (I believe) HUNTER says. neither fox. nor wolf wag their tails, &
322c270 olumes Abercrombie on the Intellectual powers. HUNTERS Animal OEconomy. edited by Owen. read several
351d057 ltered & umbilical cord,-- Broderip alluded to HUNTER'S views on this subject.-- Monstrosities, kind
355d067 showing what organs are little fixed-- (<also> HUNTERS law of monstrosity with regard to age of foetu
367d113 - by air cells connected with cheek pouches.-- HUNTER'S Animal OEconomy p. 45 "One of the most genera
379d152 e female, she will be free Marten. <Owen.> See HUNTER'S Owen-- In the Athenaeum Numbers 406, 407, 409
383d159 s whether they are not Hermaphrodites, like J. HUNTERS. Free Marten N.B. the common mule must often h
384d161 acter very different from either parent bird-- HUNTERS Animal OEconomy. (by Owen) p. 44. Classificati
389d173 but had such aversion to it, that she was held HUNTERS Eoeconomy So with inter-breeding as told by Wi
288c159 .-- are almost useless not the least notion of HUNTING, or keeping watch. how completely «nature & in
464tf6r & S., (¿ is the peculiar. N. American form)-- ¿HUNTING leopard, how strange, anyone, would have thoug
521m007 ere is a case of Mr Anson. who told a story of HUNTING «-- habitual fits.--» which my Father thinks i
588n090 oning apply chiefly to recollection. yet a dog HUNTING for a bone shows he has recollection.-- Lamarc
595n117 <lather> <foame> & sweat, when hearing merely HUNTING horn-- association or imagination [blank] [not
624o051 one pig sooner than other, rather than change HUNTING instinct. *Our tastes in mouth by my theory ar
482z011 us aura Excessively inaccurate Saw a Chouette a HUPPE courte talks of nine terrestrial Turdus falklan
413e059 f mysteries. & has grand passage upon problem.! HURRAH.-- "intermediate causes" The Sexual system of
032r041 :--If great changes of climate have happened. HURRICANE in bowels of earth cause:--<exp> does not exp
640j167 Probably no case in world like Galapagos. no HURRICANES.-- islds never joined, nature & climate very
547m111 he room. with my trunk out & I was engaged in HURRIEDLY giving orders.-- Now what was difference betw
572n034 ith general tendency of the dream.-- It does not HURT the conscience of a Boy to swear, though reason
572n034 ear, though reason may tell him not, but it does HURT his conscience, if he has been cowardly, or has
533m061 wrong (& it was not merely morally wrong, but HURTING my character I felt it)-- this is kind of cons
565n008 sneering,-- it is then more <emblem> manner of HURTING opponent by insulting his pride & is therefore
575o045 y analogy between grief & pain-- certain ideas HURTING brain, like a wound hurts body-- tears flow fr
619o042 pain: for instance either protecting sheep or HURTING them.-- Therefore in man we should expect that
537m077 & does not act up to them, no doubt disobeys & HURTS conscience more than other.-- A Scotchman will
575n045 ain-- certain ideas hurting brain, like a wound HURTS body-- tears flow from both, as when one burns
344d042 of fixed. though she resembled a harrier & her HUSBAND was pure Harrier.-- «The peculiarities of our
523m017 of the window to kill herself from jealousy of HUSBAND connection with housemaid two years before, to
523m017 r that purpose.-- this found to be true.-- Her HUSBAND never suspected during these two years that sh

524m022 state as above described, (forgetting that her HUSBAND was dead) yet instantly perceived when my Fath
321c275 at end-- The British Aviary.--do--- & Lisle's HUSBANDRY Tuckeys voyage reread Appendix Ovington Voyag
524m022 minals before execution.-- Widows not of their HUSBANDS-- My father's test of sincerity.-- People in
033r042 pellucid envelope of Lime.--form resembles the HUSKS at Coquimbo: in that case, may not central and
629o055 books &c &c.-- <]> the "secondary passion" of HUTCHESON unfolded by D. Hartley.-- Darwin's Abstract o
056r112 separating is well adapted to use of mankind.--<HUTTON show> Earthquakes part of necessary process of
073r161 in the fragmentary jasper.--do undulations (as HUTTON says) always come from without.-- "True native
275c119 on all the general arguments-- Lamarck was the HUTTON of Geology. he had few clear facts, but so bol
451e177 Catalogue of Birds of India. -- p. 555. Lieut. HUTTON counted, the ova of a tick «in India» & found
496q05a ts grown under unfavourable circumstances, as HYACINTHS in glasses &c &c Experiments Questions concer
337d022 e is distinct Creation required if he believes «HYAENA & squirrel» seal & mouse, elephant, come from
338d025 fresh creation of new forms.-- what is range of HYAENA? Hippotamus.? Indio-African, or pure Africa?--
377d139 philosophic naturalist pregnant with interest» HYAENA. thinks, when pleased cocks his ears., when fr
420e088 nt of Hyaenodon, a fossil dog-- leading towards HYAENA. -- see Comte Rendu.-- I suspect good case of f
420e088 ot between existing series of species of dogs & HYAENA. -- but a common point, whence both may have de
458t001 im. fossils-- Ruminants. & Tortoises gigantic-- HYAENA-- bear & ruminants all of larger size.-- the l
536m071 - What an animal like taste of likes smell of,.. HYAENA likes smell of that fatty substance it scrapes
358m153 species put their necks straight out & hiss.-- [HYAENA pisses from fear so does man.-- & so dog]:CD M
565n006 seless grin of passion, is like the grin of the HYAENA from fear, no actual intention to bite at mome
574n043 would show habit.-- Take the case of Jenner's <HYAENA> Jackall.-- an animal not destined by nature t
420e088 here L'Institut do. p. 419, «long» account of HYAENODON, a fossil dog-- leading towards Hyaena.-- see
203b138 ll says that old races when mingled with newer, HYBRID variety partakes chiefly of the former Eyton's
204b139 English Breed. decidedly exceedingly prolific & HYBRID about half way. Eyton says Hybrid about half a
204b139 ly prolific & hybrid about half way. Eyton says HYBRID about half aways & result the same Indian catt
204b140 same Indian cattle & common produced very fine HYBRID offspring, much larger than the dam, from thos
204b140 than the dam, from those imported by Ld. Powis HYBRID dogs offspring seldom intermediate between par
216b180 Black & White species.-- For, says he Seeds of HYBRID lillies &c &c &, (V Herbert on hybrids) thus a
216b182 (like Englishmen) men of many countenances, as HYBRID once. Is not this contradiction to his view of
220b198 d Moreton's Mare. The fact of plants going back HYBRID plants; analogous to Men. & dogs. Now if we ta
222b203 ntroduction of species.-- When species cross & «HYBRID» breed, their offspring show tendency to retur
230b235 any animal-- Athenaeum. <Jan> «p. 154»-- 1838. HYBRID Ferns It may be argued against theory of chang
248c033 revert to parent forms, & greater fertility of HYBRID & parent stock, than between two hybrids.-- As
276c122 l resemblances, than the female. The expression HYBRID & fertile Hybrids, may be used to varieties, a
291c167 ng the sex by feeding.-- no it is developing an HYBRID female it is a wonderful relation going throug
323c269 h 16. Gardner's Music of Nature-- -- Herbert on HYBRID Mixtures: Marginal notes. -- 20th. Carlyle's F
334d009 number of Mules.-- «He recollects one hatch of HYBRID geese very fine.--» How is it with plants? Thi
336d018 each sends his own likeness, & the union makes HYBRID, in fact the parents beget child like themselv
341d032 out intermediate.-- (In Zoolog Gardens there is HYBRID of Penguin duck a variety of Muscovy) with goo
342d033 produce heterogenous offsprings.-- «are not the HYBRID pheasants & grouse different.--» (if so chines
342d033 heterogenous offspring. It appears certain that HYBRID Muscovy & Common duck have been shot wild (esc
344d040 consider this with reference to "new species & HYBRID doctrine"-- I have read there are exceptions t
357d073 iven for supposing Tetrao media or Rakkelhan is HYBRID (produced commonly in Nature. both in Sweden &
357d073 ich this genus becomes crossed. ¿is red game an HYBRID?-- When I show that island would have no plant
358d086 mmon pheasant & back like silver.-- But the hen HYBRID of this bird, has long tail figure, & some deg
365d106 eing restricted species has been Made.-- In the HYBRID grouse between Black Cock & Ptarmigan (probabl
365d106 lpina.) former has blue breast, latter reddish, HYBRID purple-- be careful, See to hybrids between Ph
380d154 hermaphrodite-- (as is seen in <fe> plumage of HYBRID birds) After animal has copulated., though no
383d160 n have been dissected Zoolog. Garden. Sept 16." HYBRID between Silver & Common Pheasant. Male bird, s
393dIBC y men they cannot tell first result., or that «HYBRID» breed is uncertain Is there any peculiarity o
405e032 lour can be brownish do Saw what was said to be HYBRID between silver & gold fish-- Octob. 26th. If.
426e106 easant, & hooded crow goes against this. & wild HYBRID plants. If many wild animals were crossed, the
430e120 e.-- Mr Lonsdale evidently inclines to think it HYBRID.!!! Ask Woodward Mr Lonsdale says Trigonia cos
455eIBC plants. where they cannot «crossed» etc.-- Make HYBRID mosses.-- Leighton or some one. Father-- disea
458t002 ller-- in Zoological Gardens. informs me that a HYBRID between ass & Zebra, crossed with pony mare &
491q002 -- Questions Regarding Plants. 1. Uniformity of HYBRID & Mongrel offspring 2. How have late varieties
493q003 ent cats.-- Good observation-- examine semen of HYBRID animal. in comparison with Weigh skeleton of T
503q012 wen to ask Mr Herbert, how many generations any HYBRID has <been> reproduced itself.-- Ask Gray to as
512q019 finches-- do they throw up pellets-- (4) About HYBRID pheasants treading-- any treadde?-- Difference
559m156 pound up inflorescent parts of mosses & see if HYBRID can be made & ferns.--» Would a sensitive plan
575n044 there be stronger analogy that the tendency to HYBRID greyhound to hunt hares. «& leave the sheep» &
633j53v mes to a damp. place.-- Kolreuter mentions some HYBRID, whose flower great tendency to break off p. 2
636j50v races of peas become intermixed & gardner have HYBRID/ seedlings} p. 333. Maccullooh. brings forward.
424e103 el does not vary «one ought not to be able to HYBRIDISE the Camel» like ass «same way some plants var
491q01v ant as discovering function of seeds-- (6) To HYBRIDISE EVERY flower on melon & see whether fruit aff
472s01r true P. bracteatum; all supposed to have been HYBRIDISED == Has tried several year to obtain seed, bu
472s01v with the P. orientale in garden & all came up HYBRIDISED. It is possible to raise them pure for Miss
505q014 ts which plant seeds? (9) Melons fruit itself HYBRIDISED (10) one had no seeds, & two had plenty of s
333d005 more than of the Common cat.-- Ch IX Mongrels HYBRIDISIM Fox has half Persian cat. which bred with un
503q012 Ask Gray to ask Mr Riley to experimentice on HYBRIDISING ferns, tying them back to back 37 Col. Sykes
173b010 ish, bones instinct &c &c &c non fertility of HYBRIDITY &c &c <assuming all> if species (<a>) «(I)».
174b013 to Decandolle. (his argument applies only to HYBRIDITY.-- genera being usually peculiar to same coun
218b191 icultural Transactions and a distinct work on HYBRIDITY under title of Amaryllidae & Narcissus. Mr Do
227b228 cs.-- it would lead to closest examination of HYBRIDITY «to what circumstances favour crossing & what
228b230 Universality of generation strongly shown by HYBRIDITY of ferns.-- hybridity showing connexion of tw
228b230 ation strongly shown by hybridity of ferns.-- HYBRIDITY showing connexion of two plants. Animals-- wh
248c030 idently is tending to same end, as the law of HYBRIDITY, namely the [not located] animals unite, all
255c052 al with skeleton of such general forms.-- The HYBRIDITY of ferns bears on my doctrine of cross-genera
285c151 nce occurring.-- It will be necessary to show HYBRIDITY from few forms, parents of all species not po
358d086 of whiteness like a Male.-- Thus castration, HYBRIDITY, & breeding in & in tend to produce same effe
384d162 not distinguished. --yet may be presumed from HYBRIDITY (of ferns) afterwards they can be seen distinc
421e090 ality of law. even in fish: \.do. p. 236-- on HYBRIDITY in ferns.-- \.do p. 250-- «speaking of» the t
171b002 into play)--See Zoonomia arguements, fails in HYBRIDS where every thing else is perfect; mothes appa
178b030 ournal. of Nat. History.-- July. 1837. Eyton of HYBRIDS propagating freely In Isld neighbouring contin
179b033 ows this.-- <If> Is there a tendency in plants HYBRIDS to go back?-- If so Men & plants together would
203b138 artakes chiefly of the former Eyton's paper on HYBRIDS Loudon's Magazine. Gould on Motacilla., «speci
216b180 Seeds of hybrid lillies &c &c &, (V Herbert on HYBRIDS) thus act.-- Now the point will be to find whe
217b189 e many Northern genera peculiar to itself-- on HYBRIDS between grouse & pheasant-- Magazine. Zoology
230b240 ill be called species & mere producing fertile HYBRIDS will not destroy that evidence, as so many pla
230b240 stroy that evidence, as so many plants produce HYBRIDS, or else whole fabric will be overturned.-- He
239cIFC . 1873 Books About amount of difference: where HYBRIDS produced have any close species ever yet faile
248c033 ity of hybrid & parent stock, than between two HYBRIDS.-- As we see external influences first affect
249c034 therefore two different varieties will produce HYBRIDS but not varieties. which are not deeply impres
276c122 an the female. The expression hybrid & fertile HYBRIDS, may be used to varieties, as well as species-
280c135 cies were excessively old, they would not make HYBRIDS, whereas two newer ones, even if more differen
280c135 great difficulty in way of ascertaining about HYBRIDS.-- & is a very remarkable fact. show influence
283c145 (even though <fertile> when compelled to breed HYBRIDS produced--)» & then all that I want is granted
290c164 t., I think so because very 3/4 bred.-- (hence HYBRIDS in this case have bred). White & common pheasa
296c184 f Zool. & Bot.-- Vol I. p. 450. 4 instances of HYBRIDS between pheasants & Black fowl.-- use as argum
296c184 lack fowl.-- use as argument possibly some few HYBRIDS in nature.-- «p. 473» Webb &. Berthelot. must
326c265 h. Linn. in 1829 has given list of Spontaneous HYBRIDS. where? Sweet. Hortus Britannicus. has remarks
331dIFC eat Marlborough St Did Eytons <intermediate>. «HYBRIDS, when» interbred. show any tendency to return
331dIFC ither parent.? Is the first cross, which makes HYBRIDS. productive like geese?-- Are the number of ki
335d013 s, because they are not new breed.-- the first HYBRIDS may be compared to animal with amputated limb.
335d015 favour the Persian side.-- Theory of abortive HYBRIDS.-- If mules did breed, the offspring would «as
335d015 e to changes which every species undergoes») & HYBRIDS between very near species (that is slight alte
335d015 terations of primitive stock) are heredetary: «HYBRIDS of» Varieties is different because not long in
340d026 ail & Common Ducks, breed one with another-- & HYBRIDS fertile inter se--No directly against Eyton's
340d026 se--No directly against Eyton's rule. ¿Are the HYBRIDS similar inter se.-- [not located] the «4» Stru

342d033 produce inter se or with parent species.-- The HYBRIDS do not vary (ie the hens all alike & Cocks all
342d034 use homogeneous? I observe Bachman calls these HYBRIDS new species. Yarrell says the bird fanciers sa
345d042 tlemen of party all thought with pigs &c, that HYBRIDS were uncertain. Mr Drinkwater thought that a <
356d070 ox says) same fruit trees is analogous to some HYBRIDS breedings-- there is tendency to reproduce in
358d085 of not having sexual plumage is very common by HYBRIDS, that are infertile.-- thus the common. pheasa
359d087 mals further apart have bred inter se.-- These HYBRIDS are very wild & take <very little> in disposit
363d103 en caught wild off coast of America.-- showing HYBRIDS can fare for themselves.‖ + + first year.-- Th
365d106 er reddish, hybrid purple-- be careful, See to HYBRIDS between Pheasant & Black Cock, & other hybrids
365d106 hybrids between Pheasant & Black Cock, & other HYBRIDS-- The fact of Egyptian animals not having chan
367d112 old Buffon should be read on Mare My view, why HYBRIDS are infertile. supposes that when foetus is fo
374d134 bites Sept 25th. In considering infertility of HYBRIDS inter se, the first cross generally brothers &
381d157 considered as really cryptandrous, & they have HYBRIDS-- this is most important support to my views--
383d159 it is not overwhelming.-- Seeing in Gardens of HYBRIDS between Common & Silver Pheasant, one like coc
385d163 to .fits.-- «there is great difference between HYBRIDS & inter se offspring in latter being unhealthy
389d177 an be bred in & in.)-- [The loss of passion in HYBRIDS, perhaps connected with this same case (& not
389d177 ndency to deformity ¿this does not happen with HYBRIDS?]CD Plants must stand much breeding in & in (t
392d180 ish weeds in Hot. Houses will they flower Make HYBRIDS with moths, where fecundation can be made arti
393dIBC re fecundation can be made artificially.-- Are HYBRIDS pintail & common ducks, similar inter se? Zool
393dIBC cks. similar inter se? Zoolog. Gardens Are the HYBRIDS of those species. which cross & are fertile he
404e024 ence might be drawn that whole infertility «of HYBRIDS receive no explanation» was consequent on mind
417e075 little tendency to vary. & hence offspring are HYBRIDS,.-- Mr G. B. Sowerby <tel> showed me many land
424e102 ne wood in Anhault. & there every year produce HYBRIDS-- now this is independent good case, but very
424e103 e in lesser degree,-- how different to dog!-- (HYBRIDS of Calceolaria.)-- «:CD[Does the Power of, «ea
425e103 he Power of, «easily» making tolerably fertile HYBRIDS, bears relation to capability of variation?? an
426e106 different species produce abundantly infertile HYBRIDS, & the fact that old varieties do not so much
426e107 reat series quite otherwise & make on[ly] true HYBRIDS.-- but this is false, [give instance of series
439e143 t is very important Mr Herberts fact about the HYBRIDS (mentioned in letter to Henslow) fertilizing e
439e143 s it tends to show my view of <i>nfertility of HYBRIDS «with parent species» false, which makes it de
454e184 t wild plants, some partly dioecious? Mushroom HYBRIDS? Any «wild» plants in England, which do not pe
467t099 growing within Kitchen Garden.-- As we see in HYBRIDS that although anther «nor filaments» shrivel,
470t176 thin few yards of each other actually produced HYBRIDS-- My Father remembered when in the gardens, he
495q005 other cabbages & see whether there will result HYBRIDS-- (6) Dust flowers of one branch of Cabbage wi
495q005 .-- This in fact always takes place in natural HYBRIDS of Cabbages (7) Sow «daisy» seeds of wild cabb
497q06v) Examine pollen of those genera of which wild HYBRIDS have been formed. (15). What is History of Vib
498q009 to each other.-- As they are dioecious, if no HYBRIDS were produced by seed, we might feel sure, tha
503q012 in India?-- does not know About Yaks. & other HYBRIDS-- Dogs &c &c 38 Does only male yak cross with
507q15v -- (48) .Where «published» list of spontaneous HYBRIDS-- to see whether any Papilionaceous plants,--
511q018 it in work} Whether <Yar> knows whether Shaws HYBRIDS between Trout & Salmon were fertile & whether
512q019 urning into fine beasts.-- For comparison with HYBRIDS, is offspring of short-horn bull & hereford co
514q021 y one pineaple Horticulturists (1) Are sterile HYBRIDS healthy: number of generations: about crossing
515q021 seeds ripening has scarcely any no relation to HYBRIDS.-- (3) As peaches sport into Nectarines (does
515q022 the geese which he crossed; especially if the HYBRIDS were recrossed with either parent.-- May. 44 T
515q022 recrossed with either parent.-- May. 44 These HYBRIDS differ in colour of beak, taking after male &
575m043 ned by nature to exist. & carrying «like other HYBRIDS» with <the> it the provision for death.-- can
044r074 re of gazes. especially the most abundant. Sulp. HYD: Carb: A. Mur: A. = (& this effect of water thus
577n049 o life & mind of man.-- & we do not suppose an HYDATID to be a cause of itself.-- [by my theory no an
485z017 es it salt water so soon putrifies?? p. 319. on HYDRA-- polypi-- <Rep> do p. 324. Polypi shorter dura
567n013 <our> the willing of the simplest animals, as HYDRA towards light. being direct effect of some law.
441e150 better be shown-- soil on colour of flowers, HYDRANGEA -- black bullfinches-- & all varieties must b
499q009 othera bush.-- (19) Theory of mock flowers in HYDRANGEA (20) As Hop is Dioecious-- seedsmen who raise
390d177 are produced by budding «only» as cryptogamia & HYDRAS,-- (this repugnance to breeding in & in seems
289c161 <in wild> of species. has nothing to do with HYDRIDITY, is simply, an instinctive impulse to keep s
508q016 n man being transmitted through females, like HYDROCELE Dr. H. thinks asthma in females takes place o
614o037 ile of liver.-- ¿ is the attraction of carbon. HYDROGEN <&c> in certain definite proportions, (differ
369d115 climate of Australia, & from Ornithorhyncus & HYDROMYS not being Marsupial. («but» also mice) & thes
079r177 on coast of Guayaquil, same as Galapagos. no HYDROPHOBIA at Quito. P 281. do do Australia, C. of Good
293c174 y diseases in common between man & animals-- HYDROPHOBIA &c cowpox, proof of common origin of Man.--
252c042 8 Reference to Rüppel. travels (what language?? HYENA «venatica» <of> Cape found in Desert of Korto &
266c092 IVETS CATS only wild animals on isld.-- Niether HYENAS; jackals monkeys-- common to either coast not
468t106 apsis, Melegethes, Leptuse-- Diptera & small HYMENOPTERA Saw Humble go from great Scarlet Poppy to Rh
543m098 rhouse says far more instincts in all of the HYMENOPTERA;. <therefore> than in other orders (study Ki
543m098 just as dog & Elephant most intellectual.-- HYMENOPTERA typical insects. ie have all parts. Waterhou
424e100 ndosi, et hepatici,-- in Chara, in Marchantia & HYPNUM «Prof:» Don would have known the Composites or
058r117 s (not mere patches as in Italy proved by Coral HYPOTH. agree with great continents). Voyage aux terr
155g065 ea hypothesis, if gullies not now formed «(Mac, HYPOTH.,)» the level during any oscillation must have
248c030 quadrupeds have existed for ages.-- ∴ The most HYPOTH: part of my theory, that «two» varieties of na
195b104 that species die &. others replace them-- two HYPOTHESES fresh creations is mere assumption, it expla
132a139 ean's bottom.-- (according to M.. Parrots own HYPOTHESIS some such explanation appears to me necessar
151g041 on line of 4th shelf.-- Even on Lauder Dicks HYPOTHESIS impossible to explain absence of lines in ce
155g065 ve straight line «only» cut deep gorge on sea HYPOTHESIS, if gullies not now formed «(Mac, hypoth,)»
159g085 no peat supply.-- Consider profoundly Boulder HYPOTHESIS Thursday, from Glen Turrit to Fort Augustus
196b110 ilosop:-- I deduce from extreme difficulty of HYPOTHESIS of connecting Mollusca & vertebrata, that th
291c168 ing destroyed at Falkland Isds++.-- Mem Lyell HYPOTHESIS of change in Scicily.-- Splendid Harmony the
302c202 l subjects of classification, I must work out HYPOTHESIS, & compare it with resuts. if I acted otherw
352d056 n. such law would explain every thing.-- PURE HYPOTHESIS be careful.-- Argument for circularity of gr
370d117 a probability by induction, & to apply it as HYPOTHESIS to other points. & see whether it will solve
565n008 o when angry sneering is in progress.<--> the HYPOTHESIS of opposite muscles will want much confirmat
582n067 to do it.-- [the means must be present on any HYPOTHESIS whatever]CD an animal may so far be said to
589n091 ith my theory-- no facts, & mingled with much HYPOTHESIS.-- see M.S. notes, where strong argument in
028r030 introduced either in note in Coral Paper or HYPOTHETICAL origin of some sandstones, as in Australia.
063r121 . rocks.--then beneath water.--together with HYPOTHETICAL case of Brazil.-- Propagation. whether ordi
578n057 convulsive actions-- (except in weak people & HYSTERICAL people inclined to convulsive actions).-- Bu
021r005 , no minute description, calls it a Fucus. P <Vol I 287> P 379. Henslow Anglesea, nodules in Clay Sla
025r017 go = The Wager's Earthquake the most Southern one I have heard of In a preface, it might be well to u
025r018 to compare whole history of Europe, with America; I might add I have drawn all my illustrations from
025r018 hole history of Europe, with America; I might add I have drawn all my illustrations from America, pur
025r019 e sea.--«certainly data insufficient, yet good» «(I suspect fragments of shells will generally be fou
025r019 ells will generally be found to be old & dead)» «(I have not kept a record)» In looking over the list
026r019 remains in De la Beche, for the older formations I must believe they «the limestones» have been form
028r031 nce at least no water is absorbed into the earth <I did not see one dike in the whole Galapagos Arch;
029r033 ot occur in the South African plains.-- Sydney no I believe the secondary? formations of Brazil, all
029r034 min. W. mid. calm and clear. Caermarthen Journal I look at the cessation northwards of the Coal in C
030r035 -- On open coast, near where Challenger was lost: I know no reason for supposing these matters are no
031r037 erally appear to have fallen into lines of faults I do not think so many faults in Cordillera, as in
031r038 was very clear at base of great lava cliffs [Fig. I] line of high tidal action {P} NB. patches of mod
032r041 «top of» tidal band of action. This case differs. I think. from Patagonian steps, because the deposit
032r042 vage lines./ possibly general symetry of world.-- I feel no doubt. respecting the brecciated white st
033r043 m St Helena-- All Trachytic.--Daubeny P. 171. Vol I. Humboldt There is long discussion on Pumice «& O
033r043 is long discussion on Pumice «& Obsidian:» in the I Vol. Humb: There is rather good abstract of Humbo
035r045 of NW. America P. 209--13 P & 444 «(Yanky Edit)» <I think> At Ascension, the laminae <...> changes in
035r046 those belonging to the same district there seems. I think, little ground for skepticism, as to the ge
036r050 is called "Bramidor"(?).--it was a strange story; I believe it was necessary to ascend the hill,--but
036r051 chirping sound produced in walking over the sand: I am nearly sure, it is necessary to ascend the hil
036r052 ence; action on land different Volney, P 351. Vol I. woody bushes, «gazelles» hares, grasshoppers & R
038r057 that there must be a central core of melted rock--I think the strongest is the consideration of the s
041r065 lephant came towns driving by the want of water.--I believe in all flat countries. years of drought a
043r070 oral reefs (imperfect in latter). <At> Lyell. Vol I. P. 316. Earthquake of 1812 affected valley of Mi
044r071 would sea remove more internally than externally--I did not see any number of dikes in the cliffs.--w

Page **(Key Word)**
045r075 ic veins solution of silex & many other phenomena I do not believe that the extraordinary fissures of
045r076 re rains are so infrequent; so as to exclaim, «as I have heard» how lucky! when they hear of a place
046r079 icles altered.-- With respect to Volcanic theory. I want to ground, that the first phenomem. is an in
046r079 lutonic rocks are generated as often as Volcanic. I consider latter as accidental on the afflux of th
047r081 watching heavy swell, sea retreats & then breaks: I e to form a wave in ocean. is not this [Fig. 3] {
047r081 in ocean. is not this [Fig. 3] {P} form present, I e a part below «mean» level before the higher par
047r082 idental retreat at Portugal & Madeira (Lyell. vol I. P. 471) is explained. also the similar fact at C
047r082 ry evident movement.--The swelling first on beach I cannot understand, without (cs <[...]> raised abo
047r083 In great Calabrian wave did not sea break first? I can imagine from local form of coast (as seen in
048r084 ls on banks of Red River Louisiana. V. Lyell. Vol I. P. 191 State at St Helena. pebbles entirely coat
049r088 olcano or as an injected mass.--From conical form I incline to <latter> former; & thus occurring in g
050r090 e fragments, nature of which is doubtful. P. 180. I think my Ascension case very doubtful. -- In Icel
050r093 f Mountain Limestone in N. Wales. was it reef. -- I remember many Corals?? Breccia--Stratification? A
051r094 In smooth seas (& even turbulent as at St Helena) I have mentioned point of greatest action; I now ha
051r094 elena) I have mentioned point of greatest action; I now having seen Pernambuco believe much is owing
051r094 ere force of waves: & action on upper tidal band, I do not see how to account for oceans power.--exce
051r095 t into play; most manifest example of degradation I ever saw on beach near Callao.--From Sir. H Davy
052r098 dings introduce this discussion.--Brazil bank: (& I believe SE coast of Madagascar. where a --40 line
056r109 nd so encircled, the one slow cause is apparent. «I confess I never see such islands whose inclinatio
056r109 rcled, the one slow cause is apparent. «I confess I never see such islands whose inclination natural
057r113 r bones washed about much at Coll. of. Surgeon's? I really should think probably that B. Blanca & M.
057r114 discussion of camel urge S. Africa productions.-- I think in Patagonia white beds having proceeded fr
058r118 reat continents). Voyage aux terres Australs Vol. I. p. 54. M. Bailly says."en effet toutes les monta
059r121 Brecciated structure of all the Pitchstone (which I have seen). is a kind of concretionary structure,
060r125 Indian habitation above regions of vegetation.--«I can find nothing.» Mem Carolines quotation from T
062r128 two sides of Cordillera, where climate similar.--I do not know botanically = but picturesquely = Bot
064r135 ides.--thinks them same as recent species. -- May I not generalize the fact glaciers most abundant in
065r140 ove began to freeze. correspond to September ¿Did I make any observations on springs at S. Cruz.???--
070r154 is Let it not be overlooked that except by trees, I could not see trace of Subsidence at Uspallata.--
070r155 n latter.-- According to Mr Brown, a person (whom I met at S. W. P.) the Cordillera extend to near Sa
077r170 rock.--this N degree 2. superimposed on N degree I. even No. 2. might be mistaken for Porphyry above
078r174 'Orbigny has described it with care to 3 species. I think I have much additional information ‖ Guanax
078r174 has described it with care to 3 species. I think I have much additional information ‖ Guanaxuato, wh
079r177 l fish purple die, marevellous statements on, Vol I, P. 168. on coast of Guayaquil, same as Galapagos
087a010 ount.-- Alluvial plains of Mississippi -- No Vol. I. p212. Cuvier Oss Foss Wide range of Mammalia rea
090a021 lay common to Guyana said to extend to Cordillera I see Brewster speculates from believing meteorolit
090a023 , that was weighed,; <but> carrying on this ratio I can count 90 stones which have fallen in the 50 y
097a043 blackish colour like a dull & poor varnish, which I conceive to be analogous to the black glazing obs
098a045 ssion on Cleavage &c Geolog Transacts. Vol III. p I. p. 86. et p 95.-- It is easy to prove. (pyrites,
098a047 ization, & therefore as a consequence aggregated (I assume the same force which draws together two pa
098a048 al with layer of flint on calc.: sandstone. (& as I believe most strata) (Hence endless passages from
103a064 eing too large & pitching against each other, is, I suspect much weakened by <vi> considering how clo
107a078 n method of cooling--Very difficult subject. PP-- I think from dislocation taking place chiefly benea
109a084 bergh on flints in chalk must be studied-- though I do not think good p. 411 When discussing concreti
114a094 ea. Lyell confounds these introduce discussion -- I see Lyell talks of different composition using di
114a094 tion using difference in metamorphic action which I give at C. of Good Hope.-- A bare hill of greenst
114a095 cause of denudation, but does not tell period.-- I cannot help suspecting that clay-slates have been
118a104 rschel's view of cause of continental elevations (I) the alternation of linear bands of movement in I
120a108 but lava shows the rocks really hot. & therefore I doubt the thermometer. Is not common salt more so
121a112 Patagonia, & many shells in parts on surface, but I saw none embedded this point would be worth exami
122a115 icker, then when fluid moved [...] {P} August 25. I saw metamorphic conglomerates on shore of Loch Lo
123a116 - if so why not metals. The theory of veins will, I suspect be greatly aided by considering space for
127a127 Beaum. Memoires of French Geolog. Cantal Vol III I? p. 246. on formation of cones beneath sea.-- wit
128a130 nsactions of the Maryland Academy (at Athenaeum.) I. Part. I Vol.-- some notices on modern Tertiary s
128a130 of the Maryland Academy (at Athenaeum.) I. Part. I Vol.-- some notices on modern Tertiary strata on
128a130 otices on modern Tertiary strata on coast of do-- I believe?? coast of North America., like the Mexic
129a132 urious account of great subsidence «20 miles long I in with.» which must have been from an axis, «20
130a133 ter springs off coast of Persia In Glen Roy paper I show crust yield easily. & if easily must be thin
130a134 ott Mem. Acad. Peters. Scienc Math. Phy-- Nat. t. I, 1831. sur le temp du globe on Volcanos &c worth
132a137 des Sciences. (Sc Math. Phys. et Naturelles. Tom I. p 501.-- shows first that data wholly insufficie
132a138 but in Ocean does not. (see resumè p. 536)-- «NB. I cannot understand the argument, that cold <oceans
133a141 molten rock.-- Voyages of Adventure & Beagle vol I. p. 2 & 3. Porphyry at St. Elena. p. 6. few «livi
134a143 rghaus Chart of do Journal of Asiatic Society Vol I. {T} p. 145. on salt mines of Punjab p. 149. on t
136a147 -- letter in drawer with important letters-- When I come to treat of the age of the Pampas Deposit, I
136a147 I come to treat of the age of the Pampas Deposit, I may properly remark on the superiority of Lyell's
145g006 r 4 seams/ 3 or 4 inches thick-- {P} 35 degree is I believe about greatest dip of sandstone in upper
146g013 reservation in situ of even imperishable pebbles/ I am nearly certain there were none on surface of a
147g015 t cleavage-- in other irregular horizontal strata I suppose these upper patches if prolonged would in
147g017 lofty, & no trace of it; to the Sea more probable I did not look carefully for Marine remains-- Some
148g022 nearly as high as highest measurement but nature I am quite doubtful of as I am of all the Alluvium.
148g022 measurement but nature I am quite doubtful of as I am of all the Alluvium. At Mouth of Caledonian Ca
148g025 ogs), but they thought the breed liable to vary-- I asked this question in many ways & received same
148g028 e side-streamlet had cut them out-- In all cases «I urge» deposition marine-- because if not chain of
155g063 t most faintly on East side of Glen Turrit, where I believe they end in upwards inclined plains, as i
155g063 in upwards inclined plains, as in Corry. & as «as I believe in side ravine above houses of Roy» Maccu
155g063 ve houses of Roy» Maccullochs supernumerary shelf I doubt, much about «50 or 60 ft» «no doubt, a moun
157g073 th <near> only usually contains many pebbles, but I believe this is chiefly caused by its being lower
160g088 we descended, it is to the west of Glen Tarf What I called Alluvium shows the ascending fringes {P} w
160g090 h contorted gneiss «narrow sharp ridge with peak» I walked all round hill. Boulder about 20 ft. below
161g097 t Isthmus appears above level of shelf certainly) I took another measurement on short buttress but no
162g100 68 Air 65 degree? <.194 372 about 267 28.75 .105 I reached> 29.090 Preservation of form of land very
173b010 hybridity &c &c <assuming all> if species (<a>) «(I)». <fr> may be derived from form (2). &c.-- <(> T
175b016 because time short & no great change has happened I look at two ostriches as strong argument of possi
175b017 as we see them in space, so might they in time As I have before said isolate species <& give even les
180b036 d, which is not case,∴ MONUCULE NOT DEFINITE LIFE I think {P} Case must be that one generation then s
180b037 l extinct forms, for if each species «an ancient (I)» is capable of making, 13 recent forms.-- Twelve
185b057 egards to other tribes in that same family.-- (NB I see Waterhouse thinks Quinary only three elements
186b058 some one character of all its family; but why so? I can see no reason for these. analogies; CD[from t
186b058 vism, where real structure obliged to be altered, I can conceive colouring retained; therefore probab
186b058 back to common ancestor of Crysom. & Heterom, but I cannot understand the universality of such law.--
186b060 tly Pachydermata, less so in Miocen & so on.-- As I have traced the <type> great quadrupeds to Siberi
187b065 ot procreate, he has no issue, so with species.-- I should expect that Bear & Foxes &c same in N. Ame
188b067 o Northern species replace <No> Southern kinds-- (I) Gnu reeaches Orange river & says so far will I g
188b067 (I) Gnu reeaches Orange river & says so far will I go and no further:-- Prof. Henslow says. that whe
189b070 each with same kind of coat.-- If we could tell, I do not doubt even colour hereditary in time as in
192b088 re eocine rocks. we can only expect some steps.-- I may ask whether the series is not more perfect by
193b092 hy of plants. Essai sur la Geographie des Plants. I Vol in 4 degree.-- I have abstracted Mr Swainson'
193b092 ur la Geographie des Plants. I Vol in 4 degree.-- I have abstracted Mr Swainson's trash.-- at beginni
193b093 stribution of animals Brown Geograph Journal. Vol I p. 174. says from Swan river long South coast all
194b095 hed it would go to show a centrum for Mammalia.-- I really think a very strong case might be made out
195b098 e believe the Creator creates by any laws. which, I think is shown by the very facts of the Zoologica
196b105 eas had existed in different Continents In plants I believe not..-- It is a very great puzzle why Mar
196b110 great Kingdoms.-- Principes de Zool: Philosop:-- I deduce from extreme difficulty of hypothesis of c
197b112 animals appear.-- yet nothing about propagation= I see nothing like grandfather of Mammalia & birds
198b113 class approach on the confines? Balanidae?-- --- I cannot understand whether. G. H. thinks developen
198b114 ws & & then leaving all to follow consequences.-- I cannot make out his ideas about propagation His w
199b118 inct." Ed. N. Phi. J. p 297, No 8 Jan-Ap. 1828 -- I take higher grounds & say life is short for this
202b131 e with humps as in India. Geograph J. Vol III. P. I. p. 17 (Lat 37 degree about) Vol IV P. I. Geograp

```
202b131 III.  P. I. p. 17 (Lat 37 degree about) Vol IV P. I. Geograp. Journal. Voyage up the Massaroony by W.
203b137 tyrant  flycatcher doing the service of a swallow I think we may conclude from Australia & S. America
205b143 . 122 A Terrestrial annelidous animal p. 347. Vol I.-- compare with my planariae Leaches out of water
209b157 tarum. relation of genera to species in France is I: 5.7 in Laponia I: 2,3: Mem. Lyell on shells.-- {
209b157 genera  to species in France is I: 5.7 in Laponia I: 2,3: Mem. Lyell on shells.-- {T} Genera In North
209b157 m. Lyell on shells.-- {T} Genera In North Africa. I: 4,2 Iles Canaries I: I,46 St. Helena I: I,15 {T}
209b157 {T}  Genera In North Africa. I: 4,2 Iles Canaries I: I,46 St. Helena I: I,15 {T} Calculate my Keeling
209b157 } Genera In North Africa. I: 4,2 Iles Canaries I: I,46 St. Helena I: I,15 {T} Calculate my Keeling Ca
209b157 h Africa. I: 4,2 Iles Canaries I: I,46 St. Helena I: I,15 {T} Calculate my Keeling Case: Juan Fernand
209b157 frica. I: 4,2 Iles Canaries I: I,46 St. Helena I: I,15 {T} Calculate my Keeling Case: Juan Fernandez
210b158 rate parts of Teneriffe, the proportion of genera I: I. I can understand in «one» small island specie
210b158 e parts of Teneriffe, the proportion of genera I: I. I can understand in «one» small island species w
210b158 arts of Teneriffe, the proportion of genera I: I. I can understand in «one» small island species woul
210b159 says from Humboldt, in Laponia. genera to species I. 2,3-- From Mackenzie Iceland there 144 genera &
210b159 enera & 365 species of plants not cryptogamic but I . 2,53.-- In know varieties. there is analogy  to
213b172 ting the cross most well worthy of observation.-- I think it is certain strata could not now accumula
217b189 rouse & pheasant-- Magazine. Zoology & Botany Vol I p. 450 There is in nature a «real» repulsion «amo
222b203 acters is attempt at returning to parent stock.-- I think we may look at it so--?? It holds good even
222b204 es between land shells of Porto Santo & Madeira-- I believe very curious-- My idea of propagation alm
222b205 which can only be adaptation to changing world:-- I cannot for a moment doubt, but what cetaceae & Ph
225b217 mediate forms.-- Opponent will say. show them me, I will answer yes, if you will show me every step b
225b217 show me every step between bull Dog & Greyhound). I should say the changes were effects of external c
225b217 during  passage of birds (where is this statement I remember. L. Jenyns. talking of it) or how to mak
225b220 liams. zoology of South Sea islds. any animals?-- I believe none.-- Canary islds.? Madeira? «Tristan
226b222 ica, the jaguar has been left & Fox, & bear.-- If I had not discovered channel of communication by wh
227b224 vora than Pachydermata If my theory true, we get (I) a horizontal history of earth «within recent tim
230b235 y Lesson & Chamisso.-- Geograph Journal Vol V. P. I. p. 67. Dr. Coulter on decrease of population  in
234b255 latch  & back.-- Frogs attempted to be introduced I isle of France p. 170. «Fish introduced» Hump bac
234b255 backed  race of cows from Madagascar-- p 173. Vol I. Voyage à France-- Par un Officier du Roi.-- Mack
239c002 t offspring would be chesnut.-- On this principle I may add, that fact of half cross with parents, go
240c013 es.-- Voyage. de L'Astrolabe Zoologie. p. 60. Vol I. Cynocephalus. niger. comes from the Moluccas «Ma
240c014 ad only been found in isle of Aroe & Solor), «Vol I» Likewise new species of Parameles, which joined
241c014 t from that of the Marianna islands & at Amboina» I fancy there is marked wild breed of oxen at Java.
246c026 s variety at Madagascar, Calcutta & Sumatra,. but I do not see how it is known that they are varietie
247c029 , of very many colours, like the cattle, <">which I say "are as variously coloured as a herd in Engla
250c037 eme both in, N & S. America, is only common cause I can conceive of destruction of Great Animals in E
250c040 ction with Latitudes!? Zoological Journal.--- Vol I. p. 81. Capromys, West Indian isld. p. 120. «ref.
252c045 cies so distinct that all analogy from each other I do not know how different Sumatra Java ----------
253c047 out close species of Birds. Zoolog. Transact. Vol I p. 165.-- <a> "an account of the MANELESS lion of
254c047 skull  very slightly different.-- Q Zoolog. T. V. I p 389. Owen remarks on Entozoa, the organs of ge
256c054 ly) yet a glance would tell from which country,-- I often disputed for a moment,-- Galapagos, S. Amer
260c067 " Study Bonapartes list In the Zoological Journal I read a curious account to show that very many bir
263c078 ry-- so analogous to what we see in bodily. that <I> it does not stagger me.-- What circumstances may
264c082 ail of some ducks aberrant from-- habits.-- Gould I see quite recognizes habits in making out classif
265c086 ferent habits-- Thus bill & nostril of Puffinuria I think we may clearly attribute to heredetary orig
266c088 much  upon knowing to what type a bird belongs.-- I conceive without knowing from which country  many
266c088 s come it would be impossible to classify them.-- I would [not located] Musalman's of the  Peninsula,
268c096 Soc. Feb. 1838.-- Annals of Natural History. «vol I» p. 159 curious account of Tit mouse feeding youn
268c096 ent habits.-- showing how blind a storge From what I see of S. American birds. genera blend into each
269c102 ce in Metamorphic action.-- Geograph Journal. vol I. p. 17 & excellent sketch of plants of New Holla
269c103 ure, their life & affinity" in Scientific Memoirs I can see that perfection may be talked of with res
272c107 its) of other, thus in Chalcidious insect, which I brought from Australia, probably live in  flowers
272c108 clinging to rushes similar.-- The question which I more immediately want are there Heteromera, which
273c112 e of its whole family where female is not dull.-- I must observe that this preeminent structure is no
274c114 lisuga. Kingii very rasorial for type.-- Now here I must observe these characters vary in degree in l
274c115 ure for vegetation.-- In conversation in Museum-- I could not discover any other clear relations besi
275c119 nius Using geograph distribution of animals, <as> I use (new step in induction) as keystone of ancien
275c120 rell.-- says my view of varieties is exactly what I state.-- &c picking varieties. unnatural circumst
275c121 ll Bald Heads Mr Yarrell will mention in his work I am sorry to find Mr Yarrell's evidence about  old
276c123 hman. did for SPECIES-- between England & France. I will do with forms.-- Mention persecution of earl
278c130 ld be some term used, when there is series. Could I not give Catalogue of Mammalia arranged according
281c138 > «all men» not all equally related to each other I cannot help thinking good analogy might be traced
281c139 be  analogous to each other-- &c &c.-- V. p. 140» I should think meaning of circular arrangement  was
283c145 "oh  turn a Buccinum into a Tiger."-- but perhaps I feel the impossibility of this more than anyone.--
283c145 led to breed hybrids produced--)» & then all that I want is granted.-- For. at Galapagos. make ten sp
286c154 him as other animal-- it is the way of mankind. & I believe those who soar above Such prejudices, yet
288c160 Madeira & <seized> ceased their migrations lost?? I conceive a bird Migrating from Falkland Isd regul
290c164 ilst tips of primaries black, by examining series I cannot doubt laws of change, Will be known.-- I
290c164 etween fowls & pheasants, is most like pheasant., I think so because very 3/4 bred.-- (hence  hybrids
290c164 red). White & common pheasants. have crossed.-- </I saw» .-- The attachment of dogs to man. not altog
292c170 n in the cause if his faith is not staggered <">= I confess. no dissertation against these views, cou
294c176 eties in wild animals-- but how to show species-- I fear argument must rest upon analogy & absence of
296c184 . & paper on genus Magazine of Zool. & Bot.-- Vol I. p. 450. 4 instances of hybrids between pheasants
298c189 so  wild a bird.-- Annals of Natural History Vol. I. p. 185 case of tit lark placing withered grass o
300c196 orthy the interposition of a deity, more humble & I believe true to consider him created from animals
302c201 ity yet often retaining a family likeness, & this I believe the case. = any animal really connecting
302c202 ubject to variation. therefore lately acquired.-- I fear «great evil» from vast opposition in opinion
302c202 ion in opinion on all subjects of classification, I must work out hypothesis, & compare it with resut
302c202 work out hypothesis, & compare it with resuts. if I acted otherwise, my premises <in di> would be dis
303c204 ow are races in This respect upper & lower, which I do not know whether it <would have> differs in pr
303c204 cissitudes of present animals.-- He-will be bold. I will venture to say unphilosophical. L'Institut 1
307c216 rgans produced in domesticated animals, in plants I presume there are.? get examples.-- for instance
307c216 rom Madagascar!!!!!! Proceedings of Geol. Soc Vol I It is capable of demonstration that all animals h
308c217 t differences carried a long way. --Case of Habit I kept my tea in right hand side for-- some  month,
308c217 that  was finished kept it in-- <right> left, but I always for a week took of cover of right side, th
308c219 would have arisen under different climates &c. Do I mean that ideosyncracy of wild animals is general
310c222 l» appears to be a puzzle against my theory.-- If I be asked by what power the creator has added thou
310c222 ought to <an> so many animals of different types. I will confess my profound ignorance.-- but  seeing
310c223 s acquired & heredetary & such definite thoughts, I will never allow that because there is a chasm be
310c225 logical chronology Annals of Natural History «Vol I??» p. 318. some remarks on Bonaparte's list of bi
312c232 possible  time. Mr Willis Long eared little dogs, I am told, go to heat, take dog. but do not  become
313c236 (lateral  with no relation to time) as in buds.-- I can scarcely doubt final cause is the  adaptation
314c236 of species to circumstances by principles, which I have given .: Those animals, which only propagate
315c241 ome Europaean & Sandwich species & some of Japan. I do not understand any new ones.-- Menoir will be
315c241 h Academy Imperial-- Paper read im 1837. semestre I suspect some valuable analogies might be drawn be
317c249 pendix (& only appendix) of Congo Expedition, NB. I met an old man--, who told me that the mules betw
318c252 from  extreme north turning white like Hares??--) I never saw more beautiful adaptation for snow-- li
321c275 Man Transactions of the Entomological Society Vol I. & 1s No of Vol II. (read remainder) when out [no
322c270 ngger &c> Mitchell's Australia Walter Scotts life I & 2d & 3rd Volumes Abercrombie on the Intellectua
324c268 taux et animaux.-- on sleep & movements of Plant/ If:4s Voyage aux terres australes. Chapt. XIX. tom
326c266 on limits of painting & poetry.-- Erasmus thincks I should lik it. The Sportsman's Repository. 4to. c
327cIBC in high points Read Volney's travels in Syria Vol I. p. 71. account of Europaean plants transported.-
331dIFC ]CD [All Selected Dec. 14-- 1856]CD Towards close I first thought of selection owing to struggle July
335d016 eneration would necessarily fail.-- In last page. I should have said, "an animal <acquires <th> any n
342d034 t imperfect.-- Are Pheasant & Grouse homogeneous? I observe Bachman calls these Hybrids new  species.
344d040 h reference to "new species & hybrid doctrine"-- I have read there are exceptions to this in some la
346d047 eedin[g] [not located] half breed liable to vary. I asked this in many ways, but received same answer
347d049 OVE on it.-- Articulate animals must articulate. <I> in vertebrates tendency to improve in intellect,
```

Page
(Key Word) I / I

348d051 le of value of difference is or can be settled,-- I believe affinity may be taken literally,, though
350d054 ces" p 23. "for Nature is the art of God" Septemb I., It has been argued Man first civilized. <note>
350d055 insects move about & see, parent «(2)», female «(I)» fixed & blind: -- Macleay observed all these fa
354d065 ility mice attach themselves to man. Sept 7th. -- I was struck looking at the Indian cattle with Bump
356d069 Humboldt). G. St. Hilaire, & Lamarck have written I pretend to no originality of idea-- (though I arr
356d069 ten I pretend to no originality of idea-- (though I arrived at them quite independently & have used t
356d071 king of facts-- Savages over whole world. (Major <I> Mitchell p. 244. vol I) spit & throw dust «accor
356d071 over whole world. (Major <I> Mitchell p. 244. vol I) spit & throw dust «according to my theory of gen
357d074 s becomes crossed. ¿is red game an hybrid?-- When I show that island would have no plants were it not
357d074 ants were it not for seeds being floated about.-- I must state that. the <p> mechanism by which seeds
358d075 a ditto Juan Fernandez do Mitchell. Australia Vol I. p. 306 "The crows were amazingly bold, always ac
359d087 parents.-- (There are some 3/4 birds «of», which I think there must be some mistake in their origin)
359d088 Penguin was crossed with hen Canadian offspring, I should say in every respect most like Penguin duc
360d092 f the Penguin duck & Herberts law of ideosyncrasy I have hitherto thought that a small difference «of
360d094 y long tail, & other in having very short tail.-- I can readily see that two first might cross easier
361d095 plumage resembled that of Cross-Beak-- In lark if I understand right, all species have same character
361d095 & female similar plumage.-- in tree sparrow, (if I understand rightly) young cock & hen, all nearly
362d100 the impossibility of discovering their origin.-- I see only some «but very strange races» of them ha
363d101 ith all the varieties.-- Habits of rock pidgeon. (I suspect Pennant has described them)-- [Study horn
365d106 of Egyptian animals not having changed is good-- I scarcely hesitate to say that if there had been c
369d114 l cause here.-- & therefore different from Hunter I should say females recede in organization from sp
369d114 in organization from specific character.-- Chapt I. Also Latent Character Hunter Animal Economy p. 4
371d128 ck, or if different dieteriorating very slowly.-- I presume most of these roses, without circumstance
372d129 e organs?]CD it is of no consequence if it does= <I do not doubt, the> Do plants loose any qualities
372d130 ers artificial division of animals, as gemmation, I consider gemmation as artificial division.-- On t
374d134 sters, & therefore somewhat unfavourable-- 28th. «I do not doubt, every one till he thinks deeply has
378d147 birds, one may be sure cock & hen will be alike-- I presume converse is not true for he says Hen & co
378d148 on ducks, the young crossed amongst themselves, & I presume with common ducks. so often, that it was
379d152 mbers 406, 407, 409, Quetelet papers are given, & I think facts there mentioned about proportion of s
383d159 sibility of becoming either sex.-- In my theory I must allude to separtion of sexes as very great d
383d160 rounded. {P} & much broader., & <more ro> «three, I believe instead of» two lines.-- «faintly edged w
385d163 sequent litters having a throw of this mongrel.-- I did not ask the question.-- His bitch will not ta
385d164 h remembering that ovarium of women (Paper in Vol I of Irish Royal Academy) have contained perfect te
385d164 eth & hair-- showing foetus has gone on growing-- I believe same has happened in boys bodies.-- Lavat
385d164 - Lavaters. Essays on Phy. transl by Holcroft Vol I. p. 195. says children resemble parents in their
386d165 because showing probably not education.-- Cannot I find some animal with definite life & split it, &
387d169 s inherit./ Annals of Natural History .p. 96. Vol I. Notice the Syngnathus, or Pipe fish the male of
389d177 ps connected with this same case (& not merely as I have stated it) it is certainly very remarkable t
391d174 have so existed.» with double union.-- At present I can only say the whole object being to acquire di
397e003 are acquainted."-- this applies to one species-- I would apply it not only to population & depopulat
398e005 - «if the change could be shown to be more rapid, I should say there was some link in our train of ge
399e009 -- with long ears-- & longer hind legs??? so that I was almost doubtful which it was.-- do hind legs
400e012 r?)-- length of tail varies, & character of fur-- I am sure a very good case, might be made out of va
401e016 slip, one is tempted to think here some anomaly-- I can fancy cowslip producing primrose return to ol
402e018 th near Canton" "Here, as in all Malay countries, I noticed a peculiarity in the cats «p 10» the join
402e019 us., & skins of galiopithecus.-- Malte Brun. Vol <I> II p.,133: at Samar SE of Luçon, many monkeys, b
405e031 in the Americas probably did not.-- Octob. 25th. I observed in Windsor Park.-- the «Fallow» Deer. wh
406e037 eir tails.-- November 1st..-- Addenda to Journal. I show erratic blocks transported far S. in Norther
407e039 t have survived this mundine change..-- Therefore I argue from this that Africa «& East Indian Archip
407e042 live & multiplied, specifically & individually.-- I see clearly from F. R. it will be highly necessar
407e042 to show that if species fall, genera must. Lesson I remember says Mariana Deer very close to a Molucc
409e047 g & a porpoise was not thougt overwhelming.-- yet I will not shirk difficulty-- I have felt some diff
409e047 overwhelming.-- yet I will not shirk difficulty-- I have felt some difficulty in conceiving how inhab
409e048 line descent. My theory gives great final cause «I do not wish to say only cause, but one great fina
409e049 ty on other» hence not social instincts, which as I hope to show is «probably» the foundation of all
409e049 dispute, [although, that it was the sole object, I will dispute, when I hear from the geologist the
409e049 that it was the sole object, I will dispute, when I hear from the geologist the history, & from the A
410e051 ws of variation of one part affecting another.-- (I from looking at all facts as inducing towards law
410e052 skins worse than useless.-- yet there is no cure «I may say all this, having myself aided in such sin
411e056 g as showing some change in habits before form.-- I have already given various examples The Pipe-fish
414e060 ged, since Cretaceous period, whether progressive I know not.» (& insects.-- Stonesfield????). Have M
414e065 ke that of animals transported by floating ice.-- I agree with Mr Lyell., man is not an intruder.-- :
415e066 ves.-- There being no fossils, the only way, that I can see to discover whether the parent of man was
415e069 een so, & another to show how it came to be so.-- I speak only of the former proposition.-- as in «ra
416e071 land animals. so there are true hermaphrodites.-- I suspect this rather effect of liquid semen: there
419e085 respect to the non-development of Mollusca, which I have sometimes speculated might be owing to absol
420e088 og-- leading towards Hyaena.-- see Comte Rendu.-- I suspect good case of fossil filling up blank.-- C
421e091 motely related to the habits & food of an animal, I have always regarded as affording very clear indi
423e096 velopement in partial classes is far from true.-- I doubt not if the simplest animals could be destro
424e100 er on geographical distribution of Crustaceae.-- (I forget whether I have already referred to it.-- a
424e100 l distribution of Crustaceae.-- (I forget whether I have already referred to it.-- also on spermatic
424e102 odd since these crows are mixed in England-- for I presume Carrion Crow is found in Edinburgh.-- Why
427e110 avelled S. From the analogy of the animal kingdom I should suppose, that the pollen of crab, would PO
428e113 this to effect of male sex on locomotive system» I am bound to insist honestly that the sudden, chan
428e113 ge from Primrose to Cowslip is great difficulty. «I should doubt if wild species ever formed like sho
429e113 n external conditions» sudden loosing of horns.-- I do not believe this Nature's plan.-- Whether we c
429e114 we hear from authors (Ramond. Hort. Transact Vol I. p. 17 Append) that in the Pyrenees, that the Rho
429e115 n of sand turns the balance.-- Hort. Transact Vol I. M. Ramond. p. 19. do says lofty Alpine plants of
430e119 Folkstone, which is exactly intermediate between I. concentricus & I. sulcatus.-- the beak of this o
430e119 is exactly intermediate between I. concentricus & I. sulcatus.-- the beak of this one has concentric
430e119 wer part rayed longitudinally (give woodcut) like I. sulcatus.-- Both species are found at Folkstone.
431e122 nstance of such facts.-- Ask Phillips.-- The more I think, the more convinced I am, that extinction p
431e122 Phillips.-- The more I think, the more convinced I am, that extinction plays greater part then trans
436e137 cause?? Seeing the beautiful seed of a Bull Rush I thought, surely no "fortuitous" growth could have
436e138 iliar History of Birds by the Rev. E. Stanley Vol I. p. 72.-- Goldfinches placed near, but not in sig
439e144 so) shows either there is not so much crossing as I think,, or that these varieties have become as fi
441e149 s would a Crab tree vary if planted in rich soil, I presume not, but its seeds, I presume, probably w
441e149 anted in rich soil, I presume not, but its seeds, I presume, probably would-- at least the experiment
443e155 dite in beings with which have fluid sperm.--]CD I utterly deny the right to argue against my theory
443e156 a planet to our lives-- Being myself a geologist, I have thus argued to myself, till I can honestly r
443e156 f a geologist, I have thus argued to myself, till I can honestly reject such false reasoning Bell Bri
448e166 ding to Brown.-- Voyage of Adventure & Beagle Vol I. p. 306 Shells, as well as plants «of Juan Fernan
448e167 anama.-- These two cases highly improbable.-- yet I can see no other way of accounting for them.-- Th
448e167 es Bennetts Wandering Vol II. p 155. By inference I imagine that there are Baboons in St Thomas on W.
451e176 no like himself said not to be common-- probably, I should think grandfather first of race & if so, f
451e177 > abound with monkeys & squirrels.-- Horsbrugh E. I. Directory. Vol II. p. 46 Carimon Java. (between
451e177 <Scientific Soci> Journal of Asiatic Society Vol I. p. 261. <J> Catalogue of Birds of India.-- p. 5
451e178 ed to its body-- Journal of the Asiatic Soc. vol. I. p. 335. Catalogue of animals of Nepal. by. B. Hod
453e182 the tip crooked.-- is the form [....] Dampier. Vol I. p. 320. says no wild (carnivora) beasts on Phill
453e182 s, but much more small & shrill".-- Humboldt. Vol I. p. 275. says Teneriffe does not countenance the
453e183 inctly, that Hollyoak reproduce each other. & yet I presume seed raised in same garden.-- now this go
458t002 ul perversion of instincts-- Beechey's Voyage Vol I. p. 499. «4to. Edit»-- Horses in Lao Choo so smal
461t037 p. 97. for Man Chapt see Yarrell Syngnathus Ch 6 I presume, from my theory, as long as any structure
463t057 r explanation will probably reject this theory-- (I must answer it by rooting out curious cases of in
466t093 in crawling out brush over anther & pistil & one I SAW IMPREGNATE by pollen with which «bees» «a bee
466t093 it generally.-- In Azalea <do> «it is so» <Though I saw no Bees «several» visiting it>.-- In yellow d
466t094 amens bend up a little In a wild purple Geranium, I see Bees visit always base {a} of upper petal fro
467t103 go to nectar at foot of upper petal standing on «I saw Bee go to two species of Lupine,» two wings.

467t103 en comes out with anthers & stigma in slit» -- As I think they do in Broom & certainly when over-ripe
467t103 room & certainly when over-ripe & half withered-- I saw Bees going to clover & once this happened.--
467t104 s of yellow pollen protrudes at sheath.-- At last I saw Bee collecting pollen from <sheath> Keel of L
468t105 r with legs & take flight-- Yet we have crosses-- I see Bees almost every flower-- Blue-bells-- wild-
468t112 <flower>' plant to plant.»-- to my grt surprise-- I found all, stamens straightened pollen profusely
470t178 er On many Papilionaceous; all wh. are in flower «I saw Bees;»-- on Monk's Hood, brushing over stamen
470t178 on Monk's Hood, brushing over stamen «Egg Tree»--I think never on the Galeum saxatile & other common
470t178 never on the Galeum saxatile & other common kind--I think not on Phlox though they examine it.--«Litt
472s02r was routed. wh. was not case, on several flowers I examined some days ago-- This Bee flew from yello
472s02v red) so whole sides of flower & stigma dusted.-- <I think> When It first alights, it cleaned sucker &
472s02v hink> When It first alights, it cleaned sucker & <I think> pollen was scraped off, which appeared lik
477z001 nariae by Johnson CXII. & CXV do Azara Voyage Vol I. p. 196. According to Charpentier de Cossigny. onl
477z002 was introduced to Mauritius. 18 Azara Voyage Vol. I. p. 279 Thinks the Moruffetes of Chile different
478z004 ennatula showing it to be one animal In Australia I was assured wild dog copulates <.> freely with ta
479z005 Purple die Marvellous stories Ulloa's Voyage Vol I, p. 168 Ceratophytes common in Northern sea. Cham
480z007 n on Berre. do-- Magazine of Zoolog & Botany. Vol I p. 358. D'.Orbigny <considers> states that young
481z009 abitant of Patagonia. Mem:-- S. Cruz. Molina Vol. I. p. 244. Baccalao. migratory fish.-- See Kings dr
481z010 rra del Fuego killed in Chile. Dobrizhoffer., Vol I. p. 310. History of the Abipones-- says «the Condo
481z010 e details appear very trifling Also Berre «p. 8» (I think Planariae) Sagittella, or Fleche «p. 8» my
483z012 le of France: <Jerrold?> Bibron Zoolog. Journ Vol I. p. 125, owls seen crossing Atlantic. fact taken
483z013 papers on Intestinal worms must be studied in Vol I, Zoolog: Transact. before writing on Planariae or
486z019 of N. America & Europe. Entomolog. Transact. Vol I. p. 130. Col Sykes on balls made by dung beetles,
490q001 -- analogous or quite distinct from recent ones-- I presume some recent not found fossil (perhaps not
493q003 ortoise-shell Cats. as only one sex so coloured = I have grey-cat «wh was female» with tinge of torto
493q003 in Persian Cat-- , in Brazilian «toothless» dog-- I St. Hilaire says length differs in different cat
494q005 collapse during sleep & do of Berberies-- (latter I think certainly not) (3) Sow seeds & place cuttin
495q05a ing magnet within & see which way they fly.-- (9) I have noticed leaves covered with Honey-dew dusted
496q006 Holly-seed & they grew» (9) Place. Snap-Dragon. (I have seen one monstrous) Fox Glove & such like in
498q007 tion in plants raised by Scions, as Elms. &c &c-- I have some reason to suspect Elms.-- & Orchidaceo
498q008 e Seedsmen select at all from the plants? If not, I am surprised <plan> such plants do not degenerate
498q008 (11) Must pollen grain be whole, to impregnate?-- I presume only stigma impregnable.-- (12) At Maer C
501q011 each other & see whether cross can be obtained-- I name these three plants. because they cannot be c
501q011 ese three plants. because they cannot be crossed, I think, I expect, except by very minute insects.--
501q011 plants. because they cannot be crossed, I think, I expect, except by very minute insects.-- (30) Get
504q13v , on account of Van Mons views-- Also PEAS-- N.B. I think very likely the Peas to cross ought to be p
513q21. ains.-- Is setting of fruit. cross Conception--(‖ I could extract nothing from him)‖ Does impregnatio
515q022 -- Cover that little Ervum in Sand-walk, on which I think I have never seen Bee visit. Experiments in
515q022 that little Ervum in Sand-walk, on which I think I have never seen Bee visit. Experiments in Garden
522m011 ber Jack Baldwin at school.-- Answered To be sure I do.-- What became of him.-- Answ Had large fortun
522m014 ly when a person is tired is akin to insanity.-- <I know the feeling also of depression, & both these
522m014 & both these give strength & comfort to the body» I know the feeling, thinking over somebody who has,
523m015 nfined him. & yet disinheriting her.-- This «N B. I have read paper somewhere on horse being insane a
523m018 overed.-- <Sup> Sometimes comes on suddenly from <I> (in one case ipecacuhan-- not acting) in others
525m023 -- How is this?-- Does memory bring in old ideas <I have elsewhere remarked do> Dogs take pleasure, w
525m024 their actions do not look like fear, but shame.-- I cannot remember instances, but I feel sure I have
525m024 ar, but shame.-- I cannot remember instances, but I feel sure I have seen a dog doing what he ought n
525m024 e.-- I cannot remember instances, but I feel sure I have seen a dog doing what he ought not to do, &
526m029 ot imagination.--» Thinking over the scenes which I first recollect, «at Zoos» they are all things, w
526m030 se qualities become heredetary.-- When a man says I will improve my powers of imagination, & does so,
527m031 ongly,-- & what is frame of mind owing to.--<)>-- I verily believe free-will & chance are synonymous.
527m034 often, but not of an inventive class.-- Now that I have a test of hardness of thought, from weakness
527m034 hardness of thought, from weakness of my stomach I observe a long castle in the air, is as hard work
529m040 eceive pleasure from thinking of the fertility.-- I a geologist have illdefined notion of land covere
529m041 cal) the botanist might so view plants & trees.-- I am sure I remember my pleasure in Kensington Gard
529m041 otanist might so view plants & trees.-- I am sure I remember my pleasure in Kensington Gardens has of
532m053 instinct Fear must be simple instinctive feeling: I have awakened in the night. being slightly unwell
532m057 - but violent action.-- To avoid stating how far, I believe, in Materialism, say only that emotions,
533m058 s know a frown very early in life, <before they> (I think I have seen same thing before they could un
533m058 frown very early in life, <before they> (I think I have seen same thing before they could understand
533m060 fully expressing momentary feelings of gratitude, I had a sort of consciousness I was not right; thou
533m060 lings of gratitude, I had a sort of consciousness I was not right; though I never realized the idea t
533m060 d a sort of consciousness I was not right; though I never realized the idea that I was tending to mak
533m060 not right; though I never realized the idea that I was tending to make myself in act less grateful.-
533m061 ot merely morally wrong, but hurting my character I felt it)-- this is kind of conscience, is obscure
533m061 thought of some such remarke as now advanced; for I caught it like a flash.--. strange if judgment re
534m062 tions of the Entomological Society of London Vol. I. p. 106. Col. Sykes on Formica indefessa placed t
535m063 ch intellect. yet habit may make it act wrong, as I have done when taking lid off <tea> side of tea c
536m070 rt beat» remember how Pincher does just the same; I noticed this by perceiving myself skipping when w
537m075 - «from difference of action of approved» Yet as, I think, the opposite side has been shown-- see Mac
538m078 here seem other cases somewhat analogous, & which I think will lead to fact of old people singing son
539m081 t well when reading article by Boz.-- now in this I was interested as was I in the other, & read so i
539m081 cle by Boz.-- now in this I was interested as was I in the other, & read so intently as to be unconsc
539m082 ging up old indistinct ideas of FitzWilliam Musm. I was amused at this after seven years interval. Au
539m083 its.-- Aug. 16th. As instance of heredetary mind. I a Darwin & take after my Father in heraldic princ
540m089 person cry for joy? 17th. August Montaigne (Vol. I) has well observed, one does not fear death from
541m089 How has this instinctive fear arisen? 19th. When I went down to Woollich I was trying to unbend my m
541m089 e fear arisen? 19th. When I went down to Woollich I was trying to unbend my mind as much as possible
541m090 k minded people are fickle & full of levity (¿ do I not confound action & thought here?) The opposite
541m090 oy paper.-- this greatest mental effort, of which I am capable-- I suspect from these facts that whol
541m090 s greatest mental effort, of which I am capable-- I suspect from these facts that whole effort consis
542m094 ing to relieve circulation after stillness.-- Now I conceive if organization were changed, I conceive
542m094 s.-- Now I conceive if organization were changed, I conceive sighing might yet remain just like sneer
545m105 e to ideas so much. as objects of interest.-- do/ I was much struck with observing how the Baboon (<M
545m106 es, at every emotion & <look> <turn> of the head. I could not perceive «any» distinct wrinkle, but su
545m106 in skin of eyebrow important analogy with man.-- I see monkeys grin with passion, that is show all t
545m106 so fast to limbs that they cannot remain still.-- I do not doubt this Baboon. knew women.-- Another l
545m106 -- Another little old American monkey «(Mycelis)» I gave nut, but held it between fingers, the peevis
545m107 monkey when touched, also another monkey to dog. I showed nut & then closed my mem. expression of fu
546m108 oes not like Burke explain pleasure. August 26th. I cannot help. thinking horses admire a wide prospe
547m111 - my servant was in the room. with my trunk out & I was engaged in hurriedly giving orders.-- Now wha
547m112 closed probably by sleep & not vica versa. anyhow I might have been quite still, & not attending to b
547m112 s connected with past circumstances.-- as whether I really was going to Shrewsbury, whether I had run
547m112 whether I really was going to Shrewsbury, whether I had rung for Covington. whether he had come & ope
547m112 ington. whether he had come & opened box, whether I had thought what clothes to take (how often one c
547m113 hem, therefore could not doubt or believe.-- When I say trains, it may be instantaneous changes in or
547m113 ct of perception by senses fail first, as whether I had pulled the bell??)-- It may be deception to s
548m115 t are in progress-- In castle of air the trouble I well recollect» is in making things somewhat prob
549m121 ament future life is almost the sole object--. -- I doubt whether the last be right. The two rules co
550m126 this rather more friendship.-- Scott's. Life. Vol I, p. 127. Talks of difficulty of his own drawing c
551m129 ns pouts. partly out displeasure (& partly out of I do not know what when it looked at the glass) whe
551m129 Drinkwater's horse jumping when word Jump said-- I saw the ourang. take up a stone & pound the earth
552m132 failed, & not to any particular muscle Sept. 8th. I am tempted to say that those actions which have b
552m132 Descent of Man Moral Sense Mitchell Australia Vol I, p 292 "Dogs learn sooner to take kangaroos than
554m139 n she could not succeed of herself.-- <The male» «I saw" Jenny untying a very difficult knot-- the sa
554m139 & give food to Tommy, or anything of any sort.-- I saw Tommy picking his nose with «a» straw.-- Jenn
554m140 n knows she has done wrong & will hide herself.-- I do not know whether fear or shame.-- When she thi
555m142 (is it not present with all associated animals?) I doubted it in Fuegians, till I remembered Bynoes
555m142 sociated animals?) I doubted it in Fuegians, till I remembered Bynoes story of the women.-- The Chill

555m142 Endeavoured to classify expressions of monkeys-- I could only perceive that the American ones, often
555m143 whole skin of head they do not move eyebrows.-- (I see some of the old world ones move skin of head
555m143 un away &c having faced death like a hero, & then I had some confused idea of showing scar behind (.i
555m144 honourable wounds.-- all this was kind of wit.-- I changed I believe from hanging to head cut off. «
555m144 e wounds.-- all this was kind of wit.-- I changed I believe from hanging to head cut off. «there was
556m144 ame one after other, without ever comparing them, I neither doubted them or believed them.-- Believin
556m145 Essays on Physiognomy translated by Holcroft «Vol I» .p. 86 "We ought never to forget-- --; that ever
556m145 man who is not deformed. is born with two eyes.." I think this cannot be disputed anymore in men. tha
557m147 fight, in which case expression resembles a fox-- I can conceive the opposite muscles would act, to w
557m147 trary. when pleased tail loose & wagging-- if as (I believe) Hunter says. neither fox. nor wolf wag t
557m148 lings, no more to be analysed than fear or anger? I should think shame would be more easily analysed
557m148 e less discoverable in animals than latter.-- Yet I think one can remonstrate with a dog, & make him
558m152 k at faces of people in different trades &c &c &c I observed the Asiatic Leopard. quarrelling. mouth
558m152 me way» generic manifestation of great passion.-- I do not think they arch their backs-- Bengal tiger
558m152 ighting, & not with dog. when fear might enter?-- I believe common Swan, arch raises neck & depresses
563n002 ompanion. (mem Cyanocephalus. Sphynx howling when I struck the Keeper) may be tempted to attack him f
564n002 sorry or have a troubled conscience.-- Therefore I say grant reason to any animal with social & sexu
564n006 pper lip., is most clearly analogous to a panther I saw in garden uncovering its teeth to bite.-- the
565n010 to say "my dear," just what smile is to laugh.-- I must be very cautious. Remember how Lavater ran a
566n010 support a wife a ruling motive.-- Book IV, Chapt I on passions of mankind, as being really useful to
567n014 rter than two. V. Whewell. Induct. Sciences-- Vol I p. 334 Does a negress blush.-- I am almost sure F
567n015 . Sciences-- Vol I p. 334 Does a negress blush.-- I am almost sure Fuegia Basket did. & Jemmy, when C
567n015 et did. & Jemmy, when Chico plagued him-- Animals I should think would not have any emotion like blus
568n019 rmony common to t[he] whole kingdom of nature. If I want some good passages against, opposition of di
569n022 d by dignity-- "no mon dieu," with a shrug-- "all I can say, I am very sorry so it is"-- does not acc
569n022 y-- "no mon dieu," with a shrug-- "all I can say, I am very sorry so it is"-- does not accompany I wi
569n022 y, I am very sorry so it is"-- does not accompany I will not. I am sorry I cannot.-- Expression leave
569n022 sorry so it is"-- does not accompany I will not. I am sorry I cannot.-- Expression leave «this» out
569n022 t is"-- does not accompany I will not. I am sorry I cannot.-- Expression leave «this» out not in Libr
570n024 ribs & how he gulps in air.-- Again a master says I will see you damned first." the man shrugs his sh
570n024 ps together & shrugs his shoulders & walks off,-- I think shrugging connected with many emotions.-- (
571n030 onnection between them, & great masses of rock.-- I was much struck with this, when viewing Windsor C
571n031 s naturally & hence sublimely from natural rise-- I was also much struck in great avenue, resemblance
572n032 ich would naturally happen.-- Reynolds Works. Vol I, p. 226-- "The general idea of showing respect is
572n033 b 30th-- Dreamt somebody gave me a book in French I read the first page & pronounced each word distin
572n033 al sense of this page.-- Now <awake> «when awake» I could not picture to myself reading French book q
574n039 s> sounds singularly adapted to subject see <A> \ I think this argument might be used to show languag
574n041 a to flow «yes, certainly»-- curious association: I have seen Nina licking her chops.-- someone has d
577n049 if so ourang outang.-- oyster & zoophyte: it is (I presume-- see p. 188 of Herschel's Treatise) a "t
580n061 no subsequent ones modify it.-- «Weak people say I know it because I was always told so in childhood
580n061 s modify it.-- «Weak people say I know it because I was always told so in childhood.-- hence the beli
581n063 the phrase "heredetary habits." very clearly, all I must do is to generalize it, & see whether applic
581n065 scrape, crack, &c, imitative of the things.-- CD[I may put the argument,, that many learned men seem
583n070 as more allied to reason than instinct.) Mr Wells I can see mentally refers by reason knowledge gaine
585n076 n which they STARTED, they cannot return.-- Hence I conclude. pidgeon taken little way, whirled, & th
587n087 et things dislike others.-- dogs dislike perfume) I should think, great principle of liking, as simpl
588n090 e consequence of stooping, as sign of humility.-- I suspect very strong argument might be advanced, t
590n094 Music of nature to show barking not natural. (Vol I. p. 234) Vol. II p 153. «do». an account of a mon
591n097 to anything or excited.-- so do young dingos, as I saw wag tail when watching anything-- Keeper does
591n099 erent animals of same species.-- The general «(as I believe)» contempt at suicide. (even when no rela
591n099 the instinct of self-preservation is disobeyed-- I often have «as a boy» wondered why all abnormal s
592n101 .» however, he seems to allow it is an instinct.» I suspect the endless round of doubts & scepticisms
592n103 to convulsions.-- «Paper» must be referred to, if I follow up this subject & a reference to Brun's wo
593n111 a necessary part of man's mind.-- At Maer. Pool. I saw many coots & waterhens feeding on grassy bank
593n111 ok flight & flappered across pool to bed of flags I was astonished & having looked round saw at consi
594n112 implanted, birds soon <dis> learn to disobey it-- I have seen hawk & sparrow in Shrewsbury garden pic
595n125 orn butting, or young crocodile snapping.-- these I think are better instances of instincts (highly u
595n125 means of communication) in man, than sucking.-- [I assume a child pouts who has never seen others po
596n184 r "Humes Dissertation on the Passions." "Hartley" I should think well worth studying-- "Thomas Brown"
599o007 ave, & in the arched & leafy forests" Very good!. I grant that the thrill, which runs throug every fi
599o007 of & & or grand chorus are utterly inexplicable-- I cannot <admit> think reason sufficient to give up
599o007 with edges & crowded with towns & thoroughfares, I grant that man, from the effects of heredetary kn
599o008 nimal-- Aimé Martin de l'Education des Mères Vol. I. p. 198.-- "Moralité, raison, beau ideal, infini
600o008 -- Has little Chapter on each faculty of Soul.-- (I) <Conscience> «Moral Sentiments» imperative sense
600o08v s right by one & are ordered to do it by other.-- I suspect conscience, an heredetary compound passio
601o08b orales et physiologiques The first of these books I daresay good. 1. Sensation is the <conse> orderin
602o11b existed but in the imagination".-- Macculloch Vol I. p. 115. Attributes of Deity. on Belief.-- you be
609o29v s of the necessity of these passions, but refers (I believe) to present day & not to ruder state of S
609o030 Bees without their instincts.-- Gives art to when I say How social instincts generated? The origin of
610o031 , but could not find out-- Directed his letter, & I observed he had written Wilson & pointed it out;
612o035 period of year as much as inflorescence.-- [LHC] I here omit the case (if such there are) of animals
612o035 ing only movements such as sensitive plants. (But I include irritability for that require will in par
612o035 e so than movement of sap. or sunflower to sun? ∵ I should think there. was direct «physical» effects
613o036 o all animals modified according to species. This I suppose he deduces from the ends in each case bei
613o036 ver-- produced as soon as brain developed, and as I have said, no soul superadded, so [RHC] 6) though
615o36v instinct, leading to touch a particular organ.-- I think Pincher shows surprise, walking home one da
615o36v for five minutes every now and then howled.-- Now I don't think this only pleasure; for it was differ
621o047 lanations from origin of man.-- [RHC] By interest I do not mean any calculated pleasure, but the sati
622o049 thus avarice. &c &c.-- [RHC] 8) in the beginning I mentioned only three instincts.-- I am far from s
622o049 the beginning I mentioned only three instincts.-- I am far from saying there are not more, or that th
622o049 are not more, or that the three are as simple as I have said.-- [LHC] instinctive fear of death: of
622o049 of series in which «special» instincts decrease, I should think they were very few & general in thei
622o049 m whom, she has never received any benefit.-- Yet I think there is much truth in doctrine, for [RHC]
623o050 instincts demand some explanation [RHC] Although I cannot pretend to say how far & minutely our inst
623o050 ut to the whole past race).-- <no one> doubts) « <I cannot> » [LHC] On the Law of Utility Nothing but
625o052 think there is distinct faculty, of conscience.-- I believe that certain feelings & actions are impla
625o052 tification & hence are forgotten-- only so far do I admit its supremacy p. 37. Whewells gives Mackint
626o052 . 194. «&c &c» Butler's view given on conscience: I cannot admit it.-- see notes to it by me..' 'p. 3
627o52v s of beneficial tendency of affections.-- If ever I write on these subjects consult <following> pages
628o53v ear it might be partly made.]CD p. 21. "Why ought I to keep my word"-- gives the problem, of ethics--
628o054 nce, but not cause gives me [3] pleasure) or that I have been taught or habituated to associatical, t
628o054 aracterize the moral sense, by its "supremacy",-- I make its supremacy, solely due to greater duratio
628o054 n other passions, or instincts.-- is this good?-- I should think some parts of the emotive part of ma
628o54v will check the consciences desire for virtue.-- [I expect there is some fallacy here.-- at least poi
628o54v passion &c. whilst the passions have no relation I think this <boshes> «nonsense»-- My theory of dur
629o55v estion whether any thing is taught instinctively; I say yes, & my explanation agrees. with last head.
632j53r ributes of God Macculloch. Attribs of Deity. Vol: I it will be better always to refer to the author i
632j53r t will be better always to refer to the author if I use these facts p. 280. adduces provision of seed
632j53r was not some provision for transportation:-- But I do not want to deny laws.-- The whole universe is
632j53r universe is full of adaptations.-- but these are, I believe, only direct consequences of still higher
632j53r only direct consequences of still higher laws.-- I do not «then» believe the pappus of <th> any one
632j53r created with reference to successive developement I admit, but the admission is probably from ignoran
635j56v ne of checks & my theory) Macculloch. Attrib. Vol I. p. 330. Mentions the many cases, as in Papiliona
636j57r rganic world.-- Macculloch. Attrib. of Deity. Vol I p. 232. gives Woodpecker as instance of beautiful
638j58v y.-- Now we know what instinct is-- consider this I look at every adaptation, as the surviving one of
638j059 (with consciousness.) <th> form these schemes.-- I see no reason, why structure of brain should not
638j28r rby's view.-- Macculloch. Attributes of Deity Vol I. p. 251-- stomach hump, kinds of foot. power of c
639j28v «(as adapted to)» because their food lies deep.-- I say it is «as» simple consequence they become lon

641j29r ger Seals-- Are Porpoises numerous in cold Oceans I think not.-- Does this bear on, the absence of th
464tf6v the existence of Molina's Pudu or goat There is IBEX of Alp Pyrenees &c-- (see Blyth's work on Rumin
172b006 posites to like each other AEgyptian cats & dogs IBIS same as formerly but separate a pair & place th
346d047 oss, on some Gulls. Flamingo-- (Spoonbill Wader. IBIS)-- laws of plumage might possibly be made out.-
057r114 ata. only in upper. in accordance in Europe with ICE theory.-- Capt Ross found in Possession Bay in 7
056r141 at head. Important in forming transverse valleys ICE Sir W. Parish says they have Earthquakes in Cord
070r153 which no trace except by trees The structure of ICE in columns. show that granite when weathering in
121a111 to use. in which great Knife formed crystals of ICE were formed-- (like my gypsum case) shows power
127a125 e must be creeping «pushing» up to «the» line of ICE.-- Hence further N. when soil frozen for greater
127a125 soil frozen for greater length of time depth of ICE ought to be less.-- Memoir of the Irish Academy
131a135 ng.-- Geograph. Journal Vol. 8. p. 402.-- ground ICE-- subterranean isothermal line Athenaeum. 1839.
139aIBC ld be well worth visiting really good account of ICE.-- C. Darwin A. Glen Roy Generally received opin
217b188 interesting, because Iceland, must have been all ICE in time of ice transported.-- This gives room to
217b188 cause Iceland, must have been all ice in time of ICE transported.-- This gives room to fine speculati
225b219 s in Stonefield.. all lands united (Falkland Fox. ICE). . Mauritius what a difficulty-- where elevatio
234b255 ys cattle in Iceland. «are very» like <those of ICE Highlands of> the largest <sort> of our highland
234b255 s" --- p. 341. Black Fox sometimes introduced by ICE <no> «only few» pigs.-- birds mentioned. but few
414e065 t unlike that of animals transported by floating ICE.-- I agree with Mr Lyell. man is not an intrude
448e166 erence> p. 251. about the drifting of animals on ICE p. 643-- very curious table of all the castes fr
052r099 g in Port Famine Mr Sorrell says that numerous ICEBERGS are commonly stranded on shores of Georgia «L
510q017 Sabine says North of Siberia, no sea-current, ICEBERGS travel by wind. Aug. St. Hilaire Bot. p. 787.
050r090 I think my Ascension case very doubtful. -- In ICELAND Bladders of Lava are described, & many minute
072r159 & S. line connects western isles of Scotland & ICELAND.--Bosh nor on Norway, or Spitzbergen.--Spitzbe
096a039 Iquique. <Ceylon>. Band of Volcanic action in ICELAND parellel to Greenland: Mem.¿ Greenland subsidi
125a120 rtical trees. Uspallata.-- do p. 473. on great ICELAND stream. the 90 miles includes opposite directi
210b159 nia. genera to species I. 2,3-- From Mackenzie ICELAND there 144 genera & 365 species of plants not c
217b188 D. wonders at these species being wanderers.-- ICELAND no species to itself, a remark common to all n
217b188 northern islds.-- This is interesting, because ICELAND, must have been all ice in time of ice transpo
225b220 - Canary islds.? Madeira? «Tristan d'Acunha?» «ICELAND?--» The Connection between Mauritius & Madagas
232b247 om insular to extreme climate, <more northern> ICELAND would have possessed a most peculiar Flora «&
234b255 oi.-- Mackenzie Travel. p. 280. says cattle in ICELAND. «are very» like <those of Ice Highlands of>
234b255 and Sort, except in one respect, that those of ICELAND. are seldom seen with horns" --- p. 341. Black
234b256 body who had made great collection of birds of ICELAND. --M. Gaimard, however, will settle this.-- Wa
296c193 ble compare it with Canary Islds. Galapagos.-- ICELAND has same uniformity Primrose & Cowslip. quite
320c276 erences at end; is so said to have Mackenzie's ICELAND Molinas Chile Falkners Patagonia Azaras Voyage
481z009 - See Kings drawings.-- for real name Birds of ICELAND. Mackenzie. p 345 for comparison with Falkland
486z019 ds of do with <T> N American & T. del. Fuego & ICELAND Spix & Martius talk of birds singing in the fo
293c173 s no corelation between individual objects as ICHNEUMON & caterpillar, though our ignorance, may make
303c205 of pidgeons perfect.-- &c &c.-- do p. 136. ICHTHYOSAURUS in the Chalk Those who say «philosphically
446e160 eding & defending their young..-- The oriolus (ICTERUS Cat.) is an instance of this, & the female of
446e160 .) is an instance of this, & the female of the ICTERUS minor is a bird of more splendid plumage than
295c183 in S. America so highly developed in North.-- ICTHIOLOGY of S. America. more peculiar than its ornith
303c205 > altogether «he» has created a perfect chain. <ICTHYO> . + + +. supra & next page It is a fact pregna
444e157 der the «32» ribs are wanting. p. 144 in the ICTHYOSAURUS 60 or 70 bones in the paddle, yet all in th
024r013 shells: nothing said about K. Georges Sound The IDEA of the water at Cauquenes. coming from the [...
028r032 ctions> same cause as no colour Sir J. Herschels IDEA of escape of Heat prevented by sedimentary rock
047r083 levation in perfect conformity with <Mr Lyell's> IDEA of an injected mass of fluid rock In Patagonia
057r115 mp of which was below freezing point!!! Remember IDEA of frozen bottom or beach of sea to explain pre
061r127 blow. if one species altered: <altered> Mem: my IDEA of Volc: islands. elevated. then peculiar plant
087a011 ally very important. harmonizes well with Lyells IDEA of intertropical land.-- Siberia rises. therefo
107a079 ubul applicable to Andes & Patagonia-- On Lyells IDEA of whole centre of earth same heat, then change
113a092 y prove possibility of metamorphic theory On the IDEA of statical equilibrium, the height of lava (ha
118a103 is now cut off by denudation it gives one grand IDEA of amount denudation.-- This may be added to an
146g007 Will not curved form of hill be explained by my IDEA-- highest part must project [Blank] {P} Path Ea
173b012 ent some large country long separated.-- On this IDEA of propagation of species we can see why a form
181b045 fitted solely for air & fishes for water. If my IDEA of origin of Quinarian system is true, it will
198b113 eous system, in other nervous developed. (Owen's IDEA) states these class approach on the confines? B
198b114 Second resumé well worth studying} CD says grand IDEA in connection with Von Buch Volcanic chart & my
211b164 in connection with Von Buch Volcanic chart & my IDEA of double line of intersection.-- Dr. Horsfield
222b204 to Santo & Madeira-- I believe very curious-- My IDEA of propagation almost infers, what we call impr
227b224 ter of any «two» ancient fauna, we may form some IDEA of <origin under> connection of those two count
255c052 irds, & so for birds. We thus obtain an abstract IDEA of a bird-- An animal with skeleton <of such gen
291c167 arrogance, it our admiration of ourselves.-- The IDEA of foetus being of one «both» sexes». is stron
293c172 ing.-- Some association in such cases recall the IDEA of those parts being most easily mostrified, wh
311c226 monsters is preeminently worthy of study on the IDEA of generation being a <slip> «bud» from parent.
336d019 or rather they would not have offspring-- On the IDEA of only animals from distant countries breeding
338d025 In introduction to Eytons Anatidae.-- recurs to IDEA from cramped imagination that God created. (war
343d036 esent time, to the future-- How far grander than IDEA Septemb. 1st. Macleay & Broderip were talking o
350d054 - Australians.-- Americans. &c After Decandolles IDEA-- (though I arrived at them quite independently
356d069 arck have written I pretend to no originality of ith <gov> symmetry «& regular laws» that baffles IDEA-- My very theory requires each f
398e006 ith <gov> symmetry «& regular laws» that baffles IDEA of revolution.-- My very theory requires each f
431e123 ine of the advantage of crossing consists in the IDEA of the male being smaller, & the female larger
471tf07 ower"?.--"of their use difficult to conceive any IDEA" Linn. Trans. 18. p. 163. "D. Dod on two new ge
522m012 a thing which had just taken place.-- as if the IDEA of time had been disturbed.-- These foregoing c
522m013 ten ends in insanity or delirium.-- In Mania all IDEA of decency & affection are lost.-- most delicat
523m015 against it, his knowledge of the untruth of the IDEA, namely his poverty.-- his manner of curing it.
524m019 tly well aware that they are insane & that their IDEA is wrong.-- (Dr Ashe, the Birmingham Doctor), i
529m042 s case must have been functional.--» He has some IDEA of a son of Dr. Prietly who was cured from a fa
530m044 previous,-- because, pain prevents repetition of IDEA.-- Mr Blakeway has mentioned in Antiquities of
533m060 ess I was not right; though I never realized the IDEA that I was tending to make myself in act less g
536m071 eory probably wrong, otherwise horses would have IDEA of beautiful forms.-- With respect to free will
539m083 habit is-- In the habitual train of thought one IDEA. calls up other, & the consciousness of double
540m088 ry like best feeling of sympathy.-- Mem: Burke's IDEA of Sympathy. being real pleasure at pain of oth
541m090 this desultory thought is following out such an IDEA, as effect of sea on coves when waters had fall
541m090 facts that whole effort consists in keeping one IDEA before your mind steadily., & not merely thinki
541m091 time.-- Then if one endeavur to keep any simple IDEA as scarlet steady before mind for period, «if t
541m092 ring fall of sea.-- Is the effort greater if the IDEA is abstract as love, (or an emotion not so) tha
541m092 t as love, (or an emotion not so) than if simple IDEA as scarlet?-- How can people dwell on pain ¿ no
541m092 et?-- How can people dwell on pain ¿ no definite IDEA. nor is an emotion.-- People who can multiply l
545n104 may some ideas.-- ie habits, which must require IDEA to order muscles to do <certain> the actions.¿
545n105 becoming very often unconscious, which makes the IDEA unconscious, if so (think of this). study what
547n112 case.-- There was memory, for it related to past IDEA.-- there was a kind of ideal consciousness for
547n112 to me, that the mind is wholly absorbed with one IDEA (hence apparent vividness) & there being no oth
548n114 d or gaining by senses.-- As sleep <is> only one IDEA is awake, when one is always many necessarily ar
548n115 re works of imagination hard work.-- Keeping one IDEA present is, perhaps, hard work-- though dreams
548n116 ffering of a dear friend-- this gives one strong IDEA of what individuality is.-- Insanity is <much>
548n116 s, as shown in the tendency to forget the insane IDEA; & ones expression of double self, though as in
548n117 s singular when looking at a table one has vague IDEA something is not there, & then when one begins
553m135 asily overturned.-- so ready is change from. our IDEA of causation, to give a cause (& no one being a
553m136 it would be difficult to prove that» this innate IDEA of God in civilized nations has not been improv
555m143 ed death like a hero, & then I had some confused IDEA of showing scar behind (.instead of front) (hav
556m144 from hanging on account of blood. but all these IDEA came one after other, without ever comparing th
558m151 Martineaus one principle of charity.-- ¿ May not IDEA of God arise from our confused idea of "ought."
558m151 -- ¿ May not idea of God arise from our confused IDEA of "ought." joined with necessary notion of "ca
566n011 & beasts».-- very necessary to explain origin of IDEA of deity.-- Animals do not know they have 'thes
566n012 y notions any more than «a» Savage M. Le Comte's IDEA of theological state of science, grand idea: as
566n012 te's idea of theological state of science, grand IDEA: as before having analogy to guide one to concl
570n026 t the VII discourse by Sir J. Reynolds.-- Is our IDEA of beauty, that which we have been most general

570n026 generally accustomed to:-- analogous case to my IDEA of conscience.-- deduction from this would be t
570n027 n with male figure]CD-- As forms change, so must IDEA of beauty.-- [Old Graecians living amongst nake
572n032 .-- Reynolds Works. Vol I, p. 226-- "The general IDEA of showing respect is by making yourself less,
574n040 et into habits.-- we probably can hardly form an IDEA of a mind so limited as Birgos to become absorb
579n060 1. 1839. Herchel's Discourse p. 35. On origin of IDEA of causation; «succession of night & day does n
580n061 eristic of one kind of intellect is that when an IDEA once take hold of the mind, no subsequent ones
587n088 redetary habit.-- A blind man might be born with IDEA of scarlet, as well as remember it.-- Why do ch
593n109 ence. some obscure picture of other men. & hence IDEA of beauty.-- the social affections of animal ta
593n109 & fowls hatching stones. in some degree is so.-- IDEA of beauty of music are great distinguishing cha
602o11v eral notion of everything applicable to the high IDEA «p. 131.» in Tragic acting-- CD [My idea. would
602o11v he high idea «p. 131.» in Tragic acting-- CD [My IDEA. would make the mind have mysterious & sublime
604o017 he Human Mind. Poor.-- on insanity.-- Prevailing IDEA. owing to loss of will.-- chiefly excited by pa
604o018 teral meaning of Sublimity is height. & with the IDEA of ascension we associate something extraordina
605o19v o the ideas excited & associated with it. as the IDEA of Deity. with vastness of Eternity. which supe
606o024 now, it is therefore the embodying of a floating IDEA.-- as statue of beauty, is of the "beau ideal",
638j58v terious consciousness superadded This is similar IDEA, to cells of bee, corresponding to «every» «one
638j059 tem, gains that knowledge, before hand. & can in IDEA (with consciousness.) <th> form these schemes.-
640j167 ot separately: NB. These views quite exclude the IDEA of domesticated animals changing.-- From these
544m103 arcely really moves.-- the willing therefore is IDEAL, as all the other perceptions.-- The mind think
547m112 it related to past idea.-- there was a kind of IDEAL consciouness for moment, implied by «presence»
550m125 t> when excited by impressions, & not mental or IDEAL ones, <which> «& these» must occupy greater pro
571n029 fe time:-- the latter correspond to fashions in IDEAL taste & the former to true taste.-- Everything
593n111] Double consciousness. only extreme step of an IDEAL argument held in one's own mind, & Dr. Hollands
595n121 ling which make us think anything ugly-- a beau- IDEAL feeling. Same effect as acting on us-- <The Bab
599o008 Mères Vol. I. p. 198.-- "Moralité, raison, beau IDEAL, infini conscience; voila l'homme separe de la
600o008 ggle in man.-- two souls in one body-- (2) Beau IDEAL, refers chiefly to moral, beau desires conscien
600o008 science & love.-- [With regard to ordinary Beau IDEAL, Mem. Negro, beau,--Jeffrey denies all Beau-- H
606o022 be represented. But what is beauty?-- it is an IDEAL standard, by which real objects are judged; & h
606o024 ng idea.-- as statue of beauty, is of the "Beau IDEAL", my instinctive impression 1) September 6th. 1
092a031 1d to 3d & 19 1827 Geograp Journ There are some IDEAS about order of injected rock being determined b
171b004 ds rendered wild <through> generations, acquire IDEAS ditto. V. Zoonomia.-- There may be unknown diff
175b018 ch species changes. does it progress. Man gains IDEAS. the simplest cannot help..-- becoming more com
198b114 to follow consequences.-- I cannot make out his IDEAS about propagation His work. Philosophie Anatomi
215b178 cule" entitled "Paleontographie" developing his IDEAS on passage of forms.-- Deshayes states Lamarck
222b204 supposed to be «most» perfect (according to our IDEAS<)> of perfection); but intermediate in characte
277c128 ould answer every purpose, & would present many IDEAS of causes of change.-- The mark of analogy woul
308c219 ay be worth glancing at, as she has no original IDEAS., it will show state of knowledge «Negroes exis
398e005 e. It is curious that geology. by giving proper IDEAS of these subjects. should be absolutely necessa
399e011 stance species of Europaean genera=.-- Hope has IDEAS about generic characters. dominant. predominant
400e011 eat genus.-- because found in all quarters: his IDEAS not clear. In Australia from approach to Asiati
523m018 here are numberless people insane of particular IDEAS, «Case of Shrewsbury gentleman, unnatural union
524m021 picion can be well understood. In insanity, the IDEAS do not go back to childhood, (but appear most c
524m022 irst?-- How is this?-- Does memory bring in old IDEAS <I have elsewhere remarked do> Dogs take pleasu
539m082 thrilled across me, bringing up old indistinct IDEAS of FitzWilliam Musm. I was amused at this after
540m087 original inventive thought is that none of the IDEAS are habitual, nor recalled by obvious associati
544m102 ime, <identity,> place & personal connections-- IDEAS are strung together in manner <they> quite diff
544m103 umber, vividness, rapidity, novelty of separate IDEAS cause fatigue to the mind,-- it is solely the c
544m103 mind.-- it is solely the comparison, with past IDEAS. which makes consciousness-- & which tells one
545m104 ressions «Hume» become unconscious. so may some IDEAS.-- ie habits, which must require idea to order
545m105 viewed very often.-- former do not give rise to IDEAS so much. as objects of interest.-- do/ I was mu
545m106 erse of intellectual, there is no comparison of IDEAS-- one follows other as in blindest memory-- als
547m112 ness} & there being no other parallel trains of IDEAS connected with past circumstances.-- as whether
547m113 ntaneous changes in order <to every> calling up IDEAS of every late impression.-- (do the ideas, dire
547m113 ng up ideas of every late impression.-- (do the IDEAS, direct effect of perception by senses fail fir
548m117 Definition of happiness the number of pleasant IDEAS passing through mind in given time.-- intensity
551m128 to «Erasmus» says in Phaedo that our "necessary IDEAS arise from the preexistence of the soul, are n
551m129 th. Lockarts life of W. Scott Vol VII p. 35 "as IDEAS come & the pulse rises, or as they flag & somet
556m144 them.-- Believing consists in the comparison of IDEAS, connected with judgment. [What is the Philosop
566n010 eautiful-- is there -- anything in these absurd IDEAS.-- do they indicate mind & body retrograding to
571n027 n mind. is to me clear evidence, of the general IDEAS of our ancestors being impressed on us.-- Surel
571n031 to gloomy aisle of Churche.-- these are Mayo's IDEAS.-- In language. the possibility of poets descri
572n035 ied &c &c Are the facts (about communication of IDEAS, &c) of expression lawless, whilst they are the
575n045 e is any analogy between grief & pain-- certain IDEAS hurting brain, like a wound hurts body-- tears
575n046 I. p. 77) remembering things of youth, when new IDEAS will not enter. is something analogous. to inst
575n046 o instinct, to the permanence of old heredetary IDEAS.-- being lower faculty than the acquirement of
575n046 being lower faculty than the acquirement of new IDEAS.-- Walter Scott «(Antiquary)» Vol II p. 126 say
580n060 anguage My father says old people first fail in IDEAS of time, & perhaps of space-- in latter respect
580n061 (As Miss Clive) have only possessed very loose IDEAS.-- Have children loose ideas of time?-- Charact
580n061 ssessed very loose ideas.-- Have children loose IDEAS of time?-- Characteristic of one kind of intell
587n089 - Mackintosh first clearly insisted on assoc of IDEAS & emotions. rather ideas & bodily actions make
587n089 y insisted on assoc of ideas & emotions. rather IDEAS & bodily actions make the emotions.-- p. 272. S
589n092 tions being only bodily actions associated with IDEAS.-- A sigh, is an abortive groan.-- more power o
599o006 4. "Architecture is a fine amplification of two IDEAS in nature; a developement of the thoughts expre
602o11v by a camera obscura &c a Poussin.-- How are my IDEAS of a general notion of everything applicable to
602o11v . would make the mind have mysterious & sublime IDEAS independent of the senses & experience p. 142 "
605o19v to the greatness of an object itself or to the IDEAS excited & associated with it. as the idea of De
614o037 s-- but on its multiplicity & the comparison of IDEAS.-- As man has so very few (in adult life) insti
640j167 turalists are prone, fortunately, to take their IDEAS, which are arbitrary & empirical, from their ow
093a033 egular mud bank at Bahia Blanca. <fl> Flustra IDENTICAL. recent & bone bed.-- November 8th 1877 (Memo
098a048 yer.-- So that layer of feldspar in gneiss is IDENTICAL with layer of flint on calc.: sandstone. (& a
242c018 e brought a tortoise & toad from S. America are IDENTICAL with those from S. Africa. «M. Bibron doubts
378d148 was impossible to say which was origin of any IDENTICAL bird-- for they were of all colours.-- they w
406e035 e. St. Helena & Ascension most species like a IDENTICAL with S. America. & many very close: see full
449e170 in» islas de Sonda as well by those which are IDENTICAL, as those which are different.-- now this is
533m058 g means) if so this is precisely analogous or IDENTICAL, with bird knowing a cat. the first it sees i
588n091 »-- gives explanation & instance of starting IDENTICAL with mine.-- Lamarck. Vol II p. 319.-- Habits
589n091 etary form the instincts of animals.-- almost IDENTICAL with my theory-- no facts, & mingled with muc
609o030 <says> unites both «& shows them to be almost IDENTICAL» + What has produced the greatest good «or ra
419e086 Scandinavia from height of 200 & 300 ft are IDENTICALLY same as those of present seas.-- now in this
617o39v pprehend the force of inanimate bodies. How we IDENTIFY the two aspects as different phases of the sa
316c247 s very remarkable.-- (connect these facts with IDENTITY of land animals.-- these however come from Si
461t041 ould transplant seeds very far.-- Sept 31. The IDENTITY <of> (or only closeness) of some species-- (e
544m102 e violation of all <rules> relations of time, <IDENTITY,> place & personal connections-- ideas are st
575n044 Jenners Jackall Have we somewhat right to deny IDENTITY of instinct.-- Habits import to Bowen No one
593n111 speaks. shows, that consciousness of personnal IDENTITY is by no means a necessary part of man's mind
308c219 under different climates &c. Do I mean that IDEOSYNCRACY of wild animals is generally different, bec
338d024 ol 9. Dr. Ferguson seems most clear that the IDEOSYNCRACY of the Negro (& partly Mulatto) prevents hi
540m087 ganization.-- Same cause as colour & shape & IDEOSYNCRACY.-- Look at the Indian in slavery & look at
360d091 inking of the Penguin duck & Herberts law of IDEOSYNCRASY I have hitherto thought that a small differ
390d178 ange.-- ‖ Without some small change in form. IDEOSYNCRASY or dispositions were added or substracted a
529m042 cord he believes in Philosoph. Transactions, of IDEOT 18 years old eating white lead. who was most vi
529m042 to be ignorant, but quite sensible & no ways an IDEOT.-- «in this case must have been functional.--»
523m016 dream.-- Insanity is produced by moral causes (IDEOTCY by fear. Chile earth quakes). in people, who,
529m042 on of Dr. Prietly who was cured from a fall of IDEOTCY.-- The story of the Corbets & big noses, quite
523m015 nsane at the sight of anything scarlet.-- dogs IDEOTIC.-- dotage.--» Doctor communicated to my grandf
538m079 characters as Allen W. & Babington, both half IDEOTIC in some respects & with store of accurate & ev
560m156 ks & gestures, other cases like D. Corbet; «do» IDEOTS form habits readily?? Do the Ourang Outang lik
572n033 nly has less reasoning powers than Europaean.-- IDEOTS. defective brains.-- Erasmus does not liken te

610o032 -- My father says he has heard of many cases of IDEOTS knowing things, which are often repeated in a
610o032 must conduce to their health & comforts.-- Both IDEOTS, old People & those of weak intellects.-- West
308c219 Testament» Domesticated animals having same IDIOSYNCRASY, cause of fertility.-- varieties not produc
028r030 n Australia.--Have Limestones all been dissolved. IF so sea would separate them from indissoluble roc
031r037 raised No Faults in Patagonia[,] enormous extent; IF lowered again & covered no sign of upheaval To C
032r039 nd of greatest action not having been worn away.--IF the level of the sea was to sink by very slow &
032r039 ff (Z). to which point the waves would not reach. IF now the ocean should suddenly fall, (3) the case
032r041 to poles. nearer the surface & to the Eastward.--IF matter proceeds from great depth. from axis to s
032r041 om axis to surface must gain a Westerly current:--IF great changes of climate have happened. hurrican
033r042 although not very intelligble is a familiar case: IF refiltered with other matter how very curious a
033r042 e shells ever casts alone in Calcareous. rocks??--IF so case precisely analogous: fragments instead P
035r046 m, as to the general truth of the proposition."-- IF such can happen in troubled England; the more mi
035r048 ace oscillated equably.-- These facts become easy IF we look at the action as a deep & extensive move
038r059 hose parts. Are not the dikes generally vertical? IF so posterior to elevations? & not sources of lav
041r066 cale» «quantity of iron» being there in excess.-- IF veins {P} are secretionary, so are all those pla
042r068 ¿how far is the distance?-- Fossil bones black as IF from peat.--yet cetaceous bones so likewise «of
042r069 shwater lakes unequally elevated, which movements IF present in the Andes, would have destroyed regul
045r075 f great Lima earthquake» In the Chili earthquakes IF rise was more <than> inland than on coast it wou
045r076 Cordillera), they may be considered as accidents (IF <[...]> part of a regular system can be called a
046r078 Metallic veins likewise must separate ingredients IF we look to a constant revolution.--Are we to con
047r082 , body of water manifestly does not travel up.--) IF these view are right the coincidental retreat at
047r082 ct at Concepcion? Read the various accounts & see IF fall is not the first very evident movement.--Th
052r099 e Verds Quartz pebbles in the Cordilleras look as IF some peaks elevated.-- Greywacke. as a general f
056r109 e the stupendous mass which has been corroded. -- IF man could raise such a bulwark to the ocean, who
058r117 o. considers these earthquakes travel in order.-- IF we look at Elevations as constantly going on we
059r120 ks which harden by themselves cannot be pure. for IF so Chalk would harden.--Climate.!? or small Prop
060r125 371. Webster Antarctic veg:-- Study Ulloa to see IF Indian habitation above regions of vegetation.--
060r125 ifference of formations of S. America & Europe.-- IF great chain of Volc. had been in action during s
061r127 --change not progressif<e>: produced at one blow. IF one species altered: <altered> Mem: my idea of V
061r127 islands. elevated. then peculiar plants created. IF for such mere points; then any mountain, one is
061r127 rprised at new creation for large.--Australia's = IF for volc. isld. then for any spot of land. -- Yet
061r127 affected by Halo of neighbouring continent: ≠ as IF any creation «taking place» over certain area mu
063r130 adual change or degeneration. from circumstances: IF one species does change into another it must be
068r147 nt.-- change of volcanic focus.-- <it is certain, IF strata can be> Problem dislocate strata without
068r147 ta without ejection of the fluid propelling mass. IF one inch can be raised then all can, for fresh l
070r154 ould not see trace of Subsidence at Uspallata.-- ¿IF crust very thick would there be undulation? woul
071r156 id Peruvian Indians use arrows or Araucanians?--> IF wood now preserved over world Dicotyledones far
071r156 served over world Dicotyledones far preponderant, IF so coniferous must formerly have been most abund
073r162 globe". p. 113 How utterly incomprehensible that IF meteoric stones simply pitched from mo«o»n, that
080r178 one animal with recent dead body of other. & see IF same effects, as with man Does Indian rubber & b
086a007 c currents?-- chiefly clearly by sun's position = IF equatorial streams of warm pole; in name of Heav
089a019 ucture,-- «He» Thought of erecting machine to see IF water fell. -- <Keys off extreme point of Flori[
090a023 stones of large numbers (& how few cases recorded IF we say «100» <5>0 lbs a year too little.-- How c
090a024 0,0,000 = 2500 = tons in fifty thousand years]CD IF world increased a tenth; would the perturbation
090a024 eased a tenth; would the perturbation be serious? IF so other cause besides thin vapour bringing plan
092a029 n how electrical currents tend to deposit metals, IF in solution. My view of metamorphic in contradis
099a048 rtically than laterally. -- <In Area of this> {P} IF surface covered with oil should shrink. film par
099a048 should shrink. film parallel to longer axis. But IF great depth NB. Prof <Henslow> Sedgwicks laminat
099a051 Then every particle would tend to meet at <B. but IF particls attract each other in some increasing r
101a058 ones in interior at Falkland Isd.-- Peron does as IF well attested.-- There is no difference between
102a061 ally with little pressure? An important question! IF water yields substances from impact, «it» would
103a063 ause not wedge-formed.-- Hence fill up fissures-- IF dikes effect of horizontal elevation excepting f
105a070 -- besides general improbability. stratification. IF chain of lake. <a> the alluvium would form a suc
105a071 vium would form a succession of flights of steps; IF one lake then we must suppose barrier in the ver
105a071 with character of mouth of valleys &c; Pampas.-- IF blocks above their parent rocks. would be prove
105a072 r & concretions on coast of Madeira.? How came it IF this powder results from «decomposed sea» shells
106a073 wing to momentum. which the water has obtained.-- IF {P} inclination be great where arrow stands the
106a073 eflected from (B) which would not have been case. IF inclination small.-- The power of widening chann
106a074 ent that deflected stream cannot retain its force IF inclination be great. There could «not» be great
107r077 anean line moves upward from effects of Elevation IF not crust much thinner beneath ocean than above
108a079 id centre would lift with it isothermal line, but IF heat from centre, then crust of solid earth woul
108a080 e between sinking & rising areas.-- In Earthquake IF Subsidence we should not expect volcanos.-- not
113a090 rata having conducted away the heat of surface. & IF conducting powers had been better then 32 degree
114a095 at C. of Good Hope.-- A bare hill of greenstone, IF we know origin of greenstone tells subsidence as
115a097 deposited specimens from Anglesea in Geolog. Soc. IF numbered compare them with my rocks. when writin
115a098 , the greater power of oscillations & currents.-- IF matter was «successively» given of every degree
115a098 degree of fineness. then most regular slope-- {P} IF not course enough flat top. ended by abrupt slop
115a098 ch stratum would thin out, both inland & seaward: IF matter too coarse, then {P} that form.-- All thi
119a104) does not explain first formation of continents, IF globe be considered as condensed vapour.-- inequ
120a108 es may throw some light.-- thin dikes not cooling IF they had travelled some hundred miles through ne
120a108 o the pool is not deep. --Hot springs &c &c--then IF so, thermometer show it cannt be ordinary heat,
120a109 in <hot> cold than hot water with <-- especially IF very hot under high pressure.--» respect to form
120a109 t.?.--??? Footsteps in New Red Sandstone. look as IF a surface deposit.-- The case of the shingle in
120a109 at Chilian valleys must be profoundly considered. IF elevation near coast more than at interior effec
120a109 resent. to spread sheet of matter over surface.-- IF elevation then went on at greater rate, not only
122a114 can Herschel consider figure of earth statical.-- IF platform of mexico owes its elevation to equilib
122a114 ot be equilibrium of fluid, but of solid. because IF of fluid, the waters of the ocean would obey tha
122a114 at equilibrium so soon from; crust being cut of-- IF part of «cold» crust under ocean, became thicker
123a116 ound of dissemination & separation in volcanos.-- IF so why not metals. The theory of veins will, I s
124a119 s most curious with respect to epigmous action.-- IF the zinc were mixed with 90 percent of lead. it
124a119 e curious to know whether it would be absorbed.-- IF so exactly parallel to limestone & volcanic rock
125a121 -- does not apply it to thickness of crust.-- {P} IF crust were metal then thinner if better conducto
125a121 of crust.-- {P} if crust were metal then thinner IF better conductor, then still thinner → The probl
126a122 lin being bad conductor is {P} against my views-- IF we had rod thus & judged by increments at, how w
126a123 that surface does not become hot?-- this looks as IF bad conductor-- III But equilibrium is not attai
126a123 onductor-- III But equilibrium is not attained, & IF cold water did not percolate surface, would beco
127a126 Academy Vol 8. p. 118 water no--. oil will freeze IF cooled in a closed globule of glass. (oil may be
128a128 sh water must be sought for below the sea mark.-- IF mountain chains are matter piled up. over crevic
130a133 ia In Glen Roy paper I show crust yield easily. & IF easily must be thin: <beside mere fracture> A El
132a138 the argument, that cold <oceans> «lakes» bottom. IF not colder than mean of place, shows earth not w
147g015 x horizontal strata I suppose these upper patches IF prolonged would intersect alley above the 300 ft
147g017 ne remains-- Some of the hills almost appeared as IF they belonged to double series Whole very obscur
147g020 han river-- No exact terraces but appearances, as IF valley had been filled with sloping bed of rubbi
148g028 sloped centre-wards which would not have happened IF the side-streamlet had cut them out-- In all cas
148g028 In all cases «I urge» deposition marine-- because IF not chain of lake & if so there would be barrier
148g028 eposition marine-- because if not chain of lake & IF so there would be barrier-- recollect the case o
150g035 as those near Bridge Roy (& other cases) but then IF gradually drained, where is barrier {P} great wa
152g047 rest road comminuted shells Important contingency IF elevation from Axis, then rivers might deposit,
154g059 ? {P} Where a buttress projects from side of hill IF line suppose continued across to {P} side remove
154g060 more {P} now that it has got to the rock of cols IF--. why should it deposits River terraces often d
154g061 er terraces often descend by flights the terraces IF the largest has hollowed out most Wednesday Shel
155g065 ght line «only» cut deep gorge on sea hypothesis, IF gullies not now formed «(Mac, hypoth,)» the leve
159g085 t between river & dry Corry Scarcely conceivable. IF Hill between Corry so much cut Granite could hav
160g091 } highest point joining this hill to others 3000? IF Ben Erin is 3500 boulder Cairn leet more Habercl
163g107 h tide ripple Near Fort Augustus hill & fringe as IF it has been filled up «at» 30 ft. higher with pe
171b004 oys the effect of accidental injuries, <on> which IF animals lived for ever would be endless (that is
172b007 s, on separate islands, ought to become different IF kept long enough.-- «apart, with slightly differ
173b010 c non fertility of hybridity &c &c <assuming all> IF species (<a>) «(I)». <fr> may be derived from fo

Page **(Key Word)**

174b014 untries when we can hardly believe necessary, but IF it was necessary to one forefather, the result,
174b015 ountries longest separated greatest differences-- IF separated from immens ages possibly two distinct
175b018 est cannot help..-- becoming more complicated,; & IF we look to first origin there must be progress.
175b018 f we look to first origin there must be progress. IF we suppose monads are «constantly» formed ¿would
176b022 ng stranger in death of species, than individuals IF we suppose monad definite existence, as we may s
176b024 bdivision <six> three more, double arrangement.-- IF each Main stem of the tree is adapted for these
179b033 ter tribes in centre of S. America shows this.-- <IF> Is there a tendency in plants hybrids to go bac
179b033 there a tendency in plants hybrids to go back?-- IF so Men & plants together would establish law. ≠a
179b035 connected with gaps in the series of connection? «IF stating from same epoch certainly» The absolute
179b035 ppear very probable:-- Mem: Horse, Llama. &c &c-- IF we <suppose> «grant» similarity of animals in on
180b037 ancient types.-- with several extinct forms, for IF each species «an ancient (I)» is capable of maki
181b040 ndividuals in a country not rapidly increasing.-- IF we thus go very far back to look to the source o
181b040 his relatives shall now exist,-- In same manner, IF we take «a man from.» any large family of 12 bro
181b042 aving progeny. after that distant period.-- Hence IF this is true, that the greater the groups the gr
181b045 ments. every part would probably be not complete, IF birds were fitted solely for air & fishes for wa
181b045 ds were fitted solely for air & fishes for water. IF my idea of origin of Quinarian system is true, i
182b046 which are in far larger Proportion terrestrial,-- IF in any in the Cryptogamic flora but not atmosphe
183b051 est species often come very remote quarters. (NB. IF Plata Partridge «or Orpheus» was introduced into
183b051 wolf & Dog. (because being believed same species) IF they do not breed readily. point in view.--¿whet
186b061 , or new species; therefore animals would perish, IF there was nothing in country to superinduce a ch
187b064 odation). Now this argument applies to species.-- IF individual cannot procreate, he has no issue, so
189b070 «&» & Megatherium. each with same kind of coat.-- IF we could tell, I do not doubt even colour heredi
189b072 Equator in former periods & then splitting off.-- IF species generate «other species», their race is
189b072 ace is not utterly cut off:-- like golden pippen. IF produced by seed go on.-- otherwise all die.== T
190b075 n Have races of Plants. ever been crossed really, IF there is any difficulty in such marriages or off
190b077 g common to Southern hemisphere, does not look as IF S. africa peopled from N. Africa An originality
192b084 les.-- born from beetles with wings.& modified.-- IF simple creation, surely would have been born wit
192b087 g & tapir, yet from some common progenitor,-- Now IF the intermediate ranks had produced infinite spe
192b089 ather of Mammalia as Heterodox as ornithorhyncus. IF this last animal bred-- might not new classes be
194b095 in Europe now confined to southern hemisphere.-- IF these facts were established it would go to show
195b098 ganization constant in the shells.-- The question IF creative power acted at Galapagos it so acted th
195b098 & not permanent. this might be made very strong. IF we believe the Creator creates by any laws. whic
195b104 ption, it explains nothing further, points gained IF any facts are connected.-- No doubt in birds, mu
196b105 transport easily.-- it would have been wonderful IF the two Rheas had existed in different Continent
198b115 als> large Mammalia not being found on all isld, (IF act of fresh creation why not produced on New Ze
198b115 f fresh creation why not produced on New Zealand; IF generated <No> «an» answer <could> «can» be give
199b120 the parents; repugnance «generally» to marriage <IF offspring not fertile> <,but producing> «before
202b133 a. or like living Edentata in Africa &c &c.-- Now IF suppose world more perfectly continental. we mig
202b134 t have wanderers. (as Peccari in N. America) then IF it is doomed that only one species of family has
205b145 he male more frequently predominates," p. 20. do "IF hornless ram be put to horned ewe almost all the
206b146 se in form, less hardy, & more liable to disease" IF population of place be constant «say 2000» and a
206b147 e continued breeding from eternity «backwards.--» IF population was increasing between each lustrum,
209b155 erest in perpetuating these particular varieties. IF species made by isolation; then their distributi
210b159 seeds of marked varieties produce no difference. IF they do.--there probably will be this relation a
213b169 cies> species. in domesticated <species> races.-- IF all men were dead then monkeys make men.-- Men m
216b181 with foetus developed in the breast.--looking as IF many ova-- impregnated at once.-- Dr. Smith cons
219b194 dstone, & granite, (that is genera near Cape) see IF there are any species same as T. del Fuego & C.
219b194 & C. of Good Hope show possibility of transport. IF some cannot be explained more philosophical to s
220b198 back hybrid plants; analogous to Men. & dogs. Now IF we take structure as criterion of species Hogs d
220b198 of species Hogs different species, dogs not, but IF we take character of offspring. Hogs not differe
222b206 rm, which has wandered least from ancestral form. IF so are present typical species most near in form
223b208 next animals in mind, more than in structures.-- IF the skeleton of a Negro-- had been found what wo
223b209 pecies,, as physical changes are gradual, is this IF after isolation (seed blown into desert) or sepa
225b217 ponent will say. show them me, I will answer yes, IF you will show me every step between bull Dog & G
225b219 g? Does the mind produce any change in offspring? IF so adaptations of species by generation explaine
226b222 any species belonging to its own genera Therefore IF in small tract we have many species, we may insu
226b222 f into question of proportion of species to genus IF on one isld several species of same genus, subsi
226b222 merica, the jaguar has been left & Fox, & bear.-- IF I had not discovered channel of communication by
227b224 t») more like present carnivora than Pachydermata IF my theory true, we get (I) a horizontal history
227b227 ecome a man.-- Does not require fresh creation!-- IF continent had sprung up round Galapagos on Pacif
228b232 superadded, animals not got it, not look forward» IF we choose to let conjecture run wild then <our>
229b234 agrees. with several mammalia being peculiar (?) IF. Henslow discusses possibility of seeds of Keeli
230b235 s It may be argued against theory of changes that IF so in approaching desert country or ascending mo
230b240 take place without impregnating each other), for IF they are different then, they will be called spe
231b243 southward would merely be specifically different IF so.-- Now this is difficult to explain by creati
232b247 om the equator to add to the mean of the other.-- IF the the world had cooled by secular refrigeratio
232b247 n".-- <The Phenomena of the S. Hemisphere look as IF heat gained> Experimentise on land shells in sal
233b252 ions. of great Fresh water lakes of North America IF Parasite different, whilst man & his domesticate
235b274 e about Indian Cattle with humps.-- ¿To be solved IF horses sent to India. & long bred in & no new on
235b275 ian genus inhabit different kind of localities.-- IF so change The GRAND QUESTION Are there races of
239c002 nt is lucidly explained.-- Mr Yarrell states that IF any odd pidgeon crossed with common pidgeon, off
241c016 t 8 to 20 of Zoologie of Coquille's Voyage to see IF Lessons' remarks on the Floras can be trusted Th
248c034 es> these therefore will be chiefly heredetary.-- IF varieties «produced by slow causes, without pick
250c037 each continent. Try amongst Europaean quadrupeds IF Africa destroyed would not then some forms be pe
255c051 altered many times, but all have had feathers.-- IF wing totally obliterated. This may account for p
256c053 ted or rather a new principle is brought to bear. IF man created as now. languages. would surely have
256c055 d tertiary formation amongst Graecian isles, ¿See IF type continued?-- See to Boblaye & Virlet.-- Whe
256c055 point in tracing history of Man.-- granted.--but IF all other animals have been so formed, then man
258c060 estruction.-- simile Man living in hot countries, IF continually crossed with people from cold, child
259c065 tation, but adaptation during earliest existence; IF whole life then real adaptation The case of here
259c066 atavism) & shows my «view of» generation right?-- IF puppy born with thick coat monstrosity, if broug
259c066 ght?-- If puppy born with thick coat monstrosity, IF brought into cold country, «& there acquired» th
260c067 e been know to assist in feeding young cuckoo; as IF there was storge, which could not be resisted, w
263c074 species-- The believing that monkey would breed (IF mankind destroyed) some intellectual being thoug
263c075 surd saying) & will never be conquered by anyone (IF has any kind of prejudices) <without> who just t
263c076 other.-- grant that one instinct to be acquired (IF the medullary point in ovum. has such organizati
263c076 imagining or more profound reasoning than other-- IF this be granted!!) & whole fabric totters & fall
264c079 ssive whine, see its intelligence when spoken; as IF it understood every word said-- see its affectio
264c080 alia &c &c-- but of course they might be blended, IF archipelago turned into continent &c &.-- There
265c087 applied.-- Therefore argument not destroyed even IF these shrivelled wings could be shown to be of s
265c087 hrivelled wings could be shown to be of some use. IF we only had Puffinuria Garrottii & no other spec
270c104 rganized.-- This paper might be worth consulting, IF any Metaphysical speculations are entered in upo
273c110 e of new World, but not found in Old-World--. + + IF in any «well developed» family (Gould says) ther
273c110 which is not true, with shells?.?.?» It looks as IF animals perished by errors.-- It is most wonderf
273c111 in structure & habits (it cannot be doubted that IF swallow perished) hawks & Milvulus &c would inst
276c123 ferently from literary men.--) must remember that IF they believe & do not openly avow their belief.
276c124 ure, then Milvulus forficatus Tyrannus Sulphureus IF compelled solely to fish. structure would alter.
276c124 cannot be gradation. See what Eytons young pigs-- IF vertebrae much lengthened &c there may be tenden
276c125 rent countries habits similar ¿law?-- probable.-- IF habits & structure similar would have blended to
277c127 s those (naming them) which are found together.-- IF two species come over to this country, without r
277c127 in physiology, tell traveller what to observe.-- IF he knows he has done least part.-- that he will
278c129 true cleft-- putting out of case the Analogys.-- IF genus does not Mean this it means nothing.--Ther
278c132 re able to escape to some more fitting country,-- IF Toxodon had been found in Africa, the wonder hav
280c135 aking.-- applies it to national character.-- N.B. IF two species were excessively old, they would not
280c135 ld not make hybrids, whereas two newer ones, even IF more different might do so.-- <whi> is this true
280c136 ndency.-- it tries to go back to grandfather, but IF too unlike its own parent this impossible-- (Hen
280c136 ent this impossible-- (Hence we might expect even IF two mules bred or two certain varieties, they wo
281c137 finely the series is graeduated.-- Dr Beck doubt IF local varieties should be remembered, therefore

```
281c137 at would be best).-- Argue the case theoretically IF animals did change excessively slowly. whether g
282c142 uch men living in same country but separated, now IF one of these races had become eminently aquatic-
282c142 because we suppose all descended from same.-- but IF two original species, each became ground then th
283c144 nk, that widow bird. replaced Birds of Paradise-- IF such fantastic «sexual» ornaments. have so intim
283c145 species. some fine & some wide. which is strange IF creator had so created them.-- People will argue
283c146 f life possible with certain preexisting laws.-- . IF only one kind of plant not so many.-- The quanti
284c149 hus speckled form of young blackbird. good remark IF general.-- Where any structure is general in all
285c151 upset it-- The space which one branch of the tree IF live occupied after its decay, will be  occupied
285c152 a thousand changes, keeping distinct from other & IF a first & last individual were put together, the
287c157 ming. Philosophical Magazine & Annals. 1830 (?)." IF she has put man on the throne (of reason), she h
288c159 in  same country. may explain greater migrations, IF American & intersected wider & wider at Rio Plat
289c161 but  until it is the animals are distinct species IF any one is staggered at feathers & scales. passi
290c163 or beautiful birds.-- heredetary ambling horses, (IF not looked at as instinctive) then must be owing
291c167 absolutely necessary to explain genera & classes. IF extinct forms were all fathers of present,  then
291c168 perfect  series or gradation.-- It is easy to see IF South America grew very much hotter, then Brazil
292c170 itually we should exclaim it was instinct.-- Even IF his faith is not staggered <">= I confess. no di
293c172 e) put note. Sir W. Scott has written about it]CD IF we saw a child do some action-- which its father
293c172 we should exclaim it was instinct.-- Even IF savage takes. & was given a Great coat & this he
293c173 nctive.-- My view of instinct explains its loss ¿ IF it explains its acquirement.-- Analogy. a bird c
293c173 remember  how to swim after having once learnt, & IF that was a regular contingency the brain would b
294c176 imals change a little with external influence-- & IF those changes permanent so would the change in a
295c178 emales by song. do they by beauty, analogy of man IF so war not [not located] Erasmus says he has see
296c185 p.  564. an abstract of Mr Swainsons views. which IF abstract true are wonderfully absurd.-- p. 565 <
298c192 its attracting all the moisture.-- Henslow thinks IF leaf of plant varies, <whole cross> «all organs»
299c194 to be same.-- one grows in marsh & other dry; yet IF I. palustris be sown in dry station it will  for
302c202 st work out hypothesis, & compare it with resuts. IF I acted otherwise, my premises <in di> would  be
302c203 , being little adapted to physical change.-- .. IF Patagonia became fertile all intermediate specie
303c205 ell us. when group is true,» there are no genera. IF Mammalia are adduced. say oh look to your fossil
303c205 lia are adduced. say oh look to your fossils, now IF extinction had gone, without creation this would
303c205 oreover what will become of the future creations, IF the list is now perfect.--]CD. the creator so cr
304c206 render our knowledge a chaos: who will doubt this IF series now existed from Man to Monads-- though p
304c206 m Man to Monads-- though physiology would profit. IF the series were believed to past into each other
304c206 applicable to apterix.-- but source of error for IF some of the ostriches were to die, then they wou
305c209 anner as Europaean species = singular coincidence IF distinct creation.-- ie.-- a mere statement noth
307c215 does  in species of Apterix This is important for IF these abortive wings in the female are allowed t
307c215 ings of the male rendered abortive in the womb.-- IF these apparently useless organs do indicate such
307c216 g Transactions» & Hunter on this subject) because IF so as she can be converted «into female», it wil
308c216 ls have never at any one time formed chain, since IF cretceous period assumed, then some perished bef
308c219 fertility.-- varieties not produced as by nature. IF so. the habits which would. have formed them, wo
310c222 rnal» appears to be a puzzle against my theory,-- IF I be asked by what power the creator has added t
312c228 being spotted, & colours of little value Dr Smith IF black & white Man crosses; children heterogenous
312c231 better, than those bred in & in.-- which looks as IF qualities were not permanent, in the new cross.-
313c234 sites cannot» Read Entomological Transactions Why IF louse created should not new genus have been mad
315c240 cryptogamic  plants same in Australia & Europe.-- IF creation be absolute thing, the creation must ta
315c243 Study Bell on Expression & the Zoonomia, for IF the former shows that a man grinning is to expos
315c243 anine teeth.-- (This may be made capital argument IF man does move muscles for uncovering canines) --
318c254 ountry, like White of Selbournes Rock Ouzels.-- ..IF the line or bands of country (These facts show t
324c268 y Buffon on Varieties of Domesticated animals see IF law's cannot be made out Find out from Statistic
335d015 the Persian side.-- Theory of abortive hybrids.-- IF mules did breed, the offspring would «as in  all
336d017 al would endeavour to return to parent stock. but IF both parents are alike, offspring must be like H
336d018 , stature, are slowly obtained & hereditary; <but IF> if the change be congenital (that is most slowl
336d018 ature, are slowly obtained & hereditary; <but if> IF the change be congenital (that is most slowly ob
336d018 t individual) it is more easily inherited.-- «but IF change be in blood long, it becomes part of anim
336d019 pearance of animals being created. it is probable IF created at once. <wd> according to ordinary laws
336d019 a of generation being a <slip> <bud> from parent. IF whole parent not entirely embued with the change
337d021 mals, but merely to classes where types exist for IF so. it will be necessary to show how the first e
337d022 who  will say there is distinct Creation required IF he believes «hyæna & squirrel» seal & mouse, el
342d033 not  the hybrid pheasants & grouse different.--» IF so chinese pigs & common must be considered as d
342d035 featured by the Austrian lip & the Bourbon nose". IF this be not imagination-- then old peculiarity
345d043 .-- Why is not Tetrao Scoticus. an american form (IF so)?.-- A Sphepherd of Glen Turret. said he lear
347d049 n vertebrates tendency to improve in int.ellect,-- IF generation is condensation of changes. then anim
349d052 ippotamus, solely owing to number of lost links. IF all species know they would be innumerable. does
349d052 ds, thinking any group good, though not circular. IF characters can be established-- clearly so.-- NB
353d060 y a few pages.-- Hence (p. 59) looking at animal, IF there be many others somewhat allied whether «li
354d065 Bump. together with Bison, at some resemblance as IF the "variation in one, was analogous to specific
355d066 several aboriginal species «or ducks» (here argue IF it be said domestic fowls are descended from sev
355d067 no new fact be elicited by these speculation even IF partly true they are of the greatest service, to
356d069 e line of proof & reducing fact to law only merit IF merit there be in following work.-- The history
356d071 st <according to my theory of generation (p. 175) IF> 8th Sept Yarrell told me he had just heard of B
357d074 on, seems «?» to imply knowledge of whole world-- IF so doubtless «part of» system of great harmony.
358d085 n & in (like «courage in dogs» EFFEMINATE men),-- IF carried much further, if by the process this wer
358d085 dogs» EFFEMINATE men),-- if carried much further, IF by the process this were possible, the organs do
359d088 ed by the vigour of their propagating powers. (as IF they were a good species, or local variety, & no
360d092 to thought that a small difference «of any kind», IF very firmly fixed from long time, made no differ
360d092 time, made no difference what its kind was.-- but IF it were opposed to the difference in other  sex,
360d092 be  much more difficult to propagate-- <now» «as» IF one bird had very bright red breast & other very
360d093 oth parents to transmit there peculiarities; that IF <one had a> both had mottled breasts, <when> of
360d094 r their brain or stomach would be different.-- Or IF one species left its type in having very long le
361d095 ng plumage resembled that of Cross-Beak-- In lark IF I understand right, all species have same charac
361d095 & female similar plumage.-- in tree sparrow, (IF I understand rightly) young cock & hen, all near
365d106 changed is good-- I scarcely hesitate to say that IF there had been considerable change, it would hav
366d108 th imperfect ones.-- now Sir J. Sebright. thought IF he had had a pair he could have produced from th
366d108 nter se., or at least show repugnance to breeding IF instincts unchanged, & if their characteristic q
366d108 repugnance  to breeding if instincts unchanged, & IF their characteristic qualities were all deeply i
366d108 that  all their peculiarities must be transmitted IF their [not located] <The> every case common to m
366d112 ariation. -- «(as Hunter supposes with Monsters)» IF armless cat can propagate, ie with the chance of
367d112 so, else avitism could hardly ever occur.--».-- b IF that cannot be formed, genetal organs by that co
370d115 S. America.-- Such difficulties will always occur IF animals are thought to have been created.-- it m
371d128 e roses & bud are produced, like parent stock, or IF different dieteriorating very slowly.-- I presum
372d129 nto generative organs?]CD it is of no consequence IF it does= <I do not doubt, the> Do plants loose a
372d129 s loose any qualities by being buds--, more than IF whole branch transplanted? +.simplest forms of b
372d131 can  produce claw. but man not arm. hard to say-- IF it were possible to support the arm of Man, when
372d131 ear on this subject? A mans arm would produce arm IF supported., <so> & in making <true» bud some suc
378d137 epeatedly seen them try to pull up petticoats.. & IF woman not afraid clasp them round waist & look i
378d147 g alike, yet young ones brown.-- Sexual Selection IF masculine character. added to species,. we can s
378d147 indicate,  species changing forms, <& loosing do> IF so domestic animals ought to show them.-- Anyhow
379d152 of producing both <male> pair of male & female.-- IF there be one female, she will be free Marten. <O
380d153 ned about proportion of sexes, at birth & causes. IF an animal breeds young her growth is immediately
380d153 his argument is VERY WEAK without knowing whether IF kept they would have wings.--).-- Seep p. 84. He
380d155 eves Gestation is always some multiple of seven-- IF woman does not menstruate in the month, she will
380d155 aken from his own field to bull a cow.-- -- a dog IF led in string will not.-- some of the  tigers.--
381d157 ere abortive traces of other «sexual» organs; for IF so, separtion of sexes very simple.-- as in plan
385d163 ot ask the question.-- His bitch will not take «& IF she did take, probably would not be fertile» wit
385d165 father,  or of the mother, or sometimes of both." IF L. can be trusted, this is Lord. Moretons law.--
386d166 dge of the part, of what is good for the whole.-- IF cut off nerves in snail. (Encyclop of Anat & Phy
386d167 s two animals out of one & heals piece of skin.-- IF the tail knows how to make a head. & head & tail
386d167 his because the great changes of nature are slow. IF animals became adapted to every minute change, t
387d168 Moretons  case & Dr. Andrew Smith,» difference.-- IF A. B. C. D. E be <offspring> «animals»: if «x» m
```

Page ***(Key Word)***
387d168 nce.-- If A. B. C. D. E be <offspring> «animals»: IF «x» male impresses ovum <of> in A, «with some pe
387d169 has a tendency to descend to several generations» IF A & B be two animals which have some peculiarity
387d169 mals which have some peculiarity for first time & IF their <D & E> «all their offspring» inherit the
387d169 r degree C. & theirs again in lesser degree.--now IF the <tw> second race both have this peculiarity
387d169 ded <that> the 3d tendency from first pair.-- Now IF two of third pair of same peculiarity breed they
388d171 very near mother, & some very near father.-- now IF one of these staid in the womb, when it came out
389d176 owing generation connected with whole system, «as IF there was, a superabundance of life, like tenden
390d178 ysical differences in almost all Male animals?]CD IF the male «in the course of some generations» has
390d178 le that breeding in & in would not be deleterous IF the relative had come from different quarters) t
390d179 ugh whole series of forms to acquire differences: IF none are added object failed, & then by that cor
391d175 parents, or external circumstances during life.-- IF the circumstances which induce «which must be ex
391d175 change are always of one nature species is formed IF not.-- the changes oscillate backwards & forward
398e005 trary islands separated with some animals, &c.-- «IF the change could be shown to be more rapid, I sh
399e006 ave lasted for its time: but we ought in same bed IF very thick to find some change in upper & lower
399e009 t it is adaptation to whole life of animal, & not IF it be solely to womb, as in monster. or solely t
401e014 fertilization from turnips or other stocks. Says IF any variety of apple be sown, all sorts come up
402e017 y be explained, by such strata being merely leaf, IF one river did pour sediment in one spot, for <wh
402e019 ear the tip of the tail are generally crooked, as IF they had been broken". are born so in all Malay
403e023 of any animal (as Toxodon) & say its relations.-- IF we know its congeners then we can.-- now on my t
403e024 have not <which> not, cows hornless, (horses not) IF they give up infertility in largest sense. <es>
404e025 have been invented all at once.-- but naturalists IF they had series perfect, would expect this struc
405e031 n spread over world as early as Elephants &c &.-- IF in next 20 years none of his remain found in the
405e031 e colours vary in same Manner as they would vary, IF in wild state; thus mark on ear of cats, colour
405e032 hybrid between silver & gold fish-- Octob. 26th. IF. hereafter. M. angustidens be found to be inhabi
407e042 om F. R. it will be highly necessary to show that IF species fall, genera must. Lesson I remember say
408e043 - Europe perhaps less so, that either Americas.-- IF species change, we see external conditions have
409e049 resent, or many anterior epochs.-- but we can see IF all species, there would not be social animals.
409e049 e moral sentiments of the animated beings.-- &c-- IF man is one great object, for which the world was
409e049 onomer that the moon probably is uninhabited]CD & IF my theory be true then the formation of sexes ri
414e063 gs at all more webbed than those of other dogs.-- IF nature had had the picking she would make <them>
414e065 cal history of man is as perfect as the Elephant, IF some genus. holding same relation as Mastodon to
415e070 story. Contains many most valuable references See IF any law can be made out, that varieties are gene
417e076 locality, all left whorled.-- He kept two to see IF they would breed, It is difficult to think of ¿¿
418e084 ble, such modifications, become as much fixed, as IF added to old individuals,, during thousands of c
423e095 o, as the simplest fish &), my answer is because, IF we begin with the simplest forms & suppose them
423e096 n partial classes is far from true.-- I doubt not IF the simplest animals could be destroyed, the mor
423e096 rface of world with life.-- for otherwise a frost IF killing the vegetable in one quarter of the worl
423e099 an form.-- has several species in my fossils-- CD[IF cases of one variety in upper part of bed-- & an
425e104 «this is very important. (Sicily exception)-- see IF this can be generalized.--» islds., have peculia
426e105 Tertiary «epoch» like Sicily not having species, IF true important on my view.-- March 9th-- Is ther
426e107 ded crow goes against this. & wild hybrid plants. IF many wild animals were crossed, there would prob
428e113 e to Cowslip is great difficulty. «I should doubt IF wild species ever formed like short-tailed cat o
429e116 iberia, or even of polar regions of N. America.-- IF true curious on my view-- because these points w
430e117 t have been direct parents of any of ours,-- even IF extinction is denied.-- it will not account for
430e117 nied.-- it will not account for all species. even IF it will for all.-- Varieties are made in two way
430e118 cross.-- » Has nature any process analogous-- -- IF so she can produce great ends-- But how.-- -- «-
430e118 ifficulty apparent by cross-questioning.-- » even IF placed on Isld-- if &c &c.-- Then give my theory
430e118 y cross-questioning.-- » even if placed on Isld-- IF &c &c.-- Then give my theory.-- excellently true
431e121 ata in the uper formations Portland Stone &c &c.--IF? so «it is» good case:-- in Sowerby Min. Conch.
434e128 peaking about the controversy on Didelphys says. "IF we cannot reason from the analogies of the exist
436e135 o head pulled out. yet feathers retain character? IF separation in horizontal direction is far more e
436e136 Shark-- Owen thinks Australia part of Old World <IF> It «may» be said, that wild animals will vary,
436e137 could have produced these innumerable seeds-- yet IF a seed were produced with infinitesimal advantag
437e140 g ones, would in all probability have escaped".-- IF it had not been shot by <some> «a» shepherds, wh
438e141 bility the Brussels Sprout was slowly formed.-- » IF it shall be difficult to show that <time> the fi
439e144 n inch further would determine his preservation-- IF killed by some other animal, then that quality w
441e149 ure of the parents.-- Thus would a Crab tree vary IF planted in rich soil, I presume not, but its see
441e149 t affected by domestication, yet offspring are,-- IF Australian Dog, could bud, analogy tells us, <be
443e154 Pliny, that it is bad to graft from top shoots.-- IF prolongation of life by gemmation <can be> being
447e163 perhaps oldest in world & hence most persistent-- IF form exceedingly difficult to vary.-- the run of
447e164 a, than between the corresponding acalepha?-- But IF Acalepha do not cross there would by my theory g
448e167 recent & yet almost wholly different, is same, as IF Isthmus of Panama.-- These two cases highly impr
451e176 bably, I should think grandfather first of race & IF so, fact for my theory Cocos Isld & Preparis bet
451e178 & do not pine visibly. p. 337. it would appear as IF p. 345. The Ceylonese Elephant [...] saul forest
454e184 s of plants, in becoming double loosing fertility IF, sometimes one, sex & sometimes. other, so as to
455eIBC -- do stamina of C. Speciosus. collapse at night. IF so irritate them, «as by an insect coming always
455eIBC «as by an insect coming always at same time» see IF by so doing can be made sensitive The function o
460t019 st show how long since present forms existed, but IF it be asked how this complexity from a few types
460t019 ghtly» «a good deal» modified <& Many Forms lost; IF> «of this old stock (which from action & reactio
461t041 difficulty of rabbits of England remaining same (IF so) with those of Spain & such facts-- This uneq
463t057 o skull., two bones of tibia into one.--) because IF the animals were taken from which these series w
465t080 - Beds of La Plata. (except close to B. Ayres).-- IF we may take this as guide, the shells preserved
466t094 ls, but these <do> do not bend up-- In Lark-spur, IF Bees put proboscis within nectary «they do» they
472s02v eartease pollen.-- the pollen appeared chaffy, as IF sucked?! opens & shuts end of sucker, after havi
473s004 th of organisms holding in America as in Britain. IF there has been «as» much subsidence as elevation
491q01v me hours afterwards of nearly related plant & see IF first pollen produces any effect, as in case of
492q002 Otaheite seedless;-- are all varieties seedless-- IF so. how have varieties been formed?-- 8. Can any
492q003 Norway No Questions regarding Breeding of Animals IF two half bred animals exactly alike be interbred
492q003 et WRITTEN on crossing of Cabbages, quoted by (as IF oral) Decandoelle in V. Vol of Hort. Transacts &
493q003 & Wild Duck, & then weigh their wing bones & see IF relation is same good, avoids effects of fatness
494q004 of globule of blood in allied species similar.-- IF not how is it in <allied> varieties (9) Cross la
495q05a wig of each in bell-glass-- sow these seeds & see IF they will come up true-- whilst others are cross
496q006 ioecious-- are its female flowers always barren-- IF not how does impregnation take place male & fema
497q006 Take Bag of soil from centre of woods «especially IF date of wood be known» & other odd places & see
497q006 are transported, or how long they remain dormant. IF kinds come up, not found in wood.-- but seeds co
497q06v are very difficult to cross.-- are there races-- IF so plant them together. & raise. seed.-- In lett
497q06v vary.-- Do Bees go to Sweet Peas, IMPORTANT, for IF so, as these can be raised true, there is no cro
498q008 d, do the Seedsmen select at all from the plants? IF not, I am surprised <plan> such plants do not de
498q009 ar> close to each other.-- As they are dioecious, IF no hybrids were produced by seed, we might feel
502q11v lower of different Cabbages most carefully to see IF variation equal in flower with leaves.-- strawbe
504q013 m ǁ. Dodecatheon ǁ. Castrate apple & pear to see IF pollen naturally carried, on account of Van Mons
510q017 mum differs greatly-- how very interesting to see IF any variation in varieties. G. St. Hilaires law
513q020 of pigeons vary in manner & perfection &c &c &c--IF so probably a variety, not specific character= C
515q021 whether good & plenty of pollen is produced. & 2d IF so, whether concepcion takes place,-- the mere f
515q022 stincts of the geese which he crossed; especially IF the hybrids were recrossed with either parent.--
515q022 weet Pea) for several generations under net & see IF get sterile-- Cover that little Ervum in Sand-wa
521m007 Father thinks is mentioned in the Zoonomia.-- Now IF memory «of a tune & words» can thus lie dormant,
522m010 wever, in conversation could catch up a new train IF early association were called up.-- My F. asked
522m011 ined he had never heard of.-- Thus in many things IF he began at one end, he knew the whole subject.-
522m011 he began at one end, he knew the whole subject.-- IF at the other nothing.-- He could repeat the alph
522m012 ioned as a thing which had just taken place.-- as IF the idea of time had been disturbed.-- These for
522m013 delicate people do most indelicate actions,-- as IF «these emotions» acquired.-- this may be doubted
523m014 isgivings of the injustness of their hatreds, as <IF> in my case.-- It must be so from the curious st
523m017 c by coming on suddenly. Ans no.-- because often, IF not generally, does not really come on suddenly.
523m018 remonstrances on him» which are never generally, IF at all discovered.-- <Sup> Sometimes comes on su
524m020 ome insane.-- now this is well worth considering, IF pride & suspicion can be well understood. In ins
528m037 pleasure of perspective. which cannot be doubted IF we look at buildings, even ugly ones.-- the plea
529m041 anner.-- There is much imagination in every view. IF one were admiring one in India. & a tiger stalke

```
531m049 see how the young got on. this nest the cat could IF cats will «ever» eat little birds, this most cur
531m050 things of great importance are easily forgotten, (IF unconnected with fear &c) because people think t
532m053 Hensleigh doubts whether young babies start.-- " IF children wink. it is instinct Fear must be simpl
533m058 efore they could understand. what frowning means? IF so this is precisely analogous or identical, wit
533m061 dvanced; for I caught it like a flash.--. strange IF judgment remains, where reason is forgotten. it
534m061 as  animal born with instinctive knowledge.-- but IF so, yet this knowledge acquired by senses.-- the
534m063 wn, with simple exertion of intellectual faculty) IF ants had at once made this leap it would have be
536m072 py playing cannot doubt that they have free will, IF so all animals., then an oyster has & a polype (
536m072 apacities its senses give it of pain or pleasure, IF so free will is to mind, what chance is to matte
536m072 ce is to matter «(M. Le Compte)»-- the free will IF so called) makes change in bodily organization o
536m073 obably some error in argument, should be grateful IF it were pointed out.-- My wish to improve my tem
537m076 ceased  to destroy their offspring--¿ yet perhaps IF they had murdered their children, this moral sen
538m080 ed as it were with a second & unreasonable man.-- IF one could remember all ones farthers actions, as
541m091 one  does with novel for a length of time.-- Then IF one endeavur to keep any simple idea as  scarlet
541m091 e idea as scarlet steady before mind for period, «IF the scarlet was before one effort less» one is o
541m092 anges during fall of sea.-- Is the effort greater IF the idea is abstract as love, (or an emotion not
541m092 is abstract as love, (or an emotion not so) than IF simple idea as scarlet?-- How can people dwell o
542m094 angry it is hard not to growl out some sound even IF it be inarticulate.-- the maniac shouts & bellow
542m094 eve circulation after stillness.-- Now I conceive IF organization were changed, I conceive sighing mi
545m105 en unconscious, which makes the idea unconscious, IF so (think of this). study what impressions becom
547m114 thinking is keeping up these trains.-- especially IF they are invented as in imagination, & in rigidl
548m115 recollected,  nor of the Botanical Somnambulist. (IF he had been unhappy)-- it is because in this sta
548m116 he when well did not recollect <it> «anything».-- IF one was subject to this disease oneself, one wou
549m119 ture of mind. they make, either in themselves, or IF recollected, such part of thoughts  innumerable,
549m120 ll think the sum total of happiness greater. even IF mixed with some pain.-- than the happiness of  a
549m121 s happiness is the object of living.-- or whether IF we obey literally New Testament future life is a
550m127 eyhound about dread of water-- innate Septemb 1-- IF one performs some actions, which are pleasant, e
550m127 easure. or pleasure or pain of association.-- now IF one has these feelings, without being aware of t
554m141 ting the Keeper is obliged to go in with a stick, IF he drops it, the bird will fly at him-- Knowledg
554m141 f association. Kenyon, & then go on to show, that IF Cart horse argued from this into a theory of fri
556m144 Omnibus instantly start when they hear ready, but IF they see anything ahead. which cad cannot see, t
556m146 kneeding when old, like kittens at the breast now IF horns were to grow on horses, they must yet cont
557m147 ust contrary. when pleased tail loose & wagging-- IF as (I believe) Hunter says. neither fox. nor wol
559m156 ake & pound up inflorescent parts of mosses & see IF Hybrid can be made & ferns.--» Would a sensitive
559m156 d can be made & ferns.--» Would a sensitive plant IF irritated very regularly at one time every day.-
563n002 ¶ncher & Nina)-- or to take away food &c &c-- Now IF dogs mind were so framed that he constantly comp
564n003 ¶ Different nations having different moral sense, IF it were proved instead of militating against the
565n007 is  mouth open to hear well «as one will perceive IF in night trys to listen to growl of hounds». <wh
565n009 n is accompanied by slight protrusion of lips, as IF going to say "my dear," just what smile is to la
567n015 ike, or be indifferent about, yet feel shy.-- not IF quite stranger.-- or less so.-- When learning fa
568n017 ght one feels pain, at not performing it, (either IF prevented, or overtempted.-- «animals have shyne
568n017 og. has pleasure in following its instinct & pain IF held.-- if tempted not to follow it, by  greater
568n017 asure in following its instinct & pain if held.-- IF tempted not to follow it, by greater temptation,
568n017 tempted  not to follow it, by greater temptation, IF memory of its own emotions. (which must be intim
568n019 Harmony  common to t[he] whole kingdom of nature. IF I want some good passages against, opposition of
570n024 " the man shrugs his shoulders & replies nothing. IF he did go to reply. he would throw back his shou
571n029 mer to true taste.-- Everything that is habitual, IF heredetary, is pleasant.-- Mental & Bodily Consi
572n034 ng» «running» over imaginary words: it appears as IF the mind had dwelt on each word separately, neg
572n034 ay tell him not, but it does hurt his conscience, IF he has been cowardly, or has injured another bad
576n046 y hesitate to jump in to save his masters life,-- IF he meditated on this, it would be conscience.  A
576n049 a  cause like a deity, as M. Cousin says. because IF so ourang outang.-- oyster & zoophyte: it is  (I
578n053 g, effect of mind on individual parts of body.== (IF you <think> «fear» you shall not have e---n, «or
578n054 iss F. A. said to Mrs. B. A. how nice it would be IF your son would marry Miss. O. B.-- Mrs. B. A. bl
578n055 ity of this being discovered by anyone, especiall IF it be a person. whose opinion he regards, <& see
578n055 s her, the blood rushed to her face,"-- One blush IF one thinks that any one suspects one of having d
579n058 they  consider profoundly,-- this will be curious IF it is so.-- frown with grief,¿ bodily pain? frow
581n063 etariness in habits. be considered. as grand step IF it can be generalize.-- The tastes of man,  same
581n065 sounds)--. Horne Tookes tenses, &c &c -- <also g> IF so & seeing how simple an explanation it  offers
581n066 ns]CD.¿ but what first caused this bodily action. IF the emotion was not first felt?-- «without «slig
582n067 l to perform an instinct that it is uncomfortable IF it does not do it.-- My theory explains how it c
584n071 es; & hence we get over an apparent anomaly,. for IF anyone has taken the Woodpecker as an example fi
585n076 ied to birds, which have been carried in hampers. IF they have not known the direction in which  they
585n077 t how, stonge henge raised, yet not instinct, but IF all men placed stones in same position, it would
587n087 it. & feels no more inclined to ask [not located] IF dislike, distaste. & disapproval. were not somet
588n090 because they have memory.-- what use this faculty IF not reason.-- or does this reasoning apply chief
588n090 ection.-- Lamarck. Phil. Zoolog.-- Vol II p. 445. IF we compare the judgments & actions of a young an
588n090 «instinct», for its character is invariability.-- IF explained by habits, useful to itself, how gaine
589n091 -- could brain make a tune on the pianoforte, yes IF every individual played a little, & something de
592n103 into  convulsions.-- «Paper» must be referred to, IF I follow up this subject & a reference to Brun's
593n109 s pleasure. This [blank] will not do for insects. IF this view holds good-- then man, a socialist, do
593n111 on  grassy bank some way from water, suddenly, as IF by word of command, they all took flight & flapp
594n112 ? «it would appear that an instinct long remains, IF no steps are taken to eradicate it.--» «Emma say
594n112 ot dangerous-- wild-ducks would have fled equally IF man had appeared-- though instinct so firmly imp
596nIBC instincts of mongrel dogs Do blubbering children, IF of. convulsive tendency easily fall into convuls
596nIBC . are involuntary movement of voluntary muscles-- IF so what is trembling palsy? Expressions N [Old a
599o005 f language.-- must presume it originates slowly-- IF these speculations are utterly valueless--  then
599o005 ons are utterly valueless-- then argument fails-- IF they have, then language was progressive.-- We c
600o08a e most imperious.-- It would indeed be wonderful, IF, mind of animal was not closely allied to that o
604o016 tiplied with the organisms structure, it looks as IF consciousness an effects of sufficient perfectio
604o016 ffects of sufficient perfection of organization & IF consciousness, individuality.-- Quotes D. Stewar
608o026 ng contingencies in the way to aid motive power.--IF incorrigably bad nothing will cure him' 3) disgu
608o028 ) These views are directly opposed & inexplicable IF we suppose that the sins of a man, are under his
612o035 h as inflorescence.-- [LHC] I here omit the case (IF such there are) of animals enjoying only movemen
614o037 lysed into steps, as species change.-- Must be so IF Lamarck's theory true [RHC] Acquired instincts a
615o36v ent way of showing it, nor was there any cause, & IF surprise was felt.-- analyse feelings. Mr  Wynne
616o039 Matter  is by a metaphor said to attract; & hence IF thought &c bore the same relation to the brain t
617o40v opposite  directions. We are satisfied therefore, IF we can trace any force in inanimate matter up to
617o40v e phenomena of gravity are manifestly the same as IF every particle of matter were an animated  being
618o041 ternal consciousness, & have no objective aspect. IF thought bore the same relation to the brain that
618o41v thought  & organization run in a parallel series: IF blueness & weight always went together. & as a t
618o41v o different as action thought & organization: But IF the weight never came untill the blueness had  a
618o41v » priority of one to other: no not only thus, for IF day was first, we should not think night an effe
619o042 RHC] The history of every race of man shows this, IF we judge him by his habits, as <if> another anim
619o042 an shows this, if we judge him by his habits, as <IF> another animal. These instincts consist of a fe
619o043 without  any regard to his own interest. likewise IF such actions were prevented by force he would fe
619o043 el pain. [.By a very slight change in association IF others injured these objects, without his  being
619o043 ble to prevent it, he would likewise feel pain.-- IF he saw another man <say go> «acting in» accordan
619o043 art of that pleasure, which the acter received.-- IF either man did not obey his instincts from inter
620o044 action, & on the results following our conduct.-- IF the temptation to disobey the conscience is extr
621o046 natural  appetites.-- he is monster, or unnatural IF malevolent, or hates his children without some p
621o046 nt, or hates his children without some passion.-- IF his passions strong & his instincts weak. he wil
621o047 onduct of others, the feeling that almost (rarely IF opposed to natural instincts) any action is eith
625o50v his system not selfish.-- explained by principles IF Mackintosh.-- p. 262. Some good remarks, on anal
626o052 aces» generations led to their formation.-- 'N.B. IF feeling or emotion rises from heredetary  action
627o52v arguments of beneficial tendency of affections.-- IF ever I write on these subjects consult «followin
627o053 ions cannot be analysed into "power" &c &c &c-- & IF termed "selfish", must be subclassed as "disinte
628o53v isapproving instinct-- which were not originally, IF the shepherd dog had no instinct to commence wit
628o054 ns of this instinct, with that line of conduct, & IF taught rightly, it will be for the general good,
```

629o55v arquesans think only of prepuce, crepitando,]CD & IF passions makes one break these artifical rules,
629o055 ied by blushing, bears some relation to others 5) IF so, it is perhaps deviation from the instinctive
632j53r I it will be better always to refer to the author IF I use these facts p. 280. adduces provision of s
632j53r must be admitted there would not be these plants, IF there was not some provision for transportation:
633j53v etary [produce some peculiarity in seed vessel]CD IF man takes care they are not detrimental.-- NB. O
633j54r e.-- Hence the mistake they are created for them.   IF we once venture to say plants created to <arrest
633j54r arth, why not created to live on alpine pinnacle? IF we once to presume that God «created plants to»
639j28v an rush for life, only two of them live to breed, IF circumstances determine that, the long legged on
640j167 as been tried.-- With respect to the six puppies, IF a hare was introduced, or <a spe> became more nu
640j167 r cause, the long legged race would prevail, even IF have afforded only 10th part before & now formed
640j167 y 10th part before & now formed eighth part.-- or IF other prey diminished, total number of dogs. wou
059r123 ects of cooling [misnumbering, no page 122] In IGNEOUS rocks.--which would have the cryst of glassy F. frac
068r146 subsequently there is a coating of solidifying IGNEOUS rocks which would be too thick to be penetrate
068r147 an be raised then all can, for fresh layers of IGNEOUS rock replace strata. & it is nothing odd to fi
271c106 t is a plastic virtue.-- it is expression for IGNORANCE Two grand classes of varieties.; one where of
283c144 > called into existence in two continents our IGNORANCE is indeed profound & such it appears.-- Is th
293c173 bjects as Ichneumon & caterpillar, though our IGNORANCE, may make us think so, but only between laws-
304c206 that all his divisions merely marked his own IGNORANCE.-- the collector who plodding at making a ser
310c222 f different types. I will confess my profound IGNORANCE.-- but seeing such passions acquired & herede
547m112 n Castle & dream No answer shows our profound IGNORANCE in so simple case.-- There was memory, for it
612o035 ng to positive knowledge. lessening amount of IGNORANCE [RHC] The radicle of plants absorb by physica
632j53r t I admit, but the admission is probably from IGNORANCE]CD Who would ever have thought that the intes
225b217 effects of external causes, of which we are as IGNORANT. as my millet seed turns a Bullfinch black,
529m042 ed, but who when he recovered. was found to be IGNORANT, but quite sensible & no ways an ideot.-- <in
568n019 ucated.--»-- Old man at Cambridge observed the IGNORANT. merely looked at picture as works of imitati
618o041 might be quite familiar with thought & yet be IGNORANT of the existence of the brain. We cannot perc
478z003 ter in Paraguay p 207 Slight notice on habit of IGUANA. not pass Lat. 28 degree North p. 239 In ocean
300c197 Insects shamming death, most difficult case to IIMAGINE how art acquired.-- They reason however on th
051r094 f Organic productions. = Yet everywhere on coast (IL Defonsos «Kelp») rocks show signs of degradation
243c019 y toad is same species Coquille Voyage p. 25 Mais IL n'y a pas jusqu'aux îles Macquarie et Campbell (
243c019 espèces; et certainment on eût été bien éloigné, IL y a peu d'années, d'admettre que ces oiseaux eus
569n022 take a savage, said his wife would be grieved-- "IL leva les epaules et dit qu'il valait mieux reste
599o008 de la matiere et du temps! voila les facultes, q'IL possede seul sur la terre. J'ai trouve son âme"
058n118 lly says."en effet toutes les montagnes de cette ILE se developpent autour d'elle comme une ceinture
058r118 tandis, au contraire, que vers le centre de l ILE, elles presentent une coupe abrupte et souvent t
096a040 ch p 359 stretched out NE & SW.-- Von Buch. Can. ILE p. 406. List of Volcanos Salomon Isld,-- New Bri
322c270 st reread the previous volume. & C. Prevost on L'ILE Julie Waterton's Essays on Natural History. Octo
209b157 on shells.-- {T} Genera In North Africa. I: 4,2 ILES Canaries I: I,46 St. Helena I: I,15 {T} Calcula
243c019 oquille Voyage p. 25 Mais il n'y a pas jusqu'aux ILES Macquarie et Campbell (52 degree S) qui n'aient
328cIBC mminck has written "Coup d'oeil sur la Faune des ILES de la sonde et de L'empire du Japon Wowett on C
409e048 dividuals, & though we may not trace out all the ILL effects. -- we see it is not the order in this p
524m019 ham Doctor), in this precisely like the passion, ILL-humour & depression, which comes on from bodily
529m040 rist, in whose mind supply of food was evasive & ILL defined thought would receive pleasure from thin
557m148 ; there is no inclination to jump away,-- it is, ILL-defined fear.-- Yet one knows oneself it is quit
567n013 d, & put them in-- like child. Tommy's face, now ILL, has expression of languor & suffering The Cyano
529m040 inking of the fertility.-- I a geologist have ILLDEFINED notion of land covered with ocean, former an
523m016 , was caused by thinking over the misery of an ILLNESS at Rome, when by accidental <was> delay of mon
530m043 what bond he could have had. yet during whole ILLNESS, he had been able to direct about his own heal
532m055 e alive.-- How is it. with people nervous from ILLNESS, <the> it must be an excited action in the in
525m025 mily) are heredetary.-- other deformities are ILLNESSES of the foetus.-- some mothers. have first dea
525m026 ice versa Walter Scotts remark how odious an ILLTEMPERED fat man looks, shows same connection between
551m127 hich have no existence.-- He compared spectral ILLUSION & insanity the connexion appears to me vague-
432e123 ffspring take more after father than mother; ILLUSTRATED by the crossing of hornless sheep with horne
533m059 & co-relation of the proper muscles. may be ILLUSTRATED by the extreme difficulty of moving muscles
371d127 ned up to the (required) offices" &c &c Owen ILLUSTRATES case of Dingo (he alludes to the dholes or w
106a074 eat deflection in a "rapid".-- is a familiar ILLUSTRATION.-- Therefore stream has no tendency to wide
201b126 of annelidae.-- <">The fact is an additional ILLUSTRATION of that axiom in Natural History that all a
418e080 being Hermaphrodite-- probably will recieve ILLUSTRATION from domestication of Monooecious plants, &
025r018 th America; I might add I have drawn all my ILLUSTRATIONS from America, purposely to show what facts
206b149 nd to consider the increasing ones.-- NB As ILLUSTRATIONS are there many anomalous lizards living; or
632j53r Abstract of John Macculloch 1837 Proofs and ILLUSTRATIONS of the Attributes of God Macculloch. Attrib
079r177 "chefs d'oeuvre de l[']industrie humaine, comme ILS traitent les loix de la nature & ses phenomenes.
293c172 tion before. is strong association recalling up IMAGE which had been past-- so great an anomaly in st
526m028 ve not imagination enough to <up> recall up the IMAGE in their own mind,-- this may be worth thinking
552m130 ed at any object<)> one Shuts ones eyes) is the IMAGE not vivid as in sleep-- (one can dream of inten
552m130 ison with perceptions, & that on[e] fancies the IMAGE more vivid? Surely the image in a dream cannot
552m130 on[e] fancies the image more vivid? Surely the IMAGE in a dream cannot truly be <more> «as» vivid, «
527m033 auses the mind to create short vivid flashes of IMAGES & thoughts.-- Poetry. the latter thoughts are
534m061 enses,-- then thinking consists of sensation of IMAGES before your eyes, or ears (language mere means
552m130 e <more> «as» vivid, «a reality» as in Spectral IMAGES-- Mem Chiloe <pi> Sow, who carried from all pa
321c270 Art of Living Several of Water Savage Landons IMAGINARY Conversations-- very poor Sir T. Browne's Rel
547m111 the air, of being compelled, from some quite IMAGINARY cause to start at once to Shrewsbury., vaguel
553m135 cause (& no one being apparent, one fixes on IMAGINARY beings, many vicarious, like ourselves) that
572n033 ench book quickly, & <running> «running» over IMAGINARY words: it appears as if the mind had dwealt o
275c121 s reduced to scarcely anything.-- almost all IMAGINATION-- He says he recollects all half Bred cattle
316c244 ursting on mind & giving rise to the wildest IMAGINATION & superstitions.-- +York's Minster story of
342d035 rian lip & the Bourbon nose". if this be not IMAGINATION.-- then old peculiarity overbears the crossi
343d037 ure-- How far grander than idea from cramped IMAGINATION that God created. (warring against those ver
364d104 terwards occasionally went back-- (Effect of IMAGINATION on mother. white peeled rods mentioned in ol
380d155 disabled.-- so Elephant in confinement, & so IMAGINATION in Man, has strange effect.-- Directly a Cap
388d171 ith respect to offspring being determined by IMAGINATION of Mother.-- We see in a litter every possib
526m028 remarks that pleasure received from works of IMAGINATION very different from the inventive power,-- t
526m028 can correct every detail, yet they have not IMAGINATION enough to <up> recall up the image in their
526m028 ll perhaps show differences between memory & IMAGINATION. «Catherine thinks that children like lookin
526m028 asure of imitation (common to monkey), & not IMAGINATION.--» Thinking over the scenes which I first r
526m030 When a man says I will improve my powers of IMAGINATION, & does so.-- is not this free will,-- he im
528m036 -- There is absolute pleasure independent of IMAGINATION, (as in hearing music), this probably arises
528m039 exercise, birds singings.-- 4th. Pleasure of IMAGINATION, which correspond to those <he> awakened dur
529m041 nderful & mysterious manner.-- There is much IMAGINATION in every view. if one were admiring one in I
547m114 ins,-- especially if they are invented as in IMAGINATION, & in rigidly comparing each step as in reas
548m115 & inventing new means,-- therefore works of IMAGINATION hard work,-- Keeping one idea present is, pe
549m119 of intellect affection excited, pleasure of IMAGINATION-- therefore do these & be happy-- & these pl
551m129 something like a snow-haze. covers my whole IMAGINATION." Septembe. 3d Why when one thinks of any ob
590n094 d to be audible. one has good ground to call IMAGINATION a faculty, a power, quite distinct from self
595n117 > considerably excited, shows their power of IMAGINATION-- for it will not be allowed they can dream,
595n117 hearing merely hunting horn-- association or IMAGINATION [blank] [not located] Ernest W. playing with
602o11b and embodying, what never existed but in the IMAGINATION".-- Macculloch Vol I. p. 115. Attributes of
606o023 e not approved by the eye itself, but by the IMAGINATION through the medium of the eye"; he will allo
625o50v Some good remarks, on analogy of pleasure of IMAGINATION «the utility part being blended & lost» & mo
040r063 l life.--Patagonia In the Chonos Islds we must IMAGINE bituminous shales have been metamorphised, as
047r083 Calabrian wave did not sea break first? I can IMAGINE from local form of coast (as seen in swell) th
282c140 alogy is the close relationship in some one.-- IMAGINE the men to have greater power of change yet, a
291c166 rly) being heredetary<)>.-- it is difficult to IMAGINE it anything but structure of brain heredetary,
291c167 cious plants in animals it may be difficult to IMAGINE how sexes were separated.-- in plants we have
298c191 ry during gradual elevation of isld.-- We must IMAGINE a considerable range of one species «on -- mou
448e167 nnetts Wandering Vol II. p 155. By inference I IMAGINE that there are Baboons in St Thomas on W. coas
463t055 ermediate structure, but it is not possible to IMAGINE what habits an animal could have had with such
263c076 of a brain capable of producing more glowing IMAGINING or more profound reasoning than other-- if th

Page **(Key Word)***
529m039 erienced people-- recall pictures & therefore IMAGINING pleasure of imitation come into play.-- the t
344d040 vae of insects-- (¿glowworms) breeding-- <beet> IMAGO state fertile at once.-- Consider this with ref
593n107 as in some <instinct> «insects» which become in IMAGO state social) by smell or looks. but it does no
366d108 their characteristic qualities were all deeply IMBUED in them from long permanence, so that all thei
383d159 ate, but quarter-grown seems to show whole body IMBUED with possibility of becoming either sex.-- ¶ I
199b120 fferences blended, rather stronger tendency to IMITATE one of the parents; repugnance «generally» to
426e108 y Man. we should remember, that even birds can IMITATE the sounds surprisingly well-- In early stages
190b078 of progressive increase of organization being IMITATED in the womb, which has been passed through to
520m001 is extremely improbably that they should have IMITATED.-- when attending Mr Dryden Corbet, he could
337d022 loves, who fears, who is curious &c &c &c who IMITATES.-- who will say there is distinct Creation re
603o11b of rock--]CD or poetry, CD[my thery says yes. IMITATING song -- two primary sources, sight, & hearing
228b231 ack man other kind?» Animals with affections, IMITATION, fear <of death>. pain. sorrow for the dead.-
526m028 y taste, of animals. they know.-- pleasure of IMITATION (common to monkey), & not imagination.--» Thi
529m040 ll pictures & therefore imagining pleasure of IMITATION come into play.-- the train of thoughts vary
532m057 fects of fear.-- the state of collapse may be IMITATION of death, which many animals put on.-- The fl
569n019 gnorant. merely looked at picture as works of IMITATION.-- Hence pleasure in the beautiful. (distinct
583n069 ts].CD all this may be true,, but relation of IMITATION & reason must be thought of.-- p. 19. animals
583n070 e gained by reason: & then these qualities of IMITATION & education may be used as argument.-- for in
583n070 ve knowledge is not gained by instruction, or IMITATION.-- p. 20. Animals may be called "creatures of
624o051 h is acquired by association from education & IMITATION, has often been perverted from want of reason
583n069 tive faculty: pure instinct is not imitative: IMITATIONS seems invariably associated with reason: [N
581n065 - thinks many words, roar, scrape, crack, &c, IMITATIVE of the things.-- CD[I may put the argument,,
583n069 nting varies.-- p. 18. Animals possess strong IMITATIVE faculty: pure instinct is not imitative: imit
583n069 trong imitative faculty: pure instinct is not IMITATIVE: imitations seems invariably associated with
603o11b p. 136. Says Architecture does not come under IMITATIVE art [my view says yes. <old> mass of rock--]C
459t009 several species of deer, which are altogether IMMACULATE when grown up". Saw at Mr Bell's at Hornsey
151g040 ly water worn) are often times of rock not in IMMEDIATE neighbourhood, (as granite or gneiss of Moel
397e003 no stone can fall, or plant rise, without the IMMEDIATE agency of the deity. But we know from experie
539m081 eavour to» remember, & to think deeply, & the IMMEDIATE manner in which my head got well when reading
552m130 ense scarlet??) is it because one then has no IMMEDIATE comparison with perceptions, & that on[e] fan
568n018 ons would first take place, whether to pursue IMMEDIATE inclination or some future pleasure.-- hence
106a073 nation be great where arrow stands the force IMMEDIATELY deflected from (B) which would not have been
272c108 rushes similar.-- The question which I more IMMEDIATELY want are there Heteromera, which have habits
367d113 t more than another, which part is that most IMMEDIATELY employed in fighting" instances thighs of co
380d153 ses. If an animal breeds young her growth is IMMEDIATELY checked-- the vis formativa goes entirely to
532m056 that she was sent back to Shrewsbury,, then IMMEDIATELY fell into her old ways & became fat! What re
539m082 l of Picture. association with much pleasure IMMEDIATELY thrilled across me, bringing up old indistin
174b015 arated greatest differences-- if separated from IMMENS ages possibly two distinct type, but each havi
180b036 is). REQUIRES extinction. Thus between A. & B. IMMENS gap of relation. C & B. the finest gradation,
023r010 tralia Dampier also repeatedly talks about the IMMENSE quantities of Cuttle fish bones floating on th
035r048 try would produce equable effects.--«though so IMMENSE to short breathed traveller» Mountains, which
055r107 go) yet no historical records of eruptions how IMMENSE the time!! How well agrees with number of Crat
055r108 y meterological table the precise periods over IMMENSE areas. (& the counterbalancing variations) of
078r176 rected NW & SE). «Vol III» Mexican Cordillera "IMMENSE variety of Porphyries which are destitute of q
117a101 e oscillations of level will alone explain the IMMENSE amount of change which must have taken place,
119a105 thermometrical scale.-- (for the temp must be IMMENSE to convert rock into gneiss &c judging from wh
252c045 tor is directed to study localities of isld.--«IMMENSE importance of local faunas foundation of all o
278c131 ated with teeth of seals and dugong, therefore IMMENSE age since breccia accumulated-- surely ask Owe
304c207 posed to so many contingences??? A Question of IMMENSE difficulty is, whether Apterix descends from s
398e005 dation stone of further inductive reasoning is IMMENSE. It is curious that geology. by giving proper
480z006 its of Tubularia. p 52. May 1836 dimensions of IMMENSE Tortoises p. 81 & p 113 of 1834 On the passere
558m153 claim, your arguments are good but look at the IMMENSE difference. between man,-- forget the use of l
570n024 probably correct, to relieve respiration when IMMENSELY immersed-- mechanic apt to sigh.-- & hence ca
058r118 developpent autour d'elle comme une ceinture d'IMMENSES remparts; toutes affectent une pente plus ou
570n024 correct, to relieve respiration when immensely IMMERSED-- mechanic apt to sigh.-- & hence carried on
465t081 d of 1839) very important. it seems owing to IMMIGRATION of other races, so it is with domestic breed
465t081 S. America «Rengger.» -- now it is this very IMMIGRATION which tends to make the destroyers vary; so
580n062 through life" p. 200.-- "The desire of glory, IMMORTAL fame, &c so common in the young are symptoms
417e076 k of ¿¿Plato & Socrates, when discussing the IMMORTALITY of the Soul as the linear descendant of <Mam
213b170 & keep similar. But this is difficulty; This IMMUTABILITY of some species. In Phillips. p. 90. it see
132a137 f the moon.-- Ask Hopkins. M. Parrot, Mem. Acad. IMP. des Sciences. (Sc Math. Phys. et Naturelles. To
102a061 rtant question! If water yields substances from IMPACT, «it» would look like it. Are greenstone dike
521m009 all» «Park», after paralytic stroke. intellect IMPAIRED. <after paralytic stroke> : . could converse
560m156 icture.-- Do female monkeys not show signs of IMPATIENCE when woman present? Do they pout, or spit, o
439e144 is much facility in crossing there comes the IMPEDIMENT of instinct-- the possibility of rearing by
292c171 actions, which Many successive generations are IMPELLED to do in same way-- The improvement of reason
274c115 in parts perfect of typical structures certain <IMPER> parts changed Have <not>. S. Africa, Australia
600o008 f Soul.-- (I) <Conscience> «Moral Sentiments» IMPERATIVE sense of duty-- which makes struggle in man.
155g062 ut most Wednesday Shelf 3d dies away almost IMPERCEPTIBLY on Glen Turrit side 2nd shelf very broad «&
036r050 y to ascend the hill,--but my recollection is IMPERFECT & was recalled by note in Daubeny. P. 438., o
043r069 alapagos.--Chiloe. M. Hermoso. & Coral reefs (IMPERFECT in latter). <At> Lyell. Vol I. P. 316. Earthq
095a038 o» of Diamante near stream of same name. with IMPERFECT crater <--> near summit,-- much pumice --. ap
230b239 only be traced geologically (& then monuments IMPERFECT) or horizontally & then cross breeding preven
334d011 ses) being deaf curious case of corelation of IMPERFECT structure.-- Fox says in «Lord» Exeter's Park
342d033 olina?) off New York. therefore instincts not IMPERFECT.-- Are Pheasant & Grouse homogeneous? I obser
366d108 egs-- -- the former of which had kittens with IMPERFECT ones.-- now Sir J. Sebright. thought if he ha
525m025 as an extra number of fingers.-- hare lip or IMPERFECT roof to the mouth «stammering in my Father fa
608o026 laws of nature chance) 2) difference is from IMPERFECT condition of mind all motives do not. come in
616o038 that when we <return> turn into angels. this IMPERFECT memory may become perfect & we may look back
638j58v fect hinge.-- reason, & not death rejects the IMPERFECT attempts. In the «Bee» Mollusca the nervous s
276c124 the only argument can be, a bird practising IMPERFECTLY some habit, which the whole rest of other fa
350d056 to threads.--» Hence the Pecten, which move IMPERFECTLY has eye-point, but Broderip added it has bee
363d102 gale which both sing their own songs, though IMPERFECTLY.-- Male birds always second their songs, the
548m114 ree» trains of thought, therefore one may be IMPERFECTLY reason -- <In a> Abercrombie's case of «in B
315c241 noir will be published St. Petersburgh Academy IMPERIAL-- Paper read in 1837. semestre I suspect some
466t094 e placed all round flower as they are in Crown-IMPERIAL Lily & many other flowers-- My view of <varie
600o08v nstinct gives most pleasure. but because most IMPERIOUS.-- It would indeed be wonderful, if, mind of
629o55v it may be thought wrong.-- & conscience will IMPERIOUSLY say so, & produce shame & remorse-- [Thus pu
600o08v rice.-- Is there not something analogous to IMPERIOUSNESS of Conscience: in Maternal instinct dominee
146g013 cidental is the preservation in situ of even IMPERISHABLE pebbles/ I am nearly certain there were non
553m136 e innate knowledge of creator <is> «has been» IMPLANTED in us (<by> ¿ individually or in race?) by a
594n112 man had appeared-- though instinct so firmly IMPLANTED, birds soon <dis> learn to disobey it-- I hav
606o022 ch real objects are judged; & how obtained.-- IMPLANTED in our bosoms.-- how comes it there? Laocoon
625o052 I believe that certain feelings & actions are IMPLANTED in us. & that doing them gives pleasure & bei
390d179 ght urges?) one with opposed characters is by IMPLIANCE to breed two which have each varied from pare
553m136 improved by culture « <was> who feel the most IMPLICIT faith that through the goodness of God knowle
136a147 81. «&83» Some remarks on thinness of crust as IMPLIED by meeting with granite every-where. Phillips
538m080 probably analogous to the double individuality IMPLIED by habit, when one acts unconsciously with res
547m112 was a kind of ideal consciousness for moment, IMPLIED by «presence» my servant, «box» my own manner
048r084 ena. pebbles entirely coated with Tosca. which IMPLIES motion in the «loose» bed of pebbles. (On a se
265c084 rmined, when child is.-- shows that generation IMPLIES more than mere child, but that child should pr
292c171 is probably first stage, & an habitual action IMPLIES want of consciousness & will & therefore may b
292c171 to do in same way-- The improvement of reason IMPLIES diversity & therefore would banish individual,
569n023 but shrugged his shoulders & went away."-- he IMPLIES negation, without violence, without assigning
119a104 quired to start with (& does not Hersche theory IMPLY tendency to equilibrium.) 3d. there are mountai
357d074 e adapted for long transportation, seems «?» to IMPLY knowledge of whole world-- if so doubtless «par
613o036 xtends instinct to plants, but surely instincts IMPLY willing, therefore word misplaced The meaning o
618o41v bjectively 6) The reason why thought &c. should IMPLY «X» the existence of something in addition to m

510q018 t, curious.-- foxes-- English animals. [Made no IMPORT. remark]CD (2) Secondary male characters.-- do
575n044 at right to deny identity of instinct.-- Habits IMPORT to Bowen No one doubts that a cross of bull do
109a083 preservation of dikes & ledges of first-rate IMPORTANCE in showing not subaqueous removal--??? the d
224b212 ach group)» whether good species, & hence the IMPORTANCE Naturalists attach to Geographical range of
252c045 rected to study localities of isld.--«immense IMPORTANCE of local faunas foundation of all our knowle
276c124 to advance cause of truth It is of the utmost IMPORTANCE to show that habits sometimes go before stru
284c149 ding to the species, it is manifestly of less IMPORTANCE, as affording natural characters than among
285c151 rable conditions. (hence rise & depression of IMPORTANCE, in each group & connection of even distant
287c158 inition of relations to settling the relative IMPORTANCE of the organs in same state. in different an
291c165 ducct of Gould, remark of D'orbigny point out IMPORTANCE of habits in classification.-- Thought (or d
292c171 st be done often «to be habitual» or of great IMPORTANCE to cause long memory.-- structure is only ga
348d050 ngement depends on the organs judged to be of IMPORTANCE in inverse ratio to their variability.-- (No
348d050 h is group. because it has such characters of IMPORTANCE, "but we say such happens to be the characte
348d050 ens to be the character, of no matter of what IMPORTANCE, which prevails throughout the group & serve
370d116 t. 19th <Are> There is no scale, according to IMPORTANCE of divisions in arrangement, of the perfecti
386d167 e outside.-- is this not owing simply to more IMPORTANCE of internal organs in animals. One «invisib
531m050 nce.-- My Father remarks that things of great IMPORTANCE are easily forgotten, (if unconnected with f
531m050 d with fear &c) because people think that the IMPORTANCE of the event by itself will make it to be re
531m050 l make it to be remembered. whereas it is the IMPORTANCE.-- people very often forget where money is p
579n060 ot give notion of cause,» do p. 135.-- on the IMPORTANCE of a name, with reference to origin of langu
041r065 that part. having lived over whole bottom is IMPORTANT; because in this latter case. we cannot judge
059r123 e lines. must now proceed from great depths.--IMPORTANT.-- Decemb 10. 1802. Earthquake at Demerara. T
063r130 is <inosculation> «representation» of species IMPORTANT, each its own limit & represented.--Chiloe cr
064r135 in interior channels. there no outer coast.-- IMPORTANT effect.--? Capt. FitzRoy. -- Limited Volcanic
066r141 th its extreme force. Yet, no outlet at head. IMPORTANT in forming transverse valleys Ice Sir W. Fari
069r150 in this strata may be older than (B).-- Most IMPORTANT view Urge curious fact felspar melted gneiss/
071r156 ic action: <most> coming so near surface most IMPORTANT There is map of Cordilleras by Humboldt in Geo
087a011 r Oss Foss Wide range of Mammalia really very IMPORTANT. harmonizes well with Lyells idea of intertro
088a014 the trees---- Are any Fish seed-eaters. This IMPORTANT in transport of Fish Let a Hawk fly at Heron.
088a016 ng found on low hills East of Cordillera very IMPORTANT V. Malte brun -- Main character of Andes Meta
094a035 ne dikes which rise through granite.-- a most IMPORTANT question with respect to my theory of changes
102a061 at pressure? equally with little pressure? An IMPORTANT question! If water yields substances from imp
103a065 generally intersect metallic dikes: It is an IMPORTANT view being subsequent to dislocation of strat
107a078 body of mass.-- The last speculation becomes IMPORTANT with respect to thickness of crust broken up.
109a084 paper by rose on Greenstone, diorite, &c most IMPORTANT.:-- must be studied.-- Scientific Memoirs Edi
113a092 lly) becomes measure of force in that part.-- IMPORTANT as explaining want of levelness Major Mitchel
135a146 e as. at Bahia Blanca-- letter in drawer with IMPORTANT letters-- When I come to treat of the age of
152g047 Grant town to Forrest road comminuted shells IMPORTANT contingency if elevation from Axis, then rive
193b091 ical range Richardson-- Fauna Borealis. It is IMPORTANT the possibility of some isld not having large
200b124 sea; also Holman: <at> Keeling these are most IMPORTANT facts.-- As soon as island large enough for l
228b229 time, we might expect confusion of species.-- IMPORTANT. For instance take Voluta & Conus (??) which
244c021 cts Indian, like. Plants.--» It would be very IMPORTANT to show wide range of fish & shells in tropic
248c033 rst> latter is most intimately connected with IMPORTANT structures. <which are less obviously affecte
250c040 ef.» Philosop. Transacts 1823. Read June 5th) IMPORTANT paper by Dillwyn, on replacement of Cephalopo
261c071 mark about rock Pidgeons.-- & the latter most IMPORTANT in obviating a great apparent difficulty-- pr
264c081 d them-- From this view habits must form most IMPORTANT element in considering to which tribe,-- stru
267c095 ammon remain after rest of animal digested.-- IMPORTANT do p. 98. on a quaternary arrangement of Cryp
278c130 ralia, & several in Van Diemen's land is most IMPORTANT as showing former connection of two continent
278c131 ave been found!! rodents old inhabitants most IMPORTANT!! like Dipus of present day??! Major Mitchell
279c133 an Islds.-- Sir J. Sebright-- pamphlet-- most IMPORTANT, showing effects of peculiarities being long
279c133 lkinson. = Milking heredetary, developemen of IMPORTANT organ (.see marks on pages).-- Crosses of dif
280c135 h would even lead to anticipate mules is very IMPORTANT for L.yell said to me, the fact of existence
291c165 & heredetary. «problem solved» habits become IMPORTANT element in classification, because structure
299c195 anuring it.-- Epigonous. Perigonous &c-- very IMPORTANT in classification. here we have generative or
307c215 , what she does in species of Apterix This is IMPORTANT for if these abortive wings in the female are
313c236 n plants have habitual actions.--» «this very IMPORTANT in considering how children come to suck or o
315c243 eth at all.-- This way of viewing the subject IMPORTANT.-- Laughing modified barking., smiling modifi
315c245 ries & many to cold.-- Hence latitude is more IMPORTANT than longitude-- But in land F W s
326c266 uch on dogs.-- Reports of Brit. Assoc-- some IMPORTANT Papers. Dr. Mayo. Pathology of Human. Mind.--
335d014 ity of first cross. as stated by Fox, is very IMPORTANT, as showing above facts as first cross being
338d025 ure Africa?-- [Fossil Elephant of Africa Most IMPORTANT under this view, & Hippotamus of Madagascar:
348d050 paribus these will be the oldest) "The most IMPORTANT characters break down in certain species & be
348d051 at characters chance to be heredetary whether IMPORTANT or not,). p. 7. "The Natural arrangement of a
352d059 which experience shows us it was for.-- Most IMPORTANT law.-- Penguins wing perhaps not abortive???.
355d069 upial. Edentata.-- Pachydermata &c &c-- It is IMPORTANT with respect to extinction of species, the ca
381d157 ptandrous, & they have hybrids-- this is most IMPORTANT support to my views-- Seeing sexes separate i
389d176 male & female too closely related: this most IMPORTANT with regard to theory, showing generation con
405e035 lite man.-- L'Institut. 1838. p. 338.«V[ide]» IMPORTANT account of cross of sheep & Moufflon of Corsi
408e042 n Zoolog. Journal. for those of Cuba.-- It is IMPORTANT to understand well the relation of passage fr
410e052 anges. & external conditions of country, most IMPORTANT & will be done to all countries,-- but naming
411e054 ce.-- the laws of variation of races, may be IMPORTANT in understanding laws of specific change.--
416e070 re, great difficulty in crossing [& this most IMPORTANT obstacle to my theory] CD without the hermaph
425e104 ranean, but chiefly mountainous «this is very IMPORTANT. (Sicily exception)-- see if this can be gene
426e105 poch» like Sicily not having species, if true IMPORTANT on my view.-- March 9th-- Is there any relati
430e117 have had VERY different dispositions: this is IMPORTANT as showing small variations in offspring of w
439e143 ing passed through many changes.-- It is very IMPORTANT Mr Herberts fact about the hybrids (mentioned
440e148 th> «generation» is only modification, though IMPORTANT one, of growth «Lamark. Vol II. p. 120. obser
441e149 ing, but not on individuals is very curious & IMPORTANT.-- The existence of "laws of organization" ha
443e154 at the two kinds of generation have some most IMPORTANT difference is forced on us.-- My theory only
447e164 -- as Cocos do mer.-- Analogy shows some most IMPORTANT end.-- Festuca vivapara F ovina-- propagated
459tflv pecies, which has undergone all the changes. [IM]PORTANT view, copie[d] Gleanings of Sciences. Vol. I
465t079 ve been confounded in the caves? It is highly IMPORTANT, to bear in mind that enormous periods may el
465t081 ds paper) (Ed. Phil. Journ. end of 1839) very IMPORTANT. it seems owing to immigration of other races
480z006 bigny Vol II, p 24 Proceedings of Zoolog Soc. IMPORTANT account of habits of Tubularia. p 52. May 183
482z011 r's work of Mammali: of Paraguay must be most IMPORTANT a discussion of geographical distribution of
491q01v chnis can be propagated by cuttings.-- Try.-- IMPORTANT as discovering function of seeds-- (6) To hyb
493q003 into A. as takes place in mules ass & horse-- IMPORTANT. {In crosses does male offspring take after m
495q005 rimentise on Primrose seeds-- it really is an IMPORTANT case-- cross with cowslip pollen.-- as these
497q06v hether such vary.-- Do Bees go to Sweet Peas, IMPORTANT, for if so, as these can be raised true, ther
497q007 of Tierra del Fuego and Entomology of.-- most IMPORTANT, as furthest removed possible point.-- ¿gener
498q007 . as being difficult to cross. (5) It is most IMPORTANT to ascertain amount of variation in plants ra
498q008 ut of dogs.» much-- or the minute Orthopt.-- IMPORTANT, as we know how readily they cross.-- (9) In
501q10a r even where few are they constant: this very IMPORTANT for it wd show that such variation is not a g
509q017 ican Animal> Knows nothing [<...>] It is very IMPORTANT to know, whether Gould's observation holds go
515q022 they grow up in other respects different?--. IMPORTANT.-- Oct. 44 Tell J. Anderson's statement of En
545m106 rinkle, but such movements in skin of eyebrow IMPORTANT analogy with man.-- I see monkeys grin with p
553m137 ieves is common to that group.-- this is very IMPORTANT as showing <connection> that expression mean
554m140 commonly using, foreign bodies, for end. most IMPORTANT step in progression.-- The male Black Swan is
575n044 ossing races of dogs on their instincts, most IMPORTANT, because they obey the same laws, as the cros
575n044 ox & wolf & dog.-- the only test this is most IMPORTANT: can there be stronger analogy that the tenda
580n062 n animals & men only in Kind.-- probably very IMPORTANT work.-- Feb. 12. 1839. Sir. H. Davy -- Consol
608o026 at effort of reason to discover them: this is IMPORTANT explanation) he thinks they have none.-- Effe
613o036 Definite instincts being acquired, a most IMPORTANT argument, to show that they result from organ
026r021 nt coincidence of line of veins & cleavage is IMPORTANTS; veins appearing a galvanic phenomenon, so p
173b011 might have been Monsoons.. when they ceased IMPORTATION ceased & changes commenced.-- or intermediat
204b140 ffspring, much larger than the dam, from those IMPORTED by Ld. Powis Hybrid dogs offspring seldom int
235b263 hy do Van Diemens land people require so many, IMPORTED animals?-- At what. part of tree of life, can
450e173 an the latter-- Forrest Voyage p. 323. Sooloo. IMPORTED elephants. wild hogs-- spotted deer, no looni

450e176 in little isld of Sooloo.-- said to have been IMPORTED: shows they will propagate get dimensions-- d
534m063 lect.-- do. p. 157. Westwood remarks that some IMPORTED plants are attacked by insects & snails of th
615o36v ogous to inflorescence of Tropical plants when IMPORTED & plants sleeping good show acquirement or ob
069r149 ea The 24 ft. elevation at Concepcion. from IMPOSSIBILITY of such change having taken place unrecorde
104a067 uld be far off In Discussion on dikes argue IMPOSSIBILITY on fissure going right through superincumbe
199b120 rhaps never female.--no offspring: physical IMPOSSIBILITY to marriage.-- ¿whether those genera which
218b189 in nature a «real» repulsion «amounting to IMPOSSIBILITY, holds good in plants» between all differen
283c145 num into a Tiger."-- but perhaps I feel the IMPOSSIBILITY of this more than anyone.-- no turn the Zeb
335d016 is distinct case,-- gradation from physical IMPOSSIBILITY to (perhaps increased). fertility.-- (but m
362d100 race the laws of change in this time.-- the IMPOSSIBILITY of discovering their origin.-- I see only s
426e107 d probably be perfect series, from physical IMPOSSIBILITY to unite to perfect prolifickness.-- (<a se
555m144 ttiment about hanging came before me showing IMPOSSIBILITY of person recovering from hanging on accoun
151g041 4th shelf.-- Even on Lauder Dicks Hypothesis IMPOSSIBLE to explain absence of lines in certain parts
162g102 wer called Glengarry (Nead Roy told me) it is IMPOSSIBLE to see my new shelf, from road: Loch Ness 30
222b206 ow replace Saurians of Secondary epoch: it is IMPOSSIBLE to suppose such an accumulation at present d
266c088 rom which country many birds come it would be IMPOSSIBLE to classify them.-- I would [not located] Mu
280c136 father, but if too unlike its own parent this IMPOSSIBLE-- (Hence we might expect even if two mules b
336d017 rm of body gained in one generation, so it is IMPOSSIBLE to transmit them, & as offspring must be lik
337d021 erve to finest vibration of sound.-- which is IMPOSSIBLE.-- Mr Spence remarks that the Fringilla dome
378d148 sume with common ducks. so often, that it was IMPOSSIBLE to say which was origin of any identical bir
388d172 ng views show that transmission of mutilation IMPOSSIBLE.-- it should be observed that transmission b
423e096 ructures:= Without enormous complexity, it is IMPOSSIBLE to cover whole surface of world with life.--
443e154 longation of life by gemmation <can be> being IMPOSSIBLE. can be overturned, then the conclusion that
616o039 hey would give in support of their view it is IMPOSSIBLE to shew satisfactorily it's erroneousness. i
498q008 hole, to impregnate?-- I presume only stigma IMPREGNATE.-- (12) At Maer Cowcumbers in frames are not
384d162 related to each other, as never to be able to IMPREGNATE themselves (this never happens in plants «on
389d176 cts relating to Generation One copulation may IMPREGNATE one or many offspring.-- it affects the subs
447e164 acterised species periwinkle wants insects to IMPREGNATE allied to Asclepias Turpin cell is individua
448e165 , because it is thought to require insects to IMPREGNATE it.-- it is allied to Asclepias, where this
466t093 ng out brush over anther & pistil & one I SAW IMPREGNATE by pollen with which <bees> «a bee» was dust
492q002 bad years from Caterpillars. 5. Whether Roses IMPREGNATE each other. when close planted together: <de
498q008 (10) Is number of pollen-grains necessary to IMPREGNATE ordinary number of seeds known?-- Linnaeus h
498q008 men <?>-- (11) Must pollen grain be whole, to IMPREGNATE?-- I presume only stigma impregnable.-- (12)
505q014 lower.--Passion Flower. (as it is required to IMPREGNATE it artificially.)-- Asclepias-- Flowers not
505q014 se =Aristolochia, plant wh require insects to IMPREGNATE it (7) History of Potato field= (8) Abortive
216b181 ped in the breast.--looking as if many ova-- IMPREGNATED at once.-- Dr. Smith considers the Caffers (
312c232 old, go to heat, take dog. but do not become IMPREGNATED & puppies delicate-- they cross sister & bro
417e079 hose which <lay> «have» their eggs, <inter>, IMPREGNATED externally; nor can it be a necessary concom
417e079 essary concomitant, with moths, which can be IMPREGNATED externally-- My view of every animal being H
491q01v e, that certain plants, like Aphides produce IMPREGNATED young ones; & that it is in these that male
497q007 e agency of man not known.-- (3) How is Iris IMPREGNATED.; which part of stigma?-- (4) As Papil. flow
498q008 er Cowcumbers in frames are not artificially IMPREGNATED. Abberly says Ants-- Enquire (13) Do any of
502q012 r. which had anthers removed, did not become IMPREGNATED. (34) Any recent information about pollen of
230b240 for we know that such can take place without IMPREGNATING each other), for if they are different then
313c235 ons are produced in succession (13?) without IMPREGNATION, therefore sexual passion must arise after
333d006 lear case of heterogenous offspring from one IMPREGNATION ¿is this one impregnation, or two impregnat
333d006 offspring from one impregnation ¿is this one IMPREGNATION, or two impregnations one giving half chara
344d041 case of insects as glowworm The case of one IMPREGNATION sufficing to several births analogous to su
401e015 great part of those varieties may be due to IMPREGNATION from other apple trees.-- now seeds of crab
491q01v whether fruit affected. Mr. B. seemed to say IMPREGNATION <caused> of some seeds, caused symmetry in
496q006 male flowers always barren-- if not how does IMPREGNATION take place male & female flower in same rec
499q009 ar's produce.-- (15) Would Yew fruit without IMPREGNATION.-- (16) Any calculation of number of grains
505q014 ? Yes «From cultivation lost their horns» is IMPREGNATION necessary to fruit--; become well shaped by
507q15v h peculiarities of structure rendering cross IMPREGNATION difficult or reverse (.49) List of seeds Ga
513q21. eference to Lobelia & Clarkia-- Peas time of IMPREGNATION.-- says many flowers are dichogamous Zoster
513q21. --(‖ I could extract nothing from him)‖ Does IMPREGNATION ever regularly take place in unopened flowe
514q21. o «hollow» spur.-- Ask about Pinks & Solanum IMPREGNATION before flower open. (An. des Sci Where is B
514q21. . (An. des Sci Where is Boerhaave's paper on IMPREGNATION of violets.= Zostera= Are dwarf plants on W
636j56v ngs} p. 333. Macculloch. brings forward. the IMPREGNATION of Dioecious Plants by foreign agency-- as
333d006 regnation ¿is this one impregnation, or two IMPREGNATIONS one giving half character & other more of E
359d088 ale of every animal certainly seems chiefly to IMPRESS the young most with its form & disposition Saw
390d178 pshot of all this is that effect of Male is to IMPRESS some difference: to make the bud of the woman,
148g024 --¿ is not Mica Slate too hard & uneven to be IMPRESSED Case of Birch Wood by Inverorum being determi
248c034 w causes, without picking» become more & more IMPRESSED in blood with time, then generation will «onl
249c034 brids but not varieties. which are not deeply IMPRESSED on blood., will cross & produce fertile offsp
313c235 as well as men-- when habits much more firmly IMPRESSED.-- we see in the Entomostraca. The sexual cur
410e050 s changes, & hence no feature would be deeply IMPRESSED on it, & hence there could not be improvement
418e083 e it alive, & how the laws of generation were IMPRESSED on it.-- Seeing that <Man> «all vertebrates,
473s07v can decide-- some peculiarities may be early IMPRESSED & others later-- All poultry with same down-f
571n027 , of the general ideas of our ancestors being IMPRESSED on us.-- Surely we have taste naturally all h
609o029 t the very beginning, & therefore most deeply IMPRESSED]. shame perhaps an exception. (does it origin
611o034 modification, peculiarities of external form IMPRESSED, & different laws of movements. [LHC] Hence t
145g001 Glen Roy Generally received opinion that male IMPRESSES offspring more indelibly than female p 367 Qu
204b140 & Dog cross? Mr Yarrel thinks oldest variety IMPRESSES the offspring most forcibly-- Esquimaux dog &
345d044 1837. Generally-- received opinion that male IMPRESSES offspring more than female, yet instances «Ir
360d091 is it local (not artificial variation) which IMPRESSES offspring most.-- «& not time» thinking of th
387d168 nected with the case of successive copulation IMPRESSES offspring more & more with the added «like Lo
387d168 C. D. E be <offspring> «animals»: if «x» male IMPRESSES ovum <of> in A, «with some peculiarity» that
401e017 ncle J. says common belief. that female plant IMPRESSES main features on offspring. & male the lesser
406e035 doubt. in distinguishing <after> which parent IMPRESSES offspring most is whether mother has had any
459t013 the plumage of drake still more.-- So Penguin IMPRESSES its form both on vars & species The <male> sw
345d044 iven on opposite side,-- «The theory of males IMPRESSING most is in harmony with their wars & rivalry
036r050 opiapò which is asserted to make a noise,--My IMPRESSION. is not very distinct, from some of the lowe
463t057 s constituted that a difficulty makes greater IMPRESSION, than the grouping of «many» facts with laws
545m105 do <certain> the actions.¿ is it the <becom> IMPRESSION becoming very often unconscious, which makes
547m113 der <to every> calling up ideas of every late IMPRESSION.-- (do the ideas, direct effect of perceptio
547m113 ertion of keeping up the memory of every late IMPRESSION. & likewise gaining new ones from senses. &
551m127 like sleep does not doubt the reality of the IMPRESSION on its senses.-- insane people believe they
578n056 is a convulsive action to remove disagreeable IMPRESSION like true convulsion. (Hence pass into convu
606o024 eauty, is of the "beau ideal", my instinctive IMPRESSION 1) September 6th. 1838 Every action whatever
628o054 supremacy, solely due to greater duration of IMPRESSION of social instincts, than other passions, or
257c058 developement of lizards.-- As we have birds IMPRESSIONS in Red Sandstone. great lizards in do.-- <Wo
524m021 these places.-- My F. says, shows that early IMPRESSIONS are most durable.-- (but Miss Cogan shows th
545m104 t is customary to die"-- August 24th. As some IMPRESSIONS «Hume» become unconscious. so may some ideas
545m105 conscious, if so (think of this). study what IMPRESSIONS become unconscious those which are viewed wi
550m125 efers to the sensations <it> when excited by IMPRESSIONS, & not mental or ideal ones, <which> «& thes
553m135 , & [not located] notion, are not effects of IMPRESSIONS long repeated, without the powers of the min
563n002 re so framed that he constantly compared his IMPRESSIONS, & wished he had done so & so for his intere
602o011 &c is made up of many partial results, & the IMPRESSIONS on them are <all> remembered, when the meani
607o025 e of others-- varied capability of receiving IMPRESSIONS-- accidental (so called like chance) circums
547m114 e of ones consciousness.-- Mayo observe no IMPROBABILITIES in a dream, effect of doubting nor believi
105a070 been no tumultuous rush.-- besides general IMPROBABILITY. stratification, If chain of lake. <a> the
278c131 Owen to see whether species same, excessive IMPROBABILITY. Mem in Clifts list a rat said to have been
339d026 all divisions.-- does not seem to think any IMPROBABILITY to animals being distributed after flood (!
152g045 Collarig, by Dicks theory lake burst in most IMPROBABLE part & not in Pass, where shallowest In Glen
181b040 mmalian type of organization; it is extremely IMPROBABLE that any of <his relatives shall likewise> t
269c102 of latter render-- spontaneous generation not IMPROBABLE.-- After reading "Carus on the Kingdoms of N
438e141 quity» prevents their variation, which is not IMPROBABLE as Mr Herbert does not seem to recognize any

321
IMPROBABLE / INCLINES

Page
(Key Word)
448e167 f Isthmus of Panama.-- These two cases highly IMPROBABLE.-- yet I can see no other way of accounting
068r146 ents would keep open long after emersion, but IMPROBABLY so long, that to be surrounded by continent.
365d105 resembles Red Grouse) it may be so-- but very IMPROBABLY, for it can hardly be thought that the cross
520m001 ter died so long before, that it is extremely IMPROBABLY that they should have imitated.-- when atten
502q012 to S. America (33) Ornithologum commonly but IMPROPERLY called Canadense-- would it grow in open air
264c079 his parent, naked, artless, not improving yet IMPROVABLE» & then let him dare to boast of his proud p
309c220 sses-- avoid the contamination of <cl> castes. IMPROVE the women. (double influence) & mankind must i
309c220 e the women. (double influence) & mankind must IMPROVE-- The areas of subsidence marked out by animal
312c231 t located] Henry Thompson tells me best way to IMPROVE cattle is to cross between a good bull & <be>
347d049 nization an animal has, it tends to multiply & IMPROVE on it.-- Articulate animals must articulate. <
347d049 ust articulate. <i> in vertebrates tendency to IMPROVE in intellect,-- if generation is condensation
347d049 nsation of changes. then animals must tend to IMPROVE.-- yet fish same as, or lower than in old days
432e124 highland shepherds said.-- p. 12. Attempts to IMPROVE the native <breed> animals of any country must
526m030 es become heredetary.-- When a man says I will IMPROVE my powers of imagination, & does so,-- is not
536m073 grateful if it were pointed out.-- My wish to IMPROVE my temper, what does it arise from but organiz
536m074 most humble, he would strive <to do good> «to IMPROVE his organization» for his children's sake & fo
466t095 .-- «Is not cantering a congenital peculiarity IMPROVED.» Probably every such «new» quality becomes a
553m136 idea of God in civilized nations has not been IMPROVED by culture « <was> who feel the most implicit
568n018 pleasure in performing it.-- As soon as memory IMPROVED, direct effect of improving organization, com
620o044 obeyed) & is sure guide.-- Hence conscience is IMPROVED by attending & reasoning on its action, & on
222b204 a of propagation almost infers, what we call IMPROVEMENT, --All Mammalia from one stock, & now that o
222b206 ame argument as father of Mammalia; but have IMPROVEMENT in system of articulation. ¿whether type of
292c171 rations are impelled to do in same way-- The IMPROVEMENT of reason implies diversity & therefore woul
327cIBC . Sir. S do. do-- Buffon Suites Cline on the IMPROVEMENT of domesticated animals Fries de plantarum p
391d174 differently of what kind, either progressive IMPROVEMENT or deteriorate» that object failing, generat
410e050 impressed on it, & hence there could not be IMPROVEMENT. «& hence not «be» higher animals»' -- it was
628o53v genius allow this moral obligation? [2] [The IMPROVEMENT of the instinct of a shepherd dog, is strict
352d058 d few.-- » hence become EXTINCT, & hence the IMPROVEMENTS of every type of organization. such law wou
224b212 tell, what is the amount of difference, which IMPROVES. & checks it.-- It does not bear any precise
526m030 tion, & does so,-- is not this free will,-- he IMPROVES the faculty according to usual method, but wh
526m030 be anything ambition, avarice, &c &c An animal IMPROVES because its appetites urges it to certain act
175b019 ching upwards different types of organization IMPROVING as Owen says simplest coming in & most perfec
264c079 age, roasting his parent, naked, artless, not IMPROVING yet improvable» & then let him dare to boast
527m034 kes a discoverer, & therefore (independent of IMPROVING powers of invention) such castles in the air
568n018 As soon as memory improved. direct effect of IMPROVING organization, comparison of sensations would
255c050 ach other.-- this bird was well known for its IMPUDENCE. «This excellent case of memory without assoc
289c161 do with hydridity,, is simply, an instinctive IMPULSE to keep separate, which no doubt be overcome,
545m106 laughing» they dance with passion, ie. nervous IMPULSE to action is sent so fast to limbs that they c
582n067 tary habit:-- there must, however, be a mental IMPULSE (though unconscious of it) to move its legs so
582n067 by which an animal performs an instinct, & its IMPULSE to do it.-- [the means must be present on any
583n070 adapt its cell to circumstances], it must have IMPULSE to make a cell in certain way, which way its o
586n082 eye, & use of limbs &c, or it result from mere IMPULSE to save wax.]CD which it instinctively exerts
609o029 ginate in a doubting feel between conscience & IMPULSE] but shame «we alas know» is far easier conque
591n099 ndered why all abnormal sexual actions or even IMPULSES. (where sensations of individual are same as
366d111 of S. America take an effect.-- would perfect IMPUNITY from muskitoes bite influence propagation of
525m025 ffspring is affected by previous marriages with IMPURE breed.» A cat had its tail cut off at Shrewsb
094a036 f rock from Costorphine at Geolog. Soc: Colonel IMRIE Transact Wern. Soc. Vol. 2. p. 35 Sir J Hall Tr
451e178 April & October & like man almost (this looks INACCURATE C. D) they will catch the Malaria & die.-- O
482z011 ndiae-- histrionicus Vultus aura Excessively INACCURATE Saw a Chouette a huppe courte talks of nine
483z012 om Capt King on birds of St of Magellan. Very INACCURATE & Vol IV p. 91.-- Vol IV p. 388. Domestic mo
025r017 f the streams of Lava in Ascencion known to be INACTIVE 300 years? No Volcanic Earthquakes or Hot Spr
023r010 mountains on the continent of S. America is INADMISSIBLE «may have happened from incipient elevation
617o39v in the way in which we apprehend the force of INAMIMATE bodies. How we identify the two aspects as di
617o040 DIRECTION, & thus when we apprehend force in INANIMATE matter we feel dissatisfied until we can poin
617o40v sfied therefore, if we can trace any force in INANIMATE matter up to the action of some animated agen
542m094 rd not to growl out some sound even if it be INARTICULATE.-- the maniac shouts & bellows with passion
576n046 Vol II p. 126 says seals knit their brows when INCENSED.-- A Dog may hesitate to jump in to save his
255c053 aim that nature conscious of the principle of INCESSANT change in her offspring. ¡has invented all ki
219b196 ays Tahitian kings. would hardly produce from INCESTUOUS intercourse. a parallel fact to Blood-Hounds
279c133 difficulty in crossing race.-- bad effects of INCESTUOUS intercourse..-- excellent observations of si
391d179 ture desire fails. Every individual except by INCESTUOUS marriage has acquired from father some diffe
068r147 ut ejection of the fluid propelling mass. If one INCH can be raised then all can, for fresh layers of
081rIBC 159929 {T} Toises Pieds Myriametre = 5130., 4. 5 INCH Kilometre 513., 0. 5 Hectometre 51. 1. 10 Metre
372d131 another man.-- That the embryo the thousandth of INCH should produce a Newton is often thought wonder
439e144 cations each possess., & that tiger springing an INCH further would determine his preservation-- if k
023r009 which he pluckt a great many teeth, 2 of them, 8 INCHES long, & as big as a mans thumb, the rest not a
117a102 0 miles E of Staten land. bringing up pebbles 2 INCHES long?-- L'Institut. 1838 p. 151. Formations of
145g004 always parallel to strata 3 or 4 seams/ 3 or 4 INCHES thick-- {P} 35 degree is I believe about great
458t001 p. 539. Dr Cantor's account of fossil frog, 40 INCHES in length--! alludes to ancient gigantic salam
505q014 length from nose over head to root of tail 28½. INCHES. From sole of foot to shoulder on line of back
506q014 in length (measured same way) 47½-- in heigt 30 INCHES Examine Keel of Common & Wild Duck-- Black Duc
462tf05 men, some more intellectual than others-- is INCIDENTALLY said that a mongrel man may lose all traces
023r010 erica is inadmissible «may have happened from INCIPIENT elevation.» The volcanos originated in the bo
556m145 ting tongue from behind upper & little between INCISORS.-- like <W[...] what> person says "what a pit
056r109 t. «I confess I never see such islands whose INCLINATION natural [...] deepest astonishment.» Perhaps
078r175 cerro) have been parallel to the direction & INCLINATION of the vein".-- at Zacatecas the veta grande
085a001 award, at mouth of S. Cruz. from ascertained INCLINATION. of plains: Lias in Shropshire. or some othe
099a050 f no sort of consequence.-- Therefore < S of INCLINATION «varies with chemical attraction &c.» become
099a050 » becomes measure of force. < ∴ where little INCLINATION, little force & varying direction.--> Theref
100a052 ity, versus some fault explaining vary dip & INCLINATION.-- which last is strong character.-- A discu
105a069 oses its cutting power. (as does it when the INCLINATION becomes less & ∴ tends to finite power) wher
105a073 -- Why is serpentine course result of little INCLINATION??----It is simply as the inclination is litt
105a073 little inclination??----It is simply as the INCLINATION is little the force required to move <it> «s
106a073 ntum. which the water has obtained.-- If {P} INCLINATION be great where arrow stands the force immedi
106a073 from (B) which would not have been case. if INCLINATION small.-- The power of widening channel depen
106a074 deflected stream cannot retain its force if INCLINATION be great. There could «not» be great deflect
106a074 stream has no tendency to widen course until INCLINATION is become comparatively small, & when that i
106a075 en we have valleys of this structure. as the INCLINATION in all probability would be greater when flo
155g063 oubt, a mound of Alluvium nearly parallel--» INCLINATION of river must constantly alter with falling
155g064 tritus» than they corrode or vice versa Same INCLINATION when serpentine might remove, what above str
460t015 others & sisters half-breed showed no sexual INCLINATION for each other-- Aug. 20th The Echnida & Hed
557m148 anner quite different from fear; there is no INCLINATION to jump away,-- it is, ill-defined fear.-- Y
568n018 irst take place, whether to pursue immediate INCLINATION or some future pleasure.-- hence judgment, w
049r088 o or as an injected mass.--From conical form I INCLINE to <latter> former; & thus occurring in groups
099a050 n.--> Therefore in PILE of mud from Trapiches. INCLINED layer!!!.-- The separation in the Ponza case
148g027 Spean Bridge many flat terraces one above much INCLINED towards river all these composed-- where side
155g063 on Turrit, where I believe they end in upwards INCLINED plains, as in Corry. & as «as I believe in si
216b179 in the Indian sea.-- Deshayes.-- Mr McClay is INCLINED to think that offspring of Negro & white will
275c121 ere most like parent Brahmin bulls-- Mr. Y. is INCLINED to think that the male communicates the exter
446e162 tion by gemmation & by seed-- which Henslow is INCLINED to think very close.-- «A fruit tree by certa
578n057 s-- (except in weak people & hysterical people INCLINED to convulsive actions).-- But, the lachyrmal
586n078 ng of kind oneself knows in walking [one feels INCLINED to stop at right number of house though one c
586n082 feels no more surprise at it & feels no more INCLINED to ask [not located] If dislike, distaste. &
607o025 condition of mind which leads to motion being INCLINED that way]CD one sees this law in man in somna
058r118 arts; toutes affectent une pente plus ou moins INCLINÉE vers le rivage de la mer, tandis, au contrair
059r119 ntagnes sont formées de couches paralleles et INCLINÉES du centre d'île, vers la mer; ces couches on
044r073 ency of soft materials being consolidated; one INCLINES to belief all strata of Europe formed near co
430e120 this intermediate one.-- Mr Lonsdale evidently INCLINES to think it Hybrid.!!! Ask Woodward Mr Lonsda

055r106 II. p. 97 at Potosi the veins run from North <INCLINING> to South. inclining a little to the West: th
055r106 he veins run from North <inclining> to South. INCLINING a little to the West: the veins which follow
222b206 pose such an accumulation at present day & not INCLUDE Mammalian remains.-- The Father of all insects
356d070 d of organization good expression but does not INCLUDE so many facts as mine. The facts about half br
499q010 eeds have contrivance for transportal, does he INCLUDE seeds good to eat. (even Nux Vomica is eaten b
612o035 nly movements such as sensitive plants. (But I INCLUDE irritability for that require will in part. ¿W
251c041 Clapperton &c on Mammalia no doubt will all be INCLUDED in Smiths work «do» Vol. IV p 273. Macleay on
090a022 described as many, (one even 3000) This ninety INCLUDES all actually counted.-- The weight «or size»
125a120 p. 473. on great Iceland stream. the 90 miles INCLUDES opposite directions. Mem. S. Cruz. Assuming f
311c226 n classification» as indication of structure (INCLUDING brain & other organs difficult to analyse) wi
550m124 little that of anger or revenge.-- they are INCOMPATIBLE & the former, the more pleasant.-- Simple h
073r162 ll parts of the globe". p. 113 How utterly INCOMPREHENSIBLE that if meteoric stones simply pitched fr
618o41v but we cannot see an atom think: they are as INCONGRUOUS as blue & weight; all that can be said that
635j55r the Deity. how it acts & whether constant or INCONSTANT like that of Man.-- the cause given we know
059r119 ou les scissures.--M. B. thinks these parts INCONTESTABLY formed the parts of one whole burning mount
609o29v e). on same principle that Malthus had shown INCONTINENCE to be a vice & especially in the female Oct
637j57v rbing Camel's stomach is puzzler p. do says INCONVENIENCE would have arisen had « <not> some» some in
637j57v ided. «with proboscis» «as bee & butterfly» INCONVENIENCE.! extinction, utter extinction! let him stu
404e024 mon mind or instinct, now this is directly INCORRECT The case of my mice is good, because it is an
608o026 ngencies in the way to aid motive power.--if INCORRIGABLY bad nothing will cure him' 3) disgusted. wi
051r093 urfs: in both latter cases become petrified, & INCREASE.-- -- In Southern regions every rock is buoyed
106a076 p. 40 or Phil Mag. Dec 1837 p. 520 Mr Fox on INCREASE of temperature at great depths. All Earthquak
126a123 ld become hotter.-- hence temperature ought to INCREASE rapidly beneath level of sea.-- deep seated s
132a137 data wholly insufficient to calculate rate of INCREASE of heats in earth's crust.-- yet heat does in
132a137 ase of heats in earth's crust.-- yet heat does INCREASE,-- but in Ocean does not. (see resumè p. 536)
132a138 f earth. at the freezing point.-- accounts for INCREASE on earth by volcanic action.-- <Why> now as w
137a151 erature of the interior & p. 142 / p. 155. the INCREASE of temperature beneath the sea, is probably m
172b007 is good for Man.-- Let a pair be introduced & INCREASE slowly, from many enemies. so as often to int
181b040 brothers & sisters «in a state which does not INCREASE» it will be chances against any one «of them»
190b078 n, throughe means of every step of progressive INCREASE of organization being imitated in the womb, w
205b145 d of animals. for in <the> proportion to their INCREASE of size they become worse in form, less hardy
252c045 fferent Sumatra Java ---------- do from Indian INCREASE of knowledge would probably tell more certain
267c094 Tahiti, in spite of every attempt to check its INCREASE. The <bush> woodlands for miles in extent are
326c266 ng quoted by Malthus.-- Heberdens Observat. on INCREASE & decrease of different diseases. 4to 1801.--
374d134 ery one till he thinks deeply has assumed that INCREASE of animals exactly proportiona[l] to the numb
375d134 g of the species as inference from Malthus.-- «INCREASE of brutes, must be prevented soley by positiv
375d134 stop desire.--» in Nature production does not INCREASE, whilst no checks prevail, but the positive c
375d135 of famine & consequently death.. population in INCREASE at geometrical ratio in FAR SHORTER time than
375d135 en a few years plenty, makes population in Men INCREASE, & an ordinary crop. causes a dearth then in
399o009 s almost doubtful which it was.-- do hind legs INCREASE in any rabbits One may strongly suspect, that
408o043 towards extermination. some other species must INCREASE in number where then is the gap, for the new
423o097 eings-- but there is the strongest possible to INCREASE them, hence the degree of developement is eit
427e108 f animals & plants to each other would rapidly INCREASE, & hence number of forms. once formed. would
575n044 owen No one doubts that a cross of bull dogs-- INCREASE the courage & staunchness of greyhounds.-- bu
590n093 on the Influence of the Passion."-- p. 37. The INCREASE of Bilary secretion attends passion p. 39. Th
634j55r abit seas.-- Were they then killed out «by the INCREASE cold», & did mammifers then take their place?
032r039 dal action. this will at length be checked by INCREASED vertical <height> thickness (DZ) of mass to b
036r049 paper a gradually curved enlargement see its INCREASED length. which will represent the dilatation,
043r070 er larger spaces of the globes & "periods" of INCREASED activity.-- such as that of 1835.-- State the
048r087 oe curvilinear strata subsidence.--The sudden INCREASED dip is not parallel case to Isle of White. bu
090a024 00 = tons in fifty thousand years]CD If world INCREASED with width. for besides more surface exposed.
105a069 ong as exposed to waves of sea. cutting power INCREASED with width. for besides more surface exposed.
126a123 asily water percolates rocks,-- when pressure INCREASED or under surface, would not the fluid matter
190b079 e though species became obscurer as knowledge INCREASED, but genera stronger,-- Mr Waterhouse says no
223b211 B. C. D. & A will soon form good species! The INCREASED fertility of slightly different species & int
250c038 read then such tribes as Marsupial & Edentata INCREASED most. Certainly Africa approaches Nearest to
297c189 . Mag of Z. & B. p. 431. Missel thrush lately INCREASED in numbers over whole of England & Ireland.--
331d1FC nipples in domesticated very fertile animals INCREASED?-- Where offspring, heterogenous, in plants a
335d016 ation from physical impossibility to (perhaps INCREASED). fertility.-- (but many animals are fertile,
435e133 th the developement of form; but that tadpole INCREASED in size now the Proteus anguiformis. he remar
176b024 part of country. repugnance to intermarriage <INCREASES it--> settles it ¿We need not think that fish
298c190 n who is habitually kind to children,--<get> «INCREASES» general instinctive feeling.-- There is grea
298c191 rivals..-- &c &c --Climate altering as island INCREASES.-- upper parts attracting all the moisture.--
381d156 strange effect.-- Directly a Capon is cut, it INCREASES in size prodigiously-- Animal OEconomy by, Hu
423o097 opement is either stationary or more probably INCREASES.-- Jan 29th. Uncle John says he feels sure, t
099a051 B. but if particls attract each other in some INCREASING ratio in proportion to proximity would they
181b039 ds; like individuals in a country not rapidly INCREASING.-- If we thus go very far back to look to th
206b147 rom eternity «backwards.--» If population was INCREASING between each lustrum, the number related at
206b149 at groups, which we are bound to consider the INCREASING ones.-- NB As Illustrations are there many a
590n093 ncope p. 42. Sighing from grief. is method of INCREASING languid circulation-- no, for <grief> sighin
126a122 nst my views-- if we had rod thus & judged by INCREMENTS at, how wrong, would our judgement be-- Does
363d102 y confounded them, yet can readily be told by INCUBATION & other peculiarities.-- (Mem.-- Goulds Will
057r113 obably that B. Blanca & M. Hermoso contemp:.--INCULCATE well that Horse at least has not perished bec
242c017 picole vert instances of American forms in East. IND: Archipelago. Raffles. Horsfield. Diard. Duvauc
243c019 have representatives (& instances given) in East IND. Arch:-- Birds of New Zealand absolutely differe
629o55v emorse-- [Thus pungency of one's feelings for INDECENCY-- preposterously so, for Marquesans think onl
035r047 ishing consult book itself. P. xx: same fact is INDEED shewn? by the parallel bands of formations on
049r089 fect. than in believing mere agency of dikes: & INDEED when do these dikes lead to a conical mass. wi
283c144 to existence in two continents our ignorance is INDEED profound & such it appears.-- Is there not som
417e077 pect, as the difference between man & woman is «INDEED» (independent of sexual differences) a variety
426e108 e of Hindoism coming out more than in mother or INDEED grandmother; what is Mr S. S. parentage?-- Won
471f008 ing of their similarity «in structure» he says "INDEED it wd be difficult to point out a family so co
526m029 are brought to mind, by memory of the scenes, (INDEED my American recollections are a collection of
584n072 lled instinct,-- migrating to one spot, this is INDEED instinct.-- Australian man, may be called inst
600o08a leasure. but because most imperious.-- It would INDEED be wonderful, if, mind of animal was not close
614o037 without further end just same argument. without INDEED we are step towards some final end.-- producti
616o039 ere are living bodies without these faculties & INDEED until we know what answer they would give in s
620o044 ct. (or conscience) is always present (which is INDEED, often felt at very time it is disobeyed) & is
534m062 London Vol. I. p. 106. Col. Sykes on Formica INDEFESSA placed table in cups of water which they wade
232b245 species which survives any change may undergo INDEFINITE change., (marking in their history an eocene
263c077 is Mammalian.-- his <has> origin has not been INDEFINITE.-- it is not a deity, his end «under present
485z016 those of England.-- or ground birds-- rather INDEFINITE letter Mem Orpheus--becoming tyrant-- flycat
526m031 ecome changed.-- appetites urge the man, but INDEFINITELY, he chooses (but what makes him fix!? <)>--
428e112 what can «however» be more striking, about INDELIBLENESS, than the number of good race-horses, which
145g001 ed opinion that male impresses offspring more INDELIBLY than female p 367 Quarterly Journal of Agricl
426e106 do not so much affect first race, as it does INDELIBLY the many subsequent ones. My views, «V <see>
427e110 «.than other apples» but probably would more INDELIBLY stain offspring-- it would not reach one appl
522m013 tion are lost.-- most delicate people do most INDELICATE actions,-- as if «these emotions» acquired.-
268c099 se of destruction of «great» animals?-- Show INDEPENDENCY of shells to external features of land by s
035r047 consider arguments for oscillation of level INDEPENDENT of mineralogical nature & dependent: & then
075r166 The wealth of the veins in most part totally INDEPENDENT of the nature of the beds they intersect". =
109a083 direct (or oblique) outwards may be granted. INDEPENDENT of currents.-- mud going out can actually be
119a105 levations) we may conclude that elevation is INDEPENDENT of spreading out matter by action of the sea
179b035 f certain form from considering. S. America (INDEPENDENT of external causes) does appear very probabl
388d170 ndependently of its parent & therefore wants INDEPENDENT supply of food.-- is real. difference-- but
417e077 difference between man & woman is «indeed» (INDEPENDENT of sexual differences) a variety. The offspr
424e102 ere every year produce hybrids-- now this is INDEPENDENT good case, but very odd since these crows ar
508q016 e.-- D. Holland (7) Is Haemorragic tendency, INDEPENDENT of heredetary cases, more common in man than

527m034 of thought makes a discoverer, & therefore (INDEPENDENT of improving powers of invention) such castl
528m036 res of scenery.-- There is absolute pleasure INDEPENDENT of imagination, (as in hearing music), this
560m156 Do they pout, or spit, or cry.-- <fe> Shame, INDEPENDENT of fear: colour of bare nails--, & of eyes.-
577n051 heory of sensitive Plants Habitual actions, (INDEPENDENT of mind) in the intestinal functions &c &c &
602o11v ake the mind have mysterious & sublime ideas INDEPENDENT of the senses & experience p. 142 "Upon the
608o027 tempted, from knowing every thing he does is INDEPENDENT of himself to do harm.-- Believer in these v
618o041 ut they are known by courses of action quite INDEPENDENT of each other. A person might be quite famil
625o052 he feeling of right & wrong.-- so far it has INDEPENDENT existence. & is supreme. because it is «a» p
204b140 maux dog & Pointer «Game-fowls have courage INDEPENDENTLY of individual force» Mr Wynne has crossed D
356d069 y of idea-- (though I arrived at them quite INDEPENDENTLY & have used them since) the line of proof &
388d170 w.-- it goes through transformation, nearly INDEPENDENTLY of its parent & therefore wants independent
417e078 child may be either like father or mother, INDEPENDENTLY of its sex, or half way between, or someway
574n042 e applicable to the formation of instincts, INDEPENDENTLY of habits.-- the limits of these two action
058r116 sive lines of sea weed-- Histoire Naturelle des INDES Acosta. p. 125. of French «?» Edition states th
177b028 is particular circumstances, to which> is it an INDEX of the point whence, two favourable points of o
501q011) Which sex in Mules generally fails-- perhaps INDEXED by secondary characters-- in double flower. do
241c015 kinds common to New Guinea & rest of isle in E. INDI: Arch: In New Zealand. a sturnus of American fo
031r038 nces in old world of symetrical structure. East INDIA Archipelago. «Aleutian Arch.--» V. Fitton. Aust
044r072 mboldts. fragmens.» Read geology of N. America. INDIA.--remembering S. Africa. Australia.. Oceanic Is
091a027 1828 Ed. N. P. J. p. 105. Oct. 1828. gneiss in INDIA (falls of Garsipa) dip 30 degree. <strike> <dir
096a041 ng with Carb. Acid.-- Mr Malcolmson in Paper on INDIA p. 503. On Indian Saline Deposits. Vol II. p. 2
134a143 . 149. on the <salt mines» «saline deposits» of INDIA p. 503. On Indian Saline Deposits. Vol II. p. 2
194b094 frica, appear to represent the semnopitheque of INDIA.-- Tooth of <Spi> of Sapajou-- NB Sapajou is S
201b126 . p. 569. 1837. Account of wonderful fossils of INDIA.-- & p. 545 «great monkey» Mr Johnston says Mag
202b131 Ghilan» wooded district cattle with humps as in INDIA. Geograph . Vol III. P. I. p. 17 (Lat 37 degre
203b136 astern Asia beyond the Ganges & perhaps even in INDIA p. 261. L. Institut. 1837 Mem. Sir F. Darwin cr
208b153 iduals than of species Paris Tertiary Shells in INDIA!? A p. 28 Dr Beck. & Lyell. most curious law of
212b164 double line of intersection.-- Dr. Horsfield At INDIA House, collection of Birds from Java-- at Leyd
225b219 .» glorious fact. of absence of quadrupeds East INDIA Archipelago very good on opposite tendency.-- S
226b220 Galapagos mouse (?) brought by canoes Ceylon & INDIA.-- Van Diemen's land Australia. England & Europ
226b223 ca, ought to present great contrast in forms.-- INDIA; intermediate, see how that is.-- ¿are shell-bo
227b224 under>' connection of those two countries Hence INDIA, Mexico & Europe. one gret sea (Coral reefs:. sh
229b234 ?)-- Some paper in Institute on range of Bos in INDIA.-- Range of Zebra?-- The Crocodile & Tortise fo
231b242 y African.-- American & African forms mingle in INDIA & East Indian islds-- Monkeys different not tra
233b250 ute 1838. p 38. account of fossils of Sewalick «INDIA» Monkeys of old World. Crocodiles. Anoplotheriu
235b274 e with humps.-- ¿To be solved if horses sent to INDIA. & long bred in & no new ones introduced would
250c038 ave been formed & spread to other Africa & East INDIA Arch.-- but where these great animals had not s
250c039 d. Mem. recent Crocodiles with Palaeotherium in INDIA--: connection with Latitudes!? Zoological Journ
251c040 n the localities of certain parrots habitations INDIA & Africa.-- NB. Any monograph like Gould on Tro
253c046 required too separate isld very long» America & INDIA deer.= Africa not.-- Africa Camels?? Africa Bea
253c046 ct of Elephant same species in Borneo. Sumatra. INDIA Ceylon-- perhaps shows great persistency of cha
266c092 se of Black cattle. Not have hump like those of INDIA & Arabia p. 202-- sheep have not the enormous t
278c132 preserved, in central S. America & yet Africa & INDIA???-- & Indian Islds.-- Sir J. Sebright-- pamphl
310c224 malayas, 13,000 & Melville Isd.-- West Africa & INDIA some plants same. --America.-- See Brown Congo
311c227 107. an Ascaris inhabits the eyes of horses in INDIA in which it may be seen swimming about. A Smith
313c235 iation (case of Elephant, which had run wild in INDIA. in Heber?) is analogous to dormant instinct.--
353d060 <are> belong to same section with with those of INDIA-- Waterhouse knows three species of Paradoxurus
362d099 n to grow.-- Mr Yarrell says the «male» Axis of INDIA, breeds at times when horns not perfect-- (is n
371d127 Dingo (he alludes to the dholes or wild dogs of INDIA) in Zoolog. Garden having coloured offspring.--
392d180 g of the several species of wild fowl <in Z> of INDIA «with our common ones» in Zoolog. Gardens; <Buf
449e173 stitution from those of Europe-- for they stand INDIA. better than the latter-- Forrest Voyage p. 323
451e177 ociety Vol I. p. 261. <J> Catalogue of Birds of INDIA.-- p. 555. Lieut. Button counted, the ova of a
451e177 5. Lieut. Button counted, the ova of a tick «in INDIA» & found there were 5,283 attached to its body-
452e181 RS probable,«?» that the Hippopotamus occurs in INDIA. in the Jungles of Borabhum & Dholbum.-- Vol do
453e182 . 650. Long attested account of fall of fish in INDIA-- Windsor Earl-- Eastern Seas p. 229. Believes
465t079 is preserved in this country-- same argument to INDIA & Europe-- & Africa!,-- any negative argument a
503q012 . Sykes fertility of men & Europaean animals in INDIA?-- about Chetah & other tame animals not breedi
503q012 & other tame animals not breeding when tame in INDIA?-- does not know About Yaks. & other Hybrids--
529m041 tion in every view. if one were admiring one in INDIA. & a tiger stalked across the plains, how ones
021r005 ting marine confervae, is very common within E. INDIAN Archipelago, no minute description, calls it a
023r009 ark's Bay. Lat 25 degree. The nearest of the E. INDIAN Islands. namely Java is 1000 miles distant! Wh
034r044 m. Lyell's fact about sulphuric vapours in East INDIAN Volcanos Gypsum Andes Mem. Beechey. account of
060r125 Webster Antarctic veg:-- Study Ulloa to see if INDIAN habitation above regions of vegetation.--«I ca
080r178 other. & see if same effects, as with man Does INDIAN rubber & black lead unite chemically like grea
118a104 the alternation of linear bands of movement in INDIAN & Pacific Oceans.-- (2d--) does not explain fi
134a143 lt mines» «saline deposits» of India p. 503. On INDIAN Saline Deposits. Vol II. p. 23. p. 77 do Vol I
188b067 according to Swainson to certain islets in East INDIAN archipelago Dr Smith considers probable two No
190b076 s Northern Eastern end & die away, & partake of INDIAN character There appears in Australia great abu
191b080 cessive up & down.-- Elephants in Ceylon-- East INDIAN archipelago.-- West Indies = Opossum & Agouti
191b080 + + for instance tertiary deposits between East INDIAN islets-- (+ + +. Ireland longer separated., Ha
191b081 & life shorter or change greater-- In the East INDIAN Archipelago it would be interesting to trace t
202b133 g Monkey in France.-- of genus peculiar to East INDIAN isles.-- Compares it to fossil Didelphis (S. A
203b136 oars were wilder than parents. which is same as INDIAN Cattle.·. tameness not heredetary', having been
204b139 says Hybrid about half aways & result the same INDIAN cattle & common produced very fine Hybrid offs
209b155 ho can find trace or history of species between INDIAN cow with hump & Common;-- between Esquimaux &
211b164 s Nonsense The distribut of big Animals in East INDIAN Archipelago--, very good in connection with Vo
216b179 e ANALOGUES uses this word «for» similar in the INDIAN sea.-- Deshayes.-- Mr McClay is inclined to th
225b217 ember. L. Jenyns. talking of it; or how to make INDIAN Cow with bump & pigs foot with cloven hoof) As
229b234 Hippopotamus do.-- Giraffe do.-- Range of East INDIAN Rhinoceros (?)-- Some paper in Institute on ra
231b241 .-- When breaking up «the primeval.» continent. INDIAN Rhinoceros. Java & Sumatra ones all different.
231b242 ss, & you will have two. Tapir existing in East INDIAN Seas. Marsupials animals all show greater conn
231b242 American & African forms mingle in India & East INDIAN islds-- Monkeys different not travellers!? Roy
233b249 yage). Waterhous remark Australia Fauna so far. INDIAN all the rest. Timor according to Mountain chai
235b273 ntries-- so on with families.-- Ask Royle about INDIAN Cattle with humps.-- ¿To be solved if horses s
235b274 ers to prevent them taking peculiar character-- INDIAN Bull?-- Do species of any genus. as American o
235b275 Bull?-- Do species of any genus. as American or INDIAN genus inhabit different kind of localities.--
236b279 s case of Sub Himalayan land emys, decidedly an INDIAN form of Tortoise.-- On other hand. fresh water
242c017 Van-Hasselt, Reinwardt «Forrest» authors on E. I«NDIAN». A«rch.»¶ Borneo & Sumatra both seem to have
244c021 insectes, in 8 degree. p. 181 «who says insects INDIAN, like. Plants.--» It would be very important t
250c040 ogical Journal.--- Vol I. p. 81. Capromys, West INDIAN isld. p. 120. «ref.» Philosop. Transacts 1823.
252c045 3 <S> Species Rhinoceros Cape town good species INDIAN species so distinct that all analogy from each
252c045 w how different Sumatra Java ---------- do from INDIAN increase of knowledge would probably tell more
267c094 ddish <brown>. ears long.-- like bull terrier-- INDIAN secured one, as they always like to cross thei
278c132 central S. America & yet Africa & India???-- & INDIAN Islds.-- Sir J. Sebright-- pamphlet-- most imp
324c268 See (p. 17) for references to authors about E. INDIAN Islands. consult Dr Horsfield Silliman's Journ
354d065 o man. Sept 7th. -- I was struck looking at the INDIAN cattle with Bump. together with Bison, at some
407e039 Therefore I argue from this that Africa «& East INDIAN Archipelago»-- formerly were not so «very» EQU
408e045 world-- -- Mem. elevation & subsidence of East INDIAN Archipelago. now rising. On a particular part
417e075 owing strength to get the day» The fertility of INDIAN & Common Oxen, which one must think deserve th
425e104 orms.-- are the Labiata nearest to American, or INDIAN groups?-- = Believes some Mediterranean, but c
450e174 of instinct by domestication.-- "Notices of the INDIAN Archipelago" Published at Singapore in 1837. b
450e175 Borneo.-- Ox & hog natives of Borneo Notices of INDIAN Arch. Singapore 1837. By J. H. ? do. p. 189. «
473s006 Orchis & other plants-- & Wallich has described INDIAN Plant.-- June /42/-- June/42/ You can select c
477z002 -- do. p. 207. La punaise was not known amongst INDIAN. introduced in Paraguay in 1769 introduce in G
493q004 on differ in different breeds of dogs. Cattle, (INDIAN & Common) &c: length of life. (5) Does my Fath
540m087 as colour & shape & ideosyncracy.-- Look at the INDIAN in slavery & look at the Negro-- look at them
592n102 invented for Deaf & dumb school & used between INDIAN tribes are Many the same.-- Philosoph. Transac
071r156 em: Olivine. Volcanic product.=> <Did Peruvian INDIANS use arrows or Araucanians?--> If wood now pres

```
490q001 edded?-- Do the Tame Parrots breed amongst the INDIANS Do the Savages select their dogs Sowerby Entom
211b162 erhouse remarked Mere length of bill does not <INDICATE affinity with snipes> indicate affinity, beca
211b162 bill   does not <indicate affinity with snipes> INDICATE affinity, because similar habits produce simi
257c057 - birds that cannot fly &c &c. seem clearly to INDICATE those very changes which at first it might be
307c215 womb.--   if these apparently useless organs do INDICATE such origin, then we are bound to consider ab
361d096 ng like some of the species-- (¿do these facts INDICATE that the change is effected through the male?
378d147 earlier type by Mother?-- do these differences INDICATE, species changing forms, <& loosing do> if so
411e055 specify types, & limits of variation., & hence INDICATE gaps.-- by this means the laws probably would
566n010 e -- anything in these absurd ideas.-- do they INDICATE mind & body retrograding to ancestral type of
615o038 virtue because acquired in past ages; seems to INDICATE that when we <return> turn into angels.  this
202b131 st 120 ft Micaceous rocks. subsidence appears INDICATED.--- p. 36.-- Geograp. Journ. Vol IV. P II. p.
055r108 rtiary plants consumed) Where slope «plainly» INDICATES former boundary. (as in other unworn islands)
152g046 nt of a terrace except at very head of valley INDICATES new terrace Ballivard 2 miles North of  Grant
334d010 ese very fine.--» How is it with plants? This INDICATES a remarkable law, that first cross <not se> p
102a062 ed out. In Cleavage discussion, state broadly INDICATION of new law acting in certain directions pred
252c045 located] The systematic naturalists get clear INDICATION of circumstances in Geography to help in dis
254c047 rgans of generation, afford the least certain INDICATION of the perfection of the Species--! How does
274c116 atenthouse & <birds> Mammalia.-- We have clear INDICATION [not located] alone, but on all the general
311c226 abits can only be used «in classification» as INDICATION of structure (including brain & other organs
580n062 he infinitie & progressive nature of intellect INDICATION of better life p. 207 March 16th.-- Is not t
421e091 have always regarded as affording very clear INDICATIONS of its true affinities. We are least  likely
422e092 an lately adapted or transformed. & hence not INDICATIVE of true affinity.-- -- -- Owen says Dugong c
483z012 allied species, therefore interlock.-- Testudo INDICUS not fossil at Isle of France: <Jerrold?> Bibro
060r124 shooting upwards" of the <ground> land in the W INDIES.--p. 200. Bollingbroke voyage to the  Demerary
110a085 oth Beck & Deshayes saw fossil shells from West INDIES & declare them to be recent species--  Lyell--
179b032 ) in the two first children How is this in West INDIES «--:Humboldt. New Spain:--» Dr. Smith always u
191b080 nts in Ceylon.--» East Indian archipelago.-- West INDIES = Opossum & Agouti same as on continent-- 3 Pa
246c025 birds & bats have certainly travelled from East INDIES, isld, as far as Oualan.-- Wide space of  sea,
414e064 efly organization: though Cont of Africa & West INDIES shows organization in Black Race there gives t
499q010 (even  Nux Vomica is eaten by a Buceros in East INDIES-- Asiatic Researches) (23) Talk about Thyme. H
616o039 ctorily it's erroneousness. it is a point of INDIFFERENCE 2) In the absence of such a guide we can on
567n015 a stranger. who one may like. dislike, or be INDIFFERENT about, yet feel shy.-- not if quite stranger
391d174 whole  object being to acquire differences «INDIFFERENTLY of what kind, either progressive improvemen
479z005 besides  Wolf at Falkland ∴ black rabbits not INDIGENOUS p 112 M Lesson--Voyage of Coquille wide limi
573n039 ders.-- analyse this.-- Miss C. quite aware & INDIGNANT with Mrs C. but had no influence over  her.--
067r143 beds of shells are found on whole coast from P. INDIO to Quilmes. & at least seven miles inland.  The
241c016 '.Urville on the Distrib of Ferns in South Sea (INDIO Polynes: <)> vegetation far East) Ann: des Scie
338d025 forms.--  what is range of Hyaena? Hippotamus.? INDIO-African, or pure Africa?-- ‖Fossil Elephant of
350d054 tra", as Sir Thomas Browne says "is the only INDISPUTABLE axiom in Philosophy<"> Religio Medici. Vol
028r030 issolved. if so sea would separate them from INDISSOLUBLE rocks? Has Chalk ever been dissolved? Singu
539m082 mediately thrilled across me, bringing up old INDISTINCT ideas of FitzWilliam Musm. I was amused at t
495q05a up  periodically & see effect-- such dioecious INDIVID--small orifice (8) Carry Bees, powdered with s
063r122 emmiparous. by nature or accident). we see an INDIVIDUAL divided either at one moment or through laps
063r132 all individuals of all species. as «each» one INDIVIDUAL «divided» by different methods, associated l
063r133 more  wonder in extinction of species than of INDIVIDUAL.-- Mr Birchell says Elephant lives on very w
170b001 kind <the> which is a longer process, the new INDIVIDUAL passing throug several stages (¿typical, <of
171b004 ere may be unknown difficulty with full grown INDIVIDUAL «with fixed organization» thus being modifie
187b064 es like generation of individuals.-- Why does INDIVIDUAL die, to perpetuate certain peculiarities, (t
187b064 ). Now this argument applies to species.-- If INDIVIDUAL cannot procreate, he has no issue, so with s
204b140 ter «Game-fowls have courage independently of INDIVIDUAL force» Mr Wynne has crossed Ducks &  Widgeon
223b210 step.  species or not. A plant submits to more INDIVIDUAL change, (as some animals do more than others
276c123 f early Astronomers.--  then add chief good of INDIVIDUAL scientific men is to push their science a fe
285c152 eping distinct from other & if a first & last INDIVIDUAL were put together, they would not  according
292c171 on implies diversity & therefore would banish INDIVIDUAL, but general ones might yet be transmitted.=
293c173 of  memory.-- [There is no corelation between INDIVIDUAL objects as Ichneumon & caterpillar, though o
301c198 easonable-- Same action may be either in same INDIVIDUAL p. 7. is not squirrel hoarding, & killing gr
312c232 e qualities, with no other means whatever.--» INDIVIDUAL Men «& animals» could only exist by habit.--
336d018 is  most slowly obtained with respect to that INDIVIDUAL) it is more easily inherited.-- «but if chan
350d055 organs  have nothing to do with perfection of INDIVIDUAL, though such relation seems common, but that
360d089 d <which seems owing «determined» by the sex> INDIVIDUAL instances trouble Yarrels law. chief trust m
372d129 her part of body.-- [in plants does not whole INDIVIDUAL change into generative organs?]CD it is of n
372d130 does  Gecko produce always different tail? An INDIVIDUAL bud may be thus produced from the growth of
372d130 rom the growth of one part, (not strictly new INDIVIDUAL), or he may produced by having undergone, th
373d132 abortive  Mammae?.-- My view would make every INDIVIDUAL a spontaneous generation: what is animalcula
384d162 man  puts primordial vivifying principle) one INDIVIDUAL secretes two substances, although organs for
390d178 ame Spath, as they have only one bud.-- Every INDIVIDUAL foetus would reproduce its kind was it not f
391d179 t corelation of structure desire fails. Every INDIVIDUAL except by incestuous marriage has acquired f
391d175 e somewhat different) should be added to each INDIVIDUAL before he can procreate. these changes may b
391d175 changes  oscillate backwards & forwards & are INDIVIDUAL differences (hence every individual is diffe
391d175 rds & are individual differences (hence every INDIVIDUAL is different). (All this agrees well with my
404e024 untary variation made by man, common to every INDIVIDUAL & therefore effect of climate.-- Octob 19th.
410e053 dividuals, will care little, <in> whether the INDIVIDUAL be species or variety, but to discover physi
411e054 ysical laws of such corelations, & changes of INDIVIDUAL organs, must know whether the individuals «f
411e055 ination of the special cases, under which the INDIVIDUAL steps in the series have been fixed, to stud
447e164 impregnate allied to Asclepias Turpin cell is INDIVIDUAL May 29th.-- -- Henslow says he has not the s
448e165 ds, so that Turpin says each cell of plant is INDIVIDUAL.-- Most plants which propagate rapidly by bu
466t096 -- all affected by cultivation during life of INDIVIDUAL.  June 1st 1841. Maer Examined the Lemon-thym
490q001 s select their dogs Sowerby Entomologist Does INDIVIDUAL Shell or insect or group vary more in one co
505q014 e to see, whether stamens will be produced in INDIVIDUAL plants 17 A dead-nettle in Hot-house. will i
512q019 fferent periods in different breeds--?? or in INDIVIDUAL case: subject to disease in youth.-- Mr Toll
515q021 ortive state either in the species, or in the INDIVIDUAL by chance & under domestication.-- N.B. Bent
539m083 calls up other, & the consciousness of double INDIVIDUAL is not awakened.-- The habitual individual r
539m083 le individual is not awakened.-- The habitual INDIVIDUAL remembers things done in the other habitual
557m149 al.-- Fear is open mouthed to hear. though in INDIVIDUAL case. nothing can be heard.-- Shame would ne
577n053 it,  as in cancer, showing, efect of mind on INDIVIDUAL parts of body.== (if you <think> «fear»  you
589n091 n make a tune on the pianoforte. yes if every INDIVIDUAL played a little, & something destroyed bad b
591n099 ctions or even impulses. (where sensations of INDIVIDUAL are same as in normal cases) are held in abh
602o011 in the result of sagacity, or intuition. when INDIVIDUAL cannot give reason, though he feels he is ri
611o034 thout action of vital laws-- According to the INDIVIDUAL forms of living beings, matter is united in
611o034 e simplest forms of living beings namely «one INDIVIDUAL» vegetables, the vital laws act definitely (
611o034 enough must be present to be able to exist as INDIVIDUAL.-- [RHC] 2) In animals, growth of body preci
614o037 generation.; than from hour to hour in <man:> INDIVIDUAL-- [LHC] Perhaps even the most complicated in
615o36v eredetary in fowls & not effect of feeling of INDIVIDUAL force in any individual.-- His Malay breed «
615o36v effect  of feeling of individual force in any INDIVIDUAL.-- His Malay breed «of fowl» totally differe
616o038 -- The difference between heredetary memory & INDIVIDUAL secretion of thought, may be no more «differ
617o040 t? capable of being traced to the body of the INDIVIDUAL to whom the action is attributed; force  (be
619o042 they consist in such active sympathy that the INDIVIDUAL forgets itself, & aids & defends & acts  for
623o050 h have a beneficial tendency, (not to any one INDIVIDUAL but to the whole past race).-- <no one> doub
300c198 er it is hiatus & not saltus.-- The greater INDIVIDUALITY of mind in man, is analogous to greater ind
300c198 ity of mind in man, is analogous to greater INDIVIDUALITY of bodies of some animals over those of oth
538m080 f mind, is probably analogous to the double INDIVIDUALITY implied by habit, when one acts unconscious
548m116 former periods so «as» to <make> «give» one INDIVIDUALITY in this case.-- But now in Mayo's «p.  140»
548m116 friend-- this gives one strong idea of what INDIVIDUALITY is.-- Insanity is «much» «somewhat» the sam
604o016 fection of organization & if consciousness, INDIVIDUALITY.-- Quotes D. Stewarts System of Emotions.--
407e041 introduced live & multiplied, specifically & INDIVIDUALLY.-- I see clearly from F. R. it will be high
553m136 ator <is> «has been» implanted in us (<by> ¿ INDIVIDUALLY or in race?) by a separate act of God, & no
048r086 vast numbers of wild animals. both species & INDIVIDUALS as in the half desert country of S.  Africa.
063r132 hich become quite separate.--Considering all INDIVIDUALS of all species. as «each» one individual «di
```

Page **(Key Word)**
170b001 Two kinds of generation the coeval kind, all INDIVIDUALS absolutely similar; for instance fruit trees
171b003 h soil, many kinds, are produced, though new INDIVIDUALS produced by buds are constant, hence we see
176b022 s nothing stranger in death of species, than INDIVIDUALS If we suppose monad definite existence, as w
181b039 species in each generation only breeds; like INDIVIDUALS in a country not rapidly increasing.-- If we
187b063 s a generation of species like generation of INDIVIDUALS.-- Why does individual die, to perpetuate ce
191b082 he Progeny of female insures often mixing of INDIVIDUALS. Here we have avitism the ordinary event. &
199b118 er says. "But we could only produce domestic INDIVIDUALS & not races, without the occurence of one of
208b153 132 There is no more wonder in extinction of INDIVIDUALS than of species Paris Tertiary Shells in Ind
291c168 south ward being ready made.-- & so destroy INDIVIDUALS, wheras in Falkland Isd they would change &
305c211 ccording to a more extended relations of the INDIVIDUALS, whereby choice with memory. or reason? is n
305c211 ous shades of <dif> separation between those INDIVIDUALS thus endowed, & the community of mind, even
309c219 imb not used The only cause of similarity in INDIVIDUALS «we know of:», is relationship, children of
337d020 sh country & breed confined. to certain best INDIVIDUALS.-- scarcely any breed but what some individu
337d020 ividuals.-- scarcely any breed but what some INDIVIDUALS are picked out.-- in a really natural breed,
354d062 , but in different parts according to age of INDIVIDUALS-- (see Mammae of Women) in different parts w
406e036 ear to hold good. with regard to marriage of INDIVIDUALS, & varieties of same species & to different
409e048 therwise, there would be as many species, as INDIVIDUALS, & though we may not trace out all the ill e
410e050 sary that Physical changes should act not on INDIVIDUALS, but on masses of individuals.-- so that the
410e050 uld act not on individuals, but on masses of INDIVIDUALS.-- so that the changes should be slow & bear
410e051 ow after discussion of crossing of <species> INDIVIDUALS. with respect to representative species., wh
410e053 he formal laws of the corelation of parts in INDIVIDUALS, will care little, <in> whether the individu
411e054 of individual organs, must know whether the INDIVIDUALS «forms» are permanet, all steps in the serie
411e056 ees in Coralline powers of multiplication of INDIVIDUALS, & yet another means for individuals (Mem: t
411e056 tion of individuals, & yet another means for INDIVIDUALS (Mem: transportation will be answered) one l
411e056 logy for cause in plants. where innumberable INDIVIDUALS can be produced. & yet sexual apparatus.-- M
413e059 ms in shells. are numerous species, numerous INDIVIDUALS, & <individuals> «species» of large size,--
413e059 e numerous species, numerous individuals, & <INDIVIDUALS> «species» of large size,-- consider this (C
417e079 Moretons law holds with different species, & INDIVIDUALS of same species.--). some races of men. D'Or
418e083 d to suspect that the birth of the species & INDIVIDUALS in their present forms, are closely related-
418e084 ns, become as much fixed, as if added to old INDIVIDUALS,, during thousands of centuries,-- each of u
419e085 ller from propagation of infinite numbers of INDIVIDUALS from one, is adverse.-- Decemb. 25th.-- Lyel
431e122 shells become less in number. (¿ species, or INDIVIDUALS) the deeper one goes-- surely is this true?-
441e148 differ from possibility of concourse of two «INDIVIDUALS» & the action always of two organs-- instead
441e149 ffect of conditions on offspring, but not on INDIVIDUALS is very curious & important.-- The existence
442e153 ion my theory would require, that <species> «INDIVIDUALS» propagated by gemmation should be absolutel
446e163 would effect them, without affecting all the INDIVIDUALS-- «-- hence there would be real gradations i
447e164 ucture does not allow of crossing with other INDIVIDUALS, «with facility»-- such as cryptogamic plant
492q002 udded. with reference to extension of age of INDIVIDUALS-- 9. Do plants in becoming double ever becom
515q021 al number, they are apt to vary in number in INDIVIDUALS of same species Eyton (1) Number of eggs-- o
583n069 t the invariability of instinctive powers in INDIVIDUALS of the same species with variability of reas
477z003 s tran?? Azara Las Vinchuca or Benchuca. "Les INDIVIDUS ailes peuvent avoir <quatre> cinq lignes de l
185b055 where it could live-- but there were causes to INDUCE great change. like the Buzzard which has chang
391d175 ances during life.-- if the circumstances which INDUCE «which must be external» change are always of
410e051 ing another.-- (I from looking at all facts as INDUCING towards law of transmutation, cannot see the
567n014 side of triangle shorter than two. V. Whewell. INDUCT. Sciences-- Vol I p. 334 Does a negress blush.
257c055 een so formed, then man may be a miracle, but INDUCTION leads to other view.-- Till we know uses of o
275c119 tribution of animals, <as> I use (new step in INDUCTION) as keystone of ancient geography species tel
370d117 y is to establish a point as a probability by INDUCTION, & to apply it as hypothesis to other points.
567n016 nger.-- or less so.-- When learning facts for INDUCTION. one is obliged carefully to separate its mem
601o08b asseur, par Desgraviers, un Vol 8vo Keratry-- INDUCTIONS morales et physiologiques The first of these
322c269 nolds Discourses Lessing's Laocoon Whewells-- INDUCTIVE History.-- References at end of each Vol Hers
398e005 ry as the «basal» foundation stone of further INDUCTIVE reasoning is immense. It is curious that geol
415e069 mber 16th. The end of each volume of Whewells INDUCTIVE History. Contains many most valuable referenc
042r068 ual action of Earthquakes. «on Chili & delta of INDUS», my belief in submarine tilting alone, must be
079r177 The Fuegians treat the "chefs d'oeuvre de l[']INDUSTRIE humaine, comme ils traitent les loix de la na
583n069 of fable, & expression as cunningness of fox, INDUSTRY of bee &c &c-- p. 15."instincts act with uner
106a074 ase force is lessened. therefore rivers very INEFFECTUAL in widening valley.-- it is essentially a de
227b225 an geologists think. it agrees with excessive INEQUALITY of numbers of species in divisions. look at
119a104 f globe be considered as condensed vapour.-- INEQULITIES are required to start with (& does not Hersc
195b101 action act according to certain laws such are INEVITABLE consequen let animal be created, then by the
397e003 eem intimately related to famines., yet very INEXPLICABLE.-- do p. 529. "It accords with the most lib
599o007 last rays of & & or grand chorus are utterly INEXPLICABLE-- I cannot <admit> think reason sufficient
608o028 ion.-- 4) These views are directly opposed & INEXPLICABLE if we suppose that the sins of a man, are u
127a125 hotter) was cooled to greater depth.-- Now the <INF> subterranean isothermal line must be creeping «
313c235 ng sex of neuters) species may have had their INFANCIES, as well as men-- when habits much more firml
520m002 ad memories for things which happened in early INFANCY-- of this fact Mr Dryden C. is good instance a
532m055 their dotage, who sing the songs «& tales» of INFANCY, it is very doubtful whether they could recoll
538m079 ong was to her like one which though learnt in INFANCY, had often been repeated: Now it is remarked t
548m115 blem., one does not care for the pains of ones INFANCY.-- one cannot bring it to one self.-- nor of a
623o050 eived pleasure from some one «person» in early INFANCY, during many generations giving love of mother
580n062 H. Davy -- Consolats: "the recollections of the INFANT likewise before two years are soon lost; yet m
113a091 h of 32 degree. being little we may confidently INFER that time has not been allowed for lower beds t
297c186 , makes the group aberrant When species rare we INFER extermination, when group few in number of kind
429e113 have been produced (look at the Dahlias. we may INFER it in animals.)-- Azara gives account of produc
618o041 for metal of another person at all, we can only INFER it from his its behaviour. Thought is only know
640j167 is thus offered.-- From these views, one would INFER that Mollusca would offer few species, or rathe
375d134 does not convey the warring of the species as INFERENCE from Malthus.-- «increase of brutes, must be
404e024 pplies to wild animals only. from which plain INFERENCE might be drawn that whole infertility «of hyb
443e155 » functions to complexity is evident, yet the INFERENCE from some plants & some mollusca being hermap
448e167 s tables Bennetts Wandering Vol II. p 155. By INFERENCE I imagine that there are Baboons in St Thomas
636j56v plants, had there been no insects. The right INFERENCE is, there were insects «¿when were Palms form
034r044 h is very strong about Trachyte being the most INFERIOR rocks--The stream at Portillo Pass example of
038r059 levated considerably. which shows an afflux of INFERIOR melted rocks to those parts. Are not the dike
319c256 ery sweet notes.-- Their soft-billed birds are INFERIOR to ours, & our lark ranks very high.-- Upon t
358d076 ale?)».-- children & women = "women recognized INFERIOR intellectually"= Opposed to these facts are e
431e121 ays the T. costatus is in England found in the INFERIOR Oolite, & the T. elongata in the uper formati
450e175 - seldom equals 13 hands-- those of Lao & Siam INFERIOR to those of Pegu-- in Sumatra two breeds both
572n032 om."-- this all applies to bodily weakness & INFERIORITY, but now we carry it on to mental inferiorit
572n032 nferiority, but now we carry it on to mental INFERIORITY-- when we do not expect any bodily harm-- ca
113a092 ms deep. perfectly still water. Major Mitchell INFERRED subsidence; Mem my remarks on coast of Austra
279c133 nature.-- Whole art of making varieties may be INFERRED from <this> facts stated.-- + +. Fully suppor
304c208 ecies <osculant> but, where much death, may be INFERRED much time elapsed & therefore descended from
366d112 e «That Mutilations will not alter form may be INFERRED from Australians knocking out teeth.-- the ac
389d176 ich wishes to throw itself off.--» as might be INFERRED from annual plant being prolonged till it has
454e184 be budded «& rendered of great age» as must be INFERRED from what Mr Knight says. Hort. Transat. V. I
537m075 reason.-- How slow habits are changed may be INFERRED from expression. "relict of bad habit." as ch
222b204 ve very curious-- My idea of propagation almost INFERS, what we call improvement, --All Mammalia from
276c122 ing & ordinary offspring:-- this gradation is INFERTILE offspring. without organs of generation?! By
335d016 (but many animals are fertile, when offspring INFERTILE,-- two considerations are here combined). In
341d031 th Pea <cock> Hen.-- offspring female, yet so INFERTILE never even in seven years produced even an eg
358d085 l plumage is very common by hybrids, that are INFERTILE.-- thus the common. pheasant & fowl when cros
358d085 silver & common pheasant crossed, has a cock (INFERTILE) <with> the breast of which is like common ph
359d087 is singular pheasant & fowl being so totally INFERTILE whereas animals further apart have bred inter
367d112 ould be read on Mare My view, why hybrids are INFERTILE. supposes that when foetus is forming the ovu
383d160 lver & Common Pheasant. Male bird, said to be INFERTILE.-- spurs rather smaller than in <ma> silver m
390d179 sex. = The highest bred Blood-hound. would be INFERTILE with highest bred of other ¿ breed.= Therefor
426e106 act that different species produce abundantly INFERTILE hybrids, & the fact that old varieties do not
493q004 detary constitution, to see whether offspring INFERTILE.-- (4) Does the number of pulse, Respiration,

Page **(Key Word)**
255c052 ears on my doctrine of cross-generation. The INFERTILITY of crosse & cross, is method of nature to pr
281c137 uld go back to grandfather, which is true) & INFERTILITY is consequence.-- The simple expression of s
364d103 ight-- has almost lost his Owl-Pidgeons from INFERTILITY.-- Yarrell says in such case they exchange b
374d134 forms & Trilobites Sept 25th. In considering INFERTILITY of hybrids inter se, the first cross general
403e024 cows hornless, (horses not) If they give up INFERTILITY in largest sense. <es> «as» test of species.
404e024 ch plain inference might be drawn that whole INFERTILITY «of hybrids receive no explanation» was cons
439e143 me flower,-- as it tends to show my view of <I>NFERTILITY of hybrids «with parent species» false, whi
137a149 sandstone «said to be» analogous to granite INFILTERING some of its constituents into chert. [blank]
599o008 ol. I. p. 198.-- "Moralité, raison, beau ideal, INFINI conscience; voila l'homme separe de la matiere
172b005 rtaking of characters of both parents, & these INFINITE in Number In Man it has been said, there is i
181b043 ides affinities from three elements, from the «INFINITE» variations, & all coming from one stock & ob
192b087 ,-- Now if the intermediate ranks had produced INFINITE species probably the series would have been m
297c186 mination.-- New forms made through probably an INFINITE number of forms.-- therefore an isolated form
389d173 s case: without the nervous matter consists of INFINITE number of globules: generally sufficient for
419e085 ality, as argued by Müller from propagation of INFINITE numbers of individuals from one, is adverse.-
460t017 s of organization (ie those laws which prevent INFINITE variation in every possible way.-- the laws w
460t019 that there is evidence in the organic world of INFINITE & growing complexity from a few types, it mus
580n062 &c so common in the young are symptoms of the INFINITE & progressive nature of intellect indication
600o008 mpounded passion gained in life time]CD 3. The INFINITE. -- lives by hopes, looks to eternity. (4) Re
073r161 constantly found mixed with lead & copper is INFINITELY rare in all parts of the globe". p. 113 How
416e071 as species-- but latter far more perfectly & INFINITELY slower.-- No domesticated animal is perfectl
600o008 onfesses these faculties of soul, treating of INFINITES not definable.-- Has little Chapter on each f
057r115 erschel. says. precip. of Sulph. B. all the INFINITESIMAL cryst. arrange themselves in planes. «Mem s
436e137 le seeds-- yet if a seed were produced with the INFINITESIMAL advantage it would have better chance of be
498d008 t degenerate,-- as the Bees will mingle the INFINITESIMAL varieties which must occur.-- ¿is «it» not
498o008 ties which must occur.-- ¿is «it» not these INFINITESIMAL varieties, which counterbalance each other?
604o018 o God the notion of living in lofty regions. 3 INFINITY eternity. darkness, power. being associated w
577n051 inge «in a bivalve shell» does to reason.-- an INFLAMED membrane from local irritation to passion.--
529m043 .-- July 22d. 1838 No Deliriums, yet in some INFLAMMATORY diseases, where there has been no cloud on
367d113 es, a high-flying bat, which has the power of INFLATING its body like balloon-- by air cells connecte
401e017 e the lesser peculiarities.-- brilliancy of INFLORESCENCE Gardeners. by chance <often> sometimes graf
612o035 willing comes on period of year as much as INFLORESCENCE.-- [LHC] I here omit the case (if such ther
615o36v ect of former tropical climate analogous to INFLORESCENCE of Tropical plants when imported & plants s
559m156 g abstract of Smith's views «Take & pound up INFLORESCENT parts of mosses & see if Hybrid can be produ
077r171 f Marfil is divided, appear to have a decided INFLUENCE on the richness of the veta madre of [misnumb
101a056 wing present form of land in Northern England INFLUENCE dispersion of Boulders.-- See Rogers for Sout
139a180 galls.-- is it not effect of superadded vital INFLUENCE?-- See End of Note Book. called R. N.-- Massa
171b003 change, temperature & all circumstances which INFLUENCE living beings.-- We see <living beings>. the
194b096 ve male & female organs together, yet receive INFLUENCE from other plants.-- Does not Lyell give some
195b098 ide in space before island existed.-- Such an INFLUENCE Must exist in such spots. We know birds do ar
201b129 tance he says wings of bat, are from external INFLUENCE.-- ...... Hence name of analogy. the structure
235b272 and & Australia [blank] Falconer's remarks on INFLUENCE of climates, situations &c on 242. Hook Smell
280c135 hybrids.-- & is a very remarkable fact. show INFLUENCE of mind It is not difficult to see that it is
294c176 ticated animals change a little with external INFLUENCE-- & if those changes permanent so would the c
309c220 on of <cl> castes. improve the women. (double INFLUENCE) & mankind must improve-- The areas of subsid
323c269 e of Napoleon.» April 5d Dr. Edwards of <ter> INFLUENCE of Physical causes: well skimmed Bartrams Tra
324c268 logy Flemming. ditto Falconers remarks on the INFLUENCE of climate White's regular gradation in Man.
347d049 ed to Taenia.-- hard so as to resist external INFLUENCE.-- 27th. August. There must be some law, that
359d087 case, & why should organic affections always INFLUENCE the sexual organs alone.-- It is singular phe
366d111 -- would perfect impunity from muskitoes bite INFLUENCE propagation of species.-- Case of Association
387d169 of same peculiarity breed they will have same INFLUENCE as first pair + tendency they inherited from
387d169 ndency they inherited from second pair, + the INFLUENCE they themselves inherit./ Annals of Natural H
404e026 tance such as Hunter, or some one mentions of INFLUENCE on parent affecting offspring.-- & as adapta
430e118 n whole mass of species are subjected to some INFLUENCE, & this would take place from changing countr
435e133 is put to it.-- April 6th "Dr. Edwards on the INFLUENCE of <external> Physical agents". «translated b
544m102 ion as flying. (No memory of past events?) or INFLUENCE on our conduct, the links which when consciou
547m111 that muscular, mental, <&> digestive nervous INFLUENCE replace each other August 29th. Went to Bed.
573m039 uite aware & indignant with Mrs C. but had no INFLUENCE over her.-- Hensleigh says. Douglas. «& Spenc
590m093 before false sneeze.-- "A Dissertation on the INFLUENCE of the Passion."-- p. 37. The increase of Bil
607o025 eredelary constitution,-- education under the INFLUENCE of others-- varied capability of receiving im
623o051 nct of pointing as a dog.-- also age has much INFLUENCE.)-- & only that which is beneficial to save,
171b003 e of generations even mind & instinct becomes INFLUENCED.-- child of savage not civilized man.--birds
193b090 es in course of ages ten thousand varieties, (INFLUENCED itself perhaps by circumstances) & those alo
216b181 mare crossing by Ld. Moreton, where mare was INFLUENCED in this cross to after births, like aphides.
579n059 - love being an emotion does it regard «is it INFLUENCED by» other emotions? When a man keeps perfect
201b129 of Lamarck, which he says depends on external INFLUENCES.-- For instance he says wings of bat, are fr
202b130 tructure A Race of domestic animals made from INFLUENCES in one country is permanent in another.-- Go
248c033 han between two hybrids.-- As we see external INFLUENCES first affect external [for]m, so will the in
135a145 835. p. 437. Tours by Benza Neilgherries-- Much INFORM. on. decomposition of granite-- Bengal. J. vol
148g025 Black faced sheep are crossed with English my INFORMANT said the lambs were nearly like each other «&
163g106 layers of coarse & fine & many Sea shells. My INFORMANT saw them himself-- Sand with tiple Ripple Near
078r174 to 3 species. I think I have much additional INFORMATION Guanaxuato, which has yielded the most meet
205b142 c, JUST COMMENCING. Kirby says (not definite INFORMATION) of Rocky Mountains Asiatic types disco
212b166 at.-- different from English. Waterhouse has INFORMATION respecting the Water Rat.-- ¿Consult Dr Smit
229b232 nt for true molluscs coupling.-- Dr. Smith's INFORMATION Long Horned (very) aboriginal at Cape: cross
249c035 species. Dr. Smith will give me some capital INFORMATION ¿Carnivora of New & Old word. do not form tw
260c068 acts,. regarding habits range & all kinds of INFORMATION-- instinct Swainson's remarks in Fauna Borea
502g012 did not become impregnated. (34) Any recent INFORMATION about pollen of Subularia Royle & Horsfield
503g012 oyle In Royle's productive Resources Book no INFORMATION & Hope about Silk worms. Varieties effects o
066r141 Kerguelen 40 by 20 leagues. dimensions: Bynoe INFORMS me that in Obstruction Sound, in the narrow pa
071r157 umboldt in Geolog. Society Sir Woodbine Parish INFORMS me that town near Tucuman and Salta. towards t
093a034 distributed to various subjects) Dr. A. Smith INFORMS me that in the year a Rhinoceros was found <em
458t002 de on them. Mr Miller-- in Zoological Gardens. INFORMS me that a hybrid between ass & Zebra, crossed
391d174 ired from father some differences. V. Supra <v. INFRA> p 179, continued from Is a flower bud produced
045r076 eneral belief in N. Chili, where rains are so INFREQUENT; so as to exclaim, «as I have heard» how luc
638j58v rish. & who will dare to say that this is an INFRINGEMENT on the wisdom or Providence. when whole roc
093a031 bility in. L Institut p 247. 1837.-- The most INFUSIBLE first injected.-- Basalt: last because it cou
021rIFC ore constant: This view supposes the simplest INFUSORIA same since commencement of world.-- [not loca
207b150 ar «(p 321)» at Tertiary epock p. 330. Fossil INFUSORIA found of unknown forms, a circumstance undisc
241c015 eur in Australia.-- p 67 ¿American forms? All INFUSORIA. not extinct species. good Resumé do/p. .62 ?
254c049 tra or Ascidia spicule in sponge. stomachs in INFUSORIA, generation in each joint of taenia worm.-- f
267c095 . 1838. p. 67. Australian rodents Abstract of INFUSORIA. <p> do p. 62. Ehrenberg Annals of Nat. Hist.
270c104 al structure in Echinodermata.-- Agassiz says INFUSORIA «are» insecta.-- G. R. Treviranus, Biologie r
282c143 covered When one reads in Ehrenbergs Paper on INFUSORIA on the the enormous production-- millions in
283c146 & L'Institut.) that there are Tertiary fossil INFUSORIA, of same forms with recent, we have nothing t
434e128 ary address 1839, p. 9.,-- talks about fossil INFUSORIA becoming extinct not so soon as other forms.-
228b229 whale, or classifyng a mite, a fungus, or an INFUSORIAN. is "What are the laws of life".-- Where we
172b007 . Mocking birds; Falkland Fox-- Chiloe, fox,-- INGLISH & Irish Hare.-- As we thus believe species var
046r078 st.--» Metallic veins likewise must separate INGREDIENTS if we look to a constant revolution.--Are we
098a045 alks of LAMINATED structure (∴ separation of INGREDIENTS) as uniting with cretionary.-- it may <cl> c
312c233 «+++ p 234», different species to different,-- INGUINAL louse African & Europaean. different.-- thora
235b275 cies of any genus. as American or Indian genus INHABIT different kind of localities.-- if so change T
276c125 r of vertebrae-- ¿Where two very close species INHABIT same country are not habits different, (Mem: G
276c125 : Gould's Willow Wren) but where close species INHABIT different countries habits similar ¿law?-- pro
294c175 ns of adaptation of animals & the country they INHABIT.--» & the first one that bred one was disease
634j55r w are we to explain this.-- Did reptiles first INHABIT seas.-- Were they then killed out «by the incr
377d140 .-- Is not puma, same colour as Lion. because INHABITANT of plain & Jaguar of woods &c like ground bi
405e032 If. hereafter. M. angustidens be found to be INHABITANT of S. America & as it is <falle> embedded wi

I have felt some difficulty in conceiving how INHABITANT of Tierra del Fuego is to be converted into
cean between Lat 56 degree and 57 degree only INHABITANT crust Entomost of the genera-- Cyclops p. 13
very curious.-- replaced by didelphidae Skunk INHABITANT of Patagonia. Mem:-- S. Cruz. Molina Vol. 1.
e of Zebra?-- The Crocodile & Tortise former INHABITANTS of Mauritius Freycinet Voyage, agrees. with
!! in stomach --or in feathers--seeds.-- Two INHABITANTS of the Tropics, (whether one fossil or not)
a rat said to have been found!! rodents old INHABITANTS most important!! like Dipus of present day??
against genera.-- How long will the wretched INHABITANTS of NW. Australia, go on blinking their eyes.
but not at Feis (near island) do p. 190. The INHABITANTS of Sunmagi, a territory in the small isld of
ean lakes, near Volcanoes. lakes of brine all INHABITED: Go steadily through all the limits of birds
ay be.-- subterranean lakes, hot spring &c &c INHABITED therefore mud wood be inhabited, then how is
spring &c &c inhabited therefore mud wood be INHABITED, then how is this effected by-- for instance,
more than shells) owing to variety of station INHABITED by them-- Timor. Australian forms amongst bir
he serpent man? about zones separated by non-INHABITED spaces: has he published? does he understand
the works of man Feeling surprise at Mastodon INHABITING plains of Patagonia is removed by reflecting
1. Horsfield on two bears very close species, INHABITING Borneo & Sumatra. differ only in form of whi
ted in S. America like other forms, but those INHABITING 3d zone of <latit> height & 3d of latitude m
ke the Siam race with long nozzle & few hairs) INHABITS Celebes & few of the larger islands.-- -- Ant
es & varieties P. 708. Columba Oceanica (Less) INHABITS Caroline < «NB. The» > isld. (.perhaps Philli
of replacement)-- Coquille Voyage The caswary, INHABITS Ceram, Bourou & especially New Guinea (replac
Edinburgh. Transact. Vol IX p. 107. an Ascaris INHABITS the eyes of horses in India in which it may b
>36.-- A crustacean animal is mentioned which INHABITS the Pinna of Rio Janeiro, (like some Mediteri
onsciousness.-- & somnambulism. Do people when INHALING Nitrous oxide, forget what they did when in t
are not recollected whether from frequency, or INHERENT structure of mind. they make, either in thems
time & if their <D & E> «all their offspring» INHERIT the same peculiarity in lesser degree C. & the
m second pair, + the influence they themselves INHERIT./ Annals of Natural History .p. 96. Vol I. Not
surprise.-- heart beginning to beat-- children INHERIT it <ins> like instinct, preeminently so-- who
oss-births, or other accidents of delivery-- INHERITABLE.?-- Bell cd ask Accouchers Is any peculiarit
Accouchers Is any peculiarity in milk teeth INHERITABLE!!! very good Any peculiarity in the males of
ntain-- In races the differences depend upon INHERITANCE & in species are only ancient & perfectly ad
respect to that individual) it is more easily INHERITED.-- «but if change be in blood long, it become
same influence as first pair + tendency they INHERITED from second pair, + the influence they themse
acquired« «character» » of characters being INHERITED at corresponding age & sex, opposed by canteri
tion of heart.-- now this is the oldest <her> INHERITED & therefore remains, when the actual movement
o varieties are crossed as in Shepherd dogs-- INHERITED Habits: Have Effect in Bones is valuable it s
ny <peculiarity> change from the form which it INHERITS from its parents <stock> without it be small
er only left him a guinea a week. yet. he was INIMITABLE chess player.-- Peacocks remark about mathem
erfect inverted «with <Mr Lyell's> idea of an INJECTED mass of fluid rock In Patagonia plains. long
T. del Fuego as nucleus of a Volcano or as an INJECTED mass.--From conical form I incline to <latter
place strata. & it is nothing odd to find them INJECTED by veins & mass[es] [Fig. 8] {P} (A. B. C, no
grap Journ There are some ideas about order of INJECTED rock being determined by fusibility in. L Ins
stitut p 247. 1837.-- The most infusible first INJECTED.-- Basalt: last because it could reach the su
1; the proportional force of crust of globe & INJECTING matter on the great rise). -- The great rains
d rise, & angular displacement, consequent of INJECTION of fluid rock.-- Try on globe. with slip pape
accompanied by horizontal elevations, so are INJECTION of mountain chains. accompanied by do.-- Give
trata {P} do p. 171. argument against lateral INJECTION. from probability of fissures being prolonged
vements effects of irritability, though means INJECTION of fluid different from contraction of fibre)
ick to be penetrated by the repeted trifeling INJECTIONS.--Old vents would keep open long after emers
to think «are» to make others happy & wrong to INJURE them without temptation.-- This probably is na
lt deposited on windows of houses. & trees all INJURED on Eastern side, far inland.-- even 70 miles f
nking over somebody who has, perhaps, slightly INJURED me, plotting speeches, yet with a sort of cons
abits, & dread of misery of future thinking of INJURED moral sense.-- Notion of deity effect of reaso
is conscience, if he has been cowardly, or has INJURED another bad, vindictive.-- or lied &c &c Are t
a very slight change in association if others INJURED these objects, without his being able to preve
nyone, who will reflect must feel, how like to INJURED conscience, is the feeling of any custom of so
, generation destroys the effect of accidental INJURIES, <on> which if animals lived for ever would b
ch cases, is that the plumage has not been so INJURIOUS to bird as to allow any other kind of animal
e can be handed down without being absolutely INJURIOUS «(or requiring nutrition)» to a certain amoun
surp its place.-- & therefore the degree of INJURIOUSNESS must have been exceedingly small.-- This is
violently shaken, by earthquake. but no serious INJURY.-- <Analysis of Atacama. Iron in Edinburgh. F
eless member.-- in fact they do it in disease & INJURY.-- The sympathy of part is probably part of sa
assion, but muscular exertion consequent on the INJURY & consequently excited action of heart.-- now
ter muscles, only on the principle like does an INJURY of the spine-- that it paralyzes all muscular
y-- Have insane people any misgivings of the INJUSTNESS of their hatreds, as <if> in my case.-- It m
n the Chili earthquakes if rise was more <than> INLAND than on coast it would be invariably discover
alf a long the coast. <"> The rocks in the most INLAND part of this ray are perforated & smoothed lik
ape Possession. Syenite¿ Andite?-- Degrading of INLAND Days. like St. Julian & Port Desire applicable
rom P. Indio to Quilmes. & at least seven miles INLAND. The Cordoba earthquake a very remarkable phen
ouses. & trees all injured on Eastern side, far INLAND.-- even 70 miles from salt water. Mr. Arrowsmi
upt slope {P} each stratum would thin out, both INLAND & seaward: if matter too coarse, then {P} that
t 26 degree S. Wafer looking for Copiapo. found INLAND a great many sea shells some miles from coast-
g to the water in their beaks, & the young one <INLAND> directly by instinct. can dive & conceal them
riginal at Cape & a thinner-tailed kind further INLAND.-- NB. There is division of snakes. with hinde
iss <& also> also near summit on Hill on side of INN BOULDER of granite above 4th Shelf a little lowe
ite or gneiss of Moel Derry) on low hill between INN & Bouhunthine the summit «doubtless worn into co
xisting species is valuable because it shows no INNATE power of change & it also shows, what enormous
depends on the number of sources of pleasure & INNATE tastes, he partakes, taste for musical sound w
land dog would Greyhound about dread of water-- INNATE Septemb 1-- If one performs some actions, whic
er to us, as does that philosopher who says the INNATE knowledge of creator <is> «has been» implanted
over «it would be difficult to prove that» this INNATE idea of God in civilized nations has not been
s. St. Helena.-- [Fig. 7] {P} effect of heat on INNER wall, hence resists degradation longer than out
looks to analogy for cause in plants. where INNUMERABLE individuals can be produced. & yet sexual a
st links. if all species know they would be INNUMERABLE does not know any difference between perman
s is stated too strongly. for there would be INNUMERABLE species. .& hence few only social there coul
idgeons stomach true milk.> < <Species. are INNUMERABLE variations>. Every structure is capable of i
e variations>. Every structure is capable of INNUMERABLE variations, as long as each shall be perfect
fortuitous" growth could have produced these INNUMERABLE seeds-- yet if a seed were produced with inf
m, to that form being «the one alone» out of INNUMERABLE other ones, <alone> «which has been» preserv
y & mental disposition.-- My father has seen INNUMERABLE cases of people taking after their parents,
all> other trains of thought, or memory from INNUMERABLE late events.-- the fatigue of thinking is ke
es, or if recollected, such part of thoughts INNUMERABLE, which past through mind.-- These thoughts a
Malthus & Decandoelle.-- The Final cause of INNUMERABLE eggs is explained by Malthus.-- [is it anoma
sition of organs of generation!!!. Mem. Agaziz. (INO Annals of Nat. Hist) spiral structure in Echinod
th. Mr Lonsdale showed me two specimens of an INOCERANUS from the Gault of Folkstone, which is exactl
ements. [LHC] Hence there are two great <worlds. INORᴾ systems of laws «in the world» the organic & i
nts.-- grafting length of life &. & Will any INORGANIC substance cause such monstrous growth as oak
as compilation of action of organic nature on INORGANIC It is very remarkable as shown by Carus how i
ntermediate plants are between animal life & "INORGANIC life'.-- animals only live on matter already
n acts on. & is acted on by «the» organic and INORGANIC agents of this earth. like every other animal
aws different from those., that govern in the INORGANIC world; life itself being, the capability of s
peculiar system of movements. different from INORGANIC movements.-- See Lamarck for this definition
systems of laws «in the world» the organic & INORGANIC-- The inorganic are probably one principle fo
«in the world» the organic & inorganic-- The INORGANIC are probably one principle for connect of ele
relation.--As in first cases distinct species INOSCULATE, so must we believe ancient ones: «.»» not gr
th America closely approaching.-- but as they INOSCULATE, we must suppose the change is effected at o
esearches.-- Athenaeum. 1838-- p. 137. Three INOSCULATING rivers in Southern America ¿ effect of subs
e per saltum-- or species may perish. = This <INOSCULATION> «representation» of species important, eac
oe creeper: Furnarius. <Caracara> Calandria: INOSCULATION alone shows not gradation;-- An argument fo
flower beautiful? ¿even to children S. Jenyn's INQUIRY into the Origin of Evil. Reviewed by Johnson i

627o52v s leads to striking blows.--' p. 224.-- Hume's INQUIRY-- good abstract of Butler & arguments of benef
312c232 ansferable., not wonderful According to my view <INS> beccause actions are constant they are instinct
567n015 - heart beginning to beat-- children inherit it <INS> like instinct, preeminently so-- who can analys
311c226 g most easily mostified, which last produced --INSANE men in civilized countries-- this is well wort
522m013 e is perfect gradation between sound people and INSANE.-- that everybody is insane. at some time. Man
522m013 an sound people and insane.-- that everybody is INSANE. at some time. Mania is quite distinct, differ
523m014 anger must be directed against somebody.-- Have INSANE people any misgivings of the injustness of the
523m015 N B. I have read paper somewhere on horse being INSANE at the sight of anything scarlet.-- dogs ideot
523m017 ousemaid two years before, to prove she was not INSANE, answered she had known it at time & had bough
523m017 pected during these two years that she had been INSANE all the time.-- There are numberless people in
523m018 ane all the time.-- There are numberless people INSANE of particular ideas, «Case of Shrewsbury gentl
524m019 People are constantly well aware that they are INSANE & that their idea is wrong.-- (Dr Ashe, the Bi
524m020 constantly present in people, likely to become INSANE.-- now this is well worth considering, if prid
531m051 ve feeling, <may not pa.> In reflecting over an INSANE feeling of anger which came over me, when list
548m116 ousness, as shown in the tendency to forget the INSANE idea; & ones expression of double self, though
551m127 the reality of the impression on its senses.-- INSANE people believe they hear as well see things wh
522m013 people in violent intoxication, often ends in INSANITY or delirium.-- In Mania all idea of decency &
522m014 nvoluntarily when a person is tired is akin to INSANITY.-- «I know the feeling also of depression, &
523m015 my grandfather his feeling of consciousness of INSANITY coming on.-- his struggles against it, his kn
523m015 & studying mathematics.-- My Father says after INSANITY is over people often think no more about it t
523m016 ften think no more about it than of a dream.-- INSANITY is produced by moral causes (ideotcy by fear.
523m017 ful than the thing itself. Asked my F. whether INSANITY is not distinguished from whims passion &c by
524m021 f pride & suspicion can be well understood. In INSANITY, the ideas do not go back to childhood, (but
525m026 hich were short term, & lastly healthy ones.-- INSANITY & Epilepsy remain many generations in familie
525m026 .-- My fathers does not know whether trains of INSANITY are heredetary in any one family.-- In Aunt--
528m035 nalogous to my Fathers positive statement that INSANITY is only cured by forgetfulness.-- & the appro
528m035 stle in the air, or dreams real again explains INSANITY.-- Analysis of pleasures of scenery.-- There
538m078 fathers doctrine about people forgetting their INSANITY» there seem other cases somewhat analogous, &
538m080 s. what one performs habitually.-- Agrees with INSANITY, as in Dr Ash's case, when he struggled as it
548m116 es one strong idea of what individuality is.-- INSANITY is «much» «somewhat» the same as double consc
549m120 ugh he sees some «intellectual» good men, from INSANITY &c unhappy-- perhaps not so much as they appe
550m127 stinctive emotions?-- Dr Holland remarked that INSANITY like sleep does not doubt the reality of the
551m127 o existence.-- He compared spectral illusion & INSANITY the connexion appears to me vague-- Delirium
604o017 Mayo-- Pathology of the Human Mind. Poor.-- on INSANITY.-- Prevailing idea. owing to loss of will.--
604o017 iefly fail.-- Notices. struggle <between> when INSANITY is coming on «Thinks clearest analogy between
604o017 g on «Thinks clearest analogy between dreams & INSANITY.» D. Stewart on the Sublime The literal meani
607o025 CD one sees this law in man in sommambulism or INSANITY. free will (as generally used) is not there p
223b208 draw line.-- -- not so great as between perfect INSECT & former hard to tell whether articulate or in
225b218 h any insects hav[e] become attached to.-- that INSECT «not» being called <Phitophagous> omniphitopha
272c107 anied by habits) of other, thus in Chalcidious INSECT, which I brought from Australia, probably live
440e145 ion, that every «budding» tree, & every buzzing INSECT & grazing animal owes its form, to that form b
455eIBC llapse at night. if so irritate them, «as by an INSECT coming always at same time» see if by so doing
470t178 lover,--Veronica--, Ranunculus in numbers =what INSECT can get honey out of long, curved nectar of Bu
472s02r (with pollen) on legs go from clump to clump, & INSECT proboscis in many flowers, on one of which pol
490q001 s Sowerby Entomologist Does individual Shell or INSECT or group vary more in one country or district
496q05a l-- or sight.-- --. touching Mr Brown theory of INSECT-like Orchis-- & final cause of beauty of flowe
499q09v uch frequented by Bees or Butterflies or little INSECT?= or is pollen excessively minute or abundant?
499q09v a is dioecious & Pollen must be carried by some INSECT-- (21) Are there many instances of single clum
520mIFC rwin Esq 36 Grt. Marlborough Str.-- (p. 64. On <INSECT> Ants getting on Table. Col. Sykes) Private Fi
270c104 Echinodermata.-- Agassiz says Infusoria «are» INSECTA.-- G. R. Treviranus, Biologie referred to,. as
383d159 enis reduced to extreme degree of abortion).-- INSECTA.-- hermaphrodite, being not only dimidiate, bu
543m099 se Study with the greater number of insects in INSECTA-- not connected with transformation because Sp
244c021 273 2d Edit Consult Latreille. Geographie des INSECTES, in 8 degree. p. 181 «who says insects Indian
324c268 t. tom IV. p 273 Latreille Geographie des INSECTES 8vo p. 181.-- See (p. 17) for references to a
311c227 Institut 1838. p. 174. Apercu very good on INSECTIFEROUS <insects> quadrupeds geographical range ver
407e041 equatorial..-- The fact of No. Mam: Placent: INSECTIVORE being in S. America & Australia. reason, why
481z009 ifferent islds.-- Waterhouse remarks that no INSECTIVORE in S. America or Australia-- very curious.--
177b027 «birds» it would only appear like circles;-- & INSECTS amongst articulata.-- but in lower classes, pe
182b047 nstance in birds? Examine «good» collection of INSECTS with this in view.-- Geogr. Journal Vol VI. P
185b056 n genus Waterhouse says he is certain, that in INSECTS, each family, however many there may be, repre
196b108 ta> «vertebrates» could exist without plants & INSECTS had been created; but on other hand creation o
196b108 abstract we may say that vegetables & mass of INSECTS & worms.-- In abstract we may say that vegetab
196b108 abstract. we may say that vegetables & mass of INSECTS could live without animals but not vice versâ.
198b113 ent might tell something) p. III G. St Hilaire INSECTS & Molluscs allowed to be wide hiatus: states i
199b116 by propagation that out of the thousand of new INSECTS all belong to same types already established.
213b171 reptiles at Silurian age-- How long back have INSECTS been known? As Gould remarked to me, the "beau
221b200 st metamorphosis Islds X Is this applicable to INSECTS &c &c?-- (p. 23 do.-- On animal - Confervae--
222b206 include Mammalian remains.-- The Father of all INSECTS gives same argument as father of Mammalia; but
223b207 tellectual Man appearing..-- the appearance of INSECTS with other senses is more wonderful. its mind
225b218 now of any case of introduced plant, which any INSECTS hav[e] become attached to.-- that insect «not»
225b218 ophagous. But it will be said there are latent INSECTS.-- as crows against man with gun. & Bustards &
236b280 such should not be general circumstance.--In. INSECTS «in England» surely it is not-- intermediate g
244c021 ie des insectes, in 8 degree. p. 181 «who says INSECTS Indian, like. Plants.--» It would be very impo
248c033 & manner of generation «& primary divisions of INSECTS» 2. Relation, of external conditions, & of suc
271c105 which it builds, a berry on which it feeds. or INSECTS it devours is same species. yet that it should
271c107 - ie. mundine groups.-- Waterhouse tells me in INSECTS there are many plenty of instances of insects
271c107 insects there are many plenty of instances of INSECTS of one tribe taking on structure (probably acc
272c107 me other part of body.-- there are hemipterous INSECTS, having spiny legs & running quick & generl ap
272c108 ble Coleoptera.-- Donacia.-- some orthopterous INSECTS & some third, have got thighs with same peculi
273c112 ts, though swallow hawk, ,milvulus,, may catch INSECTS on the wing & pratencole (¿connecte[d] with Ch
274c114 vision Gould thinks he can trace structure for INSECTS & structure for vegetation.-- In conversation
289c162 rms.=» (NB waterhouse says of affinity of many INSECTS may be told by their larvae) but the acts of c
290c162 it speaks to amount of change only & not kind» INSECTS-- & vertebrata & plants. At first classificati
293c174 f men.-- Wax of Ear, bitter perhaps to prevent INSECTS lodging there. Now these exquisite adaptations
300c197 e true to consider him created from animals.-- INSECTS shamming death, most difficult case to imagin
301c200 & collecting thousands & tens of thousands New INSECTS, perhaps scarcely one new family & no new orde
305c210 ment in instincts in the <classes> «orders» of INSECTS, so is there none of reason in order of <ver>
308c218 Dr. Black. tea & sugar» people. reverse habits INSECTS & birds are the only two tribes fitted for wat
311c227 38. p. 174. Apercu very good on insectiferous <INSECTS> quadrupeds geographical range very good.-- Bl
313c236 s. ¿are sympathetic nerves & nervous system of INSECTS analogous.?-- («Even plants have habitual acti
338d025 3. Note Book M'. for case of change in food in INSECTS entered by mistake Surely the fossil Mamalogy
344d040 there are exceptions to this in some larvae of INSECTS-- (¿glowworms») breeding-- <beet> imago state f
344d040 once.-- Consider this with reference to those INSECTS, which have fertile offspring. Entomostraca &
344d041 ce of sexes. is probably arrived at in case of INSECTS as glowworm The case of one impregnation suffi
350d055 -- young of sponges move.-- young of Cochineal INSECTS move about & see, parent <(2)», female «(I)» f
358d076 Sept. 9th. It is worthy of observation that in INSECTS where one of the sexes is little developed, it
360d091 seen when sickly tigers have first come over, INSECTS somewhat like «between» lice & fleas. sticking
381d166 as» Cephalopods, Pectinibranchiate molluscs.-- INSECTS. spider crabs.-- (all these however do not req
402e019 second «1st» Voyage. Vol II p. 344. account of INSECTS of St. Peter & St. Pauls in Lat' 53 degree yet
402e019 degree & 70 degree of Europe.-- Many Europaean INSECTS-- list given.= some peculiar <M> p. 359. At Ma
414e060 us period, whether progressive I know not.» (& INSECTS.-- Stonesfield????). Have Mammalia?? My theory
416e071 CD [lowest terrestrial animals.-- in shells?-- INSECTS?.-- all!??!?-- Worms? [Barnacles, aquatic., <y
418e080 abortion of others.-- ¿ in hemi-hermaphrodite INSECTS is it not easier to understand ¿perfect?? deve
420e090 ies).-- might these fertilise other shells, as INSECTS do flowers.-- Mem. Spallanzani's experiments s
429e114 s (opportunities of escape for foreign birds & INSECTS) which are propagated with very little care.--
447e164 -- well characterised species periwinkle wants INSECTS to impregnate allied to Asclepias Turpin cell
448e165 oduces seeds, because it is thought to require INSECTS to impregnate it.-- it is allied to Asclepias,

329

449e170 rhouse says some of the Galapagos Heteromerous INSECTS come very near to Patagonian species-- p. 18.
460t019 this refers to time.-- Marsupial in Oolite.-- INSECTS, of do orders-- cheiroptera & caetacea in Eoce
485z017 Zoolog & B. Vol. II. p. 127. List of submarine INSECTS. Staphylinidae &c &c. with reference to those
485z018 'Institut, 1838 p. 46 Macleay Horae Entomolog. INSECTS swarm in Lapland & Stizberges wherever there i
486z018 rare.-- These views clearly explain rarity of INSECTS in T. del Fuego.-- Hence it is odd that Amber
486z018 in T. del Fuego.-- Hence it is odd that Amber INSECTS of Europe have Tropical Forms See p. 256 of No
501q011 ssed, I think, I expect, except by very minute INSECTS.-- (30) Get Abberley to plant SINGLE Peas, Kid
505q014 n top of House =Aristolochia, plant wh require INSECTS to impregnate it (7) History of Potato field=
534m063 arks that some imported plants are attacked by INSECTS & snails of this country (thus Dahlias by snai
535m064 no tea do. p. 233. Mr Lewis describes case of INSECTS «a Perga» of Terebrantia, laying eggs on leave
535m064 thy-- This remarkable case may be normal. with INSECTS, but habit forgotten in all older species. The
542m097 dale's story of Snails, Fox of cows, & many of INSECTS-- they likewise must understand each other exp
543m098 phant most intellectual.-- Hymenoptera typical INSECTS. ie have all parts. Waterhouse Study well the
543m099 s. Waterhouse Study well the greater number of INSECTS in insecta-- not connected with transformation
583n069 seems invariably associated with reason: [N B. INSECTS which have never seen their parents offer best
593n107 ther animal, (especiall as in some <instinct> «INSECTS» which become in imago state social) by smell
593n109 eceives pleasure. This [blank] will not do for INSECTS. if this view holds good-- then man, a sociali
613o035 lt there must be consciousness??? ? [LHC] ¿Can INSECTS live with no more consciousness than our intes
636j56v ion of Dioecious Plants by foreign agency-- as INSECTS, as wonderful case of adaptation.! There would
636j56v e been any Dioecious plants, had there been no INSECTS. The right inference is, there were insects «¿
636j56v no insects. The right inference is, there were INSECTS «¿when were Palms formed?» as soon as Dioeciou
637j57v ience would have arisen had « <not> some» some INSECTS <not> not been provided. «with proboscis» «as
641j29v tion of fetid «& acrid» secretion in Mollusca. INSECTS «Carabids & Staphylini» & Mammalia. The eye be
641j29v rata, & Planaria, & light affecting plants. in INSECTS the end is gained by some very different metho
069r149 change having taken place unrecorded must be INSENSIBLE. Quantity of matter from Cordillera. HORIZON
070r153 vestruz two species. certainly different. not INSENSIBLE change.-- Yet one is urged to look to common
117a101 ation & depression. C. Prevost.-- My views of INSENSIBLE oscillations of level will alone explain the
192b085 from forms marking good genera-- by steps so INSENSIBLE, that each is not more change than we know v
397e004 ge of circumstances; now we know how slowly & INSENSIBLY such changes are in progress.-- we feel inte
471tf08 completely natural & one whose groups pass so INSENSIBLY into each other". Phillips (Lardner's E. vol
472s02r withered on Monday.--» alight on upper petals & INSERT proboscis, under sigma & draw it out over & ov
138a153 t level, & in many parts has extended a league INSHORE both N & S of Lima.--judges from «beds of» san
067r144 arkable phenomenon. showing line of disturbance INSIDE Cordillera: It is not therefore so wonderful t
090a021 from believing meteorolite but old Planet, that INSIDE our globe melted magnetic metals. ∴ earthy cru
270c103 bles. p. 243 radiate animals <tu> plants turned INSIDE out. have position of organs of generation!!!
386d167 g place from outside inwards. & in animals from INSIDE to the outside.-- is this not owing simply to
035r048 f sand, in this view sink into their proper INSIGNIFICANCE; as fractures, consequent on grand rise, &
044r073 n great size of ocean; especially Pacifick: INSIGNIFICANT islets--general movements of the earth;--Sc
565n008 , when ones opponent is considered as quite INSIGNIFICANT, & when pride makes person extremely self-s
037r056 ear! Mem. La Condamaine on the Amazons. Consult INSIST on the frequency of dikes in Granitic countrie
074r165 ns of clay iron stone; iron pyrite in a fossil» INSIST strongly on the grand fact of Volcanic & non V
198b115 2d Vol about monsters worth reading NB well to INSIST upon <different animals> large Mammalia not be
266c088 ome dull coloured in sterile countries.-- Gould INSIST much upon knowing to what type a bird belongs.
428e113 of male sex on locomotive system» I am bound to INSIST honestly that the sudden, change from Primrose
609o030 specially in the female October d. 1838 perhaps INSIST?? Two classes of moralists: one says our rule
066r142 l> a lake in neigbourhood of town Mr Murchison INSISTED strongly. that taking up a piece of Falkland
587n089 of parts of Brain.-- Mackintosh first clearly INSISTED on assoc of ideas & emotions. rather ideas &
100a055 into steam, would Babbage.-- Webster Phillips INSISTS of analogy between Australia & Oolitic period.
214b173 Kerguelen land.-- Phillips. Lardner Encyclop. INSISTS on analogy between Australia and «fossils of»
245c024 tions p 293-- is very strong about this Lesson INSISTS much.-- The (p. 296) Columba Kurukuru found in
556m145 escribed as partaking of <st.> made by <ret> INSPIRATION & quickly retracting tongue from behind uppe
570n023 aking something off shoulder-- or is it from INSPIRATION, which accompanies surprise.-- & why does on
584n072 ally carried to long distance in dark "it is INSPIRATION."-- this is class of so called instincts to
570n023 , which accompanies surprise.-- & why does one INSPIRE, when surprise, can one resist blow better wit
044r073 pe formed near coast. Humboldts quotation of INSTABILITY of ground at present. day.-- applied by me g
064r137 onnected <lines> «points» of eruption [.] give INSTANCE of Etna Stromboli & Vesuvius Investigate with
069r149 . HORIZONTAL movement of fluid matter not (for INSTANCE) expansion of solid matter by Heat Consider p
100a054 of Rio Concretions in Pumice bed at Ascension INSTANCE of hollow concretions & concretion filled wit
145g002 epherds dogs coloured like Magellanic fox.. an INSTANCE of Provincial breeds. [3] Veins of Segregatio
146g013 tches of Conglomerate on S. Ventana, excellent INSTANCE, how accidental is the preservation in situ o
170b001 kind, all individuals absolutely similar; for INSTANCE fruit trees, probably polypi, gemmiparous pro
175b019 erfect «& others» occasionally dying out;--for INSTANCE secondary terebratula may have propagated rec
178b031 l, & L. Jenyns. Falkland rabbit may perhaps be INSTANCE of domesticated animals having effected; a ch
181b042 f continuous structure} «between them.».-- for INSTANCE there would be great gap between birds & mamm
182b047 int;. hence the resemblances & differences for INSTANCE of finches of Europe & America. &c &c &c The
182b047 re possible the number «of steps» known). <for INSTANCE among the Carabidae.-- instance in birds> Exa
182b047 » known). <for instance among the Carabidae.-- INSTANCE in birds> Examine «good» collection of insect
185b056 many there may be, represent every other; for INSTANCE in Heteromera, you have representatives (whic
188b068 ce established so difficult to root out.-- For INSTANCE ever so many seeds of white flower. all would
191b080 nd & Europe Ireland common animals-- + + + for INSTANCE tertiary deposits between East Indian islets-
192b086 jumps strongest marked genera? Reptiles?) For INSTANCE there never may have been grade between pig &
196b108 d-- There must be progressive development; for INSTANCE none--?, of the <vebtetrata> «vertebrates» co
198b113 fitted for as many situations as possible. for INSTANCE take birds animals, reptiles fish-- Condition
199b119 it up into many isolated races. ¿are there any INSTANCE of peculiar people banished by rest?-- ∴ most
200b122 at naturalist describe animals as species, for INSTANCE Australian dog: <yet when that> or Falkland r
201b129 he says depends on external influences.-- For INSTANCE he says wings of bat, are from external influ
203b137 s destroyed animals over the whole world-- For INSTANCE gradual reduction of tempereture from geograp
206b148 accidents: yet some causes are evident, as for INSTANCE one man killing another.-- So is it with vary
210b158 in proportion to difficulty of transport. For INSTANCE the temperate parts of Teneriffe, the proport
210b159 s. there is analogy to species & genera.-- for INSTANCE three kinds of greyhound.-- In plants. do the
211b162 rence in skeletons: VERY GOOD. Apteryx. a good INSTANCE probably of rudimentary bones.-- As Waterhous
217b183 never afterwards went in heat.-- This is good INSTANCE of same fact in Mr Galtons case.-- It explain
224b213 ffer scarcely in any external character:-- For INSTANCE two wrens forced to haunt two islands one wit
228b229 expect confusion of species.-- Important. For INSTANCE take Voluta & Conus (??) which now run togeth
229b233 ated for poison channels, but not having them, INSTANCE of useless structure-- Smith thinks several s
243c020 ish of Taiti found at <New> Isle of France: xx INSTANCE of wide range, where means of wide range work
259c062 inhabited, then how is this effected by-- for INSTANCE, fish being excessively abundant & tempting t
261c070 out its parent running to the water, is a good INSTANCE of connate instinct, better than child suckin
264c081 ding habits clearly showing true affinity, for INSTANCE tail of ground woodpecker:-- -- but tail of g
265c088 so not aberrant.-- Tenioptera rufiventris, is INSTANCE of bird belonging to family with peculiar col
273c112 wonderfully to the Humming bird, which is one INSTANCE of its whole family where female is not dull.
274c113 so much on the ground has not this structure. INSTANCE of habits going before structure).-- even one
274c114 bserve these characters vary in degree in last INSTANCE hardly at all developed. Not confined to one
280c134 ll as gain instincts. Wild & tame rabbits good INSTANCE-- instincts of many kinds in dogs, is clearly
293c172 afterwards could understand «(language better INSTANCE)» he had done this without reflection or cons
293c174 where habits of people nearly similar. Curious INSTANCE of difference in races of men.-- Wax of Ear,
294c175 heep could not live for some time at New York «INSTANCE of the fine relations of adaptation of animal
306c214 trial Planariae assuming bright colours., good INSTANCE of colours dependent on localities.-- -- Hami
307c216 ants I presume there are.? get examples.-- for INSTANCE where a tendril passes into a mere stump.-- S
313c235 f (in June 1838 when young male was added good INSTANCE of instinct showing itself, not from instruct
347d049 universe so adapted!!! & not man to Planets.-- INSTANCE of arrogance!! August-- 29th.-- Macleay in A.
366d108 pair he could have produced from these.-- this INSTANCE of monstrous variety. which could not have be
378d148 have a wariness about them quite remarkable". INSTANCE of old Species transmitting so much longer it
404e026 Mexican greyhounds.-- young being habituated. INSTANCE such as Hunter, or some one mentions of influ
412e057 lready given various examples The Pipe-fish, is INSTANCE of part of the hermaphrodite structure being
421e091 to Mr Bell's statement of the tame ones.. «an INSTANCE of a trifling peculiarity not to be eradicate
424e103 fondled.-- and we see in the Australian dog an INSTANCE of a half reclaimed animal.-- The dogs, which
426e107 n[ly] true hybrids.-- but this is false, [give INSTANCE of series from wild animals & plants]CD.-- Mr

```
431e121 ubts.-- Lonsdale thinks Ammonites would afford INSTANCE of such facts.-- Ask Phillips.-- The more I t
432e126 by  Sedgwick & Murchison; which is a beautiful INSTANCE of forms, intercalated between two great dist
436e136 s «the» difference between <pi> species,-- for INSTANCE pidgeons-- : then comes question of genera It
446e160 eir young..-- The oriolus (icterus Cat.) is an INSTANCE of this, & the female of the icterus minor is
520m002 ly infancy-- of this fact Mr Dryden C. is good INSTANCE as he is very deficient, he was nearly 9 year
531m049 ill «ever» eat little birds, this most curious INSTANCE of reason & abstinence.-- My Father remarks t
539m083 ciousness?) change its habits.-- Aug. 16th. As INSTANCE of heredetary mind. I a Darwin & take after m
554m141 s definition of "reason" & "reasoning," & take INSTANCE of Dray Horse going down hill.-- (argue sophi
556m146 sion of chin, like bulls & horses.-- 1838 good INSTANCE of useless muscular tricks accompanying emoti
566n011 ing he instinctively knows distances:, is good INSTANCE of obtaining <that> «a» faculty in the form o
566n012 h law.-- as soon as any enquiry commenced, for INSTANCE probably such a thing as thunder. would be pl
574n040 reasoning powers & the fixing of habits,-- for INSTANCE the Birgos opening a Cocoa nut shell at one e
577n049 e p. 188 of Herschel's Treatise) a "travelling INSTANCE" a-- "frontier instance".-- for it can be sho
577n049 reatise) a "travelling instance" a-- "frontier INSTANCE".-- for it can be shown that the life &  will
582n067 <Horses> Colts cantering in S. America capital INSTANCE of heredetary habit:-- there must, however, b
588n091 olog. «p. 284. Vol. II» -- gives explanation & INSTANCE of starting identical with mine,-- Lamarck. V
592n105 e» cause, and origin being so is not odd.; for INSTANCE wild cattle & deer pursuing a wounded one.--
619o042 ted by <necessity> «some force» give pain: for INSTANCE either protecting sheep or hurting them.-- Th
636j57r b. of Deity. Vol I. p. 232. gives Woodpecker as INSTANCE of beautiful adaptation.- & then Chamelion,
070r153 ny instances diametrically opposite have been  INSTANCED; it is Let it not be overlooked that except b
024r015 oving <point> power ¿ Submarine currents Find  INSTANCES; The whole coast of New Holland shoals. much:
031r038 red no sign of upheaval To Cleavage add other  INSTANCES in old world of symetrical structure. East In
070r153 lied species occur in same country? In botany  INSTANCES diametrically opposite have been instanced: i
145g001 367  Quarterly Journal of Agricl Dec 1837 Yet  INSTANCES given against it-- Mere fact of many races of
192b083 lthough many of the seeds will go back.-- Get  INSTANCES of a variety of fruit tree or plant run  wild
192b085 termediate steps.-- Quote in detail some good  INSTANCES. But it is other question, whether there have
217b187 thorse two folds [not located] Mr Don gave me  INSTANCES of one species of Australian genus being foun
239cIFC are varieties produced, by picking offspring?  INSTANCES of old Breeds taking greatest effect Account
242c017 19.. Tapir, «des» couroucous et rupicole vert  INSTANCES of American forms in East. Ind: Archipelago.
243c019 All  Australian forms have representatives (&  INSTANCES given) in East Ind. Arch:-- Birds of New Zeal
261c069 tal table of extent of all species Accumulate  INSTANCES of one family sending out structures into man
261c070 ination of species is empirical. show this by  INSTANCES Once grant my theory, & the examination of sp
271c107 tells  me in insects there are many plenty of  INSTANCES of insects of one tribe taking on structure (
272c109 genera» scattered over it.-- We have abundant  INSTANCES of remarkable structure which as far so speci
277c127 ies. until, he can show range & habits-- Take  INSTANCES of most disputed shells, such as Cyrena  This
284c150 ect to Variation" Dr. A. Smith. knows lots of  INSTANCES of replacement of one species by another. sup
289c160 zels» dive instant touch the water.-- capital  INSTANCES of typical land bird, having habits of a Greb
296c184 s Magazine of Zool. & Bot.-- Vol I. p. 450. 4  INSTANCES of hybrids between pheasants & Black  fowl.--
316c246 o West coast of Africa & E. S. America.-- get  INSTANCES.-- very good anomaly in range + + What circum
345d044 ale impresses offspring more than female, yet  INSTANCES given on opposite side,-- «The theory of male
360d089 ems owing «determined» by the sex» Individual  INSTANCES trouble Yarrels law. chief trust must be in g
367d113 s that most immediately employed in fighting"  INSTANCES thighs of cock & Neck of Bull.-- is most comm
371d127 -colour is a doubtful subject, but what other  INSTANCES are there of such changes, not acquired by pa
388d172 g on old women = Stags horns & testes curious  INSTANCES of corelation in structure = Neuter bee havin
392d180 common  dogs--> Ask my father to look out for  INSTANCES of Avitism Examine English weeds in Hot. Hous
454e184 esio. Pisa 1816 p. 27. Dr. Holland. Are there  INSTANCES of plants, in becoming double loosing fertili
459q010 carried  by some insect-- (21) Are there many  INSTANCES of single clumps of plants in counties, as of
525m024 ook like fear, but shame.-- I cannot remember  INSTANCES, but I feel sure I have seen a dog doing what
539m082 sidered as actions otherwise than habitual.--  INSTANCES?? The possibility of two quite separate train
539m083 s, & so is Catherine in some respects--. good  INSTANCES.-- when education same.-- My handwriting same
563n001 seasons,  & after? he has done <g> crowing.--  INSTANCES of expression.-- Octob. 3d. Dog obeying insti
595n125 rocodile snapping.-- these I think are better  INSTANCES of instincts (highly useful as only means of
637j57v disease in world & yet talk of perfection Get  INSTANCES of adaptations in varieties.-- greyhound to h
289c160 \-- p. 390,. young <grebes» «ring ouzels» dive INSTANT touch the water.-- capital instances of typica
547m113 or  believe.-- When I say trains, it may be  INSTANTANEOUS changes in order <to every> calling up idea
605o020 nates, & by what means it becomes an almost  INSTANTANEOUS perception.-- Taste has been supposed by so
606o021 tions. & yet, like in vision, it becomes so  INSTANTANEOUS. that we cannot ever perceive the various o
622o048 wardice» «This is not connected with sense»  INSTANTANEOUS so declaring it is right or wrong.-- «[just
375d135 s of hawk decreasing in number must effect  INSTANTANEOUSLY all the rest.-- One may say there is a hot
605o20v And it is manifestly from this fact & the  INSTANTANEOUSNESS of the result, that the term taste is met
273c111 f swallow perished) hawks & Milvulus &c would  INSTANTLY fill up their place.-- Humming bird there is
524m022 d, (forgetting that her husband was dead) yet  INSTANTLY perceived when my Father to distract her atte
556m144 ows triumph.--» Sept. 23rd. Horses in Omnibus  INSTANTLY start when they hear ready, but if they see a
572n033 page  & pronounced each word distinctly. woke  INSTANTLY but could not gather general sense of this pa
615o36v alking home one day met him, with Mark riding  INSTANTLY followed, me and for five minutes every now a
033r042 s??--if so case precisely analogous: fragments INSTEAD Peak of Teneriffe. also Cotopaxi has a <[...]>
045r077 ubsiding area. we may believe the fluid matter INSTEAD of afflux (always slightly oscillating as that
047r081 fick.--excepting in number of waves & in wind, INSTEAD of sea's bottom being in motion what differenc
090a023 x  19 = 1710 ÷ 50 = 34 pounds each year.-- but INSTEAD of 90 stones in many cases there were  flights
106a075 dency would <cut> be to cut a narrower channel INSTEAD of wider.-- This applies to all vallies (excep
223b211 species  well adapted to locality A. but it is INSTEAD of a stunted & diseased form a plant, adapted to
232b247 cooled  by secular refrigeration in chief part INSTEAD of change from insular to extreme climate, «mo
255c050 f latitude more commonly are the same species, INSTEAD of analogues-- <as> in other classes this evi
263c077 ason-- but Man has reasoning powers in excess. INSTEAD of definite instincts.-- this is a replacement
282c141 To make this case perfect, we must suppose men INSTEAD of mere colour & trifling form & head &c to be
288c159 price.  as is evident to be worth introducing, INSTEAD of breeding from original Durham breed.-- Nati
383d160 like do.-- but the black lines on each feather INSTEAD of coming to point {P} are more rounded. {P} &
383d160 & much broader., & <more ro> «three, I believe INSTEAD of» two lines.-- «faintly edged with reddish b
416e075 between  races & variety? «Man picks the Male, INSTEAD of allowing strength to get the day» The ferti
441e148 dividuals» & the action always of two organs-- INSTEAD of one part «as» in producing bud.-- Fewer of
458t002 aylor Bird uses pieces of thread, picked «up-» INSTEAD of spinning-- better case than English birds,
458t002 etter case than English birds, using cotton &c INSTEAD of natural substances-- useful perversion of i
463t055 xpands, so do the bones <are created> expand-- INSTEAD of saying as brain is created &c &c Bats are a
555m143 ad some confused idea of showing scar behind (. INSTEAD of front) (having changed hanging into his hea
564n003 aving different moral sense, if it were proved INSTEAD of militating against the existence of such an
636j57r spect, except «in» quick movements. (sliminess INSTEAD of barbs)-- In all these cases it should be re
638j58v <every» «ome or any»-- brain making structure, INSTEAD of parts of body.-- Now we know what  instinct
193b093 together  Man has no heredetary prejudices «or INSTINC» to conquer or breed together:-- Man has no li
146g012 - dislike & <f> passion of hatred To fulfil an INSTINCT a pleasure; mem. Shepherd dogs The Patches of
171b003 e» «know» in course of generations even mind & INSTINCT becomes influenced.-- child of savage not civ
172b006 te in Number In Man it has been said, there is INSTINCT for opposites to like each other AEgyptian ca
173b009 ate Every character continues to vanish, bones INSTINCT &c &c &c non fertility of hybridity &c &c <as
179b034 to their type: in animals so far removed, with INSTINCT in lieu of reason, there would probably be re
194b093 :-- Man has no limits to desire, in proportion INSTINCT more. reason less. so will aversion be L. Ins
196b105 rfowl.-- scrutinize genera, & draw up tables-- INSTINCT may confine certain birds which have wide pow
199b118 ntal propenity, of an accidental habit into an INSTINCT." Ed. N. Phi. J. p 297, No 8 Jan-Ap. 1828  --
207b151 ir beaks, & the young one <inland> directly by INSTINCT. can dive & conceal themselves in the grass.--
212b165 d.-- In L.' Institut. 1837: p. 404. account of INSTINCT of dogs.-- agreement & reason Some animals co
220b197 one.  «Although in plants, you cannot say that INSTINCT perverted, yet organization «especially» conn
220b197 e of two species to each other is evidently an INSTINCT-- & this prevents breeding. now domestication
220b197 ng. now domestication depends on perversion of INSTINCT (in plants domestication on perversion of str
220b197 , whether offspring does not depend on mind or INSTINCT of parent. Mem Lord Moreton's Mare. The  fact
227b227 - the manner of adaptation (wish of parents??) INSTINCT & structure becomes full of speculation & lin
248c030 must.  thus be taken, as «in» wild state (where INSTINCT not interfered with, or generative organs aff
255c051 xcellent case of memory without association..» INSTINCT goes before structure (habits of ducklings mu
260c069 ding habits range & all kinds of information-- INSTINCT Swainson's remarks in Fauna Borealis must  be
261c070 ng to the water, is a good instance of connate INSTINCT, better than child sucking or even  ducklings
263c076 nus may pass into each other.-- grant that one INSTINCT to be acquired (if the medullary point in ovu
```

280c134 cult & other go back to either parent.-- Shows INSTINCT (Sir J. Sebright admirable essay) heredetary
283c145 et them be wild in same country with their own INSTINCT & (even though <fertile> when compelled to br
285c150 remarkable crochets about instincts. whenever INSTINCT is mentioned some definition must be given It
288c159 ng, or keeping watch. how completely «nature & INSTINCT» modified.-- The partial migrations of birds
292c171 ind.-- Reflect much over my view of particular INSTINCT being memory transmitted without consciousnes
293c172 . had done habitually we should exclaim it was INSTINCT.-- Even if savage takes. & was given a Great
293c173 on holes it would be instinctive.-- My view of INSTINCT explains its loss ¿ if it explains its acquir
298c189 often looked at.-- this most puzzling whether INSTINCT, or reason?? Gould says he believes that he h
300c198 Van Diemen's land. Study Mr Blyth's papers on INSTINCT.-- His distinction between reason & instinct
300c198 n Instinct.-- His distinction between reason & INSTINCT very just, but these faculties being viewed a
301c198 ent from that of brutes p. Hard to say what is INSTINCT in animals. « & what reason» in precisely sam
301c199 s.-- child striking a post in passion.-- Habit INSTINCT gained during life.-- do Elephants easily acq
301c199 Key to their mental powers.? p. 8. mistakes of INSTINCT are external contingencies, where the habit i
309c221 .-- Male glow-worm knowing female good case of INSTINCT. bees turning neuter into Queen, more wonderf
313c235 d in India. in Heber?) is analogous to dormant INSTINCT.-- (How wonderful a case bees developing sex
313c235 838 when young male was added good instance of INSTINCT showing itself, not from instruction Even the
313c236 of the viscera. under sympathetic nerve may be INSTINCT or habits. ¿are sympathetic nerves & nervous
323c269 glis Transla -- The Revd. A. Wells. Lecture on INSTINCT -- Cline on the Breeding of Animals -- Spalla
326c266 Science connected with Natural Theology.-- on INSTINCT & animal, intelligence.-- very good. Endliche
340d026 to explain, or asserts to be explicable every INSTINCT in animals. Heard at Zoolog Soc their Pintail
346d048 y Boethius of ancient Caledonian Cattle. Ch 3. INSTINCT L'Institut. p. 249. (1838). Eggs discovered t
371d118 the mind. the simplest transmission is direct INSTINCT. & afterwards enlarged powers to meet with co
404e024 eive no explanation» was consequent on mind or INSTINCT, now this is directly incorrect The case of m
439e144 lity in crossing there comes the impediment of INSTINCT-- the possibility of rearing by seeds Holyoak
444e158 !-- p. 203 Chaetodon squirting water at fly.-- INSTINCT, for how could experience teach distances in
450e174 amous; tame ones highly polygamous-- change of INSTINCT by domestication.-- "Notices of the Indian Ar
452e181 [Asiatic Soc] [...] p [...] -- most wonderful INSTINCT, how could it have originated-- spins thread
527m032 ion may perhaps account for our acquiring «the INSTINCT» our notion of beauty & negroes another; but
527m033 osite ends of one scale.-- former pleases from INSTINCT the ears (rhythm & pleasant sound per se) & c
532m053 oung babies start.-- ‖ If children wink. it is INSTINCT Fear must be simple instinctive feeling: I ha
534m061 here reason is forgotten. it is conscience, or INSTINCT, Hensleigh says to say. Brain per se thinks i
537m075 alum, so more slowly does animal leave off <t> INSTINCT, when attended with bad effects Martineau. Ho
537m076 &c, being social animal, & this conscience or INSTINCT may be most firmly fixed, but it will not pre
543m101 Effect in Bones is valuable it shows that new INSTINCT can originate.-- strong argument for brain br
543m101 ument for brain bringing thought, & not merely INSTINCT, a separate thing superadded.-- we can thus t
544m104 ear-- & not death for the pain.-- How was this INSTINCT gained.? by conversation-- ∴ modified in thos
546m110 es in either state.-- does this throw light on INSTINCT, showing what trains of action may be done un
550m124 - A man, who perfectly obeys his conscience or INSTINCT, would probably feel but little that of anger
555m142 oises) have not charity-- is it in former case INSTINCT to destroy contagious disease.-- (Useful to u
555m142 troy contagious disease.-- (Useful to use term INSTINCT, when origin of heredetary habit cannot be tr
558m151 many new relations from language.-- the social INSTINCT more than mere love.-- fear for others acting
563n001 tances of expression.-- Octob. 3d. Dog obeying INSTINCT of running hare is stopped by fleas, also by
564n003 rant reason to any animal with social & sexual INSTINCT «& yet with passion» he must have conscience-
564n003 l instincts different.--Mem. Bee how different INSTINCT a solitary animal still different.-- ‖ Differ
564n004 ate effects of actions.→ till at last he face «INSTINCT of» hunger, «of» death & for the satisfaction
564n004 deity effect of reason acting on (<not social INSTINCT>) but a causation. & «perhaps» an instinct of
564n004 ial instinct>) but a causation. & «perhaps» an INSTINCT of conscience, feeling in his heart those rul
566n011 ining <that> «a» faculty in the form of a true INSTINCT, which is a real instinct in the chicken, jus
566n011 n the form of a true instinct, which is a real INSTINCT in the chicken, just bursting from egg.-- Ani
567n015 nning to beat-- children inherit it <ins> like INSTINCT, preeminently so-- who can analyse the sensat
568n017 us shepherd dog. has pleasure in following its INSTINCT & pain if held.-- if tempted not to follow it
568n017 rforming heredetary habit, (or moral sense, or INSTINCT,) one feels pain, & vice versa pleasure in pe
571n029 a» & America manage;-- This shows doctrine «of INSTINCT» has been carried too far In all the foregoin
572n033 efective brains.-- Erasmus does not liken term INSTINCT to muscular movement.-- say instinctive actio
574n043 hard to define.-- Consider the acquirement of INSTINCT by dogs, would show habit.-- Take the case of
575n044 all Have we somewhat right to deny identity of INSTINCT.-- Habits import to Bowen No one doubts that
575n046 eas will not enter. is something analogous. to INSTINCT, to the permanence of old heredetary ideas.--
576n048 learn Greek, that to have it handed down as an INSTINCT.-- Instinct is a modification of bodily struc
576n048 that to have it handed down as an instinct.-- INSTINCT is a modification of bodily structure «(conne
576n048 & intellect is a modification of <intelect> « INSTINCT»-- an unfolding & generalizing of the means b
576n048 olding & generalizing of the means by which an INSTINCT is transmitted.-- Arguing from man to animals
582n067 tween the means by which an animal performs an INSTINCT, & its impulse to do it.-- [the means must be
582n067 nimal may so far be said to will to perform an INSTINCT that it is uncomfortable if it does not do it
582n068 .-- The Revd. Algernon Wells Lecture on animal INSTINCT. 1834: p. 15. "To act from instinct is to be
582n068 on animal instinct. 1834: p. 15. "To act from INSTINCT is to be guided to the performance of a numbe
583n069 Animals possess strong imitative faculty: pure INSTINCT is not imitative: imitations seems invariably
583n070 is again assumed as more allied to reason than INSTINCT.) Mr Wells I can see mentally refers by reaso
583n070 .-- p. 20. Animals may be called "creatures of INSTINCT" with some slight dash of reason so mean are
584n071 en a species is found which does not climb CD[. INSTINCT may be divided into migration,-- subsidiary t
584n072 strings «in the dark» as well might be called INSTINCT.-- migrating to one spot, this is indeed inst
584n072 tinct,-- migrating to one spot, this is indeed INSTINCT.-- Australian man, may be called instinctive:
585n077 would not find its way back.-- ?? this is not INSTINCT, but a faculty, or sense-- "We know not how,
585n077 "We know not how, stonge henge raised, yet not INSTINCT, but if all men placed stones in same positio
585n077 en placed stones in same position, it would be INSTINCT-- instinct is heredetary knowledge of things
585n077 tones in same position, it would be instinct-- INSTINCT is heredetary knowledge of things which might
585n077 instruments to make toothed wheel. he might by INSTINCT make watch, but he does it by reason & experi
586n078 bit.-- so bird migrating to certain quarter is INSTINCT, but his knowledge of that quarter, is facul
586n080 which clearly ought to be separated-- We apply INSTINCT to one part. or another-- but (an instinctus
588n090 man & child.--)» what differs-- not <reason> «INSTINCT», for its character is invariability.-- if ex
591n099 ft to lament) is owing to the feeling that the INSTINCT of self-preservation is disobeyed-- I often h
591n100 man.-- a man eating what others by habit (not INSTINCT) think not fit, as cannabalism, is held in ab
591n101 Religion.» however, he seems to allow it is an INSTINCT.» I suspect the endless round of doubts & sce
593n105 ued.-- A dog turning round & round is some old INSTINCT <perverted> handed down & down.-- mem. Nina u
593n107 asure in» other animal, (especiall as in some <INSTINCT> «insects» which become in imago state social
594n112 ze in the air at a distance.-- How can such an INSTINCT arise?? «it would appear that an instinct lon
594n112 h an instinct arise?? «it would appear that an INSTINCT long remains, if no steps are taken to eradic
594n112 rabbits were not frightened at a dog.-- » The INSTINCT against man is perhaps, as strong as against
594n112 have fled equally if man had appeared-- though INSTINCT so firmly implanted, birds soon <dis> learn t
594n115 ank] Circumstances having given to the Bee its INSTINCT is not <more> less wonderful than man his int
600o08v us to imperiousness of Conscience: in Maternal INSTINCT domineering over love of Master and sport &c
600o08v -- The Bitch does not so act, because maternal INSTINCT gives most pleasure. but because most imperio
602o011 This power of the mind, faintly approaches to INSTINCT How strange it, that Nature should have so li
602o11b ut you do not convince me.--" Belief allied to INSTINCT.-- p. 134. a painted must not a actors, or a
609o30v instincts generated? The origin of the social INSTINCT «in man & animals» must be separately conside
610o30v oral sense, is strictly analogous to change of INSTINCT amongst animals.-- Jan 13th. 1839 My father r
613o036 C] 5) Kirby thinks that <all> there is one one INSTINCT to all animals modified according to species.
613o036 --- Agrees with ONE animal [RHC] Kirby extends INSTINCT to plants, but surely instincts imply willing
614o037 ciousness, therefore reward in good life [RHC] INSTINCT appear like heredetary memory; but first memo
614o037 dual-- [LHC] Perhaps even the most complicated INSTINCT might be analysed into steps, as species cha
614o037 of these instinctive feelings. [LHC] As sexual INSTINCT comes on late in life, man almost alone in th
614o037 fe, man almost alone in this case can perceive INSTINCT boy takes delight in mammae before any reaso
615o037 old him this distinctive mark, it is downright INSTINCT, leading to touch a particular organ.-- I thi
615o038 ed.-- [LHC] NB. Two dogs having very different INSTINCT always obtain peculiarities of external confi
619o042 Moreover <the> any action in accordance to an INSTINCT gives great pleasure, & <an> «such» actions b
620o044 hat when man afterwards thinks why was such an INSTINCT not followed for a pleasure now though so tri
620o044 more wisely other time, for he knows that the INSTINCT. (or conscience) is always present (which is
620o045 lways admire the habit formed by «obediance to INSTINCT» <conscience>., or rather the strengthened in
620o045 nct» <conscience>., or rather the strengthened INSTINCT, even when our reason tells-- + us the action

Page	**	(Key Word)	**
620o045	tion.-- This shows, that our feeling, that the	INSTINCT	ought to be followed is a consequence of that
622o049	ir. J. says is so ridiculous. [RHC] the social	INSTINCT	may be combined with feeling towards one as a
623o050	ce. & yet place calls up pleasure.-- [LHC] the	INSTINCT	of sociability & sociability, doubtless grow
623o051	onditions) (for pig will not so readily attain	INSTINCT	of pointing as a dog.-- also age has much inf
624o051	er simply put it, beneficial tendency in every	INSTINCT	to the species in which it occurs. [or, more
624o051	sooner than other, rather than change hunting	INSTINCT.	*Our tastes in mouth by my theory are due to
626o052	'p. 333 «& p. 377» some remarks. showing that	INSTINCT	cannot be said to guide will. as bird buildin
627o053	is-- but the pleasure a dog has in obeying its	INSTINCT,--	as young pointer to point-- clearly shows
628o53v	moral obligation? [2] [The improvement of the	INSTINCT	of a shepherd dog, is strictly analogous to e
628o53v	associated with the approving or disapproving	INSTINCT--	which were not originally, if the shepherd
628o53v	ere not originally, if the shepherd dog had no	INSTINCT	to commence with scarcely possible to teach i
628o054	bituated to associatical, the emotions of this	INSTINCT,	with that line of conduct, & if taught right
628o054	good, that is, the same cause, which gives the	INSTINCT.--]CD	p. 22. says affections, desires, & mora
629o54v	aternal affections (in a dog. & therefore not	<INSTINCT>	«conscience») equally <prefe> destroy all wi
638j58v	, instead of parts of body.-- Now we know what	INSTINCT	is-- consider this I look at every adaptation
638j059	as well as gain it. by habit.-- New theory of	INSTINCT,	returning to Kirby's view.-- Macculloch. Att
585n074	bably not only experience,, but also «by an»	INSTINCT<ING>	«which is only present in youth» (Mem. Mr
211b161	Animals have no notions of beauty, therefore	INSTINCTIVE	feelings against other «for sexual ends» spe
286c154	as debased his Nature «& violates every best	INSTINCTIVE	feeling» by making slave of his fellow black
289c161	nothing to do with hydridity,, is simply, an	INSTINCTIVE	impulse to keep separate, which no doubt be
290c163	edetary ambling horses, (if not looked at as	INSTINCTIVE)	then must be owing to heredetary power of M
292c171	nsciousness & will & therefore may be called	INSTINCTIVE.--	But why do some actions become heredetary
292c171	But why do some actions become heredetary &	INSTINCTIVE	& not others.-- We even see they must be don
293c173	front. &c or use of button holes it would be	INSTINCTIVE.--	My view of instinct explains its loss ¿ i
298c190	kind to children,--<get> «increases» general	INSTINCTIVE	feeling.-- There is great difficulty in Maki
300c197	ing death it is but being motionless. How is	INSTINCTIVE	dread it is exceedingly doubtful whether ani
363d102	ft open to them,-- ‖. In singing birds, part	INSTINCTIVE	& part acquired,-- thus Yarrel has Lark & Ni
377d139	r monkeys make st.-- noise‖ In case of woman	INSTINCTIVE	desire may be said to be more definite than
388d173	which Frogs copulate & fish shows how simply	INSTINCTIVE	the feeling of other sex being present is--
445e158	Sir. J. Banks. p. 212.-- p. 282. Allows this	INSTINCTIVE	power in chicken, yet says it is evidently a
466t095	pursuit having PUPPIES with the same powers	INSTINCTIVE	& doubtless not confined to sex.-- «Is not c
521m008	might be compared to birds singing, or some	INSTINCTIVE	<or> sounds.-- Miss C. memory cannot be call
526m027	various knowledge which is not heredetary &	INSTINCTIVE)	& the others <are> learnt. what they teach
527m032	ate of nature.-- Singing of birds, not being	INSTINCTIVE,	is heredetary knowledge like that of man, &
527m032	same quarter for many years]CD.-- Beauty is	INSTINCTIVE	feeling, & thus cuts the Knot:-- Sir J. Reyn
531m051	nt passions they go into, shows how truly an	INSTINCTIVE	feeling, <may not pa.> In reflecting over an
531m053	t from wish to avoid some danger-- but it is	INSTINCTIVE	because Nancy tells me very young babies sta
532m053	ren wink. it is instinct Fear must be simple	INSTINCTIVE	feeling: I have awakened in the night. being
533m059	tincts which full grown men can experience--	INSTINCTIVE	walking of animals. that is the ready moveme
534m061	, cannot think-- as well as animal born with	INSTINCTIVE	knowledge.-- but if so, yet this knowledge a
534m063	ad at once made this leap it would have been	INSTINCTIVE,	seeing that time is lost & endeavours made
541m089	hich is connected with death!-- How has this	INSTINCTIVE	fear arisen? 19th. When I went down to Wooll
542m096	gh's) smile & frown, who can doubt these are	INSTINCTIVE--	child does not sneer. because no young ani
549m118	easure of such thoughts We give no credit to	INSTINCTIVE	feelings.-- for man losing his children, any
550m127	tion «ie heredetary», does one not call them	INSTINCTIVE	emotions?-- Dr Holland remarked that insanit
558m150	ting> «yet being obscurely guided» or strong	INSTINCTIVE	sexual, parental & social instincts, giving
565n007	tance, mouth is placed open.-- Hence becomes	INSTINCTIVE	to fear., as ears down to horse.-- Horse snu
568n016	otions follow as consequences on habitual or	INSTINCTIVE	assent to propositions, which are the result
572n033	en term instinct to muscular movement.-- say	INSTINCTIVE	actions. senses. notions &c Octob 30th-- Dre
576n047	uired» this sense of right (& Whether wholly	INSTINCTIVE	as in the dog, or chiefly habitual as in man
581n063	name, writing for the surname,, analogous to	INSTINCTIVE	memory, & consequently instinctive action.--
581n063	logous to instinctive memory, & consequently	INSTINCTIVE	action.-- Sir. J. Sebright. has given the ph
583n069	".-- no p. 17. Contrast the invariability of	INSTINCTIVE	powers in individuals of the same species wi
583n069	reasoning power in one species man.-- false	INSTINCTIVE	pointing varies.-- p. 18. Animals possess st
583n070	n & education may be used as argument.-- for	INSTINCTIVE	knowledge is not gained by instruction, or i
584n072	ed instinct.-- Australian man, may be called	INSTINCTIVE:	the facts of memory of roads long after onc
586n079	bit of knowledge of points of compass may be	INSTINCTIVE.	it is a test to know how much of the wonder
586n079	or afterwards.-- child sucking whole wonder	INSTINCTIVE.--	carrier pidgeon just as wonderful in old
586n081	confusion.--) Hensleigh considers breathing	INSTINCTIVE,	certainly heart beating may be considered a
586n082	this faculty <may possibly be> «probably is»	INSTINCTIVE,	namely the knowledge of size is merely judg
588n089	the Hand p. 191 Says <childr> babies have an	INSTINCTIVE	fear of falling.-- &p. 193. that they percei
591n099	incts to woman is not followed; good case of	INSTINCTIVE	conscience.-- Why does not man eating cause
591n100	ause disgust, because he does not go against	INSTINCTIVE	feeling, only does not fullfil, like contine
593n107	mell or look, & therefore there must be some	INSTINCTIVE	feeling which is pleased by other animals sm
595n125	. frowning, pouting, «smiling», just as much	INSTINCTIVE	as a bull <tr> calf, just born butting, or y
606o024	statue of beauty, is of the "beau ideal", my	INSTINCTIVE	impression 1) September 6th. 1838 Every acti
608o026	eldom analyse his motives (originally mostly	INSTINCTIVE,	& therefore now great effort of reason to d
609o29v	ociety.-- Civilization is now altering these	INSTINCTIVE	passions--, which being unnecessary we call
609o030	what was necessary for good at all» is the	«INSTINCTIVE»	moral senses: (& this alone explains why ou
612o035	. These +willings are common to every animal	INSTINCTIVE	and unavoidable.-- +Can the word willing be
614o037	sily by reason.-- Conscience is one of these	INSTINCTIVE	feelings. [LHC] As sexual instinct comes on
621o046	, & experience only will teach him, that the	INSTINCTIVE	feeling in its nature being always present.
622o048	ociated «habitual» feelings. become like the	INSTINCTIVE,	ones, <which either lead to actions or not,
622o049	three are as simple as I have said.-- [LHC]	INSTINCTIVE	fear of death: of hoarding.. Ld. Kames, whic
622o049	-- [LHC] p. 254. &c &c [RHC] But the love is	INSTINCTIVE,	& how does it apply to mother loving child,
623o050	of» animals, in which case it undoubtedly is	INSTINCTIVE.	But does not Hartley explanation apply perf
623o051	cial tendency during past races could become	INSTINCTIVE.--	[LHC] x It is probably That becomes insti
623o051	ctive.-- [LHC] x It is probably That becomes	INSTINCTIVE,	which is repeated under many generations. (
624o051	it can be shewn, what things easiest become	INSTINCTIVE,	this part of argument fails, or rather is w
624o051	o some extent.» Hence this is the law of our	INSTINCTIVE	feelings of right & wrong,-- education, of p
624o50v	ared> «perceived» the comparison between our	INSTINCTIVE	feelings & our short lived Passions' State b
626o052	ll. as bird building nest, but supplies it--	INSTINCTIVE	feelings will doubtless lead to similar acti
626o052	edetary action on body.-- This feeling, when	INSTINCTIVE	will lead to action.-- the passion rising fr
628o53v	essities «& good» of society such conduct is	INSTINCTIVE	in me (& as a consequence, but not cause giv
629o055	s 5) if so, it is perhaps deviation from the	INSTINCTIVE.	right & wrong.-- (animals excepting domesti
629o055	mesticated ones have no right & wrong except	INSTINCTIVE	ones) Perhaps my theory of greater permanenc
586n081	inding its way, which in certain species is	INSTINCTIVEDLY	«not least by experience» directed to cert
528m037	ses very weak.-- (2d) form. some forms seem	INSTINCTIVELY	beautiful «as round, ovals»;-- then there t
533m058	hat birds learn to sing & do not acquire it	INSTINCTIVELY.	may not this be connected with their power
534m063	flies from intellect. but it does it always	INSTINCTIVELY	or habitually.-- good Heavens is it dispute
566n011	owly,, but when in grown years, thinking he	INSTINCTIVELY	knows distances:, is good instance of obtai
585n077	ls, which is faculty, they use this faculty	INSTINCTIVELY;	watchmaker has faculty by his instruments
586n078	not apply to dogs.--» they may do all this	INSTINCTIVELY	«yes because power varies in breeds,» somet
586n081	earnt by experience is in other is acquired	INSTINCTIVELY"	So with <sight> sight-- so a Bee has the f
586n082	from mere impulse to save wax.]CD which it	INSTINCTIVELY	exerts in concert with others in building c
621o048	or themselves & others, (as the parents are	INSTINCTIVELY	benevolent) they will teach to be wrong or
629o55v	other question whether any thing is taught	INSTINCTIVELY;	I say yes, & my explanation agrees. with l
165g126	etrao-- not an American form The union of two	INSTINCTS	crossing most remarkable ever obsesved? Shows
189b074	as highest.-- A bee doubtless would when the	INSTINCTS	were.-- relation of type in two countries dir
211b161	or sexual ends» species, whereas Man has such	INSTINCTS	very little. in Zoolog. Proceedings. Jan 1837
227b228	parative Anatomy, & it would lead to study of	INSTINCTS,	heredetary. & mind heredetary, whole metaphy
262c072	l dispute--» says civilization heredetary; ie	INSTINCTS	of wisdom virtue? «like senses of savages» (H
263c077	tion.-- he possesses some of the same general	INSTINCTS,	<as> & <moral> feelings as animals.-- they o
263c078	asoning powers in excess. instead of definite	INSTINCTS.--	this is a replacements in mental machinery
280c134	tary Young wild ducks.-- lose as well as gain	INSTINCTS.	Wild & tame rabbits good instance-- instinct
280c134	nstincts. Wild & tame rabbits good instance--	INSTINCTS	of many kinds in dogs, is clearly applicable
280c134	n dogs, is clearly applicable to formation of	INSTINCTS	in wild animals, many species in one genus--
285c150	&c Dr. S. has some remarkable crochets about	INSTINCTS.	whenever instinct is mentioned some definiti

299c196 n to mans mind in Vertebrate Kindgdom in more INSTINCTS in rodents than in other animals & again in M
300c198 divided body. Up 3. language seems to supply INSTINCTS,-- & those powers which allow of, acquirement
301c199 ture,.. habits precedes structure,.. habitual INSTINCTS precede structure.--duckling runs to water. b
305c210 here is no progression in the developement in INSTINCTS in the <classes> <orders» of insects, so is t
306c212 us, bud & bud, polypus & germ plant & seed.-- INSTINCTS in young animals, well developed, just like,
312c232 <ins> beccause actions are constant they are INSTINCTS, & not ·: instincts, constant. ¿ whether mutil
312c232 ions are constant they are instincts, & not ·: INSTINCTS, constant. ¿ whether mutilations non-heredata
315c240 of Malay fowls p 5. (note) in some papers on INSTINCTS L.' Institut «1838» p. 184 Botany of Bonin. "
324c268 Sciences. Nat. in Geolog Soc.» F.. Cuvier on INSTINCTS L. Jenyns paper in Annals of Nat. History Pri
342d033 caped from Carolina?) off New York. therefore INSTINCTS not imperfect.-- Are Pheasant & Grouse homoge
343d036 f harmony keep perfect in these themselves.-- INSTINCTS alter, reason is formed, & the world peopled
366d108 ., or at least show repugnance to breeding if INSTINCTS unchanged, & if their characteristic qualitie
377d140 ly, measure the rapidity of change of form, & INSTINCTS in the animal kingdom.-- It is the unit of ou
377d147 re not really fitted for water, only habits & INSTINCTS-- The young of the <p> Kingfisher (.p. 169) h
380d154 this is full of interest; for it shows latent INSTINCTS even in brain of male.-- Every animal surely
384d162 y man & woman is hermaphrodite:-- ∴ developed INSTINCTS of Capon. & power of assuming male plumage in
392d180 s it with China & Common Geese «how are their INSTINCTS?» «Chineses & Common Pigs.--» Experimentize o
409e049 ing with certainty on other» hence not social INSTINCTS, which as I hope to show is «probably» the fo
414e064 y which have the best fitted organization, or INSTINCTS (ie intellect in man) to gain the day.-- In m
439e143 rossing, in plants proves how much depends on INSTINCTS in animals.-- yet the existence of wild close
458t002 of natural substances-- useful perversion of INSTINCTS-- Beechey's Voyage Vol I. p. 499. «4to. Edit»
494q004 em with other breed.-- (12) About the blended INSTINCTS Remote Experiments-- Plants Raise seedlings s
515q022 s.-- (3) Hounds-- varying-- (4) About blended INSTINCTS of the geese which she crossed; especially if
521m007 on to another, also without consciousness, as INSTINCTS are, is not so very wonderful.-- <Now is not
522m013 ted, whether rather not going against natural INSTINCTS-- My Grand F. thought the feeling of anger,
525m023 honour.-- they feel pleasure in obeying their INSTINCTS naturally.--(generority in defending a frien
533m057 ieve, in Materialism, say only that emotions, INSTINCTS degrees of talent, which are heredetary are s
533m059 -- this may be nearest approach to <the> such INSTINCTS which full grown men can experience-- Instinc
536m073 al argument fixes on heredetary disposition & INSTINCTS-.-- Put it so.-- Probably some error in argu
537m075 ore wonderful than dogs should have different INSTINCTS-- Fact most opposed to this view, where the
538m081 on the brain, renders much less wondefful the INSTINCTS of animals-- Aug. 12th. 38. At the Athenaeum
543m098 uch to each other.-- Waterhouse says far more INSTINCTS in all of the Hymenoptera;. <therefore> than
543m101 ng distinct resemblances.-- The facts of half INSTINCTS. when two varieties are crossed as in Shepher
549m122 it, will add to happiness.-- Men having some INSTINCTS as revenge «& anger», which experience shows
549m122 itioned as they are effecting a change in his INSTINCTS-- like what is happening with other animals--
549m123 ng-- the mind of man is no more perfect, than INSTINCTS of animals to all & changing contingencies, o
558m150 strong instinctive sexual, parental & social INSTINCTS, giving rise "do unto others as yourself". "1
564n003 not have been same with mans because original INSTINCTS different.--Mem. Bee how different instinct a
573n036 ted at once by special act, provided with its INSTINCTS its place in nature. its range, its-- &c &c:-
574n042 r ones, may be applicable to the formation of INSTINCTS, independently of habits.-- the limits of the
575n044 e facts about crossing races of dogs on their INSTINCTS, most important, because they obey the same l
576n048 ected with locomotion.)» «no, for plants have INSTINCTS» «either» to obtain a certain en<s>«d»: & int
583n069 ngness of fox, industry of bee &c &c-- p. 15."INSTINCTS act with unerring precision".-- no p. 17. Con
583n069 never seen their parents offer best cases of INSTINCTS].CD all this may be true,, but relation of im
584n071 some way be able to measure the cell; p. 22. INSTINCTS & structure always go together: thus woodpeck
584n071 er: thus woodpecker: but this is not so,, the INSTINCTS may vary before the structure does; & hence w
584n071 iary to food & temperature molting & breeding INSTINCTS, sexual, social, «subordinate to,» self prese
584n072 is inspiration."-- this is class of so called INSTINCTS to which my theory no way applies.-- it is th
586n082 g comb-- My faculty often will turn out to be INSTINCTS, & so in some senses, is sight--CD [The facul
589n091 to 29».-- Habits becoming heredetary form the INSTINCTS-- almost identical with my theory
589n091 trong argument in favour of brain forming the INSTINCTS,-- could brain make a tune on the pianoforte,
591n099 l cases) are held in abhorrence it is because INSTINCTS to woman is not followed; good case of instin
591n100 ctions with <&> against benevolent & parental INSTINCTS very clear.-- even to the cold or benevelo <
592n105 - hence origin of expression-- There are some INSTINCTS unintelligible, <both> in the end gained «& t
595n125 ping.-- these I think are better instances of INSTINCTS (highly useful as only means of communication
596nIBC lects easily fall into habits Get facts about INSTINCTS of mongrel dogs Do blubbering children, if of
609o030 e, any more than a hive of Bees without their INSTINCTS.-- Gives art to when I say How social instinc
609o030 stincts.-- Gives art to when I say How social INSTINCTS generated? The origin of the social instinct
610o033 his not almost a question whether we have any INSTINCTS, or rather the amount of our instincts-- sure
610o033 ve any instincts, or rather the amount of our INSTINCTS-- surely in animals according to usual defini
613o036 Kirby extends instinct to plants, but surely INSTINCTS imply willing, therefore word misplaced The m
613o036 e made out Reason Will Consciousness Definite INSTINCTS being acquired, is a most important argument,
614o037 be so if Lamarck's theory true [RHC] Acquired INSTINCTS, analogous «(& replace)» to experience gained
614o037 eas.-- As man has so very few (in adult life) INSTINCTS.-- (this loss is compensated by vast power of
614o037 ast power of memory, reason &. & many general INSTINCTS, as love of virtue, of association, parental
614o037 The very existence of mankind requires these INSTINCTS,: though very weak so as to be overcome easil
615o36v ping good show acquirement or obliteration of INSTINCTS But habits acquired even by <children> «plant
615o038 by <children> «plants»! [RHC] 7) As definite INSTINCTS modified by heredetary;-- so succession so pe
615o038 nd as all men nearly same species, so general INSTINCTS nearly same; which same argument probably app
615o038 same argument probably applies to particular INSTINCTS of animals. even in wild state; certainly to
615o038 es of external configuration. [RHC] General-- INSTINCTS, certainly appear a sort of acquired memory.
619o042 ded that he has parental, conjugal and social INSTINCTS, and perhaps others.-- [LHC] ----- p. 113. Ma
619o042 by his habits, as <if> another animal. These INSTINCTS consist of a feeling of love <and sympathy> «
619o043 er man <say go> «acting in» accordance to his INSTINCTS, <he would know that many experienced pleasur
619o043 er received.-- If either man did not obey his INSTINCTS from interference of passion. he would feel p
620o045 s should «almost» always be sacrificed to the INSTINCTS.-- One does not feel it wrong in very youn
620o045 ge, which should show the child, which of its INSTINCTS are best to be followed.-- Yet even at this t
620o045 shows a bad child.-- Hence there are certain INSTINCTS pointing out lines of conduct to other men, [
620o046 py, we <do> try to teach him & strengthen his INSTINCTS.-- so man ought to follow certain lines of co
621o046 some passion.-- If his passions strong & his INSTINCTS weak. he will have many struggles, & experien
621o047 ing that almost (rarely if opposed to natural INSTINCTS) any action is either right or wrong.-- [RHC]
622o048 in tastes of the mouth]CD» [LHC] My theory of INSTINCTS, or heredetary habits fully explains the ceme
622o048 fully explains the cementation of habits into INSTINCTS & associations.-- often feelings which do not
622o048 e parts of our nature, <sub> subject to their INSTINCTS.-- I am far from saying there are not more, o
622o049 C] 8) in the beginning I mentioned only three INSTINCTS, as we see in dogs & pidgeons.-- But as man i
622o049 s animal at head of series in which «special» INSTINCTS decrease, I should think they were very few &
622o049 » any other might he grafted.-- Origin of the INSTINCTS Hartley, (according to Sir J) explains our lo
623o050 xplanation apply perfectly to origin of these INSTINCTS.-- the having received pleasure from some one
623o050 feelings.-- [LHC] According to my theory, all INSTINCTS demand some explanation [RHC] Although I cann
623o050 cannot pretend to say how far & minutely our INSTINCTS extend, yet as they are acquired by social an
623o051 ss the beneficial tendency [RHC] 10) that the INSTINCTS of bees & beavers «& deer» have <been formed>
624o051 ation, education changes, & probably likewise INSTINCTS, for the same law effects both.-- <such> chan
624o051 icial tendency" will most readily affect. the INSTINCTS for they are in accordance with it. thus a d
625o50v ved, that our passions too strong for our INSTINCTS. to gain long-lived good, ie happiness-- yet
628o054 e to greater duration of impression of social INSTINCTS, than other passions, or instincts.-- is this
628o054 of social instincts, than other passions, or INSTINCTS.-- is this good?-- I should think some parts
629o055 aps my theory of greater permanence of social INSTINCTS explains the feeling of right & wrong.-- arri
543m098 udy Kirby with this view) therefore there is INSTINCTUAL developement in one order, as there is Intel
586n080 ly instinct to one part. or another-- but (an INSTINCTUS means stained in?). had better refer to to t
088a015 nverted into crystals of Hornblende p. 248. L. INSTITUT 1837.-- Helms remark on common salt being fou
089a020 . -- Bed, of elevated shells on the Senegal. L INSTITUT P. 192.-- (1837. Peninsula of Cape Verd. volc
091a026 able; because transportal of Minute seeds-- L. INSTITUT. p. 209. May. 1837 Paper by Humboldt on Quito
093a031 cted rock being determined by fusibility in. L INSTITUT p 247. 1837.-- The most infusible first injec
093a032 e surface. before being cooled.-- Berzelius. L'INSTITUT. [1837 p. 297]CD thinks Olivine a preexisting
093a032 . Galapagos ∴ Basalt deepest?? Marcel Serres L'INSTITUT. 1837. p 331 Considers that Mercury & Sulpure
096a039 ion to amount of 30 ft.-- Mr Bollaert (at Roy. INSTITUT) talks of quantities of shells at Iquique. <C
097a042 ther.-- on Direction of mountains in Brazil L.'INSTITUT No degree 221 Lamellar dikes like Mica Slate

117a102 n land. bringing up pebbles 2 inches long?-- L'INSTITUT. 1838 p. 151. Formations of Payta extend clos
118a103 h --subject of fine paper this would make.-- L'INSTITUT. (1838) p. 216 M. Gay on the Geology of Chile
123a115 ere lamination appeared.-- Lyells Denmark -- L'INSTITUT (1838) p. 268. Paper by Humboldt on Bogota.
131a135 temp du globe on Volcanos &c worth reading. L'INSTITUT. 1838 p. 360. on orbicular trap thought to be
131a135 orbicular trap thought to be bombs submarine L'INSTITUT 1838 p. 400. Observations on Mountains of the
194b094 inct more. reason less. so will aversion be L. INSTITUT «1837. No 246» a section of fossil "singe" i
203b135 me however might have offspring, & then «V. L. INSTITUT p 245. 1837» we should have anomalies. as Cap
203b136 the Ganges & perhaps even in India p. 261. L. INSTITUT. 1837 Mem. Sir F. Darwin cross breed boars we
206b150 rtain shell cephalopoda.-- «Read Buckland» L'.INSTITUT. «1837.» p 319 - Brongniart.-- no dicotyleden
207b151 ered by Ehrenbergh.-- «Marcel Serres p. 331. L' INSTITUT-- considers that mercu» Geo. Joun. p. 325. Vo
211b163 ed by them, whilst they eat up the dogs.-- L.' INSTITUT. Curious paper by M. Serres on Molluscous ani
212b165 o this, so was obliged to be removed.-- In L.' INSTITUT. 1837: p. 404. account of instinct of dogs.--
221b200 t no confined species in Scicily. Jan: 1838 L'.INSTITUT. Bats, in Eocene beds, very like present spec
241c015 ill hereafter know about birds of N. Zealand L'INSTITUT. 1838. A Dipus. & other rongeur in Australia.
267c095 sie breed no description given-- Ch. 2. dogs L'INSTITUT. 1838. p. 67. Australian rodents Abstract of
283c146 , there are fossil (see Scientific Memoirs & L'INSTITUT.) that there are Tertiary fossil Infusoria, o
295c183 s ornithology X p. 12 do. excepting salmons L' INSTITUT. Sorex from Mauritius. p. 112. & paper on gen
303c205 bold. I will venture to say unphilosophical. L' INSTITUT 1838. p. 128. Extraordinary genus. Mesites bi
311c227 ries-- this is well worthy of investigation.-- INSTITUT 1838. p. 174. Apercu very good on insectifero
315c241 ls p 5. (note) in some papers on instincts L.' INSTITUT «1838» p. 184 Botany of Bonin. "grande analog
319c257 ins) have black nipples to their breasts.-- L' INSTITUT, 1838, p. 230 says the Macrotherium of Europe
347d049 of ancient Caledonian Cattle. Ch 3. Instinct L'INSTITUT p. 249. (1838). Eggs discovered to Taenia.--
354d062 in such rigmaroles about analogies & number L'INSTITUT p. 275. (1838) M. Blainville has written pape
354d064 ies of dogs cross, Erasmus says it look lik[e] L'INSTITUT. 1837. p. 351. Paradoxurus Phillippensis. Phi
404e025 fect of climate.-- Octob 19th. When reading. L'INSTITUT. 1838. p 329-- Milne Edwards, description of
405e035 f antiquity of reasonable cosmopolite man.-- L'INSTITUT. 1838. p. 338.«V[ide]» Important account of c
406e035 America. & many very close: see full paper. L'INSTITUT. 1838. p 338 A most grave source of doubt. in
407e042 iana Deer very close to a Molucca species.-- L'INSTITUT. 1837. p. 253, on animals of Antilles.-- (see
412e058 ions tend to accumulate. «on any structure.» L'INSTITUT. 1838. p. 384. List of fossil Mamm: from Pola
416e072 s are only ancient & perfectly adapted races L'INSTITUT. 1838. p. 394. Rhinoceros «tichorhinus» in Par
418e085 anges, as has any species.-- Decemb. 21th.-- L'INSTITUT. 1838. p. 412. M. Eichwald has published Fauna
420e088 dations, hitherto confounded,. of geology.-- L'INSTITUT. 1838. p. 414; M. Guyon thinks Monsters more c
420e088 ope especially with Europaeans settled there L'INSTITUT do. p. 419, «long» account of Hyaenodon, a fo
424e100 ernating strata of sand & limestone &c &c.-- L'INSTITUT.-- p. 290-- admirable paper on geographi
480z006 On the passeres of S. America. D'.Orbigny. L'.INSTITUT. No.-- 221 Good account of Condor by Humboldt
480z007 ion of Bonn Society M. Edwards on Corallines L'INSTITUT 1837. No 212 Observations on the Raptores of
484z014 ad back M Edwards,--on polypi of Tubulipores L'INSTITUT-- 1838 p. 75 A detailed comparison of product
452z018 e Edwards p. 138 on Polypi.-- Berenica &c &c L'INSTITUT, 1838 p. 46 Maclway Horae Entomolog. insects
572z033 any bodily harm-- case of habitual action.-- L'INSTITUT. 1838. p. 340. Mr Carlyle says that negro cer
106a076 ica ¿ effect of subsidence-- <Is there same.> INSTITUTE. 1838 p. 40 or Phil Mag. Dec 1837. p. 520 Mr
229b234 of East Indian Rhinoceros (?)-- Some paper in INSTITUTE on range of Bos in India.-- Range of Zebra?--
233b250 ish came from closely neighbouring localities INSTITUTE 1838. p 38. account of fossils of Sewalick «I
222b207 cient; in shells alone can this comparison be INSTITUTE. People often talk of the wonderful event of
310c223 nimals that man has different origin. «Royal INSTITUTION» Dr Royle seems to think Botanical Provinces
431e122 ry. March 20th. Phillips in Lecture in Royal INSTITUTION says shells become less in number. (¿ specie
539m082 ifficult task.-- Aug. 12th. When in National INSTITUTION & not feeling much enthusiasm, happened to g
398e006 page; we find <great> sensible change in the INSTITUTIONS. & we suppose not only revolutions, but ent
313c235 nstance of instinct showing itself, not from INSTRUCTION Even the action of the viscera. under sympat
583n070 - for instinctive knowledge is not gained by INSTRUCTION, or imitation.-- p. 20. Animals may be calle
622o048 will the conscience in these cases.-- Those INSTRUCTIONS, which the child sees uniformly performed b
056r112 renovation & so is Volcano a useful chemical INSTRUMENT.--Yet neglecting these final causes.--What m
454e184 essay on Spermatic animalcule. has described INSTRUMENT for galvanize <ani> them-- Cross Irish & Com
590n093 he, even from carious tooth cured by sight of INSTRUMENT.-- Bennett's Wanderings, Australian Dog does
072r160 gments can be separated from them with steel INSTRUMENTS." In R. Brown (Collect: «of F. W.») where th
585n077 instinctively; watchmaker has faculty by his INSTRUMENTS to make toothed wheel. he might by instinct
025r019 , at the bottom of the sea.--«certainly data INSUFFICIENT, yet good» «(I suspect fragments of shells
132a137 Tom I. p 501.-- shows first that data wholly INSUFFICIENT to calculate rate of increase of heats in e
437e138 s. good case of migrating.-- shows my theory INSUFFICIENT.-- p. 120 An Eagle is said to have been see
618o41v is because our knowledge of matter is quite INSUFFICIENT to account for the phenomena of thought. (T
232b247 igeration in chief part instead of change from INSULAR to extreme climate, <more northern> Iceland wo
348d050 hich prevails throughout the group & serves to INSULATE <them> it".-- i.e what characters chance to b
541m093 hat they are so difficult to conceal.-- a man «INSULTED» may forgive his enemy & not wish to strike h
565n008 n more <emblem> manner of hurting opponent by INSULTING his pride & is therefore of the snarling orde
226b222 if in small tract we have many species, we may INSURE mass continental on many large islands.-- Henc
255c053 r offspring. ,has invented all kinds of plan to INSURE stability; but isolate your species her plan i
446e163 hout crossing.-- & propagation by buds does not INSURE constancy of form.-- is the constancy owing to
191b082 ale animal affecting all the Progeny of female INSURES often mixing of individuals. Here we have avit
553m136 y a separate act of God, & not as a necessary INTEGRANT part of his most magnificent laws. of which w
224b214 should expect most species.-- The difference INTELLECT of Man & animals not so great as between livi
303c204 ize» colour, form of head «& features» (hence INTELLECT?) & what kinds of intellect) quantity & kind
303c204 features» (hence intellect?) & what kinds of INTELLECT) quantity & kind of hair forms of legs-- henc
347d049 te. <i> in vertebrates tendency to improve in INTELLECT,-- if generation is condensation of changes.
414e064 he best fitted organization, or instincts (ie INTELLECT in man) to gain the day.-- In man chiefly int
414e064 ect in man) to gain the day.-- In man chiefly INTELLECT, in animals chiefly organization: though Cont
414e064 in Black Race there gives them preponderance. INTELLECT in Australia to the white.-- The peculiar sku
435e134 a p. 112. Man. "standing alone in the gift of INTELLECT, he resembles, other mammalia in the effects
520m002 er thinks. people of weak minds, below par in INTELLECT frequently <are> have very bad memories for t
521m009 of the <Hall> «Park», after paralytic stroke. INTELLECT impaired. <after paralytic stroke> : . could
534m063 s lost & endeavours made must be experience & INTELLECT.-- do. p. 157. Westwood remarks that some imp
534m063 ying common wasp cuts off wings of flies from INTELLECT. but it does it always instinctively or habit
535m063 vens is it disputed that a wasp has this much INTELLECT. yet habit may make it act wrong, as I have d
539m082 ly, in proportion to variableness or power of INTELLECT.-- Some complicated trades can hardly be cons
549m119 conscience free from offence» -- pleasure of INTELLECT affection excited, pleasure of imagination--
549m123 it odd he should have had them.-- with lesser INTELLECT they might be necessary & no doubt were prese
552m131 est. Pigs & Elephants, (both Pachyderms) much INTELLECT.-- mem: Yarrell's story of wheel horse in dra
576n047 ful a consequence to each man is small. Man's INTELLECT is not become superior to that of the Greeks.
576n047 nal circumstances» Look at Spain now.-- man's INTELLECT might well deteriorate. «CD[in my theory ther
576n048 cts» «either» to obtain a certain en<s><d>: & INTELLECT is a modification of <intellect» «instinct»--
576n048 n en<s><«d»: & intellect is a modification of <INTELLECT> «instinct»-- an unfolding & generalizing of
580n061 ly has observed that some people of very weak INTELLECT (As Miss Clive) have only possessed very loos
580n061 deas of time?-- Characteristic of one kind of INTELLECT is that when an idea once take hold of the mi
580n062 very profound.-- concludes that difference of INTELLECT between animals & men only in Kind.-- probabl
580n062 ptoms of the infinite & progressive nature of INTELLECT indication of better life p. 207 March 16th.--
594n119 nct is not <more> less wonderful than man his INTELLECT Lyell has seen a little dog go to the assista
600o08v t heads. of separation between soul of man. & INTELLECT of beasts, not clear.-- ¿does not Mackintosh
638j58v inct The analogy between the works of art «or INTELLECT» such as hinge, & hinge of shell, works of la
638j58v laws of organization is remarkable-- what is INTELLECT, but organization, with mysterious consciousn
638j58v existing conditions.-- An adaptation made by INTELLECT this process is shortened, but yet analogous,
543m099 not so in punning) are people of very limited INTELLECTS, & in the same way are chess Players-- A man
596mIBC y> face in pashion.?-- cry? Do people of weak INTELLECTS easily fall into habits Get facts about inst
610o032 ts.-- Both ideots, old People & those of weak INTELLECTS.-- Westminster Review. March 1840 p. 267---
189b074 onsider those, where the {cerebral structure INTELLECTUAL faculties} most developed, as highest.-- A
222b207 People often talk of the wonderful event of INTELLECTUAL Man appearing..-- the appearance of insects
224b215 races.-- When all mixed & physical changes (¿ INTELLECTUAL being acquired alters case) other species o
263c074 nkey would breed (if mankind destroyed) some INTELLECTUAL being though not MAN.-- is as difficult to
322c270 life I & 2d & 3rd Volumes Abercrombie on the INTELLECTUAL powers. Hunters Animal OEconomy. edited by
409e046 Holland, ¿Sydney? The dog being so much more INTELLECTUAL than fox, wolf &c &c-- is precisely analogo
415e069 ey probably might, with such chances be made INTELLECTUAL, but almost certainly not made into man.--

462tf05 ion to the different races of men, some more INTELLECTUAL than others-- is incidentally said that a m
534m063 n agency is unknown, with simple exertion of INTELLECTUAL faculty) if ants had at once made this leap
539m081 f all around, yet there was no strain on the INTELLECTUAL powers-- the difference is of a man wagging
543m098 ctual developement in one order, as there is INTELLECTUAL in human-- probably some genera in differen
543m098 nced than others just as dog & Elephant most INTELLECTUAL.-- Hymenoptera typical insects. ie have all
543m100 ment gives a man common sense, & the highest INTELLECTUAL powers of perceiving & classifying distinct
545m108 ssion.-- Habitual actions are the reverse of INTELLECTUAL, there is no comparison of ideas-- one foll
549m120 obtain this happiness. though he sees some «INTELLECTUAL» good men, from insanity &c unhappy-- perha
549m121 is other question.-- little is there said of INTELLECTUAL «ple [...] hope» cultivation, main source o
627o053 this can be resolved into some operation of INTELLECTUAL faculties-- Will Eugenius allow this moral
233b252 tiful savannahs & forests dare to say that INTELLECTUALITY is only aim in this world [not located] T.
233b252 alk of higher orders, we should always say, INTELLECTUALLY higher.-- But who with the face of the ear
299c196 ing unequally developed.-- ¿is not Elephant INTELLECTUALLY developed amongst Pachydermata. like Man a
358d076 ildren & women = "women recognized inferior INTELLECTUALLY"= Opposed to these facts are effects of ca
033r042 ell out of its cast which, although not very INTELLIGBLE is a familiar case: If refiltered with other
264c079 omestication, hear expressive whine, see its INTELLIGENCE when spoken; as if it understood every word
326c266 th Natural Theology.-- on instinct & animal, INTELLIGENCE.-- very good. Endlicher has published in fi
589m091 19.-- Habits more prevalent in proportion to INTELLIGENCE less.-- p. 325 «to 29».-- Habits becoming h
601o08b mé Martin Leroy Lettres. Philosophique sur l'INTELLIGENCE des Animaux-- & Le Parfait Chasseur, par De
055r106 direction are thought by the <oldest> «most INTELLIGENT» miners to be the richest Vol II 147´ Shells
430e117 s in offspring of wild animals.-- grateful & INTELLIGENT.-- The theory that all animals have sprung f
553m138 that expression mean SOMETHING.-- Hunt (the INTELLIGENT Keeper) remarked that he had never seen any
591m097 e skin-- & hair standing on end.-- July 20th INTELLIGENT Keeper... Zoolog. Garden told me. he has oft
227b227 g condensation, test of highest organization INTELLIGIBLE.--may look to first germ-- --led to compreh
436e135 then, uniformity in «geological» formation. INTELLIGIBLE.-- «-- No. but the wandering & separation o
574n041 tion in man.-- it is bare fact, on my theory INTELLIGIBLE An habitual action must some way affect the
522m012 general effect of <early> «slight habitual» INTEMPERANCE.-- often accompanied by extreme anger, at n
560mIBC eat Marlborough St Has my Father ever known <INTEMPERANCE> «disease» in grandchild, when father has n
560mIBC . but where grandfather was the cause by his INTEMPERANCE. <No.> Cannot say.-- Private. Expression M
609o029 e.-- No checks were necessary to the vice of INTEMPERANCE, circumstances made the check.-- to licenti
505q014 ¿heredetary? (15) Abberley has a hooked Pea.-- INTENDS to breed from it and large Asparagus: result?
582n068 ure performing those actions neither knows nor INTENDS the result they will effect,.--" this not whol
592n103 re said to open mouth in fright because nature INTENDS to lay open all senses: <do> Horse prick his e
432e125 water «frowns connected with pain, as well as INTENSE thought.--» No one but a practised geologist c
464tf6v migrated to these mountains, when the cold was INTENSE just like the alpine plants-- In S. America. i
539m081 e Athenaeum Club. was very much struck with an INTENSE headache «after good days work» which came on
540m087 both semi-civilized-- Perhaps one cause of the INTENSE labour of original inventive thought is that n
544m102 good account of «thinks» are recollected when INTENSE, or when so near waking. that an associated is
549m118 Iosopher-- yet the philosopher has a much more INTENSE happiness-- so is it <with an> when same man i
549m121 le [...] hope» cultivation, main source of the INTENSE happiness-- it is again another question, whe
550m124 happiness. not being so desirable as <broken> INTENSE happiness even with some pain,-- compared to w
552m130 mage not vivid as in sleep-- (one can dream of INTENSE scarlet??) is it because one then has no immed
548m117 t ideas passing through mind in given time.-- INTENSITY to degree of <happi> pleasure of such thought
551m127 ars to me vague-- Delirium of every degree of INTENSITY-- «in old man, he had just seen mind went on
618o41v never came untill the blueness had a certain INTENSITY (& the experiment was varied) then might it n
579n058 th grief,¿ bodily pain? frown shows the mind is INTENT on one object.-- With respect to my theory of
565n006 e the grin of the Hyaena from fear, no actual INTENTION to bite at moment, but mere symbol of readine
530m046 ompounded of the involuntary thoughts.-- An INTENTIONALLY recollection of anything is solely by assoc
537m074 on others. It may be doubted whether a man INTENTIONALLY can wag his finger from real caprice. it is
524m022 says, do not dream of what they think of most. INTENTLY.-- criminals before execution.-- Widows not o
539m081 as interested as was I in the other, & read so INTENTLY as to be unconscious of all around, yet there
541m091 ore your mind steadily., & not merely thinking INTENTLY; for that one does with novel for a length of
180b038 ciple that the permanent varieties produced by <INTER> confined breeding & changing circumstances are
338d023 is not perhaps so great, as separation on <be> INTER-breeding, for otherwise we could not understand
340d026 cks, breed one with another-- & hybrids fertile INTER se--No directly against Eyton's rule. ¿Are the
340d026 against Eyton's rule. ¿Are the hybrids similar INTER se.-- [not located] the «4» Struthionidae, Mr B
342d032 common males) & lay many eggs but never produce INTER se or with parent species.-- The hybrids do not
359d087 fertile whereas animals further apart have bred INTER se.-- These hybrids are very wild & take <very
366d108 domesticated animals would cease being fertile INTER se., or at least show repugnance to breeding if
374d134 ept 25th. In considering infertility of hybrids INTER se, the first cross generally brothers & sister
385d163 -- «there is great difference between hybrids & INTER se offspring in latter being unhealthy.--» male
389d173 it, that she was held Hunters Eoeconomy So with INTER-breeding as told by Willis Q Proved facts relat
393dIBC .-- Are hybrids pintail & common ducks. similar INTER se? Zoolog. Gardens Are the hybrids of those sp
417e079 eptiles & those which <lay> «have» their eggs, <INTER>, impregnated externally; nor can it be a neces
447e164 eem to have been more rapid, & the facility for INTER marriage is greater (Hence Diceceous plants hig
515q022 Eyton (1) Number of eggs-- of half-bred geese-- INTER se, & with parents & of Chinese geese. (2) Anat
235b262 here «domesticated» animals separated. & long INTERBRED <p> having great tendency to vary? Is not man
331dIFC St Did Eytons <intermediate>. «hybrids, when» INTERBRED. show any tendency to return to either parent
449e169 law & Walkers of the male giving form-- they INTERBRED. & the young kept constant. & all alike Water
492q003 als If two half bred animals exactly alike be INTERBRED will offspring be uniform.-- Mr Ford Has M. S
432e126 son; which is a beautiful instance of forms, INTERCALATED between two great distinct formations.-- pa
135a145 there little branches at {P} from each side INTERCEPTING plain & dividing it-- Hopkins fissure at {P
219b196 kings. would hardly produce from Incestuous INTERCOURSE. a parallel fact to Blood-Hounds. Before Att
279c133 n crossing race.-- bad effects of incestuous INTERCOURSE..-- excellent observations of sickly offspri
443e155 & some mollusca being hermaphrodite is, that INTERCOURSE every time is of no consequence in that degr
616o038 ght, may be no more «difference» than sexual INTERCOURSE in plants is involuntary, in man voluntary:
198b115 ould «can» be given.-- It is a point of great INTEREST to prove animals not adopted to each country.
209b155 n Esquimaux & European dog? Yet man has had no INTEREST in perpetuating these particular varieties. I
258c061 e of birds from Cuba Vigors.-- nothing of much INTEREST XX. Hence relation of analogy may chiefly be
360d089 in general knowledge of breeders, where their INTEREST is concerned. Same man had crossed Jackal & d
377d139 g, but to philosophic naturalist pregnant with INTEREST» Hyaena. thinks, when pleased cocks his ears.
380d154 often better than a female.-- this is full of INTEREST; for it shows latent instincts even in brain
397e004 nsibly such changes are in progress.-- we feel INTEREST in discovering a change of level of a few fee
410e052 except describers having some high theoretical INTEREST,-- "the great end must be the law & causes of
541m090 ough thinking «& talking» for the moments with INTEREST on each.-- ∴ my father. is right in saying de
545m105 unconscious those which are viewed with little INTEREST, & those which are viewed very often.-- forme
545m105 not give rise to ideas so much. as objects of INTEREST.-- do/ I was much struck with observing how t
563n002 ressions, & wished he had done so & so for his INTEREST, & found he disobeyed a wish which was part o
619o043 ve him pleasure, without any regard to his own INTEREST. likewise if such actions were prevented by f
621o046 resent. & his passion shortlived, it is to his INTEREST to follow the former; & likewise <that the> t
621o047 he must soon necessarily learn that it is his INTEREST to follow it. even when opposed by some natur
621o047 origin of passions too strong for our present INTEREST receive simple explanations from origin of ma
621o047 le explanations from origin of man.-- [RHC] By INTEREST I do not mean any calculated pleasure, but th
539m081 n reading article by Boz.-- now in this I was INTERESTED as was I in the other, & read so intently as
030t036 gton.-- Mem: Micaceous formation of Chonos. INTERESTING from great quantity of altered Carbonaceous
139a180 coal charged with carbonic acid [blank] Many INTERESTING experiments might be tried by comparing Zoop
162g101 yed.-- the Brook <about> Head of which is so INTERESTING. enters by old tower called Glengarry (Nead
191b081 - In the East Indian Archipelago it would be INTERESTING to trace limits of large animals-- Owls. tra
217b188 mark common to all northern islds.-- This is INTERESTING. because Iceland, must have been all ice in
333d007 rmediate between two Parents.-- this is very INTERESTING as Esquimaux dog approaches to species. Agai
411e056 of Circus cinereus of the Falklands Isld. is INTERESTING as showing some change in habits before form
484z016 ahia Blanca & Southern Hemisphere It is most INTERESTING the way Synallaxis leads into Furnarius. by
484z016 Colouring on under side of wings It would be INTERESTING comparison to find how many of the small fin
510q017 s of Hilianthemum differs greatly-- how very INTERESTING to see if any variation in varieties. G. St.
619o043 generally be anger, as he would be tempted to INTERFERE, but with respect to himself it would be remo
248o030 taken, as «in» wild state (where instinct not INTERFERED with, or generative organs affected as with
619o043 f either man did not obey his instincts from INTERFERENCE of passion. he would feel pain, which would
064r135 generalize the fact glaciers most abundant in INTERIOR channels. there no outer coast.--important ef

```
101a058 ate? Phillips p 289.-- Alludes to big bones in INTERIOR at Falkland Isd.-- Peron does as if well atte
120a109 nsidered. if elevation near coast more than at INTERIOR effect would be such as present. to spread sh
123a117 surface  in tosca.-- remnant of Megetherium in INTERIOR.--  <The theory of [...] .> <The> Geographica
125a121 oling process must go from surface towards the INTERIOR -- who knows how far that may have penetrated
137a151 s of meteoric waters on the temperature of the INTERIOR & p. 142 / p. 155. the increase of temperatur
343d038 not  loosing ground. Yet, as the tribes of the INTERIOR are pushing into each other from slave trade,
059c121 s a kind of concretionary structure, for the INTERLINEAL spaces are of diff cont: & even in one case
277c126 eties. but two ostriches good species because INTERLOCK analogy to be guide. in islands. species.-- e
482z012 o 26 degree CLOSELY allied species, therefore INTERLOCK.-- Testudo INDICUS not fossil at Isle of Fran
349d052 s subkingdom.-- -- evidently artificial, as INTERLOPEMENT of Marsupials will change all.-- & so on no
302c203 species, & not, therefore, solely owing to such INTERM: father-species, being little adapted to  some
176b024 or change in part of country. repugnance to INTERMARRIAGE <increases it--> settles it ¿We need not th
327c265 No  XIV. April 1839.-- Review on "Walker on INTERMARRIAGE" price 14s. March. 20t. 1839. Philosophy of
337d021 er branches «as in common greyhound» & much INTERMARRIAGE.-- In my speculations. Must not go back to
468t112 united  filaments.-- This flower hostile to INTERMARRIAGE!!xx In Phil Transact. about year 1778. Pape
635j56v er.-- this is against my theory of frequent INTERMARRIAGE.-- A plant is in the same predicament as a
172b005 nstant over whole country; beautiful law of INTERMARRIAGES <separating> partaking of characters of bo
179b034 er trifling differences are blended «by» by INTERMARRIAGES, then the black & white is so far gone, th
199b120 dency to propagate, as well as diseases. In INTERMARRIAGES; smallest differences blended, rather stro
216b179 rent) the mulattos themselves explain it by INTERMARRIAGES with people. either a little nearer black
172b007 ase slowly, from many enemies. so as often to INTERMARRY who will dare say what result According to t
640j28v ugh continually dragged back to old type by INTERMARRYING with ordinary race.-- «There is no way of e
069r151 conglomerate East of Portillo Where gone to? INTERMEDIATE space protected.-- Oh the vast power of the
080r178 Araucarian  tribe, with point affin of yew & INTERMEDIATE Puncture one animal with recent dead body o
173b012 importation ceased & changes commenced.-- or INTERMEDIATE land existed.-- or they may represent some
178b032 C. of Good. Hope the children cannot be made INTERMEDIATE, the first children partake more of the mot
183b053 ion of species, by saying, why not have some INTERMEDIATE forms been discovered. between palaeotheriu
186b059 step in dislike to union, offspring not well INTERMEDIATE Lyell Vol III. p. 379. Mammalian type of or
192b085 produce.-- Therefore all genera MAY have had INTERMEDIATE steps.-- Quote in detail some good instance
192b086 estion, whether there have existed all those INTERMEDIATE steps especially in those classes where spe
192b087 et from some common progenitor.-- Now if the INTERMEDIATE ranks had produced infinite species probabl
204b140 ed by Ld. Powis Hybrid dogs offspring seldom INTERMEDIATE between parents.-- How easily does Wolf & D
207b150 Holland?! p. 320. Says Coniferous structure INTERMEDIATE between vascular or cryptogam. (original Fl
208b154 eed not expect to find <species>, varieties, INTERMEDIATE between every species.-- Who can find trace
217b183 reeds are crossed, sometimes offspring quite INTERMEDIATE sometimes take strongly after either parent
219b194 s of American & African form, merely because INTERMEDIATE position.-- We cannot consider it as adapta
222b204 ccording to our ideas<)> of perfection); but INTERMEDIATE in character, the same reasoning will allow
223b211 ed fertility of slightly different species & INTERMEDIATE character of offsprings accounts for unifor
225b216 otter-- why to be sure there were a thousand INTERMEDIATE forms.-- Opponent will say. show them me, I
225b219 mote. ass & Horse. produce offspring exactly INTERMEDIATE.-- Reference to Pig & Dogs. My theory  will
226b223 to present great contrast in forms.-- India; INTERMEDIATE, see how that is.-- ¿are shell-boring Mollu
227b226 other: now on this view no one need look for INTERMEDIATE structures <between> say in brain.  between
236b280 -In. insects «in England» surely it is not-- INTERMEDIATE genera we might expect.-- Lindley Introduct
270c104 It is very remarkable as shown by Carus how INTERMEDIATE plants are between animal life & "inorganic
271c107 r» groups of birds in two countries, without INTERMEDIATE ones occurring in intermediate country-- ie
271c107 ries, without intermediate ones occurring in INTERMEDIATE country-- ie. mundine groups.-- Waterhouse
273c110 ce, you may be almost sure, that there exist INTERMEDIATE species,-- This is remarkable & would  lead
276c122 gradual,  so may we suppose., that something INTERMEDIATE, between no offspring & ordinary offspring:
299c194 generations come up so.-- there are not Many INTERMEDIATE shades in these cases,-- but absolute speci
302c201 e can be no animal at present time having an INTERMEDIATE affinity between two classes.-- there may b
302c201 sses.-- there may be some descendant of some INTERMEDIATE link.-- the only connection between two suc
302c203 dent as submersion of land containing all of INTERMEDIATE Father-species, & not, therefore, solely ow
302c203 al change.-- If Patagonia became fertile all INTERMEDIATE species living there would be destroyed,  &
304c206 It  is a fact pregnant with SOMETHING.? that INTERMEDIATE. species have generally perfect organs.  &
309c222 ssages between-- owls & hawks only external» INTERMEDIATE groups often have full structure «of one cl
319c257 ca,-- Are not some of the Australian fossils <INTERMEDIATE> between those of Van Diemen's land & Austra
331dIFC s Darwin 36 Great Marlborough St Did Eytons <INTERMEDIATE>. «hybrids, when» interbred. show any tende
333d007 sembled the wolf than dog.--> appeared to be INTERMEDIATE between two Parents.-- this is very interes
333d008 -- Fox thinks half Lion & Tigers are exactly INTERMEDIATE in character & Kittens alike each other.--
333d008 s one parent & one another & are not exactly INTERMEDIATE.-- Where two dogs line the same bitch & «pe
334d013 on & China goose are crossed the neck is not INTERMEDIATE in its peculiar long neck, but much  nearer
335d015 all other animals» be like either parent, or INTERMEDIATE within certain small limits (within which l
341d032 peacock.-- tail as long as Pea hen.-- about INTERMEDIATE.-- (In Zoolog Gardens there is hybrid of Pe
374d133 . 426 Sauroid fish in Coal, true fish, & not INTERMEDIATE between fish & reptiles-- yet osteology clo
374d133 -- p. 432 some plants in coal supposed to be INTERMEDIATE between Coniferous trees & Lycopodiums.-- p
413e059 has  grand passage upon problem.! Hurrah.-- "INTERMEDIATE causes" The Sexual system of the Cirrhipede
430e119 rom the Gault of Folkstone, which is exactly INTERMEDIATE between I. concentricus & I. sulcatus.-- th
430e120 are found at Folkstone.-- it is unnamed this INTERMEDIATE one.-- Mr Lonsdale evidently inclines to th
463t055 iculty not only are no animals known with an INTERMEDIATE structure, but it is not possible to imagin
463t057 st answer it by rooting out curious cases of INTERMEDIATE structure, & supposing much extinction. gi
463t057 ch these series were drawn they would not be INTERMEDIATE, but this is not required..-- Waterhouse sa
473s006 nt «in Lin. Transacts. it has three stamens» INTERMEDIATE between Orchis & other plants-- & Wallich h
495q005 ollen.-- as these are wild varieties. Is any INTERMEDIATE form found wild a The Leptosiphon densifoli
497q007 urthest removed possible point.-- ¿genera in INTERMEDIATE country (2) Any known changes in Flora of c
513q020 fox,  to see whether the characters are then INTERMEDIATE or «sometimes» all on one side, as in cross
527m033 he music of the poetry.-- (therefore singing INTERMEDIATE, who has not had his blood run cold by sing
575n043 for death.-- can we deny that brain would be INTERMEDIATE like rest of body? Can we deny relation of
124a118 t Herschel's theory offers no explanation of INTERMITTENT action of elevatory force-- Erasmus says he
235b275 ts run wild or nearly so, which <breed> do not INTERMIX,--any cultivated plants produced by seed.-- L
636j56v rieties was prevented Do races of peas become INTERMIXED & gardner have hybrid seedlings} p. 333. Mac
194b096 may be applied to show all plants do receive INTERMIXTURE.-- But how with «hermaphrodite» shells.!!!?
205b142 sult-- Says types most subject to vary where INTERMIXTURE precluded.-- Kirby Bridgewater Treatise The
110a085 clare them to be recent species-- Lyell-- Some INTERNAL changes are in process. connected with variat
122a114 - On my view the degrading action must prevent INTERNAL fluid arriving at equilibrium so soon from; c
212b165 g he had, used to burrow like fox.-- a sort of INTERNAL bark. would remain for long time together  in
248c033 nces first affect external [for]m, so will the INTERNAL parts be of longest [consta]nt & therefore mo
386d167 is this not owing simply to more importance of INTERNAL organs in animals. One «invisible» animalcul
566n010 passions of a face «& mind» sympathetics with INTERNAL organs, as action of heart Malthus on Pop. p
617o39v by the effort it costs us to exert force or by INTERNAL consciousness; the question by our external
617o39v ention. How do the senses affect us, except by INTERNAL consciousness 3) We must endeavour to do with
618o041 of subjective action-- they are known only by INTERNAL consciousness. & have no objective aspect. If
044r071 a great explosion, nor would sea remove more INTERNALLY than externally--I did not see any number of
300c196 nce thinks himself a great work. worthy the INTERPOSITION of a deity, more humble & I believe true to
301c199 & actions habitual. it surely is not worthy INTERPOSITION of deity to teach squirrel to kill ears of
059r119 spondance exacte, et lorsquelles se trouvent INTERROMPUES par quelque vallées ou par quelque scissure
527m035 e facility with which a castle in the air is INTERRUPTED & utterly forgotten--, so as to feel a sever
195b103 und where barriers «& what are barriers but» INTERRUPTION of communication. or when country changes.
075r166 ly independent of the nature of the beds they INTERSECT". = In the Guatemala part. (& Chiloe do) no v
103a065 ffect of horizontal movement) hence generally INTERSECT metallic dikes: It is an important view which
147g016 uppose these upper patches if prolonged would INTERSECT alley above the 300 ft Alluvium <abo> by Loch
288c199 ay explain greater migrations, if American & INTERSECTED wider & wider at Rio Plata. birds which  had
040r063 na; probably more abundant in this case from INTERSECTING a mass probably cold & not warm as sides of
076r168 ongsberg in Norway.--namely dendritic silver INTERSECTING carbonate of lime-- native silver in Mexico
088a013 ia.> S Cruz -- from terrace like structure-- INTERSECTION of veins prove, that there are at least sev
211b164 h Volcanic chart & my idea of double line of INTERSECTION.-- Dr. Horsfield At India House, collection
053r101 d cause parallel lines, but the rectangular INTERSECTIONS are singular-- M. Lesson considers the Sand
074r165 emical laws as in concretions perhaps makes INTERSECTIONS richest-- Humboldt has urged phenomena in v
089a018 e of Domingo formed of coral limestone, with INTERSTICES yet empty.-- In all the mountains of Saint
```

053r100 lle of Patagonia would become more or less INTERSTRATIFIED with sediment.--& escarpment worn away lik
057r114 ery diff. ages. at Port Desire on plain. & INTERSTRATIFIED.-- Urge fact of Boulders not in lower stra
032r041 e of not accumulation of Coral limestone in INTERTROPICAL)» hence varieties of substances ejected fro
087a011 ortant. harmonizes well with Lyells idea of INTERTROPICAL land.-- Siberia rises. therefore to the Sou
056r111 ather begin & explain how water separates.--(INTERTROPICS at present fix lime). <Also Volcanos separa
501q011 ey to plant SINGLE Peas, Kidney Bean & Bean, INTERTWINED, «without sticks»-- in reference to what Mr.
173b011 s old ones, (of which none of same kind had in INTERVAL arrived) might have grown altered Hence the t
313c235 therefore sexual passion must arise after long INTERVAL very good case.-- habit is awakened by associ
539m082 m Musm. I was amused at this after seven years INTERVAL. Augt. 15th. As child gains habit «or trick»
542m096 olds them this way, when opening mouth between INTERVAL of barking, now this is smile.¶ With respect
293c172 transmitted.= Memory springing up after long INTERVALS of forgetfulness.-- after sleep, «strong» ana
628o55v l rules, get remorse-- ((hence desires do not INTERVENE between this kind of conscience & the will, t
227b224 ort, we should then know whether former lands INTERVENED.-- (2d) By character of any «two» ancient fa
352d059 e belief of their extreme antiquity (ie much INTERVENING physical change).-- distribution especially
426e107 ss.-- (<a series might be obtained>: but the INTERVENTION of domesticated ie new varieties destroys t
223b208 t & former hard to tell whether articulate or INTESTINAL, or even a mite.-- a bee «compared with chee
482z013 rence from Rüppel travels All Owens papers on INTESTINAL worms must be studied in Vol I, Zoolog: Tran
577n051 abitual actions, (independent of mind) in the INTESTINAL functions &c &c &c.-- bears. the same relati
493q003 ge of tortoise-shell «on back.--» = Length of INTESTINE in Persian Cat--, in Brazilian «toothless» d
315c242 egree in lowest animals --habitual action, in INTESTINES subject to sympathetic nerves-- The vividnes
332d004 having passed them before.-- Hence disordered INTESTINES are not healthy to worms, (like parasites of
612o035 nscience in the lower animals, as in stomach, INTESTINES & heart of man. [LHC] ¿How near in structure
613o035 ects live with no more consciousness than our INTESTINES have? [RHC] 5) Kirby thinks that <all> there
632j53r rance]CD Who would ever have thought that the INTESTINES of a thrush were means sufficient to ensure
269c102 lia with plants, with this object in view» The INTIMATE relation of Life with laws of Chemical combin
283c144 if such fantastic «sexual» ornaments. have so INTIMATE a relation to two continents as to be <replac
571n031 es that at earliest times there must have been INTIMATE connection between sound & language.-- Chines
248c033 , & of succession: the <first> latter is most INTIMATELY connected with important structures. <which
305c210 one thinking « <& Creat> sensible» principle (INTIMATELY allied to one kind of organic matter.-- brai
397e003 d. Dec 15 1856 [not located] Epidemics-- seem INTIMATELY related to famines., yet very inexplicable.-
552m131 o hear any sound.-- attitude of attention «So INTIMATELY connected is passion with sending force to m
568n017 if memory of its own emotions. (which must be INTIMATELY united with reason) it would feel «subsequen
577n051 om local irritation to passion.-- Blushing is INTIMATELY concerned with thinking of ones appearance,-
639j28v ites C. D. Endless cases.-- Macculloch p. 260 INTIMATES canines no special use to Man. Applicable to
223b208 «compared with cheese mite» with its wonderful INTINCTS <might well say how> The difference is that t
570n023 ne resist blow better with body distended.-- INTOLERABLE to be poked behind, without ones chest, bein
522m013 anges of disposition, like people in violent INTOXICATION, often ends in insanity or delirium.-- In M
036r049 st be filled up by dikes & mountain chains.-- INTRODUCE part of the above in Patagonian paper; & part
052r098 irst case. When discussing Falkland soundings INTRODUCE this discussion.--Brazil bank: (& I believe S
104a069 ask Lyell for sentence.-- Origin of Breccia, INTRODUCE in Cordillera discussion, deep sea, fragments
105a071 spute. sea. once came to Mendoza--. Will they INTRODUCE other causes to explain «alluvi» in valleys L
114a094 --shale in shall sea. Lyell confounds these INTRODUCE discussion -- I see Lyell talks of different
477z002 mongst Indian. introduced in Paraguay in 1769 INTRODUCE in Governor's tran?? Azara Las Vinchuca or Be
509q017 en. Have talked partially with him Ask him to INTRODUCE me to some Human Anatomist. Has he dissected
511g018 ing somewhat of different appearance.-- [will INTRODUCE it in work] Whether <Yar> knows whether Shaws
028r030 n all are mingled. These reflections might be INTRODUCED either in note in Coral Paper or hypothetica
044r073 rth place.--Some general reflections might be INTRODUCED on great size of ocean; especially Pacifick:
111a088 raph. Journal Vol. VII p. 216.-- Guava trees, INTRODUCED about twenty years since (1835) from Norfolk
172b007 me end which is good for Man.-- Let a pair be INTRODUCED & increase slowly, from many enemies. so as
183b051 ers. (NB. if Plata Partridge «or Orpheus» was INTRODUCED into Chili. in present states. <they> it mig
202b132 P II. p. 160. Melville Isd.-- "The buffaloes, INTRODUCED from Timor, herded separate from the English
225b218 ntomologists whether they know of any case of INTRODUCED plant, which any insects hav[e] become attac
234b255 lift up latch & back.-- Frogs attempted to be INTRODUCED I isle of France p. 170. «Fish introduced» H
234b255 be introduced I isle of France p. 170. «Fish INTRODUCED» Hump backed race of cows from Madagascar--
234b255 n with horns" --- p. 341. Black Fox sometimes INTRODUCED by ice <no> «only few» pigs.-- birds mention
235b274 s sent to India. & long bred in & no new ones INTRODUCED would not change be superinduced-- why is ev
247c029 Good Hope V. p. 44 of this Note Book Rabbits INTRODUCED in 64, of very many colours, like the cattle
253c046 Ogleby has facts to show that Australian dog INTRODUCED by savages into Australia.-- What are they?
267c093 0 years have scarcely elapsed since the Guava INTRODUCED from Norfolk Isld-- "& it now claims all the
285c150 er « <mother> » & grandfather <Might> Must be INTRODUCED & made young.--¶ father must be left out of
399e010 se in every country, where only pair has been INTRODUCED, & have freely bred, they have not lost powe
407e041 ustralia. reason, why: Marsupiata, when first INTRODUCED live & multiplied, specifically & individual
429e114 -- we must recollect the multitudes of plants INTRODUCED into our gardens (opportunities of escape fo
470t176 there was none but English,--the Palmated was INTRODUCED about '65 years ago--& soon after mules abou
477z001 de Cossigny. only 10 years ago <no> snail was INTRODUCED to Mauritius. 18 Azara Voyage Vol. I. p. 279
477z002 207. La punaise was not known amongst Indian. INTRODUCED in Paraguay in 1769 introduce in Governor's
640j167 ith respect to the six puppies, if a hare was INTRODUCED, or <a spe> became more numerous. (from deat
288c159 <go> large price. as is evident to be worth INTRODUCING, instead of breeding from original Durham br
026r022 erica. Study Geolog: Map of Europe Conybeare. INTRODUCT XII P. silicified bones not common in Britain
035r046 ructure most easily preserved.-- Mr Conybeare INTRODUCT to Geolog:--"Between the height of same beds,
236bIBC ntermediate genera we might expect.-- Lindley INTRODUCT Dict. Science. Naturelle Geographie Botanique
477z001 ology Some excellent references in L. Jenyn's INTRODUCT to Mag of Zoology and Botany. Philosoph. Tran
215b178 - Deshayes states Lamarck priority refers to INTRODUCTION to Animaux Sans Vertèbres as latest authori
222b202 velling of climates & the backward & forward INTRODUCTION of species.-- When species cross & «hybrid»
223b207 onderful. its mind more different probably & INTRODUCTION of Man. Nothing compared to the first think
322c269 -- References at end of each VoL Herschels. INTRODUCTION to Natural Philosophy R. W. Darwin's Botany
324c268 te White's regular gradation in Man. Lindlys INTRODUCTION to the Natural system Bevan on Honey Bee Du
338d025 s of Madagascar: because. contemporaries. In INTRODUCTION to Eytons Anatidae.-- recurs to idea of onl
399e009 at whole Glacial period? [not located] Study INTRODUCTION to Cuviers Regne Animal No structure will i
446e160 ed the question of spontaneous generation.-- INTRODUCTION to Bartram's Travels p. XXIII. <Some birds>
207b151 Mimosa called Botany Bay Willow V. Dr Royle INTRODUCTORY remarks to Himalaya Mountains-- Bory St. Vi
414e065 g ice.-- I agree with Mr Lyell., man is not an INTRUDER.-- : the geological history of man is as perf
602o011 s & decisions bein the result of sagacity, or INTUITION. when individual cannot give reason, though h
614o037 C] This Materialism does not tend to Atheism. INUTILITY of so high a mind without further end just sa
583n069 erring precision".-- no p. 17. Contrast the INVARIABILITY of instinctive powers in individuals of th
588n090 t <reason> «instinct», for its character is INVARIABILITY.-- if explained by habits, useful to itself
403e022 ery» «any» character even colour is good. (ie INVARIABLE) in some classes-- it is because every part
608o025 ed by what is called free will, but by strong INVARIABLE passions-- when these passions, weak, oppose
611o034 ssue <[...]> unites matter into certain form; INVARIABLE, as long as not modified by external acciden
618o41v self to what is cause & effect: it merely is «INVARIABLE» priority of one to other: no not only thus,
045r075 more <than> inland than on coast it would be INVARIABLY discovered; this may be mentioned with gener
397e003 t we call nature, have been conducted almost! INVARIABLY according to fixed laws: And since the world
543m099 ith D'Arblay of Christ of great genius, & yet INVARIABLY used to beat him-- The son of a Fruiterer in
583n069 e instinct is not imitative: imitations seems INVARIABLY associated with reason: [N B. insects which
255c053 ple of incessant change in her offspring. ,has INVENTED all kinds of plan to insure stability; but is
404e025 one is tempted to think that it must have been INVENTED all at once.-- but naturalists if they had se
547m114 ping up these trains,-- especially if they are INVENTED as in imagination, & in rigidly comparing eac
572m035 ns. recognized-- no one can say expression was INVENTED to conceal one's thought.-- Macculloch in his
579n058 on of shame-facedness for shyness, having been INVENTED, prove of the difference, which my theory bel
592n102 illiman's Journal paper showing that the signs INVENTED for Deaf & dumb school & used between Indian
593n107 iod when men. communicated before language was INVENTED,-- were musical notes the language of passion
599o05v language is an altering element, we see words INVENTED-- we see their origin in names of People.-- S
548m115 somewhat probable. in comparing every step, & INVENTING new means,-- therefore works of imagination h
527m034 therefore (independent of improving powers of INVENTION) such castles in the air are highly advantage
526m028 works of imagination very different from the INVENTIVE power,-- this, though very odd is perhaps tru
527m033 s in the air-- Catherine often, but not of an INVENTIVE class.-- Now that I have a test of hardness o
527m034 are highly advantageous, before real train of INVENTIVE thoughts are brought into play & then perhaps
540m087 s one cause of the intense labour of original INVENTIVE thought is that none of the ideas are habitua

541m091 at fatigue.-- may explain excessive labour of INVENTIVE thought.-- Examine frame of mind in following
165g124 ar Loch Lochy would be well worth examining-- INVERNESS & waters of the Tarf-- Kilfinnan Tower where
147g019 o'clock 29.642 Temp 55 Air 50 degree? Friday. INVERORUM about 20 ft above Loch Tulla 29.804 Temp 62 d
147g021 f rubbish Friday Highest part of road between INVERORUM & King's House 28.935/82 degree A Temp of Air
148g024 uneven to be impressed Case of Birch Wood by INVERORUM being determined by sheep & not deer When Bla
348d050 ds on the organs judged to be of importance in INVERSE ratio to their variability.-- (Now caeteris pa
181b043 approach animals, & some of the vertebrata INVERTEBRATES-- Such on few on each side will yet present
065r138] give instance of Etna Stromboli & Vesuvius INVESTIGATE with greater care. vegetation & climate of T
216b180 nts do so.-- As in cacti &c &c.-- as in dogs INVESTIGATE case of pidgeons. fowls. rabbits cats &c &c.
405e032 ilar in two hemispheres.-- It might be worth INVESTIGATE whether. Megatherium & Mastodon are coembedd
132a137 vour of thin crust theory.-- What a curious INVESTIGATION it would be to compare, the time of the ear
311c226 ivilized countries-- this is well worthy of INVESTIGATION.-- Institut 1838. p. 174. Apercu very good
495q05a . Have paper ruled in squares to facilitate INVESTIGATION.-- Capital in middle of ploughed field-- on
386d107 portance of internal organs in animals]. One «INVISIBLE» animalcule in four days could form 2. cubic
617o40v nimated being pulling every other particle by INVISIBLE strings & as on this supposition the forces m
559m154 he case when we consider the formation of laws INVOKING laws. & giving rise at last even to the perce
522m014 ht the feeling of anger, which rises almost INVOLUNTARILY when a person is tired is akin to insanity.
552m131 that in my grandfather remark, a tired man. INVOLUNTARILY feels angry, when brain is pumping force to
404e024 he case of my mice is good, because it is an INVOLUNTARY variation made by man, common to every indiv
530m046 ed very often, the motion becomes habitual & INVOLUNTARY.-- when a thought is thought very often it b
530m046 is thought very often it becomes habitual & INVOLUNTARY,-- that is involuntary memory, as in sleep.
530m046 it becomes habitual & involuntary,-- that is INVOLUNTARY memory, as in sleep.- a new thought arises?
530m046 .-- a new thought arises?? compounded of the INVOLUNTARY thoughts.-- An intentionally recollection of
531m053 ls in defence, &c Starting must be habitual «INVOLUNTARY» movement from wish to avoid some danger-- b
532m054 iseases of the heart are accompanied by much INVOLUNTARY fear) In these cases probably the system is
532m055 . <the> it must be an excited action in the INVOLUNTARY mind which is startled.-- My Father says he
542m093 y most earnestly wish [not] to do it, but an INVOLUNTARY laugh will burst forth, this & yawning. (com
542m094 ittle remarkable that those sounds which are INVOLUNTARY, are common to animals.-- Curious to trace,
565m006 g, hence one can help speaking, but laughing INVOLUNTARY.-- When one fear any bad news, «though in a
573m038 t avoid it.-- curious mixture of voluntary & INVOLUNTARY movements.-- Person with sore-throat told no
573m038 e-throat told not swallow spittle. will have INVOLUNTARY flow & desire to swallow.-- tells himself no
596n1BC considered convulsive.-- is convulsion. are INVOLUNTARY movement of voluntary muscles-- if so what i
616o038 erence» than sexual intercourse in plants is INVOLUNTARY, in man voluntary: ¿ False,-- secretion in b
616o038 man voluntary: ¿ False,-- secretion in both INVOLUNTARY, <application in> «ejection only has» will:
616o038 some time governed by will in some animals, INVOLUNTARY in others. [1]) Why may it not be said that
046r079 want to ground, that the first phenomem. is an INWARD afflux of melted matter.--Volcanos perhaps may
445e159 s resulted from the necessity of supposing some INWARD progressive developing power.-- My theory leav
605o18v an from any height.-- 6 That the superiority & "INWARD glorrying, which height. by its accompanying &
605o19v , as great height, eternity, &c &c. produces an INWARD pride & glorying. (often however accompanied w
605o19v me, that we may often trace the source of this "INWARD glorying" to the greatness of an object itself
610o031 is Christian name was Wilson!!-- How curious an INWARD. unconscious memory.-- Jan 14th. 1839.-- My fa
162g104 ll angular-- rounded pebbles-- dip sideward, & INWARDS-- deposited when water stood at higher Loch Ke
386d167 latter the division taking place from outside INWARDS. & in animals from inside to the outside.-- is
225b217 as why millet seed turns a Bullfinch black, or IODINE on glands of throat, (or colour of plumage alt
523m018 times comes on suddenly from <I> (in one case IPECACUHAN-- not acting) in others from drinking cold d
095a039 fs of shells & waterworn rocks «at Cobija». At IQUIQUE of elevation to amount of 30 ft.-- Mr Bollaert
096a039 oy. Institut) talks of quantities of shells at IQUIQUE. <Ceylon>. Band of Volcanic action in Iceland
028r031 been dissolved? Singularity of fresh water at IQUIQUI. not from rain, because alluvium saline; Mem:
523m018 tinction between enthusiasm passion & madness.-- IRA furor brevis est.-- My father quite believe my g
127a126 Tullamore found Sulph of Soda in peat ashes in IRELAND dikes in mountains. «(not on continents)» prov
170b1FC p. 26 30 41 46 50 54 56 67 69 76 79 91 93 107 IRELAND 113 117 This Book was commenced about July. 18
191b080 y with Europaean formations-- England & Europe IRELAND common animals-- * + + for instance tertiary d
191b080 deposits between East Indian islets-- (+ + +. IRELAND longer separated., Hare of two countries diffe
191b080 separated., Hare of two countries different.-- IRELAND & Isle of Man possessed Elk not England. Did I
191b080 d Isle of Man possessed Elk not England. Did IRELAND possesse Mastodons?? Negative facts tell for l
196b107 . No 3. p. 207 "It is not generally known that IRELAND possesses varieties of the furze, broom, & yew
196b107 t Britain, British varieties are also found in IRELAND-- There must be progressive development; for i
218b191 nn remarks to me. that give him a species from IRELAND, England, Scotland & other localities & each o
233b249 species in different isld. (as far East as New IRELAND. see Coquilles Voyage), Waterhous remark Austr
240c013 ¶-- All the Moluccas, Waggious New Guinea. New IRELAND, have phalangista, which differ in «form & hea
241c015 «162» list of some birds of Tongatabou. & New IRELAND.-- Gould will hereafter know about birds of N.
242c017 cas & Pelew Isds.-- p. 22. New Calidonia-- New IRELAND & Britain same kind of dog, with those of New
242c018 certa vittata extends <to> from Amboina to New IRELAND p. 23 Voyage of Coquille Lesson No (p. 24) bat
244c023 North. Hemisphere.-- p. 158 Cuscus albus. New IRELAND ---- maculatus -- Waigiou Speaking of Lepus Ma
247c028 &». p 49 on all the Moluccas «New Guinea, New IRELAND» & «even» Java. & very common on Otaheite-- ac
298c189 y increased in numbers over whole of England & IRELAND.-- curious in so wild a bird.-- Annals of Natu
298c192 nts, even those on Guernsey & on West coast of IRELAND, are absolutely (& who better authority) simil
344d041 he Rev R. Jones gave an admirable harrier from IRELAND to Brighton Park--first rate bitch-- tried to
357d073 e, one or two were landed as at present in New IRELAND &continent since grown.-- This will explain.
421e091 . Hist 1838 on «a» freshwater fish peculiar to IRELAND, ¶do p. 283. on the dark ears of the wild Chil
500q10a his at Canarys Arch-- it is so at Galapagos.-- IRELAND, doubtful species-- Does any genus of Plants.
468t111 to Rhododendron-- <Loasa» «Anchusa»-- speedwell IRIS-- Azalea. Rhodendron. Fraxinella to Anchusa <ne
497q007 o.-- where agency of man not known.-- (3) How is IRIS impregnated.; which part of stigma?-- (4) As Pa
542m095 the contraction & wrinkling of the skin contract IRIS?-- same way as one lifts up eyebrows to see thi
127a126 depth of ice ought to be less.-- Memoir of the IRISH Academy Vol 8. p. 118 water no--. oil will free
172b007 birds; Falkland Fox-- Chiloe, fox,-- Inglish & IRISH Hare.-- As we thus believe species vary, <in> c
212b166 Proceedings of Zoolog. Soc June 1837 p. 53. an IRISH Rat.-- different from English. Waterhouse has i
226b221 tive system Of this we see example in English & IRISH Hare.-- Galapagos.-- shrews, & when big contine
235b262 ent countries a case in point.-- All cases like IRISH & English Hare bear upon this.-- Why do Van Die
319c257 hose of Van Diemen's land & Australia proper.-- IRISH Elk case of fossil geographical range.-- [blank
322c270 Natural History. Octob 2d Transactions of Royal IRISH Academy. ----do Lavater's Physiognomy ----- Octo
353d061 s discovery of different number of vertebrae in IRISH & English Hare.-- good case these hares compare
385d164 bering that ovarium of women (Paper in Vol I of IRISH Royal Academy) have contained perfect teeth & h
454e184 bed instrument for galvanize <ani> them-- Cross IRISH & Common Hare Decandoelle has chapter on sensit
499q010 from <other> clumps from other parts? Don says IRISH, Scotch & English plants generally distinguisha
500q10a or New Zealand? Babington about differences of IRISH & British Species & British & distant parts of
515q022 than Continental horses. {About the leaping of IRISH Horses, bred in this country. {Chinese Dog's He
538m079 about eyes.-- one botanist & great knowledge of IRISH Politics, «both bad jokers.--» the other army o
027r028 fect of Elevation. United service Journal In the IRON sand formation <would> wood converted into sili
041r066 Chiloe, full of sand.--the <scale> «quantity of IRON» being there in excess.-- If veins {P} are secr
071r156 but no serious injury.-- <Analysis of Atacama. IRON in Edinburgh. Phisoph. Transactions. = Mem: Oli
072r160 ure exhibited to the Mexicans enormous masses of IRON and Nickel, & these masses which are scattered
073r161 n says) always come from without.-- "True native IRON that to which we cannot attribute a meteoric or
074r165 ther» «(state simplest case. concretions of clay IRON stone; iron pyrite in a fossil» Insist strongly
074r165 e simplest case. concretions of clay iron stone; IRON pyrite in a fossil» Insist strongly on the gran
074r165 & non Volcanic. Then Solfataras. «Mem: Micaceous IRON ore.» N.B. To show how metals may be transporte
074r165 w & steam of salts, quite curious case of oxided IRON by Mitterschlich. Vol. II Journal of Nat. & Geo
075r167 lightly broken up, & has there not been vein «of IRON» discovered?-- Klaproth analysed silver ores fr
075r167 Peru consisted of native silver & brown oxide of IRON in Mexico. sulphuretted silver, arsenical grey
076r168 in Mexico & Peru, that those oxidated masses of IRON. which contain silver are peculiar to that part
077r172 bad conductor of Heat do of Electricity Does not IRON, combined with nickel & cobalt (meteoric) resis
089a020 Isle of Cayenne. Syenite & diorite, covered with IRON clay common to Guyana said to extend to Cordill
093a032 1837. p 331 Considers that Mercury & Sulpuret of IRON has been sublimed into the tertiary limestones
114a093 03 Ehrenberg on ferrugineous Gallionella Examine IRON stone of C. of Good Hope & Australia/ and mud o
114a094 us specimens of «transversely fibrous» quartz. & IRON stone alternating. bear on subject of cleavage
137a149 aces do/p. 84 on the effects of veins of slag in IRON furnaces affecting to some distance & blending
496q006 Try.. Nitrate of Soda-- Salt. Gypsum. Magnesium IRON Rust Carb. of Ammonia.-- Horse Urine &c &c on a
556m145 led over upper with mouth shut. expressing cool IRONY, not biting? What is Emotion analysis of expres

Page   **************************************(Key Word)*************************************
110a086 th grenish. black specks of hornblende, large IRREGULAR cryst of reddish felspar. & scales. of mica.-
119a105 sea exists there.-- But Sir John considers an IRREGULAR figure to be that of equilibrium,-- What caus
146g014 ndy soil-- These Buttresses formed vestige of IRREGULAR terrace perhaps near 300 ft above Loch.-- Fro
147g015 part pure sand in current cleavage-- in other IRREGULAR horizontal strata I suppose these upper patch
147g020 e wide valley scattered with few very small & IRREGULAR hills of alluvium-- nothing very striking yet
148g026 ty well rounded stones, mixed with some quite IRREGULAR very like rubbish at head of Loch Dochart <Ne
151g042 omed strait. connecting flat on one side with IRREGULAR gravel plain of other, which must have been w
186b057 e's representatives agrees with breeding.. in IRREGULAR trees. & extinction of forms.?? It is in simp
201b129 f Maclay &c. appears to me the same, as the IRREGULARITIES in the degradation of structure of Lamarck
119a105 uilibrium,-- What causes that of tendency to IRREGULARITY,--. Why does Sir John assume it to be const
122a115 nnel. on origin of mud with stones scattered IRREGULARLY.-- (Mem near Gregory Bay). Shropshire case w
176b021 & alkali organized beings represent a tree. IRREGULARLY branched some branches far more branched.--
315c242 ls.-- (surely in plants movements effects of IRRITABILITY, though means injection of fluid different
6110034 to be direct physical effect of touch & not IRRITABILITY, which at least shows a local will, though
612o035 nts such as sensitive plants. (But I include IRRITABILITY for that require will in part. ¿Why more so
455eIBC mina of C. Speciosus. collapse at night. if so IRRITATE them, «as by an insect coming always at same
559m156 e made & ferns.--» Would a sensitive plant if IRRITATED very regularly at one time every day.-- natur
577n051 to reason.-- an inflamed membrane from local IRRITATION to passion.-- Blushing is intimately concern
579n060 analogy with my view of blushing-- in former IRRITATION on a piece of skin cut off made the blush co
305c211 nimals.-- --Animal Magnetism-- principles of IRRITATIONS sleep walking. fits, laught &c&c Man & Man m
071r158 felt. different case from shore of Pacific.--ISABELLE'S volcano, many amygdaloids.--Boussingualt «(L
050r090 h fact stated to exist in Peru. -- Ascension At ISD. capped by Calcareous rock, following Curvature
023r012 of Legrand, (SW part). describes a Small granite ISD. & great banks. effect of Elevation. United serv
027r027 of rock & sand mixed with sea shells--about 500 ISD.--Mem St Helena-- All Trachytic.--Daubeny P. 171
033r043 iversal: could not climb up many parts, in James ISD.-- Specimens of rocks were brought home in Capt.
065r138 ion & Crozet. L. Auckland. Macqueries.--Sandwich ISD. to grow on shoals like Fucus giganteus! 24 fath
066r140 e thicker than thumb» Sea weed said at Kerguelen ISD De Lucs travels Beauforts Karamania Capt. Ross.
080r180 of coasts 93 action of sea on coast. 27. Bahama ISD Mawes travels down the Brazil.-- Did Melaspena p
081r181 Farquhar Mathison travels Brazil. Peru. Sandwich ISD. p. 351.. NB. Mackenzie talks of gravel on basal
096a039 and: Mem.¿ Greenland subsiding.) Von Buch Canary ISD. p 170.-- Mem. Cordillera Can Greenstone dikes.
097a042 Lamellar dikes like Mica Slate Von. Buch. Canary ISD.-- Peron does as if well attested.-- There is no
101a058 .-- Alludes to big bones in interior at Falkland ISD into Geograph Journal Vol VII p. 279. Carcases o
111a088 ced about twenty years since (1835) from Norfolk ISD.-- "The buffaloes, introduced from Timor, herded
202b132 - Geograp. Journ. Vol IV. P II. p. 160. Melville ISD.-- (3d) We know that structure of every organ in A
227b224 gret sea (Coral reefs:. shallow water at Melville ISD case good one of animals not soon being subjecte
230b240 treme difficulty, argument in circle.-- Falkland ISD. & New Caledonia peculiar species of cassicans:
243c019 t found in Australia only New Zealand-- Norfolk. ISD.--» very good. Study D'Orbigny. & range on West
268c099 ierra del Fuego. & forest «Parrots in Macquarrie ISD regularly to main land, proof of. land having be
288c160 lost?? I conceive a bird Migrating from Falkland ISD they would change & make new species.-- alpine s
291c168 .-- & so destroy individuals, wheras in Falkland ISD «shows where proper dampness seeds can arrive qu
296c184 (& flora of Galapagos?) same condition. Keeling ISD.? ought to agree with Java?? Terrestrial Planari
306c213 Timor 215 degree. What productions Sandal. Wood ISD.-- West Africa & India some plants same. --Ameri
310c224 ely allied» species Himalayas, 13,000 & Melville ISD of Bass' Straits The common Mush room & other cr
314c239 nd & Australia.-- The wombat (Brown) is found in ISD 8 waders. 22 palmipedes: out of the first 9.:4 r
482z011) p. 205. only 9. Terrestrial birds at Falklands ISD-- Lutke Voyage Vol III p 322 Dr Martens says onl
486z019 those from Chiloe. Amblyrhyncus de marlin James ISD-- geology. volcanic? Applies to my geology & Spe
514q21. Orchis & Asclepias &-- carnosa?-- good-- Norfolk ISD) although having heredetary superfluities Man co
528m038 mmetry, of forms-- the beauty of some as Norfolk ISD fir shows this, or sea weed, &c &c-- this gives
638j58v h step being perfect «or nearly so (except no in ISDS «nor at St. Helena.--» Humboldt. New Spain Vol.
079r177 P 281. do do Australia, C. of Good Hope.--Azores ISDS.-- p. 22. New Caledonia-- New Ireland & Britain
242c017 mmon`.-- Galiopithecus common to Moluccas & Pelew ISDS.-- p. 7. New Caledonia-- probably united by Land to
291c168 es.-- alpine species being destroyed at Falkland ISDS++.-- Mem Lyell hypothesis of change in Scicily.
502q11v uculus lucidus is.-- Ask Sulivan about Falkland ISDS.-- Snipe Migratory probably united by Land to
325c267 e l'organization des Hommes. & les animaux.-- by ISID. Geoffroy. St. Hilaires. 1832. confirms all his
065r139 of Elevation.--The longer diameter of Deception ISL is six Geographical miles and width 2 & ¼ miles
217b187 -- Araucaria. species. Brazil {P} Chile, Norfolk ISL.-- Isle of Pines-- Australia.-- A <South> Americ
056r109 t, or diluvial waves. but when we see an entire ISLAND So encircled, the one slow cause is apparent.
157g075 & around which shelf 2d «almost» forms it into ISLAND-- whole hill composed of remarkable gneiss wit
173b010 g Lyells arguments of transportal) <continents> ISLAND near continents might have some species same a
185b055 xplained on principle, of animal having come to ISLAND. where it could live-- but there were causes t
195b098 ermanent a breath cannot reside in space before ISLAND existed.-- Such an influence Must exist in suc
200b124 ng these are most important facts.-- As soon as ISLAND large enough for land birds, seeds picked from
210b158 of genera I: I. I can understand in «one» small ISLAND. species would not be manufactured, <but why th
218b189 fferent forms; therefore when from being put on ISLAND. & fresh species made. parents do not cross--
218b192 peculiarity.-- Now when we hear that the whole ISLAND. is volcanic surmounted by water & studded with
219b193 ds. these plants) [Mem Fact stated by Mr Don in ISLAND. Teneriffe, St. Helena. J. Fernandez. Galapago
219b193 ce formed trees]CD & would creator <on volcanic ISLAND.> make plants <grow closely> When this volcani
226b221 transported them to new station.-- When the new ISLAND splits & grows larger species are formed of th
232b249 Phalangista New Holland form) is found in many ISLAND Celebes «Waggiou» &c &c. (See Lyell. Vol III p
271c106 o Juan Fernandez in birds. but ¿whether to same ISLAND in plants?-- What is this halo.-- Continents a
284c148 g average equal.-- The Cocos do Mar on the Mahé ISLAND, one the higher parts & only on those, & the
298c191 fresh arrivals..-- &c &c --Climate altering as ISLAND increases.-- upper parts attracting all the mo
357d074 sed. ¿is red game an hybrid?-- When I show that ISLAND-- In Sooloo we find the elephant-- in Magindan
402e020 pread from the N. end of Borneo to the adjacent ISLAND-- «connects with Asia» between two polar lands,
403e021 is found at Pellew & Eap, but not at Feis (near ISLAND) do p. 190. The inhabitants of Summagi, a terr
407e041 sider probable form of land,-- -- S America, an ISLAND.-- » p. 175., 28 sho[r]t eared owls were count
437e140 an Eagle always procured its prey from another ISLAND. namely Java is 1000 miles distant! Where are
023r009 y. Lat 25 degree. The nearest of the E. Indian ISLANDS we take in at once the stupendous mass which
056r109 indicates former boundary. (as in other unworn ISLANDS whose inclination natural [...] deepest astoni
056r109 cause is apparent. I confess I never see such ISLANDS. elevated. then peculiar plants created. if fo
061r127 ecies altered: <altered> Mem: my idea of Volc: ISLANDS & Coral Formation Lyell's Salband p. 86 Shells
085aIFC s far as concerns "Geolog Observat on Volcanic ISLANDS. from number of craters very ancient. which ag
085a002 layas penetrated like Bolivian Chain. Volcanic ISLANDS. & generally where rock metamorphic & thicknes
104a066 ent of Albite 80/100 X 6/100 = 480 In Falkland ISLANDS, ought to become different if kept long enough
172b007 lt According to this view animals, on separate ISLANDS. so permanent a breath cannot reside in space b
195b098 ery facts of the Zoological character of these ISLANDS; "ou bien encore on pourrait au plus en conclu
209b136 em:» Juan Fernandez]CD. From study of Flora of ISLANDS.-- & Artic are in same relation. We find species
209b157 se: Juan Fernandez Galapagos =Radack Isds = .. ISLANDS are near continent: compare Siicily & Galapago
210b160 | The creative power seems to be checked when ISLANDS.-- Bear peculiar to Sumatra & not found on Jav
212b164 rds from Java.-- at Leyden series from several ISLANDS.-- A genus with species in Van Diemen's land an
217b187 atra: again another of other Genus in Sandwich ISLANDS. on one with one kind of herbage & one with other,
224b213 :-- For instance two wrens forced to haunt two ISLANDS & on Arctic shores evidently due to «the» chan
226b221 genera & few species. The number of genera on ISLANDS.-- Hence this must have been condition of Pari
226b222 , we may insure mass continental or many large ISLANDS.-- -- Antelope in Celebes, Bourou new species
240c014 ew hairs) inhabits Celebes & few of the larger ISLANDS.-- at Amboina» I fancy there is marked wild bre
241c014 ccensis is different from that of the Marianna ISLANDS.-- ¿Europe has many species but not genera dis
250c038 whole world. America might have been string of ISLANDS. species.-- each describer giving his test nam
277c127 cies because interlock analogy to be guide. in ISLANDS. nor any any other part of world.-- no other p
284c148 he islets separated at high water.-- not other ISLANDS. close species, «on these isld» &c will probabl
285c170 ot possible in some detail,-- the relations to ISLANDS-- Endeavour to find out whether African forms.
296c184 » Webb &. Berthelot. must be studied on Canary ISLANDS-- of south seas says so in preface.-- Mr Brown s
314c238 h. Endlicker (He will give sketch of botany of ISLANDS.-- p. 250 admirable table of plants of St. Jago
317c250 . which grow on the lower region of the Canary ISLANDS. consult Dr Horsfield Silliman's Journal Rengg
324c268 17) for references to authors about E. Indian ISLANDS. separated with some animals, &c.-- «if the cha
398e005 n the score of small change.-- on the contrary ISLANDS presents us, for the most part, with the same
402e020 Voyage. Vol II. p 367. "The Fauna of the Sunda ISLANDS, having been purely result of elevation,-- «al
408e044 l species of Galapagos, must be owing to these ISLANDS.-- In his work on the Labiatae, some of these s
425e104 group having species peculiar to the separate ISLANDS--

```
449e170 urse to Fauna of Japan-- that the «animals of» ISLANDS N. of Timor are allied to the «type of genera
450e175 There  are prevailing colours in the different ISLANDS.-- The horse is only found wild in the  plains
451e177 e. 50 minute S. adjoining it are several small ISLANDS. abounding with deer-- Horsburgs. Vol II. p. 5
640j167 ng.-- From these views we can deduce why small ISLANDS. should possess many peculiar species. for  as
641j29r are  in progress; (on the same principles that ISLANDS are favourable,) because it must take so  long
449e170 of  Timor are allied to the «type of genera in» ISLAS de Sonda as well by those which are  identical,
054r105 ement with Coquimbo; [not located] [not located] ISLD near coast of America not reached. Juan. Galapa
061r127 creation for large.--Australia's = if for volc. ISLD. then for any spot of land. = Yet new  creation
065r138 istan D. Acuneha. Kerguelen Land. Prince Edwards ISLD. Marion & Crozet. L. Auckland. Macqueries.--San
065r138 home in Capt. Forster expedition from <Deception ISLD.> South Shetland Cape Possession. Syenite¿ Andi
072r159 t line of Old Greenland, close to W of Jan Meyen ISLD.--Mr Barrow thinks N & S. line connects western
092a030 p.  69.-- Geograp Journal Earthquake at Melville ISLD New Holland Augus 1d to 3d & 19 1827 Geograp Jo
096a040 s of gravel on basalt of Heckla-- All the Azores ISLD. Von Buch p 359 stretched out NE & SW.-- Von Bu
096a040 Buch.  Can. Ile p. 406. List of Volcanos Salomon ISLD.-- New Britain-- &c &c In Ascension for centuri
096a041 applicable  to Auvergne??? The fact of Galapagos ISLD. steep side to windward in allusion to St. Hele
134a142 f Part Lucomia-- Phillipines there is volcano on ISLD in large lake-- Berghaus Chart ot do Journal of
149z032 e hillock with beach & channel precisely as with ISLD-- {P} do they extend round hill too low line dr
172b006 rmely but. separate a pair & place them on fresh ISLD. it is very doubtful whether they would  remain
175b017 vary  quicker Unknown causes of change. Volcanic ISLD.-- Electricity Each species changes. does it pr
178b030 ly. 1837. Eyton of Hybrids propagating freely In ISLD neighbouring continent where some species  have
193b091 orealis. It is important the possibility of some ISLD not having large quadrupeds-- Humboldt has writ
198b115 t animals> large Mammalia not being found on all ISLD, (if act of fresh creation why not produced  on
218b192 r & studded with others.-- we see a beginning to ISLD. Graham isld.-- we know many seeds, might be tr
218b192 ith others.-- we see a beginning to isld. Graham ISLD.-- we know many seeds, might be transported som
219b194 annot consider it as adaptation because volcanic ISLD. whilst <neig> Africa, sandstone, & granite, (t
226b220 ando Po & Coast of Africa. equally good.-- Small ISLD off New Guinea same fact see Coquille's Voyage.
226b222 tion of proportion of species to genus If on one ISLD several species of same genus, subsided land.--
233b249 l. Vol III p. 30) different species in different ISLD. (as far East as New Ireland. see Coquilles Voy
236b278 production  of varieties is it not per saltum.-- ISLD bordering continents same type collect cases.--
236b278 ng continents same type collect cases.-- African ISLD.-- «How is Juan Fernandez-- Humming Birds» type
244c022 y from Africa p. 122. Mus decumanus, at Caroline ISLD, & a Roussette p. 136. Isle of France.-- the Te
245c023 ers, wild pigs. said by Forrest to swim from one ISLD to another--❚ «It is a good species, with diffe
245c025 f any one plant might make conditions in any one ISLD different]CD.-- p. 414. dogs of New Zealand  of
246c025 bats  have certainly travelled from East Indies, ISLD, as far as Oualan.-- Wide space of sea, to East
246c027 Oceanica  (Less) inhabits Caroline < «NB. The» > ISLD. (.perhaps Phillippines & perhaps, Friendly Isl
247c028 d in note to p 21» to Quoy & Gaimard in Sandwich ISLD. & according to Chamisso in Radack isld.-- p. 6
247c028 Sandwich isld. & according to Chamisso in Radack ISLD.-- p. 69. Sharks very generally distributed: Me
250c040 Journal.--- Vol I. p. 81. Capromys, West Indian ISLD. p. 120. «ref.» Philosop. Transacts 1823. Read
252c045 The Collector is directed to study localities of ISLD.--«immense importance of local faunas foundatio
252c045 rtain guide ∵ because time required too separate ISLD very long» America & India deer.= Africa not.--
258c061 ght.--XX Zoolog. Journal-- Parrots in Macquarrie ISLD. vol III p 430 alluded to by Capt. King do.  p.
266c092 PHICAL JOURNAL. Vol V. p 201 Wellsted. Memoir on ISLD of Socotra. Cattle generally marked like  those
266c092 abia & Egypt.-- CIVETS CATS only wild animals on ISLD.-- Niether Hyenas; jackals monkeys-- common  to
267c093 elapsed  since the Guava introduced from Norfolk ISLD--"& it now claims all the moist & fertile land
284c148 rt of world.-- no other plants peculiar to these ISLD.-- ¿Brown can <not» bear the least salt water.-
285c151 he relations to islands close species, «on these ISLD» &c will probably upset it-- The space which on
298c191 one in lower country during gradual elevation of ISLD.-- We must imagine a considerable range of  one
303c203 f Japan near Himalaya, & European forms on that  ISLD.-- The races of men differ chiefly in <size> co
311c227 Blainville  Ovington's Voyage to Surat, floating ISLD. off coast of Africa .p 69. with tall grass.  &
317c250 lants of St. Jago showing many common to Canary. ISLD., Europe, & St Jago upper region, & some to Cap
321c275 ept Brown's Appendix & excellent table of Canary ISLD: Plants Home's History of Man Transactions of t
328cIBC Cattle-- (Waterhouse has it) shells from Barrier ISLD many relations with a living Matica & many shel
403a022 inhabitants of Summagi, a territory in the small ISLD of Eap in the Carolines, are remarkably short.-
411e056 - My account of Circus cinereus of the Falklands ISLD. is interesting as showing some change in habit
421e090 a»-- (near Oran) approach in character to Canary ISLD.-- ie Canary Isld approaches more to neighbouri
421e090 pproach in character to Canary Isld.-- ie Canary ISLD approaches more to neighbouring coast of Africa
425e104 es are described.-- Capital case,-- for Sandwich ISLD are very similar to Galapagos-- study Flora-- w
426e105 once extinct-- Lyell's argument about <Tertiary> ISLD "neighbouring" formed in the Tertiary «epoch» l
430e118 rent by cross-questioning.-- » even if placed on ISLD-- if &c &c.-- Then give my theory.-- excellentl
437e140 pherds, who was watching the scene.-- «In Shiant ISLD. it is said, that an Eagle always procured  its
449e170 the  causes which give same species to different ISLD. is the same as that which gives genera.-- <it
450e175 ppears there are shades of difference in all the ISLD, like in wild animals).-- There are  prevailing
450e175 that  probably not original there)-- shows these ISLD not fit for horse. Forrest--. (p. 270) says man
450e176 43. «& 45» the Breed of elephants <of> in little ISLD of Solomo.-- said to have been imported:  shows
451e177 first  of race if so, fact for my theory Cocos  ISLD & Preparis between Andaman & Pegu. <have> aboun
452e181 cas Archipelago they are only to be found on the ISLD of Batchian near SE. end of Gilolo.-- "-- Forre
452e181 radise were first procured from Gilolo p. 253 In ISLD of Bunwood (18 miles in circum) there are  hogs
500q10a nfined genera By my theory in volcanic or rising ISLD, there ought to be a good many races or doubtfu
514q21. s of Tahiti. Dr. Boott-- says caricas from every ISLD differs-- do they also differ in different coun
640j167 alous, that the smallest newest, & most wretched ISLD should possess species to themselves.-- Probabl
038r058 existed.  These higher portions probably formed  ISLDS from which proceeded pebbles & on which trees g
040r063 uctive to animal life.--Patagonia In the Chonos  ISLDS we must imagine bituminous shales have been met
053r101 art Are there Earthquakes in the Radack & Ralix  ISLDS? In my Cleavage paper Dr Fittons Australia case
068r145 ll suggested to me that no metals in Polynesian  ISLDS--. Volcanic plenty in S. America!! Metamorphic
115a097 re them with my rocks. when writing on Falkland  ISLDS p. 94. Von Buch's Travels account of Norway cha
209b156 on Buch,.-- Canary Isles: French Edit. Flora of  ISLDS very poor «(p. 145)» 25. plants. 36 St  Helena,
209b157 Keeling  Case: Juan Fernandez Galapagos =Radack  ISLDS = ∴ Islands & Artic are in same relation. We fi
214b173 . Helena.  -- Galapagos--Juan Fernandez Falkland ISLDS-- Kerguelen land.-- Phillips. Lardner Encyclo.
217b188 cies to itself, a remark common to all northern  ISLDS.-- This is interesting, because Iceland, must h
221b200 » those which have undergone most metamorphosis  ISLDS X Is this applicable to insects &c &c?-- (p. 23
225b220 -- Study Ellis & Williams. zoology of South Sea  ISLDS. any animals?-- I believe none.-- Canary islds.
225b220 islds.  any animals?-- I believe none.-- Canary  ISLDS.? Madeira? «Tristan d'Acunha?» «Iceland?--» The
231b242 n & African forms mingle in India & East Indian  ISLDS-- Monkeys different not travellers?? Royles cas
245c024 e Voyage Durville has written Flora of Falkland  ISLDS where is it? All the Society isles have the sa
245c025 ice. In Amboina only Cuscus & Barbyroussa «NB»  [ISLDS. Springing up more likely to <M> have different
247c028 ocodile was washed on shore at one of the Pelew  ISLDS.-- killed a woman. Chamisso p. 189 Tome III: Ko
278c132 l S. America & yet Africa & India???-- & Indian  ISLDS.-- Sir J. Sebright-- pamphlet-- most important,
294c176 ad argument will explain very close Species in  ISLDS. near continent, Must we resort to quite differ
299c193 r from any other plants of same species Channel  ISLDS (& probably Isle of Man) no plants peculiar  to
299c193 mselves. this remarkable compare it with Canary  ISLDS. Galapagos.-- Iceland has same uniformity Primr
425e104 s so.--» March 6th. Mr Bentham says in Sandwich  ISLDS. he believes, there are, many cases of genera p
425e104 exception)-- see if this can be generalized.--»  ISLDS. , have peculiar forms.-- on the southern flanks
453e182 y of polymorphous plants, abounding in volcanic  ISLDS.-- <Cocks> The possibility of different varieti
481z008 ches Compare land shells of Galapagos different  ISLDS.-- Waterhouse remarks that no insectivore in S.
500q10a very close species & see whether they come from  ISLDS. or different parts or same district.-- About <
514q21. er in different countries-- on flora of African  ISLDS-- names of Plants found on mountains of N. Amer
640j167 r more easily be changed.-- Hence the Galapagos  ISLDS are explained. On distinct Creation, how anomal
640j167 case in world like Galapagos. no hurricanes.--  ISLDS never joined, nature & climate very  different,
049r087 The sudden increased dip is not parallel case to ISLE of White. but rather to one out of a series  of
089a020 92.-- (1837. Peninsula of Cape Verd. volcanic.-- ISLE of Gory. rocks encrusted with serpula-- Isle of
089a020 -- Isle of Gory. rocks encrusted with serpula--  ISLE of Cayenne. Syenite & diorite, covered with iro
191b080 d., Hare of two countries different.-- Ireland &  ISLE of Man possessed Elk not England. Did Ireland p
215b178 latest authority. The case of the tailess cat of ISLE of Man mentioned in Loudons (analogue of  Blood
217b187 caria. species. Brazil {P} Chile, Norfolk Isl.-- ISLE of Pines-- Australia.-- A <South> American «for
218b190 to  plants Female pig apt to produce monsters in ISLE of France-- Madagascar oxen with hump.-- p 1
234b255 tch  & back.-- Frogs attempted to be introduced I ISLE of France p. 170. «Fish introduced» Hump backed
240c013 phis Brunii) which as yet had only been found in ISLE of Aroe & Solor), «Vol I» likewise new  species
241c015 y <genera>. kinds common to New Guinea & rest of ISLE in E. Indi: Arch: In New Zealand. a sturnus of
```

243c020 n form? p. 27. many fish of Taiti found at <New> ISLE of France: xx instance of wide range, where mea
244c022 cumanus, at Caroline Isld, & a Roussette p. 136. ISLE of France.-- the Tenrecs from Madagascar. Monke
247c028 .-- <in 1813, a venemous snake was> one Gecko on ISLE of France Scincus multilineatus (p 45) Moluccas
299c193 plants of same species Channel Islds (& probably ISLE of Man) no plants peculiar to themselves. this
320c276 rnal. about 13 numbers have been read Voyage a l'ISLE de Frances Voyage de l'Astrolabe Partie Zoologi
320c276 'Astrolabe Partie Zoologique Pernety. voyage a l ISLE Malouines Zoological Journal 5 Vols Voyage de l
400e013 in other «places» besides Galapagos do. p. 376. ISLE Tres Marias off Mexico with small Hares & racco
483z012 efore interlock.-- Testudo INDICUS not fossil at ISLE of France: <Jerrold?> Bibron Zoolog. Journ Vol
023r012 followed Capt. Henry's vessel from the Friendly ISLES. to Sydney; know by having been seen & from the
044r072 ia.--remembering S. Africa. Australia.. Oceanic ISLES. Geology of whole world will turn out simple.--
072r159 --Mr Barrow thinks N & S. line connects western ISLES of Scotland & Iceland.--Bosh nor on Norway, or
097a043 ks of the Orinoco".-- <but> on one of the Ponza ISLES. but no minute description is given.-- Vol II.
202b133 y in France.-- of genus peculiar to East Indian ISLES.-- Compares it to fossil Didelphis (S. American
209b156 & Lyells. to certain extent Von Buch.-- Canary ISLES: French Edit. Flora of Islds very poor «(p. 145
242c018 s. <V.> p. 123 Crocodile at New Guinea. All the ISLES of Oceania have the Scincus with golden streaks
242c018 age of Coquille Lesson No (p. 24) batrachian in ISLES of Great ocean says in conformity with Bory's V
245c024 of Falkland Islds, where is it? All the Society ISLES have the same productions p 293-- is very stron
246c027 sld. (.perhaps Phillippines & perhaps, Friendly ISLES «& Hebrides») is very closely allied to C. musc
246c027 n Van Diemen's land.-- Vol II p. 8 no snakes on ISLES of central Pacific, yet there appears to be one
247c027 on Oualan.-- «Mitchill says snakes on Friendly ISLES. p. 50. LX. Journal of Silliman» «Study Silliman
252c042 ther in Jamaica & perhaps one extant at Leeward ISLES, ¿See if type continued?-- See to Boblaye & Vir
256c055 ton-- found tertiary formation amongst Graecian ISLES. Hope says positively he has seen. a Calosoma.
399e010 ur there,, but are unknown in Hervey or Society ISLES.-general movements of the earth;--Scarcity of
044r073 ze of ocean; especially Pacifick: insignificant ISLETS in East Indian archipelago Dr Smith considers
188b067 rrots peculiar according to Swainson to certain ISLETS-- (+ + +. Ireland longer separated. Hare of t
191b080 instance tertiary deposits between East Indian ISLETS.-- Next to animals land birds.-- & life shorte
191b081 e less easily transported-- Mem plants on Coral ISLETS not well fixed.-- Peron thinks Van Diemen's la
207b152 en monde".-- Considers forms in recent volcanic ISLETS & a sub-genus in Southern Africa In same manne
232b249 bution of Lemurs in Madagascar, on neighbouring ISLETS off Arabian Coast.-- ⫴Vol VI. p. 89.-- Lieut W
266c093 t of Arabia not even antelopes though common on ISLETS separated at high water.-- not other islands,
284c148 d,, one the higher parts & only on those, & the ISLETS Mr Blyth remark that a resemblanc between some
284c148 by deaths?!)-- looks like subsidence.-- on the ISLETS have different forms it is either effects of h
405e032 In such cases as at Galapagos. where different ISLETS-- March 5th. Lyell says «fossil» shells from
424e100 - several cases of species peculiar to separate ISLETS.-- Lyell says «fossil» shells from
175b017 e, so might they in time As I have before said ISOLATE species <& give even less change» especially w
228b230 mongrels.-- It really would be worth trying to ISOLATE some plants, under glass bells & see what offs
255c053 ted all kinds of plan to insure stability; but ISOLATE your species her plan is frustrated or rather
112a089 untain masses we must look for that.-- how few ISOLATED volcanos there are. where one alone has been
175b020 tendency to change, (& to multiplications when ISOLATED, requires deaths of species to keep numbers o
199b119 ry.-- people very apt to be split up into many ISOLATED races. ¿are there any instance of peculiar pe
297c186 ly an infinite number of forms.-- therefore an ISOLATED form probably a remnant.-- Pachydermata. & Ho
304c208 duced by circumstances In ostrich which is not ISOLATED, we must suppose the changes from typical str
305c209 ostriches were to die, then they would appear ISOLATED.⫴ In my birds from S. Hemisphere there are so
318c255 seems common in Rocky Mountains & on one lofty ISOLATED spot on the Alleghanies to which it migrats e
464t065 teryx, yet it shows the Apteryx is not «quite» ISOLATED in its present locality-- there have been at
464tf6r opard, how strange, anyone, would have thought ISOLATED species Mr Blyth, however, believes in the ex
499q09v r abundant? do they seed plentifully? Look for ISOLATED females.-- Also any plants which are known ea
522m011 t straight, but did not know [Z]CD when heard ISOLATELY.-- In old people. (Aunt. B.) when they hear a
209b155 hese particular varieties. If species made by ISOLATION; then their distribution (after physical chan
223b209 hysical changes are gradual, is this if after ISOLATION (seed blown into desert) or separation by mou
258c059 ties formed with extreme slowness, even where ISOLATION, from general circumstances effecting the are
304c207 is opinion?-- EXCELLENT PRINCIPLE OF ABORTION ISOLATION of range « <far more prob> » tends to alterat
107a077 Vol VI Edinburgh. Phil: Transacts.-- Does the ISOTHERMAL {P} subterranean line moves upward from effe
108a079 ge in form of fluid centre would lift with it ISOTHERMAL line, but if heat from centre, then crust of
126a124 n in siberia.-- from water {P} thawed at + in ISOTHERMAL curve. East-clinal. West clinal. S.-clinal.
127a125 o greater depth.-- Now the <inf> subterranean ISOTHERMAL line must be creeping «pushing» up to «the»
131a135 l Vol. 8. p. 402.-- ground ice-- subterranean ISOTHERMAL line Athenaeum. 1839. p. 52. On Frozen soil
246c026 ies.-- Vol `694. King-fisher of Europe, (Alcedo ISPIDA) from Molluccas. scarcely differs at all from
342d035 s shown in osteology of young Ostrich. 16th. D ISRAELI (Cur of Literat. Vol II p 11) accidentally say
187b064 es.-- If individual cannot procreate, he has no ISSUE, so with species.-- I should expect that Bear &
134a141 m Valparaiso to Santiago p. 328. dead trees on ISTHMUS of Pen. Tres Montes.-- as by subsidence ⫴ Fitz
160g088 ore feet above station! There is long straight ISTHMUS connecting E & W connecting Glen Bought & Glen
160g091 round hill. Boulder about 20 ft. below summit <ISTHMUS> {P} highest point joining this hill to others
160g091 in is 3500 boulder Cairn leet more Haberclador ISTHMUS broad flat Loch {P} XX Barom 28.92 A 75 Air 70
160g092 at Loch {P} XX Barom 28.92 A 75 Air 70 degree? ISTHMUS broad flat peat mass-- (general character in t
160g093 or about 3/4 of mile on <one> S side of «this» ISTHMUS (which runs E & W) broad terrace «of pebbles?
161g097 flights above it-- (NB the buttress or pass at ISTHMUS appears above level of shelf certainly) I took
448e167 & yet almost wholly different, is same, as if ISTHMUS of Panama.-- These two cases highly improbable
337d020 r Europaean varieties must be very unnatural-- ITALIAN Greyhound is probably the effect of <sev> loca
416e075 without any relation to running hares.-- as in ITALIAN Greyhound not so species every part of newly a
489qqIFC bits of different caterpillar races.--Name of ITALIAN who sold eggs.-- Temporary Question 1 Where ha
058r117 .-- The great movements (not mere patches as in ITALY proved by Coral hypoth. agree with great contin
337d022 North Europe is replaced by the F. cisalpina in ITALY, which is so like that difference would not be
398e004 of a few feet during last two thousand years in ITALY, but what «changes» would such a change originat
025r016 a [lighthouse] [12 degree 35 minute S.] 9 124 ITAPICURU [R.] [11 degree 46 minute S.] 9 200 R. Real [
546m110 o himself. she was one day reading a book, with IVORY paper cutter, which she valued, & she was sudde
522m011 e heard of him).-- My F. then said you remember JACK Baldwin at school.-- Answered To be sure I do.-
305c210 or wild birds-- At Zoolog Garden there is half JACKAL & Scotch Terrier.-- certainly more like Jackal
360d090 eir interest is concerned. Same man had crossed JACKAL & dog-- (offspring did not go to heat. but par
305c210 Jackal & Scotch Terrier.-- certainly more like JACKALL in gait, size, fur.; manner in which ears droo
333d007 ing always more resembled foxes than dogs (Mem JACKALL in Zoolog Garden) He has seen in a show half W
360d090 omewhat like dog-- though it has full share of JACKALL shape of body.-- disposition wild, & fearful.
360d090 tion wild, & fearful. though it not so much as in JACKALL.-- In case where Jackall was father resemblanc
360d090 ugh not so much as in Jackall.-- In case where JACKALL was father resemblance much nearer to Jackall.
360d090 Jackall was father resemblance much nearer to JACKALL.-- This Keeper has seen when sickly tigers hav
393dIBC tion of half breed Cattle at Cinbermere? How is JACKALL & dog at Z. Gardens D E Finished July 10th 183
574n043 ow habit.-- Take the case of Jenner's <Hyaena> JACKALL.-- an animal not destined by nature to exist.
575n044 se they obey the same laws, as the crossing of JACKALL & Fox & wolf & dog.-- the only test this is mo
575n044 greyhound to hunt hares. «& leave the sheep» & JACKALL to skulk about & hunt mice-- Jenners Jackall H
575n044 & jackall to skulk about & hunt mice-- Jenners JACKALL Have we somewhat right to deny identity of ins
266c092 S only wild animals on isld.-- Niether Hyenas; JACKALS monkeys-- common to either coast not found her
436e136 circumstances <April 12th..> Cestracion, Port JACKSON Shark-- Owen thinks Australia part of Old Worl
473s05r ent lakes of N. Wales can be distinguished-- & JACKSON here (Capel-Curig) says that he can certainly
027r024 dity of germination in young corals.--vide L. JACKSON'S paper. Philosoph Transact: at R. de Janeiro.
052r099 pared to the step = formed streams of lava at St JAGO. C. de Verds Quartz pebbles in the Cordilleras
260c068 checking beautiful colours of species-- Mem. St. JAGO--solitary Halcyon bird of passage.-- M. Coronat
290c163 t appear an analogy NB Pyrrho-alauda (bird of St JAGO) of brown colour; lives on ground, colour of ha
310c224 propagating itself. In Tropical countries (as St JAGO Cape de Verds) the shells in equal periods with
317c249 "Cercopithecus saboeus" said to be monkey of St JAGO C. de Verd; same as on coast of Africa.-- Macle
317c250 Professor Smith's Journal) on the heights of St JAGO found a Euphorbia so near Piscatoria as scarcel
317c250 slands-- p. 250 admirable table of plants of St JAGO showing many common to Canary. isld., Europe, &
317c250 owing many common to Canary. isld., Europe, & St JAGO upper region, & some to Cape.-- some proper, wel
219b196 «for the creator» two quadrupeds at S. America JAGUAR & Tiger & Europe, as to produce same one. ⫴Alt
226b222 ugh the Horse has perished from S. America, the JAGUAR has been left & Fox, & bear.-- If I had not di
226b223 rope to America., How strange would presence of JAGUAR been in S. America.-- <East.> «W» Coast of Afr
259c063 fish being excessively abundant & tempting the JAGUAR to use its feet much in swimming, & every deve
377d140 e colour as Lion. because inhabitant of plain & JAGUAR of woods &c like ground birds [not located] :H
423e097 s adaptation. though perhaps difference between JAGUAR & tiger may not be so.-- Considering the Kingd

252c042 tion of Molina) (p. 277.). probably another in JAMAICA & perhaps one extant at Leeward Isles. p. 388
029r033 QUAKE AT SEA.--Extract from the log-book of the JAMES Cruikshank, Captain John Young, on her voyage f
033r043 ot universal: could not climb up many parts, in JAMES Isd.--Mem St Helena-- All Trachytic.--Daubeny P
486z019 like those from Chiloe. Amblyrhyncus de marlin JAMES Isd-- Lutke Voyage Vol III p 322 Dr Martens say
037r056 of Basalt & other Volcanic rocks. Bahia, Rio de JAN: B. Oriental? level surface not disturbed.--Whol
072r159 on on coast line of Old Greenland, close to W of JAN Meyen Isld.--Mr Barrow thinks N & S. line connec
199b118 it into an instinct." Ed. N. Phi. J. p 297, No 8 JAN-Ap. 1828 -- I take higher grounds & say life is
211b162 h instincts very little. in Zoolog. Proceedings. JAN 1837, «by Eyton» Account of three, kinds of pigs
220b198 per on Botany of Tahiti In Charlesworth Magazine JAN: 1830. most curious paper on heredetary fear (li
221b200 s remarked about no confined species in Scicily. JAN: 1838 L'.Institut. Bats, in Eocene beds, very li
230b235 ffspring: applicable to any animal-- Athenaeum. <JAN> «p. 154»-- 1838. Hybrid Ferns It may be argued
323c269 d Brownes travel in Africa; «well skimmed.» 1839 JAN 10t.-- All life of W. Scott.--, except the V Vol
420e089 common point, whence both may have descended.-- JAN. 6th The rudiment of a tail, shows man was origi
423e098 either stationary or more probably increases.-- JAN 29th. Uncle John says he feels sure, that the re
579n060 ent of surface under the will? of the animal.(-- JAN 21. 1839. Herchel's Discourse p. 35. On origin o
610o031 alogous to change of instinct amongst animals.-- JAN 13th. 1839 My father received a letter from Mr R
610o032 !-- How curious an inward. unconscious memory.-- JAN 14th. 1839.-- My father says he has heard of man
024r016 atingua SE [23 degree 22 minute S.] 5 35 R. de JANEIRO SE [23 degree 58 minute S.] 18 77 C. Frio [23
027r024 Jackson's paper. Philosoph Transact: at R. de JANEIRO. Coquimbo. Balanidae. at Concepcion. Humb: Per
420e090 l is mentioned which inhabits the Pinna of Rio JANEIRO, (like some Mediterranean species).-- might th
067r144 lcano in Pampas» Pasto Earthquake. Happened on JANUARY 20th. 1834 Mr Sowerby. younger. says that Falk
170bIFC menced about July. 1837 p. 235. was written in JANUARY 183[8]: probably ended in beginning of Februar
579n059 r.-- a man cries from grief, joy. & sublimity. JANUARY 6th.-- What passes in a man's mind. when he sa
249o036 orms being within Tropics.-- Europaean birds at JAPAN. connected with Europaean forms on Himalaya??--
303c203 survived would be few in number.-- Parallel of JAPAN near Himalaya, & Europaean forms on that isld.--
315c241 n", some Europaean & Sandwich species & some of JAPAN. I do not understand any new ones.-- Menoir wil
449e170 of Temmincks. Preliminary discourse to Fauna of JAPAN-- that the «animals of» islands N. of Timor are
315c241 any of Bonin. "grande analogie avec le flore du JAPON", some Europaean & Sandwich species & some of J
328cIBC la Faune des iles de la sonde et de L'empire du JAPON Wowett on Cattle-- (Waterhouse has it) shells f
311c225 en's land & Australia different Temminck Fauna JAPONICA (?!) 82 mammalia 293 Phalangista of Australia
297c186 s in Hemiptera do. p. 160. soft plumage of night JAR. like owls. analogy in habits adaptation to noct
328cIBC rlula Cham. Cardium. Porcellus Turbo. Cerithium JARDIN du Roi Java fossils at same time Study Botanic
312c228 half Muscovy ducks, «black cock & pheasant see JARDINES Journal.»-- consult on this point-- pigs alwa
073r161 -/ is there any flexure <fr> in the fragmentary JASPER.--do undulations (as Hutton says) always come
023r009 ee. The nearest of the E. Indian Islands. namely JAVA is 1000 miles distant! Where are Hippotami foun
064r137 419> p 428» states that Von Buch has urged that JAVA volcanoes differ from all others in quantity of
188b067 merica.-- ¿are there any? Rhinoceros peculiar to JAVA, & another to Sumatra --Mem Parrots peculiar ac
191b082 nsport mice alive? Species formed by subsidence. JAVA & Sumatra. Rhinoceros. Elevate & join keep dist
212b164 rsfield At India House, collection of Birds from JAVA.-- at Leyden series from several islands.-- Bea
212b164 lands.-- Bear peculiar to Sumatra & not found on JAVA-- Monkey peculiar to. latter not to former-- Mr
216b179 f Blood hound-- Bull. Soc. Geolog. 1834. p. 217. JAVA Fossils 10 out of twenty have ANALOGUES uses th
231b241 up «the primeval.» continent. Indian Rhinoceros. JAVA & Sumatra ones all different.-- Join Sumatra &
231b241 a & Sumatra ones all different.-- Join Sumatra & JAVA <to> together, by elevations now in Progress, &
241c014 a» I fancy there is marked wild breed of oxen at JAVA.--.p. 140, calls it Bos. leucoprymnus. does not
244c022 ance.-- the Tenrecs from Madagascar. Monkey from JAVA.-- Hairs, & deer.-- Procured two makis alive fr
246c026 auritius Lesson &c p. 620. Centropus (Coucal) of JAVA & Phillippines, has variety at Madagascar, Calc
247c028 the Moluccas «New Guinea, New Ireland» & «even» JAVA. & very common on Otaheite-- according «stated
252c045 m each other I do not know how different Sumatra JAVA --------- do from Indian increase of knowledge
264c033 by them-- Timor. Australian forms amongst birds JAVA. not so much-- Peculiarities of structure. as s
306c213 oductions Sandal. Wood Isd.? ought to agree with JAVA?? Terrestrial Planariae assuming bright colours
328cIBC ardium. Porcellus Turbo. Cerithium Jardin du Roi JAVA fossils at same time Study Botanical work on Bu
343d037 n all <nature> organic nature) the Rhinoceros of JAVA & Sumatra, that since the time of the Silurian,
344d040 given list in Linnaean Transactions of birds of JAVA Caterpillars not being fertile is same as child
450e175 se of Pegu-- in Sumatra two breeds both small -- JAVA pony occasionally reaches 13 hands.-- Phillipin
450e175 of Celebes is somewhat larger than the Sambawa, JAVA & Sumatra breeds, (.Hence it appears there are
451e177 Horsbrugh E. I. Directory. Vol II. p. 46 Carimon JAVA. (between Borneo & Java) Lat 5 degree. 50 minut
451e177 y, Vol II. p. 46 Carimon Java. (between Borneo & JAVA) Lat 5 degree. 50 minute S. adjoining it are se
462t051 ry, on the grounds of similarity in condition in JAVA & Sumatra & dissimilarity of forms-- yet how va
462t051 ion as between Mammalia in cases such as that of JAVA & Sumatra Nov 15th Waterhouse showed me the com
510q017 e-- what? What does Blume say on alpine Flora of JAVA? Has Schow written on double creations & where?
023r009 hich were still sound and not petrified, and the JAW was also firm, out of which we pluckt a great ma
641j29v gy-- There is an analogy between fang of snake, (JAW of spider?) sting of bee, sting of nettle.-- Are
030r037 ferous mountain Limestone are they allied to the JAWS of the Cocos fish Rio Shells argument for rise
574n041 omeone has described slovering «gum» «teeth»less-JAWS. as picture of disgusting lewd old man. ones te
574n041 h flow of saliva, & hence with action of mouth & JAWS.-- Lascivious women. are described as biting: s
260c068 inds of bird uniting together in pursuit of Blue-JAY, when <one> birds hears <dis> crys of distress o
378d147 pies assume the metallic tints, such as Magpie, JAY, & perhaps all the rollers-- «He says» whenever
437e138 and stationary, in southern stays only winter.-- JAYS & chaffinches sometimes migratory.-- p. 103. Tu
523m017 herself out of the window to kill herself from JEALOUSY of husband connection with housemaid two year
557m148 ack absurdly arched. & tail stiff.-- is shame, JEALOUSY, envy all primitive feelings, no more to be a
557m148 think shame would be more easily analysed than JEALOUSY, because less discoverable in animals than la
557m149 fraid the whole way. but ashamed of himself.-- JEALOUSY probably originally entirely sexual; first tr
563n002 the Keeper) may be tempted to attack him from JEALOUSY. (Pincher & Nina)-- or to take away food &c &
590n093 es on before circulation is affected. p. 44.-- JEALOUSY. causes spasm in bile duct, & throws bile in
606o025 one's finger-- one feels it in passion, love-- JEALOUSY-- «as» effect of bodily organisms-- one knows
609o29v cumstances made the check.-- to licentiousness JEALOUSY, & every one being married to keep up populat
609o29v -, which being unnecessary we call vicious.-- (JEALOUSY in a dog no one calls vice). on same principl
267c093 sted "on coast of Arabia between Ras Mohammed & JEDDAH". Sheep. numerous "of two kinds one white with
308c218 here most evident!!? examine into this case D. JEFFREY (life of Mackintosh Vol II. p. 495)-- in fact,
600o008 ard to ordinary Beau ideal, Mem. Negro, beau,--JEFFREY denies all Beau-- How does Hen determine which
023r009 est not above half so long; The maw was full of JELLY which stank extreamly."--This shark was caught
567n015 blush.-- I am almost sure Fuegia Basket did. & JEMMY, when Chico plagued him-- Animals I should thin
250c040 o Phil Transacts. (read November 20th) Paper by JENNER, on birds seen far at sea, migrations of speci
483z012 5, owls seen crossing Atlantic. fact taken from JENNER (1825) Phils: Transact.-- "on Migrations of Bi
574n043 by dogs, would show habit.-- Take the case of JENNER'S <Hyaena> Jackall.-- an animal not destined by
575n044 sheep» & jackall to skulk about & hunt mice-- JENNERS Jackall Have we somewhat right to deny identit
376d138 other sexes,-- «by taking up tail» Mem: Ourang JENNY with Tommy.-- Good evidence of knowledge of Wom
554m138 nwards to look up womens petticoats-- just like JENNY with Tommy ourang.-- Very curious.-- Mr Yarrell
554m138 y ourang.-- Very curious.-- Mr Yarrell has seen JENNY, when Keeper was away, take her chair & bang ag
554m139 ld not succeed of herself.-- <The male> «I saw» JENNY untying a very difficult knot-- the sailor on b
554m140 I saw Tommy picking his nose with «a» straw.-- JENNY will often do a thing, which she had been told
567n013 y false, may lead to something.-- October. 8th. JENNY was amusing herself--, by getting out ears of c
590n094 «do- an account of a monkey in a passion like JENNY.-- Dr. Abel has given an account of an Ourang.--
178b031 pecies from the continent look over. Bell, & L. JENYNS. Falkland rabbit may perhaps be instance of do
225b217 f birds (where is this statement I remember. L. JENYNS. talking of it) or how to make Indian Cow with
243c020 e, where means of wide range work this out-- L. JENYNS, about my fish New Zealand & New Holland fish
324c268 Nat. in Geolog Soc.» F.. Cuvier on instincts L. JENYNS paper in Annals of Nat. History Prichard.-- La
424e102 oes Fleming consider them varieties & what says JENYNS to it?-- In argument of origin of Wolf, dif
477z001 racted Zoology Some excellent references in L. JENYN'S introduct to Mag of Zoology and Botany. Philos
604o015 e-- why flower beautiful? ¿even to children S. JENYN'S Inquiry into the Origin of Evil. Reviewed by J
233b251 of old World. Crocodiles. Anoplotherium.-- <M. JERROD> & Dumeril great work on Reptiles. M. J. says
482z012 tribution of Mammalidae &c &c &c-- The French <JERROLD> «Bibrons coworker of Dumeril" who is writing
483z012 Testudo INDICUS not fossil at Isle of France: <JERROLD?> Bibron Zoolog. Journ Vol I. p. 125, owls see
559m155 ume published by Colonel in army on "Wheat." in JERSEY.-- very curious facts about early production o
345d044 -good case of Provincial Breed-- Highland Sheep JET black legs, & face & tail, just like species.--
274c113 ucker--, parrots with claw like lark (NB The La JEUNE veuve parrot though so much on the ground has n
603o012 ainly curious. Chinese, S. American. Polynesians JEWS, African all sacrifices. How completely men mus
055r107 No cliffs at Ascension (or modern streams of St JGO) yet no historical records of eruptions how imme

```
024r016 Sebastian  [23 degree 52 minute S.] 12 50 {T} JOATINGUA SE [23 degree 22 minute S.] 5 35 R. de Janeir
029r033 om the log-book of the James Cruikshank, Captain JOHN Young, on her voyage from Demerara to London:--
119a105 of  the sea.-- as no sea exists there.-- But Sir JOHN considers an irregular figure to be that of equ
119a105 hat of tendency to irregularity,--. Why does Sir JOHN assume it to be constant.-- It is to be profoun
400e013 province  of Guadalaxura-- October 11th.-- Uncle JOHN-- says Decandoelle, distributed seeds of Dahlia
401e014 de by fertilizing one plant with another-- Uncle JOHN says he has no doubt bees fertilize enormous nu
423e098 ry or more probably increases.-- Jan 29th. Uncle JOHN says he feels sure, that the reason people send
423e098 e whole fields, some for cauliflower &c.-- Uncle JOHN believes one single turnip in a garden is suffi
632j53r n unfolded by D. Hartley.-- Darwin's Abstract of JOHN Macculloch 1837 Proofs and Illustrations of the
470t177 > show no trace of palmation!!? Bees at Wild St JOHNS Wort--Scabies, Cyanoglossum--Reseda wild very m
323c269 rals by Eugenius Feb 14th. Bo«s»well's life of JOHNSON. 4. Vols 25th Phillips. Geology. Larder 2d vol
477z001 -- CXVI. P 111 do Observations on Planariae by JOHNSON CXII. & CXV do Azara Voyage Vol I p. 196. Acco
604o015 s Inquiry into the Origin of Evil. Reviewed by JOHNSON in the Literary Magazine. 1756-- Ceased in 175
201b126 fossils of India.-- & p. 545 «great monkey» Mr JOHNSTON says Mag of Zooly & Bot. p 65 Vol II  talking
289c162 rhouse) Magazine of Zooly & Bot-- Vol II p. Dr JOHNSTON <on> Entomostraca Daphnia, produce young, cap
191b082 ubsidence. Java & Sumatra. Rhinoceros. Elevate & JOIN keep distinct. two species made elevation & sub
231b241 Rhinoceros. Java & Sumatra ones all different.-- JOIN Sumatra & Java <to> together, by elevations now
240c014 Vol I» likewise new species of Parameles, which JOINED to Casoars, perroquets, establishes its «zoolo
558m151 of God arise from our confused idea of "ought." JOINED with necessary notion of "causation", in refer
640j167 ld like Galapagos. no hurricanes.-- islds never JOINED, nature & climate very different, from adjoini
160g091 0 ft. below summit <Isthmus> {P} highest point JOINING this hill to others 3000? if Ben Erin is  3500
612o035 rgid vessels; effect of heat, light or shade.) JOINING two difficulties into one common one always sa
473s05v Siabod. wh. flows into Conway by Bettws & there JOINS streams from Capel-Curig-- Mr Bunbury says Mier
254c049 onge. stomachs in infusoria, generation in each JOINT of taenia worm.-- formative energies easily exp
363d101 several have a group of white speckles on elbow JOINT-- in Bewick drawing the the rock Pidgeon has no
402e019 I noticed a peculiarity in the cats «p 10» the JOINTS near the tip of the tail are generally crooked
453e182 ll Malay countries «the» cats are born with the JOINTS near the tip crooked.-- is the form [...] Damp
469t119 one  specimen had on one foot, a toe-nail & two JOINTS-- as it is on one foot probably monstruous & n
538m079 &  great knowledge of Irish Politics, «both bad JOKERS.--» the other army officer, horticulture & rel
555m143 erson was hung & came to life, & then made many JOKES. about not having run away &c having faced deat
538m079 what  is wrong in his head. & Babington's silly JOKING The possibility of the brain having whole trai
555m144 ead cut off. «there was the feeling of banter & JOKING» because the whole train of Dr Monro experimen
344d041 sup. is sucking foetus.-- August 23d The Rev R. JONES gave an admirable harrier from Ireland to Brigh
344d042 her offspring came out one big & one small. Now JONES, before this happened from her looks thougt she
345d042 for  50 or 70? years-- now «well fixed» breed.: JONES says Sussex cattle were all white headed, but t
364d104 is  owing to old <story> return.-- The Revd. R. JONES told me precisely same story about some Souther
379d151 still  further back reptiles & Cephalopoda: Old JONES remarked to me, that one of the children of Sir
543m099 unicating knowledge to each other-- August 23d. JONES said the great calculators, from the confined n
543m100 g profound reasoners.-- all same fact-- for, as JONES observed, in playing chess however many places,
559m155 on of foreign seeds.-- many varieties.-- Rev R. JONES has it.-- very curious book.-- Hume's essay on
050r090 g Volcano of Priamang.--Marsden Sumatra. M. De. JONNES seems to think that Volcanic eruptions form fo
446e161 rather  the reverse of facts stated by Smith of JORDAN Hill.-- May 27th.-- Henslow One of the 4 speci
489qIFC turists p. 21--23 Eyton p. 22 Schomburgk.---- 1 JORDAN Smith. p 1. Sowerby Cuming. -- p 1 Owen p 17 H
490q001 described  Atavism alluded to by Dr. Holland-- <JORDAN> Smith of Jordan Hill-- character of the extin
490q001 m alluded to by Dr. Holland-- <Jordan> Smith of JORDAN Hill-- character of the extinct land-shells of
121a111 let strata from decomposed muscles.. Smith of JORDANHILL has seen same thing-- Consider profoundly Ho
207b151 p. 151. L'Institut-- considers that mercu> Geo. JOUN. p. 325. Vol. IV. Ducks on rivers in Guiana, bu
091a027 SSW & NNE dip 30 degree - 80 degree Ed. N. Phil JOURN. p. 410. 1828 Ed. N. P. J. p. 105. Oct. 1828. g
092a030 ld New Holland Augus 1d to 3d & 19 1827 Geograp JOURN There are some ideas about order of injected ro
095a037 from Hudson's Bay. 2. species Vol VI. Geograph. JOURN. Analysis of Poenig Voyage Valparaiso Dr. Gilli
129a132 k.-- M. Bichoffs Papers, in Edinburgh New Phil. JOURN. 1838. several case given of hot heads &c heat b
199b119 o much change In Number 6'.? of E.d. N. Philos. JOURN.-- Paper by Crawford on Mission to Ava. account
202b132 sidence appears indicated.--- p. 36.-- Geograp. JOURN. Vol IV. P II. p. 160. Melville Isd.-- "The buf
267c093 clotted hair resembling that of goats" Geograp. JOURN. Vol VII. p.  218. Mr Bennett Voyage round world
465t081 exterminated  (see Pritchards paper) (Ed. Phil. JOURN. end of 1839) very important. it seems owing to
465t089 epend on other forms Lyell's Paper, in Taylor's JOURN.-- Phil. Mag. May. 1840 p. 362.-- some  Mammals
483z012 il at Isle of France: <Jerrold?> Bibron Zoolog. JOURN Vol 1  p. 125, owls seen crossing Atlantic. fac
027r027 eat banks. effect of Elevation. United service JOURNAL In the Iron sand formation <would> wood conver
029r033 g. 22 min. W. mid. calm and clear. Caermarthen JOURNAL I look at the cessation northwards of the Coal
067r142 acter of fossils deception complete.= Silliman JOURNAL, year 1835 excellent account of N. American ge
068r147 a near Kosir, land appears elevated. Geograph. JOURNAL p 202 Vol IV When recollecting Gulf of Califor
074r165 case  of oxided Iron by Mitterschlich. Vol. II JOURNAL of Nat. & Geograph Siciences? -- H. says in Po
077r172 us.  M. Chladni.--on meteoric Mexican stone. JOURNAL des Mines 1809. No. 151. p. 79. [misnumbering,
089a018 of ocean boil; yet heat never reach surface.-- JOURNAL de Physique, et D Histoire Naturelle, C«o»urre
092a030 ter passing sandstone Vol II. p. 69.-- Geograp JOURNAL Earthquake at Melville Isld New Holland  Augus
110a087 the gneiss beds in the mica slate.-- Geograph. JOURNAL. Vol IV (p 321) Mr Hillhouse describes central
111a087 tions on heights of valleys in Chile Geograph JOURNAL Vol VII p. 216.-- Guava trees, introduced abo
111a088 rs since (1835) from Norfolk Isd into Geograph JOURNAL Vol VII p. 279. Carcases of birds drifting out
112a089 ats on conducting powers of rock-- -- Geograph JOURNAL Vol IV p. 36. on subsidence of the land in Gui
112a090 4. probably will be published in the Geograph. JOURNAL.--» A meeting of the Geograph Soc, April 9 183
116a099 Read  Kylau on Granite Edinburgh Philosophical JOURNAL Rapport on D'Orbigny's Voyage. good section of
119a104 hough not very analogous (see Edinburgh. Phil. JOURNAL <JCD>, no great chains like Andes or Himalayas
123a117 .-- <The theory of [...] .> <The> Geographical JOURNAL Vol VIII. ( 1838) p 212. Facts from Erman abou
128a128 old submarine orifices in Cordillera Geograph. JOURNAL Vol II. p 89. at Madras. surrounded by salt wa
129a131 ldonado enter into this case.-- Ed. New. Phil. JOURNAL Vol XXI. p. 213. Beyond the limits of Alps siz
131a135 w Sir. J. Herschel's theory wrong.-- Geograph. JOURNAL. Vol. 8. p. 402.-- ground ice-- subterranean is
134a143 o on isld in large lake-- Berghaus Chart of do JOURNAL of Asiatic Society Vol I. {T} p. 145. on  salt
134a144 do Vol 5. p. 798 do Vol 7. p. <52> 363. do {T} JOURNAL of Asiatic Soc Vol V. p. p 96. apparently qood
135a145 coast-- quote passage to show abundance Bengal JOURNAL. Vol 4. 1835. p. 437. Tours by Benza Neilgherr
145g001 ing more indelibly than female p 367 Quarterly JOURNAL of Agricl Dec 1837 Yet instances given against
178b030 ene with respect to Miocene of Europe? Loudon. JOURNAL. of Nat History.-- July. 1837. Eyton of Hybrid
182b048 lection of insects with this in view.-- Geogr. JOURNAL Vol VI. P II. p 89.-- Lieut. Wellsted obtained
183b052 ted animals like races of man.-- M. Flourens. .JOURNAL des Savants.-- April 1837. p. 243 it is said a
193b093 aphical distribution of animals Brown Geograph JOURNAL Vol I p. 174. says from Swan river long South
202b131 17 (Lat 37 degree about) Vol IV P. I. Geograp. JOURNAL Voyage up the Massaroony by W. Hillhouse.-- D
218b190 st of Africa.-- changing hair-- The Edinburgh. JOURNAL of Natural History-- Preface appeared good wit
230b235 e mention[ed] by Lesson & Chamisso.-- Geograph JOURNAL Vol V. P. I. p. 67. Dr. Coulter on decrease of
247c027 chill says snakes on Friendly isles. p 50. LX. JOURNAL of Silliman» «Study Silliman.--» Vol II. p. 10
250c040 ndia--: connection with Latitudes!? Zoological JOURNAL.---- Vol I. p. 81. Capromys, West Indian  isld.
251c041 Gould on Trogons worth studying.-- «do» Zoolog JOURNAL Vol 2. p 221. Horsfield on two bears very clos
258c061 s may, be considered in this light.--XX Zoolog. JOURNAL-- Parrots in Macquarrie isld. vol III p 430 al
260c067 erica Study Bonapartes list In the Zoological JOURNAL I read a curious account to show that very man
266c092 germinating.--!! Henslow doubts? GEOGRAPHICAL JOURNAL Vol V. p 201 Wellsted. Memoir on isld of Soco
269c102 y take place in Metamorphic action.-- Geograph JOURNAL vol I. p. 17 &c excellent sketch of plants of
278c131 Diemen's Land.-- V. ls. Number of Geographical JOURNAL to discover whether dog found at Swan river. T
312c228 ovy ducks, «black cock & pheasant see Jardines JOURNAL.»-- consult on this point-- pigs always go aga
317c250 g,) passed us. do. p. 243 ( Professor Smith's JOURNAL) on the heights of St Jago found a Euphorbia s
317c251 forms.-- Study Appendix to Tuckey's Expedition JOURNAL of the Academy of Natural Sciences of Philadel
318c254 g to remain about a fortnight, «See Silliman's JOURNAL 1837. Paper by Bachman.» that is succession of
320c276 obrizhoffer. Abipom<e>nes. Edinburgh New. Phil JOURNAL. about 13 numbers have been read Voyage a l'is
320c276 Pernety.  voyage a 1 isle Malouines Zoological JOURNAL 5 Vols Voyage de la Coquille Zoological Transa
320c276 rts published March 1838 Whole of Geographical JOURNAL Asiatic Journal to end of 1837. re«a»d-- conta
320c276 rch 1838 Whole of Geographical Journal Asiatic JOURNAL to end of 1837. re«a»d-- contains very little
325c267 ndian Islands. consult Dr Horsfield Silliman's JOURNAL Rengger on Mammalia of Paraguay. account of wi
328cIBC Nat. des Pigeons et des Gallinaces. Silliman's JOURNAL. during 1837. paper by Bachman on migration of
345d043 n mother bein[g] oldest breed?.-- -- Quarterly JOURNAL of Agriculture p. 367. Dec. 1837. Generally--
405e032 Mastodon  are coembedded in N. America. see my JOURNAL for references In such cases as at  Galapagos.
406e037 ang their tails.-- November 1st..-- Addenda to JOURNAL. I show erratic blocks transported far S. in N
```

408e042 animals of Antilles.-- (see Macleay in Zoolog. JOURNAL. for those of Cuba.-- It is important to under
432e126 n.-- Splendid Pamplet. (published in Philosop. JOURNAL <Mar> April 1st 1839) by Sedgwick & Murchison;
451e176 f Rhinoceros as well as Tapir.-- <do do p 75> «JOURNAL of Asiatic Soc.. Vol V. p. 565. in a Paper by
451e177 Horsburgs. Vol II. p. 527.-- <Scientific Soci> JOURNAL of Asiatic Society Vol I. p. 261. <J> Catalogu
451e178 found there were 5,283 attached to its body-- JOURNAL of the Asiatic Soc. vol. I. p. 335. Catalogue
452e181 e hogs & monkeys <at> near shore of Magindanao JOURNAL of [Asiatic Soc] [...] p [...] -- most wonderf
458t001 andwriting, to compare with my own.-- E Bengal JOURNAL Vol 7. p. 658-- Falconer on Sub. Him. fossils-
458t002 t prototype!!.-- Copied Vol II p. 502. «Bengal JOURNAL» The Taylor Bird uses pieces of thread, picked
473s03r Two more of the flowers withered.-- Sillimans JOURNAL <vo> 1842. p. 142-- Sus americana & Hippotamus
481z009 45 for comparison with Falkland. good also for JOURNAL.-- 18 Admirable engravings in Meyen Zoology on
592n102 In one of the six (?) first Vol of Silliman's JOURNAL paper showing that the signs invented for Deaf
530m044 - & forgetfulness after bad accidents:-- After JOURNEY, a fit of = gout, has affected his memory of e
530m044 his memory of everything in <he a [...] Mr B> JOURNEY. short time previous,-- because, pain prevents
540m084 e A dog whines, & so does man.-- dogs laughs for JOY, so does dog bark. (not shout) when opening his
540m089 ther children naughty.-- Why does person cry for JOY? 17th. August Montaigne (Vol. I) has well observ
575n045 s when one burns end of nose with a hot razor.-- JOY <p> a mental pleasure. with pleasure of senses.
578n057 me of the muscles «or twitching».-- But why does JOY & OTHER EMOTION make grown up people cry.-- What
579n059 limity, sexual ardour.- a man cries from grief, JOY. & sublimity. January 6th.-- What passes in a ma
046r080 here is a native mouse Did wave first retreat at JUAN Fernandez: the first great movement was one of
054r105 located] Isld near coast of America not reached. JUAN. Galapagos. Cocos-- Ulloas voyage North of Call
173b011 h species all different. In cases as Galapagos & JUAN Fernandez. When continent of Pacific existed mig
209b156 istance (?). & similarity of type (?).-- [«Mem:» JUAN Fernandez]CD. From study of Flora of islands;
209b157 t. Helena I: I,15 (T) Calculate my Keeling Case: JUAN Fernandez Galapagos =Radack Islds = ∴ Islands &
214b173 of plants in Beetsons St. Helena. -- Galapagos--JUAN Fernandez Falkland Islds-- Kerguelen land.-- Ph
236b278 me type collect cases.-- African isld.-- «How is JUAN Fernandez-- Humming Birds» types of former dogs
269c100 th this view) Tristan D'Acunha, St Helena &c &c. JUAN Fernandez A communication on Geograph. Soc. in
271c106 rdum. The creative American halo has extended to JUAN Fernandez in birds. but ¿whether to same island
358d074 ent geography of Atlantic Tristan D'Acunha ditto JUAN Fernandez to Mitchell. Australia Vol I. p. 306
421e091 as Madagascar does to other side of Africa.-- CD JUAN Fernandez to Chile??) Falklands to southern por
448e166 agle Vol I. p. 306 Shells, as well as plants «of JUAN Fernandez» differ from American Coast Vol II.--
041r065 portant; because in this latter case. we cannot JUDGE whether such fossils. lived in groups or not. I
054r105 ay be concluded to have been covered by the sea JUDGE from the pebbles such as those on the beach--"T
126a122 , this is ascertained by conducting powers-- we JUDGE from the surface, & say 60 ft to degree.-- but
558m153 e. between man,-- forget the use of language, & JUDGE only by what you see. compare, the Fuegian & Ou
619o042 history of every race of man shows this, if we JUDGE him by his habits, as <if> another animal. Thes
126a122 is {P} against my views-- if we had rod thus & JUDGED by increments at, how wrong, would our judgeme
278c128 t the formation of subgenera is empirical, & is JUDGED solely by comparison with other genera in othe
348d050 leays plan of arrangement depends on the organs JUDGED to be of importance in inverse ratio to their
349d053 nera, common stock.-- so that value can only be JUDGED of in each «separate» line of descent.-- <& he
350d053 n, best rule for genera, & so mount upwards.-- «JUDGED by analogy»-- Consider all this NB. How can lo
586n082 inctive, namely the knowledge of size is merely JUDGED by eye, & use of limbs &c, or it result from m
606o022 is an ideal standard, by which real objects are JUDGED; & how obtained.-- implanted in our bosoms.--
126a122 judged by increments at, how wrong, would our JUDGEMENT be-- Does condensed metal, conduct heat bette
138a153 extended a league inshore both N & S of Lima.--JUDGES from «beds of» sand & gravel & shells. p. 47.
119a106 must be immense to convert rock into gneiss &c JUDGING from what we see when trap in dike & approach
151g041 ass of Glen Collarig two little lines of Hill JUDGING from external form alluvium) descend from shel
265c085 e adaptation, other monster.-- The only way of JUDGING whether structure is owing to habits, or hered
419e086 -- now in this country we have better means of JUDGING of slowness of physical changes, than in any o
609o030 hy our moral sense points <is> to revenge). In JUDGING of <our ha> of the rule of happiness we must l
275c119 clear facts, but so bold in many such profound JUDGMENT, that he forseeing. consequence., was endowed
414e063 more easily than man,-- though man's practiced JUDGMENT. even without time can do much.-- (yet one cr
533m061 d; for I caught it like a flash.--. strange if JUDGMENT remains, where reason is forgotten. it is con
543m100 n has keep in mind. all is certain.-- there is JUDGMENT of probabilities, therefore this judgment giv
543m100 e is judgment of probabilities, therefore this JUDGMENT gives a man common sense, & the highest intel
556m144 sts in the comparison of ideas, connected with JUDGMENT. [What is the Philosophy of Shame & Blushing]
568n018 e inclination or some future pleasure.-- hence JUDGMENT, which is part of reason Octob. 19th. Did
588n090 l. Zoolog.-- Vol II p. 445. If we compare the JUDGMENTS & actions of a young animal with an old.-- <d
612o035 s absorb by physical laws of endosmic & exosmic JUICES. arms of polypus, show either local or general
057r113 of Bones. = slowness of elevation proved at St JULIAN. = do not these bones differ as much nearly as
065r139 e¿ Andite?-- Degrading of inland bays. like St. JULIAN & Port Desire applicable to Craters of Elevati
093a034 ver.-- in reference to fossil guanaco of P. St. JULIAN. -- Mr Scrope seems to consider that elevation
040r062 eaward, the alternating with such matter at St JULIANS looks like such?--destructive to animal life.-
322c270 read the previous volume. & C. Prevost on L'Ile JULIE Waterton's Essays on Natural History. Octob 2d
021rIFC   R. N up to 1 degree / JULY 1835. the excess of harbor = 180 See Daubisson
170bIFC 07 Ireland 113 117 This Book was commenced about JULY. 1837 p. 235. was written in January 183[8]: pr
178b030 ne of Europe? Loudon. Journal. of Nat History.-- JULY. 1837. Eyton of Hybrids propagating freely In I
200b125 Ed. N. Phil. J. Morse found in Virginia p. 325. JULY 1828. Animal now confined to extreme North.--
239cIFC arwin written between («beginning of» February & JULY 1838) All good References selected Dec. 13 1856
332d001 e I first thought of selection owing to struggle JULY 15th. 1838 Finished. October 2d As a proof. wha
332d002 e that time it has been rare disease.-- but now JULY 1838) he has seen more case in a month, than in
332d004 cal countries cannot endure this climate-- .) --JULY 23d. Eyton, a stone blind horse, seemed to perc
397eIFC How is Jackall & dog at Z. Gardens D E Finished JULY 10th 1839.-- Selected. Dec 15 1856 [not located
489qIFC & Horsfield Sykes p. 12 Maer. p. 13 Question &c. JULY. 1842.-- Shrewsbury p. 14 Henslow (2d time) p.
502q11v ripen in Scotland?-- to show acclimatisation.-- JULY <1842> When nettle leaf. put into spirits, pois
505q014 of these small Bees-- at Sundorne has large Bees JULY/42/ Mark has six day's puppy of small true Bull
520m001 tions on Expression -- 1838 Selected Dec 16 1856 JULY 15th 1838 My father says he thinks bodily compl
529m043 te conjectural, in Blakeways book of Sheriffs.-- JULY 22d. 1838 No Deliriums, yet in some inflammator
591n097 ar giving goose skin-- & hair standing on end.-- JULY 20th Intelligent Keeper... Zoolog. Garden told
545m107 ed nut & then closed my mem. expression of fury, JUMP to scratch my face. The ourang outang, under sa
551m129 "right." or Drinkwater's horse jumping when word JUMP said-- I saw the ourang. take up a stone & poun
557m148 different from fear; there is no inclination to JUMP away,-- it is, ill-defined fear.-- Yet one know
576m046 eir brows when incensed.-- A Dog may hesitate to JUMP in to save his masters life,-- if he meditated
606o025 knows it, when one wishes to do some action (as JUMP off a bridge to save another) & yet dare not --
613o036 are through the bar of a gate before him, & then JUMP over the gate & bring it. ---- Agrees with ONE
233b250 f cats sometimes heterogenous.-- Australian dog JUMPED into tub leaving only nose above it-- pulled b
551m129 r cad cries out "right." or Drinkwater's horse JUMPING when word Jump said-- I saw the ourang. take u
192b086 (NB in those classes with few species greatest JUMPS strongest marked genera? Reptiles?) For instanc
472s01r bt. whether there is such a thing as a species-- JUN 1. 1842 Allen W. sowed some years since gathered
110a086 t on that side-- separation DISTINCT from dike   JUNCTION mechanical: DIKE base reddish feldspathes wit
110a087 such do not occur in either dike or fragment.-- JUNCTION certainly most distinct on dike side.-- oozed
158g083 «in Upper Glen Roy» In this upper part «about   JUNCTION of Upper & from Glenroy» near the upper shelf
060r125 to the Demerary Earthquakes at St Helena. 1756. JUNE 1780, Sept. 21st. 1817.--p 371. Webster Antarct
117a102 e scene of ruin in late Natical Magazine (before JUNE 1838) that 70. F were obtained 100 miles E of S
212b166 ritius & Madagascar.? Proceedings of Zoolog. Soc JUNE 1837 p. 53. an Irish Rat.-- different from Engl
250c040 d. p. 120. «ref.» Philosop. Transacts 1823. Read JUNE 5th) important paper by Dillwyn, on replacement
310c224 n plants found in other parts of world Athenaeum JUNE 3d 1838. quotes M. Turpins assertion that globu
313c235 The sexual curiosity of the orang outang of (in JUNE 1838 when young male was added good instance of
323c269 «references at end of each Chapter» Crabbes Life JUNE Is. King & FitzRoy To be read Humbold. New Spai
325c267 . Paper on Consciousness in Brutes in Blackwood. JUNE 18. H. C. Watson on Geograph. Distrib: of Brit
449e168 & Mamm. in ornithyrhycus-- is not this right?-- JUNE 18th. Eyton tells me, that Yarrell knows of a G
450e174 no loonies, but cocatores & small green parrots. JUNE 26th-- Yarrell.-- Black Swan «in domestication
466t091 s are there in old Secondary Series-- few-- Maer JUNE/41/, observed 3 plants of Caltha Palustris alon
466t093 ilst both others were in nearly full flower Maer JUNE/41/ Rhododendrum-- nectary marked by orange fre
466t099 fected by cultivation during life of individual. JUNE 1st 1841. Maer Examined the Lemon-thyme.-- equa
467t105 on Cabbage--white Butterflies suck nectar: «Maer JUNE 41» Rhubarb. pollen very minute--not excessivel
469t135 monstruous & not a second species.-- <Saw> Maer. JUNE 15./41/. Watched plants of Fraxinella, with sev
472s02r hird considerably paler, all rest very similar-- JUNE 2. 42 Maer <Thursday> Thursday After watching 1
473s004 probably buried in the depths of the sea-- Maer. JUNE/42/ June/42/-- Mr. Bunbury says has heard the T

473s05r buried in the depths of the sea-- Maer. June/42/ JUNE/42/-- Mr. Bunbury says has heard the Trout from
473s006 plants-- & Wallich has described Indian Plant.-- JUNE /42/-- June/42/ You can select cattle & sheep f
473s07r allich has described Indian Plant.-- June /42/-- JUNE/42/ You can select cattle & sheep for horns & y
559m155 ess in Brutes & Animals. in Blackwood's Magazine JUNE. 1838. Copied Mr H. C. Watson on Geographical d
452e181 that the Hippopotamus occurs in India. in the JUNGLES of Borabhum & Dholbum.-- Vol do. p. 634, allud
118a103 f Chile.-- P p217. Pentlands Fossils & Meyens --<JURA &> Chalk When we consider parallelism of dikes
120a110 nian blocks (1200 ft??). Scotland at least 2200. JURA 4000 feet.-- The veins of segregation in Greens
243c019 ecies Coquille Voyage p. 25 Mais il n'y a pas JUSQU'AUX iles Macquarie et Campbell (52 degree S) qui
571n028 thout abstracting them & reasoning on them (as JUSTICE?? as ancients did high forehead sign of exalte
609o029 lings no doubt orginally necessary revenge was JUSTICE.-- No checks were necessary to the vice of int
317c249 rees p. 259 120 ft in length, some branches of JUSTICIA still growing,) passed us. do. p. 243 (, Prof
286c155 those who soar above Such prejudices, yet have JUSTLY excalted nature of man. like to think his orig
537m075 observe, p. 21-26. argues «with examples» very JUSTLY there is no universal moral sense.-- «from dif
403e021 Pellew to Eap.-- There is another great lizard. KALUZ. which is found at Pellew & Eap, but not at Fei
622o049 C] instinctive fear of death: of hoarding.. Ld. KAMES, which Sir. J. says is so ridiculous. [RHC] the
193b091 n vacuum to each other p. 306.--. Chamisso on KAMSCHATKA quadrupeds Kotezebues first Voyage Copied in
486z019 III p 322 Dr Martens says only one Reptile in KAMTCHATKA (Salamandra aquatica). Compare with T del Fu
429e116 way. Lapland & Greenland, but not with those KAMTSCHATKA, Siberia, or even of polar regions of N. Ame
240c013 tements to such effects false In New Guinea. a KANGAROO d'Aroe (Didelphis Brunii) which as yet had on
639j28r nts (?) p. 251. all animals run by hind legs-- KANGAROO. only a caricature; Penguin.-- Pincers in Sco
552m132 ralia Vol I, p 292 "Dogs learn sooner to take KANGAROOS than emu, although young dogs get sadly torn
134a142 s at Bahia Blanca Mr. Malcolmson says are like KANKAER South of Part Luconia-- Phillipines there is v
610o033 l of Locke, Bentham, & Hartley, &. the school of KANT. to Coleridge, is regarding the sources of know
080r181 ast. 27. Bahama Isd De Lucs travels Beauforts KARAMANIA Capt. Ross. & Scoresby deep soundings Gilbert
467t104 t last I saw Bee collecting pollen from <sheath> KEEL of Lupine-- Seen Bees on Potato & several times
506q014 sured same way) 47½-- in heigt 30 inches Examine KEEL of Common & Wild Duck-- Black Duck & Penguin He
062r129 ca. Zorilla: wide limits of Waders: Ascension. KEELING: at sea so commonly seen. at long distances; g
200b124 quantities of seeds in sea; also Holman: <at> KEELING these are most important facts.-- As soon as i
209b157 es I: I,46 St. Helena I: I,15 (T) Calculate my KEELING Case: Juan Fernandez Galapagos =Radack Islds =
229b234 If. Henslow discusses possibility of seeds of KEELING standing transport.-- <tr but> Get him to disc
268c100 eru» Henslow in talking of so many families on KEELING seemed to consider it owing to one of each, be
296c184 Helena (& flora of Galapagos?) same condition. KEELING Isd «shows where proper dampness seeds can arr
506q15v requisite.-- Brown's paper 43. Any flowers of KEELING Dioecious, or Monooecious, besides the Nettle.
068r146 e repeted trifeling injections.--Old vents would KEEP open long after emersion, but improbably so lon
175b020 ons when isolated, requires deaths of species to KEEP numbers of forms equable:-- but is there any re
179b034 ainly are according to all common language) will KEEP to their type: in animals so far removed, with
180b037 rys must have left no offspring at all, so as to KEEP number of species constant.-- With respect to e
186b059 or animals) whether, <in> races have tendency to KEEP to <each> either parent, (this is what French c
188b069 ith Dr. Smith's fact of races of men tendency to KEEP to one line Dr Smith says very. close species g
191b082 ence. Java & Sumatra. Rhinoceros. Elevate & join KEEP distinct. two species made elevation & subsiden
194b096 some argument about varieties being difficult to KEEP on account of pollen from other plants because
213b170 ll?, which have wide range and therefore cross & KEEP similar. But this is difficulty; This immutabil
239cIFC me of two pigeons,-- «with specks» which cross & KEEP colour on wing Effects of colour on parent, whi
258c060 imate.-- Descent. or true relationship, tends to KEEP to species to one form, (but is modified), the
289c161 hydridity,, is simply, an instinctive impulse to KEEP separate, which no doubt be overcome, but until
304c206 eved to past into each other-- Different classes KEEP to their types. with different degrees of close
312c228 .-- Are mules homogenious owing to no attempt to KEEP up offspring, are not half lion & tigers ditto.
343d036 ther, & their bodies, by certain laws of harmony KEEP perfect in these themselves.-- instincts alter,
366d112 count of the people on the NW. Coast blinking to KEEP out flies might be used» The wild ass has no cr
378d147 .. we can see why young & female alike Good Ch 6 KEEP Is it Male that assumes change, & is the offspr
494q004 is price of fox. otter. Badger &c &c &c.-- (11) KEEP. Tumbling pigeons. cross them with other breed.
541m091 for a length of time.-- Then if one endeavur to KEEP any simple idea as scarlet steady before mind f
542m093 ut a most distinct will, he will find it hard to KEEP his lip from stiffening over his canine teeth.-
543m100 ess however many places, & contingency a man has KEEP in mind. all is certain.-- there is judgment of
609o29v tiousness jealousy, & every one being married to KEEP up population. with the existences of so many p
628o53v might be partly made.]CD p. 21. "Why ought I to KEEP my word"-- gives the problem, of ethics-- [my a
162g105 rds-- deposited when water stood at higher Loch KEEPER tells me, that Loch Lochy is 8 ft below Loch O
360d089 er were the parents of the three little ones.-- KEEPER said in <two> crosses "twice made" between ter
360d091 ther resemblance much nearer to Jackall.-- This KEEPER has seen when sickly tigers have first come ov
545m106 -chit, quickly uncovering their teeth, this the KEEPER thinks is from pleasure, & may be compared to
553m137 t its lip, looking absurdly sulky «as» often as KEEPER spoke to it,-- but he thinks not sulkiness-- t
553m138 ression mean SOMETHING.-- Hunt (the intelligent KEEPER) remarked that he had never seen any of the Am
554m138 Very curious.-- Mr Yarrell has seen Jenny, when KEEPER was away, take her chair & bang against the do
554m139 sitor, & before eating «everytime», look up to «KEEPER» see whether, this was permitted & eat it.-- g
554m140 she had been told not to do.-- when she thinks KEEPER will not see her.-- <but is> then knows she ha
554m141 Swan is very fierce when female is sitting the KEEPER is obliged to go in with a stick, if he drops
563n002 Cyanocephalus. Sphynx howling when I struck the KEEPER) may be tempted to attack him from jealousy. (
567n014 & suffering The Cyanocephalus when fondling the KEEPER... clasping «& rubbed» his arm. & show signs of
591n097 & hair standing on end.-- July 20th Intelligent KEEPER... Zoolog. Garden told me. he has often watche
591n097 gos, as I saw wag tail when watching anything-- KEEPER does not think they drop their ears.-- -- Geor
595n115 amentable howls & & was not comforted until the KEEPER took it <her> in his arms & carried to see.--
596n184 ocephalus, macacus. Cercopithecus? very much., «KEEPER says some of the monkeys move <its> «the» ears
317c248 w facts at end of "The British Aviary" or Bird KEEPERS Companion Study Appendix (& only appendix) of
500q010 eggs in all varieties of Cabbage. (26) Do deer KEEPERS cross the breed-- desirable as in Cattle in Ch
046r080 first one would think movement. owing to water KEEPING its level whilst land rose up & down.--But fro
285c152 es May have passed through a thousand changes, KEEPING distinct from other & if a first & last indivi
288c159 st useless not the least notion of hunting, or KEEPING watch. how completely «nature & instinct» modi
382d157 animal is <evidently> actually split in two-- KEEPING sexes separate. Owen say such view worthy of a
430e118 n thus produced, but by training, & crossing & KEEPING breed pure.-- «& so in plants effectually the
523m015 ely his poverty.-- his manner of curing it. by KEEPING the sum-total of his accounts in his pocket, &
541m090 from these facts that whole effort consists in KEEPING one idea before your mind steadily., & not mer
547m113 do less work & yet do so, from the exertion of KEEPING up the memory of every late impression. & like
547m114 able late events.-- the fatigue of thinking is KEEPING up these trains,-- especially if they are inve
548m115 -- therefore works of imagination hard work,-- KEEPING one idea present is, perhaps, hard work-- thou
579n059 is it influenced by» other emotions? When a man KEEPS perfect. time in walking, to chronometer, is se
640j28v most necessary: the fertility of Man in old age KEEPS woman alive: for Man & woman are same: fertilit
037r052 ater in the Southern ocean not being buoyed with KELP.-- With respect to degradation of rocks--It may
051r093 . -- In Southern regions every rock is buoyed by KELP, now Kelp sends forth branching roots which mus
051r093 uthern regions every rock is buoyed by Kelp, now KELP sends forth branching roots which must protect
051r094 uctions. = Everywhere on coast (Il Defonsos «KELP») rocks show signs of degradation; (soft substa
212b165 .-- would pull the garden bell, & then run into KENNEL to watch who would come to the door-- would co
529m041 trees.-- I am sure I remember my pleasure in KENSINGTON Gardens has often been greatly excited by lo
554m141 ing down hill.-- (argue sophism of association. KENYON, & then go on to show, that if Cart horse argu
025r019 nerally be found to be old & dead)» «(I have not KEPT a record)» In looking over the lists of organic
035c047 wonderful level «of same beds» should have been KEPT; it shows that throughout all England, whole su
060r126 of Andes fusible? no. mad dogs. Azores. although KEPT in numbers. p. 124. Webster Consult W. Parish.
100a055 m Lardner's (p. 213) form of escarpment relation KEPT to sea coast ∴ curious exception in Wealden.--
120a110 en Roy Notebook-- & scraps on Salsisbury Craigs. KEPT amongst <old> papers read before societies.-- S
125a121 penetrated,-- lower down the temperature may be KEPT up far higher from circulation of heated fluid
172b007 n separate islands, ought to become different if KEPT long enough.-- «apart, with slightly differen c
308c217 ifferences carried a long way. --Case of Habit I KEPT my tea in right hand side for-- some month, & t
308c217 for-- some month, & then when that was finished KEPT it in-- <right> left, but I always for a week t
360d091 is effect of climate, or state in which they are KEPT?-- Is there any mistake about Yarrell's law, is
362d100 nerations) the strangest peculiarities have been KEPT perfect-- also to trace the laws of change in t
370d116 all die.-- bees breed but few, because they are KEPT in security.-- Hunter doubts about production o
371d128 , being handed down? Are not Loddiges 1279 roses KEPT in same soil. same atmosphere?-- may they produ
376d138 xamine sexes of every Has repeatedly seen one he KEPT pull up feathers of tail of Hen; which lived wi
380d153 argument is VERY WEAK without knowing whether if KEPT they would have wings.--).-- Seep p. 84. Hens «

Page
(Key Word)
```
417e076 ecies: from one locality, all left whorled.-- He KEPT two to see if they would breed, It is difficult
439e142 ant probably would do-- or be with difficulty be KEPT alive.-- Nevertheless much probably depends  on
449e169 e male giving form-- they interbred. & the young KEPT constant. & all alike Waterhouse says some of t
504q014 seeds  alone remain to be compared-- Cabbages.-- KEPT true Try experiment (30/p.11) (2) Yew Berries g
527m033 nner vivid & grand. the frame of mind being just KEPT up by the music of the poetry.-- (therefore sin
544m102 e, or when so near waking. that an associated is KEPT up with waking thought.-- Ld Brougham thinks no
580n061 this precisely the same, as the double-conscious KEPT playing so well.-- Lr. Brougham «Dissert.» on s
613o035 re other senses come into play, when relation is KEPT up with distant object. where many such objects
601o08b Parfait  Chasseur, par Desgraviers, un Vol 8vo KERATRY-- Inductions morales et physiologiques The fir
065r138 . vegetation & climate of Tristan D. Acuneha. KERGUELEN Land. Prince Edwards Isld. Marion & Crozet. L
066r140 ms, none thicker than thumb» Sea weed said at KERGUELEN Isd. to grow on shoals like Fucus giganteus!
066r141 ucus giganteus! 24 fathoms deep 24» under 50. KERGUELEN land. = the way it stands gales = very strong
066r141 strong. Stones as bigger than a man's head.-- KERGUELEN 40 by 20 leagues. dimensions: Bynoe informs m
214b173 -- Galapagos--Juan Fernandez Falkland Islds-- KERGUELEN land.-- Phillips. Lardner Encyclop. insists o
400e012 f variation analogous to specific variations.-- KERR'S Collect of Voyages Vol 8 «p. 46» Capt Davis in
358d075 solutely necessary to watch our meat, while in KETTLES on the fire, & on one occasion, not withstandi
100a053 ns and cleavage conjoined very good.-- It is the KEY to the story.-- consider stalactites.-- agate ri
298c190 to passion, becomes habitually passionate.-- the KEY to the affections might perhaps thus be  found--
301c199 - do Elephants easily acquire habits is this the KEY to their mental powers.? p. 8. mistakes of insti
531m050 s placed.-- (How often one forgets where put one KEY, where all keys are placed) Memory cannot solely
089d019 ht of erecting machine to use if water fell. -- «KEYS off extreme point of Flori[da]> Excellent paper
531m050 w often one forgets where put one key. whetr all KEYS are placed) Memory cannot solely be number of t
275c119 animals, <as> I use (new step in induction) as KEYSTONE of ancient geography species tell of Physical
556m146 ting, they put down ears, when «turning round to KICK» kicking they do the same. although it is  then
556m146 in  face of donkey, horse & zebra. when going to KICK.-- Why does dog put down ears, when pleased.--
545m107 circumstances,  threw itself down on its back & KICKED & cryed like naughty child.-- Do monkeys cry?-
585n075 rses, asses, <mi> Zebras &c &c.-- Here there is KICKER but not bite.-- Henslow remarks that Chimpanze
556m146 ey put down ears, when «turning round to kick» KICKING they do the same. although it is then quite us
556m146 they  must yet continue to put down ears, when KICKING.-- -- good case of expression showing real aff
585n075 culiar manner.-- ]CD April 3d. 1839 The Giraffe KICKS with front legs & knocks with back of Head, yet
466t093 lie  in fairway to nectary.-- Is not this so in KIDNEY Bean. How is it generally.-- In Azalea <do> «i
467t103 ees «frequent» & seem to act, something like on KIDNEY Bean, they go to nectar at foot of upper petal
501q011 ects.-- (30) Get Abberley to plant SINGLE Peas, KIDNEY Bean & Bean, intertwined, «without sticks»-- i
165g124 examining)--  Inverness & waters of the Tarf-- KILFINNAN Tower where stream enters at head of which hi
301c199 rthy interposition of deity to teach squirrel to KILL ears of corn according to my views, habits give
423e096 the  vegetable in one quarter of the world would KILL all of the one herbivorous. & its one carnivoro
494q004 s (9) Cross largest Malay with Bantam-- will egg KILL Hen Bantam.-- Cross common Fowl with Dorking (1
496q05a whether  grains flaccid, as Koelreuter describes KILL Sparrow after feeding on oats, give body to Haw
496q006 ake Duck eat Spawn, eggs of snail, row of fish & KILL them in hour or two «My Father made hens cast H
523m017 rs. C. O. who threw herself out of the window to KILL herself from jealousy of husband connection wit
048r086 to  quote Burchell. V. where the Rhinoceros was KILLED. -- In Patagonia, are all beds same age? is wh
187b062 animals lived in cold countries & therefore not KILLED by <Siberia a more> cold countries.-- Seeing h
187b063 ers.-- Will Mr Lyell say that some circumstance KILLED it over a tract from Spain to S. America.-- We
211b163 bay  breed with dogs.-- the bitches never being KILLED by them, whilst they eat up the dogs.-- L.' In
247c028 as washed on shore at one of the Pelew Islds.-- KILLED a woman. Chamisso p. 189 Tome III: Kotzebue.--
251c040 s seen far at sea, migrations of species. geese KILLED in Newfoundland, with crops full of maize. (ge
294c177 ys he has seen in old Book last Bear in England KILLED in year 1000. reference to succession of types
333d006 eared «so» very like common cat, that they were KILLED, & other two very closely resembled in form of
345d043 with» black lip) & then calf «in both cases» is KILLED. Notes from Glen Roy Note Book.-- Why is not T
364d103 inst each other, till it has been known one has KILLED itself.-- Q Sir. J. Sebright-- has almost lost
375d135 an average, every species must have same number KILLED, year with year, by hawks. by. cold &c--. eve
429m115 line..  it is other plants.-- a broad border of KILLED trees would form fringe.-- but there is a cont
439e144 h further would determine his preservation-- if KILLED by some other animal, then that quality  which
481z010 n p. 92.-- great Kingfisher of Tierra del Fuego KILLED in Chile. Dobrizhoffer., Vol I p. 310. History
516q24v at order plants would reappear after <th> being KILLED Experiments not connected with Species  Theory
634j55r d reptiles first inhabit seas.-- Were they then KILLED out «by the increase cold», & did mammifers th
206b148 me causes are evident, as for instance one man KILLING another.-- So is it with varying races of man:
301c199 e individual p. 7. is not squirrel hoarding, & KILLING grains. acquirable through hoarding from short
423e096 of world with life.-- for otherwise a frost if KILLING the vegetable in one quarter of the world woul
437n139 & hares to the young ones to exercise them in KILLING them.-- "Sometimes it seems hares, rabbits, ra
508q016 Cape  any selection of Males in «cattle» or in KILLING the worst =or in dogs= (12) Do Hottentots gene
297c185 time alive ¿therefore other species mice & only KILLS them when urged by hunger.-- p. 65. Aberrant gr
081rIBC T} Toises Pieds Myriametre = 5130., 4. 5 inch KILOMETRE 513., 0. 5 Hectometre 51. 1. 10 Metre 3. 0. 1
059r121 of  all the Pitchstone (which I have seen). is a KIND of concretionary structure, for the interlineal
062r130 xtinguished by change of circumstances: The same KIND of relation that common ostrich bears to (Petis
098a047 es of feldspar in gneiss.-- Veins in septaria. a KIND of concretionary process (analogous to layers o
153g054 er side by, opposite entrance into Glen Fintec a KIND of landing place is formed Ben Erin summit 27.8
170b001 uary Zoonomia Two kinds of generation the coeval KIND, all individuals absolutely similar; for instan
170b001 n. bisection of Planariae. &c &c.-- The ordinary KIND <the> which is a longer process, the new indivi
173b011 arrivals others old ones, (of which none of same KIND had in interval arrived) might have grown alter
189b070 n, armadilloes «&» & Megatherium. each with same KIND of coat.-- If we could tell, I do not doubt eve
202b131 0 «12» feet beneath surface forest trees fallen «KIND well known, carbonized»--; clay fifty feet, the
224b213 o wrens forced to haunt two islands one with one KIND of herbage & one with other, might change organ
224b214 marcks» Without two species will generate common KIND, which is not probable; then monkeys will never
228b231 t slave holders wish to make the black man other KIND?» Animals with affections, imitation, fear <of
228b231 have  no more reason to expect the father of man KIND. than Macrauchenia yet he may be found:-- We mu
229b233 iled sheep aboriginal at Cape & a thinner-tailed KIND further inland.-- NB. There is division of snak
235b275 s. as American or Indian genus inhabit different KIND of localities.-- if so change The GRAND QUESTIO
242c017 . 22. New Calidonia-- New Ireland & Britain same KIND of dog, with those of New S. Wales. <V.> p. 123
244c021 st of the dogs of Payta-- belong to the hairless KIND, «said» to come originally from Africa p.  122.
245c023 esticated» like in general appearance to Siamese KIND.-- but considered good species from dental char
257c057 ent under peculiar conditions-- of «nearly» same KIND country distant. <Study> The circumstance of gr
260c067 ccount to show that very many birds of different KIND have been know to assist in feeding young cucko
263c075 & will never be conquered by anyone (if has any KIND of prejudices) <without> who just takes up & la
283c146 le with certain preexisting laws-- ..If only one KIND of plant not so many.-- The quantity of life on
290c162 ed. «as it speaks to amount of change only & not KIND» insects-- & vertebrata & plants. At first clas
297c186 infer extermination, when group few in number of KIND, extermination.-- New forms made through probab
298c190 rhaps thus be found-- a person who is habitually KIND to children,--<get> «increases» general instinc
298c191 al parts become occupied by a third best adapted KIND-- lower species would then revert to  pristine
299c195 ty. wild carrot. made into biennial domesticated KIND with large root by sowing it at wrong time of y
303c204 ntellect?) & what kinds of intellect) quantity & KIND of hair forms of legs-- hence the father of man
303c204 of  hair forms of legs-- hence the father of man KIND probably possessed a structure in these  points
305c210 t> sensible» principle (intimately allied to one KIND of organic matter.-- brain. & which <prin> thin
305c211 less forms, bearing a close relation in degree & KIND to the endless forms of the living beings.-- We
307c216 ere stump.-- Shall abortive organs «of very same KIND» in these cases, have plain meaning & none in o
312c231 .-- In the Bantam clubs, they used to fix on the KIND wanted, colouring of each feather weight & size
318c254 ghtingales»-- other birds (& this seems common «KIND» migration of America) migrate singly flying fe
318c254 s in flock, these birds seem clearly directed by KIND of country; «kinds of migration quite different
318c255 ight yet determine» which way to fly.-- There is a KIND of Wren (Bebyk??) which seems common in Rocky M
347d050 Africa  -- p. 4. sticks to genus or group of any KIND not being perfect till circular. p. 5 Most clea
351d092 Hunter's views on this subject.-- Monstrosities, KIND of determined by age of foetus.-- As Larvae may
352d058 but number of species with abortive organ of any KIND few.-- » hence become EXTINCT, & hence the IMPR
360d092 hitherto thought that a small difference <of any KIND>, if very firmly fixed from long time, made  no
360d092 ixed from long time, made no difference what its KIND was.-- but if it were opposed to the difference
360d093 would allow the offspring to have some different KIND of mottle, each feather partaking of  character
370d116 aper on bees in same work. (it is said that some KIND lay <pu> up honey even for single rainy day-- &
375d135 undred thousand wedges trying force <into> every KIND of adapted structure into the gaps <of> in  the
390d178 d.-- Every individual foetus would reproduce its KIND was it not for the necessity of some  change.--
```

391d174 re should be something «each time» added to that KIND of generation, which typifies the whole course
391d174 ng to acquire differences «indifferently of what KIND, either progressive improvement or deteriorate»
392d175 be added EACH TIME. but after some time]CD What KIND of plants are Monooecious or dioecious.-- very
400e012 ablishing good groups.-- ears varying so much,-- KIND of fur-- (do tips of ears take any colour?)-- l
403e021 differs from that of S. Europe p. 189. The gaut, KIND of crocodile, sometimes wanders from Pellew to
405e031 a trace of horizontal mark on flank.; & tail. & KIND of semilunar {P} mark on each side darker,, so
407e038 rated) of the world. from the <Tropical> Equable KIND of climate to the extreme.-- Therefore species
408e015 em symmetrically.-- it is easy to get 50 of same KIND of monstrosities.-- G. B. Sowerby-- Looking ov
411e055 causes. «All Cuviers generalization. of teeth to KIND of extremities come under this head» 27th Novem
419e085 living in the salt?.]-- very few animals of any KIND-- Fauna, must be very curious.-- With respect t
427e110 n of crab, would POSSIBLY «No, for pollen of any KIND would fertilize it» fertilize an apple somewhat
440e147 been so injurious to bird as to allow any other KIND of animal to usurp its place.-- & therefore the
442e153 on stocks of other varieties & we know that the KIND of stock greatly affects the Graft.-- ¶Plants c
460t017 ermine the kinds of monstrosity, & determine the KIND of variation «& sporting» in flowers & domestic
470t178 hink never on the Galeum saxatile & other common KIND--I think not on Phlox though they examine it.--
495q05a dog, which has swum-- on pools & rivers-- every KIND of seed must be distributed.-- Examine scum of
498q009 by seed, we might feel sure, that pollen of own KIND is much more effective than of foreign-- Eyton
525m024 ar.-- My father thinks that selfishness, pride & KIND of folly like (Mr George S.) is very heredetary
528m037 of the organ of sight, which is common to every KIND of view-- as likewise is novelty of view even o
533m061 g, but hurting my character I felt it)-- this is KIND of conscience, is obscure memory of having read
536m074 views would make a man a predestinarian of a new KIND, because he would tend to be an atheist. Man th
547m112 mory, for it related to past idea.-- there was a KIND of ideal consciousness for moment, implied by «
553m136 t capable to <do> produce every effect, of every KIND which surrounds us. Moreover «it would be diffi
555m143 having changed hanging into his head cut off) as KIND of wit, showing he had honourable wounds.-- all
555m144 showing he had honourable wounds.-- all this was KIND of wit.-- I changed I believe from hanging to h
580n061 ren loose ideas of time?-- Characteristic of one KIND of intellect is that when an idea once has hol
580n062 rence of intellect between animals & men only in KIND.-- probably very important work.-- Feb. 12. 183
581n063 of better life p. 207 March 16th.-- Is not that KIND of memory. which makes you do a thing properly,
582n068 teristics;-- they are more truly sensations??. a KIND of mental pain & pleasure.-- The Revd. Algernon
586n078 es because power varies in breeds,» something of KIND oneself knows in walking [one feels inclined to
596n184 of whole forehead & 2 ears.-- emotions of every KIND.-- «[Are monkeys <are> right-handed??]CD» Cyano
600o08v oks to eternity. (4) Reason, some transcendental KIND-- (5) Conscience, not clear-- Then these last h
614o037 l than thoughts-- One organic body likes one <m> KIND more than another-- What is matter? the whole a
615o038 science This «X» memory especially «the» general KIND taking pleasure in virtue because acquired in p
616o038 d; so love of virtue enhanced by this heredetary KIND of memory.-- The difference between heredetary
629o55v -- ((hence desires do not intervene between this KIND of conscience & the will, though <this> conscie
639j28v icable to Bell's sneering-theory.-- p. 263. This KIND of doctrine runs through Macculloch, the bills
299c186 on We see gradation to mans mind in Vertebrate KINGDOM in more instincts in rodents than in other an
619o042 lence towards fellow <living> creatures, or of KINDNESS to wife [RHC] 2) and children would give him
062r130 n that common ostrich bears to (Petisse. & diff KINDS of Fourmillier): extinct Guanaco to recent: in
170b001 bly ended in beginning of February Zoonomia Two KINDS of generation the coeval kind, all individuals
171b003 ance,-- seeds of plants sown in rich soil, many KINDS, are produced, though new individuals produced
182b048 ny sheep from Arabian coast. "These were of two KINDS one «white» with a black face, & similar t
188b067 able two Northern species replace <No> Southern KINDS-- (I) Gnu reeaches Orange river & says so far w
195b100 pecies could, arrive-- did it only create those KINDS not so likely to wander. Did it create two spec
210b159 alogy to species & genera.-- for instance three KINDS of greyhound.-- In plants. do the seeds of mark
211b162 eedings. Jan 1837, «by Eyton» Account of three, KINDS of pigs. difference in skeletons: VERY GOOD. Ap
219b193 ht bring in stomach-- &c &c. (Mem discover what KINDS of seeds. these plants) [Mem Fact stated by Mr
222b205 s) Case with fish-- as some of the most perfect KINDS the shark. lived in remotest epochs.-- ¿ lizard
227b226 a successive one.-- It leads to knowledge what KINDS of structure may pass into each other: now on t
236b279 , decidedly next species to some South American KINDS.-- Are the closest allied species always from d
240c004 plants.-- All these facts clearly point out two KINDS of varieties.-- One approaching to nature of Mo
241c015 mon ¿species? with New Guinea.-- Many <genera». KINDS common to New Guinea & rest of isle in E. Indi:
255c053 sant change in her offspring. ,has invented all KINDS of plan to insure stability; but isolate your s
260c068 bird, in same way Wilson (p. 5). describes many KINDS of bird uniting together in pursuit of Blue-Jay
260c069 f valuable facts,. regarding habits range & all KINDS of information-- instinct Swainson's remarks in
267c093 Ras Mohammed & Jeddah". sheep numerous "of two KINDS one white «a» black face, & similar to tho
280c134 tame rabbits good instance-- instincts of many KINDS in dogs, is clearly applicable to formation of
303c204 ge of head «& features» (hence intellect?) & what KINDS of intellect) quantity & kind of hair forms of
315c243 ng. Barking to tell other animals in associated KINDS of good news. discovery of prey.-- arising no d
316c244 ge of shells like Cryptogamic plants. of Marine KINDS, there are some restricted genera, but then the
318c253 the birds seem to follow narrow bands, certain KINDS as gallinules taking the low country near coast
318c254 irds seem clearly directed by kind of country; «KINDS of migration quite different in species of same
356d069 pect any amount of change from eating different KINDS of food: grazing animals who eat every species
372d131 o caterpillar state; the vast difference of two KINDS of generation shown by their happening in same
402e020 ies are peculiar to them" do-- p. 368. "Several KINDS of animals have spread from the N. end of Borne
402e020 oo we find the elephant-- in Magindanao several KINDS of the large monkeys.-- Fewer <of th> «Mammalia
419e085 ublished Fauna of Caspian.-- fishes fresh water KINDS. (yet living in the salt?.)-- very few animals
439e145 like greyhound-- fox-dog-- turnspit & two other KINDS It seems absurd proposition, that every «buddin
443e154 he best way to get young trees, from worn-out KINDS, & quotes from Pliny, that it is bad to graft f
443e154 be overturned, then the conclusion that the two KINDS of generation have some most important differen
460t017 ry possible way.-- the laws which determine the KINDS of monstrosity, & determine the kind of variati
462t051 ss this objection, when one thinks of different KINDS of cattle in every part of England. &c &c NB. I
495q05a Examine scum of pond for seeds.-- 11. Soak all KINDS of seeds for week in Salt. artificial water.--
497q006 ransported, or how long they remain dormant. if KINDS come up, not found in wood.-- but seeds continu
500q010 yme. Horned Oranges. Spallanzani Essay-- Fig 2 KINDS of flower annually.-- Periwinkle. (not asclepia
515q021 duce as large capsules of seed, as the commoner KINDS-- Cattle are horned, Suffolk have <abortive> «n
612o035 kewise «does». [LHC] ¿ in Corallina are not two KINDS of life vegetable and animal strictly united? [
615o038 or under contingencies of stimulants of certain KINDS such secretion) or an association of pleasures
623o051 dency to them, as <social> animals of peculiar <KINDS> «social feelings», & living under certain cond
638j28r tributes of Deity Vol I. p. 251-- stomach hump, KINDS of foot. power of closing nostril, foot, sack.
246c026 at they are varieties & not species.-- Vol :694. KING-fisher of Europe, (Alcedo ispida) from Mollucca
258c061 cquarrie isld. vol III p 430 alluded to by Capt. KING do. p. 434. Table of birds from Cuba Vigors.--
323c269 es at end of each Chapter» Crabbes Life June 1s. KING & FitzRoy To be read Humbold. New Spain-- Much
483z012 irds".-- 18 do. Vol III p. 422. letter from Capt KING on birds of St of Magellan. Very inaccurate & V
603o0012 to many races»-- thinks action towards <man> «a KING <changed into> is carried on toward deity.-- &
603o0012 changed into> is carried on toward deity.-- & as KING might like cruel pleasure, so sacrifices cruel.
072r160 Fig. 10] {P} motion from within and without H. KINGDOM N. Spa. Vol III p. 113 "Nature exhibited to th
104a068 with him that <th> gneiss, mica-slate of whole KINGDOM of Norway was contorted yet no mountain chain
174b015 t Australia-- Will this apply to whole organic KINGDOM, when our planet first cooled.-- Countries lon
254c048 392.-- except generation & digestion in Acrite KINGDOM all organs blended together, & same organ «whe
262c073 ese «aberrant» varieties will be formed in any KINGDOM of nature, where scheme not filled up, (most f
377d140 y of change of form, & instincts in the animal KINGDOM.-- It is the unit of our calendar-- epochs &
423e097 aguar & tiger may not be so.-- Considering the KINGDOM of nature as it now is, it would not be possib
427e110 ve travelled S. From the analogy of the animal KINGDOM I should suppose, that the pollen of crab, wou
568n018 urring sounds in Harmony common to t[he] whole KINGDOM of nature. If I want some good passages agains
196b109 most dependent on vegetables. of the two great KINGDOMS.-- Principes de Zool: Philosop:-- I deduce fr
269c103 not improbable.-- After reading "Carus on the KINGDOMS of Nature, their life & affinity" in Scientif
581n064 ralize.-- The tastes of man, same as in Allied KINGDOMS-- "food, sm<e>ll. (ourang-outang), music, col
185b05i4 nd «do. <w> parrot.--» woodpecker-- a desert. KINGFISHER.-- mountain tringas.-- Upland goose.-- water
189b070 Galapagos). Little wings of Apteryx Dacelo & KINGFISHER same colours Strong odour of negroes, a poin
274c114 of habits going before structure).-- even one KINGFISHER-- Gould has seen with long tarsi.-- «Ground
377d147 nly habits & instincts-- The young of the <p> KINGFISHER (.p. 169) has the colour on its back bright
378d148 like parent, so is species old: Hence <young> KINGFISHER & pies, have long had their present plumage.
437e138 ng till they drop off their perch.-- p. 101-- KINGFISHER in northern part of England stationary, in
481z010 nularia Alcedo stellata. Meyen p. 92.-- great KINGFISHER of Tierra del Fuego killed in Chile. Dobrizh
274c114 ound woodpecker» Secretary bird.-- & Millisuga. KINGII very rasorial for type.-- Now here I must obse
147g021 Friday Highest part of road between Inverorum & KING'S House 28.935/82 degree A Temp of Air 65 degree

219b196 le from T. del Fuego.-- Ellis (?) says Tahitian KINGS. would hardly produce from Incestuous intercour
306c213 es &c Timor is connected with Australia «map to KING'S Australia» by a bank of soundings of which «th
385d163 dd loosing visible powers» in Zoolog Gardens. & KINGS at Otaheite) <Think> Last litters are considere
481z009 Vol. I. p. 244. Baccalao. migratory fish.-- See KINGS drawings.-- for real name Birds of Iceland. Mac
483z013 ntage. with Cumms collections & my own: & Capt. KING'S p 453-- Planariae velellae (Less) parasitical
522m011 large fortune left. him, took name of Child «of KINLET» & married Miss A. B.-- all the same names as
522m010 My F. asked him, did he know whom Mr Child «of KINLETT» had married.-- Answered never heard of such a
204b141 first season bred readily with common ducks.--» KIRBY all through Bridgewater errs greatly in thinkin
205b142 orders, Carnivora, rodents &c, JUST COMMENCING. KIRBY says (not definite information) West of Rocky M
205b143 subject to vary where intermixture precluded.-- KIRBY Bridgewater Treatise There are some good accoun
496q05a inal cause of beauty of flowers-- contrasted by KIRBY-- with animal reproductive system.-- -- cover f
543m098 ptera;. <therefore> than in other orders (study KIRBY with this view) therefore there is Instinctual
613o036 onsciousness than our intestines have? [RHC] 5) KIRBY thinks that <all> there is one one instinct to
613o036 e & bring it. ---- Agrees with ONE animal [RHC] KIRBY extends instinct to plants, but surely instinct
613o036 from organization of brain; «[LHC] not used by KIRBY» (:analogy:-- as races are formed or modificati
638j059 habit.-- New theory of instinct, returning to KIRBY'S view.-- Macculloch. Attributes of Deity Vol I
574n041 ure of disgusting lewd old man. ones tendency to KISS, & almost bite, that which one sexually loves i
467t099 y is equally abortive--and both growing within KITCHEN Garden.-- As we see in Hybrids that although a
573n037 re expressive than movements of fingers.-- like KITTEN with mice.-- A person with St Vitus' dance bad
331dIFC ds. productive like geese?-- Are the number of KITTENS between Lion & Tiger at litter as numerous as
333d005 <un>common cat, exact variety unknown., three KITTENS, alike each other, partaking <more> «very clos
333d005 bred with unknown common house cat.-- had four KITTENS. two appeared «so» very like common cat, that
333d008 Tigers are exactly intermediate in character & KITTENS alike each other.-- Even in children of parent
366d108 ithout front legs-- -- the former of which had KITTENS with imperfect ones.-- now Sir J. Sebright. th
432e124 therefore the children do not, (& in hairless KITTENS we see same fact) go back, & this is argument
525m025 A cat had its tail cut off at Shrewsbury & its KITTENS <h> (in number 3) had all short tails; but one
525m025 le longer than rest «they all died»:-- she had KITTENS before & afterwards with tails. My father says
556m146 n quite useless-- Cats kneeding when old, like KITTENS at the breast now if horns were to grow on hor
075r167 as there not been vein «of iron» discovered?-- KLAPROTH analysed silver ores from Peru consisted of n
486z018 a greater duration of Heat. hence musquitoes & KNATS abound during short summer far N. where this ot
556m146 same. although it is then quite useless-- Cats KNEADING when old, like kittens at the breast now if h
572n032 s, but the manner, whether by bowing the body, KNEELING, prostration «uncovering body» &c &c is matte
204b141 ngst themselves or with parent birds.-- «W. Fox. KNEW of case of male widgeon, winged & turned on poo
264c079 ery word said-- see its affection.-- to those it KNEW.-- see its passion & rage, sulkiness, & very ac
312c228 ng most like <parents> Mother.-- like dogs Smith KNEW chinese hairless dog & common spaniel crossed.-
334d010 ng guided entirely by their smell.-- Fox says he KNEW «a» carter well, who placed his stallion as sec
376d137 y baboon & monkey, big & little that ever he saw KNEW women.-- he has repeatedly seen them try to pul
376d137 curiosity to pull up trousers of men. Evidently KNEW <men> women, thinks perhaps by smell.-- but mon
470t176 s-- My Father remembered when in the gardens, he KNEW there was none but English,--the Palmated was i
522m011 - Thus in many things if he began at one end, he KNEW the whole subject.-- if at the other nothing.--
545m106 nnot remain still.-- I do not doubt this Baboon. KNEW women.-- Another little old American monkey «(M
568n019 Purdie, (beginning of Vol V) «finally» says "he KNEW no more what is pretty & what ugly than a cow-
022r008 athern sack, very thick & so tough that a sharp KNIFE could not cut it: in which we found the Head &
121a111 gh <to form> for potters to use. in which great KNIFE formed crystals of ice were formed-- (like my g
546m111 mus had almost same thing happen to him about a KNIFE. which he had hid some years before.-- was grea
417e077 ld find its place in the Systema Naturae.-- Mr. KNIGHT makes this analogy between grafting & sexual u
435e130 ange their sex,-- this effect from age, what Mr KNIGHT [not located] the stigma retains its power.--
454e184 of great age» as must be inferred from what Mr KNIGHT says. Hort. Transat. V. II p. 252. Is there an
491q01v s in case of woodpidgeon & Hen. mentioned by Mr KNIGHT. Vol IV Hort. Transact.-- 4 May we no suppose,
497q007 ains numerous spec. of Cape Heath by facility. ¿KNIGHT take opposite view. Gaertner talks of the seve
442e152 s, throws enormous difficulty in the way of Mr KNIGHTS. theory «without seeds are freshly transported
442e153 e thus characterized study Von Buch.]CD Now Mr KNIGHTS statements about fruit trees. grafted. alterin
446e162 r very rarely») here is a case in answer to Mr KNIGHTS doctrine.-- Case like Corallina-- «Does it flo
501q011 ted or cuttings taken or tuber-- talk about Mr KNIGHTS theory with Henslow.-- Dr. Fleming says yes. (
513q21. -- says many flowers are dichogamous Zostera-- KNIGHTS notion of pollen & stigma generally not being
576n046 ter Scott «(Antiquary)» Vol II p. 126 says seals KNIT their brows when incensed.-- A Dog may hesitate
073r161 form masses have layers been accumulated, round KNOBS, or pushed where soft, or redissolved soft.--/
366d112 ot alter form may be inferred from Australians KNOCKING out teeth.-- the account of the people on the
585n075 il 3d. 1839 The Giraffe kicks with front legs & KNOCKS with back of Head, yet never puts down its ear
527m032 - Beauty is instinctive feeling, & thus cuts the KNOT:-- Sir J. Reynolds explanation may perhaps acco
554m139 The male» «I saw» Jenny untying a very difficult KNOT-- the sailor on board the ship could not puzzle
023r012 nry's vessel from the Friendly Isles. to Sydney; KNOW by having been seen & from the contents of its
028r029 e Tropical oceans detained by Organic powers. We KNOW the waters of the ocean all are mingled. These
030r035 On open coast, near where Challenger was lost: I KNOW no reason for supposing these matters are not n
062r128 of Cordillera, where climate similar.--I do not KNOW botanically = but picturesquely = Both N & S. g
113a090 of heat from centre.-- But is this not wrong? we KNOW mean of surface formerly much higher, «so» that
114a095 of Good Hope.-- A bare hill of greenstone, if we KNOW origin of greenstone tells subsidence as plainl
124a119 rcent of lead. it would be still more curious to KNOW whether it would be absorbed.-- (2d) By exactly p
132a138 e on earth by volcanic action.-- <Why> now as we KNOW volcanic action prevails more beneath the sea,
165g125 dormant-- Man from Glen Turret said he learnt to KNOW the lambs because most like Mother in face-- as
171b002 ife short, Why such high object generation.-- We KNOW world subject to cycle of change, temperature &
171b003 ns to vary. or adaptation.-- Again we <believe» «KNOW» in course of generations even mind & instinct
181b044 in type, & the gradation will be sudden-- Heaven KNOW whether this agrees with Nature: Cuidado The ab
186b059 iversality of such law.-- It would be curious to KNOW in plants, (or animals) whether, <in> races hav
188b069 ferent localities, so that they become useful to KNOW what is species.-- In proof that structure is n
192b085 insensible, that each is not more change than we KNOW varieties can produce.-- Therefore all genera M
195b098 - Such an influence Must exist in such spots. We KNOW birds do arrive & seeds.-- The same remarks app
195b100 . not new at Galapagos.-- did the creative force KNOW that «these» species could, arrive-- did it onl
195b100 allied to Mus. coronata, but not coronata.-- We KNOW that domestic animals vary in countries, withou
200b125 erminating.--- It would be curious experiment to KNOW whether soaking seeds in salt water &c has any
210b159 ies of plants not cryptogamic but I . 2,53.-- In KNOW varieties. there is analogy to species & genera
216b180 hus act.-- Now the point will be to find whether KNOW varieties in plants do so.-- As in cacti &c &c.
218b192 -- we see a beginning to isld. Graham isld.-- we KNOW many seeds, might be transported some blown--fl
219b194 explained more philosophical to state we do not KNOW how transported.-- (Glaciers might have acted a
225b218 with cloven hoof) Ask Entomologists whether they KNOW of any case of introduced plant, which any inse
227b224 g ascertained means of transport, we should then KNOW whether former lands intervened.-- (2d) By char
227b224 al reefs. shallow water at Melville Isd. (3d) We KNOW that structure of every organ in A. B. C. three
227b224 > structures in two genera.-- <we then cease to KNOW the steps.> although D E F. follow close to A.
227b228 -«» & generation, causes of change «in order» to KNOW what we have come from & to what we tend.-- thi
230b240 dence of two races of plants run wild.-- (for we KNOW that such can take place without impregnating e
231b241 ers varying in wild state to show that we do not KNOW what amount of difference prevents breeding;--n
241c015 ongatabou. & New Ireland.-- Gould will hereafter KNOW about birds of N. Zealand I'Institut. 1838. A D
248c029 point, or migrated from those quarter, where we KNOW quadrupeds have existed for ages.-- ∴ The most
252c045 stinct that all analogy from each other I do not KNOW how different Sumatra Java ----------- do from I
257c056 le, but induction leads to other view.-- Till we KNOW uses of organs clearly, we cannot guess causes
260c067 that very many birds of different kind have been KNOW to assist in feeding young cuckoo; as if there
265c087 e have only ornithorhyncus, then we should never KNOW how much structure was connected with habits, &
266c091 ners they are as opposite as day & night: yet we KNOW how remote the periods at which both left the l
278c129 s & subgenera; are analogical, because we do not KNOW, whether nearest species of each might not bree
283c146 royed far remote genera. will be produced. As we KNOW from Ehrenberg, there are fossil (see Scientif
286c154 signals.-- animals understand the language, they KNOW the crys of pain, as well as we.-- It is our ar
286c155 east every nation has. done so as. yet.-- We now KNOW what is the natural arrangement, it is the clas
288c159 till circles completed Major Mitchell, does not KNOW whether breeds of oxen have deteriorated, or al
289c162 emaining through winter. «It would be curious to KNOW whether a variety could be transmitted more eas
292c169 , otherwise it would be unlimited. We absolutely KNOW that tendency is greater in Mammalia, than in s
303c204 es in This respect upper & lower, which I do not KNOW whether it <would have> differs in present race
309c219 The only cause of similarity in individuals «we KNOW of:», is relationship, children of one parent,
341d032 n an egg.-- a most curious bird, did not seem to KNOW itself,. at last associated with the ducks.-- m

Page
(Key Word)

343d039 000 years Negro probably a distinct species-- We KNOW how long a Mammal may go on as one species from
345d043 -- A Sphepherd of Glen Turret. said he learnt to KNOW lambs, because in their faces they were most li
349d052 y owing to number of lost links.\ if all species KNOW they would be innumerable\ does not know any di
349d052 species know they would be innumerable\ does not KNOW any difference between permanent variety & spec
362d100 h it appears that all the <bird> varieties «,now KNOW» were then <pr> existing.-- he has also some ve
377d139 ants lived. & when present animals-- lived.-- we KNOW the great time, necessary to form channel & (&
380d153 ffect of breeding in & in.-- Mr Yarrell does not KNOW of any case of old Male. becoming like female,
385d163 take, probably would not be fertile» without she KNOW & LIKES HIM & then is actually obliged to be he
397e003 ithout the immediate agency of the deity. But we KNOW from experience! that these operations of what
397e004 one race to some change of circumstances; now we KNOW how slowly & insensibly such changes are in pro
398e005 ave studied history of the world most closely, & KNOW the amounts of change now in progress, will be
399e006 thick & yet have same fossils. does not Lonsdale KNOW some case of change in vertical series: Look at
403e023 animal (as Toxodon) & say its relations.-- if we KNOW its congeners then we can.-- now on my theory t
408e043 xtermination becomes part of same law.-- When we KNOW what a great effect. light has in colouring pla
411e054 orelations, & changes of individual organs, must KNOW whether the individuals «forms» are permanet, a
414e060 , since Cretaceous period, whether progressive I KNOW not.» (& insects.-- Stonesfield????). Have Mamm
423e099 have most slightly modified organic forms.-- we KNOW not rate of deposition has been equal even in o
429e113 of varieties of domesticated animals, yet as we KNOW how many plants have been produced (look at the
429e115 at 1600 metres precisely & stops at 2600. & yet KNOW that plant can be cultivated with ease near Lon
442e153 er are, namely on stocks of other varieties & we KNOW that the kind of stock greatly affects the Graf
465t079 Lund more Mammals, than at present «in Europe we KNOW there has been several successions of Mammals.--
494q004 & Common) &c: length of life. (5) Does my Father KNOW any case of quick or slow pulse being heredetar
498q008 much-- or the minute Orthopt.-- important, as we KNOW how readily they cross.-- (9) In the nurseries,
499q009 is Dioecious-- seedsmen who raise Hop-seed-- may KNOW something about proportion of plants necessary
503q012 mals not breeding when tame in India?-- does not KNOW About Yaks. & other Hybrids-- Dogs &c &c 38 Doe
506q015 of my questions amongst agriculturists. whom he KNOW.-- Col. le Couteur on Wheat.-- (41) Have any mo
506q15v (44) Zostera. Has he seen it in flower? does he KNOW Botanist who does-- What is Ruppia Bennett says
507q016 les of a family-- Where one tooth aborts, do you KNOW whether any trace in germ. (2) Any more cases o
509q017 l= Knows nothing [<...>] It is very important to KNOW, whether Gould's observation holds good, that i
511q018 secondary male characters appeared.= (4) Does he KNOW any seed-raisers (5) List of qualities in birds
521m010 said, who is tha? Mr C. answered why do you not KNOW, that is A. B my gardener.-- Thus was he in eve
522m010 iation were called up.-- My F. asked him, did he KNOW whom Mr Child «of Kinlett» had married.-- Answe
522m011 could repeat the alphabet straight, but did not KNOW [Z]CD when heard isolately.-- In old people. (A
522m014 when a person is tired is akin to insanity.-- «I KNOW the feeling also of depression, & both these gi
522m014 oth these give strength & comfort to the body» I KNOW the feeling, thinking over somebody who has, pe
525m026 y generations in families.-- My fathers does not KNOW whether trains of insanity are heredetary in an
526m028 <ani> pictures, an early taste, of animals. they KNOW.-- pleasure of imitation (common to monkey), &
527m031 ngland, in Tierra del Fuego not one.-- now as we KNOW birds learn from each other «though different s
533m058 iring language.-- Hensleigh. W. says that babies KNOW a frown very early in life, <before they> (I th
535m064 ay, but bit pencil when touched with it-- do not KNOW their own larvae, but one female may be moved t
535m069 he tracing facts to laws. without any attempt to KNOW their nature.-- Reviewer considers this profoun
543m097 e.-- <That> How far they communicate not easy to KNOW,-- but this capability of understanding languag
546m108 art life of. p. 27), says <sympathy> we can only KNOW what others think by putting ourselves in their
551m129 partly out displeasure (& partly out of I do not KNOW what when it looked at the glass) when pouting
554m140 e has done wrong & will hide herself.-- I do not KNOW whether fear or shame.-- When she thinks she is
556m144 Philosophy of Shame & Blushing]CD «Does Elephant KNOW shame-- dog knows triumph.--» Sept. 23rd. Horse
566n011 xplain origin of idea of deity.-- Animals do not KNOW they have `these necessary notions any more tha
567n014 g» on principles, which even animals practically KNOW «art precedes science-- art is experience & obs
571n029 & hence become heredetary; on same principle we KNOW many tastes become acquired during life time:--
580n061 subsequent ones modify it.-- «Weak people say I KNOW it because I was always told so in childhood.--
585n074 n youth (Mem. Mr Worsley's story of chicken) to KNOW that which we touch & what [...] the same.--(th
585n075 me.--(this Hensleigh therefore problem is how we KNOW that thing is same, which touches two parts of
585n075 es one part. very quickly successively.--» [& we KNOW from experiment of crossing fingers, that we on
585n075 experiment of crossing fingers, that we only do KNOW that it is one, when applied in peculiar manner
585n077 s is not instinct, but a faculty, or sense-- "We KNOW not how, stonge henge raised, yet not instinct,
586n078 ition. with respect to pidgeons, is that they do KNOW from look of Heavens, points of compass, & they
586n078 om look of Heavens, points of compass, & they do KNOW which way they go; & so return.-- «but does not
586n079 s of compass may be instinctive. it is a test to KNOW how much of the wonder consists in the action b
593n107 state social) by smell or looks. but it does not KNOW its own smell or look, & therefore there must b
593n109 iew looks good-- then man, a socialist, does not KNOW other men by smell, but by looks. hence. some o
596nIBC ed round & round in fainting state would it then KNOW its direction.-- In slight convulsions. are the
606o024 inter or sculpture, but rather subjects which we KNOW, it is therefore the embodying of a floating id
608o027 except man who has thought very much, & he will KNOW his happiness lays in doing good & being perfec
609o029 between conscience & impulse) but shame «we alas KNOW» is far easier conquered than the deeper & wors
616o039 bodies without these faculties & indeed until we KNOW what answer they would give in support of their
618o041 vely? -- ? the brain only objectively. We do not KNOW attraction objectively 6) The reason why though
618o41v other, & <the> (or conceive it) & that is all we KNOW of attraction, but we cannot see an atom think:
619o043 cting in» accordance to his instincts, <he would KNOW that many experienced pleasure,> & by associati
620o044 Whatever the cause of this may be, everyone must KNOW, how soon the pleasure from good dinner, or fro
621o047 e he has the right & wrong in his mind.-- Now we KNOW it is easy by association to give «almost» any
633j53v y they not be accidental? We have good reason to KNOW that they would not be detrimental accidents, &
635j55r terly useless-- it foretells nothing» because we KNOW nothing of the will of the Deity. how it acts &
635j55r nconstant like that of Man.-- the cause given we KNOW not the effect [blank] 6 p. 412. Macculloch exp
638j58v ng structure, instead of parts of body.-- Now we KNOW what instinct is-- consider this I look at ever
062r129 to change of circumstances; reversed argument. KNOWING it to be a desert.-- Tempted to believe animal
096a041 Malcolmson in Paper on India gives reason for KNOWING that Mur. Soda. and Carb of lime decompose eac
199b116 pe with the great continents we get a means of KNOWING movements.-- How can we understand excepting b
266c088 in sterile countries.-- Gould insist much upon KNOWING to what type a bird belongs.-- I conceive with
266c088 what type a bird belongs.-- I conceive without KNOWING from which country many birds come it would be
288c159 ally crossed would continue to cross, means of KNOWING directions: mysterious. Were the woodcocks,. w
301c199 heory must encounter all these difficulties.-- KNOWING that animals have some reason, & actions habit
309c221 wo classes however different.-- Male glow-worm KNOWING female good case of instinct. bees turning neu
380d153 them-- (but this argument is VERY WEAK without KNOWING whether if kept they would have wings.--).-- S
418e083 sexual organs, different in each species,-- & KNOWING from analogy, that all these very animals with
533m058 is precisely analogous or identical, with bird KNOWING a cat, the first it sees it.-- it is frightene
533m058 e first it sees it.-- it is frightened without KNOWING why-- the child dislikes the frown without kno
533m058 ing why-- the child dislikes the frown without KNOWING why-- a man as in Guy. Mannering. feels, pleas
533m059 in seeing the scenes of his childhood without KNOWING why-- had not conscious of recollecting it-- t
567n013 er teeth from the straw, & just like child not KNOWING what to do with them, came several times & ope
571n029 ntal & Bodily Consider case of grazing animals KNOWING poisonous «herbs:» & man not.-- ¿no vegetable
578n055 nto his face,-- "as <she> «the» thought of his KNOWING «it», suddenly came across her, the blood rush
595n121 guaze over her head. & came near him, although KNOWING it was Snow.-- Is this part of same feeling wh
608o027 perfect, & therefore will not be tempted, from KNOWING every thing he does is independent of himself
610o032 ther says he has heard of many cases of ideots KNOWING things, which are often repeated in a wonderfu
190b079 me, that he though species became obscurer as KNOWLEDGE increased, but genera stronger,-- Mr Waterhou
195b104 ng so many thousand miles distant.-- Absolute KNOWLEDGE that species die &. others replace them-- two
227b226 of animals & a successive one.-- It leads to KNOWLEDGE what kinds of structure may pass into each ot
227b226 en extremities of any great divisions) thus a KNOWLEDGE of possible changes is discovered, for specul
252c045 ortance of local faunas foundation of all our KNOWLEDGE especially great continents» Give Specimen of
252c045 ra Java ---------- do from Indian increase of KNOWLEDGE would probably tell more certainly Get Closer
252c045 eties in N America» «some doubt, from want of KNOWLEDGE of time-- Analogy from three first will give
292c170 tion "Let anyone even with a very superficial KNOWLEDGE «like myself» of <classi> real affinities. ie
304c206 ng at making a series, which would render our KNOWLEDGE a chaos: who will doubt this if series now ex
308c219 has no original ideas., it will show state of KNOWLEDGE «Negroes existed since time of earliest Egypt
325c267 used as tendril into stump Library of useful KNOWLEDGE. Horse, Cow, Sheep.-- Very. Philosophie d'Hi
327cIBC itu et analogia commentatia Library of Useful KNOWLEDGE on Horse & Cow & Sheep Clarke's Travels.-- Te
348d051 eal relationship is doubtful.-- not till much KNOWLEDGE is elicited.-- It will rest upon the discover
357d074 d for long transportation, seems «?» to imply KNOWLEDGE of whole world-- if so doubtless «part of» sy

```
360d089 e Yarrels law. chief trust must be in general KNOWLEDGE of breeders, where their interest is concerne
364d104 , by crossing-- It seems from Lib. of Useful. KNOWLEDGE that sheep originally. black. & Yarrell think
372d130 le comprehended in itself.-- it must have the KNOWLEDGE how to grow, & therefore to repair wounds-- b
376d138 : Ourang Jenny with Tommy.-- Good evidence of KNOWLEDGE of Woman-- ▯ The noise st st. which the C. Sp
386d166 is part of the reproductive system.-- of that KNOWLEDGE of the part, of what is good for the whole.--
390d179 Suites de.-- Horse & Cattle Library of Useful KNOWLEDGE Bell's Quadrupeds the effects of breeding in,
523m015 ty coming on.-- his struggles against it, his KNOWLEDGE of the untruth of the idea, namely his povert
526m027 l & probably the only one affected by various KNOWLEDGE which is not heredetary & instinctive) & the
527m032 f birds, not being instinctive, is heredetary KNOWLEDGE like that of man, & this agrees with the stat
534m061 ink-- as well as animal born with instinctive KNOWLEDGE.-- but if so, yet this knowledge acquired by
534m061 instinctive knowledge.-- but if so, yet this KNOWLEDGE acquired by senses,-- then thinking consists
538m079 ects & with store of accurate & even profound KNOWLEDGE or other & unusual line-- both odd appearance
538m079 appearance about eyes.-- one botanist & great KNOWLEDGE of Irish Politics, «both bad jokers.--» the o
543m099 rs have many,-- great powers of communicating KNOWLEDGE to each other-- August 23d. Jones said the gr
553m136 as   does that philosopher who says the innate KNOWLEDGE of creator <is> «has been» implanted in us (<
553m136 plicit faith that through the goodness of God KNOWLEDGE has been communicat to us».-- & that it does
554m141 k, if he drops it, the bird will fly at him-- KNOWLEDGE.-- Sept. 13th It will be good to give Abercro
564n005 by   experience is supposed in man to guide to KNOWLEDGE, was transmitted perfectly to chicken so as t
568n019 against, opposition of divines to progress of KNOWLEDGE. see Lyell on Scrope, Quarterly Review. 1827?
583n070 Mr Wells I can see mentally refers by reason KNOWLEDGE gained by reason: & then these qualities of i
583n070 on may be used as argument.-- for instinctive KNOWLEDGE is not gained by instruction, or imitation.--
584n071 social, «subordinate to,» self preservation, (KNOWLEDGE of enemies). use of muscles, progression.-- u
584n071 e of muscles, progression.-- use of senses.-- KNOWLEDGE of location ducks & turtles running to water,
585n077 it would be instinct-- instinct is heredetary KNOWLEDGE of things which might be «possibly» acquired
586n078 ating to certain quarter is instinct, but his KNOWLEDGE of that quarter,, is faculty, whether by sun,
586n079 thout consciousness & by habit, such habit of KNOWLEDGE of points of compass may be instinctive. it i
586n082 bly be> «probably is» instinctive, namely the KNOWLEDGE of size is merely judged by eye, & use of lim
599o007 rant that man, from the effects of heredetary KNOWLEDGE, has produced almost → greater changes in the
608o026 his desire to do arises from motives.--& his KNOWLEDGE that it is good for him effect of Education &
610o033 nt. to Coleridge, is regarding the sources of KNOWLEDGE.-- whether <we th there> "anything can be <an
610o033 "anything can be <any> «the» object of «our» KNOWLEDGE except our experience".-- is this not almost
610o033 according to usual definition, there is much KNOWLEDGE without experience. so there may be in men--
612o035 s satisfactory, though not adding to positive KNOWLEDGE. lessening amount of ignorance [RHC] The radi
618o41v omething in addition to matter is because our KNOWLEDGE of matter is quite insufficient to account fo
638j059 llusca the nervous system is endowed with the KNOWLEDGE of trying a hundred schemes of structure, in
638j059 ep».-- in Man, the nervous system, gains that KNOWLEDGE, before hand. & can in idea (with consciousne
025r017 y freshness of the streams of Lava in Ascencion KNOWN to be inactive 300 years? No Volcanic Earthquak
036r051 ound was long & prolonged). NB, Is it generally KNOWN. the acute chirping sound produced in walking o
119a106 as hot as lava-- of which temperature is partly KNOWN-- <[...]> moreover gradation from gneiss to gra
126a122 thinner → The problem is, you have temperature KNOWN at surface,-- you have temperature known far be
126a122 rature known at surface,-- you have temperature KNOWN far below surface, say 1000-- «III but an equil
173b009 ven than in Lamarck's time. Gray's remark, best KNOWN species. (as some common land shells) Most diff
182b047 f form. (& where possible the number «of steps» KNOWN). <for instance among the Carabidae.-- instance
183b052 avants.-- April 1837. p. 243 it is said as well KNOWN fact that "serin avec le chardonneret, avec  la
192b088 malia in ages long gone past.. & still more so «KNOWN» with fishes & reptiles.-- In mere eocine rocks
194b095 logical divisions.-- Mem. species doubtful when KNOWN only by bones.-- Mem: Silurian fossils: ¿How ar
196b107 . Philosop J. No 3. p. 207 "It is not generally KNOWN that Ireland possesses varieties of the  Furze,
202b131 beneath surface forest trees fallen «kind well KNOWN, carbonized--; clay fifty feet, then forest 12
213b171 Silurian age-- How long back have insects been KNOWN? As Gould remarked to me, the "beauty of specie
213b171 auty of species is their exactness,' but do not KNOWN varieties do the same, May you not breed, ten t
214b175 gham, and supposed to be descended from a breed KNOWN to be there since the time of Charles.,-- and n
217b184 & half tim> as often one way as other.-- He has KNOWN case of good pointer & rough water spaniel «pro
227b225 foretell  species. limits of good species being KNOWN.-- It explains the blending of two genera-- It
239c001 ose varieties to be oldest which have long been KNOWN in any country; he states that Esquimaux dog wh
246c026 Calcutta & Sumatra,. but I do not see how it is KNOWN that they are varieties & not species.-- Vol :6
255c050 ues apart from each other.-- this bird was well KNOWN for its impudence. «This excellent case of memo
277c126 so   There should be mark to every species. only KNOWN by analogy genera of course distinct. analogy f
284c149 value of organs vary in different group. & Not KNOWN in single ones--. viz. Macleay letter to Flemin
290c164 g series I cannot doubt laws of change, Will be KNOWN.-- It appeared to me that half between fowls  &
364d103 valry sing against each other, till it has been KNOWN one has killed itself.-- Q Sir. J. Sebright-- h
365d106 changing  circumstances.-- More probably during KNOWN changes climate became unfit for. subalpina, or
379d152 ember 11' Generation Mr Yarrell says it is well KNOWN that in breeding very pure South Down that  the
385d163 arlborough Str, Hair dresser, assures me he has KNOWN many cases of bitch going to mongrel, & all sub
385d164 ents in their bodies "It is a fact equally well KNOWN, that we observe in the temper, especially of t
411e054 specific change▯.-- When the laws of change are KNOWN.-- -- then primary forms may be speculated on,
424e100 . in Marchantia & Hypnum «Prof:» Don would have KNOWN the Composites of Galapagos were South American
461tf03 illivray says "<A Thrush &> Blackbird have been KNOWN in <their> «its» natural state to mate with a t
463t055 are a great difficulty not only are no animals KNOWN with an intermediate structure, but it is not p
477z002 inute. Lat -- -- do. p. 207. La punaise was not KNOWN amongst Indian. introduced in Paraguay in  1769
497q006 centre of woods «especially if date of wood be KNOWN» & other odd places & see what plants will spri
497q007 oint.-- ¿genera in intermediate country (2) Any KNOWN changes in Flora of countries during last centu
497q007 last century or two.-- where agency of man not KNOWN.-- (3) How is Iris impregnated.; which part of
498q008 ecessary to impregnate ordinary number of seeds KNOWN?-- Linnaeus has shown that each pistil is conne
499q09v r isolated females.-- Also any plants which are KNOWN easily to be crossed & all monooecious plants.-
523m017 , to prove she was not insane, answered she had KNOWN it at time & had bought arsenic for that purpos
525m024 betray himself by looking ashamed before it was KNOWN he had been on the table,-- guilty conscience.-
528m039 Colour «& light» has very much to do, as may be KNOWN by autumn, on clear day.-- 3d pleasure associat
533m058 (& phrenologists state that brain alters) It is KNOWN that birds learn to sing & do not acquire it in
560mIBC rwin 36 Great Marlborough St Has my Father ever KNOWN <intemperance> «disease» in grandchild, when fa
585m076 have been carried in hampers. if they have not KNOWN the direction in which they STARTED, they canno
610o031 letter  from Mr Roberts-- «a person he had long KNOWN & directed many letters to-- could not <rememb
616o39v erceptive action by which bodily action is made KNOWN to us, revealing respectively what are called i
617o40v g round to the <subjective> aspect of action as KNOWN by the exertion of our own power & consciousnes
618o041 n &c. are modes of subjective action-- they are KNOWN only by internal consciousness, & have no objec
618o041 ty by which the brain is perceived but they are KNOWN by courses of action quite independent of  each
618o041 nfer it from his its behaviour. Thought is only KNOWN subjectively? -- ? the brain only  objectively.
125a121 ust go from surface towards the interior -- who KNOWS how far that may have penetrated,-- lower down
244c022 ed two makis alive from there.-- Mem Waterhouse KNOWS of some species which escaped there.-- p.  139.
277c127 iology, tell traveller what to observe.-- if he KNOWS he has done least part.-- that he will not have
284c150 emains less subject to Variation" Dr. A. Smith. KNOWS lots of instances of replacement of one species
289c162 ensing must alter method of generation-- Heaven KNOWS how.-- This reaction takes place in every organ
353d061 e section with with those of India-- Waterhouse KNOWS three species of Paradoxurus common to Van Diem
386d167 out of one & heals piece of skin.-- if the tail KNOWS how to make a head. & head & tail, & the  belly
449e168 ight?-- June 18th. Eyton tells me, that Yarrell KNOWS of a Gull, which has laid in domestication eggs
507q15v bing amount of Horticultural Variation? Henslow KNOWS only on Citrons 47. Ficus carica Henslow presum
509q017 aking bulls. is it Domesticated African Animal= KNOWS nothing [<...>] It is very important to know, w
511q018 ce.-- {will introduce it in work} Whether <Yar> KNOWS whether Shaws hybrids between Trout & Salmon we
528m036 oloured.-- that light is a beautiful object one KNOWS from seeing artificial lights in the night.-- f
540m088 ing a book.-- Consider this.-- "The fledge-dove KNOWS the prowlers of the air" &c &c &c so is conscie
554m140 thinks keeper will not see her.-- <but is> then KNOWS she has done wrong & will hide herself.-- I  do
556m144 e & Blushing]CD «Does Elephant know shame-- dog KNOWS triumph.--» Sept. 23rd. Horses in Omnibus insta
557m148 jump away,-- it is, ill-defined fear.-- Yet one KNOWS oneself it is quite different from that.-- like
566n011 when in grown years, thinking he instinctively KNOWS distances:. is good instance of obtaining <that
567n014 & observation.--» in balancing a body & an ass KNOWS one side of triangle shorter than two. V. Whewe
582n068 e the creature performing those actions neither KNOWS nor intends the result they will effect,.--" th
582n068 this not wholly true, for we must grant a bird KNOWS what is about when building its nest; it  knows
582n068 knows  what is about when building its nest; it KNOWS its object but not result (first time of buildi
586n078 er varies in breeds,» something of kind oneself KNOWS in walking [one feels inclined to stop at right
```

Page **(Key Word)**
606o025 ealousy-- «as» effect of bodily organisms-- one KNOWS it, when one wishes to do some action (as jump
620o044 etermines to act more wisely other time, for he KNOWS that the instinct. (or conscience) is always pr
496q05a with single & see whether grains flaccid, as KOELREUTER describes Kill Sparrow after feeding on oats
326c265 Plants. Herbert. p. 348. gives reference to KOLKREUTER'S Papers Wiegman has published German Pamphle
451e180 be of a bolder & more generous temper-- Hodgson KOLOFF. voyage through the Moluccas 1825--"No wild an
633j53v g, & splits when it comes to a damp. place.-- KOLREUTER mentions some hybrid, whose flower great tend
076r168 Is <as> there, that are found in the veins of KONGSBERG in Norway.--namely dendritic silver intersect
266c091 to extend, their dominion, armed alike with the KORAN and the sword" \\quote Whewells Bridgewater trea
252c042 of> Cape found in Desert of Korto & Steppes of KORDOFAN p. 401. Admirable letter from Macleay to Bich
252c042) Hyena «venatica» <of> Cape found in Desert of KORTO & Steppes of Kordofan p. 401. Admirable letter
068r147 8] {P} (A. B. C, now grown solid.) Red Sea near KOSIR, land appears elevated. Geograph. Journal p 202
193b091 p. 306.--. Chamisso on Kamschatka quadrupeds KOTEZEBUES first Voyage Copied into list Entomological
081r181 ena publish his travels? Bellinghausen in 1819 KOTZEBUE 1816 Constant log always additive to convert
088a015 es common in Northern seas p. 312. Chamisso in KOTZEBUE. Study Humboldt. Fragmens Asiatiques account
247c028 s.-- killed a woman. Chamisso p. 189 Tome III: KOTZEBUE.-- p 22. a Gecko on St Helena.-- <in 1813, a
479z006 ratophytes common in Northern sea. Chamisso in KOTZEBUE p. 312 Leaches on leaves in Sumatra Marsden.
322c269 Sir G. Staunton's Embassy to China. Oct. 12t KOTZEBUE'S two voyages, skimmed well. do Lutke's Voyage
402e019 ntries W. Earl Eastern Seas. p 233 Octob 12th KOTZEBUE'S second «1st» Voyage. Vol II p. 344. account
402e020 ith regard to this archipelago Octob. 13th.-- KOTZEBUES first Voyage. Vol II. p 367. "The Fauna of th
403e021 d of the group, the number is very limited.-- KOTZEBUE'S Second Voyage do Vol III p. 77. Many foreign
480z008 ze-- How does this apply to pale brown Caracara KRAUSS on Corallinae from S. Seas written in German.-
242c017 Raffles. Horsfield. Diard. Duvaucel. Leschenault KUHL. Van-Hasselt, Reinwardt «Forrest» authors on E.
306c214 ailed The dogs called Persian «greyhounds» are KURDISH & come also from Asia Minor.-- tail like sette
245c024 is Lesson insists much.-- The (p. 296) Columba KURUKURU found in all Malasia-- & oceania, offers many
116a099 tter would have same effect as too coarse. Read KYLAU on Granite Edinburgh Philosophical Journal Rapp
104a068 with salt to surface Lyell remarked to me that KYLOW (?) was astonished with him that <th> gneiss, m
596n184 anocephalus, Macacus, Niger. Cercopithecus make LABIAL st st. S. American monkeys. pull back skin fro
095a037 rc Trans Geolog Soc Vol 2. p 257 {T} The Pota: LABIATA certainly is found with the Mactra. at Buenos
425e104 -- study Flora-- what general forms.-- are the LABIATA nearest to American, or Indian groups?-- = Bel
425e104 r to the separate islands-- In his work on the LABIATAE, some of these species are described.-- Capit
467t100 --some of the stigmas of {P} shape of ordinary LABIATAE --the chief part with ordinary divisions, & a
023r012 ched over board early in the passage!!-- M. LABILLARDIERE in Bay of Legrand, (SW part). describes a S
228b232 suffering «& famine»; our slaves in the most LABORIOUS work, our companion in our amusements. they m
263c075 grapple with great effect produced, is a most LABORIOUS, & painful effort of the mind (although this
540m087 mi-civilized-- Perhaps one cause of the intense LABOUR of original inventive thought is that none of
541m091 position great fatigue.-- may explain excessive LABOUR of inventive thought.-- Examine frame of mind
530m044 rd (Hunter?) that pulse of new born babies of LABOURING classes are slower than those of gentlefolks.
504q013) Which. Rhododendrum seeds??-- Bladder-nut \\. LABURNUM \\. Dodecatheon \\. Castrate apple & pear to s
641j29r tious.-- <some others>: study these facts read LACEPEDE on Cetacea & Geographical Distrib of larger S
242c018 nia have the Scincus with golden streaks-- the LACERTA vittata extends <to> from Amboina to New Irela
578n057 e inclined to convulsive actions).-- But, the LACHYRMAL gland is «not» under voluntary power, (or onl
465t089 ent forms.-- -- How many «tertiary» estuary & LACRUSTINE formations contain fossils,-- mammals-- a fe
136a147 l II p. 73.: some remarks on veins: Phillips in LADNER Vol. II p. 80-- some remarks on dikes: applica
136a147 y meeting with granite every-where. Phillips in LADNER Vol II p. 125. Good discussion on mineral vein
546m110 Mayo compares it with Somnambulism.-- the young LADY almost equally in her senses in either state.--
546m110 is concerned.?-- Mr. Mayo told me the case of a LADY, (whose name was told me, who told the fact to
546m111 ologising to party, went out & found it there!!! LADY in perfect «mental» health.-- «Erasmus had almo
128a130 xican Gulf. is fouled by bars of sand & shallow LAGOON.-- when describing Coast of. Brazil. Maldonado
449e168 ells me, that Yarrell knows of a Gull, which has LAID in domestication eggs of two shapes & colour.--
616o039 language that the onus probandi might fairly be LAID with those who would support the propriety of t
046r080 water oscillated between 2 & 3 ft. (as in Chili LAKE). Therefore motion of sea ought to be considere
066r142 quakes in Cordoba. one of which dried up <all> a LAKE in neigbourhood of town Mr Murchison insisted s
070r155 thin.--Concepcion earthquake Draw close Analogy LAKE of Cordill: of Copiápò & Desaguadero.--three ri
071r158 Cordova.-- There the Cordova earthquake in which LAKE was absorbed.--Earthquakes felt. different case
094a035 of granites into Trachytes.-- Mention Osorno in LAKE. few Volcanos now in lakes.-- Mr Murchison. M.S
098a044 Guantajaya contains salt see Geolog. proceedings LAKE let out by steps in Central France not very con
098a044 tin de la Soc. Geolog: 1833-34. p. 35.-- Ancient LAKE Lemagne in Auvergne Proofs from Phryganea NB. S
105a070 neral improbability. stratification, If chain of LAKE. <a> the alluvium would form a succession of fl
105a071 ld form a succession of flights of steps; if one LAKE then we must suppose barrier in the very part,
134a142 -- Phillipines there is volcano on isld in large LAKE-- Berghaus Chart of do Journal of Asiatic Socie
147g016 - Rivers could not have deposited it. Barrier of LAKE very lofty, & no trace of it; to the Sea more p
147g018 s of waterworn pebbles in Alluvium which without LAKE or sea could not be placed in present position
148g028 rge» deposition marine-- because if not chain of LAKE & if so there would be barrier-- recollect the
149g030 s, composed of sand & perfectly rounded stones-- LAKE required to deposit this Remember however the g
149g032 ifficult to explain on <formation> deposition in LAKE On the summit «& on Spean side» of Meal-- Derry
150g035 ne running up great bight just as Dick shows NB. LAKE gradually draining off would form plains such a
151g042 f other, which must have been waterworn after 3d LAKE.-- 4th shelf runs up some way on great sloping
152g045 t to shelf {P} In Glen Collarig, by Dicks theory LAKE burst in most improbable part & not in Pass, wh
152g048 greater cut through, not applicable to Glen Roy LAKE, must have remained very long at 4th shelf from
153g051 per edge of cliff» Others below it--argument for LAKE «or sea» at successive levels-- {P} Shelf oppos
154g060 r to do this vast quantity when during repose of LAKE it did but little more {P} now that it has got
156g068 head of valley fragments which had fallen before LAKE drained could be told from «some of» those sinc
162g103 tium aquarum with, left of Bright.-- like bed of LAKE with trace of terraces on each side High up the
399e011 esfield slate., & a Melolonittha-- In marl from «LAKE» Constance species of Europaean genera=.-- Hope
473s05v rt» wh flows by Tremadoc. but can tell them from LAKE S. of Moel Siabod. wh. flows into Conway by Bet
040r063 ones have undergone the same process.-- Neither LAKES or Avalanches (Glaciers very rare) to cause flo
042r068 ults in mountains: to elevated strata in eocene LAKES of France, & unequal action of Earthquakes. «on
042r069 ity in S. America: in France we have freshwater LAKES unequally elevated, which movements if present
062r128 ure of climate. = Perpetual snow.--subterranean LAKES, near Volcanoes. lakes of brine all inhabited:
062r128 tual snow.--subterranean lakes, near Volcanoes. LAKES of brine all inhabited: Go steadily through all
086a006 elevation. vapour from below-- Malte Brun «Salt LAKES» Siberia must be read as well as Pallas before
094a035 .-- Mention Osorno in lake. few Volcanos now in LAKES.-- Mr Murchison. M.S. Chapter on drift.-- Beyon
114a093 of C. of Good Hope & Australia/ and mud of salt-LAKES of Rio Negro--Mr Bowerbank-- Dr. A. Smith's cur
115a098 ts> «fragments» but relative to currents. Small LAKES have power of levelling their shores where curr
117a102 ells of Cobija doubtful. Examine well shores of LAKES. to see effects of degradation, «no» tides, wat
132a138 ot understand the argument, that cold <oceans> «LAKES» bottom. if not colder than mean of place, show
132a138 Parrot suppose there is no volcanicity beneath LAKES)?» Suppose ocean represents proper <state> temp
233b251 upeds-- Study Productions. of great Fresh water LAKES of North America If Parasite different, whilst
259c062 animals.-- then argumen May be.-- subterranean LAKES, hot spring &c &c inhabited therefore mud wood
259c064 woodpecker &c.--, fresh water animals of great LAKES are American form, that one is brought to admit
473s05r Bunbury says has heard the Trout from different LAKES of N. Wales can be distinguished-- & Jackson he
473s05r tell Trout from Ogwen, Capel Curig & some other LAKES, (different waters) He cannot, however, tell th
225b216 imited view.-- With respect to how species are. LAMAKS "willing" doctrine absurd. (as equally are arg
419e087 ich may be the work of a few years as with the LAMANTIN of Steller tells much less, (though <the> it
173b009 ot Produced?-- <Granting> Species according to LAMARCK disappear as collection made perfect.-- truer
201b129 egularities in the degradation of structure of LAMARCK, which he says depends on external influences.
215b178 s ideas on passage of forms.-- Deshayes states LAMARCK priority refers to introduction to Animaux San
275c119 ted] alone, but on all the general arguments-- LAMARCK was the Hutton of Geology. he had few clear fa
287c157 not syn.-- Linn: Transact Vol XIV.--\\. p. 24. LAMARCK bears to Cuvier that relation of theoretical a
287c158 icsome saltus. «p. 19»\\ Macleay seems to limit LAMARCK definition of relations to settling the relati
292c168 Scicily.-- Splendid Harmony these views-- did LAMARCK connect extermination of some forms with his v
323c269 ains skimmed -- Pliny Nat. Hist of World do -- LAMARCK. II Vol. Philo. Zoology «references at end of
356d069 ng what Von Buch (Humboldt). G. St. Hilaire, & LAMARCK have written I pretend to no originality of id
440e145 oduction» is common to all living beings. vide LAMARCK Vol II. p. 115. 4 four laws Who can say, how m
445e159 it is evidently acquired by experience in baby LAMARCK. Vol II p. 152.-- Philosophie Zoologie. says i
588n090 unting for a bone shows he has recollection.-- LAMARCK. Phil. Zoolog.-- Vol II p. 445. If we compare
588n091 how gained. reason? or some unnamed faculty-- LAMARCK. Philosop. Zoolog. «p. 284. Vol. II» -- gives
589n091 & instance of starting identical with mine,-- LAMARCK. Vol II p. 319.-- Habits more prevalent in pro

Page **(Key Word)**
611o034 nts. different from inorganic movements.-- See LAMARCK for this definition given in full.-- [LHC] ¿Ha
382d157 exes separate. Owen say such view worthy of a LAMARCKIAN.-- Mine is much simpler.-- Hunter shows almo
173b009 collection made perfect.-- truer even than in LAMARCK'S time. Gray's remark, best known species. (as
224b214 ghts (animal). «∴ my theory very distinct from LAMARCKS» Without two species will generate common kin
614o037 nto steps, as species change.-- Must be so if LAMARCK'S theory true [RHC] Acquired instincts analogou
408e046 monstrosities.-- G. B. Sowerby.-- Looking over LAMARK surprised to see how many Tropical genera come
440e148 modification, though important one, of growth «LAMARK. Vol II. p. 120. observes it commences only, w
259c063 n with any peculiarity. or any race of plants--LAMARK'S willing absurd, ∵ not applicable to plant, Ep
437e138 20 An Eagle is said to have been seen carrying a LAMB two miles towards the Morne Mountains, it then
437e139 ed the bird. p. 124-- Mr Willoughby found a dead LAMB «& hare» by the side of Eagles nest, which show
148g025 are crossed with English my informant said the LAMBS were nearly like each other «& half between par
148g025 ion in many ways & received same answer Thought LAMBS most like MOTHER!-- the cross not so hardy but
165g125 Man from Glen Turret said he learnt to know the LAMBS because most like Mother in face-- asked stated
205b145 ornless ram be put to horned ewe almost all the LAMBS will be hornless.--" does this apply to where s
345d043 phepherd of Glen Turret. said he learnt to know LAMBS, because in their faces they were most like the
346d047 many ways, but received same answer.-- Thought LAMBS were more like father than Mother.-- The cross
364d104 rrell thinks the occasional production of black LAMBS is owing to old <story> return.-- The Revd. R.
379d152 st never be put to any other breed else all the LAMBS will deteriorate.-- Lord Moreton's Case.-- When
512q019 easants treading-- any treadèe?-- Difference in LAMBS of different breeds Is there any difference in
097a042 mountains in Brazil L.'Institut No degree 221 LAMELLAR dikes like Mica Slate Von. Buch. Canary Isd.
591n099 mpt at suicide. (even when no relatives left to LAMENT) is owing to the feeling that the instinct of
595n115 at opposite end of house. & commenced a most LAMENTABLE howls & & was not comforted until the Keeper
021r005 slate same.--longer axis in line of Cleavage. LAMINAE fold round them; Quote this. Valparaiso Granit
027r023 tution of matter in shells, like Concretions & LAMINAE, show what movements take place in semiconsoli
035r045 444 «(Yanky Edit)» <I think> At Ascension, the LAMINAE <...> changes in rocks. connected with & alter
098a046 allization. therefore concretions in this case LAMINAR. hence the thick wedges of feldspar in gneiss.
098a045 e Proofs from Phryganea NB. Sedgwick talks of LAMINATED structure (∴ separation of ingredients) as un
099a051 e of Scrope parallel to walls of dykes-- Mem. LAMINATED dikes in Cordillera.!!!-- In stratum OP. let
099a049 t if great depth NB. Prof <Henslow> Sedgwicks LAMINATION parallel to stratification evidently small s
122a115 (Mem near Gregory Bay). Shropshire case where LAMINATION appeared.-- Lyells Denmark -- L'Institut (1
207b152 speces polymorphes que toute la terre ferme de LANCIEN monde".-- Considers forms in recent volcanic i
030r034 by outlying rocks; (such as between Mocha & main LAND). <[...]> At Carelmapu.--Within Chiloe:-- On op
030r036 position» position at N. S. Wales & Van Diemen's LAND.-- Whole coast S. of Concepcion where there are
030r036 coast to see proportional cliff & low or sloping LAND What are the "palatal Tritores" found in the co
032r039 P} The action of sea A. B. will be to eat in the LAND in line of highest tidal action. this will at 1
036r051 and much subsidence: hence difference; action on LAND different Volney, P 351. Vol I. woody bushes, «
039r060 red the subject.-- S. America in the form of the LAND decidedly bears the stamp of recent elevation.
041r065 . In Rio paper. when discussing probable rise of LAND: Mention M. Gay's fact about shells: Hibernatio
046r080 ovement. owing to water keeping its level whilst LAND rose up & down.--But from above reasons, do not
055r107 toises thick.--Vol II. p. 252 Urge cliff form of LAND, in St Helena. Ascension. Azores. («sandstone f
058r115 vation. Effects of great waves to obliterate all LAND marks.--At the first it would though be easy to
059r120 s of much granite on all East side of Van Diemen LAND. All the Calcareous rocks which harden by thems
060r124 thquakes "seem to arise from some efforts in the LAND to lift itself higher & to grow upwards; for th
060r124 to lift itself higher & to grow upwards; for the LAND is constantly pushing the sea (which of course
060r124 tion with "the shooting upwards" of the <ground> LAND in the W Indies.--p. 200. Bolingbroke voyage t
061r127 alia's = if for volc. isld. then for any spot of LAND.-- ≠ New creation affected by Halo of neighbo
064r135 earthquakes & great but local elevations of the LAND in Europe-- Urge difference of plutonic rocks &
065r138 ation & climate of Tristan D. Acuneha. Kerguelen LAND. Prince Edwards Isld. Marion & Crozet. L. Auckl
065r140 observations on springs at S. Cruz.???-- Form of LAND shows subsidence in T. del Fuego, and connectio
066r141 ganteus! 24 fathoms deep 24» under 50. Kerguelen LAND, -- the way it stands gales = very strong. Stone
068r147 (A. B. C, now grown solid.) Red Sea near Kosir, LAND appears elevated. Geograph. Journal p 202 Vol I
086a004 shells.-- accustomed to such terms "fixed as the LAND, stable as the water"-- It may be worth noticin
087a008 a-- Lately elevated When Siberia went up. Arctic LAND went down.-- Probably more Arctic land would be
087a008 p. Arctic land went down.-- Probably more Arctic LAND would be required to produce climate resembling
087a011 armonizes well with Lyells idea of intertropical LAND.-- Siberia rises. therefore to the South sinks.
087a011 orresponding to Europaean risings. Pacific great LAND. -- Will use argument of proof of slow corrosio
101a056 . 270-4, good discussion showing present form of LAND in Northern England influence dispersion of Bou
105a072 plain «alluvi» in valleys Lowe in his paper says LAND shells found with calcareous matter & concretio
105a072 owder results from «decomposed sea» shells, that LAND shells should be preserved in it-- some error?
112a089 graph Journal Vol IV p. 36. on subsidence of the LAND in Guiana, worthy of consideration. When discus
117a102) Lhat 70. F were obtained 100 miles E of Staten LAND. bringing up pebbles 2 inches long?-- L'Institu
128a129 due to rivers-- By Roy Bridge, a tongue of flat LAND on one side. produce subsidence of water on oth
128a129 librium This will be only a modifying cause. {P} LAND, with terraces of each side of the two valleys
148g029 due to rivers-- By Roy Bridge, a tongue of flat LAND, with terraces of each side of the two valleys
152a046 g «on» one side ravine. Are the lip, or necks of LAND on level with shelves effect of corrosion & not
162g101 5 .105 I reached² 29.090 Preservation of form of LAND very much due to Peat & Heather When it did not
173b009 ay's remark, best known species. (as some common LAND, which were late arrivals others old ones, (of
173b010 ntinents might have some species same as nearest LAND, which were late arrivals others old ones, (of
173b012 on ceased & changes commenced.-- or intermediate LAND existed.-- or they may represent some large cou
176b023 in the tree of life owing to three elements air, LAND & water, & the endeavour of each <one> typical
182b050 istinct species between Australia & Van Diemen's LAND. & Austral & New Zealand Mr Gould says in sub-g
185b055 .-- Upland goose.-- weather chionis water rat with LAND structures; :--law of chance would cause this to
189b072 connection between great groups.-- Speculate on LAND being grouped towards centres near Equator in f
191b080 ntinent-- 3 Paradupasi in common to Van Diemen's LAND & Australia From the consideration of these arc
191b081 -- Mem plants on Coral islets.-- Next to animals LAND birds.-- & life shorter or change greater-- In
199b116 th of quadrupeds being created on small spots of LAND, of the same type with the great continents we
200b124 tant facts.-- As soon as island large enough for LAND birds, seeds picked from the beach by the birds
207b152 lets not well fixed.-- Peron thinks Van Diemen's LAND long separated from Hobart Town-- {from differe
214b173 pagos--Juan Fernandez Falkland Islds-- Kerguelen LAND.-- Phillips. Lardner Encyclop. insists on analo
215b177 lose desire. Native dog not found in V. Diemen's LAND J. de Physique. Tom 59. p 467. Peron G. St. Hil
217b187 h islands-- A genus with species in Van Diemen's LAND and Tierra del Fuego.-- Araucaria, species. Bra
222b204 ral observations-- difference of species between LAND shells of Porto Santo & Madeira-- I believe ver
225b219 nly hope.-- New Zealand «compare to Van Diemen's LAND.» glorious fact. of absence of quadrupeds East
226b220 brought by canoes Ceylon & India.-- Van Diemen's LAND Australia. England & Europe.-- It will be well
226b222 nce this must have been condition of Paris basin LAND.-- (How is this with Fernando Po.). with plants
226b222 one isld several species of same genus, subsided LAND.-- Mauritius? <In plants where do most species
232b248 isphere look as if heat gained> Experimentise on LAND shells in salt water & lizards do.-- Ask Eyton
233b251 aded Duck-- Large proportion of Water & small of LAND or few quadrupeds-- Study Productions. of great
235b263 nglish Hare bear upon this.-- Why do Van Diemens LAND people require so many, imported animals?-- At
235b264 ng to three elements How is Fauna of Van Diemens LAND & Australia [blank] Falconer's remarks on influ
236b279 --of Europe Mem. Mr Bell's case of Sub Himalayan LAND emys, decidedly an Indian form of Tortoise.-- Vo
246c027 Guinea scarcely differs more from, <Van Diemen's LAND. ℕ> Australia more than Van Diemen's land.-- Vol
246c027 iemen's land. ℕ> Australia more than Van Diemen's LAND.-- Vol II p. 8 no snakes on isles of central Pa
247c028 Taeniatole austral» «caught» Chile, Van Diemen's LAND of Tahiti, in spite of every attempt to check i
267c093 Isld-- "& it now claims all the moist & fertile LAND & Cape of Good Hope V. p. 44 of this Note Book
268c095 &c. valuable paper on quadrupeds of Van Diemen's LAND, which appear diff. from Australia (Waterhouse
268c099 w independency of shells to external features of LAND by seeing how many species common to Patagonia
274c115 ms in water birds,-- but no web forms in <water> LAND birds,.-- Grups of very different value have th
278c130 thority) in Australia, & several in Van Diemen's LAND is most important as showing former connection
278c131 oes not think that dog was found in Van Diemen's LAND.-- V. 1s. Number of Geographical Journal to dis
288c160 rd Migrating from Falkland Isd regularly to main LAND, proof of. land having been formerly nearer.--
288c160 m Falkland Isd regularly to main land, proof of. LAND, having been formerly nearer.-- «Selby» Magazine
289c160 touch the water.-- capital instances of typical LAND bird, having habits of Grebe, structures migh
296c184 rs altered much.-- these others will be plants & LAND animals. & land shells.-- all in short Extreme
296c184 -- these others will be plants & land animals. & LAND shells.-- all in short Extreme North = = to pea
300c197 of Magellan.-- Change of habits in Van Diemen's LAND. Study Mr Blyth's papers on Instinct.-- His dis
302c203 ay sometimes be due to accident as submersion of LAND containing all of intermediate Father-species,
308c218 are the only two tribes fitted for water, air, & LAND, (Macleay has this remark) Mem. number 5 here m

Page ***(Key Word)***
311c225 reat diff of Latd. p 355 Echidna of Van Diemen's LAND & Australia different Temminck Fauna Japonica (
311c225 alia 293 Phalangista of Australia & Van Diemen's LAND diff.-- Habits can only be used «in classificat
314c239 utious about Goulds case of birds of Van Diemens LAND & Australia.-- The wombat (Brown) is found in I
316c245 more important element than longitude.-- But in LAND & F W shells there is more confinement. thus th
316c247 arkable.-- (connect these facts with identity of LAND animals.-- these however come from Siberia it c
317c249 cleay tells me same thing p. 55. 40 leagues from LAND several patches of reed & trees p. 259 120 ft i
319c257 ssils intermediate between those of Van Diemen's LAND & Australia proper.-- Irish Elk case of fossil
353d061 ee species of Paradoxurus common to Van Diemen's LAND & Australia well developed <tits> «Mammae» in m
365d107 ich it must have taken to separate. Van Diemen's LAND from Australia &c &c Sept. 14th. When Macleay s
379d151 with the Black sea.-- it would be ocean, what is LAND to continent-- Original Paper, worth studying.
400e011 n part near Timor, & to Europaean in Van Diemens LAND, where there is close species of elater.-- Wher
407e041 fossil ones of Europe, Consider probable form of LAND,-- -- S America, an island, «connects with Asia
414e064 of Bolivia-- strictly fossil «& in Van Diemen's LAND»-- they have been exterminated on principles. s
416e071 ontradicted by Plants). & as there are no fixed. LAND animals. so there are true hermaphrodites.-- I
417e076 ybrids,.-- Mr G. B. Sowerby <tel> showed me many LAND shells of the common species: from one locality
432e124 rounding circumstances According to my theory no LAND animal with fluid seeds can be true hermaphrodi
436e137 of Falkland.-- Old Red Sandstone-- Van Diemen's LAND.-- Porphyries of Andes. A familiar History of B
451e180 V.-- : do. Chat XXI. Wild cattle & Hogs on Timor-LAND-- monkeys do not exist. there & it is a singula
465t079 favourable for the preservation of shells; where LAND broken, rivers entering.-- & yet no shells-- no
479z005 al-- RN. p 24 Bougainville Voyage round world no LAND animal besides Wolf at Falkland ∴ black rabbits
481z008 anca. De la Beche theoretical researches Compare LAND shells of Galapagos different islds.-- Waterhou
490q001 Smith of Jordan Hill-- character of the extinct LAND-shells of Madeira-- analogous or quite distinct
500q10a here any fine doubtful species from Van Diemen's LAND? or New Zealand? Babington about differences of
502q11v rs in short turf. for abortion. or for sterility LAND Birds Madeira Migratory-- ask Gould about N. Ze
502q11v nds Isds.-- Snipe Migratory-- probably united by LAND to S. America (33) Ornithologum commonly but im
529m040 ility.-- I a geologist have illdefined notion of LAND covered with ocean, former animals, slow force
540m085 cold regions of the North,-- the elevated table LAND of Peru the hot plains of the Amazons & Brazil-
540m086 rica, (or again the black man of <B> Van Diemens LAND & the energetic copper coloured natives of New
634j54v p. do]CD <Mac. remarks all Mammifers originally LAND--animals. as> 5 p. 314. Mac. remarks all <land>
634j55r y land--animals. as> 5 p. 314. Mac. remarks all <LAND> Mammiferous animals originally terrestrial.--
357d072 is strongly tempted to believe, one or two were LANDED as at present in New Ireland & continent since
153g054 , opposite entrance into Glen Fintec a kind of LANDING place is formed Ben Erin summit 27.813. 65 55
067r144 to old Silurian system. Apply degradation of LANDLOCKED harbors to Craters of elevation.-- Lyell sug
321c270 sophy of Art of Living Several of Water Savage LANDONS Imaginary Conversations-- very poor Sir T. Bro
205b143 tacea. Vol II. p 75 a Fish which emigrates over LANDS is a siluris, p. 123 A climbing fish. p. 122 A
225b219 uped since days of Didelphis in Stonefield∴ all LANDS united (Falkland Fox. ice). . Mauritius what a
227b224 f transport, we should then know whether former LANDS intervened.-- (2d) By character of any «two» an
266c091 w how remote the periods at which both left the LANDS of their forefathers;-- the first to escape the
407e041 island, «connects with Asia» between two polar LANDS,--; Africa not so equatorial..-- The fact of No
309c221 and farmer,-- useful could not leap fences:-- Dr LANG on Polynesian nations (quoted) p. 4.-- do. p. 1
321c275 tural History of Selbourne References at end Dr. LANG Australia «trash» skimmed Macleay's Horae Entom
179b034 ies they certainly are according to all common LANGUAGE) will keep to their type: in animals so far r
252c042 les. p. 388 Reference to Rüppel. travels (what LANGUAGE?) Hyena «venatica» <of> Cape found in Desert
260c068 istress of other parents.-- Shows community of LANGUAGE Desert country. is as effectual as a cold one
264c079 of his proud preeminence.-- «not understanding LANGUAGE of Fuegian, puts on par with Monkeys» Gould s
286c154 ground) Man signals.-- animals understand the LANGUAGE, they know the crys of pain, as well as we.--
293c172 s he put on & we afterwards could understand «(LANGUAGE better instance)» he had done this without re
300c198 . his mind is still only a divided body. ‖p 3. LANGUAGE seems to supply instincts,-- & those powers w
300c198 & those powers which allow of, acquirement of LANGUAGE. heredetary & acquirable.-- therefore mans mi
375d134 nges in nature of locality. Even the energetic LANGUAGE of <Malthus> «Decandoelle» does not convey th
434e130 y become so by suppression of one organ. (here LANGUAGE forces on us the change, which <to> seems to
450e175 only found wild in the plains of Celebes. (but LANGUAGE shows that probably not original there)-- sho
533m058 his be connected with their power of acquiring LANGUAGE.-- Hensleigh. W. says that babies know a frow
534m061 sensation of images before your eyes, or ears (LANGUAGE mere means of exciting association.)-- or of
542m096 useless.-- The distinction «as often said» of LANGUAGE in man is very great from all animals-- but d
543m097 o know,-- but this capability of understanding LANGUAGE is considerable, thus carthorse & dog.-- bird
558m151 out.-- bearing in mind many new relations from LANGUAGE.-- the social instinct more than mere love.--
558m153 e difference. between man,-- forget the use of LANGUAGE, & judge only by what you see. compare, the F
568n018 , which is part of reason Octob. 19th. Did our LANGUAGE commence with singing-- is this origin of our
569n020 that sound as perfectly as a man.-- Probably, LANGUAGE commenced in some necessary connexion between
571n031 sle of Churche.-- these are Mayo's ideas.-- In LANGUAGE. the possibility of poets describing gentle t
571n031 ty of poets describing gentle things in gentle LANGUAGE, & vice versa.-- almost proves that at earlie
571n031 have been intimate connection between sound & LANGUAGE.-- Chinese. simplest language. Much pantomimi
571n031 between sound & language.-- Chinese. simplest LANGUAGE. Much pantomimic gesture?? which would natura
574n039 ‖ I think this argument might be used to show LANGUAGE had a beginning, which my theory requires. Th
579n060 ortance of a name, with reference to origin of LANGUAGE My father says old people first fail in ideas
581n065 der there is good evidence in the structure of LANGUAGE, that it was progressively formed. (--names l
593n107 on to the period when men. communicated before LANGUAGE was invented,-- were musical notes the langua
593n107 anguage was invented,-- were musical notes the LANGUAGE of passion & hence does music now excite our
599o005 taphysic,"-- give my doctrines about origin of LANGUAGE-- & effect of reason. reason could not have e
599o005 ve existed without it.-- quotes Ld Mondobbo.-- LANGUAGE commenced in whole sentences.-- signs-- ¿ wer
599o005 t it appears all speculations of the origin of LANGUAGE.-- must presume it originates slowly-- if the
599o005 ess-- then argument fails-- if they have, then LANGUAGE was progressive.-- We cannot doubt that langu
599o05v nguage was progressive.-- We cannot doubt that LANGUAGE is an altering element, we see words invented
603o013 eity.-- H. Tooke has shown one chief object of LANGUAGE is promptness «of consequence» hence language
603o013 ves from nouns & verbs-- so that much of EVERY LANGUAGE shows traces of anterior state?? Edinburgh Re
616o039 xpression, & so foreign to the use of ordinary LANGUAGE that the onus probandi might fairly be laid w
210b158 t being crossed with others.-- Compares it to LANGUAGES But how do plants cross?-- = admirable discus
231b244 ent of remains-- ¿agrees with non-blending of LANGUAGES?-- Till man acquired reason, he would be limi
256c053 le is brought to bear. If man created as now. LANGUAGES. would surely have been more homogenious.-- T
569n020 f mouth & smell. Descent of Man Understanding LANGUAGES seem simplits case of Association.-- Elephant
603o013 language is promptness «of consequence» hence LANGUAGES become corrupt, & whole classes of words «are
590n093 2. Sighing from grief. is method of increasing LANGUID circulation-- no, for <grief> sighing comes on
567n013 hild. Tommy's face, now ill, has expression of LANGUOR & suffering The Cyanocephalus when fondling th
272c109 .-- thus the banded tarsi is common to all the LANIADAE & Muscicapidae of new World, but not found in
450e175 ay Ava Pegue-- seldom equals 13 hands-- those of LAO & Siam inferior to those of Pegu-- in Sumatra tw
458t002 's Voyage Vol I. p. 499. «4to. Edit»-- Horses in LAO Choo so small, that person with long legs can ha
326c266 y.-- Dampier. probably worth reading Lessings. LAOCCAON.-- (translated in 1837) on limits of painting
322c269 arefully read.-- Reynolds Discourses Lessing's LAOCOON Whewells-- inductive History.-- References at
606o022 nd undergoes in gaining the result.-- Lessings LAOCOON. 2d Lect-- The object of art., sculpture & pai
606o023 implanted in our bosoms.-- how comes it there? LAOCOON p. 75 "The beauties developed in a work of art
606o024 ous colours &c &c surely to be added. Lessings LAOCOON p. 125-- says new subjects are not fit for pai
429e115 nts of & Pyrennees agree with those of Norway. LAPLAND & Greenland, but not with those Kamtschatka, S
485z018 . 46 Macleay Horae Entomolog. insects swarm in LAPLAND & Stizbergen wherever there is extreme heat, t
514q21. ts found on mountains of N. America similar to LAPLAND Plants --will get answer= Is pollen of cultiva
209b157 on of genera to species in France is I: 5.7 in LAPONIA I: 2,3: Mem. Lyell on shells.-- {T} Genera In
210b159 ble discussion Von Buch says from Humboldt, in LAPONIA. genera to species I. 2,3-- From Mackenzie Ice
063r130 tion, in latter time. (or changes consequent on LAPSE) being the relation.--As in first cases distinc
063r132 ividual divided either at one moment or through LAPSE of ages.--Therefore we are not so much surprise
212b167 f S. African Cattle Phillips Geology «p 81» in LARDENS Encyclop. Proportions between fossils & recent
323c269 ife of Johnson. 4. Vols 25th Phillips. Geology. LARDER 2d vol.-- March 16. Gardner's Music of Nature-
100a054 s the best case of cleavage.-- Phillips (113) «LARDNER Encyclop.--» absolutely considers gneiss an aq
100a055 n filled with unconsolidated matter-- Phillips LARDNER p. 197. refers to salt as being produced by lo
136a147 . 13. Vol II. Lardner's-- Treatise Phillips in LARDNER Vol II p. 73.: some remarks on veins: Phillips
136a147 on dikes: applicable to Cordillera Phillips in LARDNER Vol II. p. 81. «&83» Some remarks on thinness
214b173 z Falkland Islds-- Kerguelen land.-- Phillips. LARDNER Encyclop. insists on analogy between Australia
214b173 n) found on Delaware is it Edentate? Phillips. LARDNER p. 289 It is certain, that North American foss
100a055 rison rather loose.-- perhaps worth Says from LARDNER'S (p. 213) form of escarpment relation kept to

101a056 in which most substance lies <.-->? Phillips. LARDNER'S p. 270-4, good discussion showing present for
136a147 n to that of Phillips as given p. 13. Vol II. LARDNER'S-- Treatise Phillips in Lardner Vol II p. 73.:
471tf09 ass so insensibly into each other". Phillips (LARDNER'S E. vol. II p. 18.) capital list of all the fo
037r055 torrents. «marine» Tortoise & other species of LARGE lizard.-- There would probably be no other orga
050r090 there is a pumiceous conglomerate with small & LARGE fragments, nature of which is doubtful. P. 180.
051r094 eakers or in waterfall: Excepting by removal of LARGE fragments by mere force of waves: & action on u
055r106 , a sufficient proof, that the sea formed these LARGE cavities", &c &c &c Vol II. Chapt VIII. p. 97 a
061r127 e is falsely less surprised at new creation for LARGE.--Australia's = if for volc. isld. then for any
090a023 s in many cases there were flights of stones of LARGE numbers (& how few cases recorded if we say «10
103a064 ng of mountains being formed by crust being too LARGE & pitching against each other, is, I suspect mu
110a086 athes with greenish. black specks of hornblende, LARGE irregular cryst of reddish felspar. & scales. o
110a086 cryst of reddish felspar. & scales. of mica.-- LARGE cryst of Hornblende blending into base-- Salban
123a117 in many part.-- great Dasypus near Canelones -- LARGE quadruped bigger than ox.-- at Buenos Ayres 20½
130a133 tudy Bishoofs Paper.-- Weelsted told me of some LARGE fresh Water springs off coast of Persia In Glen
134a142 conia-- Phillipines there is volcano on isld in LARGE lake-- Berghaus Chart of do Journal of Asiatic
146g014 followed up to neighbourhood of Tyndrum where a LARGE sort of <plain» space is thickly studded with r
156g069 ne, where water entered are not proportionately LARGE to those now formed in same spot by present tor
159g084 Block on 2d shelf & below it some way; several LARGE ones (one 6 ft across) on top of spit between r
173b012 iate land existed.-- or they may represent some LARGE country long separated.-- On this idea of propa
181b040 -- In same manner, if we take «a man from.» any LARGE family of 12 brothers & sisters «in a state whi
191b081 lago it would be interesting to trace limits of LARGE animals-- Owls. transport mice alive? Species f
193b091 portant the possibility of some isld not having LARGE quadrupeds-- Humboldt has written on the geogra
196b106 -- it will not do to say period unfavourable to LARGE quadrupeds--horse not large-- Ed. New. Philosop
196b106 iod unfavourable to large quadrupeds--horse not LARGE-- Ed. New. Philosop J. No 3 p. 207 "It is not
198b115 ding N§ well to insist upon <different animals> LARGE Mammalia not being found on all isld. (if act o
200b124 e are most important facts.-- As soon as island LARGE enough for land birds, seeds picked from the be
217b183 haired shepherd dog lined her & produced a very LARGE litter.-- never afterwards went in heat.-- This
224b213 .-- Definition of Species: one that remains «at LARGE» with constant characters, together with other
226b220 y terrestrial Mammalia.-- Especially moderately LARGE ones.-- Is the flora of <S. America> Tierra del
226b222 species, we may insure mass continental or many LARGE islands.-- Hence this must have been condition
233b251 s. Dodo. Apteryx Penguin-- Logger-headed Duck-- LARGE proportion of Water & small of land or few quad
246c025 different]CD.-- p. 414. dogs of New Zealand of LARGE size, resemble, chien-loup.--long, black & whit
250c039 t be argued what is not cause of destruction of LARGE quadrupeds.-- common to two types of animals Wh
265c085 g to habits, or heredetary is to see, whether a LARGE family has it, & one member of that family, hav
273c110 -; <then> & one species has small band & others LARGE, then he says from long experience, you may be
283c143 acter of analogical.-- Gould says it is only in LARGE groups. where you have representations.-- The a
288c159 that rams & bulls from England fetch very <go> LARGE price. as is evident to be worth introducing, i
292c168 n of some forms with his views.-- as genera are LARGE probably only few of extinct forms have generat
299c195 rrot. made into biennial domesticated kind with LARGE root by sowing it at wrong time of year & manur
314c239 SPECIES of Australian GENERA: good case. rather LARGE flora. (150?) Mr Brown did not observe scarcely
332d003 some fever, be attended by the transmission of LARGE number of worms the child not having passed the
333d005 in progeny peculiar/ limbs very long, eyes very LARGE, very fierce to dogs <otherwise habits not diff
337d021 ht.-- (Mem whole plant may be considered as one LARGE eye-- have they smell, do plants emit odour sol
340d029 emarked that greater difference in than in many LARGE orders of birds. The Emu & Cassowary closest.--
367d113 all four legs ringed with brown.-- animal like LARGE, heavily made cream coloured ass.-- stripe on b
379d151 Q [not located] The present age is the one for LARGE Cetacea, as the past for other Mammalia. & stil
402e020 e elephant-- in Magindanao several kinds of the LARGE monkeys-- Fewer <of th> «Mammalia» have passed
413e059 erous individuals, & <individuals» «species» of LARGE size,-- consider this (Cetaceae) with reference
423e097 rnivorous devourer.;-- it is quite clear that a LARGE part of the complexity of structure is adaptati
426e107 rance of this series & makes one think that one LARGE body of varieties are fertile & make mongrel, &
436e134 ell says that naked cuttle fish now bear a very LARGE proportion to other mollusca in cold parts of s
449e170 ra.-- <it is not transportation> now in case of LARGE [not located] Mr Greenough on his Map of the Wo
458t001 ar & ruminants all of larger size.-- the law of LARGE size established-- «Australia.. S. America-- T
468t105 owers not attractive, very small--stigma rather LARGE & rough-- flowers common many winged thrips
468t105 things, covered with pollen-- <Thrips» about as LARGE as bit of chopped horse hair with legs & take f
471tf10 apital list of all the fossil Mamm. of Europe-- LARGE Lizards in Navigatores. Williams. Narrative of
472s02r ow & purple heartease without doubt.-- Bee. not LARGE, very dusky & broad never saw such a one before
499q010 n counties, as of rare green Cotton Plant-- How LARGE «area» clump there? Distinguishable from <other
505q014 as a hooked Pea.-- intends to breed from it and LARGE Asparagus: result? = failed to germinate 16 Wil
505q014 d a hive. of these small Bees-- at Sundorne has LARGE Bees July/42/ Mark has six day's puppy of small
515q021 o the most cultivated show Heartease produce as LARGE capsules of seed, as the commoner kinds-- Cattl
522m011 be sure I do.-- What became of him.-- Answ Had LARGE fortune left. him, took name of Child «of Kinle
541m092 a. nor is an emotion.-- People who can multiply LARGE numbers in their head must have this high facul
549m120 with whom sensual enjoyments of the minute make LARGE <parts> portion of daily <happiness> «pleasure»
550m124 e pleasant.-- Simple happiness «as of child» is LARGE proportion of pleasant to unpleasant mental sen
593n111 ooked round saw at considerable distance a very LARGE hawk, which are «so» rare « <s.> » here,. that
594n115 big dog. which was fast struggling with another LARGE dog his companion. Descent --Affection & [...]
641j29r able to terrestrial Mammifers-- Marine ones «of LARGE size» <to> are best nourished by arctic regions
181b039 mstances.-- Vide two. pages back. Diagram The LARGENESS of present genera renders it probable that <t
043r070 is is contrasted to contemporaneous action over LARGER spaces of the globes & "periods" of increased
182b046 e, it will not occur in plants which are in far LARGER Proportion terrestrial,-- if in any in the Cry
204b140 ommon produced very fine Hybrid offspring, much LARGER than the dam, from those imported by Ld. Powis
226b221 w station.-- When the new island splits & grows LARGER species are formed of those genera.-- & hence
240c014 zzle & few hairs) inhabits Celebes & few of the LARGER islands.-- -- Antelope in Celebes, Bourou new
266c092 like those of the Alderney breed, but size not LARGER than those of Black cattle. Not have hump like
298c192 those over whole of country.-- some species are LARGER &c in different countries. These facts show ho
357d072 ave suffered most with respect to extinction of LARGER forms.-- From observing way the Marsupials of
359d087 tificial breed-- but Mr Miller says that breeds LARGER numbers, & rears an unusual number out of any
416e072 ion to African Species <good observations.-- >, LARGER than any living [not located] A Greyhound migh
431e123 he idea of the male being smaller, & the female LARGER than average size: (surely this is very limite
450e175 somewhat resembles that of Celebes is somewhat LARGER than the Sambawa, Java & Sumatra breeds, (.Hen
458t001 ses gigantic-- hyaena-- bear & ruminants all of LARGER size.-- the law of large size established-- «A
641j29r d Lacépède on Cetacea & Geographical Distrib of LARGER Seals-- Are Porpoises numerous in cold Oceans
154g061 s often descend by flights the terraces if the LARGEST has hollowed out most Wednesday Shelf 3d dies
234b255 are very» like <those of Ice Highlands of> the LARGEST <sort> of our highland Sort, except in one res
368d114 females (as Owen observes) in Raptorial birds LARGEST.-- p. 47. (<"> is evidently the male which rec
403e024 s, (horses not) If they give up infertility in LARGEST sense. <es> «as» test of species.-- they must
451e178 there]» flee during 8 months out of 12.-- the LARGEST mammifers in the world consistently reside & a
494q004 not how is it in <allied» varieties (9) Cross LARGEST Malay with Bantam-- will. egg kill Hen Bantam.
274c113 ully shown in long legged cuckoos with claw like LARK, (one in Australia is called swamp pheasant) Go
274c113 p pheasant) Goatsucker-- parrots with claw like LARK,(NB The La jeune veuve parrot though so much on
298c189 ls of Natural History Vol. I. p. 185 case of tit LARK placing withered grass over nest, when often lo
319c256 ir soft-billed birds are inferior to ours, & our LARK ranks very high.-- Upon the whole thinks <many>
361d095 rsling plumage resembled that of Cross-Beak-- In LARK if I understand right, all species have same ch
363d102 t instinctive & part acquired,-- thus Yarrel has LARK & Nightingale which both sing their own songs,
466t094 rs & pistils, but these <do> do not bend up-- In LARK-spur, if Bees put proboscis within nectary «the
471tf11 sionary enterprises Dr Andrew Smith says in the LARKS from S. Africa he can almost make series from e
468t111 rom great Scarlet Poppy to Rhododendron-- from LARKSPUR to Lupine two species of Larkspur -- two vari
468t111 ndron-- from Larkspur to Lupine two species of LARKSPUR -- two varieties of Cistus Speedwell to Rhodo
470t153 Tree in one minute Great Humble 17 flowers of LARKSPUR on two plants in do Humble 24 flowers of smal
496q05a wers-- also do with honey-- What is use of Bee LARKSPUR= =Toad Orchis= How many flowers in minute do
462tf4r -- Athenaeum 1839. p. 708.-- Shrew, found by M. LARTET same as existing species. We see the same obje
358d076 ays female which approaches in character to the LARVA, or less developed state.-- the female & young
444e158 e, that as the transformations from the egg, or LARVA. or foetus to perfect animal are adapted by for
535m064 on leaves of Eucalyptus, watching few days till LARVA excluded, then though not feeding them «nor hel
535m064 uded, then though not feeding them «nor helping LARVA from egg» watching them, brooding over them, pr
289c162 f affinity of many insects may be told by their LARVAE) but the acts of condensing must alter method
344d040 have read there are exceptions to this in some LARVAE of insects-- (¿glowworms) breeding-- <beet> im

```
352d057 ties, kind of determined by age of foetus.-- As LARVAE may be more perfect (as we use the word) than
352d057 many ABORTIVE organs, might be expected to have LARVAE more perfect-- this is applicable to young of
370d116 wasps,  is supposed cells properly are made for LARVAE.-- CD[(p. 451.)-- Wasps breed many females, bu
380d153 nsiders the house bug, as a <female which have> LARVAE which have bred before the vis formativa had c
535m064 il when touched with it-- do not know their own LARVAE, but one female may be moved to other  larvae,
535m064 wn larvae, but one female may be moved to other LARVAE, with two groups near. mother desert one somet
477z003 uay in 1769 introduce in Governor's tran?? Azara LAS Vinchuca or Benchuca. "Les individus ailes peuve
326c265 ks quoted by Herbert. p. 338 Schiede in 1825. & LASCH. Linn. in 1829 has given list of Spontaneous Hy
574n041 aliva, & hence with action of mouth & jaws.-- LASCIVIOUS women. are described as biting: so do stalli
244c022 > holds same relation with equator--that Vesp. LASIURUS does in North. Hemisphere.-- p. 158 Cuscus al
022c008 & storms:--«in Volney's travels also» Dampier's LAST voyage to New Holland P 127.--Caught a shark 11
024r016 [23  degree S.] 7 60 {T} Soundings about same as LAST to N. of C. Frio Except at Abrolhos. [18 degree
046r078 out together? How are they eliminated.--«Sulphur LAST.--» Metallic veins likewise must separate ingre
079r177 s coming from three directions. from W. NW & S.--LAST to Seaward partaking of the character of a Arau
093a031 .-- The most infusible first injected.-- Basalt: LAST because it could reach the surface. before bein
098a047 rtz & feldspar) within other concretion.-- state LAST page thus. point of attempted  crystallization.
100a053 fault explaining vary dip & inclination.-- which LAST is strong character.-- A discussion on concreti
103a065 crust might be considered a level.-- Dikes being LAST action. (effect of horizontal movement) hence g
107a078 use heat proceeds from great body of mass.-- The LAST speculation becomes important with respect to t
157g074 oft shelf-- 29.330 A 84 degree compare this with LAST measurement of shelf of 3d:-- granite block a y
157g078 9.316 divortium aquarium «about 12 ft higher than LAST station» 29.316 true terrace «2d» near divortiu
161g094 terrace  «station perhaps 6 ft too low» (to test LAST on Peat-Mass Divortium aquarum) Barom. 29.200 A
161g094 A.77 degree Air 70 degree? Barom. 066 lower than LAST. but A 77 station was <a few> «about 3» feet <1
161g097 this  terrace Barom. 29.264 A 82 75 degree? This LAST measurement turns out too low, (NB .260 would h
162g099 on the Lochy side of it-- the terraces of which, LAST measurements belong are so complicated, that no
162g100 e side & dies away on the steep & rocky gully of LAST stream Friday Loch Lochy near Letter Finlay Bar
192b089 Mammalia as Heterodox as ornithorhyncus. If this LAST animal bred-- might not new classes be brought
212b168 species now least in number (as cephalopods,) in LAST tertiary epochs most genera dead? --Examine int
217b183 und bitch which would never take the dog. But at LAST a rough-haired shepherd dog lined her & produce
224b212 tures Mem. Eyton's Hogs & dogs.-- The passage in LAST page explains that between Species from «modera
266c091 e first to escape the doctrines of Muhanmad, the LAST to extend, their dominion, armed alike with the
274c114 must  observe these characters vary in degree in LAST instance hardly at all developed. Not  confined
275c121 ickens of each brood.-- These bantam feathers at LAST got ducky, then took white Chinese Bantam cross
281c139 quite fatal.-- ¶Relations of analogy being those LAST obtained.-- less firmly fixed & therefore  most
285c152 nges, keeping distinct from other & if a first & LAST individual were put together, they would not ac
285c153 heredetary & fixed, certain physical changes at LAST become unfit, the animal cannot change quick en
294c177 he rest,-- Lonsdale says he has seen in old Book LAST Bear in England killed in year 1000.  reference
302c202 .-- Birds same remarks. Characters of analogy.-- LAST acquired,-- or aberrant, therefore more  easily
302c202 would be disputed.-- according to Principles of LAST page. <an> osculant groups between two  circles
306c212 got  into habit of going down stream which would LAST were the stream 1000 miles long.-- a monkey. (B
311c226 those  parts being most easily mostrified, which LAST produced --insane men in civilized  countries--
335d016 ile,-- two considerations are here combined). In LAST page, we have seen mules could have no offsprin
335d016 rgans of generation would necessarily fail.-- In LAST page. I should have said, "an animal <acquires
341d032 t curious bird, did not seem to know itself,. at LAST associated with the ducks.-- most strange voice
354d062 tant round,--each particle is placed in place of LAST by the ordering of the nerves, but in different
360d094 y see that two first might cross easier than two LAST. Sept. 11. N Mr. Blyth, at Zoolog. Meeting stat
385d163 in  Zoolog Gardens. & Kings at Otaheite) <Think> LAST litters are considered the most valuable. becau
385d165 the  son, & of the father in the daughters! This LAST remark good. because showing probably not educa
397e004 scovering a change of level of a few feet during LAST two thousand years in Italy, but what «changes»
398e005 e amounts of change now in progress, will be the LAST to object to this theory on the score of  small
399e009 uction to Cuviers Regne Animal No structure will LAST. without it is adaptation to whole life of ani
429e116 e curious on my view-- because these points were LAST connected with those northern regions-- do p. 2
467t104 mass  of yellow pollen protrudes at sheath.-- At LAST I saw Bee collecting pollen from <sheath>  Keel
494q004 quick or slow pulse being heredetary. (6) In the LAST 1000 years how many generations of man have the
497q007 ] Any known changes in Flora of countries during LAST century or two.-- where agency of man not known
500q10a of deer. Contests of <green-- Q.30) March 1842. <LAST> Year «before last» beans & peas were planted i
500q10a f sexes.-- Q.30) March 1842. <Last> Year «before LAST» beans & peas were planted in rows adjoining &
500q10a s adjoining & seeds gathered there were planted <LAST year» pell mell, without sticks & seeds gathere
549m118 ng his puppies-- This looks like free will.-- V. LAST page. A healthy child is «more» entirely  happy
549m121 almost the sole object--. -- I doubt whether the LAST be right. The two rules come very near each oth
559m154 ormation of laws invoking laws. & giving rise at LAST even to the perception of a final cause.-- Read
564n004 sensitive. ulitmate effects of actions.→ till at LAST he face «instinct of» hunger, «of» death & for
599o007 ich runs throug every fibre, when one behold the LAST rays of & & or grand chorus are utterly inexpli
600o08v al kind-- (5) Conscience, not clear-- Then these LAST heads. of separation between soul of man. & int
629o55v tively; I say yes, & my explanation agrees. with LAST head.-- (4) It is other question, how the feeli
029r033 the  ship in a most violent manner. Although it LASTED about a minute, there was no uncommon ripple o
398e006 on.-- My very theory requires each form to have LASTED for its time: but we ought in same bed if very
058r117 ee a cause for Volcanos part of same phenomena LASTING so long.-- The great movements (not mere patch
620o045 f that being part of our nature, & its effects LASTING, whilst passions although equally natural leav
620o045 ons although equally natural leave effects not LASTING. By association one gains the rule, that the p
206b149 rwise the relationship would converge sooner) & LASTLY perhaps some one single one.-- Will not this a
257c056 erved B. Tricolor in Patagonia. then in Chile & LASTLY 12,000 ft above sea in Bolivia; he  examined--
525m025 hildren, then children which were short term, & LASTLY healthy ones.-- Insanity & Epilepsy remain man
535m069 ic <virtue> natures. accounting for fossils). & LASTLY the tracing facts to laws. without any attempt
371d128 rated> continue of same variety as long as life LASTS, yet they cannot transmit through seeds these c
023r009 treamly."--This shark was caught in Shark's Bay. LAT 25 degree. The nearest of the E. Indian Islands.
052r099 ergs are commonly stranded on shores of Georgia «LAT degree ( )», he has rocks on surface. applicable
065r139 aphical miles and width 2 & ½ miles S. Shetland. LAT. 62 degree 55 minute. <only> one lichen. only pr
135a144 son-- worth reading-- Burnetts. vol 4. p. 193 in LAT 26 degree S. Wafer looking for Copiapo. found in
202b131 s as in India. Geograph J. Vol III. P. I. p. 17 (LAT 37 degree about) Vol IV P. I. Geograp. Journal.
402e019 . account of insects of St. Peter & St. Pauls in LAT' 53 degree yet fauna like that 60 degree & 70 de
451e177 II.  p. 46 Carimon Java. (between Borneo & Java) LAT 5 degree. 50 minute S. adjoining it are  several
477z002 co.) are all found south of 26 degree 30 minute. LAT ---- do. p. 207. La punaise was not known among
478z003 P 207 Slight notice on habit of Iguana. not pass LAT. 28 degree North p. 239 In ocean between Lat. 56
478z003 Lat. 28 degree North p. 239 In ocean between LAT 56 degree and 57 degree only inhabitant crust En
498g007 ld grow there (7) Where parts of fructification <LAT> retrograde into leaves-- is this ever effect of
234b255 his  own eyes. new gate. opening towards pig.-- LATCH on other side.-- Pigs put legs over, & then wit
234b255 - Pigs put legs over, & then with snout lift up LATCH & back.-- Frogs attempted to be introduced I is
311c225 to some other case of 180 degree & great diff of LATD. p 355 Echidna of Van Diemen's land & Australia
117a102 erwise the world would daily be scene of ruin in LATE Natical Magazine (before June 1838) that 70.  F
173b010 ve some species same as nearest land, which were LATE arrivals others old ones, (of which none of sam
208b153 in  Arctic. in proportion to genera. agrees with LATE production of those regions, & consequently not
469t135 y flowers-- Saw Bees frequent these flowers till LATE in evening-- On rough calc. 280 flowers-- allow
473s07v can  be concepcional, as limbs &c only appear LATE in pregnancy, & then may just as well be born a
473s07v be  born a tendency to alter or assume some form LATE in youth,-- only facts can decide-- some peculi
491q002 ormity of hybrid & Mongrel offspring 2. How have LATE varieties of Peas & been obtained? 3.. Whether
520m001 with  colour, than with form of body.-- thus the LATE Colonel Leigton resembled his father in body, b
547m113 es in order <to every> calling up ideas of every LATE impression.-- (do the ideas, direct effect of p
547m113 m the exertion of keeping up the memory of every LATE impression. & likewise gaining new ones from se
547m114 er trains of thought, or memory from innumerable LATE events.-- the fatigue of thinking is keeping up
582n067 ages state.-- N B. According to my view marrying LATE, will make average of life longer.-- for short-
614o037 tive feelings. [LHC] As sexual instinct comes on LATE in life, man almost alone in this case can perc
087a007 l mountains so cold.-- Siberia no plants to it, LATELY raised above level of the Sea. Lyells Encyclop
087a007 above  level of the Sea. Lyells Encyclopaedia-- LATELY elevated When Siberia went up. Arctic land wen
149g031 se animals subject to much variation which have LATELY acquired their peculiarities? The slope of A &
297c189 om. Vol 2. Mag of Z. & B. p. 431. Missel thrush LATELY increased in numbers over whole of England & I
299c194 y seeds.-- There are two Dandelions, which just LATELY have been shewn to be same.-- one grows in mar
302c202 RE CHARACTER VARIABLE it is (one of analogy or) LATELY ACQUIRED. In pigs number of vertebrae. subject
```

302c202 r of vertebrae. subject to variation. therefore LATELY acquired.-- I fear «great evil» from vast oppo
327c265 ce 14s. Marh. 20t. 1839. Philosophy of Blushing LATELY advertised. /6s Mrs Necker on Education preemi
401e015 ty of apple be sown, all sorts come up from it. LATELY saw a nonpareil sowed by Mr Tollet so produce.
422e092 ial ones eminently adaptive.-- does it not mean LATELY adapted or transformed. & hence not indicative
441e148 one part «as» in producing bud.-- Fewer of the LATELY acquired peculiarities are transmitted it is d
225b218 omniphitophagous. But it will be said there are LATENT insects.-- as crows against man with gun. & Bu
369d114 zation from specific character.-- Chapt I. Also LATENT Character Hunter Animal Economy p. 482 (Same b
380d154 emale.-- this is full of interest; for it shows LATENT instincts even in brain of male.-- Every anima
178b032 first children partake more of the mother, the LATER ones of the father; is not this owing to each c
189b073 e at one time, though produced either sooner or LATER.-- Prove animal like plants:-- trace gradation
473s07v e peculiarities may be early impressed & others LATER-- All poultry with same down-feathers. Zoology
504q013 rd.= At Shrewsbury one branch of Rhod. flowered LATER.-- effect of accident?? (7) Which. Rhododendrum
032r039 d & from the resistance offered to the greater LATERAL extension of the waves. by the part beneath th
046r080 mmunicated to it as well as by the vertical as LATERAL movement.--At first one would think movement.
104a066 reat, one can conceive anticlinal lines near. (LATERAL pressure would always produce it) but where gr
124a120 on such strata {P} to p. 171. argument against LATERAL injection. from probability of fissures being
313c236 on (.Ehrenberg) --not necessary to generation (LATERAL with no relation to time) as in buds.-- I can
099a048 ? Why have particles in such cases moved more LATERALLY than vertically, in concretions more vertical
099a048 rtically, in concretions more vertically than LATERALLY.-- <In Area of this> {P} If surface covered
425e105 ough these are connected with other mountains LATERALLY.-- Owen. Fossil Mammalia. p. 55. talks of Tap
215b178 rs to introduction to Animaux Sans Vertèbres as LATEST authority. The case of the tailess cat of Isle
260c068 itary Halcyon bird of passage.-- M. Coronata of LATHAM, wrong,-- Mr Yarrell says-- that some birds or
595n117 think of Eyton's horses becoming <white> with <LATHER> <foame> & sweat, when hearing merely hunting
217b187 es-- Australia.-- A <South> American «form of» LATHYRUS has one species in Europe Madagascar has seve
467t104 h ejects pollen from tip of sheath.-- «Also in LATHYRUS pratensis yellow saw stigma project» In commo
222b204 as Lowe written any other papers besides one in LATIN one <of> «on» Madeira-- any general observation
255c050 e other forms, but those inhabiting 3d zone of <LATIT> height & 3d of latitude more commonly are the
029r033 e on the water. It was quite calm at the time. LATITUDE 8 deg. 47 min. N: longitude 61 deg. 22 min. W
255c050 e inhabiting 3d zone of <latit> height & 3d of LATITUDE more commonly are the same species, instead o
316c245 fined to hot countries & many to cold.-- Hence LATITUDE is more important element than longitude.-- B
087a008 ce climate resembling S. America in Europaean LATITUDES.-- Will it be supposed that the armadilloes h
243c019 eussent leurs représentants dans de si hautes LATITUDES"., --¿translate?) All Australian forms have r
250c039 ith Palaeotherium in India--: connection with LATITUDES!? Zoological Journal.-- Vol I. p. 81. Caprom
441e151 be absurd.-- As there are plants, in northern LATITUDES, which are generated by buds alone or roots,
332d003 used to be called Worm Fever, as used much more LATELY diseased Mesenteric glands.-- My Father has se
244c021 ales Chap XXXIX tom IV p. 273 2d Edit Consult LATREILLE. Geographie des insectes, in 8 degree. p. 181
324c268 x terres australes. Chapt. XIX. tom IV. p 273 LATREILLE Geographie des insectes 8vo p. 181.-- See (p.
343d037 tly so. by the extinction of prominent ones in <LATTE> one: The latter will take place when Condition
041r065 over whole bottom is important; because in this LATTER case. we cannot judge whether such fossils. li
043r069 Chiloe. M. Hermoso. & Coral reefs (imperfect in LATTER). <At> Lyell. Vol I. P. 316. Earthquake of 181
046r079 are generated as often as Volcanic. I consider LATTER as accidental on the afflux of the former. --
049r088 injected mass.--From conical form I incline to <LATTER> former; & thus occurring in groups.--As these
051r093 ina survive, in the most violent surfs: in both LATTER cases become petrified, & increase. -- In Sout
053r100 y rocks, in Alps becomes metalliferous. Vol III LATTER Part Are there Earthquakes in the Radack & Ral
054r102 cleavage E & W! at Payta. talcose slates, do at LATTER place. sandy. sandstone with gypsum, covered b
063r130 Guanaco to recent: in former case position, in LATTER time. (or changes consequent on lapse) being t
070r155 uadero.--three ridges in Copiapo, as well as in LATTER.-- According to Mr Brown, a person (whom I met
073r183 secondary alway in primitive & transition; the LATTER rarely appear in central Cordillera. particula
077r171 o & Moran--of Guanaxuato to SW. with respect to LATTER doubts whether bed or vein (very like that of
105a072 ng a valley depends on serpentine course.-- the LATTER (it is generally said) is consequence of <rapi
126a124 & springs beneath sea-- + According to this LATTER view the rod is reversed, upper part metal «co
212b164 matra & not found on Java-- Monkey peculiar to. LATTER not to former-- Mr Martens of Zoolog Soc told
227b225 es is due to adaptation + heredetary structure. LATTER far chief element.·. little service habits in cl
231b243 viz descent.-- Hence far greater discordance in LATTER-- Have change in form.-- This probably explain
239c002 rossed with common pidgeon, offspring most like LATTER, because oldest variety.-- -- He says of two v
248c033 ternal conditions, & of succession: the <first> LATTER is most intimately connected with important st
251c040 dland, with crops full of maize. (get limits of LATTER from <Tarton> Barton.-- swifts return after ye
251c041 not mix with the Gobbah or village Gazal.-- ¿is LATTER same species domesticated, strangely contradic
261c071 s Yarrell's remark about rock Pidgeons.-- & the LATTER most important in obviating a great apparent d
269c102 of Chemical combination, & the universality of LATTER render-- spontaneous generation not improbable
271c106 ne where offspring picked, one where not.-- the LATTER made by man & Nature; but cannot be counteract
273c111 l species, & likewise perching (Gould), but the LATTER is obscure because nearly all are so.-- Thus i
286c155 e classification of <arrangement> relationship; LATTER word meaning descent.-- A tree is taken by Fle
299c194 e wild, but they affect different localities,-- LATTER on banks & in damp parts.-- both propagated by
320c276 le Macleay's letter to Dr. Fleming. & Review of LATTER in Quarterly Sir J. Sebright's Pamphlets Wilki
343d037 xtinction of prominent ones in <latte> one: The LATTER will take place when Conditions are unfavourab
364d105 tioned between Capercailzie & Black Cock.-- The LATTER has crossed with the Ptarmigan. subalpina in w
365d106 n (probably subalpina.) former has blue breast, LATTER reddish, hybrid purple-- be careful, See to hy
381d157 Heautandrous.-- How is fecundation effected in LATTER; are <it> «organs» open to water? Would not Fe
384d161 ides sexual marks into primary & secondary, the LATTER only being developed, when the first <are> bec
385d163 ference between hybrids & inter se offspring in LATTER being unhealthy.--» males «bred in & in» never
386d167 als very different from that of plants. (though LATTER does sometimes occur in animals). latter the d
386d167 though latter does sometimes occur in animals). LATTER the division taking place from outside inwards
388d170 said. that difference between bud & seed, that LATTER carries with stock of food.-- the generalizati
389d177 should produce same effect as too little.-- in (LATTER case female often takes males but does not pro
398e005 not as «without every animal preserved.».-- the LATTER pages in the history are perfect, we obtain a
402e018 t probably by Chinese>, "the breed being of the LATTER being the same as the fox-like animals. which
416e071 e made by percisely same means as species-- but LATTER far more perfectly & infinitely slower.-- No d
417e079 or the similarity at first births.-- it is the LATTER only that one refers to in speaking of resembl
424e101 Carrion crow. have in Europe different ranges-- LATTER not going North of the Elbe.-- yet they meet i
446e162 (the other three by buds & seeds «though by the LATTER very rarely") here is a case in answer to Mr K
449e173 Europe-- for they stand India. better than the LATTER-- Forrest Voyage p. 323. Sooloo. imported elep
470t153 do-- do 3 of do in about <3/4> of minute These LATTER were pollen gatherers & they seem slow= Maer 1
493q004 to preponderate; according to Mr Yarrell the LATTER ought: either in first breed or permanently.--
494q005 simus collapse during sleep & do of Berberis-- (LATTER I think certainly not) (3) Sow seeds & place c
503q012 ana & yet seedless-- no light Hemslow or Royle, LATTER says seedless-- Also about Sugar-Cane Edwards
520m001 of people taking after their parents, when the LATTER died so long before, that it is extremely impr
527m033 vid Flashes of images & thoughts.-- Poetry. the LATTER thoughts are in same manner vivid & grand. the
557m148 ousy, because less discoverable in animals than LATTER.-- Yet I think one can remonstrate with a dog,
571n029 tastes become acquired during life time:-- the LATTER correspond to fashions in ideal taste & the fo
571n030 re> habits. & heredetary habits. & perhaps even LATTER may be vitiated. or rather altered. The Reason
580n061 fail in ideas of time, & perhaps of space-- in LATTER respect he thinks he certainly has observed th
588n089 ied up or downstairs, or dangled up & down-- in LATTER case they struggle their arms.-- do. p. 306 "t
619o043 love of exercise & its love of its puppies: the LATTER generally soon conquers, & the dog [RHC] 3) pr
151g041 band of pebbles on line of 4th shelf.-- Even on LAUDER Dicks Hypothesis impossible to explain absence
541m092 in a real snarl, they are enjoying a satirical. LAUGH.-- when snarling real bitter sarcasm.-- <These>
542m093 rnestly wish [not] to do it, but an involuntary LAUGH will burst forth, this & yawning. (common to ot
565n006 he noise , the face clearly passes into smile-- LAUGH long prior to talking, hence one can help speak
565n009 f going to say "my dear," just what smile is to LAUGH.-- I must be very cautious. Remember how Lavate
579n058 y of smile. remember children smile before they LAUGH.-- Has frowning anything to do with ancient mov
594n113 y 3 months old. What is absurdity, why does one LAUGH at it-- sensation of disgust with nausea, (when
315c243 - This way of viewing the subject important.-- LAUGHING modified barking., smiling modified laughing.
315c243 - Laughing modified barking., smiling modified LAUGHING. Barking to tell other animals in associated
532m054 ell & felt so much afraid though my reason was LAUGHING & told me there was nothing, & tried to seize
545m106 thinks is from pleasure, & may be compared to LAUGHING» they dance with passion, ie. nervous impulse
556m145 see, they do not move muscle.-- reason CD[The LAUGHING noise which C. Sphynx made at Z. Gardens may
565n006 n extreme.-- Looking at ones face <&> «whilst» LAUGHING in glass. & then as one ceases, or stops the
565n006 r to talking, hence one can help speaking, but LAUGHING involuntary.-- When one fear any bad news, "t

357
LAUGHING / LAW
Page
(Key Word)
569m020 ing with the sound of letter)-- crying yawning LAUGHING being necessary sounds... not produced by wil
540m084 than Locke A dog whines, & so does man.-- dogs LAUGHS for joy, so does dog bark. (not shout) when op
305c211 principles of irritations sleep walking. fits, LAUGHT &c&c Man & Man may have some relation together
557m149 wning, & anguish.-- shyness not so.-- affected LAUGHTER.-- A dog who goes home from shooting. runs aw
557m150 t mental affection make body tremble. Why much LAUGHTER tears.-- & shaking body.-- Are those parts of
592h103 6.-- children have been tickled into excessive LAUGHTER & so into convulsions.-- «Paper» must be refe
428e111 6 generations did become, acclimatized. & says LAURELS have not been so. (which is case adduced by He
025r017 of the extraordinary freshness of the streams of LAVA in Ascencion known to be inactive 300 years? No
031r038 . This structure was very clear at base of great LAVA cliffs [Fig. I] line of high tidal action {P} N
033r043 m of conical crater: at Teneriffe Wall of Porph. LAVA with base of Pitchstone; Mem Galapagos. chiefly
038r059 if so posterior to elevations? & not sources of LAVA streams.--Urge not tilted strata. It will be
049r089 graduate into granites the <conta> passage from LAVA to Granite is much more perfect. than in believ
050r090 on case very doubtful. -- In Iceland Bladders of LAVA are described. & many minute craters as at Gala
052r099 h truth compared to the step = formed streams of LAVA at St Jago. C. de Verds Quartz pebbles in the C
067r143 ellent account of N. American geology. Conybeare LAVA in Cordillera & on Eastern plains «by Antuco».
072r159 verthrown,--no mountains Mackenzie has talked of LAVA flowing up Hill; ¿what does he mean?) Consult D
091a026 to Volcanoes & another on Mexican Trachyte <roc> LAVA called Andesite. Red Coral in the Mediterranean
113a092 the idea of statical equilibrium, the height of LAVA (habitually) becomes measure of force in that p
119a106 approach other rocks. & trap at least as hot as LAVA-- of which temperature is partly known-- <[...]
120a108 added, that which give cleavage to rocks.--, but LAVA shows the rocks really hot. & therefore I doubt
044r072 valleys.--central peak small; yet great body of LAVAS have flowed from centre-- Pisolitic balls occur
049r087 f faults. [Fig. 4] {P} In Cordill: should basal LAVAS be called Volcanic or Plutonic The cellular sta
055r108 iffs of a thousand feet in height, of the solid LAVAS.--proportionally high to age. (we do not wonder
564n006 e from.-- Octob. 4th. Seeing some drawings «in LAVATER, P. cii Vol III» of excessively cross-half fur
565n009 emotion,-- why does emotion make tears fall?? LAVATER says derision lies in wrinkles about the nose,
565n010 laugh.-- I must be very cautious. Remember how LAVATER ran away with new Lavaters.-- Ye Gods!:-- says
566n010 ng to ancestral type of consciousness &c &c.-- LAVATER. (Holcroft Translat) Vol III. p.37, quotes fro
322c270 d Transactions of Royal Irish Academy. ----do LAVATER'S Physiognomy ---- Octob 3d Malthus on Populati
385d164 I believe same has happened in boys bodies.-- LAVATERS. Essays on Phy. transl by Holcroft Vol I. p.
556m145 ike <W[...] what? person says "what a pity"-- LAVATER'S Essays on Physiognomy translated by Holcroft
565n010 utious. Remember how Lavater ran away with new LAVATERS,-- Ye Gods!:-- says fleshy lips denote sensua
585n074 25 Wrong Entry Madagascar Lemur seemed to like LAVENDAR «very much» Henslow. N.. Necker has rem
074r165 etals may be transported by complicated chemical LAW & steam of salts, quite curious case of oxided I
098a046 ed by Sir J. Hershel is all crystals obeying one LAW of crystallization. therefore concretions in thi
099a049 all fluid at once, the films vertical. Ascertain LAW of attraction of particles of same nature: then
102a062 vage discussion, state broadly indication of new LAW acting in certain directions predominantly, conn
122a114 f fluid, the waters of the ocean would obey that LAW. & lie over the platform:-- On my view the degra
172b005 cies are constant over whole country; beautiful LAW of intermarriages «separating» partaking of char
174b014 why modern animals same type as extinct which is LAW almost proved.-- We can see «why» structure is c
176b021 erous. changes not result of will of animal, but LAW of adaptation as much as acid & alkali organized
179b033 k?-- If so Men & plants together would establish LAW. «as above stated:-- no one can doubt that lesse
181b043 tions, & all coming from one stock & obeying one LAW, they may approach,-- some birds may approach an
181b045 pted preeminently for each element, but it seems LAW that such tribes, as far as compatible with such
185b055 water chionis water rat with land structures; :-LAW of chance would cause this to have happened in a
186b058 but I cannot understand the universality of such LAW.-- It would be curious to know in plants, (or an
186b061 nera of those two countries ought to be similar ¿LAW: existence definite without change, superinduced
197b112 alia & birds &c p. 32. reference to M Edwards. LAW of crustacea-- with respect to mouth those beaut
202b133 in plaster of Paris.-- Now this is exception to LAW of type. like horse in S. America. or like livin
208b153 n India!? A p. 28 Dr Beck. & Lyell. most curious LAW of species few in Arctic. in proportion to gener
208b154 many yet multiplied: NB How does this bear with LAW referred to by Richardson in Report about each g
219b196 ifficulty to account for movement of all, by one LAW. as to account for each separate one, so to say
223b210) but yet propagates varieties according to same LAW with animals?? Why are species not formed. durin
230b236 nd in same mountain).-- How is this explained by LAW of small differences producing more fertile offs
248c030 , which evidently is tending to same end, as the LAW of hybridity, namely the [not located] animals u
257c059 y heredetary??.? Does not atavism relate to this LAW.?-- Local varieties formed with extreme slowness
276c125 cies inhabit different countries habits similar ¿LAW?-- probable.-- if habits & structure similar wou
277c125 ilar would have blended together Mem Mr Herberts LAW; habits determining fertility Scheme for abolish
313c234 o very different considerations, with respect to LAW of mammals shorter duration, than molluscs, argu
313c234 pe & S. America. «very difficult case» Does this LAW of duration apply to uniter extinction or rapidit
334d010 w is it with plants? This indicates a remarkable LAW, that first cross <not se> plentiful, second abs
335d014 tes. heredetary & therefore exceptions. to above LAW.-- Study what these monsters are:-- are they «ab
335d015 y might return to either parent), then according LAW, that in proportion as things are long in blood
336d019 lse.-- The creator would thus contradict his own LAW. So far is there any appearance of animals being
347d049 al influence.-- 27th. August. There must be some LAW in nature an animal may acquire organs, but lose
352d058 applicable to young of Cochineal?? Is there some LAW would explain every thing.-- PURE HYPOTHESIS be
352d058 IMPROVEMENTS of every type of organization. such LAW.-- Penguins wing perhaps not abortive???. Apteri
352d059 experience shows us it was for.-- Most important LAW of this. Do any varieties of sheep «evidently ar
354d065 acter of other species in genus."-- Is there any LAW referred to lead to change of foetus. dis
355d066 y among> be different in species of same genus.' LAW of monstrosity not prospective, but retrospectiv
355d067 g what organs are little fixed-- (<also> Hunters LAW of monstrosity with regard to age of foetus. dis
360d089 them since) the line of proof & reducing fact to LAW only merit if merit there be in following work.-
359d088 uin duck.-- which is strange anomaly in Yarrells LAW.-- it probably is explained by the vigour of the
360d091 by the sex> Individual instances trouble Yarrels LAW, chief trust must be in general knowledge of bre
360d091 are kept?-- Is there any mistake about Yarrell's LAW, is it local (not artificial variation) which im
360d091 ot time» thinking of the Penguin duck & Herberts LAW of ideosyncrasy I have hitherto thought that a s
366d112 acted «Referred to <other> Book M.» Is there any LAW of variation. -- «(as Hunter supposes with Monst
366d112 at same time, & make breed, one would doubt any LAW.-- Yet seeing the feathers along one toe of the
366d112 long one toe of the Pouter one thinks there is a LAW.-- that there must have been a tendency for fea
385d164 h." If L. can be trusted, this is Lord. Moretons LAW.-- "How often do we find in the son, the charact
386d166 ng for great length of time.-- There is probably LAW of nature that any organ. which is not used is a
386d166 any organ. which is not used is absorbed.-- this LAW acting against heredetary tendency causes aborti
386d166 ncy causes abortive organs.-- the origin of this LAW is part of the reproductive system.-- of that kn
386d168 ympathy of part is probably part of same general LAW, which makes two animals out of one & heals piec
387d168 her. (with only difference of time) is the above LAW anyways connected with the case of successive co
391d175 triplets &c &c in» in litter. Why is there some LAW about sexes of twins in former case.-- (many mon
406e035 of Corsica. <would not>, sadly against Yarrell's LAW.-- not so much against my modification of it-- G
408e043 , & therefore extermination becomes part of same LAW.-- When we know what a great effect. light has i
410e051 (I from looking at all facts as inducing towards LAW of transmutation, cannot see the deductions whic
410e052 eoretical interest,-- "the great end must be the LAW & causes of change".-- A philosopher, would as s
415e070 ontains many most valuable references See if any LAW can be made out, that varieties are generally ad
415e070 ecessity of the «so called» progressive tendency LAW.-- In animals analogy leads one to suppose that
417e077 larity of child to parent appears to follow same LAW in two of «the» same <species>, variety, as in t
417e079 ather «What is cause of this.-- » (Lord Moretons LAW holds with different species, & individuals of s
417e079 s of children to their parents.-- Lord Moreton's LAW cannot hold with fish, «& there are mule fishe
421e090 of Syngnathus.= curious as showing generality of LAW. even in fish: .do. p. 236-- on Hybridity in fe
423e096 ?-- supposing such to be the case, it proves the LAW of developement in partial classes is far from t
428e112 ttle between strong & weak March 11th. Yarrell's LAW must be partly true, as enuntiated by him to me,
440e145 little part of the Grand Mystery is this,-- the LAW of growth, that which changes the acorn into the
440e148 se mammae in man & wings under united elytra The LAW of <growth> «generation» is only modification, t
441e149 rrot seems to show this.-- This would be curious LAW, Certainly Australian Dog is not affected by dom
441e150 st not be overlooked.-- it makes fourth cause or LAW of change.-- The weakest part of my theory is, t
444e157 strength to it.-- p. 139. Doubts altogether the LAW of balancing of organs.-- In the Batracian Order
449e169 mon goose thus contradicting (probably) Yarrells LAW & Walkers of the male giving form-- they interbr
458e001 ena-- bear & ruminants all of larger size.-- the LAW of large size established-- «Australia,. S. Amer
459tf02 this may be, perhaps. squeezed into Mr Walker's LAW Gleanings of Science Vol III. p 320. Mr Hodgson
510e017 e if any variation in varieties. G. St. Hilaires LAW of Balancement Wm Yarrell (1) About non-breeding
525m023 e, though only acquired rules by art.-- like the LAW of honour.-- they feel pleasure in obeying their
527m031 st:-- so in thoughts, one will rise according to LAW. How strange <all> «so many» birds singing in En

566n012 conclusion that any one fact was connected with LAW.-- as soon as any enquiry commenced, for instanc
567n013 hydra towards light. being direct effect of some LAW.-- have plants any notion of cause & effect, «th
607o025 motion being inclined that way]CD one sees this LAW in man in somnambulism or insanity. free will (a
622o048 & all around him, will be paramount,-- hence the LAW of honour. & the etiquettes of Society.--   [LHC]
622o048 - [LHC] Sir J. M. gives different explanation of LAW of honour from Paley [RHC] Anyone, who will refl
623o050 itions, in this world, they <will conform to the LAW,> «can only be such, as are consistent» with soc
623o050 .-- <no one> doubts) « <I cannot> » [LHC] On the LAW of Utility Nothing but that which has beneficial
624o051 ul at present to some extent.» Hence this is the LAW of our instinctive feelings of right & wrong.--
624o051 quently teach same end.-- Hence this becomes the LAW of right & wrong, though, that part, which is ac
624o051 ges, & probably likewise instincts, for the same LAW effects both.-- <such> changes «in accordance to
632j53r ransportation. it follows from some more general LAW.-- [that the laws of propagation, were created w
633j54r l such facts are merely relations of more general LAW. the plants were no more created to arrest the e
635j55r anation-- it has not the character of a physical LAW, «& is therefore utterly useless-- it foretells
572n035 bout communication of ideas, &c) of expression LAWLESS, whilst they are the only steady & universal m
546m110 e valued, & she was suddenly called to go on the LAWN to see something, on her return could not find
303c204 roduced».-- make abstract on this subject from LAWRENCE. Blumenbach & Prichard -- Now we might expect
324c268 ns paper in Annals of Nat. History Prichard.-- LAWRENCE Bory St. Vincent. Vol III p. 164. on unfixed
073r162 tudy well products of Solfataras[.] some general LAWS. association of lead & silver. Sulp. of Barytes
074r165 tain chain. there after NW <W>.-- «same chemical LAWS as in concretions perhaps makes intersections r
077r172 Mem Sir W. P. stone It is clear to me, there are LAWS of solution & deposition under great pressure.
176b022 case. their creation being dependent on definite LAWS, then those which have changed most. «owing to
195b098 strong. if we believe the Creator creates by any LAWS. which, I think is shown by the very facts of t
195b101 me power let attraction act according to certain LAWS such are inevitable consequen let animal be cre
195b101 nsequen let animal be created, then by the fixed LAWS of generation, such will be their successors.--
198b114 ll worth studying] CD says grand idea god giving LAWS & then leaving all to follow consequences.--
199b118 without the occurence of one of the most general LAWS of life, the transmission of a fortuitous modif
210b161 gin to very early stage) & which, follow certain LAWS according to species. present an analogy to pro
228b228 of <species> structure in species, might lead to LAWS of change, which would then be main object of s
228b229 te, a fungus, or an infusorian. is "What are the LAWS of life".-- Where we have near genera far back,
248c033]nt & therefore most permanent Owen [re]markable LAWS of Brain & manner of generation «& primary divi
261c070 istant countries., may give thread to conduct to LAWS of change of organization! The little turtle, w
265c084 it is direct effect, <of> according to Physical LAWS, as sulphuric acid disorganizes wood, but adapt
269c102 ject in view» The intimate relation of Life with LAWS of Chemical combination, & the universality of
270c103 multiplicity" this unity,-- this distinctness of LAWS from rest of]CD universe «which Carus considers
276c122 eneration?! By profound study of local varieties LAWS of change-- whether beak (as it appears to me).
276c122 beak (as it appears to me). colour of plumage & LAWS, which might probably be reduced «but the Frenc
282c143 n> merely determined to such points by the vital LAWS.-- so that all character originally may «must»
283c146 uantum of life possible with certain preexisting LAWS.-- .If only one kind of plant not so many.-- Th
287c158 express relationship. & by so doing discover the LAWS of change in organizaton. But the classification
290c164 maries black, by examining series I cannot doubt LAWS of change, Will. be known.-- It appeared to me t
293c173 gnorance, may make us think so, but only between LAWS--]CD. Many diseases in common between man & ani
305c210 aving acting principle» according to subordinate LAWS.-- There is one thinking « <& Creat> sensible»
324c268 fon on Varieties of Domesticated animals see if LAW'S cannot be made out Find out from Statistical So
324c268 al Society-- where M. Quetelet has published his LAWS about sexes relative to age of Marriages Brown
336d019 e if created at once. <wd> according to ordinary LAWS, the character of offspring would vary, or rath
343d036 ng affect each other, & their bodies, by certain LAWS of harmony keep perfect in these themselves.--
343d037 on that God created. (warring against those very LAWS he established in all <nature> organic nature)
346d047 some Gulls. Flamingo-- (Spoonbill Wader. Ibis)-- LAWS of plumage might possibly be made out.-- August
355d067 prediction.-- The only advantage of discovering LAWS is to foretell what will happen & to see bearin
362d100 ities have been kept perfect-- also to trace the LAWS of change in this time.-- the impossibility of
397e003 conducted almost! invariably according to fixed LAWS: And since the world began, the causes of popul
397e003 ion have been probably as constant as any of the LAWS of nature with which we are acquainted."-- this
398e006 ly revolutions, but entire obliterations & fresh LAWS created., & yet with <gov> symmetry «& regular
398e006 s created., & yet with <gov> symmetry «& regular LAWS» that baffles idea of revolution.-- My very the
406e036 ding the similarity of offspring to Parents same LAWS appear to hold good. with regard to marriage of
410e051 N.orth & South Thinking of effects of my theory, LAWS probably will be discovered. of co relation of
410e051 be discovered. of co relation of parts, from the LAWS of variation of one part affecting another.-- (
410e053 of change occurs.-- those discovering the formal LAWS of the corelation of parts in individuals, will
410e053 be species or variety, but to discover physical LAWS of such corelations, & changes of individual or
411e054 & every possible contingent circumstance.-- \the LAWS of variation of races, may be important in unde
411e054 tion of races, may be important in understanding LAWS of specific change\.-- When the laws of change
411e054 derstanding laws of specific change\.-- When the LAWS of change are known.-- -- then primary forms ma
411e054 -- -- then primary forms may be speculated on, & LAWS of life.-- the end of Natural History, will be
411e054 ill be approximated to.-- Treating of the formal LAWS of corelation of parts & organs it may serve pe
411e055 ion., & hence indicate gaps.-- by this means the LAWS probably would be. generalized, & afterwards by
417e078 r, that is both parents, for they are one.-- The LAWS, therefore, of likenesses of fathers to childre
417e078 s, when species & races are crossed.-- Now these LAWS are, that child may be either like father or mo
418e083 osition of the atom, to make it alive, & how the LAWS of generation were impressed on it.-- Seeing th
440e145 ving beings. vide Lamarck Vol II. p. 115. 4 four LAWS Who can say, how much structure is due to exter
440e146 f external circumstances. results of complicated LAWS of organization: as we see these strange plumag
441e150 is very curious & important.-- The existence of "LAWS of organization" had better be shown-- soil on
441e150 varieties must be presumed to be result of such LAWS.-- The effect of one part being greatly develop
460t017 nditions (ie. analogical structure) & partly the LAWS of organization (ie those laws which prevent in
460t017 ure) & partly the laws of organization (ie those LAWS which prevent infinite variation in every possi
460t017 infinite variation in every possible way.-- the LAWS which determine the kinds of monstrosity, & det
463t057 pression, than the grouping of «many» facts with LAWS & their explanation will probably reject this t
473o03v step lower in America-- How curious all negative LAWS of America of depth of organisms holding in Ame
535m069 ting for fossils). & lastly the tracing facts to LAWS. without any attempt to know their nature.-- Re
535m069 Now it is not a little remarkable that the fixed LAWS of nature should be «universally» thought to be
535m069 cts that our will may <be> «arise from» a good LAWS of organization.-- M. le Comte argues against a
543m101 in <our own> limits of examination.-- obeys same LAWS. as other parts of. structure. C. D.27 Can an an
553m136 necessary integrant part of his most magnificent LAWS. of which we profane «degnen» in thinking not c
559m154 nwillingness to consider Creator as governing by LAWS is probably that as long as we consider each ob
559m154 to be the case when we consider the formation of LAWS invoking laws. & giving rise at last even to th
559m154 when we consider the formation of laws invoking LAWS. & giving rise at last even to the perception o
573n036 e systems of universe <of man> to be governed by LAWS,, but the smallest insect, we wish to be create
573n036 ts-- &c &c:--must be a special act, or result of LAWS. yet we placidly believe the Astronomer, when h
575n044 ncts, most important, because they obey the same LAWS, as the crossing of jackall & Fox & wolf & dog.
591n097 located] <& other cows--> Mr. Hamilton on vital LAWS (in the Athaenaeum Library) describes effects o
608o025 M. Le Comte case of Philosophy, & savage calling LAWS of nature chance) 2) difference is from imperfe
610o034 Life in the abstract is matter united by certain LAWS different from those., that govern in the inorg
610o034 mple non-organic matter, without action of vital LAWS-- According to the individual forms of living b
610o034 iarities of external form impressed, & different LAWS of movements. [LHC] Hence there are two great <
610o034 ce there are two great <worlds, inor> systems of LAWS «in the world» the organic & inorganic-- The in
610o034 n, heat & gravity is probable.-- And the Organic LAWS probably have some unknown relation to them-- [
610o034 gs namely «one individual» vegetables, the vital LAWS act definitely (<like> «as» chemical laws,) as
610o034 vital laws act definitely (<like> «as» chemical LAWS,) as long as certain contingencies are present,
612o035 e [RHC] The radicle of plants absorb by physical LAWS of endosmic & exosmic juices. arms of polypus,
632j53r for transportation:-- But I do not want to deny LAWS.-- The whole universe is full of adaptations.--
632j53r elieve, only direct consequences of still higher LAWS.-- I do not «then» believe the pappus of <th> a
632j53r follows from some more general law.-- [that the LAWS of propagation, were created with reference to
633j54r o all situations, where in accordance to certain LAWS they can live.-- Hence the mistake they are cre
633j54r Marsh plants. All flow from some grand & simple LAWS.-- 4 «Study Cuviers Anatomie Comparé» p 308. Ir
637j57v re reduced simply statement of productiveness, & LAWS of adaptation\ p. 234. The non-absorbing Camel'
638j58v llect» such as hinge, & hinge of shell, works of LAWS of organization is remarkable-- what is simile
263c075 ind of prejudices) <without> who just takes up & LAY down the subject without long meditation-- His b
289c162 e young, capable of producing young many times & LAY two sorts of eggs-- one remaining through winter
342d032 umbers with pure bred (just like common mules) & LAY many eggs but never produce inter se or with par

(Key Word)

358d085 common. pheasant & fowl when crossed never even LAY eggs. & the men cannot «hardly» tell any sex by
370d116 on bees in same work. (it is said that some kind LAY <pu> up honey even for single rainy day-- & from
417e079 there are mule fishes» & reptiles & those which <LAY> «have» their eggs, <inter>, impregnated externa
592n103 o open mouth in fright because nature intends to LAY open all senses: <do> Horse prick his ears «& sn
098a047 lactite).-- some force crystallizes minerals in LAYER. therefore aggregates them in layer.-- So that
098a047 minerals in layer. therefore aggregates them in LAYER.-- So that layer of feldspar in gneiss is ident
098a048 . therefore aggregates them in layer.-- So that LAYER of feldspar in gneiss is identical with layer o
098a048 t layer of feldspar in gneiss is identical with LAYER of flint on calc.: sandstone. (& as I believe m
099a051 erefore in PILE of mud from Trapiches. inclined LAYER!!!.-- The separation in the Ponza case of Scrop
162g104 sloping buttresses, an[d] one alternate curved LAYER of fine sand & small angular-- rounded pebbles-
385d163 ays, there is breed of Fowls called everlasting LAYER--. or Polish breed. (he thinks half pheasant, h
049r089 ony Pearlstones alternate together in contorted LAYERS: Mem: Phillips Mineralogy some such fact state
068r147 one inch can be raised then all can, for fresh LAYERS of igneous rock replace strata. & it is nothin
073r161 «of F. W.») where the stalactiform masses have LAYERS been accumulated, round knobs, or pushed where
098a047 . a kind of concretionary process (analogous to LAYERS of quartz & feldspar) within other concretion.
099a049 ot attract strong. a third.-- & this would make LAYERS.-- (Gravity can have no effect, on particles o
102a061 nes & felspar.?? Are the great crystals, & the LAYERS first of felspar & then quartz &c, owing to se
163g106 cks now are (32 ft rise) they found alternating LAYERS of coarse & fine & many Sea shells. My informa
399e006 very thick to find some change in upper & lower LAYERS.-- good objection to my theory: a modern bed a
448e165 -- Most plants which propagate rapidly by buds, LAYERS &c &-- do not seed freely.-- The periwinkle
165g125 ad of which hill is round & not merely thoughts LAYING dormant!-- Man from Glen Turret said he learnt
535m064 ribes case of insects «a Perga» of Terebrantia, LAYING eggs on leaves of Eucalyptus, watching few day
608o027 thought very much, & he will know his happiness LAYS in doing good & being perfect, & therefore will
358d075 not withstanding our vigilance a piece of pork 3 LB was taken from a boiling pot, & carried off by o
090a023 s (& how few cases recorded if we say «100» <5>0 LBS a year too little.-- How comes it none in fossil
204b140 much larger than the dam, from those imported by LD. Powis Hybrid dogs offspring seldom intermediate
216b181 osoph. Transaction on a quagga & mare crossing by LD. Moreton, where mare was influenced in this cros
275c120 e.-- &c picking varieties. unnatural circumstance LD Orfords had breed of greyhounds fleestest in Eng
326c266 arches. on Horse in. N. America.-- Owen has it.-- LD. Brougham. Dissertations on subject of Science c
346d048 made out.-- August 25th Athenaeum (1838) p. 611. LD. Tankerville account of wild cattle of Chillingh
367d113 gs reminded me strongly of Zebra.-- Mem. Quagga & LD Moreton Mare ringed Owen says that Bell in Encyc
389d173 atured by female;<]CD> such view no ways explains LD. Moretons case: without the nervous matter consi
544m102 t an associated is kept up with waking thought.-- LD Brougham thinks no dreams except at this time. h
599o005 eason could not have existed without it.-- quotes LD Mondobbo.-- language commenced in whole sentence
622o049 -- [LHC] instinctive fear of death: of hoarding.. LD. Kames, which Sir. J. says is so ridiculous. [RH
624o50v . 152. Reason never can lead to action.-- p. 164. LD. Shatsbury under term of Reflex Senses seems to
205b143 mal p. 347. Vol I.-- compare with my planariae LEACHES out of water Does the odd Petrel of F. del F.
479z006 n in Northern sea. Chamisso in Kotzebue p. 312 LEACHES on leaves in Sumatra Marsden. p. 311 D'.Orbign
049r089 re agency of dikes: & indeed when do these dikes LEAD to a conical mass. will this conical mass be gr
073r161 ic origin & which is constantly found mixed with LEAD & copper is infinitely rare in all parts of the
073r162 Solfataras[.] some general laws. association of LEAD & silver. Sulp. of Barytes: Fluoric. Barytes:--
074r164 spar. bayte. asbestos garnets.--carb & chrom. of LEAD. orpiment. chrysop[r]ase. opal:-- Veins in Lime
075r167 n Europe. common there accompanied by molybdated LEAD & «argentiferous lead»; sulfated Barytes very «
075r167 accompanied by molybdated lead & «argentiferous LEAD»; sulfated Barytes very «un»common in Mexico. F
076r169 in New Spain, contrary to Europe. argentiferous LEAD not abundant. = considerable quantity of silver
080r178 effects, as with man Does Indian rubber & black LEAD unite chemically like grease & mercury [blank]
124a119 ion.-- if the zinc were mixed with 90 percent of LEAD. it would be still more curious to know whether
227b228 recent & Fossil Comparative Anatomy, & it would LEAD to study of instincts, heredetary. & mind hered
227b228 & mind heredetary, whole metaphysics.-- it would LEAD to closest examination of hybridity «to what ci
228b228 assages of <species> structure in species, might LEAD to laws of change, which would then be main obj
254c049 r, (surely rather parents). (NB These views must LEAD to spontaneous generation??) This whole Paper M
273c110 termediate species,-- This is remarkable & would LEAD one to suppose, that species in same group gene
273c110 e group generally contemporary «++.» «This would LEAD one to expect that fossil forms would generally
280c135 <whi> is this true?? My views, which would even LEAD one to anticipate mules is very important for L.yel
287c157 lso placed a series of animals on the steps that LEAD up to it" p. 20 ‖ +++hiatus & saltus not syn.--
426e106 ubsequent ones. My views, «V <see> p. 103» would LEAD me to think that a <do> variety of one species
529m042 Transactions, of ideot 18 years old eating white LEAD. who was most violently purged «believe worms w
538m078 r cases somewhat analogous, & which I think will LEAD to fact of old people singing songs of their ch
540m089 erwise as he remarks sympathy could be barren. & LEAD people from scenes of distress.-- see how a cro
567n013 o do many animals.-- analogy probably false, may LEAD to something.-- October. 8th. Jenny was amusing
587n089 used with regard to the senses. Reason does not LEAD to action.-- p. 248. Theory of Association. owi
622o048 become like the instinctive, ones, <which either LEAD to actions or not, as feeling of cowardice» «Th
622o048 ts & associations.-- often feelings which do not LEAD to action are repressed thus avarice. &c &c.--
624o50v . so is our moral taste p. 152. Reason never can LEAD to action.-- p. 164. Ld. Shatsbury under term o
626o052 upplies it.-- instinctive feelings will doubtless LEAD to similar actions which in prior <races> gener
626o052 n on body.-- This feeling, when instinctive will LEAD to action.-- the passion rising from weariness
334d011 second horse between <whe> shaft mare & another LEADER mare,-- this stallion though eager to all othe
622o049 t may be combined with feeling towards one as a LEADER,-- the conjugal feeling may be directed toward
264c080 is beautiful gradations of forms in Australia LEADING on one side into shrikes & at the other into i
298c189 ian.-- Curious this +ready answer, without any LEADING question.-- [+] This might be mentioned in not
333d008 aux.-- this agrees perfectly with Yarrell & no LEADING question was put.-- Fox thinks half Lion & Tig
381d157 animals <are> is there gradation of structure LEADING to supposition, that the Cryptandrous are real
420e088 9, «long» account of Hyaenodon, a fossil dog-- LEADING towards Hyaena.-- see Comte Rendu.-- I suspect
589n092 zle in what Sir. J. M. says of pure reason not LEADING to action & yet our emotions being only bodily
615o037 is distinctive mark, it is downright instinct, LEADING to touch a particular organ.-- I think Pincher
622o048 instincts. [RHC] Feelings of the mind, whether LEADING to action or not, are the parts of our nature,
038r058 erent from the Porphyries: certainly appearance LEADS me to believe mere fissures filled up.--the app
227b225 ld, & on amount changes which may happen-- ‖ It LEADS you to believe the world older than geologists
227b226 ecies in divisions. look at Articulata!!!--! It LEADS to Nature of physical change between one group
227b226 n one group of animals & a successive one.-- It LEADS to knowledge what kinds of structure may pass i
257c055 ormed, then man may be a miracle, but induction LEADS to other view.-- Till we know uses of organs cl
356d069 inary effects of different Medicines on organs, LEADS one to suspect any amount of change from eating
381d157 ng sexes separate in some of the lowest tribes, LEADS one to suppose still more that they must in eff
416e070 progressive tendency law.-- In animals analogy LEADS one to suppose that seminal fluid fluid, (& not
484z016 phere It is most interesting the way Synallaxis LEADS into Furnarius. by Patagonian Furnarius.-- into
607o025 the wax was soft,-- the condition of mind which LEADS to motion being inclined that way]CD one sees t
626o052 to action.-- the passion rising from weariness LEADS to striking blows.--' p. 224.-- Hume's Inquiry-
298c192 attracting all the moisture.-- Henslow thinks if LEAF of plant varies, <whole cross» «all organs» var
298c192 ns» vary in plant. The variation in character of LEAF of plants is remarkable what is analogous to it
402e017 n only be explained, by such strata being merely LEAF, if one river did pour sediment in one spot, fo
502q11v show acclimatisation.-- July <1842> When nettle LEAF, put into spirits, poison-drop exudes-- does no
515q021 es. (4) Do apples "sport" in fruit, or time of LEAFING (5) Do the most cultivated show Heartease prod
599o006 ts expressed in Fingals cave, & in the arched & LEAFY forests" Very good!. I grant that the thrill, w
054r105 many years since, the sea covered above half a LEAGUE of what is now Terra Firma & the extent of a l
054r105 ue of what is now Terra Firma & the extent of a LEAGUE & a half a long the coast. <"> The Rocks in th
138a153 s present level, & in many parts has extended a LEAGUE inshore both N & S of Lima.--judges from «beds
024r015 er remarks on great flats on the NW coast:-- 8 LEAGUES, from Sydney 90 fathoms La Peyrouse. South of
024r015 Brazil? where not rivers in my Coral paper {T} LEAGUES Fathoms Parallel of St Catherine [27 degree 30
025r016 r this a very shoal coast. Beyond the 10 or 12 LEAGUES sea deepens suddenly. coast of Brazil generall
054r105 Callao, the country, to the distance of 3 or 4 LEAGUES «from the coast» may be concluded to have been
054r105 is particularly observable in a bay about five LEAGUES North of Callao, called Marques, where in all
058r116 rom Chili to Quito a distance of more than 500 LEAGUES. A little time after a bad earthquake in Chili
066r141 bigger than a man's head.-- Kerguelen 40 by 20 LEAGUES. dimensions: Bynoe informs me that in Obstruct
255c050 tion of food, at the Mission of Mojos (even 20 LEAGUES apart from each other.-- this bird was well kn
317c249 frica.-- Macleay tells me same thing p. 55. 40 LEAGUES from land several patches of reed & trees p. 2
309c221 accident. New England farmer,-- useful could not LEAP fences:-- Dr Lang on Polynesian nations (quoted
534m063 ellectual faculty) if ants had at once made this LEAP it would have been instinctive, seeing that tim
515q022 e in tail, than Continental horses. {About the LEAPING of Irish Horses, bred in this country. {Chines

534m062 ttle further, they ascended about a foot & then LEAPT across. (Col Sykes compares this with pidgeons
376d136 made for it at first, but in two or three days LEARN its comfort & though could not put it on, yet t
527m031 ierra del Fuego not one.-- now as we know birds LEARN from each other «though different species» when
527m032 erent species» when in confinement, so may they LEARN in a state of nature.-- Singing of birds, not b
533m058 state that brain alters} It is known that birds LEARN to sing & do not acquire it instinctively. may
552m132 ral Sense Mitchell Australia Vol I, p 292 "Dogs LEARN sooner to take kangaroos than emu, although you
576n048 r higher & far more complicated organization to LEARN Greek, that to have it handed down as an instin
585n074 ker has remarks on the means. by which children LEARN (probably not only experience,, but also «by an
594n112 instinct so firmly implanted, birds soon <dis> LEARN to disobey it-- I have seen hawk & sparrow in S
621o047 in lines of conduct, & he must soon necessarily LEARN that it is his interest to follow it. even when
581n065 hings.-- CD[I may put the argument,, that many LEARNED men seem to consider there is good evidence in
594n112 ng as against hawk, but the birds at Maer have LEARNED that he is not dangerous-- wild-ducks would ha
567n016 -- not if quite stranger.-- or less so.-- When LEARNING facts for induction. one is obliged carefully
568n016 nes of association.-- is totally distinct from LEARNING it by heart. Do not our necessary notions fol
593n107 t may be attempted to be said that young animal LEARNS parent smell & look so by association receives
165g125 s laying dormant-- Man from Glen Turret said he LEARNT to know the lambs because most like Mother in
293c173 mory can remember how to swim after having once LEARNT, & if that was a regular contingency the brain
345d043 (if so)?.-- A Sphepherd of Glen Turret. said he LEARNT to know lambs, because in their faces they wer
526m027 ot heredetary & instinctive) & the others <are> LEARNT. what they teach by the same means & therefore
538m079 then the song was to her like one which though LEARNT in infancy, had often been repeated: Now it is
573m037 tions in her countenance-- before they can have LEARNT by experience, that movements of face are more
586n081 animal has faculty of walking. which in man is LEARNT by experience is in other is acquired instinct
022r008 Caught a shark 11 ft long. "Its maw was like a LEATHERN sack, very thick & so tough that a sharp knif
537m075 bing them with alum, so more slowly does animal LEAVE off <t> instinct, when attended with bad effect
569n022 y I will not. I am sorry I cannot.-- Expression LEAVE «this» out not in Library no good There is a Lu
575n044 tendency to hybrid greyhound to hunt hares. «& LEAVE the sheep» & jackall to skulk about & hunt mice
615o36v an. begin to prowl about in the evening «seldom LEAVE their perch till evening» crow different.-- Her
620o045 sting, whilst passions although equally natural LEAVE effects not lasting. By association one gains t
030r035 from thickly wooded mountains, probably chiefly LEAVES.--This position agrees with character of.. «in
262c074 (2.typical 3.subtypical) where power arbitrary. LEAVES door open for Quinarians to deceive himself.--
351d056 -points-- Macleay then answered, because nature LEAVES vestiges of what she does-- does not move per
433e127 has been much greater, & that systems, are only LEAVES out of whole volumes.-- The fact of tumbling p
446e160 nward progressive developing power.-- My theory LEAVES quite untouched the question of spontaneous ge
479z006 ern sea. Chamisso in Kotzebue p. 312 Leaches on LEAVES in Sumatra Marsden. p. 311 D'.Orbigny consider
495q05a & see which way they fly.-- (9) I have noticed LEAVES covered with Honey-dew dusted with pollen of n
498d007 e parts of fructification <lat> retrograde into LEAVES.-- is this ever effect of want of. nutrition.--
502q11v efully to see if variation equal in flower with LEAVES.-- strawberries How <soon> «early» do characte
528m038 c-- this gives beauty to a single tree,-- & the LEAVES of the foreground either owe their beauty to a
528m038 o the repetition of similar forms as in angular LEAVES,-- (this Rhythmical beauty is shown by Humbold
535m064 nsects «a Perga» of Terebrantia, laying eggs on LEAVES of Eucalyptus, watching few days till larva ex
577n050 actions take place every day, & closing of the LEAVES, comes on from want of stimulus, after certain
120a109 ther its own matter. but would cut wide gorge. LEAVING cliffs, on each side, such as now exist.-- cau
198b114 g] CD says grand idea god giving laws & & then LEAVING all to follow consequences.-- I cannot make ou
233b250 heterogenous.-- Australian dog jumped into tub LEAVING only nose above it-- pulled bell.---- It was
563n001 rton describes. pheasant springing from nest & LEAVING no tracks.-- My Father says pea-hens do Wood p
606o022 es in gaining the result.-- Lessings Laocoon. 2d LECT-- The object of art., sculpture & painting, is
323c269 eration. Englis Transla -- The Revd. A. Wells. LECTURE on instinct -- Cline on the Breeding of Animal
431e122 animals tend to vary. March 20th. Phillips in LECTURE in Royal Institution says shells become less i
582n068 al pain & pleasure.-- The Revd. Algernon Wells LECTURE on animal instinct. 1834: p. 15. "To act from
592n103 ke. Sir Ch. Bell on Expression «First Croonian LECTURES by Parsons.» following pages contain remarks
133a140 an rocks, within say the Tertiary period. one is LED, to look at globe as resting on film of molten r
227b227 e.> transmutation & geographical grouping we are LED to endeavour to discover causes of change.-- the
227b228 zation intelligible.--may look to first germ-- --LED to comprehend true affinities. My theory would g
293c173 ithout being web footed yet with much practice & LED on by circumstanc it becomes web footed, now Man
316c246 ood anomaly in range + + What circumstances have LED to formation of new species some few have been s
380d155 from his own field to bull a cow.-- -- a dog if LED in string will not.-- some of the tigers.-- cat,
418e083 re descended from some one single stock,--one is LED to suspect that the birth of the species & indiv
471tf11 ke series from end to end-- so that he is almost LED to doubt. whether there is such a thing as a spe
623o051 ave been formed by the circumstances, which have LED to the peculiarities, & hence «must have» «only
626o052 milar actions which in prior <races> generations LED to their formation.-- 'N.B. If feeling or emotio
109a083 tually be seen.-- ¿ The preservation of dikes & LEDGES of first-rate importance in showing not subaqu
468t106 ost every flower-- Blue-bells-- wild-raspberry--LEFKS-- Flowers which thought very unattractive-- Fou
160g090 tween one of these & Glen Tarf Hill «Cairn <taw> LEER peak» Barom 28.700. A.75 degree 75 degree? Boul
160g093 o others 3000? if Ben Erin is 3500 boulder Cairn LEET more Haberclador Isthmus broad flat Loch {P] XX
252c042 bly another in Jamaica & perhaps one extant at LEEWARD Isles. p. 388 Reference to Rüppel. travels (wh
162g103 it> Has a very long, flat divatium aquarum with, LEFT of Bright.-- like bed of lake with trace of ter
180b037 t forms.-- Twelve of the contemporarys must have LEFT no offspring at all, so as to keep number of sp
181b039 bable that <the> «many» contemporary, would have LEFT scarcely any type of their existence in the pre
196b106 uzzle why Marsupials & Edentata should only have LEFT off springs <ne> in or near South Hemisphere. W
226b222 as perished from S. America, the jaguar has been LEFT & Fox, & bear.-- If I had not discovered channe
231b243 bably explains crag & miocene.-- The descendants LEFT in cooling climate might change twice over, whe
250c038 America Some portion of the world «Africa» being LEFT more equable. yet America preeminently equable.
266c091 yet we know how remote the periods at which both LEFT the lands of their forefathers;-- the first. to
285c150 st be introduced & made young.-- ¸ father must be LEFT out of case, that difference occurring.-- It wi
291c167 of others, (which is almost proved. Elephant has LEFT no descendant in Europe<)> Toxodon in S. Americ
308c217 then when that was finished kept it in--<right> LEFT, but I always for a week took of cover of right
360d094 stomach would be different.-- Or if one species LEFT its type in having very long legs, & another in
363d102 in swamps-- (owing to barns, perhaps, not being LEFT open to them,-- ,. In singing birds, part insti
417e076 ls of the common species: from one locality, all LEFT whorled.-- He kept two to see if they would bre
522m011 .-- What became of him.-- Answ Had large fortune LEFT. him, took name of Child «of Kinlet» & married
524m022 en my Father to distract her attention took her «LEFT» hand to pretend to feel her pulse.-- What fail
533m059 ns-- the difficulty of getting on a horse on the LEFT side (not good example,) because leg is right h
543m100 ond St. was so great a fool that his Father only LEFT him a guinea a week. yet. he was inimitable che
591n099 e)» contempt as suicide. (even when no relatives LEFT to lament) is owing to the feeling that the ins
255c050 ological part of Voyage of??? A Urubu, (with one LEG) attended the distribution of food, at the Missi
403e022 common.-- without arm, <skin> hands thumb,-- one LEG, hare lip &c &c. in Vol II p 363 account of Flor
533m059 rse on the left side (not good example,) because LEG is right handed.-- In Review (Edinburgh) of Frou
601o009 relation to the primary sensation.-- man moving LEG when asleep-- «or habitual actions» perhaps poly
274c113 The Rasorial type is wonderfully shown in long LEGGED cuckoos with claw like lark, (one in Australia
309c221 se Dwights' Travels in America, speaks of short LEGGED sheep. heredetary proceeding from an accident.
639j28v reed, if circumstances determine that, the long LEGGED one shall rather oftener than any other one. s
639j28v er one. survive. in ten thousand years the long LEGGED race will get the upper hand. though continual
640j167 ath of its destroyer), or other cause, the long LEGGED race would prevail, even if have afforded only
640j167 number of dogs. would diminish, whilst the long LEGGED variety would prevail.-- Not separately: NB. T
023r012 in the passage!!-- M. Labillardiere in Bay of LEGRAND, (SW part). describes a Small granite Isd. cap
146g011 faced sheep, sometimes mottled with white black LEGS & tail like species in colouring Strike an anal
205b143 tise There are some good accounts of passages of LEGS into mouth-pieces of Crustacea. Vol II. p 75 a
234b255 g towards pig.-- latch on other side.-- Pigs put LEGS over, & then with snout lift up latch & back.--
272c107 y.-- there are hemipterous insects, having spiny LEGS & running quick & generl appearance of blattae
283c145 species of Orpheus-- one of which has very short LEGS & long tail «short much curved beak.--», other
303c204 s of intellect) quantity & kind of hair forms of LEGS-- hence the father of man kind probably possess
303c205 iting pidgeons & gallinaceous birds & parrots.-- LEGS of pidgeons perfect.-- &c &c.-- do p. 136. Icht
318c252 ion for snow-- like snow shoes. than feet & hind LEGS of these white hares, fitted for regions of sno
345d044 e of Provincial Breed-- Highland Sheep jet black LEGS, & face & tail, just like species.-- high activ
360d094 if one species left its type in having very long LEGS, & another in having very long tail, & other in
363d101 inks work on Pidgeons--. & see whether feathered LEGS.-- «Carruncles on beak & in Muscovy duck» crest
363d101 long ears of rabbits. & long fur.-- feathers on LEGS of Ptarmigan & in Bantam.-- }CD CD[In the Pidge
366d108 told me of a cat & of a dog, born without front LEGS-- -- the former of which had kittens with imper

367d113 . apparently fathered by a donkey. with all four LEGS ringed with brown.-- animal like large, heavily
367d113 ade cream coloured ass.-- stripe on back also.-- LEGS reminded me strongly of Zebra.-- Mem. Quagga &
383d160 silver pheasant. yet why green? & not purple?-- LEGS pale coloured.-- In the back feathers, we have
399e009 paler & buffer.-- with long ears-- & longer hind LEGS??? so that I was almost doubtful which it was.-
399e009 at I was almost doubtful which it was.-- do hind LEGS increase in any rabbits One may strongly suspec
445e159 ers-- says lemur.-- volans, has skin between its LEGS.-- -- strangely consider existing «long-organ»
458t002 rses in Lao Choo so small, that person with long LEGS can hardly ride on them. Mr Miller-- in Zoologi
468t105 about as large as bit of chopped horse hair with LEGS & take flight-- Yet we have crosses-- I see Bee
472s02r o day saw very odd dusky humble (with pollen) on LEGS go from clump to clump, & insect proboscis in m
484z014 anner of walking-- foot bill crest feathering on LEGS-- habits-- Does the Secretary, make noise & thr
552m131 rily feels angry, when brain is pumping force to LEGS & body, & especially, when to whole body, being
582n067 l impulse (though unconscious of it) to move its LEGS so, as much as in the young salmon to go toward
585n075]CD April 3d. 1839 The Giraffe kicks with front LEGS & knocks with back of Head, yet never puts down
639j28r et & rodents (?) p. 251. all animals run by hind LEGS-- Kangaroo. only a caricature; Penguin.-- Pince
639j28v e out of every hundred litters is born with long LEGS» & in the Malthusian rush for life, only two of
497q007 re there unusually many species in genera of LEGUMINOSAE.-- Herbert explains numerous spec. of Cape H
455eIBC cannot «crossed» etc.-- Make Hybrid mosses.-- LEIGHTON or some one. Father-- diseases common to men
520m001 han with form of body.-- thus the late Colonel LEIGHTON resembled his father in body, but his mother i
098a044 la Soc. Geolog: 1833-34. p. 35.-- Ancient Lake LEMAGNE in Auvergne Proofs from Phryganea NB. Sedgwick
446e162 .-- May 27th.-- Henslow One of the 4 species of LEMNA only reproduces itself «in England, as yet obse
446e163 forth quantities of blossoms--» The case of LEMNA, «and the vivaporous grasses, which no doubt ar
447e164 my theory does require crossing.-- The case of LEMNA shows dispersion of germs is not end of seminal
447e164 lpina vivaparous sometimes seeds All species of LEMNA sometimes though very rarely flower [bu]t the o
466t099 of individual. June 1st 1841. Maer Examined the LEMON-thyme.-- equally abortive as it was in autumn:
445e159 imals & takes series of flying mammifers-- says LEMUR.-- volans, has skin between its legs.-- -- stra
585n074 respiration V E. p. 125 Wrong Entry Madagascar LEMUR seemed to like Lavendar Water «very much» Hensl
232b249 as most curious facts about the distribution of LEMURS in Madagascar, on neighbouring islets & a sub-
423e095 uppose them to have changed, these very changes <LEN> tend to give rise to others.-- Why then has the
032r039 d in line of highest tidal action. this will at LENGTH be checked by increased vertical <height> thic
036r049 gradually curved enlargement see its increased LENGTH. which will represent the dilatation, which di
053r101 per Dr Fittons Australia case must be quoted at LENGTH. The Lines of Mountain appear to me to be effe
127a125 - Hence further N. when soil frozen for greater LENGTH of time depth of ice ought to be less.-- Memoi
139a180 ied by comparing Zoophite to plants.-- grafting LENGTH of life &c & Will any inorganic substance cau
211b162 udimentary bones.-- As Waterhouse remarked Mere LENGTH of bill does not <indicate affinity with snipe
303c204 + + form of head «& features»;. but likewise in LENGTH of extremities, how are races in This respect
317c209 everal patches of reed & trees p. 259 120 ft in LENGTH, some branches of Justicia still growing,) pas
317c251 ky Mountains have peculiar character in extreme LENGTH of ears & length of limbs, so that he first th
317c251 peculiar character in extreme length of ears & LENGTH of limbs, so that he first thougt only one spe
347d049 ing) quote Whewell as profound. because he says LENGTH of days adapted to duration of sleep of man.!!
386d165 life & split it, & see whether it retains same LENGTH of life.-- like Golden Pippen trees! How is th
386d165 ged bud being formed & one part dying for great LENGTH of time.-- There is probably law of nature tha
400e012 d of fur-- (do tips of ears take any colour?)-- LENGTH of tail varies, & character of fur-- I am sure
417e079 eny more than others.-- does this more refer to LENGTH of time that the resemblance is permanent, or
432e125 the succession of organisms tell nothing about LENGTH of time, only order of succession.-- Splendid
458t001 r Cantor's account of fossil frog, 40 inches in LENGTH--! alludes to ancient gigantic salamanders-- E
466t099 as it was in autumn: filaments united in whole LENGTH = corolla--anthers minute, distinctly doubled
493q003 le» with tinge of tortoise-shell «on back.--» = LENGTH of intestine in Persian Cat-- , in Brazilian «
493q003 Brazilian «toothless» dog-- I. St. Hilaire says LENGTH differs in different cats.-- Good observation-
493q004 t breeds of dogs. Cattle, (Indian & Common) &c: LENGTH of life. (5) Does my Father know any case of q
505q014 rk has six day's puppy of small true Bull-Dog-- LENGTH from nose over head to root of tail 28½. inche
506q014 ine of back, height 17½/. The Greyhound. was in LENGTH (measured same way) 47½-- in heigt 30 inches E
511q018 Mad> Porto Santo Rabbit. Descript. of colour «& LENGTH of ears» & skeleton, & skin= Van. Voorst often
54lm091 ng intently; for that one does with novel for a LENGTH of time.-- Then if one endeavur to keep any si
614o097 or superiority of memory does not depend on its LENGTH.: Many animals (as horses) very long & good me
273c110 «black & white» bars on wings of trogons are LENGTHENED rump feathers.--; <then> & one species has s
276c124 ee what Eytons young pigs-- if vertebrae much LENGTHENED &c there may be tendency to divide, which of
468t112 , stamens straightened pollen profusely shed; LENGTHENED & turned up «more than stamens», so that all
509q017 General Questions (1) Particulars about Sierra LEONE. cow. taking bulls , is it Domesticated African
439e144 o think it absurd. that the presence of of the LEOPARD & Tiger together depended on some nice qualifi
450e174 Mr. J. H. Moor-- -- p. 1. Elephant. Rhinoceros LEOPARD (but not Royal Tiger), &c are found but only i
464tf6r is the peculiar. N. American form)-- ¿Hunting LEOPARD, how strange, anyone, would have thought isola
558m152 fferent trades &c &c &c I observed the Asiatic LEOPARD. quarrelling. mouth wide open, each [lip] draw
448e168 of Africa Owen Linn. Soc. April 2d. 1839 The LEPIDOSIREN-- Amblyrhyncus & Toxodon, <all> equally aber
495q05a s. Is any intermediate form found wild a The LEPTOSIPHON densifolium «an annual» <sleep> «closes flow
468t106 ith small Staphylinidae-- Anapsis, Melegethes, LEPTUSE-- Diptera & small Hymenoptera Saw Humble go fr
244c023 w Ireland ---- maculatus -- Waigiou Speaking of LEPUS Magellanicus says; <after> "après un examen att
601o08b Hat de l'education des Mères par L Aimé Martin LEROY Lettres. Philosophique sur l'intelligence des A
242c017 elago. \Raffles. Horsfield. Diard. Duvaucel. LESCHENAULT Kuhl. Van-Hasselt, Reinwardt «Forrest» autho
106a074 paratively small, & when that is case force is LESSENED. therefore rivers very ineffectual in widenin
109a083 e difficulty of such preservation certainly is LESSENED.-- Coral flats. argument for Heaping up.-- ve
612o035 ory, though not adding to positive knowledge. LESSENING amount of ignorance [RHC] The radicle of plan
179b033 law. ≠as above stated:-- no one can doubt that LESSER trifling differences are blended «by» by inter
262c073 ch group absurd.-- the mere fact of division of LESSER & more power (2.typical 3.subtypical) where po
263c075 choose & firmly believe in his new faith of the LESSER of the difficulties ONce grant that «species»
387d168 , & likewise ovum in (B) <an C> that in (C) «in LESSER degree» -- Then when (C) unites with Male (X)
387d169 heir offspring» inherit the same peculiarity in LESSER degree C. & theirs again in lesser degree.--no
387d169 culiarity in lesser degree C. & theirs again in LESSER degree.--now if the <tw> second race both have
401e017 mpresses main features on offspring. & make the LESSER peculiarities.-- brilliancy of inflorescence G
424e103 y some plants vary more than others» & horse in LESSER degree,-- how different to dog!-- (Hybrids of
549m123 d nor is it odd he should have had them.-- with LESSER intellect they might be necessary & no doubt w
565m009 lubbering-- sulkiness is same as pouting, <but> LESSER in degree, no smile, no frown showing thought,
322c269 Voyage. carefully read.-- Reynolds Discourses LESSING'S Laocoon Whewells-- inductive History.-- Refer
326c266 Physiology.-- Dampier. probably worth reading LESSINGS. Laocoon.-- (translated in 1837) on limits o
606o022 ch the mind undergoes in gaining the result.-- LESSINGS Laocoon. 2d Lect-- The object of art., sculpt
606o022 simplicity with grandeur of character.-- Hence LESSINGS shows expression of pain cannot be represente
606o024 f harmonious colours &c &c surely to be added. LESSINGS Laocoon p. 125-- says new subjects are not fi
040r062 found in perfection on that side.-- Add from M. LESSON. character of Flora to New Zealand, which agre
053r101 the rectangular intersections are singular-- M. LESSON considers the Sandstone & Granite districts to
054r102 be separated by profound valley [.] Sydney. -- LESSON Zoologie Grand tertiary formation of Payta: N.
178b031 naturalists thought was species <Ascensi> Study LESSON Voyage of Coquille.-- Dr. Smith says he is cer
229b234 tr but> Get him to discuss those mention[ed] by LESSON & Chamisso.-- Geograph Journal Vol V. P. I. p.
242c018 Amboina to New Ireland p. 23 Voyage of Coquille LESSON No (p. 24) batrachian in isles of Great ocean
243c020 Zealand & New Holland fish very similar.-- NB. LESSON method of generalizing without tables or refer
245c024 e productions p 293-- is very strong about this LESSON insists much.-- The (p. 296) Columba Kurukuru
246c026 whether any birds Except. Dodo!!-- in Mauritius LESSON &c p. 620. Centropus (Coucal) of Java & Philli
407e042 sary to show that if species fall, genera must. LESSON I remember says Mariana Deer very close to a M
479z005 Falkland ∴ black rabbits not indigenous p 112 M LESSON--Voyage of Coquille wide limits of Nullipora D
480z007 D Orbigny no IV Mag. of Zoolog & Botany p. 356 LESSON on Berre. do-- Magazine of Zoolog & Botany. Vo
482z011 rld German a reference to a luminous Sertularia LESSON Zoolog. Coq: p. 120 Coati Roux. Tatous & perha
274c115 tatives;, the rasorial may be observed even in LESSONIA &c & In relations of affinity all organs chan
241o016 20 of Zoologie of Coquille's Voyage to see if LESSONS' remarks on the Floras can be trusted The chan
284c149 n group we may suppose is it oldest, & therefore LEST subject to variation.-- + <being> good for gene
070r154 iametrically opposite have been instanced: it is LET it not be overlooked that except by trees, I cou
088a014 seed-eaters. This important in transport of Fish LET a Hawk fly at Heron.-- Ceratophytes common in No
098a044 ajaya contains salt see Geolog. proceedings Lake LET out by steps in Central France not very conclusi
099a051 minated dikes in Cordillera.!!!-- In stratum OP. LET force drag particls to line {P} AB, & likewise g
172b007 alters it from some end which is good for Man.-- LET a pair be introduced & increase slowly, from man
195b101 untry, but how much more simple, & sublime power LET attraction act according to certain laws such ar

(Key Word)*
195b101 ng to certain laws such are inevitable consequen LET animal be created, then by the fixed laws of gen
195b102 of generation, such will be their successors.-- LET the powers of transportal be such & so will be t
195b102 so will be the form of one country to another.-- LET geological changes go at such a rate, so will be
228b232 ls not got it, not look forward» if we choose to LET conjecture run wild then <our> animals our fello
264c079 other animals were alive, which have perished.-- LET man visit Ourang-outang in domestication, hear e
264c079 n & rage, sulkiness, & very actions of despair; «LET him look at savage, roasting his parent, naked,
264c079 d, artless, not improving yet improvable» & then LET him dare to boast of his proud preeminence.-- «n
282c141 elapse; during time such changes had elapsed.-- LET these families take <dogs> «domestic animals» wi
283c145 anyone.-- no turn the Zebra into the Quagga.-- «LET them be wild in same country with their own inst
283c145 ved beak.--», other very long beak, with short., LET these only have progeny with species & there wil
283c145 ogeny with species & there will be two genera.-- LET short billed one, be exaggerated, & all rest des
289c161 ed at feathers & scales. passing into each other LET him look at wings & orbits of penguin & then he
292c170 66 wants to see absurdity of Quinary arrangement LET him look at abstract of Swainson on Classificati
292c170 look at abstract of Swainson on Classification "LET anyone even with a very superficial knowledge «l
292c170 eal affinities. ie structure of the whole animal LET him read Mr Swainson's on the Classification of
343d037 e dignity of him, who «is supposed to have» said LET there be light & there was light.-- «bad taste {
343d037 - «bad taste {whom it has been declared "he said LET there be light & there was light."-- » August 17
403e023 . because it has been so pronounced ex cathedrâ. LET us look at facts. considering few domestic anima
540m085 ucture.-- He who doubts about national character LET him compare the American whether in the cold reg
578n055 iss. O. B.-- Mrs. B. A. blushed. analyse this:-- LET a person have committed any «concealed» action h
578n055 ommitted any «concealed» action he should not, & LET him be thinking over it with sorrow,-- let the p
578n055 ot, & let him be thinking over it with sorrow,-- LET the possibility of this being discovered by anyo
581n063 dinary habits is this my new part of the view.-- LET the proof of heredetariness in habits. be consid
637j57v y» inconvenience.! extinction, utter extinction! LET him study Malthus & Decandoelle.-- The Final cau
637j58r icken could not have lived had it not been so.-- LET egg shells grow harder. so must those with weak
095a038 of Poenig Voyage Valparaiso Dr. Gillies in MS. LETTER in Sir. W. Parish Possession. talks of <hill>
112a090 -» A meeting of the Geograph Soc, April 9 1838. LETTER from M. Erhman stating that the mean temp at Y
123a117 wever. veins of segregation in Salisbury Craigs LETTER from M Angelis. B. Ayres. 3d. May. states rema
135a146 of Coromandel. just same as. at Bahia Blanca-- LETTER in drawer with important letters-- When I come
162g100 cky gully of last stream Friday Loch Lochy near LETTER Finlay Barom 30.267, A 68 Air 65 degree? <.194
252c042 f Korto & Steppes of Kordofan p. 401. Admirable LETTER from Macleay to Bicheno much excellent detail
284c149 oup. & Not known in single ones--. viz. Macleay LETTER to Fleming p. 32 "where it (mode of generation
287c157 ive of Man created by distinct miracle. Macleay LETTER to Dr. Fleming. Philosophical Magazine & Annal
320c276 f 1837. re«a»d-- contains very little Macleay's LETTER to Dr. Fleming. & Review of latter in Quarterl
439e143 r Herberts fact about the hybrids (mentioned in LETTER to Henslow) fertilizing each other, better tha
453e183 it with hollyoaks, flaxes &c &c? Mr Herbert in LETTER says distinctly, that Hollyoak reproduce each
483z012 Migrations of Birds".-- 18 do. Vol III p. 422. LETTER from Capt King on birds of St of Magellan. Ver
485z016 England.-- or ground birds-- rather indefinite LETTER Mem Orpheus--becoming tyrant-- flycatcher-- sh
497q06v if so plant them together. & raise. seed.-- In LETTER Mr Herbert says do about OEnothera.-- (14) Exa
565n007 ary.-- When one fear any bad news, «though in a LETTER?» why is person painted with mouth open.-- why
569n020 ers, symbol of word beginning with the sound of LETTER)-- crying yawning laughing being necessary sou
610o031 animals.-- Jan 13th. 1839 My father received a LETTER from Mr Roberts-- «a person he had long known
610o031 ion book, but could not find out-- Directed his LETTER, & I observed he had written Wilson & pointed
135a146 Bahia Blanca-- letter in drawer with important LETTERS-- When I come to treat of the age of the Pampa
569n020 lion &c &c. (in same way alphabet. arose from LETTERS, symbol of word beginning with the sound of le
610o031 -- «a person he had long known & directed many LETTERS to»-- could not <remember> <read> Christian na
601o08b l'education des Mères par L Aimé Martin Leroy LETTRES. Philosophique sur l'intelligence des Animaux-
449e169 n Doveçot are very like a Himalaya species -- LEUCONOTES-- Magazine of Nat. History. 1839. p. 106.--
241c014 eed of oxen at Java.--.p. 140. calls it Bos. LEUCOPRYMNUS. does not say whether wild or not p. 156.--
243c019 et Campbell {52 degree S} qui n'aient egalement LEUR espèces; et certainment on eût été bien éloigné
243c019 eu d'années, d'admettre que ces oiseaux eussent LEURS représentants dans de si hautes latitudes"., --
569n022 e a savage, said his wife would be grieved-- "il LEVA les epaules et dit qu'il valait mieux rester a
026r022 age.-- In Europe proofs of many oscillations of LEVEL, which in the nature of strata & Organic remain
032r039 atest action not having been worn away.--If the LEVEL of the sea was to sink by very slow & gradual m
035r047 ondary:-- consider arguments for oscillation of LEVEL independent of mineralogical nature & dependent
035r047 ogical nature & dependent: & then how wonderful LEVEL «of same beds» should have been kept; it shows
037r056 Volcanic rocks. Bahia, Rio de Jan: B. Oriental? LEVEL surface not disturbed.--Whole West coast. Chono
046r080 ould think movement. owing to water keeping its LEVEL whilst land rose up & down.--But from above rea
047r081 g. 3] {P} form present, i e a part below «mean» LEVEL before the higher part. -- Does the sea fall on
060r124 shing the sea (which of course must retain same LEVEL) to a greater distance".--Afterwards speaks of
087a007 -- Siberia no plants to it, lately raised above LEVEL of the Sea. Lyells Encyclopaedia-- Lately eleva
103a064 therefore that the crust might be considered a LEVEL.-- Dikes being last action. (effect of horizont
117a101 evost.-- My views of insensible oscillations of LEVEL will alone explain the immense amount of change
122a113 - read Herschels astronomy with oscillations of LEVEL.-- {P} will point {P} be the one which generall
126a123 e temperature ought to increase rapidly beneath LEVEL, & in many parts has extended a league inshore
138a153 erly stood three hundred feet above their present LEVEL, & in many parts has extended a league inshore
152g046 e side ravine. Are the lip, or necks of land on LEVEL with shelves effect of corrosion & not cause. M
152g049 hey cut near Loch Tring-- Tuesday Bridge of Roy LEVEL of «bed of» River 30.221/65 degree/ Temp of air
152g050 ral plain. 30.127 A 72 degree Air 65 degree? at LEVEL of upper terrace The butresses of Alluvium rise
152g050 nearly up to Glen Collarig up within 200 ft of LEVEL of 4th shelf= argument against river--compositi
153g051 e buttresses on 4th shelf: others «lines not so LEVEL because of upper edge of cliff» Others below it
154g058 above bank or terrace, from terrace of 2d shelf LEVEL of shelf of Glen Guoy form comparison with gran
155g065 if gullies not now formed «Mac, hypoth,)» the LEVEL during any oscillation must have been so carefu
156g071 at head of shelf 3d almost all granite pebbles LEVEL of plain of 4th shelf at head of Lower Glenroy
160g093 «of pebbles? & Alluvium» which appear perfectly LEVEL, <on op> dies away on gradual slope-- : on N si
161g095 sed the mouth, (deep) of above valley this road LEVEL with Peat moss most distinct then lost by slope
161g095 , then clear. this bit to eye certainly appears LEVEL with rock road, & with piece of excised rock lost at
161g096 could be followed for at least 2 miles on dead LEVEL «by eye» to moss-- on this terrace Barom. 29.26
161g097 B the buttress or pass at Isthmus appears above LEVEL of shelf certainly) I took another measurement
397e004 .-- we feel interest in discovering a change of LEVEL of a few feet during last two thousand years in
398e005 ve at right conclusion about species Changes of LEVEL &c are easily recorded, but changes of species
115a098 lative to currents. Small lakes have power of LEVELLING their shores where currents very weak??-- too
113a092 that part.-- Important as explaining want of LEVELNESS Major Mitchell showed me a river <near> W. of
153g051 ow it"--argument for lake «or sea» at successive LEVELS-- {P} Shelf opposite Glen collarig at bend & h
148g022 ium. At Mouth of Caledonian Canal opposite Loch LEVEN two terraces perhaps upper one 100 ft & other o
541m090 ct.-- ∴ weak minded people are fickle & full of LEVITY (¿ do I not confound action & thought here?) T
574n041 <gum» «teeth»less-jaws. as picture of disgusting LEWD old man. ones tendency to kiss, & almost bite,
376d138 of <a> dog«s». but did not seen to evince more LEWDNESS for bitch than dog: Monkey thus examine each
535m064 > side of tea chest, when no tea do. p. 233. Mr LEWIS describes case of insects «a Perga» of Terebran
212b164 ndia House, collection of Birds from Java.-- at LEYDEN series from several islands.-- Bear peculiar t
611o034 ee Lamarck for this definition given in full.-- [LHC] ¿Has any vegetable or animal matter been formed
611o034 form impressed, & different laws of movements. [LHC] Hence there are two great <worlds, inor> system
611o034 are present, (contingencies as heat light &c). [LHC] This is true as long as movement of sensitive p
612o035 s on period of year as much as inflorescence.-- [LHC] I here omit the case (if such there are) of ani
612o035 cal or general will, & stomach likewise «does». [LHC] in Corallina are not two kinds of life vegeta
612o035 mals, as in stomach, intestines & heart of man. [LHC] ¿How near in structure is the ganglionic system
613o035 easure is felt there must be consciousness??? ? [LHC] ¿Can insects live with no more consciousness th
613o036 thinking principle is the same in all animals. [LHC] «3)» Eyton told me that his retriever Sailor be
613o036 w that they result from organization of brain; «[LHC] not used by Kirby» (:analogy-- as races are fo
614o037 nother-- What is matter? the whole a mystery.-- [LHC] This Materialism does not tend to Atheism. inut
614o037 ; than from hour to hour in <man:> individual-- [LHC] Perhaps even the most complicated instinct. mig
614o037 conscience is one of these instinctive feelings. [LHC] As sexual instinct comes on late in life, man a
615o038 in wild state; certainly to the domesticated.-- [LHC] NB. Two dogs having very different instinct alw
618o042 on to forces & mentality because effort is felt [LHC] 1) May 5th. 1839.-- Maer Mackintosh Ethical Phi
619o042 gal and social instincts, and perhaps others.-- [LHC] ----- p. 113. Mackintosh Grotius has argued nea
620o044 on to disobey the conscience is extremely great [LHC] The cause perhaps lies in its frequency & in it
621o047 ven when opposed by some natural passion.-- (a) [LHC] The conscience rebukes malevolent feelings, as
622o048 e law of honour. & the etiquettes of Society.-- [LHC] Sir J. M. gives different explanation of law of

Page **(Key Word)**
622o048 r wrong.-- «[just as in tastes of the mouth]CD» [LHC] My theory of instincts, or heredetary habits fu
622o049 that the three are as simple as I have said.-- [LHC] instinctive fear of death: of hoarding.. Ld. Ka
622o049 om having received benefits from this person.-- [LHC] p. 254. &c &c [RHC] But the love is instinctive
623o050 in the place. & yet place calls up pleasure.-- [LHC] the instinct of sociability & sociability, doub
623o050 many generations giving the social feelings.-- [LHC] According to my theory, all instincts demand so
623o050 e past race).-- <no one> doubts) « <I cannot> » [LHC] On the Law of Utility Nothing but that which ha
623o051 y during past races could become instinctive.-- [LHC] x It is probably That becomes instinctive, whic
624o051 Eugenius says, slow growth of rule of right.-- [LHC] *for it strives to give conduct beneficial to a
625o052 -- My theory explains both, perhaps, by habit-- [LHC] 11) Whewells preface. [RHC] It appears that Sir
626o052 theory.-- see p. 349.-- remark on this point.-- [LHC] p. 194. «&c &c» Butler's view given on conscien
635j56r h to some, &c &c)» These are reasons, just as LIABILITY to accidents & any other cause.-- (& my theor
148g025 » (& not like dogs), but they thought the breed LIABLE to vary-- I asked this question in many ways &
205b145 e they become worse in form, less hardy, & more LIABLE to disease" If population of place be constant
346d047 high active breedin[g] [not located] half breed LIABLE to vary. I asked this in many ways, but receiv
085a001 . Cruz. from ascertained inclination. of plains: LIAS in Shropshire. or some other wonderful outlyer.
327c265 he Cultivation of Fruit trees in. N. America" in LIB. of Hort. Soc Mr Neil. has written good article
364d104 g the fresh feather, by crossing-- It seems from LIB. of Useful. Knowledge that sheep originally. bla
397e003 icable.-- do p. 529. "It accords with the most LIBERAL! spirit of philosophy to believe that no stone
325c267 on has ceased to be used as tendril into stump LIBRARY of useful knowledge. Horse, Cow, Sheep.-- Vere
327cIBC sentum crypt. transitu et analogia commentatia LIBRARY of Useful Knowledge on Horse & Cow & Sheep Cla
390d179 oks to read Buffon Suites de.-- Horse & Cattle LIBRARY of Useful Knowledge Bell's Quadrupeds the effe
569n022 I cannot.-- Expression leave «this» out not in LIBRARY no good There is a Lutké's Voyage autour du Mo
591n097 Mr. Hamilton on vital laws (in the Athaenaeum LIBRARY) describes effects of emotions-- fear giving g
360d091 first come over, insects somewhat like «between» LICE & fleas. sticking on them, but never in an anim
560m156 ke smells «peppermint» «& music».-- Have monkeys LICE?-- picture.-- Do female monkeys not show signs
325c266 los: Prostitution of Paris. with respect to LICENTIOUSNESS, destroying children. --it is not effect,
609o29v perance, circumstances made the check.-- to LICENTIOUSNESS jealousy, & every one being married to kee
065r139 Shetland. Lat. 62 degree 55 minute. <only> one LICHEN. only production. a body which had long been b
536m071 their nostrils when excited by love? Stallion LICKING udders of mare strictly analogous to men's aff
574n041 ainly»-- curious association: I have seen Nina LICKING her chops.-- someone has described slovering <
535m063 ay make it act wrong, as I have done when taking LID off <tea> side of tea chest, when no tea do. p.
122a114 , the waters of the ocean would obey that Law. & LIE over the platform:-- On my view the degrading ac
130a133 all across a hot.--spring.-- Hot water would not LIE. at bottom.-- Surely we here have proofs of hot
314c237 he corolla being (probably) small to allow it to LIE on one side.-- but in other species, this segmen
466t093 pistils curve upwards, so that anthers & stigma LIE in fairway to nectary.-- Is not this so in Kidne
466t094 do» they must disturb all anthers, wh otherwise LIE protected by the hairy black lip of lower divisi
521m007 ia.-- Now if memory «of a tune & words» can thus LIE dormant, during a whole life time, quite unconsc
572n034 ly, or has injured another bad, vindictive.-- or LIED &c &c Are the facts (about communication of ide
053r100 entioned by Humboldt under name of Rothe-todte-LIEGENDE is perhaps same with that of Pernambuco? Quot
101a056 elves in that direction, in which most substance LIES <.-->? Phillips. Lardner's p. 270-4, good discu
158g082 Gneiss cut smooth on sides of hill where Boulder LIES. buttresses «occur» high up on Shelf 2d «in Upp
467t104 upper petal & the curvature of <an> pistil, etc LIES in gangway= In Lotus corniculatus saw Humble pr
565n009 emotion make tears fall?? Lavater says derision LIES in wrinkles about the nose, & arrogance in uppe
620o044 ience is extremely great [LHC] The cause perhaps LIES in its frequency & in its consisting in desire
639j28v made» long «(as adapted to)» because their food LIES deep.-- I say it is «as» simple consequence the
179b034 ype: in animals so far removed, with instinct in LIEU of reason, there would probably be repugnance &
182b048 in view.-- Geogr. Journal Vol VI. P II. p 89.-- LIEUT. Wellsted obtained many sheep from Arabian coas
266c093 n islets off Arabian Coast.-- \Vol VI. p. 89.-- LIEUT Wellsted "on coast of Arabia between Ras Mohamn
451e176 l of Asiatic Soc.. Vol V. p. 565. in a Paper by LIEUT. Newbold.--» A Malayan albino described "To thi
451e177 61. <J> Catalogue of Birds of India. -- p. 555. LIEUT. Hutton counted, the ova of a tick «in India» &
025r019 soundings, from about 80 fathoms & upwards. that LIFE is exceedingly rare, at the bottom of the sea.-
040r062 Julians looks like such?--destructive to animal LIFE.--Patagonia In the Chonos Islds we must imagine
063r132 idual «divided» by different methods, associated LIFE only adds one other method where the division i
071r157 wn by earthquake with great destruction of human LIFE.--Temple mentions some earthquake at Cordova.--
139a180 mparing Zoophite to plants.-- grafting length of LIFE &c &c Will any inorganic substance cause such m
171b002 mals, where mind, & therefore relations to other LIFE has not come into play)--See Zoonomia arguement
171b002 «cut» Stallions nor nuns are longer lived Why is LIFE short, Why such high object generation.-- We kn
171b005 stem of body & universe therefore final cause of LIFE With this tendency to vary by generation, why a
176b023 s» must in each state of existence have shortest LIFE.; Hence shortness of life of Mammalia.-- Would
176b023 xistence have shortest life.; Hence shortness of LIFE of Mammalia.-- Would there not be a triple bran
176b023 d there not be a triple branching in the tree of LIFE owing to three elements air, land & water, & th
177b025 guins really pass into each other.-- The tree of LIFE should perhaps be called the coral of life, bas
177b025 ee of life should perhaps be called the coral of LIFE, base of branches dead; so that passages cannot
177b027 not. {P} We may fancy, according to shortness of LIFE of species that in perfection, the bottom of br
177b029 genera have died; have they all one determinate LIFE dependent on genus,. that genus upon another, w
179b035 fferently circumstanced.-- ¿Is this shortness of LIFE of species in certain orders connected with gap
179b035 ging from one branch, & the monucle has definite LIFE, then all die at one. period, which is not case
179b035 eriod, which is not case,∴ MONUCULE NOT DEFINITE LIFE I think {P} Case must be that one generation th
191b081 Coral islets.-- Next to animals land birds.-- & LIFE shorter or change greater-- In the East Indian
191b082 pecies.-- Man & wife being constant together for LIFE, is in accordance with The Male animal affectin
199b118 the occurence of one of the most general laws of LIFE, the transmission of a fortuitous modification,
199b118 No 8 Jan-Ap. 1828 -- I take higher grounds & say LIFE is short for this object & others, viz not too
228b229 ngus, or an infusorian. is "What are the laws of LIFE".-- Where we have near genera far back, as well
235b263 ny, imported animals?-- At what. part of tree of LIFE, can orders like birds & animals separate &c &c
257c058 o speculate «not only» about beginning of animal LIFE.: generally, but even about great division, our
259c065 t adaptation during earliest existence; if whole LIFE then real adaptation The case of heredetary dis
259c065 n the other case is <adaptation> «change» during LIFE of parent, & therefore being always necessary m
269c102 th this object in view» The intimate relation of LIFE with laws of Chemical combination, & the univer
269c103 reading "Carus on the Kingdoms of Nature, their LIFE & affinity" in Scientific Memoirs I can see tha
269c103 that perfection may be talked of with respect to LIFE generally.-- where <">unity constantly develops
270c104 Carus how intermediate plants are between animal LIFE & "inorganic life".-- animals only live on matt
270c104 iate plants are between animal life & "inorganic LIFE".-- animals only live on matter already organiz
270c104 ny Metaphysical speculations are entered in upon LIFE. Namely Carus.-- How remarkable that Turdus Mag
282c143 ally produce so great an effect.-- the spirit of LIFE must be every where ambient. & <in> merely dete
283c146 ecies & genera, is probably to add to quantum of LIFE possible with certain preexisting laws.-- .If o
284c147 one kind of plant not so many.-- The quantity of LIFE on planet at different periods, depends,-- on r
285c152 logy breed together.-- The bottom of the tree of LIFE is utterly rotten & obliterated in the course o
286c154 ntenance. «[s]hare of sickness,-- death, unequal LIFE,-- stimulated by same passions-- brought into t
301c199 post in passion.-- Habit instinct gained during LIFE.-- do Elephants easily acquire habits is this t
308c218 st evident!!? examine into this case D. Jeffrey (LIFE of Mackintosh Vol II. p. 495)-- in fact, in all
321c270 803-6 Nothing Lyells Elements of Geology Gibbons LIFE on himself Hume's do, with correspond. with Rou
322c270 <Rengger &c> Mitchell's Australia Walter Scotts LIFE I & 2d & 3rd Volumes Abercrombie on the Intelle
323c269 l in Africa; «well skimmed.» 1839 Jan 10t.-- All LIFE of W. Scott.--, except the V Volume.-- -- 19t.
323c269 ions on morals by Eugenius Feb 14th. Bo«s»well's LIFE of Johnson. 4. Vols 25th Phillips. Geology. Lar
323c269 30th Lives of Hayd & Mozart «Apri 25th Lockarts LIFE of Napoleon.» April 5d Dr. Edwards of <ter> inf
323c269 logy «references at end of each Chapter» Crabbes LIFE June Is. King & FitzRoy To be read Humbold. New
352d059 alogies-- <be> when it is considered the tree of LIFE must be erect not pressed on paper, to study th
371d128 eteriorated» continue of same variety as long as LIFE lasts, yet they cannot transmit through seeds t
380d153 annual being rendered biennial-- the hardness of LIFE in female Moth &c Mr Y. says that Macleay consi
384d161 even a fruit tree or rose degenerate during its LIFE so that successive buds do differ-- any variety
386d165 ation.-- Cannot I find some animal with definite LIFE & split it, & see whether it retains same lengt
386d165 plit it, & see whether it retains same length of LIFE.-- like Golden Pippen trees! How is this with b
386d165 is with buds of plants, does annual give buds.-- LIFE may be thus prolonged bud being formed & one pa
389d176 le system, «as if there was, a superabundance of LIFE, like tendency to budding, which wishes to thro
391d175 ces of parents, or external circumstances during LIFE.-- if the circumstances which induce «which mus
392d175 all are really dioecious..) only simple form of LIFE are monooecious. Will ova of fishes & Mollusca
399e009 ure will last. without it is adaptation to whole LIFE of animal, & not if it be solely to womb, as in
404e026 puppies case of this.-- tendency in «manner of» LIFE to be mottled + heredetary tendency determines

Page ***(Key Word)***
411e054 en primary forms may be speculated on, & laws of LIFE,-- the end of Natural History, will be approxim
416e071 rather effect of liquid semen: therefore animal LIFE commenced in the Water!» It is a beautiful part
423e096 impossible to cover whole surface of world with LIFE.-- for otherwise a frost if killing the vegetab
443e154 d to graft from top shoots.-- If prolongation of LIFE by gemmation <can be> being impossible. can be
466t096 only hereditary characters, wh. come on in after LIFE of Plants-- also goodness of flavour in fruit--
466t096 ur in fruit-- all affected by cultivation during LIFE of individual. June 1st 1841. Maer Examined the
473s07r e frame of animals can adapt itself to course of LIFE, «as in trades» there is no reason, why the pec
473s07r shd be born,-- may come in corresponding time of LIFE of offspring-- No peculiarity in external struc
493q004 of dogs. Cattle, (Indian & Common) &c: length of LIFE. (5) Does my Father know any case of quick or s
521m007 ne & words» can thus lie dormant, during a whole LIFE time, quite unconsciously of it, surely memory
528m039 ection with poetry, abundance, fertility, rustic LIFE, virtuous happiness.-- recall scraps of poetry;
533m058 . W. says that babies know a frown very early in LIFE, <before they> (I think I have seen same thing
533m060 ight handed.-- In Review (Edinburgh) of Froude's LIFE. that author remarks, that writing down his con
546m108 aculty of understanding. Adam Smith (.D. Stewart LIFE of. p. 27), says <sympathy> we can only know wh
546m109 he <small> fact that no one, looking back to his LIFE, would say how many good dinners or......... he
549m121 hether if we obey literally New Testament future LIFE is almost the sole object--. -- I doubt whether
550m126 -- is not this rather more friendship.-- Scott's LIFE. Vol I, p. 127. Talks of difficulty of his own
551m129 ang. take up a stone & pound the earth. Lockarts LIFE of W. Scott Vol VII p. 35 "as ideas come & the
555m143 manner. thought that a person was hung & came to LIFE, & then made many jokes. about not having run a
560mIBC useless sudden movement of muscle) very early in LIFE Do they wink, when anything placed before their
568n017 ss has it.-- When one is «simply» habituated «in LIFE time» to any line of action, or thought one fee
568n019 Scrope, Quarterly Review. 1827? In Water Scotts LIFE.. Tom Purdie, (beginning of Vol V) «finally» sa
571n029 ciple we know many tastes become acquired during LIFE time:-- the latter correspond to fashions in id
576n046 Dog may hesitate to jump in to save his masters LIFE,-- if he meditated on this, it would be conscie
576n047 t he would ever repent, & wished he had lost his LIFE in doing so.-- nor would he regret «having acqu
576n047 man), for it added much to the happiness of his LIFE, & the chance, of so dreadful a consequence to
577n049 ontier instance".-- for it can be shown that the LIFE & will of a conferva is not an antagonist quali
577n049 ll of a conferva is not an antagonist quality to LIFE & mind of man.-- & we do not suppose an hydatid
580n062 habits acquired in that age are retained through LIFE" p. 200.-- "The desire of glory, immortal fame,
580n062 ressive nature of intellect indication of better LIFE p. 207 March 16th.-- Is not that kind of memory
582n066 ese are sensations» <Gardner in his work> In the LIFE of Hayd & Mozart. fine music is evidently consi
582n067 g to my view marrying late, will make average of LIFE longer.-- for short-lived constitutions will th
595n184 s) [blank] "Adam Smith Moral Sentiments" much on LIFE & character "Humes Dissertation on the Passions
600o008 member.-- avarice a compounded passion gained in LIFE time]CD 3. The Infinite, -- lives by hopes, loo
608o028 of a man, are under his control, & that a future LIFE is a reward or retribution.-- it may be a conse
609o030 ? Two classes of moralists: one says our rule of LIFE is what will produce the greatest happiness.--
610o034 the reviewer seems to doubt. [RHC] 1) Effects of LIFE in the abstract is matter united by certain law
610o034 from those., that govern in the inorganic world; LIFE itself being, the capability of such matter obe
612o035 does». [LHC] ¿ in Corallina are not two kinds of LIFE vegetable and animal strictly united? [RHC] It
614o037 herefore consciousness, therefore reward in good LIFE [RHC] Instinct appear like heredetary memory; b
614o037 son of ideas.-- As man has so very few (in adult LIFE) instincts.-- this loss is compensated by vast
614o037 lings. [LHC] As sexual instinct comes on late in LIFE, man almost alone in this case can perceive ins
635j56r] 6 p. 412. Macculloch explains the shortness of LIFE (peculiar to each species) owing to the growing
636j56v s Dioecious Plants were formed. Macculloch says, LIFE, forms a broken, recurrent series, whilst the h
636j56v tion «or world» simple series.-- My theory shows LIFE equally simple series, & therefore trace of beg
639j28v orn with long legs» & in the Malthusian rush for LIFE, only two of them live to breed, if circumstanc
640j28v man are same: fertility of either sex determines LIFE:.» «With respect to whether Galapagos beings ar
614o037 s «(& replace)» to experience gained by man in LIFETIME Heredetary memory not so wonderful as at firs
624o50v redetary habit (& modified & associated during LIFETIME). so is our moral taste p. 152. Reason never
060r124 "seem to arise from some efforts in the land to LIFT itself higher & to grow upwards; for the land i
108a079 heat, then change in form of fluid centre would LIFT with it isothermal line, but if heat from centr
234b255 er side.-- Pigs put legs over, & then with snout LIFT up latch & back.-- Frogs attempted to be introd
102a059 n Undulation in open ocean. as pebbles would be LIFTED up & down. on coast itself, undertow would dra
542m095 ng of the skin contract iris?-- same way as one LIFTS up eyebrows to see things in dark. & hence is t
120a108 . study Hopkins. theory of dikes may throw some LIGHT.-- thin dikes not cooling if they had travelled
258c060 olours so all changes may be considered in this LIGHT.--XX Zoolog. Journal-- Parrots in Macquarrie is
337d021 is formed.-- how one nerve becomes sensitive to LIGHT.-- (Mem whole plant may be considered as one la
343d037 im, who «is supposed to have» said let there be LIGHT & there was light.-- «bad taste {whom it has be
343d037 ed to have» said let there be light & there was LIGHT.-- «bad taste {whom it has been declared "he sa
343d037 whom it has been declared "he said let there be LIGHT & there was light".-- » August 17th Two regions
343d037 eclared "he said let there be light & there was LIGHT".-- » August 17th Two regions may be Zoolo-geog
408e043 f same law.-- When we know what a great effect. LIGHT has in colouring plants,-- who can say. what <l
408e043 t has in colouring plants,-- who can say. what <LIGHT> «colours». acting. by a most delicate organ, o
435e133 to tadpoles. p. 210. Shows. that the action of LIGHT is concerned with the developement of form; but
503q012 Talk about races of Banana & yet seedless-- no LIGHT Henslow or Royle, latter says seedless-- Also a
528m036 real a cause as in music) from the splendour of LIGHT, especially when coloured.-- that light is a be
528m036 dour of light, especially when coloured.-- that LIGHT is a beautiful object one knows from seeing art
528m039 thm applies to the view as a whole.-- Colour «& LIGHT» has very much to do, as may be known by autumn
538m078 two people. Consider this profoundly, may throw LIGHT on consciousness, explained by Dr Dewar on prin
546m110 n her senses in either state.-- does this throw LIGHT on instinct, showing what trains of action may
567n013 ling of the simplelst animals, as hydra towards LIGHT. being direct effect of some law.-- have plants
604o016 onsciousness,, it choosing food-- crawling from LIGHT.-- Yet we can split Planaria into three animals
611o034 ntingencies are present, (contingencies as heat LIGHT &c). [LHC] This is true as long as movement of
612o035 of more or less turgid vessels; effect of heat, LIGHT or shade.) Joining two difficulties into one co
641j29v llusca, Articulata, & Vertebrata, & Planaria, & LIGHT affecting plants.-- in insects the end is gained
535m069 ery true, no doubt savage attribute thunder & LIGHTENING to Gods anger.-- (∴ more poetry in that stat
451e179 ephant [...] saul forests by having a smaller, LIGHTER head, carried more elevated & higher forequart
025r016 gree 22 minute S.] 9 120 {T} Garcia de Avila [LIGHTHOUSE] [12 degree 35 minute S.] 9 124 Itapicuru [R
536m070 f feeling.-- It is as much effort to walk then LIGHTLY as to endeavur to stop heart beating: one ceas
553m135 ges (mem York Minster) consider the thunder & LIGHTNING the direct will of the God (<thus> & hence ar
528m036 autiful object one knows from seeing artificial LIGHTS in the night.-- from the mere exercise of the
081rIBC 513., 0. 5 Hectometre 51. 1. 10 Metre 3. 0. 11 LIG[NES] Decimetre 3. 8 Centimetre 4.4 {T} C. Darwin R
477z003 Les individus ailes peuvent avoir <quatre> cinq LIGNES de long et volent. p. 208 Fleas only appear in
029r034 osition of <Coal> Forests. These thick beds of LIGNITE stratified with substances so like the Coal me
318c253 e extending their ranges. «even migratory birds, LIK swallows» -- degree/ migrations of birds he ment
326c266 of painting & poetry.-- Erasmus thincks I should LIK it. The Sportsman's Repository. 4to. contains mu
022r008 P 127.--Caught a shark 11 ft long. "Its maw was LIKE a leathern sack, very thick & so tough that a s
023r011 rely accidental apertures still open.--The fault LIKE appearance «arising from the manner of horizont
027r023 e: the strange substitution of matter in shells, LIKE Concretions & laminae, show what movements take
029r034 ck beds of Lignite stratified with substances so LIKE the Coal measures in England (Excepting Conglom
037r055 he Galapagos.--strata must be accumulating which LIKE the secondary strata of England, «besides ordin
040r062 alternating with such matter at St Julians looks LIKE such?--destructive to animal life.--Patagonia I
041r066 vein» terminated each way, which little vein was LIKE the rest of these thin veins which project outw
051r093 eat swell & turbid water) organic bodies protect LIKE peat reef of sandstone.--Corals, & Corallina su
053r100 tratified with sediment.--& escarpment worn away LIKE english escarpment The great conglomerate of th
054r105 nland part of this bay are perforated & smoothed LIKE those washed by the waves, a sufficient proof,
065r139 on. Syenite¿ Andite?-- Degrading of inland bays. LIKE St. Julian & Port Desire applicable to Craters
066r140 ea weed said at Kerguelen Isd. to grow on shoals LIKE Fucus giganteus! 24 fathoms deep 24» under 50.
069r150 d with <gneiss>.--(Mica Slate) [Fig. 9] {P} ((3) LIKE Bell of Quillota.) (A) in this strata may be ol
074r165 as urged phenomena in veins, chemical affinities LIKE in composed rock. granites syenite» «strangling
077r171 spect to latter doubts whether bed or vein (very LIKE that of Spital of Schemnitz in Hungary.) Humbol
078r174 Triptera D'Orbigny has figured animal with setae LIKE my undescribed[.] p. 140. Flèche of Quoy et Gai
080r178 Does Indian rubber & black lead unite chemically LIKE grease & mercury [blank] NB. P. 73. General ref
085a002 . Arrowsmith tells me, that Himalayas penetrated LIKE Bolivian Chain. Volcanic islands. from number o
087a012 of valley of <Patagonia.> S Cruz -- from terrace LIKE structure-- Intersection of veins prove, that t
089a019 find stone not thus composed on the NE part more LIKE marble requires polish to see structure,-- «He»
097a042 Brazil L.'Institut No degree 221 Lamellar dikes LIKE Mica Slate Von. Buch. Canary Isd. p 170.-- Mem.
097a043 y coated by a thin pellicle of a blackish colour LIKE a dull & poor varnish, which I conceive to be a

Page ***(Key Word)***

```
102a061 r yields substances from impact, «it» would look LIKE it. Are greenstone dike in Granite residual mat
110a086 es; base in part. block not crystallized Salband LIKE basalt. full of circular cryst of glassy felspa
115a097 els account of Norway chain being broken through LIKE that near-- Obstruction Sound in S. America The
116a100 remity of mountains of Cordova project on plain, LIKE <re> a reef on a sea beach-- «p. 151» first dis
119a104 Edinburgh.  Phil. Journal <]CD>, no great chains LIKE Andes or Himalayas, but great circular mountain
121a111 reat Knife formed crystals of ice except from-- (LIKE my gypsum case) shows power of segregation.-- &
121a112 38 Near Woolwich there are plains & valleys just LIKE Patagonia, & many shells in parts on surface, b
122a115 orphic conglomerates on shore of Loch Lochy very LIKE those of Andes Speculate under head of Beagle C
123a116 bes near Alps great beds of rivers which must be LIKE the Chilian ones.-- Septemb. 2d.-- Sulphur like
123a116 like  the Chilian ones.-- Septemb. 2d.-- Sulphur LIKE carbon must go round of dissemination & separat
126a124 art metal «conveying heat in one direction only, LIKE water below 39 degree» & lower part glass.-- th
128a130 ast of do-- I believe?? coast of North America., LIKE the Mexican Gulf. is fouled by bars of sand & s
133a140 , would be very high.-- M. Parrot ends his paper LIKE a fool.-- Feb 25' All facts show how slowly hea
134a142 nodules  at Bahia Blanca Mr. Malcolmson says are LIKE Kankaer South of Part Luconia-- Phillipines the
145g002 bury Craigs The Highland shepherds dogs coloured LIKE Magellanic fox. an instance of Provincial bree
146g011 , sometimes mottled with white black legs & tail LIKE species in colouring Strike an analogy between
148g025 English  my informant said the lambs were nearly LIKE each other «& half between parents» (& not like
148g025 like  each other «& half between parents» (& not LIKE dogs), but they thought the breed liable to var
148g025 y ways & received same answer Thought lambs most LIKE MOTHER!-- the cross not so hardy but more easil
148g027 ded stones, mixed with some quite irregular very LIKE rubbish at head of Loch Dochart <Nea> Above Spe
156g071 81 A 82 75 degree? From this point plain appears LIKE one uniform slope slightly bending up each main
158g079 ium aquarium Peaty Mass of this point very nearly LIKE head of Glen Guoy nor is horizontal line appare
160g093 ies want on rocky place, but narrow shelves just LIKE road of Glen Roy-- appears to lip with moss  On
162g103 g, flat divatium aquarum with, left of Bright.-- LIKE bed of lake with trace of terraces on each side
165g125 et said he learnt to know the lambs because most LIKE Mother in face-- asked stated this generally th
172b006 as been said, there is instinct for opposites to LIKE each other AEgyptian cats & dogs ibis same as f
172b008 ose the change is effected at once, -- something LIKE a variety produced-- --[every space in that cas
177b027 so that in Mammalia «birds» it would only appear LIKE circles;-- & insects amongst articulata.-- but
180b037 ell adapted, & thus perish out, or on other hand LIKE Orpheus. being favourable many might be produce
181b039 nly each species in each generation only breeds; LIKE individuals in a country not rapidly increasing
183b051 t in view.--¿whether highly domesticated animals LIKE races of man.-- M. Flourens.  .Journal des Savan
185b055 -- but there were causes to induce great change. LIKE the Buzzard which has changed into Cara cara at
187b063 America.--  Never They die; without they change; LIKE Golden Pippens, it is a generation of species l
187b063 ke Golden Pippens, it is a generation of species LIKE generation of individuals.-- Why does individua
189b072 r species», their race is not utterly cut off:-- LIKE golden pippen. if produced by seed go on.-- oth
189b073 ls <are> of same species are bound together just LIKE buds of plants, which die at one time, though p
189b073 produced  either sooner or later.-- Prove animal LIKE plants:-- trace gradation between associated  &
190b079 age between good genera-- How remarkable spines, LIKE on a porcupine on Echidna-- Good to study Regne
194b094 NB  Sapajou is S. American form. therefore it is LIKE case of great <rodent> edentate [has been doubt
195b099 pplicable to fossil animals same type, armadillo LIKE covering created.-- passage for vertebrae in ne
195b099 en of use in progenitor-- or it may be of use.-- LIKE Mammae on mens' breasts.-- How does it come wan
197b112 . -- yet nothing about propagation= I see nothing LIKE grandfather of Mammalia & birds &c ‖ p. 32. ref
202b133 f Paris.-- Now this is exception to law of type. LIKE horse in S. America. or like living Edentata in
202b133 ion to law of type. like horse in S. America. or LIKE living Edentata in Africa &c &c.-- Now if suppo
206b148 s act towards each other, and are acted on. just LIKE the two fine families «no doubt a different set
211b163 without tails: some long & some short: therefore LIKE dogs.-- Ogleby says, Wolves at Hudson bay breed
212b165 told me an Australian dog he had, used to burrow LIKE fox.-- a sort of internal bark. would remain fo
212b167 a according to periods.-- NB. Was Europe desert (LIKE S. Africa) after Coal Period.-- ¿In those divis
216b181 re was influenced in this cross to after births, LIKE aphides.-- Case of boy with foetus developed in
216b182 ated at once.-- Dr. Smith considers the Caffers (LIKE Englishmen) men of many countenances, as hybrid
217b184 od pointer & rough water spaniel «produce litter LIKE both parents» & Mr Bell has half bloodhound & g
217b184 directly  one after the other, puppies differ, & LIKE both parents.-- Fox told me of case of mare cov
220b198 an: 1830. most curious paper on heredetary fear (LIKE rooks with guns) of the Bustards in  Germany.--
221b200 an: 1838 L'.Institut. Bats, in Eocene beds, very LIKE present species. , p 8. ¿Are mundine forms, long
222b203 trifling  differences of expression -- one child LIKE father another like mother Has Lowe written any
222b203 s of expression -- one child like father another LIKE mother Has Lowe written any other papers beside
226b221 -- Is the flora of <S. America> Tierra del Fuego LIKE that of North Europe, many genera & few species
226b223 , see how that is.-- ¿are shell-boring Molluscs, LIKE Carnivorous Mammalia in their wide range & in t
226b223 s Mamm: in Paris basin «allied to present») more LIKE present carnivora than Pachydermata If my theor
228b231 Animals--  whom we have made our slaves we do not LIKE to consider our equals.-- «Do not slave holders
229b233 Cape:  crossed with English Bull. offspring very LIKE common English.-- Hottentots say great tailed s
232b246 not measure time but physical changes (we assume LIKE weather on long average tolerably»  uniform).--
232b246 .-- Comparing fossils with whole world. would be LIKE in a Meteorologic table «in comparison of tempe
234b255 vel. p. 280. says cattle in Iceland. «"are very» LIKE <those of Ice Highlands of> the largest  <sort>
235b262 different countries a case in point.-- All cases LIKE Irish & English Hare bear upon this.-- Why do V
235b263 als?-- At what. part of tree of life, can orders LIKE birds & animals separate &c &c Work out Quinary
239c002 geon crossed with common pidgeon, offspring most LIKE latter, because oldest variety.-- -- He says of
240c003 be called genera., yet retains markings of wings LIKE the wild rock pidgeon.-- fact analogous to Owen
240c003 in mule & pig being half way. Yet dogs sometimes LIKE father, sometimes like mother. The fact of grea
240c014 f way. Yet dogs sometimes like father, sometimes LIKE mother. The fact of great monstrosities being p
241c014 r wild or not p. 156.-- Parroket with stiff tail LIKE the Siam race with long nozzle & few hairs) inh
244c021 s, in 8 degree. p. 181 «who says insects Indian, LIKE Plants.--» It would be very important to  show
245c023 cte! p. 171. Sus papuensis «partly domesticated» LIKE in general appearance to Siamese kind.-- but co
247c029 Rabbits  introduced in 64, of very many colours, LIKE the cattle, <">which I say "are as variously co
249c035 ng again The Varieties of Cardoon are cases <sp> LIKE those of Primrose & Cowslip run wild, The two s
251c040 habitations  India & Africa.-- NB. Any monograph LIKE Gould on Trogons worth studying.-- «do»  Zoolog
255c050 ny. Birds of prey, are distributed in S. America LIKE other forms, but those inhabiting 3d zone of <l
257c059 duces young ones capable of producing young ones LIKE itself, but ¿whether great assumption? not sole
257c059 ¿whether  great assumption? not solely producing LIKE itself, not applicable to monsters:-- Are monst
259c066 than  there is reference to more than offspring  LIKE atavism) & shows my «view of» generation right?
261c069 amily sending out structures into many genera.-- LIKE Synallaxis or Marsupial animals of N. America H
261c071 on now existing? Mr Gould says wherever any mark LIKE red patch on wings of Furnarius, Synallaxis &c
262c072 tion heredetary; ie instincts of wisdom virtue? «LIKE senses of savages? (How come its some countries
265c084 e than mere child, but that child should produce LIKE children. ‖Lyell has story from.-- Beck about s
266c092 moir on isld of Socotra. Cattle generally marked LIKE those of the Alderney breed, but size not large
266c092 larger than those of Black cattle. Not have hump LIKE those of India & Arabia p. 202-- sheep have not
267c094 0 together» colour reddish <brown>. ears long.-- LIKE bull terrier-- Indian secured one, as they alwa
267c094 ull terrier-- Indian secured one, as they always LIKE to cross their breed p. 333-- alludes to the Ma
268c100 ¿may it not be explained by mere chance?-- or it LIKE each great class of animals having its aquatic,
272c108 e Heteromera, which have habits & part structure LIKE Curculionidae.-- Are there any Crysomelidae, wi
274c113 nderfully shown in long legged cuckoos with claw LIKE lark, (one in Australia is called swamp pheasan
274c113 swamp  pheasant) Goatsucker--, parrots with claw LIKE lark (NB The La jeune veuve parrot though so mu
275c121 cts all half Bred cattle of L'Darnleys were most LIKE parent Brahmin bulls-- Mr. Y. is inclined to th
278c131 found!! rodents cold inhabitants most important!! LIKE Dipus of present day??! Major Mitchell does not
284c148 t to crawl up an hill, then by deaths?!»-- looks LIKE subsidence.-- on the islets Mr Blyth remark tha
286c155 judices, yet have justly excalted nature of man. LIKE to think his origin godlike, at least every nat
290c163 e road at right anglles, how pleased it is, just LIKE man. emotions very similar.-- Geology. Transact
290c164 garden  at Zoology Soc. it's pale ash grey back, LIKE a black bird washed, Whilst tips of primaries b
290c164 me  that half between fowls & pheasants, is most LIKE pheasant., I think so because very 3/4 bend.--
292c170 t anyone even with a very superficial knowledge «LIKE myself» of <classi> real affinities. ie structu
295c178 ed that character -- -- female & young seem most LIKE mean characters the others assumed--+++ ‖Daines
296c184 an) on Peaks. Did Creator make all new yet forms LIKE neighbouring Continent. This fact speaks volume
297c186 Hemiptera do. p. 160. soft plumage of night jar. LIKE owls. analogy in habits adaptation to nocturnal
298c189 t he has seen half fox & dog. & that it was most LIKE fox.-- He felt sure the half breed of Australia
298c189 e half breed of Australian dogs, would be «most» LIKE Australian.-- Curious this +ready answer, witho
299c193 lusion must be arrived at, when one sees a plant LIKE Paris quadrifolium growing in one wood far from
299c196 t intellectually developed amongst Pachydermata. LIKE Man amongst Monkeys-- or dogs in Carnivora.-- M
```

Page **(Key Word)**
305c209 cking thrushes of Galapagos having tone of voice LIKE S. American.-- Have not Ruffs & Reeves a remark
305c210 s half Jackal & Scotch Terrier.-- certainly more LIKE Jackall in gait, size, fur.; manner in which ea
305c210 in gait, size, fur.; manner in which ears droop LIKE dog older character & manner of wagging tail.--
306c212 instincts in young animals, well developed, just LIKE habits easily gained in child hood.-- Young sa
306c212 aten behaved very differently from a dog.-- more LIKE man. continued long in a passion & looked out f
306c214 are Kurdish & come also from Asia Minor.-- tail LIKE setters. long ears-- colours vary, but form con
307c215 but form constant.-- The females of some moths, LIKE glowworm <are> have «These abortive organs in s
312c228 ous, he feels sure of this, first offspring most LIKE <parents> Mother.-- like dogs Smith knew chines
312c228 s, first offspring most like <parents> Mother.-- LIKE dogs Smith knew chinese hairless dog & common s
312c228 - 3 puppies PERFECTLY like chinese & 3 perfectly LIKE chinese & 3 perfectly like spaniel even when gr
312c228 common spaniel crossed.-- 3 puppies PERFECTLY LIKE spaniel even when grown up.-- Are mules homogen
316c244 dote --.Sowerby.--. Geographical range of shells LIKE Cryptogamic plants. of Marine kinds, there are
316c246 N. America & England.-- but the fossils are not LIKE, except in very few cases, those of Tertiary Eu
318c252 ase of Squirrel from extreme north turning white LIKE Hares??--) I never saw more beautiful adaptatio
318c252 I never saw more beautiful adaptation for snow-- LIKE snow shoes. than feet & hind legs of these whit
318c254 t a fortnight in one particular part of country, LIKE White of Selbournes Rock Ouzels.-- ..If the lin
318c255 ts every year,; and probably a <chance» wanderer LIKE the first pair of Pipra flycatcher.-- Bachman s
331dIFC the first cross, which makes hybrids. productive LIKE geese?-- Are the number of kittens between Lion
332d004 disordered intestines are not healthy to worms, (LIKE parasites of Tropical countries cannot endure t
333d006 cat.-- had four Kittens. two appeared «so» very LIKE common cat, that they were killed, & other two
335d013 ill not remain,-- yet offspring must be somewhat LIKE parents,-- therefore offspring will tend to go
335d014 . ¶ -- Are not dreadful monsters, abortive, just LIKE mules. Fox's half bred Persians «cat» favour th
335d015 the offspring would «as in all other animals» be LIKE either parent, or intermediate within certain s
335d015 species" will have no tendency to have offspring LIKE parent, but as they must like or there will be
335d015 to have offspring like parent, but as they must LIKE or there will be none, therefore a mule can hav
336d017 ossible to transmit them, & as offspring must be LIKE parent, therefore mule has no offspring & there
336d017 but if both parents are alike, offspring must be LIKE Hence mutilations not heredetary,, but size of
336d018 on makes hybrid, in fact the parents beget child LIKE themselves. expression of countenances, organic
337d022 placed by the F. cisalpina in Italy, which is so LIKE that difference would not be discovered by an u
338d024 aking any form of Malaria-- adaptation & species-LIKE.-- -- Says Negro-- thick skinned My hairdresser
339d026 r flood (!) according to affinities!. confounds, LIKE Whewell affinity with analogy-- Good table at e
341d029 tructure of pelvis & was not adaptive structure, LIKE little wings of Auks which does not make that b
341d030 is so anomalous among true deer, yet is spotted LIKE so many deer.-- very curious like some facts of
341d030 yet is spotted like so many deer.-- very curious LIKE some facts of Mr Blyth on birds.-- Dr. Bachman
341d032 ducks.-- most strange voice often in the night, LIKE peacock.-- tail as long as Pea hen.-- about int
342d032 duced in full equal Numbers with pure bred (just LIKE common mules) & lay many eggs but never produce
345d043 now lambs, because in their faces they were most LIKE their mothers believes this resemblance general
345d044 are made.-- The Highland Shepherd dogs, coloured LIKE Magellanic Fox.-- peculiar hair & appearance-
345d044 ghland Sheep jet black legs, & face & tail, just LIKE species.-- high active breeding[g] [not located]
346d047 received same answer.-- Thought lambs were more LIKE father than Mother.-- The cross not so hardy as
350d055 leay & Broderip were talking of some Crustacean, LIKE Trilobite. (Polirus?? female blind & of quite
353d060 if there be many others somewhat allied whether «LIKE parent stock, or not. Now wings for flight-- t
354d063 o Varieties of dogs cross, Erasmus says it look LIK[E] Institut. 1837. p. 351. Paradoxurus Philippen
354d065 sion.-- The Galapagos mouse probably transported LIKE the New Zealand one-- It should be observed wit
358d085 we have beautiful proof of the breeding in & in (LIKE «courage in dogs» EFFEMINATE men),-- if carried
358d086 a cock (infertile) <with> the breast of which is LIKE common pheasant & back like silver.-- But the h
358d086 e breast of which is like common pheasant & back LIKE silver.-- But the hen hybrid of this bird, has
358d086 has long tail figure, & some degree of whiteness LIKE a Male.-- Thus castration, hybridity, & breedin
359d088 an offspring, I should say in every respect most LIKE Penguin duck.-- which is strange anomaly in Yar
359d088 local variety, & not effect of breeding in & in LIKE our pidgeons) The male of every animal certainl
359d089 th its form & disposition Saw three young ducks, LIKE each other,-- (& not very like either either wi
359d089 hree young ducks, like each other,-- (& not very LIKE either either wild or Pintail duck) from which
359d089 k, into pintail.-- Of these there were four, two LIKE each other & two dark-coloured & different.-- -
360d090 hence general appearance of face & tail somewhat LIKE dog-- though it has full share of Jackall shape
360d091 ly tigers have first come over, insects somewhat LIKE «between» lice & fleas. sticking on them, but n
361d095 cies have same character which is mottled, & not LIKE any existing species-- [In two herons, <both> p
361d096 , all nearly similar.-- in blackbird group young LIKE some of the species-- (¿do these facts indicate
363d101 geons, trace the washing out of the forked band, LIKE in plumage of ducks.-- Mr Yarrell says in very
363d103 awn compare with fallow? deer. & Moschus &c & -- LIKE young blackbirds Dr Bachman told me that 1/2 Mu
367d113 . with all four legs ringed with brown.-- animal LIKE large, heavily made cream coloured ass.-- strip
367d113 g bat, which has the power of inflating its body LIKE balloon-- by air cells connected with cheek pou
368d114 recedes from the species all females being most LIKE offspring, Q (how is this with those females wh
368d114 Q (how is this with those females which put on (LIKE some waders) the bright plumage.-- «thinks» Hen
371d118 childs nervous system should build up its body, LIKE its parent, than that it should be provided wit
371d118 fully developed tails, & one with beak turned up LIKE Avocette. here is what [not located] that it sh
371d128 after year, successive roses & bud are produced, LIKE parent stock, or if different dieteriorating ve
372d129 e peculiarities not so a seed.-- Bud probably is LIKE cutting off tail of Planaria. the whole grown t
372d129 s added from one part of the body «(or of other <LIKE «similar» body)» to another part of body.-- [i
373d132 es of changes from the placentates, Having Hair. LIKE true Mammalia, no more wonderful. than Echidna.
375d135 n ordinary crop. causes a dearth then in Spring, LIKE food used for other purposes as wheat for makin
375d135 sly all the rest.-- One may say there is a force LIKE a hundred thousand wedges trying force <into> a
376d136 s Cyanocephalus Porcarius.-- this Monkey did not LIKE a great coat made for it at first, but in two o
377d140 because inhabitant of plain & Jaguar of woods &c LIKE ground birds [not located] :Hence, also structu
378d148 not connected with habits According as child is LIKE parent, so is species old: Hence <young> Kingfi
379d151 hat one of the children of Sir J. H. was so very LIKE Sir W. whilst Sir J. himself is not like-- now
379d151 so very like Sir W. whilst Sir J. himself is not LIKE-- now this is clear case of avitism. but then ¿
380d153 t they would have wings.--).-- Seep p. 84. Hens «LIKE»-- Cocks from effect of breeding in & in.-- Mr
380d153 does not know of any case of old Male. becoming LIKE female, though many of old female becoming like
380d154 like female, though many of old female becoming LIKE cocks.-- It is very singular. so many Gallinace
383d159 of Hybrids between Common & Silver Pheasant, one LIKE cock & other like Hen.-- one doubts whether the
383d159 Common & Silver Pheasant, one like cock & other LIKE Hen.-- one doubts whether they are not Hermaphr
383d159 one doubts whether they are not Hermaphrodites, LIKE J. Hunters. Free Marten N.B. the common mule mu
383d160 s rather smaller than in <ma> silver male-- Head LIKE silver except in not having tuft,-- back like d
383d160 ad like silver except in not having tuft,-- back LIKE do.-- but the black lines on each feather inste
383d160 on tail much <blacker> «broader.-- » Breast red LIKE Common pheasant.-- lower part of breast, each f
385d163 KES HIM & then is actually obliged to be held.-- LIKE she wolf of Hunter.-- young take distemper very
386d165 & see whether it retains same length of life.-- LIKE Golden Pippen trees! How is this with buds of p
386d167 imalcule in four days could form 2. cubic stone. LIKE that of Billin.-- <Generation--> V. p. 152 It i
387d168 impresses offspring more & more with the added «LIKE Lord Moretons case & Dr. Andrew Smith,» differe
388d171 rding to this view more semen to one child. more LIKE father.-- stuff.!-- How much opposed. the Quagg
389d176 tem, «as if there was, a superabundance of life, LIKE tendency to budding, which wishes to throw itse
389d176 nt being prolonged till it has bred.-- Offspring LIKE both father & mother, or very close to either.-
391d174 two common buds??? Amongst buds each one exactly LIKE its parent. <but these buds do not procreate> &
391d175 dioecious oldest forms) why are twin in man more LIKE «each other» than twins «or triplets &c &c in»
399e011 e says positively he has seen, a Calosoma. (very LIKE American form) in Stonesfield slate., & & Melol
402e018 ed being of the latter being the same as the fox-LIKE animals. which are met with near Canton" "Here,
402e019 t. Peter & St. Pauls in Lat' 53 degree yet fauna LIKE that 60 degree & 70 degree of Europe.-- Many Eu
406e035 f Teneriffe. St. Helena & Ascension most species LIKE & identical with S. America. & many very close:
406e036 same species & to different species-- sometimes LIKE one parent & sometimes other & sometimes ½ way.
410e052 nts, without some end.-- Respect good describers LIKE Richardson.-- The relations of numbers of speci
412e057 hrodite structure being retained in the male.-- <LIKE> «far» more than marsupial bones., & even more
412e058 nciples, will account for all (1) Grandchildren. LIKE. grandfathers (2) Tendency to small change.. we
414e063 - When two races of men meet, they act precisely LIKE two species of animals.-- they fight, eat each
415e065 the» organic and inorganic agents of this earth. LIKE every other animal.-- Would anyone raise an arg
415e066 olite, he would probably be confined in locality LIKE Ornithorhyncus,: since being cosmopolite, we do
417e078 of true <pare> hermaphrodite, would of course be LIKE either, that is both parents, for they are one.
417e078 .-- Now these laws are, that child may be either LIKE father or mother, independently of its sex, or
417e078 between, or someway different from either: & or LIKE progenitors.-- in some families all the childre

Page	**	(Key Word)	**
417e078	progenitors.-- in some families all the children	LIKE	mother & in some like father «What is cause of
417e078	families all the children like mother & in some	LIKE	father «What is cause of this.-- » (Lord Moreto
419e086	vated shells in Bayfields district are much more	LIKE	those of Scandinavia, than of the N. American s
420e089	rent of all vertebrate animals.-- must have been	LIKE	some molluscous «bisexual» animal with a verteb
420e090	tioned which inhabits the Pinna of Rio Janeiro,	(LIKE	some Mediterranean species).-- might these fert
421e091	frica, than to other parts of that continent. in	LIKE	manner as Madagascar does to other side of Afri
422e092	udes to difference between fossil & recent Bull;	LIKE	fossil & recent shells of the <new> raised beac
424e103	one ought not to be able to hybridise the Camel»	LIKE	ass «same way some plants vary more than others
426e105	ld «neighbouring» formed in the Tertiary «epoch»	LIKE	Sicily not having species, if true important on
426e108	Mr Marsh has some nephews, who are astonishingly	LIKE	to some distant cousins, the nearest blood bein
428e113	lty. «I should doubt if wild species ever formed	LIKE	short-tailed cat or dog has been without recurr
430e117	, describes his tame Hares, attacking a sick one	LIKE	Chillingham bulls are described.-- His three ha
430e119	e lower part rayed longitudinally (give woodcut)	LIKE	I. sulcatus.-- Both species are found at Folkst
431e122	hat change may be anticipated, & this would look	LIKE	fresh Creation. the gardener separates a plant
436e134	oportion to other mollusca in cold parts of sea,	LIKE	Cetaceae,-- although the Cephalopods, seem to h
439e145	III. p. 33-- They had several breeds of dogs.--	LIKE	greyhound-- fox-dog-- turnspit & two other kind
440e146	ent, or past generation.-- thus cabbages growing	LIKE	Nepenthes.-- cases of pidgeons with tufts &c &c
446e162	a case in answer to Mr Knights doctrine.-- Case	LIKE	Corallina-- «Does it flower anywhere?-- Yes on
447e164	ant end.-- Festuca vivapara F ovina-- propagated	LIKE	oni[on] Poa alpina because vivaparous. Henslow
448e168	lly aberrant-- the two former connecting classes	LIKE	Toxodon «In orders»-- Fish & reptiles in former
449e169	some of the pidgeons in common Dovecot are very	LIKE	a Himalaya species -- leuconotes-- Magazine of
450e175	there are shades of difference in all the isld,	LIKE	in wild animals).-- There are prevailing colour
451e176	ere of the usual colour. His sister is an albino	LIKE	himself said not to be common-- probably, I sho
451e178	imals into this region between April & October &	LIKE	man almost (this looks inaccurate C. D) they wi
453e182	. <">The wild, small fowls at Pulo Condore "crow	LIKE	ours, but much more small & shrill".-- Humboldt
459t009	p 320. Mr Hodgson on Musk Deer-- young spotted	<LIKE	in> "prettty much as we see in the young of the
464tf6v	these mountains, when the cold was intense just	LIKE	the alpine plants-- In S. America. it appears f
467t103	Lupine, Bees «frequent» & seem to act, something	LIKE	on Kidney Bean, they go to nectar at foot of up
472s01r	p true.-- colour of flower & foliage not «being»	LIKE	the true P. bracteatum; all supposed to have be
472s02v	<I think> pollen was scraped off, which appeared	LIKE	Heartease pollen.-- the pollen appeared chaffy,
473s03r	liocene strata! Mastodon longirostris in miocene	LIKE	in Europe-- Cuvier never found remains of Sus w
482z011	pecies! birds of passage!! sylvia macloviana, 2d	LIKE	sylvia cisticola.-- Embriza melanodera-- a linn
484z015	.-- What structure do the auks bear traces of.--	LIKE	Puffinuria does of Petrel?-- Study Birds of Eur
486z019	p. 130. Col Sykes on balls made by dung beetles,	LIKE	those from Chiloe. Amblyrhyncus de marlin James
491q01v	act.-- 4 May we no suppose, that certain plants,	LIKE	Aphides produce impregnated young ones; & that
491q002	the viviparous grasses & onion, produce flowers,	LIKE	the Oxalis from C. of Good Hope mentioned by Mr
496q05a	sight.-- --. touching Mr Brown theory of insect-	LIKE	Orchis-- & final cause of beauty of flowers-- c
496q006	on. (I have seen one monstrous) Fox Glove & such	LIKE	in very rich soil-- As they have little tendenc
502q11v	bbages.-- History of Pheasant-fowl. Hen coloured	LIKE	cock-pheasant: said not to sit on own eggs Flow
507q15v	lants growing together. under same conditions.--	LIKE	cowslip & primrose, but less strongly marked.--
508q016	urring in man being transmitted through females,	LIKE	Hydrocele Dr. H. thinks asthma in females takes
510q017	& winds in Antarctic ocean: are they from West,	LIKE	as between Australia & S. America? Sabine says
522m013	ple recognized,-- sudden changes of disposition,	LIKE	people in violent intoxication, often ends in i
523m018	from drinking cold drink.-- then brain affected	LIKE	getting suddenly into passion.-- There seems no
524m019	Ashe, the Birmingham Doctor), in this precisely	LIKE	the passion, ill-humour & depression, which com
525m023	up a stone, though only acquired rules by art.--	LIKE	the law of honour.-- they feel pleasure in obey
525m023	home.-- in these cases their actions do not look	LIKE	fear, but shame.-- I cannot remember instances,
525m024	r thinks that selfishness, pride & kind of folly	LIKE	(Mr George S.) is very heredetary.-- My father
526m028	- mem Erasmus & mine taste for music.-- Children	LIKE	hearing a story told though they remember it so
526m028	y & imagination. «Catherine thinks that children	LIKE	looking at <ani> pictures, an early taste, of a
527m032	, not being instinctive, is heredetary knowledge	LIKE	that of man, & this agrees with the stated fact
530m043	- He has seen other cases of similar nature.-- --	LIKE	FitzRoy in sleep giving directions,-- & forgetf
532m056	fat! What remarkable affection to a place.-- How	LIKE	strong feelings of Man.-- The sensation of fear
533m061	me such remarke as now advanced; for I caught it	LIKE	a flash.--. strange if judgment remains, where
536m071	ng: one ceasing, so does other.-- What an animal	LIKE	taste of likes smell of,∴ Hyaena likes smell of
537m076	-- This probably is natural. consequence of man,	LIKE	deer &c, being social animal, & this conscience
538m079	as remembered her song, then the song was to her	LIKE	one which though learnt in infancy, had often b
540m088	ears, with sensations of sorrowful delight, very	LIKE	best feeling of sympathy.-- Mem: Burke's idea o
542m093	ding to say so, his step will grow erect & stiff	LIKE	that of turkey.-- he may be amused, he need not
542m094	, & which now connected physical relations.-- CD[	LIKE	sighing to relieve circulation after stillness.
542m095	hanged, I conceive sighing might yet remain just	LIKE	sneering does.-- is yawning habitual from awaki
542m096	y with the wish to make it out?-- Seeing a Baby	(LIKE	Hensleigh's) smile & frown, who can doubt these
545m106	n, that is show all the teeth: «& make noise not	LIKE	pish, but like chit-chit-chit, quickly uncoveri
545m106	all the teeth: «& make noise not like pish, but	LIKE	chit-chit-chit, quickly uncovering their teeth,
545m107	ingers, the peevish expression was most curious	<LIKE>	«remember» the expostulatory angry look of bla
545m107	, threw itself down on its back & kicked & cryed	LIKE	naughty child.-- Do monkeys cry?-- «they whine
545m107	e naughty child.-- Do monkeys cry?-- «they whine	LIKE	children.--» Expression, is an heredetary habit
546m108	ing ourselves in their situation, & then we feel	LIKE	them-- . hence sympathy very unsatisfactory beca
546m108	ce sympathy very unsatisfactory because does not	LIKE	Burke explain pleasure. August 26th. I cannot h
546m111	years afterwards whilst at a meal, she suddenly	LIKE	a flash without any assignable cause, remembere
549m118	more than to dog losing his puppies-- This looks	LIKE	free will.-- V. last page. A healthy child is «
549m122	s they are effecting a change in his instincts--	LIKE	what is happening with other animals-- is far f
549m123	cessary & no doubt were preservative, & are now,	LIKE	all other structures slowly vanishing-- the min
550m127	ve emotions?-- Dr Holland remarked that insanity	LIKE	sleep does not doubt the reality of the impress
551m129	e & the pulse rises, or as they flag & something	LIKE	a snow-haze. covers my whole imagination." Sept
553m135	, one fixes on imaginary beings, many vicarious,	LIKE	ourselves) that savages (mem York Minster) cons
554m138	ad downwards to look up womens petticoats-- just	LIKE	Jenny with Tommy ourang.-- Very curious.-- Mr Y
555m143	ever the expression of passion with open mouths	LIKE	the old world ones.-- Though the[y] move whole
555m143	about not having run away &c having faced death	LIKE	a hero, & then I had some confused idea of show
556m145	e from behind upper & little between incisors.--	LIKE	<W[...] what> person says "what a pity"-- Lavat
556m146	ion of desire-- is there not protrusion of chin,	LIKE	bulls & horses.-- 1838 good instance of useless
556m146	is then quite useless-- Cats kneeding when old,	LIKE	kittens at the breast now if horns were to grow
557m148	knows oneself it is quite different from that.--	LIKE	«slight» passion from blood rushing in face, wi
557m150	heard.-- Shame would never make person tremble,	LIKE	fear.-- Why does any great mental affection mak
560m156	about heredetary tricks & gestures, other cases	LIKE	D. Corbet; «do» ideots form habits readily?? Do
560m156	deots form habits readily?? Do the Ourang Outang	LIKE	smells «peppermint» «& music».-- Have monkeys l
564n005	ly to chicken so as to seize small moving object	LIKE	fly.-- young partridge can run even with its sh
564n005	hey have always been studied appears to me to be	LIKE	puzzling at Astronomy without Mechanics.-- Expe
565n006	eth to bite.-- the senseless grin of passion, is	LIKE	the grin of the Hyaena from fear, no actual int
567n013	rs of corn with her teeth from the straw, & just	LIKE	child not knowing what to do with them, came se
567n013	several times & opened my hand, & put them in--	LIKE	child. Tommy's face, now ill, has expression of
567n014	ed» his arm. & show signs of affecting something	LIKE	man. Has an oyster necessary notion of space--
567n015	nimals I should think would not have any emotion	LIKE	blush.-- when extreme sensation of heat shows b
567n015	rt beginning to beat-- children inherit it <ins>	LIKE	instinct, preeminently so-- who can analyse the
567n015	sensation, when meeting a stranger. who one may	LIKE.	dislike, or be indifferent about, yet feel shy
569n019	te.-- Whilst music extremely primitive.-- almost	LIKE	tastes of mouth & smell. Descent of Man Underst
570n023	nderstanding reason.-- surprise with negation.--	LIKE	shaking something off shoulder-- or is it from
573n037	are more expressive than movements of fingers.--	LIKE	Kitten with mice.-- A person with St Vitus' dan
574n041	robably due to our distant ancestors having been	LIKE	dogs to bitches.-- How comes such an associatio
575n043	mal not destined by nature to exist. & carrying	«LIKE	other hybrids» with <the> it the provision for
575n043	.-- can we deny that brain would be intermediate	LIKE	rest of body? Can we deny relation of mind & br
575n045	ween grief & pain-- certain ideas hurting brain,	LIKE	a wound hurts body-- tears flow from both, as w
576n049	nimals is philosophical. viz. man is not a cause	LIKE	a deity, as M. Cousin says. because if so ouran
577n051	ace of man, face, neck-- «upper» bosom in woman:	LIKE	erection shyness is certainly very much connect
578n056	ose the sphincter muscles, only on the principle	LIKE	does an injury of the spine-- that it paralyzes
578n056	vulsive action to remove disagreeable impression	LIKE	true convulsion. (Hence pass into convulsions?)
579n059	ve him, or her.-- it is blind feeling, something	LIKE	sexual feelings-- love being an emotion does it
581n065	uage, that it was progressively formed. (--names	LIKE	sounds)--. Horne Tookes tenses, &c &c -- <also

Page **(Key Word)**
585n074 E. p. 125 Wrong Entry Madagascar Lemur seemed to LIKE Lavendar Water «very much» Henslow. N.. Necker
586n081 ry habit, is a part never subject to volition.-- LIKE plants going to sleep.-- "A bird has the facult
587n087 is necessarily disagreeable to organs adapted to LIKE sugar, acid, &c, which may be doubted for possi
587n087 arb: again on other hand, it is said people, who LIKE sweet things dislike others.-- dogs dislike per
587n088 -outang & chimpanze. pout.-- Former, whines just LIKE a child. Get a Dictionary & make a list of ever
589n092 power over muscles of voice than respiration.-- LIKE sigh before false sneeze.-- "A Dissertation on
590n094 p 153. «do». an account of a monkey in a passion LIKE Jenny.-- Dr. Abel has given an account of an Ou
591n097 tche tame young wolf & it never dropped its ears LIKE dog-- wagged its tail «a little» when attending
591n100 inst instinctive feeling, only does not fullfil, LIKE continent man.-- a man eating what others by ha
592n103 Philosoph. Transactions Vol 44. 1746-47. Paper. LIKE. Sir Ch. Bell on Expression «First Croonian Lec
600n08v pect conscience, an heredetary compound passion. LIKE avarice-- Is there not something analogous to
603o012 o> is carried on toward deity.-- & as king might LIKE cruel pleasure, so sacrifices cruel.-- Somethin
606o20v ong series of experiments & observations. & yet, LIKE in vision, it becomes so instantaneous. that we
607o025 of receiving impressions-- accidental (so called LIKE chance) circumstances. As man hearing Bible for
608o026 none.-- Effects.-- One must view a wrecked man, LIKE a sickly one(P)-- We cannot help loathing a dis
610o031 member» «read» Christian name; fancied it looked LIKE. W. but concluded it could not be so.--Looked a
611o034 ual» vegetables, the vital laws act definitely (<LIKE> «as» chemical laws,) as long as certain contin
614o037 refore reward in good life [RHC] Instinct appear LIKE heredetary memory; but first memory in many cas
616o39v wanting, by which thought or memory. might be in LIKE manner attributed to the brain. There are two m
622o048 ey [RHC] Anyone, who will reflect must feel, how LIKE to injured conscience, is the feeling of any cu
622o048 All these associated «habitual» feelings. become LIKE the instinctive, ones, <which either lead to ac
625o052 not> which regulates our feelings steadily & not LIKE our appetites & passion, which receive enjoymen
633j54r ume that God «created plants to» arrests earth, (LIKE a Dutchman plants them to stop the moving sand)
635j55r ty. how it acts & whether constant or inconstant LIKE that of Man.-- the cause given we know not the
638j58v th weak beaks be sifted away.-- 4 & the species, LIKE 10,000 others, perish. & who will dare to say t
639j28r lly». petrel & Whale in some respects Chamaelion LIKE power in Octopus & Chamaelion.-- C. D. Sucking
640j167 ecies to themselves.-- Probably no case in world LIKE Galapagos. no hurricanes.-- islds never joined,
554m139 ish the smell of Verbena & Pocket Handerchief & LIKED the taste of Peppermint.--» Perfect understand
047r083 plains. long periods of rest & vice versâ more LIKELY to be coincidental than single elevations alon
195b100 arrive-- did it only create those kinds not so LIKELY to wander. Did it create two species closely a
225b218 ristan D'Acunha. may be said to deceive man. as LIKELY as fossils in old rocks for same purpose.!! Ca
245c025 us & Barbyroussa «NB» [islds. Springing up more LIKELY to <M> have different species than those sinki
421e091 ndications of its true affinities. We are least LIKELY in the modifications of these organs to mistak
504q13v Van Mons views-- Also PEAS-- N.B. I think very LIKELY the Peas to cross ought to be placed far from
524m020 F says are almost constantly present in people, LIKELY to become insane.-- now this is well worth con
572n033 -- Ideots. defective brains.-- Erasmus does not LIKEN term instinct to muscular movement.-- say insti
281c138 h as they are-- My theory explains that family LIKENESS, which as in absolute human family is undescr
302c201 d become affinity yet often retaining a family LIKENESS, & this I believe the case. = any animal real
336d018 s-- When two animals cross. each sends his own LIKENESS, & the union makes hybrid, in fact the parent
372d131 facts that the skin grows over a wound.-- Does LIKENESS of twin bear on this subject? A mans arm woul
417e078 , for they are one.-- The laws, therefore, of LIKENESSES of fathers to children of mankind, no doubt
417e078 ildren of mankind, no doubt are applicable to LIKENESSES, when species & races are crossed.-- Now the
455eIBC Father-- diseases common to men & animals.--:LIKENESSES of children CD(Does any annual give buds, or
385d163 obably would not be fertile» without she know & LIKES HIM & then is actually obliged to be held.-- li
536m071 , so does other.-- What an animal like taste of LIKES smell of,∴ Hyaena likes smell of that fatty sub
536m071 an animal like taste of likes smell of,∴ Hyaena LIKES smell of that fatty substance it scrapes off it
546m109 nkeys-- smell with many animals-- see how a dog LIKES smell of Partridge--, man's taste for smell of
614o037 less wonderful than thoughts-- One organic body LIKES one <m> kind more than another-- What is matter
032r041 rought into play As in Ocean & Air; there are «LIKEWISE» differences of temperature «at equal distanc
035r047 on any Geolog Map: Quoted from Daubeny P 402: LIKEWISE, mean height of tertiary. being less than mar
042r068 black as if from peat.--yet cetaceous bones so LIKEWISE «of miocene period».--Mem Bahia blanca P. 204
046r078 eliminated.--«Sulphur last.--» Metallic veins LIKEWISE must separate ingredients if we look to a con
099a051 OP. let force drag particls to line {P} AB, & LIKEWISE gravity MN. Then every particle would tend to
114a095 apis. (now we have banished diluvial waves). & LIKEWISE «tells,» «offers a presumption» it has been e
181b040 ly improbable that any of <his relatives shall LIKEWISE> the successors of his relatives shall now ex
240c014 y been found in isle of Aroe & Solor), «Vol I» LIKEWISE new species of Parameles, which joined to Cas
254c048 modifications, typical of succeeding classes & LIKEWISE those much higher in scale. So Owen actually
273c111 rmed for flight & great movement in the air, & LIKEWISE rasorial species, & likewise perching (Gould)
273c111 ent in the air, & likewise rasorial species, & LIKEWISE perching (Gould), but the latter is obscure b
273c112 wings, but tail stiff.-- Swallow & goatsuckers LIKEWISE exaggerated.-- There is one most remarkable c
303c204 r colour + + + form of head «& features»;. but LIKEWISE in length of extremities, how are races in Th
318c253 pra Flycatcher, which is now becoming common-- LIKEWISE of the Hirundo fulva (added by Audubon in App
333d007 a show half Wolf & «half Esquimaux» dog which <LIKEWISE> more resembled the wolf than dog.--> appeared
387d168 typ that in (B) to <a slight> «some» degree, & LIKEWISE ovum in (B) <an C> that in (C) «in lesser deg
406e037 transported far S. in Northern. Hemisphere.-- LIKEWISE far North in Southern.-- Great animals. of sa
447e164 of germs is not end of seminal reproduction.-- LIKEWISE grasses. &-- very heavy seeds.-- as Cocos do
528m037 ht, which is common to every kind of view-- as LIKEWISE is novelty of view even old one. every time o
530m045 in trades (,as sailor tailor blacksmiths?) are LIKEWISE heredetary, & therefore that their children h
538m080 ciously with respect to more energetic self, & LIKEWISE one forgets. what one performs habitually.--
542m097 Snails, Fox of cows, & many of insects-- they LIKEWISE must understand each other expressions, sound
547m113 ping up the memory of every late impression. & LIKEWISE gaining new ones from senses. & <comparing th
579n057 he sublime & Beautiful there are some notes. & LIKEWISE on Wordsworth's dissertation on Poetry.-- The
580n062 -- Consolats: "the recollections of the infant LIKEWISE before two years as soon lost; yet many of t
612o035 , show either local or general will, & stomach LIKEWISE «does». [LHC] ¿ in Corallina are not two kind
619o043 asure, without any regard to his own interest. LIKEWISE if such actions were prevented by force he wo
619o043 without his being able to prevent it, he would LIKEWISE feel pain.-- If he saw another man <say go> «
621o046 it is to his interest to follow the former; & LIKEWISE <that the> then receive the moral approbation
624o051 om civilization, education changes, & probably LIKEWISE instincts, for the same law effects both.-- <
587n087 ike perfume) I should think, great principle of LIKING, as simply heredetary habit.-- A blind man mig
059r119 s de couches paralleles et inclinées du centre d'LILE, vers la mer; ces couches ont entre elles une c
207b152 a Mountains-- Bory St. Vincent Vol. III. p. 164. LILE de la Reunion presente elle seule plus d'espece
216b180 White species.-- For, says he Seeds of hybrid LILLIES &c &c &, (V Herbert on hybrids) thus act.-- No
466t093 no Bees «several» visiting it>.-- In yellow day LILY, the Bees visit base of upper petal, though not
466t094 d all round flower as they are in Crown-Imperial LILY & many other flowers-- My view of «variety acqu
045r075 hat Volcanos!! were in eruption at time of great LIMA earthquake-- In the Chili earthquakes if rise wa
058r116 e in Chili; Arequipa in 82 was overthrown, & 86. LIMA. next year Quito. considers these earthquakes t
138a153 (Lyells Book)» Observaciones sobre El Clima del LIMA par Dr. H. Unanùe says he believes the sea has
139a153 acts has extended a league inshore both N & S of LIMA.--judges from «beds of» sand & gravel & shells.
415e066 nce being cosmopolite, we do find his remains.-- LIMA.-- caves.-- There being no fossils, the only wa
448e166 rious table of all the castes from Stephenson at LIMA The same numerical relation (both in species an
483z013 uille's Voyage p 302 Vol II p. 302. Vaginulus of LIMA described "Arion" of Ascension. p. do.-- some S
309c219 ve organs to mules in their genitals & even to a LIMB not used The only cause of similarity in indivi
332d003 s.-- My Father has seen case of pleurisy, broken LIMB «in children» & other such disorders accompanie
335d014 hybrids may be compared to animal with amputated LIMB.-- Heredetary <thr> Six fingered people, <Hill>
223b210 (as some animals do more than others, & cut off LIMBS & new ones are formed) but yet propagates varie
317c251 character in extreme length of ears & length of LIMBS, so that he first thougt only one species. & al
333d005 - fur short. (tail cut off in progeny peculiar) LIMBS very long, eyes very large, very fierce to dogs
459tf02 hinnus, resembles horse in its head ears, tail LIMBS-- in the mules, these parts resemble ass. (& pa
473s07v y in external structure can be concepcional, as LIMBS &c &c only appear late in pregnancy, & then may
545m106 e. nervous impulse to action is sent so fast to LIMBS that they cannot remain still.-- I do not doubt
586n082 ledge of size is merely judged by eye, & use of LIMBS &c, or it result from mere impulse to save wax.
022r007 tendency to change their position? Carbonate of LIME disseminated through the great Plas Newydd dike
028r029 cumulating in deep seas forms slates: How is the LIME separated; is it washed from the solid rock by
028r029 nown Volcanic process? How does it come that all LIME is not accumulated in the Tropical oceans detai
033r042 n, each particles coated by pellucid envelope of LIME.--form resembles the husks at Coquimbo: in that
033r042 y not central and rather differently constituted LIME have been removed?--As shell out of its cast wh
046r078 t production of Trachyte. look at Sulphur. salt, LIME, are spread over «whole» surface; how comes it
056r111 w water separates.--(intertropics at present fix LIME). <Also Volcanos separate.> Volcanos blend all

056r111 Volcanic action separates some sulphur (perhaps LIME) salt. & metallic ores.--which mingling & separ
059r121 s are of diff cont: & even in one case contained LIME.--All bear close analogy to Obsidian, & all sho
076r168 amely dendritic silver intersecting carbonate of LIME-- native silver in Mexico is always accompanied
079r176 lv.[,] Sulph. of do.[,]galena[,]quartz, Carb. of LIME. accompany.-- Ulloa has said silver in the high
096a041 Mr Brayley says he can give me facts respecting LIME <n> being heated without parting with Carb. Aci
096a041 s reason for knowing that Mur. Soda. and Carb of LIME decompose each other.-- on Direction of mountai
098a047 e which draws together two particles of Carb. of LIME, tends to crystallize them as seen in stalactit
221b201 .-- origin of Fresh-water genera? The absence of LIME in Plutonic & Volcanic rocks. most remarkable.-
499q009 on Bees thighs (18) Place pin's heads with Bird LIME near male yew tree & see whether they catch pol
022f007 ingly rare Mem C. [Cape] Turn P. 434 & 419 As LIMESTONE passes into schist scales of chlorites--Mem.
026r023 ally petrified into a peculiar cream-coloured LIMESTONE: the strange substitution of matter in shells
029r034 land (Excepting Conglomerates?) «& absence of LIMESTONE?» have been collected on the open coast. Perh
030r036 Tritores" found in the coraliferous mountain LIMESTONE are they allied to the jaws of the Cocos fish
032r041 weak.»; «(cause of not accumulation of Coral LIMESTONE in intertropical)» hence varieties of substan
039r061 ol III, says that in N. Pliocene formation of LIMESTONE, casts of shells, as in some older formations
050r093 l [blank] What is nature of strip of Mountain LIMESTONE, in N. Wales. was it reef. -- I remember many
054r102 ace. sandy. sandstone with gypsum, covered by LIMESTONE with recent shells 200 ft, how exact agreemen
074r164 ad. orpiment. chrysop[r]ase. opal:-- Veins in LIMESTONE & Grauwacke: Silver appears far more abundant
074r164 Silver appears far more abundant in the upper LIMESTONE, which H. calls by several secondary names «S
075r166 mine of Gualgayoc or Chota & Pasco in "alpine LIMESTONE" = "The wealth of the veins in most part tota
077r171 istaken for Porphyry above ancient freestone, LIMESTONE & <many> «other secondary» rocks. Vein traver
078r176 metal in Tasco vein in Mica Slate & overlying LIMESTONE Balls of Silver core occur in do veins. At Hua
089a018 tains on west side of Domingo formed of coral LIMESTONE, with interstices yet emptty.-- In all the mo
117a101 and with vegetable remains formed near coast, LIMESTONE deep water. will bear on formations. during e
124a119 ould be absorbed.-- if so exactly parallel to LIMESTONE & volcanic rock containing magnesia Lyell. El
374d133 ny. existing genera of shells in the mountain LIMESTONE (how different from plants!) But the Cephalop
423e099 ed, much less in alternating strata of sand & LIMESTONE &c &c.-- L'Institut 1838.-- p. 290-- admirabl
433e126 to another: in each system, the changes from LIMESTONE to sandstone &c. show some great change who c
026r019 the older formations I must believe they «the LIMESTONES» have been formed in shallow water: so have
028r030 in of some sandstones, as in Australia.--Have LIMESTONES all been dissolved. if so sea would separate
093a032 t of Iron has been sublimed into the tertiary LIMESTONES of Vendarques. Mem sublimation of sulphur to
107a077 m Heat. Specific gravities of many artificial LIMESTONES produced by Sir J. Hal. End of pages. p. 157
373d133 5 genera of reptiles.-- <M> p. 417. Magnesian LIMESTONES & Zechstein oldest rock in which reptiles ha
061r128 area must have peculiar character: Contrast low LIMIT of Palms, evergreen trees, arborescent grasses,
063r130 resentation» of species important, each its own LIMIT & represented.--Chiloe creeper: Furnarius. <Car
106a075 (except mere talus «over cliffs edge» of which LIMIT cannot be great over) with very gently sloping
113a090 degree Fah. at a greater depth than 400. & the LIMIT being 400 ft. shows that the strata have very u
287c158 a frolicsome saltus. «p. 19»¶ Macleay seems to LIMIT Lamarck definition of relations to settling the
633j53v takes care they are not detrimental.-- NB. One LIMIT to the transmission of abortive organs will be
056r109 ocean, who would ever suppose that its age was LIMITED? Who could suppose such trifling means could e
064r135 coast.--important effect.--? Capt. FitzRoy. -- LIMITED Volcanic action & limited earthquakes & great
064r135 -? Capt. FitzRoy.-- Limited Volcanic action & LIMITED earthquakes & great but local elevations of th
224b216 imals with same general structure.-- miserable LIMITED view.-- With respect to how species are. Lamak
231b244 uages?-- Till man acquired reason, he would be LIMITED animal in range-- & hence probability of start
402e020 n the most northern of the group the number is LIMITED of the group, the number is very limited.-- Ko
403e021 er is limited of the group, the number is very LIMITED.-- Kotzebue's Second Voyage do Vol III p. 77.
431e123 larger than average size: (surely this is very LIMITED view, though perhaps a true element) «give exa
543m099 s (it is not so in punning) are people of very LIMITED intellects, & in the same way are chess Player
574m040 There probably is some connection between very LIMITED reasoning powers & the fixing of habits,-- for
574m040 probably can hardly form an idea of a mind so LIMITED as Birgos to become absorbed by one end of Coc
294c177 he whole object of the Work is its proof, «its LIMITING, the allowing at same time true species.» & i
022f006 , amongst the varying & dubious granites.--Wide LIMITS of this mineral in Australia. Fitton's appendi
062f128 escent grasses, parasitic plants, Cacti - & with LIMITS of no vegetation at S. Shetland = Great contra
062f128 rine all inhabited: Go steadily through all the LIMITS of birds & animals in S. America. Zorilla: wid
062f129 of birds & animals in S. America. Zorilla: wide LIMITS of Waders: Ascension. Keeling: at sea so commo
101a056 spersion of Boulders.-- See Rogers for Southern LIMITS of Boulders in N. America do/p. 280. the grave
129a131 New. Phil. Journal Vol XXI. p. 213. Beyond the LIMITS of Alps size of boulders sorted: ditto Murchis
182b047 w system of Natural History will be to describe LIMITS of form. (& where possible the number «of step
191b081 an Archipelago it would be interesting to trace LIMITS of large animals-- Owls. transport mice alive?
194b093 inc» to conquer or breed together:-- Man has no LIMITS to desire, in proportion instinct more. reason
198b113 system, or three elements p 66]CD With unknown LIMITS, every tribe appears fitted for as many situat
227b225 could pass into (D).-- We may foretell species. LIMITS of good species being known.-- It explains the
251c040 in Newfoundland, with crops full of maize. (get LIMITS of latter from <Tarton> Barton.-- swifts retur
282c140 ilar-- & constitution of each species in their LIMITS of change, would be same-- Yet each family mig
326c266 g lessings. Laoccaon.-- (translated in 1837) on LIMITS of painting & poetry.-- Erasmus thincks I shou
335d015 er parent, or intermediate within certain small LIMITS (within which limits they might return to eith
335d015 diate within certain small limits (within which LIMITS they might return to either parent), then acco
349d053 f in each «separate» line of descent.-- <& here LIMITS of varieties being constant. it would be excee
407e038 Lyell) Now «Equatorial» America from the «low» LIMITS of blocks both North & South, has probably und
411e055 gans it may serve perfectly to specify types, & LIMITS of variation., & hence indicate gaps.-- by thi
436e136 ccording to my Malthusian views, within certain LIMITS, but beyond these not.-- argue against this--
479z005 igenous p 112 M Lesson--Voyage of Coquille wide LIMITS of Nullipora Discussion good on Falklands bird
508q016 out species of Rhinoceros. becoming rare beyond LIMITS of the metropolis of each-- Cause?-- Andrew Sm
537m076 ngs of right & wrong--(& therefore in fact only LIMITS moral sense)» which she seems to think «are» t
543m101 on of thought.-- it is brought within <our own> LIMITS of examination.-- obeys same laws. as other pa
574m042 on of instincts, independently of habits.-- the LIMITS of these two actions either on form or brain v
641j29r ents have been split up,-- who can decide their LIMITS.-- To show how little we understand of the Phy
403e021 Guahon. (Mariannes), "for example the prickly LIMONIA trifoliata, which cannot now be checked".-- Ma
473s006 cribed in Linn: Transacts. a Brazilian plant «in LIN. Transacts. it has three stamens» intermediate b
470t153 on two plants in do Humble 24 flowers of small LINARIA in do Domestic do 6 Campanula (two species)--
470t177 . have abortive stamens= Many Humbles on hedge LINARIA= (Plenty of Humble Bees on Phlox Down, 1854, S
511q018 of crossing them with common pigs= [it is a LINCOLNSHIRE Breed]CD-- Sir. R. H. supposes is now extin
236bIBC s not-- intermediate genera we might expect.-- LINDLEY Introduct Dict. Science. Naturelle Geographie
500q010 annually.-- Periwinkle. (not asclepiadea. «in» LINDLEY (24) Do Bees distinguish species, they do not
514q21. Alps--; are any Europaean forms found there-- LINDLEY says that only one pineaple Horticulturists (1
324c268 e of climate White's regular gradation in Man. LINDLYS introduction to the Natural system Bevan on Ho
021r005 itution «same as» of slate same.--longer axis in LINE of Cleavage. laminae fold round them; Quote thi
026r021 lace in melted rocks The frequent coincidence of LINE of veins & cleavage is importants; veins appear
031r038 very clear at base of great lava cliffs [Fig. I] LINE of high tidal action {P} NB. patches of modern
032r039 ction of sea A. B. will be to eat in the land in LINE of highest tidal action. this will at length be
032r039 was to sink by very slow & gradual movements to LINE (2). The part (o) which was before beneath band
032r040 h changes could be shown (as represented), along LINE of coast.--[Fig. 2] Mem San. Lorenzo; Valley of
042r067 ing gales would tend to travel on a <me> central LINE of Patagonia. «NB. Mr Lyell. P. 211 Vol III. tal
042r067 atagonia. «NB. Mr Lyell P. 211 Vol III. talks of LINE of cliff marking a pause» When mentioning pumic
043r071 auses go together. add. <">" <from> "in the same LINE" to "from the epoch of Ammonite to the present
047r083 coincidental than single elevations along whole LINE of coast Darby mentions beds of marine shells o
052r097 many faults.-- Necessary form; as long as coast LINE fixed.--[Fig. 5] {P} * Slope necessary for seaw
052r098 & I believe SE coast of Madagascar. where a --40 LINE <shows> runs at equal distance?) 1st cases. --
067r144 earthquake a very remarkable phenomenon. showing LINE of disturbance inside Cordillera: It is not the
069r150 onsider profoundly the sandstone of the Portillo LINE.--connected with <gneiss>.--(Mica Slate) [Fig.
072r159 nd about bubbles. -- No Volcanic action on coast LINE of Old Greenland, close to W of Jan Meyen Isld.
072r159 to W of Jan Meyen Isld.--Mr Barrow thinks N & S. LINE connects western isles of Scotland & Iceland.--
078r175 ummits, where mountains most torn.--(¿anticlinal LINE?). -- Mines of Catorce «(Principal veins)» 25 d
088a013 analogy as continental elevations slow. so would LINE of mountain chain be Mr <Lyell> «Waterhouse» ha
099a051 .!!!-- In stratum OP. let force drag particls to LINE {P} AB, & likewise gravity MN. Then every parti
107a077 ransacts.-- Does the isothermal {P} subterranean LINE moves upward from effects of Elevation if not c
108a079 rm of fluid centre would lift with it isothermal LINE, but if heat from centre, then crust of solid e
108a079 olid earth would be thicker.-- PP Andes mark the LINE between sinking & rising areas.-- In Earthquake

Page **(Key Word)**
```
114a095 tion» it has been excessively slow because beach LINE chief cause of denudation, but does not tell pe
127a125 . S.-clinal. N-clinal & anticlinal «synclinal--» LINE.-- <ditto of synclinal> simply clinal lines. di
127a125 r depth.-- Now the <inf> subterranean isothermal LINE must be creeping «pushing» up to «the» line  of
127a125 rmal line must be creeping «pushing» up to «the» LINE of ice.-- Hence further N. when soil frozen for
130a134 littoral deposits there probably would be marked LINE of separation A Paper by Parrott Mem. Acad. Pet
131a135 . p. 402.-- ground ice-- subterranean isothermal LINE Athenaeum. 1839. p. 52. On Frozen soil of Siber
149g033 ith Isld-- {P} do they extend round hill too low LINE drawn plain red talus line on N. side of Spean
150g035 nd round hill too low line drawn plain red talus LINE on N. side of Spean most clear & upper line run
150g035 alus line on N. side of Spean most clear & upper LINE running up great bight just as Dick shows NB. L
151g040 nto coincidence» has beach or band of pebbles on LINE of 4th shelf.-- Even on Lauder Dicks Hypothesis
151g044 tion & the upper shores may correspond with some LINE subsequent to shelf {P} In Glen Collarig, by Di
154g059 } Where a buttress projects from side of hill if LINE suppose continued across to {P} side removed al
155g064 hen serpentine might remove, what above straight LINE «only» cut deep gorge on sea hypothesis, if gul
156g072 3 degree ∴ plain of 4 minute shelf slope, above «LINE of 4th» shelf This shelf at head where <granite
157g077 s unite in upper Glen Roy» very little back from LINE 2d; little action since «that shelf» formed Upp
158g079 nearly  like head of Glen Guoy nor is horizontal LINE apparently continuation of upper terrace --  on
188b069 h's fact of races of men tendency to keep to one LINE Dr Smith says very.  close species generally fre
198b113 ether. G. H. thinks developen in quite straight LINE, or branching S. H What does the expression mea
211b164 with Von Buch Volcanic chart & my idea of double LINE of intersection.-- Dr. Horsfield at India House
223b207 the  first thinking being. although hard to draw LINE.-- -- not so great as between perfect insect  &
227b227 stinct & structure becomes full of speculation & LINE of observation.-- View of generation. being con
306c213 k of soundings of which «there appears to be one LINE, in which» greatest depth «appears to be» is no
306c213 t 200 miles distant).-- directly beyond produced LINE of Timor 215 degree. What productions Sandal.  N
318c254 like White of Selbournes Rock Ouzels.-- ..If the LINE or bands of country (These facts show the Norma
333d009 & are not exactly intermediate.-- Where two dogs LINE the same bitch & «perfect» spaniels & setters a
341d029 ered as adaptive & that he did not see where the LINE could be drawn-- thus the most remarkable chara
341d031 cts of Mr Blyth on birds.-- Dr. Bachman tells me LINE of Rocky Mountains separate almost all  Mammals
341d031 represented by one not differing except by black LINE,-- A Bunting by one only differing by some perm
349d053 t value can only be judged of in each «separate» LINE of descent.-- <& here limits of varieties being
356d069 quite  independently & have used them since) the LINE of proof & reducing fact to law only merit if m
370d117 must  have degenerated. as mast spices &c &c The LINE of argument «often» pursued throughout my theor
409e048 change.--  now take greater area of water & snow LINE descent. My theory gives great final cause «I d
427e109 less  extreme. (& so account for descent of snow LINE there «& there & there only: as stated by Capt.
429e115 ltivated with ease near London.-- what makes the LINE, as trees in Beagle Channel.-- it is not elemen
429e115 n not elements.-- we cannot believe in such a LINE., it is other plants.-- a broad border of Kille
505q014 il 28½. inches. From sole of foot to shoulder on LINE of back, height 17½/. The Greyhound. was in len
538m079 ate & even profound knowledge or other & unusual LINE-- both odd appearance about eyes.-- one botanis
568n017 one is «simply» habituated «in life time» to any LINE of action, or thought one feels pain, at not pe
608o026 d how often one try to persuade person to change LINE of conduct. as being better & making him happie
628o054 atical, the emotions of this instinct, with that LINE of conduct, & if taught rightly, it will be for
092a028 on of transitions clay slate &c nearly vertical LINEAR earthquake 500 by 90.-- in Syria Geolog. Proc.
118a104 f continental elevations (I) the alternation of LINEAR bands of movement in Indian & Pacific Oceans.-
177b027 iculata.-- in lower classes, perhaps a more LINEAR arrangement.-- ¿How is it that there come aber
281c139 lar arrangement was only so far true as avoided LINEAR arrangement the central twigs dying, affinitie
309c222 ses Mr Ed Blyth does not believe in circular or LINEAR arrangement.-- Thinks passages very rare., in
417e076 n discussing the Immortality of the Soul as the LINEAR descendant of <Mammferus» «Mammiferous» <vert>
217b183 he dog. But at last a rough-haired shepherd dog LINED her & produced a very large litter.-- never aft
217b184 f bloodhound & greyhound.-- Where two dogs have LINED bitch directly one after the other, puppies dif
271c105 North  by other species.--) should build a nest LINED with mud, in forest where not a tree in which i
030r037 kes do not generally appear to have fallen into LINES of faults I do not think so many faults in Cord
032r041 f earth cause:--<exp> does not explain cleavage LINES./ possibly general symetry of world.-- I feel n
038r059 not so much aluminated. As argument in favor of LINES of anticlinal violence crossing lines of crater
038r059 favor  of lines of anticlinal violence crossing LINES of crater, <arg> state that all the great Volca
053r101 ns Australia case must be quoted at length. The LINES of Mountain appear to me to be effect of expans
053r101 ofound earthquakes), which would cause parallel LINES, but the rectangular intersections are singular
058r116 would though be easy to see on beach successive LINES of sea weed-- Histoire Naturelle des Indes Acos
059r123 from  great depths, so changes, acting in those LINES. must now proceed from great depths.--important
064r137 g structure of Cordillera it must be said, that LINES of elevation have connected <lines> «points» of
064r137 e said, that lines of elevation have connected <LINES> «points» of eruption [.] give instance of Etna
066r141 narrow  parts which break through the N & South LINES the tides form eddies with its extreme force. Y
104a066 <strata> not great, one can conceive anticlinal LINES near. (lateral pressure would always produce it
110a085 fect of» elevation & subsidence. examine these «LINES» Description of rocks in Lyells'. Capital Norwa
120a107 uakes.-- by the narrowness which the anticlinal LINES are apart-- the curvatures of the strata.¿  the
127a125 l--» line.-- <ditto of synclinal> simply clinal LINES. dipping so & so or may be used East-clinal lin
127a125 nes. dipping so & so or may be used East-clinal LINES & c & . .-- But Siberia was once thawed. & hence.
132a138 vails more beneath the sea, <than> «&» on coast LINES, than on continents. it ought, (according to M.
150g038 drainage & far more earthy than what is usual-- LINES die away where slope less., best developed on s
153g041 cks Hypothesis impossible to explain absence of LINES in certain parts.-- At the Pass of Glen Collari
151g041 arts.-- At the Pass of Glen Collarig two little LINES of Hill (judging from external form alluvium) d
153g051 us-- {P} where buttresses on 4th shelf: others «LINES not so level because of upper edge of cliff» Ot
155g065 rved as to have thrown water in & same «drainage» LINES Mound of Gneiss though wonderful-- <that they a
155g066 hat they  are preserved» how much more so, these LINES & even water-scooped rock «only decay from frag
155g066 particular  hardness no wonder that all «three» LINES «should be» EQUALLY preserved 2d or upper one m
156g067 han upper & lower but quite as perfect as those LINES in Glen Collarig, & some «other parts» Boulders
157g073 l when mica slate, only sand blow away]CD where LINES appear to cross stony parts; appearance chiefly
352d059 s group without centre but circle, two or three LINES deep-- with respect to Macleay's theory of anal
383d160 ot having tuft,-- back like do.-- but the black LINES on each feather instead of coming to point  {P}
383d160 , & <more ro» «three, I believe instead of» two LINES.-- «faintly edged with reddish brown» black mar
528m037 rived in a river from seeing how the serpentine LINES narrow in the distance.-- & even on paper two w
528m037 - & even on paper two waving perfectly parallel LINES are elegant.-- Again there is beauty in  rhythm
567n016 efully to separate its memory from all ordinary LINES of association.-- is totally distinct from lear
574n039 . «& Spencer», an old Scotch Poet, has numerous LINES. of poetry.-- <signs> sounds singularly adapted
620o045 Hence  there are certain instincts pointing out LINES of conduct to other men, [RHC] 5) which are nat
620o046 his instincts.-- so man ought to follow certain LINES of conduct, <although> even when tempted not to
621o047 have a feeling, that he ought to follow certain LINES of conduct, & he must soon necessarily learn th
302c201 here may be some descendant of some intermediate LINK.-- the only connection between two such classes
398e005 wn to be more rapid, I should say there was some LINK in our train of geological reasoning, extremely
356d071 not overwhelming} What comparative solutions & LINKING of facts-- Savages over whole world. (Major <I
349d052 s in Hippopatamus, solely owing to number of lost LINKS. ‖ if all species know they would be innumerable
352d058 re is no central radiating point, all united - (LINKS which when conscious connect past, present & fu
544m102 past  events?) or influence on our conduct, the LINKS in circle must be granted unequal, because foss
085a001 n Shropshire. or some other wonderful outlyer.-- LINN: Transact. Vol. 8. p. 288. Salt deposited on wi
219b195 ight have acted at Tristan d'Acunha-- Carmichael LINN. Transacts. Vol XII.-- The Alpine plants of the
236b1BC nique De Candoelle. Geol. Soc Horae Entomolgicae LINN. Soc. Geoffry-St. Hilaire Philosophy of Zoology
287c157 up  to it" p. 20 ‖ +++hiatus & saltus not syn.-- LINN. Transact Vol XIV.--‖. p. 24. Lamarck bears t
326c265 ted by Herbert. p. 338 Schiede in 1825. & Lasch. LINN. in 1829 has given list of Spontaneous Hybrids.
448e168 Baboons  in St Thomas on W. coast of Africa Owen LINN. Soc. April 2d. 1839 The Lepidosiren-- Amblyrhy
471tf07 rfly Orchis & Listera? Bryony saw common Bee on: LINN. Trans 18. p. 133 Westwood on the Fulgoridae en
471tf08 .--"of their use difficult to conceive any idea" LINN. Trans. 18. p 163. "D. Dod on two new genera o
473s006 l-Curig-- Mr Bunbury says Miers has described in LINN. Transacts. a Brazilian plant «in Lin. Transact
480z008 uttgart ranks these bodies amongst Vegetables in LINN. Soc.-- Mr. Donn Carmichael Linn. Transacts Vol
480z008 t Vegetables in Linn. Soc.-- Mr. Donn Carmichael LINN. Transacts Vol XII. p 496. Birds at Tristan d'A
204b139 d has written paper on affinity and analogy in LINNAEAN Transactions Mr Wynne distinctly says that th
209b155 rays--  from certain sports.-- Agrees with old LINNAEAN doctrine & Lyells. to certain extent Von Buch
268c096 fied catalogue of animals of Nepal read before LINNAEAN Soc. Feb. 1838.-- Annals of Natural  History.
286c156 review says nat: fam: of Willows contains many LINNAEAN genera.-- Now are the characters which  unite
344d040 n Suites de Buffon.-- Vigors has given list in LINNAEAN Transactions of birds of Java Caterpillars no
427e109 plants  being effect of abortion of one sex.-- LINNAEAN class Dioecia & Monooecia. ought to be preemi
```

287c157 tortoise shows hiatus-- but not saltus-- when LINNAEUS put whale between cow & hawk a frolicsome sal
459tf02 & sheep, it seems male gives form. admitted by LINNAEUS.-- seems to doubt its applicability to common
498q008 o impregnate ordinary number of seeds known?-- LINNAEUS has shown that each pistil is connected with
499q010 = What structure of seeds.-- (Paris) (22) When LINNAEUS says so great percentage of seeds have contri
502q012 nadense-- would it grow in open air in Sweden. LINNAEUS found 2 flower. which had anthers removed, di
545m105 how the Baboon (<Macaco> «Cyanocephalus Sphynx LINNAEUS») constantly moved the skin of forehead over
482z011 like sylvia cisticola.-- Embriza melanodera-- a LINNET not caught.-- Troglodytis Furnarius.-- Sturnu
361d095 Zoolog. Meeting stated, that Green-finch, all LINNETS red-pole, goldfinch, hawfinch-- in nursling pl
183b052 fact that "serin avec le chardonneret, avec la LINOTTE, avec le verdier" &, <fr> silver gold & common
261c070 s of Man.-- black men, black bull finches from LINSEED-- not solely effects of climate on some «antec
504q014 rees sexes-- (3) Get Holyhoaks. races planted & LINUM Perenne.-- Herbert's. fact.= (4) Effects of Nit
505q014 field= (8) Abortive Thyme seeds weather wet--? LINUM flavum put in Spirits which plant seeds? (9) Me
506q015 &c (36) Has not H. raised races of white & Blue LINUM-- did parent plants grow near each other.--? C
253c047 Vol I p. 165.-- <a> "an account of the MANELESS LION of Guzerat by Capt. W. Shee. considered merely
312c228 to no attempt to keep up offspring, are not half LION & tigers ditto. (see Griffith) & half Muscovy d
331dIFC like geese?-- Are the number of kittens between LION & Tiger at litter as numerous as in common lion
331dIFC Lion & Tiger at litter as numerous as in common LION? Are the number of nipples in domesticated very
333d008 & no leading question was put.-- Fox thinks half LION & Tigers are exactly intermediate in character
377d140 m in the Heaverns.-- Is not puma, same colour as LION. because inhabitant of plain & Jaguar of woods
508q016 mals breeding at the Cape.-- ᴸAbout two vars: of LION: Annales des Sciencesᴵ (4) Prolifickness of fem
569n020 connexion between things & voice, as roaring for LION &c &c. (in same way alphabet. arose from letter
591n098 not think they drop their ears.-- -- George the LION is extraordinarily cowardly.-- the other one no
512q020 nglish animals & birds breed (3) In cases where LIONS have bred, have they been raised from young one
152g046 f shelves entering «on» one side ravine. Are the LIP, or necks of land on level with shelves effect o
157g078 16 true terrace «2d» near divortium aquarium is a LIP with it-- Dick right-- Mac mistook terrace also
160g093 shelves just like road of Glen Roy-- appears to LIP with moss On this terrace «station perhaps 6 ft
161g098 degree? The River <the> of which the source is a LIP with the new shelf flows into canal between L. L
342d035 ole dynasties have been featured by the Austrian LIP & the Bourbon nose". if this be not imagination.
345d043 n with white head (,or «short-horned with» black LIP) & then calf «in both cases» is killed. Notes fr
403e022 without arm, <skin> hands thumb,-- one leg, hare LIP &c &c. in Vol II p 363 account of Flora of pacif
466t094 s, wh otherwise lie protected by the hairy black LIP of lower division of nectary: «wh. itself resemb
525m025 deformity, as an extra number of fingers.-- hare LIP or imperfect roof to the mouth «stammering in my
541m092 dog in play has his mouth open ready to bark, & LIP twisted up, in that peculiar manner they do, eve
542m093 distinct will, he will find it hard to keep his LIP from stiffening over his canine teeth.-- He may
553m137 oups. thus the pig-tailed baboon, shoved out its LIP, looking absurdly sulky «as» often as keeper spo
556m145 mals.-- In the drawings of Voltaire why is under LIP curled over upper with mouth shut. expressing co
558m152 tic Leopard. quarrelling. mouth wide open, each [LIP] drawn back & driving air out of mouth «hairs er
564n006 n or muscles are contracted between eyes & upper LIP., is most clearly analogous to a panther I saw i
565n008 extremely self-sufficient,-- the corner of lower LIP are depressed & opposite muscles used to when an
565n009 in wrinkles about the nose, & arrogance in upper LIP. <The> Children having peculiar expression is re
022r008 ound the Head & Boans of a Hippotomus; the hairy LIPS of which were still sound and not petrified, an
542m096 has canine teeth.-- A dog when he barks puts his LIPS in peculiar position, & he holds them this way,
551m129 looked at the glass) when pouting proctrudes its LIPS into point-- man, though he does not pout. push
551m129 -- man, though he does not pout. pushes out both LIPS in contempt <&> disgust & defiance.-- different
555m143 ous, angry cross look, followed by protrusion of LIPS, in which respect resembles some of the old one
565n009 affection is accompanied by slight protrusion of LIPS, as if going to say "my dear." just what smile
565n010 way with new Lavaters,-- Ye Gods!:-- says fleshy LIPS denote sensuality (p 192 Vol. III Octav. Edit)-
570n024 s determined not to say anything. he presses his LIPS together & shrugs his shoulders & walks off,--
594n113 aby-- association-- pouting child same as anger, LIPS not compressed sullen, protruded. determined to
103a065 sence of dikes. argues in favour of pressure of LIQUID rock. Andes discussion-- Albite certainly cont
416e071 rmaphrodites.-- I suspect this rather effect of LIQUID semen: therefore animal life commenced in the
046r080 all first movement (as in Peru 1746).--At great LISBON Earthquake Loch Lomond water oscillated betwee
321c270 der) when out [not located] <Rays Wisdom of> LISIANSKY'S Voyage round world. 1803-6 Nothing Lyells El
321c275 ferences at end-- The British Aviary.--do--- & LISLE'S Husbandry Tuckeys voyage reread Appendix Oving
096a040 etched out NE & SW.-- Von Buch. Can. Ile p. 406. LIST of Volcanos Salomon Isld,-- New Britain-- &c &c
103a063 e this after supposition p. 461 «of Proceedings» LIST of collections in Geological Society. Pumice at
193b091 a quadrupeds Kotezebues first Voyage Copied into LIST Entomological Magazine paper on Geographical ra
214b173 analogies to shells of living seas.?-- Roxburgh. LIST of plants in Beetsons St. Helena. -- Galapagos-
218b192 ormed by nature. Carmichael. Tristan D'Acunha, a LIST of its Flora. is given Mr Don remarked to me. t
241c015 - a Synallaxis. ¿American?). Pₒ 159. & 160 «162» LIST of some birds of Tongatabou. & New Ireland.-- G
260c067 regions-- crows in N. Americaᴵ Study Bonapartes LIST In the Zoological Journal I read a curious acco
278c131 ies same, excessive improbability. Mem in Clifts LIST a rat said to have been found!! rodents old inh
303c205 , but to place all that ever lived <on> into one LIST is unfair, [moreover what will become of the fu
303c205 what will become of the future creations, if the LIST is now perfect.--]CD. the creator so creates an
311c225 ry «Vol I??» p. 318. some remarks on Bonaparte's LIST of birds in Europe & N. America, on closely all
326c265 chiede in 1825. & Lasch. Linn. in 1829 has given LIST of Spontaneous Hybrids. where? Sweet. Hortus Br
327c265 udying in Metaphysical point of view Henslow has LIST of plants of Mauritius with locality in which e
344d040 reptiles in Suites de Buffon.-- Many Europaean insects-- LIST in Linnaean Transactions of birds of Java Cater
402e019 70 degree of Europe.-- Many Europaean insects-- LIST given,-- some peculiar <M> p. 359. At Manilla a
412e058 e. «on any structure.» L'Institut. 1838. p. 384. LIST of fossil Mamm: from Poland. &c.-- Three princi
430e119 ive my theory.-- excellentiy true theory Examine LIST of St. Helena Plants & see whether those which
471tf09 . Phillips (Lardner's E. vol. II p. 18.) capital LIST of all the fossil Mamm. of Europe-- Large Lizar
485z017 ed together Mag: of Zoolog & B. Vol. II. p. 127. LIST of submarine insects. Staphylinidae &c &c. with
501q011 t observe on this subject-- (31) Ask Henslow for LIST of annuals to place in Hot house to see effect
507q15v ics are of this class.-- (48) .Where «published» LIST of spontaneous Hybrids-- to see whether any Pap
507q15v ng cross impregnation difficult or reverse (.49) LIST of seeds Gaertner de fruct:-- for woodcut-- 1 d
511q018 appeared.= (4) Does he know any seed-raisers (5) LIST of qualities in birds & animals for prizes.= Pi
587n088 nes just like a child. Get a Dictionary & make a LIST of every word, expressing a mental «desire» «qu
565n007 well «as one will perceive if in night trys to LISTEN to growl of hounds». <when> as fear to «man as
554m139 itted & eat it.-- good case of association.-- «LISTENED with great attention to Harmonicon. & readily
531m051 ane feeling of anger which came over me, when LISTENING one evening when tired «-- how true the heart
565n007 painted with mouth open.-- why when person is LISTENING is mouth open to hear well «as one will perce
470t178 t of long, curved nectar of Butterfly Orchis & LISTERA? Bryony saw common Bee on: Linn. Trans 18. p.
025r019 (I have not kept a record)» In looking over the LISTS of organic remains in De la Beche, for the olde
604o018 ams & insanity.» D. Stewart on the Sublime The LITERAL meaning of Sublimity is height. & with the ide
348d051 be settled,-- I believe affinity may be taken LITERALLY, though how far we can ever discover the rea
549m121 the object of living.-- or whether if we obey LITERALLY New Tamarind future life is almost the sole
276c123 n advance only of their age. (differently from LITERARY men.--) must remember that if they believe &
604o015 the Origin of Evil. Reviewed by Johnson in the LITERARY Magazine. 1756-- Ceased in 1758-- Read the Re
342d035 logy of young Ostrich. 16th. D Israeli (Cur of LITERAT. Vol II p 11) accidentally says "--is distinct
502q1v asa» Anchusa «Campanula» &c & dead-nettle.-- LITHOSPERMUM. Blue Gloss. it is not possible to see orif
214b176 me so often «that» it was difficult to obtain a LITTER without this defect, Very curious case = W. D.
217b183 shepherd dog lined her & produced a very large LITTER.-- never afterwards went in heat.-- This is go
217b184 of good pointer & rough water spaniel «produce LITTER like both parents» & Mr Bell has half bloodhou
275c120 orm of grey hound.-- picking out finest of each LITTER & crossing them with finest greyhounds.-- Sir.
312c233 delicate-- they cross sister & brother of same LITTER, those. of different litters or of father & ch
331dIFC e the number of kittens between Lion & Tiger at LITTER as numerous as in common lion? Are the number
385d163 st valuable. because smallest sized dogs.-- one LITTER big & then second small & so.-- Says, there is
388d171 ermined by imagination of Mother.-- We see in a LITTER every possible variation from being very near
391d175 ach other» than twins «or triplets &c &c in» in LITTER. Why is there some law about sexes of twins in
312c233 & brother of same litter, those. of different LITTERS or of father & child are thought long breeding
385d163 es of bitch going to mongrel, & all subsequent LITTERS having a throw of this mongrel.-- I did not as
385d163 log Gardens. & Kings at Otaheite) <Think» Last LITTERS are considered the most valuable. because smal
639j28v & it so chances, that one out of every hundred LITTERS is born with long legs» & in the Malthusian ru
025r016 120 parallel of Olinda Shoaler N. of Olinda.--a LITTLE WNW of C. Rock. [5 degree 29 minute S.] still
031r038 n. Australia: cases in Europe.-- Auvergne. very LITTLE Pumice, though Trachyte. same fact in Galapago
033r043 . American Geolog. in Daubeny. P. 349 Admirable LITTLE table showing long PERIODS of great violence v
035r046 ht of same beds, deposited in different basins; LITTLE or no relation appears to <exist> be made out,

Page **(Key Word)**

035r046 ging to the same district there seems. I think, LITTLE ground for skepticism, as to the general truth
038r057 wn off; & again when in great crater. different LITTLE craters are all burning, surely there must be
041r066 ort <fissure> «vein» terminated each way, which LITTLE vein was like the rest of these thin veins whi
055r106 un from North <inclining> to South. inclining a LITTLE to the West: the veins which follow this direc
058r116 to Quito a distance of more than 500 leagues. A LITTLE time after a bad earthquake in Chili; Arequipa
059r123 t of glassy F. fractured. have been melted with LITTLE pressure. & perhaps cooled suddenly. -- As the
090a023 es recorded if we say «100» <5>0 lbs a year too LITTLE. -- How comes it none in fossil state? suppose
099a050 action &c.» becomes measure of force. < ∴ where LITTLE inclination, little force & varying direction.
099a050 measure of force. < ∴ where little inclination, LITTLE force & varying direction. --> Therefore in PIL
102a060 ion, each particle coated. &c will be aware how LITTLE common Gravity has to do with arrangement of p
102a061 nces soluble under great pressure? equally with LITTLE pressure? An important question! If water yiel
105a073 en valley. -- Why is serpentine course result of LITTLE inclination??----It is simply as the inclinati
105a073 ination??----It is simply as the inclination is LITTLE the force required to move <it> «stream» aside
109a082 a must always on actual beach act same way. -- a LITTLE further from beach action probably modified by
113a091 better or worse & the depth of 32 degree. being LITTLE we may confidently infer that time has not bee
135a145 near Caubul. parallel ranges. with here & there LITTLE branches at {P} from each side intercepting pl
149g032 ide of Inn BOULDER of granite above 4th Shelf a LITTLE lower down the hillock with beach & channel pr
151g041 rtain parts. -- At the Pass of Glen Collarig two LITTLE lines of Hill (judging from external form allu
154g060 quantity when during repose of lake it did but LITTLE more {P} now that it has got to the rock of co
157g077 here two branches unite in upper Glen Roy» very LITTLE back from line 2d; little action since «that s
157g077 upper Glen Roy» very little back from line 2d; LITTLE action since «that shelf» formed Upper terrace
157g078 pey <29.35161> 29.360? A 79 degree 75 degree? A LITTLE below Divortium on slope towards Loch Spey 29.
158g082 lly retired, hard to explain on river doctrine <LITTLE Hill with granite blocks almost encircled> <fr
158g083 f 3d runs up with buttresses on each side «very LITTLE way» in Upper Glen Roy at pass {P} River Gorge
159g086 .008 A. 75 degree Air 70 degree? This station a LITTLE way down slope of obscure terraces (& conical
163g109 hropshire Quartz what substance is collected in LITTLE spots Speculate on «under head of» Beagle Chan
189b070 reditary in time as in space. (Mem: Galapagos). LITTLE wings of Apteryx Dacelo & Kingfisher same colo
191b080 nd possesse Mastodons?? Negative facts tell for LITTLE) Geographic distribution of Mammalia more valu
199b120 cing» «before domestication, afterwards none or LITTLE with «fertile offspring» » fertile offspring;
211b161 s» species, whereas Man has such instincts very LITTLE. in Zoolog. Proceedings. Jan 1837, «by Eyton»
216b179 lain it by intermarriages with people. either a LITTLE nearer black or white as it may happen.-- Dr S
216b182 ot mingling?-- In Foxes case of Blood Hounds, a LITTLE mingling would probably have been good, namely
227b225 heredetary structure. latter far chief element.: LITTLE service habits in classification. or rather th
231b243 ange take over, Whereas those which migrated a LITTLE to the southward would merely be specifically
236b280 we might expect disseminated species to vary a LITTLE, but such should not be general circumstance.-
258c061 f nature-- Whether species may not be made by a LITTLE more vigour being given to the chance offsprin
260c067 not be resisted, when hearing crys of hunger of LITTLE bird, in same way Wilson (p. 5). describes man
261c070 conduct to laws of change of organization! The LITTLE turtle, without its parent running to the wate
263c075 of slow movements &c &c. this multiplication of LITTLE means & bringing the mind to grapple with grea
294c176 ed argument, that domesticated animals change a LITTLE with external influence-- & if those changes p
302c203 ely owing to such interm: father-species, being LITTLE adapted to some physical change.-- If Patagoni
312c227 & tigers & sharks, being spotted, & colours of LITTLE value Dr Smith if black & white Man crosses; c
312c232 in shortest possible time. Mr Willis Long eared LITTLE dogs, I am told, go to heat, take dog. but do
320c276 Journal to end of 1837. re«a»d-- contains very LITTLE Macleay's letter to Dr. Fleming. & Review of l
341d029 re of pelvis & was not adaptive structure, like LITTLE wings of Auks which does not make that bird a
341d030 ller than in other Struthios. was adaptation to LITTLE Movement.-- nocturnal crawling bird.-- Wings r
355d067 e, but retrospective as showing what organs are LITTLE fixed-- (<also> Hunters law of monstrosity wit
358d076 ation that in insects where one of the sexes is LITTLE developed, it is always female which approache
359d087 se.-- These hybrids are very wild & take <very LITTLE> in disposition after their «pheasant» parents
359d089 . -- -- the former were the parents of the three LITTLE ones.-- Keeper said in <two> crosses «twice ma
370d116 nged-- Neuters are true females, but with parts LITTLE developed.-- Sept. 19th <Are> There is no scal
376d137 front.-- Dr Smith every baboon & monkey, big & LITTLE that ever he saw knew women.-- he has repeated
385d163 & in» never lose passion. (Mem: so it was said LITTLE cock «yet very odd loosing visible powers» in
389d177 ch difference should produce same effect as too LITTLE.-- in (latter case female often takes males bu
410e053 traces of changes in forms of organs, will care LITTLE for species, except so far as wanting names to
410e053 e corelation of parts in individuals, will care LITTLE, <in> whether the individual be species or var
414e060 tion of fishes & Mollusca (& plants???) been so LITTLE progressive «!Agassiz makes it wonderfully cha
417e075 eserve the name of species, may be owing to the LITTLE fixity of organization, in the two races,. owi
417e075 estication of both.-- Now in the ass-- there is LITTLE tendency to vary. & hence offspring are hybrid
420e090 s.-- Mem. Spallanzani's experiments showing how LITTLE of the spermatic fluid fertilized spawn of fro
421e091 erely adaptive to an essential character--" How LITTLE clear meaning has this compared to what it mig
426e108 est blood being a great great-grandfather,-- -- LITTLE Miss Hibbert case of Hindoism coming out more
429e114 birds & insects) which are propagated with very LITTLE care.-- & which might spread themselves, as we
440e145 as been» preserved.-- but be it remembered. how LITTLE part of the Grand Mystery is this,-- the law o
442e152 me permanent?» because its very essence is that LITTLE change is produced.-- The fact just alluded to
450e176 x. p. 43. «& 45» the Breed of elephants <of> in LITTLE isld of Sooloo.-- said to have been imported:
458t002 crossed with pony mare & produced a very pretty LITTLE animal, showing something of Mule in its ears-
466t093 not differently coloured-- & stamens bend up a LITTLE In a wild purple Geranium, I see Bees visit al
467t099 t stigma does not, so we may feel somewhat «but LITTLE» less surprised at Henslow's remark that pisti
470t178 -I think not on Phlox though they examine it.--<LITTLE Dusty & Blue» Butterflies at Clover,--Veronica
472s01v since gave her some She means to try this year. LITTLE variation in the 60 one brighter with mere tra
481z010 hink Planariae) Sagittaria, or Fleche «p. 8» my LITTLE animal with horns. Madrepores p. 26 Nullipora
496q006 ve & suck like in very rich soil-- As they have LITTLE tendency to double; what would be effect-- (10
499q09v they much frequented by Bees or Butterflies or LITTLE insect?= or is pollen excessively minute or ab
515q022 ons under net & see if get sterile-- Cover that LITTLE Ervum in Sand-walk, on which I think I have ne
525m025 h> (in number 3) had all short tails; but one a LITTLE longer than rest «they all died»:-- she had ki
530m045 tary, & therefore that their children have some LITTLE advantage in these trades.-- Delirium seems to
531m049 this nest the cat could If cats will «ever» eat LITTLE birds, this most curious instance of reason &
534m062 ves from wall to table.-- table being removed a LITTLE further, they ascended about a foot & then lea
535m069 e.-- How is it with children.-- Now it is not a LITTLE remarkable that the fixed laws of nature shoul
538m078 p. 365. Case of double consciousness, one only «LITTLE» less perfect than other, absolutely two peopl
542m094 ac shouts & bellows with passion.-- It is not a LITTLE remarkable that those sounds which are involun
545m105 become unconscious those which are viewed with LITTLE interest, & those which are viewed very often.
545m106 do not doubt this Baboon. knew women.-- Another LITTLE old American monkey «(Mycelis)» I gave nut, bu
546m109 r.--origin of colours?-- Nothing shows one how LITTLE happiness depends on the senses.; than the <sm
549m121 with that of New Testament as in other question. LITTLE is there said of intellectual <pie [...] hope>
550m124 conscience or instinct, would probably feel but LITTLE that of anger or revenge.-- they are incompati
556m145 & quickly retracting tongue from behind upper & LITTLE between incisors.-- like <W[...] what> person
565n007 .-- Contempt look obliquely so does dog. when a LITTLE one attacks him Contempt, when there is some a
578n057 d is «not» under voluntary power, (or only very LITTLE so) & hence by association, there pour out tea
585n076 annot return.-- Hence I conclude. pidgeon taken LITTLE way, whirled, & then taken other way-- would n
589n091 he pianoforte, yes if every individual played a LITTLE, & something destroyed bad brain. see p. 90.--
591n097 dropped its ears like dog-- wagged its tail «a LITTLE» when attending to anything or excited.-- so d
594n113 nsation of disgust with nausea, (when stomach a LITTLE disordered) at thought of almost anything ugly
594n115 nderful than man his intellect Lyell has seen a LITTLE dog go to the assistance & bite a big dog. whi
596n184 American monkeys. pull back skin from head very LITTLE Does blood go in <body> face in pashion.?-- cr
600o008 oul, treating of infinites not definable.-- Has LITTLE Chapter on each faculty of Soul.-- (I) <Consci
602o11v How strange it, that Nature should have so LITTLE to do with art (p 128) R. compares a view take
605o18v r so often concomitant with sublime. adds not a LITTLE to the effect: as when we look at the vast oce
641j29r p.-- who can decide their limits.-- To show how LITTLE we understand of the Physiological relations o
122a113 hich generally yields.-- Will this not explain LITTORAL mountains & volcanos.-- Why on one coast? How
130a134 cautious. mud banks & sand. dunes.-- in these LITTORAL deposits there probably would be marked line
473s004 nts of cretaceous periods, together with their LITTORAL deposits are probably buried in the depths of
183b051 might continue & thus two species be created) & LIVE in same country. How is propagation of wolf & D
185b055 of animal having come to island. where it could LIVE-- but there were causes to induce great change.
195b099 se, such beautiful adaptations yet other animals LIVE so well.-- This view of propagation gives. <ro>
196b108 may say that vegetables & mass of insects could LIVE without animals but not vice versâ. ¿could plan
196b109 ithout animals but not vice versâ. ¿could plants LIVE without carbonic acid gaz.)-- Yet unquestionabl

225b216 are arguments against it-- namely how did otter LIVE before being made otter-- why to be sure there
257c057 possible,-- it has been asked how did the otter LIVE before it had its web-feet-- all Nature answers
270c104 n animal life & "inorganic life".-- animals only LIVE on matter already organized.-- This paper might
272c107 insect, which I brought from Australia, probably LIVE in flowers & has Elytra. formed from developme
285c151 t it-- The space which one branch of the tree if LIVE occupied after its decay, will be occupied by t
294c175 ats near Bath. Lonsdale do. says Sheep could not LIVE for some time at New York «instance of the fine
374d134 ls exactly proportiona[1] to the number that can LIVE.--» We ought to be far from wondering of change
392d180 Mollusca «& Frogs» pass through birds stomachs & LIVE? In Muscovy ducks do young take most after fath
401e017 ften> sometimes graft pears on apples. they will LIVE but not flourish-- a medlar may be Grafted on p
407e041 . reason, why: Marsupiata, when first introduced LIVE & multiplied, specifically & individually.-- I
435e134 s given of high temperature at which fish &c can LIVE.-- Lyell says that naked cuttle fish now bear a
445e159 other, by the means of the medium in which they LIVE do. "Additions". p. 454.-- does really attribut
451e178 ankind «(& yet afterwards says native tribes can LIVE there)» flee during 8 months out of 12.-- the l
451e178 re breeds of Men the Thârû & the Dhangar who can LIVE there & do not pine visibly. p. 337. it would a
512q019 he believe. Stanley's fact of Hawks distributing LIVE Mamals (3) Do most Hawks eat stomach. of finche
515q021 far north in Scotland-- do they flower-- do they LIVE healthily, or does fruit merely not ripen.-- Th
613o035 re must be consciousness??? ? [LHC] ¿Can insects LIVE with no more consciousness than our intestines
633j54r ns, where in accordance to certain laws they can LIVE.-- Hence the mistake they are created for them.
633j54r should plants require earth, why not created to LIVE on alpine pinnacle? if we once to presume that
639j28v n the Malthusian rush for life, only two of them LIVE to breed, if circumstances determine that, the
041r064 e depth of shells (which being packed. in beds) LIVED there, makes it very doubtful whether they coul
041r064 makes it very doubtful whether they could have LIVED in so deep a sea.--Perhaps agrees with formatio
187b062 ange, agrees with belief. that Siberian animals LIVED over whole bottom is important; because in this
041r065 n; or being only preserved in that part. having LIVED in groups or not. Ferruginous veins of this fig
041r065 ter case. we cannot judge whether such fossils. LIVED Why is life short, Why such high object generat
171b002 Eunuchs nor «cut» Stallions nor nuns are longer LIVED for ever would be endless (that is with our pre
171b004 t of accidental injuries, <on» which if animals LIVED in cold countries & therefore not killed by <Si
213b170 lips. p. 90. it seems the most organized fishes LIVED far back, fish approaching to reptiles at Silur
222b205 -- as some of the most perfect kinds the shark. LIVED in remotest epochs.-- ¿ lizards of secondary pe
264c079 1.--man, (rude, uncivilized man) might not have LIVED mean temp 60 degree mean, then temp at depth o
269c101 e to metamorphs theory suppose when rhinoceros LIVED <on> into one list is unfair, [moreover what wi
303c205 ould have been fair, but to place all that ever LIVED in estuaries its taste. taught it to go to <sea
306c212 ld hood.-- Young salmons. first a species which LIVED with it.-- also of <a> dog«s». but did not seen
376d138 he kept pull up feathers of tail of Hen; which LIVED & when present animals-- lived.-- we know the
377d139 England was united to Continent, when elephants LIVED.-- we know the great time, necessary to form ch
377d139 when elephants lived. & when present animals-- LIVED.-- we know the great time, necessary to form ch
582n067 , will make average of life longer.-- for short-LIVED constitutions will then be cut off.-- <Horses>
624o50v on between our instinctive feelings & our short LIVED Passions' State broadly in child or animal it i
625o50v are too strong for our instincts. to gain long-LIVED good, ie happiness-- yet this system not selfis
637j58r break egg. shells-- why chicken could not have LIVED had it not been so.-- let egg shells grow harde
614o037 be, seems as much function of organ, as bile of LIVER.-- ¿ is the attraction of carbon. hydrogen <&c
508q016 sthma in females takes place of gout.-- How are LIVERS obscure organ. no answer?-- 3 Andrew Smith, ab
048r085 e nature of the country in which the Rhinoceros LIVES in S. Africa: the same caution is applicable to
064r134 than of individual.-- Mr Birchell says Elephant LIVES on very wretched cou[n]tries thinly covered by
246c027 s very closely allied to C. muscadivora., which LIVES in the Eastern Moluccas, New Guinea.-- (Case of
290c163 yrrho-alauda (bird of St Jago) of brown colour; LIVES on ground. colour of habitation. Must have some
323c269 roduction -- Treatise on Domestic pidgeons 30th LIVES of Hayd & Mozart «Apri 25th Lockarts life of Na
325c267 H. Understanding (sometime) Du Stewart works. & LIVES of Reid, Smith & giving abstract of their views
432e125 ments refer not to revolutions of the sun & our LIVES,, but to period necessary to form heaps of pebb
435e134 in size now the Proteus anguiformis. he remarks LIVES in dark caverns of Carniola p. 112. Man. "stand
443e155 -- what relation in duration of a planet to our LIVES-- Being myself a geologist, I have thus argued
559m155 well worth reading Copied <Smith> «D. Stewart» LIVES of Adam Smith Reid, &c worth reading. as giving
600o008 sion gained in life time]CD 3. The Infinite,-- LIVES by hopes, looks to eternity. (4) Reason, some t
021rIFC > Friday Thursday 29th gale Lyell's Geology The LIVING atoms having definite existence, those that ha
041r065 titudes.-- The question of shell's concretions, LIVING only in that spot & being cause of concretion;
051r094 spray (there pale green confervae) coated with LIVING beings; In smooth seas (& even turbulent as at
057r114 Ross found in Possession Bay in 73 degree 39 N. LIVING worms in the mud which he drew up from 1,000 f
116a099 pecies non decrite de petites corbules analogue LIVING in mouth of Plate. p. 26. Geology of Arica <Sc
133a141 I. p. 2 & 3. Porphyry at St. Elena. p. 6. few «LIVING» shells. on coast of do p 8.-- soft Clay beds
171b003 temperature & all circumstances which influence LIVING beings.-- We see <living beings>. the young of
171b003 tances which influence living beings.-- We see <LIVING beings>. the young of living beings, become pe
171b003 beings.-- We see <living beings>. the young of LIVING beings, become permanently changed or subject
180b036 t be that one generation then should be as many LIVING as now To do this & to have many species in sa
181b041 hances against any one «of them» having progeny LIVING ten thousand years hence; because at present d
184b053 otherium, megalonyx mastodon, & the species now LIVING.-- Now according to my view. in S. America par
202b133 law of type. like horse in S. America. or like LIVING Edentata in Africa &c &c.-- Now if suppose wor
206b146 nstant «say 2000» and at present day, every ten LIVING souls on average are related to the (200dth ye
206b146 gree. Then 200 years ago, there were 200 people LIVING who now have successors.-- Then the chance of
206b149 Illustrations   are there many anomalous lizards LIVING; or of the tribe fish extinct. or of Pachyderm
214b172 ertiary deposits present analogies to shells of LIVING seas.?-- Roxburgh. list of plants in Beetsons
214b174 fossils bear the closest relation to those now LIVING in the sea.-- See Rogers report to Brit Assoc
221b201 Geolog. Researches. facts of salt-water shells LIVING in absolutely fresh water.-- origin of Fresh-w
224b214 ellect of Man & animals not so great as between LIVING thing without thought (plants) & living thing
224b214 between living thing without thought (plants) & LIVING thing with thoughts (animal). «∴ my theory ver
258c060 d people. would cause destruction.-- simile Man LIVING in hot countries, if continually crossed with
281c138 t be traced between relationship of all men now LIVING & the classification of animals-- talking of m
282c141 hese analogies.-- We must two races of such men LIVING in same country but separated, now if one of t
285c152 ies is only fixed thing with reference to other LIVING being.-- one species May have passed through a
302c203 tagonia became fertile all intermediate species LIVING there would be destroyed, & N & S. existing sp
305c210 Mammals.-- Mem Elephants & dog.-- There is one LIVING spirit, prevalent over this word, (subject to
305c211 on in degree & kind to the endless forms of the LIVING beings.-- We see thus Unity in thinking and ac
321c270 neaus» How to observe Mayo Philosophy of Art of LIVING Several of Water Savage Landons Imaginary Conv
328cIBC shells from Barrier isld many relations with a LIVING Matica & many shells of Genera Corlula Cham. C
347d049 pt to be extinguished.--???» Mayo (.Philosop of LIVING) quote Wheewell as profound. because he says le
369d115 "the necessity of combining observation of the LIVING habits of animals, with anatomical & Zoologica
372d129 ty + in the separated part every element of the LIVING body is present-- in generation something is a
373d132 s animalcular semen-- but this-- -- the <nerve> LIVING nerve nursed in Mould.-- Lyells Elements. p. 2
374d133 s!) But the Cephalopoda depart more widely from LIVING forms.-- p 458 Upper Silurian, fishes oldest f
409e049 social   there could not be one body of animals, LIVING with certainty on other» hence not social inst
416e072 pecies <good observations.-- >, larger than any LIVING [not located] A Greyhound might be made «almos
419e085 na of Caspian.-- fishes fresh water kinds. (yet LIVING in the salt?.)-- very few animals of any kind-
423e097 acea to--- ? &c) without reducing the number of LIVING beings-- but there is the strongest possible t
440e145 rition, growth & reproduction» is common to all LIVING beings. vide Lamarck Vol II. p. 115. 4 four la
544m102 so pain gives fear of death. Mayo Philosophy of LIVING. p. 140-- Dreams good account of «thinks» are
549m121 estion, whether this happiness is the object of LIVING.-- or whether if we obey literally New Testame
550m126 ell applicable to free will. Mayo. Philosop of LIVING. p. 293. Animals "have notion of property" -- t
570n027 hange, so must idea of beauty.-- [Old Graecians LIVING amongst naked figures, & observing powers comm
599o006 ten show traces of origin.-- Mayo Philosophy of LIVING. p. 264. "Architecture is a fine amplification
604o018 e & other reasons we apply to God the notion of LIVING in lofty regions. 3 Infinity eternity. darknes
611o034 tal laws-- According to the individual forms of LIVING beings, matter is united in different modifica
611o034 lation to them-- [RHC] In the simplest forms of LIVING beings namely «one individual» vegetables, the
616o039 iousness memory &c. have the same relation to a LIVING body (especially the cerebral portions of it)
616o039 it might with equal propriety be said that the LIVING brain perceived, thought, remembered &c. Well
616o039 <expr> question above stated, because there are LIVING bodies without these faculties & indeed until
619o042 expect that acts of benevolence towards fellow <LIVING> creatures, or of kindness to wife [RHC] 2) an
623o050 nd, yet as they are acquired by social animals, LIVING under certain conditions, in this world, they
623o051 nimals of peculiar <kinds> «social feelings», & LIVING under certain conditions; by my theory they ha
636j56v same predicament as a group of bisexual animals LIVING on the borders of a country favourable to chan

Page ***(Key Word)**

185b056 Fernando Noronha Ophyressa bilineata (Gray) new <LIZA> species, belonging to true. American genus Wat
037r055 the bones of <two graniniverous> a herbivorous LIZARD.-- from the action of torrents. «marine» Torto
037r055 nts. «marine» Tortoise & other species of large LIZARD.-- There would probably be no other organic re
372d129 own to that part.-- claw added to crab, tail to LIZARD,-- healing of wound.-- reproductive faculty +
403e021 rs from Pellew to Eap.-- There is another great LIZARD. Kaluz. which is found at Pellew & Eap, but no
206b149 - NB As Illustrations are there many anomalous LIZARDS living; or of the tribe fish extinct. or of Pa
222b205 kinds the shark. lived in remotest epochs.-- ¿ LIZARDS of secondary period in same predicament. It is
232b248 > Experimentise on land shells in salt water & LIZARDS do.-- Ask Eyton to procure me some Get Hope to
257c057 ammalia in secondary-epocks, & developement of LIZARDS.-- As we have birds impressions in Red Sandsto
257c058 have birds impressions in Red Sandstone. great LIZARDS in do.-- <Wood> <Dicot wood> Coniferous wood i
471tf10 list of all the fossil Mamm. of Europe-- Large LIZARDS in Navigators. Williams. Narrative of Mission
641j29v other bird could catch mouse by night» Sailing LIZARDS. squirrels & Opossums «& fish»: flying lizards
641j29v lizards. squirrels & Opossums «& fish»: flying LIZARDS.--Mammalia. C. D.--
062r129 des a parallel case.-- Should urge that extinct LLAMA owed its death not to change of circumstances;
179b035 auses) does appear very probable:-- Mem: Horse, LLAMA. &c &c-- If we <suppose> «grant» similarity of
468t111 arieties of Cistus Speedwell to Rhododendron-- <LOASA> «Anchusa»-- speedwell Iris-- Azalea. Rhodendro
502q11v op exudes-- does not elm. does it «in» melon-- <LOASA> Anchusa «Campanula» &c & dead-nettle.- Lithos
608o026 ked man, like a sickly one(P)-- We cannot help LOATHING a diseased offensive object, so we view wicke
467t100 f part with ordinary divisions, & a few with one LOBE again divided «Have dried some»-- some with no
513q21. ower before stigmas expanded-- in reference to LOBELIA & Clarkia-- Peas time of impregnation.-- says
639j28r f> sun-fish, in mouth of swine & in stomach of LOBSTERS-- analogy in Flamingo & Duck. Ornithorhyncus
047r083 ave did not sea break first? I can imagine from LOCAL form of coast (as seen in swell) the undertow &
064r135 lcanic action & limited earthquakes & a goat but LOCAL elevations of the land in Europe-- Urge differe
100a055 ner p. 197. refers to salt as being produced by LOCAL heat, Ask Capt. Beaufort, whether, water flashi
234b256 this.-- Waterhouse says he is certain there are LOCAL varieties «of colour & size, but not for[m]» of
252c045 udy localities of isld.--«immense importance of LOCAL faunas foundation of all our knowledge especial
258c059 ary??.? Does not atavism relate to this law.?-- LOCAL varieties formed with extreme slowness, even wh
276c122 out organs of generation?! By profound study of LOCAL varieties laws of change-- whether beak (as it
281c137 ly the series is graeduated.-- Dr Beck doubt if LOCAL varieties should be remembered, therefore do no
337d020 st be most complicated, because they are partly LOCAL & then the local ones are taken to fresh countr
337d020 cated, because they are partly local & then the LOCAL ones are taken to fresh country & breed confine
337d020 ral breed, not one is picked out, & few even of LOCAL varieties approaches quite to wild local variet
337d020 ven of local varieties approaches quite to wild LOCAL variety.-- our Europaean varieties must be very
337d020 alian Greyhound is probably the effect of <sev> LOCAL variety many times changed together with some t
338d023 ty of transport in the species itself, & in the LOCAL circumstances of the two countries in times pre
350d053 ged by analogy»-- Consider all this NB. How can LOCAL species as at Galapagos., be distinguished from
359d088 ing powers. (as if they were a good species, or LOCAL variety, & not effect of breeding in & in like
360d091 Is there any mistake about Yarrell's law, is it LOCAL (not artificial variation) which impresses offs
410e050 n to the whole changes of country, & not to the LOCAL changes. = this could only be effected by sexes
430e118 ill for all.-- Varieties are made in two ways-- LOCAL varieties, when whole mass of species are subje
577n051 ll» does to reason.-- an inflamed membrane from LOCAL irritation to passion.-- Blushing is intimately
611o034 ouch & not irritability, which at least shows a LOCAL will, though perhaps not conscious sensation. [
612o035 & exosmic juices. arms of polypus, show either LOCAL or general will, & stomach likewise shows. [LH
188b069 species generally frequet slightly different LOCALITIES, so that they become useful to know what is
218b191 ecies from Ireland, England, Scotland & other LOCALITIES & each one will have a peculiar «constant» a
233b250 to distinguish came from closely neighbouring LOCALITIES Institute 1838. p 38. account of fossils of
235b275 can or Indian genus inhabit different kind of LOCALITIES.-- if so change The GRAND QUESTION Are there
251c040 urn after years to nest Vol II. p. 49. on the LOCALITIES of certain parrots habitations India & Afric
252c045 species.-- The Collector is directed to study LOCALITIES of isld.--«immense importance of local fauna
264c080 t many species when close come from different LOCALITIES, as my Furnaria-- some genus of yellow & br
289c160 do.-- all shows how nicely adapted species to LOCALITIES¶-- p. 390,, young «grebes» «ring ouzels» div
299c194 owslip, quite wild, but they affect different LOCALITIES.-- latter on banks & in damp parts.-- both p
306c214 lours., good instance of colours dependent on LOCALITIES.-- Hamilton will give an account in his T
179b033 Spain:--» Dr. Smith always urges the distinct LOCALITY or Metropolis of every species: believes in r
223b211 of A. becoming a good species well adapted to LOCALITY A. but it is instead a stunted & diseased for
299c195 e-- The two. former produced by difference of <LOCALITY>. station & varieties chiefly produced by cul
327c265 w Henslow has list of plants of Mauritius with LOCALITY in which each one is found. very good to see
375d134 er of species, from small changes in nature of LOCALITY. Even the energetic language of <Malthus> «De
415e066 cosmopolite, he would probably be confined in LOCALITY like Ornithorhyncus,: since being cosmopolite
417e076 ny land shells of the common species: from one LOCALITY, all left whorled.-- He kept two to see if th
464t065 Apteryx is not «quite» isolated in its present LOCALITY-- there have been at least other birds, with
093a034 t forces. but they are parts of one force, one LOCALLY relieving the other. -- Is the felspar glassy
021r001 soria same since commencement of world.-- [not LOCATED] La. billardiere mentions the floating marine
027r025 of Cuba for comparison (?) with St Pauls [not LOCATED] [not located] The frequency of shells in the
027r026 omparison (?) with St Pauls [not located] [not LOCATED] The frequency of shells in the Calc. Sandston
054r103 00 ft, how exact agreement with Coquimbo; [not LOCATED] [not located] Isld near coast of America not
054r104 ct agreement with Coquimbo; [not located] [not LOCATED] Isld near coast of America not reached. Juan.
108a081 cid gaz here.-- [top portion page excised, not LOCATED] Bull:. Soc: Geolog. Tome IX 1837-8. p. 24. ro
108a082 Well described [top portion page excised, not LOCATED] -- do-- [...] Subaqueous. removal, shown by t
201b127 ave produced species, have produced fe[w] [not LOCATED] The relation of Analogy of Maclay &c. appears
217b185 ered by blood horse & Carthorse two folds [not LOCATED] Mr Don gave me instances of one species of Au
230b237 untry when change has taken place, Nature [not LOCATED] Any change suddenly acquired is with difficul
234b253 intellectuality is only aim in this world [not LOCATED] T. Carlyle, saw with his own eyes. new gate.
234b257 r London.= Dr. Smith, he says, is deeply [not LOCATED] of <all> genera, «in all classes» are not a f
240c005 up by hand.-- These facts all account for [not LOCATED] Falkner Patagonia no description of wild anim
248c031 end, as the Law of hybridity, namely the [not LOCATED] animals unite, all the change that has been a
252c043 MUST BE STUDIED: genera founded in nature [not LOCATED] The systematic naturalists get clear indicati
266c089 be impossible to classify them.-- I would [not LOCATED] Musalman's of the Peninsula, are, generally s
268c097 yrannula can we worked out into. Milvulus [not LOCATED] element geographical distribution is.-- ¿Pela
274c117 rds» Mammalia.-- We have clear indication [not LOCATED] alone, but on all the general arguments-- Lam
295c179 y by beauty, analogy of man if so war not [not LOCATED] Erasmus says he has seen old Stallion tempted
297c187 my theory-- else my theory not applicable [not LOCATED] p. 428. Ouzel sometimes builds nest without d
312c229 er of vertebrae» new acquisition, we must [not LOCATED] Henry Thompson tells me best way to improve c
321c274 s No of Vol II. (read remainder) when out [not LOCATED] <Rays Wisdom>: Lisiansky's Voyage round wor
340d027 ule. ¿Are the hybrids similar inter se.-- [not LOCATED] the «4» Struthionidae, Mr Blyth remarked that
346d045 st like species.-- high active breedin[g] [not LOCATED] half breed liable to vary. I asked this in ma
358d077 ales & of age or castration on females.-- [not LOCATED] hen freely.-- here we have beautiful proof of
361d097 N. America.-- the teal which some authors [not LOCATED] September 13th The passion of the doe to the
366d109 eculiarities must be transmitted if their [not LOCATED] <The> every case common to many good species;
371d119 eak turned up like Avocette. here is what [not LOCATED] that it shall beget young different in colour
377d141 in & Jaguar of woods &c like ground birds [not LOCATED] :Hence, also structure not really fitted for
378d149 ental peculiarities. Wildness Reversion Q [not LOCATED] The present age is the one for large Cetacea,
397e001 d July 10th 1839.-- Selected. Dec 15 1856 [not LOCATED] Epidemics-- seem intimately related to famine
399e007 cal series: Look at whole Glacial period? [not LOCATED] Study introduction to Cuviers Regne Animal No
405e027 ndency determines the puppies to be so.-- [not LOCATED] Did man spread over world as early as Elephan
405e033 ving been long separated, or having never [not LOCATED] ARGUMENT REAL of antiquity of reasonable cosm
414e061 certainly requires progression, otherwise [not LOCATED] Are the feet of water-dogs at all more webbed
416e073 observations.-- >, larger than any living [not LOCATED] A Greyhound might be made «almost» without an
418e081 double sexual developement is spread over [not LOCATED] it utterly untold,-- what is added to the com
422e093 ecent shells of the <new> raised beaches» [not LOCATED] The enormous number of animals in the world o
435e131 x,-- this effect from age, what Mr Knight [not LOCATED] the stigma retains its power.-- R. Brown foun
449e171 not transportation> now in case of large [not LOCATED] Mr Greenough on his Map of the World, has. wr
521m003 potence of habit is shown about meals, no [not LOCATED] There is a case of Mr Anson. who told a story
530m047 habit of producing a train of thought.-- [not LOCATED] Fox believe cats discover birds nests & watch
535m065 btful one of Acanthosoma grisea described [not LOCATED] as first caused by will of Gods. «or God» see
553m133 ing for a swift dog to overtake an emu, & [not LOCATED] notion, are not effects of impressions long r
586n083 ise at it & feels no more inclined to ask [not LOCATED] If dislike, distaste. & disapproval. were not
591n095 ower, quite distinct from self. «or will» [not LOCATED] <& other cows--> Mr. Hamilton on vital laws (

Page ***(Key Word)***
595n119 horn-- association or imagination [blank] [not LOCATED] Ernest W. playing with Snow. when 2½ years ol
595n123 at wild beasts in Zoolog. Garden [blank] [not LOCATED] A child crying. frowning, pouting, «smiling»,
584n071 , progression.-- use of senses.-- knowledge of LOCATION ducks & turtles running to water,-- young cro
046r080 t (as in Peru 1746).--At great Lisbon Earthquake LOCH Lomond water oscillated between 2 & 3 ft. (as i
122a115 25. I saw metamorphic conglomerates on shore of LOCH Lochy very like those of Andes Speculate under
146g013 l Thursday On side of Hill South of upper end of LOCH Dochart buttresses of Alluvium or rather mass o
146g014 e of irregular terrace perhaps near 300 ft above LOCH.-- From this point could be followed up to neig
147g016 tersect alley above the 300 ft Alluvium <abo> by LOCH Dochart-- Rivers could not have deposited it. B
147g019 r 50 degree? Friday. Inverorum about 20 ft above LOCH Tulla 29.804 Temp 62 degree Air 60 degree Below
147g020 Tulla 29.804 Temp 62 degree Air 60 degree Below LOCH Tulla whole wide valley scattered with few very
148g022 Alluvium. At Mouth of Caledonian Canal opposite LOCH Leven two terraces perhaps upper one 100 ft & o
148g027 ome quite irregular very like rubbish at head of LOCH Dochart <Nea> Above Spean Bridge many flat terr
148g029 o there would be barrier-- recollect the case of LOCH <in> <below> «by» pass of Glencoe-- the erosion
152g048 buttresses, to upper edge of which they cut near LOCH Tring-- Tuesday Bridge of Roy Level of «bed of»
157g077 ion since «that shelf» formed Upper terrace near LOCH Spey <29.35161> 29.360? A 79 degree 75 degree?
157g078 egree? A little below Divortium on slope towards LOCH Spey 29.297 A 79.½ 29.316 divortium aquarum «ab
158g080 ace 15 ft divortium my measurements here side of LOCH Spey Forms terrace about 60 feet above Loch tra
158g080 e of Loch Spey Forms terrace about 60 feet above LOCH trace of this terrace «on <will> Granite ridge
160g089 connecting Glen Bought & Glen Tarf a perfect old LOCH, making <several> two divortiums aquarum, viz t
160g091 r Cairn leet more Haberclador Isthmus broad flat LOCH {P} XX Barom 28.92 A 75 Air 70 degree? Isthmus
161g098 continuous & it was 29.200 minus .008 ---- .192 LOCH Lochy 4 ft above water Barom: 30.372 A 76 degre
162g100 on the steep & rocky gully of last stream Friday LOCH Lochy near Letter Finlay Barom 30.267, A 68 Air
162g102 it is impossible to see my new shelf, from road: LOCH Ness 30.140. A 66 degree 30.095 .0458 or 6 diff
162g102 e 30.095 .0458 or 6 difference between bedrock & LOCH Ness <30.100> <Donald Macphee> Saturday Morning
162g105 & inwards-- deposited when water stood at higher LOCH Keeper tells me, that Loch Lochy is 8 ft below
162g105 water stood at higher Loch Keeper tells me, that LOCH Lochy is 8 ft below Loch Oich wh is 92 ft above
162g105 h Keeper tells me, that Loch Lochy is 8 ft below LOCH Oich wh is 92 ft above sea-- Loch Ness 40 ft ab
162g105 is 8 ft below Loch Oich wh is 92 ft above sea-- LOCH Ness 40 ft above do. When cutting bank where Lo
164g123 29.958 - 1.17 28.788 + 28.8 30.372 29.200 1.172 LOCH Oich 92 each Loch 8 ft. The Metamorphic conglom
164g123 788 + 28.8 30.372 29.200 1.172 Loch Oich 92 each LOCH 8 ft. The Metamorphic conglomerates near Loch L
165g124 ch Loch 8 ft. The Metamorphic conglomerates near LOCH Lochy would be well worth examining-- Inverness
122a115 saw metamorphic conglomerates on shore of Loch LOCHY very like those of Andes Speculate under head o
161g098 nuous & it was 29.200 minus .008 ---- .192 Loch LOCHY 4 ft above water Barom: 30.372 A 76 degree 75 d
161g098 with the new shelf flows into canal between L. LOCHY & Oich. is a brook on the Lochy side of it-- th
162g099 anal between L. Lochy & Oich. is a brook on the LOCHY side of it-- the terraces of which, last measur
162g100 steep & rocky gully of last stream Friday Loch LOCHY near Letter Finlay Barom 30.267, A 68 Air 65 de
162g105 stood at higher Loch Keeper tells me, that Loch LOCHY is 8 ft below Loch Oich wh is 92 ft above sea--
164g122 [...] shells from [...] Wenlock Edge [blank] L. LOCHY 12 ft 96 L. Oich 12 84 {P} 29.958 - 1.17 28.788
165g124 h 8 ft. The Metamorphic conglomerates near Loch LOCHY would be well worth examining-- Inverness & wat
323c269 idgeons 30th Lives of Hayd & Mozart «Apri 25th LOCKARTS life of Napoleon.» April 5d Dr. Edwards of <t
551m129 the ourang. take up a stone & pound the earth. LOCKARTS life of W. Scott Vol VII p. 35 "as ideas come
539m084 n <will> would do more towards metaphysics than LOCKE A dog whines, & so does man.-- dogs laughs for
610o033 division amongst metaphysicians-- the school of LOCKE, Bentham, & Hartley, &. the school of Kant. to
163g106 ch Ness 40 ft above do. When cutting bank where LOCKS now are (32 ft rise) they found alternating lay
350d056 thus senses, especially sight connected with LOCOMOTION.-- «Mem. Dr. Blackwell (Abercrmbies) compari
416e071 true hermaphroditism is a consequence of non-LOCOMOTION-- (contradicted by Plants). & as there are n
576n048 fication of bodily structure «(connected with LOCOMOTION.)» «no, for plants have instincts» «either»
589n092 p. 90.-- The relation of reason to organs of LOCOMOTION-- or that our faculties have been given us t
423e118 lker attributes this to effect of male sex on LOCOMOTIVE system» I am bound to insist honestly that t
371d118 to meet with contingency.-- Sept. 23rd. Saw in LODDIGES Garden. 1279 varieties of roses!!! proof of c
371d128 acquired by parent, being handed down? Are not LODDIGES 1279 roses kept in same soil. same atmosphere
293c174 Wax of Ear, bitter perhaps to prevent insects LODGING there. Now these exquisite adaptations can har
147g016 uld not have deposited it. Barrier of lake very LOFTY, & no trace of it; to the Sea more probable I d
275c119 c spirit in science--. the highest endowment of LOFTY genius Using geograph distribution of animals,
296c184 ing oldest. & having undergone changes «no near LOFTY country»?? p. 475 NB. This bears on fossils of
318c255 which seems common in Rocky Mountains & on one LOFTY isolated spot on the Alleghanies to which it mi
429e115 Hort. Transact Vol I. M. Ramond. p. 19. do says LOFTY Alpine plants of & Pyrennees agree with those o
604o018 reasons we apply to God the notion of living in LOFTY regions. 3 Infinity eternity. darkness, power.
029r033 Pernambuco. EARTHQUAKE AT SEA.--Extract from the LOG-book of the James Cruikshank, Captain John Young
081rIBC ls? Bellinghausen in 1819 Kotzebue 1816 Constant LOG always additive to convert French Toise into Eng
233b251 Continents-- Ostriches. Dodo. Apteryx Penguin-- LOGGER-headed Duck-- Large proportion of Water & smal
294c175 the first one that bred one was diseased in its LOINS & all were so afterwards, (forgets authority).-
079r177 de l[']industrie humaine, comme ils traitent les LOIX de la nature & ses phenomenes."-- Ulloa's Voyag
046r080 in Peru 1746).--At great Lisbon Earthquake Loch LOMOND water oscillated between 2 & 3 ft. (as in Chil
029r033 tain John Young, on her voyage from Demerara to LONDON:-- "Feb. 12, 1835. At 10h. 15m. a severe shock
234b256 mon at South end, but <rare> «absent from» near LONDON. = Dr. Smith, he says, is deeply [not located]
423e098 that the reason people send for their seeds to LONDON is that people in the southern Counties have w
429e115 now that plant can be cultivated with ease near LONDON.-- what makes the line, as trees in Beagle Cha
534m062 8. Transactions of the Entomological Society of LONDON Vol. I. p. 106. Col. Sykes on Formica indefess
022r008 yage to New Holland P 127.--Caught a shark 11 ft LONG. "Its maw was like a leathern sack, very thick
023r009 e pluckt a great many teeth, 2 of them, 8 inches LONG, & as big as a mans thumb, the rest not above h
023r009 big as a mans thumb, the rest not above half so LONG; The maw was full of jelly which stank extreaml
033r043 hytic.--Daubeny P. 171. Vol I. Humboldt There is LONG discussion on Pumice «& Obsidian:» in the I Vol
033r043 n Daubeny. P. 349 Admirable little table showing LONG PERIODS of great violence volcanic. from Humbol
036r051 which occurred in a sandy place.--(the sound was LONG & prolonged). NB, Is it generally known. the ac
040r064 n oscillations in the upheaval of Andes.--but as LONG as all below water no evidence--The depth of sh
047r083 injected mass of fluid rock In Patagonia plains. LONG periods of rest & vice versâ more likely to be
052r097 ault or rather many faults.-- Necessary form; as LONG as coast line fixed.--[Fig. 5] {P} * Slope nece
054r105 Terra Firma & the extent of a league & a half a LONG the coast. <"> The rocks in the most inland par
058r117 e for Volcanos part of same phenomena lasting so LONG.-- The great movements (not mere patches as in
062r129 Ascension. Keeling: at sea so commonly seen. at LONG distances; generally first arrives:-- New Zeala
065r139 y> one lichen. only production. a body which had LONG been buried, <see> from rotten state of coffin
065r139 from rotten state of coffin «buried in a mound» LONG consigned to the earth. yet body had scarcely u
068r146 trifeling injections.--Old vents would keep open LONG after emersion, but improbably so long, that to
068r146 keep open long after emersion, but improbably so LONG, that to be surrounded by continent.-- change o
105a069 tends to finite power) whereas sea. on coast, as LONG as exposed to waves of sea, cutting power incre
117a102 s E of Staten land. bringing up pebbles 2 inches LONG?-- L'Institut. 1838 p. 151. Formations of Payta
124a117 hat Siberia must have been in same condition for LONG period Subsidence in Demarara p. 131 (B.) Wrong
129a132 st curious account of great subsidence «20 miles LONG I in with.» which must have been from an axis,
152g048 icable to Glen Roy Lake, must have remained very LONG at 4th shelf from size of buttresses, to upper
160g088 marine, 400 or more feet above station! There is LONG straight isthmus connecting E & W connecting Gl
160g090 75 degree? Boulder, much covered by turf 2ft. 8- LONG of syenite with pinkish felspar;-- whole hill d
162g103 extreme right arm of River Tarf <it> Has a very LONG, flat divatium aquarum with, left of Bright.--
172b007 arate islands, ought to become different if kept LONG enough.-- «apart, with slightly differen circum
173b012 sted.-- or they may represent some large country LONG separated.-- On this idea of propagation of spe
182b048 rought from Abyssinia, & others dark brown, with LONG clotted hair resembling that of goats." Progres
190b077 undance of species of few genera or families.-- (LONG separated.-- Proteaceae & other forms (?) being
192b088 nefied slate, the father of all Mammalia in ages LONG gone past.. & still more so «known» with fishes
193b093 raph Journal. Vol I p. 174. says from Swan river LONG South coast all the remarkable Australia genera
202b133 ould we get them to associate together" There is LONG rigmarole article by S Hilaire on wonder of fin
206b147 may conclude that there will be a period though LONG distant, when of the present men (of all races)
207b152 not well fixed.-- Peron thinks Van Diemen's land LONG separated from Hobart Town-- {from difference o
211b163 ish.-- Progeny of Manks cats without tails: some LONG & some short: therefore like dogs.-- Ogleby say
212b165 fox.-- a sort of internal bark. would remain for LONG time together in tub of water with only nose pr
213b170 n.-- Men makes angels-- Those species which have LONG remained are those ¿Lyell?, which have wide ran
213b171 sh approaching to reptiles at Silurian age-- How LONG back have insects been known? As Gould remarked
229b233 rue molluscs coupling.-- Dr. Smith's Information LONG Horned (very) aboriginal at Cape: crossed with
232b246 but physical changes (we assume like weather on LONG average tolerably> uniform).-- Comparing fossil

Page ***(Key Word)***
235b262 cases, where «domesticated» animals separated. & LONG interbred <p> having great tendency to vary? Is
235b274 umps.-- ¿To be solved if horses sent to India. & LONG bred in & no new ones introduced would not chan
239c001 resuming those varieties to be oldest which have LONG been known in any country; he states that Esqui
240c014 roussa, (when young very like the Siam race with LONG nozzle & few hairs) inhabits Celebes & few of t
246c025 ew Zealand of large size, resemble, chien-loup.--LONG, black & white, ears short & straight-- do not
252c045 e ∵ because time required too separate isld very LONG» America & India deer.= Africa not.-- Africa Ca
263c075 who just takes up & lay down the subject without LONG meditation-- His best chance is to have profoun
267c093 ought from Abyssinia; the others dark brown with LONG clotted hair resembling that of goats" Geograp.
267c094 «30 or 40 together» colour reddish <brown>. ears LONG.-- like bull terrier-- Indian secured one, as t
271c106 ted by Man.-- effect of external contingencies & LONG bred in-- Mem, <an> a statement in Mr Wynne's b
273c110 has small band & others large, then he says from LONG experience, you may be almost sure, that there
274c113 uit.-- The Rasorial type is wonderfully shown in LONG legged cuckoos with claw like lark, (one in Aus
274c114 re).-- even one kingfisher-- Gould has seen with LONG tarsi.-- «Ground woodpecker» Secretary bird.--
279c133 mportant, showing effects of peculiarities being LONG in blood.-- ++ thinks difficulty in crossing ra
280c136 This is right way of viewing it.-- Variety when LONG in blood, gets stronger & stronger, so that tho
281c138 l cock birds in Gallinaceous having tendency to LON[G] or peculiar tails strange ¿¿/Genus only natura
283c144 ant circles.-- explained by such not having been LONG in blood?-- My theory agrees with unequal dista
283c145 of Orpheus-- one of which has very short legs & LONG tail «short much curved beak.--», other very lo
283c145 ong tail «short much curved beak.--», other very LONG beak, with short., let these only have progeny
284c147 tions of desert, open ocean, &c this probably on LONG average, equal quantity, 2d on relations of hea
284c147 ion of stations & diversity--.-- this perhaps on LONG average equal.-- The Cocos do Mar on the Mahé i
292c171 «to be habitual» or of great importance to cause LONG memory.-- structure is only gained slowly.-- th
293c172 t yet be transmitted.= Memory springing up after LONG intervals of forgetfulness.-- after sleep, «str
304c207 ould not then be told whether not descended from LONG way back.-- aberrant forms produced where many
306c212 ream which would last were the stream 1000 miles LONG.-- a monkey. (Baboon) at Z. Gardens upon being
306c212 fferently from a dog.-- more like man. continued LONG in a passion & looked out for him to come again
306c214 cannot be transported from their country.-- the LONG-haired cats are supposed to come from there.--
306c214 come also from Asia Minor.-- tail like setters. LONG ears-- colours vary, but form constant.-- The f
308c217 ldren of one father.-- yet differences carried a LONG way. --Case of Habit I kept my tea in right han
308c218 s every where & then circumference no where-- as LONG as this is so-- !!Metaphysics!!! Mrs Somerville
311c225 sult p. 326 wild ass extending over 90 degree of LONG. & Col. Sykes alludes to some other case of 180
312c227 ny families, «+very often in number 5» will have LONG tail.-- in raptorial birds, & tigers & sharks,
312c232 n» produced in shortest possible time. Mr Willis LONG eared little dogs, I am told, go to heat, take
312c233 fferent litters or of father & child are thought LONG breeding in. Must not trust him Hope says that
313c235 ation, therefore sexual passion must arise after LONG interval very good case.-- habit is awakened by
314c237 tube consisting of pistils & stamens united into LONG organ, moved on being touched, so as to protect
333d005 t. (tail cut off in progeny peculiar) limbs very LONG, eyes very large, very fierce to dogs «otherwis
333d007 e that <alw> «no» domesticated ones have been so LONG as wild one under present form.-- Fox has seen
334d013 sed the neck is not intermediate in its peculiar LONG neck, but much nearer to common goose.-- What h
335d013 eck, but much nearer to common goose.-- What has LONG been in blood, will remain in blood.-- --conver
335d015 according law, that in proportion as things are LONG in blood so will they remain, a mule «being new
335d015 «Hybrids of» Varieties is different because not LONG in blood.= The case of union of perfect animals
336d018 e easily inherited.-- «but if change be in blood LONG, it becomes part of animal &» by a succession o
341d032 oice often in the night, like peacock.-- tail as LONG as Pea hen.-- about intermediate.-- (In Zoolog
343d037 at since the time of the Silurian, he has made a LONG succession of vile Molluscous animals-- How ben
343d039 Negro probably a distinct species-- We know how LONG a Mammal may go on as one species from Egyptian
355d066 rom several stock, then species are fertile; as LONG as opponents will «are» not «able to» tie thems
356d071 e observed not what comparative difficulties (as LONG as not overwhelming) What comparative solutions
357d074 the <p> mechanism by which seeds are adapted for LONG transportation, seems «?» to imply knowledge of
358d086 e silver.-- But the hen hybrid of this bird, has LONG tail figure, & some degree of whiteness like a
359d086 fference.-- it may be so, but this assumption as LONG as animals are healthy which is often the case,
360d091 icking on them, but never in an animal, that had LONG been in confinement-- is this effect of climate
360d092 ference <of any kind>, if very firmly fixed from LONG time, made no difference what its kind was.-- b
360d094 - Or if one species left its type in having very LONG legs, & another in having very long tail, & oth
360d094 having very long legs, & another in having very LONG tail, & other in having very short tail.-- I ca
363d101 Study horns of wild cattle.-- plumage of fowls-- LONG ears of rabbits. & long fur.-- feathers on legs
363d101 le.-- plumage of fowls-- long ears of rabbits. & LONG fur.-- feathers on legs of Ptarmigan & in Banta
366d108 ic qualities were all deeply imbued in them from LONG permanence, so that all their peculiarities mus
366d111 ation of species-- Will it against genera.-- How LONG will the wretched inhabitants of NW. Australia,
371d128 will <deteriorated> continue of same variety as LONG as life lasts, yet they cannot transmit through
378d148 ecies old: Hence <young> Kingfisher & pies, have LONG had their present plumage.-- How is this in Pid
391d174 rom frequency of this tendency all mammalia must LONG have so existed.» with double union.-- At prese
399e009 ame colour as a Hare, but paler & buffer.-- with LONG ears-- & longer hind legs??? so that I was almo
401e016 ssage from many varieties, & probably would take LONG before all the stain would be got out of it.--
405e032 ferent forms it is either effects of having been LONG separated, or having never [not located] ARGUME
412e057 ructure is capable of innumerable variations, as LONG as each shall be perfectly adapted to circumsta
420e088 Europaeans settled there L'Institut do. p. 419, «LONG» account of Hyaenodon, a fossil dog-- leading t
438e142 ll much varied breeds both plants & animals have LONG been subjected to domestication.-- the constitu
445e159 ween its legs.-- -- strangely consider existing «LONG-organized» forms as parent forms of existing hi
453e182 those from Rhine to the Caribs.-- Vol II p. 650. LONG attested account of fall of fish in India.-- Wi
458t001 ost strange, & shows as in shells some forms are LONG preserved.-- vol VI. p. 539. Dr Cantor's accoun
458t002 -- Horses in Lao Choo so small, that person with LONG legs can hardly ride on them. Mr Miller-- in Zo
460t019 measures.-- Shells in Cambrian & Crust show how LONG since present forms existed, but if it be asked
461t037 ll Syngnathus Ch 6 I presume, from my theory, as LONG as any structure can be handed down without bei
466t094 have no relation. In Monk's Hood, a bee entering LONG nectary, would «necessary» cross directly over
468t112 stigma after bee had brushed over the anthers of LONG stamens {P} as stamens grow old «& shed some po
470t178 lus in numbers =what insect can get honey out of LONG, curved nectar of Butterfly Orchis & Listera? B
477z003 idus ailes peuvent avoir <quatre> cinq lignes de LONG et volent. p. 208 Fleas only appear in winter i
497q006 ich will show, how seeds are transported, or how LONG they remain dormant. if kinds come up, not foun
509q017 enheres Scottish Colliers, when men & women have LONG worked, whether children, who have not worked h
512q019 - Sow cast-up-balls of Hawks or even owls.-- How LONG do seeds remain in stomach of birds-- Mem: how
520m001 ing after their parents, when the latter died so LONG before, that it is extremely improbably that th
527m034 thought, from weakness of my stomach I observe a LONG castle in the air, is as hard work (abstracting
542m095 & streching & yawning can be explained from too LONG rest of muscles.-- evidently habitual when tran
552m132 hose actions which have been found necessary for LONG generation, (as friendship to fellow animals in
552m132 pleasure, & not as Paleys rule is those that on LONG run will do good.-- alter will in all cases to
553m135 located] notion, are not effects of impressions LONG repeated, without the powers of the mind being
559m154 Creator as governing by laws is probably that as LONG as we consider each object an act of separate c
559m156 every day.-- naturally close at that time after LONG period.-- My Father about double consciousness.
565n006 ise , the face clearly passes into smile-- laugh LONG prior to talking, hence one can help speaking,
581n064 lty in expressing their want, pleasure, or pains LONG before they can speak-- or understand-- thinks
584n072 ment,-- carrier pidgeons proverbially carried to LONG distance in dark "it is inspiration."-- this is
584n072 called instinctive: the facts of memory of roads LONG after once visited by horse & dogs. (even blind
594n112 stinct arise?? «it would appear that an instinct LONG remains, if no steps are taken to eradicate it.
606o20v Although taste must necessarily be acquired by a LONG series of experiments & observations. & yet, li
610o031 ived a letter from Mr Roberts-- «a person he had LONG known & directed many letters to»-- could not <
611o034 s act definitely (<like> «as» chemical laws,) as LONG as certain contingencies are present, (continge
611o034 gencies as heat light &c). [LHC] This is true as LONG as movement of sensitive plant can be shewn to
611o034 unites matter into certain form; invariable, as LONG as not modified by external accidents, & in suc
614o037 nd on its length.: Many animals (as horses) very LONG & good memories-- but on its multiplicity & the
621o048 parents think will be good for the child on the LONG run, & for themselves & others, (as the parents
625o50v ssions are too strong for our instincts. to gain LONG-lived good, ie happiness-- yet this system not
633j53v o the transmission of abortive organs will be as LONG as they are not detrimental.-- p. 285 the seed-
633j54r wer great tendency to break off p. 292. Mac. has LONG rigmarole about plants being created to arrest
636j56v Plants would be subject to extreme variation as LONG as crossing with other varieties was prevented
637j58r nsider these barren Virgins p. 235. talks of the LONG spinous processes in Giraffe &c, as adaptations
637j58r inous processes in Giraffe &c, as adaptations to LONG necks-- why they may as well say, «long» neck i
637j58r tions to long necks-- why they may as well say, «LONG» neck is adapted to long necks.-- p. 236. Marsu

637j58r they may as well say, «long» neck is adapted to LONG necks.-- p. 236. Marsupial bones especial adapt
639j28v the bills of the Grallae <are> «have been made» LONG «(as adapted to)» because their food lies deep.
639j28v I say it is «as» simple consequence they become LONG. not at once, but by steps. of which we have ma
639j28v at one out of every hundred litters is born with LONG legs» & in the Malthusian rush for life, only t
639j28v e to breed, if circumstances determine that, the LONG legged one shall rather oftener than any other
639j28v ny other one. survive. in ten thousand years the LONG legged race will get the upper hand. though con
640j167 rom death of its destroyer, or other cause, the LONG legged race would prevail, even if have afforde
640j167 total number of dogs. would diminish, whilst the LONG legged variety would prevail.-- Not separately:
640j167 ds. should possess many peculiar species. for as LONG as physical change is in progress or is, presen
641j29r islands are favourable,) because it must take so LONG to change species-- yet this is contradicted by
021r005 nodule, constitution «same as» of slate same.--LONGER axis in line of Cleavage. laminae fold round t
056r110 f heat on inner wall, hence resists degradation LONGER than outer parts.-- The common occurrence of a
065r139 Desire applicable to Craters of Elevation.--The LONGER diameter of Deception Isl is six Geographical
099a048 overed with oil should shrink. film parallel to LONGER axis. But if great depth NB. Prof <Henslow> Se
150g036 3 2 Terrace 3 Alluvials 3a 3a 2 Mass 3 2 rather LONGER than 3a Sunday In Glen Collarig, when water up
170b001 ae. &c &c.-- The ordinary kind <the> which is a LONGER process, the new individual passing throug sev
171b002 -- Yet Eunuchs nor «cut» Stallions nor nuns are LONGER lived Why is life short, Why such high object
191b080 ts between East Indian islets-- (+ + +. Ireland LONGER separated., Hare of two countries different.--
246c026 e of Europe, but beak rather sharper., & rather LONGER in proportion, colour slightly different. Who
336d017 » without it be small & slowly obtained NB. The LONGER a thing is in the blood, the more persistent.-
378d148 ". instance of old Species transmitting so much LONGER its Mental peculiarities. Wildness Reversion Q
399e009 Hare, but paler & buffer.-- with long ears-- & LONGER hind legs??? so that I was almost doubtful whi
469t135 e on 14 days. (except some wet ones/ & wd go on LONGER-- Woodfords Marrow fat, Early frame, Groom's D
480z007 considers> states that young birds of prey have LONGER tails than old ones-- in America & sexes not o
525m025 number 3) had all short tails; but one a little LONGER than rest «they all died»:-- she had kittens b
582n067 y view marrying late, will make average of life LONGER.-- for short-lived constitutions will then be
174b015 dom, when our planet first cooled.-- Countries LONGEST separated greatest differences-- if separated
221b200 ike present species., p 8. ¿Are mundine forms, LONGEST persistent?? do.-- The most perfect Plants Com
248c033 ermal [for]m, so will the internal parts be of LONGEST [consta]nt & therefore most permanent Owen [re
255c051 -- Feathers on, Apterix because we may suppose LONGEST part of structure.-- shape of wings have alter
364d103 ciers match their birds to see which will sing LONGEST, & they in evident rivalry sing against each o
233b252 <M> Vertebrate. But how does this agree with LONGEVITY of species in Molluscs!!! When we talk of hig
185b056 ken for) Carabidae, Crysomela, Scarabadae, & LONGICORNES.-- Again taking a subdivision of Heteromera
473s03r & Mylodon» in post pliocene strata! Mastodon LONGIROSTRIS in miocene like in Europe-- Cuvier never fo
029r033 calm at the time. Latitude 8 deg. 47 min. N: LONGITUDE 61 deg. 22 min. W. mid. calm and clear. Caerm
316c245 Hence latitude is more important element than LONGITUDE.-- But in land & F W shells there is more con
430e119 concentric striae, all the lower part rayed LONGITUDINALLY (give woodcut) like I. sulcatus.-- Both sp
293c175 .-- There is breed of tailless cats near Bath. LONSDALE do. says Sheep could not live for some time a
294c176 all were so afterwards, (forgets authority).-- LONSDALE is ready to admit, permanent small alteration
294c177 & analogy will necessarily explain the rest,-- LONSDALE says he has seen in old Book last Bear in Eng
294c177 ion of types ¿different species;-- Horse-- &c <LONSDALE says. that first shee> State broadly scarcely
399e006 e very thick & yet have same fossils. does not LONSDALE know some case of change in vertical series:
430e119 n to other parts of the world-- March 16th. Mr LONSDALE showed me two specimens of an Inoceranus from
430e120 e.-- it is unnamed this intermediate one.-- Mr LONSDALE evidently inclines to think it Hybrid.!!! Ask
430e120 nclines to think it Hybrid.!!! Ask Woodward Mr LONSDALE says Trigonia costata & elongata thougt consi
431e121 nd together on coast of France.-- L. doubts.-- LONSDALE thinks Ammonites would afford instance of suc
542m097 errate-- animals communicate each other.-- LONSDALE'S story of Snails, Fox of cows, & many of inse
029r034 n. W. mid. calm and clear. Caermarsten Journal I LOOK at the cessation northwards of the Coal in Chil
035r048 illated equably.-- These facts become easy if we LOOK at the action as a deep & extensive movement of
046r078 s.--neglecting the first production of Trachyte. LOOK at Sulphur. salt. lime, are spread over «whole»
046r078 c veins likewise must separate ingredients if we LOOK to a constant revolution.--Are we to consider t
049r088 a??> Porphyry at Valparaiso; Epidote -- Must we LOOK at regular greenstone cones at S. T. del Fuego
052r099 o. C. de Verds Quartz pebbles in the Cordilleras LOOK as if some peaks elevated.-- Greywacke. as a ge
055r107 as become of the Basalt. Gone into fine sediment LOOK at St Helena!!-- There are some arguments which
058r117 iders these earthquakes travel in order.-- If we LOOK at Elevations as constantly going on we shall s
070r153 nt. not insensible change.-- Yet one is urged to LOOK to common parent? why should two of the most cl
100a054 deposit resulting from disintegrated granite!!! LOOK at gneiss of Rio Concretions in Pumice bed at A
102a061 water yields substances from impact, «it» would LOOK like it. Are greenstone dike in Granite residua
112a089 her trap.-- It is in the mountain masses we must LOOK for that.-- how few isolated volcanos there are
112a089 anos there are. where one alone has been formed-- LOOK at the now active volcanos & see what high they
113a090 surface formerly much higher, «so» that we must LOOK at the upper four hundred feet of strata having
120a109 of salt.?.--??? Footsteps in New Red Sandstone. LOOK as if a surface deposit.-- The case of the shin
133a140 , within say the Tertiary period. one is led, to LOOK at globe as resting on film of molten rock.-- V
147g017 trace of it; to the Sea more probable I did not LOOK carefully for Marine remains-- Some of the hill
175b016 ause time short & no great change has happened I LOOK at two ostriches as strong argument of possibil
175b018 not help..-- becoming more complicated,; & if we LOOK to first origin there must be progress. if we s
175b020 ent terebratula, but Megatherium nothing. We may LOOK at Megatheria, armadillos & sloths as all offsp
178b031 British» Shrews diff: species from the continent LOOK over. Bell, & L. Jenyns. Falkland rabbit may pe
181b040 dly increasing.-- If we thus go very far back to LOOK to the source of the Mammalian type of organiza
186b060 the <type> great quadrupeds to Siberia; we must LOOK to type of organization.-- extinct species of s
190b077 ?) being common to Southern hemisphere, does not LOOK as if S. africa peopled from N. Africa An origi
222b203 t at returning to parent stock.-- I think we may LOOK at it so--?? It holds good even with trifling d
225b219 ptations of species by generation explained? NB. LOOK over Bell on Quadrupeds for some facts.-- about
227b225 e inequality of numbers of species in divisions. LOOK at Articulata!!!--! It leads to Nature of physi
227b226 ss into each other: now on this view no one need LOOK for intermediate structures <between> say in br
227b227 test of highest organization intelligible.--may LOOK to first germ-- --led to comprehend true affini
228b232 nt of all is superadded, animals not got it, not LOOK forward» if we choose to let conjecture run wil
232b247 geration".-- <The Phenomena of the S. Hemisphere LOOK as if heat gained> Experimentise on land shells
263c076 be granted!!) & whole fabric totters & falls.-- LOOK abroad, study gradation. study unity of type--
264c079 , sulkiness, & very actions of despair; «Let him LOOK at savage, roasting his parent, naked, artless
274c116 & northern Asia & Northern America.-- may we not LOOK to these Northern regions as the receptacles of
286c154 our arrogance, to raise on the same shelf-- to (LOOK at common ancestor, (scarcely!)] conceivable in
289c161 athers & scales. passing into each other let him LOOK at wings & orbits of penguin & then he will cea
292c170 to see absurdity of Quinary arrangement let him LOOK at abstract of Swainson on Classification "Let
303c205 e are no genera. if Mammalia are adduced. say oh LOOK to your fossils, how few if extinction had gone, wi
304c206 r types. with different degrees of closeness. -- LOOK how close birds! look at Mammals: how wide.-- t
304c206 t degrees of closeness. -- look how close birds! LOOK at Mammals: how wide.-- therefore birds younger
354d063 hen two Varieties of dogs cross, Erasmus says it LOOK lik[e] Institut. 1837. p. 351. Paradoxurus Phil
376d137 , & if woman not afraid clasp them round waist & LOOK in their faces & Mak the st. st noise.-- The Ce
392d180 alian) dogs with common dogs--> Ask my father to LOOK out for instances of Avitism Examine English we
399e006 ale know some case of change in vertical series:> LOOK at whole Glacial period? [not located] Study in
403e023 se it has been so pronounced ex cathedrā. let us LOOK at facts. considering few domestic animals few.
429e113 t as we know how many plants have been produced (LOOK at the Dahlias. we may infer it in animals.)--
431e122 ce where any species is most common, we need not LOOK for change, because its number show it is perfe
431e122 re, that change may be anticipated, & this would LOOK like fresh Creation. the gardener separates a p
465t079 broken, rivers entering.-- & yet no shells-- now LOOK at Scotland-- coasts of Chile, excepting Concep
499q09v ly minute or abundant? do they seed plentifully? LOOK for isolated females.-- Also any plants which a
505q014 aller & more vicious. Will try to get me some to LOOK at:-- Was once offered a hive. of these small B
525m023 ning home.-- in these cases their actions do not LOOK like fear, but shame.-- I cannot remember insta
528m037 re of perspective. which cannot be doubted if we LOOK at buildings, even ugly ones.-- the pleasure fr
530m044 out big noses & name Corbet, perhaps nonsense.-- LOOK to it My father has somewhere heard (Hunter?) t
532m054 adily taking same movements, senses being on the LOOK out, & the conveying means from the senses to t
540m087 - Same cause as colour & shape & ideosyncracy.-- LOOK at the Indian in slavery & look at the Negro--
540m087 ideosyncracy.-- Look at the Indian in slavery & LOOK at the Negro-- look at them both savage-- look
540m087 k at the Indian in slavery & look at the Negro-- LOOK at them both savage-- look at them both semi-ci
540m087 & look at the Negro-- look at them both savage-- LOOK at them both semi-civilized-- Perhaps one cause
541m093 im, but he will find it far more difficult to to LOOK tranquil.-- He may despise a man & say nothing,
545m105 skin of forehead over eyes, at every emotion & <LOOK> «turn» of the head. I could not perceive «any»

Page
(Key Word)

```
545m107 urious <like> «remember» the expostulatory angry LOOK of black spider monkey when touched, also anoth
550m125 cussion-- by saying what is Happiness?-- When we LOOK back to happy days, are they not those of which
554m138 Sebe??) he has seen place its head downwards to LOOK up womens petticoats-- just like Jenny with Tom
554m139 ead from a visitor, & before eating «everytime», LOOK up to «keeper» see whether, this was permitted
555m143 position from others, slow cautious, angry cross LOOK, followed by protrusion of lips, in which respe
558m151 man & Brutes) no one can doubt this connexion.-- LOOK at faces of people in different trades &c &c
558m153 ly all will exclaim, your arguments are good but LOOK at the immense difference. between man,-- forge
565m007 rect & stiff, with head up.-- Why does suspicion LOOK obliquely.-- who can analyse suspicion-- yet wh
565m007 n analyse suspicion-- yet who does not recognise LOOK of suspicion, even child will do so.-- Contempt
565m007 of suspicion, even child will do so.-- Contempt LOOK obliquely so does dog. when a little one attack
571m030 or rather altered. The Reason why New Buildings LOOK ugly is because there is some connection betwee
576m047 dark ages.-- «effects of external circumstances» LOOK at Spain now.-- man's intellect might well dete
583m070 by means of ordinary senses & muscles, we cannot LOOK at him, as machine to make cell of certain form
586m078 h respect to pidgeons, is that they do know from LOOK of Heavens, points of compass, & they do know w
593m107 or looks. but it does not know its own smell or LOOK, & therefore there must be some instinctive fee
593m107 be said that young animal learns parent smell & LOOK so by association receives pleasure. This [blan
605o18v ime. adds not a little to the effect: as when we LOOK at the vast ocean from any height.-- 6 That the
609o030 ing of <our ha> of the rule of happiness we must LOOK far forward «& to the general action» -- certai
609o030 r our good far back.-- (much further than we can LOOK forward: however our <[...]> rule may sometimes b
616o038 his imperfect memory may become perfect & we may LOOK back to definite action or to our conscious sel
638j58v - Now we know what instinct is-- consider this I LOOK at every adaptation, as the surviving one of te
037r056 obably be no other organic remains.-- On Pampas LOOKED in vain for a pebble of any sort; not one was
170bIFC useful pages cut out Dec. 7th. /1856/ (& again LOOKED through April 21 1873) p. 26 30 41 46 50 54 56
239cIFC All good References selected Dec. 13 1856 Also LOOKED through April 23. 1873 Books About amount of d
258c061 st XX. Hence relation of analogy may chiefly be LOOKED for in the aberrant groups.-- It is having wal
259c065 is reduced by atavism) Even a deformity may be LOOKED at as the best attempt of nature under certain
290c163 ful birds.-- heredetary ambling horses, (if not LOOKED at as instinctive) then must be owing to hered
298c189 rk placing withered grass over nest, when often LOOKED at.-- this most puzzling whether instinct, or
306c212 -- more like man. continued long in a passion & LOOKED out for him to come again very differently fro
551m129 ure (& partly out of I do not know what when it LOOKED at the glass) when pouting protrudes its lips
552m130 d Why when one thinks of any object, (or having LOOKED at any object<)> one Shuts ones eyes) is the i
563nIFC ion) Selected «for Species Theory» Dec. 16 1856 LOOKED through & all other Books May 1873-- October 2
569n019 man at Cambridge observed the ignorant. merely LOOKED at picture as works of imitation.-- Hence plea
593n111 pool to bed of flags I was astonished & having LOOKED round saw at considerable distance a very larg
604o016 d the Review or the article. A Planaria must be LOOKED at as animal, with consciousness,, it choosing
610o031 ot <remember> «read» Christian name; fancied it LOOKED like. W. but concluded it could not be so.--Lo
610o031 ked like. W. but concluded it could not be so.--LOOKED at a direction book, but could not find out--
610o031 ilson into his head.-- remembered, that he had. LOOKED in direction book under head of Wilson, referr
025r019 e old & dead)» «(I have not kept a record)» In LOOKING over the lists of organic remains in De la Bec
135a144 netts. vol 4. p. 193 in Lat 26 degree S. Wafer LOOKING for Copiapo. found inland a great many sea she
206b147 a few will have successors. at present day. in LOOKING at two fine families one with successors «for»
216b181 e of boy with foetus developed in the breast.--LOOKING as if many ova-- impregnated at once.-- Dr. Sm
256c054 , or A C D E H. Very striking to see M. Bibron LOOKING over reptiles he often had difficulty in disti
353d060 .-- he get merely a few pages.-- Hence (p. 59) LOOKING at animal, if there be many others somewhat al
354d065 h themselves to man. Sept 7th. -- I was struck LOOKING at the Indian cattle with Bump. together with
387d168 veral times, that it becomes fixed in blood.-- LOOKING at ovum of mother & ovum in offspring, as simi
408e046 same kind of monstrosities.-- G. B. Sowerby.-- LOOKING over Lamark surprised to see how many Tropical
410e051 ation of one part affecting another.-- (I from LOOKING at all facts as inducing towards law of transm
417e077 this analogy between grafting & sexual union-- LOOKING at simple generation as being the action of tw
511q018 on Mr Wynne, &c Could by selection a different LOOKING animal be formed-- not caring whether good or
525m024 ve seen a dog doing what he ought not to do, & LOOKING ashamed of himself.-- Squib at Maer, used to b
525m024 lf.-- Squib at Maer, used to betray himself by LOOKING ashamed before it was known he had been on the
526m028 gination. «Catherine thinks that children like LOOKING at <ani> pictures, an early taste, of animals.
529m041 gton Gardens has often been greatly excited by LOOKING at trees at [i.e., as] great compound animals
546m109 he senses.; than the <small> fact that no one, LOOKING back to his life, would say how many good dinn
548m117 o states.-- August 30th.-- It is singular when LOOKING at a table one has vague idea something is not
553m137 hus the pig-tailed baboon, shoved out its lip, LOOKING absurdly sulky «as» often as keeper spoke to i
565n006 l of readiness, & therefore done in extreme.-- LOOKING at ones face <&> «whilst» laughing in glass. &
565n009 inkle when smile.-- Hope is the expectant eye. LOOKING to distant object, brightened & moistened by e
614o037 s, say attribute of such higher animals may be LOOKING back, ∴ therefore consciousness, therefore rew
619o042 sh Ethical Philosophy [RHC] On the Moral Sense LOOKING at Man, as a Naturalist would at any other mam
040r062 the alternating with such matter at St Julians LOOKS like such?--destructive to animal life.--Patago
126a123 years, that surface does not become hot?-- this LOOKS as if bad conductor-- III But equilibrium is no
273c110 pecies, which is not true, with shells?.?.?» It LOOKS as if animals perished by errors.-- It is most
284c148 -- but to crawl up an hill, then by deaths?!)-- LOOKS like subsidence.-- on the islets Mr Blyth remar
312c231 ed are better, than those bred in & in.-- which LOOKS as if qualities were not permanent, in the new
344d042 small. Now Jones, before this happened from her LOOKS thougt she was halfbred Beagle Staghound. «++».
359d087 Canada Goose.-- Former strange mishaped bird-- LOOKS very artificial breed-- but Mr Miller says that
411e056 uals (Mem: transportation will be answered) one LOOKS to analogy for cause in plants. where innumbera
451e178 between April & October & like man almost (this LOOKS inaccurate C. D) they will catch the Malaria &
525m026 Scotts remark how odious an illtempered fat man LOOKS, shows same connection between organization & m
528m037 is novelty of view even old one. every time one LOOKS at it.-- these two causes very weak.-- (2d) for
541m091 difficult EXPERIMENTIZE upon this effort.-- it LOOKS so analogous to muscle in one position great fa
549m118 any more than to dog losing his puppies-- This LOOKS like free will.-- V. last page. A healthy child
593n107 which become in imago state social) by smell or LOOKS. but it does not know its own smell or look, &
593n107 eling which is pleased by other animals smell & LOOKS.-- no doubt it may be attempted to be said that
593n109 alist, does not know other men by smell, but by LOOKS. hence. some obscure picture of other men. & he
600o008 ife time]CD 3. The Infinite, -- lives by hopes, LOOKS to eternity. (4) Reason, some transcendental ki
604o016 mes multiplied with the organisms structure, it LOOKS as if consciousness an effects of sufficient pe
450e173 ported elephants. wild hogs-- spotted deer, no LOONIES, but cocatores & small green parrots. June 26t
355d066 «able to» tie themselves down, they can find LOOPHOLES] "It is well worthy of examination whether va
048r084 coated with Tosca. which implies motion in the «LOOSE» bed of pebbles. (On a sea beach under a cascad
051r093 on. (where occassionally most tremendous surf & LOOSE sandy beach) deposits «calcareous» encrustation
100a055 Australia & Oolitic period.-- comparison rather LOOSE. -- perhaps worth Says from Lardner's (p. 213) f
215b176 When dogs are bred into each other, the females LOOSE desire, and it is required to give the canthair
334d011 rsing every month) & worked in the same cart in LOOSE chains, by being at first beaten from her, & al
372d129 nce if it does= <I do not doubt, the> Do plants LOOSE any qualities by being buds-- , more than if wh
557m147 . back arched. just contrary. when pleased tail LOOSE & wagging-- if as (I believe) Hunter says. neit
578o056 o thoughts of other person Decemb. 27th.-- Fear LOOSE the sphincter muscles, only on the principle li
580n061 ellect (As Miss Clive) have only possessed very LOOSE ideas.-- Have children loose ideas of time?-- C
580n061 nly possessed very loose ideas.-- Have children LOOSE ideas of time?-- Characteristic of one kind of
578n056 ?) in the most perfect fainting, sphincters are LOOSED is a convulsive action to remove disagreeable
105a069 & sea.; the former as its channel becomes wider LOOSES its cutting power. (as does it when the inclin
390d179 ill not breed, but the female at least (;male?) LOOSES all appetite.-- It is the comparison of each a
109a083 la Beches argument of low coast gaining & high LOOSING answered by this -- No one can doubt. A-B once
343d038 inct. others though the negro of Africa is not LOOSING ground. Yet, as the tribes of the interior ab
378d147 fferences indicate, species changing forms, <& LOOSING do> if so domestic animals ought to show them.
385d163 (Mem: so it was said little cock «yet very odd LOOSING visible powers» in Zoolog Gardens. & Kings at
429e113 urrent tendency in external conditions» sudden LOOSING of horns.-- I do not believe this Nature's pla
454e184 there instances of plants, in becoming double LOOSING fertility if, sometimes one, sex & sometimes.
492q002 s in becoming double ever become monooecious-- LOOSING one sex & not other: which generally fails fir
586n080 he act of crossing the sea in dark night & not LOOSING its direction, equally wonderful in young & ol
265c083 t progeny-- many dogs in England must have been LOPPED off & sheeps tails cut yet there is no record
127a126 - will it bear on central fluidity.-- do p. 137. LORD Tullamore found Sulph of Soda in peat ashes in
220b197 es not depend on mind or instinct of parent. Mem LORD Moreton's Mare. The fact of plants going back h
334d009 race was determined by male: & How completely is LORD Moreton's case opposed to this fact & views.--
334d012 orelation of imperfect structure.-- Fox says in «LORD» Exeter's Park «or in the Duke of Marlborough»
335d014 -- Heredetary <thr> Six fingered people, <Hill> «LORD Berwick» family with defective palates. heredet
```

Page **(Key Word)**
379d152 her breed else all the lambs will deteriorate.-- LORD Moreton's Case.-- When cows have twins, <one> t
385d165 ometimes of both." If L. can be trusted, this is LORD. Moretons law.-- "How often do we find in the s
387d168 esses offspring more & more with the added «like LORD Moretons case & Dr. Andrew Smith,» difference.-
417e079 in some like father «What is cause of this.-- » (LORD Moretons law holds with different species, & in
417e079 of resemblances of children to their parents.-- LORD Moreton's law cannot hold with fishes, «& there
525m025 he mouth «stammering in my Father family» (as in LORD Berwick's family) are heredetary.-- other defor
032r040 nted), along line of coast.--[Fig. 2] Mem San. LORENZO; Valley of Copiapò & parts of coast of Chile.-
059r119 nt entre elles une correspondance exacte, et LORSQUELLES se trouvent interrompues par quelque vallées
215b177 were any constant species Both males & females. LOSE desire. Native dog not found in V. Diemen's lan
280c134 admirable essay) heredetary Young wild ducks.-- LOSE as well as gain instincts. Wild & tame rabbits
294c178 asing to grow far N. becomes stunted, altered, & LOSE (mere sickness)? fertility ¿because offspring t
352d058 law in nature an animal may acquire organs, but LOSE them with more difficulty, «contradicted by abo
385d163 r being unhealthy.--» males «bred in & in» never LOSE passion. (Mem: so it was said little cock «yet
462tf05 rs-- is incidentally said that a mongrel man may LOSE all traces of his parentage in «about» seven «7
549m118 ve no credit to instinctive feelings.-- for man LOSING his children, any more than to dog losing his
549m118 r man losing his children, any more than to dog LOSING his puppies-- This looks like free will.-- V.
217b183 same fact in Mr Galtons case.-- It explains the LOSS & expense, (must probably have occurred to ever
293c173 instinctive.-- My view of instinct explains its LOSS ¿ if it explains its acquirement.-- Analogy. a
389d177 ul observing cattle can be bred in & in.)-- [The LOSS of passion in hybrids. perhaps connected with t
419e086 & yet 200-300 ft elevation & no change & even no LOSS of species.-- It must never be overlooked that
419e087 ion well grounded, on time;-- therefore the mere LOSS of species, which may be the work of a few year
512q019 th.-- Mr Tollett-- about selection for milking-- LOSS of early habits in Dorsetshire sheep migration
548m117 t stop to reason what there should be & discover LOSS Definition of happiness the number of pleasant
604o017 Poor.-- on insanity.-- Prevailing idea. owing to LOSS of will.-- chiefly excited by passive emotions.
614o037 as so very few (in adult life) instincts.-- this LOSS is compensated by vast power of memory, reason
048r085 S America we cannot dive into the causes of the LOSSES of the «species of» Mastodons. which ranged fr
056r109 scarcely a pebble might remain to tell of these LOSSES.-- Cause of chimney. to crater. as at Galapago
636j57r hout these adaptations.--fossil forms show such LOSSES.-- Consider ground Woodpecker stiff tailed cor
030r035 iloe:-- On open coast, near where Challenger was LOST: I know no reason for supposing these matters a
088a013 at least several attempts at elevation From the LOST & turned about position of strata, prooff thick
104a068 on surface of plains due to whole moisture being LOST by evaporation therefore capillary attraction w
161g095 his road level with Peat moss most distinct then LOST by slope, then concealed by fragments, then cle
161g095 rs level with road, & with piece of excised rock LOST at point of valley chiefly from rockiness When
275c120 rds had breed of greyhounds fleestest in England LOST courage-- (Bull-dogs are used because they have
288c160 came Madeira & <seized> ceased their migrations LOST?? I conceive a bird Migrating from Falkland Isd
334d012 Half Muscovy Fox says a settler near Swan river, LOST his <on> two cows entirely, changed his residen
349d052 res, as in Hippotamus, solely owing to number of LOST links.\ if all species know they would be innum
364d103 killed itself.-- Q Sir. J. Sebright-- has almost LOST his Owl-Pidgeons from infertility.-- Yarrell sa
389d176 irths, & even produce fertile offspring-- DESIRE LOST when male & female too closely related: this mo
399e010 en introduced, & have freely bred, they have not LOST power of producing. Williams. Narrative of Miss
460t019 <slightly> «a good deal» modified <& Many Forms LOST; if> «of this old stock (which from action & re
463t057 ransmut, from one organ graduating into other is LOST, <be> (as vertebrae into skull., two bones of t
498q007 nutrition.-- Horned oranges so? --Yes, my Father LOST this character in grt degree from charcoal & go
505q014 eeds produced good pollen? Yes «From cultivation LOST their horns» is impregnation necessary to fruit
522m013 .-- In Mania all idea of decency & affection are LOST.-- most delicate people do most indelicate acti
534m063 would have been instinctive, seeing that time is LOST & endeavours made must be experience & intellec
573n036 ne, but a piece» of coloured glass <&admires> is LOST in astonishment at the artificer.--» Our facult
576n047 ife.-- yet he would ever repent, & wished he had LOST his life in doing so.-- nor would he regret «ha
580n062 of the infant likewise before two years are soon LOST; yet many of the habits acquired in that age ar
625o50v of imagination «the utility part being blended & LOST» & moral sense.-- My theory explains both, perh
284c150 s less subject to Variation" Dr. A. Smith. knows LOTS of instances of replacement of one species by a
467t104 rvature of <an> pistil, etc lies in gangway= In LOTUS corniculatus saw Humble press down wings which
178b030 pe of Eocene with respect to Miocene of Europe? LOUDON. Journal. of Nat History.-- July. 1837. Eyton
203b138 lla,. «species peculiar to Continent & England» LOUDON Mag: Septemb or Octob 1837 Westwood has writte
203b138 chiefly of the former Eyton's paper on Hybrids LOUDON'S Magazine. Gould on Motacilla,. «species pecul
215b178 of the tailess cat of Isle of Man mentioned in LOUDONS (analogue of Blood hound-- Bull. Soc. Geolog.
025r017 coast of Brazil generally.-- Mrs Power at Port LOUIS talked of the extraordinary freshness of the st
048r084 s beds of marine shells on banks of Red River LOUISIANA. V. Lyell. Vol I. P. 191 State at St Helena.
183b052 pheasants & fowls..-- "On sait que le "métis" du LOUP et du chien, que celui de la chevre et du belie
246c025 gs of New Zealand of large size, resemble, chien-LOUP.--long, black & white, ears short & straight--
312c233 34», different species to different,-- inguinal LOUSE African & Europaean. different.-- thorax & head
313c234 cannot» Read Entomological Transactions Why if LOUSE created should not new genus have been made, &
146g012 en pleasures of association, & passions, such as LOVE-- dislike & <f> passion of hatred To fulfil an
291c165 ness as well as wildness-- cf Sir J. Sebright.-- LOVE. of man gained & heredetary. «problem solved» h
291c166 brain heredetary,. analogy points out to this.-- LOVE of the deity effect of organization. oh you Mat
316c244 .-- a dog snarling in play.-- Hensleigh says the LOVE of the deity & thought of him «or eternity», on
362d099 he skin off horns to fight-- is analogous to the LOVE of woman (as Mitchell remarks seen in savages)
536m071 nt orders turn up their nostrils when excited by LOVE? Stallion licking udders of mare strictly analo
541m092 Is the effort greater if the idea is abstract as LOVE, (or an emotion not so) than if simple idea as
544m101 ndition to receive pleasure) gives pleasure, ie. LOVE.-- & so pain gives fear of death. Mayo Philosop
558m150 ncts, giving rise "do unto others as yourself". "LOVE thy neighbour as thyself". Analyse this out.--
558m151 m language.-- the social instinct more than mere LOVE.-- fear for others acting in unison.-- active a
570n026 ountaineer <takes> born out of country yet would LOVE mountains, & a negro, similarly treated would t
579n059 with the habitual expressemotions, which make us LOVE him, or her.-- it is blind feeling, something l
579n059 blind feeling, something like sexual feelings-- LOVE being an emotion does it regard «is it influenc
582n068 heart is the seat of the emotions.-- but are not LOVE & hate emotions; what are their characteristics
600o008 fers chiefly to moral, beau desires conscience & LOVE.-- [With regard to ordinary Beau ideal, Mem. Ne
600o08v onscience: in Maternal instinct domineering over LOVE of Master and sport &c &c -- The Bitch does not
606o025 wagging one's finger-- one feels it in passion, LOVE-- jealousy-- «as» effect of bodily organisms--
614o037 f memory, reason &. & many general instincts, as LOVE of virtue, of association, parental affection--
615o038 ied in many countries, hence national character, LOVE of country, of association &c stronger in some
616o038 is enhanced by memory of what has been heard; so LOVE of virtue enhanced by this heredetary kind of m
619o042 animal. These instincts consist of a feeling of LOVE <and sympathy> «or benevolence» to the object i
619o043 a dog we see a struggle between its appetite, or LOVE of exercise & its love of its puppies: the latt
619o043 between its appetite, or love of exercise & its LOVE of its puppies: the latter generally soon conqu
622o049 incts Hartley, (according to Sir J) explains our LOVE of another, as pleasure arising from associatio
622o049 this person.-- [LHC] p. 254. &c &c [RHC] But the LOVE is instinctive, & how does it apply to mother l
623o050 th in doctrine, for [RHC] 9) We can thus explain LOVE of place.-- although here we have not received
623o050 in early infancy, during many generations giving LOVE of mother; the having received some advantages
628o054 part of man, may be quite artificial, as avarice LOVE of gold.-- love of fame-- Yes Hartley explains
628o054 be quite artificial, as avarice love of gold.-- LOVE of fame-- Yes Hartley explains this & Mackintos
337d022 xpression <of a> of Monkey, «when offended» who LOVES, who fears, who is curious &c &c &c who imitate
574n041 to kiss, & almost bite, that which one sexually LOVES is probably connected with flow of saliva, & he
579n059 -- What passes in a man's mind. when he says he LOVES a person-- do not the features pass before him
622o049 e is instinctive, & how does it apply to mother LOVING child, from whom, she has never received any b
030r036 of Patagonian coast to see proportional cliff & LOW or sloping land What are the "palatal Tritores"
037r052 the entire absence of any rock situated beneath LOW water in the Southern ocean not being buoyed wit
045r077 at attend Eruptions: Mr P. Scopes explanation of LOW Barometer? In a subsiding area. we may believe t
061r128 tain area must have peculiar character: Contrast LOW limit of Palms, evergreen trees, arborescent gra
088a016 37.-- Helms remark on common salt being found on LOW hills East of Cordillera very important V. Malte
096a040 it might be percieved on which side craters were LOW --¿ applicable to Auvergne??? The fact of Galapa
109a083 flat sand beach. {P} -- De la Beches argument of LOW coast gaining & high loosing answered by this --
109a083 red by this -- No one can doubt. A-B once formed LOW coast.-- Annales des Mines. a translation of pap
149g033 as with Isld-- {P} do they extend round hill too LOW line drawn plain red talus line on N. side of Sp
151g040 ourhood, (as granite or gneiss of Moel Derry) on LOW hill between Inn & Bouhunthine the summit «doubt
161g094 h moss On this terrace «station perhaps 6 ft too LOW» (to test last on Peat-Mass Divortium aquarum) B
161g097 2 75 degree? This last measurement turns out too LOW, (NB .260 would have been more correct) there we
219b195 ave been separated by short space from mountains LOW down, therefore plants common take an example fr

Page
***(Key Word)**

254c047 ow does this agree with grand fact of Marsupial, LOW Cerebral structure??-- «do» p. 390. All classes
318c253 ow bands, certain kinds as gallinules taking the LOW country near coast & others the mountains, & the
388d170 with stock of food.-- the generalization begins LOW.-- it goes through transformation, nearly indepe
407e038 ews of Lyell) Now «Equatorial» America from the «LOW» limits of blocks both North & South, has probab
408e042 climate of N. America, must have been equable & LOW-- more so than any other part of the World.-- Eu
430e119 Helena Plants & see whether those which grow in LOW grounds are those, which are common & nearest be
545m108 - one follows other as in blindest memory»-- also LOW faculty of understanding. Adam Smith (.D. Stewar
105a072 duce other causes to explain «alluvi» in valleys LOW in his paper says land shells found with calcar
222b204 -- one child like father another like mother Has LOWE written any other papers besides one in Latin o
511q018 & skeleton, & skin= Van. Voorst often writes to LOWE (7) In breeding. pointers. Bull-Dogs. Spaniels-
036r050 ression. is not very distinct, from some of the LOWER orders; it was connected with movement of sand.
057r114 interstratified.-- Urge fact of Boulders not in LOWER strata. only in upper. in accordance in Europe
066r142 e could not distinguish from stone Caradoc from LOWER of third Silurian division--Together with same
113a091 een better then 32 degree would have been found LOWER.-- We have no right to consider the conducting
113a091 dently infer that time has not been allowed for LOWER beds to cool down. & then in 50000 years the de
125a121 - who knows how far that may have penetrated,-- LOWER down the temperature may be kept up far higher
126a124 e direction only, like water below 39 degree» & LOWER part glass.-- then the high temperature would b
149g032 Inn BOULDER of granite above 4th Shelf a little LOWER down the hillock with beach & channel precisely
153g056 ather smaller block 30 ft «above» & other 50 ft LOWER & other smaller ones «these boulders are decayi
156g067 «glen» than 3d. 3(a) less perfect than upper & LOWER but quite as perfect as those lines in Glen Col
156g071 pebbles Level of plain of 4th shelf at head of LOWER Glenroy 29.581 A 82 75 degree? From this point
157g073 t I believe this is chiefly caused by its being LOWER.-- [no pebbles in parts of Beagle Channel when
159g087 sited» the slope is continued some hundred feet LOWER & begins about 60 higher-- There are however fr
161g094 m. 29.200 A.77 degree Air 70 degree? Barom. 066 LOWER than last. but A 77 station was <a few> «about
161g094 t. but A 77 station was <a few> «about 3» feet <LOWER too high about a quarter of a mile further on,
170b002 s highest office in organization (especially in LOWER animals, where mind, & therefore relations to o
177b027 rcles;-- & insects amongst articulata.-- but in LOWER classes, perhaps a more linear arrangement.-- ¿
192b085 ld have been born without them.= In some of the LOWER orders a perfect gradation can be found from fo
197b112 nimals in scale.-- In monsters «also» organs of LOWER animals appear.-- yet nothing about propagation
219b195 new formations» because snow formerly descended LOWER, therefore species of lower genera altered. or
219b195 formerly descended lower, therefore species of LOWER genera altered. or northern plants «No» CD[Mem.
289c162 , «+++ must (on my theory) =supported by foetal LOWER developed forms.»» (NB waterhouse says of affin
298c190 note.-- try to trace from simplest reasoning in LOWER animals many times produced, a general tendency
298c191 ficulty in Making an alpine species from one in LOWER country during gradual elevation of isld.-- We
298c191 become occupied by a third best adapted kind.-- LOWER species would then revert to pristine form (whi
298c191 ave been altered by crossing) with alpine form) LOWER species afterwards would probably often be dest
303c204 remities, how are races in This respect upper & LOWER, which I do not know whether it <would have> di
317c250 & several old acquaintances. which grow on the LOWER region of the Canary islands-- p. 250 admirable
347d049 ls must tend to improve.-- yet fish same as, or LOWER than in old days: «for a very old variety will
374d134 est formation highly organized.-- do. p. 461.-- LOWER Silurian-- several existing genera. Nautilus tu
383d160 broader.-- » Breast red like Common pheasant.-- LOWER part of breast, each feather is fine metallic g
399e006 ed if very thick to find some change in upper & LOWER layers.-- good objection to my theory: a modern
423e099 one variety in upper part of bed-- & another in LOWER is very rare, the conclusion will be that our g
430e119 beak of this one has concentric striae, all the LOWER part rayed longitudinally (give woodcut) like I
466t094 herwise lie protected by the hairy black lip of LOWER division of nectary: «wh. itself resembles a Bee
473s03v range no plants in our Devonian-- Fish one step LOWER in America-- How curious all negative laws of A
478z003 inets Voyage Vol p. 597 Many descriptions about LOWER animals of Falklands &c &c Bennett on Chinchill
546m110 place.-- Vide page 103, supra (by mistake) have LOWER animals these vivid thoughts In same book (p. 1
565m008 rson extremely self-sufficient,-- the corner of LOWER lip are depressed & opposite muscles used to wh
575m046 the permanence of old heredetary ideas.-- being LOWER faculty than the acquirement of new ideas.-- Wa
582n069 s scarcely a faculty in man not met with in the LOWER animals.-- hence the general aim of fable, & ex
601o009 «or habitual actions» perhaps polypi-- (so that LOWER animals are sleeping higher animals & not plant
612o035 to certain stimulants without conscience in the LOWER animals, as in stomach, intestines & heart of m
612o035 w near in structure is the ganglionic system of LOWER animals & sympathetic of man [RHC] ¿How does co
633j54r an plants them to stop the moving sand) we <do> LOWER the creator to the standard of one his weak cre
031r037 ordillera, as in English Coal field -- because LOWERED & raised--so on--but gradually & simply raised
031r037 No Faults in Patagonia[,] enormous extent; if LOWERED again & covered no sign of upheaval To Cleavag
079r176 oa has said silver in the highest & gold in the LOWEST. Humboldt states that some of the richest gold
114a095 etamorphosed than other deposits.-- NB. because LOWEST. first accumulated in bed of ocean With the ex
227b226 iate structures <between> say in brain. between LOWEST Mammal & Reptile. (or between extremities of a
254c048 ??-- «do» p. 390. All classes of Acrite exhibit LOWEST stages of animal organization, "& are analogou
315c242 l actions in plants, it allows of any degree in LOWEST animals --habitual action, in intestines subje
381d157 my views-- Seeing sexes separate in some of the LOWEST tribes, leads one to suppose still more that t
416e070 now how is it-- in Planaria, they couple-- CD [LOWEST terrestrial animals.-- in shells?-- insects?.-
509q017 n, which has abortive bone. (ask more about the LOWEST cervical vertebrae process developed into ribs
542m095 ut up their hands, & as attention would amongst LOWEST savages clearly be directed chiefly by objects
580n062 , as the double-conscious kept playing so well.-- LR. Brougham «Dissert.» on subject of science conne
400e011 this collection is particularly rich. «as in LUCANIDAE» less difficulty in establishing good gr
367d113 carnivora. Where females, are peacable-- (Mem LUCANUS & Copris &c).-- In birds singing of cocks sett
239c002 s with parents, going back to either parent is LUCIDLY explained.-- Mr Yarrell states that if any odd
502q11v atory-- ask Gould about N. Zealand, as Cuculus LUCIDUS is.-- Ask Sulivan about Falklands Isds.-- Snip
045r076 equent; so as to exclaim, «as I have heard» how LUCKY! when they hear of a place having a pretty seve
402e019 - Malte Brun. Vol <I> II p.,133: at Samar SE of LUCON, many monkeys, buffaloes &c &c-- Malte Brun. wo
402e020 <of th» «Mammalia» have passed to Paragua & in LUCON the most northern of the group the number is li
134a142 Malcolmson says are like Kankaer South of Part LUCONIA-- Phillipines there is volcano on isld in lang
080r181 sts 93 action of sea on coast. 27. Bahama Isd De LUCS travels Beauforts Karamania Capt. Ross. & Score
478z004 ray has given paper to Royal Soc on glow worm. LUMINOUS property-- Curious arrangement of animals in
481z010 en. Voyage round World German a reference to a LUMINOUS Sertularia Lesson Zoolog. Coq: p. 120 Coati R
048r084 f grass penetrating by action of Organic power a LUMP of hard clay. -- In the History of S America we
463t063 uch remark & before the conclusion of his work-- LUND makes his wonderful discoveries= negative facts
465t079 e alpine plants-- In S. America. it appears from LUND more Mammals, than at present «in Europe we kno
464tf6r lied to the Struthonidae than any other forms-- LUND'S Antilope in Brazil another point of agreement
467t103 The abortive stamen are of useful height.-- In LUPINE, Bees «frequent» & seem to act, something like
467t103 tal standing on «I saw Bee go to two species of LUPINE», two wings. & when the Lupine flower is perfe
467t103 o two species of Lupine», two wings. & when the LUPINE flower is perfectly ripe & pollen abundant fil
467t104 saw Bee collecting pollen from <sheath> Keel of LUPINE-- Seen Bees on Potato & several times on Beans
468t111 carlet Poppy to Rhododendron-- from Larkspur to LUPINE two species of Larkspur -- two varieties of Ci
501q011 - Dr. Fleming says yes. (29) Are there RACES of LUPINE, Stocks Clover, to experimentize on by sowing
058r115 cryst. arrange themselves in planes. «Mem silky LUSTRE» ask Erasmus. whether electricity would affect
206b147 .--» If population was increasing between each LUSTRUM, the number related at the first start must be
206b147 t be greater, & this number would vary at each LUSTRUM, & the calculation of chance of the relationsh
206b147 genitors would have different formula for each LUSTRUM.-- We may conclude that there will be a period
486z019 from Chiloe. Amblyrhyncus de marlin James Isd-- LUTKE Voyage Vol III p 322 Dr Martens says only one R
569n022 opposite ends of series or harmonious prose.-- LUTKE Voyage in Carolinas Vol II p. 132. offered to t
322c269 . 12t Kotzebue's two voyages, skimmed well. do LUTKE'S Voyage. carefully read.-- Reynolds Discourses
569n022 e «this» out not in Library no good There is a LUTKE'S Voyage autour du Monde (1826-9) Paris. 1835 Qu
108a080 springs near coral reefs.-- Where vegetation LUXURIANT it might be almost as well said probably much
247c027 -- «Mitchill says snakes on Friendly isles. p 50. LX. Journal of Silliman» «Study Silliman.--» Vol II
235b275 ix,--any cultivated plants produced by seed.-- LYCHNIS-- Flax.-- Read Swainson [blank] In production
434e128 e & my theory explains why it is sure guide.-- LYCHNIS April 3d.-- Henslow tells me following facts:
434e129 sme following facts: believes that «only» red LYCHNIS grows in <south> Wales & certainly <old> only
434e129 hey have been thought to be different species.-- LYCHNIS dioica, generally dioicous. yet parts only ver
491q01v p. 427-- says biennial-flowers & scarlet LYCHNIS can be propagated by cuttings.-- Try.-- Import
489q09v -- Frog Bit, Valerian-- Urtica Dioica Sorrell. LYCHNIS. Butchers Broom-- «also, Vinca,» Examine all t
505q014 (4) Effects of Nitrate of Soda under Beech.-- LYCHNIS dioica answers this question= (5) Open more Ho
374d133 o be intermediate between Coniferous trees & LYCOPODIUMS.-- p. 437. Many. existing genera of shells i
021rIFC ee Daubisson both Volumes, and Molina 1st Vol & LYELL Sailed, 27th <Friday gale 29th> Friday Thursday
036r052 ili. separate geographical ranges of plants. V. LYELL. Chap XI Vol II. Urge the entire absence of any

038r057 natic It has been common practice of geologist. LYELL considers (P 84 Vol III.) whole of Etna series
038r057 he discussion it will be better not to refer to LYELL. but merely to state these reasons, & saying th
039r060 & the more true analogy from the Galapagos-- Mr LYELL. P. 111 & 113. «seems to» considers that succes
039r061 cent elevation. which is different from what Mr LYELL supposes. Lyell P 116 Vol III, says that in N.
039r061 which is different from what Mr Lyell supposes. LYELL P 116 Vol III, says that in N. Pliocene formati
040r063 ...> Molina's Case At Vesuvius. Vol III P. 124. LYELL. dikes have a parting of pitchstone; which is d
041r065 lat countries. years of drought are common.--Mr LYELL has mentioned the drifting of carcases putrid.
042r067 el on a <me> central line of Patagonia. «NB. Mr LYELL P. 211 Vol III. talks of line of cliff marking
042r068 cene period».--Mem Bahia blanca P. 204 Vol III. LYELL Owing to «open» faults in mountains: to elevate
043r070 moso. & Coral reefs (imperfect in latter). <At> LYELL. Vol I. P. 316. Earthquake of 1812 affected val
046r079 mittance of water, through the rent strata: «Mr LYELL considers that Plutonic rocks are generated as
047r082 the coincidental retreat at Portugal & Madeira (LYELL. vol I. P. 471) is explained. also the similar
048r084 rine shells on banks of Red River Louisiana. V. LYELL. Vol I. P. 191 State at St Helena. pebbles enti
049r088 In Granite great crystals arranged on sides. V. LYELL P 355 Vol III. constitution of veins, is there
053r100 ernambuco? Quote Miers about shells at Quillota LYELL, states that contact of Granite & sedimentary r
064r137 d. neglecting Cordillera itself now remaining-- LYELL « <p 419> p 428» states that Von Buch has urged
068r145 f landlocked harbors to Craters of elevation.-- LYELL suggested to me that no metals in Polynesian Is
071r158 le's volcano, many amygdaloids.--Boussingualt <(LYELL)» cracks mountains falling in.-- Earthquakes at
088a014 ns slow. so would line of mountain chain be Mr LYELL «Waterhouse» has frequently heard that Herons
104a068 traction would bring water with salt to surface LYELL remarked to me that Kylow (?) was astonished wi
104a068 tain chain case parallel to Banda Orientel. ask LYELL for sentence.-- Origin of Breccia, introduce in
110a085 st Indies & declare them to be recent species-- LYELL-- Some internal changes are in process. connect
114a094 tamorphosed clay slate.-- --shale in shall sea. LYELL confounds these introduce discussion -- I see L
114a094 I confounds these introduce discussion -- I see LYELL talks of different composition using difference
121a111 altered, & a crystalline structure superinduced LYELL on Sweden p. 5. «& 7.» violet strata from decom
121a113 surface of Patagonia, yet none in shingle beds. LYELL on Sweden. p. 12. proofs of small rise at Stock
123a116 PAPER on fossil shells of S. America. Von Buch LYELL. (under head of Delta) describes near Alps grea
124a120 o limestone & volcanic rock containing magnesia LYELL. Elements p.119 on such strata {P} do p. 171. a
186b059 slike to union, offspring not well intermediate LYELL Vol III. p. 379. Mammalian type of organization
187b063 same time in such different quarters.-- Will Mr LYELL say that some circumstance killed it over a tra
194b096 receive influence from other plants.-- Does not LYELL give some argument about varieties being diffic
198b115 ansportal otherwise not so numerous: quote from LYELL: assuming truth of quadrupeds being created on
208b153 s Tertiary Shells in India!? A p. 28 Dr Beck. & LYELL. most curious law of species few in Arctic. in
209b157 ies in France is I: 5.7 in Laponia I: 2,3; Mem. LYELL on shells.-- {T} Genera In North Africa. I: 4,2
213b170 ose species which have long remained are those ¿LYELL?, which have wide range and therefore cross & k
220b200 l statements about mundine & confined genera.-- LYELL has remarked about no confined species in Scici
232b249 nd in many island Celebes «Waggiou» &c &c. (See LYELL. Vol III p. 30) different species in different
235b272 log. 842 White regular gradat in Man poor trash LYELL 1024 Flemings Philosophy of Zoolog Royle on Him
265c084 , but that child should produce like children. ⎰LYELL has story from.-- Beck about six fingered child
280c135 lead to anticipate mules is very important for L.YELL said to me, the fact of existence of mules app
285c153 animal cannot change quick enough & perishes.-- LYELL has show such Physical changes will be unequall
291c168 ecies being destroyed at Falkland Isds++.-- Mem LYELL hypothesis of change in Scicily.-- Splendid Har
294c176 ent small alterations in wild animals, & thinks LYELL has overlooked argument, that domesticated anim
322c270 ions-- very poor Sir T. Browne's Religio Medici LYELL Book III There are many marginal notes <Rengger
325c266 ss, destroying children. --it is not effect, as LYELL suggested, of organ being worn out as. otherwis
407e038 f S. America.-- (Explained by profound views of LYELL) Now «Equatorial» America from the «low» limits
413e059 in proportion to support of parents December 2d LYELL tells me Beck considers the characteristics of
414e065 transported by floating ice.-- I agree with Mr LYELL., man is not an intruder.-- : the geological hi
419e086 viduals from one, is adverse.-- Decemb. 25th.-- LYELL says the elevated shells in Bayfields district
424e100 ecies peculiar to separate islets.-- March 5th. LYELL says «fossil» shells from North America, Scotla
424e101 il remains, & therefore not. to be trusted.-- LYELL tells me, on authority of Beck, that Hooded cro
425e105 us American form. found in Eocene beds of Paris LYELL has remarked species never reappear when once e
436e134 f high temperature at which fish &c can live.-- LYELL says that naked cuttle fish now bear a very lar
473e63r ier never found remains of Sus with Elephants-- LYELL says New Red Sandstone of. N. America is Red Sa
551m128 RAMBLING till excited by question.» Sept. 4th. LYELL in his Principles talks of it as wonderful that
568n019 sition of divines to progress of knowledge. see LYELL on Scrope, Quarterly Review. 1827? In Water Sco
594n115 ot <more> less wonderful than man his intellect LYELL has seen a little dog go to the assistance & bi
021rIFC h <Friday gale 29th> Friday Thursday 29th gale LYELL'S Geology The living atoms having definite exist
034r044 ount of ejected granitic fragments P. 386 Mem. LYELL'S fact about sulphuric vapours in East Indian Vo
047r083 after elevation in perfect conformity with <Mr LYELL'S> idea of an injected mass of fluid rock In Pat
085aIFC Observat on Volcanic islands & Coral Formation LYELL'S Salband p. 86 Shells near Woollich p. 112 Spec
087a007 ts to it, lately raised above level of the Sea. LYELLS Encyclopaedia-- Lately elevated When Siberia w
087a011 lia really very important. harmonizes well with LYELLS idea of intertropical land.-- Siberia rises. t
107a079 ia & Tubul applicable to Andes & Patagonia-- On LYELLS idea of whole centre of earth same heat, then
110a086 . examine these «lines» Description of rocks in LYELLS'. Capital Norway case.-- The fragment. consist
112a089 of old volcanos within Cordillera-- allude to LYELLS view of not discovering dike one end granite &
122a115). Shropshire case where lamination appeared.-- LYELLS Denmark -- L'Institut (1838) p. 268. Paper by
125a121 of heated fluid or gases under pressure.-- {P} LYELLS view of transmission of heat by gases-- does n
136a147 t, I may properly remark on the superiority of LYELL'S classification to that of Phillips as given p.
138a153 . Observat. in Southern Brazil. [blank] «p. 4. (LYELLS Book)» Observaciones sobre El Clima del Lima p
163g109 on «under head of» Beagle Channel. Forchammers (LYELLS Denmark) Shrewsbury rubbish.-- Speculate on or
173b010 ived from form (2). &c.-- <{> Then (remembering LYELLS arguments of transportal) «continents» island
209b155 n sports.-- Agrees with old Linnaean doctrine & LYELLS. to certain extent Von Buch.,--- Canary Isles:
221b201 r ages as metamorphic; & therefore according to LYELLS doctrine removed?? Is the prevalence of Conife
250c039 s but not. genera distinct from rest of world?? LYELLS Principles, must be abstracted & answered Much
256c053 e homogeneous.-- There must be some sophism. in LYELLS statement that some species vary more,. than w
263c074 ugh not MAN.-- is as difficult to understand as LYELLS doctrine of slow movements &c &c. this multipl
295c178 yderm in Portland stone of Alps!!!? No) p. 15 (LYELL'S Pamphlet) Is man more hairy than woman. becaus
321c270 Lisiansky's Voyage round world. 1803-6 Nothing LYELLS Elements of Geology Gibbons life on himself Hu
352d060 perhaps not abortive??? Apterix certainly.-- LYELL'S excellent view of geology, of each formation b
373d133 -- the <nerve> living nerve nursed in Mould.-- LYELL'S Elements. p. 290. Dr. Beck on numerical propor
426e105 ked species never reappear when once extinct.-- LYELL'S argument about <Tertiary> Isld «neighbouring»
448e167 eous»-- how is this with the Eocene beds.-- see LYELL'S tables Bennetts Wandering Vol II. p 155. By in
465t089 ain shows how much forms depend on other forms LYELL'S Paper, in Taylor's Journ.-- Phil. Mag. May. 18
481z010 d in the Tucuman mountains The fourth Vol. «in LYELL'S possession» of Zoolog. of Voyage of Astrolabe
109a082 baqueous. removal, shown by the number of bones LYING at the bottom of sea. off coast of England.-- S
486z020 rchaeologia arrow=heads described in Suffolk as LYING under strata of gravel & clay about 10 feet in
547m111 rewsbury., vaguely thought of packing up.-- was LYING on my back fell to sleep for second & wakened.-
489qIFC n p 17 Hooker p. 17 {T} Mrs. Whitby. Newlands LYMINGTON Hants. Habits of different caterpillar races.
383d160 to be infertile.-- spurs rather smaller than in <MA> silver male-- Head like silver except in not ha
155g065 e on sea hypothesis, if gullies not now formed «(MAC, hypoth,)» the level during any oscillation must
157g078 ivortium aquarium is a lip with it-- Dick right-- MAC mistook terrace also right-- Granite such as bou
633j54r whose flower great tendency to break off p. 292. MAC. has long rigmarole about plants being created t
634j54v ars.-- p. 313 Many other good cases -- p. do]CD <MAC. remarks all Mammifers originally land--animals.
634j55r ammifers originally land--animals. as> 5 p. 314. MAC. remarks all <land> Mammiferous animals original
545m105 was much struck with observing how the Baboon (<MACACO> «Cyanocephalus Sphynx Linnaeus») constantly m
596m184 head,-- "scarcely able St.-- » Cyanocephalus, MACACUS. Cercopithecus? very much., «Keeper says some
596m184 ot> Chimpaze. does not gradation towards man.-- MACACUS especially pulls back skin of whole forehead &
596m184 onkeys <are> right-handed??]CD» Cyanocephalus, MACACUS, Niger. Cercopithecus make labial st st. S. Am
184b054 Water-- on what is species. very good Has not MACCULLOCH written on same changes in Fish Mem. Rabbit
572m035 vidual was invented to conceal one's thought.-- MACCULLOCH in his Chapter on the Existence of a Deity h
602o11b what never existed but in the imagination'.-- MACCULLOCH Vol I. p. 115. Attributes of Deity. on Belie
632j53r ed by D. Hartley.-- Darwin's Abstract of John MACCULLOCH 1837 Proofs and Illustrations of the Attribu
632j53r fs and Illustrations of the Attributes of God MACCULLOCH. Attribs of Deity: Vol: I it will be better
635j56r iven we know not the effect [blank] 6 p. 412. MACCULLOCH explains the shortness of life (peculiar to
635j56v true, (& the doctrine of checks & my theory) MACCULLOCH. Attrib. Vol I. p. 330. Mentions the many ca
636j56v ixed & gardner have hybrid seedlings} p. 333. MACCULLOCH. brings forward. the impregnation of Dioecio
636j56v ed?» as soon as Dioecious Plants were formed. MACCULLOCH says, life, forms a broken, recurrent series

636j57r refore trace of beginning in organic world.-- MACCULLOCH. Attrib. of Deity. Vol I p. 232. gives Woodp
638j28r ory of instinct, returning to Kirby's view.-- MACCULLOCH. Attributes of Deity Vol I. p. 251-- stomach
639j28v Echidna.. & Aphrodites C. D. Endless cases.-- MACCULLOCH. p. 260 intimates canines no special use to M
639j28v -- p. 263. This kind of doctrine runs through MACCULLOCH, the bills of the Grallae <are> «have been m
641j29v by very different process from Bats. CD]CD. «MACCULLOCH says no other bird could catch mouse by nigh
155g063 believe in side ravine above houses of Roy» MACCULLOCHS supernumerary shelf I doubt, much about <50
354d062 rated by Mountains. & & &c.-- do. p. 69. A Dr MACDONALD believes the Quaternary arrangement & not the
481tf03 raphically than others. Athenaeum p. 605 Mr. MACGILLIVRAY says "<A Thrush &> Blackbird have been know
089a019 h to see structure.-- «H» Thought of erecting MACHINE to see if water fell. -- <Keys off extreme poi
495q05a turning powers-- then carry them in Electrical MACHINE, reversing the poles test by suspending magnet
583n070 ry senses & muscles, we cannot look at him, as MACHINE to make cell of certain form. (& especially as
263c078 instincts.-- this is a replacements in mental MACHINERY-- so analogous to what we see in bodily. that
050r090 inute craters as at Galapagos. <|> Sir George MACKENZIE must be worth reading Some earthquakes of Sum
072r159 wise» Mendoza never overthrown,--no mountains MACKENZIE has talked of lava flowing up Hill; ;what doe
096a039 subsiding.) Von Buch Canary Isd. p. 351.. NB. MACKENZIE talks of gravel on basalt of Heckla-- All the
210b159 , in Laponia. genera to species I. 2,3-- From MACKENZIE Iceland there 144 genera & 365 species of pla
234b255 . Voyage à France-- Par un Officier du Roi.-- MACKENZIE Travel. p. 280. says cattle in Iceland. «"are
481z009 s drawings.-- for real name Birds of Iceland. MACKENZIE. p 345 for comparison with Falkland. good als
320c276 h have references at end; is so said to have MACKENZIE'S Iceland Molinas Chile Falkners Patagonia Aza
308c218 !? examine into this case D. Jeffrey (life of MACKINTOSH Vol II. p. 495)-- in fact, in all reasonings
325c267 Reid, Smith & giving abstract of their views MACKINTOSH Ethical Philos: Prostitution of Paris. with
537m075 think, the opposite side has been shown-- see MACKINTOSH.-- Must grant, that the conscience varies in
558m151 well as the works of the whole world.-- Read MACKINTOSH on Moral sense & emotions.-- The whole argum
587n089 expressing a mental <desire> «quality» &c &c MACKINTOSH Ethics p. 97. on Devotional feeling p. 103--
587n089 ed believe, try contiguity of parts of Brain.-- MACKINTOSH first clearly insisted on assoc of ideas & e
600o08v & intellect of beasts, not clear.-- ;does not MACKINTOSH make great difference between moral sense &
618o042 effort is felt [LHC] 1) May 5th. 1839.-- Maer MACKINTOSH Ethical Philosophy [RHC] On the Moral Sense
619o042 ts, and perhaps others.-- [LHC] ----- p. 113. MACKINTOSH Grotius has argued nearly so [RHC] The histo
625o50v tem not selfish.-- explained by principles if MACKINTOSH.-- p. 262. Some good remarks, on analogy of
627o52v sult <following> pages. <p. 231> marked in my MACKINTOSH 1) Mackintosh's Ethnical Philosophy p. 6-- "
628o054 moral sense all different.-- P. 22. Butler & MACKINTOSH characterize the moral sense, by its "suprem
628o054 -- love of fame-- Yes Hartley explains this & MACKINTOSH shows the change produced.-- 4) p 38 Conscie
629o54v sh of outward gratification,-- see what cases MACKINTOSH gives & try it.-- p. 241 (1) Any action by h
323c269 ay 18th Stanley familiar History of Birds -- MACKINTOSHS' Ethical Philosophy -- Bell's Bridgewater Tr
625o052 I admit its supremacy p. 37. Whewells gives MACKINTOSH'S theory: the remarks about "contact with wil
627o053 > pages. <p. 231> marked in my Mackintosh 1) MACKINTOSH'S Ethnical Philosophy p. 6-- "The pleasure wh
201b129 fe[w] [not located] The relation of Analogy of MACLAY &c. appears to me the same, as the irregularit
208b154 rld Is monkey. peculiar to C. de Verd's.--? NO MACLEAY Name given in Congo Expedition We need not expa
251c042 be included in Smiths work «do» Vol. IV p 273. MACLEAY on Caproms. 4 species probably in Cuba (p 271
252c042 ppes of Kordofan p. 401. Admirable letter from MACLEAY to Bicheno much excellent detail & fine, views
284c149 rent group. & Not known in single ones--. viz. MACLEAY letter to Fleming p. 32 "where it (mode of gen
287c157 lternative of Man created by distinct miracle. MACLEAY letter to Dr. Fleming. Philosophical Magazine
287c158 tween cow & hawk a frolicsome saltus.-- «p. 19»¶ MACLEAY seems to limit Lamarck definition of relations
288c158 on these same organs,-- habits, range. &c &c-- MACLEAY rests his whole groundwork of analogy on its c
302c202 acters of analogy.-- see my notes on p. 37. of MACLEAY. wonderfully accordant. with fact there stated
308c218 nly two fishes fitted for water, air, & land, (MACLEAY has this remark) Mem. number 5 here most evide
317c249 Jago C. de Verd; same as on coast of Africa.-- MACLEAY tells me same thing p. 55. 40 leagues from lan
347d050 ts.-- instance of arrogance!! August-- 29th.-- MACLEAY in A. Smith's Zoolog.-- of Africa -- p. 4. sti
349d052 rmanent variety & species!! (given in note.)-- MACLEAY <met> uses term genus when it is so many steps
350d055 icans. &c After Decandolles idea Septemb. 1st. MACLEAY & Broderip were talking of some Crustacean, li
350d055 , parent «(2)», female «(I)» fixed & blind: -- MACLEAY observed all these facts prove that perfection
351d056 ted that stationary Spondylus has eye-points-- MACLEAY then answered, because nature leaves vestiges
365d107 n's land from Australia &c &c Sept. 14th. When MACLEAY says their is no difference between <t> "perma
380d153 ness of life in female Moth &c Mr Y. says that MACLEAY considers the house bug, as a <female which ha
403e022 a of pacific, given in my coral paper Oct 14th MACLEAY says, that <every> «any» character even colour
403e023 part is under change, now one part now another MACLEAY says it is nonsense to say take a tooth of any
407e042 t 1837. p. 253, on animals of Antilles.-- (see MACLEAY in Zoolog. Journal. for those of Cuba.-- It is
485z018 olypi.-- Berenica &c & L'Institut, 1838 p. 46 MACLEAY Horae Entomolog. insects swarm in Lapland & St
320c275 to end of 1837. re«a»d-- contains very little MACLEAY'S letter to Dr. Fleming. & Review of latter in
321c275 ces at end Dr. Lang Australia «trash» skimmed MACLEAY'S Horae Entomologica Ray's Wisdom of God refere
348d050 atural (p 6) as expressing natural affinities¶ MACLEAYS plan of arrangement depends on the organs jud
352d059 le, two or three lines deep-- with respect to MACLEAY'S theory of analogies-- <be> when it is conside
284c149 uralists in their test of value of character-- MACLEYS rule is converse, «when> value of character de
482z011 & another species! birds of passage!! sylvia MACLOVIANA, 2d like sylvia cisticola-- Embriza melanod
162g102 e between bedrock & Loch Ness <30.100> <Donald MACPHEE> Saturday Morning 29.958 A 64 degree, air 60 «
243c019 Voyage p. 25 Mais il n'y a pas jusqu'aux îles MACQUARIE et Campbell (52 degree S) qui n'aient egaleme
258c061 this light.--XX Zoolog. Journal-- Parrots in MACQUARRIE isld. vol III p 430 alluded to by Capt. King
268c099 sert & Tierra del Fuego. & forest «Parrots in MACQUARRIE Isd.--» very good. Study D'Orbigny. & range
065r138 e Edwards Isld. Marion & Crozet. L. Auckland. MACQUERIES.--Sandwich Isd-- Species of rocks were bro
228b231 eason to expect the father of man kind. than MACRAUCHENIA yet he may be found:-- We must not compare
278c132 nas of these countries, greter than Toxodon, MACRAUCHENIA, &c compared to America.-- the wonder is th
319c257 reasts.-- L' Institut, 1838, p. 230 says the MACROTHERIUM of Europe is between the Anteater of C of G
095a037 } The Pota: labiata certainly is found with the MACTRA. at Buenos Ayres at the Zoolog: Soc: Terebratu
244c023 here.-- p. 158 Cuscus albus. New Ireland ---- MACULATUS -- Waigiou Speaking of Lepus Magellanicus say
156g070 e now formed in same spot by present torrents MACULLOCH wrong in saying no transported materials <int
156g069 the 3 shelves» Solid rock is much notched on MACULLOCH'S supposition;-- the old ravine, where water e
267c094 & to cross their breed p. 333-- alludes to the MACUSIE breed no description given-- Ch. 2. dogs L'Ins
060r126 ve been. The red Sandstone of Andes fusible? no. MAD dogs. Azores. although kept in numbers. p. 124.
511q018 rizes.= Pidgeons. Canary birds-- Bantams.-- (6) <MAD> Porto Santo Rabbit. Descript. of colour «& leng
052r098 ssion.--Brazil bank: (& I believe SE coast of MADAGASCAR. where a --40 line <shows> runs at equal dis
212b166 t & reason Some animals common to Mauritius & MADAGASCAR.? Proceedings of Zoolog. Soc June 1837 p. 53
217b187 «form of» Lathyrus has one species in Europe MADAGASCAR has several American forms-- The above facts
218b190 pt to produce monsters in Isle of France-- -- MADAGASCAR oxen with hump.-- p 173. Voyage par un Offic
226b220 celand?--» The Connection between Mauritius & MADAGASCAR very good.-- Fernando Po & Coast of Africa.
232b249 ous facts about the distribution of Lemurs in MADAGASCAR, on neighbouring islets & a sub-genus in Sou
233b251 M. J. says some reptiles same from Maurice & MADAGASCAR & C. of Good Hope.-- His book Probably worth
234b255 ish introduced» Hump backed race of cows from MADAGASCAR-- p 173. Vol I. Voyage à France-- Par un Off
244c022 te p. 136. Isle of France.-- the Tenrecs from MADAGASCAR. Monkey from Java.-- Hairs, & deer.-- Procur
246c026 oucal) of Java & Phillippines, has variety at MADAGASCAR, Calcutta & Sumatra,. but I do not see how i
303c205 . 128. Extraordinary genus. Mesites bird from MADAGASCAR uniting pidgeons & gallinaceous birds & parr
307c216 nt of tusk «A Molar tooth» of Hippotamus from MADAGASCAR!!!!! Proceedings of Geol. Soc Vol I It is c
338d025 st important under this view, & Hippotamus of MADAGASCAR: because. contemporaries. In introduction to
421e091 er parts of that continent. in like manner as MADAGASCAR does to other side of Africa.-- (& Juan Fern
459tf1r t of Africa-- Vol II. p. 250-- wild cattle at MADAGASCAR-- «p. 121» No beasts of Prey. any country sh
585n074 affected respiration V E. p. 125 Wrong Entry MADAGASCAR Lemur seemed to like Lavendar Water «very mu
032r040 continuing. a another sloping platform would be MADE, & so on.-- This is grounded on the belief of c
035r046 ins; little or no relation appears to <exist> be MADE out, but in those belonging to the same distric
103a065 ocation of strata. A capital discussion might be MADE between dikes & «axis of» mountain-chain in pro
145g002 Animals in Britain shows that either races soon MADE or crosses difficult Salisbury Craigs The Highl
162g099 s belong are so complicated, that nothing can be MADE out of them-- but it may be said that a mound s
173b009 ies according to Lamarck disappear as collection MADE perfect.-- truer even than in Lamarck's time. G
178b032 cross at C. of Good. Hope the children cannot be MADE intermediate, the first children partake more o
181b045 least during subsequent ages.-- The Creator has MADE tribes of animals adapted preeminently for each
191b082 ceros. Elevate & join keep distinct. two species MADE elevation & subsidence continually forming spec
194b094 o 246» a section of fossil "singe", it cannot be MADE to approach the Colobes which in south Africa,
194b095 ia.-- I really think a very strong case might be MADE out of world before zoological divisions.-- Mem
195b098 on are arbitrary, & not permanent. this might be MADE very strong. if we believe the Creator creates
197b112 ol of Fish p. 59.]CD Cuvier has said each animal MADE for itself does not agree with old & modern typ

202b130 mmon end of structure A Race of domestic animals MADE from influences in one country is permanent in
209b155 petuating these particular varieties. If species MADE by isolation; then their distribution (after ph
218b189 e when from being put on island. & fresh species MADE. parents do not cross-- we see it even in men);
219b193 volcanic point appeared in the great ocean, have MADE plants of American & African form, merely becau
225b216 inst it-- namely how did otter live before being MADE otter-- why to be sure there were a thousand in
228b231 connexion of two plants. Animals-- whom we have MADE our slaves we do not like to consider our equal
234b256 tion of 1838. (Newcastle) about somebody who had MADE great collection of birds of Iceland. --M. Gaim
248c030 (which can be shown probable,)» varieties may be MADE in wild state, there will be presumption that t
258c061 is system of nature-- Whether species may not be MADE by a little more vigour being given to the chan
263c078 at circumstances may have been necessary to have MADE man! Seclusion want &c & perhaps a train of ani
271c106 re offspring picked, one where not.-- the latter MADE by man & Nature; but cannot be counteracted by
285c150 er» » & grandfather <Might> Must be introduced & MADE young.--\ father must be left out of case, that
285c153 which other marked difference in the varieties «MADE by» of Nature & Man.-- The constitution being h
291c168 ian species would migrate south ward being ready MADE.-- & so destroy individuals, wheras in Falkland
297c186 ew in number of kind, extermination.-- New forms MADE through probably an infinite number of forms.--
299c195 seeds   produce <offspring> variety. wild carrot. MADE into biennial domesticated kind with large root
308c217 ght back memory» old habit of putting tea in pot MADE me go to tea chest almost unconsciously.-- why
313c234 if louse created should not new genus have been MADE, & only species, good argument for origin of ma
315c243 a baboon with great canine teeth.-- (This may be MADE capital argument if man does move muscles for u
324c268 s of Domesticated animals see if law's cannot be MADE out Find out from Statistical Society-- where M
337d022 he Entomological Soc> A capital passage might be MADE from comparison of Man, with expression <of a>
343d037 tra, that since the time of the Silurian, he has MADE a long succession of vile Molluscous animals--
345d044 ion in breeding. how easy races or varieties are MADE.-- The Highland Shepherd dogs, coloured like Ma
346d047 Wader. Ibis)-- laws of plumage might possibly be MADE out.-- August 25th Athenaeum (1838) p. 611. Ld.
352d058 groups.   When <species of> a group of species is MADE. father probably will be dead-- hence there is
354d065 nge-- Argue the case of Probability. has Creator MADE rat for Ascension.-- The Galapagos mouse probab
360d089 tile ones.-- Keeper said in <two> crosses «twice MADE» between terrier & hairless dogs of Africa,-- s
360d092 any kind>, if very firmly fixed from long time, MADE no difference what its kind was.-- but if it we
364d104 been thought that silver Pheasants about a house MADE other pheasants have white feathers).-- It cert
365d106 ern species, & being restricted species has been MADE.-- In the hybrid grouse between Black Cock & Pt
367d113 ringed with brown.-- animal like large, heavily MADE cream coloured ass.-- stripe on back also.-- le
370d116 om case of wasps, is supposed cells properly are MADE for larvae.-- CD[(p. 451.)-- Wasps breed many f
376d136 rcarius.-- this Monkey did not like a great coat MADE for it at first, but in two or three days leann
376d137 ugh could not put it on, yet threw it over it, & MADE it meet in front.-- Dr Smith every baboon & mon
377d138 e noise st st. which the C. Sphynx makes is also MADE by the C. porcarious., together with a grunting
392d130 ake Hybrids with moths, where fecundation can be MADE artificially.-- Are hybrids pintail & common du
400e012 er of fur-- I am sure a very good case, might be MADE out of variation analogous to specific variatio
400e013 ppose one aboriginal variety.-- for they are all MADE by fertilizing one plant with another-- Uncle J
403e023 d.-- Octob. 16th. A very strong passage might be MADE-- why seeing great variation in external form o
404e024 is good, because it is an involuntary variation MADE by man, common to every individual & therefore
415e068 attendant   organization, Hand & throat) that has MADE a man.-- CD [any monkey probably might, with su
415e069 [any monkey probably might, with such chances be MADE intellectual, but almost certainly not made int
415e069 s be made intellectual, but almost certainly not MADE into man.-- It is one thing to prove that a thi
415e070 y most valuable references See if any law can be MADE out, that varieties are generally additive, & n
416e071 that <domesticated> races. of <a> organics. are MADE by percisely same means as species-- but latter
416e075 an any living [not located] A Greyhound might be MADE «almost» without any relation to running hares.
430e118 pecies. even if it will for all.-- Varieties are MADE in two ways-- local varieties, when whole mass
432e124 he native <breed> animals of any country must be MADE with great caution; owing to its adaptation to
451e176 o is held sacred by the credulous natives, & vow MADE at it. Both his parents were of the usual colou
455e1BC g always at same time» see if by so doing can be MADE sensitive The function of sleeping someway usef
486z019 log. Transact. Vol I. p. 130. Col Sykes on balls MADE by dung beetles, like those from Chiloe. Amblyx
496q008 ow of fish & kill them in hour or two «My Father MADE hens cast Holly-seed & they grew» (9) Place. Sn
510q018 onfinement   curious.-- foxes-- English animals. [MADE no import. remark]CD (2) Secondary male charact
533m060 Capt. F. R. candour & ready confession of error MADE him less repentant.-- In making too much profes
534m063 ion of intellectual faculty] if ants had at once MADE this leap it would have been instinctive, seein
534m063 stinctive, seeing that time is lost & endeavours MADE must be experience & intellect.-- do. p. 157. W
535m069 mind: the Chileno says the mountains are as God MADE them,-- next step plastic <virtue> natures. acc
539m081 on from reading «review of» M. Comte Phil. which MADE me «endeavour to» remember, & to think deeply,
555m143 ht that a person was hung & came to life, & then MADE many jokes. about not having run away &c having
556m145 .-- reason CD[The laughing noise which C. Sphynx MADE at Z. Gardens may be described as partaking of
556m145 . Gardens may be described as partaking of <st.> MADE by <ret> inspiration & quickly retracting tongu
559m156 lorescent parts of mosses & see if Hybrid can be MADE & ferns.--» Would a sensitive plant if irritate
569n023 e grief-- "there is nothing more to be said."-- "MADE no reply, but shrugged his shoulders & went awa
579n060 in former irritation on a piece of skin cut off MADE the blush come.-- it is an excitement of surfac
602o011 he is right-- it is because each decision &c is MADE up of many partial results, & the impressions o
609o029 ssary to the vice of intemperance, circumstances MADE the check.-- to licentiousness jealousy, & ever
613o036 ore word misplaced The meaning of Words, must be MADE out Reason Will Consciousness Definite instinct
616o39v s of perceptive action by which bodily action is MADE known to us, revealing respectively what are ca
628o53v act by gusto, though by fear it might be partly MADE.]CD p. 21. "Why ought I to keep my word"-- give
638j28v to the then existing conditions.-- An adaptation MADE by intellect this process is shortened, but yet
638j58v is shortened, but yet analogous, no savage ever MADE a perfect hinge.-- reason, & not death rejects
639j28v lloch, the bills of the Grallae <are> «have been MADE» long «(as adapted to)» because their food lies
047r082 e right the coincidental retreat at Portugal & MADEIRA (Lyell. vol I. P. 471) is explained. also the
105a072 th calcareous matter & concretions on coast of MADEIRA.? How came it if this powder results from «dec
222b204 ther papers besides one in Latin one <of> «on» MADEIRA-- any general observations-- difference of spe
222b204 f species between land shells of Porto Santo & MADEIRA-- I believe very curious-- My idea of propagat
225b220 ny animals?-- I believe none.-- Canary islds.? MADEIRA? «Tristan d'Acunha?» «Iceland?-- The Connecti
288c160 s: mysterious. Were the woodcocks., which came MADEIRA & <seized> ceased their migrations lost?? I co
490q001 Hill-- character of the extinct land-shells of MADEIRA-- analogous or quite distinct from recent ones
502q11v urf. for abortion. or for sterility Land Birds MADEIRA Migratory-- ask Gould about N. Zealand, as Cuc
511q018 hem & Glöger Consul Hunt, birds from Azores or MADEIRA Mr. Blyth (1) Mentions some breeder who raises
523m018 ms no distinction between enthusiasm passion & MADNESS.-- ira furor brevis est.-- My father quite bel
523m018 and F doctrine is true, that the only cure for MADNESS is forgetfulness.-- which does appear a real d
524m019 oes appear a real difference, between oddity & MADNESS.-- but then people do not well recollect what
128a128 n Cordillera Geograph. Journal vol II. p 89. at MADRAS. surrounded by salt water. purest fresh water
077r171 a decided influence on the richness of the veta MADRE of [misnumbered page] Dr D. remarks. bad conduc
481z010 or Fleche «p. 8» my little animal with horns. MADREPORES p. 26 Nullipora p. 29-- In Meyen. Voyage rou
043r070 ake of 1812 affected valley of Missisippi & New MADRID & Caraccas.-- Is this mentioned by Humboldt in
366d111 -- Case of Association very disagreeable hearing MAED servant cleaning door outside, as often as she
317c249 ¿cut? has its offspring short tails /one born at MAER. Tuckeys voyage-- p. 36 "Cercopithecus saboeus"
319c257 ith South America.-- In Home's History of Man at MAER, it is said the Samoyed women (¿ north end of t
466t091 ations are there in old Secondary Series-- few-- MAER June/41/, observed 3 plants of Caltha Palustris
466t093 n, whilst both others were in nearly full flower MAER June/41/ Rhododendrum-- nectary marked by orang
466t099 vation during life of individual. June 1st 1841. MAER Examined the Lemon-thyme.-- equally abortive as
467t105 ther on Cabbage--white Butterflies suck nectar: «MAER June 41» Rhubarb. pollen very minute--not exces
469t135 bably monstruous & not a second species.-- <Saw> MAER. June 15./41/. Watched plants of Fraxinella, wi
470t176 e latter were pollen gatherers & they seem slow= MAER 1840 My Father formerly planted «Turkey or» Pal
472s02r erably paler, all rest very similar-- June 2. 42 MAER <Thursday> Thursday After watching 14 days. man
473s004 s are probably buried in the depths of the sea-- MAER. June/42/ June/42/-- Mr. Bunbury says has heard
489qIFC iments {T} Gowen, Royle, & Horsfield Sykes p. 12 MAER. p. 13 Question &c. July. 1842.-- Shrewsbury p.
497q006 show that such proportions not effect of Chance MAER.= (12) Take Bag of soil from centre of woods «e
498q008 ?-- I presume only stigma impregnable.-- (12) At MAER Cowcumbers in frames are not artificially impre
503q013 ie jaw cross with cow: is not reverse possible?? MAER (1) Yew Trees near Boat House «ANY male branch.
525m024 to do, & looking ashamed of himself.-- Squib at MAER, used to betray himself by looking ashamed befo
593n111 by no means a necessary part of man's mind.-- At MAER. Pool. I saw many coots & waterhens feeding on
594n112 aps, as strong as against hawk, but the birds at MAER have learned that he is not dangerous-- wild-du
618o042 because effort is felt [LHC] 1) May 5th. 1839.-- MAER Mackintosh Ethical Philosophy [RHC] On the Mora
106a076 - <Is there same.> Institute. 1838 p. 40 or Phil MAG. Dec 1837. p. 520 Mr Fox on increase of temperat

201b126 ndia.-- & p. 545 «great monkey» Mr Johnston says MAG of Zooly & Bot. p 65 Vol II talking of annelidae
203b138 «species peculiar to Continent & England» Loudon MAG: Septemb or Octob 1837 Westwood has written pape
297c186 rms. & they are remnants.-- Cephalopoda ditto.-- MAG of Zoolog. & Bot. Vol. II p. 125 Allusion to abo
297c189 Ouzel sometimes builds nest without doom. Vol 2. MAG of Z. & B. p. 431. Missel thrush lately increase
465t089 forms Lyell's Paper, in Taylor's Journ.-- Phil. MAG. May. 1840 p. 362.-- some Mammals of Norfolk Cra
477z001 excellent references in L. Jenyn's introduct to MAG of Zoology and Botany. Philosoph. Transacts. 3.
480z007 es of S. America translated from D Orbigny no IV MAG. of Zoolog & Botany p. 356 Lesson on Berre. do--
485z017 ites 4 distinct Camelidae. do not breed together MAG: of Zoolog & B. Vol. II. p. 127. List of submari
117a102 d would daily be scene of ruin in late Natical MAGAZINE (before June 1838) that 70. F were obtained 1
193b091 es First Voyage Copied into list Entomological MAGAZINE paper on Geographical range Richardson-- Faun
203b138 f the former Eyton's paper on Hybrids Loudon's MAGAZINE. Gould on Motacilla.. «species peculiar to Co
217b189 tself-- on hybrids between grouse & pheasant-- MAGAZINE. Zoology & Botany Vol I p. 450 There is in na
220b198 ces, paper on Botany of Tahiti In Charlesworth MAGAZINE Jan: 1830. most curious paper on heredetary f
267c095 2. Ehrenberg Annals of Nat. Hist. precursor of MAGAZINE??? p. 75. roe of Asterias in stomach. of Samm
287c157 . Macleay letter to Dr. Fleming. Philosophical MAGAZINE & Annals. 1830 (?)." if she has put man on th
288c160 f. land having been formerly nearer.-- «Selby» MAGAZINE of Zoology & Botany No XI p. 390. a slight ch
289c162 :Scales into Teeth in Bering Pike (Waterhouse) MAGAZINE of Zooly & Bot-- Vol II p. Dr Johnston <on> E
292c170 hells ¿ univalves or bivalves.-- Anyman No VI. MAGAZINE of Zoology & Botany p. 566 wants to see absur
296c184 Sorex from Mauritius. p. 112. & paper on genus MAGAZINE of Zool. & Bot.-- Vol I. p. 450. 4 instances
296c185 ild Cattle. breed freely with the tame Vol II. MAGAZINE of Zoology p. 56. Peregrine Falcon holds bird
313c235 her specific extinction-- In the Entomostraca (MAGAZINE of Zoology & Botany) where several generation
320c275 racted Scientific Memoirs. published by Taylor MAGAZINE of. Zoology & Botany & Continuation «Annals o
324c268 s &c Richardson's Faun. Borealis Entomological MAGAZINE (paper on Geograp. range Study Buffon on Vari
353d061 point of resemblance with man.-- September 3d MAGAZINE of Natural History. 1838 vol II p. 402. Mr Go
449e169 e very like a Himalaya species -- leuconotes-- MAGAZINE of Nat. History. 1839. p. 106.-- Waterhouse r
469t151 p in 1840 true. Shrewsbury.-- Abberley-- Early MAGAZINE-- &c. double-blossomed «& dwarf-fan Bean» bea
480z007 f Zooolog & Botany p. 356 Lesson on Berre. do-- MAGAZINE of Zoolog & Botany. Vol IV p. 358. D'.Orbigny
559m155 sciousness in Brutes & Animals. in Blackwood's MAGAZINE June. 1838. Copied Mr H. C. Watson on Geograp
604o015 n of Evil. Reviewed by Johnson in the Literary MAGAZINE. 1756-- Ceased in 1758-- Read the Review or t
048r087 ine channels. such as that in front of Sts. of MAGELLAN In Chiloe curvilinear strata subsidence.--The
300c197 pecies.-- a few eggs transported to the Str of MAGELLAN.-- Change of habits in Van Diemen's land. Stu
483z012 . 422. letter from Capt King on birds of St of MAGELLAN. Very inaccurate & Vol IV p. 91.-- Vol IV p.
145g002 igs The Highland shepherds dogs coloured like MAGELLANIC fox.-- an instance of Provincial breeds. [3]
345d044 .-- The Highland Shepherd dogs, coloured like MAGELLANIC Fox.-- peculiar hair & appearance-- good cas
244c023 ---- maculatus.-- Waigiou Speaking of Lepus MAGELLANICUS says; <after> "après un examen attentif, et
271c105 . Namely Carus.-- How remarkable that Turdus MAGELLANICUS. in the. S. Hemisphere. (replaced to the No
482z011 caught.-- Troglodytis Furnarius.-- Sturnus MAGELLANICUS.-- p. 210. Scolopax very close to ours Reng
485z017 replaced by Brazilian Species.-- Mem Turdus MAGELLANICUS.-- C, <Chingolo> Chimango-- Diuca?? See Rep
402e020 island-- In Sooloo we find the elephant-- in MAGINDANAO several kinds of the large monkeys.-- Fewer
452e181) there are hogs & monkeys <at> near shore of MAGINDANAO Journal of [Asiatic Soc] [...] p [...] -- mo
124a119 rallel to limestone & volcanic rock containing MAGNESIA Lyell. Elements p.119 on such strata {P} do p
373d133 elkalk, & 5 genera of reptiles.-- <M> p. 417. MAGNESIAN Limestones & Zechstein oldest rock in which r
496q006 .= (11) Try.. Nitrate of Soda-- Salt. Gypsum. MAGNESIUM Iron Rust Carb. of Ammonia.-- Horse Urine &c
495q05a machine, reversing the poles test by suspending MAGNET within & see which way they fly.-- (9) I have
073r162 »n, that the metals should be those which have MAGNETIC properties. Study well products of Solfataras
090a021 e but old Planet, that inside our globe melted MAGNETIC metals. ∴ earthy crust compared to those of f
585n076 when,, man went out of room.-- all theories of MAGNETIC powe in birds, seeing the sun &c are absolute
586n078 er,, is faculty, whether by sun, & heavens, or MAGNETIC virtue,-- the most probably supposition. with
102a062 ain directions predominantly, connection with MAGNETISM &c counteracting gravity.-- As volcanic erupt
305c211 es, & recurrent habits in animals.-- --Animal MAGNETISM-- principles of irritations sleep walking. fi
342d036 not thus characterized.-- 16th Aug.-- What a MAGNIFICENT view one can take of the world Astronomical
553m136 ot as a necessary integrant part of his most MAGNIFICENT laws. of which we profane «degnen» in thinki
496q05a such flowers as do not seed or seed rarely-- MAGNOLIAS. «Azaleas» & plants grown under unfavourable
378d147 of the pies assume the metallic tints, such as MAGPIE, Jay, & perhaps all the rollers-- «He says» wh
284c148 on long average equal.-- The Cocos do Mar on the MAHE island`, one the higher parts & only on those,
637j58r pecial adaptation, to «young».-- good God & yet MAILS have them. What trash p. 237. Gives as Summary
030r034 rmed by outlying rocks; (such as between Mocha & MAIN land). <[...]> At Carelmapu.--Within Chiloe:--
088a015 st of Cordillera very important V. Malte brun -- MAIN character of Andes Metamorphic action -- Mem: r
156g071 like one uniform slope slightly bending up each MAIN valley.-- & that river alone had modified it--
176b024 <six> three more, double arrangement.-- if each MAIN stem of the tree is adapted for these three are
181b044 anomaly & <the g> bearing stamp of <some> great MAIN type, & the gradation will be sudden-- Heaven k
228b228 ight lead to laws of change, which would then be MAIN object of study, to guide our <past> speculatio
233b250 sion of form in birds.-- Waterhouse thinks two MAIN divisions of cats. Tortoise shell--& grey-bande
288c160 a bird Migrating from Falkland Isd regularly to MAIN land, proof of. land having been formerly nearo
401e017 says common belief. that female plant impresses MAIN features on offspring. & male the lesser peculi
471tf08 o new genera of coniferae".-- referring to the 3 MAIN divisions & speaking of their similarity «in st
549m121 id of intellectual <ple [...] hope> cultivation, MAIN source of the intense happiness.-- it is again
521m009 er he had gardener of name A. B., &c &c. & he MAINTAINED he had never heard of such a man & had no ga
522m010 xplained who he wa & all about him, but still MAINTAINED he had never heard of him).-- My F. then sai
522m011 all the same names as a few minutes before he MAINTAINED he had never heard of.-- Thus in many things
422e092 p. 318 M. Pictet of writings of Goethe.-- who MAINTAINS, that «Alludes to difference between fossil &
531m051 hildren shows that sympathy is based as Burke MAINTAINS on pleasure in beholding the misfortunes of o
243c019 -» My toad is same species Coquille Voyage p. 25 MAIS il n'y a pas jusqu'aux iles Macquarie et Campbe
251c040 eese killed in Newfoundland, with crops full of MAIZE. (get limits of latter from <Tarton> Barton.--
021r005 P 379. Henslow Anglesea, nodules in Clay Slate. MAJOR axis 2.½ ft.-- singular structure of nodule, co
113a092 rt.-- Important as explaining want of levelness MAJOR Mitchell showed me a river <near> W. of Port Ph
113a092 k.-- 4 & 5 fathoms deep. perfectly still water. MAJOR Mitchell inferred subsidence; My remarks on th
278c131 s most important!! like Dipus of present day??! MAJOR Mitchell does not think that dog was found in V
288c159 ie, cannot be discovered till circles completed MAJOR Mitchell, does not know whether breeds of oxen
306c213 dog. perhaps being in passion chief difference MAJOR Mitchell is not aware that Australian dogs ever
356d071 & linking of facts-- Savages over whole world. (MAJOR <I> Mitchell p. 244. vol I) spit & throw dust <
376d137 d clasp them round waist & look in their faces & MAK the st. st noise.-- The Cercopithecus chinensis:
036r050 ere is a Hill. near Copiapò which is asserted to MAKE a noise,--My impression. is not very distinct,
065r140 began to freeze. correspond to September ¿Did I MAKE any observations on springs at S. Cruz.???-- Fo
069r152 ace protected.-- Oh the vast power of the ocean! MAKE a grand analogy between Wealden & Bolivia Trans
087a010 out the Megatherium.-- The Guanaco the Camel.? MAKE note about N. American bone not probably in lak
099a049 they not attract strong. a third.-- & this would MAKE layers.-- (Gravity can have no effect. on parti
118a103 s equal width --subject of fine paper this would MAKE.-- L'Institut. (1838) p. 216 M. Gay on the Geol
156g068 not scooped rock on <bend> of 3(a) Cannot <see> «MAKE out» composition of shelves: generally angular
179b034 e would probably be repugnance & art required to MAKE marriage.-- as Dr Smith remarked Man & wild ani
198b114 n leaving all to follow consequences.-- I cannot MAKE out his ideas about propagation His work. Philo
213b169 cies» races.-- If all men were dead then monkeys MAKE men.-- Men makes angels-- Those species which h
219b193 d trees]CD & would creator <on volcanic island.> MAKE plants <grow closely> When this volcanic point
223b209 red they will cross (perhaps more fertility & so MAKE that sudden step. species or not. A plant submi
225b217 I remember. L. Jenyns. talking of it) or how to MAKE Indian Cow with bump & pigs foot with cloven ho
225b219 diate.-- Reference to Pig & Dogs. My theory will MAKE me deny the creation of any new quadruped since
228b231 ider our equals.-- «Do not slave holders wish to MAKE the black man other kind?» Animals with affecti
230b236 you have stunted species, but not such as would MAKE species (except perhaps in some plants & then a
244c021 istance, makes species but barrier.-- --it would MAKE a strong contrast with southern regions.-- «it wo
245c025 sinking, because arrival of any one plant might MAKE conditions in any one isld different]CD.-- p. 4
258c060 nship of Analogy is a divellent power & tends to MAKE forms remote antagonist powers.-- Every animal
280c135 two species were excessively old, they would not MAKE hybrids, whereas two newer ones, even if more d
282c141 » with them. they might be supposed to change. & MAKE genera of bird. analogous. animals would be pos
282c141 ent races of man, yet altogether different.-- To MAKE this case perfect, we must suppose men instead
283c145 all that I want is granted.-- For. at Galapagos. MAKE ten species of Orpheus-- one of which has very
291c168 uals, wheras in Falkland Isd they would change & MAKE new species.-- alpine species being destroyed a
293c173 hneumon & caterpillar, though our ignorance, may MAKE us think so, but only between laws--]CD. Many d
296c184 s. (anyhow not Australian) on Peaks. Did Creator MAKE all new yet forms like neighbouring Continent.

Page ***(Key Word)***
303c204 n some monkeys clitoris wonderfully produced».-- MAKE abstract on this subject from Lawrence. Blumenb
341d029 ucture, like little wings of Auks which does not MAKE that bird a Penguin.-- (i.e. whether relation i
366d112 ith the chance of two being born at same time, & MAKE breed, one would doubt any law.-- Yet seeing th
367d113 general marks is the superior strength <of> «of MAKE in» the males; & another circumstance, perhaps,
373d132 ave Marsupiata abortive Mammae?.-- My view would MAKE every individual a spontaneous generation: what
377d139 mith more distant.-- But he thinks other monkeys MAKE st.-- noise‖ In case of woman instinctive desir
381d157 ecious & some monooecious-- (& cultivation might MAKE one set of organs barren in one plant & not in
386d166 f nerves in snail. (Encyclop of Anat & Phys) can MAKE a head; the other part may surely absorb a usel
386d167 heals piece of skin.-- if the tail knows how to MAKE a head. & head & tail, & the belly both head &
390d178 effect of Male is to impress some difference: to MAKE the bud of the woman, not a bud in every respec
391d174 on fails.-- How completely circumstances «alone» MAKE changes or species!! CD[The view of <In> each M
392d180 ne English weeds in Hot. Houses will they flower MAKE Hybrids with moths, where fecundation can be ma
414e063 dogs.-- if nature had had the picking she would MAKE <them> such a variety far more easily than man,
418e083 what is added to the composition of the atom, to MAKE it alive, & how the laws of generation were imp
423e098 ed of Cauliflower.-- (How curious it would be to MAKE enquiries of some of these great seed-growers--
426e107 k that one large body of varieties are fertile & MAKE mongrel, & other great series quite otherwise &
426e107 mongrel, & other great series quite otherwise & MAKE on[ly] true hybrids.-- but this is false, [give
430e118 she can produce great ends-- But how.-- -- «-- .MAKE the difficulty apparent by cross-questioning.--
455eIBC ising plants. where they cannot «crossed» etc.-- MAKE Hybrid mosses.-- Leighton or some one. Father--
465t081 - now it is this very immigration which tends to MAKE the destroyers vary; so that we here see reason
471tf11 h says in the larks from S. Africa he can almost MAKE series from end to end-- so that he is almost l
484z014 eathering on legs-- habits-- Does the Secretary, MAKE noise & throw head back M Edwards,--on polypi o
496q006 lace male & female flower in same receptacle (8) MAKE Duck eat Spawn, eggs of snail, row of fish & ki
531m050 that the importance of the event by itself will MAKE it to be remembered. whereas it is the importan
533m060 at writing down his confessions of sins. did not MAKE him more humble.-- it has obscurely occurred to
533m060 I never realized the idea that I was tending to MAKE myself in act less grateful.-- How comes this t
535m063 at a wasp has this much intellect. yet habit may MAKE it act wrong, as I have done when taking lid of
536m073 bodily organization of oyster. so may free will MAKE change in man.-- the real argument fixes on her
536m074 sited upon the children.-- The above views would MAKE a man a predestinarian of a new kind, because h
537m076 moral sense)» which she seems to think «are» to MAKE others happy & wrong to injure them without tem
538m080 ey would not be more different, & yet they would MAKE one's father & self one person-- & thus eternal
542m095 viz seeing something obscurely with the wish to MAKE it out?-- Seeing a Baby (like Hensleigh's) smil
545m106 rin with passion, that is show all the teeth: «& MAKE noise not like pish, but like chit-chit-chit, q
548m116 s does not go back to former periods so «as» to <MAKE> «give» one individuality in this case.-- But n
549m118 h an» when same man is compared to peasant.-- To MAKE greatest number of pleasant thoughts, he must h
549m119 m frequency, or inherent structure of mind. they MAKE, either in themselves, or if recollected, such
549m120 sant, with whom sensual enjoyments of the minute MAKE large <parts> portion of daily <happiness> «ple
552m131 oe <pi> Sow, who carried from all parts straw to MAKE its nest. Pigs & Elephants, (both Pachyderms) m
553m136 to M. le Comte).-- Those savages who thus argue, MAKE the same mistake, more apparent however to us,
557m147 in passion.-- Cat when pleased, erect its tail & MAKE it very stiff «& back» when savage «no» & ready
557m148 -- Yet I think one can remonstrate with a dog, & MAKE him ashamed of himself, in manner quite differe
557m150 case. nothing can be heard.-- Shame would never MAKE person tremble, like fear.-- Why does any great
557m150 like fear.-- Why does any great mental affection MAKE body tremble. Why much laughter tears.-- & shak
565n009 tened & moistened by emotion,-- why does emotion MAKE tears fall?? Lavater says derision lies in wrin
573n038 bed.-- in case spittle, effect of thought is to MAKE saliva flow, & therefore thinking of subject, e
578n053 t to do so.= = How directly personal remark will MAKE any one blush.-- Is there not some saying about
578n057 r twitching».-- But why does joy & OTHER EMOTION MAKE grown up people cry.-- What is emotion? At end
579n059 marked, with the habitual expressemotions, which MAKE us love him, or her.-- it is blind feeling, som
582n067 -- N B. According to my view marrying late, will MAKE average of life longer.-- for short-lived const
583n070 & muscles, we cannot look at him, as machine to MAKE cell of certain form. (& especially as it adapt
583n070 cell to circumstances), it must have impulse to MAKE a cell in certain way, which way its organs are
585n077 ly; watchmaker has faculty by his instruments to MAKE toothed wheel. he might by instinct make watch,
585n077 ents to make toothed wheel. he might by instinct MAKE watch, but he does it by reason & experience, o
587n088 er, whines just like a child. Get a Dictionary & MAKE a list of every word, expressing a mental <desi
587n089 ideas & emotions. rather ideas & bodily actions MAKE the emotions.-- p. 272. Some remarks applicable
589n091 ur of brain forming the instincts,-- could brain MAKE a tune on the pianoforte, yes if every individu
593n105 down & down.-- mem. Nina used to get into hay & MAKE a nest for herself.-- the object is to make sau
593n105 ay & make a nest for herself.-- the object is to MAKE saucer-shaped depression.-- [blank] Does music
595n121 t was Snow.-- Is this part of same feeling which MAKE us think anything ugly-- a beau-ideal feeling.
596n184 CD» Cyanocephalus, Macacus, Niger. Cercopithecus MAKE labial st st. S. American monkeys. pull back sk
600o08v ect of beasts, not clear.-- ¿does not Mackintosh MAKE great difference between moral sense & conscien
602o11v «p. 131.» in Tragic acting-- CD [My idea. would MAKE the mind have mysterious & sublime ideas indepe
628o056 cterize the moral sense, by its "supremacy",-- I MAKE its supremacy, solely due to greater duration o
638j059 re of brain should not be born. with tendency to MAKE animal perform some action.-- as well as gain i
041r064 ells (which being packed. in beds) lived there, MAKES it very doubtful whether they could have lived
074r165 - «same chemical laws as in concretions perhaps MAKES intersections richest-- Humboldt has urged phen
160g088 Alluvium shows the ascending fringes {P} which MAKES me think it submarine, 400 or more feet above s
165g126 rkable ever obseved? Shows that <nervous> brain MAKES thought Glen Roy B C. Darwin All useful pages c
177b026 ant succession of germs in progress.-- «no only MAKES it excessively complicated.» {P} Is it thus fis
213b169 all men were dead then monkeys make men.-- Men MAKES angels-- Those species which have long remained
218b189 ttentots coexisting. proves this-- but when Man MAKES variety these are vitiated.-- This barely appli
244c021 ical sea, it. would demonstrate.; not distance, MAKES species but barrier.-- --it would make strong c
256c053 atement that some species vary more, than what MAKES species in other animals.--? Forster on South S
291c167 wonderful relation going through all Nature.-- MAKES hermaphroditisms. one step in <scale>. Series--
297c185 ., same circumstances., which by causing death, MAKES the group aberrant When species rare we infer e
331dIFC rn to either parent.? Is the first cross, which MAKES hybrids. productive like geese?-- Are the numbe
336d018 cross. each sends his own likeness, & the union MAKES hybrid, in fact the parents beget child like th
375d135 check amongst men.-- «Even a few years plenty, MAKES population in Men increase, & an ordinary crop.
377d138 Woman-- ‖ The noise st st. which the C. Sphynx MAKES is also made by the C. porcarious., together wi
384d162 own for some generations Theory of sexes (woman MAKES, bud, man puts primordial vivifying principle)
386d166 art is probably part of same general law, which MAKES two animals out of one & heals piece of skin.--
386d167 ls of Natural History. 1838. p. 123. Ehrenberg. MAKES gemmation in animals very different from that o
414e060 plants???) been so little progressive «!Agassiz MAKES it wonderfully changed, since Cretaceous period
417e077 its place in the Systema Naturae.-- Mr. Knight MAKES this analogy between grafting & sexual union--
426e107 ieties destroys the appearance of this series & MAKES one think that one large body of varieties are
429e115 can be cultivated with ease near London.-- what MAKES the line, as trees in Beagle Channel.-- it is n
439e143 y of hybrids «with parent species» false, which MAKES it determined by a facility in returning to old
441e150 eloped on another, must not be overlooked.-- it MAKES fourth cause or law of change.-- The weakest pa
443e155 he right to argue against my theory, because it MAKES the world far older than what Geologists, think
463t057 the whole mind is constituted that a difficulty MAKES greater impression, than the grouping of «many»
463t063 to come-- Owen in his description of my fossils MAKES same such remark & before the conclusion of his
463t063 mark & before the conclusion of his work-- Lund MAKES his wonderful discoveries= negative facts are v
526m031 the man, but indefinitely, he chooses (but what MAKES him fix!? <)>-- frame of mind, though perhaps h
527m034 ght.-- the capability of such trains of thought MAKES a discoverer, & therefore (independent of impro
536m071 ff its bottom.-- it is relic of same thing that MAKES one dog smell posterior at another.-- Why do bu
536m072 «(M. Le Compte)»-- the free will (if so called) MAKES change in bodily organization of oyster. so may
540m085 t its fingers. Seeing a dog & horse & man yawn, MAKES me feel how <much> all animals <are> built on o
544m103 s solely the comparison, with past ideas. which MAKES consciousness-- & which tells one of reality--
545m105 pression becoming very often unconscious, which MAKES the idea unconscious, if so (think of this). st
564n003 ! Man moreover who reasons much on his actions, MAKES his conscience far more sensitive. ulitmate eff
565n008 considered as quite insignificant, & when pride MAKES person extremely self-sufficient,-- the corner
574n041 of Cocoa nut.-- November 27th.-- Sexual desire MAKES saliva to flow «yes, certainly»-- curious assoc
581n063 March 16th.-- Is not that kind of memory. which MAKES you do a thing properly, even when you cannot r
583n070 would be "creatures of habit."-- CD[as the bee MAKES its cells, by means of ordinary senses & muscle
591n100 cannabalism, is held in abhorrence.-- all this MAKES analogy of actions with <&> against benevolent
600o008 al Sentiments» imperative sense of duty-- which MAKES struggle in man.-- two souls in one body-- (2)
616o038 memory; you cannot call the frame of mind which MAKES music pleasant, a memory; yet that frame is enh
620o045 . [RHC] 4) as starvation, or fear of death, one MAKES allowance & either excuses the «non-» following

Page
(Key Word)
629o55v k only of prepuce, crepitando,]CD & if passions MAKES one break these artifical rules, get remorse--
633j54r ted to arrest mud &c at deltas.-- Now my theory MAKES all organic beings perfectly adapted to all sit
090a021 ¿ does this bear upon the sorting of matter. in MAKING trachyte come out before.-- What must be the e
124a119 f elevatory force-- Erasmus says he has seen in MAKING brass a piece of copper not melted absorb, zin
160g089 ing Glen Bought & Glen Tarf a perfect old Loch, MAKING <several> two divortiums aquarum, viz two bran
180b037 if each species «an ancient (I)» is capable of MAKING, 13 recent forms.-- Twelve of the contemporary
242c017 slow & offspring not picked.-- as men do. when MAKING varieties.-- Voyage of Coquille. Zoolog. p 19.
264c082 abits.-- Gould I see quite recognizes habits in MAKING out classification of birds Birds vary much (m
279c133 o that not propagated by nature.-- Whole art of MAKING varieties may be inferred from <this> facts st
286c154 «& violates every best instinctive feeling» by MAKING slave of his fellow black, often wished to con
298c191 inctive feeling.-- There is great difficulty in MAKING an alpine species from one in lower country du
304c206 own ignorance.-- the collector who plodding at MAKING a series, which would render our knowledge a c
372d131 arm would produce arm if supported., <so> & in MAKING «true» bud some such process is effected.-- a
375d135 like food used for other purposes as wheat for MAKING brandy.--» take Europe on an average, every sp
377d139 change in English species-- time no element in MAKING change, only in fixing it: only circumstances.
388d173 semen. must actually reach the ovum.-- [Why in MAKING a bud, which is to pass through all transforma
425e103 alceolaria.]-- «:CD[Does the Power of, «easily» MAKING tolerably fertile hybrids, bears relation to c
436e135 n horizontal direction is far more efficient in MAKING species, than time (as cause of change) which
533m060 nfession of error made him less repentant.-- In MAKING too much profession, or rather in only fully e
548m115 tle of air the trouble «I well recollect» is in MAKING things somewhat probable. in comparing every s
558m152 t on back» «wide open» with prodigious force.-- MAKING growling, guggling noise. Puma did same & & so
572n032 26-- "The general idea of showing respect is by MAKING yourself less, but the manner, whether by bowi
594n113 ned on the 24th of May at Cresselly by the boys MAKING faces at it, so much so that the nurse had to
608o026 on to change line of conduct. as being better & MAKING him happier.-- he agrees & yet does not.-- bec
638j58v , corresponding to <every> «one or any»-- brain MAKING structure, instead of parts of body.-- Now we
244c022 nkey from Java.-- Hairs, & deer.-- Procured two MAKIS alive from there.-- Mem Waterhouse knows of som
569n022 it qu'il valait mieux rester a Farroïlap quelque MAL qu'on y fût."-- Expression common to Savage & Fr
242c017 hant & has orangs, ∥ Tapir common to Sumatra & MALACCA∥ Borneo & Malacca «& Cochin China» are said to
242c017 ∥ Tapir common to Sumatra & Malacca∥ Borneo & MALACCA «& Cochin China» are said to have orang-utang
353d060 flight-- therefore ostrich not. The peculiar <MALACCA> «Malacca» bears, <are> belong to same section
353d060 therefore ostrich not. The peculiar <Malacca> «MALACCA» bears, <are> belong to same section with with
450e176 get dimensions-- do App. p 73 State of Muar in MALACCA.-- speaks of Rhinoceros as well as Tapir.-- <d
338d024 artly Mulatto) prevents his taking any form of MALARIA-- adaptation & species-like,-- -- Says Negro--
451e178 his looks inaccurate C. D) they will catch the MALARIA & die.-- On the other hand there are breeds of
245c024 .-- The (p. 296) Columba Kurukuru found in all MALASIA-- & oceania, offers many varieties in each pla
315c240 many spots & since disseminated See. Habits of MALAY fowls p 5. (note) in some papers on instincts L
402e018 hich are met with near Canton" "Here, as in all MALAY countries, I noticed a peculiarity in the cats
402e019 as if they had been broken". are born so in all MALAY Countries W. Earl Eastern Seas. p 233 Octob 12t
453e182 und in Borneo.-- «p. 233» There, as well in all MALAY countries «the» cats are born with the joints n
494q004 w is it in <allied> varieties (9) Cross largest MALAY with Bantam-- will egg kill Hen Bantam.-- Cross
615o58v ng of individual force in any individual.-- His MALAY breed «of fowl» totally different habits from E
451e176 V. p. 565. in a Paper by Lieut. Newbold.-- A MALAYAN albino described "To this day the tomb of his
096a041 heated without parting with Carb. Acid.-- Mr MALCOLMSON in Paper on India gives reason for knowing t
134a142 Some of the Tosca nodules at Bahia Blanca Mr. MALCOLMSON says are like Kankaer South of Part Luconia-
134a144 p. p 96. apparently good geological paper. by MALCOLMSON-- worth reading-- Burnetts. vol 4. p. 193 in
135a146 ividing it-- Hopkins fissure at {P} -- G. J. MALCOLMSON has described formation of shore of Coromand
022r007 passes into schist scales of chlorites--Mem. MALDONADO P 375 Much Chlorite in some of the dikes.--P
128a130 w lagoon.-- when describing Coast of. Brazil. MALDONADO enter into this case.-- Ed. New. Phil. Journa
290c163 olour of habitation. Must have some effect.-- MALDONADO as good forests for beautiful birds.-- herede
484z016 . by Patagonian Furnarius.-- into Oxyurus, by MALDONADO creeper of same plumage.-- general red mark o
485z016 to find how many of the small finches walk at MALDONADO & Patagonia compared with those of England.--
145g001 rwin A. Glen Roy Generally received opinion that MALE impresses offspring more indelibly than female
191b082 ant together for life, is in accordance with The MALE animal affecting all the Progeny of female insu
194b096 outh American shells? ¿Do not plants, which have MALE & female organs together, yet receive influence
204b141 or with parent birds.-- "W. Fox. knew of case of MALE widgeon, winged & turned on pool, first season
205b145 are observed in their offspring, but that of the MALE more frequently predominates," p. 20. do "If he
228b229 tree <go> producing crab. is the offspring of a MALE & female animal of one variety going back ¿whet
275c121 min bulls-- Mr. Y. is inclined to think that the MALE communicates the external resemblances, than th
307c215 are allowed to the fully organized wings of the MALE rendered abortive in the womb.-- if these appar
309c221 .-- duration in two classes however different.-- MALE glow-worm knowing female good case of instinct.
313c235 of the orang outang of (in June 1838 when young MALE: was added good instance of instinct showing its
333d009 argue the whole effect of race was determined by MALE: & How completely is Lord Moreton's case oppose
345d044 67. Dec. 1837. Generally-- received opinion that MALE impresses offspring more than female, yet insta
350d055 s??) female blind & of quite different from adult MALE with eyes!-- (are not these differences in sex
353d061 nd & Australia well developed <tits> «Mammae» in MALE ourang-outang. other point of resemblance with
358d076 mage «(that is where the female differs from the MALE?)».-- children & women = "women recognized infe
358d086 g tail figure, & some degree of whiteness like a MALE.-- Thus castration, hybridity, & breeding in &
359d088 ut of any one nest. even more than common duck-- MALE Penguin was crossed with hen Canadian offspring
359d088 ffect of breeding in & in like our pidgeons) The MALE of every animal certainly seems chiefly to impr
361d096 indicate that the change is effected through the MALE??)-- Yarrell observed that female of some water
361d096 ssume for breeding a more brilliant plumage than MALE.-- «My case of Caracara. N. Zelandiae.--» Mr Bl
362d099 resh horns begin to grow.-- Mr Yarrell says the «MALE» Axis of India, breeds at times when horns not
363d102 both sing their own songs, though imperfectly.-- MALE birds always second their songs, the ++ Cervus
368d114 s settle point.-- (do the females then fight for MALE) & are merely most attracted). -- singing best
368d114 al birds largest.-- p. 47. (<"> is evidently the MALE which recedes from the species all females bein
368d114 hinks» Hence specific character most perfect in <MALE> «hermaphrodite» (Fishes have no secondary char
373d132 an Echidna. & Hedgehog having spines.-- Does not MALE Pidgeon (yes) surely) secrete milk? from stomac
373d132 eeding young, & to abortive <organs> «mammae» in MALE Mammalia:-- ¿is not this argument, for Mammalia
378d147 ee why young & female alike Good Ch 6 Keep Is it MALE that assumes change, & is the offspring brought
379d152 e twins, <one> though capable of producing both <MALE> pair of male & female.-- if there be one femal
379d152 though capable of producing both <male> pair of MALE & female.-- if there be one female, she will be
380d153 n.-- Mr Yarrell does not know of any case of old MALE. becoming like female, though many of old femal
380d154 ; for it shows latent instincts even in brain of MALE.-- Every animal surely is hermaphrodite-- (as i
381d156 oderms. Acalephes. Polyps. Sponges Heautandrous, MALE organs formed to fecundate female (as in plants
382d158 molluscs. & neuter bee would shew this-- (Do any MALE animals give suck)--But this not distinctly sta
383d160 pt. 16." Hybrid between Silver & Common Pheasant. MALE bird, said to be infertile.-- spurs rather smal
383d160 tile.-- spurs rather smaller than in <ma> silver MALE-- Head like silver except in not having tuft,--
384d161 st <are> become of use∥ <Great characteristic of MALE greater strength, (p 45) & that strength? In sp
384d162 y in subordinate manner in the plants which have MALE & female flower on same stem.--» so that Mollus
384d162 eveloped instincts of Capon. & power of assuming MALE plumage in females., & female plumage in castra
384d162 umage in females., & female plumage in castrated MALE.-- «Men giving milk--» Sept. 25th Young man at
387d168 f A. B. C. D. E be <offspring> «animals»: if «x» MALE impresses ovum <of> in A, «with some peculiarit
387d168 «in lesser degree»-- Then when (C) unites with MALE (X) «assume that every peculiarity has a tenden
387d169 . Vol I. Notice the Syngnathus, or Pipe fish the MALE of which receives <young> «eggs» in belly.-- an
389d176 ts the subsequent offspring, <when> though other MALE may have copulated.-- two animals may unite & e
389d176 ven produce fertile offspring-- DESIRE LOST when MALE & female too closely related: this most importa
389d176 both father & mother, or very close to either.-- MALE & female as foetus one sex; & therefore both ca
390d178 ther.-- The upshot of all this is that effect of MALE is to impress some difference: to make the bud
390d178 cted with the physical differences in almost all MALE animals?]CD If the male «in the course of some
390d178 ifferences in almost all Male animals?]CD If the MALE «in the course of some generations» has gained
390d179 which will not breed, but the female at least (¿MALE?) looses all appetite.-- It is the comparison o
393dIBC her dogs renders subsequent progeny faulty. Does MALE fail in passion.-- Disposition of half bred Cat
401e017 le plant impresses main features on offspring, & MALE the lesser peculiarities.-- brilliancy of inflo
412e057 he hermaphrodite structure being retained in the MALE.-- <like> «far» more than marsupial bones., & e
416e075 fference between races & variety? «Man picks the MALE, instead of allowing strength to get the day» T
428e113 egotten <?> «Walker attributes this to effect of MALE sex on locomotive system» I am bound to insist
431e123 f Animals, p. 8. size of foetus in proportion to MALE parent p. 8. his whole doctrine of the advantag
431e123 dvantage of crossing consists in the idea of the MALE being smaller, & the female larger than average

434e129 uscus aculeatus. a dioecious plant, in which the MALE plant sometimes bears female flowers, the organ
446e160 inor is a bird of more splendid plumage than the MALE.-- Athenaeum May 18. 1839. p. 377.-- Statement
449e169 the young of two hatches «all alike» between the MALE Chinense & female common goose took after the c
449e169 dicting (probably) Yarrells law & Walkers of the MALE giving form-- they interbred. & the young kept
459tf02 - From Buffon cross of he-goat & sheep, it seems MALE gives form. admitted by Linnaeus.-- seems to do
459tf02 one case bastard of wolf a dog had more form of MALE, & another of both progenitors-- the hinnus, re
460t013 n impresses its form both on vars & species The <MALE> swan-gander with common goose produce full as
461t025 with fins, & hence do not require sac.-- but the MALE in these hatch young-- are there not some. Mars
491q01v pregnated young ones; & that it is in these that MALE organs (not being always useful). fail-- Really
492q002 sex & not other: which generally fails first?-- MAL[E] 10. Henslow says semi-doubl flowers are those
493q003 mules ass & horse-- important. {In crosses does MALE offspting take after male parent & vice versâ =
493q003 tant. {In crosses does male offspring take after MALE parent & vice versâ = History of Tortoise-shell
493q004 see result.-- According to Mr Walker the form of MALE with <old> female of old breed & see result.--
496q006 barren-- if not how does impregnation take place MALE ought to preponderate; according to Mr Yarrell
499q009 highs (18) Place pin's heads with Bird lime near MALE & female flower in same receptacle (8) Make Duc
503q012 Yaks. & other Hybrids-- Dogs &c &c 38 Does only MALE yew tree & see whether they catch pollen-- <Ne>
503q013 ssible?? Maer (1) Yew Trees near Boat House «ANY MALE yak cross with cow: is not reverse possible?? M
503q013 eds in beginning of November 1841.-- Trees above MALE? (2) Result of Edwards experiment in Cabbages g
508q016 iseases) translated. (10) About Daltonism in the MALE branch.» --¿number of seeds in beginning of Nov
511q018 nimals. [Made no import. remark]CD (2) Secondary MALE Troughtons.-- paper in Taylors Scientific Memoi
511q018 remark]CD (2) Secondary male characters.-- does MALE characters.-- does male transmit to male more o
511q018 condary male characters.-- does male transmit to MALE transmit to male more of his features-- in negr
511q018 Has since recrossed this breed.-- Have secondary MALE more of his features-- in negro & white (3) Abo
513q21. rchidiae Where does J. Hunter use expression of "MALE characters appeared.= (4) Does he know any seed
513q21. of "male principle of arrangement."-- would not "MALE principle of arrangement."-- would not male or
515q022 e Hybrids differ in colour of beak, taking after MALE or female "constructive principle" be better. o
540m085 e & womb.-- We need not feel so much surprise at MALE & female parent.-- Will they grow up in other r
540m085 e outang at Zoolog Gardens touched pud. of young MALE animals smelling vaginae of females.-- when it
554m138 d philosopher himself it is chiefly shown in old MALE & smelt its fingers. Seeing a dog & horse & man
554m139 n, when she could not succeed of herself.-- <The MALE.-- A very green monkey (from Senegal he thinks
554m141 r end. most important step in progression.-- The MALE «I saw» Jenny untying a very difficult knot--
570n027 ilarly treated would think negress beautiful,-- [MALE Black Swan is very fierce when female is sittin
570n027 less admires female. showing. no connection with MALE glow worm doubtless admires female. showing. no
215b177 n.-- There never were any constant species Both MALES figure]CD-- As forms change, so must idea of be
258c061 orious seals, hence deer victorious deer, hence MALES & females. lose desire. Native dog not found in
307c215 wworm <are> have «These abortive organs in some MALES armed & pugnacious (all order: cocks all warlik
318c254 ssion of birds.-- «in» some species «a Tanagra» MALES animals, Mammae in Men, capable of giving milk)
345d044 tances given on opposite side.-- «The theory of MALES come first & the females in flocks.-- as in Engl
358d076 sed to these facts are effects of castration on MALES & of age or castration on females.-- [not locat
367d113 is the superior strength <of> «of make in» the MALES impressing most is in harmony with their wars &
367d113 -- is most common in vegetable feeders. because MALES; & another circumstance, perhaps, equally so, i
368d114 tracted). -- singing best sign of most vigorous MALES always armed in carnivora. Where females, are p
370d116 tion of Queens.-- Neuters are bred first, «then MALES.-- «(NB. most strange cocks & hens. being eithe
373d132 secrete milk? from stomach. analogous to other MALES--» how has this been arranged-- Neuters are tru
385d163 hter se offspring in latter being unhealthy.--» MALES feeding young, & to abortive <organs» «mammae»
389d177 oo little.-- in (latter case female often takes MALES «bred in & in» never lose passion. (Mem: so it
499q009 14) Bowman female branch At What distances from MALES, will female (a) Willows or Yews some poplar's
507q016 inheritable!!! very good Any peculiarity in the MALES of a family-- Where one tooth aborts, do you kn
508q016 11) And. Smith Savages at Cape any selection of MALES in «cattle» or in Killing the worst =or in dogs
592n101 ays hen doves & the female chamaeleon court the MALES by odd gestures. In one of the six (?) first Vo
620o045 est to be followed.-- Yet even at this time, MALEVOLENCE,, when not urged to it by passion, shows a b
625o50v r to obey anger as benevolence (but not cool MALEVOLENCE). it is only after reason comes into play th
621o046 l appetites.-- he is monster, or unnatural if MALEVOLENT, or hates his children without some passion.
621o047 l passion.-- (a) [LHC] The conscience rebukes MALEVOLENT feelings, as much as actions, therefore Sir
332d001 , than for ten years he never saw one case of MALIGNANT erysipelas spreading over the head, not cause
072r160 d over the surface of the ground are fibrous. MALLEABLE & of so great tenacity, that it is with diffi
320c276 be Partie Zoologique Pernety. voyage a l isle MALOUINES Zoological Journal 5 Vols Voyage de la Coquil
086a006 nnected with its elevation. vapour from below-- MALTE Brun «Salt Lakes» Siberia must be read as well
088a016 low hills East of Cordillera very important V. MALTE Brun.-- Main character of Andes Metamorphic act
402e019 all Cercopithecus., & skins of galiopithecus.-- MALTE Brun. Vol <I> II p.,133: at Samar SE of Luçon,
402e019 ar SE of Luçon, many monkeys, buffaloes &c &c-- MALTE Brun. would be worth skimming over with regard
322c270 my. ---do Lavater's Physiognomy ---- Octob 3d MALTHUS on Population W. Earls'. Eastern Seas. .Octobl
326c266 ssy to Thibet, perhaps worth reading quoted by MALTHUS.-- Heberdens Observat. on increase & decrease
375d134 e of locality. Even the energetic language of <MALTHUS» «Decandoelle» does not convey the warring of
375d134 y the warring of the species as inference from MALTHUS.-- «increase of brutes, must be prevented sole
375d135 than 25 years-- yet until the one sentence of MALTHUS no one clearly perceived the great check among
376d135 dapt it to change.-- to do that for form which MALTHUS shows, is the final effect, (by means however
566n010 tics with internal organs, as action of heart MALTHUS on Pop. p. 32, origin of Chastity in women.--
609o29v is is encroaching on views in second volume of MALTHUS). Adam Smith also talks of the necessity of th
609o29v dog no one calls vice). on same principle that MALTHUS had shown incontinence to be a vice & especial
637j57v .! extinction, utter extinction! let him study MALTHUS & Decandoelle.-- The Final cause of innumerabl
637j58r inal cause of innumerable eggs is explained by MALTHUS.-- [is it anomaly in me to talk of Final cause
436e136 that wild animals will vary, according to my MALTHUSIAN views, within certain limits, but beyond the
639j28v dred litters is born with long legs» & in the MALTHUSIAN rush for life, only two of them live to free
534m063 stem of Hollyhock, although ordinary Habitat is MALVA sylvestris. do. p. 228 Newport says Dr Darwin m
251c041 & tame horses.-- p. 246-- Gmnura-- new genus of MAM: found in Sumatra p. 452 Append to Denham Clappe
407e041 --; Africa not so equatorial..-- The fact of No. MAM: Placent: insectivore being in S. America & Aust
338d025 n insects entered by mistake Surely the fossil MAMALOGY of Britain & Europe is African. & the only di
512q019 ieve. Stanley's fact of Hawks distributing live MAMALS (3) Do most Hawks eat stomach. of finches-- do
174b015 poses time when no Mammalia existed; Australian; MAMM were produced from propagation from different s
226b223 in their duration of species. (¿are carnivorous MAMM: in Paris basin «allied to present») more like
412e058 cture.» L'Institut. 1838. p. 384. List of fossil MAMM: from Poland. &c.-- Three principles, will acco
448e168 h & reptiles in former case-- Reptiles & Birds & MAMM. in ornithyrhycus-- is not this right?-- June
471tf09 . vol. II p. 18.) capital list of all the fossil MAMM. of Europe-- Large Lizards in Navigatores. Will
195b099 use in progenitor-- or it may be of use.-- like MAMMAE on mens' breasts.-- How does it come wandering
307c215 e «These abortive organs in some Males animals, MAMMAE in Men, capable of giving milk)» rudimentary w
353d061 iemen's land & Australia well developed <tits> «MAMMAE» in male ourang-outang. other point of resembl
354d062 nt parts according to age of individuals-- (see MAMMAE of Women) in different parts when age changes
373d132 er males feeding young, & to abortive <organs> «MAMMAE» in male Mammalia:-- ¿is not this argument, fo
373d132 ever disappearing??-- Have Marsupiata abortive MAMMAE?.-- My view would make every individual a spon
380d155 ., though no offspring, Milk sometimes comes in MAMMAE & even when bitch is in heat.-- Yarrell believ
383d159 ould be argument for developement of either.-- (MAMMAE or sheath of Horses poenis reduced to extreme
387d169 ungs> «eggs» in belly.-- analogous to men having MAMMAE.-- There is an analogy between caterpillars wi
388d172 g both sexes abortive fact of same tendency. -- MAMMAE in man. having given milk. -- testis & ovaria
412e057 r» more than marsupial bones., & even more than MAMMAE, which have given milk.-- is secretion from Pi
440e147 no one can be shocked at absence of final cause MAMMAE in man & wings under united elytra The law of
461t037 ertain amount it will be so handed down«(».. as MAMMAE of men «callosities on Camels & Horses--».-«
540m085 n are probably merely coorganic as connexion of MAMMAE & womb.-- We need not feel so much surprise at
614o037 ase can perceive instinct. boy takes delight in MAMMAE before any reason had told him this distinctiv
638j58v eredetary superfluities Man could exist without MAMMAE.» to the then existing conditions.-- An adapta
227b226 ructures <between> say in brain. between lowest MAMMAL & Reptile. (or between extremities of any grea
251c041 od case.-- p. 526. (ref) To Temminck Monograph. MAMMAL; «4to» good facts about distribution of Cats V
343d039 robably a distinct species-- We know how long a MAMMAL may go on as one species from Egyptian Mummies
482z011 Scolopax very close to ours Rengger's work of MAMMALI: of Paraguay must be most important a discussi
021rIFC t number of changes towards perfection (namely MAMMALIA) must have a shorter duration, than the more
087a011 No Vol. I. p212. Cuvier Oss Foss Wide range of MAMMALIA really very important. harmonizes well with L
174b015 as in Australia» This presupposes time when no MAMMALIA existed; Australian; Mamm were produced from

176b023 ave shortest life.; Hence shortness of life of MAMMALIA.-- Would there not be a triple branching in t
177b027 n, the bottom of branches deaden.-- so that in MAMMALIA «birds» it would only appear like circles;--
181b042 tance there would be great gap between birds & MAMMALIA, Still greater between Vertebrate and Articul
191b081 ts tell for little) Geographic distribution of MAMMALIA more valuable than any other, because less ea
192b088 l animal in Stonefied slate, the father of all MAMMALIA in ages long gone past.. & still more so «kno
192b088 is not more perfect by the discovery of fossil MAMMALIA than before & that is all that can be expecte
192b089 -- This answers Cuvier-- Perhaps the father of MAMMALIA as Heterodox as ornithorhyncus. If this last
194b095 established it would go to show a centrum for MAMMALIA.-- I really think a very strong case might be
194b097 ht to say there never was common progenitor to MAMMALIA & fish. when there now exist such strange for
197b112 propagation= I see nothing like grandfather of MAMMALIA & birds &c ∥ p. 32. reference to M Edwards. l
198b115 well to insist upon <different animals> large MAMMALIA not being found on all isld, (if act of fresh
213b169 nt state, well worth consideration--» Tabulate MAMMALIA on this principle. <Varieties> < «Races» > Ma
219b196 ount for each separate one, so to say that all MAMMALIA, were born from one stock, & since distribute
221b202 tyledenous a fact analogous to reptiles before MAMMALIA. Think about Miocene fossils some species bein
222b204 almost infers, what we call improvement, --All MAMMALIA from one stock, & now that one stock cannot b
222b205 of Zoology may not be perfecting by change of MAMMALIA for Reptiles, which can only be adaptation to
222b206 f all insects gives same argument as father of MAMMALIA; but have improvement in system of articulati
226b220 ndly the origin & history of every terrestrial MAMMALIA.-- Especially moderately large ones.-- Is the
226b223 - ¿are shell-boring Molluscs, like Carnivorous MAMMALIA in their wide range & in their duration of sp
229b234 uritius Freycinet Voyage, agrees. with several MAMMALIA being peculiar (?) If. Henslow discusses poss
236b278 s» types of former dogs. character of Miocene MAMMALIA--of Europe Mem. Mr Bell's case of Sub Himalay
251c041 matra p. 452 Append to Denham Clapperton &c on MAMMALIA no doubt will all be included in Smiths work
257c057 to the possibility.-- My views will explain no MAMMALIA in secondary-epocks, & developement of lizard
274c116 Will this not agree with Waterhouse & <birds> MAMMALIA.-- We have clear indication [not located] alo
278c130 there is series. Could I not give Catalogue of MAMMALIA arranged according to my own methods. Dasyuru
287c157 cal astronomer to plain observer∥ +++. between MAMMALIA & fishes, one penguin, one tortoise shows hia
292c169 We absolutely know that tendency is greater in MAMMALIA, than in shells ¿ univalves or bivalves.-- An
302c201 ase. = any animal really connecting the fish & MAMMALIA, must be sprung from some source anterior to
303c205 . when group is true,» there are no genera. if MAMMALIA are adduced. say oh look to your fossils, now
306c213 es of very high order. not for vertebrata, but MAMMALIA & reptiles &c Timor is connected with Austral
311c225 alia different Temminck Fauna Japonica (?!) 82 MAMMALIA 293 Phalangista of Australia & Van Diemen's l
313c233 accounts for the species changing « ∴ because MAMMALIA can subsist where parasites cannot» Read Ento
313c236 ren come to suck or other actions in foetus of MAMMALIA, or chick eat») Generation becomes necessary,
325c267 ult Dr Horsfield Silliman's Journal Rengger on MAMMALIA of Paraguay. account of wild cattle & Montagu
343d039 t may be worth adding in note than amongst the MAMMALIA of Europe the shells of do-- shells of. N. Am
348d050 k down in certain species & become worthless-- MAMMALIA Edentata∥ We do (p 6) say such is group. beca
352d059 physical change).-- distribution especially of MAMMALIA As every organ is modified by use, every abor
369d115 ave reference to the Great distances which the MAMMALIA of N. S. Wales are generally compelled to tra
370d115 of climate cause of N. Zealand not having any MAMMALIA.-- Type of geographical organization. no more
373d132 The Marsupial structure shows that they became MAMMALIA, through a different series of changes from t
373d132 s from the placentates, Having Hair. like true MAMMALIA, no more wonderful. than Echidna. & Hedgehog
373d132 young, & to abortive <organs> «mammae» in male MAMMALIA:-- ¿is not this argument, for Mammalia recent
373d132 in male Mammalia:-- ¿is not this argument, for MAMMALIA recent creation.-- why. what tendency can the
379d151 e one for large Cetacea, as the past for other MAMMALIA. & still further back reptiles & Cephalopoda:
391d174 hat way «& from frequency of this tendency all MAMMALIA must long have so existed.» with double union
391d174 s or species!! CD[The view of <In> each Man or MAMMALIA being abortive hermaphrodite simplifys case m
402e020 l kinds of the large monkeys.-- Fewer <of th> «MAMMALIA» have passed to Paragua & in Luçon the most n
407e040 temperate climate, <that> of America, that the MAMMALIA of S. America are as diferent from the existi
414e060 now not.» (& insects.-- Stonesfield????). Have MAMMALIA?? My theory certainly requires progression, o
425e105 with other mountains laterally.-- Owen. Fossil MAMMALIA. p. 55. talks of Tapirus American form. found
435e134 in the gift of intellect, he resembles, other MAMMALIA in the effects produced on organization. by p
461t025 hese hatch young-- are there not some. Marsup. MAMMALIA, which <do> have not sack,-- Most curious fac
462t051 n be no sharp division of partition as between MAMMALIA in cases such as that of Java & Sumatra Nov 1
465t080 served must be as much a casualty as, bones of MAMMALIA in caves:-- :argue first case of bones (New R
494q004 d.-- in order to guess how many generations in MAMMALIA. in group effect of crossing.-- (7) Are the E
641j29v in Mollusca. insects «Carabids & Staphylini» & MAMMALIA. The eye being formed in Mollusca, Articulata
641j29v quirrels & Opossums «& fish»: flying lizards.--MAMMALIA. C. D.--
181b040 go very far back to look to the source of the MAMMALIAN type of organization; it is extremely improba
186b059 not well intermediate Lyell Vol III. p. 379. MAMMALIAN type of organization same from one period to
222b206 an accumulation at present day & not include MAMMALIAN remains.-- The Father of all insects gives sa
263c077 sus coelum attentus" is an exception.-- He is MAMMALIAN.-- his <has> origin has not been indefinite--
568n018 union of birds voice & Laste for singing with MAMMALIAN structure. «-- American monkeys utter pleasan
482z011 a discussion of geographical distribution of MAMMALIDAE &c &o &c-- The French <Jerrold?> «Bibrons co
295c178 ontains many genera of Pachydermata &c & other MAMMALS.-- «otter; civet cat, rodents.» (Pachydern in
304c206 of closeness. -- look how close birds! look at MAMMALS: how wide.-- therefore birds younger???? or <h
305c210 . so is there none of reason in order of <ver> MAMMALS.-- Mem Elephants & dog.-- There is one living
313c234 forein considerations, with respect to law of MAMMALS shorter duration, than molluscs, argue case bo
341d031 me line of Rocky Mountains separate almost all MAMMALS of N. America & many birds.-- which however ar
357d072 rld without rule? The destruction of the great MAMMALS over whole world shows there is rule.-- S. Ame
465t079 nts-- In S. America. it appears from Lund more MAMMALS, than at present «in Europe we know there has
465t079 we know there has been several successions of MAMMALS.--» yet only two monkeys, <there are now> have
465t089 s Journ.-- Phil. Mag. May. 1840 p. 362.-- some MAMMALS of Norfolk Crag. mentioned-- allied Beaver to
465t089 ary & Lacrustine formations contain fossils,-- MAMMALS-- a few only -- & how many estuary formations
417e076 lity of the Soul as the linear descendant of <MAMMFERUS> «Mammiferous» <vert> animal, which would fin
417e076 oul as the linear descendant of <Mammferus> «MAMMIFEROUS> <vert> animal, which would find its place i
619o042 g at Man, as a Naturalist would at any other MAMMIFEROUS animal, it may be concluded that he has pare
634j55r imals. as> 5 p. 314. Mac. remarks all <land> MAMMIFEROUS animals originally terrestrial.-- for we fin
422e092 achyderm. which was the origin of the aquatic MAMMIFERS» p. 306, the Dugongs cannot be united with tr
445e159 to habits of animals & takes series of flying MAMMIFERS-- says lemur.-- volans, has skin between its
451e178 flee during 8 months out of 12.-- the largest MAMMIFERS in the world consistently reside & are bred.
461t041 y closeness) of some species-- (especially of MAMMIFERS) in old beds & existing species is valuable b
634j54v ther good cases -- p. do]CD <Mac. remarks all MAMMIFERS originally land--animals. as> 5 p. 314. Mac.
634j55r then killed out «by the increase cold», & did MAMMIFERS then take their place? Would they not first o
641j29r ntries are supposed favourable to terrestrial MAMMIFERS-- Marine ones «of large size» <to> are best n
086a003 surprise at Horse being found in America, when MAMMOTH & narrow toothed Mastodon.-- argue against the
048r085 s so arranged from the forseight of the works of MAN Feeling surprise at Mastodon inhabiting plains o
056r109 e stupendous mass which has been corroded.-- If MAN could raise such a bulwark to the ocean, who wou
080r178 ad body of other. & see if same effects, as with MAN Does Indian rubber & black lead unite chemically
148g026 cross not so hardy but more easily fatten, This MAN confirmed the account of the «YOUNG» Shepherd do
165g125 is round & not merely thoughts laying dormant-- MAN from Glen Turret said he learnt to know the lamb
171b004 omes influenced.-- child of savage not civilized MAN.--birds rendered wild <through> generations, acq
172b005 s of both parents, & these infinite in Number In MAN it has been said, there is instinct for opposite
172b006 hat is alters it from some end which is good for MAN.-- Let a pair be introduced & increase slowly, f
175b018 ctricity Each species changes. does it progress. MAN gains ideas. the simplest cannot help..-- becomi
179b034 equired to make marriage.-- as Dr Smith remarked MAN & wild animals in this respect are differently c
181b040 shall now exist,-- In same manner, if we take «a MAN from.» any large family of 12 brothers & sisters
182b048 Fries rule-- What subject has Mr Newman the (7) MAN studied The condition of every animal is partly
182b049 ves final cause for enormous periods anterior to MAN.. difficult for man to be unprejudiced about sel
182b049 enormous periods anterior to Man.. difficult for MAN to be unprejudiced about self, but considering p
183b051 hether highly domesticated animals like races of MAN-- M. Flourens. .Journal des Savants.-- April 18
190b078 has been passed through to form that species.-- <MAN is derived from Monad, each fresh--> Mr Don rema
191b080 of two countries different.-- Ireland & Isle of MAN possessed Elk not England. Did Ireland possesse
191b082 tion & subsidence continually forming species.-- MAN & wife being constant together for life, is in a
193b093 remarkable Australia genera, collected together MAN has no heredetray prejudices «or instinc» to con
194b093 ces «or instinc» to conquer or breed together:-- MAN has no limits to desire, in proportion instinct
197b111 onsiders generation as a short process, by which MAN «one animal» passes from worm to man; «highest»
197b111 s, by which man «one animal» passes from worm to MAN; «highest» as typical of <sa>. changes, which ca
199b119 by Crawford on Mission to Ava. account of HAIRY MAN. because ancestors hairy with one hairy child, a

```
****************************************************(Key Word)****************************************************
206b148 yet some causes are evident, as for instance one MAN killing another.-- So is it with varying races o
206b148 illing another.-- So is it with varying races of MAN: these races may be overlooked mere variations c
209b155 Common;-- between Esquimaux & European dog? Yet MAN has had no interest in perpetuating these partic
211b161 against other «for sexual ends» species, whereas MAN has such instincts very little. in Zoolog. Proce
213b169 malia on this principle. <Varieties> < «Races» > MAN in savage state may be called, <species> species
215b178 uthority. The case of the tailess cat of Isle of MAN mentioned in Loudons (analogue of Blood hound--
218b189 & Hottentots coexisting. proves this-- but when MAN makes variety these are vitiated.-- This barely
222b207 ften talk of the wonderful event of intellectual MAN appearing..-- the appearance of insects with oth
223b207 s mind more different probably & introduction of MAN. Nothing compared to the first thinking being. a
223b208 The difference is that there is wide gap between MAN & next animals in mind, more than in structures.
224b214 pect most species.-- The difference intellect of MAN & animals not so great as between living thing w
224b214 is not probable; then monkeys will never produce MAN. but both monkeys & man may produce other specie
224b215 nkeys will never produce man. but both monkeys & MAN may produce other species., man already has prod
224b215 t both monkeys & man may produce other species., MAN already has produced marked varieties & may some
225b218 aid there are latent insects.-- as crows against MAN with gun. & Bustards &c &c!!! An American & Afri
225b218 ound in Tristan D'Acunha. may be said to deceive MAN. as likely as fossils in old rocks for same purp
227b227 for speculating on future. !.fish never become a MAN.-- Does not require fresh creation!-- If contine
228b231 .-- «Do not slave holders wish to make the black MAN other kind?» Animals with affections, imitation,
228b231 t We have no more reason to expect the father of MAN kind. than Macrauchenia yet he may be found:-- W
228b231 -- We must not compare «chances of embedment in» MAN in present state. with what he is as former spec
231b244 eations.-- Will Dromedaries & Camels breed?-- As MAN has not had time to form good species, so cannot
231b244 s ¿agrees with non-blending of languages?-- Till MAN acquired reason, he would be limited animal in r
232b248 of the savage, they perceive the superiority of MAN over animals, without such resorts Mr Waterhouse
233b252 s of North America If Parasite different, whilst MAN & his domesticated quadrupeds are not so. greate
235b262 erbred <p> having great tendency to vary? Is not MAN thus circumstanced; varieties of dogs in differe
235b272 ie Philos of Zoolog. 842 White regular gradat in MAN poor trash Lyell 1024 Flemings Philosophy of Zoo
239cIFC ever yet failed. About trades affecting form of MAN. Could you get racehorse from Cart horse by pick
255c052 ature to prevent the picking of monstrosities as MAN does.-- One is tempted to exclaim that nature co
256c053 or rather a new principle is brought to bear. If MAN created as now. languages. would surely have bee
256c055 ersary Speech. Feb 1838 thinks gradation between MAN & animal, small point in tracing history of Man.
256c055 Man & animal, small point in tracing history of MAN.-- granted.--but if all other animals have been
257c055 t if all other animals have been so formed, then MAN may be a miracle, but induction leads to other v
258c060 asoned people. would cause destruction.-- simile MAN living in hot countries, if continually crossed
261c070 ven ducklings & fowls-- When talking of races of MAN.-- black men, black bull finches from linseed--
263c074 nd destroyed) some intellectual being though not MAN.-- is as difficult to understand as Lyells doctr
263c076 . has such organization as to <per> force in one MAN-- wonderful Man. "divino ore versus coelum at
263c077 ion of fossil with recent. the fabric falls! But MAN-- wonderful Man. "divino ore versus coelum at
263c077 recent. the fabric falls! But Man-- -- wonderful MAN. "divino ore versus coelum attentus" is an excep
263c077 s animals.-- they on other hand can reason-- but MAN has reasoning powers in excess. instead of defin
263c078 rcumstances may have been necessary to have made MAN! Seclusion want &c & perhaps a train of animals
264c079 -- the world now being fit, for such an animal.--MAN, (rude, uncivilized man) might not have lived wh
264c079 it, for such an animal.--man, (rude, uncivilized MAN) might not have lived when certain other animals
264c079 r animals were alive, which have perished.-- Let MAN visit Ourang-outang in domestication, hear expre
271c106 ring picked, one where not.-- the latter made by MAN & Nature; but cannot be counteracted by Man.-- e
271c106 e by man & Nature; but cannot be counteracted by MAN.-- effect of external contingencies & long bred
282c140 ns over whole world. similar-- & constitution of MAN originally similar limits of change, would be sa
282c141 alls would be possessed by the different races of MAN, yet altogether different.-- To make this case p
285c153 ifference in the varieties «made by» of Nature & MAN.-- The constitution being heredetary & fixed, ce
285c153 «??», & therefore Most especially under care of MAN. & external circumstances not variable.-- Animal
286c154 ances not variable.-- Animals have voice, so has MAN. Not saltus. but hiatus animals expression of co
286c154 o the world same way» they may convey much thus, MAN has expression.-- animals signals. (rabbit stamp
286c154 ion.-- animals signals. (rabbit stamping ground) MAN signals.-- animals understand the language, they
286c154 ely[)] conceivable in savages) Has not the white MAN, who has debased his Nature «& violates every be
286c155 h prejudices, yet have justly excalted nature of MAN. like to think his origin godlike, at least ever
286c157 Fleming treats subject to put in alternative of MAN created by distinct miracle. Macleay letter to D
287c157 cal Magazine & Annals. 1830 (?)." if she has put MAN on the throne (of reason), she has also placed a
290c163 en we SEE structure gained by habit Talent &c in MAN not heredetary, because crossed with women with
290c163 d at right anglles, how pleased it is, just like MAN. emotions very similar.-- Geology. Transact. Vol
290c165 rossed.-- </I saw> .-- The attachment of dogs to MAN. not altogether explained by F. Cuvier, «--.Mem
290c165 » it is more, he cuts the matter short by saying MAN cannot be companion but master.-- heredetary tam
291c165 ell as wildness-- cf Sir J. Sebright.-- love. of MAN gained & heredetary. «problem solved» habits bec
292c170 ular views there offered, & he must be a zealous MAN in the cause if his faith is not staggered <">=
293c173 led on by circumstanc it becomes web footed, now MAN by effort of Memory can remember how to swim aft
293c174 tween laws--]CD. Many diseases in common between MAN & animals-- Hydrophobia &c cowpox, proof of comm
293c174 Hydrophobia &c cowpox, proof of common origin of MAN-- different contagious diseases, where habits o
293c174 would not cause misery to other.-- else smell of MAN would be disagreeable to Mosquitoes We never may
295c178 tone of Alps!!!? No) p. 15 (Lyell's Pamphlet) Is MAN more hairy than woman. because ancestors so, or
295c178 t females by song. do they by beauty, analogy of MAN if so war not [not located] Erasmus says he has
298c190 s produced, a general tendency produced. Such as MAN getting habitually into passion, becomes habitua
299c193 f same species Channel Islds (& probably Isle of MAN) no plants peculiar to themselves. this remarkab
299c196 ellectually developed amongst Pachydermata. like MAN amongst Monkeys-- or dogs in Carnivora.-- Man in
300c196 ke Man amongst Monkeys-- or dogs in Carnivora.-- MAN in his arrogance thinks himself a great work. wo
300c198 t saltus.-- The greater individuality of mind in MAN, is analogous to greater individuality of bodies
300c198 mind of different animals less divided.-- But as MAN has heredetary tendencies. his mind is still onl
301c199 is applicable to any habitual action. even which MAN performs.-- child striking a post in passion.--
303c204 kind of hair forms of legs-- hence the father of MAN kind probably possessed a structure in those par
303c204 Now we might expect that animal halfway between MAN & monkey, would have differed in hair colour + +
304c206 : who will doubt this if series now existed from MAN to Monads-- though physiology would profit. if t
305c211 of irritations sleep walking. fits, laught &c&c MAN & Man may have some relation together, as well a
305c211 ritations sleep walking. fits, laught &c&c Man & MAN may have some relation together, as well as Man
305c211 Man may have some relation together. as well as MAN & child, polypus & polypus, bud & bud, polypus &
306c212 behaved very differently from a dog.-- more like MAN. continued long in a passion & looked out for hi
308c217 ss it were fixed what a species means» civilized MAN, May exclaim with Christian «we are all» Brother
310c223 ever allow that because there is a chasm between MAN (--<a> «&» chasm«&» <necessary to account> «cons
310c223 e of» for the scheme of nature) and animals that MAN has different origin. «Royal Institution» Dr Roy
312c228 olours of little value Dr Smith if black & white MAN crosses; children heterogenous, he feels sure of
313c234 ade, & only species, good argument for origin of MAN one.-- Is the extinction & change of species two
315c243 n & the Zoonomia, for if the former shows that a MAN grinning is to exposes his canine teeth. no doub
315c243 e teeth.-- (This may be made capital argument if MAN does move muscles for uncovering canines) --. Bl
316c244 r eternity», only difference between the mind of MAN & animals.-- yet how faint in a Fuegian or Austr
317c249 appendix) of Congo Expedition, NB. I met an old MAN--, who told me that the mules between canary bir
319c257 trast with South America.-- In Home's History of MAN at Maer, it is said the Samoyed women (¿ north e
321c275 t table of Canary isld: Plants Home's History of MAN Transactions of the Entomological Society Vol I.
324c268 nfluence of climate White's regular gradation in MAN. Lindlys introduction to the Natural system Beva
337d022 capital passage might be made from comparison of MAN, with expression <of a> of Monkey, «when offende
339d025 Zoological Provinces» according to varieties of MAN.? «In Australia. plants E & W very different.--
339d025 .? «In Australia. plants E & W very different.-- MAN not so, but N. & S. New Zealand & New +++ Caledo
346d047 Black faced, but more tendency to fatten-- This MAN confirmed my account of the Shepherd dogs.-- Aug
347d049 s length of days adapted to duration of sleep of MAN.!!! whole universe so adapted!!! & not man to Pl
347d049 ep of man.!!! whole universe so adapted!!! & not MAN to Planets.-- instance of arrogance!! August-- 2
350d054 s the art of God" Septemb I,. It has been argued MAN first civilized. <note> add this in note. ¿mere
351d057 er saltum-- yet does nothing in vain!! Foetus of MAN undergoes metamorphosis., heart altered & umbili
353d061 e ourang-outang. other point of resemblance with MAN.-- September 3d Magazine of Natural History. 183
354d062 erfly-- is not more wonderful than the body of a MAN undergoing a constant round,--each particle is p
354d065 . p. 351. Paradoxurus Philippensis. Philippines MAN have varies the range-- Argue the case of Probab
354d065 ved with what facility mice attach themselves to MAN. Sept 7th. -- I was struck looking at the Indian
355d068 - the very same must take place in copulation-- (MAN & woman separate parts of same plant<s>)-- now i
360d090 reeders, where their interest is concerned. Same MAN had crossed Jackal & dog-- (offspring did not go
```

Page
(Key Word)

```
372d130 ith generation.-- Why crab can produce claw. but MAN not man. hard to say-- if it were possible to su
372d131 say-- if it were possible to support the arm of MAN, when cut off , it would produce another  man.--
372d131 of  Man, when cut off , it would produce another MAN.-- That the embryo the thousandth of inch should
376d135 volition» of this populousness, on the energy of MAN» D.'Orbigny. Comtes Rendus p. 569. 1838 says the
380d155 so  Elephant in confinement, & so imagination in MAN, has strange effect.-- Directly a Capon is  cut,
382d158 nipples <are> «though» abortive, are so plain in MAN, & yet no trace of abortive womb, or  ovarium.--
384d162 e generations Theory of sexes (woman makes, bud, MAN puts primordial vivifying principle) one individ
384d162 d, & sexes as in Vertebrate tak place.-- ∴ Every MAN & woman is hermaphrodite:-- ∴ developed instinct
385d163 ted male.-- «Men giving milk--» Sept. 25th Young MAN at Willis «Grt. Marlborough Str, Hair dresser, a
387d170 rs with respect to moths, & monkey & men.-- each MAN passess through its caterpillar state. the monke
388d172 rent puppies out of same mother.-- The view that MAN & «or cock» pheasant &c is abortive hermaphrodit
388d172 xes abortive fact of same tendency. -- Mammae in MAN. having given milk. -- testis & ovaria The follo
391d174 but  why should it demand some further change?-- MAN properly is hermaphrodite (hence monstrosities t
391d174 ke changes or species!! CD[The view of <In> each MAN or mammalia being abortive hermaphrodite simplif
391d175 merally dioecious oldest forms) why are twin in MAN more like «each other» than twins «or triplets &
398e004 ds, that most sublime discovery of the genius of MAN Those who have studied history of the world most
404e024 , because it is an involuntary variation made by MAN, common to every individual & therefore effect o
405e031 rmines the puppies to be so.-- [not located] Did MAN.-- L'Institut. 1838. p. 338.«V[ide]» Important a
405e035 MENT REAL of antiquity of reasonable cosmopolite MAN, exceeding monkeys;-- Having proved mens & brute
409e046 fox, wolf &c &c-- is precisely analogous case to MAN.-- ask the missionaries about Australians yet sl
409e047 erra del Fuego is to be converted into civilized MAN is one great object, for which the world was bro
409e049 ral sentiments of the animated beings.-- &c-- If MAN,-- though man's practiced judgment. even without
414e063 make  <them> such a variety far more easily than MAN to gain the day.-- In man chiefly intellect, in
414e064 tted organization, or instincts (ie intellect in MAN chiefly intellect, in animals chiefly organizati
414e064 ncts (ie intellect in man) to gain the day.-- In MAN is not unlike that of animals transported by flo
414e065 ictly applicable to the universe.-- The range of MAN is not an intruder.-- : the geological history o
414e065 orted by floating ice.-- I agree with Mr Lyell.. MAN is as perfect as the Elephant, if some genus. ho
414e065 s not an intruder.-- : the geological history of MAN were to be discovered. Man acts on. & is  acted
414e065 some genus. holding same relation as Mastodon to MAN. acts on. & is acted on by «the» organic and inor
415e065 ation as Mastodon to Man. were to be discovered. MAN became cosmopolite, he would probably be confine
415e066 es» of the Ornithorhyncus be found.;-- yet until MAN was quadruped or bimanous,. is to see, what part
415e066 that I can see to discover whether the parent of MAN & monkeys have equal chance that progenitor  was
415e068 depend  on the number of the species.: therefore MAN.-- CD [any monkey probably might, with such chan
415e068 ant organization, Hand & throat) that has made a MAN.-- It is one thing to prove that a thing has bee
415e069 intellectual, but almost certainly not made into MAN December 16th. The end of each volume of Whewell
415e069 on.-- as in «races of» Dogs, so in species, & in MAN arbitrarily destroying certain forms & not other
416e071 ns.-- (hence great variation in each birth) from MAN picks the Male, instead of allowing strength  to
416e075 ected Hence difference between races & variety? «MAN & woman is «indeed» (independent of sexual diffe
417e077 this  we might expect, as the difference between MAN was originally quadru«manous» «ped.-- ». Hairy.-
418e083 generation  were impressed on it.-- Seeing that <MAN> «all vertebrates. [Müller's Physiolog. p.24.]CD
420e089 cended.-- Jan. 6th The rudiment of a tail, shows MAN we should remember, that even birds can imitate
426e108 ge?-- Wonderful as is the possession of voice by MAN probably assumes the hairy character of his fore
432e124 nimal with fluid seeds can be true hermaphrodite MAN. "standing alone in the gift of intellect, he re
435e134 emarks lives in dark caverns of Carniola p. 112. MAN can offer of changes.-- as desert «or rock» plan
438e142 -- the constitution of some may resist the means MAN & wings under united elytra The law of  <growth>
440e147 n be shocked at absence of final cause mammae in MAN almost (this looks inaccurate C. D) they will ca
451e178 into  this region between April & October & like MAN Chapt see Yarrell Syngnathus Ch 6 I presume, fro
461t026 whole order of the fish.-- Embryology p. 97. for MAN may lose all traces of his parentage in  «about»
462tf05 han others-- is incidentally said that a mongrel MAN-, valueless.-- May not several generations have b
465t079 frica!,-- any negative argument against-- monkey-MAN, valueless.-- May not several generations have b
494q004 ) In the last 1000 years how many generations of MAN have been. be on what principles calculated
497q007 es during last century or two.-- where agency of MAN not known.-- (3) How is Iris impregnated.; which
508q016 y more cases of diseases, generally occurring in MAN being transmitted through females, like Hydrocel
508q016 independent  of heredetary cases, more common in MAN than in female-- (8) In Hump-back ever heredetar
509q017 cow.-- degree of soldering of tibia & fibula: in MAN any abortive bones??? do. Wing in Apteryx ‖‖  no
510q017 hysics of Morphology. ▌-- Schelgel is he serpent MAN? about zones separated by non-inhabited  spaces:
521m009 &c. & he maintained he had never heard of such a MAN & had no gardener.-- My F. then asked Mr C. to c
522m010 t» had married.-- Answered never heard of such a MAN.-- (My Father explained who he wa & all about hi
525m026 lter Scotts remark how odious an illtempered fat MAN looks, shows same connection between organizatio
526m027 , example of others or teaching of others.-- (NB MAN much more affected by other fellow-animals, than
526m030 thus these qualities become heredetary.-- When a MAN says I will improve my powers of imagination, &
526m031 themselves  become changed.-- appetites urge the MAN, but indefinitely, he chooces (but what makes hi
527m032 nstinctive, is heredetary knowledge like that of MAN, & this agrees with the stated fact, that «birds
527m032 ; but it does not explain the feeling in any one MAN.-- Music & poetry opposite ends of one  scale.--
532m054 the mind tries to fix upon some object:-- When a MAN, child or colt has once been frightened & starte
532m056 ection to a place.-- How like strong feelings of MAN.-- The sensation of fear is accompanied by «trou
533m058 child dislikes the frown without knowing why-- a MAN as in Guy. Mannering. feels, pleasure. in seeing
536m070 trivance-- it is what my views tend to.-- When a MAN is in a passion he puts himself stiff, & walks h
536m073 ation of oyster. so may free will make change in MAN.-- the real argument fixes on heredetary disposi
536m074 pon the children.-- The above views would make a MAN a predestinarian of a new kind, because he would
536m074 ew kind, because he would tend to as an atheist. MAN thus believing, <yet> would more earnestly  pray
537m074 example  on others.▌ It may be doubted whether a MAN intentionally can wag his finger from real capri
537m076 tion.-- This probably is natural. consequence of MAN, like deer &c, being social animal, & this consc
538m080 truggled as it were with a second & unreasonable MAN.-- If one could remember all ones farthers actio
539m081 the intellectual powers-- the difference is of a MAN wagging his foot & working with his toe to perfo
539m082 gains habit «or trick» so much more easily than MAN, so may animal obtain it far more easily, in pro
539m084 igid.-- How is this? dealt with p. 241 Origin of MAN now proved.-- Metaphysic must flourish.-- He who
540m084 s metaphysics than Locke A dog whines, & so does MAN.-- dogs laughs for joy, so does dog bark. (not s
540m085 male & smelt its fingers. Seeing a dog & horse & MAN yawn, makes me feel how <much> all animals <are>
540m086 - with the negros of Africa, (or again the black MAN of <B> Van Diemens land & the energetic copper c
541m093 onder that they are so difficult to conceal.-- a MAN «insulted» may forgive his enemy & not wish to s
541m093 ifficult to to look tranquil.-- He may despise a MAN & say nothing, but without a most distinct will,
542m095 red, (also how often) to the tale of a wearisome MAN.-- is frowning, result of straining vision, as s
542m096 - The distinction «as often said» of language in MAN is very great from all animals-- but do not over
543m099 ellects, & in the same way are chess Players-- A MAN at Cambridge, during his time, almost an absolut
543m100 aying chess however many places, & contingency a MAN has keep in mind. all is certain.-- there is jud
543m100 f probabilities, therefore this judgment gives a MAN common sense, & the highest intellectual  powers
544m104 0, by mistake.» N B. Everything which happens to MAN who does not produce children. or after he has u
545m108 ements in skin of eyebrow important analogy with MAN.-- I see monkeys grin with passion, that is show
546m108 dmire a wide precept.-- The very superiority of MAN perhaps depends on the number of sources of plea
549m118 We give no credit to instinctive feelings.-- for MAN losing his children, any more than to dog losing
549m118 intense happiness-- so is it <with an> when same MAN is compared to peasant.-- To make greatest numbe
549m120 portion  of daily <happiness> «pleasure». A wise MAN will try to obtain this happiness. though he see
549m123 other structures slowly vanishing-- the mind of MAN is no more perfect, than instincts of animals to
550m124 vil under form of Baboon is our grandfather!-- A MAN, who perfectly obeys his conscience or instinct,
551m127 Delirium  of every degree of intensity-- «in old MAN, he had just seen mind went on RAMBLING till exc
551m129 ss) when pouting protrudes its lips into point-- MAN, though he does not pout. pushes out both lips i
552m131 muscles,  that in my grandfather remark, a tired MAN. involuntarily feels angry, when brain is pumpin
552m132 igin as well as rule will be given.-- Descent of MAN Moral Sense Mitchell Australia Vol I, p 292 "Dog
553m138 ucture" The monkeys understand the affinities of MAN, better than the boasted philosopher himself  it
556m145 p. 86 "We ought never to forget-- --; that every MAN is born with a portion of phsiognomical sensat
556m145 phsiognomical sensation, as certainly as every MAN who is not deformed. is born with two eyes.." I
557m147 xpression «as constant smiles, cheerful face».-- MAN when at ease has smooth brow contrary to wrinkle
558m151 mind, (to show hiatus in mind not saltus between MAN & Brutes) no one can doubt this connexion.-- loo
558m153 t out & hiss.-- [Hyaena pisses from fear so does MAN.-- & so dog]:CD Man grins & stamps with passion.
558m153 ena pisses from fear so does man.-- & so dog]:CD MAN grins & stamps with passion. can expression be u
558m153 good but look at the immense difference. between MAN,-- forget the use of language, & judge only by w
564n003 n attribute would be rather favourable to it--!! MAN moreover who reasons much on his actions,  makes
```

564n005 y which eyes, aided by experience is supposed in MAN to guide to knowledge, was transmitted perfectly
565n007 o listen to growl of hounds». <when> as fear to «MAN as» animals. comes at distance, mouth is placed
565n007 every time really wishing to smell its enemy.-- MAN & dogs show triumph (& pride) same way walk erec
566n011 ed. before my view of origin of evil passions.-- MAN getting sight slowly,, but when in grown years,
567n014 is arm. & show signs of affecting something like MAN. Has an oyster necessary notion of space-- plant
568n019 a cow--" «so it is with all uneducated.--»-- Old MAN at Cambridge observed the ignorant. merely looke
569n019 almost like tastes of mouth & smell. Descent of MAN Understanding languages seem simplits case of As
569n020 th said, recognizes that sound as perfectly as a MAN.-- Probably, language commenced in some necessar
570n024 a master says I will see you damned first." the MAN shrugs his shoulders & replies nothing. if he di
571n029 of grazing animals knowing poisonous «herbs:» & MAN not.-- ¿no vegetable good «for man» to eat poiso
571n029 us «herbs:» & man not.-- ¿no vegetable good «for MAN» to eat poisonous?-- How did animals in «Austral
573n036 uns. universe, nay whole systems of universe <of MAN> to be governed by laws,, but the smallest insec
574n041 eth»less-jaws. as picture of disgusting lewd old MAN. ones tendency to kiss, & almost bite, that whic
574n041 ed as biting: so do stallions always.. = No doubt MAN has great tendency. to exert all senses, when th
574n041 s to bitches.-- How comes such an association in MAN.-- it is bare fact, on my theory intelligible An
576n046 he meditated on this, it would be conscience. A MAN, might not <t> do so even to save a friend, or w
576n047 inctive as in the dog, or chiefly habitual as in MAN,), for it added much to the happiness of his life
576n047 the chance, of so dreadful a consequence to each MAN is small. Man's intellect is not become superior
576n049 which an instinct is transmitted.-- Arguing from MAN to animals is philosophical. viz. man is not a c
576n049 guing from man to animals is philosophical. viz. MAN is not a cause like a deity, as M. Cousin says.
577n049 a is not an antagonist quality to life & mind of MAN.-- & we do not suppose an hydatid to be a cause
577n051 thought drive blood to surface exposed, face of MAN, face, neck-- «upper» bosom in woman: like erect
578n056 e-- that it paralyzes all muscular action -- «in MAN & animals» Blubbering of a child (different in d
579n059 g anything to do with ancient movement of ears A MAN shivers, from fear, sublimity, sexual ardour.--
579n059 hivers, from fear, sublimity, sexual ardour.-- a MAN cries from grief, joy. & sublimity. January 6th.
579n059 ard «is it influenced by» other emotions? When a MAN keeps perfect. time in walking, to chronometer,
581n064 nd step if it can be generalize.-- The tastes of MAN, same as in Allied Kingdoms-- "food, sm<e>ll. (o
582n066 acceleration of pulse. or rigidity of muscles.-- MAN cannot be said to be angry.--» «He may have pain
582n066 ming it.-- p. 14. There is scarcely a faculty in MAN not met with in the lower animals.-- hence the g
583n069 th variability of reasoning power in one species MAN.-- false instinctive pointing varies.-- p. 18. A
584n072 one spot, this is indeed instinct.-- Australian MAN, may be called instinctive: the facts of memory
585n076 ow remarks that Chimpanze pouted & whined, when, MAN went out of room.-- all theories of magnetic pow
586n081 r"-- "An animal has faculty of walking. which in MAN is learnt by experience is in other is acquired
587n088 of liking, as simply heredetary habit.-- A blind MAN might be born with idea of scarlet, as well as r
588n090 difference same, but less in degree, as between MAN & child.--)» what differs-- not <reason> «instin
591n100 d case of instinctive conscience.-- Why does not MAN eating cause disgust, because he does not go aga
591n100 e feeling, only does not fullfil, like continent MAN.-- a man eating what others by habit (not instin
591n100 , only does not fullfil, like continent man.-- a MAN eating what others by habit (not instinct) think
591n100 clear.-- even to the cold or benevelo- continent MAN Hume has section (IX) on the Reason of animals E
593n109 t do for insects. if this view holds good-- then MAN, a socialist, does not know other men by smell,
593n109 beauty.-- the social affections of animal taking MAN in place of other animals is hostile «is subvers
593n109 music are great distinguishing character between MAN & animals.-- [blank] Double consciousness. only
593n111 held in one's own mind, & Dr. Hollands story of MAN in Delirium tremens hearing other man speaks. sh
593n111 s story of man in Delirium tremens hearing other MAN speaks. shows, that consciousness of personal i
594n112 ot frightened at a dog.-- » The instinct against MAN is perhaps, as strong as against hawk, but the
594n112 angerous-- wild-ducks would have fled equally if MAN had appeared-- though instinct so firmly implant
594n115 e its instinct is not <more> less wonderful than MAN his intellect Lyell has seen a little dog go to
595n125 highly useful as only means of communication) in MAN, than sucking.-- [I assume a child pouts who has
596n184 s but <not> Chimpaze. does not gradation towards MAN.» Macacus especially pulls back skin of whole fo
599o007 crowded with towns & thoroughfares, I grant that MAN, from the effects of heredetary knowledge, has p
600o008 perative sense of duty-- which makes struggle in MAN.-- two souls in one body-- (2) Beau ideal, refer
600o08v these last heads. of separation between soul of MAN. & intellect of beasts, not clear.-- ¿does not M
601o009 having some relation to the primary sensation.-- MAN moving leg when asleep-- «or habitual actions» p
601o009 ess is sensation No. 2. with memory added to it, MAN in sleep not conscious, nor child-- Evidence of
603o012 «common to many races»-- thinks action towards <MAN> «a king» <changed into> is carried on toward de
607o025 dental (so called like chance) circumstances. As MAN hearing Bible for first time, & great effect bei
607o025 being inclined that way]CD one sees this law in MAN in somnambulation or insanity. free will (as gener
608o026 eral delusion about free will obvious.-- because MAN has power of action, & he can seldom analyse his
608o026 have none.-- Effects.-- One must view a wrecked MAN, like a sickly one(P)-- We cannot help loathing
608o026 ever be more proper to pity than to hate & be †A MAN may put himself in the way of Contingencies.--bu
608o027 n be really fully convinced of its truth. except MAN who has thought very much, & he will know his ha
608o028 & inexplicable if we suppose that the sins of a MAN, are under his control, & that a future life is
609o029 > 2d. 1838 Those emotions which are strongest in MAN, are common to other animals & therefore to prog
609o30v generated? The origin of the social instinct «in MAN & animals» must be separately considered.-- The
609o30v y considered.-- The difference between civilized MAN & savage,-- is that former is endeavoring to cha
612o035 er animals, as in stomach, intestines & heart of MAN. [LHC] ¿How near in structure is the ganglionic
612o035 glionic system of lower animals & sympathetic of MAN [RHC] ¿How does consciousness commence; where ot
613o035 of body. to do such.-- All this can take place & MAN not conscious as in sleep; or in sleep is man mo
613o035 & man not conscious as in sleep; or in sleep is MAN momentarily conscious, but is memory gone?-- Whe
613o036 we see particular trains of thoughts as fear of MAN.-- crows fear gun,-- pointers method of standing
614o037 ted from generation.; than from hour to hour in <MAN:> individual-- [LHC] Perhaps even the most compl
614o037 analogous «(& replace)» to experience gained by MAN in lifetime Heredetary memory not so wonderful a
614o037 its multiplicity & the comparison of ideas.-- As MAN has so very few (in adult life) instincts.-- thi
614o037 [LHC] As sexual instinct comes on late in life, MAN almost alone in this case can perceive instinct.
615o038 place without corresponding change in «external» MAN; and as all men nearly same species, so general
616o038 y may go back to animals which were changed into MAN .. they meet their reward! X Perhaps should hardl
616o038 sexual intercourse in plants is involuntary, in MAN voluntary: ¿ False,-- secretion in both involunt
619o042 l Philosophy [RHC] On the Moral Sense Looking at MAN, as a Naturalist would at any other mammiferous
619o042 ued nearly so [RHC] The history of every race of MAN shows this, if we judge him by his habits, as <i
619o042 protecting sheep or hurting them.-- Therefore in MAN we should expect that acts of benevolence toward
619o043 he would likewise feel pain.-- If he saw another MAN <say go> «acting in» accordance to his instincts
619o043 pleasure, which the acter received.-- If either MAN did not obey his instincts from interference of
620o044 [RHC] 3) probably thinks no more of it.-- Not so MAN, from his memory & <pow> mental capacity of call
620o044 blow struck in passion fades away, so that when MAN afterwards thinks why was such an instinct not f
620o045 tells-- + us the action was superfluous, as one MAN trying to save another in desperation.-- This sh
620o046 (& which «when present» give pleasure.) & which MAN ought to follow-- it is his duty to do so.-- So
620o046 ry to teach him & strengthen his instincts.-- so MAN ought to follow certain lines of conduct, <altho
621o047 approbation of his fellow men.-- [RHC] 6) Hence MAN must have a feeling, that he ought to follow cer
621o047 erest receive simple explanations from origin of MAN.-- [RHC] By interest I do not mean any calculate
622o049 ard to discover this, for the different races of MAN may have different instincts, as we see in dogs
622o049 nstincts, as we see in dogs & pidgeons.-- But as MAN is animal at head of series in which «special» i
623o050 her [RHC] This feeling seems to vary in races of MAN. & certainly in «species of» animals, in which c
623o050 mother; the having received some advantages from MAN. during many generations giving the social feeli
628o054 I should think some parts of the emotive part of MAN, may be quite artificial, as avarice love of gol
633j53v y [produce some peculiarity in seed vessel]CD if MAN takes care they are not detrimental.-- NB. One l
634j54v d for another".-- What bosch!! Put it to case of MAN. <&> The <design> determination of a God-head.--
634j54v rate movements in the Holocentrus ruber (a fish) MAN has abortive muscles to his ears.-- p. 313 Many
635j55r ts & whether constant or inconstant like that of MAN.-- the cause given we know not the effect [blank
638j58v in isd) although having heredetary superfluities MAN could exist without Mammae.» to the then existin
638j059 ture, in the course of ages «step by step»-- in MAN, the nervous system, gains that knowledge, befor
639j28v lloch p. 260 intimates canines no special use to MAN, Applicable to Bell's sneering-theory.-- p. 263.
640j28v , but for youth most necessary: the fertility of MAN in old age keeps woman alive: for Man & woman ar
640j28v rtility of Man in old age keeps woman alive: for MAN & woman are same: fertility of either sex determ
571n029 ous?-- How did animals in «Australia» & America MANAGE;-- This shows doctrine «of instinct» has been
641j29r arctic seas. (on other hand Spermaceti Whale & MANATEE.-- Naturalists must be cautious.-- <some other
531m053 therefore brain has been accustomed to send a MANDATE to the muscles & when the noise comes it canno
307c216 ame fundamental organs even in Haustellata & MANDIBULATA.--!! --Argument, when general argument is ex
253c047 ansact. Vol I p. 165.-- <a> "an account of the MANELESS lion of Guzerat by Capt. W. Shee. considered

Page **(Key Word)**
399e010 497. Vampire bat abound in the Navigators & at MANGUIA, but are unknown Eastward of the Navigators. S
399e009 monster. or solely to childhood, or solely to MANHOOD,-- it will decrease & be driven outwards in th
522m013 sane.-- that everybody is insane. at some time. MANIA is quite distinct, different also from delirium
522m013 ation, often ends in insanity or delirium.-- In MANIA all idea of decency & affection are lost.-- mos
542m094 ut some sound even if it be inarticulate.-- the MANIAC shouts & bellows with passion.-- It is not a l
525m026 e affections «& N B affections very soon go in MANIACS» seem to have failed even more than the memory
051r095 pting when pebbles are brought into play; most MANIFEST example of degradation I ever saw on beach ne
269c103 multiplicity<"> [(his definition "constant MANIFESTATION of unity through multiplicity" this unity,-
471tf07 endless variety of form in the same> organ "MANIFESTATION of divine power"?.--"of their use difficult
558m152 air «dogs snarl much the same way» generic MANIFESTATION of great passion.-- I do not think they arc
617o040 ised by our external senses consists in the MANIFESTATION of force i.e. movement? capable of being tr
617o040 y considered in themselves consist in a force MANIFESTED in every particle of matter directed towards
617o40v e strings & as on this supposition the forces MANIFESTED would be <sat> fundamentally accounted for,
594n113 n, protruded. determined to do nothing. & so MANIFESTING sulleness. [blank] Circumstances having give
047r082 thquake wave is an oscillation, body of water MANIFESTLY does not travel up.--) If these view are rig
284c149 ation) varies according to the species, it is MANIFESTLY of less importance, as affording natural cha
605o20v inating & respecting good from bad. And it is MANIFESTLY from this fact & the instantaneousness of th
617o40v imated agent Now the phenomena of gravity are MANIFESTLY the same as if every particle of matter were
639j28v g. not at once, but by steps. of which we have MANIFOLD traces in the several genera of Grallae Suppo
402e019 ts-- list given,= some peculiar <M> p. 359. At MANILLA a small Cercopithecus., & skins of galiopithec
056r112 ingling & separating is well adapted to use of MANKIND.--<Hutton show> Earthquakes part of necessary
057r112 ese final causes.--What more awful scourges to MANKIND than the Volcano & Earthquake.--Earthquakes ac
263c074 es-- The believing that monkey would breed (if MANKIND destroyed) some intellectual being though not
286c154 onsider him as other animal-- it is the way of MANKIND. & I believe those who soar above Such prejudi
309c220 astes. improve the women. (double influence) & MANKIND must improve-- The areas of subsidence marked
334d011 her.-- case parallel to brothers & sisters in MANKIND.-- The case of all blue eyed cats (Fox has see
343d038 . America. With respect to future destinies of MANKIND, some of species or varieties are becoming ext
415e067 he Bladder.-- The numbers of fatal diseases in MANKIND, «the more valuable domesticated animals» no d
417e078 efore, of likenesses of fathers to children of MANKIND, no doubt are applicable to likenesses, when s
451e178 region (mentioned by Heber) «from» which «all» MANKIND «(& yet afterwards says native tribes can live
537m076 would have been so much, as in other races of MANKIND..-- p. 27. Mart. allows some universal feeling
566n010 ling motive.-- Book IV, Chapt I on passions of MANKIND, as being really useful to them: this must be
614o037 on, parental affection-- The very existence of MANKIND requires these instincts,: though very weak so
211b163 finity & affinity-- whales & fish.-- Progeny of MANKS cats without tails: some long & some short: the
023r011 n.--The fault like appearance «arising from the MANNER of horizontal upheaval» of the shore of the Pa
029r033 of earthquake shook the ship in a most violent MANNER. Although it lasted about a minute, there was
042r069 e of valleys.--All my observations of period «& MANNER» of elevation Volcanic action, must be more ex
063r133 lely adaptation of animals.--extinction in same MANNER may not depend.--There is no more wonder in ex
094a036 s of granite clearly effect of remodelling same MANNER. as bits of Patagonian boulders might be trans
181b040 ors of his relatives shall now exist,-- In same MANNER, if we take «a man from.» any large family of
195b101 et to move in its particular destiny.-- In same MANNER God orders each animal created with certain fo
227b227 o endeavour to discover causes of change.-- the MANNER of adaptation (wish of parents??) instinct & s
232b249 islets & a sub-genus in Southern Africa In same MANNER. Cuscus, (a sub genus of Phalangista New Holla
248c033 ost permanent Owen [re]markable laws of Brain & MANNER of generation «& primary divisions of insects»
271c105 uilds its nest in <same country> something same MANNER, much mud.-- These facts show, habits heredeta
286c156 fter all a slight one It will be necessary from MANNER Fleming treats subject to put in alternative o
305c209 ch vary in colour of plumage in same remarkable MANNER as Europaean species = singular coincidence if
305c210 ertainly more like Jackall in gait, size, fur.; MANNER in which ears droop like dog older character &
305c210 in which ears droop like dog older character & MANNER of wagging tail.-- habitual movements connecte
384d161 s has a disposition to deviate from Nature in a MANNER peculiar to itself" <Is this so> Each part <no
384d162 is never happens in plants «only in subordinate MANNER in the plants which have male & female flower
388d173 division <only> is developement of gemma.-- The MANNER in which Frogs copulate & fish shows how simpl
404e026 d coloured puppies case of this.-- tendency in «MANNER of» life to be mottled + heredetary tendency d
405e031 partly retained, therefore colours vary in same MANNER as they would vary, if in wild state; thus mar
421e091 than to other parts of that continent. in like MANNER as Madagascar does to other side of Africa.--
459t013 er the Penguin in the form of its body & in the MANNER of walking but not waddling; its colour was da
484z014 ican representative of Caracaras of Americas.-- MANNER of walking-- foot bill crest feathering on leg
501q011 peaking of Thyme doubts about stigma in similar MANNER ever failing.-- answered by Gaertner (28) Can
513q020 skeleton== Does the tumbling of pigeons vary in MANNER & perfection &c &c &c--if so probably a variet
520m002 lance was in odd twiching of muscles, & general MANNER of holding hands &c &c.-- Mr Dryden Co said he
523m015 untruth of the idea, namely his poverty.-- his MANNER of curing it. by keeping the sum-total of his
527m033 ghts.-- Poetry. the latter thoughts are in same MANNER vivid & grand. the frame of mind being just ke
529m041 mpound animals united by wonderful & mysterious MANNER.-- There is much imagination in every view. if
539m081 o» remember, & to think deeply, & the immediate MANNER in which my head got well when reading article
541m092 ady to bark, & lip twisted up, in that peculiar MANNER they do, even more than in a real snarl, they
544m102 onal connections-- ideas are strung together in MANNER <they> quite different from when awake.-- pecu
547m112 implied by «presence» my servant, «box» my own MANNER of ordering things to be done.-- The senses ar
555m143 -- Sept 21st Was witty in a dream in a confused MANNER. thought that a person was hung & came to life
557m148 e with a dog, & make him ashamed of himself, in MANNER quite different from fear; there is no inclina
564n006 described as an exaggerated habitual sneer» the MANNER in which whole skin or muscles are contracted
565n008 ovement as sneering,-- it is then more <emblem> MANNER of hurting opponent by insulting his pride & i
572n032 ing respect is by making yourself less, but the MANNER, whether by bowing the body, kneeling, prostra
574n042 tual action must some way affect the brain in a MANNER which can be transmitted.-- this is analogous
579n058 erence, which my theory believes in.-- From the MANNER short-sighted people frown, frowning must have
585n075 o know that it is one, when applied in peculiar MANNER.--]CD April 3d. 1839 The Giraffe kicks with f
605o19v uperiority we transfer to ourselves in the same MANNER as we are acted on by sympathy. D. Stewart on
610o032 things, which are often repeated in a wonderful MANNER.-- as the hour of the day &c-- All habits must
616o39v ng by which thought or memory. might be in like MANNER attributed to the brain. There are two modes o
629o055 perance, acquired by education.-- CD[In similar MANNER our desires become fixed to ambition. money, b
533m059 frown without knowing why-- a man as in Guy. MANNERING. feels, pleasure. in seeing the scenes of his
266c091 the same tracts;. & that in their appearance & MANNERS they are as opposite as day & night: yet we kn
439e145 im, would be the one encouraged-- » Wilkinsons MANNERS & Customs of the Ancient Egyptians Vol III. p.
023r009 y teeth, 2 of them, 8 inches long, & as big as a MANS thumb, the rest not above half so long; The maw
066r141 ds gales = very strong. Stones as bigger than a MAN'S head.-- Kerguelen 40 by 20 leagues. dimensions:
192b084 d in foreign country.-- When one sees nipple on MAN'S breast. one does not say some use, but <no.> se
299c196 wers unimpregnated Babington We see gradation to MANS mind in Vertebrate Kindgdom in more instincts i
299c196 ncts in rodents than in other animals & again in MANS mind, in different races, being unequally devel
300c198 f language. heredetary & acquirable.-- therefore MANS mind not so different from that of brutes p. Ha
372d131 -- Does likeness of twin bear on this subject? A MANS arm would produce arm if supported., <so> & in
414e063 ch a variety far more easily than man,-- though MAN'S practiced judgment. even without time can do mu
542m097 ments.-- some say dogs understand expression of MAN'S face.-- <That> How far they communicate not eas
546m109 als-- see how a dog likes smell of Partridge--, MAN'S taste for smell of flowers, owing to parent bei
550m125 ust occupy greater proportion of <each> «every» MAN'S time.-- Begin discussion-- by saying what is Ha
564n003 -- Dogs conscience would not have been same with MANS because original instincts different.--Mem. Bee
576n047 so dreadful a consequence to each man is small. MAN'S intellect is not become superior to that of the
576n047 of external circumstances» Look at Spain now.-- MAN'S intellect might well deteriorate. «CD[in my the
579n059 y. & sublimity. January 6th.-- What passes in a MAN'S mind. when he says he loves a person-- do not t
581n063 emember it. as my father trying to remember the MAN'S Christian name, writing for the surname,, analo
593n111 nal identity is by no means a necessary part of MAN'S mind.-- At Maer. Pool. I saw many coots & water
608o026 ling on prepared materials. From contingencies a MANS character may change-- because motive power cha
634j54v ipotent creator, exhausted & abandoned. Such is MAN'S philosophy. when he argues about his Creator! p
639j28r in.-- Pincers in Scorpion & Crust in Squilla. & MANTIS. C D woodcuts stones swallowed by birds & by A
210b158 uld not be manufactured, <but why they should be MANU> Does it not present analogy to what takes plac
210b158 d in «one» small island species would not be MANUFACTURED, <but why they should be manu> Does it not
492q003 nursery gardens? <are the> is the ground much MANURED In species of close genus do more than three p
299c195 arge root by sowing it at wrong time of year & MANURING it.-- Epigonous. Perigonous &c-- very importa
026r022 e in the Cordillera of S. America. Study Geolog: MAP of Europe Conybeare. Introduct XII P. silicified
035r047 y the parallel bands of formations on any Geolog MAP: Quoted from Daubeny P 402: likewise, mean heigh

Page
(Key Word)
036r049 t in grand discussion Consult. reconsult Geolog. MAP of Europe Consult charts for distribution of peb
071r157 > coming so near surface most important There is MAP of Cordillera by Humboldt in Geolog. Society Sir
306c213 & reptiles &c Timor is connected with Australia «MAP to King's Australia» by a bank of soundings of w
449e173 case of large [not located] Mr Greenough on his MAP of the World, has. written. Mastodon found at Ti
326c266 south Sea. Botany R. Brown. has curious coloured MAPS. by Copenhagen Botanist of range of plants Book
284c148 is perhaps on long average equal.-- The Cocos do MAR on the Mahé island,, one the higher parts & only
432e126 lendid Pamplet. (published in Philosop. Journal <MAR> April 1st 1839) by Sedgwick & Murchison; which
029r033 anitic rocks Mem. Chanticleers voyage at <[...] MARANH> Pernambuco. EARTHQUAKE AT SEA.--Extract from
089a019 tone not thus composed on the NE part more like MARBLE requires polish to see structure,-- «He» Thoug
089a018 ices yet emptty.-- In all the mountains of Saint MARC et des Gonaïves it is difficult to find stone n
093a032 ng mineral.-- Mem. Galapagos ∴ Basalt deepest?? MARCEL Serres L'Institut. 1837. p 331 Considers that
207b151 , a circumstance undiscovered by Ehrenbergh.-- <MARCEL Serres p. 331. L'Institut-- considers that mer
325c267 heep.-- Verey. Philosophie d'Histoire Naturelle MARCEL de Serres Cavernes d'Ossements 3d. Edit. Octav
065r139 t to have belonged to an Englishman.--On 8th of MARCH cove began to freeze. correspond to September ¿
067r143 is.» This work is reviewed in present Edinburgh MARCH 1835 Sir W. Parish says. that beds of shells ar
269c101 communication to Geograph. Soc. in February or MARCH 1838 on soil in Siberia being frozen to 400 ft
320c276 ical Transactions. <done> up to parts published MARCH 1838 Whole of Geographical Journal Asiatic Jour
323c269 . Vols 25th Phillips. Geology. Larder 2d vol.-- MARCH 16. Gardner's Music of Nature-- -- Herbert on H
424e100 cases of species peculiar to separate islets.-- MARCH 5th. Lyell says «fossil» shells from North Amer
425e104 capability of variation!? my theory says so.--» MARCH 6th. Mr Bentham says in Sandwich Islds. he beli
426e106 having species, if true important on my view.-- MARCH 9th-- Is there any relation between the fact th
428e112 nt, but a vigorous battle between strong & weak MARCH 11th. Yarrell's law must be partly true, as enu
429e114 of production of hornless cattle-- ¿& others?-- MARCH 12th-- It is difficult to believe in the dreadf
430e119 rest being common to other parts of the world-- MARCH 16th. Mr Lonsdale showed me two specimens of an
431e122 es to vary-- domesticated animals tend to vary. MARCH 20th. Phillips in Lecture in Royal Institution
486z020 of gravel & clay about 10 feet in thickness.-- (MARCH, 1842) Questions & Experiments (T) Gowen, Royle
500q10a arieties &c of deer. Contests of sexes.-- Q.30) MARCH 1842. <Last> Year <before last» beans & peas we
581n063 e of intellect indication of better life p. 207 MARCH 16th.-- Is not that kind of memory. which makes
581n064 kling sensation.-- in savages other tastes few. MARCH 16th. Gardner's Music of Nature. p. 31. remark
594n113 picking from same bone A child born on the 1st MARCH was frightened on the 24th of May at Cresselly
610o033 those of weak intellects.-- Westminster Review. MARCHIA 1840 p. 267--- says the great division amongst
424e100 n Musci frondosi, et hepatici,-- in Chara, in MARCHANTIA & Hypnum «Prof:» Don would have known the Co
216b181 ee paper in Philosoph. Transaction on a quagga & MARE crossing by Ld. Moreton, where mare was influen
216b181 n a quagga & mare crossing by Ld. Moreton, where MARE was influenced in this cross to after births, l
217b184 r, & like both parents.-- Fox told me of case of MARE covered by blood horse & Carthorse two folds [n
220b197 n mind or instinct of parent. Mem Lord Moreton's MARE. The fact of plants going back hybrid plants; a
239c002 for many generations back, were crossed with Bay MARE, only bay a few generations, that offspring wou
295c183 ys he has seen old Stallion tempted to cover old MARE by being shown, young one.-- Many African monke
334d010 Stallion, according to Erasmus preferring young MARE to old, explained by Stallions, (according to F
334d010 his stallion as second horse between <whe> shaft MARE & another leader mare,-- this stallion though e
334d011 horse between <whe> shaft mare & another leader MARE,-- this stallion though eager to all other mare
334d011 mare,-- this stallion though eager to all other MARE had been entirely broken from their mares, (tho
366d112 tame donkey has. CD[old Buffon should be read on MARE My view, why hybrids are infertile. supposes th
367d113 me strongly of Zebra.-- Mem. Quagga & Ld Moreton MARE ringed Owen says that Bell in Encyclop of Anat
458t002 a hybrid between ass & Zebra, crossed with pony MARE & produced a very pretty little animal, showing
459tf02 mules, these parts resemble ass. (& part of body MARE)-- -- this may be, perhaps. squeezed into Mr Wa
536m071 when excited by love? Stallion licking udders of MARE strictly analogous to men's affect for womens b
334d011 other mare had been entirely broken from their MARES, (though horsing every month) & worked in the s
079r177 s."-- Ulloa's Voyage, Shell fish purple die, MAREVELLOUS statements on, Vol I, P. 168. on coast of Gu
077r171 189. "The small ravins into which the valley of MARFIL is divided, appear to have a decided influence
322c270 s Religio Medici Lyell Book III There are many MARGINAL notes <Rengger &c> Mitchell's Australia Walte
323c269 sic of Nature-- -- Herbert on Hybrid Mixtures: MARGINAL notes. -- 20th. Carlyle's French Revolution 3
327c265 - Review on "Walker on intermarriage" price 14s. MARH. 20t. 1839. Philosophy of Blushing lately adver
107a079 ficult (better conductor) Fitz Roy's Case of S. MARIA & Tubul applicable to Andes & Patagonia-- On Ly
407e042 cies fall, genera must. Lesson I remember says MARIANA Deer very close to a Molucca species.-- L'Inst
241c014 rvus moluccensis is different from that of the MARIANNA islands & at Amboina» I fancy there is marked
573n037 nodding is less strongly marked than negation MARIANNE. says. that she has constantly observed that
403e021 reign plants have been cultivated in Guahon. (MARIANNES), "for example the prickly Limonia trifoliata
245c024 p. 372. Bourous. the Barberousa; a Cervus near MARIANNES new, & some rats & mice. In Amboina only Cusc
400e013 places)» besides Galapagos do. p. 376. Isle Tres MARIAS off Mexico with small Hares & raccoons.-- «S.
021r005 located] La. billardiere mentions the floating MARINE confervae, is very common within E. Indian Arc
037r055 secondary strata of England, «besides ordinary MARINE remains» may contains <shells few corals Torto
037r055 ivorous lizard.-- from the action of torrents. «MARINE» Tortoise & other species of large lizard.-- T
048r084 long whole line of coast Darby mentions beds of MARINE remains on banks of Red River Louisiana. V. Lye
147g017 Sea more probable I did not look carefully for MARINE remains-- Some of the hills almost appeared as
148g028 cut them out-- In all cases «I urge» deposition MARINE-- because if not chain of lake & if so there w
316c244 cal range of shells like Cryptogamic plants. of MARINE kinds, there are some restricted genera, but t
477z001 -- Part II.-- 35. Phil Trans Burrowing & boring MARINE animals-- CXVI. P 111 do Observations on Plana
641j29r supposed favourable to terrestrial Mammifers-- MARINE ones «of large size» <to> are best nourished b
065r138 . Acuneha. Kerguelen Land. Prince Edwards Isld. MARION & Crozet. L. Auckland. Macqueries.--Sandwich I
039r060 3. «seems to» considers that successive terraces MARK as many distinct elevations; hence it would app
108a079 rust of solid earth would be thicker.-- PP Andes MARK the line between sinking & rising areas.-- In E
128a128 est fresh water must be sought for below the sea MARK.-- If mountain chains are matter piled up. over
147g021 of Air 65 degree? Glenoe, 6 ft above high water MARK 30.380. 68 degree 65 degree? For comparison wit
251c041 g Borneo & Sumatra. differ only in form of white MARK on breast: p. 234.-- good case-- p. 526. (ref)
261c071 not on now existing» Mr Gould says wherever any MARK like red patch on wings of Furnarius, Synallaxi
277c126 reason now assigned for doing so There should be MARK to every species. only known by analogy genera
277c128 ld present many ideas of causes of change.-- The MARK of analogy would be empirical because as soon a
332d003 us years, two having consulted him on one day.-- MARK at Shrewsbury thinks the half bred Alderney Cow
362d100 ery strange races» of them have the forked black MARK of the Rock Pidgeon,-- several have a group of
405e031 kish brown.-- yet retained a trace of horizontal MARK on flank.; & tail. & kind of semilunar {P} mark
405e031 mark on flank.; & tail. & kind of semilunar {P} MARK on each side darker,-- so that whole colour is c
405e031 anner as they would vary, if in wild state; thus MARK on ear of cats, colour can be brownish do Saw w
469t135 is visited 30 times a day is considerably under MARK, & this has now gone on 14 days. (except some w
484z016 Maldonado creeper of same plumage.-- general red MARK on wings of all-- Spix has described Philedon.
505q014 small Bees-- at Sundorne has large Bees July/42/ MARK has six day's puppy of small true Bull-Dog-- le
615o037 before any reason had told him this distinctive MARK, it is downright instinct, leading to touch a p
615o36v ows surprise, walking home one day met him, with MARK riding instantly followed, me and for five minu
049r087 be well examined At M. Video «facts of Passages MARKED by do.» discuss quartz veins, there contemp--y
130a134 these littoral deposits there probably would be MARKED line of separation A Paper by Parrott Mem. Aca
192b086 asses with few species greatest jumps strongest MARKED genera? Reptiles?) For instance there never ma
210b159 inds of greyhound.-- In plants. do the seeds of MARKED varieties produce no difference. if they do.--
222b203 we find this same tendency (only less strongly MARKED) between what are called varieties.-- NB. one
224b215 roduce other species., man already has produced MARKED varieties & may someday produce something else
228b230 s go back again, not a monstrous plant, but any MARKED. variety.-- Strawberry produced by seed«s»??--
241c014 Marianna islands & at Amboina» I fancy there is MARKED wild breed of oxen at Java.--.p. 140, calls it
266c092 ed. Memoir on isld of Socotra. Cattle generally MARKED like those of the Alderney breed, but size not
273c110 ell developed» family (Gould says) there is any MARKED colouring of plumage (as «black & white» bars
273c111 ery family of bird., even the most <m> strongly MARKED, there is a preeminently aerial,-- formed for
273c111 p their place.-- Humming bird there is strongly MARKED variety,:.in the Tyrannidae.-- Milvulus-- Even
285c153 irmly embedded in the constitution, which other MARKED difference in the varieties «made by» of Natur
304c206 ist, who believed that all his divisions merely MARKED his own ignorance.-- the collector who ploddin
306c213 re that Australian dogs ever hunt in company -- MARKED difference with dogs of La Plata & Guyana peo
309c220 mankind must improve-- The areas of subsidence MARKED out by animals of same <spe> genera. is not eq
309c220 pe» genera. is not equal to areas of elevation: MARKED out by existence of elevated «extinct!?» genera
334d012 a breed of white-tailed squirrels, which form a MARKED wild variety. doubtful whether all are white.
342d035 Vol II p 11) accidentally says "--is distinctly MARKED as whole dynasties have been featured by the A
405e031 r,, so that whole colour is changed, these best MARKED characters are partly retained, therefore colo

466t093 ull flower Maer June/41/ Rhododendrum-- nectary MARKED by orange freckles on {a} upper petal; bees &
507q15v s.-- like cowslip & primrose, but less strongly MARKED.-- 31. Plant seeds of the Fuller's plants ,Tea
573n037 hence assertion.-- but nodding is less strongly MARKED than negation Marianne. says. that she has con
579n059 a person-- do not the features pass before him MARKED, with the habitual expressemotions, which make
627o52v se subjects consult <following> pages. <p. 231> MARKED in my Mackintosh 1) Mackintosh's Ethnical Phil
042r067 r Lyell P. 211 Vol III. talks of line of cliff MARKING a pause» When mentioning pumice of Bahia Blanc
192b085 rs a perfect gradation can be found from forms MARKING good genera-- by steps so insensible, that eac
232b245 es any change may undergo indefinite change., (MARKING in their history an eocene miocene & pliocene
240c003 that they would be called genera., yet retains MARKINGS of wings like the wild rock pidgeon.-- fact a
058r115 . Effects of great waves to obliterate all land MARKS.--At the first it would though be easy to see o
255c051 his may account for permanence in many trifling MARKS.-- such as the bands on pidgeons back.-- Accord
275c121 eons which have such different skulls, but same MARKS on wings are Blue Pouters & small Bald Heads Mr
279c133 eredetary, developemen of important organ (.see MARKS on pages).-- Crosses of diff: breeds succeed, y
367d113 Animal OEconomy p. 45 "One of the most general MARKS is the superior strength <of> «of make in» the
383d160 nes.-- «faintly edged with reddish brown» black MARKS on tail much «blacker» «broader.-- » Breast red
384d161 species not similarly subject-- ▌Divides sexual MARKS into primary & secondary, the latter only being
057r112 ake.--Earthquakes act as ploughs [,] Volcanos as MARL-pits: Consider well age of Bones. = slowness of
399e011 rm) in Stonesfield slate., & a Melolonittha-- In MARL from «Lake» Constance species of Europaean gene
331dIFC n Buds & Gemmae. C D Charles Darwin 36 Great MARLBOROUGH St Did Eytons <intermediate>. «hybrids, when
334d012 s in «Lord» Exeter's Park «or in the Duke of MARLBOROUGH» there is a breed of white-tailed squirrels,
385d163 milk--» Sept. 25th Young man at Willis «Grt. MARLBOROUGH Str, Hair dresser, assures me he has known m
520mIFC ents Expression M Charles Darwin Esq 36 Grt. MARLBOROUGH Str.-- (p. 64. On <insect> Ants getting on T
560mIBC h they first see it? Charles Darwin 36 Great MARLBOROUGH St Has my Father ever known <intemperance> «
486z019 eetles, like those from Chiloe. Amblyrhyncus de MARLIN James Isd-- Lutke Voyage Vol III p 322 Dr Mart
054r105 bay about five leagues North of Callao, called MARQUES, where in all appearances not many years since
629o55v elings for indecency-- preposterously so, for MARQUESANS think only of prepuce, crepitando,]CD & if p
179b034 probably be repugnance & art required to make MARRIAGE-- as Dr Smith remarked Man & wild animals in
199b120 one of the parents; repugnance «generally» to MARRIAGE <if offspring not fertile> <,but producing» «
199b120 with «fertile offspring» » fertile offspring; MARRIAGE never probably excepting from «strict» domest
199b120 male.--no offspring: physical impossibility to MARRIAGE.-- ¿whether those genera which unite very dif
206b148 extinct.-- Who can analyze causes, dislike to MARRIAGE, heredetary disease, effects of contagions &
391d179 e fails. Every individual except by incestuous MARRIAGE has acquired from father some differences. V.
406e036 same lines appear to hold good. with regard to MARRIAGE of individuals, & varieties of same species &
447e164 have been more rapid, & the facility for inter MARRIAGE is greater (Hence Dioecious plants highest,--
190b075 ed really, if there is any difficulty in such MARRIAGES or offspring show tendency to go back-- there
324c268 ished his laws about sexes relative to age of MARRIAGES Brown at end of Flinders & at end of the Cong
525m025 hat bitch's offspring is affected by previous MARRIAGES with impure breed.-- A cat had its tail cut o
522m010 im, did he know whom Mr Child «of Kinlett» had MARRIED.-- Answered never heard of such a man.-- (My F
522m011 ne left. him, took name of Child «of Kinlet» & MARRIED Miss A. B.-- all the same names as a few minut
609o2bv to licentioustness jealousy, & every one being MARRIED to keep up population. with the existences of
463t055 talking, constantly said as the <brain> spinal MARROW expands, so do the bones <are created> expand-
469t151 pt some wet ones; & wd go on longer-- Woodfords MARROW fat, Early frame, Groom's Dwarf. planted in ro
578n054 s. B. A. how nice it would be if your son would MARRY Miss. O. B.-- Mrs. B. A. blushed. analyse this:
172b006 hey would remain constant; is it not said that MARRYING in deteriorates a race, that is alters it fro
582n067 of savages state.-- N B. According to my view MARRYING late, will make average of life longer.-- for
050r090 tion with a neighbouring Volcano of Priamang.--MARSDEN Sumatra. M. De. Jonnes seems to think that Vol
403e021 a trifoliata, which cannot now be checked".-- MAR[S]DEN p. 94 (1st Edit) of Sumatra has given account
479z006 n Kotzebue p. 312 Leaches on leaves in Sumatra MARSDEN. p. 311 D'.Orbigny considers Dasypus villosus
299c194 tely have been shewn to be same.-- one grows in MARSH & other dry; yet if T. palustris be sown in dry
426e108 e of series from wild animals & plants]CD.-- Mr MARSH has some nephews, who are astonishingly like to
633j54r mountains upheaved by volcanic force, for these MARSH plants. All flow from some grand & simple laws.
087a010 te about N. American bone not probably in salt MARSHES Efflorescence nothing -- Study account.-- Alla
344d041 e facts-- they are eating foetuses, as young of MARSUP. is sucking foetus.-- August 23d The Rev R. Jo
461t025 male in these hatch young-- are there not some. MARSUP. Mammalia, which <do> have not sack,-- Most cu
192b087 nes in Echidna & Hedgehog]CD-- As we have one MARSUPIAL animal in Stonefied slate, the father of all
204b141 s some much higher generalization in view. In MARSUPIAL division «do» we not see a splitting in order
249c036 drupeds in Eocene period. Have the Edentata & MARSUPIAL forms been chiefly preserved,-- where shut up
250c038 at animals had not spread then such tribes as MARSUPIAL & Edentata increased most. Certainly Africa a
254c047 ies--! How does this agree with grand fact of MARSUPIAL, low Cerebral structure??-- «do» p. 390. All
261c069 ctures into many genera.-- like Synallaxis or MARSUPIAL animals of N. America Hence it is universally
355d068 ormorly one great sea, & two Polar Continents MARSUPIAL. Edentata.-- Pachydermata &c &c-- It is impor
369d115 a, & from Ornithorhyncus & Hydromys not being MARSUPIAL. («but» also mice) & these being water animal
373d132 shown by their happening in same plant.-- The MARSUPIAL structure shows that they became Mammalia, th
412e057 etained in the male.-- <like» «far» more than MARSUPIAL bones., & even more than Mammae, which have g
460t019 t not be supposed that this refers to time.-- MARSUPIAL in Oolite.-- insects, of do orders-- cheiropt
461t025 divides into two divisions, one of which are MARSUPIAL & the other have young which undergo metamorp
637j58r ongo neck is adapted to long necks.-- p. 236. MARSUPIAL bones especal adaptation, to «young».-- good
174b015 it is.-- Hence Antelopes at C. of Good Hope-- MARSUPIALS. at Australia-- Will this apply to whole org
196b106 believe not..-- It is a very great puzzle why MARSUPIALS & Edentata should only have left off springs
231b242 have two. Tapir existing in East Indian Seas. MARSUPIALS animals all show greater connexion in quadru
349d052 -- evidently artificial, as interlopement of MARSUPIALS will change all.-- & so on no one will settl
357d072 ion of larger forms.-- From observing way the MARSUPIALS of Australia have branched out into orders o
357d073 ing Mundine form., & the less developement of MARSUPIALS in S. America. from presence of Edentata-- E
357d073 erica. from presence of Edentata-- Edentata & MARSUPIALS have been almost destroyed wherever other an
639j28r Fly. Gecko &c. Prehensile tail. in Monkeys & MARSUPIALS. Harvest mouse & (Chamaelion?) C. D. Spines
373d132 for abortive organ ever disappearing??-- Have MARSUPIATA abortive Mammae?.-- My view would make every
407e041 being in S. America & Australia. reason, why: MARSUPIATA, when first introduced live & multiplied, sp
537m076 so much, as in other races of mankind..-- p. 27. MART. allows some universal feelings of right & wron
379d152 ale.-- If there be one female, she will be free MARTEN. (Owen.) See Hunter's Owen-- In the Athenaeum
383d159 y are not Hermaphrodites, like J. Hunters. Free MARTEN N.B. the common mule must often have been diss
212b165 Monkey peculiar to. latter not to former-- Mr MARTENS of Zoolog Soc told me an Australian dog he had
486z019 rlin James Isd-- Lutke Voyage Vol III p 322 Dr MARTENS says only one Reptile in Kamtchatka (Salamandr
076r129 considerable quantity of silver procured from MARTIAL pyrites; great blocks of pure silver not commo
599o008 e polity of Nature than any other animal-- Aimé MARTIN de l'Education des Mères Vol. I. p. 198.-- "Mo
601o08a r takes Hat de l'education des Mères par L Aimé MARTIN Leroy Lettres. Philosophique sur l'intelligenc
537m075 <t> instinct, when attended with bad effects MARTINEAU. How to observe, p. 21-26. argues «with examp
555m142 her more steps.-- dispute about words.-- Miss MARTINEAU (How to Observe p. 213) says charity is found
321c270 me's do, with correspond. with Rousseau Miss M«MARTINEAUS» How to observe Mayo Philosophy of Art of Li
558m151 -- active assistance. &c &c. it comes to Miss MARTINEAUS one principle of charity.-- ¿ May not idea o
486z020 T> N American & T. del. Fuego & Iceland Spix & MARTIUS talk of birds singing in the forests of Brazil
293c172 mbered in next generation. [NB what are those MARVELLOUS cases, when you feel sure you have heard con
479z005 f Quoy et Gaimard Ulloa shell fish Purple die MARVELLOUS stories Ulloa's Voyage Vol I, p. 168 Ceratop
532m056 ough so fond of her & of servant of Richard & of MARY & her bed brought from Shrewsbury) yet for a fo
128a130 form which forces dilemma. Transactions of the MARYLAND Academy (at Athenaeum.) I. Part. I Vol.-- som
332d004 perceive turn on road where No houses to Eaton MASCOTT, where he had been accustomed to turn down.--
378d147 , yet young ones brown.-- Sexual Selection If MASCULINE character. added to species., we can see why
032r039 by increased vertical <height> thickness (DZ) of MASS to be removed & from the resistance offered to
032r040 ime of rest. so would the size of the triangular MASS removed vary.--The gradual rising continuing. a
040r063 y more abundant in this case from intersecting MASS probably cold & not warm as sides of a crater a
046r078 in chemical nature, or has a subterranean fluid MASS itself changed.--No. -- Yet the fluid granitic
046r079 s itself changed.--No. -- Yet the fluid granitic MASS under <[...]> less pressure might have its «pro
047r083 conformity with <Mr Lyell's> idea of an injected MASS of fluid rock In Patagonia plains. long periods
049r088 Fuego as nucleus of a Volcano or as an injected MASS.--From conical form I incline to <latter> forme
049r089 : & indeed when do these dikes lead to a conical MASS. will this conical mass be granite? Why not mor
049r089 dikes lead to a conical mass. will this conical MASS be granite? Why not more probably greenstone? W
056r109 nworn islands) we take in at once the stupendous MASS which has been corroded. -- If man could raise
064r136 CREVICE of Andes--therefore flowed towards it. a MASS on each side 3000 ft thick & 150 broad. neglect
068r147 strata without ejection of the fluid propelling MASS. If one inch can be raised then all can, for fr

Page
**************************************(Key Word)**************************************

103a065 ain-chain in proportion to weight of super [...] MASS.-- Absence of Caverns, in Plutonic rocks argume
104a067 ty on fissure going right through superincumbent MASS (varying hardness,-- takes time to trace) from
107a077 e it no because heat proceeds from great body of MASS.-- The last speculation becomes important with
146g011 three hundred feet in vertical height-- enormous MASS thunder storm, many <hundred> thousand tuns. Bl
146g013 of Loch Dochart buttresses of Alluvium or rather MASS of well rounded pebbles in yellowish argillaceo
147g018 tain there must once have been very considerable MASS of waterworn pebbles in Alluvium which without
150g036 a 3a 3 Bouthoner 3 2 Terrace 3 Alluvials 3a 3a 2 MASS 3 2 rather longer than 3a Sunday In Glen Collar
158g079 uch as boulder on <thes> Divortium aquarum Peaty MASS of this point very nearly like head of Glen Guo
160g092 8.92 A 75 Air 70 degree? Isthmus broad flat peat MASS-- (general character in these mountains & not r
161g094 tion perhaps 6 ft too low» (to test last on Peat-MASS Divortium aquarum) Barom. 29.200 A.77 degree Ai
196b108 orms.-- In abstract we may say that vegetables & MASS of insects could live without animals but not v
226b222 small tract we have many species, we may insure MASS continental or many large islands.-- Hence this
430e118 e made in two ways-- local varieties, when whole MASS of species are subjected to some influence, & t
467t104 sing either one or both of Pea's wings, stigma & MASS of yellow pollen protrudes at sheath.-- At last
506q015 round nuts (42) How are Orchidiae fecundated, as MASS of pollen is requisite.-- Brown's paper 43. Any
603o11b ome under imitative art [my view says yes. <old> MASS of rock--]CD or poetry, CD[my thery says yes. i
139aIBC nce?-- See End of Note Book. called R. N.-- MASSAC[H]USSET would be well worth visiting really good a
129a131 repeated oscillations» Hitchcock Report on MASSACUHSSETS. p. 133 The most wonderful case of great bl
202b131 Vol IV P. I. Geograp. Journal. Voyage up the MASSAROONY by W. Hillhouse.-- Demerara. In note. Demera
028r028 of rivers currents. & sea beaches. All mineral MASSES must have a tendency. to mingle; The sea would
068r147 s nothing odd to find them injected by veins & MASS[ES] [Fig. 8] {P} (A. B. C, now grown solid.) Red
072r160 113 "Nature exhibited to the Mexicans enormous MASSES of Iron and Nickel, & these masses which are s
072r160 ans enormous masses of Iron and Nickel, & these MASSES which are scattered over the surface of the gr
073r161 wn (Collect: «of F. W.») where the stalactiform MASSES have layers been accumulated, round knobs, or
076r168 rved both in Mexico & Peru, that those oxidated MASSES of iron. which contain silver are peculiar to
112a089 d granite & other trap.-- It is in the mountain MASSES we must look for that.-- how few isolated volc
157g074 ts; appearance chiefly cause by fall of angular MASSES from above on soft shelf-- 29.330 A 84 degree
410e050 l changes should act not on individuals, but on MASSES of individuals.-- so that the changes should b
435e133 retains its power.-- R. Brown found the <poll> MASSES of pollen of Asclepias placed on Orchis (so ve
514q21.) really always hit stigma by projecting pollen-MASSES?-- = answered = Has Ophrys nectary?= Bunbury s
571n030 there is some connection between them, & great MASSES of rock.-- I was much struck with this, when v
290c165 ter short by saying man cannot be companion but MASTER.-- heredetary tameness as well as wildness-- c
406e037 ack any animal except, dog when absent from its MASTER.-- dogs when strayed many their tails.-- Novem
570n024 on on the ribs & how he gulps in air.-- Again a MASTER says I will see you damned first." the man shr
600o08v : in Maternal instinct domineering over love of MASTER and sport &c &c -- The Bitch does not so act,
601o08a n French on what dog dreams, awakes-- does when MASTER takes Hat de l'education des Mères par L Aimé
550m126 f place for nest.)-- with dogs "have notion of MASTERS property"-- is not this rather more friendship
576n046 d.-- A Dog may hesitate to jump in to save his MASTERS life,-- if he meditated on this, it would be c
048r085 seight of the works of man Feeling surprise at MASTODON inhabiting plains of Patagonia is removed by
086a005 ound in America, when Mammoth & narrow toothed MASTODON.-- argue against the prejudice of not believi
086a005 ntates & camels in deserts & rodentia In Plata MASTODON Toxodon Is the general saline tendency of Ame
184b053 n discovered. between palaeotherium, megalonyx MASTODON, & the species now living.-- Now according to
187b062 ica-- It is a wonderful fact Horse, Elephant & MASTODON dying out about same time in such different q
203b135 pecies without apparent physical cause:-- Mem: MASTODON all over S. America. Hilaire does not seem(?)
253o046 e Elephas primigenious over so wide a range, & MASTODON angustidens.-- Ogleby has facts to show that
405e032 ht be worth investigate whether. Megatherium & MASTODON are coembedded in N. America. see my Journal
414e065 phant, if some genus. holding same relation as MASTODON to Man. were to be discovered. Man acts on. &
449e173 eenough on his Map of the World, has. written. MASTODON found at Timor.-- thinks he has seen specimen
473s03r grandmother & Mylodon» in post pliocene strata! MASTODON longirostris in miocene like in Europe-- Cuvi
048r085 the causes of the losses of the «species of» MASTODONS. which ranged from Equatorial plains to S. Pa
191b080 ssessed Elk not England. Did Ireland possesse MASTODONS?? Negative facts tell for little) Geographic
462tf4v g species. We see the same object gained by the MATACO-armadillo & the woodlouse-- -- a good analogy-
477z002 gonicus les tatous (.4 pichye, pelud, mulita et MATACO.) are all found south of 26 degree 30 minute.
364d103 hemselves. || + + first year.-- The bird fanciers MATCH their birds to see which will sing longest, & t
240c013 Cynocephalus. niger. comes from the Moluccas «MATCHIAN» & Celebes.-- \ Amboina; Viverra Zibetha.--/-
461tf03 ave been known in <their> «its» natural state to MATE with a thrush"-- Athenaeum 1839. p. 708.-- Shre
524m019 from Barrier isld many relations with a living MATICA & many shells of Genera Corlula Cham. Cardium.
532m057 n.-- To avoid stating how far, I believe, in MATERIALISM. say only that emotions, instincts degrees o
614o037 is matter? the whole a mystery.-- [LHC] This MATERIALISM does not tend to Atheism. inutility of so hi
291c166 of the deity effect of organization. oh you MATERIALIST!-- Read Barclay on organization!! Avitism in
044r073 a on the other: The extreme frequency of soft MATERIALS being consolidated; one inclines to belief al
046r078 (state facts) traverse granites, are granitic MATERIALS simply altered by circumstances; & not in che
156g070 ents Maculloch wrong in saying no transported MATERIALS <into> on upper shelves granite & some other
608o026 gnize an accidental spark falling on prepared MATERIALS. From contingencies a mans character may chan
600o08v g analogous to imperiousness of Conscience: in MATERNAL instinct domineering over love of Master and
600o08v rt &c &c -- The Bitch does not so act, because MATERNAL instinct gives most pleasure. but because mos
629o54v f durableness will explain it.-- Would not the MATERNAL affections (in a dog. & therefore not <instin
130a134 ion A Paper by Parrott Mem. Acad. Peters. Scienc MATH. Phy-- Nat. t. I, 1831. sur le temp du globe on
132a137 ns. M. Parrot, Mem. Acad. Imp. des Sciences. (Sc MATH. Phys. et Naturelles. Tom I. p 501.-- shows fir
099a049 ction of particles of same nature: then get MATHEMATICIAN to when two particles <would> are aggregate
543m100 table chess player.-- Peacocks remark about MATHEMATICIANS not being profound reasoners.-- all same f
523m015 al of his accounts in his pocket, & studying MATHEMATICS.-- My Father says after insanity is over peo
081r181 ss. & Scoresby deep soundings Gilbert Farquhar MATHISON travels Brazil. Peru. Sandwich Isd Mawes trav
328cIBC from Barrier isld many relations with a living MATICA & many shells of Genera Corlula Cham. Cardium.
599o008 infini conscience; voila l'homme separe de la MATIERE et du temps! voila les facultes, q'il possede
026r023 coloured Limestone: the strange substitution of MATTER in shells, like Concretions & laminae, show wh
027r028 wood converted into siliceous pyritous & coaly MATTER. Mem: Chiloe In the endless cycle of revolutio
028r028 ea would separate quartzose sand from the finer MATTER resulting from degradation of Feldspar & other
028r028 dspar & other minerals containing Alumen.--This MATTER accumulating in deep seas forms slates: How is
028r032 commenced before any great accumulation of such MATTER.-- Dr A. Smith says. that Boulders do not occu
030r035 Every winter torrents must bring much vegetable MATTER from thickly wooded mountains, probably chiefl
032r041 tion» & a <circulation owing> rotation in fluid MATTER of globe. must there not be a circulation <how
032r041 oles. nearer the surface & to the Eastward.--If MATTER proceeds from great depth. from axis to surfac
033r042 le is a familiar case: If refiltered with other MATTER how very curious a structure: Have shells ever
040r062 eing unique, yet no quadrupeds. --- Is the white MATTER beneath pebbles. the degraded matter of such p
040r062 the white matter beneath pebbles, the degraded MATTER of such pebbles extending to seaward, the alte
040r062 extending to seaward, the alternating with such MATTER at St Julians looks like such?--destructive to
044r074 Mur: A. = (& this effect of water thus holding MATTER in solution must be great: & in the fact of bo
045r076 roportional force of crust of globe & injecting MATTER on the great rise). -- The great rains which a
045r077 ? In a subsiding area. we may believe the fluid MATTER instead of afflux (always slightly oscillating
046r079 e first phenomem. is an inward afflux of melted MATTER.--Volcanos perhaps may be admittance of water,
052r097 lope necessary for seaward transportal of drift MATTER.-- Give various cases. [Fig. 6] {P} A advancin
059r121 arden.--Climate.!? or small Proportion of Alum: MATTER.--all pale cream colour.-- The Brecciated stru
064r136 canic metalliferous-- Urge enormous quantity of MATTER from CREVICE of Andes--therefore flowed toward
069r149 lace unrecorded must be insensible. Quantity of MATTER from Cordillera. HORIZONTAL movement of fluid
069r149 r from Cordillera. HORIZONTAL movement of fluid MATTER not (for instance) expansion of solid matter b
069r149 id matter not (for instance) expansion of solid MATTER by Heat Consider profoundly the sandstone of t
090a021 g stones.--¿ does this bear upon the sorting of MATTER. in making trachyte come out before.-- What mu
100a054 retions & concretion filled with unconsolidated MATTER-- Phillips Lardner p. 197. refers to salt as b
102a061 ike it. Are greenstone dike in Granite residual MATTER of upper quartzose ones & felspar.?? Are the g
105a072 is paper says land shells found with calcareous MATTER & concretions on coast of Madeira.? How came i
115a098 greater power of oscillations & currents.-- if MATTER was «successively» given of every degree of fi
115a098 ratum would thin out, both inland & seaward: if MATTER too coarse, then {P} that form.-- All this dep
116a099 urrents very weak??-- too great an abundance of MATTER would have same effect as too coarse. Read Kyl
119a105 that elevation is independent of spreading out MATTER by action of the sea.-- as no sea exists there
120a109 ct would be such as present. to spread sheet of MATTER over surface.-- if elevation then went on at g
120a109 ate, not only river would carry further its own MATTER. but would cut wide gorge. leaving cliffs, on
120a109 .-- caution about action of rivers.-- Excess of MATTER brought down Mention absolute elevation of Pat

Page **(Key Word)**
126a122 um is supposed to have been attained.» how much MATTER separates them, this is ascertained by conduct
126a123 increased or under surface, would not the fluid MATTER be driven upwards & so conduct heat?-- How com
128a128 or below the sea mark.-- If mountain chains are MATTER piled up. over crevice from effect of general
265c085 adaptation as best attempt of nature, colouring MATTER being absent.-- again dwarf plant in alpine di
270c104 life & "inorganic life".-- animals only live on MATTER already organized.-- This paper might be worth
290c165 ensleighs objection.--» it is more, he cuts the MATTER short by saying man cannot be companion but ma
291c166 rain, more wonderful than gravity a property of MATTER? It is our arrogance, it our admiration of our
305c210 d, (subject to certain contingencies of organic MATTER & chiefly heat), which assumes a multitude of
305c211 ciple (intimately allied to one kind of organic MATTER.-- brain. & which <prin> thinking principle. s
348d050 we say such happens to be the character, of no MATTER of what importance, which prevails throughout
377d140 & forming mountain-chains, when we estimate the MATTER removed by the waves of the sea, on beaches--
389d173 explains Ld. Moretons case: without the nervous MATTER consists of infinite number of globules: gener
536m072 , if so free will is to mind, what chance is to MATTER «(M. Le Compte)»-- the free will (if so called
572n032 neeling, prostration «uncovering body» &c &c is MATTER of custom."-- this all applies to bodily weakn
610o034 bt. [RHC] 1) Effects of Life in the abstract is MATTER united by certain laws different from those.,
611o034 orld; life itself being, the capability of such MATTER obeying a certain & peculiar system of movemen
611o034 n in full.-- [LHC] ¿Has any vegetable or animal MATTER been formed by the union of simple non-organic
611o034 been formed by the union of simple non-organic MATTER, without action of vital laws-- According to t
611o034 rding to the individual forms of living beings, MATTER is united in different modification, peculiari
611o034 C] During growth <extres> tissue <[...]> unites MATTER into certain form; invariable, as long as not
614o037 likes one <m> kind more than another-- What is MATTER? the whole a mystery.-- [LHC] This Materialism
616o039 portions of it) that attraction has to ordinary MATTER. The relation of attraction to ordinary matter
616o039 matter. The relation of attraction to ordinary MATTER is that which action bears to the agent. Matte
616o039 matter is that which action bears to the agent. MATTER is by a metaphor said to attract; & hence if t
616o039 e relation to the brain that attraction does to MATTER, it might with equal propriety be said that th
616o39v ceptive action» by which we come to conceive of MATTER as attracting & shew that the groundwork <of t
617o040 phenomenon apprehended by the same faculty with MATTER & being necessarily exhibited in & by matter.
617o040 th matter & being necessarily exhibited in & by MATTER. The phenomena of gravity considered in themse
617o040 sist in a force manifested in every particle of MATTER directed towards every other particle; but FOR
617o040 xistence consists in its communication to other MATTER in the course of its DIRECTION, & thus when we
617o040 ON, & thus when we apprehend force in inanimate MATTER we feel dissatisfied until we can point out Ho
617o40v erefore, if we can trace any force in inanimate MATTER up to the action of some animated agent Now th
617o40v are manifestly the same as if every particle of MATTER were an animated being pulling every other par
618o41v y «X» the existence of something in addition to MATTER is because our knowledge of matter is quite in
618o41v addition to matter is because our knowledge of MATTER is quite insufficient to account for the pheno
030r035 was lost: I know no reason for supposing these MATTERS are not now collecting, in the bottom of an op
513q21. s notion of pollen & stigma generally not being MATURE at same time on same plant --Flora of Australi
388d173 hrough transformation, & was received into bud MATURED by female;<]CD> such view no ways explains Ld.
233b251 n Reptiles. M. J. says some reptiles same from MAURICE & Madagascar & C. of Good Hope.-- His book Pro
044r071 the epoch of Ammonite to the present day. at MAURITIUS. (consult Bory «dip of strata on East») canno
210b160 alapagos!!-- Some of the animals peculiar. to MAURITIUS are not found at Bourbond Zoolog. Proceedings
212b166 .-- agreement & reason Some animals common to MAURITIUS & Madagascar.? Proceedings of Zoolog. Soc Jun
225b219 ield∴ all lands united (Falkland Fox. ice). . MAURITIUS what a difficulty-- where elevation Subsidenc
226b220 'Acunha?» «Iceland?--» The Connection between MAURITIUS & Madagascar very good.-- Fernando Po & Coast
226b222 veral species of same genus, subsided land.-- MAURITIUS? <In plants where do most species occur.)> Al
229b234 The Crocodile & Tortise former inhabitants of MAURITIUS Freycinet Voyage, agrees. with several mammal
246c025 oubtful whether any birds Except. Dodo!!-- in MAURITIUS Lesson &c p. 620. Centropus (Coucal) of Java
295c183 do. excepting salmons L' Institut. Sorex from MAURITIUS. p. 112. & paper on genus Magazine of Zool. &
327c265 l point of view Henslow has list of plants of MAURITIUS with locality in which each one is found. ver
477z001 nly 10 years ago <no> snail was introduced to MAURITIUS. 18 Azara Voyage Vol. I. p. 279 Thinks the Mo
022r008 Holland P 127.--Caught a shark 11 ft long. "Its MAW was like a leathern sack, very thick & so tough
023r009 mans thumb, the rest not above half so long; The MAW was full of jelly which stank extreamly."--This
023r012 w by having been seen & from the contents of its MAW, amongst which were things pitched over board ea
081r181 har Mathison travels Brazil. Peru. Sandwich Isd MAWES travels down the Brazil.-- Did Melaspena publis
023r010 on the continent of S. America is inadmissible «MAY have happened from incipient elevation.» The vol
025r018 sions substantiated over S. America & Europe. we MAY believe them applicable to the world.-- My gener
033r042 m resembles the husks at Coquimbo: in that case, MAY not central and rather differently constituted l
035r047 ngland; the more minute equalities of elevation, MAY well be preserved at Patagonia. The English fact
037r053 Kelp.-- With respect to degradation of rocks--It MAY be a question. whether organic remains protect a
037r055 ta of England, «besides ordinary marine remains» MAY contains <shells few corals Tortoise> «remains o
040r064 t warm as sides of a crater as Vesuvius.-- There MAY have been oscillations in the upheaval of Andes.
041r067 ending of pebbles & the appearance of travelling MAY be owing to successive transportal from prevaili
044r074 after seeing small Bombs. without a vesicle. we MAY consider appearances of eruption at bottom.--sol
045r075 on coast it would be invariably discovered; this MAY be mentioned with general slope of the country;
045r076 th> at sea (as wells as in the Cordillera), they MAY be considered as accidents (if <[...]> part of a
045r077 nation of low Barometer? In a subsiding area. we MAY believe the fluid matter instead of afflux (alwa
046r079 nward afflux of melted matter.--Volcanos perhaps MAY be admittance of water, through the rent strata:
052r099) 1st cases. -- The terraces in Valleys of Chili MAY be with much truth compared to the step = formed
054r105 the distance of 3 or 4 leagues «from the coast» MAY be concluded to have been covered by the sea jud
056r109 rk?--In valleys one is not sure whether fissures MAY not have helped it, or diluvial waves. but when
063r130 e into another it must be per saltum--or species MAY perish. = This <inosculation> «representation» o
063r131 - An argument for the Crust of globe being thin, MAY be drawn. from. Cordillera. rocks.--When beneath
063r133 daptation of animals.--extinction in same manner MAY not depend.--There is no more wonder in extincti
064r135 f tides.--thinks them same as recent species. -- MAY I not generalize the fact glaciers most abundant
065r140 of quadrupeds.--although recent elevation, there MAY have been great subsidence previously. Mem. pebb
069r150 ((3) like Bell of Quillota.) (A) in this strata MAY be older than (B).-- Most important view Urge cu
073r163 with gold veins visible:--"Porphyries of Mexico MAY be considered for most parts as rock eminently r
074r165 em: Micaceous iron ore.» N.B. To show how metals MAY be transported by complicated chemical law & ste
086a005 ms "fixed as the land, stable as the water"-- It MAY be worth noticing edentates & camels in deserts
088a014 ons bring eels alive to their nests; & then they MAY picked up beneath the trees---- Are any Fish see
091a026 ansportal of Minute seeds-- L. Institut. p. 209. MAY. 1837 Paper by Humboldt on Quito Volcanoes & ano
098a045 of ingredients) as uniting with cretionary.-- it MAY <of> come of use in discussion on Cleavage &c Ge
109a083 effect, a tendency direct (or oblique) outwards MAY be granted. independent of currents.-- mud going
110a085 ess. connected with variation of compass & these MAY cause «or be effect of» elevation & subsidence.
113a091 worse & the depth of 32 degree. being little we MAY confidently infer that time has not been allowed
118a103 ives one grand idea of amount denudation.-- This MAY be added to any place where dikes described-- {P
119a104 untains are effect of continental elevations) we MAY conclude that elevation is independent of spread
120a107 undly considered. study Hopkins. theory of dikes MAY throw some light.-- thin dikes not cooling if th
123a117 bury Craigs Letter from M Angelis. B. Ayres. 3d. MAY. states remains found in many part.-- great Dasy
125a121 e towards the interior -- who knows how far that MAY have penetrated,-- lower down the temperature ma
125a121 ay have penetrated,-- lower down the temperature MAY be kept up far higher from circulation of heated
126a122 m the surface, & say 60 ft to degree.-- but this MAY be very wrong,-- The fact of a dumplin being bad
127a125 nclinal> simply clinal lines. dipping so & so or MAY be used East-clinal lines & c & .-- But Siberia
127a126 eze if cooled in a closed globule of glass. (oil MAY be cooled to 0 degree!)-- shows effects of press
127a127 ents)» prove elevation.-- great mountain chains. MAY be effects of subsidence Elie de Beaum. Memoires
132a139 give result less to be trusted than any others-- MAY not the cold «bottom of» ocean. (with fresh sedi
136a147 ome to treat of the age of the Pampas Deposit, I MAY properly remark on the superiority of Lyell's cl
148g029 <in> <below> «by» pass of Glencoe-- the erosion MAY often be due to rivers-- By Roy Bridge, a tongue
151g044 ponds to shelf the truncation & the upper shores MAY correspond with some line subsequent to shelf {P
162g099 d, that nothing can be made out of them-- but it MAY be said that a mound stretches along, parallel t
171b004 tions, acquire ideas ditto. V. Zoonomia.-- There MAY be unknown difficulty with full grown individual
173b010 c &c <assuming all> if species (<a>) «(I)». <fr> MAY be derived from form (2). &c.-- <(> Then (rememb
173b012 enced.-- or intermediate land existed.-- or they MAY represent some large country long separated.-- O
175b019 y dying out;--for instance secondary terebratula MAY have propagated recent terebratula, but Megather
175b020 recent terebratula, but Megatherium nothing. We MAY look at Megatheria, armadillos & sloths as all o
176b022 ls If we suppose monad definite existence, as we MAY suppose is the case. their creation being depend
177b027 wn to simple organization.-- birds-- not. {P} We MAY fancy, according to shortness of life of species
178b031 nt look over. Bell, & L. Jenyns. Falkland rabbit MAY perhaps be instance of domesticated animals havi

Page		(Key Word)	
180b037	e can easily see that variety of ostrich, Petise	MAY	not be well adapted, & thus perish out, or on ot
181b039	of their existence in the present world.-- or we	MAY	suppose only each species in each generation onl
181b043	ll coming from one stock & obeying one law, they	MAY	approach,-- some birds may approach animals, & s
181b043	obeying one law, they may approach,-- some birds	MAY	approach animals, & some of the vertebrata inver
185b055	ing on. attempt at adaptation of each element.--	MAY	this not be explained on principle, of animal ha
185b056	that in insects, each family, however many there	MAY	be, represent every other; for instance in Heter
186b058	ably in some <heteromera> colouring of crysomela	MAY	be going back to common ancestor of Crysom. & He
192b085	ow varieties can produce.-- Therefore all genera	MAY	have had intermediate steps.-- Quote in detail s
192b086	rked genera? Reptiles?) For instance there never	MAY	have been grade between pig & tapir, yet from so
192b088	eocine rocks. we can only expect some steps.-- I	MAY	ask whether the series is not more perfect by th
194b096	account of pollen from other plants because this	MAY	be applied to show all plants do receive intermi
195b099	it might have been of use in progenitor-- or it	MAY	be of use.-- like Mammae on mens' breasts.-- How
195b103	e the numbers & distribution of the species!! It	MAY	be argued representative species chiefly found w
196b105	- scrutinize genera, & draw up tables-- Instinct	MAY	confine certain birds which have wide powers of
196b108	ical nature of insects & worms.-- In abstract we	MAY	say that vegetables & mass of insects could live
198b114	an used by Cuvier, that all animals (though some	MAY	be) have not been created on the same plan. [Sec
203b137	atcher doing the service of a swallow I think we	MAY	conclude from Australia & S. America. that only
206b147	ld have different formula for each lustrum.-- We	MAY	conclude that there will be a period though long
206b148	So is it with varying races of man: these races	MAY	be overlooked mere variations consequent on clim
206b148	fferent set of causes must act in the two case,»	MAY	this not be extended to all animals first consid
213b169	ple. <Varieties> < «Races» > Man in savage state	MAY	be called, <species> species. in domesticated <s
213b171	ctness,' but do not known varieties do the same,	MAY	you not breed, ten thousand grey hounds & will t
216b180	ple. either a little nearer black or white as it	MAY	happen.-- Dr Smith says he is sure of the case a
219b196	e distributed by such means as we can recognize,	MAY	be thought to explain nothing.-- it being as eas
222b203	tempt at returning to parent stock.-- I think we	MAY	look at it so--?? It holds good even with trifli
222b205	another question, whether whole scale of Zoology	MAY	not be perfecting by change of Mammalia for Rept
222b206	tem of articulation. ¿whether type of each order	MAY	not be supposed that form, which has wandered le
224b213	> beings of very near structure.-- Hence species	MAY	be good ones & differ scarcely in any external c
224b215	s will never produce man. but both monkeys & man	MAY	produce other species., man already has produced
224b215	es., man already has produced marked varieties &	MAY	someday produce something else., but not probabl
225b218	n form of plant being found in Tristan D'Acunha.	MAY	be said to deceive man. as likely as fossils in
226b222	efore if in small tract we have many species, we	MAY	insure mass continental or many large islands.--
227b224	(2d) By character of any «two» ancient fauna, we	MAY	form some idea of <origin under> connection of t
227b225	ure that structure (C) could pass into (D).-- We	MAY	foretell species. limits of good species being k
227b225	t that they are not far the most serviceable. We	MAY	speculate of durability of succession from what
227b225	ve seen. in old world, & on amount changes which	MAY	happen-- ‖ It leads you to believe the world old
227b226	-- It leads to knowledge what kinds of structure	MAY	pass into each other: now on this view no one ne
227b227	ion, test of highest organization intelligible.--	MAY	look to first germ-- --led to comprehend true af
228b229	one variety going back ¿whether this going back	MAY	not be owing to cross from other trees.???? Do t
228b231	the father of man kind. than Macrauchenia yet he	MAY	be found:-- We must not compare «chances of embe
228b232	ious work, our companion in our amusements. they	MAY	partake, from our origin in <there> one common a
229b232	rom our origin in <there> one common ancestor we	MAY	be all netted together.-- Hermaphrodite animals
230b235	thenaeum. <Jan> «p. 154»-- 1838. Hybrid Ferns It	MAY	be argued against theory of changes that if so i
232b245	All» The <one> species which survives any change	MAY	undergo indefinite change., (marking in their hi
232b245	eocene miocene & pliocene epoch), whilst others	MAY	die out or move South ward.... species must be c
239c002	ffspring would be chesnut.-- On this principle I	MAY	add, that fact of half cross with parents, going
240c003	me being an element in the transmission of form,	MAY	explain mule & pig being half way. Yet dogs some
248c030	grant «(which can be shown probable,)» varieties	MAY	be made in wild state, there will be presumption
249c036	to special districts???? north of 30 degree.--,	MAY	be connected with, Mr Blyth's statement of birds
255c051	n propagation.-- Feathers on, Apterix because we	MAY	suppose longest part of structure.-- shape of wi
255c051	ad feathers.-- if wing totally obliterated. This	MAY	account for permanence in many trifling marks.--
257c055	all other animals have been so formed, then man	MAY	be a miracle, but induction leads to other view.
258c060	some analogy in not gaudy colours so all changes	MAY	be considered in this light.--XX Zoolog. Journal
258c061	g of much interest XX. Hence relation of analogy	MAY	chiefly be looked for in the aberrant groups.--
258c061	usion in this system of nature-- Whether species	MAY	not be made by a little more vigour being given
259c062	f of the non creation of animals.-- then argumen	MAY	be.-- subterranean lakes, hot spring &c &c inhab
259c065	species is reduced by atavism) Even a deformity	MAY	be looked at as the best attempt of nature under
259c065	fe of parent, & therefore being always necessary	MAY	be called adaptation With respect to my theory o
261c070	examination of species from distant countries.,	MAY	give thread to conduct to laws of change of orga
263c075	ous, & painful effort of the mind (although this	MAY	appear an absurd saying) & will never be conquer
263c076	difficulties Once grant that «species» one genus	MAY	pass into each other.-- grant that one instinct
263c078	<I> it does not stagger me.-- What circumstances	MAY	have been necessary to have made man! Seclusion
264c081	This must be profoundly considered.-- Structure	MAY	be obliterating, whilst habits are changing-- or
264c081	ating, whilst habits are changing-- or structure	MAY	be obtaining, whilst habits slightly preceed the
265c085	ptation.-- albino however is monster. yet albino	MAY	so far be considered an adaptation as best attem
265c086	s-- Thus bill & nostril of Puffinuria I think we	MAY	clearly attribute to heredetary origin & not ada
265c086	gs beneath soldered wing-cases-- Yet these wings	MAY	be of some use,-- Nature is never extravagant th
268c100	wing to one of each, being fitted for transport ¿	MAY	it not be explained by mere chance?-- or it like
269c101	riod & no ways assisted by fluid currents which,	MAY	take place in Metamorphic action.-- Geograph Jou
269c103	in Scientific Memoirs I can see that perfection	MAY	be talked of with respect to life generally.-- w
273c110	rs large, then he says from long experience, you	MAY	be almost sure, that there exist intermediate sp
273c112	to same habits, though swallow hawk, ,milvulus,,	MAY	catch insects on the wing & pratencole (¿connect
274c115	t value have their represetatives;, the rasorial	MAY	be observed even in Lessonia &c & In relations o
274c116	ith Europe & northern Asia & Northern America.--	MAY	we not look to these Northern regions as the rec
275c119	e forseeing. consequence., was endowed with what	MAY	be called the prophetic spirit in science--. the
276c122	female. The expression hybrid & fertile Hybrids,	MAY	be used to varieties, as well as species-- as fo
276c122	ecies-- as formation of species <of> gradual, so	MAY	we suppose., that something intermediate, betwee
276c124	ung pigs-- if vertebrae much lengthened &c there	MAY	be tendency to divide, which often enough repeat
278c129	d this will be most difficult. Sub-genera so far	MAY	be eliminated. where every species of a section
279c133	gated by nature.-- Whole art of making varieties	MAY	be inferred from <this> facts stated.-- + +. Ful
281c139	rmly fixed & therefore most subject to change.--	MAY	account for certain organs not being fixed, <whi
282c143	e vital laws.-- so that all character originally	MAY	«must» have had the character of analogical.-- G
284c149	structure is general in all species in group we	MAY	suppose it is oldest, & therefore lest subject t
285c152	h reference to other living being.-- one species	MAY	have passed through a thousand changes, keeping
286c154	passions-- brought into the world same way» they	MAY	convey much thus, Man has expression.-- animals
287c158	ations in eye of mollusca. [+] These questions	MAY	be all disputable, but the one end of classifica
288c159	The partial migrations of birds in same country.	MAY	explain greater migrations, if American & inters
289c162	culty, except as distinct creation.-- Generation	MAY	be viewed as condensor, «+++ must (on my theory)
289c162	(NB waterhouse says of affinity of many insects	MAY	be told by their larvae) but the acts of condens
291c165	cause structure has tendency to follow it, or it	MAY	be heredetary & strictly point out affinities. c
291c167	on> monoeecious & dioecious plants in animals it	MAY	be difficult to imagine how sexes were separated
292c171	implies want of consciousness & will & therefore	MAY	be called instinctive.-- But why do some actions
293c173	s Ichneumon & caterpillar, though our ignorance,	MAY	make us think so, but only between laws--]CD. Ma
293c175	Man would be disagreeable to Musquitoes We never	MAY	be able to trace the steps by which the organiza
294c176	logy & absence of varieties in a wild state-- it	MAY	be said argument will explain very close Species
301c198	habitual in men & what reasonable-- Same action	MAY	be either in same individual p. 7. is not squirr
302c201	termediate affinity between two classes.-- there	MAY	be some descendant of some intermediate link.--
302c203	most discordant groups. The formation of genera	MAY	sometimes be due to accident as submersion of la
304c207	lteration views.-- ostriches do-- but then there	MAY	have existed series between apterix & other bird
304c208	e many species <osculant> but, where much death,	MAY	be inferred much time elapsed & therefore descen
305c211	tions sleep walking. fits, laught &c&c Man & Man	MAY	have some relation together, as well as Man & ch
308c217	were fixed what a species means» civilized Man,	MAY	exclaim with Christian «we are all» Brothers in
308c219	omerville, Connection of Physical Sciences p 276	MAY	be worth glancing at, as she has no original ide
311c227	inhabits the eyes of horses in India in which it	MAY	be seen swimming about. A Smith is firmly believ
313c235	l a case bees developing sex of neuters) species	MAY	have had their infancies, as well as men-- when
313c236	e action of the viscera. under sympathetic nerve	MAY	be instinct or habits. ¿are sympathetic nerves &
314c238	lant that this part must be protected however it	MAY	be effected.-- Prodromus Florae Norfolkicae. 183
314c239	an character in Timor plants, yet it seems there	MAY	be Eucalyptus!-- (Hostile fact) Be cautious abou

Page
(Key Word)
```
315c240 n means of Transport & creation exists..-- pooh. MAY have been Created at many spots & since dissemin
315c243 being  a baboon with great canine teeth.-- (This MAY be made capital argument if man does move muscle
316c245 -- this is remarkable.-- Fish & drift sea weed-- MAY transport ova of shells.-- Conchifera. hermaphro
323c269 uses: well skimmed Bartrams Travels in N America MAY 18th Stanley familiar History of Birds -- Mackin
335d013 ause they are not new breed.-- the first hybrids MAY be compared to animal with amputated limb.-- Her
337d021 e becomes sensitive to light.-- (Mem whole plant MAY be considered as one large eye-- have they smell
343d037 & there was light".-- » August 17th Two regions MAY be Zoolo-geographically divided either by develo
343d039 y a distinct species-- We know how long a Mammal MAY go on as one species from Egyptian Mummies & fro
343d039 erent figure 19th. With respect to the Deluge it MAY be worth adding in note than amongst the Mammali
348d051 rence is or can be settled,-- I believe affinity MAY be taken literally,, though how far we can  ever
352d057 kind of determined by age of foetus.-- As Larvae MAY be more perfect (as we use the word) than parent
352d057 ore perfect (as we use the word) than parent, so MAY species retrograde, but these facts are  rare.--
352d058 ochineal?? Is there some law in nature an animal MAY acquire organs, but lose them with more difficul
354d062 ternary arrangement & not the Quinary.-- any one MAY believe anything in such rigmaroles about analog
359d086 eding in & in tend to produce same effects.-- CD[MAY it be said, that breeding in & in tends to produ
359d086 «or» to perpetuate some organic difference.-- it MAY be so, but this assumption as long as animals ar
365d105 n summer dress somewhat resembles Red Grouse) it MAY be so-- but very improbably, for it can hardly b
365d107 is  adaptation to habits (& habit second nature) MAY be more in constitutional.,-- more conformable t
366d112 grow there «That Mutilations will not alter form MAY be inferred from Australians knocking out teeth.
369d115 nected with animals being compelled to travers" "MAY have reference to the Great distances which  the
371d128 1279 roses kept in same soil. same atmosphere?-- MAY they produced not be transplanted?, & yet year a
372d129 ust be owing to their unity in one stem.-- a bud MAY be transplanted & carry all these  peculiarities
372d130 produce always different tail? An Individual bud MAY be thus produced from the growth of one part, (n
372d130 f one part, (not strictly new individual), or he MAY produced by having undergone, the endless change
375d134 soley  by positive checks, excepting that famine MAY stop desire.--» in Nature production does not in
375d135 must  effect instanteously all the rest.-- One MAY say there is a force like a hundred thousand wed
377d139 st.-- noise\ In case of woman instinctive desire MAY be said to be more definite than with bitch, for
377d139 ng must urge them to these actions. «These facts MAY be turned to ridicule, or may be thought disgus
377d139 ons. «These facts may, be turned to ridicule, or MAY be thought disgusting, but to philosophic natura
378d147 tallic brilliancy is present in Young birds, one MAY be sure cock & hen will be alike-- I presume con
381d156 no «constant» relation to separation of sexes, as MAY be seen in Monooecious & Dioecious plants.-- NB.
384d162 the  double purpose are not distinguished. --yet MAY be presumed from hybridity of ferns)  afterwards
385d164 rtile, but parent bird will never sit on them.-- MAY be just worth remembering that ovarium of  women
386d165 th buds of plants, does annual give buds.-- life MAY be thus prolonged bud being formed & one part dy
386d166 of  Anat & Phys) can make a head; the other part MAY surely absorb a useless member.-- in fact they d
389d176 oved facts relating to Generation One copulation MAY impregnate one or many offspring.-- it affects t
389d176 e subsequent offspring, <when> though other male MAY have copulated.-- two animals may unite & each h
389d176 ugh other male may have copulated.-- two animals MAY unite & each have offspring by same mother.-- on
391d175 ndividual before he can procreate. these changes MAY be effect of differences of parents, or external
399e010 was.--  do hind legs increase in any rabbits One MAY strongly suspect, that breeding in & in, produce
401e015 s it probable that great part of those varieties MAY be due to impregnation from other apple trees.--
401e017 ples. they will live but not flourish-- a medlar MAY be Grafted on pear. Mountain-ash & white  Thorn!
404e026 efore chance & unfavourable conditions to parent MAY be become favourable to offspring;\ Australian d
408e043 g. by a most delicate organ, on the whole system MAY produce-- ? When a species becomes rarer, as it
409e048 be  as many species, as individuals, & though we MAY not trace out all the ill effects. -- we see  it
410e052 ns worse than useless.-- yet there is no cure «I MAY say all this, having myself aided in such  sins»
411e054 circumstance.-- \the laws of variation of races, MAY be important in understanding laws of specific c
411e054 aws of change are known.-- -- then primary forms MAY be speculated on, & laws of life,-- the end of N
411e054 e formal laws of corelation of parts & organs it MAY serve perfectly to specify types, & limits of va
416e071 <yet>  Crustacean, & true hermaphrodites] CD «It MAY be said that true hermaphroditism is a consequen
416e071 oying certain forms & not others.-- Term variety MAY be used to gradation of changes which  gradation
417e075 hich one must think deserve the name of species, MAY be owing to the little fixity of organization, i
417e078 es are crossed.-- Now these laws are, that child MAY be either like father or mother, independently o
419e087 ime;-- therefore the mere loss of species, which MAY be the work of a few years as with the  Lamantin
420e088 dogs & Hyaena.-- but a common point, whence both MAY have descended.-- Jan. 6th The rudiment of a tai
423e095 used by the advancing complexity of others.-- It MAY be said, why should there not be at any time  as
423e097 though perhaps difference between jaguar & tiger MAY not be so.-- Considering the Kingdom of nature a
429e113 ants have been produced (look at the Dahlias. we MAY infer it in animals.)-- Azara gives account of p
431e122 apted; it where few stray ones. are, that change MAY be anticipated, & this would look like fresh Cre
436e136 Owen thinks Australia part of Old World <If> It «MAY be said, that wild animals will vary, according
437e141 a  field, where there was great swarm of mice.-- MAY 4th.-- The Brussels Sprout returning suddenly to
438e142 g between varieties & species, yet the amount of MAY depend on many circumstances, time of domesticat
438e142 ted to domestication.-- the constitution of some MAY resist the means Man can offer of changes.--  as
441e148 or the changes in fruit trees. mentioned by Mr K MAY be caused by the diversity of stocks, on which t
441e151 by buds alone or roots, & never flower, so there MAY be animals as Coralline, or others. which only g
442e151 thousand generations.-- any amount of generation MAY take place by gemmation «My theory will not admi
442e152 his theory, & the sexual reproduction of species MAY stop for any number of generations-- Gozze in No
445e159 is  really hermaphrodite. & <thinks> even oyster MAY fecundate each other, by the means of the medium
446e161 more splendid plumage than the male.-- Athenaeum MAY 18. 1839. p. 377.-- Statement that the climate i
446e162 verse of facts stated by Smith of Jordan Hill.-- MAY 27th.-- Henslow One of the 4 species of Lemna on
447e165 te allied to Asclepias Turpin cell is individual MAY 29th.-- -- Henslow says he has not the slightest
459tf02 ts resemble ass. (& part of body mare)-- -- this MAY be, perhaps. squeezed into Mr Walker's law Glean
462tf05 others-- is incidentally said that a mongrel man MAY lose all traces of his parentage in «about» seve
465t079 tive argument against-- monkey-man, valueless.-- MAY not several generations have been confounded  in
465t079 important, to bear in mind that enormous periods MAY elapse, even in situations apparently favourable
465t080 of La Plata. (except close to B. Ayres).-- If we MAY take this as guide, the shells preserved must be
465t089 s Lyell's Paper, in Taylor's Journ.-- Phil. Mag. MAY  1840 p. 362.-- some Mammals of Norfolk Crag. me
467t099 r filaments" shrivel, yet stigma does not, so we MAY feel somewhat «but little» less surprised at Hen
473s07r no  reason, why the peculiarities shd be born,-- MAY come in corresponding time of life of offspring-
473s07r imbs & &c only appear late in pregnancy, & then MAY just as well be born a tendency to alter or assu
473s07v th,-- only facts can decide-- some peculiarities MAY be early impressed & others later-- All  poultry
480z006 Important   account of habits of Tubularia. p 52. MAY 1836 dimensions of immense Tortoises p. 81 & p 1
491q01v ntioned by Mr Knight. Vol IV Hort. Transact.-- 4 MAY we no suppose, that certain plants, like Aphides
499q009 Hop is Dioecious-- seedsmen who raise Hop-seed-- MAY know something about proportion of plants necess
506q015 e pollen.-- or FEW. or bad seeds formed; badness MAY be merely not ripening= (38) Have Dioecious plan
506q15v Nettle. at Galapagos-- Dioecious.-- Carex.-- We MAY presume Nettle spreads by seeds= (44) Zostera. H
515q022 the hybrids were recrossed with either parent.-- MAY  44 These Hybrids differ in colour of beak, taki
522m013 tions,-- as if «these emotions» acquired.-- this MAY be doubted, whether rather not going against nat
524m019 rame of mind, analogous to those feelings. which MAY be considered as truly spritual.-- a person twit
526m027 me means & therefore properly no free will.-- we MAY easily fancy there is, as we fancy there is such
526m028 p> recall up the image in their own mind,-- this MAY be worth thinking over.-- it. will perhaps  show
526m030 but what urges him,-- absolute free will, motive MAY be anything ambition, avarice, &c &c An animal i
527m031 hough different species» when in confinement, so MAY they learn in a state of nature.-- Singing of bi
527m032 hus cuts the Knot:-- Sir J. Reynolds explanation MAY perhaps account for our acquiring «the instinct»
528m039 hole.-- Colour «& light» has very much to do, as MAY be known by autumn, on clear day.-- 3d  pleasure
531m051 o into, shows how truly an instinctive feeling, <MAY not pa.> In reflecting over an insane feeling of
531m052 f excited action, accompanying violent movement; MAY not passion be the feeling «consequent on the vi
532m057 n usual effects of fear.-- the state of collapse MAY be imitation of death, which many animals put on
533m058 learn to sing & do not acquire it instinctively. MAY not this be connected with their power of acquir
533m059 hy-- had not conscious of recollecting it-- this MAY be nearest approach to <the> such instincts whic
533m059 dy movement & co-relation of the proper muscles. MAY be illustrated by the extreme difficulty of movi
535m063 d that a wasp has this much intellect. yet habit MAY make it act wrong, as I have done when taking li
535m064 t-- do not know their own larvae, but one female MAY be moved to other larvae, when two groups  near.
535m064 d broods appeared healthy-- This remarkable case MAY be normal, with insects, but habit forgotten  in
535m070 . When one sees this, one suspects that our will MAY <be> «arise from» as fixed laws of organization.
536m073 akes change in bodily organization of oyster. so MAY free will make change in man.-- the real argumen
536m073 t arise from but organization. that organization MAY have been affected by circumstances & education,
537m074 e & for the effect of his example on others.\ It MAY be doubted whether a man intentionally can wag h
```

537m075 ettled by reason.--╫ How slow habits are changed MAY be inferred from expression. "relict of bad habi
537m076 ing social animal, & this conscience or instinct MAY be most firmly fixed, but it will not prevent ot
537m077 er.-- A Scotchman will his country or Swis.-- it MAY be answered effects of education, may be opposed
537m077 Swis.-- it may be answered effects of education, MAY be opposed undoubted cases of heredetary pride &
538m078 absolutely two people. Consider this profoundly, MAY throw light on consciousness, explained by Dr De
539m082 abit «or trick» so much more easily than man, so MAY animal obtain it far more easily, in proportion
539m083 going on in the mind as in double consciousness MAY really explain what habit is-- In the habitual t
541m091 ogous to muscle in one position great fatigue.-- MAY explain excessive labour of inventive thought.--
541m093 are so difficult to conceal.-- a man «insulted» MAY forgive his enemy & not wish to strike him, but
541m093 it far more difficult to to look tranquil.-- He MAY despise a man & say nothing, but without a most
542m093 lip from stiffening over his canine teeth.-- He MAY feel satisfied with himself, & though dreading t
542m093 ill grow erect & stiff like that of turkey.-- he MAY be amused, he need not express it, he may most e
542m093 .-- he may be amused, he need not express it, he MAY most earnestly wish [not] to do it, but an invol
544m103 The mind thinks with extraordinary rapidity-- we MAY conclude that neither number, vividness, rapidit
545m104 s some impressions «Hume» become unconscious. so MAY some ideas.-- ie habits, which must require idea
545m106 eeth, this the Keeper thinks is from pleasure, & MAY be compared to laughing» they dance with passion
546m110 light on instinct, showing what trains of action MAY be done unconsciously as far as the ordinary sta
547m113 ld not doubt or believe.-- When I say trains, it MAY be instantaneous changes in order <to every> cal
547m113 first, as whether I had pulled the bell??)-- It MAY be deception to say the mind <thinks> quicker in
547m113 on to say the mind <thinks> quicker in sleep, it MAY do less work & yet do so, from the exertion of k
548m114 est. though. most vivid & rapid thought.-- There MAY be some «two or three» trains of thought, theref
548m114 «two or three» trains of thought, therefore one MAY be imperfectly reason -- «In a» Abercrombie's ca
550m124 t other people experience.--» But then sensation MAY be more or less pleasant & unpleasant, in same t
551m129 from sneer-- How easily. horses associate sounds MAY be seen by omnibuss Horses starting, when door s
556m145 aughing noise which C. Sphynx made at Z. Gardens MAY be described as partaking of <st.> made by <ret>
558m150 most subject to habit, as being less so will.-- MAY not moral sense arise from our enlarged capacity
558m151 to Miss Martineaus one principle of charity.-- ¿ MAY not idea of God arise from our confused idea of
563nIFC y» Dec. 16 1856 Looked through & all other Books MAY 1873-- October 2d.. 1838 Essays on Natural Histo
563n022 phalus. Sphynx howling when I struck the Keeper) MAY be tempted to attack him from jealousy. (Pincher
564n006 » of excessively cross-half furious faces «which MAY be described as an exaggerated habitual sneer» t
567n013 k, so so many animals.-- analogy probably false, MAY lead to something.-- October. 8th. Jenny was amu
567n015 the sensation, when meeting a stranger. who one MAY like. dislike, or be indifferent about, yet feel
570n025 , & criminal,-- who has stolen. neither. or both MAY be said to have fear, both both have shame-- Ani
571n028 high forehead sign of exalted character???) Why MAY not our heredetary nature thus acquire some gene
571n030 bits. & heredetary habits. & perhaps even latter MAY be vitiated. or rather altered. The Reason why N
572n034 the conscience of a Boy to swear, though reason MAY tell him not, but it does hurt his conscience, i
574n042 ced with strong arms, outliving the weaker ones, MAY be applicable to the formation of instincts, ind
576n046 ys seals knit their brows when incensed.-- A Dog MAY hesitate to jump in to save his masters life,--
581n065 rape, crack, &c, imitative of the things.-- CD[I MAY put the argument,, that many learned men seem to
582n066 uscles.-- man cannot be said to be angry.--» «He MAY have pain or pleasure these are sensations» <Gar
582n067 present on any hypothesis whatever]CD an animal MAY so far be said to will to perform an instinct th
583n069 rents offer best cases of instincts].CD all this MAY be true,, but relation of imitation & reason mus
583n070 & then these qualities of imitation & education MAY be used as argument.-- for instinctive knowledge
583n070 d by instruction, or imitation.-- p. 20. Animals MAY be called "creatures of instinct" with some slig
584n071 s woodpecker: but this is not so,, the instincts MAY vary before the structure does; & hence we get o
584n071 ecies is found which does not climb CD[.instinct MAY be divided into migration,-- subsidiary to food
584n072 spot, this is indeed instinct.-- Australian man, MAY be called instinctive: the facts of memory of ro
586n078 or return.-- «but does not apply to dogs.--» they MAY do all this instinctively «yes because power var
586n079 it, such habit of knowledge of points of compass MAY be instinctive. it is a test to know how much of
586n081 s breathing instinctive, certainly heart beating MAY be considered also such.-- heredetary habit, is
586n082 of building «regular» cells-- [but this faculty <MAY possibly be» «probably is» instinctive, namely t
587n087 to organs adapted to like sugar, acid, &c, which MAY be doubted for possibly even taste of senna. mig
593n107 ed by other animals smell & looks.-- no doubt it MAY be attempted to be said that young animal learns
594n113 n on the 1st March was frightened on the 24th of MAY at Cresselly by the boys making faces at it, so
601o009 sensation is conveyed over whole body (which it MAY be in first case. as when the excised heart is p
605o19v ed. we call sublime.-- It appears to me, that we MAY often trace the source of this "inward glorying"
608o026 f mind all motives do not. come into play.-- †It MAY be urged how often one try to persuade person to
608o026 d materials. From contingencies a mans character MAY change-- because motive power changes with organ
608o026 be more proper to pity than to hate & be †A man MAY put himself in the way of Contingencies.--but hi
608o028 t a future life is a reward or retribution.-- it MAY be a consequence but nothing further.-- October
609o030 than we can look forward: hence our <[...]> rule MAY sometimes be hard to tell) + + Society could not
610o033 e is much knowledge without experience. so there MAY be in men-- which the reviewer seems to doubt. [
614o037 , so [RHC] 6) thought, however unintelligible it MAY be, seems as much function of organ, as bile of
614o037 -- perhaps, say attribute of such higher animals MAY be looking back, ∴ therefore consciousness, ther
616o038 <return> turn into angels. this imperfect memory MAY become perfect & we may look back to definite ac
616o038 s. this imperfect memory may become perfect & we MAY look back to definite action or to our conscious
616o038 action or to our conscious selves.-- Such memory MAY go back to animals which were changed into man ∴
616o038 detary memory & individual secretion of thought, MAY be no more «difference» than sexual intercourse
616o039 in some animals, involuntary in others. [1)] Why MAY it not be said that thought perceptions will, co
618o042 rces & mentality because effort is felt [LHC] 1) MAY 5th. 1839.-- Maer Mackintosh Ethical Philosophy
620o044 ralist would at any other mammiferous animal, it MAY be concluded that has parental, conjugal and
620o044 twards think of it.-- Whatever the cause of this MAY be, everyone must know, how soon the pleasure fr
621o047 from some trifling circumstance.-- Thus a child MAY be taught to think almost anything nasty. (<& ac
621o047 in taste, & tight & wrong in action, so a child MAY be taught, or will acquire from seeing conduct o
621o048 y will teach to be wrong or right; this teaching MAY be curiously modified by circumstances of countr
622o049 says is so ridiculous. [RHC] the social instinct MAY be combined with feeling towards one as a leader
622o049 towards one as a leader,-- the conjugal feeling MAY be directed towards one or more.-- It will be ha
622o049 to discover this, for the different races of man MAY have different instincts, as we see in dogs & pi
624o051 , for they are in accordance with it. thus a dog MAY be trained to hunt one pig sooner than other, ra
628o054 uld think some parts of the emotive part of man, MAY be quite artificial, as avarice love of gold.--
629o55v gives & try it.-- p. 241 (1) Any action by habit MAY be thought wrong.-- & conscience will imperiousl
633j53v certain that these are necessary adaptations.-- MAY they not be accidental? We have good reason to k
633j53v nts, & domesticated variations show us accidents MAY become heredetary [produce some peculiarity in s
637j58r affe &c, as adaptations to long necks-- why they MAY as well say, «long» neck is adapted to long neck
321c270 . with Rousseau Miss M«artineau» How to observe MAY Philosophy of Art of Living Several of Water Sa
322c269 sophy R. W. Darwin's Botany.-- References at end MAYO Pathology of the Human. Mind Evelyns Sylva. ski
326c266 rts of Brit. Assoc.-- some important Papers. Dr. MAYO. Pathology of Human. Mind.-- Audubons. Ornithol
347d049 , & therefore more apt to be extinguished.--???» MAYO (.Philosor of Living) quote Whewell as profound
544m102 sure, ie. love.-- & so pain gives fear of death. MAYO Philosophy of Living. p. 140-- Dreams good acco
546m110) wonderful case of perfect double consciousness MAYO compares it with Somnambulism.-- the young lady
546m110 s far as the ordinary state is concerned.?-- Mr. MAYO told me the case of a lady, (whose name was tol
546m110 (whose name was told me, who told the fact to Mr MAYO himself. she was one day reading a book, with i
547m111 the time. & could trace no chain of association» MAYO Philos. seems certain that muscular, mental, <&
547m114 p» old ones, to be sure of ones consciousness.-- MAYO observe no improbabilities in a dream, effect o
550m126 are confounded.-- well applicable to free will. MAYO. Philosop. of Living p. 293. Animals "have noti
599o006 .-- declension &c often show traces of origin.-- MAYO Philosophy of Living. p. 264. "Architecture is
604o017 y.-- Quotes D. Stewarts System of Emotions.-- T. MAYO-- Pathology of the Human Mind. Poor.-- on insan
548m116 » one individuality in this case.-- But now in MAYO'S «p. 140» case of double consciousness, one wou
571n031 mblance to gloomy aisle of Churche.-- these are MAYO'S ideas.-- In language. the possibility of poets
026r020 mon opinion The Tertiary formation South of the MAYPO at one period of elevation must in its configur
477z002 of La Plata or Paraguay.-- do. p. 365. 3 cats (MBARA caya. le negro, et le pajero) l'yaguaré «the zo
216b179 or» similar in the Indian sea.-- Deshayes.-- Mr MCCLAY is inclined to think that offspring of Negro &
216b180 Dr Smith says he is sure of the case at Cape.-- MCCLAY argues from it Black & White species.-- For, s
038r058 t from the Porphyries: certainly appearance leads ME to believe mere fissures filled up.--the appeara
042r067 nerally & during gales would tend to travel on a <ME> central line of Patagonia. «NB. Mr Lyell P. 211
044r073 stability of ground at present. day. applied by ME geologically to vertical movements. In Cord: aft
053r101 quoted at length. The Lines of Mountain appear to ME to be effect of expansions acting at great depth
066r141 uelen 40 by 20 leagues. dimensions: Bynoe informs ME that in Obstruction Sound, in the narrow parts w

068r145 ors to Craters of elevation.-- Lyell suggested to ME that no metals in Polynesian Islds--. Volcanic p
070r155 s. from the Cordovise range. Signor Rozales tells ME at seven oclock Novem <5th> Concepcion most viol
071r157 dt in Geolog. Society Sir Woodbine Parish informs ME that town near Tucuman and Salta. towards the Ve
077r172 t oxidation?-- Mem Sir W. P. stone It is clear to ME, there are laws of solution & deposition under g
085a002 en 70 miles from salt water. Mr. Arrowsmith tells ME, that Himalayas penetrated like Bolivian Chain.
093a034 ributed to various subjects) Dr. A. Smith informs ME that in the year a Rhinoceros was found <emb> in
096a041 t. Helena discussion. Mr Brayley says he can give ME facts respecting lime <n> being heated without p
097a044 -- Vol II. 2d Series. p. 221.-- Mr Bollaert tells ME, that the upper strata alone at Guantajaya conta
104a068 ring water with salt to surface Lyell remarked to ME that Kylow (?) was astonished with him that <th>
108a080 hocks directly after great shock -- It appears to ME unphilosophical to think calcareous springs near
113a092 xplaining want of levelness Major Mitchell showed ME a river <near> W. of Port Philip. which had bar
124a118 ountain Chains in N. America Erasmus suggested to ME that Herschel's theory offers no explanation of
130a133 t bottom.-- Study Bishoofs Paper.-- Weelsted told ME of some large fresh Water springs off coast of P
132a139 s own hypothesis some such explanation appears to ME necessary) as M. Parrots shows from variation in
160g088 uvium shows the ascending fringes {P} which makes ME think it submarine, 400 or more feet above stati
162g102 ters by old tower called Glengarry (Nead Roy told ME) it is impossible to see my new shelf, from road
162g105 ited when water stood at higher Loch Keeper tells ME, that Loch Lochy is 8 ft below Loch Oich wh is 9
190b079 ived from Monad, each fresh--> Mr Don remarked to ME, that he though species became obscurer as knowl
192b083 Smith.-- «to state that most clearly». Fox tells ME, that beyond all doubt seeds of Ribston Pippin,
201b129 The relation of Analogy of Maclay &c. appears to ME the same, as the irregularities in the degradati
210b161 roceedings▯. <p> 1832.p. III Mr Owen suggested to ME, that the <cas> production «of monsters» (which,
212b165 ter not to former-- Mr Martens of Zoolog Soc told ME, an Australian dog he had, used to burrow like fo
213b171 ack have insects been known? As Gould remarked to ME, the "beauty of species is their exactness,' wan
214b173 nd «fossils of» Oolitic Series does not appear to ME very strong What is Osteopora platycephalus. (Ha
215b177 ired to give the cantharides and milk--Fox tells ME that it is generally said.= How came first speci
217b184 , puppies differ, & like both parents.-- Fox told ME of case of mare covered by blood horse & Carthor
217b187 e & Carthorse two folds [not located] Mr Don gave ME instances of one species of Australian genus bei
218b191 nn considers Mr H. rather wild Mr Donn remarks to ME. that give him a species from Ireland, England,
218b192 a list of its Flora. is given Mr Don remarked to ME. that some good African & some good S. American
225b217 ntermediate forms.-- Opponent will say. show them ME, I will answer yes, if you will show me every st
225b217 show them me, I will answer yes, if you will show ME every step between bull Dog & Greyhound). I shou
225b219 e.-- Reference to Pig & Dogs. My theory will make ME deny the creation of any new quadruped since day
231b242 se of Himalayan, plants ¿migratory birds, he told ME some story of crane from Holland!!! in stomach --
232b248 n salt water & lizards do.-- Ask Eyton to procure ME some Get Hope to give me an account of parasitic
232b248 .-- Ask Eyton to procure me some Get Hope to give ME an account of parasitic animals of beasts varyin
239c001 f habits Mr Yarrell «Give it as his theory» tells ME. he has no doubt that oldest variety, takes grea
249c035 s. case of replacing species. Dr. Smith will give ME some capital information ¿Carnivora of New & Old
263c078 at we see in bodily. that <I> it does not stagger ME.-- What circumstances may have been necessary to
271c107 nimals in certain countries.-- Fraser remarked to ME at Zoological Society,, that you never find two
271c107 country-- ie. mundine groups.-- Waterhouse tells ME in insects there are many plenty of instances of
276c122 s laws of change-- whether beak (as it appears to ME). colour of plumage & laws, which might probably
280c135 cipate mules is very important for L.yell said to ME, the fact of existence of mules appeared to him
290c164 t laws of change, Will be known.-- It appeared to ME that half between fowls & pheasants, is most lik
302c201 remained somewhat similar.--!!! My theory drives ME to say that there can be no animal at present ti
308c217 back memory» old habit of putting tea in pot made ME go to tea chest almost unconsciously.-- why do a
312c231 ition, we must [not located] Henry Thompson tells ME best way to improve cattle is to cross between a
314c237 by scission can not alter much.?? Mr Brown showed ME Bauer's drawings of a curious plant where a tube
317c249 ongo Expedition, NB. I met an old man--, who told ME that the mules between canary birds & goldfinche
317c249 Verd; same as on coast of Africa.-- Macleay tells ME same thing p. 55. 40 leagues from land several p
317c251 of N. America, & of the shrews.-- Dr Bachman told ME. that near Charlestown ?three species, near New
318c253 gions of snow.-- Acclimatisation.-- Bachman tells ME in Audubon there is most curious history of firs
341d031 e facts of Mr Blyth on birds.-- Dr. Bachman tells ME line of Rocky Mountains separate almost all Mamm
356d072 of generation (p. 175) if> 8th Sept Yarrell told ME he had just heard of Black game & Ptarmigan havi
363d103 hus &c & -- like young blackbirds Dr Bachman told ME that 1/2 Muscovy & common duck were often caught
364d104 to old <story> return.-- The Revd. R. Jones told ME precisely same story about some Southern «see p.
366d108 mere monstrosity propagated by art. Yarrell told ME of a cat & of a dog, born without front legs-- -
367d113 oured ass.-- stripe on back also.-- legs reminded ME strongly of Zebra.-- Mem. Quagga & Ld Moreton Ma
377d139 , as when food is offered, as much as to say give ME-- the other when Dr. Smith more distant.-- But h
379d151 ack reptiles & Cephalopoda: Old Jones remarked to ME, that one of the children of Sir J. H. was so ve
385d163 llis «Grt. Marlborough Str. Hair dresser, assures ME he has known many cases of bitch going to mongre
413e059 ion to support of parents December 2d Lyell tells ME Beck considers the characteristics of the Tropic
417e076 ing are hybrids,.-- Mr G. B. Sowerby <tel> showed ME many land shells of the common species: from one
424e101 , & therefore not to be trusted.-- Lyell tells ME, on authority of Beck, that Hooded crow & Carrio
426e106 quent ones. My views, <V <see> p. 103> would lead ME to think that a <do> variety of one species woul
428e112 law must be partly true, as enuntiated by him to ME, for otherwise breeders who only care for first
430e119 rts of the world-- March 16th. Mr Lonsdale showed ME two specimens of an Inoceranus from the Gault of
434e129 s sure guide.-- Lychnis April 3d.-- Henslow tells ME following facts: believes that «only» red Lychni
449e168 cus-- is not this right?-- June 18th. Eyton tells ME, that Yarrell knows of a Gull, which has laid in
458t002 them. Mr Miller-- in Zoological Gardens. informs ME that a hybrid between ass & Zebra, crossed with
463t055 that of Java & Sumatra Nov 15th Waterhouse showed ME the component vertebrae of the head of Snake won
505q014 Bees are smaller & more vicious. Will try to get ME some to look at:-- Was once offered a hive. of t
509q017 ve talked partially with him Ask him to introduce ME to some Human Anatomist. Has he dissected any an
510q017 understand English.-- Miguel to collect facts for ME-- what? What does Blume say on alpine Flora of J
511q018 actually rejected?? (8) Get Sir. R. Heron to give ME Pigs foot undivided, & more particulars regardin
522m014 over somebody who has, perhaps, slightly injured ME, plotting speeches, yet with a sort of conscious
531m051 g over an insane feeling of anger which came over ME, when listening one evening when tired «-- how t
531m053 anger-- but it is instinctive because Nancy tells ME very young babies start at anything they hear or
532m054 much afraid though my reason was laughing & told ME there was nothing, & tried to seize hold of obje
532m056 ill whilst their minds were sound. Caroline tells ME that Nina, when brought from Shrewsbury to Clayt
533m060 e him more humble.-- it has obscurely occurred to ME that Capt. F. R. candour & ready confession of e
536m073 & by choice which at that time organization gave ME to will-- Verily the faults of the fathers, corp
539m081 rom reading «review of» M. Comte Phil. which made ME «endeavour to» remember, & to think deeply, & th
539m082 on with much pleasure immediately thrilled across ME, bringing up old indistinct ideas of FitzWilliam
540m085 s fingers. Seeing a dog & horse & man yawn, makes ME feel how <much> all animals <are> built on one s
546m110 the ordinary state is concerned.-- Mr. Mayo told ME the case of a lady, (whose name was told me, who
546m110 told me the case of a lady, (whose name was told ME, who told the fact to Mr Mayo himself. she was o
547m112 would not have turned into dream.-- It appears to ME, that the mind is wholly absorbed with one idea
551m127 tral illusion & insanity the connexion appears to ME vague-- Delirium of every degree of intensity--
555m144 of Dr Monro experiment about hanging came before ME showing impossibility of person recovering from
564m005 ysic, as they have always been studied appears to ME to be like puzzling at Astronomy without Mechani
571m027 .CD-- The existence of taste in human mind. is to ME clear evidence, of the general ideas of our ance
572m033 ses. notions &c Octob 30th-- Dreamt somebody gave ME a book in French I read the first page & pronoun
588m090 ons of reverence & piety are felt." it appears to ME. mere consequence of stooping, as sign of humilit
591m097 ly 20th Intelligent Keeper.... Zoolog. Garden told ME. he has often watche tame young wolf & it never
602o11b ou say is perfectly true, but you do not convince ME.--" Belief allied to instinct.-- p. 134. a paint
605o19v efore mentioned. we call sublime.-- It appears to ME, that we may often trace the source of this "inw
613o036 is the same in all animals. [LHC] «3}» Eyton told ME that his retriever Sailor he has seen push a har
615o36v day met him, with Mark riding instantly followed, ME and for five minutes every now and then howled.-
626o052 ks about "contact with will" is unintelligible to ME.-- conscience regulates feelings, as of cowardic
626o052 feelings, as of cowardice.-- the whole appears to ME rather rigmarole.-- He does not say anything abo
626o052 nscience: I cannot admit it.-- see notes to it by ME..' 'p. 333 «& p. 377» some remarks. showing that
628o53v & good» of society such conduct is instinctive in ME (& as a consequence, but not cause gives me [3]
628o53v ve in me (& as a consequence, but not cause gives ME [3] pleasure) or that I have been taught or habi
637j58r eggs is explained by Malthus.-- [is it anomaly in ME (& as a consequence, but not cause gives me [3]
149g032 ition in lake On the summit «& on Spean side» of MEAL-- Derry there were perfectly rounded «base» peb
546m111 n vain for it-- ten years afterwards whilst at a MEAL, she suddenly like a flash without any assignab
520m002 died.-- The omnipotence of habit is shown about MEALS, no [not located] There is a case of Mr Anson.
035r047 Geolog Map: Quoted from Daubeny P 402: likewise, MEAN height of tertiary. being less than secondary:-

047r081 his [Fig. 3] {P} form present, i e a part below «MEAN» level before the higher part. -- Does the sea
050r091 tions form foundations for Coral reefs.--does he MEAN in contradistinction to sand?? B. Roussin state
072r159 as talked of lava flowing up Hill; ¿what does he MEAN?) Consult Dr Holland about bubbles. -- No Volca
112a090 l 9 1838. Letter from M. Erhman stating that the MEAN temp at Yakous in Siberia being -8 Reaumur.-- t
113a090 at from centre.-- But is this not wrong? we know MEAN of surface formerly much higher, «so» that we m
132a138 cold <oceans> «lakes» bottom. if not colder than MEAN of place, shows earth not with central heat.--»
198b114 line, or branching S. H What does the expression MEAN used by Cuvier, that all animals (though some m
232b246 e will take a day from the equator to add to the MEAN of the other.-- If the world had cooled by
269c101 etamorphs theory suppose when rhinocerose lived. MEAN temp 60 degree mean, then temp at depth of four
269c101 pose when rhinocerose lived. mean temp 60 degree MEAN, then temp at depth of four hundred feet would
278c129 ng out of case the Analogys.-- If genus does not MEAN this it means nothing.--There should be some te
295c178 at character -- -- female & young seem most like MEAN characters the others assumed--H+ |Daines Barr
308c219 ld have arisen under different climates &c. Do I   MEAN that ideosyncracy of wild animals is generally
393d1BC the throw of two varieties is uncertain do they MEAN they cannot tell first result., or that «hybrid
422e092 essential ones eminently adaptive.-- does it not MEAN lately adapted or transformed. & hence not indi
53m137 mportant as showing <connection> that expression   MEAN SOMETHING.-- Hunt (the intelligent Keeper) rema
583n070 of instinct" with some slight dash of reason so MEAN are called "creatures of reason", more appropri
621o047 from origin of man.-- [RHC] By interest I do not MEAN any calculated pleasure, but the satisfaction o
281c139 ach other-- &c &c.-- V. p. 140» I should think MEANING of circular arrangement was only so far true a
286c155 ion of <arrangement> relationship; latter word MEANING descent.-- A tree is taken by Fleming as emble
291c166 itism in corporeal structure are facts full of MEANING.-- Why is thought. being a secretion of brain,
307c216 «of very same kind» in these cases, have plain MEANING & none in other case! Savigny has shown same f
421e091 to an essential character--" How little clear MEANING has this compared to what it might have.-- Wha
524m021 ecessary)-- the words second childhood full of MEANING:-- Dreams do not go back to childhood-- People
602o011 essions on them are <all> remembered, when the MEANING or reasons are forgotten. Our happiness &c, ou
604o018 sanity.» D. Stewart on the Sublime The literal MEANING of Sublimity is height. & with the idea of asc
613o036 ts imply willing, therefore word misplaced The MEANING of Words, must be made out Reason Will Conscio
056r109 ge was limited? Who could suppose such trifling MEANS could efface & obliterate so grand a work?--In
171b003 constant, hence we see generation here seems a MEANS to vary. or adaptation.-- Again we <believe> «k
190b078 aptation) is given by true generation, throughe MEANS of every step of progressive increase of organi
199b116 he same type with the great continents we get a MEANS of knowing movements.-- How can we understand e
219b196 orn from one stock, & since distributed by such MEANS as we can recognize, may be thought to explain
227b224 s points of speculation; for having ascertained MEANS of transport, we should then know whether forme
231b242 in quadrupeds, <bu> plants do not follow by any MEANS.-- Ostriches.-- Hippotamus only African.-- Amer
243c020 sle of France: xx instance of wide range, where MEANS of wide range work this out-- L. Jenyns, about
263c075 movements &c &c. this multiplication of little MEANS & bringing the mind to grapple with great effec
278c129 the Analogys.-- If genus does not Mean this it MEANS nothing.--There should be some term used, when
288c159 had originally crossed would continue to cross, MEANS of knowing directions: mysterious. Were the woo
308c217 sion ustil, unless it were fixed what a species MEANS» civilized Man, May exclaim with Christian «we
312c231 erely picking opposite qualities, with no other MEANS whatever.--» Individual Men «& animals» could o
315c240 n must take place on[ly] when creator sees. the MEANS of transport fail.-- otherwise no relation betw
315c240 transport fail.-- otherwise no relation between MEANS of Transport & creation exists.-- pooh. May ha
315c242 lants movements effects of irritability, though MEANS injection of fluid different from contraction o
376d135 m which Malthus shows, is the final effect, (by MEANS however of volition) of this populousness, on t
389d177 as parturition is concerned.-- Generation being MEANS to propagate & perpetuate differences (of body,
411e055 of variation., & hence indicate gaps.-- by this MEANS the laws probably would be. generalized, & afte
411e056 of multiplication of individuals, & yet another MEANS for individuals (Mem: transportation will be an
416e071 es. of <a> organics. are made by percisely same MEANS as species-- but latter far more perfectly & in
419e086 sent seas.-- now in this country we have better MEANS of judging of slowness of physical changes, tha
422e095 he forms became complicated, they opened fresh, MEANS of adding to their complexity.-- but yet there
438e142 tion.-- the constitution of some may resist the MEANS, which wild plants never are, namely on stocks
442e153 ut on other hand, fruit trees are propagated by MEANS, which will plants never are, namely on stocks
445e159 s> even oyster may fecundate each other, by the MEANS of the medium in which they live do. "Additions
472s01v r Miss Bent three years since gave her some She MEANS to try this year. Little variation in the 60 on
521m010 very respect, no communication could be held by MEANS of hearing.-- Mr Corbet, however, in conversati
526m027 thers <are> learnt. what they teach by the same MEANS & therefore properly no free will.-- we may eas
532m054 , senses being on the look out, & the conveying MEANS from the senses to the mind being more alive.--
533m058 ing before they could understand. what frowning MEANS) if so this is precisely analogous or identical
534m061 images before your eyes, or ears (language mere MEANS of exciting association.)-- or of memory of suc
548m115 bable. in comparing every step, & inventing new MEANS,-- therefore works of imagination hard work,--
564n005 tion of axe of eyes, state of surface, or other MEANS by which eyes, aided by experience is supposed
572n035 ss, whilst they are the only steady & universal MEANS. recognized-- no one can say expression was inv
576n048 «instinct»-- an unfolding & generalizing of the MEANS by which an instinct is transmitted.-- Arguing
582n067 d.-- N B. There is wide difference, between the MEANS by which an animal performs an instinct, & its
582n067 rms an instinct, & its impulse to do it.-- [the MEANS must be present on any hypothesis whatever]CD a
582n069 to result (first time of building?), but not the MEANS of performing it.-- p. 14. There is scarcely a
583n070 of habit."-- CD[as the bee makes its cells, by MEANS of ordinary senses & muscles, we cannot look at
584n072 ung crocodile snapping-- p. 28. how curious the MEANS of guiding themselves through the air,-- waterb
585n074 ry much» Henslow. N. Necker has remarks on the MEANS. by which children learn (probably not only exp
585n077 abit. so bees in building cells, must have some MEANS of measuring cells, which is faculty, they use
586n080 ct to one part. or another-- but (an instinctus MEANS stained in?). had better refer to to the herede
593n111 at consciousness of personnal identity is by no MEANS a necessary part of man's mind.-- At Maer. Pool
595n125 r instances of instincts (highly useful as only MEANS of communication) in man, than sucking.--[I as
605o18v ensations so often gives, when excited by other MEANS, as moral excellences, brings to our recollecti
605o020 ow taste is gained how it originates, & by what MEANS it becomes an almost instantaneous perception.-
613o036 rom the ends in each case being the same, & the MEANS very similar.-- It does not appear more than sa
632j53r ve thought that the intestines of a thrush were MEANS sufficient to ensure propagation of Misseltoe?-
637j57v just as much as Woodpecker. --only we here see MEANS-- but not in the other. ¶All Bridgewater Treati
573n037 ovement negation.-- he dropped his head when he MEANT to eat, hence assertion.-- but nodding is less
099a050 «varies with chemical attraction &c.» becomes MEASURE of force. < ∴ where little inclination, little
113a092 brium, the great fallacy (habitually) becomes MEASURE of force in that part.-- Important as explaini
147g021 degree 65 degree? For comparison with all the MEASURE before There some of the half rounded gravel n
232b246 ghboring sea.-- For change of species does not MEASURE time but physical changes (we assume like weat
257c058 -- <Wood> <Dicot wood> Coniferous wood in Coal MEASURE.-- highest fish in Old Red Sandstone.-- Nautil
377d140 the waves of the sea, on beaches-- we really, MEASURE the rapidity of change of form, & instincts in
583n071 fficient for hence it must some way be able to MEASURE the cell; p. 22. instincts & structure always
506q034 ck, height 17½/. The Greyhound. was in length (MEASURED same way) 47½-- in heigt 30 inches Examine Ke
148g022 alf rounded gravel nearly as high as highest MEASUREMENT but nature I am quite doubtful of as I am of
157g074 -- 29.330 A 84 degree compare this with last MEASUREMENT of shelf of 3d:-- granite block a yard acros
161g097 race Barom. 29.264 A 82 75 degree? This last MEASUREMENT turns out too low, (NB .260 would have been
161g097 ove level of shelf certainly) I took another MEASUREMENT on short buttress but not continuous & it wa
158g079 f granite obscure terrace 15 ft divortium my MEASUREMENTS here side of Loch Spey Forms terrace about
162g099 chy side of it-- the terraces of which, last MEASUREMENTS in England (Excepting Conglomerates?) «& abse
432e125 ally comprehend how old the world is, as the MEASUREMENTS refer not to revolutions of the sun & our l
029r034 te stratified with substances so like the Coal MEASURES in England (Excepting Conglomerates?) «& abse
039r059 that all the dikes in Cornwall or in the coal MEASURES have been conduits to volcanoes.-- Talking of
460t019 a & caetacea in Eocene-- dicot. plants in coal MEASURES.-- Shells in Cambrian & Crust show how long s
585n077 es in building cells, must have some means of MEASURING cells, which is faculty, they use this facult
358d075 o camp; it was absolutely necessary to watch our MEAT, while in kettles on the fire, & on one occasio
525m023 when doing anything which is wrong.-- as eating MEAT. doing their dirt, running home.-- in these ca
570n025 erence is there between Squib after having eaten MEAT on table, & criminal,-- who has stolen. neither
570n024 relieve respiration when immensely immersed-- MECHANIC apt to sigh.-- & hence carried on as trick) «
110a086 side-- separation DISTINCT from dike junction MECHANICAL: DIKE base reddish feldspathers with grenish.
608o025 cated one calls them free will--the chance of MECHANICAL phenomena.-- (Mem: M. Le Comte case of Philo
564n005 o me to be like puzzling at Astronomy without MECHANICS.-- Experience shows the problem of the mind c
293c174 there must be some coralation. but. the whole MECHANISM is so beautiful.-- The coralations are not, h
357d074 g floated about.-- I must state that. the <p> MECHANISM by which seeds are adapted for long transport
404e025 . 329-- Milne Edwards, description of curious MECHANISM of respiration, or rather ventilation peculia

601o009 nervous filaments.-- ¿plants? yes by distinct MECHANISM 2. Sensation of higher order. where the sensa
459tf02 Sciences. Vol. III p. 83. Paper translated from MECKEL. Comp. Anat.-- From Buffon cross of he-goat &
357d073 1838. p. 654. Reason given for supposing Tetrao MEDIA or Rakkelhan is hybrid (produced commonly in Na
105a070 or filling up of valleys-- subsequent opening a MEDIAL gorge by slow erosion. but we have evidence in
326c266 ork on Bees, new Edit 1838 Harlaam. Physical & MEDICAL Researches. on Horse in. N. America.-- Owen ha
327c265 urgh. Encyclop.-- The <Edin> British & Foreign MEDICAL Review No XIV. April 1839.-- Review on "Walker
508q016 closely than Caffres.= 13 Where are there any MEDICAL Statisics, proportion of diseases (heredetary?
321c270 nversations-- very poor Sir T. Browne's Religio MEDICI Lyell Book III There are many marginal notes <
350d054 nly indisputable axiom in Philosophy<"> Religio MEDICI. Vol II. Sir T Browne's Works p. 20 There are
550m126 ur recollections are pleasant.-- Browne Religio MEDICI, p. 21-24. Curious passages showing how easily
356d069 t there be in following work.-- The history of MEDICINE, the extraordinary effects of different Medic
356d069 icine, the extraordinary effects of different MEDICINES on organs, leads one to suspect any amount of
276m046 to jump in to save his masters life,-- if he MEDITATED on this, it would be conscience. A man, might
263c075 takes up & lay down the subject without long MEDITATION-- His best chance iis to have profoundly cover
542m094 m of agony, sigh of discomfort & weariness. & MEDITATIVE tranquility. «whine of children. puppies do
091a026 roc> lava called Andesite. Red Coral in the MEDITERRANEAN 700 feet deep in some of. the twopenny peti
420e090 habits the Pinna of Rio Janeiro, (like some MEDITERRANEAN species).-- might these fertilise other she
425e104 erican, or Indian groups?-- = Believes some MEDITERRANEAN, but chiefly mountainous «this is very impo
087a011 ia rises. therefore to the South sinks.---- MEDITERRANEAN continent corresponding to Europaean rising
445e159 r may fecundate each other, by the means of the MEDIUM in which they live do. "Additions". p. 454.--
606o023 eye itself, but by the imagination through the MEDIUM of the eye"; he will allow the secondary pleas
401e017 on apples. they will live but not flourish-- a MEDLAR may be Grafted on pear. Mountain-ash & white T
263c076 rant that one instinct to be acquired (if the MEDULLARY point in ovum. has such organization as to <p
099a051 se gravity MN. Then every particle would tend to MEET at <B. but if particls attract each other in so
151g041 al form alluvium) descend from shelf 3d & almost MEET, but are separated by flat bottomed strait. con
371d118 direct instinct. & afterwards enlarged powers to MEET with contingency.-- Sept. 23rd. Saw in Loddiges
376d137 d not put it on, yet threw it over it, & made it MEET in front.-- Dr Smith every baboon & monkey, big
414e063 s breed is destroyed)-- -- When two races of men MEET, they act precisely like two species of animals
424e102 - latter not going North of the Elbe.-- yet they MEET in one wood in Anhault. & there every year prod
616o038 ck to animals which were changed into man ∴ they MEET their reward! X Perhaps should hardly be called
112a090 ll be published in the Geograph. Journal.--» A MEETING of the Geograph Soc, April 9 1838. Letter from
136a147 ome remarks on thinness of crust as implied by MEETING with granite every-where. Phillips in Ladner V
361d095 an two last. Sept. 11. N Mr. Blyth, at Zoolog. MEETING stated, that Green-finch, all linnets red-pole
567n015 ently so-- who can analyse the sensation, when MEETING a stranger. who one may like. dislike, or be i
183b053 forms been discovered. between palaeotherium, MEGALONYX mastodon, & the species now living.-- Now acc
175b020 tula, but Megatherium nothing. We may look at MEGATHERIA, armadillos & sloths as all offsprings of so
048r085 om Equatorial plains to S. Patagonia. To the MEGATHERIUM.--To the Horse. = One might fancy that it wa
087a009 osed that the armadilloes have eaten out the MEGATHERIUM. -- The Guanaco the Camel.? Make note about
175b020 may have propagated recent terebratula, but MEGATHERIUM nothing. We may look at Megatheria, armadill
184b054 arent of all armadilloes might be brother to MEGATHERIUM.-- uncle now dead. Bulletin Geologique April
189b070 is not simple adaptation, armadilloes «&» & MEGATHERIUM. each with same kind of coat.-- If we could
405e032 es.-- It might be worth investigate whether. MEGATHERIUM & Mylodon» in post pliocene strata! Mastodon
473s03r 2. p. 142-- Sus americana & Hippotamus «with MEGATHERIUM & Mastodon are coembedded in N. America. see
123a117 20 Varas from surface in tosca.-- remnant of MEGETHERIUM in interior..-- <The theory of [...] .> <The
482z011 cloviana, 2d like sylvia cisticola.-- Embriza MELANOPERA-- a linnet not caught.-- Troglogdytis Furnar
108a081 -8. p. 24. rocks of Chimborazo., & Pichincha. MELAPHYRE. = Andesite-- Albite & amphibole» Cook found
081r181 wich Isd Mawes travels down the Brazil.-- Did MELASPENA publish his travels? Bellinghausen in 1819 Ko
468t106 swarming with small Staphylinidae-- Anapsis, MELEGETHES, Leptuse-- Diptera & small Hymenoptera Saw H
467t105 gh.--green-cabbage «in flower»-- swarmed with MELEGETHES & small Staphylinidae on all their bodies po
499q009 ower (17) Catch Bees, Butterflies-- Syrphus-- MELIGETHES & see whether they are dusted with pollen--
500q10a eds gathered there were planted «last year» pell MELL, without sticks & seeds gathered & these are no
569n021 iated pleasure &c accompanying such memory.-- A MELODY on flute & Epic poem, opposite ends of series
399e011 ke American form) in Stonesfield slate., & a MELOLONITTHA-- In marl from «Lake» Constance species of
491q01v ion of seeds-- (6) To hybridise EVERY flower on MELON & see whether fruit affected. Mr. B. seemed to
502q11v poison-drop exudes-- does not elm. does it «in» MELON-- «Loasa» Anchusa «Campanula» &c & dead-nettle.
505q014 um flavum put in Spirits which plant seeds? (9) MELONS fruit itself hybridised (10) one had no seeds,
026r021 erminate planes ∴ such action can take place in MELTED rocks The frequent coincidence of line of vein
038r057 believing that there must be a central core of MELTED rocks--I think the strongest is the considerati
038r059 considerably. which shows an afflux of inferior MELTED rocks to those parts. Are not the dikes genera
046r079 that the first phenomem. is an inward afflux of MELTED matter.--Volcanos perhaps may be admittance of
059r123 ave the cryst of glassy F. fractured. have been MELTED with little pressure. & perhaps cooled suddenl
069r150 - Most important view Urge curious fact felspar MELTED gneiss/// QUARTZ!!! Analogous to Von Buch. Bas
090a021 teorolite but old Planet, that inside our globe MELTED magnetic metals. ∴ earthy crust compared to t̄h
124a119 has seen in making brass a piece of copper not MELTED absorb, zinc thrugout its thickness.-- this mo
540m088 ndered ductile by grief, or by bodily weakness, MELTS into tears, with sensations of sorrowful deligh
092a030 Vol II. p. 69.-- Geograp Journal Earthquake at MELVILLE Isld New Holland Augus 1d to 3d & 19 1827 Geo
202b132 p. 36.-- Geograp. Journ. Vol IV. P II. p. 160. MELVILLE Isd.-- "The buffaloes, introduced from Timor,
227b224 e. one gret sea (Coral reefs: shallow water at MELVILLE Isd. (3d) We know that structure of every org
310c224 > «closely allied» species Himalayas. 13,000 & MELVILLE Isd.-- West Africa & India some plants same.
022r007 isseminated through the great Plas Newydd dike.--MEM tres Montes. ((Henslow Anglesea)) great variety
022r007 w Anglesea)) great variety in nature of a dike.--MEM. at Chonos & Concepcion. P. 417 Veins of quartz
022r007 cepcion. P. 417 Veins of quartz exceedingly rare MEM C. [Cape] Turn P. 434 & 419 As Limestone passes
022r007 imestone passes into schist scales of chlorites--MEM. Maldonado P 375 Much Chlorite in some of the di
026r022 t XII P. silicified bones not common in Britain. MEM Concepcion Says Echinites. Encrinites. Asteriae,
027r024 mations are now seen in regular descending steps MEM.; rapidity of germination in young corals.--vide
027r028 onverted into siliceous pyritous & coaly matter. MEM: Chiloe In the endless cycle of revolutions. by
028r031 Iquiqui. not from rain, because alluvium saline; MEM. on coast of Northern Chili as springs become ra
029r033 iginate from the decomposition of Granitic rocks MEM. Chanticleers voyage at <[...] Maranh> Pernambuc
030r036 lus.-- «at Guacho» «on N. Chile? Washington.--» MEM. Micaceous formation of Chonos. interesting from
032r040 (as represented), along line of coast.--[Fig. 2] MEM San. Lorenzo; Valley of Copiapò & parts of coast
033r043 ffe Wall of Porph. Lava with base of Pitchstone; MEM Galapagos. chiefly red glassy scoriae.--could wa
033r043 l: could not climb up many parts, in James Isd.--MEM St Helena-- All Trachytic.--Daubeny P. 171. Vol
034r044 ood account of ejected granitic fragments P. 386 MEM. Lyell's fact about sulphuric vapours in East In
034r045 ric vapours in East Indian Volcanos Gypsum Andes MEM. Beechey. account of regular change in soundings
035r045 all old strata of England. formed near surface; MEM. Patagonian pebbles beds, most unfavourable to pr
037r056 sort; not one was found.-- Miers saw then near? MEM. La Condamaine on the Amazons. Consult Insist on
039r061 e, casts of shells, as in some older formations; MEM the envelopes at Coquimbo. the analogy is now pe
040r063 g of pitchstone; which is described as very rare MEM. St Helena; probably more abundant in this case
041r064 ells.-- the effect of Salt water of the Salado.--MEM. in Owens Africa it is mentioned that the Elepha
042r068 etaceous bones so likewise «of miocene period».--MEM Bahia blanca P. 204 Vol III. Lyell Owing to «ope
044r072 . -- Such have been seen to form in atmosphere.--MEM. Ascencion. concretions & Galapagos.-- «Humboldt
048r086 white substance triturated Porphyritic rock. s (MEM white tufas with purple Claystones of P. Desire)
049r089 rlstones alternate together in contorted layers: MEM: Phillips Mineralogy some such fact stated to ex
053r101 be effect of expansions acting at great depths (MEM: profound earthquakes), which would cause parall
057r115 m or beach of sea to explain preserved animals.--MEM: stream of water in the country.-- Sir J. Hersch
058r115 initesimal cryst. arrange themselves in planes. (MEM silky lustre» ask Erasmus. whether electricity w
060r125 ve regions of vegetation.--«I can find nothing.» MEM Carolines quotation from Temple Urge the mineral
061r127 d at one blow. if one species altered: «altered» MEM: my idea of Volc: islands. elevated. then maculi
064r137 m all others in quantity of Sulph. acid emitted: MEM: Grand gypseous formation of Cordillera In descr
065r140 there may have been great subsidence previously: MEM. pebbles of Porphyry.--Falklands.--off East Coas
069r152 of conglomerate between two ranges mysterious!-- MEM. SUBSIDENCE Uspallata of which no trace except b
071r156 ama. Iron in Edinburgh. Fhisoph. Transactions. = MEM: Olivine. Volcanic product.=> <Did Peruvian Indi
074r165 ct of Volcanic & non Volcanic. Then Solfataras. «MEM: Micaceous iron ore.» N.B. To show how metals ma
077r172 h nickel & cobalt (meteoric) resist oxidation?-- MEM Sir W. P. stone It is clear to me, there are law
088a016 -- Main character of Andes Metamorphic action?-- MEM: red sand of Europe no fossil shells --¿ action
089a017 olatilized at bottom, condensed before rising?-- MEM. granite heated.-- Metamorphic action in red san
090a025 showing schistose structure (& veins appearing): MEM. Henslows Anglesea solution of silex also shewn.
093a032 . 297]CD thinks Olivine a preexisting mineral.-- MEM. Galapagos ∴ Basalt deepest?? Marcel Serres L'In

093a032 imed into the tertiary limestones of Vendarques. MEM sublimation of sulphur to form salts of America.
096a039 olcanic action in Iceland parellel to Greenland: MEM.¿ Greenland subsiding.) Von Buch Canary Isd. p.
097a042 like Mica Slate Von. Buch. Canary Isd. p 170.-- MEM. Cordillera Can Greenstone dikes. be residue of
099a051 onza case of Scrope parallel to walls of dykes-- MEM. laminated dikes in Cordillera.!!!-- In stratum
113a092 still water. Major Mitchell inferred subsidence; MEM my remarks on coast of Australia.-- Great NW. di
120a107 cracking by vertical planes into small pieces-- MEM coal-field.-- the structure of Andes. where we b
122a115 gin of mud with stones scattered irregularly.-- (MEM near Gregory Bay). Shropshire case where laminat
123a116 idering space formed-- great vacuum-- by dike.-- MEM. however. veins of segregation in Salisbury Crai
125a120 ream. the 90 miles includes opposite directions. MEM. S. Cruz. Assuming from Sir. W. Herschel's views
126a122 ndensed metal, conduct heat better than plain?-- MEM 1000 {P} how easily water percolates rocks.-- wh
130a134 be marked line of separation A Paper by Parrott MEM. Acad. Peters. Scienc Math. Phy-- Nat. t. I, 183
132a137 e passage of the moon.-- As Hopkins. M. Parrot, MEM. Acad. Imp. des Sciences. (Sc Math. Phys. et Nat
146g012 sion of hatred To fulfil an instinct a pleasure; MEM. Shepherd dogs The Patches of Conglomerate on S.
148g024 lliam yet in Glencoe in parts no trace of them-- MEM Coast of Chile--¿ is not Mica Slate too hard & u
179b035 of external causes) does appear very probable:-- MEM: Horse, Llama. &c &c-- If we <suppose» «grant» s
184b054 s not Macculloch written on same changes in Fish MEM. Rabbit of Falklands described by Q. & G. as new
187b065 uadrupeds:-- same argument applies to England.-- MEM. Shew Mice.-- -- Animals common to South and No
188b067 noceros peculiar to Java, & another to Sumatra --MEM Parrots peculiar according to Swainson to certai
189b070 ubt even colour hereditary in time as in space. (MEM: Galapagos). Little wings of Apteryx Dacelo & Ki
191b081 han any other, because less easily transported-- MEM plants on Coral islets.-- Next to animals land b
194b095 made out of world before zoological divisions.-- MEM. species doubtful when known only by bones.-- Me
194b095 em. species doubtful when known only by bones.-- MEM: Silurian fossils. ¿How are. South American shel
203b135 th of species without apparent physical cause:-- MEM: Mastodon all over S. America. Hilaire does not
203b136 perhaps even in India p. 261. L. Institut. 1837 MEM. Sir F. Darwin cross breed boars were wilder tha
209b156 n to distance (?). & similarity of type (?).-- [«MEM:» Juan Fernandez]CD. From study of Flora of isla
209b157 o species in France is I: 5.7 in Laponia I: 2,3: MEM. Lyell on shells.-- {T} Genera In North Africa.
211b162 ause similar habits produce similar structure.-- MEM. Ornitho Rhyncus Would not relationship express,
218b190 ith hump.-- p 173. Voyage par un Officier du Roi MEM. Capt. Owen's story of cats on West coast of Afr
219b193 a Chloropus)--» might bring in stomach-- &c &c. MEM discover what kinds of seeds. these plants) [Mem
219b193 Mem discover what kinds of seeds. these plants) [MEM Fact stated by Mr Don in island, Teneriffe, St.
219b195 lower genera altered. or northern plants «No» CD[MEM. the antarctic flora must formerly have been sep
220b197 g does not depend on mind or instinct of parent. MEM Lord Moreton's Mare. The fact of plants going ba
224b212 does not bear any precise relation to structures MEM. Eyton's Hogs & dogs.-- The passage in last page
236b279 r dogs. character of Miocene Mammalia--of Europe MEM. Mr Bell's case of Sub Himalayan land emys, deci
244c022 & deer.-- Procured two makis alive from there.-- MEM Waterhouse knows of some species which escaped t
247c028 sld.-- p. 69. Sharks very generally distributed: MEM of great geological age-- Gastrobranchus «only»
250c039 h Palaeotherium in Paris quarries & at Binstead. MEM. recent Crocodiles with Palaeotherium in India--
251c042 in Cuba (p 271 Viedo says American dogs silent. MEM. contrary assertion of Molina) (p. 277.). probabl
260c068 one. in checking beautiful colours of species-- MEM. St. Jago--solitary Halcyon bird of passage.-- M
268c096 each other in very same district.-- <The same> <MEM> Tennioptera & Tyrannula (NB work out how many f
269c102 .-- great space seems to act per se as barrier-- MEM. Tartary & China.--, both coasts of New Holland.
270c103 e out. have position of organs of generation!!!. MEM. Agaziz. (INo Annals of Nat. Hist) spiral struct
271c106 ffect of external contingencies & long bred in-- MEM, <an> a statement in Mr Wynne's book, about not
276c120 inhabit same country are not habits different, MEM Gould's Willow Wren) but where close species in
277c125 & structure similar would have blended together MEM Mr Herberts law; habits determining fertility Sc
278c131 e whether species same, excessive improbability. MEM in Clifts list a rat said to have been found!! r
290c165 man. not altogether explained by F. Cuvier, «-- MEM Hensleighs objection.--» it is more, he cuts th
291c168 ne species being destroyed at Falkland Isds++.-- MEM Lyell hypothesis of change in Sicicly.-- Splendi
300c197 ired.-- They reason however on this to a degree. MEM Spider only dropping where ground thick.-- shamm
305c210 here none of reason in order of <ver> Mammals.-- MEM Elephants & dog.-- There is one living spirit, p
308c218 or water, air, & land, (Macleay has this remark) MEM. number 5 here most evident!!? examine into this
331dIFC s, in plants are the number of seeds greater.?-- MEM. for Eyton.-- Sir. R. Heron's case of breed of p
333d007 offspring always more resembled foxes than dogs MEM Jackall in Zoolog Garden) He has seen in a show
334d009 s were equally fertile with pure bred animals.-- MEM. number of Mules.-- «He recollects one hatch of
337d021 .-- how one nerve becomes sensitive to light.-- (MEM whole plant may be considered as one large eye--
338d025 only animals from distant countries breeding!. «MEM 3 species of grouse»! Has not Goldfinch & Greenf
346d047 uck with pink shade on plumage of the Pelican.-- MEM pink spots on Albatross, on some Gulls. Flamingo
350d056 , especially sight connected with locomotion.-- «MEM. Dr. Blackwell (Abercrmbies) comparison of sight
355d068 e with Secondary Species distinct-- but close.-- MEM. Von Buch on Cordillera fossils same remark. ¿wa
363d102 be told by incubation & other peculiarities.-- (MEM.-- Goulds Willow Wren.--) (Goulds story of Water
367d113 ck also.-- legs reminded me strongly of Zebra.-- MEM. Quagga & Ld Moreton Mare ringed Owen says that
367d113 med in carnivora. Where females, are peacable-- (MEM Lucanus & Copris &c).-- In birds singing of cock
376d138 examine each other sexes,-- «by taking up tail» MEM: Ourang Jenny with Tommy.-- Good evidence of kno
385d163 hy.-- males «bred in & in» never lose passion. (MEM: so it was said little cock «yet very odd loosin
408e045 t changes even in that quarter of the world-- -- (MEM: elevation & subsidence of East Indian Archipela
411e056 ndividuals, & yet another means for individuals (MEM: transportation will be answered) one looks to a
420e090 fertilise other shells, as insects do flowers.-- MEM. Spallanzani's experiments showing how little of
465t081 is case crossing has had somewhat to do with it. MEM. dogs «& pigs» in Polynesia; & dogs in S. Americ
481z008 t Voluta found in not less than 7 fathoms water. MEM Bahia Blanca. De la Beche theoretical researches
481z009 ed by didelphidae Skunk inhabitant of Patagonia. MEM:-- S. Cruz. Molina Vol. I. p. 244. Baccalao. mig
485z016 nd.-- or ground birds-- rather indefinite Letter MEM Orpheus--becoming tyrant-- flycatcher-- shown by
485z017 st.-- not being replaced by Brazilian Species.-- MEM Turdus Magellanicus.-- C, <Chingolo> Chimango--
512q019 - How long do seeds remain in stomach of birds-- MEM: how many miles they fly in few hours Zoological
526m028 ower.-- this, though very odd is perhaps true.-- MEM Erasmus & mine taste for music.-- Children like
540m088 l delight, very like best feeling of sympathy.-- MEM. Burke's idea of Sympathy. being real pleasure a
545m107 her monkey to dog. I showed nut & then closed my MEM. expression of fury, jump to scratch my face. Th
552m131 «as» vivid, «a reality» as in Spectral images-- MEM Chiloe <pi> Sow, who carried from all parts stra
552m131 & Elephants, (both Pachyderms) much intellect.-- MEM: Yarrell's story of wheel horse in drays, scrapi
553m135 s, many vicarious, like ourselves) that savages (MEM York Minster) consider the thunder & lightning t
563n002 mptation as bitch: or dogs defending companion. MEM Cyanocephalus. Sphynx howling when I struck the
564n003 with mans because original instincts different.--MEM. Bee how different instinct a solitary animal st
570n026 ons: shewn by similarity of the earliest arts.-- MEM.-- Stokes'-- arrow heads &c &c October 27th Consu
585n074 instinct<ing> «which is only present in youth» (MEM. Mr Worsley's story of chicken) to know that whi
593n105 e old instinct <perverted> handed down & down.-- MEM. Nina used to get into hay & make a nest for her
600o008 e & love.-- [With regard to ordinary Beau ideal, MEM. Negro, beau,--Jeffrey denies all Beau-- How doe
608o025 ree will--the chance of mechanical phenomena.-- (MEM: M. Le Comte case of Philosophy, & savage callin
265c085 is to see, whether a large family has it, & one MEMBER of that family, having it with very different
386d166 ead; the other part may surely absorb a useless MEMBER.-- in fact they do it in disease & injury.-- T
577n051 a bivalve shell» does to reason.-- an inflamed MEMBRANE from local irritation to passion.-- Blushing
634j54v 309. says the ribs in Draco support the flying MEMBRANE?!!-- that the phalanges have separate movemen
257c056 he examined-- all species & found "beaucoup des MEMES oiseaux. que nous avions déjà observés en Patag
257c057 logues,-- quand ce nétaient pas tout à fait les MEMES." This good case. of replacement under peculiar
114a093 range in Australian Alps.-- Taylors Scientific MEMOIR, Part IV. p. 403 Ehrenberg on ferrugineous Gal
127a126 length of time depth of ice ought to be less.-- MEMOIR of the Irish Academy Vol 8. p. 118 water no--.
266c092 s? GEOGRAPHICAL JOURNAL. Vol V. p 201 Wellsted. MEMOIR on isld of Socotra. Cattle generally marked li
295c178 ? fertility ¿because offspring too unlike.--?? MEMOIRE by Charles D'orbigny on Plastic Clay of Paris
478z004 lidae Zoolog Transacts. worth reading Cuvier's MEMOIRE 133 1803. on Pennatula showing it to be one an
089a020 a]> Excellent paper on Erratic blocks in Alps. MEMOIRES de la Soc. «de Geneva» Vol 3' P. II. -- Bed,
127a127 s. may be effects of subsidence Elie de Beaum. MEMOIRES of French Geolog. Cantal Vol III I7? p. 246. o
324c268 he Natural system Bevan on Honey Bee Dutrochet MEMOIRES sur les Vegetaux et animaux.-- on sleep & mov
109a084 st important.:-- must be studied.-- Scientific MEMOIRS Edited by Taylor Ehrenbergh on flints in chalk
269c103 f Nature, their life & affinity" in Scientific MEMOIRS I can see that perfection may be talked of wit
283c146 m Ehrenbergh. there are fossil (see Scientific MEMOIRS & L'Institut.) that there are Tertiary fossil
320c275 Wilkinson on Cattle not abstracted Scientific MEMOIRS. published by Taylor Magazine of. Zoology & Bo
322c270 y of my subjects Elie De Beaumonts. 10 Vol. of MEMOIRS of Geology of France.= on Etna. Almost reread
508q016 MALE Troughtons.-- paper in Taylors Scientific MEMOIRS (11) And. Smith Savages at Cape any selection
093a033 ical. recent & bone bed.-- November 8th 1877 (MEMORANDA so far distributed to various subjects) Dr. A
520m002 ar in intellect frequently <are> have very bad MEMORIES for things which happened in early infancy--

Page
(Key Word)
```
614o037 th.: Many animals (as horses) very long & good MEMORIES-- but on its multiplicity & the comparison of
255c050 nown for its impudence. «This excellent case of MEMORY without association..» Instinct goes before at
292c171 much  over my view of particular instinct being MEMORY transmitted without consciousness «a most poss
292c171 habitual»  or of great importance to cause long MEMORY.-- structure is only gained slowly.-- therefor
293c172 al, but general ones might yet be transmitted.= MEMORY springing up after long intervals of forgetful
293c172 fulness.-- after sleep, «strong» analogies with MEMORY in offspring.-- Some association in such cases
293c173 anc it becomes web footed, now Man by effort of MEMORY can remember how to swim after having once lea
293c173 uld become webfooted & there would be no act of MEMORY.-- [There is no corelation between  individual
305c211 lations of the individuals, whereby choice with MEMORY, or reason? is necessary.--) which is modified
308c217 sometimes vibrate-- «seeing no tea brought back MEMORY» old habit of putting tea in pot made me go to
315c241 ons of plants «when exciting cause is absent» & MEMORY of animals.-- (surely in plants movements effe
315c242 tic nerves-- The vividness of first «thoughts» «MEMORY» in children or rather their memory. very rema
315c242 «thoughts» «memory» in children or rather their MEMORY. very remarkably-- scenes in themselves accide
521m007 r thinks is mentioned in the Zoonomia.-- Now if MEMORY «of a tune & words» can thus lie dormant, duri
521m007 le life time, quite unconsciously of it, surely MEMORY from one generation to another, also without c
521m008 bitual disease of the muscles.???> Miss Cogan's MEMORY of the tune, might be compared to birds singin
521m008 ing, or some instinctive <or> sounds.-- Miss C. MEMORY cannot be called memory, because she did not r
521m008 <or>  sounds.-- Miss C. memory cannot be called MEMORY, because she did not remembered, it was an hab
524m022 ulse.-- What fails first?-- How is this?-- Does MEMORY bring in old ideas <I have elsewhere  remarked
525m026 Maniacs» seem to have failed even more than the MEMORY.-- therefore affections effect of organization
526m028 er.-- it. will perhaps show differences between MEMORY & imagination. «Catherine thinks that children
526m029 y are all things, which are brought to mind, by MEMORY of the scenes, (indeed my American recollectio
526m029 the part of page.-- one is tempted to think all MEMORY consists in a set of sketches. some real-- som
526m029 some  real-- some fancied.-- this fact of early MEMORY consisting of things seen, quite agrees with m
530m044 fter journey, a fit of = gout, has affected his MEMORY of everything in <he a [...] Mr B> journey. sh
530m046 s habitual & involuntary,-- that is involuntary MEMORY, as in sleep.-- a new thought arises?? compoun
531m050 s where put one key. where all keys are placed) MEMORY cannot solely be number of times repeated, bec
533m061 lt it)-- this is kind of conscience, is obscure MEMORY of having read or thought of some such remarke
534m062 ge means of exciting association.)-- or of such MEMORY of such sensations, & memory is repetition  of
534m062 ociation.)-- or of memory of such sensations, & MEMORY is repetition of whatever takes place in brain
544m102 when awake.-- peculiar sensation as flying. (No MEMORY of past events?) or influence on our  conduct,
545m108 son of ideas-- one follows other as in blindest MEMORY-- also low faculty of understanding. Adam Smit
547m112 ofound ignorance in so simple case.-- There was MEMORY, for it related to past idea.-- there was a ki
547m113 yet  do so, from the exertion of keeping up the MEMORY of every late impression. & likewise gaining n
567n016 of not having <all> other trains of thought, or MEMORY from innumerable late events.-- the fatigue of
567n016 ction. one is obliged carefully to separate its MEMORY from all ordinary lines of association.-- is t
568n017 ted not to follow it, by greater temptation, if MEMORY of its own emotions. (which must be intimately
568n018 e versa pleasure in performing it.-- As soon as MEMORY improved. direct effect of improving organizat
569n021 & the associated pleasure &c accompanying such MEMORY.-- A Melody on flute & Epic poem, opposite end
575n043 ucture of brain.?-- is this more wonderful than MEMORY. affected by diseases. &c &c, double conscious
577n050 ill, no one could discover he had not it.-- The MEMORY of Plants, must be association,-- a certain ro
577n051 ns &c &c &c.-- bears. the same relation to true MEMORY, that the formation of a hinge «in a bivalve s
580n061 eligions.» Emma W. says that when in playing by MEMORY. she does not think at all, whether she can or
581n063 r life p. 207 March 16th.-- Is not that kind of MEMORY. which makes you do a thing properly, even whe
581n063 ting for the surname, analogous to instinctive MEMORY, & consequently instinctive action.-- Sir.  J.
584n072 an man, may be called instinctive: the facts of MEMORY of roads long after once visited by horse & do
584n073 orses & dogs) shows it is somewhat analogous to MEMORY. Shrugging shoulders seems sign of helplessnes
588n090 d, that animals have reasons, because they have MEMORY.-- what use this faculty if not reason.-- or d
601o009 Buffon)  Consciousness is sensation No. 2. with MEMORY added to it, man in sleep not conscious, nor c
601o009 .-- Consciousness bears some relation to time & MEMORY Reynolds X discourse very curious as showing "
610o031 s Wilson!!-- How curious an inward. unconscious MEMORY.-- Jan 14th. 1839.-- My father says he has hea
613o035 r in sleep is man momentarily conscious, but is MEMORY gone?-- Where pain & pleasure is felt there mu
614o037 good life [RHC] Instinct appear like heredetary MEMORY; but first memory in many cases cannot be acqu
614o037 stinct appear like heredetary memory; but first MEMORY in many cases cannot be acquired by experience
614o037 child  sucking.-- And is it more wonderful that MEMORY should be transmitted from generation.; than f
614o037 experience gained by man in lifetime Heredetary MEMORY not so wonderful as at first appears, & no too
614o037 , & no too great advantage.; for superiority of MEMORY does not depend on its length.: Many animals (
614o037 ts.-- this loss is compensated by vast power of MEMORY, reason &. & many general instincts, as love o
615o038 Instincts,  certainly appear a sort of acquired MEMORY. a permanent secretion of thought, (or under c
615o038 performed  by your parents, conscience This «X» MEMORY especially «the» general kind taking  pleasure
616o038 en we <return> turn into angels. this imperfect MEMORY may become perfect & we may look back to defin
616o038 inite action or to our conscious selves.-- Such MEMORY may go back to animals which were changed into
616o038 their reward! X Perhaps should hardly be called MEMORY; you cannot call the frame of mind which makes
616o038 the frame of mind which makes music pleasant, a MEMORY; yet that frame is enhanced by memory of  what
616o038 easant, a memory; yet that frame is enhanced by MEMORY of what has been heard; so love of virtue enha
616o038 e of virtue enhanced by this heredetary kind of MEMORY.-- The difference between heredetary memory &
616o038 of  memory.-- The difference between heredetary MEMORY & individual secretion of thought, may be no m
616o039 id that thought perceptions will, consciousness MEMORY &c. have the same relation to a living body (e
616o039v f this> is entirely wanting by which thought or MEMORY. might be in like manner attributed to the bra
618o041 o this in the relation of thought, perceptions, MEMORY &c. either to our bodily frame or the cerebral
620o044 ly thinks no more of it.-- Not so man, from his MEMORY & <pow> mental capacity of calling up past sen
178b032 .-- Dr. Smith says he is certain that when White MEN & Hottentots or Negros cross at C. of Good. Hope
179b033 f species in wild state.-- No doubt «C. D.» wild MEN do not cross readily, distinctness of tribes  in
179b033 a tendency in plants hybrids to go back?-- If so MEN & plants together would establish law. #as above
188b068 s same bearing with Dr. Smith's fact of races of MEN tendency to keep to one line Dr Smith says very.
206b147 period  though long distant, when of the present MEN (of all races) not more than a few will have suc
207b152 from Hobart Town-- {from difference of races of MEN and animals} See R. N. p. 130 Speculations on ra
213b169 ecies. in domesticated <species> races.-- If all MEN were dead then monkeys make men.-- Men makes ang
213b169 fowls.-- If all men were dead then monkeys make MEN.-- Men makes angels-- Those species which have l
213b169 -- If all men were dead then monkeys make men.-- MEN makes angels-- Those species which have long rem
216b181 fowls.  rabbits cats &c &c.-- When black & white MEN cross some offspring black others white which is
216b182 r. Smith considers the Caffers (like Englishmen) MEN of many countenances, as hybrid once. Is not thi
218b189 s made. parents do not cross-- we see it even in MEN); the possibility of Caffers & Hottentots coexis
220b198 of plants going back hybrid plants; analogous to MEN. & dogs. Now if we take structure as criterion o
242c017 ysical change, slow & offspring not picked.-- as MEN do. when making varieties.-- Voyage of Coquille.
259c063 habit.-- perhaps in process of change.-- Are any MEN born with any peculiarity. or any race of plants
261c070 &  fowls-- When talking of races of Man.-- black MEN, black bull finches from linseed-- not solely of
276c123 .-- then add chief good of individual scientific MEN is to push their science a few years in  advance
276c123 ce only of their age. (differently from literary MEN.--) must remember that if they believe & do  not
281c138 ropagation of forms.--just same way as <we> «all MEN» not all equally related to each other I  cannot
281c138 logy might be traced between relationship of all MEN now living & the classification of animals-- tal
281c138 ing & the classification of animals-- talking of MEN as related in the third & fourth degree.-- a spe
282c140 he close relationship in some one.-- imagine the MEN to have greater power of change yet. as external
282c141 nt.-- To make this case perfect, we must suppose MEN instead of mere colour & trifling form & head &c
282c141 ave these analogies.-- We must two races of such MEN living in same country but separated, now if one
282c142 e second race would not obtain a cast of washing MEN, but might hire the preexisting race, thus  the
292c171 ithout consciousness «a most possible thing. see MEN walking in sleep».-- an action becomes  habitual
293c174 ilar. Curious instance of difference in races of MEN.-- Wax of Ear, bitter perhaps to prevent insects
301c198 ly same way not possible to say what habitual in MEN & what reasonable-- Same action may be either in
303c204 , & Europaean forms on that isld.-- The races of MEN differ chiefly in <size> colour, form of head «&
307c215 abortive organs in some Males animals, Mammae in MEN, capable of giving milk)» rudimentary wings.  so
311c226 easily  mostrified, which last produced --insane MEN in civilized countries-- this is well worthy of
312c232 ies, with no other means whatever.--» Individual MEN «& animals» could only exist by habit.-- therefo
313c235 species may have had their infancies, as well as MEN-- when habits much more firmly impressed.-- we s
339d025 S. New Zealand & New +++ Caledonia. two races of MEN, but not plants» will it hold good.-- Thinks Tem
358d085 eding in & in (like «courage in dogs» EFFEMINATE MEN).-- if carried much further, if by the process t
358d085 t & fowl when crossed never even lay eggs. & the MEN cannot «hardly» tell any sex by appearance.-- Th
362d099 n (as Mitchell remarks seen in savages) to brave MEN.-- Effect of castration horns drop off., replace
```

(Key Word)
365d107 tric> those <restr> restricted in their range by MEN & by art.-- the former only giving average of ef
375d135 no one clearly perceived the great check amongst MEN.-- «Even a few years plenty, makes population in
375d135 -- «Even a few years plenty, makes population in MEN increase, & an ordinary crop. causes a dearth th
376d137 Monkeys had no curiosity to pull up trousers of MEN. Evidently knew <men> women, thinks perhaps by s
376d137 sity to pull up trousers of men. Evidently knew <MEN> women, thinks perhaps by smell.-- but monkeys e
384d162 females., & female plumage in castrated male.-- «MEN giving milk-->> Sept. 25th Young man at Willis «G
387d169 receives <young> «eggs» in belly.-- analogous to MEN having mammae.-- There is an analogy between cat
387d170 n caterpillars with respect to moths, & monkey & MEN.-- each man passess through its caterpillar stat
406e036 ½ way. Ed. New-Phil. Transact. Rabies, common to MEN, dogs, horses cows, pigs & sheep.-- diseases com
406e036 horses cows, pigs & sheep.-- diseases common to MEN & animals cow pox.-- case in Spain of pustular d
414e063 f his breed is destroyed)-- -- When two races of MEN meet, they act precisely like two species of ani
414e064 ralia to the white.-- The peculiar skulls of the MEN on the plains of Bolivia-- strictly fossil «& in
417e079 & individuals of same species.--). some races of MEN. D'Orbigny. affect the common progeny more than
451e178 a & die.-- On the other hand there are breeds of MEN the Thârû & the Dhangar who can live there & do
455eIBC eighton or some one. Father-- diseases common to MEN & animals.--:likenesses of children CD[Does any
461t037 ount it will be so handed down«(».. as mammae of MEN «callosities on Camels & Horses--».--«(» & there
462tf05 n politics in relation to the different races of MEN, some more intellectual than others-- is inciden
465t081 ofound consideration of method by which races of MEN have been exterminated (see Pritchards paper) (E
503q012 ing them back to back 37 Col. Sykes fertility of MEN & Europaean animals in India?-- about Chetah & o
508q016 & father answered (5) About cross-bred races of MEN taking after sex. A Smith. About species of Rhin
509q017 orner. On Mr Tremenheres Scottish Colliers, when MEN & women have long worked, whether children, who
525m026 ina with her puppy.-- The common remark that fat MEN are goodnatured, & vice versa Walter Scotts rema
533m059 pproach to <the> such instincts which full grown MEN can experience-- Instinctive walking of animals.
542m094 ldren. puppies do so dogs nearly silent, so with MEN.-- How is crying-- peculiar not common?--» no ba
544m102 ept at this time. how does he account for dogs & MEN speaking in their sleep.-- Characters of dreams
549m120 ppiness. though he sees some «intellectual» good MEN, from insanity &c unhappy-- perhaps not so much
549m122 e, & then acting on it, will add to happiness.-- MEN having some instincts as revenge «& anger», whic
555m143 ld world ones move skin of head & ears,-- ∴ some MEN have this power abortive muscles) The black Spid
556m145 es.." I think this cannot be disputed anymore in MEN. than in animals.-- In the drawings of Voltaire
560m156 nails--, & of eyes.-- Do female monkeys care for MEN.-- Have we any ferns in the hothouse at home Nat
571n028 y do some children acquire it soon. & why do all MEN. agree ultimately?-- We acquire many notions unc
580n062 s that difference of intellect between animals & MEN only in Kind.-- probably very important work.--
581n065 -- CD[I may put the argument,, that many learned MEN seem to consider there is good evidence in the s
585n077 tonge henge raised, yet not instinct, but if all MEN placed stones in same position, it would be inst
587n088 ell as remember it.-- Why do children pout & not MEN-- orang-outang & chimpanze. pout.-- Former, whin
593n107 Does music bear any relation to the period when MEN. communicated before language was invented,-- we
593n109 ood-- then man, a socialist, does not know other MEN by smell, but by looks. hence. some obscure pict
593n109 t by looks. hence. some obscure picture of other MEN. & hence idea of beauty.-- the social affections
600o08a mind of animal was not closely allied to that of MEN, when the five senses were the same-- In its act
603o012 ans Jews, African all sacrifices. How completely MEN must have personified the deity.-- H. Tooke has
605o020 ptible of pleasure from these causes who are not MEN of taste & the reverse of this. taste evidently
610o033 knowledge without experience. so there may be in MEN-- which the reviewer seems to doubt. [RHC] Ef
615o038 rresponding change in «external» man; and as all MEN nearly same species, so general instincts nearly
620o045 instincts pointing out lines of conduct to other MEN, [RHC] 5) which are natural (& which «when prese
621o046 then receive the moral approbation of his fellow MEN.-- [RHC] 6) Hence man must have a feeling, that
070r155 alf across the continent.--He states plains of MENDOZA smooth. Sir W. P. states that in Helm's travel
071r158 g in.-- Earthquakes at Quito. tranquillity «at MENDOZA» exception.--«formerly perhaps otherwise» Mend
071r158 doza» exception.--«formerly perhaps otherwise» MENDOZA never overthrown,--no mountains Mackenzie has
105a071 tion -- no one will dispute. sea. once came to MENDOZA--. Will they introduce other causes to explain
315c241 e of Japan. I do not understand any new ones.-- MENOIR will be published St. Petersburg Academy Impe
195b099 ogenitor-- or it may be of use.-- like Mammae on MENS' breasts.-- How does it come wandering birds. s
409e047 case to man, exceeding monkeys;-- Having proved MENS & brutes bodies on one type: almost superfluous
536m071 on licking udders of mare strictly analogous to MEN'S affect for womens breasts.∴ Dr Darwin's theory
570n026 y of blushing solves this.--» The similarity of MEN'S reasons: shewn by similarity of the earliest ar
380d155 ys some multiple of seven-- if woman does not MENSTRUATE in the month, she will in 5 weeks.-- A Bull
263c078 definite instincts.-- this is a replacements in MENTAL machinery-- so analogous to what we see in bod
291c166 st!-- Read Barclay on organization!! Avitism in MENTAL structure or disposition. & avitism in corpore
301c199 easily acquire habits is this the Key to their MENTAL powers.? p. 8. mistakes of instinct are extern
336d018 . expression of countenances, organic diseases. MENTAL disposition, stature, are slowly obtained & he
378d148 of Old Species transmitting so much longer its MENTAL peculiarities. Wildness Reversion Q [not locat
520m001 8 My father says he thinks bodily complaints «& MENTAL disposition» oftener go with colour, than with
520m001 his father in body, but his mother in bodily & MENTAL disposition.-- My father has seen innumerable
522m012 had been disturbed.-- These foregoing cases of «MENTAL» failure very general effect of <early> «sligh
541m090 allen, as in my Glen Roy paper.-- this greatest MENTAL effort, of which I am capable-- I suspect from
546m111 , went out & found it there!!! Lady in perfect «MENTAL» health.-- «Erasmus had almost same thing happ
547m111 tion» Mayo Philos. seems certain that muscular, MENTAL, <&> digestive nervous influence replace each
548m114 h step as in reasoning-- hence delirium & sleep MENTAL rest. though. most vivid & rapid thought.-- Th
550m124 » is large proportion of pleasant to unpleasant MENTAL sensations in any given time «-- compared to w
550m125 sations <it> when excited by impressions, & not MENTAL or ideal ones, <which> «& these» must occupy g
557m150 person tremble, like fear.-- Why does any great MENTAL affection make body tremble. Why much laughter
571n029 that is habitual, if hereditary, is pleasant.-- MENTAL & Bodily Consider case of grazing animals know
572n032 akness & inferiority, but now we carry it on to MENTAL inferiority-- when we do not expect any bodily
575n045 burns end of nose with a hot razor.-- joy <p> a MENTAL pleasure. with pleasure of senses. The shudder
577n052 ve people apt to blush.-- -- The power of vivid MENTAL affection, on separate organs most curiously s
582n067 f heredetary habit:-- there must, however, be a MENTAL impulse (though unconscious of it) to move its
582n068 ;-- they are more truly sensations??. a kind of MENTAL pain & pleasure.-- The Revd. Algernon Wells Le
587n088 onary & make a list of every word, expressing a MENTAL «desire» «quality» &c &c Mackintosh Ethics p.
588n090 do. p. 306 "the eyes are rolled upwards during MENTAL agony, & whilst strong emotions of reverence &
606o20v he term taste is metaphorically applied to this MENTAL power Although taste must necessarily be acqui
608o026 e that it is good for him effect of Education & MENTAL capabilities.-- '(P) Animals do attack the wea
620o044 re of it.-- Not so man, from his memory & <pow> MENTAL capacity of calling up past sensations, he wil
618o41v JCD Cause and effect has relation to forces & MENTALITY because effort is felt [LHC] 1) May 5th. 1839
583n070 d to reason than instinct.) Mr Wells I can see MENTALLY refers by reason knowledge gained by reason:
041r064 bbles & vertical trees Grand Seco at B. Ayres; MENTION about the deer approaching the wells.-- the ef
041r065 paper. when discussing probable rise of land: MENTION M. Gay's fact about shells: Hibernation of fre
042r067 pause» When mentioning pumice of Bahia Blanca, MENTION black scoriaceous rocks of R Chupat. & fall of
048r087 g of such substances being worn into channels. MENTION submarine channels. such as that in front of S
094a035 eory of changes. of granites into Trachytes.-- MENTION Osorno in lake. few Volcanos now in lakes.-- M
120a110 ion of rivers.-- Excess of matter brought down MENTION absolute elevation of Patagonian blocks (1200
275c121 lue Pouters & small Bald Heads Mr Yarrell will MENTION in his work I am sorry to find Mr Yarrell's ev
276c123 ween England & France. I will do with forms.-- MENTION persecution of early Astronomers.-- then add c
332d001 g> «unknown» causes act upon people. My Father MENTION, than for ten years he never saw one case of m
041r064 er of the Salado.--Mem. in Owens Africa it is MENTIONED that the Elephant came towns driving by the w
041r065 s. years of drought are common.--Mr Lyell has MENTIONED the drifting of carcases putrid. In Rio paper
043r070 Mississippi & New Madrid & Caraccas.-- Is this MENTIONED by Humboldt in his account of extensive areas
045r075 t would be invariably discovered; this may be MENTIONED with general slope of the country; (perhaps g
051r094 eas (& even turbulent as at St Helena) I have MENTIONED point of greatest action; I now having seen P
053r100 e great conglomerate of the Amazons & Orinoco MENTIONED by Humboldt under name of Rothe-todte-liegend
215b178 y. The case of the tailess cat of Isle of Man MENTIONED in Loudons (analogue of Blood hound-- Bull. S
223b209 Bolivian human species?-- Small «new» animal MENTIONED from Fernando Po Zoolog. Proceedings October
229b233 extreme South coast. Elephant he believes is MENTIONED by old writers on extreme Northern Coast. Hip
229b234 ransport.-- <tr but> Get him to discuss those MENTION[ED] by Lesson & Chamisso.-- Geograph Journal Vo
234b256 troduced by ice <no> «only few» pigs.-- birds MENTIONED. but few.-- CD [There was notice in report of
285c150 rochets about instincts. whenever instinct is MENTIONED some definition must be given It would not be
298c189 out any leading question.-- [+] This might be MENTIONED in note.-- try to trace from simplest reasoni
318c253 k swallows»-- degree/ migrations of birds he MENTIONED many most curious case-- -- the birds seem to
341d030 aid in breathing.-- Animals from Hobart Town MENTIONED, it seems most of species from there now foun
364d104 t of imagination on mother. white peeled rods MENTIONED in old Testament placed before sheep-- it has

Page **(Key Word)**
364d105 - In Scandinavia besides the Rakhekna, before MENTIONED between Capercailzie & Black Cock.-- The latt
379d152 telet papers are given, & I think facts there MENTIONED about proportion of sexes, at birth & causes.
420e090 um .1839. p. <8>36.-- A crustaceous animal is MENTIONED which inhabits the Pinna of Rio Janeiro, (lik
439e143 important Mr Herberts fact about the hybrids (MENTIONED in letter to Henslow) fertilizing each other,
441e148 transmitted, for the changes in fruit trees. MENTIONED by Mr K may be caused by the diversity of sto
445e158 ly.-- "whilst the shell stuck to its tail" as MENTIONED by Sir. J. Banks. p. 212.-- p. 282. Allows th
447e165 case precisely analogous to the Canada onion MENTIONED in Hort. Transact. Aira caespitosa becomes vi
451e178 dgson. p. 336 In the most pestiferous region (MENTIONED by Heber) «from» which «all» mankind «(& yet
465t089 1840 p. 362.-- some Mammals of Norfolk Crag. MENTIONED-- allied Beaver to present forms.-- -- How ma
482z011 Yagourundi near Concepcion!!!.-- (no species MENTIONED) p. 205. only 9. Terrestrial birds at Falklan
491q01v any effect, as in case of woodpidgeon & Hen. MENTIONED by Mr Knight. Vol IV Hort. Transact.-- 4 May
491q002 flowers, like the Oxalis from C. of Good Hope MENTIONED by Mr Herbert in vol IV. Hort. Transact.-- 4.
521m007 - habitual fits.--» which my Father thinks is MENTIONED in the Zoonomia.-- Now if memory «of a tune &
522m012 p pain. or pleasure. & is often recurred to & MENTIONED as a thing which had just taken place.-- as i
530m044 revents repetition of idea.-- Mr Blakeway has MENTIONED in Antiquities of Shrewsbury something about
605o19v «this» emotion, from the associations before MENTIONED. we call sublime.-- It appears to me, that we
622o049 avarice. &c &c.-- [RHC] 8) in the beginning I MENTIONED only three instincts.-- I am far from saying
042r067 talks of line of cliff marking a pause» When MENTIONING pumice of Bahia Blanca, mention black scoria
021r005 ment of world.-- [not located] La. billardiere MENTIONS the floating marine confervae, is very common
027r023 nts take place in semiconsolidated rocks P xv. MENTIONS in what formations Conglomerates are found. --
048r084 gle elevations along whole line of coast Darby MENTIONS beds of marine shells on banks of Red River L
071r157 with great destruction of human life.--Temple MENTIONS some earthquake at Cordova.-- There the Cordo
404e026 bituated. instance such as Hunter, or some one MENTIONS of influence on parent affecting offspring.--
478z003 small shells (patella) sea weed & many pebbles MENTIONS stinging Millepora. Quoy. Freycinets Voyage V
512q019 nt, birds from Azores or Madeira Mr. Blyth (1) MENTIONS some breeder who raises many English birds--
633j53v his metaphorical pace.-- p.-- Kolreuter MENTIONS some hybrid, whose flower great tendency to b
635j56v my theory) Macculloch. Attrib. Vol I. p. 330. MENTIONS the many cases, as in Papilionaceous flower,
504q013 hose stamens move one after other to flower & MENYANTHES whose pollen bursts before flower is open--
485z017 a?? See Report <by> on D'Orbigny on species of MEPHITES 4 distinct Camelidae. do not breed together M
058r118 ente plus ou moins inclinée vers le rivage de la MER, tandis, au contraire, que vers le centre de' l
059r119 aralleles et inclinées du centre d'l'ile, vers la MER; ces couches ont entre elles une correspondance
447e164 ise grasses. &-- very heavy seeds.-- as Cocos do MER.-- Analogy shows some most important end.-- Fest
207b151 rcel Serres p. 331. L'Institut-- considers that MERCU> Geo. Joun. p. 325. Vol. IV. Ducks on rivers in
499q09v out proportion of plants necessary &c &c (a) MERCURIALIS-- Frog Bit. Valerian-- Urtica Dioica Sorrell
080r178 er & black lead unite chemically like grease & MERCURY [blank] NB. P. 73. General reflections on the
093a032 Serres L'Institut. 1837. p 331 Considers that MERCURY & Sulpuret of Iron has been sublimed into the
038r058 hyries: certainly appearance leads me to believe MERE fissures filled up.--the appearance will here b
049r089 Granite is much more perfect. than in believing MERE agency of dikes: & indeed when do these dikes l
051r094 fall: Excepting by removal of large fragments by MERE force of waves: & action on upper tidal band, I
058r117 mena lasting so long.-- The great movements (not MERE patches as in Italy proved by Coral hypoth. agr
061r127 vated. then peculiar plants created. if for such MERE points; then any mountain, one is falsely less
070r154 thick would there be undulation? would it not be MERE vibration? but walls & feeling shows undulation
106a075 of wider.-- This applies to all vallies (except MERE talus «over cliffs edge» of which limit cannot
130a133 yield easily. & if easily must be thin: <beside MERE fracture> A Elevation as in Patagonia {P} B sub
145g001 Agricl Dec 1837 Yet instances given against it-- MERE fact of many races of Animals in Britain shows
192b088 ill more so «known» with fishes & reptiles.-- In MERE eocine rocks. we can only expect some steps.--
195b104 replace them-- two hypotheses fresh creations is MERE assumption, it explains nothing further, points
206b148 ying races of man: these races may be overlooked MERE variations consequent on climate &c-- the whole
211b162 y of rudimentary bones.-- As Waterhouse remarked MERE length of bill does not <indicate affinity with
228b230 &c produce real crabs, & in each case similar or MERE mongrels.-- It really would be worth trying to
230b240 re different then, they will be called species & MERE producing fertile hybrids will not destroy that
262c073 t as for number five in each group absurd.-- the MERE fact of division of lesser & more power (2.typi
265c084 ld is.-- shows that generation implies more than MERE child, but that child should produce like child
268c100 fitted for transport ¿may it not be explained by MERE chance?-- or it like each great class of animal
282c141 his case perfect, we must suppose men instead of MERE colour & trifling form & head &c to become grea
294c178 to grow far N. becomes stunted, altered, & lose (MERE sickness)? fertility ¿because offspring too unl
302c200 rful, partly explained on my theory, = otherwise MERE fact creator chooses so to create.-- It is very
305c209 ular coincidence if distinct creation.-- ie.-- a MERE statement nothing is explained.-- this is fact
307c216 es.-- for instance where a tendril passes into a MERE stump.-- Shall abortive organs «of very same ki
350d054 d Man first civilized. <note> add this in note. ¿MERE conjecture?-- Australians.-- Americans. &c Afte
365d107 which has been adapted to former changes. than a MERE monstrosity propagated by art. Yarrell told me
404e025 s p. 330 distinctly states that the flipper is a MERE simple modification of an organ present in whol
410e052 nt & will be done to all countries,-- but naming MERE «single specimens in» skins worse than useless.
410e052 -- A philosopher, would as soon turn tailor, as, MERE describer of species, from its garments, withou
419e087 sumption well grounded, on time;-- therefore the MERE loss of species, which may be the work of a few
463t063 nd sentence about the Animaux fossiles-- being a MERE fragment of the discoveries to come-- Owen in h
465t079 ow-- -- species in S. America. -- so see what a «MERE» vestige, is preserved in this country-- same a
472s01v ar. Little variation in the 60 one brighter with MERE traces of black spot at base, one paler with te
507q15v Bull Rush-- Dandelion-- Sycamore. & seeds with «MERE» border-- & Humboldts spinning seed.-- (50) Any
515q021 2d if so, whether concepcion takes place,-- the MERE fact of seeds ripening has scarcely any no rela
528m036 eeing artificial lights in the night.-- from the MERE exercise of the organ of sight, which is common
534m061 on of images before your eyes, or ears (language MERE means of exciting association.)-- or of memory
558m151 s from language.-- the social instinct more than MERE love.-- fear for others acting in unison.-- act
565n006 fear, no actual intention to bite at moment, but MERE symbol of readiness, & therefore done in extrem
569n021 ween one or two absent things.-- reason probably MERE consequence of vividness & multiplicity of thin
586n082 ged by eye, & use of limbs &c, or it result from MERE impulse to save wax.]CD which it instinctively
588n090 of reverence & piety are felt." it appears to me MERE consequence of stooping, as sign of humility.--
617o40v his metaphorical mode of stating the <force exhibited in every> phe MERELY
023r011 an. & the present Volcanos have been said to be MERELY accidental apertures still open.--The fault li
038r057 on it will be better not to refer to Lyell. but MERELY to state these reasons, & saying that they ref
165g125 eam enters at head of which hill is round & not MERELY thoughts laying dormant-- Man from Glen Turret
219b194 n, have made plants of American & African form, MERELY because intermediate position.-- We cannot con
230b240 being subjected to change in Americas. perhaps MERELY gone back previous to fresh change Get a good
231b243 which migrated a little to the southward would MERELY be specifically different if so.-- Now this is
253c047 SS lion of Guzerat by Capt. W. Shee. considered MERELY variety.-- red form of skull very slightly dif
282c143 rit of life must be every where ambient. & <in> MERELY determined to such points by the vital laws.--
304c206 ystematist, who believed that all his divisions MERELY marked his own ignorance.-- the collector who
312c231 agreeing almost to the point in question.-- «--MERELY picking opposite qualities, with no other mean
337d021 not go back to first stock of all animals, but MERELY to classes where types exist for if so. it wil
352d060 ellent view of geology, of each formation being MERELY a page torn out of a history, & the geologist
353d060 ace the structure of animals & plants.-- he get MERELY a few pages.-- Hence (p. 59) looking at animal
368d114 t.-- (do the females then fight for male) & are MERELY most attracted).-- -- singing best sign of most
389d177 s. perhaps connected with this same case (& not MERELY as I have stated it) it is certainly very rema
390d179 uadrupeds the effects of breeding in, it is not MERELY the too close animals, which will not breed, b
402e017 ta, can only be explained, by such strata being MERELY leaf, if one river did pour sediment in one sp
421e091 the modifications of these organs to mistake a MERELY adaptive to an essential character--" How litt
430e120 rtional dimensions, must <almost> be considered MERELY varieties. & even Mr Sowerby is coming to this
434e127 ds in altering <or> form of body, or whether it MERELY has tendency (as effects of cultivation on suc
506q015 .-- or FEW. or bad seeds formed; badness may be MERELY not ripening-- (38) have Dioecious plants any s
515q021 flower-- do they live healthily, or does fruit MERELY not ripen.-- The point to attend to is whether
533m061 How did my mind feel it was wrong (& it was not MERELY morally wrong, but hurting my character I felt
540m084 Many of actions as hiccough & yawn are probably MERELY coorganic as connexion of mammae & womb.-- We
541m091 ping one idea before your mind steadily., & not MERELY thinking intently; for that one does with nove
543m101 rong argument for brain bringing thought, & not MERELY instinct, a separate thing superadded.-- we ca
551m128 n> when something good was shown him, than when MERELY ordered to do it.-- Plato «Erasmus» says in Ph
568n019 »-- Old man at Cambridge observed the ignorant. MERELY looked at picture as works of imitation.-- Hen
586n082 s» instinctive, namely the knowledge of size is MERELY judged by eye, & use of limbs &c, or it result
595n117 te> with <lather> <foame> & sweat, when hearing MERELY hunting horn-- association or imagination [bla

Page
(Key Word)

618o41v nt reduces itself to what is cause & effect: it MERELY is «invariable» priority of one to other: no n
623o050 have not received pleasure from the place, but MERELY in the place. & yet place calls up pleasure.--
633j54r of one his weak creations.-- All such facts are MERELY relations of one general law. the plants were
599o008 y other animal-- Aimé Martin de l'Education des MERES Vol. I. p. 198.-- "Moralité, raison, beau ideal
601o08a - does when Master takes Hat de l'education des MERES par L Aimé Martin Leroy Lettres. Philosophique
482z012 from Galapagos!!! Azara. Voyage dans l'Amerique MERID. Tatu noir. abundant from Paraguay to 27 degree
449e173 Paris Museum.-- Athenaeum: 1839. p. 451. Sheep MERINOS from Cape of Good Hope. has different constit
356d069) the line of proof & reducing fact to law only MERIT if merit there be in following work.-- The hist
356d069 e of proof & reducing fact to law only merit if MERIT there be in following work.-- The history of Me
538m079 , some quarrelsome as B.e on board Beagle, some MERRY goodhumoured as self.-- «When Miss Cogan has re
332d003 Worm Fever, as used much more latley diseased MESENTERIC glands.-- My Father has seen case of pleuris
303c205 L'Institut 1838. p. 128. Extraordinary genus. MESITES bird from Madagascar uniting pidgeons & gallin
070r155 atter.-- According to Mr Brown, a person (whom I MET at S. W. P.) the Cordillera extend to near Salta
317c249 dix (& only appendix) of Congo Expedition, NB. I MET an old man--, who told me that the mules between
342d034 is uncertain-- Yarrell remarks he has somewhere MET conjecture that all salt-water fish were once sa
349d052 variety & species!! (given in note.)-- Macleay <MET> uses term genus when it is so many steps from a
402e018 eing the same as the fox-like animals. which are MET with near Canton" "Here, as in all Malay countri
582n069 -- p. 14. There is scarcely a faculty in man not MET with in the lower animals.-- hence the general a
615o36v ink Pincher shows surprise, walking home one day MET him, with Mark riding instantly followed, me and
510q017 orked have any peculiar configuration.-- Hooker <META> Metaphysics of Morphology. ‖-- Schelgel is he
078r175 mation ‖ Guanaxuato, which has yielded the most METAL, where the direction of ravins, and the slope o
078r176 in hornblend & vitreous felspar".-- p. 215 Same METAL in Tasco vein in Mica Slate & overlying Limesto
125a121 y it to thickness of crust.-- {P} if crust were METAL then thinner if better conductor, then still th
126a122 wrong, would our judgement be-- Does condensed METAL, conduct heat better than plain?-- Mem 1000 {P}
126a124 his latter view the rod is reversed, upper part METAL «conveying heat in one direction only, like wat
618o044 ar & rounded.-- Fox Philosoph. Transactions on METALLIC veins. 1830 P. 399.-- Carne. Geolog. Trans: C
045r074 bubbles on the surface bespeak the changes. -- METALLIC veins solution of silex & many other phenomen
046r078 r? How are they eliminated.--«Sulphur last.--» METALLIC veins likewise must separate ingredients if w
056r111 separates some sulphur (perhaps lime) salt. & METALLIC ores.--which mingling & separating is well ad
074r165 relation)--Galapagos vein. vein of secretion.--METALLIC veins follow mountain chain. there after NW <
103a065 horizontal movement) hence generally intersect METALLIC dikes: It is an important view being subseque
137a149 & 162 [blank] Ed. New. Phil J. 1838. p. 72. on METALLIC vapours condensed from furnaces do/p. 84 on t
378d147 -- «thus young of» Many of the pies assume the METALLIC tints, such as Magpie, Jay, & perhaps all the
378d147 & perhaps all the rollers-- «He says» whenever METALLIC brilliancy is present in Young birds, one may
383d160 .-- lower part of breast, each feather is fine METALLIC green. <from> with tip & part of shaft metall
383d160 etallic green. <from> with tip & part of shaft METALLIC green.-- This green doubtless is effect of Me
383d160 lic green.-- This green doubtless is effect of METALLIC hue of silver pheasant. yet why green? & not
053r100 ranite & sedimentary rocks, in Alps becomes METALLIFEROUS. Vol III Latter Part Are there Earthquakes
064r136 rge difference of plutonic rocks & Volcanic METALLIFEROUS-- Urge enormous quantity of matter from CRE
068r145 s of elevation.-- Lyell suggested to me that no METALS in Polynesian Islds--. Volcanic plenty in S. A
073r162 ric stones simply pitched from mo«o»n, that the METALS should be those which have magnetic properties
073r163 Barytes:-- Humboldt. New Spain. Vol III. p. 130 METALS in Mexico rarely in secondary alway in primiti
074r165 as. «Mem: Micaceous iron ore.» N.B. To show how METALS may be transported by complicated chemical law
090a021 d Planet, that inside our globe melted magnetic METALS. ∴ earthy crust compared to those of falling s
092a029 s shewn how electrical currents tend to deposit METALS, if in solution. My view of metamorphic in con
123a116 ation & separation in volcanos.-- if so why not METALS. The theory of veins will, I suspect be greatl
124a118 101. On Frozen Soil of Siberia (with refer to METAMOR) wrong entrance Athenaeum. 1838. p. 652. Dr. D
146g010 The hill has been well-- denuded.-- «of hard METAMORPH» path only covering Great Slip, 10 years sinc
068r145 ian Islds--. Volcanic plenty in S. America!! METAMORPHIC Volcanos only burst out where strata in act
071r156 must formerly have been most abundant tree-- METAMORPHIC action: <most> coming so near surface most i
088a016 ant V. Malte brun -- Main character of Andes METAMORPHIC action-- Mem: red sand of Europe no fossil
089a017 nsed before rising?-- Mem. granite heated.-- METAMORPHIC action in red sandstone.-- Certainly Volcani
092a029 o deposit metals, if in solution. My view of METAMORPHIC in contradistinct to Volcanic will explain t
104a066 In Falkland islands. & generally where rock METAMORPHIC & thickness of <strata> not great, one can c
113a091 of temperature clearly prove possibility of METAMORPHIC theory On the idea of statical equilibrium,
114a094 of different composition using difference in METAMORPHIC action which I give at C. of Good Hope.-- A
119a105 nstant.-- It is to be profoundly considered, METAMORPHIC rocks at surface. & great heigth on mountain
119a106 dation from gneiss to granite shows that the METAMORPHIC rocks have just floated over the absolutely
120a107 s-- the equal movements of Glen Roy road. (¿ METAMORPHIC action at the bottom of the sea?) All this p
122a115 when fluid moved [...] {P} August 25. I saw METAMORPHIC conglomerates on shore of Loch Lochy very li
165g124 9.200 1.172 Loch Oich 92 each Loch 8 ft. The METAMORPHIC conglomerates near Loch Lochy would be well
221b201 so slow., that all have existed for ages as METAMORPHIC; & therefore according to Lyells doctrine re
269c101 d by fluid currents which, may take place in METAMORPHIC' action.-- Geograph Journal. vol I. p. 17 &c
040r063 we must imagine bituminous shales have been METAMORPHISED, as in Brazil feruginous sandy ones have un
114a094 & not ever in Secondary in Europe. gneiss-- METAMORPHOSED clay slate.-- --shale in small sea. Lyell c
114a095 that clay-slates have been more frequently METAMORPHOSED than other deposits.-- NB. because lowest.
115a096 e to have any horizontal non cleaving beds. METAMORPHOSED. The chemical action which gives polarity t
420e089 y.-- could move his ears The head being six METAMORPHOSED vertebrae, the parent of all vertebrate ani
443e154 propagated by gemmation do not now undergo METAMORPHOSES, but to arrive at their present structure t
445e159 Additions". p. 454.-- does really attribute METAMORPHOSES to habits of animals & takes series of flyi
221b200 .--!!«good» those which have undergone most METAMORPHOSIS Islds X Is this applicable to insects &c &c
351d057 s nothing in vain!! Foetus of man undergoes METAMORPHOSIS., heart altered & umbilical cord,-- Broderi
421e090 Natural History. (p 225. 1838.) account of METAMORPHOSIS in the young of Syngnathus.= curious as sho
461t025 supial & the other have young which undergo METAMORPHOSIS & are provided with fins, & hence do not re
269c101 surprise that it is not 700) is applicable to METAMORPHS theory suppose when rhinocerose lived. mean
616o039 hich action bears to the agent. Matter is by a METAPHOR said to attract; & hence if thought &c bore t
605o18v f these feelings & thus we apply to them the METAPHORICAL term sublime ? So that in this Essay. D. St
617o40v fundamentally accounted for, we prefer this METAPHORICAL mode of stating the fact to the mere statem
605o019 s causes of those sensations, which we call METAPHORICALLY sublime, but that it is through a complica
606o20v sness of the result, that the term taste is METAPHORICALLY applied to this mental power Although tast
539m084 dealt with p. 241 Origin of man now proved.-- METAPHYSIC must flourish.-- He who understands baboon -
564n005 n run even with its shell on back.-- To study METAPHYSIC, as they have always been studied appears to
599o005 henaeum "Smart-- Beginning of a new School of METAPHYSIC,"-- give my doctrines about origin of langua
270c104 This paper might be worth consulting, if any METAPHYSICAL speculations are entered in upon life. Name
327c265 Education preeminently worthy of studying in METAPHYSICAL point of view Henslow has list of plants of
535m069 «or God» secondly that these are replaced by METAPHYSICAL abstractions, such as plastic virtue, «&c»
599o004 & USELESS notes about the moral sense & some METAPHYSICAL points written about the year 1837 & Earlie
610o033 O p. 267-- says the great division amongst METAPHYSICIANS-- the school of Locke, Bentham, & Hartley,
553m135 mallest casuistical doubts.-- The history of METAPHYSICKS shows that such a view cannot be, anyhow, e
227b228 tincts, heredetary. & mind heredetary, whole METAPHYSICS.-- it would lead to closest examination of h
308c218 ference no where-- as long as this is so-- !!METAPHYSICS!!! Mrs Somerville, Connection of Physical Sc
510q017 any peculiar configuration.-- Hooker <Meta> METAPHYSICS of Morphology. ‖-- Schelgel is he serpent ma
520mIFC ivate Finished. Octob. 2d. This Book full of METAPHYSICS on Morals & Speculations on Expression -- 18
539m084 erstands baboon <will> would do more towards METAPHYSICS than Locke A dog whines, & so does man.-- do
563nIFC e in monkeys.-- Charles Darwin [Private.]CD (METAPHYSICS & Expression) Selected «for Species Theory»
073r161 ative iron that to which we cannot attribute a METEORIC origin & which is constantly found mixed with
073r162 ". p. 113 How utterly incomprehensible that if METEORIC stones simply pitched from mo«o»n, that the m
077r172 Does not iron, combined with‿nickel & cobalt (METEORIC) resist oxidation?-- Mem Sir W. P. stone It i
077r172 re. (? heat!) unknown to us. ‖ M. Chladni.--on METEORIC Mexican stone. Journal des Mines 1809. No. 15
090a022 t before.-- What must be the effect of all the METEORIC stone which must have fallen on the globe sin
137a151 8. p. 132. «& 134» Bischoff. On the effects of METEORIC waters on the temperature of the interior & p
090a021 era I see Brewster speculates from believing METEOROLITE but old Planet, that inside our globe melted
232b246 fossils with whole world. would be like in a METEOROLOGIC table «in comparison of temperature of two
022t008 ndes. In Dampier's voyage there is a mine of METEOROLOGY with respect to the discussion of winds & st
055t108 e of the great rivers prove better than any METEROLOGICAL table the precise periods over immense area
063r132 nt methods, associated life only adds one other METHOD where the division is not perfect.-- Dogs. Cat
107a078 os &c &c This view will bear much reflection on METHOD of cooling--Very difficult subject. PP-- I thi

Page
(Key Word)
```
243c020 d & New Holland fish very similar.-- NB. Lesson METHOD of generalizing without tables or references h
255c052 neration. The infertility of crosse & cross, is METHOD of nature to prevent the picking of monstrosit
289c162 r larvae) but the acts of condensing must alter METHOD of generation-- Heaven knows how.-- This react
290c162 is reaction takes place in every organ‖ «Hence «METHOD» of generation is very good «general» characte
293c174 e adaptations can hardly be accounted for by My METHOD of breeding there must be some corelation. but
372d130 arents have.-- -- Not this is effected by short METHOD in generation.-- Ehrenberg considers artificia
465t081 n go on to shells-- A profound consideration of METHOD by which races of men have been exterminated (
526m030 l,-- he improves the faculty according to usual METHOD, but what urges him,-- absolute free will, mot
590n093 rrhaea» & syncope p. 42. Sighing from grief. is METHOD of increasing languid circulation-- no, for <g
613o036 ts as fear of man,-- crows fear gun,-- pointers METHOD of standing,-- method of attacking peccari-- -
613o036 rows fear gun,-- pointers method of standing,-- METHOD of attacking peccari-- --retriever-- produced
641j29v nsects the end is gained by some very different METHOD. in pedunculated eye of Chamelion. crabs Crabs
063r132 s «each» one individual «divided» by different METHODS, associated life only adds one other method wh
278c130 logue of Mammalia arranged according to my own METHODS. Dasyurus being found fossil in Australia, & c
183b052 & common pheasants & fowls..-- "On sait que le "METIS" du loup et du chien, que celui de la chevre et
081rIBC French Toise into English ft. 0.8058372 French METRE into English ft. 0.5159929 {T} Toises Pieds Myr
081rIBC inch  Kilometre 513., 0. 5 Hectometre 5i. 1. 10 METRE 3. 0. 11 lig[nes] Decimetre 3. 8 Centimetre 4.4
429m115 at the Rhododendron ferrugineum. begins at 1600 METRES precisely & stops at 2600. & yet know that pla
179b033 . Smith always urges the distinct locality or METROPOLIS of every species: believes in repugnance in
508q016 hinoceros. becoming rare beyond limits of the METROPOLIS of each-- Cause?-- Andrew Smith. (6) What si
077r172 at!) unknown to us. ▌M. Chladni.--on meteoric MEXICAN stone. Journal des Mines 1809. No. 151. p. 79.
078r176 h some exception» directed NW & SE). «Vol III» MEXICAN Cordillera "immense variety of Porphyries whic
091a026 er by Humboldt on Quito Volcanoes & another on MEXICAN Trachyte <roc> lava called Andesite. Red Coral
128a130 I believe?? coast of North America., like the MEXICAN Gulf. is fouled by bars of sand & shallow lago
404e026 on of an organ present in whole class. Case of MEXICAN greyhounds.-- young being habituated. instance
528m038 beauty is shown by Humboldt from occurrence in MEXICAN & Graecian to be single cause) this symmetry &
072r160 . Spa. Vol III p. 113 "Nature exhibited to the MEXICANS enormous masses of Iron and Nickel, & these m
073r163 Humboldt. New Spain. Vol III. p. 130 Metals in MEXICO rarely in secondary alway in primitive & trans
073r163 gneiss with gold veins visible:--"Porphyries of MEXICO may be considered for most parts as rock emine
075r167 isted of native silver & brown oxide of Iron in MEXICO. sulphuretted silver, arsenical grey copper, a
075r167 rous lead»; sulfated Barytes very «un»common in MEXICO. Fluor spar only in certain mines. «Vol. III»
076r168 . «Vol. III» "In general it is observed both in MEXICO & Peru, that those oxidated masses of iron. wh
076r168 tersecting carbonate of lime-- native silver in MEXICO is always accompanied by Sulp. silver sometime
122a114 ider figure of earth statical.-- if platform of MEXICO owes its elevation to equilibrium.-- it cannot
227b224 connection of those two countries Hence India, MEXICO & Europe. one gret sea (Coral reefs'. shallow w
400e013 ides Galapagos do. p. 376. Isle Tres Marias off MEXICO with small Hares & raccoons.-- «S. American fo
072r159 coast line of Old Greenland, close to W of Jan MEXICO Isld.--Mr Barrow thinks N & S. line connects we
111a087 vidently dike. V. VII. p. 316 & 328 VI. p. 365. MEYEN on Chile must be studied Analysis of Voyage: ma
480z006 ccount of Condor by Humboldt Zoologie Recuiel-- MEYEN has written account of Guanaca. In transaction
481z009 also for Journal.-- 18 Admirable engravings in MEYEN Zoology on animal of Campanularia Alcedo stella
481z010 logy on animal of Campanularia Alcedo stellata. MEYEN p. 92.-- great Kingfisher of Tierra del Fuego k
481z010 th horns. Madrepores p. 26 Nullipora p. 29-- In MEYEN. Voyage round World German a reference to a lum
118a103 Geology of Chile.-- P p217. Pentlands Fossils & MEYENS --<Jura &> Chalk When we consider parallelism
585n075 wn its ear. good to contrast with horses, asses, <MI> Zebras &c &c.-- Here there is kicker but not bi
069r150 f the Portillo line.--connected with <gneiss>.--(MICA Slate) [Fig. 9] {P} ((3) like Bell of Quillota.
076r170 ne porphyrys & phonolites do. amphibole quartz & MICA very rare.-- ancient freestone & breccia is the
078r176 s felspar".-- p. 215 Same metal in Tasco vein in MICA Slate & overlying limestone Balls of Silver ore
097a042 il M.'Institut No degree 22I Lamellar dikes like MICA Slate Von. Buch. Canary Isd. p 170.-- Mem. Cord
104a068 ow (?) was astonished with him that <th> gneiss, MICA-slate of whole kingdom of Norway was contorted
110a086 irregular cryst of reddish felspar. & scales. of MICA.-- large cryst of Hornblende blending into base
110a087 rocks, most probably from the gneiss beds in the MICA slate-- Geograph. Journal. Vol IV (p 321) Mr H
115a097 very general absence of fragments «& pebbles» in MICA slate & gneiss, can only (see «supra» p 94) be
148g024 no  trace of them-- Mem Coast of Chile--¿ is not MICA Slate too hard & uneven to be impressed Case of
157g073 r,-- [no pebbles in parts of Beagle Channel when MICA slate, only sand blow away]CD where lines appea
159g084 is side (return) between 2d & 3d shelf Mountain <MICA> «composed of» Gneiss Block on 2d shelf & below
030r036 «at Guacho» «on N. Chile? Washington.--» Mem: MICACEOUS formation of Chonos. interesting from great q
074r165 lcanic & non Volcanic. Then Solfataras. «Mem: MICACEOUS iron ore.» N.B. To show how metals may be tra
202b131 nized»---; clay fifty feet, then forest 120 ft MICACEOUS rocks. subsidence appears indicated.--- p. 36
187b065 -- same argument applies to England.-- Mem. Shew MICE.-- -- Animals common to South and North Americ
191b082 trace limits of large animals-- Owls. transport MICE alive? Species formed by subsidence. Java & Sum
233b250 most curious to observe, that all the species of MICE in S. America. which were hard to distinguish c
245c024 rousa; a Cervus near Marianus new, & some rats & MICE. In Amboina only Cuscus & Barbyroussa «NB» [isl
248c029 Galapagos Mouse not the same section, with house MICE. It is wonderful how it could have been transpo
249c035 o words. have not species, generally wide range? MICE.-- Waterhouse's remarkable fact of no forms pec
252c045 ell more certainly Get Closer species-- FOX IS & MICE of America «good case on account of varieties i
297c185 rds for some time alive ¿therefore other species MICE & only kills them when urged by hunger.-- p. 65
318c252 rances-- Now this is precisely the case with the MICE of S. America, with respect.. to the Cordillera
354d065 d one-- It should be observed with what facility MICE attach themselves to man. Sept 7th. -- I was st
369d115 ncus & Hydromys not being Marsupial. («but» also MICE) & these being water animals <that this> struct
404e024 t, now this is directly incorrect The case of my MICE is good, because it is an involuntary variation
437e140 unted in a field, where there was great swarm of MICE.-- May 4th.-- The Brussels Sprout returning sud
573n037 ve than movements of fingers.-- like Kitten with MICE.-- A person with St Vitus' dance badly, told sh
575n043 s» then can we deny that the grand child dug for MICE from some peculiarity of structure of brain.--
575n044 leave the sheep» & jackall to skulk about & hunt MICE-- Jenners Jackall Have we somewhat right to. den
046r079 of the former. -- Ascension. Vegetation? Rats & MICES. At St Helena there is a native mouse Did wave
046r080 ve been owing to absolute movement of ground). MICHELL (Philos: Transacts) «seems to» considers that
467t100 pistil does not become abortive. Examined in MICROSCOPE--some of the stigmas of {P} shape of ordinar
029r033 e 8 deg. 47 min. N: longitude 61 deg. 22 min. W. MID. calm and clear. Caermarthen Journal I look at t
041r066 in modern sandstone. a circle,.{P}, had in its MIDDLE a short <fissure> «vein» terminated each way,
495q05a uares to facilitate investigation.-- Capital in MIDDLE of ploughed field-- on hills.-- 10 Shoot tame
504q013 open-- -- No (6) There is apple with branch in MIDDLE of tree with flowers near end of orchard.= At
037r056 for a pebble of any sort; not one was found.-- MIERS saw then near? Mem. La Condamaine on the Amazon
053r100 is perhaps same with that of Pernambuco? Quote MIERS about shells at Quillota Lyell, states that con
473s006 oins streams from Capel-Curig-- Mr Bunbury says MIERS has described in Linn: Transacts. a Brazilian p
569n022 eved-- "il leva les epaules et dit qu'il valait MIEUX rester a Farroïlap quelque mal qu'on y fût."--
025r018 t Southern one I have heard of In a preface, it MIGHT be well to urge, geologists to compare whole hi
025r018 ompare whole history of Europe, with America; I MIGHT add I have drawn all my illustrations from Amer
028r030 of the ocean all are mingled. These reflections MIGHT be introduced either in note in Coral Paper or
044r073 was its birth place.--Some general reflections MIGHT be introduced on great size of ocean; especiall
045r074 steam condensed.--Perhaps these mighty changes MIGHT go on. & not a bubbles on the surface bespeak t
046r079 fluid granitic mass under <[...]> less pressure MIGHT have its «proportional» particles altered.-- Wi
046r080 ovement was one of rise (any smaller prior ones MIGHT have been owing to absolute movement of ground)
048m085 gonia. To the Megatherium.--To the Horse. = One MIGHT fancy that it was so arranged from the forseigh
056r109 eepest astonishment.» Perhaps scarcely a pebble MIGHT remain to tell of these losses.-- Cause of chim
077r170 gree 2. superimposed on N degree I. even No. 2. MIGHT be mistaken for Porphyry above ancient freeston
089a017 ion in red sandstone.-- Certainly Volcanic-- CD[MIGHT not bottom of ocean boil; yet heat never reach
094a036 ing same manner. as bits of Patagonian boulders MIGHT be transported.-- {T} on grooved rocks. Specime
096a040 &c  &c In Ascension for centuries afterwards it MIGHT be perceived on which side craters were low --¿
103a064 e dislocations occur & therefore that the crust MIGHT be considered a level.-- Dikes being last actio
103a065 to dislocation of strata. A capital discussion MIGHT be made between dikes & «axis of» mountain-chai
108a080 ar coral reefs.-- Where vegetation luxuriant it MIGHT be almost as well said probably much Carbonic A
110a086 ryst of Hornblende blending into base-- Salband MIGHT have oozed out of cleavage plates: the crystals
139a180 bonic acid [blank] Many interesting experiments MIGHT be tried by comparing Zoophite to plants.-- gra
152g047 contingency if elevation from Axis, then rivers MIGHT deposit, & afterwards with greater cut through,
155g064 or vice versa Same inclination when serpentine MIGHT remove, what above straight line «only» cut dee
173b010 ransportal) <continents> island near continents MIGHT have some species same as nearest land, which w
173b011 hich none of same kind had in interval arrived) MIGHT have grown altered Hence the type would be of t
173b011 uan Fernandez. When continet of Pacific existed MIGHT have been Monsoons.. when they ceased importati
```

Page **(Key Word)**
175b017 y of such change,-- as we see them in space, so MIGHT they in time As I have before said isolate spec
180b038 other hand like Orpheus. being favourable many MIGHT be produced.-- This requires principle that the
183b051 oduced into Chili. in present states. <they> it MIGHT continue & thus two species be created) & live
184b054 y view. in S. America parent of all armadilloes MIGHT be brother to Megatherium.-- uncle now dead. Bu
192b089 x as ornithorhyncus. If this last animal bred-- MIGHT not new classes be brought into play.-- The fat
194b095 r Mammalia.-- I really think a very strong case MIGHT be made out of world before zoological division
195b098 > division are arbitrary, & not permanent. this MIGHT be made very strong. if we believe the Creator
195b099 ng place for many unintelligible structures. it MIGHT have been of use in progenitor-- or it may be o
195b101 s, without any assignable reason.-- Astronomers MIGHT formerly have said that God ordered, each plane
198b113 eration of range of capabilities past & present MIGHT tell something) p. III G. St Hilaire Insects &
202b133 if suppose world more perfectly continental. we MIGHT have wanderers. (as Peccari in N. America) then
202b134 merous there would be wanderers.-- Some however MIGHT have offspring, & then «V. L. Institut p 245. 1
206b146 ve successors.-- Then the chance of 200 people <MIGHT be> being related within 200 years backward mig
206b146 ght be> being related within 200 years backward MIGHT be calculated & this number elimanated say 150
206b146 ple, and so on backwards to one progenitor, who MIGHT have continued breeding from eternity «backward
218b192 ing to isld. Graham isld.-- we know many seeds, MIGHT be transported some blown--floating trees Thrus
219b193 (Emberiza Brasiliensis?) (Fulica Chloropus)--» MIGHT bring in stomach-- &c &c. (Mem discover what ki
219b195 ate we do not know how transported.-- (Glaciers MIGHT have acted at Tristan d'Acunha-- Carmichael Lin
219b196 ounds. Before Attract of Gravity discovered. it MIGHT have been said it was as gret a difficulty to a
223b208 with cheese mite» with its wonderful intincts <MIGHT well say how> The difference is that there is w
224b213 one with one kind of herbage & one with other, MIGHT change organization of stomach & hence remain d
226b223 hannel of communication by which great Edentate MIGHT have roamed to Europe & Pachydermata from Europ
227b227 gos on Pacific side. the Oolite order of things MIGHT have easily been formed.-- With belief of <chan
228b228 ect passages of <species> structure in species, MIGHT lead to laws of change, which would then be mai
228b229 genera far back, as well as at present time, we MIGHT expect confusion of species.-- Important. For i
231b243 cene.-- The descendants left in cooling climate MIGHT change twice over, whereas those which migrated
232b245 t, or moved towards equator.-- «or some species MIGHT then have been wanderers.--» There ought to be
236b280 Decandoelle says, no he only says sometimes we MIGHT expect disseminated species to vary a little, b
236b280 land» surely it is not-- intermediate genera we MIGHT expect.-- Lindley Introduct Dict. Science. Natu
245c025 those sinking, because arrival of any one plant MIGHT make conditions in any one isld different]CD.--
250c038 more equable. yet America preeminently equable. MIGHT have allowed fresh species to have been formed
250c038 e been condition of former whole world. America MIGHT have been string of islands.-- ¿Europe has many
250c039 Principles, must be abstracted & answered Much MIGHT be argued what is not cause of destruction of l
257c057 o indicate those very changes which at first it MIGHT be doubted were possible,-- it has been asked h
260c067 . 226 Wilson's Ornithology, D'orbigny, Spix, &c MIGHT compare birds of N. America & South-- Any how t
263c078 o produce contingents proper.-- Present monkeys MIGHT not,-- but probably would.-- the world now bein
264c079 r such an animal.--man, (rude, uncivilized man) MIGHT not have lived when certain other animals were
264c080 ed bird in Australia &c &c-- but of course they MIGHT be blended, if archipelago turned into continen
270c104 live on matter already organized.-- This paper MIGHT be worth consulting, if any Metaphysical specul
276c122 appears to me). colour of plumage & laws, which MIGHT probably be reduced What the Frenchman. did for
278c129 we do not know, whether nearest species of each MIGHT not breed:-- Genus must be a true cleft-- putti
280c135 whereas two newer ones, even if more different MIGHT do so.-- <whi> is this true?? My views, which w
280c135 mules appeared to him most strange.-- This even MIGHT be said.-- My theory thus explains a grand appa
280c136 like its own parent this impossible-- (Hence we MIGHT expect even if two mules bred or two certain va
281c138 each other I cannot help thinking good analogy MIGHT be traced between relationship of all men now l
282c140 mits of change, would be same-- Yet each family MIGHT have its own character,-- we here suppose these
282c141 take <dogs> «domestic animals» with them. they MIGHT be supposed to change. & make genera of bird. a
282c141 ture & even to certain degree in habits, yet we MIGHT have these analogies.-- We must two races of su
282c142 ce would not obtain a cast of washing men-- but MIGHT hire the preexisting race, thus the analogy wou
285c150 maphrodites» father « <mother> » & grandfather <MIGHT> Must be introduced & made young.-- father mus
289c160 land bird, having habits of a Grebe, structures MIGHT follow.-- examine structure of this bird & get
289c162 sily in those born without coitus, than with.» MIGHT be given as a hopeless difficulty, except as di
290c162 & plants. At first classification on generation MIGHT appear an analogy NB Pyrrho-alauda (bird of St
292c171 efore would banish individual, but general ones MIGHT yet be transmitted.= Memory springing up after
298c189 nswer, without any leading question.-- [+] This MIGHT be mentioned in note.-- try to trace from simpl
298c190 itually passionate.-- the Key to the affections MIGHT perhaps thus be found-- a person who is habitua
302c202 any small family. having analogous characters, MIGHT be multiplied.-- we must argue reversely. WHERE
303c204 from Lawrence. Blumenbach & Prichard -- Now we MIGHT expect that animal halfway between man & monkey
315c241 837. semestre I suspect some valuable analogies MIGHT be drawn between habitual actions of plants «wh
318c255 ion of Migration) gradually separated the birds MIGHT yet remember which way to fly.-- There is a kin
325c267 on on the culture of Dahlias Mrs. Gore on Roses MIGHT be worth consult. Paper on Consciousness in Bru
335d015 certain small limits (within which limits they MIGHT return to either parent), then according law, t
337d022 ons of the Entomological Soc> A capital passage MIGHT be made from comparison of Man, with expression
341d029 oint or many) Owen answered that all characters MIGHT be considered as adaptive & that he did not see
346d047 ngo-- (Spoonbill Wader. Ibis)-- laws of plumage MIGHT possibly be made out.-- August 25th Athenaeum (
352d057 Those animals which have many ABORTIVE organs, MIGHT be expected to have larvae more perfect-- this
360d092 bright red breast & other very bright blue, it MIGHT be harder <to tr> for both parents to transmit
360d093 o> the most aquatic & most terrestrial species, MIGHT be harder to cross than two less opposed in hab
360d094 short tail.-- I can readily see that two first MIGHT cross easier than two last. Sept. 11. N Mr. Bly
366d112 ple on the NW. Coast blinking to keep out flies MIGHT be used» The wild ass has no cross. how comes i
370d115 animals are thought to have been created.-- it MIGHT as well be attempted to be shown from peculiari
372d131 e» bud some such process is effected.-- a child MIGHT be so born. but it would be very different from
381d157 e dioecious & some monooecious-- (& cultivation MIGHT make one set of organs barren in one plant & no
388d171 f these staid in the womb, when it came out. it MIGHT partake of shade of fathers character.-- accord
389d176 udding, which wishes to throw itself off.--» as MIGHT be inferred from annual plant being prolonged t
399e006 objection to my theory: a modern bed at present MIGHT be very thick & yet have same fossils. does not
400e012 character of fur-- I am sure a very good case, MIGHT be made out of variation analogous to specific
403e023 ll of God.-- Octob. 16th. A very strong passage MIGHT be made-- why seeing great variation in externa
404e024 o wild animals only. from which plain inference MIGHT be drawn that whole infertility «of hybrids rec
404e025 s structure would become obscure & therefore it MIGHT thus have arisen, & M. Edwards p. 330 distinctl
405e032 ca is somewhat similar in two hemispheres.-- It MIGHT be worth investigate whether. Megatherium & Mas
407e039 fitted for such a preeminently equable climate. MIGHT not have been able to have survived a change, (
407e039 ough other parallel species in other continents MIGHT have survived this mundine change..-- Therefore
408e044 tion,-- «all» modern & wholly volcanic-- Azores MIGHT be prophecied to have this character.-- worth g
415e068 that has made a man.-- CD [any monkey probably MIGHT, with such chances be made intellectual, but al
416e075 arger than any living [not located] A Greyhound MIGHT be made «almost» without any relation to runnin
417e077 ecies>, variety, as in two varieties, & this we MIGHT expect, as the difference between man & woman i
419e085 of Mollusca, which I have sometimes speculated MIGHT be owing to absolute quantity of vitality «in t
420e090 o Janeiro, (like some Mediterranean species).-- MIGHT these fertilise other shells, as insects do flo
421e091 ttle clear meaning has this compared to what it MIGHT have.-- What is the difference between an essen
426e107 to unite to perfect prolifickness.-- (<a series MIGHT be obtained>: but the intervention of domestica
429e114 are propagated with very little care.-- & which MIGHT spread themselves, as well as our wild plants,
483z013 Tahiti and Chile The North & S. Range of shells MIGHT perhaps be worked out with advantage. with Cumm
498q009 ecious, if no hybrids were produced by seed, we MIGHT feel sure, that pollen of own kind is much more
507q15v as showing how simple, but beautiful adaptation MIGHT be arrived at.= Any book with drawing of Seed.
521m008 e muscles.???> Miss Cogan's memory of the tune, MIGHT be compared to birds singing, or some instincti
529m041 ently habitual to become poetical) the botanist MIGHT so view plants & trees.-- I am sure I remember
542m094 f organization were changed, I conceive sighing MIGHT yet remain just like sneering does.-- is yawnin
547m112 ed probably by sleep & not vica versa. anyhow I MIGHT have been quite still, & not attending to bodil
549m123 would have had them.-- with lesser intellect they MIGHT be necessary & no doubt were preservative, & ar
558m152 r back when fighting, & not with dog. when fear MIGHT enter?-- I believe common Swan, arch raises nec
563n002 , & constant, for a wish which was only short & MIGHT otherwise have been relieved, he would be sorry
574n039 pted to subject see <A> I think this argument MIGHT be used to show language had a beginning, which
576n046 ditated on this, it would be conscience. A man, MIGHT not <t> do so even to save a friend, or wife.--
576n047 umstances» Look at Spain now.-- man's intellect MIGHT well deteriorate. «CD[in my theory there is no
584n072 ,-- bats avoiding strings «in the dark» as well MIGHT be called instinct,-- migrating to one spot, th
585n077 nstinct is heredetary knowledge of things which MIGHT be «possibly» acquired by habit. so bees in bui
585n077 ty by his instruments to make toothed wheel. he MIGHT by instinct make watch, but he does it by reaso

```
587n087 ay be doubted for possibly even taste of senna. MIGHT be acquired. as the Turks have of Rhubarb: agai
587n088 king, as simply heredetary habit.-- A blind man MIGHT be born with idea of scarlet, as well as rememb
588n090 n of humility.-- I suspect very strong argument MIGHT be advanced, that animals have reasons, because
592n101 spect the endless round of doubts & scepticisms MIGHT be solved by considering the origin of  reason.
603o012 d into> is carried on toward deity.-- & as king MIGHT like cruel pleasure, so sacrifices cruel.-- Som
614o037 HC] Perhaps even the most complicated instinct. MIGHT be analysed into steps, as species change.-- Mu
616o039 to the brain that attraction does to matter, it MIGHT with equal propriety be said that the living br
616o039 use of ordinary language that the onus probandi MIGHT fairly be laid with those who would support the
616o39v is entirely wanting by which thought or memory. MIGHT be in like manner attributed to the brain. Ther
618o041 ction quite independent of each other. A person MIGHT be quite familiar with thought & yet be ignoran
618o41v in intensity (& the experiment was varied) then MIGHT it now be said, that blueness caused weight, be
622o049 of right & wrong.-- on which «almost» any other MIGHT be grafted.-- Origin of the instincts  Hartley,
628o53v e with scarcely possible to teach it-- all dogs MIGHT be taught, but not cat, that is not act by gust
628o53v at, that is not act by gusto, though by fear it MIGHT be partly made.]CD p. 21. "Why ought I to  keep
632j53r gn.-- perhaps they are so.-- but the coral rock MIGHT have been uninhabited as the Alpine pinnacles.-
636j56v borders of a country favourable to change.-- Mr MIGHT be concluded that Plants would be subject to ex
044r074 of of such gaz) steam condensed.--Perhaps these MIGHT changes might go on. & not a bubbles on the su
291c168 very much hotter, then Brazilian species would MIGRATE south ward being ready made.-- & so destroy in
298c184 on fossils of Europe., those species which can MIGRATE remaining constant in form, others altered muc
318c254 this seems common «kind» migration of America) MIGRATE singly flying few miles every day «generally b
431e122 greater. part then transmutation.-- Do species MIGRATE & die out.?-- In the place where any species i
231b243 e might change twice over, whereas those which MIGRATED a little to the southward would merely be spe
248c029 rupeds.: that either created in each point, or MIGRATED from those quarter, where we know  quadrupeds
464tf6v s work on Ruminants,-- these species must have MIGRATED to these mountains, when the cold was intense
318c254 ght» -- other birds which is strictly diurnal, MIGRATES singly by night.-- others in flock, these bir
288c160 sed their migrations lost?? I conceive a bird MIGRATING from Falkland Isd regularly to main land, pro
437e180 ay to the Caymans from Honduras. good case of MIGRATING.-- shows my theory insufficient.-- p. 120  An
584n072 the dark» as well might be called instinct.-- MIGRATING to one spot, this is indeed instinct.-- Austr
586n078 t by reason & experience, or habit.-- so bird MIGRATING to certain quarter is instinct, but his knowl
318c254 s» -- other birds (& this seems common «kind» MIGRATION of America) migrate singly flying few miles e
318c254 learly directed by kind of country; «kinds of MIGRATION quite different in species of same genus.» Th
318c254 try (These facts show the Normal condition of MIGRATION) gradually separated the birds might yet reme
328cIBC n's Journal. during 1837. paper by Bachman on MIGRATION of birds Temminck has written "Coup d'oeil su
512q019 g-- loss of early habits in Dorsetshire sheep MIGRATION of coots-- variation in hounds= An ugly  calf
584n071 es not climb CD[.instinct may be divided into MIGRATION,-- subsidiary to food & temperature molting &
586n079 dgeon just as wonderful in old bird as new.-- MIGRATION, <only> «only» more wonderful in young, becau
251c040 h) Paper by Jenner, on birds seen far at sea, MIGRATIONS of species, geese killed in Newfoundland, wi
288c159 y «nature & instinct» modified.-- The partial MIGRATIONS of birds in same country. may explain greate
288c159 of birds in same country. may explain greater MIGRATIONS, if American & intersected wider & wider  at
288c160 ,. which came Madeira & <seized> ceased their MIGRATIONS lost?? I conceive a bird Migrating from Falk
318c253 ven migratory birds, lik swallows» -- degree/ MIGRATIONS of birds he mentioned many most curious case
332d004 stomed to turn down.-- -- applicable to birds MIGRATIONS & Australian Savages.-- W. D. Fox has a cat.
463z012 ken from Jenner (1825) Phils: Transact.-- "on MIGRATIONS of Birds".-- 18 do. Vol III p. 422. letter f
220b199 eir representative species in each other, are MIGRATORY species from warmer countries. When will this
231b242 ravellers?? Royles case of Himalayan, plants ¿MIGRATORY birds, he told me some story of crane  from Ho
249c036 & America, which are of different forms being MIGRATORY; also with Temminks fact of forms being withi
318c253 & how birds are extending their ranges, «even MIGRATORY birds, lik swallows» -- degree/ migrations of
437e138 s only winter.-- Jays & chaffinches sometimes MIGRATORY.-- p. 103. Turtles finding their way to the C
481z009 :-- S. Cruz. Molina Vol. I. p. 244. Baccalao. MIGRATORY fish-- See Kings drawings.-- for real name B
502q11v abortion. or for sterility Land Birds Madeira MIGRATORY-- ask Gould about N. Zealand, as Cuculus luci
502q11v .-- Ask Sulivan about Falklands Isds.-- Snipe MIGRATORY-- probably united by Land to S. America  (33)
527m032 from» certain districts have the best song. [MIGRATORY birds return to same quarter for many years]C
318c255 y isolated spot on the Alleghanies to which it MIGRATS every year,; and probably a «chance» wanderer
510q017 has he published? does he understand English.-- MIGUEL to collect facts for me-- what? What does Blum
450e175 ny wild horses, bullocks, & deer South part of MILDANAO.-- Q Horse do. Appendix. p. 43. «& 45» the Br
160g093 de Barom 29.200 A 80 70 degree? for about 3/4 of MILE on <one> S side of «this» Isthmus (which runs E
161g094 ut 3» feet <lower> too high about a quarter of a MILE further on, where three [...] abutted Having cr
023r009 t of the E. Indian Islands. namely Java is 1000 MILES distant! Where are Hippotami found in that Arch
023r011 al upheaval» of the shore of the Pacifick is 60 MILES distant from the grand ancient volcanic axis «o
024r015 dney 90 fathoms La Peyrouse. South of Mocha; 19 MILES. 65 Fathoms Vide facts in Beechey. on NW  coast
024r015 of  America off Cape of Good Hope 70 fathoms 20 MILES from the shore? Beagle Coast of Brazil? where n
030r035 ry> rather deep soundings, 60-100 fathoms 2 & 3 MILES from shore. V. Chart) Every winter torrents mus
064r134 ys at Santos «M Birchels» at foot of range some MILES from shore. rock of oysters quite above reach o
065r139 r diameter of Deception Isl is six Geographical MILES and width 2 & ½ miles S. Shetland. Lat. 62 degr
065r139 n Isl is six Geographical miles and width 2 & ½ MILES S. Shetland. Lat. 62 degree 55 minute. <only> o
067r143 oast from P. Indio to Quilmes. & at least seven MILES inland. The Cordoba earthquake a very remarkabl
085a001 injured  on Eastern side, far inland.-- even 70 MILES from salt water. Mr. Arrowsmith tells me,  that
117a102 (before June 1838) that 70. F were obtained 100 MILES E of Staten land. bringing up pebbles 2  inches
119a105 ace. whether they can have been plunged so many MILES deep into the bowels of the earth, as would  be
120a108 not  cooling if they had travelled some hundred MILES through nearly cold rock.-- in volcano the pool
125a120 a.-- do p. 473. on great Iceland stream. the 90 MILES includes opposite directions. Mem. S. Cruz. Ass
129a132 -- Most curious account of great subsidence «20 MILES long I in with.» which must have been from an a
135a144 iapo. found inland a great many sea shells some MILES from coast-- quote passage to show abundance Be
152g047 end of valley indicates new terrace Ballivard 2 MILES North of Grant town to Forrest road  comminuted
161g096 stance hid it, could be followed for at least 2 MILES on dead level «by eye» to moss-- on this terrac
195b103 agos orpheus.= Put this strong so many thousand MILES distant.-- Absolute knowledge that species  die
267c094 to check its increase. The <bush> woodlands for MILES in extent are composed solely of this shrub".--
278c130 a height of more than 1000 ft. & many hundred MILES from the sea, associated with teeth of seals an
306c212 wn stream which would last were the stream 1000 MILES long.-- a monkey. (Baboon) at Z. Gardens upon b
306c213 in» the whole area,. 120 is greatest (about 200 MILES distant).-- directly beyond produced line of Ti
317c251 Charlestown ?three species, near New York. (600 MILES N.?) replaced by three other species.-- Says al
318c254 migration of America) migrate singly flying few MILES every day «generally by night» -- other birds w
334d012 ws entirely, changed his residence a great many MILES.-- yet one day <th> a cow walked in, then disap
437e138 e is said to have been seen carrying a lamb two MILES towards the Morne Mountains, it then dropped it
452e181 cured from Gilolo p. 253 In isld of Bunwood (18 MILES in circum) there are hogs & monkeys <at> near s
512q019 eeds remain in stomach of birds-- Mem: how many MILES they fly in few hours Zoological Soc (1) Do the
564n003 ent moral sense, if it were proved instead of MILITATING against the existence of such an attribute w
215b177 and  it is required to give the canthairides and MILK--Fox tells me that it is generally said.= How c
307c215 Males animals, Mammae in Men, capable of giving MILK)» rudimentary wings. so nature can produce in s
310c224 38. quotes M. Turpins assertion that globules of MILK produce a plant capable of growing!! & propagat
373d132 s.-- Does not male Pidgeon (yes) surely) secrete MILK? from stomach. analogous to other males feeding
380d155 fter animal has copulated., though no offspring, MILK sometimes comes in Mammae & even when bitch  is
384d162 female  plumage in castrated male.-- «Men giving MILK--» Sept. 25th Young man at Willis «Grt. Marlbor
388d172 of same tendency. -- Mammae in man. having given MILK. -- testis & ovaria The following views show th
412e057 ones,. & even more than Mammae, which have given MILK.-- is secretion from Pidgeons stomach true milk
412e057 milk.-- is secretion from Pidgeons stomach true MILK.-- ‖ <Species. are innumerable variations>. Eve
507q016 ?-- Bell cd ask Accouchers Is any peculiarity in MILK teeth inheritable!!! very good Any  peculiarity
279c133 ted.-- + +. Fully supported by Mr Wilkinson. = MILKING heredetary, developemen of important organ (.s
512q019 e in youth.-- Mr Tollett-- about selection for MILKING-- loss of early habits in Dorsetshire sheep mi
478z003 la) sea weed & many pebbles Mentions stinging MILLEPORA. Quoy. Freycinets Voyage Vol p. 597 Many desc
359d087 ped bird-- looks very artificial breed-- but Mr MILLER says that breeds larger numbers, & rears an un
458t002 rson with long legs can hardly ride on them. Mr MILLER-- in Zoological Gardens. informs me that a hyb
494q004 ar to those of another Duck. ¿in Pidgeon?-- Mr. MILLER said yes with regard to former (8) Is form of
512q020 raised  from young ones, bred in captivity --Mr MILLER says Wombwalls were (4) About fertility of ass
553m137 tralian savage or one of Tierra del Fuego.-- Mr MILLER (superintendent of the Zoological Gardens) rem
225b217 nal causes, of which we are as ignorant. as why MILLET seed turns a Bullfinch black, or iodine on gla
282c143 on  Infusoria on the the enormous production-- MILLIONS in few days-- one doubt that one animal can r
274c114 si.-- «Ground woodpecker» Secretary bird.-- & MILLISUGA. Kingii very rasorial for type.-- Now here  I
```

404e025 19th. When reading. L'Institut: .1838 p. 329-- MILNE Edwards, description of curious mechanism of re
485z018 pi shorter duration than cells.-- reproduced.-- MILNE Edwards p. 138 on Polypi.-- Berenica &c &c L'In
203b137 heredetary?, having been gained in short time. MILVULUS forficetus <has a wide range> is a tyrant fly
268c096 w many forms Tyrannula can we worked out into. MILVULUS [not located] element geographical distributi
273c111 t be doubted that if swallow perished) hawks & MILVULUS &c would instantly fill up their place.-- Hum
273c111 strongly marked variety,:.in the Tyrannidae.-- MILVULUS-- Even flying woodpeckers., with powerful win
273c112 of all. ¿whether in most aerial of swallows.» MILVULUS, & still more wonderfully to the Humming bird
273c112 plicable to same habits, though swallow hawk, .MILVULUS, may catch insects on the wing & pratencole
276c124 amily practise with a peculiar structure, then MILVULUS forficatus Tyrannus Sulphureus if compelled s
566m010 Vol III. p.37, quotes from Burke, who says on MIMICKING expression of emotions, he has felt the passi
207b151 succeed so well at St. Helena. as. Pineaster & MIMOSA called Botany Bay Willow V. Dr Royle introduct
454e184 Transat. V. II p. 252. Is there any very sleepy MIMOSA, nearly allied to the Sensitive Plant.-- p. 29
455eIBC these are same as trees.-- Shake some sleeping MIMOSA-- do stamina of C. Speciosus. collapse at nigh
494g005 ect? and silk caterpillars (1) Shake a sleeping MIMOSA, or half bred mimosa (a) between sensitive & s
494q005 llars (1) Shake a sleeping mimosa, or half bred MIMOSA (a) between sensitive & sleeping species, & se
029r033 t was quite calm at the time. Latitude 8 deg. 47 MIN. N: longitude 61 deg. 22 min. W. mid. calm and c
029r033 Latitude 8 deg. 47 min. N: longitude 61 deg. 22 MIN. W. mid. calm and clear. Caermarthen Journal I l
431e121 e &c &c.--if? so «it is» good case:-- in Sowerby MIN. Conch. it is however, said they have been found
055r108 na!!-- There are some arguments which strike the MIND with force.--the exact yearly rise of the great
171b002 organization (especially in lower animals, where MIND, & therefore relations to other life has not co
171b003 e <believe> «know» in course of generations even MIND & instinct becomes influenced.-- child of savag
220b197 ho can say, whether offspring does not depend on MIND or instinct of parent. Mem Lord Moreton's Mare.
223b207 insects with other senses is more wonderful. its MIND more different probably & introduction of Man.
223b208 there is wide gap between Man & next animals in MIND, more than in structures.-- If the skeleton of
225b219 ent produce any character on offspring? Does the MIND produce any change in offspring? if so adaptati
227b228 would   lead to study of instincts, heredetary. & MIND heredetary, whole metaphysics.-- it would lead
263c075 is multiplication of little means & bringing the MIND to grapple with great effect produced, is a mos
263c075 ed, is a most laborious, & painful effort of the MIND (although this may appear an absurd saying) & w
268c099 forms similar--birds??-- We must always bear in MIND proofs of most equable climate both in S. & N.
280c135 - & is a very remarkable fact. show influence of MIND It is not difficult to see that it is less repu
292c170 <to so convin> brought so much conviction to my MIND.-- Reflect much over my view of particular inst
299c196 unimpregnated Babington We see gradation to mans MIND in Vertebrate Kingdom in more instincts in rod
299c196 in rodents than in other animals & again in Mans MIND, in different races, being unequally developed.
300c198 tus & not saltus.-- The greater individuality of MIND of different animals less divided.-- But as man
300c198 dies of some animals over those of others.-- the MIND is still only a divided body. p 3. language se
300c198 ded.-- But as man has heredetary tendencies. his MIND not so different from that of brutes p. Hard to
300c198 guage. heredetary & acquirable.-- therefore mans MIND There is no progression in the developement in
305c210 agging tail.-- habitual movements connected with MIND, even in the tendency to delicate emotions betw
305c211 ose individuals thus endowed, & the community of MIND of man & animals.-- yet how faint in a Fuegian
316c244 f him «or eternity», only difference between the MIND & giving rise to the wildest imagination & supe
316c244 or bad enough for Hell.-- «glimpses bursting on MIND Evelyns Sylva. skimmed, stupid Brownes travel i
322c269 - References at end Mayo Pathology of the Human. MIND.-- Audubons. Ornithological Biography. 4. Volum
326c266 important Papers. Dr. Mayo. Pathology of Human. MIND. The simplest transmission is direct instinct.
371d118 with many contingencies how to act-- so with the MIND.-- , but both are present in every animals, bu
382d158 ite,-- but not of poenis & clitoris, shows to my MIND.-- , that both are present in every animals, bu
389d177 to propagate & perpetuate differences (of body, MIND & constitution) is the end frustrated, when nea
404e024 ybrids receive no explanation» was consequent on MIND or instinct, now this is directly incorrect The
409e047 luous to consider minds.-- as difference between MIND of a dog & a porpoise was not thougt overwhelmi
424e102 -- In argument of origin of Wolf, difference of MIND is most relied on, but Bell has some account of
463t057 sted. (vipers tooth cause a difficult), the whole MIND is constituted that a difficulty makes greater
465t079 in the caves? It is highly important, to bear in MIND that enormous periods may elapse, even in situa
524m019 old water brings on suddenly in head, a frame of MIND, analogous to those feelings. which may be cons
526m027 ks, shows same connection between organization & MIND.-- thinking over these things, one doubts exist
526m028 enough to <up> recall up the image in their own MIND,-- this may be worth thinking over.-- it. will
526m029 Zoos» they are all things, which are brought to MIND, by memory of the scenes, (indeed my American r
526m030 logists are right about habitual exercise of the MIND, altering form of head, & thus these qualities
526m031 chooses (but what makes him fix!? <)>-- frame of MIND, though perhaps he chooses wrongly,-- & what is
526m031 perhaps he chooses wrongly,-- & what is frame of MIND owing to.--<)>-- I verily believe free-will & c
527m033 rs (rhythm & pleasant sound per se) & causes the MIND to create short vivid flashes of images & thoug
527m033 s are in same manner vivid & grand. the frame of MIND being just kept up by the music of the poetry.-
529m040 in different people., an agriculturist, in whose MIND supply of food was evasive & ill defined though
529m043 y diseases, where there has been no cloud on the MIND, every occurrence for a day or two are absolute
530m054 probably the system is affected, & by habit the MIND tries to fix upon some object:-- When a man, ch
532m055 ut, & the conveying means from the senses to the MIND being more alive.-- How is it. with people nerv
532m055 it must be an excited action in the involuntary MIND which is startled.-- My Father says he should t
533m060 w comes into situation in these cases? How did my MIND feel it was wrong (& it was not merely morally
535m069 to Gods anger.-- (.. more poetry in that state of MIND: the Chileno says the mountains are as God made
536m072 ve it of pain or pleasure, if so free will is to MIND, what chance is to matter «(M. Le Compte)»-- th
538m080 perception separate, from the ordinary state of MIND is probably analogous to the double individual
539m083 ity of two quite separate trains going on in the MIND in double consciousness may really explain w
539m083 s habits.-- Aug. 16th. As instance of heredetary MIND. I a Darwin & take after my Father in heraldic
540m088 tion Fine poetry, or a strain of music, when the MIND is rendered ductile by grief, or by bodily weak
541m089 went down to Woollich I was trying to unbend my MIND as much as possible (testing success by decreas
541m089 sing headache) & found best plan was allowing my MIND to skip from subject to subject as quick as it
541m091 effort consists in keeping one idea before your MIND steadily, & not merely thinking intently; for
541m091 to keep any simple idea as scarlet steady before MIND for period, «if the scarlet was before one effo
541m091 labour of inventive thought.-- Examine frame of MIND in following changes during fall of sea.-- Is t
543m100 ver many places, & contingency a man has keep in MIND. all is certain.-- there is judgment of probabi
544m103 party dreamt of <goo> feasts of good food-- The MIND wills to do this & hears that, but yet scarcely
544m103 re is ideal, as all the other perceptions.-- The MIND thinks with extraordinary rapidity-- We may con
544m103 , novelty of separate ideas cause fatigue to the MIND,-- it is solely the comparison, with past ideas
544m103 ot real.= = dreaming appears clearly rest of the MIND, with all other faculties: «Vide page 110, by m
547m112 turned into dream.-- It appears to me, that the MIND is wholly absorbed with one idea (hence apparen
547m113 ry heirs of every action, & always running on in MIND, being absent. one could not compare the castle
547m113 led the bell??)-- It may be deception to say the MIND <thinks> quicker in sleep, it may do less work
548m117 the reality does not resemble the picture in one MIND, but does not stop to reason what there should
548m117 ess the number of pleasant ideas passing through MIND in given time.-- intensity to degree of <happi>
549m119 whether from frequency, or inherent structure of MIND. they make, either in themselves, or if recolle
549m119 part of thoughts innumerable, which past through MIND.-- These thoughts are most pleasant. when the c
549m119 re most pleasant. when the conscience tells our [MIND], good has been done-- & conscience free from
549m123 like all other structures slowly vanishing-- the MIND of man is no more perfect, than instincts of an
551m127 ree of intensity-- «in old man, he had just seen MIND went on RAMBLING till excited by question.» Sep
553m135 essions long repeated, without the powers of the MIND being EQUAL to the smallest casuistical doubts.
558m151 bour as thyself". Analyse this out.-- bearing in MIND many new relations from language.-- the social
558m151 ructure takes its value. from its connexion with MIND, (to show hiatus in mind not saltus between man
558m151 from its connexion with mind, (to show hiatus in MIND not saltus between man & Brutes) no one can dou
563n002 Nina)-- or to take away food &c &c-- Now if dogs MIND were so framed that he constantly compared his
564n005 Mechanics.-- Experience shows the problem of the MIND cannot be solved by attacking the citadel itsel
564n005 be solved by attacking the citadel itself.-- the MIND is function of body.-- we must bring some stabl
566n010 ything in these absurd ideas.-- do they indicate MIND & body retrograding to ancestral type of consci
566n010 emotions, he has felt the passions of a face «& MIND» sympathetics with internal organs, as action o
569n020 nal feelings, probably some distant power of the MIND.-- superstition & charity & prayer, or eloquent
571n027 savages???].CD-- The existence of taste in human MIND. is to me clear evidence, of the general ideas
572n034 ning» over imaginary words: it appears as if the MIND had dwelt on each word separately, neglecting
574n040 bits.-- we probably can hardly form an idea of a MIND so limited as Birgos to become absorbed by one
575n043 diate like rest of body? Can we deny relation of MIND & brain. «Do we deny the mind of a greyhound &
575n043 e deny relation of mind & brain. «Do we deny the MIND of a greyhound & spaniel. differs from their br
577n049 conferva is not an antagonist quality to life & MIND of man.-- & we do not suppose an hydatid to be

Page   ***(Key Word)***
577n051 nsitive Plants Habitual actions, (independent of MIND) in the intestinal functions &c &c &c.-- bears.
577n053 nds to give it, as in cancer, showing, effect of MIND on individual parts of body.== (if you <think>
578n056 ars. replaced & squeezed out again-- as power of MIND by habit gets more perfect over voluntary muscl
579n058 frown with grief,¿ bodily pain? frown shows the MIND is intent on one object.-- With respect to my t
579n059 sublimity. January 6th.-- What passes in a man's MIND, when he says he loves a person-- do not the fe
580n061 llect is that when an idea once take hold of the MIND, no subsequent ones modify it.-- «Weak people s
581n066 s.-- [Emotions are the heredetary effects on the MIND, accompanying certain bodily actions]CD.¿ but w
584n073 gh, commences as soon as painful thought crosses MIND, before it can have affected respiration V E. p
592n101 atural Hist. of Religion" on its origin in Human MIND.-- Andrew Smith says hen doves & the female cha
593n111 reme step of an ideal argument held in one's own MIND. & Dr. Hollands story of man in Delirium tremen
593n111 dentity is by no means a necessary part of man's MIND.-- At Maer. Pool. I saw many coots & waterhens
595n117 - think well over this;-- it shows similarity in MIND.-- think of Eyton's horses becoming <white> wit
600o08a t imperious.-- It would indeed be wonderful, if, MIND of animal was not closely allied to that of men
602o011 upon the "habitual reason".-- This power of the MIND, faintly approaches to instinct How strange it,
602o11v » in Tragic acting-- CD [My idea. would make the MIND have mysterious & sublime ideas independent of
604o017 of Emotions.-- T. Mayo-- Pathology of the Human MIND. Poor.-- on insanity.-- Prevailing idea. owing
606o021 t ever perceive the various operations which the MIND undergoes in gaining the result.-- Lessings Lao
607o025 produced.-- the wax was soft,-- the condition of MIND which leads to motion being inclined that way]C
608o026 ce) 2) difference is from imperfect condition of MIND all motives do not. come into play.-- †It may b
614o037 does not tend to Atheism. inutility of so high a MIND without further end just same argument. without
616o038 y be called memory; you cannot call the frame of MIND which makes music pleasant, a memory; yet that
621o047 calculated pleasure, but the satisfaction of the MIND, which is «much» formed by past recollections.-
621o047 ections.-- Hence he has the right & wrong in his MIND.-- Now we know it is easy by association to giv
622o048 of habits into instincts. [RHC] Feelings of the MIND, whether leading to action or not, are the part
629o55v eeling of ought, shame. right & wrong comes into MIND in first case-- seeing how shame is accompanied
541m090 rium rest-- therefore dreams thus act.-- ∴ weak MINDED people are fickle & full of levity (¿ do I not
282c142 ; «NB, aquatic. i.e relation to elements & not MINDING particular trades.--» then the second race wou
283c145 eated them.-- People will argue & fortify their MINDS with such sentences as "oh turn a Buccinum into
409o047 ies on one type: almost. superfluous to consider MINDS.-- as difference between mind of a dog & a porp
520o002 his father.-- My father thinks. people of weak MINDS, below par in intellect. frequently <are> have v
532o055 ame things from any effort of will whilst their MINDS were sound. Caroline tells me that Nina, when b
559m154 se we can compare it to the standard of our own MINDS. which ceases to be the case when we consider t
022r008 432. as in Andes. In Dampier's voyage there is a MINE of metereology with respect to the discussion o
054r102 and account of cleavage differs wonderfully from MINE: phyllade covered by quartzose sandstones: refe
054r102 rnetty: account of streams of stones agrees with MINE.--At Conception, cleavage E & W! at Payta. talc
075r166 yry, containing grenats. In Peru. on other hand, MINE of Gualgayoc or Chota & Pasco in "alpine limest
260c069 ey frequent!!?-- Wilson's American ornithology a MINE of valuable facts,. regarding habits range & al
340d026 <birds». Anatidae.-- Consult this book again.-- MINE, is a bold theory. which attempts to explain, or
356d070 expression but does not include so many facts as MINE. The facts about half breed animals being wilde
382d157 te. Owen say such view worthy of a Lamarckian.-- MINE is much simpler.-- Hunter shows almost all anim
485z017 Staphylinidae &c &c. with reference to those of MINE from T. del Fuego p. 141. How comes it salt wat
526m028 though very odd is perhaps true.-- mem Erasmus & MINE taste for music.-- Children like hearing a stor
572n035 ce of a Deity has an expression the very same as MINE about our origin of a notion of a Deity We can
588n091 xplanation & instance of starting identical with MINE,-- Lamarck. Vol II p. 319.-- Habits more preval
022r006 rying & dubious granites.--Wide limits of this MINERAL in Australia. Fitton's appendix Would Slate. &
028r028 actions of rivers currents. & sea beaches. All MINERAL masses must have a tendency. to mingle; The se
093a032 . [1837 p. 297]CD thinks Olivine a preexisting MINERAL.-- Mem. Galapagos ∴ Basalt deepest?? Marcel Se
136a147 ps in Ladner Vol II p. 125. Good discussion on MINERAL veins p. 125 to 129 & p. 135--160 & 162 [blank
035r047 nts for oscillation of level independent of MINERALOGICAL nature & dependent: & then how wonderful l e
060r125 em Carolines quotation from Temple Urge the MINERALOGICAL difference of formations of S. America & Eu
049r089 e together in contorted layers: Mem: Phillips MINERALOGY some such fact stated to exist in Peru. -- A
028r028 resulting from degradation of Feldspar & other MINERALS containing Alumen.--This matter accumulating
076r168 ilas in New Biscay. "Nature, exhibits the same MINERALS <as> there, that are found in the veins of Ko
098d047 seen in stalactite.-- some force crystallizes MINERALS in layer. therefore aggregates them in layer.
055r106 are thought by the <oldest> «most intelligent» MINERS to be the richest Vol II 147 Shells at Concepe
566n010 92 Vol. III Octav. Edit)-- certainly neither a MINERVA or Apollo would have them because not beautifu
073r163 idered for most parts as rock eminently rich in MINES of gold & silver." «p. 131» The above porphyrie
075r167 un»common in Mexico. Fluor spar only in certain MINES. «Vol. III» "In general it is observed both in
076r168 nearest to the surface of the earth."--p. 156. MINES of Batopilas in New Biscay, "Nature, exhibits t
077r172 hladni.--on meteoric Mexican stone. Journal des MINES 1809. No. 151. p. 79. [misnumbering, no page 17
078r175 e mountains most torn.--(¿anticlinal line?). -- MINES of Catorce «(Principal veins)» 25 degree to 30
079r176 . Humboldt states that some of the richest gold MINES on ridge of Cordillera near Pataz, also at Gual
109a084 doubt. A-B once formed low coast.-- Annales des MINES. a translation of paper by rose on Greenstone,
134a143 l of Asiatic Society Vol I. {T} p. 145. on salt MINES of Punjab p. 149. on the <salt mines> «saline d
134a143 5. on salt mines of Punjab p. 149. on the <salt MINES> «saline deposits» of India p. 503. On Indian S
028r028 es. All mineral masses must have a tendency. to MINGLE; The sea would separate quartzose sand from th
231b242 otamus only African.-- American & African forms MINGLE in India & East Indian islds-- Monkeys differe
498q008 ch plants do not degenerate,-- as the Bees will MINGLE the infinitesimal varieties which must occur.-
028r030 owers. We know the waters of the ocean all are MINGLED. These reflections might be introduced either
203b138 n shells-- Mr Yarrell says that old races when MINGLED with newer, hybrid variety partakes chiefly of
565n008 herefore of the snarling order.-- But contempt MINGLED with disgust, when ones opponent is considered
589n091 almost identical with my theory-- no facts, & MINGLED with much hypothesis.-- see M.S. notes, where
056r111 r (perhaps lime) salt. & metallic ores.--which MINGLING & separating is well adapted to use of mankin
216b182 ot this contradiction to his view of races not MINGLING?-- In Foxes case of Blood Hounds, a little mi
216b182 ing?-- In Foxes case of Blood Hounds, a little MINGLING would probably have been good, namely such as
181b045 as far as compatible with such structure are in MINOR degrees adapted for. other elements. every part
306c214 ton will give an account in his Travels in Asia MINOR of the domestic animals-- At Angora «centre of
306c214 he domestic animals-- At Angora «centre of Asia MINOR» are the fine-haired goats. which it is said ca
306c214 «greyhounds» are Kurdish & come also from Asia MINOR.-- tail like setters. long ears-- colours vary,
446e160 n instance of this, & the female of the icterus MINOR is a bird of more splendid plumage than the mal
316c244 wildest imagination & superstitions.-- +York's MINSTER story of storm of snow after his brothers murd
553m135 arious, like ourselves) that savages (mem York MINSTER) consider the thunder & lightning the direct w
161g098 ort buttress but not continuous & it was 29.200 MINUS .008 ----- .192 Loch Lochy 4 ft above water Baro
021r005 is very common within E. Indian Archipelago, no MINUTE description, calls it a Fucus. P «Vol I 287» P
022r006 ce in facility of conducting Electricity? Would MINUTE particles have a tendency to change their posi
024r016 Fathoms Parallel of St Catherine [27 degree 30 MINUTE S.] 18--70 Paranagua [25 degree 42 minute S.]
024r016 ee 30 minute S.] 18--70 Paranagua [25 degree 42 MINUTE S.] 12--40 St Sebastian [23 degree 52 minute S
024r016 42 minute S.] 12--40 St Sebastian [23 degree 52 MINUTE S.] 12 50 {T} Joatingua SE [23 degree 22 minut
024r016 minute S.] 12 50 {T} Joatingua SE [23 degree 22 MINUTE S.] 5 35 R. de Janeiro SE [23 degree 58 minute
024r016 minute S.] 5 35 R. de Janeiro SE [23 degree 58 MINUTE S.] 18 77 C. Frio [23 degree S.] 7 60 {T} Soun
024r016 at Abrolhos. [18 degree S.] Bahia [12 degree 57 MINUTE S.] 8 200 Morro S Paulo [13 degree 22 minute S
025r016 57 minute S.] 8 200 Morro S Paulo [13 degree 22 MINUTE S.] 9 120 {T} Garcia de Avila [lighthouse] [12
025r016 {T} Garcia de Avila [lighthouse] [12 degree 35 MINUTE S.] 9 124 Itapicuru [R.] [11 degree 46 minute
025r016 5 minute S.] 9 124 Itapicuru [R.] [11 degree 46 MINUTE S.] 9 200 R. Real [11 degree 31 minute S.] & [
025r016 egree 46 minute S.] 9 200 R. Real [11 degree 31 MINUTE S.] & [R.] Sergipe [11 degree 10 minute S.] 20
025r016 gree 31 minute S.] & [R.] Sergipe [11 degree 10 MINUTE S.] 20 190 R. San Francisco [10 degree 32 minu
025r016 inute S.] 20 190 R. San Francisco [10 degree 32 MINUTE S.] 10 50 Whole coast to Olinda [8 degree S.]
025r016 Olinda.--a little WNW of C. Rock. [5 degree 29 MINUTE S.] still shoaler, coast composed of sand dune
029r033 most violent manner. Although it lasted about a MINUTE, there was no uncommon ripple on the water. It
035r046 f such can happen in troubled England; the more MINUTE equalities of elevation, may well be preserved
050r090 Iceland Bladders of Lava are described, & many MINUTE craters as at Galapagos. <|> Sir George Macken
065r139 idth 2 & ¼ miles S. Shetland. Lat. 62 degree 55 MINUTE. <only> one lichen. only production. a body wh
091a025 far out to sea valuable; because transportal of MINUTE seeds-- L. Institut. p. 209. May. 1837 Paper b
093a033 phur to form salts of America. -- The number of MINUTE turbos in red earth with volutas. prove regula
097a043 oco".-- <but> on one of the Ponza isles. but no MINUTE description is given.-- Vol II. 2d Series. p.
156g072 er on shelf 3d. 29.455 A 83 degree ∴ plain of 4 MINUTE shelf slope, above «line of 4th» shelf This sh
290c163 with pretty faces When horse goes a round, the MINUTE gets into the road at right anglles, how pleas
386d167 re are slow. if animals became adapted to every MINUTE change, they would not be fitted to the slow g

420e089 dwriting said to be heredetary. shows well what MINUTE details of structure heredetary'-- Athenaeum .
451e177 Java. (between Borneo & Java) Lat 5 degree. 50 MINUTE S. adjoining it are several small islands. abo
466t099 ents united in whole length to corolla--anthers MINUTE, distinctly doubled, brown, but with no pollen
468t105 uck nectar: «Maer June 41» Rhubarb. pollen very MINUTE--not excessively abundant flowers not attracti
468t112 veral gloomy days. hot one, Bees almost P every MINUTE to Fraxinella «& from <flower> plant to plant.
469t135 owers-- allowing each Bee visits 10 flowers in «MINUTE» each flower will be visited in 28 minutes-- s
470t153 resown.-- Humble 22 flowers of Egg Tree in one MINUTE Great Humble 17 flowers of Larkspur on two pla
470t153 species)-- in do-- do 3 of do in about <3/4> of MINUTE These latter were pollen gatherers & they seem
477z002 et mataco.) are all found south of 26 degree 30 MINUTE. Lat -- -- do. p. 207. La punaise was not know
496q05a Bee Larkspur= =Toad Orchis= How many flowers in MINUTE do they visit?? good=!! Examine pollen of doub
498q008 ges «& Cowcumber's out of doors.» much-- or the MINUTE Orthopt.-- important, as we know how readily t
499q09v ies or little insect?= or is pollen excessively MINUTE or abundant? do they seed plentifully? Look fo
501q011 t be crossed, I think, I expect, except by very MINUTE insects.-- (30) Get Abberley to plant SINGLE P
549m119 n,-- <but the> «&» the sensual enjoyment of the MINUTE add to the happiness.-- but as they are not re
549m120 a peasant, with whom sensual enjoyments of the MINUTE make large <parts> portion of daily <happiness
623o050 HC] Although I cannot pretend to say how far & MINUTELY our instincts extend, yet as they are acquire
469t135 f Fraxinella, with seven flower stalks for ten MINUTES. it was visited by 13 Bees-- & each examined v
469t135 visited this same bunch & on this day in five MINUTES eleven Humbles came & each visited many flower
469t135 in «minute» each flower will be visited in 28 MINUTES-- say then each flower is visited 30 times a d
522m011 rried Miss A. B.-- all the same names as a few MINUTES before he maintained he had never heard of.--
615o36v ark riding instantly followed, me and for five MINUTES every now and then howled.-- Now I don't think
186b059 another, preeminently Pachydermata, less so in MIOCEN & so on.-- As I have traced the <type> great q
042r068 rom peat.--yet cetaceous bones so likewise «of MIOCENE period».--Mem Bahia blanca P. 204 Vol III. Lye
178b030 y bifurcated.-- Type of Eocene with respect to MIOCENE of Europe? Loudon. Journal. of Nat History.--
221b202 logous to reptiles before Mammalia Think about MIOCENE fossils some species being recent agreeing wit
231b243 hange in form.-- This probably explains crag & MIOCENE.-- The descendants left in cooling climate mig
232b245 e change., (marking in their history an eocene MIOCENE & pliocene epoch), whilst others may die out o
236b278 ming Birds, types of former dogs. character of MIOCENE Mammalia--of Europe Mem. Mr Bell's case of Sub
473s03r post pliocene strata! Mastodon longirostris in MIOCENE like in Europe-- Cuvier never found remains of
257c055 animals have been so formed, then man may be a MIRACLE, but induction leads to other view.-- Till we
286c156 put in alternative of Man created by distinct MIRACLE. Macleay letter to Dr. Fleming. Philosophical
635j56r D though such a scheme. would require constant MIRACLES.-- p. 420 thinks the great fecundity of germs
224b216 eating animals with same general structure.-- MISERABLE limited view.-- With respect to how species a
293c174 wever, perfect, else one animal would not cause MISERY to other.-- else smell of Man would be disagre
523m016 In Mr Hardinge, was caused by thinking over the MISERY of an illness at Rome, when by accidental <was
564n004 ollowing conscience, obeying habits, & dread of MISERY of future thinking of injured moral sense.-- N
531m051 Burke maintains on pleasure in beholding the MISFORTUNES of others.-- In young children, the violent
523m014 ed against somebody.-- Have insane people any MISGIVINGS of the injustness of their hatreds, as <if>
359d087 «from Bombay» & Canada Goose.-- Former strange MISHAPED bird-- looks very artificial breed-- but Mr M
077r170 luence on the richness of the veta madre of [MISNUMBERED page] Dr D. remarks. bad conductor of Heat d
059r121 emical action as well as effects of cooling [MISNUMBERING, no page 122] In Igneous rocks.--which have
078r172 ne. Journal des Mines 1809. No. 151. p. 79. [MISNUMBERING, no page 173] Under name of Sagitta Tripter
613o036 urely instincts imply willing, therefore word MISPLACED The meaning of Words, must be made out Reason
321c270 imself Hume's do, with correspond. with Rousseau MISS M«artineaus» How to observe Mayo Philosophy of
399e010 lost power of producing. Williams. Narrative of MISS. Enterprise, p. 497. Vampire bat abound in the
426e108 ood being a great great-grandfather.-- -- Little MISS Hibbert case of Hindoism coming out more than i
472s01v ybridised. It is possible to raise them pure for MISS Bent three years since gave her some She means
521m008 epilepsy an habitual disease of the muscles.??? MISS Cogan's memory of the tune, might be compared t
521m008 irds singing, or some instinctive <or> sounds.-- MISS C. memory cannot be called memory, because she
522m011 t. him, took name of Child «of Kinlet» & married MISS A. B.-- all the same names as a few minutes bef
524m021 that early impressions are most durable.-- (but MISS Cogan shows that repetition is not necessary)--
538m078 singing songs of their childhood. & certainly of MISS Cogan, & fully corroborates the fact of her not
538m079 Beagle, some merry goodhumoured as self.-- «when MISS Cogan has remembered her song, then the song wa
555m142 only rather more steps.-- dispute about words.-- MISS Martineau (How to Observe p. 213) says charity
558m151 unison.-- active assistance. &c &c. it comes to MISS Martineaus one principle of charity.-- ¿ May no
573n039 ing not to flow-- flow it will.-- My father told MISS. C. of the bad conduct of Mrs C. (her brother's
573n039 ng but shrugged her shoulders.-- analyse this.-- MISS C. quite aware & indignant with Mrs C. but had
578n054 anger,-- or habitual friends.-- but half & half. MISS F. A. said to Mrs. B. A. how nice it would be i
578n054 A. how nice it would be if your son would marry MISS. O. B.-- Mrs. B. A. blushed. analyse this:-- Le
580n061 rved that some people of very weak intellect (As MISS Clive) have only possessed very loose ideas.--
297c189 est without doom. Vol 2. Mag of Z. & B. p. 431. MISSEL thrush lately increased in numbers over whole
632j53r ere means sufficient to ensure propagation of MISSELTOE?-- do p. 284. it is hard on my theory of gain
199b119 E.d. N. Philos. Journ.-- Paper by Crawford on MISSION to Ava. account of HAIRY man. because ancestor
255c050 leg) attended the distribution of food, at the MISSION of Mojos (even 20 leagues apart from each othe
409e047 o be converted into civilized man.-- ask the MISSIONARIES about Australians yet slow progress has don
471tf10 izards in Navigatores. Williams. Narrative of MISSIONARY enterprises Dr Andrew Smith says in the lark
043r070 P. 316. Earthquake of 1812 affected valley of MISSISSIPPI & New Madrid & Caraccas.-- Is this mentioned
087a010 thing -- Study account.-- Alluvial plains of MISSISSIPPI -- No Vol. I. p212. Cuvier Oss Foss Wide ran
338d025 r case of change in food in insects entered by MISTAKE Surely the fossil Mamalogy of Britain & Europe
359d087 4 birds «of», which I think there must be some MISTAKE in their origin) Saw cross between Penguin Duc
360d091 r state in which they are kept?-- Is there any MISTAKE about Yarrell's law, is it local (not artifici
363d102 illow Wren.--) (Goulds story of Water-Wagtails MISTAKE Both species scattered over Europe)-- The habi
421e091 likely in the modifications of these organs to MISTAKE a merely adaptive to an essential character--"
544m103 , with all other faculties: «Vide page 110, by MISTAKE.,» N B. Everything which happens to man who doe
546m110 nt in such a place.-- Vide page 103, supra (by MISTAKE) have lower animals these vivid thoughts In sa
553m136 -- Those savages who thus argue, make the same MISTAKE, more apparent however to us, as does that phi
633j54r nce to certain laws they can live.-- Hence the MISTAKE they are created for them. If we once venture
077r170 perimposed on N degree I. even No. 2. might be MISTAKEN for Porphyry above ancient freestone, limesto
185b056 have representatives (which at first would be MISTAKEN for) Carabidae, Crysomela, Scarabadae, & long
534m063 sylvestris. do. p. 228 Newport says Dr Darwin MISTAKEN in saying common wasp cuts off wings of flies
301c199 is this the Key to their mental powers.? p. 8. MISTAKES of instinct are external contingencies, where
157g078 um aquarium is a lip with it-- Dick right-- Mac MISTOOK terrace also right-- Granite such as boulder o
113a092 mportant as explaining want of levleness Major MITCHELL showed me a river <near> W. of Port Philip. w
113a092 & 5 fathoms deep. perfectly still water. Major MITCHELL inferred subsidence; Mem my remarks on coast
278c131 important!! like Dipus of present day??! Major MITCHELL does not think that dog was found in Van Diem
288c159 not be discovered till circles completed Major MITCHELL, does not know whether breeds of oxen have de
306c213 erhaps being in passion chief difference Major MITCHELL is not aware that Australian dogs ever hunt i
356d071 f facts-- Savages over whole world. (Major <I> MITCHELL p. 244. vol I) spit & throw dust <according t
358d075 antic Tristan D'Acunha ditto Juan Fernandez do MITCHELL. Australia Vol I. p. 306 "The crows were amaz
362d099 fight-- is analogous to the love of woman (as MITCHELL remarks seen in savages) to brave men.-- Effe
552m132 le will be given.-- Descent of Man Moral Sense MITCHELL Australia Vol I, p 292 "Dogs learn sooner to
278c130 fossil in Australia, & only one tree species (MITCHELL'S authority) in Australia, & several in Van Di
322c270 II There are many marginal notes <Rengger &c> MITCHELL'S Australia Walter Scotts life I & 2d & 3rd Vo
246c027 m account of natives, & probably on Oualan.-- «MITCHILL says snakes on Friendly isles. p 50. LX. Jour
223b208 tell whether articulate or intestinal, or even a MITE.-- a bee «compared with cheese mite» with its w
223b208 l, or even a mite.-- a bee «compared with cheese MITE» with its wonderful intincts <might well say ho
228b229 re him, when dissecting a whale, or classifyng a MITE, a fungus, or an infusorian. is "What are the l
074r165 salts, quite curious case of oxided Iron by MITTERSCHLICH. Vol. II Journal of Nat. & Geograph Scienc
251c041 d that the "Asseel Gazal. (Bos Gazoeus) does not MIX with the Gobbah or village Gazal.-- ¿is latter s
027r027 an others of the Bahama consists of rock & sand MIXED with sea shells--about 500 Isd. & great. banks.
073r161 e a meteoric origin & which is constantly found MIXED with lead & copper is infintely rare in all pa
109a084 bonate soda. formed by Ca. of L. & Mur. of Soda MIXED.-- Turner's Chemistry p. 206 Both Beck & Deshay
124a119 respect to epigmous action.-- if the zinc were MIXED with 90 percent of lead. it would be still more
148g026 current cleavage, & pretty well rounded stones, MIXED with some quite irregular very like rubbish at
224b215 probable owing to mixture of races.-- When all MIXED & physical changes (¿intellectual being acquire
424e102 t good case, but very odd since these crows are MIXED in England-- for I presume Carrion Crow is foun
495q005 of generation (5) Place pollen of Red Cabbage «MIXED with own pollen» on flowers of other cabbages &
549m120 ink the sum total of happiness greater. even if MIXED with some pain.-- than the happiness of a peasa

570n025 gging aroused acting» Octob 25. Why is modesty, MIXED with triumphant feeling so similar to shame aft
191b082 fecting all the Progeny of female insures often MIXING of individuals. Here we have avitism the ordin
204b139 Transactions Mr Wynne distinctly says that the MIXTURE between Chinese & English Breed. decidedly exc
224b215 uce something else., but not probable owing to MIXTURE of races.-- When all mixed & physical changes
573n038 t touching table.-- cannot avoid it.-- curious MIXTURE of voluntary & involuntary movements.-- Person
323c269 rdner's Music of Nature-- -- Herbert on Hybrid MIXTURES: Marginal notes. -- 20th. Carlyle's French Re
099a051 drag particls to line {P} AB, & likewise gravity MN. Then every particle would tend to meet at <B. b
451e180 e through the Moluccas 1825--"No wild animals in MOA.--" Chapt.-- V.-- : do. Chat XXI. Wild cattle &
024r015 s, from Sydney 90 fathoms La Peyrouse. South of MOCHA; 19 miles. 65 Fathoms Vide facts in Beechey. co
030r034 asin formed by outlying rocks; (such as between MOCHA & main land). <[...]> At Carelmapu.--Within Chi
499c009 ollen-- <Ne> In Oenothera bush.-- (19) Theory of MOCK flowers in Hydrangea (20) As Hop is Dioecious--
172b007 ren circumstances.--» Now Galapagos Tortoises, MOCKING birds; Falkland Fox-- Chiloe, fox,-- Inglish &
305c209 hing is explained.-- this is fact analogous to MOCKING thrushes of Galapagos having tone of voice lik
319c255 Pipra flycatcher.-- Bachman says he thinks the MOCKING thrush beats all English birds in song.-- one
319c256 thrushes of N. America. singing so well. & the MOCKING thrush being so very beautiful gret contrast w
284c149 viz. Macleay letter to Fleming p. 32 "where it MODE (of generation) varies according to the species,
616o39v of such a guide we can only <shew> point out the MODE «of perceptive action» by which we come to conc
617o40v tally accounted for, we prefer this metaphorical MODE of stating the fact to the mere statement of th
149g030 have deposited much-- On other hand remember MODELLING power of sea N of Valparaiso are those animal
224b212 last page explains that between Species from «MODERATELY» distant countries. there is no test but gen
226b220 y of every terrestrial Mammalia.-- Especially MODERATELY large ones.-- Is the flora of <S. America> T
031r038 I] line of high tidal action {P} NB. patches of MODERN Conglomerates [Fig. 2] {P} The action of sea A
041r066 es in Australia. New Red Sandstone. at Bahia in MODERN sandstone. a circle,.{P}, had in its middle a
055r107 h number of craters. No cliffs at Ascension (or MODERN streams of St Jgo) yet no historical records o
077r170 of plains of Amazon, no relation--there is more MODERN breccia, chiefly owing to destruction of porph
114a094 p «& therefore extensive» water ∴ not formed in MODERN formation & not ever in Secondary in Europe. g
117a102 ormations of Payta extend close to Guayaquil.-- MODERN shells of Cobija doubtful. Examine well shores
128a130 t Athenaeum.) I. Part. I Vol.-- some notices on MODERN Tertiary strata on coast of do-- I believe?? c
174b014 a different countries. Propagation explains why MODERN animals same type as extinct which is law almo
197b112 nimal made for itself does not agree with old & MODERN types being constant. Cuvier's theory of Condi
231b244 so cannot the domesticated animals with him.!-- MODERN origin shown by only one species, far more tha
278c132 ngland from Rhinoceros Elephants &c in the most MODERN period, compared to Faunas of these countries,
303c204 eeing that Tertiary geology has obeyed rules of MODERN causes. & considering over the viccissitudes o
399e006 lower layers.-- good objection to my theory: a MODERN bed at present might be very thick & yet have
408e044 having been purely result of elevation,-- «all» MODERN & wholly volcanic-- Azores might be prophecied
424e101 are there any cases of union of two regions in MODERN times.-- this would depend on negative evidenc
616o39v e manner attributed to the brain. There are two MODES of perceptive action by which bodily action is
618o041 bral portion of it Thoughts, perception &c. are MODES of subjective action-- they are known only by i
578n053 about a person even blushing in the dark-- «so MODEST a person.» A person who blushes in the dark is
578n053 who blushes in the dark is proverbially a most MODEST person one carries on, by association, the que
570n025 fear. both both have shame-- Animals have not MODESTY. analyse this.-- «Excellent-- my theory of blu
570n025 k) «Shrugging aroused acting» Octob 25. Why is MODESTY, mixed with triumphant feeling so similar to s
199b118 ws of life, the transmission of a fortuitous MODIFICATION, into a durable form, of a fugitive want in
404e025 tly states that the flipper is a mere simple MODIFICATION of an organ present in whole class. Case of
406e035 inst Yarrell's law.-- not so much against my MODIFICATION of it-- Goat & Moufflon will not breed.-- p
440e148 tra The Law of <growth> «generation» is only MODIFICATION, though important one, of growth «Lamark. V
576n048 handed down as an instinct.-- Instinct is a MODIFICATION of bodily structure «(connected with locomo
576n048 obtain a certain en<s>«d»: & intellect is a MODIFICATION of <intellect> «instinct»-- an unfolding &
611o034 living beings, matter is united in different MODIFICATION, peculiarities of external form impressed,
613o036 y Kirby» (:analogy:-- as races are formed or MODIFICATION of external form. so modifications of brain
615o038 ristian over Heathen race.-- But as no great MODIFICATION in brain would probably take place without
254c048 ind species each class successively present MODIFICATIONS, typical of succeeding classes & likewise t
418e084 losely related-- By birth the the succesive MODIFICATIONS of structure being added to the germ, at a
418e084 ood) when the organization is pliable, such MODIFICATIONS, become as much fixed, as if added to old i
421e091 true affinities. We are least likely in the MODIFICATIONS of these organs to mistake a merely adaptiv
611o034 fied by external accidents, & in such cases MODIFICATIONS bear fixed relation to such accidents. But
613o036 formed or modification of external form. so MODIFICATIONS of brain) As in animals no prejudices about
042r068 my belief in submarine tilting alone, must be MODIFIED. «Moreover, the Volcanos from sea there burst
109a082 -- a little further from beach action probably MODIFIED by form of waves & currents.-- but this must
156g072 up each main valley.-- & that river alone had MODIFIED it-- perhaps however sea also,-- Barometer on
158g080 of this terrace «on <will> Granite ridge or a MODIFIED Granite ridge» at head of Glen Roy on same si
171b004 ndividual «with fixed organization» thus being MODIFIED,-- therefore generation to adapt & alter the
192b084 a of beetles.-- born from beetles with wings.& MODIFIED.-- if simple creation, surely would have been
258c060 tends to keep to species to one form, (but is MODIFIED), the relationship of Analogy is a divellent
288c159 ping watch. how completely «nature & instinct» MODIFIED.-- The partial migrations of birds in same co
302c202 acquired,-- or aberrant, therefore more easily MODIFIED-- = this is not easily told, for any small fa
305c211 h memory. or reason? is necessary.--) which is MODIFIED into endless forms, bearing a close relation
315c243 y of viewing the subject important.-- Laughing MODIFIED barking., smiling modified Laughing. Barking
315c243 portant.-- Laughing modified barking., smiling MODIFIED laughing. Barking to tell other animals in as
342d036 of the world Astronomical <& unknown> causes, MODIFIED by unknown ones. cause changes in geography &
352d059 ution especially of Mammalia As every organ is MODIFIED by use, every abortive organ must have been o
382d158 nction in both sexes.-- are never double, only MODIFIED. those which perform very different, are both
423e099 which is sufficient only to have most slightly MODIFIED organic forms.-- we know not rate of depositi
460t019 are the descendants, <slightly> «a good deal» MODIFIED <& Many Forms lost; if> «of this old stock (w
526m030 petites urges it to certain actions, which are MODIFIED by circumstances, & thus the appetites themse
544m104 was this instinct gained.? by conversation-- ∴ MODIFIED in those races, where it is customary to die-
611o034 into certain form; invariable, as long as not MODIFIED by external accidents, & in such cases modifi
613o036 <all> there is one one instinct to all animals MODIFIED according to species. This I suppose he deduc
615o038 dren> «plants»! [RHC] 7) As definite instincts MODIFIED by heredetary;-- so succession so perhaps gen
615o038 fear of death in Hindoo population.-- Slightly MODIFIED in many countries, hence national character,
621o048 wrong or right; this teaching may be curiously MODIFIED by circumstances of country, so will the cons
624o50v theory are due to <habit> heredetary habit (& MODIFIED & associated during lifetime). so is our mora
580n061 once take hold of the mind, no subsequent ones MODIFY it.-- «Weak people say I know it because I was
128a129 gain statical equilibrium This will be only a MODIFYING cause. {P} land protuberant water to counterb
277c126 from every country & class tells us that ¡O. MODULATOR & O. Patagonicus. till neutral ground ascerta
151g040 mmediate neighbourhood, (as granite or gneiss of MOEL Derry) on low hill between Inn & Bouhunthine th
473s05v s by Tremadoc. but can tell them from lake S. of MOEL Siabod. wh. flows into Conway by Bettws & there
266c093 Lieut Wellsted "on coast of Arabia between Ras MOHAMNED & Jeddah". sheep numerous "of two kinds one w
058r118 es remparts; toutes affectent une pente plus ou MOINS inclinée vers le rivage de la mer, tandis, au c
257c056 e nous avions déjà observés en Patagonie. ou au MOINS des espèces tres-analogues,-- quand ce nétaient
267c093 ed from Norfolk Isld-- "& it now claims all the MOIST & fertile land of Tahiti, in spite of every att
565m009 eye. looking to distant object, brightened & MOISTENED by emotion,-- why does emotion make tears fal
104a068 essure? Salt on surface of plains due to whole MOISTURE being lost by evaporation therefore capillary
288c191 nd increases.-- upper parts attracting all the MOISTURE.-- Henslow thinks if leaf of plant varies, <w
255c050 ded the distribution of food, at the Mission of MOJOS (even 20 leagues apart from each other.-- this
307c216 o genera & classes. p. 479. fragment of tusk «& MOLAR tooth» of Hippotamus from Madagascar!!!!!! Proc
444e157 alludes to the resemblances of the snout of the MOLE & Pig in having two additional bones to give st
639j28r s in the old world!]CD p. 252 analogy of hand in MOLE, & Mole cricket & rodents (?) p. 251. all anima
639j28r old world!]CD p. 252 analogy of hand in mole, & MOLE cricket & rodents (?) p. 251. all animals run b
115a097 (see <supra> p 94) be accounted for by great MOLECULAR attraction of every atom in rock On a coast,
170b001 > or shortened repetition of what the original MOLECULE has done).-- This appears highest office in o
021rIFC of harbor = 180 See Daubisson both Volumes, and MOLINA 1st Vol & Lyell Sailed, 27th <Friday gale 29th
251c042 American dogs silent. Mem contrary assertion of MOLINA) (p. 277.). probably another in Jamaica & perh
481z009 Skunk inhabitant of Patagonia. Mem:-- S. Cruz. MOLINA N.p. 244. Baccalao. migratory fish.-- Se
040r063 hich must aid in preserving the terraces <...> MOLINA'S Case At Vesuvius. Vol III P. 124. Lyell. dike
320c276 at end; is so said to have Mackenzie's Iceland MOLINAS Chile Falkners Patagonia Azaras Voyages & Quad
464tf6r r Blyth, however, believes in the existence of MOLINA'S Pudu or goat There is ibex of Alp Pyrenees &c
246c026 . King-fisher of Europe, (Alcedo ispida) from MOLLUCCAS. scarcely differs at all from those of Europe

Page **(Key Word)**
196b110 extreme difficulty of hypothesis of connecting MOLLUSCA & vertebrata, that there must be very great g
212b167 recent shells between herbivorous & zoophagous MOLLUSCA according to periods.-- NB. Was Europe desert
287c158 rd better character, than variations in eye of MOLLUSCA. [+] These questions may be all disputable,
381d156 matode & cestoid Entozoa Allotriandrous <or M> MOLLUSCA, with pectinibranchiate order-- the Annelida.
381d157 t in effect be so in all.-- 2 NB. In Pectinibr MOLLUSCA.-- «or Cephalopoda» are there abortive traces
392d180 of life are monooecious. Will ova of fishes & MOLLUSCA «& Frogs» pass through birds stomachs & live?
405e032 shells.-- shows that progression of change in MOLLUSCA is somewhat similar in two hemispheres.-- It
414e060 b. 4th.-- Why has the organization of fishes & MOLLUSCA (& plants???) been so little progressive «!Ag
419e085 ious.-- With respect to the non-development of MOLLUSCA, which I have sometimes speculated might be o
421e090 -- \.do p. 250-- «speaking of» the terrestrial MOLLUSCA of Morocco «Mr Forbes says the Fauna»-- (near
436e134 fish now bear a very large proportion to other MOLLUSCA in cold parts of sea, like Cetaceae,-- althou
443e155 ent, yet the inference from some plants & some MOLLUSCA being hermaphrodite is, that intercourse ever
447e164 uch as cryptogamic plants & true hermaphrodite MOLLUSCA, & probably corals.-- these forms then ought
447e164 resemblance between carboniferous. «& recent» MOLLUSCA, than between the corresponding acalepha?-- B
638j059 h rejects the imperfect attempts. In the «Bee» MOLLUSCA the nervous system is endowed with the knowle
640j167 ered.-- From these views, one would infer that MOLLUSCA would offer few species, or rather be very sl
641j29v ls-- Exudation of fetid «& acrid» secretion in MOLLUSCA. insects «Carabids & Staphylini» & Mammalia.
641j29v taphylini» & Mammalia. The eye being formed in MOLLUSCA, Articulata, & Vertebrata, & Planaria, & ligh
641j29v n pedunculated eye of Chamelion. crabs Crabs & MOLLUSCA we have analogues The stillness p. 276) of fl
211b163 - L.' Institut. Curious paper by M. Serres on MOLLUSCOUS animals representing foetuses of Vertebrata,
343d037 lurian, he has made a long succession of vile MOLLUSCOUS animals-- How beneath the dignity of him, wh
384d162 male & female flower on same stem.--» so that MOLLUSCOUS hermaphroditism takes place.-- thus one orga
420e089 ertebrate animals.-- must have been like some MOLLUSCOUS «bisexual» animal with a vertebra only & no
198b113 tell something) p. III G. St Hilaire Insects & MOLLUSCS allowed to be wide hiatus: states in one the
212b168 a) after Coal Period.-- ¿In those divisions of MOLLUSCS. where species now least in number (as cephal
226b223 rmediate, see how that is.-- ¿are shell-boring MOLLUSCS, like Carnivorous Mammalia in their wide rang
229b232 ermaphrodite animals couple: argument for true MOLLUSCS coupling.-- Dr. Smith's Information Long Horn
233b252 w does this agree with longevity of species in MOLLUSCS!!! When we talk of higher orders, we should a
250c040 , on replacement of Cephalopods & Trachilidous MOLLUSCS. by each other in secondary & Tertiary period
313c234 spect to law of mammals shorter duration, than MOLLUSCS, argue case both in Europe & S. America. «ver
379d151 ted to salt water?-- peculiar species, crabs & MOLLUSCS few.-- ¿are not some same-- what is the allia
381d156 e Dioeecous as» Cephalopods, Pectinibranchiate MOLLUSCS.-- insects. spider crabs.-- (all these howeve
382d158 , but unequally developed.-- surely analogy of MOLLUSCS. & neuter bee would shew this-- (Do any male
479z005 hout shells.! p 442.-- Planariae p 451.-- many MOLLUSCS Under the name of Sagitta Triptera M. D'Orbig
133a140 is led, to look at globe as resting on film of MOLTEN rock.-- Voyages of Adventure & Beagle vol I. p
584n071 migration,-- subsidiary to food & temperature MOLTING & breeding instincts, sexual, social, «subordi
407e042 n I remember says Mariana Deer very close to a MOLUCCA species.-- L'Institut 1837. p. 253, on animals
240c013 60. Vol I. Cynocephalus. niger. comes from the MOLUCCAS «Matchian» & Celebes.-- \ Amboina; Viverra Zi
240c013 bes.-- \ Amboina; Viverra Zibetha. \-- All the MOLUCCAS, Waggious New Guinea. New Ireland, have phala
242c017 g & Pongo in common\-- Galiopithecus common to MOLUCCAS & Pelew Isds.-- p. 22. New Calidonia-- New Ir
246c027 to C. muscadivora., which lives in the Eastern MOLUCCAS, New Guinea.-- (Case of replacement)-- Coquil
247c028 on Isle of France Scincus multilineatus (p 45) MOLUCCAS & New S. Wales Scincus Cyanurus «p 8 &». p 49
247c028 ales Scincus Cyanurus «p 8 &». p 49 on all the MOLUCCAS «New Guinea, New Ireland» & «even» Java. & ve
451e180 us temper-- Hodgson Koloff. voyage through the MOLUCCAS 1825--"No wild animals in Moa.--" Chapt.-- V.
452e181 e & it is a singular thing that throughout the MOLUCCAS Archipelago they are only to be found on the
241c014 lebes, Bourou new species of Axis.-- «Cervus MOLUCCENSIS is different from that of the Marianna islan
075r167 o rare in Europe. common there accompanied by MOLYBDATED lead & «argentiferous lead»; sulfated Baryte
063r132 nt). we see an individual divided either at one MOMENT or through lapse of ages.--Therefore we are no
206b149 as simile In a decreasing population at any one MOMENT fewer closely related;∴ (few species of genera
222b206 adaptation to changing world:-- I cannot for a MOMENT doubt, but what cetaceae & Phocae now replace
256c054 ll from which country,-- I often disputed for a MOMENT,-- Galapagos, S. American-- -- genus.-- The ci
547m112 .-- there was a kind of ideal consciousness for MOMENT, implied by «presence» my servant, «box» my ow
565n006 yaena from fear, no actual intention to bite at MOMENT, but mere symbol of readiness, & therefore don
613o035 ot conscious as in sleep; or in sleep is man MOMENTARILY conscious, but is memory gone?-- Where pain
533m060 rofession, or rather in only fully expressing MOMENTARY feelings of gratitude, I had a sort of consci
541m090 chose.-- although thinking «& talking» for the MOMENTS with interest on each.-- ∴ my father. is right
106a073 there more degradation at first angle owing to MOMENTUM. which the water has obtained.-- If {P} incli
291c167 ale>. Series-- in plants we have a step between <MON> monoeecious & dioecious plants in animals it ma
569n022 vage & Frenchman, unaccompanied by dignity-- "no MON dieu," with a shrug-- "all I can say, I am very
176b022 eath of species, than individuals If we suppose MONAD definite existence, as we may suppose is the ca
177b029 er, whole class would die out, therefore not.-- MONAD has not definite existence.-- There does appear
190b078 gh to form that species.-- <Man is derived from MONAD, each fresh--> Mr Don remarked to me, that he t
175b018 st origin there must be progress. if we suppose MONADS are «constantly» formed ¿would they not be pre
304c206 ll doubt this if series now existed from Man to MONADS-- though physiology would profit. if the serie
152g046 l with shelves effect of corrosion & not cause. MONDAY a rapid descent of a terrace except at very he
472s02r before-- Saw Fly 21 «this Heartease withered on MONDAY.--» alight on upper petals & insert proboscis,
207b152 polymorphes que toute la terre ferme de lancien MONDE".-- Considers forms in recent volcanic islets n
569n022 ary no good There is a Lutké's Voyage autour du MONDE (1826-9) Paris. 1835 Quoted repeatedly by Waitz
599o005 could not have existed without it.-- quotes Ld MONDOBBO.-- language commenced in whole sentences.-- s
523m016 ness at Rome, when by accidental <was> delay of MONEY, he was «only» NEARLY thrown into a hospital.--
531m050 he importance.-- people very often forget where MONEY is placed.-- (How often one forgets where put o
629o055 ar manner our desires become fixed to ambition. MONEY, books &c &c.-- <]> the "secondary passion" of
385d163 s me he has known many cases of bitch going to MONGREL, & all subsequent litters having a throw of th
385d163 all subsequent litters having a throw of this MONGREL.-- I did not ask the question.-- His bitch wil
426e107 one large body of varieties are fertile & make MONGREL, & other great series quite otherwise & make o
462tf05 tual than others-- is incidentally said that a MONGREL man may lose all traces of his parentage in «a
491q002 ns Regarding Plants. 1. Uniformity of hybrid & MONGREL offspring 2. How have late varieties of Peas &
596nIBC fall into habits Get facts about instincts of MONGREL dogs Do blubbering children, if of. convulsive
491q01v affected.-- does one branch of Cabbage being MONGRELIZED affect other branches-- The French Apple tre
179b032 purely bred owing to <first> having once borne MONGRELS he has thus seen the black blood come out fro
228b230 uce real crabs, & in each case similar or mere MONGRELS.-- It really would be worth trying to isolate
333d005 f mother: more than of the Common cat.-- Ch IX MONGRELS Hybridisim Fox has half Persian cat. which br
201b126 f wonderful fossils of India.-- & p. 545 «great MONKEY» Mr Johnston says Mag of Zooly & Bot. p 65 Vol
202b133 arole article by S Hilaire on wonder of finding MONKEY in France.-- of genus peculiar to East Indian
203b135 erica. Hilaire does not seem(?) to consider the MONKEY as a wanderer, but as produced by climate?-- M
208b154 ing its parent type in hotter parts of world Is MONKEY. peculiar to C. de Verd's.--? NO Macleay Name
212b164 Bear peculiar to Sumatra & not found on Java-- MONKEY peculiar to. latter not to former-- Mr Martens
244c012 Isle of France.-- the Tenrecs from Madagascar. MONKEY from Java.-- Hairs, & deer.-- Procured two mak
263c074 rix, many genera & species-- The believing that MONKEY would breed (if mankind destroyed) some intell
303c204 might expect that animal halfway between man & MONKEY, would have differed in hair colour + + + form
306c212 would last were the stream 1000 miles long.-- a MONKEY. (Baboon) at Z. Gardens upon being beaten beha
317c249 yage-- p. 36 "Cercopithecus saboeus" said to be MONKEY of St Jago C. de Verd; same as on coast of Afr
337d022 om comparison of Man, with expression <of a> of MONKEY, «when offended» who loves, who fears, who is
376d136 ve been told is Cyanocephalus Porcarius.-- this MONKEY did not like a great coat made for it at first
376d137 ade it meet in front.-- Dr Smith every baboon & MONKEY, big & little that ever he saw knew women.-- h
376d137 - The Cercopithecus chinensis: (or bonnet faced MONKEY he has seen do this.-- These Monkeys had no cu
376d138 een to evince more lewdness for bitch than dog: MONKEY thus examine each other sexes,-- «by taking up
387d170 y between caterpillars with respect to moths, & MONKEY & men.-- each man passess through its caterpil
387d170 man passess through its caterpillar state. the MONKEY represents this state.-- When it is said. that
402e018 d.-- G. W. Earl's Eastern seas. p. 206-- shot a MONKEY, ceased their cries. "many of them descending
415e068 , Hand & throat) that has made a man.-- CD [any MONKEY probably might, with such chances be made inte
465t079 -- & Africa!,-- any negative argument against-- MONKEY-man, valueless.-- May not several generations
526m028 . they know.-- pleasure of imitation (common to MONKEY), & not imagination.--» Thinking over the scen
545m106 boon. knew women.-- Another little old American MONKEY «(Mycelis)» I gave nut, but held it between fi
545m107 r» the expostulatory angry look of black spider MONKEY when touched, also another monkey to dog. I sh
545m107 black spider monkey when touched, also another MONKEY to dog. I showed nut & then closed my mem. exp
553m138 rked that he had never seen any of the American MONKEY show any desire for women-- «very curious. as
554m138 it is chiefly shown in old male.-- A very green MONKEY (from Senegal he thinks Callitrix Sebe??) he h

555m143 e this power abortive muscles) The black Spider MONKEY, very different disposition from others, slow
590n094 I. p. 234) Vol. II p 153. «do». an account of a MONKEY in a passion like Jenny.-- Dr. Abel has given
595n115 een dead about two months. saw a «black» spider MONKEY brought it at opposite end of house. & commenc
213b169 d <species> races.-- If all men were dead then MONKEYS make men.-- Men makes angels-- Those species w
224b214 erate common kind, which is not probable; then MONKEYS will never produce man. but both monkeys & man
224b215 then monkeys will never produce man. but both MONKEYS & man may produce other species., man already
231b242 an forms mingle in India & East Indian islds-- MONKEYS different not travellers?? Royles case of Hima
233b250 . p 38. account of fossils of Sewalick «India» MONKEYS of old World. Crocodiles. Anoplotherium.-- <M.
253c046 vota??-- <compare rodents of two countries» & «MONKEYS.» Fact of Elephant same species in Borneo. Sum
263c078 ecies to produce contingents proper.-- Present MONKEYS might not,-- but probably would.-- the world n
264c079 standing language of Fuegian, puts on par with MONKEYS-- Gould seems to think that many species when c
266c092 ild animals on isld.-- Niether Hyenas; jackals MONKEYS-- common to either coast not found here not ev
295c183 mare by being shown, young one.-- Many African MONKEYS in Fernando Po-- no new forms only species!! N
299c196 veloped amongst Pachydermata. like Man amongst MONKEYS-- or dogs in Carnivora.-- Man in his arrogance
303c204 other. points. female genital organs «in some MONKEYS clitoris wonderfully produced».-- make abstrac
331dIFC et.-- 1838 [In this Book some curious notes on MONKEYS recognising Sexes of animals:]CD [All Selected
376d136 . Smith «Remarks on extraordinary curiosity of MONKEYS». The Baboon of which anecdotes have been told
376d137 nnet faced monkey he has seen do this.-- These MONKEYS had no curiosity to pull up trousers of men. E
376d137 ew <men> women, thinks perhaps by smell.-- but MONKEYS examine sexes of every has repeatedly seen one
377d139 Dr. Smith more distant.-- But he thinks other MONKEYS make st.-- noise In case of woman instinctive
402e018 , small bears or badgers, deer. apes, baboons, MONKEYS & an animal probably a tapir p. 233. dogs in B
402e019 Vol <I> II p.,133: at Samar SE of Luçon, many MONKEYS, buffaloes &c &c-- Malte Brun. would be worth
402e020 ant-- in Magindanao several kinds of the large MONKEYS.-- Fewer <of th> «Mammalia» have passed to Par
409e046 is precisely analogous case to man, exceeding MONKEYS;-- Having proved mens & brutes bodies on one t
415e068 on the number of the species.: therefore Man & MONKEYS have equal chance that progenitor was bimanous
451e177 ris between Andaman & Pegu. <have> abound with MONKEYS & squirrels.-- Horsbrugh E. I. Directory. Vol
451e180 . Chat XXI. Wild cattle & Hogs on Timor-land-- MONKEYS do not exist. there & it is a singular thing t
452e181 Bunwood (18 miles in circum) there are hogs & MONKEYS <at> near shore of Magindanao Journal of [Asia
464t063 ful discoveries= negative facts are valueless= MONKEYS» Owen has described a greatt Struthonidous Bir
465t079 everal successions of Mammals.--» yet only two MONKEYS, <there are now> have been found fossil in S.
542m094 uliar not common?--» no bark of anger nor have MONKEYS & many other animals,-- but yet when angry it
543m097 ble, thus carthorse & dog.-- birds many cries. MONKEYS communicate much to each other.-- Waterhouse s
545m106 of eyebrow important analogy with man.-- I see MONKEYS grin with passion, that is show all the teeth:
545m107 back & kicked & cryed like naughty child.-- Do MONKEYS cry?-- «they whine like children.--» Expressio
546m109 aste for musical sound with birds. & ¿ howling MONKEYS-- smell with many animals-- see how a dog like
551m128 oul, are not derivable from experience.-- read MONKEYS for preexistence <">-- The young Ourang in <Zo
553m137 remarked that <exp> the expression & noises of MONKEYS go in groups. thus the pig-tailed baboon, shov
553m138 very curious. as they depart in structure» The MONKEYS understand the affinities of man, better than
555m142 rdens-- Endeavoured to classify expressions of MONKEYS-- I could only perceive that the American ones
560m156 ang like smells «peppermint» «& music».-- Have MONKEYS lice?-- picture.-- Do female monkeys not show
560m156 ».-- Have monkeys lice?-- picture.-- Do female MONKEYS not show signs of impatience when woman presen
560m156 colour of bare nails--, & of eyes.-- Do female MONKEYS care for men.-- Have we any forms in the hotho
563nIFC n M Expression N What are sexual difference in MONKEYS.-- Charles Darwin [Private.]CD (Metaphysics &
568n018 - is this origin of our pleasure in music-- do MONKEYS howl in harmony-- frogs chirp in do-- union of
568n018 singing with Mammalian structure. «-- American MONKEYS utter pleasant plaintive cry--» The taste of r
594n115 dog his companion. Descent --Affection & [...] MONKEYS «Ogleby» seen Zool. Soc-- 1838 remember with d
596n184 pithecus? very much., «Keeper says some of the MONKEYS move <its> «the» ears but <not> Chimpaze. does
596n184 ad & 2 ears.-- emotions of every kind.-- «[Are MONKEYS <are> right-handed??]CD» Cyanocephalus, Macacu
596n184 . Cercopithecus make labial st. S. American MONKEYS. pull back skin from head very little Does blo
639j28r og. Walrus. Fly. Gecko &c. Prehensile tail. in MONKEYS & Marsupials. Harvest mouse & (Chamaelion?) C.
466t094 to which stamen & pistils have no relation. In MONK'S Hood, a bee entering long nectary, would «nece
470t178 ceous; all wh. are in flower «I saw Bees;»-- on MONK'S Hood, brushing over stamen «Egg Tree»--I think
507q15v hether any Papilionaceous plants,-- whether many MONO or Dioeious plants, & any with peculiarities of
423e098 of these great seed-growers--).-- Feb. 24th. MONOCEROS, which Sowerby says, is an American form.-- h
206b150 - Brongniart.-- no dicotyledenous plants & few MONOCOT in Coal formation? p. 320 <Think> States Crypt
390d177 e developed forms) Study buds-- gemmae-- & MONOCOTYLEDENOUS, do those which are Monocotyledenous have
390d177 e-- & monocotyledenous, do those which are MONOCOTYLEDENOUS have many flowers same Spath, as they hav
291c167 e separated.-- in plants we have some flowers MONOCOTYLEDENOUS & others dioecious. some flowers hermaphrodi
291c167 ies-- in plants we have a step between <mon> MONOECIOUS & dioecious plants in animals it may be diff
450e174 ack Swan «in domestication & nature» strictly MONOGAMOUS-- geese polygamous (;when wild) but only som
450e174 only some birds are so when wild-- wild ducks MONOGAMOUS-- tame ones highly polygamous-- change of ins
251c040 parrots habitations India & Africa.-- NB. Any MONOGRAPH like Gould on Trogons worth studying.-- «do»
251c041 234.-- good case.-- p. 526. (ref) To Temminck MONOGRAPH. Mammal; «4to» good facts about distribution
427e109 ortion of one sex.-- Linnaean class Dioecia & MONOECIA. ought to be preeminently artificial.-- Would
435e130 abortive organs, even more so than Polygamia: MONOECIA & Dioecia, preeminiently artificial, so that
381d157 ion to separtion of sexes, as may be seen in MONOECIOUS & Dioecious plants.-- NB. in Heautandrous an
381d157 nts even in same genus some dioecious & some MONOECIOUS-- (& cultivation might make one set of organ
392d175 t after some time]CD What kind of plants are MONOECIOUS or dioecious.-- very curious how this was su
392d175 ly dioecious..) only simple form of life are MONOECIOUS. Will ova of fishes & Mollusca «& Frogs» pas
418e080 l recieve illustration from domestication of MONOECIOUS plants, & abortion of others.-- ¿ in hemi-he
434e130 o have taken place.-- Almost all Dioecious & MONOECIOUS plants have rudimentary abortive organs, eve
454e184 ex & sometimes. other, so as to become <all> MONOECIOUS.-- Are there not wild plants, some partly di
492q002 9. Do plants in becoming double ever become MONOECIOUS-- loosing one sex & not other: which general
499q09v s which are known easily to be crossed & all MONOECIOUS plants.-- Hooker says Rafflesia is dioecious
506q015 eir smell-- form of flowers-- Nectaries-- In MONOECIOUS «order» flower occupy particular position.--
506q015 -- Col. le Couteur on Wheat.-- (41) Have any MONOECIOUS or dioecious plants the Papilionaceous struc
506q15v per 43. Any flowers of Keeling Dioecious, or MONOECIOUS, besides the Nettle. at Galapagos-- Dioeciou
213b168 rmerly there would have been many genera of MONOTREMATOUS animals.-- p. 82 «There are many tables in
555m144 banter & joking» because the whole train of Dr MONRO experiment about hanging came before me showing
504q013 e if pollen naturally carried, on account of Van MONS views-- Also PEAS-- N.B. I think very likely th
173b011 en continet of Pacific existed might have been MONSOONS.. when they ceased importation ceased & chang
240c004 s of varieties.-- One approaching to nature of MONSTER, heredetary. other adaptation.-- Mr Yarrell sa
265c085 izes wood, but adaptation.-- albino however is MONSTER. yet albino may so far be considered an adapta
265c085 dwarf plants from seed, one adaptation, other MONSTER.-- The only way of judging whether structure i
391d175 w about sexes of twins in former case.-- (many MONSTER are really twins.)-- It is absolutely necessar
399e009 f animal, & not if it be solely to womb., as in MONSTER, or solely to childhood, or solely to manhood,
621o046 to do so, by other natural appetites.-- he is MONSTER, or unnatural if malevolent, or hates his chil
197b111 same organ in different animals in scale.-- In MONSTERS «also» organs of lower animals appear.-- yet
198b114 His work. Philosophie Anatomique. 2d Vol about MONSTERS worth reading NB well to insist upon <differe
210b161 suggested to me, that the <cas> production «of MONSTERS» (which, Hunter says owe their origin to very
218b190 ly applies to plants Female pig apt to produce MONSTERS in Isle of France-- -- Madagascar oxen with h
257c059 olely producing like itself, not applicable to MONSTERS:-- Are monstrosity heredetary??.? Does not at
311c226 facts about abortive organs &c The doctrine of MONSTERS is preeminently worthy of study on the idea o
335d014 & exceptions. to above law.-- Study what these MONSTERS are:-- are they «abortive» twins.-- The fer
335d014 cross being new species, -- Are not dreadful MONSTERS, abortive, just like mules. Fox's half bred P
366d112 law of variation. -- «(as Hunter supposes with MONSTERS)» if armless cat can propagate, ie with the c
384d161 l OEconomy. (by Owen) p. 44. Classification of MONSTERS. (1) From praeternatural situation of parts (
420e088 gy.-- L'Institut 1838. p. 414; M. Guyon thinks MONSTERS more common in Africa than in Europe especial
240c004 r, sometimes like mother. The fact of great MONSTROSITIES being produced, & handed down, with ease, i
255c052 method of nature to prevent the picking of MONSTROSITIES as Man does.-- One is tempted to exclaim th
351d057 lluded to Hunter's views on this subject.-- As MONSTROSITIES, kind of determined by age of foetus.-- As
391d174 nge?-- Man properly is hermaphrodite (hence MONSTROSITIES tend that way «& from frequency of this ten
408e045 lly.-- it is easy to get 50 of same kind of MONSTROSITIES-- G. B. Sowerby.-- Looking over Lamark sur
257c059 ke itself, not applicable to monsters:-- Are MONSTROSITY heredetary??.? Does not atavism relate to th
259c066 ation right?-- If puppy born with thick coat MONSTROSITY, if brought into cold country, «& there acqu
355d066 different in species of same genus." Law of MONSTROSITY not prospective, but retrospective as showin
355d067 ns are little fixed-- (<also> Hunters law of MONSTROSITY with regard to age of foetus. distinct consi
365d107 giving average of effects of country, (& no MONSTROSITY, or adaptation to unhealthy state of womb).-

```
365d107 been   adapted to former changes. than a mere MONSTROSITY propagated by art. Yarrell told me of a  cat
460t017 way.-- the laws which determine the kinds of MONSTROSITY, & determine the kind of variation «& sporti
139a180 &c &c Will any inorganic substance cause such MONSTROUS growth as oak galls or rose <buds> galls.-- i
199b119 of peculiar people banished by rest?-- ∴ most MONSTROUS form has tendency to propagate, as well as di
228b230 some   plant, whose seeds go back again, not a MONSTROUS plant, but any marked. variety.--   Strawberry
366d108 have   produced from these.-- this instance of MONSTROUS variety. which could not have been persistent
441e151 o escape it «in any case» we must draw such a MONSTROUS conclusion, that every organ is become fixed.
490q001 haracter of shells of Sandwich group {Sowerby MONSTROUS Cardium-- does it remind him of other species
492q002 emi-doubl flowers are those whose stamens are MONSTROUS, how then are seeds ever raised? 11. Is not n
496q006 rew» (9) Place. Snap-Dragon. (I have seen one MONSTROUS) Fox Glove & such like in very rich soil-- As
469t119 & two joints-- as it is on one foot probably MONSTROUOS & not a second species.-- <Saw> Maer. June 1
058r118 I. p. 54. M. Bailly says."en effet toutes les MONTAGNES de cette ile se developpent autour d'elle com
058r118 e abrupte et souvent taillée pic. Toutes ces MONTAGNES sont formées de couches paralleles et incliné
059r119 auteurs communes sur le revers de chacune des MONTAGNES qui forment les vallées ou les scissures.--M.
253c046 vages into Australia.-- What are they? Colonel MONTAGU probably contains some facts about close speci
325c267 Mammalia of Paraguay. account of wild cattle & MONTAGU on birds (facts about close species)  Wilson's
540m089 .-- Why does person cry for joy? 17th. August MONTAIGNE (Vol. I) has well observed, one does not fear
544m104 nt for early education.-- fear of death!!! as MONTAINE observes. distinct from pain, for one hates p
078r175 to  NE. vein of Moran 84 degree NE. of Real del MONTE 85 degree to S. // Tasco 40 degree to NW (after
022r007 d through the great Plas Newydd dike.--Mem tres MONTES. ((Henslow Anglesea)) great variety in  nature
049r088 aries in nature, -- Does not granite at C. Tres MONTES become more siliceous in close contact? -- «Co
134a141 iago p. 328. dead trees on Isthmus of Pen. Tres MONTES.-- as by subsidence ‖ Fitz Roy refers to ‖ & R
308c217 bit I kept my tea in right hand side for-- some MONTH, & then when that was finished kept it in-- <ri
332d002 but   now (July 1838) he has seen more case in a MONTH, than in several previous years, two having con
334d011 broken   from their mares, (though horsing every MONTH) & worked in the same cart in loose chains,  by
380d155 of   seven-- if woman does not menstruate in the MONTH, she will in 5 weeks.-- A Bull is never taken f
220b198 different.--  Henslow says. (Feb 1838) that few MONTHS since in Annales des Sciences, paper on Botany
451e178 ys native tribes can live there)» flee during 8 MONTHS out of 12.-- the largest mammifers in the worl
594n113 nurse had to carry it out of the room. nearly 3 MONTHS old. What is absurdity, why does one laugh  at
595n115 whose   companion had S[...] been dead about two MONTHS. saw a «black» spider monkey brought it at opp
179b035 ntry owing to springing from one branch, & the MONUCULE has definite life, then all die at one. period
179b035 en all die at one. period, which is not case,∴ MONUCULE NOT DEFINITE LIFE I think {P} Case must be th
230b239 hange can only be traced geologically (& then MONUMENTS imperfect) or horizontally & then cross breed
073r162 ble that if meteoric stones simply pitched from MO«O»N, that the metals should be those which have ma
119a104 to   equilibrium.) 3d. there are mountains in the MOON, which though not very analogous (see Edinburgh
131a135 ut 1838 p. 400. Observations on Mountains of the MOON. by Dr. Nichol-- adduces the case to show  Sir.
132a137 hquake of Chile, with that of the passage of the MOON.-- Ask Hopkins. M. Parrot, Mem. Acad. Imp.  des
138a153 two   years.-- will serve for comparison with the MOON at some future time [blank] Sir. J. Halls Paper
409e049 gist the history, & from the Astronomer that the MOON probably is uninhabited]CD & if my theory be tr
450e174 go" Published at Singapore in 1837. by Mr. J. H. MOOR-- -- p. 1. Elephant. Rhinoceros Leopard (but no
450e175 East of y Burrampooster & S of Tropic-- By J. H. MOORE after quitting Bengal this fact is noticed in C
263c077 ses some of the same general instincts, <as> & <MORAL> feelings as animals.-- they on other hand  can
385d165 son, the character, constitution, & most of the MORAL qualities of the father!. In how many daughters
409e049 foundation of all that is most beautiful in the MORAL sentiments of the animated beings.-- &c-- If ma
523m016 t it than of a dream.-- Insanity is produced by MORAL causes (ideotcy by fear. Chile earth quakes). i
537m075 ith examples» very justly there is no universal MORAL sense.-- «from difference of action of approved
537m076 ts.-- Fact most opposed to this view, where the MORAL sense seems to have changed suddenly-- but  are
537m076 rhaps if they had murdered their children, this MORAL sense, would have been so much, as in other rac
537m076 right & wrong «(& therefore in fact only limits MORAL sense)» which she seems to think «are» to  make
552m132 as well as rule will be given.-- Descent of Man MORAL Sense Mitchell Australia Vol I, p 292 "Dogs lea
558m150 ject to habit, as being less so will.-- May not MORAL sense arise from our enlarged capacity «acting»
558m151 works   of the whole world.-- Read Mackintosh on MORAL sense & emotions.-- The whole argument of expre
564n003 fferent.--  ‖ Different nations having different MORAL sense, if it were proved instead of  militating
564n004 & dread of misery of future thinking of injured MORAL sense.-- Notion of deity effect of reason actin
568n017 e is prevented performing heredetary habit, (or MORAL sense, or instinct,) one feels pain, & vice ver
578n054 , especially a women think of my face,"? to one MORAL conduct.-- either good or bad. either giving  a
595n184 pressive, (Vol. 4 of Works) [blank] "Adam Smith MORAL Sentiments" much on life & character "Humes Dis
599o004 y? Expressions N [Old & USELESS notes about the MORAL sense & some metaphysical points written  about
600o008 er on each faculty of Soul.-- (I) <Conscience» «MORAL Sentiments» imperative sense of duty-- which ma
600o008 in one body-- (2) Beau ideal, refers chiefly to MORAL, beau desires conscience & love.-- [With regard
600o08v es not Mackintosh make great difference between MORAL sense & conscience? we admire what is right  by
605o18v so often gives, when excited by other means, as MORAL excellences, brings to our recollection the ori
609o030 greatest   happiness.-- The other says we have a MORAL sense.-- But my view <says» unites both «& show
609o030 necessary for good at all» is the «instinctive» MORAL senses: (& this alone explains why our moral se
609o030 »» moral senses: (& this alone explains why our MORAL sense, any more than a hive of Bees without the
609o30l ell) + + Society could not go on except for the MORAL sense which experience (education is the experi
609o30v ormer is endeavoring to change that part of the MORAL sense, is strictly analogous to change of insti
610o30v ertain degree <of> right.-- The change <of> our MORAL sense, Looking at Man, as a Naturalist would  at
619o042 Maer Mackintosh Ethical Philosophy [RHC] On the MORAL Sense p. 152. Reason never can lead to action.-
619o043 orse as will be presently shown.-- This then is MORAL approbation, as far as it goes.]CD But  should
621o046 former;   & likewise <that the> then receive the MORAL approbation of his fellow men.-- [RHC] 6) Hence
624o50v dified & associated during lifetime). so is our MORAL taste p. 152. Reason never can lead to action.-
625o50v ation «the utility part being blended & lost» & MORAL sense.-- My theory explains both, perhaps, by h
627o053 . 14. It is allowed, that we have conception of MORAL obligation «when grown up???» & the question is
627o053 tellectual faculties-- Will Eugenius allow this MORAL obligation? [2] [The improvement of the instinc
628o054 stinct.--]CD p. 22. says affections, desires, & MORAL sense all different.-- P. 22. Butler & Mackinto
628o054 .-- P. 22. Butler & Mackintosh characterize the MORAL sense, by its "supremacy",-- I make its suprema
601o08b r Desgraviers, un Vol 8vo Keratry-- Inductions MORALES et physiologiques The first of these books I d
609o030 tober d. 1838 perhaps insist? Two classes of MORALISTS: one says our rule of life is what will produ
599o008 in de l'Education des Mères Vol. I. p. 198.-- "MORALITÉ, raison, beau ideal, infini conscience; voila
533m061 my mind feel it was wrong (& it was not merely MORALLY wrong, but hurting my character I felt it)-- t
323c269 Davy Consolations in Travels -- Observations on MORALS by Eugenius Feb 14th. Bo«s»well's life of John
520mIFC ed. Octob. 2d. This Book full of Metaphysics on MORALS & Speculations on Expression -- 1838  Selected
077r171 he veta grande of Zacatecas, & veins of Tasco & MORAN--of Guanaxuato to SW. with respect to latter do
078r175 1 veins)» 25 degree to 30 degree to NE. vein of MORAN 84 degree NE of Real del Monte 85 degree to S.
521m008 thought-secreting   organs, brought into play by MORBID action.-- Old Elspeth's «in Antiquary» power o
042r068 in submarine tilting alone, must be modified. «MOREOVER, the Volcanos from sea there burst out, after
047r081 bouring sea must partake in absolute movement» MOREOVER wave «with same general character» reaches fa
055r106 of  them being packed clean. & without earth.-- MOREOVER that such do not occur on the beaches. Perhap
119a106 of which temperature is partly known-- <[...]> MOREOVER gradation from gneiss to granite shows that t
303c205 that ever lived <on> into one list is unfair, [MOREOVER what will become of the future creations,  if
553m136 very effect, of every kind which surrounds us. MOREOVER «it would be difficult to prove that» this in
564n003 ibute would be rather favourable to it--!! Man MOREOVER who reasons much on his actions, makes his co
619o042 efends & acts for others at its own expense.-- MOREOVER <the> any action in accordance to an instinct
216b181 Transaction on a quagga & mare crossing by Ld. MORETON, where mare was influenced in this cross to af
367d113 inded me strongly of Zebra.-- Mem. Quagga & Ld MORETON Mare ringed Owen says that Bell in Encyclop of
220b197 epend on mind or instinct of parent. Mem Lord MORETON'S Mare. The fact of plants going back hybrid pl
334d009 determined   by male: & How completely is Lord MORETON'S case opposed to this fact & views.-- Fox says
379d152 d else all the lambs will deteriorate.-- Lord MORETON'S Case.-- "When cows have twins, <one> though ca
385d165 of both." If L. can be trusted, this is Lord MORETON law.-- "How often do we find in the son,  the
387d168 ffspring more & more with the added «like Lord MORETONS case & Dr. Andrew Smith,» difference.-- If A.
389d173 by female;<]CD> such view no ways explains Ld. MORETONS case: without the nervous matter consists  of
417e079 like father «What is cause of this.-- » (Lord MORETONS law holds with different species, & individua
417e079 mblances of children to their parents.-- Lord MORETON'S law cannot hold with fishes, «& there are mul
514q21. nstructive action on germ." '=?? answered Does MORMODES (one of the Catasetums) really always hit sti
437e138 been seen carrying a lamb two miles towards the MORNE Mountains, it then dropped it & was found alive
147g019 past   8 Tyndrum 29.<625> «636» Temp. 62 Friday MORNING ¼ past seven o'clock 29.642 Temp 55 Air 50 deg
162g103 & Loch Ness <30.100> <Donald Macphee> Saturday MORNING 29.958 A 64 degree, air 60 «Evening do» The ex
```

Page **(Key Word)**
421e090 50-- «speaking of» the terrestrial mollusca of MOROCCO «Mr Forbes says the Fauna»-- (near Oran) appro
510q017 configuration.-- Hooker <Meta> Metaphysics of MORPHOLOGY. ¶-- Schelgel is he serpent man? about zones
025r016 degree S.] Bahia [12 degree 57 minute S.] 8 200 MORRO S Paulo [13 degree 22 minute S.] 9 120 {T} Garc
200b125 any tendency to form varieties? Ed. N. Phil. J. MORSE found in Virginia p. 325. July 1828. Animal now
549m122 rules come very near each other.→ The rules to MORTIFY yourself do not tend to this-- though believin
477z002 us. 18 Azara Voyage Vol. I. p. 279 Thinks the MORUFFETES of Chile different from those of La Plata or
341d030 om there now found in Australia New species of MOSCHUS, characterized by Ogleby, who observed that th
363d103 white when a fawn compare with fallow? deer. & MOSCHUS &c & -- like young blackbirds Dr Bachman told
160g093 just like road of Glen Roy-- appears to lip with MOSS On this terrace «station perhaps 6 ft too low»
161g095 (deep) of above valley this road level with Peat MOSS most distinct then lost by slope, then conceale
161g096 d for at least 2 miles on dead level «by eye» to MOSS-- on this terrace Bacom. 29.264 A 82 75 degree?
431e123 ost strange.-- Does not spermatic animalcule in MOSSES, render my view of the crossing of mosses & al
431e123 le in Mosses, render my view of the crossing of MOSSES & all others by action of wind difficult.-- Cl
455eIBC where they cannot «crossed» etc.-- Make Hybrid MOSSES.-- Leighton or some one. Father-- diseases com
559m156 's views «Take & pound up inflorescent parts of MOSSES & see if Hybrid can be made & ferns.--» Would
161g096 ther side {P} Shelf A Shelf A at head of Gentle MOSSY slope, which from a distance hid it, could be f
608o026 & he can seldom analyse his motives (originally MOSTLY INSTINCTIVE, & therefore now great effort of r
311c226 on the idea of those parts being most easily MOSTRIFIED, which last produced --insane men in civiliz
203b138 paper on Hybrids Loudon's Magazine. Gould on MOTACILLA,. «species peculiar to Continent & England» L
380d153 ndered biennial-- the hardness of life in female MOTH &c Mr Y. says that Macleay considers the house
148g025 & received same answer Thought lambs most like MOTHER!-- the cross not so hardy but more easily fatt
165g125 d he learnt to know the lambs because most like MOTHER in face-- asked stated this generally the case
178b032 mediate, the first children partake more of the MOTHER, the later ones of the father; is not this owi
179b032 blood come out from the grandfather, (when the MOTHER was nearly quite white) in the two first child
222b203 d) between what are called varieties.-- NB. one MOTHER bringing forth young having very different cha
222b203 xpression -- one child like father another like MOTHER Has Lowe written any other papers besides one
240c003 Yet dogs sometimes like father, sometimes like MOTHER. The fact of great monstrosities being produce
285c150 nts in a circle,-- & «hermaphrodites» father « <MOTHER> » & grandfather <Might> Must be introduced &
312c228 re of this, first offspring most like <parents> MOTHER.-- like dogs Smith knew chinese hairless dog &
333d005 her, partaking <more> «very closely» of form of MOTHER: more than of the Common cat.-- Ch IX Mongrels
345d043 elieves this resemblance general. ¿depends upon MOTHER bein[g] oldest breed?.-- -- Quarterly Journal
346d047 wer.-- Thought lambs were more like father than MOTHER.-- The cross not so hardy as Black faced, but
364d104 asionally went back-- (Effect of imagination on MOTHER. white peeled rods mentioned in old Testament
378d147 s the offspring brought back to earlier type by MOTHER?-- do these differences indicate, species chan
385d165 milarity to the temper of the father, or of the MOTHER, or sometimes of both." If L. can be trusted,
385d165 In how many daughters does the character of the MOTHER revive! Or the character of the mother in the
385d165 r of the mother revive! Or the character of the MOTHER in the son, & of the father in the daughters!
387d168 it becomes fixed in blood.-- Looking at ovum of MOTHER & ovum in offspring, as similar to the several
387d168 in offspring, as similar to the several ova in MOTHER, (with only difference of time) is the above l
388d171 to offspring being determined by imagination of MOTHER.-- We see in a litter every possible variation
388d171 r every possible variation from being very near MOTHER, & some very near father.-- now if one of thes
388d172 «2» dog begetting different puppies out of same MOTHER.-- The view that man & «or cock» pheasant &c i
389d176 animals may unite & each have offspring by same MOTHER.-- one animal will fecundate female for severa
389d176 ill it has bred.-- Offspring like both father & MOTHER, or very close to either.-- Male & female as f
392d180 uscovy ducks do young take most after father or MOTHER according as they are crossed? & How is it wit
406e035 hich parent impresses offspring most is whether MOTHER has had any offspring before.-- -- now this is
417e078 ws are, that child may be either like father or MOTHER, independently of its sex, or half way between
417e078 itors.-- in some families all the children like MOTHER & in some like father «What is cause of this.-
426e108 ibbert case of Hindoism coming out more than in MOTHER or indeed grandmother; what is Mr S. S. parent
432e123 &c» p. 10 offspring take more after father than MOTHER; illustrated by the crossing of hornless sheep
520m001 l Leigton resembled his father in body, but his MOTHER in bodily & mental disposition.-- My father ha
535m064 be moved to other larvae, when two groups near. MOTHER desert one sometimes & go to other, so that tw
622o049 the love is instinctive, & how does it apply to MOTHER loving child, from whom, she has never receive
623o050 infancy, during many generations giving love of MOTHER; the having received some advantages from man.
206b148 s of cats.-- <& other tribes>.-- &c &c Exclude MOTHERS & then try this as simile In a decreasing popu
345d043 cause in their faces they were most like their MOTHERS believes this resemblance general. ¿depends up
525m025 eformities are illnesses of the foetus.-- some MOTHERS. have first dead children, then children which
535m064 esert one sometimes & go to other, so that two MOTHERS to one group.-- (as in birds blind storge-- Th
537m076 ch «sudden» changes rare,-- as when Polynesian MOTHERS ceased to destroy their offspring--¿ yet perha
171b002 s in hybrids where every thing else is perfect; MOTHERS apparently only born to breed.-- annuals tende
307c215 vary, but form constant.-- The females of some MOTHS, like glowworm <are> have «These abortive organ
387d170 an analogy between caterpillars with respect to MOTHS, & monkey & men.-- each man passess through its
392d180 Hot. Houses will they flower Make Hybrids with MOTHS, where fecundation can be made artificially.--
417e079 ly; nor can it be a necessary concomitant, with MOTHS, which can be impregnated externally-- My view
046r080 between 2 & 3 ft. (as in Chili lake). Therefore MOTION of sea ought to be considered as a plain movem
047r081 ves & in wind, instead of sea's bottom being in MOTION What difference? In watching heavy swell, sea
048r084 bbles entirely coated with Tosca. which implies MOTION in the «loose» bed of pebbles. (On a sea beach
048r084 e, one can understand pebbles thus coated.--The MOTION is most wonderful, from chemical attraction, a
072r160 towards space tend to form ring. [Fig. 10] {P} MOTION from within and without. H. Kingdom N. Spa. Vol
191b080 Good to study Regne Animal for Geography.-- The MOTION of the earth must be excessive up & down.-- El
530m046 sinks.-- When a muscle is moved very often, the MOTION becomes habitual & involuntary.-- when a thoug
607o025 as soft,-- the condition of mind which leads to MOTION being inclined that way]CD one sees this law i
300c197 round thick.-- shamming dealt it but being MOTIONLESS. How is instinctive dread it is exceedingly
526m030 thod, but what urges him,-- absolute free will, MOTIVE may be anything ambition, avarice, &c An an
566m010 ained.-- on the wish to support a wife a ruling MOTIVE.-- Book IV, Chapt I on passions of mankind, as
606o025 . 1838 Every action whatever is the effect of a MOTIVE.-- [-- must be so, analyse (a) ones feelings w
608o026 happier.-- he agrees & yet does not.-- because MOTIVE power not in proper state.-- When the admoniti
608o026 ingencies a mans character may change-- because MOTIVE power changes with organization The general de
608o026 cate by putting contingencies in the way to aid MOTIVE power.--if incorrigably bad nothing will cure
606o025) & yet dare not -- one could do it, but other MOTIVES prevent the action see Abercrombie conclusive
607o025 bercrombie conclusive remarks p. 205 & 206.]CD MOTIVES are units in the universe. [Effect of heredata
607o025 y used) is not there present, but he acts from MOTIVES, nearly as usual (a) one well feels how many a
608o026 erence is from imperfect condition of mind all MOTIVES do not. come into play.-- †It may be urged how
608o026 s power of action, & he can seldom analyse his MOTIVES (originally mostly INSTINCTIVE, & therefore no
608o026 ntingencies.--but his desire to do arises from MOTIVES.-& his knowledge that it is good for him refle
360d093 ow the offspring to have some different kind of MOTTLE, each feather partaking of character of other.
146g101 d> thousand tuns. Black faced sheep, sometimes MOTTLED with white black legs & tail like species in c
360d093 re peculiarities; that if <one had a> both had MOTTLED breasts, <when> of a sort that would allow the
361d095 ight, all species have same character which is MOTTLED, & not like any existing species-- [In two har
404e026 favourable to offspring: Australian dogs have MOTTLED coloured puppies case of this.-- tendency in «
404e026 of this.-- tendency in «manner of» life to be MOTTLED + heredatary tendency determines the puppies t
405e035 «V[ide]» Important account of cross of sheep & MOUFFLON of Corsica. <would not>, sadly against Yarrel
406e035 so much against my modification of it-- Goat & MOUFFLON will not breed.-- p. do.-- Fish of Teneriffe.
373d132 ut this-- «the <nerve> living nerve nursed in MOULD.-- Lyells Elements. p. 290. Dr. Beck on numeric
065r139 <see> from rotten state of coffin «buried in a MOUND» long consigned to the earth. yet body had scar
155g063 I doubt, much about «50 or 60 ft» «no doubt, a MOUND of Alluvium nearly parallel--» Inclination of r
155g066 s to have thrown water in same «drainage» lines MOUND of Gneiss though wonderful-- <that they are pre
162g099 be made out of them-- but it may be said that a MOUND stretches along, parallel to Shelf on opposite
350d053 us,.,--> Propagation, best rule for genera, & so MOUNT upwards.-- «judged by analogy»-- Consider all t
030r036 e "palatal Tritores" found in the coraliferous MOUNTAIN Limestone are they allied to the jaws of the
036r049 ch dilated cracks must be filled up by dikes & MOUNTAIN chains.-- Introduce part of the above in Pata
038r057 state at a grand eruption when whole summit of MOUNTAIN is blown off; & again when in great crater. d
050r093 s at Brazil [blank] What is nature of strip of MOUNTAIN Limestone in N. Wales. was it reef. -- I reme
053r101 ia case must be quoted at length. The Lines of MOUNTAIN appear to me to be effect of expansions actin
059r120 testably formed the parts of one whole burning MOUNTAIN, & that the central part fell in.--Says poste
061r127 nts created. if for such mere points; then any MOUNTAIN, one is falsely less outspread at new creatio
074r165 ein. vein of secretion.--metallic veins follow MOUNTAIN chain. there after NW <W>.-- «same chemical l
088a013 continental elevations slow. so would line of MOUNTAIN chain be Mr <Lyell> «Waterhouse» has frequent

Page ***(Key Word)***
101a058 ested.-- There is no difference between dike & MOUNTAIN axis. except in relative <strata> size with s
101a058 trata. where they have yielded conical axis of MOUNTAIN.-- only when dikes reach near the surface. th
102a062 by horizontal elevations, so are injection of MOUNTAIN chains. accompanied by do.-- Give this after
103a065 ussion might be made between dikes & «axis of» MOUNTAIN-chain in proportion to weight of super [...]
104a068 f whole kingdom of Norway was contorted yet no MOUNTAIN chain case parallel to Banda Orientel. ask Ly
112a089 e one end granite & other trap.-- It is in the MOUNTAIN masses we must look for that.-- how few isola
124a118 trance Athenaeum. 1838. p. 652. Dr. Daubeny on MOUNTAIN Chains in N. America Erasmus suggested to me
127a127 «(not on continents)» prove elevation.-- great MOUNTAIN chains. may be effects of subsidence Elie de
128a128 er must be sought for below the sea mark.-- If MOUNTAIN chains are matter piled up. over crevice from
128a128 al elevation,-- when subsidence takes place.-- MOUNTAIN will first fall-- the problem will be falling
135a145 osition of granite-- Bengal. J. vol 7. p. 522. MOUNTAIN c near Caubul. parallel ranges. with here & t
159g084 ks on this side (return) between 2d & 3d shelf MOUNTAIN <Mica> «composed of» Gneiss Block on 2d shelf
185b055 arrot.--» woodpecker-- a desert. Kingfisher.-- MOUNTAIN tringas.-- Upland goose.-- water chionis wate
223b209 tion (seed blown into desert) or separation by MOUNTAIN chains &c the species have not been much alte
223b210 ? Why are species not formed. during ascent of MOUNTAIN or approach of desert?-- because the crossing
230b235 so in approaching desert country or ascending MOUNTAIN you ought to have a gradation of species, now
230b236 lants & then a chain of steps is found in same MOUNTAIN).-- How is this explained by law of small dif
233b249 o far. Indian all the rest. Timor according to MOUNTAIN chain ought to be Australian.?-- Mr Gould has
284c148 have travelled by <dead> «each» trees dying & MOUNTAIN torrents.-- but to crawl up an hill, then by
298c191 ine a considerable range of one species «on -- MOUNTAIN side» of which the central parts become occup
374d133 p. 437. Many. existing genera of shells in the MOUNTAIN limestone (how different from plants!) But th
377d140 vating forces in raising continents, & forming MOUNTAIN-chains, when we estimate the matter removed b
401e017 ot flourish-- a medlar may be Grafted on pear. MOUNTAIN-ash & white Thorn! Species not being observed
416e072 in external conditions.-- -- as in plant up a MOUNTAIN-- In races the differences depend upon inheri
514q21. lets.= Zostera= Are dwarf plants on Wellington MOUNTAIN described in Flinders= Alpine Australia Flora
552m131 m 1838. p. 652. Dr Daubeny on the direction of MOUNTAIN chains in N. America Fear probably is connect
570n026 ience.-- deduction from this would be that a MOUNTAINEER <takes> born out of country yet would love m
425e104 - = Believes some Mediterranean, but chiefly MOUNTAINOUS «this is very important. (Sicily exception)-
023r010 fissures, produced by the elevations of those MOUNTAINS on the continent of S. America is inadmissibl
030r035 ing much vegetable matter from thickly wooded MOUNTAINS, probably chiefly leaves.--This position agre
035r048 hough so immense to short breathed traveller» MOUNTAINS, which in size are grains of sand, in this vi
042r068 204 Vol III. Lyell Owing to «open» faults in MOUNTAINS: to elevated strata in eocene lakes of France
071r158 y amygdaloids.--Boussingualt «(Lyell)» cracks MOUNTAINS falling in.-- Earthquakes at Quito. tranquill
071r158 haps otherwise» Mendoza never overthrown,--no MOUNTAINS Mackenzie has talked of lava flowing up Hill;
078r175 the direction of ravins, and the slope of the MOUNTAINS (flaqueza del cerro) have been parallel to th
078r175 opes. but on the most elevated summits, where MOUNTAINS most torn.--(¿anticlinal line?). -- Mines of
086a004 g recent elevation, yet sea shells at tops of MOUNTAINS we ought to sympathize with. old doubters of
086a007 in name of Heaven why are tops of Equatorial MOUNTAINS so cold.-- Siberia no plants to it, lately ra
089a018 tes secousses."-- Tom 54. p. 106 do-- p. 110. MOUNTAINS on west side of Domingo formed of coral limes
089a018 ne, with interstices yet emptty.-- In all the MOUNTAINS of Saint Marc et des Gonaïves it is difficult
097a042 lime decompose each other.-- on Direction of MOUNTAINS in Brazil L.'Institut No degree 221 Lamellar
103a064 difficulty.-- All De la Beche's reasoning of MOUNTAINS being formed by crust being too large & pitch
116a100 rmations. p. 150. at Portezuelo, extremity of MOUNTAINS of Cordova project on plain, like <re> a reef
119a104 imply tendency to equilibrium.) 3d. there are MOUNTAINS in the moon, which though not very analogous
119a104 s like Andes or Himalayas, but great circular MOUNTAINS, yet so analogous, that as we see mountains f
119a104 r mountains, yet so analogous, that as we see MOUNTAINS formed (& mountains are effect of continental
119a104 analogous, that as we see mountains formed (& MOUNTAINS are effect of continental elevations) we may
119a105 tamorphic rocks at surface. & great heigth on MOUNTAINS.-- consist of rocks with fossils,, therefore
122a113 ally yields.-- Will this not explain littoral MOUNTAINS & volcanos.-- Why on one coast? How can Hersc
127a127 lph of Soda in peat ashes in Ireland dikes in MOUNTAINS. «(not on continents)» prove elevation.-- gre
131a135 arine L'Institut 1838 p. 400. Observations on MOUNTAINS of the Moon. by Dr. Nichol-- adduces the case
160g092 flat peat mass-- (general character in these MOUNTAINS & not ridges) between arm of Glen Bright flow
205b142 says (not definite information) West of Rocky MOUNTAINS Asiatic types discoverable.-- Bridgewater Tre
207b151 V. Dr Royle introductory remarks to Himalaya MOUNTAINS-- Bory St. Vincent Vol. III. p. 164. Lîle de
219b195 merly have been separated by short space from MOUNTAINS low down, therefore plants common take an exa
317c251 ies.-- Says all the hares West of <all> Rocky MOUNTAINS have peculiar character in extreme length of
318c253 aking the low country near coast & others the MOUNTAINS, & these appearing to remain about a fortnigh
318c255 of Wren (Bebyk??) which seems common in Rocky MOUNTAINS & on one lofty isolated spot on the Alleghani
319c257 d the Samoyed women (¿ north end of the Oural MOUNTAINS) have black nipples to their breasts.-- L' In
341d031 n birds.-- Dr. Bachman tells me line of Rocky MOUNTAINS separate almost all Mammals of N. America & m
353d061 th American hares. Many species, separated by MOUNTAINS. & &c.-- do. p. 69. A Dr Macdonald believes
425e105 nks of Alps.-- many peculiar plants on single MOUNTAINS, though these are connected with other mounta
425e105 ntains, though these are connected with other MOUNTAINS laterally.-- Owen. Fossil Mammalia. p. 55. ta
437e138 n carrying a lamb two miles towards the Morne MOUNTAINS, it then dropped it & was found alive.-- Stan
447e165 ansact. Aira caespitosa becomes vivaparous on MOUNTAINS & yet can be raised in gardens.-- Poa alpina,
464tf6v ,-- these species must have migrated to these MOUNTAINS, when the cold was intense just like the alpi
481z010 ays «the Condor» <it> is found in the Tucuman MOUNTAINS The fourth Vol. «in Lyell's possession» of Zo
513q21. same time on same plant --Flora of Australian MOUNTAINS.-- Is setting of fruit. cross Conception--(
514q21. ia Flora= Banana's seedless-- 20 varieties in MOUNTAINS of Tahiti. Dr. Boott-- says caricas from ever
514q21. a of African Islds-- names of Plants found on MOUNTAINS of N. America similar to Lapland Plants --wil
535m069 y in that state of mind: the Chileno says the MOUNTAINS are as God made them,-- next step plastic <vi
570n026 er <takes> born out of country yet would love MOUNTAINS, & a negro, similarly treated would think neg
633j54r ves to form rain to wash down earth, from the MOUNTAINS upheaved by volcanic force, for these Marsh p
638j58v sdom or Providence. when whole rocks nay very MOUNTAINS are formed of such dead & extinct forms.-- th
046r079 n? Rats & Mices. At St Helena there is a native MOUSE Did wave first retreat at Juan Fernandez: the f
226b220 ea same fact see Coquille's Voyage.-- Galapagos MOUSE (?) brought by canoes Ceylon & India.-- Van Die
248c029 thus handed down for nearly 70 year. Galapagos MOUSE not the same section, with house mice. It is wo
268c096 History. «vol I» p. 159 curious account of Tit MOUSE feeding young of redstart & actually driving aw
337d022 uired if he believes «hyaena & squirrel» seal & MOUSE, elephant, come from one stock.-- Theory of Geo
354d065 Creator made rat for Ascension.-- The Galapagos MOUSE probably transported like the New Zealand one--
483z012 urate & Vol IV p. 91.-- Vol IV p. 388. Domestic MOUSE of Egypt is Mus Cahirimus. of Geof.-- reference
639j28r ehensile tail. in Monkeys & Marsupials. Harvest MOUSE & (Chamaelion?) C. D. Spines in Hedge Hog & Ech
641j29v]CD. «Macculloch says no other bird could catch MOUSE by night» Sailing lizards. squirrels & Opossums
085a001 ulate on the extension of Patagonia seaward, at MOUTH of S. Cruz. from ascertained inclination. of pl
105a071 le.-- The sea harmonizes well with character of MOUTH of valleys &c; Pampas.-- If blocks above their
112a088 ting out to sea do p. 358. changed soundings in MOUTH of S. Cruz in connection with Fitz Roys fact of
113a092 iver <near> W. of Port Philip. which had bar at MOUTH excavated in solid rock.-- 4 & 5 fathoms deep.
116a099 decrite de petites corbules analogue living in MOUTH of Plate. p. 26. Geology of Arica <Schit> Schmi
129a132 en from an axis, «20 ft at least in depth» near MOUTH of Columbia river-- Read Mr Parker's Book.-- M.
148g022 ite doubtful of as I am of all the Alluvium. At MOUTH of Caledonian Canal opposite Loch Leven two ter
161g095 n, where three [...] abutted Having crossed the MOUTH, (deep) of above valley this road level with Pe
197b112 o M Edwards. law of crustacea-- with respect to MOUTH those beautiful passages from one to other orga
205b143 are some good accounts of passages of legs into MOUTH-pieces of Crustacea. Vol II. p 75 a Fish which
525m025 of fingers.-- hare lip or imperfect roof to the MOUTH «stammering in my Father family» (as in Lord Be
540m084 so does dog bark. (not shout) when opening his MOUTH in romps, <so> he smiles. Many of actions as hi
541m092 eople. Aug. 21st. 38 When a dog in play has his MOUTH open ready to bark, & lip twisted up, in that p
542m096 osition, & he holds them this way, when opening MOUTH between interval of barking, now this is smile.
554m139 nicon. & readily put it. when guided to her own MOUTH.-- seemed to relish the smell of Verbena & Pock
556m145 oltaire why is under lip curled over upper with MOUTH shut. expressing cool irony, not biting? What i
558m152 &c I observed the Asiatic Leopard. quarrelling. MOUTH wide open, each [lip] drawn back & driving air
558m152 pen, each [lip] drawn back & driving air out of MOUTH «hairs erect on back» «wide open» with prodigio
565n007 «though in a letter» why is person painted with MOUTH open.-- why when person is listening is mouth o
565n007 h mouth open.-- why when person is listening is MOUTH open to hear well «as one will perceive if in n
565n007 as fear to «man as» animals. comes at distance, MOUTH is placed open.-- Hence becomes instinctive to
565n009 le, no frown showing thought, no compression of MOUTH showing action,-- sulkiness all negative expres
569n019 ic extremely primitive.-- almost like tastes of MOUTH & smell. Descent of Man Understanding languages
569n020 n.-- Elephant often given food & word open your MOUTH said, recognizes that sound as perfectly as a m
571n029 general notions, which are taste? Real taste in MOUTH, according to my theory must be acquired, by ce

Page **(Key Word)**

574n041 ted with flow of saliva, & hence with action of MOUTH & jaws.-- Lascivious women. are described as bi
592n103 5, 25. 40. 61. CD[a person is here said to open MOUTH in fright because nature intends to lay open al
622o048 is right or wrong.-- «[just as in tastes of the MOUTH]CD» [LHC] My theory of instincts, or heredetary
624o50v er than change hunting instinct. *Our tastes in MOUTH by my theory are due to <habit> heredetary habi
639j28r . «grinding» teeth in <stomach of> sun-fish, in MOUTH of swine & in stomach of lobsters-- analogy in
557m149 in failing to drive away rival.-- Fear is open MOUTHED to hear. though in individual case. nothing ca
115a096 ere there are cliffs there ought to be creeks & MOUTHS of rivers ought to be deep.-- Henslow has depo
254c049 organ «where eliminated is» often repeated, as MOUTHS in Polypi, Surely not correct view of Flustra
555m143 hardly ever the expression of passion with open MOUTHS like the old world ones.-- Though the[y] move
105a073 the inclination is little the force required to MOVE <it> «stream» aside is not great.-- Is there mo
195b101 merly have said that God ordered, each planet to MOVE in its particular destiny.-- In same manner God
232b245 & pliocene epoch), whilst others may die out or MOVE South ward.... species must be compared to neig
315c243 - (This may be made capital argument if man does MOVE muscles for uncovering canines) --. Blend this
350d055 nnulosa?) Remarked that young of Cirrhipedes can MOVE & see, parent fixed,-- young of sponges move.--
350d055 can move & see, parent fixed,-- young of sponges MOVE.-- young of Cochineal insects move about & see,
350d055 ng of sponges move.-- young of Cochineal insects MOVE about & see, parent «(2)», female «(I)» fixed &
350d055 of sight to threads.--» Hence the Pecten, which MOVE imperfectly has eye-point, but Broderip added i
351d056 ture leaves vestiges of what she does-- does not MOVE per saltum-- yet does nothing in vain!! Foetus
420e089 ginally quadru<mamous> «ped.-- ». Hairy.-- could MOVE his ears The head being six metamorphosed verte
485z017 r to some of the Fluvicolae?-- The Birds seem to MOVE much further North on West coast of S. America.
504q013 //Yes// (5) Examine the Parnassia whose stamens MOVE one after other to flower & Menyanthes whose po
555m143 mouths like the old world ones.-- Though the[y] MOVE whole skin of head they do not move eyebrows.--
555m143 hough the[y] move whole skin of head they do not MOVE eyebrows.-- (I see some of the old world ones m
555m143 ve eyebrows.-- (I see some of the old world ones MOVE skin of head & ears.-- ∴ some men have this pow
556m144 nything ahead. which cad cannot see, they do not MOVE muscle.-- reason CD[The laughing noise which C.
582n067 e a mental impulse (though unconscious of it) to MOVE its legs so, as much as in the young salmon to
596n184 co> --full of practical observations Ourang do not MOVE eyebrows.-- or skin of head,-- «scarcely able V
596n184 us? very much., «Keeper says some of the monkeys MOVE <its> «the» ears but <not> Chimpaze. does not g
618o41v t have no reference to place. [We see a particle MOVED one to another, & <the> (or conceive it) & that
098a048 ot horizontal? Why have particles in such cases MOVED more laterally than vertically, in concretions
122a114 st under ocean, became thicker, then when fluid MOVED [...] {P} August 25. I saw metamorphic conglome
129a131 The most wonderful case of great block of rock MOVED by gale-- When writing on Valleys. «Tertiary st
232b245 ed to colder climate whilst others died out, or MOVED towards equator.-- «or some species might then
314c237 ng of pistils & stamens united into long organ, MOVED on being touched, so as to protect itself, one
530m046 & afterwards patient sinks.-- When a muscle is MOVED very often, the motion becomes habitual & invol
535m064 ot know their own larvae, but one female may be MOVED to other larvae, when two groups near. mother d
545m105 co> «Cyanocephalus Sphynx Linnaeus») constantly MOVED the skin of forehead over eyes, at every emotio
035r048 if we look at the action as a deep & extensive MOVEMENT of viscid nucleus, which in any one country w
036r050 ome of the lower orders; it was connected with MOVEMENT of sand.--it is called "Bramidor"(?).--it was
036r051 submarine Volc: in harmony with the prevailing MOVEMENT being one of elevation alone.--In England muc
046r080 rst retreat at Juan Fernandez: the first great MOVEMENT was one of rise (any smaller prior ones might
046r080 r prior ones might have been owing to absolute MOVEMENT of ground). Michell (Philos: Transacts) «seem
046r080 ransacts) «seems to» considers that fall first MOVEMENT (as in Peru 1746).--At great Lisbon Earthquak
046r080 otion of sea ought to be considered as a plain MOVEMENT communicated to it as well as by the vertical
046r080 ed to it as well as by the vertical as lateral MOVEMENT.--At first one would think movement. owing to
046r080 as lateral movement.--At first one would think MOVEMENT. owing to water keeping its level whilst land
047r081 also neighbouring sea must partake in absolute MOVEMENT.» Moreover wave «with same general character»
047r082 ts & see if fall is not the first very evident MOVEMENT.--The swelling first on beach I cannot unders
069r149 Quantity of matter from Cordillera. HORIZONTAL MOVEMENT of fluid matter not (for instance) expansion
103a065 Dikes being last action. (effect of horizontal MOVEMENT) hence generally intersect metallic dikes: It
118a104 vations (I) the alternation of linear bands of MOVEMENT in Indian & Pacific Oceans.-- (2d--) does not
219b196 aid it was as gret a difficulty to account for MOVEMENT of all, by one law. as to account for each se
273c111 eeminently aerial,-- formed for flight & great MOVEMENT in the air, & likewise rasorial species, & li
314c237 s converted into hood which possesses power of MOVEMENT. & not the organ itself How except by direct
341d030 n in other Struthios. was adaptation to little MOVEMENT.-- nocturnal crawling bird.-- Wings reduced t
423e096 others.-- Why then has there been a retrograde MOVEMENT in Cephalopoda & fish & reptiles.?-- supposin
531m052 art as of excited action, accompanying violent MOVEMENT; may not passion be the feeling «consequent o
531m052 inherited & therefore remains, when the actual MOVEMENT does not take place.-- A start is HABITUAL mo
531m052 ent does not take place.-- A start is HABITUAL MOVEMENT to avoid any danger-- Fear, shamming death, o
531m053 ce, &c Starting must be habitual «involuntary» MOVEMENT from wish to avoid some danger-- but it is in
533m059 tinctive walking of animals. that is the ready MOVEMENT & co-relation of the proper muscles. may be i
542m096 ct to sneering the very essence of an habitual MOVEMENT is continuing it when useless.-- <&> therefor
545m107 dren.--» Expression, is an heredetary habitual MOVEMENT consequent on some action, which the progenit
556m146 put down ears, when pleased.-- is it opposite MOVEMENT to drawing them close on head, when going to
557m149 ss action of the heart.-- tendency to muscular MOVEMENT, hence shy people (shame of ridicule) are sin
560mIBC f Babies-- Do babies start, (ie useless sudden MOVEMENT of muscle) very early in life Do they wink, w
565n008 nger «& respect to opponent» is showed by same MOVEMENT as sneering,-- it is then more <emblem> manne
572n033 asmus does not liken term instinct to muscular MOVEMENT.-- say instinctive actions. senses. notions &
573n037 to one side & then to other. & hence rotatory MOVEMENT-- he dropped his head when he meant
579n058 gh.-- Has frowning anything to do with ancient MOVEMENT of ears A man shivers, from fear, sublimity,
579n059 alking, to chronometer, is seen to be muscular MOVEMENT. The Blushing of Camelion & Octopus; strong a
596nIBC d convulsive.-- is convulsion. are involuntary MOVEMENT of voluntary muscles-- if so what is tremblin
611o034 heat light &c). [LHC] This is true as long as MOVEMENT of sensitive plant can be shewn to be direct
612o035 r that require will in part. ¿Why more so than MOVEMENT of sap. or sunflower to sun? ∵ I should think
617o040 es consists in the manifestation of force i.e. MOVEMENT? capable of being traced to the body of the i
617o040 rce be recognized by our external senses--only MOVEMENT can.-- 4) the source from which it arises. Bu
027r023 shells, like Concretions & laminae, show what MOVEMENTS take place in semiconsolidated rocks P xv. me
032r039 of the sea was to sink by very slow & gradual MOVEMENTS to line (2). The part (o) which was before be
039r061 tinent> We are more abound to take analogy of MOVEMENTS of W coast in explaining plains because such
042r069 ve freshwater lakes unequally elevated, which MOVEMENTS if present in the Andes, would have destroyed
044r073 ially Pacifick: insignificant islets--general MOVEMENTS of the earth;--Scarcity of Organic remains.--
044r073 day.-- applied by me geologically to vertical MOVEMENTS. In Cord: after seeing small Bombs. without a
058r117 f same phenomena lasting so long.-- The great MOVEMENTS (not mere patches as in Italy proved by Coral
068r146 ions & eruptions can only happen during first MOVEMENTS, and therefore beneath ocean, for subsequentl
086a003 artz.» unmixed is very pleasing; owing to the MOVEMENTS being of one order.-- There should not be su
120a107 es of what were fluid undulations-- the equal MOVEMENTS of Glen Roy road. (¿ metamorphic action at th
199b116 he great continents we get a means of knowing MOVEMENTS.-- How can we understand excepting by propaga
263c074 cult to understand as Lyells doctrine of slow MOVEMENTS connected with mind There is no progression i
305c210 haracter & manner of wagging tail.-- habitual MOVEMENTS effects of irritability, though means injecti
315c242 ent» & memory of animals.-- (surely in plants MOVEMENTS of Plant/ I£:4s Voyage aux terres australes.
324c268 res sur les Vegetaux et animaux.-- on sleep & MOVEMENTS, senses being on the look out, & the conveyin
532m054 is partly owing to heart? readily taking same MOVEMENTS.-- some say dogs understand expression of man
542m097 hand each other expressions, sounds, & signal MOVEMENTS.-- some say dogs understand expression of man
545m106 not perceive «any» distinct wrinkle, but such MOVEMENTS in skin of eyebrow important analogy with man
573n037 fore they can have learnt by experience, that MOVEMENTS of face are more expressive than
573n037 at movements of face are more expressive than MOVEMENTS of fingers.-- like Kitten with mice-- A pers
573n038 -- curious mixture of voluntary & involuntary MOVEMENTS.-- Person with sore-throat told not swallow s
592n104 o opening eyes in fear The effect of habitual MOVEMENTS in muscles of face, is well seen in shortsigh
601o009 s, nor child-- Evidence of consciousness, <t> MOVEMENTS i) anterior to any direct sensation, in orde
611o034 matter obeying a certain & peculiar system of MOVEMENTS. different from inorganic movements.-- See La
611o034 system of movements. different from inorganic MOVEMENTS.-- See Lamarck for this definition given in f
611o034 external form impressed, & different laws of MOVEMENTS. [LHC] Hence there are two great <worlds, ino
612o035 (if such there are) of animals enjoying only MOVEMENTS such as sensitive plants. (But I include irri
612o035 tly united? [RHC] It is easy to conceive such MOVEMENTS & choice, & obedience to certain stimulants w
634j54v embrane?!!-- that the phalanges have separate MOVEMENTS in the Holocentrus ruber (a fish) Man has abo
636j57r , differs in every respect, except «in» quick MOVEMENTS. (sliminess instead of barbs)-- In all these
045r077 lways slightly oscillating as that of a spring) MOVES away.--Will geology ever succeed in showing a d
107a077 ts.-- Does the isothermal {P} subterranean line MOVES upward from effects of Elevation if not crust m

531m053 see. which frightens. them.-- Now every animal MOVES quickly away from any sudden sound or noise, &
544m103 o do this & hears that, but yet scarcely really MOVES.-- the willing therefore is ideal, as all the o
567n014 ter necessary notion of space-- plant though it MOVES doubtless has not.-- Turkey cock in passion & s
024r014 than 100 fathoms. proves the existence of some MOVING <point> power ¿ Submarine currents Find instan
103a063 d. Geological Society-- Dikes have not been the MOVING agents, because not wedge-formed.-- Hence fill
533m059 may be illustrated by the extreme difficulty of MOVING muscles in different way from what they have b
564m005 itted perfectly to chicken so as to seize small MOVING object like fly.-- young partridge can run eve
601o009 g some relation to the primary sensation.-- man MOVING leg when asleep-- <or habitual actions> perhap
617o40v an opposition of forces balancing each other & MOVING in opposite directions. We are satisfied there
633j54r earth, (like a Dutchman plants them to stop the MOVING sand) we <do> lower the creator to the standar
323c269 atise on Domestic pidgeons 30th Lives of Hayd & MOZART «Aprl 25th Lockarts life of Napoleon.» April 5
582n066 ns» «Gardner in his work» In the life of Hayd & MOZART. fine music is evidently considered as analogo
450e176 propagate get dimensions-- do App. p 73 State of MUAR in Malacca.-- speaks of Rhinoceros as well as T
055r107 irst gives» half demolished craters).--worn into MUD & dust.--connection with age, & agreement with n
057r114 ssion Bay in 73 degree 39 N. living worms in the MUD which he drew up from 1,000 f[athoms], & the tem
093a033 turbos in red earth with volutas. prove regular MUD bank at Bahia Blanca. <fl> Flustra identical. re
093a034 in the year a Rhinoceros was found <emb> in the MUD, of the Salt river.-- in reference to fossil gua
099a050 rce & varying direction.--> Therefore in PILE of MUD from Trapiches. inclined layer!!!.-- The separat
109a083 wards may be granted. independent of currents.-- MUD going out can actually be seen.-- ¿ The preserva
114a093 e Iron stone of C. of Good Hope & Australia/ and MUD of salt-lakes of Rio Negro--Mr Bowerbank-- Dr. A
122a115 ulate under head of Beagle Channel. on origin of MUD with stones scattered irregularly.-- (Mem near G
130a134 Patagonia {P} B subsidence; <as in> be cautious. MUD banks & sand. dunes.-- in these littoral deposit
163g109 origin pebbles brought by different cause: from MUD.-- [blank] {P} muddy nodular strata coral upside
259c062 nean lakes, hot spring &c &c inhabited therefore MUD wood be inhabited, then how is this effected by-
271c105 other species.--) should build a nest lined with MUD, in forest where not a tree in which it builds,
271c105 st in <same country> something same manner, much MUD.-- These facts show, habits heredetary whilst sp
463t055 Could anyone. have foreseen, sailing, climbing & MUD-walking fish? difficult-- yet suggested. (vipers
633j54r g rigmarole about plants being created to arrest MUD &c at deltas.-- Now my theory makes all organic
164g119 ght by different cause: from mud.-- [blank] {P} MUDDY nodular strata coral upside down strata coarse
266c091 athers;-- the first to escape the doctrines of MUHAMMAD, the last to extend, their dominion, armed al
338d024 r that the ideosyncracy of the Negro (& partly MULATTO) prevents his taking any form of Malaria-- ada
216b179 e cross often whiter<)> than white parent) the MULATTOS themselves explain it by intermarriages with
240c003 element in the transmission of form, may explain MULE & pig being half way. Yet dogs sometimes like f
335d015 things are long in blood so will they remain, a MULE «between new species» will have no tendency to ha
335d015 hey must like or there will be none, therefore a MULE can have no offspring.= but as «badly» deformed
335d015 lar offsprings, so will the worst mules (as real MULE) have offspring,-- slight deformities «as super
336d017 m, & as offspring must be like parent, therefore MULE has no offspring & therefore no generative orga
336d019 multiplied, & great change be effected, but in a MULE these conditions are not fullfilled.-- «{My gra
367d113 f parts, will not be produced.-- Sept. 17th. Saw MULE. apparently fathered by a donkey. with all four
383d159 es, like J. Hunters. Free Marten N.B. the common MULE must often have been dissected Zoolog. Garden.
417e079 eton's law cannot hold with fishes, «& there are MULE fishes» & reptiles & those which <lay> <have» t
458t002 very pretty little animal, showing something of MULE in its ears-- ((this is good case as showing gr
459tf02 us.-- seems to doubt its applicability to common MULE & hinnus-- in one case bastard of wolf & dog ha
280c135 ? My views, which would even lead to anticipate MULES is very important for L.yell said to me, the fa
280c135 for L.yell said to me, the fact of existence of MULES appeared to him most strange.-- This even might
280c136 impossible-- (Hence we might expect even if two MULES bred or two certain varieties, they would go ba
309c219 ews.-- Argue <argue> case of abortive organs to MULES in their genitals & even to a limb not used The
312c228 erfectly like spaniel even when grown up.-- Are MULES homogenious owing to no attempt to keep up offs
317c249 n, NB. I met an old man--, who told me that the MULES between canary birds & goldfinches differed con
334d009 ertile with pure bred animals.-- Mem. number of MULES.-- «He recollects one hatch of hybrid geese ver
335d014 Are not dreadful monsters, abortive, just like MULES. Fox's half bred Persians «cat» favour the Pers
335d015 ersian side.-- Theory of abortive hybrids.-- If MULES did breed, the offspring would «as in all other
335d015 imes have similar offsprings, so will the worst MULES (as real mule) have offspring,-- slight deformi
335d016 are here combined). In last page, we have seen MULES could have no offspring, & this being case, owi
335d016 ing, which have been gained slowly, now all the MULES have their whole <body> form of body gained in
336d019 not fullfilled.-- «{My grandfather's theory of MULES not hereditary, because generation -- highest p
342d032 equal Numbers with pure bred (just like common MULES) & lay many eggs but never produce inter se or
459tf02 les horse in its head ears, tail limbs-- in the MULES, these parts resemble ass. (& part of body mare
470t176 as introduced about '65 years ago--& soon after MULES abounded--so that palmated has now nearly disap
470t176 w nearly disappeared. <& old English> But these MULES <in our garden> show no trace of palmation!!? B
493g003 om A. into B. from B. into A. as takes place in MULES ass & horse-- important. {In crosses does male
501g011 pe & less in another region-- (27) Which sex in MULES generally fails-- perhaps indexed by secondary
477g002 -- A. Patagonicus les tatous (.4 pichye, pelud, MULITA et mataco.) are all found south of 26 degree 3
482z012 . abundant from Paraguay to 27 degree, then the MULITA from 41 degree to 26 degree CLOSELY allied spe
498g08v seen group of different species growing White MULLEIN good plant to sow & try to get other species <
419e085 ld»,-- the production of vitality, as argued by MULLER from propagation of infinite numbers of indivi
434e127 tered, that is the Nisus formativus. (what does MULLER call it) succeeds in altering <or> form of bod
418e083 on it.-- Seeing that <Man> «all vertebrates. [MULLER'S Physiolog. p.24.]CD» can be traced to a germ,
247c028 ke was> one Gecko on Isle of France Scincus MULTILINEATUS (p 45) Moluccas & New S. Wales Scincus Cyan
380d155 t.-- Yarrell believes Gestation is always some MULTIPLE of seven-- if woman does not menstruate in th
222b202 one recent, yet genera same.-- Speculate on MULTIPLICATION of species by travelling of climates & the
263c075 ells doctrine of slow movements &c &c. this MULTIPLICATION of little means & bringing the mind to gra
388d172 Athenaeum 1838. p 653. Ehrenberg<h> thinks MULTIPLICATION by division <only> is developement of gemm
411e056 oot.-- When one sees in Coralline powers of MULTIPLICATION of individuals, & yet another means for in
175b020 out.-- With this tendency to change, (& to MULTIPLICATIONS when isolated, requires deaths of species
269c103 erally.-- where <">unity constantly develops MULTIPLICITY<"> [(his definition "constant manifestation
269c103 ion "constant manifestation of unity through MULTIPLICITY" this unity,-- this distinctness of laws fr
569n021 son probably mere consequence of vividness & MULTIPLICITY of things remembered & the associated pleas
614o037 rses) very long & good memories»-- but on its MULTIPLICITY & the comparison of ideas.-- As man has so
208b154 of those regions, & consequently not many yet MULTIPLIED: NB How does this bear with law referred to
302c202 family. having analogous characters, might be MULTIPLIED.-- we must argue reversely. WHERE CHARACTER
336d018 nges> generations, these small changes become MULTIPLIED, & great change be effected, but in a mule t
407e041 why: Marsupiata, when first introduced live & MULTIPLIED, specifically & individually.-- I see clearl
604o016 o three animals, & this consciousness becomes MULTIPLIED with the organisms structure, it looks as if
302c201 be those of analogy, which when sufficiently MULTILIIED become affinity yet often retaining a family
347d049 aterer organization an animal has, it tends to MULTIPLY & IMPROVE on it.-- Articulate animals must ar
377d140 circumstances. a contingency of time. When we MULTIPLY the effects of <earthquakes>, elevating force
541m092 nite idea. nor is an emotion.-- People who can MULTIPLY large numbers in their head must have this hi
398e005 asoning, extremely faulty» The difficulty of MULTIPLYING effects & to <ponder> conceive the result wi
231b240 to explain by creation-- or we must suppose a MULTITUDE of small creations.-- Will Dromedaries & Came
305c210 ganic matter & chiefly heat), which assumes a MULTITUDE of forms «each having acting principle» accor
041r065 ut shells: Hibernation of fresh water Shells. MULTITUDES.-- The question of shell's concretions, livi
429e114 ds. & smiling fields.-- we must recollect the MULTITUDES of plants introduced into our gardens (oppor
343d039 Mammal may go on as one species from Egyptian MUMMIES & from the existing animals found fossil when
438e142 ion [see Wikinson on dogs of Egypt & Cuvier on MUMMIES]CD [NB TIME is element in change, as in Dahlia
266c092 Art. 1831. p 160. account of Bulbous root from MUMMY: after 2000 years, germinating.--!! Henslow dou
500q10a Birds in New Zealand, plants so few-- Range of MUNDANE genera, «in Birds» in accordance with range of
509g017 er Gould's observation holds good, that in the MUNDANE genera, the species <are> have wide range-- Ho
182b050 hey undoubtedly come from same countries.-- In MUNDANE genera, the nearest species often come very re
196b104 any facts are connected.-- No doubt in birds, MUNDANE genera, Bats .Foxes. Mus are birds that are ap
196b105 de powers of flight; but are there any genera, MUNDANE, which cannot transport easily.-- it would hav
203b137 de from Australia & S. America. that only some MUNDANE cause has destroyed animals over the whole wor
220b199 ll be curious.-- Some general statements about MUNDANE & confined genera.-- Lyell has remarked about
221b200 ne beds, very like present species., p 8. ¿Are MUNDANE forms, longest persistent?? do.-- The most per
271c107 e ones occurring in intermediate country-- ie. MUNDANE groups.-- Waterhouse tells me in insects there
310c225 -- Here is an element of extreme difficulty in MUNDANE geological chronology Annals of Natural Histor
357d073 ll explain. S. American case & Didelphis being MUNDANE form., & the less developement of Marsupials i
407e039 s in other continents might have survived this MUNDANE change..-- Therefore I argue from this that Af

Page
(Key Word)*
323c269 of W. Scott.--, except the V Volume.-- -- 19t. MUNGO Park.-- travels Feb 12. Sir. H. Davy Consolation
044r074 specially the most abundant. Sulp. Hyd: Carb: A. MUR: A. = (& this effect of water thus holding matte
079r176 occur in do veins. At Huantajaia. Humboldt says, MUR of Silv.[,] Sulph. of do.[,]galena[,]quartz, Car
096a041 in Paper on India gives reason for knowing that MUR. Soda. and Carb of lime decompose each other.--
109a084 oncretions Carbonate soda. formed by Ca. of L. & MUR. of Soda mixed.-- Turner's Chemistry p. 206 Both
094a036 7 Dr Buckland Reliquiae Diluvianae p. 201. & seq MURC Trans Geolog Soc Vol 2.'p 257 {T} The Pota: lab
066r142 ed up <all> a lake in neigbourhood of town Mr MURCHISON insisted strongly. that taking up a piece of
094a036 orno in lake. few Volcanos now in lakes.-- Mr MURCHISON. M.S. Chapter on drift.-- Beyond region of gr
236b279 fresh water. tortoise from Germany. (where Mr MURCHISON fox was found), decidedly next species to som
432e126 . Journal <Mar> April 1st 1839) by Sedgwick & MURCHISON: which is a beautiful instance of forms, inte
129a131 limits of Alps size of boulders sorted: ditto MURCHISONS case.--¿ does it bear on Patagonia? «Facts a
316c244 nster story of storm of snow after his brothers MURDER.--» good anecdote --.Sowerby.--. Geographical
537m076 roy their offspring--¿ yet perhaps if they had MURDERED their children, this moral sense, would have
075r167 me native silver because not very abundant.--‖ MURIATED silver. which is so rare in Europe. common th
478z004 ely with tame: comes to houses on purpose Mr J. MURRAY has given paper to Royal Soc on glow worm. lum
195b100 der. Did it create two species closely allied to MUS. coronata, but not coronata.-- We know that dome
196b104 No doubt in birds, mundine genera, Bats .Foxes. MUS are birds that are apt to wander & of easy trans
244c022 d, «said» to come originally from Africa p. 122. MUS decumanus, at Caroline Isld, & a Roussette p. 13
483z012 91.-- Vol IV p. 388. Domestic mouse of Egypt is MUS Cahirimus. of Geof.-- reference from Rüppel trav
266c091 ble to classify them.-- I would [not located] MUSALMAN'S of the Peninsula, are, generally speaking, a
246c027 s «& Hebrides») is very closely allied to C. MUSCADIVORA., which lives in the Eastern Moluccas, New G
373d133 en. oldest birds. p. 411 -- Decapod Crust in MUSCHELKALK, & 5 genera of reptiles.-- <M> p. 417. Magne
424e100 erred to it.-- also on spermatic animalcules in MUSCI frondosi, et hepatici,-- in Chara, in Marchanti
599o005 e sentences.-- signs-- ¿ were signs originally MUSICAL!!!??-- At least it appears all speculations o
318c254 uite different in species of same genus.» The MUSCICAPA solitaria stay about a fortnight in one parti
272c109 banded tarsi is common to all the Laniadae & MUSCICAPIDAE of New World, but not found in Old-World--.
530m046 y stimulus & afterwards patient sinks.-- When a MUSCLE is moved very often, the motion becomes habitu
541m091 ZE upon this effort.-- it looks so analogous to MUSCLE in one position great fatigue.-- may explain e
552m131 ole body, being failed, & not to any particular MUSCLE Sept. 8th. I am tempted to say that those acti
556m144 g ahead. which cad cannot see, they do not move MUSCLE.-- reason CD[The laughing noise which C. Sphyn
560mIBC Do babies start, (ie useless sudden movement of MUSCLE) very early in life Do they wink, when anythin
121a111 den p. 5. «& 7.» violet strata from decomposed MUSCLES.. Smith of Jordanhill has seen same thing-- Co
290c163 ive) then must be owing to heredetary power of MUSCLES.-- then we SEE structure gained by habit Talen
315c243 may be made capital argument if man does move MUSCLES for uncovering canines) --. Blend this argumen
336d018 ations not heredetary,, but size of particular MUSCLES-- When two animals cross. each sends his own l
432e125 n to our present wants. or structure, than the MUSCLES of the ears to our hearing powers E. frowns pr
515q022 ith parents & of Chinese geese. (2) Anatomy of MUSCLES of stumps of tailess dogs & cats.-- (3) Hounds
520m002 fact.-- the resemblance was in odd twiching of MUSCLES, & general manner of holding hands &c &c.-- Mr
521m007 Now is not epilepsy an habitual disease of the MUSCLES.???> Miss Cogan's memory of the tune, might be
524m020 ives sensation of pain, emits its power on the MUSCLES in the twitching. Pride & suspicion are qualit
531m053 n has been accustomed to send a mandate to the MUSCLES & when the noise comes it cannot help doing it
532m057 roubled» beating of heart, sweat, trembling of MUSCLES, are not these effects of violent running away
533m059 the ready movement & co-relation of the proper MUSCLES. may be illustrated by the extreme difficulty
533m059 llustrated by the extreme difficulty of moving MUSCLES in different way from what they have been accu
536m070 .-- «He cannot avoid sending will of action to MUSCLES, any more than «prevent» heart beat» remember
539m084 as pale & trembling. & not as flushing & with MUSCLES rigid.-- How is this? dealt with p. 241 Origin
542m095 yawning can be explained from too long rest of MUSCLES.-- evidently habitual when transferred, (also
545m104 -- ie habits, which must require idea to order MUSCLES to do <certain> the actions.¿ is it the <becom
552m131 ely connected is passion with sending force to MUSCLES, that in my grandfather remark, a tired man. i
555m143 & ears,-- ∴ some men have this power abortive MUSCLES) The black Spider Monkey, very different dispo
557m147 resembles a fox-- I can conceive the opposite MUSCLES would act, to when in a passion.-- dog tail cu
564m006 itual sneer» the manner in which whole skin or MUSCLES are contracted between eyes & upper lip., is m
565m008 e corner of lower lip are depressed & opposite MUSCLES used to when angry sneering is in progress.<--
565m008 is in progress.<--> the hypothesis of opposite MUSCLES will want much confirmation. A grave person cl
565m008 much confirmation. A grave person close those MUSCLES, which wrinkle when smile.-- Hope is the expec
578m056 erson Decemb. 27th.-- Fear loose the sphincter MUSCLES, only on the principle like does an injury of
578m056 mind by habit gets more perfect over voluntary MUSCLES. these convulsive actions-- (except in weak pe
578m057 is slight convulsive wrinkling of some of the MUSCLES «or twitching».-- But why does joy & OTHER EMO
582m066 » flush, acceleration of pulse. or rigidity of MUSCLES.-- man cannot be said to be angry.--» «He may
583m070 makes its cells, by means of ordinary senses & MUSCLES, we cannot look at him, as machine to make cel
584m071 f preservation, (knowledge of enemies). use of MUSCLES, progression.-- use of senses-- knowledge of
589m092 sigh, is an abortive groan.-- more power over MUSCLES of voice than respiration.-- like sigh before
592m104 es in fear The effect of habitual movements in MUSCLES of face, is well seen in shortsighted people.-
596mIBC ts direction.-- In slight convulsions. are the MUSCLES of the face first affected?-- Can shivering &
596mIBC vulsion. are involuntary movement of voluntary MUSCLES-- if so what is trembling palsy? Expressions N
634j54v he Holocentrus ruber (a fish) Man has abortive MUSCLES to his ears.-- p. 313 Many other good cases --
312c228 alf lion & tigers ditto. (see Griffith) & half MUSCOVY ducks, «black cock & pheasant see Jardines Jou
334d009 of Mr Strutt» of his used to breed to Common & MUSCOVY Ducks.-- English. <Common> «China» & Canada Ge
334d012 btful whether all are white. Fox says the Half MUSCOVY Fox says a settler near Swan river, lost his <
341d032 s there is hybrid of Penguin duck a variety of MUSCOVY) with goose!!) Dr. Bachman regularly breeds «i
342d032 n regularly breeds «in Carolina» for his table MUSCOVY & common ducks-- they are produced in full equ
342d033 nous offspring. It appears certain that hybrid MUSCOVY & Common duck have been shot wild (escaped fro
363d101 her feathered legs.-- «Carruncles on beak & in MUSCOVY duck» crested feather, pouters, fan tails are
363d103 e young blackbirds Dr Bachman told me that 1/2 MUSCOVY & common duck were often caught wild off coast
392d180 Frogs» pass through birds stomachs & live? In MUSCOVY ducks do young take most after father or Mothe
531m052 sion be the feeling «consequent on the violent MUSCULAR exertion» which accompanies violent attack,--
531m052 e probably there is no feeling of passion, but MUSCULAR exertion consequent on the injury & consequen
531m053 ath, or running away. accompanied with want of MUSCULAR exertion, palpitation, voiding urine because
547m111 f association» Mayo Philos. seems certain that MUSCULAR, mental, <&> digestive nervous influence repl
556m146 ulls & horses.-- 1838 good instance of useless MUSCULAR tricks accompanying emotion.-- when horses fi
557m149 , with less action of the heart.-- tendency to MUSCULAR movement, hence shy people (shame of ridicule
572n033 ins.-- Erasmus does not liken term instinct to MUSCULAR movement.-- say instinctive actions. senses.
578m056 an injury of the spine-- that it paralyzes all MUSCULAR action -- «in man & animals» Blubbering of a
579n059 time in walking, to chronometer, is seen to be MUSCULAR movement. The Blushing of Camelion & Octopus;
274c115 structure for vegetation.-- In conversation in MUSEUM-- I could not discover any other clear relatio
449e173 t Timor.-- thinks he has seen specimen at Paris MUSEUM-- Athenaeum: 1839. p. 451. Sheep Merinos from
315c240 own) is found in lsd of Bass' Straits The common MUSH room & other cryptogamic plants same in Austral
454e184 there not wild plants, some partly dioecious? MUSHROOM Hybrids? Any «wild» plants in England, which
323c269 . Geology. Larder 2d vol.-- March 16. Gardner's MUSIC of Nature-- -- Herbert on Hybrid Mixtures: Marg
526m028 is perhaps true.-- mem Erasmus & mine taste for MUSIC.-- Children like hearing a story told though th
527m033 does not explain the feeling in any one man.-- MUSIC & poetry opposite ends of one scale.-- former p
527m033 nd. the frame of mind being just kept up by the MUSIC of the poetry.-- (therefore singing intermediat
528m036 sure independent of imagination, (as in hearing MUSIC), this probably arises from (1) harmony of colo
528m036 bsolute beauty. (which is as real a cause as in MUSIC) from the splendour of light, especially when c
528m039 which correspond to those <he> awakened during MUSIC.-- connection with poetry, abundance, fertility
540m088 , Galignani Edition Fine poetry, or a strain of MUSIC, when the mind is rendered ductile by grief, or
560m156 o the Ourang Outang like smells «peppermint» «& MUSIC».-- Have monkeys lice?-- picture.-- Do female m
568m018 ith singing-- is this origin of our pleasure in MUSIC-- do monkeys howl in harmony-- frogs chirp in d
569n019 from sexual beauty) is acquired taste.-- Whilst MUSIC extremely primitive.-- almost like tastes of mo
574n041 ulated, smell, as Sir. Ch. Bell says, & hearing MUSIC. to certain degree sexual.-- The association of
575n045 nses. The shudder of pleasure. from pleasure of MUSIC Audubon IV Vol of Ornith. Biog. case of Newfoun
581n064 ied Kingdoms-- "food, sm<e>ll. (ourang-outang), MUSIC, colours we must suppose <we> «Pea-hens» admire
581n064 avages other tastes few. March 16th. Gardiner's MUSIC of Nature. p. 31. remarks chicken have no diff
582n066 in his work» In the life of Hayd & Mozart. fine MUSIC is evidently considered as analogous to glowing
590n094 Australian Dog does not Bark-- quotes Gardner's MUSIC of nature to show barking not natural. (Vol I.
593n107 o make saucer-shaped depression.-- [blank] Does MUSIC bear any relation to the period when men. commu
593n107 ical notes the language of passion & hence does MUSIC now excite our feelings.-- How does Social anim
593n109 ones. in some degree is so.-- idea of beauty of MUSIC are great distinguishing character between man

Page **(Key Word)**

603o014 , with good expression-- statues not painted-- <MUSIC> very good article-- why flower beautiful? ¿eve
616o038 ; you cannot call the frame of mind which makes MUSIC pleasant, a memory; yet that frame is enhanced
546m109 easure & innate tastes, he partakes, taste for MUSICAL sound with birds. & ¿ howling monkeys-- smell
593n107 municated before language was invented,-- were MUSICAL notes the language of passion & hence does mus
271c105 o strictly <f> agree in habits with the Turdus MUSICUS «not found in N. America» whose Southern range
459t009 eanings of Science Vol III. p 320. Mr Hodgson on MUSK Deer-- young spotted <like in> "prettty much as
366d111 take an effect.-- would perfect impunity from MUSKITOES bite influence propagation of species.-- Case
539m082 bringing up old indistinct ideas of FitzWilliam MUSM. I was amused at this after seven years interva
293c174 -- else smell of Man would be disagreeable to MUSQUITOES We never may be able to trace the steps by w
366d111 ation, & change of structure.-- When will the MUSQUITOES of S. America take an effect.-- would perfec
486z018 lly require a greater duration of Heat. hence MUSQUITOES & knats abound during short summer far N. wh
021rIFC of changes towards perfection (namely mammalia) MUST have a shorter duration, than the more constant
026r019 mains in De la Beche, for the older formations I MUST believe they «the limestones» have been formed
026r020 on South of the Maypo at one period of elevation MUST in its configuration have resembled Chiloe In D
028r028 vers currents. & sea beaches. All mineral masses MUST have a tendency. to mingle; The sea would separ
030r035 iles from shore. V. Chart) Every winter torrents MUST bring much vegetable matter from thickly wooded
032r040 ; Valley of Copiapò & parts of coast of Chile.-- MUST first explain «top of» tidal band of action. Th
032r041 lation owing> rotation in fluid matter of globe. MUST there not be a circulation «however slow & weak
032r041 proceeds from great depth. from axis to surface MUST gain a Westerly current:--If great changes of c
035r045 ocks. connected with & alternating with obsidian MUST clearly be chemical differences. & not those of
036r049 l represent the dilatation, which dilated cracks MUST be filled up by dikes & mountain chains.-- Intr
037r055 ions? Strong currents off the Galapagos.--strata MUST be accumulating which like the secondary strata
038r057 to state all arguments for believing that there MUST be a central core of melted rock--I think the s
038r057 ent little craters are all burning, surely there MUST be «somewhere» below a field of fluid rock.--In
040r063 o animal life.--Patagonia In the Chonos Islds we MUST imagine bituminous shales have been metamorphis
040r063 ers very rare) to cause floods in valleys, which MUST aid in preserving the terraces <...> Molina's C
042r068 of Indus», my belief in submarine tilting alone, MUST be modified. «Moreover, the Volcanos from sea t
043r069 period «& manner» of elevation Volcanic action, MUST be more exclusively confined to that country. R
044r074 effect of water thus holding matter in solution MUST be great: & in the fact of bombs in tufa there
046r078 connection of Pacifick & S. America. -- Volcanos MUST be considered as chemical retorts.--neglecting
046r078 ated.--«Sulphur last.--» Metallic veins likewise MUST separate ingredients if we look to a constant r
047r081 great wave on record. -- «also neighbouring sea MUST partake in absolute movement» Moreover wave «wi
047r081 hes far beyond coast, which has been raised.--It MUST be considered as an oscillation, from violence.
047r083 coast (as seen in swell) the undertow & overfall MUST vary proportionally Partial shrinking after ele
048r086 ame caution is applicable to the Siberia case We MUST not think alluvial plains «always» most favoura
049r087 he cellular state of all the Porphyry specimens, MUST be well examined At M. Video «facts of Passages
049r088 ordillera???» Porphyry at Valparaiso; Epidote -- MUST we look at regular greenstone cones at S. T. de
050r090 raters as at Galapagos. <|> Sir George Mackenzie MUST be worth reading Some earthquakes of Sumatra no
051r094 Kelp, now Kelp sends forth branching roots which MUST protect surface; On «hard» exposed rocks near B
053r101 ? In my Cleavage paper Dr Fittons Australia case MUST be quoted at length. The Lines of Mountain appe
059r123 great depths, so changes, acting in those lines. MUST now proceed from great depths.--important.-- De
060r124 d is constantly pushing the sea (which of course MUST retain same level) to a greater distance".--Aft
061r128 if any creation «taking place» over certain area MUST have peculiar character: Contrast low limit of
063r130 s in first cases distinct species inosculate, so MUST we believe ancient ones: «∴» not gradual change
063r130 nces: if one species does change into another it MUST be per saltum--or species may perish. = This <i
064r137 dillera In describing structure of Cordillera it MUST be said, that lines of elevation have connected
069r149 ity of such change having taken place unrecorded MUST be insensible. Quantity of matter from Cordille
069r151 alt where Basalt. trachyte where trachyte. There MUST have been as much conglomerate on West of Peuqu
069r151 of Peuquenes as on East. Where gone to.?-- There MUST have been some conglomerate East of Portillo Wh
070r153 s. show that granite when weathering into balls. MUST exhibit orbicular structure.--When we recollect
071r156 Dicotyledones far preponderant, if so coniferous MUST formerly have been most abundant tree-- Metamor
086a006 our from below-- Malte Brun «Salt Lakes» Siberia MUST be read as well as Pallas before Geology is wri
090a022 tter. in making trachyte come out before.-- What MUST be the effect of all the meteoric stone which m
090a022 st be the effect of all the meteoric stone which MUST have fallen on the globe since the Cambrian sys
104a067 w dikes which have given rise to eruptions.-- We MUST suppose everywhere--, in granitic areas &c &c v
105a071 cession of flights of steps; if one lake then we MUST suppose barrier in the very part, where barrier
106a076 depths. All Earthquake unaccompanied by Volcanos MUST be sought after proofs of sinking.-- No Sweden!
107a078 ng place chiefly beneath water & volcanos. crust MUST be thinner «under water» but cause most difficu
109a082 t the bottom of sea. off coast of England.-- Sea MUST always on actual beach act same way.-- a little
109a082 modified by form of waves & currents.-- but this MUST be continued. no currents & elevation have same
109a084 se on Greenstone, diorite, &c most important.:-- MUST be studied.-- Scientific Memoirs Edited by Tayl
109a084 s Edited by Taylor Ehrenbergh on flints in chalk MUST be studied-- though I do not think good p. 411
110a087 have oozed out of cleavage plates: the crystals MUST have recrystallized, as such do not occur in ei
111a087 V. VII. p. 316 & 328 VI. p. 365. Meyen on Chile MUST be studied Analysis of Voyage: many observation
112a089 & other trap.-- It is in the mountain masses we MUST look for that.-- how few isolated volcanos ther
113a090 an of surface formerly much higher, «so» that we MUST look at the upper four hundred feet of strata h
115a096 cleave, & which unite the homogenious crystals., MUST aid in adding effects to common heat.-- Where t
117a101 alone explain the immense amount of change which MUST have taken place, otherwise the world would dai
119a105 equired by thermometrical scale.-- (for the temp MUST be immense to convert rock into gneiss &c judgi
120a109 case of the shingle in the great Chilian valleys MUST be profoundly considered. if elevation near coa
122a114 the platform:-- On my view the degrading action MUST prevent internal fluid arriving at equilibrium
123a116) describes near Alps great beds of rivers which MUST be like the Chilian ones.-- Septemb. 2d.-- Sulp
123a116 ilian ones.-- Septemb. 2d.-- Sulphur like carbon MUST go round of dissemination & separation in volca
124a117 frozen soil. p. 211 Consider proved that Siberia MUST have been in same condition for long period Sub
125a121 ews earth originally fluid, then cooling process MUST go from surface towards the interior -- who kno
127a125 th.-- Now the <inf> subterranean isothermal line MUST be creeping «pushing» up to «the» line of ice.-
128a128 as. surrounded by salt water. purest fresh water MUST be sought for below the sea mark.-- If mountain
129a132 reat subsidence «20 miles long I in with.» which MUST have been from an axis, «20 ft at least in dept
130a133 Roy paper I show crust yield easily. & if easily MUST be thin: <beside mere fracture> A Elevation as
133a140 re the abysses where fluid rock has been ejected MUST remain fluid for an enormous period: now when w
146g007 m of hill be explained by my idea-- highest part MUST project [Blank] {P} Path East End near Holyrood
147g018 eries Whole very obscure but it is certain there MUST once have been very considerable mass of waterw
149g030 member however the great Chilian valley Acongua, MUST there have deposited much-- On other hand remem
151g039 rely united.-- die away also, without any cause, MUST be tides. &c. {P} <notch> roads very much this
151g042 side with irregular gravel plain of other, which MUST have been waterworn after 3d lake.-- 4th shelf
152g048 er cut through, not applicable to Glen Roy Lake, MUST have remained very long at 4th shelf from size
155g063 Alluvium nearly parallel--» Inclination of river MUST constantly alter with falling sea & so corrode
155g064 errace as regressed What <alter> a balance there MUST be in power of rivers either bringing more «det
155g065 (Mac, hypoth,)» the level during any oscillation MUST have been so carefully preserved as to have thr
157g076 uartz, & garnets.-- Boulders as before certainly MUST have <come> «been drifted» here: on very summit
163g107 er with pebbles now worn away-- The above shells MUST have been about 60 ft above sea-- soon decayed
172b008 losely approaching.-- but as they inosculate, we MUST suppose the change is effected at once, -- some
175b018 complicated,; & if we look to first origin there MUST be progress. if we suppose monads are «constant
176b022 anged most. «owing to the accident of positions» MUST in each state of existence have shortest life.;
180b036 se,∴ MONUCULE NOT DEFINITE LIFE I think {P} Case MUST be that one generation then should be as many i
180b037 , 13 recent forms.-- Twelve of the contemporarys MUST have left no offspring at all, so as to keep nu
183b053 of great quadrupeds declining as great reptiles MUST have once declined.-- Read his theory of the Ea
186b060 raced the <type> great quadrupeds to Siberia; we MUST look to type of organization.-- extinct species
191b080 Animal for Geography.-- The motion of the earth MUST be excessive up & down.-- Elephants in Ceylon--
195b098 space before island existed.-- Such an influence MUST exist in such spots. We know birds do arrive &
196b108 tish varieties are also found in Ireland-- There MUST be progressive development; for instance none--
196b108 ted; but on other hand creation of small animals MUST have gone on since from parasitical nature of i
196b110 of connecting Mollusca & vertebrata, that there MUST be very great gaps.-- yet some analogy The exis
197b111 less differences:-- does not say propagated, but MUST have concluded so= Evidently «or hints» conside
206b147 h lustrum, the number related at the first start MUST be greater, & this number would vary at each lu
206b148 ine families «no doubt a different set of causes MUST act in the two case,» May this not be extended
217b183 Galtons case.-- It explains the loss & expense, (MUST probably have occurred to every one) of rare br
217b188 n islds.-- This is interesting, because Iceland, MUST have been all ice in time of ice transported.--

Page **(Key Word)**

219b195 sacts. Vol XII.-- The Alpine plants of the Alps. MUST be <Alpine> «new formations» because snow forme
219b195 northern plants «No» CD[Mem. the antarctic flora MUST formerly have been separated by short space fro
223b211 fsprings accounts for uniformity of species & we MUST confess. that we canot tell, what is the amount
226b221 by same chance few representative species. this MUST happen. & thus acquired will explain representa
226b222 continental or many large islands.-- Hence this MUST have been condition of Paris basin land.-- (How
228b231 ind. than Macrauchenia yet he may be found:-- We MUST not compare «chances of embedment in» man in pr
231b243 this is difficult to explain by creation-- or we MUST suppose a multitude of small creations.-- Will
232b245 thers may die out or move South ward.... species MUST be compared to neighboring sea.-- For change of
242c017 the Floras can be trusted The changes in species MUST be very slow, owing to physical change, slow &
248c030 g, will not readily breed together: The argument MUST thus be taken, as «in» wild state (where instin
250c039 istinct from rest of world??? Lyells Principles, MUST be abstracted & answered Much might be argued w
252c042 h excellent detail & fine, views about Species-- MUST BE STUDIED: genera founded in nature [not locat
254c049 other, (surely rather parents). (NB These views MUST lead to spontaneous generation??) This whole Pa
254c049 ad to spontaneous generation??) This whole Paper MUST be studied.-- <Three p. 7. Am.> D'orbigny. Bird
256c053 would surely have been more homogenious.-- There MUST be some sophism. in Lyells statement that some
258c062 nother-- (according as parent types are present) MUST follow after there is proof of the non creation
259c063 ct on offspring-- but WHOLE race of that species MUST take to that particular habit.-- All structures
261c069 -- instinct Swainson's remarks in Fauna Borealis MUST be studied. There is capital table of extent of
264c081 ubt how far structure & habits go together. This MUST be profoundly considered.-- Structure may be ob
264c081 ts slightly preceed them-- From this view habits MUST form most important element in considering to w
265c083 rth do not effect progeny-- many dogs in England MUST have been lopped off & sheeps tails cut yet the
268c099 ution is.-- ¿Pelagic forms similar--birds??-- We MUST always bear in mind proofs of most equable clim
272c108 conditions the Staphylinidae on St. Pauls Rocks MUST be placed under Gould says most subgenera confi
273c112 f its whole family where female is not dull.-- I MUST observe that this preeminent structure is not a
274c114 uga. Kingii very rasorial for type.-- Now here I MUST observe these characters vary in degree in last
276c123 of their age. (differently from literary men.--) MUST remember that if they believe & do not openly a
277c128 This is reform which probably will be slow. but MUST take place-- such a classification would answer
278c129 nearest species of each might not breed:-- Genus MUST be a true cleft-- putting out of case the Analo
281c139 elated in the third & fourth degree.-- a species MUST be compared to family entirely separated from a
282c141 ether different.-- To make this case perfect, we MUST suppose men instead of mere colour & trifling f
282c141 habits, yet we might have these analogies.-- We MUST two races of such men living in same country bu
282c143 produce so great an effect.-- the spirit of life MUST be every where ambient. & <in> merely determine
282c143 al laws.-- so that all character originally may «MUST» have had the character of analogical.-- Gould
284c148 sly heavy (where trees of such Nature far apart. MUST have travelled by <dead> «each» trees dying & m
285c150 . whenever instinct is mentioned some definition MUST be given It would not be difficult to arrange c
285c150 dites» father « <mother> » & grandfather <Might> MUST be introduced & made young.--‖ father must be l
285c150 ight> Must be introduced & made young.--‖ father MUST be left out of case, that difference occurring.
285c152 thing with regard to contemporaries-- fertility MUST settle it.-- Changes in structure being «necess
287c158 of change in organizaton. But the classfication MUST chiefly rest on these same organs,-- habits, ra
289c162 n.-- Generation may be viewed as condensor, «+++ MUST (on my theory) =supported by foetal lower devel
289c162 told by their larvae) but the acts of condensing MUST alter method of generation-- Heaven knows how.-
290c163 n colour; lives on ground, colour of habitation. MUST have some effect.-- Maldonado as good forests f
290c163 g horses, (if not looked at as instinctive) then MUST be owing to heredetary power of Muscles.-- then
292c170 ffered of the singular views there offered, & he MUST be a zealous man in the cause if his faith is n
292c171 y & instinctive & not others.-- We even see they MUST be done often «to be habitual» or of great impo
293c174 be accounted for by My method of breeding there MUST be some corelation. but. the whole Mechanism is
294c176 mals-- but how to show species-- I fear argument MUST rest upon analogy & absence of varieties in a w
294c176 ain very close Species in islds. near continent, MUST we resort to quite different origin when specie
296c184 hybrids in nature.-- «p. 473» Webb &. Berthelot. MUST be studied on Canary islands-- Endeavour to fin
297c186 s adaptation to nocturnal habits-- to cats &c.-- MUST be acquired by my theory-- else my theory not a
298c191 r country during gradual elevation of isld.-- We MUST imagine a considerable range of one species «on
298c191 pecies would then revert to pristine form (which MUST have been altered by crossing) with alpine form
299c193 how very permanent plants are. & this conclusion MUST be arrived at, when one sees a plant like Paris
301c199 le through hoarding from short time.-- My theory MUST encounter all these difficulties.-- Knowing tha
302c201 ny animal really connecting the fish & Mammalia, MUST be sprung from some source anterior to giving o
302c202 analogous characters, might be multiplied.-- we MUST argue reversely. WHERE CHARACTER VARIABLE it is
302c202 in opinion on all subjects of classification, I MUST work out hypothesis, & compare it with resuts.
302c202 lant groups between two circles «of equal value» MUST be so from characters of analogy.-- see my note
304c208 rcumstances In ostrich which is not isolated, we MUST suppose the changes from typical structure have
308c217 e, then there always have been gaps, & there now MUST be, ∴extinction of species bears relation to ex
308c218 omplex nature itself with which our speculations MUST end as well as begin" &c &c then centre is ever
309c220 improve the women. (double influence) & mankind MUST improve-- The areas of subsidence marked out by
312c228 ithout «number of vertebrae» new acquisition, we MUST [not located] Henry Thompson tells me best way
312c233 of father & child are thought long breeding in. MUST not trust him Hope says that genus of parasite
313c235) without impregnation, therefore sexual passion MUST arise after long interval very good case.-- hab
314c238 -- the consciousness of the plant that this part MUST be protected however it may be effected.-- Prod
315c240 e.-- if creation be absolute thing, the creation MUST take place on[ly] when creator sees. the means
335d013 t has not been, will not remain,-- yet offspring MUST be somewhat like parents,-- therefore offspring
335d015 dency to have offspring like parent, but as they MUST like or there will be none, therefore a mule ca
336d017 t is impossible to transmit them, & as offspring MUST be like parent, therefore mule has no offspring
336d017 stock. but if both parents are alike, offspring MUST be like Hence mutilations not heredetary,, but
337d020 ish.-- The Varieties of the domesticated animals MUST be most complicated, because they are partly lo
337d020 to wild local variety.-- our Europaean varieties MUST be very unnatural-- Italian Greyhound is probab
337d021 und» & much intermarriage.-- In my speculations. MUST not go back to first stock of all animals, but
342d033 rouse different.--» (if so chinese pigs & common MUST be considered as distant species?? or is time t
342d034 -water fish were once salt water (as they almost MUST have been on elevation of continents) but Ogleb
343d038 rom slave trade, & colonization of S. Africa, so MUST the tribes become blended & prevent that strong
343d039 om the existing animals found fossil when Europe MUST have worn a quite different figure 19th. With r
344d042 pure Harrier.-- «The peculiarities of our breeds MUST have been acquired, & hence this is then case o
345d042 ght that a <pure blooded> «"first blood"» animal MUST have gone on for many years, before deserves <n
347d049 resist external influence.-- 27th. August. There MUST be some law, that whatever organization an anim
347d049 o multiply & IMPROVE on it.-- Articulate animals MUST articulate. <i> in vertebrates tendency to impr
347d049 eration is condensation of changes. then animals MUST tend to improve.-- yet fish same as, or lower t
352d058 l radiating point, all united . (links in circle MUST be granted unequal, because fossil) Now what is
352d059 es-- <be> when it is considered the tree of life MUST be erect not pressed on paper, to study the cor
352d059 y organ is modified by use, every abortive organ MUST have been once changed.-- what is abortive? whe
355d068 place in the formation of a bud-- the very same MUST take place in copulation-- (Man & woman separat
357d074 s were it not for seeds being floated about.-- I MUST state that. the <p> mechanism by which seeds ar
359d087 ere are some 3/4 birds «of», which I think there MUST be some mistake in their origin) Saw cross betw
360d089 idual instances trouble Yarrels law. chief trust MUST be in general knowledge of breeders, when thei
365d107 zle, than none, for the «e»normous time which it MUST have taken to separate. Van Diemen's land from
366d108 long permanence, so that all their peculiarities MUST be transmitted if their [not located] <The> eve
366d112 Pouter one thinks there is a law.,-- that there MUST have been a tendency for feathers to grow there
367d112 s forming the ovum within it, is forming «& this MUST be so, else avitism could hardly ever occur.--»
370d117 s in all parts of world, thus tea tree in Brazil MUST have degenerated. as must spices &c &c The line
370d117 hus tea tree in Brazil must have degenerated. as MUST spices &c &c The line of argument «often» pursu
371d128 nsmitting them with such facility to bud.-- this MUST be owing to their unity in one stem.-- a bud ma
372d130 division.-- On this view each particle of animal MUST have structure of whole comprehended in itself.
372d130 structure of whole comprehended in itself.-- it MUST have the knowledge how to grow, & therefore to
375d134 s inference from Malthus.-- «increase of brutes, MUST be prevented soley by positive checks, exceptin
375d135 ndy.--» take Europe on an average, every species MUST have same number killed, year with year, by haw
375d135 .. even one species of hawk decreasing in number MUST effect instantaneously all the rest.-- One may
375d135 ker ones. «The final cause of all this wedgings, MUST be to sort out proper structure & adapt it to c
377d139 more definite than with bitch, for some feeling MUST urge them to these actions. «These facts may, b
379d152 at in breeding very pure South Down that the ewe MUST never be put to any other breed else all the la
381d157 ribes, leads one to suppose still more that they MUST in effect be so in all.-- 2 NB. In Pectinibr Mo
383d159 ility of becoming either sex.-- ‖ In my theory I MUST allude to separtion of sexes as very great diff
383d159 ike J. Hunters. Free Marten N.B. the common mule MUST often have been dissected Zoolog. Garden. Sept

Page **(Key Word)**
388d173 sex being present is-- it also shows that semen. MUST actually reach the ovum.-- [Why in making a bud
389d177 ty ¿this does not happen with hybrids?]CD Plants MUST stand much breeding in & in (those which have s
390d177 cs brought from foreign country. (<annuals> & so MUST those forms which are produced by budding «only
391d174 «& from frequency of this tendency all mammalia MUST long have so existed.» with double union.-- At
391d175 life.-- if the circumstances which induce «which MUST be external» change are always of one nature sp
403e024 largest sense. <es> «as» test of species.-- they MUST deny species which is absurd.-- their only esca
404e025 s» of crustacea, one is tempted to think that it MUST have been invented all at once.-- but naturalis
407e042 y necessary to show that if species fall, genera MUST. Lesson I remember says Mariana Deer very close
408e042 to S. American forms. The climate of N. America, MUST have been equable & low-- more so than any othe
408e043 resses towards extermination. some other species MUST increase in number where then is the gap, for t
408e044 to enter?-- The wonderful species of Galapagos, MUST be owing to these islands, having been purely r
410e052 some high theoretical interest,-- "the great end MUST be the law & causes of change".-- A philosopher
410e053 -- but then permanent varieties in same country, MUST be distinguished, from permanent varieties not
411e054 uch corelations, & changes of individual organs, MUST know whether the individuals «forms» are perman
417e075 The fertility of Indian & Common Oxen, which one MUST think deserve the name of species, may be owing
419e085 salt?.)-- very few animals of any kind-- Fauna, MUST be very curious.-- With respect to the non-deve
419e087 tion & no change & even no loss of species.-- It MUST never be overlooked that the chronology of geol
419e087 of change) than a slow gradation in form, «which MUST be effect of slow change + & Therefore preclude
419e087 erefore precludes effects of catastrophes, which MUST serve to confound our chronology» «CONSIDER ALL
420e089 rtebrae, the parent of all vertebrate animals.-- MUST have been like some molluscous «bisexual» anima
423e096 heir places.-- The Geologico-geographico changes MUST tend sometimes to augment & sometimes to simpli
428e112 between strong & weak March 11th. Yarrell's law MUST be partly true, as enuntiated by him to me, for
429e114 ng on the peaceful woods. & smiling fields.-- we MUST recollect the multitudes of plants introduced i
430e120 siderably different, in proportional dimensions, MUST <almost> be considered merely varieties. & even
432e124 mprove the native <breed> animals of any country MUST be made with great caution; owing to its adapta
433e126 during the deposition of the beds-- The argument MUST be thus put, shall we give up whole system, of
440e146 tufts &c &c here there is no final cause yet it MUST be effect of some condition of external circums
440e147 place.-- & therefore the degree of injuriousness MUST have been exceedingly small.-- This is far more
441e150 Hydrangea -- black bullfinches-- & all varieties MUST be presumed to be result of such laws.-- The ef
441e150 of one part being greatly developed on another, MUST not be overlooked.-- it makes fourth cause or l
441e151 ss with another.-- to escape it «in any case» we MUST draw such a monstrous conclusion, that every or
443e153 the Graft.-- ‖Plants circumstanced as the Gorze MUST be propagated by its roots: now it is curious M
443e154 s, but to arrive at their present structure they MUST have <done> been propagated by sexual commerce
444e158 perfect animal are adapted by foreknowledge, so MUST the mutations of species.!!-- p. 203 Chaetodon
454e184 nuals can be budded «& rendered of great age» as MUST be inferred from what Mr Knight says. Hort. Tra
460t019 finite & growing complexity from a few types, it MUST not be supposed that this refers to time.-- Mar
460t019 this complexity from a few types originated, we MUST go to the first origin of the world.-- our pres
463t057 xplanation will probably reject this theory-- (I MUST answer it by rooting out curious cases of inter
464tf6v (see Blyth's work on Ruminants,-- these species MUST have migrated to these mountains, when the cold
465t080 we may take this as guide, the shells preserved MUST be as much a casualty as, bones of Mammalia in
466t094 Bees put proboscis within nectary «they do» they MUST disturb all anthers, wh otherwise lie protected
481z010 's possession» of Zoolog. of Voyage of Astrolabe MUST be studied for anatomy. of. corals.-- neverthel
482z011 e to ours Rengger's work of Mammali: of Paraguay MUST be most important a discussion of geographical
483z013 pel travels All Owens papers on Intestinal worms MUST be studied in Vol I, Zoolog: Transact. before w
484z014 ?? in N. America?? Wilson N. American Ornitholog MUST be studied before writing my general account--
495q05a as swum-- on pools & rivers-- every kind of seed MUST be distributed.-- Examine scum of pond for seed
498q008 es will mingle the infinitesimal varieties which MUST occur.-- ¿is «it» not these infinitesimal varie
498q008 cted with separate division of germen <?>-- (11) MUST pollen grain be whole, to impregnate?-- I presu
499q09v s.-- Hooker says Rafflesia is dioecious & Pollen MUST be carried by some insect-- (21) Are there many
523m014 ness not just.-- From habit the feeling of anger MUST be directed against somebody.-- Have insane peo
523m014 tness of their hatreds, as <if> in my case.-- It MUST be so from the curious story of the Birmingham
529m042 ite sensible & no ways an ideot.-- «in this case MUST have been functional.--» He has some idea of a
531m053 use done by some animals in defence, &c Starting MUST be habitual «involuntary» movement from wish to
532m053 start.-- ‖ If children wink. it is instinct Fear MUST be simple instinctive feeling: I have awakened
532m055 it. with people nervous from illness., <the> it MUST be an excited action in the involuntary mind wh
532m057 are not these effects of violent running away, & MUST not <this> «running away» have been usual effec
534m063 tive, seeing that time is lost & endeavours made MUST be experience & intellect.-- do. p. 157. Westwo
537m075 opposite side has been shown-- see Mackintosh.-- MUST grant, that the conscience varies in different
539m084 th p. 241 Origin of man now proved.-- Metaphysic MUST flourish.-- He who understands baboon <will> wo
541m092 ple who can multiply large numbers in their head MUST have this high faculty, yet not clever people.
542m097 , Fox of cows, & many of insects-- they likewise MUST understand each other expressions, sounds, & si
545m104 conscious. so may some ideas.-- ie habits, which MUST require idea to order muscles to do <certain> t
548m115 hen one is deeply reasoning besides these (which MUST be present, though one is not conscious of them
549m118 To make greatest number of pleasant thoughts, he MUST have contingency of good food, no pain,-- <but
549m122 as revenge «& anger», which experience shows it MUST for his happiness to check-- that is external c
550m125 s, & not mental or ideal ones, <which> «& these» MUST occupy greater proportion of <each> «every» man
556m146 breast now if horns were to grow on horses, they MUST yet continue to put down ears, when kicking.--
564n003 social & sexual instinct «& yet with passion» he MUST have conscience-- this is capital view.-- Dogs
564n005 del itself.-- the mind is function of body.-- we MUST bring some stable foundation to argue from.-- O
565n010 say "my dear," just what smile is to laugh.-- I MUST be very cautious. Remember how Lavater ran away
566n010 of mankind, as being really useful to them: this MUST be studied. before my view of origin of evil pa
568n017 emptation, if memory of its own emotions. (which MUST be intimately united with reason) it would feel
570n027 ection with male figure]CD-- As forms change, so MUST idea of beauty.-- [Old Graecians living amongst
571n029 ste? Real taste in mouth, according to my theory MUST be acquired, by certain foods being habitual--
571n031 sa.-- almost proves that at earliest times there MUST have been intimate connection between sound & l
573n036 ts its place in nature. its range, its-- &c &c:--MUST be a special act, or result of laws. yet we pla
574n042 ct, on my theory intelligible An habitual action MUST some way affect the brain in a manner which can
576n048 on, excepting from favourable circumstances!» We MUST believe, that it require a far higher & far mor
577n050 discover he had not it.-- The memory of Plants, MUST be association,-- a certain round of actions ta
579n058 the manner short-sighted people frown, frowning MUST have some relation to short-sightedness.-- do n
581n063 phrase "heredetary habits." very clearly, all I MUST do is to generalize it, & see whether applicabl
581n064 ood, sm<e>ll. (ourang-outang), music, colours we MUST suppose <we> «Pea-hens» admire peacock's tail,
581n065 re they can speak-- or understand-- thinks so it MUST have been in the dawn of civilization-- thinks
582n067 ca capital instance of heredetary habit:-- there MUST, however, be a mental impulse (though unconscio
582n067 n instinct, & its impulse to do it.-- [the means MUST be present on any hypothesis whatever]CD an ani
582n068 ey will effect,.--" this not wholly true, for we MUST grant a bird knows what is about when building
583n069 may be true,, but relation of imitation & reason MUST be thought of.-- p. 19. animals capable of educ
583n070 ially as it adapt its cell to circumstances), it MUST have impulse to make a cell in certain way, whi
583n071 which way its organs are sufficient for hence it MUST some way be able to measure the cell; p. 22. in
585n077 y» acquired by habit. so bees in building cells, MUST have some means of measuring cells, which is fa
592n103 essive laughter & so into convulsions.-- «Paper» MUST be referred to, if I follow up this subject & a
593n107 ot know its own smell or look, & therefore there MUST be some instinctive feeling which is pleased by
599o005 rs all speculations of the origin of language.-- MUST presume it originates slowly-- if these specula
603o11b " Belief allied to instinct.-- p. 134. a painted MUST not a actors, or a scene in garden.-- yet both
603o012 Jews, African all sacrifices. How completely men MUST have personified the deity.-- H. Tooke has show
604o016 758-- Read the Review or the article. A Planaria MUST be looked at as animal, with consciousness,, it
606o20v ally applied to this mental power Although taste MUST necessarily be acquired by a long series of exp
606o025 action whatever is the effect of a motive.-- [-- MUST be so, analyse (a) ones feelings when wagging o
608o026 tion) he thinks they have none.-- Effects.-- One MUST view a wrecked man, like a sickly one(P)-- We c
609o030 judging of <our ha> of the rule of happiness we MUST look far forward «& to the general action» -- c
609o30v origin of the social instinct «in man & animals» MUST be separately considered.-- The difference betw
610o032 manner.-- as the hour of the day &c-- All habits MUST conduce to their health & comforts.-- Both ideo
611o034 xed relation to such accidents. But such tissue <MUST> bears relation to whole, that is enough must b
611o034 e <must> bears relation to whole, that is enough MUST be present to be able to exist as individual.--
613o035 mory gone?-- Where pain & pleasure is felt there MUST be consciousness??? ? [LHC] ¿Can insects live w
613o036 , therefore word misplaced The meaning of Words, MUST be made out Reason Will Consciousness Definite
614o037 ght be analysed into steps, as species change.-- MUST be so if Lamarck's theory true [RHC] Acquired i
616o038 <application in> «ejection only has» will: there MUST be cases of secretion being some time governed

Page
(Key Word)
617o0040 ffect us, except by internal consciousness 3) We MUST endeavour to do without it as well as we can. <
620o0044 t.-- Whatever the cause of this may be, everyone MUST know, how soon the pleasure from good dinner, o
621o0047 robation of his fellow men.-- [RHC] 6) Hence man MUST have a feeling, that he ought to follow certain
621o0047 e ought to follow certain lines of conduct, & he MUST soon necessarily learn that it is his interest
622o0048 honour from Paley [RHC] Anyone, who will reflect MUST feel, how like to injured conscience, is the fe
623o0050 ain from our reason, that all which <has> (as we MUST admit) has been acquired, does possess the bene
623o0051 s, which have led to the peculiarities, & hence <MUST have» «only that which» had a beneficial tenden
624o0051 s* to same end.-- & general actions of community MUST frequently teach same end.-- Hence this becomes
627o0053 d into "power" &c &c &c-- & if termed "selfish", MUST be subclassed as "disinterested" p. 14. It is a
632j53r uninhabited as the Alpine pinnacles.-- One thing MUST be admitted there would not be these plants, if
634j54v ign, pursued to its utmost exhaustion, & till it MUST be abandoned for another".-- What bosch!! Put i
637j58r it not been so.-- let egg shells grow harder. so MUST those with weak beaks be sifted away.-- 4 & the
639j28r Camels? all good cases of corelations.-- [There MUST have been deserts in the old world!]CD p. 252 a
641j29r hat they most widely differ» 3 A very wide range MUST be destructive to species, when physical change
641j29r nciples that islands are favourable,) because it MUST take so long to change species-- yet this is co
641j29r r hand Spermaceti Whale & Manatee.-- Naturalists MUST be cautious.-- <some others>: study these facts
444e158 mal are adapted by foreknowledge, so must the MUTATIONS of species.!!-- p. 203 Chaetodon squirting wa
602o0010 he source of part of the highest enjoyment in MUTILATED statues In Elliotson's Physiology much about
312o232 ced in short time in some extent counterpart, MUTILATION being «variation» produced in shortest possi
388d172 The following views show that transmission of MUTILATION impossible.-- it should be observed that tra
312o232 ncts, & not '; instincts, constant. ¿ whether MUTILATIONS non-heredetary & variations produced in shor
335d015 spring.= but. as «badly» deformed people & as MUTILATIONS «(produced very quickly)» sometimes have sim
336d018 ents are alike, offspring must be like Hence MUTILATIONS not heredetary, . but size of particular Musc
366d112 a tendency for feathers to grow there «That MUTILATIONS will not alter form may be inferred from Aus
391d174 harmony with all its previous changes, which MUTILATIONS are not). but why should it demand some furt
416e070 le to my theory] CD without the hermaphrodites MUTUALLY couple,-- now how is it-- in Planaria, they c
380d155 though caterwhalling. & put into female, when MUZZLED, he is disabled.-- so Elephant in confinement,
024r015 hore? Beagle Coast of Brazil? where not rivers in MY Coral paper {T} leagues Fathoms Parallel of St C
025r018 urope, with America; I might add I have drawn all MY illustrations from America, purposely to show wh
025r019 e. we may believe them applicable to the world.-- MY general opinion from the examination of sounding
035r045 l differences. & not those of rapid cooling &c &c MY results go to believe that much of all old strat
036r050 near Copiapò which is asserted to make a noise,--MY impression. is not very distinct, from some of t
036r050 believe it was necessary to ascend the hill,--but MY recollection is imperfect & was recalled by note
042r068 tion of Earthquakes. «on Chili & delta of Indus», MY belief in submarine tilting alone, must be modif
042r069 ve destroyed regularity of slope of valleys.--All MY observations of period «& manner» of elevation V
050r090 nts, nature of which is doubtful. P. 180. I think MY Ascension case very doubtful. -- In Iceland Blad
053r101 there Earthquakes in the Radack & Ralix Islds? In MY Cleavage paper Dr Fittons Australia case must be
061r127 one blow. if one species altered: <altered> Mem: MY idea of Volc: islands. elevated. then peculiar p
078r174 tera D'Orbigny has figured animal with setae like MY undescribed[.] p. 140. Flèche of Quoy et Gaimard
092a029 currents tend to deposit metals, if in solution. MY view of metamorphic in contradistinct to Volcani
094a035 nite.-- a most important question with respect to MY theory of changes. of granites into Trachytes.--
107a078 with respect to thickness of crust broken up. ----MY view of Volcanos &c &c This view will bear much
108a081 = Cook found Granite at Christmas Sound Vol XIV. (MY Edition) p 500. Well described [top portion page
113a092 ll water. Major Mitchell inferred subsidence; Mem MY remarks on coast of Australia.-- Great NW. dip i
115a097 sea in Geolog. Soc. if numbered compare them with MY rocks. when writing on Falkland Islds p. 94. Von
117a101 ons. during elevation & depression. C. Prevost.-- MY views of insensible oscillations of level will a
119a106 the softness & curvature of quartz rock?) also by MY phenomena of earthquakes.-- by the narrowness wh
121a111 Knife formed crystals of ice were formed-- (like MY gypsum case) shows power of segregation.-- & has
121a113 proofs of small rise at Stockholm.-- analogous to MY Valparaiso case. Consider profoundly all consequ
122a114 ould obey that Law. & lile over the platform:-- On MY view the degrading action must prevent internal
126a122 t of a dumplin being bad conductor is {P} against MY views-- if we had rod thus & judged by increment
146g007 on-- Will not curved form of hill be explained by MY idea-- highest part must project [Blank] {P} Pat
148g025 r When Black faced sheep are crossed with English MY informant said the lambs were nearly like each o
158g079 resses of granite obscure terrace 15 ft divortium MY measurements here side of Loch Spey Forms terrac
162g102 ngarry (Nead Roy told me) it is impossible to see MY new shelf, from road: Loch Ness 30.140. A 66 deg
163g106 nating layers of coarse & fine & many Sea shells. MY informant saw them himself-- Sand with tide ripp
181b045 were fitted solely for air & fishes for water. If MY idea of origin of Quinarian system is true, it w
184b054 don, & the species now living.-- Now according to MY view. in S. America parent of all armadilloes mi
205b143 l annelidous animal p. 347. Vol I.-- compare with MY planariae Leaches out of water Does the odd Petr
209b157 Canaries I: 1,46 St. Helena I: 1,15 {T} Calculate MY Keeling Case: Juan Fernandez Galapagos =Radack I
211b164 good in connection with Von Buch Volcanic chart & MY idea of double line of intersection.-- Dr. Horsf
222b204 Porto Santo & Madeira-- I believe very curious-- MY idea of propagation almost infers, what we call
224b214 plants) & living thing with thoughts (animal). «: MY theory very distinct from Lamarcks» Without two
225b219 exactly intermediate.-- Reference to Pig & Dogs. MY theory will make me deny the creation of any new
227b224 more like present carnivora than Pachydermata If MY theory true, we get (I) a horizontal history of
227b228 first germ-- --led to comprehend true affinities. MY theory would give zest to recent & Fossil Compar
232b248 ng in different climates Those will not object to MY theory, those the philosophers who soar above th
242c018 those from S. Africa. «M. Bibron doubts fact.-'» MY toad is same species Coquille Voyage p. 25 Mais
243c020 ns of wide range work this out-- L. Jenyns, about MY fish New Zealand & New Holland fish very similar
248c030 ve existed for ages.-- .'. The most hypoth: part of MY theory, that «two» varieties of many ages standi
255c052 general forms.-- The hybridity of ferns bears on MY doctrine of cross-generation. The infertility of
257c057 b-feet-- all Nature answers to the possibility.-- MY views will explain no Mammalia in secondary-epoc
259c066 ecessary may be called adaptation With respect to MY theory of generation, fact of armless parent not
259c066 nce to more than offspring (like atavism) & shows MY «view of» generation right?-- If puppy born with
261c070 s is empirical. show this by instances Once grant MY theory, & the examination of species from distan
264c080 ies when close come from different localities, as MY Furnarii.-- some genus of yellow & brown-breaste
275c120 tells of horizontal barriers!-- Mr Yarrell.-- says MY view of varieties is exactly what I state.-- &c
278c130 give Catalogue of Mammalia arranged according to MY own methods. Dasyurus being found fossil in Aust
280c135 ore different might do so.-- <whi> is this true?? MY views, which would even lead to anticipate mules
280c135 to him most strange.-- This even might be said.-- MY theory thus explains a grand apparent anomaly in
281c138 logists would not find fossils such as they are-- MY theory explains that family likeness, which as i
283c145 xplained by such not having been long in blood?-- MY theory agrees with unequal distances between spe
289c161 ne structure of this bird & get account of habits MY definition <in wild> of species. has nothing to
289c162 eration may be viewed as condensor, «+++ must (on MY theory) =supported by foetal lower developed for
292c170 had <so so convin> brought so much conviction to MY mind.-- Reflect much over my view of particular
292c171 o much conviction to my mind.-- Reflect much over MY view of particular instinct being memory transmi
293c173 or use of button holes it would be instinctive.-- MY view of instinct explains its loss ¿ if it expla
293c174 uisite adaptations can hardly be accounted for by MY method of breeding there must be some corelation
294c177 first sheep> State broadly scarcely any novelty in MY theory, only slight differences. «the opinion of
296c184 eaks volumes. 2 Chapters. translated by Hooker.-- MY theory explains this. but no other will.-- St. H
297c186 turnal habits-- to cats &c.-- must be acquired by MY theory-- else my theory not applicable [not loca
297c186 o cats &c.-- must be acquired by my theory-- else MY theory not applicable [not located] p. 428. Ouze
301c199 s. acquirable through hoarding from short time.-- MY theory must encounter all these difficulties.--
301c199 teach squirrel to kill ears of corn according to MY views, habits give structure,.. habits precedes
302c200 & no new orders,-- Wonderful, partly explained on MY theory, = otherwise mere fact creator chooses so
302c201 to be sure, have remained somewhat similar.--!!! MY theory drives me to say that there can be no ani
302c202 , & compare it with results. if I acted otherwise, MY premises <in di> would be disputed.-- according
302c202 lue» must be so from characters of analogy.-- see MY notes on p. 37. of Macleay. wonderfully accordan
305c209 were to die, then they would appear isolated.\[In MY birds from S. Hemisphere there are some godwits
308c217 rences carried a long way. --Case of Habit I kept MY tea in right hand side for-- some month, & then
308c217 ys for a week took of cover of right side, though MY hand <vibrate> would sometimes vibrate-- «seeing
308c222 l-grebe. external' appears to be a puzzle against MY theory.-- If I be asked by what power the creato
310c222 inciple transferable. not wonderful According to MY view <ins> beccause actions are constant they ar
312c233 y remarkably- scenes in themselves accidental-- MY first thought of sea side-- Study Bell on Expres
322c270 n. read several papers-- all, that bear on any of MY subjects Elie De Beaumonts. 10 Vol. of Memoirs o
325c267 can Ornithology Read Aristotle to see whether any MY views very ancient? Study with profound care, ab
332d001 what <trifling> «unknown» causes act upon people. MY Father mention, than for ten years he never saw

332d003 dest breed-- He believes all pretty much alike.-- MY Father Water-in the hair a century since used to
332d003 ed much more latley diseased Mesenteric glands.-- MY Father has seen case of pleurisy, broken limb «i
334d010 <not se> plentiful, second absolutely sterile.-- MY case of Stallion, according to Erasmus preferrin
336d019 n a mule these conditions are not fulfilled.-- «[MY grandfather's theory of Mules not hereditary, be
337d021 s in common greyhound> & much intermarriage.-- In MY speculations. Must not go back to first stock of
338d024 n & species-like.-- -- Says Negro-- thick skinned MY hairdresser (Willis) says <black> «that strength
346d047 but more tendency to fatten-- This man confirmed MY account of the Shepherd dogs.-- Aug. 24th. Was s
349d052 is paper worth referring to again.-- According to MY theory, every species in any sub-genus will be.
356d070 se, but something prevents the completion.-- ⟨Say MY Grandfathers expression of generat. being highes
356d071 en acquired & heredetary tameness.-- In comparing MY theory with any other. it should be observed not
356d071 ll p. 244. vol I) spit & throw dust <according to MY theory of generation (p. 175) if> 8th Sept Yarre
361d096 r breeding a more brilliant plumage than male.-- «MY case of Caracara. N. Zelandiae.--» Mr Blyth stat
362d099 s so in S. America with C. Campestris<)> refer to MY notes) & Mr Yarrell supposes this a consequence
366d108 ot have been persistent in nature.-- According to MY view, the domesticated animals would cease being
367d112 donkey has. CD[old Buffon should be read on Mare MY view, why hybrids are infertile. supposes that w
370d117 c The line of argument «often» pursued throughout MY theory is to establish a point as a probability
373d132 appearing??-- Have Marsupiata abortive Mammae?.-- MY view would make every individual a spontaneous g
381d157 have hybrids-- this is most important support to MY views-- Seeing sexes separate in some of the low
382d158 hrodite,-- but not of poenis & clitoris, shows to MY mind.-- , that both are present in every animals
383d159 d with possibility of becoming either sex.-- ⟨ In MY theory I must allude to separtion of sexes as ve
384d161 deficiency of parts, as in Hermaphrodites, (shows MY doctrine of Hermaphrodite differs from Hunter)--
391d175 ividual is different). (All this agrees well with MY view of those «forms» slightly favoured, getting
392d180 imaux (& Australian) dogs with common dogs--> Ask MY father to look out for instances of Avitism Exam
397e004 corelations Octob. 4th. It cannot be objected to MY theory, that the amount of change within histori
398e006 regular laws» that baffles idea of revolution.-- MY very theory requires each form to have lasted fo
399e006 ange in upper & lower layers.-- good objection to MY theory: a modern bed at present might be very th
401e015 parating apple tree entirely from all others-- so MY experiment of strawberry not so absurd.-- Thinks
403e022 ol II p 363 account of Flora of pacific, given in MY coral paper Oct 14th Macleay says, that «every»
403e023 -- if we know its congeners then we can-- now on MY theory this «certainly» can be accounted for, on
404e024 tinct, now this is directly incorrect The case of MY mice is good, because it is an involuntary varia
405e032 rium & Mastodon are coembedded in N. America. see MY Journal for references In such cases as at Galap
406e035 adly against Yarrell's law.-- not so much against MY modification of it-- Goat & Moufflon will not br
409e048 w take greater area of water & snow line descent. MY theory gives great final cause «I do not wish to
409e049 mer that the moon probably is uninhabited]CD & if MY theory be true then the formation of sexes rigid
410e051 when going N.orth & South Thinking of effects of MY theory, laws probably will be discovered. of co
411e055 d» 27th November When summing up argument against MY theory, doubtless, the presence of animals in <o
411e056 iduals can be produced. & yet sexual apparatus.-- MY account of Circus cinereus of the Falklands Isld
413e059 ize,-- consider this (Cetaceae) with reference to MY theory Babbage 2d Edit, p. 226.-- Herschel calls
414e060 » (& insects.-- Stonesfield????). Have Mammalia?? MY theory certainly requires progression, otherwise
415e066 animal.-- Would anyone raise an argument against, MY theory, should no fossil «very distinct species»
415e067 see, what parts of structure abortive.-- Remember MY fathers remark about the Bladder.-- The numbers
416e070 ty in crossing [& this most important obstacle to MY theory] CD without the hermaphrodites mutually c
416e071 ommenced in the Water!» It is a beautiful part of MY theory, that «domesticated» races. of <a> organi
418e080 with moths, which can be impregnated externally-- MY view of every animal being Hermaphrodite-- proba
423e095 bly always have done so, as the simplest fish &), MY answer is because, if we begin with the simplest
423e098 ys, is an American form.-- has several species in MY fossils-- CD[If cases of one variety in upper pa
425e103 rids, bears relation to capability of variation?? MY theory says so.--» March 6th. Mr Bentham says in
426e105 e Sicily not having species, if true important on MY view.-- March 9th-- Is there any relation betwee
426e106 e, as it does indelibly the many subsequent ones. MY views, «V <see> p. 103» would lead me to think t
427e109 hence all present types are ancient. According to MY views of <Dioecious p> all plants, being occasio
428e111 y Herbert) because not reared by seedlings.-- Now MY principle does not apply to any plant reared art
428e111 partially to the Zizanias in in Sir. J's ponds-- MY principle being the destruction of all the less
429e116 polar regions of N. America.-- if true curious on MY view-- because these points were last connected
430e118 » even if placed on Isld-- if &c &c.-- Then give MY theory-- excellently true theory Examine list o
431e123 - Does not spermatic animalcule in Mosses, render MY view of the crossing of mosses & all others by a
432e124 ion to the surrounding circumstances According to MY theory no land animal with fluid seeds can be tr
434e128 e has shown we can & that analogy is sure guide & MY theory explains why it is sure guide.-- Lychnis
436e136 e said, that wild animals will vary, according to MY Malthusian views, within certain limits, but bey
436e137 ng propagated & so &c. The greatest difficulty to MY theory, is same type of shells in oldest formati
437e138 ns from Honduras. good case of migrating.-- shows MY theory insufficient.-- p. 120 An Eagle is said t
439e143 the pollen of same flower,-- as it tends to show MY view of <i>nfertility of hybrids «with parent sp
441e150 rth cause or law of change.-- The weakest part of MY theory is, the absolute necessity, that every <a
442e151 amount of generation may take place by gemmation «MY theory will not admit this, now that tulips brea
442e153 there? whether> According to the above suggestion MY theory would require, that «species» «individual
443e154 some most important difference is forced on us.-- MY theory only requires that organic beings propaga
443e155 .--]CD I utterly deny the right to argue against MY theory, because it makes the world far older tha
446e160 osing some inward progressive developing power.-- MY theory leaves quite untouched the question of sp
446e163 ds arriving.»-- throws a very great difficulty in MY theory, here we have a plant remaining constant,
447e164 ha?-- But if Acalepha do not cross there would by MY theory gradation of form from one species to oth
447e164 tion of form from one species to other: therefore MY theory does require crossing.-- The case of Lemn
451e176 think grandfather first of race & if so, fact for MY theory Cocos Isld & Preparis between Andaman & P
455eIBC seless. Grandfather's Handwriting, to compare with MY own.-- E Bengal Journal Vol 7. p. 658-- Falconer
461t037 Chapt see Yarrell Syngnathus Ch 6 I presume, from MY theory, as long as any structure can be handed d
463t063 discoveries to come-- Owen in his description of MY fossils makes same such remark & before the conc
466t095 are in Crown-Imperial Lily & many other flowers-- MY view of <variety acquired> « <character> » of ch
468t112 Fraxinella «& from <flower> plant to plant.»-- to MY grt surprise-- I found all, stamens straightened
470t176 were pollen gatherers & they seem slow= Maer 1840 MY Father formerly planted «Turkey or» Palmated and
470t176 w grands of each other actually produced hybrids-- MY Father remembered when in the gardens, he knew t
479z005 me of Sagitta Triptera M. D'Orbigny has described MY animal with teeth {P} p. 140. Fléche of Quoy et
481z010 (I think Planariae) Sagittella, or Fleche «p. 8» MY little animal with horns. Madrepores p. 26 Nulli
483z013 rked out with advantage. with Cumms collections & MY own: & Capt. King's p 453-- Planariae velellae (
484z014 merican Ornitholog must be studied before writing MY general account-- ¿ Do not the Penguins replace
484z016 -- Spix has described Philedon. allied to some of MY birds-- These groups strictly American. Colourin
490z001 Ranunculus in the nectaries. The former best for MY experiment on Selection. Experiments in crossing
494z004 e. (Indian & Common) &c: length of life. (5) Does MY Father know any case of quick or slow pulse bein
496z006 of snail, row of fish & kill them in hour (or two «MY Father made hens cast Holly-seed & they grew» (9
498z007 of want of nutrition.-- Horned oranges so? --Yes, MY Father lost this character in grt degree from ch
500q10a ndemic &> wandering species of confined genera By MY theory in volcanic or rising isld, there ought t
506q015 uttings.-- (40) Ask Henslow to distribute some of MY questions amongst agriculturists. whom he know.-
507q016 eeds of the Fuller's plants ,Teazle Dr. Holland ; MY Father. Andrew Smith (1) Are cross-births, or ot
514q21. ood-- Norfolk Isd-- geology. volcanic? Applies to MY geology & Species theory-- peculiar Fauna?.- {Aus
520m001 ssion -- 1838 Selected Dec 16 1856 July 15th 1838 MY father says he thinks bodily complaints «& menta
520m001 but his mother in bodily & mental disposition.-- MY father has seen innumerable cases of people taki
520m002 ryden Co said he could not remember his father.-- MY father thinks. people of weak minds, below par i
521m007 ld a story of hunting «-- habitual fits.--» which MY Father thinks is mentioned in the Zoonomia.-- No
521m009 on which he exclaimed, why it is dinner time.-- » MY father asked him whether he had gardener of name
521m009 ad never heard of such a man & had no gardener.-- MY F. then asked Mr C. to come to the window & poin
521m010 Mr C. answered why do you not know, that is A. B MY gardener.-- Thus was he in every respect, no com
522m010 new train if early association were called up.-- MY F. asked him, did he know whom Mr Child «of Kinl
522m010 married.-- Answered never heard of such a man.-- MY Father explained who he wa & all about him, but
522m010 ut still maintained he had never heard of him).-- MY F. then said you remember Jack Baldwin at school
522m013 nied by extreme anger, at not being understood.-- MY F. says there is perfect gradation between sound
522m013 her rather not going against natural instincts.-- MY Grand F. thought the feeling of anger, which ris
523m014 gs of the injustness of their hatreds, as <if> in MY case.-- It must be so from the curious story of
523m015 dogs ideotic.-- dotage.--» Doctor communicated to MY grandfather his feeling of consciousness of insa
523m015 accounts in his pocket, & studying mathematics.-- MY Father says after insanity is over people often
523m016 y, he was «only» NEARLY thrown into a hospital.-- MY father was nearly drowned at High Ercall, the th
523m017 was far more painful than the thing itself. Asked MY F. whether insanity is not distinguished from wh

Page   **(Key Word)**
523m018 iasm passion & madness.-- ira furor brevis est.-- MY father quite believe my grand F doctrine is true
523m018 - ira furor brevis est.-- My father quite believe MY grand F doctrine is true, that the only cure for
524m020 twitching. Pride & suspicion are qualities, which MY F says are almost constantly present in people,
524m021 ugh she never naturally talked of these places.-- MY F. says, shows that early impressions are most d
524m021 g:-- Dreams do not go back to childhood-- People, MY Father says, do not dream of what they think of
524m022 efore execution.-- Widows not of their husbands-- MY father's test of sincerity.-- People in old age.
524m022 er husband was dead) yet instantly perceived when MY Father to distract her attention took her «left»
525m024 - Not probable in Squib's case any direct fear.-- MY father thinks that selfishness, pride & kind of
525m025 f folly like (Mr George S.) is very heredetary.-- MY father says on authority of Mr Wynne that bitch'
525m025 - she had kittens before & afterwards with tails. MY father says, perfect deformity, as an extra numb
525m025 lip or imperfect roof to the mouth «stammering in MY Father family' (as in Lord Berwick's family) are
525m026 & Epilepsy remain many generations in families.-- MY fathers does not know whether trains of insanity
526m029 brought to mind, by memory of the scenes, (indeed MY American recollections are a collection of pictu
526m029 mory consisting of things seen, quite agrees with MY Fathers case of Mr Corbet of the Hall understand
526m030 come heredetary.-- When a man says I will improve MY powers of imagination, & does so,-- is not this
527m034 e a test of hardness of thought, from weakness of MY stomach I observe a long castle in the air, is a
528m035 in cannot be discovered-- is closely analogous to MY Fathers positive statement that insanity is only
529m041 ht so view plants & trees.-- I am sure I remember MY pleasure in Kensington Gardens has often been gr
529m042 ow the scenery would rise. Deer in Parks ditto.-- MY Father says there is case on record he believes
530m043 nce for a day or two are absoluteley forgotten.-- MY father signed a bond, yet when he paid the Attor
530m044 ses & name Corbet, perhaps nonsense.-- look to it MY father has somewhere heard (Hunter?) that pulse
531m050 s most curious instance of reason & abstinence.-- MY Father remarks that things of great importance a
532m054 eing slightly unwell & felt so much afraid though MY reason was laughing & told me there was nothing,
532m055 tion in the involuntary mind which is startled.-- MY Father says he should think that in old people,
533m060 - How comes this tendency in these cases? How did MY mind feel it was wrong (& it was not merely mora
533m061 g (& it was not merely morally wrong, but hurting MY character I felt it)-- this is kind of conscienc
535m070 Comte argues against all contrivance-- it is what MY views tend to.-- When a man is in a passion he p
536m073 ent, should be grateful if it were pointed out.-- MY wish to improve my temper, what does it arise fr
536m073 eful if it were pointed out.-- My wish to improve MY temper, what does it arise from but organization
538m078 ar on principle of association.--«fully bears out MY fathers doctrine about people forgetting their i
539m081 to think deeply, & the immediate manner in which MY head got well when reading article by Boz.-- now
539m083 tance of heredetary mind. I a Darwin & take after MY Father in heraldic principle. & Eras a Wedgwood
539m083 pects--. good instances.-- when education same.-- MY handwriting same as Grandfather. Aug. 16th Anger
541m089 en I went down to Woollich I was trying to unbend MY mind as much as possible (testing success by dec
541m089 creasing headache) & found best plan was allowing MY mind to skip from subject to subject as quick as
541m090 lking» for the moments with interest on each.-- ∴ MY father. is right in saying delirium rest-- there
541m090 ect of sea on coves when waters had fallen, as in MY Glen Roy paper.-- this greatest mental effort, o
545m107 another monkey to dog. I showed nut & then closed MY mem. expression of fury, jump to scratch my face
545m107 losed my mem. expression of fury, jump to scratch MY face. The ourang outang, under same circumstance
547m111 y., vaguely thought of packing up.-- was lying on MY back fell to sleep for second & wakened.-- had v
547m111 rized» dream, in continuation of waking thought-- MY servant was in the room. with my trunk out & I w
547m111 waking thought-- my servant was in the room. with MY trunk out & I was engaged in hurriedly giving or
547m112 l consciousness for moment, implied by «presence» MY servant, «box» my own manner of ordering things
547m112 r moment, implied by «presence» my servant, «box» MY own manner of ordering things to be done.-- The
551m129 as they flag & something like a snow-haze. covers MY whole imagination." Septembe. 3d Why when one th
552m131 is passion with sending force to muscles, that in MY grandfather remark, a tired man. involuntarily f
557m149 y apt to catch tricks.-- so are people in passion MY F. rubbing hands.-- stamping. grinding teeth.--
559m156 naturally close at that time after long period.-- MY Father about double consciousness.-- & somnambul
560m1BC ee it? Charles Darwin 36 Great Marlborough St Has MY Father ever known <intemperance> <disease» in gr
563o001 easant springing from nest & leaving no tracks.-- MY Father says pea-hens do Wood pidgeons building n
565n009 by slight protrusion of lips, as if going to say "MY dear,' just what smile is to laugh.-- I must be
566n011 ally useful to them: this must be studied. before MY view of origin of evil passions.-- Man getting s
567n013 what to do with them, came several times & opened MY hand, & put them in-- like child. Tommy's face,
570n025 ls have not modesty. analyse this.-- «Excellent-- MY theory of blushing solves this.--» The similarit
570n026 most generally accustomed to:-- analogous case to MY idea of conscience.-- deduction from this would
571n029 hich are taste? Real taste in mouth, according to MY theory must be acquired, by certain foods being
573n039 , even when wishing not to flow-- flow it will.-- MY father told Miss. C. of the bad conduct of Mrs C
574n039 t be used to show language had a beginning, which MY theory requires. There probably is some connecti
574n041 such an association in man.-- it is bare fact, on MY theory intelligible An habitual action must some
576n047 -- man's intellect might well deteriorate. «CD[in MY theory there is no absolute tendency to progress
577n049 suppose an hydatid to be a cause of itself.-- [by MY theory no animal. as now existing can be cause o
577n049 re is great probability against free action.-- on MY view of free will, no one could discover he had
577n050 shment of this principle of Association will help MY theory of sensitive Plants Habitual actions, (in
577n053 sudden cures of tooth ache before being drawn,-- MY father «even» believes that the general talking
578n054 on, "one will anyone, especially a women think of MY face,"? to one moral conduct.-- either good or b
579n058 ing been invented, prove of the difference, which MY theory believes in.-- From the manner short-sigh
579n058 e mind is intent on one object.-- With respect to MY theory of smile. remember children smile before
579n060 ushing of Camelion & Octopus; strong analogy with MY view of blushing-- in former irritation on a pie
580n060 e of a name, with reference to origin of language MY father says old people first fail in ideas of ti
581n063 ng properly, even when you cannot remember it. as MY father trying to remember the man's Christian na
581n063 es.-- & analogize it with ordinary habits that is MY new part of the view.-- let the proof of heredet
582n067 ary remains of savages state.-- N B. According to MY view marrying late, will make average of life lo
582n068 that it is uncomfortable if it does not do it.-- MY theory explains how it comes that the heart is t
584n072 "-- this is class of so called instincts to which MY theory no way applies.-- it is the acquirement o
586n082 exerts in concert with others in building comb-- MY faculty offen will turn out to be instincts. & s
588n089 he emotions.-- p. 272. Some remarks applicable to MY theory of happiness.-- Bell on the Hand p. 191 S
589n091 the instincts of animals.-- almost identical with MY theory-- no facts, & mingled with much hypothesi
590o005 Beginning of a new School of Metaphysic,"-- give MY doctrines about origin of language-- & effect of
590o007 cannot <admit> think reason sufficient to give up MY theory-- Viewing from eminence. the wide expanse
602o11v taken by a camera obscura &c a Poussin.-- How are MY ideas of a general notion of everything applicab
602o11v to the high idea «p. 131.» in Tragic acting-- CD [MY idea. would make the mind have mysterious & subl
603o11b s Architecture does not come under imitative art [MY view says yes. <old> mass of rock--]CD or poetry
603o11b w says yes. <old> mass of rock--]CD or poetry, CD[MY thery says yes. imitating song -- two primary so
606o024 a.-- as statue of beauty, is of the "beau ideal", MY instinctive impression 1) September 6th. 1838 Ev
609o030 ss.-- The other says we have a moral sense.-- But MY view <says> unites both «& shows them to be alike
610o031 nge of instinct amongst animals.-- Jan 13th. 1839 MY father received a letter from Mr Roberts-- «a pe
610o032 n inward. unconscious memory.-- Jan 14th. 1839.-- MY father says he has heard of many cases of ideots
622o048 ong.-- «[just as in tastes of the mouth]CD» [LHC] MY theory of instincts, or heredetary habits fully
623o050 giving the social feelings.-- [LHC] According to MY theory, all instincts demand some explanation [R
623o051 feelings», & living under certain conditions; by MY theory they have been formed by the circumstance
624o50v change hunting instinct. *Our tastes in mouth by MY theory are due to <habit> heredetary habit (& mo
625o50v ility part being blended & lost» & moral sense.-- MY theory explains both, perhaps, by habit-- [LHC]
626o052 ny principles born in us.-- Great difference with MY theory.-- see p. 349.-- remark on this point.--
627o52v cts consult <following> pages. <p. 231> marked in MY Mackintosh 1) Mackintosh's Ethnical Philosophy p
628o53v ht be partly made.]CD p. 21. "Why ought I to keep MY word"-- gives the problem, of ethics-- [my answe
628o53v o keep my word"-- gives the problem, of ethics-- [MY answer would be to all such cases-- either, that
628o54v ve no relation I think this <boshes> «nonsense»-- MY theory of durableness will explain it.-- Would n
629o55v r any thing is taught instinctively; I say yes, & MY explanation agrees. with last head.-- (4) It is
629o055 no right & wrong except instinctive ones) Perhaps MY theory of greater permanence of social instincts
633j53v pagation of Misseltoe?-- do p. 284. it is hard on MY theory of gain of small advantages thus to expla
633j54r s being created to arrest mud &c at deltas.-- Now MY theory makes all organic beings perfectly adapte
635j56r as liability to accidents & any other cause.-- (& MY theory [ALL PARTS OF ONE GREAT SYSTEM. C. D]CD [
635j56r other beings.-- true, (& the doctrine of checks & MY theory) Macculloch. Attrib. Vol I. p. 330. Menti
635j56v ould not be exposed to weather.-- this is against MY theory of frequent intermarriage.-- A plant is i
636j56v whilst the habitation «or world» simple series.-- MY theory shows life equally simple series, & there
545m106 women.-- Another little old American monkey «(MYCELIS)» I gave nut, but held it between fingers, the
473s03r Sus americana & Hippotamus «with Megatherium & MYLODON» in post pliocene strata! Mastodon longirostri
173b012 o continents; all bred in from one parent why. MYOTHERA several species in S. America why, 2 of ostri

343d036 r, reason is formed, & the world peopled «with MYRIADS of distinct forms» from a period short of eter
081rIBC e into English ft. 0.5159929 {T} Toises Pieds MYRIAMETRE = 5130., 4. 5 inch Kilometre 513., 0. 5 Hect
292c170 ne even with a very superficial knowledge «like MYSELF» of <classi> real affinities. ie structure of
410e052 et there is no cure «I may say all this, having MYSELF aided in such sins» (do not add name, without
443e156 on in duration of a planet to our lives-- Being MYSELF a geologist, I have thus argued to myself, til
443e156 Being myself a geologist, I have thus argued to MYSELF, till I can honestly reject such false reasoni
533m060 er realized the idea that I was tending to make MYSELF in act less grateful.-- How comes this tendenc
536m070 oes just the same; I noticed this by perceiving MYSELF skipping when wanting not to feel angry-- such
572n033 Now <awake> «when awake» I could not picture to MYSELF reading French book quickly, & <running> «runn
413e059 the appearance of new species. the mystery of MYSTERIES. & has grand passage upon problem.! Hurrah.--
069r152 ransportal of conglomerate between two ranges MYSTERIOUS!-- Mem. SUBSIDENCE Uspallata of which no tra
288c159 ntinue to cross, means of knowing directions: MYSTERIOUS. Were the woodcocks., which came Madeira & <
404e026 ting offspring.--\ & as adaptation,-- however MYSTERIOUS such is case\. therefore chance & unfavourab
529m041 great compound animals united by wonderful & MYSTERIOUS manner.-- There is much imagination in every
602o11v cting-- CD [My idea. would make the mind have MYSTERIOUS & sublime ideas independent of the senses &
638j58v e-- what is intellect, but organization, with MYSTERIOUS consciousness superadded This is similar ide
413e059 schel calls the appearance of new species. the MYSTERY of mysteries. & has grand passage upon problem
440e145 be it remembered. how little part of the Grand MYSTERY is this-- the law of growth, that which chang
614o037 ore than another-- What is matter? the whole a MYSTERY. -- [LHC] This Materialism does not tend to Ath
553m137 races -- whether in Ancient Greeks, with their MYSTICAL but sublime views, or the wretched fears & st
030r036 ta there is Coal-- ¿ No shells in all cases. «.MYTILUS.--» «at Guacho» «on N. Chile? Washington.--» M
316c245 F W shells there is more confinement. thus the NAIADS (study De Ferrussac) are confined to <S.> Amer
469t119 mined 7 says one specimen had on one foot, a toe-NAIL & two joints-- as it is on one foot probably mo
303c204 er of negro probably was first black at base of NAILS & over white of eyes,-- + + +, Will he say crea
560m156 <fe> Shame, independent of fear: colour of bare NAILS--, & of eyes.-- Do female monkeys care for men.
264c079 ; «let him look at savage, roasting his parent, NAKED, artless, not improving yet improvable» & then
436e134 re at which fish &c can live.-- Lyell says that NAKED cuttle fish now bear a very large proportion to
570n027 idea of beauty.-- [Old Graecians living amongst NAKED figures, & observing powers common to savages??
053r100 he Amazons & Orinoco mentioned by Humboldt under NAME of Rothe-todte-liegende is perhaps same with th
075r167 , horn silver, black silver & red silver, do not NAME native silver because not very abundant.--\ mur
078r174 o. 151. p. 79. [misnumbering, no page 173] Under NAME of Sagitta Triptera D'Orbigny has figured anima
086a007 osition = If equatorial streams of warm pole; in NAME of Heaven why are tops of Equatorial mountains
095a038 f «hill» «cerro» of Diamante near stream of same NAME. with imperfect crater <--> near summit,-- much
159g087 No Granite blocks in higher parts?? Bought Glen NAME of Glen by which we descended, it is to the wes
202b130 bat, are from external influence.--...... Hence NAME of analogy, the structures in the two animals b
208b154 monkey. peculiar to C. de Verd's.--? NO Macleay NAME given in Congo Expedition We need not expect to
239cIFC t. Hilaire Philosophy of Zoology Waterhouse B C N[A]ME of two pigeons,-- «with specks» which cross &
345d042 st have gone on for many years, before deserves <NAME> «to be so called»,-- the short horned cattle h
410e052 s, having myself aided in such sins» (do not add NAME, without reference to description), except desc
417e075 & Common Oxen, which one must think deserve the NAME of species, may be owing to the little fixity o
479z005 42.-- Planariae p 451.-- many molluscs Under the NAME of Sagitta Triptera M. D'Orbigny has described
481z098 migratory fish.-- See Kings drawings.-- for real NAME Birds of Iceland. Mackenzie. p 345 for comparis
489qIFC flants. Habits of different caterpillar races. --NAME of Italian who sold eggs.-- Temporary Question
501q011 ch other & see whether cross can be obtained-- I NAME these three plants. because they cannot be cros
521m009 » My father asked him whether he had gardener of NAME A. B., &c &c. & he maintained he had never hear
522m011 of him.-- Answ Had large fortune left. him, took NAME of Child «of Kinleto» & married Miss A. B.-- al
530m044 uities of Shrewsbury something about big noses & NAME Corbet, perhaps nonsense.-- look to it My fathe
546m110 .?-- Mr. Mayo told me the case of a lady, (whose NAME was told me, who told the fact to Mr Mayo himse
579n060 n of cause,» do p. 135.-- on the importance of a NAME, with reference to origin of language My father
581n063 my father trying to remember the man's Christian NAME, writing for the surname., analogous to instinc
610o031 ters to»-- could not <remember> «read» Christian NAME; fancied it looked like. W. but concluded it co
610o031 Wilson, referred to Robert & found his Christian NAME was Wilson!!-- How curious an inward. unconscio
021rIFC greatest number of changes towards perfection (NAMELY mammalia) must have a shorter duration, than t
023r009 5 degree. The nearest of the E. Indian Islands. NAMELY Java is 1000 miles distant! Where are Hippotam
076r168 are found in the veins of Kongsberg in Norway.--NAMELY dendritic silver intersecting carbonate of lim
216b182 little mingling would probably have been good, NAMELY such as blood hounds from other parts of Engla
225b216 absurd. (as equally are arguments against it-- NAMELY how did otter live before being made otter-- w
248c030 s tending to same end, as the law of hybridity, NAMELY the [not located] animals unite, all the chang
270c104 physical speculations are entered in upon life. NAMELY Carus.-- How remarkable that Turdus Magellanic
277c127 ands. species.-- each describer giving his test NAMELY differ as much as those (naming them) which ar
355d067 e greatest service, towards the end of science. NAMELY prediction.-- till facts are grouped. & called
414e063 r &c, but then comes the more deadly struggle,, NAMELY which have the best fitted organization, or in
442e153 opagated by means, which wild plants never are, NAMELY on stocks of other varieties & we know that th
523m015 t it, his knowledge of the untruth of the idea, NAMELY his poverty.-- his manner of curing it. by kee
586m082 ty <may possibly be> «probably is» instinctive, NAMELY the knowledge of size is merely judged by eye,
611o034 -- [RHC] In the simplest forms of living beings NAMELY «one individual» vegetables, the vital laws ac
074r164 limestone, which H. calls by several secondary NAMES «Study Hoffmans account of steam acting on trac
277c126 mining fertility Scheme for abolishing specific NAMES & giving subgenera. true value.-- as in Opetior
410e053 re little for species, except so far as wanting NAMES to refer to, to those forms. where the terminat
514q21. fferent countries-- on flora of African Islds-- NAMES of Plants found on mountains of N. America simi
522m011 «of Kinlet» & married Miss A. B.-- all the same NAMES as a few minutes before he maintained he had ne
581n065 language, that it was progressively formed. --NAMES like sounds)--. Horne Tookes tenses, &c &c --<
599o05v we see words invented-- we see their origin in NAMES of People.-- Sound of words-- argument of origi
277c127 giving his test namely differ as much as those (NAMING them) which are found together.-- If two speci
410e052 ossible.)-- Ascertainment of closest species (& NAMING them) with relation to habits, ranges. & exter
410e052 mportant & will be done to all countries,-- but NAMING mere «single specimens in» skins worse than us
531m053 old some danger-- but it is instinctive because NANCY tells me very young babies start at anything th
323c269 s of Hayd & Mozart «Apri 25th Lockarts life of NAPOLEON.» April 5d Dr. Edwards of <ter> influence of
218b191 ork on Hybridity under title of Amaryllidae & NARCISSUS. Mr Donn considers Mr H. rather wild Mr Donn
399e010 y have not lost power of producing. Williams. NARRATIVE of Miss. Enterprise. p. 497. Vampire bat abou
459tflr is good case as showing gradations, Boteler's NARRATIVE Voyage East coast of Africa-- Vol II. p. 256-
471tfl0 ope-- Large Lizards in Navigatores. Williams. NARRATIVE of Missionary enterprises Dr Andrew Smith say
066r141 oe informs me that in Obstruction Sound, in the NARROW parts which break through the N & South lines
086a003 at Horse being found in America, when Mammoth & NARROW toothed Mastodon.-- argue against the prejudic
160g090 dark grey fine grained. Much contorted gneiss «NARROW sharp ridge with peak» I walked all round hill
160g093 e-- : on N side.. dies away on rocky place, but NARROW shelves just like road of Glen Roy-- appears t
318c253 most curious case-- -- the birds seem to follow NARROW bands, certain kinds as gallinules taking the
528m037 in a river from seeing how the serpentine lines NARROW in the distance.-- & even on paper two waving
106a075 en at (C) its tendency would <cut> be to cut a NARROWER channel instead of wider.-- This applies to a
119a106 also by my phenomena of earthquakes.-- by the NARROWS which the anticlinal lines are apart-- the c
641j29r re best nourished by arctic regions-- Whales. «NARWHAL» Polar bear. Walrus, great Seals of Antarctic
621o047 a child may be taught to think almost anything NASTY. (<& accidentally» «by odd association» comes t
621o047 it). It will be only rarely that it thinks that NASTY, which the natural tastes say is good. yet hors
621o047 ven this is possible.-- So that as there nice & NASTY in taste, & right & wrong in action, so a child
074r165 oxided Iron by Mitterschlich. Vol. II Journal of NAT. & Geograph Siciences? -- H. says in Potosi the
130a134 by Parrott Mem. Acad. Peters. Scienc Math. Phy-- NAT. t. I, 1831. sur le temp du globe on Volcanos &c
178b030 espect to Miocene of Europe? Loudon. Journal. of NAT History.-- July. 1837. Eyton of Hybrids propagat
267c095 of Infusoria. <p> do p. 62. Ehrenberg Annals of NAT. Hist. precursor of magazine??? p. 75. roe of As
270c103 ns of generation!!!. Mem. Agaziz. [No Annals of NAT. Hist) spiral structure in Echinodermata.-- Agas
286c156 ought in one plane Fleming Quarterly review says NAT: fam: of Willows contains many Linnaean genera.--
323c269 -- Wilkinsons Egyptian remains skimmed -- Pliny NAT. Hist of World do -- Lamarck. II Vol. Philo. Zoo
324c268 ie. or Geographical distrib. «in Dict. Sciences. NAT. in Geolog Soc.» F.. Cuvier on instincts L. Jeny
324c268 Cuvier on instincts L. Jenyns paper in Annals of NAT. History Prichard.-- Lawrence Bory St. Vincent.
327cIBC & Cow & Sheep Clarke's Travels.-- Temminck Hist. NAT. des Pigeons et des Gallinaces. Silliman's Journ
421e091 ds to southern portion.-- \.do p. 269. Annals of NAT. Hist 1838 on «a» freshwater fish peculiar to Ir
449e169 e a Himalaya species -- leuconotes-- Magazine of NAT. History. 1839. p. 106.-- Waterhouse refers to f
580n062 «Dissert.» on subject of science connected with NAT. Theology.-- says animals have abstraction becau
117a102 the world would daily be scene of ruin in late NATICAL Magazine (before June 1838) that 70. F were ob

286c155 ike to think his origin godlike, at least every NATION has. done so as. yet.-- We now know what is th
553m135 arises   the theological age of science in every NATION according to M. le Comte).-- Those savages who
280c134 cause written on dog-Breaking.-- applies it to NATIONAL character.-- N.B. If two species were excessi
539m082 form some difficult task.-- Aug. 12th. When in NATIONAL Institution & not feeling much enthusiasm, ha
540m085 built   on one structure.-- He who doubts about NATIONAL character let him compare the American whethe
615o038 .-- Slightly modified in many countries, hence NATIONAL character, love of country, of association &c
309c221 could not leap fences:-- Dr Lang on Polynesian NATIONS (quoted) p. 4.-- do. p. 186. quotes Burkhardt
553m136 ove that» this innate idea of God in civilized NATIONS has not been improved by culture « <was> who f
564n003 solitary animal still different.-- ▯ Different NATIONS having different moral sense, if it were prove
046r079 getation? Rats & Mices. At St Helena there is a NATIVE mouse Did wave first retreat at Juan Fernandez
073r161 Hutton says) always come from without.-- "True NATIVE iron that to which we cannot attribute a meteo
073r164 n offer zeolite. stilbite. grammalite. pyenite. NATIVE sulphur.. fluor spar. bayte. asbestos garnets.
075r167 oth analysed silver ores from Peru consisted of NATIVE silver & brown oxide of Iron in Mexico. sulphu
075r167 silver, black silver & red silver, do not name NATIVE silver because not very abundant.--▯ muriated
076r168 ndritic silver intersecting carbonate of lime-- NATIVE silver in Mexico is always accompanied by Sulp
205b145 h same female p. 28 "It <is> wrong to enlarge a NATIVE breed of animals. for in <the> proportion to t
215b177 tant species Both males & females. lose desire. NATIVE dog not found in V. Diemen's land J. de Physiq
216b179 that   offspring of Negro & white will return to NATIVE stock (the cross often whiter<)> than white pa
288c159 stead of breeding from original Durham breed.-- NATIVE dogs & English cross readily-- think about hal
288c159 pearance.-- bark about half way «in tone»-- the NATIVE dogs howl most dismally, very rarely bark.-- a
432e124 hepherds said.-- p. 12. Attempts to improve the NATIVE <breed> animals of any country must be made wi
451e178 om» which «all» mankind «(& yet afterwards says NATIVE tribes can live there)» flee during 8 months o
246c027 e appears to be one at Botouma from account of NATIVES, & probably on Oualan.-- «Mitchill says snakes
402e020 t, with the same families and genera, that are NATIVES of S. Asia, but many of the species are peculi
450e174 t the northern peninsula of Borneo.-- Ox & hog NATIVES of Borneo Notices of Indian Arch. Singapore 18
451e176 also an albino is held sacred by the credulous NATIVES, & vow made at it. Both his parents were of th
540m086 n Diemens land & the energetic copper coloured NATIVES of New Zealand)-- the American in Brazil is un
350d054 al species as in two formations? by no way.?-- "NATURA nihil agit frustra", as Sir Thomas Browne says
417e079 mal, which would find its place in the Systema NATURAE.-- Mr. Knight makes this analogy between graft
030r035 with   character of.. «in Basins from rivers. & NATURAL position» position at N. S. Wales & Van Diemen
056r109 ess I never see such islands whose inclination NATURAL [...] deepest astonishment.» Perhaps scarcely
182b047 f Europe & America. &c &c &c The new system of NATURAL History will be to describe limits of form. (&
201b126 is an additional illustration of that axiom in NATURAL History that all aberrant & osculant groups ar
218b190 a.-- changing hair-- The Edinburgh. Journal of NATURAL History-- Preface appeared good with facts abo
256c054 nimals in those regions Species so far are not NATURAL, that they are either A. B. C. D E., or A C D
268c096 ad before Linnaean Soc. Feb. 1838.-- Annals of NATURAL History. «vol I» p. 159 curious account of Tit
281c138 lon[g] or peculiar tails strange ¿¿/Genus only NATURAL from death or slow propagation of forms.--just
284c149 is manifestly of less importance, as affording NATURAL characters than among those groups, where it r
286c155 as. done so as. yet.-- We now know what is the NATURAL arrangement, it is the classification of «arra
298c189 land.-- curious in so wild a bird.-- Annals of NATURAL History Vol. I. p. 185 case of tit lark placin
308c218 human nature is the object, there is really no NATURAL starting place, because there is nothing more
310c225 lty in mundine geological chronology Annals of NATURAL History «Vol I??» p. 318. some remarks on Bona
317c251 Tuckey's Expedition Journal of the Academy of NATURAL Sciences of Philadelphia Vol VII. Part II. 18
320c275 of. Zoology & Botany & Continuation «Annals of NATURAL History» Skimmed Von Buch Travels Whites Natur
321c275 tural History» Skimmed Von Buch Travels Whites NATURAL History f Selbourne References at end Dr. Lan
322c270 C. Prevost on L'Ile Julie Waterton's Essays on NATURAL History. Octob 2d Transactions of Royal Irish
322c269 at   end of each Vol Herschels. introduction to NATURAL Philosophy R. W. Darwin's Botany.-- References
324c268 gradation in Man. Lindlys introduction to the NATURAL system Bevan on Honey Bee Dutrochet Memoires s
326c266 ertations on subject of Science connected with NATURAL Theology.-- on instinct & animal, intelligence
337d020 some individuals are picked out.-- in a really NATURAL breed, not one is picked out, & few even of lo
348d050 es as now used almost any group.-- ▯all groups NATURAL (p 6) as expressing natural affinities▯ Maclea
348d050 oup.-- ▯all groups natural (p 6) as expressing NATURAL affinities▯ Macleays plan of arrangement depen
348d051 edetary whether important or not,). p. 7. "The NATURAL arrangement of animals themselves is the quest
348d051 mselves is the question in point." Now what is NATURAL arrangement,-- affinities, what is that, amoun
353d061 semblance with man.-- September 3d Magazine of NATURAL History. 1838 vol II p. 402. Mr Gould on Austr
365d107 ealthy state of womb).-- One can perceive that NATURAL varieties or species., all the structure of wh
379d151 that case ceases to be true avitism Annals of. NATURAL. History. p. 135. Natural History of the Caspi
379d151 e avitism Annals of. Natural. History. p. 135. NATURAL History of the Caspian. Fresh Water Fish!! ¿ad
386d167 great   changes really in progress.-- Annals of NATURAL History. 1838. p. 123. Ehrenberg. makes gemmat
387d169 influence   they themselves inherit./ Annals of NATURAL History .p. 96. Vol I. Notice the Syngnathus,
411e054 be speculated on, & laws of life,-- the end of NATURAL History, will be approximated to.-- Treating o
421e090 c fluid fertilized spawn of frogs.-- Annals of NATURAL History. (p 225. 1838.) account of metamorphos
458t002 than English birds, using cotton &c instead of NATURAL substances-- useful perversion of instincts--
461t025 ed more complex & some simplified.-- Annals of NATURAL History. <no. XII.> Vol. 2. p. 96 & p. 451. 18
461tf03 &> Blackbird have been known in <their> «its» NATURAL state to mate with a thrush"-- Athenaeum 1839.
471tf08 difficult   to point out a family so completely NATURAL & one whose groups pass so insensibly into eac
495q005 up true.-- This in fact always takes place in NATURAL Hybrids of Cabbages (7) Sow «daisy» seeds of w
497q007 te view. Gaertner talks of the several great & NATURAL Families, as being difficult to cross. (5) (1)
513q020 with another & see result, for comparison with NATURAL species, as dioecious plants, when crossed R.
522m013 y be doubted, whether rather not going against NATURAL instincts.-- My Grand F. thought the feeling o
537m076 re them without temptation.-- This probably is NATURAL consequence of man, like deer &c, being socia
560mIBC n.-- Have we any ferns in the hothouse at home NATURAL History of Babies-- Do babies start. (ie usele
563n001 r Books May 1873-- October 2d. 1838 Essays on NATURAL History Waterton describes. pheasant springing
571n031 e which rises naturally & hence sublimely from NATURAL rise-- I was also much struck in great avenue,
590n094 Gardner's Music of nature to show barking not NATURAL, (Vol I. p. 234) Vol. II p 153. «do». an accou
591n101 heism, at p. 424 Vol. II «Sect XV. Dialogue on NATURAL Religion» however, he seems to allow it is an
592n101 ume on Sceptical Philosophy. Hume has written "NATURAL Hist. of Religion" on its origin in Human mind
620o045 ects lasting, whilst passions although equally NATURAL leave effects not lasting. By association one
620o046 es of conduct to other men, [RHC] 5) which are NATURAL (& which «when present» give pleasure.) & whic
620o046 ough» even when tempted not to do so, by other NATURAL appetites.-- he is monster, or unnatural if ma
621o047 terest to follow it. even when opposed by some NATURAL passion.-- (a) [LHC] The conscience rebukes ma
621o047 ly rarely that it thinks that nasty, which the NATURAL tastes say is good. yet horseflesh show that e
621o047 the feeling that almost (rarely if opposed to NATURAL instincts) any action is either right or wrong
200b122 p. 410 <Nov> 1828 It is daily happening; that NATURALIST describe animals as species, for instance Au
228b229 ast & future. The Grand Question, which every NATURALIST ought to have before him, when dissecting a
281c137 onsequence.-- The simple expression of such a NATURALIST "splitting up his species & genera very fine
377d139 may be thought disgusting, but to philosophic NATURALIST pregnant with interest» Hyaena. thinks, when
619o042 [RHC] On the Moral Sense Looking at Man, as a NATURALIST would at any other mammiferous animal, it ma
178b031 ls having effected; a change whi[ch] the Fr. NATURALISTS thought was species <Ascensi> Study Lesson V
224b212 whether good species, & hence the importance NATURALISTS attach to Geographical range of species.-- D
245c024 ffers many varieties in each place to puzzle NATURALISTS.-- p. 372. Bourous. the Barbyrousa; a Cervus
252c045 unded in nature [not located] The systematic NATURALISTS get clear indication of circumstances in Geo
284c149 now some such «characters» rule are used by NATURALISTS in their test of value of character-- Macley
404e025 t must have been invented all at once.-- but NATURALISTS if they had series perfect, would expect thi
443e157 Ostrich to that of the Camel has not escaped NATURALISTS." Before he alludes to the resemblances of t
640j167 gs are species. it should be remembered that NATURALISTS are prone, fortunately, to take their ideas,
641j29r (on other hand Spermaceti Whale & Manatee.-- NATURALISTS must be cautious.-- <some others>: study the
504q013 on ▯ Castrate apple & pear to see if pollen NATURALLY carried, on account of Van Mons views-- Also
524m021 rnh[l]l, (where she was born) though she never NATURALLY talked of these places.-- My F. says, shows t
525m023 they feel pleasure in obeying their instincts NATURALLY.-- (generosity in defending a friendly dog.)-
547m113 en one recollects circumstances were such one NATURALLY would so so!) Now all these parallel trains o
559m156 tated very regularly at one time every day.-- NATURALLY close at that time after long period.-- My Fa
571n027 being impressed on us.-- Surely we have taste NATURALLY all has not been acquired by education. else
571n030 this, when viewing Windsor Castle which rises NATURALLY & hence sublimely from natural rise-- I was a
571n031 nguage. Much pantomimic gesture?? which would NATURALLY happen.-- Reynolds Works. Vol I, p. 226-- "Th
022r007 s Montes. ((Henslow Anglesea)) great variety in NATURE of a dike.--Mem. at Chonos & Concepcion. P. 41
026r022 ofs of many oscillations of level, which in the NATURE of strata & Organic remains does not appear to
035r047 cillation of level independent of mineralogical NATURE & dependent: & then how wonderful level «of sa

Page **(Key Word)**
046r078 ply altered by circumstances; & not in chemical NATURE, or has a subterranean fluid mass itself chang
048r085 ns of Patagonia is removed by reflecting on the NATURE of the country in which the Rhinoceros lives i
049r088 s there said granite in close contact varies in NATURE, -- Does not granite at C. Tres Montes become
050r090 eous conglomerate with small & large fragments, NATURE of which is doubtful. P. 180. I think my Ascen
050r093 . coast of Africa. as at Brazil [blank] What is NATURE of strip of Mountain Limestone in N. Wales. wa
062r128 icturesquely = Both N & S. great contrast. from NATURE of climate. = Perpetual snow.--subterranean la
063r132 r by cutting an animal in two. (gemmiparous. by NATURE or accident). we see an individual divided eit
072r160 and without H. Kingdom N. Spa. Vol III p. 113 "NATURE exhibited to the Mexicans enormous masses of I
075r166 e veins in most part totally independent of the NATURE of the beds they intersect". = In the Guatemal
076r168 h."--p. 156. Mines of Batopilas in New Biscay, "NATURE, exhibits the same minerals <as> there, that a
079r177 trie humaine, comme ils traitent les loix de la NATURE & ses phenomenes."-- Ulloa's Voyage, Shell fis
099a049 scertain law of attraction of particles of same NATURE: then get mathematician to when two particles
148g022 ravel nearly as high as highest measurement but NATURE I am quite doubtful of as I am of all the Allu
181b044 e sudden-- Heaven know whether this agrees with NATURE: Cuidado The above speculations are applicable
196b108 nimals must have gone on since from parasitical NATURE of insects & worms.-- In abstract we may say t
218b189 zine. Zoology & Botany Vol I p. 450 There is in NATURE a «real» repulsion «amounting to impossibility
218b191 rieties, though of trifling order are formed by NATURE. Carmichael. Tristan D'Acunha, a list of its F
227b226 divisions. look at Articulata!!!--! It leads to NATURE of physical change between one group of animal
230b236 ed to that country when change has taken place, NATURE [not located] Any change suddenly acquired is
240c004 ut two kinds of varieties.-- One approaching to NATURE of Monster, heredetary. other adaptation.-- Mr
252c042 ut Species-- MUST BE STUDIED: genera founded in NATURE [not located] The systematic naturalists get c
255c052 The infertility of crosse & cross, is method of NATURE to prevent the picking of monstrosities as Man
255c053 s as Man does.-- One is tempted to exclaim that NATURE conscious of the principle of incessant change
257c057 the otter live before it had its web-feet-- all NATURE answers to the possibility.-- My views will ex
258c061 & which causes the confusion in this system of NATURE-- Whether species may not be made by a little
259c065 formity may be looked at as the best attempt of NATURE under certain very unfavourable conditions.--
262c073 ant» varieties will be formed in any kingdom of NATURE, where scheme not filled up, (most false to sa
262c073 not filled up, (most false to say no passages; NATURE is full off them.-- wading birds partially web
263c074 & elevate & enlarge New Zealand; a division of NATURE of Apterix, many genera & species-- The believ
265c083 these not adaptations «they are counteracted by NATURE by crossing with other varieties»-- but <accid
265c085 be considered an adaptation as best attempt of NATURE, colouring matter being absent.-- again dwarf
265c086 g-cases-- Yet these wings may be of some use,-- NATURE is never extravagant though clearly not of the
269c103 able.-- After reading "Carus on the Kingdoms of NATURE, their life & affinity" in Scientific Memoirs
270c104 ferred to,. as compilation of action of organic NATURE on inorganic It is very remarkable as shown by
271c106 cked, one where not.-- the latter made by man & NATURE; but cannot be counteracted by Man.-- effect o
279c133 pring being cut off-- so that not propagated by NATURE.-- Whole art of making varieties may be inferr
280c135 heory thus explains a grand apparent anomaly in NATURE. <t>---- Many animals not breeding at all in d
280c136 t difficult to see that it is less repugnant to NATURE to produce one offspring unlike itself, than t
284c148 -- Nuts prodigiously heavy (where trees of such NATURE far apart. Must have travelled by <dead> «each
285c153 marked difference in the varieties «made by» of NATURE & Man.-- The constitution being heredetary & f
286c154 ges) Has not the white Man, who has debased his NATURE «& violates every best instinctive feeling» by
286c155 above Such prejudices, yet have justly excalted NATURE of man. like to think his origin godlike, at l
288c159 n of hunting, or keeping watch. how completely «NATURE & instinct» modified.-- The partial migrations
291c167 le it is a wonderful relation going through all NATURE.-- Makes hermaphroditisms. one step in <scale>
296c184 -- use as argument possibly some few hybrids in NATURE.-- «p. 473» Webb &. Berthelot. must be studied
307c215 capable of giving milk)» rudimentary wings. so NATURE can produce in sex, what she does in species o
308c218 5)-- in fact, in all reasonings, of which human NATURE is the object, there is really no natural star
308c218 re is nothing more elementary than that complex NATURE itself with which our speculations must end as
308c219 se of fertility.-- varieties not produced as by NATURE. if so. the habits which would. have formed th
310c223 to account> «consequence of» for the scheme of NATURE) and animals that man has different origin. «R
323c269 . Larder 2d vol.-- March 16. Gardner's Music of NATURE-- -- Herbert on Hybrid Mixtures: Marginal note
343d037 against those very laws he established in all <NATURE> organic nature) the Rhinoceros of Java & Suma
343d037 ery laws he established in all <nature> organic NATURE) the Rhinoceros of Java & Sumatra, that since
350d054 Browne's Works p. 20 There are no grotesques in NATURE; not anything framed to fill up empty cantons,
350d054 empty cantons, & unnecessary spaces" p 23. "for NATURE is the art of God" Septemb I,. It has been arg
350d056 reproduce Here there is some error-- Observed, NATURE does nothing in vain, therfore organ fitted to
351d056 has eye-points-- Macleay then answered, because NATURE leaves vestiges of what she does-- does not mo
352d058 le to young of Cochineal?? Is there some law in NATURE an animal may acquire organs, but lose them wi
357d073 ia or Rakkelhan is hybrid (produced commonly in NATURE. both in Sweden & anciently in Britain) betwee
365d107 f which is adaptation to habits (& habit second NATURE) may be more in constitutional.,-- more confor
366d108 ariety. which could not have been persistent in NATURE.-- According to my view, the domesticated anim
369d115 , in order to establish entirely their place in NATURE, as well as fully to understand their oeconomy
375d134 ges in number of species, from small changes in NATURE of locality. Even the energetic language of <M
375d134 s, excepting that famine may stop desire.--» in NATURE production does not increase, whilst no checks
375d135 structure into the gaps <of> in the oeconomy of NATURE, or rather forming gaps by thrusting out weake
384d161 every species has a disposition to deviate from NATURE in a manner peculiar to itself" <Is this so> E
386d166 reat length of time.-- There is probably law of NATURE that any organ. which is not used is absorbed.
386d167 inal cause of this because the great changes of NATURE are slow. if animals became adapted to every m
391d175 hich must be external» change are always of one NATURE species is formed if not.-- the changes oscill
397e003 perience! that these operations of what we call NATURE, have been conducted almost! invariably accord
397e003 been probably as constant as any of the laws of NATURE with which we are acquainted."-- this applies
414e063 all more webbed than those of other dogs.-- if NATURE had had the picking she would make <them> such
423e097 ger may not be so.-- Considering the Kingdom of NATURE as it now is, it would not be possible to simp
429e114 es, as well as our wild plants, we see how full NATURE. how firmly each holds its place.-- When we he
430e118 ring are picked & not allowed to cross.-- » Has NATURE any process analogous-- -- if so she can produ
450e174 th-- Yarrell.:-- Black Swan «in domestication & NATURE» strictly monogamous-- geese polygamous (¿when
527m032 in confinement, so may they learn in a state of NATURE.-- Singing of birds, not being instinctive, is
530m043 ead> Neck.-- He has seen other cases of similar NATURE.-- --like FitzRoy in sleep giving directions,-
535m069 acts to laws. without any attempt to know their NATURE.-- Reviewer considers this profoundly true.--
535m069 not a little remarkable that the fixed laws of NATURE should be «universally» thought to be the will
543m099 s said the great calculators, from the confined NATURE of their associations (it is not so in punning
568n018 nds in Harmony common to t[he] whole kingdom of NATURE. If I want some good passages against, opposit
571n028 xalted character???) Why may not our heredetary NATURE thus acquire some general notions, which are t
573n036 l act, provided with its instincts its place in NATURE. its range, its-- &c &c:--must be a special ac
575n043 s <Hyaena> Jackall.-- an animal not destined by NATURE to exist. & carrying «like other hybrids» with
580n062 oung are symptoms of the infinite & progressive NATURE of intellect indication of better life p. 207
581n064 her tastes few. March 16th. Gardiner's Music of NATURE. p. 31. remarks children have no difficulty in
590n094 n Dog does not Bark-- quotes Gardner's Music of NATURE to show barking not natural. (Vol I. p. 234) V
592n103 on is here said to open mouth in fright because NATURE intends to lay open all senses: <do> Horse pri
599o006 tecture is a fine amplification of two ideas in NATURE; a developement of the thoughts expressed in F
599o07v duced almost → greater changes in the polity of NATURE than any other animal-- Aimé Martin de l'Educa
602o11v tly approaches to instinct How strange it, that NATURE should have so little to do with art (p 128) R
605o020 from Blair receiving pleasures from beauties of NATURE & art." But as we often see people who are sus
608o025 te case of Philosophy, & savage calling laws of NATURE chance) 2) difference is from imperfect condit
620o044 ified gives only short pleasure. passion in its NATURE is only temporary, & we do not afterwards thin
620o045 owed is a consequence of that being part of our NATURE, & its effects lasting, whilst passions althou
620o046 o watch the house.-- it is part of <duty> their NATURE.-- When a pointer spring his bird. one says fo
621o046 teach him, that the instinctive feeling in its NATURE being always present. & his passion shortlived
622o048 leading to action or not, are the parts of our NATURE, <sub> subject to their instincts & associatio
622o049 uld think they were very few & general in their NATURE.-- So that we have some, it is sufficient to g
625o052 ce. & is supreme. because it is «a» part of our NATURE, <not> which regulates our feelings steadily &
640j167 Galapagos. no hurricanes.-- islds never joined, NATURE & climate very different, from adjoining coast
058r116 beach successive lines of sea weed-- Histoire NATURELLE des Indes Acosta. p. 125. of French «?» Editi
089a018 surface.-- Journal de Physique, et D Histoire NATURELLE, C«o»urrejolles. 11th Observ.-- Les grands tr
236bIBC ht expect.-- Lindley Introduct Dict. Science. NATURELLE Geographie Botanique De Candoelle. Geol. Soc
325c267 , Cow, Sheep.-- Verey. Philosophie d'Histoire NATURELLE Marcel de Serres Cavernes d'Ossements 3d. Edi
132a137 . Acad. Imp. des Sciences. (Sc Math. Phys. et NATURELLES. Tom I. p 501.-- shows first that data wholl
429e113 dden loosing of horns.-- I do not believe this NATURE'S plan.-- Whether we can or not trace history o

535m069 as God made them,-- next step plastic <virtue> NATURES. accounting for fossils). & lastly the tracing
535m069 ught to be the will of a superior being; whose NATURES can only be rudely traced out. When one sees t
379d151 iginal Paper, worth studying. Archiv. fur. NATURGESCHICHTE. September 11ᵗ Generation Mr Yarrell says
531m051 -- The extreme pleasure children show in the NAUGHTINESS of brothers children shows that sympathy is
540m089 at an accident,-- children with other children NAUGHTY.-- Why does person cry for joy? 17th. August M
545m107 itself down on its back & kicked & cryed like NAUGHTY child.-- Do monkeys cry?-- <they whine like ch
594m113 oes one laugh at it-- sensation of disgust with NAUSEA, (when stomach & little disordered) at thought
257c058 easure.-- highest fish in Old Red Sandstone.-- NAUTILI in----. it is useless to speculate <not only>
374d134 1.-- Lower Silurian-- several existing genera. NAUTILUS turbo. buccinum. turritella. terebratula, orb
471tℓ10 he fossil Mamm. of Europe-- Large Lizards in NAVIGATORES. Williams. Narrative of Missionary enterpris
399e010 Enterprise, p. 497. Vampire bat abound in the NAVIGATORS & at Manguia, but are unknown Eastward of th
399e010 & at Manguia, but are unknown Eastward of the NAVIGATORS. Snakes occur there,, but are unknown in Her
573n036 can allow «satellites», planets, suns. universe, NAY whole systems of universe <of man> to be governe
638j58v nt on the wisdom or Providence. when whole rocks NAY very mountains are formed of such dead & extinct
023r010 ean, before arriving at the Abrolhos shoals ⫝̸ -- N.B. The view of the Volcanos of the chain of the Co
031r038 ava cliffs [Fig. I] line of high tidal action {P} NB. patches of modern Conglomerates [Fig. 2] {P} Th
036r051 a sandy place.--(the sound was long & prolonged). NB. Is it generally known. the acute chirping sound
042r067 d to travel on a <me> central line of Patagonia. «NB. Mr Lyell P. 211 Vol III. talks of line of cliff
047r082 the sea fall on banks as a Bore wave rushes up? (NB. Earthquake wave in an oscillation, body of wate
068r146 nly burst out where strata in act of dislocation (NB. dislocation connected with fluidity of rock ∴ «
074r165 nic. Then Solfataras. «Mem: Micaceous iron ore.» N.B. To show how metals may be transported by compli
080r180 ad unite chemically like grease & mercury [blank] NB. P. 73. General reflections on the geology of th
096a039 eenland subsiding.) Von Buch Canary Isd. p. 351.. NB. Mackenzie talks of gravel on basalt of Heckla--
098a045 nt Lake Lemagne in Auvergne Proofs from Phryganea NB. Sedgwick talks of LAMINATED structure (∴ separa
099a049 film parallel to longer axis. But if great depth NB. Prof <Henslow> Sedgwicks lamination parallel to
114a095 e frequently metamorphosed than other deposits.-- NB. because lowest. first accumulated in bed of oce
132a138 .-- but in Ocean does not. (see resumé p. 536)-- «NB. I cannot understand the argument, that cold <oc
147g015 dded with ridges & flat topped hill/ do alluvium. NB In one part pure sand in current cleavage-- in o
150g035 er line running up great bight just as Dick shows NB. Lake gradually draining off would form plains s
154g057 m aquarium-- tidal channel-- 12ft obscure obscure NB In Glen Collarig tidal channel, sides <alm> 15 f
161g097 degree? This last measurement turns out too low, (NB .260 would have been more correct) there were se
161g097 bscure but not far continuous flights above it-- (NB the buttress or pass at Isthmus appears above le
183b051 nearest species often come very remote quarters,. NB. if Plata Partridge <or Orpheus> was introduced
185b057 h regards to other tribes in that same family-- (NB I see Waterhouse thinks Quinary only three eleme
192b086 lly in those classes where species not numerous. (NB in those classes with few species greatest jumps
194b094 opitheque of India.-- Tooth of <Spi> of Sapajou-- NB Sapajou is S. American form. therefore it is lik
198b115 e Anatomique. 2d Vol about monsters worth reading NB well to insist upon <different animals> large Ma
206b149 h we are bound to consider the increasing ones.-- NB As Illustrations are there many anomalous lizard
208b154 regions, & consequently not many yet multiplied: NB How does this bear with law referred to by Richa
212b167 ous & zoophagous Mollusca according to periods.-- NB. Was Europe desert (like S. Africa) after Coal P
222b203 ngly marked) between what are called varieties.-- NB. one mother bringing forth young having very dif
225b219 o adaptations of species by generation explained? NB. Look over Bell on Quadrupeds for some facts.--
225b219 on Quadrupeds for some facts.-- about dogs & &c NB. Animals very remote. ass & Horse. produce offsp
229b233 at Cape & a thinner-tailed kind further inland.-- NB. There is division of snakes. with hinder teeth
243c020 sh New Zealand & New Holland fish very similar.-- NB. Lesson method of generalizing without tables or
245c025 ats & mice. In Amboina only Cuscus & Barbyroussa «NB» [islds. Springing up more likely to <M> have di
246c027 708. Columba Oceanica (Less) inhabits Caroline < «NB. The» > isld. (.perhaps Phillippines & perhaps,
251c040 of certain parrots habitations India & Africa.-- NB. Any monograph like Gould on Trogons worth study
254c049 ta,-- typical of other, (surely rather parents). (NB These views must lead to spontaneous generation?
268c096 rict.-- <The same> <Mem> Tennioptera & Tyrannula (NB. work out how many forms Tyrannula can we worked
269c100 consequence, because applicable to N. Hemisphere (NB.-- Examine Abrolhos Flora with this view) Trista
274c113 asant) Goatsucker--, parrots with claw like lark (NB The La jeune veuve parrot though so much on the
280c135 -Breaking.-- applies it to national character.-- N.B. If two species were excessively old, they would
282c142 e of these races had become eminently aquatic--; «NB, aquatic, i.e relation to elements & not minding
289c162 y) =supported by foetal lower developed forms.=» (NB waterhouse says of affinity of many insects may
290c163 ssification on generation might appear an analogy NB Pyrrho-alauda (bird of St Jago) of brown colour;
293c172 that it should be remembered in next generation. [NB what are those Marvellous cases, when you feel s
296c184 ndergone changes «no near lofty country»?? p. 475 NB. This bears on fossils of Europe., those species
309c221 urkhardt to show black colour of certain Arabs.-- NB avoid quoting these hackneyed cases Mr Ed Blyth
317c249 y Appendix (& only appendix) of Congo Expedition, NB. I met an old man--, who told me that the mules
336d017 nts <stock> without. it be small & slowly obtained NB. The longer a thing is in the blood, the more pe
349d052 if characters can be established-- clearly so.-- NB. This paper worth referring to again.-- Accordin
350d053 pwards.-- «judged by analogy»-- Consider all this NB. How can local species as at Galapagos., be dist
368d114 -- singing best sign of most vigorous males.-- «(NB most strange cocks & hens. being either alike o
381d157 may be seen in Monooecious & Dioecious plants.-- NB. in Heautandrous animals <are> is there gradatio
381d157 1 more that they must in effect be so in all.-- 2 NB. In Pectinibr Mollusca.-- «or Cephalopoda» are t
383d159 not Hermaphrodites, like J. Hunters. Free Marten N.B. the common mule must often have been dissected
438e142 Wikinson on dogs of Egypt & Cuvier on Mummies]CD [NB TIME is element in change, as in Dahlias]CD all
462t051 t kinds of cattle in every part of England. &c &c NB. In botanical geography, there can be no sharp d
504q13v ried, on account of Van Mons views-- Also PEAS-- N.B. I think very likely the Peas to cross ought to
515q021 he individual by chance & under domestication.-- NB. Benthams remarks, where parts of flower are red
526m027 tion, example of others or teaching of others.-- (NB man much more affected by other fellow-animals,
615o038 ild state; certainly to the domesticated.-- [LHC] NB. Two dogs having very different instinct always
624o051 ich is beneficial to race, will have reoccured'. NB. Until, it can be shewn, what things easiest bec
626o052 r <races> generations led to their formation.-- 'N.B. If feeling or emotion rises from heredetary act
628o54v wish to gratify <it>-- anything contrary to it]CD NB. the very end of conscience is stop to wishes of
633j53v]CD if man takes care they are not detrimental.-- NB. One limit to the transmission of abortive organ
634j55r ancient.-- Are Cetaceae found in Paris Basin?-- NB) The explanation of types of structure in classe
640j167 g legged variety would prevail.-- Not separately: NB. These views quite exclude the idea of domestica
078r175 rce «(Principal veins)» 25 degree to 30 degree to NE. vein of Moran 84 degree NE. of Real del Monte 8
078r175 egree to 30 degree to 40 NE. vein of Moran 84 degree NE. of Real del Monte 85 degree to S. // Tasco 40 d
089a019 difficult to find stone not thus composed on the NE part more like marble requires polish to see str
096a040 All the Azores Isld. Von Buch p 359 stretched out NE & SW.-- Von Buch. Can. IIe p. 406. List of Volca
196b106 als & Edentata should only have left off springs <NE> in or near South Hemisphere. Were they produced
244c023 forts surtout de l'opinion du baron Cuvier, nous NE balançons pas a la regarder comme une espèce dis
499q009 male very tree & see whether they catch pollen-- <NE> In Oenothera bush.-- (19) Theory of mock flower
148g027 gular very like rubbish at head of Loch Dochart <NEA> Above Spean Bridge many flat terraces one above
162g102 teresting. enters by old tower called Glengarry (NEAD Roy told me) it is impossible to see my new she
032r041 nt downwards & to West.--From Equator to poles. NEARER the surface & to the Eastward.--If matter proc
050r091 s & gravel more common. the shoaler the water & NEARER the Banks Is there not a sudden deepening on E
126a124 lass.-- then the high temperature would be much NEARER the surface. especially at bottom of great oce
216b179 by intermarriages with people. either a little NEARER black or white as it may happen.-- Dr Smith sa
239c001 en crossed with pointer produces offspring much NEARER Esquimaux than Pointer.-- He has no doubt that
288c160 main land, proof of. land having been formerly NEARER.-- «Selby» Magazine of Zoology & Botany No XI
334d013 ntermediate in its peculiar long neck, but much NEARER to common goose.-- What has long been in blood
360d090 case where Jackall was father resemblance much NEARER to Jackall.-- This Keeper has seen when sickly
023r009 was caught in Shark's Bay. Lat 25 degree. the NEAREST of the E. Indian Islands. namely Java is 1000
076r168 silver are peculiar to that part of the veins, NEAREST to the surface of the earth."--p. 156. Mines o
173b010 ear continents might have some species same as NEAREST land, which were late arrivals others old ones
183b051 from same countries.-- In mundine genera, the NEAREST species often come very remote quarters. (NB.
209b156 s. 36 St Helena, without ferns.-- analogous to the NEAREST continent: poorness in exact proportion to dis
250c038 ta increased most. Certainly Africa approaches NEAREST to what is supposed to have been peopled--
278c129 re analogical, because we do not know, whether NEAREST species of each might not breed:-- Genus must
358d074 uds Volume on the Botany of the Pacific.--» to NEAREST continent.-- With respect to ancient geography
425e104 Flora-- what general forms.-- are the Labiata NEAREST to American, or Indian groups?-- = Believes so
426e108 stonishingly like to some distant cousins, the NEAREST blood being a great great-grandfather.-- -- 1L
430e119 w in low grounds are those, which are common & NEAREST being common to other parts of the world-- Mar
533m059 not conscious of recollecting it-- this may be NEAREST approach to <the> such instincts which full gr

Page ***(Key Word)***
036r051 g sound produced in walking over the sand: I am NEARLY sure, it is necessary to ascend the hill.-- Th
057r113 St Julian. = do not these bones differ as much NEARLY as the Eocene. = Should Mr Owen consider bones
077r171 rses both Clay slate, Porphyry North 52 W, & is NEARLY the same with that of the veta grande of Zacat
090a023 average when several are given), this will give NEARLY 19 pounds average for each stone. that fell, t
091a027 rata SE. direction of transitions clay slate &c NEARLY vertical Linear earthquake 500 by 90.-- in Syr
120a108 f they had travelled some hundred miles through NEARLY cold rock.-- in volcano the pool is not deep.
146g013 tion in situ of even imperishable pebbles/ I am NEARLY certain there were none on surface of any hill
148g022 re before There some of the half rounded gravel NEARLY as high as highest measurement but nature I am
148g025 d with English my informant said the lambs were NEARLY like each other «& half between parents» (& no
150g037 ill of Bohunthine upper road (2) extends as far NEARLY as house, the 3d below them opposite to where
152g050 of upper terrace The butresses of Alluvium rise NEARLY up to Glen Collarig up within 200 ft of level
155g063 ut «50 or 60 ft» «no doubt, a mound of Alluvium NEARLY parallel--» Inclination of river must constant
158g079 Divortium aquarum Peaty Mass of this point very NEARLY like head of Glen Guoy nor is horizontal line
179b032 out from the grandfather, (when the mother was NEARLY quite white) in the two first children How is
207b150 am. (original Flora) and Dicotyledenous, which «NEARLY» first appear «(p 321)» at Tertiary epock p. 3
235b275 QUESTION Are there races of plants run wild or NEARLY so, which <breed> do not intermix,--any cultiv
247c029 Grey varieties of rabbits thus handed down for NEARLY 70 year. Galapagos Mouse not the same section,
257c057 of replacement under peculiar conditions-- of «NEARLY» same kind country distant. <Study> The circum
273c111 hing (Gould), but the latter is obscure because NEARLY all are so.-- Thus in Hawks, there is a swallo
293c174 ent contagious diseases, where habits of people NEARLY similar. Curious instance of difference in rac
310c223 to think Botanical Provinces will turn out not NEARLY so confined as now thought.-- N. American, Eur
342d034 ion of continents) but Ogleby well answers that NEARLY all F. W. Fish are Abdominals. ∴that order fir
346d048 - Nonsense a flock of more than 100.-- Agrees, «NEARLY» with. the account. given by Boethius of ancie
361d096 (if I understand rightly) young cock & hen, all NEARLY similar.-- in blackbird group young like some
382d158 o Hermaphroditism,-- those organs which perform NEARLY same function in both sexes.-- are never doubl
388d170 n begins low.-- it goes through transformation, NEARLY independently of its parent & therefore wants
405e031 dsor Park.-- the «Fallow» Deer. which were of a NEARLY uniform <dusky> blackish brown.-- yet retained
454e184 V. II p. 252. Is there any very sleepy Mimosa, NEARLY allied to the Sensitive Plant.-- p. 290. Dr. E
458t001 Sivatherium & Anoplotherium, with existing, or NEARLY existing forms of aquatic reptiles most strang
466t091 -pods turning brown, whilst both others were in NEARLY full flower Maer June/41/ Rhododendrum-- necta
470t176 after mules abounded--so that palmated has now NEARLY disappeared. <& old English> But these mules <
491q01v different genus & then some hours afterwards of NEARLY related plant & see if first pollen produces a
520m002 s good instance as he is very deficient, he was NEARLY 9 years old. when his father died.-- The omnip
523m016 accidental <was> delay of money, he was «only» NEARLY thrown into a hospital.-- My father was nearly
523m016 NEARLY thrown into a hospital.-- My father was NEARLY drowned at High Ercall, the thoughts of it, fo
542m094 quility. «whine of children. puppies do so dogs NEARLY silent, so with men.-- How is crying-- peculia
555m142 nes, often put on a peevish expression, but not NEARLY so often <that> hardly ever the expression of
558m153 art of nostril, when passion commences.-- <All> NEARLY all will exclaim, your arguments are good but
594n113 that the nurse had to carry it out of the room. NEARLY 3 months old. What is absurdity, why does one
607o025 is not there present, but he acts from motives, NEARLY as usual (a) one well feels how many actions a
615o038 onding change in «external» man; and as all men NEARLY same species, so general instincts nearly same
615o038 l men nearly same species, so general instincts NEARLY same; which same argument probably applies to
619o042 HC] ----- p. 113. Mackintosh Grotius has argued NEARLY so [RHC] The history of every race of man show
624o051 n, « <then> each himself» & parents, & hence to NEARLY all the world.-- As conditions change, from ci
638j58v thousand trials.-- each step being perfect «or NEARLY so (except no in isd) although having heredeta
234b256 an at once tell by general colouring a group of NEBRIA complanata from. Devonshire, from another from
285c153 ust settle it.-- Changes in structure being «NECESSARILY» excessively slow, they become firmly embedd
294c177 wo of the willow wrens &c &c. & analogy will NECESSARILY explain the rest,-- Lonsdale says he has see
335d016 ns of system, the organs of generation would NECESSARILY fail.-- In last page. I should have said, "a
548m115 ly one idea is awake, when one is awake many NECESSARILY are., when one is deeply reasoning besides t
570n087 adapted to other objects. (as that senna is NECESSARILY disagreeable to organs adapted to like sugar
606o20v ied to this mental power Although taste must NECESSARILY be acquired by a long series of experiments
617o040 nded by the same faculty with matter & being NECESSARILY exhibited in & by matter. The phenomena of g
621o047 low certain lines of conduct, & he must soon NECESSARILY learn that it is his interest to follow it.
036r050 ?).--it was a strange story; I believe it was NECESSARY to ascend the hill,--but my recollection is i
036r051 alking over the sand: I am nearly sure, it is NECESSARY to ascend the hill.-- The absence of Second f
038r057 of Etna series of coatings; hence it will be NECESSARY to state all arguments for believing that the
052r097 deep: a great fault or rather many faults.-- NECESSARY form; as long as coast line fixed.--[Fig. 5]
052r097 ng as coast line fixed.--[Fig. 5] {P} * Slope NECESSARY for seaward transportal of drift matter.-- Gi
056r112 f mankind.--<Hutton show> Earthquakes part of NECESSARY process of terrestrial renovation & so is Vol
132a139 ypothesis some such explanation appears to me NECESSARY) as M. Parrots shows from variation in strata
174b014 certain countries when we can hardly believe NECESSARY, but if it was necessary to one forefather, t
174b014 e can hardly believe necessary, but if it was NECESSARY to one forefather, the result, would be as it
259c065 ring life of parent, & therefore being always NECESSARY may be called adaptation With respect to my t
263c078 tagger me.-- What circumstances may have been NECESSARY to have made man! Seclusion want &c & perhaps
285c151 case, that difference occurring.-- It will be NECESSARY to show hybridity from few forms, parents of
286c156 iculty, but after all a slight one It will be NECESSARY from manner Fleming treats subject to put in
291c167 urope<)> Toxodon in S. America) is absolutely NECESSARY to explain genera & classes. if extinct forms
305c211 ls, whereby choice with memory. or reason? is NECESSARY.--) which is modified into endless forms, bea
310c223 e is a chasm between Man (--<a> «&» chasm«s» <NECESSARY to account> «consequence of» for the scheme o
313c236 f Mammalia, or chick eat») Generation becomes NECESSARY, when organs of parent are concentrated in di
313c236 l cases probably gemmation (.Ehrenberg) --not NECESSARY to generation (lateral with no relation to ti
337d021 asses where types exist for if so. it will be NECESSARY to show how the first eye is formed.-- how on
358d075 nying us from camp to camp; it was absolutely NECESSARY to watch our meat, while in kettles on the fi
377d139 nt animals-- lived.-- we know the great time, NECESSARY to form channel & (& Basses St) yet no change
391d175 monster are really twins.)-- It is absolutely NECESSARY that some «but not great» difference (for eve
398e005 with that clearness of conviction, absolutely NECESSARY as the «basal» foundation stone of further in
398e005 ideas of these subjects. should be absolutely NECESSARY to arrive at right conclusion about species C
407e042 -- I see clearly from F. R. it will be highly NECESSARY to show that if species fall, genera must. Le
409e049 y be true then the formation of sexes rigidly NECESSARY.-- Without sexual crossing, there would be en
410e050 not «be» higher animals» -- it was absolutely NECESSARY that Physical changes should act not on indiv
417e079 ter», impregnated externally); nor can it be a NECESSARY concomitant, with moths, which can be impregn
422e095 g to their complexity.-- but yet there is no «NECESSARY» tendency in the simple animals to become com
432e125 utions of the sun & our lives,, but to period NECESSARY to form heaps of pebbles &c &c: the successio
466t094 k's Hood, a bee entering long nectary, would «NECESSARY» cross directly over the bunch of anthers & p
495q05a garden on gravel walk» will drift many seeds= NECESSARY to answer Wiessenborns doctrine of Equivocal
498q008 e each other? (10) Is number of pollen-grains NECESSARY to impregnate ordinary number of seeds known?
499q009 may know something about proportion of plants NECESSARY &c &c (a) Mercurialis-- Frog Bit, Valerian--
505q014 cultivation lost their horns» is impregnation NECESSARY to fruit--; become well shaped by care 13 Aru
524m021 (but Miss Cogan shows that repetition is not NECESSARY)-- the words second childhood full of meaning
547m113 so!) Now all these parallel trains of thought NECESSARY heirs of every action, & always running on in
549m123 d them.-- with lesser intellect they might be NECESSARY & no doubt were preservative, & are now, like
551m128 t.-- Plato «Erasmus» says in Phaedo that our "NECESSARY ideas" arise from the preexistence of the sou
552m132 say that those actions which have been found NECESSARY for long generation, (as friendship to fellow
553m136 n race?) by a separate act of God, & not as a NECESSARY integrant part of his most magnificent laws.
558m151 rom our confused idea of "ought." joined with NECESSARY notion of "causation", in reference to this "
566n011 icken, just bursting from egg.-- Animals have NECESSARY notions. which of them? & curiosity «strongly
566n011 ous artifices to take birds & beasts».-- very NECESSARY to explain origin of idea of deity.-- Animals
566n011 deity.-- Animals do not know they have 'these NECESSARY notions any more than «a» Savage M. Le Comte'
567n012 eological.-- Origin of cause & effect being a NECESSARY notion is it connected with <our> the willing
567n014 f affecting something like man. Has an oyster NECESSARY notion of space-- plant though it moves doubt
568n016 istinct from learning it by heart. Do not our NECESSARY notions follow as consequences on habitual or
568n016 des of a triangle shorter than third. is this NECESSARY notion, ass has it.-- When one is «simply» ha
569n020 a man.-- Probably, language commenced in some NECESSARY connexion between things & voice, as roaring
569n020 nd of letter)-- crying yawning laughing being NECESSARY sounds... not produced by will <by> but by co
593n111 usness of personnal identity is by no means a NECESSARY part of man's mind.-- At Maer. Pool. I saw ma
608o027 - It is not more strange that there should be NECESSARY. wickedness than disease. This view should te
609o029 elings. These bad feelings no doubt orginally NECESSARY revenge was justice.-- No checks were necessa

```
Page    ******************************************(Key Word)**********************************************
609o029 cessary revenge was justice.-- No checks were NECESSARY to the vice of intemperance, circumstances ma
609o030 roduced the greatest good «or rather what was NECESSARY for good at all» is the «instinctive» moral s
633j53v er seeds.-- But are we certain that these are NECESSARY adaptations.-- May they not be accidental? We
640j28v n, or gaining adaptations, but for youth most NECESSARY: the fertility of Man in old age keeps woman
306c212 taught it to go to <sea> salter water (& its NECESSITIES teach it taste, but that is much more genera
628o53v be to all such cases-- either, that from the NECESSITIES «& good» of society such conduct is instinct
369d115 mal Economy p. 482 (Same book) Owen says "the NECESSITY of combining observation of the living habits
389d173 r]CD \It should be observed that the constant NECESSITY for change. in process by generation applies
390d178 s would reproduce its kind was it not for the NECESSITY of some change.-- \ Without some small change
391d174 ot in other trees.-- -- Why should there be a NECESSITY that there should be something «each time» ad
415e070 ve, & not abortive: with reference to the non-NECESSITY of the «so called» progressive tendency law.-
417e077 wo bodies, <th> we can as well understand the NECESSITY of a relation between the fluids of the two a
441e150 he weakest part of my theory is, the absolute NECESSITY, that every <animal> «organic being» should c
445e159 ghly organized forms-- this resulted from the NECESSITY of supposing some inward progressive developi
447e164 rms then ought to be very persistent,, & then NECESSITY of crossing is much less,-- now certainly in
609o29v ume of Malthus). Adam Smith also talks of the NECESSITY of these passions, but refers (I believe) to
619o042 re, & <an> «such» actions being prevented by <NECESSITY> «some force» give pain: for instance either
195b099 ike covering created.-- passage for vertebrae in NECK same cause, such beautiful adaptations yet othe
334d013 x says when common & China goose are crossed the NECK is not intermediate in its peculiar long neck,
334d013 he neck is not intermediate in its peculiar long NECK, but much nearer to common goose.-- What has lo
367d113 employed in fighting" instances thighs of cock & NECK of Bull.-- is most common in vegetable feeders.
530m043 health.-- his complaint was carbbuncl on <Head> NECK.-- He has seen other cases of similar nature.--
558m152 ight enter?-- I believe common Swan, arch raises NECK & depresses chin-- strikes with wing arches win
577m051 ive blood to surface exposed, face of man, face, NECK-- «upper» bosom in woman: like erection shyness
637j58r to long necks-- why they may as well say, «long» NECK is adapted to long necks.-- p. 236. Marsupial b
327c265 ilosophy of Blushing lately advertised. /6s Mrs NECKER on Education preeminently worthy of studying i
585n074 to like Lavendar Water «very much» Henslow. N.. NECKER has remarks on the means. by which children le
152g046 entering «on» one side ravine. Are the lip, or NECKS of land on level with shelves effect of corrosi
558m153 es black Swan.-- Goose do all species put their NECKS straight out & hiss.-- [Hyaena pisses from fear
637j58r processes in Giraffe &c, as adaptations to long NECKS-- why they may as well say, «long» neck is adap
637j58r may as well say, «long» neck is adapted to long NECKS.-- p. 236. Marsupial bones especial adaptation,
467t103 act, something like on Kidney Bean, they go to NECTAR at foot of upper petal standing on «I saw Bee
467t104 beautifully to protect sheath {a} In all these NECTAR seems to be at base of upper petal & the curva
467t105 saw another on Cabbage--white Butterflies suck NECTAR: «Maer June 41» Rhubarb. pollen very minute--n
468t111 Humble alighted on base of filaments & reached NECTAR =again= between them, hence quite below stigma
470t178 =what insect can get honey out of long, curved NECTAR of Butterfly Orchis & Listera? Bryony saw comm
466t094 s not prevent bees visiting it.» In Columbine NECTARIES are placed all round flower as they are in Cr
490q001 vary much in their spurs & Ranunculus in the NECTARIES. The former best for my experiment on Selecti
506q015 of flowers-- Their smell-- form of flowers-- NECTARIES-- In Monooecious «order» flower occupy partic
515q021 ation to hybrids.-- (3) As peaches sport into NECTARINES (does reverse happen?) what is effect of cro
515q021 happen?) what is effect of crossing peaches & NECTARINES: same question with regard to Primroses. (4)
466t093 early full flower Maer June/41/ Rhododendrum-- NECTARY marked by orange freckles on {a} upper petal;
466t093 ds, so that anthers & stigma lie in fairway to NECTARY.-- Is not this so in Kidney Bean. How is it ge
466t094 relation. In Monk's Hood, a bee entering long NECTARY, would «necessary» cross directly over the bun
466t094 p-- In Lark-spur, if Bees put proboscis within NECTARY «they do» they must disturb all anthers, wh of
466t094 ed by the hairy black lip of lower division of NECTARY: «wh. itself resembles a Bee, but does not pre
468t112 - So that, finally Fraxinella. with respect to NECTARY is same case as Azalea or Rhododendron xx afte
514q21. cting pollen-masses?-- = answered = Has Ophrys NECTARY?= Bunbury says no «hollow» spur.-- Ask about P
068t148 lecting Gulf of California. Beagle Channel.--One NEED never be afraid of speculating on the sea The 2
177b025 to intermarriage «increases it--> settles it ;We NEED not think that fish & penguins really pass into
208b154 --? N Macleay Name given in Congo Expedition he NEED not expect to find <species>, varieties, interm
227b226 ay pass into each other: now on this view no one NEED look for intermediate structures <between> say
230b239 nge at once. but afterwards will not alter. This NEED not apply to very slow changes. without crossin
388d173 o pass through all transformations, should there NEED two organs; whilst in common bud there is no su
388d173 wo organs; whilst in common bud there is no such NEED.-- one would <one> suppose that the vital porti
392d175 r addition of differences. shows that difference NEED not be added EACH TIME. but after some time]CD
431e122 n the place where any species is most common, we NEED not look for change, because its number show it
540m085 ly coorganic as connexion of mammae & womb.-- We NEED not feel so much surprise at male animals smell
542m093 tiff like that of turkey.-- he may be amused, he NEED not express it, he may most earnestly wish [not
472s02r out over & over again & wipe off pollen. (as a NEEDLE becomes covered) so whole sides of flower & st
570s023 ugged his shoulders & went away."-- he implies NEGATION, without violence, without assigning or under
570s023 gning or understanding reason.-- surprise with NEGATION.-- like shaking something off shoulder-- or i
573s037 ide & then to other. & hence rotatory movement NEGATION.-- he dropped his head when he meant to eat,
573s037 on.-- but nodding is less strongly marked than NEGATION Marianne. says. that she has constantly obser
191b080 not England. Did Ireland possesse Mastodons?? NEGATIVE facts tell for little) Geographic distributio
424e101 egions in modern times.-- this would depend on NEGATIVE evidence of fossil remains, & therefore not t
464t063 s work-- Lund makes his wonderful discoveries-- NEGATIVE facts are valueless= monkeys= Owen has descri
465t079 argument to India & Europe-- & Africa!,-- any NEGATIVE argument against-- monkey-man, valueless.-- W
473s03v sh one step lower in America-- How curious all NEGATIVE laws of America of depth of organisms holding
491q01v th abortive stamens» answers first question in NEGATIVE.-- Questions Regarding Plants. 1. Uniformity
565n009 ssion of mouth showing action,-- sulkiness all NEGATIVE expression? Expression of affection is accomp
040r078 anos must be considered as chemical retorts.--NEGLECTING the first production of Trachyte. look at Su
057r112 is Volcano a useful chemical instrument.--Yet NEGLECTING these final causes.--What more awful scourge
064r136 mass on each side 3000 ft thick & 150 broad, NEGLECTING Cordillera itself now remaining- Lyell <p
572n034 the mind had dwelt on each word separately, NEGLECTING time, & general sense, anymore than connecte
567n015 hewell. Induct. Sciences-- Vol I p. 334 Does a NEGRESS blush.-- I am almost sure Fuegia Basket did. &
570n027 ains, & a negro, similarly treated would think NEGRESS beautiful,-- [male glow worm doubtless admires
114a093 Hope & Australia/ and mud of salt-lakes of Rio NEGRO--Mr Bowerbank-- Dr. A. Smith's curious specimen
116a099 port on D'Orbigny's Voyage. good section of Rio NEGRO beds.-- -- refers to species non decrite de pet
216b179 r McClay is inclined to think that offspring of NEGRO & white will return to native stock (the cross
223b208 more than in structures.-- If the skeleton of a NEGRO had been found what would Anatomists have sai
303c204 ve> differs in present races, & form of feet.= (NEGRO or father of negro probably was first black at
303c204 ent races, & form of feet.= (Negro or father of NEGRO probably was first black at base of nails & ove
338d024 n seems most clear that the ideosyncracy of the NEGRO (& partly Mulatto) prevents his taking any form
338d024 Malaria-- adaptation & species-like,-- -- Says NEGRO-- thick skinned My hairdresser (Willis) says <b
343d038 rieties are becoming extinct. others though the NEGRO of Africa is not loosing ground. Yet, as the tr
343d039 uld have taken place. otherwise in 10,000 years NEGRO probably a distinct species-- We know how long
452e182 rnal that skulls found near Vienna appoximat to NEGRO form; those from Rhine to the Caribs.-- Vol II
477z002 Paraguay.-- do. p. 365. 3 cats (mbara caya. le NEGRO, et le pajero) l'yaguaré «the zorilla-arskink»
511g018 male transmit to male more of his features-- in NEGRO & white (3) About the Bantams at Zoolog Soc.--
540m086 American in Brazil is under same conditions as NEGRO on the other side of the Atlantic. Why then is
540m087 .-- Look at the Indian in slavery & look at the NEGRO-- look at them both savage-- look at them both
570n026 rn out of country yet would love mountains, & a NEGRO, similarly treated would think negress beautifu
572n033 L'Institut. 1838. p. 340. Mr Carlyle says that NEGRO certainly has less reasoning powers than Europa
600o008 ve.-- [With regard to ordinary Beau ideal, Mem. NEGRO, beau,--Jeffrey denies all Beau-- How does Hen
189b071 celo & Kingfisher same colours Strong odour of NEGROES, a point of real repugnance.-- Waterhouse says
205b142 able.-- Bridgewater Treatise p 85. Parasite of NEGROES different from European.-- Horse & Ox have dif
308c219 ginal ideas., it will show state of knowledge «NEGROES existed since time of earliest Egyptian drawin
527m032 cquiring «the instinct» our notion of beauty & NEGROES another; but it does not explain the feeling i
178b032 is certain that when White Men & Hottentots or NEGROS cross at C. of Good. Hope the children cannot
540m086 e hot plains of the Amazons & Brazil-- with the NEGROS of Africa, (or again the black man of <B> Van
219b194 it as adaptation because volcanic isld. whilst <NEIG> Africa, sandstone, & granite, (that is genera
066r142 rdoba. one of which dried up <all> a lake in NEIGHBOURHOOD of town Mr Murchison insisted strongly. tha
043r070 nsive areas. -- P. 322 In any archipelago. & NEIGBOURING Volcanos. eruption from «more than» one orif
153g056 smaller ones «these boulders are decaying.» NEIGHBORING rock gneiss & [...] sandstone actually resti
232b246 e South ward.... species must be compared to NEIGHBORING sea.-- For change of species does not measur
558m150 rise "do unto others as yourself". "love thy NEIGHBOUR as thyself". Analyse this out.-- bearing in m
146g014 .-- From this point could be followed up to NEIGHBOURHOOD of Tyndrum where a large sort of <plain> sp
```

150g038 h extend below the Houses The Hills in this NEIGHBOURHOOD appear very round-topped with much drainage
151g040 n) are often times of rock not in immediate NEIGHBOURHOOD, (as granite or gneiss of Moel Derry) on lo
047r081 so. (1822) no great wave on record. -- «also NEIGHBOURING sea must partake in absolute movement» More
050r090 earthquakes of Sumatra no connection with a NEIGHBOURING Volcano of Priamang.--Marsden Sumatra. M. D
061r127 land. = Yet new creation affected by Halo of NEIGHBOURING continent: ≠ as if any creation «taking pla
178b030 Eyton of Hybrids propagating freely In Isld NEIGHBOURING continent where some species have passed ov
232b249 the distribution of Lemurs in Madagascar, on NEIGHBOURING islets & a sub-genus in Southern Africa In
233b250 h were hard to distinguish came from closely NEIGHBOURING localities Institute 1838. p 38. account of
296c184 aks. Did Creator make all new yet forms like NEIGHBOURING Continent. This fact speaks volumes. 2 Chap
421e090 ry Isld.-- ie Canary Isld approaches more to NEIGHBOURING coast of Africa, than to other parts of tha
426e105 ct-- Lyell's argument about <Tertiary> Isld «NEIGHBOURING» formed in the Tertiary «epoch» like Sicily
448e166 tween the Crag & Touraine beds, the one with NEIGHBOURING & Arctic sea, & the other with neighbouring
448e166 neighbouring & Arctic sea, & the other with NEIGHBOURING & Senegal as sea.-- is remarkable.-- Again
495q05a covered with Honey-dew dusted with pollen of NEIGHBOURING grass= Spread sheets of Paper. covered with
327c265 it trees in. N. America" in Lib. of Hort. Soc Mr NEIL. has written good article on Horticulture in Ed
135a145 Journal. Vol 4. 1835. p. 437. Tours by Benza NEILGHERRIES-- Much inform. on. decomposition of granite
364d105 with the Ptarmigan. subalpina in wild state.-- NEILSON has given figure of it.-- In England no doubt
381d156 quire coition every generation)-- Epizoa & the NEMATOID Entozoa-- Therefore highness in scale has no
268c096 - by Wiegman Classified catalogue of animals of NEPAL read before Linnaean Soc. Feb. 1838.-- Annals o
451e178 ic Soc. vol. I. p. 335. Catalogue of animals of NEPAL by. B. Hodgson. p. 336 In the most pestiferous
440e146 past generation.-- thus cabbages growing like NEPENTHES.-- cases of pidgeons with tufts &c & here th
426e108 wild animals & plants]CD.-- Mr Marsh has some NEPHEWS, who are astonishingly like to some distant co
313c236 on the action of the viscera. under sympathetic NERVE may be instinct or habits. ¿are sympathetic ner
337d021 to show how the first eye is formed.-- how one NERVE becomes sensitive to light.-- (Mem whole plant
337d021 solely for others parts of creation) & another NERVE to finest vibration of sound.-- which is imposs
373d132 what is animalcular semen-- but this-- -- the <NERVE> living nerve nursed in Mould.-- Lyells Element
373d132 lcular semen-- but this-- -- the <nerve> living NERVE nursed in Mould.-- Lyells Elements. p. 290. Dr.
558m150 sobbing) which are most under great sympathetic NERVE. most subject to habit, as being less so will.-
313c236 rve may be instinct or habits. ¿are sympathetic NERVES & nervous system of insects analogous.?-- («Ev
315c242 al action, in intestines subject to sympathetic NERVES-- The vividness of first <thoughts> «memory» i
354d062 placed in place of last by the ordering of the NERVES, but in different parts according to age of in
386d166 rt, of what is good for the whole.-- if cut off NERVES in snail. (Encyclop of Anat & Phys) can make a
388d173 one would <one> suppose that the vital portion ¿NERVES? passed through transformation, & was received
602o10v es In Elliotson's Physiology much about sleep-- NERVES.-- Volition &c Reynold XIII Discourse (p 115)
165g126 sing most remarkable ever obseved? Shows that <NERVOUS> brain makes thought Glen Roy B C. Darwin All
198b113 states in one the sanguineous system, in other NERVOUS developed. (Owen's idea) states these class ap
313c236 instinct or habits. ¿are sympathetic nerves & NERVOUS system of insects analogous.?-- («Even plants
371d118 solve them.-- It is less wonderful that childs NERVOUS system should build up its body, like its pare
389d173 o ways explains Ld. Moretons case: without the NERVOUS matter consists of infinite number of globules
532m055 ind being more alive.-- How is it. with people NERVOUS from illness., <the> it must be an excited act
545m106 ared to laughing» they dance with passion, ie. NERVOUS impulse to action is sent so fast to limbs tha
547m111 s certain that muscular, mental, <&> digestive NERVOUS influence replace each other August 29th. Went
601o009 consciousness is absent) in fibres united with NERVOUS filaments.-- ¿plants? yes by distinct mechanis
638j059 imperfect attempts. In the «Bee» Mollusca the NERVOUS system is endowed with the knowledge of trying
638j059 e course of ages «step by step».-- in Man, the NERVOUS system, gains that knowledge, before hand. & c
162g102 impossible to see my new shelf, from road: Loch NESS 30.140. A 66 degree 30.095 .0458 or 6 differenc
162g102 095 .0458 or 6 difference between bedrock & Loch NESS <30.100> <Donald Macphee> Saturday Morning 29.9
162g105 ft below Loch Oich wh is 92 ft above sea-- Loch NESS 40 ft above do. When cutting bank where Locks n
251c040 <Tarton> Barton.-- swifts return after years to NEST Vol II. p. 49. on the localities of certain par
271c105 to the North by other species.--) should build a NEST lined with mud, in forest where not a tree in w
271c105 ne- The black & white thrush of Azara builds its NEST in <same country> something same manner, much m
297c189 ble [not located] p. 428. Ouzel sometimes builds NEST without doom. Vol 2. Mag of Z. & B. p. 431. Mis
298c189 185 case of tit lark placing withered grass over NEST, when often looked at.-- this most puzzling whe
359d088 umbers, & rears an unusual number out of any one NEST, even more than common duck-- Male Penguin was
437e139 found a dead lamb «& hare» by the side of Eagles NEST, which shows power of carrying great weight. p.
446e160 ciprocally assist in domestic cares, as building NEST, sitting on eggs, & feeding & defending their y
530m049 ng are big enough to eat.-- There was blackbirds NEST, near hot-house at Shrewsbury, which the cat wa
531m049 to visit daily to see how the young got on. this NEST the cat could If cats will «ever» eat little bi
550m126 perty. (--regarding food & in birds of place for NEST.)-- with dogs "have notion of masters property"
552m131 ow, who carried from all parts straw to make its NEST. Pigs & Elephants, (both Pachyderms) much intel
563n001 tory Waterton describes. pheasant springing from NEST & leaving no tracks.-- My Father says pea-hens
582n068 ant a bird knows what is about when building its NEST; it knows its object but not result (first time
584n072 es through the air,-- waterbirds, the bee to its NEST,-- cats when carried in confinement,-- carrier
593n105 down.-- mem. Nina used to get into hay & make a NEST for herself.-- the object is to make saucer-sha
626o052 t cannot be said to guide will. as bird building NEST, but supplies it-- instinctive feelings will do
088a014 tly heard that Herons bring eels alive to their NESTS; & then they may picked up beneath the trees---
530m049 - [not located] Fox believe cats discover birds NESTS & watch them till the young are big enough to e
515q022 Pea (& Sweet Pea) for several generations under NET & see if get sterile-- Cover that little Ervum i
257c056 u moins des espèces tres-analogues,-- quand ce NÉTAIENT pas tout à fait les mêmes." This good case. o
229b232 in in <there> one common ancestor we may be all NETTED together.-- Hermaphrodite animals couple: argu
599o007 ing from eminence. the wide expanse, of county, NETTED with edges & crowded with towns & thoroughfare
502q11v ?-- to show acclimatisation.-- July <1842> When NETTLE leaf. put into spirits, poison-drop exudes-- d
502q11v v melon-- «Loasa» Anchusa «Campanula» &c & dead-NETTLE.-- Lithospernum. Blue Gloss. it is not possibl
505q014 will be produced in individual plants 17 A dead-NETTLE in Hot-house. will it seed?-- (Skim through Pe
506q15v Keeling Dioecious, or Monooecious, besides the NETTLE. at Galapagos-- Dioecious.-- Carex.-- We may p
506q15v alapagos-- Dioecious.-- Carex.-- We may presume NETTLE spreads by seeds= (44) Zostera. Has he seen it
641j29v snake, (jaw of spider?) sting of bee, sting of NETTLE.-- Are there any other analogies-- prickly pla
307c216 el argument.-- Are there any abortive organs in NEUTER bee, (There is paper by Yarrell in Zoolog Tra
309c221 wing female good case of instinct. bees turning NEUTER into Queen, more wonderful case Dwights' Trave
382d158 ally developed.-- surely analogy of molluscs. & NEUTER bee would shew this-- (Do any male animals giv
388d172 curious instances of corelation in structure = NEUTER bee having both sexes abortive fact of same te
313c235 - (How wonderful a case bees developing sex of NEUTERS) species may have had their infancies, as well
370d116 -- Hunter doubts about production of Queens.-- NEUTERS are bred first, «then males--» how has this be
370d116 t, «then males--» how has this been arranged-- NEUTERS are true females, but with parts little develo
061r127 t Bra in Brine Springs. (Henslow) Speculate on NEUTRAL ground of 2. ostriches; bigger one encroaches
277c126 s us that ¿O. Modulator & O. Patagonicus. till NEUTRAL ground ascertained, call them varieties. but t
023r009 Hippotami found in that Archipelago? Such have NEVER been observed in Australia Dampier also repeate
056r109 d, the one slow cause is apparent. «I confess I NEVER see such islands whose inclination natural [...
068r148 g Gulf of California. Beagle Channel.--One need NEVER be afraid of speculating on the sea The 24 ft.
071r158 xception.--«formerly perhaps otherwise» Mendoza NEVER overthrown,--no mountains Mackenzie has talked
089a017 c-- CD[Might not bottom of ocean boil; yet heat NEVER reach surface.-- Journal de Physique, et D Hist
187b063 led it over a tract from Spain to S. America.-- NEVER They die; without they change; like Golden Pipp
192b086 st marked genera? Reptiles?) For instance there NEVER may have been grade between pig & tapir, yet fr
194b097 !? We have not the slightest right to say there NEVER was common progenitor to Mammalia & fish. when
199b120 ertile offspring> » fertile offspring; marriage NEVER probably excepting from «strict» domestication
199b120 not fertile. or at least most rarely & perhaps NEVER female.--no offspring: physical impossibility t
211b163 es at Hudson bay breed with dogs.-- the bitches NEVER being killed by them, whilst they eat up the do
215b177 said.= How came first species to go on.-- There NEVER were any constant species Both males & females.
216b183 . had a very fine blood hound bitch which would NEVER take the dog. But at last a rough-haired shephe
217b183 dog lined her & produced a very large litter.-- NEVER afterwards went in heat.-- This is good instanc
224b214 kind, which is not probable; then monkeys will NEVER produce man. but both monkeys & man may produce
227b227 s discovered, for speculating on future. !.fish NEVER become a man.-- Does not require fresh creation
263c075 though this may appear an absurd saying) & will NEVER be conquered by anyone (if has any kind of prej
265c087 Yet these wings may be of some use,-- Nature is NEVER extravagant though clearly not of the use to wh
265c087 as we have only ornithorhyncus, we should NEVER know how much structure was connected with habi
271c107 remarked to me at Zoological Society,, that you NEVER find two «similar» groups of birds in two count
282c142 oos. would not be affinity, but the truth would NEVER be discovered When one reads in Ehrenbergs Pape
288c160 shed the Grasshopper Warbler-- --Yellow Wagtail NEVER seen in one district, though common on another,

Page	**	(Key Word)	**
293c175	l of Man would be disagreeable to Musquitoes We	NEVER	may be able to trace the steps by which the org
308c216	capable of demonstration that all animals have	NEVER	at any one time formed chain, since if cretceou
310c223	d & heredetary & such definite thoughts, I will	NEVER	allow that because there is a chasm between Man
318c252	m extreme north turning white like Hares??--) I	NEVER	saw more beautiful adaptation for snow-- like s
332d001	eople. My Father mention, than for ten years he	NEVER	saw one case of malignant erysipelas spreading
341d031	cock> Hen.-- offspring female, yet so infertile	NEVER	even in seven years produced even an egg.-- a m
342d032	ed (just like common mules) & lay many eggs but	NEVER	produce inter se or with parent species.-- The
358d085	- thus the common. pheasant & fowl when crossed	NEVER	even lay eggs. & the men cannot «hardly» tell a
360d091	e «between» lice & fleas. sticking on them, but	NEVER	in an animal, that had long been in confinement
362d099	horns drop off., replaced by hairy ones. which	NEVER	«dry up &» peel off their skin (not being wante
362d099	f their skin (not being wanted for war) & hence	NEVER	fall off.| Curious the rapidity of the change i
379d152	breeding very pure South Down that the ewe must	NEVER	be put to any other breed else all the lambs wi
380d155	in the month, she will in 5 weeks.-- A Bull is	NEVER	taken from his own field to bull a cow.-- -- a
382d158	rform nearly same function in both sexes.-- are	NEVER	double, only modified. those which perform very
384d162	: they then become so related to each other, as	NEVER	to be able to impregnate themselves (this never
384d162	never to be able to impregnate themselves (this	NEVER	happens in plants «only in subordinate manner i
385d163	latter being unhealthy.--» males «bred in & in»	NEVER	lose passion. (Mem: so it was said little cock
385d164	alf fowls.-- eggs fertile, but parent bird will	NEVER	sit on them.-- May be just worth remembering th
401e015	s is produced.-- Thinks probably experiment was	NEVER	tried of separating apple tree entirely from al
405e032	ffects of having been long separated, or having	NEVER	[not located] ARGUMENT REAL of antiquity of rea
406e035	has had any offspring before.-- -- now this is	NEVER	stated.-- Regarding the similarity of offspring
410e053	tions of numbers of species to genera &c &c can	NEVER	be told, without species being described.-- but
419e087	no change & even no loss of species.-- It must	NEVER	be overlooked that the chronology of geology re
425e105	Eocene beds of Paris Lyell has remarked species	NEVER	reappear when once extinct-- Lyell's argument a
441e151	, which are generated by buds alone or roots, &	NEVER	flower, so there may be animals as Coralline, o
442e152	number of generations-- Gorze in Norway, which	NEVER	flowers!!-- <How did it get there? whether> Acc
442e153	rees are propagated by means, which wild plants	NEVER	are, namely on stocks of other varieties & we k
444e158	experience teach distances in air, in which it	NEVER	touches objects.-- far better case than chicken
468t111	is-- Azalea. Rhodendron. Fraxinella to Anchusa	<NEVER>	«once» P on Fraxinella <Heartease> «small. Hum
470t178	Hood, brushing over stamen «Egg Tree»--I think	NEVER	on the Galeum saxatile & other common kind--I t
472s02r	ny times every day. many clumps of heartseases,	NEVER	saw any Bee go to them. Yesterday remarked that
472s02r	out doubt.-- Bee, not large, very dusky & broad	NEVER	saw such a one before-- Saw Fly 21 «this Hearte
473s03r	longirostris in miocene like in Europe-- Cuvier	NEVER	found remains of Sus with Elephants-- Lyell say
505q014	ey has planted seeds of pale green Cynoglossum.	NEVER	germinated 12 Does the horned orange. wh. never
505q014	never germinated 12 Does the horned orange. wh.	NEVER	has seeds produced good pollen? Yes «From culti
515q022	tle Ervum in Sand-walk, on which I think I have	NEVER	seen Bee visit. Experiments in Garden Sow stone
521m009	er of name A. B., &c &c. & he maintained he had	NEVER	heard of such a man & had no gardener.-- My F.
522m010	m Mr Child «of Kinlett» had married.-- Answered	NEVER	heard of such a man.-- (My Father explained who
522m010	wa & all about him, but still maintained he had	NEVER	heard of him).-- My F. then said you remember J
522m011	es as a few minutes before he maintained he had	NEVER	heard of.-- Thus in many things if he began at
523m017	purpose.-- this found to be true.-- Her Husband	NEVER	suspected during these two years that she had b
523m018	s restrained by remonstrances on him» which are	NEVER	generally, if at all discovered.-- <Sup> Someti
524m021	ton & Ternhill, (where she was born) though she	NEVER	naturally talked of these places.-- My F. says,
527m033	s blood run cold by singing).-- Granny says she	NEVER	builds castles in the air-- Catherine often, bu
533m060	sort of consciousness I was not right; though I	NEVER	realized the idea that I was tending to make my
544m103	astle in the air, is more prolonged than dream.	NEVER	fatiguing,-- else it is only our consciousness,
553m138	t (the intelligent Keeper) remarked that he had	NEVER	seen any of the American Monkey show any desire
556m145	translated by Holcroft «Vol I» .p. 86 "We ought	NEVER	to forget-- --; that every man is born with a p
557m150	idual case. nothing can be heard.-- Shame would	NEVER	make person tremble, like fear.-- Why does any
559m153	o great... "Ay Sir there is much in analogy, we	NEVER	find out." This unwillingness to consider Creat
583n069	ssociated with reason: [N B. insects which have	NEVER	seen their parents offer best cases of instinct
585n075	with front legs & knocks with back of Head, yet	NEVER	puts down its ear. good to contrast with horses
586n081	idered also such.-- heredetary habit, is a part	NEVER	subject to volition.-- like plants going to sle
591n097	ld me. he has often watche tame young wolf & it	NEVER	dropped its ears like dog-- wagged its tail «a
595n125	than sucking.-- [I assume a child pouts who has	NEVER	seen others pout]CD [blank] Goldsmiths Essays N
602o11b	«all» art is the realizing and embodying, what	NEVER	existed but in the imagination".-- Macculloch V
618o41v	ction thought & organization: But if the weight	NEVER	came untill the blueness had a certain intensit
622o049	pply to mother loving child, from whom, she has	NEVER	received any benefit.-- Yet I think there is mu
624o50v	lifetime). so is our moral taste p. 152. Reason	NEVER	can lead to action.-- p. 164. Ld. Shatsbury und
640j167	in world like Galapagos. no hurricanes.-- islds	NEVER	joined, nature & climate very different, from a
201b126	s» there are many gaps. & those forms which	«NEVERTHELESS»	have produced species, have produced fe[w]
439e142	do-- or be with difficulty be kept alive.--	NEVERTHELESS	much probably depends on circumstances favo
481z010	e must be studied for anatomy. of. corals.--	NEVERTHELESS	the details appear very trifling Also Berre
022r008	Volney's travels also» Dampier's last voyage to	NEW	Holland P 127.--Caught a shark 11 ft long. "Its
024r015	rine currents Find instances; The whole coast of	NEW	Holland shoals much: Dampier remarks on great fl
027r027	cted with frequency of shells in flints in Chalk	NEW	Providence more hilly than others of the Bahama
031r038	in Galapagos. Daubeny P 24 V. back of page 1 of	NEW	Zealand Geological Notes. at St. Helena. This st
040r062	ide.-- Add from M. Lesson. character of Flora to	NEW	Zealand, which agrees with St Helena in being un
041r066	retionary, so are all those plates in Australia.	NEW	Red Sandstone. at Bahia in modern sandstone. a c
043r070	rthquake of 1812 affected valley of Missisippi &	NEW	Madrid & Caraccas.-- Is this mentioned by Humbol
054r102	ie Grand tertiary formation of Payta: N. part of	NEW	Zeeland entirely volcanic!! New Zeeland rich in
054r102	ayta: N. part of New Zeeland entirely volcanic!!	NEW	Zeeland rich in particular genera of plants: All
061r127	n any mountain, one is falsely less surprised at	NEW	creation for large.--Australia's = if for volc.
061r127	for volc. isld. then for any spot of land. = Yet	NEW	creation affected by Halo of neighbouring contin
062r129	n. at long distances; generally first arrives:--	NEW	Zealand rats offering in the history of rats, in
073r163	Sulp. of Barytes: Fluoric. Barytes:-- Humboldt.	NEW	Spain. Vol III. p. 130 Metals in Mexico rarely i
076r168	ce of the earth."--p. 156. Mines of Batopilas in	NEW	Biscay, "Nature, exhibits the same minerals <as>
076r169	nied by Sulp. silver sometimes by selenite.-- in	NEW	Spain, contrary to Europe. argentiferous lead no
079r177	pe.--Azores Isds «nor at St Helena.--» Humboldt.	NEW	Spain Vol. IV. «p. 58» At Acapulco earthquakes a
092a030	9.-- Geograp Journal Earthquake at Melville Isld	NEW	Holland Augus 1d to 3d & 19 1827 Geograp Journ T
096a040	an. Ile p. 406. List of Volcanos Salomon Isld,--	NEW	Britain-- &c &c In Ascension for centuries after
102a062	Cleavage discussion, state broadly indication of	NEW	law acting in certain directions predominantly,
120a109	espect to formation of salt.?.--??? Footsteps in	NEW	Red Sandstone. look as if a surface deposit.-- T
129a131	f. Brazil. Maldonado enter into this case.-- Ed.	NEW.	Phil. Journal Vol XXI. p. 213. Beyond the limit
129a132	arker's Book.-- M. Bichoffs Papers, in Edinburgh	NEW	Phil. Journ 1838. several case given of hot head
137a149	ns p. 125 to 129 & p. 135--160 & 162 [blank] Ed.	NEW.	Phil J. 1838. p. 72. on metallic vapours conden
137a151	some of its constituents into chert. [blank] Ed:	NEW.	Phil J. 1838. p. 132. «& 134» Bischoff. On the
152g046	terrace except at very head of valley indicates	NEW	terrace Ballivard 2 miles North of Grant town to
161g098	iver <the> of which the source is a lip with the	NEW	shelf flows into canal between L. Lochy & Oich.
162g102	ry (Nead Roy told me) it is impossible to see my	NEW	shelf, from road: Loch Ness 30.140. A 66 degree
170b001	dinary kind <the> which is a longer process, the	NEW	individual passing throug several stages (¿typic
171b003	n in rich soil, many kinds, are produced, though	NEW	individuals produced by buds are constant, hence
176b021	Hence Genera.-- «as many terminal buds dying, as	NEW	ones generated» There is nothing stranger in dea
179b032	hildren How is this in West Indies «--:Humboldt.	NEW	Spain:--» Dr. Smith always urges the distinct lo
182b047	nce of finches of Europe & America. &c &c &c The	NEW	system of Natural History will be to describe li
182b050	tween Australia & Van Diemen's land. & Austral &	NEW	Zealand Mr Gould says in sub-genera, they undoub
184b054	Mem. Rabbit of Falklands described by Q. & G. as	NEW	Species. Cuvier examined it. There certainly app
185b056	gos. Fernando Noronha Ophyressa bilineata (Gray)	NEW	<liza> species, belonging to true. American genu
186b061	stence definite without change, superinduced, or	NEW	species; therefore animals would perish, if ther
190b076	peculiarities of Flora. on East & West. ends of	NEW	Holland. diminishing towards centre (p. 586)-- P
192b089	thorhyncus. If this last animal bred-- might not	NEW	classes be brought into play.-- The father being
195b100	es it come wandering birds. such sandpipers. not	NEW	at Galapagos.-- did the creative force know that
196b107	rable to large quadrupeds--horse not large-- Ed.	NEW.	Philosop J. No 3. p. 207 "It is not generally k
198b115	d, (if act of fresh creation why not produced on	NEW	Zealand; if generated <No> «an» answer <could> «
199b116	pting by propagation that out of the thousand of	NEW	insects all belong to same types already establi
206b150	hink> States Cryptogam. Flora formerly common to	NEW	Holland?! p. 320. Says Coniferous structure inte
219b195	The Alpine plants of the Alps. must be <Alpine>	«NEW	formations» because snow formerly descended lowe
223b209	nd's account of Bolivian human species?-- Small	«NEW»	animal mentioned from Fernando Po Zoolog. Proce

Page
(Key Word)*
223b209 Po Zoolog. Proceedings October (?) 1837 Contrast NEW Zealand with Tasmania The reason why there is no
223b210 e animals do more than others, & cut off limbs & NEW ones are formed) but yet propagates varieties ac
225b219 My theory will make me deny the creation of any NEW quadruped since days of Didelphis in Stonefield:.
225b219 s what a difficulty-- where elevation Subsidence NEW is only hope.-- New Zealand «compare to Van Diem
225b219 - where elevation Subsidence New is only hope.-- NEW Zealand «compare to Van Diemen's land.» glorious
226b220 Coast of Africa. equally good.-- Small isld off NEW Guinea same fact see Coquille's Voyage.-- Galapa
226b221 to survive or chance having transported them to NEW station.-- When the new island splits & grows la
226b221 ving transported then to new station.-- When the NEW island splits & grows larger species are formed
232b249 same manner. Cuscus, (a sub genus of Phalangista NEW Holland form) is found in many island Celebes «W
233b249 erent species in different isld. (as far East as NEW Ireland. see Coquilles Voyage), Waterhous remark
234b255 [not located] T. Carlyle, saw with his own eyes. NEW gate. opening towards pig.-- latch on other side
235b274 ved if horses sent to India. & long bred in & no NEW ones introduced would not change be superinduced
240c013 Viverra Zibetha. ꟿ-- All the Moluccas, Waggious NEW Guinea. New Ireland, have phalangista, which dif
240c013 etha. ꟿ-- All the Moluccas, Waggious New Guinea. NEW Ireland, have phalangista, which differ in «form
240c013 h differ in «form & head &» colour from those of NEW Holland.-- The New Holland species are not found
240c013 head &» colour from those of New Holland.-- The NEW Holland species are not found in the Archipelago
240c013 ago-- Former statements to such effects false In NEW Guinea. a Kangaroo d'Aroe (Didelphis Brunii) whi
240c014 found in isle of Aroe & Solor) «Vol I» likewise NEW species of Parameles, which joined to Casoars, p
240c014 rroquets, establishes its «zoolog» alliance with NEW Holland. The Barbaroussa, (when young very like
241c014 larger islands.-- -- Antelope in Celebes, Bourou NEW species of Axis.-- «Cervus moluccensis is differ
241c015 irds of Australia. Many in common ¿species? with NEW Guinea.-- Many <genera>. kinds common to New Gui
241c015 ith New Guinea.-- Many <genera>. kinds common to NEW Guinea & rest of isle in E. Indi: Arch: In New Z
241c015 o New Guinea & rest of isle in E. Indi: Arch: In NEW Zealand. a sturnus of American form-- a Synallax
241c015 & 160 «162» list of some birds of Tongataboo. & NEW Ireland.-- Gould will hereafter know about birds
242c017 thecus common to Moluccas & Pelew Isds.-- p. 22. NEW Calidonia-- New Ireland & Britain same kind of d
242c017 Moluccas & Pelew Isds.-- p. 22. New Calidonia-- NEW Ireland & Britain same kind of dog, with those o
242c017 reland & Britain same kind of dog, with those of NEW S. Wales. <V.> p. 123 Crocodile at New Guinea. A
242c018 those of New S. Wales. <V.> p. 123 Crocodile at NEW Guinea. All the isles of Oceania have the Scincu
242c018 the lacerta vittata extends <to> from Amboina to NEW Ireland p. 23 Voyage of Coquille Lesson No (p. 2
243c019 & instances given) in East Ind. Arch:-- Birds of NEW Zealand absolutely different.-- --Philedon circi
243c019 Philedon circinnatus not found in Australia only NEW Zealand-- Norfolk. Isd. & New Caledonia peculiar
243c019 in Australia only New Zealand-- Norfolk. Isd. & NEW Caledonia peculiar species of cassicans: (¿cassi
243c020 ralian form? p. 27. many fish of Taiti found at <NEW> Isle of France: xx instance of wide range, wher
243c020 e range work this out-- L. Jenyns, about my fish NEW Zealand & New Holland fish very similar.-- NB. L
243c020 his out-- L. Jenyns, about my fish New Zealand & NEW Holland fish very similar.-- NB. Lesson method o
244c023 oes in North. Hemisphere.-- p. 158 Cuscus albus. NEW Ireland ---- maculatus -- Waigiou Speaking of Le
245c024 Bourous. the Barbyrousa; a Cervus near Marianus NEW, & some rats & mice. In Amboina only Cuscus & Ba
246c025 in any one isld different]CD.-- p. 414. dogs of NEW Zealand of large size, resemble, chien-loup.--lo
246c027 scadivora., which lives in the Eastern Moluccas, NEW Guinea.-- (Case of replacement)-- Coquille Voyag
246c027 The caswary, inhabits Ceram, Bourou & especially NEW Guinea (replaces, Emeu) in North of New Holland.
246c027 pecially New Guinea (replaces, Emeu) in North of NEW Holland.-- New Guinea scarcely differs more from
246c027 inea (replaces, Emeu) in North of New Holland.-- NEW Guinea scarcely differs more from, <Van Diemen's
247c028 f France Scincus multilineatus (p 45) Moluccas & NEW S. Wales Scincus Cyanurus «p 8 &». p 49 on all t
247c028 ncus Cyanurus «p 8 &». p 49 on all the Moluccas «NEW Guinea, New Ireland» & «even» Java. & very commo
247c028 s «p 8 &». p 49 on all the Moluccas «New Guinea, NEW Ireland» & «even» Java. & very common on Otaheit
248c029 ld have been transported. ¿What section does the NEW Zealand Rat belong to There is this great advant
249c035 l give me some capital information ¿Carnivora of NEW & Old word. do not form two sections is this foun
251c041 nduct of wild & tame horses.-- p. 246-- Gmnura-- NEW genus of Mam: found in Sumatra p. 452 Append to
255c053 your species her plan is frustrated or rather a NEW principle is brought to bear. If man created as
263c074 of Apterix-- split, depress & elevate & enlarge NEW Zealand; a division of nature of Apterix, many g
263c075 ion; then he will choose & firmly believe in his NEW faith of the lesser of the difficulties Once gra
265c083 ails cut yet there is no record of any effect.-- NEW Hollanders have gone on boring their noses. &c &
269c102 l. vol I. p. 17 &c excellent sketch of plants of NEW Holland, supplementary to Appendix to Flinders V
269c102 arrier-- Mem. Tartary & China.--, both coasts of NEW Holland.--«Compare birds of Australia with plan
272c109 is common to all the Laniadae & Muscicapidae of NEW World, but not found in Old-World--. + + If in a
275c119 ng geograph distribution of animals, <as> I use (NEW step in induction) as keystone of ancient geogra
277c127 least part.-- that he will not have brought home NEW species. until, he can show range & habits-- Tak
291c168 wheras in Falkland Isd they would change & make NEW species.-- alpine species being destroyed at Fal
294c175 e do. says Sheep could not live for some time at NEW York «instance of the fine relations of adaptati
295c183 one.-- Many African monkeys in Fernando Po-- no NEW forms only species!! No salamanders (D'orbigny R
296c184 w not Australian) on Peaks. Did Creator make all NEW yet forms like neighbouring Continent. This fact
297c186 en group few in number of kind, extermination.-- NEW forms made through probably an infinite number o
301c200 untry & collecting thousands & tens of thousands NEW insects, perhaps scarcely one new family & no ne
301c200 s of thousands New insects, perhaps scarcely one NEW family & no new orders,-- Wonderful, partly expl
301c200 ew insects, perhaps scarcely one new family & no NEW orders.-- Wonderful, partly explained on my theo
309c221 d sheep. heredetary proceeding from an accident. NEW England farmer,-- useful could not leap fences:-
312c228 s go against this, without «number of vertebrae» NEW acquisition, we must [not located] Henry Thompso
312c231 looks as if qualities were not permanent, in the NEW cross.-- In the Bantam clubs, they used to fix o
313c234 cal Transactions Why if louse created should not NEW genus have been made, & only species, good argum
314c238 Flora, N. Zealand & N. Caledonia with a dash of NEW Holland. As in N. Zealand-- Some species of Aust
314c239 Some species same (Palm & Phormium tenax) as in NEW Zealand & Australia, some SPECIES of Australian
315c241 species & some of Japan. I do not understand any NEW ones.-- Menoir will be published St. Petersburgh
316c246 + + What circumstances have led to formation of NEW species some few have been scattered over whole
317c251 d me. that near Charlestown ?three species, near NEW York. (600 miles N.?) replaced by three other sp
320c276 f Paraguay Dobrizhoffer. Abipom<e>nes. Edinburgh NEW. Phil Journal. about 13 numbers have been read V
324c268 Life June Is. King & FitzRoy To be read Humbold. NEW Spain-- Much about Castes &c Richardson's Faun.
326c266 Volumes well worth reading Bevans work on Bees, NEW Edit 1838 Harlaam. Physical & Medical Researches
335b013 not apply to first parents, because they are not NEW breed.-- the first hybrids may be compared to an
335b014 ant, as showing above facts as first cross being NEW species, ꟿ -- Are not dreadful monsters, abortiv
335b015 long in blood so will they remain, a mule «being NEW species» will have no tendency to have offspring
335b016 should have said, "an animal <acquires <th> any NEW> is <only> able to transmit «only» those peculia
338d025 ms from Northern part & not by fresh creation of NEW forms.-- what is range of Hyaena? Hippotamus.? I
339d025 E & W very different.-- Man not so, but N. & S. NEW Zealand & New +++ Caledonia. two races of Men, b
339d025 fferent.-- Man not so, but N. & S. New Zealand & NEW +++ Caledonia. two races of Men, but not plants»
341d030 ost of species from there now found in Australia NEW species of Moschus, characterized by Ogleby, who
342d033 have been shot wild (escaped from Carolina?) off NEW York. therefore instincts not imperfect.-- Are P
342d034 mogeneous? I obseeve Bachman calls these Hybrids NEW species. Yarrell says the bird fanciers say the
343d037 geographically divided either by developement of NEW forms in one., or apparently so. by the extincti
344d040 not being so.-- consider this with reference to "NEW species & hybrid doctrine"-- I have read there a
354d065 he Galapagos mouse probably transported like the NEW Zealand one-- It should be observed with what fa
355d067 vary? «See Cuvier Ossemens Fossiles» Although no NEW fact be elicited by these speculation even if pa
356d069 s of food: grazing animals who eat every species NEW.-- Sept. 8'. A Golden Pippen or Ribston do produ
357d073 believe, one or two were landed as at present in NEW Ireland & continent since grown.-- This will exp
364d104 of variation is soon reached-- as in pidgeons no NEW races.-- In Scandinavia besides the Rakhekna, be
369d115 traverse in order to quench their thirst"-- But NEW Guinea.--!! S. America.-- Such difficulties will
372d130 duced from the growth of one part, (not strictly NEW individual), or he may produced by having underg
397e003 depopulation, but extermination & production of NEW forms.-- their number & corelations Octob. 4th.
406e036 parent & sometimes other & sometimes ½ way. Ed. NEW-Phil. Transact. Rabies, common to men, dogs, hor
408e043 ncrease in number where then is the gap, for the NEW one to enter?-- The wonderful species of Galapag
408e046 prised to see how many Tropical genera come from NEW Holland, ¿Sydney? The dog being so much more int
413e059 Edit, p. 226.-- Herschel calls the appearance of NEW species. the mystery of mysteries. & has grand p
422e092 recent Bull; like fossil & recent shells of the <NEW> raised beaches» [not located] The enormous numb
422e095 although all perhaps will have done so from the NEW relations caused by the advancing complexity of
426e107 tained>: but the intervention of domesticated ie NEW varieties destroys the appearance of this series
436e135 w, probably would be most efficient in producing NEW species; also one being reduced in numbers, but
461t037 any structure would rather become accomodated to NEW circumstances than it would be eliminated, & hen
464t065 n has described a greatt Struthonidous Bird from NEW Zealand-- <so> not an Apteryx, yet it shows the
465t080 Mammalia in caves:-- :argue first case of bones (NEW Red Sandstone) & then go on to shells-- A profou

466t095 ital peculiarity improved.» Probably every such «NEW» quality becomes associated with some other, as
471tf08 ny idea" Linn. Trans. 18. p. 163. "D. Dod on two NEW genera of coniferae".-- referring to the 3 main
473s03r found remains of Sus with Elephants-- Lyell says NEW Red Sandstone of. N. America is Red Sandstone. &
500q10a ar copied Gould.-- Number of species of Birds in NEW Zealand, plants so few-- Range of mundane genera
500q10a fine doubtful species from Van Diemen's Land? or NEW Zealand? Babington about differences of Irish &
521m009 any subject when once started,-- could receive a NEW train through eyesight, though, not through hear
522m010 orbet, however, in conversation could catch up a NEW train if early association were called up.-- My
530m044 ther has somewhere heard (Hunter?) that pulse of NEW born babies of labouring classes are slower than
530m046 ,-- that is involuntary memory, as in sleep.-- a NEW thought arises?? compounded of the involuntary t
536m074 ove views would make a man a predestinarian of a NEW kind, because he would tend to be an atheist. Ma
540m086 land & the energetic copper coloured natives of NEW Zealand)-- the American in Brazil is under same
543m101 : Have Effect in Bones is valuable it shows that NEW instinct can originate.-- strong argument for br
547m113 ory of every late impression. & likewise gaining NEW ones from senses. & <comparing their> «calling u
548m115 t probable. in comparing every step, & inventing NEW means,-- therefore works of imagination hard wor
549m121 ether this rule of happiness agrees with that of NEW Testament is other question.-- little is there s
549m121 ect of living.-- or whether if we obey literally NEW Testament future life is almost the sole object-
558m151 yself". Analyse this out.-- bearing in mind many NEW relations from language.-- the social instinct m
565n010 ery cautious. Remember how Lavater ran away with NEW Lavaters,-- Ye Gods!:-- says fleshy lips denote
571n030 y be vitiated. or rather altered. The Reason why NEW Buildings look ugly is because there is some con
575n046 Vol II. p. 77) remembering things of youth, when NEW ideas will not enter. is something analogous. to
575n046 s.-- being lower faculty than the acquirement of NEW ideas.-- Walter Scott «(Antiquary)» Vol II p. 12
581n063 - & analogize it with ordinary habits that is my NEW part of the view.-- let the proof of heredetarin
584n072 ory no way applies.-- it is the acquirement of a NEW sense,-- bats avoiding strings «in the dark» as
586n079 carrier pidgeon just as wonderful in old bird as NEW.-- migration, <only> «only» more wonderful in yo
599o005 rlier--]CD in Athenaeum "Smart-- Beginning of a NEW School of Metaphysic,"-- give my doctrines about
606o024 rely to be added. Lessings Laocoon p. 125-- says NEW subjects are not fit for painter or sculpture, b
638j059 m some action.-- as well as gain it. by habit.-- NEW theory of instinct. returning to Kirby's view.--
640j167 ge is in progress or is, present with respect to NEW arrivers. the small body of species would far mo
451e176 atic Soc.. Vol V. p. 565. in a Paper by Lieut. NEWBOLD.--» A Malayan albino described "To this day th
234b256 ce in report of British Association of 1838. (NEWCASTLE) about somebody who had made great collection
203b138 r Yarrell says that old races when mingled with NEWER, hybrid variety partakes chiefly of the former
280c135 y old, they would not make hybrids, whereas two NEWER ones, even if more different might do so.-- <wh
640j167 inct Creation, how anomalous, that the smallest NEWEST, & most wretched isld should possess species t
251c040 sea, migrations of species, geese killed in NEWFOUNDLAND, with crops full of maize. (get limits of l
550m126 family can draw-- says friend viewed him as NEWFOUNDLAND dog would Greyhound about dread of water--
575n045 usic Audubon IV Vol of Ornith. Biog. case of NEWFOUNDLAND dogs. who will not enter water, till he see
489qIFC -- p 1 Owen p 17 Hooker p. 17 {T} Mrs. Whitby. NEWLANDS Lymington Hants. Habits of different caterpil
416e075 Italian Greyhound not so species every part of NEWLY acquired structure is fully practised & perfect
182b046 r, is not this Fries rule-- What subject has Mr NEWMAN the (7) Man studied The condition of every ani
534m063 dinary Habitat is Malva sylvestris. do. p. 228 NEWPORT says Dr Darwin mistaken in saying common wasp
315c243 o tell other animals in associated kinds of good NEWS. discovery of prey.-- arising no doubt from wan
565n007 ut laughing involuntary.-- When one fear any bad NEWS, «though in a letter» why is person painted wit
045r075 ions of Volcanos. (where there are no country NEWSPAPERS)--At the Calabrian earthquake things pitched
372d131 embryo the thousandth of inch should produce a NEWTON is often thought wonderful. it is part of same
022r007 ate of Lime disseminated through the great Plas NEWYDD do.--Mem tres Montes. ((Henslow Anglesea)) g
058r116 hili; Arequipa in 82 was overthrown, & 86. Lima. NEXT year Quito. considers these earthquakes travel
102a061 ly, first the more fusible substance, & then the NEXT being sucked out. In Cleavage discussion, state
191b081 sily transported-- Mem plants on Coral islets.-- NEXT to animals land birds.-- & life shorter or chan
223b208 fference is that there is wide gap between Man & NEXT animals in mind, more than in structures.-- If
236b279 y. (where Mr Murchison fox was found), decidedly NEXT species to some South American kinds.-- Are the
286c156 an constant number of stamens.-- in order, or in NEXT family? In considering fossil animals, what rel
293c172 more wonderful, that it should be remembered in NEXT generation. [NB what are those Marvellous cases
303c205 reated a perfect chain. <Icthyo> .+ + + supra & NEXT page It is a fact pregnant with SOMETHING.? tha
405e031 d over world as early as Elephants &c &.-- if in NEXT 20 years none of his remain found in the Americ
535m069 ileno says the mountains are as God made them,-- NEXT step plastic <virtue> natures. accounting for f
439e144 of the Leopard & Tiger together depended on some NICE qualifications each possess., & that tiger spri
578n054 t half & half. Miss F. A. said to Mrs. B. A. how NICE it would be if your son would marry Miss. O. B.
621o047 w that even this is possible.-- So that as there NICE & nasty in taste, & right & wrong in action, so
262c073 ing overrun-- Tahiti. thistle. Pampas. show how NICELY things adapted--.-- These «aberrant» varieties
289c160 ome countries-- nightingale do.-- all shows how NICELY adapted species to localities¶-- p. 390,. youn
131a135 . Observations on Mountains of the Moon. by Dr. NICHOL-- adduces the case to show Sir. J. Herschel's
072r160 ted to the Mexicans enormous masses of Iron and NICKEL, & these masses which are scattered over the s
077r172 do of Electricity Does not iron, combined with NICKEL & cobalt (meteoric) resist oxidation?-- Mem Si
266c092 pt.-- CIVETS CATS only wild animals on isld.-- NIETHER Hyenas; jackals monkeys-- common to either coa
240c013 Astrolabe Zoologie. p. 60. Vol I. Cynocephalus. NIGER. comes from the Moluccas «Matchian» & Celebes.-
596m184 are> right-handed??]CD> Cyanocephalus, Macacus. NIGER. Cercopithecus make labial st st. S. American m
266c091 earance & manners they are as opposite as day & NIGHT: yet we know how remote the periods at which bo
297c186 racles in Hemiptera do. p. 160. soft plumage of NIGHT jar. like owls. analogy in habits adaptation to
318c254 singly flying few miles every day «generally by NIGHT»-- other birds which is strictly diurnal, migr
318c254 s which is strictly diurnal, migrates singly by NIGHT.-- others in flock, these birds seem clearly di
341d032 th the ducks.-- most strange voice often in the NIGHT, like peacock.-- tail as long as Pea hen.-- abo
455eIBC imosa-- do stamina of C. Speciosus. collapse at NIGHT. if so irritate them, «as by an insect coming a
480z008 s some of the «species of» smaller petrels, are NIGHT birds agree. with <pe> nocturnal habits of Crus
528m036 one knows from seeing artificial lights in the NIGHT.-- from the mere exercise of the organ of sight
532m054 ple instinctive feeling: I have awakened in the NIGHT. being slightly unwell & felt so much afraid th
532m056 ot remain quiet in any room, would not sleep at NIGHT even when in bed room-- grew very thin, would n
565n007 h open to hear well «as one will perceive if in NIGHT trys to listen to growl of hounds». <when> as f
579n060 On origin of idea of causation; «succession of NIGHT & day does not give notion of cause,» do p. 135
586n080 ere to go-- the act of crossing the sea in dark NIGHT & not loosing its direction, equally wonderful
618o41v thus, for if day was first, we should not think NIGHT an effect.]CD Cause and effect has relation to
641j29v culloch says no other bird could catch mouse by NIGHT» Sailing lizards. squirrels & Opossums «& fish»
289c160 lden creted wren so rare in some countries-- NIGHTINGALE do.-- all shows how nicely adapted species t
363d102 ve & part acquired.-- thus Yarrel has Lark & NIGHTINGALE which both sing their own songs, though impe
318c254 irst & the females in flocks. «as in English NIGHTINGALES»-- other birds («& this seems common «kind»
350d054 ies as in two formations? by no way.?-- "Natura NIHIL agit frustra", as Sir Thomas Browne says "is th
525m026 ization which can hardly be doubted, when seeing NINA with her puppy.-- The common remark that fat me
532m056 t their minds were sound. Caroline tells me that NINA, when brought from Shrewsbury to Clayton, (thou
563n002 tempted to attack him from jealousy. (Pincher & NINA)-- or to take away food &c &c-- Now if dogs min
574n041 s, certainly»-- curious association: I have seen NINA licking her chops.-- someone has described slov
593n105 instinct <perverted> handed down & down.-- mem. NINA used to get into hay & make a nest for herself.
090a022 but are described as many, (one even 3000) This NINETY includes all actually counted.-- The weight «o
192b084 nt run wild in foreign country.-- When one sees NIPPLE on man's breast. one does not say some use, bu
319c257 ¿ north end of the Oural mountains) have black NIPPLES to their breasts.-- L' Institut, 1838, p. 230
331dIFC numerous as in common lion? Are the number of NIPPLES in domesticated very fertile animals increased
382d158 ent in every shade of perfection --How came it NIPPLES <are> «though» abortive, are so plain in Man,
434e127 ther the body of parent be altered, that is the NISUS formativus. (what does Müller call it) succeeds
496q006 double Crows-foot. or Ranunculus.= (11) Try.. NITRATE of Soda-- Salt. Gypsum. Magnesium Iron Rust Ca
505q014 um Perenne.-- Herbert's. fact.= (4) Effects of NITRATE of Soda under Beech.-- Lychnis dioica answers
560m156 ess.-- & sommambulism. Do people when inhaling NITROUS oxide, forget what they did when in this state
091a027 ast of Africa. Clay Slate & Quartz. strike SSW & NNE dip 30 degree - 80 degree Ed. N. Phil Journ. p.
200b121 cra, Coast of Africa. Clay slate. strike. SSW. & NNE. dip 30 degree - 80 degree (?).-- Ed. Phi.l.. N.
039r059 kes certainly have not been points of eruption. NOBODY supposes that all the dikes in Cornwall or in
297c186 r. like owls. analogy in habits adaptation to NOCTURNAL habits-- to cats &c.-- must be acquired by my
341d030 ruthios. was adaptation to little Movement.-- NOCTURNAL crawling bird.-- Wings reduced to rudiment.--
481z008 ler petrels, are night birds agree. with <pe> NOCTURNAL habits of Crustaceae Mr Broderip says that Vo
573n037 d when he meant to eat, hence assertion.-- but NODDING is less strongly marked than negation Marianne
164g119 different cause: from mud.-- [blank] {P} muddy NODULAR strata coral upside down strata coarse agglome
021r005 ate. major axis 2.¼ ft.-- singular structure of NODULE, constitution «same as» of slate same.--longer

Page **(Key Word)**
021r005 Fucus. P «Vol I 287» P 379. Henslow Anglesea, NODULES in Clay Slate. major axis 2.½ ft.-- singular s
021r005 ld round them; Quote this. Valparaiso Granitic NODULES in Gneiss. Epidote seems commonly to occur whe
134a142 on whale bones in Falklands Some of the Tosca NODULES at Bahia Blanca Mr. Malcolmson says are like K
482z012 gos!!! Azara. Voyage dans l'Amerique Merid. Tatu NOIR. abundant from Paraguay to 27 degree, then the
036r050 Hill. near Copiapò which is asserted to make a NOISE,--My impression. is not very distinct, from som
376d137 nd waist & look in their faces & Mak the st. st NOISE.-- The Cercopithecus chinensis: (or bonnet face
377d138 .-- Good evidence of knowledge of Woman-- ‖ The NOISE st st. which the C. Sphynx makes is also made b
377d138 by the C. porcarious., together with a grunting NOISE, the former signifies recognition with pleasure
377d139 stant.-- But he thinks other monkeys make st.-- NOISE‖ In case of woman instinctive desire may be sai
484z014 ing on legs-- habits-- Does the Secretary, make NOISE & throw head back M Edwards,--on polypi of Tubu
531m053 mal moves quickly away from any sudden sound or NOISE, & therefore brain has been accustomed to send
531m053 med to send a mandate to the muscles & when the NOISE comes it cannot help doing it.-- Fanny Hensleig
545m106 th passion, that is show all the teeth: «& make NOISE not like pish, but like chit-chit-chit, quickly
556m145 ey do not move muscle.-- reason CD[The laughing NOISE which C. Sphynx made at Z. Gardens may be descr
558m152 h prodigious force.-- making growling, guggling NOISE. Puma did same & & some others-- Thus <sudden>
565n006 ng in glass. & then as one ceases, or stops the NOISE , the face clearly passes into smile-- laugh lo
553m137 l Gardens) remarked that <exp> the expression & NOISES of monkeys go in groups. thus the pig-tailed b
063r133 & bred. no doubt with perfect success.--showing NON Creation does not bear upon solely adaptation of
074r165 Insist strongly on the grand fact of Volcanic & NON Volcanic. Then Solfataras. «Mem: Micaceous iron
115a096 ception of sandstone rare to have any horizontal NON cleaving beds. metamorphosed. The chemical actio
116a099 ection of Rio Negro beds.-- -- refers to species NON decrite de petites corbules analogue living in m
173b010 ter continues to vanish, bones instinct &c &c &c NON fertility of hybridity &c &c <assuming all> if s
180b039 (contrary to what would appear from America) of NON adaptation of circumstances.-- Vide two. pages b
181b044 Cuidado The above speculations are applicable to NON progressive development which certainly is the c
189b073 e plants:-- trace gradation between associated & NON associated animals.-- & the story will be comple
231b244 igin shown by only one species, far more than by NON-embedment of remains-- ¿agrees with non-blending
231b244 than by non-embedment of remains-- ¿agrees with NON-blending of languages?-- Till man acquired reaso
258c062 present) must follow after there is proof of the NON creation of animals.-- then argumen May be.-- su
284c149 s converse, <when> value of character depends on NON-variation, & not on extension ¿these go together
312c232 not ∵ instincts, constant. ¿ whether mutilations NON-heredetary & variations produced in short time i
415e070 additive, & not abortive: with reference to the NON-necessity of the «so called» progressive tendenc
416e071 id that true hermaphroditism is a consequence of NON-locomotion-- (contradicted by Plants). & as ther
419e085 una, must be very curious.-- With respect to the NON-development of Mollusca, which I have sometimes
492q002 rous, how then are seeds ever raised? 11. Is not NON-flowering gorze common in Norway No Questions re
510q017 lgel is he serpent man? about zones separated by NON-inhabited spaces: has he published? does he unde
510q018 Hilaires law of Balancement Wm Yarrell (1) About NON-breeding of animals in confinement, curious.-- f
611o034 animal matter been formed by the union of simple NON-organic matter, without action of vital laws-- A
620o045 death, one makes allowance & either excuses the «NON-» following of ones conscience. & palliates the
637j57v roductiveness, & laws of adaptation‖ p. 234. The NON-absorbing Camel's stomach is puzzler p. do says
066r140 ff. Christmas sound. -- «(Think some 60 fathoms, NONE thicker than thumb» Sea weed said at Kerguelen
090a024 «100» <5>0 lbs a year too little.-- How comes it NONE in fossil state? suppose «100» <5>0£ x 50,000 x
121a112 ia, & many shells in parts on surface, but I saw NONE embedded this point would be worth examining. t
121a112 to support. shells on surface of Patagonia, yet NONE in shingle beds. Lyell on Sweden. p. 12. proofs
146g013 rishable pebbles/ I am nearly certain there were NONE on surface of any hill Thursday On side of Hill
173b011 ch were late arrivals others old ones, (of which NONE of same kind had in interval arrived) might hav
196b108 re must be progressive development; for instance NONE--?, of the <vebtetrata> «vertebrates» could exi
199b120 but producing> «before domestication, afterwards NONE or little with <fertile offspring> » fertile of
222b202 whilst Crag <agrees with> according to Beck has NONE recent, yet genera same.-- Speculate on multipl
225b220 ogy of South Sea islds. any animals?-- I believe NONE.-- Canary islds.? Madeira? «Tristan d'Acunha?»
305c210 n the <classes> «orders» of insects, so is there NONE of reason in order of <ver> Mammals.-- Mem Elep
307c216 same kind» in these cases, have plain meaning & NONE in other case! Savigny has shown same fundament
335d013 herefore offspring will tend to go back, or have NONE-- the argument does not apply to first parents,
335d015 e parent, but as they must like or there will be NONE, therefore a mule can have no offspring.= but a
365d106 change, it would have been greater puzzle, than NONE, for the «e»normous time which it must have tak
390d179 whole series of forms to acquire differences: if NONE are added object failed, & then by that corelat
405e031 s early as Elephants &c &.-- if in next 20 years NONE of his remain found in the Americas probably di
411e055 so in S. America, however) is very remarkable & NONE discovered before them in any part of World.--
470t176 embered when in the gardens, he knew there was NONE but English,--the Palmated was introduced about
538m079 is remarked that A. Bessy repeated things, which NONE about her had EVER before heard, so very probab
540m087 nse labour of original inventive thought is that NONE of the ideas are habitual, nor recalled by obvi
608o026 is is important explanation) he thinks they have NONE.-- Effects.-- One must view a wrecked man, like
401e015 sown, all sorts come up from it. lately saw a NONPAREIL sowed by Mr Tollet so produce.-- thinks it pr
211b163 etuses of Vertebrata, &c 1837 p. 370 Owen says NONSENSE The distribut of big Animals in East Indian A
346d048 heir preserving character & breeding in & in-- NONSENSE a flock of more than 100.-- Agrees, «nearly»
403e023 e, now one part now another Macleay says it is NONSENSE to say take a tooth of any animal (as Toxodon
530m044 mething about big noses & name Corbet, perhaps NONSENSE.-- look to it My father has somewhere heard (
534m061 Hensleigh says to say. Brain per se thinks is NONSENSE; yet who will venture to say germ within eg,
628o54v ssions have no relation I think this <boshes> «NONSENSE»-- My theory of durableness will explain it.-
044r071 a on East») cannot believe in a great explosion, NOR would sea remove more internally than externally
072r159 nects western isles of Scotland & Iceland.--Bosh NOR on Norway, or Spitzbergen.--Spitzbergen animals
079r177 do do Australia, C. of Good Hope.--Azores Isds «NOR at St Helena.--» Humboldt. New Spain Vol. IV. «p
100a052 plicable. it does not explain CLEAVAGE of rock-- NOR the Falkland case, nor. the arrangement of parti
100a052 xplain CLEAVAGE of rock-- nor the Falkland case, NOR. the arrangement of particles of granite in Hens
158g079 of this point very nearly like head of Glen Guoy NOR is horizontal line apparently continuation of up
171b002 annuals rendered perennial. &c &c.-- Yet Eunuchs NOR «cut» Stallions nor nuns are longer lived Why is
171b002 ennial. &c &c.-- Yet Eunuchs nor «cut» Stallions NOR nuns are longer lived Why is life short, Why suc
202b132 Timor, herded separate from the English cattle, NOR could we get them to associate together" There i
240c013 alkner Patagonia no description of wild animals, NOR in Dobrizhoffer Abipones.-- Voyage. de L'Astrola
284c148 ts separated at high water.-- not other islands, NOR any any other part of world.-- no other plants p
343d039 appearance of sudden termination of existence.-- NOR is there in the Tertiary <older> geological epoc
417e079 ve» their eggs, <inter>, impregnated externally; NOR can it be a necessary concomitant, with moths, w
467t099 en.-- As we see in Hybrids that although anther «NOR filaments» shrivel, yet stigma does not, so we m
535m064 ll larva excluded, then though not feeding them «NOR helping larva from egg» watching them, brooding
540m087 thought is that none of the ideas are habitual, NOR recalled by obvious associations. as by reading
541m092 How can people dwell on pain ¿ no definite idea. NOR is an emotion.-- People who can multiply large n
542m094 rying-- peculiar not common?--» no bark of anger NOR have monkeys & many other animals,-- but yet whe
547m114 o improbabilities in a dream, effect of doubting NOR believing, effect of not reasoning. effect of no
548m115 es infancy.-- one cannot bring it to one self.-- NOR of a bad dream, when that is not recollected, no
548m115 or of a bad dream, when that is not recollected, NOR of the Botanical Somnambulist. (if he had been u
549m123 s happening with other animals-- is far from odd NOR is it odd he should have had them.-- with lesser
557m147 ng-- if as (I believe) Hunter says. neither fox. NOR wolf wag their tails, &c. it is very curious, re
576n047 nt, & wished he had lost his life in doing so.-- NOR would he regret «having acquired» this sense of
582n068 creature performing those actions neither knows NOR intends the result they will effect,.--" this no
601o009 memory added to it, man in sleep not conscious, NOR child-- Evidence of consciousness, <t>" movements
608o027 ng. (yet one takes it for beauty & good temper), NOR ought one to blame others.-- This view will not
615o36v leasure; for it was different way of showing it, NOR was there any cause, & if surprise was felt.-- a
111a088 ntroduced about twenty years since (1835) from NORFOLK Isd into Geograph Journal Vol VII p. 279. Carc
217b187 Fuego.-- Araucaria, species. Brazil {P} Chile, NORFOLK Isl.-- Isle of Pines-- Australia.-- A <South>
243c019 atus not found in Australia only New Zealand-- NORFOLK. Isd. & New Caledonia peculiar species of cass
267c093 arcely elapsed since the Guava introduced from NORFOLK Isld-- "& it now claims all the moist & fertil
465t089 Phil. Mag. May. 1840 p. 362.-- some Mammals of NORFOLK Crag. mentioned-- allied Beaver to present for
514q21. vated Orchis & Asclepias &-- carnosa?-- good-- NORFOLK Isd-- geology. volcanic? Applies to my geology
528m038 m & symmetry, of forms-- the beauty of some as NORFOLK Isd fir shows this, or sea weed, &c &c-- this
314c238 wever it may be effected.-- Prodromus Florae NORFOLKICAE. 1833 Steph. Endlicker (He will give sketch
153g052 umulations At gentler bends roads disappear The NORMAL condition of 4th shelf, some way below House o
292c169 es or two animals:-- «When» sexes <being in one NORMAL-- when so the> are united (which probably is f
318c254 line or bands of country (These facts show the NORMAL condition of Migration) gradually separated th
515q021 remarks, where parts of flower are reduced from NORMAL number, they are apt to vary in number in indi

```
535m064 appeared healthy-- This remarkable case may be NORMAL. with insects, but habit forgotten in all olde
591n099 (where  sensations of individual are same as in NORMAL cases) are held in abhorrence it is because in
185b056 nged into Cara cara at the Galapagos. Fernando NORONHA Ophyressa bilineata (Gray) new <liza> species,
463t059 -- Waterhouse says perhaps animals of Fernando NORONHA are found unknown coast in front of it.-- Cuvi
050r091 n to sand?? B. Roussin states that generally in NORTH part of Brazil. <gravel becomes> sand less & gr
054r105 reached. Juan. Galapagos. Cocos-- Ulloas voyage NORTH of Callao, the country, to the distance of 3 or
054r105 icularly observable in a bay about five leagues NORTH of Callao, called Marques, where in all appeara
055r106 Chapt  VIII. p. 97 at Potosi the veins run from NORTH <inclining> to South. inclining a little to the
077r171 rocks. Vein traverses both Clay slate, Porphyry NORTH 52 W, & is nearly the same with that of the vet
128a130 ry strata on coast of do-- I believe?? coast of NORTH America., like the Mexican Gulf. is fouled by b
152g047 valley  indicates new terrace Ballivard 2 miles NORTH of Grant town to Forrest road comminuted shells
188b066 m. Shew Mice.-- --- Animals common to South and NORTH America.-- ¿are there any? Rhinoceros peculiar
195b098 with plumage «&» tone of voice partly American NORTH & South.-- (& geographical <distri> division ar
200b125 325.   July 1828. Animal now confined to extreme NORTH.-- ||.do p. 326. 2 Fossil species of ox in N. Am
209b157 a I; 2,3: Mem. Lyell on shells.-- {T} Genera In NORTH Africa. I: 4,2 Iles Canaries I: I,46 St. Helena
214b174 e? Phillips. Lardner p. 289 It is certain, that NORTH American fossils bear the closest relation to t
226b221 a of <S. America> Tierra del Fuego like that of NORTH Europe, many genera & few species. The number o
232b247 it would have possessed a most peculiar Flora «& NORTH of Europe»-- As European forms have travelled
232b247 Equator,  <th> so would the plants from extreme NORTH, which according to all analogy would have been
233b251 tudy Productions. of great Fresh water lakes of NORTH America If Parasite different, whilst man & his
244c022 ation with equator--that Vesp. lasiurus does in NORTH. Hemisphere.-- p. 158 Cuscus alsia. New Ireland
246c027 rou & especially New Guinea (replaces, Emeu) in NORTH of New Holland.-- New Guinea scarcely differs m
249c036 forms peculiar to word to special districts???? NORTH of 30 degree.--, may be connected with, Mr Blyt
271c105 anicus. in the. S. Hemisphere. (replaced to the NORTH by other species.--) should build a nest  lined
295c183 rt. p. 11) in S. America so highly developed in NORTH.-- Ichthiology of S. America. more peculiar than
296c184 animals.  & land shells.-- all in short Extreme NORTH = = to peak of Teyde in relation to surrounding
318c252 of adaptation.-- (case of Squirrel from extreme NORTH turning white like Hares??--) I never saw  more
319c257 of Man at Maer, it is said the Samoyed women (¿ NORTH end of the Oural mountains) have black  nipples
337d022 Spence  remarks that the Fringilla domestica of NORTH Europe is replaced by the F. cisalpina in Italy
353d061 glish Hare.-- good case these hares compared to NORTH American hares. Many species, separated by Moun
363d102 ttered over Europe)-- The habits of some «same» NORTH American & Europaean birds «slight» different--
406e037 far  S. in Northern. Hemisphere.-- likewise far NORTH in Southern.-- Great animals. of same two great
406e037 e two great orders destroyed about same time in NORTH & South. America.-- Whole wor[1]d, formerly pos
407e037 at  present days,, which S. America now does to NORTH. America & Europe.-- S. America favourable to T
407e038 l» America from the «low» limits of blocks both NORTH & South, has probably undergone a greater chang
410e051 respect  to representative species., when going N.ORTH & South Thinking of effects of my theory, laws
424e100 s.-- March 5th. Lyell says «fossil» shells from NORTH America, Scotland, Uddevalla. Many species same
424e102 e in Europe different ranges-- latter not going NORTH of the Elbe.-- yet they meet in one wood in Anh
427e109 etreat: effect on snow of arctic climate in far NORTH regions? Arctic forms have travelled S. From th
478z003 ice on habit of Iguana. not pass Lat. 28 degree NORTH p. 239 In ocean between Lat 56 degree and 57 de
483z013 are  described Shells from Tahiti and Chile The NORTH & S. Range of shells might perhaps be worked ou
482z017 uvicolae?-- The Birds seem to move much further NORTH on West coast of S. America. than on East.-- no
485z018 extreme heat, the tropical forms extend further NORTH, because during winter they can bear the cold w
510q017 as  between Australia & S. America? Sabine says NORTH of Siberia, no sea-current, icebergs travel  by
515q021 lionaceous order (2) History of fruit trees far NORTH in Scotland-- do they flower-- do they live hea
540m085 the American whether in the cold regions of the NORTH,-- the elevated table land of Peru the hot plai
028r031 ain, because alluvium saline; Mem: on coast of NORTHERN Chili as springs become rarer. so does the ra
088a015 a  Hawk fly at Heron.-- Ceratophytes common in NORTHERN seas p. 312. Chamisso in Kotzebue. Study Humb
101a056 ood discussion showing present form of land in NORTHERN England influence dispersion of Boulders.-- S
188b067 an archipelago Dr Smith considers probable two NORTHERN species replace <No> Southern kinds-- (I) Gnu
190b076 ree-35 degree, source of forms. reduce towards NORTHERN Eastern end & die away, & partake of Indian c
217b188 d no species to itself, a remark common to all NORTHERN islds.-- This is interesting, because Iceland
217b188 ves room to fine speculation.-- Are there many NORTHERN genera peculiar to itself-- on hybrids betwee
219b195 therefore  species of lower genera altered. or NORTHERN plants «No» CD[Mem. the antarctic flora must
229b233 elieves is mentioned by old writers on extreme NORTHERN Coast. Hippopotamus do.-- Giraffe do.-- Range
232b247 change  from insular to extreme climate, <more NORTHERN> Iceland would have possessed a most peculiar
247c028 l age-- Gastrobranchus «only» 2 species one in NORTHERN Hemisphere 2d in southern --p. 71 Chimera-- A
274c116 s in common.,-- but each several with Europe & NORTHERN Asia & Northern America.-- may we not look to
274c116 but each several with Europe & northern Asia & NORTHERN America.-- may we not look to these  Northern
274c116 & Northern America.-- may we not look to these NORTHERN regions as the receptacles of the wanderers o
338d025 nce is by the extinction of certain forms from NORTHERN part & not by fresh creation of new  forms.--
365d106 s climate became unfit for. subalpina, or some NORTHERN species, & being restricted species has  been
402e020 ia» have passed to Paragua & in Luçon the most NORTHERN. of the group the number is limited of the gro
406e037 l. I show erratic blocks transported far S. in NORTHERN. Hemisphere.-- likewise far North in Southern
422e092 nsult this passage, when considering origin of NORTHERN Cetaceae).-- -- ||.do. p. 318 M. Pictet of wri
424e100 ica, Scotland, Uddevalla. Many species same. & NORTHERN forms-- & American ones & Europaean-- agree v
429e116 se these points were last connected with those NORTHERN regions-- do p. 21 says. many plants skirt ea
437e138 drop off their perch.-- p. 101-- Kingfisher in NORTHERN part of England stationary, in southern stays
441e151 s show to be absurd.-- As there are plants, in NORTHERN latitudes, which are generated by buds  alone
442e152 nge is produced.-- The fact just alluded to of NORTHERN flowers, throws enormous difficulty in the wa
446e161 ar as vegetation is concerned, in parts of the NORTHERN «French» expedition,-- rather the reverse of
450e174 Tiger),  &c are found but only in one part the NORTHERN peninsula of Borneo.-- Ox & hog natives of Bo
479z006 's Voyage Vol I, p. 168 Ceratophytes common in NORTHERN sea. Chamisso in Kotzebue p. 312 Leaches on l
484z014 ed comparison of production without Tropics in NORTHERN & Southern America-- valuable & practicable d
484z015 e Penguins replace the <Auk> Guillemost of the NORTHERN Hemisphere, & the Puffinuria, the Awks.-- Wha
029r034 . Caermarthen Journal I look at the cessation NORTHWARDS of the Coal in Chili as clearly bearing a re
316c247 here continent between N. America & Europe?-- ||.NORTON has written on fossils of N. America.-- At the
072r159 stern isles of Scotland & Iceland.--Bosh nor on NORWAY, or Spitzbergen.--Spitzbergen animals (?). ≠ T
076r168 re, that are found in the veins of Kongsberg in NORWAY.--rain dendritic silver intersecting carbona
104a068 hat <th> gneiss, mica-slate of whole kingdom of NORWAY was contorted yet no mountain chain case paral
110a086 lines» Description of rocks in Lyells'. Capital NORWAY case.-- The fragment. consisted of hornblende
115a097 land Islds p. 94. Von Buch's Travels account of NORWAY chain being broken through like that near-- Ob
429e153 lpine plants of & Pyrennees agree with those of NORWAY. Lapland & Greenland, but not with those Kamts
442e152 y stop for any number of generations-- Gorze in NORWAY, which never flowers!!-- <How did it get there
442e153 should be absolutely similar; [all the gorze in NORWAY ought to be thus characterized study Von Buch.
492q002 aised? 11. Is not non-flowering gorze common in NORWAY No Questions regarding Breeding of Animals  If
212b165 for long time together in tub of water with only NOSE projecting.-- would pull the garden bell, & the
233b250 s.-- Australian dog jumped into tub leaving only NOSE above it-- pulled bell.-- -- It was most curiou
342d035 been  featured by the Austrian lip & the Bourbon NOSE". if this be not imagination.-- then old peculi
505q014 day's puppy of small true Bull-Dog-- length from NOSE over head to root of tail 28½. inches. From sol
554m140 anything  of any sort.-- I saw Tommy picking his NOSE with «a» straw.-- Jenny will often do a  thing,
565n009 Lavater says derision lies in wrinkles about the NOSE, & arrogance in upper lip. <The> Children havin
575n045 - tears flow from both, as when one burns end of NOSE with a hot razor.-- joy <p> a mental  pleasure.
265c083 ect.-- New Hollanders have gone on boring their NOSES. &c & This congenital changes show that grandso
529n042 ll of ideotcy.-- The story of the Corbets & big NOSES, quite conjectural, in Blakeways book of Sherif
530m044 n Antiquities of Shrewsbury something about big NOSES & name Corbet, perhaps nonsense.-- look to it M
265c086 ng it with very different habits-- Thus bill & NOSTRIL of Puffinuria I think we may clearly attribute
558m153 e is a curious drawing out of the side part of NOSTRIL, when passion commences.-- <All> Nearly all wi
638j281 stomach  hump, kinds of foot. power of closing NOSTRIL, foot, sack. power of endurance &c &c  Camels?
536m071 ses, animals of different orders turn up their NOSTRILS when excited by love? Stallion licking udders
592n103 ses: <do> Horse prick his ears «& snort clears NOSTRILS» when frightened, does not hair & rabbit depr
151g039 lso, without any cause, must be tides. &c. {P} <NOTCH> roads very much this character.-- The boulders
156g069 fallen.  «on the 3 shelves} Solid rock is much NOTCHED on Maculloch's supposition;-- the old  ravine,
028r030 These  reflections might be introduced either in NOTE in Coral Paper or hypothetical origin of some s
036r050 t my recollection is imperfect & was recalled by NOTE in Daubeny. P. 438., of similar fact near the R
085a FC nge of Sharks Nothing For any Purpose A. Geology NOTE on Woolwich Nothing on any Subject As far as p.
087a010 the Megatherium. -- The Guanaco the Camel.? Make NOTE about N. American bone not probably in salt mar
112a088 Caldcleughs collection of facts See page 101. in NOTE Book (C) for some speculats on conducting power
```

Page   **************************************(Key Word)**

139a1BC fect of superadded vital influence?-- See End of NOTE Book. called R. N.-- Massac[h]usset would be we
202b131 p the Massaroony by W. Hillhouse.-- Demerara. In NOTE. Demerara. 10 «12» feet beneath surface forest
247c028 & very common on Otaheite-- according «stated in NOTE to p 21» to Quoy & Gaimard in Sandwich isld. &
247c028 emen's land & Cape of Good Hope V. p. 44 of this NOTE Book Rabbits introduced in 64, of very many col
293c172 anomaly in structure of brain not probable) put NOTE. Sir W. Scott has written about it]CD If we saw
298c189 ading question.-- [+] This might be mentioned in NOTE.-- try to trace from simplest reasoning in lowe
315c240 ce disseminated See. Habits of Malay fowls p 5. (NOTE) in some papers on instincts L.' Institut «1838
338d025 with colour. black being strongest.-- V. p. 63. NOTE Book M'. for case of change in food in insects
343d039 respect to the Deluge it may be worth adding in NOTE than amongst the Mammalia of Europe the shells
345d043 f «in both cases» is killed. Notes from Glen Roy NOTE Book.-- Why is not Tetrao Scoticus. an american
349d052 between permanent variety & species!! (given in NOTE.)-- Macleay <met> uses term genus when it is so
350d054 emb l,. It has been argued Man first civilized. <NOTE> add this in note. ¿mere conjecture?-- Australi
350d054 n argued Man first civilized. <note> add this in NOTE. ¿mere conjecture?-- Australians.-- Americans.
472s01r e year (1827), always been empty.-- See separate NOTE-- Elizabeth says several years ago seeds were p
486z019 ects of Europe have Tropical Forms See p. 256 of NOTE Book (C) for comparison of singing powers of bi
120a110 raigs well worthy of attention-- rear Glen Roy NOTEBOOK-- & scraps on Salsisbury Craigs. Kept amongst
031r038 24 V. back of page 1 of New Zealand Geological NOTES. at St. Helena. This structure was very clear a
302c202 must be so from characters of analogy.-- see my NOTES on p. 37. of Macleay. wonderfully accordant. wi
319c255 er thrushes-- yet they have one with very sweet NOTES.-- Their soft-billed birds are inferior to ours
322c270 o Medici Lyell Book III There are many marginal NOTES <Rengger &c> Mitchell's Australia Walter Scotts
323c269 ature-- -- Herbert on Hybrid Mixtures: Marginal NOTES. -- 20th. Carlyle's French Revolution 3? vols.
331d1FC h solid feet.-- 1838 [In this Book some curious NOTES on Monkeys recognising Sexes of animals:]CD [Al
345d043 ack lip) & then calf «in both cases» is killed. NOTES from Glen Roy Note Book.-- Why is not Tetrao Sc
362d099 in S. America with C. Campestris <)> refer to my NOTES) & Mr Yarrell supposes this a consequence of th
493q003 lation with species-- answered «by Henslow» see NOTES In varieties is there any difference in off spr
578n057 essay on the sublime & Beautiful there are some NOTES. & likewise on Wordsworth's dissertation on Poe
589n091 cts, & mingled with much hypothesis.-- see M.S. NOTES, where strong argument in favour of brain formi
593n107 ed before language was invented,-- were musical NOTES the language of passion & hence does music now
599o004 s trembling palsy? Expressions N [Old & USELESS NOTES about the moral sense & some metaphysical point
626o052 w given on conscience: I cannot admit it.-- see NOTES to it by me..' 'p. 333 «& p. 377» some remarks.
023r013 ture of hill; states could discover no shells: NOTHING said about K. Georges Sound The idea of the wa
060r125 tion above regions of vegetation.--«I can find NOTHING odd to find them injected by veins & mass[es]
068r147 layers of igneous rock replace strata. & it is NOTHING For any Purpose A. Geology Note on Woolwich No
081r BC imetre 4.4 {T} C. Darwin R. N. Range of Sharks NOTHING For any Purpose A. Geology Note on Woolwich
085a FC ng For any Purpose A. Geology Note on Woolwich NOTHING on any Subject As far as p. 33. distributed to
087a010 one not probably in salt marshes Efflorescence NOTHING -- Study account.-- Alluvial plains of Mississ
123a115 68. Paper by Humboldt on Bogota. Cordillera,-- NOTHING.-- salt & coal near Bogota; p 270.-- SPLENDID
147g020 few very small & irregular hills of alluvium-- NOTHING very striking yet possibly sea more probably t
162g099 t measurements belong are so complicated, that NOTHING can be made out of them-- but it may be said t
175b020 propagated recent terebratula, but Megatherium NOTHING. We may look at Megatheria, armadillos & sloth
176b022 al buds dying, as new ones generated» There is NOTHING stranger in death of species, than individuals
186b061 ; therefore animals would perish, if there was NOTHING in country to superinduce a change? Seeing ani
195b104 resh creations is mere assumption, it explains NOTHING further, points gained if any facts are connec
197b112 s «also» organs of lower animals appear.-- yet NOTHING about propagation= I see nothing like grandfat
197b112 appear.-- yet nothing about propagation= I see NOTHING like grandfather of Mammalia & birds &c 〚 p. 3
219b196 as we can recognize, may be thought to explain NOTHING.-- it being as easy to produce «for the creato
223b207 more different probably & introduction of Man. NOTHING compared to the first thinking being. although
258c061 do. p. 434. Table of birds from Cuba Vigors.-- NOTHING of much interest XX. Hence relation of analogy
278c129 alogys.-- If genus does not Mean this it means NOTHING.--There should be some term used, when there i
283c146 Infusoria, of same forms with recent, we have NOTHING to do with CREATION.-- <On> The end of formati
289c161 habits My definition <in wild> of species. has NOTHING to do with hydridity,, is simply, an instincti
303c205 who say «philosophically to a certain extent,-- NOTHING but experience. will, tell us. when group is t
305c209 if distinct creation.-- ie.-- a mere statement NOTHING is explained.-- this is fact analogous to mock
308c218 ly no natural starting place, because there is NOTHING more elementary than that complex nature itsel
321c270 dom of> Lisiansky's Voyage round world. 1803-6 NOTHING Lyells Elements of Geology Gibbons life on him
350d055 ese facts prove that perfection of organs have NOTHING to do with perfection of individual, though su
350d056 re there is some error-- Observed, nature does NOTHING in vain, therfore organ fitted to animals plac
351d056 does-- does not move per saltum-- yet does NOTHING in vain!! Foetus of man undergoes metamorphosi
372d130 w, & therefore to repair wounds-- but this has NOTHING to do with generation.-- Why crab can produce
409e048 o say only cause, but one great final cause,-- NOTHING probably exists for one cause» of sexes «in se
432e125 ebbles &c &c: the succession of organisms tell NOTHING about length of time, only order of succession
509q017 ulls. is it Domesticated African Animal= Knows NOTHING [<...>] It is very important to know, whether
513q21. of fruit. cross Conception--(〚 I could extract NOTHING from him)〛 Does impregnation ever regularly ta
522m011 , he knew the whole subject.-" if at the other NOTHING.-- He could repeat the alphabet straight, but
532m054 ugh my reason was laughing & told me there was NOTHING, & tried to seize hold of objects to be fright
541m093 to look tranquil.-- He may despise a man & say NOTHING, but without a most distinct will, he will fin
546m109 rent being fruit eater.-- origin of colours?-- NOTHING shows one how little happiness depends on the
557m149 en mouthed to hear. though in individual case. NOTHING can be heard.-- Shame would never make person
569n023 thout, however, very sincere grief-- "there is NOTHING more to be said."-- "made no reply, but shrugg
570n024 first." the man shrugs his shoulders & replies NOTHING. if he did go to reply. he would throw back hi
573n039 uct of Mrs C. (her brother's wife). & she said NOTHING but shrugged her shoulders.-- analyse this.--
591n098 n is extraordinarily cowardly.-- the other one NOTHING will frighten-- hence variation in character i
594n113 compressed sullen, protruded. determined to do NOTHING. & so manifesting sulleness. [blank] Circumsta
608o026 way to aid motive power.--if incorrigably bad NOTHING will cure him' 3) disgusted. with them. Yet it
608o028 or retribution.-- it may be a consequence but NOTHING further.-- October <8> 2d. 1838 Those emotions
618o041 a actually apprehensible by sense. 5) There is NOTHING analogous to this in the relation of thought,
623o050 ts) « <I cannot> » [LHC] On the Law of Utility NOTHING but that which has beneficial tendency through
635j55r «& is therefore utterly useless-- it foretells NOTHING because we know nothing of the will of the De
635j55r seless-- it foretells nothing» because we know NOTHING of the will of the Deity. how it acts & whethe
234b256 gs.-- birds mentioned. but few.-- CD [There was NOTICE in report of British Association of 1838. (New
387d169 erit./ Annals of Natural History .p. 96. Vol I. NOTICE the Syngnathus, or Pipe fish the male of which
478z003 only appear in winter in Paraguay p 207 Slight NOTICE on habit of Iguana. not pass Lat. 28 degree No
402e018 ar Canton" "Here, as in all Malay countries, I NOTICED a peculiarity in the cats «p 10» the joints ne
450e175 J. H. Moore after quitting Bengal this fact is NOTICED in Cassay Ava Pegue-- seldom equals 13 hands--
495q05a within & see which way they fly.-- (9) I have NOTICED leaves covered with Honey-dew dusted with poll
536m070 at» remember how Pincher does just the same; I NOTICED this by perceiving myself skipping when wantin
128a130 Academy (at Athenaeum.) I. Part. I Vol.-- some NOTICES on modern Tertiary strata on coast of do-- I b
450e174 mous-- change of instinct by domestication.-- "NOTICES of the Indian Archipelago" Published at Singap
450e175 insula of Borneo.-- Ox & hog vertues of Borneo NOTICES of Indian Arch. Singapore 1837. By J. H. ? do.
604o017 Book.-- Sympathy & affections chiefly fail.-- NOTICES. struggle <between> when insanity is coming on
026r021 g. Trans: Cornwall «Vol II» It is a fact worth NOTICING that cryst of glassy felspar in Phonolite arr
086a005 e land, stable as the water"-- It may be worth NOTICING edentates & camels in deserts & rodentia In P
261c071 n form has changed. Can be said that animals no NOTION of beauty, when does prefer most powerful buck
288c159 rarely bark.-- are almost useless not the least NOTION of hunting, or keeping watch. how completely «
513q21. many flowers are dichogamous Zostera-- Knights NOTION of pollen & stigma generally not being mature
527m032 ps account for our acquiring «the instinct» our NOTION of beauty & negroes another; but it does not e
529m040 the fertility.-- I a geologist have illdefined NOTION of land covered with ocean, former animals, sl
550m126 Mayo. Philosop. of Living p. 293. Animals "have NOTION of property" -- their own property. (--regardi
550m126 in birds of place for nest.)-- with dogs "have NOTION of masters property"-- is not this rather more
553m135 a swift dog to overtake an emu, & [not located] NOTION, are not effects of impressions long repeated,
558m151 confused idea of "ought." joined with necessary NOTION of "causation", in reference to this "ought,"
564m004 ry of future thinking of injured moral sense.-- NOTION of deity effect of reason acting on (<not soci
567n012 l.-- Origin of cause & effect being a necessary NOTION is it connected with <our> the willing of the
567n013 ng direct effect of some law.-- have plants any NOTION of cause & effect, «they have habitual action.
567n013 hich depends on such confidence" when does such NOTION commence?-- Children understand before they ca
568n016 ing something like man. Has an oyster necessary NOTION of space-- plant though it moves doubtless has
568n016 triangle shorter than third. is this necessary NOTION, ass has it.-- When one is «simply» habituated
572n035 ion the very same as mine about our origin of a NOTION of a Deity We can allow «satellites», planets,

579n060 ation; «succession of night & day does not give NOTION of cause,» do p. 135.-- on the importance of a
602o11v a &c a Poussin.-- How are my ideas of a general NOTION of everything applicable to the high idea «p.
604o018 From these & other reasons we apply to God the NOTION of living in lofty regions. 3 Infinity eternit
211b161 gy to production of species.-- Animals have no NOTIONS of beauty, therefore instinctive feelings agai
566n011 st bursting from egg.-- Animals have necessary NOTIONS. which of them? & curiosity «strongly shewn in
566n011 Animals do not know they have 'these necessary NOTIONS any more than «a» Savage M. Le Comte's idea of
568n016 rom learning it by heart. Do not our necessary NOTIONS follow as consequences on habitual or instinct
571n028 o all men. agree ultimately?-- We acquire many NOTIONS unconsciously, without abstracting them & reas
571n028 ur heredetary nature thus acquire some general NOTIONS, which are taste? Real taste in mouth, accordi
572n033 r movement.-- say instinctive actions. senses. NOTIONS.&c Octob 30th-- Dreamt somebody gave me a book
230b235 ght to have a gradation of species, now this NOTORIOUSLY is not the case, you have stunted species, b
603o013 words «are abbreviations» he thus derives from NOUNS & verbs-- so that much of EVERY language shows
641j29r s-- Marine ones «of large size» <to> are best NOURISHED by arctic regions-- Whales. «Narwhal» Polar b
244c023 , et forts surtout de l'opinion du baron Cuvier, NOUS ne balançons pas a la regarder comme une espèce
257c056 species & found "beaucoup des mêmes oiseaux. que NOUS avions déjà observés en Patagonie. ou au moins
200b121 ree - 80 degree (?).-- Ed. Phi.l.. N. J. p. 410 <NOV> 1828 It is daily happening; that naturalist des
463t105 Mammalia in cases such as that of Java & Sumatra NOV 15th Waterhouse showed me the component vertebra
338d024 m. p. 505. some (very poor account) of plants of NOVA Zenbla -- in review of Baers work Edinburgh. Ro
307c215 of same tendency in species, this is capital & NOVEL argument.-- Are there any abortive organs in ne
541m091 erely thinking intently; for that one does with NOVEL for a length of time.-- Then if one endeavur to
294c171 s. that first shee» State broadly scarcely any NOVELTY in my theory, only slight differences. «the op
528m037 common to every kind of view-- as likewise is NOVELTY of view even old one. every time one looks at
544m103 lude that neither number, vividness, rapidity, NOVELTY of separate ideas cause fatigue to the mind,--
070r155 range. Signor Rozales tells me at seven oclock NOVEM <5th> Concepcion most violently shaken, by eart
093a033 . <fl> Flustra identical. recent & bone bed.-- NOVEMBER 8th 1877 (Memoranda so far distributed to var
250c040 riods.-- p. 125. ref. to Phil Transacts. (read NOVEMBER 20th) Paper by Jenner, on birds seen far at s
406e037 aster.-- dogs when strayed hang their tails.-- NOVEMBER 1st..-- Addenda to Journal. I show erratic d
411e055 kind of extremities come under this head» 27th NOVEMBER When summing up argument against my theory, d
503q013 le branch.» --¿number of seeds in beginning of NOVEMBER 1841.-- Trees above male? (2) Result of Edwar
573n037 erful structure of a beetle than a Universe.-- NOVEMBER 20th Saw the youngest child of H. W. constant
574n041 to become absorbed by one end of Cocoa nut.-- NOVEMBER 27th.-- Sexual desire makes saliva to flow «y
575n045 g preferred from not having any smell.-- 27th. NOVEMBER.-- Think, whether there is any analogy betwee
482z011 the first 9.:4 raptores. Falco poliosoma -- NOVOZELANDIAE -- histrionicus Vultus aura Excessively ins
240c014 , (when young very like the Siam race with long NOZZLE & few hairs) inhabits Celebes & few of the lar
023r010 lleras as arising from «the expulsion of fluid NUCLEUS through» faults or fissures, produced by the e
035r048 ction as a deep & extensive movement of viscid NUCLEUS, which in any one country would produce equabl
038r058 e reasons, & saying that they refer to CENTRAL NUCLEUS & that envelopes no doubt existed. These highe
049r088 regular greenstone cones at S. T. del Fuego as NUCLEUS of a Volcano or as an injected mass.--From con
056r111 of granite: In discussing circulation of fluid NUCLEUS,--the similarity of Volcanic products «over wh
112a089 ana, worthy of consideration. When discussing NUCLEUSE'S of old volcanos within Cordillera-- allude t
479z005 2 M Lesson--Voyage of Coquille wide limits of NULLIPORA Discussion good on Falklands birds Discussion
481z010 my little animal with horns. Madrepores p. 26 NULLIPORA p. 29-- In Meyen. Voyage round World German a
021rIFC istence, those that have undergone the greatest NUMBER of changes towards perfection (namely mammalia
044r071 e internally than externally--I did not see any NUMBER of dikes in the cliffs.--wide valleys.--centra
047r081 swell travelling across Pacifick.--excepting in NUMBER of waves & in wind, instead of sea's bottom be
055r107 & dust.--connection with age, & agreement with NUMBER of craters. No cliffs at Ascension (or modern
055r107 ons how immense the time!! How well agrees with NUMBER of Craters!--At S. Cruz. there is no occasion
064r134 Rhinoceros quite in deserts.--Much struck with NUMBER of animal[s] at Cape of Good Hope Says at Sant
085a002 ted like Bolivian Chain. Volcanic islands. from NUMBER of craters very ancient. which agrees. with pe
093a033 ion of sulphur to form salts of America. -- The NUMBER of minute turbos in red earth with volutas. pr
109a082 -- do-- [...] Subaqueous. removal, shown by the NUMBER of bones lying at the bottom of sea. off coast
172b005 characters of both parents, & these infinite in NUMBER In Man it has been said, there is instinct for
176b021 quable:-- but is there any reason for supposing NUMBER of forms equable: this being due to subdivisio
180b037 st have left no offspring at all, so as to keep NUMBER of species constant.-- With respect to extinct
182b047 describe limits of form. (& where possible the NUMBER «of steps» known). <for instance among the Car
199b119 his object & others, viz not too much change In NUMBER 6'.? of E.d. N. Philos. Journ.-- Paper by Craw
206b146 n 200 years backward might be calculated & this NUMBER elimanated say 150 people four hundred years s
206b147 lation was increasing between each lustrum, the NUMBER related at the first must be greater, &
206b147 ated at the first start must be greater, & this NUMBER would vary at each lustrum, & the calculation
212b168 visions of molluscs. where species now least in NUMBER (as cephalopods,) in last tertiary epochs most
226b221 of North Europe, many genera & few species. The NUMBER of genera on islands & on Arctic shores eviden
245c023 nother--\ «It is a good species, with different NUMBER of teats» Coquille Voyage Durville has written
249c036 a?--- This is very remarkable, when we consider NUMBER of quadrupeds in Eocene period. Have the Edent
262c073 ly will have some aberrant groups.-- but as for NUMBER five in each group absurd.-- the mere fact of
276c124 uld alter.-- It is a difficulty how a different NUMBER of vertebrae are produced, where, (& in all su
276c125 ch often enough repeated would cause an unequal NUMBER of vertebrae-- ¿Where two very close species i
278c131 at dog was found in Van Diemen's Land.-- V. ls. NUMBER of Geographical Journal to discover whether do
284c147 erefore probably fewer now than formerly.-- The NUMBER of forms depends on the external relations (a
286c156 ich unite these of older standing than constant NUMBER of stamens.-- in order, or in next family? In
292c169 d species. & of 100 extinct species the greater NUMBER probably have no descendants on earth.-- +++.
297c186 rare we infer extermination, when group few in NUMBER of kind, extermination.-- New forms made throu
297c186 .-- New forms made through probably an infinite NUMBER of forms.-- therefore an isolated form probabl
302c202 is (one of analogy or) LATELY ACQUIRED. In pigs NUMBER of vertebrae. subject to variation. therefore
303c203 sculant species. which survived would be few in NUMBER.-- Parallel of Japan near Himalaya, & Europaea
308c218 er, air, & land, (Macleay has this remark) Mem. NUMBER 5 here most evident!!? examine into this case
312c227 certain birds in many families, «†very often in NUMBER 5» will have long tail.-- in raptorial birds,
312c228 s point-- pigs always go against this, without «NUMBER of vertebrae» new acquisition, we must [not lo
312c231 each feather weight & size & they would produce NUMBER agreeing almost to the point in question.-- «-
331dIFC makes hybrids. productive like geese?-- Are the NUMBER of kittens between Lion & Tiger at litter as n
331dIFC t litter as numerous as in common lion? Are the NUMBER of nipples in domesticated very fertile animal
331dIFC here offspring, heterogenous, in plants are the NUMBER of seeds greater.?-- Mem. for Eyton.-- Sir. R.
332d003 fever, be attended by the transmission of large NUMBER of worms the child not having passed them befo
334d009 equally fertile with pure bred animals.-- Mem. NUMBER of Mules.-- «He recollects one hatch of hybrid
338d023 for otherwise we could not understand the vast NUMBER of domesticated races.-- Athenaeum. p. 505. so
349d052 s structures, as in Hippotamus, solely owing to NUMBER of lost links.\ if all species know they would
349d052 s will change all.-- & so on no one will settle NUMBER of primary divisions.-- Complains (p. 53) of M
352d058 fficulty «contradicted by abortive organs, but NUMBER of species with abortive organ of any kind few
353d061 71»). alludes to Eyton's discovery of different NUMBER of vertebrae in Irish & English Hare.-- good c
354d062 e anything in such grimaroles about analogies & NUMBER L'Institut p. 275. (1838) M. Blainville has wr
359d088 that breeds larger numbers, & rears an unusual NUMBER out of any one nest. even more than common duc
374d134 crease of animals exactly proportional[l] to the NUMBER that can live.--» We ought to be far from wond
374d134 We ought to be far from wondering of changes in NUMBER of species, from small changes in nature of lo
375d135 op on an average, every species must have same NUMBER killed, year with year, by hawks. by. cold &c-
375d135 d &c--. even one species of hawk decreasing in NUMBER must effect instantaneously all the rest.-- On
389d173 without the nervous matter consists of infinite NUMBER of globules: generally sufficient for one birt
397e003 xtermination & production of new forms.-- their NUMBER & corelations Octob. 4th. It cannot be objecte
401e014 hn says he has no doubt bees fertilize enormous NUMBER of plants-- it is scarcely possible to purchas
402e020 a & in Luçon the most northern of the group the NUMBER is limited of the group, the number is very li
403e021 e group the number is limited of the group, the NUMBER is very limited.-- Kotzebue's Second Voyage do
403e023 rieties, do we suppose bones will not change in NUMBER. (even species do not this). because it has be
408e043 ermination. some other species must increase in NUMBER where then is the gap, for the new one to ente
415e068 .-- The Value of a group does not depend on the NUMBER of the species.: therefore Man & monkeys have
422e095 new» raised beaches» [not located] The enormous NUMBER of animals in the world depends, of their vari
423e097 tus) Crustacea to--- ? &c) without reducing the NUMBER of living beings-- but there is the strongest
427e108 s to each other would rapidly increase, & hence NUMBER of forms. once formed. would remain stationary
428e113 be more striking, about indelibleness, than the NUMBER of good race-horses, which Eclipse? has begott
431e122 ommon, we need not look for change, because its NUMBER show it is perfectly adapted; it where few str
431e122 in Royal Institution says shells become less in NUMBER. (¿ species, or individuals) the deeper one go

(Key Word)

```
442e152 sexual reproduction of species may stop for any NUMBER of generations-- Gorze in Norway, which never
493q004 see whether offspring infertile.-- (4) Does the NUMBER of pulse, Respiration, period of gestation dif
496q006 e &c &c on associated plants. when proportional NUMBER appear equal-- & see whether proportions will
498q008 eties, which counterbalance each other? (10) Is NUMBER of pollen-grains necessary to impregnate ordin
498q008 pollen-grains necessary to impregnate ordinary  NUMBER of seeds known?-- Linnaeus has shown that each
499q009 without impregnation.-- (16) Any calculation of NUMBER of grains of pollen in any one flower (17) Cat
500q10a are now to be planted this year copied Gould.-- NUMBER of species of Birds in New Zealand, plants  so
503q013 Yew Trees near Boat House «ANY male branch.» --¿ NUMBER of seeds in beginning of November 1841.-- Tree
514q021 orticulturists (1) Are sterile hybrids healthy:  NUMBER of generations: about crossing of plants; espe
515q021 . where parts of Clover are reduced from normal NUMBER, they are apt to vary in number in individuals
515q021 ced from normal number, they are apt to vary in NUMBER in individuals of same species Eyton (1) Numbe
515q022 number in individuals of same species Eyton (1) NUMBER of eggs-- of half-bred geese-- inter se, & wit
525m025 ail cut off at Shrewsbury & its kittens <h> (in NUMBER 3) had all short tails; but one a little longe
525m025 My  father says, perfect deformity, as an extra NUMBER of fingers.-- hare lip or imperfect roof to th
531m050 re all keys are placed) Memory cannot solely be NUMBER of times repeated, because some people can rem
543m099 ve all parts. Waterhouse Study well the greater NUMBER of insects in insecta-- not connected with tra
544m103 rdinary rapidity-- We may conclude that neither NUMBER, vividness, rapidity, novelty of separate idea
546m108 very  superiority of man perhaps depends on the NUMBER of sources of pleasure & innate tastes, he par
548m117 be  & discover loss Definition of happiness the NUMBER of pleasant ideas passing through mind in give
549m118 man  is compared to peasant.-- To make greatest NUMBER of pleasant thoughts, he must have contingency
582n068 nstinct is to be guided to the performance of a NUMBER of prearranged actions, which will bring about
586n078 in walking [one feels inclined to stop at right NUMBER of house though one cannot remember it.]CD bac
640j167 ghth part.-- or if other prey diminished, total NUMBER of dogs. would diminish, whilst the long legge
115a097 ted specimens from Anglesea in Geolog. Soc. if NUMBERED compare them with my rocks. when writing on F
523m018 she had been insane all the time.-- There are NUMBERLESS people insane of particular ideas, «Case of
048r086 In  what part of the globe are there such vast NUMBERS of wild animals. both species & individuals as
060r126 usible? no. mad dogs. Azores. although kept in NUMBERS. p. 124. Webster Consult W. Parish. & Azara ab
090a023 ny cases there were flights of stones of large NUMBERS (& how few cases recorded if we say «100» <5>0
175b020 n isolated, requires deaths of species to keep NUMBERS of forms equable:-- but is there any reason fo
195b102 ical changes go at such a rate, so will be the NUMBERS & distribution of the species!! It may be argu
227b225 think.  it agrees with excessive inequality of NUMBERS of species in divisions. look at Articulata!!!
297c185 ged by hunger.-- p. 65. Aberrant groups few in NUMBERS & vary much in character, divided into many sm
297c139 & B. p. 431. Missel thrush lately increased in NUMBERS over whole of England & Ireland.-- curious  in
320c276 m<e>nes. Edinburgh New. Phil Journal. about 13 NUMBERS have been read Voyage a l'isle de Frances Voya
342d032 common ducks-- they are produced in full equal NUMBERS with pure bred (just like common mules) & lay
343d037 take place when Conditions are unfavourable to NUMBERS of animals. as in changing from <hot>. Warm to
359d087 breed-- but Mr Miller says that breeds larger NUMBERS, & rears an unusual number out of any one nest
379d152 . <Owen.> See Hunter's Owen-- In the Athenaeum NUMBERS 406, 407, 409, Quetelet papers are given, & I
410e053 describes like Richardson.-- The relations of NUMBERS of species to genera &c &c can never be  told,
415e067 ber my fathers remark about the Bladder.-- The NUMBERS of fatal diseases in mankind, «the more valuab
419e085 argued by Müller from propagation of infinite NUMBERS of individuals from one, is adverse.-- Decemb.
436e135 oducing new species; also one being reduced in NUMBERS, but not so much these, because circumstances»
436e136 tainly appears that swallows have decreased in NUMBERS, what cause?? Seeing the beautiful seed of a B
470t178 terflies at Clover,--Veronica--, Ranunculus in NUMBERS =what insect can get honey out of long, curved
541m092 is an emotion.-- People who can multiply large NUMBERS in their head must have this high faculty, yet
373d133 Mould.-- Lyells Elements. p. 290. Dr. Beck on NUMERICAL proportion in shells in Arctic Ocean. p.  350
448e166 l the castes from Stephenson at Lima The same NUMERICAL relation (both in species and subgenera) betw
448e166 the resemblance between the Superga & Paris, NUMERICALLY the same with recent & yet almost wholly dif
049r089 tone? What probable origin can be given to the NUMEROUS hills of greenstone? -- Daubeny. P 95. Glassy
052r099 excepting  in Port Famine Mr Sorrell says that NUMEROUS icebergs are commonly stranded on shores of G
110a087 as NW / SE. Vol VI. p. 247. Mr. Schombürgk NW. NUMEROUS boulders of GRANITE" "direction of strata  on
176b021 f differences, so forms would be about equally NUMEROUS. changes not result of will of animal, but la
192b086 especially  in those classes where species not NUMEROUS. (NB in those classes with few species greate
198b115 .-- Provision for transportal otherwise not so NUMEROUS. quote from Lyell: assuming truth of quadrupe
202b134 erers would not, but where original forms most NUMEROUS there would be wanderers.-- Some however migh
213b168 -- p. 82 «There are many tables in Phillips of NUMEROUS genera in fossil & recent state, well worth c
267c093 f Arabia between Ras Mohammed & Jeddah". sheep NUMEROUS "of two kinds one white with «a» black  face,
331d1FC r of kittens between Lion & Tiger at litter as NUMEROUS as in common lion? Are the number of  nipples
413c059 teristics of the Tropical Forms in shells. are NUMEROUS species, numerous individuals, & <individuals
413e059 ropical Forms in shells. are numerous species, NUMEROUS individuals, & <individuals> «species» of lar
469t135 -- /during several succeeding days <many> «most NUMEROUS» bees visited this same bunch & on this day i
497q007 s in genera of Leguminosae.-- Herbert explains NUMEROUS spec. of Cape Heath by facility. ¿Knight take
501q10a , in other countries, where species are either NUMEROUS or even where few are they constant: this ver
566n011 ch of them? & curiosity «strongly shewn in the NUMEROUS artifices to take birds & beasts».-- very nec
574n039 Douglas. «& Spencer», an old Scotch Poet, has NUMEROUS lines. of poetry.-- <signs> sounds singularly
640j167 a  hare was introduced, or <a spe> became more NUMEROUS. (from death of its destroyer), or other caus
641j29r phical Distrib of larger Seals-- Are Porpoises NUMEROUS in cold Oceans I think not.-- Does this  bear
171b002 al. &c &c.-- Yet Eunuchs nor «cut» Stallions nor NUNS are longer lived Why is life short, Why such hi
594n113 he boys making faces at it, so much so that the NURSE had to carry it out of the room. nearly 3 month
373d132 semen--  but this-- -- the <nerve> living nerve NURSED in Mould.-- Lyells Elements. p. 290. Dr. Beck
498q008 we  know how readily they cross.-- (9) In the NURSERIES, when «seed of» the varieties of Cabbages, pe
492q003 doubt by Herbert Do forest-trees sport much in NURSERY gardens? <are the> is the ground much  manured
361d095 all linnets red-pole, goldfinch, hawfinch-- in NURSLING plumage resembled that of Cross-Beak-- In lar
361d095 cies-- [In two herons, <both> plumage of both (NURSLING) quite similar.-- one species retained this c
504q013 dent?? (7) Which. Rhododendrum seeds??-- Bladder-NUT ℕ. Laburnum ℕ. Dodecatheon ℕ . Castrate apple  &
545m106 er little old American monkey «(Mycelis)» I gave NUT, but held it between fingers, the peevish expres
545m107 en touched, also another monkey to dog. I showed NUT & then closed my mem. expression of fury, jump t
574n040 abits,-- for instance the Birgos opening a Cocoa NUT shell at one end.-- Children & old people get in
574n040 as Birgos to become absorbed by one end of Cocoa NUT.-- November 27th.-- Sexual desire makes saliva t
632j53r eeds for transportation through the air.-- cocoa NUT by water «fucus for adhesion».-- as examples of
440e145 the acorn into the oak.-- In short all which «NUTRITION, growth & reproduction» is common to all livi
461t037 out being absolutely injurious «(or requiring NUTRITION)» to a certain amount it will be so handed do
498q007 into  leaves-- is this ever effect of want of NUTRITION.-- Horned oranges so? --Yes, my Father lost t
284c148 -- ¿Brown can «not» bear the least salt water.-- NUTS prodigiously heavy (where trees of such Nature
506q015 the  Papilionaceous structure of flower-- Ground NUTS (42) How are Orchidiae fecundated, as mass of p
499q010 portal, does he include seeds good to eat. (even NUX Vomica is eaten by a Buceros in East Indies-- As
024r015 hoals much: Dampier remarks on great flats on the NW coast:-- 8 leagues, from Sydney 90 fathoms La Pe
024r015 a; 19 miles. 65 Fathoms Vide facts in Beechey. on NW coast of America off Cape of Good Hope 70 fathom
034r045 change  in soundings. on approaching the coast of NW. America P. 209--13 P & 444 «(Yanky Edit)» <I th
051r097 any bodies [blank] Beechey.--changes in bottom in NW coast of America. from shingle to sand &c &c. <V
074r165 metallic veins follow mountain chain. there after NW <W>.-- «same chemical laws as in concretions per
078r175 l del Monte 85 degree to S. // Tasco 40 degree to NW (afterwards said to be «all with some exception»
078r175 rds said to be «all with some exception» directed NW & SE). «Vol III» Mexican Cordillera "immense var
079r177 cognized as coming from three directions. from W. NW & S.--last to Seaward partaking of the character
110a087 se describes central granitic ridge of Guayana as NW / SE. Vol VI. p. 247. Mr. Schombürgk NW. numerou
110a087 uayana as NW / SE. Vol VI. p. 247. Mr. Schombürgk NW. numerous boulders of GRANITE" "direction of str
111a087 of  strata on the Berbice N. 35 degree. E. dip to NW to 80 degree faults with red wacke contorted evi
113a092 ce; Mem my remarks on coast of Australia.-- Great NW. dip in SE part of Australia.-- Probably a  case
366d111 nera-- How long will the wretched inhabitants of NW. Australia, go on blinking their eyes. without e
366d112 ing out teeth.-- the account of the people on the NW. Coast blinking to keep out flies might be used»
139a180 organic substance cause such monstrous growth as OAK galls or rose <buds> galls.-- is it not effect o
440e145 of growth, that which changes the acorn into the OAK.-- In short all which «Nutrition, growth & repro
326c265 iegman has published German Pamphlet on crossing OATS, &c «Horticultural Transacts.--» Mr Coxe "Views
496q05a elreuter describes Kill Sparrow after feeding on OATS, give body to Hawk & sow pellet. ejected.  done
506q15v ate. of flower 45. Charlsworth. vol II. p. 670-- OATS cut down turning into Rye.-- 46). Book describi
620o045 fence; one always admire the habit formed by «OBEDIANCE to instinct» <conscience>. , or rather the str
612o035 s easy to conceive such movements & choice, & OBEDIENCE to certain stimulants without conscience in t
122a114 cause if of fluid, the waters of the ocean would OBEY that Law. & lie over the platform:-- On my view
```

Page   **(Key Word)**
549m121 ness is the object of living.-- or whether if we OBEY literally New Testament future life is almost t
575n044 on their instincts, most important, because they OBEY the same laws, as the crossing of jackall & Fox
619o043 hich the acter received.-- If either man did not OBEY his instincts from interference of passion. he
625o50v oadly in child or animal it is equally proper to OBEY anger as benevolence (but not cool malevolence)
303c204 ion is at end, seeing that Tertiary geology has OBEYED rules of modern causes. & considering over the
098a046 as suggested by Sir J. Hershel is all crystals OBEYING one law of crystallization. therefore concreti
181b043 ite» variations, & all coming from one stock & OBEYING one law, they may approach,-- some birds may a
525m023 ike the law of honour.-- they feel pleasure in OBEYING their instincts naturally.-- (generosity in de
563n001 g.-- instances of expression.-- Octob. 3d. Dog OBEYING instinct of running hare is stopped by fleas,
564n004 for the satisfaction of following conscience, OBEYING habits, & dread of misery of future thinking o
611o034 fe itself being, the capability of such matter OBEYING a certain & peculiar system of movements. diff
627o053 d against this-- but the pleasure a dog has in OBEYING its instinct,-- as young pointer to point-- cl
543m101 ought within <our own> limits of examination.-- OBEYS same laws. as other parts of structure. C. D.27
550m124 boon is our grandfather!-- A man, who perfectly OBEYS his conscience or instinct, would probably feel
171b002 e longer lived Why is life short, Why such high OBJECT generation.-- We know world subject to cycle o
199b118 ake higher grounds & say life is short for this OBJECT & others, viz not too much change In Number 6'
228b228 ead to laws of change, which would then be main OBJECT of study, to guide our <past> speculations wit
232b248 ts varying in different climates Those will not OBJECT to my theory, those the philosophers who
269c102 mpare birds of Australia with plants, with this OBJECT in view» The intimate relation of Life with la
294c177 nion of many people in conversation.» the whole OBJECT of the Work is its proof, <its limiting, the a
308c218 in all reasonings, of which human nature is the OBJECT, there is really no natural starting place, be
390d178 ocess would be similar to budding. which is not OBJECT of generation.-- therefore passions fail.-- In
390d179 forms to acquire differences: if none are added OBJECT failed, & then by that correlation of structure
391d174 le union.-- At present I can only say the whole OBJECT being to acquire differences «indifferently of
391d174 er progressive improvement or deteriorate» that OBJECT failing, generation fails.-- How completely ci
398e005 of change now in progress, will be the last to OBJECT to this theory on the score of small change.--
409e049 the animated beings.-- &c-- If man is one great OBJECT, for which the world was brought into present
409e049 w will dispute, [although, that it was the sole OBJECT, I will dispute, when I hear from the geologis
462tf4v artet same as existing species. We see the same OBJECT gained by the Mataco-armadillo & the woodlouse
528m036 ally when coloured.-- that light is a beautiful OBJECT one knows from seeing artificial lights in the
532m054 ted, & by habit the mind tries to fix upon some OBJECT:-- When a man, child or colt has once been fri
549m121 another question, whether this happiness is the OBJECT of living.-- or whether if we obey literally N
549m121 ly New Testament future life is almost the sole OBJECT--. -- I doubt whether the last be right. The i
552m130 ation." Septembe. 3d Why when one thinks of any OBJECT, (or having looked at any object<)> one Shuts
552m130 thinks of any object, (or having looked at any OBJECT<)> one Shuts ones eyes) is the image not vivid
557m149 rely sexual; first try «to» attract female, (or OBJECT of attachment) & then failing to drive away ri
559m154 ws is probably that as long as we consider each OBJECT an act of separate creation. we admire it more
564n005 erfectly to chicken so as to seize small moving OBJECT like fly.-- young partridge can run even with
565n009 - Hope is the expectant eye. looking to distant OBJECT, brightened & moistened by emotion,-- why does
579n058 ily pain? frown shows the mind is intent on one OBJECT.-- With respect to my theory of smile. remembe
582n068 t is about when building its nest; it knows its OBJECT but not result (first time of building?), but
593n105 o get into hay & make a nest for herself.-- the OBJECT is to make saucer-shaped depression.-- [blank]
601o009 aoid it-- beetles feigning death upon seeing an OBJECT.-- are Planariae conscious.-- Consciousness be
602o11b e p. 142 "Upon the whole it seems."-- "that the OBJECT of «all» art is the realizing and embodying, w
603o013 ified the deity.-- H. Tooke has shown one chief OBJECT of language is promptness <of consequence» hen
605o19v f this "inward glorying" to the greatness of an OBJECT itself or to the ideas excited & associated wi
605o020 e acted on by sympathy. D. Stewart on taste The OBJECT of this essay is to show how taste is gained h
606o022 g the result.--Lessings Laocoon. 2d Lect-- The OBJECT of art., sculpture & painting, is beauty.-- wh
608o026 -- We cannot help loathing a diseased offensive OBJECT, so we view wickedness.-- it would however be
610o033 ther <we th there>» "anything can be <any> «the» OBJECT of <our» knowledge except our experience".-- i
613o035 nto play, when relation is kept up with distant OBJECT. where many such objects are present, & where
617o39v the two aspects as different phases of the same OBJECT of thought is a question which ought to be cle
619o042 of love <and sympathy» «or benevolence» to the OBJECT in question. Without regarding their origin, w
627o053 phy p. 6-- "The pleasure which results when the OBJECT is attained (the gratification of one's offspr
397ж004 number & corelations Octob. 4th. It cannot be OBJECTED to my theory, that the amount of change withi
290c165 explained by F. Cuvier, «-- .Mem. Hensleighs OBJECTION.--» it is more, he cuts the matter short by s
399e006 d some change in upper & lower layers.-- good OBJECTION to my theory: a modern bed at present might b
462t051 ssimilarity of forms-- yet how valueless this OBJECTION, when one thinks of different kinds of cattle
616o39v espectively what are called it's subjective & OBJECTIVE aspect. The subjective aspect of bodily actio
617o39v exert force or by internal consciousness; the OBJECTIVE, by our external what senses in the way in wh
617o040 vour to do without it as well as we can. <The OBJECTIVE aspect of> bodily action as recognised by our
618o041 own only by internal consciousness, & have no OBJECTIVE aspect. If thought bore the same relation to
617o040 ed towards every other particle; but FORCE, <OBJECTIVELY> considered, by our external senses is a phe
618o041 only known subjectively? -- ? the brain only OBJECTIVELY. We do not know attraction objectively 6) Th
618o041 only objectively. We do not know attraction OBJECTIVELY 6) The reason why thought &c. should imply «
183b053 ead his theory of the Earth attentively Cuvier OBJECTS to <tran» propagation of species, by saying, w
293c173 .-- [There is no corelation between individual OBJECTS as Ichneumon & caterpillar, though our ignoran
444e158 ch distances in air, in which it never touches OBJECTS.-- far better case than chicken pecking fly.--
462t051 > seven «7» generations.-- so many!! Hensleigh OBJECTS to transmut. theory, on the grounds of similar
532m054 me there was nothing, & tried to seize hold of OBJECTS to be frightened at.-- (again diseases of the
542m095 lowest savages clearly be directed chiefly by OBJECTS of vision.-- Does the contraction & wrinkling
545m105 - former do not give rise to ideas so much. as OBJECTS of interest.-- do/ I was much struck with obse
569m021 ably is single comparison by senses of any two OBJECTS-- they by VIVID power of conception between on
587n087 e not something more than the unfitness of the OBJECTS then viewed. to organs adapted to other object
587n087 bjects then viewed. to organs adapted to other OBJECTS. (as that senna is necessarily disagreeable to
606o022 auty?-- it is an ideal standard, by which real OBJECTS are judged; & how obtained.-- implanted in our
613o035 s kept up with distant object. where many such OBJECTS are present, & where will directs <to> other p
618o41v to account for the phenomena of thought. (The OBJECTS of thought have no reference to place. [We see
619o043 change in association if others injured these OBJECTS, without his being able to prevent it, he woul
627o053 is allowed, that we have conception of moral OBLIGATION «when grown up??» & the question is, whethe
627o053 al faculties-- Will Eugenius allow this moral OBLIGATION? [2] [The improvement of the instinct of a s
186b058 the principle of atavism, where real structure OBLIGED to be altered, I can conceive colouring retain
212b165 to the door-- would constantly do this, so was OBLIGED to be removed.-- In L.' Institut. 1837: p. 404
352d060 e torn out of a history, & the geologist being OBLIGED to fill up the gaps.-- is possibly the same wi
385d163 ithout she know & LIKES HIM & then is actually OBLIGED to be held.-- like she wolf of Hunter.-- young
541m091 the scarlet was before one effort less» one is OBLIGED to repeat the word, & think of qualities as fl
554m141 ry fierce when female is sitting the Keeper is OBLIGED to go in with a stick, if he drops it, the bir
567n016 o.-- When learning facts for induction. one is OBLIGED carefully to separate its memory from all ordi
109a083 vation have same effect, a tendency direct (or OBLIQUE) outwards may be granted. independent of cause
146g010 d be traced Grey in front on wall perhaps wall OBLIQUE The hill has been well-- denuded.-- «of hard m
565n007 tiff, with head up.-- Why does suspicion look OBLIQUELY.-- who can analyse suspicion-- yet who does n
565n007 icion, even child will do so.-- Contempt look OBLIQUELY so does dog. when a little one attacks him Co
056r109 ld suppose such trifling means could efface & OBLITERATE so grand a work?--In valleys one is not sure
058r115 sign of elevation. Effects of great waves to OBLITERATE all land marks.--At the first it would thoug
187b064 n peculiarities, (therefore adaptation), & to OBLITERATE accidental varieties, & to accomodate itself
386d167 frame.-- «one of» The final cause of sexes to OBLITERATE differences. final cause of this because the
255c051 but all have had feathers.-- if wing totally OBLITERATED. This may account for permanence in many tri
285c152 ttom of the tree of life is utterly rotten & OBLITERATED in the course of ages.-- As species is <cert
384d162 akes place.-- thus one organ in each becomes OBLITERATED, & sexes as in Vertebrate tak place.-- ∴ Eve
407e038 urope. in which all Tropical forms have been OBLITERATED) of the world. from the <Tropical> Equable b
264c081 be profoundly considered.-- Structure may be OBLITERATING, whilst habits are changing-- or structure
364d103 thout fresh feather, & consequent trouble in OBLITERATING the fresh feather, by crossing-- It seems f
615o36v d & plants sleeping good show acquitement or OBLITERATION of instincts But habits acquired even by <c
398e006 we suppose not only revolutions, but entire OBLITERATIONS & fresh laws created., & yet with <gov> sym
067r143 tuco» Athenaeum April 1836 (p302) Coleccion de OBRAS. 2 Vols fol: Buenos Ayres 1836: W. Parish?? «by
602o11v t (p 128) R. compares a view taken by a camera OBSCURA &c a Poussin.-- How are my ideas of a general
147g017 s if they belonged to double series Whole very OBSCURE but it is certain there must once have been ve
154g057 lain divortium aquarium-- tidal channel-- 12ft OBSCURE obscure NB In Glen Collarig tidal channel, sid

(Key Word)

```
154g057 ortium aquarium-- tidal channel-- 12ft obscure OBSCURE NB In Glen Collarig tidal channel, sides <alm>
158g079 nd {P} rounded waterworn buttresses of granite OBSCURE terrace 15 ft divortium my measurements here s
159g086 egree? This station a little way down slope of OBSCURE terraces (& conical hills on same) of «semi» w
160g092 f L. Oich, & waters flowing into west end with OBSCURE terraces on one side Barom 29.200 A 80 70 degr
161g097 uld have been more correct) there were several OBSCURE but not far continuous flights above it-- (NB
200b123 ome cause of change,, yet such causes are most OBSCURE. without doubt:-- Vide cattle: The grand fact
273c111 & likewise perching (Gould), but the latter is OBSCURE because nearly all are so.-- Thus in Hawks, th
404e025 fect, would expect this structure would become OBSCURE & therefore it might thus have arisen, & M. Ed
508q016 females  takes place of gout.-- How are livers OBSCURE organ. no answer?-- 3 Andrew Smith, about tame
534m061 et I felt it)-- this is kind of conscience, is OBSCURE memory of having read or thought of some  such
593h109 other  men by smell, but by looks. hence. some OBSCURE picture of other men. & hence idea of beauty.-
533m060 sins.  did not make him more humble.-- it has OBSCURELY occurred to me that Capt. F. R. candour & rea
542m095 expression of surprise-- viz seeing something OBSCURELY with the wish to make it out?-- Seeing a Baby
558m150 rom our enlarged capacity <acting> «yet being OBSCURELY guided» or strong instinctive sexual, parenta
190b079 remarked  to me, that he though species became OBSCURER as knowledge increased, but genera stronger,-
089a018 , et D Histoire Naturelle, C«o»urrejolles. 11th OBSERV.-- Les grands tremblemens de terre sont presqu
054r105 as  those on the beach--"This is particularly OBSERVABLE in a bay about five leagues North of Callao,
138a153 thern Brazil. [blank] «p. 4. (Lyells Book)» OBSERVACIONES sobre El Clima del Lima par Dr. H. Unanùe s
085aIFC p 140-- abstracted as far as concerns "Geolog OBSERVAT on Volcanic islands & Coral Formation Lyell's
138a151 . or 3?) there is an account of Sellow Geolog. OBSERVAT. in Southern Brazil. [blank] «p. 4. (Lyells B
326c266 s worth reading quoted by Malthus.-- Heberdens OBSERVAT. on increase & decrease of different diseases
190b076 pecies.-- «Brown Appendix» A most remarkable OBSERVATION of Mr Brown, about peculiarities of Flora. o
213b171 ties affecting the cross most worthy of OBSERVATION.-- I think it is certain strata could not no
227b227 ucture becomes full of speculation & line of OBSERVATION.-- View of generation. being condensation, t
285c150 t savages are not born with any capacity for OBSERVATION of tracks &c &c Dr. S. has some remarkable c
358d076 this  Caracara.-- Sept. 9th. It is worthy of OBSERVATION that in insects where one of the sexes is li
369d115 book)  Owen says "the necessity of combining OBSERVATION of the living habits of animals, with anatom
493q003 ays length differs in different cats.-- Good OBSERVATION-- examine semen of Hybrid animal. in compari
509q017 t is very important to know, whether Gould's OBSERVATION holds good, that in the mundane genera,  the
567n014 «art  precedes science-- art is experience & OBSERVATION.--» in balancing a body & an ass knows one s
042r069 oyed regularity of slope of valleys.--All my OBSERVATIONS of period «& manner» of elevation  Volcanic
065r140 eze. correspond to September ¿Did I make any OBSERVATIONS on springs at S. Cruz.???-- Form of land sh
111a087 ile must be studied Analysis of Voyage: many OBSERVATIONS on heights of valleys in Chile Geograph. Jo
131a135 o be bombs submarine L'Institut 1838 p. 400. OBSERVATIONS on Mountains of the Moon. by Dr. Nichol-- a
222b204 in Latin one <of> «on» Madeira-- any general OBSERVATIONS-- difference of species between land shells
279c133 ects of incestuous intercourse.-- excellent OBSERVATIONS of sickly offspring being cut off-- so that
323c269 12.  Sir. H. Davy Consolations in Travels -- OBSERVATIONS on morals by Eugenius Feb 14th. Bo«s»well's
416e072 sin.-- its relation to African Species <good OBSERVATIONS.-- >, larger than any living [not  located)
477z001 ing & boring marine animals-- CXVI. P 111 do OBSERVATIONS on Planarias by Johnson CXII. & CXV do Azar
480z007 dwards on Corallines L'Institut 1837. No 212 OBSERVATIONS on the Raptores of S. America translated fr
596m184 n worthy of close study.-- full of practical OBSERVATIONS Ourang do not move eyebrows.-- or skin of h
606o20v e acquired by a long series of experiments & OBSERVATIONS. & yet, like in vision, it becomes so insta
233b250 it--  pulled bell.-- -- It was most curious to OBSERVE, that all the species of mice in S. America. w
273c112 hole family where female is not dull.-- I must OBSERVE that this preeminent structure is not always a
274c114 ngii very rasorial for type.-- Now here I must OBSERVE these characters vary in degree in last instan
277c127 preventing the chaos.-- will point out what to OBSERVE.-- will aid us in physiology, tell traveller w
277c127 l aid us in physiology, tell traveller what to OBSERVE.-- if he knows he has done least part.-- that
292c170 Swainson's on the Classification of animals. & OBSERVE the character of the demonstrations offered of
314c239 e. rather large flora. (150?) Mr Brown did not OBSERVE scarcely any Australian character in Timor pla
321c270 espond. with Rousseau Miss M«artineaus» How to OBSERVE Mayo Philosophy of Art of Living Several of Wa
342d034 erfect.-- Are Pheasant & Grouse homogeneous? I OBSERVE Bachman calls these Hybrids new species. Yarre
385d164 dies "It is a fact equally well known, that we OBSERVE in the temper, especially of the youngest chil
501q011 out sticks»-- in reference to what Mr. Herbert OBSERVE on this subject-- (31) Ask Henslow for list of
527m034 ness of thought, from weakness of my stomach I OBSERVE a long castle in the air, as hard work (abs
537m075 en attended with bad effects Martineau. How to OBSERVE, p. 21-26. argues «with examples» very  justly
547m114 ones, to be sure of ones consciousness.-- Mayo OBSERVE no improbabilities in a dream, effect of doubt
555m142 dispute  about words.-- Miss Martineau (How to OBSERVE p. 213) says charity is found everywhere (is i
323r009 ound in that Archipelago? Such have never been OBSERVED in Australia Dampier also repeatedly talks ab
076r168 in certain mines. «Vol. III» "In general it is OBSERVED both in Mexico & Peru, that those oxidated ma
097a043 conceive  to be analogous to the black glazing OBSERVED by Humboldt on the granitic rocks of the Orin
205b145 by Mr Cline "The character of both parents are OBSERVED in their offspring, but that of the male more
257c056 of cow!!-- &c &c D'orbigny (p 108) says having OBSERVED B. Tricolor in Patagonia. then in Chile & las
274c115 ave their represetatives;, the rasorial may be OBSERVED even in Lessonia &c & In relations of affinit
341d030 ecies of Moschus, characterized by Ogleby, who OBSERVED that the young of this animal, which is so an
350d055 «(2)», female «(I)» fixed & blind: -- Macleay OBSERVED all these facts prove that perfection of orga
350d056 g able to reproduce Here there is some error-- OBSERVED, nature does nothing in vain, therfore  organ
354d065 ported like the New Zealand one-- It should be OBSERVED with what facility mice attach themselves  to
356d071 mparing my theory with any other. it should be OBSERVED not what comparative difficulties (as long as
361d096 ange is effected through the male??)-- Yarrell OBSERVED that female of some water birds, (as Phalarop
388d172 ssion of mutilation impossible.-- it should be OBSERVED that transmission bears no relation to utilit
389d173 cient for one birth or rather]CD ▮It should be OBSERVED that the constant necessity for change. in pr
390d178 eways different from himself, for it should be OBSERVED that from Books to read Buffon Suites de.-- H
402e017 Mountain-ash  & white Thorn! Species not being OBSERVED to change «is very great difficulty» in thick
402e017 or <whole» many epochs-- such changes would be OBSERVED.-- G. W. Earl's Eastern seas. p. 206-- shot a
405e031 he Americas probably did not.-- Octob. 25th. I OBSERVED in Windsor Park.-- the "Fallow" Deer. which w
435e130 this  structure.-- Some willow trees have been OBSERVED to change their sex,-- this effect from  age,
443e153 ated by its roots: now it is curious Mr K. has OBSERVED that to graft from the roots is the best  way
446e162 mna only reproduces itself «in England, as yet OBSERVED» by buds-- (the other three by buds & seeds «
449e168 ation eggs of two shapes & colour.-- Eyton has OBSERVED same thing in Brent Goose. Eyton says some of
466t091 in old Secondary Series-- few-- Maer June/41/. OBSERVED 3 plants of Caltha Palustris alone together.
540m089 joy? 17th. August Montaigne (Vol. I) has well OBSERVED, one does not fear death from its pain, but o
543m100 und reasoners.-- all same fact-- for, as Jones OBSERVED, in playing chess however many places, & cont
558m152 faces of people in different trades &c &c &c I OBSERVED the Asiatic Leopard. quarrelling. mouth wide
568n019 with all uneducated.--»-- Old man at Cambridge OBSERVED the ignorant. merely looked at picture as wor
573n037 gation Marianne. says. that she has constantly OBSERVED that very young children. express the greates
580n061 - in latter respect he thinks he certainly has OBSERVED that some people of very weak intellect (As M
610o031 could  not find out-- Directed his letter, & I OBSERVED he had written Wilson & pointed it out; he wa
287c157 at relation of theoretical astronomer to plain OBSERVER +++. between Mammalia & fishes, one penguin,
337d022 nce would not be discovered by an unscientific OBSERVER-- «Transactions of the Entomological Soc» A
257c056 aucoup des mêmes oiseaux. que nous avions déjà OBSERVES en Patagonie. ou au moins des espèces t;es-an
368d114 lay beauty of plumage.-- (The females (as Owen OBSERVES) in Raptorial birds largest.-- p. 47, (<"> is
384d161 phrodite differs from Hunter)-- Hunter (p. 45) OBSERVES "every species has a disposition to deviate f
440e148 ortant one, of growth «Lamark. Vol II. p. 120. OBSERVES it commences only, when growth stops».-- Spal
544m104 rly education.-- fear of death!!! as Montaigne OBSERVES. distinct from pain, for one hates pain  from
357d072 respect to extinction of larger forms.-- From OBSERVING way the Marsupials of Australia have branched
389d177 ow connected (This seems case, for by careful OBSERVING cattle can be bred in & in.)-- [The loss of p
536m070 feel  angry-- such efforts prevent anger, but OBSERVING eyes thus unconsciously discover struggle of
545m105 cts of interest.-- do/ I was much struck with OBSERVING how the Baboon (<Macaco> «Cyanocephalus Sphyn
570n027 Old Graecians living amongst naked figures, & OBSERVING powers common to savages???].CD-- The existen
165g126 of two instincts crossing most remarkable ever OBSEVED? Shows that <nervous> brain makes thought Glen
033r043 Humboldt There is long discussion on Pumice «& OBSIDIAN:» in the I Vol. Humb: There is rather good ab
035r045 es in rocks. connected with & alternating with OBSIDIAN must clearly be chemical differences. & not t
059r121 ase contained lime.--All bear close analogy to OBSIDIAN, & all show chemical action as well as effect
416e070 difficulty  in crossing [& this most important OBSTACLE to my theory] CD without the hermaphrodites m
066r141 eagues. dimensions: Bynoe informs me that in OBSTRUCTION Sound, in the narrow parts which break throu
115a097 chain  being broken through like that near-- OBSTRUCTION Sound in S. America The very general absence
214b176 - This came so often «that» it was difficult to OBTAIN a litter without this defect, Very curious cas
255c052 ancestor  of all birds, & so for birds. We thus OBTAIN an abstract idea of a bird-- An animal with sk
```

Page **(Key Word)**
282c142 cular trades.--» then the second race would not OBTAIN a cast of washing men-- but might hire the pre
398e006 the latter pages in the history are perfect, we OBTAIN a glimpse only of the changes which the govern
398e006 the government is subject to.-- further back we OBTAIN here & there in order a scattered page; we fin
472s01r ve been hybridised == Has tried several year to OBTAIN seed, but the pods have (except this one year
539m082 ck» so much more easily than man, so may animal OBTAIN it far more easily, in proportion to variablen
549m120 <happiness> «pleasure». A wise man will try to OBTAIN this happiness. though he sees some «intellect
576n048 .)» «no, for plants have instincts» «either» to OBTAIN a certain en<s>«d»: & intellect is a modificat
615o038 Two dogs having very different instinct always OBTAIN peculiarities of external configuration. [RHC]
106a073 t angle owing to momentum. which the water has OBTAINED.-- If {P} inclination be great where arrow st
117a102 al Magazine (before June 1838) that 70. F were OBTAINED 100 miles E of Staten land. bringing up pebbl
182b048 Journal Vol VI. P II. p 89.-- Lieut. Wellsted OBTAINED many sheep from Arabian coast. "These were of
281c139 atal.-- ‖Relations of analogy being those last OBTAINED.-- less firmly fixed & therefore most subject
336d017 s parents «stock» without it be small & slowly OBTAINED NB. The longer a thing is in the blood, the m
336d018 eases, mental disposition, stature, are slowly OBTAINED & hereditary; <but if> if the change be conge
336d018 the change be congenital (that is most slowly OBTAINED with respect to that individual) it is more e
426e107 o perfect prolifickness.-- (<a series might be OBTAINED>: but the intervention of domesticated ie new
491q002 ing 2. How have late varieties of Peas &c been OBTAINED? 3.. Whether the viviparous grasses & onion,
501q011 ing near each other & see whether cross can be OBTAINED-- I name these three plants. because they can
606o022 ndard, by which real objects are judged; & how OBTAINED.-- implanted in our bosoms.-- how comes it th
620o045 t is owing to some <subsequent> power (reason) OBTAINED by age, which should show the child, which of
264c081 1st habits are changing-- or structure may be OBTAINING, whilst habits slightly preceed them-- From t
566n011 ctively knows distances:, is good instance of OBTAINING <that> «a» faculty in the form of a true inst
261c071 ck Pidgeons.-- & the latter most important in OBVIATING a great apparent difficulty-- preservation of
540m087 one of the ideas are habitual, nor recalled by OBVIOUS associations. as by reading a book.-- Consider
608o026 anization The general delusion about free will OBVIOUS.-- because man has power of action, & he can s
248c033 ed with important structures. <which are less OBVIOUSLY affected by external circumstances> these the
055r107 th number of Craters!--At S. Cruz. there is no OCCASION to wonder what has become of the Basalt. Gone
358d075 r meat, while in kettles on the fire, & on one OCCASION, not withstanding our vigilance a piece of po
364d104 sheep originally. black. & Yarrell thinks the OCCASIONAL production of black lambs is owing to old <s
175b019 simplest coming in & most perfect «& others» OCCASIONALLY dying out;--for instance secondary terebrat
356d070 8'. A Golden Pippen or Ribston do producing OCCASIONALLY (as Fox says) same fruit trees is analogous
364d104 tle with white heads; which years afterwards OCCASIONALLY went back-- (Effect of imagination on mothe
427e109 my views of <Dioecious p> all plants, being OCCASIONALLY dioecious; & really dioecious plants being
450e175 n Sumatra two breeds both small -- Java pony OCCASIONALLY reaches 13 hands.-- Phillipines Pony somewh
051r093 lous action of ocean.--at Ascension. (where OCCASSIONALLY most tremendous surf & loose sandy beach) d
285c151 The space which one branch of the tree if live OCCUPIED after its decay, will be occupied by the vigo
285c151 tree if live occupied after its decay, will be OCCUPIED by the vigorous shoots from each branch No: b
298c191 untain side» of which the central parts become OCCUPIED by a third best adapted kind.-- lower species
459tf1v should during [...] conditions-- every spot is OCCUPIED & has been occupied [...] species, which has
459tf1v conditions-- every spot is occupied & has been OCCUPIED [...] species, which has undergone all the ch
383d159 r.-- Do testes, & ovaria when they first appear OCCUPY their proper positions,-- this would be argume
506q015 ers-- Nectaries-- In Monooecious «order» flower OCCUPY particular position.-- (39) What does he think
550m125 ot mental or ideal ones, <which> «& these» must OCCUPY greater proportion of <each> «every» man's tim
634j55r ers then take their place? Would they not first OCCUPY the Poles? Is this origin of Polar attributes
022r006 ic nodules in Gneiss. Epidote seems commonly to OCCUR where rocks have undergone action of heat. it i
029r032 atter.-- Dr A. Smith says. that Boulders do not OCCUR in the South African plains.-- Sydney no I beli
043r070 ion from «more than» one orifice <...> does not OCCUR at same time: this is contrasted to contemporan
044r072 lavas have flowed from centre-- Pisolitic balls OCCUR in the Ashes which fill up theatre of Pompeei (
055r106 an. & without earth.--Moreover that such do not OCCUR on the beaches. Perhaps these facts attest a <m
070r153 y should two of the most closely allied species OCCUR in same country? In botany instances diametrica
079r176 Slate & overlying Limestone Balls of Silver ore OCCUR in do veins. At Huantajaia. Humboldt says, mur
103a064 by <vi> considering how close the dislocations OCCUR & therefore that the crust might be considered
110a087 ystals must have recrystallized, as such do not OCCUR in either dike or fragment. junction certainly
158g082 n sides of hill where Boulder lies. buttresses «OCCUR» high up on Shelf 2d «in Upper Glen Roy» In thi
182b046 origin of Quinarian system is true, it will not OCCUR in plants which are in far larger Proportion te
226b222 .-- Mauritius? <In plants where do most species OCCUR.)> Although the Horse has perished from S. Amer
284c149 t vary in the species of it.-- «where does such OCCUR?» now some such «characters» rule are used by N
367d112 this must be so, else avitism could hardly ever OCCUR.--».-- & if that cannot be formed, genetal orga
370d115 -!! S. America.-- Such difficulties will always OCCUR if animals are thought to have been created.--
386d167 m that of plants. (though latter does sometimes OCCUR in animals). latter the division taking place f
399e010 are unknown Eastward of the Navigators. Snakes OCCUR there,, but are unknown in Hervey or Society is
493q003 close genus do more than three primary colours OCCUR in relation with species-- answered «by Henslow
498q008 l mingle the infinitesimal varieties which must OCCUR.-- ¿is «it» not these infinitesimal varieties,
199b118 domestic individuals & not races, without the OCCURENCE of one of the most general laws of life, the
036r051 438., of similar fact near the Red Sea.--which OCCURRED in a sandy place.--(the sound was long & prol
217b183 plains the loss & expense, (must probably have OCCURRED to every one) of rare breeds of dogs, from ow
533m060 d not make him more humble.-- it has obscurely OCCURRED to me that Capt. F. R. candour & ready confes
056r110 adation longer than outer parts.-- The common OCCURRENCE of a breccia of primitive rocks between that
528m038 s Rhythmical beauty is shown by Humboldt from OCCURRENCE in Mexican & Graecian to be single cause) th
529m043 re there has been no cloud on the mind, every OCCURRENCE for a day or two are absoluteley forgotten.-
049r088 cal form I incline to <latter> former; & thus OCCURRING in groups.--As these greenstone rocks are see
271c107 s in two countries, without intermediate ones OCCURRING in intermediate country-- ie. mundine groups.
285c150 her must be left out of case, that difference OCCURRING.-- It will be necessary to show hybridity fro
508q016 rm. (2) Any more cases of diseases, generally OCCURRING in man being transmitted through females, lik
116a100 r. formed of rounded pebbles-- it is clear gold OCCURS in submarine alluvium, or sublittoral formatio
156g072 head where <granite &> «veined» gneiss <unite> «OCCURS» abundantly with perfectly rounded pebbles of
185b057 taking a subdivision of Heteromera same. thing OCCURS with regards to other tribes in that same fami
240c004 & handed down, with ease, is analogous to what OCCURS in plants.-- All these facts clearly point out
410e053 to those forms. where the termination of change OCCURS.-- those discovering the formal laws of the co
452e181 , It APPEARS probable,«?» that the Hippopotamus OCCURS in India. in the Jungles of Borabhum & Dholbum
524m020 a person twitching when a disagreeable thought OCCURS, is closely analogous to Epilepsy & convulsion
624o051 cy in every instinct to the species in which it OCCURS. [or, more correctly «in which it» has been so
023r010 uttle fish bones floating on the surface of the OCEAN, before arriving at the Abrolhos shoals ‖ -- N.
023r011 .» The volcanos originated in the bottom of the OCEAN. & the present Volcanos have been said to be me
028r030 ed by Organic powers. We know the waters of the OCEAN all are mingled. These reflections might be int
032r039 ich point the waves would not reach. If now the OCEAN should suddenly fall, (3) the case would be as
032r041 ition & accumulation is brought into play As in OCEAN & Air; there are «likewise» differences of temp
037r052 rock situated beneath low water in the Southern OCEAN not being buoyed with Kelp.-- With respect to d
044r073 eflections might be introduced on great size of OCEAN; especially Pacifick: insignificant islets--gen
047r081 a retreats & then breaks: i e to form a wave in OCEAN. is not this [Fig. 3] {P} form present, i e a p
051r093 ?? Breccia--Stratification? Anomalous action of OCEAN.--at Ascension. (where occassionally most treme
056r109 ed. -- If man could raise such a bulwark to the OCEAN, who would ever suppose that its age was limite
068r146 k ∴ «in earliest stage» when covered up beneath OCEAN).--The first dislocations & eruptions can only
068r146 n during first movements, and therefore beneath OCEAN, for subsequently there is a coating of solidif
069r151 ate space protected.-- Oh the vast power of the OCEAN! Make a grand analogy between Wealden & Bolivia
089a017 .-- Certainly Volcanic-- CD[Might not bottom of OCEAN boil; yet heat never reach surface.-- Journal d
102a059 ace. that strata yield.-- In Undulation in open OCEAN. as pebbles would be lifted up & down. on coast
107a077 of Elevation if not crust much thinner beneath OCEAN than above it no because heat proceeds from gre
114a095 NB. because lowest. first accumulated in bed of OCEAN With the exception of sandstone rare to have an
122a114 f solid. because if of fluid, the waters of the OCEAN would obey that Law. & lie over the platform:--
122a114 st being cut of-- if part of «cold» crust under OCEAN, became thicker, then when fluid moved [...] {P
126a124 arer the surface. especially at bottom of great OCEAN, where the circulations from surface can take p
132a137 rth's crust.-- yet heat does increase,-- but in OCEAN does not. (see resumè p. 536)-- «NB. I cannot u
132a138 here is no volcanicity beneath lakes)?» Suppose OCEAN represents proper <state> temperature of earth.
132a138 rrots argument against central heat to warm the OCEAN).-- and M. Parrot does conjecture that in Score
132a138 case volcanicity has warmed it. Is not cold of OCEAN accounted for, by the circulation being greater
132a139 than any others-- may not the cold «bottom of» OCEAN. (with fresh sediment added to bottom) be cause
219b193 When this volcanic point appeared in the great OCEAN, have made plants of American & African form, m

242c018 Lesson No (p. 24) batrachian in isles of Great OCEAN says in conformity with Bory's Views.-- <Says>
284c147 eriods, depends.-- on relations of desert, open OCEAN, &c this probably on long average, equal quanti
373d133 eck on numerical proportion in shells in Arctic OCEAN. p. 350 Grallae in Wealden. oldest birds. p. 41
379d151 the alliance with the Black sea.-- it would be OCEAN, what is land to continent-- Original Paper, wo
478z003 Iguana. not pass Lat. 28 degree North p. 239 In OCEAN between Lat 56 degree and 57 degree only inhabi
483z013 lae (Less) parasitical on Vellellae in Atlantic OCEAN Gould agrees with D'Orbigny, that Serpent Eater
510q017 s & where? How are current & winds in Antarctic OCEAN: are they from West, like as between Australia
529m040 ist have illdefined notion of land covered with OCEAN, former animals, slow force cracking surface &c
605o18v ttle to the effect: as when we look at the vast OCEAN from any height.-- 6 That the superiority & "in
242c018 123 Crocodile at New Guinea. All the isles of OCEANIA have the Scincus with golden streaks-- the lac
245c024 296) Columba Kurukuru found in all Malasia-- & OCEANIA, offers many varieties in each place to puzzle
044c072 ca. India.--remembering S. Africa. Australia. . OCEANIC Isles. Geology of whole world will turn out si
246c026 ay whether species & varieties P. 708. Columba OCEANICA (Less) inhabits Caroline < «NB. The» > isld.
023r011 fissure in a deep & therefore weak part of the OCEAN'S bottom. With respect to Sharks distributing fo
028r029 hat all Lime is not accumulated in the Tropical OCEANS detained by Organic powers. We know the waters
051r095 per tidal band, I do not see how to account for OCEANS power.--excepting when pebbles are brought int
118a104 of linear bands of movement in Indian & Pacific OCEANS.-- (2d--) does not explain first formation of
132a138 B. I cannot understand the argument, that cold <OCEANS> «lakes» bottom. if not colder than mean of pl
132a139 tion being greater, than the transmission from OCEAN'S bottom.-- (according to M. . Parrots own hypoth
641j28r f larger Seals-- Are Porpoises numerous in cold OCEANS I think not.-- Does this bear on, the absence
070r155 rdovise range. Signor Rozales tells me at seven OCLOCK Novem <5th> Concepcion most violently shaken,
091a027 N. Phil Journ. p. 410. 1828 Ed. N. P. J. p. 105. OCT. 1828. gneiss in India (falls of Garsipa) dip 30
322c269 Octob12th.-- Sir G. Staunton's Embassy to China. OCT. 12t Kotzebue's two voyages, skimmed well. do Lu
323c269 s. -- 20th. Carlyle's French Revolution 3? vols. OCT: -- 26th Blumenbach's Essay on Generation. Engli
403c022 unt of Flora of pacific, given in my coral paper OCT 14th Macleay says, that <every> «any» character
515q022 up in other respects different?--. Important.-- OCT. 44 Tell J. Anderson's statement of English Hors
325c267 Marcel de Serres Cavernes d'Ossements 3d. Edit. OCTAV. (good to trace Europèan forms compared with Af
566n010 s fleshy lips denote sensuality (p 192 Vol. III OCTAV. Edit)-- certainly neither a Minerva or Apollo
203b138 to Continent & England» Loudon Mag: Septemb or OCTOB 1837 Westwood has written paper on affinity and
322c270 Ile Julie Waterton's Essays on Natural History. OCTOB 2d Transactions of Royal Irish Academy. ----do
322c270 rish Academy. ----do Lavater's Physiognomy ---- OCTOB 3d Malthus on Population W. Earls'. Eastern Sea
322c270 Malthus on Population W. Earls'. Eastern Seas. .OCTOB12th.-- Sir G. Staunton's Embassy to China. Oct.
397e004 tion of new forms.-- their number & corelations OCTOB. 4th. It cannot be objected to my theory, that
399e009 en outwards in the grand crush of population.-- OCTOB 10th. Saw. two undoubtedly rabbits in poulterer
402e019 all Malay Countries W. Earl Eastern Seas. p 233 OCTOB 12th Kotzebue's second «1st» Voyage. Vol II p.
402e020 h skimming over with regard to this archipelago OCTOB. 13th.-- Kotzebues first Voyage. Vol II. p 367.
403e023 nted for, on any other it is the will of God.-- OCTOB. 16th. A very strong passage might be made-- wh
404e025 ery individual & therefore effect of climate.-- OCTOB 19th. When reading. L'Institut: .1838 p. 329--
405e031 emain found in the Americas probably did not.-- OCTOB. 25th. I observed in Windsor Park.-- the «Fallo
405e032 said to be hybrid between silver & gold fish-- OCTOB. 26th. If. hereafter. M. angustidens be found t
520mIFC getting on Table. Col. Sykes) Private Finished. OCTOB. 2d. This Book full of Metaphysics on Morals &
563n001 done <g> crowing.-- instances of expression.-- OCTOB. 3d. Dog obeying instinct of running hare is st
564n004 hose rules, which he wills to give his child.-- OCTOB 3d. Was told by W of Downing. Coll. that he had
564n006 t bring some stable foundation to argue from.-- OCTOB. 4th. Seeing some drawings «in Lavater, P. cii
568n018 sure.-- hence judgment, which is part of reason OCTOB 19th. Did our language commence with singing--
570n025 carried on as trick) «Shrugging aroused acting» OCTOB 25. Why is modesty, mixed with triumphant feeli
572n033 -- say instinctive actions. senses. notions &c OCTOB 30th-- Dreamt somebody gave me a book in French
223b209 mentioned from Fernando Po Zoolog. Proceedings OCTOBER (?) 1837 Contrast New Zealand with Tasmania Th
332d001 on owing to struggle July 15th. 1838 Finished. OCTOBER 2d As a proof. what <trifling> «unknown» cause
400e013 merican form.--» off province of Guadalaxura-- OCTOBER 11th.-- Uncle John-- says Decandoelle, distrib
451e178 tame animals into this region between April & OCTOBER & like man almost (this looks inaccurate C. D)
563n001 56 Looked through & all other Books May 1873-- OCTOBER 2d.. 1838 Essays on Natural History Waterton d
567n013 alogy probably false, may lead to something.-- OCTOBER. 8th. Jenny was amusing herself--, by trying
570n026 iest arts.-- Mem.-- Stokes-- arrow heads &c &c OCTOBER 27th Consult the VII discourse by Sir J. Reyno
609o029 it may be a consequence but nothing further.-- OCTOBER <8> 2d. 1838 Those emotions which are stronges
609o030 inence to be a vice & especially in the female OCTOBER d. 1838 perhaps insist?? Two classes of morali
579n060 muscular movement. The Blushing of Camelion & OCTOPUS; strong analogy with my view of blushing-- in
639j28r hale in some respects Chamaelion like power in OCTOPUS & Chamaelion.-- C. D. Sucking feet in Frog. Wa
068r147 of igneous rock replace strata. & it is nothing ODD to find them injected by veins & mass[es] [Fig.
205b144 with my planariae Leaches out of water Does the ODD Petrel of F. del F. take form of awk, because th
206b149 one single one.-- Will not this account for the ODD genera with few species which stand between grea
239c002 cidly explained.-- Mr Yarrell states that if any ODD pigeon crossed with common pidgeon, offspring m
358d085 le, the organs doubtless would shrivel up.-- Yet ODD they should have so much sexual character as the
385d163 sion. (Mem: so it was said little cock «yet very ODD loosing visible powers» in Zoolog Gardens. & Kin
424e102 ds-- now this is independent good case, but very ODD since these crows are mixed in England-- for I p
453e183 fferent varieties being raised by seed is highly ODD-- as it is not so with the esculent vegetables--
472s02r flowers had suddenly withered, & to day saw very ODD dusky humble (with pollen) on legs go from clump
472s02v ithdrawn it.-- Saw 4 more Bees at work-- another ODD genus-- & a small common Humble»- & more of same
486z018 rarity of insects in T. del Fuego.-- Hence it is ODD that Amber insects of Europe have Tropical Forms
497q006 ds «especially if date of wood be known» & other ODD places & see what plants will spring up which wi
520m002 struck with this fact.-- the resemblance was in ODD twiching of muscles, & general manner of holding
526m028 nt from the inventive power,-- this, though very ODD is perhaps true.-- mem Erasmus & mine taste for
538m079 rofound knowledge or other & unusual line-- both ODD appearance about eyes.-- one botanist & great kn
549m122 at is happening with other animals-- is far from ODD nor is it odd he should have had them.-- with le
549m123 g with other animals-- is far from odd nor is it ODD he should have had them.-- with lesser intellect
592n101 choes & the female chamaeleon court the males by ODD gestures. In one of the six (?) first Vol of Sil
592n105 therefore the» cause, and origin being so is not ODD. ; for instance wild cattle & deer pursuing a wou
610o031 inted it out; he was astonished, & said how very ODD.-- --could not think what had put Wilson into hi
621o047 ink almost anything nasty. (<& accidentally> «by ODD association» comes to this conclusion, not owing
524m019 -- which does appear a real difference, between ODDITY & madness.-- but then people do not well recol
525m026 dnatured, & vice versa Walter Scotts remark how ODIOUS an illtempered fat man looks, shows same conne
189b071 Apteryx Dacelo & Kingfisher same colours Strong ODOUR of negroes, a point of real repugnance.-- Water
337d021 one large eye-- have they smell, do plants emit ODOUR solely for others parts of creation) & another
322c270 bie on the Intellectual powers. Hunters Animal OECONOMY. edited by Owen. read several papers-- all, t
367d113 onnected with cheek pouches.-- Hunter's Animal OECONOMY p. 45 "One of the most general marks is the s
369d115 n nature, as well as fully to understand their OECONOMY, is now universally admitted.""-- p. 483. Owe
375d135 of adapted structure into the gaps <of> in the OECONOMY of Nature, or rather forming gaps by thrustin
381d156 ut, it increases in size prodigiously-- Animal OECONOMY by, Hunter. (edited by Owen) p. 34.-- Owen ol
384d151 erent from either parent bird-- Hunters Animal OECONOMY. (by Owen) p. 44. Classification of Monsters.
328eIBC migration of birds Temminck has written "Coup d'OEIL sur-la Faune des iles de la sonde et de L'empir
497q06v e. . seed.-- In letter Mr Herbert says do about OENOTHERA-- (14) Examine pollen of those genera of whi
499q009 ree & see whether they catch pollen-- <Ne> In OENOTHERA bush.-- (19) Theory of mock flowers in Hydran
079r177 e says P 291.-- The Fuegians trace the "chefs d'OEUVRE de l[']industrie humaine, comme ils traitent l
024r015 ms Vide facts in Beechey. on NW coast of America OFF Cape of Good Hope 70 fathoms 20 miles from the s
036r050 ult charts for distribution of pebbles.--Plains. OFF coast of Patagonia.-- British channel &c &c. The
037r055 o the Bay of Bengal. dimensions? Strong currents OFF the Galapagos.--strata must be accumulating whic
038r057 eruption when whole summit of mountain is blown OFF; & again when in great crater. different little
045r075 ers)--At the Calabrian earthquake things pitched OFF the ground. «Ulloa states that Volcanos!! were i
065r140 eviously. Mem. pebbles of Porphyry.--Falklands.--OFF East Coast. -- Capt. Cook found soundings. (end
065r140 (end of 2d voyage outside coast of T. del Fuego. OFF. Christmas sound. -- «(Think some 60 fathoms, no
089a019 erecting machine to see if water fell. -- <Keys OFF extreme point of Flori[da]> Excellent paper on E
104a066 e great thickness is affected, they would be far OFF In Discussion on dikes argue impossibility on fi
104a069 Cordillera discussion, deep sea, fragments fall OFF cliffs. but then how spread abroad?-- There is t
109a082 the number of bones lying at the bottom of sea. OFF coast of England.-- Sea must always on actual be
118a103 come scoriform. has thinned upwards & is now cut OFF by denudation it gives one grand idea of amount
130a134 elsted told me of some large fresh Water springs OFF coast of Persia In Glen Roy paper I show crust y
150g035 t just as Dick shows NB. Lake gradually draining OFF would form plains such as those near Bridge Roy
189b072 near Equator in former periods & then splitting OFF.-- If species generate «other species», their ra

Page **(Key Word)**
189b072 e «other species», their race is not utterly cut OFF:-- like golden pippen. if produced by seed go on
196b106 why Marsupials & Edentata should only have left OFF springs <ne> in or near South Hemisphere. Were t
196b106 ere. Were they produced in several places & died OFF in some? Why did not fossil horse breed in S. Am
200b121 do not show the possibility of common branching OFF.? Accra, Coast of Africa. Clay slate. strike. SS
223b210 nge, (as some animals do more than others, & cut OFF limbs & new ones are formed) but yet propagates
226b220 Po & Coast of Africa. equally good.-- Small isld OFF New Guinea same fact see Coquille's Voyage.-- Ga
259c065 is on the same principle that, cut a sheeps tail OFF plenty of times & you will have no tail (example
262c073 , (most false to say no passages; nature is full OFF them.-- wading birds partially webbed. &c &c.--)
265c083 eny-- many dogs in England must have been lopped OFF & sheeps tails cut yet there is no record of any
266c093 rabia not even antelopes though common on islets OFF Arabian Coast.-- ‖Vol VI. p. 89.-- Lieut Wellste
275c120 cock «is not that old variety» & then recrossing OFF spring. till size diminished, but feathers conti
279c133 llent observations of sickly offspring being cut OFF-- so that not propagated by nature.-- Whole art
302c201 st be sprung from some source anterior to giving OFF of these two families, but we see analogies betw
304c207 from same parent with other birds,, or branched OFF anteriorly think what principles are there to gu
311c227 ville Ovington's Voyage to Surat, floating isld. OFF coast of Africa .p 69. with tall grass. & p 72 h
332d005 re. skin black & wrinkled-- fur short. (tail cut OFF in progeny peculiar) limbs very long, eyes very
342d033 uck have been shot wild (escaped from Carolina?) OFF New York. therefore instincts not imperfect.-- A
358d075 ork 3 lb was taken from a boiling pot, & carried OFF by one of these birds" Case of bird of different
362d099 he doe to the victorious stag. who rubs the skin OFF horns to fight-- is analogous to the love of wom
362d099 to brave men.-- Effect of castration horns drop OFF., replaced by hairy ones. which never «dry up &»
362d099 laced by hairy ones. which never «dry up &» peel OFF their skin (not being wanted for war) & hence ne
362d099 in (not being wanted for war) & hence never fall OFF.‖ Curious the rapidity of the change in 5 or 6 w
363d103 1/2 Muscovy & common duck were often caught wild OFF coast of America.-- showing hybrids can fare for
372d129 es not so a seed.-- Bud probably is like cutting OFF tail of Planaria, the whole grown to that part.-
372d131 ere possible to support the arm of Man, when cut OFF , it would produce another man.-- That the embry
386d166 he part, of what is good for the whole.-- if cut OFF nerves in snail. (Encyclop of Anat & Phys) can m
389d176 endency to budding, which wishes to throw itself OFF.--» as might be inferred from annual plant being
400e013 » besides Galapagos do. p. 376. Isle Tres Marias OFF Mexico with small Hares & raccoons.-- «S. Americ
400e013 small Hares & raccoons.-- «S. American form.--» OFF province of Guadalaxura-- October 11th.-- Uncle
415e067 iseases, & such constitutions only being cleared OFF by fatal diseases.-- The Value of a group does n
436e138 in sight of each other will sing till they drop OFF their perch.-- p. 101-- Kingfisher in northern p
437e139 & not being sufficiently weakened by wounds got OFF from the young ones while they were amusing them
472s02r der sigma & draw it out over & over again & wipe OFF pollen. (as a needle becomes covered) so whole s
472s02v it cleaned sucker & <I think> pollen was scraped OFF, which appeared like Heartease pollen.-- the pol
493q003 ee notes In varieties is there any difference in OFF spring from A. into B. from B. into A. as takes
525m025 ages with impure breed.-- A cat had its tail cut OFF at Shrewsbury & its kittens <h> (in number 3) ha
529m042 most violently purged «believe worms were passed OFF.» & vomited, but who when he recovered. was foun
534m063 ys Dr Darwin mistaken in saying common wasp cuts OFF wings of flies from intellect. but it does it al
535m063 ake it act wrong, as I have done when taking lid OFF <tea> side of tea chest, when no tea do. p. 233.
536m071 a likes smell of that fatty substance it scrapes OFF its bottom.-- it is relic of same thing that mak
537m075 them with alum, so more slowly does animal leave OFF <t> instinct, when attended with bad effects Mar
555m143 front) (having changed hanging into his head cut OFF) as kind of wit, showing he had honourable wound
555m144 .-- I changed I believe from hanging to head cut OFF. «there was the feeling of banter & joking» beca
570n023 surprise with negation.-- like shaking something OFF shoulder-- or is it from inspiration, which acco
570n024 his lips together & shrugs his shoulders & walks OFF,-- I think shrugging connected with many emotion
579n060 ng-- in former irritation on a piece of skin cut OFF made the blush come.-- it is an excitement of su
582n067 - for short-lived constitutions will then be cut OFF.-- <Horses> Colts cantering in S. America capita
606o025 s it, when one wishes to do some action (as jump OFF a bridge to save another) & yet dare not -- one
633j53v ome hybrid, whose flower great tendency to break OFF p. 292. Mac. has long rigmarole about plants bei
549m119 , good has been done-- «& conscience free from OFFENCE» -- pleasure of intellect affection excited, p
620o045 following of ones conscience. & palliates the OFFENCE; one always admire the habit formed by «obedia
337d022 f Man, with expression <of a> of Monkey, «when OFFENDED» who loves, who fears, who is curious &c &c &
608o026 y one(P)-- We cannot help loathing a diseased OFFENSIVE object, so we view wickedness.-- it would how
073r164 in great analogy to Hungary. = Veins of Zimapan OFFER zeolite. stilbite. grammalite. pyenite. native
438e142 nstitution of some may resist the means Man can OFFER of changes.-- as desert «or rock» plant probabl
583n069 B. insects which have never seen their parents OFFER best cases of instincts].CD all this may be tru
640j167 hese views, one would infer that Mollusca would OFFER few species, or rather be very slowly changed &
640j167 tebrata much so.-- so far true, but do not fish OFFER a most striking anomaly to this. Have they wide
032r039 Z) of mass to be removed & from the resistance OFFERED to the greater lateral extension of the waves.
292c170 & observe the character of the demonstrations OFFERED of the singular views there offered, & he must
292c170 nstrations offered of the singular views there OFFERED, & he must be a zealous man in the cause if hi
377d138 ies recognition with pleasure, as when food is OFFERED, as much as to say give me-- the other when Dr
505q014 Will try to get me some to look at:-- Was once OFFERED a hive. of these small Bees-- at Sundorne has
569n022 se.-- Lutké Voyage in Carolinas Vol II p. 132. OFFERED to take a savage, said his wife would be griev
640j167 adjoining coast. Admirable explanation is thus OFFERED.-- From these views, one would infer that Moll
062r129 s; generally first arrives:-- New Zealand rats OFFERING in the history of rats, in the antipodes a pa
114a095 banished diluvial waves). & likewise <tells,> «OFFERS a presumption» it has been excessively slow be
124a118 Erasmus suggested to me that Herschel's theory OFFERS no explanation of intermittent action of eleva
177b025 ; so that passages cannot be seen.-- this again OFFERS contradiction to constant succession of germs
245c024 umba Kurukuru found in all Malasia-- & oceania, OFFERS many varieties in each place to puzzle natural
299c194 absolute species formed. The Anagallis perhaps, OFFERS another case of permanent varieties in wild st
429e116 ere else not even in branch valleys-- M. Ramond OFFERS no explanation.-- Poet Cowper, describes his t
581n065 g> if so & seeing how simple an explanation it OFFERS of radical diversity of tongues.-- [Emotions a
170b002 inal molecule has done).-- This appears highest OFFICE in organization (especially in lower animals,
538m079 Politics, «both bad jokers.--» the other army OFFICER, horticulture & religious sects.-- yet Allen.
095a039 e Planchon talks of very much of Gypsum.-- The OFFICERS of the Bonite. French discovery ship, found c
371d127 to be more easily trained up to the (required) OFFICES" &c &c Owen illustrates case of Dingo (he allu
218b190 agascar oxen with hump.-- p 173. Voyage par un OFFICIER du Roi Mem. Capt. Owen's story of cats on Wes
234b255 ascar-- p 173. Vol I. Voyage à France-- Par un OFFICIER du Roi.-- Mackenzie Travel. p. 280. says catt
478z004 Curious arrangement of animals in rays Par un OFFICIER du Roi Rapid growth of Coral-- RN. p 24 Bouga
145g001 enerally received opinion that male impresses OFFSPRING more indelibly than female p 367 Quarterly Jo
180b037 Twelve of the contemporarys must have left no OFFSPRING at all, so as to keep number of species const
186b059 bably this is first step in dislike to union, OFFSPRING not well intermediate Lyell Vol III. p. 379.
190b075 there is any difficulty in such marriages or OFFSPRING show tendency to go back-- there is an end to
199b120 rents; repugnance «generally» to marriage <if OFFSPRING not fertile> <,but producing> «before domesti
199b120 tion, afterwards none or little with <fertile OFFSPRING> » fertile offspring; marriage never probably
199b120 or little with <fertile offspring> » fertile OFFSPRING; marriage never probably excepting from «stri
199b120 robably excepting from «strict» domestication OFFSPRING not fertile. or at least most rarely & perhap
199b120 least most rarely & perhaps never female.--no OFFSPRING: physical impossibility to marriage.-- ¿wheth
202b134 is doomed that only one species of family has OFFSPRING the chance is that these wanderers would not,
202b134 would be wanderers.-- Some however might have OFFSPRING, & then «V. L. Institut p 245. 1837» we shoul
204b140 ian cattle & common produced very fine Hybrid OFFSPRING, much larger than the dam, from those importe
204b140 from those imported by Ld. Powis Hybrid dogs OFFSPRING seldom intermediate between parents.-- How ea
204b140 Mr Yarrel thinks oldest variety impresses the OFFSPRING most forcibly-- Esquimaux dog & Pointer «Game
204b141 force» Mr Wynne has crossed Ducks & Widgeon & OFFSPRING either amongst themselves or with parent bird
205b145 aracter of both parents are observed in their OFFSPRING, but that of the male more frequently predomi
216b179 shayes.-- Mr McClay is inclined to think that OFFSPRING of Negro & white will return to native stock
216b181 ts &c &c.-- When black & white men cross some OFFSPRING black others white which is more closely alli
217b183 ogs of opposite breeds are crossed, sometimes OFFSPRING quite intermediate sometimes take strongly af
220b197 on & not production. But who can say, whether OFFSPRING does not depend on mind or instinct of parent
220b198 pecies, dogs not, but if we take character of OFFSPRING. Hogs not different. some dogs different.-- H
222b203 -- When species cross & «hybrid» breed, their OFFSPRING show tendency to return to one parent, this i
223b211 ith (B) (& B having crossed with (C) prevents OFFSPRING of A. becoming a good species well adapted to
225b219 ishing of the Parent produce any character on OFFSPRING? Does the mind produce any change in offsprin
225b219 ffspring? Does the mind produce any change in OFFSPRING? if so adaptations of species by generation e
225b219 NB. Animals very remote. ass & Horse. produce OFFSPRING exactly intermediate.-- Reference to Pig & Do
228b229 bston Pippin tree <go> producing crab. is the OFFSPRING of a male & female animal of one variety goin
228b230 ate some plants, under glass bells & see what OFFSPRING would come from them. Ask Henslow for some pl

(Key Word)

```
229b233 boriginal at Cape: crossed with English Bull. OFFSPRING very like common English.-- Hottentots say gr
230b235 population  in California cessation of female OFFSPRING: applicable to any animal-- Athenaeum.  <Jan>
230b236 w of small differences producing more fertile OFFSPRING.-- 1st. All variation of animal is either eff
233b250 oise shell--& grey-banded. ¿species?-- thinks OFFSPRING of cats sometimes heterogenous.--  Australian
239cIFC e room How are varieties produced, by picking OFFSPRING? Instances of old Breeds taking greatest effe
239c001 that oldest variety, takes greatest effect on OFFSPRING. Thus presuming those varieties to be  oldest
239c001 uimaux dog when crossed with pointer produces OFFSPRING much nearer Esquimaux than Pointer.-- He  has
239c002 th Bay mare, only bay a few generations, that OFFSPRING would be chesnut.-- On this principle I may a
239c002 any  odd pidgeon crossed with common pidgeon, OFFSPRING most like latter, because oldest variety.-- -
242c017 e very slow, owing to physical change, slow & OFFSPRING not picked.-- as men do. when making varietie
249c034 time,  then generation will «only» produce an OFFSPRING capable of producing «two» such as itself.--
249c034 essed on blood., will cross & produce fertile OFFSPRING in the first case it will either produce no o
249c034 in  the first case it will either produce no OF[FSPRI]NG or such as not capable of producing again Th
255c053 s of the principle of incessant change in her OFFSPRING, ,has invented all kinds of plan to insure st
258c061 little  more vigour being given to the chance OFFSPRING who have any slight peculiarity of structure.
259c063 to the parent so tending to produce effect on OFFSPRING-- but WHOLE race of that species must take to
259c066 d, shows than there is reference to more than OFFSPRING (like atavism) & shows my «view of» generatio
260c068 ls are placed in white rooms to give tinge to OFFSPRING.-- Darkness effect on human offspring.-- whit
260c068 inge to offspring.-- Darkness effect on human OFFSPRING.-- white, snow.-- the fine green of vegetatio
271c106 ce Two grand classes of varieties.; one where OFFSPRING picked, one where not.-- the latter made by m
276c122 ose., that something intermediate, between no OFFSPRING & ordinary offspring:-- this gradation is inf
276c122 intermediate, between no offspring & ordinary OFFSPRING:-- this gradation is infertile offspring. wit
276c122 nary offspring:-- this gradation is infertile OFFSPRING. without organs of generation?! By profound s
279c133 ercourse.-- excellent observations of sickly OFFSPRING being cut off-- so that not propagated by nat
280c136 it is less repugnant to nature to produce one OFFSPRING unlike itself, than to produce that capable o
293c172 fter sleep, «strong» analogies with memory in OFFSPRING.-- Some association in such cases recall  the
294c178 d, & lose (mere sickness)? fertility ¿because OFFSPRING too unlike.--?? Memoire by Charles  D'orbigny
299c195 g parent in rich soils & their seeds produce <OFFSPRING> variety. wild carrot. made into biennial dom
312c228 en heterogenous, he feels sure of this, first OFFSPRING most like «parents» Mother.-- like dogs Smith
312c228 es homogenious owing to no attempt to keep up OFFSPRING, are not half lion & tigers ditto. (see Griff
312c231 ull & <be» the provincial breed, & that first OFFSPRING thus produced are better, than those bred  in
317c249 y now & then a short-tailed cat.¿cut? has its OFFSPRING short tails /one born at Maer. Tuckeys voyage
331dIFC cated very fertile animals increased?-- Where OFFSPRING, heterogenous, in plants are the number of se
333d006 Here  then we have clear case of heterogenous OFFSPRING from one impregnation ¿is this one impregnati
333d007 that when a wild animal is crossed with tame, OFFSPRING always take most after wild.-- i.e that «alw»
333d007 s seen several cases of foxes & dogs crossed, OFFSPRING always more resembled foxes than dogs (Mem Ja
335d013 se, what has not been, will not remain,-- yet OFFSPRING must be somewhat like parents,-- therefore of
335d013 ng must be somewhat like parents,-- therefore OFFSPRING will tend to go back, or have none-- the argu
335d015 f abortive hybrids.-- If mules did breed, the OFFSPRING would «as in all other animals» be like eithe
335d015 ng new species» will have no tendency to have OFFSPRING like parent, but as they must like or there w
335d015 re will be none, therefore a mule can have no OFFSPRING.= but as «badly» deformed people & as mutilat
335d015 , so will the worst mules (as real mule) have OFFSPRING,-- slight deformities «as supernumerary finge
335d016 tility.-- (but many animals are fertile, when OFFSPRING infertile,-- two considerations are here comb
335d016 n last page, we have seen mules could have no OFFSPRING, & this being case, owing to the corelations
335d016 o transmit «only» those peculiarities, to its OFFSPRING, which have been gained slowly, now all the m
336d017 n, so it is impossible to transmit them, & as OFFSPRING must be like parent, therefore mule has no of
336d017 ng must be like parent, therefore mule has no OFFSPRING & therefore no generative organ.-- Same Prop.
336d017 n animal Either parent cannot transmit to its OFFSPRING any <peculiarity> change from the form  which
336d017 parent  stock. but if both parents are alike, OFFSPRING must be like Hence mutilations not heredetary
336d019 according  to ordinary laws, the character of OFFSPRING would vary, or rather they would not have off
336d019 ing would vary, or rather they would not have OFFSPRING-- On the idea of generation being a <slip> «b
341d031 ossed cock Guinea Fowl with Pea <cock» Hen.-- OFFSPRING female, yet so infertile never even in  seven
342d033 breed most freely. & produce somewhat fertile OFFSPRING produce heterogenous offspring. It appears ce
342d033 mewhat fertile offspring produce heterogenous OFFSPRING. It appears certain that hybrid Muscovy & Com
344d040 eference to those insects, which have fertile OFFSPRING. Entomostraca & Aphides. The extreme differen
344d041 us to superfoetation, & to successive fertile OFFSPRING in Entomostraca & Aphides Developement of sex
344d042 rate bitch-- tried to breed from her, but her OFFSPRING came out one big & one small. Now Jones, befo
345d044 erally-- received opinion that male impresses OFFSPRING more than female, yet instances given on oppo
359d088 -- Male Penguin was crossed with hen Canadian OFFSPRING, I should say in every respect most like Peng
360d090 ncerned. Same man had crossed Jackal & dog-- (OFFSPRING did not go to heat. but parts swelled, though
360d091 al (not artificial variation) which impresses OFFSPRING most.--«& not time» thinking of the Penguin
360d093 reasts, <when» of a sort that would allow the OFFSPRING to have some different kind of mottle, each f
368d114 from  the species all females being most like OFFSPRING, Q (how is this with those females which  put
371d127 cked-- the vis formativa goes entirely to the OFFSPRING.-- but surely in all these cases an unseen ch
378d147 Keep Is it Male that assumes change, & is the OFFSPRING brought back to earlier type by Mother?--  do
380d153 the vis formativa goes entirely to the female OFFSPRING-- this is clearly the converse of annual bein
380d155 birds) After animal has copulated., though no OFFSPRING, Milk sometimes comes in Mammae & even when b
385d163 s great difference between hybrids & inter se OFFSPRING in latter being unhealthy.--» males «bred  in
387d168 blood.--  Looking at ovum of mother & ovum in OFFSPRING, as similar to the several ova in mother. (wi
387d168 h the case of successive copulation impresses OFFSPRING more & more with the added «like Lord Moreton
387d168 ew Smith,» difference.-- If A. B. C. D. E be <OFFSPRING> «animals»: if «x» male impresses ovum <of> i
387d169 for  first time & if their <D & E» «all their OFFSPRING» inherit the same peculiarity in lesser degre
388d171 s does not apply to potato.-- With respect to OFFSPRING being determined by imagination of  Mother.--
389d176 ion One copulation may impregnate one or many OFFSPRING.-- it affects the subsequent offspring, <when
389d176 r many offspring.-- it affects the subsequent OFFSPRING, <when> though other male may have copulated.
389d176 opulated.-- two animals may unite & each have OFFSPRING by same mother.-- one animal will fecundate f
389d176 le for several births, & even produce fertile OFFSPRING like both father & mother, or very close to e
389d176 ual plant being prolonged till it has bred.-- OFFSPRING without coition or addition of differences. s
392d175 & forming species)-- [Aphides having fertile OFFSPRING. &,male the lesser peculiarities.-- brilliann
401e017 that  female plant impresses main features on OFFSPRING.--‖ & as adaptation,-- however mysterious suc
404e026 one mentions of influence on parent affecting OFFSPRING:‖ Australian dogs have mottled coloured puppi
404e026 ditions to parent may be become favourable to OFFSPRING before.-- -- now this is never stated.-- Rega
406e035 distinguishing <after> which parent impresses OFFSPRING most is whether mother has had any  offspring
406e035 offspring  most is whether mother has had any OFFSPRING before.-- -- now this is never stated.-- Rega
406e036 s never stated.-- Regarding the similarity of OFFSPRING to Parents same laws appear to hold good. wit
417e075 s-- there is little tendency to vary. & hence OFFSPRING are hybrids.-- Mr G. B. Sowerby <tell> showed
417e077 pendent of sexual differences) a variety. The OFFSPRING of true <pare> hermaphrodite, would of course
427e110 ples» but probably would more indelibly stain OFFSPRING-- it would not reach one apple sooner than <o
427e110 prefer  other. but produces greater effect on OFFSPRING-- Mr. Herbert says «p 347. Amyyralidae» Plant
430e117 s is important as showing small variations in OFFSPRING of wild animals.-- grateful & intelligent.--
430e118 breed pure.-- «& so in plants effectually the OFFSPRING are picked & not allowed to cross.-- » Has na
432e123 igs, with small chinese boars &c &c &c» p. 10 OFFSPRING take more after father than mother; illustrat
434e127 ) to do so, the effects are equally handed to OFFSPRING.-- Whewell's anniversary address 1839, p. 9.,
441e149 ian Dog is not affected by domestication, yet OFFSPRING are,-- if Australian Dog, could bud,  analogy
441e149 alian Dog, analogy tells us, <be» OFFSPRING would be similar to <f> first form.-- The gr
441e149 rst form.-- The great effect of conditions on OFFSPRING, but not on individuals is very curious & imp
459t013 en grown up". Saw at Mr Bell's at Hornsey the OFFSPRING of a black & white <duck of pecu» <drake» wit
460t013 . crossed with common goose <to has» «produce OFFSPRING with» so much of the swan-goose in appearance
473s07r -- may come in corresponding time of life of OFFSPRING-- No peculiarity in external structure can be
491q002 ing Plants. 1. Uniformity of hybrid & Mongrel OFFSPRING 2. How have late varieties of Peas &c been ob
492q003 animals exactly alike be interbred will OFFSPRING be uniform.-- Mr Ford Has M. Sageret WRITTEN
493q003 ss & horse-- important. {In crosses does male OFFSPRING take after male parent & vice versâ = History
493q004 erent heredetary constitution, to see whether OFFSPRING infertile.-- (4) Does the number of pulse, Re
512q019 s> sometimes turns into fine beast. would its OFFSPRING have ugly calves. also turning into fine beas
512q019 ine beasts.-- For comparison with hybrids, is OFFSPRING of short-horn bull & hereford cow similar  to
525m025 er says on authority of Mr Wynne that bitch's OFFSPRING is affected by previous marriages with impure
537m076 en Polynesian mothers ceased to destroy their OFFSPRING--¿ yet perhaps if they had murdered their chi
627o053 bject is attained (the gratification of one's OFFSPRING) is not the aim of the agent, for it does not
```

Page **(Key Word)**
175b020 ook at Megatheria, armadillos & sloths as all OFFSPRINGS of some still older type some of the branche
223b211 different species & intermediate character of OFFSPRINGS accounts for uniformity of species & we Must
335d015 roduced very quickly)» sometimes have similar OFFSPRINGS, so will the worst mules (as real mule) have
342d033 ear species or varieties produce heterogenous OFFSPRINGS.-- «are not the hybrid pheasants & grouse di
046r079 considers   that Plutonic rocks are generated as OFTEN as Volcanic. I consider latter as accidental on
148g029 <below> «by» pass of Glencoe-- the erosion may OFTEN be due to rivers-- By Roy Bridge, a tongue of f
151g040 lders (one of Gnelss remarkably water worn) are OFTEN times of rock not in immediate neighbourhood, (
154g061 ols if--. why should it deposits River terraces OFTEN descend by flights the terraces if the largest
172b007 ced & increase slowly, from many enemies. so as OFTEN to intermarry who will dare say what result Acc
183b051 tries.-- In mundine genera, the nearest species OFTEN come very remote quarters. (NB. if Plata Partri
191b082 mal affecting all the Progeny of female insures OFTEN mixing of individuals. Here we have avitism the
205b145 --" does this apply to where same animal breeds OFTEN with same female p. 28 "It <is> wrong to enlarg
214b176 on would say the tail was broken-- This came so OFTEN «that» it was difficult to obtain a litter with
216b179 & white will return to native stock (the cross OFTEN whiter<)> than white parent) the mulattos thems
217b184 after either parent. about <half & half tim» as OFTEN one way as other.-- He has known case of good p
222b207 alone can this comparison be instituted. People OFTEN talk of the wonderful event of intellectual Man
254c049 ed together, & same organ «where eliminated is» OFTEN repeated, as mouths in Polypi, Surely not corre
256c054 iking to see M. Bibron looking over reptiles he OFTEN had difficulty in distinguishing which were spe
256c054 yet a glance would tell from which country,-- I OFTEN disputed for a moment,-- Galapagos, S. American
276c124 hened &c there may be tendency to divide, which OFTEN enough repeated would cause an unequal number o
286c154 e feeling» by making slave of his fellow black, OFTEN wished to consider him as other animal-- it is
292c171 e & not others.-- We even see they must be done OFTEN «to be habitual» or of great importance to caus
298c189 tit lark placing withered grass over nest, when OFTEN looked at.-- this most puzzling whether instinc
298c191 e form) lower species afterwards would probably OFTEN be destroyed.-- or regrafted with fresh arrival
302c201 en sufficiently Multipliied become affinity yet OFTEN retaining a family likeness, & this I believe t
309c222 owls & hawks only external» intermediate groups OFTEN have full structure «of one class» & full of se
312c227 ntation. certain birds in many families, «+very OFTEN in number 5» will have long tail.-- in raptoria
317c248 eferences very good. also "Rays Wisdom of God." OFTEN refer to these.-- Also some few facts at end of
341d032 associated with the ducks.-- most strange voice OFTEN in the night, like peacock.-- tail as long as P
359d087 umption as long as animals are healthy which is OFTEN the case, & why should organic affections alway
363d103 man told me that 1/2 Muscovy & common duck were OFTEN caught wild off coast of America.-- showing hyb
366d111 hearing maed servant cleaning door outside, as OFTEN as she touched handle, though really fully awar
370d117 ted. as must spices &c &c The line of argument «OFTEN» pursued throughout my theory is to establish a
372d131 e a thousandth of inch should produce a Newton is OFTEN thought wonderful. it is part of same class of
378d148 t themselves, & I presume with common ducks. so OFTEN, that it was impossible to say which was origin
380d154 ed.-- A capon will sit upon eggs, as well as, & OFTEN better than a female.-- this is full of interes
383d159 Hunters.   Free Marten N.B. the common mule must OFTEN have been dissected Zoolog. Garden. Sept 16." H
385d165 be trusted, this is Lord. Moretons law.-- "How OFTEN do we find in the son, the character, constitut
389d177 effect as too little.-- in (latter case female OFTEN takes males but does not produce) tendency to d
401e017 illiancy of inflorescence Gardeners. by chance <OFTEN> sometimes graft pears on apples. they will liv
509q017 me Human Anatomist. Has he dissected any animal OFTEN, which has abortive bone. (ask more about the l
511q018 length of ears» & skeleton, & skin= Van. Voorst OFTEN writes to Lowe (7) In breeding. pointers. Bull-
520m001 , said, although constantly seeing him, she was OFTEN struck with this fact.-- the resemblance was in
522m012 ld people. (Aunt. B.) when they hear a thing it OFTEN does not take any effect at the time, but some
522m012 afterwards it calls up pain. or pleasure. & is OFTEN recurred to & mentioned as a thing which had ju
522m012 ct of <early> «slight habitual» intemperance.-- OFTEN accompanied by extreme anger, at not being unde
522m013 sposition, like people in violent intoxication, OFTEN ends in insanity or delirium.-- In Mania all id
523m015 -- My Father says after insanity is over people OFTEN think no more about it than of a dream.-- Insan
523m017 ion &c by coming on suddenly. Ans no.-- because OFTEN, if not generally, does not really come on sudd
527m033 she never builds castles in the air-- Catherine OFTEN, but not of an inventive class.-- Now that I ha
529m041 remember my pleasure in Kensington Gardens has OFTEN been greatly excited by looking at trees at [i.
530m046 ds patient sinks.-- When a muscle is moved very OFTEN, the motion becomes habitual & involuntary.-- w
530m046 & involuntary.-- when a thought is thought very OFTEN it becomes habitual & involuntary,-- that is in
531m050 ed. whereas it is the importance.-- people very OFTEN forget where money is placed.-- (How often one
531m050 very often forget where money is placed.-- (How OFTEN one forgets where put one key. where all keys a
538m079 er like one which though learnt in infancy, had OFTEN been repeated: Now it is remarked that A. Bessy
542m095 evidently habitual when transferred, (also how OFTEN) to the tale of a wearisome man.-- Is frowning,
542m096 ering the canine useless.-- The distinction «as OFTEN said» of language in man is very great from all
545m105 ns.¿ it is the <becom> impression becoming very OFTEN unconscious, which makes the idea unconscious,
545m105 little interest, & those which are viewed very OFTEN.-- former do not give rise to ideas so much. as
547m112 whether I had thought what clothes to take (how OFTEN one cannot tell whether one has rung the bell.,
553m137 shoved out its lip, looking absurdly sulky «as» OFTEN as keeper spoke to it,-- but he thinks not sulk
554m140 y picking his nose with «a» straw.-- Jenny will OFTEN do a thing, which she had been told not to do.-
555m142 - I could only perceive that the American ones, OFTEN put on a peevish expression, but not nearly so
555m142 put on a peevish expression, but not nearly so OFTEN <that> hardly ever the expression of passion wi
569n020 s seem simplits case of Association.-- Elephant OFTEN given food & word open your mouth said, recogni
586n082 ncert with others in building comb-- My faculty OFTEN will turn out to be instincts, & so in some sen
591n097 ligent Keeper... Zoolog. Garden told me. he has OFTEN watche tame young wolf & it never dropped its e
591n099 instinct of self-preservation is disobeyed-- I OFTEN have «as a boy» wondered why all abnormal sexua
599o05v argument of original formation.-- declension &c OFTEN show traces of origin.-- Mayo Philosophy of Liv
602o11b belief things you can give no proof for, & one OFTEN replies "what you say is perfectly true, but yo
604o018 the association of power &c &c with height, we OFTEN apply the term sublime, where there is no real
605o18v sublimity   5 The emotions of terror & wonder so OFTEN concomitant with sublime. adds not a little to
605o18v by its accompanying & associated sensations so OFTEN gives, when excited by other means, as moral ex
605o19v y, &c &c. produces an inward pride & glorying. (OFTEN however accompanied with terror & wonderment) <
605o19v e call sublime.-- It appears to me, that we may OFTEN trace the source of this "inward glorying" to t
605o020 sures from beauties of nature & art." But as we OFTEN see people who are susceptible of pleasure from
608o026 do not. come into play.-- †It may be urged how OFTEN one try to persuade person to change line of co
610o032 many cases of ideots knowing things, which are OFTEN repeated in a wonderful manner.-- as the hour o
620o044 conscience) is always present (which is indeed, OFTEN felt at very time it is disobeyed) & is sure gu
621o047 eculiarity of organ of taste, for when grown up OFTEN conquers it). It will be only rarely that it th
622o048 b> subject to their instincts & associations.-- OFTEN feelings which do not lead to action are repres
640o051 by association from education & imitation, has OFTEN been perverted from want of reason.-- Hence as
520m001 hinks bodily complaints «& mental disposition» OFTENER go with colour, than with form of body.-- thus
639j28v termine that, the long legged one shall rather OFTENER than any other one. survive. in ten thousand y
211b163 some long & some short: therefore like dogs.-- OGLEBY says, Wolves at Hudson bay breed with dogs.--
253c046 over so wide a range, & Mastodon angustidens.-- OGLEBY has facts to show that Australian dog introduc
341d030 tralia New species of Moschus, characterized by OGLEBY, who observed that the young of this animal, w
342d034 must have been on elevation of continents) but OGLEBY well answers that nearly all F. W. Fish are Ab
594n115 companion. Descent --Affection & [...] Monkeys «OGLEBY» seen Zool. Soc-- 1838 remember with distress
473s05r rig) says that he can certainly tell Trout from OGWEN, Capel Curig & some other lakes, (different wat
069r151 lo Where gone to? Intermediate space protected.-- OH the vast power of the ocean! Make a grand analog
213b172 in every sea, from Equatorial to extreme poles.-- OH. Wealden.-- Wealden. Do the N. American. Tertiar
232b246 two countries» finding a very hot day, «in one», OH we will take a day from the equator to add to th
283c145 gue & fortify their minds with such sentences as "OH turn a Buccinum into a Tiger."-- but perhaps I f
291c166 this.-- love of the deity effect of organization. OH you Materialist!-- Read Barclay on organization!
303c205 there are no genera. if Mammalia are adduced. say OH look to your fossils, now if extinction had gone
160g092 een arm of Glen Bright flowing into E. end of L. OICH, & waters flowing into west end with obscure te
161g098 he new shelf flows into canal between L. Lochy & OICH. is a brook on the Lochy side of it-- the terra
162g105 per tells me, that Loch Lochy is 8 ft below Loch OICH wh is 92 ft above sea-- Loch Ness 40 ft above d
164g122 [...] Wenlock Edge [blank] L. Lochy 12 ft 96 L. OICH 12 84 {P} 29.958 - 1.17 28.788 + 28.8 30.372 29
164g123 58 - 1.17 28.788 + 28.8 30.372 29.200 1.172 Loch OICH 92 each Loch 8 ft. The Metamorphic conglomerate
099a048 <In Area of this> {P} If surface covered with OIL should shrink. film parallel to longer axis. But
127a126 r of the Irish Academy Vol 8. p. 118 water no--. OIL will freeze if cooled in a closed globule of gla
127a126 freeze if cooled in a closed globule of glass. (OIL may be cooled to 0 degree!)-- shows effects of p
243c019 oigné, il y a peu d'années, d'admettre que ces OISEAUX eussent leurs représentants dans de si hautes
257c056 ined-- all species & found "beaucoup des mêmes OISEAUX. que nous avions déjà observés en Patagonie. o
025r019 ragments of shells will generally be found to be OLD & dead)» «(I have not kept a record)» In looking

```
031r038 n of upheaval To Cleavage add other instances in OLD world of symetrical structure. East India Archip
035r045 &c  &c My results go to believe that much of all OLD strata of England. formed near surface: Mem Pata
067r144  - says that Falkland fossils decidedly belong to OLD Silurian system. Apply degradation of landlocked
068r146 penetrated by the repeted trifeling injections.--OLD vents would keep open long after emersion, but i
072r159 bubbles.  -- No Volcanic action on coast line of OLD Greenland, close to W of Jan Meyen Isld.--Mr Bar
086a004 t tops of mountains we ought to sympathize with. OLD doubters of what are fossil shells.-- accustomed
090a021 ewster speculates from believing meteorolite but OLD Planet, that inside our globe melted magnetic me
112a089 of  consideration. When discussing nucleuse's of OLD volcanos within Cordillera-- allude to Lyell's v
120a110 k-- & scraps on Salsisbury Craigs. Kept amongst <OLD> papers read before societies.-- Sir. J Hall Vol
127a127 mation of cones beneath sea.-- with reference to OLD submarine orifices in Cordillera Geograph. Journ
156g069 s much notched on Maculloch's supposition;-- the OLD ravine, where water entered are not proportionat
160g089 & W connecting Glen Bought & Glen Tarf a perfect OLD Loch, making <several> two divortiums aquarum, v
162g101 bout> Head of which is so interesting. enters by OLD tower called Glengarry (Nead Roy told me) it  is
173b011 as nearest land, which were late arrivals others OLD ones, (of which none of same kind had in interva
197b112 each  animal made for itself does not agree with OLD & modern types being constant. Cuvier's theory o
203b138 l heat.-- But then shells-- Mr Yarrell says that OLD leaves when mingled with newer, hybrid variety pa
209b155 be  in rays-- from certain sports.-- Agrees with OLD Linnaean doctrine & Lyells. to certain extent Vo
213b171 they not be greyhounds?-- Yarrell's remark about OLD varieties affecting the cross most well worthy o
225b218 be  said to deceive man. as likely as fossils to OLD rocks for same purpose.!! Can the wishing of the
227b225 ability of succession from what we have seen. in OLD world, & on amount changes which may happen--
229b233 outh coast. Elephant he believes is mentioned by OLD writers on extreme Northern Coast. Hippopotamus
233b250 ccount of fossils of Sewalick «India» Monkeys of OLD World. Crocodiles. Anoplotherium.-- <M. Jerrod>
239cIFC ies produced, by picking offspring? Instances of OLD Breeds taking greatest effect Account of the [..
249c035 me  some capital information ¿Carnivora of New & OLD word. do not form two sections is this not conne
257c058 niferous wood in Coal Measure.-- highest fish in OLD Red Sandstone.-- Nautili n----. it is useless t
272c109 ae & Muscicapidae of new World, but not found in OLD-World--. + + If in any «well developed» family (
275c120 by crossing with common Polish cock «is not that OLD variety» & then recrossing off spring. till size
275c121 yellow one & crossing with duck bantams procured OLD variety.-- The pidgeons which have such differen
275c121 I am sorry to find Mr Yarrell's evidence about OLD varieties is reduced to scarcely anything.-- alm
277c126 e.. rupestris -- good species it is reverting to OLD plan, but reason now assigned for doing so There
278c131 fts list a rat said to have been found!! rodents OLD inhabitants most important!! like Dipus of prese
280c135 haracter.-- N.B. If two species were excessively OLD, they would not make hybrids, whereas two  newer
284c149 rm in birds is visible, when young, but not when OLD.-- thus speckled form of young blackbird. good r
294c177 explain the rest,-- Lonsdale says he has seen in OLD Book last Bear in England killed in year 1000. r
295c183 o war not [not located] Erasmus says he has seen OLD Stallion tempted to cover old mare by being show
295c183 s says he has seen old Stallion tempted to cover OLD mare by being shown, young one.-- Many African m
307c216 rted «into female», it will be splendid argument OLD female, turning into cock, abortive spurs. growi
308c217 es vibrate-- «seeing no tea brought back memory» OLD habit of putting tea in pot made me go to tea ch
308c219 isted since time of earliest Egyptian drawings & OLD Testament» Domesticated animals having same idio
317c249 only appendix) of Congo Expedition, NB. I met an OLD man--, who told me that the mules between canary
317c250 carcely to be distinguished from it.-- & several OLD acquaintances. which grow on the lower region of
325c266 suggested, of organ being worn out as. otherwise OLD whores would not have children Turners embassy t
334d010 n, according to Erasmus preferring young mare to OLD, explained by Stallions, (according to Fox) bein
342d034 dominals. ..that order first converted-- is it an OLD order Geologically? Owen says relation of Osteol
342d035 ourbon nose". if this be not imagination.-- then OLD peculiarity overbears the crossing with  females
346d048 lingham,-- habits peculiar,-- young one 203 days OLD butted violently. & fell.-- gore to death the ol
346d048 ld butted violently. & fell.-- gore to death the OLD & wounded,-- see Annals. vol. 2. 1839.-- are bad
347d049 to improve.-- yet fish same as, or lower than in OLD days: «for a very old variety will be harder  to
347d049 same  as, or lower than in old days: «for a very OLD variety will be harder to vary, & therefore more
362d100 ncy to breed at particular times. Mr Yarrell has OLD book 1765? Treatise on Domestic Pidgeon, in whic
364d104 occasional production of black lambs is owing to OLD «story» return.-- The Revd. R. Jones told me pre
364d104 nation on mother. white peeled rods mentioned in OLD Testament placed before sheep-- it has been thou
366d112 cross. how comes it that the tame donkey has. CD[OLD Buffon should be read on Mare My view, why hybri
378d148 According as child is like parent, so is species OLD: Hence <young> Kingfisher & pies, have long  had
378d148 riness about them quite remarkable". instance of OLD Species transmitting so much longer its Mental p
379d151 ia. & still further back reptiles & Cephalopoda: OLD Jones remarked to me, that one of the children o
380d153 & in.-- Mr Yarrell does not know of any case of OLD Male. becoming like female, though many of old f
380d154 f old Male. becoming like female, though many of OLD female becoming like cocks.-- It is very singula
388d172 dite is supported by change which takes place in OLD age of female assuming plumage of cock, & beards
388d172 le assuming plumage of cock, & beards growing on OLD women = Stags horns & testes curious instances o
401e016 I can fancy cowslip producing primrose return to OLD stock, but not primrose producing cowslip  Uncle
418e084 ifications, become as much fixed, as if added to OLD individuals,, during thousands of centuries,-- e
418e084 housands of centuries,-- each of us, then <is as OLD, as the oldest animal>, have passed through as m
426e106 ce abundantly infertile hybrids, & the fact that OLD varieties do not so much affect first race, as i
432e125 a practised geologist can really comprehend how OLD the world is, as the measurements refer not to r
434e129 red  Lychnis grows in <south> Wales & certainly <OLD> only white in Cambridge, in some counties somet
436e136 rt Jackson Shark-- Owen thinks Australia part of OLD World <If> It «may» be said, that wild animals w
436e137 ver, extend round world.-- Quartz of Falkland.-- OLD Red Sandstone-- Van Diemen's land.-- Porphyries
437e140 one  day a rabbit escaped into a hole, where the OLD Eagle could not find it..-- The parent bird anot
439e143 akes it determined by a facility in returning to OLD type Mr Herbert showing the extreme facility  of
460t019 deal»  modified <& Many Forms lost; if> «of this OLD stock (which from action & reaction grew more co
461t041 ) of some species-- (especially of mammifers) in OLD beds & existing species is valuable because it s
465t089 ly -- & how many estuary formations are there in OLD Secondary Series-- few-- Maer June/41/, observed
468t112 the  anthers of long stamens {P} as stamens grow OLD «& shed some pollen». they turn upwards & bend o
470t176 -so that palmated has now nearly disappeared. <& OLD English» But these mules <in our garden> show no
480z077 that  young birds of prey have longer tails than OLD ones-- in America & sexes not of different size-
493q004 ls.-- &c (1) To cross some artificial male with <OLD> female of old breed & see result.-- According t
493q004 cross  some artificial male with <old> female of OLD breed & see result.-- According to Mr Walker the
505q014 pollen from other flowers? Can flys' escape from OLD flower» (14) Has planted seeds of Geranium pyren
509q017 more plain in young Rhinoceros or Whale, than in OLD?? Falconer says all in cases. Owen. Have  talked
520m001 elp thinking, he was prescribing to his father & OLD Mrs Harrison, said, although constantly seeing h
520m002 e as he is very deficient, he was nearly 9 years OLD. when his father died.-- The omnipotence of habi
521m008 ng organs, brought into play by morbid action.-- OLD Elspeth's «in Antiquary» power of repeating poet
522m012 ut did not know [Z]CD when heard isolately.-- In OLD people. (Aunt. B.) when they hear a thing it oft
524m021 delirium after epilepsy, but in the failing from OLD age, they constantly do.-- In Mrs P. <...> of B.
524m022 nds-- My father's test of sincerity.-- People in OLD age. exceedingly sharp in some things, though so
524m022 ils first?-- How is this?-- Does memory bring in OLD ideas <I have elsewhere remarked do> Dogs take p
526m029 Mr Corbet of the Hall understanding. (on hearing OLD association brought up) by sight & not by hearin
528m037 nd of view-- as likewise is novelty of view even OLD one. every time one looks at it.-- these two cau
529m042 es in Philosoph. Transactions, of ideot 18 years OLD eating white lead. who was most violently purged
532m055 artled.-- My Father says he should think that in OLD people, in their dotage, who sing the songs «& t
532m056 k to Shrewsbury,-- then immediately fell into her OLD ways & became fat! What remarkable affection  to
538m078 analogous,  & which I think will lead to fact of OLD people singing songs of their childhood. & certa
539m082 sure immediately thrilled across me, bringing up OLD indistinct ideas of FitzWilliam Musm. I was amus
545m106 doubt  this Baboon. knew women.-- Another little OLD American monkey «(Mycelis)» I gave nut, but held
547m113 es from senses. & <comparing their> «calling up» OLD ones, to be sure of ones consciousness.-- Mayo o
551m127 ue-- Delirium of every degree of intensity-- «in OLD man, he has just seen mind went on RAMBLING till
554m138 asted philosopher himself is chiefly shown in in OLD male.-- A very green monkey (from Senegal he thi
555m143 expression of passion with open mouths like the OLD world ones.-- Though the[y] move whole skin of h
555m143 they  do not move eyebrows.-- (I see some of the OLD world ones move skin of head & ears,-- ∴ some me
555m143 of  lips, in which respect resembles some of the OLD ones.-- -- S. American group sneer.-- Sept 21st
556m146 gh it is then quite useless-- Cats kneeding when OLD, like kittens at the breast now if horns were to
568n019 han a cow--" «so it is with all uneducated.--»-- OLD man at Cambridge observed the ignorant. merely l
570n027 CD-- As forms change, so must idea of beauty.-- [OLD Graecians living amongst naked figures, & observ
574n039 her.-- Hensleigh says. Douglas. «& Spencer», an OLD Scotch Poet, has numerous lines. of poetry.--<s
574n040 ening a Cocoa nut shell at one end.-- Children of OLD people get into habits.-- we probably can hardly
574n041 «teeth»less-jaws. as picture of disgusting lewd OLD man. ones tendency to kiss, & almost bite, that
575n046 only  winged.-- fetches two birds out at once.-- OLD People-- (Antiquary Vol II. p. 77) remembering t
```

(Key Word)

575n046 ing analogous. to instinct, to the permanence of OLD heredetary ideas.-- being lower faculty than the
580n060 h reference to origin of language My father says OLD people first fail in ideas of time, & perhaps of
586n079 tinctive.-- carrier pidgeon just as wonderful in OLD bird as new.-- migration, <only> «only» more won
586n080 sing its direction. equally wonderful in young & OLD.-- These facts point out some essential differen
588n090 he judgments & actions of a young animal with an OLD.-- (dog horse, sow) we perceive great difference
593n105 g pursued.-- A dog turning round & round is some OLD instinct <perverted> handed down & down.-- mem.
594n113 had to carry it out of the room. nearly 3 months OLD. What is absurdity, why does one laugh at it-- s
595n121 ated] Ernest W. playing with Snow. when 2½ years OLD. was frightened when Snow put a guaze over her h
599o004 -- if so what is trembling palsy? Expressions N [OLD & USELESS notes about the moral sense & some met
603o11b not come under imitative art [my view says yes. <OLD> mass of rock--]CD or poetry, CD[my thery says y
610o032 nduce to their health & comforts.-- Both ideots, OLD People & those of weak intellects.-- Westminster
620o046 nter spring his bird. one says for shame (& the «OLD» dog really feels ashamed?) not so puppy, we <do
639j28r elations.-- [There must have been deserts in the OLD world!]CD p. 252 analogy of hand in mole, & Mole
639j28v e upper hand. though continually dragged back to OLD type by intermarrying with ordinary race.-- «The
640j28v .-- «There is no way of eliminating the evils of OLD age, after breeding season, or gaining adaptatio
640j28v or youth most necessary: the fertility of Man in OLD age keeps woman alive: for Man & woman are same:
025r019 ists of organic remains in De la Beche, for the OLDER formations I must believe they «the limestones»
026r021 ere is a resemblance at Hobart town between the OLDER strata & the bottom of sea near T. del Fuego.--
039r061 ation of Limestone, casts of shells, as in some OLDER formations: Mem the envelopes at Coquimbo. the
069r150 ke Bell of Quillota.) (A) in this strata may be OLDER than (B).-- Most important view Urge curious fa
175b020 dillos & sloths as all offsprings of some still OLDER type some of the branches dying out.-- with thi
227b225 ay happen-- ‖ It leads you to believe the world OLDER than geologists think. it agrees with excessive
286c156 .-- Now are the characters which unite these of OLDER standing than constant number of stamens.-- in
305c210 with which they have been crossed--is Alderney OLDER character & manner of wagging tail.-- habitual
332d003 ainst my theory, because it makes the world far OLDER breed-- He believes all pretty much alike.-- M
343d039 n of existence.-- nor is there in the Tertiary <OLDER> geological epochs.-- There are some admirable
443e155 ut fifty years since to geologists.-- & what is OLDER than what Geologists, think: it would be doing,
443e156 ormal. which follow this direction are thought OLDER-- what relation in duration of a planet to our
535m064 om living forms.-- p 458 Upper Silurian, fishes OLDER species. The earwig & a doubtful one of Acantho
055r106 easily does Wolf & Dog cross? Mr Yarrel thinks <OLDEST> «most intelligent» miners to be the richest V
204b140 t as his theory» tells me. he has no doubt that OLDEST variety impresses the offspring most forcibly-
239c001 offspring. Thus presuming those varieties to be OLDEST variety, takes greatest effect on offspring. T
239c001 on pidgeon, offspring most like latter, because OLDEST which have long been known in any country; he
239c002 domesticated varieties a tendency to go back to OLDEST variety.-- -- He says of two varieties of pidg
248c030 al in all species in group we may suppose it is OLDEST race, which evidently is tending to same end,
284c149 n of peak-- altogether original. owing to being OLDEST, & therefore lest subject to variation.-- + <b
296c184 with which they have been crossed--is Alderney OLDEST. & having undergone changes «no near lofty cou
345d043 semblance general. ¿depends upon mother bein[g] OLDEST breed-- He believes all pretty much alike.-- M
348d050 lity.-- (Now caeteris paribus these will be the OLDEST) ‖"The most important characters break down in
373d133 lls in Arctic Ocean. p. 350 Grallae in Wealden. OLDEST birds. p. 411 -- Decapod Crust in Muschelkalk,
374d133 -- <M» p. 417. Magnesian Limestones & Zechstein OLDEST rock in which reptiles have been found. p. 426
374d133 om living forms.-- p 458 Upper Silurian, fishes OLDEST formation highly organized.-- do. p. 461.-- Lo
391d174 e (Are not Coniferous trees generally dioecious OLDEST forms) why are twin in man more like «each oth
418e084 enturies,-- each of us, then <is as old, as the OLDEST animal>, have passed through as many changes,
436e137 ficulty to my theory, is same type of shells in OLDEST formations:-- The Cambrian formations do not h
447e163 on to another.--» -- these simple forms perhaps OLDEST in world & hence most persistent-- if form exc
531m052 ntly excited action of heart.-- now this is the OLDEST <her> inherited & therefore remains, when the
025r016 o [10 degree 32 minute S.] 10 50 Whole coast to OLINDA [8 degree S.] 9-10 = 30-40 {T} at twice or 18-
025r016 {T} at twice or 18-20 <60>--80 120 parallel of OLINDA Shoaler N. of Olinda.--a little WNW of C. Rock
025r016 0 <60>--80 120 parallel of Olinda Shoaler N. of OLINDA. --a little WNW of C. Rock. [5 degree 29 minute
071r156 on in Edinburgh. Phisoph. Transactions.-- Mem: OLIVINE. Volcanic product.=> <Did Peruvian Indians use
093a032 Berzelius. L'Institut. [1837 p. 297]CD thinks OLIVINE a preexisting mineral.-- Mem. Galapagos ∴ Basa
612o035 of year as much as inflorescence.-- [LHC] I here OMIT the case (if such there are) of animals enjoyin
556m144 -- dog knows triumph.--» Sept. 23rd. Horses in OMNIBUS instantly start when they hear ready, but if t
551m129 easily. horses associate sounds may be seen by OMNIBUSS Horses starting, when door shut or cad cries
225b218 t insect «moto being called <Phitophagous> OMNIPHITOPHAGOUS. But it will be said there are latent ins
520m002 rly 9 years old. when his father died.-- The OMNIPOTENCE of habit is shown about meals, no [not locat
634j54v ermination of a God-head.-- the designs of an OMNIPOTENT creator, exhausted & abandoned. Such is Man'
056r109 dary. (as in other unworn islands) we take in at ONCE the stupendous mass which has been corroded. --
099a049 small scale of concretionary action all fluid at ONCE, the films vertical. Ascertain law of attractio
105a071 sea-- beach action -- no one will dispute. sea. ONCE came to Mendoza--. Will they introduce other ca
109a083 oosing answered by this -- No one can doubt. A B ONCE formed low coast.-- Annales des Mines. a transl
118a103 lera. St Helena &c &c.-- in Cordillera, it is at ONCE evident only small proportion of dikes have rea
127a125 used East-clinal lines & c & .-- But Siberia was ONCE thawed. & hence. (when climate hotter) was cold
147g018 Whole very obscure but it is certain there must ONCE have been very considerable mass of waterworn p
172b008 ulate, we must suppose the change is effected at ONCE, -- something like a variety produced-- --[ever
176b024 of affinity in each branch A species as soon as ONCE formed by separation or change in part of count
179b032 ies are less purely bred owing to <first> having ONCE borne Mongrels he has thus seen the black blood
183b053 quadrupeds declining as great reptiles must have ONCE declined.-- Read his theory of the Earth attent
188b068 no further:-- Prof. Henslow says. that when race ONCE established so difficult to root out.-- For ins
216b182 breast.--looking as if many ova-- impregnated at ONCE.-- Dr. Smith considers the Caffers (like Englis
216b182 Englishmen) men of many countenances, as hybrid ONCE. Is not this contradiction to his view of races
230b239 nt will admit of a certain quantity of change at ONCE. but afterwards will not alter. This need not a
234b256 m]» of animals.-- He says Stephens say he can at ONCE tell by general colouring a group of Nebria com
261c070 of species is empirical. show this by instances ONCE grant my theory, & the examination of species f
263c076 his new faith of the lesser of the difficulties ONCE grant that «species» one genus may pass into ea
293c173 of Memory can remember how to swim after having ONCE learnt, & if that was a regular contingency the
294c176 different origin when species rather further.-- ONCE grant good species as carrion crow & rook forme
336d019 mals being created. it is probable if created at ONCE. <wd> according to ordinary laws, the character
342d034 ere met conjecture that all salt-water fish were ONCE salt water (as they almost must have been on el
344d040 wworms) breeding-- <beet> imago state fertile at ONCE.-- Consider this with reference to those insect
352d059 fied by use, every abortive organ must have been ONCE changed.-- what is abortive? when it does not p
404e025 to think that it must have been invented all at ONCE.-- but naturalists if they had series. perfect.
425e105 s Lyell has remarked species never reappear when ONCE extinct-- Lyell's argument about <Tertiary> Isl
427e108 would rapidly increase, & hence number of forms. ONCE formed. would remain stationary, hence all pres
441e151 als as Coralline, or others. which only generate ONCE in a thousand generations.-- any amount of gene
467k103 e & half withered-- I saw Bees going to clover & ONCE this happened.-- And in common Beans it is wond
468t111 alea. Rhodendron. Fraxinella to Anchusa <never> «ONCE» P on Fraxinella <Heartease» «small. Humble ali
505q014 cious. Will try to get me some to look at:-- Was ONCE offered a hive. of these small Bees-- at Sundor
521m009 oke> : . could converse well on any subject when ONCE started,-- could receive a new train through ey
531m050 ed, because some people can remember poetry when ONCE read over.-- The extreme pleasure children show
532m054 pon some object:-- When a man, child or colt has ONCE been frightened & started more much apt, this p
534m063 exertion of intellectual faculty) if ants had at ONCE made this leap it would have been instinctive,
547m111 led, from some quite imaginary cause to start at ONCE to Shrewsbury., vaguely thought of packing up.-
573n036 ut the smallest insect, we wish to be created at ONCE by special act, provided with its instincts its
575n045 nded, or only winged.-- fetches two birds out at ONCE.-- Old People-- (Antiquary Vol II. p. 77) remem
580n061 ic of one kind of intellect is that when an idea ONCE take hold of the mind, no subsequent ones modif
584n072 inctive: the facts of memory of roads long after ONCE visited by horse & dogs. (even blind horses & d
633j54r nce the mistake they are created for them. If we ONCE venture to say plants created to <arrest> «prev
633j54r hy not created to live on alpine pinnacle? if we ONCE to presume that God «created plants to» arrests
634j54v ing "It is the determination to adhere to a plan ONCE adopted; & it is from these very circumstances,
639j28v «as» simple consequence they become long. not at ONCE, but by steps. of which we have manifold traces
472s01r the seeds of Papaver bracteatum, & the Papaver ONCITATE was growing in same garden. & out of 60 seedl
548m116 nything».-- if one was subject to this disease ONESELF, one would only feel sympathy. as for for the
557m148 ay,-- it is, ill-defined fear.-- Yet one knows ONESELF it is quite different from that.-- like «sligh
577n052 certainly very much connected with thinking of ONESELF.-- «blushing» is connected with sexual, becaus
586n078 use power varies in breeds,» something of kind ONESELF knows in walking [one feels inclined to stop a
447e164 d.-- Festuca vivapara F ovina-- propagated like ONI[ON] Poa alpina because vivaparous. Henslow has se

447e165 which is case precisely analogous to the Canada ONION mentioned in Hort. Transact. Aira caespitosa be
491q002 obtained? 3.. Whether the viviparous grasses & ONION, produce flowers, like the Oxalis from C. of Go
059r119 inées du centre d'Ile, vers la mer; ces couches ONT entre elles une correspondance exacte, et lorsqu
618o039 foreign to the use of ordinary language that the ONUS probandi might fairly be laid with those who wo
227b227 sprung up round Galapagos on Pacific side. the OOLITE order of things might have easily been formed.
431e121 T. costatus is in England found in the Inferior OOLITE, & the T. elongata in the uper formations Port
460t019 pposed that this refers to time.-- Marsupial in OOLITE.-- insects, of do orders-- cheiroptera & caeta
100a055 hillips insists of analogy between Australia & OOLITIC period.-- comparison rather loose.-- perhaps w
214b173 on analogy between Australia and «fossils of» OOLITIC Series does not appear to me very strong What
110a086 nblende blending into base-- Salband might have OOZED out of cleavage plates: the crystals must have
110a087 unction certainly most distinct on dike side.-- OOZED from one of the true rocks, most probably from
099a051 m. laminated dikes in Cordillera.!!!-- In stratum OP. let force drag particls to line {P} AB, & likew
160g093 es? & Alluvium» which appear perfectly level, <on OP> dies away on gradual slope-- : on N side.. dies
074r164 -carb & chrom. of lead. oripment. chrysop[r]ase. OPAL:-- Veins in Limestone & Grauwacke: Silver appea
023r011 een said to be merely accidental apertures still OPEN.--The fault like appearance «arising from the m
029r034 bsence of limestone?» have been collected on the OPEN coast. Perhaps as at Concepcion. favoured by ba
030r035 and). <[...]> At Carelmapu.--Within Chiloe:-- On OPEN coast, near where Challenger was lost: I know n
030r035 ters are not now collecting, in the bottom of an OPEN & not deep sea.--(Character of coast regular &
042r068 Mem Bahia blanca P. 204 Vol III. Lyell Owing to «OPEN» faults in mountains: to elevated strata in eoc
068r146 eted trifeling injections.--Old vents would keep OPEN long after emersion, but improbably so long, th
102a059 e surface. that strata yield.-- In Undulation in OPEN ocean. as pebbles would be lifted up & down. on
105a069 idth. for besides more surface exposed. bay more OPEN to turbulence. Bull. Soc. Geolog «1837» p. 320.
185b055 situations as possible.-- Why should we have in OPEN country a ground «do. <w> parrot.--» woodpecker
262c074 3.subtypical) where power arbitrary. leaves door OPEN for Quinarians to deceive himself.-- Give the c
284c147 rent periods, depends,-- on relations of desert, OPEN ocean, &c this probably on long average, equal
363d102 wamps-- (owing to barns, perhaps, not being left OPEN to them,-- . In singing birds, part instinctiv
381d157 ecundation effected in latter; are <it> «organs» OPEN to water? Would not Ferns according to this doc
502q012 t improperly called Canadense-- would it grow in OPEN air in Sweden. Linnaeus found 2 flower. which h
504q013 Menyanthes whose pollen bursts before flower is OPEN--- No (6) There is apple with branch in middl
505q014 ech.-- Lychnis dioica answers this question:- (5) OPEN more Horned oranges.= (6) Figs, flower.--Passio
514q21. about Pinks & Solanum impregnation before flower OPEN. (An. des Sci Where is Boerhaave's paper on imp
527m034 r, is as hard work (abstracting it being done in OPEN air, with exercise &c no organs of sense being
541m092 . Aug. 21st. 38 When a dog in play has his mouth OPEN ready to bark, & lip twisted up, in that peculi
554m138 ke her chair & bang against the door to force it OPEN, when she could not succeed of herself.-- <The
555m143 that> hardly ever the expression of passion with OPEN mouths like the old world ones.-- Though the[y]
557m149 t) & then failing to drive away rival.-- Fear is OPEN mouthed to hear. though in individual case. not
558m152 ved the Asiatic Leopard. quarrelling. mouth wide OPEN, each [lip] drawn back & driving air out of mou
558m152 ing air out of mouth «hairs erect on back» «wide OPEN» with prodigious force.-- making growling, gugg
565n007 in a letter» why is person painted with mouth OPEN.-- why when person is listening is mouth open t
565n007 th open.-- why when person is listening is mouth OPEN to hear well «as one will perceive if in night
565n007 as» animals. comes at distance, mouth is placed OPEN.-- Hence becomes instinctive to fear., as ears
569n020 Association.-- Elephant often given food & word OPEN your mouth said, recognizes that sound as perfe
592n103 n p. 15, 25. 40. 61. CD[a person is here said to OPEN mouth in fright because nature intends to lay o
592n103 en mouth in fright because nature intends to lay OPEN all senses; <do> Horse prick his ears «& snort
637j57v air to water-- bull dog to bulls.-- primrose to <OPEN fields> banks-- cowslip to <banks> fields-- the
422e095 .-- hence as the forms became complicated, they OPENED fresh, means of adding to their complexity.--
547m112 I had rung for Covington. whether he had come & OPENED box, whether I had thought what clothes to tak
567n013 wing what to do with them, came several times & OPENED my hand, & put them in-- like child. Tommy's f
105a070 account for filling up of valleys-- subsequent OPENING a medial gorge by slow erosion. but we have ev
234b255] T. Carlyle, saw with his own eyes. new gate. OPENING towards pig.-- latch on other side.-- Pigs put
309c219 ldren of one parent, races of animals-- argue «OPENING» case. «thus» Educate all classes-- avoid the
540m084 hs for joy, so does dog bark. (not shout) when OPENING his mouth in romps, <so> he smiles. Many of ac
542m096 uliar position, & he holds them this way, when OPENING mouth between interval of barking, now this is
574n040 he fixing of habits,-- for instance the Birgos OPENING a Cocoa nut shell at one end.-- Children & old
592n103 Shutting eyes in contempt opposite action to OPENING eyes in fear The effect of habitual movements
276c123 -) must remember that if they believe & do not OPENLY avow their belief. they do as much to retard,
472s02v n.-- the pollen appeared chaffy, as if sucked?! OPENS & shuts end of sucker, after having withdrawn i
281c137 ra very finely" show how arbitrary & optional OPERATION it is.-- show how finely the series is graedu
627o053 on is, whether this can be resolved into some OPERATION of intellectual faculties-- Will Eugenius all
397e003 eity. But we know from experience! that these OPERATIONS of what we call nature, have been conducted
606o021 ous. that we cannot ever perceive the various OPERATIONS which the mind undergoes in gaining the resu
277c126 mes & giving subgenera. true value.-- as in OPETIORHYNCUS. fulginosus. (a) Falklands (b). F. del Fueg
514q21. by projecting pollen-masses?-- = answered = Has OPHRYS nectary?= Bunbury says no «hollow» spur.-- Ask
185b056 Cara cara at the Galapagos. Fernando Noronha OPHYRESSA bilineata (Gray) new <liza> species, belongin
025r019 eve them applicable to the world.-- My general OPINION from the examination of soundings, from about
026r019 s: Yet this view is directly opposed to common OPINION The Tertiary formation South of the Maypo at o
145g001 ce.-- C. Darwin A. Glen Roy Generally received OPINION that male impresses offspring more indelibly t
244c023 près un examen attentif, et forts surtout de l'OPINION du baron Cuvier, nous ne balançons pas a la re
276c123 ef. they do as much to retard, as those, whose OPINION they believe have endeavoured to advance cause
294c177 ty in my theory, only slight differences. «the OPINION of many people in conversation.» the whole obj
302c202 -- I fear «great evil» from vast opposition in OPINION on all subjects of classification, I must work
304c207 ink what principles are there to guide in this OPINION?-- EXCELLENT PRINCIPLE OF ABORTION ISOLATION o
345d044 ulture p. 367. Dec. 1837. Generally-- received OPINION that male impresses offspring more than female
578n055 by anyone, especiall if it be a person. whose OPINION he regards, <& see how> feel how the blood gus
191b080 lon-- East Indian archipelago.-- West Indies = OPOSSUM & Agouti same as on continent-- 3 Paradupasi i
194b095 reat <rodent> edentate [has been doubted?]CD & OPOSSUM found in Europe now confined to southern hemis
641j29v h mouse by night» Sailing lizards. squirrels & OPOSSUMS «& fish»: flying lizards.--Mammalia. C. D.--
225b217 re there were a thousand intermediate forms.-- OPPONENT will say. show them me, I will answer yes, if
565n008 ntempt, when there is some anger «& respect to OPPONENT» is showed by same movement as sneering,-- it
565n008 .-- it is then more <emblem> manner of hurting OPPONENT by insulting his pride & is therefore of the
565n008 - But contempt mingled with disgust, when ones OPPONENT is considered as quite insignificant, & when
355d068 stock, then species are fertile!; as long as OPPONENTS will «are» not «able to» tie themselves down,
429e114 udes of plants introduced into our gardens (OPPORTUNITIES of escape for foreign birds & insects) whic
026r019 e the Conglomerates: Yet this view is directly OPPOSED to common opinion The Tertiary formation South
334d009 male: & How completely is Lord Moreton's case OPPOSED to this fact & views.-- Fox says a cousin «one
358d076 = "women recognized inferior intellectually"= OPPOSED to these facts are effects of castration on ma
360d092 difference what its kind was.-- but if it were OPPOSED to the difference in other sex, it would be mu
360d093 pecies, might be harder to cross than two less OPPOSED in habits, though externally similar.-- this h
388d172 child. more like father.-- stuff.!-- How much OPPOSED. the Quagga case appears to that of «2» dog be
390d179 to breed (as Sir J. Sebright urges?) one with OPPOSED characters is by impliance to breed two which
466k095 ers being inherited at corresponding age & sex, OPPOSED by cantering horses having colts which can can
537m075 s should have different instincts.-- Fact most OPPOSED to this view, where the moral sense seems to h
537m077 t may be answered effects of education, may be OPPOSED undoubted cases of heredetary pride & in singl
576n047 superior to that of the Greeks.-- (which seems OPPOSED to progressive. developement) on account of da
608o025 variable passions-- when these passions, weak, OPPOSED & complicated one calls them free will--the ch
608o028 on to Education.-- 4) These views are directly OPPOSED & inexplicable if we suppose that the sins of
621o047 then it is his interest to follow it. even when OPPOSED by some natural passion.-- (a) [LHC] The consc
621o047 of others, the feeling that almost (rarely if OPPOSED to natural instincts) any action is either rig
070r153 ame country? In botany instances diametrically OPPOSITE have been instanced: it is Let it not be over
125a120 on great Iceland stream. the 90 miles includes OPPOSITE directions. Mem. S. Cruz. Assuming from Sir.
148g022 all the Alluvium. At Mouth of Caledonian Canal OPPOSITE Loch Leven two terraces perhaps upper one 100
150g038 ends as far nearly as house, the 3d below them OPPOSITE to where side ravine enters On opposite side
150g038 w them opposite to where side ravine enters On OPPOSITE side of valley both extend below the Houses T
153g051 lake «or sea» at successive levels-- {P} Shelf OPPOSITE Glen collarig at bend & here most accumulatio
153g054 tion of sediment? Where ravines enter side by, OPPOSITE entrance into Glen Fintec a kind of landing p
162g100 a mound stretches along, parallel to Shelf on OPPOSITE side & dies away on the steep & rocky gully o
197b112 passages from one to other organ.-- Cuvier on OPPOSITE side; Is Vol of Fish p. 59.]CD Cuvier has sai
200b123 fact is to establish whether in crossing very OPPOSITE races, whether you would expect equal fertili

Page **(Key Word)**
217b183 great care of them. Fox says when two dogs of OPPOSITE breeds are crossed, sometimes offspring quite
225b219 quadrupeds East India Archipelago very good on OPPOSITE tendency.-- Study Ellis & Williams. zoology o
266c091 that in their appearance & manners they are as OPPOSITE as day & night: yet we know how remote the pe
312c231 to the point in question.-- «--merely picking OPPOSITE qualities, with no other means whatever.--» I
345d044 pring more than female, yet instances given on OPPOSITE side,-- «The theory of males impressing most
497q007 spec. of Cape Heath by facility. ¿Knight take OPPOSITE view. Gaertner talks of the several great & n
527m033 n the feeling in any one man.-- Music & poetry OPPOSITE ends of one scale.-- former pleases from inst
537m075 ce of action of approved» Yet as, I think, the OPPOSITE side has been shown-- see Mackintosh.-- Must
541m090 do I not confound action & thought here?) The OPPOSITE extreme of this desultory thought is followin
556m146 does dog put down ears, when pleased.-- is it OPPOSITE movement to drawing them close on head, when
557m147 xpression resembles a fox-- I can conceive the OPPOSITE muscles would act, to when in a passion.-- do
565n008 ent.-- the corner of lower lip are depressed & OPPOSITE muscles used to when angry sneering is in pro
565n008 sneering is in progress.<--> the hypothesis of OPPOSITE muscles will want much confirmation. A grave
569n021 such memory.-- A Melody on flute & Epic poem, OPPOSITE ends of series or harmonious prose.-- Lutké V
592n103 ce to Brun's work.-- Shutting eyes in contempt OPPOSITE action to opening eyes in fear The effect of
595n115 ths. saw a «black» spider monkey brought it at OPPOSITE end of house. & commenced a most lamentable h
617o40v ion of forces balancing each other & moving in OPPOSITE directions. We are satisfied therefore, if we
172b006 n Man it has been said, there is instinct for OPPOSITES to like each other AEgyptian cats & dogs ibis
302c202 ely acquired.-- I fear «great evil» from vast OPPOSITION in opinion on all subjects of classification
568n019 nature. If I want some good passages against, OPPOSITION of divines to progress of knowledge. see Lye
617o40v at we ourselves can originate in any point an OPPOSITION of forces balancing each other & moving in o
281c137 ies & genera very finely" show how arbitrary & OPTIONAL operation it is.-- show how finely the series
215b178 m 59. p 467. Peron G. St. Hilaire has written "OPUSCULE" entitled "Paleontographie" developing his id
492q003 RITTEN on crossing of Cabbages, quoted by (as if ORAL) Decandoelle in V. Vol of Hort. Transacts & M.
421e090 ca of Morocco «Mr Forbes says the Fauna»-- (near ORAN) approach in character ʇo Canary Isld.-- ie Can
242c017 neo & Malacca «& Cochin China» are said to have ORANG-utang & Pongo in common-- Galiopithecus common
313c235 n the Entomostraca. The sexual curiosity of the ORANG outang of (in June 1838 when young male was add
587n088 remember it.-- Why do children pout & not men-- ORANG-outang & chimpanze. pout.-- Former, whines just
188b067 replace <No> Southern kinds-- (I) Gnu reeaches ORANGE river & says so far will I go and no further:-
466t093 Maer June/41/ Rhododendrum-- nectary marked by ORANGE freckles on {a} upper petal; bees & flies seen
505q014 ynoglossum. never germinated 12 Does the horned ORANGE. wh. never has seeds produced good pollen? Yes
498q007 his ever effect of want of nutrition.-- Horned ORANGES so? --Yes, my Father lost this character in gr
500q010 atic Researches) (23) Talk about Thyme. Horned ORANGES. Spallanzani Essay-- Figs 2 kinds of flower an
505q014 ca answers this question-- (5) Open more Horned ORANGES.= (6) Figs, flower.--Passion Flower. (as it is
242c017 rneo & Sumatra both seem to have elephant & has ORANGS, Tapir common to Sumatra & Malacca Borneo &
070o153 nite when weathering into balls. must exhibit ORBICULAR structure.--When we recollect connection of c
070o153 .--When we recollect connection of columnar & ORBICULAR in basalt.-- When we see Avestruz two species
131a135 &c worth reading. L'Institut. 1838 p. 360. on ORBICULAR trap thought to be bombs submarine L'Institut
374d134 lus turbo. buccinum. turritella. terebratula, ORBICULAS, with many extinct forms & Trilobites Sept 25
078b174 no page 173] Under name of Sagitta Triptera D'ORBIGNY has figured animal with setae like my undescri
078i174 ribed[.] p. 140. Flèche of Quoy et Gaimard.--D'ORBIGNY has described it with care to 3 species. I thi
242c018 ys in conformity with Bory's Views.-- <Says> D'ORBIGNY is said to have brought a tortoise & toad from
255c050 e Paper Must be studied.-- <Three p. 7. Am.> D'ORBIGNY. Birds of prey, are distributed in S. America
257c056 s of change.-- hump on back of cow!!-- &c &c D'ORBIGNY (p 108) says having observed B. Tricolor in Pa
260c067 ology, Vol III. p. 226 Wilson's Ornithology, D'ORBIGNY, Spix, &c might compare birds of N. America &
268c099 rrots in Macquarrie Isd.--» very good. Study D'ORBIGNY, & range on West Coast «Guayaquil & Peru» Hens
278c129 an Andrew Smith, «Richardson» a Vaillant, a D'ORBIGNY has travelled this will be most difficult. Sub
291c165 out affinities. conducct of Gould, remark of D'ORBIGNY point out importance of habits in classificati
295c178 offspring too unlike.--?? Memoire by Charles D'ORBIGNY on Plastic Clay of Paris contains many genera
295c183 no new forms only species!! No salamanders (D'ORBIGNY) Rapport. p. 11) in S. America so highly develo
376d136 of this populousness, on the energy of Man» D. 'ORBIGNY. Comtes Rendus p. 569. 1838 says the cross bet
417e079 uals of same species.--). some races of men. D. 'ORBIGNY. affect the common progeny more than others.--
479z005 lluscs Under the name of Sagitta Triptera M. D'ORBIGNY has described my animal with teeth {P} p. 140.
479z006 eaches on leaves in Sumatra Marsden. p. 311 D'ORBIGNY considers Dasypus villosus is true Peludo Cavi
480z006 113 of 1834 On the passeres of S. America. D'. ORBIGNY. L'.Institut. No.-- 221 Good account of Condor
480z007 n the Raptores of S. America translated from D ORBIGNY no IV Mag. of Zoolog & Botany p. 356 Lesson on
480z007 Magazine of Zoolog & Botany. Vol I p. 358. D'.ORBIGNY <considers> states that young birds of prey ha
484z014 ellellae in Atlantic Ocean Gould agrees with D'ORBIGNY, that Serpent Eater-- or Secretary is S. Afric
485z017 ngolo> Chimango-- Diuca?? See Report <by> on D'ORBIGNY on species of Mephites 4 distinct Camelidae. d
116a099 Edinburgh Philosophical Journal Rapport on D'ORBIGNY'S Voyage. good section of Rio Negro beds.-- --
289c161 passing into each other let him look at wings & ORBITS of penguin & then he will cease to doubt :Scal
504q013 nch in middle of tree with flowers near end of ORCHARD.= At Shrewsbury one branch of Rhod. flowered l
498q007 c-- I have some reason to suspect Elms.-- & ORCHIDACEAE plants no other case.-- (6) Will plant accu
506q015 tructure of flower-- Ground nuts (42) How are ORCHIDAE fecundated, as mass of pollen is requisite.--
513q21. r-- <doubt> disbelieve this in Bauers case of ORCHIDIAE Where does J. Hunter use expression of "male
435e133 <poll> masses of pollen of Asclepias placed on ORCHIS (so very different) that the granules exserted
470t178 t honey out of long, curved nectar of Butterfly ORCHIS & Listera? Bryony saw common Bee on: Linn. Tra
473s006 cts. it has three stamens> intermediate between ORCHIS & other plants-- & Wallich has described India
496q05a -- --. touching Mr Brown theory of insect-like ORCHIS-- & final cause of beauty of flowers-- contras
496q05a with honey-- What is use of Bee Larkspur-= =Toad ORCHIS-- How many flowers in minute do they visit?? go
514q21. ants --will get answer= Is pollen of cultivated ORCHIS & Asclepias &-- carnosa?-- good-- Norfolk Isd-
058r116 ar Quito. considers these earthquakes travel in ORDER.-- If we look at Elevations as constantly going
086a003 y pleasing; owing to the movements being of one ORDER. -- There should not be surprise at Horse being
092a031 9 1827 Geograp Journ There are some ideas about ORDER of injected rock being determined by fusibility
218b191 » aspect. That is varieties, though of trifling ORDER are formed by nature. Carmichael. Tristan D'Acu
222b206 n system of articulation. ¿whether type of each ORDER may not be supposed that form, which has wander
227b227 up round Galapagos on Pacific side. the Oolite ORDER of things might have easily been formed.-- With
227b228 events it--» & generation, causes of change «in ORDER» to know what we have come from & to what we te
258c061 rious deer, hence males armed & pugnacious (all ORDER; cocks all warlike)» «thiis wars against in any
258c062 the discussion <after> about affinity & how one ORDER first becomes developed & then another-- (accor
286c156 standing than constant number of stamens.-- in ORDER, or in next family? In considering fossil anima
305c210 ders» of insects, so is there none of reason in ORDER of <ver> Mammals.-- Mem Elephants & dog.-- Ther
306c213 & grandest divisions. but for ones of very high ORDER. not for vertebrata, but mammalia & reptiles &c
342d034 hat nearly all F. W. Fish are Abdominals. .:that ORDER first converted-- is it an old order Geological
342d034 als. .:that order first converted-- is it an old ORDER Geologically? Owen says relation of Osteology o
369d115 mals. with anatomical & Zoological research, in ORDER to establish entirely their place in nature, as
369d115 S. Wales are generally compelled to traverse in ORDER to quench their thirst"-- But New Guinea.--!! S
381d156 androus <or M> Mollusca, with pectinibranchiate ORDER-- the Annelida. All others, <animals,> «are Dio
398e006 ct to.-- further back we obtain here & there in ORDER a scattered page; we find <great> sensible chan
407e038 orld formerly much more so. yet climate of same ORDER as that of S. America.-- (Explained by profound
409e048 ut all the ill effects. -- we see it is not the ORDER in this perfect <uni> world, either at the pres
432e125 ganisms tell nothing about length of time, only ORDER of succession.-- Splendid Pamplet. (published i
444e157 law of balancing of organs.-- In the Batracian ORDER the «32» ribs are wanting. p. 144 in the Icthyo
458t001 alludes to ancient gigantic salamanders-- Every ORDER (except whales) have great prototype!!.-- Copie
461t025 facts & this paper deserves fresh study & whole ORDER of the fish.-- Embryology p. 97. for Man Chapt
486z018 und during short summer far N. where this other ORDER is comparatively rare.-- These views clearly ex
494q004 ere been.-- on what principles calculated.-- in ORDER to guess how many generations in Mammalia. in g
506q015 - form of flowers-- Nectaries-- In Monooecious «ORDER» flower occupy particular position.-- (39) What
514q021 t crossing of plants; especially Papilionaceous ORDER (2) History of fruit trees far north in Scotlan
516q24v d, with different salts & poisons & see in what ORDER plants would reappear after <th> being killed E
543m098 refore there is Instinctual developement in one ORDER, as there is Intellectual in human-- probably s
545m104 ideas.-- ie habits, which must require idea to ORDER muscles to do <certain> the actions.¿ is the ab
547m113 say trains, it may be instantaneous changes in ORDER <to every> calling up ideas of every late impre
565n008 ulting his pride & is therefore of the snarling ORDER.-- But contempt mingled with disgust, when ones
601o009 es by distinct mechanism 2. Sensation of higher ORDER. where the sensation is conveyed over whole bod
601o009 ements «¿» anterior to any direct sensation, in ORDER to avaoid it-- beetles feigning death upon seei
195b101 Astronomers might formerly have said that God ORDERED, each planet to move in its particular destiny

```
551m128 something good was shown him, than when merely ORDERED to do it.-- Plato «Erasmus» says in Phaedo tha
600o08v nscience? we admire what is right by one & are ORDERED to do it by other.-- I suspect conscience,  an
354d062 ach particle is placed in place of last by the ORDERING of the nerves, but in different parts accordi
547m112 «presence»  my servant, «box» my own manner of ORDERING things to be done.-- The senses are closed pr
601o009 ks I daresay good. 1. Sensation is the <conse> ORDERING contraction (that is the only evidence.  when
036r050 n. is not very distinct, from some of the lower ORDERS; it was connected with movement of sand.--it i
179b035 Is this shortness of life of species in certain ORDERS connected with gaps in the series of connectio
192b085 e been born without them.= In some of the lower ORDERS a perfect gradation can be found from forms ma
195b101 in its particular destiny.-- In same manner God ORDERS each animal created with certain form in certa
204b141 rsupial division «do» we not see a splitting in ORDERS, Carnivora, rodents &c, JUST COMMENCING. Kirby
226b221 ue to «the» chance of some one of the different ORDERS being able to survive or chance having transpo
233b252 f species in Molluscs!!! When we talk of higher ORDERS, we should always say, intellectually higher.--
235b263 d animals?-- At what. part of tree of life, can ORDERS like birds & animals separate &c &c Work out Q
301c200 sects, perhaps scarcely one new family & no new ORDERS,-- Wonderful, partly explained on my theory, =
305c210 the developement in instincts in the <classes> «ORDERS» of insects, so is there none of reason in ord
340d029 d that greater difference in than in many large ORDERS of birds. The Emu & Cassowary closest.-- Ostri
357d072 Marsupials  of Australia have branched out into ORDERS one is strongly tempted to believe, one or two
404e025 ather ventilation peculiar to <the class> «some ORDERS» of crustacea, one is tempted to think that it
406e037 in Southern.-- Great animals. of same two great ORDERS destroyed about same time in North & South. Am
407e040 of S. America are as different from the existing ORDERS, as the Eocene of Paris! (Great Edentata at th
411e055 the  presence of animals in <own> «the present» ORDERS (not so in S. America, however) is very remark
448e168 two  former connecting classes like Toxodon «In ORDERS»-- Fish & reptiles in former case-- Reptiles &
460t019 o time.-- Marsupial in Oolite.-- insects, of do ORDERS-- cheiroptera & caetacea in Eocene-- dicot. pl
536m071 .-- Why do bulls & horses, animals of different ORDERS turn up their nostrils when excited by love? S
543m098 of the Hymenoptera;. <therefore> than in other ORDERS (study Kirby with this view) therefore there i
543m098 al in human-- probably some genera in different ORDERS more advanced than others just as dog & Elepha
547m111 y trunk out & I was engaged in hurriedly giving ORDERS.-- Now what was difference between Castle & dr
037r055 like the secondary strata of England, «besides ORDINARY marine remains» may contains <shells few cora
063r132 hetical case of Brazil.-- Propagation. whether ORDINARY. hermaphrodite. or by cutting an animal in tw
120a108 c &c--then if so, thermometer show it cannt be ORDINARY heat, then there is something superadded, tha
170b001 pagation. bisection of Planariae. &c &c.-- The ORDINARY kind <the> which is a longer process, the new
191b083 ixing of individuals. Here we have avitism the ORDINARY event. & succession the extraordinary South A
276c122 something intermediate, between no offspring & ORDINARY offspring:-- this gradation is infertile offs
336d019 probable if created at once. <wd> according to ORDINARY laws, the character of offspring would  vary,
375d135 plenty, makes population in Men increase, & an ORDINARY crop. causes a dearth then in Spring, like fo
467t100 icroscope--some of the stigmas of {P} shape of ORDINARY Labiatae --the chief part with ordinary divis
467t100 ape of ordinary Labiatae --the chief part with ORDINARY divisions, & a few with one lobe again divide
498q008 umber of pollen-grains necessary to impregnate ORDINARY number of seeds known?-- Linnaeus has shown t
534m063 nsformation in the stem of Hollyhock, although ORDINARY Habitat is Malva sylvestris. do. p. 228 Newpo
538m080 ughts, feeling & perception separate, from the ORDINARY state of mind, is probably analogous to the d
546m110 action may be done unconsciously as far as the ORDINARY state is concerned.?-- Mr. Mayo told me the c
567n016 iged carefully to separate its memory from all ORDINARY lines of association.-- is totally distinct f
581n063 applicable to all cases.-- & analogize it with ORDINARY habits that is my new part of the view.-- let
583n070 "-- CD[as the bee makes its cells. by means of ORDINARY senses & muscles. we cannot look at him, as m
600o008 u desires conscience & love.-- [With regard to ORDINARY Beau ideal. Mem. Negro, beau,--Jeffrey denies
616o039 erebral portions of it) that attraction has to ORDINARY matter. The relation of attraction to ordinar
616o039 ordinary matter. The relation of attraction to ORDINARY matter is that which action bears to the agen
616o039 artling expression, & so foreign to the use of ORDINARY language that the onus probandi might  fairly
640j28v dragged back to old type by intermarrying with ORDINARY race.-- «There is no way of eliminating the e
074r165 Volcanic.  Then Solfataras. «Mem: Micaceous iron ORE.» N.B. To show how metals may be transported  by
079r176 Mica Slate & overlying Limestone Balls of Silver ORE occur in do veins. At Huantajaia. Humboldt says,
263c077 abric falls! But Man-- -- wonderful Man. "divino ORE versus coelum attentus" is an exception.-- He is
056r111 tes some sulphur (perhaps lime) salt. & metallic ORES.--which mingling & separating is well adapted t
075r167 «of iron» discovered?-- Klaproth analysed silver ORES from Peru consisted of native silver & brown ox
275c120 c picking varieties. unnatural circumstance Ld ORFORDS had breed of greyhounds fleestest in England l
197b111 ies. States there is but one animal: one set of ORGAN-- the others «animals» created with endless di
197b111 l of <sa>. changes, which can be traced in same ORGAN in different animals in scale.-- In monsters «a
197b112 outh those beautiful passages from one to other ORGAN.-- Cuvier on opposite side; Is Vol of Fish p. 5
227b224 ville Isd. (3d) We know that structure of every ORGAN in A. B. C. three species «of one genus» can pa
254c049 ite Kingdom all organs blended together, & same ORGAN «where eliminated is» often repeated, as mouths
279c133 = Milking heredetary, developemen of important ORGAN (.see marks on pages).-- Crosses of diff: breed
290c162 knows how.-- This reaction takes place in every ORGAN‖ «Hence «method» of generation is very good «ge
314c237 onsisting of pistils & stamens united into long ORGAN, moved on being touched, so as to protect itsel
314c237 od which possesses power of movement. & not the ORGAN itself How except by direct adaptation has such
325c266 ren. --it is not effect, as Lyell suggested, of ORGAN being worn out as. otherwise old whores would n
336d017 mule has no offspring & therefore no generative ORGAN.-- Same Prop. better enunciated.-- "An animal E
350d056 Observed, nature does nothing in vain, therfore ORGAN fitted to animals place in creation.-- thus sen
352d058 ive organs, but number of species with abortive ORGAN of any kind few.-- » hence become EXTINCT, & he
352d059 -- distribution especially of Mammalia As every ORGAN is modified by use, every abortive organ must h
352d059 every  organ is modified by use, every abortive ORGAN must have been once changed.-- what is abortive
373d132 -- why. what tendency can there be for abortive ORGAN ever disappearing??-- Have Marsupiata  abortive
384d162 luscous hermaphroditism takes place.-- thus one ORGAN in each becomes obliterated, & sexes as in Vert
386d166 ime.-- There is probably law of nature that any ORGAN. which is not used is absorbed.-- this law acti
404e025 the flipper is a mere simple modification of an ORGAN, on the whole system may. produce-- ? When a sp
408e043 t <light> «colours». acting. by a most delicate ORGAN, in which case the hermaphroditism would not be
418e080 one sex on one side, than the addition of other ORGAN. (here language forces on us the change,  which
434e130 e, so that they become so by suppression of one ORGAN. is become fixed. & cannot vary.-- which all fac
441e151 st draw such a monstrous conclusion, that every ORGAN graduating into other is lost, <be> (as vertebr
463t057 arked, that any argument for transmut, from one ORGAN "manifestation of divine power"?.--"of their us
471tf07 nd such an endless variety of form in the same> ORGAN. no answer?-- 3 Andrew Smith, about tamed  wild
508q016 s take place of gout.-- How are livers obscure ORGAN. is absent by abortion, but appears in  abortive
515q021 - Is there any genus of plants, «in» which some ORGAN of sight, which is common to every kind of view
528m037 s in the night.-- from the mere exercise of an ORGAN, as bile of liver.-- ¿ is the attraction of car
614o037 telligible it may be, seems as much function of ORGAN.-- I think Pincher shows surprise, walking home
615o037 wnright instinct, leading to touch a particular ORGAN. of taste, for when grown up often conquers it).
621o047 to this conclusion, not owing to peculiarity of ORGANIC remains in De la Beche, for the older formatio
025r019 t kept a record)» In looking over the lists of ORGANIC remains does not appear to have taken place in
026r022 ions of level, which in the nature of strata & ORGANIC powers. We know the waters of the ocean all ar
028r029 accumulated in the Tropical oceans detained by ORGANIC structure most easily preserved.-- Mr Conybear
035r046 cified wood. Cordilleras, Chiloe. &c seems the ORGANIC remains protect a rock, or that the rock not w
037r053 dation of rocks--It may be a question. whether ORGANIC remains.-- On Pampas looked in vain for a pebb
037r055 rge lizard.-- There would probably be no other ORGANIC remains.--Unequal distribution of Volcanic act
044r073 --general movements of the earth;--Scarcity of ORGANIC power a lump of hard clay. -- In the History o
048r084 , as a blade of grass penetrating by action of ORGANIC bodies protect like peat reef of sandstone.--C
051r093 s.--At Pernambuco (great swell & turbid water) ORGANIC productions. = Yet everywhere on coast (Il Def
051r094 nambuco believe much is owing to protection of ORGANIC kingdom, when our planet first cooled.-- Count
174b015 pials. at Australia-- Will this apply to whole ORGANIC being.-- the fact of guavas having overrun- T
262c073 powers of reasoning &c &c.-- Study the wars of ORGANIC nature on inorganic It is very remarkable as
270c104 ogie referred to,. as compilation of action of ORGANIC matter & chiefly heat), which assumes a multit
305c210 his word, (subject to certain contingencies of ORGANIC matter.-- brain. & which «prin> thinking princ
305c211 e» principle (intimately allied to one kind of ORGANIC diseases, mental disposition, stature, are slo
336d018 d like themselves. expression of countenances, ORGANIC beings.-- Animals «of same classes» differ  in
337d023 stock.-- Theory of Geograph. Distrib: of <ani> ORGANIC world, as adaptation. & these changing affect
343d036 es.-- these superinduce changes of form in the ORGANIC nature) the Rhinoceros of Java & Sumatra, that
343d037 those very laws he established in all <nature> ORGANIC difference.-- it may be so, but this assumptio
359d086 oduce unhealthiness,-- «or» to perpetuate some ORGANIC affections always influence the sexual  organs
359d087 healthy  which is often the case, & why should ORGANIC affections always influence the sexual  organs
398e004 physical changes & oscillations, not affecting ORGANIC forms, that the whole value of the  geological
```

415e065 iscovered. Man acts on. & is acted on by «the» ORGANIC and inorganic agents of this earth. like every
423e099 sufficient only to have most slightly modified ORGANIC forms.-- we know not rate of deposition has be
429e114 to believe in the dreadful «but quiet» war of ORGANIC beings. going on the peaceful woods. & smiling
441e150 , the absolute necessity, that every <animal> «ORGANIC being» should cross with another.-- to escape
443e154 s forced on us.-- My theory only requires that ORGANIC beings propagated by gemmation do not now unde
460t019 When it is said that there is evidence in the ORGANIC world of infinite & growing complexity from a
460t019 o the first origin of the world.-- our present ORGANIC beings are the descendants, <slightly> «a good
611o034 matter been formed by the union of simple non-ORGANIC matter, without action of vital laws-- Accordi
611o034 rlds, inor> systems of laws «in the world» the ORGANIC & inorganic-- The inorganic are probably one p
611o034 raction, heat & gravity is probable.-- And the ORGANIC laws probably have some unknown relation to th
614o037 ies] really less wonderful than thoughts-- One ORGANIC body likes one <m> kind more than another-- Wh
633j54r st mud &c at deltas.-- Now my theory makes all ORGANIC beings perfectly adapted to all situations, wh
636j56v mple series, & therefore trace of beginning in ORGANIC world.-- Macculloch. Attrib. of Deity. Vol I p
416e071 f my theory, that «domesticated» races. of <a> ORGANICS. are made by percisely same means as species-
432e123 orm heaps of pebbles &c &c: the succession of ORGANISMS tell nothing about length of time, only order
473s03v ious all negative laws of America of depth of ORGANISMS holding in America as in Britain. If there ha
604o016 his consciousness becomes multiplied with the ORGANISMS structure, it looks as if consciousness an ef
606o025 sion, love-- jealousy-- «as» effect of bodily ORGANISMS-- one knows it, when one wishes to do some ac
170b002 has done).-- This appears highest office in ORGANIZATION (especially in lower animals, where mind, &
171b004 culty with full grown individual «with fixed ORGANIZATION» thus being modified,-- therefore generatio
175b019 imal is branching upwards different types of ORGANIZATION improving as Owen says simplest coming in &
177b026 thus fish can be traced right down to simple ORGANIZATION.-- birds-- not. {P} We may fancy, according
177b028 f the point whence, two favourable points of ORGANIZATION commenced branching.-- As all the species o
181b040 look to the source of the Mammalian type of ORGANIZATION; it is extremely improbable that any of <hi
186b059 ate Lyell Vol III. p. 379. Mammalian type of ORGANIZATION same from one period to another, preeminent
186b060 adrupeds to Siberia; we must look to type of ORGANIZATION.-- extinct species of that country parents
190b078 ans of every step of progressive increase of ORGANIZATION being imitated in the womb, which has been
192b087 because in each there is possibility of such ORGANIZATION. [Spines in Echidna & Hedgehog]CD-- As we h
195b097 strange forms as ornithorhyncus The type of ORGANIZATION constant in the shells.-- The question if c
220b197 you cannot say that instinct perverted, yet ORGANIZATION «especially» connected with generation cert
224b213 nd of herbage & one with other, might change ORGANIZATION of stomach & hence remain distinct. Where c
227b227 eration. being condensation, test of highest ORGANIZATION intelligible.--may look to first germ-- --1
254c048 es of Acrite exhibit lowest stages of animal ORGANIZATION, '& are analogous to the earliest condition
261c070 give thread to conduct to laws of change of ORGANIZATION! The little turtle, without its parent runn
263c076 ed (if the medullary point in ovum. has such ORGANIZATION as to <per> force in one man the developeme
291c166 s out to this.-- love of the deity effect of ORGANIZATION. oh you Materialist!-- Read Barclay on orga
291c168 ation. oh you Materialist!-- Read Barclay on ORGANIZATION!! Avitism in mental structure or dispositio
293c175 may be able to trace the steps by which the ORGANIZATION of the eye, passed from simpler stage to mo
293c175 - the wonderful power of adaptation given to ORGANIZATION.-- This really perhaps greatest difficulty
325c267 Generalle et particuliere des Anomalies de l'ORGANIZATION des Hommes. & les animaux.-- by Isid. Geoff
336d019 tary, because generation -- highest point of ORGANIZATION] CD» false.-- The creator would thus contra
347d049 ugust. There must be some law, that whatever ORGANIZATION an animal has, it tends to multiply & IMPRO
352d058 T, & hence the IMPROVEMENTS of every type of ORGANIZATION. such law would explain every thing.-- PURE
356d070 expression of generat. being highest end of ORGANIZATION good expression but does not include so man
369d114 t from Hunter I should say females recede in ORGANIZATION from specific character.-- Chapt I. Also La
370d115 having any Mammalia.-- Type of geographical ORGANIZATION. no more can be said.... In paper on bees i
414e060 -- Read this Work-- Decb. 4th.-- Why has the ORGANIZATION of fishes & Mollusca (& plants???) been so
414e064 , namely which have the best fitted ORGANIZATION, or instincts (ie intellect in man) to gain
414e064 In man chiefly intellect, in animals chiefly ORGANIZATION: though Cont of Africa & West Indies shows
414e064 n: though Cont of Africa & West Indies shows ORGANIZATION in Black Race there gives them preponderanc
415e068 a chance it, has been, (with what attendant ORGANIZATION, Hand & throat) that has made a man.-- CD [
417e075 pecies, may be owing to the little fixity of ORGANIZATION, in the two races,, owing to the domesticat
418e084 , at a time, (as even in childhood) when the ORGANIZATION is pliable, such modifications, become as m
423e097 is, it would not be possible to simplify the ORGANIZATION of the different beings, (all fishes to the
435e134 s, other mammalia in the effects produced on ORGANIZATION. by physical agents." p. 466. Many facts gi
440e146 ircumstances. results of complicated laws of ORGANIZATION: as we see these strange plumage in pidgeon
441e150 ous & important.-- The existence of "laws of ORGANIZATION" had better be shown-- soil on colour of fl
460t017 . analogical structure) & partly the laws of ORGANIZATION (ie those laws which prevent infinite varia
525m026 the memory.-- therefore affections effect of ORGANIZATION which can hardly be doubted, when seeing Ni
525m026 fat man looks, shows same connection between ORGANIZATION & mind.-- thinking over these things, one d
535m070 will may <be> «arise from» as fixed laws of ORGANIZATION.-- M. le Comte argues against all contrivan
536m072 oyster, one can fancy to be direct effect of ORGANIZATION, by the capacities its senses give it of pa
536m073 e will (if so called) makes change in bodily ORGANIZATION of oyster. so may free will make change in
536m073 prove my temper, what does it arise from but ORGANIZATION. that organization may have been affected b
536m073 at does it arise from but organization. that ORGANIZATION may have been affected by circumstances & e
536m074 & education, & by choice which at that time ORGANIZATION gave me to will-- Verily the faults of the
536m074 he would strive <to do good» «to improve his ORGANIZATION» for his children's sake & for the effect o
540m086 e Atlantic. Why then is he so different-- in ORGANIZATION.-- Same cause as colour & shape & ideosyncr
542m094 ulation after stillness.-- Now I conceive if ORGANIZATION were changed, I conceive sighing might yet
568m018 memory improved. direct effect of improving ORGANIZATION, comparison of sensations would first take
576m048 require a far higher & far more complicated ORGANIZATION to learn Greek, that to have it handed down
604o016 sness an effects of sufficient perfection of ORGANIZATION & if consciousness, individuality.-- Quotes
608o026 y change-- because motive power changes with ORGANIZATION The general Delusion about free will obviou
613o036 tant argument, to show that they result from ORGANIZATION of brain; «[LHC] not used by Kirby» {:analo
618o41v weight; all that can be said that thought & ORGANIZATION run in a parallel series: if blueness & wei
618o41v n> «things» so different as action thought & ORGANIZATION: But if the weight never came untill the bl
638j58v as hinge, & hinge of itself, works of laws of ORGANIZATION is remarkable-- what is intellect, but orga
638j58v ation is remarkable-- what is intellect, but ORGANIZATION, with mysterious consciousness superadded T
287c158 & by so doing discover the laws of change in ORGANIZATON. But the classfication must chiefly rest on
176b021 ut law of adaptation as much as acid & alkali ORGANIZED beings represent a tree. irregularly branched
213b170 . In Phillips. p. 90. it seems the most ORGANIZED fishes lived far back, fish approaching to re
270c104 life".-- animals only live on matter already ORGANIZED.-- This paper might be worth consulting, if a
307c215 wings in the female are allowed to the fully ORGANIZED wings of the male rendered abortive in the wo
374d133 pper Silurian, fishes oldest formation highly ORGANIZED.-- do. p. 461.-- Lower Silurian-- several exi
423e096 t animals could be destroyed, the more highly ORGANIZED ones. would soon be disorganized to fill thei
445e159 legs.-- -- strangely consider existing «long-ORGANIZED» forms as parent forms of existing highly org
445e159 zed» forms as parent forms of existing highly ORGANIZED forms-- this resulted from the necessity of s
194b096 hells? ¿Do not plants, which have male & female ORGANS together, yet receive influence from other pla
197b111 ifferent animals in scale.-- In monsters «also» ORGANS of lower animals appear.-- yet nothing about p
220b197 erversion of structures especially reproductive ORGANS) & therefore the one distinction of species wo
248c030 ere instinct not interfered with, or generative ORGANS affected as with plants) no animals VERY diffe
254c047 g. T. V. I. p 389. Owen remarks on Entozoa, the ORGANS of generation, afford the least certain indica
254c049 pt generation & digestion in Acrite Kingdom all ORGANS blended together, & same organ «where eliminat
257c056 ion leads to other view.-- Till we know uses of ORGANS clearly, we cannot guess causes of change.-- h
270c103 <tu> plants turned inside out. have position of ORGANS of generation!!!. Mem. Agaziz. (INo Annals of
274c115 n in Lessonia &c & In relations of affinity all ORGANS change together, in analogy certain parts perf
276c122 this gradation is infertile offspring. without ORGANS of generation?! By profound study of local re
281c139 st subject to change.-- may account for certain ORGANS not being fixed, <whi> in some genera, which a
284c149 xtension ¿these go together? Therefore value of ORGANS vary in different group. & Not known in single
287c158 ions to settling the relative importance of the ORGANS in same state. in different animals, & the val
287c158 ate. in different animals, & the value of those ORGANS, when changed in different animals.-- + whethe
287c158 e classfication must chiefly rest on these same ORGANS.-- habits, range. &c &c-- Macleay rests his wh
298c192 nks if leaf of plant varies, <whole cross> «all ORGANS» vary in plant. The variation in character of
299c195 tant in classification. here we have generative ORGANS. first character.-- In dioecious plants many o
301c200 all animals of same class being about equal.-- ORGANS of generation about equally complicated.-- An
303c204 a less time than other. points. female genital ORGANS «in some monkeys clitoris wonderfully produced
304c208 at intermediate. species have generally perfect ORGANS. do changes of habit affect particular organs.
304c206 t organs. do changes of habit affect particular ORGANS.-- of two adjoining families & not all organs

Page ***(Key Word)***
304c206 r organs.-- of two adjoining families & not all ORGANS blending away.-- +++ Hopeless work to systemat
306c213 Plata & Guyana‖ people will say. not species.-- ORGANS of generation a captial character. (Owen) not
307c215 moths, like glowworm <are> have «These abortive ORGANS in some Males animals, Mammae in Men, capable
307c215 tive in the womb.-- if these apparently useless ORGANS do indicate such origin, then we are bound to
307c215 origin, then we are bound to consider abortive ORGANS of same tendency in species, this is capital &
307c216 ital & novel argument.-- Are there any abortive ORGANS in neuter bee, (There is paper by Yarrell «in
307c216 ortive spurs. growing.-- Are there any abortive ORGANS produced in domesticated animals, in plants I
307c216 dril passes into a mere stump.-- Shall abortive ORGANS «of very same kind» in these cases, have plain
307c216 other case! Savigny has shown same fundamental ORGANS even in Haustellata & mandibulata.--!! --Argum
309c219 rbert's views.-- Argue <argue> case of abortive ORGANS to mules in their genitals & even to a limb no
311c226 ndication of structure (including brain & other ORGANS difficult to analyse) will not this separate f
311c226 se) will not this separate facts about abortive ORGANS &c The doctrine of monsters is preeminently wo
313c236 chick eat») Generation becomes necessary, when ORGANS of parent are concentrated in different parts,
325c267 ery ancient? Study with profound care, abortive ORGANS produced in domesticated plants; where functio
335d016 g case, owing to the corelations of system, the ORGANS of generation would necessarily fail.-- In las
348d050 es‖ Macleays plan of arrangement depends on the ORGANS judged to be of importance in inverse ratio to
350d055 served all these facts prove that perfection of ORGANS have nothing to do with perfection of individu
352d057 2d Sept Those animals which have many ABORTIVE ORGANS, might be expected to have larvae more perfect
352d058 there some law in nature an animal may acquire ORGANS, but lose them with more difficulty, «contradi
352d058 with more difficulty, «contradicted by abortive ORGANS, but number of species with abortive organ of
355d067 prospective, but retrospective as showing what ORGANS are little fixed-- (<also> Hunters law of mons
356d069 extraordinary effects of different Medicines on ORGANS, leads one to suspect any amount of change fro
358d085 ther, if by the process this were possible, the ORGANS doubtless would shrivel up.-- Yet odd they sho
359d087 organic affections always influence the sexual ORGANS alone.-- It is singular pheasant & fowl being
367d112 ccur.--».-- & if that cannot be formed, genetal ORGANS by that co-relation of parts, will not be prod
372d129 oes not whole individual change into generative ORGANS?]CD it is of no consequence if it does= <I do
373d132 us to other males feeding young, & to abortive <ORGANS> «mammae» in male Mammalia:-- ¿is not this arg
381d156 fies Hermaphrodites. Cryptandrous. (only female ORGANS visible). Oyster. cystic Entozoa. Echinoderms.
381d156 . Acalephes. Polyps. Sponges Heautandrous, male ORGANS formed to fecundate female (as in plants) Cirr
381d157 ow is fecundation effected in latter; are <it> «ORGANS» open to water? Would not Ferns according to t
381d157 da» are there abortive traces of other «sexual» ORGANS; for if so, separtion of sexes very simple.--
381d157 ooecious-- (& cultivation might make one set of ORGANS barren in one plant & not in other), Hunter <a
382d158 all animals subject to Hermaphroditism,-- those ORGANS which perform nearly same function in both sex
384d162 ne individual secretes two substances, although ORGANS for the double purpose are not distinguished.
384d162 (in dioecious plants are there abortive sexual ORGANS?): they then become so related to each other,
386d166 ing against heredetary tendency causes abortive ORGANS.-- the origin of this law is part of the repro
386d167 not owing simply to more importance of internal ORGANS in animals‖. One «invisible» animalcule in fou
388d173 ough all transformations, should there need two ORGANS; whilst in common bud there is no such need.--
410e053 me country.-- The traces of changes in forms of ORGANS, will care little for species, except so far a
411e054 ws of such corelations, & changes of individual ORGANS, must know whether the individuals «forms» are
411e054 ing of the formal laws of corelation of parts & ORGANS it may serve perfectly to specify types, & lim
417e077 at simple generation as being the action of two ORGANS in one body.-- or in two bodies, <th> we can a
418e083 vital principle, which gives rise to the sexual ORGANS, different in each species,-- & knowing from a
421e091 ct» in his paper on the Dugong, "The generative ORGANS being those which are most remotely related to
421e091 are least likely in the modifications of these ORGANS to mistake a merely adaptive to an essential c
434e130 Male plant sometimes bears female flowers, the ORGANS are most clearly abortive, so that they become
434e130 & Monooecious plants have rudimentary abortive ORGANS, even more so than Polygamia: Monooecia & Dioe
441e148 of two «individuals» & the action always of two ORGANS-- instead of one part «as» in producing bud.--
444e157 139. Doubts altogether the law of balancing of ORGANS.-- In the Batracian Order the «32» ribs are wa
491q01v ted young ones; & that it is in these that male ORGANS (not being always useful). fail-- Really good
495q005 English plants in Hothouse & see what effect on ORGANS of generation (5) Place pollen of Red Cabbage
501q011 place in Hot house to see effect on generative ORGANS of great Heat (32) Can Henslow ask question of
509q017 range-- How is this in «Plants??» Are abortive ORGANS as <young> teeth, more plain in young Rhinocer
521m008 it was an habitual action of thought-secreting ORGANS, brought into play by morbid action.-- Old Els
524m020 epsy & convulsion.-- affections of the thinking ORGANS ‖ -- the action of brain which gives sensation
527m034 it being done in open air, with exercise &c no ORGANS of sense being required) as the closest train
566n010 s of a face «& mind» sympathetics with internal ORGANS, as action of heart‖ Malthus on Pop. p. 32, or
577n052 he power of vivid mental affection, on separate ORGANS most curiously shown in the sudden cures of to
583n071 se to make a cell in certain way, which way its ORGANS are sufficient for hence it must some way be a
587n087 an the unfitness of the objects then viewed. to ORGANS adapted to other objects. (as that senna is ne
587n087 . (as that senna is necessarily disagreeable to ORGANS adapted to like sugar, acid, &c, which may be
589n092 d brain. see p. 90.-- The relation of reason to ORGANS of locomotion-- or that our faculties have bee
633j53v - NB. One limit to the transmission of abortive ORGANS will be as long as they are not detrimental.--
609o029 worser feelings. These bad feelings no doubt ORGINALLY necessary revenge was justice.-- No checks we
037r056 & other Volcanic rocks. Bahia, Rio de Jan: B. ORIENTAL? level surface not disturbed.--Whole West coa
472s01v ral years ago seeds were procured with the P. ORIENTALE in garden & all came up hybridised. It is pos
104a068 d yet no mountain chain case parallel to Banda ORIENTEL. ask Lyell for sentence.-- Origin of Breccia,
043r070 ouring Volcanos. eruption from «more than» one ORIFICE <...> does not occur at same time: this is con
495q05a y & see effect-- such dioecious individ--small ORIFICE (8) Carry Bees, powdered with starch & Carmine
502q11v spernum. Blue Gloss. it is not possible to see ORIFICE of poison-tube-- so put carmine in spirits & t
127a127 beneath sea.-- with reference to old submarine ORIFICES in Cordillera Geograph. Journal vol II. p 89.
028r030 d either in note in Coral Paper or hypothetical ORIGIN of some sandstones, as in Australia.--Have Lim
049r089 Why not more probably greenstone? What probable ORIGIN can be given to the numerous hills of greensto
073r161 on that to which we cannot attribute a meteoric ORIGIN & which is constantly found mixed with lead &
104a069 el to Banda Orientel. ask Lyell for sentence.-- ORIGIN of Breccia, introduce in Cordillera discussion
114a095 d Hope.-- A bare hill of greenstone, if we know ORIGIN of greenstone tells subsidence as plainly as T
122a115 ndes Speculate under head of Beagle Channel. on ORIGIN of mud with stones scattered irregularly.-- (M
163g109 lls Denmark) Shrewsbury rubbish.-- Speculate on ORIGIN pebbles brought by different cause: from mud.-
175b018 coming more complicated,; & if we look to first ORIGIN there must be progress. if we suppose monads a
181b045 olely for air & fishes for water. If my idea of ORIGIN of Quinarian system is true, it will not occur
210b161 ion «of monsters» (which, Hunter says owe their ORIGIN to very early stage) & which, follow certain l
221b201 ater shells living in absolutely fresh water.-- ORIGIN of Fresh-water genera? The absence of lime in
226b220 ill be well worth while to study profoundly the ORIGIN & history of every terrestrial Mammalia.-- Esp
227b224 «two» ancient fauna, we may form some idea of <ORIGIN under> connection of those two countries Hence
229b232 n in our amusements. they may partake, from our ORIGIN in <there> one common ancestor we may be all n
231b244 ot the domesticated animals with him.!-- Modern ORIGIN shown by only one species, far more than by no
263c077 is an exception.-- He is Mammalian.-- his <has> ORIGIN has not been indefinite-- he is not a deity, h
265c086 I think we may clearly attribute to heredetary ORIGIN & not adaptation. to its habits.-- Few will di
286c155 ustly excalted nature of man. like to think his ORIGIN godlike, at least every nation has. done so as
293c174 nimals-- Hydrophobia &c cowpox, proof of common ORIGIN of Man.-- different contagious diseases, where
294c176 ar continent, Must we resort to quite different ORIGIN when species rather further.-- once grant good
307c215 hese apparently useless organs do indicate such ORIGIN, then we are bound to consider abortive organs
310c223 e of nature) and animals that man has different ORIGIN. «Royal Institution» Dr Royle seems to think B
313c234 ve been made, & only species, good argument for ORIGIN of man one.-- Is the extinction & change of sp
359d087 ich I think there must be some mistake in their ORIGIN) Saw cross between Penguin Duck «from Bombay»
362d100 time.-- the impossibility of discovering their ORIGIN.-- I see only some «but very strange races» of
378d148 often, that it was impossible to say which was ORIGIN of any identical bird-- for they were of all c
386d166 redetary tendency causes abortive organs.-- the ORIGIN of this law is part of the reproductive system
422e092 hydermata.-- <it was a Pachyderm. which was the ORIGIN of the aquatic Mammifers> p. 306, the Dugongs
422e092 seal.-- (Consult this passage, when considering ORIGIN of northern Cetaceae).-- -- ‖.do. p. 318 M. Pi
424e102 s & what says Jenyns to it?-- -- In argument of ORIGIN of Wolf, difference of mind is most relied on,
460t019 a few types originated, we must go to the first ORIGIN of the world.-- our present organic beings are
539m084 muscles rigid.-- How is this? dealt with p. 241 ORIGIN of man now proved.-- Metaphysic must flourish.
546m109 f flowers, owing to parent being fruit eater.-- ORIGIN of colours?-- Nothing shows one how little hap
550m123 r bodies of either.-- Our descent, then, is the ORIGIN of our evil passions!!-- The Devil under form
552m132 ll do good.-- alter will in all cases to have & ORIGIN as well as rule will be given.-- Descent of Ma
555m142 s disease.-- (Useful to use term instinct, when ORIGIN of heredetary habit cannot be traced) V. D. p.
566n010 ans, as action of heart‖ Malthus on Pop. p. 32, ORIGIN of Chastity in women.-- rationally explained.-

566n011 o them: this must be studied. before my view of ORIGIN of evil passions.-- Man getting sight slowly,,
566n011 ke birds & beasts».-- very necessary to explain ORIGIN of idea of deity.-- Animals do not know they h
567n012 od. Zoology itself is now purely theological.-- ORIGIN of cause & effect being a necessary notion is
568n018 id our language commence with singing-- is this ORIGIN of our pleasure in music-- do monkeys howl in
572n035 s an expression the very same as mine about our ORIGIN of a notion of a Deity We can allow «satellite
579n060 (-- Jan 21. 1839. Herchel's Discourse p. 35. On ORIGIN of idea of causation; «succession of night & d
579n060 on the importance of a name, with reference to ORIGIN of language My father says old people first fa
591n101 the Reason of animals Essays Vol 2.-- «also on ORIGIN of religion or polytheism, at p. 424 Vol. II «
592n101 scepticisms might be solved by considering the ORIGIN of reason. as gradually developed. see Hume on
592n101 has written "Natural Hist. of Religion" on its ORIGIN in Human mind.-- Andrew Smith says hen doves &
592n104 e, is well seen in shortsighted people.-- hence ORIGIN of expression-- There are some instincts unint
592n105 in the end gained «& therefore the» cause, and ORIGIN being so is not odd.; for instance wild cattle
599o005 chool of Metaphysic,"-- give my doctrines about ORIGIN of language-- & effect of reason. reason could
599o005 ?-- At least it appears all speculations of the ORIGIN of language.-- must presume it originates slow
599o005v g element, we see words invented-- we see their ORIGIN in names of People.-- Sound of words-- argume
599o05v formation.-- declension &c often show traces of ORIGIN.-- Mayo Philosophy of Living. p. 264. "Archive
603o012 Staunton Embassy Vol II p. 405.-- Speculates on ORIGIN of sacrifices. «common to many races»-- thinks
603o012 so sacrifices cruel.-- Something wrong here.-- ORIGIN is certainly curious. Chinese, S. American. Po
604o015 ? ¿even to children S. Jenyn's Inquiry into the ORIGIN of Evil. Reviewed by Johnson in the Literary M
609o30v when I say How social instincts generated? The ORIGIN of the social instinct «in man & animals» must
619o042 the object in question. Without regarding their ORIGIN, we see in other animals they consist in such
621o047 too much about the contiguity to will. (a) The ORIGIN of passions too strong for our present interes
621o047 esent interest receive simple explanations from ORIGIN of man.-- [RHC] By interest I do not mean any
622o049 on which «almost» any other might be grafted.-- ORIGIN of the instincts Hartley, (according to Sir J)
623o050 does not Hartley explanation apply perfectly to ORIGIN of these instincts.-- the having received plea
634j55r Would they not first occupy the Poles? Is this ORIGIN of Polar attributes of the Cetaceae.-- How cam
170b001 , <of the» or shortened repetition of what the ORIGINAL molecule has done).-- This appears highest of
202b134 e is that these wanderers would not, but where ORIGINAL forms most numerous there would be wanderers.
207b150 e intermediate between vascular or cryptogam. (ORIGINAL Flora) and Dicotyledenous, which «nearly» fir
282c142 suppose all descended from same.-- but if two ORIGINAL species, each became ground then the relation
288c159 be worth introducing, instead of breeding from ORIGINAL Durham breed.-- Native dogs & English cross r
296c184 quick enough» Vegetation of peak-- altogether ORIGINAL. owing to being oldest. & having undergone ch
308c219 p 276 May be worth glancing at, as she has no ORIGINAL ideas., it will show state of knowledge «Negr
379d151 it would be ocean, what is Land to continent-- ORIGINAL Paper, worth studying. Archiv. fur. Naturgesc
450e175 Celebes. (but language shows that probably not ORIGINAL there)-- shows these isld not fit for horse.
540m087 d-- Perhaps one cause of the intense labour of ORIGINAL inventive thought is that none of the ideas a
564n003 nce would not have been same with mans because ORIGINAL instincts different.--Mem. Bee how different
599o05v ames of People.-- Sound of words-- argument of ORIGINAL formation.-- declension &c often show traces
605o18v al excellences, brings to our recollection the ORIGINAL cause of these feelings & thus we apply to th
634j54v tances, that we become satisfied respecting an ORIGINAL thought, or design, pursued to its utmost exh
190b078 ok as if S. africa peopled from N. Africa An ORIGINALITY is given (& power of adaptation) is given by
356d069 aire, & Lamarck have written I pretend to no ORIGINALITY of idea-- (though I arrived at them quite in
125a121 Assuming from Sir. W. Herschel's views earth ORIGINALLY fluid, then cooling process must go from sur
244c021 - belong to the hairless kind, «said» to come ORIGINALLY from Africa p. 122. Mus decumanus, at Caroli
282c140 whole world. similar-- & constitution of man ORIGINALLY similar limits of change, would be same-- Ye
282c143 ts by the vital laws.-- so that all character ORIGINALLY may «must» have had the character of analogi
288c159 d wider & wider at Rio Plata. birds which had ORIGINALLY crossed would continue to cross, means of kn
364d104 ems from Lib. of Useful. Knowledge that sheep ORIGINALLY. black. & Yarrell thinks the occasional prod
391d174 abortive hermaphrodite simplifys case much; & ORIGINALLY <her> each hermaphrodite being simple (Are n
420e089 an. 6th The rudiment of a tail, shows man was ORIGINALLY quadri<manous> «ped.-- ». Hairy.-- could mov
557m149 . but ashamed of himself.-- Jealousy probably ORIGINALLY entirely sexual; first try «to» attract fema
599o005 ed in whole sentences.-- signs-- ¿ were signs ORIGINALLY musical!!??-- At least it appears all spec
608o026 action, & he can seldom analyse his motives (ORIGINALLY mostly INSTINCTIVE, & therefore now great ef
628o53v ing or disapproving instinct-- which were not ORIGINALLY, if the shepherd dog had no instinct to comm
634j54v cases -- p. do]CD <Mac. remarks all Mammifers ORIGINALLY land--animals. as> 5 p. 314. Mac. remarks al
634j55r . Mac. remarks all <land> Mammiferous animals ORIGINALLY terrestrial.-- for we find even in Cetaceae
029r033 ieve the secondary? formations of Brazil, all ORIGINATE from the decomposition of Granitic rocks Mem.
543m101 es is valuable it shows that new instinct can ORIGINATE in a doubting feel between conscience & impul
609o029 ressed). shame perhaps an exception. (does it ORIGINATE in any point an opposition of forces balancin
617o40v of it we are conscious that we ourselves can ORIGINATE in any point an opposition of forces balancin
023r010 pened from incipient elevation.» The volcanos ORIGINATED in the bottom of the ocean. & the present Vo
452e181 -- most wonderful instinct, how could it have ORIGINATED-- spins thread of cotton.-- do p. 583, It AP
460t019 be asked how this complexity from a few types ORIGINATED, we must go to the first origin of the world
599o005 of the origin of language.-- must presume it ORIGINATES slowly-- if these speculations are utterly v
605o020 s essay is to show how taste is gained how it ORIGINATES, & by what means it becomes an almost instan
053r100 rpment The great conglomerate of the Amazons & ORINOCO mentioned by Humboldt under name of Rothe-todt
097a043 erved by Humboldt on the granitic rocks of the ORINOCO".-- <but> on one of the Ponza isles. but no mi
446e160 ggs. & feeding & defending their young..-- The ORIOLUS (icterus Cat.) is an instance of this, & the f
283c144 irds of Paradise-- if such fantastic «sexual» ORNAMENTS. have so intimate a relation to two continent
575o045 asure. from pleasure of music Audubon IV Vol of ORNITH. Biog. case of Newfoundland dogs. who will not
211b162 milar habits produce similar structure.-- Mem. ORNITHO Rhyncus Would not relationship express, a real
484z014 erers.--?? in N. America?? Wilson N. American ORNITHOLOG must be studied before writing my general ac
255c050 es to greater range of such forms.-- p. 56: ORNITHOLOGICAL part of Voyage of??? A Urubu, (with one la
326c266 Mayo. Pathology of Human. Mind.-- Audubons. ORNITHOLOGICAL Biography. 4. Volumes well worth reading B
511q018 fertile & whether homogeneous {About German ORNITHOLOGISTS, Bhem & Glöger Consul Hunt, birds from Azo
502q012 - probably united by Land to S. America (33) ORNITHOLOGUM commonly but improperly called Canadense--
260c067 - No Carrion Vultures in Australia!! Wilsons ORNITHOLOGY, Vol III. p. 226 Wilson's Ornithology, D'orb
260c067 ilsons Ornithology, Vol III. p. 226 Wilson's ORNITHOLOGY, D'orbigny, Spix, &c might compare birds of
260c069 . which they frequent!!?-- Wilson's American ORNITHOLOGY a mine of valuable facts,. regarding habits
295c183 iology of S. America. more peculiar than its ORNITHOLOGY X p. 12 do. excepting salmons L' Institut. S
325c267 facts about close species) Wilson's American ORNITHOLOGY Read Aristotle to see whether any my views v
349d051 tionidae> «in the Cetoniadae»,-- when will ORNITHORHYNCHUS come in circle?!!! p. 8-- Anomalous struct
192b089 haps the father of Mammalia as Heterodox as ORNITHORHYNCUS. If this last animal bred-- might not new
194b097 when there now exist such strange forms as ORNITHORHYNCUS The type of organization constant in the s
265c087 rottii & no other species-- as we have only ORNITHORHYNCUS, then we should never know how much struct
369d115 en thinks from climate of Australia, & from ORNITHORHYNCUS & Hydromys not being Marsupial. («but» als
415e066 ld no fossil «very distinct species» of the ORNITHORHYNCUS be found.;-- yet until man became cosmopol
415e066 would probably be confined in locality like ORNITHORHYNCUS,: since being cosmopolite, we do find his
639j28r h of lobsters-- analogy in Flamingo & Duck, ORNITHORHYNCUS «externally». petrel & Whale in some respe
448e168 n former case-- Reptiles & Birds & Mamm. in ORNITYHYRHYCUS-- is not this right?-- June 18th. Eyton te
180b037 pted, & thus perish out, or on other hand like ORPHEUS. being favourable many might be produced.-- Th
183b051 y remote quarters. (NB. if Plata Partridge «or ORPHEUS» was introduced into Chili. in present states.
195b103 cies so close as Patagonian <Chat> & Galapagos ORPHEUS.= Put this strong so many thousand miles dista
283c145 nted.-- For. at Galapagos. make ten species of ORPHEUS-- one of which has very short legs & long tail
485z016 or ground birds-- rather indefinite letter Mem ORPHEUS--becoming tyrant-- flycatcher-- shown by habit
074r164 yte. asbestos garnets.--carb & chrom. of lead. ORPIMENT. chrysop[r]ase. opal:-- Veins in Limestone &
498q008 owcumber's out of doors.» much-- or the minute ORTHOPT.-- important, as we know how readily they cros
272c108 ingly resemble Coleoptera.-- Donacia.-- some ORTHOPTEROUS insects & some third, have got thighs with
509q017 ny abortive bones??? do. Wing in Apteryx ‖‖ no as OS Coccygis-- Turbinated bones? False ribs Wings of
391d175 ature species is formed if not.-- the changes OSCILLATE backwards & forwards & are individual differe
035r047 ws that throughout all England, whole surface OSCILLATED equably.-- These facts become easy if we loo
046r080 -At great Lisbon Earthquake Loch Lomond water OSCILLATED between 2 & 3 ft. (as in Chili lake). Theref
045r077 id matter instead of afflux (always slightly OSCILLATING as that of a spring) moves away.--Will geolo
035r047 ess than secondary:-- consider arguments for OSCILLATION of level independent of mineralogical nature
047r081 as been raised.--It must be considered as an OSCILLATION, from violence. Is it not same as swell trav
047r082 e wave rushes up? (NB. Earthquake wave is an OSCILLATION, body of water manifestly does not travel up
108a080 ot expect volcanos.-- not so much horizontal OSCILLATION. or so many shocks directly after great shoc
155g065 formed «(Mac, hypoth,)» the level during any OSCILLATION must have been so carefully preserved as to

026r022 Roussin's voyage.-- In Europe proofs of many OSCILLATIONS of level, which in the nature of strata & O
027r023 ations Conglomerates are found. -- The above OSCILLATIONS remarkable because the formations are now s
040r064 a crater as Vesuvius.-- There may have been OSCILLATIONS in the upheaval of Andes.--but as long as a
115a098 he shallower the water, the greater power of OSCILLATIONS & currents.-- if matter was «successively»
117a101 ession. C. Prevost.-- My views of insensible OSCILLATIONS of level will alone explain the immense amo
122a113 hown by arc.-- read Herschels astronomy with OSCILLATIONS of level.-- {P} will point {P} be the one w
129a131 acts about subsided forests.-- Many repeated OSCILLATIONS» Hitchcock Report on Massacuhssets. p. 133
398e004 the circumstance of small physical changes & OSCILLATIONS, not affecting organic forms, that the whol
201b126 t axiom in Natural History that all aberrant & OSCULANT groups are not only few in species, but every
302c202 .-- according to Principles of last page. <an> OSCULANT groups between two circles «of equal value» m
303c203 r of genera-- whatever the cause is. <the> any OSCULANT species. which survived would be few in numbe
304c208 -- aberrant forms produced where many species <OSCULANT> but, where much death, may be inferred much
094a035 changes. of granites into Trachytes.-- Mention OSORNO in lake. few Volcanos now in lakes.-- Mr Murch
087a010 plains of Mississippi -- No Vol. I. p212. Cuvier OSS Foss Wide range of Mammalia really very importan
355d067 S of genus Sus. do vertebrae vary? «See Cuvier OSSEMENS Fossiles» Although no new fact be elicited by
325c267 istoire Naturelle Marcel de Serres Cavernes d'OSSEMENTS 3d. Edit. Octav. (good to trace Européan form
342d035 old order Geologically? Owen says relation of OSTEOLOGY of birds to Reptiles shown in osteology of yo
342d035 on of Osteology of birds to Reptiles shown in OSTEOLOGY of young Ostrich. 16th. D Israeli (Cur of Lit
374d133 ot intermediate between fish & reptiles-- yet OSTEOLOGY closely resembles reptiles.-- p. 432 some pla
101a507 England different from Boulder beds-- What is OSTEOPORA platycephalus (Harlan) found on the Delaware.
214b173 ies does not appear to me very strong What is OSTEOPORA platycephalus. (Harlan) found on Delaware is
062r130 stances: The same kind of relation that common OSTRICH bears to (Petisse. & diff kinds of Fourmillier
180b037 o extinction we can easily see that variety of OSTRICH, Petise may not be well adapted, & thus perish
304c208 heredetary & not produced by circumstances In OSTRICH which is not isolated, we must suppose the cha
340d029 rders of birds. The Emu & Cassowary closest.-- OSTRICH & Rhea closest.-- (& two Rheas still closer).-
342d035 birds to Reptiles shown in osteology of young OSTRICH. 16th. D Israeli (Cur of Literat. Vol II p 11)
353d060 tock, or not. Now wings for flight-- therefore OSTRICH not. The peculiar <Malacca> «Malacca» bears, <
443e157 d.-- p. 94.-- "The resemblance of the foot the OSTRICH to that of the Camel has not escaped Naturalis
061r127 . (Henslow) Speculate on neutral ground of 2. OSTRICHES; bigger one encroaches on smaller.--change no
174b013 thera several species in S. America why, 2 of OSTRICHES in. S. America-- This is answer to Decandoell
175b016 & no great change has happened I look at two OSTRICHES as strong argument of possibility of such cha
231b242 ds, <bu> plants do not follow by any means.-- OSTRICHES.-- Hippotamus only African.-- American & Afri
233b251 th studying.-- Wingless birds S. Continents-- OSTRICHES. Dodo. Apteryx Penguin-- Logger-headed Duck--
277c126 und ascertained, call them varieties. but two OSTRICHES good species because interlock analogy to be
304c207 far more prob> » tends to alteration views.-- OSTRICHES do-- but then there may have existed series b
305c209 rix.-- but source of error for if some of the OSTRICHES were to die, then they would appear isolated.
247c028 . New Ireland» & «even» Java. & very common on OTAHEITE-- according «stated in note to p 21» to Quoy
358d074 uliar character of St. Helena.-- contrast with OTAHEITE in relation «See Gaudichauds Volume on the Bo
385d163 visible powers» in Zoolog Gardens. & Kings at OTAHEITE) <Think> Last litters are considered the most
492q002 written on them? 7... Are the wild Bananas of OTAHEITE seedless;-- are all varieties seedless-- if s
615o038 general ones.-- Parental feelings weakened in OTAHIATI; fear of death in Hindoo population.-- Slight
071r158 ty «at Mendoza» exception.--«formerly perhaps OTHERWISE» Mendoza never overthrown,--no mountains Mack
117a101 amount of change which must have taken place, OTHERWISE the world would daily be scene of ruin in lat
189b072 e golden pippen. if produced by seed go on.-- OTHERWISE all die.== The fossil horse, generated in S.
198b115 to each country.-- Provision for transportal OTHERWISE not so numerous: quote from Lyell: assuming t
206b149 species of genera) ultimately few genera (for OTHERWISE the relationship would converge sooner) & las
292c169 tage) the tendency to change cannot be great, OTHERWISE it would be unlimited. We absolutely know tha
302c200 - Wonderful, partly explained on my theory, = OTHERWISE mere fact creator chooses so to create.-- It
302c202 othesis, & compare it with resuts. if I acted OTHERWISE, my premises <in di> would be disputed.-- acc
315c240 creator sees. the means of transport fail.-- OTHERWISE no relation between means of Transport & crea
325c266 Lyell suggested, of organ being worn out as. OTHERWISE old whores would not have children Turners em
333d005 y long, eyes very large, very fierce to dogs «OTHERWISE habits not different; tone of voice. perhaps
338d023 at, as separation on <be> inter-breeding, for OTHERWISE we could not understand the vast number of do
343d039 lended & prevent that strong separation which OTHERWISE would have taken place. otherwise in 10,000 y
343d039 ation which otherwise would have taken place. OTHERWISE in 10,000 years Negro probably a distinct spe
409e048 cause» of sexes «in separate» «animals»: for OTHERWISE, there would be as many species, as individua
414e060 a?? My theory certainly requires progression, OTHERWISE [not located] Are the feet of water-dogs at a
423e096 cover whole surface of world with life.-- for OTHERWISE a frost if killing the vegetable in one quart
428e107 le & make mongrel, & other great series quite OTHERWISE & make on[ly] true hybrids.-- but this is fal
428e112 partly true, as enuntiated by him to me, for OTHERWISE breeders who only care for first generations,
466t094 v «they do» they must disturb all anthers, wh OTHERWISE lie protected by the hairy black lip of lower
523m016 Chile earth quakes). in people, who, probably OTHERWISE would not have been so.-- In Mr Hardinge, was
536m071 breasts... Dr Darwin's theory probably wrong, OTHERWISE horses would have idea of beautiful forms.--
539m082 ed trades can hardly be considered as actions OTHERWISE than habitual.-- instances?? The possibility
540m089 thers, with rational desire to assist them,-- OTHERWISE as he remarks sympathy could be barren. & lea
563n002 tant, for a wish which was only short & might OTHERWISE have been relieved, he would be sorry or have
225b216 ually are arguments against it-- namely how did OTTER live before being made otter-- why to be sure t
225b216 t-- namely how did otter live before being made OTTER-- why to be sure there were a thousand intermed
257c057 were possible,-- it has been asked how did the OTTER live before it had its web-feet-- all Nature an
295c178 y genera of Pachydermata &c & other Mammals.-- «OTTER; civet cat, rodents.» (Pachyderm in Portland st
494q004 as   show. & curiosities!! What is price of fox. OTTER. Badger &c &c &c.-- (11) Keep. Tumbling pigeons
246c025 nly travelled from East Indies, isld, as far as OUALAN.-- Wide space of sea, to East of America. woul
246c027 Botouma   from account of natives, & probably on OUALAN.-- «Mitchill says snakes on Friendly isles. p
046r080 ft. (as in Chili lake). Therefore motion of sea OUGHT to be considered as a plain movement communicat
086a004 evation, yet sea shells at tops of mountains we OUGHT to sympathize with. old doubters of what are fo
112a090 at Yakous in Siberia being -8 Reaumur.-- there OUGHT to be 32 degree Fah. at a greater depth than 40
115a096 to   common heat.-- Where there are cliffs there OUGHT to be creeks & mouths of rivers ought to be dee
115a096 ffs there ought to be creeks & mouths of rivers OUGHT to be deep.-- Henslow has deposited specimens f
126a123 rface, would become hotter.-- hence temperature OUGHT to increase rapidly beneath level of sea.-- dee
127a125 frozen for greater length of time depth of ice OUGHT to be less.-- Memoir of the Irish Academy Vol 8
132a138 han» «&» on coast lines, than on continents. it OUGHT, (according to M. Parrots argument against cent
172b007 ding to this view animals, on separate islands, OUGHT to become different if kept long enough.-- «apa
172b008 believe species vary, <in> changing climate we OUGHT to find representative species; this we do in S
186b060 f American.-- Now Genera of those two countries OUGHT to be similar ¿Law: existence definite without
226b223 » «W» Coast of Africa & <West.> «E» of America, OUGHT to present great contrast in forms.-- India; in
228b229 ure. The Grand Question, which every naturalist OUGHT to have before him, when dissecting a whale, or
230b235 aching desert country or ascending mountain you OUGHT to have a gradation of species, now this notori
232b245 pecies might then have been wanderers.-- There OUGHT to be fewer species in proportion to genera tha
233b249 all the rest. Timor according to Mountain chain OUGHT to be Australian.?-- Mr Gould has been struck w
284c149 ion.-- + <being> good for generic divisions [+] OUGHT genus to be founded on such characters, as do n
286c156 mals, what relation in classification in books, OUGHT they to hold,-- Birds having web-feet, where we
306c213 215 degree. What productions Sandal. Wood Isd.? OUGHT to agree with Java?? Terrestrial Planariae assu
374d134 oportiona[l] to the number that can live.--» We OUGHT to be far from wondering of changes in number o
378d147 ng forms, <& loosing do> if so domestic animals OUGHT to show them.-- Anyhow not connected with habit
390d178 fails.-- therefore «each» seedling of one apple OUGHT to differ from those of other.-- The upshot of
398e006 s each form to have lasted for its time: but we OUGHT in same bed if very thick to find some change i
407e041 of vegetation, & conchology,-- shells of Africa OUGHT most to resemble fossil ones of Europe, Conside
424e103 in hot countries.-- CD[Camel does not vary «one OUGHT not to be able to hybridise the Camel» like ass
427e109 one sex.-- Linnaean class Dioecia & Monooecia. OUGHT to be preeminently artificial.-- Would not subs
428e112 by Herbert] to sift out the weaker ones: there OUGHT to be no weeding or encouragement, but a vigoro
442e153 be absolutely similar; [all the gorze in Norway OUGHT to be thus characterized study Von Buch.]CD Now
447e164 Mollusca, & probably corals.-- these forms then OUGHT to be very persistent, & then necessity of cro
493q004 sult.-- According to Mr Walker the form of male OUGHT to preponderate; according to Mr Yarrell the la
493q004 reponderate; according to Mr Yarrell the latter OUGHT: either in first breed or permanently.-- (2) Cr
500q10a By my theory in volcanic or rising isld, there OUGHT to be a good many races or doubtful species; ho
504q13v AS-- N.B. I think very likely the Peas to cross OUGHT to be placed far from all other Peas, from Wieg
525m024 but I feel sure I have seen a dog doing what he OUGHT not to do, & looking ashamed of himself.-- Squi
556m145 gnomy translated by Holcroft «Vol I» .p. 86 "We OUGHT never to forget-- --; that every man is born wi

558m151 ot idea of God arise from our confused idea of "OUGHT." joined with necessary notion of "causation",
558m151 ry notion of "causation", in reference to this "OUGHT," as well as the works of the whole world.-- Re
586n080 nt out some essential difference, which clearly OUGHT to be separated-- We apply instinct to one part
608o026 tack the weak & sickly as we do the wicked.--we OUGHT to pity & assist & educate by putting contingen
608o027 yet one takes it for beauty & good temper), nor OUGHT one to blame others.-- This view will not do ha
617o39v the same object of thought is a question which OUGHT to be clearly comprehended by anyone who wishes
620o045 This shows, that our feeling, that the instinct OUGHT to be followed is a consequence of that being p
620o046 hich «when present» give pleasure.) & which man OUGHT to follow-- it is his duty to do so.-- So we sa
620o046 it is his duty to do so.-- So we say a pointer OUGHT to stand a <spaniels> «housedog's» duty is to w
620o046 teach him & strengthen his instincts.-- so man OUGHT to follow certain lines of conduct, <although>
621o047 [RHC] 6) Hence man must have a feeling, that he OUGHT to follow certain lines of conduct, & he must s
628o53v by fear it might be partly made.]CD p. 21. "Why OUGHT I to keep my word"-- gives the problem, of ethi
629o55v -- (4) It is other question, how the feeling of OUGHT, shame. right & wrong comes into mind in first
319c257 t is said the Samoyed women (¿ north end of the OURAL mountains) have black nipples to their breasts.
264c079 ere alive, which have perished.-- Let man visit OURANG-outang in domestication, hear expressive whine
353d061 ustralia well developed <tits> «Mammae» in male OURANG-outang. other point of resemblance with man.--
376d138 ne each other sexes,-- «by taking up tail» Mem: OURANG Jenny with Tommy.-- Good evidence of knowledge
469t119 l Transact. about year 1778. Paper by Camper on OURANG-outang, has examined 7 says one specimen had o
540m085 ted that smell of ones own pud. not disagree.-- OURANG outang at Zoolog Gardens touched pud. of young
545m107 xpression of fury, jump to scratch my face. The OURANG outang, under same circumstances, threw itself
551m129 s horse jumping when word Jump said-- I saw the OURANG. take up a stone & pound the earth. Lockarts l
551m129 - read monkeys for preexistence <">-- The young OURANG in <Zoolog> Gardens pouts. partly out. displea
554m138 womens petticoats-- just like Jenny with Tommy OURANG.-- Very curious.-- Mr Yarrell has seen Jenny,
559m153 ge only by what you see. compare, the Fuegian & OURANG & outang, & dare to say difference so great...
560m156 orbet; «do» ideots form habits readily?? Do the OURANG Outang like smells «peppermint» «& music».-- H
576n049 like a deity, as M. Cousin says. because if so OURANG outang.-- oyster & zoophyte: it is (I presume-
581n064 . same as in Allied Kingdoms-- "food, sm<e>ll. (OURANG-outang), music, colours we must suppose <we> «
581n064 k's tail, as much as we do.-- touch apparently. OURANG-outang very fond of soft, silk-handkerchief--
590n094 ke Jenny.-- Dr. Abel has given an account of an OURANG.-- see his Travels.-- When one sees in Cowper,
596n184 f close study.-- full of practical observations OURANG do not move eyebrows.-- or skin of head,-- «sc
319c256 notes.-- Their soft-billed birds are inferior to OURS, & our lark ranks very high.-- Upon the whole t
430e117 iod could not have been direct parents of any of OURS,-- even if extinction is denied.-- it will not
453e182 The wild, small fowls at Pulo Condore "crow like OURS, but much more small & shrill".-- Humboldt. Vol
482z011 s Magellanicus.-- p. 210. Scolopax very close to OURS Rengger's work of Mammali: of Paraguay must be
291c166 er? It is our arrogance, it our admiration of OURSELVES.-- The idea of foetus being of one «both» sex
546m108 we can only know what others think by putting OURSELVES in their situation, & then we feel like them-
553m135 xes on imaginary beings, many vicarious, like OURSELVES) that savages (mem York Minster) consider the
605o19v of Eternity. which superiority we transfer to OURSELVES in the same manner as we are acted on by symp
617o40v consciousness of it we are conscious that we OURSELVES can originate in any point an opposition of f
264c079 ve, which have perished.-- Let man visit Ourang-OUTANG in domestication, hear expressive whine, see i
313c235 Entomostraca. The sexual curiosity of the orang OUTANG of (in June 1838 when young male was added goo
353d061 a well developed <tits> «Mammae» in male ourang-OUTANG. other point of resemblance with man.-- Septem
469t119 act. about year 1778. Paper by Camper on Ourang-OUTANG, has examined 7 says one specimen had on one f
540m085 t smell of ones own pud. not disagree.-- Ourang OUTANG at Zoolog Gardens touched pud. of young male &
545m107 on of fury, jump to scratch my face. The ourang OUTANG, under same circumstances, threw itself down o
559m153 y what you see. compare, the Fuegian & Ourang & OUTANG, & dare to say difference so great... "Ay Sir
560m156 «do» ideots form habits readily?? Do the Ourang OUTANG 'like smells «peppermint» «& music».-- Have mon
576n049 deity, as M. Cousin says. because if so ourang OUTANG.-- oyster & zoophyte: it is (I presume-- see p
581n064 as in Allied Kingdoms-- "food, sm<e>ll. (ourang-OUTANG), music, colours we must suppose <we> «Pea-hen
581n064 l, as much as we do.-- touch apparently. ourang OUTANG very fond of soft, silk-handkerchief-- cats &
587n088 er it.-- Why do children pout & not men--orang-OUTANG & chimpanze. pout.-- Former, whines just like
056r110 ner wall, hence resists degradation longer than OUTER parts.-- The common occurrence of a breccia of
064r135 rs most abundant in interior channels. there no OUTER coast.--important effect.--? Capt. FitzRoy. --
134a141 refers to ‖ & Rocks p. 375. on the soundings on OUTER coast of T. del. Fuego.-- p 385 Rocks of S. Wes
066r141 des form eddies with its extreme force. Yet, no OUTLET at head. Important in forming transverse valle
120a107 re of Andes. where we believe we can trace the OUTLINES of what were fluid undulations-- the equal mo
574n042 ren. which chance? produced with strong arms, OUTLIVING the weaker ones, may be applicable to the for
085a001 s: Lias in Shropshire. or some other wonderful OUTLYER.-- Linn: Transact. Vol. 8. p. 288. Salt deposi
030r034 as at Concepcion. favoured by basin formed by OUTLYING rocks; (such as between Mocha & main land). <
065r140 Capt. Cook found soundings. (end of 2d voyage OUTSIDE coast of T. del Fuego. off. Christmas sound. -
095a038 > near summit.-- much pumice --. appears to be OUTSIDE of the Cordillera-- Near the Planchon talks of
366d111 isagreeable hearing maed servant cleaning door OUTSIDE, as often as she touched handle, though really
386d167 nimals). latter the division taking place from OUTSIDE inwards. & in animals from inside to the outsi
386d167 tside inwards. & in animals from inside to the OUTSIDE.-- is this not owing simply to more importance
628o54v 4) p 38 Conscience checks the wish to <other> OUTWARD gratification, whilst <the> no desire of grati
629o54v nscience») equally <prefe> destroy all wish of OUTWARD gratification,-- see what cases Mackintosh gi
041r066 ike the rest of these thin veins which project OUTWARDS.-- In Patagonia. the blending of pebbles & th
102a059 down. on coast itself, undertow would draw it OUTWARDS.-- form of breaker affected some way out to s
399e009 ely to manhood,-- it will decrease & be driven OUTWARDS in the grand crush of population.-- Octob 10t
297c180 my theory not applicable [not located] p. 428. OUZEL sometimes builds nest without doom. Vol 2. Mag
255c051 (habits of ducklings and chickens) Young water OUZELS hence aversion to generation, before great dif
289c160 to localities"-- p. 390,. young <grebes> "ring OUZELS» dive instant touch the water.-- capital insta
318c254 part of country, like White of Selbournes Rock OUZELS.-- ..If the line or bands of country (These fa
216b181 lus developed in the breast.--looking as if many OVA-- impregnated at once.-- Dr. Smith considers the
316c245 arkable.-- Fish & drift sea weed-- may transport OVA of shells.-- Conchifera. hermaphrodites-- eggs i
387d168 r & ovum in offspring, as similar to the several OVA in mother. (with only difference of time) is the
392d180) only simple form of life are monocecious. Will OVA of fishes & Mollusca «& Frogs» pass through bird
451e177 of India. -- p. 555. Lieut. Hutton counted, the OVA of a tick «in India» & found there were 5,283 at
528m037 e forms seem instinctively beautiful «as round, OVALS»;-- then there the pleasure of perspective. whi
382d158 n female.-- the <add> presence of both testes & OVARIA in Hermaphrodite,-- but not of poenis & clitor
382d158 not distinctly stated by Hunter.-- Do testes, & OVARIA when they first appear occupy their proper pos
388d172 - Mammae in man. having given milk. -- testis & OVARIA The following views show that transmission of
382d158 in in Man, & yet no trace of abortive womb, or OVARIUM.-- or testicles in female.-- the <add> presenc
385d164 on them.-- May be just worth remembering that OVARIUM of women (Paper in Vol I of Irish Royal Academ
342d035 is be not imagination.-- then old peculiarity OVERBEARS the crossing with females not thus characteri
289c161 ve impulse to keep separate, which no doubt be OVERCOME, but until it is the animals are distinct spe
614o037 these instincts,: though very weak so as to be OVERCOME easily by reason.-- Conscience is one of thes
047r083 orm of coast (as seen in swell) the undertow & OVERFALL must vary proportionally Partial shrinking of
070r154 site have been instanced: it is Let it not be OVERLOOKED that except by trees, I could not see trace
206b148 with varying races of man: these races may be OVERLOOKED mere variations consequent on climate &c-- t
294c176 terations in wild animals, & thinks Lyell has OVERLOOKED argument, that domesticated animals change a
419e087 & even no loss of species.-- It must never be OVERLOOKED that the chronology of geology rests upon am
441e150 ing greatly developed on another, must not be OVERLOOKED.-- it makes fourth cause or law of change.--
365d107 tween <t> "permanent varieties" & species. he OVERLOOKS-- <restric> those <restr> restricted in thir
077r171 mboldt says fragments from roof & penetrating OVERLYING beds tells the secret.-- p. 189. "The small r
078r176 215 Same metal in Tasco vein in Mica Slate & OVERLYING Limestone Balls of Silver ore occur in do vei
542m097 an is very great from all animals-- but do not OVERRATE-- animals communicate to each other.-- Lonsda
262c073 of organic being.-- the fact of guavas having OVERRUN-- Tahiti. thistle. Pampas. show how nicely thi
553m132 former. But it is one thing for a swift dog to OVERTAKE an emu, & [not located] notion, are not effec
568n017 not performing it, (either if prevented, or OVERTEMPTED.-- "animals have shyness with strangers" «as
058r118 a bad earthquake in Chili; Arequipa in 82 was OVERTHROWN, & 86. Lima. next year Quito. considers thes
071r157 an and Salta. towards the Vermejo was utterly OVERTHROWN by earthquake with great destruction of huma
071r158 --«formerly perhaps otherwise» Mendoza near OVERTHROWN,--no mountains Mackenzie has talked of lava
230b240 produce hybrids, or else whole fabric will be OVERTURNED.-- Hence extreme difficulty, argument in cir
443e154 y gemmation <can be> being impossible. can be OVERTURNED, then the conclusion that the two kinds of g
553m135 ws that such a view cannot be, anyhow, easily OVERTURNED.-- so ready is change from. our idea of caus
356d071 hat comparative difficulties (as long as not OVERWHELMING) What comparative solutions & linking of fa

```
383d159 then give speculation to show that it is not OVERWHELMING.-- Seeing in Gardens of Hybrids between Com
409e047 en mind of a dog & a porpoise was not thougt OVERWHELMING.-- yet I will not shirk difficulty-- I have
447e164 s some most important end.-- Festuca vivapara F OVINA-- propagated like oni[on] Poa alpina because vi
447e165 at Festuca vivapara is the same species with F. OVINA, <& this> rendered vivaparous by growing on hei
321c275 sle's Husbandry Tuckeys voyage reread Appendix OVINGTON Voyage to Surinam. Voyage Congo expedition: Z
311c227 ds geographical range very good.-- Blainville OVINGTON'S Voyage to Surat, floating isld. off coast of
355d068 w in some Polypi we see young bud changing into OVULES.-- Captain Grants. Himalaya. shells (see Paper
254c048 higher  classes, during which the changes of the OVUM or embryo succeed each other with the greatest
263c076 stinct to be acquired (if the medullary point in OVUM. has such organization as to <per> force in one
367d112 ertile. supposes that when foetus is forming the OVUM within it, is forming «& this must be so,  else
387d168 es, that it becomes fixed in blood.-- Looking at OVUM of mother & ovum in offspring, as similar to th
387d168 es fixed in blood.-- Looking at ovum of mother & OVUM in offspring, as similar to the several ova  in
387d168 be  <offspring> «animals»: if «x» male impresses OVUM <of> in A, «with some peculiarity» that in  (B)
387d168 t in (B) to <a slight> «some» degree, & likewise OVUM in (B) <an C> that in (C) «in lesser degree» --
388d173 t also shows that semen. must actually reach the OVUM.-- [Why in making a bud, which is to pass throu
210b161 as> production «of monsters» (which, Hunter says OWE their origin to very early stage) & which, follo
528m038 le tree,-- & the leaves of the foreground either OWE their beauty to absolute forms or to the repetit
062r129 parallel case.-- Should urge that extinct Llama OWED its death not to change of circumstances; rever
057r113 differ as much nearly as the Eocene. = Should Mr OWEN consider bones washed about much at Coll. of. S
175b019 rds different types of organization improving as OWEN says simplest coming in & most perfect «& other
210b161 ourbond Zoolog. Proceedings\. <p> 1832.p. III Mr OWEN suggested to me, that the <cas> production «of
211b163 resenting foetuses of Vertebrata, &c 1837 p. 370 OWEN says Nonsense The distribut of big Animals in E
248c033 of longest [consta]nt & therefore most permanent OWEN [re]markable laws of Brain & manner of generati
254c047 slightly different.-- Q Zoolog. T. V. I. p 389. OWEN remarks on Entozoa, the organs of generation, a
254c048 lasses & likewise those much higher in scale. So OWEN actually believes in this view!!! p. 392.-- exc
261c072 n of beauty, when does prefer most powerful buck OWEN talking of Plesiossaurus Plesiosaurus.  alludes
278c131 mense age since breccia accumulated-- surely ask OWEN to see whether species same, excessive improbab
306c213 es.-- organs of generation a captial character. (OWEN) not for first & grandest divisions. but for on
322c270 ctual powers. Hunters Animal OEconomy. edited by OWEN. read several papers-- all, that bear on any of
325c267 832. contains all his fathers views.-- Quoted by OWEN.-- Hunter has written quarto. work on Physiolog
325c267 rk on Physiology besides the papers collected by OWEN. (at Shrewsbury) Yarrells Paper on change of pl
326c266 ere appears to be good art.. on <Etn> Entozoa by OWEN in Encyclop. of Anat. & Physiology.--  Dampier.
326c266 & Medical Researches. on Horse in. N. America.-- OWEN has it.-- Ld. Brougham. Dissertations on subjec
341d029 .-- (i.e. whether relation in one point or many) OWEN answered that all characters might be considere
342d035 rst converted-- is it an old order Geologically? OWEN says relation of Osteology of birds to Reptiles
367d113 of Zebra.-- Mem. Quagga & Ld Moreton Mare ringed OWEN says that Bell in Encyclop of Anat & Phys. desc
368d114 ds display beauty of plumage.-- (The females (as OWEN observes) in Raptorial birds largest.-- p.  47.
369d115 aracter Hunter Animal Economy p. 482 (Same book) OWEN says "the necessity of combining observation of
369d115 conomy, is now universally admitted.""-- p. 483. OWEN thinks from climate of Australia, & from Ornith
371d127 sily trained up to the (required) offices" &c &c OWEN illustrates case of Dingo (he alludes to the dh
379d152 f there be one female, she will be free Marten. <OWEN.> See Hunter's Own-- In the Athenaeum  Numbers
379d152 e, she will be free Marten. <Owen.> See Hunter's OWEN-- In the Athenaeum Numbers 406, 407, 409, Quete
381d156 giously-- Animal OEconomy by, Hunter. (edited by OWEN) p. 34.-- Owen classifies Hermaphrodites. Crypt
381d156 I OEconomy by, Hunter. (edited by Owen) p. 34.-- OWEN classifies Hermaphrodites. Cryptandrous.  (only
382d157 not in other), Hunter <asks> p. 36 is thought by OWEN to ask. whether a Heautandrous animal is <evide
382d157 actually  split in two-- keeping sexes separate. OWEN say such view worthy of a Lamarckian.-- Mine is
384d161 ither parent bird-- Hunters Animal OEconomy. (by OWEN) p. 44. Classification of Monsters. (1) From pr
421e091 arity not to be eradicated.-- \.do. p. 305.-- Mr OWEN says <tha> «in abstract» in his paper on the Du
422e092 & hence not indicative of true affinity.-- -- -- OWEN says Dugong connected with Pachydermata.-- <it
425e105 are  connected with other mountains laterally.-- OWEN. Fossil Mammalia. p. 55. talks of Tapirus Ameri
436e136 <April 12th..> Cestracion, Port Jackson Shark-- OWEN thinks Australia part of Old World <If> It «may
448e148 e are Baboons in St Thomas on W. coast of Africa OWEN Linn. Soc. April 2d. 1839 The Lepidosiren-- Amb
463t063 ing a new fragment of the discoveries to commer- OWEN in his description of my fossils makes same suc
464t065 coveries- negative facts are valueless= monkeys= OWEN has described a great Struthoniidous Bird  from
489qIFC ---- l Jordan Smith. p l. Sowerby Cuming. -- p l OWEN p 17 Hooker p. 17 (T) Mrs. Whitby. Newlands Lym
509q017 Whale, than in old?? Falconer says all in cases. OWEN.  Have talked partially with him Ask him to intr
041r064 he effect of Salt water of the Salado.--Mem. in OWENS Africa it is mentioned that the Elephant came t
061r127 expedition.  -- ¿ Another one in 1816 (?).-- Mr OWEN'S curious fact about Crust Bra in Brine Springs.
198b113 anguineous system, in other nervous developed. (OWEN'S idea) states these class approach on the confi
218b190 p 173. Voyage par un Officier du Roi Mem. Capt. OWEN'S story of cats on West coast of Africa.-- chang
240c003 like the wild rock pidgeon.-- fact analogous to OWEN'S Phil: remark of Apteryx having feathers.--  It
483z013 s. of Geof.-- reference from Rüppel travels All OWENS papers on Intestinal worms must be studied in V
122a114 igure of earth statical.-- if platform of mexico OWES its elevation to equilibrium.-- it cannot be eq
440e145 g» tree, & every buzzing insect & grazing animal OWES its form, to that from being «the one alone» ou
032r041 ances from centre of rotation» & a <circulation OWING rotation in fluid matter of globe. must  there
041r067 f pebbles & the appearance of travelling may be OWING to successive transportal from prevailing swell
042r068 eriod».--Mem Bahia blanca P. 204 Vol III. Lyell OWING to «open» faults in mountains: to elevated stra
046r080 of rise (any smaller prior ones might have been OWING to absolute movement of ground). Michell (Philo
046r080 l movement.--At first one would think movement. OWING to water keeping its level whilst land rose  up
051r094 n; I now having seen Pernambuco believe much is OWING to protection of Organic productions. = Yet eve
077r170 relation--there is more modern breccia, chiefly OWING to destruction of porphyries. whereas other  to
086a003 f Ventana's «Quartz.» unmixed is very pleasing; OWING to the movements being of one order. -- There s
102a061 & the layers first of felspar & then quartz &c, OWING to separation having taken place most gradually
106a073 eat.-- Is there more degradation at first angle OWING to momentum. which the water has obtained.-- If
176b022 nite laws, then those which have changed most. «OWING to the accident of positions» must in each stat
176b023 e not be a triple branching in the tree of life OWING to three elements air, land & water, & the ende
179b032 ther, the later ones of the father; is not this OWING to each copulation producing its effect; as whe
179b032 t; as when bitches puppies are less purely bred OWING to <first> having once borne Mongrels he has th
179b035 e> «grant» similarity of animals in one country OWING to springing from one branch, & the monucle has
224b229 meday produce something else., but not probable OWING to mixture of races.-- When all mixed & physica
228b229 going  back ¿whether this going back may not be OWING to cross from other trees.???? Do the seeds  of
242c017 usted The changes in species must be very slow, OWING to physical change, slow & offspring not picked
258c060 als having wide range, by preventing adaptation OWING to crossing, with unseasoned people. would caus
264c082 ion of birds Birds vary much (more than shells) OWING to variety of station inhabited by them-- Timor
265c085 -- The only way of judging whether structure is OWING to habits, or heredetary is to see, whether a l
267c095 quaternary  arrangement of Cryptogamic plants.--OWING to plants not being adapted to Air!! p. 11--&c.
268c100 many  families on Keeling seemed to consider it OWING to one of each, being fitted for transport ¿may
290c163 (if  not looked at as instinctive) then must be OWING to heredetary power of Muscles.-- then we SEE s
296c184 ough» Vegetation of peak-- altogether original. OWING to being oldest. & having undergone changes «no
302c203 ediate Father-species, & not, therefore, solely OWING to such interm: father-species, being little ad
312c228 iel even when grown up.-- Are mules homogenious OWING to no attempt to keep up offspring, are not hal
331dIFC 6]CD Towards close I first thought of selection OWING to struggle July 15th. 1838 Finished. October 2
335d016 les could have no offspring, & this being case, OWING to the corelations of system, the organs of gen
349d052 Anomalous  structures, as in Hippotamus, solely OWING to number of lost links.\ if all species know t
360d089 e hairy-- the former preponderated <which seems OWING «determined» by the sex> Individual instances t
363d102 eeds in <flags> «thick vegetation» in swamps-- (OWING to barns, perhaps, not being left open to them,
364d104 nks the occasional production of black lambs is OWING to old <story> return.-- The Revd. R. Jones tol
365d105 oubt the cross between Pheasant & Black game is OWING to their rarity., a single female in wood  with
371d128 them  with such facility to bud.-- this must be OWING to their unity in one stem.-- a bud may be tran
386d167 imals from inside to the outside.-- is this not OWING simply to more importance of internal organs in
408e044 ?-- The wonderful species of Galapagos, must be OWING to these islands, having been purely result  of
412e057 to  circumstances of times. & from persistency «OWING to their slow formation» these variations  tend
415e067 more valuable domesticated animals" no doubt is OWING to the rearing up of every heredetary tendency
417e075 must   think deserve the name of species, may be OWING to the little fixity of organization, in the tw
417e075 ttle fixity of organization, in the two races,, OWING to the domestication of both.-- Now in the ass»
419e085 sca, which I have sometimes speculated might be OWING to absolute quantity of vitality «in the World»
432e124 of any country must be made with great caution; OWING to its adaptation to the surrounding circumstan
446e163 ot insure constancy of form.-- is the constancy OWING to similarity of conditions-- & that no  change
```

Page
(Key Word)
465t081 1. Journ. end of 1839) very important. it seems OWING to immigration of other races, so it is with do
526m031 s he chooses wrongly,-- & what is frame of mind OWING to.--<)>-- I verily believe free-will & chance
532m054 frightened & started much more apt, this partly OWING to heart? readily taking same movements, senses
546m109 Partridge--, man's taste for smell of flowers, OWING to parent being fruit eater.-- origin of colour
587n089 ead to action.-- p. 248. Theory of Association. OWING to time when entered brain, try contiguity of p
591n099 ide. (even when no relatives left to lament) is OWING to the feeling that the instinct of self-preser
604o017 n Mind. Poor.-- on insanity.-- Prevailing idea. OWING to loss of will.-- chiefly excited by passive e
620o045 re than in an animal.-- which shows that. it is OWING to some <subsequent> power (reason) obtained by
621o047 odd association» comes to this conclusion, not OWING to peculiarity of organ of taste, for when grow
635j56r he shortness of life (peculiar to each species) OWING to the growing size of the world? & the physica
363d102 ican & Europaean birds «slight» different-- Barn OWL <the> in the former place breeds in <flags> «thi
364d103 self.-- Q Sir. J. Sebright-- has almost lost his OWL-Pidgeons from infertility,-- Yarrell says in suc
641j29v ave analogues The stillness p. 276) of flight of OWL remarkable, [gained by very different process fr
191b082 e interesting to trace limits of large animals-- OWLS. transport mice alive? Species formed by subsid
297c186 tera do. p. 160. soft plumage of night jar. like OWLS. analogy in habits adaptation to nocturnal habi
309c222 anatomical structure.-- «the passages between-- OWLS & hawks only external» intermediate groups ofte
437e140 om another island.-- » p. 175., 28 sho[r]t eared OWLS were counted in a field, where there was great
483z012 : <Jerrold?> Bibron Zoolog. Journ Vol I. p. 125, OWLS seen crossing Atlantic. fact taken from Jenner
512q019 erse cross.-- Sow cast-up-balls of Hawks or even OWLS.-- How long do seeds remain in stomach of birds
217b183 rred to every one) of rare breeds of dogs, from OWNERS great care of them. Fox says when two dogs of
123a117 pus near Canelones -- large quadruped bigger than OX.-- at Buenos Ayres 20½ quadras from river; 20 va
200b125 extreme North.-- \.do p. 326. 2 Fossil species of OX in N. America: as well as 2 recent See Geolog. P
205b142 ite of Negroes different from European-- Horse & OX have different parasite in different climates.--
418e080 se the hermaphroditism would not be perfect as in OX. the amount of double sexual developement is spr
450e174 y in one part the northern peninsula of Borneo.-- OX & hog natives of Borneo Notices of Indian Arch.
491q002 rous grasses & onion, produce flowers, like the OXALIS from C. of Good Hope mentioned by Mr Herbert i
218b190 oduce monsters in Isle of France-- -- Madagascar OXEN with hump.-- p 173. Voyage par un Officier du R
241c014 t Amboina» I fancy there is marked wild breed of OXEN at Java.--.p. 140, calls it Bos. leucoprymnus.
288c159 Major Mitchell, does not know whether breeds of OXEN have deteriorated, or altered, but it is certai
417e075 to get the day» The fertility of Indian & Common OXEN, which one must think deserve the name of speci
216b183 hounds from other parts of England. Mr Bell of OXFORD St'-. had a very fine blood hound bitch which
076t168 is observed both in Mexico & Peru, that those OXIDATED masses of iron. which contain silver are pecu
077t172 mbined with nickel & cobalt (meteoric) resist OXIDATION?-- Mem Sir W. P. stone It is clear to me, the
075t167 es from Peru consisted of native silver & brown OXIDE of Iron in Mexico. sulphuretted silver, arsenic
560m156 & sommambulism. Do people when inhaling Nitrous OXIDE, forget what they did when in this state, or re
074r165 cal law & steam of salts, quite curious case of OXIDED Iron by Mitterschlich. Vol. II Journal of Nat.
516q23v 2) Athenaeum 1840 p. 777. Decaying wood absorbs OXYGEN & forms Carbonic Acid. will this bear on Petri
484z016 nto Furnarius. by Patagonian Furnarius.-- into OXYURUS, by Maldonado creeper of same plumage.-- gener
381d156 es. Cryptandrous. (only female organs visible). OYSTER. cystic Entozoa. Echinoderms. Acalephes. Polyp
445e159 l fish is really hermaphrodite. & <thinks> even OYSTER may fecundate each other, by the means of the
536m072 hey have free will, if so all animals., then an OYSTER has & a polype (& a plant in some senses, perh
536m072 only to be changed by habits). now free will of OYSTER, one can fancy to be direct effect of organiza
536m073 called) makes change in bodily organization of OYSTER. so may free will make change in man.-- the re
567n014 w signs of affecting something like man. Has an OYSTER necessary notion of space-- plant though it mo
576n049 M. Cousin says. because if so ourang outang.-- OYSTER & zoophyte: it is (I presume-- see p. 188 of H
064r134 t foot of range some miles from shore. rock of OYSTERS quite above reach of tides.--thinks them same
501q10a ntingent on country.-- How is it in Patella or OYSTERS or Helix. Or does any «one» species of plant,
531m051 shows how truly an instinctive feeling, <may not PA.> In reflecting over an insane feeling of anger
250c037 on.-- Whatever destroyed great <quadrupeds>; PACHYDERM in S. America destroyed great Edentata or Ame
295c178 ther Mammals.-- «otter; civet cat, rodents.» (PACHYDERM in Portland stone of Alps!!!? No) p. 15 (Lyel
422e092 ame from one period to another, pre-eminently PACHYDERMATA, less so in Miocen & so on.-- As I have tra
186b059 living; or of the tribe fish extinct. or of PACHYDERMATA, or of coniferous trees; or in certain shel
206b149 great Edentate might have roamed to Europe & PACHYDERMATA from Europe to America., How strange would
226b223 o present») more like present carnivora than PACHYDERMATA If my theory true, we get (I) a horizontal
226b223 lastic Clay of Paris contains many genera of PACHYDERMATA. &c & other Mammals.-- «otter; civet cat, ro
295c178 efore an isolated form probably a remnant.-- PACHYDERMATA. & Horses few forms. & they are remnants.--
297c186 ot Elephant intellectually developed amongst PACHYDERMATA. like Man amongst Monkeys-- or dogs. in Carn
299c196 two Polar Continents Marsupial. Edentata.-- PACHYDERMATA &c &c-- It is important with respect to ext
355d068 ity.-- Owen says Dugong connected with PACHYDERMATA.-- <it was a Pachyderm. which was the origi
422e092 ith true Cetacea or whales.-- but are aquatic PACHYDERMS. & Walrus-- aquatic seal.-- (Consult this pa
422e092 raw to make its nest. Pigs & Elephants, (both PACHYDERMS) much intellect.-- mem: Yarrell's story of w
552m131 Earthquakes felt. different case from shore of PACIFIC.--Isabelle's volcano, many amygdaloids.--Bouss
071r158 continent corresponding to Europaean risings. PACIFIC great land. -- Will use argument of proof of s
087a011 nation of linear bands of movement in Indian & PACIFIC Oceans.-- (2d--) does not explain first format
118a104 s Galapagos & Juan Fernandez. When continet of PACIFIC existed might have been Monsoons.. when they c
173b011 If continent had sprung up round Galapagos on PACIFIC side. the Oolite order of things might have ea
227b227 Vol II p. 8 no snakes on isles of central PACIFIC, yet there appears to be one at Botouma from a
246c027 n «See Gaudichauds Volume on the Botany of the PACIFIC.--» to nearest continent.-- With respect to an
358d074 lip &c &c. in Vol II p 363 account of Flora of PACIFIC, given in my coral paper Oct 14th Macleay says
403e022 er of horizontal upheaval» of the shore of the PACIFICK is 60 miles distant from the grand ancient vo
023r011 introduced on great size of ocean; especially PACIFICK: insignificant islets--general movements of t
044r073 another falls.--When discussing connection of PACIFICK & S. America.-- Volcanos must be considered
045r077 nce. Is it not same as swell travelling across PACIFICK.--excepting in number of waves & in wind, ins
047r081 r no evidence--The depth of shells (which being PACKED. in beds) lived there, makes it very doubtful
040r064 50 toises above the sea. = talks of them being PACKED clean. & without earth.--Moreover that such do
055r106 art at once to Shrewsbury., vaguely thought of PACKING up.-- was lying on my back fell to sleep for s
547m111 I. p. 325-- Wild dogs of Guayana always hunt in PACKS <30 or 40 together> colour reddish <brown>. ear
267c094 same fact in Galapagos. Daubeny P 24 V. back of PADDLE, yet all in the arm are perfect.-- p. 144.-- A
444e157 . 144 in the Icthyosaurus 60 or 70 bones in the PAGE 1 of New Zealand Geological Notes. at St. Helen
031r038 as well as effects of cooling [misnumbered, no PAGE 122] In Igneous rocks.--which have the cryst of
059r121 n the richness of the veta madre of [misnumbered, no PAGE] Dr D. remarks. bad conductor of Heat do of Ele
077r170 es Mines 1809. No. 151. p. 79. [misnumbered, no PAGE 173] Under name of Sagitta Triptera D'Orbigny t
078r172 feldspar) within other concretion.-- state last PAGE thus. point of attempted crystallization, & the
098a047 ably much Carbonic Acid gaz here.-- [top portion PAGE excised, not located] Bull:. Soc: Geolog. Tome
108a081 (My Edition) p 500. Well described [top portion PAGE excised, not located] -- do-- [...] Subaqueous.
108a082 f stone.-- & Caldcleughs collection of facts See PAGE 101. in Note Book (C) for some speculats on con
112a088 Mem. Eyton's Hogs & dogs.-- The passage in last PAGE explains that between Species from «moderately»
224b212 d be disputed.-- according to Principles of last PAGE. <an> osculant groups between two circles «of e
302c202 d a perfect chain. <Icthyo> .+ + +. supra & next PAGE It is a fact pregnant with SOMETHING.? that int
303c205 - two considerations are here combined). In last PAGE, we have seen mules could have no offspring, &
335d016 of generation would necessarily fail.-- In last PAGE. I should have said, "an animal <acquires <th>
352d060 iew of geology, of each formation being merely a PAGE torn out of a history, & the geologist being ob
398e006 back we obtain here & there in order a scattered PAGE; we find <great> sensible change in the institu
526m029 bers a thing in a book, one remember the part of PAGE.-- one is tempted to think all memory consists
544m103 est of the mind, with all other faculties: «Vide PAGE 110, by mistake.» N B. Everything which happens
546m110 many happy days he spent in such a place.-- Vide PAGE 103, supra (by mistake) have lower animals thes
549m118 s puppies-- This looks like free will.-- V. last PAGE. A healthy child is «more» entirely happy (cont
572m033 mebody gave me a book in French I read the first PAGE & pronounced each word distinctly. woke instant
572m033 antly but could not gather general sense of this PAGE.-- Now <awake> «when awake» I could not picture
107a077 icial limestones produced by Sir J. Hal. End of PAGES. p. 157. Vol VI Edinburgh. Phil: Transacts.-- D
170bIFC n makes thought Glen Roy B C. Darwin All useful PAGES cut out Dec. 7th. /1856/ (& again looked throug
180b039 of non adaptation of circumstances.-- Vide two. PAGES back. Diagram The largeness of present genera r
241c016 es. Septmemb. 1825 Get Henslow to read over the PAGES from about 8 to 20 of Zoologie of Coquille's Vo
279c133 , developemen of important organ (.see marks on PAGES).-- Crosses of diff: breeds succeed, yet seems
353d060 ture of animals & plants.-- he get merely a few PAGES.-- Hence (p. 59) looking at animal, if there be
398e005 «without every animal preserved.».-- the latter PAGES in the history are perfect, we obtain a glimpse

(Key Word)

592n103 «First Croonian Lectures by Parsons.» following PAGES contain remarks worthy of attention p. 15, 25.
627o52v r I write on these subjects consult <following> PAGES. <p. 231> marked in my Mackintosh 1) Mackintosh
530m043 orgotten.-- My father signed a bond, yet when he PAID the Attorneys bill, he asked what bond he could
228b231 als with affections, imitation, fear <of death>. PAIN. sorrow for the dead.-- respect We have no more
228b232 n wild then <our> animals our fellow brethren in PAIN, disease death & suffering «& famine»; our slav
286c154 s understand the language, they know the crys of PAIN, as well as we.-- It is our arrogance, to raise
300c197 whether animals have any fear of death, only of PAIN, of death acquired?. The S. American dung beetl
432e125 drinking very cold water «frowns connected with PAIN, as well as intense thought.--"» No one but a pr
522m012 t the time, but some time afterwards it calls up PAIN. or pleasure. & is often recurred to & mentione
524m020 -- the action of brain which gives sensation of PAIN, emits its power on the muscles in the twitchin
530m044] Mr B> journey. short time previous,-- because, PAIN prevents repetition of idea.-- Mr Blakeway has
536m072 in some senses, perhaps, though from not having PAIN or pleasure actions unavoidable & only to be ch
536m072 ization, by the capacities its senses give it of PAIN or pleasure, if so free will is to mind, what c
540m088 Burke's idea of Sympathy. being real pleasure at PAIN of others, with rational desire to assist them,
541m089 well observed, one does not fear death from its PAIN, but one only fears that pain, which is connect
541m089 ear death from its pain, but one only fears that PAIN, which is connected with death!-- How has this
541m092 imple idea as scarlet?-- How can people dwell on PAIN ¿ no definite idea. nor is an emotion.-- People
544m101 ceive pleasure) gives pleasure, ie. love.-- & so PAIN gives fear of death. Mayo Philosophy of Living.
544m104 of death!!! as Montaigne observes. distinct from PAIN, for one hates pain from this fear-- & not deat
544m104 igne observes. distinct from pain, for one hates PAIN from this fear-- & not death for the pain.-- Ho
544m104 hates pain from this fear-- & not death for the PAIN.-- How was this instinct gained.? by conversati
549m118 ughts, he must have contingency of good food, no PAIN.-- «but the» «&» the sensual enjoyment of the m
549m120 al of happiness greater. even if mixed with some PAIN.-- than the happiness of a peasant, with whom s
550m125 ble as <broken> intense happiness even with some PAIN.-- compared to what others experience in same t
550m127 t circumstance calls up pleasure. or pleasure or PAIN of association.-- now if one has these feelings
568n017 ime» to any line of action, or thought one feels PAIN, at not performing it, (either if prevented, or
568n017 rd dog. has pleasure in following its instinct & PAIN if held.-- if tempted not to follow it, by grea
568n017 habit, (or moral sense, or instinct,) one feels PAIN, & vice versa pleasure in performing it.-- As s
575n045 nk, whether there is any analogy between grief & PAIN-- certain ideas hurting brain, like a wound hur
579n058 curious if it is so.-- frown with grief,¿ bodily PAIN? frown shows the mind is intent on one object.-
582n066 man cannot be said to be angry.--» «He may have PAIN or pleasure these are sensations» <Gardner in h
582n068 ey are more truly sensations??. a kind of mental PAIN & pleasure.-- The Revd. Algernon Wells Lecture
606o022 character.-- Hence Lessings shows expression of PAIN cannot be represented. But what is beauty?-- it
613o035 mentarily conscious, but is memory gone?-- Where PAIN & pleasure is felt there must be consciousness?
619o042 being prevented by <necessity> «some force» give PAIN: for instance either protecting sheep or hurtin
619o043 ch actions were prevented by force he would feel PAIN. [.By a very slight change in association if ot
619o043 being able to prevent it, he would likewise feel PAIN.-- If he saw another man <say go> «acting in» a
619o043 ncts from interference of passion. he would feel PAIN, which would generally be anger, as he would be
636j57r nsider ground Woodpecker stiff tailed cormorant: PAIN & disease in world & yet talk of perfection Get
263c075 great effect produced, is a most laborious, & PAINFUL effort of the mind (although this may appear a
523m016 ghts of it, for some years after, was far more PAINFUL than the thing itself. Asked my F. whether ins
584n073 ys she can perceive sigh, commences as soon as PAINFUL thought crosses mind, before it can have affec
544m101 awn between «heredetary» associated pleasures & PAINS & emotions-- such as child sucking, gives pleas
548m115 is curious problem., one does not care for the PAINS of ones infancy.-- one cannot bring it to one s
581n064 fficulty in expressing their want, pleasure, or PAINS long before they can speak-- or understand-- th
565n007 y bad news, «though in a letter» why is person PAINTED with mouth open.-- why when person is listenin
603o11b me.--" Belief allied to instinct.-- p. 134. a PAINTED must not a actors, or a scene in garden.-- yet
603o014 oung woman, with good expression-- statues not PAINTED-- <music> very good article-- why flower beaut
606o024 oon p. 125-- says new subjects are not fit for PAINTER or sculpture, but rather subjects which we kno
326c266 Laoccaon.-- (translated in 1837) on limits of PAINTING & poetry.-- Erasmus thincks I should lik it.
606o022 oon. 2d Lect-- The object of art., sculpture & PAINTING, is beauty.-- which he thinks is a better def
172b006 cats & dogs ibis same as formerly but separate a PAIR & place them on fresh isld. it is very doubtful
172b007 it from some end which is good for Man.-- Let a PAIR be introduced & increase slowly, from many enem
318c255 and probably a «chance» wanderer like the first PAIR of Pipra flycatcher.-- Bachman says he thinks t
366d108 .-- now Sir J. Sebright. thought if he had had a PAIR he could have produced from these.-- this insta
379d152 s, <one> though capable of producing both <male> PAIR of male & female.-- if there be one female, she
387d169 strongly; they transmit with same force as first PAIR, but to this tendency is added <that> the 3d te
387d169 dency is added <that> the 3d tendency from first PAIR.-- Now if two of third pair of same peculiarity
387d169 tendency from first pair.-- Now if two of third PAIR of same peculiarity breed they will have same i
387d169 ity breed they will have same influence as first PAIR + tendency they inherited from second pair, + t
387d169 first pair + tendency they inherited from second PAIR, + the influence they themselves inherit./ Anna
399e010 similarity, because in every country, where only PAIR has been introduced, & have freely bred, they h
477z002 do. p. 365. 3 cats (mbara caya. le negro, et le PAJERO) l'yaguaré «the zorilla-arskink» le quiyá (Coi
146g009 project [Blank] {P} Path East End near Holyrood PALACE In same way at top the trap could be traced Gr
183b053 intermediate forms been discovered. between PALAEOTHERIUM, megalonyx mastodon, & the species now livi
250c039 pes of animals What reptiles coexisted with PALAEOTHERIUM in Paris quarries & at Binstead. Mem. recen
250c039 & at Binstead. Mem. recent Crocodiles with PALAEOTHERIUM in India--: connection with Latitudes!? Zoo
030r036 onal cliff & low or sloping land What are the "PALATAL Tritores" found in the coraliferous mountain L
335d014 e, <Hill> «Lord Berwick» family with defective PALATES. heredetary & therefore exceptions. to above l
051r094 hia, whole surface to where highest spray (there PALE green confervae) coated with living beings: In
059r121 ate.!? or small Proportion of Alum: matter.--all PALE cream colour.-- The Brecciated structure of all
290c164 eeing common gull in garden at Zoology Soc. it's PALE ash grey back, like a black bird washed, Whilst
383d160 er pheasant. yet why green? & not purple?-- legs PALE coloured.-- In the back feathers, we have chara
480z007 s not of different size-- How does this apply to PALE brown Caracara Krauss on Corallinae from S. Sea
505q014 they seed.?-- (11) Abberley has planted seeds of PALE green Cynoglossum. never germinated 12 Does the
539m084 Descript of Queen) «O» of Hell Cant IV or V.) as PALE & trembling. & not as flushing & with muscles r
215b178 . Hilaire has written "opuscule" entitled "PALEONTOGRAPHIE" developing his ideas on passage of forms.
399e009 poulterer shops., of same colour as a Hare, but PALER & buffer.-- with long ears-- & longer hind legs
472s01v ter with mere traces of black spot at base, one PALER with less riged foliage & no black spot & a thi
472s01v foliage & no black spot & a third considerably PALER, all rest very similar-- June 2. 42 Maer <Thurs
622o048 ves different explanation of law of honour from PALEY [RHC] Anyone, who will reflect must feel, how l
552m132 are good & consequently give pleasure, & not as PALEYS rule is those that on long run will do good.--
086a006 «Salt Lakes» Siberia must be read as well as PALLAS before Geology is written Cuvier. Europe posse
620o045 es the «non-» following of ones conscience. & PALLIATES the offence; one always admire the habit form
314c239 species of Australian Genera Some species same (PALM & Phormium tenax) as in New Zealand & Australia
470t176 er 1840 My Father formerly planted «Turkey or» PALMATED and English, planted within few yards of each
470t176 dens, he knew there was none but English,--the PALMATED was introduced about '65 years ago-& soon af
470t176 ears ago-& soon after mules abounded--so that PALMATED has now nearly disappeared. <& old English> B
470t176 these mules <in our garden> show no trace of PALMATION!!? Bees at Wild St Johns Wort--Scabies, Cyano
482z011 rrestrial birds at Falklands Isd 8 waders. 22 PALMIPEDES: out of the first 9.:4 raptores. Falco polio
061r128 have peculiar character: Contrast low limit of PALMS, evergreen trees, arborescent grasses, parasiti
447e164 e is greater (Hence Dioecious plants highest,-- PALMS &c &c)-- Is there greater resemblance between c
636j56v ht inference is, there were insects «¿when were PALMS formed?» as soon as Dioecious Plants were forme
531m053 accompanied with want of muscular exertion, PALPITATION, voiding urine because done by some animals
596nIBC of voluntary muscles-- if so what is trembling PALSY? Expressions N [Old & USELESS notes about the m
299c194 .-- one grows in marsh & other dry; yet if T. PALUSTRIS be sown in dry station it will for some gener
466t091 -- Maer June/41/, observed 3 plants of Caltha PALUSTRIS alone together. one had seed-pods turning bro
037r056 ould probably be no other organic remains.-- On PAMPAS looked in vain for a pebble of any sort; not o
067r144 ful that volcanic rocks at M. Video «Volcano in PAMPAS» Pasto Earthquake. Happened on January 20th. 1
105a071 zes well with character of mouth of valleys &c; PAMPAS.-- If blocks above their parent rocks. would b
136a147 etters-- When I come to treat of the age of the PAMPAS Deposit, I may properly remark on the superior
262c073 act of guavas having overrun-- Tahiti. thistle. PAMPAS. show how nicely things adapted--.-- These «ab
279c133 India???-- & Indian Islds.-- Sir J. Sebright-- PAMPHLET-- most important, showing effects of peculiar
295c178 Portland stone of Alps!!!? No) p. 15 (Lyell's PAMPHLET) Is man more hairy than woman. because ancest
326c265 lkreuter's Papers Wiegman has published German PAMPHLET on crossing Oats, &c «Horticultural Transacts
320c275 view of latter in Quarterly Sir J. Sebright on PAMPHLETS Wilkinson on Cattle not abstracted Scientific
432e126 of time, only order of succession.-- Splendid PAMPLET. (published in Philosop. Journal <Mar> April 1
448e167 ost wholly different, is same, as if Isthmus of PANAMA.-- These two cases highly improbable.-- yet I

Page **(Key Word)**
564n006 s & upper lip., is most clearly analogous to a PANTHER I saw in garden uncovering its teeth to bite.-
571n031 language.-- Chinese. simplest language. Much PANTOMIMIC gesture?? which would naturally happen.-- Re
472s01r . sowed some years since gathered the seeds of PAPAVER bracteatum, & the Papaver oncitate was growing
472s01r athered the seeds of Papaver bracteatum, & the PAPAVER oncitate was growing in same garden. & out of
024r015 e Coast of Brazil? where not rivers in my Coral PAPER {T} leagues Fathoms Parallel of St Catherine [2
027r024 germination in young corals.--vide L. Jackson's PAPER. Philosoph Transact: at R. de Janeiro. Coquimbo
028r030 ons might be introduced either in note in Coral PAPER or hypothetical origin of some sandstones, as i
035r048 ection of fluid rock.-- Try on globe. with slip PAPER a gradually curved enlargement see its increase
036r049 ns.-- Introduce part of the above in Patagonian PAPER; & part in grand discussion Consult. reconsult
041r065 ntioned the drifting of carcases putrid. In Rio PAPER. when discussing probable rise of land: Mention
053r101 kes in the Radack & Ralix Islds? In my Cleavage PAPER Dr Fittons Australia case must be quoted at len
089a020 <Keys off extreme point of Flori[da]> Excellent PAPER on Erratic blocks in Alps. Memoires de la Soc.
091a026 f Minute seeds-- L. Institut. p. 209. May. 1837 PAPER by Humboldt on Quito Volcanoes & another on Mex
092a029 species same as Paris. 1500 ft high Mr Bird in PAPER to Brit. Assoc: has shewn how electrical curren
096a041 out parting with Carb. Acid.-- Mr Malcolmson on PAPER on India gives reason for knowing that Mur. Sod
105a070 to turbulence. Bull. Soc. Geolog «1837» p. 320. PAPER on shrinking of Clay. applicable to Cleavage. C
105a072 uses to explain «alluvi» in valleys Lowe in his PAPER says land shells found with calcareous matter &
109a084 ow coast.-- Annales des Mines. a translation of PAPER by rose on Greenstone, dicrite, &c most importa
118a103 al deposit always equal width --subject of fine PAPER this would make.-- L'Institut. (1838) p. 216 M.
123a115 -- Lyells Denmark -- L'Institut (1838) p. 268. PAPER by Humboldt on Bogota. Cordillera,-- nothing.--
123a115 g.-- salt & coal near Bogota; p 270.-- SPLENDID PAPER on fossil shells of S. America. Von Buch Lyell.
130a133 ers have proofs of hot bottom.-- Study Bishoofs PAPER.-- Weelsted told me of some large fresh Water s
130a133 h Water springs off coast of Persia In Glen Roy PAPER I show crust yield easily. & if easily must be
130a134 e probably would be marked line of separation A PAPER by Parrott Mem. Acad. Peters. Scienc Math. Phy-
133a140 esert, would be very high.-- M. Parrot ends his PAPER like a fool.-- Feb 25' All facts show how slowl
134a144 Soc Vol V. p. p 96. apparently good geological PAPER. by Malcolmson-- worth reading-- Burnetts. vol
138a155 moon at some future time [blank] Sir. J. Halls PAPER on the consolidation of strata-- he heated sand
193b091 Voyage Copied into list Entomological Magazine PAPER on Geographical range Richardson-- Fauna Boreal
199b119 ange In Number 6'.? of E.d. N. Philos. Journ.-- PAPER by Crawford on Mission to Ava. account of HAIRY
203b138 variety partakes chiefly of the former Eyton's PAPER on Hybrids Loudon's Magazine. Gould on Motacill
204b139 Mag: Septemb or Octob 1837 Westwood has written PAPER on affinity and analogy in Linnaean Transaction
211b163 t they eat up the dogs.-- L.' Institut. Curious PAPER by M. Serres on Molluscous animals representing
216b181 closely allied to case of cross of dogs.-- See PAPER in Philosoph. Transaction on a quagga & mare cr
220b198 that few months since in Annales des Sciences, PAPER on Botany of Tahiti In Charlesworth Magazine Ja
220b198 n Charlesworth Magazine Jan: 1830. most curious PAPER on heredetary fear (like rooks with guns) of th
220b199 y species from warmer countries. When will this PAPER be published it will be curious.-- Some general
229b234 o.-- Range of East Indian Rhinoceros (?)-- Some PAPER in Institute on range of Bos in India.-- Range
250c040 losop. Transacts 1823. Read June 5th) important PAPER by Dillwyn, on replacement of Cephalopods & Tra
250c040 5. ref. to Phil Transacts. (read November 20th) PAPER by Jenner, on birds seen far at sea, migrations
254c049 st lead to spontaneous generation??) This whole PAPER Must be studied.-- <Three p. 7. Am.> D'orbigny.
268c095 not being adapted to Air!! p. 11--&c. valuable PAPER on quadrupeds of Van Diemen's land, which appea
270c104 s only live on matter already organized.-- This PAPER might be worth consulting, if any Metaphysical
282c143 ever be discovered When one reads in Ehrenbergs PAPER on Infusoria on the the enormous production-- m
286c155 here is same difficulty in arranging animals in PAPER as drying plant, all brought in one plane Flemi
295c183 ns L' Institut. Sorex from Mauritius. p. 112. & PAPER on genus Magazine of Zool. & Bot.-- Vol I p. 4
307c216 re any abortive organs in neuter bee. (There is PAPER by Yarrell «in Zoolog Transactions» & Hunter on
315c241 be published St. Petersburgh Academy Imperial-- PAPER read in 1837. semestre I suspect some valuable
318c254 bout a fortnight, «See Silliman's Journal 1837. PAPER by Bachman.» that is succession of births.-- «in
324c268 ardson's Faun. Borealis Entomological Magazine (PAPER on Geograp. range Study Buffon on Varieties of
324c268 Geolog Soc,» F.. Cuvier on instincts L. Jenyns PAPER in Annals of Nat. History Prichard.-- Lawrence
325c267 ers collected by Owen. (at Shrewsbury) Yarrells PAPER on change of plumage in Hen Pheasants <Zoologic
325c267 lias Mrs. Gore on Roses might be worth consult. PAPER on Consciousness in Brutes in Blackwood. June 1
328cIBC es Gallinaceas. Silliman's Journal. during 1837. PAPER by Bachman on migration of birds Temminck has w
349d052 ers can be established-- clearly so.-- NB. This PAPER worth referring to again.-- According to my the
352d059 d the tree of life must be erect not pressed on PAPER, to study the corresponding points.-- The prese
354d062 stitut p. 275. (1838) M. Blainville has written PAPER to show Stonesfield Didelphis not Didelphis «An
355d068 ovules.-- Captain Grants. Himalaya. shells (see PAPER in Geolog Transacts) same appearance with Secon
370d116 phical organization. no more can be said.... In PAPER on bees in same work. (it is said that some kin
379d151 be ocean, what is land to continent-- Original PAPER, worth studying. Archiv. fur. Naturgeschichte.
385d164 e just worth remembering that ovarium of women (PAPER in Vol I of Irish Royal Academy) have contained
403e022 account of Flora of pacific, given in my coral PAPER Oct 14th Macleay says, that <every> «any» chara
406e035 al with S. America. & many very close: see full PAPER. L'Institut 1838. p. 338 A most grave source of
421e091 305.-- Mr Owen says <tha> «in abstract» in his PAPER on the Dugong, "The generative organs being tho
424e100 &c &c.-- L'Institut 1838.-- p. 290-- admirable PAPER on geographical distribution of Crustaceae.-- (
451e176 > «Journal of Asiatic Soc.. Vol V. p. 565. in a PAPER by Lieut. Newbold.--» A Malayan albino describe
459tf02 copie[d] Gleanings of Sciences. Vol. III p. 83. PAPER translated from Meckel. Comp. Anat.-- From Buff
461t025 51. 1839-- Translation of P. Fries most curious PAPER on the Pipe-fish-- which he divides into two di
461t025 <do> have not sack,-- Most curious facts & this PAPER deserves fresh study & whole order of the fish.
465t081 s of men have been exterminated (see Pritchards PAPER) (Ed. Phil. Journ. end of 1839) very important.
465t089 ws much forms depend on other forms Lyell's PAPER, in Taylor's Journ.-- Phil. Mag. May. 1840 p. 3
469t119 marriage!!xx In Phil Transact. about year 1778. PAPER by Camper on Ourang-outang. has examined 7 says
478z004 mes to houses on purpose Mr J. Murray has given PAPER to Royal Soc on glow worm. luminous property--
483z013 ng on Planariae or Polypi & is especially grand PAPER. p. 387. "on Classification of such animals.."-
495q05a pollen of neighbouring grass- Spread sheets of PAPER. covered with some sticky stuff in flat places
495q05a of Equivocal Generation Charlworth p. 377. Have PAPER ruled in squares to facilitate investigation.--
506g015 ated, as mass of pollen is requisite.-- Brown's PAPER 43. Any flowers of Keeling Dioecious, or Monooe
508q016 (10) About Daltonism in the MALE Troughtons.-- PAPER in Taylors Scientific Memoirs (11) And. Smith S
514q21. flower open. (An. des Sci Where is Boerhaave's PAPER on impregnation of violets.= Zostera= Are dwarf
523m015 yet disinheriting her.-- This «N B. I have read PAPER somewhere on horse being insane at the sight of
528m037 ntine lines narrow in the distance.-- & even on PAPER two waving perfectly parallel lines are elegant
541m090 coves when waters had fallen, as in my Glen Roy PAPER.-- this greatest mental effort, of which I am c
546m110 elf. she was one day reading a book, with ivory PAPER cutter, which she valued, & she was suddenly ca
546m110 to see something, on her return could not find PAPER cutter, hunted in vain for it-- ten years after
559m155 ven to the perception of a final cause.-- Read. PAPER on consciousness in Brutes & Animals. in Blackw
592n102 of the six (?) first Vol of Silliman's Journal PAPER showing that the signs invented for Deaf & dumb
592n103 ame.-- Philosoph. Transactions Vol 44. 1746-47. PAPER. like. Sir Ch. Bell on Expression «First Crooni
592n103 to excessive laughter & so into convulsions.-- «PAPER» must be referred to, if I follow up this subje
120a110 scraps on Salsisbury Craigs. Kept amongst <old> PAPERS read before societies.-- Sir. J Hall Vol VI. p
126a124 this view.-- however it is said in some of the PAPERS that there are springs even in siberia.-- from
129a132 ia river-- Read Mr Parker's Book.-- M. Bichoffs PAPERS, in Edinburgh New Phil. Journ 1838. several ca
218b191 changes when animals transported.) Mr Herbert's PAPERS are in the Horticultural Transactions and a di
222b204 another like mother Has Lowe written any other PAPERS besides one in Latin one <of> «on» Madeira-- a
300c198 f habits in Van Diemen's land. Study Mr Blyth's PAPERS on Instinct.-- His distinction between reason
315c240 See. Habits of Malay fowls p 5. (note) in some PAPERS on instincts L.' Institut «1838» p. 184 Botany
322c270 s Animal OEconomy. edited by Owen. read several PAPERS--, all, that bear on any of my subjects Elie De
325c267 written quarto. work on Physiology besides the PAPERS collected by Owen. (at Shrewsbury) Yarrells Pa
326c265 ogs.-- Reports of Brit. Assoc.-- some important PAPERS. Dr. Mayo. Pathology of Human. Mind.-- Audubon
326c265 erbert. p. 348. gives reference to Kolkreuter's PAPERS Wiegman has published German Pamphlet on cross
379d152 n the Athenaeum Numbers 406, 407, 409, Quetelet PAPERS are given, & I think facts there mentioned abo
477z001 of Zoology and Botany. Philosoph. Transacts. 3. PAPERS connected with transform of Crust-- Westwood &
483z013 Geof.-- reference from Rüppel travels All Owens PAPERS on Intestinal worms must be studied in Vol I.
527m035 train of thought this does not happen. because PAPERS, &c &c round one. one recalls the castle by go
497q007 is impregnated.; which part of stigma?-- (4) As PAPIL. flowers appear difficult to cross, are there u
470t178 is ejected with violence in shower On many PAPILIONACEOUS; all wh. are in flower «I saw Bees;»-- on
506g015 ave any monooecious or dioecious plants the PAPILIONACEOUS structure of flower-- Ground nuts (42) How
507q15v of spontaneous Hybrids-- to see whether any PAPILIONACEOUS plants,-- whether many mono or Dioeious pl
514q021 tions: about crossing of plants; especially PAPILIONACEOUS order (2) History of fruit trees far north
635j56v l I. p. 330. Mentions the many cases, as in PAPILIONACEOUS flower, where such care seems to be taken

Page
(Key Word)
632j53r till higher laws.-- I do not «then» believe the PAPPUS of <th> any one seed. (all have not it) was DI
245c023 arder comme une espèce distincte! p. 171. Sus PAPUENSIS «partly domesticated» like in general appeara
272c109 e tail in cock peacock,. widowbird.-- Birds of PARADISE. Trogons.-- the one feather in wing the curio
283c144 d to think, that widow bird. replaced Birds of PARADISE-- if such fantastic «sexual» ornaments. have
440e147 so no <cause> corresponding change in Birds of PARADISE.-- All that we can say in such cases, is that
452e181 no wild animals in Gilolo.-- p. 134: Birds of PARADISE were first procured from Gilolo p. 253 In isl
353d061 of India-- Waterhouse knows three species of PARADOXURUS common to Van Diemen's land & Australia well
354d064 says it look lik[e] Institut. 1837. p. 351. PARADOXURUS Phillippensis. Philippines Man have varies t
191b080 s = Opossum & Agouti same as on continent-- 3 PARADUPASI in common to Van Diemen's Land & Australia F
402e020 eys.-- Fewer <of th> «Mammalia» have passed to PARAGUA in Luçon the most northern of the group the
320c276 kners Patagonia Azaras Voyages & Quadrupeds of PARAGUAY Dobrizhoffer. Abipom<e>nes. Edinburgh New. Ph
325c267 ield Silliman's Journal Rengger on Mammalia of PARAGUAY. account of wild cattle & Montagu on birds (f
477z002 s of Chile different from those of La Plata or PARAGUAY.-- do. p. 365. 3 cats (mbara caya. le negro,
477z002 se was not known amongst Indian. introduced in PARAGUAY in 1769 introduce in Governor's tran?? Azara
478z003 volent. p. 208 Fleas only appear in winter in PARAGUAY p 207 Slight notice on habit of Iguana. not p
482z011 ry close to ours Rengger's work of Mammali: of PARAGUAY must be most important a discussion of geogra
482z012 ans l'Amerique Merid. Tatu noir. abundant from PARAGUAY to 27 degree, then the Mulita from 41 degree
024r016 t rivers in my Coral paper {T} leagues Fathoms PARALLEL of St Catherine [27 degree 30 minute S.] 18--
025r016 -10 = 30-40 {T} at twice or 18-20 <60>--80 120 PARALLEL of Olinda Shoaler N. of Olinda.--a little WNW
035r047 self. P. xx: same fact is indeed shewn? by the PARALLEL bands of formations on any Geolog Map: Quoted
049r087 a subsidence.--The sudden increased dip is not PARALLEL case to Isle of White. but rather to one out
053r101 (mem: profound earthquakes), which would cause PARALLEL lines, but the rectangular intersections are
062r129 ing in the history of rats, in the antipodes a PARALLEL case.-- Should urge that extinct Llama owed i
078r175 f the mountains (flaquexa del cerro) have been PARALLEL to the direction & inclination of the vein".-
099a048 f surface covered with oil should shrink. film PARALLEL to longer axis. But if great depth NB. Prof <
099a049 depth NB. Prof <Henslow> Sedgwicks lamination PARALLEL to stratification evidently small scale of co
099a051 .-- The separation in the Ponza case of Scrope PARALLEL to walls of dykes-- Mem. laminated dikes in C
104a068 orway was contorted yet no mountain chain case PARALLEL to Banda Orientel. ask Lyell for sentence.--
124a119 whether it may be absorbed.-- if so exactly PARALLEL to limestone & volcanic rock containing magne
135a145 gal. J. vol 7. p. 522. Mountain c near Caubul. PARALLEL ranges. with here & there little branches at
145g004 ins, amygdaloidal-- as well as base not always PARALLEL to strata 3 or 4 seams/ 3 or 4 inches thick--
155g063 r 60 ft» «no doubt, a mound of Alluvium nearly PARALLEL--» Inclination of river must constantly alter
162g099 t it may be said that a mound stretches along, PARALLEL to Shelf on opposite side & dies away on the
190b076 Holland. diminishing towards centre (p. 586)-- PARALLEL 33 degree-35 degree, source of forms. reduce
219b196 hardly produce from Incestuous intercourse. a PARALLEL fact to Blood-Hounds. Before Attract of Gravi
288c158 le groundwork of analogy on its concurrence in PARALLEL parts of his series, ie, cannot be discovered
303c203 cies. which survived would be few in number.-- PARALLEL of Japan near Himalaya, & Europaean forms on
334d011 n from her, & always accustomed to her.-- case PARALLEL to brothers & sisters in Mankind.-- The case
407e039 change, (& become transmuted), although other PARALLEL species in other continents might have surviv
408e045 Himmalaya-- Humboldt bones at 7800 in Andes-- PARALLEL & curious facts.-- The Himmalaya. case, bears
463t057 tructure,, & supposing much extinction. give a PARALLEL case) Waterhouse remarked, that any argument
528m037 stance.-- & even on paper two waving perfectly PARALLEL lines are elegant.-- Again there is beauty in
547m112 nce apparent vividness) & there being no other PARALLEL trains of ideas connected with past circumsta
547m113 such one naturally would so so!) Now all these PARALLEL trains of thought necessary heirs of every ac
618o41v n be said that thought & organization run in a PARALLEL series: if blueness & weight always went toge
059r119 Toutes ces montagnes sont formées de couches PARALLELES et inclinées du centre d'lile, vers la mer;
118a103 s & Meyens --<Jura &> Chalk When we consider PARALLELISM of dikes (Hopkins) & that every dike. which
521m000 Case of Mr Corbet of the <Hall> «Park», after PARALYTIC stroke. intellect impaired. <after paralytic
521m009 paralytic stroke. intellect impaired. <after PARALYTIC stroke> : . could converse well on any subjec
578m056 le like does an injury of the spine-- that it PARALYZES all muscular action -- «in man & animals» Blu
240c014 roe & Solor), «Vol I» likewise new species of PARAMELES, which joined to Casoars, perroquets, establi
622o048 med by the teachers & all around him, will be PARAMOUNT,-- hence the law of honour. & the etiquettes
024r016 St Catherine [27 degree 30 minute S.] 18--70 PARANAGUA [25 degree 42 minute S.] 12--40 St Sebastian
205b142 pes discoverable.-- Bridgewater Treatise p 85. PARASITE of Negroes different from European.-- Horse &
205b142 ent from European.-- Horse & Ox have different PARASITE in different climates.-- Hunbt. Humboldt? Vol
233b252 of great Fresh water lakes of North America If PARASITE different, whilst man & his domesticated quad
312c233 in. Must not trust him Hope says that genus of PARASITE to genus of animals different «+++ p 234», di
313c233 rent.-- thorax & head differ Africa Australia PARASITES die, when brought over on tropical animals wh
313c233 anging « ∴ because mammalia can subsist where PARASITES cannot» Read Entomological Transactions Why i
332d004 ed intestines are not healthy to worms, (like PARASITES of Tropical countries cannot endure this clim
061r128 Palms, evergreen trees, arborescent grasses, PARASITIC plants, Cacti: & with limits of no vegetation
232b248 ure me some Get Hope to give me an account of PARASITIC animals of beasts varying in different climat
196b108 f small animals must have gone on since from PARASITICAL nature of insects & worms.-- In abstract we
483z013 pt. King's p 453-- Planariae velellae (Less) PARASITICAL on Vellellae in Atlantic Ocean Gould agrees
417e078 l differences) a variety. The offspring of true <PARE> hermaphrodite, would of course be like either,
096a039 . <Ceylon>. Band of Volcanic action in Iceland PARELLEL to Greenland: Mem.¿ Greenland subsiding.) Von
070r153 le change.-- Yet one is urged to look to common PARENT? why should two of the most closely allied spe
105a071 of valleys &c; Pampas.-- If blocks above their PARENT rocks. would be prove of subsidence.-- removal
173b012 rm peculiar to continents; all bred in from one PARENT why. Myothera several species in S. America wh
184b054 ving.-- Now according to my view. in S. America PARENT of all armadillos might be brother to Megathe
186b059 n> races have tendency to keep to <each> either PARENT, (this is what French call (atavism) Probably
204b141 n & offspring either amongst themselves or with PARENT birds.-- «W. Fox. knew of case of male widgeon
208b154 ichardson in Report about each genus having its PARENT type in hotter parts of world Is monkey. pecul
216b179 ive stock (the cross often whiter<)> than white PARENT) the mulattos themselves explain it by interma
217b184 termediate sometimes take strongly after either PARENT. about <half & half tim> as often one way as o
220b197 ffspring does not depend on mind or instinct of PARENT. Mem Lord Moreton's Mare. The fact of plants g
222b203 their offspring show tendency to return to one PARENT, this is only character. & yet we find this s
222b203 different characters is attempt at returning to PARENT stock.-- I think we may look at it so--?? It h
225b219 ocks for same purpose.!! Can the wishing of the PARENT produce any character on offspring? Does the m
239cIFC ross & keep colour on wing Effects of colour on PARENT, white room How are varieties produced, by pic
239c002 f half cross with parents, going back to either PARENT is lucidly explained.-- Mr Yarrell states that
248c033 transmitted;.-- hence the tendency to revert to PARENT forms, & greater fertility of hybrid & parent
248c033 o parent forms, & greater fertility of hybrid & PARENT stock, than between two hybrids.-- As we see e
258c062 ecomes developed & then another-- (according as PARENT types are present) must follow after there is
259c063 every developement giving greater vigour to the PARENT so tending to produce effect on offspring-- bu
259c065 er case is <adaptation> «change» during life of PARENT, & therefore being always necessary may be cal
259c066 ect to my theory of generation, fact of armless PARENT not having armless child, shows than there is
261c070 of organization! The little turtle, without its PARENT running to the water, is a good instance of co
264c079 despair; «let him look at savage, roasting his PARENT, naked, artless, not improving yet improvable»
268c096 eding young of redstart & actually driving away PARENT birds.-- showing how blind a storge From what
275c121 l half Bred cattle of L'Darnleys were most like PARENT Brahmin bulls-- Mr. Y. is inclined to think th
280c133 grant, that difficult & other go back to either PARENT.-- Shows instinct (Sir J. Sebright admirable e
280c136 back to grandfather, but if too unlike its own PARENT this impossible-- (Hence we might expect even
299c195 ion & varieties chiefly produced by cultivating PARENT in rich soils & their seeds produce <offspring
304c207 fficulty is, whether Apterix descends from same PARENT with other birds., or branched off anteriorly
304c207 racters as red band on wing show to be from one PARENT.-- same forms of beak &c without these trifles
309c219 «we know of:», is relationship, children of one PARENT, races of animals-- argue «opening» case. «thu
313c236 ») Generation becomes necessary, when organs of PARENT are concentrated in different parts, & scissio
331dIFC nterbred. show any tendency to return to either PARENT.? Is the first cross, which makes hybrids. pro
333d008 n of parents <some> one sometimes resembles one PARENT & one another & are not exactly intermediate.-
335d015 would «as in all other animals» be like either PARENT, or intermediate within certain small limits (
335d015 within which limits they might return to either PARENT), then according law, that in proportion as th
335d015 s» will have no tendency to have offspring like PARENT, but as they must like or there will be none,
336d017 e to transmit them, & as offspring must be like PARENT, therefore mule has no offspring & therefore n
336d017 me Prop. better enunciated.-- "An animal Either PARENT cannot transmit to its offspring any «peculiar
336d017 his is that animal would endeavour to return to PARENT stock. but if both parents are alike, offsprin
336d019 he idea of generation being a <slip> «bud» from PARENT. if whole parent not entirely embued with the
336d019 tion being a <slip> «bud» from parent. if whole PARENT not entirely embued with the change, a bud cou

342d033 ay many eggs but never produce inter se or with PARENT species.-- The hybrids do not vary (ie the hen
342d033 the hens all alike & Cocks all alike) More than PARENT species-- Mr Blyth remarked only near species
350d055 arked that young of Cirrhipedes can move & see, PARENT fixed,-- young of sponges move.-- young of Coc
350d055 -- young of Cochineal insects move about & see, PARENT «(2)», female «(I)» fixed & blind: -- Macleay
352d057 e may be more perfect (as we use the word) than PARENT, so may species retrograde, but these facts ar
353d060 e be many others somewhat allied whether «like» PARENT stock, or not. Now wings for flight-- therefor
371d118 rvous system should build up its body, like its PARENT, than that it should be provided with many con
371d127 nces are there of such changes, not acquired by PARENT, being handed down? Are not Loddiges 1279 rose
371d128 year, successive roses & bud are produced, like PARENT stock, or if different dieteriorating very slo
378d148 onnected with habits According as child is like PARENT, so is species old: Hence «young» Kingfisher &
383d160 s, we have character very different from either PARENT bird-- Hunters Animal OEconomy. (by Owen) p. 4
385d164 half pheasant, half fowls.-- eggs fertile, but PARENT bird will never sit on them.-- May be just wor
388d170 ugh transformation, nearly independently of its PARENT & therefore wants independent supply of food.-
390d178 s tendency to propagate the whole difference of PARENT, tree, but it fails.-- therefore «each» seedli
390d179 liance to breed two which have each varied from PARENT stock.-- The very theory of generation being t
391d174 buds??? Amongst buds each one exactly like its   PARENT. <but these buds do not procreate> & all alike
391d174 these buds do not procreate> & all alike in one PARENT or tree, (but not in other trees.-- -- Why sho
404e026 as Hunter, or some one mentions of influence on PARENT affecting offspring.--। & as adaptation,-- how
404e026 . therefore chance & unfavourable conditions to PARENT may be become favourable to offspring:। Austra
406e035 ource of doubt. in distinguishing «after» which PARENT impresses offspring most is whether mother has
406e036 ies & to different species-- sometimes like one PARENT & sometimes other & sometimes ½ way. Ed. New-P
415e066 nly way, that I can see to discover whether the PARENT of man was quadruped or bimanous,, is to see,
417e077 g of trees.--]CD CD[The similarity of child to PARENT appears to follow same law in two of «the» sam
420e089 The head being six metamorphosed vertebrae, the PARENT of all vertebrate animals.-- must have been li
431e123 als, p. 8. size of foetus in proportion to male PARENT p. 8. his whole doctrine of the advantage of c
434e127 heredetary ambling horses. Whether the body of PARENT be altered, that is the Nisus formativus. (wha
437e140 , where the old Eagle could not find it..-- The PARENT bird another day brought to her young ones the
439e143 show my view of <i>nfertility of hybrids «with PARENT species» false, which makes it determined by a
445e159 ely consider existing «long-organized» forms as PARENT forms of existing highly organized forms-- thi
493q003 {In crosses does male offspring take after male PARENT & vice versâ = History of Tortoise-shell Cats.
506q015 not H. raised races of white & Blue Linum-- did PARENT plants grow near each other.-- ? Cannot rememb
515q022 ially if the hybrids were recrossed with either PARENT.-- May. 44 These Hybrids differ in colour of b
515q022 r in colour of beak, taking after male & female PARENT.-- Will they grow up in other respects differe
533m057 edetary are so because brain of child resemble, PARENT stock.-- (& phrenologists state that brain alt
546m109 e--, man's taste for smell of flowers, owing to PARENT being fruit eater.-- origin of colours?-- Noth
593n107 e attempted to be said that young animal learns PARENT smell & look so by association receives pleasu
426e108 other or indeed grandmother; what is Mr S. S.   PARENTAGE?-- Wonderful as is the possession of voice by
462tf05 that a mongrel man may lose all traces of his   PARENTAGE in «about» seven «7» generations.-- so many!!
387d168 2 It is very singular the same difference from PARENTAL stock having been repeated several times, tha
558m150 bscurely guided» or strong instinctive sexual, PARENTAL & social instincts, giving rise "do unto othe
591n100 alogy of actions with <&> against benevolent & PARENTAL instincts very clear.-- even to the cold or b
614o037 instincts, as love of virtue, of association, PARENTAL affection-- The very existence of mankind req
615o038 ry-- so succession so perhaps general ones.-- PARENTAL feelings weakened in Otahiati; fear of death
619o042 ferous animal, it may be concluded that he has PARENTAL, conjugal and social instincts, and perhaps o
148g025 bs were nearly like each other «& half between PARENTS» (& not like dogs), but they thought the breed
172b005 s «separating» partaking of characters of both PARENTS, & these infinite in Number In Man it has been
186b060 rganization.-- extinct species of that country PARENTS of American.-- Now Genera of those two countri
199b120 rather stronger tendency to imitate one of the PARENTS; repugnance «generally» to marriage <if offspr
203b136 r F. Darwin cross breed boars were wilder than PARENTS. which is same as Indian Cattle.∴ tameness not
204b140 rid dogs offspring seldom intermediate between PARENTS.-- How easily does Wolf & Dog cross? Mr Yarrel
205b145 of Animals by Mr Cline "The character of both PARENTS are observed in their offspring, but that of t
217b184 rough water spaniel «produce litter like both PARENTS»-- & Mr Bell has half bloodhound & greyhound.--
217b184 e after the other, puppies differ, & like both PARENTS.-- Fox told me of case of mare covered by bloo
218b189 rom being put on island. & fresh species made. PARENTS do not cross-- we see it even in men); the pos
227b227 of change.-- the manner of adaptation (wish of PARENTS??) instinct & structure becomes full of specul
239c002 nciple I may add, that fact of half cross with PARENTS, going back to either parent is lucidly explai
254c049 in Acrita,-- typical of other, (surely rather PARENTS. (NB These views must lead to spontaneous gen
259c065 example probably not true).-- or again healthy PARENTS have healthy children the other case is <adapt
260c068 e> birds hears <dis> crys of distress of other PARENTS.-- Shows community of language Desert country.
285c150 d not be difficult to arrange children of same PARENTS in a circle,-- & «hermaphrodites» father < «mo
285c151 be necessary to show hybridity from few forms, PARENTS of all species not possible in some detail,--
312c228 feels sure of this, first offspring most like <PARENTS» Mother.-- like dogs Smith knew chinese hairle
333d007 og.--> appeared to be intermediate between two PARENTS.-- this is very interesting as Esquimaux dog a
333d008 ittens alike each other.-- Even in children of PARENTS <some> one sometimes resembles one parent & on
335d013 remain,-- yet offspring must be somewhat like PARENTS,-- therefore offspring will tend to go back, o
335d013 ve none-- the argument does not apply to first PARENTS, because they are not new breed.-- the first h
336d017 hange from the form which it inherits from its PARENTS «stock» without it be small & slowly obtained
336d017 devour to return to parent stock. but if both PARENTS are alike, offspring must be like Hence mutila
336d018 ikeness, & the union makes hybrid, in fact the PARENTS beget child like themselves. expression of cou
356d071 cts about. half breed animals being wilder than PARENTS is very curious as pointing out difference bet
359d087 little> in disposition after their «pheasant» PARENTS.-- (There are some 3/4 birds «of», which I thi
359d089 coloured & different.-- -- the former were the PARENTS of the three little ones.-- Keeper said in <tw
360d093 ight blue, it might be harder <to tr> for both PARENTS to transmit there peculiarities; that if <one
371d127 ll these cases an unseen change is produced in PARENTS--colour is a doubtful subject, but what other
372d130 ving undergone, the endless changes, which its PARENTS have.-- -- Not this is effected by short metho
385d164 Holcroft Vol I. p. 195. says children resemble PARENTS in their bodies "It is a fact equally well kno
391d175 these changes may be effect of differences of PARENTS, or external circumstances during life.-- if t
406e036 ed.-- Regarding the similarity of offspring to PARENTS same laws appear to hold good. with regard to
413e058 3) Great fertility in proportion to support of PARENTS December 2d Lyell tells me Beck considers the
417e078 , would of course be like either, that is both PARENTS, for they are one.-- The laws, therefore, of l
417e079 speaking of resemblances of children to their PARENTS.-- Lord Moreton's law cannot hold with fishes,
430e117 ls in Eocene period could not have been direct PARENTS of any of ours,-- even if extinction is denied
439e142 y possessed-- <that is animals> «or rather the PARENTS» having passed through many changes.-- It is v
441e148 gating» constitution. but not structure of the PARENTS.-- Thus would a Crab tree vary if planted in r
451e176 credulous natives, & now made at it. Both his PARENTS were of the usual colour. His sister is an alb
515q022 f eggs-- of half-bred geese-- inter se, & with PARENTS & of Chinese geese. (2) Anatomy of muscles of
520m001 innumerable cases of people taking after their PARENTS, when the latter died so long before, that it
583n069 son: [N B. insects which have never seen their PARENTS offer best cases of instincts].CD all this may
615o038 easures with certain actions performed by your PARENTS, conscience This «X» memory especially «the» g
621o048 either right or wrong.-- [RHC] 7) Hence, what PARENTS think will be good for the child on the long r
621o048 e long run, & for themselves & others, (as the PARENTS are instinctively benevolent) they will teach
624o051 ive feelings of right & wrong,-- education, of PARENTS strives* to same end.-- & general actions of c
624o051 to all the children, « <then> each himself> & PARENTS, & hence to nearly all the world.-- As conditi
344d042 und. «++».∴the grandchildren went back to either PARET & breed not fixed. though she resembled a harr
601o08b sophique sur l'intelligence des Animaux-- & Le PARFAIT Chasseur, par Desgraviers,,, un Vol 8vo Keratry-
348d050 se ratio to their variability.-- (Now caeteris PARIBUS these will be the oldest) ¶"The most important
092a028 Tertiary <bea> formation twenty species same as PARIS. 1500 ft high Mr Bird in paper to Brit. Assoc:
202b133 sil Didelphis (S. American genus) in plaster of PARIS.-- Now this is exception to law of type. like h
208b153 er in extinction of individuals than of species PARIS Tertiary Shells in India!? A p. 28 Dr Beck. & L
226b222 lands.-- Hence this must have been condition of PARIS basin land.-- (How is this with Fernando Po.).
226b223 duration of species. (¿are carnivorous Mamm: in PARIS basin «allied to present») more like present ca
250c039 s What reptiles coexisted with Palaeotherium in PARIS quarries & at Binstead. Mem. recent Crocodiles
295c178 Memoire by Charles D'orbigny on Plastic Clay of PARIS contains many genera of Pachydermata &c & other
299c193 must be arrived at, when one sees a plant like PARIS quadrifolium growing in one wood far from any o
325c266 iews Mackintosh Ethical Philos: Prostitution of PARIS. with respect to licentiousness, destroying chi
407e040 rent from the existing orders, as the fossor of PARIS! (Great Edentata at that period) Analyse this,-
416e072 titut 1838. p. 394. Rhinoceros «tichorhinus» in PARIS basin.-- its relation to African Species <good
425e105 Tapirus American form. found in Eocene beds of PARIS Lyell has remarked species never reappear when

Page **(Key Word)**
448e166 .-- Again the resemblance between the Superga & PARIS, numerically the same with recent & yet almost
449e173 ound at Timor.-- thinks he has seen specimen at PARIS Museum.-- Athenaeum: 1839. p. 451. Sheep Merino
499q010 y distinguishable.= What structure of seeds.-- (PARIS) (22) When Linnaeus says so great percentage of
569n022 re is a Lutké's Voyage autour du Monde (1826-9) PARIS. 1835 Quoted repeatedly by Waitz (In Theil. V)
634j55r irds? They are ancient.-- Are Cetaceae found in PARIS Basin?.-- NB) The explanation of types of struc
060r126 ugh kept in numbers. p. 124. Webster Consult W. PARISH. & Azara about dry season[.] 1791. seen common
066r142 ortant in forming transverse valleys Ice Sir W. PARISH says they have Earthquakes in Cordoba. one of
067r143 ion de obras. 2 Vols fol: Buenos Ayres 1836: W. PARISH?? «by Pedro de Angelis.» This work is reviewed
067r143 reviewed in present Edinburgh March 1835 Sir W. PARISH says. that beds of shells are found on whole c
071r157 era by Humboldt in Geolog. Society Sir Woodbine PARISH informs me that town near Tucuman and Salta. t
095a038 Valparaiso Dr. Gillies in MS. letter in Sir. W. PARISH Possession. talks of <hill> «cerro» of Diamant
323c269 . Scott.--, except the V Volume.-- -- 19t. Mungo PARK-- travels Feb 12. Sir. H. Davy Consolations in
334d012 perfect structure.-- Fox says in «Lord» Exeter's PARK «or in the Duke of Marlborough» there is a bree
344d041 ve an admirable harrier from Ireland to Brighton PARK--first rate bitch-- tried to breed from her, bu
405e031 ly did not.-- Octob. 25th. I observed in Windsor PARK.-- the «Fallow» Deer. which were of a nearly un
500q010 he breed-- desirable as in Cattle in Chillingham PARK-- What Book on varieties &c of deer. Contests o
521m009 nt. B. ditto.-- Case of Mr Corbet of the <Hall> «PARK», after paralytic stroke. intellect impaired. <
129a132 depth» near mouth of Columbia river-- Read Mr PARKER'S Book.-- M. Bichoffs Papers, in Edinburgh New
529m041 excited, & how the scenery would rise. Deer in PARKS ditto.-- My Father says there is case on record
504q013 ring. & within garden //Yes// (5) Examine the PARNASSIA whose stamens move one after other to flower
241c014 us. does not say whether wild or not p. 156.-- PARROKET with stiff tail like woodpecker.-- Birds of A
132a137 t of the passage of the moon.-- Ask Hopkins. M. PARROT, Mem. Acad. Imp. des Sciences. (Sc Math. Phys.
132a138 shows earth not with central heat.--» «(does M. PARROT suppose there is no volcanicity beneath lakes)
132a138 ainst central heat to warm the ocean).-- and M. PARROT does conjecture that in Scoresby's case volcan
133a140 hara de> a dry desert, would be very high.-- M. PARROT ends his paper like a fool.-- Feb 25' All fact
185b055 hould we have in open country a ground «do. <w> PARROT.--» woodpecker-- a desert. Kingfisher.-- mount
274c113 rots with claw like lark (NB The La jeune veuve PARROT though so much on the ground has not this stru
132a138 than on continents. it ought, (according to M. PARROTS argument against central heat to warm the ocea
132a139 ssion from ocean's bottom.-- (according to M.. PARROTS own hypothesis some such explanation appears t
132a139 uch explanation appears to me necessary) as M. PARROTS shows from variation in strata earth a very ba
188b067 s peculiar to Java, & another to Sumatra --Mem PARROTS peculiar according to Swainson to certain isle
251c040 st Vol II. p. 49. on the localities of certain PARROTS habitations India & Africa.-- NB. Any monograp
258c061 onsidered in this light.--XX Zoolog. Journal-- PARROTS in Macquarrie isld. vol III p 430 alluded to b
268c099 Patagonia desert & Tierra del Fuego. & forest «PARROTS in Macquarrie Isd.--» very good. Study D'Orbig
274c113 tralia is called swamp pheasant) Goatsucker--, PARROTS with claw like lark (NB The La jeune veuve par
274c114 erally small genus. ¿are there not many ground PARROTS? are there not many ground woodpeckers?-- In e
303c205 gascar uniting pidgeons & gallinaceous birds & PARROTS.-- legs of pidgeons perfect.-- &c &c.-- do p.
450e173 deer, no loonies, but cocatores & small green PARROTS. June 26th-- Yarrell.:-- Black Swan «in domest
490q001 common recent ones not embedded?-- Do the Tame PARROTS breed amongst the Indians Do the Savages selec
557m147 is expression of anger in species of swans, in PARROTS &c &c -- -- peacock & turkey cock in passion.-
130a134 would be marked line of separation A Paper by PARROTT Mem. Acad. Peters. Scienc Math. Phy-- Nat. t.
502q11v ucture Compare flowers of wild & tame carrot-- PARSLEY & Fennel. Verbena Compare flower of different
592n103 Bell on Expression «First Croonian Lectures by PARSONS.» following pages contain remarks worthy of at
047r081 wave on record. -- «also neighbouring sea must PARTAKE in absolute movement» Moreover wave «with same
178b032 annot be made intermediate, the first children PARTAKE more of the mother, the later ones of the fath
190b076 uce towards Northern Eastern end & die away, & PARTAKE of Indian character There appears in Australia
228b232 ork, our companion in our amusements. they may PARTAKE, from our origin in <there> one common ancesto
388d171 staid in the womb, when it came out. it might PARTAKE of shade of fathers character.-- according to
203b158 races when mingled with newer, hybrid variety PARTAKES chiefly of the former Eyton's paper on Hybrid
546m109 ber of sources of pleasure & innate tastes, he PARTAKES, taste for musical sound with birds. & ¿ howl
080r178 directions. from W. NW & S.--last to Seaward PARTAKING of the character of a Araucarian tribe, with
172b005 beautiful law of intermarriages <separating> PARTAKING of characters of both parents, & these infini
333d005 ty unknown., three kittens, alike each other, PARTAKING <more> «very closely» of form of mother: more
360d093 e some different kind of mottle, each feather PARTAKING of character of other.-- <so> the most aquati
556m145 Sphynx made at Z. Gardens may be described as PARTAKING of <st.> made by <ret> inspiration & quickly
047r083 e undertow & overfall must vary proportionally PARTIAL shrinking after elevation in perfect conformit
288c159 completely «nature & instinct» modified.-- The PARTIAL migrations of birds in same country. may expla
423e096 the case, it proves the law of developement in PARTIAL classes is far from true.-- I doubt not if the
602o011 is because each decision &c is made up of many PARTIAL results, & the impressions on them are <all> r
262c073 ages; nature is full off them.-- wading birds PARTIALLY webbed. &c &c.--)-- & in round of chances eve
428e111 to any plant reared artificially, & only very PARTIALLY to the Zizanias in in Sir. J's ponds-- my pri
509q017 Falconer says all in cases. Owen. Have talked PARTIALLY with him Ask him to introduce me to some Huma
072r160 ncretions somewhat analogous to septa.-- would PARTICLE attracted towards space tend to form ring. [F
099a051 line {P} AB, & likewise gravity MN. Then every PARTICLE would tend to meet at <B. but if particls att
102a060 mygdaloid. calcareous rocks of Ascension, each PARTICLE coated. &c will be aware how little common Gr
354d062 dy of a man undergoing a constant round,--each PARTICLE is placed in place of last by the ordering of
372d130 on as artificial division.-- On this view each PARTICLE of animal must have structure of whole compre
617o040 mselves consist in a force manifested in every PARTICLE of matter directed towards every other partic
617o040 article of matter directed towards every other PARTICLE; but FORCE, <objectively> considered, by our
617o40v of gravity are manifestly the same as if every PARTICLE of matter were an animated being pulling ever
617o40v ter were an animated being pulling every other PARTICLE by invisible strings & as on this supposition
618o41v thought have no reference to place. [We see a PARTICLE move one to another, & <the> (or conceive it)
022r006 ility of conducting Electricity? Would minute PARTICLES have a tendency to change their position? Car
033r042 cension In Calc: sandstone at Ascension, each PARTICLES coated by pellucid envelope of Lime.--form re
046r079 > less pressure might have its «proportional» PARTICLES altered.-- With respect to Volcanic theory. I
098a047 ssume the same force which draws together two PARTICLES of Carb. of Lime, tends to crystallize them a
098a048 iss to granite): Why not horizontal? Why have PARTICLES in such cases moved more laterally than verti
099a049 ilms vertical. Ascertain law of attraction of PARTICLES of same nature: then get mathematician to whe
099a049 me nature: then get mathematician to when two PARTICLES <would> are aggregated, would they not attrac
099a049 ake layers.-- (Gravity can have no effect, on PARTICLES of equal weight.--)¿ cleavage not vertical ∵
100a052 or the Falkland case, nor. the arrangement of PARTICLES of granite in Henslow's Grit, yet it is worth
102a060 common Gravity has to do with arrangement of PARTICLES in rock. This applies to cleavage & concretio
099a051 Cordillera.!!!-- In stratum OP. let force drag PARTICLS to line {P} AB, & likewise gravity MN. Then e
099a051 very particle would tend to meet at <B. but if PARTICLS attract each other in some increasing ratio i
054r102 eland entirely volcanic!! New Zeeland rich in PARTICULAR genera of plants: All St. Catherine & coast
155g066 rock «only decay from fragment falling» of no PARTICULAR hardness no wonder that all «three» lines «s
177b028 into many species, so have the awks, there is PARTICULAR circumstances, to which> is it an index of t
195b101 that God ordered, each planet to move in its PARTICULAR destiny.-- In same manner God orders each an
209b155 man has had no interest in perpetuating these PARTICULAR varieties. If species made by isolation; the
259c063 WHOLE race of that species must take to that PARTICULAR habit.-- All structures either direct effect
282c142 uatic, i.e relation to elements & not minding PARTICULAR trades.--» then the second race would not ob
292c171 on to my mind.-- Reflect much over my view of PARTICULAR instinct being memory transmitted without co
304c206 ly perfect organs. do changes of habit affect PARTICULAR organs.-- of two adjoining families & not al
318c254 icapa solitaria stay about a fortnight in one PARTICULAR part of country, like White of Selbournes Ro
336d018 ence mutilations not heredetary,, but size of PARTICULAR Muscles-- When two animals cross. each sends
362d100 domesticated cattle have tendency to breed at PARTICULAR times. Mr Yarrell has old book 1765? Treatis
408e045 of East Indian Archipelago. now rising. On a PARTICULAR part of coast of Somersetshire the Cockles a
433e126 ated between two great distinct formations.-- PARTICULAR air given. p. 246.-- 248 & p. 258 A beautifu
506q015 taries-- In Monooecious «order» flower occupy PARTICULAR position.-- (39) What does he think of Dr. F
523m018 time.-- There are numberless people insane of PARTICULAR ideas, «Case of Shrewsbury gentleman, unnatu
552m131 hen to whole body, being failed, & not to any PARTICULAR muscle Sept. 8th. I am tempted to say that t
613o036 in animals no prejudices about souls, we see PARTICULAR trains of thoughts as fear of man,-- crows f
615o037 it is downright instinct, leading to touch a PARTICULAR organ.-- I think Pincher shows surprise, wal
615o038 same; which same argument probably applies to PARTICULAR instincts of animals. even in wild state; ce
054r105 pebbles such as those on the beach--"This is PARTICULARLY observable in a bay about five leagues Nort
073r163 latter rarely appear in central Cordillera. PARTICULARLY between 18 degree & 22 degree N. = formatio
400e011 pecies of elater.-- Where this collection is PARTICULARLY rich. «as in Lucanidae» <no> less difficult
403e022 , are remarkably short.-- & Deformations are PARTICULARLY common.-- without arm, <skin> hands thumb,-

578n053 you wont. ==)== No surer way to blush, than PARTICULARLY to wish not to do so.= = How directly perso
509q017 in same races Mr. Gray General Questions (1) PARTICULARS about Sierra Leone. cow. taking bulls. is it
511q018 Heron to give me Pigs foot undivided, & more PARTICULARS regarding effects of crossing them with comm
325c267 ith African <A> Annals Histoire Generalle et PARTICULIERE des Anomalies de l'organization des Hommes.
320c276 oyage a l'isle de Frances Voyage de l'Astrolabe PARTIE Zoologique Pernety. voyage a l isle Malouines
040r063 Vesuvius. Vol III P. 124. Lyell. dikes have a PARTING of pitchstone; which is described as very rare
096a041 facts respecting lime <n> being heated without PARTING with Carb. Acid.-- Mr Malcolmson in Paper on I
462t051 geography, there can be no sharp division of PARTITION as between Mammalia in cases such as that of
063r132 eing Zoophite producing distinct animals. still PARTLY united. & eggs which become quite separate.--C
106a075 with very gently sloping sides This argument is PARTLY taken from Delabechs Theoretical Researches.--
119a106 least as hot as lava-- of which temperature is PARTLY known-- <[...]> moreover gradation from gneiss
159g086 nical hills on same) of «semi» waterwork & some PARTLY well worn pebbles-- «which river could not hav
159g087 of valley much more gentle than in Glen Roy, & PARTLY shut in No Granite blocks in higher parts?? Bo
182b046 7) Man studied The condition of every animal is PARTLY due to direct adaptation & partly to heredatar
192b046 ery animal is partly due to direct adaptation & PARTLY to heredetary taint;. hence the resemblances &
195b098 ted that bi[r]ds with plumage «&» tone of voice PARTLY American North & South.-- (& geographical <dis
245c023 me une espèce distincte! p. 171. Sus papuensis «PARTLY domesticated» like in general appearance to Si
302c200 ly one new family & no new orders.-- Wonderful, PARTLY explained on my theory, = otherwise mere fact
337d020 mals must be most complicated, because they are PARTLY local & then the local ones are taken to fresh
338d024 ost clear that the ideosyncracy of the Negro (& PARTLY Mulatto) prevents his taking any form of Malar
355d067 w fact be elicited by these speculation even if PARTLY true they are of the greatest service, towards
405e031 ur is changed, these best marked characters are PARTLY retained, therefore colours vary in same Manne
428e112 strong & weak March 11th. Yarrell's law must be PARTLY true, as enuntiated by him to me, for otherwis
454e184 monooecious.--   Are there not wild plants, some PARTLY dioecious? Mushroom Hybrids? Any «wild» plants
460t017 gehog Tenrec both having spines, is the effect, PARTLY of the same external conditions (ie. analogica
460t017 xternal conditions (ie. analogical structure) & PARTLY the laws of organization (ie those laws which
532m054 e been frightened & started much more apt, this PARTLY owing to heart? readily taking same movements,
549m120 -- perhaps not so much as they appear & perhaps PARTLY their fault.-- Whether this rule of happiness
551m129 >-- The young Ourang in <Zoolog» Gardens pouts. PARTLY out displeasure (& partly out of I do not know
551m129 oolog» Gardens pouts. partly out displeasure (& PARTLY out of I do not know what when it looked at th
584n071 s an example fitted for climbing, his arguments PARTLY fall, when a species is found which does not c
628o53v is not act by gusto, though by fear it might be PARTLY made.]CD p. 21. "Why ought I to keep my word"-
183b051 ften come very remote quarters. (NB. if Plata PARTRIDGE «or Orpheus» was introduced into Chili. in pr
546m109 h many animals-- see how a dog likes smell of PARTRIDGE--, man's taste for smell of flowers, owing to
564n005 o seize small moving object like fly.-- young PARTRIDGE can run even with its shell on back.-- To stu
389d177 ting, but one is rendered abortive as far as PARTURITION is concerned.-- Generation being means to pr
345d042 is then case of avitism.++» Three gentlemen of PARTY all thought with pigs &c, that hybrids were unc
544m103 are broken-- Sir J. Franklin when starved, all PARTY dreamt of <goo» feasts of good food-- The mind
546m111 had put it in branch of tree, & apologising to PARTY, went out & found it there!!! Lady in perfect «
243c019 same species Coquille Voyage p. 25 Mais il n'y a PAS jusqu'aux îles Macquarie & Campbell (52 degree
244c023 de l'opinion du baron Cuvier, nous ne balançons PAS a la regarder comme une espèce distincte! p. 171
257c057 des espèces tres-analogues,-- quand ce nétaint PAS tout à fait les mêmes. This good case, of repla
075r166 ru. on other hand, mine of Gualgayoc or Chota & PASCO in "alpine limestone" = "The wealth of the vein
596n184 ad very little Does blood go in <body> face in PASHION.?-- cry? Do people of weak intellects easily f
034r044 the most inferior rocks--The stream at Portillo PASS example of do? <Poor> Daubeny good account of e
148g029 r-- recollect the case of loch <in> <below> «by» PASS of Glencoe-- the erosion may often be due to ri
151g041 lain absence of lines in certain parts.-- At the PASS of Glen Collarig two little lines of Hill (judg
152g045 eory lake burst in most improbable part & not in PASS, where shallowest In Glen Collarig good case of
158g083 each side «very little way» in Upper Glen Roy at PASS {P} River Gorge 4th Sh side of valley Granite b
161g097 ontinuous flights above it-- (NB the buttress or PASS at Isthmus appears above level of shelf certain
177b025 t ¿We need not think that fish & penguins really PASS into each other.-- The tree of life should perh
227b224 gan in A. B. C. three species «of one genus» can PASS into each other «by steps we see»: but this can
227b225 B. C. we cannot be sure that structure (C) could PASS into (D).-- We may foretell species. limits of
227b226 t leads to knowledge what kinds of structure may PASS into each other: now on this view no one need l
263c076 iculties Once grant that «species» one genus may PASS into each other.-- grant that one instinct to b
388d173 ch the ovum.-- [Why in making a bud, which is to PASS through all transformations, should there need
392d180 oecious. Will ova of fishes & Mollusca «& Frogs» PASS through birds stomachs & live? In Muscovy ducks
471tf08 family so completely natural & one whose groups PASS so insensibly into each other". Phillips (Lardn
478z003 guay p 207 Slight notice on habit of Iguana. not PASS Lat. 28 degree North p. 239 In ocean between La
578n056 greeable impression like true convulsion. (Hence PASS into convulsions?)-- squeeze out tears. replace
579n059 he says he loves a person-- do not the features PASS before him marked, with the habitual expressemo
023r012 ch were things pitched over board early in the PASSAGE!!-- M. Labillardiere in Bay of Legrand, (SW pa
049r089 are seen to graduate into granites the <conta> PASSAGE from lava to Granite is much more perfect. tha
132a137 e of the earthquake of Chile, with that of the PASSAGE of the moon.-- Ask Hopkins. M. Parrot, Mem. Ac
135a144 many sea shells some miles from coast-- quote PASSAGE to show abundance Bengal Journal. Vol 4. 1835.
190b079 stronger,-- Mr Waterhouse says no real <separ> PASSAGE between good genera-- How remarkable spines, l
195b099 same type, armadillo like covering created.-- PASSAGE for vertebrae in neck same cause, such beautif
196b110 some analogy The existence of plants, & their PASSAGE to animals appears greatest argument against t
215b178 tled 'Paleontographie' developing his ideas on PASSAGE of forms.-- Deshayes states Lamarck priority r
224b212 to structures Mem. Eyton's Hogs & dogs.-- The PASSAGE in last page explains that between Species Fro
225b217 f throat, (or colour of plumage altered during PASSAGE of birds (where is this statement I remember.
260c068 cies-- Mem. St. Jago--solitary Halcyon bird of PASSAGE.-- M. Coronata of Latham, wrong.-- Mr Yarrell
286c156 web-feet, where we see scarcely any traces of PASSAGE a difficulty, but after all a slight one It wi
337d022 ansactions of the Entomological Soc> A capital PASSAGE might be made from comparison of Man, with exp
401e016 that such variety as red cabbage produced from PASSAGE from many varieties, & probably would take lon
403e023 the will of God.-- Octob. 16th. A very strong PASSAGE might be made-- why seeing great variation in
408e042 s important to understand well the relation of PASSAGE from N. to S. American forms. The climate of N
413e059 species. the mystery of mysteries. & has grand PASSAGE upon problem.! Hurrah.-- "intermediate causes"
422e092 erms. & Walrus-- aquatic seal.-- (Consult this PASSAGE, when considering origin of northern Cetaceae)
482z011 s! Says the thrush & another species! birds of PASSAGE!! sylvia macloviana, 2d like sylvia cisticola.
601o08a -- In its action-- emotions.-- p 176 & 177 good PASSAGE in French on what dog dreams, awakes-- does wh
602o011 &c Reynold XIII Discourse (p 115) a very good PASSAGE. about actions & decisions bein the result of
049r087 s, must be well examined At M. Video «facts of PASSAGES marked by do.» discuss quartz veins, there co
098a048 e. (& as I believe most strata) (Hence endless PASSAGES from gneiss to granite). Why not horizontal?
177b025 coral of life, base of branches dead; so that PASSAGES cannot be seen.-- this again offers contradic
197b112 stacea-- with respect to mouth those beautiful PASSAGES from one to other organ.-- Cuvier on opposite
205b143 water Treatise There are some good accounts of PASSAGES of legs into mouth-pieces of Crustacea. Vol I
228b228 tend.-- this & <direct» examination of direct PASSAGES of <species> structure in species, might lead
262c073 re scheme not filled up, (most false to say no PASSAGES; nature is full off them.-- wading birds part
309c222 ve in circular or linear arrangement.-- Thinks PASSAGES very rare., in anatomical structure.-- «the p
309c222 es very rare., in anatomical structure.-- «the PASSAGES between-- owls & hawks only external» interme
403t055 inct!!-- He would not allow such series showed PASSAGES-- yet in talking, constantly said as the <bra
550m126 nt.-- Browne Religio Medici, p. 21-24. Curious PASSAGES showing how easily chance & will of Deity are
568n019] whole kingdom of nature. If I want some good PASSAGES against, opposition of divines to progress of
178b030 neighbouring continent where some species have PASSED over, & where other species have <come> <'air"
190b078 tion being imitated in the womb, which has been PASSED through to form that species.--<Man is derive
285c152 e to other living being.-- one species May have PASSED through a thousand changes, keeping distinct f
293c175 the steps by which the organization of the eye, PASSED from simpler stage to more perfect. preserving
317c249 ngth, some branches of Justicia still growing,) PASSED us. do. p. 243 (, Professor Smith's Journal) o
332d004 n of large number of worms the child not having PASSED them before.-- Hence disordered intestines may
388d173 d <one> suppose that the vital portion ¿nerves? PASSED through transformation, & was received into bu
402e020 large monkeys.-- Fewer <of th> «Mammalia» have PASSED to Paragua & in Luçon the most northern of the
418e084 s, then <is as old, as the oldest animal>, have PASSED through as many changes, as has any species.--
439e142 that is animals» «or rather the parents» having PASSED through many changes.-- It is very important M
529m042 o was most violently purged «believe worms were PASSED off.» & vomited, but who when he recovered. wa
480z006 immense Tortoises p. 81 & p 113 of 1834 On the PASSERES of S. America. D'.Orbigny. L'.Institut. No.--
482z011 f nine terrestrial Turdus falklandii & then 9. PASSERES! Says the thrush & another species! birds of
022r007 re Mem C. [Cape] Turn P. 434 & 419 As Limestone PASSES into schist scales of chlorites--Mem. Maldonad

(Key Word)

197b111 n as a short process, by which man «one animal» PASSES from worm to man; «highest» as typical of <sa>
307c216 .? get examples.-- for instance where a tendril PASSES into a mere stump.-- Shall abortive organs «of
565n006 e ceases, or stops the noise , the face clearly PASSES into smile-- laugh long prior to talking, henc
579n059 om grief, joy. & sublimity. January 6th.-- What PASSES in a man's mind. when he says he loves a perso
387d170 h respect to moths, & monkey & men.-- each man PASSESS through its caterpillar state. the monkey repr
076r170 America. Geology of Guanuaxuato.--Clay slate. PASSING into talcose & chloritic slate. with beds of s
092a030 High up the Essequibo, granite & quartz, after PASSING sandstone Vol II. p. 69.-- Geograp Journal Ear
170b001 which is a longer process, the new individual PASSING throug several stages (¿typical, <of the> or s
186b057 h in all excepting specific character); and in PASSING from species to genera, each retains some one
289c161 If any one is staggered at feathers & scales. PASSING into each other let him look at wings & orbits
390d179 ock.-- The very theory of generation being the PASSING through whole series of forms to acquire diffe
548m117 tion of happiness the number of pleasant ideas PASSING through mind in given time.-- intensity to deg
146g012 tion, & passions, such as love-- dislike & <f> PASSION of hatred To fulfil an instinct a pleasure; me
264c079 e its affection.-- to those it knew.-- see its PASSION & rage, sulkiness, & very actions of despair;
298c190 produced. Such as man getting habitually into PASSION, becomes habitually passionate.-- the Key to t
301c199 which Man performs.-- child striking a post in PASSION.-- Habit instinct gained during life.-- do Ele
306c212 om a dog.-- more like man. continued long in a PASSION & looked out for him to come again very differ
306c212 in very differently from dog. perhaps being in PASSION chief difference Major Mitchell is not aware t
313c235 n (13?) without impregnation, therefore sexual PASSION must arise after long interval very good case.
362d099 some authors [not located] September 13th The PASSION of the doe to the victorious stag. who rubs th
385d163 unhealthy.--» males «bred in & in» never lose PASSION. (Mem: so it was said little cock «yet very od
389d177 ng cattle can be bred in & in.).-- [The loss of PASSION in hybrids. perhaps connected with this same c
393dIBC s subsequent progeny faulty. Does male fail in PASSION.-- Disposition of half bred Cattle at Cinberme
505q014 Open more Horned oranges.= (6) Figs, flower.--PASSION Flower. (as it is required to impregnate it ar
523m017 ether insanity is not distinguished from whims PASSION.-- There seems no distinction between enthusia
523m018 then brain affected like getting suddenly into PASSION.-- There seems no distinction between enthusia
523m018 There seems no distinction between enthusiasm PASSION & madness.-- ira furor brevis est.-- My father
524m019 e do not well recollect what they have done in PASSION.-- People are constantly well aware that they
524m019 Birmingham Doctor), in this precisely like the PASSION, ill-humour & depression, which comes on from
531m052 action, accompanying violent movement; may not PASSION be the feeling «consequent on the violent musc
531m052 turneth,, here probably there is no feeling of PASSION, but muscular exertion consequent on the injur
532m057 animals put on.-- The flush which accompanies PASSION & not sweat is the <state> effect of short --
536m070 is what my views tend to.-- When a man is in a PASSION he puts himself stiff, & walks hard.-- «He can
542m094 articulate.-- the maniac shouts & bellows with PASSION.-- It is not a little remarkable that those so
545m106 nt analogy with man.-- I see monkeys grin with PASSION, that is show all the teeth: «& make noise not
545m106 & may be compared to laughing» they dance with PASSION, ie. nervous impulse to action is sent so fast
552m131 itude of attention «So intimately connected is PASSION with sending force to muscles, that in my gran
555m143 so often <that> hardly ever the expression of PASSION with open mouths like the old world ones.-- Th
557m147 e the opposite muscles would act, to when in a PASSION.-- dog tail curled when angry & very stiff. ba
557m147 n parrots &c &c -- -- peacock & turkey cock in PASSION.-- Cat when pleased, erect its tail & make it
557m148 is quite different from that.-- like «slight» PASSION from blood rushing in face, with less action o
557m149 ularly apt to catch tricks.-- so are people in PASSION my F. rubbing hands.-- stamping. grinding teet
558m152 h the same way» generic manifestation of great PASSION.-- I do not think they arch their backs-- Beng
558m153 es man.-- & so dog]:CD Man grins & stamps with PASSION. can expression be used more correctly than th
558m153 drawing out of the side part of nostril, when PASSION commences.-- <All> Nearly all will exclaim, yo
564n003 imal with social & sexual instinct «& yet with PASSION» he must have conscience-- this is capital vie
565n006 ing its teeth to bite.-- the senseless grin of PASSION, is like the grin of the Hyaena from fear, no
567n014 h it moves doubtless has not.-- Turkey cock in PASSION & sends blood to its breast &c &c All Science
577n051 an inflamed membrane from local irritation to PASSION.-- Blushing is intimately concerned with think
590n093 eze.-- "A Dissertation on the Influence of the PASSION."-- p. 37. The increase of Bilary secretion at
590n093 . 37. The increase of Bilary secretion attends PASSION p. 39. The sweat that accompanies fear is the
590n094 1. II p 153. «do». an account of a monkey in a PASSION like Jenny.-- Dr. Abel has given an account of
593n107 invented,-- were musical notes the language of PASSION & hence does music now excite our feelings.--
600o008 best singer-- Remember.-- avarice a compounded PASSION-- gained in life time]CD 3. The Infinite, -- liv
600o09v - I suspect conscience, an heredetary compound PASSION. like avarice.-- Is there not something analog
606o025 gs when wagging one's finger-- one feels it in PASSION, love-- jealousy-- «as» effect of bodily organ
619o043 id not obey his instincts from interference of PASSION. he would feel pain, which would generally be
619o043 s it goes.).CD But should be prevented by some PASSION or appetite, what would be the result? In a do
620o044 appetite gratified gives only short pleasure. PASSION in its nature is only temporary, & we do not a
620o045 ure from good dinner, or from a blow struck in PASSION fades away, so that when man afterwards thinks
620o045 not feel it wrong in very young child to be in PASSION, any more than in an animal.-- which shows tha
620o045 is time, malevolence,, when not urged to it by PASSION, shows a bad child.-- Hence there are certain
621o046 malevolent, or hates his children without some PASSION.-- If his passions strong & his instincts weak
621o046 ling in its nature being always present. & his PASSION shortlived, it is to his interest to follow th
621o047 o follow it. even when opposed by some natural PASSION.-- (a) [LHC] The conscience rebukes malevolent
625o052 r feelings steadily & not like our appetites & PASSION, which receive enjoyment from gratification &
626o052 g, when instinctive will lead to action.-- the PASSION rising from weariness leads to striking blows.
628o54v he very end of conscience is stop to wishes of PASSION &c. whilst the passions have no relation I thi
629o055 tion. money, books &c &c.-- <]> the "secondary PASSION" of Hutcheson unfolded by D. Hartley.-- Darwin
298c190 g habitually into passion, becomes habitually PASSIONATE.-- the Key to the affections might perhaps t
146g012 an analogy between pleasures of association, & PASSIONS, such as love-- dislike & <f> passion of hatr
286c154 s,-- death, unequal life,-- stimulated by same PASSIONS-- brought into the world same way» they may c
310c222 nfess my profound ignorance.-- but seeing such PASSIONS acquired & heredetary & such definite thought
390d178 which is not object of generation.-- therefore PASSIONS fail.-- In fruit trees no doubt there is tend
531m051 es of others.-- In young children, the violent PASSIONS they go into, shows how truly an instinctive
550m123 - Our descent, then, is the origin of our evil PASSIONS!!-- The Devil under form of Baboon is our gra
566n010 icking expression of emotions, he has felt the PASSIONS of a face «& mind» sympathetics with internal
566n010 a wife a ruling motive.-- Book IV, Chapt I on PASSIONS of mankind, as being really useful to them: t
566n011 t be studied. before my view of origin of evil PASSIONS.-- Man getting sight slowly,, but when in gro
596n184 on life & character "Humes Dissertation on the PASSIONS." "Hartley" I should think well worth studyin
608o025 is called free will, but by strong invariable PASSIONS-- when these passions, weak, opposed & compli
608o025 but by strong invariable passions-- when these PASSIONS, weak, opposed & complicated one calls them f
609o29v dam Smith also talks of the necessity of these PASSIONS, but refers (I believe) to present day & not
609o29v Civilization is now altering these instinctive PASSIONS--. which being unnecessary we call vicious.--
620o045 t of our nature, & its effects lasting, whilst PASSIONS-- although equally natural leave effects not la
620o045 g. By association one gains the rule, that the PASSIONS & appetites should «almost» always be sacrifi
621o046 es his children without some passion.-- If his PASSIONS strong & his instincts weak. he will have man
621o047 out the continguity to will. (a) The origin of PASSIONS too strong for our present interest receive s
624o051 which it» has been so in some past time, hence PASSIONS]CD «although perhaps useful at present to som
624o50v een our instinctive feelings & our short lived PASSIONS' State broadly in child or animal it is equal
625o50v ng.--. for then only is it perceived, that our PASSIONS are too strong for our instincts. to gain lon
628o054 of impression of social instincts, than other PASSIONS, or instincts.-- is this good?-- I should thi
628o54v ce is stop to wishes of passion &c. whilst the PASSIONS have no relation I think this <boshes> «nonse
629o55v ans think only of prepuce, crepitando,]CD & if PASSIVE emotions.-- Cannot quite perceive drift of Boo
604o017 a. owing to loss of will.-- chiefly excited by PASSIVE emotions.-- Cannot quite perceive drift of Boo
147g019 be placed in present position Thursday Evening ½ PAST 8 Tyndrum 29.<625> «636» Temp. 62 Friday mornin
147g019 Tyndrum 29.<625> «636» Temp. 62 Friday morning ½ PAST seven o'clock 29.642 Temp 55 Air 50 degree? Fri
192b088 te, the father of all Mammalia in ages long gone PAST.. & still more so «known» with fishes & reptile
198b113 (Perhaps consideration of range of capabilities PAST & present might tell something) p. III G. St Hi
228b228 ould then be main object of study, to guide our <PAST> speculations with respect to past & future. Th
228b229 to guide our <past> speculations with respect to PAST & future. The Grand Question, which every natur
293c172 ng association recalling up image which had been PAST-- so great an anomaly in structure of brain not
304c206 ogy would profit. if the series were believed to PAST into each other-- Different classes Keep to the
338d023 mstances of the two countries in times present & PAST. The effect of physical conditions of country i
379d151 present age is the one for large Cetacea, as the PAST for other Mammalia. & still further back reptil
434e128 e analogies of the existing to the events of the PAST world, we have no foundation for our science".-
440e146 ency, without final cause. either in present, or PAST generation.-- thus cabbages growing like Nepent
544m102 e.-- peculiar sensation as flying. (No memory of PAST events?) or influence on our conduct, the links

Page **(Key Word)**
544m102 conduct, the links which when conscious connect PAST, present & future thoughts are broken-- Sir J.
544m103 to the mind,-- it is solely the comparison, with PAST ideas. which makes consciousness-- & which tell
547m112 mple case.-- There was memory, for it related to PAST idea.-- there was a kind of ideal consciousness
547m112 no other parallel trains of ideas connected with PAST circumstances.-- as whether I really was going
549m119 lected, such part of thoughts innumerable, which PAST through mind.-- These thoughts are most pleasan
564n004 hatched few hours placed on table & when fly ran PAST it. cocked its head, & picked it-- Here then, t
615o038 nd taking pleasure in virtue because acquired in PAST ages; seems to indicate that when we <return> t
620o044 his memory & <pow> mental capacity of calling up PAST sensations, he will be forced to reflect on his
621o047 isfaction of the mind, which is «much» formed by PAST recollections.-- Hence he has the right & wrong
623o050 ncy, (not to any one individual but to the whole PAST race).-- <no one> doubts) « <I cannot> » [LHC]
623o051 nly that which» had a beneficial tendency during PAST races could become instinctive.-- [LHC] x It is
624o051 more correctly «in which it» has been so in some PAST time, hence passions]CD «although perhaps usefu
067r144 volcanic rocks at M. Video «Volcano in Pampas» PASTO Earthquake. Happened on January 20th. 1834 Mr S
031r037 n--but gradually & simply raised No Faults in PATAGONIA[,] enormous extent; if lowered again & covere
035r047 lities of elevation, may well be preserved in PATAGONIA. The English fact is astonishing consult book
036r050 istribution of pebbles.--Plains. off coast of PATAGONIA.-- British channel &c &c. There is a Hill. ne
036r052 tic of the deserts of Syria <chara> ditto for PATAGONIA, especially rocky parts of central Patagonia
036r052 Patagonia, especially rocky parts of central PATAGONIA Does Andes in Chili. separate geographical ra
040r062 ooks like such?--destructive to animal life.--PATAGONIA In the Chonos Islds we must imagine bituminou
041r067 these thin veins which project outwards.-- In PATAGONIA. the blending of pebbles & the appearance of
042r067 ould tend to travel on a <me> central line of PATAGONIA. «NB. Mr Lyell P. 211 Vol III. talks of line
047r083 's> idea of an injected mass of fluid rock In PATAGONIA plains. long periods of rest & vice versâ mor
048r085 ns. which ranged from Equatorial plains to S. PATAGONIA. To the Megatherium.--To the Horse. = One mig
048r085 ing surprise at Mastodon inhabiting plains of PATAGONIA is removed by reflecting on the nature of the
048r086 ll. V. where the Rhinoceros was killed. -- In PATAGONIA, are all beds same age? is white substance tr
052r099 ( )», he has rocks on surface. applicable to PATAGONIA. During a period of subsidence the shinglle o
053r100 During a period of subsidence the shinglle of PATAGONIA would become more or less interstratified wit
057r114 amel urge S. Africa productions.-- I think in PATAGONIA white beds having proceeded from gravel prove
085a001 Woollich p. 112 Speculate on the extension of PATAGONIA seaward, at mouth of S. Cruz. from ascertaine
087a012 ment of proof of slow corrosion of valley of <PATAGONIA.> S Cruz -- from terrace like structure-- Int
107a079 ase of S. Maria & Tubul applicable to Andes & PATAGONIA-- On Lyells idea of whole centre of earth sam
121a112 Woolwich there are plains & valleys just like PATAGONIA, & many shells in parts on surface, but I saw
121a112 h examining. to support. shells on surface of PATAGONIA, yet none in shingle beds. Lyell on Sweden. p
129a131 ed: ditto Murchisons case.--¿ does it bear on PATAGONIA? «Facts about subsided forests.-- Many repeat
130a134 hin: <beside mere fracture> A Elevation as in PATAGONIA {P} B subsidence; <as in> be cautious. mud ba
240c013 e facts all account for [not located] Falkner PATAGONIA no description of wild animals, nor in Dobriz
257c056 y (p 108) says having observed B. Tricolor in PATAGONIA. then in Chile & lastly 12,000 ft above sea i
268c099 of land by seeing how many species common to PATAGONIA desert & Tierra del Fuego. & forest «Parrots
302c203 little adapted to some physical change.-- If PATAGONIA became fertile all intermediate species livin
320c276 ve Mackenzie's Iceland Molinas Chile Falkners PATAGONIA Azaras Voyages & Quadrupeds of Paraguay Dobri
465t080 and-- coasts of Chile, excepting Concepcion-- PATAGONIA-- Beds of La Plata. (except close to B. Ayres
481z009 - replaced by didelphidae Skunk inhabitant of PATAGONIA. Mem:-- S. Cruz. Molina Vol. I. p. 244. Bacca
485z016 many of the small finches walk at Maldonado & PATAGONIA compared with those of England.-- or ground b
030r036 altered Carbonaceous shales Examine chart of PATAGONIAN coast to see proportional cliff & low or slo
032r041 d of action. This case differs. I think. from PATAGONIAN steps, because the deposition & accumulation
035r045 d strata of England. formed near surface: Mem PATAGONIAN pebbles beds, most unfavourable to preservat
036r049 tain chains.-- Introduce part of the above in PATAGONIAN paper; & part in grand discussion Consult. r
094a036 effect of remodelling same manner. as bits of PATAGONIAN boulders might be transported.-- {T} On groo
120a110 er brought down Mention absolute elevation of PATAGONIAN blocks (1200 ft??). Scotland at least 2200.
195b103 conditions would produce species so close as PATAGONIAN <Chat> & Galapagos orpheus.= Put this strong
449e170 apagos Heteromerous insects come very near to PATAGONIAN species-- p. 18. of Temmincks. Preliminary d
484z016 g the way Synallaxis leads into Furnarius. by PATAGONIAN Furnarius.-- into Oxyurus, by Maldonado cree
277c126 try & class tells us that ¿O. Modulator & O. PATAGONICUS. till neutral ground ascertained, call them
477z002 lla-arskink» le quiyá (Coipu) viscacha.-- A. PATAGONICUS les tatous (.4 pichye, pelud, mulita et mata
257c056 mes oiseaux. que nous avions déjà observés en PATAGONIE. ou au moins des espèces tres-analogues,-- qu
079r176 richest gold mines on ridge of Cordillera near PATAZ, also at Gualgayoc. where many petrified shells
261c071 sting» Mr Gould says wherever any mark like red PATCH on wings of Furnarius, Synallaxis &c &c. sure t
516q24v comes up.-- [Unnumbered blank] Experiment Cover PATCH of ground, with different salts & poisons & see
031r038 ffs [Fig. I] line of high tidal action {P} NB. PATCHES of modern Conglomerates [Fig. 2] {P} The actio
058r117 sting so long.-- The great movements (not mere PATCHES as in Italy proved by Coral hypoth. agree with
146g013 an instinct a pleasure; mem. Shepherd dogs The PATCHES of Conglomerate on S. Ventana, excellent insta
147g015 egular horizontal strata I suppose these upper PATCHES if prolonged would intersect alley above the 3
317c249 same thing p. 55. 40 leagues from land several PATCHES of reed & trees p. 259 120 ft in length, some
360d089 gs of Africa,-- some puppies hairless. some in PATCHES, & some hairy-- the former preponderated <whic
478z003 In white Cape Pidgeon's stomach small shells (PATELLA) sea weed & many pebbles Mentions stinging Mil
501q10a er,, but contingent on country.-- How is it in PATELLA or Oysters or Helix. Or does any «one» species
146g009 my idea-- highest part must project [Blank] {P} PATH East End near Holyrood Palace In same way at to
146g010 l has been well-- denuded.-- «of hard metamorph» PATH only covering Great Slip, 10 years since three
322c269 W. Darwin's Botany.-- References at end Mayo PATHOLOGY of the Human. Mind Evelyns Sylva. skimmed, st
326c266 it. Assoc.-- some important Papers. Dr. Mayo. PATHOLOGY of Human. Mind.-- Audubons. Ornithological Bi
604o017 s D. Stewarts System of Emotions.-- T. Mayo-- PATHOLOGY of the Human Mind. Poor.-- on insanity.-- Pre
530m045 some doctors care it, by stimulus & afterwards PATIENT sinks.-- When a muscle is moved very often, th
332d001 nd, when suddenly during one time he had three PATIENTS at very distant quarters of the county, who h
262c072 nses of savages» (How come its some countries PATRIOTIC?)-- but more especially the powers of reasoni
537m077 revent other being engrafted.-- No one doubts PATRIOTISM & family pride are heredetary.. & therefore
025r016 .] Bahia [12 degree 57 minute S.] 8 200 Morro S PAULO [13 degree 22 minute S.] 9 120 {T} Garcia de Av
027r024 entine form: of Cuba for comparison (?) with St PAULS [not located] [not located] The frequency of sh
272c108 at peculiar conditions the Staphylinidae on St. PAULS Rocks must be placed under Gould says most subg
402e019 I p. 344. account of insects of St. Peter & St. PAULS in Lat' 53 degree yet fauna like that 60 degree
042r067 . 211 Vol III. talks of line of cliff marking a PAUSE» When mentioning pumice of Bahia Blanca, mentio
595n117 st» dreaming, growling. & yelpings. «& twitching PAWS» which they only do when <great> considerably e
325c267 sants <Zoological> Philosop. Transactions. 1827 PAXTON on the culture of Dahlias Mrs. Gore on Roses m
608o027 mself to do harm.-- Believer in these views will PAY great attention to Education.-- 4) These views a
054r102 -- Lesson Zoologie Grand tertiary formation of PAYTA: N. part of New Zeeland entirely volcanic!! New
054r102 s with mine.--At Conception, cleavage E & W! at PAYTA. talcose slates, do at latter place. sandy. san
117a102 long?-- L'Institut. 1838 p. 151. Formations of PAYTA extend close to Guayaquil.-- modern shells of C
244c021 n. some tatous!!! p. 120.-- Most of the dogs of PAYTA-- belong to the hairless kind, «said» to come o
480z008 of» smaller petrels, are night birds agree. with <PE> nocturnal habits of Crustaceae Mr Broderip says
341d031 &c Dr. Bachman has crossed cock Guinea Fowl with PEA <cock> Hen.-- offspring female, yet so infertile
341d032 en in the night, like peacock.-- tail as long as PEA hen.-- about intermediate.-- (In Zoolog Gardens
369d114 condary characters.--)p. 49. (wonderful case of PEA hen. taking feathers of Peacock & spurs-- no fin
467t104 s pratensis yellow saw stigma project» In common PEA saw Humble so press down sheath, that stigma cov
505q014 ossing & ¿heredetary? (15) Abberley has a hooked PEA.-- intends to breed from it and large Asparagus:
515q022 ountry. {Chinese Dog's Head to send Cover common PEA (& Sweet Pea) for several generations under net
515q022 ese Dog's Head to send Cover common Pea (& Sweet PEA) for several generations under net & see if get
563n001 from nest & leaving no tracks.-- My Father says PEA-hens do Wood pidgeons building near houses. yet
581n064 ng-outang), music, colours we must suppose <we> «PEA-hens» admire peacock's tail, as much as we do.--
367d113 always armed in carnivora. Where females, are PEACABLE-- (Mem Lucanus & Copris &c).-- In birds singi
368d114 ent in recently altered genera. Guinea Fowl & PEACCOCKS.!!» other birds display beauty of plumage.--
429e114 but quiet» war of organic beings. going on the PEACEFUL woods. & smiling fields.-- we must recollect
515q021 scarcely any no relation to hybrids.-- (3) As PEACHES sport into Nectarines (does reverse happen?) w
515q021 es reverse happen?) what is effect of crossing PEACHES & nectarines: same question with regard to Pri
272c109 is concerned superabundant,-- the tail in cock PEACOCK,. widowbird.-- Birds of Paradise. Trogons.-- t
341d032 -- most strange voice often in the night, like PEACOCK.-- tail as long as Pea hen.-- about intermedia
369d114 (wonderful case of Pea hen. taking feathers of PEACOCK & spurs-- no final cause here.-- & therefore d
557m147 er in species of swans, in parrots &c &c -- -- PEACOCK & turkey cock in passion.-- Cat when pleased,
513q020 ty of Bantams from different countries= Do the PEACOCKS cross.= Young Chinese or Penguin Duck in very
543m100 a week. yet. he was inimitable chess player.-- PEACOCKS remark about mathematicians not being profoun

(Key Word)

```
581n064 olours we must suppose <we> «Pea-hens» admire PEACOCK'S tail, as much as we do.-- touch apparently. o
033r043 f so case precisely analogous: fragments instead PEAK of Teneriffe. also Cotopaxi has a <[...]> cylin
064g072 of  dikes in the cliffs.--wide valleys.--central PEAK small; yet great body of lavas have flowed from
064g090 one  of these & Glen Tarf Hill «Cairn <taw> leer PEAK» Barom 28.700. A.75 degree 75 degree? Boulder,
160g090 . Much contorted gneiss «narrow sharp ridge with PEAK» I walked all round hill. Boulder about 20  ft.
296n184 ess seeds can arrive quick enough» Vegetation of PEAK-- altogether original. owing to being oldest. &
296n184 land shells.-- all in short Extreme North = = to PEAK of Teyde in relation to surrounding countries &
052r099 artz pebbles in the Cordilleras look as if some PEAK elevated.-- Greywacke. as a general fact absent
296n184 ether African forms. (anyhow not Australian) on PEAKS. Did Creator make all new yet forms like neighb
401e017 ye but not flourish-- a medlar may be Grafted on PEAR. Mountain-ash & white Thorn! Species not  being
504q013 ᴎ. Laburnum ᴎ. Dodecatheon ᴎ. Castrate apple & PEAR to see if pollen naturally carried, on  account
049r089 greenstone? -" Daubeny. P 95. Glassy & Stony PEARLSTONES alternate together in contorted layers: Mem:
401e017 ce Gardeners. by chance <often> sometimes graft PEARS on apples. they will live but not flourish--
467t104 ole breast-- {b} pressing either one or both of PEA'S wings, stigma & mass of yellow pollen protrudes
491q002 Mongrel  offspring 2. How have late varieties of PEAS &c been obtained? 3.. Whether the viviparous gr
497q06v parts,  whether such vary.-- Do Bees go to Sweet PEAS, IMPORTANT, for if so, as these can be raised t
498q008 eries, when «seed of» the varieties of Cabbages, PEAS, beans, as raised, do the Seedsmen select at al
500q10a 0) March 1842. <Last> Year «before last» beans & PEAS were planted in rows adjoining & seeds gathered
501q011 ute insects.-- (30) Get Abberley to plant SINGLE PEAS, Kidney Bean & Bean, intertwined, «without stic
504q013 lly carried, on account of Van Mons views-- Also PEAS-- N.B. I think very likely the Peas to cross ou
504q13v views-- Also PEAS-- N.B. I think very likely PEAS to cross ought to be placed far from all  other
504q13v s to cross ought to be placed far from all other PEAS, from Wiegman Shrewsbury (1) Peas.-- Beans seed
504q014 from all other Peas, from Wiegman Shrewsbury (1) PEAS.-- Beans seeds alone remain to be compared-- Ca
513q21. s expanded-- in reference to Lobelia & Clarkia-- PEAS time of impregnation.-- says many flowers are d
636j56v g with other varieties was prevented Do races of PEAS become intermixed & gardner have hybrid seedlin
549m118 o is it <with an> when same man is compared to PEASANT.-- To make greatest number of pleasant thought
549m120 ixed with some pain.-- than the happiness of a PEASANT, with whom sensual enjoyments of the minute ma
042r068 is the distance?-- Fossil bones black as if from PEAT.--yet cetaceous bones so likewise «of miocene p
051r093 well & turbid water} organic bodies protect like PEAT reef of sandstone.--Corals, & Corallina survive
127a126 do p. 137. Lord Tullamore found Sulph of Soda in PEAT ashes in Ireland dikes in mountains. «(not on c
154g057 te N W W of Ben Erin {P} Shelf of Glen Guoy flat PEAT plain divortium aquarium-- tidal channel-- 12ft
158g081 r Glen Roy great plain about 60 ft beneath shelf PEAT on pebbles tidal plain as sea gradually retired
159g085 orry so much cut Granite could have remained, no PEAT supply.-- Consider profoundly Boulder hypothesi
160g092 rom 28.92 A 75 Air 70 degree? Isthmus broad flat PEAT mass-- (general character in these mountains  &
161g094 «station  perhaps 6 ft too low» (to test last on PEAT-Mass Divortium aquarum) Barom. 29.200 A.77 degr
161g095 uth, (deep) of above valley this road level with PEAT moss most distinct then lost by slope, then con
162g101 90 Preservation of form of land very much due to PEAT & Heather When it did not grow at first-- relic
516q23v ected with Species Theory (1) Will an extract of PEAT do to preserve fungi or animal substances-- (At
516q23v es-- (Athenaeum (40) p. 823 chemical analysis of PEAT (2) Athenaeum 1840 p. 777. Decaying wood absorb
158g079 ite such as boulder on <thes> Divortium aquarium PEATY Mass of this point very nearly like head of Gle
037r056 ganic remains.-- On Pampas looked in vain for a PEBBLE of any sort; not one was found.-- Miers saw th
056r109 [...] deepest astonishment.» Perhaps scarcely a PEBBLE might remain to tell of these losses.-- Cause
035r045 f England. formed near surface: Mem Patagonian PEBBLES beds, most unfavourable to preservation of bon
036r050 p of Europe Consult charts for distribution of PEBBLES.--Plains. off coast of Patagonia.-- British ch
038r058 ons probably formed Islds from which proceeded PEBBLES & on which trees grew.--? Are not the dikes in
040r062 no  quadrupeds. -- Is the white matter beneath PEBBLES. the degraded matter of such pebbles extending
040r062 r beneath pebbles. the degraded matter of such PEBBLES extending to seaward, the alternating with suc
041r064 deep  a sea.--Perhaps agrees with formation of PEBBLES & vertical trees Grand Seco at B. Ayres; menti
041r067 ject outwards.-- In Patagonia. the blending of PEBBLES & the appearance of travelling may be owing to
048r084 a. V. Lyell. Vol I. P. 191 State at St Helena. PEBBLES entirely coated with Tosca. which implies moti
048r084 ca. which implies motion in the «loose» bed of PEBBLES. (On a sea beach under a cascade, one can unde
048r084 sea beach under a cascade, one can understand PEBBLES thus coated.--The motion is most wonderful, fr
051r095 w to account for oceans power.--excepting when PEBBLES are brought into play; most manifest example o
052r099 streams of lava at St Jago. C. de Verds Quartz PEBBLES in the Cordilleras look as if some peaks eleva
054r105 to have been covered by the sea judge from the PEBBLES such as those on the beach--"This is particula
065r140 ay have been great subsidence previously. Mem. PEBBLES of Porphyry.-Falklands.--off East Coast. -- C
094a036 r on drift.-- Beyond region of great boulders, PEBBLES of granite clearly effect of remodelling  same
102a059 trata yield.-- In Undulation in open ocean. as PEBBLES» would be lifted up & down. on coast itself, un
115a097 erica The very general absence of fragments «& PEBBLES» in mica slate & gneiss, can only (see «supra»
116a100 ft or so above bed of river. formed of rounded PEBBLES-- it is clear gold occurs in submarine alluviu
117a102 tained 100 miles E of Staten land. bringing up PEBBLES 2 inches long?-- L'Institut. 1838 p. 151. Form
146g013 the  preservation in situ of even imperishable PEBBLES/ I am nearly certain there were none on surfac
146g013 ses of Alluvium or rather mass of well rounded PEBBLES in yellowish argillaceous or sandy soil-- Thes
147g018 have  been very considerable mass of waterworn PEBBLES in Alluvium which without lake or sea could no
149g032 al-- Derry there were perfectly rounded «base» PEBBLES of quartz & other rocks not apparently in situ
151g040 ss worn into coincidence» has beach or band of PEBBLES on line of 4th shelf.-- Even on Lauder Dicks H
156g070 r rocks at head of shelf 3d almost all granite PEBBLES Level of plain of 4th shelf at head of Lower G
156g072 te» «occurs» abundantly with perfectly rounded PEBBLES of granite & forming «sloping» buttresses  Yet
157g073 ly shelf 4th <near> only usually contains many PEBBLES, but I believe this is chiefly caused by its b
157g073 is is chiefly caused by its being lower,-- [no PEBBLES in parts of Beagle Channel when mica slate, on
157g075 front  of which shelf 3d form beach of granite PEBBLES, & around which shelf 2d «almost» forms it int
158g081 great  plain about 60 ft beneath shelf peat on PEBBLES tidal plain as sea gradually retired, hard to
158g083 roy» near the upper shelfs ground strewed with PEBBLES Shelf 3d runs up with buttresses on each  side
159g086 e) of «semi» waterwork & some partly well worn PEBBLES--«which river could not have deposited» the s
160g093 » Isthmus (which runs E & W) broad terrace «of PEBBLES? & Alluvium» which appear perfectly level, <on
162g104 d layer of fine sand & small angular-- rounded PEBBLES-- dip sideward, & inwards-- deposited when wat
163g107 it  has been filled up «at» 30 ft. higher with PEBBLES now worn away-- The above shells must have bee
163g109 ark) Shrewsbury rubbish.-- Speculate on origin PEBBLES brought by different cause: from mud.-- [blank
432e125 ves,, but to period necessary to form heaps of PEBBLES &c &c: the succession of organisms tell nothin
478z003 stomach small shells (patella) sea weed & many PEBBLES Mentions stinging Millepora. Quoy. Freycinets
202b134 ctly continental. we might have wanderers. as PECCARI in N. America) then if it is doomed that  only
613o036 ters method of standing,-- method of attacking PECCARI-- --retriever-- produced as soon as brain deve
445e158 ouches objects.-- far better case than chicken PECKING fly.-- "whilst the shell stuck to its tail" as
350d056 s} comparison of sight to threads.--» Hence the PECTEN, which move imperfectly has eye-point, but Bro
381d157 they  must in effect be so in all.-- 2 NB. In PECTINIBR Mollusca.-- «or Cephalopoda» are there aborti
381d156 ozoa Allotriandrous <or M> Mollusca, with PECTINIBRANCHIATE order-- the Annelida. All others, <animal
381d156 animals,> «are Dioeecous as» Cephalopods, PECTINIBRANCHIATE molluscs.-- insects. spider crabs.-- (all
459t013 ornsey the offspring of a black & white «duck of PECU» «drake» with the penguin duck. it took after t
026r023 Encrinites. Asteriae, usually petrified into a PECULIAR cream-coloured Limestone: the strange substit
061r127 Mem:  my idea of Volc: islands. elevated. then PECULIAR plants created. if for such mere points; then
061r128 ion «taking place» over certain area must have PECULIAR character: Contrast low limit of Palms, everg
076r168 dated masses of iron. which contain silver are PECULIAR to that part of the veins, nearest to the sur
085a002 er of craters very ancient. which agrees. with PECULIAR character of Vegetation. -- So accustomed to
173b012 f propagation of species we can see why a form PECULIAR to continents; all bred in from one parent wh
174b013 lies only to hybridity.-- genera being usually PECULIAR to same country, different genera different c
188b067 nd North America.-- ¿are there any? Rhinoceros PECULIAR to Java, & another to Sumatra --Mem Parrots p
188b067 ar to Java, & another to Sumatra --Mem Parrots PECULIAR according to Swainson to certain islets in Ea
199b119 any isolated races. ¿are there any instance of PECULIAR people banished by rest?-- ∴ most monstrous f
202b133 wonder of finding Monkey in France.-- of genus PECULIAR to East Indian isles.-- Compares it to fossil
203b138 udon's Magazine. Gould on Motacilla,. «species PECULIAR to Continent & England» Loudon Mag: Septemb o
208b154 arent type in hotter parts of world Is monkey. PECULIAR to C. de Verd's.--? NO Macleay Name given  in
210b160 re Siicily & Galapagos!!-- Some of the animals PECULIAR. to Mauritius are not found at Bourbond Zoolo
212b164 at  Leyden series from several islands.-- Bear PECULIAR to Sumatra & not found on Java-- Monkey pecul
212b164 culiar to Sumatra & not found on Java-- Monkey PECULIAR to. latter not to former-- Mr Martens of Zool
217b188 speculation.-- Are there many Northern genera PECULIAR to itself-- on hybrids between grouse & pheas
218b191 land & other localities & each one will have a PECULIAR «constant» aspect. That is varieties, though
229b234 et Voyage, agrees. with several mammalia being PECULIAR (?) If. Henslow discusses possibility of seed
232b247 northern> Iceland would have possessed a most PECULIAR Flora «& north of Europe»-- As Europaean form
```

Page **(Key Word)**
234b261 » are not a few only cosmopolites, & in genera PECULIAR to any one country do not species generally a
235b274 from different quarters to prevent them taking PECULIAR character-- Indian Bull?-- Do species of any
243c020 ly New Zealand-- Norfolk. Isd. & New Caledonia PECULIAR species of cassicans: (¿cassicans Australian
243c020 s shells towards extremities of the continents PECULIAR to the different points.-- Consult Voyage aux
249c036 ce.-- Waterhouse's remarkable fact of no forms PECULIAR to word to special districts???? north of 30
250c037 Africa destroyed would not then some forms be PECULIAR to it, so on, & so on.-- Whatever destroyed g
257c057 s mêmes." This good case. of replacement under PECULIAR conditions-- of «nearly» same kind country di
265c088 , is instance of bird belonging to family with PECULIAR coloured plumage, where colours have changed
272c108 nsects & some third, have got thighs with same PECULIAR structure & habits of clinging to rushes simi
272c108 the Horae Entomologicae will tell this.-- What PECULIAR conditions the Staphylinidae on St. Pauls Roc
276c124 the whole rest of other family practise with a PECULIAR structure, then Milvulus forficatus Tyrannus
281c138 s in Gallinaceous having tendency to lon[g] or PECULIAR tails strange ¿¿/Genus only natural from deat
284c148 any any other part of world.-- no other plants PECULIAR to these isld.-- ¿Brown can «not» bear the le
295c183 ped in North.-- Icthiology of S. America. more PECULIAR than its ornithology X p. 12 do. excepting sa
299c193 annel Islds (& probably Isle of Man) no plants PECULIAR to themselves. this remarkable compare it wit
317c251 l the hares West of <all> Rocky Mountains have PECULIAR character in extreme length of ears & length
318c252 e species. & all hares on East side have other PECULIAR appearances-- Now this is precisely the case
327c265 ch each one is found. very good to see whether PECULIAR plants-- in high points Read Volney's travels
333d005 wrinkled-- fur short. (tail cut off in progeny PECULIAR) limbs very long, eyes very large, very fierc
334d013 re crossed the neck is not intermediate in its PECULIAR long neck, but much nearer to common goose.--
345d044 Shepherd dogs, coloured like Magellanic Fox.-- PECULIAR hair & appearance-- good case of Provincial B
346d048 ccount of wild cattle of Chillingham,-- habits PECULIAR,-- young one 203 days old butted violently. &
353d060 wings for flight-- therefore ostrich not. The PECULIAR <Malacca> «Malacca» bears, <are> belong to sa
358d074 ubtless «part of» system of great harmony. The PECULIAR character of St. Helena.-- contrast with otah
379d151 . Fresh Water Fish!! ¿adapted to salt water?-- PECULIAR species, crabs & molluscs few.-- ¿are not som
384d161 disposition to deviate from Nature in a manner PECULIAR to itself" <Is this so> Each part <not> of ea
402e019 .-- Many Europaean insects-- list given,= some PECULIAR <M> p. 359. At Manilla a small Cercopithecus.
402e020 atives of S. Asia, but many of the species are PECULIAR to them" do-- p. 368. "Several kinds of anima
404e025 echanism of respiration, or rather ventilation PECULIAR to <the class> «some orders» of crustacea, on
414e064 ce. intellect in Australia to the white.-- The PECULIAR skulls of the men on the plains of Bolivia--
421e091 nnals of Nat. Hist 1838 on «a» freshwater fish PECULIAR to Ireland. ‖do p. 283. on the dark ears of t
424e100 ere South American.-- several cases of species PECULIAR to separate islets.-- March 5th. Lyell says «
425e104 . he believes, there are, many cases of genera PECULIAR to the group having species peculiar to the s
425e104 of genera peculiar to the group having species PECULIAR to the separate islands-- In his work on the
425e104 ee if this can be generalized.--» islds., have PECULIAR forms.-- on the southern flanks of Alps.-- ma
425e105 orms.-- on the southern flanks of Alps.-- many PECULIAR plants on single mountains, though these are
464tf6r of agreement with. N. America & S., (¿ is the PECULIAR. N. American form)-- ¿Hunting leopard, how st
509q017 whether children, who have not worked have any PECULIAR configuration.-- Hooker <Meta> Metaphysics of
514q21. anic? Applies to my geology & Species theory-- PECULIAR Fauna?. {Australian Alps--; are any Europaean
522m013 uite distinct, different also from delirium, a PECULIAR complaint stomach not acted upon by Emetics.-
539m082 siasm, happened to go close to one & smelt the PECULIAR smell of Picture. association with much pleas
541m092 open ready to bark, & lip twisted up, in that PECULIAR manner they do, even more than in a real snar
542m094 nearly silent, so with men.-- How is crying-- PECULIAR not common?--» no bark of anger nor have monk
542m096 teeth.-- A dog when he barks puts his lips in PECULIAR position, & he holds them this way, when open
544m102 nner <they> quite different from when awake.-- PECULIAR sensation as flying. (No memory of past event
565n009 arrogance in upper lip. <The> Children having PECULIAR expression is remarkable. the pouting, & blub
585n075 e only do know that it is one, when applied in PECULIAR manner.--]CD April 3d. 1839 The Giraffe kick
611o034 capability of such matter obeying a certain & PECULIAR system of movements. different from inorganic
623o051 icial tendency to them, as <social> animals of PECULIAR <kinds> «social feelings», & living under cer
635j56r 12. Macculloch explains the shortness of life (PECULIAR to each species) owing to the growing size of
640j167 deduce why small islands. should possess many PECULIAR species. for as long as physical change is in
149g031 variation which have lately acquired their PECULIARITIES? The slope of A & B regular & even towards
187b064 does individual die, to perpetuate certain PECULIARITIES, (therefore adaptation), & to obliterate ac
190b076 t remarkable observation of Mr Brown, about PECULIARITIES of Flora. on East & West. ends of New Holla
264c083 ian forms amongst birds Java. not so much-- PECULIARITIES of structure. as six fingered people are so
279c133 mphlet-- most important, showing effects of PECULIARITIES being long in blood.-- ++ thinks difficulty
335d016 ew> is <only> able to transmit «only» those PECULIARITIES, to its offspring, which have been gained s
344d042 rier & her husband was pure Harrier.-- «The PECULIARITIES of our breeds must have been acquired, & he
360d093 <to tr> for both parents to transmit there PECULIARITIES; that if <one had a> both had mottled breas
362d100 ears-- (how many generations) the strangest PECULIARITIES have been kept perfect-- also to trace the
363d102 t can readily be told by incubation & other PECULIARITIES.-- (Mem.-- Goulds Willow Wren.--) (Goulds s
366d108 hem from long permanence, so that all their PECULIARITIES must be transmitted if their [not located]
370d115 might as well be attempted to be shown from PECULIARITIES of climate cause of N. Zealand not having a
372d129 a bud may be transplanted & carry all these PECULIARITIES not so a seed.-- Bud probably is like cutti
378d148 cies transmitting so much longer its Mental PECULIARITIES. Wildness Reversion Q [not located] The pre
401e017 in features on offspring. & male the lesser PECULIARITIES.-- brilliancy of inflorescence Gardeners. b
441e148 oducing bud.-- Fewer of the lately acquired PECULIARITIES are transmitted it is doubtful whether any
473s07r «as in trades» there is no reason, why the PECULIARITIES shd be born,-- may come in corresponding ti
473s07v te in youth,-- only facts can decide-- some PECULIARITIES may be early impressed & others later-- All
507q15v er many mono or Dioeious plants, & any with PECULIARITIES of structure rendering cross impregnation d
530m045 re slower than those of gentlefolks. & that PECULIARITIES of form in trades (,as sailor tailor blacks
611o034 matter is united in different modification, PECULIARITIES of external form impressed, & different law
615o038 aving very different instinct always obtain PECULIARITIES of external configuration. [RHC] General--
623o051 by the circumstances, which have led to the PECULIARITIES, & hence <must have> «only that which» had
218b192 these <African forms> forms would have some PECULIARITY.-- Now when we hear that the whole island is
258c061 to the chance offspring who have any slight PECULIARITY of structure. «hence seals take victorious s
259c063 ocess of change.-- Are any men born with any PECULIARITY. or any race of plants--Lamark's willing abs
336d017 parent cannot transmit to its offspring any <PECULIARITY> change from the form which it inherits from
342d035 ose". if this be not imagination.-- then old PECULIARITY overbears the crossing with females not thus
387d168 x» male impresses ovum <of> in A, «with some PECULIARITY» that in (B) to <a slight> «some» degree, &
387d168 (C) unites with Male (X) «assume that every PECULIARITY has a tendency to descend to several generat
387d169 ons» If A & B be two animals which have some PECULIARITY for first time & if their <D & E> «all their
387d169 & E> «all their offspring» inherit the same PECULIARITY in lesser degree C. & theirs again in lesser
387d169 --now if the <tw> second race both have this PECULIARITY strongly; they transmit with same force as f
387d169 rst pair.-- Now if two of third pair of same PECULIARITY breed they will have same influence as first
393dIBC hat «hybrid» breed is uncertain Is there any PECULIARITY or variation common to any zoophyte «born in
402e018 Here, as in all Malay countries, I noticed a PECULIARITY in the cats «p 10» the joints near the tip o
421e091 the tame ones.-- ‖an instance of a trifling PECULIARITY not to be eradicated.-- ‖.do. p. 305.-- Mr O
453e183 cribed Atavism.-- ask Dr Holland cases where PECULIARITY has first appeared.-- "Storia della Riproduz
466t095 ned to sex.-- «Is not cantering a congenital PECULIARITY improved.» Probably every such «new» quality
473s07r corresponding time of life of offspring-- No PECULIARITY in external structure can be concepcional, a
507q016 nheritable.?-- Bell cd ask Accouchers Is any PECULIARITY in milk teeth inheritable!!! very good Any p
507q016 y in milk teeth inheritable!!! very good Any PECULIARITY in the males of a family-- Where one tooth a
575n043 that the grand child dug for mice from some PECULIARITY of structure of brain.?-- is this more wonde
621o047 tion» comes to this conclusion, not owing to PECULIARITY of organ of taste, for when grown up often c
633j53v ccidents may become heredetary [produce some PECULIARITY in seed vessel]CD if man takes care they are
420e089 a tail, shows man was originally quadru<manous> «PED.-- ». Hairy.-- could move his ears The head bein
067r143 2 Vols fol: Buenos Ayres 1836: W. Parish?? «by PEDRO de Angelis.» This work is reviewed in present E
641j29v is gained by some very different method. in PEDUNCULATED eye of Chamelion. crabs Crabs & Mollusca we
362d099 , replaced by hairy ones. which never «dry up &» PEEL off their skin (not being wanted for war) & hen
364d104 back-- (Effect of imagination on mother. white PEELED rods mentioned in old Testament placed before
545m106 » I gave nut, but held it between fingers, the PEEVISH expression was most curious <like> «remember»
555m142 erceive that the American ones, often put on a PEEVISH expression, but not nearly so often <that> har
450e175 hands-- those of Lao & Siam inferior to those of PEGU-- in Sumatra two breeds both small -- Java pony
451e177 y theory Cocos Isld & Preparis between Andaman & PEGU. <have> abound with monkeys & squirrels.-- Hors
450e175 tting Bengal this fact is noticed in Cassay Ava PEGUE-- seldom equals 13 hands-- those of Lao & Siam
268c099 ated] element geographical distribution is.-- ¿PELAGIC forms similar--birds??-- We must always bear i
242c017 in common‖-- Galiopithecus common to Moluccas & PELEW Isds.-- p. 22. New Calidonia-- New Ireland & Br

```
*********************************************(Key Word)*********************************************
247c028 hat Crocodile was washed on shore at one of the PELEW Islds.-- killed a woman. Chamisso p. 189 Tome I
346d047 . Was struck with pink shade on plumage of the PELICAN.-- Mem pink spots on Albatross, on some Gulls.
500q10a & seeds gathered there were planted «last year» PELL mell, without sticks & seeds gathered & these a
496g05a after feeding on oats, give body to Hawk & sow PELLET. ejected. done Examine pollen of such flowers
512q019 wks eat stomach. of finches-- do they throw up PELLETS-- (4) About hybrid pheasants treading-- any tr
403e021 gaut. Kind of crocodile, sometimes wanders from PELLEW to Eap.-- There is another great lizard. Kaluz
403e021 another great lizard. Kaluz. which is found at PELLEW & Eap, but not at Feis (near island) do p. 190
097a043 s of Trachyte, "superficially coated by a thin PELLICLE of a blackish colour like a dull & poor varni
033r042 ndstone at Ascension, each particles coated by a PELLUCID envelope of Lime.--form resembles the husks a
477z002 scacha.-- A. Patagonicus les tatous (.4 pichye, PELUD, mulita et mataco.) are all found south of 26 d
479z006 1 D'.Orbigny considers Dasypus villosus is true PELUDO Cavia Australis. Dorbigny Vol II, p 24 Proceed
341d029 closer).-- Mr Blyth asked whether structure of PELVIS & was not adaptive structure, like little wing
134a141 iso to Santiago p. 328. dead trees on Isthmus of PEN. Tres Montes.-- as by subsidence ‖ Fitz Roy refe
535m064 he» eat & enemies-- would not fly away, but bit PENCIL when touched with it-- do not know their own l
429e116 each side of the great N & S. valleys, which PENETRATE Pyrenees but are found no where else not even
068r146 igneous rocks which would be too thick to be PENETRATED by the repeted trifeling injections.--Old v
085a002 ater. Mr. Arrowsmith tells me, that Himalayas PENETRATED like Bolivian Chain. Volcanic islands. from
125a121 e interior -- who knows how far that may have PENETRATED,-- lower down the temperature may be kept up
133a140 od: now when we see how many points have been PENETRATED by volcanic & trappean rocks, within say the
048r084 rom chemical attraction, as a blade of grass PENETRATING by action of Organic power a lump of hard cl
077r171 ungary.) Humboldt says fragments from roof & PENETRATING overlying beds tells the secret.-- p. 189. "
113a092 -- Probably a case of rivers turning round & PENETRATING their own range in Australian Alps.-- Taylor
233b251 birds S. Continents-- Ostriches. Dodo. Apteryx PENGUIN-- Logger-headed Duck-- Large proportion of Wat
287c157 observer‖ +++. between Mammalia & fishes, one PENGUIN, one tortoise shows hiatus-- but not saltus--
289c161 o each other let him look at wings & orbits of PENGUIN & then he will cease to doubt :Scales into Tee
341d029 wings of Auks which does not make that bird a PENGUIN.-- (i.e. whether relation in one point or many
341d032 diate.-- (In Zoolog Gardens there is hybrid of PENGUIN duck a variety of Muscovy) with goose!!) Dr. B
359d087 ome mistake in their origin) Saw cross between PENGUIN Duck «from Bombay» & Canada Goose.-- Former st
359d088 ny one nest. even more than common duck-- Male PENGUIN was crossed with hen Canadian offspring, I sho
359d088 pring, I should say in every respect most like PENGUIN duck.-- which is strange anomaly in Yarrells l
360d091 offspring most.-- «& not time» thinking of the PENGUIN duck & Herberts law of ideosyncrasy I have hit
459t013 black & white <duck of pecu» «drake» with the PENGUIN duck. it took after the Penguin in the form of
459t013 rake» with the penguin duck. it took after the PENGUIN in the form of its body & in the manner of wal
459t013 t not waddling; its colour was darker than the PENGUIN & the bright feathers on its wing resembled th
459t013 esembled the plumage of drake still more.-- So PENGUIN impresses its form both on vars & species The
494q004 effect of crossing.-- (7) Are the Eggs of the PENGUIN Duck quite similar to those of another Duck. ¿
506q014 mine Keel of Common & Wild Duck-- Black Duck & PENGUIN Henslow &c (36) Has not H. raised races of whi
513q020 ries= Do the Peacocks cross.-- Young Chinese or PENGUIN Duck in very young state for skeleton== Does t
639j28r un by hind legs-- Kangaroo. only a caricature; PENGUIN.-- Pincers in Scorpion & Crust in Squilla. & M
177b025 t--> settles it ¿We need not think that fish & PENGUINS really pass into each other.-- The tree of li
352d059 e shows us it was for.-- Most important law.-- PENGUINS wing perhaps not abortive???. Apterix certain
484z015 fore writing my general account-- ¿ Do not the PENGUINS replace the <Auk> Guillemost of the northern
089a020 s on the Senegal. L Institut p. 192.-- (1837. PENINSULA of Cape Verd. volcanic.-- Isle of Gory. rocks
266c091 em.-- I would [not located] Musalman's of the PENINSULA, are, generally speaking, a much fairer race
450e174 c are found but only in one part the northern PENINSULA of Borneo.-- Ox & hog natives of Borneo Notic
363d101 arieties.-- Habits of rock pidgeon. (I suspect PENNANT has described them)-- [Study horns of wild cat
478z004 . worth reading Cuvier's Memoire 133 1803. on PENNATULA showing it to be one animal In Australia I wa
505q014 tle in Hot-house. will it seed?-- (Skim through PENNY Cyclopaedia) Abberley says that some Bees are s
058r118 nture d'immenses remparts; toutes affectent une PENTE plus ou moins inclinée vers le rivage de la mer
118a103 216 M. Gay on the Geology of Chile.-- P p217. PENTLANDS Fossils & Meyens --<Jura &> Chalk When we con
223b208 what would Anatomists have said.-- ¿where is PENTLAND'S account of Bolivian human species?-- Small «
497q06v eeds being sown= See in Cultivated Plants, as PENTSTEMON, which have abortive parts, whether such var
199b119 uguese. priest.--In first settling a country.-- PEOPLE very apt to be split up into many isolated rac
199b119 ated races. ¿are there any peculiar instance of PEOPLE banished by rest?-- ∴ most monstrous form has
206b146 ear) degree. Then 200 years ago, there were 200 PEOPLE living who now have successors.-- Then the cha
206b146 o now have successors.-- Then the chance of 200 PEOPLE <might be> being related within 200 years back
206b146 be calculated & this treatment say 150 PEOPLE four hundred years since were progenitors of
206b146 hundred years since were progenitors of present PEOPLE, and so on backwards to one progenitor, who mi
216b179 os themselves explain it by intermarriages with PEOPLE. either a little nearer black or white as it m
222b207 shells alone can this comparison be instituted. PEOPLE often talk of the wonderful event of intellect
235b263 Hare bear upon this.-- Why do Van Diemens land PEOPLE require so many, imported animals?-- At what.
258c060 g adaptation owing to crossing, with unseasoned PEOPLE. would cause destruction.-- simile Man living
258c060 g in hot countries, if continually crossed with PEOPLE from cold, children would not become adapted t
264c083 h-- Peculiarities of structure. as six fingered PEOPLE are sometimes heredetary,-- yet these not adap
283c145 ch is strange if creator had so created them.-- PEOPLE will argue & fortify their minds with such sen
293c172 recall the idea. «or simple structure in brain PEOPLE in fevers recollecting things utterly forgotte
293c174 different contagious diseases, where habits of PEOPLE nearly similar. Curious instance of difference
294c177 , only slight differences. «the opinion of many PEOPLE in conversation.» the whole object of the Work
306c213 rked difference with dogs of La Plata & Guyana‖ PEOPLE will say. not species.-- organs of generation
308c217 ously.-- why do absent «Dr. Black. tea & sugar» PEOPLE. reverse habits Insects & birds are the only t
332d001 roof. what <trifling» «unknown» causes act upon PEOPLE. My Father mention, than for ten years he neve
332d001 t see other cases.-- He thinks apoplexy affects PEOPLE all over England at same periods When he began
335d014 amputated limb.-- Heredetary <thr> Six fingered PEOPLE, <Hill» «Lord Berwick» family with defective p
335d015 can have no offspring.-- but as «badly» deformed PEOPLE & as mutilations «(produced very quickly)» som
366d112 alians knocking out teeth.-- the account of the PEOPLE on the NW. Coast blinking to keep out flies mi
423e098 Uncle John says he feels sure, that the reason PEOPLE send for their seeds to London is that people
423e098 n people send for their seeds to London is that PEOPLE in the southern Counties have whole fields, so
520m001 tion.-- My father has seen innumerable cases of PEOPLE taking after their parents, when the latter di
520m002 ld not remember his father.-- My father thinks. PEOPLE of weak minds, below par in intellect frequent
522m012 d not know [Z]CD when heard isolately.-- In old PEOPLE. (Aunt. B.) when they hear a thing it often do
522m013 . says there is perfect gradation between sound PEOPLE and insane.-- that everybody is insane. at som
522m013 complaint. stomach not acted upon by Emetics.-- PEOPLE recognized,-- sudden changes of disposition, l
522m013 cognized,-- sudden changes of disposition. like PEOPLE in violent intoxication, often ends in insanit
522m013 f decency & affection are lost.-- most delicate PEOPLE do most indelicate actions,-- as if «these emo
523m014 ust be directed against somebody.-- Have insane PEOPLE any misgivings of the injustness of their hatr
523m015 matics.-- My Father says after insanity is over PEOPLE often think no more about it than of a dream.-
523m016 auses (ideotcy by fear. Chile earth quakes). in PEOPLE, who, probably otherwise would not have been s
523m018 een insane all the time.-- There are numberless PEOPLE insane of particular ideas, «Case of Shrewsbur
524m019 ifference, between oddity & madness.-- but then PEOPLE do not well recollect what they have done in p
524m019 hell recollect what they have done in passion.-- PEOPLE are constantly well aware that they are insane
524m020 hich my F says are almost constantly present in PEOPLE, likely to become insane.-- now this is well w
524m021 meaning:-- Dreams do not go back to childhood-- PEOPLE, my Father says, do not dream of what they thi
524m022 eir husbands-- My father's test of sincerity.-- PEOPLE in old age. exceedingly sharp in some things,
529m039 of poetry;-- former thoughts, & in experienced PEOPLE-- recall pictures & therefore imagining pleasu
529m040 he train of thoughts vary no doubt in different PEOPLE., an agriculturist, in whose mind supply of fo
531m050 orgotten, (if unconnected with fear &c) because PEOPLE think that the importance of the event by itse
531m050 be remembered. whereas it is the importance.-- PEOPLE very often forget where money is placed.-- (Ho
531m050 olely be number of times repeated, because some PEOPLE can remember poetry when once read over.-- The
532m055 to the mind being more alive.-- How is it. with PEOPLE nervous from illness., <the> it must be an exc
532m055 d.-- My Father says he should think that in old PEOPLE, in their dotage, who sing the songs «& tales»
538m078 little» less perfect than other, absolutely two PEOPLE. Consider this profoundly, may throw light on
538m078 on.--«fully bears out my fathers doctrine about PEOPLE forgetting their insanity» there seem other ca
538m078 ogous, & which I think will lead to fact of old PEOPLE singing songs of their childhood.-- & certainly
538m079 ecurs, such as -- -- of Trinity always thinking PEOPLE were calling him a bastard.-- when drunk.-- ha
540m089 as he remarks sympathy could be barren. & lead PEOPLE from scenes of distress.-- see how a crowd col
541m090 st-- therefore dreams thus act.-- ∴ weak minded PEOPLE are fickle & full of levity (¿ do I not confou
541m092 t so) than if simple idea as scarlet?-- How can PEOPLE dwell on pain ¿ no definite idea. nor is an em
541m092 n pain ¿ no definite idea. nor is an emotion.-- PEOPLE who can multiply large numbers in their head m
```

Page **(Key Word)***
541m092 ead must have this high faculty, yet not clever PEOPLE. Aug. 21st. 38 When a dog in play has his mout
543m099 heir associations (it is not so in punning) are PEOPLE of very limited intellects, & in the same way
550m124 ns in any given time «-- compared to what other PEOPLE experience.--» But then sensation may be more
551m127 ality of the impression on its senses.-- insane PEOPLE believe they hear as well see things which hav
557m149 art.-- tendency to muscular movement, hence shy PEOPLE (shame of ridicule) are singularly apt to catc
557m149 e) are singularly apt to catch tricks.-- so are PEOPLE in passion my F. rubbing hands.-- stamping. gr
558m151 ne can doubt this connexion.-- look at faces of PEOPLE in different trades &c &c &c I observed the As
560m156 bout double consciousness.-- & somnambulism. Do PEOPLE when inhaling Nitrous oxide, forget what they
574n040 a Cocoa nut shell at one end.-- Children & old PEOPLE get into habits.-- we probably can hardly form
575n046 winged.-- fetches two birds out at once.-- Old PEOPLE-- (Antiquary Vol II. p. 77) remembering things
577n052 sex.-- Hence, animals. not being such thinking PEOPLE. do not blush.-- sensitive people apt to blush
577n052 such thinking people. do not blush.-- sensitive PEOPLE apt to blush.-- -- The power of vivid mental a
578n057 les, these convulsive actions-- (except in weak PEOPLE & hysterical people inclined to convulsive act
578n057 e actions-- (except in weak people & hysterical PEOPLE inclined to convulsive actions).-- But, the la
578n057 But why does joy & OTHER EMOTION make grown up PEOPLE cry.-- What is emotion? At end of Burke's essa
579n058 ry believes in.-- From the manner short-sighted PEOPLE frown, frowning must have some relation to sh
579n058 on to short-sightedness.-- do not short sighted PEOPLE squinny-- when they consider profoundly,-- thi
580n060 erence to origin of language My father says old PEOPLE first fail in ideas of time, & perhaps of spac
580n061 t he thinks he certainly has observed that some PEOPLE of very weak intellect (As Miss Clive) have on
580n061 the mind, no subsequent ones modify it.-- «Weak PEOPLE say I know it because I was always told so in
582n066 as analogous to glowing conversation of several PEOPLE.-- Children have an uncommon pleasure in hidin
582n066 elves & skulking about in shrubbery. when other PEOPLE are about: this is analogous to young pigs hid
587n087 ave of Rhubarb: again on other hand, it is said PEOPLE, who like sweet things dislike others.-- dogs
592n104 n muscles of face, is well seen in shortsighted PEOPLE.-- hence origin of expression-- There are some
596nIBC blood go in <body> face in pashion.?-- cry? Do PEOPLE of weak intellects easily fall into habits Get
599o05v ords invented-- we see their origin in names of PEOPLE.-- Sound of words-- argument of original forma
605o020 beauties of nature & art." But as we often see PEOPLE who are susceptible of pleasure from these cau
610o032 to their health & comforts.-- Both ideots, old PEOPLE & those of weak intellects.-- Westminster Revi
190b077 hern hemisphere, does not look as if S. africa PEOPLED from N. Africa An originality is given (& powe
343d036 instincts alter, reason is formed, & the world PEOPLED «with Myriads of distinct forms» from a period
554m139 ena & Pocket Handerchief & liked the taste of PEPPERMINT.--» Perfect understand voice.-- will do anyt
560m156 s readily?? Do the Ourang Outang like smells «PEPPERMINT» «& music».-- Have monkeys lice?-- picture.-
063r130 one species does change into another it must be PER saltum--or species may perish. = This <inosculat
104a066 . Andes discussion-- Albite certainly contains 6 PER cent more silica than common felspar therefore o
104a066 on axis of Cordillera, in Andite - containing 80 PER cent of Albite 80/100 X 6/100 = 480 In Falkland
236b278 son [blank] In production of varieties is it not PER saltum.-- Isld bordering continents same type co
263c076 lary point in ovum. has such organization as to <PER> force in one man the developement of a brain ca
269c102 ders Voyage by Brown.-- great space seems to act PER se as barrier-- Mem. Tartary & China.--, both co
351d056 leaves vestiges of what she does-- does not move PER saltum-- yet does nothing in vain!! Foetus of ma
527m033 from instinct the ears (rhythm & pleasant sound PER se) & causes the mind to create short vivid flas
534m061 ience, or instinct. Hensleigh says to say. Brain PER se thinks is nonsense; yet who will venture to s
232b248 s who soar above the pride of the savage, they PERCEIVE the superiority of man over animals, without
332d004 uly 23d. Eyton, a stone blind horse, seemed to PERCEIVE turn on road where No houses to Eaton Mascott
365d107 ptation to unhealthy state of womb).-- One can PERCEIVE that Natural varieties or species., all the s
545m106 otion & <look> «turn» of the head. I could not PERCEIVE «any» distinct wrinkle, but such movements in
555m142 classify expressions of monkeys-- I could only PERCEIVE that the American ones, often put on a peevis
565n007 tening is mouth open to hear well «as one will PERCEIVE if in night trys to listen to growl of hounds
584n073 ers seems sign of helplessness E. says she can PERCEIVE sigh, commences as soon as painful thought cr
588n089 tinctive fear of falling.-- &p. 193. that they PERCEIVE the difference on being carried up or downsta
588n090 oung animal with an old.-- (dog horse, sow) we PERCEIVE great difference.-- «(& is not this differenc
604o017 ly excited by passive emotions.-- Cannot quite PERCEIVE drift of Book.-- Sympathy & affections chiefl
606o021 becomes so instantaneous. that we cannot ever PERCEIVE the various operations which the mind undergo
614o037 ate in life, man almost alone in this case can PERCEIVE instinct. boy takes delight in mammae before
618o041 orant of the existence of the brain. We cannot PERCEIVE the thought attraction of sulphuric acid for
375d135 il the one sentence of Malthus no one clearly PERCEIVED the great check amongst men.-- «Even a few ye
524m022 ting that her husband was dead) yet instantly PERCEIVED when my Father to distract her attention took
534m062 tever takes place in brain. when sensation is PERCEIVED.= = Aug. 7th--38. Transactions of the Entomol
616o039 equal propriety be said that the living brain PERCEIVED, thought, remembered &c. Well the heart is sa
618o041 force does to the bodily frame, they could be PERCEIVED by the faculty by which the brain is perceive
618o041 erceived by the faculty by which the brain is PERCEIVED but they are known by courses of action quite
624o50v rm of Reflex Senses seems to have <compared> «PERCEIVED» the comparison between our instinctive feeli
625o50v n be said to be wrong.--. for then only is it PERCEIVED, that our passions are too strong for our ins
548m117 not there, & then when one begins eating one PERCEIVES butter or salt is not there.-- the reality do
536m070 Pincher does just the same; I noticed this by PERCEIVING myself skipping when wanting not to feel ang
543m100 n sense, & the highest intellectual powers of PERCEIVING & classifying distinct resemblances.-- The f
124a119 gmous action.-- if the zinc were mixed with 90 PERCENT of lead. it would be still more curious to kno
181b041 he <descen> fathers would be reduced to small PERCENTAGE.-- & <in> therefore the chances are excessiv
499q010 s.-- (Paris) (22) When Linnaeus says so great PERCENTAGE of seeds have contrivance for transportal, d
538m080 ain having whole train of thoughts, feeling & PERCEPTION separate, from the ordinary state of mind, i
547m113 impression.-- (do the ideas, direct effect of PERCEPTION by senses fail first, as whether I had pulle
559m154 oking laws. & giving rise at last even to the PERCEPTION of a final cause.-- Read. Paper on conscious
587n089 votional feeling p. 103-- Abstraction p. 152. PERCEPTION very different from emotion.-- The former is
605o020 what means it becomes an almost instantaneous PERCEPTION.-- Taste has been supposed by some to consis
618o041 frame or the cerebral portion of it Thoughts, PERCEPTION &c. are modes of subjective action-- they ar
544m103 willing therefore is ideal, as all the other PERCEPTIONS.-- The mind thinks with extraordinary rapidi
552m130 se one then has no immediate comparison with PERCEPTIONS, & that on[e] fancies the image more vivid?
616o039 rs. [1)] Why may it not be said that thought PERCEPTIONS will, consciousness memory &c. have the same
618o041 nalogous to this in the relation of thought, PERCEPTIONS, memory &c. either to our bodily frame or th
616o39v ide we can only <shew> point out the mode «of PERCEPTIVE action» by which we come to conceive of matt
616o39v tributed to the brain. There are two modes of PERCEPTIVE action by which bodily action is made known
436e138 f each other will sing till they drop off their PERCH.-- p. 101-- Kingfisher in northern part of Engl
615o36v prowl about in the evening «seldom leave their PERCH till evening» crow different.-- Heredetary effe
273c111 e air, & likewise rasorial species, & likewise PERCHING (Gould), but the latter is obscure because ne
096a040 scension for centuries afterwards it might be PERCIEVED on which side craters were low --¿ applicable
416e071 sticated» races. of <a> organics. are made by PERCISELY same means as species-- but latter far more p
126a123 rium is not attained, & if cold water did not PERCOLATE surface, would become hotter.-- hence tempera
126a123 olumn.--» of cold water show, that water does PERCOLATE, & springs beneath sea-- → According to this
126a123 r than plain?-- Mem 1000 {P} how easily water PERCOLATES rocks,-- when pressure increased or under su
296c185 h the tame Vol II. Magazine of Zoology p. 56. PEREGRINE Falcon holds birds for some time alive ¿there
504q014 xes-- (3) Get Holyhoaks. races planted & Linum PERENNE.-- Herbert's. fact.= (4) Effects of Nitrate of
171b002 rently only born to breed.-- annuals rendered PERENNIAL. &c &c.-- Yet Eunuchs nor «cut» Stallions nor
039r061 the envelopes at Coquimbo. the analogy is now PERFECT <The grand propulsion of fluid rock, which ele
047r083 rtionally Partial shrinking after elevation in PERFECT conformity with <Mr Lyell's> idea of an inject
049r089 nta> passage from lava to Granite is much more PERFECT. than in believing mere agency of dikes: & ind
063r132 dds one other method where the division is not PERFECT.-- Dogs. Cats. Horses. Cattle. Goat. Asses. ha
063r133 Asses. have all run wild & bred. no doubt with PERFECT success.--showing non Creation does not bear u
155g066 uld be» EQUALLY preserved 2d or upper one more PERFECT in this <part> «glen» than 3d. 3(a) less perfe
156g067 rfect in this <part> «glen» than 3d. 3(a) less PERFECT than upper & lower but quite as perfect as tho
156g067) less perfect than upper & lower but quite as PERFECT as those lines in Glen Collarig, & some «other
160g089 ing E & W connecting Glen Bought & Glen Tarf a PERFECT old Loch, making <several> two divortiums aqua
171b002 ts, fails in hybrids where every thing else is PERFECT; mothes apparently only born to breed.-- annua
173b009 ording to Lamarck disappear as collection made PERFECT.-- truer even than in Lamarck's time. Gray's r
175b019 proving as Owen says simplest coming in & most PERFECT «& others» occasionally dying out;--for instan
192b085 n without them.= In some of the lower orders a PERFECT gradation can be found from forms marking good
192b087 ecies probably the series would have been more PERFECT. because in each there is possibility of such
192b088 ps.-- I may ask whether the series is not more PERFECT by the discovery of fossil Mammalia than befor
221b200 ine forms, longest persistent?? do.-- The most PERFECT Plants Composites.--!!«good» those which have
222b204 that one stock cannot be supposed to be «most» PERFECT (according to our ideas<)> of perfection); but

Page
(Key Word)

```
222b205 rhaps is) Case with fish-- as some of the most PERFECT kinds the shark. lived in remotest epochs.-- ¿
223b208 ard to draw line.-- -- not so great as between PERFECT insect & former hard to tell whether articulat
223b209 land with Tasmania The reason why there is not PERFECT gradation of change in species, as physical c
230b239 or horizontally & then cross breeding prevents PERFECT change.-- It is scarcely possible to get evide
274c115 gans change together, in analogy certain parts PERFECT of typical structures certain <imper> parts ch
282c141 yet altogether different.-- To make this case PERFECT, we must suppose men instead of mere colour  &
291c168 re all fathers of present, then there would be PERFECT series or gradation.-- It is easy to see if So
293c174 beautiful.-- The corelations are not, however, PERFECT, else one animal would not cause misery to oth
293c175 of the eye, passed from simpler stage to more PERFECT. preserving its relations.-- the wonderful pow
303c205 llinaceous birds & parrots.-- legs of pidgeons PERFECT.-- &c &c.-- do p. 136. Ichthyosaurus in the Ch
303c205 me of the future creations, if the list is now PERFECT.--]CD. the creator so creates animals, «, it w
303c205 gaps. yet <what> altogether «he» has created a PERFECT chain. <Icthyo> .+ + +. supra & next page It i
304c206 NG.? that intermediate. species have generally PERFECT organs. do changes of habit affect  particular
316c244 o greater difficulty for Deity to choose, when PERFECT enough for future state, that when good enough
333d009 diate.-- Where two dogs line the same bitch & «PERFECT» spaniels & setters are produced. one would ar
335d015 cause not long in blood.= The case of union of PERFECT animals is distinct case,-- gradation from phy
336d019 ken, without it either went back, or not being PERFECT would perish.-- The Varieties of the domestica
343d036 their bodies, by certain laws of harmony keep PERFECT in these themselves.-- instincts alter, reason
347d050 sticks to genus or group of any kind not being PERFECT till circular. p. 5 Most clearly shows that ge
352d057 ined by age of foetus.-- As Larvae may be more PERFECT (as we use the word) than parent, so may speci
352d057 organs, might be expected to have larvae more PERFECT-- this is applicable to young of Cochineal?? I
362d099 Axis of India, breeds at times when horns not PERFECT-- (is not this so in S. America with C. Campes
362d100 ns) the strangest peculiarities have been kept PERFECT-- also to trace the laws of change in this tim
366d111 squitoes of S. America take an effect.-- would PERFECT impunity from muskitoes bite influence propaga
368d114 mage.-- «thinks» Hence specific character most PERFECT in <male> «hermaphrodite» (Fishes have no seco
385d164 n Vol I of Irish Royal Academy) have contained PERFECT teeth & hair-- showing foetus has gone on grow
398e005 erved.».-- the latter pages in the history are PERFECT, we obtain a glimpse only of the changes which
404e025 at  once.-- but naturalists if they had series PERFECT, would expect this structure would become obsc
409e048 effects. -- we see it is not the order in this PERFECT <uni> world, either at the present, or many an
414e065 ruder.-- : the geological history of man is as PERFECT as the Elephant, if some genus. holding same r
418e080 rodite insects is it not easier to understand ¿PERFECT?? development of one sex on one side, than th
418e080 in which case the hermaphroditism would not be PERFECT as in Ox. the amount of double sexual develope
426e106 s would cross easier with 2d species, than two PERFECT species; but facts of grouse, & pheasant, & ho
426e107 animals  were crossed, there would probably be PERFECT series, from physical impossibility to unite t
426e107 eries, from physical impossibility to unite to PERFECT prolificness.-- (<a series might be obtained>
436e135 ased since earliest times-- Apteryx has a most PERFECT Struthio head pulled out. yet feathers  retain
444e157 70 bones in the paddle, yet all in the arm are PERFECT.-- p. 144.-- Alludes to two theories;-- that s
444e158 ormations from the egg, or larva. or foetus to PERFECT animal are adapted by foreknowledge, so must t
454e184 ds? Any «wild» plants in England, which do not PERFECT their seed?-- What annuals can be budded «& re
465t081 ers vary; so that we here see reasons-- why no PERFECT gradation can be expected in any one country.--
522m013 at not being understood.-- My F. says there is PERFECT gradation between sound people and insane.-- t
525m025 efore & afterwards with tails. My father says, PERFECT deformity, as an extra number of fingers.-- ha
538m078 f double consciousness, one only «little» less PERFECT than other, absolutely two people. Consider th
538m078 repeating song» when she had recollected it in PERFECT senses.-- These things, & drunkedness, show wh
546m110 oughts In same book (p. 143) wonderful case of PERFECT double consciousness Mayo compares it with Som
546m111 to party, went out & found it there!!! Lady in PERFECT «mental» health.-- «Erasmus had almost same th
549m123 slowly  vanishing-- the mind of man is no more PERFECT, than instincts of animals to all & changing c
554m139 anderchief & liked the taste of Peppermint.--» PERFECT understand voice.-- will do anything.-- will t
578n056 ild (different in different ones?) in the most PERFECT fainting, sphincters are loosed is a convulsiv
578n056 ut again-- as power of mind by habit gets more PERFECT over voluntary muscles, these convulsive actio
579n059 nfluenced by» other emotions? When a man keeps PERFECT. time in walking. to chronometer, is seen to b
608o027 know  his happiness lays in doing good & being PERFECT, & therefore will not be tempted, from knowing
616o038 into  angels. this imperfect memory may become PERFECT & we may look back to definite action or to ou
638j58v one of ten, thousand trials.-- each step being PERFECT «or nearly so (except no in isd) although havi
638j58v ened, but yet analogous, no savage ever made a PERFECT hinge.-- reason, & not death rejects the imper
416e075 newly acquired structure is fully practised & PERFECTED Hence difference between races & variety? «Ma
222b205 on, whether whole scale of Zoology may not be PERFECTING by change of Mammalia for Reptiles, which ca
021rIFC ergone the greatest number of changes towards PERFECTION (namely mammalia) must have a shorter durati
039r061 n explaining plains because such are found in PERFECTION on that side.-- Add from M. Lesson. Characte
177b027 rding to shortness of life of species that in PERFECTION, the bottom of branches dead.-- so that in
177b029 some connection shortness of existence, <in» PERFECT<ION>, «species from many» <therefore> changes an
222b204 «most»  perfect (according to our ideas<)> of PERFECTION); but intermediate in character, the same re
254c047 n, afford the least certain indication of the PERFECTION of the Species--! How does this agree with g
269c103 ffinity" in Scientific Memoirs I can see that PERFECTION may be talked of with respect to life genera
350d055 - Macleay observed all these facts prove that PERFECTION of organs have nothing to do with perfection
350d055 perfection  of organs have nothing to do with PERFECTION of individual, though such relation seems co
350d055 , though such relation seems common, but that PERFECTION consists in being able to reproduce Here the
370d116 mportance of divisions in arrangement, of the PERFECTION of their separation.-- thus Vertebrate blend
382d158 different, are both present in every shade of PERFECTION --How came it nipples <are> «though» abortiv
513q020 Does the tumbling of pigeons vary in manner & PERFECTION &c &c &c--if so probably a variety, not spec
602o010 olds X discourse very curious as showing "the PERFECTION of this science of abstract form" is the sou
604o016 as  if consciousness an effects of sufficient PERFECTION of organization & if consciousness, individu
636j57r morant: pain & disease in world & yet talk of PERFECTION Get instances of adaptations in varieties.--
113a092 xcavated in solid rock.-- 4 & 5 fathoms deep. PERFECTLY still water. Major Mitchell inferred subsiden
149g030 corresponding as in Andes, composed of sand & PERFECTLY rounded stones-- lake required to deposit thi
149g032 «& on Chean side» of Meal-- Derry there were PERFECTLY rounded «base» pebbles of quartz & other rock
156g072 ined» gneiss <unite» «occurs» abundantly with PERFECTLY rounded pebbles of granite & forming «sloping
160g093 terrace «of pebbles? & Alluvium» which appear PERFECTLY level, <on op> dies away on gradual slope-- :
202b133 in  Africa &c &c.-- Now if suppose world more PERFECTLY continental. we might have wanderers. (as Pec
312c228 ess dog & common spaniel crossed.-- 3 puppies PERFECTLY like chinese & 3 perfectly like spaniel even
312c228 ossed.-- 3 puppies PERFECTLY like chinese & 3 PERFECTLY like spaniel even when grown up.-- Are  mules
333d008 take much most after Esquimaux.-- this agrees PERFECTLY with Yarrell & no leading question was put.--
366d111 e she was not coming in, could not help being PERFECTLY distracted «Referred to <other> Book M.» Is t
411e054 of  correlation of parts & organs it may serve PERFECTLY to specify types, & limits of variation., & h
412e057 umerable variations, as long as each shall be PERFECTLY adapted to circumstances of times. & from per
416e071 y same means as species-- but latter far more PERFECTLY & infinitely slower.-- No domesticated animal
416e071 nfinitely slower.-- No domesticated animal is PERFECTLY adapted to external conditions.-- (hence grea
416e072 n inheritance & in species are only ancient & PERFECTLY adapted races I'Institut 1838. p. 394. Rhinoc
431e122 ook for change, because its number show it is PERFECTLY adapted; it where few stray ones. are, that c
467t103 pine,» two wings. & when the Lupine flower is PERFECTLY ripe & pollen abundant filaments & stamens al
528m037 in the distance.-- & even on paper two waving PERFECTLY parallel lines are elegant.-- Again there  is
547m111 & wakened.-- had very clear & pretty vivid «& PERFECTLY characterized» dream, in continuation of waki
550m124 rm of Baboon is our grandfather!-- A man, who PERFECTLY obeys his conscience or instinct, would proba
564n005 in man to guide to knowledge, was transmitted PERFECTLY to chicken so as to seize small moving object
569n020 pen your mouth said, recognizes that sound as PERFECTLY as a man.-- Probably, language commenced in s
602o11b oof for, & one often replies "what you say is PERFECTLY true, but you do not convince me.--" Belief a
623o050 ctive. But does not Hartley explanation apply PERFECTLY to origin of these instincts.-- the having re
633j54r tas.-- Now my theory makes all organic beings PERFECTLY adapted to all situations, where in accordanc
054r105 rocks in the most inland part of this bay are PERFORATED & smoothed like those washed by the waves, a
229b233 here is division of snakes. with hinder teeth PERFORATED for poison channels, but not having them, in
352d059 changed.-- what is abortive? when it does not PERFORM that function which experience shows us it was
382d158 bject to Hermaphroditism,-- those organs which PERFORM nearly same function in both sexes.-- are neve
382d158 - are never double, only modified. those which PERFORM very different, are both present in every shad
539n081 man wagging his foot & working with his toe to PERFORM some difficult task.-- Aug. 12th. When in Nati
582n067 ver]CD an animal may so far be said to will to PERFORM an instinct that it is uncomfortable if it doe
638j059 ould not be born. with tendency to make animal PERFORM some action.-- as well as gain it. by habit.--
582n068 "To act from instinct is to be guided to the PERFORMANCE of a number of prearranged actions, which wi
586n079 ch of the wonder consists in the action being PERFORMED or emotion felt in early childhood (before ex
```

615o038 association of pleasures with certain actions PERFORMED by your parents, conscience This «X» memory e
622o048 instructions, which the child sees uniformly PERFORMED by the teachers & all around him, will be par
568n017 of action, or thought one feels pain, at not PERFORMING it, (either if prevented, or overtempted.--
568n017 se had been]CD-- «Also» When one is prevented PERFORMING heredetary habit, (or moral sense, or instin
568n017 ct,) one feels pain, & vice versa pleasure in PERFORMING it.-- As soon as memory improved. direct eff
582n068 ng about a certain result, while the creature PERFORMING those actions neither knows nor intends the
582o069 irst time of building?), but not the means of PERFORMING it.-- p. 14. There is scarcely a faculty in
301c199 licable to any habitual action. even which Man PERFORMS.-- child striking a post in passion.-- Habit
538m080 ergetic self, & likewise one forgets. what one PERFORMS habitually.-- Agrees with insanity, as in Dr
550m127 out dread of water-- innate Septemb 1-- If one PERFORMS some actions, which are pleasant, every conco
582o067 fference, between the means by which an animal PERFORMS an instinct, & its impulse to do it.-- [the m
587n087 ke sweet things dislike others.-- dogs dislike PERFUME) I should think, great principle of liking, as
535m064 . p. 233. Mr Lewis describes case of insects «a PERGA» of Terebrantia, laying eggs on leaves of Eucal
030r034 stone?» have been collected on the open coast. PERHAPS as at Concepcion. favoured by basin formed by
041r064 ether they could have lived in so deep a sea.--PERHAPS agrees with formation of pebbles & vertical tr
044r074 there is proof of such gaz) steam condensed.--PERHAPS these mighty changes might go on. & not a bubb
045o075 mentioned with general slope of the country; (PERHAPS generally over whole world) Yet eruptions <bot
046r079 s an inward afflux of melted matter.--Volcanos PERHAPS may be admittance of water, through the rent s
053r100 Humboldt under name of Rothe-todte-liegende is PERHAPS same with that of Permambuco? Quote Miers abou
055r106 oreover that such do not occur on the beaches. PERHAPS these facts attest a <more> decided elevation
056r109 clination natural [...] deepest astonishment.» (PERHAPS scarcely a pebble might remain to tell of thes
056r111 n. But Volcanic action separates some sulphur (PERHAPS lime) salt. & metallic ores.--which mingling &
059r123 ured. have been melted with little pressure. & PERHAPS cooled suddenly.-- As the rude symmetry of the
071r158 ranquillity «at Mendoza» exception.-«formerly   PERHAPS otherwise» Mendoza never overthrown,--no mount
074r165 W <W>.-- «same chemical laws as in concretions PERHAPS makes intersections richest-- Humboldt has urg
100a055 & Oolitic period.-- comparison rather loose.-- PERHAPS worth Says from Lardner's (p. 213) form of esc
145g006 in upper part «of Salisbury Craigs» 25 degree PERHAPS most common-- Will not curved form of hill be
146g010 the trap could be traced Grey in front on wall PERHAPS wall oblique The hill has been well-- denuded.
146g014 Buttresses formed vestige of irregular terrace PERHAPS near 300 ft above Loch.-- From this point coul
148g023 edonian Canal opposite Loch Leven two terraces PERHAPS upper one 100 ft & other one 40-- \ traces of
156g072 valley.-- & that river alone had modified it-- PERHAPS however sea also,-- Barometer on shelf 3d. 29.
161g094 ears to lip with moss On this terrace «station PERHAPS 6 ft too low» (to test last on Peat-Mass Divor
177b025 ass into each other.-- The tree of life should PERHAPS be called the coral of life, base of branches
177b027 ts amongst articulata.-- but in lower classes, PERHAPS a more linear arrangement.-- ¿How is it that t
178b031 k over. Bell, & L. Jenyns. Falkland rabbit may PERHAPS be instance of domesticated animals having eff
192b089 l that can be expected-- This answers Cuvier-- PERHAPS the father of Mammalia as Heterodox as ornitho
193b090 ges ten thousand varieties, (influenced itself PERHAPS by circumstances) & those alone preserved whic
198b113 les fish-- Conditions will not explain status (PERHAPS consideration of range of capabilities past &
199b120 fspring not fertile. or at least most rarely & PERHAPS never female.--no offspring: physical impossib
203b136 in central & Eastern Asia beyond the Ganges & PERHAPS even in India p. 261. L. Institut. 1837 Mem. S
206b149 e relationship would converge sooner) & lastly PERHAPS some one single one.-- Will not this account f
222b205 ng will allow of decrease in character. (which PERHAPS is) Case with fish-- as some of the most perfe
223b209 es have not been much altered they will cross (PERHAPS more fertility & so make that sudden step. spe
230b236 es, but not such as would make species (except PERHAPS in some plants & then a chain of steps is found
230b240 ot soon being subjected to fresh change in Americas. PERHAPS merely gone back previous to fresh change Get
246c027 (Less) inhabits Caroline < «NB. The» > isld. (. PERHAPS Phillippines & perhaps, Friendly Isles «& Hebr
246c027 e < «NB. The» > isld. (.perhaps Phillippines & PERHAPS, Friendly Isles «& Hebrides») is very closely
252c042 lina) (p. 277.). probably another in Jamaica & PERHAPS one extant at Leeward Isles.-- p. 388 Reference
253c046 ame species in Borneo. Sumatra. India Ceylon-- PERHAPS shows great persistency of character. Hence El
259c063 or heredetary «& combined» effect of habit.-- PERHAPS in process of change.-- Are any men born with
261c070 effects of climate on some «antecedent races, PERHAPS not on now existing» Mr Gould says wherever an
263c078 ecessary to have made man! Seclusion want &c & PERHAPS a train of animals of hundred generations of s
283c145 es as "oh turn a Buccinum into a Tiger."-- but PERHAPS I feel the impossibility of this more than any
284c147 subdivision of stations & diversity--.-- this PERHAPS on long average equal.-- The Cocos do Mar on t
293c174 fference in races of men.-- Wax of Ear, bitter PERHAPS to prevent insects lodging there. Now these ex
293c175 daptation given to organization.-- This really PERHAPS greatest difficulty to whole theory.-- There i
298c190 passionate.-- the Key to the affections might PERHAPS thus be found-- a person who is habitually kin
299c194 ,-- but absolute species formed. The Anagallis PERHAPS, offers another case of permanent varieties in
301c200 ing thousands & tens of thousands New insects, PERHAPS scarcely one new family & no new orders,-- Won
306c212 r him to come again very differently from dog. PERHAPS being in passion chief difference Major Mitche
326c266 d not have children Turners embassy to Thibet, PERHAPS worth reading quoted by Malthus.-- Heberdens O
333d005 otherwise habits not different; tone of voice. (PERHAPS rather different».-- crossed with <un>common c
335d016 e,-- gradation from physical impossibility to (PERHAPS increased). fertility.-- (but many animals are
338d023 ffect of physical conditions of country is not PERHAPS so great, as separation on <be> inter-breeding
352d059 was for.-- Most important law.-- Penguins wing PERHAPS not abortive???. Apterix certainly-- Lyell's
363d102 thick vegetation» in swamps-- (owing to barns, PERHAPS, not being left open to them,-- \. In singing
367d113 of make in» the males; & another circumstance, PERHAPS, equally so, is this strength being directed t
376d137 ers of men. Evidently knew <men> women, thinks PERHAPS by smell.-- but monkeys examine sexes of every
378d147 ume the metallic tints, such as Magpie, Jay, & PERHAPS all the rollers-- «He says» whenever metallic
389d177 d in & in.)-- [The loss of passion in hybrids. PERHAPS connected with this same case (& not merely as
408e042 so than any other part of the World.-- Europe PERHAPS less so, that either Americas.-- If species ch
422e095 ple animals to become complicated although all PERHAPS will have done so from the new relations cause
423e097 complexity of structure is adaptation. though PERHAPS difference between jaguar & tiger may not be s
431e123 ize: (surely this is very limited view, though PERHAPS a true element) «give examples, pigs, with sma
447e163 ne region to another.--» -- these simple forms PERHAPS oldest in world & hence most persistent-- if f
459tf02 e ass. (& part of body mare)-- -- this may be, PERHAPS. squeezed into Mr Walker's Law Gleanings of Sc
460t019 rom action & reaction grew more complex)» some PERHAPS rendered more complex & some simplified.-- Ann
463t055 n animal could have had with such structure.-- PERHAPS greatest Could anyone. have foreseen, sailing,
463t059 , but this is not required.-- Waterhouse says   PERHAPS animals of Fernando Noronha are found unknown
482z011 esson Zoolog. Coq: p. 120 Coati Roux. Tatous & PERHAPS Yagourundi near Concepcion!!!.-- (no species m
483z013 and Chile The North & S. Range of shells might PERHAPS be worked out with advantage. with Cumms colle
490q001 ones-- I presume some recent not found fossil (PERHAPS not embedded ¿ are there any very common recen
501q011 on-- (27) Which sex in Mules generally fails-- PERHAPS indexed by secondary characters-- in double fl
522m014 w the feeling, thinking over somebody who has, PERHAPS, slightly injured me, plotting speeches, yet w
526m028 he inventive power,-- this, though very odd is PERHAPS true.-- mem Erasmus & mine taste for music.--
526m028 -- this may be worth thinking over.-- it. will PERHAPS show differences between memory & imagination.
526m031 at makes him fix!? <)>-- frame of mind, though PERHAPS he chooses wrongly,-- & what is frame of mind
527m032 ts the Knot:-- Sir J. Reynolds explanation may PERHAPS account for our acquiring «the instinct» our n
527m035 nventive thoughts are brought into play & then PERHAPS the sooner castles in the air are banished the
530m044 sbury something about big noses & name Corbet, PERHAPS nonsense.-- look to trick My father has somehow
536m072 ster has & a polype (& a plant in some senses, PERHAPS, though from not having pain or pleasure actio
537m076 thers ceased to destroy their offspring--¿ yet PERHAPS if they had murdered their children, this mora
540m087 th savage-- look at them both semi-civilized-- PERHAPS one cause of the intense labour of original in
546m108 a wide prospect.-- The very superiority of man PERHAPS depends on the number of sources of pleasure &
548m115 tion hard work,-- Keeping one idea present is, PERHAPS, hard work-- though dreams do that One Reflect
549m118 different it refers to wishes for future) than PERHAPS well «regulated» philosopher-- yet the philoso
549m120 llectual» good men, from insanity &c unhappy-- PERHAPS not so much as they appear & perhaps partly th
549m120 unhappy-- perhaps not so much as they appear & PERHAPS partly their fault.-- Whether this rule of hap
564n004 on <not social instinct>) but a causation. & «PERHAPS» an instinct of conscience, feeling in his hea
571n030 uth from <here> habits. & heredetary habits.   «PERHAPS» of space-- in latter respect he thinks he cert
580n060 says old people first fail in ideas of time, & PERHAPS of space-- in latter respect he thinks he cert
594n112 ened at a dog.-- » The instinct against man is PERHAPS as strong as against hawk, but the birds at M
601o009 moving leg when asleep» «or habitual actions» PERHAPS polypi-- (so that lower animals are sleeping h
609o029 ing, & therefore most deeply impressed). shame PERHAPS an exception. (does it originate in a doubting
609o030 ice & especially in the female October d. 1838 PERHAPS insist?? Two classes of moralists: one says ou
611o034 ity, which at least shows a local will, though PERHAPS not conscious sensation. [RHC] During growth <
614o037 me final end.-- production of higher animals-- PERHAPS, say attribute of such higher animals may be l
614o037 from hour to hour in <man:> individual-- [LHC] PERHAPS even the most complicated instinct. might be a

```
615o038 cts modified by heredetary;-- so succession so PERHAPS general ones.-- Parental feelings weakened in
616o038 e changed into man ∴ they meet their reward! X PERHAPS should hardly be called memory; you cannot cal
619o042 s parental, conjugal and social instincts, and PERHAPS others.-- [LHC] ----- p. 113. Mackintosh Groti
620o044 conscience  is extremely great [LHC] The cause PERHAPS lies in its frequency & in its consisting in d
624o051 in some past time, hence passions]CD «although PERHAPS useful at present to some extent.» Hence  this
625o50v ost» & moral sense.-- My theory explains both, PERHAPS, by habit-- [LHC] 11) Whewells preface.   [RHC]
629o055 bears  some relation to others 5) if so, it is PERHAPS deviation from the instinctive. right & wrong.
629o055 have no right & wrong except instinctive ones) PERHAPS my theory of greater permanence of social inst
632j53r ucus for adhesion».-- as examples of design.-- PERHAPS they are so.-- but the coral rock might have b
299c195 rong time of year & manuring it.-- Epigonous.   PERIGONOUS &c-- very important in classification.  here
026r020 he Tertiary formation South of the Maypo at one PERIOD of elevation must in its configuration have re
042r068 t.--yet cetaceous bones so likewise «of miocene PERIOD».--Mem Bahia blanca P. 204 Vol III. Lyell Owin
042r069 ty of slope of valleys.--All my observations of PERIOD «& manner» of elevation Volcanic action,  must
053r100 s on surface. applicable to Patagonia. During a PERIOD of subsidence the shinglle of Patagonia  would
060r125 in of Volc. had been in action during secondary PERIOD how diff. would the rocks have been. The red S
100a055 insists  of analogy between Australia & Oolitic PERIOD.-- comparison rather loose.-- perhaps worth Sa
114a095 ne chief cause of denudation, but does not tell PERIOD.-- I cannot help suspecting that clay-slates h
124a117 beria must have been in same condition for long PERIOD Subsidence in Demarara p. 131 (B.) Wrong Entra
133a140 been  ejected must remain fluid for an enormous PERIOD: now when we see how many points have been pen
133a140 canic & trappean rocks, within say the Tertiary PERIOD. one is led, to look at globe as resting on fi
179b035 monucle has definite life, then all die at one. PERIOD, which is not case,∴ MONUCLE NOT DEFINITE LIF
181b041 o of the 12. having progeny. after that distant PERIOD.-- Hence if this is true, that the greater the
183b053 some  great system acting over whole world, the PERIOD of great quadrupeds declining as great reptile
186b059 9. Mammalian type of organization same from one PERIOD to another, preeminently Pachydermata, less so
196b106 rse breed in S. America-- it will not do to say PERIOD unfavourable to large quadrupeds--horse not la
206b147 lustrum.-- We may conclude that there will be a PERIOD though long distant, when of the present men (
212b167 . Was Europe desert (like S. Africa) after Coal PERIOD.-- ¿In those divisions of molluscs. where spec
222b205 ed in remotest epochs.-- ¿ lizards of secondary PERIOD in same predicament. It is another question, w
249c036 when we consider number of quadrupeds in Eocene PERIOD. Have the Edentata & Marsupial forms been chie
269c101 of change have travelled that thickness in that PERIOD & no ways assisted by fluid currents which, ma
278c132 from Rhinoceros Elephants &c in the most modern PERIOD, compared to Faunas of these countries, greter
308c216 t any one time formed chain, since if cretceous PERIOD assumed, then some perished before, Carbonifer
343d036 peopled «with Myriads of distinct forms» from a PERIOD short of eternity to the present time, to the
399e006 hange in vertical series: Look at whole Glacial PERIOD? [not located] Study introduction to Cuviers R
407e040 as the Eocene of Paris! (Great Edentata at that PERIOD) Analyse this,-- consider state of vegetation,
414e060 makes  it wonderfully changed, since Cretaceous PERIOD, whether progressive I know not.» & insects.-
419e086 navia, than of the N. American species--Glacial PERIOD Dr. Beck says the shells in Scandinavia from h
423e099 atest formations <are> have been deposited in a PERIOD (say 10,000 years) which is sufficient only to
430e117 ancient  generic forms.-- the animals in Eocene PERIOD could not have been direct parents of any of o
432e125 to  revolutions of the sun & our lives,, but to PERIOD necessary to form heaps of pebbles &c &c:  the
446e163 s-- & that no change would affect them in short PERIOD & hence no change would effect them, without a
493q004 e.-- (4) Does the number of pulse, Respiration, PERIOD of gestation differ in different breeds of dog
541m091 y simple idea as scarlet steady before mind for PERIOD, «if the scarlet was before one effort less» o
559m156 day.-- naturally close at that time after long PERIOD.-- My Father about double consciousness.-- & s
593n107 .-- [blank] Does music bear any relation to the PERIOD when men. communicated before language was inv
612o035 ubject to accident; the sexual willing comes on PERIOD of year as much as inflorescence.-- [LHC] I he
091a026 ranean 700 feet deep in some of. the twopenny PERIODICAL said so. «Campbell the Poet» Accra. Coast of
495q05a ronella also sleeps on ditto-- Cover them up PERIODICALLY & see effect-- such dioecious individ--smal
032r040 the  belief of constant rising with successive PERIODS of greater activity & rest.--Such changes coul
033r043 ny. P. 349 Admirable little table showing long PERIODS of great violence volcanic. from Humboldt: Com
043r070 ous action over larger spaces of the globes & "PERIODS" of increased activity.-- such as that of 1835
047r083 d mass of fluid rock In Patagonia plains. long PERIODS of rest & vice versã more likely to be coincid
055r108 etter than any meterological table the precise PERIODS over immense areas. (& the counterbalancing va
182b049 ive development gives final cause for enormous PERIODS anterior to Man.. difficult for man to be unpr
189b072 grouped towards centres near Equator in former PERIODS & then splitting off.-- If species generate «o
212b167 herbivorous & zoophagous Mollusca according to PERIODS.-- NB. Was Europe desert (like S. Africa) afte
250c040 olluscs. by each other in secondary & Tertiary PERIODS.-- p. 125. ref. to Phil Transacts. (read Novem
266c091 ite as day & night: yet we know how remote the PERIODS at which both left the lands of their forefath
266c091 ontinuing for some time to flower at their own PERIODS.-- Arcana of Science & Art. 1831. p 160. accou
284c147 -- The quantity of life on planet at different PERIODS, depends,-- on relations of desert, open ocean
310c224 (as St Jago Cape de Verds) the shells in equal PERIODS with Europe would probably have changed much l
332d001 oplexy affects people all over England at same PERIODS When he began practice, he remember during a y
465t079 ighly important, to bear in mind that enormous PERIODS may elapse, even in situations apparently favo
473s004 as elevation then all continents of cretaceous PERIODS, together with their littoral deposits are pro
512q019 ep in the sprouting of the horns. at different PERIODS in different breeds--?? or in individual case:
548m116 , the consciousness does not go back to former PERIODS so «as» to «make» «give» one individuality  in
063r130 o another it must be per saltum--or species may PERISH. = This <inoculation» «representation» of spe
180b037 ostrich, Petise may not be well adapted, & thus PERISH out, or on other hand like Orpheus. being favo
186b061 nduced, or new species; therefore animals would PERISH, if there was nothing in country to superinduc
336d019 it either went back, or not being perfect would PERISH.-- The Varieties of the domesticated animals m
463j58v ed away.-- 4 & the species, like 10,000 others, PERISH. & who will dare to say that this is an infrin
057r113 ::.--Inculcate well that Horse at least has not PERISHED because too cold:--With discussion of camel u
189b072 generated in S. Africa Zebra.-- & continued.-- PERISHED in America All animals <are> of same  species
226b222 o most species occur.)> Although the Horse has PERISHED from S. America, the jaguar has been left & P
264c079 n certain other animals were alive, which have PERISHED.-- Let man visit Ourang-outang in domesticati
273c110 true, with shells?.?.?» It looks as if animals PERISHED by errors.-- It is most wonderful how in ever
273c111 & habits (it cannot be doubted that if swallow PERISHED) hawks & Milvulus &c would instantly fill  up
308c216 , since if cretceous period assumed, then some PERISHED before, Carboniferous some perished before, t
308c216 then  some perished before, Carboniferous some PERISHED before, then there always have been gaps, & t
285c153 unfit, the animal cannot change quick enough & PERISHES.-- Lyell has show such Physical changes  will
447e164 on the continent-- well characterised species PERIWINKLE wants insects to impregnate allied to Asclep
448e165 ds, layers &c & &-- do not seed freely.-- The PERIWINKLE seldom produces seeds, because it is thought
500q010 ni Essay-- Figs 2 kinds of flower annually.-- PERIWINKLE. (not asclepiadae. «in» Lindley) (24) Do Bee
255c051 ing totally obliterated. This may account for PERMANENCE in many trifling marks.-- such as the  bands
366d108 ties were all deeply imbued in them from long PERMANENCE, so that all their peculiarities must be tra
414e063 out time can do much.-- (yet one cross, & the PERMANENCE of his breed is destroyed)-- -- When two rac
575n046 . is something analogous. to instinct, to the PERMANENCE of old heredetary ideas.-- being lower facul
629o055 nstinctive ones) Perhaps my theory of greater PERMANENCE of social instincts explains the feeling of
180b038 produced.-- This requires principle that the PERMANENT varieties produced by <inter> confined breedi
195b098 phical <distri> division are arbitrary, & not PERMANENT. this might be made very strong. if we believ
195b098 the  Zoological character of these islands so PERMANENT a breath cannot reside in space before island
200b123 1 breed.-- But what a character is this? Race PERMANENT, because every trifle heredetary, without som
202b130 nimals made from influences in one country is PERMANENT in another.-- Good argument for species not b
210b158 ce from time? Von Buch distinctly states that PERMANENT varieties. become species. p.147 «p. 150»-- n
248c033 rts be of longest [consta]nt & therefore most PERMANENT Owen [re]markable laws of Brain & manner of g
294c176 ets authority).-- Lonsdale is ready to admit, PERMANENT small alterations in wild animals, & thinks L
294c176 with  external influence-- & if those changes PERMANENT so would the change in animal be permanent.--
294c176 es permanent so would the change in animal be PERMANENT.-- It will be easy to prove persistent Variet
298c192 ifferent countries. These facts show how very PERMANENT plants are. & this conclusion must be arrived
299c194 The Anagallis perhaps, offers another case of PERMANENT varieties in wild state-- The two. former pro
312c231 & in.-- which looks as if qualities were not PERMANENT, in the new cross.-- In the Bantam clubs, the
341d031 ne,-- A Bunting by one only differing by some PERMANENT white streaks.-- &c &c Dr. Bachman has crosse
349d052 merable] does not know any difference between PERMANENT variety & species! (given in note.)-- Maclea
365d107 leay says their is no difference between <t> "PERMANENT varieties" & species. he overlooks-- <restric
410e053 , without species being described.-- but then PERMANENT varieties in same country, must be distinguis
410e053 in  same country, must be distinguished, from PERMANENT varieties not in same country.-- The traces o
417e079 fer to length of time that the resemblance is PERMANENT, or the similarity at first births.-- is t
442e151 tulips break by cultivation can a form become PERMANENT?» because its very essence is that little cha
```

615o038 certainly appear a sort of acquired memory. a PERMANENT secretion of thought, (or under contingencies
209b156 nt le plus aisément en espèces distinctes et PERMANENTES." p. 145. In Humboldt great work de distribu
171b003 beings>. the young of living beings, become PERMANENTLY changed or subject to variety, according «to
230b239 change suddenly acquired is with difficulty PERMANENTLY transmitted.-- <It will admit> a plant will
493q004 l the latter ought: either in first breed or PERMANENTLY.-- (2) Cross two half-bred animals. which ar
411e054 must know whether the individuals «forms» are PERMANET, all steps in the series, their relation to t
554m139 e», look up to «keeper» see whether, this was PERMITTED & eat it.-- good case of association.-- «List
029r033 ks Mem. Chanticleers voyage at <[...] Maranh> PERNAMBUCO. EARTHQUAKE AT SEA.--Extract from the log-bo
051r093 ous» encrustations; At Bahia ferruginous.--At PERNAMBUCO (great swell & turbid water) organic bodies
051r094 d point of greatest action; I now having seen PERNAMBUCO believe much is owing to protection of Organ
053r100 e-todte-liegende is perhaps same with that of PERNAMBUCO? Quote Miers about shells at Quillota Lyell,
054r102 sandstones: refers to broken hill described by PERNETTY: account of streams of stones agrees with min
320c276 rances Voyage de l'Astrolabe Partie Zoologique PERNETY. voyage a l isle Malouines Zoological Journal
101a058 des to big bones in interior at Falkland Isd.-- PERON does as if well attested.-- There is no differe
207b152 rms in recent volcanic islets not well fixed.-- PERON thinks Van Diemen's land long separated from Ho
215b177 V. Diemen's land J. de Physique. Tom 59. p 467. PERON G. St. Hilaire has written "opuscule" entitled
062c128 S. great contrast. from nature of climate. -- PERPETUAL snow.--subterranean Lakes. near Volcanoes. la
187b064 of individuals.-- Why does individual die, to PERPETUATE certain peculiarities, (therefore adaptation
359d086 in tends to produce unhealthiness,-- «or» to PERPETUATE some organic difference.-- it may be so, but
389d177 rned.-- Generation being means to propagate & PERPETUATE differences (of body, mind & constitution) i
209b155 European dog? Yet man has had no interest in PERPETUATING these particular varieties. If species made
045r077 pretty severe shock). are much more curious & PERPLEXING. than those that attend Eruptions: Mr P. Sco
240c014 pecies of Parameles, which joined to Casoars, PERROQUETS, establishes its «zoolog» alliance with New
027r024 neiro. Coquimbo. Balanidae. at Concepcion. Humb: PERS. N. vii P. 56 Serpentine form: of Cuba for comp
276c123 nd & France. I will do more.-- Mention PERSECUTION of early Astronomers.-- then add chief good
130a133 of some large fresh Water springs off coast of PERSIA In Glen Roy paper I show crust yield easily. &
306c214 All the sheep are thick-tailed The dogs called PERSIAN «greyhounds» are Kurdish & come also from Asia
333d005 cat.-- Ch IX Mongrels Hybridism Fox has half PERSIAN cat. which bred with unknown common house cat.
333d006 mbled in form of tail, fur &c to the half bred PERSIAN.-- Here then we have clear case of heterogenou
335d015 les. Fox's half bred Persians «cat» favour the PERSIAN side.-- Theory of abortive hybrids.-- If mules
493q003 se-shell «on back.--» = Length of intestine in PERSIAN Cat-- , in Brazilian «toothless» dog-- I. St.
493q004 Esquimaux dog. with the hairless Brazilian or PERSIAN animals of different heredetary constitution,
335d015 rs, abortive, just like mules. Fox's half bred PERSIANS «cat» favour the Persian side.-- Theory of ab
253c046 Sumatra. India Ceylon-- perhaps shows great PERSISTENCY of character. Hence Elephas primigenious ove
412e057 ly adapted to circumstances of times. & from PERSISTENCY «owing to their slow formation» these variat
221b200 nt species., p 8. ¿Are mundine forms, longest PERSISTENT?? do.-- The most perfect Plants Composites.-
294c176 imal be permanent.-- It will be easy to prove PERSISTENT Varieties in wild animals-- but how to show
336d017 The longer a thing is in the blood, the more PERSISTENT.-- «any amount of change» shorter time less
366d108 monstrous variety. which could not have been PERSISTENT in nature.-- According to my view, the domes
447e163 le forms perhaps oldest in world & hence most PERSISTENT-- if form exceedingly difficult to vary.-- t
447e164 y corals.-- these forms then ought to be very PERSISTENT, , & then necessity of crossing is most ness
070r155 s well as in latter.-- According to Mr Brown, a PERSON (whom I met at S. W. P.) the Cordillera extend
214b176 il, enlarged two very considerably, so that any PERSON would say the tail was broken-- This came so o
208c190 the affections might perhaps thus be found-- a PERSON who is habitually kind to children.--<get> «in
458t002 «4to. Edits-- Horses in Lao Choo so small, that PERSON with long legs can hardly ride on them. Mr Mil
522m014 anger, which rises almost involuntarily when a PERSON is tired is akin to insanity.-- «I know the fe
524m020 which may be considered as truly spiritual.-- a PERSON twitching when a disagreeable thought occurs,
538m080 , & yet they would make one's father & self one PERSON-- & thus eternal punishment explained. These f
540m089 hildren with other children naughty.-- Why does PERSON cry for joy? 17th. August Montaigne (Vol. I) h
555m143 in a dream in a confused manner. thought that a PERSON was hung & came to life, & then made many joke
555m144 hanging came before me showing impossibility of PERSON recovering from hanging on account of blood. b
556m145 & little between incisors.-- like <W[...] what> PERSON says "what a pity"-- Lavater's Essays on Physi
557m150 nothing can be heard.-- Shame would never make PERSON tremble, like fear.-- Why does any great menta
565m007 fear any bad news, «though in a letter» why is PERSON painted with mouth open.-- why when person is
565m007 y is person painted with mouth open.-- why when PERSON is listening is mouth open to hear well «as on
565m008 ered as quite insignificant, & when pride makes PERSON extremely self-sufficient,-- the corner of low
565m008 te muscles will want much confirmation. A grave PERSON close those muscles, which wrinkle when smile.
570m023 d, without ones chest, being distended. touch a PERSON on the ribs & how he gulps in air.-- Again a m
573m038 ements of fingers.-- like Kitten with mice.-- A PERSON with St Vitus' dance badly, told should have s
573m038 mixture of voluntary & involuntary movements.-- PERSON with sore-throat told not swallow spittle. wil
578m053 y one blush.-- Is there not some saying about a PERSON even blushing in the dark-- «so modest a perso
578m053 person even blushing in the dark-- «so modest a PERSON.» A person who blushes in the dark is proverbi
578m053 blushing in the dark-- «so modest a person.» A PERSON who blushes in the dark is proverbially a most
578m053 ushes in the dark is proverbially a most modest PERSON one carries on, by association, the question,
578m055 B.-- Mrs. B. A. blushed. analyse this:-- Let a PERSON have committed any «concealed» action he shoul
578m055 eing discovered by anyone, especiall if it be a PERSON. whose opinion he regards, <& see how> feel ho
578m055 lways bear some references to thoughts of other PERSON Decemb. 27th.-- Fear loose the sphincter muscl
578m059 passes in a man's mind. when he says he loves a PERSON-- do not the features pass before him marked,
588m116 rks worthy of attention p. 15, 25. 40. 61. CD[a PERSON is here said to open mouth in fright because n
608o026 †It. may be urged how often one try to persuade PERSON to change line of conduct. as being better & m
610o031 y father received a letter from Mr Roberts-- «a PERSON he had long known & directed many letters to»-
618o041 es of action quite independent of each other. A PERSON might be quite familiar with thought & yet be
618o041 traction of sulphuric acid for metal of another PERSON at all, we can only infer it from his its beha
621o047 sociation to give «almost» any taste to a young PERSON. or it is accidentally acquired from some trif
622o049 ciation from having received benefits from this PERSON.-- [LHC] P. 254. &c &c [RHC] But the love is i
623o050 .-- the having received pleasure from some one «PERSON» in early infancy, during many generations giv
544m102 <rules> relations of time, <identity,> place & PERSONAL connections-- ideas are strung together in ma
578m053 icularly to wish not to do so.= = How directly PERSONAL remark will make any one blush.-- Is there no
603o012 all sacrifices. How completely men must have PERSONIFIED the deity.-- H. Tooke has shown one chief ob
593m111 ther man speaks. shows, that consciousness of PERSONNAL identity is by no means a necessary part of m
528m037 round, ovals»;-- then there the pleasure of PERSPECTIVE. which cannot be doubted if we look at build
528m037 ildings, even ugly ones.-- the pleasure from PERSPECTIVE is derived in a river from seeing how the se
608o026 play.-- †It may be urged how often one try to PERSUADE person to change line of conduct. as being be
090a024 ars]CD If world increased a tenth; would the PERTURBATION be serious? if so other cause besides thin
046r080 ms to» considers that fall first movement (as in PERU 1746).--At great Lisbon Earthquake Loch Lomond
049r089 ips Mineralogy some such fact stated to exist in PERU. -- Ascension At Ischia there is a pumiceous co
075r166 red by a clayey porphyry, containing grenats. In PERU. on other hand, mine of Gualgayoc or Chota & Pa
075r167 discovered?-- Klaproth analysed silver ores in PERU consisted of native silver & brown oxide of Iro
076r168 III» "In general it is observed both in Mexico & PERU, that those oxidated masses of iron. which cont
081r181 ndings Gilbert Farquhar Mathison travels Brazil. PERU. Sandwich Isd Mawes travels down the Brazil.--
268c099 dy D'Orbigny. & range on West Coast «Guayaquil & PERU» Henslow in talking of so many families on Keel
540m085 gions of the North,-- the elevated table land of PERU the hot plains of the Amazons & Brazil-- with t
071r156 ions. = Mem: Olivine. Volcanic product.=> <Did PERUVIAN Indians use arrows or Araucanians?--> If wood
220b197 events breeding. now domestication depends on PERVERSION of instinct (in plants domestication on perv
220b197 rsion of instinct (in plants domestication on PERVERSION of structures especially reproductive organs
458t002 ton &c instead of natural substances-- useful PERVERSION of instincts-- Beechey's Voyage Vol I. p. 49
220b197 hough in plants, you cannot say that instinct PERVERTED, yet organization «especially» connected with
593m105 g turning round & round is some old instinct <PERVERTED> handed down & down.-- mem. Nina used to get
624o051 on from education & imitation, has often been PERVERTED from want of reason.-- Hence as Eugenius says
451e178 of Nepal by. B. Hodgson. p. 336 In the most PESTIFEROUS region (mentioned by Heber) «from» which «al
466t093 nectary marked by orange freckles on {a} upper PETAL; bees & flies seen directed to it-- The Humbles
466t093 n yellow abar lily. the Bees visit base of upper PETAL, though not differently coloured-- & stamens be
466t094 nium, I see Bees visit always base {a} of upper PETAL from facility of alighting? which is not differ
467t103 Kidney Bean, they go to nectar at foot of upper PETAL standing on «I saw Bee go to two species of Lup
467t104 n all these nectar seems to be at base of upper PETAL & the curvature of <an> pistil, etc lies in gan
472s02r eartease withered on Monday.--» alight on upper PETALS & insert proboscis, under sigma & draw it out
402e019 oyage. Vol II p. 344. account of insects of St. PETER & St. Pauls in Lat' 53 degree yet fauna like th
130a134 ine of separation A Paper by Parrott Mem. Acad. PETERS. Scienc Math. Phy-- Nat. t. I, 1831. sur le te

315c241 any new ones.-- Menoir will be published St. PETERSBURGH Academy Imperial-- Paper read in 1837. semes
180b037 tion we can easily see that variety of ostrich, PETISE may not be well adapted, & thus perish out, or
062r130 kind of relation that common ostrich bears to (PETISSE. & diff kinds of Fourmillier): extinct Guanaco
089a018 et suivis, queque temps avant et apres, par de PETITES secousses."-- Tom 54. p. 106 do-- p. 110. Moun
116a099 ro beds.-- -- refers to species non decrite de PETITES corbules analogue living in mouth of Plate. p.
200b121 genera which unite very different structure as. PETREL & alk. do not show the possibility of common b
205b144 my planariae Leaches out of water Does the odd PETREL of F. del F. take form of awk, because there i
310c222 ll of second--this class of facts «analogous to PETREL-grebe. external» appears to be a puzzle agains
484z015 auks bear traces of.-- like Puffinuria does of PETREL?-- Study Birds of Europe for other representat
639j28r n Flamingo & Duck, Ornithorhyncus «externally». PETREL & Whale in some respects Chamaelion like power
177b028 arts belonging to each)» approaching another. <PETRELS have divided themselves into many species, so
480z008 loropus. says some of the «species of» smaller PETRELS, are night birds agree. with <pe> nocturnal ha
516q23v gen & forms Carbonic Acid. will this bear on PETRIFACTION?-- [blank] Questions & Experiments Expressi
022r008 hairy lips of which were still sound and not PETRIFIED, and the jaw was also firm, out of which we p
026r022 Says Echinites. Encrinites. Asteriae, usually PETRIFIED into a peculiar cream-coloured Limestone: the
051r093 st violent surfs: in both latter cases become PETRIFIED, & increase.-- In Southern regions every roc
079r176 era near Pataz, also at Gualgayoc. where many PETRIFIED shells Bougainville says P 291.-- The Fuegian
376d137 -- he has repeatedly seen them try to pull up PETTICOATS., & if woman not afraid clasp them round wai
554m138 en place its head downwards to look up womens PETTICOATS-- just like Jenny with Tommy ourang.-- Very
243c019 ; et certainment on eût été bien éloigné, il y a PEU d'années, d'admettre que ces oiseaux eussent leu
069r151 ust have been as much conglomerate on West of PEUQUENES as on East. Where gone to.?-- There must have
477z003 Las Vinchuca or Benchuca. "Les individus ailes PEUVENT avoir <quatre> cinq lignes de long et volent.
024r015 coast:-- 8 leagues, from Sydney 90 fathoms La PEYROUSE. South of Mocha; 19 miles. 65 Fathoms Vide fa
551m128 ely ordered to do it.-- Plato «Erasmus» says in PHAEDO that our "necessary ideas" arise from the pree
634j54v aco support the flying membrane?!!-- that the PHALANGES have separate movements in the Holocentrus ru
232b249 rica In same manner, Cuscus, (a sub genus of PHALANGISTA New Holland form) is found in many island Ce
240c013 ccas, Waggious New Guinea. New Ireland, have PHALANGISTA, which differ in «form & head &» colour from
311c225 Temminck Fauna Japonica (?!) 82 mammalia 293 PHALANGISTA of Australia & Van Diemen's land diff.-- Hab
361d096 observed that female of some water birds, (as PHALAROPE) assume for breeding a more brilliant plumage
617o39v s. How we identify the two aspects as different PHASES of the same object of thought is a question wh
217b189 culiar to itself-- on hybrids between grouse & PHEASANT-- Magazine. Zoology & Botany Vol I p. 450 The
274c113 w like lark, (one in Australia is called swamp PHEASANT) Goatsucker--, parrots with claw like lark (N
290c164 t half between fowls & pheasants, is most like PHEASANT., I think so because very 3/4 bred.-- (hence
312c228 Griffith) & half Muscovy ducks, «black cock & PHEASANT see Jardines Journal.»-- consult on this poin
342d033 York. therefore instincts not imperfect.-- Are PHEASANT & Grouse homogeneous? I observe Bachman calls
358d085 ybrids, that are infertile.-- thus the common. PHEASANT & fowl when crossed never even lay eggs. & th
358d085 l any sex by appearance.-- The silver & common PHEASANT crossed, has a cock (infertile) <with> the br
358d086 ile) <with> the breast of which is like common PHEASANT & back like silver.-- But the hen hybrid of t
359d087 ence the sexual organs alone.-- It is singular PHEASANT & fowl being so totally infertile whereas ani
359d087 take <very little> in disposition after their «PHEASANT» parents.-- (There are some 3/4 birds «of», w
365d105 of it.-- In England no doubt the cross between PHEASANT & Black game is owing to their rarity., a sin
365d106 id purple-- be careful, See to hybrids between PHEASANT & Black Cock, & other hybrids-- The fact of E
383d159 in Gardens of Hybrids between Common & Silver PHEASANT, one like cock & other like Hen.-- one doubts
383d160 rden. Sept 16." Hybrid between Silver & Common PHEASANT. Male bird, said to be infertile.-- spurs rat
383d160 <blacker» «broader.-- » Breast red like Common PHEASANT.-- lower part of breast, each feather is fine
383d160 doubtless is effect of Metallic hue of silver PHEASANT. yet why green? & not purple?-- legs pale col
385d164 ting layer--. or Polish breed. (he thinks half PHEASANT, half fowls.-- eggs fertile, but parent bird
388d172 f same mother.-- The view that man & «or cock» PHEASANT &c is abortive hermaphrodite is supported by
426e106 an two perfect species; but facts of grouse, & PHEASANT, & hooded crow goes against this. & wild hybr
502q1lv : Compare young. beans. cabbages.-- History of PHEASANT-fowl. Hen coloured like cock-pheasant: said n
502q1lv story of Pheasant-fowl. Hen coloured like cock- PHEASANT: said not to sit on own eggs Flowers in short
563n001 Essays on Natural History Waterton describes. PHEASANT springing from nest & leaving no tracks.-- My
563n001 rth Hill shows it is evergreens they seek Cock PHEASANT claps this wings before? crowing & only in bre
183b052 avec le verdier" &, <fr> silver gold & common PHEASANTS & fowls.-- "On sait que le "métis" du loup e
290c164 - It appeared to me that half between fowls & PHEASANTS, is most like pheasant., I think so because v
290c164 brids in this case have bred). White & common PHEASANTS. have crossed.-- </I saw> .-- The attachment
296c184 Vol I. p. 450. 4 instances of hybrids between PHEASANTS & Black fowl.-- use as argument possibly some
325c267 y) Yarrells Paper on change of plumage in Hen PHEASANTS <Zoological> Philosop. Transactions. 1827 Pax
342d033 eterogenous offsprings.-- «are not the hybrid PHEASANTS & grouse different.--» (if so chinese pigs &
364d104 efore sheep-- it has been thought that silver PHEASANTS about a house made other pheasants have white
364d104 hat silver Pheasants about a house made other PHEASANTS have white feathers).-- It certainly appears
365d105 o their rarity., a single female in wood with PHEASANTS would sure to be trod, & in many parts of Sca
512q019 - do they throw up pellets-- (4) About hybrid PHEASANTS treading-- any treadèe?-- Difference in lambs
046r079 canic theory. I want to ground, that the first PHENOMEM. is an inward afflux of melted matter.--Volca
045r074 metallic veins solution of silex & many other PHENOMENA I do not believe that the extraordinary fissu
058r117 e shall see a cause for Volcanos part of same PHENOMENA lasting so long.-- The great movements (not m
060r124 greater distance".--Afterwards speaks of this PHENOMENA in connection with "the shooting upwards" of
074r165 es intersections richest-- Humboldt has urged PHENOMENA in veins, chemical affinities like in compose
119a106 tness & curvature of quartz rock?) also by my PHENOMENA of earthquakes.-- by the narrowness which the
232b247 tion played on secular refrigeration".-- <The PHENOMENA of the S. Hemisphere look as if heat gained?
604o018 ness, power. being associated with God. these PHENOMENA we (feel & ?) call sublime.-- 4 From the asso
608o025 alls them free will--the chance of mechanical PHENOMENA.-- (Mem: M. Le Comte case of Philosophy, & sa
617o040 ing necessarily exhibited in & by matter. The PHENOMENA of gravity considered in themselves consist i
617o40v to the action of some animated agent Now the PHENOMENA of gravity are manifestly the same as if ever
617o40v e statement of the «force exhibited in every» PHENOMENA actually apprehensible by sense. 5) There is
618o41v tter is quite insufficient to account for the PHENOMENA of thought. (The objects of thought have no r
079r177 omme ils traitent les loix de la nature a ses PHENOMENES."-- Ulloa's Voyage, Shell fish purple die, m
026r021 age is important; veins appearing a galvanic PHENOMENON, so probably will the Cleavage be There is a
067r144 and. The Cordoba earthquake a very remarkable PHENOMENON. showing line of disturbance inside Cordille
617o040 attributed; force (be it remembered) being a PHENOMENON apprehended by the same faculty with matter
617o040 vely> considered, by our external senses is a PHENOMENON the essence of whose existence consists in i
199b118 of an accidental habit into an instinct." Ed. N. PHI. J. p 297, No 8 Jan-Ap. 1828 -- I take higher gr
091a025 ea solution of silex also shewn. No 3d of Ed. N. PHIL. J. p 194. Fact of dust blown far out to sea va
091a027 trike SSW & NNE dip 30 degree - 80 degree En. PHIL Journ. p. 410. 1828 Ed. N. P. J. p. 105. Oct. 1
106a076 ence-- <Is there same.> Institute. 3d. p. 40 or PHIL Mag. Dec 1837. p. 520 Mr Fox on increase of tem
107a077 J. Hal. End of pages. p. 157. Vol VI Edinburgh. PHIL: Transacts.-- Does the isothermal {P} subterran
119a104 which though not very analogous (see Edinburgh. PHIL. Journal <JCD>, no great chains like Andes or H
129a131 azil. Maldonado enter into this case.-- Ed. New. PHIL. Journal Vol XXI. p. 213. Beyond the limits of
129a132 r's Book.-- M. Bichoffs Papers, in Edinburgh New PHIL. Journ 1838. several case given of hot heads &c
137a149 125 to 129 & p. 135--160 & 162 [blank] Ed. New. PHIL J. 1838. p. 72. on metallic vapours condensed f
137a151 of its constituents into chert. [blank] Ed. New. PHIL J. 1838. p. 132. «& 134» Bischoff. On the effec
200b121 SW. & NNE. dip 30 degree - 80 degree (?).-- Ed. PHI.L.. N. J. p. 410 <Nov> 1828 It is daily happening
200b125 er &c has any tendency to form varieties? Ed. N. PHIL. J. Morse found in Virginia p. 325. July 1828.
240c003 he wild rock pidgeon.-- fact analogous to Owen's PHIL: remark of Apteryx having feathers.-- It is pos
250c040 secondary & Tertiary periods.-- p. 125. ref. to PHIL Transacts. (read November 20th) Paper by Jenner
320c276 aguay Dobrizhoffer. Abipom<e>nes. Edinburgh New. PHIL Journal. about 13 numbers have been read Voyage
406e036 ent & sometimes other & sometimes ½ way. Ed. New-PHIL. Transact. Rabies, common to men, dogs, horses
465t081 ve been exterminated (see Pritchards paper] (Ed. PHIL. Journ. end of 1839) very important. it seems o
465t089 other forms Lyell's Paper, in Taylor's Journ.-- PHIL. Mag. May. 1840 p. 362.-- some Mammals of Norfo
469t119 s.-- This flower hostile to intermarriage!!xx In PHIL Transact. about year 1778. Paper by Camper on O
477z001 of Crust-- Westwood & Thompsons-- Part II.-- 35. PHIL Trans Burrowing & boring marine animals-- CXVI.
538m078 eredetary pride & in single families. Edinburgh. PHIL. Transact. p. 365. Case of double consciousness
539m081 which came on from reading «review of» M. Comte PHIL. which made me «endeavour to» remember, & to th
588m090 for a bone shows he has recollection.-- Lamarck. PHIL. Zoolog.-- Vol II p. 445. If we compare the jud
317c251 ournal of the Academy of Natural Sciences of PHILADELPHIA Vol VII. Part II/. 1837 account of the vari
243c019 Birds of New Zealand absolutely different.-- --PHILEDON circinnatus not found in Australia only New Z
484z016 red mark on wings of all-- Spix has described PHILEDON. allied to some of my birds-- These groups st
113a092 or Mitchell showed me a river <near> W. of Port PHILIP. which had bar at mouth excavated in solid roc

354d064 ut. 1837. p. 351. Paradoxurus Phillippensis. PHILIPPINES Man have varies the range-- Argue the case o
134a142 ays are like Kankaer South of Part Luconia-- PHILLIPINES there is volcano on isld in large lake-- Ber
450e175 - Java pony occasionally reaches 13 hands.-- PHILLIPINES Pony somewhat resembles that of Celebes is s
453e182 . p. 320. says no wild (carnivora) beasts on PHILLIPINES. Forrest somewhere says same.-- do p. 393. <
354d064 lik[e] Institut. 1837. p. 351. Paradoxurus PHILLIPINES. Philippines Man have varies the range-- Ar
246c026 sson &c p. 620. Centropus (Coucal) of Java & PHILLIPINES, has variety at Madagascar, Calcutta & Suma
246c027 abits Caroline < «NB. The» > isld. (.perhaps PHILLIPINES & perhaps, Friendly Isles «& Hebrides») is
316c245 other countries «++ p. 246» as Cyclostoma in PHILLIPINES & Anphidesma in S. America-- yet there are
049r089 s alternate together in contorted layers: Mem: PHILLIPS Mineralogy some such fact stated to exist in
061r126 world. «(Was it so in Sydney, consult history? PHILLIPS.» 1826.27.28. grt. drought at Sydney. which c
100a054 ite of Gypsum, is the best case of cleavage.-- PHILLIPS (113) «Lardner Encyclop.--» absolutely consid
100a055 concretion filled with unconsolidated matter-- PHILLIPS Lardner p. 197. refers to salt as being produ
100a055 flashing into steam, would Babbage.-- Webster PHILLIPS insists of analogy between Australia & Ooliti
101a056 direction, in which most substance lies <.-->? PHILLIPS. Lardner's p. 270-4, good discussion showing
101a057 Harlan) found on the Delaware. is it Edentate? PHILLIPS p 289.-- Alludes to big bones in interior at
136a147 periority of Lyell's classification to that of PHILLIPS as given p. 13. Vol II. Lardner's-- Treatise
136a147 s as given p. 13. Vol II. Lardner's-- Treatise PHILLIPS in Lardner Vol II p. 73.: some remarks on vei
136a147 Lardner Vol II p. 73.: some remarks on veins: PHILLIPS in Ladner Vol. II p. 80-- some remarks on dik
136a147 ome remarks on dikes: applicable to Cordillera PHILLIPS in Lardner Vol II. p. 81. «&83» Some remarks
136a147 s implied by meeting with granite every-where. PHILLIPS in Ladner Vol II p. 125. Good discussion on m
212b167 ¿Consult Dr Smith History of S. African Cattle PHILLIPS Geology «p 81» in Lardens Encyclop. Proportio
212b168 ochs most genera dead? --Examine into this «in PHILLIPS».-- According to this, formerly there would h
213b168 ous animals.-- p. 82 «There are many tables in PHILLIPS of numerous genera in fossil & recent state,
213b170 ficulty; This immutability of some species. In PHILLIPS. p. 90. it seems the most organized fishes li
214b173 n Fernandez Falkland Islds-- Kerguelen land.-- PHILLIPS. Lardner Encyclop. insists on analogy between
214b173 us. (Harlan) found on Delaware is it Edentate? PHILLIPS. Lardner p. 289 It is certain, that North Ame
323c269 4th. Bo«s»well's life of Johnson. 4. Vols 25th PHILLIPS. Geology. Larder 2d vol.-- March 16. Gardner'
431e121 tes would afford instance of such facts.-- Ask PHILLIPS.-- The more I think, the more convinced I am,
431e122 domesticated animals tend to vary. March 20th. PHILLIPS in Lecture in Royal Institution says shells b
471tf09 se groups pass so insensibly into each other". PHILLIPS (Lardner's E. vol. II p. 18.) capital list of
323c269 Pliny Nat. Hist of World do -- Lamarck. II Vol. PHILO. Zoology «references at end of each Chapter» Cr
046r080 owing to absolute movement of ground). Michell (PHILOS: Transacts) «seems to» considers that fall fir
199b117 a thought to more uniform than existing. Ed. N. PHILOS J. p. <191> «p 191» No. 5. Ap 1827 F. Cuvier s
199b119 z not too much change In Number 6'.? of E.d. N. PHILOS. Journ.-- Paper by Crawford on Mission to Ava.
235b272 of climates, situations &c on 242. Hook Smellie PHILOS of Zoolog. 842 White regular gradat in Man poo
325c267 ving abstract of their views Mackintosh Ethical PHILOS: Prostitution of Paris. with respect to licent
547m111 me. & could trace no chain of association" Mayo PHILOS. seems certain that muscular, mental, <&> dige
196b107 o large quadrupeds--horse not large-- Ed. New. PHILOSOP J. No 3. p. 207 "It is not generally known th
196b110 f the two great Kingdoms.-- Principes de Zool: PHILOSOP:-- I deduce from extreme difficulty of hypoth
250c040 81. Capromys, West Indian isld. p. 120. «ref.» PHILOSOP. Transacts 1823. Read June 5th) important pap
325c267 hange of plumage in Hen Pheasants <Zoological> PHILOSOP. Transactions. 1827 Paxton on the culture of
347d049 fore more apt to be extinguished.--???» Mayo (.PHILOSOP of Living) quote Whewell as profound. because
432e126 succession.-- Splendid Pamplet. (published in PHILOSOP. Journal <Mar> April 1st 1839) by Sedgwick &
491q01v & only allow <few> one flower (5) Dr Fleming. PHILOSOP. of Zoolog. vol 1. p. 427-- says biennial-wal
550m126 founded.-- well applicable to free will. Mayo. PHILOSOP. of Living p. 293. Animals "have notion of pr
588n091 ed. reason? or some unnamed faculty-- Lamarck. PHILOSOP. Zoolog. «p. 284. Vol. II» -- gives explanati
026r020 tinction is given to angular & rounded.-- Fox PHILOSOPH. Transactions on metallic veins. 1830 P. 399.
027r024 on in young corals.--vide L. Jackson's paper. PHILOSOPH Transact: at R. de Janeiro. Coquimbo. Baланid
216b181 lied to case of cross of dogs.-- See paper in PHILOSOPH. Transaction on a quagga & mare crossing by L
477z001 nyn's introduct to Mag of Zoology and Botany. PHILOSOPH. Transacts. 3. papers connected with transfor
529m042 r says there is case on record he believes in PHILOSOPH. Transactions, of ideot 18 years old eating w
592n103 ed between Indian tribes are Many the same.-- PHILOSOPH. Transactions Vol 44. 1746-47. Paper. like. S
353d060 - is possibly the same with the <Zoologist> «PHILOSOPHER», who has trace the structure of animals & p
410e052 end must be the law & causes of change".-- A PHILOSOPHER, would as soon turn tailor, as, mere describ
549m118 es for future) than perhaps well «regulated» PHILOSOPHER-- yet the philosopher has a much more intens
549m118 rhaps well «regulated» philosopher-- yet the PHILOSOPHER has a much more intense happiness-- so is it
553m136 e, more apparent however to us, as does that PHILOSOPHER who says the innate knowledge of creator <is
554m138 e affinities of man, better than the boasted PHILOSOPHER himself it is chiefly shown in old male.-- A
232b248 hose will not object to my theory, those the PHILOSOPHERS who soar above the pride of the savage, the
377d139 dicule, or may be thought disgusting, but to PHILOSOPHIC naturalist pregnant with interest» Hyaena. t
116a099 too coarse. Read Kylau on Granite Edinburgh PHILOSOPHICAL Journal Rapport on D'Orbigny's Voyage. good
219b194 transport. If some cannot be explained more PHILOSOPHICAL to state we do not know how transported.--
287c157 nct miracle. Macleay letter to Dr. Fleming. PHILOSOPHICAL Magazine & Annals. 1830 (?)." if she has pu
576n049 ansmitted.-- Arguing from man to animals is PHILOSOPHICAL. viz. man is not a cause like a deity, as M
198b114 ke out his ideas about propagation His work. PHILOSOPHIE Anatomique. 2d Vol about monsters worth read
324c268 rs & at end of the Congo Voyage Decandoelle. PHILOSOPHIE. or Geographical distrib. «in Dict. Sciences
325c267 seful knowledge. Horse, Cow, Sheep.-- Verey. PHILOSOPHIE d'Histoire Naturelle Marcel de Serres Cavern
445e159 experience in baby Lamarck. Vol II p. 152.-- PHILOSOPHIE Zoologie. says it is not sufficiently proved
601o08b des Mères par L Aimé Martin Leroy Lettres. PHILOSOPHIQUE sur L'intelligence des Animaux-- & Le Parfa
235b272 gradat in Man poor trash Lyell 1024 Flemings PHILOSOPHY of Zoolog Royle on Himalaya Plants-- Would i
236bIBC e Entomolgicae Linn: Soc. Geoffry-St. Hilaire PHILOSOPHY of Zoology Waterhouse B C N[a]me of two pige
321c270 ousseau Miss M«artineau» How to observe Mayo PHILOSOPHY of Art of Living Several of Water Savage Lan
322c269 f each Vol Herschels. introduction to Natural PHILOSOPHY R. W. Darwin's Botany.-- References at end M
324c268 orms. Dr. Royle on Himalayan. types. Smellie. PHILOSOPHY of Zoology Flemming. ditto Falconers remarks
327c265 on intermarriage" price 14s. Marh. 20t. 1839. PHILOSOPHY of Blushing lately advertised.-- /6s Mrs Necke
350d054 rowne says "is the only indisputable axiom in PHILOSOPHY<"> Religio Medici. Vol II. Sir T Browne's Wo
397e003 "It accords with the most liberal! spirit of PHILOSOPHY to believe that no stone can fall, or plant
544m102 . love.-- & so pain gives fear of death. Mayo PHILOSOPHY of Living. p. 140-- Dreams good account of «
556m144 ideas, connected with judgment. [What is the PHILOSOPHY of Shame & Blushing]CD «Does Elephant know s
592n101 as gradually developed. see Hume on Sceptical PHILOSOPHY. Hume has written "Natural Hist. of Religion
599o006 ension &c often show traces of origin.-- Mayo PHILOSOPHY of Living. p. 264. "Architecture is a fine a
608o025 anical phenomena.-- (Mem: M. Le Comte case of PHILOSOPHY, & savage calling laws of nature chance) 2)
618o042 1) May 5th. 1839.-- Maer Mackintosh Ethical PHILOSOPHY [RHC] On the Moral Sense Looking at Man, as
627o053 ked in my Mackintosh 1) Mackintosh's Ethical PHILOSOPHY. p. 6-- "The pleasure which results when the
634j54v creator, exhausted & abandoned. Those who say «PHILOSOPHICALLY to a certain extent.-- nothing but experie
303c205 . Ichthyosaurus in the Chalk Those who say «PHILOSOPHICALLY to a certain extent.-- nothing but experie
323c269 liar History of Birds -- Mackintoshs' Ethical PHILOSPOHY -- Bell's Bridgewater Treatise -- Wilkinsons
500q10a arts of Europe.-- Gocould-- go over the Pigeons, PHILOTIS, Dacelo. Alcyone, where there are very close
094a036 sact Wern. Soc. Vol. 2. p. 35 Sir J Hall Trans. PHILS Royal Ed. Vol 7 Dr Buckland Reliquiae Diluviana
483z012 rossing Atlantic. fact taken from Jenner (1825) PHILS: Transact.-- "on Migrations of Birds".-- 18 do.
071r156 y. -- <Analysis of Atacama. Iron in Edinburgh. PHISOPH. Transactions. = Mem: Olivine. Volcanic produc
225b218 tached to.-- that insect «not» being called <PHITOPHAGOUS> omniphitophagous. But it will be said ther
470t177 les on hedge Linaria» (Plenty of Humble Bees on PHLOX Down, 1854, Sept.) In Spanish Broom by pulling
470t178 um saxatile & other common kind--I think not on PHLOX though they examine it.--«Little Dusty & Blue»
222b206 cannot for a moment doubt, but what cetaceae & PHOCAE now replace Saurians of Secondary epoch: it is
026r021 orth noticing that cryst of glassy felspar in PHONOLITE arrange themselves in determinate planes ∴ su
076r170 covered by conformable greenstone porphyrys & PHONOLITE do. amphibole quartz & mica very rare.-- anc
073r164 ntly only vitreous felspar: = gold veins in a PHONOLITIC porphyry. = several parts of N. Spain great
314c239 of Australian Genera Some species same (Palm & PHORMIUM tenax) as in New Zealand & Australia, some SP
581n063 ctive action.-- Sir. J. Sebright. has given the PHRASE "heredetary habits." very clearly, all I must
526m030 & not by hearing One is tempted to believe PHRENOLOGISTS are right about habitual exercise of the mi
533m057 brain of child resemble, parent stock.-- (& PHRENOLOGISTS state that brain alters) It is known that b
098a044 Ancient Lake Lemagne in Auvergne Proofs from PHRYGANEA NB. Sedgwick talks of LAMINATED structure (∴
556m145 ; that every man is born with a portion of PHSIOGNOMINICAL sensation, as certainly as every man who i
130a134 Paper by Parrott Mem. Acad. Peters. Scienc Math. PHY-- Nat. t. I, 1831. sur le temp du globe on Volca
385d164 s mentioned in boys bodies.-- Lavaters. Essays on PHY. transl by Holcroft Vol I. p. 195. says children
054r102 unt of cleavage differs wonderfully from mine: PHYLLADE covered by quartzose sandstones: refers to br
132a137 Parrot, Mem. Acad. Imp. des Sciences. (Sc Math. PHYS. et Naturelles. Tom I. p 501.-- shows first tha

367d113 ringed Owen says that Bell in Encyclop of Anat & PHYS. describes, a high-flying bat, which has the po
386d166 if cut off nerves in snail. (Encyclop of Anat & PHYS) can make a head; the other part may surely abs
199b120 rarely & perhaps never female.--no offspring: PHYSICAL impossibility to marriage.-- ¿whether those g
203b135 rt of theory death of species without apparent PHYSICAL cause:-- Mem: Mastodon all over S. America. H
209b155 e by isolation; then their distribution (after PHYSICAL changes) would be in rays-- from certain spor
223b209 ot perfect gradation of change in species,, as PHYSICAL changes are gradual, is this if after isolati
224b215 owing to mixture of races.-- When all mixed & PHYSICAL changes (¿intellectual being acquired alters
227b226 look at Articulata!!!--! It leads to Nature of PHYSICAL change between one group of animals & a succe
232b246 or change of species does not measure time but PHYSICAL changes (we assume like weather on long avera
242c017 changes in species must be very slow, owing to PHYSICAL change, slow & offspring not picked.-- as men
265c084 ill say it is direct effect, <of> according to PHYSICAL laws, as sulphuric acid disorganizes wood, bu
275c119 keystone of ancient geography species tell of PHYSICAL relations in time «& forms» distribution tell
285c153 constitution being heredetary & fixed, certain PHYSICAL changes at last become unfit, the animal cann
285c153 quick enough & perishes.-- Lyell has show such PHYSICAL changes will be unequally rapid, with respect
302c203 : father-species, being little adapted to some PHYSICAL change.-- If Patagonia became fertile all int
308c219 !!Metaphysics!!! Mrs Somerville, Connection of PHYSICAL Sciences p 276 May be worth glancing at, as s
323c269 n.» April 5d Dr. Edwards of <ter> influence of PHYSICAL causes: well skimmed Bartrams Travels in N Am
326c266 ng Bevans work on Bees, new Edit 1838 Harlaam. PHYSICAL & Medical Researches. on Horse in. N. America
335d016 ect animals is distinct case,-- gradation from PHYSICAL impossibility to (perhaps increased). fertili
337d023 e time they have been separated; together with PHYSICAL differences of country: the time of separatio
338d023 untries in times present & past. The effect of PHYSICAL conditions of country is not perhaps so great
343d036 f climate superadded to change of climate from PHYSICAL causes.-- these superinduce changes of form i
352d059 f their extreme antiquity (ie much intervening PHYSICAL change).-- distribution especially of Mammali
390d178 n every respect.-- [Is this connected with the PHYSICAL differences in almost all Male animals?]CD If
398e004 getation &c.-- It is the circumstance of small PHYSICAL changes & oscillations, not affecting organic
410e050 r animals» -- it was absolutely necessary that PHYSICAL changes should act not on individuals, but on
410e053 ividual be species or variety, but to discover PHYSICAL laws of such corelations, & changes of indivi
411e055 ps in the series have been fixed, to study the PHYSICAL causes. «All Cuviers generalization. of teeth
413e058 2) Tendency to small change.. «especially with PHYSICAL change» (3) Great fertility in proportion to
419e086 we have better means of judging of slowness of PHYSICAL changes, than in any other, & yet 200-300 ft
419e087 the chronology of geology rests upon amount of PHYSICAL change <affecting whole bodies of species>, &
426e107 , there would probably be perfect series, from PHYSICAL impossibility to unite to perfect prolifickne
435e133 th "Dr. Edwards on the Influence of <external> PHYSICAL agents". «translated by Dr. Hodgkins» p. 54.
435e134 ia in the effects produced on organization. by PHYSICAL agents." p. 466. Many facts given of high tem
530m046 ly by association, & association is probably a PHYSICAL effect of brain the «similar remark» thoughts
538m081 a train of though[t] action &c will arise from PHYSICAL action on the brain, renders much less wondef
542m094 se actions are habitual, & which now connected PHYSICAL relations.-- CD[like sighing to relieve circu
611o034 t of sensitive plant can be shewn to be direct PHYSICAL effect of touch & not irritability, which at
612o035 er to sun? ‍ I should think there. was direct «PHYSICAL» effects of more or less turgid vessels; effe
612o035 gnorance [RHC] The radicle of plants absorb by PHYSICAL laws of endosmic & exosmic juices. arms of po
635j55r no explanation-- it has not the character of a PHYSICAL law, «& is therefore utterly useless-- it for
635j56r owing to the growing size of the world? & the PHYSICAL changes it was to undergo «animals feeding on
640j167 possess many peculiar states. for as long as PHYSICAL change is in progress or is, present with res
641j29r ide range must be destructive to species, when PHYSICAL changes are in progress; (on the same princip
322c270 ons of Royal Irish Academy. ----do Lavater's PHYSIOGNOMY ---- Octob 3d Malthus on Population W. Earls
556m145 son says "what a pity"-- Lavater's Essays on PHYSIOGNOMY translated by Holcroft «Vol I» .p. 86 "We ou
418e083 Seeing that <Man> «all vertebrates. [Müller's PHYSIOLOG. p.24.]CD» can be traced to a germ, endowed w
641j29r .-- To show however little we understand of the PHYSIOLOGICAL relations of animals. equatorial countries
601o08b un Vol 8vo Keratry-- Inductions morales et PHYSIOLOGIQUES The first of these books I daresay good. 1
277c127 l point out what to observe.-- will aid us in PHYSIOLOGY, tell traveller what to observe.-- if he kno
304c206 eries now existed from Man to Monads-- though PHYSIOLOGY would profit. if the series were believed to
325c267 by Owen.-- Hunter has written quarto. work on PHYSIOLOGY besides the papers collected by Owen. (at Sh
326c266 <Etn> Entozoa by Owen in Encyclop. of Anat. & PHYSIOLOGY.-- Dampier. probably worth reading Lessings.
454e184 Decandoelle has chapter on sensitive plants; PHYSIOLOGY Get Habberley to try experiments. about rais
602o10v enjoyment in mutilated statues In Elliotson's PHYSIOLOGY much about sleep-- Nerves.-- Volition &c Rey
089a018 il; yet heat never reach surface.-- Journal de PHYSIQUE, et D Histoire Naturelle, C«o»urrejolles. 11t
215b177 Native dog not found in V. Diemen's land J. de PHYSIQUE. Tom 59. p 467. Peron G. St. Hilaire has writ
436e136 ow variation as much as «the» difference between <PI> species,-- for instance pidgeons-- : then comes
552m131 , «a reality» as in Spectral images-- Mem Chiloe <PI> Sow, who carried from all parts straw to make i
531m010 true the heart the scene of anger.--» to the PIANOFORTE, it seemed solely to be feelings of discomfo
589n091 e instincts,-- could brain make a tune on the PIANOFORTE, yes if every individual played a little, &
058r118 presentent une coupe abrupte et souvent tailée a PIC. Toutes ces montagnes sont formées de couches pa
108a081 ome IX 1837-8. p. 24. rocks of Chimborazo., & PICHINCHA. Melaphyre. = Andesite-- Albite & amphibole=
472c002 oipu) viscacha.-- A. Patagonicus les tatous (.4 PICHYE, pelud, mulita et mataco.) are all found south
088a014 ring eels alive to their nests; & then they may PICKED up beneath the trees---- Are any Fish seed-eat
200b124 on as island large enough for land birds, seeds PICKED from the beach by the birds; most seeds germin
242c017 owing to physical change, slow & offspring not PICKED.-- as men do. when making varieties.-- Voyage
271c106 rand classes of varieties.; one where offspring PICKED, one where not.-- the latter made by man & Nat
337d020 carcely any breed but what some individuals are PICKED out.-- in a really natural breed, not one is p
337d020 ed out.-- in a really natural breed, not one is PICKED out, & few even of local varieties approaches
430e118 - «& so in plants effectually the offspring are PICKED & not allowed to cross.-- » Has nature any pro
458t002 Journal» The Taylor Bird uses pieces of thread, PICKED «up-» instead of spinning-- better case than E
564n004 able & when fly ran past it. cocked its head, & PICKED it-- Here then, that faculty, whether for posi
239cIFC ent, white room How are varieties produced, by PICKING offspring? Instances of old Breeds taking grea
239cIFC an. Could you get racehorse from Cart horse by PICKING without change of habits Mr Yarrell «Give it a
248c034 If varieties «produced by slow causes, without PICKING» become more & more impressed in blood with ti
255c052 se & cross, is method of nature to prevent the PICKING of monstrosities as Man does.-- One is tempted
275c120 iew of varieties is exactly what I state.-- &c PICKING varieties. unnatural circumstance Ld Orfords h
275c120 ash of blood, with whole form of grey hound.-- PICKING out finest of each litter & crossing them with
275c120 ill size diminished, but feathers continued by PICKING chickens of each brood.-- These bantam feather
275c121 yellow & others yellowish & white varieties by PICKING the yellow one & crossing with duck bantams pr
312c231 g almost to the point in question.-- «--merely PICKING opposite qualities, with no other means whatev
414e063 n those of other dogs.-- if nature had had the PICKING she would make <them> such a variety far more
525m023 - as carrying a basket, bringing back game, or PICKING up a stone, though only acquired rules by art.
554m139 Tommy, or anything of any sort.-- I saw Tommy PICKING his nose with «a» straw.-- Jenny will often do
594n112 have seen hawk & sparrow in Shrewsbury garden PICKING from same bone A child born on the 1st March w
416e075 Hence difference between races & variety? «Man PICKS the Male, instead of allowing strength to get t
422e092 gin of northern Cetaceae).-- » ‍.do. p. 318 M. PICTET of writings of Goethe.-- who maintains, that «
539m082 go close to one & smelt the peculiar smell of PICTURE. association with much pleasure immediately th
548m117 not there.-- the reality does not resemble the PICTURE. in one mind, but does not stop to reason what
560m156 «peppermint» «& music».-- Have monkeys lice?-- PICTURE.-- Do female monkeys not show signs of impatie
569m019 bridge observed the ignorant. merely looked at PICTURE as works of imitation.-- Hence pleasure in the
572n033 s page.-- Now «awake» «when awake» I could not PICTURE to myself reading French book quickly, & <runn
574n041 described slovening «gum» «teeth«less-jaws. as PICTURE of disgusting lewd old man. ones tendency to k
593n109 en by smell, but by looks. hence, some obscure PICTURE of other men. & hence idea of beauty.-- the so
526m028 ine thinks that children like looking at <ami> PICTURES, an early taste, of animals. they know.-- ple
526m029 my American recollections are a collection of PICTURES).-- when one remembers a thing in a book, one
529m039 mer thoughts, & in experienced people-- recall PICTURES & therefore imagining pleasure of imitation c
062r128 e similar.--I do not know botanically - but PICTURESQUELY = Both N & S. great contrast. from nature o
239c002 explained.-- Mr Yarrell states that if any odd PIDGEON crossed with common pidgeon, offspring most li
239c002 es that if any odd pidgeon crossed with common PIDGEON, offspring most like latter, because oldest va
240c003 dest variety.-- -- He says of two varieties of PIDGEON, although having skulls so different, that the
240c003 t retains markings of wings like the wild rock PIDGEON.-- fact analogous to Owen's Phil: remark of Ap
362d100 arrell has old book 1765? Treatise on Domestic PIDGEON, in which it appears that all the <bird> varie
362d100 of them have the forked black mark of the Rock PIDGEON,-- several have a group of white speckles on e
363d101 n elbow joint-- in Bewick drawing the the rock PIDGEON has not: now how many wild pidgeons have spang
363d101 found in any colours of plumage &c &c «Pouting PIDGEON exaggeration of cooing.--» & compare them with
363d101 them with all the varieties.-- Habits of rock PIDGEON. (I suspect Pennant has described them)-- [Stu

```
373d132 dna. & Hedgehog having spines.-- Does not male PIDGEON (yes) surely) secrete milk? from stomach. anal
494q004 ck quite similar to those of another Duck. ¿in PIDGEON?-- Mr. Miller said yes with regard to former (
585n076 ARTED, they cannot return.-- Hence I conclude. PIDGEON taken little way, whirled, & then taken  other
586n079 ld sucking whole wonder instinctive.-- carrier PIDGEON just as wonderful in old bird as new.-- migrat
596nIBC endency easily fall into convulsions A carrier PIDGEON carried & turned round & round in fainting sta
216b180 cacti  &c &c.-- as in dogs investigate case of PIDGEONS. fowls. rabbits cats &c.-- When black & wh
240c004 ion.-- Mr Yarrell says, that after breeding in PIDGEONS with very much care that, it requires the gre
255c051 in many trifling marks.-- such as the bands on PIDGEONS back.-- According to this description of clas
261c071 p.-- it is same as Yarrell's remark about rock PIDGEONS.-- & the latter most important in obviating a
275c121 with  duck bantams procured old variety.-- The PIDGEONS which have such different skulls, but same ma
303c205 ry genus. Mesites bird from Madagascar uniting PIDGEONS & gallinaceous birds & parrots.-- legs of pid
303c205 eons & gallinaceous birds & parrots.-- legs of PIDGEONS perfect.-- &c &c.-- do p. 136.  ichthyosaurus
323c269 on Animal Reproduction -- Treatise on Domestic PIDGEONS 30th Lives of Hayd & Mozart «April 25th Lockar
359d088 ety, & not effect of breeding in & in like our PIDGEONS) The male of avery animal certainly seems chi
362d100 s also some very fine recent drawing «of prize PIDGEONS» in 1834-- now this would be most curious  to
363d101 he the rock Pidgeon has not: now how many wild PIDGEONS have spangles on this part: this will be well
363d101 ll worth working out.-- Study Temminks work on PIDGEONS--, & see whether feathered legs.-- «Carruncle
363d101 legs of Ptarmigan & in Bantam.-- ]CD CD[In the PIDGEONS, trace the washing out of the forked band, li
364d103 - Q Sir. J. Sebright-- has almost lost his Owl-PIDGEONS from infertility.-- Yarrell says in such case
364d104 he amount of variation is soon reached-- as in PIDGEONS no new races.-- In Scandinavia besides the Ra
378d148 ng had their present plumage.-- How is this in PIDGEONS & fowls.--??? Wate[r]ton «p. 197» put 12 wild
412e057 ae, which have given milk.-- is secretion from PIDGEONS stomach true milk.-- | <Species. are innumera
430e118 rom changing country: but greyhound. & poutter PIDGEONS «race-horse». have not been thus produced, bu
433e127 s out of whole volumes.-- The fact of tumbling PIDGEONS-- : then comes question of genera It certainl
436e136 ifference between <pi> species,-- for instance PIDGEONS-- : flying high all together & then tumbling, fa
440e146 us cabbages growing like Nepenthes.-- cases of PIDGEONS with tufts & &c here there is no final cause
440e146 ganization: as we see these strange plumage in PIDGEONS yet no change of habits, so no <cause> corres
449e169 e thing in Brent Goose. Eyton says some of the PIDGEONS in common Dovecot are very like a Himalaya sp
478z003 ra-- Cyclops p. 134. and p. 115 In white Cape PIDGEON'S stomach small shells (patella) sea weed & man
511q018 t of qualities in birds & animals for prizes.= PIDGEONS. Canary birds-- Bantams.-- (6) <Mad» Porto Sa
534m062 en leapt across. (Col Sykes compares this with PIDGEONS finding their way home-- there is something w
563n001 g no tracks.-- My Father says pea-hens do Wood PIDGEONS building near houses. yet so shy at all other
584n072 -- cats when carried in confinement,-- carrier PIDGEONS proverbially carried to long distance in dark
586n078 the most probably supposition. with respect to PIDGEONS, is that they do know from look of Heavens, p
622o049 have  different instincts, as we see in dogs & PIDGEONS.-- But as man is animal at head of series  in
066r142 r Murchison insisted strongly. that taking up a PIECE of Falkland Sandstone. he could not distinguish
088a013 strata, prooff thickness not very great; where PIECE turned over axis or hinge no doubt fluid.-- ana
124a119 ce-- Erasmus says he has seen in making brass a PIECE of copper not melted absorb, zinc thrugout  its
138a176 ting wood by yielding Carbonic Acid unite with «PIECE of cabbage» alklali & precipitate silica / or c
161g095 o eye certainly appears level with road, & with PIECE of excised rock lost at point of valley chiefly
358d075 one  occasion, not withstanding our vigilance a PIECE of pork 3 lb was taken from a boiling pot, & ca
386d167 Law, which makes two animals out of one & heals PIECE of skin.-- if the tail knows how to make a head
573n036 c «The Savage admires not a steam engine, but a PIECE» of coloured glass <&admires> is lost in astoni
579n060 my view of blushing-- in former irritation on a PIECE of skin cut off made the blush come.-- it is an
580n061 ink at all, whether she can or can not play the PIECE, she plays <f> better than when she tries is no
120a107 earth is cracking by vertical planes into small PIECES-- mem coal-field.-- the structure of Andes. wh
205b143 me good accounts of passages of legs into mouth-PIECES of Crustacea. Vol II. p 75 a Fish which emigra
458t002 I p. 502. «Bengal Journal» The Taylor Bird uses PIECES of thread, picked «up» instead of spinning--
081rIBC nch metre into English ft. 0.5159929 {T} Toises PIEDS Myriametre = 5130., 4. 5 inch Kilometre 513., 0
378d147 back  bright blue.-- «thus young of» Many of the PIES assume the metallic tints, such as Magpie, Jay,
378d148 t, so is species old: Hence <young> Kingfisher & PIES, have long had their present plumage.-- How  is
588n090 agony,  & whilst strong emotions of reverence & PIETY are felt." it appears to me mere consequence of
192b086 instance there never may have been grade between PIG & tapir, yet from some common progenitor,--  Now
218b190 vitiated.-- This barely applies to plants Female PIG apt to produce monsters in Isle of France-- -- M
225b219 e offspring exactly intermediate.-- Reference to PIG & Dogs. My theory will make me deny the creation
234b255 saw with his own eyes. new gate. opening towards PIG.-- latch on other side.-- Pigs put legs over, &
240c003 in  the transmission of form, may explain mule & PIG being half way. Yet dogs sometimes like  father,
444e157 s to the resemblances of the snout of the mole & PIG in having two additional bones to give  strength
553m137 ssion & noises of monkeys go in groups. thus the PIG-tailed baboon, shoved out its lip, looking absur
623o051 y generations. (& under unknown conditions) for PIG will not so readily attain instinct of  pointing
624o051 e with it. thus a dog may be trained to hunt one PIG sooner than other, rather than change hunting in
239cIFC osophy of Zoology Waterhouse B C N[a]me of two PIGEONS,-- «with specks» which cross & keep colour  on
327cIBC ep Clarke's Travels.-- Temminck Hist. Nat. des PIGEONS et des Gallinaces. Silliman's Journal.  during
473s07r no  difference in calves--how is this in young PIGEONS--dogs--cattle? As we see the frame of  animals
494q004 otter.  Badger &c &c &c.-- (11) Keep. Tumbling PIGEONS. cross them with other breed.-- (12) About the
500g10a distant parts of Europe.-- Gould-- go over the PIGEONS, Philotis, Dacelo. Alcyone, where there are ve
513q020 oung state for skeleton== Does the tumbling of PIGEONS vary in manner & perfection &c &c--if so pr
211b162 Jan  1837, «by Eyton» Account of these, kinds of PIGS. difference in skeletons: VERY GOOD. Apteryx. a
225b217 ing of it) or how to make Indian Cow with bump & PIGS foot with cloven hoof) Ask Entomologists whethe
234b255 e. opening towards pig.-- latch on other side.-- PIGS put legs over, & then with snout lift up latch
234b256 Fox  sometimes introduced by ice <no> «only few» PIGS.-- birds mentioned. but few.-- CD [There was no
244c021 «has»  taken place with quadrupeds» p. 118. wild PIGS of Falklands, generally "red of brick" hair, ve
245c023 idered good species from dental characters, wild PIGS. said by Forrest to swim from one isld to anoth
276c124 there cannot be gradation. See what Eytons young PIGS-- if vertebrae much lengthened &c there may  be
302c202 LE it is (one of analogy or) LATELY ACQUIRED. In PIGS number of vertebrae. subject to variation. ther
312c228 see Jardines Journal.»-- consult on this point-- PIGS always go against this, without «number of vert
331dIFC m. for Eyton.-- Sir. R. Heron's case instead of PIGS with solid feet.-- 1838 [In this Book some curi
342d033 pheasants  & grouse different.--» (if so chinese PIGS &c, that hybrids were uncertain. Mr Drinkwater
345d042 sm.++» Three gentlemen of party all thought with PIGS &c.-- Experimentatio on crossing of the several sp
392d184 se «how are their instincts?» «Chineses & Common PIGS.--» Experimentatio on crossing of the several sp
402e018 n".-- p. 229. Borneo.-- only animals he heard of PIGS, small bears or badgers, deer, apes, baboons, m
406e036 nsact. Rabies, common to men, dogs, horses cows, PIGS' & sheep.-- diseases common to men & animals cow
431e123 , though perhaps a true element) «give examples, PIGS, with small chinese boars &c &c &c» p. 10 offsp
465t081 ing has had somewhat to do with it. mem. dogs «& PIGS» in Polynesia; & dogs in S. America «Rengger.»
511q018 ally rejected?? (8) Get Sir. R. Heron to give me PIGS foot undivided, & more particulars regarding ef
511q018 s regarding effects of crossing them with common PIGS= [it is a Lincolnshire Breed]CD-- Sir. R. H. su
552m131 o carried from all parts straw to make its nest. PIGS & Elephants, (both Pachyderms) much intellect.-
582n066 her people are about: this is analogous to young PIGS hiding themselves; & heredetary remains of sava
289c161 will cease to doubt :Scales into Teeth in Bering PIKE (Waterhouse) Magazine of Zooly & Bot-- Vol II p
099a050 ittle force & varying direction.--> Therefore in PILE of mud from Trapiches. inclined layer!!!.-- The
128a128 w the sea mark.-- If mountain chains are matter PILED up. over crevice from effect of general elevati
639j28z legs-- Kangaroo. only a caricature; Penguin.-- PINCERS in Scorpion & Crust in Squilla. & Mantis. C D
536m070 y more than «prevent» heart beat- remember how PINCHER does just the same; I noticed this by perceivi
563n002 ] may be tempted to attack him from jealousy. (PINCHER & Nina)-- or to take away food &c &c-- Now  if
615o36v leading to touch a particular organ.-- I think PINCHER shows surprise, walking home one day met  him,
451e178 Thârû & the Dhangar who can live there & do not PINE visibly. p. 337. it would appear as if p.  345.
514q21. forms found there-- Lindley says that only one PINEAPLE Horticulturists (1) Are sterile hybris healt
207b151 s no trees succeed so well, at St. Helena, as. PINEASTER & Mimosa called Botany Bay Willow V. Dr Royle
217b187 ecies. Brazil {P} Chile, Norfolk Isl.-- Isle of PINES-- Australia.-- A <South» American «form of» Lat
346d047 the  Shepherd dogs.-- Aug. 24th. Was struck with PINK shade on plumage of the Pelican.-- Mem pink spo
346d047 with pink shade on plumage of the Pelican.-- Mem PINK spots on Albatross, on some Gulls. Flamingo-- (
160g090 h covered by turf 2ft. 8- long of syenite with PINKISH felspar;-- whole hill dark grey fine  grained.
514q21. ry?= Bunbury says no «hollow» spur.-- Ask about PINKS & Solanum impregnation before flower open. (An.
420e090 staceous animal is mentioned which inhabits the PINNA of Rio Janeiro, (like some Mediterranean specie
633j54r quire earth, why not created to live on alpine PINNACLE? if we once to presume that God «created plan
632j53r ock might have been uninhabited as the Alpine PINNACLES.-- One thing must be admitted there would not
499q009 en) is ball of pollen on Bees thighs (18) Place PIN'S heads with Bird lime near male yew tree & see w
339d025 finch & Greenfinch bred, & surely wild Duck & «PINTAIL» Widgeon!-- Divides animals «world into Zoolog
```

```
340d026 instinct in animals. Heard at Zoolog Soc their PINTAIL & Common Ducks, breed one with another-- & hyb
359d089 ther,-- (& not very like either either wild or PINTAIL duck) from which they were descended-- they de
359d089 ch they were descended-- they descend from 1/2 PINTAIL <into> «by» duck, into pintail.-- Of these the
359d089 escend from 1/2 pintail <into> «by» duck, into PINTAIL.-- Of these there were four, two like each oth
393dIBC dation can be made artificially.-- Are hybrids PINTAIL & common ducks. similar inter se? Zoolog. Gard
387d169 History .p. 96. Vol I. Notice the Syngnathus, or PIPE fish the male of which receives <young> «eggs»
412e057 orm.-- I have already given various examples The PIPE-fish is instance of part of the hermaphrodite s
461t025 ranslation of P. Fries most curious paper on the PIPE-fish-- which he divides into two divisions, one
189b072 heir race is not utterly cut off:-- like golden PIPPEN. if produced by seed go on.-- otherwise all di
192b083 ston Pippin, produce Ribstone Pippins, & Golden PIPPEN, goldens-- hence-- sub-varieties & hence possi
356d070 who eat every species new.-- Sept. 8'. A Golden PIPPEN or Ribston do producing occasionally (as Fox s
386d165 r it retains same length of life.-- like Golden PIPPEN trees! How is this with buds of plants, does a
187b063 ver They die; without they change; like Golden PIPPENS, it is a generation of species like generation
192b083 ells me, that beyond all doubt seeds of Ribstone PIPPIN, produce Ribstone Pippins, & Golden Pippin, go
228b229 both genera formerly abundant. Seed of Ribston PIPPIN tree <go> producing crab. is the offspring of
228b230 s from other trees.???? Do the seeds of Ribston PIPPIN & Golden Pippin &c produce real crabs, & in ea
228b230 es.???? Do the seeds of Ribston Pippin & Golden PIPPIN &c produce real crabs, & in each case similar
192b083 oubt seeds of Ribston Pippin, produce Ribstone PIPPINS, & Golden Pippen, goldens-- hence-- sub-variet
318c253 history of first appearance of the S. American PIPRA Flycatcher, which is now becoming common-- like
318c255 ably a «chance» wanderer like the first pair of PIPRA flycatcher.-- Bachman says he thinks the Mockin
453e183 toria della Riproduzione Vegetale". by Gallesio. PISA 1816 p. 27. Dr. Holland. Are there instances of
317c250 heights of St Jago found a Euphorbia so near PISCATORIA as scarcely to be distinguished from it.-- &
545m106 at is show all the teeth: «& make noise not like PISH, but like chit-chit-chit, quickly uncovering th
044r072 great body of lavas have flowed from centre-- PISOLITIC balls occur in the Ashes which fill up theatr
558m153 put their necks straight out & hiss.-- (Hyaena PISSES from fear so does man.-- & so dog]:CD Man grin
466t093 the Humbles in crawling out brush over anther & PISTIL & one I SAW IMPREGNATE by pollen with which <b
467t099 little» less surprised at Henslow's remark that PISTIL does not become abortive. Examined in microsco
467t104 at base of upper petal & the curvature of <an> PISTIL, etc lies in gangway= In Lotus corniculatus sa
498q008 of seeds known?-- Linnaeus has shown that each PISTIL is connected with separate division of germen
314c237 of a curious plant where a tube consisting of PISTILS & stamens united into long organ, moved on bei
466t093 <bees> «a bee» was dusted over. {P} Stamens & PISTILS curve upwards, so that anthers & stigma lie in
466t094 s not differently coloured & to which stamen & PISTILS have no relation. In Monk's Hood, a bee enteri
466t094 ry» cross directly over the bunch of anthers & PISTILS, but these <do> do not bend up-- In Lark-spur,
023r012 contents of its maw, amongst which were things PITCHED over board early in the passage!!-- M. Labilla
045r075 ewspapers)--At the Calabrian earthquake things PITCHED off the ground. «Ulloa states that Volcanos!!
073r162 ncomprehensible that if meteoric stones simply PITCHED from mo«o»n, that the metals should be those w
103a064 ntains being formed by crust being too large & PITCHING against each other, is, I suspect much weaken
033r043 at Teneriffe Wall of Porph. Lava with base of PITCHSTONE; Mem Galapagos. chiefly red glassy scoriae.-
040r063 ol III P. 124. Lyell. dikes have a parting of PITCHSTONE; which is described as very rare Mem. St Hel
059r121 colour.-- The Brecciated structure of all the PITCHSTONE (which I have seen). is a kind of concretion
057r112 -Earthquakes act as ploughs [,] Volcanos as Marl-PITS: Consider well age of Bones. = slowness of elev
548m116 «p. 140» case of double consciousness, one would PITY suffering in one state almost as much as in the
556m145 cisors.-- like <W[...]» person says "what a PITY"-- Lavater's Essays on Physiognomy translated b
608o026 wickedness.-- it would however be more proper to PITY than to hate & be †A man may put himself in the
608o026 weak & sickly as we do the wicked.--we ought to PITY & assist & educate by putting contingencies in
026r021 es in determinate planes ∴ such action can take PLACE in melted rocks The frequent coincidence of lin
026r022 & Organic remains does not appear to have taken PLACE in the Cordillera of S. America. Study Geolog:
027r023 Concretions & laminae, show what movements take PLACE in semiconsolidated rocks P xv. mentions in wha
038r051 ct near the Red Sea.--which occurred in a sandy PLACE.--(the sound was long & prolonged). NB, Is it g
044r073 ate for this science. that Europe was its birth PLACE.--Some general reflections might be introduced
045r076 as I have heards how lucky! when they hear of a PLACE having a pretty severe shock), are much more cu
054r102 e E & W! at Payta. talcose slates, do at latter PLACE. sandy. sandstone with gypsum, covered by limes
061r128 bouring continent: ≠ as if any creation «taking PLACE» over certain area must have peculiar character
069r149 from impossibility of such change having taken PLACE unrecorded must be insensible. Quantity of matt
102a061 hen quartz &c, owing to separation having taken PLACE most gradually, first the more fusible substanc
107a078 t subject. PP-- I think from dislocation taking PLACE chiefly beneath water & volcanos. crust must be
117a101 immense amount of change which must have taken PLACE, otherwise the world would daily be scene of ru
118a103 f amount denudation.-- This may be added to any PLACE where dikes described-- {P} Cordillera. St Hele
126a124 n, where the circulations from surface can take PLACE.-- ‖ the depth of frozen soil is against this v
128a128 t of general elevation.-- when subsidence takes PLACE.-- Mountain will first fall-- the problem will
132a138 ans» «lakes» bottom. if not colder than mean of PLACE, shows earth not with central heat.--» «(does M
153g054 ite entrance into Glen Fintec a kind of landing PLACE is formed Ben Erin summit 27.813. 65 55 degree?
160g093 radual slope-- : on N side.. dies away on rocky PLACE, but narrow shelves just like road of Glen Roy-
172b006 ogs ibis same as formerly but separate a pair & PLACE them on fresh isld. it is very doubtful whether
178b030 where other species have <come» «"air" of that PLACE. Will it be said, those have been there created
195b099 .-- This view of propagation gives.-- <ro» hiding PLACE for many unintelligible structures. it might ha
206b146 rdy, & more liable to disease" If population of PLACE be constant «say 2000» and at present day, ever
210b158 manu» Does it not present analogy to what takes PLACE from time? Von Buch distinctly states that perm
230b236 st fitted to that country when change has taken PLACE, Nature [not located] Any change suddenly acqui
230b240 nts turn wild.-- (for we know that such can take PLACE without impregnating each other), for if they a
244c021 d now represent. what <actually is» «has» taken PLACE with quadrupeds» p. 118. wild pigs of Falklands
245c024 asia-- & oceania, offers many varieties in each PLACE to puzzle naturalists.-- p. 372. Bourous. the B
269c101 ways assisted by fluid currents which, may take PLACE in Metamorphic action.-- Geograph Journal. vol
273c111 wks & Milvulus &c would instantly fill up their PLACE.-- Humming bird there is strongly marked variet
277c128 form which probably will be slow. but must take PLACE-- such a classification would answer every purp
284c150 f replacement of one species by another. supply PLACE in each others‖ economy Dr. S. showed that sava
290c162 ation-- Heaven knows how.-- This reaction takes PLACE in every organ‖ «Hence «method» of generation i
303c205 hout creation this would have been fair, but to PLACE all that» ever lived <on» into one list is unfai
308c218 the object, there is really no natural starting PLACE, because there is nothing more elementary than
315c240 ation be absolute thing, the creation must take PLACE on[ly] when creator sees. the means of transpor
318c253 on in Appendix) showing WHAT CHANGES are taking PLACE & how birds are extending their ranges. «even m
343d037 inent ones in <latte» one: The latter will take PLACE when Conditions are unfavourable to numbers of
343d039 ong separation which otherwise would have taken PLACE. otherwise in 10,000 years Negro probably a dis
350d056 thing in vain, therfore organ fitted to animals PLACE in creation.-- thus senses, especially sight co
354d062 g a constant round,--each particle is placed in PLACE of last by the ordering of the nerves, but in d
355d068 o see bearing of scattered facts..-- What takes PLACE in the formation of a bud-- the very same must
355d068 he formation of a bud-- the very same must take PLACE in copulation-- (Man & woman separate parts of
363d102 light» different-- Barn Owl <the» in the former PLACE breeds in <flags» «thick vegetation» in swamps-
369d115 research, in order to establish entirely their PLACE in nature, as well as fully to understand their
384d162 em.--» so that Molluscous hermaphroditism takes PLACE.-- thus one organ in each becomes obliterated,
384d162 comes obliterated, & sexes as in Vertebrate tak PLACE.-- ∴ Every man & woman is hermaphrodite:-- ∴ de
386d167 s occur in animals). latter the division taking PLACE from outside inwards. & in animals from inside
388d172 ermaphrodite is supported by change which takes PLACE in old age of female assuming plumage of cock,
413e060 ual. Buckland's Reliqu: Diluv. says Africa only PLACE, where, Elephant, Rhinoceros, Hippot, Haena &c
417e076 ammiferous» <vert» animal, which would find its PLACE in the Systema Naturae.-- Mr. Knight makes this
429e114 see how full nature. how firmly each holds its PLACE.-- When we hear from authors (Ramond. Hort. Tra
430e118 subjected to some influence, & this would take PLACE from changing country: but greyhound. & poutter
431e122 tion.-- Do species migrate & die out.?-- In the PLACE where any species is most common, we need not l
434e130 n us the change, which <to» seems to have taken PLACE.-- Almost all Dioecious & Monooecious plants ha
440e147 to allow any other kind of animal to usurp its PLACE.-- & therefore the degree of injuriousness must
442e151 enerations.-- any amount of generation may take PLACE by gemmation «My theory will not admit this, no
493q003 spring from A. into B. from B. into A. as takes PLACE in mules ass & horse-- important. (In crosses d
495q005 (latter I think certainly not) (3) Sow seeds & PLACE cuttings or bulbs in several different soils &
495q005 e & see what effect on organs of generation (5) PLACE pollen of Red Cabbage «mixed with own pollen» o
495q005 on springs up true.-- This in fact always takes PLACE in natural Hybrids of Cabbages (7) Sow <daisy»
496q006 ways barren-- if not how does impregnation take PLACE male & female flower in same receptacle (8) Mak
496q006 ther made hens cast Holly-seed & they grew» (9) PLACE. Snap-Dragon. (I have seen one monstrous) Fox G
499q009 r broken) is ball of pollen on Bees thighs (18) PLACE pin's heads with Bird lime near male yew tree &
```

Page ***(Key Word)***

501q011 bject-- (31) Ask Henslow for list of annuals to PLACE in Hot house to see effect on generative organs
508q016 Hydrocele Dr. H. thinks asthma in females takes PLACE of gout.-- How are livers obscure organ. no ans
513q21. rom him)‖ Does impregnation ever regularly take PLACE in unopened flower-- <doubt> disbelieve this in
515q021 produced. & 2d if so, whether concepcion takes PLACE,-- the mere fact of seeds ripening has scarcely
522m012 to & mentioned as a thing which had just taken PLACE.-- as if the idea of time had been disturbed.--
531m052 remains, when the actual movement does not take PLACE.-- A start is HABITUAL movement to avoid any da
532m056 ys & became fat! What remarkable affection to a PLACE.-- How like strong feelings of Man.-- The sensa
534m062 tions, & memory is repetition of whatever takes PLACE in brain. when sensation is perceived.= = Aug.
544m102 n of all <rules> relations of time, <identity,> PLACE & personal connections-- ideas are strung toget
546m109 ould say how many happy days he spent in such a PLACE.-- Vide page 103, supra (by mistake) have lower
550m126 r own property. (--regarding food & in birds of PLACE for nest.)-- with dogs "have notion of masters
554m138 Senegal he thinks Callitrix Sebe??) he has seen PLACE its head downwards to look up womens petticoats
568m018 tion, comparison of sensations would first take PLACE, whether to pursue immediate inclination or som
573m036 by special act, provided with its instincts its PLACE in nature. its range, its-- &c &c:--must be a s
577m050 association,-- a certain round of actions take PLACE every day, & closing of the leaves, comes on fr
593m109 - the social affections of animal taking man in PLACE of other animals is hostile «is subversive of»
613o033 parts of body. to do such.-- All this can take PLACE & man not conscious as in sleep; or in sleep is
614o037 afinite proportions. (different from what takes PLACE out of bodies) really less wonderful than thoug
615o038 great modification in brain would probably take PLACE without corresponding change in «external» man;
618o41v t. (The objects of thought have no reference to PLACE. [We see a particle move one to another, & <the
623o050 trine, for [RHC] 9) We can thus explain love of PLACE.-- although here we have not received pleasure
623o050 ugh here we have not received pleasure from the PLACE, but merely in the place. & yet place calls up
623o050 ived pleasure from the place, but merely in the PLACE. & yet place calls up pleasure.-- [LHC] the ins
623o050 from the place, but merely in the place. & yet PLACE calls up pleasure.-- [LHC] the instinct of soci
633j53v rolled along, & splits when it comes to a damp. PLACE.-- Kolreuter mentions some hybrid, whose flower
634j55r increase cold», & did mammifers then take their PLACE? Would they not first occupy the Poles? Is this
033r043 Teneriffe. also Cotopaxi has a <[...]> cylinder PLACED on the rim of conical crater: at Teneriffe Wal
147g018 Alluvium which without lake or sea could not be PLACED in present position Thursday Evening ½ past 8
200b122 descendant, which of course most rare, or when PLACED together they will breed.-- But what a charact
260c068 r Yarrell says-- that some birds or animals are PLACED in white rooms to give tinge to offspring.-- D
272c108 ns the Staphylinidae on St. Pauls Rocks must be PLACED under Gould says most subgenera confined to co
278c128 e empirical because as soon as two species were PLACED in different subgenera, then it would be usele
287c157 put man on the throne (of reason), she has also PLACED a series of animals on the steps that lead up
334d010 smell.-- Fox says he knew «a» carter well, who PLACED his stallion as second horse between <whe> sha
354d062 undergoing a constant round,--each particle is PLACED in place of last by the ordering of the nerves
364d104 r. white peeled rods mentioned in old Testament PLACED before sheep-- it has been thought that silver
430e118 ulty apparent by cross-questioning.-- » even if PLACED on Isld-- if &c &c.-- Then give my theory.-- e
435e133 found the <poll> masses of pollen of Asclepias PLACED on Orchis (so very different) that the granule
436e138 the Rev. E. Stanley Vol I. p. 72.-- Goldfinches PLACED near, but not in sight of each other will sing
466t094 t bees visiting it.» In Columbine nectaries are PLACED all round flower as they are in Crown-Imperial
498q007 . «& 374» Will plant accustomed to rich soil, when PLACED in very poor flower, but not fruit-- -- Do not
504q13v think very likely the Peas to cross ought to be PLACED far from all other Peas, from Wiegman Shrewsbu
531m050 ance.-- people very often forget where money is PLACED.-- (How often one forgets where put one key. w
531m050 e forgets where put one key. where all keys are PLACED) Memory cannot solely be number of times repea
534m062 Vol. I. p. 106. Col. Sykes on Formica indefessa PLACED table in cups of water which they waded. or sw
560mIBC very early in life Do they wink, when anything PLACED before their eyes, very young, before experien
564n004 that he had seen chicken only hatched few hours PLACED on table & when fly ran past it. cocked its he
565n007 o «man as» animals. comes at distance, mouth is PLACED open.-- Hence becomes instinctive to fear., as
566n012 ance probably such a thing as thunder. would be PLACED to the will of God. Zoology itself is now pure
585n077 henge raised, yet not instinct, but if all men PLACED stones in same position, it would be instinct-
407e041 ica not so equatorial..-- The fact of No. Mam: PLACENT: insectivore being in S. America & Australia.
373d132 rough a different series of changes from the PLACENTATES, Having Hair. like true Mammalia, no more wo
196b106 South Hemisphere. Were they produced in several PLACES & died off in some? Why did not fossil horse b
400e012 , «& 374» Spaniards says no Tortoises in other «PLACES» besides Galapagos do. p. 376. Isle Tres Maria
423e096 ones. would soon be disorganized to fill their PLACES.-- The Geologico-geographic changes must tend
495q05a f Paper. covered with some sticky stuff in flat PLACES & see whether wind, on «dry» windy day, «flowe
497q006 specially if date of wood be known» & other odd PLACES & see what plants will spring up which will sh
524m021 orn) though she never naturally talked of these PLACES.-- My F. says, shows that early impressions ar
543m100 s Jones observed, in playing chess however many PLACES, & contingency a man has keep in mind. all is
573n036 st be a special act, or result of laws. yet we PLACIDLY believe the Astronomer, when he tells us sate
298c189 atural History Vol. I. p. 185 case of tit lark PLACING withered grass over nest, when often looked at
567n015 st sure Fuegia Basket did. & Jemmy, when Chico PLAGUED him-- Animals I should think would not have an
046r080 efore motion of sea ought to be considered as a PLAIN movement communicated to it as well as by the v
057r114 of rocks of very diff. ages. at Port Desire on PLAIN. & interstratified.-- Urge fact of Boulders not
116a100 o, extremity of mountains of Cordova project on PLAIN, like <re> a reef on a sea beach-- «p. 151» fir
126a122 Does condensed metal, conduct heat better than PLAIN?-- Mem 1000 {P} how easily water percolates roc
135a145 tle branches at {P} from each side intercepting PLAIN & dividing it:-- Hopkins fissure at {P}.-- G. J
146g014 neighbourhood of Tyndrum where a large sort of <PLAIN> space is thickly studded with ridges & flat to
149g033 P} do they extend round hill too low line drawn PLAIN red talus line on N. side of Spean most clear &
151g042 nnecting flat on one side with irregular gravel PLAIN of other, which must have been waterworn after
151g042 .-- 4th shelf runs up some way on great sloping PLAIN of alluvium (much corroded by rivers) & not to
151g043 vium (much corroded by rivers) & not to head of PLAIN.-- but below houses where rivulet enters two gr
152g049 iver & bed of river about 40 ft beneath general PLAIN. 30.127 A 72 degree Air 65 degree? at level of
154g057 W of Ben Erin {P} Shelf of Glen Guoy flat peat PLAIN divortium aquarium-- tidal channel-- 12ft obscu
155g062 uced» from same «cause» as «great» spit <is> or PLAIN <now> formed on shelf 4th 2d & 3d can be traced
155g064 constantly alter with falling sea & so corrode PLAIN into terrace as regressed What <alter> a balanc
156g071 of shelf 3d almost all granite pebbles Level of PLAIN of 4th shelf at head of Lower Glenroy 29.581 A
156g071 Glenroy 29.581 A 82 75 degree? From this point PLAIN appears like one uniform slope slightly bending
156g072 ,-- Barometer on shelf 3d. 29.455 A 83 degree ∴ PLAIN of 4 minute shelf slope, above «line of 4th» sh
158g081 where two rivers unite in Upper Glen Roy great PLAIN about 60 ft beneath shelf peat on pebbles tidal
158g081 about 60 ft beneath shelf peat on pebbles tidal PLAIN as sea gradually retired, hard to explain on ri
287c157 vier that relation of theoretical astronomer to PLAIN observer‖ +++. between Mammalia & fishes, one p
307c216 organs «of very same kind» in these cases, have PLAIN meaning & none in other case! Savigny has shown
377d140 uma, same colour as Lion. because inhabitant of PLAIN & Jaguar of woods &c like ground birds [not loc
382d158 came it nipples <are> «though» abortive, are so PLAIN in Man, & yet no trace of abortive womb, or ova
404e024 t rule applies to wild animals only. from which PLAIN inference might be drawn that whole infertility
460t015 e & common goose.-- the stripe down back pretty PLAIN in in these «half» «3/4» bred ones-- The brothe
509q017 s??» Are abortive organs as «young» teeth, more PLAIN in young Rhinoceros or Whale, than in old?? Fal
055r108 to see tertiary places consumed) Where slope PLAINLY indicates former boundary. (as in other unwor
114a095 know origin of greenstone tells subsidence as PLAINLY as Temple of Serapis. (now we have banished di
029r032 that Boulders do not occur in the South African PLAINLY.-- Sydney no I believe the secondary? formatio
036r050 pe Consult charts for distribution of pebbles.--PLAINS. off coast of Patagonia.-- British channel &c
039r061 e analogy of movements of W coast in explaining PLAINS because such are found in perfection on that s
047r083 of an injected mass of fluid rock In Patagonia PLAINS. long periods of rest & vice versâ more likely
048r085 ies of» Mastodons. which ranged from Equatorial PLAINS to S. Patagonia. To the Megatherium.--To the H
048r085 of man Feeling surprise at Mastodon inhabiting PLAINS of Patagonia is removed by reflecting on the n
048r086 to the Sibiria case We must not think alluvial PLAINS «always» most favourable; In what part of the
055r108 high to age. (we do not wonder to see tertiary PLAINS consumed) Where slope «plainly» indicates form
067r143 logy. Conybeare Lava in Cordillera & on Eastern PLAINS «by Antuco». Athenaeum April 1836 (p302) Colec
070r155 uquisaca. half across the continent.--He states PLAINS of Mendoza smooth. Sir W. P. states that in He
077r170 e & breccia is the same with that on surface of PLAINS of Amazon, no relation--there is more modern li
085a001 th of S. Cruz. from ascertained inclination. of PLAINS: Lias in Shropshire. or some other wonderful o
087a010 lorescence nothing -- Study account.-- Alluvial PLAINS of Mississippi -- No Vol. I. p212. Cuvier Oss
104a068 temperature ¿ with pressure? Salt on surface of PLAINS due to whole moisture being lost by evaporatio
121a112 racks 11th August. 1838 Near Woolwich there are PLAINS & valleys just like Patagonia, & many shells i
148g027 ine entered terraces formed successive bays but PLAINS sloped centre-wards which would not have taken
150g035 hows NB. Lake gradually draining off would form PLAINS such as those near Bridge Roy (& other cases)
155g063 t, where I believe they end in upwards inclined PLAINS, as in Corry. & as «as I believe in side ravin

414e064 white.-- The peculiar skulls of the men on the PLAINS of Bolivia-- strictly fossil «& in Van Diemen'
450a175 islands.-- The horse is only found wild in the PLAINS of Celebes. (but language shows that probably
529m041 ring one in India. & a tiger stalked across the PLAINS, how ones feelings would be excited, & how the
540m086 orth,-- the elevated table land of Peru the hot PLAINS of the Amazons & Brazil-- with the negros of A
568n018 tructure. «-- American monkeys utter pleasant PLAINTIVE cry--» The taste of recurring sounds in Harmo
198b114 h some may be) have not been created on the same PLAN. [Second resumé well worth studying] CD says ge
255c053 nge in her offspring. ,has invented all kinds of PLAN to insure stability; but isolate your species h
255c053 o insure stability; but isolate your species her PLAN is frustrated or rather a new principle is brou
277c126 rupestris -- good species it is reverting to old PLAN, but reason now assigned for doing so There sho
348d050 (p 6) as expressing natural affinities\ Macleays PLAN of arrangement depends on the organs judged to
429e113 osing of horns.-- I do not believe this Nature's PLAN.-- Whether we can or not trace history of first
498q008 at all from the plants? If not, I am surprised <PLAN> such plants do not degenerate,-- as the Bees w
541m089 ing success by decreasing headache] & found best PLAN was allowing my mind to skip from subject to su
634j54v saying "It is the determination to adhere to a PLAN once adopted; & it is from these very circumsta
372d129 ed.-- Bud probably is like cutting off tail of PLANARIA, the whole grown to that part.-- claw added t
416e070 hrodites mutually couple,-- now how is it-- in PLANARIA, they couple-- CD [lowest terrestrial animals
6040016 ed in 1758-- Read the Review or the article. A PLANARIA must be looked at as animal, with consciousne
6040016 food-- crawling from light.-- Yet we can split PLANARIA into three animals, & this consciousness beco
641j29v ormed in Mollusca, Articulata, & Vertebrata, & PLANARIA, & light affecting plants. in insects the end
170b001 polypi, gemmiparous propagation. bisection of PLANARIAE. &c &c.-- The ordinary kind <the> which is a
205b143 idous animal p. 347. Vol I.-- compare with my PLANARIAE Leaches out of water Does the odd Petrel of F
306c214 Isd.? ought to agree with Java?? Terrestrial PLANARIAE assuming bright colours., good instance of co
477z001 rine animals-- CXVI. P 111 do Observations on PLANARIAE by Johnson CXII. & CXV do Azara Voyage Vol I
479z005 la,-- Salpa Anatifs without shells.! p 442.-- PLANARIAE p 451.-- many molluscs Under the name of Sagi
481z010 pear very trifling Also Berre «p. 8» (I think PLANARIAE) Sagittella, or Fleche «p. 8» my little anima
483z013 in Vol I, Zoolog: Transact. before writing on PLANARIAE or Polypi & is especially grand paper. p. 387
483z013 collections & my own: & Capt. King's p 453-- PLANARIAE velellae (Less) parasitical on Vellellae in A
601o009 s feigning death upon seeing an object.-- are PLANARIAE conscious.-- Consciousness bears some relatio
095a038 ars to be outside of the Cordillera-- Near the PLANCHON talks of very much of Gypsum.-- The officers
258c064 ermination of species-- Epidemic amongst trees. PLANE trees all died certain year. Extreme difficulty
286c155 ls in paper as drying plant, all brought in one PLANE Fleming Quarterly review says nat: fam: of Will
026r021 in Phonolite arrange themselves in determinate PLANES ∴ such action can take place in melted rocks T
058r115 the infinitesimal cryst. arrange themselves in PLANES. «Mem silky lustre» ask Erasmus. whether elect
102a060 ncretions.-- Septaria in concretion arranged in PLANES, case of separation.-- the branching cracks--
120a107 ty with which the earth is cracking by vertical PLANES into small pieces-- mem coal-field.-- the stru
090a021 r speculates from believing meteorolite but old PLANET, that inside our globe melted magnetic metals.
174b015 l this apply to whole organic kingdom, when our PLANET first cooled.-- Countries longest separated gr
195b101 might formerly have said that God ordered, each PLANET to move in its particular destiny.-- In same m
284c147 of plant not so many.-- The quantity of life on PLANET at different periods, depends,-- on relations
443e156 what is older-- what relation in duration of a PLANET to our lives-- Being myself a geologist, I hav
090a024 if so other cause besides thin vapour bringing PLANETS to an end? Fragmentary granite showing schisto
347d049 .!!! whole universe so adapted!!! & not man to PLANETS.-- instance of arrogance!! August-- 29th.-- Ma
573n036 a notion of a Deity We can allow «satellites», PLANETS, suns. universe, nay whole systems of universe
634j55r will of the deity, to create animals on certain PLANS.-- is no explanation-- it has not the character
192b083 .-- Get instances of a variety of fruit tree or PLANT run wild in foreign country.-- When one sees ni
223b210 y & so make that sudden step. species or not. A PLANT, submits to more individual change, (as some ani
223b211 . but it is instead a stunted & diseased form a PLANT, adapted to A. B. C. D.-- Destroy plants B. C.
225b218 sts whether they know of any case of introduced PLANT, which any insects hav[e] become attached to.--
225b218 Bustards & &c!!!! An American & African form of PLANT being found in Tristan D'Acunha. may be said to
228b230 ring would come from them. Ask Henslow for some PLANT, whose seeds go back again, not a monstrous pla
228b230 ant, whose seeds go back again, not a monstrous PLANT, but any marked. variety.-- Strawberry produced
230b239 ty permanently transmitted.-- <It will admit> a PLANT will admit of a certain quantity of change at o
245c025 than those sinking, because arrival of any one PLANT might make conditions in any one isld different
259c063 s--Lamark's willing absurd, ∴ not applicable to PLANT, Epidemics of South Sea, wonderful case of exte
265c085 e, colouring matter being absurd.-- again dwarf PLANT in alpine district & dwarf plants from seed, on
283c146 ertain preexisting laws.-- .If only one kind of PLANT not so many.-- The quantity of life on planet a
286c155 ficulty in arranging animals in paper as drying PLANT, all brought in one plane Fleming Quarterly rev
298c192 g all the moisture.-- Henslow thinks if leaf of PLANT varies, <whole cross> «all organs» vary in plan
298c192 lant varies, <whole cross> «all organs» vary in PLANT. The variation in character of leaf of plants i
299c193 conclusion must be arrived at, when one sees a PLANT like Paris quadrifolium growing in one wood far
305c211 d, polypus & polypus, bud & bud, polypus & germ PLANT & seed.-- instincts in young animals, well deve
310c224 rpins assertion that globules of milk produce a PLANT capable of growing!! & propagating itself. In T
314c237 r Brown showed me Bauer's drawings of a curious PLANT where a tube consisting of pistils & stamens un
314c238 hange been effected.-- the consciousness of the PLANT that this part must be protected however it may
324c268 Vegetaux et animaux.-- on sleep & movements of PLANT/ If.:4s Voyage aux terres australes. Chapt. XIX.
337d021 nerve becomes sensitive to light.-- (Mem whole PLANT may be considered as one large eye-- have they
372d131 of generation shown by their happening in same PLANT.-- The Marsupial structure shows that they beca
381d157 tion might make one set of organs barren in one PLANT & not in other), Hunter <asks> p. 36 is thought
389d176 itself off.--» as might be inferred from annual PLANT being prolonged till it has bred.-- Offspring l
397e003 hilosophy to believe that no stone can fall, or PLANT rise, without the immediate agency of the deity
401e014 ety.-- for they are all made by fertilizing one PLANT with another-- Uncle John says he has no doubt
401e017 owslip Uncle J. says common belief. that female PLANT impresses main features on offspring. & male th
416e072 n difference in external conditions.-- -- as in PLANT up a mountain-- In races the differences depend
428e111 lings.-- Now my principle does not apply to any PLANT reared artificially, & only very partially to t
429e115 tres precisely & stops at 2600. & yet know that PLANT can be cultivated with ease near London.-- what
431e122 k like fresh Creation. the gardener separates a PLANT he wishes to vary-- domesticated animals tend t
434e129 a few seeds,-- -- Ruscus aculeatus. a dioecious PLANT, in which the Male plant sometimes bears female
434e129 aculeatus. a dioecious plant, in which the Male PLANT sometimes bears female flowers, the organs are
438e142 Man can offer of changes.-- as desert «or rock» PLANT probably would do-- or be with difficulty be ke
446e163 y great difficulty in my theory, here we have a PLANT remaining constant, without crossing.-- & propa
447e164 the run of chances, would prevent it varying. A PLANT <producing> propagating itself by buds is in sa
448e165 e are endless curious facts about every part of PLANT producing buds, so that Turpin says each cell o
448e165 roducing buds, so that Turpin says each cell of PLANT is individual.-- Most plants which propagate ra
454e184 y sleepy Mimosa, nearly allied to the Sensitive PLANT.-- p. 290. Dr. Edwards in his essay on Spermati
468t112 t P every minute to Fraxinella «& from <flower> PLANT to plant.»-- to my grt surprise-- I found all,
468t112 minute to Fraxinella «& from <flower> plant to PLANT.»-- to my grt surprise-- I found all, stamens s
473s006 s has described in Linn: Transacts. a Brazilian PLANT «in Lin. Transacts. it has three stamens» inter
473s006 & other plants-- & Wallich has described Indian PLANT.»-- June /42/-- June/42/ You can select cattle &
491q01v & then some hours afterwards of nearly related PLANT & see if first pollen produces any effect, as i
495q05a seeds for week in Salt. artificial water.-- 12. PLANT two races of Cabbages near each other-- & enclo
497q06v y difficult to cross.-- are there races-- if so PLANT them together. & raise. seed.-- In letter Mr He
498q007 Orchidaceous plants no other case.-- (6) Will PLANT accustomed to rich soil, when placed in very po
498q08v of different species growing White Mullein good PLANT to sow & try to get other species <near> close
499q010 of plants in counties, as of rare green Cotton PLANT-- How large «area» clump there? Distinguishable
501q10a Oysters or Helix. Or does any «one» species of PLANT, vary in one region of Europe & less in another
501q011 by very minute insects.-- (30) Get Abberley to PLANT SINGLE Peas, Kidney Bean & Bean, intertwined, «
505q014 of boiled earth on top of House =Aristolochia, PLANT wh require insects to impregnate it (7) History
505q014 eather wet--? Linum flavum put in Spirits which PLANT seeds? (9) Melons fruit itself hybridised (10)
505q014 sparagus: result? = failed to germinate 16 Will PLANT some of the Thyme with abortive stamens by Terr
507q15v lip & primrose, but less strongly marked.-- 31. PLANT seeds of the Fuller's plants ,Teazle Dr. Hollan
513q21. generally not being mature at same time on same PLANT --Flora of Australian Mountains.-- Is setting o
536m072 ll animals., then an oyster has a polype (& a PLANT in some senses, perhaps, though from not having
559m156 ybrid can be made & ferns.--» Would a sensitive PLANT if irritated very regularly at one time every d
567n014 man. Has an oyster necessary notion of space-- PLANT though it moves doubtless has not.-- Turkey coc
611o034] This is true as long as movement of sensitive PLANT can be shewn to be direct physical effect of to
633j53v detrimental.-- p. 285 the seed-pod of a desert PLANT (Anastatica) is rolled along, & splits when it
636j56v gainst my theory of frequent intermarriage.-- A PLANT is in the same predicament as a group of bisexu
209b157 p. 145. In Humboldt great work de distribut. PLANTARUM. relation of genera to species in France is I

Page **(Key Word)**
327cIBC improvement of domesticated animals Fries de PLANTARUM proesentum crypt. transitu et analogia commen
188b068 white flower. all would come up white, though PLANTED in same soil with blue. Now this is same beari
441e149 the parents.-- Thus would a Crab tree vary if PLANTED in rich soil, I presume not, but its seeds, I
469t151 dfords Marrow fat, Early frame, Groom's Dwarf. PLANTED in rows «close to each other» & seeds gathered
469t151 double-blossomed «& dwarf-fan Bean» bean, were PLANTED in rows, & seeds gathered same year came up tr
470t176 & they seem slow= Maer 1840 My Father formerly PLANTED «Turkey or» Palmated and English, planted with
470t176 erly planted «Turkey or» Palmated and English, PLANTED within few yards of each other actually produc
492q002 hether Roses impregnate each other. when close PLANTED together: <do> Can Holyoak be raised distinct
500q10a 2. <Last> Year «before last» beans & peas were PLANTED in rows adjoining & seeds gathered there were
500q10a in rows adjoining & seeds gathered there were PLANTED «last year» pell mell, without sticks & seeds
500q10a sticks & seeds gathered & these are now to be PLANTED this year copied Gould.-- Number of species of
504q014 ?-- Yew trees sexes-- (3) Get Holyhoaks. races PLANTED & Linum Perenne.-- Herbert's. fact.= (4) Effec
505q014 owcumbers will they seed.?-- (11) Abberley has PLANTED seeds of pale green Cynoglossum. never germina
505q014 rs? Can flys' escape from old flower= (14) Has PLANTED seeds of Geranium pyrenaicum. small white-flow
253c046 rica not.-- Africa Camels?? Africa Bears??-- PLANTIGRADE Carnivora??-- «compare rodents of two countr
036r052 Andes in Chili. separate geographical ranges of PLANTS. V. Lyell. Chap XI Vol II. Urge the entire abs
054r102 anic!! New Zeeland rich in particular genera of PLANTS: All St. Catherine & coast Granite: P. 199; Fa
061r127 idea of Volc: islands. elevated. then peculiar PLANTS created. if for such mere points; then any mou
062r128 evergreen trees, arborescent grasses, parasitic PLANTS, Cacti: & with limits of no vegetation at S. S
087a007 s of Equatorial mountains so cold.-- Siberia no PLANTS to it, lately raised above level of the Sea. L
139a180 riments might be tried by comparing Zoophite to PLANTS.-- grafting length of life &c &c Will any inor
163g108 C. Watson Geographical distribution of British PLANTS Shropshire Quartz what substance is collected
171b003 ariety, according «to» circumstance,-- seeds of PLANTS sown in rich soil, many kinds, are produced, t
179b033 erica shows this.-- <If> Is there a tendency in PLANTS hybrids to go back?-- If so Men & plants toget
179b033 ncy in plants hybrids to go back?-- If so Men & PLANTS together would establish law. ≠as above stated
181b043 and Articulata. still greater between animals & PLANTS But yet besides affinities from three elements
182b046 Quinarian system is true, it will not occur in PLANTS which are in far larger Proportion terrestrial
186b059 y of such law.-- It would be curious to know in PLANTS, (or animals) whether, <in> races have tendenc
189b073 me species are bound together just like buds of PLANTS, which die at one time, though produced either
189b073 ced either sooner or later.-- Prove animal like PLANTS:-- trace gradation between associated & non as
190b075 on to facilities of communication Have races of PLANTS. ever been crossed really, if there is any dif
191b081 ny other, because less easily transported-- Mem PLANTS on Coral islets.-- Next to animals land birds.
193b092 peds-- Humboldt has written on the geography of PLANTS. Essai sur la Geographie des Plants. I Vol in
193b092 eography of plants. Essai sur la Geographie des PLANTS. I Vol in 4 degree.-- I have abstracted Mr Swa
194b096 ssils: ¿How are. South American shells? ¿Do not PLANTS, which have male & female organs together, yet
194b096 gans together, yet receive influence from other PLANTS.-- Does not Lyell give some argument about var
194b096 fficult to keep on account of pollen from other PLANTS because this may be applied to show all plants
194b096 plants because this may be applied to show all PLANTS do receive intermixture.-- But how with «herma
196b105 wo Rheas had existed in different Continents In PLANTS I believe not..-- It is a very great puzzle wh
196b108 <vebtetrata> «vertebrates» could exist without PLANTS & insects had been created; but on other hand
196b109 live without animals but not vice versâ. ¿could PLANTS live without carbonic acid gaz.)-- Yet unquest
196b110 great gaps.-- yet some analogy The existence of PLANTS, & their passage to animals appears greatest a
200b123 her you would expect equal fertility-- ditto in PLANTS.<==> It will be well to refer to Chamisso Vol
206b150 1837.» p 319 -- Brongniart.-- no dicotyledenous PLANTS & few Monocot in Coal formation? p. 320 <Think
209b156 h Edit. Flora of Islds very poor «(p. 145)» 25. PLANTS. 36 St Helena, without ferns.-- analogous to n
210b158 h others.-- Compares it to languages But how do PLANTS cross?-- = admirable discussion Von Buch says
210b159 enzie Iceland there 144 genera & 365 species of PLANTS not cryptogamic but I . 2,53.-- In know variet
210b159 .-- for instance three kinds of greyhound.-- In PLANTS. do the seeds of marked varieties produce no d
214b173 to shells of living seas.?-- Roxburgh. list of PLANTS in Beetsons St. Helena. -- Galapagos--Juan Fer
216b180 point will be to find whether know varieties in PLANTS do so.-- As in cacti &c &c.-- as in dogs inves
218b189 sion «amounting to impossibility, holds good in PLANTS» between all different forms; therefore when f
218b189 ty these are vitiated.-- This barely applies to PLANTS Female pig apt to produce monsters in Isle of
219b193 &c &c. (Mem discover what kinds of seeds. these PLANTS) [Mem Fact stated by Mr Don in island, Tenerif
219b193 s]CD & would creator <on volcanic island.> make PLANTS <grow closely> When this volcanic point appear
219b194 ic point appeared in the great ocean, have made PLANTS of American & African form, merely because int
219b195 rmichael Linn. Transacts. Vol XII.-- The Alpine PLANTS of the Alps. must be <Alpine> «new formations»
219b195 re species of lower genera altered. or northern PLANTS «No» CD[Mem. the antarctic flora must formerly
219b195 short space from mountains low down, therefore PLANTS common take an example from T. del Fuego.-- El
220b197 & Europe, as to produce same one. \Although in PLANTS, you cannot say that instinct perverted, yet o
220b197 stication depends on perversion of instinct (in PLANTS domestication on perversion of structures espe
220b197 of parent. Mem Lord Moreton's Mare. The fact of PLANTS going back hybrid plants; analogous to Men. &
220b198 on's Mare. The fact of plants going back hybrid PLANTS; analogous to Men. & dogs. Now if we take stru
221b200 ms, longest persistent?? do.-- The most perfect PLANTS Composites.--!!«good» those which have undergo
223b211 form a plant, adapted to A. B. C. D.-- Destroy PLANTS B. C. D. & A will soon form good species! The
224b214 great as between living thing without thought (PLANTS) & living thing with thoughts (animal). «∴ my
226b222 n land.-- (How is this with Fernando Po.). with PLANTS of St. Helena & Tristan D'Acunha, resolves its
226b222 of same genus, subsided land.-- Mauritius? <In PLANTS where do most species occur.)> Although the Ho
228b230 It really would be worth trying to isolate some PLANTS, under glass bells & see what offspring would
228b230 of ferns.-- hybridity showing connexion of two PLANTS. Animals-- whom we have made our slaves we do
230b236 h as would make species (except perhaps in some PLANTS & then a chain of steps is found in same mount
230b240 arcely possible to get evidence of two races of PLANTS run wild.-- (for we know that such can take pl
230b240 rids will not destroy that evidence, as so many PLANTS produce hybrids, or else whole fabric will be
230b241 change Get a good many examples of animals «or PLANTS» very close (take Europaean birds. Mr Goulds'
231b242 all show greater connexion in quadrupeds, <bu> PLANTS do not follow by any means.-- Ostriches.-- Hip
231b242 rent not travellers?? Royles case of Himalayan, PLANTS ¿migratory birds, he told me some story of cra
232b247 ve travelled towards Equator, <th> so would the PLANTS from extreme north, which according to all ana
235b272 Flemings Philosophy of Zoolog Royle on Himalaya PLANTS-- Would it not be possible to work through all
235b275 so change The GRAND QUESTION Are there races of PLANTS run wild or nearly so, which <breed> do not in
235b275 which <breed> do not intermix,--any cultivated PLANTS produced by seed.-- Lychnis.-- Flax.-- Read Sw
240c004 down, with ease, is analogous to what occurs in PLANTS.-- All these facts clearly point out two kinds
244c021 degree. p. 181 «who says insects Indian, like. PLANTS.--» It would be very important to show wide ra
248c030 red with, or generative organs affected as with PLANTS) no animals VERY different will breed together
259c063 y men born with any peculiarity. or any race of PLANTS--Lamark's willing absurd, ∵ not applicable to
265c085 -- again dwarf plant in alpine district & dwarf PLANTS from seed, one adaptation, other monster.-- Th
266c091 e Whewells Bridgewater treatise, (p..26). about PLANTS from Cape of Good Hope continuing for some tim
267c095 98. on a quaternary arrangement of Cryptogamic PLANTS.--Owing to plants not being adapted to Air!! p
267c095 ry arrangement of Cryptogamic plants.--Owing to PLANTS not being adapted to Air!! p. 11--&c. valuable
269c102 ph Journal. vol I. p. 17 &c excellent sketch of PLANTS of New Holland, supplementary to Appendix to F
269c102 New Holland.-- «Compare birds of Australia with PLANTS, with this object in view» The intimate relati
270c103 than in vegetables. p. 243 radiate animals <tu> PLANTS turned inside out. have position of organs of
270c104 y remarkable as shown by Carus how intermediate PLANTS are between animal life & "inorganic life".--
271c106 nandez in birds. but ¿whether to same island in PLANTS?-- What is this halo.-- Continents are not sta
284c148 ds, nor any other part of world.-- no other PLANTS peculiar to these isld.-- ¿Brown can «not» bea
290c162 hange only & not kind» insects-- & vertebrata & PLANTS. At first classification on generation might a
291c167 maphroditisms. one step in <scale>. Series-- in PLANTS we have a step between <mon> monoeecious & dio
291c167 ve a step between <mon> monoeecious & dioecious PLANTS in animals it will be difficult to imagine how
291c167 icult to imagine how sexes were separated.-- in PLANTS we have some flowers monoecious & others dioec
296c184 rm, others altered much.-- these others will be PLANTS & land animals. & land shells.-- all in short
298c192 in plant. The variation in character of leaf of PLANTS is remarkable what is analogous to it in anima
298c192 gous to it in animals?-- Babington says in most PLANTS, even those on Guernsey & on West coast of Ire
298c192 countries. These facts show how very permanent PLANTS are. & this conclusion must be arrived at, whe
299c193 rifolium growing in one wood far from any other PLANTS of same species Channel Islds (& probably Isle
299c193 ecies Channel Islds (& probably Isle of Man) no PLANTS peculiar to themselves. this remarkable compar
299c195 erative organs. first character.-- In dioecious PLANTS many of the Female flowers unimpregnated Babin
307c216 ive organs produced in domesticated animals, in PLANTS I presume there are.? get examples.-- for inst
310c224 ,000 & Melville Isd.-- West Africa & India some PLANTS same. --America.-- See Brown Congo Expedition:
310c224 a.-- See Brown Congo Expedition: 400 Australian PLANTS found in other parts of world Athenaeum June 3
313c236 nervous system of insects analogous.?-- («Even PLANTS have habitual actions.--» «this very important

314c239 erve scarcely any Australian character in Timor PLANTS, yet it seems there may be Eucalyptus!-- (Host
315c240 traits The common Mush room & other cryptogamic PLANTS same in Australia & Europe.-- if creation be a
315c241 gies might be drawn between habitual actions of PLANTS «when exciting cause is absent» & memory of an
315c241 se is absent» & memory of animals.-- (surely in PLANTS movements effects of irritability, though mean
315c241 re)-- it is most remarkable habitual actions in PLANTS, it allows of any degree in lowest animals --h
316c244 . Geographical range of shells like Cryptogamic PLANTS. of Marine kinds, there are some restricted ge
316c245 ermaphrodites-- eggs in groups.. Have Dioecious PLANTS more restricted ranges than other plants.-- Ma
316c245 ecious plants more restricted ranges than other PLANTS.-- Many «Some» genera confined to hot countrie
317c250 the Canary islands-- p. 250 admirable table of PLANTS of St. Jago showing many common to Canary. isl
321c275 wn's Appendix & excellent table of Canary isld: PLANTS Home's History of Man Transactions of the Ento
325c267 care, abortive organs produced in domesticated PLANTS; where function has ceased to be used as tendr
325c267 8 H. C. Watson on Geograph. Distrib: of British PLANTS. Humes Essay on H. Understanding (sometime) Du
326c266 loured maps. by Copenhagen Botanist of range of PLANTS Books quoted by Herbert. p. 338 Schiede in 182
326c265 us Britannicus. has remarks on acclimatizing of PLANTS. Herbert. p. 348. gives reference to Kolkreute
327c265 Metaphysical point of view Henslow has list of PLANTS-- in high points Read Volney's travels in Syri
327c265 one is found. very good to see whether peculiar PLANTS transported.-- Crawford. Eastern Archipelago.
327cIBC els in Syria Vol I. p. 71. account of Europaean PLANTS are the number of seeds greater.?-- Mem. for E
331dIFC increased?-- Where offspring, heterogenous, in PLANTS are in those which grow in low grounds
334d010 ch of hybrid geese very fine.--» How is it with PLANTS? This indicates a remarkable law, that first c
337d021 nsidered as one large eye-- have they smell, do PLANTS emit odour solely for others parts of creation
338d024 Athenaeum. p. 505. some (very poor account) of PLANTS of Nova Zenbla -- in review of Baers work Edin
339d025 according to varieties of Man.? «In Australia. PLANTS E & W very different.-- Man not so, but N. & S
339d025 & New +++ Caledonia. two races of Men, but not PLANTS» will it hold good.-- Thinks Temmink doubtful
353d060 pher», who has trace the structure of animals & PLANTS.-- he get merely a few pages.-- Hence (p. 59)
355d068 pulation-- (Man & woman separate parts of same PLANT<S>)-- now in some Polypi we see young bud changi
357d074 hybrid?-- When I show that island would have no PLANTS were it not for seeds being floated about.-- I
370d117 tinct.-- Collect cases of difficulty of growing PLANTS in all parts of world, thus tea tree in Brazil
372d129 «similar» body)» to another part of body.-- [in PLANTS does not whole individual change into generati
372d129 ppreessance if it does= <I do not doubt, the> Do PLANTS loose any qualities by being buds-- , more tha
374d133 ology closely resembles reptiles.-- p. 432 some PLANTS in coal supposed to be intermediate between Co
374d133 s in the mountain limestone (how different from PLANTS!) But the Cephalopoda depart more widely from
381d156 , male organs formed to fecundate female (as in PLANTS) Cirrhipeds rotifers, trematode & cestoid Ento
381d157 exes, as may be seen in Monooecious & Dioecious PLANTS.-- NB. in Heautandrous animals <are> is there
381d157 if so, separtion of sexes very simple.-- as in PLANTS even in same genus some dioecious & some monoo
384d162 rwards they can be seen distinct. (in dioecious PLANTS are there abortive sexual organs?): they then
384d162 to impregnate themselves (this never happens in PLANTS «only in subordinate manner in the plants whic
384d162 ns in plants «only in subordinate manner in the PLANTS which have male & female flower on same stem.-
386d165 e Golden Pippen trees! How is this with buds of PLANTS, does annual give buds.-- life may be thus pro
386d167 emmation in animals very different from that of PLANTS. (though latter does sometimes occur in animal
389d177 eformity ¿this does not happen with hybrids?]CD PLANTS must stand much breeding in & in (those which
392d175 EACH TIME. but after some time]CD What kind of PLANTS are Monooecious or dioecious.-- very curious h
401e014 has no doubt bees fertilize enormous number of PLANTS-- it is scarcely possible to purchase seeds of
403e021 's Second Voyage do Vol III p. 77. Many foreign PLANTS have been cultivated in Guahon. (Mariannes), "
408e043 now what a great effect. light has in colouring PLANTS,-- who can say. what <light> «colours». acting
411e056 be answered) one looks to analogy for cause in PLANTS. where innumberable individuals can be produce
414e060 hy has the organization of fishes & Mollusca (& PLANTS???) been so little progressive «!Agassiz makes
416e070 pose that seminal fluid fluid, (& not dry as in PLANTS) therefore, great difficulty in crossing [& th
416e071 onsequence of non-locomotion-- (contradicted by PLANTS). & as there are no fixed. land animals. so th
418e080 illustration from domestication of Monooecious PLANTS, & abortion of others.-- ¿ in hemi-hermaphrodi
424e103 to hybridise the Camel» like ass «same way some PLANTS vary more than others» & horse in lesser degre
425e105 on the southern flanks of Alps.-- many peculiar PLANTS on single mountains, though these are connecte
426e108 & hooded crow goes against this. & wild hybrid PLANTS. If many wild animals were crossed, there woul
426e107 s, [give instance of series from wild animals & PLANTS]CD.-- Mr Marsh has some nephews, who are aston
427e108 s of transmutations, the relations of animals & PLANTS to each other would rapidly increase, & hence
427e109 ent. According to my views of <Dioecious p> all PLANTS, being occasionally dioecious; & really dioeci
427e109 eing occasionally dioecious; & really dioecious PLANTS being effect of abortion of one sex.-- Linnaea
427e111 fspring-- Mr. Herbert says «p 347. Amyyralidae» PLANTS do not become acclimatised by crossing, or by
429e113 f domesticated animals, yet as we know how many PLANTS have been produced (look at the Dahlias. we ma
429e114 g fields.-- we must recollect the multitudes of PLANTS introduced into our gardens (opportunities of
429e114 ch might spread themselves, as well as our wild PLANTS, we see how full nature. how firmly each holds
429e115 we cannot believe in such a line., it is other PLANTS.-- a broad border of Killed trees would form f
429e115 t Vol I. M. Ramond. p. 19. do says lofty Alpine PLANTS of & Pyrennees agree with those of Norway. Lap
429e116 th those northern regions-- do p. 21 says. many PLANTS skirt each side of the great N & S. valleys, w
430e118 ng, & crossing & keeping breed pure.-- «& so in PLANTS effectually the offspring are picked & not all
430e119 ellently true theory Examine list of St. Helena PLANTS & see whether those which grow in low grounds
434e127 cts of cultivation on successive generations of PLANTS) to do so, the effects are equally handed to o
434e130 ken place.-- Almost all Dioecious & Monooecious PLANTS have rudimentary abortive organs, even more so
438e142 e, as in Dahlias]CD all much varied breeds both PLANTS & animals have long been subjected to domestic
439e143 rt showing the extreme facility of crossing, in PLANTS proves how much depends on instincts in animal
439e143 s.-- yet the existence of wild close species of PLANTS shows there is tendency to prevent the crossin
441e151 ich all facts show to be absurd.-- As there are PLANTS, in northern latitudes, which are generated by
442e153 fruit trees are propagated by means, which wild PLANTS never are, namely on stocks of other varieties
443e153 the kind of stock greatly affects the Graft.-- PLANTS circumstanced as the Gorze must be propagated
443e155 plexity is evident, yet the inference from some PLANTS & some mollusca being hermaphrodite is, that i
447e164 ividuals, «with facility»-- such as cryptogamic PLANTS & true hermaphrodite Mollusca, & probably cora
447e164 for inter marriage is greater (Hence Dioecious PLANTS highest,-- Palms &c)-- Is there greater res
448e165 n says each cell of plant is individual.-- Most PLANTS which propagate rapidly by buds, layers &c &
448e166 nture & Beagle Vol I. p. 306 Shells, as well as PLANTS «of Juan Fernandez» differ from American Coast
453e182 does not countenance the theory of polymorphous PLANTS, abounding in volcanic islds.-- «Cocks» The po
454e184 1816 p. 27. Dr. Holland. Are there instances of PLANTS, in becoming double loosing fertility if, some
454e184 become <all> monooecious.-- Are there not wild PLANTS, some partly dioecious? Mushroom Hybrids? Any
454e184 partly dioecious? Mushroom Hybrids? Any «wild» PLANTS in England, which do not perfect their seed?--
454e184 ommon Hare Decandoelle has chapter on sensitive PLANTS; Physiology Get Habberley to try experiments.
455eIBC Get Habberley to try experiments. about raising PLANTS. where they cannot «crossed» etc.-- Make Hybri
460t019 ers-- cheiroptera & caetacea in Eocene-- dicot. PLANTS in coal measures.-- Shells in Cambrian & Crust
464tf6v when the cold was intense just like the alpine PLANTS-- In S. America. it appears from Lund more Mam
466t091 ondary Series-- few-- Maer June/41/, observed 3 PLANTS of Caltha Palustris alone together. one had se
466t096 ditary characters, wh. come on in after life of PLANTS-- also goodness of flavour in fruit-- all affe
469t135 nd species.-- «Saw» Maer. June 15./41/. Watched PLANTS of Fraxinella, with seven flower stalks for te
469t151 together blossomed together The seeds of these PLANTS will be collected & resown.-- Humble 22 flower
470t153 nute Great Humble 17 flowers of Larkspur on two PLANTS in do Humble 24 flowers of small Linaria in do
470t177 & Humbles-- «Humbles & common» On silene, many PLANTS of wh. have abortive stamens= Many Humbles on
473s03v of. N. America is Red Sandstone. & Birds true! PLANTS in Devonian-- How strange no plants in our Dev
473s03v Birds true! Plants in Devonian-- How strange no PLANTS in our Devonian-- Fish one step lower in Ameri
473s006 ee stamens» intermediate between Orchis & other PLANTS-- & Wallich has described Indian Plant.-- June
490q01v riment on Selection. Experiments in crossing &c PLANTS 1 Repeat the French experiment of Carrot 2 {al
491q01v . Transact.-- 4 May we no suppose, that certain PLANTS, like Aphides produce impregnated young ones;
491q01v ent.--«to repeat Spallanzani» Raise only single PLANTS & only allow <few> one flower (5) Dr Fleming.
491q002 rst question in negative.-- Questions Regarding PLANTS. 1. Uniformity of hybrid & Mongrel offspring 2
492q004 ence to extension of age of individuals-- 9. Do PLANTS in becoming double ever become monooecious-- l
494q005 bout the blended instincts Remote Experiments-- PLANTS Raise seedlings surrounded by various bright c
495q005 cuttings &c (4) Raise annuals or common English PLANTS in Hothouse & see what effect on organs of gen
495q005 > seeds of wild cabbage in VERY rich soil, will PLANTS abort?, does it require successive generations
496q05a ot seed or seed rarely-- Magnolias. «Azaleas». PLANTS grown under unfavourable circumstances, as Hya
496q006 glasses &c &c Experiments Questions concerning PLANTS Is the common Fig Dioecious-- are its female f
496q006 . of Ammonia.-- Horse Urine &c &c on associated PLANTS. when proportional number appear equal-- & see
497q006 of wood be known» & other odd places & see what PLANTS will spring up which will show, how seeds are
497q06v result from seeds being sown= See in Cultivated PLANTS, as Pentstemon, which have abortive parts, whe

498q007 t important to ascertain amount of variation in PLANTS raised by Scions, as Elms. &c &c-- I have some
498q007 some reason to suspect Elms.-- & Orchidaceous PLANTS no other case.-- (6) Will plant accustomed to
498q008 raised, do the Seedsmen select at all from the PLANTS? If not, I am surprised <plan> such plants do
498q008 the plants? If not, I am surprised <plan> such PLANTS do not degenerate,-- as the Bees will mingle t
499q008 p-seed-- may know something about proportion of PLANTS necessary &c &c (a) Mercurialis-- Frog Bit, Va
499q09v ntifully? Look for isolated females.-- Also any PLANTS which are known easily to be crossed & all mon
499q09v re known easily to be crossed & all monooecious PLANTS.-- Hooker says Rafflesia is dioecious & Pollen
499q010 1) Are there many instances of single clumps of PLANTS in counties, as of rare green Cotton Plant-- H
499q010 m other parts? Don says Irish, Scotch & English PLANTS generally distinguishable.= What structure of
500q10a d.-- Number of species of Birds in New Zealand, PLANTS so few-- Range of mundane genera, «in Birds» i
501q10a - Ireland, doubtful species-- Does any genus of PLANTS. vary & hard to separate specifically in one c
501q011 ther cross can be obtained-- I name these three PLANTS. because they cannot be crossed, I think, I ex
505q014 whether stamens will be produced in individual PLANTS 17 A dead-nettle in Hot-house. will it seed?--
506q015 raised races of white & Blue Linum-- did parent PLANTS grow near each other.-- ? Cannot remember at a
506q015 .-- ? Cannot remember at all. (37) Any cases of PLANTS. which will not produce seed in this country--
506q015 may be merely not ripening= (38) Have Dioecious PLANTS any secondary, sexual characters.-- Stature, p
506q015 Wheat.-- (41) Have any monooecious or dioecious PLANTS the Papilionaceous structure of flower-- Groun
507q15v ous Hybrids-- to see whether any Papilionaceous PLANTS,-- whether many mono or Dioeious plants, & any
507q15v naceous plants,-- whether many mono or Dioeious PLANTS, & any with peculiarities of structure renderi
507q15v inning seed.-- (50) Any cases of wild varieties PLANTS growing together. under same conditions.-- lik
507q15v ongly marked.-- 31. Plant seeds of the Fuller's PLANTS, Teazle Dr. Holland ; My Father. Andrew Smith
509q017 species <are> have wide range-- How is this in «PLANTS??» Are abortive organs as <young> teeth, more
513q020 r comparison with natural species, as dioecious PLANTS, when crossed R. BROWN-- will pollen act on an
514q21. on impregnation of violets.= Zostera= Are dwarf PLANTS on Wellington Mountain described in Flinders=
514q21. ountries-- on flora of African Islds-- names of PLANTS found on mountains of N. America similar to La
514q21. d on mountains of N. America similar to Lapland PLANTS --will get answer= Is pollen of cultivated Orc
514q021 althy: number of generations: about crossing of PLANTS; especially Papilionaceous order (2) History o
515q021 mes have dangling ones.-- Is there any genus of PLANTS, «in» which some organ is absent by abortion,
516q24v h different salts & poisons & see in what order PLANTS would reappear after <th> being killed Experim
529m041 to become poetical) the botanist might so view PLANTS & trees.-- I am sure I remember my pleasure in
534m063 do. p. 157. Westwood remarks that some imported PLANTS are attacked by insects & snails of this count
559m155 Watson on Geographical distribution of British PLANTS A Volume published by Colonel in army on "Whea
567n013 light. being direct effect of some law.-- have PLANTS any notion of cause & effect, «they have habit
576n048 ructure «(connected with locomotion.)» «no, for PLANTS have instincts» «either» to obtain a certain e
577n050 e could discover he had not it.-- The memory of PLANTS, must be association,-- a certain round of act
577n050 of Association will help my theory of sensitive PLANTS Habitual actions, (independent of mind) in the
586n081 it, is a part never subject to volition.-- like PLANTS going to sleep.-- "A bird has the faculty of f
601o009 nt) in fibres united with nervous filaments.-- ¿PLANTS? yes by distinct mechanism 2. Sensation of hig
601o009 lower animals are sleeping higher animals & not PLANTS as supposed by Buffon) Consciousness is sensat
612o035 In animals, growth of body precisely same as in PLANTS, but as animals bear relation to less simple b
612o035 imals enjoying only movements such as sensitive PLANTS. (But I include irritability for that require
612o035 sening amount of ignorance [RHC] The radicle of PLANTS absorb by physical laws of endosmic & exosmic
613o036 with ONE animal [RHC] Kirby extends instinct to PLANTS, but surely instincts imply willing, therefore
615o36v climate analogous to inflorescence of Tropical PLANTS when imported & plants sleeping good show accu
615o36v nflorescence of Tropical plants when imported & PLANTS sleeping good show acquirement or obliteration
615o36v stincts But habits acquired even by <children> «PLANTS»! [RHC] 7) As definite instincts modified by h
616o038 no more «difference» than sexual intercourse in PLANTS is involuntary, in man voluntary: ¿ False,-- s
632j53r thing must be admitted there would not be these PLANTS, if there was not some provision for transport
633j54r break off p. 292. Mac. has long rigmarole about PLANTS being created to arrest mud &c at deltas.-- No
633j54r are created for them. If we once venture to say PLANTS created to <arrest> «prevent» the valuable soi
633j54r s why should the earth have drifted; why should PLANTS require earth, why not created to live on alpi
633j54r nnacle? if we once to presume that God «created PLANTS to» arrests earth, (like a Dutchman plants the
633j54r ated plants to» arrests earth, (like a Dutchman PLANTS them to stop the moving sand) we <do> lower th
633j54r ts are merely relations of one general law. the PLANTS were no more created to arrest the earth, than
633j54r ins upheaved by volcanic force, for these Marsh PLANTS. All flow from some grand & simple laws.-- 4 «
636j56v ourable to change.-- It might be concluded that PLANTS would be subject to extreme variation as long
636j56v . brings forward. the impregnation of Dioecious PLANTS by foreign agency-- as insects, as wonderful c
636j56v ation.! There would not have been any Dioecious PLANTS, had there been no insects. The right inferenc
636j56v «¿when were Palms formed?» as soon as Dioecious PLANTS were formed. Macculloch says, life, forms a br
641j29v ttle.-- Are there any other analogies-- prickly PLANTS or animals-- Exudation of fetid «& acrid» secr
641j29v ta, & Vertebrata, & Planaria, & light affecting PLANTS. in insects the end is gained by some very dif
022r007 Carbonate of Lime disseminated through the great PLAS Newydd dike.--Mem tres Montes. ((Henslow Angles
202b133 it to fossil Didelphis (S. American genus) in PLASTER of Paris.-- Now this is exception to law of ty
271c106 rring proofs not always continents».-- it is a PLASTIC virtue.-- it is expression for ignorance Two g
295c178 oo unlike.--?? Memoire by Charles D'orbigny on PLASTIC Clay of Paris contains many genera of Pachyder
535m069 replaced by metaphysical abstractions, such as PLASTIC virtue, «&c» (Very true, no doubt savage attri
535m069 he mountains are as God made them,-- next step PLASTIC <virtue> natures. accounting for fossils). & l
037r054 allows such Compare the elevated estuary of the PLATA. to the Bay of Bengal. dimensions? Strong curre
056r110 ot <so> great«er» for all Europe, than from the PLATA to Caraccas, which is all of granite: In discus
068a005 ing edentates & camels in deserts & rodentia In PLATA Mastodon Toxodon Is the general saline tendency
183b051 pecies often come very remote quarters. (NB. if PLATA Partridge «or Orpheus» was introduced into Chil
288c159 if American & intersected wider & wider at Rio PLATA. birds which had originally crossed would cont
306c213 in company -- marked difference with dogs of La PLATA & Guyana] people will say. not species.-- organ
465t080 , excepting Concepcion-- Patagonia- Beds of La PLATA. (except close to B. Ayres).-- If we may take t
477z002 Moruffetes of Chile different from those of La PLATA or Paraguay.-- do. p. 365. 3 cats (mbara caya.
116a099 de petites corbules analogue living in mouth of PLATE. p. 26. Geology of Arica <Schit> Schmidtmeyer t
041r066 If veins {P} are secretionary, so are all those PLATES in Australia. New Red Sandstone. at Bahia in m
110a086 base-- Salband might have oozed out of cleavage PLATES: the crystals must have recrystallized, as suc
032r040 e gradual rising continuing. a another sloping PLATFORM would be made, & so on.-- This is grounded on
122a114 rschel consider figure of earth statical.-- if PLATFORM of mexico owes its elevation to equilibrium.-
122a114 the ocean would obey that Law. & lie over the PLATFORM-- On my view the degrading action must preve
417e076 they would breed, It is difficult to think of ¿¿PLATO & Socrates, when discussing the Immortality of
551m128 shown him, than when merely ordered to do it.-- PLATO «Erasmus» says in Phaedo that our "necessary id
101a057 erent from Boulder beds-- What is Osteopora PLATYCEPHALUS (Harlan) found on the Delaware. is it Edent
214b113 appear to me very strong What is Osteopora PLATYCEPHALUS. (Harlan) found on Delaware is it Edentate?
032r041 se the deposition & accumulation is brought into PLAY As in Ocean & Air; there are «likewise» differe
051r095 power.--excepting when pebbles are brought into PLAY; most manifest example of degradation I ever sa
171b002 refore relations to other life has not come into PLAY)--See Zoonomia arguements, fails in hybrids whe
192b089 mal bred-- might not new classes be brought into PLAY.-- The father being climatized, climatizes the
315c243 imals» becomes very curious.-- a dog snarling in PLAY-- Hensleigh says the love of the deity & thoug
521m008 action of thought-secreting organs, brought into PLAY by morbid action.-- Old Elspeth's «in Antiquary
527m035 eal train of inventive thoughts are brought into PLAY & then perhaps the sooner castles in the air ar
529m040 refore imagining pleasure of imitation come into PLAY.-- the train of thoughts vary no doubt in diffe
541m092 t not clever people. Aug. 21st. 38 When a dog in PLAY has his mouth open ready to bark, & lip twisted
543m099 during his time, almost an absolute fool used to PLAY regularly with D'Arblay of Christ of great geni
580n061 oes not think at all, whether she can or can not PLAY the piece, she plays <f> better than when she t
608o026 condition of mind all motives do not. come into PLAY.-- †It may be urged how often one try to persua
613o035 ciousness commence; where other senses come into PLAY, when relation is kept up with distant object.
625o50v malevolence). it is only after reason comes into PLAY that anger can be said to be wrong.-- for then
232b247 unlike Southern Europaean ones.-- "a variation PLAYED on secular refrigeration".-- <The Phenomena of
589n091 tune on the pianoforte, yes if every individual PLAYED a little, & something destroyed had brain. see
543m100 m a guinea a week. yet. he was inimitable chess PLAYER.-- Peacocks remark about mathematicians not be
543m069 imited intellects, & in the same way are chess PLAYERS-- A man at Cambridge, during his time, almost
056r110 en that formation and the secondary (stated in PLAYFAIR to be the case p. 51). presupposes an elevate
536m072 s.-- With respect to free will, seeing a puppy PLAYING cannot doubt that they have free will, if so a
543m100 .-- all same fact-- for, as Jones observed, in PLAYING chess however many places, & contingency a man
580n061 strange religions.» Emma W. says that when in PLAYING by memory. she does not think at all, whether
580n061 ecisely the same, as the double-conscious kept PLAYING so well.-- Lr. Brougham «Dissert.» on subject

595n121 or imagination [blank] [not located] Ernest W. PLAYING with Snow. when 2½ years old. was frightened w
431e122 think, the more convinced I am, that extinction PLAYS greater part then transmutation.-- Do species m
580n061 whether she can or can not play the piece, she PLAYS <f> better than when she tries is not this prec
549m121 uestion.-- little is there said of intellectual <PLE [...] hope> cultivation, main source of the inte
527m033 ormer pleases from instinct the ears (rhythm & PLEASANT sound per se) & causes the mind to create sho
548m117 ver loss Definition of happiness the number of PLEASANT ideas passing through mind in given time.-- i
549m118 pared to peasant.-- To make greatest number of PLEASANT thoughts, he must have contingency of good fo
549m119 h past through mind.-- These thoughts are most PLEASANT. when the conscience tells our [mind], good h
550m124 - they are incompatible & the former, the more PLEASANT.-- Simple happiness «as of child» is large pr
550m124 happiness «as of child» is large proportion of PLEASANT to unpleasant mental sensations in any given
550m124 nce.--» But then sensation may be more or less PLEASANT & unpleasant, in same time,-- therefore degre
550m125 y not those of which all our recollections are PLEASANT.-- Browne Religio Medici, p. 21-24. Curious p
550m127 mb 1-- If one performs some actions, which are PLEASANT, every concomitant circumstance calls up plea
568n018 ammalian structure. «-- American monkeys utter PLEASANT plaintive cry--» The taste of recurring sound
571n029 Everything that is habitual, if heredetary, is PLEASANT.-- Mental & Bodily Consider case of grazing a
616o038 annot call the frame of mind which makes music PLEASANT, a memory; yet that frame is enhanced by memo
290c163 inute gets into the road at right angles, how PLEASED it is, just like man. emotions very similar.--
377d139 t pregnant with interest> Hyaena. thinks, when PLEASED cocks his ears., when frighten depresses them.
556m146 ng to kick.-- Why does dog put down ears, when PLEASED.-- is it opposite movement to drawing them clo
557m147 & very stiff. back arched. just contrary. when PLEASED tail loose & wagging-- if as (I believe) Hunte
557m147 ontrary to wrinkled: (a horse when winnowing & PLEASED pricks his ears?--).-- How is expression of an
557m147 - peacock & turkey cock in passion.-- Cat when PLEASED, erect its tail & make it very stiff «& back»
593n107 here must be some instinctive feeling which is PLEASED by other animals smell & looks.-- no doubt it
527m033 & poetry opposite ends of one scale.-- former PLEASES from instinct the ears (rhythm & pleasant soun
086a003 plicity of Ventana's «Quartz.» unmixed is very PLEASING; owing to the movements being of one order. -
146g012 <f> passion of hatred To fulfil an instinct a PLEASURE; mem. Shepherd dogs The Patches of Conglomera
377d138 g noise, the former signifies recognition with PLEASURE, as when food is offered, as much as to say g
522m012 but some time afterwards it calls up pain. or PLEASURE. & is often recurred to & mentioned as a thin
525m023 ideas <I have elsewhere remarked do> Dogs take PLEASURE, when doing. what they consider their duty.--
525m023 by art.-- like the law of honour.-- they feel PLEASURE in obeying their instincts naturally.-- (gene
526m028 e the two statements.-- Catherine remarks that PLEASURE received from works of imagination very diffe
526m028 ures, an early taste, of animals. they know.-- PLEASURE of imitation (common to monkey), & not imagin
528m036 s of pleasures of scenery.-- There is absolute PLEASURE independent of imagination, (as in hearing mu
528m037 beautiful «as round, ovals»;-- then there the PLEASURE of perspective. which cannot be doubted if we
528m037 if we look at buildings, even ugly ones.-- the PLEASURE from perspective is derived in a river from s
528m039 as may be known by autumn, on clear day.-- 3d PLEASURE association warmth, exercise, birds singings.
528m039 ation warmth, exercise, birds singings.-- 4th. PLEASURE of imagination, which correspond to those «he
529m039 people-- recall pictures & therefore imagining PLEASURE of imitation come into play.-- the train of t
529m040 as evasive & ill defined thought would receive PLEASURE from thinking of the fertility.-- I a geologi
529m041 view plants & trees.-- I am sure I remember my PLEASURE in Kensington Gardens has often been greatly
531m051 mber poetry when once read over.-- The extreme PLEASURE children show in the naughtiness of brothers
531m051 s that sympathy is based as Burke maintains on PLEASURE in beholding the misfortunes of others.-- In
533m059 owing why-- a man as in Guy. Mannering. feels, PLEASURE. in seeing the scenes of his childhood withou
536m072 enses, perhaps, though from not having pain or PLEASURE actions unavoidable & only to be changed by h
536m072 y the capacities its senses give it of pain or PLEASURE, if so free will is to mind, what chance is t
539m082 culiar smell of Picture. association with much PLEASURE immediately thrilled across me, bringing up o
540m088 y.-- Mem: Burke's idea of Sympathy. being real PLEASURE, at pain of others, with rational desire to as
544m101 ains & emotions-- such as child sucking, gives PLEASURE, & always has done therefore sight of own chi
544m101 own child. (when frame in condition to receive PLEASURE) gives pleasure, ie. love.-- & so pain gives
544m101 frame in condition to receive pleasure) gives PLEASURE, ie. love.-- & so pain gives fear of death. M
545m106 ng their teeth, this the Keeper thinks is from PLEASURE, & may be compared to laughing» they dance wi
546m108 tisfactory because does not like Burke explain PLEASURE. August 26th. I cannot help. thinking horses
546m108 an perhaps depends on the number of sources of PLEASURE & innate tastes, he partakes, taste for music
548m117 n given time.-- intensity to degree of <happi> PLEASURE of such thoughts We give no credit to instinc
549m119 een done-- «& conscience free from offence»-- PLEASURE of intellect affection excited, pleasure of i
549m119 e»-- pleasure of intellect affection excited, PLEASURE of imagination-- therefore do these & be happ
549m120 ke large <parts> portion of daily <happyness> «PLEASURE». A wise man will try to obtain this happines
550m125 ared to what others experience in same time.-- PLEASURE more usually refers to the sensations <it> wh
550m127 asant, every concomitant circumstance calls up PLEASURE. or pleasure or pain of association.-- now if
550m127 concomitant circumstance calls up pleasure. or PLEASURE or pain of association.-- now if one has thes
552m132) are those which are good & consequently give PLEASURE, & not as Paleys rule is those that on long r
557m147 r tails, &c. it is very curious, recurrence of PLEASURE so teaching expression «as constant smiles, c
568n017 se, or instinct,) one feels pain, & vice versa PLEASURE in following its instinct & pain if held.-- i
568n017 to pursue immediate inclination or some future PLEASURE in performing it.-- As soon as memory improve
568n018 commence with singing-- is this origin of our PLEASURE in music-- do monkeys howl in harmony-- frogs
569n019 oked at picture as works of imitation.-- Hence PLEASURE in the beautiful. (distinct from sexual beaut
569n021 iplicity of things remembered & the associated PLEASURE &c accompanying such memory.-- A Melody on fl
575o045 d of nose with a hot razor.-- joy <p> a mental PLEASURE. with pleasure of senses. The shudder of plea
575o045 a hot razor.-- joy <p> a mental pleasure. with PLEASURE of senses. The shudder of pleasure. from plea
575o045 asure. with pleasure of senses. The shudder of PLEASURE. from pleasure of music Audubon IV Vol of Orn
575o045 asure of senses. The shudder of pleasure. from PLEASURE of music Audubon IV Vol of Ornith. Biog. case
581n064 n have no difficulty in expressing their want, PLEASURE, or pains long before they can speak-- or und
582n066 t be said to be angry.--» «He may have pain or PLEASURE these are sensations» <Gardner in his work> I
582n066 of several people.-- Children have an uncommon PLEASURE in hiding themselves & skulking about in shru
582n068 re truly sensations??. a kind of mental pain & PLEASURE.-- The Revd. Algernon Wells Lecture on animal
593n107 gs.-- How does Social animal recognize «& take PLEASURE in» other animal, (especiall as in some <inst
593n107 parent smell & look so by association receives PLEASURE. This [blank] will not do for insects. if thi
600o08v t so act, because maternal instinct gives most PLEASURE. but because most imperious.-- It would indee
603o012 on toward deity.-- & as king might like cruel PLEASURE, so sacrifices cruel.-- Something wrong here.
605o020 as we often see people who are susceptible of PLEASURE from these causes who are not men of taste &
613o035 conscious, but is memory gone?-- Where pain & PLEASURE is felt there must be consciousness??? ? [LHC
615o36v and then howled.-- Now I don't think this only PLEASURE; for it was different way of showing it, nor
615o038 X» memory especially «the» general kind taking PLEASURE, & <an» «such» actions being·prevented by <ne
619o042 ction in accordance to an instinct gives great PLEASURE, & <an» «such» actions being·prevented by <ne
619o043 s to wife [RHC] 2) and children would give him PLEASURE, without any regard to his own interest. like
619o043 nstincts, <he would know that many experienced PLEASURE,> & by association he would feel part of that
619o043 ,> & by association he would feel part of that PLEASURE, which the acter received.-- If either man it
620o044 choice: an appetite gratified gives only short PLEASURE. passion in its nature is only temporary, & w
620o044 this may be, everyone must know, how soon the PLEASURE from good dinner, or from a blow struck in pa
620o044 ks which was such an instinct not followed for a PLEASURE now though so trifling he feels remorse.-- He
620o044 d & therefore as soon as desire is fullfilled, PLEASURE forgotten. [RHC] 4) as starvation, or fear of
620o046 which are natural (& which «when present» give PLEASURE.) & which man ought to follow-- it is his dut
621o047 [RHC] By interest I do not mean any calculated PLEASURE, but the satisfaction of the mind, which is «
622o049 ing to Sir J) explains our love of another, as PLEASURE arising from association from having received
623o050 of place.-- although here we have not received PLEASURE from the place, but merely in the place. & ye
623o050 but merely in the place. & yet place calls up PLEASURE.-- [LHC] the instinct of sociability & sociab
623o050 igin of these instincts.-- the having received PLEASURE from some one «person» in early infancy, duri
625o50v sh.-- p. 262. Some good remarks, on analogy of PLEASURE of imagination «the utility part being blende
625o052 s are implanted in us. & that doing them gives PLEASURE & being prevented uneasiness, & that this is
627o053) Mackintosh's Ethnical Philosophy p. 6-- "The PLEASURE which results when the object is attained (th
627o053 Eugenius would contend against this-- but the PLEASURE a dog has in obeying its instinct,-- as young
628o054 & as a consequence, but not cause gives me [3] PLEASURE) or that I have been taught or habituated to
146g012 pecies in colouring Strike an analogy between PLEASURES of association, & passions, such as love-- di
528m036 s real again explains insanity.-- Analysis of PLEASURES of scenery.-- There is absolute pleasure inde
543m101 logy be drawn between «heredetary» associated PLEASURES & pains & emotions-- such as child sucking, g
549m119 ion-- therefore do these & be happy-- & these PLEASURES are so very great, that every one who has tas
605o020 exquisite susceptibility from Blair receiving PLEASURES from beauties of nature & art." But as we oft

Page **(Key Word)**
606o023 dium of the eye"; he will allow the secondary PLEASURES of harmonious colours &c &c surely to be adde
615o038 in kinds such secretion) or an association of PLEASURES with certain actions performed by your parent
334d010 s a remarkable law, that first cross <not se> PLENTIFUL, second absolutely sterile.-- My case of Stal
499q09v excessively minute or abundant? do they seed PLENTIFULLY? Look for isolated females.-- Also any plant
068r145 that no metals in Polynesian Islds--. Volcanic PLENTY in S. America!! Metamorphic Volcanos only burs
259c065 the same principle that, cut a sheeps tail off PLENTY of times & you will have no tail (example prob
271c107 - Waterhouse tells me in insects there are many PLENTY of instances of insects of one tribe taking on
375d135 he great check amongst men.-- «Even a few years PLENTY, makes population in Men increase, & an ordina
470t177 ortive stamens= Many Humbles on hedge Linaria= (PLENTY of Humble Bees on Phlox Down, 1854, Sept.) In
505q014 elf hybridised (10) one had no seeds, & two had PLENTY of seed & these Seeds of unimpregnated Cowcumb
515q021 pen.-- The point to attend to is whether good & PLENTY of pollen is produced. & 2d if so, whether con
261c072 powerful buck Owen talking of Plesiossaurus PLESIOSAURUS. alludes to some structure in head, which h
261c072 s prefer most powerful buck Owen talking of PLESIOSSAURUS Plesiosaurus. alludes to some structure in
332d003 esenteric glands.-- My Father has seen case of PLEURISY, broken limb «in children» & other such disor
418e084 as even in childhood) when the organization is PLIABLE, such modifications, become as much fixed, as
323c269 atise -- Wilkinsons Egyptian remains skimmed -- PLINY Nat. Hist of World do -- Lamarck. II Vol. Philo
443e154 young trees, from worn-out kinds, & quotes from PLINY, that it is bad to graft from top shoots.-- If
039r061 supposes. Lyell P 116 Vol III, says that in N. PLIOCENE formation of Limestone, casts of shells, as i
232b245 (marking in their history an eocene miocene & PLIOCENE epoch), whilst others may die out or move Sou
473s03r ippotamus «with Megatherium & Mylodon» in post PLIOCENE strata! Mastodon longirostris in miocene like
304c206 marked his own ignorance.-- the collector who PLODDING at making a series, which would render our kn
522m014 omebody who has, perhaps, slightly injured me, PLOTTING speeches, yet with a sort of consciousness no
495q05a cilitate investigation.-- Capital in middle of PLOUGHED field-- on hills.-- 10 Shoot tame duck on pon
057r112 the Volcano & Earthquake.--Earthquakes act as PLOUGHS [,] Volcanos as Marl-pits: Consider well age o
023r009 ied, and the jaw was also firm, out of which we PLUCKT a great many teeth, 2 of them, 8 inches long,
195b098 ted at Galapagos it so acted that bi[r]ds with PLUMAGE «&» tone of voice partly American North & Sout
225b217 , or iodine on glands of throat, (or colour of PLUMAGE altered during passage of birds (where is this
265c088 ird belonging to family with peculiar coloured PLUMAGE, where colours have changed in accordance to h
273c110 (Gould says) there is any marked colouring of PLUMAGE (as «black & white» bars on wings of trogons a
273c112 erent families.-- that sexes <are> «have» same PLUMAGE.-- <no> this is applicable to swallow-hawk, «t
276c122 whether beak (as it appears to me). colour of PLUMAGE & laws, which might probably be reduced What t
297c186 ortive spiracles in Hemiptera do. p. 160. soft PLUMAGE of night jar. like owls. analogy in habits ada
305c209 cies, and the sexes of which vary in colour of PLUMAGE in same remarkable manner as Europaean species
305c209 - Have not Ruffs & Reeves a remarkably varying PLUMAGE for wild birds-- At Zoolog Gardens there is ha
325c267 n. (at Shrewsbury) Yarrells Paper on change of PLUMAGE in Hen Pheasants <Zoological> Philosop. Transa
346d047 gs.-- Aug. 24th. Was struck with pink shade on PLUMAGE of the Pelican.-- Mem pink spots on Albatross,
346d047 . Flamingo-- (Spoonbill Wader. Ibis)-- laws of PLUMAGE might possibly be made out.-- August 25th Athe
358d076 le & young of all birds resemble each other in PLUMAGE «(that is where the female differs from the ma
358d085 they have This character of not having sexual PLUMAGE is very common by hybrids, that are infertile.
361d095 ts red-pole, goldfinch, hawfinch-- in nursling PLUMAGE resembled that of Cross-Beak-- In lark if I un
361d095 any existing species-- [In two herons, <both> PLUMAGE of both (nursling) quite similar.-- one specie
361d095 ly]CD In common sparrow young & female similar PLUMAGE.-- in tree sparrow, (if I understand rightly)
361d096 halarope) assume for breeding a more brilliant PLUMAGE than male.-- «My case of Caracara. N. Zelandia
363d101 pouters, fan tails are found in any colours of PLUMAGE &c &c «Pouting pidgeon exaggeration of cooing.
363d101 scribed them)-- [Study horns of wild cattle.-- PLUMAGE of fowls-- long ears of rabbits. & long fur.--
363d101 ce the washing out of the forked band, like in PLUMAGE of ducks.-- Mr Yarrell says in very close spec
368d114 & Peaccocks.!!» other birds display beauty of PLUMAGE.-- (The females (as Owen observes) in Raptoria
368d114 les which put on (like some waders) the bright PLUMAGE.-- «thinks» Hence specific character most perf
378d148 Kingfisher & pies, have long had their present PLUMAGE.-- How is this in Pidgeons & fowls.--??? Wate[
380d154 ar. so many Gallinaceous birds have cock & hen PLUMAGE so different, yet the Cassowary & Guinea Fowl
380d154 surely is hermaphrodite-- (as is seen in <fe> PLUMAGE of hybrid birds) After animal has copulated.,
384d162 d instincts of Capon. & power of assuming male PLUMAGE in females., & female plumage in castrated mal
384d162 of assuming male plumage in females., & female PLUMAGE in castrated male.-- «Men giving milk--» Sept.
388d172 hich takes place in old age of female assuming PLUMAGE of cock, & beards growing on old women = Stags
440e146 laws of organization: as we see these strange PLUMAGE in pidgeons yet no change of habits, so no <ca
440e147 All that we can say in such cases, is that the PLUMAGE has not been so injurious to bird as to allow
446e160 f the icterus minor is a bird of more splendid PLUMAGE than the male.-- Athenaeum May 18. 1839. p. 37
459t013 ke.-- another of same half breed resembled the PLUMAGE of drake still more.-- So Penguin impresses it
484z016 .-- into Oxyurus, by Maldonado creeper of same PLUMAGE.-- general red mark on wings of all-- Spix has
485z016 coming tyrant-- flycatcher-- shown by habits & PLUMAGE so very similar to some of the Fluvicolae?-- T
119a105 ormed near surface. whether they can have been PLUNGED so many miles deep into the bowels of the eart
046r079 ough the rent strata: «Mr Lyell considers that PLUTONIC rocks are generated as often as Volcanic. I c
049r087 dill: should basal lavas be called Volcanic or PLUTONIC The cellular state of all the Porphyry specim
064r136 ons of the land in Europe-- Urge difference of PLUTONIC rocks & Volcanic metalliferous-- Urge enormou
103a065 of super [...] mass.-- Absence of Caverns, in PLUTONIC rocks argument against great bodies of vapour
221b201 of Fresh-water genera? The absence of lime in PLUTONIC & Volcanic rocks. most remarkable.--¿ Have th
223b209 ies?-- Small «new» animal mentioned from Fernando PO Zoolog. Proceedings October (?) 1837 Contrast Ne
226b220 ween Mauritius & Madagascar very good.-- Fernando PO & Coast of Africa. equally good.-- Small isld of
226b222 of Paris basin land.-- (How is this with Fernando PO.). with plants of St. Helena & Tristan D'Acunha,
295c183 wn, young one.-- Many African monkeys in Fernando PO-- no new forms only species!! No salamanders (D'
447e164 stuca vivapara F ovina-- propagated like oni[on] POA alpina because vivaparous. Henslow has seen this
447e164 ina because vivaparous. Henslow has seen this-- POA alpina vivaparous sometimes seeds All species of
447e165 s on mountains & yet can be raised in gardens.-- POA alpina, thougt generally vivaparous sometimes se
523m015 by keeping the sum-total of his accounts in his POCKET, & studying mathematics.-- My Father says afte
554m139 outh.-- seemed to relish the smell of Verbena & POCKET Handerchief & liked the taste of Peppermint.--
633j53v g as they are not detrimental.-- p. 285 the seed-POD of a desert plant (Anastatica) is rolled along,
466t091 of Caltha Palustris alone together. one had seed-PODS turning brown, whilst both others were in nearl
472s01r = Has tried several year to obtain seed, but the PODS have (except this one year (1827), always been
569n021 mpanying such memory.-- A Melody on flute & Epic POEM, opposite ends of series or harmonious prose.--
095a037 2. species Vol VI. Geograph. Journ. Analysis of POENIG Voyage Valparaiso Dr. Gillies in MS. letter in
382d158 testes & ovaria in Hermaphrodite,-- but not of POENIS & clitoris, shows to my mind.-- , that both ar
383d159 pement of either.-- (Mammae or sheath of Horses POENIS reduced to extreme degree of abortion).-- Inse
091a026 . the twopenny periodical said so. «Campbell the POET» Accra. Coast of Africa. Clay Slate & Quartz. s
430e117 nch valleys-- M. Ramond offers no explanation.-- POET Cowper, describes his tame Hares, attacking a s
574n039 sleigh says. Douglas. «& Spencer», an old Scotch POET, has numerous lines. of poetry.-- <signs> sound
529m040 animals, slow force cracking surface &c truly POETICAL. (V. Wordsworth about science being sufficien
529m040 science being sufficiently habitual to become POETICAL) the botanist might so view plants & trees.--
326c266 -- (translated in 1837) on limits of painting & POETRY.-- Erasmus thincks I should lik it. The Sports
521m008 Old Elspeth's «in Antiquary» power of repeating POETRY in her dotage is fact of same sort. Aunt. B. d
527m033 t explain the feeling in any one man.-- Music & POETRY opposite ends of one scale.-- former pleases f
527m033 ate short vivid flashes of images & thoughts.-- POETRY. the latter thoughts are in same manner vivid
527m033 of mind being just kept up by the music of the POETRY.-- (therefore singing intermediate, who has no
528m039 e <he> awakened during music.-- connection with POETRY, abundance, fertility, rustic life, virtuous h
529m039 ic life, virtuous happiness.-- recall scraps of POETRY;-- former thoughts, & in experienced people--
531m050 imes repeated, because some people can remember POETRY when once read over.-- The extreme pleasure ch
535m069 e thunder & lightening to Gods anger.-- (∴ more POETRY in that state of mind: the Chileno says the mo
540m088 ridge,-- Zapoyla p. 117, Galignani Edition Fine POETRY, or a strain of music, when the mind is render
574n039 er», an old Scotch Poet, has numerous lines. of POETRY.-- <signs> sounds singularly adapted to subjec
579n057 tes. & likewise on Wordsworth's dissertation on POETRY.-- The expression of shame-facedness for shyne
603o11b t [my view says yes. <old> mass of rock--]CD or POETRY, CD[my thery says yes. imitating song -- two p
571n031 Mayo's ideas.-- In language. the possibility of POETS describing gentle things in gentle language, &
024r014 0 fathoms. proves the existence of some moving <POINT> power ¿ Submarine currents Find instances; The
032r039 a uniform slope to base of cliff (Z). to which POINT the waves would not reach. If now the ocean sho
032r041 hence varieties of substances ejected from same POINT. & changes. «(changes in variation?)» as in Cor
042r068 many» In the Valle del Yeso it is probable that POINT of Porphyry has been upheaved in a dry form It
051r094 ven turbulent as at St Helena) I have mentioned POINT of greatest action; I now having seen Pernambuc
057r114 athoms], & the temp of which was below freezing POINT!!! Remember idea of frozen bottom or beach of s
080r178 ng of the character of a Araucarian tribe, with POINT affin of yew & intermediate Puncture one animal

Page
(Key Word)

```
089a019 hine to see if water fell. -- <Keys off extreme POINT of Flori[da]> Excellent paper on Erratic blocks
098a047 ithin other concretion.-- state last page thus. POINT of attempted crystallization, & therefore as a
100a053 on transverse.-- or rather radiating to central POINT. can cleavage be radiation from some grand cent
121a112 parts on surface, but I saw none embedded in this POINT would be worth examining. to support. shells on
122a113 stronomy with oscillations of level.-- {P} will POINT {P} be the one which generally yields.-- Will t
132a138 r <state> temperature of earth. at the freezing POINT.-- accounts for increase on earth by volcanic a
146g014 ace perhaps near 300 ft above Loch.-- From this POINT could be followed up to neighbourhood of Tyndru
156g068 these three shelves soil is <the> usually slaty POINT of rounded not scooped rock on <bend> of 3(a) C
156g071 Lower  Glenroy 29.581 A 82 75 degree? From this POINT plain appears like one uniform slope slightly b
158g079 on  <thes> Divortium aquarum Peaty Mass of this POINT very nearly like head of Glen Guoy nor is horiz
160g091 about 20 ft. below summit <Isthmus> {P} highest POINT joining this hill to others 3000? if Ben Erin i
161g095 with road, & with piece of excised rock lost at POINT of valley chiefly from rockiness When on other
177b028 circumstances,  to which> is it an index of the POINT whence, two favourable points of organization c
183b051 ved same species) if they do not breed readily. POINT in view.--¿whether highly domesticated animals
186b057 in genus resembles each other, (at least in one POINT, in truth in all excepting specific character);
189b071 gfisher same colours Strong odour of negroes, a POINT of real repugnance.-- Waterhouse says there is
198b115 > «an» answer <could> «can» be given.-- It is a POINT of great interest to prove animals not adopted
216b180 c &, (V Herbert on hybrids) thus act.-- Now the POINT will be to find whether know varieties in plant
219b193 > make plants <grow closely> When this volcanic POINT appeared in the great ocean, have made plants o
231b244 ange-- & hence probability of starting from one POINT.-- In the crag we see the process of change of
235b262 ieties of dogs in different countries a case in POINT.-- All cases like Irish & English Hare bear upo
240c004 hat occurs in plants.-- All these facts clearly POINT out two kinds of varieties.-- One approaching t
248c029 nge of quadrupeds.: that either created in each POINT, or migrated from those quarter, where we  know
256c055 38 thinks gradation between Man & animal, small POINT in tracing history of Man.-- granted.--but if a
263c076 t one instinct to be acquired (if the medullary POINT in ovum. has such organization as to <per> forc
275c120 st greyhounds.-- Sir. J. Sebright first got {P} POINT on hackles on Bantams by crossing with common P
277c127 - This will aid in preventing the chaos.-- will POINT out what to observe.-- will aid us in physiolog
291c165 o follow it, or it may be heredetary & strictly POINT out affinities. conducct of Gould, remark of D'
291c165 inities. conducct of Gould, remark of D'orbigny POINT out importance of habits in classification.-- T
312c228 easant see Jardines Journal.»-- consult on this POINT--pigs always go against this, without «number
312c231 hey would produce number agreeing almost to the POINT in question.-- «--merely picking opposite quali
315c243 assistance.--  crying is a puzzler-- Under this POINT of view. expression «of all animals» becomes ve
327c265 preeminently worthy of studying in Metaphysical POINT of view Henslow has list of plants of Mauritius
336d019 s not heredatary, because generation -- highest POINT of organization] CD» false.-- The creator would
341d029 bird a Penguin.-- (i.e. whether relation in one POINT or many) Owen answered that all characters migh
348d051 gement of animals themselves is the question in POINT." Now what is natural arrangement,-- affinities
351d056 ence the Pecten, which move imperfectly has eye-POINT, but Broderip added it has been stated that sta
352d058 l be dead-- hence there is no central radiating POINT, all united . (links in circle must be granted
353d061 ed <tits> «Mammae» in male ourang-outang. other POINT of resemblance with man.-- September 3d Magazin
368d114 Copris  &c).-- In birds singing of cocks settle POINT.-- (do the females then fight for male) & are m
370d117 pursued  throughout my theory is to establish a POINT as a probability by induction, & to apply it as
383d160 lack lines on each feather instead of coming to POINT {P} are more rounded. {P} & much broader., & <m
420e088 ies of species of dogs & Hyaena.-- but a common POINT, whence both may have descended.-- Jan. 6th The
443e155 ce «The fact of Corallina & Halimeda is case in POINT».-- The relation of these «sexual» functions to
464tf6r other forms-- Lund's Antilope in Brazil another POINT of agreement with. N. America & S., (¿ is the p
471tf08 tructure" he says "indeed it wd be difficult to POINT out a family so completely natural & one whose
497q007 -- most important, as furthest removed possible POINT.-- ¿genera in intermediate country (2) Any know
515q021 ealthily, or does fruit merely ripen.-- The POINT-- man, though he does not pout. pushes out both
551m129 the glass) when pouting protrudes its lips into POINT-- man, though he does not pout. pushes out both
558m151 hole argument of expression more than any other POINT of structure takes its value. from its connexio
586n080 equally wonderful in young & old.-- These facts POINT out some essential difference, which clearly ou
616o039 shew satisfactorily it's erroneousness. it is a POINT of indifference 2) In the absence of such a gui
616o39v the  absence of such a guide we can only <shew> POINT out the mode «of perceptive action» by which we
617o040 nimate matter we feel dissatisfied until we can POINT out How can force be recognized by our external
617o40v onscious that we ourselves can originate in any POINT an opposition of forces balancing each other  &
626o052 with  my theory.-- see p. 349.-- remark on this POINT.-- [LHC] p. 194. «&c &c» Butler's view given on
627o053 n  obeying its instinct,-- as young pointer to POINT-- clearly shows this is true. p. 13. Affections
628o54v I expect there is some fallacy here.-- at least POINT of «false» honour will stop all wish to gratify
637j58r h p. 237. Gives as Summary of adaptations Horny POINT to chickens beak. to break egg. shells-- why ch
363d102 be  told apart, so that after differences were POINTED out Selby confounded them, yet can readily be
521m009 My F. then asked Mr C. to come to the window & POINTED out the Gardener & said, who is tha? Mr C. ans
536m073 ror in argument, should be grateful if it were POINTED out.-- My wish to improve my temper, what does
610o031 s letter, & I observed he had written Wilson & POINTED it out; he was astonished, & said how very odd
204b140 the  offspring most forcibly-- Esquimaux dog & POINTER a rough water spaniel «produce litter like bot
217b184 one  way as other.-- He has known case of good POINTER & rough water spaniel «produce litter like bot
239c001 he states that Esquimaux dog when crossed with POINTER.-- He has no doubt that same thing would happe
239c001 produces  offspring much nearer Esquimaux than POINTER produces offspring much nearer Esquimaux than
620o046 ollow-- it is his duty to do so.-- So we say a POINTER ought to stand a <spaniels> «housedog's» duty
620o046 -- it is part of <duty> their nature.-- When a POINTER spring his bird. one says for shame (& the «ol
627o053 a  dog has in obeying its instinct,-- as young POINTER to point-- clearly shows this is true. p.  13.
511q018 . Voorst often writes to Lowe (7) In breeding. POINTERS. Bull-Dogs. Spaniels-- Grey-hounds-- is there
613o036 f thoughts as fear of man,-- crows fear gun,-- POINTERS method of standing,-- method of attacking pec
356d071 s being wilder than parents is very curious as POINTING out difference between acquired & heredetary
466t095 quality becomes associated with some other, as POINTING with smell.= These qualities have been given
583n069 power  in one species man.-- false instinctive POINTING varies.-- p. 18. Animals possess strong imita
620o045 bad child.-- Hence there are certain instincts POINTING out lines of conduct to other men, [RHC] 5) w
623o051 for pig will not so readily attain instinct of POINTING as a dog.-- also age has much influence.)-- &
039o59v St Helena, where dikes certainly have not been POINTS of eruption. Nobody supposes that all the dike
061r127 then  peculiar plants created. if for such mere POINTS; then any mountain, one is falsely less surpri
064r137 or an enormous period: now when we see how many POINTS of eruption [.] give instance of Etna Strombo
133a140 r these three elements, there will be certainly POINTS have been penetrated by volcanic & trappean ro
176b024 it an index of the point whence, two favourable POINTS of affinity in each branch A species as soon a
177b028 s mere assumption, it explains nothing further, POINTS of organization commenced branching.-- As all
195b104 of earth «within recent times.». & many curious POINTS gained if any facts are connected.-- No doubt
227b224 ies of the continents peculiar to the different POINTS.-- Consult Voyage aux terres Australes Chap XX
243c020 s! but they were not shut up!! Extreme southern POINTS of S. Hemisphere fully characterized, of  each
250c037 like)» «this wars against in any class:, those POINTS which are different from each other, & resembl
258c061 where ambient. & <in> merely determined to such POINTS by the vital laws.-- so that all character ori
282c143 ing but structure of brain heredetary,. analogy POINTS out to this.-- love of the deity effect of org
291c166 an kind probably possessed a structure in these POINTS for <t> a less time than other. points. female
303c204 in these points for <t> a less time than other. POINTS. female genital organs «in some monkeys clitor
303c204 y good to see whether peculiar plants-- in high POINTS Read Volney's travels in Syria Vol I. p. 71. a
327c265 s been stated that stationary Spondylus has eye-POINTS-- Macleay then answered, because nature leaves
351d056 ot pressed on paper, to study the corresponding POINTS.-- The present geographical distribution of an
352d059 induction, & to apply it as hypothesis to other POINTS. & see whether it will solve them.-- It is les
370d117 a.-- if true curious on my view-- because these POINTS were last connected with those northern region
429e116 ons, is that they do know from look of Heavens, POINTS of compass, & they do know which way they  go;
586n078 iousness & by habit, such habit of knowledge of POINTS of compass may be instinctive. it is a test to
586n079 notes about the moral sense & some metaphysical POINTS written about the year 1837 & Earlier-- ]CD in
599o004 ses: (& this alone explains why our moral sense POINTS <is> to revenge). In judging of <our ha> of th
609o030 ion of snakes. with hinder teeth perforated for POISON channels, but not having them, instance of use
229b233 July <1842> When nettle leaf. put into spirits, POISON-drop exudes-- does not elm. does it «in» melon
502q11v lue Gloss. it is not possible to see orifice of POISON-tube-- so put carmine in spirits & then experi
571n029 dily Consider case of grazing animals knowing POISONOUS «herbs:» & man not.-- ¿no vegetable good fr
571n029 an not.-- ¿no vegetable good «for man» to eat POISONOUS?-- How did animals in «Australia» & America m
516q24v Cover  patch of ground, with different salts & POISONS & see in what order plants would reappear afte
570n023 better with body distended.-- intolerable to be POKED behind, without ones chest, being distended. to
```

(Key Word)
412e058 stitut. 1838. p. 384. List of fossil Mamm: from POLAND. &c.-- Three principles, will account for all
355d068 emark. ¿was there formerly one great sea, & two POLAR Continents Marsupial. Edentata.-- Pachydermata
407e041 ca, an island, «connects with Asia» between two POLAR lands,--; Africa not so equatorial..-- The fact
429e116 not with those Kamtschatka, Siberia, or even of POLAR regions of N. America.-- if true curious on my
634j55r y not first occupy the Poles? Is this origin of POLAR attributes of the Cetaceae.-- How came Bats als
641j29r nourished by arctic regions-- Whales. «Narwhal» POLAR bear. Walrus, great Seals of Antarctic seas. (o
115a096 metamorphosed. The chemical action which gives POLARITY to atoms in slates that cleave, & which unite
086a007 y sun's position = If equatorial streams of warm POLE; in name of Heaven why are tops of Equatorial m
361d095 eeting stated, that Green-finch, all linnets red-POLE, goldfinch, hawfinch-- in nursling plumage rese
032r041 hanges in variation?)» as in Cordillera.-- From POLES to Equator current downwards & to West.--From E
032r041 r current downwards & to West.--From Equator to POLES. nearer the surface & to the Eastward.--If matt
213b172 found in every sea, from Equatorial to extreme POLES.-- Oh. Wealden.-- Wealden. Do the N. American.
495q05a carry them in Electrical machine, reversing the POLES test by suspending magnet within & see which wa
634j55r ke their place? Would they not first occupy the POLES? Is this origin of Polar attributes of the Ceta
482z011 ipedes: out of the first 9.:4 raptores. Falco POLIOSOMA -- novozelandiae -- histrionicus Vultus aura
350d055 e talking of some Crustacean, like Trilobite. (POLIRUS??) female blind & of quite different form from
089a019 mposed on the NE part more like marble requires POLISH to see structure,-- «He» Thought of erecting m
275c120 t on hackles on Bantams by crossing with common POLISH cock «is not that old variety» & then recrossi
385d163 s breed of Fowls called everlasting layer--. or POLISH breed. (he thinks half pheasant, half fowls.--
462tf05 -- A curious theoretical French book review on POLITICS in relation to the different races of men, so
538m079 yes.-- one botanist & great knowledge of Irish POLITICS, «both bad jokers.--» the other army officer,
599o07v e, has produced almost + greater changes in the POLITY of Nature than any other animal-- Aimé Martin
435e133 stigma retains its power.-- R. Brown found the <POLL> masses of pollen of Asclepias placed on Orchis
194b096 varieties being difficult to keep on account of POLLEN from other plants because this may be applied
427e110 f the animal kingdom I should suppose, that the POLLEN of crab, would POSSIBLY «No, for pollen of any
427e110 hat the pollen of crab, would POSSIBLY «No, for POLLEN of any kind would fertilize it» fertilize an a
435e133 ts power.-- R. Brown found the <poll> masses of POLLEN of Asclepias placed on Orchis (so very differe
435e133 w Mr Herbert has shown that stigma swells, when POLLEN even most remote is put to it.-- April 6th "Dr
439e143 enslow) fertilizing each other, better than the POLLEN of same flower,-- as it tends to show my view
439e144 become as fixed as species, & prefer their own POLLEN to that of other variety.-- «Elizabeth & Hensl
468t093 over anther & pistil & one I SAW IMPREGNATE by POLLEN with which <bees> «a bee» was dusted over. {P}
467t099 minute, distinctly doubled, brown, but with no POLLEN.-- Common Thyme growing close by is equally ab
467t103 s. & when the Lupine flower is perfectly ripe & POLLEN abundant filaments & stamens all protrude «the
467t103 end of stigma, which forces out from extremity POLLEN, or pollen comes out with anthers & stigma in
467t103 gma, which forces out from extremity pollen, or POLLEN comes out with anthers & stigma in slit»-- As
467t104 ulatus saw Humble press down wings which ejects POLLEN from tip of sheath.-- «Also in Lathyrus praten
467t104 so press down sheath, that stigma covered with POLLEN was pressed & rubbed along whole breast-- {b}
467t104 or both of Pea's wings, stigma & mass of yellow POLLEN protrudes at sheath.-- At last I saw Bee colle
467t104 rudes at sheath.-- At last I saw Bee collecting POLLEN from <sheath> Keel of Lupine-- Seen Bees on Po
467t105 ethes & small Staphylinidae on all their bodies POLLEN-- on a sulphur Broccoli not many do-- pollen n
467t105 es pollen-- on a sulphur Broccoli not many do-- POLLEN not very abundant. not very small-- Saw one sm
468t105 utterflies suck nectar: «Maer June 41» Rhubarb. POLLEN very minute--not excessively abundant flowers
468t105 owers common-- many winged thrips, covered with POLLEN-- «Thrips» about as large as bit of chopped ho
468t112 rt surprise-- I found all, stamens straightened POLLEN profusely shed; lengthened & turned up «more t
468t112 ng stamens {P} as stamens grow old «& shed some POLLEN». they turn upwards & bend over stigma:-- but
470t153 f do in about <3/4> of minute These latter were POLLEN gatherers & they seem slow= Maer 1840 My Fathe
470t178 Sept.) In Spanish Broom by pulling back Wings, POLLEN is ejected with violence in shower On many Pap
472s02r hered, & to day saw very odd dusky humble (with POLLEN) on legs go from clump to clump, & insect prob
472s02r sect proboscis in many flowers, on one of which POLLEN was routed. wh. was not case, on several flowe
472s02r igma & draw it out over & over again & wipe off POLLEN. (as a needle becomes covered) so whole sides
472s02v It first alights, it cleaned sucker & <I think> POLLEN was scraped off, which appeared like Heartease
472s02v was scraped off, which appeared like Heartease POLLEN.-- the pollen appeared chaffy, as if sucked?!
472s02v ff, which appeared like Heartease pollen.-- the POLLEN appeared chaffy, as if sucked?! opens & shuts
491q01v ch soil & propagate from their seed 3. To apply POLLEN of different genus & then some hours afterward
491q01v terwards of nearly related plant & see if first POLLEN produces any effect, as in case of woodpidgeon
495q005 e what effect on organs of generation (5) Place POLLEN of Red Cabbage «mixed with own pollen» on flow
495q005 (5) Place pollen of Red Cabbage «mixed with own POLLEN» on flowers of other cabbages & see whether th
495q005 (6). Dust flowers of one branch of Cabbage with POLLEN of other, count seeds, & see how great a propo
495q005 eally is an important case-- cross with cowslip POLLEN-- as these are wild varieties. Is any interme
495q05a ticed leaves covered with Honey-dew dusted with POLLEN of neighbouring grass= Spread sheets of Paper.
496q05a owers in minute do they visit?? good=!! Examine POLLEN of double flowers. compared with single & see
496q05a ody to Hawk & sow pellet. ejected. done Examine POLLEN of such flowers as do not seed or seed rarely-
497q06v Herbert says do about OEnothera.-- (14) Examine POLLEN of those genera of which wild hybrids have bee
498q008 ch counterbalance each other? (10) Is number of POLLEN-grains necessary to impregnate ordinary number
498q008 ith separate division of germen <?>-- (11) Must POLLEN grain be whole, to impregnate?-- I presume onl
498q009 were produced by seed, we might feel sure, that POLLEN of own kind is much more effective than of for
499q009 .-- (16) Any calculation of number of grains of POLLEN in any one flower (17) Catch Bees, Butterflies
499q009 - Meligethes & see whether they are dusted with POLLEN-- in what state (whole or broken) is ball of p
499q009 en-- in what state (whole or broken) is ball of POLLEN on Bees thighs (18) Place pin's heads with Bir
499q09v ime near male yew tree & see whether they catch POLLEN-- <Ne> In Oenothera bush.-- (19) Theory of moc
499q09v by Bees or Butterflies or little insect?= or is POLLEN excessively minute or abundant? do they seed p
502q012 plants.-- Hooker says Rafflesia is dioecious & POLLEN must be carried by some insect-- (21) Are ther
504q013 ve one after_other to flower & Menyanthes about POLLEN bursts before flower is open-- -- No (6) There
504q013 Dodecatheon Ʌ . Castrate apple & pear to see if POLLEN naturally carried, on account of Van Mons view
505q014 orned orange. wh. never has seeds produced good POLLEN? Yes «From cultivation lost their horns» is im
505q014 it--; become well shaped by care 13 Arum before POLLEN is shed can you find flys dusted with pollen f
505q014 re pollen is shed can you find flys dusted with POLLEN from other flowers? Can flys' escape from old
506q015 country-- where cause not apparent-- Any where POLLEN is not produced or small in quantity -- Any un
506q015 where germen does not swell, although there be POLLEN.-- or FEW. or bad seeds formed; badness may be
506q015 s (42) How are Orchidiae fecundated, as mass of POLLEN is requisite-- Brown's paper 43. Any flowers
513q21. dioecious plants, when crossed R. BROWN-- will POLLEN act on any flower before stigmas expanded-- in
513q21. ers are dichogamous Zostera-- Knights notion of POLLEN & stigma generally not being mature at same ti
514q21. asetums) really always hit stigma by projecting POLLEN-masses?-- = answered = Has Ophrys nectary?= Bu
514q21. similar to Lapland Plants --will get answer= Is POLLEN of cultivated Orchis & Asclepias &-- carnosa?-
515q021 point to attend to is whether good & plenty of POLLEN is produced, & 2d if so, whether concepcion ta
507q15v Ficus carica Henslow presumes females produce. POLYGAM. trioecia. (are female flowers ever productive
435e130 udimentary abortive organs, even more so than POLYGAMIA: Monoecia & Dioecia, preeminiently artificia
450e174 ication & nature» strictly monogamous-- geese POLYGAMOUS (¿when wild) but only some birds are so when
450e174 ild-- wild ducks monogamous; tame ones highly POLYGAMOUS-- change of instinct by domestication.-- "No
207b152 a Reunion presente elle seule plus d'especes POLYMORPHES que toute la terre ferme de lancien monde".--
453e182 Teneriffe does not countenance the theory of POLYMORPHOUS plants, abounding in volcanic islds.-- <Coc
241c016 le on the Distrib of Ferns in South Sea (Indio POLYNES: <)> vegetation far East) Ann: des Sciences. S
465t081 somewhat to do with it. mem. dogs «& pigs» in POLYNESIA; & dogs in S. America «Rengger.» -- now it is
068r145 on.-- Lyell suggested to me that no metals in POLYNESIAN Islds--. Volcanic plenty in S. America!! Met
309c221 -- useful could not leap fences:-- Dr Lang on POLYNESIAN nations (quoted) p. 4.-- do. p. 186. quotes
537m076 are not such «sudden» changes rare,-- as when POLYNESIAN mothers ceased to destroy their offspring--¿
603o012 is certainly curious. Chinese, S. American. POLYNESIANS Jews, African all sacrifices. How completely
536m072 ill, if so all animals., then an oyster has & a POLYPE (& a plant in some senses, perhaps, though fro
170b001 ely similar; for instance fruit trees, probably POLYPI, gemmiparous propagation. bisection of Planari
254c049 ere eliminated is» often repeated, as mouths in POLYPI, Surely not correct view of Flustra or Ascidia
355d068 separate parts of same plant<s>)-- now in some POLYPI we see young bud changing into ovules.-- Capta
483z013 oolog: Transact. before writing on Planariae or POLYPI & is especially grand paper. p. 387. "on Class
484z014 ry, make noise & throw head back M Edwards,--on POLYPI of Tubulipores L'Institut-- 1838 p. 75 A detai
485z017 lt water so soon putrifies?? p. 319. on Hydra-- POLYPI-- <Rep> do p. 324. Polypi shorter duration tha
485z018 ?? p. 319. on Hydra-- polypi-- <Rep> do p. 324. POLYPI shorter duration than cells.-- reproduced.-- M
485z018 cells.-- reproduced.-- Milne Edwards p. 138 on POLYPI.-- Berenica &c &c L'Institut, 1838 p. 46 Macle
601o009 leg when asleep-- «or habitual actions» perhaps POLYPI-- (so that lower animals are sleeping higher a

381d156 Oyster. cystic Entozoa. Echinoderms. Acalephes. POLYPS. Sponges Heautandrous, male organs formed to f
305c211 ome relation together, as well as Man & child, POLYPUS & polypus, bud & bud, polypus & germ plant & s
305c211 on together, as well as Man & child, polypus & POLYPUS, bud & bud, polypus & germ plant & seed.-- ins
305c211 as Man & child, polypus & polypus, bud & bud, POLYPUS & germ plant & seed.-- instincts in young anim
612o035 cal laws of endosmic & exosmic juices. arms of POLYPUS, show either local or general will, & stomach
591n101 ssays Vol 2.-- «also on origin of religion or POLYTHEISM, at p. 424 Vol. II «Sect XV. Dialogue on Nat
044r072 ls occur in the Ashes which fill up theatre of POMPEEI (?). -- Such have been seen to form in atmosph
495q05a oughed field-- on hills.-- 10 Shoot tame duck on POND with Duck-weed-- coots-- waterhens-- examine do
495q05a d of seed must be distributed.-- Examine scum of POND for seeds.-- 11. Soak all kinds of seeds for we
398e005 ty» The difficulty of multiplying effects & to <PONDER> conceive the result with that clearness of co
428e111 y very partially to the Zizanias in in Sir. J's PONDS-- my principle being the destruction of all the
242c017 «& Cochin China» are said to have orang-utang & PONGO in common»-- Galiopithecus common to Moluccas &
450e175 Pegu-- in Sumatra two breeds both small -- Java PONY occasionally reaches 13 hands.-- Phillipines Po
450e175 ony occasionally reaches 13 hands.-- Phillipines PONY somewhat resembles that of Celebes is somewhat
458t002 that a hybrid between ass & Zebra, crossed with PONY mare & produced a very pretty little animal, sh
097a043 ic rocks of the Orinoco".-- <but> on one of the PONZA isles. but no minute description is given.-- Vo
099a051 hes. inclined layer!!!.-- The separation in the PONZA case of Scrope parallel to walls of dykes-- Mem
315c240 between means of Transport & creation exists.-- POOH. May have been Created at many spots & since di
119a106 ocks have just floated over the absolutely fluid POOL.-- (this is shown by the softness & curvature o
120a108 miles through nearly cold rock.-- in volcano the POOL is not deep. --Hot springs &c &c--then if so, t
204b141 knew of case of male widgeon, winged & turned on POOL, first season bred readily with common ducks.--
593n111 means a necessary part of man's mind.-- At Maer. POOL. I saw many coots & waterhens feeding on grassy
593n111 command, they all took flight & flappered across POOL to bed of flags I was astonished & having looke
495q05a -- waterhens-- examine dog, which has swum-- on POOLS & rivers-- every kind of seed must be distribut
034r044 cks--The stream at Portillo Pass example of do? <POOR> Daubeny good account of ejected granitic fragm
097a043 thin pellicle of a blackish colour like a dull & POOR varnish, which I conceive to be analogous to th
129a131 erica» read parts of this work, though it is but POOR. Athenaeum. 1838 p. 791 -- Most curious account
209b156 - Canary Isles: French Edit. Flora of Islds very POOR «(p. 145)» 25. plants. 36 St Helena, without fe
235b272 hilos of Zoolog. 842 White regular gradat in Man POOR trash Lyell 1024 Flemings Philosophy of Zoolog
321c270 er Savage Landons Imaginary Conversations-- very POOR Sir T. Browne's Religio Medici Lyell Book III T
338d024 esticated races.-- Athenaeum. p. 505. some (very POOR account) of plants of Nova Zenbla -- in review
498q007 ant accustomed to rich soil, when placed in very POOR flower, but not fruit-- -- Do not orchards beco
604o017 otions.-- T. Mayo-- Pathology of the Human Mind. POOR.-- on insanity.-- Prevailing idea. owing to los
209b156 thout ferns.-- analogous to nearest continent: POORNESS in exact proportion to distance (?). & simila
498q007 -- -- Do not orchards become unproductive from POORNESS of soil.-- yet crabs probably would grow ther
566n010 internal organs, as action of heart Malthus on POP. p. 32, origin of Chastity in women.-- rationall
499q009 om males, will female (a) Willows or Yews some POPLAR'S produce.-- (15) Would Yew fruit without impre
468t111 ll Hymenoptera Saw Humble go from great Scarlet POPPY to Rhododendron-- from Larkspur to Lupine two s
468t111 ertain flowers, to day early, the great scarlet POPPY-- So that, finally Fraxinella. with respect to
206b146 orm, less hardy, & more liable to disease" If POPULATION of place be constant «say 2000» and at prese
206b147 nued breeding from eternity «backwards.--» If POPULATION was increasing between each lustrum, the num
206b149 ers & then try this as simile In a decreasing POPULATION at any one moment fewer closely related;∴ (f
230b235 ol V. P. I. p. 67. Dr. Coulter on decrease of POPULATION in California cessation of female offspring:
322c270 avater's Physiogmony ---- Octob 3d Malthus on POPULATION W. Earls'. Eastern Seas. .Octob12th.-- Sir G
375d135 sitive check of famine & consequently death.. POPULATION in increase at geometrical ratio in FAR SHOR
375d135 mongst men.-- «Even a few years plenty, makes POPULATION in Men increase, & an ordinary crop. causes
397e003 aws: And since the world began, the causes of POPULATION & depopulation have been probably as constan
397e003 to one species-- I would apply it not only to POPULATION & depopulation, but extermination & producti
399e009 se & be driven outwards in the grand crush of POPULATION.-- Octob 10th. Saw. two undoubtedly rabbits
609o29r ealousy, & every one being married to keep up POPULATION. with the existences of so many positive che
615o038 weakened in Otahiati; fear of death in Hindoo POPULATION.-- Slightly modified in many countries, henc
376d135 fect, (by means however of volition) of this POPULOUSNESS, on the energy of Man» D.'Orbigny. Comtes R
377d138 ch the C. Sphynx makes is also made by the C. PORCARIOUS., together with a grunting noise, the former
376d136 many anecdotes have been told is Cyanocephalus PORCARIUS.-- this Monkey did not like a great coat made
328cIBC many shells of Genera Corlula Cham. Cardium. PORCELLUS Turbo. Cerithium Jardin du Roi Java fossils s
191b079 ood genera-- How remarkable spines, like on a PORCUPINE on Echidna-- Good to study Regne Animal for G
358d075 asion, not withstanding our vigilance a piece of PORK 3 lb was taken from a boiling pot, & carried of
033r043 the rim of conical crater: at Teneriffe Wall of PORPH. Lava with base of Pitchstone; Mem Galapagos. c
038r058 kes in upper strata. quite different from the PORPHYRIES: certainly appearance leads me to believe me
073r163 e granits & gneiss with gold veins visible:--"PORPHYRIES of Mexico may be considered for most parts a
073r164 n mines of gold & silver." «p. 131» The above PORPHYRIES characterized by no quartz & amphibole frequ
077r170 dern breccia, chiefly owing to destruction of PORPHYRIES. whereas other to ancient rock.--this N degr
078r176 l III» Mexican Cordillera "immense variety of PORPHYRIES which are destitute of quartz, & wh abound b
436e137 d.-- Old Red Sandstone-- Van Diemen's land.-- PORPHYRIES of Andes. A familiar History of Birds by the
048r086 beds same age? is white substance triturated PORPHYRITIC rock. s (mem white tufas with purple Claysto
042r068 he Valle del Yeso it is probable that point of PORPHYRY has been upheaved in a dry form It is clear t
049r087 anic or Plutonic The cellular state of all the PORPHYRY specimens, must be well examined At M. Video
049r088 siliceous in close contact? -- «Cordillera???» PORPHYRY at Valparaiso; Epidote -- Must we look at neg
065r140 n great subsidence previously. Mem. pebbles of PORPHYRY.--Falklands.--off East Coast. -- Capt. Cook f
073r163 18 degree & 22 degree N. = formations of amph: PORPHYRY. greenstone[,] amygdaloid. basalt & other tra
075r165 vitreous felspar: = gold veins in a phonolitic PORPHYRY. = several parts of N. Spain great analogy to
075r166 ined in a primitive slate, covered by a clayey PORPHYRY, containing grenats. In Peru. on other hand,
077r170 N degree I. even No. 2. might be mistaken for PORPHYRY above ancient freestone, limestone & <many> «
077r171 ondary» rocks. Vein traverses both Clay slate, PORPHYRY North 52 W, & is nearly the same with that of
133a141 Voyages of Adventure & Beagle vol I. p. 2 & 3. PORPHYRY at St. Elena. p. 6. few «living» shells. on o
076r170 50 degree-- covered by conformable greenstone PORPHYRYS & phonolites do. amphibole quartz & mica very
409e047 nds.-- as difference between mind of a dog & a PORPOISE was not thougt overwhelming.-- yet I will not
555m142 ory of the women.-- The Chillingham cattle (& PORPOISES) have not charity-- is it in former case inst
592n105 wild cattle & deer pursuing a wounded one.-- PORPOISES a ditto-- it is probably some secondary one--
641j29r & Geographical Distrib of larger Seals-- Are PORPOISES numerous in cold Oceans I think not.-- Does t
025r017 denly. coast of Brazil generally.-- Mrs Power at PORT Louis talked of the extraordinary freshness of
052r099 eneral fact absent in T. del Fuego, excepting at PORT Famine Mr Sorrell says that numerous icebergs a
057r114 rious similarity of rocks of very diff. ages. at PORT Desire on plain. & interstratified.-- Urge fact
065r139 e?-- Degrading of inland bays. like St. Julian & PORT Desire applicable to Craters of Elevation.--The
113a092 ss Major Mitchell showed me a river <near> W. of PORT Philip. which had bar at mouth excavated in sol
436e136 ecause circumstances» <April 12th..> Cestracion, PORT Jackson Shark-- Owen thinks Australia part of O
116a100 luvium, or sublittoral formations. p. 150. at PORTEZUELO, extremity of mountains of Cordova project o
116a100 all» bits of red granite between 40 & 50 from PORTEZUELO. Bull: Soc. Geolog. 1837. December. p. 91. a
034r044 e being the most inferior rocks--The stream at PORTILLO Pass example of do? <Poor> Daubeny good accou
069r150 Heat Consider profoundly the sandstone of the PORTILLO line.--connected with <gneiss>.--(Mica Slate)
069r151 There must have been some conglomerate East of PORTILLO Where gone to? Intermediate space protected.-
108a081 d probably much Carbonic Acid gaz here.-- [top PORTION page excised, not located] Bull:. Soc: Geolog.
108a082 l XIV. (My Edition) p 500. Well described [top PORTION page excised, not located] -- do-- [...] Subaq
250c038 tion of Great Animals in Europe & America Some PORTION of the world «Africa» being left more equable.
388d173 need.-- one would <one> suppose that the vital PORTION ¿nerves? passed through transformation, & was
421e101 an Fernandez to Chile??) Falklands to southern PORTION.-- ℕ.do p. 269. Annals of Nat. Hist 1838 on «a
549m120 al enjoyments of the minute make large <parts> PORTION of daily «happiness» «pleasure». A wise man wi
556m145 to forget-- --; that every man is born with a PORTION of phsiognominical sensation, as certainly as
618o041 &c. either to our bodily frame or the cerebral PORTION of it Thoughts, perception &c. are modes of su
038r058 that envelopes no doubt existed. These higher PORTIONS probably formed Islds from which proceeded pe
616o039 tion to a living body (especially the cerebral PORTIONS of it) that attraction has to ordinary matter
295c178 .-- «otter; civet cat, rodents.» (Pachyderm in PORTLAND stone of Alps!!!? No) p. 15 (Lyell's Pamphlet
431e121 lite, & the T. elongata in the upper formations PORTLAND Stone &c p.? No) so «it is» good case:-- -- n
222b204 -- difference of species between land shells of PORTO Santo & Madeira-- I believe very curious-- My i
511q018 = Pidgeons. Canary birds-- Bantams.-- (6) <Mad> PORTO Santo Rabbit. Descript. of colour «& length of
332d005 es.-- W. D. Fox has a cat. which he bought in PORTSMOUTH, said to come from coast of Guinea, -- ears
047r082 ese view are right the coincidental retreat at PORTUGAL & Madeira (Lyell. vol I. P. 471) is explained
199b119 of albino DISEASE being banished, & given to PORTUGUESE. priest.--In first settling a country.-- peo

Page **(Key Word)**
022r006 nute particles have a tendency to change their POSITION? Carbonate of Lime disseminated through the g
029r034 Chili as clearly bearing a relation to present POSITION of <Coal> Forests. These thick beds of Lignit
030r035 oded mountains, probably chiefly leaves.--This POSITION agrees with character of.. «in Basins from ri
030r035 aracter of.. «in Basins from rivers. & natural POSITION» position at N. S. Wales & Van Diemen's land.
030r036 .. «in Basins from rivers. & natural position» POSITION at N. S. Wales & Van Diemen's land.-- Whole c
062r130 er): extinct Guanaco to recent: in former case POSITION, in latter time. (or changes consequent on la
086a007 mospheric currents?-- chiefly clearly by sun's POSITION = If equatorial streams of warm pole; in name
088a013 mpts at elevation From the lost & turned about POSITION of strata, prooff thickness not very great; w
147g018 out lake or sea could not be placed in present POSITION Thursday Evening ½ past 8 Tyndrum 29.<625> «6
219b194 an & African form, merely because intermediate POSITION.-- We cannot consider it as adaptation becaus
270c103 te animals <tu> plants turned inside out. have POSITION of organs of generation!!!. Mem. Agaziz. (INo
506q015 s any secondary, sexual characters.-- Stature, POSITION of flowers-- Their smell-- form of flowers--
506q015 n Monooecious «order» flower occupy particular POSITION.-- (39) What does he think of Dr. Flemings st
510q017 travel by wind. Aug. St. Hilaire Bot. p. 787. POSITION of embryo in close species of Hilianthemum di
541m091 fort.-- it looks so analogous to muscle in one POSITION great fatigue.-- may explain excessive labour
542m096 A dog when he barks puts his lips in peculiar POSITION, & he holds them this way, when opening mouth
564n005 cked it-- Here then, that faculty, whether for POSITION of axe of eyes, state of surface, or other me
585n077 instinct, but if all men placed stones in same POSITION, it would be instinct-- instinct is heredetar
176b022 have changed most. «owing to the accident of POSITIONS» must in each state of existence have shortes
383d159 ia when they first appear occupy their proper POSITIONS,-- this would be argument for developement of
375d134 increase of brutes, must be prevented soley by POSITIVE checks, excepting that famine may stop desire
375d134 ot increase, whilst no checks prevail, but the POSITIVE check of famine & consequently death.. popula
528m035 iscovered-- is closely analogous to my Fathers POSITIVE statement that insanity is only cured by forg
609o29v up population. with the existences of so many POSITIVE checks.-- (This is encroaching on views in se
612o035 one always satisfactory, though not adding to POSITIVE knowledge. lessening amount of ignorance [RHC
399e011 unknown in Hervey or Society isles. Hope says POSITIVELY he has seen. a Calosoma. (very like American
599o008 matiere et du temps! voila les facultes, q'il POSSEDE seul sur la terre. J'ai trouve son âme" &c-- -
439e144 ther depended on some nice qualifications each POSSESS., & that tiger springing an inch further would
583n069 instinctive pointing varies.-- p. 18. Animals POSSESS strong imitative faculty: pure instinct is not
623o050 as> (as we must admit) has been acquired, does POSSESS the beneficial tendency [RHC] 10) that the ins
640j167 views we can deduce why small islands. should POSSESS many peculiar species. for as long as physical
640j167 e smallest newest, & most wretched isld should POSSESS species to themselves.-- Probably no case in w
191b080 of Man possessed Elk not England. Did Ireland POSSESSE Mastodons?? Negative facts tell for little) G
086a006 llas before Geology is written Cuvier. Europe POSSESSED a great edentata. -- How much is temperature
191b080 countries different.-- Ireland & Isle of Man POSSESSED Elk not England. Did Ireland possesse Mastodo
232b247 e climate, <more northern> Iceland would have POSSESSED a most peculiar Flora «& north of Europe»-- A
282c141 e genera of bird. analogous. animals would be POSSESSED by the different races of man, yet altogether
303c204 legs-- hence the father of man kind probably POSSESSED a structure in these points for <t> a less ti
406e037 h & South. America.-- Whole wor[l]d, formerly POSSESSED a climate compared to S. America at present d
439e142 ring the reappearance of characters, formerly POSSESSED-- <that is animals> «or rather the parents» h
580n061 very weak intellect (As Miss Clive) have only POSSESSED very loose ideas.-- Have children loose ideas
196b107 . 207 "It is not generally known that Ireland POSSESSES varieties of the furze, broom, & yew very dif
263c077 e are deceived) then he is no exception.-- he POSSESSES some of the same general instincts, <as> & <m
314c237 es, this segment is converted into hood which POSSESSES power of movement. & not the organ itself How
057r114 Europe with ice theory.-- Capt Ross found in POSSESSION Bay in 73 degree 39 N. living worms in the m
065r138 on from <Deception Isld.> South Shetland Cape POSSESSION. Syenite¿ Andite?-- Degrading of inland bays
095a038 o Dr. Gillies in MS. letter in Sir. W. Parish POSSESSION. talks of <hill> «cerro» of Diamante near st
214b175 since the time of Charles.,-- and now in the POSSESSION of Mr Howard Galton, have one of the vertebr
426e108 is Mr S. S. parentage?-- Wonderful as is the POSSESSION of voice by Man. we should remember, that ev
481z010 Tucuman mountains The fourth Vol. «in Lyell's POSSESSION» of Zoolog. of Voyage of Astrolabe must be s
113a091 ve transmission of temperature clearly prove POSSIBILITY of metamorphic theory On the idea of statica
175b016 look at two ostriches as strong argument of POSSIBILITY of such change,-- as we see them in space, s
192b083 pen, goldens-- hence-- sub-varieties & hence POSSIBILITY of reproducing any variety, although many of
192b087 been more perfect. because in each there is POSSIBILITY of such organization. [Spines in Echidna & H
193b091 ardson-- Fauna Borealis. It is important the POSSIBILITY of some isld not having large quadrupeds-- H
200b121 structure as. petrel & alk. do not show the POSSIBILITY of common branching off.? Accra, Coast of Af
218b189 s do not cross-- we see it even in men); the POSSIBILITY of Caffers & Hottentots coexisting. proves t
219b194 same as T. del Fuego & C. of Good Hope show POSSIBILITY of transport. If some cannot be explained mo
229b234 lia being peculiar (?) If. Henslow discusses POSSIBILITY of seeds of Keeling standing transport.-- <t
257c057 had its web-feet-- all Nature answers to the POSSIBILITY.-- My views will explain no Mammalia in seco
259c064 rican form, that one is brought to admit the POSSIBILITY (any great change in species is reduced by a
383d159 r-grown seems to show whole body imbued with POSSIBILITY of becoming either sex.-- ‖ In my theory I m
439e144 there comes the impediment of instinct-- the POSSIBILITY of rearing by seeds Holyoaks-- (how far is t
441e148 in connection with buds.-- They differ from POSSIBILITY of concourse of two «individuals» & the acti
453e183 , abounding in volcanic islds.-- <Cocks> The POSSIBILITY of different varieties being raised by seed
538m080 in his head. & Babington's silly joking The POSSIBILITY of the brain having whole train of thoughts,
539m083 s otherwise than habitual.-- instances?? The POSSIBILITY of two quite separate trains going on in the
571n031 - these are Mayo's ideas.-- In language. the POSSIBILITY of poets describing gentle things in gentle
578n055 m be thinking over it with sorrow,-- let the POSSIBILITY of this being discovered by anyone, especial
603o014 «EXCELLENT». Deficient in not explaining the POSSIBILITY of <handsome> «UGLY healthy» young woman, wi
182b047 y will be to describe limits of form. (& where POSSIBLE the number «of steps» known). <for instance a
185b055 to accomodate itself to as many situations as POSSIBLE.-- Why should we have in open country a groun
198b113 tribe appears fitted for as many situations as POSSIBLE. for instance take birds animals, reptiles fi
227b226 es of any great divisions) thus a knowledge of POSSIBLE changes is discovered, for speculating on fut
230b240 ding prevents perfect change.-- It is scarcely POSSIBLE to get evidence of two races of plants run wi
235b273 log Royle on Himalaya Plants-- Would it not be POSSIBLE to work through all genera, & see how many co
240c003 il: remark of Apteryx having feathers.-- It is POSSIBLE, time being an element in the transmission of
257c057 hanges which at first it might be doubted were POSSIBLE,-- it has been asked how did the otter live b
265c086 . to its habits.-- Few will dispute that it is POSSIBLE to have structure without habits-- after seei
283c146 genera, is probably to add to quantum of life POSSIBLE with certain preexisting laws.-- .If only one
285c151 ity from few forms, parents of all species not POSSIBLE in some detail,-- the relations to islands cl
292c171 mory transmitted without consciousness «a most POSSIBLE thing. see men walking in sleep».-- an action
301c198 ls. « & what reason» in precisely same way not POSSIBLE to say what habitual in men & what reasonable
312c232 ilation being «variation» produced in shortest POSSIBLE time. Mr Willis Long eared little dogs, I am
358d085 ried much further, if by the process this were POSSIBLE, the organs doubtless would shrivel up.-- Yet
372d131 law. but man not arm. hard to say-- if it were POSSIBLE to support the arm of Man, when cut off , it
388d171 gination of Mother.-- We see in a litter every POSSIBLE variation from being very near mother, & some
401e014 ize enormous number of plants-- it is scarcely POSSIBLE to purchase seeds of any cabbage, where a gre
410e051 smutation, cannot see the deductions which are POSSIBLE.)-- Ascertainment of closest species (& namin
411e054 their relation to the external world, & every POSSIBLE contingent circumstance.-- ‖the laws of varia
423e097 ingdom of nature as it now is, it would not be POSSIBLE to simplify the organization of the different
423e097 of living beings-- but there is the strongest POSSIBLE to increase them, hence the degree of develop
460t017 laws which prevent infinite variation in every POSSIBLE way.-- the laws which determine the kinds of
463t055 with an intermediate structure, but it is not POSSIBLE to imagine what habits an animal could have h
472s01v tale in garden & all came up hybridised. It is POSSIBLE to raise them pure for Miss Bent three years
497q007 logy of.-- most important, as furthest removed POSSIBLE point.-- ¿genera in intermediate country (2)
502q11v -nettle.-- Lithospernum. Blue Gloss. it is not POSSIBLE to see orifice of poison-tube-- so put carmin
503q012 s only male yak cross with cow: is not reverse POSSIBLE?? Maer (1) Yew Trees near Boat House «ANY mal
541m089 lich I was trying to unbend my mind as much as POSSIBLE (testing success by decreasing headache) & fo
621o047 is good. yet horseflesh show that even this is POSSIBLE.-- So that as there nice & nasty in taste, &
628o53v dog had no instinct to commence with scarcely POSSIBLE to teach it-- all dogs might be taught, but n
032r041 ause:--<exp> does not explain cleavage lines./ POSSIBLY general symetry of world.-- I feel no doubt.
147g020 hills of alluvium-- nothing very striking yet POSSIBLY sea more probably than river-- No exact terra
174b015 st differences-- if separated from immens ages POSSIBLY two distinct type, but each having its repres
292c170 ss. no dissertation against these views, could POSSIBLY have had <to so convin> brought so much convi
296c184 ween pheasants & Black fowl.-- use as argument POSSIBLY some few hybrids in nature.-- «p. 473» Webb &
346d047 Spoonbill Wader. Ibis)-- laws of plumage might POSSIBLY be made out.-- August 25th Athenaeum (1838) p
352d060 logist being obliged to fill up the gaps.-- is POSSIBLY the same with the <Zoologist> «philosopher»,

Page
(Key Word)*
427e110 should suppose, that the pollen of crab, would POSSIBLY «No, for pollen of any kind would fertilize i
585n077 heredetary knowledge of things which might be «POSSIBLY» acquired by habit. so bees in building cells
586n082 lding «regular» cells-- [but this faculty «may POSSIBLY be» «probably is» instinctive, namely the kno
587n087 like sugar, acid, &c, which may be doubted for POSSIBLY even taste of senna. might be acquired. as th
301c199 ion. even which Man performs.-- child striking a POST in passion.-- Habit instinct gained during life
473s03r ana & Hippotamus «with Megatherium & Mylodon» in POST pliocene strata! Mastodon longirostris in mioce
038c059 . Are not the dikes generally vertical? if so POSTERIOR to elevations? & not sources of lava streams.
059r120 ntain, & that the central part fell in.--Says POSTERIOR craters in centre:-- Bailly talks of much gra
536m071 relic of same thing that makes one dog smell POSTERIOR at another.-- Why do bulls & horses, animals
308c217 brought back memory» old habit of putting tea in POT made me go to tea chest almost unconsciously.--
358d075 ce a piece of pork 3 lb was taken from a boiling POT, & carried off by one of these birds" Case of bi
505q014 ically.)-- Asclepias-- Flowers not seeding= Put POT of boiled earth on top of House =Aristolochia, p
095a037 & seq Murc Trans Geolog Soc Vol 2. p 257 {T} The POTA: labiata certainly is found with the Mactra. at
388d170 s real. difference-- but this does not apply to POTATO.-- With respect to offspring being determined
467t104 len from <sheath> Keel of Lupine-- Seen Bees on POTATO & several times on Beans Rough.--green-cabbage
505q014 require insects to impregnate it (7) History of POTATO field= (8) Abortive Thyme seeds weather wet--?
055r106 avities", &c &c &c Vol II. Chapt VIII. p. 97 at POTOSI the veins run from North <inclining> to South.
075r166 nal of Nat. & Geograph Siciences? -- H. says in POTOSI the silver is contained in a primitive slate,
121a111 sact) has seen clay stiff enough <to form> for POTTERS to use. in which great Knife formed crystals o
367d113 ke balloon-- by air cells connected with cheek POUCHES.-- Hunter's Animal OEconomy p. 45 "One of the
097a043 vein in higher parts? & felspathic veins?-- Mr POULETT Scrope. talks of Trachyte, "superficially coat
399e009 - Octob 10th. Saw. two undoubtedly rabbits in POULTERER shops., of same colour as a Hare, but paler &
473s07r es may be early impressed & others later-- All POULTRY with same down-feathers. Zoology 1856 Skimmed
509q017 s of Apterix: clavicle in-? Combs in combless POULTRY-- Teeth in foetal state: Mr. Horner. On Mr Tre
551m129 Jump said-- I saw the ourang. take up a stone & FOUND the earth. Lockarts life of W. Scott Vol VII p.
559m156 ng. as giving abstract of Smith's views «Take & FOUND up inflorescence parts of mosses & see if Hybrid
090a023 25 stones.-- The total weight recorded is 473. POUNDS (taking about average when several are given),
090a023 en several are given), this will give nearly 19 POUNDS average for each stone. that fell, that was we
090a023 len in the 50 years. ∴ 90 x 19 = 1710 ÷ 50 = 34 POUNDS each year.-- but instead of 90 stones in many
402e017 such strata being merely leaf, if one river did POUR sediment in one spot, for <whole> many epochs--
578n057 ly very little so) & hence by association, there POUR out tears, & there is slight convulsive wrinkli
209b156 study of Flora of islands; "ou bien encore on POURRAIT au plus en conclure quels sont les genres qui
602o11v compares a view taken by a camera obscura &c a POUSSIN.-- How are my ideas of a general notion of eve
551m129 es its lips into point-- man, though he does not POUT. pushes out both lips in contempt <&> disgust &
560m156 signs of impatience when woman present? Do they POUT, or spit, or cry.-- <fe> Shame, independent of
587n088 arlet, as well as remember it.-- Why do children POUT & not men-- orang-outang & chimpanze. pout.-- F
587n088 ldren pout & not men-- orang-outang & chimpanze. POUT.-- Former, whines just like a child. Get a Dict
595m125 I assume a child pouts who has never seen others POUT]CD [blank] Goldsmiths Essays No XV,. on sounds
585n076 but not bite.-- Henslow remarks that Chimpanze POUTED & whined, when, man went out of room.-- all th
366d112 -- Yet seeing the feathers along one toe of the POUTER one thinks there is a law.,-- that there must
275c121 erent skulls, but same marks on wings are Blue POUTERS & small Bald Heads Mr Yarrell will mention in
363d101 es on beak & in Muscovy duck" crested feather, POUTERS, fan tails are found in any colours of plumage
363d101 ils are found in any colours of plumage &c &c «POUTING pidgeon exaggeration of cooing.--» & compare t
551m129 ot know what when it looked at the glass) when POUTING protrudes its lips into point-- man, though he
565n009 having peculiar expression is remarkable. the POUTING, & blubbering-- sulkiness is same as pouting,
565n009 e pouting, & blubbering-- sulkiness is same as POUTING, <but> lesser in degree, no smile, no frown sh
594n113 of almost anything ugly. baby-- association-- POUTING child same as anger, lips not compressed sulle
595m125 blank] [not located] A child crying. frowning, POUTING, «smiling», just as much instinctive as a bull
551m129 ence <">-- The young Ourang in <Zoolog> Gardens POUTS. partly out displeasure (& partly out of I do n
595m125 tion) in man, than sucking.-- [I assume a child POUTS who has never seen others pout]CD [blank] Golds
430e118 place from changing country: but greyhound. & POUTER Pidgeons «race-horse». have not been thus prod
523m015 owledge of the untruth of the thing, namely his POVERTY.-- his manner of curing it. by keeping the sum
620o044 no more of it.-- Not so man, from his memory & «POW» mental capacity of calling up past sensations,
105a072 tions on coast of Madeira.? How came it if this POWDER results from «decomposed sea» shells, that lan
495q05a oecious individ--small orifice (8) Carry Bees, POWDERED with starch & Carmine & experimentia on thei
585n076 man went out of room.-- all theories of magnetic POWER in birds, seeing the sun &c are absolutely unal
024r014 ms. proves the existence of some moving <point> POWER ¿ Submarine currents Find instances; The whole
025r017 pens suddenly. coast of Brazil generally.-- Mrs POWER at Port Louis talked of the extraordinary fresh
048r084 blade of grass penetrating by action of Organic POWER a lump of hard clay. -- In the History of S Ame
051r095 al band, I do not see how to account for oceans POWER.--excepting when pebbles are brought into play;
069r151 to? Intermediate space protected.-- Oh the vast POWER of the ocean! Make a grand analogy between Weal
105a069 - There is thus wide difference between erosive POWER of river & sea.; the former as its channel beco
105a069 as its channel becomes wider looses its cutting POWER. (as does it when the inclination becomes less
105a069 he inclination becomes less & ∴ tends to finite POWER) whereas sea. on coast, as long as exposed to w
105a069 st, as long as exposed to waves of sea, cutting POWER increased with width. for besides more surface
105a073 > slow course, & with slow course small erosive POWER. therefore tendency of running water to deepen
106a074 not have been case. if inclination small.-- The POWER of widening channel depends on power of deflect
106a074 all.-- The power of widening channel depends on POWER of deflection with stream retaining its force,
113a090 ws that the strata have very unusual conducting POWER of heat from centre.-- But is this not wrong? w
115a098 n a coast, the shallower the water, the greater POWER of oscillations & currents.-- if matter was «su
115a098 nts" but relative to currents. Small lakes have POWER of levelling their shores where currents very w
121a111 f ice were formed-- (like my gypsum case) shows POWER of segregation.-- & has heated angular fragment
149g030 posited much-- On other hand remember modelling POWER of sea N of Valparaiso are those animals subjec
159g064 gressed What <alter> a balance there must be in POWER of rivers either bringing more «detritus» than
182b049 to be unprejudiced about self, but considering POWER, extending range, reason & futurity. it does as
190b078 opled from N. Africa An originality is given (& POWER of adaptation) is given by true generation, thr
195b098 stant in the shells.-- The question if creative POWER acted at Galapagos it so acted that bi[r]ds wit
195b101 in country, but how much more simple, & sublime POWER let attraction act according to certain laws su
210b160 e this relation also.‖Yes « Fox» ‖ The creative POWER seems to be checked when islands are near conti
258c060 ed), the relationship of Analogy is a divellent POWER & tends to make forms remote antagonist powers.
262c073 d.-- the mere fact of division of lesser & more POWER (2.typical 3.subtypical) where power arbitrary.
262c074 ser & more power (2.typical 3.subtypical) where POWER arbitrary. leaves door open for Quinarians to d
282c140 in some one.-- imagine the men to have greater POWER of change yet, as external conditions over whol
290c163 s instinctive) then must be owing to heredetary POWER of Muscles.-- then we SEE structure gained by h
293c175 fect. preserving its relations.-- the wonderful POWER of adaptation given to organization.-- This rea
310c222 zzle against my theory.-- If I be asked by what POWER the creator has added thought to <an> so many a
314c237 segment is converted into hood which possesses POWER of movement. & not the organ itself How except
367d113 ys. describes, a high-flying bat, which has the POWER of inflating its body like balloon-- by air cel
384d162 maphrodite:-- ∴ developed instincts of Capon. & POWER of assuming male plumage in females., & female
399e010 roduced, & have freely bred, they have not lost POWER of producing. Williams. Narrative of Miss. Ente
425e103 og!-- (Hybrids of Calceolaria.)-- «:CD[Does the POWER of, «easily» making tolerably fertile hybrids,
435e133 Mr Knight [not located] the stigma retains its POWER.-- R. Brown found the <poll> masses of pollen o
437e139 & hare» by the side of Eagles nest, which shows POWER of carrying great weight. p. 125 is said that E
445e158 anks. p. 212.-- p. 282. Allows this instinctive POWER in chicken, yet says it is evidently acquired b
445e159 of supposing some inward progressive developing POWER.-- My theory leaves quite untouched the questio
461t041 species is valuable because it shows no innate POWER of change & it also shows, what enormous change
471tf07 orm in the same> organ "manifestation of divine POWER"?.--"of their use difficult to conceive any ide
521m008 y morbid action.-- Old Elspeth's «in Antiquary» POWER of repeating poetry in her dotage is fact of sa
524m020 brain which gives sensation of pain, emits its POWER on the muscles in the twitching. Pride & suspic
526m028 f imagination very different from the inventive POWER,-- this, though very odd is perhaps true.-- mem
533m058 inctively. may not this be connected with their POWER of acquiring language.-- Hensleigh. W. says tha
539m082 r more easily, in proportion to variableness or POWER of intellect.-- Some complicated trades can har
555m143 ove skin of head & ears,-- ∴ some men have this POWER abortive muscles) The black Spider Monkey, very
569m020 e.-- Devotional feelings, probably some distant POWER of the mind-- superstition & charity & prayer,
569n021 on by senses of any two objects-- they by VIVID POWER of conception between one or two absent things.
577n052 lush.-- sensitive people apt to blush.-- -- The POWER of vivid mental affection, on separate organs m
578n056 e out tears. replaced & squeezed out again-- as POWER of mind by habit gets more perfect over volunta
578n057 t, the lachyrmal gland is «not» under voluntary POWER, (or only very little so) & hence by associatio

583n069 the same species with variability of reasoning POWER in one species man.-- false instinctive pointin
586n078 they may do all this instinctively «yes because POWER varies in breeds,» something of kind oneself kn
589n092 h ideas.-- A sigh, is an abortive groan.-- more POWER over muscles of voice than respiration.-- like
590n094 as good ground to call imagination a faculty, a POWER, quite distinct from self. «or will» [not locat
595n117 when <great> considerably excited, shows their POWER of imagination-- for it will not be allowed the
602o011 eing depends upon the "habitual reason".-- This POWER of the mind, faintly approaches to instinct How
604o018 we associate something extraordinary & of great POWER-- -- 2 From these & other reasons we apply to G
604o018 n lofty regions. 3 Infinity eternity. darkness, POWER. being associated with God. these phenomena we
604o018 & ?) call sublime.-- 4 From the association of POWER &c &c with height, we often apply the term subl
605o20v tly does not consist of this. but rather in the POWER of discriminating & respecting good from bad. A
606o20v taste is metaphorically applied to this mental POWER Although taste must necessarily be acquired by
608o026 r.-- he agrees & yet does not.-- because motive POWER not in proper state.-- When the admonition succ
608o026 es a mans character may change-- because motive POWER changes with organization The general delusion
608o026 sion about free will obvious.-- because man has POWER of action, & he can seldom analyse his motives
608o026 putting contingencies in the way to aid motive POWER.--if incorrigably bad nothing will cure him' 3)
612o035 es communication from without, & gives wondrous POWER of willing. These +willings are common to every
614o037) instincts.-- this loss is compensated by vast POWER of memory, reason &. & many general instincts,
617o40v t of action as known by the exertion of our own POWER & consciousness of it we are conscious that we
620o045 ch shows that. it is owing to some <subsequent> POWER (reason) obtained by age, which should show the
627o053 rue. p. 13. Affections cannot be analysed into "POWER" &c &c &c-- & if termed "selfish", must be subc
638j28r ty Vol I. p. 251-- stomach hump, kinds of foot. POWER of closing nostril, foot, sack. power of endura
638j28r of foot. power of closing nostril, foot, sack. POWER of endurance &c &c Camels? all good cases of co
639j28r petrel & Whale in some respects Chamaelion like POWER in Octopus & Chamaelion.-- C. D. Sucking feet i
261c071 als no notion of beauty, when does prefer most POWERFUL buck Owen talking of Plesiosaurus Plesiosaur
273c111 e.-- Milvulus-- Even flying woodpeckers,. with POWERFUL wings, but tail stiff.-- Swallow & goatsucker
028r029 ated in the Tropical oceans detained by Organic POWERS. We know the waters of the ocean all are mingl
059r123 enly.-- As the rude symmetry of the globe shows POWERS have acted from great depths, so changes, acti
112a088 Note Book (C) for some speculats on conducting POWERS of rock-- -- Geograph Journal Vol IV p. 36. on
113a090 ucted away the heat of surface. & if conducting POWERS had been better then 32 degree would have been
113a091 .-- We have no right to consider the conducting POWERS either better or worse & the depth of 32 degre
126a122 parates them, this is ascertained by conducting POWERS-- we judge from the surface, & say 60 ft to de
195b102 ation, such will be their successors.-- let the POWERS of transportal be such & so will be the form o
196b105 tinct may confine certain birds which have wide POWERS of flight; but are there any genera, mundine,
258c060 t power & tends to make forms remote antagonist POWERS.-- Every animal in cold country has some analo
262c072 countries patriotic?)-- but more especially the POWERS of reasoning &c &c.-- Study the wars of organi
263c077 n other hand can reason-- but Man has reasoning POWERS in excess. instead of definite instincts.-- th
300c198 . language seems to supply instincts,-- & those POWERS which allow of, acquirement of language. hered
301c199 acquire habits is this the Key to their mental POWERS.? p. 8. mistakes of instinct are external cont
322c270 d & 3rd Volumes Abercrombie on the Intellectual POWERS. Hunters Animal OEconomy. edited by Owen. read
359d088 is explained by the vigour of their propagating POWERS. (as if they were a good species, or local var
371d118 ssion is direct instinct. & afterwards enlarged POWERS to meet with contingency.-- Sept. 23rd. Saw in
385d163 said little cock «yet very odd loosing visible POWERS» in Zoolog Gardens. & Kings at Otaheite) <Thin
411e095 te, than the muscles of-- When one sees in Coralline POWERS of multiplication of individuals, & yet anothe
432e125 trained to pursuit having PUPPIES with the same POWERS E. frowns prodigiously when drinking very cold
466t095 trained to pursuit having PUPPIES with the same POWERS instinctive & doubtless not confined to sex.--
486z019 256 of Note Book (C) for comparison of singing POWERS of birds of N. America & Europe. Entomolog. Tr
495q05a ch & Carmine & experimentise on their returning POWERS-- then carry them in Electrical machine, rever
526m030 heredatary.-- When a man says I will improve my POWERS of imagination, & does so,-- is not this free
527m034 scoverer, & therefore (independent of improving POWERS of invention) such castles in the air are high
539m081 nd, yet there was no strain on the intellectual POWERS-- the difference is of a man wagging his foot
543m099 ansformation because Spiders have many,-- great POWERS of communicating knowledge to each other-- Aug
543m100 a man common sense, & the highest intellectual POWERS of perceiving & classifying distinct resemblan
553m135 fects of impressions long repeated, without the POWERS of the mind being EQUAL to the smallest casuis
570n027 cians living amongst naked figures, & observing POWERS common to savages???].CD-- The existence of ta
572n033 le says that negro certainly has less reasoning POWERS than Europaean.-- Ideots. defective brains.--
574n040 some connection between very limited reasoning POWERS & the fixing of habits,-- for instance the Bir
583n069 . 17. Contrast the invariability of instinctive POWERS in individuals of the same species with variab
589n092 ty of a tree having reason: or dog, having high POWERS without hand or voice.-- there is some great p
612o035 simple bodies, and to more extended space, such POWERS of relation required to be extended. Hence a s
204b140 larger than the dam, from those imported by Ld. POWIS Hybrid dogs offspring seldom intermediate betwe
406e036 & sheep.-- diseases common to men & animals cow POX.-- case in Spain of pustular disease following h
280c136 onger & stronger, so that though by great effort <PR> one unlike can be produced, yet to produce whol
362d100 t all the <bird> varieties «,now know» were then <PR> existing.-- he has also some very fine recent d
484z014 in Northern & Southern America-- valuable & PRACTICABLE deed Caricaridae wanderers.--?? in N. Americ
596n184 Association worthy of close study.-- full of PRACTICAL observations Ourang do not move eyebrows.-- o
557n014 ematizing» on principles, which even animals PRACTICALLY know «art precedes science-- art is experien
038r057 C. of Good Hope.--Carnatic It has been common PRACTICE of geologist. Lyell considers (P 84 Vol III.)
293c173 an swim without being web footed yet with much PRACTICE & led on by circumstanc it becomes web footed
332d002 all over England at same periods When he began PRACTICE, he remember during a year or two he saw many
414e063 iety far more easily than man,-- though man's PRACTICED judgment. even without time can do much.-- (y
276c124 me habit, which the whole rest of other family PRACTISE with a peculiar structure, then Milvulus forf
416e075 ery part of newly acquired structure is fully PRACTISED & perfected Hence difference between races &
432e125 , as well as intense thought.--» No one but a PRACTISED geologist can really comprehend how old the w
276c124 tructures.-- the only argument can be. a bird PRACTISING imperfectly some habit, which the whole rest
384d161 p. 44. Classification of Monsters. (1) From PRAETERNATURAL situation of parts (2) addition of parts,
523m014 rom the curious story of the Birmingham Doctor PRAISING his sister who confined him. & yet disinherit
273c112 ,milvulus,. may catch insects on the wing & PRATENCOLE (¿connecte[d] with Chionis), the Tropic
467t104 ollen from tip of sheath.-- «Also in Lathyrus PRATENSIS yellow saw stigma project» In common Pea saw
191b083 Africa. proof of subsidence. & recent elevation: PRAY ask Dr. Smith.-- «to state that most clearly».
536m074 . Man thus believing, <yet> would more earnestly PRAY "deliver us from temptation,' he would be most
569n020 nt power of the mind-- superstition & charity & PRAYER, or eloquent request. Reason in simplets form
582n068 be guided to the performance of a number of PREARRANGED actions, which will bring about a certain re
301c199 abits precedes structure,.. habitual instincts PRECEDE structure.--duckling runs to water. before it
089a018 nds tremblemens de terre sont presque toujours PRECEDES et suivis, queque temps avant et apres, par d
301c199 g to my views, habits give structure,.. habits PRECEDES structure,.. habitual instincts precede struc
567n014 ples, which even animals practically know «art PRECEDES science-- art is experience & observation.--»
264c081 cture may be obtaining, whilst habits slightly PRECEED them-- From this view habits must form most im
057r115 water in the country.-- Sir J. Herschel. says. PRECIP. of Sulph. B. all the infinitesimal cryst. arr
138a176 Acid unite with «piece of cabbage» alklali & PRECIPITATE silica / or charcoal charged with carbonic a
055r108 prove better than any meterological table the PRECISE periods over immense areas. (& the counterbala
224b212 improves. & checks it.-- It does not bear any PRECISE relation to structures Mem. Eyton's Hogs & dog
033r042 asts alone in Calcareous. rocks??--if so case PRECISELY analogous: fragments instead Peak of Teneriff
149g032 e lower down the hillock with beach & channel PRECISELY as with Isld-- [F) do they extend round hill
301c198 t is instinct in animals. « & what reason» in PRECISELY same way not possible to say what habitual in
318c252 have other peculiar appearances-- Now this is PRECISELY the case with the mice of S. America, with re
364d104 <story> return.-- The Revd. R. Jones told me PRECISELY same story about some Southern «see p. 43 sup
409e046 h more intellectual than fox, wolf &c &c-- is PRECISELY analogous case to man, exceeding monkeys;-- H
414e063 ed).-- -- When two races of men meet, they act PRECISELY like two species of animals.-- they fight, ea
429e115 ododendron ferrugineum. begins at 1600 metres PRECISELY & stops at 2600. & yet know that plant can be
447e165 seen it propagated in a garden, which is case PRECISELY analogous to the Canada onion mentioned in Ho
524m019 .-- (Dr Ashe, the Birmingham Doctor), in this PRECISELY like the passion, ill-humour & depression, wh
533m058 nderstand. what frowning means) if so this is PRECISELY analogous or identical, with bird knowing a c
580n061 ys <f> better than when she tries is not this PRECISELY the same, as the double-conscious kept playin
612o035 vidual.-- [RHC] 2) In animals, growth of body PRECISELY same as in plants, but as animals bear relati
583n069 ee &c &c-- p. 15."instincts act with unerring PRECISION.'-- no p. 17. Contrast the invariability of i
205b142 types most subject to vary where intermixture PRECLUDED.-- Kirby Bridgewater Treatise There are some
419e087 h must be effect of slow change + & Therefore PRECLUDES effects of catastrophes, which must serve to
267c095 <p> do p. 62. Ehrenberg Annals of Nat. Hist. PRECURSOR of magazine??? p. 75. roe of Asterias in stom

536m074 ldren.-- The above views would make a man a PREDESTINARIAN of a new kind, because he would tend to be
222b205 chs.-- ¿ lizards of secondary period in same PREDICAMENT. It is another question, whether whole scale
365d105 birds are very far from common.-- Under this PREDICAMENT, probably, alone would species cross in wild
447e164 ucing» propagating itself by buds is in same PREDICAMENT, as one, in which structure does not allow o
636j56v uent intermarriage.-- A plant is in the same PREDICAMENT as a group of bisexual animals living on the
227b224 h other «by steps we see»: but this cannot be PREDICATED of <genus.> structures in two genera.-- <we
355d067 t service, towards the end of science. namely PREDICTION.-- till facts are grouped. & called. there c
355d067 facts are grouped. & called. there can be no PREDICTION.-- The only advantage of discovering laws is
400e011 as ideas about generic characters. dominant. PREDOMINANT &c having relation to geographical distribut
102a062 ion of new law acting in certain directions PREDOMINANTLY, connection with magnetism &c counteracting
205b145 spring, but that of the male more frequently PREDOMINATES," p. 20. do "If hornless ram be put to horn
264c079 e» & then let him dare to boast of his proud PREEMINENCE.-- «not understanding language of Fuegian, p
273c112 emale is not dull.-- I must observe that this PREEMINENT structure is not always applicable to same h
181b045 e Creator has made tribes of animals adapted PREEMINENTLY for each element, but it seems law that suc
186b059 rganization same from one period to another, PREEMINENTLY Pachydermata, less so in Miocen & so on.--
250c038 Africa» being left more equable. yet America PREEMINENTLY equable. might have allowed fresh species t
273c111 ven the most <m> strongly marked, there is a PREEMINENTLY aerial,-- formed for flight & great movemen
274c113 opic bird, has very different habits, though PREEMINENTLY belonging to this type, ¿the Humming bird?
311c226 ortive organs &c The doctrine of monsters is PREEMINENTLY worthy of study on the idea of those parts
327c265 tely advertised. /6s Mrs Necker on Education PREEMINENTLY worthy of studying in Metaphysical point of
407e039 refore species, which were fitted for such a PREEMINENTLY equable climate. might not have been able t
407e040 productions.-- Hence it is, from the ancient PREEMINENTLY equable & temperate climate, <that> of Amer
427e109 naean class Dioecia & Monooecia. ought to be PREEMINENTLY artificial.-- Would not subsidence of Green
567n015 t-- children inherit it <ins> like instinct, PREEMINENTLY so-- who can analyse the sensation, when me
435e130 ore so than Polygamia: Monooecia & Dioecia, PREEMINENTLY artificial, so that even some species only
551m128 do that our "necessary ideas" arise from the PREEXISTENCE of the soul, are not derivable from experie
551m128 erivable from experience.-- read monkeys for PREEXISTENCE <">-- The young Ourang in <Zoolog> Gardens
083a032 L'Institut. [1837 p. 297]CD thinks Olivine a PREEXISTING mineral.-- Mem. Galapagos ∴ Basalt deepest??
282c142 n a cast of washing men-- but might hire the PREEXISTING race, thus the analogy would not in all case
283c146 add to quantum of life possible with certain PREEXISTING laws.-- .If only one kind of plant not so ma
025r018 ake the most Southern one I have heard of In a PREFACE, it might be well to urge, geologists to compa
218b190 -- The Edinburgh. Journal of Natural History-- PREFACE appeared good with facts about changes when an
314c238 of botany of islands of south seas says so in PREFACE.-- Mr Brown says character of Flora, N. Zealan
625o052 s both, perhaps, by habit-- [LHC] 11) Whewells PREFACE. [RHC] It appears that Sir. J. & others think
629o54v therefore not «instinct» «conscience"» equally <PREFE> destroy all wish of outward gratification,-- s
261c071 aid that animals no notion of beauty, when does PREFER most powerful buck Owen talking of Plesiossaur
427e110 - same way one variety of <animal> dog does not PREFER other. but produces greater effect on offsprin
439e144 se varieties have become as fixed as species, & PREFER their own pollen to that of other variety.-- «
617o40v would be <sat> fundamentally accounted for, we PREFER this metaphorical mode of stating the fact to
575n044 staunchness of greyhounds.-- bull-dogs being PREFERRED from not having any smell.-- 27th. November.-
334d010 .-- My case of Stallion, according to Erasmus PREFERRING young mare to old, explained by Stallions, (
473s07v cepcional, as limbs &c &c only appear late in PREGNANCY, & then may just as well be born a tendency t
304c206 Icthyo> .+ + +. supra & next page It is a fact PREGNANT with SOMETHING.? that intermediate. species h
377d139 ught disgusting, but to philosophic naturalist PREGNANT with interest» Hyaena. thinks, when pleased c
639j28r Sucking feet in Frog. Walrus. Fly. Gecko &c. PREHENSILE tail. in Monkeys & Marsupials. Harvest mouse
086a004 narrow toothed Mastodon.-- argue against the PREJUDICE of not believing recent elevation, yet sea sh
193b093 era, collected together Man has no heredetray PREJUDICES «or instinc» to conquer or breed together:--
263c075 er be conquered by anyone (if has any kind of PREJUDICES) <without> who just takes up & lay down the
286c154 ankind. & I believe those who soar above Such PREJUDICES, yet have justly excalted nature of man. lik
571n030 cases most difficult to distinguish. between PREJUDICES of youth from <here> habits. & heredetary ha
613o036 . so modifications of brain) As in animals no PREJUDICES about souls, we see particulat trains of tho
449e170 to Patagonian species-- p. 18. of Temnincks. PRELIMINARY discourse to Fauna of Japan-- that the «anim
183b052 et du belier, cessent d être feconds. dès les PREMIERES generations" go back to type of either animal
302c202 mpare it with resuts. if I acted otherwise, my PREMISES <in di> would be disputed.-- according to Pri
608o026 s not recognize an accidental spark falling on PREPARED materials. From contingencies a mans characte
451e177 race & if so, fact for my theory Cocos Isld & PREPARIS between Andaman & Pegu. <have> abound with mo
414e064 organization in Black Race there gives them PREPONDERANCE. intellect in Australia to the white.-- The
071r156 d now preserved over world Dicotyledones far PREPONDERANT, if so coniferous must formerly have been m
493q004 rding to Mr Walker the form of male ought to PREPONDERATE; according to Mr Yarrell the latter ought:
360d089 some in patches, & some hairy-- the former PREPONDERATED <which seems owing «determined» by the sex>
629o55v pungency of one's feelings for indecency-- PREPOSTEROUSLY so, for Marquesans think only of prepuce,
629o55v reposterously so, for Marquesans think only of PREPUCE, crepitando,]CD & if passions makes one break
520m001 n Corbet, he could not help thinking, he was PRESCRIBING to his father & old Mrs Harrison, said, alth
103a065 vapour. according to Hopkins theory.-- general PRESENCE of dikes. argues in favour of liq
226b223 ata from Europe to America. How strange would PRESENCE of Jaguar been in S. America.-- <East.> «W» C
357d073 developement of Marsupials in S. America. from PRESENCE of Edentata-- Edentata & Marsupials have been
382d158 ovarium.-- or testicles in female.-- the <add> PRESENCE of both testes & ovaria in Hermaphrodite,-- b
411e055 up argument against my theory, doubtless, the PRESENCE of animals in <own> «the present» orders (not
439e144 Hensleigh. seemed to think it absurd. that the PRESENCE of of the Leopard & Tiger together depended o
547m112 of ideal consciousness for moment, implied by «PRESENCE» my servant, «box» my own manner of ordering
023r011 s originated in the bottom of the ocean. & the PRESENT Volcanos have been said to be merely accidenta
029r034 Coal in Chili as clearly bearing a relation to PRESENT position of <Coal> Forests. These thick beds o
042r069 r lakes unequally elevated, which movements if PRESENT in the Andes, would have destroyed regularity
043r071 me line" to "from the epoch of Ammonite to the PRESENT day. at Mauritius. (consult Bory «dip of strat
044r073 umboldts quotation of instability of ground at PRESENT. day.-- applied by me geologically to vertical
045r075 dinary fissures of the ground at Calabria when PRESENT at the Concepcion earthquake.--expatiate on di
047r081 a wave in ocean. is not this [Fig. 3] {P} form PRESENT, i e a part below «mean» level before the high
056r111 explain how water separates.--(intertropics at PRESENT fix lime). <Also Volcanos separate.> Volcanos
067r143 by Pedro de Angelis.» This work is reviewed in PRESENT Edinburgh March 1835 Sir W. Parish says. that
101a056 s. Lardner's p. 270-4, good discussion showing PRESENT form of land in Northern England influence dis
120a109 more than at interior effect would be such as PRESENT. to spread sheet of matter over surface.-- if
138a153 as formerly stood three hundred feet above its PRESENT level, & in many parts has extended a league i
147g018 ich without lake or sea could not be placed in PRESENT position Thursday Evening ½ past 8 Tyndrum 29.
156g070 tely large to those now formed in same spot by PRESENT torrents Maculloch wrong in saying no transpor
171b005 ed for ever would be endless (that is with our PRESENT system of body & universe therefore final caus
181b039 Vide two. pages back. Diagram The largeness of PRESENT genera renders it probable that <the> «many» c
181b039 ft scarcely any type of their existence in the PRESENT world.-- or we may suppose only each species i
181b041 ny living ten thousand years hence; because at PRESENT day many are relatives, so that by tracing bac
181b043 ertebrates-- Such on few on each side will yet PRESENT some anomaly & <the g> bearing stamp of <some>
183b051 dge «or Orpheus» was introduced into Chili. in PRESENT states. <they> it might continue & thus two sp
198b113 consideration of range of capabilities past & PRESENT might tell something) p. III G. St Hilaire Ins
206b146 ulation of place be constant «say 2000» and at PRESENT day, every ten living souls on average are rel
206b146 e four hundred years since were progenitors of PRESENT people, and so on backwards to one progenitor,
206b147 l be a period though long distant, when of the PRESENT men (of all races) not more than a few will ha
206b147) not more than a few will have successors. at PRESENT day. in looking at two fine families one with
210b158 red, <but why they should be manu> Does it not PRESENT analogy to what takes place from time? Von Buc
210b161 ich, follow certain laws according to species. PRESENT an analogy to production of species.-- Animals
214b172 Wealden. Do the N. American. Tertiary deposits PRESENT analogies to shells of living seas.?-- Roxburg
221b200 8 L'.Institut. Bats, in Eocene beds, very like PRESENT species., p 8. ¿Are mundine forms, longest per
222b206 impossible to suppose such an accumulation at PRESENT day & not include Mammalian remains.-- The Fat
222b206 wandered least from ancestral form. If so are PRESENT typical species most near in form to ancient;
226b223 t of Africa & <West.> «E» of America, ought to PRESENT great contrast in forms.-- India; intermediate
226b223 re carnivorous Mamm: in Paris basin «allied to PRESENT"») more like present carnivora than Pachydermat
226b223 in Paris basin «allied to present"») more like PRESENT carnivora than Pachydermata If my theory true,
228b229 re we have near genera far back, as well as at PRESENT time, we might expect confusion of species.--
228b231 t not compare «chances of embedment in» man in PRESENT state. with what he is as former species. His
232b245 fewer species in proportion to genera than in PRESENT seas, «All» The <one> species which survives a
254c048 "-- so we find species each class successively PRESENT modifications, typical of succeeding classes &

Page **(Key Word)**
257c058 there come to be, many genera of fish &c &c at PRESENT day.-- It is ASSUMPTION to say generation prod
258c062 then another-- (according as parent types are PRESENT) must follow after there is proof of the non c
263c077 indefinite-- he is not a deity, his end «under PRESENT form» will come, (or how dredfully we are dece
263c078 ns of species to produce contingents proper.-- PRESENT monkeys might not,-- but probably would.-- the
268c099 te both in S. & N. Hemisphere just anterior to PRESENT. ¿cause of destruction of «great» animals?-- S
277c128 sification would answer every purpose, & would PRESENT many ideas of causes of change.-- The mark of
278c131 old inhabitants most important!! like Dipus of PRESENT day??! Major Mitchell does not think that dog
291c167 classes. if extinct forms were all fathers of PRESENT, then there would be perfect series or gradati
296c184 f Teyde in relation to surrounding countries & PRESENT tropical countries. p. 564. an abstract of Mr
302c201 rives me to say that there can be no animal at PRESENT time having an intermediate affinity between t
303c204 do not know whether it <would have> differs in PRESENT races, & form of feet.= (Negro or father of ne
303c204 auses. & considering over the viccissitudes of PRESENT animals.-- He-will be bold. I will venture to
316c246 been scattered over whole world Many shells at PRESENT day same (or according to Sowerby fine species
333d007 cated ones have been so long as wild one under PRESENT form.-- Fox has seen several cases of foxes &
338d023 al circumstances of the two countries in times PRESENT & past. The effect of physical conditions of c
343d036 forms» from a period short of eternity to the PRESENT time, to the future-- How far grander than ide
352d059 aper, to study the corresponding points.-- The PRESENT geographical distribution of animals countenan
357d073 mpted to believe, one or two were landed as at PRESENT in New Ireland & continent since grown.-- This
372d129 rated part every element of the living body is PRESENT-- in generation something is added from one pa
377d139 ted to Continent, when elephants lived. & when PRESENT animals-- lived.-- we know the great time, nec
378d147 rs-- «He says» whenever metallic brilliancy is PRESENT in Young birds, one may be sure cock & hen wil
378d148 <young> Kingfisher & pies, have long had their PRESENT plumage.-- How is this in Pidgeons & fowls.--?
379d151 rities. Wildness Reversion Q [not located] The PRESENT age is the one for large Cetacea, as the past
382d158 . those which perform very different, are both PRESENT in every shade of perfection --How came it nip
382d158 clitoris, shows to my mind.-- , that both are PRESENT in every animals, but unequally developed.-- s
388d173 ply instinctive the feeling of other sex being PRESENT is-- it also shows that semen. must actually r
391d174 long have so existed.» with double union.-- At PRESENT I can only say the whole object being to acqui
399e006 - good objection to my theory: a modern bed at PRESENT might be very thick & yet have same fossils. d
400e013 generations before they began to double.-- at PRESENT time Uncle J. does not suppose one aboriginal
404e025 pper is a mere simple modification of an organ PRESENT in whole class. Case of Mexican greyhounds.--
407e037 possessed a climate compared to S. America at PRESENT days,, which S. America now does to North. Ame
407e039 so «very» EQUABLE. or so tropical, & therefore PRESENT state of world is not so different, with <d> r
409e049 der in this perfect <uni> world, either at the PRESENT, or many anterior epochs.-- but we can see if
409e049 t object, for which the world was brought into PRESENT state.-- «whether he was or not. He is present
409e049 present state.-- «whether he was or not. He is PRESENT a social animal» a fact few will dispute, [alt
411e055 ubtless, the presence of animals in <own> «the PRESENT» orders (not so in S. America, however) is ver
418e083 he birth of the species & individuals in their PRESENT forms, are closely related-- By birth the the
419e086 200 & 300 ft are identically same as those of PRESENT seas.-- now in this country we have better mea
424e101 & Europaean-- agree very much closer, than the PRESENT ones., which according to Beck are different.-
427e108 nce formed. would remain stationary, hence all PRESENT types are ancient. According to my views of <D
432e125 eeth, or sneering, has no more relation to our PRESENT wants. or structure, than the muscles of the e
440e146 xternal agency, without final cause. either in PRESENT, or past generation.-- thus cabbages growing l
443e154 undergo metamorphoses, but to arrive at their PRESENT structure they must have <done> been propagate
460t019 Shells in Cambrian & Crust show how long since PRESENT forms existed, but if it be asked how this com
460t019 ust go to the first origin of the world.-- our PRESENT organic beings are the descendants, <slightly>
464t065 ows the Apteryx is not «quite» isolated in its PRESENT locality-- there have been at least other bird
465t079 ca. it appears from Lund more Mammals, than at PRESENT «in Europe we know there has been several succ
465t089 of Norfolk Crag. mentioned-- allied Beaver to PRESENT forms.-- -- How many «tertiary» estuary & Lacr
524m020 alities, which my F says are almost constantly PRESENT in people, likely to become insane.-- now this
544m102 , the links which when conscious connect past, PRESENT & future thoughts are broken-- Sir J. Franklin
548m115 deeply reasoning besides these (which must be PRESENT, though one is not conscious of them, else one
548m115 s of imagination hard work,-- Keeping one idea PRESENT is, perhaps, hard work-- though dreams do that
555m142 3) says charity is found everywhere (is it not PRESENT with all associated animals?) I doubted it in
560m156 onkeys not show signs of impatience when woman PRESENT? Do they pout, or spit, or cry.-- <fe> Shame,
582n067 , & its impulse to do it.-- [the means must be PRESENT on any hypothesis whatever]CD an animal may so
585n074 but also «by an» instinct<ing> «which is only PRESENT in youth» (Mem. Mr Worsley's story of chicken)
607o025 ty. free will (as generally used) is not there PRESENT, but he acts from motives, nearly as usual (a)
609o29v y of these passions, but refers (I believe) to PRESENT day & not to ruder state of Society.-- Civiliz
611o034 al laws,) as long as certain contingencies are PRESENT, (contingencies as heat light &c). [LHC] This
611o034 ears relation to whole, that is enough must be PRESENT to be able to exist as individual.-- [RHC] 2)
613o035 th distant object. where many such objects are PRESENT, & where will directs <to> other parts of body
620o044 s that the instinct. (or conscience) is always PRESENT (which is indeed, often felt at very time it i
620o046 men, [RHC] 5) which are natural (& which «when PRESENT» give pleasure.) & which man ought to follow--
621o046 instinctive feeling in its nature being always PRESENT. & his passion shortlived, it is to his intere
621o047 (a) The origin of passions too strong for our PRESENT interest receive simple explanations from orig
624o051 hence passions]CD «although perhaps useful at PRESENT to some extent.» Hence this is the law of our
640j167 long as physical change is in progress or is, PRESENT with respect to new arrivers, the small body o
207b152 . Vincent Vol. III. p. 164. Lîle de la Reunion PRESENTE elle seule plus d'especes polymorphes que tou
058r118 ontraire, que vers le centre de' l île, elles PRESENTENT une coupe abrupte et souvent tailée a pic. T
619o043 ect to himself it would be remorse as will be PRESENTLY shown.-- This then is moral approbation, as f
402e020 Vol II. p 367. "The Fauna of the Sunda islands PRESENTS us, for the most part, with the same families
035r045 atagonian pebbles beds, most unfavourable to PRESERVATION of bones &c &c--Yet <silicified> turn over
109a083 mud going out can actually be seen.-- ¿ The PRESERVATION of dikes & ledges of first-rate importance
109a083 baqueous removal--??? the difficulty of such PRESERVATION certainly is lessened.-- Coral flats. argum
146g013 a, excellent instance, how accidental is the PRESERVATION in situ of even imperishable pebbles/ I am
162g101 4 372 about 267 28.75 .105 I reached> 29.090 PRESERVATION of form of land very much due to Peat & Hea
261c071 t in obviating a great apparent difficulty-- PRESERVATION of colouring, when form has changed. Can be
428e112 all the less hardy ones. & the <accidental> PRESERVATION of accidental hardy seedlings: (which are c
439e144 pringing an inch further would determine his PRESERVATION-- if killed by some other animal, then that
465t079 in situations apparently favourable for the PRESERVATION of shells; where land broken, rivers enteri
584n071 ncts, sexual, social, «subordinate to,» self PRESERVATION, (knowledge of enemies). use of muscles, pr
591n099 ing to the feeling that the instinct of self-PRESERVATION is disobeyed-- I often have «as a boy» wond
549m123 lect they might be necessary & no doubt were PRESERVATIVE, & are now, like all other structures slowl
516q23v ecies Theory (1) Will an extract of peat do to PRESERVE fungi or animal substances-- (Athenaeum (40)
035r046 e. &c seems the organic structure most easily PRESERVED.-- Mr Conybeare introduct to Geolog--"Between
035r047 e minute equalities of elevation, may well be PRESERVED at Patagonia. The English fact is astonishing
041r065 ot & being cause of concretion; or being only PRESERVED in that part. having lived over whole bottom
057r115 a of frozen bottom or beach of sea to explain PRESERVED animals.--Mem: stream of water in the country
065r139 ergone any decomposition: countenance so well PRESERVED. that it was thought not to have belonged to
071r156 ans use arrows or Araucanians?--> If wood now PRESERVED over world Dicotyledones far preponderant, if
105a072 posed sea» shells, that land shells should be PRESERVED in it-- some error? (because more recent) ---
155g065 g any oscillation must have been so carefully PRESERVED as to have thrown water in same «drainage» li
155g066 d of Gneiss though wonderful-- <that they are PRESERVED> how much more so, these lines & even water-s
155g066 er that all «three» lines «should be» EQUALLY PRESERVED 2d or upper one more perfect in this <part> «
193b090 tself perhaps by circumstances) & those alone PRESERVED which are well adapted, This would account fo
249c036 e the Edentata & Marsupial forms been chiefly PRESERVED,-- where shut up by themselves without other
278c132 .-- The difficulty is how came it animals not PRESERVED,», in central S. America & yet Africa & India??
398e005 of species «are» not as «without every animal PRESERVED.».-- the latter pages in the history are perf
440e145 umerable other ones, <alone> «which has been» PRESERVED.-- but be it remembered. how little part of t
458t001 nge, & shows as in shells some forms are long PRESERVED.-- vol VI. p. 539. Dr Cantor's account of fos
465t079 America. -- so see what a «mere» vestige, is PRESERVED in this country-- same argument to India & Eu
465t080).-- If we may take this as guide, the shells PRESERVED must be as much a casualty as, bones of Mamma
040r063 to cause floods in valleys, which must aid in PRESERVING the terraces <...> Molina's Case At Vesuvius
293c175 e, passed from simpler stage to more perfect. PRESERVING its relations.-- the wonderful power of adap
346d048 remarked that it was against all rules their PRESERVING character & breeding in & in-- Nonsense a fl
535m064 from egg» watching them, brooding over them, PRESERVING them from «the» sun & enemies-- would not fl
089a018 Observ.-- Les grands tremblemens de terre sont PRESQUE toujours precedes et suivis, queque temps avan
467t104 es in gangway= In Lotus corniculatus saw Humble PRESS down wings which ejects pollen from tip of shea

Page **(Key Word)**
467t104 saw stigma project» In common Pea saw Humble so PRESS down sheath, that stigma covered with pollen wa
352d059 considered the tree of life must be erect not PRESSED on paper, to study the corresponding points.--
467t104 wn sheath, that stigma covered with pollen was PRESSED & rubbed along whole breast-- {b} pressing eit
570n024 show, he is determined not to say anything. he PRESSES his lips together & shrugs his shoulders & wal
467t104 was pressed & rubbed along whole breast-- {b} PRESSING either one or both of Pea's wings, stigma & m
044r074 es of eruption at bottom.--solution under high PRESSURE of gazes. especially the most abundant. Sulp.
046r079 Yet the fluid granitic mass under <[...]> less PRESSURE might have its «proportional» particles alter
059r123 ssy F. fractured. have been melted with little PRESSURE. & perhaps cooled suddenly.-- As the rude sym
077r172 are laws of solution & deposition under great PRESSURE. (? heat!) unknown to us. ‖M. Chladni.--on m
102a061 ive rocks-- Are substances soluble under great PRESSURE? equally with little pressure? An important q
102a061 uble under great pressure? equally with little PRESSURE? An important question! If water yields subst
103a065 general presence of dikes. argues in favour of PRESSURE of liquid rock. Andes discussion-- Albite cer
104a066 e can conceive anticlinal lines near. (lateral PRESSURE would always produce it) but where great thic
104a068 ility of fluids varies with temperature ¿ with PRESSURE? Salt on surface of plains due to whole moist
120a109 ter with «-- especially if very hot under high PRESSURE.--» respect to formation of salt.?.--??? Foot
125a121 rom circulation of heated fluid or gases under PRESSURE.-- {P} Lyells view of transmission of heat by
126a123 {P} how easily water percolates rocks,-- when PRESSURE increased or under surface, would not the flu
127a126 may be cooled to 0 degree!)-- shows effects of PRESSURE in change of form as the result of heat.-- wi
307c216 produced in domesticated animals, in plants I PRESUME there are.? get examples.-- for instance where
371d128 r if different dieteriorating very slowly.-- I PRESUME most of these roses, without circumstances ver
378d147 , one may be sure cock & hen will be alike-- I PRESUME converse is not true for he says Hen & cock St
378d148 cks, the young crossed amongst themselves, & I PRESUME with common ducks. so often, that it was impos
424e102 since these crows are mixed in England-- for I PRESUME Carrion Crow is found in Edinburgh.-- Why does
441e149 ld a Crab tree vary if planted in rich soil, I PRESUME not, but its seeds, I presume, probably would-
441e149 in rich soil, I presume not, but its seeds, I PRESUME, probably would-- at least the experiment of t
453e183 y, that Hollyoak reproduce each other. & yet I PRESUME seed raised in same garden.-- now this good qu
461t037 7. for Man Chapt see Yarrell Syngnathus Ch 6 I PRESUME, from my theory, as long as any structure can
490q001 alogous or quite distinct from recent ones-- I PRESUME some recent not found fossil (perhaps not embe
498q008 Must pollen grain be whole, to impregnate?-- I PRESUME only stigma impregnable.-- (12) At Maer Cowcum
506q15v e. at Galapagos-- Dioecious.-- Carex.-- We may PRESUME Nettle spreads by seeds= (44) Zostera. Has he
577n049 o ourang outang.-- oyster & zoophyte: it is (I PRESUME-- see p. 188 of Herschel's Treatise) a "travel
599o005 speculations of the origin of language.-- must PRESUME it originates slowly-- if these speculations a
633j54r ated to live on alpine pinnacle? if we once to PRESUME that God «created plants to» arrests earth, (l
384d162 le purpose are not distinguished. --yet may be PRESUMED from hybridity of ferns) afterwards they can
441e150 -- black bullfinches-- & all varieties must be PRESUMED to be result of such laws.-- The effect of on
507q15v knows only on Citrons 47. Ficus carica Henslow PRESUMES females produce. Polygam. trioecia. (are fema
239c001 ety, takes greatest effect on offspring. Thus PRESUMING those varieties to be oldest which have long
114a095 luvial waves). & likewise <tells,> «offers a PRESUMPTION» it has been excessively slow because beach
248c030 ies may be made in wild state, there will be PRESUMPTION that they would not. breed together.-- We se
056r110 y (stated in Playfair to be the case p. 51). PRESUPPOSES an elevated country of granite, not <so> gre
174b015 its representatives-- «as in Australia» This PRESUPPOSES time when no Mammalia existed; Australian; M
356d069 ldt). G. St. Hilaire, & Lamarck have written I PRETEND to no originality of idea-- (though I arrived
524m022 distract her attention took her «left» hand to PRETEND to feel her pulse.-- What fails first?-- How i
623o050 emand some explanation [RHC] Although I cannot PRETEND to say how far & minutely our instincts extend
459t009 odgson on Musk Deer-- young spotted <like in> "PRETTTY much as we see in the young of the wild hog &
045r076 » how lucky! when they hear of a place having a PRETTY severe shock). are much more curious & perplex
148g026 an, hills of «sea» gravel, current cleavage, & PRETTY well rounded stones, mixed with some quite irr
175b018 nads are «constantly» formed ¿would they not be PRETTY similar over whole world under similar climate
290c163 not heredetary, because crossed with women with PRETTY faces When horse goes a round, the minute gets
332d003 sed--is Alderney oldest breed-- He believes all PRETTY much alike.-- My Father Water-in the hair a ce
361d096 h stated «that there are» two ducks, which have PRETTY close representative species in England & N. Am
458t002 Zebra, crossed with pony mare & produced a very PRETTY little animal, showing something of Mule in it
460t015 an-goose & common goose.-- the stripe down back PRETTY plain in in these <half> «3/4» bred ones-- The
547m111 sleep for second & wakened.-- had very clear & PRETTY vivid «& perfectly characterized» dream, in co
568n019 Vol V) «finally» says "he knew no more what was PRETTY & what ugly than a cow--" «so it is with all u
375d134 production does not increase, whilst no checks PREVAIL, but the positive check of famine & consequent
640j167 r), or other cause, the long legged race would PREVAIL, even if have afforded only 10th part before &
640j167 diminish, whilst the long legged variety would PREVAIL.-- Not separately: NB. These views quite exclu
036r051 cept near submarine Volc: in harmony with the PREVAILING movement being one of elevation alone.--In E
041r067 g may be owing to successive transportal from PREVAILING swell, (as Shingle travels on the Chesil ban
450e175 the isld, like in wild animals).-- There are PREVAILING colours in the different islands.-- The hors
604o017 ogy of the Human Mind. Poor.-- on insanity.-- PREVAILING idea. owing to loss of will.-- chiefly excit
132a138 action.-- <Why> now as we know volcanic action PREVAILS more beneath the sea, <than> «&» on coast lin
272c109 us.-- Remarkable how small detail in structure PREVAILS amongst the same species & subgenera in famil
348d050 racter, of no matter of what importance, which PREVAILS throughout the group & serves to insulate <th
221b202 according to Lyells doctrine removed?? Is the PREVALENCE of Coniferous Woods before Dicotyledenous a
305c210 lephants & dog.-- There is one living spirit, PREVALENT over this word, (subject to certain contingen
589n091 mine,-- Lamarck. Vol II p. 319.-- Habits more PREVALENT in proportion to intelligence less.-- p. 325
122a114 atform:-- On my view the degrading action must PREVENT internal fluid arriving at equilibrium so soon
235b274 s to cross. animals from different quarters to PREVENT them taking peculiar character-- Indian Bull?-
255c052 lity of crosse & cross, is method of nature to PREVENT the picking of monstrosities as Man does.-- On
293c174 races of men.-- Wax of Ear, bitter perhaps to PREVENT insects lodging there. Now these exquisite ada
343d038 S. Africa, so must the tribes become blended & PREVENT that strong separation which otherwise would h
401e014 ly be reared without greatest care be taken to PREVENT fertilization from turnips or other stocks. Sa
439e143 e species of plants shows there is tendency to PREVENT the crossing.-- in animals where there is much
447e163 difficult to vary.-- the run of chances, would PREVENT it varying. A plant <producing> propagating it
460t017 the laws of organization (ie those laws which PREVENT infinite variation in every possible way.-- th
466t094 ary: «wh. itself resembles a Bee, but does not PREVENT bees visiting it.» In Columbine nectaries are
536m070 ding will of action to muscles, any more than «PREVENT» heart beat» remember how Pincher does just th
536m070 when wanting not to feel angry-- such efforts PREVENT anger, but observing eyes thus unconsciously d
537m077 inct may be most firmly fixed, but it will not PREVENT other being engrafted.-- No one doubts patriot
606o025 dare not -- one could do it, but other motives PREVENT the action see Abercrombie conclusive remarks
619o043 jured these objects, without his being able to PREVENT it, he would likewise feel pain.-- If he saw a
633j54r nce venture to say plants created to <arrest> «PREVENT» the valuable soil in its seaward course,-- we
028r032 olour Sir J. Herschels idea of escape of Heat PREVENTED by sedimentary rocks, & hence Volcanic action
375d134 from Malthus.-- «increase of brutes, must be PREVENTED soley by positive checks, excepting that fami
568n017 feels pain, at not performing it, (either if PREVENTED, or overtempted.-- «animals have shyness with
568n017 er the cause had been]CD-- «Also» When one is PREVENTED performing heredetary habit, (or moral sense,
619o042 s great pleasure, & <an> <such> actions being PREVENTED by <necessity> «some force» give pain: for in
619o043 s own interest. likewise if such actions were PREVENTED by force he would feel pain. [.By a very slig
619o043 obation, as far as it goes.].CD But should he PREVENTED by some passion or appetite, what would be th
625o052 us. & that doing them gives pleasure & being PREVENTED uneasiness, & that this is the feeling of rig
636j56v as long as crossing with other varieties was PREVENTED Do races of peas become intermixed & gardner
258c060 area equably.-- Animals having wide range, by PREVENTING adaptation owing to crossing, with unseasone
277c127 . (b) until data be given.-- This will aid in PREVENTING the chaos.-- will point out what to observe.
220b197 o each other is evidently an instinct-- & this PREVENTS breeding. now domestication depends on perver
223b210 - because the crossing of species less altered PREVENTS the complete adaptation which would ensue A.
223b211 crossing with (B) (& B having crossed with (C) PREVENTS offspring of A. becoming a good species well
227b228 «to what circumstances favour crossing & what PREVENTS it--» & generation, causes of change «in orde
230b239 perfect) or horizontally & then cross breeding PREVENTS perfect change.-- It is scarcely possible to
230b241 that we do not know what amount of difference PREVENTS breeding;--or as others would exjudge it amou
338d024 e ideosyncracy of the Negro (& partly Mulatto) PREVENTS his taking any form of Malaria-- adaptation &
356d070 dency to reproduce in each case, but something PREVENTS the completion.-- ‖Say my Grandfathers expres
468e141 ime> the fixity of characters «from antiquity» PREVENTS their variation, which is not improbable as M
530m044 journey. short time previous,-- because, pain PREVENTS repetition of idea.-- Mr Blakeway has mention
230b240 o change in Americas. perhaps merely gone back PREVIOUS to fresh change Get a good many examples of a
322c270 Geology of France.= on Etna. Almost reread the PREVIOUS volume. & C. Prevost on L'Ile Julie Waterton'
332d002 has seen more case in a month, than in several PREVIOUS years, two having consulted him on one day.--

391d174 differences which are in harmony with all its PREVIOUS changes, which mutilations are not). but why
525m025 Mr Wynne that bitch's offspring is affected by PREVIOUS marriages with impure breed.-- A cat had its
530m044 thing in <he a [...] Mr B> journey. short time PREVIOUS.-- because, pain prevents repetition of idea.
065r140 evation, there may have been great subsidence PREVIOUSLY. Mem. pebbles of Porphyry.--Falklands.--off
105a070 shrinking of Clay. applicable to Cleavage. C. PREVOST.-- In Cordillera. a rush of water will account
117a101 formations. during elevation & depression. C. PREVOST.-- My views of insensible oscillations of leve
322c270 Etna. Almost reread the previous volume. & C. PREVOST on L'Ile Julie Waterton's Essays on Natural Hi
255c050 studied.-- <Three p. 7. Am.> D'orbigny. Birds of PREY, are distributed in S. America like other forms
315c243 s in associated kinds of good news. discovery of PREY.-- arising no doubt from want of assistance.--
437e140 d. it is said, that an Eagle always procured its PREY from another island.-- » p. 175., 28 sho[r]t ea
459tf1r ild cattle at Madagascar-- «p. 121» No beasts of PREY. any country should during [...] conditions-- e
480z007 '.Orbigny <considers> states that young birds of PREY have longer tails than old ones-- in America &
557m147 iff «& back» when savage «no» & ready to dash at PREY streched out & flaccid, when furious «with frig
640j167 t before & now formed eighth part.-- or if other PREY diminished, total number of dogs. would diminis
050r090 a no connection with a neighbouring Volcano of FRAMANG.--Marsden Sumatra. M. De. Johnnes seems to thi
288c159 rams & bulls from England fetch very «go» large PRICE. as is evident to be worth introducing, instead
327c265 ril 1839.-- Review on "Walker on intermarriage" PRICE 14s. Marh. 20t. 1839. Philosophy of Blushing la
494q004 hough valuable as show. & curiosities!! What is PRICE of fox. otter. Badger &c &c &c.-- (11) Keep. Tu
303c204 ct on this subject from Lawrence. Blumenbach & PRICHARD -- Now we might expect that animal halfway be
324c268 ncts L. Jenyns paper in Annals of Nat. History PRICHARD.-- Lawrence Bory St. Vincent. Vol III p. 164.
592n103 ture intends to lay open all senses: <do> Horse PRICK his ears «& snort clears nostrils» when frighte
601o009 be in first case. as when the excised heart is PRICKED) and certain action. (only evidence. when not
403e021 vated in Guahon. (Marianness), "for example the PRICKLY Limonia trifoliata, which cannot now be checke
641j29v g of nettle.-- Are there any other analogies-- PRICKLY plants or animals-- Exudation of fetid «& acri
557m147 to wrinkled: (a horse when winnowing & pleased PRICKS his ears?--).-- How is expression of anger in
565n007 nuffs «& snorts», the air «& raises its head, & PRICKS its ears» when afraid, though not every time r
232b248 eory, those the philosophers who soar above the PRIDE of the savage, they perceive the superiority of
524m020 mits its power on the muscles in the twitching. PRIDE & suspicion are qualities, which my F says are
524m020 nsane.-- now this is well worth considering, if PRIDE & suspicion can be well understood. In insanity
525m024 rect fear.-- My father thinks that selfishness, PRIDE & kind of folly like (Mr George S.) is very her
537m077 engrafted.-- No one doubts patriotism & family PRIDE are heredetary., & therefore he has these stron
537m077 n, may be opposed undoubted cases of heredetary PRIDE & in single families. Edinburgh. Phil. Transact
565n007 o smell its enemy.-- Man & dogs show triumph (& PRIDE) same way walk erect & stiff, with head up.-- W
565n008 em» manner of hurting opponent by insulting his PRIDE & is therefore of the snarling order.-- But con
565n008 nt is considered as quite insignificant, & when PRIDE makes person extremely self-sufficient,-- the c
605o19v eat height, eternity, &c &c. produces an inward PRIDE & glorying. (often however accompanied with ter
199b119 DISEASE being banished, & given to Portuguese. PRIEST.-- In first settling a country.-- people very a
529m042 unctional.--» He has some idea of a son of Dr. PRIETLY who was cured from a fall of ideotcy.-- The st
290c164 ack, like a black bird washed, Whilst tips of PRIMARIES black, by examining series I cannot doubt law
248c034 rkable laws of Brain & manner of generation «& PRIMARY divisions of insects» 2. Relation, of external
349d052 ge all.-- & so on no, one will settle number of PRIMARY divisions.-- Complains (p. 53) of M. Edwards,
384d161 similarly subject-- {Divides sexual marks into PRIMARY & secondary, the latter only being developed,
411e054 - When the laws of change are known.-- - then PRIMARY forms may be speculated on, & laws of life,--
493q003 d In species of close genus do more than three PRIMARY colours occur in relation with species-- answe
601o009 ced in consequence having some relation to the PRIMARY sensation.-- man moving leg when asleep-- «or
603o11b y, CD[my thery says yes. imitating song -- two PRIMARY sources, sight, & hearing-- Staunton Embassy V
231b241 varying in wild state.-- When breaking up «the PRIMEVAL.» continent. Indian Rhinoceros. Java & Sumatr
253c046 reat persistency of character. Hence Elephas PRIMIGENIOUS over so wide a range, & Mastodon angustiden
056r110 arts.-- The common occurrence of a breccia of PRIMITIVE rocks between that formation and the secondar
073r163 Metals in Mexico rarely in secondary alway in PRIMITIVE & transition; the latter rarely appear in cen
075r166 . says in Potosi the silver is contained in a PRIMITIVE slate, covered by a clayey porphyry, containi
102a060 hing cracks-- only bear relations to VEINS in PRIMITIVE rocks-- Are substances soluble under great pr
335d015 erary fingers» (that is slight alterations of PRIMITIVE stocks «relative to changes which every speci
335d015 y near species (that is slight alterations of PRIMITIVE stock) are heredetary: «Hybrids of» Varieties
557m148 & tail stiff.-- is shame, jealousy, envy all PRIMITIVE feelings, no more to be analysed than fear or
569n019) is acquired taste.-- Whilst music extremely PRIMITIVE.-- almost like tastes of mouth & smell. Desce
384d162 s Theory of sexes (woman makes, bud, man puts PRIMORDIAL vivifying principle) one individual secretes
249c035 ieties of Cardoon are cases <sp> like those of PRIMROSE & Cowslip run wild, The two species of Clenom
299c194 slds. Galapagos.-- Iceland has same uniformity PRIMROSE & Cowslip, quite wild, but they affect differ
401e016 of it.-- Now this is curiously different from PRIMROSE suddenly produce cowslip, one is tempted to t
401e016 e some anomaly-- I can fancy cowslip producing PRIMROSE return to old stock, but not primrose produci
401e016 roducing primrose return to old stock, but not PRIMROSE producing cowslip Uncle J. says common belief
428e113 o insist honestly that the sudden, change from PRIMROSE to Cowslip is great difficulty. «I should dou
437e141 at the French experiment of Carrot 2 {also try PRIMROSE & Cowslip suddenly changing into each other,
495q005 w weeds in such soil.-- 7 (a) Experimentise on PRIMROSE seeds-- it really is an important case-- cros
507q15v ether. under same conditions.-- like cowslip & PRIMROSE, but less strongly marked.-- 31. Plant seeds
637j57v waterdog hair to water-- bull dog to bulls.-- PRIMROSE to <open fields> banks-- cowslip to <bankings
515q021 es & nectarines: same question with regard to PRIMROSES. (^) Do apples "sport" in fruit, or time of l
305c211 to one kind of organic matter.-- brain. & which <PRIN> thinking principle. seems to be given or assum
065r138 climate of Tristan D. Acuneha. Kerguelen Land. PRINCE Edwards Isld. Marion & Crozet. L. Auckland. Ma
078r175 .--(¿anticlinal line?). -- Mines of Catorce «(PRINCIPAL veins)» 25 degree to 30 degree to NE. vein of
196b110 t on vegetables. of the two great Kingdoms.-- PRINCIPES de Zool: Philosop:-- I deduce from extreme di
180b038 rable many might be produced.-- This requires PRINCIPLE that the permanent varieties produced by <int
185b055 each element.-- May this not be explained on PRINCIPLE, of animal having come to island. where it co
186b058 e no reason for these. analogies; CD[from the PRINCIPLE of atavism, where real structure obliged to b
213b169 th consideration--» Tabulate Mammalia on this PRINCIPLE. <Varieties> < «Races» > Man in savage state
239c002 s, that offspring would be chesnut.-- On this PRINCIPLE I may add, that fact of half cross with paren
255c053 mpted to exclaim that nature conscious of the PRINCIPLE of incessant change in her offspring. ,has in
255c053 pecies her plan is frustrated or rather a new PRINCIPLE is brought to bear. If man created as now. la
259c065 he case of heredetary disease, is on the same PRINCIPLE that, cut a sheeps tail off plenty of times &
304c207 e there to guide in this opinion?-- EXCELLENT PRINCIPLE OF ABORTION ISOLATION of range « <far more pr
305c210 umes a multitude of forms «each having acting PRINCIPLE»» according to subordinate laws.-- There is on
305c210 - There is one thinking « <& Creat> sensible» PRINCIPLE (intimately allied to one kind of organic mat
305c211 anic matter.-- brain. & which <prin> thinking PRINCIPLE. seems to be given or assumed according to a
305c211 s.-- We see thus Unity in thinking and acting PRINCIPLE in the various shades of <dif> separation bet
312c232 » could only exist by habit.-- therefore same PRINCIPLE transferable., not wonderful According to my
384d162 man makes, bud, man puts primordial vivifying PRINCIPLE) one individual secretes two substances, alth
418e083 n be traced to a germ, endowed with the vital PRINCIPLE, which gives rise to the sexual organs, diffe
428e111 lly to the Zizanias in in Sir. J's ponds-- my PRINCIPLE does not apply to any plant reared artificial
428e111 they can bear the cold when torpid.-- On this PRINCIPLE being the destruction of all the less hardy o
485z018 they can bear the cold when torpid.-- On this PRINCIPLE tropical forms in N. America extend much furt
513q21. Where does J. Hunter use expression of "male PRINCIPLE of arrangement."-- would not male or female "
514q21. nt.'-- would not male or female "constructive PRINCIPLE" be better. or "constructive action on germ."
538m078 ht on consciousness, explained by Dr Dewar on PRINCIPLE of association.--fully bears out my fathers
539m083 I a Darwin & take after my Father in heraldic PRINCIPLE. & Eras a Wedgwood in many respects & some of
558m151 tance. &c &c. it comes to Miss Martineaus one PRINCIPLE of charity.-- ¿ May not idea of God arise fro
571n029 habitual-- & hence become heredetary; on same PRINCIPLE we know many tastes become acquired during li
574n042 having children with strong arms.-- The other PRINCIPLE of those children. which chance? produced wit
577n050 ciated with them.-- The establishment of this PRINCIPLE of Association will help my theory of sensiti
578n056 Fear loose the sphincter muscles, only on the PRINCIPLE like does an injury of the spine-- that it pa
587n087 - dogs dislike perfume] I should think, great PRINCIPLE» of liking, as simply heredetary habit.-- A bl
605o019 y. D. Stewart does not attempt «by one common PRINCIPLE» to explain the various causes of those sensa
609o29v jealousy in a dog no one calls vice). on same PRINCIPLE that Malthus had shown incontinence to be a v
610o034 & inorganic-- The inorganic are probably one PRINCIPLE for connect of electricity chemical attractio
613o036 not appear more than saying that the thinking PRINCIPLE is the same in all animals. [LHC] «3)» Eyton
250c039 genera distinct from rest of world??? Lyells PRINCIPLES, must be abstracted & answered Much might be
302c202 ses <in di> would be disputed.-- according to PRINCIPLES of last page. <an> osculant groups between t

304c207 birds,, or branched off anteriorly think what PRINCIPLES are there to guide in this opinion?-- EXCELL
305c211 ent habits in animals.-- --Animal Magnetism-- PRINCIPLES of irritations sleep walking. fits, laught &
314c236 the adaptation of species to circumstances by PRINCIPLES, which I have given ∴ Those animals, which o
412e058 List of fossil Mamm: from Poland. &c.-- Three PRINCIPLES, will account for all (1) Grandchildren. lik
414e064 emen's land»-- they have been exterminated on PRINCIPLES. strictly applicable to the universe.-- The
494q004 generations of man have there been.-- on what PRINCIPLES calculated.-- in order to guess how many gen
551m128 excited by question.» Sept. 4th. Lyell in his PRINCIPLES talks of it as wonderful that Elephants unde
567n014 l Science is reason acting «systematizing» on PRINCIPLES, which even animals practically know «art pr
625o50v - yet this system not selfish.-- explained by PRINCIPLES if Mackintosh.-- p. 262. Some good remarks,
626o052 gmarole.-- He does not say anything about any PRINCIPLES born in us.-- Great difference with my theor
641j29r hysical changes are in progress; (on the same PRINCIPLES that islands are favourable,) because it mus
046r080 rst great movement was one of rise (any smaller PRIOR ones might have been owing to absolute movement
565n006 the face clearly passes into smile-- laugh long PRIOR to talking, hence one can help speaking, but la
626o052 will doubtless lead to similar actions which in PRIOR <races> generations led to their formation.-- '
215b178 on passage of forms.-- Deshayes states Lamarck PRIORITY refers to introduction to Animaux Sans Vertèb
618o41v t is cause & effect: it merely is «invariable» PRIORITY of one to other: no not only thus, for if day
298c191 ted kind.-- lower species would then revert to PRISTINE form (which must have been altered by crossin
465t081 hich races of men have been exterminated (see PRITCHARDS paper] (Ed. Phil. Journ. end of 1839) very i
520mIFC On <insect> Ants getting on Table. Col. Sykes) PRIVATE Finished. Octob. 2d. This Book full of Metaphy
560mIBC cause by his intemperance. <No.> Cannot say.-- PRIVATE. Expression M Expression N What are sexual dif
563nIFC exual difference in monkeys.-- Charles Darwin [PRIVATE.]CD (Metaphysics & Expression) Selected «for S
362d100 - he has also some very fine recent drawing «of PRIZE pidgeons» in 1834-- now this would be most curi
511q018 rs (5) List of qualities in birds & animals for PRIZES.= Pidgeons. Canary birds-- Bantams.-- (6) <Mad
304c207 CIPLE OF ABORTION ISOLATION of range « <far more PROB> » tends to alteration views.-- ostriches do--
304c208 efore descended from branch high up.-- Such PROBABILITIES only guides.-- Yet trifles are produced by
543m100 ind. all is certain.-- there is judgment of PROBABILITY, therefore this judgment gives a man common
106a075 of this structure. as the inclination in all PROBABILITY would be greater when flowing over (B) than
124a120 71. argument against lateral injection. from PROBABILITY of fissures being prolonged to surface. see
231b244 e would be limited animal in range-- & hence PROBABILITY of starting from one point.-- In the crag we
354d065 an have varies the range-- Argue the case of PROBABILITY. has Creator made rat for Ascension.-- The G
370d117 ghout my theory is to establish a point as a PROBABILITY by induction, & to apply it as hypothesis to
437e140 perately bitten the young ones, would in all PROBABILITY have escaped".-- if it had not been shot by
438e141 aracter of antecedent races.-- «& yet in all PROBABILITY the Brussels Sprout was slowly formed.-- » i
577n049 e cause of itself.]CD & hence there is great PROBABILITY against free action.-- on my view of free wi
041r065 carcases putrid. In Rio paper. when discussing PROBABLE rise of land: Mention M. Gay's fact about she
042r068 lcanos in Germany» In the Valle del Yeso it is PROBABLE that point of Porphyry has been upheaved in a
049r089 ranite? Why not more probably greenstone? What PROBABLE origin can be given to the numerous hills of
098a044 ance not very conclusive proofs, but certainly PROBABLE. Bulletin de la Soc. Geolog: 1833-34. p. 35.-
105a071 barrier in the very part, where barrier least PROBABLE.-- The sea harmonizes well with character of
147g017 very lofty, & no trace of it; to the Sea more PROBABLE I did not look carefully for Marine remains--
179b035 dependent of external causes) does appear very PROBABLE:-- Mem: Horse, Llama. &c &c-- If we <suppose>
181b039 ram The largeness of present genera renders it PROBABLE that <the> «many» contemporary, would have le
188b067 in East Indian archipelago Dr Smith considers PROBABLE two Northern species replace <No> Southern ki
224b214 pecies will generate common kind, which is not PROBABLE; then monkeys will never produce man. but bot
224b215 & may someday produce something else., but not PROBABLE owing to mixture of races.-- When all mixed &
248c030 ogether, so when we grant «(which can be shown PROBABLE,)» varieties may be made in wild state, there
276c125 bit different countries habits similar ¿law?-- PROBABLE.-- if habits & structure similar would have b
293c172 so great an anomaly in structure of brain not PROBABLE) put note. Sir W. Scott has written about it]
336d019 any appearance of animals being created. it is PROBABLE if created at once. <wd> according to ordinar
390d178 difference «from what it received» (for it is PROBABLE that breeding in & in would not be deletereou
401e015 eil sowed by Mr Tollet so produce.-- thinks it PROBABLE that great part of those varieties may be due
407e041 st to resemble fossil ones of Europe, Consider PROBABLE form of land,-- -- S America, an island, «con
440e147 ave been exceedingly small.-- This is far more PROBABLE way of explaining, much structure, than attem
452e181 pins thread of cotton.-- do p. 583, It APPEARS PROBABLE,«?» that the Hippopotamus occurs in India. in
525m024 men on the table,-- guilty conscience.-- Not PROBABLE in Squib's case any direct fear.-- My father
548m115 I well recollect» is in making things somewhat PROBABLE. in comparing every step, & inventing new mea
611o034 tricity chemical attraction, heat & gravity is PROBABLE.-- And the Organic Laws probably have some un
026r021 nts; veins appearing a galvanic phenomenon, so PROBABLY will the Cleavage be There is a resemblance a
028r029 e solid rock by the actions of Springs or more PROBABLY by some unknown Volcanic process? How does it
030r035 egetable matter from thickly wooded mountains, PROBABLY chiefly leaves.--This position agrees with ch
037r055 & other species of large lizard.-- There would PROBABLY be no other organic remains.-- On Pampas look
038r058 elopes no doubt existed. These higher portions PROBABLY formed Islds from which proceeded pebbles & o
038r058 so much; or no rapilli; & from action of water PROBABLY not so much aluminated. As argument in favor
040r063 hich is described as very rare Mem. St Helena; PROBABLY more abundant in this case from intersecting
040r063 abundant in this case from intersecting a mass PROBABLY cold & not warm as sides of a crater as Vesuv
049r089 ill this conical mass be granite? Why not more PROBABLY greenstone? What probable origin can be given
057r113 at Coll. of. Surgeon's? I really should think PROBABLY that B. Blanca & M. Hermoso contemp:.--Inculc
087a008 When Siberia went up. Arctic land went down.-- PROBABLY more Arctic land would be required to produce
087a010 e Camel.? Make note about N. American bone not PROBABLY in salt marshes Efflorescence nothing -- Stud
108a080 tion luxuriant it might be almost as well said PROBABLY much Carbonic Acid gaz here.-- [top portion p
109a082 same way.-- a little further from beach action PROBABLY modified by form of waves & currents.-- but t
110a087 side.-- oozed from one of the true rocks, most PROBABLY from the gneiss beds in the mica slate.-- Geo
112a090 hat high they are «See Athenaeum. 1838. p 274. PROBABLY will be published in the Geograph. Journal.--
113a092 ia.-- Great NW. dip in SE part of Australia.-- PROBABLY a case of rivers turning round & penetrating
130a134 and. dunes.-- in these littoral deposits there PROBABLY would be marked line of separation A Paper by
137a151 he increase of temperature beneath the sea, is PROBABLY much more rapid than beneath continents In Be
147g020 -- nothing very striking yet possibly sea more PROBABLY than river-- No exact terraces but appearance
170bIFC y. 1837 p. 235. was written in January 183[8]: PROBABLY ended in beginning of February Zoonomia Two k
170b001 absolutely similar; for instance fruit trees, PROBABLY polypi, gemmiparous propagation. bisection of
175b017 even less change> especially with some change PROBABLY <change> vary quicker Unknown causes of chang
179b034 , with instinct in lieu of reason, there would PROBABLY be repugnance & art required to make marriage
181b045 adapted for. other elements. every part would PROBABLY be not complete, if birds were fitted solely
182b048 ptogamic flora but not atmospheric type. Hence PROBABLY only four, is not this Fries rule-- What subj
186b058 . I can conceive colouring retained; therefore PROBABLY in some <heteromera> colouring of crysomela m
186b059 er parent, (this is what French call (atavism) PROBABLY this is first step in dislike to union, offsp
192b087 termediate ranks had produced infinite species PROBABLY the series would have been more perfect. beca
199b120 offspring> » fertile offspring; marriage never PROBABLY excepting from «strict» domestication offspri
210b159 ties produce no difference. if they do.--there PROBABLY will be this relation also. |Yes « Fox» ‖ The
211b162 skeletons: VERY GOOD. Apteryx. a good instance PROBABLY of rudimentary bones.-- As Waterhouse remarke
216b182 case of Blood Hounds, a little mingling would PROBABLY have been good, namely such as blood hounds f
217b183 case.-- It explains the loss & expense, (must PROBABLY have occurred to every one) of rare breeds of
223b207 ses is more wonderful. its mind more different PROBABLY & introduction of Man. Nothing compared to th
231b243 rdance in latter-- Have change in form.-- This PROBABLY explains crag & miocene.-- The descendants le
233b251 ice & Madagascar & C. of Good Hope.-- His book PROBABLY worth studying.-- Wingless birds S. Continent
246c027 o be one at Botouma from account of natives, & PROBABLY on Oualan.-- «Mitchill says snakes on Friendl
251c042 Vol. IV p 273. Macleay on Capromys. 4 species PROBABLY in Cuba (p 271 Viedo says American dogs silen
252c042 . Mem contrary assertion of Molina) (p. 277.). PROBABLY another in Jamaica & perhaps one extant at Le
252c045 --- do from Indian increase of knowledge would PROBABLY tell more certainly Get Closer species-- FOX
253c046 to Australia.-- What are they? Colonel Montagu PROBABLY contains some facts about close species of Bi
256c054 n other animals.--? Forster on South Sea, will PROBABLY contain descriptions of domesticated animals
259c065 enty of times & you will have no tail (example PROBABLY not true).-- or again healthy parents have hr
263c078 nts proper.-- Present monkeys might not,-- but PROBABLY would.-- the world now being fit, for such an
271c107 s of insects of one tribe taking on structure (PROBABLY accompanied by habits) of other, thus in Chal
272c107 didous insect, which I brought from Australia, PROBABLY live in flowers & has Elytra. formed from dev
276c122 to me). colour of plumage & laws, which might PROBABLY be reduced What the Frenchman. did for SPECIE
277c128 ed shells, such as Cyrena This is reform which PROBABLY will be slow. but must take place-- such a cl
283c146 > The end of formation of species & genera, is PROBABLY to add to quantum of life possible with certa
284c147 -- on relations of desert, open ocean, &c this PROBABLY on long average, equal quantity, 2d on relati

Page		(Key Word)	
284c147	ity, 2d on relations of heat & cold. therefore	PROBABLY	fewer now than formerly.-- The number of form
285c151	islands close species, «on these isld» &c will	PROBABLY	upset it-- The space which one branch of the
292c168	me forms with his views.-- as genera are large	PROBABLY	only few of extinct forms have generated spec
292c169	s. & of 100 extinct species the greater number	PROBABLY	have no descendants on earth.-- +++. Even at
292c169	ndants on earth.-- +++. Even at Falklands some	PROBABLY	would stand change better than others.-- The
292c169	in one normal-- when so the> are united (which	PROBABLY	is first stage) the tendency to change cannot
292c171	ing in sleep».-- an action becomes habitual is	PROBABLY	first stage, & an habitual action implies wan
297c186	kind, extermination.-- New forms made through	PROBABLY	an infinite number of forms.-- therefore an i
297c186	number of forms.-- therefore an isolated form	PROBABLY	a remnant.-- Pachydermata. & Horses few forms
298c191	th alpine form) lower species afterwards would	PROBABLY	often be destroyed.-- or regrafted with fresh
299c193	other plants of same species Channel Islds (&	PROBABLY	Isle of Man) no plants peculiar to themselves
303c204	r forms of legs-- hence the father of man kind	PROBABLY	possessed a structure in these points for <t>
303c204	es, & form of feet.= (Negro or father of negro	PROBABLY	was first black at base of nails & over white
304c208	a.-- when it can be traced through series then	PROBABLY	heredetary & not produced by circumstances In
310c224	the shells in equal periods with Europe would	PROBABLY	have changed much less.-- Here is an element
313c236	ess.-- but why two sexes scission in all cases	PROBABLY	gemmation (.Ehrenberg) --not necessary to gen
314c237	tect itself, one segment of the corolla being	(PROBABLY)	small to allow it to lie on one side.-- but
318c255	leghanies to which it migrats every year,; and	PROBABLY	a «chance» wanderer like the first pair of Pi
326c266	in Encyclop. of Anat. & Physiology.-- Dampier.	PROBABLY	worth reading Lessings. Laoccaon.-- (translat
327cIBC	transported.-- Crawford. Eastern Archipelago.	PROBABLY	some account Raffles. Sir. S do. do-- Buffon
337d020	must be very unnatural-- Italian Greyhound is	PROBABLY	the effect of <sev> local variety many times
343d039	e taken place. otherwise in 10,000 years Negro	PROBABLY	a distinct species-- We know how long a Mamma
344d041	& Aphides. The extreme difference of sexes. is	PROBABLY	arrived at in case of insects as glowworm The
352d058	species of> a group of species is made. father	PROBABLY	will be dead-- hence there is no central radi
354d065	made rat for Ascension.-- The Galapagos mouse	PROBABLY	transported like the New Zealand one-- It sho
359d088	which is strange anomaly in Yarrells law.-- it	PROBABLY	is explained by the vigour of their propagati
365d105	ery far from common.-- Under this predicament,	PROBABLY,	alone would species cross in wild state.-- I
365d106	e adapted it to changing circumstances.-- More	PROBABLY	during known changes climate became unfit for
365d106	hybrid grouse between Black Cock & Ptarmigan	(PROBABLY	subalpina.) former has blue breast, latter re
372d129	y all these peculiarities not so a seed.-- Bud	PROBABLY	is like cutting off tail of Planaria, the who
385d163	-- His bitch will not take «& if she did take,	PROBABLY	would not be fertile» without she know & LIKE
385d165	ghters! This last remark good. because showing	PROBABLY	not education.-- Cannot I find some animal wi
386d166	art dying for great length of time.-- There is	PROBABLY	law of nature that any organ. which is not us
386d166	in disease & injury.-- The sympathy of part is	PROBABLY	part of same general law, which makes two ani
397e003	causes of population & depopulation have been	PROBABLY	as constant as any of the laws of nature with
401e015	effect from apple trees is produced.-- Thinks	PROBABLY	experiment was never tried of separating appl
401e016	e produced from passage from many varieties, &	PROBABLY	would take long before all the stain would be
402e018	gers, deer, apes, baboons, monkeys & an animal	PROBABLY	a tapir p. 233. dogs in Borneo-- <brought pro
402e018	ably a tapir p. 233. dogs in Borneo-- <brought	PROBABLY	by Chinese>, "the breed being of the latter b
405e031	years none of his remain found in the Americas	PROBABLY	did not.-- Octob. 25th. I observed in Windsor
407e038	«low» limits of blocks both North & South, has	PROBABLY	undergone a greater change, than any part, (e
409e048	ly cause, but one great final cause,-- nothing	PROBABLY	exists for one cause» of sexes «in separate»
409e049	social instincts, which as I hope to show is	«PROBABLY»	the foundation of all that is most beautiful
409e049	e history, & from the Astronomer that the moon	PROBABLY	is uninhabited]CD & if my theory be true then
410e051	& South Thinking of effects of my theory, laws	PROBABLY	will be discovered. of co relation of parts,
411e055	hence indicate gaps.-- by this means the laws	PROBABLY	would be. generalized, & afterwards by the ex
415e066	;-- yet until man became cosmopolite, he would	PROBABLY	be confined in locality like Ornithorhyncus,:
415e068	throat) that has made a man.-- CD [any monkey	PROBABLY	might, with such chances be made intellectual
418e080	My view of every animal being Hermaphrodite--	PROBABLY	will recieve illustration from domestication
423e095	many species tending to dis-developement (some	PROBABLY	always have done so, as the simplest fish &),
423e097	e of developement is either stationary or more	PROBABLY	increases.-- Jan 29th. Uncle John says he fee
426e107	If many wild animals were crossed, there would	PROBABLY	be perfect series, from physical impossibilit
427e110	somewhat more readily, «than other apples» but	PROBABLY	would more indelibly stain offspring-- it wou
432e124	with fluid seeds can be true hermaphrodite Man	PROBABLY	assumes the hairy character of his forefather
433e126	s elapsed between each of these gaps, far more	PROBABLY	than during the deposition of the beds-- The
436e135	- No. but the wandering & separation of a few,	PROBABLY	would be most efficient in producing new spec
438e142	offer of changes.-- as desert «or rock» plant	PROBABLY	would do-- or be with difficulty be kept aliv
439e142	difficulty be kept alive.-- Nevertheless much	PROBABLY	depends on circumstances favouring the reappe
441e149	soil, I presume not, but its seeds, I presume,	PROBABLY	would-- at least the experiment of the carrot
447e164	ogamic plants & true hermaphrodite Mollusca, &	PROBABLY	corals.-- these forms then ought to be very p
449e169	ook after the common goose thus contradicting	(PROBABLY)	Yarrells law & Walkers of the male giving fo
450e175	he plains of Celebes. (but language shows that	PROBABLY	not original there)-- shows these isld not fi
451e176	an albino like himself said not to be common--	PROBABLY,	I should think grandfather first of race & i
461t037	sities on Camels & Horses--».--«)» & therefore	PROBABLY	any structure would rather become accomodated
463t057	many» facts with laws & their explanation will	PROBABLY	reject this theory-- (I must answer it by roo
466t095	cantering a congenital peculiarity improved.»	PROBABLY	every such «new» quality becomes associated w
469t119	a toe-nail & two joints-- as it is on one foot	PROBABLY	monstruous & not a second species.-- <Saw> Ma
473s004	ods, together with their littoral deposits are	PROBABLY	buried in the depths of the sea-- Maer. June/
498q007	nproductive from poorness of soil.-- yet crabs	PROBABLY	would grow there (7) Where parts of fructific
502q11v	ivan about Falklands Isds.-- Snipe Migratory--	PROBABLY	united by Land to S. America (33) Ornithologu
513q020	ns vary in manner & perfection &c &c &c--if so	PROBABLY	a variety, not specific character= Cross Rump
523m016	by fear. Chile earth quakes). in people, who,	PROBABLY	otherwise would not have been so.-- In Mr Har
526m027	other fellow-animals, than any other animal &	PROBABLY	the only one affected by various knowledge wh
528m036	nt of imagination, (as in hearing music), this	PROBABLY	arises from (1) harmony of colours, <whi> & t
530m046	ing is solely by association, & association is	PROBABLY	a physical effect of brain the «similar remar
531m052	-- Even the worm when trod upon turneth,, here	PROBABLY	there is no feeling of passion, but muscular
532m054	anied by much involuntary fear) In these cases	PROBABLY	the system is affected, & by habit the mind t
536m071	affect for womens breasts.∴ Dr Darwin's theory	PROBABLY	wrong, otherwise horses would have idea of be
536m073	tary disposition & instincts--.-- Put it so.--	PROBABLY	some error in argument, should be grateful if
537m076	rong to injure them without temptation.-- This	PROBABLY	is natural. consequence of man, like deer &c,
538m079	none about her had EVER before heard, so very	PROBABLY	forgotten.» Such facts bear on such character
538m080	separate, from the ordinary state of mind, is	PROBABLY	analogous to the double individuality implied
540m084	smiles. Many of actions as hiccough & yawn are	PROBABLY	merely coorganic as connexion of mammae & wom
543m098	one order, as there is Intellectual in human--	PROBABLY	some genera in different orders more advanced
547m112	ing things to be done.-- The senses are closed	PROBABLY	by sleep & not vica versa. anyhow I might hav
550m124	fectly obeys his conscience or instinct, would	PROBABLY	feel but little that of anger or revenge.-- t
552m131	irection of mountain chains in N. America Fear	PROBABLY	is connected with habitual stopping of breath
557m149	whole way. but ashamed of himself.-- Jealousy	PROBABLY	originally entirely sexual; first try «to» at
559m154	ss to consider Creator as governing by laws is	PROBABLY	that as long as we consider each object an ac
566n012	as soon as any enquiry commenced, for instance	PROBABLY	such a thing as thunder. would be placed to t
567n013	e they can talk, so do many animals.-- analogy	PROBABLY	false, may lead to something.-- October. 8th.
568n017	ral good, or in case of any fantastic custom»	«PROBABLY	bashfulness is connected with some disturbed
569n020	recognizes that sound as perfectly as a man.--	PROBABLY,	language commenced in some necessary connexi
569n020	by corporeal structure.-- Devotional feelings,	PROBABLY	some distant power of the mind-- superstition
569n021	, or eloquent request. Reason in simplets form	PROBABLY	is single comparison by senses of any two obj
569n021	ion between one or two absent things.-- reason	PROBABLY	mere consequence of vividness & multiplicity
570n024	th many emotions.-- (Explanation of sighing is	PROBABLY	correct, to relieve respiration when immensel
574n040	d a beginning, which my theory requires. There	PROBABLY	is some connection between very limited reaso
574n040	-- Children & old people get into habits.-- we	PROBABLY	can hardly form an idea of a mind so limited
574n041	almost bite, that which one sexually loves is	PROBABLY	connected with flow of saliva, & hence with a
574n041	degree sexual.-- The association of saliva, is	PROBABLY	due to our distant ancestors having been like
580n062	ntellect between animals & men only in Kind.--	PROBABLY	very important work.-- Feb. 12. 1839. Sir. H.
585n074	remarks on the means. by which children learn	(PROBABLY	not only experience,, but also «by an» instin
586n078	sun, & heavens, or magnetic virtue,-- the most	PROBABLY	supposition. with respect to pidgeons, is tha
586n082	» cells-- [but this faculty <may possibly be>	«PROBABLY	is» instinctive, namely the knowledge of size
592n105	ing a wounded one.-- porpoises a ditto-- it is	PROBABLY	some secondary one-- blood being disagreeable
594n112	hawk, which are «so» rare « <s.> » here,, that	PROBABLY	few had ever before seen one, yet all-- flew
611o034	d» the organic & inorganic-- The inorganic are	PROBABLY	one principle for connect of electricity chem

Page ***(Key Word)***
611o034 & gravity is probable.-- And the Organic laws PROBABLY have some unknown relation to them-- [RHC] In
615o038 -- But as no great modification in brain would PROBABLY take place without corresponding change in «e
615o038 ral instincts nearly same; which same argument PROBABLY applies to particular instincts of animals. e
620o044 er generally soon conquers, & the dog [RHC] 3) PROBABLY thinks no more of it.-- Not so man, from his
623o051 aces could become instinctive.-- [LHC] x It is PROBABLY That becomes instinctive, which is repeated u
624o051 hange, from civilization, education changes, & PROBABLY likewise instincts, for the same law effects
632j53r ive developement I admit, but the admission is PROBABLY from ignorance]CD Who would ever have thought
640j167 d isld should possess species to themselves.-- PROBABLY no case in world like Galapagos. no hurricane
616o039 to the use of ordinary language that the onus PROBANDI might fairly be laid with those who would sup
068r147 nic focus.-- <it is certain, if strata can be> PROBLEM dislocate strata without ejection of the fluid
126a122 if better conductor, then still thinner → The PROBLEM is, you have temperature known at surface,-- y
128a128 takes place.-- Mountain will first fall-- the PROBLEM will be falling of an arch weighted in its cen
291c165 Sebright.-- love. of man gained & heredetary. «PROBLEM solved» habits become important element in cla
413e059 mystery of mysteries. & has grand passage upon PROBLEM.! Hurrah.-- "intermediate causes" The Sexual s
548m115 o that One Reflective Consciousness is curious PROBLEM., one does not care for the pains of ones infa
564n005 nomy without Mechanics.-- Experience shows the PROBLEM of the mind cannot be solved by attacking the
585n075 hat [...] the same.--(this Hensleigh therefore PROBLEM is how we know that thing is same, which touch
628o53v 21. "Why ought I to keep my word"-- gives the PROBLEM, of ethics-- [my answer would be to all such c
466t094 o> do not bend up-- In Lark-spur, if Bees put PROBOSCIS within nectary «they do» they must disturb al
472s02r len) on legs go from clump to clump, & insect PROBOSCIS in many flowers, on one of which pollen was r
472s02r on Monday.--» alight on upper petals & insert PROBOSCIS, under sigma & draw it out over & over again
637j57v » some insects <not> not been provided. «with PROBOSCIS» «as bee & butterfly» inconvenience.! extinct
092a028 Linear earthquake 500 by 90.-- in Syria Geolog. PROC. p. 541. year 1837 In Upper Assam. Geolog Proc
092a028 . Proc. p. 541. year 1837 In Upper Assam. Geolog PROC p. 566 1837.-- Tertiary <bea> formation twenty
201b126 x in N. America: as well as 2 recent See Geolog. PROC. p. 569. 1837. Account of wonderful fossils of
059r123 s, so changes, acting in those lines. must now PROCEED from great depths.--important.-- Decemb 10. 18
038r058 her portions probably formed Islds from which PROCEEDED pebbles & on which trees grew.--? Are not the
057r114 ons.-- I think in Patagonia white beds having PROCEEDED from gravel proved.-- curious similarity of r
309c221 ica, speaks of short legged sheep. heredetary PROCEEDING from an accident. New England farmer,-- usef
097a044 lone at Guantajaya contains salt see Geolog. PROCEEDINGS Lake let out by steps in Central France not
103a063 do.-- Give this after supposition p. 461 «of PROCEEDINGS» List of collections in Geological Society.
210b160 Mauritius are not found at Bourbond Zoolog. PROCEEDINGS‖. <p> 1832.p. III Mr Owen suggested to me, t
211b162 n has such instincts very little. in Zoolog. PROCEEDINGS. Jan 1837, «by Eyton» Account of three, kind
212b166 e animals common to Mauritius & Madagascar.? PROCEEDINGS of Zoolog. Soc June 1837 p. 53. an Irish Rat
223b209 w» animal mentioned from Fernando Po Zoolog. PROCEEDINGS October (?) 1837 Contrast New Zealand with T
307c216 r tooth» of Hippotamus from Madagascar!!!!!! PROCEEDINGS of Geol. Soc Vol I It is capable of demonstr
480z006 eludo Cavia Australis. Dorbigny Vol II, p 24 PROCEEDINGS of Zoolog Soc. Important account of habits o
032r041 arer the surface & to the Eastward.--If matter PROCEEDS from great depth. from axis to surface must g
107a077 er beneath ocean than above it no because heat PROCEEDS from great body of mass.-- The last speculati
028r029 ings or more probably by some unknown Volcanic PROCESS? How does it come that all Lime is not accumul
040r063 feruginous sandy ones have undergone the same PROCESS.-- Neither lakes or Avalanches (Glaciers very
056r112 .--<Hutton show> Earthquakes part of necessary PROCESS of terrestrial renovation & so is Volcano a us
098a046 t concretions are connected with a crystalline PROCESS.-- now cleavage as suggested by Sir J. Hershel
098a047 .-- Veins in septaria. a kind of concretionary PROCESS (analogous to layers of quartz & feldspar) wit
110a085 species-- Lyell-- Some internal changes are in PROCESS. connected with variation of compass & these m
125a121 l's views earth originally fluid, then cooling PROCESS must go from surface towards the interior -- w
170b001 c.-- The ordinary kind <the> which is a longer PROCESS, the new individual passing throug several sta
197b111 tly «or hints» considers generation as a short PROCESS, by which man «one animal» passes from worm to
232b245 rting from one point.-- In the crag we see the PROCESS of change of those forms, which have succeeded
259c063 ary «& combined» effect of habit.-- perhaps in PROCESS of change.-- Are any men born with any peculia
313c236 different parts, & scission cannot effect the PROCESS.-- but why two sexes scission in all cases pro
358d085 ATE men),-- if carried much further, if by the PROCESS this were possible, the organs doubtless would
372d131 ported., <so> & in making «true» bud some such PROCESS is effected.-- a child might be so born. but i
389d173 ved that the constant necessity for change. in PROCESS by generation applies only the more complicate
390d178 racted at each, or in several generations, the PROCESS would be similar to budding. which is not obje
391d174 change from simplest form.-- (Because by this PROCESS it separates those differences which are in ha
430e118 ked & not allowed to cross.-- » Has nature any PROCESS analogous-- -- if so she can produce great end
509q017 (ask more about the lowest cervical vertebrae PROCESS developed into ribs.) & does its abortion vary
638j58v itions.-- An adaptation made by intellect this PROCESS is shortened, but yet analogous, no savage eve
641j29v t of Owl remarkable, [gained by very different PROCESS from Bats. CD]CD. «Macculloch says no other bi
637j58r ren Virgins p. 235. talks of the long spinous PROCESSES in Giraffe &c, as adaptations to long necks--
187b064 nt applies to species.-- If individual cannot PROCREATE, he has no issue, so with species.-- I should
391d174 actly like its parent. <but these buds do not PROCREATE> & all alike in one parent or tree, (but not
391d175 uld be added to each individual before he can PROCREATE. these changes may be effect of differences o
232b248 lls in salt water & lizards do.-- Ask Eyton to PROCURE me some Get Hope to give me an account of para
076r169 ot abundant. = considerable quantity of silver PROCURED from martial pyrites; great blocks of pure si
244c022 dagascar. Monkey from Java.-- Hairs, & deer.-- PROCURED two makis alive from there.-- Mem Waterhouse
275c121 ng the yellow one & crossing with duck bantams PROCURED old variety.-- The pidgeons which have such d
437e140 Shiant Isld. it is said, that an Eagle always PROCURED its prey from another island.-- » p. 175., 28
452e181 Gilolo.-- p. 134: Birds of Paradise were first PROCURED from Gilolo p. 253 In isld of Bunwood (18 mil
472s01v -- Elizabeth says several years ago seeds were PROCURED with the P. orientale in garden & all came up
268c096 this)> says differently) do. do. on the genus PROCYON.-- by Wiegman Classified catalogue of animals
558m152 mouth «hairs erect on back» «wide open» with PRODIGIOUS force.-- making growling, guggling noise. Pu
284c148 can «not» bear the least salt water.-- Nuts PRODIGIOUSLY heavy (where trees of such Nature far apart
381d156 irectly a Capon is cut, it increases in size PRODIGIOUSLY-- Animal OEconomy by, Hunter. (edited by Ow
432e125 of the ears to our hearing powers E. frowns PRODIGIOUSLY when drinking very cold water «frowns conne
314c238 st be protected however it may be effected.-- PRODROMUS Florae Norfolkicae. 1833 Steph. Endlicker (He
035r048 viscid nucleus, which in any one country would PRODUCE equable effects.--«though so immense to short
087a008 Probably more Arctic land would be required to PRODUCE climate resembling S. America in Europaean lat
104a066 nal lines near. (lateral pressure would always PRODUCE it) but where great thickness is affected, the
128a129 ntre.-- Will not abrasion of land on one side. PRODUCE subsidence of water on other. from tendency to
180b038 eding & changing circumstances are continued & PRODUCE according to the adaptation of such circumstan
192b083 that beyond all doubt seeds of Ribston Pippin, PRODUCE Ribstone Pippins, & Golden Pippen, goldens-- h
192b085 is not more change than we know varieties can PRODUCE.-- Therefore all genera MAY have had intermedi
195b103 s under equator that external conditions would PRODUCE species so close as Patagonian <Chat> & Galapa
199b118 5. Ap 1827 F. Cuvier says. "But we could only PRODUCE domestic individuals & not races, without the
210b159 -- In plants. do the seeds of marked varieties PRODUCE no difference. if they do.--there probably wil
211b162 pes> indicate affinity, because similar habits PRODUCE similar structure.-- Mem. Ornitho Rhyncus Woul
217b184 wn case of good pointer & rough water spaniel «PRODUCE litter like both parents» & Mr Bell has half b
218b190 his barely applies to plants Female pig apt to PRODUCE monsters in Isle of France-- -- Madagascar oxe
219b196 -- Ellis (?) says Tahitian kings. would hardly PRODUCE from Incestuous intercourse. a parallel fact t
219b196 ught to explain nothing.-- it being as easy to PRODUCE «for the creator» two quadrupeds at S. America
220b197 s at S. America Jaguar & Tiger & Europe, as to PRODUCE same one. ‖Although in plants, you cannot say
224b214 which is not probable; then monkeys will never PRODUCE man. but both monkeys & man may produce other
224b215 never produce man. but both monkeys & man may PRODUCE other species., man already has produced marke
224b215 dy has produced marked varieties & may someday PRODUCE something else., but not probable owing to mix
225b219 same purpose.!! Can the wishing of the Parent PRODUCE any character on offspring? Does the mind prod
225b219 duce any character on offspring? Does the mind PRODUCE any change in offspring? if so adaptations of
225b219 gs &c &c NB. Animals very remote. ass & Horse. PRODUCE offspring exactly intermediate.-- Reference to
228b230 the seeds of Ribston Pippin & Golden Pippin &c PRODUCE real crabs, & in each case similar or mere mon
230b240 l not destroy that evidence, as so many plants PRODUCE hybrids, or else whole fabric will be overturn
248c034 n blood with time, then generation will «only» PRODUCE an offspring capable of producing <two> such a
249c034 self.-- therefore two different varieties will PRODUCE hybrids but not varieties. which are not deepl
249c034 e not deeply impressed on blood., will cross & PRODUCE fertile offspring in the first case it will ei
249c034 ile offspring in the first case it will either PRODUCE no of[fspri]ng or such as not capable of produ
259c063 ing greater vigour to the parent so tending to PRODUCE effect on offspring-- but WHOLE race of that s
263c078 f animals of hundred generations of species to PRODUCE contingents proper.-- Present monkeys might no
265c084 es more than mere child, but that child should PRODUCE like children. ‖Lyell has story from.-- Beck a

Page	**	(Key Word)	**
280c136	to see that it is less repugnant to nature to	PRODUCE	one offspring unlike itself, than to produce t
280c136	o produce one offspring unlike itself, than to	PRODUCE	that capable of producing itself alike.-- in o
280c136	effort <pr> one unlike can be produced, yet to	PRODUCE	whole generation unlike would go against the t
282c143	ew days-- one doubt that one animal can really	PRODUCE	so great an effect.-- the spirit of life must
289c162	l II p. Dr Johnston <on> Entomostraca Daphnia,	PRODUCE	young, capable of producing young many times &
299c195	cultivating parent in rich soils & their seeds	PRODUCE	<offspring> variety. wild carrot. made into bi
307c215	giving milk)» rudimentary wings. so nature can	PRODUCE	in sex, what she does in species of Apterix Th
310c224	tes M. Turpins assertion that globules of milk	PRODUCE	a plant capable of growing!! & propagating its
312c231	ing of each feather weight & size & they would	PRODUCE	number agreeing almost to the point in questio
342d032	t like common mules) & lay many eggs but never	PRODUCE	inter se or with parent species.-- The hybrids
342d033	Blyth remarked only near species or varieties	PRODUCE	heterogenous offsprings.-- «are not the hybrid
342d033	en do those SPECIES which breed most freely. &	PRODUCE	somewhat fertile offspring produce heterogenou
342d033	t freely. & produce somewhat fertile offspring	PRODUCE	heterogenous offspring. It appears certain tha
358d086	tration, hybridity, & breeding in & in tend to	PRODUCE	same effects.-- CD[May it be said, that breedi
359d086	May it be said, that breeding in & in tends to	PRODUCE	unhealthiness,-- «or» to perpetuate some organ
372d129	d? +.simplest forms of budding. Why does Gecko	PRODUCE	always different tail? An Individual bud may b
372d130	nothing to do with generation.-- Why crab can	PRODUCE	claw. but man not arm. hard to say-- if it wer
372d131	upport the arm of Man, when cut off , it would	PRODUCE	another man.-- That the embryo the thousandth
372d131	That the embryo the thousandth of inch should	PRODUCE	a Newton is often thought wonderful. it is par
372d131	of twin bear on this subject? A mans arm would	PRODUCE	arm if supported., <so> & in making «true» bud
389d176	ll fecundate female for several births, & even	PRODUCE	fertile offspring-- DESIRE LOST when male & fe
389d177	ery remarkable that too much difference should	PRODUCE	same effect as too little.-- in (latter case f
389d177	ter case female often takes males but does not	PRODUCE)	tendency to deformity ¿this does not happen w
398e004	Italy, but what «changes» would such a change	PRODUCE	in climate vegetation &c.-- It is the circumst
401e015	. lately saw a nonpareil sowed by Mr Tollet so	PRODUCE.	-- thinks it probable that great part of those
401e015	on from other apple trees.-- now seeds of crab	PRODUCE	crab, so that some effect from apple trees is
401e016	is curiously different from primrose suddenly	PRODUCE	cowslip, one is tempted to think here some ano
408e043	most delicate organ, on the whole system may.	PRODUCE-	- ? When a species becomes rarer, as it progre
424e102	eet in one wood in Anhault. & there every year	PRODUCE	hybrids-- now this is independent good case, b
426e106	lation between the fact that different species	PRODUCE	abundantly infertile hybrids, & the fact that
430e118	ature any process analogous-- -- if so she can	PRODUCE	great ends-- But how.-- -- «-- .Make the diffi
434e129	bortive & bed of female flowers will sometimes	PRODUCE	a few seeds,-- -- Ruscus aculeatus. a dioeciou
460t013	ecies The <male> swan-gander with common goose	PRODUCE	full as many eggs as pure bred common.-- the h
460t013	d ganders. crossed with common goose <to has> «	PRODUCE	offspring with» so much of the swan-goose in a
491q01v	no suppose, that certain plants, like Aphides	PRODUCE	impregnated young ones; & that it is in these
491q002	d? 3.. Whether the viviparous grasses & onion,	PRODUCE	flowers, like the Oxalis from C. of Good Hope
499q009	will female (a) Willows or Yews some poplar's	PRODUCE.	-- (15) Would Yew fruit without impregnation.-
506q015	all. (37) Any cases of plants. which will not	PRODUCE	seed in this country-- where cause not apparen
507q15v	rons 47. Ficus carica Henslow presumes females	PRODUCE.	Polygam. trioecia. (are female flowers ever p
515q021	fing (5) Do the most cultivated show Heartease	PRODUCE	as large capsules of seed, as the commoner kin
544m104	. Everything which happens to man who does not	PRODUCE	children. or after he has useless. does not af
553m136	ofane «degnen» in thinking not capable to <do>	PRODUCE	every effect, of every kind which surrounds us
609o030	alists: one says our rule of life is what will	PRODUCE	the greatest happiness.-- The other says we ha
629o55v	ong.-- & conscience will imperiously say so, &	PRODUCE	shame & remorse-- [Thus pungency of one's feel
633j53v	tions show us accidents may become heredetary [	PRODUCE	some peculiarity in seed vessel]CD if man take
023r010	of fluid nucleus through» faults or fissures,	PRODUCED	by the elevations of those mountains on the c
023r011	ined side of volcanic activity.» That axis was	PRODUCED,	from a fissure in a deep & therefore weak pa
036r051	s it generally known. the acute chirping sound	PRODUCED	in walking over the sand: I am nearly sure, i
061r127	roaches on smaller.--change not progressif<e>:	PRODUCED	at one blow. if one species altered: <altered
100a055	illips Lardner p. 197. refers to salt as being	PRODUCED	by local heat, Ask Capt. Beaufort, whether, w
107a077	ecific gravities of many artificial limestones	PRODUCED	by Sir J. Hal. End of pages. p. 157. Vol VI E
155g062	n Turrit side 2nd shelf very broad «& cut out,	PRODUCED»	from same «cause» as «great» spit <is> or pl
171b003	s of plants sown in rich soil, many kinds, are	PRODUCED,	though new individuals produced by buds are
171b003	ny kinds, are produced, though new individuals	PRODUCED	by buds are constant, hence we see generation
172b008	effected at once, -- something like a variety	PRODUCED-	- --[every grade in that case surely is not P
173b009	ed-- --[every grade in that case surely is not	PRODUCED?	-- <Granting> Species according to Lamarck di
174b015	hen no Mammalia existed; Australian; Mamm were	PRODUCED	from propagation from different set, as the r
180b038	d like Orpheus. being favourable many might be	PRODUCED.	-- This requires principle that the permanent
180b038	equires principle that the permanent varieties	PRODUCED	by <inter> confined breeding & changing circu
189b072	not utterly cut off:-- like golden pippen. if	PRODUCED	by seed go on.-- otherwise all die.== The fos
189b073	buds of plants, which die at one time, though	PRODUCED	either sooner or later.-- Prove animal like p
192b087	rogenitor,-- Now if the intermediate ranks had	PRODUCED	infinite species probably the series would ha
196b106	gs <ne> in or near South Hemisphere. Were they	PRODUCED	in several places & died off in some? Why did
198b115	on all isld, (if act of fresh creation why not	PRODUCED	on New Zealand; if generated <No> «an» answer
201b126	gaps. & those forms which «nevertheless» have	PRODUCED	species, have produced fe[w] [not located] Th
201b126	ich «nevertheless» have produced species, have	PRODUCED	fe[w] [not located] The relation of Analogy o
203b135	) to consider the monkey as a wanderer, but as	PRODUCED	by climate?-- M. Baer (thinks) the Aurock, wa
204b139	aways & result the same Indian cattle & common	PRODUCED	very fine Hybrid offspring, much larger than
217b183	t last a rough-haired shepherd dog lined her &	PRODUCED	a very large litter.-- never afterwards went
224b215	an may produce other species., man already has	PRODUCED	marked varieties & may someday produce someth
224b215	acquired alters case) other species or angels.	PRODUCED	«&» Has the Creator since the Cambrian format
228b230	s plant, but any marked. variety.-- Strawberry	PRODUCED	by seed«s»??-- Universality of generation str
235b275	breed> do not intermix,--any cultivated plants	PRODUCED	by seed.-- Lychnis.-- Flax.-- Read Swainson [
239cIFC	colour on parent, white room How are varieties	PRODUCED,	by picking offspring? Instances of old Breed
239cIFC	ooks About amount of difference: where hybrids	PRODUCED	have any close species ever yet failed. About
240c004	mother. The fact of great monstrosities being	PRODUCED,	& handed down, with ease, is analogous to wh
248c034	re will be chiefly heredetary.-- If varieties «	PRODUCED	by slow causes, without picking» become more
263c075	bringing the mind to grapple with great effect	PRODUCED,	is a most laborious, & painful effort of the
276c124	iculty how a different number of vertebrae are	PRODUCED,	where, (& in all such structure) there canno
280c136	though by great effort <pr> one unlike can be	PRODUCED,	yet to produce whole generation unlike would
282c142	ce, thus the analogy would not in all cases be	PRODUCED,	but would depend upon exclusion.-- The same
283c145	ough <fertile> when compelled to breed hybrids	PRODUCED-	-)» & then all that I want is granted.-- For.
283c145	all rest destroyed far remote genera. will be	PRODUCED.	As we know from Ehrenbergh, there are fossil
298c190	simplest reasoning in lower animals many times	PRODUCED,	a general tendency produced. Such as man get
298c190	nimals many times produced, a general tendency	PRODUCED.	Such as man getting habitually into passion,
299c195	nent varieties in wild state-- The two. former	PRODUCED	by difference of <locality>. station & variet
299c195	nce of <locality>. station & varieties chiefly	PRODUCED	by cultivating parent in rich soils & their s
303c204	l organs «in some monkeys clitoris wonderfully	PRODUCED»	.-- make abstract on this subject from Lawren
304c207	descended from long way back.-- aberrant forms	PRODUCED	where many species <osculant> but, where much
304c208	h probabilities only guides.-- Yet trifles are	PRODUCED	by circumstances. Spines on Echidna.-- when i
304c208	through series then probably heredetary & not	PRODUCED	by circumstances In ostrich which is not isol
306c213	t (about 200 miles distant).-- directly beyond	PRODUCED	line of Timor 215 degree. What productions Sa
307c216	purs. growing.-- Are there any abortive organs	PRODUCED	in domesticated animals, in plants I presume
308c219	osyncrasy, cause of fertility.-- varieties not	PRODUCED	as by nature. if so. the habits which would.
311c226	parts being most easily mostrified, which last	PRODUCED	--insane men in civilized countries-- this is
312c231	provincial breed, & that first offspring thus	PRODUCED	are better, than those bred in & in.-- which
312c232	hether mutilations non-heredetary & variations	PRODUCED	in short time in some extent counterpart, mut
312c232	tent counterpart, mutilation being «variation»	PRODUCED	in shortest possible time. Mr Willis Long ear
313c235	oology & Botany) where several generations are	PRODUCED	in succession (13?) without impregnation, the
325c267	ent? Study with profound care, abortive organs	PRODUCED	in domesticated plants; where function has ce
333d009	same bitch & «perfect» spaniels & setters are	PRODUCED.	one would argue the whole effect of race was
335d015	as «badly» deformed people & as mutilations «(	PRODUCED	very quickly)» sometimes have similar offspri
341d031	le, yet so infertile never even in seven years	PRODUCED	even an egg.-- a most curious bird, did not s
342d032	or his table Muscovy & common ducks-- they are	PRODUCED	in full equal Numbers with pure bred (just li
355d066	l worthy of examination whether variations are	PRODUCED	only in those character which are seen to <va
357d073	supposing Tetrao media or Rakkelhan is hybrid (	PRODUCED	commonly in Nature. both in Sweden & ancientl
366d108	ht. thought if he had had a pair he could have	PRODUCED	from these.-- this instance of monstrous vari
367d112	gans by that co-relation of parts, will not be	PRODUCED.	-- Sept. 17th. Saw mule. apparently fathered

371d127 surely in all these cases an unseen change is PRODUCED in parents--colour is a doubtful subject, but
371d128 kept in same soil. same atmosphere?-- may they PRODUCED, not be transplanted?, & yet year after year,
371d128 et year after year, successive roses & bud are PRODUCED, like parent stock, or if different dieterior
372d130 different tail? An Individual bud may be thus PRODUCED from the growth of one part, (not strictly ne
372d130 part, (not strictly new individual), or he may PRODUCED by having undergone, the endless changes, whi
390d177 ry. (<annuals> & so must those forms which are PRODUCED by budding «only» as cryptogamia & hydras,--
391d174 . infra» p 179, continued from Is a flower bud PRODUCED by union of two common buds??? Amongst buds e
401e015 crab, so that some effect from apple trees is PRODUCED.-- Thinks probably experiment was never tried
401e016 d.-- Thinks-- that such variety as red cabbage PRODUCED from passage from many varieties, & probably
411e056 plants. where innumberable individuals can be PRODUCED. & yet sexual apparatus.-- My account of Circ
427e110 <other:> «that of another» apple. only effect PRODUCED would be different.-- same way one variety of
429e134 mals, yet as we know how many plants have been PRODUCED (look at the Dahlias. we may infer it in anim
430e118 tter Pidgeons «race-horse». have not been thus PRODUCED, but by training, & crossing & keeping breed
435e134 t, he resembles, other mammalia in the effects PRODUCED on organization. by physical agents." p. 466.
436e137 ught, surely no "fortuitous" growth could have PRODUCED these innumerable seeds-- yet if a seed were
436e137 d these innumerable seeds-- yet if a seed were PRODUCED with infinitesimal advantage it would have be
442e152 ause its very essence is that little change is PRODUCED.-- The fact just alluded to of Northern flowe
458t002 between ass & Zebra, crossed with pony mare & PRODUCED a very pretty little animal, showing somethin
470t176 lanted within few yards of each other actually PRODUCED hybrids-- My Father remembered when in the ga
498q009 r.-- As they are dioecious, if no hybrids were PRODUCED by seed, we might feel sure, that pollen of o
505q014 12 Does the horned orange. wh. never has seeds PRODUCED good pollen? Yes «From cultivation lost their
505q014 ens by Terrace to see, whether stamens will be PRODUCED in individual plants 17 A dead-nettle in Hot-
506q015 e cause not apparent-- Any where pollen is not PRODUCED or small in quantity -- Any unproductive, whe
515q021 ttend to is whether good & plenty of pollen is PRODUCED. & 2d if so, whether concepcion takes place,-
523m016 o more about it than of a dream.-- Insanity is PRODUCED by moral causes (ideotcy by fear. Chile earth
569n020 yawning laughing being necessary sounds... not PRODUCED by will <by> but by corporeal structure.-- De
574n042 her principle of those children. which chance? PRODUCED with strong arms, outliving the weaker ones,
599o007 from the effects of heredetary knowledge, has PRODUCED almost → greater changes in the polity of Nat
601o009 n. (only evidence. when not consciousness) are PRODUCED in consequence having some relation to the pr
607o025 ing Bible for first time, & great effect being PRODUCED.-- the wax was soft,-- the condition of mind
609o030 shows them to be almost identical» + What has PRODUCED the greatest good «or rather what was necessa
613o036 -- method of attacking peccari-- --retriever-- PRODUCED as soon as brain developed, and as I have sai
628o054 ey explains this & Mackintosh shows the change PRODUCED.-- 4) p 38 Conscience checks the wish to <oth
193b090 climatizes the child. ¿-- whether every animal PRODUCES in course of ages ten thousand varieties, (in
239c001 s that Esquimaux dog when crossed with pointer PRODUCES offspring much nearer Esquimaux than Pointer.
257c059 sent day.-- It is ASSUMPTION to say generation PRODUCES young ones capable of producing young ones li
399e010 e may strongly suspect, that breeding in & in, PRODUCES bad effects solely, because of similarity, be
427e110 ety of <animal> dog does not prefer other. but PRODUCES greater effect on offspring-- Mr. Herbert say
448e165 -- do not seed freely.-- The periwinkle seldom PRODUCES seeds, because it is thought to require insec
491q01v of nearly related plant & see if first pollen PRODUCES any effect, as in case of woodpidgeon & Hen.
605o19v tain causes, as great height, eternity, &c &c. PRODUCES an inward pride & glorying. (often however ac
063r132 are not so much surprised at seeing Zoophite PRODUCING distinct animals. still partly united. & eggs
179b032 father; is not this owing to each copulation PRODUCING its effect; as when bitches puppies are less
199b120 to marriage <if offspring not fertile> <,but PRODUCING> «before domestication, afterwards none or li
228b229 ly abundant. Seed of Ribston Pippin tree <go> PRODUCING crab. is the offspring of a male & female ani
230b236 is this explained by law of small differences PRODUCING more fertile offspring.-- 1st. All variation
230b240 rent then, they will be called species & mere PRODUCING fertile hybrids will not destroy that evidenc
249c034 n will «only» produce an offspring capable of PRODUCING <two> such as itself.-- therefore two differe
249c034 duce no of[fspri]ng or such as not capable of PRODUCING again The Varieties of Cardoon are cases <sp>
257c059 say generation produces young ones capable of PRODUCING young ones like itself, but ¿whether great as
257c059 lf, but ¿whether great assumption? not solely PRODUCING like itself, not applicable to monsters:-- Ar
263c076 ne man the developement of a brain capable of PRODUCING more glowing imagining or more profound reaso
280c136 nlike itself, than to produce that capable of PRODUCING itself alike.-- in one case it changes one, i
289c162 tomostraca Daphnia, produce young, capable of PRODUCING young many times & lay two sorts of eggs-- on
356d070 ew.-- Sept. 8'. A Golden Pippen or Ribston do PRODUCING occasionally (as Fox says) same fruit trees i
379d152 When cows have twins, <one> though capable of PRODUCING both «male» pair of male & female.-- if there
399e010 have freely bred, they have not lost power of PRODUCING Williams. Narrative of Miss. Enterprise, p.
401e016 think here some anomaly-- I can fancy cowslip PRODUCING primrose return to old stock, but not primros
401e016 rimrose return to old stock, but not primrose PRODUCING cowslip Uncle J. says common belief. that fem
436e135 of a few, probably would be most efficient in PRODUCING new species; also one being reduced in number
441e148 s of two organs-- instead of one part «as» in PRODUCING bud.-- Fewer of the lately acquired peculiari
447e164 f chances, would prevent it varying. A plant <PRODUCING> propagating itself by buds is in same predic
448e165 dless curious facts about every part of plant PRODUCING buds, so that Turpin says each cell of plant
530m046 me part of brain, or the tendency to habit of PRODUCING a train of thought.-- [not located] Fox belie
071r156 hisoph. Transactions. = Mem: Olivine. Volcanic PRODUCT.=> <Did Peruvian Indians use arrows or Araucan
446e162 & then they break-- -- each tulip is the <of> PRODUCT of fresh bud-- here then is case of change ana
046r078 ed as chemical retorts.--neglecting the first PRODUCTION of Trachyte. look at Sulphur. salt. lime, ar
065r139 62 degree 55 minute. <only> one lichen. only PRODUCTION. a body which had long been buried, <see> fr
208b153 ic. in proportion to genera. agrees with late PRODUCTION of those regions, & consequently not many ye
210b161 . III Mr Owen suggested to me, that the <cas> PRODUCTION «of monsters» (which, Hunter says owe their
210b161 s according to species. present an analogy to PRODUCTION of species.-- Animals have no notions of bea
220b197 fail. But this applies only to coition & not PRODUCTION. But who can say, whether offspring does not
236b278 - Lychnis.-- Flax.-- Read Swainson [blank] In PRODUCTION of varieties is it not per saltum.-- Isld bo
282c143 nbergs Paper on Infusoria on the the enormous PRODUCTION-- millions in few days-- one doubt that one
364d104 nally. black. & Yarrell thinks the occasional PRODUCTION of black lambs is owing to old <story> retur
370d116 y are kept in security.-- Hunter doubts about PRODUCTION of Queens.-- Neuters are bred first, «then m
375d134 ing that famine may stop desire.--» in Nature PRODUCTION does not increase, whilst no checks prevail,
397e003 opulation & depopulation, but extermination & PRODUCTION of new forms.-- their number & corelations O
419e085 te quantity of vitality «in the World».-- the PRODUCTION of vitality, as argued by Müller from propag
427e111 me acclimatised by crossing, or by accidental PRODUCTION of seedling with hardier constitution.-- Now
429e113 nfer it in animals.)-- Azara gives account of PRODUCTION of hornless cattle-- ¿& others?-- March 12th
446e162 his case».-- & strong case showing analogy of PRODUCTION by gemmation & by seed-- which Henslow is in
484z014 nstitut-- 1838 p. 75 A detailed comparison of PRODUCTION without Tropics in Northern & Southern Ameri
559m155 " in Jersey.-- very curious facts about early PRODUCTION of foreign seeds.-- many varieties.-- Rev R
614o037 indeed we are step towards some final end.-- PRODUCTION of higher animals-- perhaps, say attribute o
051r094 lieve much is owing to protection of Organic PRODUCTIONS. = Yet everywhere on coast (Il Defonsos «Kel
057r113 ld:--With discussion of camel urge S. Africa PRODUCTIONS.-- I think in Patagonia white beds having pr
233b251 er & small of land or few quadrupeds-- Study PRODUCTIONS. of great Fresh water lakes of North America
245c024 e is it? All the Society isles have the same PRODUCTIONS p 293-- is very strong about this Lesson ins
306c213 yond produced line of Timor 215 degree. What PRODUCTIONS Sandal. Wood Isd.? ought to agree with Java?
407e037 Europe.-- S. America favourable to Tropical PRODUCTIONS. The world formerly much more so. yet climat
407e040 s not so different, with <d> regard to their PRODUCTIONS.-- Hence it is, from the ancient preeminentl
331dIFC nt.? Is the first cross, which makes hybrids. PRODUCTIVE like geese?-- Are the number of kittens betw
503q012 not seed-- «Bruce says does» Royle In Royle's PRODUCTIVE Resources Book no information & Hope about S
507q15v . Polygam. trioecia. (are female flowers ever PRODUCTIVE) Smith says many trees in Tropics are of Chi
637j57v Treatises. are reduced simply statement of PRODUCTIVENESS, & laws of adaptation| p. 234. The non-abs
056r111 of fluid nucleus,--the similarity of Volcanic PRODUCTS «over whole world» argument, as well as separ
056r111 te.> Volcanos blend all substances together; & PRODUCTS being similar over whole world, general circu
073r162 ose which have magnetic properties. Study well PRODUCTS of Solfataras[.] some general laws. associati
327cIBC nt of domesticated animals Fries de plantarum PROESENTUM crypt. transitu et analogia commentatia Libr
099a049 parallel to longer axis. But if great depth NB. PROF <Henslow> Sedgwicks lamination parallel to stra
188b068 river & says so far will I go and no further:-- PROF. Henslow says. that when race once established
424e100 et hepatici,-- in Chara, in Marchantia & Hypnum «PROF:» Don would have known the Composites of Galapa
553m136 part of his most magnificent laws. of which we PROFANE «degnen» in thinking not capable to <do> produ
533m060 made him less repentant.-- In making too much PROFESSION, or rather in only fully expressing momentar
317c250 icia still growing,) passed us. do. p. 243 (, PROFESSOR Smith's Journal) on the heights of St Jago fo
304c206 ed from Man to Monads-- though physiology would PROFIT. if the series were believed to past into each
059r119 par quelque vallées ou par quelque scissures PROFONDES, on les voit se reproduire a des hauteurs com

Page   **(Key Word)**
053r101 ect of expansions acting at great depths (mem: PROFOUND earthquakes), which would cause parallel line
053r101 ndstone & Granite districts to be separated by PROFOUND valley [.] Sydney. -- Lesson Zoologie Grand t
263c076 le of producing more glowing imagining or more PROFOUND reasoning than other-- if this be granted!!)
275c119 had few clear facts, but so bold in many such PROFOUND judgment, that he forseeing. consequence., wa
276c122 e offspring. without organs of generation?! By PROFOUND study of local varieties laws of change-- whe
283c144 ence in two continents our ignorance is indeed PROFOUND & such it appears.-- Is there not some statem
310c222 animals of different types. I will confess my PROFOUND ignorance.-- but seeing such passions acquire
325c267 whether any my views very ancient? Study with PROFOUND care, abortive organs produced in domesticate
347d049 ?» Mayo (.Philosop of Living) quote Whewell as PROFOUND. because he says length of days adapted to du
407e038 e order as that of S. America.-- (Explained by PROFOUND views of Lyell) Now «Equatorial» America from
465t081 (New Red Sandstone) & then go on to shells-- A PROFOUND consideration of method by which races of men
538m079 some respects & with store of accurate & even PROFOUND knowledge or other & unusual line-- both odd
543m100 Peacocks remark about mathematicians not being PROFOUND reasoners.-- all same fact-- for, as Jones ob
547m112 nce between Castle & dream No answer shows our PROFOUND ignorance in so simple case.-- There was memo
580n062 traction because they understand signs.-- very PROFOUND.-- concludes that difference of intellect bet
608o027 dness than disease. This view should teach one PROFOUND humility, one deserves no credit for anything
051r097 o whole E. America. <East> Africa. Australia. PROFOUNDLY deep: a great fault or rather many faults.--
069r150 e) expansion of solid matter by Heat Consider PROFOUNDLY the sandstone of the Portillo line.--connect
119a105 John assume it to be constant.-- It is to be PROFOUNDLY considered, metamorphic rocks at surface. &
120a107 ic action at the bottom of the sea?) All this PROFOUNDLY considered. study Hopkins. theory of dikes m
120a109 shingle in the great Chilian valleys must be PROFOUNDLY considered. if elevation near coast more tha
121a112 of Jordanhill has seen same thing-- Consider PROFOUNDLY How came it. that Glen Roy district could ha
122a113 .-- analogous to my Valparaiso case. Consider PROFOUNDLY all consequences of EXTREME FLUIDITY of eart
159g085 uld have remained, no peat supply.-- Consider PROFOUNDLY Boulder hypothesis Thursday, from Glen Turri
226b220 urope.-- It will be well worth while to study PROFOUNDLY the origin & history of every terrestrial Ma
263c075 long meditation-- His best chance is to have PROFOUNDLY over the enormous difficulty of reproduction
264c081 structure & habits go together. This must be PROFOUNDLY considered.-- Structure may be obliterating,
484z015 annulae-- replaces warblers of Europe-- Study PROFOUNDLY shells of Bahia Blanca & Southern Hemisphere
535m069 know their nature.-- Reviewer considers this PROFOUNDLY true.-- How is it with children.-- Now it is
538m078 n other, absolutely two people. Consider this PROFOUNDLY, may throw light on consciousness, explained
579n058 t sighted people squinny-- when they consider PROFOUNDLY-- this will be curious if it is so.-- frown
468t112 se-- I found all, stamens straightened pollen PROFUSELY shed; lengthened & turned up «more than stame
192b087 ade between pig & tapir, yet from some common PROGENITOR,-- Now if the intermediate ranks had produce
194b097 slightest right to say there never was common PROGENITOR to Mammalia & fish. when there now exist suc
195b099 ible structures. it might have been of use in PROGENITOR-- or it may be of use.-- like Mammae on mens
206b146 of present people, and so on backwards to one PROGENITOR, who might have continued breeding from eter
415e068 herefore Man & monkeys have equal chance that PROGENITOR was bimanous, or quadrumanous.-- What a clan
545m107 movement consequent on some action, which the PROGENITOR did, when excited or disturbed by the same c
609o029 n, are common to other animals & therefore to PROGENITOR far back, (anger <to> at the very beginning,
206b146 say 150 people four hundred years since were PROGENITORS of present people, and so on backwards to on
206b147 ulation of chance of the relationship of the PROGENITORS would have different formula for each lustru
417e078 or someway different from either: & or like PROGENITORS.-- in some families all the children like mo
459tf02 dog had more form of male, & another of both PROGENITORS-- the hinnus, resembles horse in its head ea
181b041 ll be chances against any one «of them» having PROGENY living ten thousand years hence; because at pr
181b041 ively great against, any two of the 12. having PROGENY. after that distant period.-- Hence if this is
191b082 ordance with The Male animal affecting all the PROGENY of female insures often mixing of individuals.
211b163 a real affinity & affinity-- whales & fish.-- PROGENY of Manks cats without tails: some long & some
265c083 «accidental» changes after birth do not effect PROGENY-- many dogs in England must have been lopped o
283c145 ry long beak, with short., let these only have PROGENY with species & there will be two genera.-- let
333d005 black & wrinkled-- fur short. (tail cut off in PROGENY peculiar) limbs very long, eyes very large, ve
393dIBC copulation with other dogs renders subsequent PROGENY faulty. Does male fail in passion.-- Dispositi
417e079 ome races of men. D'Orbigny. affect the common PROGENY more than others.-- does this more refer to le
175b018 d.-- Electricity Each species changes. does it PROGRESS. Man gains ideas. the simplest cannot help.--
175b018 d,; & if we look to first origin there must be PROGRESS. if we suppose monads are «constantly» formed
177b026 ntradiction to constant succession of germs in PROGRESS.-- «no only makes it excessively complicated.
231b241 tra & Java <to> together, by elevations now in PROGRESS, & you will have two. Tapir existing in East
386d167 be fitted to the slow great changes really in PROGRESS.-- Annals of Natural History. 1838. p. 123. E
397e004 ow how slowly & insensibly such changes are in PROGRESS.-- we feel interest in discovering a change o
398e005 t closely, & know the amounts of change now in PROGRESS, will be the last to object to this theory on
409e047 sk the missionaries about Australians yet slow PROGRESS has done so.-- Show a savage a dog, & ask him
548m115 and) a crowd of other trains of thought are in PROGRESS-- In castle of air the trouble «I well recoll
565n008 site muscles used to when angry sneering is in PROGRESS.<--> the hypothesis of opposite muscles will
568n019 ood passages against, opposition of divines to PROGRESS of knowledge. see Lyell on Scrope, Quarterly
640j167 species. for as long as physical change is in PROGRESS or is, present with respect to new arrivers,
641j29r ctive to species, when physical changes are in PROGRESS: (on the same principles that islands are fav
408e043 oduce-- ? When a species becomes rarer, as it PROGRESSES towards extermination. some other species mu
061r127 igger one encroaches on smaller.--change not PROGRESSIF<E>: produced at one blow. if one species alte
305c210 al movements connected with mind There is no PROGRESSION in the developement in instincts in the <cla
405e032 dded with almost recent shells.-- shows that PROGRESSION of change in Mollusca is somewhat similar in
414e060 Have Mammalia?? My theory certainly requires PROGRESSION, otherwise [not located] Are the feet of wat
554m140 eign bodies, for end. most important step in PROGRESSION.-- The male Black Swan is very fierce when f
576n047 n my theory there is no absolute tendency to PROGRESSION, excepting from favourable circumstances!» W
584n071 ion, (knowledge of enemies). use of muscles, PROGRESSION.-- use of senses.-- knowledge of location du
181b044 The above speculations are applicable to non PROGRESSIVE development which certainly is the case at l
182b049 long clotted hair resembling that of goats-- PROGRESSIVE development gives final cause for enormous p
190b078 generation, throughe means of every step of PROGRESSIVE increase of organization being imitated in t
196b108 es are also found in Ireland-- There must be PROGRESSIVE development; for instance none--?, of the <v
391d174 ferences «indifferently of what kind, either PROGRESSIVE improvement or deteriorate» that object fail
414e060 shes & Mollusca (& plants???) been so little PROGRESSIVE «!Agassiz makes it wonderfully changed, sinc
414e060 ly changed, since Cretaceous period, whether PROGRESSIVE I know not.» (& insects.-- Stonesfield????).
415e070 ence to the non-necessity of the «so called» PROGRESSIVE tendency law.-- In animals analogy leads one
445e159 from the necessity of supposing some inward PROGRESSIVE developing power.-- My theory leaves quite u
576n047 hat of the Greeks.-- (which seems opposed to PROGRESSIVE. developement) on account of dark ages.-- «e
580n062 in the young are symptoms of the infinite & PROGRESSIVE nature of intellect indication of better lif
599o005 ment fails-- if they have, then language was PROGRESSIVE.-- We cannot doubt that language is an alter
581n065 e in the structure of language, that it was PROGRESSIVELY formed. (--names like sounds)--. Horne Took
041r066 in was like the rest of these thin veins which PROJECT outwards.-- In Patagonia. the blending of pebb
116a100 Portezuelo, extremity of mountains of Cordova PROJECT on plain, like <re> a reef on a sea beach-- «p
146g007 ll be explained by my idea-- highest part must PROJECT [Blank] {P} Path East End near Holyrood Palace
467t104 «Also in Lathyrus pratensis yellow saw stigma PROJECT» In common Pea saw Humble so press down sheath
151g043 t below houses where rivulet enters two great PROJECTING butresses, upper slope of which corresponds
212b165 time together in tub of water with only nose PROJECTING.-- would pull the garden bell, & then run in
514q21. f the Catasetums) really always hit stigma by PROJECTING pollen-masses?-- = answered = Has Ophrys nec
154g059 87. 72 degree Air 65? 70? {P} Where a buttress PROJECTS from side of hill if line suppose continued a
204b139 Chinese & English Breed. decidedly exceedingly PROLIFIC & hybrid about half way. Eyton says Hybrid ab
426e107 physical impossibility to unite to perfect PROLIFICKNESS.-- (<a series might be obtained>: but the i
508q016 wo vars: of Lion: Annales des Sciences (4) PROLIFICKNESS of female, relation to healthiness? & fathe
443e154 hat it is bad to graft from top shoots.-- If PROLONGATION of life by gemmation <can be> being impossi
036r051 rred in a sandy place.--(the sound was long & PROLONGED). NB, Is it generally known. the acute chirpi
124a120 injection. from probability of fissures being PROLONGED to surface. see p. 181 on do subject do p. 44
147g015 ontal strata I suppose these upper patches if PROLONGED would intersect alley above the 300 ft Alluvi
386d155 ts, does annual give buds.-- life may be thus PROLONGED bud being formed & one part dying for great l
389d176 as might be inferred from annual plant being PROLONGED till it has bred.-- Offspring like both fathe
544m103 s one of reality-- castle in the air, is more PROLONGED than dream. never fatiguing,-- else it is onl
558m152 ame & & some others-- Thus <sudden> «forcible PROLONGED» expulsion of air «dogs snarl much the same w
343d037 one., or apparently so. by the extinction of PROMINENT ones in <latte> one: The latter will take pla
603o013 oke has shown one chief object of language is PROMPTNESS «of consequence» hence languages become corr
640j167 s. it should be remembered that Naturalists are PRONE, fortunately, to take their ideas, which are ar

403e023 species   do not this). because it has been so PRONOUNCED ex cathedrâ. let us look at facts. consideri
572n033 e me a book in French I read the first page & PRONOUNCED each word distinctly. woke instantly but cou
044r074 great: & in the fact of bombs in tufa there is PROOF of such gaz) steam condensed.--Perhaps these mi
055r106 ed like those washed by the waves, a sufficient PROOF, that the sea formed these large cavities", &c
087a012 gs. Pacific great land. -- Will use argument of PROOF of slow corrosion of valley of <Patagonia.> S C
189b069 hey become useful to know what is species.-- In PROOF that structure is not simple adaptation, armadi
191b083 t. & succession the extraordinary South Africa. PROOF of subsidence. & recent elevation: Pray ask Dr.
258c062 t types are present) must follow after there is PROOF of the non creation of animals.-- then argumen
288c160 ating from Falkland Isd regularly to main land, PROOF of. land having been formerly nearer.-- «Selby»
293c174 between   man & animals-- Hydrophobia &c cowpox, PROOF of common origin of Man.-- different contagious
294c177 versation.» the whole object of the Work is its PROOF, «its limiting, the allowing at same time   true
332d001 uggle July 15th. 1838 Finished. October 2d As a PROOF. what <trifling> «unknown» causes act upon peop
356d069 dependently & have used them since) the line of PROOF & reducing fact to law only merit if merit ther
358d051 t located] hen freely.-- here we have beautiful PROOF of the breeding in & in (like «courage in dogs»
371d118 in Loddiges Garden. 1279 varieties of roses!!! PROOF of capability of variation.-- Saw his collectio
581n063 bits that is my new part of the view.-- let the PROOF of heredetariness in habits. be considered. as
602o11b on Belief.-- you belief things you can give no PROOF for, & one often replies "what you say is perfe
088a013 rom the lost & turned about position of strata, PROOFF thickness not very great; where piece turned o
026r022 account of Baron Roussin's voyage.-- In Europe PROOFS of many oscillations of level, which in the na
095a039 the   Bonite. French discovery ship, found clear PROOFS of shells & waterworn rocks «at Cobija.» At Iq
098a044 by   steps in Central France not very conclusive PROOFS, but certainly probable. Bulletin de la Soc. G
098a044 3-34. p. 35.-- Ancient Lake Lemagne in Auvergne PROOFS from Phryganea NB. Sedgwick talks of LAMINATED
106a076 unaccompanied by Volcanos must be sought after PROOFS of sinking.-- No Sweden!! swelling of rock fro
121a113 t none in shingle beds. Lyell on Sweden. p. 12. PROOFS of small rise at Stockholm.-- analogous to my
130a133 would not lie. at bottom.-- Surely we here have PROOFS of hot bottom.-- Study Bishoofs Paper.-- Weels
268c099 similar--birds??-- We must always bear in mind PROOFS of most equable climate both in S. & N. Hemisp
271c166 halo.-- Continents are not stationary «unerring PROOFS not always continents».-- it is a plastic virt
632j53r ey.-- Darwin's Abstract of John Macculloch 1837 PROOFS and Illustrations of the Attributes of God Mac
336d017 ffspring & therefore no generative organ.-- Same PROP. better enunciated.-- "An animal Either parent
199b119 rest?-- ∴ most monstrous form has tendency to PROPAGATE, as well as diseases. In intermarriages; smal
314c237 hich I have given ∴ Those animals, which only PROPAGATE by scission can not alter much.?? Mr Brown sh
360d092 other sex, it would be much more difficult to PROPAGATE--<now> «as» if one bird had very bright   red
366d112 r supposes with Monsters)» if armless cat can PROPAGATE, ie with the chance of two being born at same
389d177 ion is concerned.-- Generation being means to PROPAGATE & perpetuate differences (of body, mind & con
390d178 In fruit trees no doubt there is tendency to PROPAGATE the whole difference of parent, tree, but it
448e165 l of plant is individual.-- Most plants which PROPAGATE rapidly by buds, layers &c & &-- do not seed
450e176 - said to have been imported: shows they will PROPAGATE get dimensions-- do App. p 73 State of Muar i
491q01v 2 {also try Primrose & Cowslip in rich soil & PROPAGATE from their seed 3. To apply pollen of differe
175b020 --for instance secondary terebratula may have PROPAGATED recent terebratula, but Megatherium nothing.
197b111 ated with endless differences:-- does not say PROPAGATED, but must have concluded so= Evidently «or h
279c133 sickly offspring being cut off-- so that not PROPAGATED by nature.-- Whole art of making varieties m
299c194 es,-- latter on banks & in damp parts.-- both PROPAGATED by seeds.-- There are two Dandelions, which
365d107 ed to former changes. than a mere monstrosity PROPAGATED by art. Yarrell told me of a cat & of a dog,
429e114 escape for foreign birds & insects) which are PROPAGATED with very little care.-- & which might sprea
436e137 dvantage it would have better chance of being PROPAGATED & so &c. The greatest difficulty to my theor
442e153 y would require, that <species> «individuals» PROPAGATED by gemmation should be absolutely similar; [
442e153 e to this: but on other hand, fruit trees are PROPAGATED by means, which wild plants never are, namel
443e153 -- ¶Plants circumstanced as the Gorze must be PROPAGATED by its roots: now it is curious Mr K. has ob
443e153 - My theory only requires that organic beings PROPAGATED by gemmation do not now undergo metamorphose
443e154 present   structure they must have <done> been PROPAGATED by sexual commerce «The fact of Corallina &
446e163 nd the vivaparous grasses, which no doubt are PROPAGATED during hundreds of years, without fresh seed
447e164 t important end.-- Festuca vivapara F ovina-- PROPAGATED like oni[on] Poa alpina because vivaparous.
447e165 s by growing on heights.-- yet he has seen it PROPAGATED in a garden, which is case precisely analogo
491q01v iennial-wall-flowers & scarlet Lychnis can be PROPAGATED by cuttings.-- Try.-- Important as discoveri
494q004 's, Bears Badgers,-- How few wild animals are PROPAGATED,, though valuable as show. & curiosities!! W
506q015 s statement of Sweet Williams & Stocks, being PROPAGATED many years, by cuttings.-- (40) Ask Henslow
223b210 cut off limbs & new ones are formed) but yet PROPAGATES varieties according to same law with animals
178b030 Nat History.-- July. 1837. Eyton of Hybrids PROPAGATING freely In Isld neighbouring continent. where
310c224 milk produce a plant capable of growing!! & PROPAGATING itself. In Tropical countries (as St Jago Ca
359d088 probably is explained by the vigour of their PROPAGATING powers. (as if they were a good species, or
389d176 foetus one sex; & therefore both capable of PROPAGATING, but one is rendered abortive as far as part
441e148 & more of the effects of conditions on the «PROPAGATING» constitution. but not structure of the para
447e164 ould prevent it varying. A plant <producing> PROPAGATING itself by buds is in same predicament, as on
063r132 together with hypothetical case of Brazil.-- PROPAGATION. whether ordinary. hermaphrodite. or by cutt
170b001 ce fruit trees, probably polypi, gemmiparous PROPAGATION. bisection of Planariae. &c &c.-- The ordina
173b012 ge country long separated.-- On this idea of PROPAGATION of species we can see why a form peculiar to
174b014 untry, different genera different countries. PROPAGATION explains why modern animals same type as ext
174b015 existed; Australian; Mamm were produced from PROPAGATION from different set, as the rest of the world
183b051 s be created) & live in same country. How is PROPAGATION of wolf & Dog. (because being believed same
183b053 e Earth attentively Cuvier objects to <tran> PROPAGATION of species, by saying, why not have some int
195b099 t other animals live so well.-- This view of PROPAGATION gives. <ro> hiding place for many unintellig
197b112 of lower animals appear.-- yet nothing about PROPAGATION= I see nothing like grandfather of Mammalia
198b114 quences.-- I cannot make out his ideas about PROPAGATION His work. Philosophie Anatomique. 2d Vol wou
199b116 ements.-- How can we understand excepting by PROPAGATION that out of the thousand of new insects all
199b117 s of forms, should they all be classified.-- PROPAGATION explains this.---- Ancient Flora thought to
223b204 adeira-- I believe very curious-- My idea of PROPAGATION almost infers, what we call improvement, --A
234b261 tations;-- this would be strong argument for PROPAGATION of species-- Again is there not similarity e
255c011 on to generation, before great difficulty in PROPAGATION.-- Feathers on, Apterix because we may suppo
281c138 nge ¿¿/Genus only natural from death or slow PROPAGATION, best rule for genera, & so mount upwards.--
350d053 call,, one group genus & other subgenus,,--> PROPAGATION of forms.--just same way as <we> «all men» n
419e085 uction of vitality, as argued by Müller from PROPAGATION of infinite numbers of individuals from one.
446e163 nt remaining constant, without crossing.-- & PROPAGATION by buds does not insure constancy of form.--
632j53r m some more general law.-- [that the laws of PROPAGATION, were created with reference to successive d
632j53r of a thrush were means sufficient to ensure PROPAGATION of Misseltoe?-- do p. 284. it is hard on my
068r142 islocate strata without ejection of the fluid PROPELLING mass. If one inch can be raised then all can
199b118 e form, of a fugitive want into a fundamental PROPENITY, of an accidental habit into an instinct." Ed
035r048 re grains of sand, in this view sink into their PROPER insignificance; as fractures, consequent on gr
132a138 icity beneath lakes)?» Suppose ocean represents PROPER <state> temperature of earth. at the freezing
263c078 d generations of species to produce contingents PROPER.-- Present monkeys might not,-- but probably w
296c184 agos?) same condition. Keeling Isd «shows where PROPER dampness seeds can arrive quick enough» Vegeta
317c250 & St Jago upper region, & some to Cape.-- some PROPER well-worth studying, with respect to forms.--
319c257 between   those of Van Diemen's land & Australia PROPER.-- Irish Elk case of fossil geographical range
376d135 cause of all this wedgings, must be to sort out PROPER structure & adapt it to change.-- to do that f
383d159 s, & ovaria when they first appear occupy their PROPER positions,-- this would be argument for develo
398e005 immense.   It is curious that geology. by giving PROPER ideas of these subjects. should be absolutely
533m059 that is the ready movement & co-relation of the PROPER muscles. may be illustrated by the extreme dif
608o026 s & yet does not.-- because motive power not in PROPER state.-- When the admonition succeeds who does
608o026 we view wickedness.-- it would however be more PROPER to pity than to hate & be †A man may put himse
625o50v State broadly in child or animal it is equally PROPER to obey anger as benevolence (but not cool mal
136a147 treat of the age of the Pampas Deposit, I may PROPERLY remark on the superiority of Lyell's classifi
291c166 in classification.-- Thought (or desires more PROPERLY) being heredetary<)>.-- it is difficult to im
370d116 day-- & from case of wasps, is supposed cells PROPERLY are made for larvae.-- CD[(p. 451.)-- Wasps b
391d174 hy should it demand some further change?-- Man PROPERLY is hermaphrodite (hence monstrosities tend th
526m027 what they teach by the same means & therefore PROPERLY no free will.-- we may easily fancy there is,
581n063 hat kind of memory. which makes you do a thing PROPERLY, even when you cannot remember it. as my fath
073r162 he metals should be those which have magnetic PROPERTIES. Study well products of Solfataras[.] some g
291c166 retion of brain, more wonderful than gravity a PROPERTY of matter? It is our arrogance, it our admira

478z004 iven paper to Royal Soc on glow worm. luminous PROPERTY-- Curious arrangement of animals in rays Par
550m126 sop. of Living p. 293. Animals "have notion of PROPERTY" -- their own property. (--regarding food & i
550m126 Animals "have notion of property" -- their own PROPERTY. (--regarding food & in birds of place for ne
550m126 for nest.)-- with dogs "have notion of masters PROPERTY"-- is not this rather more friendship.-- Scot
408e044 l» modern & wholly volcanic-- Azores might be PROPHECIED to have this character.-- worth going there
275c119 nce., was endowed with what may be called the PROPHETIC spirit in science--. the highest endowment of
059r121 f so Chalk would harden.--Climate.!? or small PROPORTION of Alum: matter.--all pale cream colour.-- T
099a051 ttract each other in some increasing ratio in PROPORTION to proximity would they not unite in B. K.>
103a065 e between dikes & «axis of» mountain-chain in PROPORTION to weight of super [...] mass.-- Absence of
118a103 Cordillera, it is at once evident only small PROPORTION of dikes have reached the surface Arguments
182b046 l not occur in plants which are in far larger PROPORTION terrestrial,-- if in any in the Cryptogamic
194b093 d together:-- Man has no limits to desire, in PROPORTION instinct more. reason less. so will aversion
205b145 large a Native breed of animals. for in <the> PROPORTION to their increase of size they become worse
208b153 most curious law of species few in Arctic. in PROPORTION to genera. agrees with late production of th
209b156 ogous to nearest continent: poorness in exact PROPORTION to distance (?). & similarity of type (?).--
210b158 are in same relation. We find species few in PROPORTION to difficulty of transport. For instance the
210b158 nstance the temperate parts of Teneriffe, the PROPORTION of genera I: I. I can understand in «one» sm
226b222 an D'Acunha, resolves itself into question of PROPORTION of species to genus If on one isld several s
232b245 derers--» There ought to be fewer species in PROPORTION to genera than in present seas, «All» The <o
233b251 Apteryx Penguin-- Logger-headed Duck-- Large PROPORTION of Water & small of land or few quadrupeds--
246c026 but beak rather sharper., & rather longer in PROPORTION, colour slightly different. Who can say whet
335d015 o either parent), then according law, that in PROPORTION as things are long in blood so will they rem
337d023 asses» differ in different countries in exact PROPORTION to the time they have been separated; togeth
373d133 yells Elements. p. 290. Dr. Beck on numerical PROPORTION in shells in Arctic Ocean. p. 350 Grallae in
379d152 given, & I think facts there mentioned about PROPORTION of sexes, at birth & causes. If an animal br
413e058 with physical change» (3) Great fertility in PROPORTION to support of parents December 2d Lyell tell
431e123 breeding of Animals, p. 8. size of foetus in PROPORTION to male parent p. 8. his whole doctrine of t
436e134 that naked cuttle fish now bear a very large PROPORTION to other mollusca in cold parts of sea, like
495q005 llen of other, count seeds, & see how great a PROPORTION springs up true.-- This in fact always takes
499q009 who raise Hop-seed-- may know something about PROPORTION of plants necessary &c &c (a) Mercurialis--
508q016 s.= 13 Where are there any medical Statisics, PROPORTION of diseases (heredetary?) in diff. countries
539m082 , so may animal obtain it far more easily, in PROPORTION to variableness or power of intellect.-- Som
550m124 nt.-- Simple happiness «as of child» is large PROPORTION of pleasant to unpleasant mental sensations
550m125 l ones, <which> «& these» must occupy greater PROPORTION of <each» «every» man's time.-- Begin discus
589n091 ck. Vol II p. 319.-- Habits more prevalent in PROPORTION to intelligence less.-- p. 325 «to 29».-- Ha
030r036 les Examine chart of Patagonian coast to see PROPORTIONAL cliff & low or sloping land What are the "p
045r076 regular system can be called accidental; the PROPORTIONAL force of crust of globe & injecting matter
046r079 under <[...]> less pressure might have its «PROPORTIONAL» particles altered.-- With respect to Volca
374d134 as assumed that increase of animals exactly PROPORTIONA[L] to the number that can live.--» We ought t
430e120 & elongata thougt considerably different, in PROPORTIONAL dimensions, must <almost> be considered mer
496q006 Horse Urine &c &c on associated plants. when PROPORTIONAL number appear equal-- & see whether proport
047r083 in swell) the undertow & overfall must vary PROPORTIONALLY Partial shrinking after elevation in perfe
055r108 ousand feet in height, of the solid lavas.--PROPORTIONALLY high to age. (we do not wonder to see tert
156g069 he old ravine, where water entered are not PROPORTIONATELY large to those now formed in same spot by
212b167 Phillips Geology «p 81» in Lardens Encyclop. PROPORTIONS between fossils & recent shells between herb
496q006 ortional number appear equal-- & see whether PROPORTIONS will vary, which will show that such proport
496q006 ortions will vary, which will show that such PROPORTIONS not effect of Chance Maer.= (12) Take Bag of
614c037 of carbon. hydrogen <&c> in certain definite PROPORTIONS, (different from what takes place out of bod
035r046 r skepticism, as to the general truth of the PROPOSITION."-- If such can happen in troubled England;
415e069 came to be so.-- I speak only of the former PROPOSITION.-- as in «races of» Dogs, so in species, & i
440e145 - turnspit & two other kinds It seems absurd PROPOSITION, that every «budding» tree, & every buzzing
588n016 quences on habitual or instinctive assent to PROPOSITIONS, which are the result of our senses, or our
616o039 ttraction does to matter, it might with equal PROPRIETY be said that the living brain perceived, thou
616o039 irly be laid with those who would support the PROPRIETY of the expression. They would do well to ask
039c061 quimbo. the analogy is now perfect <the grand PROPULSION of fluid rock. which elevates a continent> W
569n021 pic poem, opposite ends of series or harmonious PROSE.-- Lutké Voyage in Carolinas Vol II p. 132. off
546m108 . I cannot help. thinking horses admire a wide PROSPECT.-- The very superiority of man perhaps depend
355d066 ecies of same genus." Law of monstrosity not PROSPECTIVE, but retrospective as showing what organs ar
325c266 ct of their views Mackintosh Ethical Philos: PROSTITUTION of Paris. with respect to licentiousness, d
572n032 anner, whether by bowing the body, kneeling, PROSTRATION «uncovering body» &c &c is matter of custom.
190b077 few genera or families.-- (long separated.-- PROTEACEAE & other forms (?) being common to Southern h
037r053 -It may be a question. whether organic remains PROTECT a rock, or that the rock not weathering allows
051r093 co (great swell & turbid water) organic bodies PROTECT like peat reef of sandstone.--Corals, & Corall
051r094 ow Kelp sends forth branching roots which must PROTECT surface; On «hard» exposed rocks near Bahia, w
314c237 o long organ, moved on being touched, so as to PROTECT itself, one segment of the corolla being (prob
467t103 lently: in Beans the wings seem beautifully to PROTECT sheath {a} In all these nectar seems to be at
069r151 of Portillo Where gone to? Intermediate space PROTECTED.-- Oh the vast power of the ocean! Make a gra
314c238 ciousness of the plant that this part must be PROTECTED however it may be effected.-- Prodromus Flora
466t094 ey must disturb all anthers, wh otherwise lie PROTECTED by the hairy black lip of lower division of n
619o042 > «some force» give pain: for instance either PROTECTING sheep or hurting them.-- Therefore in man we
051r094 ving seen Pernambuco believe much is owing to PROTECTION of Organic productions. = Yet everywhere on
435e133 by Dr. Hodgkins» p. 54. The axolotl, siren, & PROTEUS, affinity to tadpoles. p. 210. Shows. that the
435e134 rm; but that tadpole increased in size now the PROTEUS anguiformis. he remarks lives in dark caverns
458t001 ders-- Every order (except whales) have great PROTOTYPE!!.-- Copied Vol II p. 502. «Bengal Journal» T
467t103 ripe & pollen abundant filaments & stamens all PROTRUDE «there is a brush at end of stigma, which for
594n113 ld same as anger, lips not compressed sullen, PROTRUDED. determined to do nothing. & so manifesting s
467t104 f Pea's wings, stigma & mass of yellow pollen PROTRUDES at sheath.-- At last I saw Bee collecting pol
551m129 hat when it looked at the glass) when pouting PROTRUDES its lips into point-- man, though he does not
555m143 slow cautious, angry cross look, followed by PROTRUSION of lips, in which respect resembles some of
556m146 alysis of expression of desire-- is there not PROTRUSION of chin, like bulls & horses.-- 1838 good in
565n009 ression of affection is accompanied by slight PROTRUSION of lips, as if going to say "my dear," just
128a129 his will be only a modifying cause. {P} land PROTUBERANT water to counterbalance How strongly the Gle
264c079 improvable» & then let him dare to boast of his PROUD preeminence.-- «not understanding language of F
055r108 rce.--the exact yearly rise of the great rivers PROVE better than any meterological table the precise
088a013 terrace like structure-- Intersection of veins PROVE, that there are at least several attempts at el
093a033 ber of minute turbos in red earth with volutas. PROVE regular mud bank at Bahia Blanca. <fl> Flustra
098a046 . Vol III. p I . p. 86. et p 95.-- It is easy to PROVE. (pyrites, agates, calcareous balls) that concr
105a071 -- If blocks above their parent rocks. would be PROVE of subsidence.-- removal downwards by successiv
113a091 successive transmission of temperature clearly PROVE possibility of metamorphic theory On the idea o
127a127 eland dikes in mountains. «(not on continents)» PROVE elevation.-- great mountain chains. may be effe
175b016 mal has tendency to change.-- this difficult to PROVE cats &c from Egypt no answer because time short
189b073 time, though produced either sooner or later.-- PROVE animal like plants:-- trace gradation between a
198b115 be given.-- It is a point of great interest to PROVE animals not adopted to each country.-- Provisio
294c176 ge in animal be permanent.-- It will be easy to PROVE persistent Varieties in wild animals-- but how
350d055 ed & blind: -- Macleay observed all these facts PROVE that perfection of organs have nothing to do wi
415e069 rtainly not made into man.-- It is one thing to PROVE that a thing has been so, & another to show how
523m017 connection with housemaid two years before, to PROVE she was not insane, answered she had known it a
553m136 urrounds us. Moreover «it would be difficult to PROVE that»" this innate idea of God in civilized nati
579n058 me-facedness for shyness, having been invented, PROVE of the difference, which my theory believes in.
057r113 ider well age of Bones. = slowness of elevation PROVED at St Julian. = do not these bones differ as m
057r114 tagonia white beds having proceeded from gravel PROVED.-- curious similarity of rocks of very diff. a
058r117 e great movements (not mere patches as in Italy PROVED by Coral hypoth. agree with great continents).
124a117 ut great depths of frozen soil. p. 211 Consider PROVED that Siberia must have been in same condition
174b014 nimals same type as extinct which is law almost PROVED.-- We can see «why» structure is common in cer
200b122 of proving to them it is not; one when they can PROVED descendant, which of course most rare, or when
281c137 be remembered, therefore do not consider it as PROVED that they are varieties, (though that would be
291c167 forms & succession of others, (which is almost PROVED. Elephant has left no descendant in Europe<)>
389d176 nomy So with inter-breeding as told by Willis Q PROVED facts relating to Generation One copulation ma

409e047 logous case to man, exceeding monkeys;-- Having PROVED mens & brutes bodies on one type: almost super
445a159 ilosophie Zoologie. says it is not sufficiently PROVED that any shell fish is really hermaphrodite. &
539m084 ow is this? dealt with p. 241 Origin of man now PROVED.-- Metaphysic must flourish.-- He who understa
564n003 ations having different moral sense, if it were PROVED instead of militating against the existence of
578n053 person.» A person who blushes in the dark is PROVERBIALLY a most modest person one carries on, by ass
584n072 n carried in confinement,-- carrier pidgeons PROVERBIALLY carried to long distance in dark "it is ins
024r014 shoaling of the water to more than 100 fathoms. PROVES the existence of some moving <point> power ¿ S
055r108 nt «daily» yearly brought down by every torrent PROVES the decay atmospheric of the most solid rocks.
218b189 possibility of Caffers & Hottentots coexisting. PROVES this-- but when Man makes variety these are vi
423e096 reptiles.?-- supposing such to be the case, it PROVES the law of developement in partial classes is
439e143 ing the extreme facility of crossing, in plants PROVES how much depends on instincts in animals.-- ye
571n031 ings in gentle language, & vice versa.-- almost PROVES that at earliest times there must have been in
371d118 body, like its parent, than that it should be PROVIDED with many contingencies how to act-- so with
461t025 r have young which undergo metamorphosis & are PROVIDED with fins, & hence do not require sac.-- but
573n036 we wish to be created at once by special act, PROVIDED with its instincts its place in nature. its r
637j57v had < <not> some» some insects <not> not been PROVIDED. «with proboscis» «as bee & butterfly» inconv
027r027 th frequency of shells in flints in Chalk New PROVIDENCE more hilly than others of the Bahama consist
638j58v that this is an infringement on the wisdom or PROVIDENCE. when whole rocks nay very mountains are for
202b131 ot being so closely adapted. Near the Caspian «PROVINCE of Ghilan» wooded district cattle with humps
400e013 Hares & raccoons.-- «S. American form.--» off PROVINCE of Guadalaxura-- October 11th.-- Uncle John--
203b135 - This supposes world divided into Zoological PROVINCES-- united-- & now divided again-- Weakest part
310c223 nstitution? Dr Royle seems to think Botanical PROVINCES will turn out not nearly so confined as now t
339d025 eon!-- Divides animals «world into Zoological PROVINCES» according to varieties of Man.? «In Australi
145g002 coloured like Magellanic fox.. an instance of PROVINCIAL breeds. [3] Veins of Segregation in Salisbur
312c231 le is to cross between a good bull & <be> the PROVINCIAL breed, & that first offspring thus produced
345d044 .. peculiar hair & appearance-- good case of PROVINCIAL Breed-- HighLand Sheep jet black legs, & fac
200b122 r Falkland rabbit.-- There is only two ways of PROVING to them it is not; one when they can proved de
198b115 prove animals not adopted to each country.-- PROVISION for transportal otherwise not so numerous: qu
575n043 rrying «like other hybrids» with <the> it the PROVISION for death.-- can we deny that brain would be
632j53r e author if I use these facts p. 280. adduces PROVISION of seeds for transportation through the air.-
632j53r ld not be these plants, if there was not some PROVISION for transportation:-- But I do not want to de
615o36v tally different habits from Europaean. begin to PROWL about in the evening «seldom leave their perch
540m088 -- Consider this.-- "The fledge-dove knows the PROWLERS of the air" &c &c &c so is conscience &c &c C
099a051 her in some increasing ratio in proportion to PROXIMITY would they not unite in B. K.> {P} on the dia
265c084 ect to question what is adaptation.-- Ermine, PTARMIGAN hare becoming white in winter of Arctic count
356d072 ell told me he had just heard of Black game & PTARMIGAN having crossed in wild state-- & the English
363d101 of rabbits. & long fur.-- feathers on legs of PTARMIGAN & in Bantam.--]CD CD[In the Pidgeons, trace
364d105 Black Cock.-- The latter has crossed with the PTARMIGAN. subalpina in wild state.-- Neilson has given
365d106 .-- In the hybrid grouse between Black Cock & PTARMIGAN (probably subalpina.) former has blue breast,
370d116 ees in same work. (it is said that some kind lay <PU> up honey even for single rainy day-- & from cas
081r181 Mawes travels down the Brazil.-- Did Melaspena PUBLISH his travels? Bellinghausen in 1819 Kotzebue 18
112a090 «See Athenaeum. 1838. p 274. probably will be PUBLISHED in the Geograph. Journal.--» A meeting of the
220b199 rom warmer countries. When will this paper be PUBLISHED it will be curious.-- Some general statements
315c241 not understand any new ones.-- Menoir will be PUBLISHED St. Petersburgh Academy Imperial-- Paper read
320c276 e Zoological Transactions. <done> up to parts PUBLISHED March 1838 Whole of Geographical Journal Asia
320c275 on Cattle not abstracted Scientific Memoirs. PUBLISHED by Taylor Magazine of. Zoology & Botany & Con
324c268 m Statistical Society-- where M. Quetelet has PUBLISHED his laws about sexes relative to age of Marri
326c266 mal, intelligence.-- very good. Endlicher has PUBLISHED in first volume of Annales of Vienna. sketch
326c265 reference to Kolkreuter's Papers Wiegman has PUBLISHED German Pamphlet on crossing Oats, &c «Horticu
418e085 h.-- L'Institut 1838. p. 412. M. Eichwald has PUBLISHED Fauna of Caspian.-- fishes fresh water kinds.
432e126 nly order of succession.-- Splendid Pamplet. (PUBLISHED in Philosop. Journal <Mar> April 1st 1839) to
450e174 cation.-- "Notices of the Indian Archipelago" PUBLISHED at Singapore in 1837. by Mr. J. H. Moor-- --
507q15v in Tropics are of this class.-- (48) Where «PUBLISHED» list of spontaneous Hybrids-- to see whether
510q017 nes separated by non-inhabited spaces: has he PUBLISHED? does he understand English.-- Miguel to coll
559m155 hical distribution of British Plants A Volume PUBLISHED by Colonel in army on "Wheat." in Jersey.-- v
540m085 -- when it is recollected that smell of ones own PUD. not disagree.-- Ourang outang at Zoolog Gardens
540m085 agree.-- Ourang outang at Zoolog Gardens touched PUD. of young male & smelt its fingers. Seeing a dog
464tf6r , however, believes in the existence of Molina's PUDU or goat There is ibex of Alp Pyrenees &c-- (see
265c086 ery different habits-- Thus bill & nostril of PUFFINURIA I think we may clearly attribute to heredeta
265c087 ld be shown to be of some use. If we only had PUFFINURIA Garrottii &c in other species-- as we have on
484z015 Guillemot of the northern Hemisphere, & the PUFFINURIA, the Awks.-- What structure do the auks bear
484z015 structure do the auks bear traces of.-- like PUFFINURIA does of Petrel?-- Study Birds of Europe for
258c061 nce deer victorious deer, hence males armed & PUGNACIOUS (all order; cocks all warlike)» «thiis wars
212b165 tub of water with only nose projecting.-- would PULL the garden bell, & then run into Kennel to watc
376d137 knew women.-- he has repeatedly seen them try to PULL up petticoats, & if woman not afraid clasp the
376d137 een do this.-- These Monkeys had no curiosity to PULL up trousers of men. Evidently knew <men> women,
376d138 e sexes of every Has repeatedly seen one he kept PULL up feathers of tail of Hen; which lived with it
596n184 pithecus make labial st st. S. American monkeys. PULL back skin from head very little Does blood go i
233b250 og jumped into tub leaving only nose above it-- PULLED bell.-- -- It was most curious to observe, tha
436e135 imes-- Apteryx has a most perfect Struthio head PULLED out. yet feathers retain character? If separat
547m113 rception by senses fail first, as whether I had PULLED the bell??)-- It may be deception to say the m
470t178 n Phlox Down, 1854, Sept.) In Spanish Broom by PULLING back Wings, pollen is ejected with violence in
617o40v very particle of matter were an animated being PULLING every other particle by invisible strings & as
596n184 not gradation towards man.» Macacus especially PULLS back skin of whole forehead & 2 ears.-- emotion
453e182 s same.-- do p. 393. <">The wild, small fowls at PULO Condore "crow like ours, but much more small &
493q004 r offspring infertile.-- (4) Does the number of PULSE, Respiration, period of gestation differ in dif
494q004) Does my Father know any case of quick or slow PULSE being heredetary. (6) In the last 1000 years ho
524m022 ion took her <left» hand to pretend to feel her PULSE.-- What fails first?-- How is this?-- Does memo
530m044 it My father has somewhere heard (Hunter?) that PULSE of new born babies of labouring classes are slo
551m129 of W. Scott Vol VII p. 35 "as ideas come & the PULSE rises, or as they flag & something like a snow-
581n066 elt?-- «without «slight» flush, acceleration of PULSE. or rigidity of muscles.-- man cannot be said t
377d140 olutions of our system in the Heavens.-- Is not PUMA, same colour as Lion. because inhabitant of pla
558m152 igious force.-- making growling, guggling noise. PUMA did same & & some others-- Thus <sudden> «forci
031r038 ralia: cases in Europe.-- Auvergne. very little PUMICE, though Trachyte. same fact in Galapagos. Daub
032r042 of Chiloe, after having examined the changes of PUMICE at Ascension In Calc: sandstone at Ascension,
033r043 71. Vol I. Humboldt There is long discussion on PUMICE «& Obsidian:» in the I Vol. Humb: There is rat
042r067 line of cliff marking a pause» When mentioning PUMICE of Bahia Blanca, mention black scoriaceous roc
095a038 with imperfect crater <--> near summit,-- much PUMICE --. appears to be outside of the Cordillera--
100a054 granite!!! Look at gneiss of Rio Concretions in PUMICE bed at Ascension instance of hollow concretion
103a063 ngs» List of collections in Geological Society. PUMICE at South Shetland. Geological Society-- Dikes
050r090 st in Peru.-- Ascension At Ischia there is a PUMICEOUS conglomerate with small & large fragments, na
567n015 - when extreme sensation of heat shows blood is PUMPED over whole body.-- is it connected with surpri
552m131 man. involuntarily feels angry, when brain is PUMPING force to legs & body, & especially, when to wh
477z002 26 degree 30 minute. Lat -- -- do. p. 207. La PUNAISE was not known amongst Indian. introduced in Pa
080r178 tribe, with point affin of yew & intermediate PUNCTURE one animal with recent dead body of other. &
629o55v usly say so, & produce shame & remorse-- [Thus PUNGENCY of one's feelings for indecency-- preposterou
608o027 im' 3) disgusted. with them. Yet it is right to PUNISH criminals; but solely to deter others.-- It is
538m080 e's father & self one person-- & thus eternal PUNISHMENT explained. These facts showing what a train
134a143 tic Society Vol I. {T} p. 145. on salt mines of PUNJAB p. 149. on the "salt mines» «saline deposits»
543m099 nature of their associations (it is not so in PUNNING) are people of very limited intellects, & in t
179b032 pulation producing its effect; as when bitches PUPPIES are less purely bred owing to <first> having o
217b184 have lined bitch directly one after the other, PUPPIES differ, & like both parents.-- Fox told me of
312c228 ese hairless dog & common spaniel crossed.-- 3 PUPPIES PERFECTLY like chinese & 3 perfectly like span
312c233 eat, take dog. but do not become impregnated & PUPPIES delicate-- they cross sister & brother of same
360d089 ween terrier & hairless dogs of Africa,-- some PUPPIES hairless. some in patches, & some hairy-- the
388d172 appears, to that of «2» dog begetting different PUPPIES out of same mother.-- The view that man & «or
404e026 spring:｜ Australian dogs have mottled coloured PUPPIES case of this.-- tendency in «manner of» life t
404e026 e mottled + heredetary tendency determines the PUPPIES to be so.-- [not located] Did man spread over

```
424e103 of  wolf in Zoolog. Gardens, which brought its PUPPIES to be fondled.-- and we see in the  Australian
466t095 can  canter-- & DOGS trained to pursuit having PUPPIES with the same powers instinctive & doubtless n
542m094 &  meditative tranquility. «whine of children. PUPPIES do so dogs nearly silent, so with men.-- How i
549m118 his  children, any more than to dog losing his PUPPIES-- This looks like free will.-- V. last page. A
619o043 ppetite, or love of exercise & its love of its PUPPIES: the latter generally soon conquers, & the dog
639j28v s in the several genera of Grallae Suppose six PUPPIES are born «& it so chances, that one out of eve
640j167 ther has been tried.-- With respect to the six PUPPIES, if a hare was introduced, or <a spe> became m
259c066 sm) & shows my «view of» generation right?-- If PUPPY born with thick coat monstrosity, if brought in
505q014 orne has large Bees July/42/ Mark has six day's PUPPY of small true Bull-Dog-- length from nose  over
525m026 an hardly be doubted, when seeing Nina with her PUPPY.-- The common remark that fat men are goodnatur
536m072 ul forms.-- With respect to free will, seeing a PUPPY playing cannot doubt that they have free  will,
6200046 (& the «old» dog really feels ashamed?) not so PUPPY, we <do> try to teach him & strengthen his inst
401e014 number  of plants-- it is scarcely possible to PURCHASE seeds of any cabbage, where a great many will
568n019 rterly Review.  1827? In Water Scotts life.. Tom PURDIE, (beginning of Vol V) «finally» says "he  knew
059r120 reous rocks which harden by themselves cannot be PURE. for if so Chalk would harden.--Climate.!? or s
076r169 r procured from martial pyrites; great blocks of PURE silver not common in <S.> America: In all clima
076r169 ]> sometimes concentrated: wonderful quantity of PURE silver in S. America. Geology of Guanuaxuato.--
147g015 &  flat topped hill/ do alluvium. NB In one part PURE sand in current cleavage-- in other irregular h
334d009 they  this first cross were equally fertile with PURE bred animals.-- Mem. number of Mules.-- «He rec
338d025 range  of Hyaena? Hippotamus.? Indio-African, or PURE Africa?-- ‖Fossil Elephant of Africa Most impor
342d032 s-- they are produced in full equal Numbers with PURE bred (just like common mules) & lay many eggs b
344d042 though she resembled a harrier & her husband was PURE Harrier.-- «The peculiarities of our breeds mus
345d042 ds were uncertain. Mr Drinkwater thought that a <PURE blooded> «"first blood"» animal must have  gone
345d043 hite headed, but this was bred out & now all are PURE red, yet calf every now & then born with  white
352d058 anization. such law would explain every thing.-- PURE HYPOTHESIS be careful.-- Argument for circulari
379d152 rell says it is well known that in breeding very PURE South Down that the ewe must never be put to an
430e118 ced, but by training, & crossing & keeping breed PURE.-- «& so in plants effectually the offspring ar
460t013 r with common goose produce full as many eggs as PURE bred common.-- the half of the cross, as above,
472s01v came up hybridised. It is possible to raise them PURE for Miss Bent three years since gave her some S
583n069 p. 18. Animals possess strong imitative faculty- PURE instinct is not imitative: imitations seems inv
589n092 is  some great puzzle in what Sir. J. M. says of PURE reason not leading to action & yet our emotions
179b032 ng its effect; as when bitches puppies are less PURELY bred owing to <first> having once borne Mongre
408e044 os, must be owing to these islands, having been PURELY result of elevation,-- «all» modern & wholly v
567n012 laced to the will of God. Zoology itself is now PURELY theological.-- Origin of cause & effect  being
128a128 II.  p 89. at Madras. surrounded by salt water. PUREST fresh water must be sought for below the sea m
529m042 s old eating white lead. who was most violently PURGED «believe worms were passed off.» & vomited, bu
048r086 rated Porphyritic rock. s (mem white tufas with PURPLE Claystones of P. Desire). = Where talking of s
079r177 & ses phenomenes."-- Ulloa's Voyage, Shell fish PURPLE die, marevellous statements on, Vol I, P. 168.
365d106 former  has blue breast, latter reddish, hybrid PURPLE-- be careful, See to hybrids between  Pheasant
383d160 ic hue of silver pheasant. yet why green? & not PURPLE?-- legs pale coloured.-- In the back feathers,
466t094 coloured-- & stamens bend up a little In a wild PURPLE Geranium, I see Bees visit always base {a}  of
472s02r ays ago-- This Bee flew from yellow to yellow & PURPLE heartease without doubt.-- Bee, not large, ver
479z005 140. Flèche of Quoy et Gaimard Ulloa shell fish PURPLE die Marvellous stories Ulloa's Voyage Vol I, p
081r BC . Darwin R. N. Range of Sharks Nothing For any PURPOSE A. Geology Note on Woolwich Nothing on any Sub
225b218 an. as likely as fossils in old rocks for same PURPOSE.!! Can the wishing of the Parent produce any c
277c128 ace-- such a classification would answer every PURPOSE, & would present many ideas of causes of chang
384d162 two substances, although organs for the double PURPOSE are not distinguished. --yet may be presumed f
461t037 ated, & hence, the application of structure to PURPOSE after purpose would tend to render complex the
461t037 the  application of structure to purpose after PURPOSE would tend to render complex the series.--  Ch
478z004 lates <.> freely with tame: comes to houses on PURPOSE Mr J. Murray has given paper to Royal Soc on g
523m017 known it at time & had bought arsenic for that PURPOSE.-- this found to be true.-- Her Husband  never
025r018 have drawn all my illustrations from America, PURPOSELY to show what facts can be supported from that
375d135 earth then in Spring, like food used for other PURPOSES as wheat for making brandy.--» take Europe on
568m018 f sensations would first take place, whether to PURSUE immediate inclination or some future pleasure.
370d117 must spices &c &c The line of argument «often» PURSUED throughout my theory is to establish a point a
593n105 ing disagreeable & anything disagreeable being PURSUED.-- A dog turning round & round is some old ins
634j54v ied respecting an original thought, or design, PURSUED to its utmost exhaustion, & till it must be ab
592n105 o is not odd.; for instance wild cattle & deer PURSUING a wounded one.-- porpoises a ditto-- it is pr
260c068 scribes many kinds of bird uniting together in PURSUIT of Blue-Jay, when <one> birds hears <dis> crys
466t095 ing colts which can canter-- & DOGS trained to PURSUIT having PUPPIES with the same powers instinctiv
276c123 dd chief good of individual scientific men is to PUSH their science a few years in advance only of th
613o036 on told me that his retriever Sailor ho has seen PUSH a hare through the bar of a gate before him,  &
073r161 s have layers been accumulated, round knobs, or PUSHED where soft, or redissolved soft.--/ is there a
551m129 lips into point-- man, though he does not pout. PUSHES out both lips in contempt <&> disgust & defian
060r124 & to grow upwards; for the land is constantly PUSHING the sea (which of course must retain same leve
127a125 subterranean isothermal line must be creeping «PUSHING» up to «the» line of ice.-- Hence further N. w
343d038 ground. Yet, as the tribes of the interior are PUSHING into each other from slave trade, & colonizati
406e036 n to men & animals cow pox.-- case in Spain of PUSTULAR disease following handling sheep-- all cases:
195b103 close as Patagonian <Chat> & Galapagos orpheus.= PUT this strong so many thousand miles distant.-- Ab
205b145 tly predominates." p. 20. do "If hornless ram be PUT to horned ewe almost all the lambs will be hornl
218b189 n all different forms; therefore when from being PUT on island. & fresh species made. parents do  not
234b255 ening towards pig.-- latch on other side.-- Pigs PUT legs over, & then with snout lift up latch & bac
277c127 country,  without range, or habits ascertained-- PUT them as (a). (b) until data be given.-- This wil
285c152 t from other & if a first & last individual were PUT together, they would not according to all analog
286c156 necessary  from manner Fleming treats subject to PUT in alternative of Man created by distinct miracl
287c157 ophical Magazine & Annals. 1830 (?)." if she has PUT man on the throne (of reason), she has also plac
287c157 se shows hiatus-- but not saltus-- when Linnaeus PUT whale between cow & hawk a frolicsome saltus. «p
293c172 t an anomaly in structure of brain not probable) PUT note. Sir W. Scott has written about it]CD If we
293c172 savage takes. & was given a Great coat & this he PUT on & we afterwards could understand «(language b
333d008 perfectly with Yarrell & no leading question was PUT.-- Fox thinks half Lion & Tigers are exactly int
368d114 fspring, Q (how is this with those females which PUT on (like some waders) the bright plumage.-- «thi
376d136 three  days learn its comfort & though could not PUT it on, yet threw it over it, & made it meet in f
378d148 is in Pidgeons & fowls.--??? Wate[r]ton «p. 197» PUT 12 wild duck's eggs under common ducks, the youn
379d152 very  pure South Down that the ewe must never be PUT to any other breed else all the lambs will deter
380d155 me of the tigers.-- cat, though caterwhalling. & PUT into female, when muzzled, he is disabled.--  so
384d161 & that strength"? In speaking of generation alway PUT female first Will not even a fruit tree or  rose
433e127 position of the beds-- The argument must be thus PUT, shall we give up whole system, of transmut., or
435e133 t stigma swells, when pollen even most remote is PUT to it.-- April 6th "Dr. Edwards on the Influence
466t094 hese <do> not bend up-- In Lark-spur, if Bees PUT proboscis within nectary «they do» they must dis
496q05a animal  reproductive system.-- -- cover flower-- PUT artificial flowers-- also do with honey-- What i
502q11v acclimatisation.-- July <1842> When nettle leaf. PUT into spirits, poison-drop exudes-- does not elm.
502q11v not  possible to see orifice of poison-tube-- so PUT carmine in spirits & then experimentise: for gra
505q014 rtificially.)-- Asclepias-- Flowers not seeding-- PUT pot of boiled earth on top of House =Aristolochi
505q014 Abortive Thyme seeds weather wet--? Linum flavum PUT in Spirits which plant seeds? (9) Melons fruit i
531m050 money  is placed.-- (How often one forgets where PUT one key. where all keys are placed) Memory canno
532m057 se may be imitation of death, which many animals PUT on.-- The flush which accompanies passion &  not
536m073 fixes on heredetary disposition & instincts--.-- PUT it so.-- Probably some error in argument, should
542m095 ult of straining vision, as savages without hats PUT their hands, & as attention would amongst low
546m111 without any assignable cause, remembered she had PUT it in branch of tree, & apologising to party, we
554m139 ed with great attention to Harmonicon. & readily PUT it. when guided to her own mouth.-- seemed to re
555m142 ould only perceive that the American ones, often PUT on a peevish expression, but not nearly so often
556m146 companying emotion.-- when horses fighting, they PUT down ears, when «turning round to kick» kicking
556m146 ere to grow on horses, they must yet continue to PUT down ears, when kicking.-- -- good case of expre
556m146 orse & zebra. when going to kick.-- Why does dog PUT down ears, when pleased.-- is it opposite moveme
558m153 ngs-- as does black Swan.-- Goose do all species PUT their necks straight out & hiss.-- [Hyaena pisse
567n013 ith them, came several times & opened my hand, & PUT them in-- like child. Tommy's face, now ill, has
581n065 , crack, &c, imitative of the things.-- CD[I may PUT the argument,, that many learned men seem to con
595n121 now. when 2½ years old. was frightened when Snow PUT a guaze over her head. & came near him, although
```

Page
(Key Word)
608o026 more proper to pity than to hate & be †A man may PUT himself in the way of Contingencies.--but his de
610o031 said `how very odd.-- --could not think what had PUT Wilson into his head.-- remembered, that he had.
624o051 fails, or rather is weak.-- [RHC] Better simply PUT it, beneficial tendency in every instinct to the
634j54v t must be abandoned for another".-- What bosch!! PUT it to case of man. <&> The <design> determinatio
041r065 Mr Lyell has mentioned the drifting of carcases PUTRID. In Rio paper. when discussing probable rise o
485z017 Fuego p. 141. How comes it salt water so soon PUTRIFIES?? p. 319. on Hydra-- polypi-- <Rep> do p. 324
264c079 nence.-- «not understanding language of Fuegian, PUTS on par with Monkeys» Gould seems to think that
384d162 nerations Theory of sexes (woman makes, bud, man PUTS primordial vivifying principle) one individual
536m070 y views tend to.-- When a man is in a passion he PUTS himself stiff, & walks hard.-- «He cannot avoid
542m096 g animal has canine teeth.-- A dog when he barks PUTS his lips in peculiar position, & he holds them
585n075 front legs & knocks with back of Head, yet never PUTS down its ear. good to contrast with horses, ass
278c129 ight not breed:-- Genus must be a true cleft-- PUTTING out of case the Analogys.-- If genus does not
308c217 eeing no tea brought back memory» old habit of PUTTING tea in pot made me go to tea chest almost unco
546m108 ympathy» we can only know what others think by PUTTING ourselves in their situation, & then we feel 1
608o026 icked.--we ought to pity & assist & educate by PUTTING contingencies in the way to aid motive power.-
196b106 In plants I believe not..-- It is a very great PUZZLE why Marsupials & Edentata should only have lef
245c024 oceania, offers many varieties in each place to PUZZLE naturalists.-- p. 372. Bourous. the Barbyrousa
310c222 gous to petrel-grebe. external» appears to be a PUZZLE against my theory,-- If I be asked by what pow
365d106 considerable change, it would have been greater PUZZLE, than none, for the «e»normous time which it m
554m139 t knot-- the sailor on board the ship could not PUZZLE her-- with aid of teeth & hands.-- Descent 183
589n092 rs without hand or voice.-- there is some great PUZZLE in what Sir. J. M. says of pure reason not lea
315c243 o doubt from want of assistance.-- crying is a PUZZLER-- Under this point of view. expression «of all
637j57v ℕ p. 234. The non-absorbing Camel's stomach is PUZZLER p. do says inconvenience would have arisen had
298c189 s over nest, when often looked at.-- this most PUZZLING whether instinct, or reason?? Gould says he b
564n005 e always been studied appears to me to be like PUZZLING at Astronomy without Mechanics.-- Experience
073c164 f Zimapan offer zeolite. stilbite. grammalite. PYENITE. native sulphur.. fluor spar. bayte. asbestos
505q014 ld flower= (14) Has planted seeds of Geranium PYRENAICUM. small white-flowered var. with abortive sta
429e114 ort. Transact Vol I. p. 17 Append) that in the PYRENEES, that the Rhododendron ferrugineum. begins at
429e116 e of the great N & S. valleys, which penetrate PYRENEES but are found no where else not even in branc
464tf6v of Molina's Pudu or goat There is ibex of Alp PYRENEES &c-- (see Blyth's work on Ruminants,-- these
429e115 mond. p. 19. do says lofty Alpine plants of & PYRENNEES agree with those of Norway. Lapland & Greenla
074r165 lest case. concretions of clay iron stone; iron PYRITE in a fossil» Insist strongly on the grand fact
076r169 rable quantity of silver procured from martial PYRITES; great blocks of pure silver not common in <S.
098a046 . p I. p. 86. et p 95.-- It is easy to prove. (PYRITES, agates, calcareous balls) that concretions ar
110a086 crystals being red) «with» cleavage, veins of PYRITES, few curious fissures; base in part. block not
027r028 ormation <would> wood converted into siliceous PYRITOUS & coaly matter. Mem: Chiloe In the endless cy
484z015 e for other representatives of this class--. PYROCEPHALUS & many Tyrannulae-- replaces warblers of Eu
290c163 cation on generation might appear an analogy NB PYRRHO-alauda (bird of St Jago) of brown colour; live
123a117 uadruped bigger than ox.-- at Buenos Ayres 20½ QUADRAS from river; 20 varas from surface in tosca.--
299c193 arrived at, when one sees a plant like Paris QUADRIFOLIUM growing in one wood far from any other plan
415e068 qual chance that progenitor was bimanous, or QUADRUMANOUS.-- What a chance it, has been, (with what a
420e089 udiment of a tail, shows man was originally QUADRU<MANOUS> «ped.-- ». Hairy.-- could move his ears Th
123a117 rt-- great Dasypus near Canelones -- large QUADRUPED bigger than ox.-- at Buenos Ayres 20½ quadras
225b219 ory will make me deny the creation of any new QUADRUPED since days of Didelphis in Stonefield. all la
415e066 see to discover whether the parent of man was QUADRUPED or bimanous,, is to see, what parts of struct
040r062 agrees with St Helena in being unique, yet no QUADRUPEDS. -- Is the white matter beneath pebbles. the
065r140 subsidence in T. del Fuego, and connection of QUADRUPEDS.--although recent elevation, there may have
183b053 t.-- ¿Whether extinction of great S. American QUADRUPEDS. part of some great system acting over whole
183b053 acting over whole world, the period of great QUADRUPEDS declining as great reptiles must have once d
186b060 & so on.-- As I have traced the <type> great QUADRUPEDS to Siberia; we must look to type of organiza
187b065 cause separated since time of extinct QUADRUPEDS:-- same argument applies to England.-- Mem.
183b091 each other p. 306.--. Chamisso on Kamschatka QUADRUPEDS Kotzebues first Voyage Copied into list Ent
193b091 the possibility of some isld not having large QUADRUPEDS-- Humboldt has written on the geography of p
196b106 ll not do to say period unfavourable to large QUADRUPEDS--horse not large-- Ed. New. Philosop J. No 5
199b116 numerous: quote from Lyell;: assuming truth of QUADRUPEDS being created on small spots of land, of the
219b196 eing as easy to produce «for the creator» two QUADRUPEDS at S. America Jaguar & Tiger & Europe, as to
225b219 y generation explained? NB. Look over Bell on QUADRUPEDS for some facts.-- about dogs &c &c NB. Anima
225b219 Diemen's land.» glorious fact. of absence of QUADRUPEDS East India Archipelago very good on opposite
231b242 supials animals all show greater connexion in QUADRUPEDS,-- <bu> plants do not follow by any means.-- O
233b251 ge proportion of Water & small of land or few QUADRUPEDS-- Study Productions. of great Fresh water la
233b252 site different, whilst man & his domesticated QUADRUPEDS are not so. greater facilities of change in
244c021 nt. what <actually is> «has» taken place with QUADRUPEDS» p. 118. wild pigs of Falklands, generally "
248c029 eat advantage in studying. Geograph. range of QUADRUPEDS.: that either created in each point, or migr
248c029 migrated from those quarter, where we know QUADRUPEDS have existed for ages.-- ∴ The most hypoth:
249c036 s very remarkable, when we consider number of QUADRUPEDS in Eocene period. Have the Edentata & Marsup
250c037 zed, of each continent. Try amongst Europaean QUADRUPEDS if Africa destroyed would not then some form
250c037 , so on, & so on.-- Whatever destroyed great <QUADRUPEDS>; Pachyderm in S. America destroyed great Ed
250c039 ued what is not cause of destruction of large QUADRUPEDS.-- common to two types of animals What repti
257c058 uestion is not, how there come to be fishes & QUADRUPEDS, but how there come to be, many genera of fi
268c095 adapted to Air!! p. 11--&c. valuable paper on QUADRUPEDS of Van Diemen's land, which appear diff. fro
311c227 . Apercu very good on insectiferous <insects> QUADRUPEDS geographical range very good.-- Blainville O
320c276 nas Chile Falkners Patagonia Azaras Voyages & QUADRUPEDS of Paraguay Dobrizhoffer. Abipom<e>nes. Edin
356d072 can dove.-- The extinction of the S. American QUADRUPEDS is difficulty on any theory-- without God is
390d179 e & Cattle Library of Useful Knowledge Bell's QUADRUPEDS the effects of breeding in, it is not merely
216b181 ogs.-- See paper in Philosoph. Transaction on a QUAGGA & mare crossing by Ld. Moreton, where mare was
283c145 more than anyone.-- no turn the Zebra into the QUAGGA.-- «let them be wild in same country with their
367d113 o.-- legs reminded me strongly of Zebra.-- Mem. QUAGGA & Ld Moreton Mare ringed Owen says that Bell i
388d172 like father.-- stuff.!-- How much opposed. the QUAGGA case appears to that of «2» dog begetting diff
523m016 d by moral causes (ideotcy by fear. Chile earth QUAKES). in people, who, probably otherwise would not
439e144 pard & Tiger together depended on some nice QUALIFICATIONS each possess., & that tiger springing an i
312c231 than those bred in & in.-- which looks as if QUALITIES were not permanent, in the new cross.-- In th
312c231 int in question.-- «--merely picking opposite QUALITIES, with no other means whatever.--»» Individual
366d108 nstincts unchanged, & if their characteristic QUALITIES were all deeply imbued in them from long perm
372d129 es= <I do not doubt, the> Do plants loose any QUALITIES by being buds-- , more than if whole branch t
385d165 character, constitution, & most of the moral QUALITIES of the father!. In how many daughters does th
466t095 th some other, as pointing with smell.-- These QUALITIES have been given to foetus <fr> before sex dev
511q018 (4) Does he know any seed-raisers (5) List of QUALITIES in birds & animals for prizes.= Pidgeons. Can
524m020 scles in the twitching. Pride & suspicion are QUALITIES, which my F says are almost constantly presen
526m030 the mind, altering form of head, & thus these QUALITIES become heredetary.-- When a man says I will i
541m091 one is obliged to repeat the word, & think of QUALITIES as flowers, cloth &c & with all this difficul
583n070 ason knowledge gained by reason: & then these QUALITIES of imitation & education may be used as argum
439e144 on-- if killed by some other animal, then that QUALITY which saved him, would be the one encouraged--
466t095 culiarity improved.» Probably every such «new» QUALITY becomes associated with some other, as pointin
577n049 life & will of a conferva is not an antagonist QUALITY to life & mind of man.-- & we do not suppose a
587n088 t of every word, expressing a mental «desire» «QUALITY» &c &c Mackintosh Ethics p. 97. on Devotional
257c056 onie. ou au moins des espèces tres-analogues,-- QUAND ce nétaient pas tout à fait les mêmes." This go
023r010 mpier also repeatedly talks about the immense QUANTITIES of Cuttle fish bones floating on the surface
096a039 ft.-- Mr Bollaert (at Roy. Institut) talks of QUANTITIES of shells at Iquique. <Ceylon>. Band of Volc
200b124 ll to refer to Chamisso Vol III p. 155. about QUANTITIES of seeds in sea; also Holman: <at> Keeling t
446e162 by certain treatment will suddenly send forth QUANTITIES of blossoms--» The case of the Lemna, «and t
030r036 us formation of Chonos. interesting from great QUANTITY of altered Carbonaceous shales Examine chart
041r066 e balls at Chiloe, full of sand.--the <scale> «QUANTITY of iron» being there in excess.-- If veins {P
064r136 rocks & Volcanic metalliferous--' Urge enormous QUANTITY of matter from CREVICE of Andes--therefore fl
064r137 d that Java volcanos differ from all others in QUANTITY of Sulph. acid emitted: mem: Grand gypseous f
069r149 ing taken place unrecorded must be insensible. QUANTITY of matter from Cordillera. HORIZONTAL movemen
076r169 rgentiferous lead not abundant. = considerable QUANTITY of silver procured from martial pyrites; grea
076r169 ated <[...]> sometimes concentrated: wonderful QUANTITY of pure silver in S. America. Geology of Guan
154g060 ell & good, but how came river to do this vast QUANTITY when during repose of lake it did but little

230b239 It will admit> a plant will admit of a certain QUANTITY of change at once. but afterwards will not al
284c147 .If only one kind of plant not so many.-- The QUANTITY of life on planet at different periods, depen
284c147 ocean, &c this probably on long average, equal QUANTITY, 2d on relations of heat & cold. therefore pr
284c147 rms depends on the external relations (a fixed QUANTITY) & on subdivision of stations & diversity--.-
303c204 (hence intellect?) & what kinds of intellect) QUANTITY & kind of hair forms of legs-- hence the fath
419e085 ometimes speculated might be owing to absolute QUANTITY of vitality «in the World»,-- the production
506q015 - Any where pollen is not produced or small in QUANTITY -- Any unproductive, where germen does not sw
283c146 ion of species & genera, is probably to add to QUANTUM of life possible with certain preexisting laws
558m152 des &c &c &c I observed the Asiatic Leopard. QUARRELLING. mouth wide open, each [lip] drawn back & dr
538m079 lly been so.-- some always sentimental, some QUARRELSOME as B.e on board Beagle, some merry goodhumou
250c039 reptiles coexisted with Palaeotherium in Paris QUARRIES & at Binstead. Mem. recent Crocodiles with Pa
161g094 a few> «about 3» feet <lower> too high about a QUARTER of a mile further on, where three [...] abutte
248c029 created in each point, or migrated from those QUARTER, where we know quadrupeds have existed for age
383d159 - hermaphrodite, being not only dimidiate, but QUARTER-grown seems to show whole body imbued with pos
408c045 . case, bears on the vast changes even in that QUARTER of the world-- -- Mem. elevation & subsidence
423e096 erwise a frost if killing the vegetable in one QUARTER of the world would kill all of the one herbivo
527m032 the best song. [Migratory birds return to same QUARTER for many years]CD.-- Beauty is instinctive fee
586m078 ence, or habit.-- so bird migrating to certain QUARTER is instinct, but his knowledge of that quarter
586m078 quarter is instinct, but his knowledge of that QUARTER, is faculty, whether by sun, & heavens, or ma
586m081 «not least by experience» directed to certain QUARTER"-- "An animal has faculty of walking. which in
145g001 es offspring more indelibly than female p 367 QUARTERLY Journal of Agricl Dec 1837 Yet instances give
286c156 rying plant, all brought in one plane Fleming QUARTERLY review says nat: fam: of Willows contains man
320c276 letter to Dr. Fleming. & Review of latter in QUARTERLY Sir J. Sebright's Pamphlets Wilkinson on Catt
345d043 pends upon mother bein[g] oldest breed?.-- -- QUARTERLY Journal of Agriculture p. 367. Dec. 1837. Gen
568n019 o progress of knowledge. see Lyell on Scrope, QUARTERLY Review. 1827? In Water Scotts life.. Tom Purd
183b051 ra, the nearest species often come very remote QUARTERS. (NB. if Plata Partridge «or Orpheus» was int
187b063 on dying out about same time in such different QUARTERS.-- Will Mr Lyell say that some circumstance k
235b274 ne so anxious to cross. animals from different QUARTERS to prevent them taking peculiar character-- I
332d001 one time he had three patients at very distant QUARTERS of the county, who had had no sort of communi
390d178 ereous if the relative had come from different QUARTERS) then it causes <to> a secretion of something
400e011 Hattica is great genus.-- because found in all QUARTERS: his ideas not clear. In Australia from appro
325c267 s views.-- Quoted by Owen.-- Hunter has written QUARTO. work on Physiology besides the papers collect
022r007 .--Mem. at Chonos & Concepcion. P. 417 Veins of QUARTZ exceedingly rare Mem C. [Cape] Turn P. 434 & 4
049r087 Video «facts of Passages marked by do.» discuss QUARTZ veins, there contemp--yet similar ones in Clay
052r099 formed streams of lava at St Jago. C. de Verds QUARTZ pebbles in the Cordilleras look as if some pea
069r150 view Urge curious fact felspar melted gneiss/// QUARTZ!!! Analogous to Von Buch. Basalt where Basalt.
073r164 . 131» The above porphyries characterized by no QUARTZ & amphibole frequently only vitreous felspar:
076r170 greenstone porphyrys & phonolites do. amphibole QUARTZ & mica very rare.-- ancient freestone & brecci
078r176 se variety of Porphyries which are destitute of QUARTZ, & wh abound both in hornblend & vitreous fels
079r176 says, mur of Silv.[,] Sulph. of do.[,]galena[,]QUARTZ, Carb. of Lime. accompany.-- Ulloa has said si
086a003 on in Europe, that the simplicity of Ventana's «QUARTZ.» unmixed is very pleasing; owing to the movem
091e027 the Poet» Accra. Coast of Africa. Clay Slate & «QUARTZ, strike SSW & NNE dip 30 degree - 80 degree Ed
092a030 un M. 516 1837 High up the Essequibo, granite & QUARTZ, after passing sandstone Vol II. p. 69.-- Geog
098a047 f conceationary process (analogous to layers of QUARTZ & feldspar) within other concretion.-- state l
102a061 crystalls, & the layers first of felspar & then QUARTZ &c, owing to separation having taken place mos
114a094 h's curious specimens of «transversely fibrous» QUARTZ. & iron stone alternating. bear on subject of
119a106 - (this is shown by the softness & curvature of QUARTZ rock?) also by my phenomena of earthquakes.--
149g032 there were perfectly rounded «base» pebbles of QUARTZ & other rocks not apparently in situ <& in> hi
157g075 d of remarkable gneiss with red granite veins of QUARTZ, & garnets.-- Boulders as before certainly mus
163g109 hical distribution of British Plants Shropshire QUARTZ what substance is collected in little spots Sp
436e137 ormations do not however, extend round world.-- QUARTZ of Falkland.-- Old Red Sandstone-- Van Diemen'
028r028 a tendency. to mingle; The sea would separate QUARTZOSE sand from the finer matter resulting from deg
054r102 rs wonderfully from mine: phyllade covered by QUARTZOSE sandstones: refers to broken hill described b
097a042 ordillera Can Greenstone dikes. be residue of QUARTZOSE vein in higher parts? & felspathic veins?-- M
102a061 tone dike in Granite residual matter of upper QUARTZOSE ones & felspar.?? Are the great crystalls, &
267c095 f animal digested.-- Important do p. 98. on a QUATERNARY arrangement of Cryptogamic plants.--Owing to
354d062 &c.-- do. p. 69. A Dr Macdonald believes the QUATERNARY arrangement & not the Quinary.-- any one may
477z003 r Benchuca. "Les individus ailes peuvent avoir <QUATRE> cinq lignes de long et volent. p. 208 Fleas o
309c221 good case of instinct. bees turning neuter into QUEEN, more wonderful case Dwights' Travels in Americ
539m084 ed by Spenser (Faery Queene. CD 25 (Descript of QUEEN) «O» of Hell Cant IV or V.) as pale & trembling
539m084 e» in worst form is described by Spenser (Faery QUEENE. CD 25 (Descript of Queen) «O» of Hell Cant IV
370d116 n security.-- Hunter doubts about production of QUEENS.-- Neuters are bred first, «then males--» how
059r119 e, et lorsquelles se trouvent interrompues par QUELQUE vallées ou par quelque scissures profondes, on
059r119 ouvent interrompues par quelque vallées ou par QUELQUE scissures profondes, on les voit se reproduire
569n022 s et dit qu'il valait mieux rester a Farroïlap QUELQUE mal qu'on y fût."-- Expression common to Savag
209b156 "ou bien encore on pourrait au plus en conclure QUELS sont les genres qui, sous ce climat, se divisen
369d115 are generally compelled to traverse in order to QUENCH their thirst"-- But New Guinea.--!! S. America
089a018 terre sont presque toujours precedes et suivis, QUEQUE temps avant et apres, par de petites secousses
633j54r award course.-- we sink into such contemptible QUERIES as why should the earth have drifted; why sho
037r053 h respect to degradation of rocks--It may be a QUESTION. whether organic remains protect a rock, or t
041r065 ation of fresh water Shells. multitudes.-- The QUESTION of shell's concretions, living only in that s
051r095 imal will adhere to a certain part. Apropos to QUESTION does animal adhere to rock because it does no
094a035 which rise through granite.-- a most important QUESTION with respect to my theory of changes. of gran
102a061 re? equally with little pressure? An important QUESTION! If water yields substances from impact, «it»
148g025 hought the breed liable to vary-- I asked this QUESTION in many ways & received same answer Thought l
192b085 in detail some good instances. But it is other QUESTION, whether there have existed all those interme
195b098 of organization constant in the shells.-- The QUESTION if creative power acted at Galapagos it so ac
222b205 dary period in same predicament. It is another QUESTION, whether whole scale of Zoology may not be pe
226b222 elena & Tristan D'Acunha, resolves itself into QUESTION of proportion of species to genus If on one i
228b229 tions with respect to past & future. The Grand QUESTION, which every naturalist ought to have before
235b275 t kind of localities.-- if so change The GRAND QUESTION Are there races of plants run wild or nearly
257c058 lly, but even about great division, our <only> QUESTION is not, how there come to be fishes & quadrup
265c084 fingered children heredetary¶ With respect to QUESTION what is adaptation.-- Ermine, ptarmigan hare
272c108 & habits of clinging to rushes similar.-- The QUESTION which I more immediately want are there Heter
298c189 urious this †ready answer, without any leading QUESTION.-- [+] This might be mentioned in note.-- try
304c207 ot «been» exposed to so many contingencies??? A QUESTION of immense difficulty is, whether Apterix des
312c231 produce number agreeing almost to the point in QUESTION.-- «--merely picking opposite qualities, with
333d008 his agrees perfectly with Yarrell & no leading QUESTION was put.-- Fox thinks half Lion & Tigers are
348d051 tural arrangement of animals themselves is the QUESTION in point." Now what is natural arrangement,--
385d163 g a throw of this mongrel.-- I did not ask the QUESTION.-- His bitch will not take «& if she did take
436e136 pecies,-- for instance pidgeons-- : then comes QUESTION of genera It certainly appears that swallows
446e160 power.-- My theory leaves quite untouched this QUESTION of spontaneous generation.-- Introduction to
453e183 me seed raised in same garden.-- now this good QUESTION-- single, or half double.-- anyhow fertile be
489qIFC en, Royle, & Horsfield Sykes p. 12 Maer. p. 13 QUESTION &c. July. 1842.-- Shrewsbury p. 14 Henslow (2
490q001 .--Name of Italian who sold eggs.-- Temporary QUESTION 1 Where has Duchesne described Atavism allude
491q01v ple tree «with abortive stamens» answers first QUESTION in negative.-- Questions Regarding Plants. 1.
501q011 tive organs of great Heat (32) Can Henslow ask QUESTION of Col. Le. Couteur about Wheat-- Change of S
505q014 oda under Beech.-- Lychnis dioica answers this QUESTION= (5) Open more Horned oranges-- (6) Figs, flo
515q021 effect of crossing peaches & nectarines: same QUESTION with regard to Primroses. (4) Do apples "spor
549m121 ess agrees with that of New Testament is other QUESTION.-- little is there said of intellectual <ple
549m121 f the intense happiness.-- it is again another QUESTION, whether this happiness is the object of livi
551m127 ust seen mind went on RAMBLING till excited by QUESTION.» Sept. 4th. Lyell in his Principles talks of
578n054 est person one carries on, by association, the QUESTION, "one will anyone, especially a women think o
610o033 except our experience".-- is this not almost a QUESTION whether we have any instincts, or rather the
616o039 l to ask themselves the converse of the <expr> QUESTION above stated, because there are living bodies
617o39v rent phases of the same object of thought is a QUESTION which ought to be clearly comprehended by any
619o042 nd sympathy> «or benevolence» to the object in QUESTION. Without regarding their origin, we see in ot
627o053 n of moral obligation «when grown up???» & the QUESTION is, whether this can be resolved into some op

Page
(Key Word)

629o55v between the desires & will?)) (2) It is other QUESTION what it is desirable to be taught,-- all are
629o55v all are agreed general utility (3) It is other QUESTION whether any thing is taught instinctively; I
629o55v tion agrees. with last head.-- (4) It is other QUESTION, how the feeling of ought, shame. right & wro
430e118 - «-- .Make the difficulty apparent by cross-QUESTIONING.-- » even if placed on Isld-- if &c &c.-- Th
287c158 an variations in eye of mollusca. [+] These QUESTIONS may be all disputable, but the one end of cla
489q FC y about 10 feet in thickness.-- (March, 1842) QUESTIONS & Experiments {T} Gowen, Royle, & Horsfield S
491q002 tamens» answers first question in negative.-- QUESTIONS Regarding Plants. 1. Uniformity of hybrid & M
492q003 s not non-flowering gorze common in Norway No QUESTIONS regarding Breeding of Animals If two half bre
496q006 es, as Hyacinths in glasses &c &c Experiments QUESTIONS concerning Plants Is the common Fig Dioecious
506q015 .-- (40) Ask Henslow to distribute some of my QUESTIONS amongst agriculturists. whom he know.-- Col.
509q017 iff. countries in same races Mr. Gray General QUESTIONS (1) Particulars about Sierra Leone. cow. taki
516q BC id. will this bear on Petrifaction?-- [blank] QUESTIONS & Experiments Expression M Charles Darwin Esq
324c268 t Find out from Statistical Society-- where M. QUETELET has published his laws about sexes relative t
379d152 Owen-- In the Athenaeum Numbers 406, 407, 409, QUETELET papers are given, & I think facts there menti
376d136 almost White from first generation., that with QUICHUAS the American character is more tenacious. & d
272c107 emipterous insects, having spiny legs & running QUICK & generl appearance of blattae other Hemiptera
285c153 at last become unfit, the animal cannot change QUICK enough & perishes.-- Lyell has show such Physic
296c184 d «shows where proper dampness seeds can arrive QUICK enough!» Vegetation of peak-- altogether origina
494q004 th of life. (5) Does my Father know any case of QUICK or slow pulse being heredetary. (6) In the last
541m090 wing my mind to skip from subject to subject as QUICK as it chose.-- although thinking «& talking» fo
636j57r ame food, differs in every respect, except «in» QUICK movements. (sliminess instead of barbs)-- In al
175b017 cially with some change probably <change> vary QUICKER Unknown causes of change. Volcanic isld.-- Ele
547m113 - It may be deception to say the mind <thinks> QUICKER in sleep, it may do less work & yet do so, fro
335d015 formed people & as mutilations «(produced vary QUICKLY)» sometimes have similar offsprings, so will t
531m053 hich frightens. them.-- Now every animal moves QUICKLY away from any sudden sound or noise, & therefo
545m106 noise not like pish, but like chit-chit-chit, QUICKLY uncovering their teeth, this the Keeper thinks
556m145 partaking of <st.> made by <ret> inspiration & QUICKLY retracting tongue from behind upper & little b
572o033 ould not picture to myself reading French book QUICKLY, & <running» «running» over imaginary words: i
585o075 arts of our bodies, «or touches one part. very QUICKLY successively.--» [& we know from experiment of
429e114 It is difficult to believe in the dreadful «but QUIET» war of organic beings. going on the peaceful w
532m056 ly unhappy, constantly whimed, would not remain QUIET in any room, would not sleep at night even when
053r100 hat of Pernambuco? Quote Miers about shells at QUILLOTA Lyell, states that contact of Granite & sedim
069r150 .--(Mica Slate) [Fig. 9] {P} ((3) like Bell of QUILLOTA.) A) in this strata may be older than (B).--
067r143 ells are found on whole coast from P. Indio to QUILMES. & at least seven miles inland. The Cordoba ea
182b046 r & fishes for water. If my idea of origin of QUINARIAN system is true, it will not occur in plants w
262c074) where power arbitrary. leaves door open for QUINARIANS to deceive himself.-- Give the case of Apter
185b057 hat same family.-- (NB I see Waterhouse thinks QUINARY only three elements) How far Does Waterhouse's
197b112 xistence is thought to account resemblances &·. QUINARY system, or three elements p 66]CD With unknown
235b263 s like birds & animals separate &c &c Work out QUINARY system according to three elements How is Faun
292c170 logy & Botany p. 566 wants to see absurdity of QUINARY arrangement let him look at abstract of Swains
354d062 believes the Quaternary arrangement & not the QUINARY.-- any one may believe anything in such rigmar
029r033 ere was no uncommon ripple on the water. It was QUITE calm at the time. Latitude 8 deg. 47 min. N: lo
038r058 ees grew.--? Are not the dikes in upper strata. QUITE different from the Porphyries: certainly appear
063r132 imals. still partly united. & eggs which become QUITE separate.--Considering all individuals of all s
064r134]tries thinly covered by vegetation. Rhinoceros QUITE in deserts.--Much struck with number of animal[
064r134 of range some miles from shore. rock of oysters QUITE above reach of tides.--thinks them same as rece
074r165 d by complicated chemical law & steam of salts, QUITE curious case of oxided Iron by Mitterschlich. V
148g022 as high as highest measurement but nature I am QUITE doubtful of as I am of all the Alluvium. At Mou
148g026 , & pretty well rounded stones, mixed with some QUITE irregular very like rubbish at head of Loch Doc
156g067 an 3d. 3(a) less perfect than upper & lower but QUITE as perfect as those lines in Glen Collarig, & s
179b032 om the grandfather, (when the mother was nearly QUITE white) in the two first children How is this in
198b113 understand whether. G. H. thinks development in QUITE straight line, or branching S. H What does the
217b183 pposite breeds are crossed, sometimes offspring QUITE intermediate sometimes take strongly after eith
234b261 species-- Again is there not similarity even in QUITE distinct countries in same hemisphere. more tha
264c082 one ducks aberrant from-- habits.-- Gould I see QUITE recognizes habits in making out classification
281c139 r>, in broken circles.-- which in each group is QUITE fatal.-- ¶Relations of analogy being those last
294c176 ies in islds. near continent, Must we resort to QUITE different origin when species rather further.--
299c194 Iceland has same uniformity Primrose & Cowslip, QUITE wild, but they affect different localities,-- l
318c254 irected by kind of country; «kinds of migration QUITE different in species of same genus.» The Muscic
319c256 than in America, but the few of N. America are QUITE as beautiful. The thrushes of N. America. singi
337d020 d out, & few even of local varieties approaches QUITE to wild local variety.-- our Europaean varietie
343d039 imals found fossil when Europe must have worn a QUITE different figure 19th. With respect to the Delu
350d055 , like Trilobite. (Polirus??) female blind & of QUITE different form from male with eyes!-- (are not
356d069 originality of idea-- (though I arrived at them QUITE independently & have used them since) the line
361d095 n two herons, <both> plumage of both (nursling) QUITE similar.-- one species retained this character
370d117 blend with Annelidae by some fish.-- But birds QUITE distinct.-- Collect cases of difficulty of grow
378d148 fed, but still they have a wariness about them QUITE remarkable". instance of old Species transmitti
423e097 orous. & its one carnivorous devourer.;-- it is QUITE clear that a large part of the complexity of st
426e107 re fertile & make mongrel, & other great series QUITE otherwise & make on[ly] true hybrids.-- but thi
446e160 rogressive developing power.-- My theory leaves QUITE untouched the question of spontaneous generatio
464t065 ot an Apteryx, yet it shows the Apteryx is not «QUITE» isolated in its present locality-- there have
468t111 ts & reached nectar =again= between them, hence QUITE below stigma. & so avoided it.» On certain days
490q001 e extinct land-shells of Madeira-- analogous or QUITE distinct from recent ones-- I presume some rece
494q004 rossing.-- (7) Are the Eggs of the Penguin Duck QUITE similar to those of another Duck. ¿in Pidgeon?-
521m007 can thus lie dormant, during a whole life time, QUITE unconsciously of it, surely memory from one gen
522m013 hat everybody is insane. at some time. Mania is QUITE distinct, different also from delirium, a pecul
523m018 & madness:-- ira furor brevis est.-- My father QUITE believe my grand F doctrine is true, that the o
526m029 fact of early memory consisting of things seen, QUITE agrees with my Fathers case of Mr Corbet of the
529m042 hen he recovered. was found to be ignorant, but QUITE sensible & no ways an ideot.-- «in this case mu
529m042 deotcy.-- The story of the Corbets & big noses, QUITE conjectural, in Blakeways book of Sheriffs.-- J
532m056 ortnight. continued to grow thin & did not seem QUITE happy. in five weeks was so thin, that she was
539m083 habitual.-- instances?? The possibility of two QUITE separate trains going on in the mind as in doub
544m102 ns-- ideas are strung together in manner <they> QUITE different from when awake.-- peculiar sensation
547m111 astle in the air, of being compelled, from some QUITE imaginary cause to start at once to Shrewsbury.
547m112 leep & not vica versa. anyhow I might have been QUITE still, & not attending to bodily sensation & ye
556m146 » kicking they do the same. although it is then QUITE useless-- Cats kneeding when old, like kittens
557m148 a dog, & make him ashamed of himself, in manner QUITE different from fear; there is no inclination to
557m148 ill-defined fear.-- Yet one knows oneself it is QUITE different from that.-- like «slight» passion fr
565n008 th disgust, when ones opponent is considered as QUITE insignificant, & when pride makes person extrem
567n015 or be indifferent about, yet feel shy.-- not if QUITE stranger.-- or less so.-- When learning facts f
573n039 rugged her shoulders.-- analyse this.-- Miss C. QUITE aware & indignant with Mrs C. but had no influe
590n094 ground to call imagination a faculty, a power, QUITE distinct from self. «or will» [not located] <&
6040o17 - chiefly excited by passive emotions.-- Cannot QUITE perceive drift of Book.-- Sympathy & affections
618o041 rceived but they are known by courses of action QUITE independent of each other. A person might be qu
618o041 te independent of each other. A person might be QUITE familiar with thought & yet be ignorant of the
618o41v to matter is because our knowledge of matter is QUITE insufficient to account for the phenomena of th
628o054 k some parts of the emotive part of man, may be QUITE artificial, as avarice love of gold.-- love of
640j167 ould prevail.-- Not separately: NB. These views QUITE exclude the idea of domesticated animals changi
058r116 that the same earthquake has run from Chili to QUITO a distance of more than 500 leagues. A little t
058r116 ipa in 82 was overthrown, & 86. Lima. next year QUITO. considers these earthquakes travel in order.--
071r158 » cracks mountains falling in.-- Earthquakes at QUITO. tranquillity «at Mendoza» exception.--«formerl
079r177 Guayaquil, same as Galapagos. no Hydrophobia at QUITO. P 281. do do Australia, C. of Good Hope.--Azor
091a026 nstitut. p. 209. May. 1837 Paper by Humboldt on QUITO Volcanoes & another on Mexican Trachyte <roc> 1
450e175 ramposter & S of Tropic-- By J. H. Moore after QUITTING Bengal this fact is noticed in Cassay Ava Peg
477z002 t le pajero) l'yaguaré «the zorilla-arskink» le QUIYA (Coipu) viscacha.-- A. Patagonicus les tatous (
044b073 strata of Europe formed near coast. Humboldts QUOTATION of instability of ground at present. day.-- a
060r125 etation.--«I can find nothing.» Mem Carolines QUOTATION from Temple Urge the mineralogical difference
021r005 s in line of Cleavage. laminae fold round them; QUOTE this. Valparaiso Granitic nodules in Gneiss. Ep

Page **(Key Word)**
048r086 esert country of S. Africa. It would be well to QUOTE Burchell. V. where the Rhinoceros was killed. -
053r100 egende is perhaps same with that of Pernambuco? QUOTE Miers about shells at Quillota Lyell, states th
135a144 a great many sea shells some miles from coast-- QUOTE passage to show abundance Bengal Journal. Vol 4
192b085 e all genera MAY have had intermediate steps.-- QUOTE in detail some good instances. But it is other
198b115 sion for transportal otherwise not so numerous: QUOTE from Lyell: assuming truth of quadrupeds being
266c091 ion, armed alike with the Koran and the sword" \|QUOTE Whewells Bridgewater treatise, (p..26). about p
347d049 extinguished.--???» Mayo (.Philosop of Living) QUOTE Whewell as profound. because he says length of
035r047 parallel bands of formations on any Geolog Map: QUOTED from Daubeny P 402: likewise, mean height of t
053r101 leavage paper Dr Fittons Australia case must be QUOTED at length. The Lines of Mountain appear to me
309c221 t leap fences:-- Dr Lang on Polynesian nations (QUOTED) p. 4.-- do. p. 186. quotes Burkhardt to show
325c267 laires. 1832. contains all his fathers views.-- QUOTED by Owen.-- Hunter has written quarto. work on
326c266 urners embassy to Thibet, perhaps worth reading QUOTED by Malthus.-- Heberdens Observat. on increase
326c266 e & decrease of different diseases. 4to 1801.-- QUOTED by do.-- There appears to be good art.. on <Et
326c265 by Copenhagen Botanist of range of plants Books QUOTED by Herbert. p. 338 Schiede in 1825. & Lasch. L
492q003 Has M. Sageret WRITTEN on crossing of Cabbages, QUOTED by (as if oral) Decandoelle in V. Vol of Hort.
569n022 é's Voyage autour du Monde (1826-9) Paris. 1835 QUOTED repeatedly by Waitz (In Theil. V) in describin
309c221 Polynesian nations (quoted) p. 4.-- do. p. 186. QUOTES Burkhardt to show black colour of certain Arab
310c224 in other parts of world Athenaeum June 3d 1838. QUOTES M. Turpins assertion that globules of milk pro
443e154 way to get young trees, from worn-out kinds, & QUOTES from Pliny, that it is bad to graft from top s
566n010 .-- Lavater. (Holcroft Translat) Vol III. p.37, QUOTES from Burke, who says on mimicking expression o
590n094 tt's Wanderings, Australian Dog does not Bark-- QUOTES Gardner's Music of nature to show barking not
599o005 on. reason could not have existed without it.-- QUOTES Ld Mondobbo.-- language commenced in whole sen
604o017 ganization & if consciousness, individuality.-- QUOTES D. Stewarts System of Emotions.-- T. Mayo-- Pa
309c221 show black colour of certain Arabs.-- NB avoid QUOTING these hackneyed cases Mr Ed Blyth does not bel
078r174 h setae like my undescribed[.] p. 140. Flèche of QUOY et Gaimard.--D'Orbigny has described it with ca
247c028 Otaheite-- according «stated in note to p 21» to QUOY & Gaimard in Sandwich isld. & according to Cham
478z003 weed & many pebbles Mentions stinging Millepora. QUOY. Freycinets Voyage Vol p. 597 Many descriptions
479z005 ribed my animal with teeth {P} p. 140. Fléche of QUOY et Gaimard Ulloa shell fish Purple die Marvello
178b031 ontinent look over. Bell, & L. Jenyns. Falkland RABBIT may perhaps be instance of domesticated animal
184b054 Macculloch written on same changes in Fish Mem. RABBIT of Falklands described by Q. & G. as new Speci
200b122 nce Australian dog: <yet when that> or Falkland RABBIT.-- There is only two ways of proving to them i
286c154 h thus, Man has expression.-- animals signals. (RABBIT stamping ground) Man signals.-- animals unders
437e139 ere amusing themselves with them, and one day a RABBIT escaped into a hole, where the old Eagle could
511q018 Canary birds-- Bantams.-- (6) <Mad> Porto Santo RABBIT. Descript. of colour «& length of ears» & skel
592n103 ears nostrils» when frightened, does not hair & RABBIT depress. them from squatting.-- p. 64 closing
216b180 s in dogs investigate case of pidgeons. fowls. RABBITS cats &c &c.-- When black & white men cross som
247c029 & Cape of Good Hope V. p. 44 of this Note Book RABBITS introduced in 64, of very many colours, like t
247c029 a herd in England"-- Black & Grey varieties of RABBITS thus handed down for nearly 70 year. Galapagos
280c134 -- lose as well as gain instincts. Wild & tame RABBITS good instance-- instincts of many kinds in dog
363d101 wild cattle.-- plumage of fowls-- long ears of RABBITS. & long fur.-- feathers on legs of Ptarmigan &
399e009 population.-- Octob 10th. Saw. two undoubtedly RABBITS in poulterer shops., of same colour as a Hare,
399e009 l which it was.-- do hind legs increase in any RABBITS One may strongly suspect, that breeding in & i
437e139 great weight. p. 125 is said that Eagles bring RABBITS & hares to the young ones to exercise them in
437e139 in killing them.-- "Sometimes it seems hares, RABBITS, rats & not being sufficiently weakened by wou
461t041 ain adapted.-- it does away with difficulty of RABBITS of England remaining same (if so) with those o
479z005 o land animal besides Wolf at Falkland ∴ black RABBITS not indigenous p 112 M Lesson--Voyage of Coqui
594n112 ken to eradicate it.--» «Emma says, «her» tame RABBITS were not frightened at a dog.-- » The instinct
406e036 ther & sometimes ½ way. Ed. New-Phil. Transact. RABIES, common to men, dogs, horses cows, pigs & shee
400e013 Isle Tres Marias off Mexico with small Hares & RACCOONS.-- «S. American form.--» off province of Guad
171b004 ied,-- therefore generation to adapt & alter the RACE to changing world.-- On other hand, generation
172b006 ; is it not said that marrying in deteriorates a RACE, that is alters it from some end which is good
188b068 and no further:-- Prof. Henslow says. that when RACE once established so difficult to root out.-- Fo
189b072 ff.-- If species generate «other species», their RACE is not utterly cut off:-- like golden pippen. i
200b123 they will breed.-- But what a character is this? RACE permanent, because every trifle heredetary, wit
202b130 s to a third body., or common end of structure A RACE of domestic animals made from influences in one
234b255 of France p. 170. «Fish introduced» Hump backed RACE of cows from Madagascar-- p 173. Vol I. Voyage
240c014 The Barbaroussa, (when young very like the Siam RACE with long nozzle & few hairs) inhabits Celebes
248c030 icated varieties a tendency to go back to oldest RACE, which evidently is tending to same end, as the
259c063 nding to produce effect on offspring-- but WHOLE RACE of that species must take to that particular ha
259c063 -- Are any men born with any peculiarity. or any RACE of plants--Lamark's willing absurd, ∵ not appli
266c091 eninsula, are, generally speaking, a much fairer RACE than the Hindu's, in the same tracts;. & that i
279c133 ong in blood.-- ++ thinks difficulty in crossing RACE.-- bad effects of incestuous intercourse..-- ex
282c142 ot minding particular trades.--» then the second RACE would not obtain a cast of washing men-- but mi
282c142 of washing men-- but might hire the preexisting RACE, thus the analogy would not in all cases be pro
333d009 re produced. one would argue the whole effect of RACE was determined by male: & How completely is Lor
387d169 again in lesser degree.--now if the <tw> second RACE both have this peculiarity strongly; they trans
397e004 forms is <al> solely adaptation of whole of one RACE to some change of circumstances; now we know ho
414e064 Africa & West Indies shows organization in Black RACE there gives them preponderance. intellect in Au
426e106 t that old varieties do not so much affect first RACE, as it does indelibly the many subsequent ones.
428e113 ng, about indelibleness, than the number of good RACE-horses, which Eclipse? has begotten <?> «Walker
430e118 ging country: but greyhound. & poutter Pidgeons «RACE-horse». have not been thus produced, but by tra
451e176 -- probably, I should think grandfather first of RACE & if so, fact for my theory Cocos Isld & Prepar
544m104 ildren. or after he has useless. does not affect RACE. argument for early education.-- fear of death!
553m136 been» implanted in us (<by> ¿ individually or in RACE?) by a separate act of God, & not as a necessar
615o038 rs-- Hence superiority of Christian over Heathen RACE.-- But as no great modification in brain would
619o042 has argued nearly so [RHC] The history of every RACE of man shows this, if we judge him by his habit
623o050 (not to any one individual but to the whole past RACE).-- <no one> doubts) « <I cannot> » [LHC] On th
623o051 influence.)-- & only that which is beneficial to RACE, will have reoccurred'. NB. Until, it can be sh
639j28v . survive. in ten thousand years the long legged RACE will get the upper hand. though continually dra
640j28v back to old type by intermarrying with ordinary RACE.-- «There is no way of eliminating the evils of
640j167 its destroyer), or other cause, the long legged RACE would prevail, even if have afforded only 10th
239cIFC t trades affecting form of man. Could you get RACEHORSE from Cart horse by picking without change of
145g001 instances given against it-- Mere fact of many RACES of Animals in Britain shows that either races s
145g002 y races of Animals in Britain shows that either RACES soon made or crosses difficult Salisbury Craigs
183b051 iew.--¿whether highly domesticated animals like RACES of man.-- M. Flourens. .Journal des Savants.--
186b059 s to know in plants, (or animals) whether, <in> RACES have tendency to keep to <each> either parent,
188b068 w this is same bearing with Dr. Smith's fact of RACES of men tendency to keep to one line Dr Smith sa
190b075 ct relation to facilities of communication Have RACES of Plants. ever been crossed really, if there i
199b118 e could only produce domestic individuals & not RACES, without the occurence of one of the most gener
199b119 ople very apt to be split up into many isolated RACES. ¿are there any instance of peculiar people ban
200b123 to establish whether in crossing very opposite RACES, whether you would expect equal fertility-- dit
203b138 t.-- But then shells-- Mr Yarrell says that old RACES when mingled with newer, hybrid variety partake
206b147 h long distant, when of the present men (of all RACES) not more than a few will have successors. at p
206b148 ne man killing another.-- So is it with varying RACES of man: these races may be overlooked mere vari
206b148 er.-- So is it with varying races of man: these RACES may be overlooked mere variations consequent on
206b148 variations consequent on climate &c-- the whole RACES act towards each other, and are acted on, just
207b152 eparated from Hobart Town-- {from difference of RACES of men and animals} See R. N. p. 130 Speculatio
213b169 late Mammalia on this principle. <Varieties> < «RACES» > Man in savage state may be called, <species>
213b169 d, <species> species. in domesticated <species> RACES.-- If all men were dead then monkeys make men.-
216b182 once. Is not this contradiction to his view of RACES not mingling?-- In Foxes case of Blood Hounds,
224b215 ing else., but not probable owing to mixture of RACES.-- When all mixed & physical changes (¿intellec
230b240 It is scarcely possible to get evidence of two RACES of plants run wild.-- (for we know that such ca
235b275 es.-- if so change The GRAND QUESTION Are there RACES of plants run wild or nearly so, which <breed>
261c070 ing or even ducklings & fowls-- When talking of RACES of Man.-- black men, black bull finches from li
261c070 t solely effects of climate on some «antecedent RACES, perhaps not on now existing» Mr Gould says whe
282c141 us. animals would be possessed by the different RACES of man, yet altogether different.-- To make thi
282c141 et we might have these analogies.-- We must two RACES of such men living in same country but separate
282c142 same country but separated, now if one of these RACES had become eminently aquatic--; «NB, aquatic, i

293c174 arly similar. Curious instance of difference in RACES of men.-- Wax of Ear, bitter perhaps to prevent
299c196 ther animals & again in Mans mind, in different RACES, being unequally developed.-- ¿is not Elephant
303c204 Himalaya, & Europaean forms on that isld.-- The RACES of men differ chiefly in <size> colour, form of
303c204 but likewise in length of extremities, how are RACES in This respect upper & lower, which I do not k
303c204 know whether it <would have> differs in present RACES, & form of feet.= (Negro or father of negro pro
305c211 en in the tendency to delicate emotions between RACES, & recurrent habits in animals.-- --Animal Magn
309c219 of:», is relationship, children of one parent, RACES of animals-- argue «opening» case. «thus» Educa
338d024 not understand the vast number of domesticated RACES.-- Athenaeum. p. 505. some (very poor account)
339d025 ut N. & S. New Zealand & New +++ Caledonia. two RACES of Men, but not plants» will it hold good.-- Th
345d044 with the aid of seclusion in breeding. how easy RACES or varieties are made.-- The Highland Shepherd
362d100 eir origin.-- I see only some «but very strange RACES» of them have the forked black mark of the Rock
364d104 riation is soon reached-- as in pidgeons no new RACES.-- In Scandinavia besides the Rakhekna, before
411e054 ngent circumstance.-- ‖the laws of variation of RACES, may be important in understanding laws of spec
414e063 anence of his breed is destroyed)-- -- When two RACES of men meet, they act precisely like two specie
415e069 speak only of the former proposition.-- as in «RACES of» Dogs, so in species, & in Man December 16th
416e071 eautiful part of my theory, that «domesticated» RACES. of <a> organics. are made by percisely same me
416e072 conditions.-- -- as in plant up a mountain-- In RACES the differences depend upon inheritance & in sp
416e072 in species are only ancient & perfectly adapted RACES L'Institut 1838. p. 394. Rhinoceros «tichorhinu
416e075 practised & perfected Hence difference between RACES & variety? «Man picks the Male, instead of allo
417e075 o the little fixity of organization, in the two RACES,. owing to the domestication of both.-- Now in
417e078 bt are applicable to likenesses, when species & RACES are crossed.-- Now these laws are, that child m
417e079 pecies, & individuals of same species--). some RACES of men. D'Orbigny. affect the common progeny mo
437e141 ach other, & depends on character of antecedent RACES.-- «& yet in all probability the Brussels Sprou
462tf05 review on politics in relation to the different RACES of men, some more intellectual than others-- is
465t081 s-- A profound consideration of method by which RACES of men have been exterminated (see Pritchards p
465t081 portant. it seems owing to immigration of other RACES, so it is with domestic breeds. (though in this
489qIFC ymington Hants. Habits of different caterpillar RACES. --Name of Italian who sold eggs.-- Temporary Q
495q05a week in Salt. artificial water.-- 12. Plant two RACES of Cabbages near each other-- & enclose one twi
497q06v ocuses are very difficult to cross.-- are there RACES-- if so plant them together. & raise. seed.-- I
500q10a c or rising isld, there ought to be a good many RACES or doubtful species; how is this at Canarys Arc
501q011 Henslow.-- Dr. Fleming says yes. (29) Are there RACES of Lupine, Stocks Clover, to experimentize on b
502q11v trawberries How «soon» «early» do characters of RACES of different vegetables & animals come on.-- Co
503q012 of Subularia Royle & Horsfield (35) Talk about RACES of Banana & yet seedless-- no light Henslow or
504q014 rminate?-- Yew trees sexes-- (3) Get Holyhoaks. RACES planted & Linum Perenne.-- Herbert's. fact.= (4
506q015 at & Penguin Henslow &c (36) Has not H. raised RACES of white & Blue Linum-- did parent plants grow
508q016 thiness? & father answered (5) About cross-bred RACES of men taking after sex. A Smith. About species
508q016 seases (heredatary?) in diff. countries in same RACES Mr. Gray General Questions (1) Particulars abou
537m075 grant, that the conscience varies in different RACES.-- no more wonderful than dogs should have diff
537m076 ral sense, would have been so much, as in other RACES of mankind..-- p. 27. Mart. allows some univers
544m104 gained.? by conversation-- ∴ modified in those RACES, where it is customary to die-- August 24th. As
553m136 -- & that it does exist in different degrees in RACES.-- whether in Ancient Greeks, with their mystic
575n044 suppose some essence. The facts about crossing RACES of dogs on their instincts, most important, bec
603o012 ulates on origin of sacrifices. «common to many RACES»-- thinks action towards <man> «a king» <change
613o036 rain; «[LHC] not used by Kirby» (:analogy:-- as RACES are formed or modification of external form. so
622o049 ill be hard to discover this, for the different RACES of man may have different instincts, as we see
623o050 ow together [RHC] This feeling seems to vary in RACES of man. & certainly in «species of» animals, in
623o051 at which» had a beneficial tendency during past RACES could become instinctive.-- [LHC] x It is proba
626o052 ubtless lead to similar actions which in prior <RACES> generations led to their formation.-- 'N.B. If
636j56v crossing with other varieties was prevented Do RACES of peas become intermixed & gardner have hybrid
053r101 ol III Latter Part Are there Earthquakes in the RADACK & Ralix Islds? In my Cleavage paper Dr Fittons
209b157 late my Keeling Case: Juan Fernandez Galapagos «RADACK Islds-- ∴ Islands & Artic are in same relation
247c028 rd in Sandwich isld. & according to Chamisso in RADACK isld.-- p. 69. Sharks very generally distribut
270c103 d in higher animals than in vegetables. p. 243 RADIATE animals <tu> plants turned inside out. have po
100a053 ings, crystallization transverse.-- or rather RADIATING to central point. can cleavage be radiation f
352d058 ably will be dead-- hence there is no central RADIATING point, all united . (links in circle must be
100a533 r radiating to central point. can cleavage be RADIATION from some grand centre.-- A Stalactite of Gyp
581n065 seeing how simple an explanation it offers of RADICAL diversity of tongues.-- [Emotions are the here
612o035 ledge. lessening amount of ignorance [RHC] The RADICLE of plants absorb by physical laws of endosmic
534m063 country (thus Dahlias by snails)-- <The> Apion RADIOLUM undergoes transformation in the stem of Holly
242c017 of American forms in East. Ind: Archipelago. ‖RAFFLES. Horsfield. Diard. Duvaucel. Leschenault Kuhl.
327cIBC rd. Eastern Archipelago. probably some account RAFFLES. Sir. S do. do-- Buffon Suites Cline on the im
499q09v ossed & all monooecious plants.-- Hooker says RAFFLESIA is dioecious & Pollen must be carried by some
264c079 fection.-- to those it knew.-- see its passion & RAGE, sulkiness, & very actions of despair; «let him
539m084 andwriting same as Grandfather. Aug. 16th Anger «RAGE» in worst form is described by Spenser (Faery Q
028r031 Singularity of fresh water at Iquiqui. not from RAIN, because alluvium saline; Mem: on coast of Nort
028r031 hern Chili as springs become rarer, so does the RAIN, therefore such «rain» is cause, hence at least
028r031 become rarer, so does the rain, therefore such «RAIN» is cause, hence at least no water is absorbed
055r108 se areas. (& the counterbalancing variations) of RAIN. = The Bulk of sediment «daily» yearly brought
633j54r rrest the earth, than the earth revolves to form RAIN to wash down earth, from the mountains upheaved
045r076 jecting matter on the great rise). -- The great RAINS which attend severe Earthquakes «1822¿ 1835?» a
045r076 alone, (& the general belief in N. Chili, where RAINS are so infrequent; so as to exclaim, «as I have
370d116 hat some kind lay «pu» up honey even for single RAINY day-- & from case of wasps, is supposed cells p
056r109 s mass which has been corroded. -- If man could RAISE such a bulwark to the ocean, who would ever sup
286c154 pain, as well as we.-- It is our arrogance, to RAISE on the same shelf-- to (look at common ancestor
415e066 earth. like every other animal.-- Would anyone RAISE an argument against, my theory, should no fossi
472s01v den & all came up hybridised. It is possible to RAISE them pure for Miss Bent three years since gave
491q01v ubject for experiment.--«to repeat Spallanzanin RAISE only single Plants & only allow <few> one flowe
494q005 e blended instincts Remote Experiments-- Plants RAISE seedlings surrounded by various bright colours,
495q005 l seedlings vary much more than cuttings &c (4) RAISE annuals or common English plants in Hothouse &
497q06v are there races-- if so plant them together. & RAISE. seed.-- In letter Mr Herbert says do about OEn
499q009 drangea (20) As Hop is Dioecious-- seedsmen who RAISE Hop-seed-- may know something about proportion
031r037 , as in English Coal field-- because lowered & RAISED--so on--but gradually & simply raised No Fault
031r037 lowered & raised--so on--but gradually & simply RAISED No Faults in Patagonia[,] enormous extent; if
047r081 acter» reaches far beyond coast, which has been RAISED.--It must be considered as an oscillation, fro
047r082 beach I cannot understand, without (cs <[...]> RAISED above as). -- In great Calabrian wave did not
068r147 f the fluid propelling mass. If one inch can be RAISED then all can, for fresh layers of igneous rock
087a007 ains so cold.-- Siberia no plants to it, lately RAISED above level of the Sea. Lyells Encyclopaedia--
422e092 Bull; like fossil & recent shells of the <new> RAISED beaches» [not located] The enormous number of
447e165 sa becomes vivaparous on mountains & yet can be RAISED in gardens.-- Poa alpina, thougt generally viv
453e183 s> The possibility of different varieties being RAISED by seed is highly odd-- as it is not so with t
453e183 yoak reproduce each other. & yet I presume seed RAISED in same garden.-- now this good question-- sin
453e183 alf double.-- anyhow fertile because they «are» RAISED by seed.-- Where has Duchesne described Atavis
492q002 hen close planted together: <do> Can Holyoak be RAISED distinct by seed-- Heartease. 6. -- Do not spe
492q002 stamens are monstrous, how then are seeds ever RAISED? 11. Is not non-flowering gorze common in Norw
497q06v eet Peas, IMPORTANT, for if so, as these can be RAISED true, there is no crossing by Bees.-- Henslow.
498q007 tant to ascertain amount of variation in plants RAISED by Scions, as Elms. &c &c-- I have some reason
498q008 of» the varieties of Cabbages, peas, beans, as RAISED, do the Seedsmen select at all from the plants
501q011 Wheat-- Change of Soil-- crossing-- when seeds RAISED.-- His Book.-- 32. Would wheat from AEgypt rip
506q015 Black Duck & Penguin Henslow &c (36) Has not H. RAISED races of white & Blue Linum-- did parent plant
512q020 In cases where Lions have bred, have they been RAISED from young ones, bred in captivity --Mr Miller
585n077 ulty, or sense-- "We know not how, stonge henge RAISED, yet not instinct, but if all men placed stone
511q018 haracters appeared.= (4) Does he know any seed-RAISERS (5) List of qualities in birds & animals for p
512q019 Madeira Mr. Blyth (1) Mentions some breeder who RAISES many English birds-- will young wild ones bree
558m152 fear might enter?-- I believe common Swan, arch RAISES neck & depresses chin-- strikes with wing arch
565n017 to horse.-- Horse snuffs «& snorts», the air «& RAISES its head, & pricks its ears» when afraid, thou
377d140 effects of <earthquakes>, elevating forces in RAISING continents, & forming mountain-chains, when we
455eIBC iology Get Habberley to try experiments. about RAISING plants. where they cannot «crossed» etc.-- Mak
599o008 ducation des Mères Vol. I. p. 198.-- "Moralité, RAISON, beau ideal, infini conscience; voila l'homme

364d105 ons no new races.-- In Scandinavia besides the RAKHEKNA, before mentioned between Capercailzie & Blac
357d073 4. Reason given for supposing Tetrao media or RAKKELHAN is hybrid (produced commonly in Nature. both
053r101 tter Part Are there Earthquakes in the Radack & RALIX Islds? In my Cleavage paper Dr Fittons Australi
205b145 frequently predominates," p. 20. do "If hornless RAM be put to horned ewe almost all the lambs will b
551m127 y-- «in old man, he had just seen mind went on RAMBLING till excited by question.» Sept. 4th. Lyell i
429e114 h holds its place.-- When we hear from authors (RAMOND. Hort. Transact Vol I. p. 17 Append) that in t
429e115 d turns the balance.-- Hort. Transact Vol I. M. RAMOND. p. 19. do says lofty Alpine plants of & Pyren
429e116 d no where else not even in branch valleys-- M. RAMOND offers no explanation.-- Poet Cowper, describe
288c159 deteriorated, or altered, but it is certain that RAMS & bulls from England fetch very <go> large pric
564n004 nly hatched few hours placed on table & when fly RAN past it. cocked its head, & picked it-- Here the
565n010 -- I must be very cautious. Remember how Lavater RAN away with new Lavaters,-- Ye Gods!:-- says flesh
064r134 ood Hope Says at Santos «M Birchels» at foot of RANGE some miles from shore. rock of oysters quite ab
070r155 ounts of travelled boulders. from the Cordovise RANGE. Signor Rozales tells me at seven oclock Novem
081r BC cimetre 3. 8 Centimetre 4.4 {T} C. Darwin R. N. RANGE of Sharks Nothing For any Purpose A. Geology No
087a011 ssippi -- No Vol. I. p212. Cuvier Oss Foss Wide RANGE of Mammalia really very important. harmonizes w
114a093 of rivers turning round & penetrating their own RANGE in Australian Alps.-- Taylors Scientific Memoir
182b049 ed about self, but considering power, extending RANGE, reason & futurity. it does as yet appear clim
193b091 st Entomological Magazine paper on Geographical RANGE Richardson-- Fauna Borealis. It is important th
198b113 ll not explain status (Perhaps consideration of RANGE of capabilities past & present might tell somet
203b137 in short time. Milvulus forficetus «has a wide RANGE> is a tyrant flycatcher doing the service of a
208b153 n and animals} See R. N. p. 130 Speculations on RANGE of allied species. p. 127. p. 132 There is no m
213b170 ong remained are those ¿Lyell?, which have wide RANGE and therefore cross & keep similar. But this is
224b212 e importance Naturalists attach to Geographical RANGE of species.-- Definition of Species: one that r
226b223 lluscs, like Carnivorous Mammalia in their wide RANGE & in their duration of species. (¿are carnivoro
229b233 re-- Smith thinks several species of Rhinoceros RANGE from Abyssinia to extreme South coast. Elephant
229b234 orthern Coast. Hippopotamus do.-- Giraffe do.-- RANGE of East Indian Rhinoceros (?)-- Some paper in I
229b234 ian Rhinoceros (?)-- Some paper in Institute on RANGE of Bos in India.-- Range of Zebra?-- The Crocod
229b234 paper in Institute on range of Bos in India.-- RANGE of Zebra?-- The Crocodile & Tortise former inha
231b244 acquired reason, he would be limited animal in RANGE-- & hence probability of starting from one poin
243c020 nd at <New> Isle of France: xx instance of wide RANGE, where means of wide range work this out-- L. J
243c020 xx instance of wide range, where means of wide RANGE work this out-- L. Jenyns, about my fish New Ze
244c021 nts.--» It would be very important to show wide RANGE of fish & shells in tropical sea, it. would dem
248c029 is this great advantage in studying. Geograph. RANGE of quadrupeds.: that either created in each poi
249c035 rm two sections is this not connected with wide RANGE of animals. Follow this out, where species of s
249c035 in two words. have not species, generally wide RANGE? Mice.-- Waterhouse's remarkable fact of no for
253c046 cter. Hence Elephas primigenious over so wide a RANGE, & Mastodon angustidens.-- Ogleby has facts to
255c050 other classes this evidently relates to greater RANGE of such forms.-- p. 56: Ornithological part of
258c060 fecting the area equally.-- Animals having wide RANGE & all kinds of information-- instinct Swainson'
260c069 ogy a mine of valuable facts,. regarding habits RANGE & & all kinds of information-- instinct Swainson'
268c099 acquarrie Isd.--» very good. Study D'Orbigny. & RANGE on West Coast «Guayaquil & Peru» Henslow in tal
271c105 usicus «not found in N. America» whose Southern RANGE is? <One> The black & white thrush of Azara bui
277c127 two species come over to this country, without RANGE, or habits ascertained-- put them as (a). (b) u
277c127 ve brought home new species. until, he can show RANGE & habits-- Take instances of most disputed shel
287c158 st chiefly rest on these same organs,-- habits, RANGE. &c &c-- Macleay rests his whole groundwork of
298c191 ation of isld.-- We must imagine a cosiderable RANGE of one species «on-- mountain side» of which t
304c207 -- EXCELLENT PRINCIPLE OF ABORTION ISOLATION of RANGE « <far more prob>» » tends to alteration views.--
311c227 insectiferous <insects> quadrupeds geographical RANGE very good.-- Blainville Ovington's Voyage to Su
316c244 r.--» good anecdote --.Sowerby.-- Geographical RANGE of shells like Cryptogamic plants. of Marine ki
316c246 merica.-- get instances.-- very good anomaly in RANGE + + What circumstances have led to formation of
319c257 proper.-- Irish Elk case of fossil geographical RANGE.-- [blank] Books examined: with ref: to Species
324c268 ealis Entomological Magazine (paper on Geograp. RANGE Study Buffon on Varieties of Domesticated anima
326c266 urious coloured maps. by Copenhagen Botanist of RANGE of plants Books quoted by Herbert. p. 338 Schie
338d025 & not by fresh creation of new forms.-- what is RANGE of Hyaena? Hippotamus.? Indio-African, or pure
354d065 Phillippensis. Philippines Man have varies the RANGE-- Argue the case of Probability. has Creator ma
365d107 s-- <restric> those <restr> restricted in their RANGE by men & by art.-- the former only giving avera
414d065 les. strictly applicable to the universe.-- The RANGE of man is not unlike that of animals transporte
483z013 bed Shells from Tahiti and Chile The North & S. RANGE of shells might perhaps be worked out with adva
500q10a pecies of Birds in New Zealand, plants so few-- RANGE of mundane genera, «in Birds» in accordance wi
500q10a f mundane genera, «in Birds» in accordance with RANGE of species?-- Are there any fine doubtful speci
509q017 the mundane genera, the species <are> have wide RANGE-- How is this in «Plants??» Are abortive organs
573n036 ded with its instincts its place in nature. its RANGE, its-- &c &c:--must be a special act, or result
641j29r ewn that they most widely differ» 3 A very wide RANGE must be destructive to species, when physical c
048r085 the losses of the «species of» Mastodons. which RANGED from Equatorial plains to S. Patagonia. To the
036r052 onia Does Andes in Chili. separate geographical RANGES of plants. V. Lyell. Chap XI Vol II. Urge the
069r152 Bolivia Transportal of conglomerate between two RANGES mysterious!-- Mem. SUBSIDENCE Uspallata of whi
135a145 vol 7. p. 522. Mountain c near Caubul. parallel RANGES. with here & there little branches at {P} from
316c245 groups.. Have Dioecious plants more restricted RANGES than other plants.-- Many «Some» genera confin
318c253 re taking place & how birds are extending their RANGES. «even migratory birds, lik swallows» -- degre
410e052 pecies (& naming them) with relation to habits, RANGES. & external conditions of country, most import
424e101 d crow & Carrion crow. have in Europe different RANGES-- latter not going North of the Elbe.-- yet th
640j167 a most striking anomaly to this. Have they wide RANGES? Agassiz has shewn that they most widely diffe
192b087 me common progenitor.-- Now if the intermediate RANKS had produced infinite species probably the seri
319c256 t-billed birds are inferior to ours, & our lark RANKS very high.-- Upon the whole thinks <many> more
480z008 nae from S. Seas written in German.-- Stuttgart RANKS these bodies amongst Vegetables in Linn. Soc.--
470t178 y & Blue» Butterflies at Clover,--Veronica--, RANUNCULUS in numbers =what insect can get honey out of
490q001 ecies of Aquilegia vary much in their spurs & RANUNCULUS in the nectaries. The former best for my exp
496q006 me double.-- There is a double Crows-foot. or RANUNCULUS.-- (11) Try.. Nitrate of Soda-- Salt. Gypsum.
035r045 clearly be chemical differences. & not those of RAPID cooling &c &c My results go to believe that muc
105a072 atter (it is generally said) is consequence of <RAPID> slow course, & with slow course small erosive
106a074 at. There could «not» be great deflection in a "RAPID".-- is a familiar illustration.-- Therefore str
137a151 perature beneath the sea, is probably much more RAPID than beneath continents In Berlin Transactions
152g046 elves effect of corrosion & not cause. Monday a RAPID descent of a terrace except at very head of val
285c153 as show such Physical changes will be unequally RAPID, with respect to their effects The AEgyptian an
304c208 es from typical structure have either been more RAPID than in all other birds, or that it sprung from
398e005 &c.-- «if the change could be shown to be more RAPID, I should say there was some link in our train
447e164 higher animals; changes seem to have been more RAPID, & the facility for inter marriage is greater (
478z004 ement of animals in rays Par un officier du Roi RAPID growth of Coral-- RN. p 24 Bougainville Voyage
548m114 irium & sleep mental rest. though. most vivid & RAPID thought.-- There may be some «two or three» tra
027r024 are now seen in regular descending steps Mem.; RAPIDITY of germination in young corals.--vide L. Jack
254c048 or embryo succeed each other with the greatest RAPIDITY"-- so we find species each class successively
313c234 s law of duration apply to utter extinction or RAPIDITY of specific change.? <One> the first would be
362d099 for war} & hence never fall off.{ Curious the RAPIDITY of the change in 5 or 6 weeks after castratio
377d140 f the sea, on beaches-- we really, measure the RAPIDITY of change of form, & instincts in the animal
544m108 ceptions.-- The mind thinks with extraordinary RAPIDITY-- We may conclude that neither number, vividn
544m103 e may conclude that neither number, vividness, RAPIDITY, novelty of separate ideas cause fatigue to t
126a123 hotter.-- hence temperature ought to increase RAPIDLY beneath level of sea.-- deep seated springs «s
181b039 only breeds; like individuals in a country not RAPIDLY increasing.-- If we thus go very far back to l
224b214 & hence remain distinct. Where country changes RAPIDLY, we should expect most species.-- The differen
427e108 ations of animals & plants to each other would RAPIDLY increase, & hence number of forms. once formed
448e165 t is individual.-- Most plants which propagate RAPIDLY by buds, layers &c & &-- do not seed freely.--
038r058 m subaerial one?--In former not so much; or no RAPILLI; & from action of water probably not so much a
116a099 lau on Granite Edinburgh Philosophical Journal RAPPORT on D'Orbigny's Voyage. good section of Rio Neg
295c183 forms only species!! No salamanders (D'orbigny RAPPORT., 11) in S. America so highly developed in N
480z007 es L'Institut 1837. No 212 Observations on the RAPTORES of S. America translated from D Orbigny no IV
482z011 8 waders. 22 palmipedes: out of the first 9.:4 RAPTORES. Falco poliosoma -- novozelandiae -- histrion
312c227 often in number 5» will have long tail.-- in RAPTORIAL birds, & tigers & sharks, being spotted, & co
368d114 plumage.-- (The females (as Owen observes) in RAPTORIAL birds largest.-- p. 47. (<"> is evidently the
022r007 & Concepcion. P. 417 Veins of quartz exceedingly RARE Mem C. [Cape] Turn P. 434 & 419 As Limestone pa

(Key Word)

```
025r019 t 80 fathoms & upwards. that life is exceedingly RARE, at the bottom of the sea.--«certainly data ins
040r063 ss.-- Neither lakes or Avalanches (Glaciers very RARE) to cause floods in valleys, which must aid  in
040r063 arting of pitchstone; which is described as very RARE Mem. St Helena; probably more abundant in  this
073r161 tly found mixed with lead & copper is infinitely RARE in all parts of the globe". p. 113 How  utterly
075r167 t very abundant.--\ muriated silver. which is so RARE in Europe. common there accompanied by molybdat
076r170 ys & phonolites do. amphibole quartz & mica very RARE.-- ancient freestone & breccia is the same with
115a096 in  bed of ocean With the exception of sandstone RARE to have any horizontal non cleaving beds. metam
200b122 they can proved descendant, which of course most RARE, or when placed together they will breed.-- But
217b183 e, (must probably have occurred to every one) of RARE breeds of dogs, from owners great care of them.
234b256 rieties of a Harpalus. common at South end, but <RARE> «absent from» near London. = Dr. Smith, he say
289c160 though common on another, (golden creted wren so RARE in some countries-- nightingale do.-- all shows
297c186 ing death, makes the group aberrant When species RARE we infer extermination, when group few in numbe
309c222 ar or linear arrangement.-- Thinks passages very RARE., in anatomical structure.-- «the passages betw
332d002 t cancer in women, & since that time it has been RARE disease.-- but now (July 1838) he has seen more
352d057 , so may species retrograde, but these facts are RARE.-- 2d Sept Those animals which have many ABORTI
423e099 n upper part of bed-- & another in lower is very RARE, the conclusion will be that our greatest forma
486z018 r far N. where this other order is comparatively RARE.-- These views clearly explain rarity of insect
499q010 es of single clumps of plants in counties, as of RARE green Cotton Plant-- How large «area» clump the
508q016 . A Smith. About species of Rhinoceros. becoming RARE beyond limits of the metropolis of each-- Cause
537m076 ged suddenly-- but are not such «sudden» changes RARE,-- as when Polynesian mothers ceased to destroy
594n112 rable distance a very large hawk, which are «so» RARE « <s.> » here,, that probably few had ever befo
073r163 dt. New Spain. Vol III. p. 130 Metals in Mexico RARELY in secondary alway in primitive &  transition;
073r163 ary alway in primitive & transition; the latter RARELY appear in central Cordillera. particularly bet
151g039 eloped on steep earthy slope, two circumstances RARELY united.-- die away also, without any cause, mu
199b120 ication offspring not fertile. or at least most RARELY & perhaps never female.--no offspring: physica
288c159 one»-- the native dogs howl most dismally, very RARELY bark.-- are almost useless not the least notio
446e162 hree by buds & seeds «though by the latter very RARELY»} here is a case in answer to Mr Knights doctr
447e164 eeds All species of Lemna sometimes though very RARELY flower [bu]t the one does on the continent-- w
496q05a e pollen of such flowers as do not seed or seed RARELY-- Magnolias. «Azaleas» & plants grown under un
621o047 en grown up often conquers it). It will be only RARELY that it thinks that nasty, which the natural t
621o047 ing conduct of others, the feeling that almost (RARELY if opposed to natural instincts) any action is
028r031 m: on coast of Northern Chili as springs become RARER, so does the rain, therefore such «rain» is cau
408e043 system may. produce-- ? When a species becomes RARER, as it progresses towards extermination. some o
365d105 between Pheasant & Black game is owing to their RARITY., a single female in wood with Pheasants would
486z018 mparatively rare.-- These views clearly explain RARITY of insects in T. del Fuego.-- Hence it is  odd
266c093 89.-- Lieut Wellstec "on coast of Arabia between RAS Mohamned & Jeddah". sheep numerous "of two kinds
273c111 flight & great movement in the air, & likewise RASORIAL species, & likewise perching (Gould), but the
274c113 ys, he believes does. but also on fruit.-- The RASORIAL type is wonderfully shown in long legged cuck
274c114 er» Secretary bird.-- & Millisuga. Kingii very RASORIAL for type.-- Now here I must observe these cha
274c115 ifferent value have their represetatives;, the RASORIAL may be observed even in Lessonia &c & In rela
468t106 Bees  almost every flower-- Blue-bells-- wild-RASPBERRY--leeks-- Flowers which thought very unattract
185b055 n tringas.-- Upland goose.-- water chionis water RAT with land structures; :¬law of chance would caus
212b166 eedings of Zoolog. Soc June 1837 p. 53. an Irish RAT.-- different from English. Waterhouse has inform
212b166 Waterhouse  has information respecting the Water RAT.-- ¿Consult Dr Smith History of S. African Cattl
248c029 transported.  ¿What section does the New Zealand RAT belong to There is this great advantage in study
278c131 e, excessive improbability. Mem in Clifts list a RAT said to have been found!! rodents old inhabitant
354d065 Argue  the case of Probability. has Creator made RAT for Ascension.-- The Galapagos mouse probably tr
109a083 -- ¿ The preservation of dikes & ledges of first-RATE importance in showing not subaqueous removal--?
120a109 surface.--  if elevation then went on at greater RATE, not only river would carry further its own mat
132a137 first that data wholly insufficient to calculate RATE of increase of heats in earth's crust.-- yet he
195b102 o another.-- let geological changes go at such a RATE, so will be the numbers & distribution of the s
344d041 ble harrier from Ireland to Brighton Park--first RATE bitch-- tried to breed from her, but her offspr
423e099 t slightly modified organic forms.-- we know not RATE of deposition has been equal even in one bed, m
030r035 sea.--(Character  of coast regular & <not very> RATHER deep soundings, 60-100 fathoms 2 & 3 miles fro
033r042 at  Coquimbo: in that case, may not central and RATHER differently constituted lime have been removed
033r043 mice «& Obsidian:» in the I Vol. Humb: There is RATHER good abstract of Humboldt. S. American Geolog.
049r087 dip  is not parallel case to Isle of White. but RATHER to one out of a series of faults. [Fig. 4] {P}
051r097 a. Australia. profoundly deep: a great fault or RATHER many faults.-- Necessary form; as long as coas
056r111 ent, as well as separating causes by water.--Or RATHER begin & explain how water separates.--(intertr
100a053 - agate rings, crystallization transverse.-- or RATHER radiating to central point. can cleavage be ra
100a055 etween Australia & Oolitic period.-- comparison RATHER loose.-- perhaps worth Says from Lardner's (p.
146g013 r end of Loch Dochart buttresses of Alluvium or RATHER mass of well rounded pebbles in yellowish argi
150g036 thoner 3 2 Terrace 3 Alluvials 3a 3a 2 Mass 3 2 RATHER longer than 3a Sunday In Glen Collarig, when w
153g056 anite--«band» 4 X 3 X 2 «feet» & 2 deep Another RATHER smaller block 30 ft «above» & other 50 ft lowe
159g086 om Glen Turrit to Fort Augustus Barom on upper (RATHER above)? shelf 29.290 A. 69 degree Air 68 degre
180b036 of relation. C & B. the finest gradation, B & D RATHER greater distinction Thus genera would be forme
199b120 n intermarriages; smallest differences blended, RATHER stronger tendency to imitate one of the parent
218b191 maryllimide & Narcissus. Mr Donn considers Mr H. RATHER wild Mr Donn remarks to me. that give him a sp
227b225 nt.∴ little service habits in classification. or RATHER the fact that they are not far the most servic
246c026 y differs at all from those of Europe, but beak RATHER sharper., & rather longer in proportion, colou
246c026 om those of Europe, but beak rather sharper., & RATHER longer in proportion, colour slightly differen
254c049 ny forms in Acrita,-- typical of other, (surely RATHER parents). (NB These views must lead to spontan
255c053 isolate  your species her plan is frustrated or RATHER a new principle is brought to bear. If man cre
294c176 e resort to quite different origin when species RATHER further.-- once grant good species as  carrion
314c239 , some SPECIES of Australian GENERA. good case. RATHER large flora. (150?) Mr Brown did not observe s
315c242 ies of first <thoughts> «memory» in children or RATHER their memory. very remarkably-- scenes in them
333d005 se habits not different; tone of voice. perhaps RATHER different».-- crossed with <un>common cat, exa
336d019 laws, the character of offspring would vary, or RATHER they would not have offspring-- On the idea of
375d135 nto the gaps <of> in the oeconomy of Nature, or RATHER forming gaps by thrusting out weaker ones. «Th
383d160 asant. Male bird, said to be infertile.-- spurs RATHER smaller than in <ma> silver male-- Head like s
389d173 globules: generally sufficient for one birth or RATHER]CD \It should be observed that the constant ne
404e025 ription of curious mechanism of respiration, or RATHER ventilation peculiar to <the class» «some orde
416e071 there are true hermaphrodites.-- I suspect this RATHER effect of liquid semen: therefore animal  life
439e142 ers, formerly possessed--'that is animals> «or RATHER the parents» having passed through many change
446e161 in parts of the Northern «French» expedition,-- RATHER the reverse of facts stated by Smith of Jordan
461t037 .--«)» & therefore probably any structure would RATHER become accomodated to new circumstances than i
468t105 dant flowers not attractive, very small--stigma RATHER large & rough-- flowers common-- many winged t
485z016 ared with those of England.-- or ground birds-- RATHER indefinite letter Mem Orpheus--becoming tyrant
522m013 tions» acquired.-- this may be doubted, whether RATHER not going against natural instincts.-- My Gran
533m060 repentant.--  In making too much profession, or RATHER in only fully expressing momentary feelings of
550m126 "have notion of masters property"-- is not this RATHER more friendship.-- Scott's Life. Vol I, p. 127
555m141 be discoverer "reasoning" or "reasoning"-- only RATHER more steps.-- dispute about words.-- Miss Mart
564n003 nst the existence of such an attribute would be RATHER favourable to it--!! Man moreover who  reasons
571n030 bits. & perhaps even latter may be vitiated. or RATHER altered. The Reason why New Buildings look ugl
587n089 clearly  insisted on assoc of ideas & emotions. RATHER ideas & bodily actions make the emotions.-- p.
605o20v . taste evidently does not consist of this. but RATHER in the power of discriminating & respecting go
606o024 jects are not fit for painter or sculpture, but RATHER subjects which we know, it is therefore the em
609o030 ical» + What has produced the greatest good «or RATHER what was necessary for good at all» is the «in
610o033 st a question whether we have any instincts, or RATHER the amount of our instincts-- surely in animal
620o045 ed by «obeadiance to instinct» <conscience>., or RATHER the strengthened instinct, even when our reaso
624o051 me instinctive, this part of argument fails, or RATHER is weak.-- [RHC] Better simply put it, benefic
624o051 y be trained to hunt one pig sooner than other, RATHER than change hunting instinct. *Our tastes in m
626o052 ngs, as of cowardice.-- the whole appears to me RATHER rigmarole.-- He does not say anything about an
639j28v ances determine that, the long legged one shall RATHER oftener than any other one. survive. in ten th
640j167 infer that Mollusca would offer few species, or RATHER be very slowly changed & vertebrata much so.--
090a023 fell, that was weighed,; <but> carrying on this RATIO I can count 90 stones which have fallen in  the
099a051 particls  attract each other in some increasing RATIO in proportion to proximity would they not unite
348d050 he organs judged to be of importance in inverse RATIO to their variability.-- (Now caeteris paribus t
```

Page
(Key Word)
375d135 y death.. population in increase at geometrical RATIO in FAR SHORTER time than 25 years-- yet until t
540m088 y. being real pleasure at pain of others, with RATIONAL desire to assist them,-- otherwise as he rema
566m010 on Pop. p. 32, origin of Chastity in women.-- RATIONALLY explained.-- on the wish to support a wife a
6290055 feeling of right & wrong.-- arrived at first <RATIONALLY> by feeling-- reasoned on, steps forgotten,
036r052 . woody bushes, «gazelles» hares, grasshoppers & RATS. characteristic of the deserts of Syria <chara>
046r079 afflux of the former. -- Ascension. Vegetation? RATS & Mices. At St Helena there is a native mouse D
062r129 istances; generally first arrives:-- New Zealand RATS offering in the history of rats, in the antipod
062r129 s:-- New Zealand rats offering in the history of RATS, in the antipodes a parallel case.-- Should urg
245c024 e Barbyrousa; a Cervus near Marianus new, & some RATS & mice. In Amboina only Cuscus & Barbyroussa «N
437e139 ling them.-- "Sometimes it seems hares, rabbits, RATS & not being sufficiently weakened by wounds got
148g027 d towards river all these composed-- where side RAVINE entered terraces formed successive bays but pl
150g038 house, the 3d below them opposite to where side RAVINE enters On opposite side of valley both extend
152g045 rig good case of shelves entering «on» one side RAVINE. Are the lip, or necks of land on level with s
155g063 plains, as in Corry. & as «as I believe in side RAVINE above houses of Roy» Maccullochs supernumerary
156g069 h notched on Maculloch's supposition;-- the old RAVINE, where water entered are not proportionately l
153g054 earthquake cause collection of sediment? Where RAVINES enter side by, opposite entrance into Glen Fin
077r171 ing beds tells the secret.-- p. 189. "The small RAVINS into which the valley of Marfil is divided, ap
078r175 yielded the most metal, where the direction of RAVINS, and the slope of the mountains (flaqueza del
078r175 Guanax.--the other E & W.--veins richest not in RAVINS or along gentle slopes. but on the most elevat
430e119 s one has concentric striae, all the lower part RAYED longitudinally (give woodcut) like I. sulcatus.
209b155 istribution (after physical changes) would be in RAYS-- from certain sports.-- Agrees with old Linnae
317c248 e's Selbourne." many references very good. also "RAYS Wisdom of God." Often refer to these.-- Also so
321c275 ia «trash» skimmed Macleay's Horae Entomologica RAY'S Wisdom of God references at end-- The British A
321c270 Vol II. (read remainder) when out [not located] <RAYS Wisdom of> Lisiansky's Voyage round world. 1803
478z004 ous property-- Curious arrangement of animals in RAYS Far un officier du Roi Rapid growth of Coral--
599o007 uns throug every fibre, when one behold the last RAYS of & & on grand chorus are utterly inexplicable
575n045 both, as when one burns end of nose with a hot RAZOR.-- joy <p> a mental pleasure. with pleasure of
116a100 y of mountains of Cordova project on plain, like <RE> a reef on a sea beach-- «p. 151» first discover
032r039 f cliff (Z). to which point the waves would not REACH. If now the ocean should suddenly fall, (3) the
064r134 e miles from shore. rock of oysters quite above REACH of tides.--thinks them same as recent species.
089a017 [Might not bottom of ocean boil; yet heat never REACH surface.-- Journal de Physique, et D Histoire N
093a031 first injected.-- Basalt: last because it could REACH the surface. before being cooled.-- Berzelius.
102a059 ded conical axis of mountain.-- only when dikes REACH near the surface. that strata yield.-- In Undul
388d173 nt is-- it also shows that semen. must actually REACH the ovum.-- [Why in making a bud, which is to p
427e110 d more indelibly stain offspring-- it would not REACH one apple sooner than <other:> «that of another
054r105] [not located] Isld near coast of America not REACHED. Juan. Galapagos. Cocos-- Ulloas voyage North
118a103 ce evident only small proportion of dikes have REACHED the surface Arguments against Herschel's view
162g100 ir 65 degree? <.194 372 about 267 28.75 .105 I REACHED> 29.090 Preservation of form of land very much
187b062 cold countries.-- Seeing how horse & Elephant REACHED S. America.-- explains how Zebras reached Sout
187b062 hant reached S. America.-- explains how Zebras REACHED South Africa-- It is a wonderful fact Horse, E
364d104 animals, that the amount of variation is soon REACHED-- as in pidgeons no new races.-- In Scandinavi
468t111 «small. Humble alighted on base of filaments & REACHED nectar =again= between them, hence quite below
047r081 t» Moreover wave «with same general character» REACHES far beyond coast, which has been raised.--It m
450e175 wo breeds but small -- Java pony occasionally REACHES 13 hands.-- Phillipines Pony somewhat resemble
289c162 ethod of generation-- Heaven knows how.-- This REACTION takes place in every organ‖ «Hence «method» o
460t019 t; if» «of this old stock (which from action & REACTION grew more complex)» some perhaps rendered mor
043r069 st be more exclusively confined to that country. READ description of channels or grooves in rocks at
044r072 oncretions & Galapagos.-- «Humboldts. fragmens.» READ geology of N. America. India.--remembering S. A
047r082 explained. also the similar fact at Concepcion? READ the various accounts & see if fall is not the f
086a006 below-- Malte Brun «Salt Lakes» Siberia must be READ as well as Pallas before Geology is written Cuv
116a099 of matter would have same effect as too coarse. READ Kylau on Granite Edinburgh Philosophical Journa
120a110 on Salsisbury Craigs. Kept amongst <old> papers READ before societies.-- Sir. J Hall Vol VI. p. 173.
122a113 tudy different forms of earth as shown by arc.-- READ Herschels astronomy with oscillations of level.
129a131 iliting on Valleys. «Tertiary strata of S America» READ parts of this work, though it is but poor. Athe
129a132 t least in depth» near mouth of Columbia river-- READ Mr Parker's Book.-- M. Bichoffs Papers, in Edin
183b053 ing as great reptiles must have once declined.-- READ his theory of the Earth attentively Cuvier obje
206b149 erous trees; or in certain shell cephalopoda.-- «READ Buckland» L'..Institut «1837.» p 319 -- Brongni
236b276 ed plants produced by seed.-- Lychnis.-- Flax.-- READ Swainson [blank] In production of varieties is
241c016 Ann: des Sciences. Semptemb. 1825 Get Henslow to READ over the pages from about 8 to 20 of Zoologie o
250c040 n isld. p. 120. «ref.» Philosop. Transacts 1823. READ June 5th) important paper by Dillwyn, on replac
250c040 iary periods.-- p. 125. ref. to Phil Transacts. (READ November 20th) Paper by Jenner, on birds seen f
260c067 tudy Bonapartes list In the Zoological Journal I READ a curious account to show that very many birds
268c096 Wiegman Classified catalogue of animals of Nepal READ before Linnaean Soc. Feb. 1838.-- Annals of Nat
291c166 ty effect of organization. oh you Materialist!-- READ Barclay on organization!! Avitism in mental str
292c170 nities. ie structure of the whole animal let him READ Mr Swainson's on the Classification of animals.
313c233 use mammalia can subsist where parasites cannot» READ Entomological Transactions Why if louse created
315c241 blished St. Petersburgh Academy Imperial-- Paper READ in 1837. semestre I suspect some valuable analo
320c276 gh New. Phil Journal. about 13 numbers have been READ Voyage a l'isle de Frances Voyage de l'Astrolab
320c276 aphical Journal Asiatic Journal to end of 1837. RE«A»D-- contains very little Macleay's letter to Dr.
321c275 Entomological Society Vol I. & 1s No of Vol II. (READ remainder) when out [not located] <Rays Wisdom
322c270 powers. Hunters Animal OEconomy. edited by Owen. READ several papers-- all, that bear on any of my su
322c269 ages, skimmed well. do Lutke's Voyage. carefully READ.-- Reynolds Discourses Lessing's Laocoon Whewel
324c268 pter» Crabbes Life June Is. King & FitzRoy To be READ Humbolt. New Spain-- Much about Castes &c Richa
325c267 out close species) Wilson's American Ornithology READ Aristotle to see whether any my views very anci
327cIBC to see whether peculiar plants-- in high points READ Volney's travels in Syria Vol I. p. 71. account
344d040 ence to "new species & hybrid doctrine"-- I have READ there are exceptions to this in some larvae of
366d112 hat the tame donkey has. CD[old Buffon should be READ on Mare My view, why hybrids are infertile. sup
390d179 lf, for it should be observed that from Books to READ Buffon Suites de.-- Horse & Cattle Library of U
413e060 intoceros, Hippot, Haena &c are found together.-- READ this Work-- Dech. 4th.-- Why has the organizati
523m015 im. & yet disinheriting her.-- This «N B. I have READ paper somewhere on horse being insane at the si
531m050 ecause some people can remember poetry when once READ over.-- The extreme pleasure children show in t
533m061 kind of conscience, is obscure memory of having READ or thought of some such remarke as now advanced
539m081 n this I was interested as was I in the other, & READ so intently as to be unconscious of all around,
551m128 f the soul, are not derivable from experience.-- READ monkeys for preexistence <">-- The young Ourang
558m151 ght," as well as the works of the whole world.-- READ Mackintosh on Moral sense & emotions.-- The who
559m155 last even to the perception of a final cause.-- READ. Paper on consciousness in Brutes & Animals. in
572n033 0th-- Dreamt somebody gave me a book in French I READ the first page & pronounced each word distinctl
604o015 n the Literary Magazine. 1756-- Ceased in 1758-- READ the Review or the article. A Planaria must be l
610o031 irected many letters to»-- could not <remember> «READ» Christian name; fancied it looked like. W. but
641j29r be cautious.-- <some others>: study these facts READ Lacépède on Cetacea & Geographical Distrib of l
179b033 tate.-- No doubt «C. D.» wild men do not cross READILY, distinctness of tribes in T. del Fuego. the e
183b051 ng believed same species) if they do not breed READILY. point in view.--¿whether highly domesticated
204b141 on, winged & turned on pool, first season bred READILY with common ducks.--» Kirby all through Bridge
248c030 two» varieties of many ages standing, will not READILY breed together: The argument must thus be take
288c159 al Durham breed.-- Native dogs & English cross READILY-- think about half way in appearance.-- bark a
360d094 il, & other in having very short tail.-- I can READILY see that two first might cross easier than two
363d102 ere pointed out Selby confounded them, yet can READILY be told by incubation & other peculiarities.--
385d163 he wolf of Hunter.-- young take distemper very READILY & are subject to fits.-- «there is great diffe
427e110 fertilize it» fertilize an apple somewhat more READILY, «than other apples» but probably would more i
498q008 he minute Orthopt.-- important, as we know how READILY they cross.-- (9) In the nurseries, when «seed
532m054 ted much more apt, this partly owing to heart? READILY taking same movements, senses being on the loo
554m139 Listened with great attention to Harmonicon. & READILY put it. when guided to her own mouth.-- seemed
560m156 cases like D. Corbet; «do» ideots form habits READILY?? Do the Ourang Outang like smells «peppermint
623o051 under unknown conditions) (for pig will not so READILY attain instinct of pointing as a dog.-- also a
624o051 n accordance to beneficial tendency» will most READILY affect. the instincts, for they are in accorda
357d073 Caperalkie & cock Black-cock.-- (Curious the READINESS with which this genus becomes crossed. ¿is re
565n006 tention to bite at moment, but mere symbol of READINESS, & therefore done in extreme.-- Looking at on
050r090 apagos. <|> Sir George Mackenzie must be worth READING Some earthquakes of Sumatra no connection with

130a134 831. sur le temp du globe on Volcanos &c worth READING. L'Institut. 1838 p. 360. on orbicular trap th
135a144 y good geological paper. by Malcolmson-- worth READING-- Burnetts. vol 4. p. 193 in Lat 26 degree S.
198b114 sophie Anatomique. 2d Vol about monsters worth READING NB well to insist upon <different animals> lar
269c103 spontaneous generation not improbable.-- After READING "Carus on the Kingdoms of Nature, their life &
326c266 ldren Turners embassy to Thibet, perhaps worth READING quoted by Malthus.-- Heberdens Observat. on in
326c266 Anat. & Physiology.-- Dampier. probably worth READING Lessings. Laoccaon.-- (translated in 1837) on
326c266 rnithological Biography. 4. Volumes well worth READING Bevans work on Bees, new Edit 1838 Harlaam. Ph
404e025 herefore effect of climate.-- Octob 19th. When READING. L'Institut: .1838 p. 329-- Milne Edwards, des
478z004 nnett on Chinchillidae Zoolog Transacts. worth READING Cuvier's Memoire 133 1803. on Pennatula showin
539m081 ache «after good days work» which came on from READING «review of» M. Comte Phil. which made me «ende
539m081 mmediate manner in which my head got well when READING article by Boz.-- now in this I was interested
540m087 l, nor recalled by obvious associations. as by READING a book.-- Consider this.-- "The fledge-dove kn
546m110 d the fact to Mr Mayo himself. she was one day READING a book, with ivory paper cutter, which she val
559m155 's essay on the Human Understanding well worth READING Copied <Smith> «D. Stewart» lives of Adam Smit
559m155 D. Stewart» lives of Adam Smith Reid, &c worth READING. as giving abstract of Smith's views «Take & p
572n033 ke» «when awake» I could not picture to myself READING French book quickly, & <running> «running» ove
282c143 ut the truth would never be discovered When one READS in Ehrenbergs Paper on Infusoria on the the eno
291c168 razilian species would migrate south ward being READY made.-- & so destroy individuals, wheras in Fal
294c176 afterwards, (forgets authority).-- Lonsdale is READY to admit, permanent small alterations in wild a
298c189 ould be «most» like Australian.-- Curious this +READY answer, without any leading question.-- [+] Thi
521m009 > the servant showed him watch & said dinner is READY, what, what.-- then showed the watch upon which
533m059 e-- Instinctive walking of animals. that is the READY movement & co-relation of the proper muscles. m
533m060 urely occurred to me that Capt. F. R. candour & READY confession of error made him less repentant.--
541m092 21st. 38 When a dog in play has his mouth open READY to bark, & lip twisted up, in that peculiar man
553m135 view cannot be, anyhow, easily overturned.-- so READY is change from. our idea of causation, to give
556m144 orses in Omnibus instantly start when they hear READY, but if they see anything ahead. which ted cann
557m147 make it very stiff «& back» when savage «no» & READY to dash at prey streched out & flaccid, when fu
025r016 Itapicuru [R.] [11 degree 46 minute S.] 9 200 R. REAL [11 degree 31 minute S.] & [R.] Sergipe [11 deg
078r175 30 degree to NE. vein of Moran 84 degree NE. of REAL del Monte 85 degree to S. // Tasco 40 degree to
186b058 alogies; CD[from the principle of atavism, where REAL structure obliged to be altered, I can conceive
189b071 same colours Strong odour of negroes, a point of REAL repugnance.-- Waterhouse says there is no TRUE
190b079 ed, but genera stronger.-- Mr Waterhouse says no REAL <separ> passage between good genera-- How remar
211b162 rnitho Rhyncus Would not relationship express, a REAL affinity & affinity-- whales & fish.-- Progeny
218b189 logy & Botany Vol I p. 450 There is in nature a «REAL» repulsion «amounting to impossibility, holds g
228b230 eds of Ribston Pippin & Golden Pippin &c produce REAL crabs, & in each case similar or mere mongrels.
231b243 pics, (whether one fossil or not) are related by REAL relationship. as well as effect of similar temp
259c065 on during earliest existence; if whole life then REAL adaptation The case of heredetary disease, is o
281c138 is undescribable, yet holds good, so does it in REAL classification The relation of all cock birds i
285c152 in the course of ages.-- As species is <certain> REAL thing with regard to contemporaries-- fertility
292c170 superficial knowledge «like myself» of <classi» REAL affinities. ie structure of the whole animal le
335d015 similar offsprings, so will the worst mules (as REAL mule) have offspring,-- slight deformities «as
348d051 erally,, though how far we can ever discover the REAL relationship is doubtful.-- not till much knowl
388d170 therefore wants independent supply of food.-- is REAL. difference-- but this does not apply to potato
405e035 eparated, or having never [not located] ARGUMENT REAL of antiquity of reasonable cosmopolite man.-- L
446e163 g all the individuals-- «-- hence there would be REAL gradations in species from one region to anothe
481z009 lao. migratory fish.-- See Kings drawings.-- for REAL name Birds of Iceland. Mackenzie. p 345 for com
524m019 madness is forgetfulness.-- which does appear a REAL difference, between oddity & madness.-- but the
526m029 k all memory consists in a set of sketches. some REAL-- some fancied.-- this fact of early memory con
527m034 stles in the air are highly advantageous, before REAL train of inventive thoughts are brought into pl
527m035 ten--, so as to feel a severe disappointment «in REAL train of thought this does not happen. because
528m035 o believing a vivid castle in the air, or dreams REAL again explains insanity.-- Analysis of pleasure
528m036 urs, <whi> & their absolute beauty. (which is as REAL a cause as in music) from the splendour of ligh
536m073 ster. so may free will make change in man.-- the REAL argument fixes on heredetary disposition & inst
537m074 ther a man intentionally can wag his finger from REAL caprice. it is chance, which way it will be, bu
540m088 sympathy.-- Mem: Burke's idea of Sympathy. being REAL pleasure at pain of others, with rational desir
541m092 hat peculiar manner they do, even more than in a REAL snarl, they are enjoying a satirical. laugh.--
541m092 are enjoying a satirical. laugh.-- when snarling REAL bitter sarcasm.-- <These> Seeing how ancient th
544m103 ly our consciousness, & senses tell us it is not REAL. = = dreaming appears clearly rest of the mind,
556m146 en kicking.-- -- good case of expression showing REAL affinity in face of donkey, horse & zebra. when
566m011 culty in the form of a true instinct, which is a REAL instinct in the chicken, just bursting from egg
568m017 ss with strangers» «as in case of temperance, or REAL virtue, that is action which experience shows w
571m029 s acquire some general notions, which are taste? REAL taste in mouth, according to my theory must be
604o018 often apply the term sublime, where there is no REAL sublimity 5 The emotions of terror & wonder so
606o022 it is beauty?-- it is an ideal standard, by which REAL objects are judged; & how obtained.-- implanted
544m103 ich makes consciousness-- & which tells one of REALITY-- castle in the air, is more prolonged than dr
548m117 e perceives butter or salt is not there.-- the REALITY does not resemble the picture in one mind, but
550m127 ed that insanity like sleep does not doubt the REALITY of the impression on its senses.-- insane peop
552m130 a dream cannot truly be <more> «as» vivid, «a REALITY» as in Spectral images-- Mem Chiloe <pi> Sow,
533m060 consciousness I was not right; though I never REALIZED the idea that I was tending to make myself in
602o11b eems."-- "that the object of «all» art is the REALIZING and embodying, what never existed but in the
057r113 nes washed about much at Coll. of. Surgeon's? I REALLY should think probably that B. Blanca & M. Herm
087a011 I. p212. Cuvier Oss Foss Wide range of Mammalia REALLY very important. harmonizes well with Lyells id
120a108 cleavage to rocks.--, but lava shows the rocks REALLY hot. & therefore I doubt the thermometer. Is n
139aIBC .-- Massac[h]usset would be well worth visiting REALLY good account of ice.-- C. Darwin A. Glen Roy G
177b025 tles it ¡We need not think that fish & penguins REALLY pass into each other.-- The tree of life shoul
190b075 ication have races of Plants. ever been crossed REALLY, if there is any difficulty in such marriages
194b095 it would go to show a centrum for Mammalia.-- I REALLY think a very strong case might be made out of
228b230 , & in each case similar or mere mongrels.-- It REALLY would be worth trying to isolate some plants,
282c143 ons in few days-- one doubt that one animal can REALLY produce so great an effect.-- the spirit of li
293c175 wer of adaptation given to organization.-- This REALLY perhaps greatest difficulty to whole theory.--
302c201 keness, & this I believe the case. = any animal REALLY connecting the fish & Mammalia, must be sprung
308c218 , of which human nature is the object, there is REALLY no natural starting place, because there is no
337d020 ut what some individuals are picked out.-- in a REALLY natural breed, not one is picked out, & few ev
366d111 outside, as often as she touched handle, though REALLY fully aware she was not coming in, could not h
377d140 emoved by the waves of the sea, on beaches-- we REALLY, measure the rapidity of change of form, & ins
377d147 birds [not located] :Hence, also structure not REALLY fitted for water, only habits & instincts-- Th
381d157 ading to supposition, that the Cryptandrous are REALLY, Heautandrous.-- How is fecundation effected i
381d157 rns according to this doctrine be considered as REALLY cryptandrous, & they have hybrids-- this is mo
386d167 y would not be fitted to the slow great changes REALLY in progress.-- Annals of Natural History. 1838
390d175 est bred of other ¿ breed.= Therefore it is not REALLY breeding in & in, but « <on» » breeding animal
391d175 es of twins in former case.-- (many monster are REALLY twins.)-- It is absolutely necessary that some
392d175 ious how this was superinduced? (Surely all are REALLY dioecious..) only simple form of life are mono
427e109 p» all plants, being occasionally dioecious; & REALLY dioecious plants being effect of abortion of o
432e125 hought.--» No one but a practised geologist can REALLY comprehend how old the world is, as the measur
445e159 not sufficiently proved that any shell fish is REALLY hermaphrodite. & <thinks> even oyster may fecu
445e159 which they live do. "Additions". p. 454.-- does REALLY attribute metamorphoses to habits of animals &
491q01v t male organs (not being always useful). fail-- REALLY good subject for experiment.--«to repeat Spall
495q005 1.-- 7 (a) Experimentaion on Primrose seeds-- it REALLY is an important case-- cross with cowslip poll
514q21. answered Does Mormodes (one of the Catasetums) REALLY always hit stigma by projecting pollen-masses?
523m017 no.-- because often, if not generally, does not REALLY come on suddenly.-- Case of Mrs. C. O. who thr
538m079 e calling him a bastard.-- when drunk.-- having REALLY been so.-- some always sentimental, some quarr
539m083 g on in the mind as in double consciousness may REALLY explain what habit is-- In the habitual train
544m103 wills to do this & hears that, but yet scarcely REALLY moves.-- the willing therefore is ideal, as al
547m112 nnected with past circumstances.-- as whether I REALLY was going to Shrewsbury, whether I had rung fo
565n007 ks its ears» when afraid, though not every time REALLY wishing to smell its enemy.-- Man & dogs show
566n010 ok IV, Chapt I on passions of mankind, as being REALLY useful to them: this must be studied. before m
608o027 is view will not do harm, because no one can be REALLY fully convinced of its truth. except man who h
614o037 (different from what takes place out of bodies) REALLY less wonderful than thoughts-- One organic bod

Page ***(Key Word)***

```
620o046 g his bird. one says for shame (& the «old» dog REALLY feels ashamed?) not so puppy, we <do> try to t
622o048 n.-- & how far more «feelin» acute the feeling REALLY is.-- All these associated «habitual» feelings
425e105 beds of Paris Lyell has remarked species never REAPPEAR when once extinct-- Lyell's argument about <T
516q24v lts & poisons & see in what order plants would REAPPEAR after <th> being killed Experiments not conne
439e142 bably depends on circumstances favouring the REAPPEARANCE of characters, formerly possessed-- <that i
120a110 e of Salisbury Craigs well worthy of attention-- REAR Glen Roy Notebook-- & scraps on Salsisbury Crai
240c004 are that, it requires the greatest difficulty to REAR them, eggs hatched under other birds & brought
401e014 tes to crossing.-- Cape Broccolli can hardly be REARED without greatest care be taken to prevent fert
428e111 (which  is case adduced by Herbert) because not REARED by seedlings.-- Now my principle does not appl
428e111 -- Now my principle does not apply to any plant REARED artificially, & only very partially to the Ziz
415e067 domesticated animals» no doubt is owing to the REARING up of every hereditary tendency towards fatal
439e144 he impediment of instinct-- the possibility of REARING by seeds Holyoaks-- (how far is this so) shows
359d087 ut Mr Miller says that breeds larger numbers, & REARS an unusual number out of any one nest. even mor
030r035 oast, near where Challenger was lost: I know no REASON for supposing these matters are not now collec
096a041 . Acid.-- Mr Malcolmson in Paper on India gives REASON for knowing that Mur. Soda. and Carb of lime d
176b021 ep numbers of forms equable:-- but is there any REASON for supposing number of forms equable: this be
179b034 nimals so far removed, with instinct in lieu of REASON, there would probably be repugnance & art requ
182b049 t self, but considering power, extending range, REASON & futurity. it does as yet appear clim In Mr G
186b058 ter of all its family; but why so? I can see no REASON for these. analogies; CD[from the principle of
194b093 limits  to desire, in proportion instinct more. REASON less. so will aversion be L. Institut «1837. N
195b100 imals vary in countries, without any assignable REASON.-- Astronomers might formerly have said that G
212b165 404. account of instinct of dogs.-- agreement & REASON Some animals common to Mauritius & Madagascar.
223b209 (?) 1837 Contrast New Zealand with Tasmania The REASON why there is not perfect gradation of change i
228b231 sorrow  for the dead.-- respect We have no more REASON to expect the father of man kind. than Macrauc
231b244 non-blending  of languages?-- Till man acquired REASON, he would be limited animal in range-- & hence
263c077 > feelings as animals.-- they on other hand can REASON-- but Man has reasoning powers in excess. inst
277c126 - good species it is reverting to old plan, but REASON now assigned for doing so There should be mark
287c157 1830 (?)." if she has put man on the throne (of REASON), she has also placed a series of animals on t
292c171 impelled to do in same way-- The improvement of REASON implies diversity & therefore would banish ind
298c189 d at.-- this most puzzling whether instinct, or REASON?? Gould says he believes that he has seen half
300c197 icult case to iimagine how art acquired.-- They REASON however on this to a degree. Mem Spider only d
300c198 s papers on Instinct.-- His distinction between REASON & instinct very just, but these faculties bein
301c198 rd to say what is instinct in animals. « & what REASON» in precisely same way not possible to say wha
301c199 difficulties.--  Knowing that animals have some REASON, & actions habitual. it surely is not worthy i
301c199 . 7. Mr Blyths arguments against squirrel using REASON in hiding its food is applicable to any habitu
305c210 asses» «orders» of insects, so is there none of REASON in order of <ver> Mammals.-- Mem Elephants & d
305c211 the individuals, whereby choice with memory. or REASON? is necessary.--) which is modified into endle
343d036 perfect in these themselves.-- instincts alter, REASON is formed, & the world peopled «with Myriads o
357d073 ther animals existed.-- Athenaeum 1838. p. 654. REASON given for supposing Tetrao media or Rakkelhan
407e041 t: insectivore being in S. America & Australia. REASON, why: Marsupiata, when first introduced live &
423e098 n 29th. Uncle John says he feels sure, that the REASON people send for their seeds to London is that
434e128 he controversy on Didelphys says. "If we cannot REASON from the analogies of the existing to the even
473s07r f to course of life, «as in trades» there is no REASON, why the peculiarities shd be born.-- may come
498q007 raised by Scions, as Elms. &c &c-- I have some REASON to suspect Elms.-- & Orchidaceous plants no o
531m049 eat little birds, this most curious instance of REASON & abstinence.-- My Father remarks that things
532m054 slightly unwell & felt so much afraid though my REASON was laughing & told me there was nothing, & tr
534m061 a flash.--. strange if judgment remains, where REASON is forgotten. it is conscience, or instinct. H
537m074 which  way it will be, but yet it is settled by REASON.--╢ How slow habits are changed may be inferre
548m114 ns of thought, therefore one may be imperfectly REASON-- <In a> Abercrombie's case of «in Botanical
548m114 case of «in Botanical Student» somnabulism, did REASON about himself-- but not about, facts gained or
548m117 e the picture in one mind, but does not stop to REASON what there should be & discover loss Definitio
554m141 ll be good to give Abercrombie's definition of "REASON" & "reasoning," & take instance of Dray  Horse
556m144 hich cad cannot see, they do not move muscle.-- REASON CD[The laughing noise which C. Sphynx made  at
564n002 a  troubled conscience.-- Therefore I say grant REASON to any animal with social & sexual instinct «&
564n004 njured moral sense.-- Notion of deity effect of REASON acting on (<not social instinct>) but a causat
567n014 sends  blood to its breast &c &c All Science is REASON acting «systematizing» on principles, which ev
568n017 emotions. (which must be intimately united with REASON) it would feel «subsequent» sorrow, whatever t
568n018 re pleasure.-- hence judgment, which is part of REASON Octob. 19th. Did our language commence with si
569n021 tition & charity & prayer, or eloquent request. REASON in simplets form probably is single comparison
569n021 conception  between one or two absent things.-- REASON probably mere consequence of vividness & multi
570n023 ut violence, without assigning or understanding REASON.-- surprise with negation.-- like shaking some
571n030 latter  may be vitiated. or rather altered. The REASON why New Buildings look ugly is because there i
572n034 t hurt the conscience of a Boy to swear, though REASON may tell him not, but it does hurt his conscie
577n051 rmation of a hinge «in a bivalve shell» does to REASON.-- an inflamed membrane from local  irritation
583n069 ve: imitations seems invariably associated with REASON: [N B. insects which have never seen their par
583n069 this  may be true, but relation of imitation & REASON must be thought of.-- p. 19. animals capable o
583n070 ation; (this is again assumed as more allied to REASON than innate.) Mr Wells I can see mentally re
583n070 nstinct.) Mr Wells I can see mentally refers by REASON knowledge gained by reason: & then these quali
583n070 e mentally refers by reason knowledge gained by REASON: & then these qualities of imitation & educati
583n070 dash of reason so mean are called "creatures of REASON", more appropriately they would be "creatures
586n078 might by instinct make watch, but he does it by REASON & experience, or habit.-- so bird migrating to
587n089 - The former is used with regard to the senses. REASON does not lead to action.-- p. 248. Theory of A
588n090 hey have memory.-- what use this faculty if not REASON.-- or does this reasoning apply chiefly to rec
588n090 as between man & child.--)» what differs-- not <REASON> «instinct», for its character is invariabilit
588n090 lained by habits, useful to itself, how gained. REASON? or some unnamed faculty-- Lamarck.  Philosop.
589n092 stroyed bad brain. see p. 90.-- The relation of REASON to organs of locomotion-- or that our facultie
589n092 clearly seen. in the absurdity of a tree having REASON: or dog, having high powers without hand or vo
589n092 me great puzzle in what Sir. J. M. says of pure REASON not leading to action & yet our emotions being
591n101 elo- continent man Hume has section (IX) on the REASON of animals Essays Vol 2.-- «also on origin  of
592n101 ms might be solved by considering the origin of REASON. as gradually developed. see Hume on Sceptical
599o005 octrines about origin of language-- & effect of REASON. reason could not have existed without it.-- q
599o005 about  origin of language-- & effect of reason. REASON could not have existed without it.-- quotes Ld
599o007 e utterly inexplicable-- I cannot <admit> think REASON sufficient to give up my theory-- Viewing from
600o08v nite, -- lives by hopes, looks to eternity. (4) REASON, some transcendental kind-- (5) Conscience, no
602o011 city, or intuition. when individual cannot give REASON, though he feels he is right-- it is because e
602o011 s &c, our well-being depends upon the "habitual REASON".-- This power of the mind, faintly approaches
608o026 ly INSTINCTIVE, & therefore now great effort of REASON to discover them: this is important explanatio
613o036 isplaced The meaning of Words, must be made out REASON Will Consciousness Definite instincts being ac
614o037 is loss is compensated by vast power of memory, REASON &. & many general instincts, as love of virtue
614o037 though very weak so as to be overcome easily by REASON.-- Conscience is one of these instinctive feel
614o037 nstinct. boy takes delight in mammae before any REASON had told him this distinctive mark, it is down
618o41v y. We do not know attraction objectively 6) The REASON why thought &c. should imply «X» the existence
620o045 rather the strengthened instinct, even when our REASON tells-- + us the action was superfluous, as on
620o045 s that. it is owing to some <subsequent> power (REASON) obtained by age, which should show the child,
623o050 s. could be acquired, & we are certain from our REASON, that. all which <has> (as we must admit) has b
624o051 mitation, has often been perverted from want of REASON.-- Hence as Eugenius says, slow growth of rule
624o50v during lifetime). so is our moral taste p. 152. REASON never can lead to action.-- p. 164. Ld. Shatsb
625o50v ce (but not cool malevolence). it is only after REASON comes into play that anger can be said to be w
633j53v ons.-- May they not be accidental? We have good REASON to know that they would not be detrimental acc
638j58v alogous, no savage ever made a perfect hinge.-- REASON, & not death rejects the imperfect attempts. I
638j059 ciousness.) <th> form these schemes.-- I see no REASON, why structure of brain should not be born. wi
301c198 t possible to say what habitual in men & what REASONABLE-- Same action may be either in same individu
405e035 r [not located] ARGUMENT REAL of antiquity of REASONABLE cosmopolite man.-- L'Institut. 1838. p. 338.
629o055 .-- arrived at first <rationally> by feeling-- REASONED on, steps forgotten, habit formed,-- & such h
543m100 emark about mathematicians of being profound REASONERS.-- all same fact-- for, as Jones observed, in
103a064 ch explains a difficulty.-- All De la Beche's REASONING of mountains being formed by crust being  too
222b205 ion); but intermediate in character, the same REASONING will allow of decrease in character. (which p
```

262c072 triotic?)-- but more especially the powers of REASONING &c &c.-- Study the wars of organic being.-- t
263c076 ucing more glowing imagining or more profound REASONING than other-- if this be granted!!) & whole fa
263c077 - they on other hand can reason-- but Man has REASONING powers in excess. instead of definite instinc
293c172 e this without reflection or consciousness of REASONING to tell back from front. &c or use of button
298c190 ntioned in note.-- try to trace from simplest REASONING in lower animals many times produced, a gener
398e005 here was some link in our train of geological REASONING, extremely faulty» The difficulty of multiply
398e005 «basal» foundation stone of further inductive REASONING is immense. It is curious that geology. by gi
443e156 myself, till I can honestly reject such false REASONING Bell Bridgewater's Treatise on the Hand.-- p.
547m114 fect of doubting nor believing, effect of not REASONING. effect of not having <all> other trains of t
548m114 ation, & in rigidly comparing each step as in REASONING-- hence delirium & sleep mental rest. though.
548m115 ake many necessarily are., when one is deeply REASONING besides these (which must be present, though
554m141 give Abercrombie's definition of "reason" & "REASONING," & take instance of Dray Horse going down hi
555m141 f friction & gravity. it would be discoverer "REASONING" or "reasoning"-- only rather more steps.-- d
555m141 avity. it would be discoverer "reasoning" or "REASONING"-- only rather more steps.-- dispute about wo
571n028 ons unconsciously, without abstracting them & REASONING on them (as justice?? as ancients did high fo
572n033 Mr Carlyle says that negro certainly has less REASONING powers than Europaean.-- Ideots. defective br
574n040 bably is some connection between very limited REASONING powers & the fixing of habits,-- for instance
583n069 duals of the same species with variability of REASONING power in one species man.-- false instinctive
588n090 se this faculty if not reason.-- or does this REASONING apply chiefly to recollection. yet a dog hunt
620o044 - Hence conscience is improved by attending & REASONING on its action, & on the results following our
308c218 Mackintosh Vol II. p. 495)-- in fact, in all REASONINGS, of which human nature is the object, there
038r058 t to refer to Lyell. but merely to state these REASONS, & saying that they refer to CENTRAL nucleus &
046r080 el whilst land rose up & down.--But from above REASONS, do not think so also elevating Earthquake of
465t081 make the destroyers vary; so that we here see REASONS-- why no perfect gradation can be expected in
564n003 e rather favourable to it--!! Man moreover who REASONS much on his actions, makes his conscience far
570n026 ushing solves this.--» The similarity of men's REASONS: shewn by similarity of the earliest arts.-- M
588n090 argument might be advanced, that animals have REASONS, because they have memory.-- what use this fac
602o011 them are <all> remembered, when the meaning or REASONS are forgotten. Our happiness &c, our well-bein
604o018 ary & of great power-- -- 2 From these & other REASONS we apply to God the notion of living in lofty
620o044 now though so trifling he feels remorse.-- He REASONS on it & determines to act more wisely other ti
635j56r .-- «(Causing death to some, &c &c)» These are REASONS, just as liability to accidents & any other ca
112a090 at the mean temp at Yakous in Siberia being -8 REAUMUR.-- there ought to be 32 degree Fah. at a great
621o047 me natural passion.-- (a) [LHC] The conscience REBUKES malevolent feelings, as much as actions, there
293c172 in offspring.-- Some association in such cases RECALL the idea. «or simple structure in brain people
526m028 l, yet they have not imagination enough to <up> RECALL up the image in their own mind,-- this may be
529m039 , fertility, rustic life, virtuous happiness.-- RECALL scraps of poetry;-- former thoughts, & in expe
529m039 y;-- former thoughts, & in experienced people-- RECALL pictures & therefore imagining pleasure of imi
036r050 hill,--but my recollection is imperfect & was RECALLED by note in Daubeny. P. 438., of similar fact
540m087 ht is that none of the ideas are habitual, nor RECALLED by obvious associations. as by reading a book
293c172 rd conversation before. is strong association RECALLING up image which had been past-- so great an an
527m035 t happen. because papers, &c &c round one. one RECALLS the castle by going to beginning of castle» he
369d114 fore different from Hunter I should say females RECEDE in organization from specific character.-- Cha
368d114 est.-- p. 47. (<"> is evidently the male which RECEDES from the species all females being most like o
194b096 which have male & female organs together, yet RECEIVE influence from other plants.-- Does not Lyell
194b096 ause this may be applied to show all plants do RECEIVE intermixture.-- But how with «hermaphrodite» s
404e024 ht be drawn that whole infertility «of hybrids RECEIVE no explanation» was consequent on mind or inst
521m009 well on any subject when once started,-- could RECEIVE a new train through eyesight, though, not thro
529m040 f food was evasive & ill defined thought would RECEIVE pleasure from thinking of the fertility.-- I a
544m101 ight of own child. (when frame in condition to RECEIVE pleasure) gives pleasure, ie. love.-- & so pai
621o046 follow the former; & likewise <that the> then RECEIVE the moral approbation of his fellow men.-- [RH
621o047 f passions too strong for our present interest RECEIVE simple explanations from origin of man.-- [RHC
625o052 dily & not like our appetites & passion, which RECEIVE enjoyment from gratification & hence are forgo
145g001 ount of ice.-- C. Darwin A. Glen Roy Generally RECEIVED opinion that male impresses offspring more in
148g025 to vary-- I asked this question in many ways & RECEIVED same answer Thought lambs most like MOTHER!--
345d043 of Agriculture p. 367. Dec. 1837. Generally-- RECEIVED opinion that male impresses offspring more th
346d047 liable to vary. I asked this in many ways, but RECEIVED same answer.-- Thought lambs were more like f
379d151 «not» the expression of <father> Sir W. itself RECEIVED from his father so that case ceases to be tru
388d173 ¿nerves? passed through transformation, & was RECEIVED into bud matured by female;<]CD> such view no
390d178 ions» has gained some difference «from what it RECEIVED» (for it is probable that breeding in & in wo
526m028 statements.-- Catherine remarks that pleasure RECEIVED from works of imagination very different from
610o031 ct amongst animals.-- Jan 13th. 1839 My father RECEIVED a letter from Mr Roberts-- «a person he had l
619o043 ld feel part of that pleasure, which the acter RECEIVED.-- If either man did not obey his instincts f
622o049 pleasure arising from association from having RECEIVED benefits from this person.-- [LHC] p. 254. &c
622o049 mother loving child, from whom, she has never RECEIVED any benefit.-- Yet I think there is much trut
623o050 ain love of place.-- although here we have not RECEIVED pleasure from the place, but merely in the pl
623o050 tly to origin of these instincts.-- the having RECEIVED pleasure from some one «person» in early infa
623o050 generations giving love of mother; the having RECEIVED some advantages from man. during many generat
387d169 the Syngnathus, or Pipe fish the male of which RECEIVES <young> «eggs» in belly.-- analogous to men h
593n107 l learns parent smell & look so by association RECEIVES pleasure. This [blank] will not do for insect
612o035 uired to be extended. Hence a sensorium, which RECEIVES communication from without, & gives wondrous
605o020 st of "an exquisite susceptibility from Blair RECEIVING pleasures from beauties of nature & art." But
607o025 he influence of others-- varied capability of RECEIVING impressions-- accidental (so called like chan
039r061 e form of the land decidedly bears the stamp of RECENT elevation. which is different from what Mr Lye
054r102 andstone with gypsum, covered by limestone with RECENT shells 200 ft, how exact agreement with Coquim
062r130 diff kinds of Fourmillier): extinct Guanaco to RECENT: in former case position, in latter time. (or
064r134 uite above reach of tides.--thinks them same as RECENT species. -- May I not generalize the fact glac
065r140 Fuego, and connection of quadrupeds.--although RECENT elevation, there may have been great subsidenc
080r178 of yew & intermediate Puncture one animal with RECENT dead body of other. & see if same effects, as
086a004 -- argue against the prejudice of not believing RECENT elevation, yet sea shells at tops of mountains
093a033 d bank at Bahia Blanca. <fl> Flustra identical. RECENT & bone bed.-- November 8th 1877 (Memoranda so
105a072 be preserved in it-- some error? (because more RECENT) ------ Coquimbo on. other hand?-- The widenin
110a085 il shells from West Indies & declare them to be RECENT species-- Lyell-- Some internal changes are in
175b020 tance secondary terebratula may have propagated RECENT terebratula, but Megatherium nothing. We may l
180b037 ecies «an ancient (I)» is capable of making, 13 RECENT forms.-- Twelve of the contemporarys must have
191b083 raordinary South Africa. proof of subsidence. & RECENT elevation: Pray ask Dr. Smith.-- «to state tha
200b125 ossil species of ox in N. America: as well as 2 RECENT See Geolog. Proc. p. 569. 1837. Account of won
207b152 e ferme de lancien monde".-- Considers forms in RECENT volcanic islets not well fixed.-- Peron thinks
212b167 Lardens Encyclop. Proportions between fossils & RECENT shells between herbivorous & zoophagous Mollus
213b168 bles in Phillips of numerous genera in fossil & RECENT state, well worth consideration--» Tabulate Ma
221b202 Think about Miocene fossils some species being RECENT agreeing with Senegal. whilst Crag <agrees wit
222b202 t Crag <agrees with> according to Beck has none RECENT, yet genera same.-- Speculate on multiplicatio
227b224 e get (I) a horizontal history of earth «within RECENT times.». & many curious points of speculation;
227b228 d true affinities. My theory would give zest to RECENT & Fossil Comparative Anatomy, & it would lead
250c039 eotherium in Paris quarries & at Binstead. Mem. RECENT Crocodiles with Palaeotherium in India--: conn
263c077 ical distribution study relation of fossil with RECENT. the fabric falls! But Man-- -- wonderful Man.
283c146 e Tertiary fossil Infusoria, of same forms with RECENT, we have nothing to do with CREATION.-- <On> T
316c247 le those of America than of Europe, because the RECENT ones are so close. Was there continent between
362d100 hen <pr> existing.-- he has also some very fine RECENT drawing «of prize pidgeons» in 1834-- now this
373d132 Mammalia:-- ¿is not this argument, for Mammalia RECENT creation.-- why. what tendency can there be fo
405e032 America & as it is <falle> embedded with almost RECENT shells.-- shows that progression of change in
422e092 s, that «Alludes to difference between fossil & RECENT Bull; like fossil & recent shells of the <new>
422e092 nce between fossil & recent Bull; like fossil & RECENT shells of the <new> raised beaches [not locat
447e164 e greater resemblance between carboniferous. «& RECENT» mollusca, than between the corresponding acal
448e167 the Superga & Paris, numerically the same with RECENT & yet almost wholly different, is same, as if
490q001 s of Madeira-- analogous or quite distinct from RECENT ones-- I presume some recent not found fossil
490q001 uite distinct from recent ones-- I presume some RECENT not found fossil (perhaps not embedded ¿ are t
490q001 erhaps not embedded ¿ are there any very common RECENT ones not embedded?-- Do the Tame Parrots breed
502q012 s removed, did not become impregnated. (34) Any RECENT information about pollen of Subularia Royle &

368d114 hens. being either alike or very different in RECENTLY altered genera. Guinea Fowl & Peacockss.!!» o
496q006 ation take place male & female flower in same RECEPTACLE (8) Make Duck eat Spawn, eggs of snail, row
274c116 we not look to these Northern regions as the RECEPTACLES of the wanderers out of the rest of the worl
418e080 ery animal being Hermaphrodite-- probably will RECIEVE illustration from domestication of Monooecious
446e160 some birds sing equally well. and <in> these RECIPROCALLY assist in domestic cares, as building nest,
424e103 e in the Australian dog an instance of a half RECLAIMED animal.-- The dogs, which have run wild have,
565n007 who can analyse suspicion-- yet who does not RECOGNISE look of suspicion, even child will do so.-- C
617o040 n. <The objective aspect of> bodily action as RECOGNISED by our external senses consists in the manif
331dIFC [In this Book some curious notes on Monkeys RECOGNISING Sexes of animals:]CD [All Selected Dec. 14--
377d138 with a grunting noise, the former signifies RECOGNITION with pleasure, as when food is offered, as m
219b196 , & since distributed by such means as we can RECOGNIZE, may be thought to explain nothing.-- it bein
438e141 not improbable as Mr Herbert does not seem to RECOGNIZE any difference in crossing between varieties
573n036 rtificer.--» Our faculties are more fitted to RECOGNIZE the wonderful structure of a beetle than a Un
593n107 excite our feelings.-- How does Social animal RECOGNIZE «& take pleasure in» other animal, (especiall
608o026 .-- When the admonition succeeds who does not RECOGNIZE an accidental spark falling on prepared mater
079r177 Vol. IV. «p. 58» At Acapulco earthquakes are RECOGNIZED as coming from three directions. from W. NW
358d076 from the male?)».-- children & women = "women RECOGNIZED inferior intellectually"= Opposed to these f
522m013 t stomach not acted upon by Emetics.-- people RECOGNIZED,-- sudden changes of disposition, like peopl
572m035 t they are the only steady & universal means. RECOGNIZED-- no one can say expression was invented to
617o040 sfied until we can point out How can force be RECOGNIZED by our external senses--only movement can.--
264c082 s aberrant from-- habits.-- Gould I see quite RECOGNIZES habits in making out classification of birds
569n020 often given food & word open your mouth said, RECOGNIZES that sound as perfectly as a man.-- Probably
070r153 s. must exhibit orbicular structure.--When we RECOLLECT connection of columnar & orbicular in basalt.
148g028 hain of lake & if so there would be barrier-- RECOLLECT the case of loch <in> <below> «by» pass of Gl
429e114 e peaceful woods. & smiling fields.-- we must RECOLLECT the multitudes of plants introduced into our
524m019 dity & madness.-- but then people do not well RECOLLECT what they have done in passion.-- People are
526m029 on.--» Thinking over the scenes which I first RECOLLECT, «at Zoos» they are all things, which are bro
532m055 fancy, it is very doubtful whether they could RECOLLECT these same things from any effort of will whi
548m115 ogress-- In castle of air the trouble «I well RECOLLECT» is in making things somewhat probable. in co
548m116 in the other,-- though she when well did not RECOLLECT <it> «anything».-- if one was subject to this
538d076 mbering which> «repeating song» when she had RECOLLECTED it in perfect senses.-- These things, & drun
540m085 ls smelling vaginae of females.-- when it is RECOLLECTED that smell of ones own pud. not disagree.--
544m102 p. 140-- Dreams good account of «thinks» are RECOLLECTED when intense, or when so near waking. that a
548m115 self.-- nor of a bad dream, when that is not RECOLLECTED, nor of the Botanical Somnambulist. (if he h
549m119 add to the happiness.-- but as they are not RECOLLECTED whether from frequency, or inherent structur
549m119 mind. they make, either in themselves, or if RECOLLECTED, such part of thoughts innumerable, which pa
068r148 levated. Geograph. Journal p 202 Vol IV When RECOLLECTING Gulf of California. Beagle Channel.--One he
259c064 back to distinct creations.-- it is only be RECOLLECTING that the ground woodpecker &c.--, fresh wat
293c172 r simple structure in brain people in fevers RECOLLECTING things utterly forgotten» --it is scarcely
533m059 d without knowing why-- had not conscious of RECOLLECTING it-- this may be nearest approach to <the>
036r050 it was necessary to ascend the hill,--but my RECOLLECTION is imperfect & was recalled by note in Daub
530m046 the involuntary thoughts.-- An intentionally RECOLLECTION of anything is solely by association, & ass
588n090 n.-- or does this reasoning apply chiefly to RECOLLECTION. yet a dog hunting for a bone shows he has
588n090 n. yet a dog hunting for a bone shows he has RECOLLECTION.-- Lamarck. Phil. Zoolog.-- Vol II p. 445.
605o18v r means, as moral excellences, brings to our RECOLLECTION the original cause of these feelings & thus
526m029 y memory of the scenes, (indeed my American RECOLLECTIONS are a collection of pictures).-- when one r
550m125 y days, are they not those of which all our RECOLLECTIONS are pleasant.-- Browne Religio Medici, p. 2
580n062 . 12. 1839. Sir. H. Davy -- Consolats: "the RECOLLECTIONS of the infant likewise before two years are
621o047 of the mind, which is «much» formed by past RECOLLECTIONS.-- Hence he has the right & wrong in his mi
275c121 ything.-- almost all imagination-- He says he RECOLLECTS all half Bred cattle of L'Darnleys were most
334d009 e bred animals.-- Mem. number of Mules.-- «He RECOLLECTS one hatch of hybrid geese very fine.--» How
547m113 tell whether one has rung the bell., when one RECOLLECTS circumstances since so naturally would
036r049 an paper; & part in grand discussion Consult. RECONSULT Geolog. Map of Europe Consult charts for dist
025r019 be found to be old & dead)» «(I have not kept a RECORD)» In looking over the lists of organic remains
047r081 rthquake of Valparaiso. (1822) no great wave on RECORD. -- «also neighbouring sea must partake in abs
265c083 n lopped off & sheeps tails cut yet there is no RECORD of any effect.-- New Hollanders have gone on b
437e139 leys Familiar History of Birds several cases on RECORD of stoats being carried (p. 121) & dropped hav
529m042 Parks ditto.-- My Father says there is case on RECORD he believes in Philosoph. Transactions, of ide
090a022 at is fifty years-- 90 <showers of> stones are RECORDED as falling; many of these were not single, bu
090a023 ize» is given of 25 stones.-- The total weight RECORDED is 473. pounds (taking about average when sev
090a023 ts of stones of large numbers (& how few cases RECORDED if we say «100» <5>0 lbs a year too little.--
398e005 n about species Changes of level &c are easily RECORDED, but changes of species «are» not as «without
055r107 or modern streams of St Jgo yet no historical RECORDS of eruptions how immense the time!! How well a
529m042 were passed off.» & vomited, but who when he RECOVERED. was bound to be ignorant, but quite sensible
555m144 ame before me showing impossibility of person RECOVERING from hanging on account of blood. but all th
275c120 rough breed <greyhound>. bull-dog. & crossed & RECROSSED, till there was a dash of blood, with whole f
511q018 did result appear without his wish Has since RECROSSED this breed.-- Have secondary male characters
515q022 ch he crossed; especially if the hybrids were RECROSSED with either parent.-- May. 44 These Hybrids d
275c120 Polish cock «is not that old variety» & then RECROSSING off spring. till size diminished, but feathe
110a087 of cleavage plates: the crystals must have RECRYSTALLIZED, as such do not occur in either dike or fr
053r101), which would cause parallel lines, but the RECTANGULAR intersections are singular-- M. Lesson consi
480z006 21 Good account of Condor by Humboldt Zoologie RECUIEL-- Meyen has written account of Guanaca. In tra
522m012 ards it calls up pain. or pleasure. & is often RECURRED to & mentioned as a thing which had just take
557m147 wolf wag their tails, &c. it is very curious, RECURRENCE of pleasure so teaching expression «as const
305c211 endency to delicate emotions between races, & RECURRENT habits in animals.-- --Animal Magnetism-- pri
428e113 like short-tailed cat or dog has been without RECURRENT tendency in external conditions» sudden loosi
636j56v ormed. Macculloch says, life, forms a broken, RECURRENT series, whilst the habitation «or world» simp
568n018 utter pleasant plaintive cry-- The taste of RECURRING sounds in Harmony common to t[he] whole kingd
338d025 poraries. In introduction to Eytons Anatidae.-- RECURS to idea of only animals from distant countries
538m079 n state of turn In Drunkedness same disposition RECURS, such as -- -- of Trinity always thinking peop
033r043 with base of Pitchstone; Mem Galapagos. chiefly RED glassy scoriae.--could walk round base:--not uni
036r051 te in Daubeny. P. 438., of similar fact near the RED Sea.--which occurred in a sandy place.--(the sou
041r066 onary, so are all those plates in Australia. New RED Sandstone. at Bahia in modern sandstone. a circl
048r084 Darby mentions beds of marine shells on banks of RED River Louisiana. V. Lyell. Vol I. P. 191 State a
060r125 period how diff. would the rocks have been. The RED Sandstone of Andes fusible? no. mad dogs. Azores
068r147 ass[es] [Fig. 8] {P} (A. B. C. now grown solid.) RED Sea near Kosir, land appears elevated. Geograph.
075r167 opper, and antimony, horn silver, black silver & RED silver, do not name native silver because not ve
088a016 in character of Andes Metamorphic action -- Mem: RED sand of Europe no fossil shells --¿ action of He
089a017 ?-- Mem. granite heated.-- Metamorphic action in RED sandstone.-- Geological Volcanic-- CD[Might not b
091a026 on Mexican Trachyte <roc> lava called Andesite. RED Coral in the Mediterranean 700 feet deep in some
093a033 ts of America. -- The number of minute turbos in RED earth with volutas. prove regular mud bank at Ba
110a086 f hornblende (?) & felspar, (some crystals being RED) «with» cleavage, veins of pyrites, few curious
111a087 35 degree. E. dip to NW to 80 degree faults with RED wacke contorted evidently dike. V. VII. p. 316 &
116a100 - «p. 151» first discovered «very small» bits of RED granite between 40 & 50 from Portezuelo. Bull: S
120a109 ct to formation of salt.?.--??? Footsteps in New RED Sandstone. look as if a surface deposit.-- The c
138a155 on the consolidation of strata-- he heated sand RED hot & brine was boiling on the top-- [blank] Wou
149g033 they extend round hill too low line drawn plain RED talus line on N. side of Spean most clear & uppe
157g075 -- whole hill composed of remarkable gneiss with RED granite veins & quartz, & garnets.-- Boulders a
244c021 peds» p. 118. wild pigs of Falklands, generally RED of brick" hair, very stiff, p. 120-- Coati roux
253c047 t by Capt. W. Shee. considered merely variety.-- RED form of skull very slightly different.-- Q Zoolo
257c058 nt of lizards.-- As we have birds impressions in RED Sandstone. great lizards in do.-- <Wood> <Dicot
257c058 rous wood in Coal Measure.-- highest fish in Old RED Sandstone.-- Nautili in----. it is useless to sp
261c071 w existingn Mr Gould says wherever any mark like RED patch on wings of Furnarius, Synallaxis &c &c. s
304c207 rent forms of Synallaxis. trifling characters as RED band on wing show to be from one parent.-- same
341d031 however are most closely represented.-- Thus the RED breasted thrush is represented by one not differ
345d043 headed, but this was bred out & now all are pure RED, yet calf every now & then born with white head
357d073 iness with which this genus becomes crossed. ¿is RED game an hybrid?-- When I show that island would
360d092 opagate-- <now> «as» if one bird had very bright RED breast & other very bright blue, it might be har

Page ***(Key Word)***
361d095 g. Meeting stated, that Green-finch, all linnets RED-pole, goldfinch, hawfinch-- in nursling plumage
365d105 would   species cross in wild state.-- Is English RED Grouse. a cross between Black Game. &, the subal
365d105 weden, (which in summer dress somewhat resembles RED Grouse) it may be so-- but very improbably, for
383d160 arks on tail much <blacker> «broader.-- » Breast RED like Common pheasant.-- lower part of breast, ea
401e015 y not so absurd.-- Thinks-- that such variety as RED cabbage produced from passage from many varietie
434e129 w tells me following Facts: believes that «only» RED Lychnis grows in <south> Wales & certainly <old>
436e137 extend round world.-- Quartz of Falkland.-- Old RED Sandstone-- Van Diemen's land.-- Porphyries of A
465t080 alia in caves:-- :argue first case of bones (New RED Sandstone) & then go on to shells-- A profound c
473s03r d remains of Sus with Elephants-- Lyell says New RED Sandstone of. N. America is Red Sandstone. & Bir
473s03v - Lyell says New Red Sandstone of. N. America is RED Sandstone. & Birds true! Plants in Devonian-- Ho
484z016 by Maldonado creeper of same plumage.-- general RED mark on wings of all-- Spix has described Philed
495q005 fect on organs of generation (5) Place pollen of RED Cabbage «mixed with own pollen» on flowers of ot
110a086 TINCT from dike junction mechanical: DIKE base REDDISH feldspathes with greenish. black specks of horn
110a086 specks of hornblende, large irregular cryst of REDDISH felspar. & scales. of mica.-- large cryst of H
267c094 lways hunt in packs «30 or 40 together» colour REDDISH <brown>. ears long.-- like bull terrier-- Indi
365d106 bly subalpina.) former has blue breast, latter REDDISH, hybrid purple-- be careful, See to hybrids be
383d160 e instead of» two lines.-- «faintly edged with REDDISH brown» black marks on tail much <blacker> «bro
073r161 lated, round knobs, or pushed where soft, or REDISSOLVED soft.--/ is there any flexure <fr> in the fr
363d102 well watched always very different.-- the two REDPOLES can hardly be told apart, so that after diffe
268c096 curious   account of Tit mouse feeding young of REDSTART & actually driving away parent birds.-- showi
190b076 Parallel 33 degree-35 degree, source of forms. REDUCE towards Northern Eastern end & die away, & par
377d140 the unit of our calendar.-- epochs & creations, REDUCE themselves to the revolutions of our system in
181b041 by tracing back. the <descen> fathers would be REDUCED to small percentage.-- & <in> therefore the ch
259c065 he possibility (any great change in species is REDUCED by atavism) Even a deformity may be looked at
275c121 d Mr Yarrell's evidence about old varieties is REDUCED to scarcely anything.-- almost all imagination
276c122 our of plumage & laws, which might probably be REDUCED What the Frenchman. did for SPECIES-- between
341d030 e Movement.-- nocturnal crawling bird.-- Wings REDUCED to rudiment.-- clavicle scapula &c strongly de
383d159 f either.-- (Mammae or sheath of Horses poenis REDUCED to extreme degree of abortion).-- Insecta.-- h
436e135 cient in producing new species; also one being REDUCED in numbers, but not so much these, because cir
515q021 B. Benthams remarks, where parts of flower are REDUCED from normal number, they are apt to vary in nu
637j57v in the other. ¶All Bridgewater Treatises. are REDUCED simply statement of productiveness, & laws of
618o41v both due to some common cause:-- The argument REDUCES itself to what is cause & effect: it merely is
356d069 ly & have used them since) the line of proof & REDUCING fact to law only merit if merit there be in f
423e097 the Ammocoetus) Crustacea to--- ? &c) without REDUCING the number of living beings-- but there is th
203b137 s over the whole world-- For instance gradual REDUCTION of tempereture from geographical or central h
188b067 species replace <No> Southern kinds-- (I) Gnu REEACHES Orange river & says so far will I go and no f
317c249 g. p. 55. 40 leagues from land several patches of REED & trees p. 259 120 ft in length, some branches
050r093 strip of Mountain Limestone in N. Wales. was it REEF. -- I remember many Corals?? Breccia--Stratific
051r093 & turbid water) organic bodies protect like peat REEF of sandstone.--Corals, & Corallina survive, in
051r094 ction is anomalous; It is wonderful to see Coral REEF--or confervae in the breakers or in waterfall:
116a100 untains of Cordova project on plain, like <re> a REEF on a sea beach-- «p. 151» first discovered «ver
043r069 re with Galapagos.--Chiloe. M. Hermoso. & Coral REEFS (imperfect in latter). <At> Lyell. Vol I. P. 31
050r091 t Volcanic eruptions form foundations for Coral REEFS.--does he mean in contradistinction to sand?? B
108a080 sophical to think calcareous springs near coral REEFS.-- Where vegetation luxuriant it might be almos
227b224 nce India, Mexico & Europe. one gret sea (Coral REEFS:. shallow water at Melville Isd. (3d) We know th
305c209 e of voice like S. American.-- Have not Ruffs & REEVES a remarkably varying plumage for wild birds--
250c040 I I. p. 81. Capromys, West Indian isld. p. 120. «REF.» Philosop. Transacts 1823. Read June 5th) impor
250c040 other in secondary & Tertiary periods.-- p. 125. REF. to Phil Transacts. (read November 20th) Paper b
251c041 mark on breast: p. 234.-- good case.-- p. 526. (REF) To Temminck Monograph. Mammal; «4to» good facts
319c276 ographical range.-- [blank] Books examined: with REF: to Species Most of those which have references
038r057 ck.--In the discussion it will be better not to REFER to Lyell. but merely to state these reasons, &
038r058 rely to state these reasons, & saying that they REFER to CENTRAL nucleus & that envelopes no doubt ex
124a118 Book C. p. 101. On Frozen Soil of Siberia (with REFER to Metamor) wrong entrance Athenaeum. 1838. p.
200b124 ility-- ditto in Plants.<==> It will be well to REFER to Chamisso Vol III p. 155. about quantities of
317c248 ces very good. also "Rays Wisdom of God." Often REFER to these.-- Also some few facts at end of "The
362d099 not this so in S. America with C. Campestris<)> REFER to my notes) & Mr Yarrell supposes this a conse
410e053 for species, except so far as wanting names to REFER to, to those forms. where the termination of ch
417e079 mmon progeny more than others.-- does this more REFER to length of time that the resemblance is perma
432e125 ehend how old the world is, as the measurements REFER not to revolutions of the sun & our lives. but
586n080 t (an instinctus means stained in?). had better REFER to to the heredetary part of it,-- & faculty (f
632j53r bs of Deity. Vol: I it will be better always to REFER to the author if I use these facts p. 280. addu
093a034 und <emb> in the mud, of the Salt river.-- in REFERENCE to fossil guanaco of P. St. Julian. -- Mr Scr
127a127 46. on formation of cones beneath sea.-- with REFERENCE to old submarine orifices in Cordillera Geogr
197b112 e grandfather of Mammalia & birds &c ¶ p. 32. REFERENCE to M Edwards. law of crustacea-- with respect
225b219 se. produce offspring exactly intermediate.-- REFERENCE to Pig & Dogs. My theory will make me deny th
252c042 & perhaps one extant at Leeward Isles.-- p. 388 REFERENCE to Rüppel. travels (what language?) Hyena «ve
259c066 not having armless child, shows than there is REFERENCE to more than offspring (like atavism) & shows
285c152 y doubtful A species is only fixed thing with REFERENCE to other living being.-- one species May have
294c177 ook last Bear in England killed in year 1000. REFERENCE to succession of types ¿different species;--
326c265 climatizing of Plants. Herbert. p. 348. gives REFERENCE to Kolkreuter's Papers Wiegman has published
344d040 s children not being so.-- consider this with REFERENCE to "new species & hybrid doctrine"-- I have r
344d040 o state fertile at once.-- Consider this with REFERENCE to those insects, which have fertile offsprin
369d115 animals being compelled to travers) "May have REFERENCE to the Great distances which the Mammalia of
410e052 aided in such sins" (do not add name, without REFERENCE to description), except describers having som
413e059 f large size,-- consider this (Cetacean) with REFERENCE to my theory Babbage 2d Edit, p. 226.-- Hersc
415e070 are generally additive, & not abortive: with REFERENCE to the non-necessity of the «so called» progr
421e091 ark ears of the wild Chillingham Cattle. with REFERENCE to Mr Bell's statement of the tame ones.-- ¶a
448e166 nandez» differ from American Coast Vol II.-- <REFERENCE» p. 251. about the drifting of animals on ice
481z010 p. 29-- In Meyen. Voyage round World German a REFERENCE to a luminous Sertularia Lesson Zoolog. Coq:
483z012 c mouse of Egypt is Mus Cahirimus. of Geof.-- REFERENCE from Rüppel travels All Owens papers on Intes
485z017 submarine insects. Staphylinidae &c &c. with REFERENCE to those of mine from T. del Fuego p. 141. Ho
492q002 formed?-- 8. Can any annuals be budded. with REFERENCE to extension of age of individuals-- 9. Do pl
501q011 an & Bean, intertwined, «without sticks»-- in REFERENCE to what Mr. Herbert observe on this subject--
513q21. ct on any flower before stigmas expanded-- in REFERENCE to Lobelia & Clarkia-- Peas time of impregnat
558m151 ined with necessary notion of "causation", in REFERENCE to this "ought," as well as the works of the
579n060 o p. 135.-- on the importance of a name, with REFERENCE to origin of language My father says old peop
592n103 referred to, if I follow up this subject & a REFERENCE to Brun's work.-- Shutting eyes in contempt o
618o41v a of thought. (The objects of thought have no REFERENCE to place. [We see a particle move one to anot
632j53r at the laws of propagation, were created with REFERENCE to successive developement I admit, but the a
239cIFC «beginning of» February & July 1838) All good REFERENCES selected Dec. 13 1856 Also looked through Ap
243c020 sson method of generalizing without tables or REFERENCES highly unphilosophical xx Says same remark w
317c248 ca.-- At the end of "White's Selbourne." many REFERENCES very good. also "Rays Wisdom of God." Often
319c276 with ref: to Species Most of those which have REFERENCES at end; is so said to have Mackenzie's Icela
321c275 h Travels Whites Natural History of Selbourne REFERENCES at end Dr. Lang Australia «trash» skimmed Ma
321c275 leay's Horae Entomologica Ray's Wisdom of God REFERENCES at end-- The British Aviary.--do-- & Lisle'
322c269 ing's Laocoon Whewells-- inductive History.-- REFERENCES at end of each Vol Herschels. introduction t
322c269 o Natural Philosophy R. W. Darwin's Botany.-- REFERENCES at end Mayo Pathology of the Human. Mind Eve
323c269 World do -- Lamarck. II Vol. Philo. Zoology «REFERENCES at end of each Chapter» Crabbes Life June 18
324c268 ie des insectes 8vo p. 181.-- See (p. 17) for REFERENCES to authors about E. Indian Islands. consult
405e032 coembedded in N. America. see my Journal for REFERENCES In such cases as at Galapagos. where differe
415e069 nductive History. Contains many most valuable REFERENCES See if any law can be made out, that varieti
477z001 d through & abstracted Zoology Some excellent REFERENCES in L. Jenyn's introduct to Mag of Zoology an
578n055 ither good or bad action, it always bear some REFERENCES to thoughts of other person Decemb. 27th.--
208b154 yet multiplied: NB How does this bear with law REFERRED to by Richardson in Report about each genus h
270c104 ia «are» insecta.-- G. R. Treviranus, Biologie REFERRED to,. as compilation of action of organic natu
366d111 in, could not help being perfectly distracted «REFERRED to <other> Book M.» Is there any law of varia
424e100 Crustaceae.-- (I forget whether I have already REFERRED to it.-- also on spermatic animalcules in Mus
492q003 e in V. Vol of Hort. Transacts & M. Sageret is REFERRED to with doubt by Herbert Do forest-trees spor

```
592n103 ghter & so into convulsions.-- «Paper» must be REFERRED to, if I follow up this subject & a reference
610o031 looked in direction book under head of Wilson, REFERRED to Robert & found his Christian name was Wils
349d052 ablished-- clearly so.-- NB. This paper worth REFERRING to again.-- According to my theory, every spe
471tf08 3. "D. Dod on two new genera of coniferae".-- REFERRING to the 3 main divisions & speaking of their s
054r102 mine: phyllade covered by quartzose sandstones: REFERS to broken hill described by Pernetty: account
100a055 nconsolidated matter-- Phillips Lardner p. 197. REFERS to salt as being produced by local heat, Ask C
116a099 's Voyage. good section of Rio Negro beds.-- -- REFERS to species non decrite de petites corbules ana
134a141 Pen. Tres Montes.-- as by subsidence ‖ Fitz Roy REFERS to ‖ & Rocks p. 375. on the soundings on outer
215b178 ge of forms.-- Deshayes states Lamarck priority REFERS to introduction to Animaux Sans Vertèbres as l
417e079 first  births.-- it is the latter only that one REFERS to in speaking of resemblances of children to
449e169 ine of Nat. History. 1839. p. 106.-- Waterhouse REFERS to fossil remains of the Hamster.-- is not thi
460t019 a  few types, it must not be supposed that this REFERS to time.-- Marsupial in Oolite.-- insects, th
549m118 more» entirely happy (contentmt is different it REFERS to wishes for future) than perhaps well «regul
550m125 xperience in same time.-- Pleasure more usually REFERS to the sensations <it> when excited by impress
583n070 son than instinct.) Mr Wells I can see mentally REFERS by reason knowledge gained by reason: & then t
600o008 man.--  two souls in one body-- (2) Beau ideal, REFERS chiefly to moral, beau desires conscience & lo
609o29v o talks of the necessity of these passions, but REFERS (I believe) to present day & not to ruder stat
033r042 h not very intelligble is a familiar case: If REFILTERED with other matter how very curious a structu
232c171 nvin> brought so much conviction to my mind.-- REFLECT much over my view of particular instinct being
620o044 lling up past sensations, he will be forced to REFLECT on his choice: an appetite gratified gives onl
622o048 aw of honour from Paley [RHC] Anyone, who will REFLECT must feel, how like to injured conscience,  is
048r085 inhabiting  plains of Patagonia is removed by REFLECTING on the nature of the country in which the Rh
531m051 ruly an instinctive feeling, «may not pa.> In REFLECTING over an insane feeling of anger which came o
107a078 ew of Volcanos &c &c This view will bear much REFLECTION on method of cooling--Very difficult subject
293c122 ge better instance)» he had done this without REFLECTION or consciousness of reasoning to tell back f
028r030 e waters of the ocean all are mingled. These REFLECTIONS might be introduced either in note in  Coral
044r073 at Europe was its birth place.--Some general REFLECTIONS might be introduced on great scale of  ocean;
080r180 grease  & mercury [blank] NB. P. 73. General REFLECTIONS on the geology of the world P. 14-91. gradua
548m115 erhaps, hard work-- though dreams do that One REFLECTIVE Consciousness is curious problem., one  does
624o50v o action.-- p. 164. Ld. Shatsbury under term of REFLEX Senses seems to have <compared> «perceived» th
277c128 of most disputed shells, such as Cyrena This is REFORM which probably will be slow. but must take pla
232b247 r.-- If the the world had cooled by secular REFRIGERATION in chief part instead of change from insula
232b247 aean ones.-- "a variation played on secular REFRIGERATION".-- <The Phenomena of the S. Hemisphere loo
573n037 w the youngest child of H. W. constantly. when REFUSING food, turn his head first to one side &  then
128a129 subsidence of water on other. from tendency to REGAIN statical equilibrium This will be only a modif
243c020 highly unphilosophical xx Says same remark with REGARD to shells.-- But he says shells towards extrem
285c152 ages.-- As species is <certain> real thing with REGARD to contemporaries-- fertility must settle it.-
355d067 fixed-- (<also> Hunters law of monstrosity with REGARD to age of foetus. distinct consideration) Now
389d176 e too closely related: this most important with REGARD to theory, showing generation connected with ins
402e019 - Malte Brun. would be worth skimming over with REGARD to this archipelago Octob. 13th.-- Kotzebues f
406e036 to  Parents same laws appear to hold good. with REGARD to marriage of individuals, & varieties of sam
407e040 nt state of world is not so different, with <d> REGARD to their productions.-- Hence it is, from the
494q004 r Duck. ¿in Pidgeon?-- Mr. Miller said yes with REGARD to former (8) Is form of globule of blood in a
515q021 ossing peaches & nectarines: same question with REGARD to Primroses. (4) Do apples "sport" in  fruit,
579n059 sexual feelings-- love being an emotion does it REGARD «is it influenced by» other emotions? When a m
587n089 fferent from emotion.-- The former is used with REGARD to the senses. Reason does not lead to action.
600o008 moral.  beau desires conscience & love.-- [With REGARD to ordinary Beau ideal, Mem. Negro, beau,--Jef
619o043 d children would give him pleasure, without any REGARD to his own interest. likewise if such  actions
421e091 the  habits & food of an animal, I have always REGARDED as affording very clear indications of its tr
244c020 on du baron Cuvier, nous ne balançons pas a la REGARDER comme une espèce distincte! p. 171. Sus papue
260c069 erican ornithology a mine of valuable facts., REGARDING habits range & all kinds of information-- ins
406e036 ring before.-- -- now this is never stated.-- REGARDING the similarity of offspring to Parents same l
491q002 swers first question in negative.-- Questions REGARDING Plants. 1. Uniformity of hybrid & Mongrel off
492q003 flowering gorze common in Norway No Questions REGARDING Breeding of Animals If two half bred  animals
511q018 ve me Pigs foot undivided, & more particulars REGARDING effects of crossing them with common pigs= [i
550m126 notion of property" -- their own property. (--REGARDING food & in birds of place for nest.)-- with do
610o033 tley, &. the school of Kant. to Coleridge, is REGARDING the sources of knowledge.-- whether <we th th
619o042 nevolence» to the object in question. Without REGARDING their origin, we see in other animals they co
185b057 division of Heteromera same. thing occurs with REGARDS to other tribes in that same family.-- (NB I s
578n055 especiall  if it be a proven. whose opinion he REGARDS, <& see how> feel how the blood gushed into hi
094a036 - Mr Murchison. M.S. Chapter on drift.-- Beyond REGION of great boulders, pebbles of granite  clearly
317c250 eral old acquaintances. which grow on the lower REGION of the Canary islands-- p. 250 admirable table
317c250 ommon to Canary. isld., Europe, & St Jago upper REGION, & some to Cape.-- some proper well-worth stud
447e163 re would be real gradations in species from one REGION to another.--» -- these simple forms perhaps o
451e178 by.  B. Hodgson. p. 336 In the most pestiferous REGION (mentioned by Heber) «from» which «all» mankin
451e178 reside & are bred. "take tame animals into this REGION between April & October & like man almost (thi
501q10a Or does any «one» species of plant, vary in one REGION of Europe & less in another region-- (27) Whic
501q10a vary  in one region of Europe & less in another REGION-- (27) Which sex in Mules generally fails-- pe
051f093 s  become petrified, & increase. -- In Southern REGIONS every rock is buoyed by Kelp, now Kelp sends f
060r125 Study  Ulloa to see if Indian habitation above REGIONS of vegetation.--«I can find nothing.» Mem Caro
208b153 o genera. agrees with late production of those REGIONS, & consequently not many yet multiplied: NB Ho
231b243 similar  temperature.-- now those of temperate REGIONS & tropics are only related by one connection.-
244c021 --it  would make strong contrast with southern REGIONS.-- «it would now represent. what <actually is>
256c054 descriptions  of domesticated animals in those REGIONS Species so far are not natural, that they  are
260c067 irds of N. America & South-- Any how temperate REGIONS-- crows in N. America, Study Bonapartes list I
274c116 n America.-- may we not look to these Northern REGIONS as the receptacles of the wanderers out of the
318c252 t & hind legs of these white hares, fitted for REGIONS of snow.-- Acclimatisation.-- Bachman tells me
343d037 light  & there was light".-- » August 17th Two REGIONS may be Zoolo-geographically divided either  by
424e101 idence of Greenland-- case of splitting of two REGIONS-- -- are there any cases of union of two regio
424e101 gions-- -- are there any cases of union of two REGIONS in modern times.-- this would depend on negati
427e109 effect  on snow of arctic climate in far north REGIONS? Arctic forms have travelled S. From the analo
429e116 h those Kamtschatka, Siberia, or even of polar REGIONS-- do p. 21 says. many plants skirt each side o
429e116 points were last connected with those northern REGIONS-- do p. 21 says. many plants skirt each side o
540m085 t him compare the American whether in the cold REGIONS of the North,-- the elevated table land of Per
604o018 we  apply to God the notion of living in lofty REGIONS. 3 Infinity eternity. darkness, power. being a
641j29r large  size» <to> are best nourished by arctic REGIONS-- Whales. «Narwhal» Polar bear. Walrus,  great
191b079 like  on a porcupine on Echidna-- Good to study REGNE Animal for Geography.-- The motion of the earth
399e009 od? [not located] Study introduction to Cuviers REGNE Animal No structure will last. without it is ad
298c191 wards would probably often be destroyed.-- or REGRAFTED with fresh arrivals..-- &c &c --Climate alter
155g064 alling sea & so corrode plain into terrace as REGRESSED What <alter> a balance there must be in power
576n047 e had lost his life in doing so.-- nor would he REGRET «having acquired» this sense of right (& Wheth
027r023 arkable because the formations are now seen in REGULAR descending steps Mem.; rapidity of germination
030r035 f an open & not deep sea.--(Character of coast REGULAR & <not very> rather deep soundings, 60-100 fat
034r045 Volcanos Gypsum Andes Mem. Beechey. account of REGULAR change in soundings. on approaching the coast
045r076 considered  as accidents (if <[...]> part of a REGULAR system can be called accidental; the proportio
049r088 hyry at Valparaiso; Epidote -- Must we look at REGULAR greenstone cones at S. T. del Fuego as nucleus
093a033 minute turbos in red earth with volutas. prove REGULAR mud bank at Bahia Blanca. <fl> Flustra identic
115a098 » given of every degree of fineness. then most REGULAR slope-- {P} if not course enough flat top. end
149g031 quired their peculiarities? The slope of A & B REGULAR & even towards {P} Spean Roy double terrace ri
235b272 242. Hook Smellie Philos of Zoolog. 842 White REGULAR gradat in Man poor trash Lyell 1024 Flemings P
293c173 swim after having once learnt, & if that was a REGULAR contingency the brain would become webfooted &
324c268 rs remarks on the influence of climate White's REGULAR gradation in Man. Lindlys introduction to  the
398e006 sh laws created.. & yet with <gov> symmetry «& REGULAR laws" That baffles idea of revolution.-- My ve
586n082 sight--  so a Bee has the faculty of building «REGULAR» cells-- [but this faculty «may possibly be» «
042r068 is  clear the forces have acted with far more REGULARITY in S. America: in France we have  freshwater
042r069 if present in the Andes, would have destroyed REGULARITY of slope of valleys.--All my observations co
288c160 I conceive a bird Migrating from Falkland Isd REGULARLY to main land, proof of. land having been form
342d032 variety of Muscovy!) with goose!!) Dr. Bachman REGULARLY breeds «in Carolina» for his table Muscovy &
```

Page **(Key Word)**
513q21. act nothing from him)‖ Does impregnation ever REGULARLY take place in unopened flower-- <doubt> disbe
543m099 is time, almost an absolute fool used to play REGULARLY with D'Arblay of Christ of great genius, & ye
559m156 --» Would a sensitive plant if irritated very REGULARLY at one time every day.-- naturally close at t
086a007 edentata. -- How much is temperature of world REGULATED by atmospheric currents?-- chiefly clearly by
549m118 fers to wishes for future) than perhaps well «REGULATED» philosopher-- yet the philosopher has a much
625o052 use it is «a» part of our nature, <not> which REGULATES our feelings steadily & not like our appetite
626o052 h will" is unintelligible to me.-- conscience REGULATES feelings, as of cowardice.-- the whole appear
325c267 standing (sometime) Du Stewart works. & lives of REID, Smith & giving abstract of their views Mackint
559m155 Copied <Smith> «D. Stewart» lives of Adam Smith REID, &c worth reading. as giving abstract of Smith'
242c017 ard. Duvaucel. Leschenault Kuhl. Van-Hasselt, REINWARDT «Forrest» authors on E. I«ndian». A«rch.»‖ Bo
443e156 have thus argued to myself, till I can honestly REJECT such false reasoning Bell Bridgewater's Treati
463t057 cts with laws & their explanation will probably REJECT this theory-- (I must answer it by rooting out
511q018 caring whether good or bad.-- are any actually REJECTED?? (8) Get Sir. R. Heron to give me Pigs foot
638j58v er made a perfect hinge.-- reason, & not death REJECTS the imperfect attempts. In the «Bee» Mollusca
257c059 Are monstrosity heredetary??.? Does not atavism RELATE to this law.?-- Local varieties formed with ex
206b146 ent day, every ten living souls on average are RELATED to the (200dth year) degree. Then 200 years ag
206b146 Then the chance of 200 people <might be> being RELATED within 200 years backward might be calculated
206b147 as increasing between each lustrum, the number RELATED at the first start must be greater, & this num
206b149 ing population at any one moment fewer closely RELATED;∴ (few species of genera) ultimately few gener
231b243 f the Tropics, (whether one fossil or not) are RELATED by real relationship. as well as effect of sim
231b243 those of temperate regions & tropics are only RELATED by one connection.-- viz descent.-- Hence far
281c138 ust same way as <we> «all men» not all equally RELATED to each other I cannot help thinking good anal
281c138 classification of animals-- talking of men as RELATED in the third & fourth degree.-- a species must
384d162 abortive sexual organs?): they then become so RELATED to each other, as never to be able to impregna
389d176 g-- DESIRE LOST when male & female too closely RELATED: this most important with regard to theory, sh
397e003 1856 [not located] Epidemics-- seem intimately RELATED to famines., yet very inexplicable.-- do p. 52
418e083 ndividuals in their present forms, are closely RELATED-- By birth the the succesive modifications of
421e091 ive organs being those which are most remotely RELATED to the habits & food of an animal, I have alwa
491q01v t genus & then some hours afterwards of nearly RELATED plant & see if first pollen produces any effec
547m112 in so simple case.-- There was memory, for it RELATED to past idea.-- there was a kind of ideal cons
255c050 logues.-- <as> in other classes this evidently RELATES to greater range of such forms.-- p. 56: Ornit
389d176 nter-breeding as told by Willis Q Proved facts RELATING to Generation One copulation may impregnate o
029r034 ards of the Coal in Chili as clearly bearing a RELATION to present position of <Coal> Forests. These
035r046 s, deposited in different basins; little or no RELATION appears to <exist> be made out, but in those
045r077 -Will geology ever succeed in showing a direct RELATION of a part of globe rising, when another falls
062r130 d by change of circumstances: The same kind of RELATION that common ostrich bears to (Petisse. & diff
063r130 me. (or changes consequent on lapse) being the RELATION.--As in first cases distinct species inoscula
077r170 e with that on surface of plains of Amazon, no RELATION--there is more modern breccia, chiefly owing
100a055 ays from Lardner's (p. 213) form of escarpment RELATION kept to sea coast ∴ curious exception in Weal
180b036 extinction. Thus between A. & B. immens gap of RELATION. C & B. the finest gradation, B & D rather gr
180b036 inction Thus genera would be formed.-- bearing RELATION to ancient types.-- with several extinct form
190b075 bee doubtless would when the instincts were.-- RELATION of type in two countries direct relation to f
190b075 re.-- relation of type in two countries direct RELATION to facilities of communication Have races of
201b129 species, have produced fe[w] [not located] The RELATION of Analogy of Maclay &c. appears to me the sa
209b157 n Humboldt great work de distribut. Plantarum. RELATION of genera to species in France is I: 5.7 in L
209b157 =Radack Islds = ∴ Islands & Artic are in same RELATION. We find species few in proportion to difficu
210b159 ence. if they do.--there probably will be this RELATION also.‖Yes « Fox» ‖ The creative power seems t
214b174 , that North American fossils bear the closest RELATION to those now living in the sea.-- See Rogers
224b212 s. & checks it.-- It does not bear any precise RELATION to structures Mem. Eyton's Hogs & dogs.-- The
244c022 from Buenos Ayres) replaces <Vesp.> holds same RELATION with equator--that Vesp. lasiurus does in Nor
248c033 generation «& primary divisions of insects» 2. RELATION, of external conditions, & of succession: the
258c061 a Vigors.-- nothing of much interest XX. Hence RELATION of analogy may chiefly be looked for in the a
263c077 f type-- Study geographical distribution study RELATION of fossil with recent. the fabric falls! But
269c102 plants, with this object in view» The intimate RELATION of Life with laws of Chemical combination, &
281c138 ds good, so does it in real classification The RELATION of all cock birds in Gallinaceous having tend
282c140 most fixed in others. In analogy it is not the RELATION to bear to each other, but to some external c
282c142 become eminently aquatic--; «NB, aquatic, i.e RELATION to elements & not minding particular trades.-
282c142 original species, each became ground then the RELATION of all the ground cuckoos. would not be affin
283c143 entations.-- The aerial type in each family is RELATION to elements & not habits as shown by frigate
283c144 ntastic «sexual» ornaments. have so intimate a RELATION to two continents as to be <replaced> called
286c156 xt family? In considering fossil animals, what RELATION in classification in books, ought they to hol
287c157 l XIV.--‖. p. 24. Lamarck bears to Cuvier that RELATION of theoretical astronomer to plain observer‖
291c167 developing an hybrid female it is a wonderful RELATION going through all Nature.-- Makes hermaphrodi
296c184 in short Extreme North = = to peak of Teyde in RELATION to surrounding countries & present tropical c
305c211 s modified into endless forms, bearing a close RELATION in degree & kind to the endless forms of the
305c211 ing. fits, laught &c&c Man & Man may have some RELATION together, as well as Man & child, polypus & p
308c217 here now must be, ∴extinction of species bears RELATION to existence of genera &c &c Two savages, two
313c236 --not necessary to generation (lateral with no RELATION to time) as in buds.-- I can scarcely doubt f
315c240 s. the means of transport fail.-- otherwise no RELATION between means of Transport & creation exists.
341d029 not make that bird a Penguin.-- (i.e. whether RELATION in one point or many) Owen answered that all
342d035 d-- is it an old order Geologically? Owen says RELATION of Osteology of birds to Reptiles shown in os
350d055 do with perfection of individual, though such RELATION seems common, but that perfection consists in
358d074 ter of St. Helena.-- contrast with otaheite in RELATION «See Gaudichauds Volume on the Botany of the
367d112 at cannot be formed, genetal organs by that co-RELATION of parts, will not be produced.-- Sept. 17th.
381d156 Therefore highness in scale has no «constant» RELATION to separtion of sexes, as may be seen in Mono
388d172 should be observed that transmission bears no RELATION to utility of change-- hence harelips heredet
400e011 ic characters. dominant. predominant &c having RELATION to geographical distribution-- Thus Hattica i
408e042 Cuba.-- It is important to understand well the RELATION of passage from N. to S. American forms. The
410e050 s.-- so that the changes should be slow & bear RELATION to the whole changes of country, & not to the
410e051 heory, laws probably will be discovered. of co RELATION of parts, from the laws of variation of one p
410e052 inment of closest species (& naming them) with RELATION to habits, ranges. & external conditions of c
411e054 » are permanet, all steps in the series, their RELATION to the external world, & every possible conti
414e065 t as the Elephant, if some genus. holding same RELATION as Mastodon to Man. were to be discovered. Ma
416e072 Rhinoceros «tichorhinus» in Paris basin.-- its RELATION to African Species <good observations.-- >, l
416e075 A Greyhound might be made «almost» without any RELATION to running hares.-- as in Italian Greyhound n
417e077 > we can as well understand the necessity of a RELATION between the fluids of the two as in the graft
425e103 asily» making tolerably fertile hybrids, bears RELATION to capability of variation?? my theory says s
426e106 portant on my view.-- March 9th-- Is there any RELATION between the fact that different species produ
432e125 ing the canine teeth, or sneering, has no more RELATION to our present wants. or structure, than the
443e155 Corallina & Halimeda is case in point».-- The RELATION of these «sexual» functions to complexity is
443e156 since to geologists.-- & what is older-- what RELATION in duration of a planet to our lives-- Being
448e166 tes from Stephenson at Lima The same numerical RELATION (both in species and subgenera) between the C
462tf05 theoretical French book review on politics in RELATION to the different races of men, some more inte
466t094 y coloured & to which stamen & pistils have no RELATION. In Monk's Hood, a bee entering long nectary,
493q003 us do more than three primary colours occur in RELATION with species-- answered «by Henslow» see note
493q003 d Duck, & then weigh their wing bones & see if RELATION is same good, avoids effects of fatness.-- Ex
508q016 les des Sciences‖ (4) Prolifickness of female, RELATION to healthiness? & father answered (5) About c
515q021 ere fact of seeds ripening has scarcely any no RELATION to hybrids.-- (3) As peaches sport into Necta
533m059 ng of animals. that is the ready movement & co-RELATION of the proper muscles. may be illustrated by
575n043 be intermediate like rest of body? Can we deny RELATION of mind & brain. «Do we deny the mind of a gr
577n051 testinal functions &c &c &c.-- bears. the same RELATION to true memory, that the formation of a hinge
579n058 -sighted people frown, frowning must have some RELATION to short-sightedness.-- do not short sighted
583n069 es of instincts].CD all this may be true,, but RELATION of imitation & reason must be thought of.-- p
586n082 s, is sight--CD [The faculties bear so close a RELATION to the senses, that one feels no more surpris
589n092 omething destroyed bad brain. see p. 90.-- The RELATION of reason to organs of locomotion-- or that o
593n107 aped depression.-- [blank] Does music bear any RELATION to the period when men. communicated before l
601o009 sness) are produced in consequence having some RELATION to the primary sensation.-- man moving leg wh
601o009 lanariae conscious.-- Consciousness bears some RELATION to time & memory Reynolds X discourse very cu

611o034 nd the Organic laws probably have some unknown RELATION to them-- [RHC] In the simplest forms of livi
611o034 ents, & in such cases modifications bear fixed RELATION to such accidents. But such tissue <must> bea
611o034 o such accidents. But such tissue <must> bears RELATION to whole, that is enough must be present to b
612o035 ecisely same as in plants, but as animals bear RELATION to less simple bodies, and to more extended s
612o035 es, and to more extended space, such powers of RELATION required to be extended. Hence a sensorium, w
612o035 nimals have consciousness. These willings have RELATION to external contingencies, as much as growth
613o035 mence; where other senses come into play, when RELATION is kept up with distant object. where many su
616o039 s will, consciousness memory &c. have the same RELATION to a living body (especially the cerebral por
616o039 t) that attraction has to ordinary matter. The RELATION of attraction to ordinary matter is that whic
616o039 o attract; & hence if thought &c bore the same RELATION to the brain that attraction does to matter,
618o041 . 5) There is nothing analogous to this in the RELATION of thought, perceptions, memory &c. either to
618o041 no objective aspect. If thought bore the same RELATION to the brain that force does to the bodily fr
618o041v think night an effect.]CD Cause and effect has RELATION to forces & mentality because effort is felt
628o54v hes of passion &c. whilst the passions have no RELATION I think this <boshes> «nonsense»-- My theory
629o55v w shame is accompanied by blushing, bears some RELATION to others 5) if so, it is perhaps deviation f
102a060 eparation.-- the branching cracks-- only bear RELATIONS to VEINS in primitive rocks-- Are substances
171b002 lly in lower animals, where mind, & therefore RELATIONS to other life has not come into play)--See Zo
202b130 gy, the structures in the two animals bearing RELATIONS to a third body., or common end of structure
274c115 Museum-- I could not discover any other clear RELATIONS besides aerial, & terrestial.-- How is it in
274c115 rial may be observed even in Lessonia &c & In RELATIONS of affinity all organs change together, in an
275c119 of ancient geography species tell of Physical RELATIONS in time «& forms» distribution tells of horiz
281c139 les.-- which in each group is quite fatal.-- RELATIONS of analogy being those last obtained.-- less
282c140 contingency.-- affinity is the sum of all the RELATIONS, analogy is the close relationship in some on
284c147 on planet at different periods, depends,-- on RELATIONS of desert, open ocean, &c this probably on lo
284c147 obably on long average, equal quantity, 2d on RELATIONS of heat & cold. therefore probably fewer now
284c147 - The number of forms depends on the external RELATIONS (a fixed quantity) & on subdivision of statio
285c151 ll species not possible in some detail,-- the RELATIONS to islands close species, «on these isld» &c
287c158 Macleay seems to limit Lamarck definition of RELATIONS to settling the relative importance of the or
293c175 simpler stage to more perfect. preserving its RELATIONS.-- The wonderful power of adaptation given to
294c175 r some time at New York «instance of the fine RELATIONS of adaptation of animals & the country they i
305c211 given or assumed according to a more extended RELATIONS of the individuals, whereby choice with memor
328cIBC erhouse has it) shells from Barrier isld many RELATIONS with a living Matica & many shells of Genera
389d177 onstitution) is the end frustrated, when near RELATIONS, & therefore those very close are bred into e
403e023 a tooth of any animal (as Toxodon) & say its RELATIONS.-- if we know its congeners then we can.-- no
410e053 espect good describers like Richardson.-- The RELATIONS of numbers of species to genera &c &c can nev
422e095 gh all perhaps will have done so from the new RELATIONS caused by the advancing complexity of others.
427e108 well-- In early stages of transmutations, the RELATIONS of animals & plants to each other would rapid
542m094 are habitual, & which now connected physical RELATIONS.-- CD[like sighing to relieve circulation aft
544m102 no surprise, at the violation of all <rules> RELATIONS of time, <identity,> place & personal connect
558m151 Analyse this out.-- bearing in mind many new RELATIONS from language.-- the social instinct more tha
633j54r s weak creations.-- All such facts are merely RELATIONS of one general law. the plants were no more c
641j29r how little we understand of the Physiological RELATIONS of animals. equatorial countries are supposed
206b147 lustrum, & the calculation of chance of the RELATIONSHIP of the progenitors would have different for
206b149 ra) ultimately few genera (for otherwise the RELATIONSHIP would converge sooner) & lastly perhaps som
211b162 structure.-- Mem. Ornitho Rhyncus Would not RELATIONSHIP express, a real affinity & affinity-- whale
231b243 ether one fossil or not) are related by real RELATIONSHIP. as well as effect of similar temperature.-
258c060 ecome adapted to climate.-- Descent. or true RELATIONSHIP, tends to keep to species to one form, (but
258c060 species to one form, (but is modified), the RELATIONSHIP of Analogy is a divellent power & tends to
258c061 e some other class analogy» The resemblances RELATIONSHIP, the dissenblances analogy.-- See Abercromb
281c138 hinking good analogy might be traced between RELATIONSHIP of all men now living & the classification
282c140 m of all the relations, analogy is the close RELATIONSHIP in some one.-- imagine the men to have grea
286c155 t, it is the classification of <arrangement> RELATIONSHIP; latter word meaning descent.-- A tree is t
287c158 but the one end of classification to express RELATIONSHIP. & by so doing discover the laws of change
309c219 similarity in individuals «we know of:», is RELATIONSHIP, children of one parent, races of animals--
348d051 though how far we can ever discover the real RELATIONSHIP is doubtful.-- not till much knowledge is e
101a058 erence between dike & mountain axis. except in RELATIVE <strata> size with superincumbent strata. whe
115a098 force» «size of» of <currents> «fragments» but RELATIVE to currents. Small lakes have power of levell
287c158 amarck definition of relations to settling the RELATIVE importance of the organs in same state. in di
324c268 M. Quetelet has published his laws about sexes RELATIVE to age of Marriages Brown at end of Flinders
335d015 hat is slight alterations of primitive stocks «RELATIVE to changes which every species undergoes») &
390d178 eeding in & in would not be deletereous if the RELATIVE had come from different quarters) then it cau
181b040 ; it is extremely improbable that any of <his RELATIVES shall likewise» the successors of his relativ
181b040 iatives shall likewise» the successors of his RELATIVES shall now exist,-- In same manner, if we take
181b040 years hence; because at present day many are RELATIVES, so that by tracing back. the <descen> father
591n099 believe)» contempt at suicide. (even when no RELATIVES left to lament) is owing to the feeling that
536m071 ty substance it scrapes off its bottom.-- it is RELIC of same thing that makes one dog smell posterio
162g101 Peat & Heather When it did not grow at first-- RELICS destroyed.-- the Brook <about> Head of which i
537m075 s are changed may be inferred from expression. "RELICT of bad habit." as child is cured of sucking hi
424e102 t of origin of Wolf, difference of mind is most RELIED on, but Bell has some account of wolf in Zoolo
542m094 ected physical relations.-- CD[like sighing to RELIEVE circulation after stillness.-- Now I conceive
570n024 Explanation of sighing is probably correct, to RELIEVE respiration when immensely immersed-- mechanic
564n002 ich was only short & might otherwise have been RELIEVED, he would be sorry or have a troubled conscie
093a034 but they are parts of one force, one locally RELIEVING the other. -- Is the felspar glassy in greens
321c270 nary Conversations-- very poor Sir T. Browne's RELIGIO Medici Lyell Book III There are many marginal
350d054 s the only indisputable axiom in Philosophy«"> RELIGIO Medici. Vol II. Sir T Browne's Works p. 20 The
550m126 h all our recollections are pleasant.-- Browne RELIGIO Medici, p. 21-24. Curious passages showing how
591n101 of animals Essays Vol 2.-- «also on origin of RELIGION or polytheism, at p. 424 Vol. II «Sect XV. Di
591n101 t p. 424 Vol. II «Sect XV. Dialogue on Natural RELIGION.» however, he seems to allow it is an instinc
592n101 Philosophy. Hume has written "Natural Hist. of RELIGION" on its origin in Human mind.-- Andrew Smith
580n061 dhood.-- hence the belief in the many strange RELIGIONS.» Emma W. says that when in playing by memory
538m079 rs.--» the other army officer, horticulture & RELIGIOUS sects.-- yet Allen. W. remark about his slipp
413e060 ulata, which are all truly bisexual. Buckland's RELIQU: Diluv. says Africa only place, where, Elephan
094a036 Hall Trans. Phils Royal Ed. Vol 7 Dr Buckland RELIQUIAE Diluvianae p. 201. & seq Murc Trans Geolog So
408e044 «Gales of wind would blend species» Buckland. RELIQUIAE Diluvianae. p. 222. Bones of Horse. Bear & De
554m139 t it. when guided to her own mouth.-- seemed to RELISH the smell of Verbena & Pocket Handerchief & li
056r109 astonishment.» Perhaps scarcely a pebble might REMAIN to tell of these losses.-- Cause of chimney. t
133a140 abysses where fluid rock has been ejected must REMAIN fluid for an enormous period: now when we see
172b006 sh isld. it is very doubtful whether they would REMAIN constant; is it not said that marrying in dete
212b165 rrow like fox.-- a sort of internal bark. would REMAIN for long time together in tub of water with on
224b213 r, might change organization of stomach & hence REMAIN distinct. Where country changes rapidly, we sh
267c095 ?? p. 75. roe of Asterias in stomach. of Sammon REMAIN after rest of animal digested.-- Important do
318c254 st & others the mountains, & these appearing to REMAIN about a fortnight, «See Silliman's Journal 183
335d013 mmon goose.-- What has long been in blood, will REMAIN in blood.-- --converse, what has not been, wil
335d013 lood.-- --converse, what has not been, will not REMAIN,-- yet offspring must be somewhat like parents
335d015 ortion as things are long in blood so will they REMAIN, a mule «being new species» will have no tende
405e031 ephants &c &.-- if in next 20 years none of his REMAIN found in the Americas probably did not.-- Octo
427e108 se, & hence number of forms. once formed. would REMAIN stationary, hence all present types are ancien
461t041 of conditions, some species will undergo & yet REMAIN adapted.-- it does away with difficulty of rab
497q006 ow, how seeds are transported, or how long they REMAIN dormant. if kinds come up, not found in wood.--
504q014 iegman Shrewsbury (1) Peas.-- Beans seeds alone REMAIN to be compared-- Cabbages.-- kept true Try exp
512q019 alls of Hawks or even owls.-- How long do seeds REMAIN in stomach of birds-- Mem: how many miles they
525m026 m, & lastly healthy ones.-- Insanity & Epilepsy REMAIN many generations in families.-- My fathers doe
532m056 retchedly unhappy, constantly whined, would not REMAIN quiet in any room, would not sleep at night ev
542m059 tion were changed, I conceive sighing might yet REMAIN just like sneering does.-- is yawning habitual
545m106 ction is sent so fast to limbs that they cannot REMAIN still.-- I do not doubt this Baboon. knew wome
321c275 gical Society Vol I. & 1s No of Vol II. (read REMAINDER) when out [not located] <Rays Wisdom of> Lisi
152g048 gh, not applicable to Glen Roy Lake, must have REMAINED very long at 4th shelf from size of buttresse
159g085 l between Corry so much cut Granite could have REMAINED, no peat supply.-- Consider profoundly Boulde

Page
(Key Word)
```
213b170 n makes angels-- Those species which have long REMAINED are those ¿Lyell?, which have wide range  and
302c200 gaps.--  external conditions, to be sure, have REMAINED somewhat similar.--!!! My theory drives me to
064r136 & 150 broad. neglecting Cordillera itself now REMAINING-- Lyell « <p 419> p 428» states that Von Buch
289c162 oung many times & lay two sorts of eggs-- one REMAINING through winter. «It would be curious to  know
296c184 s of Europe., those species which can migrate REMAINING constant in form, others altered much.-- thes
446e163 difficulty in my theory, here we have a plant REMAINING constant, without crossing.-- & propagation b
461t041 es away with difficulty of rabbits of England REMAINING same (if so) with those of Spain & such facts
023r012 om. With respect to Sharks distributing fossil REMAINS: Sharks followed Capt. Henry's vessel from the
025r019 record)»  In looking over the lists of organic REMAINS in De la Beche, for the older formations I mus
026r022 level, which in the nature of strata & Organic REMAINS does not appear to have taken place in the Cor
037r053 f rocks--It may be a question. whether organic REMAINS protect a rock, or that the rock not weatherin
037r055 ry strata of England, «besides ordinary marine REMAINS» may contains <shells few corals Tortoise> «re
037r055 ns» may contains <shells few corals Tortoise> «REMAINS of Amphibia, exclusively.» & Turtle bones. & t
037r055 rd.-- There would probably be no other organic REMAINS.-- On Pampas looked in vain for a pebble of an
044r073 l movements of the earth;--Scarcity of Organic REMAINS.--Unequal distribution of Volcanic action, Aus
117a101 ding to composition thinks sand with vegetable REMAINS formed near coast, limestone deep water.  will
123a117 tter from M Angelis. B. Ayres. 3d. May. states REMAINS found in many part.-- great Dasypus near Canel
147g017 e probable I did not look carefully for Marine REMAINS-- Some of the hills almost appeared as if they
222b206 ulation at present day & not include Mammalian REMAINS.-- The Father of all insects gives same argume
224b213 of  species.-- Definition of Species: one that REMAINS «at large» with constant characters,  together
231b244 one species, far more than by non-embedment of REMAINS-- ¿agrees with non-blending of languages?-- Ti
284c149 l characters than among those groups, where it REMAINS less subject to Variation" Dr. A. Smith. knows
323c269 's Bridgewater Treatise -- Wilkinsons Egyptian REMAINS.-- Lima.-- caves.-- There being no fossils, th
415e066 ncus,: since being cosmopolite, we do find his REMAINS skimmed -- Pliny Nat. Hist of World do -- Lama
424e101 is would depend on negative evidence of fossil REMAINS, & therefore not to be trusted.-- -- Lyell tel
449e169 y. 1839. p. 106.-- Waterhouse refers to fossil REMAINS of the Hamster.-- is not this Siberian animal?
473s03r in miocene like in Europe-- Cuvier never found REMAINS of Sus with Elephants-- Lyell says New Red San
531m052 this is the oldest <her> inherited & therefore REMAINS, when the actual movement does not take place.
534m061 caught it like a flash.--. strange if judgment REMAINS, where reason is forgotten. it is conscience,
582m066 to  young pigs hiding themselves; & heredetary REMAINS of savages state.-- N B. According to my  view
594n112 arise??  «it would appear that an instinct long REMAINS, if no steps are taken to eradicate it.--» «Em
641j29r not.-- Does this bear on, the absence of their REMAINS in the Wealden? In the strongly separated Arct
088a016 of Hornblende p. 248. L. Institut 1837.-- Helms REMARK on common salt being found on low hills East o
136a147 f the age of the Pampas Deposit, I may properly REMARK on the superiority of Lyell's classification t
173b009 ct.-- truer even than in Lamarck's time. Gray's REMARK, best known species. (as some common land shel
213b171 unds & will they not be greyhounds?-- Yarrell's REMARK about old varieties affecting the cross most w
217b188 ng wanderers.-- Iceland no species to itself, a REMARK common to all northern islds.-- This is intere
233b249 s New Ireland. see Coquilles Voyage), Waterhous REMARK Australia Fauna so far. Indian all the rest. T
240c003 rock  pidgeon.-- fact analogous to Owen's Phil: REMARK of Apteryx having feathers.-- It is  possible,
243c020 references  highly unphilosophical xx Says same REMARK with regard to shells.-- But he says shells to
261c071 the birds into group.-- it is same as Yarrell's REMARK about rock Pidgeons.-- & the latter most impor
284c149 looks like subsidence.-- on the islets Mr Blyth REMARK that a resemblanc between some form in birds i
284c149 .-- thus speckled form of young blackbird. good REMARK if general.-- Where any structure is general i
291c165 rictly point out affinities. conducct of Gould, REMARK of D'orbigny point out importance of habits in
308c218 itted for water, air, & land, (Macleay has this REMARK) Mem. number 5 here most evident!!? examine in
355d068 ose.-- Mem. Von Buch on Cordillera fossils same REMARK. ¿was there formerly one great sea, & two Pola
385d165 on, & of the father in the daughters! This last REMARK good. because showing probably not education.-
415e067 ts of structure abortive.-- Remember my fathers REMARK about the Bladder.-- The numbers of fatal dise
463t063 n his description of my fossils makes same such REMARK & before the conclusion of his work-- Lund mad
467t099 mewhat «but little» less surprised at Henslow's REMARK that pistil does not become abortive. Examined
509g017 does  its abortion vary, according to Bentham's REMARK. Horse or cow.-- degree of soldering of  tibia
510q018 us.-- foxes-- English animals. [Made no import. REMARK]CD (2) Secondary male characters.-- does  male
525m026 , when seeing Nina with her puppy.-- The common REMARK that fat men are goodnatured, & vice versa Wal
525m026 men are goodnatured, & vice versa Walter Scotts REMARK how odious an illtempered fat man looks, shows
530m046 robably a physical effect of brain the «similar REMARK» thoughts, being functions of same part of bra
538m079 horticulture & religious sects.-- yet Allen. W. REMARK about his slippers bad for fires, what is wron
543m100 yet. he was inimitable chess player.-- Peacocks REMARK about mathematicians low & being profound reas
552m131 ending force to muscles, that in my grandfather REMARK, a tired man. involuntarily feels angry,  when
578n053 to  wish not to do so.= = How directly personal REMARK will make any one blush.-- Is there not some s
626o052 reat difference with my theory.-- see p. 349.-- REMARK on this point.-- [LHC] p. 194. «&c &c» Butler'
027r023 omerates are found. -- The above oscillations REMARKABLE because the formations are now seen in regul
039r060 es.-- Talking of the cricket valley «the most REMARKABLE feature in the structure of Ascension» give
067r144 n miles inland. The Cordoba earthquake a very REMARKABLE phenomenon. showing line of disturbance insi
157g075 forms it into island-- whole hill composed of REMARKABLE gneiss with red granite veins & quartz, & ga
165g126 form The union of two instincts crossing most REMARKABLE ever obseved? Shows that <nervous> brain mak
190b076 an  end to species.-- «Brown Appendix» A most REMARKABLE observation of Mr Brown, about peculiarities
190b079 eal <separ> passage between good genera-- How REMARKABLE spines, like on a porcupine on Echidna-- Goo
193b093 says from Swan river long South coast all the REMARKABLE Australia genera, collected together Man has
221b201 ce of lime in Plutonic & Volcanic rocks. most REMARKABLE.--¿ Have the changes been so slow., that all
248c033 [consta]nt  & therefore most permanent Owen [RE]MARKABLE laws of Brain & manner of generation «& prim
249c036 s, generally wide range? Mice.-- Waterhouse's REMARKABLE fact of no forms peculiar to word to special
249c058 Europaean  forms on Himalaya??-- This is very REMARKABLE, when we consider number of quadrupeds in Eu
270c104 ion of organic nature on inorganic It is very REMARKABLE as shown by Carus how intermediate plants ar
271c105 are entered in upon life. Namely Carus.-- How REMARKABLE that Turdus Magellanicus. in the. S. Hemisph
272c109 ered over it.-- We have abundant instances of REMARKABLE structure which as far so species is concern
272c109 ng the curious feathers in tail of Edolius.-- REMARKABLE how small detail in structure prevails among
273c110 t there exist intermediate species,-- This is REMARKABLE & would lead one to suppose, that species in
273c112 ers likewise exaggerated.-- There is one most REMARKABLE connection between these aerial representati
280c135 of ascertaining about hybrids.-- & is a very REMARKABLE fact. show influence of mind It is not diffi
285c150 r observation of tracks &c &c Dr. S. has some REMARKABLE crochets about instincts. whenever  instinct
298c192 e variation in character of leaf of plants is REMARKABLE what is analogous to it in animals?-- Babing
299c193 f Man) no plants peculiar to themselves. this REMARKABLE compare it with Canary Islds. Galapagos.-- I
302c200 ct creator chooses so to create.-- It is very REMARKABLE, with so much death, as has gone on,. No gre
305c209 es of which vary in colour of plumage in same REMARKABLE manner as Europaean species = singular coinc
315c242 erent from contraction of fibre)-- it is most REMARKABLE habitual actions in plants, it allows of ma
316c245 ew Cyclostomes & a few Anphidesmas.-- this is REMARKABLE.-- Fish & drift sea weed-- may transport ova
316c247 rtiary Europaean fossils-- «(so much the more REMARKABLE ': carboniferous ones similar?)» Now this is
316c247 arboniferous ones similar?)» Now this is very REMARKABLE.-- (connect these facts with identity of lan
334d010 e.--» How is it with plants? This indicates a REMARKABLE law, that first cross <not se> plentiful, se
341d029 where the line could be drawn-- thus the most REMARKABLE character in Apteryx, small respiratory syst
378d148 t still they have a wariness about them quite REMARKABLE". instance of old Species transmitting so mu
388d172 domestication  (even Elephant) not breeding-- REMARKABLE Athenaeum 1838. p 653. Ehrenberg<h> thinks m
389d177 ely as I have stated it) it is certainly very REMARKABLE that too much difference should produce same
411e055 rders (not so in S. America, however) is very REMARKABLE & none discovered before them in any part of
413e060 Sexual  system of the Cirrhipedes is the more REMARKABLE from their alliance to Articulata, which are
448e166 ther with neighbouring & Senegal as sea.-- is REMARKABLE.-- Again the resemblance between the Superga
532m056 ely fell into her old ways & became fat! What REMARKABLE affection to a place.-- How like strong feel
535m064 . The deserted broods appeared healthy-- This REMARKABLE case may be normal. with insects, but  habit
535m069 is it with children.-- Now it is not a little REMARKABLE that the fixed laws of nature should be «uni
542m094 & bellows with passion.-- It is not a little REMARKABLE that those sounds which are involuntary, are
565m009 <The> Children having peculiar expression is REMARKABLE. the pouting, & blubbering-- sulkiness is sa
638j58v ge of shell, works of laws of organization is REMARKABLE-- what is intellect, but organization,  with
641j29v logues The stillness p. 276) of flight of Owl REMARKABLE, [gained by very different process from Bats
151g040 this character.-- The boulders (one of Gneiss REMARKABLY water worn) are often times of rock not in i
305c209 like S. American.-- Have not Ruffs & Reeves a REMARKABLY varying plumage for wild birds-- At Zoolog G
315c242 ory» in children or rather their memory. very REMARKABLY-- scenes in themselves accidental-- My first
403e022 n the small isld of Eap in the Carolines, are REMARKABLY short.-- & Deformations are particularly com
533m061 memory  of having read or thought of some such REMARKE as now advanced; for I caught it like a flash.
```

```
104a068 n would bring water with salt to surface Lyell REMARKED to me that Kylow (?) was astonished with him
179b034 & art required to make marriage.-- as Dr Smith REMARKED Man & wild animals in this respect are differ
190b079 an is derived from Monad, each fresh--> Mr Don REMARKED to me, that he though species became obscurer
211b162 probably of rudimentary bones.-- As Waterhouse REMARKED Mere length of bill does not <indicate affini
213b171 ow long back have insects been known? As Gould REMARKED to me, the "beauty of species is their exactn
218b192 D'Acunha, a list of its Flora. is given Mr Don REMARKED to me. that some good African & some good S.
220b200 s about mundine & confined genera.-- Lyell has REMARKED about no confined species in Sicily. Jan: 18
271c107 reed of animals in certain countries.-- Fraser REMARKED to me at Zoological Society,, that you never
340d029 [not  located] the «4» Struthionidae, Mr Blyth REMARKED that greater difference in than in many large
342d033 all alike) More than parent species-- Mr Blyth REMARKED only near species or varieties produce hetero
346d048 file,  hide their young., bold.-- a Mr W: Hall REMARKED that it was against all rules their preservin
350d055 hese differences in sex confined to annulosa?) REMARKED that young of Cirrhipedes can move & see, par
379d151 further back reptiles & Cephalopoda: Old Jones REMARKED to me, that one of the children of Sir J. H.
425e105 form.   found in Eocene beds of Paris Lyell has REMARKED species never reappear when once extinct-- Ly
463t057 h extinction. give a parallel case) Waterhouse REMARKED, that any argument for transmut, from one org
472s02r eases, never saw any Bee go to them. Yesterday REMARKED that many flowers had suddenly withered, & to
525m023 es memory bring in old ideas <I have elsewhere REMARKED do> Dogs take pleasure, when doing. what they
538m079 in infancy, had often been repeated: Now it is REMARKED that A. Bessy repeated things, which none abo
550m127 t call them instinctive emotions?-- Dr Holland REMARKED that insanity like sleep does not doubt the r
553m137 ler (superintendent of the Zoological Gardens) REMARKED that <exp> the expression & noises of monkeys
553m138 ean SOMETHING.-- Hunt (the intelligent Keeper) REMARKED that he had never seen any of the American Mo
024r015 hole coast of New Holland shoals much: Dampier REMARKS on great flats on the NW coast:-- 8 leagues, f
077r172 of the veta madre of [misnumbered page] Dr D. REMARKS. bad conductor of Heat do of Electricity Does
113a092 er. Major Mitchell inferred subsidence; Mem my REMARKS on coast of Australia.-- Great NW. dip in SE p
136a147 eatise Phillips in Lardner Vol II p. 73.: some REMARKS on veins: Phillips in Ladner Vol. II p. 80-- s
136a147 veins: Phillips in Ladner Vol. II p. 80-- some REMARKS on dikes: applicable to Cordillera Phillips in
136a147 Phillips  in Lardner Vol II. p. 81. «&83» Some REMARKS on thinness of crust as implied by meeting wit
195b099 s. We know birds do arrive & seeds.-- The same REMARKS applicable to fossil animals same type, armadi
207b151 led Botany Bay Willow V. Dr Royle introductory REMARKS to Himalaya Mountains-- Bory St. Vincent  Vol.
218b191 s. Mr Donn considers Mr H. rather wild Mr Donn REMARKS to me. that give him a species from Ireland, E
235b272 an Diemens Land & Australia [blank] Falconer's REMARKS on influence of climates, situations &c on 242
241c016 ologie of Coquille's Voyage to see if Lessons' REMARKS on the Floras can be trusted The changes in sp
254c047 ly different.-- Q Zoolog. T. V. I. p 389. Owen REMARKS on Entozoa, the organs of generation, afford t
261c069 all kinds of information-- instinct Swainson's REMARKS in Fauna Borealis must be studied. There is ca
302c201 ut we see analogies between fish.-- Birds same REMARKS. Characters of analogy.-- last acquired,-- or
310c225 nals of Natural History «Vol I??» p. 318. some REMARKS on Bonaparte's list of birds in Europe & N. Am
324c268 hilosophy of Zoology Flemming. ditto Falconers REMARKS on the influence of climate White's regular gr
326c265 Hybrids. where? Sweet. Hortus Britannicus. has REMARKS on acclimatizing of Plants. Herbert. p. 348. g
337d022 n of sound.-- which is impossible.-- Mr Spence REMARKS that the Fringilla domestica of North Europe i
342d034 any two species crossed is uncertain-- Yarrell REMARKS he has somewhere met conjecture that all salt-
362d099 is analogous to the love of woman (as Mitchell REMARKS seen in savages) to brave men.-- Effect of cas
376d136 Many generations Sept. 29th Dr. Andrew. Smith «REMARKS on extraordinary curiosity of Monkeys». The Ba
435e134 reased in size now the Proteus anguiformis. he REMARKS lives in dark caverns of Carniola p. 112. Man.
481z009 lls of Galapagos different islds.-- Waterhouse REMARKS that no insectivore in S. America or Australia
515q021 chance  & under domestication.-- N.B. Benthams REMARKS, where parts of flower are reduced from normal
526m028 -- equall true the two statements.-- Catherine REMARKS that pleasure received from works of imaginati
531m050 s instance of reason & abstinence.-- My Father REMARKS that things of great importance are easily for
533m060 view (Edinburgh) of Froude's life. that author REMARKS, that writing down his confessions of sins. di
534m063 experience & intellect.-- do. p. 157. Westwood REMARKS that some imported plants are attacked by inse
540m089 ional desire to assist them.-- otherwise as he REMARKS sympathy could be barren. & lead people from s
581n064 March 16th. Gardiner's Music of Nature. p. 31. REMARKS children have no difficulty in expressing thei
585n074 ndar Water «very much» Henslow. N.. Necker has REMARKS on the means. by which children learn (probabl
585n076 - Here there is kicker but not bite.-- Henslow REMARKS that Chimpanze pouted & whined, when, man went
588n089 dily actions make the emotions.-- p. 272. Some REMARKS applicable to my theory of happiness.-- The bo
592n103 Lectures  by Parsons.» following pages contain REMARKS worthy of attention p. 15, 25. 40. 61. CD[a pe
607o025 prevent   the action see Abercrombie conclusive REMARKS p. 205 & 206.]CD Motives are units in the mind
625o50v principles   if Mackintosh.-- p. 262. Some good REMARKS. on analogy of pleasure of imagination «the ut
625o052 p. 37. Whewells gives Mackintosh's theory: the REMARKS about "contact with will" is unintelligible to
626o052 e notes to it by me..' 'p. 333 «& p. 377» some REMARKS. showing that instinct cannot be said to guide
634j54v p. 313 Many other good cases -- p. do]CD <Mac. REMARKS all Mammifers originally land--animals. as> 5
634j55r s originally land--animals. as> 5 p. 314. Mac. REMARKS all <land> Mammiferous animals originally terr
050r093 ntain Limestone in N. Wales. was it reef. -- I REMEMBER many Corals?? Breccia--Stratification? Anomal
057r115 the   temp of which was below freezing point!!! REMEMBER idea of frozen bottom or beach of sea to expl
149g030 rounded stones-- lake required to deposit this REMEMBER however the great Chilian valley Acongua, must
149g030 must there have deposited much-- On other hand REMEMBER modelling power of sea N of Valparaiso are th
225b217 ng passage of birds (where is this statement I REMEMBER. L. Jenyns. talking of it) or how to make Ind
276c123 r age. (differently from literary men.--) must REMEMBER that if they believe & do not openly avow the
293c173 es web footed, now Man by effort of Memory can REMEMBER how to swim after having once learnt, & if th
318c255 ation) gradually separated the birds might yet REMEMBER which way to fly.-- There is a kind of Wren (
332d002 and at same periods When he began practice, he REMEMBER during a year or two he saw many cases of vir
407e042 ow that if species fall, genera must. Lesson I REMEMBER says Mariana Deer very close to a Molucca spe
415e067 is to see, what parts of structure abortive.-- REMEMBER my fathers remark about the Bladder.-- The nu
426e108 s is the possession of voice by Man. we should REMEMBER, that even birds can imitate the sounds surpr
506q015 parent plants grow near each other.-- ? Cannot REMEMBER at all. (37) Any cases of plants. which will
520m002 hands   & &c.-- Mr Dryden Co said he could not REMEMBER his father.-- My father thinks. people of wea
522m010 had never heard of him).-- My F. then said you REMEMBER Jack Baldwin at. school.-- Answered To be sure
525m024 s do not look like fear, but shame.-- I cannot REMEMBER instances, but I feel sure I have seen a dog
526m028 Children like hearing a story told though they REMEMBER it so well that they can correct every detail
526m029 ).-- when one remembers a thing in a book, one REMEMBER the part of page.-- one is tempted to think a
529m041 st might so view plants & trees.-- I am sure I REMEMBER my pleasure in Kensington Gardens has often b
531m050 ber of times repeated, because some people can REMEMBER poetry when one read over.-- The extreme ple
536m070 o muscles, any more than «prevent» heart beat» REMEMBER how Pincher does just the same; I noticed thi
538m080 th a second & unreasonable man.-- If one could REMEMBER all ones farthers actions, as one does  those
539m081 f» M. Comte Phil. which made me «endeavour to» REMEMBER, & to think deeply, & the immediate manner in
545m107 he peevish expression was most curious <like» «REMEMBER» the expostulatory angry look of black spider
560m156 e, forget what they did when in this state, or REMEMBER what they did in former one. about heredetary
565n010 smile   is to laugh.-- I must be very cautious. REMEMBER how Lavater ran away with new Lavaters,-- Ye
579n058 object.--   With respect to my theory of smile. REMEMBER children smile before they laugh.-- Has frown
581n063 you   do a thing properly, even when you cannot REMEMBER it. as my father trying to remember the man's
581n063 you cannot remember it. as my father trying to REMEMBER the man's Christian name, writing for the sur
586n078 top at right number of house though one cannot REMEMBER it.]CD back, without consciousness & by habit
587n088 might be born with idea of scarlet, as well as REMEMBER it.-- Why do children pout & not men-- orang-
594n115 & [...] Monkeys «Ogleby» seen Zool. Soc-- 1838 REMEMBER with distress their companions-- a «blue» Gib
600o008 which most beautiful cock, which best singer-- REMEMBER.-- avarice a compounded passion gained in lif
610o031 known & directed many letters to»-- could not <REMEMBER> «read» Christian name; fancied it looked lik
281c137 -- Dr Beck doubt if local varieties should be REMEMBERED, therefore do not consider it as proved that
293c172 is scarcely more wonderful, that it should be REMEMBERED in next generation. [NB what are those Marve
440e145 lone> «which have been» preserved.-- but be it REMEMBERED. how little part of the Grand Mystery is thi
470t176 h other actually produced hybrids-- My Father REMEMBERED when in the gardens, he knew there was  none
521m008 cannot  be called memory, because she did not REMEMBERED, it was an habitual action of thought-secret
531m050 nce of the event by itself will make it to be REMEMBERED.  whereas it is the importance.-- people very
538m079 goodhumoured as self.-- «When Miss Cogan has REMEMBERED her song, then the song was to her like one
546m111 ly like a flash without any assignable cause, REMEMBERED she had put it in branch of tree, & apologis
555m142 ed animals?) I doubted it in Fuegians, till I REMEMBERED Bynoes story of the women.-- The Chillingham
569n021 equence of vividness & multiplicity of things REMEMBERED & the associated pleasure &c accompanying su
602o011 results,   & the impressions on them are <all> REMEMBERED, when the meaning or reasons are  forgotten.
610o031 ot think what had put Wilson into his head.-- REMEMBERED, that he had. looked in direction book under
616o039 aid that the living brain perceived, thought, REMEMBERED &c. Well the heart is said to feel Now  this
```

Page **(Key Word)**
617o040 o whom the action is attributed; force (be it REMEMBERED) being a phenomenon apprehended by the same
636j57r d of barbs)-- In all these cases it should be REMEMBERED, that animals could not exist without these
640j167 er Galapagos beings are species. it should be REMEMBERED that Naturalists are prone, fortunately, to
044r072 agmens.» Read geology of N. America. India.--REMEMBERING S. Africa. Australia.. Oceanic Isles. Geolog
173b010 ay be derived from form (2). &c.-- <(> Then (REMEMBERING Lyells arguments of transportal) <continents
385d164 will never sit on them.-- May be just worth REMEMBERING that ovarium of women (Paper in Vol I of Iri
538m078 n, & fully corroborates the fact of her not <REMEMBERING which> «repeating song» when she had recolle
575n046 ce.-- Old People-- (Antiquary Vol II. p. 77) REMEMBERING things of youth, when new ideas will not ent
526m029 ons are a collection of pictures).-- when one REMEMBERS a thing in a book, one remember the part of p
539m083 al is not awakened.-- The habitual individual REMEMBERS things done in the other habitual state becau
490q001 wich group {Sowerby monstrous Cardium-- does it REMIND him of other species Hooker says the species o
367d113 am coloured ass.-- stripe on back also.-- legs REMINDED me strongly of Zebra.-- Mem. Quagga & Ld More
123a117 from river; 20 varas from surface in tosca.-- REMNANT of Megetherium in interior..-- <The theory of
297c186 forms.-- therefore an isolated form probably a REMNANT.-- Pachydermata. & Horses few forms. & they ar
297c186 - Pachydermata. & Horses few forms. & they are REMNANTS.-- Cephalopoda ditto.-- Mag of Zoolog. & Bot.
094a036 ulders, pebbles of granite clearly effect of REMODELLING same manner. as bits of Patagonian boulders
523m018 union with turkey cock,-- was restrained by REMONSTRANCES on him» which are never generally, if at al
557m148 n animals than latter.-- Yet I think one can REMONSTRATE with a dog, & make him ashamed of himself, i
619o043 rfere, but with respect to himself it would be REMORSE as will be presently shown.-- This then is mor
620o044 for a pleasure now though so trifling he feels REMORSE.-- He reasons on it & determines to act more w
629o55v nce will imperiously say so, & produce shame & REMORSE-- [Thus pungency of one's feelings for indecen
629o55v ons makes one break these artifical rules, get REMORSE-- ((hence desires do not intervene between thi
183b051 ine genera, the nearest species often come very REMOTE quarters. (NB. if Plata Partridge «or Orpheus»
225b219 some facts.-- about dogs &c &c NB. Animals very REMOTE. ass & Horse. produce offspring exactly interm
258c060 logy is a divellent power & tends to make forms REMOTE antagonist powers.-- Every animal in cold coun
266c091 are as opposite as day & night: yet we know how REMOTE the periods at which both left the lands of th
283c145 d one, be exaggerated, & all rest destroyed far REMOTE genera. will be produced. As we know from Ehre
435e133 shown that stigma swells, when pollen even most REMOTE is put to it.-- April 6th "Dr. Edwards on the
494q005 other breed.-- (12) About the blended instincts REMOTE Experiments-- Plants Raise seedlings surrounde
421e091 e generative organs being those which are most REMOTELY related to the habits & food of an animal, I
222b205 of the most perfect kinds the shark. lived in REMOTEST epochs.-- ¿ lizards of secondary period in sa
051r094 in the breakers or in waterfall: Excepting by REMOVAL of large fragments by mere force of waves: & a
105a071 parent rocks. would be prove of subsidence.-- REMOVAL downwards by successive torrent spread out. by
109a082 xcised, not located] -- do-- [...] Subaqueous. REMOVAL, shown by the number of bones lying at the bot
109a083 irst-rate importance in showing not subaqueous REMOVAL--??? the difficulty of such preservation certa
044r071 not believe in a great explosion, nor would sea REMOVE more internally than externally--I did not see
155g064 ce versa Same inclination when serpentine might REMOVE, what above straight line «only» cut deep gorg
578n056 sphincters are loosed is a convulsive action to REMOVE disagreeable impression like true convulsion.
032r039 vertical <height> thickness (DZ) of mass to be REMOVED & from the resistance offered to the greater l
032r040 rest. so would the size of the triangular mass REMOVED vary.--The gradual rising continuing. a anothe
033r042 rather differently constituted lime have been REMOVED?--As shell out of its cast which, although not
048r085 at Mastodon inhabiting plains of Patagonia is REMOVED by reflecting on the nature of the country in
154g059 l if line suppose continued across to {P} side REMOVED all well & good, but how came river to do this
179b034 ge) will keep to their type: in animals so far REMOVED, with instinct in lieu of reason, there would
212b165 would constantly do this, so was obliged to be REMOVED.-- In L.' Institut. 1837: p. 404. account of i
221b201 phic; & therefore according to Lyells doctrine REMOVED?? Is the prevalence of Coniferous Woods before
377d140 g mountain-chains, when we estimate the matter REMOVED by the waves of the sea, on beaches-- we reall
497q007 d Entomology of.-- most important, as furthest REMOVED possible point.-- ¿genera in intermediate coun
502q012 en. Linnaeus found 2 flower. which had anthers REMOVED, did not become impregnated. (34) Any recent i
534m062 d themselves from wall to table.-- table being REMOVED a little further, they ascended about a foot &
058r118 nt autour d'elle comme une ceinture d'immenses REMPARTS; toutes affectent une pente plus ou moins inc
269c102 mical combination, & the universality of latter RENDER-- spontaneous generation not improbable.-- Aft
304c206 or who plodding at making a series, which would RENDER our knowledge a chaos: who will doubt this if
427e109 artificial.-- Would not subsidence of Greenland RENDER climate less extreme. (& so account for descen
431e123 nge.-- Does not spermatic animalcule in Mosses, RENDER my view of the crossing of mosses & all others
461t037 tructure to purpose after purpose would tend to RENDER complex the series.-- Ch 6 Upland geese would
171b002 othes apparently only born to breed.-- annuals RENDERED perennial. &c &c.-- Yet Eunuchs nor «cut» Sta
171b004 d.-- child of savage not civilized man.--birds RENDERED wild <through> generations, acquire ideas dit
307c215 lowed to the fully organized wings of the male RENDERED abortive in the womb.-- if these apparently u
380d153 - this is clearly the converse of annual being RENDERED biennial-- the hardness of life in female Mot
389d176 refore both capable of propagating, but one is RENDERED abortive as far as parturition is concerned.-
447e165 ra is the same species with F. ovina, <& this> RENDERED vivaparous by growing on heights.-- yet he ha
454e184 ct their seed?-- What annuals can be budded «& RENDERED of great age» as must be inferred from what M
460t019 on & reaction grew more complex)» some perhaps RENDERED more complex & some simplified.-- Annals of N
540m088 poetry, or a strain of music, when the mind is RENDERED ductile by grief, or by bodily weakness, melt
507q15v plants, & any with peculiarities of structure RENDERING cross impregnation difficult or reverse (.49)
181b039 back. Diagram The largeness of present genera RENDERS it probable that <the> «many» contemporary, wo
393dIBC d in & in) that one copulation with other dogs RENDERS subsequent progeny faulty. Does male fail in p
538m081 will arise from physical action on the brain, RENDERS much less wondefful the instincts of animals--
420e088 ossil dog-- leading towards Hyaena.-- see Comte RENDU.-- I suspect good case of fossil filling up bla
376d136 sness, on the energy of Man» D.'Orbigny. Comtes RENDUS p. 569. 1838 says the cross between the Guaran
322c270 Lyell Book III There are many marginal notes <RENGGER &c> Mitchell's Australia Walter Scotts life I
325c267 lands. consult Dr Horsfield Silliman's Journal RENGGER on Mammalia of Paraguay. account of wild cattl
465t081 s «& pigs» in Polynesia; & dogs in S. America «RENGGER.» -- now it is this very immigration which ten
482z011 anicus.-- p. 210. Scolopax very close to ours RENGGER'S work of Mammali: of Paraguay must be most imp
056r112 akes part of necessary process of terrestrial RENOVATION & so is Volcano a useful chemical instrument
046r079 perhaps may be admittance of water, through the RENT strata: «Mr Lyell considers that Plutonic rocks
623o051 y that which is beneficial to race, will have REOCCURRED'. NB. Until, it can be shewn, what things ea
485z018 so soon putrifies?? p. 319. on Hydra-- polypi-- <REP> do p. 324. Polypi shorter duration than cells.-
372d130 have the knowledge how to grow, & therefore to REPAIR wounds-- but this has nothing to do with gener
490q01v Selection. Experiments in crossing &c Plants 1 REPEAT the French experiment of Carrot 2 {also try Pr
491q01v fail-- Really good subject for experiment.--«to REPEAT Spallanzani» Raise only single Plants & only a
522m011 subject.-- if at the other nothing.-- He could REPEAT the alphabet straight, but did not know [Z]CD
541m091 t was before one effort less» one is obliged to REPEAT the word, & think of qualities as flowers, clo
129a131 tagonia? «Facts about subsided forests.-- Many REPEATED oscillations» Hitchcock Report on Massacuhsse
254c049 ther, & same organ «where eliminated is» often REPEATED, as mouths in Polypi, Surely not correct view
276c125 may be tendency to divide, which often enough REPEATED would cause an unequal number of vertebrae--
334d011 - The case of all blue eyed cats (Fox has seen REPEATED cases) being deaf curious case of corelation
387d168 ame difference from parental stock having been REPEATED several times, that it becomes fixed in blood
531m050 laced) Memory cannot solely be number of times REPEATED, because some people can remember poetry when
538m079 which though learnt in infancy, had often been REPEATED: Now it is remarked that A. Bessy repeated th
538m079 een repeated: Now it is remarked that A. Bessy REPEATED things, which none about her had EVER before
553m135 d] notion, are not effects of impressions long REPEATED, without the powers of the mind being EQUAL t
610o032 ases of ideots knowing things, which are often REPEATED in a wonderful manner.-- as the hour of the d
623o051 is probably That becomes instinctive, which is REPEATED under many generations. (& under unknown cond
023r010 never been observed in Australia Dampier also REPEATEDLY talks about the immense quantities of Cuttle
376d137 little that ever he saw knew women.-- he has REPEATEDLY seen them try to pull up petticoats., & if w
376d138 ell.-- but monkeys examine sexes of every Has REPEATEDLY seen one he kept pull up feathers of tail of
569n022 e autour du Monde (1826-9) Paris. 1835 Quoted REPEATEDLY by Waitz (In Theil. V) in describing Carolin
521m008 tion.-- Old Elspeth's «in Antiquary» power of REPEATING poetry in her dotage is fact of same sort. Au
538m078 ates the fact of her not <remembering which> «REPEATING song» when she had recollected it in perfect
576n047 to save a friend, or wife.-- yet he would ever REPENT, & wished he had lost his life in doing so.--
533m060 our & ready confession of error made him less REPENTANT.-- In making too much profession, or rather i
361d096 ere are» two ducks, which have pretty close REPESENTATIVE species in England & N. America.-- the teal
068r146 ich would be too thick to be penetrated by the REPETED trifeling injections.--Old vents would keep op
170b001 veral stages (¿typical, <of the> or shortened REPETITION of what the original molecule has done).-- T
524m021 re most durable.-- (but Miss Cogan shows that REPETITION is not necessary)-- the words second childho
528m038 owe their beauty to absolute forms or to the REPETITION of similar forms as in angular leaves,-- (th

Page ***(Key Word)***
530m044 short time previous,-- because, pain prevents REPETITION of idea.-- Mr Blakeway has mentioned in Anti
534m062 or of memory of such sensations, & memory is REPETITION of whatever takes place in brain. when sensa
068r147 then all can, for fresh layers of igneous rock REPLACE strata. & it is nothing odd to find them injec
188b067 Smith considers probable two Northern species REPLACE <No> Southern kinds-- (I) Gnu reeaches Orange
195b104 Absolute knowledge that species die &. others REPLACE them-- two hypotheses fresh creations is mere
222b206 a moment doubt, but what cetaceae & Phocae now REPLACE Saurians of Secondary epoch: it is impossible
484z015 ing my general account-- ¿ Do not the Penguins REPLACE the <Auk> Guillemost of the northern Hemispher
547m111 cular, mental, <&> digestive nervous influence REPLACE each other August 29th. Went to Bed. & built «
614o037 ry true [RHC] Acquired instincts analogous «(& REPLACE)» to experience gained by man in lifetime Here
271c105 t Turdus Magellanicus. in the. S. Hemisphere. (REPLACED to the North by other species.--) should buil
283c144 - Hawk Gould seemed to think, that widow bird. REPLACED Birds of Paradise-- if such fantastic «sexual
283c144 ntimate a relation to two continents as to be <REPLACED> called into existence in two continents our
317c251 ?three species, near New York. (600 miles N.?) REPLACED by three other species.-- Says all the hares
337d022 hat the Fringilla domestica of North Europe is REPLACED by the F. cisalpina in Italy, which is so lik
362d099 e men.-- Effect of castration horns drop off., REPLACED by hairy ones. which never «dry up &» peel of
481z009 e in S. America or Australia-- very curious.-- REPLACED by didelphidae Skunk inhabitant of Patagonia.
485z017 coast of S. America. than on East.-- not being REPLACED by Brazilian Species.-- Mem Turdus Magellanic
535m069 will of Gods. «or God» secondly that these are REPLACED by metaphysical abstractions, such as plastic
578m056 e pass into convulsions?)-- squeeze out tears. REPLACED & squeezed out again-- as power of mind by ha
246c027 the Eastern Molluccas, New Guinea.-- (Case of REPLACEMENT)-- Coquille Voyage The casward, inhabits Cer
250c040 ead June 5th) important paper by Dillwyn. on REPLACEMENT of Cephalopods & Trachilidous Molluscs. by e
257c057 s tout à fait les mêmes.º This good case. of REPLACEMENT under peculiar conditions-- of «nearly» same
284c150 on" Dr. A. Smith. knows lots of instances of REPLACEMENT of one species by another. supply place in e
263c078 . instead of definite instincts.-- this is a REPLACEMENTS in mental machinery-- so analogous to what
244c022 Vespertilio bonar«i»ensis (from Buenos Ayres) REPLACES <Vesp.> holds same relation with equator--tha
246c027 nhabits Ceram, Bourou & especially New Guinea (REPLACES, Emeu) in North of New Holland.-- New Guinea
484z015 this class--. Pyrocephalus & many Tyrannulae-- REPLACES warblers of Europe-- Study profoundly shells
249c035 un wild, The two species of Clenomys. case of REPLACING species. Dr. Smith will give me some capital
300c198 ery just, but these faculties being viewed as REPLACING each other it is hiatus & not saltus.-- The g
311c225 ope & N. America, on closely allied species. «REPLACING each other». good to consult p. 326 wild ass
570n024 damned first." the man shrugs his shoulders & REPLIES nothing. if he did go to reply. he would throw
602o11b things you can give no proof for, & one often REPLIES "what you say is perfectly true, but you do no
569n023 "there is nothing more to be said."-- "made no REPLY, but shrugged his shoulders & went away."-- he
570n024 is shoulders & replies nothing. if he did go to REPLY. he would throw back his shoulders. he wishes t
129a131 orests.-- Many repeated oscillations» Hitchcock REPORT on Massachssets. p. 133 The most wonderful ca
208b154 this bear with law referred to by Richardson in REPORT about each genus having its parent type in hot
214b174 on to those now living in the sea.-- See Rogers REPORT to Brit Assoc <to> on N. American Zoology-- A
234b256 s mentioned. but few.-- CD [There was notice in REPORT of British Association of 1838. (Newcastle) ab
485z017 lanicus.-- C, <Chingolo> Chimango-- Diuca?? See REPORT <by> on D'Orbigny on species of Mephites 4 dis
326c266 an's Repository. 4to. contains much on dogs.-- REPORTS of Brit. Assoc.-- some important Papers. Dr. M
154g060 came river to do this vast quantity when during REPOSE of lake it did but little more {P} now that it
326c266 smus thincks I should lik it. The Sportsman's REPOSITORY. 4to. contains much on dogs.-- Reports of Br
036r049 argement see its increased length. which will REPRESENT the dilatation, which dilated cracks must be
173b012 - or intermediate land existed.-- or they may REPRESENT some large country long separated.-- On this
176b021 ion as much as acid & alkali organized beings REPRESENT a tree. irregularly branched some branches fa
185b056 ects, each family, however many there may be, REPRESENT every other; for instance in Heteromera, you
194b094 the Colobes which in south Africa, appear to REPRESENT the semmoptiheque of India.-- Tooth of <Spi>
244c021 ntrast with southern regions.-- «it would now REPRESENT. what <actually is> «has» taken place with qu
243c019 s, d'admettre que ces oiseaux eussent leurs REPRESENTANTS dans de si hautes latitudes"., --¿translate
063r130 species may perish. = This <inoculation> «REPRESENTATION» of species important, each its own limit
311c227 imming about. A Smith is firmly believed in REPRESENTATION. certain birds in many families, «»very of
283c143 it is only in large groups. where you have REPRESENTATIONS.-- The aerial type in each family is relat
172b008 ary, <in> changing climate we ought to find REPRESENTATIVE species; this we do in South America close
195b103 tribution of the species!! It may be argued REPRESENTATIVE species chiefly found where barriers «& wh
220b199 «N.» America & Europe, which have not their REPRESENTATIVE species in each other, are migratory speci
226b221 those genera.-- & hence by same chance few REPRESENTATIVE species. this must happen. & thus acquired
226b221 s must happen. & thus acquired will explain REPRESENTATIVE system Of this we see example in English &
410e051 g of <species> individuals. with respect to REPRESENTATIVE species., when going N.orth & South Thinki
484z014 Serpent Eater-- or Secretary is S. African REPRESENTATIVE of Caracaras of Americas.-- manner of walk
174b015 bly two distinct type, but each having its REPRESENTATIVES-- «as in Australia» This presuppose time
185b056 ther; for instance in Heteromera, you have REPRESENTATIVES (which at first would be mistaken for) Car
186b057 three elements) How far Does Waterhouse's REPRESENTATIVES agrees with breeding.. in irregular trees.
243c019 , --¿translate?) All Australian forms have REPRESENTATIVES (& instances given) in East Ind. Arch:-- B
273c112 remarkable connection between these aerial REPRESENTATIVES of the different families.-- that sexes <a
484z015 Petrel?-- Study Birds of Europe for other REPRESENTATIVES of this class--. Pyrocephalus & many Tyran
032r040 ity & rest.--Such changes could be shown (as REPRESENTED), along line of coast.--[Fig. 2] Mem San. Lo
063r130 » of species important, each its own limit & REPRESENTED.--Chiloe creeper: Furnarius. <Caracara> Cala
341d031 many birds.-- which however are most closely REPRESENTED.-- Thus the red breasted thrush is represent
341d031 presented.-- Thus the red breasted thrush is REPRESENTED by one not differing except by black line,--
606o022 Lessings shows expression of pain cannot be REPRESENTED, But what is beauty?-- it is an ideal standa
211b163 ous paper by M. Serres on Molluscous animals REPRESENTING foetuses of Vertebrata, &c 1837 p. 370 Owen
132a138 no volcanicity beneath lakes?» Suppose ocean REPRESENTS proper <state> temperature of earth. at the
387d170 ess through its caterpillar state. the monkey REPRESENTS this state.-- When it is said. that differen
274c115 -- Grups of very different value have their REPRESETATIVES;, the rasorial may be observed even in Les
622o048 ften feelings which do not lead to action are REPRESSED thus avarice. &c &c.-- [RHC] 8) in the beginn
350d055 but that perfection consists in being able to REPRODUCE Here there is some error-- Observed, nature d
356d070 some hybrids breedings-- there is tendency to REPRODUCE in each case, but something prevents the comp
390d178 only one bud.-- Every individual foetus would REPRODUCE .its kind was it not for the necessity of some
453e183 bert in letter says distinctly, that Hollyoak REPRODUCE each other. & yet I presume seed raised in sa
485z018 p. 324. Polypi shorter duration than cells.-- REPRODUCED.-- Milne Edwards p. 138 on Polypi.-- Berenic
503q012 t, how many generations any hybrid has <been> REPRODUCED itself.-- Ask Gray to ask Mr Riley to experi
446e162 -- Henslow One of the 4 species of Lemna only REPRODUCES itself «in England, as yet observed» by buds
192b083 hence-- sub-varieties & hence possibility of REPRODUCING any variety, although many of the seeds will
323c269 of Animals -- Spallanzani's Essays on Animal REPRODUCTION -- Treatise on Domestic pidgeons 30th Lives
440e145 k.-- In short all which «Nutrition, growth & REPRODUCTION» is common to all living beings. vide Lamar
442e152 rted»-- throw over this theory, & the sexual REPRODUCTION of species may stop for any number of gener
447e164 ws dispersion of germs is not end of seminal REPRODUCTION-- likewise grasses. &-- very heavy seeds.--
635j56r . D]CD [All this does not explain death, but REPRODUCTION]CD though such a scheme. would require cons
263c075 profoundly over the enormous difficulty of REPRODUCTIONS of species & certainty of destruction; then
220b197 ation on perversion of structures especially REPRODUCTIVE organs) & therefore the one distinction of
372d129 crab, tail to lizard,-- healing of wound.-- REPRODUCTIVE faculty + in the separated part every eleme
386d166 ans.-- the origin of this law is part of the REPRODUCTIVE system.-- of that knowledge of the part, of
496q05a flowers-- contrasted by Kirby-- with animal REPRODUCTIVE system.-- cover flower-- put artificial
059r119 r quelque scissures profondes, on les voit se REPRODUIRE a des hauteurs communes sur le revers de cha
508q016 mith. (6) What size book Gallesio storia del REPRODUZIONE.-- D. Holland (7) Is Haemorragic tendency,
227b226 between» say in brain. between lowest Mammal & REPTILE. (or between extremities of any great division
486z019 Voyage Vol III p 322 Dr Martens says only one REPTILE in Kamtchatka (Salamandra aquatica). Compare w
183b053 period of great quadrupeds declining as great REPTILES must have once declined.-- Read his theory of
192b086 pecies greatest jumps strongest marked genera? REPTILES?) For instance there never may have been grad
192b088 e past.. & still more so «known» with fishes & REPTILES.-- In mere eocine rocks. we can only expect s
198b113 as possible. for instance take birds animals, REPTILES fish-- Conditions will not explain status (Pe
213b170 zed fishes lived far back, fish approaching to REPTILES at Silurian age-- How long back have insects
221b202 oods before Dicotyledenous a fact analogous to REPTILES before Mammalia Think about Miocene fossils s
222b205 ay not be perfecting by change of Mammalia for REPTILES. which can only be adaptation to changing wor
233b251 therium.-- <M. Jerrod> & Dumeril great work on REPTILES. M. J. says some reptiles same from Maurice &
233b251 umeril great work on Reptiles. M. J. says some REPTILES same from Maurice & Madagascar & C. of Good H
246c025 ld account for this.-- Coquille Voyage Says no REPTILES. p 460 & very doubtful whether any birds Exce
250c039 drupeds.-- common to two types of animals What REPTILES coexisted with Palaeotherium in Paris quarrie

256c054 H. Very striking to see M. Bibron looking over REPTILES he often had difficulty in distinguishing whi
306c213 high order. not for vertebrata, but mammalia & REPTILES &c Timor is connected with Australia «map to
342d035 y? Owen says relation of Osteology of birds to REPTILES shown in osteology of young Ostrich. 16th. D
344d040 e admirable tables on Geograph distribution of REPTILES in Suites de Buffon.-- Vigors has given list
373d133 -- Decapod Crust in Muschelkalk, & 5 genera of REPTILES.-- <M> p. 417. Magnesian Limestones & Zechste
374d133 an Limestones & Zechstein oldest rock in which REPTILES have been found. p. 426 Sauroid fish in Coal,
374d133 , true fish, & not intermediate between fish & REPTILES-- yet osteology closely resembles reptiles.--
374d133 h & reptiles-- yet osteology closely resembles REPTILES-- p. 432 some plants in coal supposed to be
379d151 past for other Mammalia. & still further back REPTILES & Cephalopoda: Old Jones remarked to me, that
417e079 hold with fishes, «& there are mule fishes» & REPTILES & those which <lay> «have» their eggs, <inter
423e096 a retrograde movement in Cephalopoda & fish & REPTILES.?-- supposing such to be the case, it proves
448e168 ting classes like Toxodon «In orders»-- Fish & REPTILES in former case-- Reptiles & Birds & Mamm. in
448e168 «In orders»-- Fish & reptiles in former case-- REPTILES & Birds & Mamm. in ornityhyrhycus-- is not th
458t001 existing, or nearly existing forms of aquatic REPTILES most strange, & shows as in shells some forms
483z013 Arion" of Ascension. p. do.-- some S. American REPTILES are described Shells from Tahiti and Chile Th
634j55r tremities.-- How are we to explain this.-- Did REPTILES first inhabit seas.-- Were they then killed o
176b024 d by separation or change in part of country. REPUGNANCE to intermarriage <increases it--> settles it
179b033 y or Metropolis of every species: believes in REPUGNANCE in crossing of species in wild state.-- No d
179b034 ct in lieu of reason, there would probably be REPUGNANCE & art required to make marriage.-- as Dr Smi
189b071 ours Strong odour of negroes, a point of real REPUGNANCE.-- Waterhouse says there is no TRUE connecti
199b120 onger tendency to imitate one of the parents,-- REPUGNANCE «generally» to marriage <if offspring not fe
366d108 ase being fertile inter se., or at least show REPUGNANCE to breeding if instincts unchanged, & if the
390d177 dding «only» as cryptogamia & hydras,-- (this REPUGNANCE to breeding in & in seems connected with mor
280c136 nd It is not difficult to see that it is less REPUGNANT to nature to produce one offspring unlike its
218b189 tany Vol I p. 450 There is in nature a «real» REPULSION «amounting to impossibility, holds good in pl
568m020 - superstition & charity & prayer, or eloquent REQUEST. Reason in simplets form probably is single co
227b227 future. !.fish never become a man.-- Does not REQUIRE fresh creation!-- If continent had sprung up r
235b263 ar upon this.-- Why do Van Diemens land people REQUIRE so many, imported animals?-- At what. part of
381d156 cts. spider crabs.-- (all these however do not REQUIRE coition every generation)-- Epizoa & the nemat
442e153 ording to the above suggestion my theory would REQUIRE, that <species> «individuals» propagated by ge
447e164 one species to other: therefore my theory does REQUIRE crossing.-- The case of Lemna shows dispersion
448e165 eldom produces seeds, because it is thought to REQUIRE insects to impregnate it.-- it is allied to As
461t025 hosis & are provided with fins, & hence do not REQUIRE sac.-- but the male in these hatch young-- are
486z018 America than in Europe-- Coleoptera especially REQUIRE a greater duration of Heat. hence musquitoes &
495q005 in VERY rich soil, will plants abort?, does it REQUIRE successive generations to accustom them to suc
503q012 . Varieties effects of domestication-- said to REQUIRE Selection (36) Ask Mr Gowen to ask Mr Herbert,
505q014 earth on top of House =Aristolochia, plant wh REQUIRE insects to impregnate it (7) History of Potato
545m104 us. so may some ideas.-- ie habits, which must REQUIRE idea to order muscles to do <certain> the acti
576n048 rable circumstances!» We must believe, that it REQUIRE a far higher & far more complicated organizati
612o035 e plants. (But I include irritability for that REQUIRE will in part. ¿Why more so than movement of sa
617o39v stand this subject, but the answer to it would REQUIRE a considerable degree of attention. How do the
633j54r ould the earth have drifted; why should plants REQUIRE earth, why not created to live on alpine pinna
635j56r ut reproduction]CD though such a scheme. would REQUIRE constant miracles.-- p. 420 thinks the great f
087a008 ent down.-- Probably more Arctic land would be REQUIRED to produce climate resembling S. America in E
105a073 simply as the inclination is little the force REQUIRED to move <it> «stream» aside is not great.-- I
119a104 sidered as condensed vapour.-- inequlities are REQUIRED to start with (& does not Hersche theory impl
119a105 deep into the bowels of the earth, as would be REQUIRED by thermometrical scale.-- (for the temp must
149g030 osed of sand & perfectly rounded stones-- lake REQUIRED to deposit this Remember however the great Ch
179b034 ason, there would probably be repugnance & art REQUIRED to make marriage.-- as Dr Smith remarked Man
215b176 ach other, the females loose desire, and it is REQUIRED to give the canthairides and milk--Fox tells
252c045 l give one almost certain guide ∴because time REQUIRED to separate isld very long» America & India
337d022 tes.-- who will say there is distinct Creation REQUIRED if he believes «hyaena & squirrel» seal & mou
371d127 ition, as to be more easily trained up to the (REQUIRED) offices" &c &c Owen illustrates case of Ding
463t057 hey would not be intermediate, but this is not REQUIRED.-- Waterhouse says perhaps animals of Fernan
505q014 = (6) Figs, flower.--Passion Flower. (as it is REQUIRED to impregnate it artificially.)-- Asclepias--
527m034 air, with exercise &c no organs of sense being REQUIRED) as the closest train of geological thought.-
612o035 o more extended space, such powers of relation REQUIRED to be extended. Hence a sensorium, which rece
089a019 thus composed on the NE part more like marble REQUIRES polish to see structure,-- «He» Thought of er
126a123 th level of sea.-- deep seated springs «spring REQUIRES connected column.--» of cold water show, that
175b020 o change, (& to multiplications when isolated, REQUIRES deaths of species to keep numbers of forms eq
180b036 d. to have many species in same genus (as is). REQUIRES extinction. Thus between A. & B. immens gap o
180b038 eing favourable many might be produced.-- This REQUIRES principle that the permanent varieties produc
240c004 eding in pidgeons with very much care that, it REQUIRES the greatest difficulty to rear them, eggs ha
398e006 t baffles idea of revolution.-- My very theory REQUIRES each form to have lasted for its time: but we
414e060 ield????). Have Mammalia?? My theory certainly REQUIRES progression, otherwise [not located] Are the
443e154 t difference is forced on us.-- My theory only REQUIRES that organic beings propagated by gemmation d
574n039 show language had a beginning, which my theory REQUIRES. There probably is some connection between ve
614o037 ntal affection-- The very existence of mankind REQUIRES these instincts,: though very weak so as to b
461t037 down without being absolutely injurious «(or REQUIRING nutrition)» to a certain amount it will be so
506q015 re Orchidiae fecundated, as mass of pollen is REQUISITE.-- Brown's paper 43. Any flowers of Keeling D
321c275 iary.--do--- & Lisle's Husbandry Tuckeys voyage REREAD Appendix Ovington Voyage to Surinam. Voyage Co
322c270 Memoirs on Geology of France.= on Etna. Almost REREAD the previous volume. & C. Prevost on L'Ile Jul
369d115 abits of animals, with anatomical & Zoological RESEARCH, in order to establish entirely their place i
106a075 nt is partly taken from Delabechs Theoretical RESEARCHES.-- Athenaeum. 1838-- p. 137. Three inosculat
221b201 Confervae-- p. 23 p. 267. Dela Beche. Geolog. RESEARCHES. facts of salt-water shells living in absolu
326c266 es, new Edit 1838 Harlaam. Physical & Medical RESEARCHES. on Horse in N. America.-- Owen has it.-- L
481z008 er. Mem Bahia Blanca. De la Beche theoretical RESEARCHES Compare land shells of Galapagos different i
499q010 s eaten by a Buceros in East Indies-- Asiatic RESEARCHES) (23) Talk about Thyme. Horned Oranges. Spal
470t177 s at Wild St Johns Wort--Scabies, Cyanoglossum--RESEDA wild very many Bees & Humbles--on Thistles man
284c149 dence.-- on the islets Mr Blyth remark that a RESEMBLANCE between some form in birds is visible, when
026r021 so probably will the Cleavage be There is a RESEMBLANCE at Hobart town between the older strata & th
345d043 y were most like their mothers believes this RESEMBLANCE general. ¿depends upon mother bein[g] oldest
348d051 ement.-- affinities, what is that, amount of RESEMBLANCE,-- how can we estimate this amount, when <va
353d061 ammae» in male ourang-outang. other point of RESEMBLANCE with man.-- September 3d Magazine of Natural
354d065 ttle with Bump. together with Bison, at some RESEMBLANCE as if the "variation in one, was analogous t
360d090 Jackall.-- In case where Jackall was father RESEMBLANCE much nearer to Jackall.-- This Keeper has se
385d164 cially of the youngest children, a striking <RESEMBLANCE> similarity to the temper of the father, or
417e079 s this more refer to length of time that the RESEMBLANCE is permanent, or the similarity at first bir
443e157 ater's Treatise on the Hand.-- p. 94.-- "The RESEMBLANCE of the foot the Ostrich to that of the Camel
447e164 s highest,-- Palms &c &c)-- Is there greater RESEMBLANCE between carboniferous. «& recent» mollusca,
448e166 Senegal as sea.-- is remarkable.-- Again the RESEMBLANCE between the Superga & Paris, numerically the
520m001 , she was often struck with this fact.-- the RESEMBLANCE was in odd twiching of muscles, & general ma
571n031 se-- I was also much struck in great avenue, RESEMBLANCE to gloomy aisle of Churche.-- these are Mayo
182b047 ion & partly to heredetary taint;. hence the RESEMBLANCES & differences for instance of finches of Eu
197b112 onditions of existence is thought to account RESEMBLANCES &:. quinary system, or three elements p 66]C
258c061 er, & resemble some other class analogy" The RESEMBLANCES relationship, the dissenblances analogy,--
275c121 hink that the male communicates the external RESEMBLANCES, than the female. The expression hybrid & f
417e079 atter only that one refers to in speaking of RESEMBLANCES of children to their parents.-- Lord Moreto
444e157 caped Naturalists." Before he alludes to the RESEMBLANCES-- The snout of the mole & Pig in having tw
543m101 powers of perceiving & classifying distinct RESEMBLANCES.-- The facts of half instincts. when two va
246c025 .-- p. 414. dogs of New Zealand of large size, RESEMBLE, chien-loup.--long, black & white, ears short
258c061 points which are different from each other, & RESEMBLE some other class analogy" The resemblances re
272c108 ppearance of blattae other Hemiptera stikingly RESEMBLE Coleoptera.-- Donacia.-- some orthopterous in
316c247 iberia it cannot be said American fossils more RESEMBLE those of America than of Europe, because the
358d076 loped state.-- the female & young of all birds RESEMBLE each other in plumage «(that is where the fem
385d164 ransl by Holcroft Vol I. p. 195. says children RESEMBLE parents in their bodies "It is a fact equally
407e041 & conchology,-- shells of Africa ought most to RESEMBLE fossil ones of Europe, Consider probable form
459tf02 d ears, tail limbs-- in the mules, these parts RESEMBLE ass. (& part of body mare)-- -- this may be,

Page **(Key Word)**
508q016 orst =or in dogs= (12) Do Hottentots generally RESEMBLE each other very closely, more closely than Ca
533m057 h are heredetary are so because brain of child RESEMBLE, parent stock.-- (& phrenologists state that
548m117 r or salt is not there.-- the reality does not RESEMBLE the picture in one mind, but does not stop to
026r020 d of elevation must in its configuration have RESEMBLED Chiloe In De La Beche, article "Erratic block
333d006 at they were killed, & other two very closely RESEMBLED in form of tail, fur &c to the half bred Pers
333d007 f foxes & dogs crossed, offspring always more RESEMBLED foxes than dogs (Mem Jackall in Zoolog Garden
333d007 f & «half Esquimaux» dog which <likewise more RESEMBLED the wolf than dog.--> appeared to be intermed
344d042 o either paret, & breed not fixed. though she RESEMBLED a harrier & her husband was pure Harrier.-- «
361d095 le, goldfinch, hawfinch-- in nursling plumage RESEMBLED that of Cross-Beak-- In lark if I understand
459t013 the penguin & the bright feathers on its wing RESEMBLED the drake.-- another of same half breed resem
459t013 mbled the drake.-- another of same half breed RESEMBLED the plumage of drake still more.-- So Penguin
520m001 form of body.-- thus the late Colonel Leigton RESEMBLED his father in body, but his mother in bodily
033r042 es coated by pellucid envelope of Lime.--form RESEMBLES the husks at Coquimbo: in that case, may not
186b057 n simplest case saying every species in genus RESEMBLES each other, (at least in one point, in truth
333d008 n in children of parents <some> one sometimes RESEMBLES one parent & one another & are not exactly in
365d105 na of Sweden, (which in summer dress somewhat RESEMBLES Red Grouse) it may be so-- but very improbabl
374d133 tween fish & reptiles-- yet osteology closely RESEMBLES reptiles.-- p. 432 some plants in coal suppos
435e134 "standing alone in the gift of intellect, he RESEMBLES, other mammalia in the effects produced on or
450e175 reaches 13 hands.-- Phillipines Pony somewhat RESEMBLES that of Celebes is somewhat larger than the S
459tf02 , & another of both progenitors-- the hinnus, RESEMBLES horse in its head ears, tail limbs-- in the m
466t094 lip of lower division of nectary: «wh. itself RESEMBLES a Bee, but does not prevent bees visiting it.
555m143 lowed by protrusion of lips, in which respect RESEMBLES some of the old ones.-- -- S. American group
556m146 when going to fight, in which case expression RESEMBLES a fox-- I can conceive the opposite muscles w
087a008 tic land would be required to produce climate RESEMBLING S. America in Europaean latitudes.-- Will it
182b048 , & others dark brown, with long clotted hair RESEMBLING that of goats." Progressive development give
267c093 the others dark brown with long clotted hair RESEMBLING that of goats" Geograp. Journ. Vol VII. p. 2
195b098 r of these islands so permanent a breath cannot RESIDE in space before island existed.-- Such an infl
451e178 the largest mammifers in the world consistently RESIDE & are bred. "take tame animals into this regio
334d012 lost his <on> two cows entirely, changed his RESIDENCE a great many miles.-- yet one day <th> a cow
102a061 d look like it. Are greenstone dike in Granite RESIDUAL matter of upper quartzose ones & felspar.?? A
097a042 70.-- Mem. Cordillera Can Greenstone dikes. be RESIDUE of quartzose vein in higher parts? & felspathi
077r172 iron, combined with nickel & cobalt (meteoric) RESIST oxidation?-- Mem Sir W. P. stone It is clear t
347d049 38). Eggs discovered to Taenia.-- hard so as to RESIST external influence.-- 27th. August. There must
438e142 o domestication.-- the constitution of some may RESIST the means Man can offer of changes.-- as deser
570n023 & why does one inspire, when surprise, can one RESIST blow better with body distended.-- intolerable
032r039 ickness (DZ) of mass to be removed & from the RESISTANCE offered to the greater lateral extension of
260c067 oo; as if there was storge, which could not be RESISTED, when hearing crys of hunger of little bird,
056r110 ig. 7] {P} effect of heat on inner wall, hence RESISTS degradation longer than outer parts.-- The com
627o053 up???» & the question is, whether this can be RESOLVED into some operation of intellectual faculties
226b222 with plants of St. Helena & Tristan D'Acunha, RESOLVES itself into question of proportion of species
294c176 close Species in islds. near continent, Must we RESORT to quite different origin when species rather
232b248 superiority of man over animals, without such RESORTS Mr Waterhouse has most curious facts about the
503q012 «Bruce says does» Royle In Royle's productive RESOURCES Book no information & Hope about Silk worms.
469t151 r The seeds of these plants will be collected & RESOWN.-- Humble 22 flowers of Egg Tree in one minute
022r008 r's voyage there is a mine of metereology with RESPECT to the discussion of winds & storms:--«in Voln
023r012 herefore weak part of the ocean's bottom. With RESPECT to Sharks distributing fossil remains: Sharks
037r053 thern ocean not being buoyed with Kelp.-- With RESPECT to degradation of rocks--It may be a question.
046r079 e its «proportional» particles altered.-- With RESPECT to Volcanic theory. I want to ground, that the
077r171 ns of Tasco & Moran--of Guanaxuato to SW. with RESPECT to latter doubts whether bed or vein (very lik
094a035 ough granite.-- a most important question with RESPECT to my theory of changes. of granites into Trac
107a078 -- The last speculation becomes important with RESPECT to thickness of crust broken up.----My view of
120a109 especially if very hot under high pressure.--» RESPECT to formation of salt.?.--??? Footsteps in New
124a119 rugout its thickness.-- this most curious with RESPECT to epigmous action.-- if the zinc were mixed w
178b030 m which they bifurcated.-- Type of Eocene with RESPECT to Miocene of Europe? Loudon. Journal. of Nat
179b034 s Dr Smith remarked Man & wild animals in this RESPECT are differently circumstanced.-- ¿Is this shor
180b037 as to keep number of species constant.-- With RESPECT to extinction we can easily see that variety o
197b112 eference to M Edwards. law of crustacea-- with RESPECT to mouth those beautiful passages from one to
225b216 al structure.-- miserable limited view.-- With RESPECT to how species are. Lamaks "willing" doctrine
228b229 f study, to guide our <past> speculations with RESPECT to past & future. The Grand Question, which ev
228b231 fear <of death>. pain. sorrow for the dead.-- RESPECT We have no more reason to expect the father of
234b255 est <sort> of our highland Sort, except in one RESPECT, that those of Iceland. are seldom seen with h
259c066 always necessary may be called adaptation With RESPECT to my theory of generation, fact of armless pa
265c084 k about six fingered children heredetary|| With RESPECT to question what is adaptation.-- Ermine, ptar
269c103 can see that perfection may be talked of with RESPECT to life generally.-- where <">unity constantly
282c142 haracters which are analogical in a genus with RESPECT to rest of its family as in ground cuckoos, is
282c142 family as in ground cuckoos, is affinity with RESPECT to species of each other, because we suppose a
285c153 Physical changes will be unequally rapid, with RESPECT to their effects The AEgyptian animals domesti
303c204 n length of extremities, how are races in This RESPECT upper & lower, which I do not know whether it
313c234 pecies two very different considerations, with RESPECT to law of mammals shorter duration, than mollu
317c250 Cape.-- some proper well-worth studying, with RESPECT to forms.-- Study Appendix to Tuckey's Expedit
318c252 ely the case with the mice of S. America, with RESPECT.. to the Cordillera,---- Bachman has seen webb
325c266 sh Ethical Philos: Prostitution of Paris. with RESPECT to licentiousness, destroying children. --it i
336d018 congenital (that is most slowly obtained with RESPECT to that individual) it is more easily inherite
343d038 f the characteristic forms of S. America. With RESPECT to future destinies of mankind, some of specie
343d039 have worn a quite different figure 19th. With RESPECT to the Deluge it may be worth adding in note t
352d059 tre but circle, two or three lines deep-- with RESPECT to Macleay's theory of analogies-- <be> when i
355d069 a.-- Pachydermata &c &c-- It is important with RESPECT to extinction of species, the capability of on
357d072 & Australia appear to have suffered most with RESPECT to extinction of larger forms.-- From observin
358d074 f the Pacific.--» to nearest continent.-- With RESPECT to ancient geography of Atlantic Tristan D'Acu
359d088 hen Canadian offspring, I should say in every RESPECT most like Penguin duck.-- which is strange ano
387d170 There is an analogy between caterpillars with RESPECT to moths, & monkey & men.-- each man passess t
388d171 ce-- but this does not apply to potato.-- With RESPECT to offspring being determined by imagination o
390d178 make the bud of the woman, not a bud in every RESPECT.-- [Is this connected with the physical differ
410e051 ion of crossing of <species> individuals. with RESPECT to representative species., when going N.orth
410e052 pecies, from its garments, without some end.-- RESPECT good describers like Richardson.-- The relatio
419e085 any kind-- Fauna, must be very curious.-- With RESPECT to the non-development of Mollusca, which I ha
468t112 rlet Poppy-- So that, finally Fraxinella. with RESPECT to nectary is same case as Azalea or Rhododend
494q004 atistics of breeding in Zoolog. Gardens-- with RESPECT to conditions of animals & their general healt
521m010 at is A. B my gardener.-- Thus was he in every RESPECT, no communication could be held by means of he
536m072 ses would have idea of beautiful forms.-- With RESPECT to free will, seeing a puppy playing cannot do
538m080 ied by habit, when one acts unconsciously with RESPECT to more energetic self, & likewise one forgets
542m096 interval of barking, now this is smile.|| With RESPECT to sneering the very essence of an habitual mo
555m143 look, followed by protrusion of lips, in which RESPECT resembles some of the old ones.-- -- S. Americ
565n008 acks him Contempt, when there is some anger «& RESPECT to opponent» is showed by same movement as sne
572n032 . Vol I, p. 226-- "The general idea of showing RESPECT is by making yourself less, but the manner, wh
579n058 shows the mind is intent on one object.-- With RESPECT to my theory of smile. remember children smile
580n060 ideas of time, & perhaps of space-- in latter RESPECT he thinks he certainly has observed that some
586n078 virtue,-- the most probably supposition. with RESPECT to pidgeons, is that they do know from look of
619o043 as he would be tempted to interfere, but with RESPECT to himself it would be remorse as will be pres
636j57r , which feeding on same food, differs in every RESPECT, except «in» quick movements. (sliminess inste
640j167 rtility of either sex determines life:.» «With RESPECT to whether Galapagos beings are species. it sh
640j167 their breeding together has been tried.-- With RESPECT to the six puppies, if a hare was introduced,
640j167 ical change is in progress or is, present with RESPECT to new arrivers, the small body of species wou
032r042 general symetry of world.-- I feel no doubt. RESPECTING the brecciated white stone of Chiloe, after
096a041 cussion. Mr Brayley says he can give me facts RESPECTING lime <n> being heated without parting with C
212b166 rent from English. Waterhouse has information RESPECTING the Water Rat.-- ¿Consult Dr Smith History o
605o20v . but rather in the power of discriminating & RESPECTING good from bad. And it is manifestly from thi
634j54v very circumstances, that we become satisfied RESPECTING an original thought, or design, pursued to i

```
**************************************************(Key Word)**************************************************
616o39v bodily action is made known to us, revealing RESPECTIVELY what are called it's subjective & objective
358d075 ferent family. having very same habits in some RESPECTS as this Caracara.-- Sept. 9th. It is worthy o
515q022 & female parent.-- Will they grow up in other RESPECTS different?-- Important.-- Oct. 44 Tell J. An
538m079 llen W. & Babington, both half ideotic in some RESPECTS & with store of accurate & even profound know
539m083 heraldic principle. & Eras a Wedgwood in many RESPECTS & some of Aunt Sarahs. cranks, & so is Cather
539m083 Aunt Sarahs. cranks, & so is Catherine in some RESPECTS--. good instances.-- when education same.-- M
639j28r horhyncus «externally». petrel & Whale in some RESPECTS Chamaelion like power in Octopus & Chamaelion
404e025 Edwards, description of curious mechanism of RESPIRATION, or rather ventilation peculiar to <the clas
493q004 g infertile.-- (4) Does the number of pulse, RESPIRATION, period of gestation differ in different bre
570n024 n of sighing is probably correct, to relieve RESPIRATION when immensely immersed-- mechanic apt to si
584n073 ht crosses mind, before it can have affected RESPIRATION V E. p. 125 Wrong Entry Madagascar Lemur see
589n092 oan.-- more power over muscles of voice than RESPIRATION.-- like sigh before false sneeze.-- "A Disse
341d029 most remarkable character in Apteryx, small RESPIRATORY system; even much smaller than in other Stru
023r009 em, 8 inches long, & as big as a mans thumb, the REST not above half so long; The maw was full of jel
032r040 irst. & according to the greater or less time of REST. so would the size of the triangular mass remov
032r040 ng with successive periods of greater activity & REST.--Such changes could be shown (as represented),
041r066 minated each way, which little vein was like the REST of these thin veins which project outwards.-- I
047r083 fluid rock In Patagonia plains. long periods of REST & vice versâ more likely to be coincidental tha
174b015 uced from propagation from different set, as the REST of the world.-- This view supposes that in cour
199b119 here any instance of peculiar people banished by REST?-- ∴ most monstrous form has tendency to propag
233b249 us remark Australia Fauna so far. Indian all the REST. Timor according to Mountain chain ought to be
241c015 a.-- Many <genera>. kinds common to New Guinea & REST of isle in E. Indi: Arch: In New Zealand. a stu
250c039 pe has many species but not genera distinct from REST of world??? Lyells Principles, must be abstract
267c095 e of Asterias in stomach. of Sammon remain after REST of animal digested.-- Important do p. 98. on a
270c103 ty" this unity,-- this distinctness of laws from REST of]CD universe «which Carus considers big anima
274c116 s as the receptacles of the wanderers out of the REST of the world?-- Will this not agree with Waterh
276c124 actising imperfectly some habit, which the whole REST of other family practise with a peculiar struct
282c142 which are analogical in a genus with respect to REST of its family as in ground cuckoos, is affinity
283c145 a.-- let short billed one, be exaggerated, & all REST destroyed far remote genera. will be produced.
287c158 organizaton. But the classification must chiefly REST on these same organs,-- habits, range. &c &c--
294c176 - but how to show species-- I fear argument must REST upon analogy & absence of varieties in a wild s
294c177 ns &c &c. & analogy will necessarily explain the REST,-- Lonsdale says he has seen in old Book last B
349d051 - not till much knowledge is elicited.-- It will REST upon the discovery what characters VARY most ea
375d135 ng in number must effect instantaneously all the REST.-- One may say there is a force like a hundred
472s01v no black spot & a third considerably paler, all REST very similar-- June 2. 42 Maer <Thursday> Thurs
525m025 ad all short tails; but one a little longer than REST «they all died»:-- she had kittens before & aft
530m045 e advantage in these trades.-- Delirium seems to REST the sensorium.-- — analogous to sleep--; some
541m090 each.-- ∴ my father. is right in saying delirium REST-- therefore dreams thus act.-- ∴ weak minded pe
542m095 reching & yawning can be explained from too long REST of muscles.-- evidently habitual when transferr
544m103 l us it is not real.= = dreaming appears clearly REST of the mind, with all other faculties: «Vide pa
548m114 as in reasoning-- hence delirium & sleep mental REST. though. most vivid & rapid thought.-- There ma
575n043 an we deny that brain would be intermediate like REST of body? Can we deny relation of mind & brain.
569n022 "il leva les epaules et dit qu'il valait mieux RESTER a Farroïlap quelque mal qu'on y fût."-- Expres
133a140 rtiary period. one is led, to look at globe as RESTING on film of molten rock.-- Voyages of Adventure
153g056 hboring rock gneiss & [...] sandstone actually RESTING on them on summit of hill rounded, site N N W
365d107 ies" & species. he overlooks-- <restric> those <RESTR> restricted in their range by men & by art.-- t
523m018 eman, unnatural union with turkey cock,-- was RESTRAINED by remonstrances on him» which are never gen
365d107 ermanent varieties" & species. he overlooks-- <RESTRIC> those <restr> restricted in their range by me
316c244 gamic plants. of Marine kinds, there are some RESTRICTED genera, but then they appears always very sm
316c245 - eggs in groups.. Have Dioecious plants more RESTRICTED ranges than other plants.-- Many «Some» gene
365d106 subalpina, or some Northern species, & being RESTRICTED species has been Made.-- In the hybrid grous
365d107 ecies. he overlooks-- <restric> those <restr> RESTRICTED in their range by men & by art.-- the former
288c158 e same organs,-- habits, range. &c &c-- Macleay RESTS his whole groundwork of analogy on its concurre
419e087 er be overlooked that the chronology of geology RESTS upon amount of physical change <affecting whole
032r039 n. would now by degrees be exposed to it, & the RESULT would [be] a uniform slope to base of cliff (Z
105a073 not to widen valley.-- Why is serpentine course RESULT of little inclination??----It is simply as the
127a126 ws effects of pressure in change of form as the RESULT of heat.-- will it bear on central fluidity.--
132a139 .-- shows p. 516 that subterranean springs give RESULT less to be trusted than any others-- may not t
172b007 o as often to intermarry who will dare say what RESULT According to this view animals, on separate is
174b014 but if it was necessary to one forefather, the RESULT, would be as it is.-- Hence Antelopes at C. of
176b021 ms would be about equally numerous. changes not RESULT of will of animal, but law of adaptation as mu
204b139 half way. Eyton says Hybrid about half aways & RESULT the same Indian cattle & common produced very
336d017 amount of change» shorter time less [s]o.-- the RESULT of this is that animal would endeavour to retu
393dIBC s uncertain do they mean they cannot tell first RESULT., or that «hybrid» breed is uncertain Is there
398e005 multiplying effects & to <ponder> conceive the RESULT with that clearness of conviction, absolutely
408e044 t be owing to these islands, having been purely RESULT of elevation,-- «all» modern & wholly volcanic
441e150 inches-- & all varieties must be presumed to be RESULT of such laws.-- The effect of one part being g
444e157 Alludes to two theories;-- that species are the RESULT of circumstances,;-- or the will of the Animal
493q004 icial male with <old> female of old breed & see RESULT.-- According to Mr Walker the form of male oug
493q004 alf-bred animals. which are exactly alike & see RESULT.-- (3) Cross the Esquimaux dog. with the hairl
495q005 wers of other cabbages & see whether there will RESULT hybrids-- (6) Dust flowers of one branch of Ca
497q06v tory of Viburnum. or snow-ball-tree. what would RESULT from seeds being sown= See in Cultivated Plant
503q013 nning of November 1841.-- Trees above male? (2) RESULT of Edwards experiment in Cabbages given (3)
505q014 - intends to breed from it and large Asparagus: RESULT? = failed to germinate 16 Will plant some of t
511q018 o destroy secondary character believe No or did RESULT appear without his wish Has since recrossed th
513q020 with abortive tail or horn, with another & see RESULT, for comparison with natural species, as dioec
542m095 to the tale of a wearisome man.-- Is frowning, RESULT of straining vision, as savages without hats p
568n016 stinctive assent to propositions, which are the RESULT of our senses, or our experience.-- Two sides
573n036 range, its-- &c &c;--must be a special act, or RESULT of laws. yet we placidly believe the Astronome
582n068 anged actions, which will bring about a certain RESULT, while the creature performing those actions n
582n068 ing those actions neither knows nor intends the RESULT they will effect,.--" this not wholly true, fo
582n068 building its nest; it knows its object but not RESULT (first time of building?), but not the means o
586n082 merely judged by eye, & use of limbs &c, or it RESULT from mere impulse to save wax.]CD which it ins
602o011 ood passage. about actions & decisions bein the RESULT of sagacity, or intuition. when individual can
605o20v y from this fact & the instantaneousness of the RESULT, that the term taste is metaphorically applied
606o021 rations which the mind undergoes in gaining the RESULT.-- Lessings Lacoon. 2d Lect-- The object of a
609o030 general action» -- certainly because it is the RESULT of what has generally been best for our good f
613o036 is a most important argument, to show that they RESULT from organization of brain; «[LHC] not used by
619o043 by some passion or appetite, what would be the RESULT? In a dog we see a struggle between its appeti
445e159 orms of existing highly organized forms-- this RESULTED from the necessity of supposing some inward p
028r028 separate quartzose sand from the finer matter RESULTING from degradation of Feldspar & other minerals
100a054 absolutely considers gneiss an aqueo deposit RESULTING from disintegrated granite!!! Look at gneiss
634j55r anation of types of structure in classes-- as RESULTING from the will of the deity, to create animals
035r045 erences. & not those of rapid cooling &c &c My RESULTS go to believe that much of all old strata of E
105a072 coast of Madeira.? How came it if this powder RESULTS from «decomposed sea» shells, that land shells
440e146 t of some condition of external circumstances. RESULTS of complicated laws of organization: as we see
602o011 se each decision &c is made up of many partial RESULTS, & the impressions on them are <all> remembere
620o044 attending & reasoning on its action, & on the RESULTS following our conduct.-- If the temptation to
627o053 Ethnical Philosophy p. 6-- "The pleasure which RESULTS when the object is attained (the gratification
132a138 at does increase,-- but in Ocean does not. (see RESUME p. 536)-- «NB. I cannot understand the argumen
198b114 have not been created on the same plan. [Second RESUME well worth studying] CD says grand idea god gi
241c015 forms? All Infusoria. not extinct species. good RESUME do/p. .62 ??? Age of Deinotherium. p. 23.. Bul
302c202 , I must work out hypothesis, & compare it with RESUTS. if I acted otherwise, my premises <in di> wou
556m145 may be described as partaking of <st.> made by <RET> inspiration & quickly retracting tongue from be
060r124 onstantly pushing the sea (which of course must RETAIN same level) to a greater distance".--Afterward
106a074 it will be evident that deflected stream cannot RETAIN its force if inclination be great. There could
436e135 perfect Struthio head pulled out. yet feathers RETAIN character? If separation in horizontal directi
121a111 & has heated angular fragments of rock, which RETAINED their angles sharp-- yet with character compl
```

Page **(Key Word)**
```
186b058 bliged to be altered, I can conceive colouring RETAINED; therefore probably in some <heteromera> colo
361d095 f both (nursling) quite similar.-- one species RETAINED this character in adult stage, other alters e
405e031 a nearly uniform <dusky> blackish brown.-- yet RETAINED a trace of horizontal mark on flank.; & tail.
405e031 anged, these best marked characters are partly RETAINED, therefore colours vary in same Manner as the
412e057 e of part of the hermaphrodite structure being RETAINED in the male.-- <like> «far» more than marsupi
580n062 et many of the habits acquired in that age are RETAINED through life" p. 200.-- "The desire of glory,
106a074 el depends on power of deflection with stream RETAINING its force, now it will be evident that deflec
302c201 ciently Multipliied become affinity yet often RETAINING a family likeness, & this I believe the case.
186b057 ); and in passing from species to genera, each RETAINS some one character of all its family; but why
240c003 ferent, that they would be called genera., yet RETAINS markings of wings like the wild rock pidgeon.-
386d165 ith definite life & split it, & see whether it RETAINS same length of life.-- like Golden Pippen tree
435e133 m age, what Mr Knight [not located] the stigma RETAINS its power.-- R. Brown found the <poll> masses
276c123 ot openly avow their belief. they do as much to RETARD, as those, whose opinion they believe have end
158g081 f peat on pebbles tidal plain as sea gradually RETIRED, hard to explain on river doctrine <Little Hil
046r078 ca. -- Volcanos must be considered as chemical RETORTS.--neglecting the first production of Trachyte.
556m145 of  <st.> made by <ret> inspiration & quickly RETRACTING tongue from behind upper & little between in
046r080 Helena  there is a native mouse Did wave first RETREAT at Juan Fernandez: the first great movement wa
047r082 p.--) If these view are right the coincidental RETREAT at Portugal & Madeira (Lyell. vol I. P. 471) i
427e109 w less extreme, than before arctic forms would RETREAT: effect on snow of arctic climate in far north
052r098 s. [Fig. 6] {P} A advancing coast to Seaward. RETREATING case in excess as first case. When discussin
047r081 what  difference? In watching heavy swell, sea RETREATS & then breaks: i e to form a wave in ocean. i
608o028 control, & that a future life is a reward or RETRIBUTION.-- it may be a consequence but nothing furth
613o036 ll animals. [LHC] «3]» Eyton told me that his RETRIEVER Sailor he has seen push a hare through the ba
613o036 f standing,-- method of attacking peccari-- --RETRIEVER-- produced as soon as brain developed, and as
352d057 we  use the word) than parent, so may species RETROGRADE, but these facts are rare.-- 2d Sept Those a
423e096 e rise to others.-- Why then has there been a RETROGRADE movement in Cephalopoda & fish & reptiles.?-
498g007 there (7) Where parts of fructification <lat> RETROGRADE into leaves-- is this ever effect of want of
568o010 absurd ideas.-- do they indicate mind & body RETROGRADING to ancestral type of consciousness &c &c.--
432e125 ent against Blyth's doctrine of young birds RETROGRESSING-- Uncovering the canine teeth, or sneering,
355d066 s." Law of monstrosity not prospective, but RETROSPECTIVE as showing what organs are little fixed-- (
159g084 Sh  side of valley Granite blocks on this side (RETURN) between 2d & 3d shelf Mountain <Mica> «compos
216b179 d to think that offspring of Negro & white will RETURN to native stock (the cross often whiter<>) tha
222b203 hybrid» breed, their offspring show tendency to RETURN to one parent, this is only character., & yet
251c040 limits of latter from <Tarton> Barton.-- swifts RETURN after years to nest Vol II. p. 49. on the loca
331dIFC «hybrids, when» interbred. show any tendency to RETURN to either parent.? Is the first cross, which m
335d015 in small limits (within which limits they might RETURN to either parent), then according law, that in
336d017 esult of this is that animal would endeavour to RETURN to parent stock. but if both parents are alike
364d104 oduction of black lambs is owing to old <story> RETURN.-- The Revd. R. Jones told me precisely same s
401e014 eds of any cabbage, where a great many will not RETURN to all sorts of varieties, which he attributes
401e016 nomaly-- I can fancy cowslip producing primrose RETURN to old stock, but not primrose producing cowsl
527m032 districts  have the best song. [Migratory birds RETURN to same quarter for many years]CD.-- Beauty is
546m110 lled to go on the lawn to see something, on her RETURN could not find paper cutter, hunted in vain fo
585n076 he direction in which they STARTED, they cannot RETURN.-- Hence I conclude. pidgeon taken little way,
586n078 compass, & they do know which way they go; & so RETURN.-- «but does not apply to dogs.--» they may do
615o038 d in past ages; seems to indicate that when we <RETURN> turn into angels. this imperfect memory may b
222b203 aving very different characters is attempt at RETURNING to parent stock.-- I think we may look at  it
437e141 arm of mice.-- May 4th.-- The Brussels Sprout RETURNING suddenly to type when brought back to home. (
439e143 e, which makes it determined by a facility in RETURNING to old type Mr Herbert showing the extreme fa
495g05a ith starch & Carmine & experimentise on their RETURNING powers-- then carry them in Electrical machin
638j059 gain  it. by habit.-- New theory of instinct, RETURNING to Kirby's view.-- Macculloch. Attributes  of
207b152 Bory  St. Vincent Vol. III. p. 164. Lile de la REUNION presente elle seule plus d'especes polymorphes
344d041 g of Marsup. is sucking foetus.-- August 23d The REV R. Jones gave an admirable harrier from  Ireland
436e138 ies of Andes. A familiar History of Birds by the REV. E. Stanley Vol I. p. 72.-- Goldfinches placed n
558m155 production of foreign seeds.-- many varieties.-- REV R. Jones has it.-- very curious book.-- Hume's e
323c269 ach's Essay on Generation. Englis Transla -- The REVD. A. Wells. Lecture on instinct -- Cline on  the
364d104 lack lambs is owing to old <story> return.-- The REVD. R. Jones told me precisely same story about so
582n068 tions??. a kind of mental pain & pleasure.-- The REVD. Algernon Wells Lecture on animal instinct. 183
304c207 trifling characters, in common with other birds REVEAL the secret.-- Now all the different forms of S
617o39v ect. The subjective aspect of bodily action is REVEALED to us by the effort it costs us to exert forc
616o39v n by which bodily action is made known to us, REVEALING respectively what are called it's  subjective
549m122 d to happiness.-- Men having some instincts as REVENGE «& anger», which experience shows it must  for
550m124 ould probably feel but little that of anger or REVENGE.-- they are incompatible & the former, the mor
609o029 hese bad feelings no doubt orginally necessary REVENGE was justice.-- No checks were necessary to the
609o030 ne explains why our moral sense points <is> to REVENGE). In judging of <our ha> of the rule of happin
588n090 ing mental agony, & whilst strong emotions of REVERENCE & piety are felt." it appears to me mere cons
059r119 it se reproduire a des hauteurs communes sur le REVERS de chacune des montagnes qui forment les vallé
308c217 why do absent «Dr. Black. tea & sugar» people. REVERSE habits Insects & birds are the only two tribes
446e161 the Northern «French» expedition,-- rather the REVERSE of facts stated by Smith of Jordan Hill.-- May
503q012 c 38 Does only male yak cross with cow: is not REVERSE possible?? Maer (1) Yew Trees near Boat  House
507q15v ture rendering cross impregnation difficult or REVERSE (.49) List of seeds Gaertner de fruct:-- for w
512q019 g of short-horn bull & hereford cow similar to REVERSE cross.-- Sow cast-up-balls of Hawks or even ow
515q021 .-- (3) As peaches sport into Nectarines (does REVERSE happen?) what is effect of crossing peaches  &
545m108 tes the expression.-- Habitual actions are the REVERSE of intellectual, there is no comparison of ide
605o020 om these causes who are not men of taste & the REVERSE of this. taste evidently does not consist of t
062r129 owed its death not to change of circumstances; REVERSED argument. knowing it to be a desert.-- Tempte
126a124 a-- → According to this latter view the rod is REVERSED, upper part metal «conveying heat in one dire
302c202 racters, might be multiplied.-- we must argue REVERSELY. WHERE CHARACTER VARIABLE it is (one of analo
495q05a wers-- then carry them in Electrical machine, REVERSING the poles test by suspending magnet within  &
378d148 uch longer its Mental peculiarities. Wildness REVERSION Q [not located] The present age is the one fo
248c033 cannot be transmitted;.-- hence the tendency to REVERT to parent forms, & greater fertility of hybrid
298c191 best adapted kind.-- lower species would then REVERT to pristine form (which must have been altered
277c126 e (E) Chile.. rupestris -- good species it is REVERTING to old plan, but reason now assigned for doin
286c156 ant, all brought in one plane Fleming Quarterly REVIEW says nat: fam: of Willows contains many Linnae
320c276 very  little Macleay's letter to Dr. Fleming. & REVIEW of latter in Quarterly Sir J. Sebright's Pamph
327c265 ncyclop.-- The <Edin> British & Foreign Medical REVIEW No XIV. April 1839.-- Review on "Walker on int
327c265 &  Foreign Medical Review No XIV. April 1839.-- REVIEW on "Walker on intermarriage" price 14s.  Marh.
338d024 ry poor account) of plants of Nova Zenbla -- in REVIEW of Baers work Edinburgh. Royal. Transact.-- p.
462tf05 1839 p. 772-- A curious theoretical French book REVIEW on politics in relation to the different races
533m060 ood example,) because leg is right handed.-- In REVIEW (Edinburgh) of Froude's life. that author rema
539m081 ter good days work" which came on from reading «REVIEW of» M. Comte Phil. which made me «endeavour to
568n019 ss of knowledge. see Lyell on Scrope, Quarterly REVIEW. 1827? In Water Scotts life.. Tom Purdie, (beg
603o014 uage shows traces of anterior state?? Edinburgh REVIEW Vol 18. (1st Article) on Taste «EXCELLENT». De
604o015 rary Magazine. 1756-- Ceased in 1758-- Read the REVIEW or the article. A Planaria must be looked at a
610o033 eople & those of weak intellects.-- Westminster REVIEW. March 1840 p. 267--- says the great  division
067r143 . Parish?? «by Pedro de Angelis.» This work is REVIEWED in present Edinburgh March 1835 Sir W. Parish
604o015 en S. Jenyn's Inquiry into the Origin of Evil. REVIEWED by Johnson in the Literary Magazine. 1756-- C
535m069 s. without any attempt to know their nature.-- REVIEWER considers this profoundly true.-- How is it w
610o033 experience..so there may be in men-- which the REVIEWER seems to doubt. [RHC] 1) Effects of Life in t
385d165 many daughters does the character of the mother REVIVE! Or the character of the mother in the son,  &
046r078 separate ingredients if we look to a constant REVOLUTION.--Are we to consider that the dikes which so
323c269 es: Marginal notes. -- 20th. Carlyle's French REVOLUTION 3? vols. oct: -- 26th Blumenbach's Essay  on
398e006 ymmetry «& regular laws» that baffles idea of REVOLUTION.-- My very theory requires each form to have
027r028 matter. Mem: Chiloe In the endless cycle of REVOLUTIONS. by actions of rivers currents. & sea beache
377d140 epochs & creations, reduce themselves to the REVOLUTIONS of our system in the Heaverns.-- Is not puma
398e006 e in the institutions. & we suppose not only REVOLUTIONS, but entire obliterations & fresh laws creat
432e125 e world is, as the measurements refer not to REVOLUTIONS of the sun & our lives,, but to period neces
633j54r re created to arrest the earth, than the earth REVOLVES to form rain to wash down earth, from the mou
608o028 re under his control, & that a future life is a REWARD or retribution.-- it may be a consequence but
```

(Key Word)
```
614o037 king back, ∴ therefore consciousness, therefore REWARD in good life [RHC] Instinct appear like herede
616o038 s which were changed into man ∴ they meet their REWARD! X Perhaps should hardly be called memory; you
602o011 ology much about sleep-- Nerves.-- Volition &c REYNOLD XIII Discourse (p 115) a very good passage. ab
322c269 med well. do Lutke's Voyage. carefully read.-- REYNOLDS Discourses Lessing's Laocoon Whewells-- induc
527m032 nctive feeling, & thus cuts the Knot:-- Sir J. REYNOLDS explanation may perhaps account for our acqui
570n026 tober 27th Consult the VII discourse by Sir J. REYNOLDS.-- Is our idea of beauty, that which we have
572n032 imic gesture?? which would naturally happen.-- REYNOLDS Works. Vol I, p. 226-- "The general idea of s
602o010 ciousness bears some relation to time & memory REYNOLDS X discourse very curious as showing "the perf
610o034 be  in men-- which the reviewer seems to doubt. [RHC] 1) Effects of Life in the abstract is matter un
611o034 s probably have some unknown relation to them-- [RHC] In the simplest forms of living beings namely «
611o034 l will, though perhaps not conscious sensation. [RHC] During growth <extres> tissue <[...]> unites ma
612o035 be present to be able to exist as individual.-- [RHC] 2) In animals, growth of body precisely same as
612o035 sitive knowledge. lessening amount of ignorance [RHC] The radicle of plants absorb by physical laws o
612o035 s of life vegetable and animal strictly united? [RHC] It is easy to conceive such movements & choice,
613o035 ic system of lower animals & sympathetic of man [RHC] ¿How does consciousness commence; where other s
613o036 no more consciousness than our intestines have? [RHC] 5) Kirby thinks that <all> there is one one ins
613o036 he gate & bring it. ---- Agrees with ONE animal [RHC] Kirby extends instinct to plants, but surely in
614o037 ped, and as I have said, no soul superadded, so [RHC] 6) thought, however unintelligible it may be, s
614o037 re consciousness, therefore reward in good life [RHC] Instinct appear like heredetary memory; but fir
614o037 s change.-- Must be so if Lamarck's theory true [RHC] Acquired instincts analogous «(& replace)» to e
615o038 ut habits acquired even by <children> «plants»! [RHC] 7) As definite instincts modified by heredetary
615o038 obtain peculiarities of external configuration. [RHC] General-- Instincts, certainly appear a sort of
619o042 5th. 1839.-- Maer Mackintosh Ethical Philosophy [RHC] On the Moral Sense Looking at Man, as a Natural
619o042 p. 113. Mackintosh Grotius has argued nearly so [RHC] The history of every race of man shows this, if
619o043 llow <living> creatures, or of kindness to wife [RHC] 2) and children would give him pleasure, withou
620o044 : the latter generally soon conquers, & the dog [RHC] 3) probably thinks no more of it.-- Not so man,
620o045 on as desire is fullfilled, pleasure forgotten. [RHC] 4) as starvation, or fear of death, one makes a
620o046 cts pointing out lines of conduct to other men, [RHC] 5) which are natural (& which «when present» gi
621o047 eive the moral approbation of his fellow men.-- [RHC] 6) Hence man must have a feeling, that he ought
621o047 ceive simple explanations from origin of man.-- [RHC] By interest I do not mean any calculated pleasu
621o048 stincts) any action is either right or wrong.-- [RHC] 7) Hence, what parents think will be good for t
622o048 fferent explanation of law of honour from Paley [RHC] Anyone, who will reflect must feel, how like to
622o048 lains the cementation of habits into instincts. [RHC] Feelings of the mind, whether leading to action
622o049 to  action are repressed thus avarice. &c &c.-- [RHC] 8) in the beginning I mentioned only three inst
622o049 Ld. Kames, which Sir. J. says is so ridiculous. [RHC] the social instinct may be combined with feelin
622o049 enefits from this person.-- [LHC] p. 254. &c &c [RHC] But the love is instinctive, & how does it appl
623o050 et I think there is much truth in doctrine, for [RHC] 9) We can thus explain love of place.-- althoug
623o050 iability & sociability, doubtless grow together [RHC] This feeling seems to vary in races of man. & c
623o050 y theory, all instincts demand some explanation [RHC] Although I cannot pretend to say how far & minu
623o051 acquired,  does possess the beneficial tendency [RHC] 10) that the instincts of bees & beavers «& dee
624o051 is part of argument fails, or rather is weak.-- [RHC] Better simply put it, beneficial tendency in ev
625o052 perhaps, by habit-- [LHC] 11) Whewells preface. [RHC] It appears that Sir. J. & others think there is
340d029 birds.  The Emu & Cassowary closest.-- Ostrich & RHEA closest.-- (& two Rheas still closer).-- Mr Bly
196b105 sily.-- it would have been wonderful if the two RHEAS had existed in different Continents In plants I
341d029 ary closest.-- Ostrich & Rhea closest.-- (& two RHEAS still closer).-- Mr Blyth asked whether structu
452e182 near Vienna appoximat to Negro form; those from RHINE to the Caribs.-- Vol II p. 650. Long attested a
048o085 ing on the nature of the country in which the RHINOCEROS lives in S. Africa: the same caution is appl
048o086 would be well to quote Burchell. V. where the RHINOCEROS was killed. -- In Patagonia, are all beds sa
064r134 hed cou[n]tries thinly covered by vegetation. RHINOCEROS quite in deserts.--Much struck with number o
093a034 s) Dr. A. Smith informs me that in the year a RHINOCEROS was found <emb> in the mud, of the Salt rive
188b067 to South and North America.-- ¿are there any? RHINOCEROS peculiar to Java, & another to Sumatra --Mem
191b082 Species formed by subsidence. Java & Sumatra. RHINOCEROS. Elevate & join keep distinct. two species m
229b233 s structure-- Smith thinks several species of RHINOCEROS range from Abyssinia to extreme South coast.
229b234 amus do.-- Giraffe do.-- Range of East Indian RHINOCEROS (?)-- Some paper in Institute on range of Bo
231b241 breaking up «the primeval.» continent. Indian RHINOCEROS. Java & Sumatra ones all different.-- Join S
252c045 » Give Specimen of arrangement, 3 <5> Species RHINOCEROS Cape town good species Indian species so dis
278c132 und at Swan river. The change in England from RHINOCEROS Elephants &c in the most modern period, comp
343d037 tablished in all <nature> organic nature) the RHINOCEROS of Java & Sumatra, that since the time of th
413e060 luv. says Africa only place, where, Elephant, RHINOCEROS, Hippot, Haena &c are found together.-- Read
416e072 fectly adapted races L'Institut 1838. p. 394. RHINOCEROS «tichorhinus» in Paris basin.-- its relation
450e174 1837.  by Mr. J. H. Moor-- -- p. 1. Elephant. RHINOCEROS Leopard (but not Royal Tiger), &c are found
450e176 p. p 73 State of Muar in Malacca.-- speaks of RHINOCEROS as well as Tapir.-- <do do p 75> «Journal of
453e182 sor Earl-- Eastern Seas p. 229. Believes the <RHINOCEROS> «Tapir» is found in Borneo.-- «p. 233» Ther
508q016 n taking after sex. A Smith. About species of RHINOCEROS. becoming rare beyond limits of the metropol
509q017 organs  as <young> teeth, more plain in young RHINOCEROS or Whale, than in old?? Falconer says all in
513q020 (5) About callosities on Camels-horses. &c &c RHINOCEROS= (6) Cross. Sus Barlyroussa with tame.-- (7)
269c101 applicable to metamorphs theory suppose when RHINOCEROSE lived. mean temp 60 degree mean, then temp a
504q013 ear end of orchard.= At Shrewsbury one branch of RHOD. flowered later.-- effect of accident?? (7) Whi
468t111 <Loasa>  «Anchusa»-- speedwell Iris-- Azalea. RHODENDION. Fraxinella to Anchusa <never> «once» P on F
429e115 p. 17 Append) that in the Pyrenees, that the RHODODENDRON ferrugineum. begins at 1600 metres precisel
468t111 ra Saw Humble go from great Scarlet Poppy to RHODODENDRON-- from Larkspur to Lupine two species of La
468t111 spur -- two varieties of Cistus Speedwell to RHODODENDRON-- <Loasa> «Anchusa»-- speedwell Iris-- Azal
468t112 respect to nectary is same case as Azalea or RHODODENDRON xx after several gloomy days. hot one, Bees
466t093 ers were in nearly full flower Maer June 14/ RHODODENDRUM-- nectary marked by orange freckles on {a}
504q013 red later.-- effect of accident?? (7) Which. RHODODENDRUM seeds??-- Bladder-nut ⅄. Laburnum ⅄. Dodeca
468t105 -white Butterflies suck nectar: "Maer June 14" RHUBARB. pollen very minute--not excessively 'abundant
468t106 lowers which thought very unattractive-- Found RHUBARB. blossom swarming with small Staphylinidae-- An
587n087 senna. might be acquired. as the Turks have of RHUBARB: again on other hand, it is said people, who l
211b162 bits produce similar structure.-- Mem. Ornitho RHYNCUS Would not relationship express, a real affinit
527m033 scale.-- former pleases from instinct the ears (RHYTHM & pleasant sound per se) & causes the mind to
528m038 l lines are elegant.-- Again there is beauty in RHYTHM & symmetry, of forms-- the beauty of some as N
528m038 & Graecian to be single cause) this symmetry & RHYTHM applies to the view as a whole.-- Colour «& li
528m038 f similar forms as in angular leaves,-- (this RHYTHMICAL beauty is shown by Humboldt from occurrence
039r060 the great subsidence at the famous eruption of RIALEJA, & the more true analogy from the Galapagos--
444e157 ing of organs.-- In the Batracian Order the «32» RIBS are wanting. p. 144 in the Icthyosaurus 60 or 7
509q017 lowest cervical vertebrae process developed into RIBS.) & does its abortion vary, according to Bentha
509q017 x ‖‖ no as Os Coccygis-- Turbinated bones? False RIBS Wings of Apterix: clavicle in--? Combs in combl
570n023 es chest, being distended. touch a person on the RIBS & how he gulps in air.-- Again a master says  I
634j54v en he argues about his Creator! p. 309. says the RIBS in Draco support the flying membrane?!!-- that
192b083 . Fox tells me, that beyond all doubt seeds of RIBSTON Pippin, produce Ribstone Pippins, & Golden Pip
228b229 ere not both genera formerly abundant. Seed of RIBSTON Pippin tree <go> producing crab. is the offspr
228b230 to cross from other trees.???? Do the seeds of RIBSTON Pippin & Golden Pippin &c produce real  crabs,
356d070 ry species new.-- Sept. 8'. A Golden Pippen or RIBSTON do producing occasionally (as Fox says) same f
192b083 ond all doubt seeds of Ribston Pippin, produce RIBSTONE Pippins, & Golden Pippen, goldens-- hence-- s
054r102 t of New Zeeland entirely volcanic!! New Zeeland RICH in particular genera of plants: All St. Catheri
073r163 y be considered for most parts as rock eminently RICH in mines of gold & silver." «p. 131» The  above
171b003 ing «to» circumstance,-- seeds of plants sown in RICH soil, many kinds, are produced, though new indi
299c195 ieties chiefly produced by cultivating parent in RICH soils & their seeds produce <offspring> variety
400e011 elater.-- Where this collection is particularly RICH. «as in Lucanidae» <no> less difficulty in esta
441e149 nts.-- Thus would a Crab tree vary if planted in RICH soil, I presume not, but its seeds, I  presume,
491q01v ment of Carrot 2 {also try Primrose & Cowslip in RICH soil & propagate from their seed 3. To apply po
495q005 es (7) Sow <daisy> seeds of wild cabbage in VERY RICH soil, will plants abort?, does it require succe
496q006 een one monstrous} Fox Glove & such like in very RICH soil-- As they have little tendency to  double;
498q007 ts no other case.-- (6) Will plant accustomed to RICH soil, when placed in very poor flower, but  not
532m056 layson, (though so fond of her & of servant of RICHARD & of Mary & her bed brought from Shrewsbury) y
131a136 flowing  from beneath frozen crust in America RICHARDSON.-- From strata being not only vertical, but
193b091 ological Magazine paper on Geographical range RICHARDSON-- Fauna Borealis. It is important the possib
208b154 NB How does this bear with law referred to by RICHARDSON in Report about each genus having its parent
```

278c129 his test.-- Excepting where an Andrew Smith, «RICHARDSON» a Vaillant, a D'orbigny has travelled this
410e052 hout some end.-- Respect good describers like RICHARDSON.-- The relations of numbers of species to ge
324c268 ad Humbold. New Spain-- Much about Castes &c RICHARDSON'S Faun. Borealis Entomological Magazine (pape
055r106 e <oldest> «most intelligent» miners to be the RICHEST Vol II 147 Shells at Concepecion 50 toises abo
074r165 as in concretions perhaps makes intersections RICHEST-- Humboldt has urged phenomena in veins, chemi
078r175 direction as Guanax.--the other E & W.--veins RICHEST not in ravins or along gentle slopes. but on t
079r176 n the lowest. Humboldt states that some of the RICHEST gold mines on ridge of Cordillera near Pataz,
077r171 ded, appear to have a decided influence on the RICHNESS of the veta madre of [misnumbered page] Dr D.
458t002 so small, that person with long legs can hardly RIDE on them. Mr Miller-- in Zoological Gardens. inf
079r176 t states that some of the richest gold mines on RIDGE of Cordillera near Pataz, also at Gualgayoc. wh
110a087 (p 321) Mr Hillhouse describes central granitic RIDGE of Guayana as NW / SE. Vol VI. p. 247. Mr. Scho
158g080 e Loch trace of this terrace «on <will> Granite RIDGE» or a modified Granite ridge» at head of Glen Ro
158g080 «on <will> Granite ridge or a modified Granite RIDGE» at head of Glen Roy on same side where two riv
160g090 ne grained. Much contorted gneiss «narrow sharp RIDGE with peak» I walked all round hill. Boulder abo
070r155 ke of Cordill: of Copiápô & Desaguadero.--three RIDGES in Copiapo, as well as in latter.-- According
147g015 e sort of <plain> space is thickly studded with RIDGES & flat topped hill/ do alluvium. NB In one par
160g092 s-- (general character in these mountains & not RIDGES) between arm of Glen Bright flowing into E. en
377d139 these actions. «These facts may, be turned to RIDICULE, or may be thought disgusting, but to philoso
557m149 muscular movement, hence shy people (shame of RIDICULE) are singularly apt to catch tricks.-- so are
622o049 oarding.. Ld. Kames, which Sir. J. says is so RIDICULOUS. [RHC] The social instinct may be combined w
615o36v rprise, walking home one day met him, with Mark RIDING instantly followed, me and for five minutes ev
472s01v aces of black spot at base, one paler with less RIGED foliage & no black spot & a third considerably
047r082 festly does not travel up.--) If these view are RIGHT the coincidental retreat at Portugal & Madeira
104a067 n on dikes argue impossibility on fissure going RIGHT through superincumbent mass (varying hardness,-
113a091 egree would have been found lower.-- We have no RIGHT to consider the conducting powers either better
157g077 lders» {P} cory stream hill with boulders river RIGHT Hand Cascade has <cut> «where two branches unit
157g078 near divortium aquarium is a lip with it-- Dick RIGHT-- Mac mistook terrace also right-- Granite such
157g078 with it-- Dick right-- Mac mistook terrace also RIGHT-- Granite such as boulder on <thes> Divortium a
158g079 apparently continuation of upper terrace -- on RIGHT hand {P} rounded waterworn buttresses of granit
162g103 58 A 64 degree, air 60 «Evening do» The extreme RIGHT arm of River Tarf <it> Has a very long, flat di
177b026 complicated.» {P} Is it thus fish can be traced RIGHT down to simple organization.-- birds-- not. {P}
194b097 phrodite» shells.!!!? We have not the slightest RIGHT to say there never was common progenitor to Mam
259c066 (like atavism) & shows my «view of» generation RIGHT?-- If puppy born with thick coat monstrosity, i
280c136 her it changes thousands in futurity.-- This is RIGHT way of viewing it.-- Variety when long in blood
290c163 goes a round, the minute gets into the road at RIGHT anglles, how pleased it is, just like man. emot
308c217 ed a long way. --Case of Habit I kept my tea in RIGHT hand side for-- some month, & then when that wa
308c217 th, & then when that was finished kept it in-- <RIGHT> left, but I always for a week took of cover of
308c217 left, but I always for a week took of cover of RIGHT side, though my hand <vibrate> would sometimes
361d095 ed that of Cross-Beak-- In lark if I understand RIGHT, all species have same character which is mottl
398e005 ts. should be absolutely necessary to arrive at RIGHT conclusion about species Changes of level &c ar
443e155 hich have fluid sperm.--]CD I utterly deny the RIGHT to argue against my theory, because it makes th
448e168 & Birds & Mamm. in ornityhyrhycus-- is not this RIGHT?-- June 18th. Eyton tells me, that Yarrell know
526m030 ing One is tempted to believe phrenologists are RIGHT about habitual exercise of the mind, altering f
533m059 he left side (not good example,) because leg is RIGHT handed.-- In Review (Edinburgh) of Froude's lif
533m060 titude, I had a sort of consciousness I was not RIGHT; though I never realized the idea that I was te
537m076 p. 27. Mart. allows some universal feelings of RIGHT & wrong «(& therefore in fact only limits moral
541m090 oments with interest on each.-- ∴ my father. is RIGHT in saying delirium rest-- therefore dreams thus
549m121 e sole object--. -- I doubt whether the last be RIGHT. The two views come very near each other.→ The
551m129 rses starting, when door shut or cad cries out "RIGHT." or Drinkwater's horse jumping when word Jump
575n044 & hunt mice-- Jenners Jackall Have we somewhat RIGHT to deny identity of instinct.-- Habits import t
576n047 would he regret «having acquired» this sense of RIGHT (& Whether wholly instinctive as in the dog, or
586n078 knows in walking [one feels inclined to stop at RIGHT number of house though one cannot remember it.]
596n184 - emotions of every kind.-- «[Are monkeys <are> RIGHT-handed??]CD» Cyanocephalus, Macacus, Niger. Cer
600o08v een moral sense & conscience? we admire what is RIGHT by one & are ordered to do it by other.-- I sus
602o011 idual cannot give reason, though he feels he is RIGHT-- it is because each decision &c is made up of
608o027 ll cure him' 3) disgusted. with them. Yet it is RIGHT to punish criminals; but solely to deter others
610o30v ore rule of happiness is to certain degree <of> RIGHT.-- The change <of> our moral sense, is strictly
621o047 ormed by past recollections.-- Hence he has the RIGHT & wrong in his mind.-- Now we know it is easy b
621o047 le.-- So that as there nice & nasty in taste, & RIGHT & wrong in action, so a child may be taught, or
621o047 osed to natural instincts) any action is either RIGHT or wrong.-- [RHC] 7) Hence, what parents think
622o048 vely benevolent) they will teach to be wrong or RIGHT; this teaching may be curiously modified by cir
622o048 ed with senses» instantaneous so declaring it is RIGHT or wrong.-- «[just as in tastes of the mouth]CD
622o049 it is sufficient to give rise to the feeling of RIGHT & wrong.-- on which «almost» any other might be
624o051 this is the law of our instinctive feelings of RIGHT & wrong.-- education, of parents strives» to sa
624o051 teach same end.-- Hence this becomes the law of RIGHT & wrong, though. that part, which is acquired b
624o051 Hence as Eugenius says, slow growth of rule of RIGHT.-- [LHC] *for it strives to give conduct benefi
625o052 ented uneasiness, & that this is the feeling of RIGHT & wrong.-- so far it has independent existence.
628o53v child,-- causing many actions to be considered RIGHT & wrong.-- to be associated with the approving
629o55v ther question, how the feeling of ought, shame. RIGHT & wrong comes into mind in first case-- seeing
629o055 , it is perhaps deviation from the instinctive. RIGHT & wrong.-- (animals excepting domesticated ones
629o055 -- (animals excepting domesticated ones have no RIGHT & wrong except instinctive ones) Perhaps my the
629o055 nce of social instincts explains the feeling of RIGHT & wrong.-- arrived at first <rationally> by fee
636j56v ioecious plants, had there been no insects. The RIGHT inference is, there were insects «;when were Pa
361d095 r plumage.-- in tree sparrow, (if I understand RIGHTLY) young cock & hen, all nearly similar.-- in bl
628o054 stinct, with that line of conduct, & if taught RIGHTLY, it will be for the general good, that is, the
539m084 e & trembling. & not as flushing & with muscles RIGID.-- How is this? dealt with p. 241 Origin of man
581n066 hout «slight» flush, acceleration of pulse. or RIGIDITY of muscles.-- man cannot be said to be angry.
409e049 my theory be true then the formation of sexes RIGIDLY necessary.-- Without sexual crossing, there wo
548m114 y if they are invented as in imagination, & in RIGIDLY comparing each step as in reasoning-- hence de
202b133 get them to associate together" There is long RIGMAROLE article by S Hilaire on wonder of finding Mon
626o052 f cowardice.-- the whole appears to me rather RIGMAROLE.-- He does not say anything about any princip
633j54r t tendency to break off p. 292. Mac. has long RIGMAROLES about analogies & number L'Institut p. 275.
354d062 inary.-- any one may believe anything in such RIGMAROLES about analogies & number L'Institut p. 275.
503q012 <been> reproduced itself.-- Ask Gray to ask Mr RILEY to experimentise on hybridising ferns, tying th
033r043 so Cotopaxi has a <[....]> cylinder placed on the RIM of conical crater: at Teneriffe Wall of Porph. L
072r160 ld particle attracted towards space tend to form RING. [Fig. 10] {P} motion from within and without H
289c160 pecies to localities‖-- p. 390,. young <grebes> «RING ouzels» dive instant touch the water.-- capital
367d113 rently fathered by a donkey. with all four legs RINGED with brown.-- animal like large, heavily made
367d113 ongly of Zebra.-- Mem. Quagga & Ld Moreton Mare RINGED Owen says that Bell in Encyclop of Anat & Phys
100a053 y to the mine-- consider stalactites.-- agate RINGS, crystallization transverse.-- or rather radiat
030r037 ne are they allied to the jaws of the Cocos fish RIO Shells argument for rise In Cordillera, the dike
037r056 , hills of Basalt & other Volcanic rocks. Bahia, RIO de Jan: B. Oriental? level surface not disturbed
041r065 as mentioned the drifting of carcases putrid. In RIO paper. when discussing probable rise of land: Me
100a054 from disintegrated granite!!! Look at gneiss of RIO Concretions in Pumice bed at Ascension instance
114a093 Good Hope & Australia/ and mud of salt-lakes of RIO Negro--Mr Bowerbank-- Dr. A. Smith's curious spe
116a099 l Rapport on D'Orbigny's Voyage. good section of RIO Negro beds.-- -- refers to species non decrite d
288c159 ions, if American & intersected wider & wider at RIO Plata. birds which had originally crossed would
420e090 animal is mentioned which inhabits the Pinna of RIO Janeiro, (like some Mediterranean species).-- mi
467t103 two wings. & when the Lupine flower is perfectly RIPE & pollen abundant filaments & stamens all protr
467t103 s I think they do in Broom & certainly when over-RIPE & half withered-- I saw Bees going to clover &
502q11v ised.-- His Book.-- 32. Would wheat from AEgypt RIPEN in Scotland?-- to show acclimatisation.-- July
515q021 o they live healthily, or does fruit merely not RIPEN?-- The point to attend to is whether good & ple
506q015 or bad seeds formed; badness may be merely not RIPENING= (38) Have Dioecious plants any secondary, se
515q021 ncepcion takes place,-- the mere fact of seeds RIPENING has scarcely any no relation to hybrids.-- (3
029r033 it lasted about a minute, there was no uncommon RIPPLE on the water. It was quite calm at the time. L
163g107 My informant saw them himself-- Sand with tide RIPPLE Near Fort Augustus hill & fringe as if it has
453e183 uliarity has first appeared.-- "Storia della RIPRODUZIONE Vegetale". by Gallesio. Pisa 1816 p. 27. Dr
030r037 e jaws of the Cocos fish Rio Shells argument for RISE In Cordillera, the dikes do not generally appea

035r048 nsignificance; as fractures, consequent on grand RISE, & angular displacement, consequent of injectio
041r065 s putrid. In Rio paper. when discussing probable RISE of land: Mention M. Gay's fact about shells: Hi
042r068 er, the Volcanos from sea there burst out, after RISE from sea: <As did> as did those aerial Volcanos
045r075 eat Lima earthquake» In the Chili earthquakes if RISE was more <than> inland than on coast it would b
045r076 f crust of globe & injecting matter on the great RISE). -- The great rains which attend severe Earthq
046r080 n Fernandez: the first great movement was one of RISE (any smaller prior ones might have been owing t
055r108 ch strike the mind with force.--the exact yearly RISE of the great rivers prove better than any meter
094a035 Is the felspar glassy in greenstone dikes which RISE through granite.-- a most important question wi
104a067 s time to trace) from few dikes which have given RISE to eruptions.-- We must suppose everywhere--, i
121a113 le beds. Lyell on Sweden. p. 12. proofs of small RISE at Stockholm.-- analogous to my Valparaiso case
152g050 level of upper terrace The butresses of Alluvium RISE nearly up to Glen Collarig up within 200 ft of
163g106 do. When cutting bank where Locks now are (32 ft RISE) they found alternating layers of coarse & fine
285c151 s> is effects of unfavourable conditions. (hence RISE & depression of importance, in each group & con
316c244 for Hell.-- «+glimpses bursting on mind & giving RISE to the wildest imagination & superstitions.-- +
397e003 ophy to believe that no stone can fall, or plant RISE, without the immediate agency of the deity. But
418e083 m, endowed with the vital principle, which gives RISE to the sexual organs, different in each species
423e096 e changed, these very changes <len> tend to give RISE to others.-- Why then has there been a retrogra
527m031 ne will be uppermost:-- so in thoughts, one will RISE according to law. How strange <all> «so many» b
529m041 alings would be excited, & how the scenery would RISE. Deer in Parks ditto.-- My Father says there is
545m105 hich are viewed very often.-- former do not give RISE to ideas so much. as objects of interest.-- do/
558m150 tive sexual, parental & social instincts, giving RISE "do unto others as yourself". "love thy neighbo
559m154 er the formation of laws invoking laws. & giving RISE-- at last even to the perception of a final cause
571n031 h rises naturally & hence sublimely from natural RISE-- I was also much struck in great avenue, resem
622o049 - So that we have some, it is sufficient to give RISE to the feeling of right & wrong.-- on which «al
087a011 th Lyells idea of intertropical land.-- Siberia RISES. therefore to the South sinks.---- Mediterranea
522m013 My Grand F. thought the feeling of anger, which RISES almost involuntarily when a person is tired is
551m129 Scott Vol VII p. 35 "as ideas come & the pulse RISES, or as they flag & something like a snow-haze.
571n030 ck with this, when viewing Windsor Castle which RISES naturally & hence sublimely from natural rise--
626o052 their formation.-- 'N.B. If feeling or emotion RISES from heredetary action on body.-- This feeling,
032r040 the triangular mass removed vary.--The gradual RISING continuing. a another sloping platform would b
032r040 n.-- This is grounded on the belief of constant RISING with successive periods of greater activity &
045r077 in showing a direct relation of a part of globe RISING, when another falls.--When discussing connecti
089a017 bubbles volatilized at bottom, condensed before RISING?-- Mem. granite heated.-- Metamorphic action i
108a079 ker.-- PP Andes mark the line between sinking & RISING areas.-- In Earthquake if Subsidence we should
117a102 no» tides, water always falling or at least not RISING are there cliffs. Sir L. Dick says (.p 52) fri
408o045 on & subsidence of East Indian Archipelago. now RISING. On a particular part of coast of Somerset·shir
500q104 of confined genera By my theory in volcanic or RISING isld, there ought to be a good many races or d
626o052 instinctive will Lead to action.-- the passion RISING from weariness leads to striking blows.--' p.
087a011 tteranean continent corresponding to Europaean RISINGS. Pacific great land. -- Will use argument of p
058r118 ectent une pente plus ou moins inclinée vers le RIVAGE de la mer, tandis, au contraire, que vers le c
557m149 ect of attachment) & then failing to drive away RIVAL.-- Fear is open mouthed to hear. though in indi
345o044 mpressing most is in harmony with their wars & RIVALRY.--» The very many breeds of animals in Britain
364d103 see which will sing longest, & they in evident RIVALRY sing against each other, till it has been know
048r084 mentions beds of marine shells on banks of Red RIVER Louisiana. V. Lyell. Vol I. P. 191 State at St
093a034 noceros was found <emb> in the mud, of the Salt RIVER.-- in reference to fossil guanaco of P. St. Jul
105a069 s thus wide difference between erosive power of RIVER & sea.; the former as its channel becomes wider
113a092 ng want of leveliness Major Mitchell showed me a RIVER <near> W. of Port Philip. which had bar at mout
116a100 29. gold is not sought for in Chile in beds of RIVER, but in shelving «successive» banks <above> 30
116a100 cessive» banks <above> 30 ft or so above bed of RIVER. formed of rounded pebbles-- it is clear gold o
120a109 levation then went on at greater rate, not only RIVER would carry further its own matter. but would c
123a117 ger than ox.-- at Buenos Ayres 20½ quadras from RIVER; 20 varas from surface in tosca.-- remnant of M
129a132 20 ft at least in depth» near mouth of Columbia RIVER-- Read Mr Parker's Book.-- M. Bichoffs Papers,
147g020 ry striking yet possibly sea more probably than RIVER-- No exact terraces but appearances, as if vall
148g027 y flat terraces one above much inclined towards RIVER all these composed-- where side ravine entered
149g032 lar & even towards {P} Spean Roy double terrace RIVER «& to West of Spean» difficult to explain on <f
150g037 r up to shelf very shallow channel 50 ft wide & RIVER get formed in centre In Glen Collarig, on side
152g049 Tring-- Tuesday Bridge of Roy Level of «bed of» RIVER 30.221/65 degree/ Temp of air 65 degree? There
152g049 ree? There are two terraces on the East side of RIVER & bed of river about 40 ft beneath general plai
152g049 two terraces on the East side of river & bed of RIVER about 40 ft beneath general plain. 30.127 A 72
152g050 200 ft of level of 4th shelf= argument against RIVER--composition &-- stratification argument detrit
153g053 which higher up on is corroded {P} 4th [shelf] RIVER 4th [shelf] Could earthquake cause collection o
154g060 {P} side removed all well & good, but how came RIVER to do this vast quantity when during repose of
154g061 o the rock of cols if--. why should it deposits RIVER terraces often descend by flights the terraces
155g063 d of Alluvium nearly parallel--» Inclination of RIVER must constantly alter with falling sea & so cor
156g071 slightly bending up each main valley.-- & that RIVER alone had modified it-- perhaps however sea als
157g076) «boulders» {P} cory stream hill with boulders RIVER Right Hand Cascade has <cut> «where two branche
158g081 in as sea gradually retired, hard to explain on RIVER doctrine <Little Hill with granite blocks almos
158g083 «very little way» in Upper Glen Roy at pass {P} RIVER Gorge 4th Sh side of valley Granite blocks on t
159g084 e ones (one 6 ft across) on top of spit between RIVER & dry Corry Scarcely conceivable. if Hill betwe
159g086 erworn & some partly well worn pebbles-- «which RIVER could not have deposited» the slope is continue
160g089 al> two divortiums aquarum, viz two branches of RIVER Bought & between one of these & Glen Tarf Hill
161g098 water Barom: 30.372 A 76 degree 75 degree? The RIVER <the> of which the source is a lip with the new
162g103 e, air 60 «Evening do» The extreme right arm of RIVER Tarf <it> Has a very long, flat divatium aquaru
188b067 e <No> Southern kinds-- (I) Gnu reeaches Orange RIVER & says so far will I go and no further:-- Prof.
193b093 Geograph Journal. Vol I p. 174. says from Swan RIVER long South coast all the remarkable Australia g
278c131 l Journal to discover whether dog found at Swan RIVER. The change in England from Rhinoceros Elephant
334d012 s the Half Muscovy Fox says a settler near Swan RIVER, lost his <on> two cows entirely, changed his r
402e017 ained, by such strata being merely leaf, if one RIVER did pour sediment in one spot, for <whole> many
528m037 -- the pleasure from perspective is derived in a RIVER from seeing how the serpentine lines narrow in
024r015 om the shore? Beagle Coast of Brazil? where not RIVERS in my Coral paper {T} leagues Fathoms Parallel
027r028 the endless cycle of revolutions. by actions of RIVERS currents. & sea beaches. All mineral masses mu
030r035 tion agrees with character of.. «in Basins from RIVERS. & natural position» position at N. S. Wales &
055r108 with force.--the exact yearly rise of the great RIVERS prove better than any meterological table the
106a074 when that is case force is lessened. therefore RIVERS very ineffectual in widening valley.-- it is e
106a076 -- Athenaeum. 1838--. p. 137. Three inosculating RIVERS in Southern America ¿ effect of subsidence-- <
113a092 p in SE part of Australia.-- Probably a case of RIVERS turning round & penetrating their own range in
115a096 are cliffs there ought to be creeks & mouths of RIVERS ought to be deep.-- Henslow has deposited spec
120a109 e, such as now exist.-- caution about action of RIVERS.-- Excess of matter brought down Mention absol
123a116 ead of Delta) describes near Alps great beds of RIVERS which must be like the Chilian ones.-- Septemb
147g016 ove the 300 ft Alluvium <abo» by Loch Dochart-- RIVERS could not have deposited it. Barrier of lake v
148g029 ss of Glencoe-- the erosion may often be due to RIVERS-- By Roy Bridge, a tongue of flat land, with t
151g043 eat sloping plain of alluvium (much corroded by RIVERS) & not to head of plain.-- but below houses wh
152g047 ortant contingency if elevation from Axis, then RIVERS might deposit, & afterwards with greater cut t
155g064 hat <alter> a balance there must be in power of RIVERS either bringing more «detritus» than they corr
158g080 dge» at head of Glen Roy on same side where two RIVERS unite in Upper Glen Roy great plain about 60 f
207b151 hat mercu> Geo. Joun. p. 325. Vol. IV. Ducks on RIVERS in Guiana. build top of trees carry duckling t
465t079 the preservation of shells; where land broken, RIVERS entering.-- & yet no shells-- now look at Scot
495q05a hens-- examine dog, which has swum-- on pools & RIVERS-- every kind of seed must be distributed.-- Ex
151g043 not to head of plain.-- but below houses where RIVULET enters two great projecting butresses, upper s
478z004 ys Par un officier du Roi Rapid growth of Coral-- RN. p 24 Bougainville Voyage round world no land an
195b099 live so well.-- This view of propagation gives. <RO> hiding place for many unintelligible structures
383d160 P} are more rounded. {P} & much broader., & <more RO> «three, I believe instead of two lines.-- <fai
120a107 id undulations-- the equal movements of Glen Roy ROAD. (¿ metamorphic action at the bottom of the sea
134a141 on Western Coast p. 204 do. do p. 210. Height on ROAD from Valparaiso to Santiago p. 328. dead trees
147g021 th sloping bed of rubbish Friday Highest part of ROAD between Inverorum & King's House 28.935/82 degr
150g037 en Collarig, on side of Hill of Bohunthine upper ROAD (2) extends as far nearly as house, the 3d belo
152g047 Ballivrat 2 miles North of Grant town to Forrest ROAD comminuted shells Important contingency if elev
160g093 way on rocky place, but narrow shelves just like ROAD of Glen Roy-- appears to lip with moss On this

Page
(Key Word)
161g095 g crossed the mouth, (deep) of above valley this ROAD level with Peat moss most distinct then lost by
161g095 ar. this bit to eye certainly appears level with ROAD, & with piece of excised rock lost at point of
162g102 d me) it is impossible to see my new shelf, from ROAD: Loch Ness 30.140. A 66 degree 30.095 .0458 or
290c163 hen horse goes a round, the minute gets into the ROAD at right anglles, how pleased it is, just like
332d004 a stone blind horse, seemed to perceive turn on ROAD where No houses to Eaton Mascott, where he had
473s05r annot, however, tell them from L. Groznerat, «on ROAD to Bethgellert» wh flows by Tremadoc. but can t
151g039 thout any cause, must be tides. &c. {P} <notch> ROADS very much this character.-- The boulders (one o
153g052 bend & here most accumulations At gentler bends ROADS disappear The normal condition of 4th shelf, so
584n072 y be called instinctive: the facts of memory of ROADS long after once visited by horse & dogs. (even
226b223 ommunication by which great Edentate might have ROAMED to Europe & Pachydermata from Europe to Americ
581n065 in the dawn of civilization-- thinks many words, ROAR, scrape, crack, &c, imitative of the things.--
569n020 necessary connexion between things & voice, as ROARING for lion &c &c. (in same way alphabet. arose f
264c079 y actions of despair; «let him look at savage, ROASTING his parent, naked, artless, not improving yet
610o031 irection book under head of Wilson, referred to ROBERT & found his Christian name was Wilson!!-- How
610o031 13th. 1839 My father received a letter from Mr ROBERTS-- «a person he had long known & directed many
091a026 n Quito Volcanoes & another on Mexican Trachyte <ROC> lava called Andesite. Red Coral in the Mediterr
023r012 cribes a Small granite Isd. capped by Calcareous ROCK; following Curvature of hill; states could disc
025r016 Olinda Shoaler N. of Olinda.--a little WNW of C. ROCK. [5 degree 29 minute S.] still shoaler, coast c
027r027 more hilly than others of the Bahama consists of ROCK & sand mixed with sea shells--about 500 Isd. &
028r029 the Lime separated; is it washed from the solid ROCK by the actions of Springs or more probably by s
035r048 r displacement, consequent of injection of fluid ROCK.-- Try on globe, with slip paper a gradually cu
037r052 . Chap XI Vol II. Urge the entire absence of any ROCK situated beneath low water in the Southern ocea
037r053 be a question. whether organic remains protect a ROCK, or that the rock not weathering allows such Co
037r053 ther organic remains protect a rock, or that the ROCK not weathering allows such Compare the elevated
038r057 ving that there must be a central core of melted ROCK--I think the strongest is the consideration of
038r057 there must be «somewhere» below a field of fluid ROCK.--In the discussion it will be better not to re
039r061 gy is now perfect <The grand propulsion of fluid ROCK, which elevates a continent> We are more abound
047r083 h <Mr Lyell's> idea of an injected mass of fluid ROCK In Patagonia plains. long periods of rest & vic
048r086 e age! is white substance triturated Porphyritic ROCK. s (mem white tufas with purple Claystones of P
051r093 rified, & increase. -- In Southern regions every ROCK is buoyed by Kelp, now Kelp sends forth branchi
051r095 part. Apropos to question does animal adhere to ROCK because it does not decompose. or vice versâ. C
064r134 irchels» at foot of range some miles from shore. ROCK of oysters quite above reach of tides.--thinks
068r146 tion (NB. dislocation connected with fluidity of ROCK ∴ «in earliest stage» when covered up beneath o
068r147 raised then all can, for fresh layers of igneous ROCK replace strata. & it is nothing odd to find the
073r163 to great thickness. = Coast of Acapulco granitic ROCK.--in parts of table granits & gneiss with gold
073r163 of Mexico may be considered for most parts as ROCK eminently rich in mines of gold & silver." «p.
074r165 res. We here have case of such vapours washing a ROCK» Veins concretionary; concretions <dt> determin
074r165 a in veins, chemical affinities like in composed ROCK. granites syenite» «strangling &c of veins can
077r170 truction of porphyries. whereas other to ancient ROCK.--this N degree 2. superimposed on N degree I.
092a031 urn There are some ideas about order of injected ROCK being determined by fusibility in. L Institut p
094a036 transported.-- {T} On grooved rocks. Specimen of ROCK from Costorphine at Geolog. Soc: Colonel Imrie
100a052 not applicable. it does not explain CLEAVAGE of ROCK-- nor the Falkland case, nor. the arrangement o
102a060 avity has to do with arrangement of particles in ROCK. This applies to cleavage & concretions.-- Sept
103a065 of dikes. argues in favour of pressure of liquid ROCK. Andes discussion-- Albite certainly contains 6
104a066 100 = 480 In Falkland islands. & generally where ROCK metamorphic & thickness of «strata» not great,
106a076 ter proofs of sinking.-- No Sweden!! swelling of ROCK from Heat. Specific gravities of many artificia
112a088 k (C) for some speculats on conducting powers of ROCK-- -- Geograph Journal Vol IV p. 36. on subsiden
113a092 hilip. which had bar at mouth excavated in solid ROCK-- 4 & 5 fathoms deep, perfectly still water. M
115a097 r by great molecular attraction of every atom in ROCK On a coast, the shallower the water, the greate
119a105 cale.-- (for the temp must be immense to convert ROCK into gneiss & judging from what we see when tr
119a106 s is shown by the softness & curvature of quartz ROCK?) also by my phenomena of earthquakes.-- by the
120a108 travelled some hundred miles through nearly cold ROCK.-- in volcano the pool is not deep. --Hot sprin
121a111 segregation.-- & has heated angular fragments of ROCK, which retained their angles sharp-- yet with c
124a119 - if so exactly parallel to limestone & volcanic ROCK containing magnesia Lyell. Elements p.119 on su
129a131 p. 133 The most wonderful case of great block of ROCK moved by gale-- When writing on Valleys. «Terti
133a140 eat travels; & therefore the abysses where fluid ROCK has been ejected must remain fluid for an enorm
133a140 d, to look at globe as resting on film of molten ROCK.-- Voyages of Adventure & Beagle vol I. p. 2 &
151g040 Gneiss remarkably water worn) are often times of ROCK not in immediate neighbourhood, (as granite or
153g056 ones «these boulders are decaying.» neighboring ROCK gneiss & [...] sandstone actually resting on th
154g060 d but little more {P} now that it has got to the ROCK of cols if--. why should it deposits River terr
155g066 w much more so, these lines & even water-scooped ROCK «only decay from fragment falling» of no partic
156g068 <the> usually slaty Point of rounded not scooped ROCK on «bend» of 3(a) Cannot <see> «make out» compo
156g069 of» those since fallen. «on the 3 shelves» Solid ROCK is much notched on Maculloch's supposition;-- t
161g095 appears level with road, & with piece of excised ROCK lost at point of valley chiefly from rockiness
240c003 ra., yet retains markings of wings like the wild ROCK pidgeon.-- fact analogous to Owen's Phil: remar
261c071 to group.-- it is same as Yarrell's remark about ROCK Pidgeons.-- & the latter most important in obvi
318c254 icular part of country, like White of Selbournes ROCK Ouzels.-- ..If the line or bands of country (Th
362d100 races» of them have the forked black mark of the ROCK Pidgeon,-- several have a group of white speckl
363d101 ckles on elbow joint-- in Bewick drawing the the ROCK Pidgeon has not: now how many wild pidgeons hav
363d101 compare them with all the varieties.-- Habits of ROCK pidgeon. (I suspect Pennant has described them)
374d133 p. 417. Magnesian Limestones & Zechstein oldest ROCK in which reptiles have been found. p. 426 Sauro
438e142 means Man can offer of changes.-- as desert «or ROCK-- plant probably would do-- or be with difficult
571n030 some connection between them, & great masses of ROCK.-- I was much struck with this, when viewing Wi
603o11b r imitative art [my view says yes. <old> mass of ROCK--}CD or poetry, CDjmy thecry says yes. imitating
632j53r f design.-- perhaps they are so.-- but the coral ROCK might have been uninhabited as the Alpine pinna
161g095 sed rock lost at point of valley chiefly from ROCKINESS When on other side {P} Shelf A Shelf A at hea
022r006 n Gneiss. Epidote seems commonly to occur where ROCKS have undergone action of heat. it is so found i
022r006 . Fitton's appendix Would Slate. & unstratified ROCKS show any difference in facility of conducting E
026r021 e planes.: such action can take place in melted ROCKS The frequent coincidence of line of veins & cle
027r023 w what movements take place in semiconsolidated ROCKS P xv. mentions in what formations Conglomerates
028r030 if so sea would separate them from indissoluble ROCKS? Has Chalk ever been dissolved? Singularity of
028r032 idea of escape of heat prevented by sedimentary ROCKS, & hence Volcanic action, contradicted by Cordi
029r033 ll originate from the decomposition of Granitic ROCKS Mem. Chanticleers voyage at <[...] Maranh> Pern
030r034 oncepcion. favoured by basin formed by outlying ROCKS; (such as between Mocha & main land). <[...]> A
033r042 re: Have shells ever casts alone in Calcareous. ROCKS??--if so case precisely analogous: fragments in
034r044 y strong about Trachyte being the most inferior ROCKS--The stream at Portillo Pass example of do? <Po
035r045 ink> At Ascension, the laminae <...> changes in ROCKS. connected with & alternating with obsidian mus
037r053 yed with Kelp.-- With respect to degradation of ROCKS--It may be a question. whether organic remains
037r056 ut even there, hills of Basalt & other Volcanic ROCKS. Bahia, Rio de Jan: B. Oriental? level surface
038r059 rably. which shows an afflux of inferior melted ROCKS to those parts. Are not the dikes generally ver
042r067 mice of Bahia Blanca, mention black scoriaceous ROCKS of R Chupat. & fall of Ashes of Falkner, ¿how f
043r069 try. Read description of channels or grooves in ROCKS at Costorphine hills. to compare with Galapagos
046r079 rent strata: «Mr Lyell considers that Plutonic ROCKS are generated as often as Volcanic. I consider
049r088 thus occurring in groups.--As these greenstone ROCKS are seen to graduate into granites the <conta>
051r094 s which must protect surface; On «hard» exposed ROCKS near Bahia, whole surface to where highest spra
051r094 = Yet everywhere on coast (Il Defonsos «Kelp») ROCKS show signs of degradation; (soft substances wor
052r099 d on shores of Georgia «Lat degree ( )», he has ROCKS on surface. applicable to Patagonia. During a p
053r100 l, states that contact of Granite & sedimentary ROCKS, in Alps becomes metalliferous. Vol III Latter
054r105 of a league & a half a long the coast. <"> The ROCKS in the most inland part of this bay are perfora
055r108 proves the decay atmospheric of the most solid ROCKS.--The grand cliffs of a thousand feet in height
056r110 The common occurrence of a breccia of primitive ROCKS between that formation and the secondary (state
057r114 ded from gravel proved.-- curious similarity of ROCKS of very diff. ages. at Port Desire on plain. &
059r120 ast side of Van Diemen Land. All the Calcareous ROCKS which harden by themselves cannot be pure. for
059r123 cooling [misnumbering, no page 122] In Igneous ROCKS.--which have the cryst of glassy F. fractured.
060r125 ion during secondary period how diff. would the ROCKS have been. The red Sandstone of Andes fusible?
063r131 obe being thin, may be drawn. from. Cordillera. ROCKS.--When beneath water.--together with hypothetic
064r136 he land in Europe-- Urge difference of plutonic ROCKS & Volcanic metalliferous-- Urge enormous quanti
065r138 kland. Macqueries.--Sandwich Isd-- Specimens of ROCKS were brought home in Capt. Forster expedition f

067r144 It is not therefore so wonderful that volcanic ROCKS at M. Video «Volcano in Pampas" Pasto Earthquak
068r146 ently there is a coating of solidifying igneous ROCKS which would be too thick to be penetrated by th
075r166 vered. Humboldt suggests covered up by volcanic ROCKS. //St Helena has been slightly broken up, & has
077r171 freestone, limestone & <many> «other secondary» ROCKS. Vein traverses both Clay slate, Porphyry North
089a020 ninsula of Cape Verd. volcanic.-- Isle of Gory. ROCKS encrusted with serpula-- Isle of Cayenne. Syeni
094a036 boulders might be transported.-- {T} On grooved ROCKS. Specimen of rock from Costorphine at Geolog. S
095a039 ship, found clear proofs of shells & waterworn ROCKS «at Cobija.» At Iquigue of elevation to amount
097a043 ck glazing observed by Humboldt on the granitic ROCKS of the Orinoco".-- <but> on one of the Ponza is
102a060 on coast of England-- Any one. who has studied ROCKS in detail as amygdaloid. calcareous rocks of As
102a060 udied rocks in detail as amygdaloid. calcareous ROCKS of Ascension, each particle coated. &c will be
102a060 cks-- only bear relations to VEINS in primitive ROCKS-- Are substances soluble under great pressure?
103a065 r [...] mass.-- Absence of Caverns, in Plutonic ROCKS. argument against great bodies of vapour. accord
105a071 leys &c; Pampas.-- If blocks above their parent ROCKS. would be prove of subsidence.-- removal downwa
108a081 ted] Bull:. Soc: Geolog. Tome IX 1837-8. p. 24. ROCKS of Chimborazo., & Pichincha. Melaphyre. = Andes
110a086 ubsidence. examine these «lines» Description of ROCKS in Lyells'. Capital Norway case.-- The fragment
110a087 inct on dike side.-- oozed from one of the true ROCKS, most probably from the gneiss beds in the mica
115a097 n Geolog. Soc. if numbered compare them with my ROCKS. when writing on Falkland Islds p. 94. Von Buch
119a105 It is to be profoundly considered, metamorphic ROCKS at surface. & great heigth on mountains.-- cons
119a105 face. & great heigth on mountains.-- consist of ROCKS with fossils,, therefore formed near surface. w
119a106 what we see when trap in dike & approach other ROCKS. & trap at least as hot as lava-- of which temp
119a106 om gneiss to granite shows that the metamorphic ROCKS have just floated over the absolutely fluid poo
120a108 mething superadded, that which give cleavage to ROCKS.--, but lava shows the rocks really hot. & ther
120a108 h give cleavage to rocks.--, but lava shows the ROCKS really hot. & therefore I doubt the thermometer
126a123 ain?-- Mem 1000 {P} how easily water percolates ROCKS,-- when pressure increased or under surface, wo
133a140 nts have been penetrated by volcanic & trappean ROCKS, within say the Tertiary period. one is led, to
134a141 al Fuego Admiralty Sound. SE dip. much p. 136. ROCKS on Western Coast p. 204 do. do p. 210. Height o
134a141 es.-- as by subsidence || Fitz Roy refers to || & ROCKS p. 375. on the soundings on outer coast of T. d
134a141 ndings on outer coast of T. del. Fuego.-- p 385 ROCKS of S. Western Coast Vol II p. 277. on whale bon
149g032 fectly rounded «base» pebbles of quartz & other ROCKS not apparently in situ <& in> hill being gneiss
156g070 ls <into> on upper shelves granite & some other ROCKS at head of shelf 3d almost all granite pebbles
192b088 known» with fishes & reptiles.-- In mere eocine ROCKS. we can only expect some steps.-- I may ask whe
202b131 ; clay fifty feet, then forest 120 ft Micaceous ROCKS. subsidence appears indicated.--- p. 36.-- Geog
221b201 era? The absence of lime in Plutonic & Volcanic ROCKS. most remarkable.--¿ Have the changes been so s
225b218 aid to deceive man. as likely as fossils in old ROCKS for same purpose.!! Can the wishing of the Pare
272c108 uliar conditions the Staphylinidae on St. Pauls ROCKS must be placed under Gould says most subgenera
638j58v ngement on the wisdom or Providence. when whole ROCKS nay very mountains are formed of such dead & ex
036r052 f Syria <chara> ditto for Patagonia, especially ROCKY parts of central Patagonia Does Andes in Chili.
160g093 y on gradual slope-- : on N side.. dies away on ROCKY place, but narrow shelves just like road of Gle
162g100 elf on opposite side & dies away on the steep & ROCKY gully of last stream Friday Loch Lochy near Let
205b142. Kirby says (not definite information) West of ROCKY Mountains Asiatic types discoverable.-- Bridgew
317c251 her species.-- Says all the hares West of <all> ROCKY Mountains have peculiar character in extreme le
318c255 a kind of Wren (Bebyk??) which seems common in ROCKY Mountains & on one lofty isolated spot on the A
341d031 Blyth on birds.-- Dr. Bachman tells me line of ROCKY Mountains separate almost all Mammals of N. Ame
126a122 ad conductor is {P} against my views-- if we had ROD thus & judged by increments at, how wrong, would
126a124 eneath sea-- ? According to this latter view the ROD is reversed, upper part metal «conveying heat in
194b094 rican form. therefore it is like case of great <RODENT> edentate [has been doubted?]CD & opossum foun
086a005 worth noticing edentates & camels in deserts & RODENTIA In Plata Mastodon Toxodon Is the general sali
204b141 * we not see a splitting in orders, Carnivora, RODENTS &c, JUST COMMENCING. Kirby says (not definite
253c046 a Bears??-- Plantigrade Carnivora??-- «compare RODENTS of two countries» & «Monkeys.» Fact of Elephan
267c095 h. 2. dogs L'Institut. 1838. p. 67. Australian RODENTS Abstract of Infusoria. <p> do p. 62. Ehrenbra
278c131 in Clifts list a rat said to have been found!! RODENTS old inhabitants most important!! like Dipus of
295c178 rmata &c & other Mammals.-- «otter; civet cat, RODENTS.» (Pachyderm in Portland stone of Alps!!!? No)
299c196 nd in Vertebrate Kingdom in more instincts in RODENTS than in other animals & again in Mans mind, in
639j28r 252. analogy of hand in mole, & Mole cricket & RODENTS (?) p. 251. all animals run by hind legs-- Kan
364d104 - (Effect of imagination on mother. white peeled RODS mentioned in old Testament placed before sheep-
267c095 ls of Nat. Hist. precursor of magazine??? p. 75. ROE of Asterias in stomach. of Sammon remain after r
101a056 England influence dispersion of Boulders.-- See ROGERS for Southern limits of Boulders in N. America
214b174 relation to those now living in the sea.-- See ROGERS report to Brit Assoc <to> on N. American Zoolo
218b190 en with hump.-- p 173. Voyage par un Officier du ROI Mem. Capt. Owen's story of cats on West coast of
234b255 173. Vol I. Voyage à France-- Par un Officier du ROI.-- Mackenzie Travel. p. 280. says cattle in Icel
328cIBC m. Cardium. Porcellus Turbo. Cerithium Jardin du ROI Java fossils at same time Study Botanical work o
478z004 rrangement of animals in rays Par un officier du ROI Rapid growth of Coral-- RN. p 24 Bougainville Vo
588m090 struggle their arms.-- do. p. 306 "the eyes are ROLLED upwards during mental agony, & whilst strong e
633j53v the seed-pod of a desert plant (Anastatica) is ROLLED along, & splits when it comes to a damp. place
102a059 ting some way from coast is driven on to it.-- ROLLERS at Tristan d'.Acunha.-- silting up. channels o
378d147 tints, such as Magpie, Jay, & perhaps all the ROLLERS-- He says» whenever metallic brilliancy is pr
523m016 sed by thinking over the misery of an illness at ROME, when by accidental <was> delay of money, he wa
540m084 dog bark. (not shout) when opening his mouth in ROMPS, <so> he smiles. Many of actions as hiccough &
241c015 N. Zealand L'Institut. 1838. A Dipus. & other RONGEUR in Australia.-- p 67 ¿American forms? All Infu
077r171 emnitz in Hungary.) Humboldt says fragments from ROOF & penetrating overlying beds tells the secret.-
525m025 extra number of fingers.-- hare lip or imperfect ROOF to the mouth «stammering in my Father family» (
468t112 ds & bend over stigma:-- but stigma «is» almost ROOFED by united filaments.-- This flower hostile to
294c177 her.-- once grant good species as carrion crow & ROOK formed by descent. or two of the willow wrens &
220b198 30. most curious paper on heredetary fear (like ROOKS with guns) of the Bustards in Germany.-- Athena
217b188 all ice in time of ice transported.-- This gives ROOM to fine speculation.-- Are there many Northern
239cIFC olour on wing Effects of colour on parent, white ROOM How are varieties produced, by picking offsprin
315c240 is found in Isd of Bass' Straits The common Mush ROOM & other cryptogamic plants same in Australia &
532m056 constantly whined, would not remain quiet in any ROOM, would not sleep at night even when in bed room
532m056 room, would not sleep at night even when in bed ROOM-- grew very thin, would not go out of house exc
547m111 uation of waking thought-- my servant was in the ROOM. with my trunk out & I was engaged in hurriedly
585n076 Chimpanzee pouted & whined, when, man went out of ROOM.-- all theories of magnetic powe in birds, seei
594n113 uch so that the nurse had to carry it out of the ROOM. nearly 3 months old. What is absurdity, why do
260c068 that some birds or animals are placed in white ROOMS to give tinge to offspring.-- Darkness effect o
563n001 & on the ground.-- Cock fowl. on the ground, at ROOST, in all seasons, & after? he has done <g> crowi
188b068 that when race once established so difficult to ROOT out.-- For instance ever so many seeds of white
266c092 f Science & Art. 1831. p 160. account of Bulbous ROOT from Mummy: after 2000 years, germinating.--!!
299c195 made into biennial domesticated kind with large ROOT by sowing it at wrong time of year & manuring i
505q014 ll true Bull-Dog-- length from nose over head to ROOT of tail 28½. inches. From sole of foot to shoul
463t057 ably reject this theory-- (I must answer it by ROOTING out curious cases of intermediate structure,,
051r094 buoyed by Kelp, now Kelp sends forth branching ROOTS which must protect surface; On «hard» exposed r
441e151 latitudes, which are generated by buds alone or ROOTS, & never flower, so there may be animals as Cor
443e153 mstanced as the Gorze must be propagated by its ROOTS; now it is curious Mr K. has observed that to g
443e153 rious Mr K. has observed that to graft from the ROOTS is the best way to get young trees. from worn-o
501q10a rate specifically in one country & not in other; ROSA is hard in Europe, Walnut in America.-- Heaths
046r080 nt. owing to water keeping its level whilst land ROSE up & down.--But from above reasons, do not thin
109a084 .-- Annales des Mines. a translation of paper by ROSE on Greenstone, diorite, &c most important.:-- m
139a180 ance cause such monstrous growth as oak galls or ROSE <buds> galls.-- is it not effect of superadded
384d161 y put female first Will not even a fruit tree or ROSE degenerate during its life so that successive b
325c267 7 Paxton on the culture of Dahlias Mrs. Gore on ROSE might be worth consult. Paper on Consciousness
371d118 23rd. Saw in Loddiges Garden. 1279 varieties of ROSES!!! proof of capability of variation.-- Saw his
371d128 arent, being handed down? Are not Loddiges 1279 ROSES kept in same soil. same atmosphere?-- may they
371d128 ransplanted?, & yet year after year, successive ROSES & bud are produced, like parent stock, or if di
371d128 iorating very slowly.-- I presume most of these ROSES, without circumstances very unfavourable, will
492q002 cked in bad years from Caterpillars. 5. Whether ROSES impregnate each other. when close planted toget
492q002 seed-- Heartease. 6. -- Do not species of wild ROSES run into each other very much.-- Has not some o
057r114 in accordance in Europe with ice theory.-- Capt ROSS found in Possession Bay in 73 degree 39 N. livi
080r181 ma Isd De Lucs travels Beauforts Karamania Capt. ROSS. & Scoresby deep soundings Gilbert Farquhar Mat
032r041 temperature «at equal distances from centre of ROTATION» & a <circulation owing> rotation in fluid ma

032r041 om centre of rotation» & a <circulation owing> ROTATION in fluid matter of globe. must there not be a
573m037 ead first to one side & then to other. & hence ROTATORY movement negation.-- he dropped his head when
053r100 s & Orinoco mentioned by Humboldt under name of ROTHE-todte-liegende is perhaps same with that of Per
381d156 to fecundate female (as in plants) Cirrhipeds ROTIFERS, trematode & cestoid Entozoa Allotriandrous <
065r139 . a body which had long been buried, <see> from ROTTEN state of coffin «buried in a mound» long consi
285c152 er.-- The bottom of the tree of life is utterly ROTTEN & obliterated in the course of ages.-- As spec
138a176 & brine was boiling on the top-- [blank] Would ROTTING wood by yielding Carbonic Acid unite with «pie
217b183 h which would never take the dog. But at last a ROUGH-haired shepherd dog lined her & produced a very
217b184 as other.-- He has known case of good pointer & ROUGH water spaniel «produce litter like both parents
467t105 -- Seen Bees on Potato & several times on Beans ROUGH.--green-cabbage «in flower»-- swarmed with meli
468t105 t attractive, very small--stigma rather large & ROUGH-- flowers common-- many winged thrips, covered
469t135 requent these flowers till late in evening-- On ROUGH calc. 280 flowers-- allowing each Bee visits 10
021r005 --longer axis in line of Cleavage. laminae fold ROUND them; Quote this. Valparaiso Granitic nodules i
033r043 apagos. chiefly red glassy scoriae.--could walk ROUND base:--not universal: could not climb up many p
073r161 alactiform masses have layers been accumulated, ROUND knobs, or pushed where soft, or redissolved sof
113a092 Australia.-- Probably a case of rivers turning ROUND & penetrating their own range in Australian Alp
123a116 s.-- Septemb. 2d.-- Sulphur like carbon must go ROUND of dissemination & separation in volcanos.-- if
149g033 nel precisely as with Isld-- {P} do they extend ROUND hill too low line drawn plain red talus line on
150g038 ses The Hills in this neighbourhood appear very ROUND-topped with much drainage & far more earthy tha
160g090 iss «narrow sharp ridge with peak» I walked all ROUND hill. Boulder about 20 ft. below summit <Isthmu
165g124 er where stream enters at head of which hill is ROUND & not merely thoughts laying dormant-- Man from
227b227 re fresh creation!-- If continent had sprung up ROUND Galapagos on Pacific side. the Oolite order of
262c073 wading birds partially webbed. &c &c.--}-- & in ROUND of chances every family will have some aberrant
267c093 grap. Journ. Vol VII. p. 216. Mr Bennett Voyage ROUND world, 20 years have scarcely elapsed since the
290c163 with women with pretty faces When horse goes a ROUND, the minute gets into the road at right angles
321c270 ot located] <Rays Wisdom of> Lisiansky's Voyage ROUND world. 1803-6 Nothing Lyells Elements of Geolog
354d062 ul than the body of a man undergoing a constant ROUND,--each particle is placed in place of last by t
362d099 consequence of the female breeding all the year ROUND. ask Colonel Sykes.-- Even our domesticated cat
376d137 p petticoats., & if woman not afraid clasp them ROUND waist & look in their faces & Mak the st. st no
436e137 The Cambrian formations do not however, extend ROUND world.-- Quartz of Falkland.-- Old Red Sandston
466t094 ting it.» In Columbine nectaries are placed all ROUND flower as they are in Crown-Imperial Lily & man
479z005 growth of Coral-- RN. p 24 Bougainville Voyage ROUND world no land animal besides Wolf at Falkland ∴
481z010 epores p. 26 Nullipora p. 29-- In Meyen. Voyage ROUND World German a reference to a luminous Sertular
527m035 ght this does not happen. because papers, &c &c ROUND one. one recalls the castle by going to beginni
528m037 rm. some forms seem instinctively beautiful «as ROUND, ovals»;-- then there the pleasure of perspecti
556m146 ses fighting, they put down ears, when «turning ROUND to kick» kicking they do the same. although it
577n050 ory of Plants, must be association,-- a certain ROUND of actions take place every day, & closing of t
592n101 allow it is an instinct.» I suspect the endless ROUND of doubts & scepticisms might be solved by cons
593n105 ing disagreeable being pursued.-- A dog turning ROUND & round is some old instinct <perverted> handed
593n105 greeable being pursued.-- A dog turning round & ROUND is some old instinct <perverted> handed down &
593n111 o bed of flags I was astonished & having looked ROUND saw at considerable distance a very large hawk,
596nIBC convulsions A carrier pidgeon carried & turned ROUND & round in fainting state would it then know it
596nIBC ions A carrier pidgeon carried & turned round & ROUND in fainting state would it then know its direct
617o40v 4) the source from which it arises. But coming ROUND to the <subjective> aspect of action as known b
026r020 t sufficient distinction is given to angular & ROUNDED.-- Fox Philosoph. Transactions on metallic vei
116a100 ove» 30 ft or so above bed of river. formed of ROUNDED pebbles-- it is clear gold occurs in submarine
146g013 buttresses of Alluvium or rather mass of well ROUNDED pebbles in yellowish argillaceous or sandy soi
148g022 all the measure before There some of the half ROUNDED gravel nearly as high as highest measurement b
148g026 «sea», gravel, current cleavage, & pretty well ROUNDED stones, mixed with some quite irregular very l
149g030 ding as in Andes, composed of sand & perfectly ROUNDED stones-- lake required to deposit this Remembe
149g032 ean side of Meal-- Derry there were perfectly ROUNDED «base» pebbles of quartz & other rocks not app
153g056 one actually resting on them on summit of hill ROUNDED, site N N W of Ben Erin {P} Shelf of Glen Guoy
156g068 e shelves soil is <the> usually slaty Point of ROUNDED not scooped rock on <bend> of 3(a) Cannot <see
158g079 iss <unite> <occurs> abundantly with perfectly ROUNDED pebbles of granite & forming <sloping> buttres
158g079 inuation of upper terrace -- on right hand {P} ROUNDED waterworn buttresses of granite obscure terrac
162g104 te curved layer of fine sand & small angular-- ROUNDED pebbles-- dip sideward. & inwards-- deposited
383d160 eather instead of coming to point (P) are more ROUNDED. (P) & much broader., & <more ro> <three, I be
321c270 fe on himself Hume's do, with correspond. with ROUSSEAU Miss M«artineau» How to observe Mayo Philoso
244c022 p. 122. Mus decumanus, at Caroline Isld, & a ROUSSETTE p. 136. Isle of France.-- the Tenrecs from Ma
050r091 does he mean in contradistinction to sand?? B. ROUSSIN states that generally in North part of Brazil.
026r022 ear T. del Fuego.-- Is there account of Baron ROUSSIN'S voyage.-- In Europe proofs of many oscillatio
472s02r cis in many flowers, on one of which pollen was ROUTED. wh. was not case, on several flowers I examin
244c021 "red of brick" hair, very stiff, p. 120-- Coati ROUX common. near Concepcion. some tatous!!! p. 120.
482z011 nous Sertularia Lesson Zoolog. Coq: p. 120 Coati ROUX. Tatous & perhaps Yagourundi near Concepcion!!!
496g006 ceptacle (8) Make Duck eat Spawn, eggs of snail, ROW of fish & kill them in hour or two «My Father ma
469t151 rrow fat, Early frame, Groom's Dwarf. planted in ROWS «close to each other» & seeds gathered «all» ca
469t151 ossomed «& dwarf-fan Bean» bean, were planted in ROWS, & seeds gathered same year came up true «in 18
500q10a Year «before last» beans & peas were planted in ROWS adjoining & seeds gathered there were planted «
214b173 present analogies to shells of living seas.?-- ROXBURGH. list of plants in Beetsons St. Helena. -- Ga
096a039 elevation to amount of 30 ft.-- Mr Bollaert (at ROY. Institut) talks of quantities of shells at Iqui
120a107 fluid undulations-- the equal movements of Glen ROY road. (¿ metamorphic action at the bottom of the
120a110 bury Craigs well worthy of attention-- rear Glen ROY Notebook-- & scraps on Salsisbury Craigs. Kept a
121a112 ing-- Consider profoundly How came it. that Glen ROY district could have been elevated without fissur
128a129 nt water to counterbalance How strongly the Glen ROY case shows that the figure of the world has just
130a133 fresh Water springs off coast of Persia In Glen ROY paper I show crust yield easily. & if easily mus
134a141 s of Pen. Tres Montes.-- as by subsidence ‖ Fitz ROY refers to ‖ & Rocks p. 375. on the soundings on
145g FC really good account of ice.-- C. Darwin A. Glen ROY Generally received opinion that male impresses o
148g029 oe-- the erosion may often be due to rivers-- By ROY Bridge, a tongue of flat land, with terraces of
149g031 slope of A & B regular & even towards {P} Spean ROY double terrace river «& to West of Spean» diffic
150g035 off would form plains such as those near Bridge ROY (& other cases) but then if gradually drained, w
152g048 with greater cut through, not applicable to Glen ROY Lake. must have remained very long at 4th shelf
152g049 ich they cut near Loch Tring-- Tuesday Bridge of ROY Level of «bed of» River 30.221/65 degree/ Temp o
153g053 ition of 4th shelf. some way below House of Glen ROY, seems to be which higher up on is corroded {P}
155g063 as «as I believe in side ravine above houses of ROY» Maccullochs supernumerary shelf I doubt, much a
157g077 as <cut> «where two branches unite in upper Glen ROY» very little back from line 2d; little action si
158g080 dge or a modified Granite ridge» at head of Glen ROY on same side where two rivers unite in Upper Gle
158g080 n same side where two rivers unite in Upper Glen ROY great plain about 60 ft beneath shelf peat on pe
158g082 esses «occur» high up on Shelf 2d «in Upper Glen ROY» In this upper part «about junction of Upper & f
158g083 ses on each side «very little way» in Upper Glen ROY» at pass {P} River Gorge 4th Sh side of valley Gr
159g087 er Slope of valley much more gentle than in Glen ROY, & partly shut in No Granite blocks in higher pa
160g093 place, but narrow shelves just like road of Glen ROY-- appears to lip with moss On this terrace «stat
162g102 ting. enters by old tower called Glengarry (Nead ROY told me) it is impossible to see my new shelf, f
165g BC d? Shows that <nervous> brain makes thought Glen ROY B C. Darwin All useful pages cut out Dec. 7th. /
345d043 calf «in both cases» is killed. Notes from Glen ROY Note Book-- Why is not Tetrao Scoticus. an amer
541m090 a on coves when waters had fallen, as in my Glen ROY paper.-- this greatest mental effort, of which I
094a036 ern. Soc. Vol. 2. p. 35 Sir J Hall Trans. Phils ROYAL Ed. Vol 7 Dr Buckland Reliquiae Diluvianae p. 2
310c223 re) and animals that man has different origin. «ROYAL Institution» Dr Royle seems to think Botanical
322c270 ys on Natural History. Octob 2d Transactions of ROYAL Irish Academy. ----do Lavater's Physiognomy ---
338d024 va Zenbla -- in review of Baers work Edinburgh. ROYAL. Transact.-- p. 297. Vol 9. Dr. Ferguson seems
385d164 that ovarium of women (Paper in Vol I of Irish ROYAL Academy) have contained perfect teeth & hair--
431e122 end to vary. March 20th. Phillips in Lecture in ROYAL Institution says shells become less in number.
450e174 -- p. 1. Elephant. Rhinoceros Leopard (but not ROYAL Tiger), &c are found but only in one part the n
478z004 uses on purpose Mr J. Murray has given paper to ROYAL Soc on glow worm. luminous property-- Curious a
207b151 neaster & Mimosa called Botany Bay Willow V. Dr ROYLE introductory remarks to Himalaya Mountains-- Bo
235b272 trash Lyell 1024 Flemings Philosophy of Zoolog ROYLE on Himalaya Plants-- Would it not be possible t
235b273 certain countries-- so on with families.-- Ask ROYLE about Indian Cattle with humps.-- ¿To be solved
310c223 an has different origin. «Royal Institution» Dr ROYLE seems to think Botanical Provinces will turn ou
324c268 Vincent. Vol III p. 164. on unfixed forms. Dr. ROYLE on Himalayan. types. Smellie. Philosophy of Zoo

489qIFC March, 1842) Questions & Experiments {T} Gowen, ROYLE, & Horsfield Sykes p. 12 Maer. p. 13 Question &
503q012 ny recent information about pollen of Subularia ROYLE & Horsfield (35) Talk about races of Banana & y
503q012 of Banana & yet seedless-- no light Henslow or ROYLE, latter says seedless-- Also about Sugar-Cane E
503q012 Edwards says does not seed-- «Bruce says does» ROYLE In Royle's productive Resources Book no informa
231b242 dian islds-- Monkeys different not travellers?? ROYLES case of Himalayan, plants ¿migratory birds, he
503q012 ays does not seed-- «Bruce says does» Royle In ROYLE'S productive Resources Book no information & Hop
107a079 ut cause most difficult (better conductor) Fitz ROY'S Case of S. Maria & Tubul applicable to Andes &
112a088 ings in Mouth of S. Cruz in connection with Fitz ROYS fact of elevated block of stone.-- & Caldcleugh
070r155 led boulders. from the Cordovise range. Signor ROZALES tells me at seven oclock Novem <5th> Concepcio
467t104 , that stigma covered with pollen was pressed & RUBBED along whole breast-- {b} pressing either one o
567n014 cephalus when fondling the keeper., clasping «& RUBBED» his arm. & show signs of affecting something
080r178 & see if same effects, as with man Does Indian RUBBER & black lead unite chemically like grease & me
537m075 t." as child is cured of sucking his finger by RUBBING them with alum, so more slowly does animal lea
557m149 catch tricks.-- so are people in passion by F. RUBBING hands.-- stamping. grinding teeth.-- in shame
147g020 if valley had been filled with sloping bed of RUBBISH Friday Highest part of road between Inverorum
148g027 nes, mixed with some quite irregular very like RUBBISH at head of Loch Dochart <Nea> Above Spean Brid
163g109 annel. Forchammers (Lyells Denmark) Shrewsbury RUBBISH.-- Speculate on origin pebbles brought by diff
634i54v nges have separate movements in the Holocentrus RUBER (a fish) Man has abortive muscles to his ears.-
362d098 e passion of the doe to the victorious stag. who RUBS the skin off horns to fight-- is analogous to le
059r123 le pressure. & perhaps cooled suddenly.-- As the RUDE symmetry of the globe shows powers have acted f
264c079 world now being fit, for such an animal.--man, (RUDE, uncivilized man) might not have lived when cer
535m069 of a superior being; whose natures can only be RUDELY traced out. When one sees this, one suspects t
609o29v but refers (I believe) to present day & not to RUDER state of Society.-- Civilization is now alterin
341d030 -- nocturnal crawling bird.-- Wings reduced to RUDIMENT.-- clavicle scapula &c strongly developed to
420e089 whence both may have descended.-- Jan. 6th The RUDIMENT of a tail, shows man was originally quadru<ma
211b162 Y GOOD. Apteryx. a good instance probably of RUDIMENTARY bones.-- As Waterhouse remarked Mere length
307c215 als, Mammae in Men, capable of giving milk)» RUDIMENTARY wings. so nature can produce in sex, what sh
434e130 most all Dioecious & Monooecious plants have RUDIMENTARY abortive organs, even more so than Polygamia
305c209 ving tone of voice like S. American.-- Have not RUFFS & Reeves a remarkably varying plumage for wild
265c088 sm. for it not so not aberrant.-- Tenioptera RUFIVENTRIS, is instance of bird belonging to family wit
117a101 ace, otherwise the world would daily be scene of RUIN in late Natical Magazine (before June 1838) tha
182b046 ype. Hence probably only four, is not this Fries RULE-- What subject has Mr Newman the (7) Man studie
205b144 here is no awk in Southern hemisphere. does this RULE apply? A Treatise on Form of Animals by Mr Clin
284c149 ere does such occur?» now some such «characters» RULE are used by Naturalists in their test of value
284c149 ts in their test of value of character-- Macleys RULE is converse, <when> value of character depends
340d026 ds fertile inter se--No directly against Eyton's RULE. ¿Are the hybrids similar inter se.-- [not loca
350d053 up genus & other subgenus,,--> Propagation, best RULE for genera, & so mount upwards.-- «judged by an
356d072 hout God is supposed to create & destroy without RULE-- But what does he in this world without rule?
357d072 ut rule-- But what does he in this world without RULE? The destruction of the great Mammals over whol
357d072 he great Mammals over whole world shows there is RULE.-- S. America & Australia appear to have suffer
403e024 ies which is absurd.-- their only escape is that RULE applies to wild animals only. from which plain
549m120 ear & perhaps partly their fault.-- Whether this RULE of happiness agrees with that of New Testament
552m132 od & consequently give pleasure, & not as Paleys RULE is those that on long run will do good.-- Alter
552m132 er will in all cases to have & origin as well as RULE will be given.-- Descent of Man Moral Sense Mit
609o030 insist?? Two classes of moralists: one says our RULE of life is what will produce the greatest happi
609o030 <is> to revenge). In judging of <our ha> of the RULE of happiness we must look far forward «& to the
609o30v ther than we can look forward: hence our <[...]> RULE may sometimes be hard to tell) + + Society coul
610o30v hows does not tend to greatest good.-- Therefore RULE of happiness is to certain degree <of> right.--
620o045 ffects not lasting. By association one gains the RULE, that the passions & appetites should «almost»
624o051 reason.-- Hence as Eugenius says, slow growth of RULE of right.-- [LHC] *for it strives to give condu
495q05a ivocal Generation Charlworth p. 377. Have paper RULED in squares to facilitate investigation.-- Capit
303c204 at end, seeing that Tertiary geology has obeyed RULES of modern causes. & considering over the viccis
346d048 - a Mr W: Hall remarked that it was against all RULES their preserving character & breeding in & in--
525m023 me, or picking up a stone, though only acquired RULES by art.-- like the law of honour.-- they feel p
544m102 of dreams no surprise, at the violation of all <RULES> relations of time, <identity,> place & persona
549m121 . -- I doubt whether the last be right. The two RULES come very near each other.-> The rules to mortif
549m122 . The two rules come very near each other.-> The RULES to mortify yourself do not tend to this-- thoug
564n004 tinct of conscience, feeling in his heart those RULES, which he wills to give his child.-- Octob 3d.
629o55v D & if passions makes one break these artifical RULES, get remorse-- ((hence desires do not intervene
566n010 ly explained.-- on the wish to support a wife a RULING motive.-- Book IV, Chapt I on passions of mank
458t001 1 7. p. 658-- Falconer on Sub. Him. fossils-- RUMINANTS. & Tortoises gigantic-- hyaena-- bear & rumin
458t001 nants. & Tortoises gigantic-- hyaena-- bear & RUMINANTS all of larger size.-- the law of large size e
464tf6v bex of Alp Pyrenees &c-- (see Blyth's work on RUMINANTS,-- these species must have migrated to these
273c110 & white» bars on wings of trogons are lengthened RUMP feathers.--; <then> & one species has small ban
513q020 bably a variety, not specific character= Cross RUMPLESS fowls & Dorking fowls,-- or tailless dogs & f
055r106 &c Vol II. Chapt VIII. p. 97 at Potosi the veins RUN from North <inclining> to South. inclining a lit
058r116 «?» Edition states that the same earthquake has RUN from Chili to Quito a distance of more than 500
063r133 ogs. Cats. Horses. Cattle. Goat. Asses. have all RUN wild & bred. no doubt with perfect success.--sho
192b083 et instances of a variety of fruit tree or plant RUN wild in foreign country.-- When one sees nipple
212b165 projecting.-- would pull the garden bell, & then RUN into Kennel to watch who would come to the door-
228b229 For instance take Voluta & Conus (??) which now RUN together, were not both genera formerly abundant
228b232 not look forward= if we choose to let conjecture RUN wild then <our> animals our fellow brethren in p
230b240 possible to get evidence of two races of plants RUN wild.-- (for we know that such can take place wi
235b275 nge The GRAND QUESTION Are there races of plants RUN wild or nearly so, which <breed> do not intermix
249c035 are cases <sp> like those of Primrose & Cowslip RUN wild. The two species of Clenomys. case of repla
275c120 e no scent!) Mr Wynne) at end of chase would not RUN up hill-- he took thorough bred <greyhound>. bul
313c235 ened by association (case of Elephant, which had RUN wild in India. in Heber?) is analogous to dorman
424e103 a half reclaimed animal.-- The dogs, which have RUN wild have, have done so in hot countries.-- CD[C
447e163 t-- if form exceedingly difficult to vary.-- the RUN of chances, would prevent it varying. A plant <p
492q002 -- Heartease. 6. -- Do not species of wild Roses RUN into each other very much.-- Has not some one wr
527m033 singing intermediate, who has not had his blood RUN cold by singing).-- Granny says she never builds
552m132 sure, & not as Paleys rule is those that on long RUN will do good.-- alter will in all cases to have
555m143 o life, & then made many jokes. about not having RUN away &c having faced death like a hero, & then I
564n005 ll moving object like fly.-- young partridge can RUN even with its shell on back.-- To study Metaphys
618o41v all that can be said that thought & organization RUN in a parallel series: if blueness & weight alway
621o048 nts think will be good for the child on the long RUN, & for themselves & others, (as the parents are
639j28r & Mole cricket & rodents (?) p. 251. all animals RUN by hind legs-- Kangaroo. only a caricature; Peng
547m112 I really was going to Shrewsbury, whether I had RUNG for Covington. whether he had come & opened box
547m113 take (how often one cannot tell whether one has RUNG the bell., when one recollects circumstances we
105a073 rse small erosive power. therefore tendency of RUNNING water to deepen not to widen valley.-- Why is
150g035 ne on N. side of Spean most clear & upper line RUNNING up great bight just as Dick shows NB. Lake gra
261c070 ization! The little turtle, without its parent RUNNING to the water, is a good instance of connate in
272c107 e are hemipterous insects, having spiny legs & RUNNING quick & generl appearance of blattae other Hem
410e075 might be made «almost» without any relation to RUNNING hares.-- as in Italian Greyhound not so specie
525m023 is wrong.-- as eating meat., doing their dirt, RUNNING home.-- in these cases their actions do not lo
531m052 to avoid any danger-- Fear, shamming death, or RUNNING away. accompanied with want of muscular exerti
532m057 g of muscles, are not these effects of violent RUNNING away, & must not <this> «running away» have be
532m057 ts of violent running away, & must not <this> «RUNNING away» have been usual effects of fear.-- the s
547m113 ught necessary heirs of every action, & always RUNNING on in mind, being absent. one could not compar
563n001 pression.-- Octob. 3d. Dog obeying instinct of RUNNING hare is stopped by fleas, also by greater temp
572n033 ture to myself reading French book quickly, & <RUNNING> «running» over imaginary words: it appears as
572n033 self reading French book quickly, & <running» «RUNNING» over imaginary words: it appears as if the mi
584n071 enses.-- knowledge of location ducks & turtles RUNNING to water.-- young crocodile snapping-- p. 28.
052r098 E coast of Madagascar. where a --40 line <shows> RUNS at equal distance?) 1st cases. -- The terraces
151g042 t have been waterworn after 3d lake.-- 4th shelf RUNS up some way on great sloping plain of alluvium
158g083 pper shelfs ground strewed with pebbles Shelf 3d RUNS up with buttresses on each side <very little wa
160g093 of mile on <one> S side of «this» Isthmus (which RUNS E & W) broad terrace <of pebbles? & Alluvium» w
301c199 habitual instincts precede structure.--duckling RUNS to water. before it is conscious of web. feet.-

542

557m149 d laughter.-- A dog who goes home from shooting. RUNS away. is not afraid the whole way. but ashamed
599o007 ests" Very good!. I grant that the thrill, which RUNS throug every fibre, when one behold the last ra
639j28v sneering-theory.-- p. 263. This kind of doctrine RUNS through Macculloch, the bills of the Grallae <a
277c126 del Fuego differ from (C) Chiloe (E) Chile.. RUPESTRIS -- good species it is reverting to old plan,
242c017 lle. Zoolog. p 19.. Tapir, «des» couroucous et RUPICOLE vert instances of American forms in East. Ind
252c042 ne extant at Leeward Isles. p. 388 Reference to RÜPPEL. travels (what language?) Hyena «venatica» <of
483z012 ypt is Mus Cahirimus. of Geol.-- reference from RÜPPEL. travels All Owens papers on intestinal worms m
506q15v lower? does he know Botanist who does-- What is RUPPIA Bennett says in same state. of flower 45. Char
434e129 lowers will sometimes produce a few seeds.-- -- RUSCUS aculeatus. a dioecious plant, in which the Mal
105a070 able to Cleavage. C. Prevost.-- In Cordillera. a RUSH of water will account for filling up of valleys
105a070 ion of blocks, that there has been no tumultuous RUSH.-- besides general improbability. stratificatio
346d048 ol. 2. 1839.-- are bad breeders & subject to the RUSH as all animals which breed, in & in are-- colou
436e137 what cause?? Seeing the beautiful seed of a Bull RUSH I thought, surely no "fortuitous" growth could
507q15v with drawing of Seed. Anemone with, tuft-- Bull RUSH-- Dandelion-- Sycamore. & seeds with «mere» bor
639j28v ters is born with long legs» & in the Malthusian RUSH for life, only two of them live to breed, if ci
578n055 owing «it», suddenly came across her, the blood RUSHED to her face,"-- One blush if one thinks that a
047r082 t. -- Does the sea fall on banks as a Bore wave RUSHES up? (NB. Earthquake wave is an oscillation, bo
272c108 same peculiar structure & habits of clinging to RUSHES similar.-- The question which I more immediate
557m148 from that.-- like «slight» passion from blood RUSHING in face, with less action of the heart.-- tend
496q006 . Nitrate of Soda-- Salt. Gypsum. Magnesium Iron RUST Carb. of Ammonia.-- Horse Urine &c &c on associ
528m039 - connection with poetry, abundance, fertility, RUSTIC life, virtuous happiness.-- recall scraps of p
506q15v rth. vol II. p. 670-- oats cut down turning into RYE.-- 46). Book describing amount of Horticultural
197b111 passes from worm to man; «highest» as typical of <SA>. changes, which can be traced in same organ in
510q017 m West, like as between Australia & S. America? SABINE says North of Siberia, no sea-current, iceberg
317c249 at Maer. Tuckeys voyage-- p. 36 "Cercopithecus SABOEUS" said to be monkey of St Jago C. de Verd; same
461t025 & are provided with fins, & hence do not require SAC.-- but the male in these hatch young-- are there
022r008 a shark 11 ft long. "Its maw was like a leathern SACK, very thick & so tough that a sharp knife could
461t025 not some. Marsup. Mammalia, which <do> have not SACK,-- Most curious facts & this paper deserves fre
638j28r , kinds of foot. power of closing nostril, foot, SACK. power of endurance &c &c Camels? all good case
451e176 his grandfather, who was also an albino is held SACRED by the credulous natives, & vow made at it. Bo
620o045 assions & appetites should «almost» always be SACRIFICED to the instincts.-- -- One does not feel it
603o012 assy Vol II p. 405.-- Speculates on origin of SACRIFICES. «common to many races»-- thinks action towa
603o012 ty.-- & as king might like cruel pleasure, so SACRIFICES cruel.-- Something wrong here.-- Origin is
603o012 e, S. American. Polynesians Jews, African all SACRIFICES. How completely men must have personified th
406e035 ss of sheep & Mouflon of Corsica. <would not>, SADLY against Yarrell's law.-- not so much against my
553m132 ake kangaroos than emu, although young dogs get SADLY torn in conflicts with the former. But it is on
602o011 . about actions & decisions bein the result of SAGACITY, or intuition. when individual cannot give re
492q003 ed will offspring be uniform.-- Mr Ford Has M. SAGERET WRITTEN on crossing of Cabbages, quoted by (as
492q003 Decandoelle in V. Vol of Hort. Transacts & M. SAGERET is referred to with doubt by Herbert Do forest
078r174 79. [misnumbering, no page 173] Under name of SAGITTA Triptera D'Orbigny has figured animal with set
479z005 ariae p 451.-- many molluscs Under the name of SAGITTA Triptera M. D'Orbigny has described my animal
481z010 rifling Also Berre «p. 8» (I think Planariae) SAGITTELLA, or Fleche «p. 8» my little animal with horn
133a139 therefore that temperature of earth beneath <of SAHARA de> a dry desert, would be very high.-- M. Par
021rIFC bisson both Volumes, and Molina 1st Vol & Lyell SAILED, 27th <Friday gale 29th> Friday Thursday 29th
463t055 perhaps greatest Could anyone. have foreseen, SAILING, climbing & mud-walking fish? difficult-- yet
641j29v says no other bird could catch mouse by night» SAILING lizards. squirrels & Opossums «& fish»: flying
530m045 ks. & that peculiarities of form in trades (,as SAILOR tailor blacksmiths?) are likewise heredetary,
554m139 saw» Jenny untying a very difficult knot-- the SAILOR on board the ship could not puzzle her-- with
613o036 ls. [LHC] «3»" Eyton told me that his retriever SAILOR he has seen push a hare through the bar of a g
089a018 tersticies yet emptty.-- In all the mountains of SAINT Marc et des Gonaîves it is difficult to find st
183b052 > silver gold & common pheasants & fowls..-- "On SAIT que le "métis" du loup et du chien, que celui d
537m074 «to improve his organization» for his children's SAKE & for the effect of his example on others.⎸ It
041r064 ng the wells.-- the effect of Salt water of the SALADO.--Mem. in Owens Africa it is mentioned that th
295c183 Fernando Po-- no new forms only species!! No SALAMANDERS (D'orbigny Rapport. p. 11) in S. America so
458t001 hes in length--! alludes to ancient gigantic SALAMANDERS-- Every order (except whales) have great pro
486z019 Martens says only one Reptile in Kamtchatka (SALAMANDRA aquatica). Compare with T del Fuego Compare
085aIFC on Volcanic islands & Coral Formation Lyell's SALBAND p. 86 Shells near Woollich p. 112 Speculate on
110a086 fissures; base in part. block not crystallized SALBAND like basalt. full of circular cryst of glassy
110a086 als from fragment disseminated on that side of SALBAND. gradually becoming finer grained & more compa
110a086 large cryst of Hornblende blending into base-- SALBAND might have oozed out of cleavage plates: the c
028r031 ter at Iquiqui. not from rain, because alluvium SALINE; Mem: on coast of Northern Chili as springs be
086a005 dentia In Plata Mastodon Toxodon Is the general SALINE tendency of America connected with its elevati
134a143 lt mines of Punjab p. 149. on the «salt mines» «SALINE deposits» of India p. 503. On Indian Saline De
134a143 s> «saline deposits» of India p. 503. On Indian SALINE Deposits. Vol II. p. 23. p. 77 do Vol III p. 3
120a110 .-- The veins of segregation in Greenstone of SALISBURY Craigs well worthy of attention-- rear Glen R
123a116 dike.-- Mem. however. veins of segregation in SALISBURY Craigs Letter from M Angelis. B. Ayres. 3d. M
145g002 t either races soon made or crosses difficult SALISBURY Craigs The Highland shepherds dogs coloured l
145g003 rovincial breeds. [3] Veins of Segregation in SALISBURY Craigs [4] Salisbury Craigs V. Specimens-- {P
145g004 Veins of Segregation in Salisbury Craigs [4] SALISBURY Craigs V. Specimens-- {P} Veins, amygdaloidal
145g006 t greatest dip of sandstone in upper part «of SALISBURY Craigs» 25 degree perhaps most common-- Will
573n038 - in case spittle, effect of thought is to make SALIVA flow, & therefore thinking of subject, even wh
574n041 coa nut.-- November 27th.-- Sexual desire makes SALIVA to flow «yes, certainly»-- curious association
574n041 xually loves is probably connected with flow of SALIVA, & hence with action of mouth & jaws.-- Lasciv
574n041 to certain degree sexual.-- The association of SALIVA, is probably due to our distant ancestors havi
511q018 ar> knows whether Shaws hybrids between Trout & SALMON were fertile & whether homogeneous {Abour Germ
582n067 t) to move its legs so, as much as in the young SALMON to go towards the sea. or down the stream; whi
295c183 iar than its ornithology X p. 12 do. excepting SALMONS L' Institut. Sorex from Mauritius. p. 112. & p
306c212 e, habits easily gained in child hood.-- Young SALMONS. first a species which lived in estuaries its
096a040 -- Von Buch. Can. Ile p. 406. List of Volcanos SALOMON Isld,-- New Britain-- &c &c In Ascension for c
479z005 good on Falklands birds Discussion of Firola,-- SALPA Anatifs without shells.! p 442.-- Planariae p 4
120a110 ention-- rear Glen Roy Notebook-- & scraps on SALSISBURY Craigs. Kept amongst <old> papers read befor
041r064 the deer approaching the wells.-- the effect of SALT water of the Salado.--Mem. in Owens Africa it i
046r078 e first production of Trachyte. look at Sulphur. SALT. lime, are spread over «whole» surface; how com
056r111 nic action separates some sulphur (perhaps lime) SALT. & metallic ores.--which mingling & separating
085a001 erful outlyer.-- Linn: Transact. Vol. 8. p. 288. SALT deposited on windows of houses. & trees all inj
085a001 n Eastern side, far inland.-- even 70 miles from SALT water. Mr. Arrowsmith tells me, that Himalayas
086a006 h its elevation. vapour from below-- Malte Brun «SALT Lakes» Siberia must be read as well as Pallas b
087a010 Make note about N. American bone not probably in SALT marshes Efflorescence nothing -- Study account.
088a016 248. L. Institut 1837.-- Helms remark on common SALT being found on low hills East of Cordillera ver
093a034 a Rhinoceros was found <emb> in the mud, of the SALT river.-- in reference to fossil guanaco of P. S
097a044 at the upper strata alone at Guantajaya contains SALT see Geolog. proceedings Lake let out by steps i
100a055 ated matter-- Phillips Lardner p. 197. refers to SALT as being produced by local heat, Ask Capt. Beau
104a068 fluids varies with temperature ¿ with pressure? SALT on surface of plains due to whole moisture bein
104a068 fore capillary attraction would bring water with SALT to surface Lyell remarked to me that Kylow (?)
114a093 stone of C. of Good Hope & Australia/ and mud of SALT-lakes of Rio Negro--Mr Bowerbank-- Dr. A. Smith
120a109 therefore I doubt the thermometer. Is not common SALT above more soluble in <hot> cold than hot water with
120a109 under high pressure.--» respect to formation of SALT.?.--??? Footsteps in New Red Sandstone. look as
123a115 by Humboldt on Bogota. Cordillera,-- nothing.-- SALT & coal near Bogota; p 270.-- SPLENDID PAPER on
128a128 . Journal vol II. p 89. at Madras. surrounded by SALT water. purest fresh water must be sought for be
134a143 Journal of Asiatic Society Vol I. {T} p. 145. on SALT mines of Punjab p. 149. on the «salt mines» «sa
134a143 p. 145. on salt mines of Punjab p. 149. on the <SALT mines» «saline deposits» of India p. 503. On In
184b054 1 1837. p. 216 Deshayes on change in shells from SALT & F. Water-- on what is species. very good Has
200b125 ious experiment to know whether soaking seeds in SALT water &c has any tendency to form varieties? Ed
221b201 p. 267. Dela Beche. Geolog. Researches. facts of SALT-water shells living in absolutely fresh water.--
232b248 if heat gained? Experimentise on land shells in SALT water & lizards do.-- Ask Eyton to procure me s
284c148 to these isld.-- ¿Brown can «not» bear the least SALT water.-- Nuts prodigiously heavy (where trees o
342d034 remarks he has somewhere ment conjecture that all SALT-water fish were once salt water (as they almost
342d034 et conjecture that all salt-water fish were once SALT water (as they almost must have been on elevati

379d151 y of the Caspian. Fresh Water Fish!! ¿adapted to SALT water?-- peculiar species, crabs & molluscs few
419e085 .-- fishes fresh water kinds. (yet living in the SALT?.)-- very few animals of any kind-- Fauna, must
485z017 e of mine from T. del Fuego p. 141. How comes it SALT water so soon putrifies?? p. 319. on Hydra-- po
495q05a seeds.-- 11. Soak all kinds of seeds for week in SALT. artificial water.-- 12. Plant two races of Cab
496q006 ot. or Ranunculus.= (11) Try.. Nitrate of Soda-- SALT. Gypsum. Magnesium Iron Rust Carb. of Ammonia.-
548m117 n when one begins eating one perceives butter or SALT is not there.-- the reality does not resemble t
070r155 met   at S. W. P.) the Cordillera extend to near SALTA. & not far from Tucama[n]. & at Chuquisaca. hal
071r157 ne Parish informs me that town near Tucuman and SALTA. towards the Vermejo was utterly overthrown by
306c212 n estuaries its taste. taught it to go to <sea> SALTER water (& its necessities teach it taste, but t
074r165 nsported by complicated chemical law & steam of SALTS, quite curious case of oxided Iron by Mittersch
093a032 Vendarques. Mem sublimation of sulphur to form SALTS of America. -- The number of minute turbos in r
516q24v xperiment Cover patch of ground, with different SALTS & poisons & see in what order plants would reap
063r130 species does change into another it must be per SALTUM--or species may perish. = This <inosculation>
236b278 blank] In production of varieties is it not per SALTUM.-- Isld bordering continents same type collect
351d056 s vestiges of what she does-- does not move per SALTUM-- yet does nothing in vain!! Foetus of man und
286c154 variable.-- Animals have voice, so has man. Not SALTUS. but hiatus animals expression of countenance.
287c157 e steps that lead up to it" p. 20 ‖ +++hiatus or SALTUS not syn.-- Linn: Transact Vol XIV.--‖. p. 24.
287c157 ne penguin, one tortoise shows hiatus-- but not SALTUS-- when Linnaeus put whale between cow & hawk a
287c157 naeus put whale between cow & hawk a frolicsome SALTUS. «p. 19»‖ Macleay seems to limit Lamarck defin
300c198 ewed as replacing each other it is hiatus & not SALTUS.-- The greater individuality of mind in man, i
558m151 onnexion with mind, (to show hiatus in mind not SALTUS between man & Brutes) no one can doubt this co
402e019 iopitheous.-- Malte Brun. Vol <I> II p.,133: at SAMAR SE of Luçon, many monkeys, buffaloes &c &c-- Ma
450e175 es that of Celebes is somewhat larger than the SAMBANA, Java & Sumatra breeds, (.Hence it appears the
021rIFC stant: This view supposes the simplest infusoria SAME since commencement of world.-- [not located] La
021r005 t.-- singular structure of nodule, constitution «SAME as» of slate same.--longer axis in line of Clea
021r005 cture of nodule, constitution «same as» of slate SAME.--longer axis in line of Cleavage. laminae fold
024r016 C. Frio [23 degree S.] 7 60 {T} Soundings about SAME as last to N. of C. Frio Except at Abrolhos. [1
028r031 n the whole Galapagos Arch; because no sections> SAME cause as no colour Sir J. Herschels idea of esc
031r038 - Auvergne. very little Pumice, though Trachyte. SAME fact in Galapagos. Daubeny P 24 V. back of page
032r041 cal)» hence varieties of substances ejected from SAME point. & changes. «(changes in variation?)» as
035r046 eare introduct to Geolog--"Between the height of SAME beds, deposited in different basins; little   or
035r046 xist> be made out, but in those belonging to the SAME district there seems. I think, little ground fo
035r047 fact is astonishing consult book itself. P. xx: SAME fact is indeed shewn? by the parallel bands of
035r047 ture & dependent: & then how wonderful level «of SAME beds» should have been kept; it shows that thro
040r063 Brazil feruginous sandy ones have undergone the SAME process.-- Neither lakes or Avalanches (Glacier
043r070 «more than» one orifice <...> does not occur at SAME time: this is contrasted to contemporaneous act
043r071 lar causes go together. add. <">" <from> "in the SAME line" to "from the epoch of Ammonite to the pre
047r081 artake in absolute movement» Moreover wave «with SAME general character» reaches far beyond coast, wh
047r081 ered as an oscillation, from violence. Is it not SAME as swell travelling across Pacifick.--excepting
048r085 in which the Rhinoceros lives in S. Africa: the SAME caution is applicable to the Siberia case We mu
048r086 oceros was killed. -- In Patagonia, are all beds SAME age? is white substance triturated Porphyritic
053r104 dt under name of Rothe-todte-liegende is perhaps SAME with that of Pernambuco? Quote Miers about shel
058r116 a. p. 125. of French «?» Edition states that the SAME earthquake has run from Chili to Quito a distan
058r117 ing on we shall see a cause for Volcanos part of SAME phenomena lasting so long.-- The great movement
060r124 tly pushing the sea (which of course must retain SAME level) to a greater distance".--Afterwards spea
062r130 not extinguished by change of circumstances: The SAME kind of relation that common ostrich bears to (
063r133 pon solely adaptation of animals.--extinction in SAME manner may not depend.--There is no more wonder
064r134 oysters quite above reach of tides.--thinks them SAME as recent species. -- May I not generalize the
066r142 lower of third Silurian division--Together with SAME general character of fossils deception complete
070r153 two of the most closely allied species occur in SAME country? In botany instances diametrically oppo
074r165 ns follow mountain chain. there after NW «W».-- SAME chemical laws as in concretions perhaps makes i
076r168 f Batopilas in New Biscay, "Nature, exhibits the SAME minerals <as> there, that are found in the vein
076r170 very rare.-- ancient freestone & breccia is the SAME with that on surface of plains of Amazon, no re
077r171 Clay slate, Porphyry North 52 W, & is nearly the SAME with that of the veta grande of Zacatecas, & ve
078r175 of the vein".-- at Zacatecas the veta grande has SAME direction as Guanax.--the other E & W.--veins r
078r176 both in hornblend & vitreous felspar".-- p. 215 SAME metal in Tasco vein in Mica Slate & overlying L
079r177 ements on, Vol I, P. 168. on coast of Guayaquil, SAME as Galapagos. no Hydrophobia at Quito. P 281. d
080r178 animal with recent dead body of other. & see if SAME effects, as with man Does Indian rubber & black
092a028 1837.-- Tertiary <bea> formation twenty species SAME as Paris. 1500 ft high Mr Bird in paper to Brit
094a036 pebbles of granite clearly effect of remodeling SAME manner. as bits of Patagonian boulders might be
095a038 lks of <hill> «cerro» of Diamante near stream of SAME name. with imperfect crater <--> near summit,--
098a047 refore as a consequence aggregated (I assume the SAME force which draws together two particles of Car
099a049 cal. Ascertain law of attraction of particles of SAME nature: then get mathematician to when two part
106a076 thern America ¿ effect of subsidence-- <Is there SAME.> Institute. 1838 p. 40 or Phil Mag. Dec 1837.
107a079 agonia-- On Lyells idea of whole centre of earth SAME heat, then change in form of fluid centre would
109a082 f England.-- Sea must always on actual beach act SAME way.-- a little further from beach action proba
109a082 must be continued. no currents & elevation have SAME effect, a tendency direct (or oblique) outwards
116a099 ??-- too great an abundance of matter would have SAME effect as too coarse. Read Kylau on Granite Edi
121a111 ecomposed muscles.. Smith of Jordanhill has seen SAME thing-- Consider profoundly How came it. that G
124a117 I Consider proved that Siberia must have been in SAME condition for long period Subsidence in Demarar
135a146 described formation of shore of Coromandel. just SAME as. at Bahia Blanca-- letter in drawer with imp
146g010 Blank] {P} Path East End near Holyrood Palace In SAME way at top the trap could be traced Grey in fro
148g025 -- I asked this question in many ways & received SAME answer Thought lambs most like MOTHER!-- the cr
155g062 2nd shelf very broad «& cut out, produced» from SAME «cause» as «great» spit <is> or plain <now> for
155g064 more «detritus» than they corrode or vice versa SAME inclination when serpentine might remove, what
155g065 o carefully preserved as to have thrown water in SAME «drainage» lines Mound of Gneiss though wonderf
156g067 Glen Collarig, & some «other parts» Boulders of SAME granite, all on these three shelves soil is <th
156g069 not proportionately large to those now formed in SAME spot by present torrents Maculloch wrong in say
158g080 a modified Granite ridge» at head of Glen Roy on SAME side where two rivers unite in Upper Glen Roy g
159g086 wn slope of obscure terraces (& conical hills on SAME) of «semi» waterworn & some partly well worn pe
172b006 es to like each other AEgyptian cats & dogs ibis SAME as formerly but separate a pair & place them on
173b010 > island near continents might have some species SAME as nearest land, which were late arrivals other
173b011 late arrivals others old ones, (of which none of SAME kind had in interval arrived) might have grown
174b013 to hybridity.-- genera being usually peculiar to SAME country, different genera different countries.
174b014 untries. Propagation explains why modern animals SAME type as extinct which is law almost proved.-- W
179b035 ps in the series of connection? «if stating from SAME epoch certainly» The absolute end of certain fo
180b036 ving as now To do this & to have many species in SAME genus (as is). REQUIRES extinction. Thus betwee
181b040 uccessors of his relatives shall now exist,-- In SAME manner, if we take «a man from.» any large fami
182b050 d says in sub-genera, they undoubtedly come from SAME countries.-- In mundine genera, the nearest spe
183b051 ontinue & thus two species be created) & live in SAME country. How is propagation of wolf & Dog. (bec
183b051 opagation of wolf & Dog. (because being believed SAME species) if they do not breed readily. point in
184b054 species. very good Has not Macculloch written on SAME changes in Fish Mem. Rabbit of Falklands descri
185b057 rnes.-- Again taking a subdivision of Heteromera SAME. thing occurs with regards to other tribes in t
185b057 hing occurs with regards to other tribes in that SAME family.-- (NB I see Waterhouse thinks Quinary o
186b059 Vol III. p. 379. Mammalian type of organization SAME from one period to another, preeminently Pachyd
187b063 fact Horse, Elephant & Mastodon dying out about SAME time in such different quarters.-- Will Mr Lyel
187b065 species.-- I should expect that Bear & Foxes &c SAME in N. America & Asia, but many species closely
187b065 ry separated since time of extinct quadrupeds:-- SAME argument applies to England.-- Mem. Shew Mice.-
188b068 ower. all would come up white, though planted in SAME soil with blue. Now this is same bearing with D
188b068 ough planted in same soil with blue. Now this is SAME bearing with Dr. Smith's fact of races of men t
189b070 tation, armadillos «&» Megatherium. each with SAME kind of coat.-- If we could tell, I do not doub
189b070 os). Little wings of Apteryx Dacelo & Kingfisher SAME colours Strong odour of negroes, a point of rea
189b073 nued.-- perished in America All animals <are> of SAME species are bound together just like buds of pl
191b080 an archipelago.-- West Indies = Opossum & Agouti SAME as on continent-- 3 Paradupasi in common to Van
195b080 ch spots. We know birds do arrive & seeds.-- The SAME remarks applicable to fossil animals same type,
195b099 -- The same remarks applicable to fossil animals SAME type, armadillo like covering created.-- passag
195b099 overing created.-- passage for vertebrae in neck SAME cause, such beautiful adaptations yet other ani
195b101 h planet to move in its particular destiny.-- In SAME manner God orders each animal created with cert

Page		(Key Word)	
197b111	typical of <sa>. changes, which can be traced in	SAME	organ in different animals in scale.-- In monst
198b114	though some may be) have not been created on the	SAME	plan. [Second resumé well worth studying] CD sa
199b116	eds being created on small spots of land, of the	SAME	type with the great continents we get a means o
199b116	out of the thousand of new insects all belong to	SAME	types already established. why out of the thous
201b129	ation of Analogy of Maclay &c. appears to me the	SAME,	as the irregularities in the degradation of st
203b136	s breed boars were wilder than parents. which is	SAME	as Indian Cattle.∴ tameness not heredetary?, ha
204b139	Eyton says Hybrid about half aways & result the	SAME	Indian cattle & common produced very fine Hybri
205b145	bs will be hornless.--" does this apply to where	SAME	animal breeds often with same female p. 28 "It
205b145	his apply to where same animal breeds often with	SAME	female p. 28 "It <is> wrong to enlarge a Native
209b157	lapagos =Radack Islds = ∴ Islands & Artic are in	SAME	relation. We find species few in proportion to
213b171	ir exactness,' but do not known varieties do the	SAME,	May you not breed, ten thousand grey hounds &
217b183	erwards went in heat.-- This is good instance of	SAME	fact in Mr Galtons case.-- It explains the loss
219b194	s genera near Cape) see if there are any species	SAME	as T. del Fuego & C. of Good Hope show possibil
220b197	. America Jaguar & Tiger & Europe, as to produce	SAME	one. ‖Although in plants, you cannot say that i
222b202	h> according to Beck has none recent, yet genera	SAME.--	Speculate on multiplication of species by tr
222b203	ent, this is only character., & yet we find this	SAME	tendency (only less strongly marked) between wh
222b205	perfection); but intermediate in character, the	SAME	reasoning will allow of decrease in character.
222b205	otest epochs.-- ¿ lizards of secondary period in	SAME	predicament. It is another question, whether wh
222b206	alian remains.-- The Father of all insects gives	SAME	argument as father of Mammalia; but have improv
223b210	ormed) but yet propagates varieties according to	SAME	law with animals?? Why are species not formed.
224b216	ambrian formations gone on creating animals with	SAME	general structure.-- miserable limited view.--
225b218	ceive man. as likely as fossils in old rocks for	SAME	purpose.!! Can the wishing of the Parent produc
226b220	frica. equally good.-- Small isld off New Guinea	SAME	fact see Coquille's Voyage.-- Galapagos mouse (
226b221	species are formed of those genera.-- & hence by	SAME	chance few representative species. this must ha
226b222	ecies to genus If on one isld several species of	SAME	genus, subsided land.-- Mauritius? <In plants w
230b236	some plants & then a chain of steps is found in	SAME	mountain).-- How is this explained by law of sm
232b249	uring islets & a sub-genus in Southern Africa In	SAME	manner. Cuscus, (a sub genus of Phalangista New
233b251	great work on Reptiles. M. J. says some reptiles	SAME	from Maurice & Madagascar & C. of Good Hope.--
234b261	t similarity even in quite distinct countries in	SAME	hemisphere. more than in other. Are there any c
236b278	s it not per saltum.-- Isld bordering continents	SAME	type collect cases.-- African isld.-- «How is J
239c001	r Esquimaux than Pointer.-- He has no doubt that	SAME	thing would happen with Australian dog & any of
242c017	.-- p. 22. New Calidonia-- New Ireland & Britain	SAME	kind of dog, with those of New S. Wales. <V.> p
242c018	S. Africa. «M. Bibron doubts fact.--» My toad is	SAME	species Coquille Voyage p. 25 Mais il n'y a pas
243c020	les or references highly unphilosophical xx Says	SAME	remark with regard to shells.-- But he says she
244c022	ensis (from Buenos Ayres) replaces <Vesp.> holds	SAME	relation with equator--that Vesp. lasiurus does
245c024	lds, where is it? All the Society isles have the	SAME	productions p 293-- is very strong about this L
248c029	down for nearly 70 year. Galapagos Mouse not the	SAME	section, with house mice. It is wonderful how i
248c030	ck to oldest race, which evidently is tending to	SAME	end, as the law of hybridity, namely the [not l
249c035	ge of animals. Follow this out, where species of	SAME	genera in two words. have not species, generall
251c041	x with the Gobbah or village Gazal.-- ¿is latter	SAME	species domesticated, strangely contradictory t
253c046	of two countries» & «Monkeys.» Fact of Elephant	SAME	species in Borneo. Sumatra. India Ceylon-- perh
254c049	in Acrite Kingdom all organs blended together, &	SAME	organ «where eliminated is» often repeated, as
255c050	t> height & 3d of latitude more commonly are the	SAME	species, instead of analogues.-- <as> in other
257c057	lacement under peculiar conditions-- of «nearly»	SAME	kind country distant. <Study> The circumstance
259c065	tation The case of heredetary disease, is on the	SAME	principle that, cut a sheeps tail off plenty of
260c067	, when hearing crys of hunger of little bird, in	SAME	way Wilson (p. 5). describes many kinds of bird
261c071	&c. sure to unite the birds into group.-- it is	SAME	as Yarrell's remark about rock Pidgeons.-- & th
263c077	n he is no exception.-- he possesses some of the	SAME	general instincts, <as> & <moral> feelings as a
266c091	ing, a much fairer race than the Hindu's, in the	SAME	tracts;. & that in their appearance & manners t
268c096	ican birds. genera blend into each other in very	SAME	district.-- <The same> <Mem> Tennioptera & Tyra
268c096	nd into each other in very same district.-- <The	SAME>	<Mem> Tennioptera & Tyrannula (NB work out how
271c105	erry on which it feeds. or insects it devours is	SAME	species. yet that it should so strictly <f> agr
271c105	lack & white thrush of Azara builds its nest in	<SAME	country> something same manner, much mud.-- The
271c105	zara builds its nest in <same country> something	SAME	manner, much mud.-- These facts show, habits he
271c106	nded to Juan Fernandez in birds. but ¿whether to	SAME	island in plants?-- What is this halo.-- Contin
272c108	erous insects & some third, have got thighs with	SAME	peculiar structure & habits of clinging to rush
272c109	w small detail in structure prevails amongst the	SAME	species & subgenera in families.-- thus the ban
273c110	ble & would lead one to suppose, that species in	SAME	group generally contemporary «++.» «This would
273c112	he different families.-- that sexes <are> «have»	SAME	plumage.-- <no> this is applicable to swallow-h
273c112	preeminent structure is not always applicable to	SAME	habits, though swallow hawk, ,milvulus,, may ca
275c121	e pidgeons which have such different skulls, but	SAME	marks on wings are Blue Pouters & small Bald He
276c125	ertebrae-- ¿Where two very close species inhabit	SAME	country are not habits different, (Mem: Gould's
278c131	mulated-- surely ask Owen to see whether species	SAME,	excessive improbability. Mem in Clifts list a
278c132	n had been found in Africa, the wonder have been	SAME	for S. America & Europe.-- The difficulty is ho
281c138	l from death or slow propagation of forms.--just	SAME	way as <we> «all men» not all equally related t
282c140	an originally similar limits of change, would be	SAME--	Yet each family might have its own character,
282c141	ogies.-- We must two races of such men living in	SAME	country but separated, now if one of these race
282c142	produced, but would depend upon exclusion.-- The	SAME	characters which are analogical in a genus with
282c142	ach other, because we suppose all descended from	SAME.--	but if two original species, each became gro
283c145	he Zebra into the Quagga.-- «let them be wild in	SAME	country with their own instinct & (even though
283c146	t.) that there are Tertiary fossil Infusoria, of	SAME	forms with recent, we have nothing to do with C
285c150	It would not be difficult to arrange children of	SAME	parents in a circle,-- & «hermaphrodites» fathe
286c154	sickness,-- death, unequal life,-- stimulated by	SAME	passions-- brought into the world same way» the
286c154	ulated by same passions-- brought into the world	SAME	way» they may convey much thus, Man has express
286c154	ll as we.-- It is our arrogance, to raise on the	SAME	shelf-- to (look at common ancestor, (scarcely[
286c155	dichotomous arrangement which is false There is	SAME	difficulty in arranging animals in paper as dry
287c158	ettling the relative importance of the organs in	SAME	state. in different animals, & the value of tho
287c158	But the classfication must chiefly rest on these	SAME	organs,-- habits, range. &c &c-- Macleay rests
288c159	» modified.-- The partial migrations of birds in	SAME	country. may explain greater migrations, if Ame
292c171	any successive generations are impelled to do in	SAME	way-- The improvement of reason implies diversi
294c177	ork is its proof, «its limiting, the allowing at	SAME	time true species.» & its adaption to classific
296c184	other will.-- St. Helena (& flora of Galapagos?)	SAME	condition. Keeling Isd «shows where proper damp
297c185	∵circumstances not favourable to many species.,	SAME	circumstances., which by causing death, makes t
299c193	growing in one wood far from any other plants of	SAME	species Channel Islds (& probably Isle of Man)
299c193	e it with Canary Islds. Galapagos.-- Iceland has	SAME	uniformity Primrose & Cowslip, quite wild, but
299c194	delions, which just lately have been shewn to be	SAME.--	one grows in marsh & other dry; yet if T. pa
301c198	stinct in animals. « & what reason» in precisely	SAME	way not possible to say what habitual in men &
301c198	to say what habitual in men & what reasonable--	SAME	action may be either in same individual p. 7. i
301c198	& what reasonable-- Same action may be either in	SAME	individual p. 7. is not squirrel hoarding, & ki
301c200	ble. The degree of development of all animals of	SAME	class being about equal.-- organs of generation
302c201	lies, but we see analogies between fish.-- Birds	SAME	remarks. Characters of analogy.-- last acquired
304c207	nse difficulty is, whether Apterix descends from	SAME	parent with other birds,, or branched off anter
304c207	s red band on wing show to be from one parent.--	SAME	forms of beak &c without these trifles. it woul
305c209	the sexes of which vary in colour of plumage in	SAME	remarkable manner as Europaean species = singul
307c215	then we are bound to consider abortive organs of	SAME	tendency in species, this is capital & novel ar
307c216	o a mere stump.-- Shall abortive organs «of very	SAME	kind» in these cases, have plain meaning & none
307c216	meaning & none in other case! Savigny has shown	SAME	fundamental organs even in Haustellata & mandib
308c219	ngs & Old Testament» Domesticated animals having	SAME	idiosyncrasy, cause of fertility.-- varieties n
309c220	The areas of subsidence marked out by animals of	SAME	<spe> genera. is not equal to areas of elevatio
310c224	lose The most curious case is Saxifrage, almost	<SAME>	«closely allied» species Himalayas, 13,000 & M
310c224	Melville Isd.-- West Africa & India some plants	SAME.	--America.-- See Brown Congo Expedition: 400 A
312c232	animals» could only exist by habit.-- therefore	SAME	principle transferable, not wonderful Accordin
312c233	uppies delicate-- they cross sister & brother of	SAME	litter, those. of different litters or of fathe
314c239	- Some species of Australian Genera Some species	SAME	(Palm & Phormium tenax) as in New Zealand & Aus
315c240	The common Mush room & other cryptogamic plants	SAME	in Australia & Europe.-- if creation be absolut
316c246	ered over whole world Many shells at present day	SAME	(or according to Sowerby fine species) on coast
317c249	aboeus" said to be monkey of St Jago C. de Verd;	SAME	as on coast of Africa.-- Macleay tells me same
317c249	; same as on coast of Africa.-- Macleay tells me	SAME	thing p. 55. 40 leagues from land several patch

(Key Word)

```
318c254 kinds of migration quite different in species of SAME genus.» The Muscicapa solitaria stay about a fo
328c1BC s Turbo. Cerithium Jardin du Roi Java fossils ke SAME time Study Botanical work on Buds & Gemmae. C D
332d001 inks appoplexy affects people all over England at SAME periods When he began practice, he remember dur
333d006 r & other more of English, but the effect is the SAME.-- Fox thinks that when a wild animal is crosse
333d009 exactly intermediate.-- Where two dogs line the SAME bitch & «perfect» spaniels & setters are produc
334d011 es, (though horsing every month) & worked in the SAME cart in loose chains, by being at first beaten
336d017 no offspring & therefore no generative organ.-- SAME Prop. better enunciated.-- "An animal Either pa
337d023 Distrib: of <ani> organic beings.-- Animals «of SAME classes» differ in different countries in exact
344d040 birds of Java Caterpillars not being fertile is SAME as children not being so.-- consider this with
346d047 to vary. I asked this in many ways, but received SAME answer.-- Thought lambs were more like father t
347d049 s. then animals must tend to improve.-- yet fish SAME as, or lower than in old days: «for a very old
353d060 g obliged to fill up the gaps.-- is possibly the SAME with the <Zoologist> «philosopher», who has tra
353d060 uliar <Malacca> «Malacca» bears, <are> belong to SAME section with with those of India-- Waterhouse k
355d066 seen to <vary among> be different in species of SAME genus." Law of monstrosity not prospective, but
355d068 takes place in the formation of a bud-- the very SAME must take place in copulation-- (Man & woman se
355d068 e in copulation-- (Man & woman separate parts of SAME plant<s>)-- now in some Polypi we see young bud
355d068 Himalaya. shells (see Paper in Geolog Transacts) SAME appearance with Secondary Species distinct-- bu
355d068 but close.-- Mem. Von Buch on Cordillera fossils SAME remark. ¿was there formerly one great sea, & tw
356d070 Ribston do producing occasionally (as Fox says) SAME fruit trees is analogous to some hybrids breedi
358d075 s" Case of bird of different family. having very SAME habits in some respects as this Caracara.-- Sep
358d086 n, hybridity, & breeding in & in tend to produce SAME effects.-- CD[May it be said, that breeding in
360d090 of breeders, where their interest is concerned. SAME man had crossed Jackal & dog-- (offspring did n
361d095 In lark if I understand right, all species have SAME character which is mottled, & not like any exis
363d102 ies scattered over Europe)-- The habits of some «SAME» North American & Europaean birds «slight» diff
364d104 > return.-- The Revd. R. Jones told me precisely SAME story about some Southern «see p. 43 supra» bre
366d112 opagate, ie with the chance of two being born at SAME time, & make breed, one would doubt any law.--
369d115 o Latent Character Hunter Animal Economy p. 482 SAME book) Owen says "the necessity of combining obs
370d116 ion. no more can be said.... In paper on bees in SAME work. (it is said that some kind lay <pu> up ho
371d128 handed down? Are not Loddiges 1279 roses kept in SAME soil. same atmosphere?-- may they produced not
371d128 ? Are not Loddiges 1279 roses kept in same soil. SAME atmosphere?-- may they produced not be transpla
371d128 ry unfavourable, will <deteriorated> continue of SAME variety as long as life lasts, yet they cannot
372d131 Newton is often thought wonderful. it is part of SAME class of facts that the skin grows over a wound
372d131 kinds of generation shown by their happening in SAME plant.-- The Marsupial structure shows that the
375d135 ke Europe on an average, every species must have SAME number killed, year with year, by hawks. by. co
377d140 ns of our system in the Heaverns.-- In not puma, SAME colour as Lion. because inhabitant of plain & J
379d151 r species, crabs & molluscs few.-- ¿are not some SAME-- what is the alliance with the Black sea.-- it
381d157 ion of sexes very simple.-- as in plants even in SAME genus some dioecious & some monooecious-- (& cu
382d158 aphroditism,-- those organs which perform nearly SAME function in both sexes.-- are never double, onl
384d162 in the plants which have male & female flower on SAME stem.--» so that Molluscous hermaphroditism tak
385d164 - showing foetus has gone on growing-- I believe SAME has happened in boys bodies.-- Lavaters. Essays
386d165 finite life & split it, & see whether it retains SAME length of life.-- like Golden Pippen trees! How
386d166 jury.-- The sympathy of part is probably part of SAME general law, which makes two animals out of one
387d168 <Generation--> V. p. 152 It is very singular the SAME difference from parental stock having been repe
387d169 their <D & E> «all their offspring» inherit the SAME peculiarity in lesser degree C. & theirs again
387d169 ve this peculiarity strongly; they transmit with SAME force as first pair, but to this tendency is ad
387d169 y from first pair.-- Now if two of third pair of SAME peculiarity breed they will have same influence
387d169 rd pair of same peculiarity breed they will have SAME influence as first pair + tendency they inherit
388d172 at of «2» dog begetting different puppies out of SAME mother.-- The view that man & «or cock» pheasan
388d172 = Number hase having both sexes abortive fact of SAME tendency.-- Mammae in man. having given milk.
389d176 - two animals may unite & each have offspring by SAME mother.-- one animal will fecundate female for
389d177 passion in hybrids. perhaps connected with this SAME case (& not merely as I have stated it) it is c
389d177 markable that too much difference should produce SAME effect as too little.-- in (latter case female
390d177 ose which are Monocotyledenous have many flowers SAME Spath, as they have only one bud.-- Every indiv
399e006 om to have lasted for its time: but we ought in SAME bed if very thick to find some change in upper
399e006 rn bed at present might be very thick & yet have SAME fossils. does not Lonsdale know some case of ch
399e009 two undoubtedly rabbits in poulterer shops., of SAME colour as a Hare, but paler & buffer.-- with lo
400e013 lle, distributed seeds of Dahlia all over Europe SAME year.-- he sowed them for four generation befor
402e018 inese", "the breed being of the latter being the SAME as the fox-like animals. which are met with nea
402e020 islands presents us, for the most part, with the SAME families and genera, that are natives of S. Asi
405e031 s are partly retained, therefore colours vary in SAME Manner as they would vary, if in wild state; th
406e036 Regarding the similarity of offspring to Parents SAME laws appear to hold good. with regard to marria
406e036 egard to marriage of individuals, & varieties of SAME species & to different species-- sometimes like
406e037 ewise far North in Southern.-- Great animals. of SAME two great orders destroyed about same time in N
406e037 nimals. of same two great orders destroyed about SAME time in North & South. America.-- Whole wor[l]d
407e038 The world formerly much more so. yet climate of SAME order as that of S. America.-- (Explained by pr
408e043 them, & therefore extermination becomes part of SAME law.-- When we know what a great effect. light
408e045 of them symmetrically.-- it is easy to get 50 of SAME kind of monstrosities.-- G. B. Sowerby.-- Looki
410e053 ing described.-- but then permanent varieties in SAME country, must be distinguished, from permanent
410e053 e distinguished, from permanent varieties not in SAME country.-- The traces of changes in forms of or
414e065 perfect. as the Elephant, if some genus. holding SAME relation as Mastodon to Man. were to be discuss
416e071 d» races. of <a> organics. are made by percisely SAME means as species-- but latter far more perfectl
417e077 similarity of child to parent appears to follow SAME law in two of «the» same <species>, variety, as
417e077 arent appears to follow same law in two of «the» SAME <species>, variety, as in two varieties, & this
417e079 w holds with different species, & individuals of SAME species.--). some races of men. D'Orbigny. affe
419e086 avia from height of 200 & 300 ft are identically SAME as those of present seas.-- now in this country
424e100 North America, Scotland, Uddevalla. Many species SAME. & Northern forms-- & American ones & Euroaean
424e103 not to be able to hybridise the Camel» like ass «SAME way some plants vary more than others» & horse
427e110 pple. only effect produced would be different.-- SAME way one variety of «animal» dog does not prefer
432e24 e children do not, (& in hairless kittens we see SAME fact) go back, & this is argument against Blyth
436e137 so &c. The greatest difficulty to my theory, is SAME type of shells in oldest formations:-- The Camb
439e143 ertilizing each other, better than the pollen of SAME flower.-- as it tends to show my view of <i>nfe
447e164 ant <producing> propagating itself by buds is in SAME predicament, as one, in which structure does no
447e165 the slightest doubt that Festuca vivapara is the SAME species with F. ovina, <& this> rendered vivapa
448e166 le of all the castes from Stephenson at Lima The SAME numerical relation (both in species and subgene
448e167 nce between the Superga & Paris, numerically the SAME with recent & yet almost wholly different, is s
448e167 me with recent & yet almost wholly different, is SAME, as if Isthmus of Panama.-- These two cases hig
449e170 ggs of two shapes & colour.-- Eyton has observed SAME thing in Brent Goose. Eyton says some of the pi
449e170 cal, as those which are different.-- now this is SAME, as Galapagos facts & &c.-- & it shows the cau
449e170 facts &c &c.-- & it shows the causes which give SAME species to different isld. is the same as that
449e170 hich give same species to different isld. is the SAME as that which gives genera.-- <it is not transp
453e182 a) beasts on Phillipines. Forrest somewhere says SAME.-- do p. 393. <">The wild, small fowls at Pulo
453e183 oduce each other. & yet I presume seed raised in SAME garden.-- now this good question-- single, or h
455eIBC annual give buds, or tubers.-- but these are the SAME as trees.-- Shake some sleeping mimosa-- do sta
455eIBC irritate them, «as by an insect coming always at SAME time» see if by so doing can be made sensitive
459t013 rs on its wing resembled the drake.-- another of SAME half breed resembled the plumage of drake still
460t017 both having spines, is the effect, partly of the SAME external conditions (ie. analogical structure)
461t041 with difficulty of England remaining SAME (if so) with those of Spain & such facts-- This
461t041 & such facts-- This unequal duration is exactly SAME as some species extending much further geograph
462tf4r enaeum 1839. p. 708.-- Shrew, found by M. Lartet SAME as existing species. We see the same object gai
462tf4v y M. Lartet same as existing species. We see the SAME object gained by the Mataco-armadillo & the woo
463t063 me-- Owen in his description of my fossils makes SAME such remark & before the conclusion of his work
465t079 a «mere» vestige, is preserved in this country-- SAME argument to India & Europe-- & Africa!.-- any n
466t095 DOGS trained to pursuit having PUPPIES with the SAME powers instinctive & doubtless not confined to
468t112 , finally Fraxinella. with respect to nectary is SAME case as Azalea or Rhododendron xx after several
469t135 ng days «many» «most numerous» bees visited this SAME bunch & on this day in five minutes eleven Humb
469t151 an» bean, were planted in rows, & seeds gathered SAME year came up true «in 1840»: All in together bl
471tf07 n we find such an endless variety of form in the SAME> organ "manifestation of divine power"?.--"of t
472s01r racteatum, & the Papaver oncitate was growing in SAME garden. & out of 60 seedlings not one came up t
```

Page **(Key Word)**
472s02v odd genus-- & a small common Humble-- & more of SAME fly Two more of the flowers withered.-- Sillima
473s07v arly impressed & others later-- All poultry with SAME down-feathers. Zoology 1856 Skimmed through & a
484z016 rnarius.-- into Oxyurus, by Maldonado creeper of SAME plumage.-- general red mark on wings of all-- S
493q003 then weigh their wing bones & see if relation is SAME good, avoids effects of fatness.-- Experiment i
496q006 impregnation take place male & female flower in SAME receptacle (8) Make Duck eat Spawn, eggs of sna
498q008 ted. Abberley says Ants-- Enquire (13) Do any of SAME species of Willows grow in same situation & flo
498q008 e (13) Do any of same species of Willows grow in SAME situation & flower at same time. Has H. seen gr
498q008 es of Willows grow in same situation & flower at SAME time. Has H. seen group of different species gr
500q10a ther they come from islds. or different parts or SAME district.-- About <endemic &> wandering species
506q014 ght 17½/. The Greyhound. was in length (measured SAME way) 47½-- in heigt 30 inches Examine Keel of C
506q15v tanist who does-- What is Ruppia Bennett says in SAME state. of flower 45. Charlsworth. vol II. p. 67
507q15v of wild varieties plants growing together. under SAME conditions.-- like cowslip & primrose, but less
508q016 of diseases (heredetary?) in diff. countries in SAME races Mr. Gray General Questions (1) Particular
513q21. of pollen & stigma generally not being mature at SAME time on same plant --Flora of Australian Mounta
513q21. tigma generally not being mature at same time on SAME plant --Flora of Australian Mountains.-- Is set
515q021 what is effect of crossing peaches & nectarines: SAME question with regard to Primroses. (4) Do apple
515q021 they are apt to vary in number in individuals of SAME species Eyton (1) Number of eggs-- of half-bred
521m008 wer of repeating poetry in her dotage is fact of SAME sort. Aunt. B. ditto.-- Case of Mr Corbet of th
522m011 Child «of Kinlet» & married Miss A. B.-- all the SAME names as a few minutes before he maintained he
525m026 k how odious an illtempered fat man looks, shows SAME connection between organization & mind.-- think
526m027 the others <are> learnt. what they teach by the SAME means & therefore properly no free will.-- we m
527m032 s have the best song. [Migratory birds return to SAME quarter for many years]CD.-- Beauty is instinct
527m033 & thoughts.-- Poetry. the latter thoughts are in SAME manner vivid & grand. the frame of mind being j
530m046 he «similar remark» thoughts, being functions of SAME part of brain, or the tendency to habit of prod
532m054 apt, this partly owing to heart? readily taking SAME movements, senses being on the look out, & the
532m055 very doubtful whether they could recollect these SAME things from any effort of will whilst their min
533m058 arly in life, <before they> (I think I have seen SAME thing before they could understand. what frowni
536m070 » heart beat» remember how Pincher does just the SAME; I noticed this by perceiving myself skipping w
536m071 ance it scrapes off its bottom.-- it is relic of SAME thing that makes one dog smell posterior at ano
538m079 f thought depend on state of turn In drunkedness SAME disposition recurs, such as -- -- of Trinity al
539m083 ome respects--. good instances.-- when education SAME.-- My handwriting same as Grandfather. Aug. 16t
539m083 stances.-- when education same.-- My handwriting SAME as Grandfather. Aug. 16th Anger «Rage» in worst
540m086 f New Zealand)-- the American in Brazil is under SAME conditions as Negro on the other side of the At
540m087 Why then is he so different-- in organization.-- SAME cause as colour & shape & ideosyncracy.-- Look
542m095 raction & wrinkling of the skin contract iris?-- SAME way as one lifts up eyebrows to see things in d
543m099 are people of very limited intellects, & in the SAME way are chess Players-- A man at Cambridge, dur
543m100 thematicians not being profound reasoners.-- all SAME fact-- for, as Jones observed, in playing chess
543m101 within <our own> limits of examination.-- obeys SAME laws. as other parts of structure. C. D.27 Can
545m107 ump to scratch my face. The ourang outang, under SAME circumstances, threw itself down on its back &
545m107 progenitor did, when excited or disturbed by the SAME cause, which «now» excites the expression.-- Ha
546m110 take) have lower animals these vivid thoughts In SAME book (p. 143) wonderful case of perfect double
546m111 n perfect «mental» health.-- «Erasmus had almost SAME thing happen to him about a knife. which he had
548m116 iduality is.-- Insanity is <much> «somewhat» the SAME as double consciousness, as shown in the tenden
549m118 more intense happiness-- so is it <with an> when SAME man is compared to peasant.-- To make greatest
550m124 on may be more or less pleasant & unpleasant, in SAME time,-- therefore degrees of happiness-- Entire
550m125 me pain,-- compared to what others experience in SAME time.-- Pleasure more usually refers to the sen
553m136 Comte).-- Those savages who thus argue, make the SAME mistake, more apparent however to us, as does t
556m146 when «turning round to kick» kicking they do the SAME. although it is then quite useless-- Cats kneed
558m152 rce.-- making growling, guggling noise. Puma did SAME & & some others-- Thus <sudden> «forcible prolo
558m152 prolonged» expulsion of air «dogs snarl much the SAME way» generic manifestation of great passion.--
564n003 ital view.-- Dogs conscience would not have been SAME with mans because original instincts different.
565n007 l its enemy.-- Man & dogs show triumph (& pride) SAME way walk erect & stiff, with head up.-- Why doe
565n008 some anger «& respect to opponent» is showed by SAME movement as sneering,-- it is then more <emblem
565n009 rkable. the pouting, & blubbering-- sulkiness is SAME as pouting, <but> lesser in degree, no smile, n
569n020 n things & voice, as roaring for lion &c &c. (in SAME way alphabet. arose from letters, symbol of wor
571n029 s being habitual-- & hence become heredetary; on SAME principle we know many tastes become acquired d
572n035 Existence of a Deity has an expression the very SAME as mine about our origin of a notion of a Deity
575n044 instincts, most important, because they obey the SAME laws, as the crossing of jackall & Fox & wolf &
577n051 the intestinal functions &c &c &c.-- bears. the SAME relation to true memory, that the formation of
580n061 er than when she tries is not this precisely the SAME, as the double-conscious kept playing so well.-
581n064 ep if it can be generalize.-- The tastes of man, SAME as in Allied Kingdoms-- "food, sm<e>ll. (ourang
583n069 lity of instinctive powers in individuals of the SAME species with variability of reasoning power in
585n074 en) to know that which we touch & what [...] the SAME.--(this Hensleigh therefore problem is how we k
585n075 h therefore problem is how we know that thing is SAME, which touches two parts of our bodies, «or tou
585n077 et not instinct, but if all men placed stones in SAME position, it would be instinct-- instinct is he
588n090 e great difference.-- «(& is not this difference SAME, but less in degree, as between man & child.--)
590n093 on p. 39. The sweat that accompanies fear is the SAME, as that which attends great weakness.-- <Diarr
591n098 e variation in character in different animals of SAME species.-- The general «(as I believe)» contemp
591n099 en impulses. (where sensations of individual are SAME as in normal cases) are held in abhorrence it i
592n102 school & used between Indian tribes are Many the SAME.-- Philosoph. Transactions Vol 44. 1746-47. Pap
594n112 hawk & sparrow in Shrewsburg garden picking from SAME bone A child born on the 1st March was frighten
594n113 nything ugly. baby-- association-- pouting child SAME as anger, lips not compressed sullen, protruded
595n121 although knowing it was Snow.-- Is this part of SAME feeling which make us think anything ugly-- a b
595n121 e us think anything ugly-- a beau-ideal feeling. SAME effect as acting on us-- <The Baby> «Effie Wedg
600o08a ed to that of men, when the five senses were the SAME-- In its action-- emotions-- p 176 & 177 good p
605o019 ssociations that we apply to such emotions. this SAME term.-- Hence it appears, that when certain cau
605o19v hich superiority we transfer to ourselves in the SAME manner as we are acted on by sympathy. D. Stewa
609o29v ous.-- (jealousy in a dog no one calls vice). on SAME principle that Malthus had shown incontinence t
612o035 -- [RHC] 2) In animals, growth of body precisely SAME as in plants, but as animals bear relation to l
613o036 he deduces from the ends in each case being the SAME, & the means very similar.-- It does not appear
613o036 e than saying that the thinking principle is the SAME in all animals. [LHC] «3)» Eyton told me that h
614o037 ility of so high a mind without further end just SAME argument. without indeed we are step towards so
615o038 change in «external» man; and as all men nearly SAME species, so general instincts nearly same; whic
615o038 nearly same species, so general instincts nearly SAME; which same argument probably applies to partic
615o038 species, so general instincts nearly same; which SAME argument probably applies to particular instinc
616o039 ceptions will, consciousness memory &c. have the SAME relation to a living body (especially the cereb
616o039 said to attract; & hence if thought &c bore the SAME relation to the brain that attraction does to m
617o39v ntify the two aspects as different phases of the SAME object of thought is a question which ought to
617o040 emembered) being a phenomenon apprehended by the SAME faculty with matter & being necessarily exhibit
617o40v Now the phenomena of gravity are manifestly the SAME as if every particle of matter were an animated
618o041 & have no objective aspect. If thought bore the SAME relation to the brain that force does to the bo
624o051 ght & wrong,-- education, of parents strives* to SAME end.-- & general actions of community must freq
624o051 neral actions of community must frequently teach SAME end.-- Hence this becomes the law of right & wr
624o051 changes, & probably likewise instincts, for the SAME law effects both.-- <such> changes «in accordan
628o054 y, it will be for the general good, that is, the SAME cause, which gives the instinct.--]CD p. 22. sa
636j56v y of frequent intermarriage.-- A plant is in the SAME predicament as a group of bisexual animals livi
636j57r adaptation.-- & then Chamelion, which feeding on SAME food, differs in every respect, except «in» qui
640j28v n old age keeps woman alive: for Man & woman are SAME: fertility of either sex determines life:.» «Wi
641j29r , when physical changes are in progress; (on the SAME principles that islands are favourable,) becaus
267c095 gazine??? p. 75. roe of Asterias in stomach. of SAMMON remain after rest of animal digested.-- Import
319c257 Home's History of Man at Maer, it is said the SAMOYED women (¿ north end of the Oural mountains) hav
025r016 [R.] Sergipe [11 degree 10 minute S.] 20 190 R. SAN Francisco [10 degree 32 minute S.] 10 50 Whole c
032r040 represented), along line of coast.--[Fig. 2] Mem SAN. Lorenzo; Valley of Copiapò & parts of coast of
025r016 e 29 minute S.] still shoaler, coast composed of SAND dunes. 15--15 Does not seem to consider this a
027r027 lly than others of the Bahama consists of rock & SAND mixed with sea shells--about 500 Isd. & great b
027r028 of Elevation. United service Journal In the Iron SAND formation <would> wood converted into siliceous
028r028 ncy. to mingle; The sea would separate quartzose SAND from the finer matter resulting from degradatio
035r048 raveller» Mountains, which in size are grains of SAND, in this view sink into their proper insignific

036r050 lower   orders; it was connected with movement of SAND.--it is called "Bramidor"(?).--it was a strange
036r051 cute chirping sound produced in walking over the SAND: I am nearly sure, it is necessary to ascend th
041r066 in alliance with those balls at Chiloe, full of SAND.--the <scale> «quantity of iron» being there in
050r091 ral reefs.--does he mean in contradistinction to SAND?? B. Roussin states that generally in North par
050r091 erally in North part of Brazil. <gravel becomes> SAND less & gravel more common. the shoaler the wate
051r097 n bottom in NW coast of America. from shingle to SAND &c &c. <Vol II> P. 209. 211. 213. 444 «Yanky ed
088a016 haracter of Andes Metamorphic action -- Mem: red SAND of Europe no fossil shells --¿ action of Heat b
109a083 his will show effects.-- analogous to broad flat SAND beach. {P} -- De la Beches argument of low coas
117a101 Europaean strata according to composition thinks SAND with vegetable remains formed near coast, limes
128a130 ca., like the Mexican Gulf. is fouled by bars of SAND & shallow lagoon.-- when describing Coast of. B
130a134 } B subsidence; <as in> be cautious. mud banks & SAND. dunes.-- in these littoral deposits there prob
138a153 shore both N & S of Lima.--judges from <beds of> SAND & gravel & shells. p. 47. do has table of every
138a155 Paper on the consolidation of strata-- he heated SAND red hot & brine was boiling on the top-- [blank
147g015 at topped hill/ do alluvium. NB In one part pure SAND in current cleavage-- in other irregular horizo
149g030 o valleys corresponding as in Andes, composed of SAND & perfectly rounded stones-- lake required to d
157g073 in parts of Beagle Channel when mica slate, only SAND blow away]CD where lines appear to cross stony
162g104 resses, an[d] one alternate curved layer of fine SAND & small angular-- rounded pebbles-- dip sidewar
163g107 many Sea shells. My informant saw them himself-- SAND with tide ripple Near Fort Augustus hill & frin
423e099 n in one bed, much less in alternating strata of SAND & limestone &c &c.-- L'Institut 1838.-- p. 290-
429e115 n fringe.-- but there is a contest. & a grain of SAND turns the balance.-- Hort. Transact Vol I. M. R
515q022 see if get sterile-- Cover that little Ervum in SAND-walk, on which I think I have never seen Bee vi
527m031 e are synonymous.-- Shake ten thousand grains of SAND together & one will be uppermost:-- so in thoug
633j54r (like a Dutchman plants them to stop the moving SAND) we <do> lower the creator to the standard of o
306c213 uced line of Timor 215 degree. What productions SANDAL. Wood Isd.? ought to agree with Java?? Terrest
195b100 sts.-- How does it come wandering birds. such SANDPIPERS. not new at Galapagos.-- did the creative fo
027r027 located] The frequency of shells in the Calc. SANDSTONE Concret, is connected with frequency of shell
033r042 d the composition of pumice at Ascension In Calc: SANDSTONE at Ascension, each particles coated by pelluc
041r066 not. Ferruginous veins of this figure {P} in SANDSTONE: evidently depend on a concretionary contract
041r066 so are all those plates in Australia. New Red SANDSTONE. at Bahia in modern sandstone. a circle,.{P},
041r066 tralia. New Red Sandstone. at Bahia in modern SANDSTONE. a circle,.{P}, had in its middle a short <fi
051r093 ter) organic bodies protect like peat reef of SANDSTONE.--Corals, & Corallina survive, in the most vi
053r101 ctions are singular-- M. Lesson considers the SANDSTONE & Granite districts to be separated by profou
054r102 a. talcose slates, do at latter place. sandy SANDSTONE with gypsum, covered by limestone with recent
055r107 m of land, in St Helena. Ascension. Azores. («SANDSTONE first gives» half demolished craters).--worn
060r125 how diff. would the rocks have been. The red SANDSTONE of Andes fusible? no. mad dogs. Azores. altho
066r142 strongly. that taking up a piece of Falkland SANDSTONE. he could not distinguish from stone Caradoc
069r150 solid matter by Heat Consider profoundly the SANDSTONE of the Portillo line.--connected with <gneiss
089a017 . granite heated.-- Metamorphic action in red SANDSTONE.-- Certainly Volcanic-- CD[Might not bottom o
092a030 he Essequibo, granite & quartz, after passing SANDSTONE Vol II. p. 69.-- Geograp Journal Earthquake a
098a048 ss is identical with layer of flint on calc.: SANDSTONE. (& as I believe most strata) (Hence endless
115a096 mulated in bed of ocean With the exception of SANDSTONE rare to have any horizontal non cleaving beds
120a109 ormation of salt.?.--??? Footsteps in New Red SANDSTONE. look as if a surface deposit.-- The case of
137a149 re affecting to some distance & blending with SANDSTONE «said to be» analogous to granite infiltering
145g006 35 degree is I believe about greatest dip of SANDSTONE in upper part «of Salisbury Craigs» 25 degree
153g056 re decaying.» neighboring rock gneiss & [...] SANDSTONE actually resting on them on summit of hill ro
219b194 because volcanic isld. whilst <neig> Africa, SANDSTONE, & granite, (that is genera near Cape) see it
257c058 izards.-- As we have birds impressions in Red SANDSTONE: great lizards in do.-- <Wood> <Dicot wood> C
257c058 od in Coal Measure.-- highest fish in Old Red SANDSTONE.-- Nautili in----. it is useless to speculate
433e126 in each system, the changes from limestone to SANDSTONE &c. show some great change who can say how ma
436e137 round world.-- Quartz of Falkland.-- Old Red SANDSTONE-- Van Diemen's land.-- Porphyries of Andes. A
465t080 caves:-- :argue first case of bones (New Red SANDSTONE) & then go on to shells-- A profound consider
473s03r ns of Sus with Elephants-- Lyell says New Red SANDSTONE of. N. America is Red Sandstone. & Birds true
473s03v says New Red Sandstone of. N. America is Red SANDSTONE. & Birds true! Plants in Devonian-- How stran
028r030 in Coral Paper or hypothetical origin of some SANDSTONES, as in Australia.--Have Limestones all been
054r102 ully from mine: phyllade covered by quartzose SANDSTONES: refers to broken hill described by Pernetty
065r138 ld. Marion & Crozet. L. Alexand. Macqueries.--SANDWICH Isd-- Specimens of rocks were brought home in
081r181 ilbert Farquhar Mathison travels Brazil. Peru. SANDWICH Isd Mawes travels down the Brazil.-- Did Mela
217b187 nd in Sumatra; again another of other Genus in SANDWICH islands-- A genus with species in Van Diemen'
247c028 «stated in note to p 21» to Quoy & Gaimard in SANDWICH isld. & according to Chamisso in Radack isld.
315c241 ogie avec le flore du Japon", some Europaean & SANDWICH species & some of Japan. I do not understand
425e104 eory says so.--» March 6th. Mr Bentham says in SANDWICH Islds. he believes, there are, many cases of
425e104 e species are described.-- Capital case,-- for SANDWICH Isld are very similar to Galapagos-- study Fl
490q001 strict than in another? Character of shells of SANDWICH group {Sowerby monstrous Cardium-- does it re
036r051 lar fact near the Red Sea.--which occurred in a SANDY place.--(the sound was long & prolonged). NB, I
040r063 ave been metamorphised, as in Brazil feruginous SANDY ones have undergone the same process.-- Neither
051r093 here occasionally most tremendous surf & loose SANDY beach) deposits «calcareous» encrustations; At
054r102 ! at Payta. talcose slates, do at latter place. SANDY. sandstone with gypsum, covered by limestone wi
146g014 ll rounded pebbles in yellowish argillaceous or SANDY soil-- These Buttresses formed vestige of irreg
198b113 allowed to be wide hiatus: states in one the SANGUINEOUS system, in other nervous developed. (Owen's
215b178 man>k priority refers to introduction to Animaux SANS Vertèbres as latest authority. The case of the
134a141 . do p. 210. Height on road from Valparaiso to SANTIAGO p. 328. dead trees on Isthmus of Pen. Tres Mo
222b204 ference of species between land shells of Porto SANTO & Madeira-- I believe very curious-- My idea of
511q018 eons. Canary birds-- Bantams.-- (6) <Mad> Porto SANTO Rabbit. Descript. of colour <& length of ears»
064r134 umber of animal[s] at Cape of Good Hope Says at SANTOS «M Birchels» at foot of range some miles from
612o035 uire will in part. ¿Why more so than movement of SAP. or sunflower to sun? ∵ I should think there. wa
194b094 he semnopitheque of India.-- Tooth of <Spi> of SAPAJOU-- NB Sapajou is S. American form. therefore it
194b094 ause of India.-- Tooth of <Spi> of Sapajou-- NB SAPAJOU is S. American form. therefore it is like case
539m083 Eras a Wedgwood in many respects & some of Aunt SARAHS. cranks, & so is Catherine in some respects--.
541m092 satirical. laugh.-- when snarling real bitter SARCASM.-- <These> Seeing how ancient this expression
617o40v this supposition the forces manifested would be <SAT> fundamentally accounted for, we prefer this met
573n036 r origin of a notion of a Deity We can allow «SATELLITES», planets, suns. universe, may whole systems
573n036 idly believe the Astronomer, when he tells us SATELLITES &c &c «The Savage admires not a steam engine
541m092 ore than in a real snarl, they are enjoying a SATIRICAL. laugh.-- when snarling real bitter sarcasm.»
564n004 e «instinct of» hunger, «of» death & for the SATISFACTION of following conscience, obeying habits, &
621o047 do not mean any calculated pleasure, but the SATISFACTION of the mind, which is «much» formed by past
354d062 onesfield Didelphis not Didelphis «Answered SATISFACTORILY by. Valenciennes.» The change from caterpi
616o039 port of their view it is impossible to shew SATISFACTORILY it's erroneousness. it is a point of indif
612o035 two difficulties into one common one always SATISFACTORY, though not adding to positive knowledge. l
542m093 iffening over his canine teeth.-- He may feel SATISFIED with himself, & though dreading to say so, hi
617o40v other & moving in opposite directions. We are SATISFIED therefore, if we can trace any force in inani
634j54v from these very circumstances, that we become SATISFIED respecting an original thought, or design, pu
148g026 irmed the account of the «YOUNG» Shepherd dogs SATURDAY. Before coming to Bridge of Spean, hills of «
162g103 bedrock & Loch Ness <30.100> <Donald Macphee> SATURDAY Morning 29.958 A 64 degree, air 60 «Evening d
593n105 ake a nest for herself.-- the object is to make SAUCER-shaped depression.-- [blank] Does music bear a
451e179 ppear as if p. 345. The Ceylonese Elephant [...] SAUL forests by having a smaller, lighter head, carr
222b206 doubt, but what cetaceae & Phocae now replace SAURIANS of Secondary epoch: it is impossible to suppo
374d133 rock in which reptiles have been found. p. 426 SAUROID fish in Coal, true fish, & not intermediate be
171b004 mind & instinct becomes influenced.-- child of SAVAGE not civilized man.--birds rendered wild <throu
213b169 this principle. <Varieties> < <Races> > Man in SAVAGE state may be called, <species> species. in dom
232b248 he philosophers who soar above the pride of the SAVAGE, they perceive the superiority of man over ani
264c079 ss, & very actions of despair; <let him look at SAVAGE, roasting his parent, naked, artless, not impr
293c172 ly we should exclaim it was instinct.-- Even if SAVAGE takes. & was given a Great coat & this he put
321c270 yo Philosophy of Art of Living Several of Water SAVAGE Landons Imaginary Conversations-- very poor Si
409e047 ralians yet slow progress has done so.-- Show a SAVAGE a dog, & ask him, how wolf was so changed. Whe
535m069 ch as plastic virtue, «&c» (Very true, no doubt SAVAGE attribute thunder & lightening to Gods anger.-
540m087 slavery & look at the Negro-- look at them both SAVAGE-- look at them both semi-civilized-- Perhaps o
553m137 fears & strange superstitions of an Australian SAVAGE or one of Tierra del Fuego.-- Mr Miller (super
557m147 ect its tail & make it very stiff «& back» when SAVAGE «no» & ready to dash at prey streched out & fl

566n011 have 'these necessary notions any more than «a» SAVAGE M. Le Comte's idea of theological state of sci
569n022 e in Carolinas Vol II p. 132. offered to take a SAVAGE, said his wife would be grieved-- "il leva les
569n022 uelque mal qu'on y fût."-- Expression common to SAVAGE & Frenchman, unaccompanied by dignity-- "no mo
573n036 ronomer, when he tells us satellites &c &c «The SAVAGE admires not a steam engine, but a piece» of co
608o025 mena.-- (Mem: M. Le Comte case of Philosophy, & SAVAGE calling laws of nature chance) 2) difference i
609o30v dered.-- The difference between civilized man & SAVAGE,-- is that former is endeavoring to change tha
638j58v his process is shortened, but yet analogous, no SAVAGE ever made a perfect hinge.-- reason, & not dea
253c046 acts to show that Australian dog introduced by SAVAGES into Australia.-- What are they? Colonel Monta
262c072 ie instincts of wisdom virtue? «like senses of SAVAGES» (How come its some countries patribiotic?)-- bu
285c150 ace in each others' economy Dr. S. showed that SAVAGES are not born with any capacity for observation
286c154 t common ancestor, [scarcely?] conceivable in SAVAGES) Has not the white Man, who has debased his Na
308c217 ears relation to existence of genera &c &c Two SAVAGES, two species.-- «discussion ustil, unless it w
332d004 -- applicable to birds migrations & Australian SAVAGES.-- W. D. Fox has a cat. which he bought in Por
356d071 hat comparative solutions & linking of facts-- SAVAGES over whole world. (Major <I> Mitchell p. 244.
362d099 the love of woman (as Mitchell remarks seen in SAVAGES) to brave men.-- Effect of castration horns dr
490q001 Tame Parrots breed amongst the Indians Do the SAVAGES select their dogs Sowerby Entomologist Does in
508q016 in Taylors Scientific Memoirs (11) And. Smith SAVAGES at Cape any selection of Males in «cattle» or
542m095 -- Is frowning, result of straining vision, as SAVAGES without hats put up their hands, & as attentio
542m095 eir hands, & as attention would amongst lowest SAVAGES clearly be directed chiefly by objects of visi
553m135 y beings, many vicarious, like ourselves) that SAVAGES (mem York Minster) consider the thunder & ligh
553m135 very nation according to M. le Comte).-- Those SAVAGES who thus argue, make the same mistake, more ap
571n027 st naked figures, & observing powers common to SAVAGES???].CD-- The existence of taste in human mind.
581n064 & dogs fond of slight tickling sensation.-- in SAVAGES other tastes few. March 16th. Gardiner's Music
582n066 igs hiding themselves; & heredetary remains of SAVAGES state.-- N B. According to my view marrying la
233b252 of the earth covered with the most beautiful SAVANNAHS & forests dare to say that intellectuality is
183b052 like races of man.-- M. Flourens. .Journal des SAVANTS.-- April 1837. p. 243 it is said as well known
576n046 hen incensed.-- A Dog may hesitate to jump in to SAVE his masters life,-- if he meditated on this, it
576n047 e conscience. A man, might not <t> do so even to SAVE a friend, or wife.-- yet he would ever repent,
586n082 e of limbs &c, or it result from mere impulse to SAVE wax.]CD which it instinctively exerts in concer
606o025 ishes to do some action (as jump off a bridge to SAVE another) & yet dare not -- one could do it, but
620o045 the action was superfluous, as one man trying to SAVE another in desperation.-- This shows, that our
439e144 d by some other animal, then that quality which SAVED him, would be the one encouraged-- » Wilkinsons
307c216 ases, have plain meaning & none in other case! SAVIGNY has shawn same fundamental organs even in Haus
470t178 stamen «Egg Tree»--I think never on the Galeum SAXATILE & other common kind--I think not on Phlox tho
310c224 s & a fox most close The most curious case is SAXIFRAGE, almost <same» «closely allied» species Himal
132a137 opkins. M. Parrot, Mem. Acad. Imp. des Sciences. (SC Math. Phys. et. Naturelles. Tom I. p. 501.-- shows
470t177 ce of palmation!!? Bees at Wild St Johns Wort--SCABIES, Cyanoglossum--Reseda wild very many Bees & Hu
041r066 with those balls at Chiloe, full of sand.--the <SCALE> «quantity of iron» being there in excess.-- If
099a049 tion parallel to stratification evidently small SCALE of concretionary action all fluid at once, the
119a105 e earth, as would be required by thermometrical SCALE.-- (for the temp must be immense to convert roc
197b111 be traced in same organ in different animals in SCALE.-- In monsters «also» organs of lower animals a
222b205 dicament. It is another question, whether whole SCALE of Zoology may not be perfecting by change of M
254c048 ceeding classes & likewise those much higher in SCALE. So Owen actually believes in this view!!! p. 3
291c167 Nature.-- Makes hermaphroditisms. one step in <SCALE>. Series-- in plants we have a step between «mo
348d051 ow can we estimate this amount, when <value» no SCALE of value of difference is or can be settled,--
370d116 ittle developed.-- Sept. 19th <Are> There is no SCALE, according to importance of divisions in arrang
381d156 & the nematoid Entozoa-- Therefore highness in SCALE has no «constant» relation to separtion of sexe
527m033 one man.-- Music & poetry opposite ends of one SCALE.-- former pleases from instinct the ears (rhyth
022r007 rn P. 434 & 419 As Limestone passes into schist SCALES of chlorites--Mem. Maldonado P 375 Much Chlori
110a086 de, large irregular cryst of reddish felspar. & SCALES. of mica.-- large cryst of Hornblende blending
289c161 t species If any one is staggered at feathers & SCALES. passing into each other let him look at wings
289c161 rbits of penguin & then he will cease to doubt :SCALES into Teeth in Bering Pike (Waterhouse) Magazin
364d105 reached-- as in pidgeons no new races.-- In SCANDINAVIA besides the Rakhekna, before mentioned betwe
365d105 ts would sure to be trod, & in many parts of SCANDINAVIA these birds are very far from common.-- Unde
419e086 yfields district are much more like those of SCANDINAVIA, than of the N. American species--Glacial pe
419e086 --Glacial period Dr. Beck says the shells in SCANDINAVIA from height of 200 & 300 ft are identically
341d030 bird.-- Wings reduced to rudiment.-- clavicle SCAPULA &c strongly developed to aid in breathing.-- A
555m143 hero, & then I had some confused idea of showing SCAR behind (.instead of front) (having changed hang
185b056 would be mistaken for) Carabidae, Crysomela, SCARABADAE, & longicornes.-- Again taking a subdivision
056r109 n natural [...] deepest astonishment.» Perhaps SCARCELY a palpable might remain to tell of these losses
065r139 und) long consigned to the earth. yet body had SCARCELY undergone any decomposition: countenance so w
159g085 ross) on top of spit between river & dry Corry SCARCELY conceivable. if Hill between Corry so much cu
181b039 hat <the» «many» contemporary, would have left SCARCELY any type of their existence in the present wo
224b213 ure.-- Hence species may be good ones & differ SCARCELY in any external character:-- For instance two
230b240 ross breeding prevents perfect change.-- It is SCARCELY possible to get evidence of two races of plan
246c026 her of Europe, (Alcedo ispida) from Molluccas. SCARCELY differs at all from those of Europe, but beak
246c027 s, Emeu) in North of New Holland.-- New Guinea SCARCELY differs more from, <Van Diemen's land.Ṉ> Aust
267c093 . Mr Bennett Voyage round world, 20 years have SCARCELY elapsed since the Guava introduced from Norfo
275c121 l's evidence about old varieties is reduced to SCARCELY anything.-- almost all imagination-- He says
286c154 the same shelf-- to (look at common ancestor, (SCARCELY?)] conceivable in savages) Has not the white
286c156 to hold,-- Birds having web-feet, where we see SCARCELY any traces of passage a difficulty, but after
293c172 recollecting things utterly forgotten» --it is SCARCELY more wonderful, that it should be remembered
294c177 <Lonsdale says. that first shee> State broadly SCARCELY any novelty in my theory, only slight differe
301c200 sands & tens of thousands New insects, perhaps SCARCELY one new family & no new orders,-- Wonderful,
313c236 with no relation to time) as in buds.-- I can SCARCELY doubt final cause is the adaptation of specie
314c239 r large flora. (150?) Mr Brown did not observe SCARCELY any Australian character in Timor plants, yet
317c250 t Jago found a Euphorbia so near Piscatoria as SCARCELY to be distinguished from it.-- & several old
337d020 breed confined. to certain best individuals.-- SCARCELY any breed but what some individuals are picke
365d106 gyptian animals not having changed is good-- I SCARCELY hesitate to say that if there had been consid
401e014 es fertilize enormous number of plants-- it is SCARCELY possible to purchase seeds of any cabbage, wh
515q021 s place,-- the mere fact of seeds ripening has SCARCELY any no relation to hybrids.-- (3) As peaches
544m103 he mind wills to do this & hears that, but yet SCARCELY really moves.-- the willing therefore is idea
582n069 the means of performing it.-- p. 14. There is SCARCELY a faculty in man not met with in the lower an
596n184 ng do not move eyebrows.-- or skin of head,-- «SCARCELY able St.-- » Cyanocephalus, macacus. Cercopit
628o53v icant islets--general movements of the earth;--SCARCITY of Organic remains.--Unequal distribution of
044r073 icant islets--general movements of the earth;--SCARCITY of Organic remains.--Unequal distribution of
468t111 a & small Hymenoptera Saw Humble go from great SCARLET Poppy to Rhododendron-- from Larkspur to Lupin
468t111 quent certain flowers, to day early. the great SCARLET Poppy-- So that, finally Fraxinella. with resp
491q01v . vol 1. p. 427-- says biennial-wall-flowers & SCARLET Lychnis can be propagated by cuttings.-- Try.-
523m015 on horse being insane at the sight of anything SCARLET.-- dogs ideotic.-- dotage.--» Doctor communica
541m091 hen if one endeavur to keep any simple idea as SCARLET steady before mind for period, «if the scarlet
541m091 scarlet steady before mind for period, «if the SCARLET was before one effort less» one is obliged to
541m092 (or an emotion not so) than if simple idea as SCARLET?-- How can people dwell on pain ¿ no definite
552m130 vivid as in sleep-- (one can dream of intense SCARLET??) is it because one then has no immediate con
587n088 abit.-- A blind man might be born with idea of SCARLET, as well as remember it.-- Why do children pou
072r160 of Iron and Nickel, & these masses which are SCATTERED over the surface of the ground are fibrous. m
122a115 Beagle Channel. on origin of mud with stones SCATTERED irregularly.-- (Mem near Gregory Bay). Shrops
147g020 60 degree Below Loch Tulla whole wide valley SCATTERED with few very small & irregular hills of allu
272c109 t, though we have seen species «of subgenera» SCATTERED over it.-- We have abundant instances of rema
316c246 o formation of new species some few have been SCATTERED over whole world Many shells at present day s
355d067 foretell what will happen & to see bearing of SCATTERED facts.-- What takes place in the formation o
363d102 story of Water-Wagtails mistake both species SCATTERED over Europe)-- The habits of some «same» Nort
398e006 urther back we obtain here & there in order a SCATTERED page; we find <great> sensible change in the
117a101 taken place, otherwise the world would daily be SCENE of ruin in late Natical Magazine (before June 1
437e140 t by <some> «a» shepherds, who was watching the SCENE.-- «In Shiant Isld. it is said, that an Eagle a
531m051 e evening when tired «-- how true the heart the SCENE of anger.--» to the pianoforte, it seemed solel
603o11b ct.-- p. 134. a painted must not a actors, or a SCENE in garden.-- yet both beautiful! p. 136. Says A
528m036 explains insanity.-- Analysis of pleasures of SCENERY.-- There is absolute pleasure independent of i

529m041 how ones feelings would be excited, & how the SCENERY would rise. Deer in Parks ditto.-- My Father s
315c242 ldren or rather their memory. very remarkably-- SCENES in themselves accidental-- My first thought of
526m029 onkey), & not imagination.--» Thinking over the SCENES which I first recollect, «at Zoos» they are al
526m029 gs, which are brought to mind, by memory of the SCENES. (indeed my American recollections are a colle
533m059 Guy. Mannering. feels, pleasure. in seeing the SCENES of his childhood without knowing why-- had not
540m089 ks sympathy could be barren. & lead people from SCENES of distress.-- see how a crowd collects at an
275c120 rage-- (Bull-dogs are used because they have no SCENT!) Mr Wynne) at end of chase would not run up hi
592n101 f reason. as gradually developed. see Hume on SCEPTICAL Philosophy. Hume has written "Natural Hist. o
592n101 ct.» I suspect the endless round of doubts & SCEPTICISMS might be solved by considering the origin of
510q017 - Hooker <Meta> Metaphysics of Morphology. ▌-- SCHELGEL is he serpent man? about zones separated by n
262c073 will be formed in any kingdom of nature, where SCHEME not filled up, (most false to say no passages;
277c126 m Mr Herberts law; habits determining fertility SCHEME for abolishing specific names & giving subgene
310c223 <necessary to account> «consequence of» for the SCHEME of nature) and animals that man has different
635j56r xplain death, but reproduction]CD though such a SCHEME. would require constant miracles.-- p. 420 thi
638j059 endowed with the knowledge of trying a hundred SCHEMES. of structure, in the course of ages «step by s
638j059 in idea (with consciousness.) <th> form these SCHEMES.-- I see no reason, why structure of brain sho
077r171 ther bed or vein (very like that of Spital of SCHEMNITZ in Hungary.) Humboldt says fragments from roo
326c265 ange of plants Books quoted by Herbert. p. 338 SCHIEDE in 1825. & Lasch. Linn. in 1829 has given list
022r007 ape] Turn P. 434 & 419 As Limestone passes into SCHIST scales of chlorites--Mem. Maldonado P 375 Much
090a025 lanets to an end? Fragmentary granite showing SCHISTOSE structure (& veins appearing): mem. Henslows
116a100 ing in mouth of Plate. p. 26. Geology of Arica <SCHIT> Schmidtmeyer travels into Chile p 29. gold is
116a100 th of Plate. p. 26. Geology of Arica <Schit> SCHMIDTMEYER travels into Chile p 29. gold is not sought
110a087 ge of Guayana as NW / SE. Vol VI. p. 247. Mr. SCHOMBURGK NW. numerous boulders of GRANITE" "direction
489qIFC n p. 21 Horticulturists p. 21--23 Eyton p. 22 SCHOMBURGK.---- 1 Jordan Smith. p 1. Sowerby Cuming. --
522m011 -- My F. then said you remember Jack Baldwin at SCHOOL.-- Answered To be sure I do.-- What became of
592n102 showing that the signs invented for Deaf & dumb SCHOOL & used between Indian tribes are Many the same
599o005 --]CD in Athenaeum "Smart-- Beginning of a new SCHOOL of Metaphysic,"-- give my doctrines about orig
610o033 the great division amongst metaphysicians-- the SCHOOL of Locke, Bentham, & Hartley, &. the school of
610o033 the school of Locke, Bentham, & Hartley, &. the SCHOOL of Kant. to Coleridge, is regarding the source
510q017 hat does Blume say on alpine Flora of Java? Has SCHOW written on double creations & where? How are cu
514q21. olanum impregnation before flower open. (An. des SCILY Where is Boerhaave's paper on impregnation of vi
220b200 yell has remarked about no confined species in SICILY. Jan: 1838 L'.Institut. Bats, in Eocene beds,
291c168 nd Isds++.-- Mem Lyell hypothesis of change in SICILY.-- Splendid Harmony these views-- did Lamarck
130a134 eparation A Paper by Parrott Mem. Acad. Peters. SCIENC Math. Phy-- Nat. t. I, 1831. sur le temp du gl
044r073 rld will turn out simple.-- Fortunate for this SCIENCE. that Europe was its birth place.--Some genera
236bIBC era we might expect.-- Lindley Introduct Dict. SCIENCE. Naturelle Geographie Botanique De Candoelle.
266c092 me to flower at their own periods.-- Arcana of SCIENCE & Art. 1831. p 160. account of Bulbous root fr
275c119 ith what may be called the prophetic spirit in SCIENCE--. the highest endowment of lofty genius Using
276c123 of individual scientific men is to push their SCIENCE a few years in advance only of their age. (dif
326c266 t.-- Ld. Brougham. Dissertations on subject of SCIENCE connected with Natural Theology.-- on instinct
355d067 re of the greatest service, towards the end of SCIENCE. namely prediction.-- till facts are grouped.
434e128 the past world, we have no foundation for our SCIENCE".-- <it is only analogy.> but experience has s
459t009 ps. squeezed into Mr Walker's law Gleanings of SCIENCE Vol III. p 320. Mr Hodgson on Musk Deer-- youn
529m040 urface &c truly poetical. (V. Wordsworth about SCIENCE being sufficiently habitual to become poetical
553m135 (<thus> &. hence arises the theological age of SCIENCE in every nation according to M. le Comte).-- T
566n012 age M. Le Comte's idea of theological state of SCIENCE, grand idea: as before having analogy to guide
567n014 passion & sends blood to its breast &c &c All SCIENCE is reason acting «systematizing» on principles
567n014 ch even animals practically know «art precedes SCIENCE-- art is experience & observation.--» in balan
580n062 well.-- Lr. Brougham «Dissert.» on subject of SCIENCE connected with Nat. Theology.-- says animals h
602o010 ery curious as showing "the perfection of this SCIENCE of abstract form" is the source of part of the
132a137 -- Ask Hopkins. M. Parrot, Mem. Acad. Imp. des SCIENCES. (Sc Math. Phys. et Naturelles. Tom I. p 501.
220b198 Feb 1838) that few months since in Annales des SCIENCES. paper on Botany of Tahiti In Charlesworth Ma
241c016 dio Polynes: <)> vegetation far East) Ann: des SCIENCES. Septemb. 1825 Get Henslow to read over the
308c219 sics!!! Mrs Somerville, Connection of Physical SCIENCES p 276 May be worth glancing at. as she has no
317c251 s Expedition Journal of the Academy of Natural SCIENCES of Philadelphia Vol VII. Part II). 1837 accou
324c268 hilosophie. or Geographical distrib. «in Dict. SCIENCES. Nat. in Geolog Soc.» F.. Cuvier on instincts
459tf02 anges. [important view, copie[d] Gleanings of SCIENCES. Vol. III p. 83. Paper translated from Meckel
508q016 Cape.-- ¶About two vars: of Lion: Annales des SCIENCES¶ (4) Prolifixness of female, relation to hea
567n014 triangle shorter than two. V. Whewell. Induct. SCIENCES-- Vol I p. 334 Does a negress blush.-- I am a
109a084 ite, &c most important.:-- must be studied.-- SCIENTIFIC Memoirs Edited by Taylor Ehrenbergh on flint
114a093 their own range in Australian Alps.-- Taylors SCIENTIFIC Memoir, Part IV. p. 403 Ehrenberg on ferrugi
269c103 Kingdoms of Nature, their life & affinity" in SCIENTIFIC Memoirs I can see that perfection may be tal
276c123 ronomers.-- then add chief good of individual SCIENTIFIC men is to push their science a few years in
283c146 e know from Ehrenbergh, there are fossil (see SCIENTIFIC Memoirs & L'Institut.) that there are Tertia
320c275 Pamphlets Wilkinson on Cattle not abstracted SCIENTIFIC Memoirs. published by Taylor Magazine of. Zo
451e177 ing with deer-- Horsburgs. Vol II. p. 527.-- <SCIENTIFIC Soci> Journal of Asiatic Society Vol I. p. 2
508q016 sm in the MALE Troughtons.-- paper in Taylors <SCIENTIFIC Memoirs (11) And. Smith Savages at Cape any
242c018 New Guinea. All the isles of Oceania have the SCINCUS with golden streaks-- the lacerta vittata exte
247c028 enemous snake was> one Gecko on Isle of France SCINCUS multilineatus (p 45) Moluccas & New S. Wales S
247c028 s multilineatus (p 45) Moluccas & New S. Wales SCINCUS Cyanurus «p 8 &». p 49 on all the Moluccas «Ne
498q007 certain amount of variation in plants raised by SCIONS, as Elms. &c &c-- I have some reason to suspec
313c236 parent are concentrated in different parts, & SCISSION cannot effect the process.-- but why two sexe
313c236 cannot effect the process.-- but why have sexes SCISSION in all cases probably gemmation (.Ehrenberg)
314c237 given ∴ Those animals, which only propagate by SCISSION can not alter much.?? Mr Brown showed me Baue
059r119 terrompues par quelque vallées ou par quelque SCISSURES profondes, on les voit se reproduire a des ha
059r119 des montagnes qui forment les vallées ou les SCISSURES.--M. B. thinks these parts incontestably form
482z011 s Furnarius.-- Sturnus Magellanicus.-- p. 210. SCOLOPAX very close to ours Rengger's work of Mammali:
155g066 ed> how much more so, these lines & even water-SCOOPED rock «only decay from fragment falling» of no
156g068 il is <the> usually slaty Point of rounded not SCOOPED rock on <bend> of 3(a) Cannot <see> «make out»
045r077 lexing. than those that attend Eruptions: Mr P. SCOPES explanation of low Barometer? In a subsiding a
398o005 ill be the last to object to this theory on the SCORE of small change.-- on the contrary islands sepa
080r181 Lucs travels Beauforts Karamania Capt. Ross. & SCORESBY deep soundings Gilbert Farquhar Mathison trav
132a138 ean).-- and M. Parrot does conjecture that in SCORESBY'S case volcanicity has warmed it. Is not cold
042r067 ioning pumice of Bahia Blanca, mention black SCORIACEOUS rocks of R Chupat. & fall of Ashes of Falkne
033r043 Pitchstone; Mem Galapagos. chiefly red glassy SCORIAE.--could walk round base:--not universal: could
118a103 ike. which has not formed volcanos. or become SCORIFORM. has thinned upwards & is now cut off by denu
639j28r aroo. only a caricature; Penguin.-- Pincers in SCORPION & Crust in Squilla. & Mantis. C D woodcuts st
296c185 t true are wonderfully absurd.-- p. 565 <breed> SCOTCH wild Cattle. breed freely with the tame Vol II
305c210 irds-- At Zoolog Gardens there is half Jackal & SCOTCH Terrier.-- certainly more like Jackall in gait
499q010 other> clumps from other parts? Don says Irish, SCOTCH & English plants generally distinguishable.= W
574n039 -- Hensleigh says. Douglas. «& Spencer», an old SCOTCH Poet, has numerous lines. of poetry.-- <signs>
537m077 obeys & hurts conscience more than other.-- A SCOTCHMAN will his country or Swis.-- it may be answere
165g126 the case Wednesday 12/ & 3/ Why is the Tetrao SCOTICUS & Tetrao-- not an American form The union of
345d043 s from Glen Roy Note Book.-- Why is not Tetrao SCOTICUS. an american form (if so)?.-- A Sphepherd of
072r159 w thinks N & S. line connects western isles of SCOTLAND & Iceland.--Bosh nor on Norway, or Spitzberge
120a110 te elevation of Patagonian blocks (1200 ft??). SCOTLAND at least 2200. Jura 4000 feet.-- The veins of
218b191 that give him a species from Ireland, England, SCOTLAND & other localities & each one will have a pec
424e100 Lyell says «fossil» shells from North America, SCOTLAND, Uddevalla. Many species same. & Northern for
465t080 vers entering.-- & yet no shells-- now look at SCOTLAND-- coasts of Chile, excepting Concepcion-- Pat
502q11v s Book.-- 32. Would wheat from AEgypt ripen in SCOTLAND?-- to show acclimatisation.-- July <1842> Whe
515q021 order (2) History of fruit trees far north in SCOTLAND-- do they flower-- do they live healthily, or
293c172 ructure of brain not probable) put note. Sir W. SCOTT has written about it]CD If we saw a child do so
323c269 «well skimmed.» 1839 Jan 10t.-- All life of W. SCOTT.-- except the V Volume.-- -- 19t. Mungo Park--
551m129 a stone & pound the earth. Lockarts life of W. SCOTT Vol VII p. 35 "as ideas come & the pulse rises,
576n046 lty than the acquirement of new ideas.-- Walter SCOTT «(Antiquary)» Vol II p. 126 says seals knit the
509q017 in foetal state: Mr. Horner. On Mr Tremenheres SCOTTISH Colliers, when men & women have long worked,
322c270 notes <Rengger &c> Mitchell's Australia Walter SCOTTS life I & 2d & 3rd Volumes Abercrombie on the I
525m026 at fat men are goodnatured, & vice versa Walter SCOTTS remark how odious an illtempered fat man looks

Page **************************************(Key Word)**
550m126 perty"-- is not this rather more friendship.-- SCOTT'S Life. Vol I, p. 127. Talks of difficulty of hi
568n019 ell on Scrope, Quarterly Review. 1827? In Water SCOTTS life.. Tom Purdie, (beginning of Vol V) <final
057r112 eglecting these final causes.--What more awful SCOURGES to mankind than the Volcano & Earthquake.--Ea
581n065 dawn of civilization-- thinks many words, roar, SCRAPE, crack, &c, imitative of the things.-- CD[I ma
472s02v ghts, it cleaned sucker & <I think> pollen was SCRAPED off, which appeared like Heartease pollen.-- t
536m071 Hyaena likes smell of that fatty substance it SCRAPES off its bottom.-- it is relic of same thing th
552m131 mem: Yarrell's story of wheel horse in drays, SCRAPING against cornice stone to cause friction Athen
120a110 orthy of attention-- rear Glen Roy Notebook-- & SCRAPS on Salsisbury Craigs. Kept amongst <old> paper
529m039 lity, rustic life, virtuous happiness.-- recall SCRAPS of poetry;-- former thoughts, & in experienced
545m107 hen closed my mem. expression of fury, jump to SCRATCH my face. The ourang outang, under same circums
542m094 orth, this & yawning. (common to other animals) SCREAM of agony, sigh of discomfort & weariness. & me
093a034 rence to fossil guanaco of P. St. Julian. -- Mr SCROPE seems to consider that elevation & eruptions a
097a043 higher parts? & felspathic veins?-- Mr Poulett SCROPE. talks of Trachyte, "superficially coated by a
099a051 layer!!!.-- The separation in the Ponza case of SCROPE parallel to walls of dykes-- Mem. laminated di
568n019 divines to progress of knowledge. see Lyell on SCROPE, Quarterly Review. 1827? In Water Scotts life.
196b105 of easy transportal.-- Waders & Waterfowl.-- SCRUTINIZE genera, & draw up tables-- Instinct may conf
606o022 ssings Laocoon. 2d Lect-- The object of art., SCULPTURE & painting, is beauty.-- which he thinks is a
606o024 says new subjects are not fit for painter or SCULPTURE, but rather subjects which we know, it is the
495q05a very kind of seed must be distributed.-- Examine SCUM of pond for seeds.-- 11. Soak all kinds of seed
025r016 a very shoal coast. Beyond the 10 or 12 leagues SEA deepens suddenly. coast of Brazil generally.-- M
025r019 t life is exceedingly rare, at the bottom of the SEA.--«certainly data insufficient, yet good» «(I su
026r021 rt town between the older strata & the bottom of SEA near T. del Fuego.-- Is there account of Baron R
027r027 of the Bahama consists of rock & sand mixed with SEA shells--about 500 Isd. & great banks. effect of
027r028 of revolutions. by actions of rivers currents. & SEA beaches. All mineral masses must have a tendency
028r028 eral masses must have a tendency. to mingle; The SEA would separate quartzose sand from the finer mat
028r030 alia.--Have Limestones all been dissolved. if so SEA would separate them from indissoluble rocks? Has
029r033 yage at <[...] Maranh> Pernambuco. EARTHQUAKE AT SEA.--Extract from the log-book of the James Cruiksh
030r035 collecting, in the bottom of an open & not deep SEA.--(Character of coast regular & <not very> rathe
032r039 modern Conglomerates [Fig. 2] {P} The action of SEA A. B. will be to eat in the land in line of high
032r039 not having been worn away.--If the level of the SEA was to sink by very slow & gradual movements to
036r051 n Daubeny. P. 438., of similar fact near the Red SEA.--which occurred in a sandy place.--(the sound w
041r064 btful whether they could have lived in so deep a SEA.--Perhaps agrees with formation of pebbles & ver
042r068 must be modified.«Moreover, the Volcanos from SEA there burst out, after rise from sea: <As did> a
042r068 lcanos from sea there burst out, after rise from SEA: <As did> as did those aerial Volcanos in German
044r071) cannot believe in a great explosion, nor would SEA remove more internally than externally--I did no
045r076 erally over whole world) Yet eruptions <both> at SEA (as wells as in the Cordillera), they may be con
046r080 & 3 ft. (as in Chili lake). Therefore motion of SEA ought to be considered as a plain movement commu
047r081) no great wave on record. -- «also neighbouring SEA must partake in absolute movement» Moreover wave
047r081 motion what difference? In watching heavy swell, SEA retreats & then breaks: i e to form a wave in oc
047r082 "mean" level before the higher part. -- Does the SEA fall on banks as a Bore wave rushes up? (NB. Ear
047r083 ed above as). -- In great Calabrian wave did not SEA break first? I can imagine from local form of co
048r084 lies motion in the «loose» bed of pebbles. (On a SEA beach under a cascade, one can understand pebble
054r105 st» may be concluded to have been covered by the SEA judge from the pebbles such as those on the beac
054r105 ere in all appearances not many years since, the SEA covered above half a league of what is now Terra
055r106 ashed by the waves, a sufficient proof, that the SEA formed these large cavities", &c &c &c Vol II. C
055r106 II 147 Shells at Concepcion 50 toises above the SEA. = talks of them being packed clean. & without e
057r115 nt!!! Remember idea of frozen bottom or beach of SEA to explain preserved animals.--Mem: stream of wa
058r116 ough be easy to see on beach successive lines of SEA weed-- Histoire Naturelle des Indes Acosta. p. 1
060r124 upwards; for the land is constantly pushing the SEA (which of course must retain same level) to a gr
062r129 a: wide limits of Waders: Ascension. Keeling: at SEA so commonly seen. at long distances; generally f
066r140 (Think some 60 fathoms, none thicker than thumb» SEA weed said at Kerguelen Isd. to grow on shoals li
068r147 es] [Fig. 8] {P} (A. B. C, now grown solid.) Red SEA near Kosir, land appears elevated. Geograph. Jou
068r148 --One need never be afraid of speculating on the SEA The 24 ft. elevation at Concepcion. from impossi
080r180 . 14-91. gradual shoaling of coasts 93 action of SEA on coast. 27. Bahama Isd De Lucs travels Beaufor
086a004 prejudice of not believing recent elevation, yet SEA shells at tops of mountains we ought to sympathi
087a007 o plants to it, lately raised above level of the SEA. Lyells Encyclopaedia-- Lately elevated When Sib
091a025 N. Phil. J. p 194. Fact of dust blown far out to SEA valuable; because transportal of Minute seeds--
100a055 r's (p. 213) form of escarpment relation kept to SEA coast ∴ curious exception in Wealden.-- Would cr
102a059 ards.-- form of breaker affected some way out to SEA.--¿ effects on bottom a thing floating some way
104a069 reccia, introduce in Cordillera discussion, deep SEA, fragments fall off cliffs. but then how spread
105a069 wide difference between erosive power of river & SEA.; the former as its channel becomes wider looses
105a069 becomes less &. ∴ tends to finite power) whereas SEA. on coast, as long as exposed to waves of sea, c
105a069 as sea. on coast, as long as exposed to waves of SEA, cutting power increased with width. for besides
105a071 e very part, where barrier least probable.-- The SEA harmonizes well with character of mouth of valle
105a071 l downwards by successive torrent spread out. by SEA-- beach action -- no one will dispute. sea. once
105a071 t. by sea-- beach action -- no one will dispute. SEA. once came to Mendoza--). Will they introduce oth
105a072 came it if this powder results from «decomposed SEA» shells, that land shells should be preserved in
109a082 wn by the number of bones lying at the bottom of SEA. off coast of England.-- Sea must always on actu
109a082 ng at the bottom of sea. off coast of England.-- SEA must always on actual beach act same way.-- a li
111a088 ol VII p. 279. Carcases of birds drifting out to SEA do p. 358. changed soundings in Mouth of S. Cruz
114a094 s-- metamorphosed clay slate.-- --shale in shall SEA. Lyell confounds these introduce discussion -- I
116a100 Cordova project on plain, like <re> a reef on a SEA beach--«p. 151» first discovered «very small» b
119a105 pendent of spreading out matter by action of the SEA.-- as no sea exists there.-- But Sir John consid
119a105 reading out matter by action of the sea.-- as no SEA exists there.-- But Sir John considers an irregu
120a107 road. (¿ metamorphic action at the bottom of the SEA?) All this profoundly considered. study Hopkins.
126a123 ature ought to increase rapidly beneath level of SEA.-- deep seated springs «spring requires connecte
126a123 ow, that water does percolate, & springs beneath SEA-- → According to this latter view the rod is rev
127a127 Vol III? p. 246. on formation of cones beneath SEA-- with reference to old submarine orifices in C
128a128 purest fresh water must be sought for below the SEA mark.-- If mountain chains are matter piled up.
129a132 eral case given of hot heads &c heat beneath the SEA.-- CD[did not Beechy have some such case]CD what
132a138 e know volcanic action prevails more beneath the SEA, <than> «&» on coast lines, than on continents.
135a144 r looking for Copiapo. found inland a great many SEA shells some miles from coast-- quote passage to
137a151 p. 155. the increase of temperature beneath the SEA, is probably much more rapid than beneath contin
138a153 del Lima par Dr. H. Unanùe says he believes the SEA has formerly stood three hundred feet above its
147o017 ier of lake very lofty, & no trace of it; to the SEA more probable I did not look carefully for Marin
147o042 erworn pebbles in Alluvium which without lake or SEA could not be placed in present position Thursday
147o020 of alluvium-- nothing very striking yet possibly SEA more probably than river-- No exact terraces but
148o026 day. Before coming to Bridge of Speam, hills of «SEA», gravel, current cleavage, & pretty well rounde
149o030 much-- On other hand remember modelling power of SEA N of Valparaiso are those animals subject to muc
153o051 of cliff" Others below it--argument for lake «or SEA» at successive levels-- {P} Shelf opposite Glen
155o063 tion of river must constantly alter with falling SEA & so corrode plain into terrace as regressed Wha
155o065 hat above straight line «only» cut deep gorge on SEA hypothesis, if gullies not now formed «(Mac, hyp
156o072 at river alone had modified it-- perhaps however SEA also,-- Barometer on shelf 3d. 29.455 A 83 degre
158o081 ft beneath shelf peat on pebbles tidal plain as SEA gradually retired, hard to explain on river doct
162o105 Lochy is 8 ft below Loch Oich wh is 92 ft above SEA-- Loch Ness 40 ft above do. When cutting bank wh
163o106 found alternating layers of coarse & fine & many SEA shells. My informant saw them himself-- Sand wit
163o108 he above shells must have been about 60 ft above SEA-- soon decayed on exposure Mr H. C. Watson Geogr
200b124 sso Vol III p. 155. about quantities of seeds in SEA; also Holman: <at> Keeling these are most import
213b172 ut seal-bones & cetaceans.-- both found in every SEA, from Equatorial to extreme poles.-- Oh. Wealden
214b174 the closest relation to those now living in the SEA.-- See Rogers report to Brit Assoc <to> on N. Am
216b179 OGUES sums this word «for» similar in the Indian SEA.-- Deshayes.-- Mr McClay is inclined to think th
225b220 ency.-- Study Ellis & Williams. zoology of South SEA islds. any animals?-- I believe none.-- Canary i
227b224 countries Inner India, Mexico & Europe. one gret SEA (Coral reefs∴ shallow water at Melville Isd. (3d
232b246 ward.... species must be compared to neighboring SEA-- For change of species does not measure time b
241o016 - M. D'.Urville on the Distrib of Ferns in South SEA (Indio Polynes: <)> vegetation far East) Ann: de
244o021 to show wide range of fish & shells in tropical SEA, it. would demonstrate.; not distance, makes spe
246o025 Indies, isld, as far as Oualan.-- Wide space of SEA, to East of America. would account for this.-- C
251o040 mber 20th) Paper by Jenner, on birds seen far at SEA, migrations of species, geese killed in Newfound

(Key Word)

256c054 es species in other animals.--? Forster on South SEA, will probably contain descriptions of domestica
257c056 atagonia. then in Chile & lastly 12,000 ft above SEA in Bolivia; he examined-- all species & found "b
259c064 d, ".' not applicable to plant. Epidemics of South SEA. wonderful case of extermination of species-- Ep
267c094 ub".-- p. 229. carcases of birds drifting out to SEA Vol VII. p. 325-- Wild dogs of Guayana always hu
278c131 more than 1000 ft. & many hundred miles from the SEA, associated with teeth of seals and dugong, ther
306c212 ived in estuaries its taste. taught it to go to <SEA> salter water (& its necessities teach it taste,
315c242 s in themselves accidental-- My first thought of SEA side-- Study Bell on Expression & the Zoonomia,
316c245 nphidemas.-- this is remarkable.-- Fish & drift SEA weed-- may transport ova of shells.-- Conchifera
326c266 rst volume of Annales of Vienna. sketch of south SEA. Botany R. Brown. has curious coloured maps. by
355d068 ssils same remark. ¿was there formerly one great SEA, & two Polar Continents Marsupial. Edentata.-- P
377d140 estimate the matter removed by the waves of the SEA, on beaches-- we really, measure the rapidity of
379d151 some same-- what is the alliance with the Black SEA.-- it would be ocean, what is land to continent-
436e134 ge proportion to other mollusca in cold parts of SEA, like Cetaceae,-- although the Cephalopods, seem
448e166 ouraine beds, the one with neighbouring & Arctic SEA, & the other with neighbouring & Senegal as sea.
448e166 sea, & the other with neighbouring & Senegal as SEA.-- is remarkable.-- Again the resemblance betwee
462tf4v -armadillo & the woodlouse-- -- a good analogy-- SEA-Crustacea-- Tullus. Athenaeum 1839 p. 772-- A cu
473s004 eposits are probably buried in the depths of the SEA-- Maer. June/42/ June/42/-- Mr. Bunbury says has
478z003 te Cape Pidgeon's stomach small shells (patella) SEA weed & many pebbles Mentions stinging Millepora.
479z006 ge Vol I, p. 168 Ceratophytes common in Northern SEA. Chamisso in Kotzebue p. 312 Leaches on leaves i
510q017 a & S. America? Sabine says North of Siberia, no SEA-current, icebergs travel by wind. Aug. St. Hilai
528m038 beauty of some as Norfolk Isd fir shows this, or SEA weed, &c &c-- this gives beauty to a single tree
541m090 ught is following out such an idea, as effect of SEA on coves when waters had fallen, as in my Glen R
541m091 rame of mind in following changes during fall of SEA.-- Is the effort greater if the idea is abstract
582n067 as much as in the young salmon to go towards the SEA. or down the stream; which it does unconsciously
586n080 en taught, where to go-- the act of crossing the SEA in dark night & not loosing its direction, equal
213b172 certain strata could now not accumulate without SEAL-bones & cetaceans.-- both found in every sea, f
337d022 tion required if he believes «hyaena & squirrel» SEAL & mouse, elephant, come from one stock.-- Theor
422e092 - but are aquatic Pachyderms. & Walrus-- aquatic SEAL.-- (Consult this passage, when considering orig
258c061 ave any slight peculiarity of structure. «hence SEALS take victorious seals, hence deer victorious de
258c061 rity of structure. «hence seals take victorious SEALS, hence deer victorious deer, hence males armed
278c131 ed miles from the sea, associated with teeth of SEALS and dugong, therefore immense age since breccia
576n046 - Walter Scott «(Antiquary)» Vol II p. 126 says SEALS knit their brows when incensed.-- A Dog may hes
641j29r s-- Whales. «Narwhal» Polar bear. Walrus, great SEALS of Antarctic seas. (on other hand Spermaceti Wh
641j29r ède on Cetacea & Geographical Distrib of larger SEALS-- Are Porpoises numerous in cold Oceans I think
145g004 ll as base not always parallel to strata 3 or 4 SEAMS/ 3 or 4 inches thick-- {P} 35 degree is I belie
028r029 aining Alumen.--This matter accumulating in deep SEAS forms slates: How is the Lime separated; is it
047r081 epting in number of waves & in wind, instead of SEA'S bottom being in motion what difference? In watc
051r094 confervae) coated with living beings; In smooth SEA'S (& even turbulent as at St Helena) I have menti
055r106 hese facts attest a <more> decided elevation of SEA'S bottom. beds of shells. 2 - 3 toises thick.--Vo
088a015 fly at Heron.-- Ceratophytes common in Northern SEAS. p. 312. Chamisso in Kotzebue. Study Humboldt. F
214b172 y deposits present analogies to shells of living SEAS.?-- Roxburgh. list of plants in Beetsons St. He
231b242 you will have two. Tapir existing in East Indian SEAS. Marsupials animals all show greater connexion
232b245 species in proportion to genera than in present SEAS, «All» The <one> species which survives any cha
314c238 e will give sketch of botany of islands of south SEAS says so in preface.-- Mr Brown says character o
322c270 ctob 3d Malthus on Population W. Earls'. Eastern SEAS. .Octob12th.-- Sir G. Staunton's Embassy to Chi
402e018 hanges would be observed.-- G. W. Earl's Eastern SEAS. p. 206-- shot a monkey, ceased their cries. "m
402e019 e born so in all Malay Countries W. Earl Eastern SEAS. p 233 Octob 12th Kotzebue's second «1st» Voyag
419e086 300 ft are identically same as those of present SEAS.-- now in this country we have better means of
453e182 fall of fish in India.-- Windsor Earl-- Eastern SEAS p. 229. Believes the «Rhinoceros» «Tapir» is fo
480z008 pale brown Caracara Krauss on Corallinae from S. SEAS written in German.-- Stuttgart ranks these bodi
634j55r we to explain this.-- Did reptiles first inhabit SEAS.-- Were they then killed out «by the increase c
641j29r al» Polar bear. Walrus, great Seals of Antarctic SEAS. (on other hand Spermaceti Whale & Manatee.-- N
060r126 4. Webster Consult W. Parish. & Azara about dry SEASON[.] 1791. seen commonly bad over whole world. «
204b141 of male widgeon, winged & turned on pool, first SEASON bred readily with common ducks.--» Kirby all t
563n001 ps his wings before? crowing & only in breeding SEASON & on the ground.-- Cock fowl. on the ground, a
640j28v liminating the evils of old age, after breeding SEASON, or gaining adaptations, but for youth most ne
563n001 .-- Cock fowl. on the ground, at roost, in all SEASONS, & after? he has done <₽> crowing.-- instances
582n068 eory explains how it comes that the heart is the SEAT of the emotions.-- but are not love & hate emot
126a123 o increase rapidly beneath level of sea.-- deep SEATED springs «spring requires connected column.--»
040r062 e degraded matter of such pebbles extending to SEAWARD, the alternating with such matter at St Julian
052r097 ine fixed.--[Fig. 5] {P} * Slope necessary for SEAWARD transportal of drift matter.-- Give various ca
052r098 rious cases. [Fig. 6] {P} A advancing coast is SEAWARD. Retreating case in excess as first case. When
079r177 rom three directions. from W. NW & S.--last to SEAWARD partaking of the character of a Araucarian tri
085a001 p. 112 Speculate on the extension of Patagonia SEAWARD, at mouth of S. Cruz. from ascertained inclina
115a098 {P} each stratum would thin out, both inland & SEAWARD: if matter too coarse, then {P} that form.-- A
633j54r to <arrest> «prevent» the valuable soil in its SEAWARD course,-- we sink into such contemptible queri
024r016 Paranagua [25 degree 42 minute S.] 12--40 St SEBASTIAN [23 degree 52 minute S.] 12 50 {T} Joatingua
554m138 y green monkey (from Senegal he thinks Callitrix SEBE??) he has seen place its head downwards to look
275c120 rossing them with finest greyhounds.-- Sir. J. SEBRIGHT first got {P} point on hackles on Bantams by
279c133 t Africa & India???-- & Indian Islds.-- Sir J. SEBRIGHT-- pamphlet-- most important, showing effects
280c134 ack to either parent.-- Shows instinct (Sir J. SEBRIGHT admirable essay) heredetary Young wild ducks.
280c134 cumstances in both cases effect. it.-- Sir J.. SEBRIGHT excellent authority because written on dog-Br
291c165 etary tameness as well as wildness-- cf Sir J. SEBRIGHT.-- love. of man gained & heredetary. «problem
364d103 been known one has killed itself.-- Q Sir J. SEBRIGHT-- has almost lost his Owl-Pidgeons from infer
366d108 had kittens with imperfect ones.-- now Sir J. SEBRIGHT. thought if he had had a pair he could have p
390d179 ried from their stock, for to breed (as Sir J. SEBRIGHT urges?) one with opposed characters is by imp
511q018 About the Bantams at Zoolog Soc.-- did Sir J. SEBRIGHT select to destroy secondary character believe
581n063 , & consequently instinctive action.-- Sir J. SEBRIGHT. has given the phrase "heredetary habits." ve
320c275 eming. & Review of latter in Quarterly Sir J. SEBRIGHT'S Pamphlets Wilkinson on Cattle not abstracted
263c070 ces may have been necessary to have made man! SECLUSION want &c & perhaps a train of animals of hundr
345d044 of animals in Britain shows, with the aid of SECLUSION in breeding. how easy races or varieties are
041r064 with formation of pebbles & vertical trees Grand SECO at B. Ayres; mention about the deer approaching
036r051 necessary to ascend the hill.-- The absence of SECOND form, except near submarine Volc: in harmony w
198b114 ay be) have not been created on the same plan. [SECOND resumé well worth studying] CD says grand idea
282c142 ts & not minding particular trades.--» then the SECOND race would not obtain a cast of washing men--
310c222 en have full structure «of one class» & full of SECOND--this class of facts «analogous to petrel-greb
334d010 kable law, that first cross <not se> plentiful, SECOND absolutely sterile.-- My case of Stallion, acc
334d010 new «a» carter well, who placed his stallion as SECOND horse between <whe> shaft mare & another leade
363d102 songs, though imperfectly.-- Male birds always SECOND their songs, the ++ Cervus Campestris spotted
365d107 cture of which is adaptation to habits (& habit SECOND nature) may be more in constitutional.,-- more
385d163 se smallest sized dogs.-- one litter big & then SECOND small & so.-- Says, there is breed of Fowls ca
387d169 theirs again in lesser degree.--now if the <Lw> SECOND race both have this peculiarity strongly; they
387d169 ce as first pair + tendency they inherited from SECOND pair, + the influence they themselves inherit.
402e019 Earl Eastern Seas. p 233 Octob 12th Kotzebue's SECOND «1st» Voyage. Vol II p. 344. account of insect
403e021 group, the number is very limited.-- Kotzebue's SECOND Voyage do Vol III p. 77. Many foreign plants h
469t119 s it is on one foot probably monstruous & not a SECOND species.-- <Saw> Maer. June 15./41/. Watched p
524m021 s that repetition is not necessary)-- the words SECOND childhood full of meaning:-- Dreams do not go
538m080 Ash's case, when he struggled as it were with a SECOND & unreasonable man.-- If one could remember al
538m080 all ones farthers actions, as one does those in SECOND childhood, <they> or when drunk they would ver
547m111 ng up.-- was lying on my back fell to sleep for SECOND & wakened.-- had very clear & pretty vivid «&
609o29v tive checks.-- (This is encroaching on views in SECOND volume of Malthus). Adam Smith also talks of t
419e087 <affecting whole bodies of species», & only SECONDARILY,, by assumption well grounded, on time;-- th
029r033 uth African plains.-- Sydney no I believe the SECONDARY? formations of Brazil, all originate from the
035r047 ise, mean height of tertiary. being less than SECONDARY:-- consider arguments for oscillation of leve
037r055 --strata must be accumulating which like the SECONDARY strata of England, «besides ordinary marine r
056r110 rimitive rocks between that formation and the SECONDARY (stated in Playfair to be the case p. 51). pr
060r125 reat chain of Volc. had been in action during SECONDARY period how diff. would the rocks have been. T
073r163 n. Vol III. p. 130 Metals in Mexico rarely in SECONDARY alway in primitive & transition; the latter r

```
074r164 he upper limestone, which H. calls by several SECONDARY names «Study Hoffmans account of steam acting
077r171 ancient  freestone, limestone & <many> «other SECONDARY» rocks. Vein traverses both Clay slate, Porph
114a094 not  formed in modern formation & not ever in SECONDARY in Europe. gneiss-- metamorphosed clay slate.
175b019 others» occasionally dying out;--for instance SECONDARY terebratula may have propagated recent terebr
222b205 ark. lived in remotest epochs.-- ¿ lizards of SECONDARY period in same predicament. It is another que
222b206 hat setaceae & Phocae now replace Saurians of SECONDARY epoch: it is impossible to suppose such an ac
250c040 ods & Trachilidous Molluscs. by each other in SECONDARY & Tertiary periods.-- p. 125. ref. to Phil Tr
257c057 ility.-- My views will explain no Mammalia in SECONDARY-spocks, & developement of lizards.-- As we ha
355d068 per in Geolog Transacts) same appearance with SECONDARY Species distinct-- but close.-- Mem. Von Buch
369d114 ect in <male> «hermaphrodite» (Fishes have no SECONDARY characters.-- )p. 49. (wonderful case of Pea
384d161 ubject-- ⫿Divides sexual marks into primary & SECONDARY, the latter only being developed, when the fi
465t089 how  many estuary formations are there in old SECONDARY Series-- few-- Maer June/41/, observed 3 plan
501q011 in Mules generally fails-- perhaps indexed by SECONDARY characters-- in double flower. do Henslow Spe
506q015 not  ripening= (38) Have Dioecious plants any SECONDARY, sexual characters.-- Stature, position of fl
511q018 glish animals. [Made no import. remark]CD (2) SECONDARY male characters.-- does male transmit to male
511q018 Soc.-- did Sir. J. Sebright select to destroy SECONDARY character believe No or did result appear wit
511q018 s wish Has since recrossed this breed.-- Have  SECONDARY male characters appeared.= (4) Does he know a
593n105 ne.-- porpoises a ditto-- it is probably some SECONDARY one-- blood being disagreeable & anything dis
606o023 ugh the medium of the eye"; he will allow the SECONDARY pleasures of harmonious colours &c &c  surely
629o055 d to ambition. money, books &c &c.-- <]> the "SECONDARY passion" of Hutcheson unfolded by D. Hartley.
535m069 ted] as first caused by will of Gods. «or God» SECONDLY that these are replaced by metaphysical abstr
089a018 , queque temps avant et apres, par de petites SECOUSSES."-- Tom 54. p. 106 do-- p. 110. Mountains on
077r171 rom roof & penetrating overlying beds tells the SECRET.-- p. 189. "The small ravins into which the va
304c207 aracters, in common with other birds reveal the SECRET.-- Now all the different forms of  Synallaxis.
274c114 s seen with long tarsi.-- «Ground woodpecker» SECRETARY bird.-- & Millisuga. Kingii very rasorial for
484o014 grees with D'Orbigny, that Serpent Eater-- or SECRETARY is S. African representative of Caracaras of
484o014 crest  feathering on legs-- habits-- Does the SECRETARY, make noise & throw head back M  Edwards,--on
373d132 spines.--  Does not male Pidgeon (yes) surely) SECRETE milk? from stomach. analogous to other males f
384d162 primordial vivifying principle) one individual SECRETES two substances, although organs for the doubl
521m008 embered, it was an habitual action of thought-SECRETING organs, brought into play by morbid action.--
074r165 at least corelation)--Galapagos vein. vein of SECRETION.--metallic veins follow mountain chain. there
291c166 ts full of meaning.-- Why is thought. being a SECRETION of brain, more wonderful than gravity a prope
390d178 rom different quarters) then it causes <to> a SECRETION of something someways different from himself,
412eo597 more than Mammae, which have given milk.-- is SECRETION from Pidgeons stomach true milk.-- ⫿ <Species
590n093 the Passion."-- p. 37. The increase of Bilary SECRETION attends passion p. 39. The sweat that accompa
615o038 appear a sort of acquired memory. a permanent SECRETION of thought, (or under contingencies of stimul
615o038 ingencies of stimulants of certain kinds such SECRETION) or an association of pleasures with  certain
616o038 erence between heredetary memory & individual SECRETION of thought, may be no more «difference»  than
616o038 is  involuntary, in man voluntary: ¿ False,-- SECRETION in both involuntary, <application in> «ejecti
616o038 ection only has» will: there must be cases of SECRETION being some time governed by will in some anim
641j29v nts or animals-- Exudation of fetid «& acrid» SECRETION in Mollusca. insects «Carabids & Staphylini»
041r066 n» being there in excess.-- If veins {P} are SECRETIONARY, so are all those plates in Australia.  New
591n101 in of religion or polytheism, at p. 424 Vol. II «SECT XV. Dialogue on Natural Religion.» however,  he
116a099 al Journal Rapport on D'Orbigny's Voyage. good SECTION of Rio Negro beds.-- -- refers to species  non
194b094 will  aversion be L. Institut «1837. No 246» a SECTION of fossil "singe", it cannot be made to approa
248c029 r nearly 70 year. Galapagos Mouse not the same SECTION, with house mice. It is wonderful how it could
248c029 rful how it could have been transported. ¿What SECTION does the New Zealand Rat belong to There is th
278c129 ar may be eliminated. where every species of a SECTION is confined to one continent & every species t
353d060 Malacca> «Malacca» bears, <are> belong to same SECTION with with those of India-- Waterhouse knows th
591n101 o the cold or benevelo- continent man Hume has SECTION (IX) on the Reason of animals Essays Vol  2.--
028r031 e dike in the whole Galapagos Arch; because no SECTIONS> same cause as no colour Sir J. Herschels beli
249o035 ¿Carnivora  of New & Old word. do not form two SECTIONS is this not connected with wide range of anim
278c129 ontinent & every species to another then those SECTIONS & subgenera; are analogical, because we do no
538m079 he other army officer, horticulture & religious SECTS.-- yet Allen. W. remark about his slippers  bad
232b247 of the other.-- If the the world had cooled by SECULAR refrigeration in chief part instead of  change
232b247 thern Europaean ones.-- "a variation played on SECULAR refrigeration".-- <The Phenomena of the S. Hem
267c094 rown>.. ears long.-- like bull terrier-- Indian SECURED one, as they always like to cross their breed
370d116 - bees breed but few, because they are kept in SECURITY.-- Hunter doubts about production of Queens.--
098a045 Lemagne  in Auvergne Proofs from Phryganea NB. SEDGWICK talks of LAMINATED structure (∴ separation of
432e126 in  Philosop. Journal <Mar> April 1st 1839) by SEDGWICK & Murchison; which is a beautiful instance of
099a049 r axis. But if great depth NB. Prof <Henslow> SEDGWICKS lamination parallel to stratification evident
053r100 would become more or less interstratified with SEDIMENT.--& escarpment worn away like english escarpm
055r107 what  has become of the Basalt. Gone into fine SEDIMENT Look at St Helena!!-- There are some argument
055r108 erbalancing variations) of rain. = The Bulk of SEDIMENT «daily» yearly brought down by every torrent
132a139 ay not the cold <bottom of> ocean. (with fresh SEDIMENT added to bottom) be caused, by absence of cir
153g053 h [shelf] Could earthquake cause collection of SEDIMENT? Where ravines enter side by, opposite entran
402e017 trata being merely leaf, if one river did pour SEDIMENT in one spot, for <whole> many epochs-- such c
028r032 erschels idea of escape of Heat prevented by SEDIMENTARY rocks, & hence Volcanic action, contradicted
053r100 lota Lyell, states that contact of Granite & SEDIMENTARY rocks, in Alps becomes metalliferous. Vol II
088a014 may picked up beneath the trees---- Are any Fish SEED-eaters. This important in transport of Fish Let
189b072 ly cut off:-- like golden pippen. if produced by SEED go on.-- otherwise all die.== The fossil horse,
223b209 changes are gradual, is this if after isolation (SEED blown into desert) or separation by mountain ch
225b217 uses, of which we are as ignorant. as why millet SEED turns a Bullfinch black, or iodine on glands of
228b229 ogether, were not both genera formerly abundant. SEED of Ribston Pippin tree <go> producing crab.  is
235b275 not intermix,--any cultivated plants produced by SEED.-- Lychnis.-- Flax.-- Read Swainson [blank]  In
265c085 arf plant in alpine district & dwarf plants from SEED, one adaptation, other monster.-- The only  way
305c211 pus & polypus, bud & bud, polypus & germ plant & SEED.-- instincts in young animals, well  developed,
372d129 planted & carry all these peculiarities not so a SEED.-- Bud probably is like cutting off tail of Pla
388d170 - When it is said. that difference between bud & SEED, that latter carries with stock of food.-- the
423e098 ould be to make enquiries of some of these great SEED-growers-- ).-- Feb. 24th. Monoceros, which Sowe
436e137 ed in numbers, what cause?? Seeing the beautiful SEED of a Bull Rush I thought, surely no "fortuitous
436e137 have produced these innumerable seeds-- yet if a SEED were produced with infinitesimal advantage it w
446e162 showing  analogy of production by gemmation & by SEED-- which Henslow is inclined to think very close
448e165 ropagate rapidly by buds, layers &c & &-- do not SEED freely.-- The periwinkle seldom produces seeds,
453e183 ssibility of different varieties being raised by SEED is highly odd-- as it is not so with the escule
453e183 t Hollyoak reproduce each other. & yet I presume SEED raised in same garden.-- now this good question
453e183 e.-- anyhow fertile because they <are> raised by SEED,-- Where has Duchesne described Atavism.--  ask
454e184 d» plants in England, which do not perfect their SEED?-- What annuals can be budded «& rendered of gr
466t091 ants of Caltha Palustris alone together. one had SEED-pods turning brown, whilst both others were in
472s01r n hybridised == Has tried several year to obtain SEED, but the pods have (except this one year (1827)
491q01v se & Cowslip in rich soil & propagate from their SEED 3. To apply pollen of different genus & then so
492q002 together: <do> Can Holyoak be raised distinct by SEED-- Heartease. 6. -- Do not species of wild Roses
495q05a ich has swum-- on pools & rivers-- every kind of SEED must be distributed.-- Examine scum of pond for
496q05a d. done Examine pollen of such flowers as do not SEED or seed rarely-- Magnolias. «Azaleas» & plants
496q05a Examine pollen of such flowers as do not seed or SEED rarely-- Magnolias. «Azaleas & plants grown un
496q006 m in hour or two «My Father made hens cast Holly-SEED & they grew» (9) Place. Snap-Dragon. I have se
497q06v here races-- if so plant them together. & raise. SEED.-- In letter Mr Herbert says do about OEnothera
498q008 eadily they cross.-- (9) In the nurseries, when «SEED of» the varieties of Cabbages, peas, beans, are
498q009 ey are dioecious, if no hybrids were produced by SEED, we might feel sure, that pollen of own kind is
499q009 20) As Hop is Dioecious-- seedsmen who raise Hop-SEED-- may know something about proportion of plants
499q09v s pollen excessively minute or abundant? do they SEED plentifully? Look for isolated females.--  Also
503q010 ss-- Also about Sugar-Cane Edwards says does not SEED-- «Bruce says does» Royle In Royle's productive
505q014 dised (10) one had no seeds, & two had plenty of SEED & these Seeds of unimpregnated Cowcumbers  will
505q014 hese Seeds of unimpregnated Cowcumbers will they SEED.?-- (11) Abberley has planted seeds of pale gre
505q014 al plants 17 A dead-nettle in Hot-house. will it SEED?-- (Skim through Penny Cyclopaedia) Abberley sa
506q015 (37) Any cases of plants. which will not produce SEED in this country-- where cause not apparent-- An
507q15v single  hook; curved spines-- simple spines-- or SEED-cases with similar structure.= good case as sho
507q15v n might be arrived at.= Any book with drawing of SEED. Anemone with, tuft-- Bull Rush-- Dandelion-- S
```

Page **(Key Word)***

```
507q15v seeds  with «mere» border-- & Humboldts spinning SEED.-- (50) Any cases of wild varieties plants grow
511q018 male  characters appeared.= (4) Does he know any SEED-raisers (5) List of qualities in birds & animal
515q021 ated show Heartease produce as large capsules of SEED, as the commoner kinds-- Cattle are horned, Suf
632j53r do not «then» believe the pappus of <th> any one SEED. (all have not it) was DIRECTLY created. for tr
633j53v y become heredetary [produce some peculiarity in SEED vessel]CD if man takes care they are not detrim
633j53v s long as they are not detrimental.-- p. 285 the SEED-pod of a desert plant (Anastatica) is rolled al
505q014 te it artificially.)-- Asclepias-- Flowers not SEEDING= Put pot of boiled earth on top of House =Aris
492q002 on them? 7... Are the wild Bananas of Otaheite SEEDLESS;-- are all varieties seedless-- if so. how ha
492q002 anas of Otaheite seedless;-- are all varieties SEEDLESS-- if so. how have varieties been formed?-- 8.
503q012 orsfield (35) Talk about races of Banana & yet SEEDLESS-- no light Henslow or Royle, latter says seed
503q012 dless-- no light Henslow or Royle, latter says SEEDLESS-- Also about Sugar-Cane Edwards says does not
514q21. in  Flinders= Alpine Australia Flora= Banana's SEEDLESS-- 20 varieties in mountains of Tahiti. Dr. Bo
390d178 parent, tree, but it fails.-- therefore «each» SEEDLING of one apple ought to differ from those of ot
427e111 ed by crossing, or by accidental production of SEEDLING with hardier constitution.-- Now Sir. J. Bank
428e111 ase adduced by Herbert) because not reared by SEEDLINGS.-- Now my principle does not apply to any pla
428e112 <accidental> preservation of accidental hardy SEEDLINGS: (which are confessed to by Herbert) to  sift
472s01r itate was growing in same garden. & out of 60 SEEDLINGS not one came up true.-- colour of flower & fo
494q005 d instincts Remote Experiments-- Plants Raise SEEDLINGS surrounded by various bright colours, any eff
495q005 ratures & see what the effect will be.-- will SEEDLINGS vary much more than cuttings &c (4) Raise ann
636j56v peas  become intermixed & garden have hybrid SEEDLINGS?} p. 333. Macculloch. brings forward. the impr
091a025 to sea valuable; because transportal of Minute SEEDS-- L. Institut. p. 209. May. 1837 Paper by Humbo
171b003 ject to variety, according «to» circumstance,-- SEEDS of plants sown in rich soil, many kinds, are pr
188b068 ficult to root out.-- For instance ever so many SEEDS of white flower. all would come up white, thoug
192b083 t clearly». Fox tells me, that beyond all doubt SEEDS of Ribston Pippin, produce Ribstone Pippins, &
192b083 f reproducing any variety, although many of the SEEDS will go back.-- Get instances of a variety of f
195b098 exist  in such spots. We know birds do arrive & SEEDS.-- The same remarks applicable to fossil animal
200b124 to Chamisso Vol III p. 155. about quantities of SEEDS in sea; also Holman: <at> Keeling these are mos
200b124 As  soon as island large enough for land birds, SEEDS picked from the beach by the birds; most  seeds
200b124 seeds  picked from the beach by the birds; most SEEDS germinating.--- It would be curious  experiment
210b159 d be curious experiment to know whether soaking SEEDS in salt water &c has any tendency to form varie
216b180 e three kinds of greyhound.-- In plants. do the SEEDS of marked varieties produce no difference. if t
218b192 s from it Black & White species.-- For, says he SEEDS of hybrid lillies &c &c &, (V Herbert on hybrid
219b193 beginning  to isld. Graham isld.-- we know many SEEDS. might be transported some blown--floating tree
219b193 in stomach-- &c &c. (Mem discover what kinds of SEEDS. these plants) [Mem Fact stated by Mr Don in is
228b230 ndez. Galapagos. Many trees Compositae, because SEEDS first arrived «Ferns ditto.--» & hence formed t
228b230 be  owing to cross from other trees.???? Do the SEEDS of Ribston Pippin & Golden Pippin &c produce re
228b230 me from them. Ask Henslow for some plant, whose SEEDS go back again, not a monstrous plant, but any m
229b234 culiar (?) If. Henslow discusses possibility of SEED«S»??-- Universality of generation strongly shown
231b242 ne from Holland!!! in stomach --or in feathers-- SEEDS.-- Two inhabitants of the Tropics, (whether one
296c184 ition. Keeling Isd «shows where proper dampness SEEDS can arrive quick enough» Vegetation of peak-- a
299c194 on banks & in damp parts.-- both propagated by SEEDS.-- There are two Dandelions, which just lately
299c195 ced by cultivating parent in rich soils & their SEEDS produce «offspring» variety. wild carrot. made
331dIFC ring, heterogenous, in plants are the number of SEEDS greater.?-- Mem. for Eyton.-- Sir. R. Heron's c
357d074 hat island would have no plants were it not for SEEDS being floated about.-- I must state that. the <
357d074 - I must state that. the <p> mechanism by which SEEDS are adapted for long transportation, seems  «?»
371d128 as life lasts, yet they cannot transmit through SEEDS these characters though transmitting them  with
400e013 h.-- Uncle John-- says Decandoelle, distributed SEEDS of Dahlia all over Europe same year.-- he sowed
401e014 of plants-- it is scarcely possible to purchase SEEDS of any cabbage, where a great many will not ret
401e015 e to impregnation from other apple trees.-- now SEEDS of crab produce crab, so that some effect  from
423e098 els  sure, that the reason people send for their SEEDS to London is that people in the southern Counti
432e124 ccording to my theory no land animal with fluid SEEDS can be true hermaphrodite Man probably assumes
434e129 of  female flowers will sometimes produce a few SEEDS,---- Ruscus aculeatus. a dioecious plant, in w
436e137 s" growth could have produced these innumerable SEEDS-- yet if a seed were produced with infinitesima
439e144 ent of instinct-- the possibility of rearing by SEEDS Holyoaks-- (how far is this so) shows either th
441e149 if planted in rich soil, I presume not, but its SEEDS, I presume, probably would-- at least the exper
442e152 culty in the way of Mr Knights. theory «without SEEDS are freshly transported»-- throw over this theo
446e162 observed» by buds-- (the other three by buds & SEEDS «though by the latter very rarely») here is a c
446e163 pagated during hundreds of years, without fresh SEEDS arriving.»-- throws a very great difficulty  in
447e164 eproduction.-- likewise grasses. &-- very heavy SEEDS.-- as Cocos do mer.-- Analogy shows some most i
447e164 as seen this-- (Poa alpina vivaparous sometimes SEEDS All species of Lemma sometimes though very rare
447e165 a alpina, thougt generally vivaparous sometimes SEEDS.-- There are endless curious facts about every
448e165 t seed freely.-- The periwinkle seldom produces SEEDS, because it is thought to require insects to im
461t039 he series.-- Ch 6 Upland geese would transplant SEEDS very far.-- Sept 31. The identity <of> (or only
469t151 Dwarf.  planted in rows «close to each other» & SEEDS gathered «all» came up in 1840 true. Shrewsbury
469t151 & dwarf-fan Bean», were planted in rows. & SEEDS gathered same year came up true «in 1840»:  All
469t151 n 1840»: All in together blossomed together The SEEDS of these plants will be collected & resown.-- H
472s01r 42 Allen W. sowed some years since gathered the SEEDS of Papaver bracteatum, & the Papaver oncitate w
472s01v eparate note-- Elizabeth says several years ago SEEDS were procured with the P. orientale in garden &
491q01v .-- Try.-- Important as discovering function of SEEDS-- (6) To hybridise EVERY flower on melon &  see
491q01v B.  seemed to say impregnation <caused> of some SEEDS, caused symmetry in cone-- The «above Exper» ex
492q002 those whose stamens are monstrous, hen then are SEEDS ever raised? 11. Is not non-flowering gorze com
495q005 tberis-- (latter I think certainly not) (3) Sow SEEDS & place cuttings or bulbs in several  different
495q005 e branch of Cabbage with pollen of other, count SEEDS, & see how great a proportion springs up true.-
495q005 in  natural Hybrids of Cabbages (7) Sow <daisy> SEEDS of wild cabbage in VERY rich soil, will plants
495q005 in such soil.-- 7 (a) Experimentise on Primrose SEEDS: it really is an important case-- cross with c
495q05a «flower  garden on gravel walk» will drift many SEEDS-- Necessary to answer Wiessenborns doctrine of E
495q05a must be distributed.-- Examine scum of pond for SEEDS.-- 11. Soak all kinds of seeds for week in Salt
495q05a scum of pond for seeds.-- 11. Soak all kinds of SEEDS for week in Salt. artificial water.-- 12. Plant
495q05a lose one twig of each in bell-glass-- sow these SEEDS & see if they will come up true-- whilst others
497q006 what plants will spring up which will show, how SEEDS are transported, or how long they remain dorman
497q006 ant. if kinds come up, not found in wood.-- but SEEDS continually dropping in woods. by birds 13. Mr.
497q06v rnum. or snow-ball-tree. what would result from SEEDS being sown= See in Cultivated Plants, as Pentst
498q008 ains necessary to impregnate ordinary number of SEEDS known?-- Linnaeus has shown that each pistil is
499q010 s generally distinguishable.= What structure of SEEDS.-- (Paris) (22) When Linnaeus says so great per
499q010 (22)  When Linnaeus says so great percentage of SEEDS have contrivance for transportal, does he inclu
499q010 ve contrivance for transportal, does he include SEEDS good to eat. (even Nux Vomica is eaten by a Buc
500q10a » beans & peas were planted in rows adjoining & SEEDS gathered there were planted «last year» pell me
500q10a planted «last year» pell mell, without sticks & SEEDS gathered & these are now to be planted this yea
501q011 about. Wheat-- Change of Soil--" crossing-- when SEEDS raised.-- His Book.-- 32. Would wheat from AEgy
503q013 near Boat House «ANY male branch.» --¿number of SEEDS in beginning of November 1841.-- Trees above ma
504q013 -- effect of accident?? (7) Which. Rhododendrum SEEDS??-- Bladder-nut \. Laburnum \. Dodecatheon \ .
504q013 Peas, from Wiegman Shrewsbury (1) Peas.-- Beans SEEDS alone remain to be compared-- Cabbages.-- Kept
505q014 (7) History of Potato field= (8) Abortive Thyme SEEDS weather wet--? Linum flavum put in Spirits whic
505q014 wet--? Linum flavum put in Spirits which plant SEEDS? (9) Melons fruit itself hybridised (10) one ha
505q014 Melons  fruit itself hybridised (10) one had no SEEDS, & two had plenty of seed & these Seeds of unim
505q014 had  no seeds, & two had plenty of seed & these SEEDS of unimpregnated Cowcumbers will they  seed.?--
505q014 rs will they seed.?-- (11) Abberley has planted SEEDS of pale green Cynoglossum. never germinated  12
505q014 inated 12 Does the horned orange. wh. never has SEEDS produced good pollen? Yes «From cultivation los
505q014 flys'  escape from old flower-- (14) Has planted SEEDS of Geranium pyrenaicum. small white-flowered va
506q015 ell, although there be pollen.-- or FEW. or bad SEEDS formed; badness may be merely not ripening= (38
506q15v us.-- Carex.-- We may presume Nettle spreads by SEEDS= (44) Zostera. Has he seen it in flower? does h
507q15v impregnation difficult or reverse (.49) List of SEEDS Gaertner de fruct:-- for woodcut-- 1 double hoo
507q15v ith, tuft-- Bull Rush-- Dandelion-- Sycamore. & SEEDS with «mere» border-- & Humboldts spinning seed.
507q15v primrose, but less strongly marked.-- 31. Plant SEEDS of the Fuller's plants ,Teazle Dr. Holland ; My
512q019 t-up-balls of Hawks or even owls.-- How long do SEEDS remain in stomach of birds-- Mem: how many mile
515q021 ther concepcion takes place,-- the mere fact of SEEDS ripening has scarcely any no relation to hybrid
559m155 curious facts about early production of foreign SEEDS.-- many varieties.-- Rev R. Jones has it.-- ver
```

Page
632j53r I use these facts p. 280. adduces provision of SEEDS for transportation through the air.-- cocoa nut
633j53v alves of the broom.-- or the springing of other SEEDS.-- But are we certain that these are necessary
498q008 es of Cabbages, peas, beans, as raised, do the SEEDSMEN select at all from the plants? If not, I am s
499q009 lowers in Hydrangea (20) As Hop is Dioecious-- SEEDSMEN who raise Hop-seed-- may know something about
563n001 times.-- Birth Hill shows it is evergreens they SEEK Cock Pheasant claps his wings before? crowing &
025r016 r, coast composed of sand dunes. 15--15 Does not SEEK to consider this a very shoal coast. Beyond the
060r124 . 1802. Earthquake at Demerara. The earthquakes "SEEM to arise from some efforts in the land to lift
203b135 : Mastodon all over S. America. Hilaire does not SEEM(?) to consider the monkey as a wanderer, but as
242c017 s on E. I«ndian». A«rch.» ‖ Borneo & Sumatra both SEEM to have elephant & has orangs, ‖ Tapir common t
257c057 ound woodpeckers.-- birds that cannot fly &c &c. SEEM clearly to indicate those very changes which at
295c178 s he assumed that character -- -- female & young SEEM most like mean characters the others assumed--+
318c253 mentioned many most curious case-- -- the birds SEEM to follow narrow bands, certain kinds as gallin
318c254 singly by night.-- others in flock, these birds SEEM clearly directed by kind of country; «kinds of
339d026 hinks there are some small divisions.-- does not SEEM to think any improbability to animals being dis
341d032 uced even an egg.-- a most curious bird, did not SEEM to know itself,. at last associated with the du
397e003 Selected. Dec 15 1856 [not located] Epidemics-- SEEM intimately related to famines., yet very inexpl
436e135 sea, like Cetaceae,-- although the Cephalopods, SEEM to have decreased since earliest times-- Apteri
438e141 , which is not improbable as Mr Herbert does not SEEM to recognize any difference in crossing between
447e164 ,-- now certainly in the higher animals; changes SEEM to have been more rapid, & the facility for int
467t103 of useful height.-- In Lupine, Bees «frequent» & SEEM to act, something like on Kidney Bean, they go
467t103 own the wings most violently: in Beans the wings SEEM beautifully to protect sheath {a} In all these
468t111 stigma. & so avoided it.» On certain days Humble SEEM to frequent certain flowers, to day early, the
470t153 minute These latter were pollen gatherers & they SEEM slow= Maer 1840 My Father formerly planted «Tur
485z017 y similar to some of the Fluvicolae?-- The Birds SEEM to move much further North on West coast of S.
525m026 tions «& N B affections very soon go in Maniacs» SEEM to have failed even more than the memory.-- the
528m037 se two causes very weak.-- (2d) form. some forms SEEM instinctively beautiful «as round, ovals»;-- th
532m056 fter fortnight. continued to grow think & did not SEEM quite happy. in five weeks was so thin, that sh
538m078 ne about people forgetting their insanity» there SEEM other cases somewhat analogous, & which I think
569m020 & smell. Descent of Man Understanding languages SEEM simplits case of Association.-- Elephant conser
581n065 D[I may put the argument,, that many learned men SEEM to consider there is good evidence in the struc
268c100 nslow in talking of so many families on Keeling SEEMED to consider it owing to one of each, being fit
283c144 wn by frigate Bird & flying eagle.-- Hawk Gould SEEMED to think, that widow bird. replaced Birds of P
332d004 e-- .) -- July 23d. Eyton, a stone blind horse, SEEMED to perceive turn on road where No houses to Ea
439e144 hat of other variety.-- «Elizabeth & Hensleigh. SEEMED to think it absurd. that the presence of of th
491q01v r on melon & see whether fruit affected. Mr. B. SEEMED to say impregnation <caused> of some seeds, ca
531m051 rt the scene of anger.--» to the pianoforte, it SEEMED solely to be feelings of discomfort, especiall
554m139 readily put it. when guided to her own mouth.-- SEEMED to relish the smell of Verbena & Pocket Hander
585n074 ration V E. p. 125 Wrong Entry Madagascar Lemur SEEMED to like Lavendar Water «very much» Henslow. N.
022r006 Valparaiso Granitic nodules in Gneiss. Epidote SEEMS commonly to occur where rocks have undergone ac
035r046 n over silicified wood. Cordilleras, Chiloe. &c SEEMS the organic structure most easily preserved.--
035r046 t in those belonging to the same district there SEEMS. I think, little ground for skepticism, as to t
039r060 y from the Galapagos-- Mr Lyell. P. 111 & 113. «SEEMS to» considers that successive terraces mark as
046r080 vement of ground). Michell (Philos: Transacts) «SEEMS to» considers that fall first movement (as in P
050r090 no of Priamang.--Marsden Sumatra. M. De. Jonnes SEEMS to think that Volcanic eruptions form foundatio
093a034 o fossil guanaco of P. St. Julian. -- Mr Scrope SEEMS to consider that elevation & eruptions are anta
153g053 of 4th shelf, some way below House of Glen Roy, SEEMS to be which higher up on is corroded {P} 4th [s
171b003 buds are constant, hence we see generation here SEEMS a means to vary. or adaptation.-- Again we <bel
181b045 s adapted preeminently for each element, but it SEEMS law that such tribes, as far as compatible with
210b160 relation also. ‖Yes « Fox» ‖ The creative power SEEMS to be checked when islands are near continent:
213b170 ability of some species. In Phillips. p. 90. it SEEMS the most organized fishes lived far back, fish
247c028 Silliman» «Study Silliman.--» Vol II. p. 10. it SEEMS that Crocodile was washed on shore at one of th
264c080 age of Fuegian, puts on par with Monkeys» Gould SEEMS to think that many species when close come from
264c081 1 forming. according to Gould, good genus Gould SEEMS to doubt how far structure & habits go together
269c102 ndix to Flinders Voyage by Brown.-- great space SEEMS to act per se as barrier-- Mem. Tartary & China
280c133 pages).-- Crosses of diff: breeds succeed, yet SEEMS to grant, that difficult & other go back to eit
287c158 ow & hawk a frolicsome saltus. «p. 19» ‖ Macleay SEEMS to limit Lamarck definition of relations to set
288c160 I p. 390. a slight change in enclosing a common SEEMS in part of-- to have almost banished the Grassh
300c198 nd is still only a divided body. ‖p 3. language SEEMS to supply instincts,-- & those powers which all
305c211 er.-- brain. & which <prin> thinking principle. SEEMS to be given or assumed according to a more exte
310c223 different origin. «Royal Institution» Dr Royle SEEMS to think Botanical Provinces will turn out not
314c239 ny Australian character in Timor plants, yet it SEEMS there may be Eucalyptus!-- (Hostile fact) Be ca
318c254 in English Nightingales» -- other birds (& this SEEMS common «kind» migration of America) migrate sin
318c255 fly.-- There is a kind of Wren (Bebyk??) which SEEMS common in Rocky Mountains & on one lofty isolat
338d024 Royal. Transact.-- p. 297. Vol 9. Dr. Ferguson SEEMS most clear that the ideosyncracy of the Negro (
341d030 thing.-- Animals from Hobart Town mentioned, it SEEMS most of species from there now found in Austral
350d055 perfection of individual, though such relation SEEMS common, but that perfection consists in being a
357d074 hich seeds are adapted for long transportation, SEEMS «?» to imply knowledge of whole world-- if so d
359d088 ur pidgeons) The male of every animal certainly SEEMS chiefly to impress the young most with its form
360d089 & some hairy-- the former preponderated <which SEEMS owing «determined» by the sex» Individual insta
364d104 bliterating the fresh feather, by crossing-- It SEEMS from Lib. of Useful. Knowledge that sheep origi
383d159 te, being not only dimidiate, but quarter-grown SEEMS to show whole body imbued with possibility of b
389d177 o each other.-- This is somehow connected (This SEEMS case, for by careful observing cattle can be br
390d177 hydras,-- (this repugnance to breeding in & in SEEMS connected with more developed forms) Study buds
434e130 re language forces on us the change, which <to> SEEMS to have taken place.-- Almost all Dioecious & M
437e139 exercise them in killing them.-- "Sometimes it SEEMS hares, rabbits, rats & not being sufficiently w
440e145 hound-- fox-dog-- turnspit & two other kinds It SEEMS absurd proposition, that every «budding» tree,
441e149 y would-- at least the experiment of the carrot SEEMS to show this.-- This would be curious law, Cert
444e158 mstances,;-- or the will of the Animal. p. 145. SEEMS to argue, that as the transformations from the
459tf02 nat.-- From Buffon cross of he-goat & sheep, it SEEMS male gives form. admitted by Linnaeus.-- seems
459tf02 seems male gives form. admitted by Linnaeus.-- SEEMS to doubt its applicability to common mule & hin
465t081 d. Phil. Journ. end of 1839] very important. it SEEMS owing to immigration of other races, so it is w
467t104 fully to protect sheath {a} In all these nectar SEEMS to be at base of upper petal & the curvature of
523m018 ted like getting suddenly into passion.-- There SEEMS no distinction between enthusiasm passion & mad
530m045 me little advantage in these trades.-- Delirium SEEMS to rest the sensorium.-- -- analogous to sleep-
537m076 ost opposed to this view, where the moral sense SEEMS to have changed suddenly-- but are not such «su
537m076 ore in fact only limits moral sense)» which she SEEMS to think «are» to make others happy & wrong to
547m111 uld trace no chain of association» Mayo Philos. SEEMS certain that muscular, mental, <&> digestive ne
576n047 become superior to that of the Greeks.-- (which SEEMS opposed to progressive. developement) on accoun
583n069 lty: pure instinct is not imitative: imitations SEEMS invariably associated with reason: [N B. insect
584n073 mewhat analogous to memory. Shrugging shoulders SEEMS sign of helplessness E. says she can perceive s
591n101 XV. Dialogue on Natural Religion.» however, he SEEMS to allow it is an instinct.» I suspect the endl
602o11b e senses & experience p. 142 "Upon the whole it SEEMS.-- "that the object of «all» art is the realiz
610o033 ce. so there may be in men-- which the reviewer SEEMS to doubt. [RHC] 1) Effects of Life in the abstr
614o037] 6) thought, however unintelligible it may be, SEEMS as much function of organ, as bile of liver.--
615o038 easure in virtue because acquired in past ages; SEEMS to indicate that when we <return> turn into ang
623o050 ity, doubtless grow together [RHC] This feeling SEEMS to vary in races of man. & certainly in «specie
624o50v 164. Ld. Shatsbury under term of Reflex Senses SEEMS to have «compared» «perceived» the comparison b
635j56v s, as in Papilionaceous flower, where such care SEEMS to be taken that the anthers should not be expo
380d153 ing whether if kept they would have wings.--).-- SEEP p. 84. Hens «like»-- Cocks from effect of breed
192b084 or plant run wild in foreign country.-- When one SEES nipple on man's breast. one does not say some u
299c193 . & this conclusion must be arrived at, when one SEES a plant like Paris quadrifolium growing in one
315c240 the creation must take place on[ly] when creator SEES. the means of transport fail.-- otherwise no re
411e056 any part of World.-- Wealden to boot.-- When one SEES in Coralline powers of multiplication of indivi
533m058 identical, with bird knowing a cat, the first it SEES it.-- it is frightened without knowing why-- th
535m066 natures can only be rudely traced out. When one SEES this, one suspects that our will may <be> «aris
549m120 man will try to obtain this happiness. though he SEES some «intellectual» good men, from insanity &c
575n045 oundland dogs. who will not enter water, till he SEES. whether birds badly wounded, or only winged.--
590n094 ount of an Ourang.-- see his Travels.-- When one SEES in Cowper, whole sentences spoken & believed to

607o025 h leads to motion being inclined that way]CD one SEES this law in man in somnambulism or insanity. fr
622o048 ese cases.-- Those instructions, which the child SEES uniformly performed by the teachers & all aroun
314c237 on being touched, so as to protect itself, one SEGMENT of the corolla being (probably) small to allow
314c237 lie on one side.-- but in other species, this SEGMENT is converted into hood which possesses power o
120a110 t least 2200. Jura 4000 feet.-- The veins of SEGREGATION in Greenstone of Salisbury Craigs well worth
121a111 ormed-- (like my gypsum case) shows power of SEGREGATION.-- & has heated angular fragments of rock, a
123a116 t vacuum-- by dike.-- Mem. however. veins of SEGREGATION in Salisbury Craigs Letter from M Angelis. B
145g003 instance of Provincial breeds. [3] Veins of SEGREGATION in Salisbury Craigs [4] Salisbury Craigs V.
532m054 aughing & told me there was nothing, & tried to SEIZE hold of objects to be frightened at.-- (again d
584n005 , was transmitted perfectly to chicken so as to SEIZE small moving object like fly.-- young partridge
288c160 ous. Were the woodcocks,. which came Madeira & <SEIZED> ceased their migrations lost?? I conceive a b
332d001 nty, who had had no sort of communication, were SEIZED with it, & for ten years afterwards, he then d
317c248 ssils of N. America.-- At the end of "White's SELBOURNE." many references very good. also "Rays Wisdo
321c275 ed Von Buch Travels Whites Natural History of SELBOURNE References at end Dr. Lang Australia «trash»
318c254 one particular part of country, like White of SELBOURNES Rock Ouzels.-- ..If the line or bands of cou
288c160 proof of. land having been formerly nearer.-- «SELBY» Magazine of Zoology & Botany No XI p. 390. a s
363d102 art, so that after differences were pointed out SELBY confounded them, yet can readily be told by inc
204b140 ose imported by Ld. Powis Hybrid dogs offspring SELDOM intermediate between parents.-- How easily doe
234b255 cept in one respect, that those of Iceland. are SELDOM seen with horns" --- p. 341. Black Fox sometim
448e165 s &c & &-- do not seed freely.-- The periwinkle SELDOM produces seeds, because it is thought to requi
450e175 ngal this fact is noticed in Cassay Ava Pegue-- SELDOM equals 13 hands-- those of Lao & Siam inferior
608o026 us.-- because man has power of action, & he can SELDOM analyse his motives (originally mostly INSTINC
615o36v Europaean. begin to prowl about in the evening «SELDOM leave their perch till evening» crow different
473s07r ed Indian Plant.-- June /42/-- June/42/ You can SELECT cattle & sheep for horns & yet no difference i
490q001 arrots breed amongst the Indians Do the Savages SELECT their dogs Sowerby Entomologist Does individua
498q008 bbages, peas, beans, as raised, do the Seedsmen SELECT at all from the plants? If not, I am surprised
511q018 e Bantams at Zoolog Soc.-- did Sir. J. Sebright SELECT to destroy secondary character believe No or d
239cIFC of» February & July 1838) All good References SELECTED Dec. 13 1856 Also looked through April 23. 18
331dIFC Monkeys recognising Sexes of animals:]CD [All SELECTED Dec. 14-- 1856]CD Towards close I first thoug
397eIFC g at Z. Gardens D E Finished July 10th 1839.-- SELECTED. Dec 15 1856 [not located] Epidemics-- seem i
520mIFC on Morals & Speculations on Expression -- 1838 SELECTED Dec 16 1856 July 15th 1838 My father says he
563n1FC Darwin [Private.]CD (Metaphysics & Expression) SELECTED «for Species Theory» Dec. 16 1856 Looked thro
331dIFC 14-- 1856]CD Towards close I first thought of SELECTION owing to struggle July 15th. 1838 Finished. O
378d147 tarling alike, yet young ones brown.-- Sexual SELECTION If masculine character. added to species.. we
490q001 ctaries. The former best for my experiment on SELECTION. Experiments in crossing &c Plants 1 Repeat t
503g012 es effects of domestication-- said to require SELECTION (36) Ask Mr Gowen to ask Mr Herbert, how many
508q016 c Memoirs (11) And. Smith Savages at Cape any SELECTION of Males in «cattle» or in Killing the worst
511q018 eration?? HOUNDS. Eyton Mr Wynne, &c Could by SELECTION a different looking animal be formed-- not ca
512q019 ect to disease in youth.-- Mr Toilett-- about SELECTION for milking-- loss of early habits in Dorsets
076r169 lways accompanied by Sulp. silver sometimes by SELENITE.-- in New Spain, contrary to Europe. argentif
182b049 Man.. difficult for man to be unprejudiced about SELF, but considering power, extending range, reason
538m079 B.e on board Beagle, some merry goodhumoured as SELF.-- «When Miss Cogan has remembered her song, th
538m080 cts unconsciously with respect to more energetic SELF, & likewise one forgets. what one performs habi
538m080 different, & yet they would make one's father & SELF one person-- & thus eternal punishment explaine
548m115 ns of ones infancy.-- one cannot bring it to one SELF.-- nor of a bad dream, when that is not recolle
548m117 get the insane idea; & ones expression of double SELF, though as in Dr Ashe's case, one here was cons
565n008 significant, & when pride makes person extremely SELF-sufficient,-- the corner of lower lip are depre
584n071 ing instincts, sexual, social, «subordinate to,» SELF preservation, (knowledge of enemies). use of mu
590n094 gination a faculty, a power, quite distinct from SELF. «or will» [not located] <& other cows--» Mr. H
591n099 nt) is owing to the feeling that the instinct of SELF-preservation is disobeyed-- I often have «as a
625o50v lived good, ie happiness-- yet this system not SELFISH.-- explained by principles if Mackintosh.-- p.
627o053 analysed into "power" &c &c &c-- & if termed "SELFISH", must be subclassed as "disinterested" p. 14.
525m024 ase any direct fear.-- My father thinks that SELFISHNESS, pride & kind of folly like (Mr George S.) i
138a151 ansactions (1832. or 3?) there is an account of SELLOW Geolog. Observat. in Southern Brazil. [blank]
616o038 ook back to definite action or to our conscious SELVES.-- Such memory may go back to animals which we
373d132 l a spontaneous generation: what is animalcular SEMEN-- but this-- -- the <nerve> living nerve nursed
388d171 athers character.-- according to this view more SEMEN to one child. more like father.-- stuff.!-- How
388d173 other sex being present is-- it also shows that SEMEN. must actually reach the ovum.-- [Why in making
416e071 dites.-- I suspect this rather effect of liquid SEMEN: therefore animal life commenced in the Water!»
493g003 in different cats.-- Good observation-- examine SEMEN of Hybrid animal. in comparison with Weigh skel
315c241 rsburgh Academy Imperial- Paper read in 1837. SEMESTRE I suspect some valuable analogies might be dr
159g086 f obscure terraces (& conical hills on same) of «SEMI» waterworn & some partly well worn pebbles-- «w
492q002 generally fails first?-- Mal[e] 10. Henslow says SEMI-doubl flowers are those whose stamens are monst
540m087 o-- look at them both savage-- look at them both SEMI-civilized-- Perhaps one cause of the intense la
027r023 laminae, show what movements take place in SEMICONSOLIDATED rocks P xv. mentions in what formations C
405e031 horizontal mark on flank.; & tail. & kind of SEMILUNAR {P} mark on each side darker,, so that whole
416e070 - In animals analogy leads one to suppose that SEMINAL fluid fluid, (& not dry as in plants) therefor
447e164 Lemna shows dispersion of germs is not end of SEMINAL reproduction-- likewise grasses. &-- very hea
194b094 ch in south Africa, appear to represent the SEMNOPITHEQUE of India.-- Tooth of <Spi> of Sapajou-- NB
241c016 s: <>) vegetation far East) Ann: des Sciences. SEMPTEMB. 1825 Get Henslow to read over the pages from
423e098 John says he feels sure, that the reason people SEND for their seeds to London is that people in the
446e162 «A fruit tree by certain treatment will suddenly SEND forth quantities of blossoms--» The case of the
515q022 es, bred in this country. {Chinese Dog's Head to SEND Cover common Pea (& Sweet Pea) for several gene
531m053 noise, & therefore brain has been accustomed to SEND a mandate to the muscles & when the noise comes
261c069 all species Accumulate instances of one family SENDING out structures into many genera.-- like Synall
536m070 imself stiff, & walks hard.-- «He cannot avoid SENDING will of action to muscles, any more than «prev
552m131 ntion «So intimately connected is passion with SENDING force to muscles, that in my grandfather remar
051r093 regions every rock is buoyed by Kelp, now Kelp SENDS forth branching roots which must protect surfac
336d018 rticular Muscles-- When two animals cross. each SENDS his own likeness, & the union makes hybrid, in
567n014 s doubtless has not.-- Turkey cock in passion & SENDS blood to its breast &c &c All Science is reason
089a020 ol 3' P. II. -- Bed, of elevated shells on the SENEGAL. L Institut p. 192.-- (1837. Peninsula of Cape
221b202 ossils some species being recent agreeing with SENEGAL. whilst Crag <agrees with> according to Beck h
448e166 & Arctic sea, & the other with neighbouring & SENEGAL as sea.-- is remarkable.-- Again the resemblan
554m138 shown in old male.-- A very green monkey (from SENEGAL he thinks Callitrix Sebe??) he has seen place
587n087 d. to organs adapted to other objects. (as that SENNA is necessarily disagreeable to organs adapted t
587n087 which may be doubted for possibly even taste of SENNA. might be acquired. as the Turks have of Rhubar
524m020 g organs ¶ -- the action of brain which gives SENSATION of pain, emits its power on the muscles in th
532m057 ace.-- How like strong feelings of Man.-- The SENSATION of fear is accompanied by «troubled» beating
534m061 quired by senses,-- then thinking consists of SENSATION of images before your eyes, or ears (language
534m062 tition of whatever takes place in brain. when SENSATION is perceived. = Aug. 7th--38. Transactions o
544m102 > quite different from when awake.-- peculiar SENSATION as flying. (No memory of past. events?) or inf
547m112 e been quite still, & not attending to bodily SENSATION & yet the Castle would not have turned into d
550m124 to what other people experience.--» But then SENSATION may be more or less pleasant & unpleasant, in
556m145 man is born with a portion of phsiognominical SENSATION, as certainly as every man who is not deforme
567n015 t have any emotion like blush.-- when extreme SENSATION of heat shows blood is pumped over whole body
567n015 stinct, preeminently so-- who can analyse the SENSATION, when meeting a stranger. who one may like. d
581n064 erchief-- cats & dogs fond of slight tickling SENSATION.-- in savages other tastes few. March 16th. G
594n113 What is absurdity, why does one laugh at it-- SENSATION of disgust with nausea, (when stomach a littl
601o009 s The first of these books I daresay good. 1. SENSATION is the <conse> ordering contraction (that is
601o009 ents.-- ¿plants? yes by distinct mechanism 2. SENSATION of higher order. where the sensation is conve
601o009 anism 2. Sensation of higher order. where the SENSATION is conveyed over whole body (which it may be
601o009 nsequence having some relation to the primary SENSATION.-- man moving leg when asleep-- «or habitual
601o009 lants as supposed by Buffon) Consciousness is SENSATION No. 2. with memory added to it, man in sleep
601o009 ess, <t> movements «¿» anterior to any direct SENSATION, in order to avaoid it-- beetles feigning dea
611o034 ws a local will, though perhaps not conscious SENSATION. [RHC] During growth <extres> tissue <[...]>
534m062 exciting association.)-- or of memory of such SENSATIONS, & memory is repetition of whatever takes pl
540m088 or by bodily weakness, melts into tears, with SENSATIONS of sorrowful delight, very like best feeling
550m124 e proportion of pleasant to unpleasant mental SENSATIONS in any given time «-- compared to what other

550m125 e time.-- Pleasure more usually refers to the SENSATIONS <it> when excited by impressions, & not ment
568n018 fect of improving organization, comparison of SENSATIONS would first take place, whether to pursue im
582n066 y.--» «He may have pain or pleasure these are SENSATIONS» <Gardner in his work> In the life of Hayd &
582n068 their characteristics;-- they are more truly SENSATIONS??. a kind of mental pain & pleasure.-- The R
591n099 ormal sexual actions or even impulses. (where SENSATIONS of individual are same as in normal cases) a
605o18v hich height. by its accompanying & associated SENSATIONS so often gives, when excited by other means,
605o019 ciple» to explain the various causes of those SENSATIONS, which we call metaphorically sublime, but t
620o044 ry & <pow> mental capacity of calling up past SENSATIONS, he will be forced to reflect on his choice:
403e024 ses not) If they give up infertility in largest SENSE. <es> «as» test of species.-- they must deny sp
527m034 done in open air, with exercise &c no organs of SENSE being required) as the closest train of geologi
537m075 amples» very justly there is no universal moral SENSE.-- «from difference of action of approved» Yet
537m076 Fact most opposed to this view, where the moral SENSE seems to have changed suddenly-- but are not su
537m076 if they had murdered their children, this moral SENSE, would have been so much, as in other races of
537m076 & wrong «(& therefore in fact only limits moral SENSE)» which she seems to think «are» to make others
543m100 ies, therefore this judgment gives a man common SENSE, & the highest intellectual powers of perceivin
552m132 l as rule will be given.-- Descent of Man Moral SENSE Mitchell Australia Vol I, p 292 "Dogs learn soo
558m150 o habit, as being less so will.-- May not moral SENSE arise from our enlarged capacity <acting» «yet
558m151 of the whole world.-- Read Mackintosh on Moral SENSE & emotions.-- The whole argument of expression
564n003 t.-- Ɪ Different nations having different moral SENSE, if it were proved instead of militating agains
564n004 d of misery of future thinking of injured moral SENSE.-- Notion of deity effect of reason acting on (
568n017 revented performing heredetary habit, (or moral SENSE, or instinct.) one feels pain, & vice versa ple
572n033 v wise instantly but could not gather general SENSE of this page.-- Now <awake> «when awake» I coul
572o034 ach word separately, neglecting time, & general SENSE, anymore than connected with general tendency o
576n047 o.-- nor would we regret «having acquired» this SENSE of right (& Whether wholly instinctive as in th
584n072 o way applies.-- it is the acquirement of a new SENSE,-- bats avoiding strings «in the dark» as well
585n077 k.-- ?? this is not instinct, but a faculty, or SENSE-- "We know not how, stonge henge raised, yet no
599o004 ressions N [Old & USELESS notes about the moral SENSE & some metaphysical points written about the ye
600o008 (I) <Conscience> «Moral Sentiments» imperative SENSE of duty-- which makes struggle in man.-- two so
600o08v Mackintosh make great difference between moral SENSE & conscience? we admire what is right by one &
609o030 est happiness.-- The other says we have a moral SENSE.-- But my view <says> unites both «& shows them
609o030 al senses: (& this alone explains why our moral SENSE points <is> to revenge). In judging of <our ha>
609o030 + Society could not go on except for the moral SENSE, any more than a hive of Bees without their ins
609o30v is endeavoring to change that part of the moral SENSE which experience (education is the experience o
610o30v degree <of> right.-- The change <of> our moral SENSE, is strictly analogous to change of instinct am
617o40v d in every> phenomena actually apprehensible by SENSE. 5) There is nothing analogous to this in the r
619o042 ackintosh Ethical Philosophy [RHC] On the Moral SENSE Looking at Man, as a Naturalist would at any ot
622o048 eling of cowardice? «This is not connected with SENSE» instantaneous so declaring it is right or wron
625o50v «the utility part being blended & lost» & moral SENSE.-- My theory explains both, perhaps, by habit--
628o054 .--]CD p. 22. says affections, desires, & moral SENSE all different.-- P. 22. Butler & Mackintosh cha
628o054 22. Butler & Mackintosh characterize the moral SENSE, by its "supremacy",-- I make its supremacy, so
565n006 in garden uncovering its teeth to bite.-- the SENSELESS grin of passion, is like the grin of the Hyae
223b207 earing..-- the appearance of insects with other SENSES is more wonderful. its mind more different pro
262c072 eredetary; ie instincts of wisdom virtue? «like SENSES of savages» (How come its some countries patri
350d056 gan fitted to animals place in creation.-- thus SENSES, especially sight connected with locomotion.--
532m054 owing to heart? readily taking same movements, SENSES being on the look out, & the conveying means f
532m055 on the look out, & the conveying means from the SENSES to the mind being more alive.-- How is it. wit
534m061 ge.-- but if so, yet this knowledge acquired by SENSES,-- then thinking consists of sensation of imag
536m072 hen an oyster has a & a polype (& a plant in some SENSES, perhaps, though from not having pain or pleas
536m072 t effect of organization, by the capacities its SENSES give it of pain or pleasure, if so free will i
538m078 ng song» when she had recollected it in perfect SENSES.-- These things, & drunkedness, show what trai
544m103 tiguing,-- else it is only our consciousness, & SENSES tell us it is not real. = = dreaming appears cl
546m109 g shows one how little happiness depends on the SENSES.; than the <small> fact that no one, looking b
546m110 mbulism.-- the young lady almost equally in her SENSES in either state.-- does this throw light on in
547m112 own manner of ordering things to be done.-- The SENSES are closed probably by sleep & not vica versa.
547m113 - (do the ideas, direct effect of perception by SENSES fail first, as whether I had pulled the bell??
547m113 te impression. & likewise gaining new ones from SENSES. & <comparing their> «calling up» old ones, to
548m114 elf-- but not about, facts gained or gaining by SENSES.-- As sleep <is> only one idea is awake, when
551m127 not doubt the reality of the impression on its SENSES.-- insane people believe they hear as well see
568n016 nt to propositions, which are the result of our SENSES, or our experience.-- Two sides of a triangle
569n021 simplets form probably is single comparison by SENSES of any two objects-- they by VIVID power of co
572n033 o muscular movement.-- say instinctive actions. SENSES. notions &c Octob 30th-- Dreamt somebody gave
574n041 = No doubt man has great tendency. to exert all SENSES, when thus stimulated, smell, as Sir. Ch. Bell
575n045 .-- joy <p> a mental pleasure. with pleasure of SENSES & muscles, we cannot look at him, as machine t
583n070 s the bee makes its cells, by means of ordinary SENSES & muscles, progression.-- the source from which
584n071 enemies). use of muscles, progression.-- in some SENSES,-- knowledge of location ducks & turtles runni
586n082 ten will turn out to be instincts, & so in some SENSES, is sight--CD [The faculties bear so close a r
586n082 [The faculties bear so close a relation to the SENSES, that one feels no more surprise at it & feels
587n089 motion.-- The former is used with regard to the SENSES. Reason does not lead to action.-- p. 248. The
592n103 n fright because nature intends to lay open all SENSES: <do> Horse prick his ears «& snort clears nos
600o08a ot closely allied to that of men, when the five SENSES were the same-- In its action-- emotions-- p 1
602o11v e mysterious & sublime ideas independent of the SENSES & experience p. 142 "Upon the whole it seems."
609o030 ary for good at all» is the «instinctive» moral SENSES: (& this alone explains why our moral sense po
613o035] ¿How does consciousness commence; where other SENSES come into play, when relation is kept up with
617o39v sciousness; the objective, by our external what SENSES in the way in which we apprehend the force of
617o39v a considerable degree of attention. How do the SENSES affect us, except by internal consciousness 3)
617o040 of> bodily action as recognised by our external SENSES consists in the manifestation of force i.e. mo
617o040 ORCE, <objectively> considered, by our external SENSES is a phenomenon the essence of whose existence
617o040 out How can force be recognised by our external SENSES--only movement can.-- 4) the source from which
624o50v n.-- p. 34. Ld. Shatsbury under term of Reflex SENSES seems to have <compared> «perceived» the compa
305c210 nate laws.-- There is one thinking « <& Creat> SENSIBLE" principle (intimately allied to one kind of
398e006 ere in order a scattered page; we find <great> SENSIBLE change in the institutions. & we suppose not
529m042 recovered. was found to be ignorant, but quite SENSIBLE & no ways an ideot.-- «in this case must have
337d021 first eye is formed.-- how one nerve becomes SENSITIVE to light.-- (Mem whole plant may be considere
454e184 any very sleepy Mimosa, nearly allied to the SENSITIVE Plant.-- p. 290. Dr. Edwards in his essay on
454e184 rish & Common Hare Decandoelle has chapter on SENSITIVE plants; Physiology Get Habberley to try exper
455eIBC at same time» see if by so doing can be made SENSITIVE The function of sleeping someway useful.-- it
494q005 eping mimosa, or half bred mimosa (a) between SENSITIVE & sleeping species, & see whether association
559m156 see if Hybrid can be made & ferns.--» Would a SENSITIVE plant if irritated very regularly at one time
564n003 on his actions, makes his conscience far more SENSITIVE.» ulitmate effects of actions.→ till at last h
577n050 inciple of Association will help my theory of SENSITIVE Plants Habitual actions, (independent of mind
577n052 t being such thinking people. do not blush.-- SENSITIVE people apt to blush.-- -- The power of vivid
610o034 c). [LHC] This is true as long as movement of SENSITIVE plant can be shewn to be direct physical effe
612o035 e) of animals enjoying only movements such as SENSITIVE plants. (But I include irritability for that
530m045 in these trades.-- Delirium seems to rest the SENSORIUM.-- -- analogous to sleep--; some doctors care
612o035 of relation required to be extended. Hence a SENSORIUM, which receives communication from without, &
549m118 ncy of good food, no pain,-- <but the> «&» the SENSUAL enjoyment of the minute add to the happiness.-
549m120 .-- than the happiness of a peasant, with whom SENSUAL enjoyments of the minute make large <parts> po
565n010 vaters,-- Ye Gods!:-- says fleshy lips denote SENSUALITY (p 192 Vol. III Octav. Edit)-- certainly nei
235b274 ian Cattle with humps.-- ¿To be solved if horses SENT to India. a long bred in & no new ones introduc
532m056 e happy. in five weeks was so thin, that she was SENT back to Shrewsbury,, then immediately fell into
545m106 e with passion, ie. nervous impulse to action is SENT so fast to limbs that they cannot remain still.
104a068 case parallel to Banda Oriental. ask Lyell for SENTENCE.-- Origin of Breccia, introduce in Cordillera
375d135 SHORTER time than 25 years-- yet until the one SENTENCE of Malthus no more clearly perceived the great
463t063 known coast in front of it.-- Cuvier has grand SENTENCE about the Animaux fossiles-- being a mere fra
283c145 le will argue & fortify their minds with such SENTENCES as "oh turn a Buccinum into a Tiger."-- but g
590o094 his Travels.-- When one sees in Cowper, whole SENTENCES spoken & believed to be audible. one has good
599o005 es Ld Mondobbo.-- language commenced in whole SENTENCES.-- signs-- ¿ were signs originally muscical!!
538m079 runk.-- having really been so.-- some always SENTIMENTAL, some quarrelsome as B.e on board Beagle, so

409e049 on of all that is most beautiful in the moral SENTIMENTS of the animated beings.-- &c-- If man is one
595n184 , (Vol. 4 of Works) [blank] "Adam Smith Moral SENTIMENTS" much on life & character "Humes Dissertatio
600o008 ch faculty of Soul.-- (I) <Conscience> «Moral SENTIMENTS» imperative sense of duty-- which makes stru
072r160 .--Spitzbergen animals (?). ≠ The Hollowness of <SEP> Chiloe concretions somewhat analogous to septa.
076r170 lcose & chloritic slate. with beds of syenite & <SEP> serpentine dipping to SW at 45 degree to 50 deg
190b079 genera stronger,-- Mr Waterhouse says no real <SEPAR> passage between good genera-- How remarkable s
028r028 must have a tendency. to mingle; The sea would SEPARATE quartzose sand from the finer matter resultin
028r030 Limestones all been dissolved. if so sea would SEPARATE them from indissoluble rocks? Has Chalk ever
036r052 arts of central Patagonia Does Andes in Chili. SEPARATE geographical ranges of plants. V. Lyell. Chap
046r078 «Sulphur last.--» Metallic veins likewise must SEPARATE ingredients if we look to a constant revoluti
056r111 ertropics at present fix lime). <Also Volcanos SEPARATE.> Volcanos blend all substances together; & p
063r132 still partly united. & eggs which become quite SEPARATE.--Considering all individuals of all species.
172b006 Egyptian cats & dogs ibis same as formerly but SEPARATE a pair & place them on fresh isld. it is very
172b007 what result According to this view animals, on SEPARATE islands, ought to become different if kept lo
173b009 (as some common land shells) Most difficult to SEPARATE Every character continues to vanish, bones in
202b132 "The buffaloes, introduced from Timor, herded SEPARATE from the English cattle, nor could we get the
219b196 ent of all, by one law. as to account for each SEPARATE one, so to say that all Mammalia, were born f
235b263 tree of life, can orders like birds & animals SEPARATE &c &c Work out Quinary system according to th
252c045 most certain guide ·: because time required too SEPARATE isld very long» America & India deer.= Africa
289c161 ty,, is simply, an instinctive impulse to keep SEPARATE, which no doubt be overcome, but until it is
311c226 her organs difficult to analyse) will not this SEPARATE facts about abortive organs &c The doctrine o
341d031 - Dr. Bachman tells me line of Rocky Mountains SEPARATE almost all Mammals of N. America & many birds
349d053 - so that value can only be judged of in each «SEPARATE» line of descent.-- -<& here limits of varieti
355d068 e must take place in copulation-- (Man & woman SEPARATE parts of same plant<s>)-- now in some Polypi
365d107 he «e»normous time which it must have taken to SEPARATE. Van Diemen's land from Australia &c & Sept.
381d157 t important support to my views-- Seeing sexes SEPARATE in some of the lowest tribes, leads one to su
382d157 idently> actually split in two-- keeping sexes SEPARATE. Owen say such view worthy of a Lamarckian.--
409e048 ng probably exists for one cause» of sexes «in SEPARATE» «animals»: for otherwise, there would be as
424e100 erican.-- several cases of species peculiar to SEPARATE islets.-- March 5th. Lyell says «fossil» shel
425e104 ar to the group having species peculiar to the SEPARATE islands-- In his work on the Labiatae, some o
472s01r this one year (1827), always been empty.-- See SEPARATE note-- Elizabeth says several years ago seeds
498g008 s has shown that each pistil is connected with SEPARATE division of germen <?>-- (11) Must pollen gra
501q10a ies-- Does any genus of Plants. vary & hard to SEPARATE specifically in one country & not in other: R
538m080 whole train of thoughts, feeling & perception SEPARATE, from the ordinary state of mind, is probably
539m083 al.-- instances?? The possibility of two quite SEPARATE trains going on in the mind as in double cons
543m101 ain bringing thought, & not merely instinct, a SEPARATE thing superadded.-- we can thus trace causati
544m103 either manner, vividness, rapidity, novelty of SEPARATE ideas cause fatigue to the mind,-- it is sole
553m136 d in us (<by> ¿ individually or in race?) by a SEPARATE act of God, & not as a necessary integrant pa
559m154 t as long as we consider each object an act of SEPARATE creation. we admire it more. because we can c
567n016 cts for induction. one is obliged carefully to SEPARATE its memory from all ordinary lines of associa
577n052 .-- -- The power of vivid mental affection, on SEPARATE organs most curiously shown in the sudden cur
634j54v e flying membrane?!!-- that the phalanges have SEPARATE movements in the Holocentrus ruber (a fish) M
028r029 ng in deep seas forms slates: How is the Lime SEPARATED; is it washed from the solid rock by the acti
053r101 iders the Sandstone & Granite districts to be SEPARATED by profound valley [.] Sydney. -- Lesson Zool
072r160 s with difficulty that a few fragments can be SEPARATED from them with steel instruments." In R. Brow
151g042 descend from shelf 3d & almost meet, but are SEPARATED by flat bottomed strait. connecting flat on o
173b012 or they may represent some large country long SEPARATED.-- On this idea of propagation of species we
174b015 our planet first cooled.-- Countries longest SEPARATED greatest differences-- if separated from imme
174b015 s longest separated greatest differences-- if SEPARATED from immens ages possibly two distinct type,
187b065 closely allied but different, because country SEPARATED, since time of extinct quadrupeds:-- same argu
190b077 of species of few genera or families.-- (long SEPARATED.-- Proteaceae & other forms (?) being common
191b080 n East Indian islets-- (+ + +. Ireland longer SEPARATED. . Hare of two countries different.-- Ireland
207b152 fixed.-- Peron thinks Van Diemen's land long SEPARATED from Hobart Town-- (from difference of races
219b195 . the antarctic flora must formerly have been SEPARATED by short space from mountains low down, there
235b262 there any cases, where «domesticated» animals SEPARATED. & long interbred <p> having great tendency t
281c139 a species must be compared to family entirely SEPARATED from any degree; the tailors-- «in each branc
282c141 races of such men living in same country but SEPARATED, now if one of these races had become eminent
284c148 he higher parts & only on those, & the islets SEPARATED at high water.-- not other islands, nor any a
291c167 it may be difficult to imagine how sexes were SEPARATED.-- in plants we have some flowers monoecious
318c255 the Normal condition of Migration) gradually SEPARATED the birds might yet remember which way to fly
337d023 n exact proportion to the time they have been SEPARATED; together with physical differences of countr
353d061 mpared to North American hares. Many species, SEPARATED by Mountains. & & &c.-- do. p. 69. A Dr Macdo
372d129 ing of wound.-- reproductive faculty + in the SEPARATED part every element of the living body is pres
398e005 re of small change.-- on the contrary islands SEPARATED with some animals, &c.-- «if the change could
405e032 orms it is either effects of having been long SEPARATED, or having never [not located] ARGUMENT REAL
510q017 . -- Schelgel is he serpent man? about zones SEPARATED by non-inhabited spaces: has he published? do
586n080 sential difference, which clearly ought to be SEPARATED-- We apply instinct to one part. or another--
641j29v their remains in the Wealden? In the strongly SEPARATED Arctic genera, there is evidence of antiquity
572n034 ppears as if the mind had dwelt on each word SEPARATELY, neglecting time, & general sense, anymore t
609o30v he social instinct «in man & animals» must be SEPARATELY considered.-- The difference between civiliz
640j167 the long legged variety would prevail.-- Not SEPARATELY: NB. These views quite exclude the idea of d
056r111 y water.--Or rather begin & explain how water SEPARATES.--(intertropics at present fix lime). <Also V
056r111 rld, general circulation. But Volcanic action SEPARATES some sulphur (perhaps lime) salt. & metallic
126a122 posed to have been attained.» how much matter SEPARATES them, this is ascertained by conducting power
391d174 simplest form.-- (Because by this process it SEPARATES those differences which are in harmony with a
431e122 would   look like fresh Creation. the gardener SEPARATES a plant he wishes to vary-- domesticated anim
056r111 ducts «over whole world» argument, as well as SEPARATING causes by water.--Or rather begin & explain
172b005 ole country; beautiful law of intermarriages <SEPARATING> partaking of characters of both parents, &
401e015 Thinks probably experiment was never tried of SEPARATING apple tree entirely from all others-- so my
098a045 NB. Sedgwick talks of LAMINATED structure (∴ SEPARATION of ingredients) as uniting with cretionary.-
099a051 mud from Trapiches. inclined layer!!!.-- The SEPARATION in the Ponza case of Scrope parallel to wall
102a060 ria in concretion arranged in planes, case of SEPARATION.-- the branching cracks-- only bear relation
102a061 s first of felspar & then quartz &c, owing to SEPARATION having taken place most gradually, first the
110a086 g finer grained & more compact on that side-- SEPARATION DISTINCT from dike junction mechanical: DIKE
123a116 like carbon must go round of dissemination & SEPARATION in volcanos.-- if so why not metals. The the
130a134 posits there probably would be marked line of SEPARATION A Paper by Parrott Mem. Acad. Peters. Scienc
176b024 ch branch A species as soon as once formed by SEPARATION or change in part of country. repugnance to
223b209 f after isolation (seed blown into desert) or SEPARATION by mountain chains &c the species have not b
305c211 ting principle in the various shades of <dif> SEPARATION between those individuals thus endowed, & th
338d023 physical differences of country: the time of SEPARATION depends on facility of transport in the spec
338d023 itions of country is not perhaps so great, as SEPARATION on <be> inter-breeding, for otherwise we cou
343d038 e tribes become blended & prevent that strong SEPARATION which otherwise would have taken place. othe
370d117 ns in arrangement, of the perfection of their SEPARATION.-- thus Vertebrate blend with Annelidae by s
436e135 pulled out. yet feathers retain character? If SEPARATION in horizontal direction is far more efficien
436e135 . intelligible.-- «-- No. but the wandering & SEPARATION of a few, probably would be most efficient i
600o08v cience, not clear-- Then these last heads. of SEPARATION between soul of man. & intellect of beasts,
599o008 n, beau ideal, infini conscience; voila l'homme SEPARE de la matiere et du temps! voila les facultes,
381d156 ghness in scale has no «constant» relation to SEPARTION of sexes, as may be seen in Monooecious & Dio
381d157 e traces of other «sexual» organs; for if so, SEPARTION of sexes very simple.-- as in plants even in
383d159 either sex.-- In my theory I must allude to SEPARTION of sexes as very great difficulty, then give
060r125 erary Earthquakes at St Helena. 1756. June 1780, SEPT. 21st. 1817.--p 371. Webster Antarctic veg:-- S
352d057 ecies retrograde, but these facts are rare.-- 2d SEPT Those animals which have many ABORTIVE organs,
354d065 ith what facility mice attach themselves to man. SEPT 7th. -- I was struck looking at the Indian catt
356d070 od: grazing animals who eat every species new.-- SEPT. 8'. A Golden Pippen or Ribston do producing oc
356d072 ding to my theory of generation (p. 175) if> 8th SEPT Yarrell told me he had just heard of Black game
358d076 same habits in some respects as this Caracara.-- SEPT. 9th. It is worthy of observation that in insec
361d095 that two first might cross easier than two last. SEPT. 11. N Mr. Blyth, at Zoolog. Meeting stated, th

Page ***(Key Word)***
365d107 separate. Van Diemen's land from Australia &c &c SEPT. 14th. When Macleay says their is no difference
367d113 at co-relation of parts, will not be produced.-- SEPT. 17th. Saw mule. apparently fathered by a donke
370d116 true females, but with parts little developed.-- SEPT. 19th <Are> There is no scale, according to imp
371d118 ards enlarged powers to meet with contingency.-- SEPT. 23rd. Saw in Loddiges Garden. 1279 varieties o
374d134 orbiculas, with many extinct forms & Trilobites SEPT 25th. In considering infertility of hybrids int
376d136 cious. & does not disappear for Many generations SEPT. 29th Dr. Andrew. Smith <Remarks on extraordina
383d160 e must often have been dissected Zoolog. Garden. SEPT 16." Hybrid between Silver & Common Pheasant. M
385d163 plumage in castrated male.-- «Men giving milk--» SEPT. 25th Young man at Willis «Grt. Marlborough Str
461t041 Upland geese would transplant seeds very far.-- SEPT 31. The identity <of> (or only closeness) of so
470t177 ria= (Plenty of Humble Bees on Phlox Down, 1854, SEPT.) In Spanish Broom by pulling back Wings, polle
551m128 mind went on RAMBLING till excited by question.» SEPT. 4th. Lyell in his Principles talks of it as wo
552m132 dy, being failed, & not to any particular muscle SEPT. 8th. I am tempted to say that those actions wh
554m141 rops it. the bird will fly at him-- Knowladge.-- SEPT. 13th It will be good to give Abercrombie's def
555m142 ot be traced) V. D. p. 111, case of Association. SEPT. 16th Zoological Gardens-- Endeavoured to class
555m143 of the old ones.-- -- S. American group sneer.-- SEPT 21st Was witty in a dream in a confused manner.
556m144 Does Elephant know shame-- dog knows triumph.--» SEPT. 23rd. Horses in Omnibus instantly start when t
072r160 <sep> Chiloe concretions somewhat analogous to SEPTA.-- would particle attracted towards space tend
074r165 concretions <dt> determined by fissures as in SEPTARIA. (& Chiloe case, at least corelation)--Galapa
098a047 thick wedges of feldspar in gneiss.-- Veins in SEPTARIA. a kind of concretionary process (analogous t
102a060 ock. This applies to cleavage & concretions.-- SEPTARIA in concretion arranged in planes, case of sep
123a116 rivers which must be like the Chilian ones.-- SEPTEMB. 2d.-- Sulphur like carbon must go round of di
203b138 s peculiar to Continent & England» Loudon Mag: SEPTEMB or Octob 1837 Westwood has written paper on af
350d054 y spaces" p 23. "for Nature is the art of God" SEPTEMB I,. It has been argued Man first civilized. <n
350d055 alians.-- Americans. &c After Decandolles idea SEPTEMB. 1st. Macleay & Broderip were talking of some
550m127 would Greyhound about dread of water-- innate SEPTEMB 1-- If one performs some actions, which are pl
552m130 ike a snow-haze. covers my whole imagination." SEPTEMBE. 3d Why when one thinks of any object, (or ha
065r139 of March cove began to freeze. correspond to SEPTEMBER ¿Did I make any observations on springs at S.
353d061 utang. other point of resemblance with man.-- SEPTEMBER 3d Magazine of Natural History. 1838 vol II p
362d099 .-- the teal which some authors [not located] SEPTEMBER 13th The passion of the doe to the victorious
379d152 worth studying. Archiv. fur. Naturgeschichte. SEPTEMBER 11' Generation Mr Yarrell says it is well kno
606o025 he "beau ideal", my instinctive impression 1) SEPTEMBER 6th. 1838 Every action whatever is the effect
094a036 Vol 7 Dr Buckland Reliquiae Diluvianae p. 201. & SEQ Murc Trans Geolog Soc Vol 2. p 257 {T} The Pota:
114a095 stone tells subsidence as plainly as Temple of SERAPIS. (now we have banished diluvial waves). & like
025r016 9 200 R. Real [11 degree 31 minute S.] & [R.] SERGIPE [11 degree 10 minute S.] 20 190 R. San Francis
038r057 . Lyell considers (P 84 Vol III.) whole of Etna SERIES of coatings; hence it will be necessary to sta
049r087 se to Isle of White. but rather to one out of a SERIES of faults. [Fig. 4] {P} In Cordill: should bas
097a043 ut no minute description is given.-- Vol II. 2d SERIES. p. 221.-- Mr Bollaert tells me, that the uppe
147g017 s almost appeared as if they belonged to double SERIES Whole very obscure but it is certain there mus
179b035 es in certain orders connected with gaps in the SERIES of connection? «if stating from same epoch cer
192b087 anks had produced infinite species probably the SERIES would have been more perfect. because in each
192b088 only expect some steps.-- I may ask whether the SERIES is not more perfect by the discovery of fossil
212b164 use, collection of Birds from Java.-- at Leyden SERIES from several islands.-- Bear peculiar to Sumat
214b173 logy between Australia and «fossils of» Oolitic SERIES does not appear to me very strong What is Oste
278c129 --There should be some term used, when there is SERIES. Could I not give Catalogue of Mammalia arrang
281c137 optional operation it is.-- show how finely the SERIES is graeduated.-- Dr Beck doubt if local variet
287c157 n the throne (of reason), she has also placed a SERIES of animals on the steps that lead up to it" p.
288c158 ogy on its concurrence in parallel parts of his SERIES, ie, cannot be discovered till circles complet
290c164 d, Whilst tips of primaries black, by examining SERIES I cannot doubt laws of change, Will be known.-
291c167 -- Makes hermaphrodituses. one step in <scale>. SERIES-- in plants we have a step between <mon> monoe
291c168 fathers of present, then there would be perfect SERIES or gradation.-- It is easy to see if South Ame
304c206 rance.-- the collector who plodding at making a SERIES, which would render our knowledge a chaos: who
304c206 r our knowledge a chaos: who will doubt this if SERIES now existed from Man to Monads-- though physio
304c206 Monads-- though physiology would profit. if the SERIES were believed to past into each other-- Differ
304c207 ostriches do-- but then there may have existed SERIES between apterix & other birds.-- will having m
304c208 nes on Echidna.-- when it can be traced through SERIES then probably heredetary & not produced by cir
373d132 that they became Mammalia, through a different SERIES of changes from the placentates, Having Hair.
390d179 y of generation being the passing through whole SERIES of forms to acquire differences: if none are a
399e006 t Lonsdale know some case of change in vertical SERIES: Look at whole Glacial period? [not located] S
404e025 nted all at once.-- but naturalists if they had SERIES, their relation to the external world, & every
411e054 ividuals «forms» are permanet, all steps in the SERIES, would expect this structure would bec
411e055 cases, under which the individual steps in the SERIES have been fixed, to study the physical causes.
420e088 sil filling up blank.-- CD[not between existing SERIES of species of dogs & Hyaena.-- but a common po
426e107 s were crossed, there would probably be perfect SERIES, from physical impossibility to unite to perfe
426e107 bility to unite to perfect prolifickness.-- (<a SERIES might be obtained>: but the intervention of ne
426e107 e new varieties destroys the appearance of this SERIES & makes one think that one large body of varie
426e107 eties are fertile & make mongrel, & other great SERIES quite otherwise & make on[ly] true hybrids.--
426e107 hybrids.-- but this is false, [give instance of SERIES from wild animals & plants]CD.-- Mr Marsh has
445e159 bute metamorphoses to habits of animals & takes SERIES.-- Ch 6 Upland geese would transplant seeds ve
461t037 after purpose would tend to render complex the SERIES,-- volans, ha
463t055 onderful!! distinct!!-- He would not allow such SERIES showed passages-- yet in talking, constantly s
463t057 ause if the animals were taken from which these SERIES were drawn they would not be intermediate, but
465t081 expected in any one country.-- in a descending SERIES of strata This again shows how much forms depe
465t089 y estuary formations are there in old Secondary SERIES-- few-- Maer June/41/, observed 3 plants of Ca
471tf11 in the larks from S. Africa he can almost make SERIES from end to end-- so that he is almost led to
569n021 A Melody on flute & Epic poem, opposite ends of SERIES or harmonious prose.-- Lutké Voyage in Carolin
605o019 y sublime, but that it is through a complicated SERIES of associations that we apply to such emotions
606o20v gh taste must necessarily be acquired by a long SERIES of experiments & observations. & yet, like in
618o41v d that thought & organization run in a parallel SERIES: if blueness & weight always went together. &
622o049 s & pidgeons.-- But as man is animal at head of SERIES in which «special» instincts decrease, I shoul
636j56v acculloch says, life, forms a broken, recurrent SERIES, whilst the habitation «or world» simple serie
636j56v series, whilst the habitation «or world» simple SERIES.-- My theory shows life equally simple series,
636j56v e series.-- My theory shows life equally simple SERIES, & therefore trace of beginning in organic wor
183b052 837. p. 243 it is said as well known fact that "SERIN avec le chardonneret, avec la linotte, avec le
070r155 n most violently shaken, by earthquake. but no SERIOUS injury.-- <Analysis of Atacama. Iron in Edinb
090a024 d increased a tenth; would the perturbation be SERIOUS? if so other cause besides thin vapour bringin
484z014 lantic Ocean Gould agrees with D'Orbigny, that SERPENT Eater-- or Secretary is S. African representat
510q017 Metaphysics of Morphology. ▮--Schelgel is he SERPENT man? about zones separated by non-inhabited sp
027r024 idae. at Concepcion. Humb: Pers. N. vii P. 56 SERPENTINE form: of Cuba for comparison (?) with St Pau
076r170 chloritic slate. with beds of syenite & <sep> SERPENTINE dipping to SW at 45 degree to 50 degree-- co
105a072 ther hand?-- The widening a valley depends on SERPENTINE course.-- the latter (it is generally said)
105a073 water to deepen not to widen valley.-- Why is SERPENTINE course result of little inclination??-----Th
155g064 y corrode or vice versa Same inclination when SERPENTINE might remove, what above straight line «only
528m037 ive is derived in a river from seeing how the SERPENTINE lines narrow in the distance.-- & even on pa
089a020 volcanic.-- Isle of Gory. rocks encrusted with SERPULA-- Isle of Cayenne. Syenite & diorite, covered
093a032 ral.-- Mem. Galapagos ∴ Basalt deepest?? Marcel SERRES L'Institut. 1837. p 331 Considers that Mercury
207b151 cumstance undiscovered by Ehrenbergh.-- «Marcel SERRES p. 331. L'Institut-- considers that mercu> Geo
211b163 p the dogs.-- L.' Institut. Curious paper by M. SERRES on Molluscous animals representing foetuses of
325c267 rey. Philosophie d'Histoire Naturelle Marcel de SERRES Cavernes d'Ossements 3d. Edit. Octav. (good to
481z010 round World German a reference to a luminous SERTULARIA Lesson Zoolog. Coq: p. 120 Coati Roux. Tatou
366d111 of Association very disagreeable hearing maed SERVANT cleaning door outside, as often as she touched
521m009 e watch was <seen> shown him.-- «<Mr Corb» SERVANT showed him watch & said dinner is ready, what,
532m056 wsbury to Clayton, (though so fond of her & of SERVANT of Richard & of Mary & her bed brought from Sh
547m111 dream, in continuation of waking thought-- my SERVANT was in the room. with my trunk out & I was eng
547m112 ciousness for moment, implied by «presence» my SERVANT, «box» my own manner of ordering things to be
138a153 e of every earthquake, during two years.-- will SERVE for comparison with the moon at some future tim
411e054 mal laws of corelation of parts & organs it may SERVE perfectly to specify types, & limits of variati
419e087 e precludes effects of catastrophes, which must SERVE to confound our chronology» «CONSIDER ALL THIS»
348d050 portance, which prevails throughout the group & SERVES to insulate <them> it".-- i.e what characters

```
027r027 sd. & great banks. effect of Elevation. United SERVICE Journal In the Iron sand formation <would> woo
203b137 a wide range> is a tyrant flycatcher doing the SERVICE of a swallow I think we may conclude from Aust
227b225 ry structure. latter far chief element.̱ little SERVICE habits in classification. or rather the fact t
355d067 n even if partly true they are of the greatest SERVICE, towards the end of science. namely prediction
227b225 ther the fact that they are not far the most SERVICEABLE. We may speculate of durability of successio
079r177 aine, comme ils traitent les loix de la nature & SES phenomenes."-- Ulloa's Voyage, Shell fish purple
174b015 mm were produced from propagation from different SET, as the rest of the world.-- This view  supposes
197b111 f analogies. States there is but one animal: one SET of organ:-- the others «animals» created with en
206b148 like the two fine families «no doubt a different SET of causes must act in the two case,» May this no
381d157 some monooecious-- (& cultivation might make one SET of organs barren in one plant & not in other), H
526m029 one is tempted to think all memory consists in a SET of sketches. some real-- some fancied.-- this fa
078r174 itta Triptera D'Orbigny has figured animal with SETAE like my undescribed[.] p. 140. Flèche of Quoy e
306c214 rdish & come also from Asia Minor.-- tail like SETTERS. long ears-- colours vary, but form constant.-
333d009 ogs line the same bitch & «perfect» spaniels & SETTERS are produced. one would argue the whole effect
513q21. me plant --Flora of Australian Mountains.-- Is SETTING of fruit. cross Conception--(Ị I could extract
234b256 f birds of Iceland. --M. Gaimard, however, will SETTLE this.-- Waterhouse says he is certain there ar
285c152 with regard to contemporaries-- fertility must SETTLE it.-- Changes in structure being «necessarily»
349d052 rsupials will change all.-- & so on no one will SETTLE number of primary divisions.-- Complains (p. 5
368d114 canus & Copris &c).-- In birds singing of cocks SETTLE point.-- (do the females then fight for male)
348d051 > no scale of value of difference is or can be SETTLED,-- I believe affinity may be taken literally,,
420e088 rica than in Europe especially with Europaeans SETTLED there L'Institut do. p. 419, «long» account of
537m074 is chance, which way it will be, but yet it is SETTLED by reason.-- How slow habits are changed may
334d012 re white. Fox says the Half Muscovy Fox says a SETTLER near Swan river, lost his <on> two cows entire
176b024 . repugnance to intermarriage «increases it--> SETTLES it ¿We need not think that fish & penguins rea
199b119 shed, & given to Portuguese. priest.--In first SETTLING a country.-- people very apt to be split up i
287c158 ms to limit Lamarck definition of relations to SETTLING the relative importance of the organs in same
599o008 re et du temps! voila les facultes, q'il possede SEUL sur la terre. J'ai trouve son âme" &c-- -- Conf
207b152 . III. p. 164. Lile de la Reunion presente elle SEULE plus d'especes polymorphes que toute la terre f
337d020 l-- Italian Greyhound is probably the effect of <SEV> local variety many times changed together with
073r164 spar: = gold veins in a phonolitic porphyry. = SEVERAL parts of N. Spain great analogy to Hungary. =
074r164 dant in the upper limestone, which H. calls by SEVERAL secondary names «Study Hoffmans account of ste
085aIFC on any Subject As far as p. 33. distributed to SEVERAL subjects. Feb 24th 1839 As far as p 140-- abst
088a013 ection of veins prove, that there are at least SEVERAL attempts at elevation From the lost & turned a
090a023 rded is 473. pounds (taking about average when SEVERAL are given), this will give nearly 19 pounds av
129a132 ffs Papers, in Edinburgh New Phil. Journ 1838. SEVERAL case given of hot heads &c heat beneath the se
159p084 Gneiss  Block on 2d shelf & below it some way; SEVERAL large ones (one 6 ft across) on top of spit be
160g089 Bought & Glen Tarf a perfect old Loch, making <SEVERAL> two divortiums aquarium, viz two branches of R
161g097 .260  would have been more correct) there were SEVERAL obscure but not far continuous flights above i
170b001 ger process, the new individual passing throug SEVERAL stages (¿typical, <of the> or shortened repeti
173b012 nts; all bred in from one parent why. Myothera SEVERAL species in S. America why, 2 of ostriches  in.
180b037 d.-- bearing relation to ancient types.-- with SEVERAL extinct forms, for if each species «an ancient
196b106 r near South Hemisphere. Were they produced in SEVERAL places & died off in some? Why did not  fossil
212b164 on of Birds from Java.-- at Leyden series from SEVERAL islands.-- Bear peculiar to Sumatra & not foun
217b187 hyrus has one species in Europe Madagascar has SEVERAL American forms-- The above facts evidently sho
226b222 proportion  of species to genus If on one isld SEVERAL species of same genus, subsided land.-- Maurit
229b233 , instance of useless structure-- Smith thinks SEVERAL species of Rhinoceros range from Abyssinia  to
229b234 ts of Mauritius Freycinet Voyage, agrees. with SEVERAL mammalia being peculiar (?) If. Henslow discus
274c116 America  very few forms in common.,-- but each SEVERAL with Europe & northern Asia & Northern America
278c130 species (Mitchell's authority) in Australia, & SEVERAL in Van Diemen's land is most important as show
313c235 omostraca (Magazine of Zoology & Botany) where SEVERAL generations are produced in succession (13?) w
317c249 ells me same thing p. 55. 40 leagues from land SEVERAL patches of reed & trees p. 259 120 ft in lengt
317c250 a as scarcely to be distinguished from it.-- & SEVERAL old acquaintances. which grow on the lower reg
321c270 ow to observe Mayo Philosophy of Art of Living SEVERAL of Water Savage Landons Imaginary Conversation
322c270 Hunters  Animal OEconomy. edited by Owen. read SEVERAL papers-- all, that bear on any of my  subjects
332d002 838) he has seen more case in a month, than in SEVERAL previous years. two having consulted him on on
333d007 as wild one under present form.-- Fox has seen SEVERAL cases of Foxes & dogs crossed, offspring alway
333d007 x dog approaches to species. Again he has seen SEVERAL crosses between Esquimaux dog & common dogs &
344d041 worm The case of one impregnation sufficing to SEVERAL births analogous to superfoetation, & to succe
355d066 ith stripe approaches to ass.» or fowls to the SEVERAL aboriginal species «or ducks» (here argue if i
355d066 f it be said domestic fowls are descended from SEVERAL stock, then species are fertile[; as long as o
362d100 e the forked black mark of the Rock Pidgeon,-- SEVERAL have a group of white speckles on elbow joint-
371d118 .-- Saw his collection of «Humming» birds, saw SEVERAL fully developed tails, & one with beak  turned
374d134 ly organized.-- do. p. 461.-- Lower Silurian-- SEVERAL existing genera. Nautilus turbo. buccinum. tur
387d168 rence from parental stock having been repeated SEVERAL times, that it becomes fixed in blood.-- Looki
387d168 mother  & ovum in offspring, as similar to the SEVERAL ova in mother. (with only difference of  time)
387d168 every peculiarity has a tendency to descend to SEVERAL generations) If A & B be two animals which hav
389d176 mother.-- one animal will fecundate female for SEVERAL births, & even produce fertile offspring-- DES
390d178 tions were added or substracted at each, or in SEVERAL generations, the process would be similar to b
392d180 mmon Pigs.--» Experimentize on crossing of the SEVERAL species of wild fowl <in Z> of India «with our
402e020 he species are peculiar to them" do-- p. 368. "SEVERAL kinds of animals have spread from the N. end o
402e020 In Sooloo we find the elephant-- in Magindanao SEVERAL kinds of the large monkeys.-- Fewer <of th> «M
423e098 which Sowerby says, is an American form.-- has SEVERAL species in my fossils-- CD[If cases of one var
424e100 Composites of Galapagos were South American.-- SEVERAL cases of species peculiar to separate islets.-
437e139 nd alive.-- Stanleys Familiar History of Birds SEVERAL cases on record of stoats being carried (p. 12
439e145 he Ancient Egyptians Vol III. p. 33-- They had SEVERAL breeds of dogs.-- like greyhound-- fox-dog-- t
446e162 characterized.-- Tulips are cultivated during SEVERAL years & then they break-- -- each tulip is the
451e177 a) Lat 5 degree. 50 minute S. adjoining it are SEVERAL small islands. abounding with deer-- Horsburgs
459t009 ch as we see in the young of the wild hog & of SEVERAL species of deer, which are altogether immacula
465t079 n at present «in Europe we know there has been SEVERAL successions of Mammals.--» yet only two monkey
465t079 ent against-- monkey-man, valueless.-- May not SEVERAL generations have been confounded in the caves?
466t093 Azalea  <do> <it is so> <Though I saw no Bees <SEVERAL> visiting it>-- In yellow day lily, the Bees
467t104 sheath> Keel of Lupine-- Seen Bees on Potato & SEVERAL times on Beans Rough.--"green-cabbage «in flowe
468t112 s same case as Azalea or Rhododendron xx after SEVERAL gloomy days. hot one, Bees almost P every minu
469t135 ch examined very many flowers.= 22d.-- /during SEVERAL succeeding days <many> «most numerous» bees vi
472s01r supposed  to have been hybridised == Has tried SEVERAL year to obtain seed, but the pods have (except
472s01v en empty.-- See separate note-- Elizabeth says SEVERAL years ago seeds were procured with the P. orie
472s02r which  pollen was routed. wh. was not case, on SEVERAL flowers I examined some days ago-- This Bee fl
495q005 ot) (3) Sow seeds & place cuttings or bulbs in SEVERAL different soils & temperatures & see what  the
497q007 ight take opposite view. Gaertner talks of the SEVERAL great & natural Families, as being difficult t
515q022 ead to send Cover common Pea (& Sweet Pea) for SEVERAL generations under net & see if get sterile-- C
567n013 e child not knowing what to do with them, came SEVERAL times & opened my hand, & put them in-- like c
582n066 idered as analogous to glowing conversation of SEVERAL people.-- Children have an uncommon pleasure i
639j28v steps. of which we have manifold traces in the SEVERAL genera of Grallae Suppose six puppies are born
029r033 ara to London:-- "Feb. 12, 1835. At 10h. 15m. a SEVERE shock of earthquake shook the ship in a most v
045r076 he great rise). -- The great rains which attend SEVERE Earthquakes «1822¿ 1835?» alone, (& the genera
045r076 ucky! when they hear of a place having a pretty SEVERE shock). are much more curious & perplexing. th
527m035 errupted & utterly forgotten--, so as to feel a SEVERE disappointment «in real train of thought  this
233b250 es Institute 1838. p 38. account of fossils of SEWALICK «India» Monkeys of old World. Crocodiles. Ano
192b084 n's breast. one does not say some use, but <no.> SEX not having been determined.-- so with useless wi
291c167 supported by wonderful fact of bees changing the SEX by feeding.-- no it is developing an hybrid fema
307c215 lk)» rudimentary wings. so nature can produce in SEX, what she does in species of Apterix This is imp
313c235 nstinct.-- (How wonderful a case bees developing SEX of neuters) species may have had their infancies
350d055 male  with eyes!-- (are not these differences in SEX confined to annulosa?) Remarked that young of Ci
358d085 ven lay eggs. & the men cannot «hardly» tell any SEX by appearance.-- The silver & common pheasant cr
360d089 onderated <which seems owing «determined» by the SEX> Individual instances trouble Yarrels law. chief
360d092 ut if it were opposed to the difference in other SEX, it would be much more difficult to propagate--
383d159 body  imbued with possibility of becoming either SEX.-- Ŋ In my theory I must allude to separation of
388d173 hows how simply instinctive the feeling of other SEX being present is-- it also shows that semen. mus
```

389d176 y close to either.-- Male & female as foetus one SEX; & therefore both capable of propagating, but on
390d179 & not its comparison «of difference» with other SEX. = The highest bred Blood-hound. would be infert
417e078 ther like father or mother, independently of its SEX, or half way between, or someway different from
418e080 ier to understand ¿perfect?? developement of one SEX on one side, than the addition of other organ, i
427e109 dioecious plants being effect of abortion of one SEX.-- Linnaean class Dioecia & Monooecia. ought to
428e113 en <?> «Walker attributes this to effect of male SEX on locomotive system» I am bound to insist hones
435e130 willow trees have been observed to change their SEX,-- this effect from age, what Mr Knight [not loc
454e184 ming double loosing fertility if, sometimes one, SEX & sometimes. other, so as to become <all> monooe
466t095 characters being inherited at corresponding age & SEX, opposed by cantering horses having colts which
466t095 e powers instinctive & doubtless not confined to SEX.-- «Is not cantering a congenital peculiarity in
466t095 qualities have been given to foetus <fr> before SEX developed-- Double flowers & colours breaking on
492q002 ing double ever become monooecious-- loosing one SEX & not other: which generally fails first?-- Mal[
493q003 sā = History of Tortoise-shell Cats. as only one SEX so coloured = I have grey-cat «wh was female» wi
501q011 of Europe & less in another region-- (27) Which SEX in Mules generally fails-- perhaps indexed by se
508q016 d (5) About cross-bred races of men taking after SEX. A Smith. About species of Rhinoceros. becoming
577n052 blushing» is connected with sexual, because each SEX thinks more of what another thinks of him, than
577n052 nother thinks of him, than of any one of his own SEX.-- Hence, animals. not being such thinking peopl
640j28v e: for Man & woman are same: fertility of either SEX determines life:.» «With respect to whether Gala
256c055 an--- -- genus.-- The circumstance of having two SEXES is the check to distribution of birds & animals
273c112 presentatives of the different families.-- that SEXES <are> «have» same plumage.-- <no> this is appli
291c167 elves.-- The idea of foetus being of one «both» SEX«ES». is strongly supported by wonderful fact of b
291c167 s in animals it may be difficult to imagine how SEXES were separated.-- in plants we have some flower
292c169 animal the more subject to variation. therefore SEXES or two animals:-- «When» sexes «being in one no
292c169 ation. therefore sexes or two animals:-- «When» SEXES <being in one normal-- when so they» are united
305c209 ts which are close to European species, and the SEXES of which vary in colour of plumage in same rema
313c236 ission cannot effect the process.-- but why two SEXES scission in all cases probably gemmation (.Ehre
324c268 where M. Quetelet has published his laws about SEXES relative to age of Marriages Brown at end of Fl
331dIFC Book some curious notes on Monkeys recognising SEXES of animals:]CD [All Selected Dec. 14-- 1856]CD
344d041 tomostraca & Aphides. The extreme difference of SEXES. is probably arrived at in case of insects as g
344d041 pring in Entomostraca & Aphides Developement of SEXES in Caterpillar. very valuable facts-- they are
358d076 of observation that in insects where one of the SEXES is little developed, it is always female which
376d137 thinks perhaps by smell.-- but monkeys examine SEXES of every Has repeatedly seen one he kept pull u
376d138 bitch than dog: Monkey thus examine each other SEXES,-- «by taking up tail» Mem: Ourang Jenny with T
379d152 think facts there mentioned about proportion of SEXES, at birth & causes. If an animal breeds young h
381d156 cale has no «constant» relation to separation of SEXES, as may be seen in Monooecious & Dioecious plan
381d157 is most important support to my views-- Seeing SEXES separate in some of the lowest tribes, leads on
381d157 other «sexual» organs; for if so, separation of SEXES very simple.-- as in plants even in same genus
382d157 is <evidently> actually split in two-- keeping SEXES separate. Owen say such view worthy of a Lamarc
382d158 gans which perform nearly same function in both SEXES.-- are never double, only modified. those which
383d159 -- ⟨ In my theory I must allude to separation of SEXES as very great difficulty, then give speculation
384d162 t is handed down for some generations Theory of SEXES (woman makes, bud, man puts primordial vivifyin
384d162 - thus one organ in each becomes obliterated, & SEXES as in Vertebrate tak place.-- ∴ Every man & wom
386d167 y in human frame.-- «one of» The final cause of SEXES to obliterate differences. final cause of this
388d172 orelation in structure = Neuter bee having both SEXES abortive fact of same tendency. -- Mammae in ma
391d175 & &c in» in litter. Why is there some law about SEXES of twins in former case.-- (many monster are re
409e048 se,-- nothing probably exists for one cause» of SEXES «in separate» «animals»: for otherwise, there w
409e049 CD & if my theory be true then the formation of SEXES rigidly necessary.-- Without sexual crossing, t
410e051 local changes. = this could only be effected by SEXES: All the above should follow after discussion o
446e160 o Bartram's Travels p. XXIII. <Some birds> Both SEXES of some birds sing equally well. and <in> these
480z007 have longer tails than old ones-- in America & SEXES not of different size-- How does this apply to
500q010 What Book on varieties &c of deer. Contests of SEXES.-- Q.30) March 1842. <Last> Year «before last»
504q014 30/p.11) (2) Yew Berries germinate?-- Yew trees SEXES-- (3) Get Holyhoaks. races planted & Linum Pere
211b161 erefore instinctive feelings against other «for SEXUAL» ends» species, whereas Man has such instincts
283c144 replaced Birds of Paradise»-- if such fantastic «SEXUAL» ornaments. have so intimate a relation to two
313c235 uccession (13?) without impregnation, therefore SEXUAL passion must arise after long interval very go
313c235 ly impressed.-- we see in the Entomostraca. The SEXUAL curiosity of the orang outang of (in June 1838
358d085 shrivel up.-- Yet odd they should have so much SEXUAL character as they have This character of not h
358d085 acter as they have This character of not having SEXUAL plumage is very common by hybrids, that are in
359d087 should organic affections always influence the SEXUAL organs alone.-- It is singular pheasant & fowl
378d147 & cock Starling alike, yet young ones brown.-- SEXUAL Selection If masculine character. added to spe
381d157 ephalopoda» are there abortive traces of other «SEXUAL» organs; for if so, separation of sexes very si
384d161 f each species not similarly subject-- ⟨Divides SEXUAL marks into primary & secondary, the latter onl
384d162 stinct. (in dioecious plants are there abortive SEXUAL organs?): they then become so related to each
410e050 formation of sexes rigidly necessary.-- Without SEXUAL crossing, there would be endless changes, & he
411e056 innumerable individuals can be produced. & yet SEXUAL apparatus.-- My account of Circus cinereus of
413e060 n problem.! Hurrah.-- "intermediate causes" The SEXUAL system of the Cirrhipedes is the more remarkab
417e077 r. Knight makes this analogy between grafting & SEXUAL union-- Looking at simple generation as being
417e077 between man & woman is «indeed» (independent of SEXUAL differences) a variety. The offspring of true
418e080 d not be perfect as in Ox. the amount of double SEXUAL developement is spread over [not located] it u
418e083 th the vital principle, which gives rise to the SEXUAL organs, different in each species,-- & knowing
442e152 ly transported»-- throw over this theory, & the SEXUAL reproduction of species may stop for any numbe
443e155 ucture they must have <done> been propagated by SEXUAL commerce <The fact of Corallina & Halimeda is
443e155 eda is case in point».-- The relation of these «SEXUAL» functions to complexity is evident, yet the i
460t015 s-- The brothers & sisters half-breed showed no SEXUAL inclination for each other-- Aug. 20th The Ech
506q015 ning= (38) Have Dioecious plants any secondary, SEXUAL characters.-- Stature, position of flowers-- T
557m149 imself.-- Jealousy probably originally entirely SEXUAL; first try «to» attract female, (or object of
558m150 t being obscurely guided» or strong instinctive SEXUAL, parental & social instincts, giving rise "to
563nIFC .-- Private. Expression M Expression N What are SEXUAL difference in monkeys.-- Charles Darwin [Priva
564n003 I say grant reason to any animal with social & SEXUAL instinct «& yet with passion» he must have con
569n019 Hence pleasure in the beautiful. (distinct from SEXUAL beauty) is acquired taste.-- Whilst music extr
574n041 bed by one end of Cocoa nut.-- November 27th.-- SEXUAL desire makes saliva to flow «yes, certainly»--
574n041 . Bell says, & hearing music. to certain degree SEXUAL.-- The association of saliva, is probably due
577n052 king of oneself.-- «blushing» is connected with SEXUAL, because each sex thinks more of what another
579n059 nt of ears A man shivers, from fear, sublimity, SEXUAL ardour.-- a man cries from grief, joy. & subli
579n059 , or her.-- it is blind feeling, something like SEXUAL feelings-- love being an emotion does it regar
584n071 ood & temperature molting & breeding instincts, SEXUAL, social, «subordinate to,» self preservation,
591n099 often have «as a boy» wondered why all abnormal SEXUAL actions or even impulses. (where sensations of
612o035 owth of tissue and are subject to accident; the SEXUAL willing comes on period of year as much as inf
614o037 is one of these instinctive feelings. [LHC] As SEXUAL instinct comes on late in life, man almost alo
616o038 on of thought, may be no more «difference» than SEXUAL intercourse in plants is involuntary, in man v
574n041 endency to kiss, & almost bite, that which one SEXUALLY loves is probably connected with flow of sali
158g083 ay» in Upper Glen Roy at pass {P} River Gorge 4th SH side of valley Granite blocks on this side (retu
346d047 hepherd dogs.-- Aug. 24th. Was struck with pink SHADE on plumage of the Pelican.-- Mem pink spots on
382d158 rIform very different, are both present in every SHADE of perfection --How came it nipples <are> «thou
388d171 the womb, when it came out. it might partake of SHADE of fathers character.-- according to this view
612o035 r less turgid vessels; effect of heat, light or SHADE.) Joining two difficulties into one common one
299c194 s come up so.-- there are not Many intermediate SHADES in these cases,-- but absolute species formed.
305c211 in thinking and acting principle in the various SHADES of <dif> separation between those individuals
450e175 & Sumatra breeds, (.Hence it appears there are SHADES of difference in all the isld, like in wild an
334d010 aced his stallion as second horse between <whe> SHAFT mare & another leader mare,-- this stallion tho
383d160 fine metallic green. <from> with tip & part of SHAFT metallic green.-- This green doubtless is effec
455eIBC or tubers. Yes-- but these are same as trees.-- SHAKE some sleeping mimosa-- do stamina of C. Specios
494q005 colours, any effect? and silk caterpillars (1) SHAKE a sleeping mimosa, or half bred mimosa (a) betw
527m031 ly believe free-will & chance are synonymous.-- SHAKE ten thousand grains of sand together & one will
070r155 en oclock Novem <5th> Concepcion most violently SHAKEN, by earthquake. but no serious injury. -- <Ana
557m150 ake body tremble. Why much laughter tears.-- & SHAKING body.-- Are those parts of body, as heart, & c
570n023 nding reason.-- surprise with negation.-- like SHAKING something off shoulder-- or is it from inspira
114a094 Europe. gneiss-- metamorphosed clay slate.-- --SHALE in shall sea. Lyell confounds these introduce d

```
030r036  ing from great quantity of altered Carbonaceous SHALES Examine chart of Patagonian coast to see propo
040r063  In  the Chonos Islds we must imagine bituminous SHALES have been metamorphised, as in Brazil ferugino
058r117  we look at Elevations as constantly going on we SHALL see a cause for Volcanos part of same phenomena
114a094  gneiss-- metamorphosed clay slate.-- --shale in SHALL sea. Lyell confounds these introduce discussion
181b040  extremely improbable that any of <his relatives SHALL likewise> the successors of his relatives shall
181b040  shall likewise> the successors of his relatives SHALL now exist,-- In same manner, if we take «a  man
307c216  nce where a tendril passes into a mere stump.-- SHALL abortive organs «of very same kind» in these ca
371d127  ke Avocette. here is what [not located] that it SHALL beget young different in colour, form, & so alt
412e057  able of innumerable variations, as long as each SHALL be perfectly adapted to circumstances of times.
433e127  on of the beds-- The argument must be thus put, SHALL we give up whole system, of transmut., or belie
438e141  he Brussels Sprout was slowly formed.-- » if it SHALL be difficult to show that <time> the fixity of
578n053  ual parts of body.== (if you <think> «fear» you SHALL not have e---n, «or wish extraordinarily to hav
639j28v  rcumstances determine that, the long legged one SHALL rather oftener than any other one. survive.  in
026r019  ieve they «the limestones» have been formed in SHALLOW water: so have the Conglomerates: Yet this vie
128a130  the  Mexican Gulf. is fouled by bars of sand & SHALLOW lagoon.-- when describing Coast of. Brazil. Ma
150g037  In  Glen Collarig, when water up to shelf very SHALLOW channel 50 ft wide & river get formed in centr
227b224  a, Mexico & Europe. one gret sea (Coral reefs:. SHALLOW water at Melville Isd. (3d) We know that struc
115a098  raction of every atom in rock On a coast, the SHALLOWER the water, the greater power of oscillations
152a045  in  most improbable part & not in Pass, where SHALLOWEST In Glen Collarig good case of shelves enteri
525m023  osity in defending a friendly dog).-- they feel SHAME, when doing anything which is wrong.-- as eatin
525m024  cases  their actions do not look like fear, but SHAME.-- I cannot remember instances, but I feel sure
554m140  I hide herself.-- I do not know whether fear or SHAME.-- When she thinks she is going to be whipped.
556m144  ected with judgment. [What is the Philosophy of SHAME & Blushing]CD «Does Elephant know shame-- dog k
556m144  ophy of Shame & Blushing]CD «Does Elephant know SHAME-- dog knows triumph.--» Sept. 23rd. Horses in O
557m148  right» back absurdly arched. & tail stiff.-- is SHAME, jealousy, envy all primitive feelings, no more
557m148  be  analysed than fear or anger? I should think SHAME would be more easily analysed than jealousy, be
557m149  endency to muscular movement, hence shy people (SHAME of ridicule) are singularly apt to catch tricks
557m149  rubbing hands.-- stamping. grinding teeth.-- in SHAME frowning, & anguish,-- shyness not so.-- affect
557m150  ugh in individual case. nothing can be heard.-- SHAME would never make person tremble, like fear.-- W
560m156  present?  Do they pout, or spit, or cry.-- <fe> SHAME, independent of fear: colour of bare nails--, &
570n025  ty, mixed with triumphant feeling so similar to SHAME after asinine.-- both accompanied by depending
570n025  r both may be said to have fear, both both have SHAME-- Animals have not modesty. analyse this.-- «Ex
579n058  h's dissertation on Poetry.-- The expression of SHAME-facedness for shyness, having been invented, pr
609o029  beginning,  & therefore most deeply impressed). SHAME perhaps an exception. (does it originate in a d
609o029  doubting feel between conscience & impulse) but SHAME «we alas know» is far easier conquered than the
620o046  -- When a pointer spring his bird. one says for SHAME (& the «old» dog really feels ashamed?) not  so
629o55v  & conscience will imperiously say so, & produce SHAME & remorse-- [Thus pungency of one's feelings fo
629o55v  It is other question, how the feeling of ought, SHAME. right & wrong comes into mind in first  case--
629o55v  rong comes into mind in first case-- seeing how SHAME is accompanied by blushing, bears some relation
300c197  o consider him created from animals.-- Insects SHAMMING death, most difficult case to iimagine how ar
300c197  Mem Spider only dropping where ground thick.-- SHAMMING death it is but being motionless. How is inst
531m052  HABITUAL  movement to avoid any danger-- Fear, SHAMMING death, or running away. accompanied with want
255c051  use we may suppose longest part of structure.-- SHAPE of wings have altered many times, but all  have
360d090  like  dog-- though it has full share of Jackall SHAPE of body.-- disposition wild, & fearful. though
467t100  mined in microscope--some of the stigmas of {P} SHAPE of ordinary Labiatae --the chief part with ordi
540m087  ent-- in organization.-- Same cause as colour & SHAPE & ideosyncracy.-- Look at the Indian in slavery
505q014  impregnation necessary to fruit--; become well SHAPED by care 13 Arum before pollen is shed can  you
593n105  est for herself.-- the object is to make saucer- SHAPED depression.-- [blank] Does music bear any rela
449e168  ll, which has laid in domestication eggs of two SHAPES & colour.-- Eyton has observed same thing in B
286c154  but hiatus animals expression of countenance. «S[SHAPE of sickness,-- death, unequal life,-- stimulat
360d090  e & tail somewhat like dog-- though it has full SHARE of Jackall shape of body.-- disposition wild, &
022r008  r's last voyage to New Holland P 127.--Caught a SHARK 11 ft long. "Its maw was like a leathern  sack,
023r009  was full of jelly which stank extreamly."--This SHARK was caught in Shark's Bay. Lat 25 degree. The n
222b205  th fish-- as some of the most perfect kinds the SHARK. lived in remotest epochs.-- ¿ lizards of secon
436e136  tances» <April 12th..> Cestracion, Port Jackson SHARK-- Owen thinks Australia part of Old World  <If>
023r009  ch stank extreamly."--This shark was caught in SHARK'S Bay. Lat 25 degree. The nearest of the E. Indi
023r012  eak part of the ocean's bottom. With respect to SHARKS distributing fossil remains: Sharks followed C
023r012  respect  to Sharks distributing fossil remains: SHARKS followed Capt. Henry's vessel from the Friendl
081r BC  . 8 Centimetre 4.4 {T} C. Darwin R. N. Range of SHARKS Nothing For any Purpose A. Geology Note on Woo
247c028  according  to Chamisso in Radack isld.-- p. 69. SHARKS very generally distributed: Mem of great geolo
312c227  ave long tail.-- in raptorial birds, & tigers & SHARKS, being spotted, & colours of little value Dr S
022r008  e a leathern sack, very thick & so tough that a SHARP knife could not cut it: in which we found the H
121a111  fragments  of rock, which retained their angles SHARP-- yet with character completely altered, & a cr
160g090  rey fine grained. Much contorted gneiss «narrow SHARP ridge with peak» I walked all round hill. Bould
462t051  &c  NB. In botanical geography, there can be no SHARP division of partition as between Mammalia in ca
524m022  of  sincerity.-- People in old age. exceedingly SHARP in some things, though so confused in others.--
246c026  s at all from those of Europe, but beak rather SHARPER., & rather longer in proportion, colour slight
624o50v  eason never can lead to action.-- p. 164. Ld. SHATSBURY under term of Reflex Senses seems to have <co
511q018  troduce it in work} Whether <Yar> knows whether SHAWS hybrids between Trout & Salmon were fertile & w
473d07r  rades» there is no reason, why the peculiarities SHD be born,-- may come in corresponding time of lif
383d159  gument for developement of either.-- (Mammae or SHEATH of Horses poenis reduced to extreme degree  of
467t103  in  Beans the wings seem beautifully to protect SHEATH {a} the less nectar seems to be at base of
467t104  ress down wings which ejects pollen from tip of SHEATH.-- «Also in Lathyrus pratensis yellow saw stig
467t104  project-- In common Pea saw Humble so press down SHEATH, that stigma covered with pollen was pressed &
467t104  gs, stigma & mass of yellow pollen protrudes at SHEATH.-- At last I saw Bee collecting pollen from <s
467t104  th.-- At last I saw Bee collecting pollen from <SHEATH» Keel of Lupine-- Seen Bees on Potato & severa
468t112  found all, stamens straightened pollen profusely SHED; lengthened & turned up «more than stamens», so
468t112  thers of long stamens {P} as stamens grow old «& SHED some pollen». they turn upwards & bend over sti
505q014  ome well shaped by care 13 Arum before pollen is SHED can you find flys dusted with pollen from other
253c047  ount of the MANELESS lion of Guzerat by Capt. W. SHEE. considered merely variety.-- red form of skull
294c177  species;-- Horse-- &c <Lonsdale says. that first SHEEP State broadly scarcely any novelty in my theor
146g011  torm, many <hundred> thousand tuns. Black faced SHEEP, sometimes mottled with white black legs & tail
148g024  of  Birch Wood by Inverorum being determined by SHEEP & not deer When Black faced sheep are crossed w
148g025  determined by sheep & not deer When Black faced SHEEP are crossed with English my informant said  the
182b048  VI. P II. p 89.-- Lieut. Wellstead obtained many SHEEP from Arabian coast. "These were of two kinds on
229b233  e common English.-- Hottentots say great tailed SHEEP aboriginal at Cape & a thinner-tailed kind furt
266c092  have hump like those of India & Arabia p. 202-- SHEEP have not the enormous tails, which disfigure th
267c093  coast of Arabia between Ras Mohammed & Jeddah". SHEEP numerous "of two kinds one white with «a» black
294c175  d of tailless cats near Bath. Lonsdale do. says SHEEP could not live for some time at New York «insta
306c214  cats are supposed to come from there.-- All the SHEEP are thick-tailed The dogs called Persian «greyh
309c221  hts' Travels in America, speaks of short legged SHEEP. heredetary proceeding from an accident. New En
311c227  of  Africa .p 69. with tall grass. & p 72 hairy SHEEP-- Edinburgh. Transact. Vol IX p. 107. an Ascari
325c267  stump  Library of useful knowledge. Horse, Cow, SHEEP.-- Verey. Philosophie d'Histoire Naturelle Marc
327c1BC  ia Library of Useful Knowledge on Horse & Cow & SHEEP Clarke's Travels.-- Temminck Hist. Nat. des Pig
345d044  ance-- good case of Provincial Breed-- Highland SHEEP jet black legs, & face & tail, just like specie
354d065  - Is there any law of this. Do any varieties of SHEEP «evidently artificial» approach in character to
364d104  -- It seems from Lib. of Useful. Knowledge that SHEEP originally. black. & Yarrell thinks the occasio
364d104  d rods mentioned in old Testament placed before SHEEP-- it has been thought that silver Pheasants abo
400e012  able Bay with Humps on their backs & big tailed SHEEP do Vol 10. p. 373, «& 374» Spaniards says no To
405e035  . p. 338.<V[ide]» Important account of cross of SHEEP & Moufflon of Corsica. <would not>, sadly again
406e036  ables, common to men, dogs, horses cows, pigs & SHEEP.-- diseases common to men & animals cow  pox.--
406e036  in Spain of pustular disease following handling SHEEP-- all cases: d degree p. 354-- The most vicious
432e123  mother; illustrated by the crossing of hornless SHEEP with horned.-- compare this with what highland
449e173  men at Paris Museum.-- Athenaeum: 1839. p. 451. SHEEP Merinos from Cape of Good Hope., . has different
459tf02  1. Comp. Anat.-- From Buffon cross of he-goat & SHEEP, it seems male gives form. admitted by Linnaeus
473s07r  -- June /42/-- June/42/ You can select cattle & SHEEP for horns & yet no difference in calves--how is
512q019  s Is there any difference in breeds of Cattle & SHEEP in the sprouting of the horns. at different per
512q019  r milking-- loss of early habits in Dorsetshire SHEEP migration of coots-- variation in hounds= An ug
```

(Key Word)

575n044 to hybrid greyhound to hunt hares. «& leave the SHEEP» & jackall to skulk about & hunt mice-- Jenners
619o042 orce» give pain: for instance either protecting SHEEP or hurting them. -- Therefore in man we should e
259c065 y disease, is on the same principle that, cut a SHEEPS tail off plenty of times & you will have no ta
265c083 any dogs in England must have been lopped off & SHEEPS tails cut yet there is no record of any effect
120a109 rior effect would be such as present. to spread SHEET of matter over surface.-- if elevation then wen
495q05a usted with pollen of neighbouring grass~ Spread SHEETS of Paper. covered with some sticky stuff in fl
149g032 ill on side of Inn BOULDER of granite above 4th SHELF a little lower down the hillock with beach & ch
150g037 an 3a Sunday In Glen Collarig, when water up to SHELF very shallow channel 50 ft wide & river get for
151g040 ce» has beach or band of pebbles on line of 4th SHELF.-- Even on Lauder Dicks Hypothesis impossible t
151g041 dging from external form alluvium) descend from SHELF 3d & almost meet, but are separated by flat bot
151g042 h must have been waterworn after 3d lake.-- 4th SHELF runs up some way on great sloping plain of allu
151g043 butresses, upper slope of which corresponds to SHELF the truncation & the upper shores may correspon
151g044 res may correspond with some line subsequent to SHELF {P} In Glen Collarig, by Dicks theory lake burs
152g048 n Roy Lake, must have remained very long at 4th SHELF from size of buttresses, to upper edge of which
152g050 Glen Collarig up within 200 ft of level of 4th SHELF= argument against river--composition &-- strati
153g051 argument detritus-- {P} where buttresses on 4th SHELF: others «lines not so level because of upper ed
153g051 nt for lake «or sea» at successive levels-- {P} SHELF opposite Glen collarig at bend & here most accu
153g052 nds roads disappear The normal condition of 4th SHELF, some way below House of Glen Roy, seems to be
153g053 s to be which higher up on is corroded {P} 4th [SHELF] river 4th [shelf] Could earthquake cause colle
153g053 er up on is corroded {P} 4th [shelf] river 4th [SHELF] Could earthquake cause collection of sediment?
154g057 mit of hill rounded, site N N W of Ben Erin {P} SHELF of Glen Guoy flat peat plain divortium aquarium
154g057 15 ft above bank or terrace, from terrace of 2d SHELF Level of shelf of Glen Guoy form comparison wit
154g058 k or terrace, from terrace of 2d shelf Level of SHELF of Glen Guoy form comparison with granite block
155g061 if the largest has hollowed out most Wednesday SHELF 3d dies away almost imperceptibly on Glen Turri
155g062 ay almost imperceptibly on Glen Turrit side 2nd SHELF very broad «& cut out, produced» from same «cau
155g062 » as «great» spit <is> or plain <now> formed on SHELF 4th 2d & 3d can be traced some way up, but most
155g063 above houses of Roy» Maccullochs supernumerary SHELF I doubt, much about «50 or 60 ft» «no doubt, a
156g070 r shelves granite & some other rocks at head of SHELF 3d almost all granite pebbles Level of plain of
156g071 lmost all granite pebbles Level of plain of 4th SHELF at head of Lower Glenroy 29.581 A 82 75 degree?
156g072 d it-- perhaps however sea also,-- Barometer on SHELF 3d. 29.455 A 83 degree ∴ plain of 4 minute shel
156g072 helf 3d. 29.455 A 83 degree ∴ plain of 4 minute SHELF slope, above «line of 4th» shelf This shelf at
156g072 in of 4 minute shelf slope, above «line of 4th» SHELF This shelf at head where <granite &> «veined» g
156g072 ute shelf slope, above «line of 4th» shelf This SHELF at head where <granite &> «veined» gneiss <unit
157g073 te & forming «sloping» buttresses Yet certainly SHELF 4th <near> only usually contains many pebbles,
157g074 se by fall of angular masses from above on soft SHELF-- 29.330 A 84 degree compare this with last mea
157g074 84 degree compare this with last measurement of SHELF of 3d:-- granite block a yard across. On side o
157g075 ross. On side of «that» hill, in front of which SHELF 3d form beach of granite pebbles, & around whic
157g075 d form beach of granite pebbles, & around which SHELF 2d «almost» forms it into island-- whole hill c
157g077 le back from line 2d; little action since «that SHELF» formed Upper terrace near Loch Spey <29.35161>
158g081 Upper Glen Roy great plain about 60 ft beneath SHELF peat on pebbles tidal plain as sea gradually re
158g082 ere Boulder lies. buttresses «occur» high up on SHELF 2d «in Upper Glen Roy» In this upper part «abou
158g083 ar the upper shelfs ground strewed with pebbles SHELF 3d runs up with buttresses on each side «very 1
159g084 te blocks on this side (return) between 2d & 3d SHELF Mountain <Mica> «composed of» Gneiss Block on 2
159g084 ountain <Mica> «composed of» Gneiss Block on 2d SHELF & below it some way; several large ones (one 6
159g086 to Fort Augustus Barom on upper (rather above)? SHELF 29.290 A. 69 degree Air 68 degree? Barom 29.008
161g096 y chiefly from rockiness When on other side {P} SHELF A Shelf A at head of Gentle mossy slope, which
161g096 y from rockiness When on other side {P} Shelf A SHELF A at head of Gentle mossy slope, which from a d
161g097 tress or pass at Isthmus appears above level of SHELF certainly) I took another measurement on short
161g098 <the> of which the source is a lip with the new SHELF flows into canal between L. Lochy & Oich. is a
162g100 said that a mound stretches along, parallel to SHELF on opposite side & dies away on the steep & roc
162g102 ead Roy told me) it is impossible to see my new SHELF, from road: Loch Ness 30.140. A 66 degree 30.09
286c154 we.-- It is our arrogance, to raise on the same SHELF-- to (look at common ancestor, (scarcely[)] con
158g083 unction of Upper & from Glenroy) near the upper SHELFS ground strewed with pebbles Shelf 3d runs up w
033r042 erently constituted lime have been removed?--As SHELL out of its cast which, although not very intell
079r177 la nature & ses phenomenes."-- Ulloa's Voyage, SHELL fish purple die, marevellous statements on, Vol
206b149 ydermata, or of coniferous trees; or in certain SHELL cephalopoda.-- «Read Buckland» L'..Institut «18
226b223 -- India; intermediate, see how that is.-- ¿are SHELL-boring Molluscs, like Carnivorous Mammalia in t
233b250 use thinks two main divisions of cats. Tortoise SHELL--& grey-banded. ¿species?-- thinks offspring of
445e158 er case than chicken pecking fly.-- "whilst the SHELL stuck to its tail" as mentioned by Sir. J. Bank
445e159 ie. says it is not sufficiently proved that any SHELL fish is really hermaphrodite. & <thinks> even o
479z005 eth {P} p. 140. Fléche of Quoy et Gaimard Ulloa SHELL fish Purple die Marvellous stories Ulloa's Voya
490q001 their dogs Sowerby Entomologist Does individual SHELL or insect or group vary more in one country or
493q003 male parent & vice versâ = History of Tortoise-SHELL Cats. as only one sex so coloured = I have grey
493q003 grey-cat «which was female» with tinge of tortoise-SHELL «on back.--» = Length of intestine in Persian C
564n005 ke fly.-- young partridge can run even with its SHELL on back.-- To study Metaphysic, as they have al
574n040 ,-- for instance the Birgos opening a Cocoa nut SHELL at one end.-- Children & old people get into ha
577n051 ry, that the formation of a hinge «in a bivalve SHELL» does to reason.-- an inflamed membrane from lo
638j58v of art «or intellect» such as hinge, & hinge of SHELL, works of laws of organization is remarkable-
023r013 ing Curvature of hill; states could discover no SHELLS: nothing said about K. Georges Sound The idea
025r010 mestone: the strange substitution of matter in SHELLS will generally be found to be old & dead» «(I
027r023 imestone: the strange substitution of matter in SHELLS, like Concretions & laminae, show what movemen
027r027 ls [not located] [not located] The frequency of SHELLS in the Calc. Sandstone Concret, is connected w
027r027 ndstone Concret, is connected with frequency of SHELLS in flints in Chalk New Providence more hilly t
027r027 e Bahama consists of rock & sand mixed with sea SHELLS--about 500 Isd. & great banks. effect of Eleva
030r036 there are Tertiary strata there is Coal-- ¿ No SHELLS in all cases. «.Mytilus.--» «at Guacho» «on N.
030r037 e they allied to the jaws of the Cocos fish Rio SHELLS argument for rise In Cordillera, the dikes co
033r042 other matter how very curious a structure: Have SHELLS ever casts alone in Calcareous. rocks??--if so
037r055 «besides ordinary marine remains» may contains <SHELLS few corals Tortoise> «remains of Amphibia, exc
039r061 in N. Pliocene formation of Limestone, casts of SHELLS, as in some older formations: Mem the envelope
040r064 ng as all below water no evidence--The depth of SHELLS (which being packed. in beds) lived there, mak
041r065 bable rise of land: Mention M. Gay's fact about SHELLS: Hibernation of fresh water Shells. multitudes
041r065 s fact about shells: Hibernation of fresh water SHELLS. multitudes.-- The question of shell's concret
041r065 sh water Shells. multitudes.-- The question of SHELL'S concretions, living only in that spot & being
048r084 ole line of coast Darby mentions beds of marine SHELLS on banks of Red River Louisiana. V. Lyell. Vol
053r100 same with that of Pernambuco? Quote Miers about SHELLS at Quillota Lyell, states that contact of Gran
054r102 e with gypsum, covered by limestone with recent SHELLS 200 ft, how exact agreement with Coquimbo; [no
055r106 ntelligent» miners to be the richest Vol II 147 SHELLS at Concepecion 50 toises above the sea. = talk
055r106 ore? decided elevation of sea's bottom. beds of SHELLS. 2 - 3 toises thick.--Vol II. p. 252 Urge clif
067r143 rgh March 1835 Sir W. Parish says. that beds of SHELLS are found on whole coast from P. Indio to Quil
079r176 Pataz, also at Gualgayoc. where many petrified SHELLS Bougainville says P 291.-- The Fuegians treat
085aIFC islands & Coral Formation Lyell's Salband p. 86 SHELLS near Woollich p. 112 Speculate on the extensio
086a004 dice of not believing recent elevation, yet sea SHELLS at tops of mountains we ought to sympathize wi
086a004 ympathize with. old doubters of what are fossil SHELLS.-- accustomed to such terms "fixed as the land
088a016 hic action -- Mem: red sand of Europe no fossil SHELLS --¿ action of Heat bubbles volatilized at bott
089a020 . «de Geneva» Vol 3' P. II. -- Bed, of elevated SHELLS on the Senegal. L Institut p. 192.-- (1837. Pe
095a039 e. French discovery ship, found clear proofs of SHELLS & waterworn rocks «at Cobija.» At Iquique of e
096a039 laert (at Roy. Institut) talks of quantities of SHELLS at Iquique. <Ceylon>. Band of Volcanic action
105a072 «alluvi» in valleys Lowe in his paper says land SHELLS found with calcareous matter & concretions on
105a072 it if this powder results from «decomposed sea» SHELLS, that land shells should be preserved in it--
105a072 results from «decomposed sea» shells, that land SHELLS should be preserved in it-- some error? (becau
110a085 hemistry p. 206 Both Beck & Deshayes saw fossil SHELLS from West Indies & declare them to be recent s
117a102 ns of Payta extend close to Guayaquil.-- modern SHELLS of Cobija doubtful. Examine well shores of lak
121a112 re plains & valleys just like Patagonia, & many SHELLS in parts on surface, but I saw none embedded t
121a112 his point would be worth examining. to support. SHELLS on surface of Patagonia, yet none in shingle b
123a115 near Bogota; p 270.-- SPLENDID PAPER on fossil SHELLS of S. America. Von Buch Lyell. (under head of
133a141 & 3. Porphyry at St. Elena. p. 6. few «living» SHELLS. on coast of do p 8.-- soft Clay beds near C.
135a44g king for Copiapo. found inland a great many sea SHELLS some miles from coast-- quote passage to show
138a153 of Lima.--judges from «beds of» sand & gravel & SHELLS. p. 47. do has table of every earthquake, duri

Page **(Key Word)**
152g047 North of Grant town to Forrest road comminuted SHELLS Important contingency if elevation from Axis,
163g106 alternating layers of coarse & fine & many Sea SHELLS. My informant saw them himself-- Sand with tid
163g107 . higher with pebbles now worn away-- The above SHELLS must have been about 60 ft above sea-- soon de
164g119 ral upside down strata coarse agglomerate [...] SHELLS from [...] Wenlock Edge [blank] L. Lochy 12 ft
173b009 emark, best known species. (as some common land SHELLS) Most difficult to separate Every character co
184b054 ogique April 1837. p. 216 Deshayes on change in SHELLS from salt & F. Water-- on what is species. ver
194b095 Mem: Silurian fossils: ¿How are. South American SHELLS? ¿Do not plants, which have male & female orga
194b096 ve intermixture.-- But how with «hermaphrodite» SHELLS.!!!? We have not the slightest right to say th
195b097 hyncus The type of organization constant in the SHELLS.-- The question if creative power acted at Gal
203b137 e from geographical or central heat.-- But then SHELLS-- Mr Yarrell says that old races when mingled
208b153 n of individuals than of species Paris Tertiary SHELLS in India!? A p. 28 Dr Beck. & Lyell. most curi
209b157 ance is I: 5.7 in Laponia I: 2,3: Mem. Lyell on SHELLS.-- {T} Genera In North Africa. I: 4,2 Iles Can
212b167 Encyclop. Proportions between fossils & recent SHELLS between herbivorous & zoophagous Mollusca acco
214b172 merican. Tertiary deposits present analogies to SHELLS of living seas.?-- Roxburgh. list of plants in
221b201 Beche. Geolog. Researches. facts of salt-water SHELLS living in absolutely fresh water.-- origin of
222b204 servations-- difference of species between land SHELLS of Porto Santo & Madeira-- I believe very curi
222b207 ypical species most near in form to ancient; in SHELLS alone can this comparison be instituted. Peopl
232b248 e look as if heat gained> Experimentise on land SHELLS in salt water & lizards do.-- Ask Eyton to pro
243c020 hilosophical xx Says same remark with regard to SHELLS.-- But he says shells towards extremities of t
243c020 ame remark with regard to shells.-- But he says SHELLS towards extremities of the continents peculiar
244c021 be very important to show wide range of fish & SHELLS in tropical sea, it. would demonstrate.; not d
264c082 ssification of birds Birds vary much (more than SHELLS) owing to variety of station inhabited by them
268c099 tion of «great» animals?-- Show independency of SHELLS to external features of land by seeing how man
273c110 p genera & not species, which is not true, with SHELLS?.?.?» It looks as if animals perished by error
277c127 ange & habits-- Take instances of most disputed SHELLS, such as Cyrena This is reform which probably
292c169 w that tendency is greater in Mammalia, than in SHELLS ¿ univalves or bivalves.-- Anyman No VI. Magaz
296c184 se others will be plants & land animals. & land SHELLS.-- all in short Extreme North = = to peak of T
309c220 t by existence of elevated «extinct?» genera of SHELLS.-- duration in two classes however different.-
310c224 opical countries (as St Jago Cape de Verds) the SHELLS in equal periods with Europe would probably ha
316c244 d anecdote --.Sowerby.--. Geographical range of SHELLS like Cryptogamic plants. of Marine kinds, ther
316c245 -- Fish & drift sea weed-- may transport ova of SHELLS.-- Conchifera. hermaphrodites-- eggs in groups
316c245 ant element than longitude.-- But in land & F W SHELLS there is more confinement. thus the Naiads (st
316c246 <S.> America.-- Mr Sowerby says there are some SHELLS common to West coast of Africa & E. S. America
316c246 e few have been scattered over whole world Many SHELLS at present day same (or according to Sowerby f
328cIBC du Japon Wowett on Cattle-- (Waterhouse has it) SHELLS from Barrier isld many relations with a living
328cIBC isld many relations with a living Matica & many SHELLS of Genera Corlula Cham. Cardium. Porcellus Tur
343d039 in note than amongst the Mammalia of Europe the SHELLS of do-- shells of. N. America.-- shells of S.
343d039 ongst the Mammalia of Europe the shells of do-- SHELLS of. N. America.-- shells of S. America.-- ther
343d039 ope the shells of do-- shells of. N. America.-- SHELLS of S. America.-- there is no appearance of sud
355d068 anging into ovules.-- Captain Grants. Himalaya. SHELLS (see Paper in Geolog Transacts) same appearanc
373d133 ts. p. 290. Dr. Beck on numerical proportion in SHELLS in Arctic Ocean. p. 350 Grallae in Wealden. ol
374d133 Lycopodiums.-- p. 437. Many. existing genera of SHELLS in the mountain limestone (how different from
405e032 & as it is <falle> embedded with almost recent SHELLS.-- shows that progression of change in Mollusc
407e041 - consider state of vegetation, & conchology,-- SHELLS of Africa ought most to resemble fossil ones o
413e059 rs the characteristics of the Tropical Forms in SHELLS. are numerous species, numerous individuals, &
416e070 y couple-- CD [lowest terrestrial animals.-- in SHELLS?-- insects?.-- all!??!?-- Worms? [Barnacles, a
417e076 ,.-- Mr G. B. Sowerby <tel> showed me many land SHELLS of the common species: from one locality, all
419e086 erse.-- Decemb. 25th.-- Lyell says the elevated SHELLS in Bayfields district are much more like those
419e086 rican species--Glacial period Dr. Beck says the SHELLS in Scandinavia from height of 200 & 300 ft are
420e090 rranean species).-- might these fertilise other SHELLS, as insects do flowers.-- Mem. Spallanzani's e
422e092 ween fossil & recent Bull; like fossil & recent SHELLS of the <new> raised beaches» [not located] The
424e100 parate islets.-- March 5th. Lyell says «fossil» SHELLS from North America, Scotland, Uddevalla. Many
431e122 . Phillips in Lecture in Royal Institution says SHELLS become less in number. (¿ species, or individu
436e137 eatest difficulty to my theory, is same type of SHELLS in oldest formations:-- The Cambrian formation
448e166 n.-- Voyage of Adventure & Beagle Vol I. p. 306 SHELLS, as well as plants «of Juan Fernandez» differ
448e167 -- Think over this-- The Superga beds have many SHELLS in common «& are not far distant» with Tourain
458t001 of aquatic reptiles most strange, & shows as in SHELLS some forms are long preserved.-- vol VI. p. 53
460t019 a in Eocene-- dicot. plants in coal measures.-- SHELLS in Cambrian & Crust show how long since presen
465t079 s apparently favourable for the preservation of SHELLS; where land broken, rivers entering.-- & yet n
465t079 where land broken, rivers entering.-- & yet no SHELLS-- now look at Scotland-- coasts of Chile, exce
465t080 B. Ayres).-- If we may take this as guide, the SHELLS preserved must be as much a casualty as, bones
465t080 se of bones (New Red Sandstone) & then go on to SHELLS-- A profound consideration of method by which
478z003 nd p. 115 In white Cape Pidgeon's stomach small SHELLS (patella) sea weed & many pebbles Mentions sti
479z005 s Discussion of Firola,-- Salpa Anatifs without SHELLS.! p 442.-- Planariae p 451.-- many molluscs Un
481z008 De la Beche theoretical researches Compare land SHELLS of Galapagos different islds.-- Waterhouse rem
483z013 . do.-- some S. American Reptiles are described SHELLS from Tahiti and Chile The North & S. Range of
483z013 s from Tahiti and Chile The North & S. Range of SHELLS might perhaps be worked out with advantage. wi
484z015 replaces warblers of Europe-- Study profoundly SHELLS of Bahia Blanca & Southern Hemisphere It is mo
490q001 of Jordan Hill-- character of the extinct land-SHELLS of Madeira-- analogous or quite distinct from
490q001 untry or district than in another? Character of SHELLS of Sandwich group {Sowerby monstrous Cardium--
637j58r ons Horny point to chickens beak, to break egg. SHELLS-- why chicken could not have lived had it not
637j58r ld not have lived had it not been so.-- let egg SHELLS grow harder. so must those with weak beaks be
152g045 where shallowest In Glen Collarig good case of SHELVES entering «on» one side ravine. Are the lip, or
152g046 e. Are the lip, or necks of land on level with SHELVES effect of corrosion & not cause. Monday a rapi
156g067 » Boulders of same granite, all on these three SHELVES soil is <the> usually slaty Point of rounded n
156g068 of 3(a) Cannot <see> «make out» composition of SHELVES: generally angular except near head of valley
156g069 d from «some of» those since fallen. «on the 3 SHELVES» Solid rock is much notched on Maculloch's sup
156g070 aying no transported materials <into> on upper SHELVES granite & some other rocks at head of shelf 3d
160g093 N side.. dies away on rocky place, but narrow SHELVES just like road of Glen Roy-- appears to lip wi
116a100 t sought for in Chile in beds of river, but in SHELVING «successive» banks <above> 30 ft or so above
146g012 hatred To fulfil an instinct a pleasure; mem. SHEPHERD dogs The Patches of Conglomerate on S. Ventan
148g026 This man confirmed the account of the «YOUNG» SHEPHERD dogs Saturday. Before coming to Bridge of Spe
217b183 never take the dog. But at last a rough-haired SHEPHERD dog lined her & produced a very large litter.
345d044 sy races or varieties are made.-- The Highland SHEPHERD dogs, coloured like Magellanic Fox.-- peculia
346d047 fatten-- This man confirmed my account of the SHEPHERD dogs.-- Aug. 24th. Was struck with pink shade
543m101 nstincts. when two varieties are crossed as in SHEPHERD dogs-- Inherited Habits: Have Effect in Bones
568n017 is connected with some disturbed habit» [Thus SHEPHERD dog. has pleasure in following its instinct &
628o53v ion? [2] [The improvement of the instinct of a SHEPHERD dog, is strictly analogous to education of ch
628o53v g instinct-- which were not originally, if the SHEPHERD dog had no instinct to commence with scarcely
145g002 osses difficult Salisbury Craigs The Highland SHEPHERDS dogs coloured like Magellanic fox.. an instan
432e123 ith horned.-- compare this with what highland SHEPHERDS said.-- p. 12. Attempts to improve the native
437e140 ped".-- if it had not been shot by <some> «a» SHEPHERDS, who was watching the scene.-- «In Shiant Isl
529m042 noses, quite conjectural, in Blakeways book of SHERIFFS.-- July 22d. 1838 No Deliriums, yet in some i
062r128 s, Cacti: & with limits of no vegetation at S. SHETLAND = Great contrast of two sides of Cordillera,
065r138 orster expedition from <Deception Isld.> South SHETLAND Cape Possession. Syenite¿ Andite?-- Degrading
065r139 ix Geographical miles and width 2 & ½ miles S. SHETLAND. Lat. 62 degree 55 minute. <only> one lichen.
103a063 ections in Geological Society. Pumice at South SHETLAND. Geological Society-- Dikes have not been the
187b065 peds:-- same argument applies to England.-- Mem. SHEW Mice.-- --- Animals common to South and North A
382d158 - surely analogy of molluscs. & neuter bee would SHEW this-- (Do any male animals give suck)--But thi
616o039 ive in support of their view it is impossible to SHEW satisfactorily it's erroneousness. it is a poin
616o39v e 2) In the absence of such a guide we can only <SHEW> point out the mode «of perceptive action» by w
616o39v ch we come to conceive of matter as attracting & SHEW that the groundwork <of this> is entirely wanti
035r047 consult book itself. P. xx: same fact is indeed SHEWN? by the parallel bands of formations on any Geo
091a025 : mem. Henslows Anglesea solution of silex also SHEWN. No 3d of Ed. N. Phil. J. p 194. Fact of dust b
092a029 00 ft high Mr Bird in paper to Brit. Assoc: has SHEWN how electrical currents tend to deposit metals,
299c194 are two Dandelions, which just lately have been SHEWN to be same.-- one grows in marsh & other dry; y
566n011 y notions. which of them? & curiosity «strongly SHEWN in the numerous artifices to take birds & beast
570n026 olves this.--» The similarity of men's reasons: SHEWN by similarity of the earliest arts.-- Mem.-- St

Page **(Key Word)***
611o034 e as long as movement of sensitive plant can be SHEWN to be direct physical effect of touch & not irr
624o051 ce, will have reoccurred'. NB. Until, it can be SHEWN, what things easiest become instinctive, this p
640j167 aly to this. Have they wide ranges? Agassiz has SHEWN that they most widely differ» 3 A very wide ran
437e140 a» shepherds, who was watching the scene.-- «In SHIANT Isld. it is said, that an Eagle always procure
573n038 n with St Vitus' dance badly, told should have SHILLING to walk to door without touching table.-- can
041r067 cessive transportal from prevailing swell, (as SHINGLE travels on the Chesil bank. V. De la Beche). A
051r097 changes in bottom in NW coast of America. from SHINGLE to sand &c &c. <Vol II> P. 209. 211. 213. 444
120a109 ook as if a surface deposit.-- The case of the SHINGLE in the great Chilian valleys must be profoundl
121a112 t. shells on surface of Patagonia, yet none in SHINGLE beds. Lyell on Sweden. p. 12. proofs of small
053r100 o Patagonia. During a period of subsidence the SHINGLLE of Patagonia would become more or less inters
029r033 10h. 15m. a severe shock of earthquake shook the SHIP in a most violent manner. Although it lasted ab
095a039 .-- The officers of the Bonite. French discovery SHIP, found clear proofs of shells & waterworn rocks
554m139 a very difficult knot-- the sailor on board the SHIP could not puzzle her-- with aid of teeth & hand
409e047 e was not thougt overwhelming.-- yet I will not SHIRK difficulty-- I have felt some difficulty in con
596nIBC the muscles of the face first affected?-- Can SHIVERING & trembling be considered convulsive.-- is co
579n059 hing to do with ancient movement of ears A man SHIVERS, from fear, sublimity, sexual ardour.-- a man
025r016 s. 15--15 Does not seem to consider this a very SHOAL coast. Beyond the 10 or 12 leagues sea deepens
025r016 twice or 18-20 <60>--80 120 parallel of Olinda SHOALER N. of Olinda.--a little WNW of C. Rock. [5 deg
025r016 WNW of C. Rock. [5 degree 29 minute S.] still SHOALER, coast composed of sand dunes. 15--15 Does not
050r091 l becomes> sand less & gravel more common. the SHOALER the water & nearer the Banks Is there not a su
024r014 rom the [...] Cordillera & flowing The gradual SHOALING of the water to more than 100 fathoms. proves
080r180 on the geology of the world P. 14-91. gradual SHOALING of coasts 93 action of sea on coast. 27. Baha
023r010 e of the ocean, before arriving at the Abrolhos SHOALS ‖ -- N.B. The view of the Volcanos of the chai
024r015 Find instances; The whole coast of New Holland SHOALS much: Dampier remarks on great flats on the NW
066r140 umb» Sea weed said at Kerguelen Isd. to grow on SHOALS like Fucus giganteus! 24 fathoms deep 24» unde
029r033 London:-- "Feb. 12, 1835. At 10h. 15m. a severe SHOCK of earthquake shook the ship in a most violent
043r071 sco). yet whole territory vibrates from any one SHOCK-- In S. America--continuity of space in formati
045r076 hen they hear of a place having a pretty severe SHOCK). are much more curious & perplexing. than thos
108a080 llation. or so many shocks directly after great SHOCK -- It appears to me unphilosophical to think ca
440e147 tempting anything about habits-- no one can be SHOCKED at absence of final cause mammae in man & wing
108a080 not so much horizontal oscillation. or so many SHOCKS directly after great shock -- It appears to me
318c252 more beautiful adaptation for snow-- like snow SHOES. than feet & hind legs of these white hares, fi
029r033 1835. At 10h. 15m. a severe shock of earthquake SHOOK the ship in a most violent manner. Although it
495q05a al in middle of ploughed field-- on hills.-- 10 SHOOT tame duck on pond with Duck-weed-- coots-- wate
060r124 eaks of this phenomena in connection with "the SHOOTING upwards" of the <ground> land in the W Indies
557m149 affected laughter.-- A dog who goes home from SHOOTING. runs away. is not afraid the whole way. but
285c151 ter its decay, will be occupied by the vigorous SHOOTS from each branch No: because decay in that spe
443e154 es from Pliny, that it is bad to graft from top SHOOTS.-- If prolongation of life by gemmation <can b
399e009 10th. Saw. two undoubtedly rabbits in poulterer SHOPS., of same colour as a Hare, but paler & buffer.
023r011 from the manner of horizontal upheaval» of the SHORE of the Pacifick is 60 miles distant from the gr
024r015 Cape of Good Hope 70 fathoms 20 miles from the SHORE? Beagle Coast of Brazil? where not rivers in my
030r035 deep soundings, 60-100 fathoms 2 & 3 miles from SHORE. V. Chart) Every winter torrents must bring muc
064r134 s «M Birchels» at foot of range some miles from SHORE. rock of oysters quite above reach of tides.--t
071r158 bsorbed.--Earthquakes felt. different case from SHORE of Pacific.--Isabelle's volcano, many amygdaloi
122a115 } August 25. I saw metamorphic conglomerates on SHORE of Loch Lochy very like those of Andes Speculat
135a146 .-- G. J. Malcolmson has described formation of SHORE of Coromandel. just same as. at Bahia Blanca--
247c028 I. p. 10. it seems that Crocodile was washed on SHORE at one of the Pelew Islds.-- killed a woman. Ch
452e181 s in circum) there are hogs & monkeys <at> near SHORE of Magindanao Journal of [Asiatic Soc] [...] p
051r097 <Vol II> P. 209. 211. 213. 444 «Yanky edition» SHORES of Pacifick, as compared to whole E. America.
052r099 that numerous icebergs are commonly stranded on SHORES of Georgia «Lat degree ( )», he has rocks on s
115a098 ents. Small lakes have power of levelling their SHORES where currents very weak??-- too great an abun
117a102 modern shells of Cobija doubtful. Examine well SHORES of lakes. to see effects of degradation, «no»
151g044 corresponds to shelf the truncation & the upper SHORES may correspond with some line subsequent to sh
226b221 es. The number of genera on islands & on Arctic SHORES evidently due to «the» chance of some one of t
035r048 produce equable effects.--«though so immense to SHORT breathed traveller» Mountains, which in size ar
041r066 n sandstone. a circle,.{P}, had in its middle a SHORT <fissure> «vein» terminated each way, which lit
161g097 shelf certainly) I took another measurement on SHORT buttress but not continuous & it was 29.200 min
171b002 Stallions nor nuns are longer lived Why is life SHORT, Why such high object generation.-- We know wor
175b016 prove cats &c from Egypt no answer because time SHORT & no great change has happened I look at two os
197b111 Evidently «or hints» considers generation as a SHORT process, by which man «one animal» passes from
199b118 Ap. 1828 -- I take higher grounds & say life is SHORT for this object & others, viz not too much chan
203b136 tameness not heredetary?, having been gained in SHORT time. Milvulus forficetus <has a wide range> is
211b163 y of Manks cats without tails: some long & some SHORT: therefore like dogs.-- Ogleby says, Wolves at
219b195 ctic flora must formerly have been separated by SHORT space from mountains low down, therefore plants
246c025 esemble, chien-loup.--long, black & white, ears SHORT & straight-- do not bark p. 433. birds & bats h
283c145 ten species of Orpheus-- one of which has very SHORT legs & long tail «short much curved beak.--», o
283c145 - one of which has very short legs & long tail «SHORT much curved beak.--», other very long beak, wit
283c145 uch curved beak.--», other very long beak, with SHORT., let these only have progeny with species & th
283c145 with species & there will be two genera.-- let SHORT billed one, be exaggerated, & all rest destroye
290c165 hs objection.--» it is more, he cuts the matter SHORT by saying man cannot be companion but master.--
296c184 plants & land animals. & land shells.-- all in SHORT Extreme North = = to peak of Teyde in relation
301c199 illing grains. acquirable through hoarding from SHORT time.-- My theory must encounter all these diff
309c221 ful case Dwights' Travels in America, speaks of SHORT legged sheep. heredetary proceeding from an acc
312c232 lations non-heredetary & variations produced in SHORT time in some extent counterpart, mutilation bei
317c249 in their colour & appearance Every now & then a SHORT-tailed cat.¿cut? has its offspring short tails
317c249 then a short-tailed cat.¿cut? has its offspring SHORT tails /one born at Maer. Tuckeys voyage-- p. 36
332d005 inea, -- ears bare. skin black & wrinkled-- fur SHORT. (tail cut off in progeny peculiar) limbs very
343d036 «with Myriads of distinct forms» from a period SHORT of eternity to the present time, to the future-
345d042 before deserves <name> «to be so called»,-- the SHORT horned cattle have gone on for 50 or 70? years-
345d043 alf every now & then born with white head (,or «SHORT-horned with» black lip) & then calf «in both ca
360d094 n having very long tail, & other in having very SHORT tail.-- I can readily see that two first might
372d130 its parents have.-- -- Not this is effected by SHORT method in generation.-- Ehrenberg considers art
403e022 ll isld of Eap in the Carolines, are remarkably SHORT.-- & Deformations are particularly common.-- wi
428e113 I should doubt if wild species ever formed like SHORT-tailed cat or dog has been without recurrent te
437e140 its prey from another island.-- » p. 175., 28 SHO[R]T eared owls were counted in a field, where ther
440e145 that which changes the acorn into the oak.-- In SHORT all which «Nutrition, growth & reproduction» is
446e163 ditions-- & that no change would affect them in SHORT period & hence no change would effect them, wit
486z018 of Heat. hence musquitoes & knats abound during SHORT summer far N. where this other order is compara
502q11v heasant: said not to sit on own eggs Flowers in SHORT turf. for abortion. or for sterility Land Birds
512q019 -- For comparison with hybrids, is offspring of SHORT-horn bull & hereford cow similar to reverse cro
525m025 ewsbury & its kittens <h> (in number 3) had all SHORT tails; but one a little longer than rest «they
525m025 e first dead children, then children which were SHORT term, & lastly healthy ones.-- Insanity & Epile
527m033 asant sound per se) & causes the mind to create SHORT vivid flashes of images & thoughts.-- Poetry. t
530m044 ory of everything in <he a [...] Mr B> journey. SHORT time previous,-- because, pain prevents repetit
532m057 es passion & not sweat is the <state> effect of SHORT -- but violent action.-- To avoid stating how f
563n002 s system, & constant, for a wish which was only SHORT & might otherwise have been relieved, he would
579n058 which my theory believes in.-- From the manner SHORT-sighted people frown, frowning must have some r
579n058 ople frown, frowning must have some relation to SHORT-sightedness.-- do not short sighted people squi
579n058 ve some relation to short-sightedness.-- do not SHORT sighted people squinny-- when they consider pro
582n067 g late, will make average of life longer.-- for SHORT-lived constitutions will then be cut off.-- <Ho
620o044 on his choice: an appetite gratified gives only SHORT pleasure. passion in its nature is only tempora
624o50v mparison between our instinctive feelings & our SHORT lived Passions' State broadly in child or anima
170b001 throug several stages (¿typical, <of the> or SHORTENED repetition of what the original molecule has
638j58v adaptation made by intellect this process is SHORTENED, but yet analogous, no savage ever made a per
021rIFC wards perfection (namely mammalia) must have a SHORTER duration, than the more constant: This view su
191b081 islets.-- Next to animals land birds.-- & life SHORTER or change greater-- In the East Indian Archipe
313c234 considerations, with respect to law of mammals SHORTER duration, than molluscs, argue case both in Eu
336d017 the more persistent.-- «any amount of change» SHORTER time less [s]o.-- the result of this is that a

375d135 lation in increase at geometrical ratio in FAR SHORTER time than 25 years-- yet until the one sentenc
485z018 9. on Hydra-- polypi-- <Rep> do p. 324. Polypi SHORTER duration than cells.-- reproduced.-- Milne Edw
567n014 ing a body & an ass knows one side of triangle SHORTER than two. V. Whewell. Induct. Sciences-- Vol I
568n016 , or our experience.-- Two sides of a triangle SHORTER than third. is this necessary notion, ass has
176b022 ositions" must in each state of existence have SHORTEST life.; Hence shortness of life of Mammalia.--
312c232 part, mutilation being «variation» produced in SHORTEST possible time. Mr Willis Long eared little do
621o046 ts nature being always present. & his passion SHORTLIVED, it is to his interest to follow the former;
176b023 state of existence have shortest life.; Hence SHORTNESS of life of Mammalia.-- Would there not be a t
177b027 - birds-- not. {P} We may fancy, according to SHORTNESS of life of species that in perfection, the bo
177b029 xistence.-- There does appear some connection SHORTNESS of existence, «in» perfect<lon>, «species fro
179b035 ect are differently circumstanced.-- ¿Is this SHORTNESS of life of species in certain orders connecte
635j56r ect [blank] 6 p. 412. Macculloch explains the SHORTNESS of life (peculiar to each species) owing to t
592n104 ovements in muscles of face, is well seen in SHORTSIGHTED people.-- hence origin of expression-- Ther
342d033 tain that hybrid Muscovy & Common duck have been SHOT wild (escaped from Carolina?) off New York. the
402e018 observed.-- G. W. Earl's Eastern seas. p. 206-- SHOT a monkey, ceased their cries. "many of them des
437e140 probability have escaped".-- if it had not been SHOT by «some» «a» shepherds, who was watching the s
032r039 int the waves would not reach. If now the ocean SHOULD suddenly fall, (3) the case would be as at fir
035r047 dent: & then how wonderful level «of same beds» SHOULD have been kept; it shows that throughout all E
049r087 of a series of faults. [Fig. 4] {P} In Cordill: SHOULD basal lavas be called Volcanic or Plutonic The
057r113 se bones differ as much nearly as the Eocene. = SHOULD Mr Owen consider bones washed about much at Co
057r113 hed about much at Coll. of. Surgeon's? I really SHOULD think probably that B. Blanca & M. Hermoso con
062r129 ry of rats, in the antipodes a parallel case.-- SHOULD urge that extinct Llama owed its death not to
070r153 Yet one is urged to look to common parent? why SHOULD two of the most closely allied species occur i
073r162 nes simply pitched from mo«o»n, that the metals SHOULD be those which have magnetic properties. Study
098a048 h Area of this> {P} If surface covered with oil SHOULD not be surprise at Horse being found in Americ
105a072 from «decomposed sea» shells, that land shells SHOULD be preserved in it-- some error? (because more
108a080 rising areas.-- In Earthquake if Subsidence we SHOULD not expect volcanos.-- not so much horizontal
154g060 w that it has got to the rock of cols if--. why SHOULD it deposits River terraces often descend by fl
155g066 ular hardness no wonder that all «three» lines «SHOULD be» EQUALLY preserved 2d or upper one more per
177b025 really pass into each other.-- The tree of life SHOULD perhaps be called the coral of life, base of b
180b036 think {P} Case must be that one generation then SHOULD be as many living as now To do this & to have
185b055 itself to as many situations as possible.-- Why SHOULD we have in open country a ground «do. <w> parr
187b065 rocreate, he has no issue, so with species.-- I SHOULD expect that Bear & Foxes &c same in N. America
196b106 s a very great puzzle why Marsupials & Edentata SHOULD only have left off springs <ne> in or near Sou
199b117 established. why out of the thousands of forms, SHOULD they all be classified.-- Propagation explains
203b135 fspring, & then «V. L. Institut p 245. 1837» we SHOULD have anomalies. as Cape Anteater,.-- This supp
210b158 pecies would not be manufactured, <but why they SHOULD be manu> Does it not present analogy to what t
224b214 ain distinct. Where country changes rapidly, we SHOULD expect most species.-- The difference intellec
225b217 me every step between bull Dog & Greyhound). I SHOULD say the changes were effects of external cause
227b224 ; for having ascertained means of transport, we SHOULD then know whether former lands intervened.-- (
233b252 n Molluscs!!! When we talk of higher orders, we SHOULD always say, intellectually higher.-- But who w
236b280 disseminated species to vary a little, but such SHOULD not be general circumstance.-- In. insects «in
265c084 on implies more than mere child, but that child SHOULD produce like children. ⟨Lyell has story from.-
265c087 ecies-- as we have only ornithorhyncus, then we SHOULD never know how much structure was connected wi
271c105 re. (replaced to the North by other species.--) SHOULD build a nest lined with mud, in forest where n
271c105 insects it devours is same species. yet that it SHOULD so strictly <f> agree in habits with the Turdu
277c126 lan, but reason now assigned for doing so There SHOULD be mark to every species. only known by analog
278c129 nus does not Mean this it means nothing.--There SHOULD be some term used, when there is series. Could
281c137 graeduated.-- Dr Beck doubt if local varieties SHOULD be remembered, therefore do not consider it as
281c139 analogous to each other-- &c &c.-- V. p. 140» I SHOULD think meaning of circular arrangement was only
293c172 otten» --it is scarcely more wonderful, that it SHOULD be remembered in next generation. [NB what are
293c172 tion-- which its father. had done habitually we SHOULD exclain it was instinct.-- Even if savage take
313c234 Entomological Transactions Why if louse created SHOULD not new genus have been made, & only species,
326c266 imits of painting & poetry.-- Erasmus thincks I SHOULD lik it. The Sportsman's Repository. 4to. conta
335d016 ation would necessarily fail.-- In last page. I SHOULD have said, "an animal <acquires <th> any new>
354d065 bably transported like the New Zealand one-- It SHOULD be observed with what facility mice attach the
356d071 ss.-- In comparing my theory with any other. it SHOULD be observed not what comparative difficulties
358d085 gans doubtless would shrivel up.-- Yet odd they SHOULD have so much sexual character as they have Thi
359d087 mals are healthy which is often the case, & why SHOULD organic affections always influence the sexual
359d088 guin was crossed with hen Canadian offspring, I SHOULD say in every respect most like Penguin duck.--
366d112 omes it that the tame donkey has. CD[old Buffon SHOULD be read on Mare My view, why hybrids are infer
369d114 use here.-- & therefore different from Hunter I SHOULD say females recede in organization from specif
371d118 It is less wonderful that childs nervous system SHOULD build up its body, like its parent, than that
371d118 uild up its body, like its parent, than that it SHOULD be provided with many contingencies how to act
372d131 r man-- That the embryo the thousandth of inch SHOULD produce a Newton is often thought wonderful. i
386d167 & the belly both head & tail,--no wonder there SHOULD be sympathy in human frame.-- «one of» The fin
388d172 hat transmission of mutilation impossible.-- it SHOULD be observed that transmission bears no relatio
388d173 , which is to pass through all transformations, SHOULD there need two organs; whilst in common bud th
389d173 rally sufficient for one birth or rather⟩CD ⟨It SHOULD be observed that the constant necessity for ch
389d177 tainly very remarkable that too much difference SHOULD produce same effect as too little.-- in (latte
390d178 mething someways different from himself, for it SHOULD be observed that from Books to read Buffon Sui
391d174 rent or tree, (but not in other trees.-- -- Why SHOULD there be a necessity that there should be some
391d174 - -- Why should there be a necessity that there SHOULD be something «each time» added to that kind of
391d174 us changes, which mutilations are not). but why SHOULD it demand some further change?-- Man properly
391d175 t even brother & sister are somewhat different) SHOULD be added to each individual before he can proc
398e005 f the change could be shown to be more rapid, I SHOULD say there was some link in our train of geolog
398e005 logy. by giving proper ideas of these subjects. SHOULD be absolutely necessary to arrive at right con
410e050 was absolutely necessary that Physical changes SHOULD act not on individuals, but on masses of indiv
410e050 on masses of individuals.-- so that the changes SHOULD be slow & bear relation to the whole changes o
410e051 could only be effected by sexes: All the above SHOULD follow after discussion of crossing of <specie
415e066 ld anyone raise an argument against, my theory, SHOULD no fossil «very distinct species» of the Ornit
423e095 ing complexity of others.-- It may be said, why SHOULD there not be at any time as many species tendi
426e108 derful as is the possession of voice by Man. we SHOULD remember, that even birds can imitate the soun
427e110 led S. From the analogy of the animal kingdom I SHOULD suppose, that the pollen of crab, would POSSIB
428e113 rom Primrose to Cowslip is great difficulty. «I SHOULD doubt if wild species ever formed like short-t
441e150 necessity, that every <animal> «organic being» SHOULD cross with another.-- to escape it «in any cas
442e153 <species» «individuals» propagated by gemmation SHOULD be absolutely similar; [all the gorze in Norwa
451e176 ike himself said not to be common-- probably, I SHOULD think grandfather first of race & if so, fact
459tf1v ascar-- «p. 121» No beasts of Prey. any country SHOULD during [...] conditions-- every spot is occupi
520m001 fore, that it is extremely improbably that they SHOULD have imitated.-- when attending Mr Dryden Corb
532m055 ary mind which is startled.-- My Father says he SHOULD think that in old people, in their dotage, who
535m069 little remarkable that the fixed laws of nature SHOULD be «universally» thought to be the will of a s
536m073 - Put it so.-- Probably some error in argument, SHOULD be grateful if it were pointed out.-- My wish
537m075 different races.-- no more wonderful than dogs SHOULD have different instincts.-- Fact most opposed
548m117 ne mind, but does not stop to reason what there SHOULD be & discover loss Definition of happiness the
549m123 ther animals-- is far from odd nor is it odd he SHOULD have had them.-- with lesser intellect they mi
557m148 s, no more to be analysed than fear or anger? I SHOULD think shame would be more easily analysed than
567n015 id. & Jemmy, when Chico plagued him-- Animals I SHOULD think would not have any emotion like blush.--
573n038 ce.-- A person with St Vitus' dance badly, told SHOULD have shilling to walk to door without touching
578n055 person have committed any «concealed» action he SHOULD not, & let him be thinking over it with sorrow
587n087 hings dislike others.-- dogs dislike perfume) I SHOULD think, great principle of liking, as simply he
596n144 umes Dissertation on the Passions." "Hartley" I SHOULD think well worth studying-- "Thomas Brown" on
602o11v roaches to instinct How strange it, that Nature SHOULD have so little to do with art (p 128) R. compa
608o027 ter others.-- It is not more strange that there SHOULD be necessary. wickedness than disease. This vi
608o027 e necessary. wickedness than disease. This view SHOULD teach one profound humility, one deserves no c
612o035 than movement of sap. or sunflower to sun? ∵ I SHOULD think there. was direct «physical» effects of
614o037 sucking.-- And is it more wonderful that memory SHOULD be transmitted from generation.; than from hou

Page		(Key Word)	
616o038	ed into man ∴ they meet their reward! X Perhaps	SHOULD	hardly be called memory; you cannot call the f
618o41v	ction objectively 6) The reason why thought &c.	SHOULD	imply «X» the existence of something in additi
618o41v	her: no not only thus, for if day was first, we	SHOULD	not think night an effect.]CD Cause and effect
619o042	ng sheep or hurting them.-- Therefore in man we	SHOULD	expect that acts of benevolence towards fellow
619o043	s moral approbation, as far as it goes.].CD But	SHOULD	he prevented by some passion or appetite, what
620o045	e gains the rule, that the passions & appetites	SHOULD	«almost» always be sacrificed to the instincts
620o045	bsequent> power (reason) obtained by age, which	SHOULD	show the child, which of its instincts are bes
622o049	series in which «special» instincts decrease, I	SHOULD	think they were very few & general in their na
628o054	her passions, or instincts.-- is this good?-- I	SHOULD	think some parts of the emotive part of man, m
633j54r	we sink into such contemptible queries, as why	SHOULD	the earth have drifted; why should plants requ
633j54r	ries, as why should the earth have drifted; why	SHOULD	plants require earth, why not created to live
635j56v	re such care seems to be taken that the anthers	SHOULD	not be exposed to weather.-- this is against m
636j57r	iness instead of barbs)-- In all these cases it	SHOULD	be remembered, that animals could not exist wi
638j059	emes.-- I see no reason, why structure of brain	SHOULD	not be born. with tendency to make animal perf
640j167	ect to whether Galapagos beings are species. it	SHOULD	be remembered that Naturalists are prone, fort
640j167	om these views we can deduce why small islands.	SHOULD	possess many peculiar species. for as long as
640j167	that the smallest newest, & most wretched isld	SHOULD	possess species to themselves.-- Probably no c
505q014	root of tail 28½. inches. From sole of foot to	SHOULDER	on line of back, height 17⅓/. The Greyhound.
570n023	se with negation.-- like shaking something off	SHOULDER	-- or is it from inspiration, which accompanie
569n023	be said."-- "made no reply, but shrugged his	SHOULDERS	& went away."-- he implies negation, without
570n024	ill see you damned first." the man shrugs his	SHOULDERS	& replies nothing. if he did go to reply. he
570n024	f he did go to reply. he would throw back his	SHOULDERS	. he wishes to show, he is determined not to s
570n024	ng. he presses his lips together & shrugs his	SHOULDERS	& walks off,-- I think shrugging connected wi
573n039	's wife). & she said nothing but shrugged her	SHOULDERS	.-- analyse this.-- Miss C. quite aware & indi
584n073	it is somewhat analogous to memory. Shrugging	SHOULDERS	seems sign of helplessness E. says she can pe
540m084	.-- dogs laughs for joy, so does dog bark. (not	SHOUT)	when opening his mouth in romps, <so> he smile
542m094	sound even if it be inarticulate.-- the maniac	SHOUTS	& bellows with passion.-- It is not a little r
553m137	nkeys go in groups. thus the pig-tailed baboon,	SHOVED	out its lip, looking absurdly sulky «as» often
022r006	ton's appendix Would Slate. & unstratified rocks	SHOW	any difference in facility of conducting Electr
025r018	all my illustrations from America, purposely to	SHOW	what facts can be supported from that part of t
027r023	of matter in shells, like Concretions & laminae,	SHOW	what movements take place in semiconsolidated r
051r094	t everywhere on coast (Il Defonsos «Kelp») rocks	SHOW	signs of degradation; (soft substances worn int
056r112	ting is well adapted to use of mankind.--<Hutton	SHOW>	Earthquakes part of necessary process of terre
059r121	lime.--All bear close analogy to Obsidian, & all	SHOW	chemical action as well as effects of cooling [
070r153	except by trees The structure of ice in columns.	SHOW	that granite when weathering into balls. must e
074r165	n Solfataras. «Mem: Micaceous iron ore.» N.B. To	SHOW	how metals may be transported by complicated ch
109a083	. argument for Heaping up.-- very good this will	SHOW	effects.-- analogous to broad flat sand beach.
120a108	ep. --Hot springs &c &c--then if so, thermometer	SHOW	it cannt be ordinary heat, then there is someth
126a123	ring requires connected column.--» of cold water	SHOW	, that water does percolate, & springs beneath s
130a133	springs off coast of Persia In Glen Roy paper I	SHOW	crust yield easily. & if easily must be thin: <
131a135	of the Moon. by Dr. Nichol-- adduces the case to	SHOW	Sir. J. Herschel's theory wrong.-- Geograph. Jo
133a140	ends his paper like a fool.-- Feb 25' All facts	SHOW	how slowly heat travels; & therefore the abysse
135a144	shells some miles from coast-- quote passage to	SHOW	abundance Bengal Journal. Vol 4. 1835. p. 437.
190b075	is any difficulty in such marriages or offspring	SHOW	tendency to go back-- there is an end to specie
194b095	- If these facts were established it would go to	SHOW	a centrum for Mammalia.-- I really think a very
194b096	from other plants because this may be applied to	SHOW	all plants do receive intermixture.-- But how w
200b121	ery different structure as. petrel & alk. do not	SHOW	the possibility of common branching off.? Accra
217b187	veral American forms-- The above facts evidently	SHOW	that Mr D. wonders at these species being wande
219b194	y species same as T. del Fuego & C. of Good Hope	SHOW	possibility of transport. If some cannot be exp
222b203	species cross & «hybrid» breed, their offspring	SHOW	tendency to return to one parent, this is only
225b217	housand intermediate forms.-- Opponent will say.	SHOW	them me, I will answer yes, if you will show me
225b217	ay. show them me, I will answer yes, if you will	SHOW	me every step between bull Dog & Greyhound). I
231b241	f Willow wren) & others varying in wild state to	SHOW	that we do not know what amount of difference p
231b242	ting in East Indian Seas. Marsupials animals all	SHOW	greater connexion in quadrupeds, <bu> plants do
244c021	, like. Plants.--» It would be very important to	SHOW	wide range of fish & shells in tropical sea, it
253c046	e, & Mastodon angustidens.-- Ogleby has facts to	SHOW	that Australian dog introduced by savages into
260c067	e Zoological Journal I read a curious account to	SHOW	that very many birds of different kind have bee
261c070	that the discrimination of species is empirical.	SHOW	this by instances Once grant my theory, & the e
262c073	guavas having overrun-- Tahiti. thistle. Pampas.	SHOW	how nicely things adapted--.-- These «aberrant»
265c083	boring their noses. &c & This congenital changes	SHOW	that grandson is determined, when child is.-- s
268c099	ent. ¿cause of destruction of «great» animals?--	SHOW	independency of shells to external features of
271c105	> something same manner, much mud.-- These facts	SHOW	, habits heredetary whilst species have changed
276c124	cause of truth It is of the utmost importance to	SHOW	that habits sometimes go before structures.-- t
277c127	not have brought home new species. until, he can	SHOW	range & habits-- Take instances of most dispute
280c135	ng about hybrids.-- & is a very remarkable fact.	SHOW	influence of mind It is not difficult to see th
281c137	"splitting up his species & genera very finely"	SHOW	how arbitrary & optional operation it is.-- sho
281c137	show how arbitrary & optional operation it is.--	SHOW	how finely the series is graeduated.-- Dr Beck
285c151	difference occurring.-- It will be necessary to	SHOW	hybridity from few forms, parents of all specie
285c153	nnot change quick enough & perishes.-- Lyell has	SHOW	such Physical changes will be unequally rapid,
294c176	ersistent Varieties in wild animals-- but how to	SHOW	species-- I fear argument must rest upon analog
298c192	re larger &c in different countries. These facts	SHOW	how very permanent plants are. & this conclusio
304c207	allaxis. trifling characters as red band on wing	SHOW	to be from one parent.-- same forms of beak &c
308c219	ncing at, as she has no original ideas., it will	SHOW	state of knowledge «Negroes existed since time
309c221	(quoted) p. 4.-- do. p. 186. quotes Burkhardt to	SHOW	black colour of certain Arabs.-- NB avoid quoti
318c254	- ..If the line or bands of country (These facts	SHOW	the Normal condition of Migration) gradually se
331dIFC	ytons <intermediate>. «hybrids, when» interbred.	SHOW	any tendency to return to either parent.? Is th
333d007	(Mem Jackall in Zoolog Garden) He has seen in a	SHOW	half Wolf & «half Esquimaux» dog which <likewis
337d021	e types exist for if so. it will be necessary to	SHOW	how the first eye is formed.-- how one nerve be
354d062	. 275. (1838) M. Blainville has written paper to	SHOW	Stonesfield Didelphis not Didelphis «Answered s
357d074	ecomes crossed. ¿is red game an hybrid?-- When I	SHOW	that island would have no plants were it not fo
362d100	ons» in 1834-- now this would be most curious to	SHOW	that in sixty years-- (how many generations) th
366d108	would cease being fertile inter se., or at least	SHOW	repugnance to breeding if instincts unchanged,
378d147	, <& loosing do> if so domestic animals ought to	SHOW	them.-- Anyhow not connected with habits Accord
383d159	g not only dimidiate, but quarter-grown seems to	SHOW	whole body imbued with possibility of becoming
383d159	very great difficulty, then give speculation to	SHOW	that it is not overwhelming.-- Seeing in Garden
388d172	ven milk. -- testis & ovaria The following views	SHOW	that transmission of mutilation impossible.-- i
406e037	tails.-- November 1st..-- Addenda to Journal. I	SHOW	erratic blocks transported far S. in Northern.
407e042	learly from F. R. it will be highly necessary to	SHOW	that if species fall, genera must. Lesson I rem
409e047	out Australians yet slow progress has done so.--	SHOW	a savage a dog, & ask him, how wolf was so chan
409e049	» hence not social instincts, which as I hope to	SHOW	is «probably» the foundation of all that is mos
415e069	to prove that a thing has been so, & another to	SHOW	how it came to be so.-- I speak only of the for
431e122	we need not look for change, because its number	SHOW	it is perfectly adapted; it where few stray one
433e126	tem, the changes from limestone to sandstone &c.	SHOW	some great change who can say how many centurie
438e141	s slowly formed.-- » if it shall be difficult to	SHOW	that <time> the fixity of characters «from anti
439e143	than the pollen of same flower,-- as it tends to	SHOW	my view of <i>nfertility of hybrids «with paren
441e149	- at least the experiment of the carrot seems to	SHOW	this.-- This would be curious law, Certainly Au
441e151	s become fixed. & cannot vary.-- which all facts	SHOW	to be absurd.-- As there are plants, in norther
460t019	s in coal measures.-- Shells in Cambrian & Crust	SHOW	how long since present forms existed, but if it
470t176	<& old English> But these mules <in our garden>	SHOW	no trace of palmation!!? Bees at Wild St Johns
494q004	wild animals are propagated,, though valuable as	SHOW	. & curiosities!! What is price of fox. otter. B
496q006	& see whether proportions will vary, which will	SHOW	that such proportions not effect of Chance Maer
497q006	aces & see what plants will spring up which will	SHOW	, how seeds are transported, or how long they re
501q10a	are they constant: this very important for it wd	SHOW	that such variation is not a generic or specifi
502q11v	Would wheat from AEgypt ripen in Scotland?-- to	SHOW	acclimatisation.-- July <1842> When nettle leaf
505q014	all white-flowered var. with abortive stamens.--	SHOW	crossing & ¿heredetary? (15) Abberley has a hoo
515q021	t, or time of leafing (5) Do the most cultivated	SHOW	Heartease produce as large capsules of seed, as
526m028	s may be worth thinking over.-- it. will perhaps	SHOW	differences between memory & imagination. «Cath
531m051	once read over.-- The extreme pleasure children	SHOW	in the naughtiness of brothers children shows t

```
538m078 n perfect senses.-- These things, & drunkedness, SHOW what trains of thought depend on state of  turn
545m106 man.--  I see monkeys grin with passion, that is SHOW all the teeth: «& make noise not like pish, but
553m138 hat he had never seen any of the American Monkey SHOW any desire for women-- «very curious. as they d
554m141 sophism  of association. Kenyon, & then go on to SHOW, that if Cart horse argued from this into a the
558m151 kes its value. from its connexion with mind, (to SHOW hiatus in mind not saltus between man & Brutes)
560m156 monkeys lice?-- picture.-- Do female monkeys not SHOW signs of impatience when woman present? Do they
565n007 really  wishing to smell its enemy.-- Man & dogs SHOW triumph (& pride) same way walk erect &  stiff,
567n014 ling the keeper., clasping «& rubbed» his arm. & SHOW signs of affecting something like man. Has an o
570n024 he  would throw back his shoulders. he wishes to SHOW, he is determined not to say anything. he press
574n039 see <A> ‖ I think this argument might be used to SHOW language had a beginning, which my theory requi
574n043 sider the acquirement of instinct by dogs, would SHOW habit.-- Take the case of Jenner's <Hyaena> Jac
590m094 s not Bark-- quotes Gardner's Music of nature to SHOW barking not natural. (Vol I. p. 234) Vol. II p
599o05v ent of original formation.-- declension & often SHOW traces of origin.-- Mayo Philosophy of  Living.
605o020 Stewart  on taste The object of this essay is to SHOW how taste is gained how it originates, & by wha
612o035 s of endosmic & exosmic juices. arms of polypus, SHOW either local or general will, & stomach likewis
613o036 being acquired, is a most important argument, to SHOW that they result from organization of brain; «[
615o36v ical plants when imported & plants sleeping good SHOW acquirement or obliteration of instincts But ha
620o045 nt> power (reason) obtained by age, which should SHOW the child, which of its instincts are best to b
621o047 h the natural tastes say is good. yet horseflesh SHOW that even this is possible.-- So that as  there
633j53v detrimental accidents, & domesticated variations SHOW us accidents may become heredetary [produce som
636j57r t exist without these adaptations.--fossil forms SHOW such losses.-- Consider ground Woodpecker stiff
641j29r en split up.-- who can decide their limits.-- To SHOW how little we understand of the Physiological r
113a092 as  explaining want of levelness Major Mitchell SHOWED me a river <near> W. of Port Philip. which had
285c150 er. supply place in each others' economy Dr. S. SHOWED that savages are not born with any capacity fo
314c237 gate by scission can not alter man.?? Mr Brown SHOWED me Bauer's drawings of a curious plant where a
417e076 ffspring are hybrids,.-- Mr G. B. Sowerby <tel> SHOWED me many land shells of the common species: fro
430e119 er parts of the world-- March 16th. Mr Lonsdale SHOWED me two specimens of an Inoceranus from the Gau
460t015 » bred ones-- The brothers & sisters half-breed SHOWED no sexual inclination for each other-- Aug. 20
463t055 h as that of Java & Sumatra Nov 15th Waterhouse SHOWED me the component vertebrae of the head of Snak
463t055 l!! distinct!!-- He would not allow such series SHOWED passages-- yet in talking, constantly said  as
521m009 was  <seen> shown him.-- «<Mr Corb> the servant SHOWED him watch & said dinner is ready, what, what.-
521m009 atch & said dinner is ready, what, what.-- then SHOWED the watch upon which he exclaimed, why it is d
545m107 key when touched, also another monkey to dog. I SHOWED nut & then closed my mem. expression of fury,
565n008 there  is some anger «& respect to opponent» is SHOWED by same movement as sneering,-- it is then mor
470t178 back  Wings, pollen is ejected with violence in SHOWED On many Papilionaceous; all wh. are in  flower
090a022 between 1768 & 1818. that is fifty years-- 90 <SHOWERS of> stones are recorded as falling; many of th
033r043 log. in Daubeny. P. 349 Admirable little table SHOWING long PERIODS of great violence volcanic.  from
045r077 ing) moves away.--Will geology ever succeed in SHOWING a direct relation of a part of globe rising, w
063r133 n wild & bred. no doubt with perfect success.--SHOWING non Creation does not bear upon solely adaptat
067r144 rdoba earthquake a very remarkable phenomenon. SHOWING line of disturbance inside Cordillera: It is m
090a025 ringing planets to an end? Fragmentary granite SHOWING schistose structure (& veins appearing): mem.
101a056 Phillips.  Lardner's  p. 270-4, good discussion SHOWING present form of land in Northern England influ
109a083 of  dikes & ledges of first-rate importance in SHOWING not subaqueous removal--??? the difficulty of
228b230 ongly shown by hybridity of ferns.-- hybridity SHOWING connexion of two plants. Animals-- whom we hav
264c081 structure without corresponding habits clearly SHOWING true affinity, for instance tail of ground woo
268c096 dstart & actually driving away parent birds.-- SHOWING how blind a storge From what I see of S. Ameri
278c130 eral in Van Diemen's land is most important as SHOWING former connection of two continents & death of
279c133 - Sir J. Sebright-- pamphlet-- most important, SHOWING effects of peculiarities being long in blood.--
313c235 young male was added good instance of instinct SHOWING itself, not from instruction Even the action o
317c250 - p. 250 admirable table of plants of St. Jago SHOWING many common to Canary. isld., Europe, & St Jag
318c253 e Hirunda fulva (added by Audubon in Appendix) SHOWING WHAT CHANGES are taking place & how birds  are
335d014 cross, as stated by Fox, is very important, as SHOWING above facts as first rate bone new  species,
355d066 strosity not prospective, but retrospective as SHOWING what organs are little fixed-- (<also> Hunters
360d090 rts swelled, though no fluid came from them.-- SHOWING how gradually every <thing> «change» is effect
363d103 were often caught wild off coast of America.-- SHOWING hybrids can fare for themselves.‖ + + first ye
385d164 Academy) have contained perfect teeth & hair-- SHOWING foetus has gone on growing-- I believe same ha
385d165 the  daughters! This last remark good. because SHOWING probably not education.-- Cannot I find some a
389d176 ed: this most important with regard to theory, SHOWING generation connected with whole system, «as if
400e013 them for four generation before they broke.--, SHOWING effects of cultivation gradually adding up.  &
411e056 ereus of the Falklands Isld. is interesting as SHOWING some change in habits before form.-- I have al
420e090 s do flowers.-- Mem. Spallanzani's experiments SHOWING how little of the spermatic fluid fertilized s
421e090 phosis in the young of Syngnathus.= curious as SHOWING generality of law. even in fish: ‖.do. p. 236-
430e117 Y different dispositions: this is important as SHOWING small variations in offspring of wild animals.
433e126 iven. p. 246.-- 248 & p. 258 A beautiful case, SHOWING the gradation from one grand system to another
439e143 a facility in returning to old type Mr Herbert SHOWING the extreme facility of crossing, in plants pr
446e162 ifferent stocks in this case».-- & strong case SHOWING analogy of production by gemmation & by seed--
458t002 y mare & produced a very pretty little animal, SHOWING something of Mule in its ears-- ((this is good
458t002 g of Mule in its ears-- ((this is good case as SHOWING gradations, Boteler's Narrative Voyage East co
478z004 eading Cuvier's Memoire 133 1803. on Pennatula SHOWING it to be one animal In Australia I was assured
507q15v ed-cases with similar structure.= good case as SHOWING how simple, but beautiful adaptation might  be
538m081 thus eternal punishment explained. These facts SHOWING what a train of thought[] action &c will arise
546m110 er state.-- does this throw light on instinct, SHOWING what trains of action may be done unconsciousl
550m126 wne Religio Medici, p. 21-24. Curious passages SHOWING how easily chance & will of Deity are confound
553m137 mon to that group.-- this is very important as SHOWING <connection> that expression mean SOMETHING.--
555m143 ike a hero, & then I had some confused idea of SHOWING scar behind (.instead of front) (having change
555m143 hanging into his head cut off) as kind of wit, SHOWING he had honourable wounds.-- all this was  kind
555m144 Monro  experiment about hanging came before me SHOWING impossibility of person recovering from hangin
556m146 rs, when kicking.-- -- good case of expression SHOWING real affinity in face of donkey, horse & zebra
565n009 ng, <but> lesser in degree, no smile, no frown SHOWING thought, no compression of mouth showing actio
565n009 frown showing thought, no compression of mouth SHOWING action,-- sulkiness all negative expression? E
570n027 l,-- [male alone worm doubtless admires female. SHOWING. no connection with male figure]CD-- As forms
572n032 ds Works. Vol I, p. 226-- "The general idea of SHOWING respect is by making yourself less, but the ma
577n053 ut any disease tends to give it, as in cancer, SHOWING, effect of mind on individual parts of body.==
592n102 six  (?) first Vol of Silliman's Journal paper SHOWING that the signs invented for Deaf & dumb school
602o010 & memory Reynolds X discourse very curious as SHOWING "the perfection of this science of abstract fo
615o36v his only pleasure; for it was different way of SHOWING it, nor was there any cause, & if surprise was
626o052 o it by me..' 'p. 333 «& p. 377» some remarks. SHOWING that instinct cannot be said to guide will. as
032r040 greater activity & rest.--Such changes could be SHOWN (as represented), along line of coast.--[Fig. 2
109a082 not located] -- do-- [...] Subaqueous. removal, SHOWN by the number of bones lying at the bottom of s
119a106 ated over the absolutely fluid pool.-- (this is SHOWN by the softness & curvature of quartz rock?) al
122a113 Y of earth.-- study different forms of earth as SHOWN by arc.-- read Herschels astronomy with oscilla
195b098 Creator  creates by any laws. which, I think is SHOWN by the very facts of the Zoological character o
231b234 seed«s»??-- Universality of generation strongly SHOWN by hybridity of ferns.-- hybridity showing conn
231b234 domesticated animals with him.!-- Modern origin SHOWN by only one species, far more than by non-embed
248c030 breed together, so when we grant «(which can be SHOWN probable,)» varieties may be made in wild state
265c087 stroyed even if these shrivelled wings could be SHOWN to be of some use. If we only had Puffinuria Ga
270c104 ic nature on inorganic It is very remarkable as SHOWN by Carus how intermediate plants are between an
274c113 so on fruit.-- The Rasorial type is wonderfully SHOWN in long legged cuckoos with claw like lark, (on
283c143 family  is relation to elements & not habits as SHOWN by frigate Bird & flying eagle,-- Hawk Gould se
295c183 old Stallion tempted to cover old mare by being SHOWN, young one.-- Many African monkeys in  Fernando
307c216 plain meaning & none in other case! Savigny has SHOWN same fundamental organs even in Haustellata & m
342d035 says relation of Osteology of birds to Reptiles SHOWN in osteology of young Ostrich. 16th. D  Israeli
370d115 created.--  it might as well be attempted to be SHOWN from peculiarities of climate cause of N. Zeala
372d131 the  vast difference of two kinds of generation SHOWN by their happening in same plant.-- The Marsupi
398e005 ith some animals, &c.-- «if the change could be SHOWN to be more rapid, I should say there was some l
420e089 determined by most complicated circumstances, as SHOWN by difficulty in forging, yet handwriting  said
434e128 ce".-- <it is only analogy.> but experience has SHOWN we can & that analogy is sure guide & my theory
435e133 anules exserted their tubes: now Mr Herbert has SHOWN that stigma swells, when pollen even most remot
```

Page
(Key Word)
441e150 istence of "laws of organization" had better be SHOWN-- soil on colour of flowers, Hydrangea -- black
485z016 ter Mem Orpheus--becoming tyrant-- flycatcher-- SHOWN by habits & plumage so very similar to some of
498q008 ordinary number of seeds known?-- Linnaeus has SHOWN that each pistil is connected with separate div
520m002 his father died.-- The omnipotence of habit is SHOWN about meals, no [not located] There is a case o
521m098 uld not understand it, but the watch was <seen> SHOWN him.-- «<Mr Corb» the servant showed him watch
526m038 in angular leaves,-- (this Rhythmical beauty is SHOWN by Humboldt from occurrence in Mexican & Graeci
537m075 ed» Yet as, I think, the opposite side has been SHOWN-- see Mackintosh.-- Must grant, that the consci
548m116 «somewhat» the same as double consciousness, as SHOWN in the tendency to forget the insane idea; & on
551m128 ar more alacrity <than> when something good was SHOWN him, than when merely ordered to do it.-- Plato
554m138 n the boasted philosopher himself it is chiefly SHOWN in old male.-- A very green monkey (from Senega
577n049 tance" a-- "frontier instance".-- for it can be SHOWN that the life & will of a conferva is not an an
577n052 al affection, on separate organs most curiously SHOWN in the sudden cures of tooth ache before being
603o013 must have personified the deity.-- H. Tooke has SHOWN one chief object of language is promptness «of
609o29v calls vice). on same principle that Malthus had SHOWN incontinence to be a vice & especially in the f
619o043 imself it would be remorse as will be presently SHOWN.-- This then is moral approbation, as far as it
035r047 level «of same beds» should have been kept; it SHOWS that throughout all England, whole surface osci
038r059 lcanoes. have been elevated considerably. which SHOWS an afflux of inferior melted rocks to those par
052r098 ieve SE coast of Madagascar. where a --40 line <SHOWS> runs at equal distance?) 1st cases. -- The ter
059r123 d suddenly.-- As the rude symmetry of the globe SHOWS powers have acted from great depths, so changes
063r130 arius. <Caracara> Calandria: inosculation alone SHOWS not gradation;-- An argument for the Crust of g
065r140 ations on springs at S. Cruz.???-- Form of land SHOWS subsidence in T. del Fuego, and connection of q
070r154 d it not be mere vibration? but walls & feeling SHOWS undulation.: crust thin.--Concepcion earthquake
113a090 eater depth than 400. & the limit being 400 ft. SHOWS that the strata have very unusual conducting po
119a106 ...]> moreover gradation from gneiss to granite SHOWS that the metamorphic rocks have just floated ov
120a108 that which gave cleavage to rocks.--, but lava SHOWS the rocks really hot. & therefore I doubt the t
121a111 tals of ice were formed-- (like my gypsum case) SHOWS power of segregation.-- & has heated angular fr
127a126 le of glass. (oil may be cooled to 0 degree!)-- SHOWS effects of pressure in change of form as the re
128a129 o counterbalance How strongly the Glen Roy case SHOWS that the figure of the world has just that form
132a137 (Sc Math. Phys. et Naturelles. Tom I. p 501.-- SHOWS first that data wholly insufficient to calculat
132a138 akes» bottom. if not colder than mean of place, SHOWS earth not with central heat.--» «(does M. Parro
132a139 lanation appears to me necessary) as M. Parrots SHOWS from variation in strata earth a very bad condu
132a139 riation in strata earth a very bad conductor.-- SHOWS p. 516 that subterranean springs give result le
145g002 - Mere fact of many races of Animals in Britain SHOWS that either races soon made or crosses difficul
150g035 upper line running up great bight just as Dick SHOWS NB. Lake gradually draining off would form plai
160g088 to the west of Glen Tarf What I called Alluvium SHOWS the ascending fringes {P} which makes me think
165g126 nstincts crossing most remarkable ever observed? SHOWS that <nervous> brain makes thought Glen Roy B C
179b033 stence of whiter tribes in centre of S. America SHOWS this.-- <If> Is there a tendency in plants hybr
253c046 cies in Borneo. Sumatra. India Ceylon-- perhaps SHOWS great persistency of character. Hence Elephas p
259c066 act of armless parent not having armless child, SHOWS than there is reference to more than offspring
259c066 ference to more than offspring (like atavism) & SHOWS my «view of» generation right?-- If puppy born
260c068 ears <dis> crys of distress of other parents.-- SHOWS community of language Desert country. is as eff
265c084 w that grandson is determined, when child is.-- SHOWS that generation implies more than mere child, b
280c134 t difficult & other go back to either parent.-- SHOWS instinct (Sir J. Sebright admirable essay) here
285c151 each branch No: because decay in that species <SHOWS> is effects of unfavourable conditions. (hence
287c157 en Mammalia & fishes, one penguin, one tortoise SHOWS hiatus-- but not saltus-- when Linnaeus put wha
289c160 rare in some countries-- nightingale do.-- all SHOWS how nicely adapted species to localities¶-- p.
296c184 ora of Galapagos?) same condition. Keeling Isd «SHOWS where proper dampness seeds can arrive quick en
315c243 on Expression & the Zoonomia, for if the former SHOWS that a man grinning is to exposes his canine te
345d044 .--» The very many breeds of animals in Britain SHOWS, with the aid of seclusion in breeding. how eas
347d050 being perfect till circular. p. 5 Most clearly SHOWS that genus expresses as now used almost any gro
352d059 does not perform that function which experience SHOWS us it was for.-- Most important law.-- Penguins
357d072 struction of the great Mammals over whole world SHOWS there is rule.-- S. America & Australia appear
373d132 pening in same plant.-- The Marsupial structure SHOWS that they became Mammalia, through a different
376d135 to change.-- to do that for form which Malthus SHOWS, is the final effect, (by means however of voli
380d154 an a female.-- this is full of interest; for it SHOWS latent instincts even in brain of male.-- Every
382d158 a Lamarckian.-- Mine is much simpler.-- Hunter SHOWS almost all animals subject to Hermaphroditism,-
382d158 Hermaphrodite,-- but not of poenis & clitoris, SHOWS to my mind.--, that both are present in every
384d161 n & deficiency of parts, as in Hermaphrodites, (SHOWS my doctrine of Hermaphrodite differs from Hunte
388d173 ma.-- The manner in which Frogs copulate & fish SHOWS how simply instinctive the feeling of other sex
388d173 feeling of other sex being present is-- it also SHOWS that semen. must actually reach the ovum.-- [Wh
392d175 ing without coition or addition of differences. SHOWS that difference need not be added EACH TIME. bu
405e032 s <false» embedded with almost concept, though SHOWS that progression of change in Mollusca is some
414e064 ganization: though Cont of Africa & West Indies SHOWS organization in Black Race there gives them pre
416e072 be used to gradation of changes which gradation SHOWS it to be the effect of a gradation in differenc
420e089 e descended.-- Jan. 6th The rudiment of a tail, SHOWS man was originally quadru<manous> <ped.-->. Ha
420e089 forging, yet handwriting said to be hereditary. SHOWS well what minute details of structure heredetar
435e133 siren, & Proteus, affinity to tadpoles. p. 210. SHOWS. that the action of light is concerned with the
437e138 aymans from Honduras. good case of migrating.-- SHOWS my theory insufficient.-- p. 120 An Eagle is sa
437e139 lamb «& hare» by the side of Eagles nest, which SHOWS power of carrying great weight. p. 125 is said
439e143 t the existence of wild close species of plants SHOWS there is tendency to prevent the crossing.-- in
439e144 earing by seeds Holyoaks-- (how far is this so) SHOWS either there is not so much crossing as I think
447e164 eory does require crossing.-- The case of Lemna SHOWS dispersion of germs is not end of seminal repro
447e164 very heavy seeds.-- as Cocos do men.-- Analogy SHOWS some most important end.-- Festuca vivapara F o
449e170 this is same, as Galapagos facts &c &c.-- & it SHOWS the causes which give same species to different
450e175 nd wild in the plains of Celebes. (but language SHOWS that probably not original there)-- shows these
450e175 guage shows that probably not original there)-- SHOWS these isld not fit for horse. Forrest--. (p. 27
450e176 e isld of Sooloo.-- said to have been imported: SHOWS they will propagate get dimensions-- do App. p
458t001 sting forms of aquatic reptiles most strange, & SHOWS as in shells some forms are long preserved.-- v
461t041 beds & existing species is valuable because it SHOWS no innate power of change & it also shows, what
461t041 se it shows no innate power of change & it also SHOWS, what enormous changes of conditions, some spec
464t065 from New Zealand-- <so> not an Apteryx, yet it SHOWS, the Apteryx is not «quite» isolated in its pres
465t081 .-- in a descending series of strata This again SHOWS how much forms depend on other forms Lyell's Pa
524m021 naturally talked of these places.-- My F. says, SHOWS that early impressions are most durable.-- (but
524m021 impressions are most durable.-- (but Miss Cogan SHOWS that repetition is not necessary)-- the words s
525m026 remark how odious an illtempered fat man looks, SHOWS same connection between organization & mind.--
528m038 f forms-- the beauty of some a Norfolk Isd fir SHOWS this, or sea weed, &c &c-- this gives beauty to
531m051 en show in the naughtiness of brothers children SHOWS that sympathy is based as Burke maintains on pl
531m051 ng children, the violent passions they go into, SHOWS how truly an instinctive feeling, <may not pa.>
543m101 ted Habits: Have Effect in Bones is valuable it SHOWS that new instinct can originate.-- strong argum
546m109 ing fruit eater.-- origin of colours?-- Nothing SHOWS one how little happiness depends on the senses.
547m112 was difference between Castle & dream No answer SHOWS our profound ignorance in so simple case.-- The
549m122 nstincts as revenge «& anger», which experience SHOWS it must for his happiness to check-- that is ex
553m135 suistical doubts.-- The history of Metaphysicks SHOWS that such a view cannot be, anyhow, easily over
563n001 es. yet so shy at all other times.-- Birth Hill SHOWS it is evergreens they seek Cock Pheasant claps
564n005 ng at Astronomy without Mechanics.-- Experience SHOWS the problem of the mind cannot be solved by att
567n015 on like blush.-- when extreme sensation of heat SHOWS blood is pumped over whole body.-- is it connec
568n017 or real virtue, that is action which experience SHOWS will be for general good, or in case of any fan
571n029 animals in «Australia» & America manage;-- This SHOWS doctrine «of instinct» has been carried too far
579n058 is so.-- frown with grief,¿ bodily pain? frown SHOWS the mind is intent on one object.-- With respec
584n072 ted by horse & dogs. (even blind horses & dogs) SHOWS it is somewhat analogous to memory. Shrugging s
588n090 y to recollection. yet a dog hunting for a bone SHOWS he has recollection.-- Lamarck. Phil. Zoolog.--
593n111 n in Delirium tremens hearing other man speaks. SHOWS, that consciousness of personnal identity is by
595n117 they only do when <great> considerably excited, SHOWS their power of imagination-- for it will not be
595n117 ot have day-dreams-- think well over this;-- it SHOWS similarity in mind.-- think of Eyton's horses b
603o013 nouns & verbs-- so that much of EVERY language SHOWS traces of anterior state?? Edinburgh Review Vol
606o022 ty with grandeur of character.-- Hence Lessings SHOWS expression of pain cannot be represented. But w
609o030 oral sense.-- But my view <says> unites both «& SHOWS them to be almost identical» + What has produce
609o30v erience (education is the experience of others) SHOWS does not tend to greatest good.-- Therefore rul

611o034 ect of touch & not irritability, which at least SHOWS a local will, though perhaps not conscious sens
615o36v to touch a particular organ.-- I think Pincher SHOWS surprise, walking home one day met him, with Ma
619o042 early so [RHC] The history of every race of man SHOWS this, if we judge him by his habits, as <if> an
620o045 n trying to save another in desperation.-- This SHOWS, that our feeling, that the instinct ought to b
620o045 in passion, any more than in an animal.-- which SHOWS that. it is owing to some <subsequent> power (r
620o045 malevolence,-- when not urged to it by passion, SHOWS a bad child.-- Hence there are certain instinct
627o053 instinct,-- as young pointer to point-- clearly SHOWS this is true. p. 13. Affections cannot be analy
628o054 f fame-- Yes Hartley explains this & Mackintosh SHOWS the change produced.-- 4) p 38 Conscience check
636j56v abitation «or world» simple series.-- My theory SHOWS life equally simple series, & therefore trace o
462tf4r mate with a thrush"-- Athenaeum 1839. p. 708.-- SHREW, found by M. Lartet same as existing species. W
178b031 ere created there;--» Are not all our «British» SHREWS diff: species from the continent look over. Be
226b221 example in English & Irish Hare.-- Galapagos.-- SHREWS, & when big continent many species belonging t
317c251 «some since discovered» of N. America, & of the SHREWS.-- Dr Bachman told me. that near Charlestown ?
318c252 to the Cordillera,-- Bachman has seen webbed SHREWS. case of adaptation. (case of Squirrel from
163g109 Beagle Channel. Forchammers (Lyells Denmark) SHREWSBURY) rubbish.-- Speculate on origin pebbles brcug
325c267 ogy besides the papers collected by Owen. (at SHREWSBURY) Yarrells Paper on change of plumage in Hen
332d003 wo having consulted him on one day.-- Mark at SHREWSBURY thinks the half bred Alderney Cows take more
469t151 & seeds gathered «all» came up in 1840 true. SHREWSBURY.-- Abberley-- Early Magazine-- &c. double-bl
489qIFC p. 12 Maer. p. 13 Question &c. July. 1842.-- SHREWSBURY p. 14 Henslow (2d time) p. 14.-- Father. And
504q013 of tree with flowers near end of orchard.= At SHREWSBURY one branch of Rhod. flowered later.-- effect
504q014 placed far from all other Peas, from Wiegman SHREWSBURY (1) Peas.-- Beans seeds alone remain to be c
523m018 s people insane of particular ideas, «Case of SHREWSBURY gentleman, unnatural union with turkey cock,
525m025 impure breed.-- A cat had its tail cut off at SHREWSBURY & its kittens <h> (in number 3) had all shor
530m044 - Mr Blakeway has mentioned in Antiquities of SHREWSBURY something about big noses & name Corbet, per
530m049 There was blackbirds nest, near hot-house at SHREWSBURY, which the cat was seen by Hubberley to visi
532m056 aroline tells me that Nina, when brought from SHREWSBURY to Clayton, (though so fond of her & of serv
532m056 t of Richard & of Mary & her bed brought from SHREWSBURY) yet for a fortnight continued wretchedly un
532m056 weeks was so thin, that she was sent back to SHREWSBURY., then immediately fell into her old ways &
547m111 ome quite imaginary cause to start at once to SHREWSBURY., vaguely thought of packing up.-- was lying
547m112 umstances.-- as whether I really was going to SHREWSBURY, whether I had rung for Covington. whether h
594n112 to disobey it-- I have seen hawk & sparrow in SHREWSBURY garden picking from same bone A child born o
264c080 of forms in Australia leading on one side into SHRIKES & at the other into into Crows. yet all formin
453e182 Condore "crow like ours, but much more small & SHRILL".-- Humboldt. Vol I. p. 275. says Teneriffe do
099a048 of this? [P] If surface covered with oil should SHRINK. film parallel to longer axis. But if great de
047r083 w & overfall must vary proportionally Partial SHRINKING after elevation in perfect conformity with <M
105a070 ce. Bull. Soc. Geolog «1837» p. 320. paper on SHRINKING of Clay. applicable to Cleavage. C. Prevost.-
358d085 this were possible, the organs doubtless would SHRIVEL up.-- Yet odd they should have so much sexual
467c099 n Hybrids that although anther «nor filaments» SHRIVEL, yet stigma does not, so we may feel somewhat
265c087 herefore argument not destroyed even if these SHRIVELLED wings could be shown to be of some use. If w
085a001 m ascertained inclination. of plains: Lias in SHROPSHIRE. or some other wonderful outlyer.- Linn: Tr
122a115 ttered irregularly.-- (Mem near Gregory Bay). SHROPSHIRE case where lamination appeared.-- Lyells Den
163g109 n Geographical distribution of British Plants SHROPSHIRE Quartz what substance is collected in little
267c094 for miles in extent are composed solely of this SHRUB".-- p. 229. carcases of birds drifting out to s
582n066 sure in hiding themselves & skulking about in SHRUBBERY. when other people are about: this is analogo
569n022 naccompanied by dignity-- "no mon dieu," with a SHRUG-- "all I can say, I am very sorry so it is"-- d
569n023 othing more to be said."-- "made no reply, but SHRUGGED his shoulders & went away."-- he implies nega
573n039 . (her brother's wife). & she said nothing but SHRUGGED her shoulders.-- analyse this.-- Miss C. quit
570n024 & shrugs his shoulders & walks off,-- I think SHRUGGING connected with many emotions.-- (Explanation
570n024 apt to sigh.-- & hence carried on as trick) «SHRUGGING aroused acting» Octob 25. Why is modesty, mix
584n073 gs) shows it is somewhat analogous to memory. SHRUGGING shoulders seems sign of helplessness E. says
570n024 ster says I will see you damned first." the man SHRUGS his shoulders & replies nothing. if he did go
570n024 to say anything. he presses his lips together & SHRUGS his shoulders & walks off,-- I think shrugging
278c128 ra in other families.-- it will however be much <SHU> surer, when false species banished by this test
575n045 mental pleasure. with pleasure of senses. The SHUDDER of pleasure. from pleasure of music Audubon IV
159g087 lley much more gentle than in Glen Roy, & partly SHUT in No Granite blocks in higher parts?? Bought G
249c036 Marsupial forms been chiefly preserved,-- where SHUT up by themselves without other animals? but the
249c036 mselves without other animals? but they were not SHUT up!! Extreme southern points of S. Hemisphere f
470t177 omposed) Asparagus very small flowers & as much SHUT up, frequented by «many» Bees & Humbles-- «Humb
551m128 ts understand contracts.-- but W. Fox's dog that SHUT the door evidently did, for it did with far mor
551m129 y be seen by omnibuss Horses starting, when door SHUT or cad cries out "right." or Drinkwater's horse
556m145 re why is under lip curled over upper with mouth SHUT. expressing cool irony, not biting? What is Emo
472s02v pollen appeared chaffy, as if sucked?! opens & SHUTS end of sucker, after having withdrawn it.-- Saw
552m130 object, (or having looked at any object<>? one SHUTS ones eyes) is the image not vivid as in sleep--
592n103 p this subject & a reference to Brun's work.-- SHUTTING eyes in contempt opposite action to opening e
557m149 he heart.-- tendency to muscular movement, hence SHY people (shame of ridicule) are singularly apt to
563n001 ns do Wood pidgeons building near houses. yet so SHY at all other times.-- Birth Hill shows it is eve
567n015 like. dislike, or be indifferent about, yet feel SHY.-- not if quite stranger.-- or less so.-- When l
557m149 nding teeth.-- in shame frowning, & anguish,-- SHYNESS not so.-- affected laughter.-- A dog who goes
568n017 if prevented, or overtempted.-- «animals have SHYNESS with strangers» «as in case of temperance, or
577n052 , neck-- «upper» bosom in woman: like erection SHYNESS is certainly very much connected with thinking
579n058 oetry.-- The expression of shame-facedness for SHYNESS, having been invented, prove of the difference
243c019 e ces oiseaux eussent leurs représentants dans de SI hautes latitudes".. --¿translate?) All Australia
473s05v remadoc. but can tell them from lake S. of Moel SIABOD. wh. flows into Conway by Bettws & there joins
240c014 land. The Barbaroussa, (when young very like the SIAM race with long nozzle & few hairs) inhabits Cel
450e175 Pegue-- seldom equals 13 hands-- those of Lao & SIAM inferior to those of Pegu-- in Sumatra two bree
245c023 ly domesticated? Like in general appearance to SIAMESE kind.-- but considered good species from denta
048r085 Africa: the same caution is applicable to the SIBERIA case We must not think alluvial plains «always
086a006 n. vapour from below-- Malte Brun «Salt Lakes» SIBERIA must be read as well as Pallas before Geology
087a007 hy are tops of Equatorial mountains so cold.-- SIBERIA no plants to it, easily raised above level of
087a008 a. Lyells Encyclopaedia-- Lately elevated When SIBERIA went up. Arctic land went down.-- Probably mor
087a011 well with Lyells idea of intertropical land.-- SIBERIA rises. therefore to the South sinks.---- Medit
112a090 Erhman stating that the mean temp at Yakous in SIBERIA being -8 Reaumur.-- there ought to be 32 degre
124a117 hs of frozen soil. p. 211 Consider proved that SIBERIA must have been in same condition for long peri
124a118 ng Entrance. Book C. p. 101. On Frozen Soil of SIBERIA (with refer to Metamor) wrong entrance Athenae
126a124 e of the papers that there are springs even in SIBERIA.-- from water {P} thawed at + in isothermal cu
127a125 or may be used East-clinal lines & c & .-- But SIBERIA was once thawed. & hence. (when climate hotter
131a136 line Athenaeum. 1839. p. 52. On Frozen soil of SIBERIA.-- facts of water flowing from beneath frozen
186b060 s I have traced the <type> great quadrupeds to SIBERIA; we must look to type of organization.-- extin
187b062 d in cold countries & therefore not killed by <SIBERIA a more> cold countries.-- Seeing how horse & E
269c101 aph. Soc. in February or March 1838 on soil in SIBERIA being frozen to 400 ft in depth, (& Erman's su
316c247 ity of land animals.-- these however come from SIBERIA it cannot be said American fossils more resemb
429e116 d & Greenland, but not with those Kamtschatka, SIBERIA, or even of polar regions of N. America.-- if
510q017 n Australia & S. America? Sabine says North of SIBERIA, no sea-current, icebergs travel by wind. Aug.
187b062 rica; with no change, agrees with belief. that SIBERIAN animals lived in cold countries & therefore n
449e169 o fossil remains of the Hamster.-- is not this SIBERIAN animal?-- Eyton says that the young of two ha
074t165 erschlich. Vol. II Journal of Nat. & Geograph SICIENCES? -- H. says in Potosi the silver is contained
425e104 t chiefly mountainous «this is very important. (SICILY exception)-- see if this can be generalized.--
426e105 ighbouring" formed in the Tertiary «epoch» like SICILY not having species, if true important on my vi
430e117 et Cowper, describes his tame Hares, attacking a SICK one like Chillingham bulls are described.-- His
279c133 tuous intercourse. .-- excellent observations of SICKLY offspring being cut off-- so that not propagat
360d091 nearer to Jackall.-- This Keeper has seen when SICKLY tigers have first come over, insects somewhat
608o026 Effects.-- One must view a wrecked man, like a SICKLY one(P)-- We cannot help loathing a diseased of
608o026 pabilities.-- '(P) Animals do attack the weak & SICKLY as we do the wicked.--we ought to pity & assis
286c154 animals expression of countenance. «[s]hare of SICKNESS,-- death, unequal life,-- stimulated by same
294c178 far N. becomes stunted, altered, & lose (mere SICKNESS)? fertility ¿because offspring too unlike.--?
023r011 xis «of the Andes».-- «Has this fault determined SIDE of volcanic activity.» That axis was produced,
039r061 ins because such are found in perfection on that SIDE.-- Add from M. Lesson. character of Flora to Ne
044r073 of Volcanic action, Australia S. Africa-- on one SIDE. S. America on the other: The extreme frequency

Page ***(Key Word)***
059r120 ntre:-- Bailly talks of much granite on all East SIDE of Van Diemen Land. All the Calcareous rocks wh
064r136 des--therefore flowed towards it. a mass on each SIDE 3000 ft thick & 150 broad. neglecting Cordiller
085a001 indows of houses. & trees all injured on Eastern SIDE, far inland.-- even 70 miles from salt water. M
089a018 -- Tom 54. p. 106 do-- p. 110. Mountains on west SIDE of Domingo formed of coral limestone, with inte
096a040 turies afterwards it might be percieved on which SIDE craters were low --¿ applicable to Auvergne???
096a041 to Auvergne??? The fact of Galapagos Isld. steep SIDE to windward in allusion to St. Helena discussio
110a086 ase. crystals from fragment disseminated on that SIDE of salband. gradually becoming finer grained &
110a086 ly becoming finer grained & more compact on that SIDE-- separation DISTINCT from dike junction mechan
110a087 agment. junction certainly most distinct on dike SIDE.-- oozed from one of the true rocks, most proba
120a109 ut would cut wide gorge. leaving cliffs, on each SIDE, such as now exist.-- caution about action of r
128a129 n its centre.-- Will not abrasion of land on one SIDE. produce subsidence of water on other. from ten
135a145 th here & there little branches at {P} from each SIDE intercepting plain & dividing it-- Hopkins fiss
146g013 ere were none on surface of any hill Thursday On SIDE of Hill South of upper end of Loch Dochart butt
148g027 nclined towards river all these composed-- where SIDE ravine entered terraces formed successive bays
148g028 entre-wards which would not have happened if the SIDE-streamlet had cut them out-- In all cases «I ur
148g029 ge, a tongue of flat land, with terraces of each SIDE of the two valleys corresponding as in Andes, c
149g032 on> deposition in lake On the summit «& on Spean SIDE» of Meal-- Derry there were perfectly rounded «
149g032 eing gneiss <& also> also near summit on Hill on SIDE of Inn BOULDER of granite above 4th Shelf a lit
150g035 ll too low line drawn plain red talus line on N. SIDE of Spean most clear & upper line running up gre
150g037 river get formed in centre In Glen Collarig, on SIDE of Hill of Bohunthine upper road (2) extends as
150g038 ly as house, the 3d below them opposite to where SIDE ravine enters On opposite side of valley both e
150g038 opposite to where side ravine enters On opposite SIDE of valley both extend below the Houses The Hill
151g042 by flat bottomed strait. connecting flat on one SIDE with irregular gravel plain of other, which mus
152g045 Collarig good case of shelves entering «on» one SIDE ravine. Are the lip, or necks of land on level
152g049 ir 65 degree? There are two terraces on the East SIDE of river & bed of river about 40 ft beneath gen
153g054 ause collection of sediment? Where ravines enter SIDE by, opposite entrance into Glen Fintec a kind o
154g059 e Air 65? 70? {P} Where a buttress projects from SIDE of hill if line suppose continued across to {P}
154g059 of hill if line suppose continued across to {P} SIDE removed all well & good, but how came river to
155g062 3d dies away almost imperceptibly on Glen Turrit SIDE 2nd shelf very broad «& cut out, produced» from
155g062 be traced some way up, but most faintly on East SIDE of Glen Turrit, where I believe they end in upw
155g063 lined plains, as in Corry. & as «as I believe in SIDE ravine above houses of Roy» Maccullochs supernu
157g075 f shelf of 3d:-- granite block a yard across. On SIDE of «that» hill, in front of which shelf 3d form
158g080 ure terrace 15 ft divortium my measurements here SIDE of Loch Spey Forms terrace about 60 feet above
158g080 ified Granite ridge» at head of Glen Roy on same SIDE where two rivers unite in Upper Glen Roy great
158g083 pebbles Shelf 3d runs up with buttresses on each SIDE «very little way» in Upper Glen Roy at pass {P}
158g083 in Upper Glen Roy at pass {P} River Gorge 4th Sh SIDE of valley Granite blocks on this side (return)
159g084 rge 4th Sh side of valley Granite blocks on this SIDE (return) between 2d & 3d shelf Mountain <Mica>
160g092 owing into west end with obscure terraces on one SIDE Barom 29.200 A 80 70 degree? for about 3/4 of m
160g093 A 80 70 degree? for about 3/4 of mile on <one> S SIDE of «this» Isthmus (which runs E & W) broad terr
160g093 vel, <on op> dies away on gradual slope-- : on N SIDE.. dies away on rocky place, but narrow shelves
161g096 t of valley chiefly from rockiness When on other SIDE {P} Shelf A Shelf A at head of Gentle mossy slo
162g099 between L. Lochy & Oich. is a brook on the Lochy SIDE of it-- the terraces of which, last measurement
162g100 d stretches along, parallel to Shelf on opposite SIDE & dies away on the steep & rocky gully of last
162g103 like bed of lake with trace of terraces on each SIDE High up the Tarf (a Granite (boulder), sloping
181b043 e vertebrata invertebrates-- Such on few on each SIDE will yet present some anomaly & <the g> bearing
197b112 es from one to other organ.-- Cuvier on opposite SIDE; Is Vol of Fish p. 59.]CD Cuvier has said each
227b227 ntinent had sprung up round Galapagos on Pacific SIDE. the Oolite order of things might have easily b
234b255 new gate. opening towards pig.-- latch on other SIDE.-- Pigs put legs over, & then with snout lift u
264c080 gradations of forms in Australia leading on one SIDE into shrikes & at the other into into Crows. ye
298c191 onsiderable range of one species «on -- mountain SIDE» of which the central parts become occupied by
308c217 way. --Case of Habit I kept my tea in right hand SIDE for-- some month, & then when that was finished
308c217 , but I always for a week took of cover of right SIDE, though my hand <vibrate> would sometimes vibra
314c237 being (probably) small to allow it to lie on one SIDE.-- but in other species, this segment is conver
315c242 themselves accidental-- My first thought of sea SIDE-- Study Bell on Expression & the Zoonomia, for
317c251 rst thougt only one species. & all hares on East SIDE have other peculiar appearances-- Now this is p
335d015 ox's half bred Persians «cat» favour the Persian SIDE.-- Theory of abortive hybrids.-- If mules did b
345d044 ore than female, yet instances given on opposite SIDE,-- «The theory of males impressing most is in h
405e031 k.; & tail. & kind of semilunar {P} mark on each SIDE darker,, so that whole colour is changed, these
418e080 rstand ¿perfect?? developement of one sex on one SIDE, than the addition of other organ, in which cas
421e091 nent. in like manner as Madagascar does to other SIDE of Africa.-- (& Juan Fernandez to Chile??) Falk
429e116 regions-- do p. 21 says. many plants skirt each SIDE of the great N & S. valleys, which penetrate Py
437e139 Mr Willoughby found a dead lamb «& hare» by the SIDE of Eagles nest, which shows power of carrying g
484z016 ese groups strictly American. Colouring on under SIDE of wings It would be interesting comparison to
491q01v y in cone-- The «above Exper» explains apples on SIDE near other tree being affected.-- does one bran
513q020 are then intermediate or «sometimes» all on one SIDE, as in crossing varieties Amongst varieties cro
533m059 the difficulty of getting on a horse on the left SIDE (not good example,) because leg is right handed
535m063 wrong, as I have done when taking lid off <tea> SIDE of tea chest, when no tea do. p. 233. Mr Lewis
537m075 ction of approved» Yet as, I think, the opposite SIDE has been shown-- see Mackintosh.-- Must grant,
540m086 l is under same conditions as Negro on the other SIDE of the Atlantic. Why then is he so different--
558m153 e wild ass there is a curious drawing out of the SIDE part of nostril, when passion commences.-- <All
567n014 ation.--» in balancing a body & an ass knows one SIDE of triangle shorter than two. V. Whewell. Induc
573n037 . when refusing food, turn his head first to one SIDE & then to other. & hence rotatory movement nega
040r063 intersecting a mass probably cold & not warm as SIDES of a crater as Vesuvius.-- There may have been
049r088 in dikes. In Granite great crystals arranged on SIDES. V. Lyell P 355 Vol III. constitution of veins,
062r128 getation at S. Shetland = Great contrast of two SIDES of Cordillera, where climate similar.--I do not
106a075 cannot be great over) with very gently sloping SIDES This argument is partly taken from Delabechs Th
154g057 cure obscure NB In Glen Collarig tidal channel, SIDES <alm> 15 ft above bank or terrace, from terrace
158g082 ks almost encircled> <fre> Gneiss cut smooth on SIDES of hill where Boulder lies. buttresses «occur»
472s02v pollen. (as a needle becomes covered) so whole SIDES of flower & stigma dusted.-- <I think> When It
568n016 result of our senses, or our experience.-- Two SIDES of a triangle shorter than third. is this neces
162g104 e sand & small angular-- rounded pebbles-- dip SIDEWARD, & inwards-- deposited when water stood at hi
509q017 r. Gray General Questions (1) Particulars about SIERRA Leone. cow. taking bulls. is it Domesticated A
428e112 eedlings: (which are confessed to by Herbert) to SIFT out the weaker ones: there ought to be no weedi
637j58r s grow harder. so must those with weak beaks be SIFTED away.-- 4 & the species, like 10,000 others, p
542m094 ning. (common to other animals) scream of agony, SIGH of discomfort & weariness. & meditative tranqui
570n024 ration when immensely immersed-- mechanic apt to SIGH.-- & hence carried on as trick) «Shrugging arou
584n073 ms sign of helplessness E. says she can perceive SIGH, commences as soon as painful thought crosses m
589n092 g only bodily actions associated with ideas.-- A SIGH, is an abortive groan.-- more power over muscle
589n092 r over muscles of voice than respiration.-- like SIGH before false sneeze.-- "A Dissertation on the I
542m094 ch now connected physical relations.-- CD[like SIGHING to relieve circulation after stillness.-- Now
542m094 ceive if organization were changed, I conceive SIGHING might yet remain just like sneering does.-- is
570n024 onnected with many emotions.-- (Explanation of SIGHING is probably correct, to relieve respiration wh
590n093 great weakness.-- <Diarrhaea> & syncope p. 42. SIGHING from grief. is method of increasing languid ci
590n093 creasing languid circulation-- no, for <grief> SIGHING comes on before circulation is affected. p. 44
350d056 ls place in creation.-- thus senses, especially SIGHT connected with locomotion.-- «Mem. Dr. Blackwel
350d056 «Mem. Dr. Blackwell (Abercrmbies) comparison of SIGHT to threads.--» Hence the Pecten, which move imp
436e138 I. p. 72.-- Goldfinches placed near, but not in SIGHT of each other will sing till they drop off thei
496q05a rs are crossed.-- Are Bees guided by smell-- or SIGHT.-- --. touching Mr Brown theory of insect-like
523m015 ad paper somewhere on horse being insane at the SIGHT of anything scarlet.-- dogs ideotic.-- dotage.-
526m029 ing. (on hearing old association brought up) by SIGHT & not by hearing One is tempted to believe phre
528m037 night.-- from the mere exercise of the organ of SIGHT, which is common to every kind of view-- as lik
544m101 ng, gives pleasure, & always has done therefore SIGHT of own child. (when frame in condition to recei
566n011 view of origin of evil passions.-- Man getting SIGHT slowly,, but when in grown years, thinking he i
586n081 is in other is acquired instinctively" So with <SIGHT> sight-- so a Bee has the faculty of building «
586n081 ther is acquired instinctively" So with <sight> SIGHT-- so a Bee has the faculty of building «regular
586n082 rn out to be instincts, & so in some senses, is SIGHT--CD [The faculties bear so close a relation to
590n093 ys tooth ache, even from carious tooth cured by SIGHT of instrument.-- Bennett's Wanderings, Australi
603o11b ays yes. imitating song -- two primary sources, SIGHT, & hearing-- Staunton Embassy Vol II p. 405.--

(Key Word)*
579n058 my theory believes in.-- From the manner short-SIGHTED people frown, frowning must have some relation
579n058 relation to short-sightedness.-- do not short SIGHTED people squinny-- when they consider profoundly
579n058 n, frowning must have some relation to short-SIGHTEDNESS.-- do not short sighted people squinny-- whe
472s02r light on upper petals & insert proboscis, under SIGMA & draw it out over & over again & wipe off poll
031r037] enormous extent; if lowered again & covered no SIGN of upheaval To Cleavage add other instances in
058r115 circumstances of appearance at Concepcion [.] no SIGN of elevation. Effects of great waves to obliter
368d114 e) & are merely most attracted). -- singing best SIGN of most vigorous males.-- «(NB. most strange co
571n028 them (as justice?? as ancients did high forehead SIGN of exalted character???) Why may not our herede
584n073 t analogous to memory. Shrugging shoulders seems SIGN of helplessness E. says she can perceive sigh,
588n090 t appears to me mere consequence of stooping, as SIGN of humility.-- I suspect very strong argument m
542m097 st understand each other expressions, sounds, & SIGNAL movements.-- some say dogs understand expressi
286c154 onvey much thus, Man has expression.-- animals SIGNALS. (rabbit stamping ground) Man signals.-- anima
286c154 animals signals. (rabbit stamping ground) Man SIGNALS.-- animals understand the language, they know
530m043 y or two are absolutely forgotten.-- My father SIGNED a bond, yet when he paid the Attorneys bill, h
377d138 ., together with a grunting noise, the former SIGNIFIES recognition with pleasure, as when food is of
070r155 f travelled boulders. from the Cordovise range. SIGNOR Rozales tells me at seven oclock Novem «5th» C
051r094 ywhere on coast (Il Defonsos «Kelp») rocks show SIGNS of degradation; (soft substances worn into bare
560m156 s lice?-- picture.-- Do female monkeys not show SIGNS of impatience when woman present? Do they pout,
567n014 he keeper. , clasping «& rubbed» his arm. & show SIGNS of affecting something like man. Has an origin
574n039 Scotch Poet, has numerous lines. of poetry.-- <SIGNS> sounds singularly adapted to subject see <A>
580n062 nimals have abstraction because they understand SIGNS.-- very profound.-- concludes that difference o
592n102 ol of Silliman's Journal paper showing that the SIGNS invented for Deaf & dumb school & used between
599o005 bbo.-- language commenced in whole sentences.-- SIGNS-- ¿ were signs originally musical!!!??-- At le
599o005 commenced in whole sentences.-- signs-- ¿ were SIGNS originally musical!!!!??-- At least it appears
210b160 ecked when islands are near continent: compare SIICILY & Galapagos!!-- Some of the animals peculiar.
470t177 y «many» Bees & Humbles:- «Humbles & common» On SILENE, many plants of wh. have abortive stamens:= Man
251c042 robably in Cuba (p 271 Viedo says American dogs SILENT. Mem contrary assertion of Molina) (p. 277.).
542m094 . «whine of children. puppies do so dogs nearly SILENT, so with men.-- How is crying-- peculiar not c
045r074 peak the changes. -- metallic veins solution of SILEX & many other phenomena I do not believe that th
091a025 appearing): mem. Henslows Anglesea solution of SILEX also shewn. No 3d of Ed. N. Phil. J. p 194. Fac
104a066 ion-- Albite certainly contains 6 per cent more SILICA than common felspar therefore on axis of Cordi
138a176 e with «piece of cabbage» alklali & precipitate SILICA / or charcoal charged with carbonic acid [blan
027r028 on sand formation <would> wood converted into SILICEOUS pyritous & coaly matter. Mem: Chiloe In the e
049r088 oes not granite at C. Tres Montes become more SILICEOUS in close contact? -- «Cordillera???» Porphyry
026r022 og: Map of Europe Conybeare. Introduct XII P. SILICIFIED bones not common in Britain. Mem Concepcion
035r045 vourable to preservation of bones & &c--Yet <SILICIFIED> turn over silicified wood. Cordilleras, Chi
035r046 on of bones &c &c--Yet <silicified> turn over SILICIFIED wood. Cordilleras, Chiloe. &c seems the orga
494q005 unded by various bright colours, any effect? and SILK caterpillars (1) Shake a sleeping mimosa, or ha
503q012 ctive Resources Book no information & Hope about SILK worms. Varieties effects of domestication-- sai
581n064 uch apparently. ourang outang very fond of soft, SILK-handkerchief-- cats & dogs fond of slight tickl
058r115 imal cryst. arrange themselves in planes. «Mem SILKY lustre» ask Erasmus. whether electricity would
067r142 eral character of fossils deception complete.= SILLIMAN Journal. year 1835 excellent account of N. Am
247c027 snakes on Friendly isles. p 50. LX. Journal of SILLIMAN» «Study Silliman.--» Vol II. p. 10. it seems
247c027 y isles. p 50. LX. Journal of Silliman «Study SILLIMAN.--» Vol II. p. 10. it seems that Crocodile wa
318c254 e appearing to remain about a fortnight, «See SILLIMAN'S Journal 1837. Paper by Bachman.» that is suc
325c267 about E. Indian Islands. consult Dr Horsfield SILLIMAN'S Journal Rengger on Mammalia of Paraguay. acc
328cIBC nck Hist. Nat. des Pigeons et des Gallinaces. SILLIMAN'S Journal. during 1837. paper by Bachman on mi
473s03r same fly Two more of the flowers withered.-- SILLIMANS Journal <vo> 1842. p. 142-- Sus americana & H
592n102 gestures. In one of the six (?) first Vol of SILLIMAN'S Journal paper showing that the signs invente
538m079 fires, what is wrong in his head. & Babington's SILLY joking The possibility of the brain having whol
102a059 en on to it.-- rollers at Tristan d'.Acunha.-- SILTING up. channels on coast of England-- Any one. wh
066r142 inguish from stone Caradoc from lower of third SILURIAN division--Together with same general characte
067r144 that Falkland fossils decidedly belong to old SILURIAN system. Apply degradation of landlocked harbo
194b095 cies doubtful when known only by bones.-- Mem: SILURIAN fossils: ¿How are. South American shells? ¿Do
213b170 ived far back, fish approaching to reptiles at SILURIAN age-- How long back have insects been known?
343d037 of Java & Sumatra, that since the time of the SILURIAN, he has made a long succession of vile Mollus
374d133 t more widely from living forms.-- p 458 Upper SILURIAN, fishes oldest formation highly organized.--
374d134 mation highly organized.-- do. p. 461.-- Lower SILURIAN-- several existing genera. Nautilus turbo. bu
205b143 I. p 75 a Fish which emigrates over lands is a SILURIS p. 123 A climbing fish. p. 122 A Terrestrial
079r176 n do veins. At Huantajaia. Humboldt says, mur of SILV.[,] Sulph. of do.[,]galena[,]quartz, Carb. of L
073r162 ras[.] some general laws. association of lead & SILVER. Sulp.of Barytes: Fluoric. Barytes:-- Humbold
073r163 parts as rock eminently rich in mines of gold & SILVER." «p. 131» The above porphyries characterized
074r164 [r]ase. opal:-- Veins in Limestone & Grauwacke: SILVER appears far more abundant in the upper limesto
075r168 & Geograph Sciences? -- H. says in Potosi the SILVER is contained in a primitive slate, covered by
075r167 vein «of iron» discovered?-- Klaproth analysed SILVER ores from Peru consisted of native silver & br
075r167 lysed silver ores from Peru consisted of native SILVER & brown oxide of Iron in Mexico. sulphuretted
075r167 r & brown oxide of Iron in Mexico. sulphuretted SILVER, arsenical grey copper, and antimony, horn sil
075r167 lver, arsenical grey copper, and antimony, horn SILVER, black silver & red silver, do not name native
075r167 l grey copper, and antimony, horn silver, black SILVER & red silver, do not name native silver becaus
075r167 , and antimony, horn silver, black silver & red SILVER, do not name native silver because not very ab
075r167 , black silver & red silver, do not name native SILVER because not very abundant.-- muriated silver.
075r167 e silver because not very abundant.-- muriated SILVER. which is so rare in Europe,. common there acco
076r168 at those oxidated masses of iron. which contain SILVER are peculiar to that part of the veins, neares
076r168 veins of Kongsberg in Norway.--namely dendritic SILVER intersecting carbonate of lime-- native silver
076r168 silver intersecting carbonate of lime-- native SILVER in Mexico is always accompanied by Sulp. silve
076r169 silver in Mexico is always accompanied by Sulp. SILVER sometimes by selenite.-- in New Spain, contrar
076r169 s lead not abundant. = considerable quantity of SILVER procured from martial pyrites; great blocks of
076r169 ured from martial pyrites; great blocks of pure SILVER not common in <S.> America: In all climates di
076r169 n <S.> America: In all climates distribution of SILVER «in veins» very unequal sometimes disseminated
076r169 etimes concentrated: wonderful quantity of pure SILVER in S. America. Geology of Guanuaxuato.--Clay s
079r176 in in Mica Slate & overlying Limestone Balls of SILVER ore occur in do veins. At Huantajaia. Humboldt
079r176 rtz, Carb. of Lime. accompany.-- Ulloa has said SILVER in the highest & gold in the lowest. Humboldt
183b052 eret, avec la linotte, avec le verdier" &, <fr> SILVER gold & common pheasants & fowls..-- "On sait q
358d085 nnot «hardly» tell any sex by appearance.-- The SILVER & common pheasant crossed, has a cock (inferti
358d086 st of which is like common pheasant & back like SILVER.-- But the hen hybrid of this bird, has long t
364d104 placed before sheep-- it has been thought that SILVER Pheasants about a house made other pheasants h
383d159 - Seeing in Gardens of Hybrids between Common & SILVER Pheasant, one like cock & other like Hen.-- on
383d160 sected Zoolog. Garden. Sept 16." Hybrid between SILVER & Common Pheasant. Male bird, said to be infer
383d160 infertile.-- spurs rather smaller than in <ma> SILVER male-- Head like silver except in not having t
383d160 er smaller than in <ma> silver male-- Head like SILVER except in not having tuft,-- back like do.-- b
383d160 is green doubtless is effect of Metallic hue of SILVER pheasant. yet why green? & not purple?-- legs
405e032 wnish do Saw what was said to be hybrid between SILVER & gold fish-- Octob. 26th. If. hereafter. M. a
036r051 & was recalled by note in Daubeny. P. 438., of SIMILAR fact near the Red Sea.--which occurred in a se
043r071 tinuity of space in formations & durability of SIMILAR causes go together. add. <">" <from> "in the s
047r082 (Lyell. vol I. P. 471) is explained. also the SIMILAR fact at Concepcion? Read the various accounts
049r087 do.» discuss quartz veins, there contemp--yet SIMILAR ones in Clay. Slates contemporaneous others su
056r111 lend all substances together; & products being SIMILAR over whole world, general circulation. But Vol
062r128 rast of two sides of Cordillera, where climate SIMILAR.--I do not know botanically = but picturesque
170b001 on the coeval kind, all individuals absolutely SIMILAR; for instance fruit trees, probably polypi, s
175b018 «constantly» formed ¿would they not be pretty SIMILAR over whole world under similar climates & as f
175b019 y not be pretty similar over whole world under SIMILAR climates & as far as world has been uniform, a
182b048 wo kinds one <with> white with a black face, & SIMILAR to those brought from Abyssinia, & others dark
186b060 Now Genera of those two countries ought to be SIMILAR ¿Law: existence definite without change, super
211b162 finity with snipes» indicate affinity, because SIMILAR habits produce similar structure.-- Mem. Ornit
211b162 icate affinity, because similar habits produce SIMILAR structure.-- Mem. Ornitho Rhyncus Would not re
213b170 ich have wide range and therefore cross & keep SIMILAR. But this is difficulty; This immutability of
216b179 of twenty have ANALOGUES uses this word «for» SIMILAR in the Indian sea.-- Deshayes.-- Mr McClay is
228b230 n Pippin &c produce real crabs, & in each case SIMILAR or mere mongrels.-- It really would be worth t

Page   ***(Key Word)***
```
231b243 ted by real relationship. as well as effect of SIMILAR temperature.-- now those of temperate  regions
233b249 e Australian.?-- Mr Gould has been struck with SIMILAR extension of form in birds.-- ⌐Waterhouse thi
243c020 ut my fish New Zealand & New Holland fish very SIMILAR.-- NB. Lesson method of generalizing without t
267c093 "of two kinds one white with «a» black face, & SIMILAR to those brought from Abyssinia; the others da
268c099 geographical distribution is.-- ¿Pelagic forms SIMILAR--birds??-- We must always bear in mind  proofs
271c107 Zoological  Society,, that you never find two «SIMILAR» groups of birds in two countries, without int
272c108 uliar structure & habits of clinging to rushes SIMILAR.-- The question which I more immediately  want
272c108 culionidae.-- Are there any Crysomelidae, with SIMILAR habits. But the Horae Entomologicae will  tell
276c125 ose species inhabit different countries habits  SIMILAR ¿law?-- probable.-- if habits & structure simi
276c125 ilar ¿law?-- probable. -- if habits & structure SIMILAR would have blended together Mem Mr Herberts la
282c140 yet,  as external conditions over whole world.  SIMILAR-- & constitution of man originally similar lim
282c140 ld. similar-- & constitution of man originally  SIMILAR limits of change, would be same-- Yet each fam
290c163 ow pleased it is, just like man. emotions very  SIMILAR.-- Geology. Transact. Vol V. Birds bones-- in
293c174 agious diseases, where habits of people nearly  SIMILAR. Curious instance of difference in races of me
298c192 eland, are absolutely (& who better authority)   SIMILAR with those over whole of country.-- some speci
302c200 conditions, to be sure, have remained somewhat  SIMILAR.--!!! My theory drives me to say that there ca
316c247 much  the more remarkable ∵ carboniferous ones  SIMILAR?)» Now this is very remarkable.-- (connect the
335d015 tions «(produced very quickly)» sometimes have  SIMILAR offsprings; so will the worst mules (as real m
340d026 irectly against Eyton's rule. ¿Are the hybrids  SIMILAR inter se.-- [not located] the «4» Struthionida
360d093 two  less opposed in habits, though externally  SIMILAR.-- this however is a sophism for their brain o
361d095 erons. <both> plumage of both (nursling) quite  SIMILAR.-- one species retained this character in adul
361d095 s entirely]CD In common sparrow young & female  SIMILAR plumage.-- in tree sparrow, (if I understand r
361d096 derstand rightly) young cock & hen, all nearly  SIMILAR.-- in blackbird group young like some of the s
372d129 rom one part of the body «(or of other <like>  «SIMILAR» body)» to another part of body.-- [in  plants
387d168 king at ovum of mother & ovum in offspring, as  SIMILAR to the several ova in mother. (with only diffe
390d178 r in several generations, the process would be  SIMILAR to budding. which is not object of generation.
393dIBC icially.-- Are hybrids pintail & common ducks.  SIMILAR inter se? Zoolog. Gardens Are the hybrids of t
405e032 progression  of change in Mollusca is somewhat  SIMILAR in two hemispheres.-- It might be worth invest
425e104 .-- Capital case,-- for Sandwich Isld are very  SIMILAR to Galapagos-- study Flora-- what general form
441e149 d, analogy tells us, <be> «offspring» would be  SIMILAR to <f> first form.-- The great effect of condi
442e153 » propagated by gemmation should be absolutely  SIMILAR; [all the gorze in Norway ought to be thus cha
472s01v ot & a third considerably paler, all rest very  SIMILAR-- June 2. 42 Maer <Thursday> Thursday After wa
485z016 flycatcher-- shown by habits & plumage so very  SIMILAR to some of the Fluvicolae?-- The Birds seem to
494q004 .-- (7) Are the Eggs of the Penguin Duck quite  SIMILAR to those of another Duck. ¿in Pidgeon?-- Mr. M
494q004 Is  form of globule of blood in allied species  SIMILAR.-- if not how is it in <allied> varieties  (9)
501q011 nslow Speaking of Thyme doubts about stigma in  SIMILAR manner ever failing.-- answered by Gaertner (2
507q15v ed spines-- simple spines-- or seed-cases with  SIMILAR structure.= good case as showing how simple, b
512q019 is offspring of short-horn bull & hereford cow  SIMILAR to reverse cross.-- Sow cast-up-balls of Hawks
514q21. mes of Plants found on mountains of N. America  SIMILAR to Lapland Plants --will get answer= Is pollen
528m038 auty to absolute forms or to the repetition of  SIMILAR forms as in angular leaves,-- (this Rhythmical
530m043 l on <Head> Neck.-- He has seen other cases of  SIMILAR nature.-- --like FitzRoy in sleep giving direc
530m046 on is probably a physical effect of brain the  «SIMILAR remark» thoughts, being functions of same part
570n025 y is modesty, mixed with triumphant feeling so  SIMILAR to shame after asinine.-- both accompanied  by
613o036 in  each case being the same, & the means very  SIMILAR.-- It does not appear more than saying that th
626o052 -- instinctive feelings will doubtless lead to  SIMILAR actions which in prior <races> generations led
629o055 as  temperance, acquired by education.-- CD[In  SIMILAR manner our desires become fixed to ambition. me
638j58v th mysterious consciousness superadded This is  SIMILAR idea, to cells of bee, corresponding to <every
056j111 discussing circulation of fluid nucleus,--the  SIMILARITY of Volcanic products «over whole world» argu
057r114 aving proceeded from gravel proved.-- curious  SIMILARITY of rocks of very diff. ages. at Port  Desire
179b035 Horse, Llama. &c &c-- If we <suppose> «grant»  SIMILARITY of animals in one country owing to springing
209b156 orness in exact proportion to distance (?). &  SIMILARITY of type (?).-- [«Mem:» Juan Fernandez]CD. Fr
234b261 r propagation of species-- Again is there not  SIMILARITY even in quite distinct countries in same hem
309c219 s & even to a limb not used The only cause of  SIMILARITY in individuals «we know of:», is relationsh
385d164 e youngest children, a striking <resemblance>  SIMILARITY to the temper of the father, or of the mothe
399e010 & in, produces bad effects solely, because of  SIMILARITY, because in every country, where only pair h
406e036 --  now this is never stated.-- Regarding the  SIMILARITY of offspring to Parents same laws appear  to
417e077 two as in the grafting of trees.-- ]CD CD[The  SIMILARITY of child to parent appears to follow same la
417e079 ime that the resemblance is permanent, or the  SIMILARITY at first births.-- it is the latter only tha
446e163 onstancy of form.-- is the constancy owing to  SIMILARITY of conditions-- & that no change would affec
462t051 bjects to transmut. theory, on the grounds of  SIMILARITY in condition in Java & Sumatra & dissimilari
471tf08 g to the 3 main divisions & speaking of their  SIMILARITY «in structure» he says "indeed it wd be diff
570n026 t-- my theory of blushing solves this.--» The  SIMILARITY of men's reasons: shewn by similarity of the
570n026 --» The similarity of men's reasons: shewn by  SIMILARITY of the earliest arts.-- Mem.-- Stokes-- arro
595n117 day-dreams-- think well over this;-- it shows  SIMILARITY in mind.-- think of Eyton's horses  becoming
384d161 this  so? Each part <not> of each species not  SIMILARLY subject-- ⌐Divides sexual marks into  primary
570n026 country  yet would love mountains, & a negro,  SIMILARLY treated would think negress beautiful,-- [mal
206b148 es>.-- &c &c Exclude mothers & then try this as  SIMILE In a decreasing population at any one moment f
258c060 nic Isles. Geology of whole world will turn out  SIMILE Man living in hot countries, if continually cr
044c072 (P) Is it thus fish can be traced right down to  SIMPLE organization.-- birds-- not. (P) We may Fancy,
177b026 at is species.-- In proof that structure is not  SIMPLE adaptation, armadillios «&» & Megatherium. eac
189b069 - born from beetles with wings.& modified.-- if  SIMPLE creation, surely would have been born  without
192b084 tain form in certain country, but how much more  SIMPLE, & sublime power let attraction act  according
195b101 ch is true) & infertility is consequence.-- The  SIMPLE expression of such a naturalist "splitting  up
281c137 association  in such cases recall the idea. «or  SIMPLE structure in brain people in fevers recollecti
293c172 ual» organs; for if so, separtion of sexes very  SIMPLE.-- as in plants even in same genus some dioeci
381d157 ch; & originally <her> each hermaphrodite being  SIMPLE (Are not Coniferous trees generally  dioecious
391d174 duced? (Surely all are really dioecious..) only  SIMPLE form of life are monooecious. Will ova of fish
392d175 30 distinctly states that the flipper is a mere  SIMPLE modification of an organ present in whole fish
404e025 gy between grafting & sexual union-- Looking at  SIMPLE generation as being the action of two organs i
417e077 but yet there is no «NECESSARY» tendency in the  SIMPLE animals to become complicated although all per
422e095 species from one region to another.--» -- these  SIMPLE forms perhaps oldest in world & hence most per
447e163 um. Galium Burrh ≡ simple hook; curved spines--  SIMPLE spines-- or seed-cases with similar structure.
507q15v th similar structure.= good case as showing how  SIMPLE, but beautiful adaptation might be arrived at.
532m053 ⌐ If children wink. it is instinct Fear must be  SIMPLE instinctive feeling: I have awakened in the ni
534m062 aring these cases, when agency is unknown, with  SIMPLE exertion of intellectual faculty) if ants  had
541m091 gth of time.-- Then if one endeavour to keep any  SIMPLE idea as scarlet steady before mind for period,
541m092 bstract as love, (or an emotion not so) than if  SIMPLE idea as scarlet?-- How can people dwell on pai
547m112 am No answer shows our profound ignorance in so  SIMPLE case.-- There was memory, for it related to pa
550m124 incompatible & the former, the more pleasant.--  SIMPLE happiness «as of child» is large proportion of
581n065 es tenses, &c &c-- <also g> if so & seeing how  SIMPLE an explanation it offers of radical  diversity
610o034 le or animal matter been formed by the union of  SIMPLE non-organic matter, without action of vital la
612o035 in plants, but as animals bear relation to less  SIMPLE bodies, and to more extended space, such power
621o047 ons too strong for our present interest receive  SIMPLE explanations from origin of man.-- [RHC] By in
622o049 ng there are not more, or that the three are as  SIMPLE as I have said.-- [LHC] instinctive fear of de
633j54r these  Marsh plants. All flow from some grand &  SIMPLE laws.-- 4 <Study Cuviers Anatomie Comparé» p 3
636j56v urrent series, whilst the habitation «or world»  SIMPLE series.-- My theory shows life equally  simple
636j56v » simple series-- My theory shows life equally  SIMPLE series, & therefore trace of beginning in orga
639j28v ecause their food lies deep.-- I say it is «as»  SIMPLE consequence they become long. not at once, but
567n013 is it connected with <our> the willing of the  SIMPLELST animals, as hydra towards light. being direct
293c175 which the organization of the eye, passed from  SIMPLER stage to more perfect. preserving its relation
382d157 ch view worthy of a Lamarckian.-- Mine is much  SIMPLER.-- Hunter shows almost all animals subject  to
021rIFC than the more constant: This view supposes the  SIMPLEST infusoria same since commencement of world.--
074r165 etionary action, conjoined with other» «(state  SIMPLEST case. concretions of clay iron stone; iron py
175b018 hanges. does it progress. Man gains ideas. the  SIMPLEST cannot help..-- becoming more complicated,; &
175b019 t types of organization improving as Owen says  SIMPLEST coming in & most perfect «& others» occasiona
186b057 gular trees. & extinction of forms.?? It is in  SIMPLEST case saying every species in genus  resembles
298c190 ight be mentioned in note.-- try to trace from  SIMPLEST reasoning in lower animals many times produce
```

Page
(Key Word)
371d118 ntingencies how to act-- so with the mind. the SIMPLEST transmission is direct instinct. & afterwards
372d129 -- , more than if whole branch transplanted? +.SIMPLEST forms of budding. Why does Gecko produce alwa
391d174 which typifies the whole course of change from SIMPLEST form.-- (Because by this process it separates
423e095 ent (some probably always have done so, as the SIMPLEST fish &), my answer is because, if we begin wi
423e095 &), my answer is because, if we begin with the SIMPLEST forms & suppose them to have changed, these v
423e096 classes is far from true.-- I doubt not if the SIMPLEST animals could be destroyed, the more highly o
571n031 onnection between sound & language.-- Chinese. SIMPLEST language. Much pantomimic gesture?? which wou
611o034 e some unknown relation to them-- [RHC] In the SIMPLEST forms of living beings namely «one individual
569n021 arity & prayer, or eloquent request. Reason in SIMPLETS form probably is single comparison by senses
086a003 stomed to utter confusion in Europe, that the SIMPLICITY of Ventana's «Quartz.» unmixed is very pleas
606o022 r definition than Winkleman's. who says it is SIMPLICITY with grandeur of character.-- Hence Lessings
460t019 x)» some perhaps rendered more complex & some SIMPLIFIED.-- Annals of Natural History. <no. XII.> Vol
423e096 must tend sometimes to augment & sometimes to SIMPLIFY structures:= Without enormous complexity, it
423e097 ture as it now is, it would not be possible to SIMPLIFY the organization of the different beings, (al
391d174 Man or mammalia being abortive hermaphrodite SIMPLIFYS case much; & originally <her> each hermaphrod
569n020 l. Descent of Man Understanding languages seem SIMPLITS case of Association.-- Elephant often given f
031r037 ecause lowered & raised--so on--but gradually & SIMPLY raised No Faults in Patagonia[,] enormous exte
046r078 acts) traverse granites, are granitic materials SIMPLY altered by circumstances; & not in chemical na
073r162 tterly incomprehensible that if meteoric stones SIMPLY pitched from mo«o»n, that the metals should be
105a073 course result of little inclination??----It is SIMPLY as the inclination is little the force require
127a125 inal «synclinal--» line.-- <ditto of synclinal> SIMPLY clinal lines. dipping so & so or may be used E
289c161 species. has nothing to do with hydridity,, is SIMPLY, an instinctive impulse to keep separate, whic
386d167 from inside to the outside.-- is this not owing SIMPLY to more importance of internal organs in anima
388d173 manner in which Frogs copulate & fish shows how SIMPLY instinctive the feeling of other sex being pre
568n017 is necessary notion, ass has it.-- When one is «SIMPLY» habituated «in life time» to any line of acti
587n087) I should think, great principle of liking, as SIMPLY heredetary habit.-- A blind man might be born
624o051 gument fails, or rather is weak.-- [RHC] Better SIMPLY put it, beneficial tendency in every instinct
637j57v other. ¶All Bridgewater Treatises. are reduced SIMPLY statement of productiveness, & laws of adaptat
021rIFC This view supposes the simplest infusoria same SINCE commencement of world.-- [not located] La. bill
054r105 argues, where in all appearances not many years SINCE, the sea covered above half a league of what is
090a022 eoric stone which must have fallen on the globe SINCE the Cambrian system In Ures dictionary between
111a088 6.-- Guava trees, introduced about twenty years SINCE (1835) from Norfolk Isd into Geograph Journal V
146g011 amorph» path only covering Great Slip, 10 years SINCE three hundred feet in vertical height-- enormou
156g068 lake drained could be told from «some of» those SINCE fallen. «on the 3 shelves» Solid rock is much n
157g077 y» very little back from line 2d; little action SINCE «that shelf» formed Upper terrace near Loch Spe
187b065 allied but different, because country separated SINCE time of extinct quadrupeds:-- same argument app
196b108 and creation of small animals must have gone on SINCE from parasitical nature of insects & worms.-- I
206b146 er elimanated say 150 people four hundred years SINCE were progenitors of present people, and so on b
214b175 to be descended from a breed known to be there SINCE the time of Charles.,-- and now in the possessi
219b196 that all Mammalia, were born from one stock, & SINCE distributed by such means as we can recognize,
220b198 ent.-- Henslow says. (Feb 1838) that few months SINCE in Annales des Sciences, paper on Botany of Tah
224b216 species or angels. produced «&» Has the Creator SINCE the Cambrian formations gone on creating animal
225b219 make me deny the creation of any new quadruped SINCE days of Didelphis in Stonefield:. all lands unit
267c093 age round world, 20 years have scarcely elapsed SINCE the Guava introduced from Norfolk Isld-- "& it
278c131 eeth of seals and dugong, therefore immense age SINCE breccia accumulated-- surely ask Owen to see wh
308c216 nimals have never at any one time formed chain, SINCE if crteceous period assumed, then some perished
308c219 t will show state of knowledge «Negroes existed SINCE time of earliest Egyptian drawings & Old Testam
315c240 .-- pooh. May have been Created at many spots & SINCE disseminated See. Habits of Malay fowls p 5. (n
317c251 rt II/. 1837 account of the various hares «some SINCE discovered» of N. America, & of the shrews.-- D
332d002 e saw many cases of virulent cancer in women, & SINCE that time it has been rare disease.-- but now (
332d003 alike.-- My Father Water-in the hair a century SINCE used to be called Worm Fever, as used much more
343d037 nature) & the Rhinoceros of Java & Sumatra, that SINCE the time of the Silurian, he has made a long su
356d069 ed at them quite independently & have used them SINCE) the line of proof & reducing fact to law only
357d073 landed as at present in New Ireland & continent SINCE grown.-- This will explain. S. American case &
397e003 almost! invariably according to fixed laws: And SINCE the world began, the causes of population & dep
414e060 ressive --!Agassiz makes it wonderfully changed, SINCE Cretaceous period, whether progressive I know n
415e066 y be confined in locality like Ornithorhyncus,: SINCE being cosmopolite, we do find his remains.-- Li
424e102 now this is independent good case, but very odd SINCE these crows are mixed in England-- for I presum
436e135 lthough the Cephalopods, seem to have decreased SINCE earliest times-- Apterix has a most perfect Str
443e156 it would be doing, what others but fifty years SINCE to geologists.-- & what is older-- what relatio
460t019 res.-- Shells in Cambrian & Crust show how long SINCE present forms existed, but if it be asked how t
472s01r species-- Jun 1. 1842 Allen W. sowed some years SINCE gathered the seeds of Papaver bracteatum, & the
472s01v le to raise them pure for Miss Bent three years SINCE gave her some She means to try this year. Littl
511q018 ve No or did result appear without his wish Has SINCE recrossed this breed.-- Have secondary male cha
569n023 elago. Dn 75 cf., p 268 without, however, very SINCERE grief-- "there is nothing more to be said."--
524m022 s not of their husbands-- My family's test of SINCERITY.-- People in old age. exceedingly sharp in so
319c256 high.-- Upon the whole thinks <many> more birds SING in England than in America, but the few of N. A
363d102 -- thus Yarrel has Lark & Nightingale which both SING their own songs, though imperfectly.-- Male bir
364d103 ird fanciers match their birds to see which will SING longest, & they in evident rivalry sing against
364d103 ich will sing longest, & they in evident rivalry SING against each other, till it has been known one
436e138 placed near, but not in sight of each other will SING till they drop off their perch.-- p. 101-- King
446e160 p. XXIII. <Some birds> Both sexes of some birds SING equally well. and <in> these reciprocally assis
532m055 d think that in old people, in their dotage, who SING the songs «& tales» of infancy, it is very doub
533m058 at brain alters) It is known that birds learn to SING & do not acquire it instinctively. may not this
450e174 tices of the Indian Archipelago" Published at SINGAPORE in 1837. by Mr. J. H. Moor---- p. 1. Elephan
450e175 L. Institut «1837. No 246» a section of fossil "SINGE", it cannot be made to approach the Colobes whi
194N094 SINGAPORE 1837. By J. H. ? do. p. 189. «19D» No full si
600o008 determine which most beautiful cock, which best SINGER-- Remember.-- avarice a compounded passion gai
319c256 uite as beautiful. The thrushes of N. America. SINGING so well. & the mocking thrush being so very be
363d102 perhaps, not being left open to them,-- ¶. In SINGING birds, part instinctive & part acquired,-- thu
367d113 acable-- (Mem Lucanus & Copris &c).-- In Birds SINGING of cocks settle point.-- (do the females then
368d114 ght for male) & are merely most attracted). -- SINGING best sign of most vigorous males.-- «(NB. most
486z019 See p. 256 of Note Book (C) for comparison of SINGING powers of birds of N. America & Europe. Entomo
486z020 . Fuego & Iceland Spix & Martius talk of birds SINGING in the forests of Brazil H. Wedgwood says in <
521m008 memory of the tune, might be compared to birds SINGING, or some instinctive <or> sounds.-- Miss C. me
527m031 ding to law. How strange <all> «so many» birds SINGING in England, in Tierra del Fuego not one.-- now
527m032 ent, so may they learn in a state of nature.-- SINGING of birds, not being instinctive, is heredetary
527m033 pt up by the music of the poetry.-- (therefore SINGING intermediate, who has not had his blood run co
527m033 mediate, who has not had his blood run cold by SINGING).-- Granny says she never builds castles in th
538m078 which I think will lead to fact of old people SINGING songs of their childhood. & certainly of Miss
568n018 on Octob. 19th. Did our language commence with SINGING-- is this origin of our pleasure in music-- do
568n018 chirp in do-- union of birds voice & taste for SINGING with Mammalian structure. «-- American monkeys
528m039 d pleasure association warmth, exercise, birds SINGING.-- 4th. Pleasure of imagination, which corres
047r083 vice versâ more likely to be coincidental than SINGLE elevations along whole line of coast Darby men
090a022 are recorded as falling; many of these were not SINGLE, but are described as many, (one even 3000) Th
206b149 ould converge sooner) & lastly perhaps some one SINGLE one.-- Will not this account for the odd gener
284c149 organs vary in different group. & Not known in SINGLE ones--. viz. Macleay letter to Fleming p. 32 "
365d105 asant & Black game is owing to their rarity. a SINGLE female in wood with Pheasants would sure to be
370d116 said that some kind lay <pu> up honey even for SINGLE rainy day-- & for case of wasps, is supposed
410e052 ll be done to all countries,-- but naming mere «SINGLE specimens in» skins worse than useless.-- yet
418e083 these very animals are descended from some one SINGLE stock,--one is led to suspect that the birth o
423e098 e for cauliflower &c.-- Uncle John believes one SINGLE turnip in a garden is sufficent to spoil a bed
425e105 thern flanks of Alps.-- many peculiar plants on SINGLE mountains, though these are connected with oth
435e183 ised in same garden.-- now this good question-- SINGLE, or half double.-- anyhow fertile because they
491q01v experiment.--«to repeat Spallanzani» Raise only SINGLE Plants & only allow <few> one flower (5) Dr Fl
496q05a Examine pollen of double flowers. compared with SINGLE & see whether grains flaccid, as Koelreuter de
499q010 some insect-- (21) Are there many instances of SINGLE clumps of plants in counties, as of rare green
501q011 ry minute insects.-- (30) Get Abberley to plant SINGLE Peas, Kidney Bean & Bean, intertwined, «withou

Page ***(Key Word)***
507q15v odcut-- 1 double hook-- -- Geum. Galium Burrh ≡ SINGLE hook; curved spines-- simple spines-- or seed-
528m038 is, or sea weed, &c &c-- this gives beauty to a SINGLE tree,-- & the leaves of the foreground either
528m038 ldt from occurrence in Mexican & Graecian to be SINGLE cause) this symmetry & rhythm applies to the v
537m077 pposed undoubted cases of heredetary pride & in SINGLE families. Edinburgh. Phil. Transact. p. 365. C
569n021 nt request. Reason in simplets form probably is SINGLE comparison by senses of any two objects-- they
318c254 ems common «kind» migration of America) migrate SINGLY flying few miles every day «generally by night
318c254 other birds which is strictly diurnal, migrates SINGLY by night.-- others in flock, these birds seem
021r005 a, nodules in Clay Slate. major axis 2.½ ft.-- SINGULAR structure of nodule, constitution «same as» o
053r101 l lines, but the rectangular intersections are SINGULAR-- M. Lesson considers the Sandstone & Granite
292c170 character of the demonstrations offered of the SINGULAR views there offered, & he must be a zealous m
305c209 same remarkable manner as Europaean species = SINGULAR coincidence if distinct creation.-- ie.-- a m
359d087 ays influence the sexual organs alone.-- It is SINGULAR pheasant & fowl being so totally infertile wh
380d154 f old female becoming like cocks.-- It is very SINGULAR. so many Gallinaceous birds have cock & hen p
387d168 Billin.--   <Generation--> V. p. 152 It is very SINGULAR the same difference from parental stock havin
443e155 uence in that degree of developement.-- [It is SINGULAR there is no true hermaphrodite in beings with
451e180 r-land-- monkeys do not exist. there & it is a SINGULAR thing that throughout the Moluccas Archipelag
548m117 ious of the two states.-- August 30th.-- It is SINGULAR when looking at a table one has vague idea so
028r031 oluble rocks? Has Chalk ever been dissolved? SINGULARITY of fresh water at Iquiqui. not from rain, be
557m149 ent, hence shy people (shame of ridicule) are SINGULARLY apt to catch tricks.-- so are people in pass
574n039 s numerous lines. of poetry.-- <signs> sounds SINGULARLY adapted to subject see <A> ¶ I think this ar
032r039 been worn away.--If the level of the sea was to SINK by very slow & gradual movements to line (2). T
035r048 , which in size are grains of sand, in this view SINK into their proper insignificance; as fractures,
633j54r t» the valuable soil in its seaward course.-- we SINK into such contemptible queries, as why should t
106a076 ied by Volcanos must be sought after proofs of SINKING.-- No Sweden!! swelling of rock from Heat. Spe
108a079 d be thicker.-- PP Andes mark the line between SINKING & rising areas.-- In Earthquake if Subsidence
245c025 ikely to <M> have different species than those SINKING, because arrival of any one plant might make c
087a011 l land.-- Siberia rises. therefore to the South SINKS.---- Meditteranean continent corresponding to E
530m045 ctors care it, by stimulus & afterwards patient SINKS.-- When a muscle is moved very often, the motio
410e052 «I may say all this, having myself aided in such SINS» (do not add name, without reference to descrip
533m060 or remarks, that writing down his confessions of SINS. did not make him more humble.-- it has obscure
608o028 ly opposed & inexplicable if we suppose that the SINS of a man, are under his control, & that a futur
435e133 translated by Dr. Hodgkins» p. 54. The axolotl, SIREN, & Proteus, affinity to tadpoles. p. 210. Shows
312c233 ome impregnated & puppies delicate-- they cross SISTER & brother of same litter, those. of different
391d175 «but not great» difference (for even brother & SISTER are somewhat different) should be added to eac
451e176 Both his parents were of the usual colour. His SISTER is an albino like himself said not to be commo
523m014 ous story of the Birmingham Doctor praising his SISTER who confined him. & yet disinheriting her.-- T
181b040 a man from.» any large family of 12 brothers & SISTERS «in a state which does not increase» it will b
334d011 customed to her.-- case parallel to brothers & SISTERS in Mankind.-- The case of all blue eyed cats (
374d134 inter se, the first cross generally brothers & SISTERS, & therefore somewhat unfavourable-- 28th. «I
460t015 these <half> «3/4» bred ones-- The brothers & SISTERS half-breed showed no sexual inclination for ea
380d154 nea Fowl cannot be distinguished.-- A capon will SIT upon eggs, as well as, & often better than a fem
385d164 owls.-- eggs fertile, but parent bird will never SIT on them.-- May be just worth remembering that ov
502q11v wl. Hen coloured like cock-pheasant: said not to SIT on own eggs Flowers in short turf. for abortion.
153g056 ually resting on them on summit of hill rounded, SITE N N W of Ben Erin {P} Shelf of Glen Guoy flat p
446e160 ly assist in domestic cares, as building nest, SITTING on eggs. & feeding & defending their young..--
554m141 male Black Swan is very fierce when female is SITTING the Keeper is obliged to go in with a stick, i
146g013 instance. how accidental is the preservation in SITU of even imperishable pebbles/ I am nearly certa
149g032 ebbles of quartz & other rocks not apparently in SITU <& in> hill being gneiss <& also> also near sum
037r052 XI Vol II. Urge the entire absence of any rock SITUATED beneath low water in the Southern ocean not b
384d161 fication of Monsters. (1) From praeternatural SITUATION of parts (2) addition of parts, (3) deficienc
498q008 o any of same species of Willows grow in same SITUATION & flower at same time. Has H. seen group of d
546m108 at others think by putting ourselves in their SITUATION, & then we feel like them--. hence sympathy v
185b055 ant structure to accomodate itself to as many SITUATIONS as possible.-- Why should we have in open co
198b113 imits, every tribe appears fitted for as many SITUATIONS as possible. for instance take birds animals
235b272 Falconer's remarks on influence of climates, SITUATIONS &c on 242. Hook Smellie Philos of Zoolog. 84
465t079 ind that enormous periods may elapse, even in SITUATIONS apparently favourable for the preservation o
633j54r s all organic beings perfectly adapted to all SITUATIONS, where in accordance to certain laws they ca
458t001 a--» These strange forms., camels, giraffes. SIVATHERIUM & Anoplotherium, with existing, or nearly ex
362d100 now this would be most curious to show that in SIXTY years-- (how many generations) the strangest ge
032r040 o the greater or less time of rest. so would the SIZE of the triangular mass removed vary.--The gradu
035r048 to short breathed traveller» Mountains, which in SIZE are grains of sand, in this view sink into thei
044r073 general reflections might be introduced on great SIZE of ocean; especially Pacifick: insignificant is
090a022 includes all actually counted.-- The weight «or SIZE» is given of 25 stones.-- The total weight reco
101a058 ike & mountain axis. except in relative <strata> SIZE with superincumbent strata. where they have yie
115a098 m.-- All this depending not on absolute <force> «SIZE of» of <currents» «fragments» but relative to c
129a131 urnal Vol XXI. p. 213. Beyond the limits of Alps SIZE of boulders sorted: ditto Murchisons case.--¿ d
152g048 , must have remained very long at 4th shelf from SIZE of buttresses, to upper edge of which they cut
205b145 ls. for in <the> proportion to their increase of SIZE they become worse in form, less hardy, & more l
234b256 s certain there are local varieties «of colour & SIZE, but not for[m]» of animals.-- He says Stephens
246c025 erent]CD.-- p. 414. dogs of New Zealand of large SIZE, resemble, chien-loup.--long, black & white, ea
266c092 lly marked like those of the Alderney breed, but SIZE not larger than those of Black cattle. Not have
275c120 old variety» & then recrossing off spring. till SIZE diminished, but feathers continued by picking c
303c204 that isld.-- The races of men differ chiefly in <SIZE> colour, form of head «& features» (hence intel
305c210 Terri.-- certainly more like Jackall in gait, SIZE, fur.; manner in which ears droop like dog olde
312c231 kind wanted, colouring of each feather weight & SIZE & they would produce number agreeing almost to
336d018 t be like Hence mutilations not heredetary, but SIZE of particular Muscles-- When two animals cross.
381d156 fect.-- Directly a Capon is cut, it increases in SIZE prodigiously-- Animal OEconomy by, Hunter. (edi
413e059 individuals, & <individuals» <species of large SIZE,-- consider this (Cetacea) with reference to m
431e123 icult.-- Cline on the breeding of Animals, p. 8. SIZE of foetus in proportion to male parent p. 8. hi
431e123 being smaller, & the female larger than average SIZE: (surely this is very limited view, though perh
435e133 elopement of form; but that tadpole increased in SIZE now the Proteus anguiformis. he remarks lives i
458t001 gantic-- hyaena-- bear & ruminants all of larger SIZE.-- the law of large size established-- «Austral
458t001 ruminants all of larger size.-- the law of large SIZE established-- «Australia, S. America»-- These
480z007 n old ones-- in America & sexes not of different SIZE-- How does this apply to pale brown Caracara Kr
508q016 opolis of each-- Cause?-- Andrew Smith. (6) What SIZE book Gallesio storia del Reproduzione.-- D. Hol
586n082 robably is» instinctive, namely the knowledge of SIZE is merely judged by eye, & use of limbs &c, or
594n112 o bed of flags. hernes are common. not unlike in SIZE in the air at a distance.-- How can such an ins
635j56r (peculiar to each species) owing to the growing SIZE of the world? & the physical changes it was to
641j29r to terrestrial Mammifers-- Marine ones «of large SIZE» <to> are best nourished by arctic regions-- Wh
385d163 considered the most valuable. because smallest SIZED dogs.-- one litter big & then second small & so
450e175 pore 1837. By J. H. ? do. p. 189. «190» No full SIZED horse is found East of y Burramposter & S of Tr
223b208 als in mind, more than in structures.-- If the SKELETON of a Negro-- had been found what would Anatom
255c052 in an abstract idea of a bird-- An animal with SKELETON of such general forms.-- The hybridity of fer
493q003 men of Hybrid animal. in comparison with Weigh SKELETON of Tame Duck & Wild Duck, & then weigh their
511q018 bbit. Descript. of colour «& length of ears» & SKELETON, & skin= Van. Voorst often writes to Lowe (7)
513q020 hinese or Penguin Duck in very young state for SKELETON== Does the tumbling of pigeons vary in manner
633j54v atomie Compare» p 308. Traces the gradation of SKELETON in Vertebrates & constantly alludes «(& at p.
211b162 ccount of three, kinds of pigs. difference in SKELETONS: VERY GOOD. Apteryx. a good instance probably
035r046 trict there seems. I think, little ground for SKEPTICISM, as to the general truth of the proposition.
269c102 .-- Geograph Journal. vol I. p. 17 &c excellent SKETCH of plants of New Holland, supplementary to App
314c238 orfolkice. 1833 Steph. Endlicker (He will give SKETCH of botany of islands of south seas says so in
326c266 published in first volume of Annales of Vienna. SKETCH of south Sea. Botany R. Brown. has curious col
526m029 mpted to think all memory consists in a set of SKETCHES. some real-- some fancied.-- this fact of ear
505q014 17 A dead-nettle in Hot-house. will it seed?-- (SKIM through Penny Cyclopaedia) Abberley says that s
320c275 any & Continuation «Annals of Natural History» SKIMMED Von Buch Travels Whites Natural History of Sel
321c275 e References at end Dr. Lang Australia «trash» SKIMMED Macleay's Horae Entomologica Ray's Wisdom of G
322c269 ssy to China. Oct. 12t Kotzebue's two voyages, SKIMMED well. do Lutke's Voyage. carefully read.-- Rey
322c269 yo Pathology of the Human. Mind Evelyns Sylva. SKIMMED, stupid Brownes travel in Africa; «well skimme

(Key Word)
322c269 kimmed, stupid Brownes travel in Africa; «well SKIMMED.» 1839 Jan 10t.-- All life of W. Scott.--, exc
323c269 ds of <ter> influence of Physical causes: well SKIMMED Bartrams Travels in N America May 18th Stanley
323c269 ewater Treatise -- Wilkinsons Egyptian remains SKIMMED -- Pliny Nat. Hist of World do -- Lamarck. II
477zIFC poultry with same down-feathers. Zoology 1856 SKIMMED through & abstracted Zoology Some excellent me
402e019 , buffaloes &c &c-- Malte Brun. would be worth SKIMMING over with regard to this archipelago Octob. 1
332d005 said to come from coast of Guinea, -- ears bare. SKIN black & wrinkled-- fur short. (tail cut off in
362d099 of the doe to the victorious stag. who rubs the SKIN off horns to fight-- is analogous to the love o
362d099 airy ones. which never «dry up &» peel off their SKIN (not being wanted for war) & hence never fall o
372d131 rful. it is part of same class of facts that the SKIN grows over a wound.-- Does likeness of twin bea
386d167 ch makes two animals out of one & heals piece of SKIN.-- if the tail knows how to make a head. & head
403e022 mations are particularly common.-- without arm, <SKIN> hands thumb,-- one leg, hare lip &c &c. in Vol
445e159 of flying mammifers-- says lemur.-- volans, has SKIN between its legs.-- -- strangely consider exist
511q018 ript. of colour «& length of ears» & skeleton, & SKIN-- Van. Voorst often writes to Lowe (7) In breedi
542m095 ision.-- Does the contraction & wrinkling of the SKIN contract iris?-- same way as one lifts up eyebr
545m105 ocephalus Sphynx Linnaeus») constantly moved the SKIN of forehead over eyes, at every emotion & <look
545m106 ve «any» distinct wrinkle, but such movements in SKIN of eyebrow important analogy with man.-- I see
555m143 e the old world ones.-- Though the[y] move whole SKIN of head they do not move eyebrows.-- (I see som
555m143 ebrows.-- (I see some of the old world ones move SKIN of head & ears,-- ∴ some men have this power ab
564n006 erated habitual sneer» the manner in which whole SKIN or muscles are contracted between eyes & upper
570n025 ompanied by depending head., & active vessels of SKIN.-- What difference is there between Squib after
579n060 of blushing-- in former irritation on a piece of SKIN cut off made the blush come.-- it is an excitem
591n097 escribes effects of emotions-- fear giving goose SKIN-- & hair standing on end.-- July 20th Intellige
596n184 l observations Ourang do not move eyebrows.-- or SKIN of head,-- «scarcely able St.-- » Cyanocephalus
596n184 tion towards man.» Macacus especially pulls back SKIN of whole forehead & 2 ears.-- emotions of every
596n184 ake labial st st. S. American monkeys. pull back SKIN from head very little Does blood go in <body> f
338d024 tation & species-like,-- -- Says Negro-- thick SKINNED My hairdresser (Willis) says <black> «that str
402e019 M» p. 359. At Manilla a small Cercopithecus., & SKINS of galiopithecus.-- Malte Brun. Vol <I> II p.,l
410e052 ntries,-- but naming mere «single specimens in» SKINS worse than useless.-- yet there is no cure «I'm
541m089 dache) & found best plan was allowing my mind to SKIP from subject to subject as quick as it chose.--
536m070 the same; I noticed this by perceiving myself SKIPPING when wanting not to feel angry-- such efforts
429e116 e northern regions-- do p. 21 says. many plants SKIRT each side of the great N & S. valleys, which pe
575n044 to hunt hares. «& leave the sheep» & jackall to SKULK about & hunt mice-- Jenners Jackall Have we som
582n066 ve an uncommon pleasure in hiding themselves & SKULKING about in shrubbery. when other people are abo
253c047 Shee. considered merely variety.-- red form of SKULL very slightly different.-- Q Zoolog. T. V. I. p
463t057 ing into other is lost, <be> (as vertebrae into SKULL., two bones of tibia into one.--) because if th
240c003 ys of two varieties of pidgeon, although having SKULLS so different, that they would be called genera
275c121 riety.-- The pidgeons which have such different SKULLS, but same marks on wings are Blue Pouters & sm
414e064 llect in Australia to the white.-- The peculiar SKULLS of the men on the plains of Bolivia-- strictly
452e182 634, alludes to fact stated by M. Tournal that SKULLS found near Vienna appoximat to Negro form; tho
481z009 ralia-- very curious.-- replaced by didelphidae SKUNK inhabitant of Patagonia. Mem:-- S. Cruz. Molina
137a149 rom furnaces do/p. 84 on the effects of veins of SLAG in iron furnaces affecting to some distance & b
021r005 I 287» P 379. Henslow Anglesea, nodules in Clay SLATE. major axis 2.½ ft.-- singular structure of nod
021r005 structure of nodule, constitution «same as» of SLATE same.--longer axis in line of Cleavage. laminae
022r006 s mineral in Australia. Fitton's appendix Would SLATE. & unstratified rocks show any difference in fa
069r150 Portillo line--connected with <gneiss>.--(Mica SLATE. [Fig. 9] {P} ((3) like Bell of Quillota.) (A)
075r166 n Potosi the silver is contained in a primitive SLATE, covered by a clayey porphyry, containing grena
076r170 er in S. America. Geology of Guanuaxuato.--Clay SLATE. passing into talcose & chloritic slate. with b
076r170 .--Clay slate. passing into talcose & chloritic SLATE. with beds of syenite & <sep> serpentine dippin
077r171 ther secondary» rocks. Vein traverses both Clay SLATE, Porphyry North 52 W, & is nearly the same with
078r176 par".-- p. 215 Same metal in Tasco vein in Mica SLATE & overlying Limestone Balls of Silver ore occur
088a015 nt of American Volcanic action. -- Fragments of SLATE converted into crystals of Hornblende p. 248. L
091a027 Campbell the Poet» Accra. Coast of Africa. Clay SLATE & Quartz. strike SSW & NNE dip 30 degree - 80 d
091a027 ills & strata SE. direction of transitions clay SLATE &c nearly vertical Linear earthquake 500 by 90.
097a042 Institut No degree 221 Lamellar dikes like Mica SLATE Von. Buch. Canary Isd. p 170.-- Mem. Cordillera
104a068 was astonished with him that <th> gneiss, mica-SLATE of whole kingdom of Norway was contorted yet no
110a087 most probably from the gneiss beds in the mica SLATE.-- Geograph. Journal. Vol IV (p 321) Mr Hillhou
114a094 e alternating. bear on subject of cleavage Clay SLATE. a distinct formation deep «& therefore extensi
114a094 econdary in Europe. gneiss-- metamorphose clay SLATE.-- --shale in shall sea. Lyell confounds these
115a097 eneral absence of fragments «& pebbles» in mica SLATE & gneiss, can only (see <supra» p. 94) be accoun
134a141 oft Clay beds near C. Virgin p. 59. dip of Clay SLATE in T del Fuego Admiralty Sound. SE dip. much p.
148g024 ace of them-- Mem Coast of Chile--¿ is not Mica SLATE too hard & uneven to be impressed Case of Birch
157g073 no pebbles in parts of Beagle Channel when mica SLATE, only sand blow away]CD where lines appear to c
192b087 -- As we have one Marsupial animal in Stonefied SLATE, the father of all Mammalia in ages long gone p
200b121 on branching off.? Accra. Coast of Africa. Clay SLATE. strike. SSW. & NNE. dip 30 degree - 80 degree
399e011 osoma. (very like American form) in Stonesfield SLATE., & a Melolonittha-- In marl from «Lake» Consta
028r029 n.--This matter accumulating in deep seas forms SLATES: How is the Lime separated; is it washed from
049r087 veins, there contemp--yet similar ones in Clay. SLATES contemporaneous others subsequent. as in dikes
051r095 ause it does not decompose. or vice versâ. Clay SLATES unfavourable to attachment of many bodies [bla
054r102 t Conception, cleavage E & W! at Payta. talcose SLATES, do at latter place. sandy. sandstone with gyp
114a095 ll period.-- I cannot help suspecting that clay-SLATES have been more frequently metamorphosed than o
115a096 hemical action which gives polarity to atoms in SLATES that cleave, & which unite the homogenious cry
156g068 ll on these three shelves soil is <the> usually SLATY Point of rounded not scooped rock on <bend> of
228b231 e do not like to consider our equals.-- «Do not SLAVE holders wish to make the black man other kind?»
286c154 lates every best instinctive feeling» by making SLAVE of his fellow black, often wished to consider h
343d038 f the interior are pushing into each other from SLAVE trade, & colonization of S. Africa, so must the
540m087 shape & ideosyncracy.-- Look at the Indian in SLAVERY & look at the Negro-- look at them both savage
228b231 of two plants. Animals-- whom we have made our SLAVES we do not like to consider our equals.-- «Do n
228b232 pain, disease death & suffering «& famine»; our SLAVES in the most laborious work, our companion in o
292c171 ness «a most possible thing. see men walking in SLEEP».-- an action becomes habitual is probably firs
293c172 after long intervals of forgetfulness.-- after SLEEP, «strong» analogies with memory in offspring.--
305c211 --Animal Magnetism-- principles of irritations SLEEP walking. fits, laught &c&c Man & Man may have s
324c268 chet Memoires sur les Vegetaux et animaux.-- on SLEEP & movements of Plant/ I£:4s Voyage aux terres a
347d049 e he says length of days adapted to duration of SLEEP of man.!!! whole universe so adapted!!! & not m
494q005 he stamina of C. Speciossissimus collapse during SLEEP & do of Berberis-- (latter I think certainly no
495q05a wild a The Leptosiphon densifolium «an annual» <SLEEP> «closes flower» on all gloomy days.-- The «gar
530m043 or cases of similar nature.-- --like FitzRoy in SLEEP giving directions,-- & forgetfulness after bad
530m045 seems to rest the sensorium.-- --- analogous to SLEEP--; some doctors care it, by stimulus & afterwar
530m046 nvoluntary.-- that is involuntary memory, as in SLEEP.-- a new thought arises?? compounded of the inv
532m056 , would not remain quiet in any room, would not SLEEP at night even when in bed room-- grew very thin
542m095 g does.-- is yawning habitual from awaking from SLEEP see how a dog yawns when he awakes. & streching
544m102 oes he account for dogs & men speaking in their SLEEP.-- Characters of dreams no surprise, at the vio
547m111 t of packing up.-- was lying on my back fell to SLEEP for second & wakened.-- had very clear & pretty
547m112 to be done.-- The senses are closed probably by SLEEP & not vica versa. anyhow I might have been quit
547m113 e deception to say the mind <thinks> quicker in SLEEP, it may do less work & yet do so, from the exer
548m114 ng each step as in reasoning-- hence delirium & SLEEP mental rest. though. most vivid & rapid thought
548m114 about, facts gained or gaining by senses.-- As SLEEP <is> only one idea is awake, when one is awake
550m127 tions?-- Dr Holland remarked that insanity like SLEEP does not doubt the reality of the impression on
552m130 e Shuts ones eyes) is the image not vivid as in SLEEP-- (one can dream of intense scarlet??) is it be
586n081 ver subject to volition.-- like plants going to SLEEP.-- "A bird has the faculty of finding its way,
601o009 ensation No. 2. with memory added to it, man in SLEEP not conscious, nor child-- Evidence of consciou
602o10v ed statues In Elliotson's Physiology much about SLEEP-- Nerves.-- Volition & Reynold XIII Discourse
613o035 l this can take place & man not conscious as in SLEEP; or in sleep is man momentarily conscious, but
613o035 ke place & man not conscious as in sleep; or in SLEEP is man momentarily conscious, but is memory gon
455eIBC es-- but these are same as trees.-- Shake some SLEEPING mimosa-- do stamina of C. Speciosus. collapse
455eIBC So doing can be made sensitive The function of SLEEPING someway useful.-- it is only the association
494q005 any effect? and silk caterpillars (1) Shake a SLEEPING mimosa, or half bred mimosa (a) between sensi
494q005 a, or half bred mimosa (a) between sensitive & SLEEPING species, & see whether association can be giv
601o009 s» perhaps polypi-- (so that lower animals are SLEEPING higher animals & not plants as supposed by Bu

```
615o36v ence of Tropical plants when imported & plants SLEEPING good show acquirement or obliteration of inst
495q05a all  gloomy days.-- The «garden» Coronella also SLEEPS on ditto-- Cover them up periodically & see ef
454e184 Hort.  Transat. V. II p. 252. Is there any very SLEEPY Mimosa, nearly allied to the Sensitive Plant.-
258c061 eing given to the chance offspring who have any SLIGHT peculiarity of structure. «hence seals take vi
286c156 traces of passage a difficulty, but after all a SLIGHT one It will be necessary from manner Fleming t
288c160 y» Magazine of Zoology & Botany No XI p. 390. a SLIGHT change in enclosing a common seems in part of-
294c177 broadly scarcely any novelty in my theory, only SLIGHT differences. «the opinion of many people in co
335d015 he worst mules (as real mule) have offspring,-- SLIGHT deformities «as supernumerary fingers» (that i
335d015 deformities «as supernumerary fingers» (that is SLIGHT alterations of primitive stocks «relative to c
335d015 ») & hybrids between very near species (that is SLIGHT alterations of primitive stock) are heredetary
363d102 f some «same» North American & Europaean birds «SLIGHT» different-- Barn Owl <the> in the former plac
387d168 in A, «with some peculiarity» that in (B) to <a SLIGHT» «some» degree, & likewise ovum in (B) <an  C>
478z003 8 Fleas only appear in winter in Paraguay p 207 SLIGHT notice on habit of Iguana. not pass Lat. 28 de
522m012 mental» failure very general effect of <early» «SLIGHT» habitual» intemperance.-- often accompanied b
557m148 neself it is quite different from that.-- like «SLIGHT» passion from blood rushing in face, with less
565n009 sion? Expression of affection is accompanied by SLIGHT protrusion of lips, as if going to say "my dea
578n057 y association, there pour out tears, & there is SLIGHT convulsive wrinkling of some of the muscles «o
581n064 f soft, silk-handkerchief-- cats & dogs fond of SLIGHT tickling sensation.-- in savages other  tastes
581n066 if  the emotion was not first felt?-- «without «SLIGHT» flush, acceleration of pulse. or rigidity of
583n070 may be called "creatures of instinct" with some SLIGHT dash of reason so mean are called "creatures o
596nIBC ng state would it then know its direction.-- In SLIGHT convulsions. are the muscles of the face first
619o043 vented by force he would feel pain. [.By a very SLIGHT change in association if others injured  these
194b097 h «hermaphrodite» shells.!!!? We have not the SLIGHTEST right to say there never was common progenito
447e165 al May 28th.-- -- Henslow says he has not the SLIGHTEST doubt that Festuca vivapara is the same speci
045r077 eve the fluid matter instead of afflux (always SLIGHTLY oscillating as that of a spring) moves away.-
075r167 red up by volcanic rocks. //St Helena has been SLIGHTLY broken up, & has there not been vein «of iron
156g071 his point plain appears like one uniform slope SLIGHTLY bending up each main valley.-- & that river a
172g007 different  if kept long enough.-- «apart, with SLIGHTLY different circumstances.--» Now Galapagos Tort
188b069 ith says very. close species generally frequet SLIGHTLY different localities, so that they become use
223b211 form  good species! The increased fertility of SLIGHTLY different species & intermediate character of
246c026 harper., & rather longer in proportion, colour SLIGHTLY different. Who can say whether species & vari
253c047 dered merely variety.-- red form of skull very SLIGHTLY different.-- Q Zoolog. T. V. I. p 389. Owen r
264c081 - or structure may be obtaining, whilst habits SLIGHTLY preceed them-- From this view habits must for
391d175 this agrees well with my view of those «forms» SLIGHTLY favoured, getting the upper hand. <]CD> & for
423e099 0 years) which is sufficient only to have most SLIGHTLY modified organic forms.-- we know not rate of
434e129 ioica, generally dioicous. yet parts only very SLIGHTLY abortive & bed of female flowers will sometim
460t019 r present organic beings are the descendants, <SLIGHTLY» «a good deal» modified <& Many Forms lost; i
522m014 ling, thinking over somebody who has, perhaps, SLIGHTLY injured me, plotting speeches, yet with a sor
532m054 e feeling: I have awakened in the night. being SLIGHTLY unwell & felt so much afraid though my reason
558m152 ink they arch their backs-- Bengal tiger. when SLIGHTLY angry. curls tip of tail.-- do two cats  arch
615o038 tahiati; fear of death in Hindoo population.-- SLIGHTLY modified in many countries, hence national ch
636j57r every  respect, except «in» quick movements. (SLIMINESS instead of barbs)-- In all these cases it sho
035r048 of injection of fluid rock.-- Try on globe. with SLIP paper a gradually curved enlargement see its in
146g011 .-- «of hard metamorph» path only covering Great SLIP, 10 years since three hundred feet in  vertical
336d019 e offspring-- On the idea of generation being a <SLIP> «bud» from parent. if whole parent not entirel
538m079 igious sects.-- yet Allen. W. remark about his SLIPPERS bad for fires, what is wrong in his head. & B
467t103 en, or pollen comes out with anthers & stigma in SLIT» -- As I think they do in Broom & certainly whe
032r039 xposed to it, & the result would [be] a uniform SLOPE to base of cliff (Z). to which point the waves
042r069 n the Andes, would have destroyed regularity of SLOPE of valleys.--All my observations of period «& m
045r075 discovered;  this may be mentioned with general SLOPE of the country; (perhaps generally over whole w
052r097 m; as long as coast line fixed.--[Fig. 5] {P} * SLOPE necessary for seaward transportal of drift matt
055r108 t wonder to see tertiary plains consumed) Where SLOPE «plainly» indicates former boundary. (as in oth
078r175 t metal, where the direction of ravins, and the SLOPE of the mountains (flaqueza del cerro) have been
115a098 of  every degree of fineness. then most regular SLOPE-- {P} if not course enough flat top. ended by a
115a098 if  not course enough flat top. ended by abrupt SLOPE {P} each stratum would thin out, both inland  &
149g031 h have lately acquired their peculiarities? The SLOPE of A & B regular & even towards {P} Spean Roy d
150g038 arthy than what is usual-- Lines die away where SLOPE less., best developed on steep earthy slope, tw
151g039 ere slope less., best developed on steep earthy SLOPE, two circumstances rarely united.-- die away al
151g043 et enters two great projecting butresses, upper SLOPE of which corresponds to shelf the truncation  &
156g071 From  this point plain appears like one uniform SLOPE slightly bending up each main valley.-- & that
156g072 d. 29.455 A 83 degree ∴ plain of 4 minute shelf SLOPE, above «line of 4th» shelf This shelf at head w
157g078 9 degree? A little below Divortium on  SLOPE towards Loch Spey 29.297 A 79.½ 29.316 divortiu
159g086 e Air 70 degree? This station a little way down SLOPE of obscure terraces (& conical hills on same) o
159g086 es-- «which river could not have deposited» the SLOPE is continued some hundred feet lower & begins a
159g087 re however fringes of alluvium (?) still higher SLOPE of valley much more gentle than in Glen Roy,  &
160g093 r perfectly level, <on op> dies away on gradual SLOPE-- : on N side.. dies away on rocky place, but n
161g095 level with Peat moss most distinct then lost by SLOPE, then concealed by fragments, then clear.  this
161g096 ide {P} Shelf A Shelf A at head of Gentle mossy SLOPE, which from a distance hid it, could be followe
148g027 ered terraces formed successive bays but plains SLOPED centre-wards which would not have happened  if
078r175 W.--veins richest not in ravins or along gentle SLOPES. but on the most elevated summits, where mount
030r036 onian coast to see proportional cliff & low or SLOPING land What are the "palatal Tritores" found  in
032r040 ary.--The gradual rising continuing. a another SLOPING platform would be made, & so on.-- This is gro
106a075 h limit cannot be great over) with very gently SLOPING sides This argument is partly taken from Delab
147g020 appearances, as if valley had been filled with SLOPING bed of rubbish Friday Highest part of road bet
151g042 3d lake.-- 4th shelf runs up some way on great SLOPING plain of alluvium (much corroded by rivers)  &
156g072 erfectly rounded pebbles of granite & forming <SLOPING» buttresses Yet certainly shelf 4th <near> onl
162g104 ch side High up the Tarf (a Granite (boulder), SLOPING buttresses, an[d] one alternate curved layer o
175b020 othing. We may look at Megatheria, armadillos & SLOTHS as all offsprings of some still older type som
574n041 na licking her chops.-- someone has described SLOVERING «gum» «teeth»less-jaws. as picture of disgust
032r039 ay.--If the level of the sea was to sink by very SLOW & gradual movements to line (2). The part (o) w
032r041 globe.  must there not be a circulation «however SLOW & weak.»; «(cause of not accumulation of  Coral
056r109 en we see an entire island so encircled, the one SLOW cause is apparent. «I confess I never see  such
087a012 fic great land. -- Will use argument of proof of SLOW corrosion of valley of <Patagonia.> S Cruz -- f
088a013 doubt fluid.-- analogy as continental elevations SLOW. so would line of mountain chain be Mr  <Lyell>
105a070 f valleys-- subsequent opening a medial gorge by SLOW erosion. but we have evidence in distribution o
105a072 (it is generally said) is consequence of <rapid> SLOW course, & with slow course small erosive power.
105a073 d) is consequence of <rapid> slow course, & with SLOW course small erosive power. therefore  tendency
113a091 ll be greater than <5000.> 400.-- These facts of SLOW but successive transmission of temperature clea
114a095 > «offers a presumption» it has been excessively SLOW because beach line chief cause of denudation, b
221b201 ks. most remarkable.--¿ Have the changes been so SLOW. , that all have existed for ages as metamorphic
230b239 ards will not alter. This need not apply to very SLOW changes. without crossing.-- Now a gradual chan
242c017 n be trusted The changes in species must be very SLOW, owing to physical change, slow & offspring not
242c017 ies must be very slow, owing to physical change, SLOW & offspring not picked.-- as men do. when makin
248c034 chiefly  heredetary.-- If varieties «produced by SLOW causes, without picking» become more & more imp
263c074 as difficult to understand as Lyells doctrine of SLOW movements &c &c. this multiplication of  little
277c128 as  Cyrena This is reform which probably will be SLOW. but must take place-- such a classification wo
281c138 ails strange ¿¿/Genus only natural from death or SLOW propagation of forms.--just same way as <we> «a
285c153 ges in structure being «necessarily» excessively SLOW, they become firmly embedded in the constitutio
386d167 of  this because the great changes of nature are SLOW. if animals became adapted to every minute chan
386d167 y minute change, they would not be fitted to the SLOW great changes really in progress.-- Annals of N
409e047 an.-- ask the missionaries about Australians yet SLOW progress has done so.-- Show a savage a dog,  &
410e050 of  individuals.-- so that the changes should be SLOW & bear relation to the whole changes of country
412e057 ces of times. & from persistency «owing to their SLOW formation» these variations tend to accumulate.
419e087 hough <the> it also the effect of change) than a SLOW gradation in form, «which must be effect of slo
419e087 slow gradation in form, «which must be effect of SLOW change + & Therefore precludes effects of catas
470t153 e These latter were pollen gatherers & they seem SLOW= Maer 1840 My Father formerly planted «Turkey o
494q004 fe. (5) Does my Father know any case of quick or SLOW pulse being heredetary. (6) In the last 1000 ye
529m040 tion of land covered with ocean, former animals, SLOW force cracking surface &c truly poetical. (V. W
```

(Key Word)

537m075 will be, but yet it is settled by reason.--‖ How SLOW habits are changed may be inferred from express
555m143 Monkey, very different disposition from others, SLOW cautious, angry cross look, followed by protrus
624o051 d from want of reason.-- Hence as Eugenius says, SLOW growth of rule of right.-- [LHC] "for it strive
416e071 es-- but latter far more perfectly & infinitely SLOWER.-- No domesticated animal is perfectly adapted
530m044 lse of new born babies of labouring classes are SLOWER than those of gentlefolks. & that peculiaritie
133a140 paper like a fool.-- Feb 25' All facts show how SLOWLY heat travels; & therefore the abysses where fl
172b007 for Man.-- Let a pair be introduced & increase SLOWLY. from many enemies. so as often to intermarry
281c137 theoretically if animals did change excessively SLOWLY. whether geologists would not find fossils suc
292c171 o cause long memory.-- structure is only gained SLOWLY.-- therefore it can only be those actions, whi
335d016 ities, to its offspring, which have been gained SLOWLY, now all the mules have their whole <body> for
336d017 from its parents «stock» without it be small & SLOWLY obtained NB. The longer a thing is in the bloo
336d018 anic diseases, mental disposition, stature, are SLOWLY obtained & hereditary; <but if> if the change
336d018 t if> if the change be congenital (that is most SLOWLY obtained with respect to that individual) it i
371d128 rent state, or if different dieteriorating very SLOWLY.-- I presume most of these roses, without circ
397e004 o some change of circumstances; now we know how SLOWLY & insensibly such changes are in progress.-- w
438e141 yet in all probability the Brussels Sprout was SLOWLY formed.-- » if it shall be difficult to show t
537m075 g his finger by rubbing them with alum, so more SLOWLY does animal leave off <t> instinct, when atten
549m123 servative, & are now, like all other structures SLOWLY vanishing-- the mind of man is no more perfect
566n011 of origin of evil passions.-- Man getting sight SLOWLY, but when in grown years, thinking he instinc
599o005 rigin of language.-- must presume it originates SLOWLY-- if these speculations are utterly valueless-
640j167 usca would offer few species, or rather be very SLOWLY changed & vertebrata much so.-- so far true, b
057r113 os as Marl-pits: Consider well age of Bones. = SLOWNESS of elevation proved at St Julian. = do not th
258c059 is law.?-- Local varieties formed with extreme SLOWNESS, even where isolation, from general circumsta
419e086 his country we have better means of judging of SLOWNESS of physical changes, than in any other, & yet
023r012 diere in Bay of Legrand, (SW part). describes a SMALL granite Isd. capped by Calcareous rock; followi
044r012 kes in the cliffs.--wide valleys.--central peak SMALL; yet great body of lavas have flowed from centr
044r074 ly to vertical movements. In Cord: after seeing SMALL Bombs. without a vesicle. we may consider appea
050r090 t Ischia there is a pumiceous conglomerate with SMALL & large fragments, nature of which is doubtful.
059r121 e. for if so Chalk would harden.--Climate.!? or SMALL Proportion of Alum: matter.--all pale cream col
077r171 overlying beds tells the secret.-- p. 189. "The SMALL ravins into which the valley of Marfil is divid
099a049 lamination parallel to stratification evidently SMALL scale of concretionary action all fluid at once
105a073 ence of <rapid> slow course, & with slow course SMALL erosive power. therefore tendency of running wa
106a073 which would not have been case. if inclination SMALL.-- The power of widening channel depends on pow
106a074 ourse until inclination is become comparatively SMALL, & when that is case force is lessened. therefo
115a098 currents> <fragments> but relative to currents. SMALL lakes have power of levelling their shores wher
116a100 n a sea beach-- «p. 151» first discovered «very SMALL» bits of red granite between 40 & 50 from Porte
118a103 &c.-- in Cordillera, it is at once evident only SMALL proportion of dikes have reached the surface Ar
120a107 h the earth is cracking by vertical planes into SMALL pieces-- mem coal-field.-- the structure of And
121a113 shingle beds. Lyell on Sweden. p. 12. proofs of SMALL rise at Stockholm.-- analogous to my Valparaiso
147g020 Tulla whole wide valley scattered with few very SMALL & irregular hills of alluvium-- nothing very st
162g104 an[d] one alternate curved layer of fine sand & SMALL angular-- rounded pebbles-- dip sideward, & inw
181b041 back. the <descen> fathers would be reduced to SMALL percentage.-- & <in> therefore the chances are
196b108 had been created; but on other hand creation of SMALL animals must have gone on since from parasitica
199b116 : assuming truth of quadrupeds being created on SMALL spots of land, of the same type with the great
210b158 rtion of genera I: I. I can understand in «one» SMALL island species would not be manufactured, <but
223b209 Pentland's account of Bolivian human species?-- SMALL «new» animal mentioned from Fernando Po Zoolog.
226b220 Fernando Po & Coast of Africa. equally good.-- SMALL isld off New Guinea same fact see Coquille's Vo
226b222 ies belonging to its own genera Therefore if in SMALL tract we have many species, we may insure mass
230b236 me mountain).-- How is this explained by law of SMALL differences producing more fertile offspring.--
231b243 by creation-- or we must suppose a multitude of SMALL creations.-- Will Dromedaries & Camels breed?--
233b251 ogger-headed Duck-- Large proportion of Water & SMALL of land or few quadrupeds-- Study Productions.
256c055 Feb 1838 thinks gradation between Man & animal, SMALL point in tracing history of Man.-- granted.--bu
265c087 tary. The circumstance of aberrant groups being SMALL it is truism. for it not so not aberrant.-- Ten
272c109 s feathers in tail of Edolius.-- Remarkable how SMALL detail in structure prevails amongst the same s
273c110 ened rump feathers.--; <then> & one species has SMALL band & others large, then he says from long exp
274c114 ped. Not confined to one species, but generally SMALL genus. ¿are there not many ground parrots? are
275c121 lls, but same marks on wings are Blue Pouters & SMALL Bald Heads Mr Yarrell will mention in his work
294c176 ority).-- Lonsdale is ready to admit, permanent SMALL alterations in wild animals, & thinks Lyell has
297c185 ers & vary much in character, divided into many SMALL genera 'circumstances not favourable to many sp
302c202 y modified = = this is not easily told, for any SMALL family. having analogous characters, might be m
314c237 lf, one segment of the corolla being (probably) SMALL to allow it to lie on one side.-- but in other
316c244 icted genera, but then they appears always very SMALL ones as Trigonia in Australia or Concholepas in
335d015 e either parent, or intermediate within certain SMALL limits (within which limits they might return t
336d017 inherits from its parents «stock» without it be SMALL & slowly obtained NB. The longer a thing is in
336d018 succession of <such changes> generations, these SMALL changes become multiplied, & great change be ef
339d026 when he says No genera.-- thinks there are some SMALL divisions.-- does not seem to think any improba
341d029 thus the most remarkable character in Apteryx, SMALL respiratory system; even much smaller than in o
344d042 m her, but her offspring came out one big & one SMALL. Now Jones, before this happened from her looks
355d069 o extinction of species, the capability of only SMALL amount of change at any one time Seeing what Vo
360d092 of ideosyncrasy I have hitherto thought that a SMALL difference <of any kind>, if very firmly fixed
374d134 wondering of changes in number of species, from SMALL changes in nature of locality. Even the energet
385d163 lest sized dogs.-- one litter big & then second SMALL & so.-- Says, there is breed of Fowls called ev
390d178 the necessity of some change.-- ‖ Without some SMALL change in form. ideosyncrasy or dispositions we
397e004 ount of change within historical times has been SMALL-- because change in forms is <al> solely adapta
398e004 mate vegetation &c.-- It is the circumstance of SMALL physical changes & oscillations, not affecting
398e005 e last to object to this theory on the score of SMALL change.-- on the contrary islands separated wit
400e013 os do. p. 376. Isle Tres Marias off Mexico with SMALL Hares & raccoons.-- «S. American form.--» off p
402e018 . 229. Borneo.-- only animals he heard of pigs, SMALL bears or badgers, deer, apes, baboons, monkeys
402e019 given,= some peculiar <M> p. 359. At Manilla a SMALL Cercopithecus., & skins of galiopithecus.-- Mal
403e022 The inhabitants of Summagi, a territory in the SMALL isld of Eap in the Carolines, are remarkably sh
413e058 andchildren. like. grandfathers (2) Tendency to SMALL change.. «especially with physical change» (3)
430e117 rent dispositions: this is important as showing SMALL variations in offspring of wild animals.-- and
431e123 haps a true element) «give examples, pigs, with SMALL chinese boars &c &c &c» p. 10 offspring take mo
440e147 ree of injuriousness must have been exceedingly SMALL.-- This is far more probable way of explaining,
450e173 ogs-- spotted deer, no loomies, but cocatores & SMALL green parrots. June 26th-- Yarrell.-- Black Sw
450e175 r to those of Pegu-- in Sumatra two breeds both SMALL-- Java pony occasionally reaches 13 hands.-- P
451e177 5 degree. 50 minute S. adjoining it are several SMALL islands. abounding with deer-- Horsburgs. Vol I
453e182 somewhere says same.-- do p. 393. <">The wild, SMALL fowls at Pulo Condore "crow like ours, but much
453e182 at Pulo Condore "crow like ours, but much more SMALL & shrill".-- Humboldt. Vol I. p. 275. says Tene
458t002 I. p. 499. «4to. Edit»-- Horses in Lao Choo so SMALL, that person with long legs can hardly ride on
464t065 ty-- there have been at least other birds, with SMALL wings, & surely the Apteryx is more closely all
467t105 cabbage «in flower»-- swarmed with meligethes & SMALL Staphylinidae on all their bodies pollen-- on a
467t105 ot many do-- pollen not very abundant. not very SMALL-- Saw one small Bee; saw another on Cabbage--wh
467t105 len not very abundant. not very small-- Saw one SMALL Bee; saw another on Cabbage--white Butterflies
468t105 cessively abundant frequent not attractive, very SMALL--stigma rather large & rough-- flowers common--
468t106 ttractive-- Found Rhubarb blossom swarming with SMALL Staphylinidae-- Anapsis, Melegethes, Leptuse--
468t106 idae-- Anapsis, Melegethes, Leptuse-- Diptera & SMALL Hymenoptera Saw Humble go from great Scarlet Po
468t111 usa <never> «once» P on Fraxinella <Heartease» «SMALL. Humble alighted on base of filaments & reached
470t153 rkspur on two plants in do Humble 24 flowers of SMALL Linaria in do Domestic do 6 Campanula (two spec
470t177 ny (curious because a Composite) Asparagus very SMALL flowers & as much shut up, frequented by «many»
472s02v w 4 more Bees at work-- another odd genus-- & a SMALL common Humble-- & more of same fly Two more of
478z003 134. and p. 115 In white Cape Pidgeon's stomach SMALL shells (patella) sea weed & many pebbles Mentio
484z016 interesting comparison to find how many of the SMALL finches walk at Maldonado & Patagonia compared
495q05a odically & see effect-- such dioecious individ--SMALL orifice (8) Carry Bees, powdered with starch &
505q014 (14) Has planted seeds of Geranium pyrenaicum. SMALL white-flowered var. with abortive stamens.-- sh
505q014 to look at:-- Was once offered a hive. of these SMALL Bees-- at Sundorne has large Bees July/42/ Mark
505q014 large Bees July/42/ Mark has six day's puppy of SMALL true Bull-Dog-- length from nose over head to r
506q015 apparent-- Any where pollen is not produced or SMALL in quantity -- Any unproductive, where germen d

SMALL / SMITH 578
Page **(Key Word)**

546m109 tle happiness depends on the senses.; than the <SMALL> fact that no one, looking back to his life, wo
564n005 transmitted perfectly to chicken so as to seize SMALL moving object like fly.-- young partridge can r
576n047 ce, of so dreadful a consequence to each man is SMALL. Man's intellect is not become superior to that
633j53v - do p. 284. it is hard on my theory of gain of SMALL advantages thus to explain the curling of the v
640j167 changing.-- From these views we can deduce why SMALL islands. should possess many peculiar species.
640j167 r is, present with respect to new arrivers, the SMALL body of species would far more easily be change
046r080 the first great movement was one of rise (any SMALLER prior ones might have been owing to absolute m
061r127 ound of 2. ostriches; bigger one encroaches on SMALLER.--change not progressif<e>: produced at one bl
153g056 band» 4 X 3 X 2 «feet» & 2 deep Another rather SMALLER block 30 ft «above» & other 50 ft lower & othe
153g056 lock 30 ft «above» & other 50 ft lower & other SMALLER ones «these boulders are decaying.» neighborin
341d029 n Apteryx, small respiratory system; even much SMALLER than in other Struthios. was adaptation to lit
383d160 ale bird, said to be infertile.-- spurs rather SMALLER than in <ma> silver male-- Head like silver ex
431e123 rossing consists in the idea of the male being SMALLER, & the female larger than average size: (surel
451e179 lonese Elephant [...] saul forests by having a SMALLER, lighter head, carried more elevated & higher
480z008 ulica Chloropus. says some of the «species of» SMALLER petrels, are night birds agree. with <pe> noct
505q014 Cyclopaedia) Abberley says that some Bees are SMALLER & more vicious. Will try to get me some to loo
199b120 agate, as well as diseases. In intermarriages; SMALLEST differences blended, rather stronger tendency
385d163 ters are considered the most valuable. because SMALLEST sized dogs.-- one litter big & then second sm
553m135 hout the powers of the mind being EQUAL to the SMALLEST casuistical doubts.-- The history of Metaphys
573n036 erse <of man> to be governed by laws,, but the SMALLEST insect, we wish to be created at once by spec
640j167 On distinct Creation, how anomalous, that the SMALLEST newest, & most wretched isld should possess s
599o005 out the year 1837 & Earlier--]CD in Athenaeum "SMART-- Beginning of a new School of Metaphysic,"-- g
293c174 animal would not cause misery to other.-- else SMELL of Man would be disagreeable to Musquitoes We n
334d010 ccording to Fox) being guided entirely by their SMELL.-- Fox says he knew «a» carter well, who placed
337d021 may be considered as one large eye-- have they SMELL, do plants emit odour solely for others parts o
376d137 . Evidently knew <men> women, thinks perhaps by SMELL.-- but monkeys examine sexes of every Has repea
466t095 es associated with some other, as pointing with SMELL.= These qualities have been given to foetus <fr
496q05a whilst others are crossed.-- Are Bees guided by SMELL-- or sight.-- --. touching Mr Brown theory of i
506q015 racters.-- Stature, position of flowers-- Their SMELL-- form of flowers-- Nectaries-- In Monooecious
536m071 oes other.-- What an animal like taste of likes SMELL of,∴ Hyaena likes smell of that fatty substance
536m071 mal like taste of likes smell of,∴ Hyaena likes SMELL of that fatty substance it scrapes off its bott
536m071 -- it is relic of same thing that makes one dog SMELL posterior at another.-- Why do bulls & horses,
539m082 appened to go close to one & smelt the peculiar SMELL of Picture. association with much pleasure imme
540m085 ginae of females.-- when it is recollected that SMELL of ones own pud. not disagree.-- Ourang outang
546m109 musical sound with birds. & ¿ howling monkeys-- SMELL with many animals-- see how a dog likes smell o
546m109 - smell with many animals-- see how a dog likes SMELL of Partridge--, man's taste for smell of flower
546m109 dog likes smell of Partridge--, man's taste for SMELL of flowers, owing to parent being fruit eater.-
554m139 guided to her own mouth.-- seemed to relish the SMELL of Verbena & Pocket Handerchief & liked the tas
565n007 afraid, though not every time really wishing to SMELL its enemy.-- Man & dogs show triumph (& pride)
569n019 mely primitive.-- almost like tastes of mouth & SMELL. Descent of Man Understanding languages seem si
574n041 ncy. to exert all senses, when thus stimulated, SMELL, as Sir. Ch. Bell says, & hearing music. to cer
575n044 - bull-dogs being preferred from not having any SMELL.-- 27th. November.-- Think, whether there is an
581n064 es of man, same as in Allied Kingdoms-- "food, SM<E>LL. (ourang-outang), music, colours we must suppo
593n107 insects» which become in imago state social) by SMELL or looks. but it does not know its own smell or
593n107 by smell or looks. but it does not know its own SMELL or look, & therefore there must be some instinc
593n107 ctive feeling which is pleased by other animals SMELL & looks.-- no doubt it may be attempted to be s
593n107 pted to be said that young animal learns parent SMELL & look so by association receives pleasure. Thi
593n109 en man, a socialist, does not know other men by SMELL, but by looks. hence. some obscure picture of o
235b272 luence of climates, situations &c on 242. Hook SMELLIE Philos of Zoolog. 842 White regular gradat in
324c268 unfixed forms. Dr. Royle on Himalayan. types. SMELLIE. Philosophy of Zoology Flemming. ditto Falcone
540m085 need not feel so much surprise at male animals SMELLING vaginae of females.-- when it is recollected
560m156 form habits readily?? Do the Ourang Outang like SMELLS «peppermint» «& music».-- Have monkeys lice?--
539m082 much enthusiasm, happened to go close to one & SMELT the peculiar smell of Picture. association with
540m085 at Zoolog Gardens touched pud. of young male & SMELT its fingers. Seeing a dog & horse & man yawn, m
542m096 make it out?-- Seeing a Baby (like Hensleigh's) SMILE & frown, who can doubt these are instinctive--
542m096 mouth between interval of barking, now this is SMILE.‖ With respect to sneering the very essence of
565n006 stops the noise , the face clearly passes into SMILE-- laugh long prior to talking, hence one can he
565n009 person close those muscles, which wrinkle when SMILE.-- Hope is the expectant eye. looking to distan
565n009 is same as pouting, <but> lesser in degree, no SMILE, no frown showing thought, no compression of mo
565n009 f lips, as if going to say "my dear," just what SMILE is to laugh.-- I must be very cautious. Remembe
579n058 t on one object.-- With respect to my theory of SMILE. remember children smile before they laugh.-- H
579n058 espect to my theory of smile. remember children SMILE before they laugh.-- Has frowning anything to d
540m084 shout) when opening his mouth in romps, <so> he SMILES. Many of actions as hiccough & yawn are probab
557m147 of pleasure so teaching expression «as constant SMILES, cheerful face».-- Man when at ease has smooth
315c243 bject important.-- Laughing modified barking., SMILING modified laughing. Barking to tell other anima
429e114 organic beings. going on the peaceful woods. & SMILING fields.-- we must recollect the multitudes of
595n125 t located] A child crying. frowning, pouting, «SMILING», just as much instinctive as a bull <tr> calf
029r032 any great accumulation of such matter.-- Dr A. SMITH says. that Boulders do not occur in the South A
093a034 so far distributed to various subjects) Dr. A. SMITH informs me that in the year a Rhinoceros was fo
121a111 «& 7.» violet strata from decomposed muscles.. SMITH of Jordanhill has seen same thing-- Consider pr
178b032 Ascensi> Study Lesson Voyage of Coquille.-- Dr. SMITH says he is certain that when White Men & Hotten
179b033 in West Indies «--:Humboldt. New Spain:--» Dr. SMITH always urges the distinct locality or Metropoli
179b034 gnance & art required to make marriage.-- as Dr SMITH remarked Man & wild animals in this respect are
188b067 to certain islets in East Indian archipelago Dr SMITH considers probable two Northern species replace
188b069 of races of men tendency to keep to one line Dr SMITH says very. close species generally frequet slig
191b083 of subsidence. & recent elevation: Pray ask Dr. SMITH.-- «to state that most clearly». Fox tells me,
212b167 rmation respecting the Water Rat.-- ¿Consult Dr SMITH History of S. African Cattle Phillips Geology «
216b180 le nearer black or white as it may happen.-- Dr SMITH says he is sure of the case at Cape.-- McClay a
216b182 ing as if many ova-- impregnated at once.-- Dr. SMITH considers the Caffers (like Englishmen) men of
229b233 ot having them, instance of useless structure-- SMITH thinks several species of Rhinoceros range from
234b256 nd, but <rare> «absent from» near London. = Dr. SMITH, he says, is deeply [not located] of <all> gene
249c035 ies of Clenomys. case of replacing species. Dr. SMITH will give me some capital information ¿Carnivor
278c129 ished by this test.-- Excepting where an Andrew SMITH, «Richardson» a Vaillant, a D'orbigny has trave
284c150 re it remains less subject to Variation" Dr. A. SMITH. knows lots of instances of replacement of one
311c227 India in which it may be seen swimming about. A SMITH is firmly believed in representation. certain b
312c228 ks, being spotted, & colours of little value Dr SMITH if black & white Man crosses; children heteroge
312c228 fspring most like <parents> Mother.-- like dogs SMITH knew chinese hairless dog & common spaniel cros
325c267 g (sometime) Du Stewart works. & lives of Reid, SMITH & giving abstract of their views Mackintosh Eth
376d136 ear for Many generations Sept. 29th Dr. Andrew. SMITH «Remarks on extraordinary curiosity of Monkeys»
376d137 threw it over it, & made it meet in front.-- Dr SMITH every baboon & monkey, big & little that ever h
377d139 as much as to say give me-- the other when Dr. SMITH more distant.-- But he thinks other monkeys mak
387d168 the added «like Lord Moretons case & Dr. Andrew SMITH,» difference.-- If A. B. C. D. E be <offspring>
446e161 dition,-- rather the reverse of facts stated by SMITH of Jordan Hill.-- May 27th.-- Henslow One of th
471tf11 . Narrative of Missionary enterprises Dr Andrew SMITH says in the larks from S. Africa he can almost
489qIFC y p. 14 Henslow (2d time) p. 14.-- Father. And. SMITH Dr. Holland p. 16 Babington-- Gould ---- 10.(a)
489qIFC p. 21--23 Eyton p. 22 Schomburgk.---- 1 Jordan SMITH. p 1. Sowerby Cuming. -- p 1 Owen p 17 Hooker p
490q001 ed Atavism alluded to by Dr. Holland-- <Jordan> SMITH of Jordan Hill-- character of the extinct land-
507q15v trioecia. (are female flowers ever productive) SMITH says many trees in Tropics are of this class.--
507q016 plants ,Teazle Dr. Holland ; My Father. Andrew SMITH (1) Are cross-births, or other accidents of del
508q016 are livers obscure organ. no answer?-- 3 Andrew SMITH, about tamed wild animals breeding at the Cape.
508q016 out cross-bred races of men taking after sex. A SMITH. About species of Rhinoceros. becoming rare bey
508q016 its of the metropolis of each-- Cause?-- Andrew SMITH. (6) What size book Gallesio storia del Reprodu
508q016 - paper in Taylors Scientific Memoirs (11) And. SMITH Savages at Cape any selection of Males in «catt
546m108 emory-- also low faculty of understanding. Adam SMITH (.D. Stewart life of. p. 27), says <sympathy> w
559m155 Human Understanding well worth reading Copied <SMITH> «D. Stewart» lives of Adam Smith Reid, &c wort
559m155 ading Copied <Smith> «D. Stewart» lives of Adam SMITH Reid, &c worth reading. as giving abstract of S
592n101 Religion" on its origin in Human mind.-- Andrew SMITH says hen doves & the female chamaeleon court th
595n184 ing expressive, (Vol. 4 of Works) [blank] "Adam SMITH Moral Sentiments" much on life & character "Hum

Page **(Key Word)**
609o29v ing on views in second volume of Malthus). Adam SMITH also talks of the necessity of these passions,
114a094 salt-lakes of Rio Negro--Mr Bowerbank-- Dr. A. SMITH'S curious specimens of «transversely fibrous» qu
188b068 l with blue. Now this is same bearing with Dr. SMITH'S fact of races of men tendency to keep to one l
229b232 le: argument for true molluscs coupling.-- Dr. SMITH'S Information Long Horned (very) aboriginal at C
251c041 &c on Mammalia no doubt will all be included in SMITHS work «do» Vol. IV p 273. Macleay on Capromys.
317c250 l growing,) passed us. do. p. 243 (, Professor SMITH'S Journal) on the heights of St Jago found a Eup
347d050 of arrogance!! August-- 29th.-- Macleay in A. SMITH'S Zoolog.-- of Africa -- p. 4. sticks to genus o
559m155 Reid, &c worth reading. as giving abstract of SMITH'S views «Take & pound up inflorescent parts of m
051r094 green confervae) coated with living beings; In SMOOTH seas (& even turbulent as at St Helena) I have
070r155 oss the continent.--He states plains of Mendoza SMOOTH. Sir W. P. states that in Helm's travels accou
158g082 anite blocks almost encircled> <fre> Gneiss cut SMOOTH on sides of hill where Boulder lies. buttresse
557m147 smiles, cheerful face».-- Man when at ease has SMOOTH brow contrary to wrinkled: (a horse when winno
054r105 most inland part of this bay are perforated & SMOOTHED like those washed by the waves, a sufficient
386d166 t is good for the whole.-- if cut off nerves in SNAIL. (Encyclop of Anat & Phys) can make a head; the
477z001 Charpentier de Cossigny. only 10 years ago <no> SNAIL was introduced to Mauritius. 18 Azara Voyage Vo
496q006 ame receptacle (8) Make Duck eat Spawn, eggs of SNAIL, row of fish & kill them in hour or two «My Fat
534m063 some imported plants are attacked by insects & SNAILS of this country (thus Dahlias by snails)-- <Th
534m063 sects & snails of this country (thus Dahlias by SNAILS)-- <The> Apion radiolum undergoes transformati
542m097 ommunicate to each other.-- Lonsdale's story of SNAILS, Fox of cows, & many of insects-- they likewis
247c028 2. a Gecko on St Helena.-- <in 1813, a venemous SNAKE was> one Gecko on Isle of France Scincus multil
463t055 howed me the component vertebrae of the head of SNAKE wonderful!! distinct!!-- He would not allow suc
641j29v n geology-- There is an analogy between fang of SNAKE, (jaw of spider?) sting of bee, sting of nettle
229b233 kind further inland.-- NB. There is division of SNAKES. with hinder teeth perforated for poison chann
246c027 a more than Van Diemen's land.-- Vol II p. 8 no SNAKES on isles of central Pacific, yet there appears
247c027 natives, & probably on Oualan.-- «Mitchill says SNAKES on Friendly isles. p 50. LX. Journal of Sillim
399e010 ia, but are unknown Eastward of the Navigators. SNAKES occur there,, but are unknown in Hervey or Soc
496q006 ade hens cast Holly-seed & they grew» (9) Place. SNAP-Dragon. (I have seen one monstrous) Fox Glove &
584n071 & turtles running to water,-- young crocodile SNAPPING-- p. 28. how curious the means of guiding the
595n125 r> calf, just born butting, or young crocodile SNAPPING.-- these I think are better instances of inst
541m092 culiar manner they do, even more than in a real SNARL, they are enjoying a satirical. laugh.-- when s
558m152 en> «forcible prolonged» expulsion of air «dogs SNARL much the same way» generic manifestation of gre
315c243 «of all animals» becomes very curious.-- a dog SNARLING in play.-- Hensleigh says the love of the dei
541m092 , they are enjoying a satirical. laugh.-- when SNARLING real bitter sarcasm.-- <These> Seeing how anc
565n008 t by insulting his pride & is therefore of the SNARLING order.-- But contempt mingled with disgust, w
542m096 an doubt these are instinctive-- child does not SNEER. because no young animal has canine teeth.-- A
551m129 ntempt <&> disgust & defiance.-- different from SNEER-- How easily. horses associate sounds may be se
555m143 es some of the old ones.-- -- S. American group SNEER.-- Sept 21st Was witty in a dream in a confused
564n006 ich may be described as an exaggerated habitual SNEER» the manner in which whole skin or muscles are
432e125 etrogressing-- Uncovering the canine teeth, or SNEERING, has no more relation to our present wants. o
542m095 I conceive sighing might yet remain just like SNEERING does.-- is yawning habitual from awaking from
542m096 f barking, now this is smile.|| With respect to SNEERING the very essence of an habitual movement is c
565n008 ect to opponent» is showed by same movement as SNEERING,-- it is then more <emblem> manner of hurting
565n008 epressed & opposite muscles used to when angry SNEERING is in progress.<--> the hypothesis of opposit
639j28v es no special use to Man. Applicable to Bell's SNEERING-theory.-- p. 263. This kind of doctrine runs
589n092 oice than respiration.-- like sigh before false SNEEZE.-- "A Dissertation on the Influence of the Pas
502q11v cidus is.-- Ask Sulivan about Falklands Isds.-- SNIPE Migratory-- probably united by Land to S. Ameri
211b162 length of bill does not <indicate affinity with SNIPES> indicate affinity, because similar habits pro
592n103 y open all senses: <do> Horse prick his ears «& SNORT clears nostrils» when frightened, does not hair
565n007 fear., as ears down to horse.-- Horse snuffs «& SNORTS», the air «& raises its head, & pricks its ear
234b255 n other side.-- Pigs put legs over, & then with SNOUT lift up latch & back.-- Frogs attempted to be i
444e157 ." Before he alludes to the resemblances of the SNOUT of the mole & Pig in having two additional bone
062r128 at contrast. from nature of climate. = Perpetual SNOW.--subterranean lakes, near Volcanoes. lakes of
219b195 Alps. must be <Alpine> «new formations» because SNOW formerly descended lower, therefore species of
260c069 .-- Darkness effect on human offspring.-- white, SNOW.-- the fine green of vegetation,-- ¿account for
316c244 perstitions.-- +York's Minster story of storm of SNOW after his brothers murder.--» good anecdote --.
318c252 s??--) I never saw more beautiful adaptation for SNOW-- like snow shoes. than feet & hind legs of the
318c252 er saw more beautiful adaptation for snow-- like SNOW shoes. than feet & hind legs of these white har
318c252 legs of these white hares, fitted for regions of SNOW.-- Acclimatisation.-- Bachman tells me in Audub
408e044 2. Bones of Horse. Bear & Deer at 16000 ft. with SNOW on Himmalaya-- Humboldt bones at 7800 in Andes-
409e048 nt of change.-- now take greater area of water & SNOW line descent. My theory gives great final cause
427e109 imate less extreme. (& so account for descent of SNOW line there «& there & there only: as stated by
427e109 han before arctic forms would retreat: effect on SNOW of arctic climate in far north regions? Arctic
497q06v en formed. (15). What is History of Viburnum. or SNOW-ball-tree. what would result from seeds being s
551m129 pulse rises, or as they flag & something like a SNOW-haze. covers my whole imagination." Septembe. 3
595n121 ion [blank] [not located] Ernest W. playing with SNOW. when 2½ years old. was frightened when Snow pu
595n121 ith Snow. when 2½ years old. was frightened when SNOW put a guaze over her head. & came near him, alt
595n121 r head. & came near him, although knowing it was SNOW.-- Is this part of same feeling which make us t
565n007 nctive to fear., as ears down to horse.-- Horse SNUFFS «& snorts», the air «& raises its head, & pric
495q05a ributed.-- Examine scum of pond for seeds.-- 11. SOAK all kinds of seeds for week in Salt. artificial
200b125 It would be curious experiment to know whether SOAKING seeds in salt water &c has any tendency to for
232b248 object to my theory, those the philosophers who SOAR above the pride of the savage, they perceive th
286c154 it is the way of mankind. & I believe those who SOAR above Such prejudices, yet have justly excalted
557m150 -- Are those parts of body, as heart, & chest (SOBBING) which are most under great sympathetic nerve.
138a153 il. [blank] «p. 4. (Lyells Book)» Observaciones SOBRE El Clima del Lima par Dr. H. Unanùe says he bel
089a020 paper on Erratic blocks in Alps. Memoires de la SOC. «de Geneva» Vol 3' P. II. -- Bed, of elevated s
094a036 ks. Specimen of rock from Costorphine at Geolog. SOC: Colonel Imrie Transact Wern. Soc. Vol. 2. p. 35
094a036 ine at Geolog. Soc: Colonel Imrie Transact Wern. SOC. Vol. 2. p. 35 Sir J Hall Trans. Phils Royal Ed.
094a036 quiae Diluvianae p. 201. & seq Murc Trans Geolog SOC Vol 2. p 257 {T} The Pota: labiata certainly is
095a037 with the Mactra. at Buenos Ayres at the Zoolog: SOC: Terebratula from Hudson's Bay. 2. species Vol V
098a044 e proofs, but certainly probable. Bulletin de la SOC. Geolog: 1833-34. p. 35.-- Ancient Lake Lemagne
105a070 face exposed. bay more open to turbulence. Bull. SOC. Geolog «1837» p. 320. paper on shrinking of Cla
108a081 - [top portion page excised, not located] Bull:. SOC. Geolog. Tome IX 1837-8. p. 24. rocks of Chimbor
112a090 Geograph. Journal.--» A meeting of the Geograph SOC, April 9 1838. Letter from M. Erhman stating tha
115a097 has deposited specimens from Anglesea in Geolog. SOC. if numbered compare them with my rocks. when wr
117a101 d granite between 40 & 50 from Portezuelo. Bull: SOC. Geolog. 1837. December. p. 91. a classification
134a144 do Vol 7. p. <52> 363. do {T} Journal of Asiatic SOC Vol V. p. p 96. apparently good geological paper
212b165 to. latter not to former-- Mr Martens of Zoolog SOC told me an Australian dog he had, used to burrow
212b166 Mauritius & Madagascar.? Proceedings of Zoolog. SOC June 1837 p. 53. an Irish Rat.-- different from
216b179 oned in Loudons (analogue of Blood hound-- Bull. SOC. Geolog. 1834. p. 217. Java Fossils 10 out of tw
236bIBC turelle Geographie Botanique De Candoelle. Geol. SOC Horae Entomolgicae Linn: Soc. Geoffry-St. Hilair
236bIBC De Candoelle. Geol. Soc Horae Entomolgicae Linn: SOC. Geoffry-St. Hilaire Philosophy of Zoology Water
241c016 do/p. .62 ??? Age of Deinotherium. p. 23.. Bull: SOC. Geolog. 1837-8. Tom: IX.-- M. D'.Urville on the
268c096 talogue of animals of Nepal read before Linnaean SOC. Feb. 1838.-- Annals of Natural History. «vol I»
269c101 &c. Juan Fernandez A communication to Geograph. SOC. in February or March 1838 on soil in Siberia be
290c164 e forest Seeing common gull in garden at Zoology SOC. it's pale ash grey back, like a black bird wash
307c216 tamus from Madagascar!!!!!! Proceedings of Geol. SOC Vol I It is capable of demonstration that all an
324c268 ical distrib. «in Dict. Sciences. Nat. in Geolog SOC.» F.. Cuvier on instincts L. Jenyns paper in Ann
327c265 of Fruit trees in. N. America" in Lib. of Hort. SOC Mr Neil. has written good article on Horticultur
337d022 c observer.-- <Transactions of the Entomological SOC> A capital passage might be made from comparison
340d026 cable every instinct in animals. Heard at Zoolog SOC their Pintail & Common Ducks, breed one with ano
448e168 ns in St Thomas on W. coast of Africa Owen Linn. SOC. April 2d. 1839 The Lepidosiren-- Amblyrhyncus &
451e176 ell as Tapir.-- <do do p 75> «Journal of Asiatic SOC.. Vol V. p. 565. in a Paper by Lieut. Newbold.--
451e178 83 attached to its body-- Journal of the Asiatic SOC. vol. I. p. 335. Catalogue of animals of Nepal b
452e181 at> near shore of Magindanao Journal of [Asiatic SOC] [...] p [...] -- most wonderful instinct, how c
478z004 on purpose Mr J. Murray has given paper to Royal SOC on glow worm. luminous property-- Curious arrang
480z006 lis. Dorbigny Vol II, p 24 Proceedings of Zoolog SOC. Important account of habits of Tubularia. p 52.
480z008 t ranks these bodies amongst Vegetables in Linn. SOC.-- Mr. Donn Carmichael Linn. Transacts Vol XII.
489qIFC ell---- 18 Blyth---- 19-- Mr. Tollett {T} Zoolog SOC «Gardens» ---- . . 20 & Breeders Dr. Boott: R. B

Page **(Key Word)**
511q018 in negro & white (3) About the Bantams at Zoolog SOC.-- did Sir. J. Sebright select to destroy second
512q020 how many miles they fly in few hours Zoological SOC (1) Do the animals there, sometimes couple but n
594n115 --Affection & [...] Monkeys «Ogleby» seen Zool. SOC-- 1838 remember with distress their companions--
451e177 deer-- Horsburgs. Vol II. p. 527.-- <Scientific SOCI> Journal of Asiatic Society Vol I. p. 261. <J>
623o050 e calls up pleasure.-- [LHC] the instinct of SOCIABILITY & sociability, doubtless grow together [RHC]
623o050 asure.-- [LHC] the instinct of sociability & SOCIABILITY, doubtless grow together [RHC] This feeling
409e049 t we can see if all species, there would not be SOCIAL animals. «this is stated too strongly. for the
409e049 would be innumerable species. .& hence few only SOCIAL there could not be one body of animals, living
409e049 mals, living with certainty on other» hence not SOCIAL instincts, which as I hope to show is «probabl
409e049 tate.-- «whether he was or not. He is present a SOCIAL animal» a fact few will dispute, [although, th
537m076 atural. consequence of man, like deer &c, being SOCIAL animal, & this conscience or instinct may be m
552m132 generation, (as friendship to fellow animals in SOCIAL animals) are those which are good & consequent
558m150 uided» or strong instinctive sexual, parental & SOCIAL instincts, giving rise "do unto others as your
558m151 in mind many new relations from language.-- the SOCIAL instinct more than mere love.-- fear for other
564n003 Therefore I say grant reason to any animal with SOCIAL & sexual instinct «& yet with passion» he must
564n004 otion of deity effect of reason acting on (<not SOCIAL instinct>) but a causation. & «perhaps» an ins
584n071 mperature molting & breeding instincts, sexual, SOCIAL, «subordinate to,» self preservation, (knowled
593n107 does music now excite our feelings.-- How does SOCIAL animal recognize «& take pleasure in» other an
593n107 instinct> «insects» which become in imago state SOCIAL) by smell or looks. but it does not know its o
593n109 ure of other men. & hence idea of beauty.-- the SOCIAL affections of animal taking man in place of ot
609o030 their instincts.-- Gives art to when I say How SOCIAL instincts generated? The origin of the social
609o30v w social instincts generated? The origin of the SOCIAL instinct «in man & animals» must be separately
619o042 be concluded that he has parental, conjugal and SOCIAL instincts, and perhaps others.-- [LHC] ----- p
622o049 which Sir. J. says is so ridiculous. [RHC] the SOCIAL instinct may be combined with feeling towards
623o050 es from man. during many generations giving the SOCIAL feelings.-- [LHC] According to my theory, all
623o050 r instincts extend, yet as they are acquired by SOCIAL animals, living under certain conditions, in t
623o050 aw,> «can only be such, as are consistent» with SOCIAL animals, that in which have a beneficial tende
623o051 been formed> a beneficial tendency to them, as <SOCIAL> animals of peculiar <kinds> «social feelings»
623o051 them, as <social> animals of peculiar <kinds> «SOCIAL feelings», & living under certain conditions;
628o054 solely due to greater duration of impression of SOCIAL instincts, than other passions, or instincts.-
629o055 nes) Perhaps my theory of greater permanence of SOCIAL instincts explains the feeling of right & wron
593n109 nsects. if this view holds good-- then man, a SOCIALIST, does not know other men by smell, but by loo
120a110 Craigs. Kept amongst <old> papers read before SOCIETIES.-- Sir. J Hall Vol VI. p 173. (Ed. Transact)
071r157 re is map of Cordillera by Humboldt in Geolog. SOCIETY Sir Woodbine Parish informs me that town near
103a063 Proceedings» List of collections in Geological SOCIETY. Pumice at South Shetland. Geological Society-
103a063 Society. Pumice at South Shetland. Geological SOCIETY-- Dikes have not been the moving agents, becau
134a143 lake-- Berghaus Chart of do Journal of Asiatic SOCIETY Vol I. {T} p. 145. on salt mines of Punjab p.
245c024 Flora of Falkland Islds, where is it? All the SOCIETY isles have the same productions p 293-- is ver
271c107 untries.-- Fraser remarked to me at Zoological SOCIETY,, that you never find two «similar» groups of
321c275 story of Man Transactions of the Entomological SOCIETY Vol I. & 1s No of Vol II. (read remainder) whe
324c268 s cannot be made out Find out from Statistical SOCIETY-- where M. Quetelet has published his laws abo
399e010 kes occur there,, but are unknown in Hervey or SOCIETY isles. Hope says positively he has seen. a Cal
451e177 p. 527.-- <Scientific Soci> Journal of Asiatic SOCIETY Vol I. p. 261. <J> Catalogue of Birds of India
480z006 ten account of Guanaca. In transaction of Bonn SOCIETY M. Edwards on Corallines L'Institut 1837. No 2
534m062 ug. 7th--38. Transactions of the Entomological SOCIETY of London Vol. I. p. 106. Col. Sykes on Formic
609o29v elieve) to present day & not to ruder state of SOCIETY.-- Civilization is now altering these instinct
609o030 [...]> rule may sometimes be hard to tell) + + SOCIETY could not go on except for the moral sense, an
622o048 - hence the law of honour. & the etiquettes of SOCIETY.-- [LHC] Sir J. M. gives different explanation
622o048 ed conscience, is the feeling of any custom of SOCIETY broken..-- & how far more <feelin> acute the f
628o53v either, that from the necessities «& good» of SOCIETY such conduct is instinctive in me (& as a cons
266c092 RNAL. Vol V. p 201 Wellsted. Memoir on isld of SOCOTRA. Cattle generally marked like those of the Ald
417e076 d breed, It is difficult to think of ¿¿Plato & SOCRATES, when discussing the Immortality of the Soul
096a041 aper on India gives reason for knowing that Mur. SODA. and Carb of lime decompose each other.-- on Di
109a084 ood p. 411 When discussing concretions Carbonate SODA. formed by Ca. of L. & Mur. of Soda mixed.-- Tu
109a084 ns Carbonate soda. formed by Ca. of L. & Mur. of SODA mixed.-- Turner's Chemistry p. 206 Both Beck &
127a126 dity.-- do p. 137. Lord Tullamore found Sulph of SODA in peat ashes in Ireland dikes in mountains. «(
496q006 rows-foot. or Ranunculus.= (11) Try.. Nitrate of SODA-- Salt. Gypsum. Magnesium Iron Rust Carb. of Am
505q014 e.-- Herbert's. fact.= (4) Effects of Nitrate of SODA under Beech.-- Lychnis dioica answers this ques
044r073 . America on the other: The extreme frequency of SOFT materials being consolidated; one inclines to b
051r094 fonsos «Kelp») rocks show signs of degradation; (SOFT substances worn into bare cliffs evident); the
073r161 s been accumulated, round knobs, or pushed where SOFT, or redissolved soft.--/ is there any flexure <
073r161 ound knobs, or pushed where soft, or redissolved SOFT.--/ is there any flexure <fr> in the fragmentar
134a141 p. 6. few «living» shells. on coast of do p 8.-- SOFT Clay beds near C. Virgin p. 59. dip of Clay sla
157g074 ly cause by fall of angular masses from above on SOFT shelf-- 29.330 A 84 degree compare this with la
297c186 n to abortive spiracles in Hemiptera do. p. 160. SOFT plumage of night jar. like owls. analogy in hab
319c256 yet they have one with very sweet notes.-- Their SOFT-billed birds are inferior to ours, & our lark r
581n064 .-- touch apparently. ourang outang very fond of SOFT, silk-handkerchief-- cats & dogs fond of slight
607o025 me, & great effect being produced.-- the wax was SOFT;-- the condition of mind which leads to motion
119a106 absolutely fluid pool.-- (this is shown by the SOFTNESS & curvature of quartz rock?) also by my pheno
124a117 2. Facts from Erman about great depths of frozen SOIL. p. 211 Consider proved that Siberia must have
124a118 1 (B.) Wrong Entrance. Book C. p. 101. On Frozen SOIL of Siberia (with refer to Metamor) wrong entran
126a124 surface can take place.-- | the depth of frozen SOIL is against this view.-- however it is said in s
127a125 up to «the» line of ice.-- Hence further N. when SOIL frozen for greater length of time depth of ice
131a136 sothermal line Athenaeum. 1839. p. 52. On Frozen SOIL of Siberia.-- facts of water flowing from benea
146g014 unded pebbles in yellowish argillaceous or sandy SOIL-- These Buttresses formed vestige of irregular
156g068 ders of same granite, all on these three shelves SOIL is <the> usually slaty Point of rounded not sco
171b003 to» circumstance,-- seeds of plants sown in rich SOIL, many kinds, are produced, though new individua
188b068 all would come up white, though planted in same SOIL with blue. Now this is same bearing with Dr. Sm
195b103 when country changes. Will it said that Volcanic SOIL of Galapagos under equator that external condit
269c101 n to Geograph. Soc. in February or March 1838 on SOIL in Siberia being frozen to 400 ft in depth, (&
371d128 d down? Are not Loddiges 1279 roses kept in same SOIL. same atmosphere?-- may they produced not be tr
441e149 - Thus would a Crab tree vary if planted in rich SOIL, I presume not, but its seeds, I presume, proba
441e150 of "laws of organization" had better be shown-- SOIL on colour of flowers, Hydrangea -- black bullfi
491q01v of Carrot 2 {also try Primrose & Cowslip in rich SOIL & propagate from their seed 3. To apply pollen
495q005) Sow <daisy> seeds of wild cabbage in VERY rich SOIL, will plants abort?, does it require successive
495q005 successive generations to accustom them to such SOIL.-- Sow weeds in such soil.-- 7 (a) Experimentis
495q005 accustom them to such soil.-- Sow weeds in such SOIL.-- 7 (a) Experimentise on Primrose seeds-- it r
496q006 ne monstrous) Fox Glove & such like in very rich SOIL-- As they have little tendency to double; what
497q006 ons not effect of Chance Maer.= (12) Take Bag of SOIL from centre of woods «especially if date of woo
498q007 other case.-- (6) Will plant accustomed to rich SOIL, when placed in very poor flower, but not fruit
498q007 ot orchards become unproductive from poorness of SOIL.-- yet crabs probably would grow there (7) Wher
501q011 tion of Col. Le. Couteur about Wheat-- Change of SOIL-- crossing-- when seeds raised.-- His Book.-- 3
633j54r lants created to <arrest> «prevent» the valuable SOIL in its seaward course,-- we sink into such cont
299c195 chiefly produced by cultivating parent in rich SOILS & their seeds produce <offspring> variety. wild
495q005 & place cuttings or bulbs in several different SOILS & temperatures & see what the effect will be.--
514q21. ury says no «hollow» spur.-- Ask about Pinks & SOLANUM impregnation before flower open. (An. des Sci
489qIFC fferent caterpillar races. --Name of Italian who SOLD eggs.-- Temporary Question 1 Where has Duchesne
265c086 abits-- after seeing beetle with wings beneath SOLDERED wing-cases-- Yet these wings may be of some u
509q017 o Bentham's Remark. Horse or cow.-- degree of SOLDERING of tibia & fibula: in Man any abortive bones?
409e049 act few will dispute, [although, that it was the SOLE object, I will dispute, when I hear from the ge
505q014 nose over head to root of tail 28½. inches. From SOLE of foot to shoulder on line of back, height 17½
549m121 iterally New Testament future life is almost the SOLE object--. -- I doubt whether the last be right.
063r133 ccess.--showing non Creation does not bear upon SOLELY adaptation of animals.--extinction in same man
181b045 probably be not complete, if birds were fitted SOLELY for air & fishes for water. If my idea of orig
257c059 like itself, but ¿whether great assumption? not SOLELY producing like itself, not applicable to monst
261c070 lack men, black bull finches from linseed-- not SOLELY effects of climate on some «antecedent races,
267c094 ush> woodlands for miles in extent are composed SOLELY of this shrub".-- p. 229. carcases of birds dr
276c124 lus forficatus Tyrannus Sulphureus if compelled SOLELY to fish. structure would alter.-- It is a diff

Page **(Key Word)**
278c128 ormation of subgenera is empirical, & is judged SOLELY by comparison with other genera in other famil
302c203 intermediate Father-species, & not, therefore, SOLELY owing to such interm: father-species, being li
337d021 rge eye-- have they smell, do plants emit odour SOLELY for others parts of creation) & another nerve
349d052 p. 8-- Anomalous structures, as in Hippotamus, SOLELY owing to number of lost links.‖ if all species
397e004 as been small-- because change in forms is <al> SOLELY adaptation of whole of one race to some change
399e009 ptation to whole life of animal, & not if it be SOLELY to womb, as in monster. or solely to childhood
399e009 not if it be solely to womb, as in monster. or SOLELY to childhood, or solely to manhood,-- it will
399e009 womb, as in monster. or solely to childhood, or SOLELY to manhood,-- it will decrease & be driven out
399e010 ct, that breeding in & in, produces bad effects SOLELY, because of similarity, because in every count
530m046 -- An intentionally recollection of anything is SOLELY by association, & association is probably a ph
531m050 e key. where all keys are placed) Memory cannot SOLELY be number of times repeated, because some peop
531m051 scene of anger.--» to the pianoforte, it seemed SOLELY to be feelings of discomfort, especially about
544m103 parate ideas cause fatigue to the mind,-- it is SOLELY the comparison, with past ideas. which makes c
608o027 them. Yet it is right to punish criminals; but SOLELY to deter others.-- It is not more strange that
628o054 se, by its "supremacy",-- I make its supremacy, SOLELY due to greater duration of impression of socia
375d134 lthus.-- «increase of brutes, must be prevented SOLEY by positive checks, excepting that famine may s
073r162 e magnetic properties. Study well products of SOLFATARAS[.] some general laws. association of lead &
074r165 e grand fact of Volcanic & non Volcanic. Then SOLFATARAS. «Mem: Micaceous iron ore.» N.B. To show how
028r029 ow is the Lime separated; is it washed from the SOLID rock by the actions of Springs or more probably
055r108 orrent proves the decay atmospheric of the most SOLID rocks.--The grand cliffs of a thousand feet in
055r108 and cliffs of a thousand feet in height, of the SOLID lavas.--proportionally high to age. (we do not
068r147 ins & mass[es] [Fig. 8] {P} (A. B. C, now grown SOLID.) Red Sea near Kosir, land appears elevated. Ge
069r149 of fluid matter not (for instance) expansion of SOLID matter by Heat Consider profoundly the sandston
108a079 al line, but if heat from centre, then crust of SOLID earth would be thicker.-- PP Andes mark the lin
113a092 ort Philip. which had bar at mouth excavated in SOLID rock.-- 4 & 5 fathoms deep. perfectly still wat
122a114 um.-- it cannot be equilibrium of fluid, but of SOLID. because if of fluid, the waters of the ocean w
156g069 some of» those since fallen. «on the 3 shelves» SOLID rock is much notched on Maculloch's supposition
331dIFC n.-- Sir. R. Heron's case of breed of pigs with SOLID feet.-- 1838 [In this Book some curious notes o
068r146 cean, for subsequently there is a coating of SOLIDIFYING igneous rocks which would be too thick to be
318c254 rent in species of same genus.» The Muscicapa SOLITARIA stay about a fortnight in one particular part
260c068 beautiful colours of species-- Mem. St. Jago--SOLITARY Halcyon bird of passage.-- M. Coronata of Lat
389d177 stand much breeding in & in (those which have SOLITARY flower) exotics brought from foreign country.
564n003 different.--Mem. Bee how different instinct a SOLITARY animal still different.-- ‖ Different nations
240c013 ch as yet had only been found in isle of Aroe & SOLOR), «Vol I» likewise new species of Parameles, wh
104a068 nos {P} fissure dike.-- thus dikes terminated SOLUBILITY of fluids varies with temperature ¿ with pre
102a061 s to VEINS in primitive rocks-- Are substances SOLUBLE under great pressure? equally with little pres
120a109 doubt the thermometer. Is not common salt more SOLUBLE in <hot> cold than hot water with «-- especial
044r074 y consider appearances of eruption at bottom.--SOLUTION under high pressure of gazes. especially the
044r074 (& this effect of water thus holding matter in SOLUTION must be great: & in the fact of bombs in tufa
045r074 surface bespeak the changes. -- metallic veins SOLUTION of silex & many other phenomena I do not beli
077r172 P. stone It is clear to me, there are laws of SOLUTION & deposition under great pressure. (? heat!)
091a025 re (& veins appearing): mem. Henslows Anglesea SOLUTION of silex also shewn. No 3d of Ed. N. Phil. J.
092a029 ctrical currents tend to deposit metals, if in SOLUTION. My view of metamorphic in contradistinct to
092a029 contradistinct to Volcanic will explain their SOLUTION. Athenaeum M. 516 1837 High up the Essequibo,
181b042 e greater the groups the greater the gaps (or SOLUTIONS of continuous structure) «between them.».-- f
356d071 as long as not overwhelming) What comparative SOLUTIONS & linking of facts-- Savages over whole world
370d117 pothesis to other points. & see whether it will SOLVE them.-- It is less wonderful that childs nervou
235b274 Royle about Indian Cattle with humps.-- ¿To be SOLVED if horses sent to India. & long bred in & no n
291c165 t.-- love. of man gained & heredetary. «problem SOLVED» habits become important element in classifica
564n005 erience shows the problem of the mind cannot be SOLVED by attacking the citadel itself.-- the mind is
592n101 endless round of doubts & scepticisms might be SOLVED by considering the origin of reason. as gradua
570n025 lyse this.-- «Excellent-- my theory of blushing SOLVES this.--» The similarity of men's reasons: shew
234b256 British Association of 1838. (Newcastle) about SOMEBODY who had made great collection of birds of Ice
522m014 to the body» I know the feeling, thinking over SOMEBODY who has, perhaps, slightly injured me, plotti
523m014 the feeling of anger must be directed against SOMEBODY.-- Have insane people any misgivings of the i
572n033 ctions. senses. notions &c Octob 30th-- Dreamt SOMEBODY gave me a book in French I read the first pag
224b215 an already has produced marked varieties & may SOMEDAY produce something else., but not probable owin
389d177 very close are bred into each other.-- This is SOMEHOW connected (This seems case, for by careful obs
574n041 ciation: I have seen Nina licking her chops.-- SOMEONE has described slovering <gum> «teeth»less-jaws
408e065 ow rising. On a particular part of coast of SOMERSETSHIRE the Cockles are all apt to be diseased., &
308c219 as long as this is so-- !!Metaphysics!!! Mrs SOMERVILLE, Connection of Physical Sciences p 276 May b
120a108 show it cannt be ordinary heat, then there is SOMETHING superadded, that which give cleavage to rocks
172b008 st suppose the change is effected at once, -- SOMETHING like a variety produced-- --[every grade in t
198b113 nge of capabilities past & present might tell SOMETHING) p. III G. St Hilaire Insects & Molluscs allo
224b215 oduced marked varieties & may someday produce SOMETHING else., but not probable owing to mixture of r
271c105 sh of Azara builds its nest in <same country> SOMETHING same manner, much mud.-- These facts show, ha
276c122 pecies <of> gradual, so may we suppose., that SOMETHING intermediate, between no offspring & ordinary
304c206 supra & next page It is a fact pregnant with SOMETHING.? that intermediate. species have generally p
356d070 re is tendency to reproduce in each case, but SOMETHING prevents the completion.-- ‖Say my Grandfathe
372d129 of the living body is present-- in generation SOMETHING is added from one part of the body «(or of ot
390d178 quarters) then it causes <to> a secretion of SOMETHING someways different from himself, for it shoul
391d174 uld there be a necessity that there should be SOMETHING «each time» added to that kind of generation,
458t002 produced a very pretty little animal, showing SOMETHING of Mule in its ears-- ((this is good case as
467t103 .-- In Lupine, Bees «frequent» & seem to act, SOMETHING like on Kidney Bean, they go to nectar at foo
499q009 ious-- seedsmen who raise Hop-seed-- may know SOMETHING about proportion of plants necessary &c &c (a
530m044 ay has mentioned in Antiquities of Shrewsbury SOMETHING about big noses & name Corbet, perhaps nonsen
534m062 th pidgeons finding their way home-- there is SOMETHING wrong in comparing these cases, when agency i
542m095 cause of expression of surprise-- viz seeing SOMETHING obscurely with the wish to make it out?-- See
546m110 was suddenly called to go on the lawn to see SOMETHING, on her return could not find paper cutter, h
548m117 ar when looking at a table one has vague idea SOMETHING is not there, & then when one begins eating o
551m128 for it did with far more alacrity <than> when SOMETHING good was shown him, than when merely ordered
551m129 eas come & the pulse rises, or as they flag & SOMETHING like a snow-haze. covers my whole imagination
553m137 as showing <connection> that expression mean SOMETHING.-- Hunt (the intelligent Keeper) remarked tha
567n013 nimals.-- analogy probably false, may lead to SOMETHING.-- October. 8th. Jenny was amusing herself--,
567n014 «& rubbed» his arm. & show signs of affecting SOMETHING like man. Has an oyster necessary notion of s
570n023 son.-- surprise with negation.-- like shaking SOMETHING off shoulder-- or is it from inspiration, whi
575n046 s of youth, when new ideas will not enter. is SOMETHING analogous. to instinct, to the permanence of
579n059 e us love him, or her.-- it is blind feeling, SOMETHING like sexual feelings-- love being an emotion
586n078 ctively «yes because power varies in breeds,» SOMETHING of kind oneself knows in walking [one feels i
587n087 If dislike, distaste. & disapproval. were not SOMETHING more than the unfitness of the objects then v
589n091 e, yes if every individual played a little, & SOMETHING destroyed bad brain. see p. 90.-- The relatio
600o08v ompound passion. like avarice.-- Is there not SOMETHING analogous to imperiousness of Conscience: in
603o012 t like cruel pleasure, so sacrifices cruel.-- SOMETHING wrong here.-- Origin is certainly curious. Ch
604o018 ht. & with the idea of ascension we associate SOMETHING extraordinary & of great power-- -- 2 From th
618o41v thought &c. should imply «X» the existence of SOMETHING in addition to matter is because our knowledg
325c267 itish Plants. Humes Essay on H. Understanding (SOMETIME) Du Stewart works. & lives of Reid, Smith & g
076r169 Mexico is always accompanied by Sulp. silver SOMETIMES by selenite.-- in New Spain, contrary to Euro
076r169 istribution of silver «in veins» very unequal SOMETIMES disseminated <[...]> sometimes concentrated:
076r169 » very unequal sometimes disseminated <[...]> SOMETIMES concentrated: wonderful quantity of pure silv
146g011 y <hundred> thousand tuns. Black faced sheep, SOMETIMES mottled with white black legs & tail like spe
217b183 when two dogs of opposite breeds are crossed, SOMETIMES offspring quite intermediate sometimes take s
217b184 ossed, sometimes offspring quite intermediate SOMETIMES take strongly after either parent. about <hal
233b250 -banded. ¿species?-- thinks offspring of cats SOMETIMES heterogenous.-- Australian dog jumped into tu
234b255 seldom seen with horns" --- p. 341. Black Fox SOMETIMES introduced by ice <no> «only few» pigs.-- bir
236b280 untries, as Decandoelle says, no he only says SOMETIMES we might expect disseminated species to vary
240c003 y explain mule & pig being half way. Yet dogs SOMETIMES like father, sometimes like mother. The fact
240c003 ing half way. Yet dogs sometimes like father, SOMETIMES like mother. The fact of great monstrosities
264c083 ties of structure. as six fingered people are SOMETIMES heredetary,-- yet these not adaptations «they

```
276c124 of  the utmost importance to show that habits  SOMETIMES go before structures.-- the only argument can
297c189 ry not applicable [not located] p. 428. Ouzel  SOMETIMES builds nest without doom. Vol 2. Mag of Z. &
302c203 iscordant groups. The formation of genera may  SOMETIMES be due to accident as submersion of land cont
308c217 of right side, though my hand <vibrate> would   SOMETIMES vibrate-- «seeing no tea brought back memory»
333d008 her.-- Even in children of parents <some> one   SOMETIMES resembles one parent & one another & are  not
335d015 le & as mutilations «(produced very quickly)»   SOMETIMES have similar offsprings, so will the worst mu
380d155 mal has copulated., though no offspring, Milk    SOMETIMES comes in Mammae & even when bitch is in heat.
385d165 he temper of the father, or of the mother, or   SOMETIMES of both." If L. can be trusted, this is Lord.
386d167 rent from that of plants. (though latter does   SOMETIMES occur in animals). latter the division taking
401e017 of inflorescence Gardeners. by chance <often>   SOMETIMES graft pears on apples. they will live but not
403e021 . Europe p. 189. The gaut, kind of crocodile,   SOMETIMES wanders from Pellew to Eap.-- There is anothe
406e036 ties of same species & to different species--   SOMETIMES like one parent & sometimes other & sometimes
406e036 fferent species-- sometimes like one parent &   SOMETIMES other & sometimes ½ way. Ed. New-Phil. Transa
406e036 sometimes like one parent & sometimes other &   SOMETIMES ½ way. Ed. New-Phil. Transact. Rabies, common
419e085 the non-development of Mollusca, which I have   SOMETIMES speculated might be owing to absolute quantit
423e096 - The Geologico-geographico changes must tend   SOMETIMES to augment & sometimes to simplify structures
423e096 hico changes must tend sometimes to augment &   SOMETIMES to simplify structures:= Without enormous com
434e129 ld> only white in Cambridge, in some counties   SOMETIMES one & sometimes other.-- there is some differ
434e129 n Cambridge, in some counties sometimes one &   SOMETIMES other.-- there is some difference of habit be
434e129 lightly abortive & bed of female flowers will   SOMETIMES produce a few seeds,-- -- Ruscus aculeatus. a
434e129 s. a dioecious plant, in which the Male plant   SOMETIMES bears female flowers, the organs are most cle
437e138 thern stays only winter.-- Jays & chaffinches   SOMETIMES migratory.-- p. 103. Turtles finding their wa
437e139 ung ones to exercise them in killing them.--   "SOMETIMES it seems hares, rabbits, rats & not being suf
447e164 enslow has seen this-- (Poa alpina vivaparous  SOMETIMES seeds All species of Lemna sometimes though v
447e164 vaparous sometimes seeds All species of Lemna  SOMETIMES though very rarely flower [bu]t the one  does
447e165 ns.-- Poa alpina, thougt generally vivaparous  SOMETIMES seeds.-- ¶There are endless curious facts abo
454e184 nts, in becoming double loosing fertility if,  SOMETIMES one, sex & sometimes. other, so as to  become
454e184 le loosing fertility if, sometimes one, sex &  SOMETIMES. other, so as to become <all> monooecious.--
512q019 s-- variation in hounds= An ugly calf <turns>  SOMETIMES turns into fine beast. would its offspring ha
512q020 ours Zoological Soc (1) Do the animals there,  SOMETIMES couple but not conceive :Bears /Yes/ (2) Foxe
513q020 ther the characters are then intermediate or  «SOMETIMES» all on one side, as in crossing varieties Am
515q021 k have <abortive> «no» horns by abortion, but  SOMETIMES have dangling ones.-- Is there any genus of p
523m018 ever generally, if at all discovered.-- <Sup>  SOMETIMES comes on suddenly from <I> (in one case ipeca
535m064 rvae, when two groups near. mother desert one  SOMETIMES & go to other, so that two mothers to one gro
609o030 can  look forward: hence our <[...]> rule may  SOMETIMES be hard to tell) + + Society could not go  on
417e078 ependently of its sex, or half way between, or  SOMEWAY different from either: & or like progenitors.-
455eIBC can be made sensitive The function of sleeping  SOMEWAY useful.-- it is only the association which  is
390d178 ) then it causes <to> a secretion of something  SOMEWAYS different from himself, for it should be obse
072r160 . ≠ The Hollowness of <sep> Chiloe concretions  SOMEWHAT analogous to septa.-- would particle attracte
302c200 external conditions, to be sure, have remained  SOMEWHAT similar.--!!! My theory drives me to say that
335d013 been,  will not remain,-- yet offspring must be  SOMEWHAT like parents,-- therefore offspring will tend
342d033 ose SPECIES which breed most freely. & produce  SOMEWHAT fertile offspring produce heterogenous offspr
353d060 59) looking at animal, if there be many others  SOMEWHAT allied whether <like» parent stock, or not. N
360d090 dog.. & hence general appearance of face & tail  SOMEWHAT like dog-- though it has full share of Jackal
360d091 en sickly tigers have first come over, insects  SOMEWHAT like «between» lice & fleas. sticking on them
365d105 he subalpina of Sweden, (which in summer dress  SOMEWHAT resembles Red Grouse) it may be so-- but very
374d134 ross generally brothers & sisters, & therefore  SOMEWHAT unfavourable-- 28th. «I do not doubt, every o
391d175 eat» difference (for even brother & sister are  SOMEWHAT different) should be added to each individual
405e032 hows that progression of change in Mollusca is  SOMEWHAT similar in two hemispheres.-- It might be wor
427e110 ny kind would fertilize it» fertilize an apple  SOMEWHAT more readily, «than other apples» but probabl
450e175 asionally reaches 13 hands.-- Phillipines Pony  SOMEWHAT resembles that of Celebes is somewhat  larger
450e175 nes Pony somewhat resembles that of Celebes is  SOMEWHAT larger than the Sambawa, Java & Sumatra breed
465t081 breeds.  (though in this case crossing has had  SOMEWHAT to do with it. mem. dogs «& pigs» in Polynesi
467t099 » shrivel, yet stigma does not, so we may feel  SOMEWHAT «but little» less surprised at Henslow's rema
511q018 t. American & Europaean common species, having  SOMEWHAT of different appearance.-- {will introduce it
538m078 getting their insanity» there seem other cases  SOMEWHAT analogous, & which I think will lead to  fact
548m115 trouble «I well recollect» is in making things  SOMEWHAT probable. in comparing every step, & inventin
548m116 f what individuality is.-- Insanity is <much>  «SOMEWHAT» the same as double consciousness, as shown i
575n044 lk about & hunt mice-- Jenners Jackall Have we  «SOMEWHAT» right to deny identity of instinct.-- Habits
584n072 & dogs. (even blind horses & dogs) shows it is  SOMEWHAT analogous to memory. Shrugging shoulders seem
038r057 raters are all burning, surely there must be  «SOMEWHERE» below a field of fluid rock.--In the discuss
342d034 crossed is uncertain-- Yarrell remarks he has   SOMEWHERE met conjecture that all salt-water fish  were
453e182 ld (carnivora) beasts on Phillipines. Forrest   SOMEWHERE says same.-- do p. 393. <">The wild, small fo
523m019 nheriting he-- This «N B. I have read paper     SOMEWHERE on horse being insane at the sight of anythin
530m044 perhaps  nonsense.-- look to it My father has   SOMEWHERE heard (Hunter?) that pulse of new born babies
548m114 Abercrombie's case of «in Botanical Student»   SOMNAMBULISM, did reason about himself-- but not about, f
546m110 t double consciousness Mayo compares it with   SOMNAMBULISM.-- the young lady almost equally in her sen
559m156 -- My Father about double consciousness.-- &   SOMNAMBULISM. Do people when inhaling Nitrous oxide, for
607o023 ined that way]CD one sees this law in man in   SOMNAMBULISM or insanity. free will (as generally used)
548m115 hat is not recollected, nor of the Botanical   SOMNAMBULIST. (if he had been unhappy)-- it is because i
449e170 are  allied to the «type of genera in» islas de SONDA as well by those which are identical, as  those
328cIBC ritten "Coup d'oeil sur la Faune des iles de la SONDE et de L'empire du Japon Wowett on Cattle-- (Wat
295c178 es Barrington says cock birds attract females by SONG. do they by beauty, analogy of man if so war no
319c255 ks the Mocking thrush beats all English birds in SONG.-- one of their thrushes exceeds our blackbird,
527m032 hat «birds from» certain districts have the best SONG. [Migratory birds return to same quarter for ma
538m078 e fact of her not <remembering which» «repeating SONG" when she had recollected it in perfect senses.
538m078 d as self.-- «When Miss Cogan has remembered her SONG, then the song was to her like one which though
538m079 hen Miss Cogan has remembered her song, then the SONG was to her like one which though learnt in infa
603o11b ]CD or poetry, CD[my thery says yes. imitating  SONG -- two primary sources, sight, & hearing-- Stau
363d102 as Lark & Nightingale which both sing their own  SONGS, though imperfectly.-- Male birds always second
363d102 h imperfectly.-- Male birds always second their  SONGS, the ++ Cervus Campestris spotted white when  a
532m055 at in old people, in their dotage, who sing the  SONGS «& tales» of infancy, it is very doubtful wheth
538m078 I think will lead to fact of old people singing  SONGS of their childhood. & certainly of Miss  Cogan,
059r119 te et souvent taillée a pic. Toutes ces montagnes SONT formées de couches paralleles & inclinées du c
089a018 . 11th Observ.-- Les grands tremblemens de terre SONT presque toujours precedes et suivis, queque tem
209b156 ien encore on pourrait au plus en conclure quels SONT les genres qui, sous ce climat, se divisent  le
402e020 he N. end of Borneo to the adjacent island-- In  SOOLOO we find the elephant-- in Magindanao several k
450e173 better than the latter-- Forrest Voyage p. 323.  SOOLOO. imported elephants. wild hogs-- spotted deer,
450e176 » the Breed of elephants <of> in little isld of  SOOLOO.-- said to have been imported: shows they will
122a114 revent internal fluid arriving at equilibrium so SOON from; crust being cut of-- if part of «cold» cr
145g002 es of Animals in Britain shows that either races  SOON made or crosses difficult Salisbury Craigs  The
163g108 ve shells must have been about 60 ft above sea-- SOON decayed on exposure Mr H. C. Watson Geographica
176o024 y points of affinity in each branch A species as SOON as once formed by separation or change in  part
200b124 at» Keeling these are most important facts.-- As SOON as island large enough for land birds, seeds pi
223b211 o A. B. C. D.-- Destroy plants B. C. D. & A will SOON form good species! The increased fertility of s
230b240 cle.-- Falkland Isd case good one of animals not SOON being subjected to change in Americas.  perhaps
277c128 he mark of analogy would be empirical because as SOON as two species were placed in different subgene
364d104 ticated animals, that the amount of variation is SOON reached-- as in pidgeons no new races.-- In Sca
410e052 w & causes of change".-- A philosopher, would as SOON turn tailor. as, mere describer of species, fro
423e096 destroyed, the more highly organized ones. would SOON be disorganized to fill their places.-- The Geo
434e128 s about fossil Infusoria becoming extinct not so SOON as other forms.-- p. 36.. speaking about the co
470t176 & Palmated was introduced about '65 years ago--& SOON after mules abounded--so that palmated has  now
485z017 T. del Fuego p. 141. How comes it salt water so SOON putrifies?? p. 319. on Hydra-- polypi-- <Rep> d
502q11v equal in flower with leaves.-- strawberries How <SOON> «early» do characters of races of different ve
525m026 Aunt-- B. the affections «& N B affections very  SOON go in Maniacs» seem to have failed even more th
566n012 n that any one fact was connected with law.-- As SOON as any enquiry commenced, for instance probably
568n018 in, & vice versa pleasure in performing it.-- As SOON as memory improved. direct effect of  improving
571n028 education.  else why do some children acquire it SOON. & why do all men. agree ultimately?-- We acqui
```

```
580n062 ions of the infant likewise before two years are SOON lost; yet many of the habits acquired in that a
584n073 ness E. says she can perceive sigh, commences as SOON as painful thought crosses mind, before it  can
594n112 red-- though instinct so firmly implanted, birds SOON <dis> learn to disobey it-- I have seen hawk  &
613o036 of attacking peccari-- --retriever-- produced as SOON as brain developed, and as I have said, no soul
619o043 & its love of its puppies: the latter generally SOON conquers, & the dog [RHC] 3) probably thinks no
620o044 he cause of this may be, everyone must know, how SOON the pleasure from good dinner, or from a blow s
620o044 ts consisting in desire gratified & therefore as SOON as desire is fullfilled, pleasure forgotten. [R
621o047 ht to follow certain lines of conduct, & he must SOON necessarily learn that it is his interest to fo
636j56v there were insects «¿when were Palms formed?» as SOON as Dioecious Plants were formed. Macculloch say
189b073 , which die at one time, though produced either SOONER or later.-- Prove animal like plants:-- trace
206b149 (for otherwise the relationship would converge SOONER) & lastly perhaps some one single one.-- Will
427e110 stain offspring-- it would not reach one apple SOONER than <other:> «that of another» apple. only ef
527m035 oughts are brought into play & then perhaps the SOONER castles in the air are banished the better.--
552m132 nse Mitchell Australia Vol I, p 292 "Dogs learn SOONER to take kangaroos than emu, although young dog
624o051 h it. thus a dog may be trained to hunt one pig SOONER than other, rather than change hunting instinc
256c053 ve been more homogenious.-- There must be some SOPHISM. In Lyells statement that some species vary mo
360d093 though externally similar.-- this however is a SOPHISM for their brain or stomach would be different.
554m141 stance of Dray Horse going down hill.-- (argue SOPHISM of association. Kenyon, & then go on to show,
573n038 voluntary & involuntary movements.-- Person with SORE-throat told not swallow spittle. will have invo
295c183 logy X p. 12 do. excepting salmons L' Institut. SOREX from Mauritius. p. 112. & paper on genus Magazi
052r099 t in T. del Fuego, excepting in Port Famine Mr SORRELL says that numerous icebergs are commonly stran
499q09v rcurialis-- Frog Bit, Valerian-- Urtica Dioica SORRELL. Lychnis. Butchers Broom-- «also, Vinca,» Exam
228b231 h affections, imitation, fear <of death>. pain. SORROW for the dead.-- respect We have no more reason
568n017 united with reason) it would feel «subsequent» SORROW, whatever the cause had been]CD-- «Also» When
578n055 should not, & let him be thinking over it with SORROW,-- let the possibility of this being discovere
540m088 eakness, melts into tears, with sensations of SORROWFUL delight, very like best feeling of sympathy.-
275c121 Heads  Mr Yarrell will mention in his work I am SORRY to find Mr Yarrell's evidence about old varieti
564n002 might otherwise have been relieved, he would be SORRY or have a troubled conscience.-- Therefore I sa
569n022 dieu," with a shrug-- "all I can say, I am very SORRY so it is"-- does not accompany I will not.  I am
569n022 so it is"-- does not accompany I will not. I am SORRY I cannot.-- Expression leave «this» out not  in
037r056 .-- On Pampas looked in vain for a pebble of any SORT; not one was found.-- Miers saw them near? Mem.
099a050 bined with gravity.-- hence changes in dip of no SORT of consequence.-- Therefore < S of  inclination
146g014 wed up to neighbourhood of Tyndrum where a large SORT of <plain> space is thickly studded with ridges
212b165 tralian dog he had, used to burrow like fox.-- a SORT of internal bark. would remain for long time to
234b255 y» like <those of Ice Highlands of> the largest <SORT> of our highland Sort, except in one respect, t
234b255 Highlands of> the largest <sort> of our highland SORT, except in one respect, that those of  Iceland.
332d001 y distant quarters of the county, who had had no SORT of communication, were seized with it, & for te
360d093 one had a> both had mottled breasts, <whem> of a SORT that would allow the offspring to have some dif
376d135 The final cause of all this wedgings, must be to SORT out proper structure & adapt it to change.-- to
521m008 f repeating poetry in her dotage is fact of same SORT. Aunt. B. ditto.-- Case of Mr Corbet of the <Ha
522m014 ightly injured me, plotting speeches, yet with a SORT of consciousness not just.-- From habit the fee
533m060 ressing momentary feelings of gratitude, I had a SORT of consciousness I was not right; though I have
554m139 ll take & give food to Tommy, or anything of any SORT.-- I saw Tommy picking his nose with «a» straw.
615o038 n. [RHC] General-- Instincts, certainly appear a SORT of acquired memory. a permanent secretion of th
129a131 213. Beyond the limits of Alps size of boulders SORTED: ditto Murchisons case.--¿ does it bear on Pat
090a021 of falling stones.--¿ does this bear upon the SORTING of matter. in making trachyte come out before.
289c162 capable of producing young many times & lay two SORTS of eggs-- one remaining through winter. «It wou
401e014 bage, where a great many will not return to all SORTS of varieties, which he attributes to crossing.-
401e015 ocks. Says if any variety of apple be sown, all SORTS come up from it. lately saw a nonpareil sowed b
106a076 ll Earthquake unaccompanied by Volcanos must be SOUGHT after proofs of sinking.-- No Sweden!! swellin
116a100 hmidtmeyer travels into Chile p 29. gold is not SOUGHT for in Chile in beds of river, but in shelving
128a128 unded by salt water. purest fresh water must be SOUGHT for below the sea mark.-- If mountain chains a
228b232 t then have taken him over whole world.-- «--the SOUL by consent of all is superadded, animals not go
417e076 Socrates, when discussing the Immortality of the SOUL as the linear descendant of <Mammferus> «Mammif
551m128 essary ideas" arise from the preexistence of the SOUL, are not derivable from experience.-- read monk
600o008 ve son âme" &c-- -- Confesses these faculties of SOUL.-- (I) <Conscience> «Moral Sentiments» imperati
600o008 finable.-- Has little Chapter on each faculty of SOUL.-- (I) <Conscience> «Moral Sentiments» imperati
600o08v r-- Then these last heads. of separation between SOUL of man. & intellect of beasts, not clear.-- ¿do
613o036 soon as brain developed, and as I have said, no SOUL superadded, so [RHC] 6) thought, however uninte
206b146 «say 2000» and at present day, every ten living SOULS on average are related to the (200dth year) deg
600o008 se of duty-- which makes struggle in man.-- two SOULS in one body-- (2) Beau ideal, refers chiefly to
613o036 ons of brain) As in animals no prejudices about SOULS, we see particular trains of thoughts as fear o
022r008 Hippotomus;  the hairy lips of which were still SOUND and not petrified, and the jaw was also firm, o
023r013 scover no shells: nothing said about K. Georges SOUND The idea of the water at Cauquenes. coming from
036r051 ed Sea.--which occurred in a sandy place.--(the SOUND was long & prolonged). NB, Is it generally know
036r051 . NB, Is it generally known. the acute chirping SOUND produced in walking over the sand: I am  nearly
037r056 e West coast. Chonos to Copiapo.--Sydney. K. G. SOUND. C. of Good Hope.--Carnatic It has been  common
065r140 e outside coast of T. del Fuego. off. Christmas SOUND. -- «(Think some 60 fathoms, none thicker  than
066r141 imensions: Bynoe informs me that in Obstruction SOUND, in the narrow parts which break through the  N
108a081 te & amphibole= Cook found Granite at Christmas SOUND Vol XIV. (My Edition) p 500. Well described [to
115a097 ing broken through like that near-- Obstruction SOUND in S. America The very general absence of fragm
134a141 59.  dip of Clay slate in T del Fuego Admiralty SOUND. SE dip. much p. 136. Rocks on Western Coast p.
337d021 reation) & another nerve to finest vibration of SOUND.-- which is impossible.-- Mr Spence remarks tha
522m013 - My F. says there is perfect gradation between SOUND people and insane.-- that everybody is  insane.
527m033 eases from instinct the ears (rhythm & pleasant SOUND per se) & causes the mind to create short vivid
531m053 every animal moves quickly away from any sudden SOUND or noise, & therefore brain has been accustomed
532m055 from any effort of will whilst their minds were SOUND. Caroline tells me that Nina, when brought from
542m094 yet when angry it is hard not to growl out some SOUND even if it be inarticulate.-- the maniac shouts
546m109 & innate tastes, he partakes, taste for musical SOUND with birds. & ¿ howling monkeys-- smell with ma
552m131 ed with habitual stopping of breath to hear any SOUND.-- attitude of attention «So intimately connect
569n020 od & word open your mouth said, recognizes that SOUND as perfectly as a man.-- Probably, language com
569n020 from letters, symbol of word beginning with the SOUND of letter)-- crying yawning laughing being nece
571n031 here must have been intimate connection between SOUND & language.-- Chinese. simplest language.  Much
599o05v ted-- we see their origin in names of People.-- SOUND of words-- argument of original formation.-- se
130a133 some  such case]CD what would be the chance in SOUNDING over a continent to fall across a hot.--sprin
024r016 ute S.] 18 77 C. Frio [23 degree S.] 7 60 {T} SOUNDINGS about same as last to N. of C. Frio Except a
025r019 -- My general opinion from the examination of SOUNDINGS, from about 80 fathoms & upwards. that life i
030r035 ter of coast regular & <not very> rather deep SOUNDINGS, 60-100 fathoms 2 & 3 miles from shore. V. Ch
034r045 es Mem. Beechey. account of regular change in SOUNDINGS. on approaching the coast of NW. America P. 2
052r098 xcess as first case. When discussing Falkland SOUNDINGS introduce this discussion.-- Brazil bank: (& I
065r140 lklands.--off East Coast. -- Capt. Cook found SOUNDINGS. (end of 2d voyage outside coast of T. del Fu
080r181 auforts Karamania Capt. Ross. & Scoresby deep SOUNDINGS Gilbert Farquhar Mathison travels Brazil. Per
112a088 birds drifting out to sea do p. 358. changed SOUNDINGS in Mouth of S. Cruz in connection with Fitz R
134a141 || Fitz Roy refers to || & Rocks p. 375. on the SOUNDINGS on outer coast of T. del. Fuego.-- p 385 Rock
306c213 tralia «map to King's Australia» by a bank of SOUNDINGS of which «there appears to be one line, in wh
426e108 hould remember, that even birds can imitate the SOUNDS surprisingly well-- In early stages of transmu
521m008 ared to birds singing, or some instinctive <or> SOUNDS.-- Miss C. memory cannot be called memory, bec
542m094 ion.-- It is not a little remarkable that those SOUNDS which are involuntary, are common to animals.-
542m097 ikewise must understand each other expressions, SOUNDS, & signal movements.-- some say dogs understan
551m129 erent from sneer-- How easily. horses associate SOUNDS may be seen by omnibous Horses starting,  when
568n018 leasant plaintive cry--» The taste of recurring SOUNDS in Harmony common to t[he] whole kingdom of na
569n020 tter)-- crying yawning laughing being necessary SOUNDS... not produced by will <by> but by  corporeal
574n039 Poet,  has numerous lines. of poetry.-- <signs> SOUNDS singularly adapted to subject see <A> || I thin
581n065 that it was progressively formed. (--names like SOUNDS)--. Horne Tookes tenses, &c &c -- <also g> if
595n127 rs pout]CD [blank] Goldsmiths Essays No XV,, on SOUNDS of words being expressive, (Vol. 4 of Works) [
161g098 degree  75 degree? The River <the> of which the SOURCE is a lip with the new shelf flows into canal b
181b040 g.-- If we thus go very far back to look to the SOURCE of the Mammalian type of organization; it is e
190b076 centre (p. 586)-- Parallel 33 degree-35 degree, SOURCE of forms. reduce towards Northern Eastern  end
```

Page		(Key Word)	
302c201	g the fish & Mammalia, must be sprung from some	SOURCE	anterior to giving off of these two families,
304c208	this argument not applicable to apterix.-- but	SOURCE	of error for if some of the ostriches were to
406e035	ull paper. L'Institut 1838. p. 338 A most grave	SOURCE	of doubt. in distinguishing <after> which pare
549m121	intellectual <ple [...] hope> cultivation, main	SOURCE	of the intense happiness.-- it is again anothe
602o010	ection of this science of abstract form" is the	SOURCE	of part of the highest enjoyment in mutilated
605o19v	- It appears to me, that we may often trace the	SOURCE	of this "inward glorying" to the greatness of
617o40v	ur external senses--only movement can.-- 4) the	SOURCE	from which it arises. But coming round to the
038r059	vertical? if so posterior to elevations? & not	SOURCES	of lava streams.--Urge not tilted strata.-- It
546m108	iority of man perhaps depends on the number of	SOURCES	of pleasure & innate tastes, he partakes, tast
603o11b	thery says yes. imitating song -- two primary	SOURCES,	sight, & hearing-- Staunton Embassy Vol II p.
610o033	school of Kant. to Coleridge, is regarding the	SOURCES	of knowledge.-- whether <we th there> "anythin
209b156	t au plus en conclure quels sont les genres qui,	SOUS	ce climat, se divisent le plus aisément en espè
024r015	8 leagues, from Sydney 90 fathoms La Peyrouse.	SOUTH	of Mocha; 19 miles. 65 Fathoms Vide facts in Be
026r020	pposed to common opinion The Tertiary formation	SOUTH	of the Maypo at one period of elevation must in
029r032	. Smith says. that Boulders do not occur in the	SOUTH	African plains.-- Sydney no I believe the secon
055r106	Potosi the veins run from North <inclining> to	SOUTH.	inclining a little to the West: the veins whic
065r138	Capt. Forster expedition from <Deception Isld.>	SOUTH	Shetland Cape Possession. Syenite¿ Andite?-- De
066r141	in the narrow parts which break through the N &	SOUTH	lines the tides form eddies with its extreme fo
087a011	ropical land.-- Siberia rises. therefore to the	SOUTH	sinks.---- Mediterranean continent correspondin
103a063	of collections in Geological Society. Pumice at	SOUTH	Shetland. Geological Society-- Dikes have not b
134a142	hia Blanca Mr. Malcolmson says are like Kankaer	SOUTH	of Part Luconia-- Phillipines there is volcano
146g013	on surface of any hill Thursday On side of Hill	SOUTH	of upper end of Loch Dochart buttresses of Allu
172b008	t to find representative species; this we do in	SOUTH	America closely approaching.-- but as they inos
187b062	ached S. America.-- explains how Zebras reached	SOUTH	Africa-- It is a wonderful fact Horse, Elephant
188b066	land.-- Mem. Shew Mice.-- --- Animals common to	SOUTH	and North America.-- ¿are there any? Rhinoceros
191b083	ordinary event. & succession the extraordinary	SOUTH	Africa. proof of subsidence. & recent elevation
193b093	ournal. Vol I p. 174. says from Swan river long	SOUTH	coast all the remarkable Australia genera, coll
194b094	cannot be made to approach the Colobes which in	SOUTH	Africa, appear to represent the semnopitheque o
194b095	ly by bones.-- Mem: Silurian fossils: ¿How are.	SOUTH	American shells? ¿Do not plants, which have mal
195b098	umage «&» tone of voice partly American North &	SOUTH.	-- (& geographical <distri> division are arbitr
196b106	ould only have left off springs <ne> in or near	SOUTH	Hemisphere. Were they produced in several place
217b187	Norfolk Isl.-- Isle of Pines-- Australia.-- A	<SOUTH>	American «form of» Lathyrus has one species in
225b220	tendency.-- Study Ellis & Williams. zoology of	SOUTH	Sea islds. any animals?-- I believe none.-- Can
229b233	s of Rhinoceros range from Abyssinia to extreme	SOUTH	coast. Elephant he believes is mentioned by old
232b245	ocene epoch), whilst others may die out or move	SOUTH	ward.... species must be compared to neighborin
234b256	inds certain varieties of a Harpalus. common at	SOUTH	end, but <rare> «absent from» near London. = Dr
236b279	fox was found), decidedly next species to some	SOUTH	American kinds.-- Are the closest allied specie
241c016	IX.-- M. D'.Urville on the Distrib of Ferns in	SOUTH	Sea (Indio Polynes: <)> vegetation far East) An
256c054	t makes species in other animals.--? Forster on	SOUTH	Sea, will probably contain descriptions of dome
259c064	absurd, ∵ not applicable to plant, Epidemics of	SOUTH	Sea, wonderful case of extermination of species
260c067	y, Spix, &c might compare birds of N. America &	SOUTH-	- Any how temperate regions-- crows in N. Ameri
291c168	ect series or gradation.-- It is easy to see if	SOUTH	America grew very much hotter, then Brazilian s
291c168	ch hotter, then Brazilian species would migrate	SOUTH	ward being ready made.-- & so destroy individua
314c238	er (He will give sketch of botany of islands of	SOUTH	seas says so in preface.-- Mr Brown says charac
319c256	rush being so very beautiful gret contrast with	SOUTH	America.-- In Home's History of Man at Maer, it
326c266	in first volume of Annales of Vienna. sketch of	SOUTH	Sea. Botany R. Brown. has curious coloured maps
379d152	ays it is well known that in breeding very pure	SOUTH	Down that the ewe must never be put to any othe
406e037	eat orders destroyed about same time in North &	SOUTH.	America.-- Whole wor[l]d, formerly possessed a
407e038	ca from the «low» limits of blocks both North &	SOUTH,	has probably undergone a greater change, than
410e051	to representative species., when going N.orth &	SOUTH	Thinking of effects of my theory, laws probably
424e100	uld have known the Composites of Galapagos were	SOUTH	American.-- several cases of species peculiar t
434e129	cts: believes that «only» red Lychnis grows in	<SOUTH>	Wales & certainly <old> only white in Cambridg
450e175	p. 270) says many wild horses, bullocks, & deer	SOUTH	part of Mildanao.-- Q Horse do. Appendix. p. 43
477z002	pichye, pelud, mulita et mataco.) are all found	SOUTH	of 26 degree 30 minute. Lat -- -- do. p. 207. L
025r017	T. del Fuego = The Wager's Earthquake the most	SOUTHERN	one I have heard of In a preface, it might be
037r052	of any rock situated beneath low water in the	SOUTHERN	ocean not being buoyed with Kelp.-- With resp
051r093	tter cases become petrified, & increase. -- In	SOUTHERN	regions every rock is buoyed by Kelp, now Kel
101a056	uence dispersion of Boulders.-- See Rogers for	SOUTHERN	limits of Boulders in N. America do/p. 280. t
106a076	m. 1838-- p. 137. Three inosculating rivers in	SOUTHERN	America ¿ effect of subsidence-- <Is there sa
138a151	e is an account of Sellow Geolog. Observat. in	SOUTHERN	Brazil. [blank] «p. 4. (Lyells Book)» Observa
188b067	ers probable two Northern species replace <No>	SOUTHERN	kinds-- (I) Gnu reeaches Orange river & says
190b077	- Proteaceae & other forms (?) being common to	SOUTHERN	hemisphere, does not look as if S. africa peo
194b095	?]CD & opossum found in Europe now confined to	SOUTHERN	hemisphere.-- If these facts were established
205b144	. take form of awk, because there is no awk in	SOUTHERN	hemisphere. does this rule apply? A Treatise
232b247	ing to all analogy would have been very unlike	SOUTHERN	Europaean ones.-- "a variation played on secu
232b249	ascar, on neighbouring islets & a sub-genus in	SOUTHERN	Africa In same manner. Cuscus, (a sub genus o
244c021	arrier.-- --it would make strong contrast with	SOUTHERN	regions.-- «it would now represent. what <act
247c028	ly» 2 species one in Northern Hemisphere 2d in	SOUTHERN	--p. 71 Chimera-- Antarctica «also Taeniatole
250c037	r animals? but they were not shut up!! Extreme	SOUTHERN	points of S. Hemisphere fully characterized,
271c105	Turdus Musicus «not found in N. America» whose	SOUTHERN	range is? <One> The black & white thrush of A
364d104	Jones told me precisely same story about some	SOUTHERN	«see p. 43 supra» breed of cattle with white
406e037	Northern. Hemisphere.-- likewise far North in	SOUTHERN.	-- Great animals. of same two great orders de
421e091	.-- (& Juan Fernandez to Chile??) Falklands to	SOUTHERN	portion.-- N.do p. 269. Annals of Nat. Hist 1
423e098	or their seeds to London is that people in the	SOUTHERN	Counties have whole fields, some for cauliflo
425e105	ized.--» islds., have peculiar forms.-- on the	SOUTHERN	flanks of Alps.-- many peculiar plants on sin
437e138	her in northern part of England stationary, in	SOUTHERN	stays only winter.-- Jays & chaffinches somet
484z014	on of production without Tropics in Northern &	SOUTHERN	America-- valuable & practicable deed Caricar
484z015	pe-- Study profoundly shells of Bahia Blanca &	SOUTHERN	Hemisphere It is most interesting the way Syn
231b243	whereas those which migrated a little to the	SOUTHWARD	would merely be specifically different if so.
058r118	' l île, elles presentent une coupe abrupte et	SOUVENT	tailée a pic. Toutes ces montagnes sont formée
495q005	of Berberis-- (latter I think certainly not) (3)	SOW	seeds & place cuttings or bulbs in several diffe
495q005	s takes place in natural Hybrids of Cabbages (7)	SOW	<daisy> seeds of wild cabbage in VERY rich soil,
495q005	ive generations to accustom them to such soil.--	SOW	weeds in such soil.-- 7 (a) Experimentise on Pri
495q05a	her-- & enclose one twig of each in bell-glass--	SOW	these seeds & see if they will come up true-- wh
496q05a	arrow after feeding on oats, give body to Hawk &	SOW	pellet. ejected. done Examine pollen of such flo
498q08v	rent species growing White Mullein good plant to	SOW	& try to get other species <near> close to each
512q019	bull & hereford cow similar to reverse cross.--	SOW	cast-up-balls of Hawks or even owls.-- How long
516q023	have never seen Bee visit. Experiments in Garden	SOW	stones of Standard Apricot grafted on what, & se
552m131	reality» as in Spectral images-- Mem Chiloe <pi>	SOW,	who carried from all parts straw to make its ne
588n090	ons of a young animal with an old.-- (dog horse,	SOW)	we perceive great difference.-- «(& is not this
400e013	seeds of Dahlia all over Europe same year.-- he	SOWED	them for four generation before they broke.--,
401e015	l sorts come up from it. lately saw a nonpareil	SOWED	by Mr Tollet so produce.-- thinks it probable t
472s01r	uch a thing as a species-- Jun 1. 1842 Allen W.	SOWED	some years since gathered the seeds of Papaver
067r144	Earthquake. Happened on January 20th. 1834 Mr	SOWERBY.	younger. says that Falkland fossils decidedly
316c244	after his brothers murder.--» good anecdote --.	SOWERBY.	--. Geographical range of shells like Cryptoga
316c245	Ferrussac) are confined to <S.> America.-- Mr	SOWERBY	says there are some shells common to West coas
316c246	ny shells at present day same (or according to	SOWERBY	fine species) on coasts of N. America & Englan
408e045	get 50 of same kind of monstrosities.-- G. B.	SOWERBY.	-- Looking over Lamark surprised to see how ma
417e076	ry. & hence offspring are hybrids,.-- Mr G. B.	SOWERBY	<tel> showed me many land shells of the common
423e098	eed-growers--).-- Feb. 24th. Monoceros, which	SOWERBY	says, is an American form.-- has several speci
431e120	ost> be considered merely varieties. & even Mr	SOWERBY	is coming to this conclusion, from specimens i
431e121	d Stone &c &c.--if? so «it is» good case:-- in	SOWERBY	Min. Conch. it is however, said they have been
489qIFC	ton p. 22 Schomburgk.---- 1 Jordan Smith. p 1.	SOWERBY	Cuming. -- p 1 Owen p 17 Hooker p. 17 {T} Mrs.
490q001	t the Indians Do the Savages select their dogs	SOWERBY	Entomologist Does individual Shell or insect o
490q001	nother? Character of shells of Sandwich group	{SOWERBY	monstrous Cardium-- does it remind him of othe
299c195	o biennial domesticated kind with large root by	SOWING	it at wrong time of year & manuring it.-- Epig
501q011	f Lupine, Stocks Clover, to experimentize on by	SOWING	near each other & see whether cross can be obt
171b003	, according «to» circumstance,-- seeds of plants	SOWN	in rich soil, many kinds, are produced, though

585
Page
(Key Word)
299c194 ows in marsh & other dry; yet if T. palustris be SOWN in dry station it will for some generations com
401e014 or other stocks. Says if any variety of apple be SOWN, all sorts come up from it. lately saw a nonpar
497q06v ow-ball-tree. what would result from seeds being SOWN= See in Cultivated Plants, as Pentstemon, which
249c035 oducing again The Varieties of Cardoon are cases <SP> like those of Primrose & Cowslip run wild, The
072r160 {P} motion from within and without H. Kingdom N. SPA. Vol III p. 113 "Nature exhibited to the Mexican
043r071 om any one shock-- In S. America--continuity of SPACE in formations & durability of similar causes go
069r151 te East of Portillo Where gone to? Intermediate SPACE protected.-- Oh the vast power of the ocean! Ma
072r160 u to septa.-- would particle attracted towards SPACE tend to form ring. [Fig. 10] {P} motion from wi
123a116 will, I suspect greatly aided by considering SPACE formed-- great vacuum-- by dike.-- Mem. however
146g014 urhood of Tyndrum where a large sort of <plain> SPACE is thickly studded with ridges & flat topped hi
175b017 possibility of such change,-- as we see them in SPACE, so might they in time As I have before said is
189b070 not doubt even colour hereditary in time as in SPACE. (Mem: Galapagos). Little wings of Apteryx Dace
195b098 islands so permanent a breath cannot reside in SPACE before island existed.-- Such an influence Must
219b195 lora must formerly have been separated by short SPACE from mountains low down, therefore plants commo
246c025 rom East Indies, isld, as far as Oualan.-- Wide SPACE of sea, to East of America. would account for t
269c102 o Appendix to Flinders Voyage by Brown.-- great SPACE seems to act per se as barrier-- Mem. Tartary &
285c151 «on these isld» &c will probably upset it-- The SPACE which one branch of the tree if live occupied a
567n014 ing like man. Has an oyster necessary notion of SPACE-- plant though it moves doubtless has not.-- Tu
580n060 eople first fail in ideas of time, & perhaps of SPACE-- in latter respect he thinks he certainly has
612o035 ion to less simple bodies, and to more extended SPACE, such powers of relation required to be extende
043r070 ontrasted to contemporaneous action over larger SPACES of the globes & "periods" of increased activit
059r121 of concretionary structure, for the interlineal SPACES are of diff cont: & even in one case contained
350d054 framed to fill up empty cantons, & unnecessary SPACES" p 23. "for Nature is the art of God" Septemb
510q017 ent man? about zones separated by non-inhabited SPACES: has he published? does he understand English.
073r163 . of Barytes: Fluoric. Barytes:-- Humboldt. New SPAIN. Vol III. p. 130 Metals in Mexico rarely in se
073r164 in a phonolitic porphyry. = several parts of N. SPAIN great analogy to Hungary. = Veins of Zimapam of
076r169 by Sulp. silver sometimes by selenite.-- in New SPAIN, contrary to Europe. argentifeous lead not abu
079r177 Azores Isds «nor at St Helena.--» Humboldt. New SPAIN Vol. IV. «p. 58» At Acapulco earthquakes are re
179b032 en How is this in West Indies «--:Humboldt. New SPAIN:--» Dr. Smith always urges the distinct localit
187b063 t some circumstance killed it over a tract from SPAIN to S. America.-- Never They die; without they c
324c268 June Is. King & FitzRoy To be read Humbold. New SPAIN-- Much about Castes &c Richardson's Faun. Borea
406e036 ases common to men & animals cow pox.-- case in SPAIN of pustular disease following handling sheep--
461t041 of England remaining same (if so) with those of SPAIN & such facts-- This unequal duration is exactly
576n047 .-- «effects of external circumstances» Look at SPAIN now.-- man's intellect might well deteriorate.
491q01v lly good subject for experiment.--«to repeat SPALLANZANI» Raise only single Plants & only allow <few>
500q010 ches) (23) Talk about Thyme. Horned Oranges. SPALLANZANI Essay-- Figs 2 kinds of flower annually.-- P
323c269 inct -- Cline on the Breeding of Animals -- SPALLANZANI'S Essays on Animal Reproduction -- Treatise o
420e090 other shells, as insects do flowers.-- Mem. SPALLANZANI'S experiments showing how little of the sperm
440e148 es it commences only, when growth stops».-- SPALLANZANI'S facts in connection with buds.-- They diffe
363d101 dgeon has not: now how many wild pidgeons have SPANGLES on this part: this will be well worth working
376d136 9. 1838 says the cross between the Guaranis & SPANIARDS are almost White from first generation., that
400e012 & big tailed sheep do Vol 10. p. 373, «& 374» SPANIARDS says no Tortoises in other «places» besides G
217b184 e has known case of good pointer & rough water SPANIEL «produce litter like both parents» & Mr Bell h
312c228 dogs Smith knew chinese hairless dog & common SPANIEL crossed.-- 3 puppies PERFECTLY like chinese &
312c228 pies PERFECTLY like chinese & 3 perfectly like SPANIEL even when grown up.-- Are mules homogenious ow
575n043 & brain. «Do we deny the mind of a greyhound » SPANIEL. differs from their brains» then can we deny t
333d009 Where two dogs line the same bitch & «perfect» SPANIELS & setters are produced. one would argue the w
511q018 to Lowe (7) In breeding. pointers. Bull-Dogs. SPANIELS-- Grey-hounds-- is there ever any degeneratio
620o046 do so.-- So we say a pointer ought to stand a <SPANIELS> «housedog's» duty is to watch the house.-- i
470t178 of Humble Bees on Phlox Down, 1854, Sept.) In SPANISH Broom by pulling back Wings, pollen is ejected
073r164 ite. grammalite. pyenite. native sulphur.. fluor SPAR. bayte. asbestos garnets.--carb & chrom. of lea
075r167 ulfated Barytes very «un»common in Mexico. Fluor SPAR only in certain mines. «Vol. III» "In general i
608o026 n succeeds who does not recognize an accidental SPARK falling on prepared materials. Fire contingenci
361d095 dult stage, other alters entirely]CD In common SPARROW young & female similar plumage.-- in tree spar
361d095 rrow young & female similar plumage.-- in tree SPARROW, (if I understand rightly) young cock & hen, a
496q05a r grains flaccid, as Koelreuter describes Kill SPARROW after feeding on oats, give body to Hawk & sow
594n112 <dis> learn to disobey it-- I have seen hawk & SPARROW in Shrewsbury garden picking from same bone A
590n093 culation is affected. p. 44.-- Jealousy. causes SPASM in bile duct, & throws bile in circulation p. 7
390d177 ich are Monocotyledenous have many flowers same SPATH. as they have only one bud.-- Every individual
420e090 ng how little of the spermatic fluid fertilized SPAWN of frogs.-- Annals of Natural History. (p 225.
496q006 ale flower in same receptacle (8) Make Duck eat SPAWN, eggs of snail, row of fish & kill them in hour
309c220 eas of subsidence marked out by animals of same <SPE> genera. is not equal to areas of elevation: Mar
640j167 the six puppies, if a hare was introduced, or <a SPE> became more numerous. (from death of its destro
415e069 so, & another to show how it came to be so.-- I SPEAK only of the former proposition.-- as in «races
581n064 r want, pleasure, or pains long before they can SPEAK-- or understand-- thinks so it must have been i
244c023 s albus. New Ireland ---- maculatus -- Waigiou SPEAKING of Lepus Magellanicus says; called "après un
266c091 d] Musalman's of the Peninsula, are, generally SPEAKING, a much fairer race than the Hindu's, in the
384d161 e greater strength, (p 45) & that strength> In SPEAKING of generation alway put female first Will not
417e079 -- it is the latter only that one refers to in SPEAKING of resemblances of children to their parents.
421e090 236-- on Hybridity in ferns.-- ₦.do p. 250-- «SPEAKING» of the terrestrial mollusca of Morocco «Mr F
434e128 extinct not so soon as other forms.-- p. 36.. SPEAKING about the controversy on Didelphys says. "If
471t08 iferae".-- referring to the 3 main divisions & SPEAKING of their similarity «in structure» he says "i
501q011 dary characters-- in double flower. do Henslow SPEAKING of Thyme doubts about stigma in similar manne
544m102 this time. how does he account for dogs & men SPEAKING in their sleep.-- Characters of dreams no sur
565n006 augh long prior to talking, hence one can help SPEAKING, but laughing involuntary.-- When one fear an
060r124 same level") to a greater distance".--Afterwards SPEAKS of this phenomena in connection with "the shoo
290c162 imals, where much change has been added. «as it SPEAKS to amount of change only & not kind» insects--
296c184 et forms like neighbouring Continent. This fact SPEAKS volumes. 2 Chapters. translated by Hooker.-- m
309c221 ore wonderful case Dwights' Travels in America, SPEAKS of short legged sheep. hereditary proceeding p
450e176 ions-- do App. p 73 State of Muar in Malacca.-- SPEAKS of Rhinoceros as well as Tapir.-- <do do p 75>
593n111 ry of man in Delirium tremens hearing other man SPEAKS. shows, that consciousness of personnal identi
148g026 pherd dogs Saturday. Before coming to Bridge of SPEAN, hills of «sea», gravel, current cleavage, & pr
148g027 ike rubbish at head of Loch Dochart <Nea> Above SPEAN Bridge many flat terraces one above much inclin
149g031 ? The slope of A & B regular & even towards {P} SPEAN Roy double terrace river «& to West of Spean» d
149g032 P] Spean Roy double terrace river «& to West of SPEAN» difficult to explain on <formation> deposition
149g032 rmation» deposition in lake On the summit «& on SPEAN side» of Meal-- Derry there were perfectly roun
150g035 w line drawn plain red talus line on N. side of SPEAN most clear & upper line running up great bight
497q007 nera of Leguminosae.-- Herbert explains numerous SPEC. of Cape Heath by facility. ¿Knight take opposi
249c036 emarkable fact of no forms peculiar to word to SPECIAL districts???? north of 30 degree.--, may be co
411e055 alized, & afterwards by the examination of the SPECIAL cases, under which the individual steps in the
573n036 llest insect, we wish to be created at once by SPECIAL act, provided with its instincts its place in
573n036 in nature. its range, its-- &c &c:--must be a SPECIAL act, or result of laws. yet we placidly believ
622o049 t as man is animal at head of series in which «SPECIAL» instincts decrease, I should think they were
639j28v ases.-- Macculloch p. 260 intimates canines no SPECIAL use to Man. Applicable to Bell's sneering-theo
037r055 action of torrents. «marine» Tortoise & other SPECIAL of large lizard.-- There would probably be no
048r085 not dive into the causes of the losses of the «SPECIES of» Mastodons. which ranged from Equatorial pl
048r086 ore such vast numbers of wild animals. both SPECIES & individuals as in the half desert country of
061r127 ot progressif<e>: produced at one blow. if one SPECIES altered: <altered> Mem: my idea of Volc: islan
063r130 eing the relation.--As in first cases distinct SPECIES inosculate, so must we believe ancient ones: «
063r130 ge or degeneration. from circumstances: if one SPECIES does change into another it must be per saltum
063r130 change into another it must be per saltum--or SPECIES may perish. = This <inosculation> «representat
063r132 ish. = This <inosculation> «representation» of SPECIES important, each its own limit & represented.--
063r132 separate.--Considering all individuals of all SPECIES. as «each» one individual «divided» by differe
063r133 end.--"There is no more wonder in extinction of SPECIES than of individual.-- Mr Birchell says Elephan
064r134 ve reach of tidents.--thinks them same as recent SPECIES. -- May I not generalize the fact glaciers mos
070r153 rbicular in basalt.-- When we see Avestruz two SPECIES. certainly different. not insensible change.--
070r153 ent? why should two of the most closely allied SPECIES occur in same country? In botany instances dia
078r174 rd.--D'Orbigny has described it with care to 3 SPECIES. I think I have much additional information ₦

Page **(Key Word)**
092a028 p. 566 1837.-- Tertiary <bea> formation twenty SPECIES same as Paris. 1500 ft high Mr Bird in paper t
095a037 Zoolog: Soc: Terebratula from Hudson's Bay. 2. SPECIES Vol VI. Geograph. Journ. Analysis of Poenig Vo
110a085 s from West Indies & declare them to be recent SPECIES-- Lyell-- Some internal changes are in process
116a099 good section of Rio Negro beds.-- -- refers to SPECIES non decrite de petites corbules analogue livin
146g011 imes mottled with white black legs & tail like SPECIES in colouring Strike an analogy between pleasur
171b005 this tendency to vary by generation, why are. SPECIES are constant over whole country; beautiful law
172b008 ,-- Inglish & Irish Hare.-- As we thus believe SPECIES vary, <in> changing climate we ought to find r
172b008 anging climate we ought to find representative SPECIES; this we do in South America closely approachi
173b009 that case surely is not Produced?-- <Granting> SPECIES according to Lamarck disappear as collection m
173b009 n in Lamarck's time. Gray's remark, best known SPECIES. (as some common land shells) Most difficult t
173b010 fertility of hybridity &c &c <assuming all> if SPECIES (<a>) «(I)». <fr> may be derived from form (2)
173b010 inents> island near continents might have some SPECIES same as nearest land, which were late arrivals
173b011 ence the type would be of the continent though SPECIES all different. In cases as Galapagos & Juan Fe
173b012 ng separated.-- On this idea of propagation of SPECIES we can see why a form peculiar to continents;
174b013 bred in from one parent why. Myothera several SPECIES in S. America why, 2 of ostriches in. S. Ameri
175b017 ght they in time As I have before said isolate SPECIES <& give even less change> especially with some
175b018 s of change. Volcanic isld.-- Electricity Each SPECIES changes. does it progress. Man gains ideas. th
175b020 tiplications when isolated, requires deaths of SPECIES to keep numbers of forms equable:-- but is the
176b022 nerated» There is nothing stranger in death of SPECIES, than individuals If we suppose monad definite
176b024 certainly points of affinity in each branch A SPECIES as soon as once formed by separation or change
177b027 e may fancy, according to shortness of life of SPECIES that in perfection, the bottom of branches dea
177b028 angement.-- ¿How is it that there come aberant SPECIES in each genus «(with well characterized parts
177b028 er. <Petrels have divided themselves into many SPECIES, so have the awks, there is particular circums
177b029 organization commenced branching.-- As all the SPECIES of some genera have died; have they all one de
177b029 on shortness of existence, «in» perfect<ion>, «SPECIES from many» <therefore> changes and base of bra
178b030 eely In Isld neighbouring continent where some SPECIES have passed over, & where other species have <
178b030 e some species have passed over, & where other SPECIES have <come> «"air" of that place. Will it be s
178b031 ere;--» Are not all our «British» Shrews diff: SPECIES from the continent look over. Bell, & L. Jenyn
178b031 change whi[ch] the Fr. naturalists thought was SPECIES <Ascensi> Study Lesson Voyage of Coquille.-- D
179b033 s the distinct locality or Metropolis of every SPECIES: believes in repugnance in crossing of species
179b033 species: believes in repugnance in crossing of SPECIES in wild state.-- No doubt «C. D.» wild men do
179b034 hen the black & white is so far gone, that the SPECIES (for species they certainly are according to a
179b034 & white is so far gone, that the species (for SPECIES they certainly are according to all common lan
179b035 circumstanced.-- ¿Is this shortness of life of SPECIES in certain orders connected with gaps in the s
180b036 s many living as now To do this & to have many SPECIES in same genus (as is). REQUIRES extinction. Th
180b037 pes.-- with several extinct forms, for if each SPECIES «an ancient (I)» is capable of making, 13 rece
180b037 t no offspring at all, so as to keep number of SPECIES constant.-- With respect to extinction we can
180b038 f such circumstances & therefore that death of SPECIES is a consequence (contrary to what would appea
181b039 e present world.-- or we may suppose only each SPECIES in each generation only breeds; like individua
182b050 curious cases. of close but certainly distinct SPECIES between Australia & Van Diemen's land. & Austr
183b051 me countries.-- In mundine genera, the nearest SPECIES often come very remote quarters. (NB. if Plata
183b051 nt states. <they> it might continue & thus two SPECIES be created) & live in same country. How is pro
183b051 on of wolf & Dog. (because being believed same SPECIES) if they do not breed readily. point in view.-
183b053 tively Cuvier objects to <tran> propagation of SPECIES, by saying, why not have some intermediate for
184b053 tween palaeotherium, megalonyx mastodon, & the SPECIES now living.-- Now according to my view. in S.
184b054 ge in shells from salt & F. Water-- on what is SPECIES. very good Has not Macculloch written on same
184b054 abbit of Falklands described by Q. & G. as new SPECIES. Cuvier examined it. There certainly appears a
185b056 Noronha Ophyressa bilineata (Gray) new <liza> SPECIES, belonging to true. American genus Waterhouse
186b057 f forms.?? It is in simplest case saying every SPECIES in genus resembles each other, (at least in on
186b057 pting specific character); and in passing from SPECIES to genera, each retains some one character of
186b060 e must look to type of organization.-- extinct SPECIES of that country parents of American.-- Now Gen
186b061 definite without change, superinduced, or new SPECIES; therefore animals would perish, if there was
187b063 ge; like Golden Pippens, it is a generation of SPECIES like generation of individuals.-- Why does ind
187b064 is accomodation). Now this argument applies to SPECIES.-- If individual cannot procreate, he has no i
187b064 ual cannot procreate, he has no issue, so with SPECIES.-- I should expect that Bear & Foxes &c same i
187b065 & Foxes &c same in N. America & Asia, but many SPECIES closely allied but different, because country
188b067 elago Dr Smith considers probable two Northern SPECIES replace <No> Southern kinds-- (I) Gnu reeaches
188b069 to keep to one line Dr Smith says very. close SPECIES generally frequet slightly different localitie
188b069 es, so that they become useful to know what is SPECIES.-- In proof that structure is not simple adapt
189b072 r in former periods & then splitting off.-- If SPECIES generate «other species», their race is not ut
189b072 en splitting off.-- If species generate «other SPECIES», their race is not utterly cut off:-- like go
189b073 perished in America All animals <are> of same SPECIES are bound together just like buds of plants, w
190b075 show tendency to go back-- there is an end to SPECIES.-- «Brown Appendix» A most remarkable observat
190b077 There appears in Australia great abundance of SPECIES of few genera or families.-- (long separated.-
190b078 mb, which has been passed through to form that SPECIES.-- <Man is derived from Monad, each fresh--> M
190b079 fresh--> Mr Don remarked to me, that he though SPECIES became obscurer as knowledge increased, but ge
191b082 of large animals-- Owls. transport mice alive? SPECIES formed by subsidence. Java & Sumatra. Rhinocer
191b082 Rhinoceros. Elevate & join keep distinct. two SPECIES made elevation & subsidence continually formin
191b082 ade elevation & subsidence continually forming SPECIES.-- Man & wife being constant together for life
192b086 ediate steps especially in those classes where SPECIES not numerous. (NB in those classes with few sp
192b086 es not numerous. (NB in those classes with few SPECIES greatest jumps strongest marked genera? Reptil
192b087 f the intermediate ranks had produced infinite SPECIES probably the series would have been more perfe
194b095 t of world before zoological divisions.-- Mem. SPECIES doubtful when known only by bones.-- Mem: Silu
195b100 os.-- did the creative force know that «these» SPECIES could, arrive-- did it only create those kinds
195b100 nds not so likely to wander. Did it create two SPECIES closely allied to Mus. coronata, but not coron
195b102 , so will be the numbers & distribution of the SPECIES!! It may be argued representative species chie
195b103 the species!! It may be argued representative SPECIES chiefly found where barriers «& what are barri
195b103 equator that external conditions would produce SPECIES so close as Patagonian <Chat> & Galapagos orph
195b104 usand miles distant.-- Absolute knowledge that SPECIES die &. others replace them-- two hypotheses fr
200b122 happening; that naturalist describe animals as SPECIES, for instance Australian dog: <yet when that>
200b125 ined to extreme North.-- N.do p. 326. 2 Fossil SPECIES of ox in N. America: as well as 2 recent See G
201b126 aberrant & osculant groups are not only few in SPECIES, but every two or three these form genera-- th
201b126 those forms which «nevertheless» have produced SPECIES, have produced fe[w] [not located] The relatio
202b130 y is permanent in another.-- Good argument for SPECIES not being so closely adapted. Near the Caspian
202b134 N. America) then if it is doomed that only one SPECIES of family has offspring the chance is that the
203b135 ivided again-- Weakest part of theory death of SPECIES without apparent physical cause:-- Mem: Mastod
203b138 brids Loudon's Magazine. Gould on Motacilla,. «SPECIES peculiar to Continent & England» Loudon Mag: S
206b148 not be extended to all animals first consider SPECIES of cats.-- <& other tribes>.-- &c &c Exclude m
206b149 at any one moment fewer closely related;∴ (few SPECIES of genera) ultimately few genera (for otherwis
206b149 l not this account for the odd genera with few SPECIES which stand between great groups, which we are
208b153 e R. N. p. 130 Speculations on range of allied SPECIES. p. 127. p. 132 There is no more wonder in ext
208b153 re wonder in extinction of individuals than of SPECIES Paris Tertiary Shells in India!? A p. 28 Dr Be
208b153 A p. 28 Dr Beck. & Lyell. most curious law of SPECIES few in Arctic. in proportion to genera. agrees
208b154 n Congo Expedition We need not expect to find <SPECIES>, varieties, intermediate between every specie
208b154 pecies>, varieties, intermediate between every SPECIES.-- Who can find trace or history of species be
208b154 ry species.-- Who can find trace or history of SPECIES between Indian cow with hump & Common;-- betwe
209b155 in perpetuating these particular varieties. If SPECIES made by isolation; then their distribution (af
209b157 de distribut. Plantarum. relation of genera to SPECIES in France is I: 5.7 in Laponia I: 2,3: Mem. Ly
210b158 Islands & Artic are in same relation. We find SPECIES few in proportion to difficulty of transport.
210b158 a I: I. I can understand in «one» small island SPECIES would not be manufactured, <but why they shoul
210b158 inctly states that permanent varieties. become SPECIES. p.147 «p. 150»-- not being crossed with other
210b159 Buch says from Humboldt, in Laponia. genera to SPECIES I. 2,3-- From Mackenzie Iceland there 144 gene
210b159 From Mackenzie Iceland there 144 genera & 365 SPECIES of plants not cryptogamic but I . 2,53.-- In k
210b159 2,53.-- In know varieties. there is analogy to SPECIES & genera.-- for instance three kinds of greyho
210b161 age) & which, follow certain laws according to SPECIES. present an analogy to production of species.-
210b161 o species. present an analogy to production of SPECIES.-- Animals have no notions of beauty, therefor
211b161 ctive feelings against other «for sexual ends» SPECIES, whereas Man has such instincts very little. i
212b168 riod.-- ¿In those divisions of molluscs. where SPECIES now least in number (as cephalopods,) in last

213b169 «Races» > Man in savage state may be called, <SPECIES> species. in domesticated <species> races.-- I
213b169 > Man in savage state may be called, <species> SPECIES. in domesticated <species> races.-- If all men
213b169 be called, <species> species. in domesticated <SPECIES> races.-- If all men were dead then monkeys ma
213b170 n monkeys make men.-- Men makes angels-- Those SPECIES which have long remained are those ¿Lyell?, wh
213b170 this is difficult; This immutability of some SPECIES. In Phillips. p. 90. it seems the most organiz
213b171 known? As Gould remarked to me, the "beauty of SPECIES is their exactness,' but do not known varietie
215b177 me that it is generally said.= How came first SPECIES to go on.-- There never were any constant spec
215b177 cies to go on.-- There never were any constant SPECIES Both males & females. lose desire. Native dog
216b180 at Cape.-- McClay argues from it Black & White SPECIES.-- For, says he Seeds of hybrid lillies &c &c
217b187 [not located] Mr Don gave me instances of one SPECIES of Australian genus being found in Sumatra; ag
217b187 other Genus in Sandwich islands-- A genus with SPECIES in Van Diemen's land and Tierra del Fuego.-- A
217b187 emen's land and Tierra del Fuego.-- Araucaria, SPECIES. Brazil {P} Chile, Norfolk Isl.-- Isle of Pine
217b187 A <South> American «form of» Lathyrus has one SPECIES in Europe Madagascar has several American form
217b188 cts evidently show that Mr D. wonders at these SPECIES being wanderers.-- Iceland no species to itsel
217b188 at these species being wanderers.-- Iceland no SPECIES to itself, a remark common to all northern isl
218b189 erefore when from being put on island. & fresh SPECIES made. parents do not cross-- we see it even in
218b191 er wild Mr Donn remarks to me. that give him a SPECIES from Ireland, England, Scotland & other locali
219b194 that is genera near Cape) see if there are any SPECIES same as T. del Fuego & C. of Good Hope show po
219b195 cause snow formerly descended lower, therefore SPECIES of lower genera altered. or northern plants «N
220b197 h generation certainly is.= The dislike of two SPECIES to each other is evidently an instinct-- & thi
220b197 ive organs) & therefore the one distinction of SPECIES would fail. But this applies only to coition &
220b198 dogs. Now if we take structure as criterion of SPECIES Hogs different species, dogs not, but if we ta
220b198 ructure as criterion of species Hogs different SPECIES, dogs not, but if we take character of offspri
220b199 & Europe, which have not their representative SPECIES in each other, are migratory species from warm
220b199 sentative species in each other, are migratory SPECIES from warmer countries. When will this paper be
220b200 genera.-- Lyell has remarked about no confined SPECIES in Sicily. Jan: 1838 L'.Institut. Bats, in Eo
221b201 titut. Bats, in Eocene beds, very like present SPECIES., p 8. ¿Are mundine forms, longest persistent?
221b202 fore Mammalia Think about Miocene fossils some SPECIES being recent agreeing with Senegal. whilst Cra
222b202 genera same.-- Speculate on multiplication of SPECIES by travelling of climates & the backward & for
222b202 mates & the backward & forward introduction of SPECIES.-- When species cross & «hybrid» breed, their
222b203 ward & forward introduction of species.-- When SPECIES cross & «hybrid» breed, their offspring show t
222b204 ira-- any general observations-- difference of SPECIES between land shells of Porto Santo & Madeira--
222b207 from ancestral form. If so are present typical SPECIES most near in form to ancient; in shells alone
223b209 ¿where is Pentland's account of Bolivian human SPECIES?-- Small «new» animal mentioned from Fernando
223b209 hy there is not perfect gradation of change in SPECIES,, as physical changes are gradual, is this if
223b209 esert) or separation by mountain chains & the SPECIES have not been much altered they will cross (pe
223b210 aps more fertility & so make that sudden step. SPECIES or not. A plant submits to more individual cha
223b210 s according to same law with animals?? Why are SPECIES not formed. during ascent of mountain & appro
223b210 approach of desert?-- because the crossing of SPECIES less altered prevents the complete adaptation
223b211 h (C) prevents offspring of A. becoming a good SPECIES well adapted to locality A. but it is instead
223b211 estroy plants B. C. D. & A will soon form good SPECIES! The increased fertility of slightly different
223b211 The increased fertility of slightly different SPECIES & intermediate character of offsprings account
223b211 acter of offsprings accounts for uniformity of SPECIES & we Must confess. that we canot tell, what is
224b212 The passage in last page explains that between SPECIES, from «moderately» distant countries. there is
224b212 erience according to each group)» whether good SPECIES, & hence the importance Naturalists attach to
224b212 ce Naturalists attach to Geographical range of SPECIES.-- Definition of Species: one that remains «at
224b213 Geographical range of species.-- Definition of SPECIES: one that remains «at large» with constant cha
224b213 nimals> beings of very near structure.-- Hence SPECIES may be good ones & differ scarcely in any exte
224b214 country changes rapidly, we should expect most SPECIES.-- The difference intellect of Man & animals n
224b214 heory very distinct from Lamarcks» Without two SPECIES will generate common kind, which is not probab
224b215 man. but both monkeys & man may produce other SPECIES., man already has produced marked varieties &
224b215 intellectual being acquired alters case) other SPECIES or angels. produced «&» Has the Creator since
225b216 miserable limited view.-- With respect to how SPECIES are. Lamaks "willing" doctrine absurd. (as equ
225b219 any change in offspring? if so adaptations of SPECIES by generation explained? NB. Look over Bell on
226b221 o like that of North Europe, many genera & few SPECIES. The number of genera on islands & on Arctic s
226b221 n.-- When the new island splits & grows larger SPECIES are formed of those genera.-- & hence by same
226b221 a.-- & hence by same chance few representative SPECIES. this must happen. & thus acquired will explai
226b221 Galapagos.-- shrews, & when big continent many SPECIES belonging to its own genera Therefore if in sm
226b222 enera Therefore if in small tract we have many SPECIES, we may insure mass continental or many large
226b222 resolves itself into question of proportion of SPECIES to genus If on one isld several species of sam
226b222 ion of species to genus If on one isld several SPECIES of same genus, subsided land.-- Mauritius? <In
226b222 ed land.-- Mauritius? <In plants where do most SPECIES occur.)> Although the Horse has perished from
226b223 lia in their wide range & in their duration of SPECIES. (¿are carnivorous Mamm: in Paris basin «allie
227b224 hat structure of every organ in A. B. C. three SPECIES «of one genus» can pass into each other «by st
227b225 ure (C) could pass into (D).-- We may foretell SPECIES. limits of good species being known.-- It expl
227b225 (D).-- We may foretell species. limits of good SPECIES being known.-- It explains the blending of two
227b225 enera-- It explains typical structure.-- Every SPECIES is due to adaptation + heredetary structure. l
227b225 agrees with excessive inequality of numbers of SPECIES in divisions. look at Articulata!!!--! It lead
228b228 & «direct» examination of direct passages of <SPECIES> structure in species, might lead to laws of c
228b228 n of direct passages of <species> structure in SPECIES, might lead to laws of change, which would the
228b229 at present time, we might expect confusion of SPECIES.-- Important. For instance take Voluta & Conus
228b231 an in present state. with what he is as former SPECIES. His arts would not then have taken him over w
229b233 ce of useless structure-- Smith thinks several SPECIES of Rhinoceros range from Abyssinia to extreme
230b235 ding mountain you ought to have a gradation of SPECIES, now this notoriously is not the case, you hav
230b236 notoriously is not the case, you have stunted SPECIES, but not such as would make species (except pe
230b236 ve stunted species, but not such as would make SPECIES (except perhaps in some plants & then a chain
230b240 f they are different then, they will be called SPECIES & mere producing fertile hybrids will not dest
231b244 breed?-- As man has not had time to form good SPECIES, so cannot the domesticated animals with him.!
231b244 s with him.!-- Modern origin shown by only one SPECIES, far more than by non-embedment of remains-- ¿
232b245 died out, or moved towards equator.-- «or some SPECIES might then have been wanderers.-» There ought
232b245 ave been wanderers.-» There ought to be fewer SPECIES in proportion to genera than in present seas,
232b245 o genera than in present seas, «All» The <one> SPECIES which survives any change may undergo indefini
232b246 ilst others may die out or move South ward.... SPECIES must be compared to neighboring sea.-- For cha
232b246 e compared to neighboring sea.-- For change of SPECIES does not measure time but physical changes (we
233b249 u» &c &c. (See Lyell. Vol III p. 30) different SPECIES in different isld. (as far East as New Ireland
233b250 sions of cats. Tortoise shell--& grey-banded. ¿SPECIES?-- thinks offspring of cats sometimes heteroge
233b250 - It was most curious to observe, that all the SPECIES of mice in S. America. which were hard to dist
233b252 ate. But how does this agree with longevity of SPECIES in Molluscs!!! When we talk of higher orders,
234b261 & in genera peculiar to any one country do not SPECIES generally affect different stations;-- this wo
234b261 is would be strong argument for propagation of SPECIES-- Again is there not similarity even in quite
235b275 taking peculiar character-- Indian Bull?-- Do SPECIES of any genus. as American or Indian genus inha
236b279 re Mr Murchison fox was found), decidedly next SPECIES to some South American kinds.-- Are the closes
236b280 South American kinds.-- Are the closest allied SPECIES always from distant countries, as Decandoelle
236b280 ly says sometimes we might expect disseminated SPECIES to vary a little, but such should not be gener
239cIFC ference: where hybrids produced have any close SPECIES ever yet failed. About trades affecting form o
240c013 r from those of New Holland.-- The New Holland SPECIES are not found in the Archipelago-- Former stat
240c014 in isle of Aroe & Solor), «Vol I» likewise new SPECIES of Parameles, which joined to Casoars, perroqu
241c014 islands.-- -- Antelope in Celebes, Bourou new SPECIES of Axis.-- «Cervus moluccensis is different fr
241c015 dpecker.-- Birds of Australia. Many in common ¿SPECIES? with New Guinea.-- Many <genera>. kinds commo
241c015 67 ¿American forms? All Infusoria. not extinct SPECIES. good Resumé do/p. .62 ??? Age of Deinotherium
242c017 ks on the Floras can be trusted The changes in SPECIES must be very slow, owing to physical change, s
242c018 ca. «M. Bibron doubts fact.--» My toad is same SPECIES Coquille Voyage p. 25 Mais il n'y a pas jusqu'
243c020 aland-- Norfolk. Isd. & New Caledonia peculiar SPECIES of cassicans: (¿cassicans Australian form? p.
244c021 a, it. would demonstrate.; not distance, makes SPECIES but barrier.-- --it would make strong contrast
244c022 ive from there-- Mem Waterhouse knows of some SPECIES which escaped there.-- p. 139. Vespertilio bon
245c023 earance to Siamese kind.-- but considered good SPECIES from dental characters, wild pigs. said by For
245c023 swim from one isld to another--॥ «It is a good SPECIES, with different number of teats» Coquille Voya
245c025 Springing up more likely to <M> have different SPECIES than those sinking, because arrival of any one

```
Page    *************************************(Key Word)*************************************
246c026 how  it is known that they are varieties & not SPECIES.-- Vol :694. King-fisher of Europe, (Alcedo is
246c026 colour slightly different. Who can say whether SPECIES & varieties P. 708. Columba Oceanica (Less) in
247c028 great geological age-- Gastrobranchus «only» 2 SPECIES one in Northern Hemisphere 2d in southern --p.
249c035 those  of Primrose & Cowslip run wild, The two SPECIES of Clenomys. case of replacing species. Dr. Sm
249c035 The two species of Clenomys. case of replacing SPECIES. Dr. Smith will give me some capital informati
249c035 wide  range of animals. Follow this out, where SPECIES of same genera in two words. have not species,
249c035 species  of same genera in two words. have not SPECIES, generally wide range? Mice.-- Waterhouse's re
250c038 preeminently equable. might have allowed fresh SPECIES to have been formed & spread to other Africa &
250c039 ave been string of islands.-- ¿Europe has many SPECIES but not genera distinct from rest of  world???
251c040 enner, on birds seen far at sea, migrations of SPECIES, geese killed in Newfoundland, with crops full
251c041 ol 2. p 221. Horsfield on two bears very close SPECIES, inhabiting Borneo & Sumatra. differ only in f
251c041 the Gobbah or village Gazal.-- ¿is latter same SPECIES domesticated, strangely contradictory to Azara
251c042 ork «do» Vol. IV p 273. Macleay on Capromys. 4 SPECIES probably in Cuba (p 271 Viedo says American do
252c042 heno much excellent detail & fine, views about SPECIES-- MUST BE STUDIED: genera founded in nature [n
252c045 to  help in distinguishing empirically what is SPECIES.-- The Collector is directed to study localiti
252c045 ontinents» Give Specimen of arrangement, 3 <5> SPECIES Rhinoceros Cape town good species Indian speci
252c045 ement, 3 <5> Species Rhinoceros Cape town good SPECIES Indian species so distinct that all analogy fr
252c045 ecies Rhinoceros Cape town good species Indian SPECIES so distinct that all analogy from each other I
252c045 would  probably tell more certainly Get Closer SPECIES-- FOX IS & Mice of America «good case on accou
253c046 countries»  & «Monkeys.» Fact of Elephant same SPECIES in Borneo. Sumatra. India Ceylon-- perhaps sho
253c046 ntagu probably contains some facts about close SPECIES of Birds. Zoolog. Transact. Vol I p. 165.-- «a
254c047 st certain indication of the perfection of the SPECIES--! How does this agree with grand fact of Mars
254c048 other with the greatest rapidity"-- so we find SPECIES each class successively present modifications,
255c050 ht & 3d of latitude more commonly are the same SPECIES, instead of analogues.-- «as» in other classes
255c053 of  plan to insure stability; but isolate your SPECIES her plan is frustrated or rather a new princip
256c053 be some sophism. in Lyells statement that some SPECIES vary more,, than what makes species in other a
256c053 that  some species vary more,, than what makes SPECIES in other animals.--? Forster on South Sea, wil
256c054 tions of domesticated animals in those regions SPECIES so far are not natural, that they are either A
256c054 en had difficulty in distinguishing which were SPECIES, (theory admirably) yet a glance would tell fr
257c056 000 ft above sea in Bolivia; he examined-- all SPECIES & found "beaucoup des mêmes oiseaux. que  nous
258c060 escent. or true relationship, tends to keep to SPECIES to one form, (but is modified), the relationsh
258c061 e confusion in this system of nature-- Whether SPECIES may not be made by a little more vigour  being
259c063 e effect on offspring-- but WHOLE race of that SPECIES must take to that particular habit.-- All stru
259c064 South  Sea, wonderful case of extermination of SPECIES-- Epidemic amongst trees. Plane trees all died
259c064 year.  Extreme difficulty of TRACING change of SPECIES to species «although we see it affected» tempt
259c064 eme difficulty of TRACING change of species to SPECIES «although we see it affected» tempts one to br
259c065 to  admit the possibility (any great change in SPECIES is reduced by atavism) Even a deformity may be
260c068 s a cold one. in checking beautiful colours of SPECIES-- Mem. St. Jago--solitary Halcyon bird of pass
261c069 udied. There is capital table of extent of all SPECIES Accumulate instances of one family sending out
261c070 universally allowed that the discrimination of SPECIES is empirical. show this by instances Once gran
261c070 ces Once grant my theory, & the examination of SPECIES from distant countries., may give thread to co
263c074 a division of nature of Apteryx, many genera & SPECIES-- The believing that monkey would breed (if ma
263c075 er the enormous difficulty of reproductions of SPECIES & certainty of destruction; then he will choos
263c076 he lesser of the difficulties Once grant that «SPECIES» one genus may pass into each other.-- grant t
263c078 s a train of animals of hundred generations of SPECIES to produce contingents proper. -- Present monke
264c080 r with Monkeys» Gould seems to think that many SPECIES when close come from different localities, and
265c087 If we only had Puffinuria Garrottii & no other SPECIES-- as we have only ornithorhyncus, then we shou
268c099 o external features of land by seeing how many SPECIES common to Patagonia desert & Tierra del Fuego.
271c105 S. Hemisphere. (replaced to the North by other SPECIES.--) should build a nest lined with mud, in for
271c105 which  it feeds. or insects it devours is same SPECIES. yet that it should so strictly <f> agree in h
271c105 .-- These facts show, habits heredetary whilst SPECIES have changed Argumentum ad absurdum. The creat
272c109 era confined to continent, though we have seen SPECIES «of subgenera» scattered over it.-- We have ab
272c109 tances of remarkable structure which as far so SPECIES is concerned superabundant,-- the tail in cock
272c109 detail  in structure prevails amongst the same SPECIES & subgenera in families.-- thus the banded tar
273c110 are  lengthened rump feathers.--; <then> & one SPECIES has small band & others large, then he says fr
273c110 be  almost sure, that there exist intermediate SPECIES,-- This is remarkable & would lead one to supp
273c110 s remarkable & would lead one to suppose, that SPECIES in same group generally contemporary «++.» «Th
273c110 sil forms would generally fill up genera & not SPECIES, which is not true, with shells?.?.?» It looks
273c111 great movement in the air, & likewise rasorial SPECIES, & likewise perching (Gould), but the latter i
274c114 e hardly at all developed. Not confined to one SPECIES, but generally small genus. ¿are there not man
275c119 in induction) as keystone of ancient geography SPECIES tell of Physical relations in time «& forms»
276c122 Hybrids,  may be used to varieties, as well as SPECIES-- as formation of species <of> gradual, so may
276c122 arieties, as well as species-- as formation of SPECIES <of> gradual, so may we suppose., that somethi
276c123 robably be reduced What the Frenchman. did for SPECIES-- between England & France. I will do with for
276c125 al number of vertebrae-- ¿Where two very close SPECIES inhabit same country are not habits different,
276c125 nt, (Mem: Gould's Willow Wren) but where close SPECIES inhabit different countries habits similar ¿la
277c126 from  (C) Chiloe (E) Chile.. rupestris -- good SPECIES it is reverting to old plan, but reason now as
277c126 ned for doing so There should be mark to every SPECIES. only known by analogy genera of course distin
277c127 d, call them varieties. but two ostriches good SPECIES because interlock analogy to be guide. in isla
277c127 use interlock analogy to be guide. in islands. SPECIES.-- each describer giving his test namely diffe
277c127 aming them) which are found together.-- If two SPECIES come over to this country, without range, or h
277c127 part.-- that he will not have brought home new SPECIES.-- until, he can show range & habits-- Take inst
278c128 logy would be empirical because as soon as two SPECIES were placed in different subgenera, then it wo
278c128 t will however be much <shu> surer, when false SPECIES banished by this test.-- Excepting where an An
278c129 b-genera so far may be eliminated. where every SPECIES of a section is confined to one continent & ev
278c129 a section is confined to one continent & every SPECIES to another then those sections & subgenera; ar
278c129 gical, because we do not know, whether nearest SPECIES of each might not breed:-- Genus must be a tru
278c130 ing found fossil in Australia, & only one tree SPECIES (Mitchell's authority) in Australia, & several
278c131 a accumulated-- surely ask Owen to see whether SPECIES same, excessive improbability. Mem in Clifts 1
280c134 o formation of instincts in wild animals, many SPECIES in one genus-- external circumstances in  both
280c135 pplies it to national character.-- N.B. If two SPECIES were excessively old, they would not make hybr
281c137 ression of such a naturalist "splitting up his SPECIES & genera very finely" show how arbitrary & opt
281c139 n as related in the third & fourth degree.-- a SPECIES must be compared to family entirely  separated
282c142 in ground cuckoos, is affinity with respect to SPECIES of each other, because we suppose all descende
282c142 all descended from same.-- but if two original SPECIES, each became ground then the relation of all t
283c145 y theory agrees with unequal distances between SPECIES. some fine & some wide. which is strange if cr
283c145 want is granted.-- For. at Galapagos. make ten SPECIES of Orpheus-- one of which has very short  legs
283c145 with  short., let these only have progeny with SPECIES & there will be two genera.-- let short billed
283c146 with  CREATION.-- «On» The end of formation of SPECIES & genera, is probably to add to quantum of lif
284c149 neral.-- Where any structure is general in all SPECIES in group we may suppose it is oldest, & theref
284c149 nded on such characters, as do not vary in the SPECIES of it.-- «where does such occur?» now some suc
284c149 t (mode of generation) varies according to the SPECIES, it is manifestly of less importance, as affor
284c150 knows  lots of instances of replacement of one SPECIES by another. supply place in each others' econo
285c151 show  hybridity from few forms, parents of all SPECIES not possible in some detail,-- the relations t
285c151 some  detail,-- the relations to islands close SPECIES, «on these isld» &c will probably upset it-- T
285c151 ots from each branch No: because decay in that SPECIES <shows> is effects of unfavourable conditions.
285c152 hether ever arrive at true affinity doubtful A SPECIES is only fixed thing with reference to other li
285c152 tten & obliterated in the course of ages.-- As SPECIES is <certain> real thing with regard to contemp
289c160 nightingale do.-- all shows how nicely adapted SPECIES to localities-- p. 390,, young <grebes> «ring
289c161 t account of habits My definition <in wild> of SPECIES. has nothing to do with hydridity,, is simply,
289c161 come, but until it is the animals are distinct SPECIES. If any one is staggered at feathers & scales.
291c168 America  grew very much hotter, then Brazilian SPECIES would migrate south ward being ready made.-- &
291c168 s in Falkland Isd they would change & make new SPECIES.-- alpine species being destroyed at Falkland
291c168 they would change & make new species.-- alpine SPECIES being destroyed at Falkland Isds++.-- Mem Lyel
292c169 bably only few of extinct forms have generated SPECIES. & of 100 extinct species the greater number p
292c169 forms have generated species. & of 100 extinct SPECIES the greater number probably have no descendant
294c176 nt Varieties in wild animals-- but how to show SPECIES-- I fear argument must rest upon analogy & abs
```

Page ***(Key Word)**
294c176 t may be said argument will explain very close SPECIES in islds. near continent, Must we resort to qu
294c176 Must we resort to quite different origin when SPECIES rather further.-- once grant good species as c
294c176 when species rather further.-- once grant good SPECIES as carrion crow & rook formed by descent. or t
294c177 0. reference to succession of types ¿different SPECIES;-- Horse-- &c <Lonsdale says. that first shee>
294c177 «its limiting, the allowing at same time true SPECIES.» & its adaption to classification & affinitie
295c183 can monkeys in Fernando Po-- no new forms only SPECIES!! No salamanders (D'orbigny Rapport. p. 11) in
296c184 75 NB. This bears on fossils of Europe., those SPECIES which can migrate remaining constant in form,
297c185 lds birds for some time alive ¿therefore other SPECIES mice & only kills them when urged by hunger.--
297c185 l genera ∵circumstances not favourable to many SPECIES., same circumstances., which by causing death,
297c186 y causing death, makes the group aberrant When SPECIES rare we infer extermination, when group few in
298c191 There is great difficulty in Making an alpine SPECIES from one in lower country during gradual eleva
298c191 -- We must imagine a considerable range of one SPECIES «on -- mountain side» of which the central par
298c191 occupied by a third best adapted kind.-- lower SPECIES would then revert to pristine form (which must
298c191 n altered by crossing) with alpine form) lower SPECIES afterwards would probably often be destroyed.-
298c192 milar with those over whole of country.-- some SPECIES are larger &c in different countries. These fa
299c193 in one wood far from any other plants of same SPECIES Channel Islds (& probably Isle of Man) no plan
299c194 rmediate shades in these cases,-- but absolute SPECIES formed. The Anagallis perhaps, offers another
300c197 g beetles will each become the fathers of many SPECIES.-- a few eggs transported to the Str of Magell
302c203 of land containing all of intermediate Father-SPECIES, & not, therefore, solely owing to such interm
302c203 therefore, solely owing to such interm: father-SPECIES, being little adapted to some physical change.
302c203 - If Patagonia became fertile all intermediate SPECIES living there would be destroyed, & N & S. exis
302c203 ng there would be destroyed, & N & S. existing SPECIES becomes father of genera-- whatever the cause
303c203 ra-- whatever the cause is. <the> any osculant SPECIES. which survived would be few in number.-- Para
304c206 t pregnant with SOMETHING.? that intermediate. SPECIES have generally perfect organs. do changes of h
304c207 way back.-- aberrant forms produced where many SPECIES <osculant> but, where much death, may be infer
305c209 e are some godwits which are close to European SPECIES, and the sexes of which vary in colour of plum
305c209 plumage in same remarkable manner as Europaean SPECIES = singular coincidence if distinct creation.--
306c212 gained in child hood.-- Young salmons. first a SPECIES which lived in estuaries its taste. taught it
306c213 ogs of La Plata & Guyana‖ people will say. not SPECIES.-- organs of generation a captial character. (
307c215 so nature can produce in sex, what she does in SPECIES of Apterix This is important for if these abor
307c215 o consider abortive organs of same tendency in SPECIES, this is capital & novel argument.-- Are there
307c216 gument, when general argument is extended from SPECIES to genera & classes. p. 479. fragment of tusk
308c217 been gaps, & there now must be, ∴extinction of SPECIES bears relation to existence of genera &c &c Tw
308c217 to existence of genera &c &c Two savages, two SPECIES.-- «discussion ustil, unless it were fixed wha
308c217 «discussion ustil, unless it were fixed what a SPECIES means» civilized Man, May exclaim with Christi
310c223 N. American, European & Chinese Genera & some SPECIES on Himalaya.-- some English beetles, birds & a
310c224 e is Saxifrage, almost <same> «closely allied» SPECIES Himalayas, 13,000 & Melville Isd.-- West Afric
311c225 irds in Europe & N. America, on closely allied SPECIES. «replacing each other». good to consult p. 32
312c233 us of animals different «+++ p 234», different SPECIES to different,-- inguinal louse African & Europ
313c233 ver on tropical animals which accounts for the SPECIES changing which accounts for the species changi
313c233 or the species changing which accounts for the SPECIES changing « ∴ because mammalia can subsist wher
313c234 ed should not new genus have been made, & only SPECIES, good argument for origin of man one.-- Is the
313c234 in of man one.-- Is the extinction & change of SPECIES two very different considerations, with respec
313c235 nderful a case bees developing sex of neuters) SPECIES may have had their infancies, as well as men--
314c236 carcely doubt final cause is the adaptation of SPECIES to circumstances by principles, which I have g
314c237 to allow it to lie on one side.-- but in other SPECIES, this segment is converted into hood which pos
314c238 a dash of New Holland. As in N. Zealand-- Some SPECIES of Australian Genera Some species same (Palm &
314c238 aland-- Some species of Australian Genera Some SPECIES same (Palm & Phormium tenax) as in New Zealand
314c239 ium tenax) as in New Zealand & Australia, some SPECIES of Australian GENERA∴ good case. rather large
315c241 le flore du Japon", some Europaean & Sandwich SPECIES & some of Japan. I do not understand any new o
316c244 in America,-- yet many countries have far more SPECIES than other countries «++ p. 246» as Cyclostoma
316c246 hat circumstances have led to formation of new SPECIES some few have been scattered over whole world
316c246 present day same (or according to Sowerby fine SPECIES) on coasts of N. America & England.-- but the
317c251 Bachman told me. that near Charlestown ?three SPECIES, near New York. (600 miles N.?) replaced by th
317c251 York. (600 miles N.?) replaced by three other SPECIES.-- Says all the hares West of <all> Rocky Moun
317c251 gth of limbs, so that he first thougt only one SPECIES. & all hares on East side have other peculiar
318c254 man.» that is succession of birds.-- «in» some SPECIES «a Tanagra» Males come first & the females in
318c254 ountry; «kinds of migration quite different in SPECIES of same genus.» The Muscicapa solitaria stay a
319c276 range.-- [blank] Books examined: with ref: to SPECIES Most of those which have references at end; is
325c267 d cattle & Montagu on birds (facts about close SPECIES) Wilson's American Ornithology Read Aristotle
333d007 ery interesting as Esquimaux dog approaches to SPECIES. Again he has seen several crosses between Esq
335d014 s showing above facts as first cross being new SPECIES, ‖ -- Are not dreadful monsters, abortive, jus
335d015 n blood so will they remain, a mule «being new SPECIES» will have no tendency to have offspring like
335d015 mitive stocks «relative to changes which every SPECIES undergoes») & hybrids between very near specie
335d015 pecies undergoes») & hybrids between very near SPECIES (that is slight alterations of primitive stock
338d023 ration depends on facility of transport in the SPECIES itself, & in the local circumstances of the tw
338d024 his taking any form of Malaria-- adaptation & SPECIES-like,-- -- Says Negro-- thick skinned My haird
338d025 imals from distant countries breeding!. «Mem 3 SPECIES of grouse»! Has not Goldfinch & Greenfinch bre
341d030 s from Hobart Town mentioned, it seems most of SPECIES from there now found in Australia New species
341d030 species from there now found in Australia New SPECIES of Moschus, characterized by Ogleby, who obser
342d033 eggs but never produce inter se or with parent SPECIES.-- The hybrids do not vary (ie the hens all al
342d033 all alike & Cocks all alike) More than parent SPECIES-- Mr Blyth remarked only near species or varie
342d033 n parent species-- Mr Blyth remarked only near SPECIES or varieties produce heterogenous offsprings.-
342d033 se pigs & common must be considered as distant SPECIES?? or is time the varying element). Then do tho
342d033 or is time the varying element). Then do those SPECIES which breed most freely. & produce somewhat fe
342d034 ous? I observe Bachman calls these Hybrids new SPECIES. Yarrell says the bird fanciers say the throw
342d034 ays the bird fanciers say the throw of any two SPECIES crossed is uncertain-- Yarrell remarks he has
343d038 espect to future destinies of mankind, some of SPECIES or varieties are becoming extinct. others thou
343d039 wise in 10,000 years Negro probably a distinct SPECIES-- We know how long a Mammal may go on as one s
343d039 s-- We know how long a Mammal may go on as one SPECIES from Egyptian Mummies & from the existing anim
344d040 ing so.-- consider this with reference to "new SPECIES & hybrid doctrine"-- I have read there are exc
345d044 Sheep jet black legs, & face & tail, just like SPECIES.-- high active breedin[g] [not located] half b
348d050 ost important characters break down in certain SPECIES & become worthless-- Mammalia Edentata‖ We do
349d052 solely owing to number of lost links.‖ if all SPECIES know they would be innumerable‖ does not know
349d052 now any difference between permanent variety & SPECIES!! (given in note.)-- Macleay <met> uses term g
349d052 ring to again.-- According to my theory, every SPECIES in any sub-genus will be. descended from one s
350d053 analogy»-- Consider all this NB. How can local SPECIES as at Galapagos., be distinguished from tempor
350d053 at Galapagos., be distinguished from temporal SPECIES as in two formations? by no way.?-- "Natura ni
352d057 rfect (as we use the word) than parent, so may SPECIES retrograde, but these facts are rare.-- 2d Sep
352d058 contradicted by abortive organs, but number of SPECIES with abortive organ of any kind few.-- » hence
352d058 l.-- Argument for circularity of groups. When <SPECIES of> a group of species is made. father probabl
352d058 larity of groups. When <species of> a group of SPECIES is made. father probably will be dead-- hence
353d061 h with those of India-- Waterhouse knows three SPECIES of Paradoxurus common to Van Diemen's land & A
353d061 e hares compared to North American hares. Many SPECIES, separated by Mountains. & & &c.-- do. p. 69.
354d065 , was analogous to specific character of other SPECIES in genus."-- Is there any law of this. Do any
355d066 es to ass.» or fowls to the several aboriginal SPECIES «or ducks» (here argue if it be said domestic
355d066 c fowls are descended from several stock, then SPECIES are fertile‖; as long as opponents will «are»
355d066 which are seen to <vary among> be different in SPECIES of same genus." Law of monstrosity not prospec
355d067 etus. distinct consideration) Now in different SPECIES of genus Sus. do vertebrae vary? «See Cuvier O
355d068 olog Transacts) same appearance with Secondary SPECIES distinct-- but close.-- Mem. Von Buch on Cordi
355d069 It is important with respect to extinction of SPECIES, the capability of only small amount of change
356d069 t kinds of food: grazing animals who eat every SPECIES new.-- Sept. 8'. A Golden Pippen or Ribston do
359d088 ir propagating powers. (as if they were a good SPECIES, or local variety, & not effect of breeding in
360d093 er.-- <so> the most aquatic & most terrestrial SPECIES, might be harder to cross than two less oppose
360d094 ain or stomach would be different.-- Or if one SPECIES left its type in having very long legs, & anot
361d095 ross-Beak-- In lark if I understand right, all SPECIES have same character which is mottled, & not li
361d095 cter which is mottled, & not like any existing SPECIES-- [In two herons, <both> plumage of both (nurs
361d095 lumage of both (nursling) quite similar.-- one SPECIES retained this character in adult stage, other

361d096 r.-- in blackbird group young like some of the SPECIES-- (¿do these facts indicate that the change is
361d096 o ducks, which have pretty close repesentative SPECIES in England & N. America.-- the teal which some
363d102 mage of ducks.-- Mr Yarrell says in very close SPECIES, of birds, habits when well watched always ver
363d102) (Goulds story of Water-Wagtails mistake both SPECIES scattered over Europe)-- The habits of some «s
365d105 Under this predicament, probably, alone would SPECIES cross in wild state.-- Is English red Grouse.
365d106 became unfit for. subalpina, or some Northern SPECIES, & being restricted species has been Made.-- I
365d106 , or some Northern species, & being restricted SPECIES has been Made.-- In the hybrid grouse between
365d107 difference between <t> "permanent varieties" & SPECIES. he overlooks-- <restric> those <restr> restri
365d107 .-- One can perceive that Natural varieties or SPECIES., all the structure of which is adaptation to
366d111 located] <The> every case common to many good SPECIES; & therefore to genera (& the uncles & aunts)
366d111 erefore does not tell against transmutation of SPECIES-- Will it against genera.-- How long will the
366d111 y from muskitoes bite influence propagation of SPECIES.-- Case of Association very disagreeable heari
368d114 > is evidently the male which recedes from the SPECIES all females being most like offspring, Q (how
374d134 be far from wondering of changes in number of SPECIES, from small changes in nature of locality. Eve
375d134 ecandoelle» does not convey the warring of the SPECIES as inference from Malthus.-- «increase of brut
375d135 ng brandy.--» take Europe on an average, every SPECIES must have same number killed, year with year,
375d135 with year, by hawks. by. cold &c--.. even one SPECIES of hawk decreasing in number must effect insta
377d139 annel & (& Basses St) yet no change in English SPECIES-- time no element in making change, only in fi
378d147 ual Selection If masculine character. added to SPECIES. we can see why young & female alike Good Ch
378d147 pe by Mother?-- do these differences indicate, SPECIES changing forms, <& loosing do> if so domestic
378d148 abits According as child is like parent, so is SPECIES old: Hence <young> Kingfisher & pies, have lon
378d148 about them quite remarkable". instance of old SPECIES transmitting so much longer its Mental peculia
379d151 ater Fish!! ¿adapted to salt water?-- peculiar SPECIES, crabs & molluscs few.-- ¿are not some same--
384d161 from Hunter)-- Hunter (p. 45) observes "every SPECIES has a disposition to deviate from Nature in a
384d161 o itself" <Is this so> Each part <not> of each SPECIES not similarly subject-- ▯Divides sexual marks
391d174 mpletely circumstances «alone» make changes or SPECIES!! CD[The view of <In> each Man or mammalia bei
391d175 t be external» change are always of one nature SPECIES is formed if not.-- the changes oscillate back
392d175 oured, getting the upper hand. <]CD> & forming SPECIES)-- [Aphides having fertile offspring without c
392d180 s.--» Experimentize on crossing of the several SPECIES of wild fowl <in Z> of India «with our common
393dIBC r se? Zoolog. Gardens Are the hybrids of those SPECIES. which cross & are fertile heterogenous? When
397e003 hich we are acquainted."-- this applies to one SPECIES-- I would apply it not only to population & de
398e005 necessary to arrive at right conclusion about SPECIES Changes of level &c are easily recorded, but c
398e005 f level &c are easily recorded, but changes of SPECIES «are» not as «without every animal preserved.»
399e011 a Melolonittha-- In marl from «Lake» Constance SPECIES of Europaean genera=.-- Hope has ideas about g
400e011 aean in Van Diemens land, where there is close SPECIES of elater.-- Where this collection is particul
402e017 e Grafted on pear. Mountain-ash & white Thorn! SPECIES not being observed to change «is very great di
402e020 , that are natives of S. Asia, but many of the SPECIES are peculiar to them" do-- p. 368. "Several ki
403e023 suppose bones will not change in number. (even SPECIES do not this). because it has been so pronounce
403e024 nfertility in largest sense. <es> «as» test of SPECIES.-- they must deny species which is absurd.-- t
403e024 e. <es> «as» test of species.-- they must deny SPECIES which is absurd.-- their only escape is that r
406e035 Fish of Teneriffe. St. Helena & Ascension most SPECIES like & identical with S. America. & many very
406e036 o marriage of individuals, & varieties of same SPECIES & to different species-- sometimes like one pa
406e036 ls, & varieties of same species & to different SPECIES-- sometimes like one parent & sometimes other
407e039 le kind of climate to the extreme.-- Therefore SPECIES, which were fitted for such a preeminently equ
407e039 (& become transmuted), although other parallel SPECIES in other continents might have survived this m
407e042 R. it will be highly necessary to show that if SPECIES fall, genera must. Lesson I remember says Mati
407e042 mber says Mariana Deer very close to a Molucca SPECIES.-- L'Institut 1837. p. 253, on animals of Anti
408e043 pe perhaps less so, that either Americas.-- If SPECIES change, we see external conditions have great
408e043 n, on the whole system may. produce-- ? When a SPECIES becomes rarer, as it progresses towards exterm
408e043 t progresses towards extermination. some other SPECIES must increase in number where then is the gap,
408e044 gap, for the new one to enter?-- The wonderful SPECIES of Galapagos, must be owing to these islands,
408e044 going there for.-- «Gales of wind would blend SPECIES» Buckland. Reliquiae Diluvianae. p. 222. Bones
409e048 nimals»: for otherwise, there would be as many SPECIES, as individuals, & though we may not trace out
409e049 many anterior epochs.-- but we can see if all SPECIES, there would not be social animals. «this is s
409e049 d too strongly. for there would be innumerable SPECIES. .& hence few only social there could not be o
410e051 should follow after discussion of crossing of <SPECIES> individuals. with respect to representative s
410e051 s> individuals. with respect to representative SPECIES., when going N.orth & South Thinking of effect
410e052 hich are possible.)-- Ascertainment of closest SPECIES (& naming them) with relation to habits, range
410e052 uld as soon turn tailor, as, mere describer of SPECIES, from its garments, without some end.-- Respec
410e053 like Richardson.-- The relations of numbers of SPECIES to genera &c &c can never be told, without spe
410e053 ies to genera &c &c can never be told, without SPECIES being described.-- but then permanent varietie
410e053 anges in forms of organs, will care little for SPECIES, except so far as wanting names to refer to, t
410e053 ll care little, <in> whether the individual ▯e SPECIES or variety, but to discover physical laws of s
412e057 ecretion from Pidgeons stomach true milk.-- ▯ <SPECIES. are innumerable variations». Every structure
413e059 of the Tropical Forms in shells. are numerous SPECIES, numerous individuals, & <individuals> «specie
413e059 pecies, numerous individuals, & <individuals> «SPECIES» of large size.-- consider this (Cetaceae) wit
413e059 p. 226.-- Herschel calls the appearance of new SPECIES. the mystery of mysteries. & has grand passage
414e063 races of men meet, they act precisely like two SPECIES of animals.-- they fight, eat each other, the
415e066 st, my theory, should no fossil «very distinct SPECIES» of the Ornithorhyncus be found.;-- yet until
415e068 f a group does not depend on the number of the SPECIES.: therefore Man & monkeys have equal chance th
415e069 er proposition.-- as in «races of» Dogs, so in SPECIES, & in Man December 16th. The end of each volum
416e071 organics. are made by percisely same means as SPECIES-- but latter far more perfectly & infinitely s
416e072 s the differences depend upon inheritance & in SPECIES are only ancient & perfectly adapted races L'I
416e072 nus» in Paris basin.-- its relation to African SPECIES good observations.-- larger than any livin
416e075 unning hares.-- as in Italian Greyhound not so SPECIES every part of newly acquired structure is full
417e075 Oxen, which one must think deserve the name of SPECIES, may be owing to the little fixity of organiza
417e076 <tel> showed me many land shells of the common SPECIES: from one locality, all left whorled.-- He kep
417e077 pears to follow same law in two of «the» same <SPECIES>, variety, as in two varieties, & this we migh
417e078 d, no doubt are applicable to likenesses, when SPECIES & races are crossed.-- Now these laws are, tha
417e079 s.-- » (Lord Moretons law holds with different SPECIES, & individuals of same species.--). some races
417e079 with different species, & individuals of same SPECIES.--). some races of men. D'Orbigny. affect the
418e083 s rise to the sexual organs, different in each SPECIES & knowing from analogy, that all these very
418e083 ,--one is led to suspect that the birth of the SPECIES & individuals in their present forms, are clos
418e084 ave passed through as many changes, as has any SPECIES.-- Decemb. 21th.-- L'Institut 1838. p. 412. M.
419e086 those of Scandinavia, than of the N. American SPECIES--Glacial period Dr. Beck says the shells in Sc
419e086 300 ft elevation & no change & even no loss of SPECIES.-- It must never be overlooked that the chrono
419e087 of physical change «affecting whole bodies of SPECIES», & only secondarily,, by assumption well grou
419e087 rounded, on time;-- therefore the mere loss of SPECIES, which may be the work of a few years as with
420e088 up blank.-- CD[not between existing series of SPECIES of dogs & Hyaena.-- but a common point, whence
420e090 Pinna of Rio Janeiro, (like some Mediterranean SPECIES).-- might these fertilise other shells, as ins
423e095 d, why should there not be at any time as many SPECIES tending to dis-developement (some probably alw
423e098 werby says, is an American form.-- has several SPECIES in my fossils-- CD[If cases of one variety in
424e100 apagos were South American.-- several cases of SPECIES peculiar to separate islets.-- March 5th. Lyel
424e100 from North America, Scotland, Uddevalla. Many SPECIES same. & Northern forms-- & American ones & Eur
425e104 y cases of genera peculiar to the group having SPECIES peculiar to the separate islands-- In his work
425e104 s-- In his work on the Labiatae, some of these SPECIES are described.-- Capital case,-- for Sandwich
425e105 und in Eocene beds of Paris Lyell has remarked SPECIES never reappear when once extinct-- Lyell's arg
426e105 in the Tertiary «epoch» like Sicily not having SPECIES, if true important on my view.-- March 9th-- I
426e106 e any relation between the fact that different SPECIES produce abundantly infertile hybrids, & the fa
426e106 ld lead me to think that a <do> variety of one SPECIES would cross easier with 2d species, than two p
426e106 iety of one species would cross easier with 2d SPECIES, than two perfect species; but facts of grouse
428e113 p is great difficulty. «I should doubt if wild SPECIES ever formed like short-tailed cat or dog has b
430e117 ction is denied.-- it will not account for all SPECIES. even if it will for all.-- Varieties are made
430e118 two ways-- local varieties, whose mass of one SPECIES are subjected to some influence, & this would
430e119 inally (give woodcut) like I. sulcatus.-- Both SPECIES are found at Folkstone.-- it is unnamed this i
431e122 on plays greater part then transmutation.-- Do SPECIES migrate & die out.?-- In the place where any s
431e122 s migrate & die out.?-- In the place where any SPECIES is most common, we need not look for change, b

```
431e122 titution says shells become less in number. (¿ SPECIES, or individuals) the deeper one goes-- surely
434e129 so that they have been thought to be different SPECIES. Lychnis dioica, generally dioicous. yet parts
435e130 a, preeminiently artificial, so that even some SPECIES only in genera <are> have this structure.-- So
436e135 ntal direction is far more efficient in making SPECIES, than time (as cause of change) which can hard
436e135 bably would be most efficient in producing new SPECIES; also one being reduced in numbers, but not so
436e136 ation as much as «the» difference between <pi> SPECIES,-- for instance pidgeons-- : then comes questi
438e141 any difference in crossing between varieties & SPECIES, yet the amount of may depend on many circumst
439e143 view of <i>nfertility of hybrids «with parent SPECIES» false, which makes it determined by a facilit
439e143 in animals.-- yet the existence of wild close SPECIES of plants shows there is tendency to prevent t
439e144 r that these varieties have become as fixed as SPECIES, & prefer their own pollen to that of other va
442e152 over this theory, & the sexual reproduction of SPECIES may stop for any number of generations-- Gorze
442e153 bove suggestion my theory would require, that <SPECIES> «individuals» propagated by gemmation should
444e157 t.-- p. 144.-- Alludes to two theories;-- that SPECIES are the result of circumstances,;-- or the wil
444e158 ted by foreknowledge, so must the mutations of SPECIES.!!-- p. 203 Chaetodon squirting water at fly.-
446e162 ordan Hill.-- May 27th.-- Henslow One of the 4 SPECIES of Lemna only reproduces itself «in England, a
447e163 -- «-- hence there would be real gradations in SPECIES from one region to another.--» -- this simple
447e164 would  by my theory gradation of form from one SPECIES to other: therefore my theory does require cro
447e164 s-- (Poa alpina vivaparous sometimes seeds All SPECIES of Lemna sometimes though very rarely flower [
447e164 one does on the continent-- well characterised SPECIES periwinkle wants insects to impregnate  allied
447e165 ghtest doubt that Festuca vivapara is the same SPECIES with F. ovina, <& this> rendered vivaparous by
448e166 n at Lima The same numerical relation (both in SPECIES and subgenera) between the Crag & Touraine bed
449e169 ons in common Dovecot are very like a Himalaya SPECIES -- leuconotes-- Magazine of Nat. History. 1839
449e170 eromerous insects come very near to Patagonian SPECIES-- p. 18. of Temmincks. Preliminary discourse t
449e170 &c &c.-- & it shows the causes which give same SPECIES to different isld. is the same as that which g
459tflv ery spot is occupied & has been occupied [...] SPECIES, which has undergone all the changes. [im]port
459t009 see  in the young of the wild hog & of several SPECIES of deer, which are altogether immaculate  when
459t013 - So Penguin impresses its form both on vars & SPECIES The <male> swan-gander with common goose produ
461t041 The  identity <of> (or only closeness) of some SPECIES-- (especially of mammifers) in old beds & exis
461t041 specially of mammifers) in old beds & existing SPECIES is valuable because it shows no innate power o
461t041 ows, what enormous changes of conditions, some SPECIES will undergo & yet remain adapted.-- it does a
461t041 This  unequal duration is exactly same as some SPECIES extending much further geographically then th
462tf4r .-- Shrew, found by M. Lartet same as existing SPECIES. We see the same object gained by the Mataco-a
464tf6r w strange, anyone, would have thought isolated SPECIES Mr Blyth, however, believes in the existence o
464tf6v s &c-- (see Blyth's work on Ruminants,-- these SPECIES must have migrated to these mountains, when th
465t079 found fossil in S. America, there are now-- -- SPECIES in S. America.-- so see what a «mere» vestige
467t103 f upper petal standing on «I saw Bee go to two SPECIES of Lupine,» two wings. & when the Lupine flowe
468t111 to  Rhododendron-- from Larkspur to Lupine two SPECIES of Larkspur -- two varieties of Cistus Speedwe
469t119 on one foot probably monstruous & not a second SPECIES.-- <Saw> Maer. June 15./41/. Watched plants of
470t153 all Linaria in do Domestic do 6 Campanula (two SPECIES)-- in do-- do 3 of do in about <3/4> of minute
471tf11 d to doubt. whether there is such a thing as a SPECIES-- Jun 1. 1842 Allen W. sowed some years  since
480z008 ensis (??)) Fulica Chloropus. says some of the «SPECIES of» smaller petrels, are night birds agree. wi
482z011 & perhaps Yagourundi near Concepcion!!!.-- (no SPECIES mentioned) p. 205. only 9. Terrestrial birds a
482z011 &  then 9. passeres! Says the thrush & another SPECIES! birds of passage!! sylvia macloviana, 2d like
482z012 ril» who is writing with Dumeril says that two SPECIES of Tortoises come from Galapagos!!! Azara. Voy
482z012 ita from 41 degree to 26 degree CLOSELY allied SPECIES, therefore interlock.-- Testudo INDICUS not fo
485z017 han on East.-- not being replaced by Brazilian SPECIES.-- Mem Turdus Magellanicus.-- C, <Chingolo> Ch
485z017 ango-- Diuca?? See Report <by> on D'Orbigny on SPECIES of Mephites 4 distinct Camelidae. do not breed
490q001 onstrous Cardium-- does it remind him of other SPECIES Hooker says the species of Aquilegia vary much
490q001 it remind him of other species Hooker says the SPECIES of Aquilegia vary much in their spurs & Ranunc
492q002 sed distinct by seed-- Heartease. 6. -- Do not SPECIES of wild Roses run into each other very much.--
493q003 rdens? <are the> is the ground much manured In SPECIES of close genus do more than three primary colo
493q003 n three primary colours occur in relation with SPECIES-- answered «by Henslow» see notes In varieties
494q004 rmer (8) Is form of globule of blood in allied SPECIES similar.-- if not how is it in <allied> variet
494q005 f bred mimosa (a) between sensitive & sleeping SPECIES, & see whether association can be given (2) do
497q007 d difficult to cross  are there unusually many SPECIES in genera of Leguminosae.-- Herbert explains n
498q008 berley says Ants-- Enquire (13) Do any of same SPECIES of Willows grow in same situation & flower  at
498q008 r at same time. Has H. seen group of different SPECIES growing White Mullein good plant to sow &  try
498q08v e Mullein good plant to sow & try to get other SPECIES <near> close to each other.-- As they are dioe
500q010 piadae. «in» Lindley! (24) Do Bees distinguish SPECIES, they do not varieties.-- (25) Does the yellow
500q10a be planted this year copied Gould.-- Number of SPECIES of Birds in New Zealand, plants so few-- Range
500q10a genera, «in Birds» in accordance with range of SPECIES?-- Are there any fine doubtful species from Va
500q10a ange of species?-- Are there any fine doubtful SPECIES from Van Diemen's Land? or New Zealand? Babing
500q10a Babington about differences of Irish & British SPECIES & British & distant parts of Europe.-- Gould--
500q10a s, Dacelo. Alcyone, where there are very close SPECIES & see whether they come from islds. or differe
500q10a r same district.-- About <endemic &> wandering SPECIES of confined genera By my theory in volcanic or
500q10a here ought to be a good many races or doubtful SPECIES; how is this at Canarys Arch-- it is so at Gal
500q10a h-- it is so at Galapagos.-- Ireland, doubtful SPECIES-- Does any genus of Plants. vary & hard to sep
501q10a nera less difficult, in other countries, where SPECIES are either numerous or even where few are they
501q10a Patella or Oysters or Helix. Or does any «one» SPECIES of plant, vary in one region of Europe &  less
508q016 races  of men taking after sex. A Smith. About SPECIES of Rhinoceros. becoming rare beyond limits  of
509q017 on holds good, that in the mundane genera, the SPECIES <are> have wide range-- How is this in «Plants
510q017 laire Bot. p. 787. position of embryo in close SPECIES of Hilianthemum differs greatly-- how very int
511q018 xtinct= (9) About. American & Europaean common SPECIES, having somewhat of different appearance.-- {w
513q020 ther & see result, for comparison with natural SPECIES, as dioecious plants, when crossed R.  BROWN--
514q21. d-- geology. volcanic? Applies to my geology & SPECIES theory-- peculiar Fauna?. {Australian Alps--;
515q021 n, but appears in abortive state either in the SPECIES, or in the individual by chance & under domest
515q021 e apt to vary in number in individuals of same SPECIES, Eyton (1) Number of eggs-- of half-bred geese-
516q23v h- being killed Experiments not connected with SPECIES Theory (1) Will an extract of peat. do to prese
527m031 birds  learn from each other «though different SPECIES» when in confinement, so may they learn in a s
535m064 with insects. but habit forgotten in all older SPECIES. The earwig & a doubtful one of Acanthosoma gr
557m147 his  ears?--).-- How is expression of anger in SPECIES of swans, in parrots &c -- -- peacock & tur
558m153 hes wings-- as does black Swan.-- Goose do all SPECIES put their necks straight out & hiss.-- [Hyaena
563nIFC e.]CD (Metaphysics & Expression) Selected «for SPECIES Theory» Dec. 16 1856 Looked through & all othe
583n069 instinctive  powers in individuals of the same SPECIES with variability of reasoning power in one spe
583n069 ies with variability of reasoning power in one SPECIES man.-- false instinctive pointing varies.-- p.
584n071 or climbing, his arguments partly fall, when a SPECIES is found which does not climb CD[.instinct may
586n081 e faculty of finding its way, which in certain SPECIES is instinctively «not least by experience» di
591n098 tion in character in different animals of same SPECIES.-- The general «(as I believe)» contempt at su
613o036 instinct  to all animals modified according to SPECIES. This I suppose he deduces from the ends in ea
614o037 ted instinct. might be analysed into steps, as SPECIES change.-- Must be so if Lamarck's theory  true
615o038 in  «external» man; and as all men nearly same SPECIES, so general instincts nearly same; which  same
623o050 seems to vary in races of man. & certainly in «SPECIES of» animals, in which case it undoubtedly is i
624o051 , beneficial tendency in every instinct to the SPECIES in which it occurs. [or, more correctly «in wh
635j56r plains the shortness of life (peculiar to each SPECIES) owing to the growing size of the world? & th
638j58v hose with weak beaks be sifted away.-- 4 & the SPECIES, like 10,000 others, perish. & who will dare t
640j167 «With  respect to whether Galapagos beings are SPECIES. it should be remembered that Naturalists  are
640j167 y unphilosophical to assert, that they are not SPECIES, until their breeding together has been tried.
640j167 hy small islands. should possess many peculiar SPECIES. for as long as physical change is in progress
640j167 ith respect to new arrivers, the small body of SPECIES would far more easily be changed.-- Hence the
640j167 st newest, & most wretched isld should possess SPECIES to themselves.-- Probably no case in world lik
640j167 nould infer that Mollusca would offer few SPECIES, or rather be very slowly changed & vertebrata
641j29r er» 3 A very wide range must be destructive to SPECIES, when physical changes are in progress; (on th
641j29r rable,) because it must take so long to change SPECIES-- yet this is contradicted by continents <bri>
641j29r ontradicted by continents <bri> abounding with SPECIES-- there will be a balance, continents have bee
107a077 ing.-- No Sweden!! swelling of rock from Heat. SPECIFIC gravities of many artificial limestones produ
186b057 least  in one point, in truth in all excepting SPECIFIC character); and in passing from species to ge
277c126 ts determining fertility Scheme for abolishing SPECIFIC names & giving subgenera. true value.-- as in
313c234 ation apply to utter extinction or rapidity of SPECIFIC change.? <One> the first would be called. gen
```

Page **(Key Word)***
```
313c234 ne> the first would be called. generic & other SPECIFIC extinction-- In the Entomostraca (Magazine of
354d065 as if the "variation in one, was analogous to SPECIFIC character of other species in genus."-- Is th
368d114 e waders) waders the bright plumage.-- «thinks» Hence SPECIFIC character most perfect in <male> «hermaphrodi
369d114 should say females recede in organization from SPECIFIC character.-- Chapt I. Also Latent Character H
400e012 e, might be made out of variation analogous to SPECIFIC variations.-- Kerr's Collect of Voyages Vol 8
411e054 ces, may be important in understanding laws of SPECIFIC change.-- When the laws of change are known.
501q10a d show that such variation is not a generic or SPECIFIC character,, but contingent on country.-- How
513q020 ection &c &c &c--if so probably a variety, not SPECIFIC character» Cross Rumpless fowls & Dorking fow
231b243 ed a little to the southward would merely be SPECIFICALLY different if so.-- Now this is difficult to
407e041 ta, when first introduced live & multiplied, SPECIFICALLY & individually.-- I see clearly from F. R.
501q10a any genus of Plants. vary & hard to separate SPECIFICALLY in one country & not in other: Rosa is hard
411e055 on of parts & organs it may serve perfectly to SPECIFY types, & limits of variation., & hence indicat
094a036 might be transported.-- {T} On grooved rocks. SPECIMEN of rock from Costorphine at Geolog. Soc: Colo
252c045 ur knowledge especially great continents» Give SPECIMEN of arrangement, 3 <5> Species Rhinoceros Cape
449e173 Mastodon found at Timor.-- thinks he has seen SPECIMEN at Paris Museum.-- Athenaeum: 1839. p. 451. S
469t119 mper on Ourang-outang, has examined 7 says one SPECIMEN had on one foot, a toe-nail & two joints-- as
049r087 utonic The cellular state of all the Porphyry SPECIMENS, must be well examined at M. Video «facts of
065r138 zet. L. Auckland. Macqueries.--Sandwich Isd-- SPECIMENS of rocks were brought home in Capt. Forster e
114a094 Negro--Mr Bowerbank-- Dr. A. Smith's curious SPECIMENS of «transversely fibrous» quartz. & iron ston
115a097 ers ought to be deep.-- Henslow has deposited SPECIMENS from Anglesea in Geolog. Soc. if numbered com
145g004 n in Salisbury Craigs [4] Salisbury Craigs V. SPECIMENS-- {P} Veins, amygdaloidal-- as well as base n
410e052 e to all countries,-- but naming mere «single SPECIMENS in» skins worse than useless.-- yet there is
430e119 world-- March 16th. Mr Lonsdale showed me two SPECIMENS of an Inoceranus from the Gault of Folkstone,
431e120 Mr Sowerby is coming to this conclusion, from SPECIMENS in grades, now L. says the T. costatus is in
494q005 ation can be given (2) do the stamina of C. SPECIOSISSIMUS collapse during sleep & do of Berberis-- (
455eIBC Shake some sleeping mimosa-- do stamina of C. SPECIOSUS. collapse at night. if so irritate them, «as
284c149 visible, when young, but not when old.-- thus SPECKLED form of young blackbird. good remark if gener
363d101 Rock Pidgeon,-- several have a group of white SPECKLES on elbow joint-- in Bewick drawing the the ro
110a086 KE base reddish feldspathes with grenish. black SPECKS of hornblende. large irregular cryst of reddis
239cIFC y Waterhouse B C N[a]me of two pigeons,-- «with SPECKS» which cross & keep colour on wing Effects of
551m127 things which have no existence.-- He compared SPECTRAL illusion & insanity the connexion appears to
552m130 truly be <more> «as» vivid, «a reality» as in SPECTRAL images-- Mem Chiloe <pi> Sow, who carried fro
061r127 t about Crust Bra in Brine Springs. (Henslow) SPECULATE on neutral ground of 2. ostriches; bigger one
085a001 l's Salband p. 86 Shells near Woollich p. 112 SPECULATE on the extension of Patagonia seaward, at mou
122a115 shore of Loch Lochy very like those of Andes SPECULATE under head of Beagle Channel. on origin of mu
163g109 z what substance is collected in little spots SPECULATE on «under head of» Beagle Channel. Forchammer
163g109 ammers (Lyells Denmark) Shrewsbury rubbish.-- SPECULATE on origin pebbles brought by different cause:
189b072 is no TRUE connection between great groups.-- SPECULATE on land being grouped towards centres near Eq
222b202 g to Beck has none recent, yet genera same.-- SPECULATE on multiplication of species by travelling of
227b225 they are not far the most serviceable. We may SPECULATE of durability of succession from what we have
257c058 Sandstone.-- Nautili in----. it is useless to SPECULATE «not only» about beginning of animal life.: g
411e054 nge are known.-- -- then primary forms may be SPECULATED on, & laws of life,-- the end of Natural His
419e085 velopment of Mollusca, which I have sometimes SPECULATED might be owing to absolute quantity of vital
090a021 a said to extend to Cordillera I see Brewster SPECULATES from believing meteorolite but old Planet, t
603o012 & hearing-- Staunton Embassy Vol II p. 405.-- SPECULATES on origin of sacrifices. «common to many rac
068r148 Beagle Channel.--The need never be afraid of SPECULATING on the sea The 24 ft. elevation at Concepcio
227b226 ledge of possible changes is discovered, for SPECULATING on future. !.fish never become a man.-- Does
107a078 proceeds from great body of mass.-- The last SPECULATION becomes important with respect to thickness
217b188 f ice transported.-- This gives room to fine SPECULATION.-- Are there many Northern genera peculiar t
227b224 hin recent times.». & many curious points of SPECULATION; for having ascertained means of transport,
227b227 ents??) instinct & structure becomes full of SPECULATION & line of observation.-- View of generation.
355d067 s» Although no new fact be elicited by these SPECULATION even if partly true they are of the greatest
383d159 of sexes as very great difficulty, then give SPECULATION to show that it is not overwhelming.-- Seein
181b044 r this agrees with Nature: Cuidado The above SPECULATIONS are applicable to non progressive developme
208b153 f races of men and animals) See R. N. p. 130 SPECULATIONS on range of allied species. p. 127. p. 132
228b228 be main object of study, to guide our <past> SPECULATIONS with respect to past & future. The Grand Qu
270c104 ght be worth consulting, if any Metaphysical SPECULATIONS are entered in upon life. Namely Carus.-- H
308c218 an that complex nature itself with which our SPECULATIONS must end as well as begin" &c &c then centr
337d021 mon greyhound» & much intermarriage.-- In my SPECULATIONS. Must not go back to first stock of all ani
520mIFC d. This Book full of Metaphysics on Morals & SPECULATIONS on Expression -- 1838 Selected Dec 16 1856
599o005 ally musical!!!??-- At least it appears all SPECULATIONS of the origin of language.-- must presume i
599o005 must presume it originates slowly--if these SPECULATIONS are utterly valueless-- then argument fails
112a088 facts See page 101. in Note Book (C) for some SPECULATS on conducting powers of rock-- -- Geograph Jo
256c055 & Virlet.-- Whewell thinks (p 642) anniversary SPEECH. Feb 1838 thinks gradation between Man & anima
522m014 ho has, perhaps, slightly injured me, plotting SPEECHES, yet with a sort of consciousness not just.--
468t111 pecies of Larkspur -- two varieties of Cistus SPEEDWELL to Rhododendron-- <Loasa> «Anchusa»-- speedwe
468t111 eedwell to Rhododendron-- <Loasa> «Anchusa»-- SPEEDWELL Iris-- Azalea. Rhodendron. Fraxinella to Arch
337d022 vibration of sound.-- which is impossible.-- Mr SPENCE remarks that the Fringilla domestica of North
574t039 luence over her.-- Hensleigh says. Douglas. «& SPENCER», an old Scotch Poet, has numerous lines. of p
539m084 6th Anger «Rage» in worst form is described by SPENSER (Faery Queene. CD 25 (Descript of Queen) «O» o
545m109 he had had, he would say how many happy days he SPENT in such a place.-- Vide page 103, supra (by mis
443e155 e hermaphrodite in beings with which have fluid SPERM.-- JCD I utterly deny the right to argue agains
641j29r great Seals of Antarctic seas. (on other hand SPERMACETI Whale & Manatee.-- Naturalists must be cauti
420e090 nzani's experiments showing how little of the SPERMATIC fluid fertilized spawn of frogs.-- Annals of
424e100 ther I have already referred to it.-- also on SPERMATIC animalcules in Musci frondosi, et hepatici,--
431e123 rely is this true?-- most strange.-- Does not SPERMATIC animalcule in Mosses, render my view of the c
454e184 Plant.-- p. 290. Dr. Edwards in his essay on SPERMATIC animalcule. has described instrument for galv
157g077 ince «that shelf» formed Upper terrace near Loch SPEY <29.35161> 29.360? A 79 degree 75 degree? A lit
157g078 ? A little below Divortium on slope towards Loch SPEY 29.297 A 79.½ 29.316 divortium aquarium «about 1
158g080 5 ft divortium my measurements here side of Loch SPEY Forms terrace about 60 feet above Loch trace of
345d043 trao Scoticus. an american form (if so)?.-- A SPHEPHERD of Glen Turret. said he learnt to know lambs,
578n056 f other person Decemb. 27th.-- Fear loose the SPHINCTER muscles, only on the principle like does an i
578n056 ifferent ones?) in the most perfect fainting, SPHINCTERS are loosed is a convulsive action to remove
377d138 edge of Woman-- The noise st st. which the C. SPHINX makes is also made by the C. porcarious., toge
545m105 serving how the Baboon (<Macaco> «Cyanocephalus SPHINX Linnaeus») constantly moved the skin of forehe
556m145 muscle.-- reason CD[The laughing noise which C. SPHINX made at Z. Gardens may be described as partaki
558m153 ression be used more correctly than the C. SPHINX.-- In the wild ass there is a curious drawing
563n002 r dogs defending companion. (mem Cyanocephalus. SPHINX howling when I struck the Keeper) may be tempt
194b094 epresent the semnopitheque of India.-- Tooth of <SPI> of Sapajou-- NB Sapajou is S. American form. th
370d117 a tree in Brazil must have degenerated. as must SPICES &c &c The line of argument «often» pursued thr
254c049 Surely not correct view of Flustra or Ascidia SPICULE in sponge. stomachs in infusoria, generation i
300c197 -- They reason however on this to a degree. Mem SPIDER only dropping where ground thick.-- shamming d
381d156 alopods, Pectinibranchiate molluscs.-- insects. SPIDER crabs.-- (all these however do not require coi
545m107 remember» the expostulatory angry look of black SPIDER monkey when touched, also another monkey to do
555m143 men have this power abortive muscles) The black SPIDER Monkey, very different disposition from others
595n115 [...] been dead about two months. saw a «black» SPIDER monkey brought it at opposite end of house. &
641j29v re is an analogy between fang of snake, (jaw of SPIDER?) sting of bee, sting of nettle.-- Are there a
543m099 ta-- not connected with transformation because SPIDERS have many,-- great powers of communicating kno
463t055 yet in talking, constantly said as the <brain> SPINAL marrow expands, so do the bones <are created>
578n056 nly on the principle like does an injury of the SPINE-- that it paralyzes all muscular action -- «in
190b079 r> passage between good genera-- How remarkable SPINES, like on a porcupine on Echidna-- Good to stud
192b087 ach there is possibility of such organization. [SPINES in Echidna & Hedgehog]CD-- As we have one Mars
304c208 s.-- Yet trifles are produced by circumstances. SPINES on Echidna.-- when it can be traced through se
373d132 more wonderful. than Echidna. & Hedgehog having SPINES.-- Does not male Pidgeon (yes) surely) secrete
460t017 20th The Echnida & Hedgehog Tenrec both having SPINES. is the effect, partly of the same external co
507q15v k-- -- Geum. Galium Burrh = single hook; curved SPINES-- simple spines-- or seed-cases with similar s
507q15v ium Burrh = single hook; curved spines-- simple SPINES-- or seed-cases with similar structure.= good
639j28r Marsupials. Harvest mouse & (Chamaelion?) C. D. SPINES in Hedge Hog & Echidna. & Aphrodites C. D. En
```

(Key Word)

458t002 uses pieces of thread, picked «up-» instead of SPINNING-- better case than English birds, using cotto
507q15v more. & seeds with «mere» border-- & Humboldts SPINNING seed.-- (50) Any cases of wild varieties plan
637j58r these barren Virgins p. 235. talks of the long SPINOUS processes in Giraffe &c, as adaptations to lon
452e181 derful instinct, how could it have originated-- SPINS thread of cotton.-- do p. 583, It APPEARS proba
272c107 f body.-- there are hemipterous insects, having SPINY legs & running quick & generl appearance of bla
297c186 g. & Bot. Vol. II p. 125 Allusion to abortive SPIRACLES in Hemiptera do. p. 160. soft plumage of nigh
270c103 tion!!!. Mem. Agaziz. (INo Annals of Nat. Hist) SPIRAL structure in Echinodermata.-- Agassiz says Inf
275c119 s endowed with what may be called the prophetic SPIRIT in science--. the highest endowment of lofty g
282c143 al can really produce so great an effect.-- the SPIRIT of life must be every where ambient. & <in> me
305c210 s.-- Mem Elephants & dog.-- There is one living SPIRIT, prevalent over this word, (subject to certain
308c217 exclaim with Christian «we are all» Brothers in SPIRIT-- all children of one father.-- yet difference
397e003 - do p. 529. "It accords with the most liberal! SPIRIT of philosophy to believe that no stone can fal
502q11v tion.-- July «1842» When nettle leaf. put into SPIRITS, poison-drop exudes-- does not elm. does it «i
502q11v see orifice of poison-tube-- so put carmine in SPIRITS & then experimentise: for gradation in structu
505q014 Thyme seeds weather wet--? Linum flavum put in SPIRITS which plant seeds? (9) Melons fruit itself hyb
155g062 cut out. produced» from same «cause» as «great» SPIT <is> or plain <now> formed on shelf 4th 2d & 3d
159g084 ; several large ones (one 6 ft across) on top of SPIT between river & dry Corry Scarcely conceivable.
356d071 whole world. (Major <I> Mitchell p. 244. vol I) SPIT & throw dust <according to my theory of generat
560m156 impatience when woman present? Do they pout, or SPIT, or cry.-- <fe> Shame, independent of fear: col
077r171 r doubts whether bed or vein (very like that of SPITAL of Schemnitz in Hungary.) Humboldt says fragme
267c093 aims all the moist & fertile land of Tahiti, in SPITE of every attempt to check its increase. The <bu
573n038 ts.-- Person with sore-throat told not swallow SPITTLE. will have involuntary flow & desire to swallo
573n038 ot to turn in bed. will turn in bed.-- in case SPITTLE, effect of thought is to make saliva flow, & t
072r159 Scotland & Iceland.--Bosh nor on Norway, or SPITZBERGEN.--Spitzbergen animals (?). ≠ The Hollowness
072r159 eland.--Bosh nor on Norway, or Spitzbergen.--SPITZBERGEN animals (?). ≠ The Hollowness of <sep> Chilo
260c067 Vol III. p. 226 Wilson's Ornithology, D'orbigny, SPIX, &c might compare birds of N. America & South--
484z016 me plumage.-- general red mark on wings of all-- SPIX has described Philedon. allied to some of my bi
486z020 do with <T> N American & T. del. Fuego & Iceland SPIX & Martius talk of birds singing in the forests
123a115 - nothing.-- salt & coal near Bogota; p 270.-- SPLENDID PAPER on fossil shells of S. America. Von Buc
291c168 - Mem Lyell hypothesis of change in Scicily.-- SPLENDID Harmony these views-- did Lamarck connect ext
307c216 she can be converted «into female», it will be SPLENDID argument old female, turning into cock, abort
432e160 ut length of time, only order of succession.-- SPLENDID Pamplet. (published in Philosop. Journal <Mar
446e160 female of the icterus minor is a bird of more SPLENDID plumage than the male.-- Athenaeum May 18. 18
528m036 hich is as real a cause as in music) from the SPLENDOUR of light, especially when coloured.-- that li
199b119 rst settling a country.-- people very apt to be SPLIT up into many isolated races. ¿are there any ins
263c074 o deceive himself.-- Give the case of Apterix-- SPLIT, depress & elevate & enlarge New Zealand; a div
382d157 r a Heautandrous animal is <evidently> actually SPLIT in two-- keeping sexes separate. Owen say such
386d165 Cannot I find some animal with definite life & SPLIT it, & see whether it retains same length of lif
604o016 oosing food-- crawling from light.-- Yet we can SPLIT Planaria into three animals, & this consciousne
641j29r - there will be a balance, continents have been SPLIT up-- who can decide their limits.-- To show ho
226b221 rted them to new station.-- When the new island SPLITS & grows larger species are formed of those gen
633j53v a desert plant (Anastatica) is rolled along, & SPLITS when it comes to a damp. place.-- Kolreuter me
189b072 centres near Equator in former periods & then SPLITTING off.-- If species generate «other species», t
204b141 view. In Marsupial division «do» we not see a SPLITTING in orders, Carnivora, rodents &c, JUST COMMEN
281c137 - The simple expression of such a naturalist "SPLITTING up his species & genera very finely" show how
424e101 ifferent.-- Subsidence of Greenland-- case of SPLITTING of two regions-- -- are there any cases of un
423e098 s one single turnip in a garden is sufficent to SPOIL a bed of Cauliflower.-- (How curious it would b
553m137 ip, looking absurdly sulky «as» often as keeper SPOKE to it,-- but he thinks not sulkiness-- this exp
264c079 ear expressive whine, see its intelligence when SPOKEN; as if it understood every word said-- see its
590n094 els.-- When one sees in Cowper, whole sentences SPOKEN & believed to be audible. one has good ground
351d056 erip added it has been stated that stationary SPONDYLUS has eye-points-- Macleay then answered, becau
254c049 t correct view of Flustra or Ascidia spicule in SPONGE. stomachs in infusoria, generation in each joi
350d055 pedes can move & see, parent fixed,-- young of SPONGES move.-- young of Cochineal insects move about
381d156 ystic Entozoa. Echinoderms. Acalephes. Polyps. SPONGES Heautandrous, male organs formed to fecundate
254c049 ather parents). (NB These views must lead to SPONTANEOUS generation??) This whole Paper Must be studi
269c102 ation, & the universality of latter render-- SPONTANEOUS generation not improbable.-- After reading "
326c265 25. & Lasch. Linn. in 1829 has given list of SPONTANEOUS Hybrids. where? Sweet. Hortus Britannicus. h
373d132 ae?.-- My view would make every individual a SPONTANEOUS generation: what is animalcular semen-- but
446e160 heory leaves quite untouched the question of SPONTANEOUS generation.-- Introduction to Bartram's Trav
507q15v his class.-- (48) .Where «published» list of SPONTANEOUS Hybrids-- to see whether any Papilionaceous
346d047 pots on Albatross, on some Gulls. Flamingo-- (SPOONBILL Wader. Ibis)-- laws of plumage might possibly
492q003 ferred to with doubt by Herbert Do forest-trees SPORT much in nursery gardens? <are the> is the groun
515q021 ly any no relation to hybrids.-- (3) As peaches SPORT into Nectarines (does reverse happen?) what is
515q021 estion with regard to Primroses. (4) Do apples SPORT in fruit, or time of leafing (5) Do the most c
600o08v al instinct domineering over love of Master and SPORT &c &c -- The Bitch does not so act, because mat
460t017 strosity, & determine the kind of variation «& SPORTING» in flowers & domestication of animals Aug. 2
209b155 ysical changes) would be in rays-- from certain SPORTS.-- Agrees with old Linnaean doctrine & Lyells.
326c266 etry.-- Erasmus thinks I should lik it. The SPORTSMAN'S Repository. 4to. contains much on dogs.-- Re
041r065 tion of shell's concretions, living only in that SPOT & being cause of concretion; or being only pres
061r127 .--Australia's = if for volc. isld. then for any SPOT of land. = Yet new creation affected by Halo of
156g070 roportionately large to those now formed in same SPOT by present torrents Maculloch wrong in saying n
318c255 ommon in Rocky Mountains & on one lofty isolated SPOT on the Alleghanies to which it migrats every ye
402e017 rely leaf, if one river did pour sediment in one SPOT, for <whole> many epochs-- such changes would b
459tf1v y country should during [...] conditions-- every SPOT is occupied & has been occupied [...] species,
472s01v in the 60 one brighter with mere traces of black SPOT at base, one paler with less riged foliage & no
472s01v se, one paler with less riged foliage & no black SPOT & a third considerably paler, all rest very sim
584n072 ell might be called instinct,-- migrating to one SPOT, this is indeed instinct.-- Australian man, may
163g109 re Quartz what substance is collected in little SPOTS Speculate on «under head of» Beagle Channel. Fo
195b098 existed.-- Such an influence Must exist in such SPOTS. We know birds do arrive & seeds.-- The same re
199b116 ming truth of quadrupeds being created on small SPOTS of land, of the same type with the great contin
315c240 exists..-- pooh. May have been Created at many SPOTS & since disseminated See. Habits of Malay fowls
346d047 ink shade on plumage of the Pelican.-- Mem pink SPOTS on Albatross, on some Gulls. Flamingo-- (Spoonb
312c227 - in raptorial birds, & tigers & sharks, being SPOTTED, & colours of little value Dr Smith if black &
341d030 which is so anomalous among true deer, yet is SPOTTED like so many deer.-- very curious like some fa
363d103 s second their songs, the +1* Cervus Campestris SPOTTED white when a fawn compare with fallow? deer. &
450e173 . 323. Sooloo. imported elephants. wild hogs-- SPOTTED deer, no loonies, but cocatores & small green
459t009 ol III. p 320. Mr Hodgson on Musk Deer-- young SPOTTED <like in> "prettty much as we see in the young
051r094 ocks near Bahia, whole surface to where highest SPRAY (there pale green confervae) coated with living
037r078 n of Trachyte. look at Sulphur. salt. lime, are SPREAD over «whole» surface; how comes it they do not
104a069 ep sea, fragments fall off cliffs. but then how SPREAD abroad?-- There is thus wide difference betwe
105a071 ence.-- removal downwards by successive torrent SPREAD out. by sea-- beach action -- no one will disp
120a109 at interior effect would be such as present. to SPREAD sheet of matter over surface.-- if elevation t
250c038 ave allowed fresh species to have been formed & SPREAD to other Africa & East India Arch.-- but where
250c038 a Arch.-- but where these great animals had not SPREAD then such tribes as Marsupial & Edentata incre
402e020 em" do-- p. 368. "Several kinds of animals have SPREAD from the N. end of Borneo to the adjacent isla
405e031 s the puppies to be so.-- [not located] Did man SPREAD over world as early as Elephants &c &.-- if in
418e080 Ox. the amount of double sexual developement is SPREAD over [not located] it utterly untold,-- what i
429e114 opagated with very little care.-- & which might SPREAD themselves, as well as our wild plants, we see
495q05a y-dew dusted with pollen of neighbouring grass-- SPREAD sheets of Paper. covered with some sticky stuf
119a105 may conclude that elevation is independent of SPREADING out matter by action of the sea.-- as no sea
332d001 he never saw one case of malignant erysipelas SPREADING over the head, not caused by a wound, when su
506q15v -- Dioecious.-- Carex.-- We may presume Nettle SPREADS by seeds-- (44) Zostera. Has he seen it in flow
045r077 fflux (always slightly oscillating as that of a SPRING) moves away.--Will geology ever succeed in sho
126a123 ly beneath level of sea.-- deep seated springs «SPRING requires connected column.--» of cold water sh
130a133 ounding over a continent to fall across a hot.--SPRING.-- Hot water would not lie. at bottom.-- Surel
259c062 then argumen May be.-- subterranean lakes, hot SPRING &c &c inhabited therefore mud wood be inhabite
275c120 «is not that old variety» & then recrossing off SPRING. till size diminished, but feathers continued
375d135 se, & an ordinary crop. causes a dearth then in SPRING, like food used for other purposes as wheat fo

Page ***(Key Word)**
493q003 tes In varieties is there any difference in off SPRING from A. into B. from B. into A. as takes place
497q006 nown» & other odd places & see what plants will SPRING up which will show, how seeds are transported,
504q013 es the Thyme bear abortive stamens every year & SPRING. & within garden //Yes// (5) Examine the Parna
620o046 s part of <duty> their nature.-- When a pointer SPRING his bird. one says for shame (& the «old» dog
179b035 similarity of animals in one country owing to SPRINGING from one branch, & the monucle has definite l
245c025 mboina only Cuscus & Barbyroussa «NB» [islds. SPRINGING up more likely to <M> have different species
293c172 eneral ones might yet be transmitted.= Memory SPRINGING up after long intervals of forgetfulness.-- a
439e144 ce qualifications each possess., & that tiger SPRINGING an inch further would determine his preservat
563n001 Natural History Waterton describes. pheasant SPRINGING from nest & leaving no tracks.-- My Father sa
633j53v curling of the valves of the broom.-- or the SPRINGING of other seeds.-- But are we certain that the
025r017 tive 300 years? No Volcanic Earthquakes or Hot SPRINGS in T. del Fuego = The Wager's Earthquake the m
028r029 t washed from the solid rock by the actions of SPRINGS or more probably by some unknown Volcanic proc
028r031 ium saline; Mem: on coast of Northern Chili as SPRINGS become rarer, so does the rain, therefore such
061r127 r Owen's curious fact about Crust Bra in Brine SPRINGS. (Henslow) Speculate on neutral ground of 2. o
065r140 d to September ¿Did I make any observations on SPRINGS at S. Cruz.???-- Form of land shows subsidence
108a080 ears to me unphilosophical to think calcareous SPRINGS near coral reefs.-- Where vegetation luxuriant
120a108 rock.-- in volcano the pool is not deep. --Hot SPRINGS &c &c--then if so, thermometer show it cannt b
126a123 se rapidly beneath level of sea.-- deep seated SPRINGS «spring requires connected column.--» of cold
126a123 cold water show, that water does percolate, & SPRINGS beneath sea-- → According to this latter view
126a124 t is said in some of the papers that there are SPRINGS even in siberia.-- from water {P} thawed at +
130a133 .-- Weelsted told me of some large fresh Water SPRINGS off coast of Persia In Glen Roy paper I show c
132a139 ad conductor.-- shows p. 516 that subterranean SPRINGS give result less to be trusted than any others
196b106 arsupials & Edentata should only have left off SPRINGS <ne> in or near South Hemisphere. Were they pr
495q005 her, count seeds, & see how great a proportion SPRINGS up true.-- This in fact always takes place in
524m019 ose feelings. which may be considered as truly SPRITUAL.-- a person twitching when a disagreeable tho
437e141 great swarm of mice.-- May 4th.-- The Brussels SPROUT returning suddenly to type when brought back t
438e141 races.-- «& yet in all probability the Brussels SPROUT was slowly formed.-- » if it shall be difficul
512q019 difference in breeds of Cattle & sheep in the SPROUTING of the horns. at different periods in differe
437e141 es of Brassica certainly not becoming Brussels SPROUTS) is analogous to Primrose & Cowslip suddenly c
227b227 not require fresh creation!-- If continent had SPRUNG up round Galapagos on Pacific side. the Oolite
302c201 really connecting the fish & Mammalia, must be SPRUNG from some source anterior to giving off of the
304c208 more rapid than in all other birds, or that it SPRUNG from a branch high up.-- this argument not app
430e117 intelligent.-- The theory that all animals have SPRUNG from few stocks. does not bear, the least on a
466t094 pistils, but these <do> do not bend up-- In Lark-SPUR, if Bees put proboscis within nectary «they do»
514q21. = Has Ophrys nectary?= Bunbury says no «hollow» SPUR.-- Ask about Pinks & Solanum impregnation befor
307c216 rgument old female, turning into cock, abortive SPURS. growing.-- Are there any abortive organs produ
369d114 l case of Pea hen. taking feathers of Peacock & SPURS-- no final cause here.-- & therefore different
383d160 on Pheasant. Male bird, said to be infertile.-- SPURS rather smaller than in <ma> silver male-- Head
490q001 ays the species of Aquilegia vary much in their SPURS & Ranunculus in the nectaries. The former best
495q05a eration Charlworth p. 377. Have paper ruled in SQUARES to facilitate investigation.-- Capital in midd
592n103 ed, does not hair & rabbit depress. them from SQUATTING.-- p. 64 closing both eyelids express contemp
578n056 e convulsion. (Hence pass into convulsions?)-- SQUEEZE out tears. replaced & squeezed out again-- as
459tf02 part of body mare)-- -- this may be, perhaps. SQUEEZED into Mr Walker's law Gleanings of Science Vol
578n056 convulsions?)-- squeeze out tears. replaced & SQUEEZED out again-- as power of mind by habit gets mo
525m024 ught not to do, & looking ashamed of himself.-- SQUIB at Maer, used to betray himself by looking asha
570n025 els of skin.-- What difference is there between SQUIB after having eaten meat on table, & criminal,--
525m024 table,-- guilty conscience.-- Not probable in SQUIB'S case any direct fear.-- My father thinks that
639j28r ure; Penguin.-- Pincers in Scorpion & Crust in SQUILLA. & Mantis. C D woodcuts stones swallowed by bi
579n058 ort-sightedness.-- do not short sighted people SQUINNY-- when they consider profoundly,-- this will b
301c199 may be either in same individual p. 7. is not SQUIRREL hoarding, & killing grains. acquirable throug
301c199 is not worthy interposition of deity to teach SQUIRREL to kill ears of corn according to my views, h
301c199 web. feet.-- p. 7. Mr Blyths arguments against SQUIRREL using reason in hiding its food is applicable
318c252 webbed Shrews. case of adaptation.-- (case of SQUIRREL from extreme north turning white like Hares??
337d022 nct Creation required if he believes «hyaena & SQUIRREL» seal & mouse, elephant, come from one stock.
334d012 Marlborough» there is a breed of white-tailed SQUIRRELS, which form a marked wild variety. doubtful w
451e177 Andaman & Pegu. <have> abound with monkeys & SQUIRRELS.-- Horsbrugh E. I. Directory. Vol II. p. 46 C
641j29v could catch mouse by night» Sailing lizards. SQUIRRELS & Opossums «& fish»: flying lizards.--Mammali
444e158 he mutations of species.!!-- p. 203 Chaetodon SQUIRTING water at fly.-- instinct, for how could exper
091a027 ra. Coast of Africa. Clay Slate & Quartz. strike SSW & NNE dip 30 degree - 80 degree Ed. N. Phil Jour
200b121 ff.? Accra, Coast of Africa. Clay slate. strike. SSW. & NNE. dip 30 degree - 80 degree (?).-- Ed. Phi
255c053 ng. ,has invented all kinds of plan to insure STABILITY; but isolate your species her plan is frustra
086a004 -- accustomed to such terms "fixed as the land, STABLE as the water"-- It may be worth noticing edent
564n005 mind is function of body.-- we must bring some STABLE foundation to argue from.-- Octob. 4th. Seeing
362d099 er 13th The passion of the doe to the victorious STAG. who rubs the skin off horns to fight-- is anal
068r146 connected with fluidity of rock ∴ «in earliest STAGE» when covered up beneath ocean).--The first dis
210b161 ich, Hunter says owe their origin to very early STAGE) & which, follow certain laws according to spec
292c169 hen so the> are united (which probably is first STAGE) the tendency to change cannot be great, otherw
292c171 -- an action becomes habitual is probably first STAGE, & an habitual action implies want of conscious
293c175 he organization of the eye, passed from simpler STAGE to more perfect. preserving its relations.-- th
361d095 -- one species retained this character in adult STAGE, other alters entirely]CD In common sparrow you
170b001 cess, the new individual passing throug several STAGES (¿typical, <of the> or shortened repetition of
254c048 o» p. 390. All classes of Acrite exhibit lowest STAGES of animal organization, "& are analogous to th
427e108 imitate the sounds surprisingly well-- In early STAGES of transmutations, the relations of animals &
263c078 to what we see in bodily. that <I> it does not STAGGER me.-- What circumstances may have been necessa
289c161 he animals are distinct species If any one is STAGGERED at feathers & scales. passing into each other
292c170 zealous man in the cause if his faith is not STAGGERED <">= I confess. no dissertation against these
344d042 from her looks thougt she was halfbred Beagle STAGHOUND. «++»∴the grandchildren went back to either p
388d172 lumage of cock, & beards growing on old women = STAGS horns & testes curious instances of corelation
388d171 & some very near father.-- now if one of these STAID in the womb, when it came out. it might partake
401e016 ties, & probably would take long before all the STAIN would be got out of it.-- Now this is curiously
427e110 other apples» but probably would more indelibly STAIN offspring-- it would not reach one apple sooner
586n080 ne part. or another-- but (an instinctus means STAINED in?). had better refer to to the heredetary pa
073r161 In R. Brown (Collect: «of F. W.») where the STALACTIFORM masses have layers been accumulated, round
098a047 of Lime, tends to crystallize them as seen in STALACTITE).-- some force crystallizes minerals in laye
100a053 vage be radiation from some grand centre.-- A STALACTITE of Gypsum, is the best case of cleavage.-- P
100a053 od.-- It is the Key to the story.-- consider STALACTITES.-- agate rings, crystallization transverse.-
529m041 . if one were admiring one in India. & a tiger STALKED across the plains, how ones feelings would be
469t135 Watched plants of Fraxinella, with seven flower STALKS for ten minutes. it was visited by 13 Bees-- &
295c183 not [not located] Erasmus says he has seen old STALLION tempted to cover old mare by being shown, you
334d010 tiful, second absolutely sterile.-- My case of STALLION, according to Erasmus preferring young mare t
334d010 x says he knew «a» carter well, who placed his STALLION as second horse between <whe> shaft mare & an
334d011 <whe> shaft mare & another leader mare,-- this STALLION though eager to all other mare had been entir
536m071 s turn up their nostrils when excited by love? STALLION licking udders of mare strictly analogous to
171b002 red perennial. &c &c.-- Yet Eunuchs nor «cut» STALLIONS nor nuns are longer lived Why is life short,
334d010 us preferring young mare to old, explained by STALLIONS, (according to Fox) being guided entirely by
574n041 civious women. are described as biting: so do STALLIONS always..= No doubt man has great tendency. to
466t094 g? which is not differently coloured & to which STAMEN & pistils have no relation. In Monk's Hood, a
467t100 with no division in young flowers. The abortive STAMEN are of useful height.-- In Lupine, Bees «frequ
470t178 r «I saw Bees;»-- on Monk's Hood, brushing over STAMEN «Egg Tree»--I think never on the Galeum saxati
286c156 hese of older standing than constant number of STAMENS.-- in order, or in next family? In considering
314c237 ous plant where a tube consisting of pistils & STAMENS united into long organ, moved on being touched
466t093 with which <bees> «a bee» was dusted over. {P} STAMENS & pistils curve upwards, so that anthers & sti
466t093 per petal, though not differently coloured-- & STAMENS bend up a little In a wild purple Geranium, I
467t103 s perfectly ripe & pollen abundant filaments & STAMENS all protrude «there is a brush at end of stigm
468t112 to plant.»-- to my grt surprise-- I found all, STAMENS straightened pollen profusely shed; lengthened
468t112 fusely shed; lengthened & turned up «more than STAMENS», so that all were brushed by Bees & especiall
468t112 after bee had brushed over the anthers of long STAMENS {P} as stamens grow old «& shed some pollen».
468t112 rushed over the anthers of long stamens {P} as STAMENS grow old «& shed some pollen». they turn upwar

470t177 n» On silene, many plants of wh. have abortive STAMENS= Many Humbles on hedge Linaria= (Plenty of Hum
473s006 azilian plant «in Lin. Transacts. it has three STAMENS» intermediate between Orchis & other plants--
491q01v ranches-- The French Apple tree «with abortive STAMENS» answers first question in negative.-- Questio
492q002 enslow says semi-doubl flowers are those whose STAMENS are monstrous, how then are seeds ever raised?
504q013 in Heartease (4) Does the Thyme bear abortive STAMENS every year & Spring. & within garden //Yes// (
504q013 garden //Yes// (5) Examine the Parnassia whose STAMENS move one after other to flower & Menyanthes wh
505q014 aicum. small white-flowered var. with abortive STAMENS.-- show crossing & ¿heredetary? (15) Abberley
505q014 16 Will plant some of the Thyme with abortive STAMENS by Terrace to see, whether stamens will be pro
505q014 th abortive stamens by Terrace to see, whether STAMENS will be produced in individual plants 17 A dea
455eIBC me as trees.-- Shake some sleeping mimosa-- do STAMINA of C. Speciosus. collapse at night. if so irri
494q005 ee whether association can be given (2) do the STAMINA of C. Speciosissimus collapse during sleep & d
525m025 s.-- hare lip or imperfect roof to the mouth «STAMMERING in my Father family» (as in Lord Berwick's f
039r061 ica in the form of the land decidedly bears the STAMP of recent elevation. which is different from wh
181b044 will yet present some anomaly & <the g> bearing STAMP of <some> great main type, & the gradation will
286c154 Man has expression.-- animals signals. (rabbit STAMPING ground) Man signals.-- animals understand the
557m149 o are people in passion my F. rubbing hands.-- STAMPING. grinding teeth.-- in shame frowning, & angui
558m153 om fear so does man.-- & so dog]:CD Man grins & STAMPS with passion. can expression be used more corr
206b149 count for the odd genera with few species which STAND between great groups, which we are bound to con
292c169 h.-- +++. Even at Falklands some probably would STAND change better than others.-- The more complicat
389d177 is does not happen with hybrids?]CD Plants must STAND much breeding & in in (those which have solitar
449e173 nt constitution from those of Europe-- for they STAND India. better than the latter-- Forrest Voyage
548m115 ne is not conscious of them, else one would not STAND) a crowd of other trains of thought are in prog
620o046 s duty to do so.-- So we say a pointer ought to STAND a <spaniels> «housedog's» duty is to watch the
516q023 Bee visit. Experiments in Garden Sow stones of STANDARD Apricot grafted on what, & see what comes up.
559m154 mire it more. because we can compare it to the STANDARD of our own minds. which ceases to be the case
606o022 resented. But what is beauty?-- it is an ideal STANDARD, by which real objects are judged; & how obta
633j54r moving sand) we <do> lower the creator to the STANDARD of one his weak creations.-- All such facts a
229b234 slow discusses possibility of seeds of Keeling STANDING transport.-- <tr but> Get him to discuss thos
248c030 f my theory, that «two» varieties of many ages STANDING, will not readily breed together: The argumen
286c156 are the characters which unite these of older STANDING than constant number of stamens.-- in order,
435e134 ives in dark caverns of Carniola p. 112. Man. "STANDING alone in the gift of intellect, he resembles,
467t103 Bean, they go to nectar at foot of upper petal STANDING on «I saw Bee go to two species of Lupine,» t
591n097 of emotions-- fear giving goose skin-- & hair STANDING on end.-- July 20th Intelligent Keeper... Zoo
613o036 of man,-- crows fear gun,-- pointers method of STANDING,-- method of attacking peccari-- --retriever-
066r141 deep 24» under 50. Kerguelen Land, = the way it STANDS gales = very strong. Stones as bigger than a m
106a073 ined.-- If {P} inclination be great where arrow STANDS the force immediately deflected from (B) which
023r009 e half so long; The maw was full of jelly which STANK extreamly."--This shark was caught in Shark's B
323c269 skimmed Bartrams Travels in N America May 18th STANLEY familiar History of Birds -- Mackintoshs' Ethi
436e138 es. A familiar History of Birds by the Rev. E. STANLEY Vol 1. p. 72.-- Goldfinches placed near. but n
437e139 tains, it then dropped it & was found alive.-- STANLEYS Familiar History of Birds several cases on re
512q019 Get direction write to-- (2) Does he believe. STANLEY'S fact of Hawks distributing live Mamals (3) Do
641j29v d» secretion in Mollusca. insects «Carabids & STAPHYLINI» & Mammalia. The eye being formed in Mollusc
272c108 l tell this.-- What peculiar conditions the STAPHYLINIDAE on St. Pauls Rocks must be placed under Gou
467t105 n flowers-- swarmed with meligethes & small STAPHYLINIDAE on all their bodies pollen-- on a sulphur B
468t106 - Found Rhubarb blossom swarming with small STAPHYLINIDAE-- Anapsis, Melegethes, Leptuse-- Diptera &
485z017 Vol. II. p. 127. List of submarine insects. STAPHYLINIDAE &c &c. with reference to those of mine from
495q05a id--small orifice (8) Carry Bees, powdered with STARCH & Carmine & experimentise on their returning p
378d147 me converse is not true for he says Hen & cock STARLING alike, yet young ones brown.-- Sexual Selecti
119a104 condensed vapour.-- inequlities are required to START with (& does not Hersche theory imply tendency
206b147 n each lustrum, the number related at the first START must be greater, & this number would vary at ea
531m052 en the actual movement does not take place.-- A START is HABITUAL movement to avoid any danger-- Fear
531m053 nctive because Nancy tells me very young babies START at anything they hear or see. which frightens.
532m053 .-- Fanny Hensleigh doubts whether young babies START.-- ⁋ If children wink. it is instinct Fear must
547m111 g compelled, from some quite imaginary cause to START at once to Shrewsbury., vaguely thought of pack
556m144 mph.--» Sept. 23rd. Horses in Omnibus instantly START when they hear ready, but if they see anything
560mIBC e at home Natural History of Babies-- Do babies START, (ie useless sudden movement of muscle) very ea
521m009 . could converse well on any subject when once STARTED,-- could receive a new train through eyesight,
532m054 man, child or colt has once been frightened & STARTED much more apt, this partly owing to heart? rea
585n076 hey have not known the direction in which they STARTED, they cannot return.-- Hence I conclude. pidge
231b244 mited animal in range-- & hence probability of STARTING from one point.-- In the crag we see the proc
308c218 ture is the object, there is really no natural STARTING place, because there is nothing more elementa
531m053 ne because done by some animals in defence, &c STARTING must be habitual «involuntary» movement from
551m129 ssociate sounds may be seen by omnibuss Horses STARTING, when door shut or cad cries out "right." or
588n091 4. Vol. II» -- gives explanation & instance of STARTING identical with mine,-- Lamarck. Vol II p. 319
532m055 xcited action in the involuntary mind which is STARTLED.-- My Father says he should think that in old
616o039 s said to feel Now this would certainly be a STARTLING expression, & so foreign to the use of ordina
620o045 s fullfilled. pleasure forgotten. [RHC] 4} as STARVATION, or fear of death, one makes allowance & eit
544m103 ure thoughts are broken- Sir J. Franklin when STARVED, all party dreamt of <goo> feasts of good food
038r057 ries of coatings; hence it will be necessary to STATE all arguments for believing that there must be
038r057 think the strongest is the consideration of the STATE at a grand eruption when whole summit of mounta
038r058 be better not to refer to Lyell. but merely to STATE these reasons, & saying that they refer to CENT
038r059 clinal violence crossing lines of crater, <arg> STATE that all the great Volcanoes. have been elevate
043r070 f increased activity.-- such as that of 1835.-- STATE the three «or 4» fields of Earthquakes in Chili
046r078 e to consider that the dikes which so commonly (STATE facts) traverse granites, are granitic material
048r084 of Red River Louisiana. V. Lyell. Vol I. P. 191 STATE at St Helena. pebbles entirely coated with Tosc
049r087 vas be called Volcanic or Plutonic The cellular STATE of all the Porphyry specimens, must be well exa
058r115 smus. whether electricity would affect this. -- STATE the circumstances of appearance at Concepcion [
065r139 y which had long been buried, <see> from rotten STATE of coffin «buried in a mound» long consigned to
074r165 y concretionary action, conjoined with other» «(STATE simplest case. concretions of clay iron stone;
090a024 year too little.-- How comes it none in fossil STATE? suppose «100» <5>0£ x 50,000 x <50 = 250 0000>
098a047 f quartz & feldspar) within other concretion.-- STATE last page thus. point of attempted crystallizat
102a062 next being sucked out. In Cleavage discussion, STATE broadly indication of new law acting in certain
132a138 neath lakes)?» Suppose ocean represents proper <STATE> temperature of earth. at the freezing point.--
176b022 wing to the accident of positions» must in each STATE of existence have shortest life.; Hence shortne
179b033 es in repugnance in crossing of species in wild STATE.-- No doubt «C. D.» wild men do not cross readi
181b040 any large family of 12 brothers & sisters «in a STATE which does not increase» it will be chances aga
191b083 . & recent elevation: Pray ask Dr. Smith.-- «to STATE that most clearly». Fox tells me, that beyond a
213b168 Phillips of numerous genera in fossil & recent STATE, well worth consideration--» Tabulate Mammalia
213b169 rinciple. <Varieties> < «Races» > Man in savage STATE may be called, <species> species. in domesticat
219b194 some cannot be explained more philosophical to STATE we do not know how transported.-- (Glaciers mig
228b231 ompare «chances of embedment in» man in present STATE. with what he is as former species. His arts wo
230b241 ' case of Willow wren) & others varying in wild STATE to show that we do not know what amount of diff
231b241 hers would exjudge it amount of varying in wild STATE.-- When breaking up «the primeval.» continent.
248c030 : The argument must thus be taken, as «in» wild STATE (where instinct not interfered with, or generat
248c030 shown probable,)» varieties may be made in wild STATE, there will be presumption that they would not.
275c120 .-- says my view of varieties is exactly what I STATE.-- & picking varieties. unnatural circumstance
287c158 g the relative importance of the organs in same STATE. in different animals, & the value of those org
294c176 t upon analogy & absence of varieties in a wild STATE-- it may be said argument will explain very clo
294c177 ;-- Horse-- &c <Lonsdale says. that first shee> STATE broadly scarcely any novelty in my theory, only
299c194 ers another case of permanent varieties in wild STATE-- The two. former produced by difference of <lo
308c219 at, as she has no original ideas., it will show STATE of knowledge «Negroes existed since time of ear
316c244 Deity to chose, when perfect enough for future STATE, that when good enough for Heaven or bad enough
344d040 insects-- (¿glowworms) breeding-- <beet> imago STATE-- & the English & Some African dove.-- The exti
356d072 f Black game & Ptarmigan having crossed in wild STATE-- for seeds being floated about.-- I must
357d074 it not for seeds being floated about.-- I must STATE that. the <p> mechanism by which seeds are adap
358d076 es in character to the larva, or less developed STATE.-- the female & young of all birds resemble eac
360d091 in confinement-- is this effect of climate, or STATE in which they are kept?-- Is there any mistake
364d105 s crossed with the Ptarmigan. subalpina in wild STATE.-- Neilson has given figure of it.-- In England

Page ***(Key Word)***
365d105 nt, probably, alone would species cross in wild STATE.-- Is English red Grouse. a cross between Black
365d107 , (& no monstrosity, or adaptation to unhealthy STATE of womb).-- One can perceive that Natural varie
372d131 from true generation.-- there is no caterpillar STATE; the vast difference of two kinds of generation
387d170 men.-- each man passess through its caterpillar STATE. the monkey represents this state.-- When it is
387d170 s caterpillar state. the monkey represents this STATE.-- When it is said. that difference between bud
405e031 y in same Manner as they would vary, if in wild STATE; thus mark on ear of cats, colour can be browni
407e039 y» EQUABLE. or so tropical, & therefore present STATE of world is not so different, with <d> regard t
407e040 entata at that period) Analyse this,-- consider STATE of vegetation, & conchology,-- shells of Africa
409e049 t, for which the world was brought into present STATE.-- «whether he was or not. He is present a soci
423e097 ion of the different beings, (all fishes to the STATE of the Ammocoetus) Crustacea to--- ? &c) withou
450e176 ey will propagate get dimensions-- do App. p 73 STATE of Muar in Malacca.-- speaks of Rhinoceros as w
461tf03 ckbird have been known in <their> «its» natural STATE to mate with a thrush"-- Athenaeum 1839. p. 708
499q009 e whether they are dusted with pollen-- in what STATE (whole or broken) is ball of pollen on Bees thi
506q15v who does-- What is Ruppia Bennett says in same STATE. of flower 45. Charlsworth. vol II. p. 670-- oa
509q017 --? Combs in combless Poultry-- Teeth in foetal STATE: Mr. Horner. On Mr Tremenheres Scottish Collier
513q020 s.= Young Chinese or Penguin Duck in very young STATE for skeleton== Does the tumbling of pigeons var
515q021 is absent by abortion, but appears in abortive STATE either in the species, or in the individual by
524m022 though so confused in others.-- Mrs P. when in STATE as above described, (forgetting that her husban
527m032 es» when in confinement, so may they learn in a STATE of nature.-- Singing of birds, not being instin
532m057 ng away» have been usual effects of fear.-- the STATE of collapse may be imitation of death, which ma
532m057 h which accompanies passion & not sweat is the <STATE> effect of short -- but violent action.-- To av
533m057 hild resemble, parent stock.-- (& phrenologists STATE that brain alters) It is known that birds learn
535m069 htening to Gods anger.-- (∴ more poetry in that STATE of mind: the Chileno says the mountains are as
538m078 nkedness, show what trains of thought depend on STATE of turn In drunkedness same disposition recurs,
538m080 eeling & perception separate, from the ordinary STATE of mind, is probably analogous to the double in
539m083 ual remembers things done in the other habitual STATE because it will (without direct consciousness?)
546m110 ung lady almost equally in her senses in either STATE.-- does this throw light on instinct, showing w
546m110 ay be done unconsciously as far as the ordinary STATE is concerned.?-- Mr. Mayo told me the case of a
548m116 if he had been unhappy)-- it is because in this STATE, the consciousness does not go back to former p
548m116 consciousness, one would pity suffering in one STATE almost as much as in the other,-- though she wh
560m156 itrous oxide, forget what they did when in this STATE, or remember what they did in former one. about
564n005 t faculty, whether for position of axe of eyes, STATE of surface, or other means by which eyes, aided
566n012 an «a» Savage M. Le Comte's idea of theological STATE of science, grand idea: as before having analog
582n066 ing themselves; & heredetary remains of savages STATE.-- N B. According to my view marrying late, wil
593n107 some <instinct> «insects» which become in imago STATE social) by smell or looks. but it does not know
596nIBC geon carried & turned round & round in fainting STATE would it then know its direction.-- In slight c
603o013 much of EVERY language shows traces of anterior STATE?? Edinburgh Review Vol 18. (1st Article) on Tas
608o026 does not.-- because motive power not in proper STATE.-- When the admonition succeeds who does not re
609o29v efers (I believe) to present day & not to ruder STATE of Society.-- Civilization is now altering thes
615o038 o particular instincts of animals. even in wild STATE; certainly to the domesticated.-- [LHC] NB. Two
625o50v nstinctive feelings & our short lived Passions' STATE broadly in child or animal it is equally proper
049r089 layers: Mem: Phillips Mineralogy some such fact STATED to exist in Peru. -- Ascension At Ischia there
056r110 rocks between that formation and the secondary (STATED in Playfair to be the case p. 51). presupposes
165g125 lambs because most like Mother in face-- asked STATED this generally the case Wednesday 12/ & 3/ Why
179b033 plants together would establish law. ≠as above STATED:-- no one can doubt that lesser trifling diffe
219b193 er what kinds of seeds. these plants) [Mem Fact STATED by Mr Don in island, Teneriffe, St. Helena. J.
247c028 n» Java. & very common on Otaheite-- according «STATED in note to p 21» to Quoy & Gaimard in Sandwich
251c041 about distribution of Cats Vol III. <p.> p 233, STATED that the "Asseel Gazal. (Bos Gazoeus) does not
279c133 ing varieties may be inferred from <this> facts STATED.-- + +. Fully supported by Mr Wilkinson. = Mil
302c202 Macleay. wonderfully accordant. with fact there STATED, only in most discordant groups. The formation
335d014 ve» twins.-- ‖ The fertility of first cross, as STATED by Fox, is very important, as showing above fa
351d056 y has eye-point, but Broderip added it has been STATED that stationary Spondylus has eye-points-- Mac
361d095 last. Sept. 11. N Mr. Blyth, at Zoolog. Meeting STATED, that Green-finch, all linnets red-pole, goldf
361d096 «My case of Caracara. N. Zelandiae.--» Mr Blyth STATED «that there are» two ducks, which have pretty
382d158 ale animals give suck)--But this not distinctly STATED by Hunter.-- Do testes, & ovaria when they fir
389d177 ted with this same case (& not merely as I have STATED it) it is certainly very remarkable that too m
406e035 ad any offspring before.-- -- now this is never STATED.-- Regarding the similarity of offspring to Pa
409e049 es, there would not be social animals. «this is STATED too strongly. for there would be innumerable s
427e109 nt of snow line there «& there & there only: as STATED by Capt. Graah») & break up. N. American Conch
446e161 ench» expedition,-- rather the reverse of facts STATED by Smith of Jordan Hill.-- May 27th.-- Henslow
452e182 um & Dholbum.-- Vol do. p. 634, alludes to fact STATED by M. Tournal that skulls found near Vienna ap
527m032 wledge like that of man, & this agrees with the STATED fact, that «birds from» certain districts have
616o039 elves the converse of the <expr> question above STATED, because there are living bodies without these
225b217 ltered during passage of birds (where is this STATEMENT I remember. L. Jenyns. talking of it) or how
249c036 degree.--, may be connected with, Mr Blyth's STATEMENT of birds of Europe & America, which are of di
256c053 ious.-- There must be some sophism. in Lyells STATEMENT that some species vary more,, than what makes
271c106 al contingencies & long bred in-- Mem, <an> a STATEMENT in Mr Wynne's book, about not altering breed
283c144 ofound & such it appears.-- Is there not some STATEMENT about diversity of forms in aberrant circles.
305c209 ncidence if distinct creation.-- ie.-- a mere STATEMENT nothing is explained.-- this is fact analogou
421e091 illingham Cattle, with reference to Mr Bell's STATEMENT of the tame ones.-- ‖an instance of a triflin
446e161 the male.-- Athenaeum May 18. 1839. p. 377.-- STATEMENT that the climate is on the decline, as far as
506q015 on.-- (39) What does he think of Dr. Flemings STATEMENT of Sweet Williams & Stocks, being propagated
515q022 t?--. Important.-- Oct. 44 Tell J. Anderson's STATEMENT of English Horses having fewer vertebrae in t
528m035 - is closely analogous to my Fathers positive STATEMENT that insanity is only cured by forgetfulness.
617o40v phorical mode of stating the fact to the mere STATEMENT of the <force exhibited in every> phenomena a
637j57v All Bridgewater Treatises. are reduced simply STATEMENT of productiveness, & laws of adaptation‖ p. 2
079r177 's Voyage, Shell fish purple die, marevellous STATEMENTS on, Vol I, P. 168. on coast of Guayaquil, sa
220b199 published it will be curious.-- Some general STATEMENTS about mundine & confined genera.-- Lyell has
240c013 ies are not found in the Archipelago-- Former STATEMENTS to such effects false In New Guinea. a Kanga
442e153 aracterized study Von Buch.]CD Now Mr Knights STATEMENTS about fruit trees. grafted. altering is host
526m027 nes our throwing it up.-- equall true the two STATEMENTS.-- Catherine remarks that pleasure received
117a102 e 1838) that 70. F were obtained 100 miles E of STATEN land. bringing up pebbles 2 inches long?-- L'I
023r013 y Calcareous rock; following Curvature of hill; STATES could discover no shells: nothing said about K
045r075 arthquake things pitched off the ground. «Ulloa STATES that Volcanos!! were in eruption at time of gr
050r091 mean in contradistinction to sand?? B. Roussin STATES that generally in North part of Brazil. <grave
053r100 co? Quote Miers about shells at Quillota Lyell, STATES that contact of Granite & sedimentary rocks, i
058r116 des Indes Acosta. p. 125. of French «?» Edition STATES that the same earthquake has run from Chili to
064r137 a itself now remaining-- Lyell « <p 419> p 428» STATES that Von Buch has urged that Java volcanos dif
070r155 & at Chuquisaca. half across the continent.--He STATES plains of Mendoza smooth. Sir W. P. states tha
070r155 --He states plains of Mendoza smooth. Sir W. P. STATES that in Helm's travels accounts of travelled b
079r176 r in the highest & gold in the lowest. Humboldt STATES that some of the richest gold mines on ridge o
123a117 raigs Letter from M Angelis. B. Ayres. 3d. May. STATES remains found in many part.-- great Dasypus ne
183b051 Orpheus» was introduced into Chili. in present STATES. <they> it might continue & thus two species b
197b111 greatest argument against theory of analogies. STATES there is but one animal: one set of organ:-- t
198b113 e Insects & Molluscs allowed to be wide hiatus: STATES in one the sanguineous system, in other nervou
198b113 stem, in other nervous developed. (Owen's idea) STATES these class approach on the confines? Balanida
206b150 & few Monocot in Coal formation? p. 320 <Think> STATES Cryptogam. Flora formerly common to New Hollan
210b158 what takes place from time? Von Buch distinctly STATES that permanent varieties. become species. p.14
215b178 oping his ideas on passage of forms.-- Deshayes STATES Lamarck priority refers to introduction to Ani
220b199 Athenaeum. No. 537. Feb. 1838. p. 107. Mr Blyth STATES that all «genera of» birds in «N.» America & E
239c001 t which have long been known in any country; he STATES that Esquimaux dog when crossed with pointer p
239c002 ither parent is lucidly explained.-- Mr Yarrell STATES that if any odd pidgeon crossed with common pi
404e025 hus have arisen, & M. Edwards p. 330 distinctly STATES that the flipper is a mere simple modification
480z007 & Botany. Vol I p. 358. D'.Orbigny <considers> STATES that young birds of prey have longer tails tha
548m117 Ashe's case, one here was conscious of the two STATES.-- August 30th.-- It is singular when looking
113a092 ssibility of metamorphic theory On the idea of STATICAL equilibrium, the height of lava (habitually)
122a114 ast? How can Herschel consider figure of earth STATICAL.-- if platform of mexico owes its elevation t
128a129 nce of water on other. from tendency to regain STATICAL equilibrium This will be only a modifying cau

Page **(Key Word)**
112a090 graph Soc, April 9 1838. Letter from M. Erhman STATING that the mean temp at Yakous in Siberia being
179b035 ted with gaps in the series of connection? «if STATING from same epoch certainly» The absolute end of
532m057 ect of short -- but violent action.-- To avoid STATING how far, I believe, in Materialism, say only t
617o40v unted for, we prefer this metaphorical mode of STATING the fact to the mere statement of the <force e
157g078 ivortium aquarum «about 12 ft higher than last STATION» 29.316 true terrace «2d» near divortium aquar
159g086 Barom 29.008 A. 75 degree Air 70 degree? This STATION a little way down slope of obscure terraces (&
160g088 me think it submarine, 400 or more feet above STATION! There is long straight isthmus connecting E &
161g094 oy-- appears to lip with moss On this terrace «STATION perhaps 6 ft too low» (to test last on Peat-Ma
161g094 0 degree? Barom. 066 lower than last. but A 77 STATION was <a few> «about 3» feet <lower> too high ab
226b221 rvive or chance having transported them to new STATION.-- When the new island splits & grows larger s
264c082 ry much (more than shells) owing to variety of STATION inhabited by them-- Timor. Australian forms am
299c194 other dry; yet if T. palustris be sown in dry STATION it will for some generations come up so.-- the
299c195 . former produced by difference of <locality>. STATION & varieties chiefly produced by cultivating pa
271c106 ts?-- What is this halo.-- Continents are not STATIONARY «unerring proofs not always continents».-- i
351d056 t, but Broderip added it has been stated that STATIONARY Spondylus has eye-points-- Macleay then answ
423e097 m, hence the degree of developement is either STATIONARY or more probably increases.-- Jan 29th. Uncl
427e108 ce number of forms. once formed. would remain STATIONARY, hence all present types are ancient. Accord
437e138 101-- Kingfisher in northern part of England STATIONARY, in southern stays only winter.-- Jays & cha
234b261 ntry do not species generally affect different STATIONS;-- this would be strong argument for propagat
284c147 lations (a fixed quantity) & on subdivision of STATIONS & diversity--.-- this perhaps on long average
508q016 than Caffres.= 13 Where are there any medical STATISICS, proportion of diseases (heredetary?) in diff
324c268 ee if law's cannot be made out Find out from STATISTICAL Society-- where M. Quetelet has published hi
494q004 Bantam.-- Cross common Fowl with Dorking (10) STATISTICS of breeding in Zoolog. Gardens-- with respec
606o024 herefore the embodying of a floating idea.-- as STATUE of beauty, is of the "beau ideal", my instinct
602o010 of part of the highest enjoyment in mutilated STATUES In Elliotson's Physiology much about sleep-- N
603o014 Y healthy» young woman, with good expression-- STATUES not painted-- <music> very good article-- why
336d018 enances, organic diseases, mental disposition, STATURE, are slowly obtained & hereditary; <but if> if
506q015 ous plants any secondary, sexual characters.-- STATURE, position of flowers-- Their smell-- form of f
198b113 ls, reptiles fish-- Conditions will not explain STATUS (Perhaps consideration of range of capabilitie
575n044 cross of bull dogs-- increase the courage & STAUNCHNESS of greyhounds.-- bull-dogs being preferred f
603o012 ong -- two primary sources, sight, & hearing-- STAUNTON Embassy Vol II p. 405.-- Speculates on origin
322c269 W. Earls'. Eastern Seas. .Octob12th.-- Sir G. STAUNTON'S Embassy to China. Oct. 12t Kotzebue's two vo
318c254 species of same genus.» The Muscicapa solitaria STAY about a fortnight in one particular part of cou
437e138 orthern part of England stationary, in southern STAYS only winter.-- Jays & chaffinches sometimes mig
062r128 ar Volcanoes. lakes of brine all inhabited: Go STEADILY through all the limits of birds & animals in
541m091 consists in keeping one idea before your mind STEADILY., & not merely thinking intently; for that on
625o052 our nature, <not> which regulates our feelings STEADILY & not like our appetites & passion, which rec
541m091 one endeavur to keep any simple idea as scarlet STEADY before mind for period, «if the scarlet was be
572n035 of expression lawless, whilst they are the only STEADY & universal means. recognized-- no one can say
044r074 ct of bombs in tufa there is proof of such gaz) STEAM condensed.--Perhaps these mighty changes might
074r165 eral secondary names «Study Hoffmans account of STEAM acting on trachytes. also Azores. We here have
074r165 ay be transported by complicated chemical law & STEAM of salts, quite curious case of oxided Iron by
100a055 sk Capt. Beaufort, whether, water flashing into STEAM, would Babbage.-- Webster Phillips insists of a
573n036 s us satellites &c &c «The Savage admires not a STEAM engine, but a piece» of coloured glass <&admire
072r160 a few fragments can be separated from them with STEEL instruments." In R. Brown (Collect: «of F. W.»)
096a041 able to Auvergne??? The fact of Galapagos Isld. STEEP side to windward in allusion to St. Helena disc
151g039 s die away where slope less., best developed on STEEP earthy slope, two circumstances rarely united.-
162g100 el to Shelf on opposite side & dies away on the STEEP & rocky gully of last stream Friday Loch Lochy
481z010 Meyen Zoology on animal of Campanularia Alcedo STELLATA. Meyen p. 92.-- great Kingfisher of Tierra de
419e087 he work of a few years as with the Lamantin of STELLER tells much less, (though <the> it also the eff
176b024 > three more, double arrangement.-- if each Main STEM of the tree is adapted for these three elements
371d128 bud.-- this must be owing to their unity in one STEM.-- a bud may be transplanted & carry all these
384d162 e plants which have male & female flower on same STEM.--» so that Molluscous hermaphroditism takes pl
534m063 > Apion radiolum undergoes transformation in the STEM of Hollyhock, although ordinary Habitat is Malv
052r099 of Chili may be with much truth compared to the STEP = formed streams of lava at St Jago. C. de Verd
186b059 hat French call (atavism) Probably this is first STEP in dislike to union, offspring not well interme
190b078 iven by true generation, throughe means of every STEP of progressive increase of organization being i
223b209 ss (perhaps more fertility & so make that sudden STEP. species or not. A plant submits to more indivi
225b217 me, I will answer yes, if you will show me every STEP between bull Dog & Greyhound). I should say the
275c119 eograph distribution of animals, <as> I use (new STEP in induction) as keystone of ancient geography
291c167 hrough all Nature.-- Makes hermaphroditisms. one STEP in <scale>. Series-- in plants we have a step b
291c167 ne step in <scale>. Series-- in plants we have a STEP between <mon> monoeecious & dioecious plants in
473s03v How strange no plants in our Devonian-- Fish one STEP lower in America-- How curious all negative law
535m069 says the mountains are as God made them,-- next STEP plastic <virtue> natures. accounting for fossil
542m093 d with himself, & though dreading to say so, his STEP will grow erect & stiff like that of turkey.--
548m114 d as in imagination, & in rigidly comparing each STEP as in reasoning-- hence delirium & sleep mental
548m115 ing things somewhat probable. in comparing every STEP, & inventing new means,-- therefore works of im
554m140 y using, foreign bodies, for end. most important STEP in progression.-- The male Black Swan is very f
581n063 eredetariness in habits. be considered. as grand STEP if it can be generalize.-- The tastes of man, s
593n111 ls.-- [blank] Double consciousness. only extreme STEP of an ideal argument held in one's own mind, &
614o037 er end just same argument. without indeed we are STEP towards some final end.-- production of higher
638j58v he surviving one of ten, thousand trials.-- each STEP being perfect «or nearly so (except no in isd)
638j059 red schemes of structure, in the course of ages «STEP by step».-- in Man, the nervous system, gains t
638j059 mes of structure, in the course of ages «step by STEP».-- in Man, the nervous system, gains that know
314c238 effected.-- Prodromus Florae Norfolkicae. 1833 STEPH. Endlicker (He will give sketch of botany of is
234b256 & size, but not for[m]» of animals.-- He says STEPHENS say he can at once tell by general colouring
448e166 3-- very curious table of all the castes from STEPHENSON at Lima The same numerical relation (both in
252c042 venatica» <of> Cape found in Desert of Korto & STEPPES of Kordofan p. 401. Admirable letter from Macl
027r023 e formations are now seen in regular descending STEPS Mem.; rapidity of germination in young corals.-
032r041 on. This case differs. I think. from Patagonian STEPS, because the deposition & accumulation is broug
098a044 ns salt see Geolog. proceedings Lake let out by STEPS in Central France not very conclusive proofs, b
105a071 alluvium would form a succession of flights of STEPS; if one lake then we must suppose barrier in th
182b047 imits of form. (& where possible the number «of STEPS» known). <for instance among the Carabidae.-- i
192b085 an be found from forms marking good genera-- by STEPS so insensible, that each is not more change tha
192b085 Therefore all genera MAY have had intermediate STEPS.-- Quote in detail some good instances. But it
192b086 ether there have existed all those intermediate STEPS especially in those classes where species not n
192b088 - In mere eocine rocks. we can only expect some STEPS.-- I may ask whether the series is not more per
227b224 ies «of one genus» can pass into each other «by STEPS we see»: but this cannot be predicated of <genu
227b224 res in two genera.-- <we then cease to know the STEPS.> although D E F. follow close to A. B. C. we c
230b236 except perhaps in some plants & then a chain of STEPS is found in same mountain).-- How is this expla
287c157 she has also placed a series of animals on the STEPS that lead up to it" p. 20 +++hiatus & saltus
293c175 to Musquitoes We never may be able to trace the STEPS by which the organization of the eye, passed fr
349d052 acleay <met> uses term genus when it is so many STEPS from a head, as subkingdom.-- -- evidently arti
411e054 ether the individuals «forms» are permanet, all STEPS in the series, their relation to the external w
411e055 f the special cases, under which the individual STEPS in the series have been fixed, to study the phy
555m141 r "reasoning" or "reasoning"-- only rather more STEPS.-- dispute about words.-- Miss Martineau (How t
594n112 uld appear that an instinct long remains, if no STEPS are taken to eradicate it.--» «Emma says, «her»
614o037 st complicated instinct. might be analysed into STEPS, as species change.-- Must be so if Lamarck's t
629o055 at first <rationally> by feeling-- reasoned on, STEPS forgotten, habit formed,-- & such habits carrie
639j28v nsequence they become long. not at once, but by STEPS. of which we have manifold traces in the severa
266c088 beetles & birds & <f> become dull coloured in STERILE countries.-- Gould insist much upon knowing to
334d010 st cross <not se> plentiful, second absolutely STERILE.-- My case of Stallion, according to Erasmus p
514q021 that only one pineaple Horticulturists (1) Are STERILE hybrids healthy: number of generations: about
515q022 for several generations under net & see if get STERILE-- Cover that little Ervum in Sand-walk, on whi
502q11v s Flowers in short turf. for abortion. or for STERILITY Land Birds Madeira Migratory-- ask Gould abou
325c267 Humes Essay on H. Understanding (sometime) Du STEWART works. & lives of Reid, Smith & giving abstrac
546m108 low faculty of understanding. Adam Smith (.D. STEWART life of. p. 27), says <sympathy> we can only k
559m155 standing well worth reading Copied <Smith> «D. STEWART» lives of Adam Smith Reid, &c worth reading. a

604o018 learest analogy between dreams & insanity.» D. STEWART on the Sublime The literal meaning of Sublimit
605o019 rical term sublime 7 So that in this Essay. D. STEWART does not attempt «by one common principle» to
605o020 same manner as we are acted on by sympathy. D. STEWART on taste The object of this essay is to show h
604o017 & if consciousness, individuality.-- Quotes D. STEWARTS System of Emotions.-- T. Mayo-- Pathology of
554m141 s sitting the Keeper is obliged to go in with a STICK, if he drops it, the bird will fly at him-- Kno
360d091 insects somewhat like «between» lice & fleas. STICKING on them, but never in an animal, that had Lon
347d050 leay in A. Smith's Zoolog.-- of Africa -- p. 4. STICKS to genus or group of any kind not being perfec
500q10a ere were planted «last year» pell mell, without STICKS & seeds gathered & these are now to be planted
501q011 Peas, Kidney Bean & Bean, intertwined, «without STICKS»-- in reference to what Mr. Herbert observe on
495q05a rass= Spread sheets of Paper. covered with some STICKY stuff in flat places & see whether wind, on «d
121a111 all Vol VI. p 173. (Ed. Transact) has seen clay STIFF enough <to form> for potters to use. in which g
241c014 say whether wild or not p. 156.-- Parroket with STIFF tail like woodpecker.-- Birds of Australia. Man
244c021 Falklands, generally "red of brick" hair, very STIFF, p. 120-- Coati roux common. near Concepcion. s
273c112 ing woodpeckers., with powerful wings, but tail STIFF.-- Swallow & goatsuckers likewise exaggerated.-
536m070 o.-- When a man is in a passion he puts himself STIFF, & walks hard.-- «He cannot avoid sending will
542m093 dreading to say so, his step will grow erect & STIFF like that of turkey.-- he may be amused, he nee
557m147 a passion.-- dog tail curled when angry & very STIFF. back arched. just contrary. when pleased tail
557m147 Cat when pleased, erect its tail & make it very STIFF «& back» when savage «no» & ready to dash at pr
557m147 ious «with fright» back absurdly arched. & tail STIFF.-- is shame, jealousy, envy all primitive feeli
565n007 gs show triumph (& pride) same way walk erect & STIFF, with head up.-- Why does suspicion look obliqu
636j57r show such losses.-- Consider ground Woodpecker STIFF tailed cormorant: pain & disease in world & yet
542m093 ll, he will find it hard to keep his lip from STIFFENING over his canine teeth.-- He may feel satisfi
435e133 fect from age, what Mr Knight [not located] the STIGMA retains its power.-- R. Brown found the <poll>
435e133 rted their tubes: now Mr Herbert has shown that STIGMA swells, when pollen even most remote is put to
466t093 mens & pistils curve upwards, so that anthers & STIGMA lie in fairway to nectary.-- Is not this so in
467t099 at although anther «nor filaments» shrivel, yet STIGMA does not, so we may feel somewhat «but little»
467t103 tamens all protrude «there is a brush at end of STIGMA, which forces out from extremity pollen, or po
467t103 mity pollen, or pollen comes out with anthers & STIGMA in slit» -- As I think they do in Broom & cert
467t104 heath.-- «Also in Lathyrus pratensis yellow saw STIGMA project» In common Pea saw Humble so press dow
467t104 ommon Pea saw Humble so press down sheath, that STIGMA covered with pollen was pressed & rubbed along
467t104 {b} pressing either one or both of Pea's wings, STIGMA & mass of yellow pollen protrudes at sheath.--
468t105 ly abundant flowers not attractive, very small-STIGMA rather large & rough-- flowers common-- many w
468t111 nectar =again= between them, hence quite below STIGMA. & so avoided it.» On certain days Humble seem
468t112 , so that all were brushed by Bees & especially STIGMA after bee had brushed over the anthers of long
468t112 hed some pollen». they turn upwards & bend over STIGMA:-- but stigma «is» almost roofed by united fil
468t112 n». they turn upwards & bend over stigma:-- but STIGMA «is» almost roofed by united filaments.-- This
472s02v dle becomes covered) so whole sides of flower & STIGMA dusted.-- <I think> When It first alights, it
497q007 .-- (3) How is Iris impregnated.; which part of STIGMA?-- (4) As Papil. flowers appear difficult to c
498q008 grain be whole, to impregnate?-- I presume only STIGMA impregnable.-- (12) At Maer Cowcumbers in fram
501q011 ower. do Henslow Speaking of Thyme doubts about STIGMA in similar manner ever failing.-- answered by
513q21. ichogamous Zostera-- Knights notion of pollen & STIGMA generally not being mature at same time on sam
514q21. modes (one of the Catasetums) really always hit STIGMA by projecting pollen-masses?-- = answered = Ha
467t100 abortive. Examined in microscope--some of the STIGMAS of {P} shape of ordinary Labiatae --the chief
513q21. . BROWN-- will pollen act on any flower before STIGMAS expanded-- in reference to Lobelia & Clarkia-
272c108 generl appearance of blattae other Hemiptera STIKINGLY resemble Coleoptera.-- Donacia.-- some orthop
023r164 to Hungary. = Veins of Zimapan offer zeolite. STILBITE. grammalite. pyenite. native sulphur.. fluor
022r008 s of a Hippotomus; the hairy lips of which were STILL sound and not petrified, and the jaw was also f
023r011 ave been said to be merely accidental apertures STILL open.--The fault like appearance «arising from
025r016 little WNW of C. Rock. [5 degree 29 minute S.] STILL shoaler, coast composed of sand dunes. 15--15 D
063r132 at seeing Zoophite producing distinct animals. STILL partly united. & eggs which become quite separa
113a092 in solid rock.-- 4 & 5 fathoms deep. perfectly STILL water. Major Mitchell inferred subsidence; Mem
124a119 were mixed with 90 percent of lead. it would be STILL more curious to know whether it would be absorb
125a121 re metal then thinner if better conductor, then STILL thinner → The problem is, you have temperature
159g087 her-- There are however fringes of alluvium (?) STILL higher Slope of valley much more gentle than in
175b020 , armadillos & sloths as all offsprings of some STILL older type some of the branches dying out.-- wi
181b042 re would be great gap between birds & mammalia, STILL greater between Vertebrate and Articulata. stil
181b043 till greater between Vertebrate and Articulata. STILL greater between animals & Plants But yet beside
192b088 ther of all Mammalia in ages long gone past.. & STILL more so «known» with fishes & reptiles.-- In me
273c112 hether in most aerial of swallows.» Milvulus, & STILL more wonderfully to the Humming bird, which is
300c198 t as man has heredetary tendencies. his mind is STILL only a divided body. {p 3. language seems to su
317c249 259 120 ft in length, some branches of Justicia STILL growing,) passed us. do. p. 243 (, Professor Sm
341d029 osest.-- Ostrich & Rhea closest.-- (& two Rheas STILL closer).-- Mr Blyth asked whether structure of
378d148 f tame, they came to the windows to be fed, but STILL they have a wariness about them quite remarkabl
379d151 arge Cetacea, as the past for other Mammalia. & STILL further back reptiles & Cephalopoda: Old Jones
381d157 some of the lowest tribes, leads one to suppose STILL more that they must in effect be so in all.-- 2
459t013 same half breed resembled the plumage of drake STILL more.-- So Penguin impresses its form both on v
522m010 Father explained who he wa & all about him, but STILL maintained he had never heard of him).-- My F.
545m106 s sent so fast to limbs that they cannot remain STILL.-- I do not doubt this Baboon. knew women.-- An
547m112 not vica versa. anyhow I might have been quite STILL, & not attending to bodily sensation & yet the
564m003 m. Bee how different instinct a solitary animal STILL different.-- Different nations having differe
618o41v the weight, anymore that weight, the blueness STILL less between ²action» «things» so different as
632j53r ese are, I believe, only direct consequences of STILL higher Laws.-- I do not «then» believe the papp
542m094 CD[like sighing to relieve circulation after STILLNESS.-- Now I conceive if organization were change
641j29v crabs Crabs & Mollusca we have analogues The STILLNESS p. 276) of flight of Owl remarkable, [gained
612o035 ch movements & choice, & obedience to certain STIMULANTS without conscience in the lower animals, as
615o038 retion of thought, (or under contingencies of STIMULANTS of certain kinds such secretion) or an assoc
286c154 [s]hare of sickness,-- death, unequal life,-- STIMULATED by same passions-- brought into the world sa
574m041 reat tendency. to exert all senses, when thus STIMULATED, smell, as Sir. Ch. Bell says, & hearing mus
530m045 analogous to sleep--; some doctors care it, by STIMULUS & afterwards patient sinks.-- When a muscle i
577n050 & closing of the leaves, comes on from want of STIMULUS, after certain other actions, & hence becomes
641j29v analogy between fang of snake, (jaw of spider?) STING of bee, sting of nettle.-- Are there any other
641j29v n fang of snake, (jaw of spider?) sting of bee, STING of nettle.-- Are there any other analogies-- pr
478z003 lls (patella) sea weed & many pebbles Mentions STINGING Millepora. Quoy. Freycinets Voyage Vol p. 597
485z018 y Horae Entomolog. insects swarm in Lapland & STIZBERGEN wherever there is extreme heat, the tropical
437e139 iar History of Birds several cases on record of STOATS being carried (p. 121) & dropped having wounde
181b043 he «infinite» variations, & all coming from one STOCK & obeying one law, they may approach,-- some bi
216b179 ffspring of Negro & white will return to native STOCK (the cross often whiter<)> than white parent) i
219b196 so to say that all Mammalia, were born from one STOCK, & since distributed by such means as we can re
222b203 nt characters is attempt at returning to parent STOCK.-- I think we may look at it so--?? It holds go
222b204 at we call improvement, --All Mammalia from one STOCK, & now that one stock cannot be supposed to be
222b204 , --All Mammalia from one stock, & now that one STOCK cannot be supposed to be «most» perfect (accord
248c033 t forms, & greater fertility of hybrid & parent STOCK, than between two hybrids.-- As we see external
335d015 pecies (that is slight alterations of primitive STOCK) are heredetary: «Hybrids of» Varieties is diff
336d017 om the form which it inherits from its parents «STOCK» without it be small & slowly obtained NB. The
336d017 that animal would endeavour to return to parent STOCK. but if both parents are alike, offspring. must
337d021 - In my speculations. Must not go back to first STOCK of all animals, but merely to classes where typ
337d022 squirrel» seal & mouse. elephant, come from one STOCK.-- Theory of Geograph. Distrib: of <ani> organi
349d052 es in any sub-genus will be. descended from one STOCK, & that stock with other subgenera will come fr
349d052 genus will be. descended from one stock, & that STOCK with other subgenera will come from. common sto
349d053 ock with other subgenera will come from. common STOCK.-- all genera, common stock.-- so that value ca
349d053 I come from. common stock.-- all genera, common STOCK, or not. Now wings for flight-- therefore ostri
353d060 ny others somewhat allied whether «like» parent STOCK, or not. Now wings for flight-- therefore ostri
355d066 said domestic fowls are descended from several STOCK, then species are fertile²; as long as opponent
371d128 uccessive roses & bud are produced, like parent STOCK, or if different dieteriorating very slowly.--
387d168 very singular the same difference from parental STOCK having been repeated several times, that it bec
388d170 ce between bud & seed, that latter carries with STOCK of food.-- the generalization begins low.-- tha
390d179 ing animals that have neither varied from their STOCK, for to breed (as Sir J. Sebright urges?) one w
390d179 to breed two which have each varied from parent STOCK.-- The very theory of generation being the pass

Page
(Key Word)
```
401e016 fancy  cowslip producing primrose return to old STOCK, but not primrose producing cowslip Uncle J. sa
418e083 very animals are descended from some one single STOCK,--one is led to suspect that the birth of the s
443e153 s of other varieties & we know that the kind of STOCK greatly affects the Graft.-- ‖Plants circumstan
460t019 » modified <& Many Forms lost; if> «of this old STOCK (which from action & reaction grew more complex
533m057 are  so because brain of child resemble, parent STOCK.-- (& phrenologists state that brain alters) It
121a113 ell on Sweden. p. 12. proofs of small rise at STOCKHOLM.-- analogous to my Valparaiso case.  Consider
335d015 ngers» (that is slight alterations of primitive STOCKS «relative to changes which every species under
401e014 to  prevent fertilization from turnips or other STOCKS. Says if any variety of apple be sown, all sor
430e117 he theory that all animals have sprung from few STOCKS. does not bear, the least on ancient generic f
441e148 ioned by Mr K may be caused by the diversity of STOCKS, on which they bear & are grafted.-- No than by growt
442e153 y means, which wild plants never are, namely on STOCKS of other varieties & we know that the kind  of
446e162 n grafted trees «:so is not effect of different STOCKS in this case».-- & strong case showing analogy
501q011 eming says yes. (29) Are there RACES of Lupine, STOCKS Clover, to experimentize on by sowing near eac
506q015 k of Dr. Flemings statement of Sweet Williams & STOCKS, being propagated many years, by cuttings.-- (
570n026 wn by similarity of the earliest arts.-- Mem.-- STOKES-- arrow heads &c &c October 27th Consult the V
570n025 ving eaten meat on table, & criminal.-- who has STOLEN. neither, or both may be said to have fear, bo
218b193 liensis?) (Fulica Chloropus)--» might bring in STOMACH-- &c &c. (Mem discover what kinds of seeds. th
224b213 & one with other, might change organization of STOMACH & hence remain distinct. Where country changes
231b242 told me some story of crane from Holland!!! in STOMACH --or in feathers--seeds.-- Two inhabitants of
267c095 ursor of magazine??? p. 75. roe of Asterias in STOMACH. of Sammon remain after rest of animal digeste
360d094 - this however is a sophism for their brain or STOMACH would be different.-- Or if one species left i
373d132 male Pidgeon (yes) surely) secrete milk? from STOMACH. analogous to other males feeding young, & to
412e057 have  given milk.-- is secretion from Pidgeons STOMACH true milk.-- ‖ <Species. are innumerable varia
478z003 ops p. 134. and p. 115 In white Cape Pidgeon's STOMACH small shells (patella) sea weed & many pebbles
512q019 distributing live Mamals (3) Do most Hawks eat STOMACH. of finches-- do they throw up pellets-- (4) A
512q019 ks or even owls.-- How long do seeds remain in STOMACH of birds-- Mem: how many miles they fly in few
522m013 erent also from delirium, a peculiar complaint STOMACH not acted upon by Emetics.-- people recognized
527m034 st of hardness of thought, from weakness of my STOMACH I observe a long castle in the air, is as hard
594n113 t it-- sensation of disgust with nausea, (when STOMACH a little disordered) at thought of almost anyt
612o035 polypus,  show either local or general will, & STOMACH likewise «does». [LHC] ¿ in Corallina are  not
612o035 without conscience in the lower animals, as in STOMACH, intestines & heart of man. [LHC] ¿How near in
637j57v adaptation‖ p. 234. The non-absorbing Camel's STOMACH is puzzler p. do says inconvenience would have
638j28r acculloch. Attributes of Deity Vol I. p. 251-- STOMACH hump, kinds of foot. power of closing nostril,
639j28r by  Aphysia. C. D p. 258. «grinding» teeth in <STOMACH of> sun-fish, in mouth of swine & in stomach o
639j28r <stomach of> sun-fish, in mouth of swine & in STOMACH of lobsters-- analogy in Flamingo & Duck, Orni
254c049 view of Flustra or Ascidia spicule in sponge. STOMACHS in infusoria, generation in each joint of tae
392d180 fishes & Mollusca «& Frogs» pass through birds STOMACHS & live? In Muscovy ducks do young take most a
032r042 feel  no doubt. respecting the brecciated white STONE of Chiloe, after having examined the changes of
066r142 lkland Sandstone. he could not distinguish from STONE Caradoc from lower of third Silurian division--
074r165 «(state simplest case. concretions of clay iron STONE; iron pyrite in a fossil» Insist strongly on th
077r172 lt (meteoric) resist oxidation?-- Mem Sir W. P. STONE It is clear to me, there are laws of solution &
077r172 known to us. ‖ M. Chladni.--on meteoric Mexican STONE. Journal des Mines 1809. No. 151. p. 79. [misnu
089a019 nt Marc et des Gonaïves it is difficult to find STONE not thus composed on the NE part more like marb
090a022 .-- What must be the effect of all the meteoric STONE which must have fallen on the globe since the C
090a023 his will give nearly 19 pounds average for each STONE. that fell, that was weighed,; <but> carrying o
112a088 ection with Fitz Roys fact of elevated block of STONE.-- & Caldcleughs collection of facts See page 1
114a093 enberg on ferrugineous Gallionella Examine Iron STONE of C. of Good Hope & Australia/ and mud of salt
114a094 cimens of «transversely fibrous» quartz. & iron STONE alternating. bear on subject of cleavage Clay s
295c178 er; civet cat, rodents.» (Pachyderm in Portland STONE of Alps!!!? No) p. 15 (Lyell's Pamphlet) Is man
332d004 endure this climate-- .) -- July 23d. Eyton, a STONE blind horse, seemed to perceive turn on road wh
386d167 le» animalcule in four days could form 2. cubic STONE. like that of Billin.-- <Generation--> V. p. 15
397e003 iberal! spirit of philosophy to believe that no STONE can fall, or plant rise, without the  immediate
398e005 absolutely  necessary as the «basal» foundation STONE of further inductive reasoning is immense. It i
431e121 the T. elongata in the uper formations Portland STONE &c &c.--if? so «it is» good case:-- in  Sowerby
525m023 g a basket, bringing back game, or picking up a STONE, though only acquired rules by art.-- like  the
551m129 en word Jump said-- I saw the ourang. take up a STONE & pound the earth. Lockarts life of W. Scott Vo
552m131 wheel  horse in drays, scraping against cornice STONE to cause friction Athenaeum 1838. p. 652. Dr Da
192b087 gehog]CD-- As we have one Marsupial animal in STONEFIELD slate, the father of all Mammalia in ages lon
225b219 any  new quadruped since days of Didelphis in STONEFIELD/. all lands united (Falkland Fox. ice). . Mau
054r102 ll described by Pernetty: account of streams of STONES agrees with mine.--At Conception, cleavage E &
066r141 Land,  = the way it stands gales = very strong. STONES as bigger than a man's head.-- Kerguelen 40 by
073r162 3 How utterly incomprehensible that if meteoric STONES simply pitched from mo«o»n, that the metals sh
090a021 ls. ∴ earthy crust compared to those of falling STONES.--¿ does this bear upon the sorting of matter.
090a022 8 & 1818. that is fifty years-- 90 «showers of> STONES are recorded as falling; many of these were no
090a022 counted.-- The weight «or size» is given of 25 STONES.-- The total weight recorded is 473. pounds (t
090a023 d,; <but> carrying on this ratio I can count 90 STONES which have fallen in the 50 years. ∴ 90 x 19 =
090a023 ÷ 50 = 34 pounds each year.-- but instead of 90 STONES in many cases there were flights of stones of
090a023 f 90 stones in many cases there were flights of STONES of large numbers (& how few cases recorded if
122a115 r head of Beagle Channel. on origin of mud with STONES, scattered irregularly.-- (Mem near Gregory Bay
148g026 gravel, current cleavage, & pretty well rounded STONES, mixed with some quite irregular very like rub
149g030 in  Andes, composed of sand & perfectly rounded STONES-- lake required to deposit this Remember howev
516q023 never seen Bee visit. Experiments in Garden Sow STONES of Standard Apricot grafted on what, & see wha
585n077 raised, yet not instinct, but if all men placed STONES in same position, it would be instinct-- insti
593n109 s subversive of» to this view, & fowls hatching STONES. in some degree is so.-- idea of beauty of mus
639j28r pion & Crust in Squilla. & Mantis. C D woodcuts STONES swallowed by birds & by Aphysia. C. D p.  258.
354d062 838) M. Blainville has written paper to show STONESFIELD Didelphis not Didelphis «Answered satisfacto
399e011 en. a Calosoma. (very like American form) in STONESFIELD slate., & a Melolonittha-- In marl from «Lak
414e060 ether progressive I know not.» (& insects.-- STONESFIELD????). Have Mammalia?? My theory certainly re
585n077 ct, but a faculty, or sense-- "We know not how STONGE henge raised, yet not instinct, but if all men
049r089 hills of greenstone? -- Daubeny. P 95. Glassy & STONY Pearlstones alternate together in contorted lay
157g074 y sand blow away]CD where lines appear to cross STONY parts; appearance chiefly cause by fall of angu
138a153 H. Unanüe says he believes the sea has formerly STOOD three hundred feet above its present level, & i
162g104 dip  sideward, & inwards-- deposited when water STOOD at higher Loch Keeper tells me, that Loch Lochy
588n090 re felt." it appears to me mere consequence of STOOPING, as sign of humility.-- I suspect very strong
375d134 ey by positive checks, excepting that famine may STOP desire.--» in Nature production does not increa
442e152 theory, & the sexual reproduction of species may STOP for any number of generations-- Gorze in Norway
536m070 ch effort to walk then lightly as to endeavur to STOP heart beating: one ceasing, so does other.-- Wh
548m117 t resemble the picture in one mind, but does not STOP to reason what there should be & discover  loss
586n078 oneself  knows in walking [one feels inclined to STOP at right number of house though one cannot reme
628o54v cy here.-- at least point of «false» honour will STOP all wish to gratify <it>-- anything contrary to
628o54v trary to it]CD NB. the very end of conscience is STOP to wishes of passion &c. whilst the passions ha
633j54r » arrests earth, (like a Dutchman plants them to STOP the moving sand) we <do> lower the creator to t
563n001 b. 3d. Dog obeying instinct of running hare is STOPPED by fleas, also by greater temptation as bitch:
552m131 erica Fear probably is connected with habitual STOPPING of breath to hear any sound.-- attitude of at
429e115 ferrugineum. begins at 1600 metres precisely & STOPS at 2600. & yet know that plant can be cultivate
440e148 p. 120. observes it commences only, when growth STOPS».-- Spallanzani's facts in connection with buds
565m006 st» laughing in glass. & then as one ceases, or STOPS the noise , the face clearly passes into smile-
538m079 gton, both half ideotic in some respects & with STORE of accurate & even profound knowledge or other
260c067 assist in feeding young cuckoo; as if there was STORGE, which could not be resisted, when hearing cry
268c096 riving away parent birds.-- showing how blind a STORGE? From what I see of S. American birds. genera b
535m064 two mothers to one group.-- (as in birds blind STORGE-- They continue till death, thus acting 4 to 6
453e183 cases where peculiarity has first appeared.-- "STORIA della Riproduzione Vegetale". by Gallesio. Pis
508q016 se?-- Andrew Smith. (6) What size book Gallesio STORIA del Reproduzione.-- D. Holland (7) Is Haemorra
479z005 Gaimard Ulloa shell fish Purple die Marvellous STORIES Ulloa's Voyage Vol I, p. 168 Ceratophytes comm
146g011 feet in vertical height-- enormous mass thunder STORM, many <hundred> thousand tuns. Black faced shee
316c244 ion & superstitions.-- +York's Minster story of STORM of snow after his brothers murder.--» good anec
022r008 ology with respect to the discussion of winds & STORMS:--«in Volney's travels also» Dampier's last vo
036r050 --it is called "Bramidor"(?).--it was a strange STORY; I believe it was necessary to ascend the hill,
```

Page		(Key Word)	
100a053	age conjoined very good.-- It is the Key to the	STORY.--	consider stalactites.-- agate rings, crystal
189b073	en associated & non associated animals.-- & the	STORY	will be complete.-- It is absurd to talk of one
218b190	Voyage par un Officier du Roi Mem. Capt. Owen's	STORY	of cats on West coast of Africa.-- changing hai
231b242	layan, plants ¿migratory birds, he told me some	STORY	of crane from Holland!!! in stomach --or in fea
265c084	child should produce like children. Lyell has	STORY	from.-- Beck about six fingered children herede
316c244	imagination & superstitions.-- +York's Minster	STORY	of storm of snow after his brothers murder.--»
363d102	ities.-- (Mem.-- Goulds Willow Wren.--) (Goulds	STORY	of Water-Wagtails mistake both species scattere
364d104	onal production of black lambs is owing to old	<STORY>	return.-- The Revd. R. Jones told me precisely
364d104	rn.-- The Revd. R. Jones told me precisely same	STORY	about some Southern «see p. 43 supra» breed of
521m007	ocated] There is a case of Mr Anson. who told a	STORY	of hunting «-- habitual fits.--» which my Fathe
523m014	f> in my case.-- It must be so from the curious	STORY	of the Birmingham Doctor praising his sister wh
526m028	mine taste for music.-- Children like hearing a	STORY	told though they remember it so well that they
529m042	tly who was cured from a fall of ideotcy.-- The	STORY	of the Corbets & big noses, quite conjectural,
542m097	animals communicate to each other.-- Lonsdale's	STORY	of Snails, Fox of cows, & many of insects-- the
552m131	th Pachyderms) much intellect.-- mem: Yarrell's	STORY	of wheel horse in drays, scraping against corni
555m142	oubted it in Fuegians, till I remembered Bynoes	STORY	of the women.-- The Chillingham cattle (& Porpo
585n074	ch is only present in youth» (Mem. Mr Worsley's	STORY	of chicken) to know that which we touch & what
593n111	argument held in one's own mind, & Dr. Hollands	STORY	of man in Delirium tremens hearing other man sp
300c197	of many species.-- a few eggs transported to the	STR	of Magellan.-- Change of habits in Van Diemen's
385d163	Sept. 25th Young man at Willis «Grt. Marlborough	STR,	Hair dresser, assures me he has known many case
520mIFC	ression M Charles Darwin Esq 36 Grt. Marlborough	STR.--	(p. 64. On <insect> Ants getting on Table. Co
155g064	ation when serpentine might remove, what above	STRAIGHT	line «only» cut deep gorge on sea hypothesis,
160g088	400 or more feet above station! There is long	STRAIGHT	isthmus connecting E & W connecting Glen Boug
198b113	tand whether. G. H. thinks developent in quite	STRAIGHT	line, or branching S. H What does the express
246c025	chien-loup.--long, black & white, ears short &	STRAIGHT--	do not bark p. 433. birds & bats have certa
522m011	other nothing.-- He could repeat the alphabet	STRAIGHT,	but did not know [Z]CD when heard isolately.
558m153	k Swan.-- Goose do all species put their necks	STRAIGHT	out & hiss.-- [Hyaena pisses from fear so doe
468t112	-- to my grt surprise-- I found all, stamens	STRAIGHTENED	pollen profusely shed; lengthened & turned
539m081	be unconscious of all around, yet there was no	STRAIN	on the intellectual powers-- the difference is
540m088	yla p. 117, Galignani Edition Fine poetry, or a	STRAIN	of music, when the mind is rendered ductile by
542m095	of a wearisome man.-- Is frowning, result of	STRAINING	vision, as savages without hats put up their
151g042	almost meet, but are separated by flat bottomed	STRAIT.	connecting flat on one side with irregular gr
314c239	-- The wombat (Brown) is found in Isd of Bass'	STRAITS	The common Mush room & other cryptogamic plant
052r099	rrell says that numerous icebergs are commonly	STRANDED	on shores of Georgia «Lat degree ( )», he has
026r023	into a peculiar cream-coloured Limestone: the	STRANGE	substitution of matter in shells, like Concret
036r050	f sand.--it is called "Bramidor"(?).--it was a	STRANGE	story; I believe it was necessary to ascend th
194b097	to Mammalia & fish. when there now exist such	STRANGE	forms as ornithorhyncus The type of organizati
226b223	pe & Pachydermata from Europe to America., How	STRANGE	would presence of Jaguar been in S. America.--
280c135	act of existence of mules appeared to him most	STRANGE.--	This even might be said.-- My theory thus e
281c138	us having tendency to lon[g] or peculiar tails	STRANGE	¿¿/Genus only natural from death or slow propa
283c145	tween species. some fine & some wide. which is	STRANGE	if creator had so created them.-- People will
341d032	lf,. at last associated with the ducks.-- most	STRANGE	voice often in the night, like peacock.-- tail
359d087	in Duck «from Bombay» & Canada Goose.-- Former	STRANGE	mishaped bird-- looks very artificial breed--
359d088	ery respect most like Penguin duck.-- which is	STRANGE	anomaly in Yarrells law.-- it probably is expl
362d100	ring their origin.-- I see only some «but very	STRANGE	races» of them have the forked black mark of t
368d114	best sign of most vigorous males.-- «(NB. most	STRANGE	cocks & hens. being either alike or very diffe
380d155	t in confinement, & so imagination in Man, has	STRANGE	effect.-- Directly a Capon is cut, it increase
431e122	deeper one goes-- surely is this true?-- most	STRANGE.--	Does not spermatic animalcule in Mosses, re
440e146	plicated laws of organization: as we see these	STRANGE	plumage in pidgeons yet no change of habits, s
458t001	established-- «Australia,. S. America--» These	STRANGE	forms., camels, giraffes. Sivatherium & Anoplo
458t001	nearly existing forms of aquatic reptiles most	STRANGE,	& shows as in shells some forms are long pres
464tf6r	iar. N. American form)-- ¿Hunting leopard, how	STRANGE,	anyone, would have thought isolated species M
471tf07	133 Westwood on the Fulgoridae enumerates the	STRANGE	forms which the thorax & head displays.-- most
473s03v	dstone. & Birds true! Plants in Devonian-- How	STRANGE	no plants in our Devonian-- Fish one step lowe
527m031	thoughts, one will rise according to law. How	STRANGE	<all> «so many» birds singing in England, in T
533m061	now advanced; for I caught it like a flash.--.	STRANGE	if judgment remains, where reason is forgotten
553m137	cal but sublime views, or the wretched fears &	STRANGE	superstitions of an Australian savage or one o
580n061	o in childhood.-- hence the belief in the many	STRANGE	religions.» Emma W. says that when in playing
602o11v	f the mind, faintly approaches to instinct How	STRANGE	it, that Nature should have so little to do wi
608o027	; but solely to deter others.-- It is not more	STRANGE	that there should be necessary. wickedness tha
251c041	azal.-- ¿is latter same species domesticated,	STRANGELY	contradictory to Azaras fact of conduct of wi
445e159	ur.-- volans, has skin between its legs.-- --	STRANGELY	consider existing «long-organized» forms as p
176b022	dying, as new ones generated» There is nothing	STRANGER	in death of species, than individuals If we s
567n015	who can analyse the sensation, when meeting a	STRANGER.	who one may like. dislike, or be indifferent
567n015	ndifferent about, yet feel shy.-- not if quite	STRANGER.--	or less so.-- When learning facts for indu
578n054	or cheating.-- one does not blush before utter	STRANGER,--	or habitual friends.-- but half & half. Mi
568n017	or overtempted.-- «animals have shyness with	STRANGERS»	«as in case of temperance, or real virtue, t
362d100	t in sixty years-- (how many generations) the	STRANGEST	peculiarities have been kept perfect-- also t
074r165	ies like in composed rock. granites syenite»	«STRANGLING	&c of veins can only be accounted for by con
026r021	a resemblance at Hobart town between the older	STRATA	& the bottom of sea near T. del Fuego.-- Is th
026r022	y oscillations of level, which in the nature of	STRATA	& Organic remains does not appear to have take
030r036	coast S. of Concepcion where there are Tertiary	STRATA	there is Coal-- ¿ No shells in all cases. «.My
035r045	c My results go to believe that much of all old	STRATA	of England. formed near surface: Mem Patagonia
037r055	dimensions? Strong currents off the Galapagos.--	STRATA	must be accumulating which like the secondary
037r055	a must be accumulating which like the secondary	STRATA	of England, «besides ordinary marine remains»
038r058	which trees grew.--? Are not the dikes in upper	STRATA.	quite different from the Porphyries: certainl
038r059	& not sources of lava streams.--Urge not tilted	STRATA.--	It will be well to urge the case of St Hele
042r068	wing to «open» faults in mountains: to elevated	STRATA	in eocene lakes of France, & unequal action of
044r071	resent day. at Mauritius. (consult Bory «dip of	STRATA	on East») cannot believe in a great explosion,
044r073	being consolidated; one inclines to belief all	STRATA	of Europe formed near coast. Humboldts quotati
046r079	ps may be admittance of water, through the rent	STRATA:	«Mr Lyell considers that Plutonic rocks are g
048r087	front of Sts. of Magellan In Chiloe curvilinear	STRATA	subsidence.--The sudden increased dip is not p
057r114	tratified.-- Urge fact of Boulders not in lower	STRATA.	only in upper. in accordance in Europe with i
068r146	ica!! Metamorphic Volcanos only burst out where	STRATA	in act of dislocation (NB. dislocation connect
068r147	change of volcanic focus.-- <it is certain, if	STRATA	can be> Problem dislocate strata without eject
068r147	is certain, if strata can be> Problem dislocate	STRATA	without ejection of the fluid propelling mass.
068r147	l can, for fresh layers of igneous rock replace	STRATA.	& it is nothing odd to find them injected by
069r150	9] {P} ((3) like Bell of Quillota.) (A) in this	STRATA	may be older than (B).-- Most important view U
088a013	vation From the lost & turned about position of	STRATA,	prooff thickness not very great; where piece
091a027	<?>»ESE-- CD [In the Darwar. transition Hills &	STRATA	SE. direction of transitions clay slate &c nea
097a044	p. 221.-- Mr Bollaert tells me, that the upper	STRATA	alone at Guantajaya contains salt see Geolog.
098a048	flint on calc.: sandstone. (& as I believe most	STRATA)	(Hence endless passages from gneiss to granit
101a058	tween dike & mountain axis. except in relative	<STRATA>	size with superincumbent strata. where they h
101a058	t in relative <strata> size with superincumbent	STRATA.	where they have yielded conical axis of mount
102a059	-- only when dikes reach near the surface. that	STRATA	yield.-- In Undulation in open ocean. as pebbl
103a065	portant view being subsequent to dislocation of	STRATA.	A capital discussion might be made between di
104a066	enerally where rock metamorphic & thickness of	<STRATA>	not great, one can conceive anticlinal lines
111a087	NW. numerous boulders of GRANITE" "direction of	STRATA	on the Berbice N. 35 degree. E. dip to NW to 8
113a090	n 400. & the limit being 400 ft. shows that the	STRATA	have very unusual conducting power of heat fro
113a090	we must look at the upper four hundred feet of	STRATA	having conducted away the heat of surface. & i
117a101	December. p. 91. a classification of Europaean	STRATA	according to composition thinks sand with vege
120a107	iclinal lines are apart-- the curvatures of the	STRATA.¿	the enormous faults & facility with which th
121a111	uperinduced Lyell on Sweden p. 5. «& 7.» violet	STRATA	from decomposed muscles.. Smith of Jordanhill
124a120	ntaining magnesia Lyell. Elements p.119 on such	STRATA	{P} do p. 171. argument against lateral inject
128a130	Part. I Vol.-- some notices on modern Tertiary	STRATA	on coast of do-- I believe?? coast of North Am
129a131	ed by gale-- When writing on Valleys. «Tertiary	STRATA	of S America» read parts of this work, though
131a136	eath frozen crust in America Richardson.-- From	STRATA	being not only vertical, but turned over in ma
132a139	ecessary) as M. Parrots shows from variation in	STRATA	earth a very bad conductor.-- shows p. 516 tha

138a155 nk] Sir. J. Halls Paper on the consolidation of STRATA-- he heated sand red hot & brine was boiling o
145g004 loidal-- as well as base not always parallel to STRATA 3 or 4 seams/ 3 or 4 inches thick-- {P} 35 deg
147g015 urrent cleavage-- in other irregular horizontal STRATA I suppose these upper patches if prolonged wou
164g119 nt cause: from mud.-- [blank] {P} muddy nodular STRATA coral upside down strata coarse agglomerate [.
164g119 ank] {P} muddy nodular strata coral upside down STRATA coarse agglomerate [...] shells from [...] Wen
213b172 worthy of observation.-- I think it is certain STRATA could not now accumulate without seal-bones &
290c163 r.-- Geology. Transact. Vol V. Birds bones-- in STRATA of Tilgate forest Seeing common gull in garden
402e017 d to change «is very great difficulty» in thick STRATA, can only be explained, by such strata being m
402e017 in thick strata, can only be explained, by such STRATA being merely leaf, if one river did pour sedim
423e099 equal even in one bed, much less in alternating STRATA of sand & limestone &c &c.-- L'Institut 1838.-
465t081 in any one country.-- in a descending series of STRATA This again shows how much forms depend on othe
473s03r s «with Megatherium & Mylodon» in post pliocene STRATA! Mastodon longirostris in miocene like in Euro
486z020 arrow=heads described in Suffolk as lying under STRATA of gravel & clay about 10 feet in thickness.--
050r093 reef. --- I remember many Corals?? Breccia--STRATIFICATION? Anomalous action of ocean.--at Ascension.
099a043 <Henslow> Sedgwicks lamination parallel to STRATIFICATION evidently small scale of concretionary act
105a070 uous rush.-- besides general improbability. STRATIFICATION, If chain of lake. <a> the alluvium would
152g051 lf= argument against river--composition &-- STRATIFICATION argument detritus-- {P} where buttresses o
029c034 f <Coal> Forests. These thick beds of Lignite STRATIFIED with substances so like the Coal measures in
099a051 -- Mem. laminated dikes in Cordillera.!!!-- In STRATUM OP. let force drag particls to line {P} AB, &
115a098 nough flat top. ended by abrupt slope {P} each STRATUM would thin out, both inland & seaward: if matt
552m131 Mem Chiloe <pi> Sow, who carried from all parts STRAW to make its nest. Pigs & Elephants, (both Pachy
554m140 y sort.-- I saw Tommy picking his nose with «a» STRAW.-- Jenny will often do a thing, which she had b
554m140 is going to be whipped. will cover herself with STRAW, or with a blanket.-- these cases of commonly u
567n013 etting out ears of corn with her teeth from the STRAW, & just like child not knowing what to do with
502q11v if variation equal in flower with leaves.-- STRAWBERRIES How <soon> «early» do characters of races o
228b230 a monstrous plant, but any marked. variety.-- STRAWBERRY produced by seed«s»??-- Universality of gene
401e015 ntirely from all others-- so my experiment of STRAWBERRY not so absurd.-- Thinks-- that such variety
431e122 mber show it is perfectly adapted; it where few STRAY ones. are, that change may be anticipated, & th
406e037 , dog when absent from its master.-- dogs when STRAYED hang their tails.-- November 1st..-- Addenda t
242c018 isles of Oceania have the Scincus with golden STREAKS-- the lacerta vittata extends <to> from Amboin
341d031 by one only differing by some permanent white STREAKS.-- &c &c Dr. Bachman has crossed cock Guinea F
034r044 out Trachyte being the most inferior rocks--The STREAM at Portillo Pass example of do? <Poor> Daubeny
057r115 each of sea to explain preserved animals.--Mem: STREAM of water in the country.-- Sir J. Herschel. sa
095a038 ssion. talks of <hill> «cerro» of Diamante near STREAM of same name. with imperfect crater <--> near
105a073 tion is little the force required to move «it» «STREAM» aside is not great.-- Is there more degradati
106a074 ing channel depends on power of deflection with STREAM retaining its force, now it will be evident th
106a074 ts force, now it will be evident that deflected STREAM cannot retain its force if inclination be grea
106a074 pid".-- is a familiar illustration.-- Therefore STREAM has no tendency to widen course until inclinat
125a120 trees. Uspallata.-- do p. 473. on great Iceland STREAM. The 90 miles includes opposite directions. Me
157g076 valley «there are» granite) «boulders» {P} cory STREAM hill with boulders river Right Hand Cascade ha
162g100 & dies away on the steep & rocky gully of last STREAM Friday Loch Lochy near Letter Finlay Barom 30.
165g124 ss & waters of the Tarf-- Kilfinnan Tower where STREAM enters at head of which hill is round & not me
306g212 uch more general argument) & therefore down the STREAM followed ebb tide, therefore got into habit of
306g212 bb tide, therefore got into habit of going down STREAM which would last were the stream 1000 miles lo
306g212 of going down stream which would last were the STREAM 1000 miles long.-- a monkey. (Baboon) at Z. Ga
582n067 young salmon to go towards the sea. or down the STREAM; which it does unconsciously of any end.-- N B
148g028 rds which would not have happened if the side-STREAMLET had cut them out-- In all cases «I urge» depo
025r017 s talked of the extraordinary freshness of the STREAMS of Lava in Ascencion known to be inactive   300
038r059 posterior to elevations? & not sources of lava STREAMS.--Urge not tilted strata.-- It will be well to
052r099 with much truth compared to the step = formed STREAMS of lava at St Jago. C. de Verds Quartz pebbles
054r102 broken hill described by Pernetty: account of STREAMS of stones agrees with mine.--At Conception, cl
055r107 of . craters. At Ascension (or modern STREAMS of St Jgo) yet no historical records of erupti
086a007 efly clearly by sun's position = If equatorial STREAMS of warm pole; in name of Heaven why are tops o
473s03v wh. flows into Conway by Bettws & there joins STREAMS from Capel-Curig-- Mr Bunbury says Miers has d
557m147 back» when savage «no» & ready to dash at prey STRECHED out & flaccid, when furious «with fright» bac
542m095 m sleep see how a dog yawns when he awakes. & STRECHING & yawning can be explained from too long rest
338d024 ned My hairdresser (Willis) says <black> «that STRENGTH of» hair goes with colour. black being strong
367d113 "One of the most general marks is the superior STRENGTH <of> «of make in» the males; & another circum
367d113 her circumstance, perhaps, equally so, is this STRENGTH being directed to one part more than another,
384d161 of use <Great characteristic of male greater STRENGTH, (p 45) & that strength> In speaking of gener
384d161 ristic of male greater strength, (p 45) & that STRENGTH> In speaking of generation alway put female f
416e075 iety? «Man picks the Male, instead of allowing STRENGTH to get the day» The fertility of Indian & Com
444e157 e & Pig in having two additional bones to give STRENGTH to it.-- p. 139. Doubts altogether the law of
522m014 feeling also of depression, & both these give STRENGTH & comfort to the body» I know the feeling, th
620o046 ed?) not so puppy, we <do> try to teach him & STRENGTHEN his instincts.-- so man ought to follow cert
620o045 ce to instinct» <conscience>., or rather the STRENGTHENED instinct, even when our reason tells-- + us
096a040 Hecka-- All the Azores Isld. Von Buch p 359 STRETCHED out NE & SW.-- Von Buch. Can. Ile p. 406. Lis
534m062 which they waded. or swam across.-- they then STRETCHED themselves from wall to table.-- table being
162g099 out of them-- but it may be said that a mound STRETCHES along, parallel to Shelf on opposite side & d
158g083 r & from Glenroy» near the upper shelfs ground STREWED with pebbles Shelf 3d runs up with lithological
430e119 sulcatus.-- the beak of this one has concentric STRIAE, all the lower part largest longitudinally (give
256c055 e check to distribution of birds & animals Mr STRICKLAND & Hamilton-- found tertiary formation amongs
199b120 spring; marriage never probably excepting from «STRICT» domestication offspring not. fertile. or at le
271c105 devours is same species. yet that it should so STRICTLY <f> agree in habits with the Turdus Musicus «
291c165 ndency to follow it, or it may be heredetary & STRICTLY point out affinities. conducct of Gould, rema
318c254 y «generally by night» -- other birds which is STRICTLY diurnal, migrates singly by night.-- others i
372d130 hus produced from the growth of one part, (not STRICTLY new individual), or he may produced by having
414e064 r skulls of the men on the plains of Bolivia-- STRICTLY fossil «& in Van Diemen's land»-- they have b
414e064 »-- they have been exterminated on principles. STRICTLY applicable to the universe.-- The range of ma
450e174 ell.:-- Black Swan «in domestication & nature» STRICTLY monogamous-- geese polygamous (¿when wild) bu
484z016 don. allied to some of my birds-- These groups STRICTLY American. Colouring on under side of wings It
536m071 cited by love? Stallion licking udders of mare STRICTLY analogous to men's affect for womens breasts.
610o30v > right.-- The change <of> our moral sense, is STRICTLY analogous to change of instinct amongst anima
612o035 are not two kinds of life vegetable and animal STRICTLY united? [RHC] It is easy to conceive such mov
628o53v rovement of the instinct of a shepherd dog, is STRICTLY analogous to education of child,-- causing ma
055r108 at St Helena!!-- There are some arguments which STRIKE the mind with force.--the exact yearly rise of
091a027 t» Accra. Coast of Africa. Clay Slate & Quartz. STRIKE SSW & NNE dip 30 degree - 80 degree Ed. N. Phi
091a027 iss in India (falls of Garsipa) dip 30 degree. <STRIKE> «direction<?>»ESE-- CD [In the Darwar. transi
146g012 ite black legs & tail like species in colouring STRIKE an analogy between pleasures of association, &
200b121 ching off.? Accra, Coast of Africa. Clay slate. STRIKE. SSW. & NNE. dip 30 degree - 80 degree (?).--
541m093 «insulted» may forgive his enemy & not wish to STRIKE him, but he will find it far more difficult to
558m152 mmon Swan, arch raises neck & depresses chin-- STRIKES with wing arches wings-- as does black Swan.--
147g020 l & irregular hills of alluvium-- nothing very STRIKING yet possibly sea more probably than river-- N
256c054 y are either A. B. C. D E., or A C D E H. Very STRIKING to see M. Bibron looking over reptiles he oft
301c199 itual action. even which Man performs.-- child STRIKING a post in passion.-- Habit instinct gained du
385d164 temper, especially of the youngest children, a STRIKING <resemblance> similarity to the temper of the
428e112 much about breed.-- what can «however» be more STRIKING, about indelibleness, than the number of good
626o052 .-- the passion rising from weariness leads to STRIKING blows.--' p. 224.-- Hume's Inquiry-- good abs
640j167 o.-- so far true, but do not fish offer a most STRIKING anomaly to this. Have they wide ranges? Agass
250c038 of former whole world. America might have been STRING of islands.-- ¿Europe has many species but not
380d155 s own field to bull a cow.-- - a dog if led in STRING will not.-- some of the tigers.-- cat, though
584n072 he acquirement of a new sense,-- bats avoiding STRINGS «in the dark» as well might be called instinct
617o40v eing pulling every other particle by invisible STRINGS & as on this supposition the forces manifested
050r093 Africa. as at Brazil [blank] What is nature of STRIP of Mountain Limestone in N. Wales. was it reef.
355d066 Mr Herberts variety of horse, dun-coloured with STRIPE approaches to ass.» or fowls to the several ab
367d113 like large, heavily made cream coloured ass.-- STRIPE on back also.-- legs reminded me strongly of Z
460t015 lf way between swan-goose & common goose.-- the STRIPE down back pretty plain in in these <half> «3/4
536m074 temptation,' he would be most humble, he would STRIVE <to do good> «to improve his organization» for

Page **(Key Word)**
```
6240051 ings of right & wrong,-- education, of parents STRIVES* to same end.-- & general actions of community
6240051 slow   growth of rule of right.-- [LHC] *for it STRIVES to give conduct beneficial to all the children
521m009 Mr Corbet of the «Hall» «Park», after paralytic STROKE. intellect impaired. <after paralytic stroke>
521m009 ic stroke. intellect impaired. <after paralytic STROKE> : . could converse well on any subject when o
064r137 points» of eruption [.] give instance of Etna STROMBOLI & Vesuvius Investigate with greater care. veg
034r044 ldt: Comparison P 361. Daubeny Von Buch is very STRONG about Trachyte being the most inferior rocks--
037r055 of the Plata. to the Bay of Bengal. dimensions? STRONG currents off the Galapagos.--strata must be ac
066r141 erguelen Land, = the way it stands gales = very STRONG. Stones as bigger than a man's head.-- Kerguel
099a049 <would>  are aggregated, would they not attract STRONG. a third.-- & this would make layers.-- (Gravi
100a053 laining vary dip & inclination.-- which last is STRONG character.-- A discussion on concretions and c
131a136 ned over in many parts of the world.-- argument STRONG in favour of thin crust theory.-- What a curio
175b016 change  has happened I look at two ostriches as STRONG argument of possibility of such change,-- as w
189b071 ngs of Apteryx Dacelo & Kingfisher same colours STRONG odour of negroes, a point of real repugnance.-
194b095 a centrum for Mammalia.-- I really think a very STRONG case might be made out of world before zoologi
195b098 trary, & not permanent. this might be made very STRONG. if we believe the Creator creates by any laws
195b103 atagonian <Chat> & Galapagos orpheus.= Put this STRONG so many thousand miles distant.-- Absolute kno
214b173 s of» Oolitic Series does not appear to me very STRONG What is Osteopora platycephalus. (Harlan) foun
234b261 ally affect different stations;-- this would be STRONG argument for propagation of species-- Again is
244c021 e, makes species but barrier.-- --it would make STRONG contrast with southern regions.-- «it would no
245c024 isles have the same productions p 293-- is very STRONG about this Lesson insists much.-- The (p. 296)
293c172 ong intervals of forgetfulness.-- after sleep, «STRONG» analogies with memory in offspring.-- Some as
293c172 eel sure you have heard conversation before. is STRONG association recalling up image which had been
343d038 o must the tribes become blended & prevent that STRONG separation which otherwise would have taken pl
403e023 er it is the will of God.-- Octob. 16th. A very STRONG passage might be made-- why seeing great varia
428e112 or encouragement, but a vigorous battle between STRONG & weak March 11th. Yarrell's law must be partl
446e162 t effect of different stocks in this case».-- & STRONG case showing analogy of production by gemmatio
448e167 far distant» with Touraine «which as L. says is STRONG argument for their contemporaneous»-- how is t
532m056 hat remarkable affection to a place.-- How like STRONG feelings of Man.-- The sensation of fear is ac
537m077 pride are heredetary., & therefore he has these STRONG, & does not act up to them, no doubt disobeys
543m101 ble it shows that new instinct can originate.-- STRONG argument for brain bringing thought, & not mer
548m116 ard suffering of a dear friend-- this gives one STRONG idea of what individuality is.-- Insanity is <
558m150 pacity «acting» «yet being obscurely guided» or STRONG instinctive sexual, parental & social instinct
574n042 analogous  to a blacksmith having children with STRONG arms.-- The other principle of those children.
574n042 of  those children. which chance? produced with STRONG arms, outliving the weaker ones, may be applic
579n060 r movement. The Blushing of Camelion & Octopus; STRONG analogy with my view of blushing-- in former i
583n069 ctive pointing varies.-- p. 18. Animals possess STRONG imitative faculty: pure instinct is not imitat
588n090 re rolled upwards during mental agony, & whilst STRONG emotions of reverence & piety are felt." it ap
588n090 stooping, as sign of humility.-- I suspect very STRONG argument might be advanced, that animals  have
589n091 d with much hypothesis.-- see M.S. notes, where STRONG argument in favour of brain forming the instin
594n112 og.-- » The instinct against man is perhaps, as STRONG as against hawk, but the birds at Maer have le
608o025 determined  by what is called free will, but by STRONG invariable passions-- when these passions, wea
621o046 hildren without some passion.-- If his passions STRONG & his instincts weak. he will have many strugg
621o047 inguity to will. (a) The origin of passions too STRONG for our present interest receive simple explan
625o50v only is it perceived, that our passions are too STRONG for our instincts. to gain long-lived good, ie
190b079 me obscurer as knowledge increased, but genera STRONGER,-- Mr Waterhouse says no real <separ> passage
199b120 arriages; smallest differences blended, rather STRONGER tendency to imitate one of the parents; repug
280c136 viewing it.-- Variety when long in blood, gets STRONGER & stronger, so that though by great effort <p
280c136 -- Variety when long in blood, gets stronger & STRONGER, so that though by great effort <pr> one unli
575n044 only test this is most important: can there be STRONGER analogy that the tendency to hybrid greyhound
615o038 character,   love of country, of association &c STRONGER in some than others-- Hence superiority of Ch
038r057 be a central core of melted rock--I think the STRONGEST is the consideration of the state at a  grand
038r058 s filled up.--the appearance will here be the STRONGEST argument:--¿ Consider causes for subaqueous c
192b086 those classes with few species greatest jumps STRONGEST marked genera? Reptiles?) For instance  there
338d024 rength of» hair goes with colour. black being STRONGEST.-- V. p. 63. Note Book M'. for case of change
423e097 he number of living beings-- but there is the STRONGEST possible to increase them, hence the degree o
609o029 October <8> 2d. 1838 Those emotions which are STRONGEST in man, are common to other animals & therefo
066r142 in  neigbourhood of town Mr Murchison insisted STRONGLY. that taking up a piece of Falkland Sandstone
074r165 ay iron stone; iron pyrite in a fossil» Insist STRONGLY on the grand fact of Volcanic & non Volcanic.
128a129 } land protuberant water to counterbalance How STRONGLY the Glen Roy case shows that the figure of th
217b184 es offspring quite intermediate sometimes take STRONGLY after either parent. about <half & half  tim>
222b203 ., & yet we find this same tendency (only less STRONGLY marked) between what are called  varieties.--
228b230 uced by seed«s»??-- Universality of generation STRONGLY shown by hybridity of ferns.-- hybridity show
273c111 ow in every family of bird,, even the most <m> STRONGLY marked, there is a preeminently aerial,-- for
273c111 y fill up their place.-- Humming bird there is STRONGLY marked variety,:.in the Tyrannidae.-- Milvulu
291c167 idea of foetus being of one «both» sex«es». is STRONGLY supported by wonderful fact of bees  changing
341d030 ngs reduced to rudiment.-- clavicle scapula &c STRONGLY developed to aid in breathing.-- Animals from
357d072 Australia have branched out into orders one is STRONGLY tempted to believe, one or two were landed as
367d113 ass.-- stripe on back also.-- legs reminded me STRONGLY of Zebra.-- Mem. Quagga & Ld Moreton Mare rin
387d169 he <tw> second race both have this peculiarity STRONGLY; they transmit with same force as first pair,
399e010 - do hind legs increase in any rabbits One may STRONGLY suspect, that breeding in & in, produces  bad
409e049 uld not be social animals. «this is stated too STRONGLY. for there would be innumerable species. .& h
507q15v onditions.-- like cowslip & primrose, but less STRONGLY marked.-- 31. Plant seeds of the Fuller's pla
566n011 necessary notions. which of them? & curiosity «STRONGLY shewn in the numerous artifices to take birds
573n037 to eat, hence assertion.-- but nodding is less STRONGLY marked than negation Marianne. says. that she
641j29v bsence of their remains in the Wealden? In the STRONGLY separated Arctic genera, there is evidence of
064r134 vegetation.  Rhinoceros quite in deserts.--Much STRUCK with number of animal[s] at Cape of Good  Hope
233b249 in ought to be Australian.?-- Mr Gould has been STRUCK with similar extension of form in birds.-- ] W
346d047 account  of the Shepherd dogs.-- Aug. 24th. Was STRUCK with pink shade on plumage of the Pelican.-- M
354d065 ce attach themselves to man. Sept 7th. -- I was STRUCK looking at the Indian cattle with Bump. togeth
520m001 , although constantly seeing him, she was often STRUCK with this fact.-- the resemblance was in odd t
539m081 12th.  38. At the Athenaeum Club. was very much STRUCK with an intense headache «after good days work
545m105 much.  as objects of interest.-- do/ I was much STRUCK with observing how the Baboon (<Macaco> «Cyano
563n002 nion. (mem Cyanocephalus. Sphynx howling when I STRUCK the Keeper) may be tempted to attack him  from
571n030 ween them, & great masses of rock.-- I was much STRUCK with this, when viewing Windsor Castle which r
571n031 e sublimely from natural rise-- I was also much STRUCK in great avenue, resemblance to gloomy aisle o
6200044 n the pleasure from good dinner, or from a blow STRUCK in passion fades away, so that when man afterw
021r005 in  Clay Slate. major axis 2.¼ ft.-- singular STRUCTURE of nodule, constitution «same as» of slate sa
031r038 dd other instances in old world of symetrical STRUCTURE. East India Archipelago. «Aleutian Arch.--» V
031r038 Zealand Geological Notes. at St. Helena. This STRUCTURE was very clear at base of great lava cliffs [
033r042 filtered with other matter how very curious a STRUCTURE: Have shells ever casts alone in  Calcareous.
035r046 od. Cordilleras, Chiloe. &c seems the organic STRUCTURE most easily preserved.-- Mr Conybeare introdu
039r060 et valley «the most remarkable feature in the STRUCTURE of Ascension» give as an example the great su
059r121 ter.--all pale cream colour.-- The Brecciated STRUCTURE of all the Pitchstone (which I have seen). is
059r121 hich I have seen). is a kind of concretionary STRUCTURE, for the interlineal spaces are of diff cont:
064r137 ypseous formation of Cordillera In describing STRUCTURE of Cordillera it must be said, that lines of
070r153 pallata of which no trace except by trees The STRUCTURE of ice in columns. show that granite when wea
070r153 weathering into balls. must exhibit orbicular STRUCTURE.--When we recollect connection of columnar  &
087a012 y of <Patagonia.> S Cruz -- from terrace like STRUCTURE-- Intersection of veins prove, that there are
089a019 part  more like marble requires polish to see STRUCTURE,-- «New Thought of erecting machine to see if
090a025 an end? Fragmentary granite showing schistose STRUCTURE (& veins appearing): mem. Henslows Anglesea s
098a045 rom Phryganea NB. Sedgwick talks of LAMINATED STRUCTURE (∴ separation of ingredients) as uniting with
106a075 nt {P} Therefore when we have valleys of this STRUCTURE. as the inclination in all probability  would
120a107 nes into small pieces-- mem coal-field.-- the STRUCTURE of Andes. where we believe we can trace the o
121a111 character completely altered, & a crystalline STRUCTURE superinduced Lyell on Sweden p. 5. «& 7.» vio
174b014 hich is law proved.-- We can see «why» STRUCTURE is common in certain countries when we can ha
181b042 greater  the gaps (or solutions of continuous STRUCTURE) «between them.».-- for instance there  would
181b045 t such tribes, as far as compatible with such STRUCTURE are in minor degrees adapted for. other eleme
185b055 re certainly appears attempt in each dominant STRUCTURE to accomodate itself to as many situations as
```

186b058 CD[from the principle of atavism, where real STRUCTURE obliged to be altered, I can conceive colouri
189b069 eful to know what is species.-- In proof that STRUCTURE is not simple adaptation, armadilloes «&» & M
189b074 her.-- We consider those, where the {cerebral STRUCTURE intellectual faculties} most developed, as hi
200b121 ether those genera which unite very different STRUCTURE as. petrel & alk. do not show the possibility
201b129 , as the irregularities in the degradation of STRUCTURE of Lamarck, which he says depends on external
202b130 relations to a third body., or common end of STRUCTURE A Race of domestic animals made from influenc
207b150 mmon to New Holland?! p. 320. Says Coniferous STRUCTURE intermediate between vascular or cryptogam. (
211b162 inity, because similar habits produce similar STRUCTURE.-- Mem. Ornitho Rhyncus Would not relationshi
220b198 nts; analogous to Men. & dogs. Now if we take STRUCTURE as criterion of species Hogs different specie
224b213 ther with other <animals> beings of very near STRUCTURE.-- Hence species may be good ones & differ sc
224b216 ns gone on creating animals with same general STRUCTURE.-- miserable limited view.-- With respect to
227b224 llow water at Melville Isd. (3d) We know that STRUCTURE of every organ in A. B. C. three species «of
227b225 llow close to A. B. C. we cannot be sure that STRUCTURE (C) could pass into (D).-- We may foretell sp
227b225 blending of two genera-- It explains typical STRUCTURE.-- Every species is due to adaptation + hered
227b225 ery species is due to adaptation + heredatary STRUCTURE. latter far chief element.·. little service hab
227b226 ve one.-- It leads to knowledge what kinds of STRUCTURE may pass into each other: now on this view no
227b227 of adaptation (wish of parents??) instinct & STRUCTURE becomes full of speculation & line of observa
228b228 » examination of direct passages of <species> STRUCTURE in species, might lead to laws of change, whi
229b233 els, but not having them, instance of useless STRUCTURE-- Smith thinks several species of Rhinoceros
254c047 ee with grand fact of Marsupial, low Cerebral STRUCTURE??-- «do» p. 390. All classes of Acrite exhibi
255c113 y without association..» Instinct goes before STRUCTURE (habits of ducklings and chickens) Young wate
255c053 pterix because we may suppose longest part of STRUCTURE.-- shape of wings have altered many times, bu
258c061 offspring who have any slight peculiarity of STRUCTURE. «hence seals take victorious seals, hence da
261c072 f Plesiossaurus Plesiosaurus. alludes to some STRUCTURE in head, which he says (evidently is an excep
261c072 n to animals wants & not as change in typical STRUCTURE?!! Whewell «in Comment/ few will dispute--» s
264c081 ould, good genus Gould seems to doubt how far STRUCTURE & habits go together. This must be profoundly
264c081 gether. This must be profoundly considered.-- STRUCTURE may be obliterating, whilst habits are changi
264c081 obliterating, whilst habits are changing-- or STRUCTURE may be obtaining, whilst habits slightly prec
264c081 tant element in considering to which tribe,-- STRUCTURE without corresponding habits clearly showing
264c083 st birds Java. not so much-- Peculiarities of STRUCTURE. as six fingered people are sometimes heredet
265c085 er monster.-- The only way of judging whether STRUCTURE is owing to habits, or heredetary is to see,
265c086 Few will dispute that it is possible to have STRUCTURE without habits-- after seeing beetle with win
265c087 horhyncus, then we should never know how much STRUCTURE was connected with habits, & how much heredet
270c103 Mem. Agaziz. (INo Annals of Nat. Hist) spiral STRUCTURE in Echinodermata.-- Agassiz says Infusoria «a
271c107 f instances of insects of one tribe taking on STRUCTURE (probably accompanied by habits) of other, th
272c108 ome third, have got thighs with same peculiar STRUCTURE & habits of clinging to rushes similar.-- The
272c108 re there Heteromera, which have habits & part STRUCTURE like Curculionidae.-- Are there any Crysomeli
272c109 t.-- We have abundant instances of remarkable STRUCTURE which as far so species is concerned superabu
272c109 of Edolius.-- Remarkable how small detail in STRUCTURE prevails amongst the same species & subgenera
273c111 -- Thus in Hawks, there is a swallow, both in STRUCTURE & habits (it cannot be doubted that if swallo
273c112 t dull.-- I must observe that this preeminent STRUCTURE is not always applicable to same habits, thou
274c113 rot though so much on the ground has not this STRUCTURE., instance of habits going before structure).
274c113 s structure., instance of habits going before STRUCTURE).-- even one kingfisher-- Gould has seen with
274c114 -- In each division Gould thinks he can trace STRUCTURE for insects & structure for vegetation.-- In
274c114 d thinks he can trace structure for insects & STRUCTURE for vegetation.-- In conversation in Museum--
276c124 rest of other family practise with a peculiar STRUCTURE, then Milvulus forficatus Tyrannus Sulphureus
276c124 annus Sulphureus if compelled solely to fish. STRUCTURE would alter.-- It is a difficulty how a diffe
276c124 vertebrae are produced, where, (& in all such STRUCTURE) there cannot be gradation. See what Eytons y
276c125 abits similar ¿law?-- probable.-- if habits & STRUCTURE similar would have blended together Mem Mr He
282c141 form & head &c to become greatly changed. in STRUCTURE & even to certain degree in habits, yet we mi
284c149 lackbird. good remark if general.-- Where any STRUCTURE is general in all species in group we may sup
285c153 ries-- fertility must settle it.-- Changes in STRUCTURE being «necessarily» excessively slow, they ma
289c161 f a Grebe, structures might follow.-- examine STRUCTURE of this bird & get account of habits My defin
290c163 to heredetary power of Muscles.-- then we SEE STRUCTURE gained by habit Talent &c in man not heredeta
291c165 important element in classification, because STRUCTURE has tendency to follow it, or it may be hered
291c166 -- it is difficult to imagine it anything but STRUCTURE of brain heredetary., analogy points out to t
291c166 d Barclay on organization!! Avitism in mental STRUCTURE or disposition. & avitism in corporeal struct
291c166 ucture or disposition. & avitism in corporeal STRUCTURE are facts full of meaning.-- Why is thought.
292c170 «like myself» of <classi> real affinities. ie STRUCTURE of the whole animal let him read Mr Swainson'
292c171 r of great importance to cause long memory.-- STRUCTURE is only gained slowly.-- therefore it can onl
293c172 ion in such cases recall the idea. «or simple STRUCTURE in brain people in fevers recollecting things
293c172 which had been past-- so great an anomaly in STRUCTURE of brain not probable} put note. Sir W. Scott
301c199 rs of corn according to my views, habits give STRUCTURE... habits precedes structure... habitual inst
301c199 ews, habits give structure... habits precedes STRUCTURE,.. habitual instincts precede structure.--duc
301c199 cedes structure,.. habitual instincts precede STRUCTURE.--duckling runs to water. before it is consci
303c204 e the father of man kind probably possessed a STRUCTURE in these points for <t> a less time than othe
304c208 ted, we must suppose the changes from typical STRUCTURE have either been more rapid than in all other
309c222 .-- Thinks passages very rare., in anatomical STRUCTURE.-- «the passages between-- owls & hawks only
309c222 external» intermediate groups often have full STRUCTURE «of one class» & full of second--this class o
311c226 be used «in classification» as indication of STRUCTURE (including brain & other organs difficult to
334d011 deaf curious case of corelation of imperfect STRUCTURE.-- Fox says in «Lord» Exeter's Park «or in th
341d029 Rheas still closer).-- Mr Blyth asked whether STRUCTURE of pelvis & was not adaptive structure, like
341d029 hether structure of pelvis & was not adaptive STRUCTURE, like little wings of Auks which does not mak
353d060 <Zoologist> «philosopher», who has trace the STRUCTURE of animals & plants.-- he get merely a few pa
365d107 e that Natural varieties or species.. all the STRUCTURE of which is adaptation to habits (& habit sec
365d107 in constitutional.,-- more conformable to the STRUCTURE which has been adapted to former changes. tha
366d111 heir eyes. without extermination, & change of STRUCTURE.-- When will the musquitoes of S. America tak
369d115 mice) & these being water animals <that this> STRUCTURE <connected with animals being compelled to tr
372d130 n this view each particle of animal must have STRUCTURE of whole comprehended in itself.-- it must ha
373d132 heir happening in same plant.-- The Marsupial STRUCTURE shows that they became Mammalia, through a di
375d135 ges trying force <into> every kind of adapted STRUCTURE into the gaps <of> in the oeconomy of Nature,
376d135 all this wedgings, must be to sort out proper STRUCTURE & adapt it to change.-- to do that for form w
377d147 like ground birds [not located] :Hence, also STRUCTURE not really fitted for water, only habits & in
381d157 utandrous animals <are> is there gradation of STRUCTURE leading to supposition, that the Cryptandrous
388d172 s & testes curious instances of corelation in STRUCTURE = Neuter bee having both sexes abortive fact
391d179 d object failed, & then by that corelation of STRUCTURE desire fails. Every individual except by ince
399e009 Study introduction to Cuviers Regne Animal No STRUCTURE will last. without it is adaptation to whole
404e025 if they had series perfect, would expect this STRUCTURE would become obscure & therefore it might thu
412e057 fish is instance of part of the hermaphrodite STRUCTURE being retained in the male.-- <like» «far» mo
412e057 <Species. are innumerable variations>. Every STRUCTURE is capable of innumerable variations, as long
412e057 these variations tend to accumulate. «on any STRUCTURE.» L'Institut. 1838. p. 384. List of fossil Ma
415e067 druped or bimanous,, is to see, what parts of STRUCTURE abortive.-- Remember my fathers remark about
416e075 d not so species every part of newly acquired STRUCTURE is fully practised & perfected Hence differen
418e084 - By birth the the succesive modifications of STRUCTURE being added to the germ, at a time, (as even
420e089 heredetary. shows well what minute details of STRUCTURE heredetary'-- Athenaeum .1839. p. <8>36.-- A
422e095 animals in the world depends, of their varied STRUCTURE & complexity.-- hence as the forms became com
423e097 clear that a large part of the complexity of STRUCTURE is adaptation. though perhaps difference betw
432e125 has no more relation to our present wants. or STRUCTURE, than the muscles of the ears to our hearing
435e130 n some species only in genera <are> have this STRUCTURE.-- Some willow trees have been observed to ch
440e146 II. p. 115. 4 four laws Who can say, how much STRUCTURE is due to external agency, without final caus
440e147 is far more probable way of explaining, much STRUCTURE, than attempting anything about habits-- no o
441e148 ns on the «propagating» constitution. but not STRUCTURE of the parents.-- Thus would a Crab tree vary
443e154 metamorphosis, but to arrive at their present STRUCTURE they must have <done> been propagated by sexu
447e164 buds in same predicament, as one, in which STRUCTURE does not allow of crossing with other individ
460t017 the same external conditions (ie. analogical STRUCTURE) & partly the laws of organization (ie those
461t037 h 6 I presume, from my theory, as long as any STRUCTURE can be handed down without being absolutely i
461t037 ls & Horses--».--«)» & therefore probably any STRUCTURE would rather become accomodated to new circum
461t037 ld be eliminated, & hence, the application of STRUCTURE to purpose after purpose would tend to render

```
463t055 nly are no animals known with an intermediate STRUCTURE, but it is not possible to imagine what habit
463t055 hat habits an animal could have had with such STRUCTURE.-- perhaps greatest Could anyone. have forese
463t057 by   rooting out curious cases of intermediate STRUCTURE., & supposing much extinction. give a paralle
471tf08 divisions & speaking of their similarity «in   STRUCTURE» he says "indeed it wd be difficult to  point
473s07v ife of offspring-- No peculiarity in external  STRUCTURE can be concepcional, as limbs &c &c only appe
484z015 emisphere, & the Puffinuria, the Awks.-- What   STRUCTURE do the auks bear traces of.-- like Puffinuria
499q010 glish plants generally distinguishable.= What   STRUCTURE of seeds.-- (Paris) (22) When Linnaeus says s
502q11v pirits & then experimentise: for gradation in   STRUCTURE Compare flowers of wild & tame carrot-- Parsl
506q015 ecious or dioecious plants the Papilionaceous   STRUCTURE of flower-- Ground nuts (42) How are Orchidia
507q15v Dioeious  plants, & any with peculiarities of   STRUCTURE rendering cross impregnation difficult or rev
507q15v -- simple spines-- or seed-cases with similar   STRUCTURE.= good case as showing how simple, but beauti
540m085 eel how <much> all animals <are> built on one   STRUCTURE.-- He who doubts about national character let
543m101 ination.-- obeys same laws. as other parts of   STRUCTURE. C. D.27 Can an analogy be drawn between «her
549m119 collected whether from frequency, or inherent   STRUCTURE of mind. they make, either in themselves,  or
553m138 for   women-- «very curious. as they depart in  STRUCTURE» The monkeys understand the affinities of man
558m151 nt of expression more than any other point of   STRUCTURE takes its value. from its connexion with mind
568n018 irds voice & taste for singing with Mammalian   STRUCTURE. «-- American monkeys utter pleasant plaintiv
569n020 .. not produced by will <by> but by corporeal   STRUCTURE.-- Devotional feelings, probably some distant
573n036 es are more fitted to recognize the wonderful   STRUCTURE of a beetle than a Universe.-- November  20th
575n043 d child dog for mice from some peculiarity of   STRUCTURE of brain.?-- is this more wonderful than memo
576n048 tinct.-- Instinct is a modification of bodily   STRUCTURE «(connected with locomotion.)» <no, for plant
581n065 eem to consider there is good evidence in the   STRUCTURE of language, that it was progressively formed
584n071 able to measure the cell; p. 22. instincts &   STRUCTURE always go together: thus woodpecker: but this
584n071 is not so,, the instincts may vary before the   STRUCTURE does; & hence we get over an apparent anomaly
604o016 ousness becomes multiplied with the organism   STRUCTURE, it looks as if consciousness an effects of s
612o035 intestines & heart of man. [LHC] ¿How near in  STRUCTURE is the ganglionic system of lower animals & s
634j55r ris  Basin?.-- NB) The explanation of types of STRUCTURE in classes-- as resulting from the will of Ma
638j58v onding to «every» «one or any»-- brain making   STRUCTURE, instead of parts of body.-- Now we know what
638j059 the  knowledge of trying a hundred schemes of   STRUCTURE, in the course of ages «step by step».-- in M
638j059 h> form these schemes.-- I see no reason, why   STRUCTURE of brain should not be born. with tendency to
185b055 nd goose.-- water chionis water rat with land   STRUCTURE; :→law of chance would cause this to have ha
195b099 es. <ro> hiding place for many unintelligible   STRUCTURES. it might have been of use in progenitor-- o
202b130 influence.--...... Hence name of analogy, the   STRUCTURES in the two animals bearing relations to a th
220b197 nct (in plants domestication on perversion of  STRUCTURES especially reproductive organs) & therefore
223b208 ween Man & next animals in mind, more than in   STRUCTURES.-- If the skeleton of a Negro-- had been fou
224b212 t.-- It does not bear any precise relation to   STRUCTURES Mem. Eyton's Hogs & dogs.-- The passage in l
227b224 e»: but this cannot be predicated of <genus.>  STRUCTURES in two genera.-- <we then cease to know  the
227b226 n this view no one need look for intermediate   STRUCTURES <between> say in brain. between lowest Mamma
248c033 r is most intimately connected with important  STRUCTURES. <which are less obviously affected by exter
259c063 ies must take to that particular habit.-- All   STRUCTURES either direct effect of habit, or heredetary
261c069 ccumulate instances of one family sending out  STRUCTURES into many genera.-- like Synallaxis or Marsu
274c115 , in analogy certain parts perfect of typical  STRUCTURES certain <imper> parts changed Have <not>. S.
276c124 tance to show that habits sometimes go before  STRUCTURES.-- the only argument can be, a bird practisi
289c160 typical  land bird, having habits of a Grebe,  STRUCTURES might follow.-- examine structure of this bi
349d052 orhynchus come in circle?!!! p. 8-- Anomalous  STRUCTURES, as in Hippotamus, solely owing to number of
423e096 sometimes   to augment & sometimes to simplify STRUCTURES:= Without enormous complexity, it is impossi
549m123 were   preservative, & are now, like all other STRUCTURES slowly vanishing-- the mind of man is no mor
331dIFC ds close I first thought of selection owing to STRUGGLE July 15th. 1838 Finished. October 2d As a pro
414e063 each   other &c, but then comes the more deadly STRUGGLE,, namely which have the best fitted organizat
536m070 but observing eyes thus unconsciously discover STRUGGLE of feeling.-- It is as much effort to walk th
588m089 rs, or dangled up & down-- in latter case they STRUGGLE their arms.-- do. p. 306 "the eyes are rolled
600o008 iments> imperative sense of duty-- which makes STRUGGLE in man.-- two souls in one body-- (2) Beau id
604o017 Sympathy & affections chiefly fail.-- Notices. STRUGGLE <between> when insanity is coming on  «Thinks
619o043 e, what would be the result? In a dog we see a  STRUGGLE between its appetite, or love of exercise & i
538m080 s with insanity, as in Dr Ash's case, when he  STRUGGLED as it were with a second & unreasonable man.-
523m015 of consciousness of insanity coming on.-- his  STRUGGLES against it, his knowledge of the untruth of t
621o046 trong & his instincts weak. he will have many  STRUGGLES, & experience only will teach him, that the i
594n115 e assistance & bite a big dog. which was fast  STRUGGLING with another large dog his companion. Descen
544m102 tity,> place & personal connections-- ideas are STRUNG together in manner <they> quite different from
436e135 ce earliest times-- Apterix has a most perfect STRUTHIO head pulled out. yet feathers retain characte
340d029 s similar inter se.-- [not located] the «4» STRUTHNIONIDAE, Mr Blyth remarked that greater  difference
341d030 atory system; even much smaller than in other  STRUTHIOS. was adaptation to little Movement.-- nocturn
464t065 ly the Apteryx is more closely allied to the   STRUTHONIDAE than any other forms-- Lund's Antilope in B
464t065 eless= monkeys= Owen has described a greatt   STRUTHONIDOUS Bird from New Zealand-- <so> not an Apteryx
334d009 is fact & views.-- Fox says a cousin «one of Mr STRUTT» of his used to breed to Common & Muscovy Duck
048r087 ion submarine channels. such as that in front of STS. of Magellan In Chiloe curvilinear strata subsid
445e158 e than chicken pecking fly.-- "whilst the shell STUCK to its tail" as mentioned by Sir. J. Banks.  p.
147g015 where a large sort of <plain> space is thickly STUDDED with ridges & flat topped hill/ do alluvium. T
218b192 whole island is volcanic surmounted by water & STUDDED with others.-- we see a beginning to isld. Gra
548m114 --  <In a> Abercrombie's case of «in Botanical STUDENT» somnabulism, did reason about himself-- but n
102a060 hannels on coast of England-- Any one. who has STUDIED rocks in detail as amygdaloid. calcareous rock
109a084 nstone, diorite, &c most important.::-- must be STUDIED.-- Scientific Memoirs Edited by Taylor Ehrenbe
109a084 y Taylor Ehrenbergh on flints in chalk must be STUDIED-- though I do not think good p. 411 When discu
111a087 . 316 & 328 VI. p. 365. Meyen on Chile must be  STUDIED Analysis of Voyage: many observations on Volca
182b046 rule--  What subject has Mr Newman the (7) Man  STUDIED The condition of every animal is partly due to
252c042 t detail & fine, views about Species-- MUST BE  STUDIED: genera founded in nature [not located] The sy
254c049 taneous generation??) This whole Paper Must be  STUDIED.-- «Three p. 7. Am.» D'orbigny. Birds of prey,
261c069 t Swainson's remarks in Fauna Borealis must be  STUDIED. There is capital table of extent of all speci
296c184 nature.-- <p. 473» Webb &. Berthelot. must be   STUDIED on Canary islands-- Endeavour to find out whet
398e005 discovery  of the genius of man Those who have STUDIED history of the world most closely, & know  the
481z010 ion» of Zoolog. of Voyage of Astrolabe must be STUDIED for anatomy. of. corals.-- nevertheless the de
483z013 s All Owens papers on Intestinal worms must be STUDIED in Vol I, Zoolog: Transact. before writing on
484z014 merica?? Wilson N. American Ornitholog must be STUDIED before writing my general account-- ¿ Do not t
564n005 To   study Metaphysic, as they have always been STUDIED appears to me to be like puzzling at Astronomy
566n011 , as being really useful to them: this must be  STUDIED. before my view of origin of evil  passions.--
026r022 ve taken place in the Cordillera of S. America. STUDY Geolog: Map of Europe Conybeare. Introduct XII
060r125 t. 21st. 1817.--p 371. Webster Antarctic veg:-- STUDY Ulloa to see if Indian habitation above regions
073r162 should be those which have magnetic properties. STUDY well products of Solfataras[.] some general law
074r165 one, which H. calls by several secondary names «STUDY Hoffmans account of steam acting on  trachytes.
087a010 obably in salt marshes Efflorescence nothing -- STUDY account.-- Alluvial plains of Mississippi -- No
088a015 in Northern seas p. 312. Chamisso in Kotzebue.  STUDY Humboldt. Fragmens Asiatiques account of Americ
120a107 om of the sea?) All this profoundly considered.  STUDY Hopkins. theory of dikes may throw some light.-
122a113 ll consequences of EXTREME FLUIDITY of earth.--  STUDY different forms of earth as shown by arc.-- rea
130a133 .-- Surely we here have proofs of hot bottom.--  STUDY Bishoofs Paper.-- Weelsted told me of some larg
178b031 e Fr. naturalists thought was species <Ascensi> STUDY Lesson Voyage of Coquille.-- Dr. Smith says  he
191b079 pines, like on a porcupine on Echidna-- Good to STUDY Regne Animal for Geography.-- The motion of the
209b156 of  type (?).-- [«Mem:» Juan Fernandez]CD. From STUDY of Flora of islands; "ou bien encore on pourrai
225b220 a Archipelago very good on opposite tendency.-- STUDY Ellis & Williams. zoology of South Sea islds. a
226b220 land & Europe.-- It will be well worth while to STUDY profoundly the origin & history of every terres
227b228 Fossil  Comparative Anatomy, & it would lead to STUDY of instincts, heredetary. & mind heredetary, wh
228b228 s of change, which would then be main object of STUDY, to guide our <past> speculations with respect
233b251 on of Water & small of land or few quadrupeds.- STUDY Productions. of great Fresh water lakes of Nort
247c027 Friendly isles. p 50. LX. Journal of Silliman» «STUDY Silliman.--» Vol II. p. 10. it seems that Croco
252c045 what is species.-- The Collector is directed to STUDY localities of isld.--«immense importance of loc
257c057 tions-- of «nearly» same kind country distant. «STUDY> The circumstance of ground woodpeckers.-- bird
260c067 ny how temperate regions-- crows in N. America] STUDY Bonapartes list In the Zoological Journal I rea
262c073 ore especially the powers of reasoning &c &c.-- STUDY the wars of organic being.-- the fact of guavas
263c076 & whole fabric totters & falls.-- look abroad,  STUDY gradation. study unity of type-- Study geograph
```

263c076 otters & falls.-- look abroad, study gradation. STUDY unity of type-- Study geographical distribution
263c076 abroad, study gradation. study unity of type-- STUDY geographical distribution study relation of fos
263c077 unity of type-- Study geographical distribution STUDY relation of fossil with recent. the fabric fall
268c099 orest «Parrots in Macquarrie Isd.--» very good. STUDY D'Orbigny. & range on West Coast «Guayaquil & P
276c122 ing. without organs of generation?! By profound STUDY of local varieties laws of change-- whether bea
300c198 ellan.-- Change of habits in Van Diemen's land. STUDY Mr Blyth's papers on Instinct.-- His distinctio
311c226 doctrine of monsters is preeminently worthy of STUDY on the idea of those parts being most easily mo
315c243 ves accidental-- My first thought of sea side-- STUDY Bell on Expression & the Zoonomia, for if the f
316c245 lls there is more confinement. thus the Naiads (STUDY De Ferrussac) are confined to <S.> America.-- M
317c248 "The British Aviary" or Bird Keepers Companion STUDY Appendix (& only appendix) of Congo Expedition,
317c251 r well-worth studying, with respect to forms.-- STUDY Appendix to Tuckey's Expedition Journal of the
324c268 Entomological Magazine (paper on Geograp. range STUDY Buffon on Varieties of Domesticated animals see
325c267 totle to see whether any my views very ancient? STUDY with profound care. abortive organs produced in
328c1BC rithium Jardin du Roi Java fossils at same time STUDY Botanical work on Buds & Gemmae. C D Charles Da
335d014 edefary & therefore exceptions. to above law.-- STUDY what these monsters are:-- are they «abortive»
352d050 of life must be erect not pressed on paper, to STUDY the corresponding points.-- The present geograp
363d101 is part: this will be well worth working out.-- STUDY Temmincks work on Pidgeons--, & see whether feat
363d101 geon. (I suspect Pennant has described them)-- (STUDY horns of wild cattle.-- plumage of fowls-- long
390d177 & in seems connected with more developed forms) STUDY buds-- gemmae-- & monocotyledenous, do those wh
399e009 es: Look at whole Glacial period? [not located] STUDY introduction to Cuviers Regne Animal No structu
411e055 ividual steps in the series have been fixed, to STUDY the physical causes. «All Cuviers generalizatio
425e104 r Sandwich Isld are very similar to Galapagos-- STUDY Flora-- what general forms.-- are the Labiata n
442e153 gorze in Norway ought to be thus characterized STUDY Von Buch.]CD Now Mr Knights statements about fr
461t025 Most curious facts & this paper deserves fresh STUDY & whole order of the fish.-- Embryology p. 97.
484z015 traces of.-- like Puffinuria does of Petrel?-- STUDY Birds of Europe for other representatives of th
484z015 many Tyrannulae-- replaces warblers of Europe-- (STUDY profoundly shells of Bahia Blanca & Southern He
543m098 Hymenoptera;. <therefore> than in other orders (STUDY Kirby with this view) therefore there is Instin
543n099 typical insects. ie have all parts. Waterhouse STUDY well the greater number of insects in insecta--
545m105 es the idea unconscious, if so (think of this). STUDY what impressions become unconscious those which
564n005 ridge can run even with its shell on back.-- To STUDY Metaphysic, as they have always been studied ap
596n184 - "Thomas Brown" on Association worthy of close STUDY.-- full of practical observations Ourang do not
633j54v s. All flow from some grand & simple laws.-- 4 «STUDY Cuviers Anatomie Comparé» p 308. Traces the gra
637j57v enience.! extinction, utter extinction! let him STUDY Malthus & Decandoelle.-- The Final cause of inn
641j29r Naturalists must be cautious.-- <some others>- STUDY these facts read Lacépède on Cetacea & Geograph
198b114 ed on the same plan. [Second resumé well worth STUDYING] CD says grand idea god giving laws & & then
233b251 r & C. of Good Hope.-- His book Probably worth STUDYING.-- Wingless birds S. Continents-- Ostriches.
248c029 Rat belong to There is this great advantage in STUDYING. Geograph. range of quadrupeds.: that either
251c040 NB. Any monograph like Gould on Trogons worth STUDYING.-- «do» Zoolog Journal Vol 2. p 221. Horsfiel
317c250 gion, & some to Cape.-- some proper well-worth STUDYING, with respect to forms.-- Study Appendix to T
327c265 Mrs Necker on Education preeminently worthy of STUDYING in Metaphysical point of view Henslow has lis
379d151 t is land to continent-- Original Paper, worth STUDYING. Archiv. fur. Naturgeschichte. September 11
523m015 the sum-total of his accounts in his pocket, & STUDYING mathematics.-- My Father says after insanity
596n184 Passions.-- "Hartley" I should think well worth STUDYING-- "Thomas Brown" on Association worthy of clo
388d171 ew more semen to one child. more like father.-- STUFF.!-- How much opposed. the Quagga case appears t
495q05a pread sheets of Paper. covered with some sticky STUFF in flat places & see whether wind, on «dry» win
307c216 for instance where a tendril passes into a mere STUMP.-- Shall abortive organs «of very same kind» in
325c267 function has ceased to be used as tendril into STUMP Library of useful knowledge. Horse, Cow, Sheep.
515q022 s & of Chinese geese. (2) Anatomy of muscles of STUMPS of tailess dogs & cats.-- (3) Hounds-- varying
223b211 ell adapted to locality A. but it is instead a STUNTED & diseased form a plant, adapted to A. B. C. D
230b236 now this notoriously is not the case, you have STUNTED species, but not such as would make species (e
294c178 ccount of trees ceasing to grow far N. becomes STUNTED, altered, & lose (mere sickness)? fertility ¿b
056r109 other unworn islands) we take in at once the STUPENDOUS mass which has been corroded. -- If man coul
322c269 logy of the Human. Mind Evelyns Sylva. skimmed, STUPID Brownes travel in Africa; «well skimmed.» 1839
241c015 st of isle in E. Indi: Arch: In New Zealand. a STURNUS of American form-- a Synallaxis. ¿American?).
482z011 linnet not caught.-- Troglodytis Furnarius.-- STURNUS Magellanicus.-- p. 210. Scolopax very close to
061r126 .28. grt. drought at Sydney. which caused Capt. STURT expedition. -- ¿ Another one in 1816 (?).-- Mr
480z008 Corallinae from S. Seas written in German.-- STUTTGART ranks these bodies amongst Vegetables in Linn
182b050 s land. & Austral & New Zealand Mr Gould says in SUB-genera, they undoubtedly come from same countrie
192b083 tone Pippins, & Golden Pippen, goldens-- hence-- SUB-varieties & hence possibility of reproducing any
232b249 Lemurs in Madagascar, on neighbouring islets & a SUB-genus in Southern Africa In same manner. Cuscus,
232b249 us in Southern Africa In same manner. Cuscus, (a SUB genus of Phalangista New Holland form) is found
236b279 ocene Mammalia--of Europe Mem. Mr Bell's case of SUB Himalayan land emys, decidedly an Indian form of
278c129 bigny has travelled this will be most difficult. SUB-genera so far may be eliminated. where every spe
34vd052 .-- According to my theory, every species in any SUB-genus will be. descended from one stock, & that
458t001 .-- E Bengal Journal Vol 7. p. 658-- Falconer on SUB. Him. fossils-- Ruminants. & Tortoises gigantic
622o048 to action or not, are the parts of our nature (SUB> subject to their instincts & associations.-- or
038c058 ses for subaqueous crater being of diff: form SUBAERIAL one?--In former not so much; or no rapilli; &
364d105 -- The latter has crossed with the Ptarmigan. SUBALPINA in wild state.-- Neilson has given figure of
365d105 ed Grouse. a cross between Black Game. &, the SUBALPINA of Sweden, (which in summer dress somewhat re
365d105 uring known changes climate became unfit for. SUBALPINA, or some Northern species, & being restricted
365d106 ouse between Black Cock & Ptarmigan (probably SUBALPINA.) former has blue breast, latter reddish, hyb
038r058 he strongest argument:--¿ Consider causes for SUBAQUEOUS crater being of diff: form subaerial one?--I
109a082 tion page excised, not located] -- do-- [...] SUBAQUEOUS. removal, shown by the number of bones lying
109a083 edges of first-rate importance in showing not SUBAQUEOUS removal--??? the difficulty of such preserva
627o053 er" &c &c &c-- & if termed "selfish", must be SUBCLASSED as "disinterested" p. 14. It is allowed, tha
176b023 extend his domain into the other domains. & SUBDIVISION <six> three more, double arrangement.-- if e
185b056 Scarabadae, & longicornes.-- Again taking a SUBDIVISION of Heteromera same. thing occurs with regard
284c147 e external relations (a fixed quantity) & on SUBDIVISION of stations & diversity--.-- this perhaps on
176b021 g number of forms equable: this being due to SUBDIVISIONS & amount of differences, so forms would be
272c109 ls Rocks must be placed under Gould says most SUBGENERA confined to continent, though we have seen sp
272c109 to continent, though we have seen species «of SUBGENERA» scattered over it.-- We have abundant instan
272c109 structure prevails amongst the same species & SUBGENERA in families.-- thus the banded tarsi is commo
277c126 Scheme for abolishing specific names & giving SUBGENERA. true value.-- as in Opetiorhyncus. fulginosu
278c128 soon as two species were placed in different SUBGENERA, then it would be useless, but the formation
278c128 hen it would be useless, but the formation of SUBGENERA is empirical, & is judged solely by compariso
278c129 very species to another then those sections & SUBGENERA; are analogical, because we do not know, whet
349d051 ot be discovered «un»till <in> «we ascend to» SUBGENERA & families, <even in Cetionidae» «in the Ceto
349d052 ended from one stock, & that stock with other SUBGENERA will come from. common source.-- all genera, c
448e166 same numerical relation (both in species and SUBGENERA) between the Crag & Touraine beds, the one wi
350d053 edingl wrong to call, one group genus & other SUBGENUS, --> Propagation, best rule for genera, & so
039r060 t would appear he has not fully considered the SUBJECT.-- S. America in the form of the land decidedl
085a FC ose A. Geology Note on Woolwich Nothing on any SUBJECT As far as p. 33. distributed to several subjec
107a078 eflection on method of cooling--Very difficult SUBJECT. PP-- I think from dislocation taking place ch
114a094 ous- quartz. & iron stone alternating. bear on SUBJECT of cleavage Clay slate. a distinct formation d
118a103 ge of sublittoral deposit always equal width --SUBJECT of fine paper this would make.-- l'Institut. (
124a120 s being prolonged to surface. see p. 181 on do SUBJECT do p. 447 & 449. «& 450». On Vertical trees. U
149g031 power of sea N of Valparaiso are those animals SUBJECT to much variation which have lately acquired t
171b002 y such high object generation.-- We know world SUBJECT to cycle of change, temperature & all circumst
171b003 f living beings, become permanently changed or SUBJECT to variety, according «to» circumstance,-- see
182b046 bably only four, is not this Fries rule-- What SUBJECT has Mr Newman the (7) Man studied The conditio
205b042 Vol V. P II. p 565. Consult-- Says types most SUBJECT to vary where intermixture precluded.-- Kirby
263c075 es) <without> who just takes up & lay down the SUBJECT without long meditation-- His best chance is t
281c054 obtained.-- less firmly fixed & therefore most SUBJECT to change.-- may account for certain organs no
284c149 we may suppose it is oldest, & therefore lest SUBJECT to Variation.-- + <being> good for generic div
284c149 than among those groups, where it remains less SUBJECT to Variation-- Dr. A. Smith. knows lots of inst
286c156 t will be necessary from manner Fleming treats SUBJECT to put in alternative of Man created by distin
292c169 rs.-- The more complicated the animal the more SUBJECT to variation. therefore sexes or two animals:-
302c202 LATELY ACQUIRED. In pigs number of vertebrae. SUBJECT to variation. therefore lately acquired.-- I f

```
303c204 wonderfully produced».-- make abstract on this SUBJECT from Lawrence. Blumenbach & Prichard -- Now we
305c210 one  living spirit, prevalent over this word, (SUBJECT to certain contingencies of organic matter & c
307c216 rell «in Zoolog Transactions» & Hunter on this SUBJECT) because if so as she can be converted «into f
315c242 owest animals --habitual action, in intestines SUBJECT to sympathetic nerves-- The vividness of first
315c243 canine teeth at all.-- This way of viewing the SUBJECT important.-- Laughing modified barking., smili
326c266 Owen  has it.-- Ld. Brougham. Dissertations on SUBJECT of Science connected with Natural  Theology.--
346d048 see Annals. vol. 2. 1839.-- are bad breeders & SUBJECT to the rush as all animals which breed, in & i
351d057 ,-- Broderip alluded to Hunter's views on this SUBJECT.--Monstrosities, kind of determined by age of
371d127 e is produced in parents--colour is a doubtful SUBJECT, but what other instances are there of such ch
372d131 a  wound.-- Does likeness of twin bear on this SUBJECT? A mans arm would produce arm if supported.. <
382d158 uch simpler.-- Hunter shows almost all animals SUBJECT to Hermaphroditism,-- those organs which perfo
384d161 Each  part <not> of each species not similarly SUBJECT-- \Divides sexual marks into primary & seconda
385d163 ter.-- young take distemper very readily & are SUBJECT to fits.-- «there is great difference between
398e006 se only of the changes which the government is SUBJECT to.-- further back we obtain here & there in o
491q01v (not  being always useful). fail-- Really good SUBJECT for experiment.--«to repeat Spallanzani» Raise
501q011 reference  to what Mr. Herbert observe on this SUBJECT-- (31) Ask Henslow for list of annuals to plac
512q019 in different breeds--?? or in individual case: SUBJECT to disease in youth.-- Mr Tollett-- about sele
521m009 ralytic stroke> :. could converse well on any SUBJECT when once started,-- could receive a new train
522m011 ings if he began at one end, he knew the whole SUBJECT.-- if at the other nothing.-- He could  repeat
541m089 nd best plan was allowing my mind to skip from SUBJECT to subject as quick as it chose.-- although th
541m090 n was allowing my mind to skip from subject to SUBJECT as quick as it chose.-- although thinking «& t
548m116 id not recollect <it> «anything».-- if one was SUBJECT to this disease oneself, one would only feel s
558m150 h are most under great sympathetic nerve. most SUBJECT to habit, as being less so will.-- May not mor
573n038 s to make saliva flow, & therefore thinking of SUBJECT, even when wishing not to flow-- flow it will.
574n039 poetry.-- <signs> sounds singularly adapted to SUBJECT see <A> |\ I think this argument might be used
580n062 playing  so well.-- Lr. Brougham «Dissert.» on SUBJECT of science connected with Nat. Theology.-- say
586n081 also such.-- heredetary habit, is a part never SUBJECT to volition.-- like plants going to sleep.-- "
592n103 aper» must be referred to, if I follow up this SUBJECT & a reference to Brun's work.-- Shutting  eyes
612o035 ingencies, as much as growth of tissue and are SUBJECT to accident; the sexual willing comes on perio
617o39v by  anyone who wishes to fully understand this SUBJECT, but the answer to it would require a consider
622o048 ion or not, are the parts of our nature, <sub> SUBJECT to their instincts & associations.-- often fee
636j56v .-- It might be concluded that Plants would be SUBJECT to extreme variation as long as crossing  with
230b240 d Isd case good one of animals not soon being SUBJECTED to change in Americas. perhaps merely gone ba
430e118 cal varieties, when whole mass of species are SUBJECTED to some influence, & this would take place fr
438e142 d breeds both plants & animals have long been SUBJECTED to domestication.-- the constitution of  some
616o39v , revealing respectively what are called it's SUBJECTIVE & objective aspect. The subjective aspect of
617o39v alled it's subjective & objective aspect. The SUBJECTIVE aspect of bodily action is revealed to us by
617o40v rom which it arises. But coming round to the <SUBJECTIVE> aspect of action as known by the exertion o
618o041 n of it Thoughts, perception &c. are modes of SUBJECTIVE action-- they are known only by internal con
618o041 rom his its behaviour. Thought is only known SUBJECTIVELY? -- ? the brain only objectively. We do not
085aIFC ubject As far as p. 33. distributed to several SUBJECTS. Feb 24th 1839 As far as p 140-- abstracted a
093a033 1877  (Memoranda so far distributed to various SUBJECTS) Dr. A. Smith informs me that in the year a R
302c202 t evil» from vast opposition in opinion on all SUBJECTS of classification, I must work out hypothesis
322c270 d several papers-- all, that bear on any of my SUBJECTS Elie De Beaumonts. 10 Vol. of Memoirs on Geol
398e005 that  geology. by giving proper ideas of these SUBJECTS. should be absolutely necessary to arrive  at
606o024 o be added. Lessings Laocoon p. 125-- says new SUBJECTS are not fit for painter or sculpture, but rat
606o024 e not fit for painter or sculpture, but rather SUBJECTS which we know, it is therefore the embodying
627o52v ency of affections.-- If ever I write on these SUBJECTS consult <following> pages. <p. 231> marked in
349d052 enus when it is so many steps from a head, as SUBKINGDOM.-- -- evidently artificial, as interloperment
093a032 o the tertiary limestones of Vendarques. Mem SUBLIMATION of sulphur to form salts of America. --  The
195b101 n certain country, but how much more simple, & SUBLIME power let attraction act according to  certain
398e004 f the geological chronology depends, that most SUBLIME discovery of the genius of man Those who  have
553m137 her in Ancient Greeks, with their mystical but SUBLIME views, or the wretched fears & strange superst
578n057 hat is emotion? At end of Burke's essay on the SUBLIME & Beautiful there are some notes. & likewise o
602o11v My idea. would make the mind have mysterious & SUBLIME ideas independent of the senses & experience p
604o018 between  dreams & insanity.» D. Stewart on the SUBLIME The literal meaning of Sublimity is height.  &
604o018 d with God. these phenomena we (feel & ?) call SUBLIME.-- 4 From the association of power &c &c  with
604o018 wer &c &c with height, we often apply the term SUBLIME, where there is no real sublimity 5 The emotio
605o18v s of terror & wonder so often concomitant with SUBLIME. adds not a little to the effect: as when we l
605o18v &  thus we apply to them the metaphorical term SUBLIME 7 So that in this Essay. D. Stewart does not a
605o019 those sensations, which we call metaphorically SUBLIME, but that it is through a complicated series o
605o19v rom the associations before mentioned. we call SUBLIME.-- It appears to me, that we may often trace t
093a032 iders that Mercury & Sulpuret of Iron has been SUBLIMED into the tertiary limestones of Vendarques. M
571n031 Windsor  Castle which rises naturally & hence SUBLIMITY from natural rise-- I was also much struck in
579n059 nt movement of ears A man shivers, from fear, SUBLIMITY, sexual ardour.-- a man cries from grief, joy
579n059 xual ardour.-- a man cries from grief, joy. & SUBLIMITY. January 6th.-- What passes in a man's  mind.
604o018 Stewart on the Sublime The literal meaning of SUBLIMITY is height. & with the idea of ascension we as
604o018 pply the term sublime, where there is no real SUBLIMITY 5 The emotions of terror & wonder so often co
116a100 clear  gold occurs in submarine alluvium, or SUBLITTORAL formations. p. 150. at Portezuelo, extremity
118a103 e cliffs. Sir L. Dick says (.p 52) fringe of SUBLITTORAL deposit always equal width --subject of fine
024r014 the  existence of some moving <point> power ¿ SUBMARINE currents Find instances; The whole coast of N
036r051 ll.-- The absence of Second form, except near SUBMARINE Volc: in harmony with the prevailing movement
042r068 es. «on Chili & delta of Indus», my belief in SUBMARINE tilting alone, must be modified. «Moreover, t
048r087 substances  being worn into channels. mention SUBMARINE channels. such as that in front of Sts. of Ma
116a100 rounded  pebbles-- it is clear gold occurs in SUBMARINE alluvium, or sublittoral formations. p.  150.
127a127 of cones beneath sea.-- with reference to old SUBMARINE orifices in Cordillera Geograph. Journal  vol
131a135 p. 360. on orbicular trap thought to be bombs SUBMARINE L'Institut 1838 p. 400. Observations on Mount
160g088 ascending fringes {P} which makes me think i SUBMARINE, 400 or more feet above station! There is lon
485z017v Mag: of Zoolog & B. Vol. II. p. 127. List of SUBMARINE insects. Staphylinidae &c &c. with  reference
302c203 of genera may sometimes be due to accident as SUBMERSION of land containing all of intermediate Fathe
223b210 make that sudden step. species or not. A plant SUBMITS to more individual change, (as some animals do
305c210 «each  having acting principle» according to SUBORDINATE laws.-- There is one thinking « <& Creat» se
384d162 elves (this never happens in plants «only in SUBORDINATE manner in the plants which have male & femal
584n071 lting & breeding instincts, sexual, social, «SUBORDINATE to,» self preservation, (knowledge of enemie
049r087 r ones in Clay. Slates contemporaneous others SUBSEQUENT. as in dikes. In Granite great crystals arra
103a065 metallic dikes: It is an important view being SUBSEQUENT to dislocation of strata. A capital discussi
105a070 ater will account for filling up of valleys-- SUBSEQUENT opening a medial gorge by slow erosion.  but
151g044 he upper shores may correspond with some line SUBSEQUENT to shelf {P} In Glen Collarig, by Dicks theo
181b044 t which certainly is the case at least during SUBSEQUENT ages.-- The Creator has made tribes of anima
385d163 n many cases of bitch going to mongrel, & all SUBSEQUENT litters having a throw of this mongrel.--  I
389d176 gnate one or many offspring.-- it affects the SUBSEQUENT offspring, <when> though other male may have
393dIBC ) that  one copulation with other dogs renders SUBSEQUENT progeny faulty. Does male fail in passion.--
426e106 ect first race, as it does indelibly the many SUBSEQUENT ones. My views, «V <see> p. 103» would  lead
568n017 intimately united with reason) it would feel «SUBSEQUENT» sorrow, whatever the cause had been]CD-- «A
580n061 t when an idea once take hold of the mind, no SUBSEQUENT ones modify it.-- «Weak people say I know it
620o045 mal.-- which shows that. it is owing to some <SUBSEQUENT» power (reason) obtained by age. which shoul
068r146 movements,  and therefore beneath ocean, for SUBSEQUENTLY there is a coating of solidifying igneous r
129a131 se.--¿ does it bear on Patagonia? «Facts about SUBSIDED forests.-- Many repeated oscillations» Hitchc
226b222 If  on one isld several species of same genus, SUBSIDED land.-- Mauritius? «In plants where do most s
036r051 eing one of elevation alone.--In England much SUBSIDENCE: hence difference; action on land  different
039r060 re of Ascension» give as an example the great SUBSIDENCE at the famous eruption of Rialeja, & the mor
048r087 Sts. of Magellan In Chiloe curvilinear strata SUBSIDENCE.--The sudden increased dip is not parallel c
053r100 . applicable to Patagonia. During a period of SUBSIDENCE the shinglle of Patagonia would become  more
065r140 n springs at S. Cruz.???-- Form of land shows SUBSIDENCE in T. del Fuego, and connection of quadruped
065r140 h recent elevation, there may have been great SUBSIDENCE previously. Mem. pebbles of Porphyry.--Falkl
069r152 omerate between two ranges mysterious!-- Mem. SUBSIDENCE Uspallata of which no trace except by  trees
070r154 hat except by trees, I could not see trace of SUBSIDENCE at Uspallata.-- ¿If crust very thick would t
105a071 s above their parent rocks. would be prove of SUBSIDENCE.-- removal downwards by successive torrent s
```

Page **(Key Word)**
106a076 lating rivers in Southern America ¿ effect of SUBSIDENCE-- <Is there same.> Institute. 1838 p. 40 or
108a080 en sinking & rising areas.-- In Earthquake if SUBSIDENCE we should not expect volcanos.-- not so much
110a085 these may cause «or be effect of» elevation & SUBSIDENCE. examine these «lines» Description of rocks
112a089 f rock-- -- Geograph Journal Vol IV p. 36. on SUBSIDENCE of the land in Guiana, worthy of considerati
113a092 erfectly still water. Major Mitchell inferred SUBSIDENCE; Mem my remarks on coast of Australia.-- Gre
114a095 nstone, if we know origin of greenstone tells SUBSIDENCE as plainly as Temple of Serapis. (now we hav
124a118 t have been in same condition for long period SUBSIDENCE in Demarara p. 131 (B.) Wrong Entrance. Book
127a127 n.-- great mountain chains. may be effects of SUBSIDENCE Elie de Beaum. Memoires of French Geolog. Ca
128a128 vice from effect of general elevation,-- when SUBSIDENCE takes place.-- Mountain will first fall-- th
128a129 ill not abrasion of land on one side. produce SUBSIDENCE of water on other. from tendency to regain s
129a132 1838 p. 791 -- Most curious account of great SUBSIDENCE «20 miles long I in with.» which must have b
130a134 e fracture> A Elevation as in Patagonia {P} B SUBSIDENCE; <as in> be cautious. mud banks & sand. dune
134a141 trees on Isthmus of Pen. Tres Montes.-- as by SUBSIDENCE ‖ Fitz Roy refers to ‖ & Rocks p. 375. on th
191b082 Owls. transport mice alive? Species formed by SUBSIDENCE. Java & Sumatra. Rhinoceros. Elevate & join
191b082 n keep distinct. two species made elevation & SUBSIDENCE continually forming species.-- Man & wife be
191b083 sion the extraordinary South Africa. proof of SUBSIDENCE. & recent elevation: Pray ask Dr. Smith.-- «
202b131 fty feet, then forest 120 ft Micaceous rocks. SUBSIDENCE appears indicated.--- p. 36.-- Geograp. Jour
225b219 Mauritius what a difficulty-- where elevation SUBSIDENCE New is only hope.-- New Zealand «compare to
284c148 wl up an hill, then by deaths?!)-- looks like SUBSIDENCE.-- on the islets Mr Blyth remark that a rese
309c220 luence) & mankind must improve-- The areas of SUBSIDENCE marked out by animals of same <spe> genera.
408e045 at quarter of the world-- -- Mem. elevation & SUBSIDENCE of East Indian Archipelago. now rising. On a
424e101 es., which according to Beck are different.-- SUBSIDENCE of Greenland-- case of splitting of two regi
427e109 ht to be preeminently artificial.-- Would not SUBSIDENCE of Greenland render climate less extreme. (&
473s004 ca as in Britain. If there has been «as» much SUBSIDENCE as elevation then all continents of cretaceo
584n071 CD[.instinct may be divided into migration,-- SUBSIDIARY to food & temperature molting & breeding ins
045r077 P. Scopes explanation of low Barometer? In a SUBSIDING area. we may believe the fluid matter instead
096a039 celand parellel to Greenland: Mem.¿ Greenland SUBSIDING.) Von Buch Canary Isd. p. 351.. NB. Mackenzie
313c233 the species changing « ∴ because mammalia can SUBSIST where parasites cannot» Read Entomological Tra
048r086 In Patagonia, are all beds same age? is white SUBSTANCE triturated Porphyritic rock. s (mem white tuf
101a056 e themselves in that direction, in which most SUBSTANCE lies <.-->? Phillips. Lardner's p. 270-4, goo
102a061 place most gradually, first the more fusible SUBSTANCE, & then the next being sucked out. In Cleavag
139a180 fting length of life &c &c Will any inorganic SUBSTANCE cause such monstrous growth as oak galls or r
163g109 tion of British Plants Shropshire Quartz what SUBSTANCE is collected in little spots Speculate on «un
536m071 s smell of,∴ Hyaena likes smell of that fatty SUBSTANCE it scrapes off its bottom.-- it is relic of s
029r034 . These thick beds of Lignite stratified with SUBSTANCES so like the Coal measures in England (Except
032r041 mestone in intertropical)» hence varieties of SUBSTANCES ejected from same point. & changes. «(change
048r086 stones of P. Desire). = Where talking of such SUBSTANCES being worn into channels. mention submarine
051r094 Kelp») rocks show signs of degradation; (soft SUBSTANCES worn into bare cliffs evident); the action i
056r111 <Also Volcanos separate.> Volcanos blend all SUBSTANCES together; & products being similar over whol
102a061 r relations to VEINS in primitive rocks-- Are SUBSTANCES soluble under great pressure? equally with l
102a061 ssure? An important question! If water yields SUBSTANCES from impact, «it» would look like it. Are gr
384d162 ifying principle) one individual secretes two SUBSTANCES, although organs for the double purpose are
458t002 ish birds, using cotton &c instead of natural SUBSTANCES-- useful perversion of instincts-- Beechey's
516q23v xtract of peat do to preserve fungi or animal SUBSTANCES-- (Athenaeum (40) p. 823 chemical analysis o
025r018 art of the globe: & when we see conclusions SUBSTANTIATED over S. America & Europe. we may believe th
026r023 culiar cream-coloured Limestone: the strange SUBSTITUTION of matter in shells, like Concretions & lam
390d178 . ideosyncrasy or dispositions were added or SUBSTRACTED at each, or in several generations, the proc
046r078 mstances; & not in chemical nature, or has a SUBTERRANEAN fluid mass itself changed.--No. -- Yet the
062r128 . from nature of climate. = Perpetual snow.--SUBTERRANEAN lakes, near Volcanoes. lakes of brine all i
107a077 . Phil: Transacts.-- Does the isothermal {P} SUBTERRANEAN line moves upward from effects of Elevation
127a125 was cooled to greater depth.-- Now the <inf> SUBTERRANEAN isothermal line must be creeping «pushing»
131a135 raph. Journal Vol. 8. p. 402.-- ground ice-- SUBTERRANEAN isothermal line Athenaeum. 1839. p. 52. On
132a139 th a very bad conductor.-- shows p. 516 that SUBTERRANEAN springs give result less to be trusted than
259c062 reation of animals.-- then argumen May be.-- SUBTERRANEAN lakes, hot spring &c &c inhabited therefore
262c073 division of lesser & more power (2.typical 3.SUBTYPICAL) where power arbitrary. leaves door open for
502q012 . (34) Any recent information about pollen of SUBULARIA Royle & Horsfield (35) Talk about races of Ba
593n109 man in place of other animals is hostile «is SUBVERSIVE of» to this view, & fowls hatching stones. i
045r077 at of a spring) moves away.--Will geology ever SUCCEED in showing a direct relation of a part of glob
207b151 the grass.-- Beatson St. Helena says no trees SUCCEED so well at St. Helena. as. Pineaster & Mimosa
254c048 during which the changes of the ovum or embryo SUCCEED each other with the greatest rapidity"-- so we
279c133 see marks on pages).-- Crosses of diff: breeds SUCCEED, yet seems to grant, that difficult & other go
554m138 the door to force it open, when she could not SUCCEED of herself.-- <The male> «I saw» Jenny untying
232b245 process of change of those forms, which have SUCCEEDED in becoming habituated to colder climate whil
254c048 uccessively present modifications, typical of SUCCEEDING classes & likewise those much higher in scal
469t135 ed very many flowers.= 22d.-- /during several SUCCEEDING days <many> «most numerous» bees visited thi
434e127 e Nisus formativus. (what does Muller call it) SUCCEEDS in altering <or> form of body, or whether it
608o026 wer not in proper state.-- When the admonition SUCCEEDS who does not recognize an accidental spark fa
418e084 forms, are closely related-- By birth the the SUCCESIVE modifications of structure being added to the
063r133 ave all run wild & bred. no doubt with perfect SUCCESS.--showing non Creation does not bear upon sole
541m089 to unbend my mind as much as possible (testing SUCCESS by decreasing headache) & found best plan was
105a071 chain of lake. <a> the alluvium would form a SUCCESSION of flights of steps; if one lake then we mus
177b026 - this again offers contradiction to constant SUCCESSION of germs in progress.-- «no only makes it ex
191b083 s. Here we have avitism the ordinary event. & SUCCESSION the extraordinary South Africa. proof of sub
227b225 erviceable. We may speculate of durability of SUCCESSION from what we have seen. in old world, & on a
248c033 ts» 2. Relation, of external conditions, & of SUCCESSION: the <first> latter is most intimately conne
291c167 s & others not??? ‖ The death of some forms & SUCCESSION of others, (which is almost proved. Elephant
294c177 in England killed in year 1000. reference to SUCCESSION of types ¿different species;-- Horse-- &c <L
313c235 ny) where several generations are produced in SUCCESSION (13?) without impregnation, therefore sexual
318c254 an's Journal 1837. Paper by Bachman.» that is SUCCESSION of birds.-- «in» some species «a Tanagra» Ma
336d018 blood long, it becomes part of animal &» by a SUCCESSION of <such changes> generations, these small c
343d037 the time of the Silurian, he has made a long SUCCESSION of vile Molluscous animals-- How beneath the
393dIBC or variation common to any zoophyte «born in SUCCESSION» which is not transmitted by generation?? Is
432e125 necessary to form heaps of pebbles &c &c: the SUCCESSION of organisms tell nothing about length of ti
432e125 l nothing about length of time, only order of SUCCESSION.-- Splendid Pamplet. (published in Philosop.
579n060 ourse p. 35. On origin of idea of causation; «SUCCESSION of night & day does not give notion of cause
615o038 finite instincts modified by heredetary;-- so SUCCESSION so perhaps general ones.-- Parental feelings
465t079 nt «in Europe we know there has been several SUCCESSIONS of Mammals.--» yet only two monkeys, <there
032r040 rounded on the belief of constant rising with SUCCESSIVE periods of greater activity & rest.--Such ch
039r060 yell. P. 111 & 113. «seems to» considers that SUCCESSIVE terraces mark as many distinct elevations; h
041r067 the appearance of travelling may be owing to SUCCESSIVE transportal from prevailing swell, (as Shing
058r116 first it would though be easy to see on beach SUCCESSIVE lines of sea weed-- Histoire Naturelle des I
105a071 e prove of subsidence.-- removal downwards by SUCCESSIVE torrent spread out. by sea-- beach action --
113a091 r than <5000.> 400.-- These facts of SLOW but SUCCESSIVE transmission of temperature clearly prove po
116a100 r in Chile in beds of river, but in shelving «SUCCESSIVE» banks <above> 30 ft or so above bed of rive
148g027 d-- where side ravine entered terraces formed SUCCESSIVE bays but plains sloped centre-wards which wo
153g051 thers below it--argument for lake «or sea» at SUCCESSIVE levels-- {P} Shelf opposite Glen collarig at
175b019 t former epoch-- How is this Ehrenberg? every SUCCESSIVE animal is branching upwards different types
227b226 sical change between one group of animals & a SUCCESSIVE one.-- It leads to knowledge what kinds of s
292c171 fore it can only be those actions, which Many SUCCESSIVE generations are impelled to do in same way--
344d041 eral births analogous to superfoetation, & to SUCCESSIVE fertile offspring in Entomostraca & Aphides
371d128 not be transplanted?, & yet year after year, SUCCESSIVE roses & bud are produced, like parent stock,
384d161 ee or rose degenerate during its life so that SUCCESSIVE buds do differ-- any variety is not handed d
387d168 above law anyways connected with the case of SUCCESSIVE copulation impresses offspring more & more w
434e127 ly has tendency (as effects of cultivation on SUCCESSIVE generations of plants) to do so, the effects
495q005 ich soil, will plants abort?, does it require SUCCESSIVE generations to accustom them to such soil.--
632j53r f propagation, were created with reference to SUCCESSIVE developement I admit, but the admission is p
115a098 of oscillations & currents.-- if matter was «SUCCESSIVELY» given of every degree of fineness. then mo
254c048 st rapidity"-- so we find species each class SUCCESSIVELY present modifications, typical of succeedin

Page
(Key Word)
585n075 r bodies, «or touches one part. very quickly SUCCESSIVELY.--» [& we know from experiment of crossing
181b040 hat any of <his relatives shall likewise> the SUCCESSORS of his relatives shall now exist,-- In same
195b101 fixed laws of generation, such will be their SUCCESSORS.-- let the powers of transportal be such & s
206b146 go, there were 200 people living who now have SUCCESSORS.-- Then the chance of 200 people <might be>
206b147 (of all races) not more than a few will have SUCCESSORS. at present day. in looking at two fine fami
206b148 day. in looking at two fine families one with SUCCESSORS <for> centuries, the other will become extin
313c236 ry important in considering how children come to SUCK or other actions in foetus of Mammalia, or chic
382d158 bee would shew this.-- (Do any male animals give SUCK)--But this not distinctly stated by Hunter.-- D
467t105 l Bee; saw another on Cabbage--white Butterflies SUCK nectar: «Maer June 41» Rhubarb. pollen very min
102a061 e more fusible substance, & then the next being SUCKED out. In Cleavage discussion, state broadly ind
472s02v ase pollen.-- the pollen appeared chaffy, as if SUCKED?! opens & shuts end of sucker, after having wi
472s02v .-- <I think> When It first alights, it cleaned SUCKER & <I think> pollen was scraped off, which appe
472s02v red chaffy, as if sucked?! opens & shuts end of SUCKER, after having withdrawn it.-- Saw 4 more Bees
261c070 nstance of connate instinct, better than child SUCKING or even ducklings & fowls-- When talking of ra
344d041 ey are eating foetuses, as young of Marsup. is SUCKING foetus.-- August 23d The Rev R. Jones gave an
537m075 n. "relict of bad habit." as child is cured of SUCKING his finger by rubbing them with alum, so more
544m101 d pleasures & pains & emotions-- such as child SUCKING, gives pleasure, & always has done therefore s
586n079 habit) could be formed or afterwards.-- child SUCKING whole wonder instinctive.-- carrier pidgeon ju
595n125 l as only means of communication) in man, than SUCKING.-- [I assume a child pouts who has never seen
614o037 ses cannot be acquired by experience for child SUCKING.-- And is it more wonderful that memory should
639j28r on like power in Octopus & Chamaelion.-- C. D. SUCKING feet in Frog. Walrus. Fly. Gecko &c. Prehensil
048r087 n In Chiloe curvilinear strata subsidence.--The SUDDEN increased dip is not parallel case to Isle of
050r091 ler the water & nearer the Banks Is there not a SUDDEN deepening on E. coast of Africa. as at Brazil
181b044 <some> great main type, & the gradation will be SUDDEN-- Heaven know whether this agrees with Nature:
223b209 ll cross (perhaps more fertility & so make that SUDDEN step. species or not. A plant submits to more
343d039 ells of S. America.-- there is no appearance of SUDDEN termination of existence.-- nor is there in th
428e113 system» I am bound to insist honestly that the SUDDEN, change from Primrose to Cowslip is great diff
429e113 hout recurrent tendency in external conditions» SUDDEN loosing of horns.-- I do not believe this Natu
522m013 t acted upon by Emetics.-- people recognized,-- SUDDEN changes of disposition, like people in violent
531m053 -- Now every animal moves quickly away from any <SUDDEN> sound or noise, & therefore brain has been acc
537m076 ms to have changed suddenly-- but are not such <SUDDEN> changes rare,-- as when Polynesian mothers ce
558m152 ng noise. Puma did same & & some others-- Thus <SUDDEN> «forcible prolonged» expulsion of air «dogs s
560mIBC istory of Babies-- Do babies start, (ie useless SUDDEN movement of muscle) very early in life Do they
577n052 on separate organs most curiously shown in the SUDDEN cures of tooth ache before being drawn,-- My f
025r016 coast. Beyond the 10 or 12 leagues sea deepens SUDDENLY. coast of Brazil generally.-- Mrs Power at Po
032r039 waves would not reach. If now the ocean should SUDDENLY fall, (3) the case would be as at first. & ac
059r123 melted with little pressure. & perhaps cooled SUDDENLY.-- As the rude symmetry of the globe shows po
230b239 s taken place, Nature [not located] Any change SUDDENLY acquired is with difficulty permanently trans
332d001 ing over the head, not caused by a wound, when SUDDENLY during one time he had three patients at very
401e016 Now this is curiously different from primrose SUDDENLY produce cowslip, one is tempted to think have
437e141 ce.-- May 4th.-- The Brussels Sprout returning SUDDENLY to type when brought back to home. (& yet all
437e141 ls Sprouts) is analogous to Primrose & Cowslip SUDDENLY changing into each other, & depends on charac
446e162 ose.-- «A fruit tree by certain treatment will SUDDENLY send forth quantities of blossoms--» The case
472s02r them. Yesterday remarked that many flowers had SUDDENLY withered, & to day saw very odd dusky humble
523m017 stinguished from whims passion &c by coming on SUDDENLY. Ans no.-- because often, if not generally, d
523m017 ten, if not generally, does not really come on SUDDENLY.-- Case of Mrs. C. O. who threw herself out o
523m018 at all discovered.-- <Sup> Sometimes comes on SUDDENLY from <I> (in one case ipecacuhan-- not acting
523m018 cold drink.-- then brain affected like getting SUDDENLY into passion.-- There seems no distinction be
524m019 ent for materialism. that cold water brings on SUDDENLY in head, a frame of mind, analogous to those
537m076 w, where the moral sense seems to have changed SUDDENLY-- but are not such «sudden» changes rare,-- a
546m110 vory paper cutter, which she valued, & she was SUDDENLY called to go on the lawn to see something, on
546m111 t-- ten years afterwards whilst at a meal, she SUDDENLY like a flash without any assignable cause, se
578n055 - "as <she> «the» thought of his knowing «it», SUDDENLY came across her, the blood rushed to her face
593n111 ns feeding on grassy bank some way from water, SUDDENLY, as if by word of command, they all took flig
357d072 rule.-- S. America & Australia appear to have SUFFERED most with respect to extinction of larger for
228b232 our fellow brethren in pain, disease death & SUFFERING «& famine»; our slaves in the most laborious
548m116 case of double consciousness, one would pity SUFFERING in one state almost as much as in the other,-
548m116 ould only feel sympathy. as for the heard deaf SUFFERING of a dear friend-- this gives one strong idea
567n013 's face, now ill, has expression of languor & SUFFERING The Cyanocephalus when fondling the keeper.,
423e098 ohn believes one single turnip in a garden is SUFFICIENT distinction is given to angular & rounded.--
026r020 In De La Beche, article "Erratic blocks" not SUFFICIENT proof, that the sea formed these large cavit
055r106 & smoothed like those washed by the waves, a SUFFICIENT for one birth or rather]CD ▯It should be obs
389d173 sts of infinite number of globules: generally SUFFICIENT only to have most slightly modified organic
423e099 sited in a period (say 10,000 years) which is SUFFICIENT for hence it must come way be able to measur
565n008 ant, & when pride makes person extremely self-SUFFICIENT,-- the corner of lower lip are depressed & o
583n071 cell in certain way, which way its organs are SUFFICIENT to give up my theory-- Viewing from eminence
599o007 inexplicable-- I cannot <admit> think reason SUFFICIENT to ensure propagation of Misseltoe?-- do p.
604o016 e, it looks as if consciousness an effects of SUFFICIENT perfection of organization & if consciousne
622o049 n their nature.-- So that we have some, it is SUFFICIENT to give rise to the feeling of right & wrong
632j53r ht that the intestines of a thrush were means SUFFICIENT to ensure propagation of Misseltoe?-- do p.
302c201 classes will be those of analogy, which when SUFFICIENTLY Multipliied become affinity yet often retai
437e139 es it seems hares, rabbits, rats & not being SUFFICIENTLY weakened by wounds got off from the young o
445e159 152.-- Philosophie Zoologie. says it is not SUFFICIENTLY proved that any shell fish is really hermap
529m040 poetical. (V. Wordsworth about science being SUFFICIENTLY habitual to become poetical) the botanist m
344d041 ects as glowworm The case of one impregnation SUFFICING to several births analogous to superfoetation
486z020 » Vol of Archaeologia arrow=heads described in SUFFOLK as lying under strata of gravel & clay about 1
515q021 ed, as the commoner kinds-- Cattle are horned, SUFFOLK have <abortive> «no» horns by abortion, but so
308c217 nconsciously.-- why do absent «Dr. Black. tea & SUGAR» people. reverse habits Insects & birds are the
503q012 low or Royle, latter says seedless-- Also about SUGAR-Cane Edwards says does not seed-- «Bruce says d
587n087 essarily disagreeable to organs adapted to like SUGAR, acid, &c, which may be doubted for possibly ev
068r145 cked harbors to Craters of elevation.-- Lyell SUGGESTED to me that no metals in Polynesian Islds--. V
098a046 with a crystalline process.-- now cleavage as SUGGESTED by Sir J. Hershel is all crystals obeying one
124a18t besy on mountain Chains in N. America Erasmus SUGGESTED to me that Herschel's theory offers no explan
210b161 Zoolog. Proceedings▯. <p> 1832.p. III Mr Owen SUGGESTED to me, that the <cas> production «of monsters
325c266 roying children. --it is not effect, as Lyell SUGGESTED, of organ being worn out as. otherwise old wh
463t057 climbing & mud-walking fish? difficult-- yet SUGGESTED, (vipers tooth also a difficult), the whole m
442e153 it get there? whether> According to the above SUGGESTION my theory would require, that <species> <wil
075r166 t. (& Chiloe do) no veins discovered. Humboldt SUGGESTS covered up by volcanic rocks. //St Helena vic
591n099 es.-- The general «(as I believe)» contempt at SUICIDE. (even when no relatives left to lament) is ow
327cIBC ly some account Raffles. Sir. S do. do-- Buffon SUITES Cline on the improvement of domesticated anima
344d040 tables on Geograph distribution of reptiles in SUITES de Buffon.-- Vigors has given list in Linnaean
390d179 ould be observed that from Books to read Buffon SUITES de.-- Horse & Cattle Library of Useful Knowled
089a018 mens de terre sont presque toujours precedes et SUIVIS, queque temps avant et apres, par de petites s
430e119 ctly intermediate between I. concentricus & I. SULCATUS.-- the beak of this one has concentric striae
430e119 rt rayed longitudinally (give woodcut) like I. SULCATUS-- Both species are found at Folkstone.-- it
075r167 ied by molybdated lead & «argentiferous lead»; SULFATED Barytes very «un»common in Mexico. Fluor spar
502q11v about N. Zealand, as Cuculus lucidus is.-- Ask SULIVAN about Falklands Isds-- Snipe Migratory-- prob
264c079 - to those it knew.-- see its passion & rage, SULKINESS, & very actions of despair; «let him look at
553m137 en as keeper spoke to it,-- but he thinks not SULKINESS-- this expression he believes is common to th
565n009 on is remarkable. the pouting, & blubbering-- SULKINESS is same as pouting, <but> lesser in degree, n
565n009 ht, no compression of mouth showing action,-- SULKINESS all negative expression? Expression of affect
553m137 ed baboon, showed out its lip, looking absurdly SULKY «as» often as keeper spoke to it,-- but he thin
594n113 outing child same as anger, lips not compressed SULLEN, protruded. determined to do nothing. & so man
594n113 d. determined to do nothing. & so manifesting SULLENESS.-- [blank] Circumstances having given to the Be
044n074 pressure of gazes. especially the most abundant. SULP. Hyd: Carb: A. Mur: A. = (& this effect of wate
073r162 some general laws. association of lead & silver. SULP. of Barytes: Fluoric. Barytes:-- Humboldt. New
076r169 native silver in Mexico is always accompanied by SULP. silver sometimes by selenite.-- in New Spain,
057r115 he country.-- Sir J. Herschel. says. precip. of SULPH. B. all the infinitesimal cryst. arrange themse

(Key Word)*
064r137 volcanos differ from all others in quantity of SULPH. acid emitted: mem: Grand gypseous formation of
079r176 . At Huantajaia. Humboldt says, mur of Silv.[,] SULPH. of do.[,]galena[,]quartz, Carb. of Lime. accom
127a126 ral fluidity.-- do p. 137. Lord Tullamore found SULPH of Soda in peat ashes in Ireland dikes in mount
046r078 ting the first production of Trachyte. look at SULPHUR. salt. lime, are spread over «whole» surface;
046r078 flow out together? How are they eliminated.--«SULPHUR last.--» Metallic veins likewise must separate
056r111 irculation. But Volcanic action separates some SULPHUR (perhaps lime) salt. & metallic ores.--which m
073r164 zeolite. stilbite. grammalite. pyenite. native SULPHUR.. fluor spar. bayte. asbestos garnets.--carb &
093a032 y limestones of Vendarques. Mem sublimation of SULPHUR to form salts of America. -- The number of min
123a116 ust be like the Chilian ones.-- Septemb. 2d.-- SULPHUR like carbon must go round of dissemination & s
467t105 taphylinidae on all their bodies pollen-- on a SULPHUR Broccoli not many do-- pollen not very abundan
075r167 tive silver & brown oxide of Iron in Mexico. SULPHURETTED silver, arsenical grey copper, and antimony
276c124 structure, then Milvulus forficatus Tyrannus SULPHUREUS if compelled solely to fish. structure would
034r044 itic fragments P. 386 Mem. Lyell's fact about SULPHURIC vapours in East Indian Volcanos Gypsum Andes
265c084 t effect, <of> according to Physical laws, as SULPHURIC acid disorganizes wood, but adaptation.-- alb
618o041 We cannot perceive the thought attraction of SULPHURIC acid for metal of another person at all, we c
093a032 Institut. 1837. p 331 Considers that Mercury & SULPURET of Iron has been sublimed into the tertiary l
282c140 to some external contingency.-- affinity is the SUM of all the relations, analogy is the close relat
523m015 verty.-- his manner of curing it. by keeping the SUM-total of his accounts in his pocket, & studying
549m120 at every one who has tasted them, will think the SUM total of happiness greater. even if mixed with s
050r090 nzie must be worth reading Some earthquakes of SUMATRA no connection with a neighbouring Volcano of P
050r090 h a neighbouring Volcano of Priamang.--Marsden SUMATRA. M. De. Jonnes seems to think that Volcanic ar
188b067 any? Rhinoceros peculiar to Java, & another to SUMATRA --Mem Parrots peculiar according to Swainson t
191b082 ce alive? Species formed by subsidence. Java & SUMATRA. Rhinoceros. Elevate & join keep distinct. two
212b164 eries from several islands.-- Bear peculiar to SUMATRA & not found on Java-- Monkey peculiar to. latt
217b187 one species of Australian genus being found in SUMATRA; again another of other Genus in Sandwich isla
231b241 rimeval.» continent. Indian Rhinoceros. Java & SUMATRA ones all different.-- Join Sumatra & Java <to>
231b241 ros. Java & Sumatra ones all different.-- Join SUMATRA & Java <to> together, by elevations now in Pro
242c017 est» authors on E. I«ndian». A«rch.» Borneo & SUMATRA both seem to have elephant & has orangs, Tap
242c017 have elephant & has orangs, Tapir common to SUMATRA & Malacca Borneo & Malacca «& Cochin China» a
246c026 ippines, has variety at Madagascar, Calcutta & SUMATRA,. but I do not see how it is known that they a
251c041 bears very close species, inhabiting Borneo & SUMATRA. differ only in form of white mark on breast:
251c041 - p. 246-- Gmnura-- new genus of Mam: found in SUMATRA p. 452 Append to Denham Clapperton &c on Mamma
252c045 gy from each other I do not know how different SUMATRA Java ---------- do from Indian increase of kno
253c046 eys.» Fact of Elephant same species in Borneo. SUMATRA. India Ceylon-- perhaps shows great persistenc
343d037 ture» organic nature) the Rhinoceros of Java & SUMATRA, that since the time of the Silurian, he has m
403e021 w be checked".--Mar[s]den p. 94 (1st Edit) of SUMATRA has given account of Buffalo of the East which
450e175 e of Lao & Siam inferior to those of Pegu-- in SUMATRA two breeds both small -- Java pony occasionall
450e175 es is somewhat larger than the Sambawa, Java & SUMATRA breeds, (.Hence it appears there are shades of
462t051 e grounds of similarity in condition in Java & SUMATRA & dissimilarity of forms-- yet how valueless t
462t051 tween Mammalia in cases such as that of Java & SUMATRA Nov 15th Waterhouse showed me the component ve
478z006 amisso in Kotzeboe p. 312 Leaches on leaves in SUMATRA Marsden. p. 311 D'.Orbigny considers Dasypus v
403e022 is (near island) do p. 190. The inhabitants of SUMMAGI, a territory in the small isld of Eap in the C
637j58r t Mails have them. What trash p. 237. Gives as SUMMARY of adaptations Horny point to chickens beak, t
365d105 ack Game. &, the subalpina of Sweden, (which in SUMMER dress somewhat resembles Red Grouse) it may be
486z018 t. hence musquitoes & knats abound during short SUMMER far N. where this other order is comparatively
411e055 ities come under this head» 27th November When SUMMING up argument against my theory, doubtless, the
038r057 ion of the state at a grand eruption when whole SUMMIT of mountain is blown off; & again when in grea
095a038 m of same name. with imperfect crater <--> near SUMMIT,-- much pumice --. appears to be outside of th
149g032 xplain on <formation> deposition in lake On the SUMMIT «& on Spean side» of Meal-- Derry there were p
149g032 itu <& in> hill being gneiss <& also> also near SUMMIT on Hill on side of Inn BOULDER of granite abov
151g040 erry) on low hill between Inn & Bouhunthine the SUMMIT «doubtless worn into coincidence» has beach or
153g055 ntec a kind of landing place is formed Ben Erin SUMMIT 27.813. 65 55 degree? Boulder of Granite 28.36
153g056 s & [...] sandstone actually resting on them on SUMMIT of hill rounded, site N N W of Ben Erin {P} Sh
157g076 y must have «come» «been drifted» here: on very SUMMIT no granite-- (in valley «there are» granite) «
160g091 lked all round hill. Boulder about 20 ft. below SUMMIT «Isthmus» {P} highest point joining this hill
078r175 along gentle slopes. but on the most elevated SUMMITS, where mountains most torn.--(¿anticlinal line
432e125 the measurements refer not to revolutions of the SUN & our lives,, but to period necessary to form he
535m064 , brooding over them, preserving them from «the» SUN & enemies-- would not fly away, but bit pencil w
585n076 l theories of magnetic powe in birds, seeing the SUN &c are absolutely useless when applied to birds,
586n078 owledge of that quarter,, is faculty, whether by SUN, & heavens, or magnetic virtue,-- the most proba
612o035 hy more so than movement of sap. or sunflower to SUN? ∴ I should think there. was direct «physical» e
639j28r a. C. D p. 258. «grinding» teeth in <stomach of> SUN-fish, in mouth of swine & in stomach of lobsters
402e020 first Voyage. Vol II. p 367. "The Fauna of the SUNDA islands presents us, for the most part, with th
150g037 lluvials 3a 3a 2 Mass 3 2 rather longer than Sa SUNDAY In Glen Collarig, when water up to shelf very
505g014 once offered a hive. of these small Bees-- at SUNDORNE has large Bees July/42/ Mark has six day's pu
612o035 n part. ¿Why more so than movement of sap. or SUNFLOWER to sun? ∴ I should think there. was direct «p
086a007 d by atmospheric currents?-- chiefly clearly by SUN's position if equatorial streams of warm pole;
573n036 n of a Deity We can allow «satellites», planets, SUNS. universe, nay whole systems of universe <of ma
523m018 ch are never generally, if at all discovered.-- <SUP> Sometimes comes on suddenly from <I> (in one ca
103a065 s of» mountain-chain in proportion to weight of SUPER [...] mass.-- Absence of Caverns, in Plutonic r
389d176 cted with whole system, «as if there was, a SUPERABUNDANCE of life, like tendency to budding, which w
272c109 ucture which as far so species is concerned SUPERADDED,-- the tail in cock peacock,. widowbird.--
120a108 nnt be ordinary heat, then there is something SUPERADDED, that which give cleavage to rocks.--, but l
139a180 s or rose «buds» galls.-- is it not effect of SUPERADDED vital influence?-- See End of Note Book. cal
228b232 ole world.-- «--the soul by consent of all is SUPERADDED, animals not got it, not look forward» if we
342d036 use changes in geography & changes of climate SUPERADDED to change of climate from physical causes.--
543m101 ught, & not merely instinct, a separate thing SUPERADDED.-- we can thus trace causation of thought.--
613o036 brain developed, and as I have said, no soul SUPERADDED, so [RHC] 6) thought, however unintelligible
638j58v t organization, with mysterious consciousness SUPERADDED This is similar idea, to cells of bee, corre
292c170 Classification "Let anyone even with a very SUPERFICIAL knowledge «like myself» of <classi> real aff
097a043 s?-- Mr Poullet Scrope. talks of Trachyte, "SUPERFICIALLY coated by a thin pellicle of a blackish col
638j58v xcept no in isd) although having heredetary SUPERFLUITIES Man could exist without Mammae.» to the the
409e047 ved mens & brutes bodies on one type: almost SUPERFLUOUS to consider minds.-- as difference between m
620o045 when our reason tells-- + us the action was SUPERFLUOUS, as one man trying to save another in desper
344d041 on sufficing to several births analogous to SUPERFOETATION, & to successive fertile offspring in Ento
448e166 emarkable.-- Again the resemblance between the SUPERGA & Paris, numerically the same with recent & ye
448e167 f accounting for them.-- Think over this-- The SUPERGA beds have many shells in common «& are not far
077r170 eas other to ancient rock.--this N degree 2. SUPERIMPOSED on N degree I. even No. 2. might be mistake
101a058 axis. except in relative <strata> size with SUPERINCUMBENT strata. where they have yielded conical ax
104a067 mpossibility on fissure going right through SUPERINCUMBENT mass (varying hardness,-- takes time to tr
186b061 d perish, if there was nothing in country to SUPERINDUCE a change? Seeing animal die out in S. Americ
343d036 nge of climate from physical causes.-- these SUPERINDUCED changes of form in the organic world, as ada
121a111 ompletely altered, & a crystalline structure SUPERINDUCED Lyell on Sweden p. 5. «& 7.» violet strata
186b061 liar ¿law: existence definite without change, SUPERINDUCED. or new species; therefore animals would pe
235b274 & no new ones introduced would not change be SUPERINDUCED-- why is every one so anxious to cross. ani
392d175 us or dioecious.-- very curious how this was SUPERINDUCED? (Surely all are really dioecious.,) only s
553m137 ge or one of Tierra del Fuego.-- Mr Miller (SUPERINTENDENT of the Zoological Gardens) remarked that <
367d113 my p. 45 "One of the most general marks is the SUPERIOR strength <of> «of make in» the males; & anoth
535m069 d be «universally» thought to be the will of a SUPERIOR being; whose natures can only be rudely trace
576n047 on man is small. Man's intellect is not become SUPERIOR to that of the Greeks.-- (which seems opposed
136a147 Pampas Deposit, I may properly remark on the SUPERIORITY of Lyell's classification to that of Phillip
232b248 e the pride of the savage, they perceive the SUPERIORITY of man over animals, without such resorts Mr
546m108 ng horses admire a wide prospect.-- The very SUPERIORITY of man perhaps depends on the number of sour
605o18v the vast ocean from any height.-- 6 That the SUPERIORITY & "inward glorrying, which height. by its ac
605o19v a of Deity. with vastness of Eternity. which SUPERIORITY we transfer to ourselves in the same manner
614o037 irst appears, & no too great advantage.; for SUPERIORITY of memory does not depend on its length.: Ma
615o038 tion &c stronger in some than others-- Hence SUPERIORITY of Christian over Heathen race.-- But as no
155g063 ide ravine above houses of Roy» Maccullochs SUPERNUMERARY shelf I doubt, much about «50 or 60 ft» «no

335d015 e) have offspring,-- slight deformities «as SUPERNUMERARY fingers» (that is slight alterations of pri
569n020 s, probably some distant power of the mind-- SUPERSTITION & charity & prayer, or eloquent request. Re
316c244 & giving rise to the wildest imagination & SUPERSTITIONS. -- +York's Minster story of storm of snow a
553m137 lime views, or the wretched fears & strange SUPERSTITIONS of an Australian savage or one of Tierra de
269c102 excellent sketch of plants of New Holland, SUPPLEMENTARY to Appendix to Flinders Voyage by Brown.--
626o052 said to guide will. as bird building nest, but SUPPLIES it-- instinctive feelings will doubtless lead
159g085 o much cut Granite could have remained, no peat SUPPLY.-- Consider profoundly Boulder hypothesis Thur
284c150 ances of replacement of one species by another. SUPPLY place in each others' economy Dr. S. showed th
300c198 ll only a divided body. ⫯p 3. language seems to SUPPLY instincts.-- & those powers which allow of, ac
388d170 tly of its parent & therefore wants independent SUPPLY of food.-- is real. difference-- but this does
529m040 ferent people., an agriculturist, in whose mind SUPPLY of food was evasive & ill defined thought woul
121a112 bedded this point would be worth examining. to SUPPORT. shells on surface of Patagonia, yet none in s
372d131 not arm. hard to say-- if it were possible to SUPPORT the arm of Man, when cut off , it would produc
381d157 & they have hybrids-- this is most important SUPPORT to my views-- Seeing sexes separate in some of
413e058 l change» (3) Great fertility in proportion to SUPPORT of parents December 2d Lyell tells me Beck con
566n010 omen.-- rationally explained.-- on the wish to SUPPORT a wife a ruling motive.-- Book IV, Chapt I on
616o039 andi might fairly be laid with those who would SUPPORT the propriety of the expression. They would do
616o039 d until we know what answer they would give in SUPPORT of their view it is impossible to shew satisfa
634j54v ut his Creator! p. 309. says the ribs in Draco SUPPORT the flying membrane?!!-- that the phalanges ha
635j56r inks the great fecundity of germs is to afford SUPPORT to other beings.-- true, (& the doctrine of ch
025r108 America, purposely to show what facts can be SUPPORTED from that part of the globe-- & when we see co
279c133 ferred from <this> facts stated.-- + +. Fully SUPPORTED by Mr Wilkinson. = Milking heredetary, develo
289c162 iewed as condensor, «+++ must (on my theory) =SUPPORTED by foetal lower developed forms.=» (NB waterh
291c167 etus being of one «both» sex«es». is strongly SUPPORTED by wonderful fact of bees changing the sex by
372d131 this subject? A mans arm would produce arm if SUPPORTED., <so> & in making «true» bud some such proce
388d172 ock» pheasant &c is abortive hermaphrodite is SUPPORTED by change which takes place in old age of fem
056r109 se such a bulwark to the ocean, who would ever SUPPOSE that its age was limited? Who could suppose su
056r109 er suppose that its age was limited? Who could SUPPOSE such trifling means could efface & obliterate
090a024 o little.-- How comes it none in fossil state? SUPPOSE «100» <5>0£ x 50,000 x <50 = 250 0000> x 100 =
104a067 which have given rise to eruptions.-- We must SUPPOSE everywhere--, in granitic areas &c &c volcanos
105a071 of flights of steps; if one lake then we must SUPPOSE barrier in the very part, where barrier least
132a138 rth not with central heat.--» «(does M. Parrot SUPPOSE there is no volcanicity beneath lakes?» Suppo
132a138 ppose there is no volcanicity beneath lakes)?» SUPPOSE ocean represents proper <state> temperature of
147g015 avage-- in other irregular horizontal strata I SUPPOSE these upper patches if prolonged would interse
154g059 a buttress projects from side of hill if line SUPPOSE continued across to {P} side removed all well
172b008 approaching.-- but as they inosculate, we must SUPPOSE the change is effected at once, -- something l
175b018 to first origin there must be progress. if we SUPPOSE monads are «constantly» formed ¿would they not
176b022 er in death of species, than individuals If we SUPPOSE monad definite existence, as we may suppose is
176b022 we suppose monad definite existence. & we must SUPPOSE is the case. their creation being dependent on
179b055 probable:-- «em: Horse, Llama. &c &c-- If we <SUPPOSE> «grant» similarity of animals in one country
181b039 ir existence in the present world.-- or we may SUPPOSE only each species in each generation only bree
202b133 like living Edentata in Africa &c &c.-- Now if SUPPOSE world more perfectly continental. we might hav
222b206 urians of Secondary epoch: it is impossible to SUPPOSE such an accumulation at present day & not incl
231b243 difficult to explain by creation-- or we must SUPPOSE a multitude of small creations.-- Will Dromeda
255c051 agation.-- Feathers on, Apteryx because we may SUPPOSE longest part of structure.-- shape of wings ha
265c088 d in accordance to habits.-- one is tempted to SUPPOSE from beholding the ground.-- why do beetles &
269c101 is not 700) is applicable to metamorphs theory SUPPOSE when rhinocerose lived. mean temp 60 degree me
273c110 cies,-- This is remarkable & would lead one to SUPPOSE, that species in same group generally contempo
276c122 s formation of species <of> gradual, so may we SUPPOSE., that something intermediate, between no offs
282c140 family might have its own character,-- we here SUPPOSE these changes of <use> adaptation greater than
282c141 ifferent.-- To make this case perfect, we must SUPPOSE men instead of mere colour & trifling form & h
282c142 h respect to species of each other, because we SUPPOSE all descended from same.-- but if two original
284c149 ture is general in all species in group we may SUPPOSE it is oldest, & therefore lest subject to vari
304c208 nces In ostrich which is not isolated, we must SUPPOSE the changes from typical structure have either
381d157 ate in some of the lowest tribes, leads one to SUPPOSE still more that they must in effect be so in a
388d173 n bud there is no such need.-- one would <one> SUPPOSE that the vital portion ¿nerves? passed through
398d006 eat> sensible change in the institutions. & we SUPPOSE not only revolutions, but entire obliterations
400e013 to double.-- at present time Uncle J. does not SUPPOSE one aboriginal variety.-- for they are all mad
403e023 variation in external form of varieties, do we SUPPOSE bones will not change in number. (even species
416e070 endency law.-- In animals analogy leads one to SUPPOSE that seminal fluid fluid, (& not dry as in pla
423e095 because, if we begin with the simplest forms & SUPPOSE them to have changed, these very changes <len>
427e110 rom the analogy of the animal kingdom I should SUPPOSE, that the pollen of crab, would POSSIBLY «No,
481q01v r Knight. Vol IV Hort. Transact.-- 4 May we no SUPPOSE, that certain plants, like Aphides produce imp
575n043 consciousness? What other explanation-- can we SUPPOSE some essence. The facts about crossing races o
577n049 t quality to life & mind of man.-- & we do not SUPPOSE an hydatid to be a cause of itself.-- [by my t
581n064 <e>ll. (ourang-outang), music. colours we must SUPPOSE <we> «Pea-hens» admire peacock's tail, as much
608o028 iews are directly opposed & inexplicable if we SUPPOSE that the sins of a man, are under his control,
613o036 animals modified according to species. This I SUPPOSE he deduces from the ends in each case being th
639j28v nifold traces in the several genera of Grallae SUPPOSE six puppies are born «& it so chances, that on
087a009 . America in Europaean latitudes.-- Will it be SUPPOSED that the armadilloes have eaten out the Megat
126a122 surface, say 1000-- «III but an equilibrium is SUPPOSED to have been attained.» how much matter separ
214b175 ounds from Aston Hall close to Birmingham, and SUPPOSED to be descended from a breed known to be ther
222b204 from one stock, & now that one stock cannot be SUPPOSED to be «most» perfect (according to our ideas<
222b206 lation. ¿whether type of each order may not be SUPPOSED that form, which has wandered least from ance
250c038 Certainly Africa approaches Nearest to what is SUPPOSED to have been condition of former whole world.
282c141 s> «domestic animals» with them. they might be SUPPOSED to change. & make genera of bird. analogous
306c214 from their country.-- the long-haired cats are SUPPOSED to come from there.-- All the sheep are thick
343d037 mals-- How beneath the dignity of him, who «is SUPPOSED to have» said let there be light & there was
356d072 s is difficulty on any theory-- without God is SUPPOSED to create & destroy without rule-- But what d
370d116 or single rainy day-- & from case of wasps, is SUPPOSED cells properly are made for larvae.-- CD[(p.
374d133 sembles reptiles.-- p. 432 some plants in coal SUPPOSED to be intermediate between Coniferous trees &
460t019 ng complexity from a few types, it must not be SUPPOSED that this refers to time.-- Marsupial in Ooli
472s01r e not «being» like the true P. bracteatum; all SUPPOSED to have been hybridised == Has tried several
564n005 er means by which eyes, aided by experience is SUPPOSED in man to guide to knowledge, was transmitted
601o009 ls are sleeping higher animals & not plants as SUPPOSED by Buffon) Consciousness is sensation No. 2.
605o020 ost instantaneous perception.-- Taste has been SUPPOSED by some to consist of "an exquisite suscepti b
641j29r relations of animals. equatorial countries are SUPPOSED favourable to terrestrial Mammifers-- Marine
021rIFC er duration, than the more constant: This view SUPPOSES the simplest infusoria same since commencemen
039c099 ainly have not been points of eruption. Nobody SUPPOSES that all the dikes in Cornwall or in the coal
039r061 evation. which is different from what Mr Lyell SUPPOSES. Lyell P 116 Vol III, says that in N. Pliocen
175b018 ent set, as the rest of the world.-- This view SUPPOSES that in course of ages. & therefore changes.
203b135 ould have anomalies. as Cape Anteater.-- This SUPPOSES world divide into Zoological provinces-- uni
362d099 Campestris>» refer to my notes) &Mr Yarrell SUPPOSES this a consequence of the female breeding all
366d112 Is there any law of variation. -- «(as Hunter SUPPOSES with Monsters)» if armless cat can propagate,
367d112 ad on Mare My view, why hybrids are infertile. SUPPOSES that when foetus is forming the ovum within i
511q018 s= [it is a Lincolnshire Breed]CD-- Sir. R. H. SUPPOSES is now extinct= (9) About. American & Europae
030r035 ere Challenger was lost: I know no reason for SUPPOSING these matters are not now collecting, in the
176b021 forms equable:-- but is there any reason for SUPPOSING number of forms equable: this being due to su
357d073 d.-- Athenaeum 1838. p. 654. Reason given for SUPPOSING Tetrao media or Rakkelhan is hybrid (produced
423e096 movement in Cephalopoda & fish & reptiles.?-- SUPPOSING such to be the case, it proves the law of dev
445e159 d forms-- this resulted from the necessity of SUPPOSING some inward progressive developing power.-- M
463t057 t curious cases of intermediate structure,, & SUPPOSING much extinction. give a parallel case) Waterh
102a062 chains. accompanied by do.-- Give this after SUPPOSITION p. 461 «of Proceedings» List of collections
156g069 s» Solid rock is much notched on Maculloch's SUPPOSITION;-- the old ravine, where water entered are n
381d157 > is there gradation of structure leading to SUPPOSITION, that the Cryptandrous are really, Heautandr
586n078 ens, or magnetic virtue,-- the most probably SUPPOSITION. with respect to pidgeons, is that they do k
617o40v r particle by invisible strings & as on this SUPPOSITION the forces manifested would be <sat> fundame
434e130 clearly abortive, so that they become so by SUPPRESSION of one organ. (here language forces on us th

Page ***(Key Word)***
```
115a097 pebbles» in mica slate & gneiss, can only (see «SUPRA» p 94) be accounted for by great molecular attr
303c205 » has created a perfect chain. <Icthyo> .+ + +. SUPRA & next page It is a fact pregnant with SOMETHIN
364d104 isely same story about some Southern «see p. 43 SUPRA» breed of cattle with white heads; which years
391d179 e has acquired from father some differences. V. SUPRA <v. infra> p 179, continued from is a flower bu
546m110 days he spent in such a place.-- Vide page 103, SUPRA (by mistake) have lower animals these vivid tho
625o052 ce are forgotten-- only so far do I admit its SUPREMACY p. 37. Whewells gives Mackintosh's theory: th
628o054 kintosh characterize the moral sense, by its "SUPREMACY",-- I make its supremacy, solely due to great
628o054 moral sense, by its "supremacy",-- I make its SUPREMACY, solely due to greater duration of impression
625o052 g.-- so far it has independent existence. & is SUPREME. because it is «a» part of our nature, <not> w
311c227 ge very good.-- Blainville Ovington's Voyage to SURAT, floating isld. off coast of Africa .p 69. with
036r051 d produced in walking over the sand: I am nearly SURE, it is necessary to ascend the hill.-- The abse
056r109 literate so grand a work?--In valleys one is not SURE whether fissures may not have helped it, or dil
216b180 or white as it may happen.-- Dr Smith says he is SURE of the case at Cape.-- McClay argues from it Bl
225b216 d otter live before being made otter--.why to be SURE there were a thousand intermediate forms.-- Opp
227b225 ugh D E F. follow close to A. B. C. we cannot be SURE that structure (C) could pass into (D).-- We ma
261c071 d patch on wings of Furnarius, Synallaxis &c &c. SURE, to unite the birds into group.-- it is same  as
273c110 he  says from long experience, you may be almost SURE, that there exist intermediate species,--  This
293c172 B what are those Marvellous cases, when you feel SURE you have heard conversation before. is strong a
298c189 ox & dog. & that it was most like fox.-- He felt SURE the half breed of Australian dogs, would be «mo
302c200 ., No greater gaps.-- external conditions, to be SURE, have remained somewhat similar.--!!! My theory
312c228 ite Man crosses; children heterogenous, he feels SURE of this, first offspring most like <parents> Mo
365d105 y., a single female in wood with Pheasants would SURE to be trod, & in many parts of Scandinavia the
378d147 brilliancy is present in Young birds, one may be SURE cock & hen will be alike-- I presume converse i
400e012 length of tail varies, & character of fur-- I am SURE a very good case, might be made out of variatio
423e098 increases.--   Jan 29th. Uncle John says he feels SURE, that the reason people send for their seeds to
434e128 ut experience has shown we can & that analogy is SURE guide & my theory explains why it is sure guide
434e128 ogy is sure guide & my theory explains why it is SURE guide.-- Lychnis April 3d.-- Henslow tells me f
498q009 no  hybrids were produced by seed, we might feel SURE, that pollen of own kind is much more effective
522m011 emember Jack Baldwin at school.-- Answered To be SURE I do.-- What became of him.-- Answ Had large fo
525m024 shame.-- I cannot remember instances, but I feel SURE I have seen a dog doing what he ought not to do
529m041 he botanist might so view plants & trees.-- I am SURE I remember my pleasure in Kensington Gardens ha
547m113 & <comparing their> «calling up» old ones, to be SURE of ones consciousness.-- Mayo observe no improb
567n015 Vol I p. 334 Does a negress blush.-- I am almost SURE Fuegia Basket did. & Jemmy, when Chico  plagued
620o044 d, often felt at very time it is disobeyed) & is SURE guide.-- Hence conscience is improved by attend
038r057 ater. different little craters are all burning, SURELY there must be «somewhere» below a field of flu
130a133 spring.-- Hot water would not lie. at bottom.-- SURELY we here have proofs of hot bottom.-- Study Bis
172b008 variety  produced-- --[every grade in that case SURELY is not Produced?-- <Granting> Species accordin
192b084 es with wings.& modified.-- if simple creation, SURELY would have been born without them.= In some of
236b280 general circumstance.--In. insects «in England» SURELY it is not-- intermediate genera we might expec
254c049 inated is» often repeated, as mouths in Polypi, SURELY not correct view of Flustra or Ascidia spicule
254c049 find many forms in Acrita,-- typical of other, (SURELY rather parents). (NB These views must lead  to
256c053 o bear. If man created as now. languages. would SURELY have been more homogenious.-- There must be so
278c131 erefore immense age since breccia accumulated-- SURELY ask Owen to see whether species same, excessiv
301c199 nimals have some reason, & actions habitual. it SURELY is not worthy interposition of deity to  teach
315c241 citing cause is absent!» & memory of animals.-- (SURELY in plants movements effects of irritability, t
338d025 of change in food in insects entered by mistake SURELY the fossil Mamalogy of Britain & Europe is Afr
339d025 grouse»! Has not Goldfinch & Greenfinch bred, & SURELY wild Duck & «pintail» Widgeon!-- Divides anima
371d127 Zoolog. Garden having coloured offspring.-- but SURELY in all these cases an unseen change is produce
373d132 og having spines.-- Does not male Pidgeon (yes) SURELY) secrete milk? from stomach. analogous to othe
380d154 instincts even in brain of male.-- Every animal SURELY is hermaphrodite-- (as is seen in <fe> plumage
382d158 nt in every animals, but unequally developed.-- SURELY analogy of molluscs. & neuter bee would shew t
386d166 nat & Phys) can make a head; the other part may SURELY absorb a useless member.-- in fact they do  it
392d175 ous.-- very curious how this was superinduced? (SURELY all are really diocecious.).. only simple form o
431e122 species,  or individuals) the deeper one goes-- SURELY is this true?-- most strange.-- Does not sperm
431e123 maller, & the female larger than average size; (SURELY this is very limited view, though perhaps a tr
436e137 ng the beautiful seed of a Bull Rush I thought, SURELY no "fortuitous" growth could have produced the
464t065 been  at least other birds, with small wings, & SURELY the Apteryx is more closely allied to the Stru
521m007 g a whole life time, quite unconsciously of it, SURELY memory from one generation to another, also wi
552m130 ons, & that on[e] fancies the image more vivid? SURELY the image in a dream cannot truly be <more> «a
571n027 ideas of our ancestors being impressed on us.-- SURELY we have taste naturally all has not been acqui
606o023 secondary pleasures of harmonious colours &c &c SURELY to be added. Lessings Laocoon p. 125-- says ne
610o033 tincts, or rather the amount of our instincts-- SURELY in animals according to usual definition, ther
613o036 mal [RHC] Kirby extends instinct to plants, but SURELY instincts imply willing, therefore word mispla
278c128 other families.-- it will however be much <shu> SURER, when false species banished by this test.-- Ex
578n053 extraordinarily to have one» you wont. ==)== No SURER way to blush, than particularly to wish not  to
051r093 Ascension.  (where occassionally most tremendous SURF & loose sandy beach) deposits «calcareous» encr
023r010 uantities of Cuttle fish bones floating on the SURFACE of the ocean, before arriving at the  Abrolhos
032r041 & to West.--From Equator to poles. nearer the SURFACE & to the Eastward.--If matter proceeds from gr
032r041 matter proceeds from great depth. from axis to SURFACE must gain a Westerly current:--If great change
035r045 much of all old strata of England. formed near SURFACE: Mem Patagonian pebbles beds, most unfavourabl
035r047 t; it shows that throughout all England, whole SURFACE oscillated equably.-- These facts become  easy
037r056 c rocks. Bahia, Rio de Jan: B. Oriental? level SURFACE not disturbed.--Whole West coast. Chonos to Co
045r074 ty changes might go on. & not a bubbles on the SURFACE; how comes it they do not flow out together? P
046r078 t Sulphur. salt. lime, are spread over «whole» SURFACE, metallic veins solutio
051r094 sends forth branching roots which must protect SURFACE; On «hard» exposed rocks near Bahia, whole sur
051r094 ace; On «hard» exposed rocks near Bahia, whole SURFACE to where highest spray (there pale green conte
052r099 s of Georgia «Lat degree ( )», he has rocks on SURFACE. applicable to Patagonia. During a period of s
071r156 ee-- Metamorphic action: <most> coming so near SURFACE most important There is map of Cordillera by H
072r160 l, & these masses which are scattered over the SURFACE of the ground are fibrous. malleable & of so g
076r168 liar to that part of the veins, nearest to the SURFACE of the earth."--p. 156. Mines of Batopilas  in
077r170 t freestone & breccia is the same with that on SURFACE of plains of Amazon, no relation--there is mor
089a017 not bottom of ocean boil; yet heat never reach SURFACE.-- Journal de Physique, et D Histoire Naturell
093a031 Led.-- Basalt: last because it could reach the SURFACE. before being cooled.-- Berzelius. L'Institut.
099a048 ly than laterally. -- <In Area of this> {P} If SURFACE covered with oil should shrink. film parallel
102a059 of  mountain.-- only when dikes reach near the SURFACE. that strata yield.-- In Undulation in open oc
104a068 ries with temperature ¿ with pressure? Salt on SURFACE of plains due to whole moisture being lost  by
104a068 lary attraction would bring water with salt to SURFACE Lyell remarked to me that Kylow (?) was astoni
105a069 g power increased with width. for besides more SURFACE exposed. bay more open to turbulence. Bull. So
113a090 ntre.-- But is this not wrong? we know mean of SURFACE formerly much higher, «so» that we must look a
113a090 et of strata having conducted away the heat of SURFACE. & if conducting powers had been better then 3
118a103 nly small proportion of dikes have reached the SURFACE Arguments against Herschel's view of cause  of
119a105 be profoundly considered, metamorphic rocks at SURFACE. & great heigth on mountains.-- consist of roc
119a105 of  rocks with fossils, therefore formed near SURFACE. whether they can have been plunged so many mi
120a109 ? Footsteps in New Red Sandstone. look as if a SURFACE deposit.-- The case of the shingle in the grea
120a109 uch as present. to spread sheet of matter over SURFACE.-- if elevation then went on at greater  rate,
121a112 just like Patagonia, & many shells in parts on SURFACE, but I saw none embedded this point would be w
121a112 could be worth examining. to. support. shells on SURFACE of Patagonia, yet none in shingle beds. Lyell
123a117 os Ayres 20½ quadras from river; 20 varas from SURFACE in tosca.-- remnant of Megetherium in interior
124a120 rom probability of fissures being prolonged to SURFACE. see p. 181 on do subject do p. 447 & 449. «&
125a121 nally fluid, then cooling process must go from SURFACE towards the interior -- who knows how far that
126a122 The  problem is, you have temperature known at SURFACE,-- you have temperature known far below surfac
126a122 urface.-- you have temperature known far below SURFACE, say 1000-- «III but an equilibrium is suppose
126a122 ained by conducting powers-- we judge from the SURFACE, & say 60 ft to degree.-- but this may be very
126a123 ates rocks,-- when pressure increased or under SURFACE, would not the fluid matter be driven  upwards
126a123 that have gone on for thousands of years, that SURFACE does not become hot?-- this looks as if bad co
126a123 ot attained, & if cold water did not percolate SURFACE, would become hotter.-- hence temperature ough
126a124 the  high temperature would be much nearer the SURFACE. especially at bottom of great ocean, where th
126a124 om of great ocean, where the circulations from SURFACE can take place.--  the depth of frozen soil i
```

Page
***(Key Word)**
```
146g013 ebbles/ I am nearly certain there were none on SURFACE of any hill Thursday On side of Hill South of
202b131 erara. In note. Demerara. 10 «12» feet beneath SURFACE forest trees fallen «kind well known, carboniz
423e096 us complexity, it is impossible to cover whole SURFACE of world with life.-- for otherwise a frost if
529m040 ith ocean, former animals, slow force cracking SURFACE &c truly poetical. (V. Wordsworth about scienc
564n005 whether  for position of axe of eyes, state of SURFACE, or other means by which eyes, aided by experi
577n051 s appearance,--does the thought drive blood to SURFACE exposed, part of man, face, neck-- «upper» bos
579n060 made  the blush come.-- it is an excitement of SURFACE under the will? of the animal.(-- Jan 21. 1839
051r093 orals, & Corallina survive, in the most violent SURFS: in both latter cases become petrified, & incre
057r113 consider bones washed about much at Coll. of. SURGEON'S? I really should think probably that B. Blanc
321c275 keys voyage reread Appendix Ovington Voyage to SURINAM. Voyage Congo expedition: Zaire except Brown's
218b192 hen we hear that the whole island is volcanic SURMOUNTED by water & studded with others.-- we see a b
581n063 mber the man's Christian name, writing for the SURNAME, analogous to instinctive memory, & consequen
048r085 from the foresight of the works of man Feeling SURPRISE at Mastodon inhabiting plains of Patagonia is
086a003 nts being of one order. -- There should not be SURPRISE at Horse being found in America, when Mammoth
269c101 ia being frozen to 400 ft in depth, (& Erman's SURPRISE that it is not 700) is applicable to metamorp
468t112 «& from <flower> plant to plant.»-- to my grt SURPRISE-- I found all, stamens straightened pollen pr
540m085 n of mammae & womb.-- We need not feel so much SURPRISE at male animals smelling vaginae of females.-
542m095 rk. & hence is this the cause of expression of SURPRISE-- viz seeing something obscurely with the wis
544m102 king in their sleep.-- Characters of dreams no SURPRISE, at the violation of all <rules> relations of
567n015 pumped over whole body.-- is it connected with SURPRISE.-- heart beginning to beat-- children inherit
570n023 , without assigning or understanding reason.-- SURPRISE with negation.-- like shaking something off s
570n023 - or is it from inspiration, which accompanies SURPRISE.-- & why does one inspire, when surprise, can
570n023 anies surprise.-- & why does one inspire, when SURPRISE, can one resist blow better with body distend
573n037 that very young children. express the greatest SURPRISE at emotions in her countenance-- before  they
586n082 relation to the senses, that one feels no more SURPRISE at it & feels no more inclined to ask [not lo
615o36v ch a particular organ.-- I think Pincher shows SURPRISE, walking home one day met him, with Mark ridi
615o36v y of showing it, nor was there any cause, & if SURPRISE was felt.-- analyse feelings. Mr Wynne  says,
061r127 oints; then any mountain, one is falsely less SURPRISED at new creation for large.--Australia's -  if
063r132 lapse of days.--Therefore we are not so much SURPRISED at seeing Zoophite producing distinct animals
408e046 ities.-- G. B. Sowerby.-- Looking over Lamark SURPRISED to see how many Tropical genera come from New
467t099 ot, so we may feel somewhat «but little» less SURPRISED at Henslow's remark that pistil does not beco
498q008 n select at all from the plants? If not, I am SURPRISED <plan> such plants do not degenerate,-- as th
254c049 y expended & no one system developed <is> not SURPRISING to find many forms in Acrita,-- typical of o
426e108 mber, that even birds can imitate the sounds SURPRISINGLY well-- In early stages of transmutations, i
068r146 emersion,  but improbably so long, that to be SURROUNDED by continent.-- change of volcanic focus.--
128a128 ra Geograph. Journal vol II. p 89. at Madras. SURROUNDED by salt water. purest fresh water must be so
494q005 s Remote Experiments-- Plants Raise seedlings SURROUNDED by various bright colours, any effect? and s
296c184 me North = = to peak of Teyde in relation to SURROUNDING countries & present tropical countries. p. 5
432e124 reat caution; owing to its adaptation to the SURROUNDING circumstances According to my theory no land
553m136 do> produce every effect, of every kind which SURROUNDS us. Moreover «it would be difficult to  prove
244c023 s; <after> "après un examen attentif, et forts SURTOUT de l'opinion du baron Cuvier, nous ne balançon
051r093 e peat reef of sandstone.--Corals, & Corallina SURVIVE, in the most violent surfs: in both latter cas
226b221 some one of the different orders being able to SURVIVE or chance having transported them to new stati
639j28r d one shall rather oftener than any other one. SURVIVE. in ten thousand years the long legged race wi
303c203 he cause is. <the> any osculant species. which SURVIVED would be few in number.-- Parallel of Japan n
407e039 able climate. might not have been able to have SURVIVED a change, (& become transmuted), although oth
407e039 arallel species in other continents might have SURVIVED this mundine change.,-- Therefore I argue fro
232b245 in present seas, «All» The <one> species which SURVIVES any change may undergo indefinite change., (m
638j58v sider this I look at every adaptation, as the SURVIVING one of ten, thousand trials.-- each step bein
245c023 la  regarder comme une espèce distincte! p. 171. SUS papuensis «partly domesticated» like in  general
355d067 consideration) Now in different SPECIES of genus SUS. do vertebrae vary? «See Cuvier Ossemens Fossile
473s03r ithered.-- Sillimans Journal <vo> 1842. p. 142-- SUS americana & Hippotamus «with Megatherium & Mylod
473s03r e like in Europe-- Cuvier never found remains of SUS with Elephants-- Lyell says New Red Sandstone of
513q020 s on Camels-horses. &c &c Rhinoceros= (6) Cross. SUS Barlyroussa with tame.-- (7) About fertility of
605o020 upposed by some to consist of "an exquisite SUSCEPTIBILITY from Blair receiving pleasures from beauti
605o020 e & art." But as we often see people who are SUSCEPTIBLE of pleasure from these causes who are not me
025r019 .--«certainly data insufficient, yet good» «(I SUSPECT fragments of shells will generally be found to
103a064 too large & pitching against each other, is, I SUSPECT much weakened by <vi> considering how close th
123a116 so why not metals. The theory of veins will, I SUSPECT be greatly aided by considering space formed--
315c241 demy Imperial-- Paper read in 1837. semestre I SUSPECT some valuable analogies might be drawn between
356d069 of different Medicines on organs, leads one to SUSPECT any amount of change from eating different kin
363d101 ll the varieties.-- Habits of rock pidgeon. (I SUSPECT Pennant has described them)-- [Study horns  of
399e010 legs  increase in any rabbits One may strongly SUSPECT, that breeding in & in, produces bad effects s
416e071 animals. so there are true hermaphrodites.-- I SUSPECT this rather effect of liquid semen:  therefore
418e083 ded from some one single stock.--one is led to SUSPECT that the birth of the species & individuals in
420e088 leading towards Hyaena.-- see Comte Rendu.-- I SUSPECT good case of fossil filling up blank.-- CD[not
498q007 Scions, as Elms. &c &c-- I have some reason to SUSPECT Elms.-- & Orchidaceous plants no other case.--
541m090 atest mental effort, of which I am capable-- I SUSPECT from these facts that whole effort consists in
588n090 sequence of stooping, as sign of humility.-- I SUSPECT very strong argument might be advanced, that a
592n101 wever, he seems to allow it is an instinct.» I SUSPECT the endless round of doubts & scepticisms migh
600o08v ht by one & are ordered to do it by other.-- I SUSPECT conscience, an heredetary compound passion. li
523m017 -- this found to be true.-- Her Husband never SUSPECTED during these two years that she had been insa
114a095 on, but does not tell period.-- I cannot help SUSPECTING that clay-slates have been more frequently m
535m070 be  rudely traced out. When one sees this, one SUSPECTS that our will may <be> «arise from» as  fixed
578n055 face,"-- One blush if one thinks that any one SUSPECTS one of having done either good or bad action,
495q05a ectrical machine, reversing the poles test by SUSPENDING magnet within & see which way they fly.-- (9
524m020 ower on the muscles in the twitching. Pride & SUSPICION are qualities, which my F says are almost con
524m020 on this is well worth considering, if pride & SUSPICION can be well understood. In insanity, the idea
565n007 walk  erect & stiff, with head up.-- Why does SUSPICION look obliquely.-- who can analyse suspicion--
565n007 s suspicion look obliquely.-- who can analyse SUSPICION-- yet who does not recognise look of suspicio
565n007 uspicion-- yet who does not recognise look of SUSPICION, even child will do so.-- Contempt look obliq
345d042 70? years-- now «well fixed» breed,: Jones says SUSSEX cattle were all white headed, but this was bre
023r012 passage!!-- M. Labillardiere in Bay of Legrand, (SW part). describes a Small granite Isd. capped by
076r170 ith beds of syenite & <sep> serpentine dipping to SW at 45 degree to 50 degree-- covered by conformab
077r171 tecas, & veins of Tasco & Moran--of Guanaxuato to SW. with respect to latter doubts whether bed or ve
096a040 he Azores Isld. Von Buch p 359 stretched out NE & SW.-- Von Buch. Can. Ile p. 406. List of Volcanos S
188b067 to Sumatra --Mem Parrots peculiar according to SWAINSON to certain islets in East Indian  archipelago
236b276 ts produced by seed.-- Lychnis.-- Flax.-- Read SWAINSON [blank] In production of varieties is it  not
292c170 uinary arrangement let him look at abstract of SWAINSON on Classification. "Let anyone even with a ver
193b092 ts. I Vol in 4 degree.-- I have abstracted Mr SWAINSON's trash.-- at beginning of Volume on Geographi
261c069 s range & all kinds of information-- instinct SWAINSON'S remarks in Fauna Borealis must be studied. T
292c170 structure of the whole animal let him read Mr SWAINSON'S on the Classification of animals. &  observe
296c185 tropical countries. p. 564. an abstract of Mr SWAINSONS views. which if abstract true are wonderfully
203b137 is  a tyrant flycatcher doing the service of a SWALLOW I think we may conclude from Australia & S. Am
273c111 nearly all are so.-- Thus in Hawks, there is a SWALLOW, both in structure & habits (it cannot be doub
273c111 ructure & habits (it cannot be doubted that if SWALLOW perished) hawks & Milvulus &c would  instantly
273c112 ckers., with powerful wings, but tail stiff.-- SWALLOW & goatsuckers likewise exaggerated.-- There is
273c112 ve» same plumage.-- <no> this is applicable to SWALLOW-hawk, «this not the case in swallow??? which i
273c112 licable to swallow-hawk, «this not the case in SWALLOW??? which is most wonderful of all. ¿whether in
273c112 s not always applicable to same habits, though SWALLOW hawk ,milvulus,, may catch insects on the win
573n038 pittle. will have involuntary flow & desire to SWALLOW.-- tells himself not to be tend. will turn
573n038 movements.-- Person with sore-throat told not SWALLOW spittle. will have involuntary flow & desire t
639j28r ust in Squilla. & Mantis. C D woodcuts stones SWALLOWED by birds & by Aphysia. C. D p. 258. «grinding
273c112 t wonderful of all. ¿whether in most aerial of SWALLOWS.» Milvulus, & still more wonderfully to the H
318c253 nding their ranges. «even migratory birds, lik SWALLOWS» -- degree/ migrations of birds he  mentioned
436e136 s question of genera It certainly appears that SWALLOWS have decreased in numbers, what cause?? Seein
534m062 aced table in cups of water which they waded. or SWAM across.-- they then stretched themselves from w
274c113 ith claw like lark, (one in Australia is called SWAMP pheasant) Goatsucker--, parrots with claw  like
```

```
363d102 r place breeds in <flags> «thick vegetation» in SWAMPS-- (owing to barns, perhaps, not being left ope
193b093 Brown Geograph Journal. Vol I p. 174. says from SWAN river long South coast all the remarkable Austr
278c131 aphical Journal to discover whether dog found at SWAN river. The change in England from Rhinoceros El
334d012 ox says the Half Muscovy Fox says a settler near SWAN river, lost his <on> two cows entirely, changed
450e174 all green parrots. June 26th-- Yarrell.-- Black SWAN «in domestication & nature» strictly monogamous
460t013 esses its form both on vars & species The <male> SWAN-gander with common goose produce full as many e
460t013 the cross, as above, take «generally» after the SWAN-gander. one of these half-bred ganders. crossed
460t015 <to has> «produce offspring with» so much of the SWAN-goose in appearance Bell at Hornsey (though onl
460t015 f blood). that it appears about half way between SWAN-goose & common goose.-- the stripe down back pr
554m141 important  step in progression.-- The male Black SWAN is very fierce when female is sitting the Keepe
558m152 h dog. when fear might enter?-- I believe common SWAN, arch raises neck & depresses chin-- strikes wi
558m153 - strikes with wing arches wings-- as does black SWAN.-- Goose do all species put their necks straigh
557m147 --).-- How is expression of anger in species of SWANS, in parrots &c &c -- -- peacock & turkey cock i
234b256 complanata from. Devonshire, from another from SWANSEA.-- Again Waterhouse finds certain varieties of
437e140 were  counted in a field, where there was great SWARM of mice.-- May 4th.-- The Brussels Sprout retur
485z018 ut, 1838 p. 46 Macleay Horae Entomolog. insects SWARM in Lapland & Stizbergen wherever there is extre
467t105 s on Beans Rough.--green-cabbage «in flower»-- SWARMED with meligethes & small Staphylinidae on all t
468t106 ught very unattractive-- Found Rhubarb blossom SWARMING with small Staphylinidae-- Anapsis, Melegethe
572n034 .-- It does not hurt the conscience of a Boy to SWEAR, though reason may tell him not, but it does hu
532m057 is  accompanied by «troubled» beating of heart, SWEAT, trembling of muscles, are not these effects of
532m057 on.-- The flush which accompanies passion & not SWEAT is the <state> effect of short -- but violent a
590n093 of  Bilary secretion attends passion p. 39. The SWEAT that accompanies fear is the same, as that whic
595n117 horses becoming <white> with <lather> <foame> & SWEAT, when hearing merely hunting horn-- association
106a076 os must be sought after proofs of sinking.-- No SWEDEN!! swelling of rock from Heat. Specific graviti
121a111 & a crystalline structure superinduced Lyell on SWEDEN p. 5. «& 7.» violet strata from decomposed mus
121a113 f Patagonia, yet none in shingle beds. Lyell on SWEDEN. p. 12. proofs of small rise at Stockholm.-- a
357d073 is hybrid (produced commonly in Nature. both on SWEDEN & anciently in Britain) between hen Caperailki
365d105 t cross between Black Game. &, the subalpina of SWEDEN, (which in summer dress somewhat resembles Red
502q012 called Canadense-- would it grow in open air in SWEDEN. Linnaeus found 2 flower. which had anthers re
319c255 ir other thrushes-- yet they have one with very SWEET notes.-- Their soft-billed birds are inferior t
326c265 9 has given list of Spontaneous Hybrids. where? SWEET. Hortus Britannicus. has remarks on acclimatizi
497q06v rtive parts, whether such vary.-- Do Bees go to SWEET Peas, IMPORTANT, for if so, as these can be rai
506q015 What does he think of Dr. Flemings statement of SWEET Williams & Stocks, being propagated many years,
515q022 {Chinese Dog's Head to send Cover common Pea (& SWEET Pea) for several generations under net & see if
587n087 gain on other hand, it is said people, who like SWEET things dislike others.-- dogs dislike  perfume)
041r067 owing to successive transportal from prevailing SWELL, (as Shingle travels on the Chesil bank. V.  De
042r067 bank. V. De la Beche). Ask Capt. F.: R: how the SWELL, generally & during gales would tend to  travel
047r081 n oscillation, from violence. Is it not same as SWELL travelling across Pacifick.--excepting in numbe
047r081 ng in motion what difference? In watching heavy SWELL, sea retreats & then breaks: i e to form a wave
047r081 an imagine from local form of coast (as seen in SWELL) the undertow & overfall must vary proportional
051r093 ns; At Bahia ferruginous.--At Pernambuco (great SWELL & turbid water) organic bodies protect like pea
506q015 tity -- Any unproductive, where germen does not SWELL, although there be pollen.-- or FEW. or bad see
360d090 dog-- (offspring did not go to heat. but parts SWELLED, though no fluid came from them.-- showing how
047r082 l is not the first very evident movement.--The SWELLING first on beach I cannot understand, without (
106a076 sought  after proofs of sinking.-- No Sweden!! SWELLING of rock from Heat. Specific gravities of many
435e133 eir tubes: now Mr Herbert has shown that stigma SWELLS, when pollen even most remote is put to  it.--
553m132 icts with the former. But it is one thing for a SWIFT dog to overtake an emu, & [not located] notion.
251c040 . (get limits of latter from <Tarton> Barton.-- SWIFTS return after years to nest Vol II. p. 49. on t
245c023 dental characters, wild pigs. said by Forrest to SWIM from one isld to another--\ «It is a good speci
293c173 explains  its acquirement.-- Analogy. a bird can SWIM without being web footed yet with much practice
293c173 now  Man by effort of Memory can remember how to SWIM after having once learnt, & if that was a regul
259c063 & tempting the Jaguar to use its feet much in SWIMMING, & every developement giving greater vigour t
311c227 yes of horses in India in which it may be seen SWIMMING about. A Smith is firmly believed in represen
639j28r ng» teeth in <stomach of> sun-fish, in mouth of SWINE & in stomach of lobsters-- analogy in  Flamingo
537m077 re than other.-- A Scotchman will his country or SWIS.-- it may be answered effects of education, may
266c091 ir dominion, armed alike with the Koran and the SWORD'' \quote Whewells Bridgewater treatise, (p..26).
495q05a eed-- coots-- waterhens-- examine dog, which has SWUM-- on pools & rivers-- every kind of seed must b
507q15v . Anemone with, tuft-- Bull Rush-- Dandelion-- SYCAMORE. & seeds with «mere» border-- & Humboldts spi
023r012 apt. Henry's vessel from the Friendly Isles. to SYDNEY; know by having been seen & from the contents
024r015 great  flats on the NW coast:-- 8 leagues. from SYDNEY 90 fathoms La Peyrouse. South of Mocha; 19 mil
029r032 ers do not occur in the South African plains.-- SYDNEY no I believe the secondary? formations of Braz
037r056 sturbed.--Whole West coast. Chonos to Copiapo.--SYDNEY. K. G. Sound. C. of Good Hope.--Carnatic It ha
053r101 istricts to be separated by profound valley [.] SYDNEY. -- Lesson Zoologie Grand tertiary formation o
061r126 n commonly bad over whole world. «(Was it so in SYDNEY, consult history? Phillips.» 1826.27.28.  grt.
061r126 history? Phillips.» 1826.27.28. grt. drought at SYDNEY. which caused Capt. Sturt expedition.-- ¿ Ano
408e046 ow many Tropical genera come from New Holland, ¿SYDNEY? The dog being so much more intellectual  than
065r138 ception Isld.> South Shetland Cape Possession. SYENITE¿ Andite?-- Degrading of inland bays. like  St.
074r165 cal affinities like in composed rock. granites SYENITE° «strangling &c of veins can only be accounted
076r170 g into talcose & chloritic slate. with beds of SYENITE & <sep> serpentine dipping to SW at 45  degree
089a020 ocks encrusted with serpula-- Isle of Cayenne. SYENITE & diorite, covered with iron clay common to Gu
160g090 Boulder,  much covered by turf 2ft. 8- long of SYENITE with pinkish felspar;-- whole hill dark grey f
311c225 ld ass extending over 90 degree of Long. & Col. SYKES alludes to some other case of 180 degree & grea
362d099 female breeding all the year round. ask Colonel SYKES.-- Even our domesticated cattle have tendency t
486z019 Europe. Entomolog. Transact. Vol I. p. 130. Col SYKES on balls made by dung beetles, like those  from
489qIFC ons & Experiments {T} Gowen, Royle, & Horsfield SYKES p. 12 Maer. p. 13 Question &c. July. 1842.-- Sh
503q012 ridising ferns, tying them back to back 37 Col. SYKES fertility of men & Europaean animals in India?-
520mIFC (p. 64. On <insect> Ants getting on Table. Col. SYKES) Private Finished. Octob. 2d. This Book full of
534m062 ological Society of London Vol. I. p. 106. Col. SYKES on Formica indefessa placed table in cups of wa
322c269 t end Mayo Pathology of the Human. Mind Evelyns SYLVA. skimmed, stupid Brownes travel in Africa; «wel
534m063 Hollyhock. although ordinary Habitat is Malva SYLVESTRIS. do. p. 228 Newport says Dr Darwin  mistaken
482z011 he thrush & another species! birds of passage!! SYLVIA maclovians, 2d like sylvia cisticola.-- Embriz
482z011 ! birds of passage!! sylvia maclovians, 2d like SYLVIA cisticola.-- Embriza melanodera-- a linnet not
565n006 no actual intention to bite at moment. but mere SYMBOL of readiness, & therefore done in extreme.-- L
569n020 &c. (in same way alphabet. arose from letters, SYMBOL of word beginning with the sound of letter)--
031r038 Cleavage  add other instances in old world of SYMETRICAL structure. East India Archipelago. «Aleutian
032r041 not  explain cleavage lines./ possibly general SYMETRY of world.-- I feel no doubt. respecting the br
408e045 are all apt to be diseased., & some of them SYMMETRICALLY.-- it is easy to get 50 of same kind of mon
059r123 sure. & perhaps cooled suddenly.-- As the rude SYMMETRY of the globe shows powers have acted from gre
398e006 ations & fresh laws created., & yet with <gov> SYMMETRY «& regular laws» that baffles idea of revolut
491q01v ay impregnation <caused> of some seeds, caused SYMMETRY in cone-- The «above Exper» explains apples o
528m038 e elegant.-- Again there is beauty in rhythm & SYMMETRY, of forms-- the beauty of some as Norfolk Isd
528m038 in Mexican & Graecian to be single cause) this SYMMETRY & rhythm applies to the view as a whole.-- Co
313c236 uction Even the action of the viscera. under SYMPATHETIC nerve may be instinct or habits. ¿are sympat
313c236 thetic nerve may be instinct or habits. ¿are SYMPATHETIC nerves & nervous system of insects analogous
315c242 --habitual  action, in intestines subject to SYMPATHETIC nerves-- The vividness of first <thoughts> «
558m150 & chest (sobbing) which are most under great SYMPATHETIC nerve. most subject to habit, as being  less
612o035 is  the ganglionic system of lower animals & SYMPATHETIC of man [RHC] ¿How does consciousness commenc
566n010 he has felt the passions of a face «& mind» SYMPATHIES with internal organs, as action of heart\ M
086a004 t sea shells at tops of mountains we ought to SYMPATHIZE with. old doubters of what are fossil shells
386d166 in  fact they do it in disease & injury.-- The SYMPATHY of part is probably part of same general law,
386d167 y both head & tail,--no wonder there should be SYMPATHY in human frame.-- «one of» The final cause of
531m051 he naughtiness of brothers children shows that SYMPATHY is based as Burke maintains on pleasure in be
540m088 f sorrowful delight, very like best feeling of SYMPATHY.-- Mem: Burke's idea of Sympathy. being real
540m088 st feeling of sympathy.-- Mem: Burke's idea of SYMPATHY. being real pleasure at pain of others,  with
540m089 sire to assist them,-- otherwise as he remarks SYMPATHY could be barren. & lead people from scenes of
546m108 Adam Smith (.D. Stewart life of. p. 27), says <SYMPATHY> we can only know what others think by puttin
546m108 r situation, & then we feel like them--. hence SYMPATHY very unsatisfactory because does not like Bur
```

```
548m116 t to this disease oneself, one would only feel SYMPATHY. as for for the heard suffering of a dear fri
604o017 ions.-- Cannot quite perceive drift of Book.-- SYMPATHY & affections chiefly fail.-- Notices. struggl
605o19v elves in the same manner as we are acted on by SYMPATHY. D. Stewart on taste The object of this essay
619o042 se instincts consist of a feeling of love <and SYMPATHY> «or benevolence» to the object in  question.
619o042 e in other animals they consist in such active SYMPATHY that the individual forgets itself, & aids &
580n062 , immortal fame, &c so common in the young are SYMPTOMS of the infinite & progressing nature of intel
287c157 at lead up to it" p. 20 ┃ +++hiatus & saltus not SYN.-- Linn: Transact Vol XIV.--┃. p. 24. Lamarck be
241c015 n New Zealand. a sturnus of American form-- a SYNALLAXIS. ¿American?). p. 159. & 160 «162» list of so
261c069 nding out structures into many genera.-- like SYNALLAXIS or Marsupial animals of N. America Hence  it
261c071 ny mark like red patch on wings of Furnarius, SYNALLAXIS &c &c. sure to unite the birds into group.--
304c207 the  secret.-- Now all the different forms of SYNALLAXIS. trifling characters as red band on wing sho
484z016 ern Hemisphere It is most interesting the way SYNALLAXIS leads into Furnarius. by Patagonian Furnariu
127a125 est clinal. S.-clinal. N-clinal & anticlinal «SYNCLINAL--» line.-- <ditto of synclinal> simply clinal
127a125 & anticlinal «synclinal--» line.-- <ditto of SYNCLINAL> simply clinal lines. dipping so & so or  may
590n093 which  attends great weakness.-- <Diarrhaea> & SYNCOPE p. 42. Sighing from grief. is method of increa
387d169 of  Natural History .p. 96. Vol I. Notice the SYNGNATHUS, or Pipe fish the male of which receives <yo
421e090 38.) account of metamorphosis in the young of SYNGNATHUS.= curious as showing generality of law. even
461t026 - Embryology p. 97. for Man Chapt see Yarrell SYNGNATHUS Ch 6 I presume, from my theory, as long as a
527m031 <>-- I verily believe free-will & chance are SYNONYMOUS.-- Shake ten thousand grains of sand togethe
036r052 oppers & Rats. characteristic of the deserts of SYRIA <chara> ditto for Patagonia, especially rocky p
092a028 arly vertical Linear earthquake 500 by 90.-- in SYRIA Geolog. Proc. p. 541. year 1837 In Upper Assam.
327cIBC lants-- in high points Read Volney's travels in SYRIA Vol I. p. 71. account of Europaean plants trans
499q009 any  one flower (17) Catch Bees, Butterflies-- SYRPHUS-- Meligethes & see whether they are dusted wit
045c076 ered as accidents (if «[...]» part of a regular SYSTEM can be called accidental; the proportional for
067r144 lkland fossils decidedly belong to old Silurian SYSTEM. Apply degradation of landlocked harbors to Cr
090a022 ust have fallen on the globe since the Cambrian SYSTEM In Ures dictionary between 1768 & 1818. that i
171b005 ever would be endless (that is with our present SYSTEM of body & universe therefore final cause of li
182b046 es for water. If my idea of origin of Quinarian SYSTEM is true, it will not occur in plants which are
182b047 f finches of Europe & America. &c &c &c The new SYSTEM of Natural History will be to describe  limits
183b053 reat S. American quadrupeds. part of some great SYSTEM acting over whole world, the period of great q
197b112 e is thought to account resemblances &'. quinary SYSTEM, or three elements p 66]CD With unknown limits
198b113 o be wide hiatus: states in one the sanguineous SYSTEM, in other nervous developed. (Owen's idea) sta
222b206 as  father of Mammalia; but have improvement in SYSTEM of articulation. ¿whether type of each order m
226b221 en. & thus acquired will explain representative SYSTEM Of this we see example in English & Irish Hare
235b263 birds & animals separate &c &c Work out Quinary SYSTEM according to three elements How is Fauna of Va
254c049 .-- formative energies easily expended & no one SYSTEM developed <is> not surprising to find many for
258c061 dpecker &c & which causes the confusion in this SYSTEM of nature-- Whether species may not be made by
313c236 ct or habits. ¿are sympathetic nerves & nervous SYSTEM of insects analogous.?-- («Even plants have ha
324c268 ion in Man. Lindlys introduction to the Natural SYSTEM Bevan on Honey Bee Dutrochet Memoires sur  les
335d016 & this being case, owing to the corelations of SYSTEM, the organs of generation would necessarily fa
341d029 arkable character in Apteryx, small respiratory SYSTEM; even much smaller than in other Struthios.
357d074 edge of whole world-- if so doubtless «part of» SYSTEM of great harmony. The peculiar character of St
371d118 hem.-- It is less wonderful that childs nervous SYSTEM should build up its body, like its parent, tha
377d140 ns, reduce themselves to the revolutions of our SYSTEM in the Heaverns.-- Is not puma, same colour as
386d166 origin  of this law is part of the reproductive SYSTEM.-- of that knowledge of the part, of what is g
389d176 theory, showing generation connected with whole SYSTEM, «as if there was, a superabundance of life, l
408e043 acting.   by a most delicate organ, on the whole SYSTEM may. produce-- ? When a species becomes rarer,
413e060 em.! Hurrah.-- "intermediate causes" The Sexual SYSTEM of the Cirrhipedes is the more remarkable from
428e113 ibutes this to effect of male sex on locomotive SYSTEM> I am bound to insist honestly that the sudden
433e126 iful case, showing the gradation from one grand SYSTEM to another: in each system, the changes from l
433e126 ation from one grand system to another: in each SYSTEM, the changes from limestone to sandstone &c. s
433e127 gument must be thus put, shall we give up whole SYSTEM, of transmut., or believe that time has been m
496q05a contrasted  by Kirby-- with animal reproductive SYSTEM.-- cover flower-- put artificial flowers--
532m054 h involuntary fear) In these cases probably the SYSTEM is affected, & by habit the mind tries to  fix
563n002 found he disobeyed a wish which was part of his SYSTEM, & constant, for a wish which was only short &
604o017 sciousness, individuality.-- Quotes D. Stewarts SYSTEM of Emotions.-- T. Mayo-- Pathology of the Huma
610o034 ity of such matter obeying a certain & peculiar SYSTEM of movements. different from inorganic movemen
612o035 [LHC]   ¿How near in structure is the ganglionic SYSTEM of lower animals & sympathetic of man [RHC] ¿H
625o50v o gain long-lived good, ie happiness-- yet this SYSTEM not selfish.-- explained by principles if Mack
635j56r r cause.-- (& my theory [ALL PARTS OF ONE GREAT SYSTEM. C. D]CD [All this does not explain death, but
638j059 ect attempts. In the «Bee» Mollusca the nervous SYSTEM is endowed with the knowledge of trying a hund
638j059 e of ages «step by step».-- in Man, the nervous SYSTEM, gains that knowledge, before hand. & can in i
417e076 ert> animal, which would find its place in the SYSTEMA Naturae.-- Mr. Knight makes this analogy betwe
252c045 D: genera founded in nature [not located] The SYSTEMATIC naturalists get clear indication of circumst
304c206 organs blending away.-- +++ Hopeless work to SYSTEMATIST, who believed that all his divisions  merely
567n014 breast  &c &c All Science is reason acting «SYSTEMATIZING» on principles, which even animals practica
433e127 elieve that time has been much greater, & that SYSTEMS, are only leaves out of whole volumes.-- The f
573n036 atellites», planets, suns. universe, nay whole SYSTEMS of universe <of man> to be governed by  laws,,
610o034 [LHC] Hence there are two great <worlds, inor> SYSTEMS of laws «in the world» the organic & inorganic
033r043 can Geolog. in Daubeny. P. 349 Admirable little TABLE showing long PERIODS of great violence volcanic
055r108 reat rivers prove better than any meterological TABLE the precise periods over immense areas. (& the
073r163 = Coast of Acapulco granitic rock.--in parts of TABLE granits & gneiss with gold veins visible:--"For
138a153 «beds of» sand & gravel & shells. p. 47. do has TABLE of every earthquake, during two years.-- will s
232b246 th whole world. would be like in a Meteorologic TABLE «in comparison of temperature of two countries»
258c061 III   p 430 alluded to by Capt. King do. p. 434. TABLE of extent of all species Accumulate instances o
261c069 auna Borealis must be studied. There is capital TABLE of birds from Cuba Vigors.-- nothing of much in
317c250 region of the Canary islands-- p. 250 admirable TABLE of all species St. Jago showing many common to Ca
321c275 tion: Zaire except Brown's Appendix & excellent TABLE of plants of St. Jago showing many common to Ca
340d026 unds, like Whewell affinity with analogy-- Good TABLE at end of distrib: of <birds>. Anatidae.-- Cons
342d032 Bachman  regularly breeds «in Carolina» for his TABLE Muscovy & common ducks-- they are produced in f
400e012 ol 8 «p. 46» Capt Davis in 1598 found cattle in TABLE Bay with Humps on their backs & big tailed shee
448e166 rifting of animals on ice p. 643-- very curious TABLE of all the castes from Stephenson at Lima The s
520mIFC ough Str.-- (p. 64. On <insect> Ants getting on TABLE. Col. Sykes) Private Finished. Octob. 2d.  This
525m024 ashamed  before it was known he had been on the TABLE,-- guilty conscience.-- Not probable in Squib's
534m062 p.  106. Col. Sykes on Formica indefessa placed TABLE in cups of water which they waded. or swam acro
534m062 .-- they then stretched themselves from wall to TABLE.-- table being removed a little further, they a
534m062 then stretched themselves from wall to table.-- TABLE being removed a little further, they ascended a
540m085 n the cold regions of the North,-- the elevated TABLE land of Peru the hot plains of the Amazons & Br
548m117 August 30th.-- It is singular when looking at a TABLE one has vague idea something is not there, & th
564n004 d seen chicken only hatched few hours placed on TABLE & when fly ran past it. cocked its head, & pick
570n025 there  between Squib after having eaten meat on TABLE, a criminal,-- who has stolen. neither, or both
573n038 have  shilling to walk to door without touching TABLE.-- cannot avoid it.-- curious mixture of volunt
196b105 ers & Waterfowl.-- scrutinize genera, & draw up TABLES-- Instinct may confine certain birds which hav
213b168 monotrematous  animals.--p. 82 «There are many TABLES in Phillips of numerous genera in fossil & rec
243c020 ar.-- NB. Lesson method of generalizing without TABLES or references highly unphilosophical xx Says s
344d040 > geological epochs.-- There are some admirable TABLES on Geograph distribution of reptiles in Suites
448e167 how  is this with the Eocene beds.-- see Lyells TABLES Bennetts Wandering Vol II. p 155. By inference
213b169 il & recent state, well worth consideration--» TABULATE Mammalia on this principle. <Varieties> < «Ra
435e133 cerned with the developement of form; but that TADPOLE increased in size now the Proteus anguiformis.
435e133 54. The axolotl, siren, & Proteus, affinity to TADPOLES. p. 210. Shows. that the action of light is c
254c049 machs in infusoria, generation in each joint of TAENIA worm.-- formative energies easily expended & n
347d049 L'Institut.   p. 249. (1838). Eggs discovered to TAENIA.-- hard so as to resist external  influence.--
247c028 n southern  -p. 71 Chimera-- Antarctica «also TAENIATOLE austral» «caught» Chile, Van Diemen's land &
220b198 nce in Annales des Sciences, paper on Botany of TAHITI In Charlesworth Magazine Jan: 1830. most curio
262c073 ic being.-- the fact of guavas having overrun-- TAHITI. thistle. Pampas. show how nicely things adapt
267c093 & it now claims all the moist & fertile land of TAHITI, in spite of every attempt to check its increa
483z013 S.  American Reptiles are described Shells from TAHITI & Chile The North & S. Range of shells might
514q21. anana's seedless-- 20 varieties in mountains of TAHITI. Dr. Boott-- says caricas from every isld diff
```

219b196 an example from T. del Fuego.-- Ellis (?) says TAHITIAN kings. would hardly produce from Incestuous i
146g011 sheep, sometimes mottled with white black legs & TAIL like species in colouring Strike an analogy bet
214b175 have one of the vertebra, about 2/3 from base of TAIL, enlarged two very considerably, so that any pe
214b176 y considerably, so that any person would say the TAIL was broken-- This came so often «that» it was d
241c014 hether wild or not p. 156.-- Parroket with stiff TAIL like woodpecker.-- Birds of Australia. Many in
259c065 ase, is on the same principle that, cut a sheeps TAIL off plenty of times & you will have no tail (ex
259c065 eeps tail off plenty of times & you will have no TAIL (example probably not true).-- or again healthy
264c082 bits clearly showing true affinity, for instance TAIL of ground woodpecker-- -- but tail of some duck
264c082 for instance tail of ground woodpecker-- -- but TAIL of some ducks aberrant from-- habits.-- Gould I
272c109 far so species is concerned superabundant,-- the TAIL in cock peacock,. widowbird.-- Birds of Paradis
272c109 the one feather in wing the curious feathers in TAIL of Edolius.-- Remarkable how small detail in st
273c112 en flying woodpeckers., with powerful wings, but TAIL stiff.-- Swallow & goatsuckers likewise exagger
283c145 rpheus-- one of which has very short legs & long TAIL «short much curved beak.-», other very long be
305c210 oop like dog older character & manner of wagging TAIL.-- habitual movements connected with mind There
306c214 unds» are Kurdish & come also from Asia Minor.-- TAIL like setters. long ears-- colours vary, but for
312c227 milies, «+very often in number 5» will have long TAIL.-- in raptorial birds, & tigers & sharks, being
332d005 - ears bare. skin black & wrinkled-- fur short. (TAIL cut off in progeny peculiar) limbs very long, e
333d006 d, & other two very closely resembled in form of TAIL, fur &c to the half bred Persian.-- Here then w
341d032 trange voice often in the night, like peacock.-- TAIL as long as Pea hen.-- about intermediate.-- (In
345d044 Breed-- Highland Sheep jet black legs, & face & TAIL, just like species.-- high active breeding[g] [n
358d086 var.-- But the hen hybrid of this bird, has long TAIL figure, & some degree of whiteness like a Male.
360d090 father dog. & hence general appearance of face & TAIL somewhat like dog-- though it has full share of
360d094 ng very long legs, & another in having very long TAIL, & other in having very short tail.-- I can rea
360d094 ing very long tail, & other in having very short TAIL.-- I can readily see that two first might cross
372d129 ot so a seed.-- Bud probably is like cutting off TAIL of Planaria, the whole grown to that part.-- cl
372d129 whole grown to that part.-- claw added to crab, TAIL to lizard,-- healing of wound.-- reproductive f
372d129 budding. Why does Gecko produce always different TAIL? An Individual bud may be thus produced from th
376d138 repeatedly seen one he kept pull up feathers of TAIL of Hen; which lived with it.-- also of <a> dog«
376d138 y thus examine each other sexes,-- «by taking up TAIL» Mem: Ourang Jenny with Tommy.-- Good evidence
383d160 faintly edged with reddish brown» black marks on TAIL much <blacker> «broader.-- » Breast red like Co
386d167 imals out of one & heals piece of skin.-- if the TAIL knows how to make a head. & head & tail, & the
386d167 - if the tail knows how to make a head. & head & TAIL, & the belly both head & tail,--no wonder there
386d167 e a head. & head & tail, & the belly both head & TAIL,--no wonder there should be sympathy in human f
400e012 - (do tips of ears take any colour?)-- length of TAIL varies, & character of fur-- I am sure a very g
402e019 n the cats «p 10» the joints near the tip of the TAIL are generally crooked, as if they had been brok
405e031 retained a trace of horizontal mark on flank.; & TAIL. & kind of semilunar {P} mark on each side dark
420e089 may have descended.-- Jan. 6th The rudiment of a TAIL, shows man was originally quadru«manous» «ped.-
445e158 en pecking fly.-- "whilst the shell stuck to its TAIL" as mentioned by Sir. J. Banks. p. 212.-- p. 28
459tf02 -- the hinnus, resembles horse in its head ears, TAIL limbs-- in the mules, these parts resemble ass.
505q014 Bull-Dog-- length from nose over head to root of TAIL 28½. inches. From sole of foot to shoulder on l
513q020 ieties Amongst varieties cross one with abortive TAIL or horn, with another & see result, for compari
515q022 ment of English Horses having fewer vertebrae in TAIL, than Continental horses. {About the leaping of
525m025 ous marriages with impure breed.-- A cat had its TAIL cut off at Shrewsbury & its kittens «h» (in num
557m147 e muscles would act, to when in a passion.-- dog TAIL curled when angry & very stiff. back arched. ju
557m147 stiff. back arched. just contrary. when pleased TAIL loose & wagging-- if as (I believe) Hunter says
557m147 y cock in passion.-- Cat when pleased, erect its TAIL & make it very stiff «& back» when savage «no»
557m147 en furious «with fright» back absurdly arched. & TAIL stiff.-- is shame, jealousy, envy all primitive
558m152 Bengal. tiger. when slightly angry. curls tip of TAIL.-- do two cats arch their back when fighting, &
581n064 we must suppose <we> «Pea-hens» admire peacock's TAIL,-- as much as we do.-- touch apparently. ourang o
591n097 it never dropped its ears like dog-- wagged its TAIL «a little» when attending to anything or excite
591n097 g or excited.-- so do young dingos, as I saw wag TAIL when watching anything-- Keeper does not think
639j28r feet in Frog. Walrus. Fly. Gecko &c. Prehensile TAIL. in Monkeys & Marsupials. Harvest mouse & (Cham
229b233 ery like common English.-- Hottentots say TAILED sheep aboriginal at Cape & a thinner-tailed ki
229b233 eat tailed sheep aboriginal at Cape & a thinner- TAILED kind further inland.-- NB. There is division o
306c214 d to come from there.-- All the sheep are thick- TAILED The dogs called Persian «greyhounds»-- are Kurdi
317c249 ir colour & appearance Every now & then a short- TAILED cat.¿cut? has its offspring short tails /one b
334d012 Duke of Marlborough» there is a breed of white- TAILED squirrels, which form a marked wild variety. d
400e012 le in Table Bay with Humps on their backs & big TAILED sheep do Vol 10. p. 373, «& 374» Spaniards say
428e113 ld doubt if wild species ever formed like short- TAILED cat or dog has been without recurrent tendency
553m137 & noises of monkeys go in groups. thus the pig- TAILED baboon, shoved out its lip, looking absurdly s
636j57r such losses.-- Consider ground Woodpecker stiff TAILED cormorant: pain & disease in world & yet talk
058r118 , elles presentent une coupe abrupte et souvent TAILEE a pic. Toutes ces montagnes sont formées de co
215b178 Vertèbres as latest authority. The case of the TAILESS cat of Isle of Man mentioned in Loudons (analo
515q022 ese geese. (2) Anatomy of muscles of stumps of TAILESS dogs & cats.-- (3) Hounds-- varying-- (4) Abou
293c175 ifficulty to whole theory.-- There is breed of TAILLESS cats near Bath. Lonsdale do. says Sheep could
513q020 er= Cross Rumpless fowls & Dorking fowls,-- or TAILLESS dogs & fox, to see whether the characters are
410e052 of change".-- A philosopher, would as soon turn TAILOR, as, mere describer of species, from its garme
530m045 hat peculiarities of form in trades (,as sailor TAILOR blacksmiths?) are likewise heredetary, & there
281c139 family entirely separated from any degree; the TAILORS-- «in each branch would be analogous to each o
211b163 Whales & fish.-- Progeny of Manks cats without TAILS: some long & some short: therefore like dogs.--
265c083 s in England must have been lopped off & sheeps TAILS cut yet there is no record of any effect.-- New
266c092 a & Arabia p. 202-- sheep have not the enormous TAILS, which disfigure those of Arabia & Egypt.-- CIV
281c138 linaceous having tendency to lon[g] or peculiar TAILS strange ¿¿/Genus only natural from death or slo
317c249 short-tailed cat.¿cut? has its offspring short TAILS /one born at Maer. Tuckeys voyage-- p. 36 "Cerc
353d061 all Eagles. of Australia characterized by wedge TAILS.-- many of the hawks <to> are analogues to «Bus
363d101 in Muscovy duck» crested feather, poutors. fan TAILS are found in any colours of plumage &c &c «Fout
371d118 of «Humming» birds, saw several fully developed TAILS, & one with beak turned up like Avocette. here
406e037 from its master.-- dogs when strayed hang their TAILS.-- November 1st. -- Addenda to Journal. I show
480z007 rs> states that young birds of prey have longer TAILS than old ones-- in America & sexes not of diffe
525m025 y & its kittens <h> (in number 3) had all short TAILS; but one a little longer than rest «they all di
525m025 ed»:-- she had kittens before & afterwards with TAILS. My father says, perfect deformity, as an extra
557m147 e) Hunter says. neither fox. nor wolf wag their TAILS, &c. it is very curious, recurrence of pleasure
182b046 due to direct adaptation & partly to heredetary TAINT; hence the resemblances & differences for inst
243c020 ¿cassicans Australian form? p. 27. many fish of TAITI found at <New> Isle of France: xx instance of w
384d162 ch becomes obliterated, & sexes as in Vertebrate TAK place.-- ∴ Every man & woman is hermaphrodite:--
054r102 ine.--At Conception, cleavage E & W! at Payta. TALCOSE slates, do at latter place. sandy. sandstone w
076r170 logy of Guanuaxuato.--Clay slate. passing into TALCOSE & chloritic slate. with beds of syenite & <sep
542m095 bitual when transferred, (also how often) to the TALE of a wearisome man.-- Is frowning, result of st
290c163 uscles.-- then we SEE structure gained by habit TALENT &c in man not heredetary, because crossed with
533m057 m, say only that emotions, instincts degrees of TALENT, which are heredetary are so because brain of
532m055 people, in their dotage, who sing the songs «& TALES» of infancy, it is very doubtful whether they c
189b074 & the story will be complete.-- It is absurd to TALK of one animal being higher than another.-- We c
222b207 can this comparison be instituted. People often TALK of the wonderful event of intellectual Man appe
233b252 with longevity of species in Molluscs!!! When we TALK of higher orders, we should always say, intelle
486z020 merican & T. del. Fuego & Iceland Spix & Martius TALK of birds singing in the forests of Brazil H. We
500q010 uceros in East Indies-- Asiatic Researches} (23) TALK about Thyme. Horned Oranges. Spallanzani Essay-
501q011 Biennial be grafted or cuttings taken or tuber-- TALK about Mr Knights theory with Henslow.-- Dr. Fle
503q012 about pollen of Subularia Royle & Horsfield (35) TALK about races of Banana & yet seedless-- no light
567n013 commence?-- Children understand before they can TALK, so do many animals.-- analogy probably false,
636j57r tailed cormorant: pain & disease in world & yet TALK of perfection Get instances of adaptations in v
637j58r explained by Malthus.-- [is it anomaly in me to TALK of Final causes: consider this!--]CD consider t
025r017 of Brazil generally.-- Mrs Power at Port Louis TALKED of the extraordinary freshness of the streams
072r159 a never overthrown,--no mountains Mackenzie has TALKED of lava flowing up Hill; {what does he mean?}
269c103 ntific Memoirs I can see that perfection may be TALKED of with respect to life generally.-- where <">
509q017 in old?? Falconer says all in cases. Owen. Have TALKED partially with him Ask him to introduce me to
524m021 e, they constantly do.-- In Mrs P. <...> of B. <TALKED of.> thought herself near Drayton & Ternhil,
524m021 (where she was born) though she never naturally TALKED of these places.-- My F. says, shows that earl
039r060 al measures have been conduits to volcanoes.-- TALKING of the cricket valley «the most remarkable fea

Page **(Key Word)**
048r086 with purple Claystones of P. Desire). = Where TALKING of such substances being worn into channels. m
201b126 Johnston says Mag of Zooly & Bot. p 65 Vol II TALKING of annelidae.-- <">The fact is an additional i
225b217 where is this statement I remember. L. Jenyns. TALKING of it) or how to make Indian Cow with bump & p
261c070 child sucking or even ducklings & fowls-- When TALKING of races of Man.-- black men, black bull finch
261c072 auty, when does prefer most powerful buck Owen TALKING of Plesiossaurus Plesiosaurus. alludes to some
268c100 ge on West Coast «Guayaquil & Peru» Henslow in TALKING of so many families on Keeling seemed to consi
281c138 n now living & the classification of animals-- TALKING of men as related in the third & fourth degree
350d055 les idea Septemb. 1st. Macleay & Broderip were TALKING of some Crustacean, like Trilobite. (Polirus??
463t055 not allow such series showed passages-- yet in TALKING, constantly said as the <brain> spinal marrow
541m090 t as quick as it chose.-- although thinking «& TALKING» for the moments with interest on each.-- ∴ my
565n006 learly passes into smile-- laugh long prior to TALKING, hence one can help speaking, but laughing inv
577n053 ,-- My father «even» believes that the general TALKING about any disease tends to give it, as in canc
023r010 n observed in Australia Dampier also repeatedly TALKS about the immense quantities of Cuttle fish bon
042r067 ine of Patagonia. «NB. Mr Lyell P. 211 Vol III. TALKS of line of cliff marking a pause» When mentioni
055r106 hells at Concepecion 50 toises above the sea. = TALKS of them being packed clean. & without earth.--M
059r120 in.--Says posterior craters in centre:-- Bailly TALKS of much granite on all East side of Van Diemen
095a038 ies in MS. letter in Sir. W. Parish Possession. TALKS of <hill> «cerro» of Diamante near stream of sa
095a038 e outside of the Cordillera-- Near the Planchon TALKS of very much of Gypsum.-- The officers of the B
096a039 ount of 30 ft.-- Mr Bollaert (at Roy. Institut) TALKS of quantities of shells at Iquique. <Ceylon>. B
096a039 g.) Von Buch Canary Isd. p. 351.. NB. Mackenzie TALKS of gravel on basalt of Heckla-- All the Azores
097a043 parts? & felspathic veins?-- Mr Poulett Scrope. TALKS of Trachyte, "superficially coated by a thin pe
098a045 in Auvergne Proofs from Phryganea NB. Sedgwick TALKS of LAMINATED structure (∴ separation of ingredi
114a094 ounds these introduce discussion -- I see Lyell TALKS of different composition using difference in me
425e105 ains laterally.-- Owen. Fossil Mammalia. p. 55. TALKS of Tapirus American form. found in Eocene beds
434e128 -- Whewell's anniversary address 1839, p. 9.,-- TALKS about fossil Infusoria becoming extinct not so
482z011 sively inaccurate Saw a Chouette a huppe courte TALKS of nine terrestrial Turdus falklandii & then 9.
497q007 facility. ¿Knight take opposite view. Gaertner TALKS of the several great & natural Families, as bei
550m126 more friendship.-- Scott's Life. Vol I, p. 127. TALKS of difficulty of his own drawing compared to a
551m128 y question.» Sept. 4th. Lyell in his Principles TALKS of it as wonderful that Elephants understand co
609o29v s in second volume of Malthus). Adam Smith also TALKS of the necessity of these passions, but refers
621o047 elings, as much as actions, therefore Sir J. M. TALKS too much about the contiguity to will. (a) The
637j58r his!--]CD consider these barren Virgins p. 235. TALKS of the long spinous processes in Giraffe &c, as
311c227 , floating isld. off coast of Africa .p 69. with TALL grass. & p 72 hairy sheep-- Edinburgh. Transact
106a075 der.-- This applies to all vallies (except mere TALUS «over cliffs edge» of which limit cannot be gre
149g033 extend round hill too low line drawn plain red TALUS line on N. side of Spean most clear & upper lin
251c041 ontradictory to Azaras fact of conduct of wild & TAME horses.-- p. 246-- Gmnura-- new genus of Mam: f
280c134 ducks.-- lose as well as gain instincts. Wild & TAME rabbits good instance-- instincts of many kinds
296c185 breed> Scotch wild Cattle. breed freely with the TAME Vol II. Magazine of Zoology p. 56. Peregrine Fa
333d007 x thinks that when a wild animal is crossed with TAME, offspring always take most after wild.-- i.e t
366d112 The wild ass has no cross. how comes it that the TAME donkey has. CD[old Buffon should be read on Mar
378d148 were of all colours.-- they were "half wild-half TAME, they came to the windows to be fed, but still
421e091 le, with reference to Mr Bell's statement of the TAME ones.-- ‖an instance of a trifling peculiarity
430e117 ers no explanation.-- Poet Cowper, describes his TAME Hares, attacking a sick one like Chillingham bu
450e174 birds are so when wild-- wild ducks monogamous; TAME ones highly polygamous-- change of instinct by
451e178 the world consistently reside & are bred. "take TAME animals into this region between April & Octobe
478z004 I was assured wild dog copulates <.> freely with TAME: comes to houses on purpose Mr J. Murray has gi
490q001 y very common recent ones not embedded?-- Do the TAME Parrots breed amongst the Indians Do the Savage
493q003 rid animal. in comparison with Weigh skeleton of TAME Duck & Wild Duck, & then weigh their wing bones
495q05a middle of ploughed field-- on hills.-- 10 Shoot TAME duck on pond with Duck-weed-- coots-- waterhens
502q11v gradation in structure Compare flowers of wild & TAME carrot-- Parsley & Fennel. Verbena Compare flow
503q012 ropaean animals in India?-- about Chetah & other TAME animals not breeding when tame in India?-- does
503q012 ut Chetah & other tame animals not breeding when TAME in India?-- does not know About Yaks. & other H
513q020 c &c Rhinoceros= (6) Cross. Sus Barlyroussa with TAME.-- (7) About fertility of Bantams from differen
591n097 r... Zoolog. Garden told me. he has often watche TAME young wolf & it never dropped its ears like dog
594n112 are taken to eradicate it.--» «Emma says, «her» TAME rabbits were not frightened at a dog.-- » The i
508q016 scure organ. no answer?-- 3 Andrew Smith, about TAMED wild animals breeding at the Cape.-- ‖About two
203b136 than parents. which is same as Indian Cattle.∴ TAMENESS not heredetary?, having been gained in short
290c165 n cannot be companion but master.-- heredetary TAMENESS as well as wildness-- cf Sir J. Sebright.-- l
356d071 g out difference between acquired & heredetary TAMENESS.-- In comparing my theory with any other. it
318c254 is succession of birds.-- «in» some species «a TANAGRA» Males come first & the females in flocks. «as
058r118 lus ou moins inclinée vers le rivage de la mer, TANDIS, au contraire, que vers le centre de' l ile, e
346d048 .-- August 25th Athenaeum (1838) p. 611. Ld. TANKERVILLE account of wild cattle of Chillingham,-- hab
192b086 e there never may have been grade between pig & TAPIR, yet from some common progenitor,-- Now if the
231b242 levations now in Progress, & you will have two. TAPIR existing in East Indian Seas. Marsupials animal
242c017 varieties.-- Voyage of Coquille. Zoolog. p 19., TAPIR, «des» couroucous et rupicole vert instances of
242c017 atra both seem to have elephant & has orangs, ‖ TAPIR common to Sumatra & Malacca‖ Borneo & Malacca «
402e018 , apes, baboons, monkeys & an animal probably a TAPIR p. 233. dogs in Borneo-- <brought probably by C
450e176 r in Malacca.-- speaks of Rhinoceros as well as TAPIR.-- <do do p 75> «Journal of Asiatic Soc.. Vol V
453e182 Eastern Seas p. 229. Believes the <Rhinoceros> «TAPIR.» is found in Borneo.-- «p. 233» There, as well
425e105 ally.-- Owen. Fossil Mammalia. p. 55. talks of TAPIRUS American form. found in Eocene beds of Paris L
160g088 by which we descended, it is to the west of Glen TARF What I called Alluvium shows the ascending frin
160g089 s connecting E & W connecting Glen Bought & Glen TARF a perfect old Loch, making <several> two divort
160g089 es of River Bought & between one of these & Glen TARF Hill «Cairn <taw> leer peak» Barom 28.700. A.75
162g103 r 60 «Evening do» The extreme right arm of River TARF <it> Has a very long, flat divatium aquarum wit
162g103 with trace of terraces on each side High up the TARF (a Granite (boulder), sloping buttresses, an[d]
165g124 well worth examining-- Inverness & waters of the TARF-- Kilfinnan Tower where stream enters at head o
272c109 cies & subgenera in families.-- thus the banded TARSI is common to all the Laniadae & Muscicapidae of
274c114 even one kingfisher-- Gould has seen with long TARSI.-- «Ground woodpecker» Secretary bird.-- & Mill
269c102 at space seems to act per se as barrier-- Mem. TARTARY & China.--, both coasts of New Holland.-- «Com
251c040 rops full of maize. (get limits of latter from <TARTON> Barton.-- swifts return after years to nest V
077r171 hat of the veta grande of Zacatecas, & veins of TASCO & Moran--of Guanaxuato to SW. with respect to l
078r175 degree NE. of Real del Monte 85 degree to S. // TASCO 40 degree to NW (afterwards said to be «all wit
078r176 end & vitreous felspar".-- p. 215 Same metal in TASCO vein in Mica Slate & overlying Limestone Balls
539m081 & working with his toe to perform some difficult TASK.-- Aug. 12th. When in National Institution & no
223b209 ngs October (?) 1837 Contrast New Zealand with TASMANIA The reason why there is not perfect gradation
306c212 s. first a species which lived in estuaries its TASTE. taught it to go to <sea> salter water (& its n
306c212 <sea> salter water (& its necessities teach it TASTE, but that is much more general argument) & ther
343d037 id let there be light & there was light.-- «bad TASTE {whom it has been declared "he said let there b
526m028 very odd is perhaps true.-- mem Erasmus & mine TASTE for music.-- Children like hearing a story told
526m028 ildren like looking at <ani> pictures, an early TASTE, of animals. they know.-- pleasure of imitation
536m071 e ceasing, so does other.-- What an animal like TASTE of likes smell of,∴ Hyaena likes smell of that
546m109 urces of pleasure & innate tastes, he partakes, TASTE for musical sound with birds. & ¿ howling monke
546m109 see how a dog likes smell of Partridge--, man's TASTE for smell of flowers, owing to parent being fru
554m139 ell of Verbena & Pocket Handerchief & liked the TASTE of Peppermint.--» Perfect understand voice.-- w
568n018 ny-- frogs chirp in do-- union of birds voice & TASTE for singing with Mammalian structure. «-- Ameri
568n018 can monkeys utter pleasant plaintive cry--» The TASTE of recurring sounds in Harmony common to t[he]
569n019 iful. (distinct from sexual beauty) is acquired TASTE.-- Whilst music extremely primitive.-- almost l
571n027 ers common to savages???].CD-- The existence of TASTE in human mind. is to me clear evidence, of the
571n027 cestors being impressed on us.-- Surely we have TASTE naturally all has not been acquired by educatio
571n028 re thus acquire some general notions, which are TASTE? Real taste in mouth, according to my theory mu
571n029 ire some general notions, which are taste? Real TASTE in mouth, according to my theory must be acquir
571n029 e:-- the latter correspond to fashions in ideal TASTE & the former to true taste.-- Everything that i
571n029 to fashions in ideal taste & the former to true TASTE.-- Everything that is habitual, if heredetary,
587n087 cid, &c, which may be doubted for possibly even TASTE of senna. might be acquired. as the Turks have
603o014 ate?? Edinburgh Review Vol 18. (1st Article) on TASTE «EXCELLENT». Deficient in not explaining the po
605o020 r as we are acted on by sympathy. D. Stewart on TASTE The object of this essay is to show how taste i
605o020 n taste The object of this essay is to show how TASTE is gained how it originates, & by what means it
605o020 t becomes an almost instantaneous perception.-- TASTE has been supposed by some to consist of "an exq

Page
(Key Word)

```
605o020 f pleasure from these causes who are not men of TASTE & the reverse of this. taste evidently does not
605o020 who are not men of taste & the reverse of this. TASTE evidently does not consist of this. but  rather
606o20v instantaneousness  of the result, that the term TASTE is metaphorically applied to this mental  power
606o20v horically applied to this mental power Although TASTE must necessarily be acquired by a long series o
621o047 it  is  easy by association to give «almost» any TASTE to a young person. or it is accidentally acquir
621o047 onclusion, not owing to peculiarity of organ of TASTE, for when grown up often conquers it). It  will
621o047 is possible.-- So that as there nice & nasty in TASTE, & right & wrong in action, so a child may be t
624o50v & associated during lifetime). so is our moral TASTE'p. 152. Reason never can lead to action.-- p. 1
549m119 sures are so very great, that every one who has TASTED them, will think the sum total of happiness gr
546m108 s on the number of sources of pleasure & innate TASTES, he partakes, taste for musical sound with bir
569n019 Whilst music extremely primitive.-- almost like TASTES of mouth & smell. Descent of Man Understanding
571n029 come heredetary; on same principle we know many TASTES become acquired during life time:-- the latter
581n064 d. as grand step if it can be generalize.-- The TASTES of man, same as in Allied Kingdoms-- "food, sm
581n064 f slight tickling sensation.-- in savages other TASTES few. March 16th. Gardiner's Music of Nature. p
621o047 ly that it thinks that nasty, which the natural TASTES say is good. yet horseflesh show that even thi
622o048 declaring  it is right or wrong.-- «[just as in TASTES of the mouth]CD» [LHC] My theory of instincts,
624o50v ther, rather than change hunting instinct. *Our TASTES in mouth by my theory are due to <habit> hered
244c021 120--  Coati roux common. near Concepcion. some TATOUS!!! p. 120.-- Most of the dogs of Payta-- belon
477z002 le quiyá (Coipu) viscacha.-- A. Patagonicus les TATOUS (.4 pichye, pelud, mulita et mataco.) are  all
482z011 rtularia Lesson Zoolog. Coq: p. 120 Coati Roux. TATOUS & perhaps Yagourundi near Concepcion!!!.-- (no
482z012 alapagos!!! Azara. Voyage dans l'Amerique Merid. TATU noir. abundant from Paraguay to 22 degree, then
306c212 t a species which lived in estuaries its taste. TAUGHT it to go to <sea> salter water (& its necessit
560mIBC ir eyes, very young, before experience can have TAUGHT them to avoid danger Do they frown, when  they
586n079 e wonderful in young, because can not have been TAUGHT, where to go-- the act of crossing the sea  in
621o047 me trifling circumstance.-- Thus a child may be TAUGHT to think almost anything nasty. (<& accidental
621o047 e, & right & wrong in action, so a child may be TAUGHT, or will acquire from seeing conduct of others
628o53v arcely possible to teach it-- all dogs might be TAUGHT, but not cat, that is not act by gusto, though
628o054 ause gives me [3] pleasure) or that I have been TAUGHT or habituated to associatical, the emotions of
628o054 this  instinct, with that line of conduct, & if TAUGHT rightly, it will be for the general good, that
629o55v It is other question what it is desirable to be TAUGHT,-- all are agreed general utility (3) It is ot
629o55v y (3) It is other question whether any thing is TAUGHT instinctively, I say yes, & my explanation agr
160g090 & between one of these & Glen Tarf Hill «Cairn  <TAW> leer peak» Barom 28.700. A.75 degree 75 degree?
109a084 must be studied.-- Scientific Memoirs Edited by TAYLOR Ehrenbergh on flints in chalk must be studied-
320c217 not abstracted Scientific Memoirs. published by TAYLOR Magazine of. Zoology & Botany & Continuation «
458t002 !.-- Copied Vol II p. 502. «Bengal Journal» The TAYLOR Bird uses pieces of thread, picked «up» inste
114a093 etrating their own range in Australian Alps.-- TAYLORS Scientific Memoir, Part IV. p. 403 Ehrenberg o
465t089 forms  depend on other forms Lyell's Paper, in TAYLOR'S Journ.-- Phil. Mag. May. 1840 p. 362.-- some
508q016 t Daltonism in the MALE Troughtons.-- paper in TAYLORS Scientific Memoirs (11) And. Smith Savages  at
308c217 es carried a long way. --Case of Habit I kept my TEA in right hand side for-- some month, & then when
308c217 d <vibrate> would sometimes vibrate-- «seeing no TEA brought back memory» old habit of putting tea in
308c217 no tea brought back memory» old habit of putting TEA in pot made me go to tea chest almost unconsciou
308c217 y» old habit of putting tea in pot made me go to TEA chest almost unconsciously.-- why so absent «Dr.
308c217 lmost unconsciously.-- why so absent «Dr. Black. TEA & sugar» people. reverse habits Insects & birds
370d117 ty of growing plants in all parts of world, thus TEA tree in Brazil must have degenerated. as must sp
535m063 t act wrong, as I have done when taking lid off <TEA> side of tea chest, when no tea do. p. 233. Mr L
535m063 as I have done when taking lid off <tea> side of TEA chest, when no tea do. p. 233. Mr Lewis describe
535m063 taking  lid off <tea> side of tea chest, when no TEA do. p. 233. Mr Lewis describes case of insects «
301c199 surely  is not worthy interposition of deity to TEACH squirrel to kill ears of corn according to my v
306c212 to go to <sea> salter water (& its necessities TEACH it taste, but that is much more general argumen
444e158 er at fly.-- instinct, for how could experience TEACH distances in air, in which it never touches obj
526m027 stinctive) & the others <are> learnt. what they TEACH by the same means & therefore properly no  free
608o027 sary. wickedness than disease. This view should TEACH one profound humility, one deserves no credit f
620o046 ly feels ashamed?) not so puppy, we <do> try to TEACH him & strengthen his instincts.-- so man  ought
621o046 ill have many struggles, & experience only will TEACH him, that the instinctive feeling in its nature
621o048 parents are instinctively benevolent) must frequently TEACH to be wrong or right; this teaching may be curi
624o051 & general actions of community must frequently TEACH same end.-- Hence this becomes the law of right
628o53v instinct  to commence with scarcely possible to TEACH it-- all dogs might be taught, but not cat, tha
622o048 hich the child sees uniformly performed by the TEACHERS & all around him, will be paramount,-- hence
526m027 heredetary constitution, example of others or TEACHING of others.-- (NB man much more affected by ot
557m147 it  is very curious, recurrence of pleasure so TEACHING expression «as constant smiles, cheerful face
621o048 nt) they will teach to be wrong or right; this TEACHING may be curiously modified by circumstances of
361d096 sentative species in England & N. America.-- The TEAL which some authors [not located] September 13th
540m088 ile by grief, or by bodily weakness, melts into TEARS, with sensations of sorrowful delight, very lik
557m150 affection  make body tremble. Why much laughter TEARS.-- & shaking body.-- Are those parts of body, a
565n009 & moistened by emotion.-- why does emotion make TEARS fall?? Lavater says derision lies in wrinkles a
575n045 ideas  hurting brain, like a wound hurts body-- TEARS flow from both, as when one burns end of nose w
578n056 n. (Hence pass into convulsions?)-- squeeze out TEARS. replaced & squeezed out again-- as power of mi
578n057 ttle so) & hence by association, there pour out TEARS, & there is slight convulsive wrinkling of some
245c023 «It is a good species, with different number of TEATS» Coquille Voyage Durville has written Flora of
507q15v rked.-- 31. Plant seeds of the Fuller's plants ,TEAZLE Dr. Holland ; My Father. Andrew Smith (1) Are
023r009 also  firm, out of which we pluckt a great many TEETH, 2 of them, 8 inches long, & as big as a mans t
229b233 -- NB. There is division of snakes. with hinder TEETH perforated for poison channels, but not  having
278c131 any hundred miles from the sea, associated with TEETH of seals and dugong, therefore immense age sinc
289c161 ain & then he will cease to doubt :Scales into TEETH in Bering Pike (Waterhouse) Magazine of Zooly &
315c243 ws that a man grinning is to exposes his canine TEETH. no doubt a habit gained by formerly being a ba
315c243 ed by formerly being a baboon with great canine TEETH.-- (This may be made capital argument if man do
315c243 --. Blend this argument with his having canine TEETH at all.-- This way of viewing the subject impor
366d112 m may be inferred from Australians knocking out TEETH.-- the account of the people on the NW. Coast b
385d164 of  Irish Royal Academy) have contained perfect TEETH & hair-- showing foetus has gone on growing-- I
411e055 hysical causes. «All Cuviers generalization. of TEETH to kind of extremities come under this head» 27
432e125 ung birds retrogressing-- Uncovering the canine TEETH, or sneering. has no more relation to our prese
479z005 ptera M. D'Orbigny has described my animal with TEETH {P} p. 140. Flèche of Quoy et Gaimard Ulloa she
507q016 ll cd ask Accouchers Is any peculiarity in milk TEETH inheritable!!! very good Any peculiarity in the
509q017 is in «Plants??» Are abortive organs as <young> TEETH, more plain in young Rhinoceros or Whale.  than
509q017 rix: clavicle in--? Combs in combless Poultry-- TEETH in foetal state: Mr. Horner. On Mr  Tremenheres
542m093 to keep his lip from stiffening over his canine TEETH.-- He may feel satisfied with himself, & though
542m096 s not sneer. because no young animal has canine TEETH.-- A dog when he barks puts his lips in peculia
545m106 monkeys grin with passion, that is show all the TEETH: «& make noise not like pish, but like chit-chi
545m106 t like chit-chit-chit, quickly uncovering their TEETH, this the Keeper thinks is from pleasure, & may
554m139 ard the ship could not puzzle her-- with aid of TEETH & hands.-- Descent 1838 It was very curious to
557m149 ssion my F. rubbing hands.-- stamping. grinding TEETH.-- in shame frowning, & anguish.-- shyness  not
565n006 ous to a panther I saw in garden uncovering its TEETH to bite.-- the senseless grin of passion, is li
567n013 herself--, by getting out ears of corn with her TEETH from the straw, & just like child not knowing w
639j28r by  birds & by Aphysia. C. D p. 258. «grinding» TEETH in <stomach of> sun-fish, in mouth of swine & i
574n041 ops.-- someone has described slovering «gum» «TEETH»LESS-jaws. as picture of disgusting lewd old man.
417e076 ence offspring are hybrids,.-- Mr G. B. Sowerby <TELL> showed me many land shells of the common specie
056r109 ment.» Perhaps scarcely a pebble might remain to TELL of these losses.-- Cause of chimney. to crater.
114a095 ach line chief cause of denudation, but does not TELL period.-- I cannot help suspecting that clay-sl
189b070 rium. each with same kind of coat.-- If we could TELL, I do not doubt even colour hereditary in  time
191b080 Did  Ireland possesse Mastodons?? Negative facts TELL for little) Geographic distribution of Mammalia
198b113 on of range of capabilities past & present might TELL something» p. III G. St Hilaire Insects & Mollu
223b208 great  as between perfect insect & former hard to TELL whether articulate or intestinal, or even a mit
223b211 mity of species & we Must confess. that we canot TELL, what is the amount of difference, which improv
234b256 f animals.-- He says Stephens say he can at once TELL by general colouring a group of Nebria complana
252c043 from Indian increase of knowledge would probably TELL more certainly Get Closer species-- FOX IS & Mi
256c054 e species, (theory admirably) yet a glance would TELL from which country.-- I often disputed for a mo
272c108 similar habits. But the Horae Entomologicae will TELL this.-- What peculiar conditions the Staphylini
275c119 uction) as keystone of ancient geography species TELL of Physical relations in time «& forms» distrib
```

Page **(Key Word)**
277c127 ut what to observe.-- will aid us in physiology, TELL traveller what to observe.-- if he knows he has
293c172 hout reflection or consciousness of reasoning to TELL back from front. &c or use of button holes it w
303c205 certain extent,-- nothing but experience. will, TELL us. when group is true,» there are no genera. i
315c243 barking., smiling modified laughing. Barking to TELL other animals in associated kinds of good news.
358d085 d never even lay eggs. & the men cannot «hardly» TELL any sex by appearance.-- The silver & common ph
366d111 nera (& the uncles & aunts) & therefore does not TELL against transmutation of species-- Will it agai
393dIBC varieties is uncertain do they mean they cannot TELL first result., or that «hybrid» breed is uncert
432e125 ps of pebbles &c &c: the succession of organisms TELL nothing about length of time, only order of suc
473s05r on here (Capel-Curig) says that he can certainly TELL Trout from Ogwen, Capel Curig & some other lake
473s05r er lakes, (different waters) He cannot, however, TELL them from L. Groznerat, «on road to Bethgellert
473s05v ad to Bethgellert» wh flows by Tremadoc. but can TELL them from lake S. of Moel Siabod. wh. flows int
515q022 ther respects different?--. Important.-- Oct. 44 TELL J. Anderson's statement of English Horses havin
544m103 g,-- else it is only our consciousness, & senses TELL us it is not real.= = dreaming appears clearly
547m113 ought what clothes to take (how often one cannot TELL whether one has rung the bell., when one recoll
572n034 conscience of a Boy to swear, though reason may TELL him not, but it does hurt his conscience, if he
609o030 hence our <[...]> rule may sometimes be hard to TELL) + + Society could not go on except for the mor
070r155 lders. from the Cordovise range. Signor Rozales TELLS me at seven oclock Novem <5th> Concepcion most
077r171 ragments from roof & penetrating overlying beds TELLS the secret.-- p. 189. "The small ravins into wh
085a002 - even 70 miles from salt water. Mr. Arrowsmith TELLS me, that Himalayas penetrated like Bolivian Cha
097a044 ven.-- Vol II. 2d Series. p. 221.-- Mr Bollaert TELLS me, that the upper strata alone at Guantajaya c
114a095 of greenstone, if we know origin of greenstone TELLS subsidence as plainly as Temple of Serapis. (no
114a095 w we have banished diluvial waves). & likewise <TELLS,> «offers a presumption» it has been excessivel
162g105 eposited when water stood at higher Loch Keeper TELLS me, that Loch Lochy is 8 ft below Loch Oich wh
192b083 Dr. Smith.-- «to state that most clearly». Fox TELLS me, that beyond all doubt seeds of Ribston Pipp
215b177 required to give the canthairides and milk--Fox TELLS me that it is generally said.= How came first s
239c001 ge of habits Mr Yarrell «Give it as his theory» TELLS me. he has no doubt that oldest variety, takes
271c107 iate country-- ie. mundine groups.-- Waterhouse TELLS me in insects there are many plenty of instance
275c119 ysical relations in time «& forms» distribution TELLS of horizontal barriers-- Mr Yarrell.-- says my
277c126 se distinct. analogy from every country & class TELLS us that ¿O. Modulator & O. Patagonicus. till ne
312c231 quisition, we must [not located] Henry Thompson TELLS me best way to improve cattle is to cross betwe
317c249 de Verd; same as on coast of Africa.-- Macleay TELLS me same thing p. 55. 40 leagues from land sever
318c253 r regions of snow.-- Acclimatisation.-- Bachman TELLS me in Audubon there is most curious history of
341d031 some facts of Mr Blyth on birds.-- Dr. Bachman TELLS me line of Rocky Mountains separate almost all
413e059 portion to support of parents December 2d Lyell TELLS me Beck considers the characteristics of the Tr
419e087 of a few years as with the Lamantin of Steller TELLS much less, (though <the> it also the effect of
424e101 ains, & therefore not to be trusted.-- -- Lyell TELLS me, on authority of Beck, that Hooded crow & Ca
434e129 it is sure guide.-- Lychnis April 3d.-- Henslow TELLS me following facts: believes that «only» red Ly
441e149 ng are,-- if Australian Dog, could bud, analogy TELLS us, <be> «offspring» would be similar to <f> fi
449e168 yrhycus-- is not this right?-- June 18th. Eyton TELLS me, that Yarrell knows of a Gull, which has lai
531m053 me danger-- but it is instinctive because Nancy TELLS me very young babies start at anything they hea
532m056 of will whilst their minds were sound. Caroline TELLS me that Nina, when brought from Shrewsbury to C
544m103 past ideas. which makes consciousness-- & which TELLS one of reality-- castle in the air, is more pro
549m119 thoughts are most pleasant. when the conscience TELLS our [mind], good has been done-- «& conscience
573n036 yet we placidly believe the Astronomer, when he TELLS us satellites &c &c «The Savage admires not a s
573n038 ll have involuntary flow & desire to swallow.-- TELLS himself not to turn in bed. will turn in bed.--
620o045 the strengthened instinct, even when our reason TELLS-- + us the action was superfluous, as one man t
251c041 reast: p. 234.-- good case.-- p. 526. (ref) To TEMMINCK Monograph. Mammal; «4to» good facts about dis
311c225 dna of Van Diemen's land & Australia different TEMMINCK Fauna Japonica (?!) 82 mammalia 293 Phalangis
327cIBC dge on Horse & Cow & Sheep Clarke's Travels.-- TEMMINCK Hist. Nat. des Pigeons et des Gallinaces. Sil
328cIBC g 1837. paper by Bachman on migration of birds TEMMINCK has written "Coup d'oeil sur la Faune des ile
449e170 e very near to Patagonian species-- p. 18. of TEMMINCKS. Preliminary discourse to Fauna of Japan-- th
339d025 n, but not plants» will it hold good.-- Thinks TEMMINK doubtful when he says No genera.-- thinks ther
249c036 of different forms being migratory; also with TEMMINKS fact of forms being within Tropics.-- Europae
363d101 : this will be well worth working out.-- Study TEMMINKS work on Pidgeons--, & see whether feathered l
057r114 mud which he drew up from 1,000 f[athoms], & the TEMP of which was below freezing point!!! Remember i
112a090 838. Letter from M. Erhman stating that the mean TEMP at Yakous in Siberia being -8 Reaumur.-- there
119a105 be required by thermometrical scale.-- (for the TEMP must be immense to convert rock into gneiss &c
130a134 ters. Scienc Math. Phy-- Nat. t. I, 1831. sur le TEMP du globe on Volcanos &c worth reading. L'Instit
147g019 Thursday Evening ½ past 8 Tyndrum 29.<625> «636» TEMP. 62 Friday morning ½ past seven o'clock 29.642
147g019 p. 62 Friday morning ½ past seven o'clock 29.642 TEMP 55 Air 50 degree? Friday. Inverorum about 20 ft
147g019 y. Inverorum about 20 ft above Loch Tulla 29.804 TEMP 62 degree Air 60 degree Below Loch Tulla whole
147g021 ween Inverorum & King's House 28.935/82 degree A TEMP of Air 65 degree? Glenoe, 6 ft above high water
152g049 of Roy Level of «bed of» River 30.221/65 degree/ TEMP of air 65 degree? There are two terraces on the
269c101 rphs theory suppose when rhinocerose lived. mean TEMP 60 degree mean, then temp at depth of four hund
269c101 hinocerose lived. mean temp 60 degree mean, then TEMP at depth of four hundred feet would be 60 degre
385d164 fact equally well known, that we observe in the TEMPER, especially of the youngest children, a striki
385d164 ren, a striking <resemblance> similarity to the TEMPER of the father, or of the mother, or sometimes
451e179 ters: is said to be of a bolder & more generous TEMPER-- Hodgson Koloff. voyage through the Moluccas
536m073 if it were pointed out.-- My wish to improve my TEMPER, what does it arise from but organization. tha
608o027 r anything. (yet one takes it for beauty & good TEMPER), nor ought one to blame others.-- This view w
568n017 s have shyness with strangers» «as in case of TEMPERANCE, or real virtue, that is action which experi
629o055 habits carried on to other feelings, such as TEMPERANCE, acquired by education.-- CD[In similar mann
210b158 to difficulty of transport. For instance the TEMPERATE parts of Teneriffe, the proportion of genera
231b243 effect of similar temperature.-- now those of TEMPERATE regions & tropics are only related by one con
260c067 compare birds of N. America & South-- Any how TEMPERATE regions-- crows in N. America] Study Bonapart
407e040 t is, from the ancient preeminently equable & TEMPERATE climate, <that> of America, that the Mammalia
032r041 n & Air; there are «likewise» differences of TEMPERATURE «at equal distances from centre of rotation»
086a007 e possessed a great edentata. -- How much is TEMPERATURE of world regulated by atmospheric currents?-
104a068 terminated Solubility of fluids varies with TEMPERATURE ¿ with pressure? Salt on surface of plains d
106a076 Mag. Dec 1837. p. 520 Mr Fox on increase of TEMPERATURE at great depths. All Earthquake unaccompanie
113a091 facts of SLOW but successive transmission of TEMPERATURE clearly prove possibility of metamorphic the
119a106 s. & trap at least as hot as lava-- of which TEMPERATURE is partly known-- <[...]> moreover gradation
125a121 r that may have penetrated,-- lower down the TEMPERATURE may be kept up far higher from circulation o
126a122 hen still thinner → The problem is, you have TEMPERATURE known at surface,-- you have temperature kno
126a122 ave temperature known at surface,-- you have TEMPERATURE known far below surface, say 1000-- «III but
126a123 colate surface, would become hotter.-- hence TEMPERATURE ought to increase rapidly beneath level of s
126a124 degree» & lower part glass.-- then the high TEMPERATURE would be much nearer the surface. especially
132a138 s)?» Suppose ocean represents proper <state> TEMPERATURE of earth. at the freezing point.-- accounts
133a139 nce of circulating water.-- & therefore that TEMPERATURE of earth beneath <of Sahara de> a dry desert
137a151 ff. On the effects of meteoric waters on the TEMPERATURE of the interior & p. 142 / p. 155. the incre
137a151 interior & p. 142 / p. 155. the increase of TEMPERATURE beneath the sea, is probably much more rapid
171b002 -- We know world subject to cycle of change, TEMPERATURE & all circumstances which influence living b
231b243 l relationship. as well as effect of similar TEMPERATURE.-- now those of temperate regions & tropics
232b246 ke in a Meteorologic table «in comparison of TEMPERATURE of two countries» finding a very hot day, «i
435e134 al agents." p. 466. Many facts given of high TEMPERATURE at which fish &c can live.-- Lyell says that
584n071 vided into migration,-- subsidiary to food & TEMPERATURE molting & breeding instincts, sexual, social
495q005 ttings or bulbs in several different soils & TEMPERATURES & see what the effect will be.-- will seedl
203b137 le world-- For instance gradual reduction of TEMPERETURE from geographical or central heat.-- But the
060r125 can find nothing.» Mem Carolines quotation from TEMPLE Urge the mineralogical difference of formation
071r157 rthquake with great destruction of human life.--TEMPLE mentions some earthquake at Cordova.-- There t
114a095 in of greenstone tells subsidence as plainly as TEMPLE of Serapis. (now we have banished diluvial wav
350d053 pecies as at Galapagos., be distinguished from TEMPORAL species as in two formations? by no way.?-- "
490q001 lar races. --Name of Italian who sold eggs.-- TEMPORARY Question 1 Where has Duchesne described Atavi
620o044 short pleasure. passion in its nature is only TEMPORARY, & we do not afterwards think of it.-- Whatev
089a018 ont presque toujours precedes et suivis, queque TEMPS avant et apres, par de petites secousses."-- To
599o008 ience; voila l'homme separe de la matiere et du TEMPS! voila les facultes, q'il possede seul sur la t
536m074 t> would more earnestly pray "deliver us from TEMPTATION,' he would be most humble, he would strive <
537m076 e others happy & wrong to injure them without TEMPTATION.-- This probably is natural. consequence of

Page
(Key Word)

563n001 ing hare is stopped by fleas, also by greater TEMPTATION as bitch: or dogs defending companion. (mem
568n017 ld.-- if tempted not to follow it, by greater TEMPTATION, if memory of its own emotions. (which must
620o044 n the results following our conduct.-- If the TEMPTATION to disobey the conscience is extremely great
062r129 eversed argument. knowing it to be a desert.-- TEMPTED to believe animals created for a definite time
255c052 picking of monstrosities as Man does.-- One is TEMPTED to exclaim that nature conscious of the princi
265c088 have changed in accordance to habits.-- one is TEMPTED to suppose from beholding the ground.-- why do
295c183 located] Erasmus says he has seen old Stallion TEMPTED to cover old mare by being shown, young one.--
357d072 have branched out into orders one is strongly TEMPTED to believe, one or two were landed as at prese
401e016 from primrose suddenly produce cowslip, one is TEMPTED to think here some anomaly-- I can fancy cowsl
404e025 <the class» «some orders» of crustacea, one is TEMPTED to think that it must have been invented all a
526m029 book, one remember the part of page.-- one is TEMPTED to think all memory consists in a set of sketc
526m030 n brought up) by sight & not by hearing One is TEMPTED to believe phrenologists are right about habit
552m132 & not to any particular muscle Sept. 8th. I am TEMPTED to say that those actions which have been foun
563n002 phynx howling when I struck the Keeper) may be TEMPTED to attack him from jealousy. (Pincher & Nina)-
568n017 in following its instinct & pain if held.-- if TEMPTED not to follow it, by greater temptation, if me
608o027 good & being perfect, & therefore will not be TEMPTED, from knowing every thing he does is independe
619o043 which would generally be anger, as he would be TEMPTED to interfere, but with respect to himself it w
620o046 certain lines of conduct, <although» even when TEMPTED not to do so, by other natural appetites.-- he
259c063 or instance, fish being excessively abundant & TEMPTING the Jaguar to use its feet much in swimming,
259c064 pecies to species «although we see it affected» TEMPTS one to bring one back to distinct creations.--
376d136 with Quichuas the American character is more TENACIOUS. & does not disappear for Many generations Se
072r160 he ground are fibrous. malleable & of so great TENACITY, that it is with difficulty that a few fragme
314c239 alian Genera Some species same (Palm & Phormium TENAX) as in New Zealand & Australia, some SPECIES of
042r067 R: how the swell, generally & during gales would TEND to travel on a «me» central line of Patagonia.
072r160 septa.-- would particle attracted towards space TEND to form ring. [Fig. 10] {P} motion from within
092a029 o Brit. Assoc: has shewn how electrical currents TEND to deposit metals, if in solution. My view of m
099a051 & likewise gravity MN. Then every particle would TEND to meet at <B. but if particls attract each oth
228b228 der.-- to know what we have come from & to what we TEND,-- this & «direct» examination of direct passag
335d013 omewhat like parents.-- therefore offspring will TEND to go back, or have none-- the argument does no
347d049 on is condensation of changes. then animals must TEND to improve.-- yet fish same as, or lower than i
358d086 - Thus castration, hybridity, & breeding in & in TEND to produce same effects.-- CD[May it be said, t
391d174 n properly is hermaphrodite (hence monstrosities TEND that way «& from frequency of this tendency all
412e057 «owing to their slow formation» these variations TEND to accumulate. «on any structure.» L'Institut.
423e095 e them to have changed, these very changes <len» TEND to give rise to others.-- Why then has there be
423e096 places.-- The Geologico-geographico changes must TEND sometimes to augment & sometimes to simplify st
431e122 a plant he wishes to vary-- domesticated animals TEND to vary. March 20th. Phillips in Lecture in Roy
461t037 tion of structure to purpose after purpose would TEND to render complex the series.-- Ch 6 Upland gee
535m070 es against all contrivance-- it is what my views TEND to.-- When a man is in a passion he puts himsel
536m074 a predestinarian of a new kind, because he would TEND to be an atheist. Man thus believing, <yet> wou
549m122 ach other.⁴ The rules to mortify yourself do not TEND to this-- though believing it to be true, & the
609o30v tion is the experience of orders) shows does not TEND to greatest good.-- Therefore rule of happiness
614o037 ole a mystery.-- [LHC] This Materialism does not TEND to Atheism. inutility of so high a mind without
575n044 ortant: can there be stronger analogy that the TENDENCY to hybrid greyhound to hunt hares. «& leave t
300c198 als less divided.-- But as man has heredetary TENDENCIES. his mind is still only a divided body. ⦗p 3
022r006 ing Electricity? Would minute particles have a TENDENCY to change their position? Carbonate of Lime d
028r028 & sea beaches. All mineral masses must have a TENDENCY. to mingle; The sea would separate quartzose
086a005 n Plata Mastodon Toxodon Is the general saline TENDENCY of America connected with its elevation. vapo
105a073 ith slow course small erosive power. therefore TENDENCY of running water to deepen not to widen valle
106a074 miliar illustration.-- Therefore stream has no TENDENCY to widen course until inclination is become c
106a075 ter when flowing over (B) than when at (C) its TENDENCY would <cut> be to cut a narrower channel inst
109a083 d. no currents & elevation have same effect, a TENDENCY direct (or oblique) outwards may be granted.
119a104 to start with (& does not Hersche theory imply TENDENCY to equilibrium.) 3d. there are mountains in t
119a105 be that of equilibrium.-- What causes that of TENDENCY to irregularity.-- Why does Sir John assume
128a129 de. produce subsidence of water on other. from TENDENCY to regain statical equilibrium This will be o
171b005 iverse therefore final cause of life With this TENDENCY to vary by generation, why are species are c
175b016 of ages. & therefore changes. every animal has TENDENCY to change.-- this difficult to prove cats &c
175b020 pe some of the branches dying out.-- with this TENDENCY to change, (& to multiplications when isolate
179b033 re of S. America shows this.-- <If> Is there a TENDENCY in plants hybrids to go back?-- If so Men & p
186b069 plants, (or animals) whether, <in> races have TENDENCY to keep to <each> either parent, (this is wha
188b069 bearing with Dr. Smith's fact of races of men TENDENCY to keep to one line Dr Smith says very. close
190b075 difficulty in such marriages or offspring show TENDENCY to go back-- there is an end to species.-- «B
199b119 banished by rest?-- ∴ most monstrous form has TENDENCY to propagate, as well as diseases. In interma
199b120 smallest differences blended, rather stronger TENDENCY to imitate one of the parents; repugnance «ge
200b125 whether soaking seeds in salt water &c has any TENDENCY to form varieties! Ed. N. Phil. J. Morse foun
222b203 s cross & «hybrid» breed, their offspring show TENDENCY to return to one parent, this is only charact
222b203 is is only character., & yet we find this same TENDENCY (only less strongly marked) between what are
225b219 s East India Archipelago very good on opposite TENDENCY.-- Study Ellis & Williams. zoology of South S
235b262 s separated. & long interbred <p> have great TENDENCY to vary? Is not man thus circumstanced; varie
248c030 her.-- We see even in domesticated varieties a TENDENCY to go back to oldest race, which evidently is
248c033 ccumulated cannot be transmitted;.-- hence the TENDENCY to revert to parent forms, & greater fertilit
276c124 - if vertebrae much lengthened &c there may be TENDENCY to divide, which often enough repeated would
280c138 e whole generation unlike would go against the TENDENCY.-- it tries to go back to grandfather, but if
281c138 ation of all cock birds in Gallinaceous having TENDENCY to lon[g] or peculiar tails strange ¿¿/Genus
281c165 ement in classification, because structure has TENDENCY to follow it, or it may be heredetary & stric
292c169 are united (which probably is first stage) the TENDENCY to change cannot be great, otherwise it would
292c169 it would be unlimited. We absolutely know that TENDENCY is greater in Mammalia, than in shells ¿ univ
298c190 in lower animals many times produced, a general TENDENCY produced. Such as man getting habitually into
305c211 endowed, & the community of mind, even in the TENDENCY to delicate emotions between races, & recurre
307c215 are bound to consider abortive organs of same TENDENCY in species, this is capital & novel argument.
331dIFC rmediate». «hybrids, when» interbred. show any TENDENCY to return to either parent.? Is the first cro
335d015 emain, a mule «being new species» will have no TENDENCY to have offspring like parent, but as they mu
346d047 he cross not so hardy as Black faced, but more TENDENCY to fatten-- This man confirmed my account of
347d049 te animals must articulate. <i> in vertebrates TENDENCY to improve in intellect,-- if generation is c
356d070 analogous to some hybrids breedings-- there is TENDENCY to reproduce in each case, but something prev
362d100 nel Sykes.-- Even our domesticated cattle have TENDENCY to breed at particular times. Mr Yarrell has
366d112 there is a law.,-- that there must have been a TENDENCY for feathers to grow there «That Mutilations
373d132 ent, for Mammalia recent creation.-- why. what TENDENCY can there be for abortive organ ever disappea
386d166 absorbed.-- this law acting against heredetary TENDENCY causes abortive organs.-- the origin of this
387d168 Male (X) «assume that every peculiarity has a TENDENCY to descend to several generations» If A & B b
387d169 mit with same force as first pair, but to this TENDENCY is added <that> the 3d tendency from first pa
387d169 r, but to this tendency is added <that> the 3d TENDENCY from first pair.-- Now if two of third pair o
387d169 they will have same influence as first pair + TENDENCY they inherited from second pair, + the influe
388d172 er bee having both sexes abortive fact of same TENDENCY. -- Mammae in man. having given milk. -- test
389d176 s if there was, a superabundance of life, like TENDENCY to budding, which wishes to throw itself off.
389d177 female often takes males but does not produce) TENDENCY to deformity ¿this does not happen with hybri
390d178 sions fail.-- In fruit trees no doubt there is TENDENCY to propagate the whole difference of parent,
391d174 sities tend that way «& from frequency of this TENDENCY all mammalia must long have so existed.» with
404e026 have mottled coloured puppies case of this.-- TENDENCY in «manner of» life to be mottled + heredetar
404e026 in «manner of» life to be mottled + heredetary TENDENCY determines the puppies to be so.-- [not locat
413e058 all (1) Grandchildren. like. grandfathers (2) TENDENCY to small change.. «especially with physical c
415e067 is owing to the rearing up of every heredetary TENDENCY towards fatal diseases, & such constitutions
415e070 e non-necessity of the «so called» progressive TENDENCY law.-- In animals analogy leads one to suppos
417e075 on of both.-- Now in the ass-- there is little TENDENCY to vary. & hence offspring are hybrids,.-- Mr
422e095 complexity.-- but yet there is no «NECESSARY» TENDENCY in the simple animals to become complicated a
428e113 t-tailed cat or dog has been without recurrent TENDENCY in external conditions» sudden loosing of hor
434e122 ng <or> form of body, or whether it merely has TENDENCY (as effects of cultivation on successive gene
439e143 of wild close species of plants shows there is TENDENCY to prevent the crossing.-- in animals where t
473s07v n pregnancy, & then may just as well be born a TENDENCY to alter or assume some form late in youth,--

```
154g061 ts River terraces often descend by flights the TERRACES if the largest has hollowed out most Wednesda
159g086 his station a little way down slope of obscure TERRACES (& conical hills on same) of «semi» waterworn
160g092 h, & waters flowing into west end with obscure TERRACES on one side Barom 29.200 A 80 70 degree?  for
162g099 Oich. is a brook on the Lochy side of it-- the TERRACES of which, last measurements belong are so com
162g103 ft of Bright.-- like bed of lake with trace of TERRACES on each side High up the Tarf (a Granite (bou
089a018 olles. 11th Observ.-- Les grands tremblemens de TERRE sont presque toujours precedes et suivis, quequ
207b152 e seule plus d'especes polymorphes que toute la TERRE ferme de lancien monde".-- Considers forms in r
599o008 s! voila les facultes, q'il possede seul sur la TERRE. J'ai trouve son âme" &c-- -- Confesses these f
058r118 ypoth. agree with great continents). Voyage aux TERRES Australs Vol. I. p. 54. M. Bailly says."en eff
243c021 r to the different points.-- Consult Voyage aux TERRES Australes Chap XXXIX tom IV p. 273 2d Edit Con
324c268 on sleep & movements of Plant/ If:4s Voyage aux TERRES australes. Chapt. XIX. tom IV. p 273 Latreille
274c115 r any other clear relations besides aerial, & TERRESTRIAL,-- How is it in water birds, there are walki
056r112 ow> Earthquakes part of necessary process of TERRESTRIAL renovation & so is Volcano a useful chemical
182b046 in plants which are in far larger Proportion TERRESTRIAL,-- if in any in the Cryptogamic flora but no
205b143 a siluris. p. 123 A climbing fish. p. 122 A TERRESTRIAL annelidous animal p. 347. Vol I.-- compare w
226b220 udy profoundly the origin & history of every TERRESTRIAL Mammalia.-- Especially moderately large ones
306c214 andal. Wood Isd.? ought to agree with Java?? TERRESTRIAL Planariae assuming bright colours., good ins
360d093 ter of other. --<so> the most aquatic & most TERRESTRIAL animals, might be harder to cross then two l
416e070 s it-- in Pianaria, they couple-- CD (lowest TERRESTRIAL animals.-- in shells?-- insects?-- all!?!!?
421e090 in ferns.-- ¶.do p. 250-- «speaking of» the TERRESTRIAL mollusca of Morocco «Mr Forbes says the Faun
482z011 !!.-- (no species mentioned) p. 205. only 9. TERRESTRIAL birds at Falklands Isd 8 waders. 22 palmiped
482z011 Saw a Chouette a huppe courte talks of nine TERRESTRIAL Turdus falklandii & then 9. passeres! Says t
634j55r ks all <land> Mammiferous animals originally TERRESTRIAL.-- for we find even in Cetaceae traces of hi
641j29r atorial countries are supposed favourable to TERRESTRIAL Mammifers-- Marine ones «of large size» <to>
267c094 colour reddish <brown>. ears long.-- like bull TERRIER-- Indian secured one, as they always like to c
305c210 t Zoolog Gardens there is half Jackal & Scotch TERRIER.-- certainly more like Jackall in gait,  size,
360d089 per said in <two> crosses «twice made» between TERRIER & hairless dogs of Africa,-- some puppies hair
043r071 ion. Valparaiso (Copiapò & Guasco). yet whole TERRITORY vibrates from any one shock-- In S. America--
403e022 and) do p. 190. The inhabitants of Summagi, a TERRITORY in the small isld of Eap in the Carolines, ar
605o18v re there is no real sublimity 5 The emotions of TERROR & wonder so often concomitant with sublime. ad
605o19v ide & glorying. (often however accompanied with TERROR & wonderment) <which> «this» emotion, from the
026r020 view is directly opposed to common opinion The TERTIARY formation South of the Maypo at one period of
030r036 - Whole coast S. of Concepcion where there are TERTIARY strata there is Coal-- ¿ No shells in all cas
035r047 d from Daubeny P 402: likewise, mean height of TERTIARY. being less than secondary:-- consider argume
054r102 nd valley [.] Sydney. -- Lesson Zoologie Grand TERTIARY formation of Payta: N. part of New Zeeland en
055r108 tionally high to age. (we do not wonder to see TERTIARY plains consumed) Where slope «plainly» indica
092a028 837 In Upper Assam. Geolog Proc p. 566 1837.-- TERTIARY <bea> formation twenty species same as Paris.
093a032 & Sulpuret of Iron has been sublimed into the TERTIARY limestones of Vendarques. Mem sublimation  of
128a130 eum.] I. Part. I Vol.-- some notices on modern TERTIARY strata on coast of do-- I believe?? coast  of
129a131 rock moved by gale-- When writing on Valleys. «TERTIARY strata of S America» read parts of this work,
133a140 d by volcanic & trappean rocks, within say the TERTIARY period. one is led, to look at globe as resti
191b080 pe Ireland common animals-- + + + for instance TERTIARY deposits between East Indian islets-- (+ + +.
207b150 hous, which «nearly» first appear «(p 321)» at TERTIARY epock p. 330. Fossil Infusoria found of unkno
208b153 xtinction of individuals than of species Paris TERTIARY Shells in India!? A p. 28 Dr Beck. & Lyell. m
212b168 now  least in number (as cephalopods,) in last TERTIARY epochs most genera dead? --Examine into  this
214b172 -- Oh. Wealden-- Wealden. Do the N. American. TERTIARY deposits present analogies to shells of livin
250c040 ilidous Molluscs. by each other in secondary & TERTIARY periods.-- p. 125. ref. to Phil Transacts. (r
256c055 rds & animals Mr Strickland & Hamilton-- found TERTIARY formation amongst Graecian isles, ¿See if typ
283c146 ientific Memoirs & L'Institut.) that there are TERTIARY fossil Infusoria, of same forms with  recent.
303c204 +, Will he say creation is at end, seeing that TERTIARY geology has obeyed rules of modern causes.  &
316c247 e not like, except in very few cases, those of TERTIARY Europaean fossils-- «(so much the more remark
343d039 ermination of existence.-- nor is there in the TERTIARY <older> geological epochs.-- There are some a
426e105 ar when once extinct-- Lyell's argument about <TERTIARY> Isld «neighbouring» formed in the Tertiary «
426e105 t <Tertiary> Isld «neighbouring» formed in the TERTIARY «epoch» like Sicily not having species, if tr
465t089 allied Beaver to present forms.-- -- How many «TERTIARY» estuary & Lacrustine formations contain foss
161g094 this  terrace «station perhaps 6 ft too low» (to TEST last on Peat-Mass Divortium aquarum) Barom. 29.
224b212 from «moderately» distant countries. there is no TEST but generation, «(but experience according to e
227b227 ation.-- View of generation. being condensation, TEST of highest organization intelligible.--may look
277c127 in islands. species.-- each describer giving his TEST namely differ as much as those (naming them) wh
278c128 <shu> surer, when false species banished by this TEST.-- Excepting where an Andrew Smith, "Richardson
284c149 haracters' rule are used by Naturalists in their TEST of value of character-- Macleys rule is convers
403e024 give  up infertility in largest sense. <es> «as» TEST of species.-- they must deny species which is a
495q05a them  in Electrical machine, reversing the poles TEST by suspending magnet within & see which way the
524m022 on.-- Widows not of their husbands-- My father's TEST of sincerity.-- People in old age.  exceedingly
527m034 t not of an inventive class.-- Now that I have a TEST of hardness of thought, from weakness of my sto
575m044 ossing of jackall & Fox & wolf & dog.-- the only TEST this is most important: can there be stronger a
586m079 of points of compass may be instinctive. it is a TEST to know how much of the wonder consists in  the
308c219 ince time of earliest Egyptian drawings & Old TESTAMENT» Domesticated animals having same idiosyncras
364d104 on mother. white peeled rods mentioned in old TESTAMENT placed before sheep-- it has been thought tha
549m121 his rule of happiness agrees with that of New TESTAMENT is other question.-- little is there said of
549m121 living.-- or whether if we obey literally New TESTAMENT future life is almost the sole object--. -- I
382d158 sticles in female.-- the <add> presence of both TESTES & ovaria in Hermaphrodite,-- but not of poenis
382d158 -But this not distinctly stated by Hunter.-- Do TESTES, & ovaria when they first appear occupy  their
388d172 , & beards growing on old women = Stags horns & TESTES curious instances of correlation in structure =
382d158 t no trace of abortive womb, or ovarium.-- or TESTICLES in female.-- the <add> presence of both teste
541m089 trying  to unbend my mind as much as possible (TESTING success by decreasing headache) & found best p
388d172 ndency.-- Mammae in man. having given milk. -- TESTIS & ovaria The following views show that transmi
483z012 CLOSELY allied species, therefore interlock.-- TESTUDO INDICUS not fossil at Isle of France: <Jerrold
165g126 enerally the case Wednesday 12/ & 3/ Why is the TETRAO scoticus & Tetrao-- not an American form The u
165g126 Wednesday 12/ & 3/ Why is the Tetrao scoticus & TETRAO-- not an American form The union of two instin
345d043 ed. Notes from Glen Roy Note Book.-- Why is not TETRAO Scoticus. an american form (if so)?.-- A Sphep
357d073 enaeum 1838. p. 654. Reason given for supposing TETRAO media or Rakkelhan is hybrid (produced commonl
296c184 ls.-- all in short Extreme North = = to peak of TEYDE in relation to surrounding countries &  present
421e091 o be eradicated.-- ¶.do p. 305.-- Mr Owen says <THA> «in abstract» in his paper on the Dugong, "The
521m009 window & pointed out the Gardener & said, who is THA? Mr C. answered why do you not know, that is  A.
451e178 - On the other hand there are breeds of Men the THARU & the Dhangar who can live there & do not  pine
126a124 e are springs even in siberia.-- from water {P} THAWED at + in isothermal curve. East-clinal. West cl
127a125 ast-clinal lines & c & .-- But Siberia was once THAWED. & hence. (when climate hotter) was cooled  to
044r072 solitic balls occur in the Ashes which fill up THEATRE of Pompeii (?). -- Such have been seen to form
569n022 6-9) Paris. 1835 Quoted repeatedly by Waitz (In THEIL. V) in describing Caroline Archipelago. Dn 75 c
387d169 erit the same peculiarity in lesser degree C. & THEIRS again in lesser degree.--now if the <tw> secon
026r021 cryst of glassy felspar in Phonolite arrange THEMSELVES in determinate planes ∴ such action can take
058r115 ulph. B. all the infinitesimal cryst. arrange THEMSELVES in planes. «Mem silky lustre» ask Erasmus. w
059r120 and. All the Calcareous rocks which harden by THEMSELVES cannot be pure. for if so Chalk would harden
101a056 xception in Wealden.-- Would crystals arrange THEMSELVES in that direction, in which most substance l
177b028 )» approaching another. «Petrels have divided THEMSELVES into many species, so have the awks, there i
204b141 ed Ducks & Widgeon & offspring either amongst THEMSELVES or with parent birds.-- «W. Fox. knew of cas
207b151 and> directly by instinct. can dive & conceal THEMSELVES in the grass.-- Beatson St. Helena says no t
216b179 ten whiter<)> than white parent) the mulattos THEMSELVES explain it by intermarriages with people. ei
249c036 ms been chiefly preserved,-- where shut up by THEMSELVES without other animals? but they were not shu
299c193 & probably Isle of Man) no plants peculiar to THEMSELVES. this remarkable compare it with Canary Isld
315c242 her their memory. very remarkably-- scenes in THEMSELVES accidental-- My first thought of sea  disease--
336d018 hybrid, in fact the parents beget child like THEMSELVES. expression of countenances, organic disease
343d036 certain laws of harmony keep perfect in these THEMSELVES.-- instincts alter, reason is formed, & the
348d051 .). p. 7. "The Natural arrangement of animals THEMSELVES is the question in point." Now what is natur
354d065 ld be observed with what facility mice attach THEMSELVES to man. Sept 7th. -- I was struck looking at
355d066 ong as opponents will «are» not «able to» tie THEMSELVES down, they can find loopholes "It is well w
363d103 st of America.-- showing hybrids can fare for THEMSELVES. ‖ + + first year.-- The bird fanciers  match
```

Page   **(Key Word)**
377d140 of our calendar.-- epochs & creations, reduce THEMSELVES to the revolutions of our system in the Heav
378d148 under common ducks, the young crossed amongst THEMSELVES, & I presume with common ducks. so often, th
384d162 each other, as never to be able to impregnate THEMSELVES (this never happens in plants «only in subor
387d169 erited from second pair, + the influence they THEMSELVES inherit./ Annals of Natural History .p. 96.
409e048 g extinction of animals in Europe. :the forms THEMSELVES have been basis of argument of change.-- now
429e114 with very little care.-- & which might spread THEMSELVES, as well as our wild plants, we see how full
437e139 f from the young ones while they were amusing THEMSELVES with them, and one day a rabbit escaped into
526m031 dified by circumstances, & thus the appetites THEMSELVES become changed.-- appetites urge the man, bu
534m062 waded. or swam across.-- they then stretched THEMSELVES from wall to table.-- table being removed a
549m119 erent structure of mind. they make, either in THEMSELVES, or if recollected, such part of thoughts in
582n066 Children have an uncommon pleasure in hiding THEMSELVES & skulking about in shrubbery. when other pe
582n066 about: this is analogous to young pigs hiding THEMSELVES; & heredetary remains of savages state.-- N
584n072 ing-- p. 28. how curious the means of guiding THEMSELVES through the air,-- waterbirds, the bee to it
616o039 of the expression. They would do well to ask THEMSELVES the converse of the <expr> question above st
617o040 atter. The phenomena of gravity considered in THEMSELVES consist in a force manifested in every parti
621o048 be good for the child on the long run, & for THEMSELVES & others, (as the parents are instinctively
640j167 most wretched isld should possess species to THEMSELVES.-- Probably no case in world like Galapagos.
553m135 t will of the God (<thus> & hence arises the THEOLOGICAL age of science in every nation according to
568m012 y more than «a» Savage M. Le Comte's idea of THEOLOGICAL state of science, grand idea: as before havi
567m012 he will of God. Zoology itself is now purely THEOLOGICAL.-- Origin of cause & effect being a necessar
326c266 s on subject of Science connected with Natural THEOLOGY.-- on instinct & animal, intelligence.-- very
580n062 rt.» on subject of science connected with Nat. THEOLOGY.-- says animals have abstraction because they
106a075 This argument is partly taken from Delabechs THEORETICAL Researches.-- Athenaeum. 1838-- p. 137. Thre
287c157 24. Lamarck bears to Cuvier that relation of THEORETICAL astronomer to plain observer| +++. between M
410e052 ription), except describers having some high THEORETICAL interest,-- "the great end must be the law &
462tf05 -- Tullus. Athenaeum 1839 p. 772-- A curious THEORETICAL French book review on politics in relation t
481z008 fathoms water. Mem Bahia Blanca. De la Beche THEORETICAL researches Compare land shells of Galapagos
281c137 hough that would be best).-- Argue the case THEORETICALLY if animals did change excessively slowly. w
444e157 he arm are perfect.-- p. 144.-- Alludes to two THEORIES;-- that species are the result of circumstanc
585n076 ed & whined, when, man went out of room.-- all THEORIES of magnetic powe in birds, seeing the sun &c
046r079 » particles altered.-- With respect to Volcanic THEORY. I want to ground, that the first phenomem. is
057r114 only in upper. in accordance in Europe with ice THEORY.-- Capt Ross found in Possession Bay in 73 deg
094a035 -- a most important question with respect to my THEORY of changes. of granites into Trachytes.-- Ment
103a065 st great bodies of vapour. according to Hopkins THEORY.-- general presence of dikes. argues in favour
113a091 rature clearly prove possibility of metamorphic THEORY On the idea of statical equilibrium, the heigh
119a104 are required to start with (& does not Hersche THEORY imply tendency to equilibrium.) 3d. there are
120a107 All this profoundly considered. study Hopkins THEORY of dikes may throw some light.-- thin dikes no
123a116 ration in volcanos.-- if so why not metals. The THEORY of veins will, I suspect be greatly aided by c
123a117 .-- remnant of Megetherium in interior..-- <The THEORY of [...] .> <The> Geographical Journal Vol VII
124a118 America Erasmus suggested to me that Herschel's THEORY offers no explanation of intermittent action o
131a135 l-- adduces the case to show Sir. J. Herschel's THEORY wrong.-- Geograph. Journal Vol. 8. p. 402.-- g
131a136 orld.-- argument strong in favour of thin crust THEORY.-- What a curious investigation it would be to
152a045 sequent to shelf (P) In Glen Collarig, by Dicks THEORY lake burst in most improbable part & not in Pa
135b053 at reptiles must have once declined.-- Read his THEORY of the Earth attentively Cuvier objects to <tr
196b110 ge to animals appears greatest argument against THEORY of analogies. States there is but one animal:
197b112 ith old & modern types being constant. Cuvier's THEORY of Conditions of existence is thought to accou
203b135 united-- & now divided again-- Weakest part of THEORY very distinct from Lamarcks» Without two speci
224b214 s) & living thing with thoughts (animal). «: my THEORY will make me deny the creation of any new quad
225b219 tly intermediate.-- Reference to Pig & Dogs. My THEORY true, we get (I) a horizontal history of earth
227b224 like present carnivora than Pachydermata If my THEORY would give zest to recent & Fossil Comparative
227b228 germ-- --led to comprehend true affinities. My THEORY of changes that if so in approaching desert co
230b235 »-- 1838. Hybrid Ferns It may be argued against THEORY, those the philosophers who soar above the pri
232b248 different climates Those will not object to my THEORY» tells me. he has no doubt that oldest variety
239c001 out change of habits Mr Yarrell «Give it as his THEORY, that «two» varieties of many ages standing, w
248c030 isted for ages.-- .'. The most hypoth: part of my THEORY, (THEORY admirably) yet a glance would tell from which
256c054 fficulty in distinguishing which were species, (THEORY admirably) yet a glance would tell from which
259c066 ary may be called adaptation With respect to my THEORY of generation, fact of armless parent not havi
261c070 empirical. show this by instances Once grant my THEORY, & the examination of species from distant cou
269c101 that it is not 700) is applicable to metamorphs THEORY suppose when rhinocerose lived. mean temp 60 d
280c135 m most strange.-- This even might be said.-- My THEORY thus explains a grand apparent anomaly in natu
281c138 ts would not find fossils such as they are-- My THEORY explains that family likeness, which as in abs
283c145 ned by such not having been long in blood?-- My THEORY agrees with unequal distances between species.
289c162 on may be viewed as condensor, «+++ must (on my THEORY) =supported by foetal lower developed forms.=»
293c175 his really perhaps greatest difficulty to whole THEORY.-- There is breed of tailless cats near Bath.
294c177 shee» State broadly scarcely any novelty in my THEORY, only slight differences. «the opinion of many
296c184 volumes. 2 Chapters. translated by Hooker.-- my THEORY explains this. but no other will.-- St. Helena
297c186 l habits-- to cats &c.-- must be acquired by my THEORY-- else my theory not applicable [not located]
297c186 s &c.-- must be acquired by my theory-- else my THEORY not applicable [not located] p. 428. Ouzel som
301c199 quirable through hoarding from short time.-- My THEORY must encounter all these difficulties.-- Knowi
302c200 new orders,-- Wonderful, partly explained on my THEORY, = otherwise mere fact creator chooses so to c
302c201 e sure, have remained somewhat similar.--!!! My THEORY drives me to say that there can be no animal a
310c222 be. external» appears to be a puzzle against my THEORY.-- If I be asked by what power the creator has
335d015 bred Persians «cat» favour the Persian side.-- THEORY of abortive hybrids.-- If mules did breed, the
336d019 itions are not fullfilled.-- «[My grandfather's THEORY of Mules not hereditary, because generation --
337d023 seal & mouse, elephant, come from one stock.-- THEORY of Geograph. Distrib: of <ani> organic beings.
340d026 ae.-- Consult this book again.-- Mine is a bold THEORY. which attempts to explain, or asserts to be e
345d044 e, yet instances given on opposite side,-- «The THEORY of males impressing most is in harmony with th
349d052 per worth referring to again.-- According to my THEORY, every species in any sub-genus will be. desce
352d059 or three lines deep-- with respect to Macleay's THEORY of analogies-- <be> when it is considered the
356d071 quired & heredetary tameness.-- In comparing my THEORY with any other. it should be observed not what
356d071 244. vol I) spit & throw dust «according to my THEORY of generation (p. 175) if> 8th Sept Yarrell o
356d072 the S. American quadrupeds is difficulty on any THEORY-- without God is supposed to create & destroy
370d117 line of argument «often» pursued throughout my THEORY is to establish a point as a probability by n
383d159 h possibility of becoming either sex.-- | In my THEORY I must allude to separtion of sexes as very gr
384d162 d down. but is handed down for some generations THEORY of sexes (woman makes, bud, man puts primordia
389d176 ely related: this most important with regard to THEORY, showing generation connected with whole syste
390d179 have each varied from parent stock.-- The very THEORY of generation being the passing through whole
397e004 lations Octob. 4th. It cannot be objected to by THEORY, that the amount of change within historical t
398e005 in progress, will be the last to object to this THEORY on the score of small change.-- on the contrar
398e006 aws» that baffles idea of revolution.-- My very THEORY requires each form to have lasted for its time
399e006 in upper & lower layers.-- good objection to my THEORY: a modern bed at present might be very thick &
403e023 we know its congeners then we can.-- now on my THEORY this «certainly» can be accounted for, on any
409e048 e greater area of water & snow line descent. My THEORY gives great final cause «I do not wish to say
409e049 hat the moon probably is uninhabited]CD & if my THEORY be true then the formation of sexes rigidly ne
410e051 going N.orth & South Thinking of effects of my THEORY, laws probably will be discovered. of co relat
411e055 th November When summing up argument against my THEORY, doubtless, the presence of animals in <own> «
413e059 -- consider this (Cetaceae) with reference to my THEORY Babbage 2d Edit, p. 226.-- Herschel calls the
414e060 insects.-- Stonesfield????). Have Mammalia?? My THEORY certainly requires progression, otherwise [not
415e066 l.-- Would anyone raise an argument against, my THEORY, should no fossil «very distinct species» of t
416e070 crossing [& this most important obstacle to my THEORY] CD without the hermaphrodites mutually couple
416e071 ced in the Water!» It is a beautiful part of my THEORY, that «domesticated» races. of <a> organics. a
425e103 bears relation to capability of variation?? my THEORY says so.--» March 6th. Mr Bentham says in Sand
430e117 f wild animals.-- grateful & intelligent.-- The THEORY that all animals have sprung from few stocks.
430e118 en if placed on Isld-- if &c &c.-- Then give my THEORY.-- excellently true theory Examine list of St.
430e118 c &c.-- Then give my theory-- excellently true THEORY Examine list of St. Helena Plants & see whethe
432e124 o the surrounding circumstances According to my THEORY no land animal with fluid seeds can be true to
434e128 shown we can & that analogy is sure guide & my THEORY explains why it is sure guide.-- Lychnis April
436e137 opagated & so &c. The greatest difficulty to my THEORY, is same type of shells in oldest formations:-

(Key Word)

```
437e138 om Honduras. good case of migrating.-- shows my THEORY insufficient.-- p. 120 An Eagle is said to hav
441e150 ause or law of change.-- The weakest part of my THEORY is, the absolute necessity, that every <animal
442e151 t of generation may take place by gemmation «My THEORY will not admit this, now that tulips break  by
442e152 s enormous difficulty in the way of Mr Knights. THEORY «without seeds are freshly transported»-- thro
442e152 eeds are freshly transported»-- throw over this THEORY, & the sexual reproduction of species may stop
442e153 ? whether? According to the above suggestion my THEORY would require, that <species> «individuals» pr
443e154 most important difference is forced on us.-- My THEORY only requires that organic beings propagated b
443e155 CD I utterly deny the right to argue against my THEORY, because it makes the world far older than wha
446e160 some  inward progressive developing power.-- My THEORY leaves quite untouched the question of spontan
446e163 riving.»-- throws a very great difficulty in my THEORY, here we have a plant remaining constant, with
447e164 But  if Acalepha do not cross there would by my THEORY gradation of form from one species to other: t
447e164 of form from one species to other: therefore my THEORY does require crossing.-- The case of Lemna sho
451e176 grandfather  first of race & if so, fact for my THEORY Cocos Isld & Preparis between Andaman & Pegu.
453e182 p. 275. says Teneriffe does not countenance the THEORY, of polymorphous plants, abounding in  volcanic
461t037 see  Yarrell Syngnathus Ch 6 I presume, from my THEORY, as long as any structure can be handed down w
462t051 ons.-- so many!! Hensleigh objects to transmut. THEORY, on the grounds of similarity in condition  in
463t057 s & their explanation will probably reject this THEORY-- (I must answer it by rooting out curious cas
496q05a ed by smell-- or sight.-- --. touching Mr Brown THEORY of insect-like Orchis-- & final cause of beaut
499q009 y catch pollen-- <Ne> In Oenothera bush.-- (19) THEORY of mock flowers in Hydrangea (20) As Hop is Di
500q10a c &> wandering species of confined genera By my THEORY in volcanic or rising isld, there ought to  be
501q011 cuttings taken or tuber-- talk about Mr Knights THEORY with Henslow.-- Dr. Fleming says yes. (29) Are
514q21. logy. volcanic? Applies to my geology & Species THEORY-- peculiar Fauna?. {Australian Alps--; are any
516q23v g killed Experiments not connected with Species THEORY (1) Will an extract of peat do to preserve fun
536m071 o men's affect for womens breasts.·. Dr Darwin's THEORY probably wrong, otherwise horses would have id
554m141 how, that if Cart horse argued then this into a THEORY of friction & gravity. it would be  discoverer
563nIFC Metaphysics & Expression) Selected «for Species THEORY» Dec. 16 1856 Looked through & all other Books
570n025 ve not modesty. analyse this.-- «Excellent-- my THEORY of blushing solves this.»-- The similarity  of
571n029 are taste? Real taste in mouth, according to my THEORY must be acquired, by certain foods being habit
574n039 used to show language had a beginning, which my THEORY requires. There probably is some connection be
574n041 an association in man.-- it is bare fact. on my THEORY intelligible An habitual action must some  way
576n047 n's intellect might well deteriorate. «CD[in my THEORY there is no absolute tendency to progression,
577n049 se an hydatid to be a cause of itself.-- [by my THEORY no animal. as now existing can be cause of its
577n050 t of this principle of Association will help my THEORY of sensitive Plants Habitual actions, (indepen
579n058 een invented, prove of the difference, which my THEORY believes in.-- From the manner short-sighted p
579n058 d is intent on one object.-- With respect to my THEORY of smile. remember children smile before  they
582n068 it  is uncomfortable if it does not do it.-- My THEORY explains how it comes that the heart is the se
584n072 his is class of so called instincts to which my THEORY no way applies.-- it is the acquirement of a n
587n089 nses. Reason does not lead to action.-- p. 248. THEORY of Association. owing to time when entered bra
588n089 otions.-- p. 272. Some remarks applicable to my THEORY of happiness.-- Bell on the Hand p. 191 Says <
589n091 nstincts  of animals.-- almost identical with my THEORY-- no facts, & mingled with much  hypothesis.--
599o007 t <admit> think reason sufficient to give up my THEORY-- Viewing from eminence. the wide expanse,  of
614o037 s, as species change.-- Must be so if Lamarck's THEORY true [RHC] Acquired instincts analogous «(& re
620o048 - «[just as in tastes of the mouth]CD» [LHC] My THEORY of instincts, or heredetary habits fully expla
623o050 ng the social feelings.-- [LHC] According to my THEORY, all instincts demand some explanation [RHC] A
623o051 ings», & living under certain conditions; by my THEORY they have been formed by the circumstances, wh
624o50v ge hunting instinct. *Our tastes in mouth by my THEORY are due to <habit> heredetary habit (& modifie
625o50v part  being blended & lost» & moral sense.-- My THEORY explains both, perhaps, by habit-- [LHC] 11) W
625o052 ts supremacy p. 37. Whewells gives Mackintosh's THEORY: the remarks about "contact with will" is unin
626o052 inciples born in us.-- Great difference with my THEORY.-- see p. 349.-- remark on this point.-- [LHC]
628o54v relation  I think this <boshes> «nonsense»-- My THEORY of durableness will explain it.-- Would not th
629o055 ght & wrong except instinctive ones) Perhaps my THEORY of greater permanence of social instincts expl
633j53v ion of Misseltoe?-- do p. 284. it is hard on my THEORY of gain of small advantages thus to explain th
633j54r ng created to arrest mud &c at deltas.-- Now my THEORY makes all organic beings perfectly adapted  to
635j56r ability to accidents & any other cause.-- (& my THEORY [ALL PARTS OF ONE GREAT SYSTEM. C. D]CD [All t
635j56r beings.-- true, (& the doctrine of checks & my THEORY) Macculloch. Attrib. Vol I.. p. 330. Mentions t
635j56v not be exposed to weather.-- this is against my THEORY of frequent intermarriage.-- A plant is in the
636j56vr t the habitation «or world» simple series.-- My THEORY shows life equally simple series, & therefore
638j059 e action.-- as well as gain it. by habit.-- New THEORY of instinct, returning to Kirby's view.-- Macc
639j28v ecial use to Man. Applicable to Bell's sneering-THEORY.-- p. 263. This kind of doctrine runs  through
023r011 axis was produced, from a fissure in a deep & THEREFORE weak part of the ocean's bottom. With respect
028r031 li as springs become rarer, so does the rain, THEREFORE such «rain» is cause, hence at least no water
046r080 illated between 2 & 3 ft. (as in Chili lake). THEREFORE motion of sea ought to be considered as a pla
063r132 ther at one moment or through lapse of ages.--THEREFORE we are not so much surprised at seeing Zoophi
064r136 ous quantity of matter from CREVICE of Andes--THEREFORE flowed towards it. a mass on each side 3000 f
067r146 e of disturbance inside Cordillera: It is not THEREFORE so wonderful that volcanic rocks at M.  Video
068r146 s can only happen during first movements, and THEREFORE beneath ocean, for subsequently there is a co
087a011 idea  of intertropical land.-- Siberia rises. THEREFORE to the South sinks.---- Mediiteranean contine
098a046 crystals  obeying one law of crystallization. THEREFORE concretions in this case laminar. hence the t
098a047 e thus. point of attempted crystallization, & THEREFORE as a consequence aggregated (I assume the sam
098a047 -- some force crystallizes minerals in layer. THEREFORE aggregates them in layer.-- So that layer  of
099a050 e changes in dip of no sort of consequence.-- THEREFORE < S of inclination «varies with chemical attr
099a050 ination, little force & varying direction.--> THEREFORE in PILE of mud from Trapiches. inclined layer
103a064 onsidering how close the dislocations occur & THEREFORE that the crust might be considered a level.--
104a066 ns 6 per cent more silica than common felspar THEREFORE on axis of Cordillera, in Andite - containing
104a068 e to whole moisture being lost by evaporation THEREFORE capillary attraction would bring water with s
105a073 urse, & with slow course small erosive power. THEREFORE tendency of running water to deepen not to wi
106a074 in a "rapid".-- is a familiar illustration.-- THEREFORE stream has no tendency to widen course  until
106a074 small, & when that is case force is lessened. THEREFORE rivers very ineffectual in widening valley.--
106a075 ey.-- it is essentially a deepening agent {P} THEREFORE when we have valleys of this structure. as th
114a094 vage Clay slate.. a distinct formation deep «& THEREFORE extensive» water ·. not formed in modern forma
119a105 mountains.--  consist of rocks with fossil., THEREFORE formed near surface. whether they can have be
120a108 ks.--, but lava shows the rocks really hot. & THEREFORE I doubt the thermometer. Is not common salt in
132a139 caused, by absence of circulating water.-- & THEREFORE that temperature of earth beneath <of  Sahara
133a140 25' All facts show how slowly heat travels; & THEREFORE the abysses where fluid rock has been ejected
171b002 n (especially in lower animals, where mind,-- THEREFORE relations to other life has not come into pla
171b004 th fixed organization» thus being modified,-- THEREFORE generation to adapt & alter the race to chang
171b005 is with our present system of body & universe THEREFORE final cause of life With this tendency to var
175b016 This  view supposes that in course of ages. & THEREFORE changes. every animal has tendency to change.
177b029 enus upon another, whole class would die out, THEREFORE not.-- Monad has not definite existence.-- Th
177b029 ence, «in» perfect<ion>, «species from many» <THEREFORE> changes and base of branches being dead from
180b038 ing to the adaptation of such circumstances & THEREFORE that death of species is a consequence (contr
181b041 ould be reduced to small percentage.-- & <in> THEREFORE the chances are excessively great against, an
186b058 e altered, I can conceive colouring retained; THEREFORE probably in some <heteromera> colouring of cr
186b061 without change, superinduced, or new species; THEREFORE animals would perish, if there was nothing in
187b062 at Siberian animals lived in cold countries & THEREFORE not killed by <Siberia a more> cold countries
187b064 al die, to perpetuate certain peculiarities. (THEREFORE adaptation), & to obliterate accidental varie
192b085 change  than we know varieties can produce.-- THEREFORE all genera MAY have had intermediate steps.--
194b094 of  Sapajou-- NB Sapajou is S. American form. THEREFORE it is like case of great <rodent> edentate [h
211b161 species.-- Animals have no notions of beauty, THEREFORE instinctive feelings against other «for sexua
211b163 s cats without tails: some long & some short: THEREFORE like dogs.-- Ogleby says, Wolves at Hudson ba
213b170 are  those ¿Lyell?, which have wide range and THEREFORE cross & keep similar. But this is difficulty;
218b189 good  in plants» between all different forms; THEREFORE when from being put on island. & fresh specie
219b195 tions» because snow formerly descended lower, THEREFORE species of lower genera altered. or  northern
219b195 rated by short space from mountains low down, THEREFORE plants common take an example from T. del Fue
220b197 structures  especially reproductive organs; & THEREFORE the one distinction of species would fail. Bu
221b201 t all have existed for ages as metamorphic; & THEREFORE according to Lyells doctrine removed?? Is the
226b222 nent many species belonging to its own genera THEREFORE if in small tract we have many species, we ma
248c033 the internal parts be of longest [consta]nt & THEREFORE most permanent Owen [re]markable laws of Brai
```

Page		(Key Word)	
248c033	sly affected by external circumstances> these	THEREFORE	will be chiefly heredetary.-- If varieties «p
249c034	capable of producing <two> such as itself.--	THEREFORE	two different varieties will produce hybrids
259c062	ubterranean lakes, hot spring &c &c inhabited	THEREFORE	mud wood be inhabited, then how is this effec
259c065	adaptation> «change» during life of parent, &	THEREFORE	being always necessary may be called adaptati
265c087	e use to which wings are generally applied.--	THEREFORE	argument not destroyed even if these shrivell
269c101	undred feet would be 60 degree + 6 degree??.,	THEREFORE	34 degree degrees of change have travelled th
278c131	a, associated with teeth of seals and dugong,	THEREFORE	immense age since breccia accumulated-- surel
281c137	oubt if local varieties should be remembered,	THEREFORE	do not consider it as proved that they are va
281c139	ng those last obtained.-- less firmly fixed &	THEREFORE	most subject to change.-- may account for cer
284c147	ual quantity, 2d on relations of heat & cold.	THEREFORE	probably fewer now than formerly.-- The numbe
284c149	ecies in group we may suppose it is oldest, &	THEREFORE	lest subject to variation.-- + <being> good f
284c149	ation, & not on extension ¿these go together?	THEREFORE	value of organs vary in different group. & No
285c153	ts The AEgyptian animals domesticated «??», &	THEREFORE	Most especially under care of Man. & external
292c169	ted the animal the more subject to variation.	THEREFORE	sexes or two animals:-- «When» sexes <being i
292c171	action implies want of consciousness & will &	THEREFORE	may be called instinctive.-- But why do some
292c171	memory.-- structure is only gained slowly.--	THEREFORE	it can only be those actions, which Many succ
292c171	The improvement of reason implies diversity &	THEREFORE	would banish individual, but general ones mig
296c185	grine Falcon holds birds for some time alive ¿	THEREFORE	other species mice & only kills them when urg
297c186	rough probably an infinite number of forms.--	THEREFORE	an isolated form probably a remnant.-- Pachyd
300c198	ement of language. heredetary & acquirable.--	THEREFORE	mans mind not so different from that of brute
302c202	s of analogy.-- last acquired,-- or aberrant,	THEREFORE	more easily modified = = this is not easily t
302c202	gs number of vertebrae. subject to variation.	THEREFORE	lately acquired.-- I fear «great evil» from v
302c203	ng all of intermediate Father-species, & not,	THEREFORE,	solely owing to such interm: father-species,
304c206	how close birds! look at Mammals: how wide.--	THEREFORE	birds younger???? or «have» not «been» expose
304c208	ch death, may be inferred much time elapsed &	THEREFORE	descended from branch high up.-- Such probabi
306c212	te, but that is much more general argument) &	THEREFORE	down the stream followed ebb tide, therefore
306c212	therefore down the stream followed ebb tide,	THEREFORE	got into habit of going down stream which wou
309c219	e. arise a good deal from climate & habits, &	THEREFORE	less fertile. according to Mr Herbert's views
312c232	Men «& animals» could only exist by habit.--	THEREFORE	same principle transferable., not wonderful A
313c235	ced in succession (13?) without impregnation,	THEREFORE	sexual passion must arise after long interval
335d013	et offspring must be somewhat like parents,--	THEREFORE	offspring will tend to go back, or have none-
335d014	» family with defective palates. heredetary &	THEREFORE	exceptions. to above law.-- Study what these
335d015	but as they must like or there will be none,	THEREFORE	a mule can have no offspring.= but as «badly»
336d017	mit them, & as offspring must be like parent,	THEREFORE	mule has no offspring & therefore no generati
336d017	ike parent, therefore mule has no offspring &	THEREFORE	no generative organ.-- Same Prop. better enun
342d033	t wild (escaped from Carolina?) off New York.	THEREFORE	instincts not imperfect.-- Are Pheasant & Gro
347d049	a very old variety will be harder to vary, &	THEREFORE	more apt to be extinguished.--???» Mayo (.Phi
353d060	parent stock, or not. Now wings for flight--	THEREFORE	ostrich not. The peculiar <Malacca> «Malacca»
366d111	he> every case common to many good species; &	THEREFORE	to genera (& the uncles & aunts) & therefore
366d111	therefore to genera (& the uncles & aunts) &	THEREFORE	does not tell against transmutation of specie
369d114	of Peacock & spurs-- no final cause here.-- &	THEREFORE	different from Hunter I should say females re
372d130	.-- it must have the knowledge how to grow, &	THEREFORE	to repair wounds-- but this has nothing to do
374d134	e first cross generally brothers & sisters, &	THEREFORE	somewhat unfavourable-- 28th. «I do not doubt
381d156	generation)-- Epizoa & the nematoid Entozoa--	THEREFORE	highness in scale has no «constant» relation
388d170	rmation, nearly independently of its parent &	THEREFORE	wants independent supply of food.-- is real.
389d176	either.-- Male & female as foetus one sex; &	THEREFORE	both capable of propagating, but one is rende
389d177	is the end frustrated, when near relations, &	THEREFORE	those very close are bred into each other.--
390d178	budding. which is not object of generation.--	THEREFORE	passions fail.-- In fruit trees no doubt ther
390d178	e difference of parent, tree, but it fails.--	THEREFORE	«each» seedling of one apple ought to differ
390d179	nfertile with highest bred of other ¿ breed.=	THEREFORE	it is not really breeding in & in, but « <on>
404e024	ion made by man, common to every individual &	THEREFORE	effect of climate.-- Octob 19th. When reading
404e025	expect this structure would become obscure &	THEREFORE	it might thus have arisen, & M. Edwards p. 33
404e026	aptation,-- however mysterious such is case.	THEREFORE	chance & unfavourable conditions to parent ma
405e031	e best marked characters are partly retained,	THEREFORE	colours vary in same Manner as they would var
407e039	al> Equable kind of climate to the extreme.--	THEREFORE	species, which were fitted for such a preemin
407e039	s might have survived this mundine change..--	THEREFORE	I argue from this that Africa «& East Indian
407e039	were not so «very» EQUABLE. or so tropical, &	THEREFORE	present state of world is not so different, w
408e043	ernal conditions have great effect on them, &	THEREFORE	extermination becomes part of same law.-- Whe
415e068	oes not depend on the number of the species.:	THEREFORE	Man & monkeys have equal chance that progenit
416e070	seminal fluid fluid, (& not dry as in plants)	THEREFORE,	great difficulty in crossing [& this most im
416e071	I suspect this rather effect of liquid semen:	THEREFORE	animal life commenced in the Water!» It is a
417e078	s both parents, for they are one.-- The laws,	THEREFORE,	of likenesses of fathers to children of mank
419e087	ily,, by assumption well grounded, on time;--	THEREFORE	the mere loss of species, which may be the wo
419e087	orm, «which must be effect of slow change + &	THEREFORE	precludes effects of catastrophes, which must
424e101	end on negative evidence of fossil remains, &	THEREFORE	not to be trusted.-- -- Lyell tells me, on au
432e124	his forefathers only when advanced in age, &	THEREFORE	the children do not, (& in hairless kittens w
440e147	other kind of animal to usurp its place.-- &	THEREFORE	the degree of injuriousness must have been ex
447e164	gradation of form from one species to other:	THEREFORE	my theory does require crossing.-- The case o
461t037	en «callosities on Camels & Horses--».--«)» &	THEREFORE	probably any structure would rather become ac
482z012	1 degree to 26 degree CLOSELY allied species,	THEREFORE	interlock.-- Testudo INDICUS not fossil at Is
525m026	m to have failed even more than the memory.--	THEREFORE	affections effect of organization which can h
526m027	> learnt. what they teach by the same means &	THEREFORE	properly no free will.-- we may easily fancy
527m033	g just kept up by the music of the poetry.-- (	THEREFORE	singing intermediate, who has not had his blo
527m034	such trains of thought makes a discoverer, &	THEREFORE	(independent of improving powers of invention
529m039	, & in experienced people-- recall pictures &	THEREFORE	imagining pleasure of imitation come into pla
530m045	ilor blacksmiths?) are likewise heredetary, &	THEREFORE	that their children have some little advantag
531m052	t.-- now this is the oldest <her> inherited &	THEREFORE	remains, when the actual movement does not ta
531m053	uickly away from any sudden sound or noise, &	THEREFORE	brain has been accustomed to send a mandate t
537m076	some universal feelings of right & wrong «(&	THEREFORE	in fact only limits moral sense)» which she s
537m077	patriotism & family pride are heredetary., &	THEREFORE	he has these strong, & does not act up to the
541m090	my father. is right in saying delirium rest--	THEREFORE	dreams thus act.-- ∴ weak minded people are f
542m096	movement is continuing it when useless.-- <&>	THEREFORE	it is here continued when the uncovering the
543m098	r more instincts in all of the Hymenoptera;. <	THEREFORE>	than in other orders (study Kirby with this
543m098	in other orders (study Kirby with this view)	THEREFORE	there is Instinctual developement in one orde
543m100	ertain.-- there is judgment of probabilities,	THEREFORE	this judgment gives a man common sense, & the
544m101	ld sucking, gives pleasure, & always has done	THEREFORE	sight of own child. (when frame in condition
544m103	but yet scarcely really moves.-- the willing	THEREFORE	is ideal, as all the other perceptions.-- The
547m113	. one could not compare the castle with them,	THEREFORE	could not doubt or believe.-- When I say trai
548m114	may be some «two or three» trains of thought,	THEREFORE	one may be imperfectly reason -- <In a> Aberc
548m115	omparing every step, & inventing new means,--	THEREFORE	works of imagination hard work,-- Keeping one
549m119	affection excited, pleasure of imagination--	THEREFORE	do these & be happy-- & these pleasures are s
550m124	r less pleasant & unpleasant, in same time,--	THEREFORE	degrees of happiness-- Entire happiness. not
564n002	uld be sorry or have a troubled conscience.--	THEREFORE	I say grant reason to any animal with social
565n006	te at moment, but mere symbol of readiness, &	THEREFORE	done in extreme.-- Looking at ones face <&> «
565n008	hurting opponent by insulting his pride & is	THEREFORE	of the snarling order.-- But contempt mingled
573n038	, effect of thought is to make saliva flow, &	THEREFORE	thinking of subject, even when wishing not to
585n075	touch & what [...] the same.--(this Hensleigh	THEREFORE	problem is how we know that thing is same, wh
592n105	s unintelligible, <both> in the end gained «&	THEREFORE	the» cause, and origin being so is not odd.;
593n107	but it does not know its own smell or look, &	THEREFORE	there must be some instinctive feeling which
606o024	ure, but rather subjects which we know, it is	THEREFORE	the embodying of a floating idea.-- as statue
608o026	his motives (originally mostly INSTINCTIVE, &	THEREFORE	now great effort of reason to discover them:
608o027	ppiness lays in doing good & being perfect, &	THEREFORE	will not be tempted, from knowing every thing
609o029	rongest in man, are common to other animals &	THEREFORE	to progenitor far back, (anger <to> at the ve
609o029	ar back, (anger <to> at the very beginning, &	THEREFORE	most deeply impressed). shame perhaps an exce
610o30v	hers) shows does not tend to greatest good.--	THEREFORE	rule of happiness is to certain degree <of> r
613o036	o plants, but surely instincts imply willing,	THEREFORE	word misplaced The meaning of Words, must be
614o037	of such higher animals may be looking back, ∴	THEREFORE	consciousness, therefore reward in good life
614o037	y be looking back, ∴ therefore consciousness,	THEREFORE	reward in good life [RHC] Instinct appear lik

```
617o40v ving in opposite directions. We are satisfied THEREFORE, if we can trace any force in inanimate matte
619o042 ce either protecting sheep or hurting them.-- THEREFORE in man we should expect that acts of benevole
620o044 ncy & in its consisting in desire gratified & THEREFORE as soon as desire is fullfilled, pleasure for
621o047 ukes malevolent feelings, as much as actions, THEREFORE Sir J. M. talks too much about the continguit
629o54v ould not the maternal affections (in a dog. & THEREFORE not <instinct> «conscience»} equally <prefe>
635j55r as not the character of a physical law, «& is THEREFORE utterly useless-- it foretells nothing» becau
636j56v My theory shows life equally simple series, & THEREFORE trace of beginning in organic world.-- Maccul
350d056 error-- Observed, nature does nothing in vain, THERFORE organ fitted to animals place in creation.--
120a108 s not deep. --Hot springs &c &c--then if so, THERMOMETER show it cannt be ordinary heat, then there i
120a108 he rocks really hot. & therefore I doubt the THERMOMETER. Is not common salt more soluble in <hot> co
119a105 owels of the earth, as would be required by THERMOMETRICAL scale.-- (for the temp must be immense to
603o11b s yes. <old> mass of rock--]CD or poetry, CD[my THERY says yes. imitating song -- two primary sources
158g079 terrace also right-- Granite such as boulder on <THES> Divortium aquarum Peaty Mass of this point ver
326c266 ores would not have children Turners embassy to THIBET, perhaps worth reading quoted by Malthus.-- He
022r008 t long. "Its maw was like a leathern sack, very THICK & so tough that a sharp knife could not cut it:
029r034 on to present position of <Coal> Forests. These THICK beds of Lignite stratified with substances so l
055r106 n of sea's bottom. beds of shells. 2 - 3 toises THICK.--Vol II. p. 252 Urge cliff form of land, in St
064r136 flowed towards it. a mass on each side 3000 ft THICK & 150 broad. neglecting Cordillera itself now r
068r146 of solidifying igneous rocks which would be too THICK to be penetrated by the repeted trifeling injec
070r154 ce of Subsidence at Uspallata.--? ¿If crust very THICK would there be undulation? would it not be mere
096a046 ore concretions in this case laminar. hence the THICK wedges of feldspar in gneiss.-- Veins in septar
145g004 parallel to strata 3 or 4 seams/ 3 or 4 inches THICK-- {P} 35 degree as I believe about greatest dip
259c066 view of> generation right?-- If puppy born with THICK coat monstrocity, if brought into cold country,
300c197 a degree. Mem Spider only dropping where ground THICK.-- shamming death it is but being motionless. H
306c214 upposed to come from there.-- All the sheep are THICK-tailed The dogs called Persian «greyhounds» are
338d024 -- adaptation & species-like,-- -- Says Negro-- THICK skinned My hairdresser (Willis) says <black> «t
363d102 wl <the> in the former place breeds in <flags> «THICK vegetation» in swamps-- (owing to barns, perhap
399e006 for its time: but we ought in same bed if very THICK to find some change in upper & lower layers.--
399e006 y theory: a modern bed at present might be very THICK & yet have same fossils. does not Lonsdale know
402e017 bserved to change «is very great difficulty» in THICK strata, can only be explained, by such strata b
066r140 istmas sound. -- --(Think some 60 fathoms, none THICKER than thumb» Sea weed said at Kerguelen Isd. to
108a079 rom centre, then crust of solid earth would be THICKER.-- PP Andes mark the line between sinking & ri
122a114 -- if part of <cold> crust under ocean, became THICKER, then when fluid moved [...] {P} August 25. I
030r035 torrents must bring much vegetable matter from THICKLY wooded mountains, probably chiefly leaves.--Th
147g015 Tyndrum where a large sort of <plain> space is THICKLY studded with ridges & flat topped hill/ do all
032r039 gth be checked by increased vertical <height> THICKNESS (DZ) of mass to be removed & from the resista
073r163 daloid. basalt & other trap cover it to great THICKNESS. = Coast of Acapulco granitic rock.--in parts
088a013 ost & turned about position of strata, prooff THICKNESS not very great; where piece turned over axis
104a066 islands. & generally where rock metamorphic & THICKNESS of <strata> not great, one can conceive antic
104a066 sure would always produce it) but where great THICKNESS is affected, they would be far off In Discuss
107a078 speculation becomes important with respect to THICKNESS of crust broken up.----My view of Volcanos &c
124a119 f copper not melted absorb, zinc thrugout its THICKNESS.-- this most curious with respect to epigmous
125a121 ssion of beat by gases-- does not apply it to THICKNESS of crust.-- {P} if crust were metal then thin
269c101 degree degrees of change have travelled that THICKNESS in that period & no ways assisted by fluid cu
486z020 nder strata of gravel & clay about 10 feet in THICKNESS.-- (March, 1842) Questions & Experiments {T}
272c108 ome orthopterous insects & some third, have got THIGHS with same peculiar structure & habits of cling
367d113 ost immediately employed in fighting" instances THIGHS of cock & Neck of Bull.-- is most common in ve
499q009 ate (whole or broken) is ball of pollen on Bees THIGHS (18) Place pin's heads with Bird lime near mal
258c061 d & pugnacious (all order; cocks all warlike)» «THIS wars against in any class:. those points which
041r066 ay, which little vein was like the rest of these THIN veins which project outwards.-- In Patagonia. t
063r131 tion;-- An argument for the Crust of globe being THIN, may be drawn. from. Cordillera. rocks.--When b
070r154 ion? but walls & feeling shows undulation:. crust THIN.--Concepcion earthquake Draw close Analogy Lake
090a024 rturbation be serious? if so other cause besides THIN vapour bringing planets to an end? Fragmentary
097a043 e. talks of Trachyte, "superficially coated by a THIN pellicle of a blackish colour like a dull & poo
103a063 above unite with those from below. would always THIN out above which explains a difficulty.-- All De
115a098 op. ended by abrupt slope {P} each stratum would THIN out, both inland & seaward: if matter too coars
120a108 Hopkins. theory of dikes may throw some light.-- THIN dikes not cooling if they had travelled some hu
130a133 r I show crust yield easily. & if easily must be THIN: <beside mere fracture> A Elevation as in Patag
131a136 rts of the world.-- argument strong in favour of THIN, crust theory.-- What a curious investigation it
532m056 sleep at night even when in bed room-- grew very THIN, would not go out of house except with Caroline
532m056 th Caroline-- After fortnight. continued to grow THIN & did not seem quite happy. in five weeks was s
532m056 & did not seem quite happy. in five weeks was so THIN, that she was sent back to Shrewsbury,. then im
326c266 837) on limits of painting & poetry.-- Erasmus THINCKS I should lik it. The Sportsman's Repository. 4
102a059 ted some way out to sea.--¿ effects on bottom a THING floating some way from coast is driven on to i
121a111 sed muscles.. Smith of Jordanhill has seen same THING-- Consider profoundly How came it. that Glen Ro
171b002 onomia arguements, fails in hybrids where every THING else is perfect; mothes apparently only born to
185b057 Again taking a subdivision of Heteromera same. THING occurs with regards to other tribes in that sam
204b141 nking every animal born to consume this or that THING.-- There is some much higher generalization in
224b214 of Man & animals not so great as between living THING without thought (plants) & living thing with th
224b214 living thing without thought (plants) & living THING with thoughts (animal). «∴ my theory very disti
239c001 imaux than Pointer.-- He has no doubt that same THING would happen with Australian dog & any of our c
285c152 true affinity doubtful A species is only fixed THING with reference to other living being.-- one spe
285c152 course of ages.-- As species is <certain> real THING with regard to contemporaries-- fertility must
292c171 nsmitted without consciousness «a most possible THING. see men walking in sleep».-- an action become
315c240 n Australia & Europe.-- if creation be absolute THING, the creation must take place only] when creat
317c249 as on coast of Africa.-- Macleay tells me same THING p. 55. 40 leagues from land several patches of
336d017 it be small & slowly obtained NB. The longer a THING is in the blood, the more persistent.-- «any am
352d058 e of organization. such law would explain every THING.-- PURE HYPOTHESIS be careful.-- Argument for c
360d090 came from them.-- showing how gradually every <THING> «change» is effected)-- the one in garden is f
415e069 almost certainly not made into man.-- It is one THING to prove that a thing has been so, & another to
415e069 ade into man.-- It is one thing to prove that a THING has been so, & another to show how it came to b
449e168 two shapes & colour.-- Eyton has observed same THING in Brent Goose. Eyton says some of the pidgeons
451e180 monkeys do not exist. there & it is a singular THING that throughout the Moluccas Archipelago they a
471tf11 is almost led to doubt. whether there is such a THING as a species-- Jun 1. 1842 Allen W. sowed some
522m012 y.-- In old people. (Aunt. B.) when they hear a THING it often does not take any effect at the time,
522m012 easure. & is often recurred to & mentioned as a THING which had just taken place.-- as if the idea of
523m016 some years after, was far more painful than the THING itself. Asked my F. whether insanity is not dis
526m027 ily fancy there is, is as we fancy there is such a THING as chance.-- chance governs the descent of a fa
526m029 collection of pictures).-- when one remembers a THING in a book, one remember the part of page.-- one
533m058 n life, <before they> (I think I have seen same THING before they could understand. what frowning mea
536m071 t scrapes off its bottom.-- it is relic of same THING that makes one dog smell posterior at another.-
543m101 ging thought, & not merely instinct, a separate THING superadded.-- we can thus trace causation of th
546m111 ect <mental> health.-- «Erasmus had almost same THING happen to him about a knife. which he had hid s
553m132 orn in conflicts with the former. But it is one THING for a swift dog to overtake an emu, & [not loca
554m140 is nose with «a» straw.-- Jenny will often do a THING, which she had been told not to do.-- when she
566n012 enquiry commenced, for instance probably such a THING as thunder. would be placed to the will of God.
581n063 s not that kind of memory. which makes you do a THING properly, even when you cannot remember it. as
585n075 Hensleigh therefore problem is how we know that THING is same, which touches two parts of our bodies,
608o027 erefore will not be tempted, from knowing every THING he does is independent of himself to do harm.--
618o41v blueness & weight always went together. & as a THING grew blue it «uniquely» grew heavier yet it cou
629o55v al utility (3) It is other question whether any THING is taught instinctively; I say yes, & my explan
632j53r been uninhabited as the Alpine pinnacles.-- One THING must be admitted there would not be these plant
023r012 rom the contents of its maw, amongst which were THINGS pitched over board early in the passage!!-- M.
045r075 ountry newspapers)--At the Calabrian earthquake THINGS pitched off the ground. «Ulloa states that Vol
227b227 Galapagos on Pacific side. the Oolite order of THINGS might have easily been formed.-- With belief o
262c073 rrun-- Tahiti. thistle. Pampas. show how nicely THINGS adapted--.-- These «aberrant» varieties will b
293c172 tructure in brain people in fevers recollecting THINGS utterly forgotten» --it is scarcely more wonde
335d015 ent), then according law, that in proportion as THINGS are long in blood so will they remain, a mule
```

Page **(Key Word)**
520m002 ect frequently <are> have very bad memories for THINGS which happened in early infancy-- of this fact
522m011 aintained he had never heard of.-- Thus in many THINGS if he began at one end, he knew the whole subj
524m022 -- People in old age. exceedingly sharp in some THINGS, though so confused in others.-- Mrs P. when i
526m027 ween organization & mind.-- thinking over these THINGS, one doubts existence of free will every actio
526m029 which I first recollect, «at Zoos» they are all THINGS, which are brought to mind, by memory of the s
526m029 cied.-- this fact of early memory consisting of THINGS seen, quite agrees with my Fathers case of Mr
531m050 f reason & abstinence.-- My Father remarks that THINGS of great importance are easily forgotten, (if
532m055 oubtful whether they could recollect these same THINGS from any effort of will whilst their minds wer
538m078 e had recollected it in perfect senses.-- These THINGS, & drunkedness, show what trains of thought de
538m079 ated: Now it is remarked that A. Bessy repeated THINGS, which none about her had EVER before heard, s
539m083 t awakened.-- The habitual individual remembers THINGS done in the other habitual state because it wi
542m095 ris?-- same way as one lifts up eyebrows to see THINGS in dark. & hence is this the cause of expressi
547m112 ce» my servant, «box» my own manner of ordering THINGS to be done.-- The senses are closed probably b
548m115 air the trouble «I well recollect» is in making THINGS somewhat probable. in comparing every step, &
551m127 .-- insane people believe they hear as well see THINGS which have no existence.-- He compared spectra
569n020 e commenced in some necessary connexion between THINGS & voice, as roaring for lion &c &c. (in same w
569n021 D power of conception between one or two absent THINGS.-- reason probably mere consequence of vividne
569n021 mere consequence of vividness & multiplicity of THINGS remembered & the associated pleasure &c accomp
571n031 age. the possibility of poets describing gentle THINGS in gentle language, & vice versa.-- almost pro
575n046 People-- (Antiquary Vol II. p. 77) remembering THINGS of youth, when new ideas will not enter. is so
581n065 ords, roar, scrape, crack, &c, imitative of the THINGS.-- CD[I may put the argument,, that many learn
585n077 instinct-- instinct is heredetary knowledge of THINGS which might be «possibly» acquired by habit. s
587n087 n other hand, it is said people, who like sweet THINGS dislike others.-- dogs dislike perfume) I shou
602o11b 5. Attributes of Deity. on Belief.-- you belief THINGS you can give no proof for, & one often replies
610o032 ys he has heard of many cases of ideots knowing THINGS, which are often repeated in a wonderful manne
618o41v ght, the blueness, still less between <action> «THINGS» so different as action thought & organization
624o051 e reoccurred'. NB. Until, it can be shewn, what THINGS easiest become instinctive, this part of argum
031r037 ar to have fallen into lines of faults I do not THINK so many faults in Cordillera, as in English Coa
032r041 of» tidal band of action. This case differs. I THINK. from Patagonian steps, because the deposition
035r045 W. America P. 209--13 P & 444 «(Yanky Edit)» <I THINK> At Ascension, the laminae <...> changes in roc
035r046 e belonging to the same district there seems. I THINK, little ground for skepticism, as to the genera
038r057 there must be a central core of melted rock--I THINK the strongest is the consideration of the state
046r080 rtical as lateral movement.--At first one would THINK movement. owing to water keeping its level whil
046r080 rose up & down.--But from above reasons, do not THINK so also elevating Earthquake of Valparaiso. (18
048r086 n is applicable to the Siberia case We must not THINK alluvial plains «always» most favourable; In wh
050r090 agments, nature of which is doubtful. P. 180. I THINK my Ascension case very doubtful. -- In Iceland
050r091 amang.--Marsden Sumatra. M. De. Jonnes seems to THINK that Volcanic eruptions form foundations for Co
057r113 ut much at Coll. of. Surgeon's? I really should THINK probably that B. Blanca & M. Hermoso contemp:.-
057r114 ussion of camel urge S. Africa productions.-- I THINK in Patagonia white beds having proceeded from g
066r140 ast of T. del Fuego. off. Christmas sound. -- «(THINK some 60 fathoms, none thicker than thumb» Sea w
078r174 igny has described it with care to 3 species. I THINK I have much additional information Guanaxuato
107a078 thod of cooling--Very difficult subject. PP-- I THINK from dislocation taking place chiefly beneath w
108a080 at shock -- It appears to me unphilosophical to THINK calcareous springs near coral reefs.-- Where ve
109a084 ints in chalk must be studied-- though I do not THINK good p. 411 When discussing concretions Carbona
160g088 shows the ascending fringes {P} which makes me THINK it submarine, 400 or more feet above station! T
177b025 rriage <increases it--> settles it ¿We need not THINK that fish & penguins really pass into each othe
180b036 hich is not case,∴ MONUCULE NOT DEFINITE LIFE I THINK {P} Case must be that one generation then shoul
194b095 d go to show a centrum for Mammalia.-- I really THINK a very strong case might be made out of world b
195b098 lieve the Creator creates by any laws. which, I THINK is shown by the very facts of the Zoological ch
203b137 ant flycatcher doing the service of a swallow I THINK we may conclude from Australia & S. America. th
206b150 plants & few Monocot in Coal formation? p. 320 <THINK> States Cryptogam. Flora formerly common to New
213b172 the cross most well worthy of observation.-- I THINK it is certain strata could not now accumulate w
216b179 ian sea.-- Deshayes.-- Mr McClay is inclined to THINK that offspring of Negro & white will return to
221b202 us a fact analogous to reptiles before Mammalia THINK about Miocene fossils some species being recent
222b203 rs is attempt at returning to parent stock.-- I THINK we may look at it so--?? It holds good even wit
227b225 you to believe the world older than geologists THINK. it agrees with excessive inequality of numbers
264c080 egian, puts on par with Monkeys» Gould seems to THINK that many species when close come from differen
265c086 nt habits-- Thus bill & nostril of Puffinuria I THINK we may clearly attribute to heredetary origin &
275c121 ke parent Brahmin bulls-- Mr. Y. is inclined to THINK that the male communicates the external resembl
278c131 Dipus of present day??! Major Mitchell does not THINK that dog was found in Van Diemen's Land.-- V. 1
281c139 us to each other-- &c &c.-- V. p. 140» I should THINK meaning of circular arrangement was only so far
283c144 ate Bird & flying eagle.-- Hawk Gould seemed to THINK, that widow bird. replaced Birds of Paradise--
286c155 yet have justly excalted nature of man. like to THINK his origin godlike, at least every nation has.
288c159 breed.-- Native dogs & English cross readily-- THINK about half way in appearance.-- bark about half
290c164 en fowls & pheasants, is most like pheasant., I THINK so because very 3/4 bred.-- (hence hybrids in t
293c173 caterpillar, though our ignorance, may make us THINK so, but only between laws--]CD. Many diseases i
304c207 t with other birds,, or branched off anteriorly THINK what principles are there to guide in this opin
310c223 t origin. «Royal Institution» Dr Royle seems to THINK Botanical Provinces will turn out not nearly so
339d026 re are some small divisions.-- does not seem to THINK any improbability to animals being distributed
359d087 ents.-- (There are some 3/4 birds «of», which I THINK there must be some mistake in their origin) Saw
379d152 s 406, 407, 409, Quetelet papers are given, & I THINK facts there mentioned about proportion of sexes
385d163 owers» in Zoolog Gardens. & Kings at Otaheite) <THINK> Last litters are considered the most valuable.
401e016 ose suddenly produce cowslip, one is tempted to THINK here some anomaly-- I can fancy cowslip produci
404e025 > «some orders» of crustacea, one is tempted to THINK that it must have been invented all at once.--
417e075 rtility of Indian & Common Oxen, which one must THINK deserve the name of species, may be owing to th
417e076 to see if they would breed, It is difficult to THINK of ¿¿Plato & Socrates, when discussing the Immo
426e106 es. My views, «V <see> p. 103» would lead me to THINK that a <do> variety of one species would cross
426e107 troys the appearance of this series & makes one THINK that one large body of varieties are fertile &
430e120 ediate one.-- Mr Lonsdale evidently inclines to THINK it Hybrid.!!! Ask Woodward Mr Lonsdale says Tri
431e122 nce of such facts.-- Ask Phillips.-- The more I THINK, the more convinced I am, that extinction plays
439e144 shows either there is not so much crossing as I THINK,, or that these varieties have become as fixed
439e144 er variety.-- «Elizabeth & Hensleigh. seemed to THINK it absurd. that the presence of of the Leopard
443e155 makes the world far older than what Geologists, THINK: it would be doing, what others but fifty years
446e162 mation & by seed-- which Henslow is inclined to THINK very close.-- «A fruit tree by certain treatmen
448e167 can see no other way of accounting for them.-- THINK over this-- The Superga beds have many shells i
451e176 self said not to be common-- probably, I should THINK grandfather first of race & if so, fact for my
467t103 omes out with anthers & stigma in slit» -- As I THINK they do in Broom & certainly when over-ripe & h
470t178 Monk's Hood, brushing over stamen «Egg Tree»--I THINK never on the Galeum saxatile & other common kin
470t178 r on the Galeum saxatile & other common kind--I THINK not on Phlox though they examine it.--«Little D
472s02v so whole sides of flower & stigma dusted.-- <I THINK> When It first alights, it cleaned sucker & <I
472s02v > When It first alights, it cleaned sucker & <I THINK> pollen was scraped off, which appeared like He
481z010 tails appear very trifling Also Berre «p. 8» (I THINK Planariae) Sagittella, or Fleche «p. 8» my litt
494q005 lapse during sleep & do of Berberis-- (latter I THINK certainly not) (3) Sow seeds & place cuttings o
501q011 three plants. because they cannot be crossed, I THINK, I expect, except by very minute insects.-- (30
504q13v account of Van Mons views-- Also PEAS-- N.B. I THINK very likely the Peas to cross ought to be place
506q015 occupy particular position.-- (39) What does he THINK of Dr. Flemings statement of Sweet Williams & S
515q022 over that little Ervum in Sand-walk, on which I THINK I have never seen Bee visit. Experiments in Gar
523m015 Father says after insanity is over people often THINK no more about it than of a dream.-- Insanity is
524m022 ople, my Father says, do not dream of what they THINK of most. intently.-- criminals before execution
526m029 remember the part of page.-- one is tempted to THINK all memory consists in a set of sketches. some
531m050 n, (if unconnected with fear &c) because people THINK that the importance of the event by itself will
532m055 d which is startled.-- My Father says he should THINK that in old people, in their dotage, who sing t
533m058 ow a frown very early in life, <before they> (I THINK I have seen same thing before they could unders
534m061 who will venture to say germ within egg, cannot THINK-- as well as animal born with instinctive knowl
537m075 rom difference of action of approved» Yet as, I THINK, the opposite side has been shown-- see Mackint
537m076 ct only limits moral sense)» which she seems to THINK «are» to make others happy & wrong to injure th
538m078 seem other cases somewhat analogous, & which I THINK will lead to fact of old people singing songs o
539m081 il. which made me «endeavour to» remember, & to THINK deeply, & the immediate manner in which my head

Page **(Key Word)**

541m091 fort less» one is obliged to repeat the word, & THINK of qualities as flowers, cloth &c & with all th
545m105 cious, which makes the idea unconscious, if so (THINK of this). study what impressions become unconsc
546m108), says <sympathy> we can only know what others THINK by putting ourselves in their situation, & then
549m119 great, that every one who has tasted them, will THINK the sum total of happiness greater. even if mix
556m145 who is not deformed. is born with two eyes.." I THINK this cannot be disputed anymore in men. than in
557m148 ore to be analysed than fear or anger? I should THINK shame would be more easily analysed than jealou
557m148 ss discoverable in animals than latter.-- Yet I THINK one can remonstrate with a dog, & make him asha
558m152 eric manifestation of great passion.-- I do not THINK they arch their backs-- Bengal tiger. when slig
567n015 emmy, when Chico plagued him-- Animals I should THINK would not have any emotion like blush.-- when e
570n024 ogether & shrugs his shoulders & walks off,-- I THINK shrugging connected with many emotions.-- (Expl
570n026 e mountains, & a negro, similarly treated would THINK negress beautiful,-- [male glow worm doubtless
574n039 ounds singularly adapted to subject see <A> ‖ I THINK this argument might be used to show language ha
575n045 from not having any smell.-- 27th. November.-- THINK, whether there is any analogy between grief & p
578n053 of mind on individual parts of body.== (if you <THINK> «fear» you shall not have e---n, «or wish extr
578n054 question, "one will anyone, especially a women THINK of my face,"? to one moral conduct.-- either go
580n061 ys that when in playing by memory. she does not THINK at all, whether she can or can not play the pie
587n087 islike others.-- dogs dislike perfume) I should THINK, great principle of liking, as simply heredetar
591n097 g tail when watching anything-- Keeper does not THINK they drop their ears.-- -- George the lion is e
591n100 man eating what others by habit (not instinct) THINK not fit, as cannabalism, is held in abhorrence.
595n117 allowed they can dream, & not have day-dreams-- THINK well over this;-- it shows similarity in mind.-
595n117 ell over this;-- it shows similarity in mind.-- THINK of Eyton's horses becoming <white> with <lather
595n121 w.-- Is this part of same feeling which make us THINK anything ugly-- a beau-ideal feeling. Same effe
595n125 butting, or young crocodile snapping.-- these I THINK are better instances of instincts (highly usefu
596n184 ssertation on the Passions." "Hartley" I should THINK well worth studying-- "Thomas Brown" on Associa
599o007 rus are utterly inexplicable-- I cannot <admit> THINK reason sufficient to give up my theory-- Viewin
610o031 astonished, & said how very odd.-- --could not THINK what had put Wilson into his head.-- remembered
612o035 ovement of sap. or sunflower to sun? ∵ I should THINK there. was direct «physical» effects of more or
615o36v tinct, leading to touch a particular organ.-- I THINK Pincher shows surprise, walking home one day me
615o36v inutes every now and then howled.-- Now I don't THINK this only pleasure; for it was different way of
618o41v e know of attraction, but we cannot see an atom THINK: they are as incongruous as blue & weight; all
618o41v only thus, for if day was first, we should not THINK night an effect.]CD Cause and effect has relati
620o044 ature is only temporary, & we do not afterwards THINK of it.-- Whatever the cause of this may be, eve
621o047 g circumstance.-- Thus a child may be taught to THINK almost anything nasty. (<& accidentally> «by od
621o048 right or wrong.-- [RHC] 7) Hence, what parents THINK will be good for the child on the long run, & f
622o049 in which «special» instincts decrease, I should THINK they were very few & general in their nature.--
622o049 om, she has never received any benefit.-- Yet I THINK there is much truth in doctrine, for [RHC] 9) W
625o052 preface. [RHC] It appears that Sir. J. & others THINK there is distinct faculty, of conscience.-- I b
628o054 sions, or instincts.-- is this good?-- I should THINK some parts of the emotive part of man, may be q
628o54v sion &c. whilst the passions have no relation I THINK this <boshes> «nonsense»-- My theory of durable
629o55v r indecency-- preposterously so, for Marquesans THINK only of prepuce, crepitando,]CD & if passions m
641j29r Seals-- Are Porpoises numerous in cold Oceans I THINK not.-- Does this bear on, the absence of their
204b141 Kirby all through Bridgewater errs greatly in THINKING every animal born to consume this or that thi
223b207 oduction of Man. Nothing compared to the first THINKING being. although hard to draw line.-- -- not s
281c138 ll equally related to each other I cannot help THINKING good analogy might be traced between relation
305c210 according to subordinate laws.-- There is one THINKING « <& Creat> sensible» principle (intimately a
305c211 ind of organic matter.-- brain. & which <prin> THINKING principle. seems to be given or assumed accor
305c211 s of the living beings.-- We see thus Unity in THINKING and acting principle in the various shades of
349d052 divisions.-- Complains (p. 53) of M. Edwards, THINKING any group good, though not circular, if chara
360d091 which impresses offspring most.-- «& not time» THINKING of the Penguin duck & Herberts law of ideosyn
410e051 esentative species., when going N.orth & South THINKING of effects of my theory, laws probably will b
520m001 attending Mr Dryden Corbet, he could not help THINKING, he was prescribing to his father & old Mrs H
522m014 gth & comfort to the body» I know the feeling, THINKING over somebody who has, perhaps, slightly inju
523m016 have been so.-- In Mr Hardinge, was caused by THINKING over the misery of an illness at Rome, when b
524m020 to Epilepsy & convulsion.-- affections of the THINKING organs ‖ -- the action of brain which gives s
526m027 same connection between organization & mind.-- THINKING over these things, one doubts existence of fr
526m028 e image in their own mind,-- this may be worth THINKING over.-- it. will perhaps show differences bet
526m029 tion (common to monkey), & not imagination.--» THINKING over the scenes which I first recollect, «at
529m040 ll defined thought would receive pleasure from THINKING of the fertility.-- I a geologist have illdef
534m061 yet this knowledge acquired by senses,-- then THINKING consists of sensation of images before your e
538m079 sition recurs, such as -- -- of Trinity always THINKING people were calling him a bastard.-- when dru
541m090 ct to subject as quick as it chose.-- although THINKING «& talking» for the moments with interest on
541m091 idea before your mind steadily., & not merely THINKING intently; for that one does with novel for a
546m108 explain pleasure. August 26th. I cannot help. THINKING horses admire a wide prospect.-- The very sup
547m114 from innumerable late events.-- the fatigue of THINKING is keeping up these trains,-- especially if t
553m136 nificent laws. of which we profane «degnen» in THINKING not capable to <do> produce every effect, of
564n004 e, obeying habits, & dread of misery of future THINKING of injured moral sense.-- Notion of deity eff
566n011 etting sight slowly,, but when in grown years, THINKING he instinctively knows distances:, is good in
573n038 of thought is to make saliva flow, & therefore THINKING of subject, even when wishing not to flow-- f
577n051 ssion.-- Blushing is intimately concerned with THINKING of ones appearance,--does the thought drive b
577n052 shyness is certainly very much connected with THINKING of oneself.-- «blushing» is connected with se
577n052 his own sex.-- Hence, animals. not being such THINKING people. do not blush.-- sensitive people apt
578n055 «concealed» action he should not, & let him be THINKING over it with sorrow,-- let the possibility of
613o036 - It does not appear more than saying that the THINKING principle is the same in all animals. [LHC] «
059r119 ui forment les vallées ou les scissures.--M. B. THINKS these parts incontestably formed the parts of
064r134 e. rock of oysters quite above reach of tides.--THINKS them same as recent species. -- May I not gene
072r159 nland, close to W of Jan Meyen Isld.--Mr Barrow THINKS N & S. line connects western isles of Scotland
093a032 ooled.-- Berzelius. L'Institut. [1837 p. 297]CD THINKS Olivine a preexisting mineral.-- Mem. Galapago
117a101 on of Europaean strata according to composition THINKS sand with vegetable remains formed near coast,
185b057 bes in that same family.-- (NB I see Waterhouse THINKS Quinary only three elements) How far Does Wate
198b113 nidae?-- --- I cannot understand whether. G. H. THINKS develepent in quite straight line, or branchin
203b136 anderer, but as produced by climate?-- M. Baer (THINKS) the Aurock, was found in Germany & thinks eve
203b136 aer (thinks) the Aurock, was found in Germany & THINKS even now in central & Eastern Asia beyond the
204b140 .-- How easily does Wolf & Dog cross? Mr Yarrel THINKS oldest variety impresses the offspring most fo
207b152 recent volcanic islets not well fixed.-- Peron THINKS Van Diemen's land long separated from Hobart T
229b233 ing them, instance of useless structure-- Smith THINKS several species of Rhinoceros range from Abyss
233b250 ilar extension of form in birds.-- | Waterhouse THINKS two main divisions of cats. Tortoise shell--&
233b250 ats. Tortoise shell--& grey-banded. ¿species?-- THINKS offspring of cats sometimes heterogenous.-- Au
256c055 continued?-- See to Boblaye & Virlet.-- Whewell THINKS (p 642) anniversary Speech. Feb 1838 thinks gr
256c055 ell thinks (p 642) anniversary Speech. Feb 1838 THINKS gradation between Man & animal, small point in
274c114 ny ground woodpeckers?-- In each division Gould THINKS he can trace structure for insects & structure
279c133 ects of peculiarities being long in blood.-- ++ THINKS difficulty in crossing race.-- bad effects of
294c176 permanent small alterations in wild animals, & THINKS Lyell has overlooked argument, that domesticat
298c192 er parts attracting all the moisture.-- Henslow THINKS if leaf of plant varies, <whole cross> «all or
300c196 -- or dogs in Carnivora.-- Man in his arrogance THINKS himself a great work. worthy the interposition
309c222 ot believe in circular or linear arrangement.-- THINKS passages very rare., in anatomical structure.-
319c255 rst pair of Pipra flycatcher.-- Bachman says he THINKS the Mocking thrush beats all English birds in
319c256 s, & our lark ranks very high.-- Upon the whole THINKS <many> more birds sing in England than in Amer
332d001 terwards, he then did not see other cases.-- He THINKS apoplexy affects people all over England at sa
332d003 consulted him on one day.-- Mark at Shrewsbury THINKS the half bred Alderney Cows take more after Al
333d006 e of English, but the effect is the same.-- Fox THINKS that when a wild animal is crossed with tame,
333d007 osses between Esquimaux dog & common dogs & Fox THINKS they decidedly take much most after Esquimaux.
333d008 th Yarrell & no leading question was put.-- Fox THINKS half Lion & Tigers are exactly intermediate in
339d025 es of Men, but not plants» will it hold good.-- THINKS Temmink doubtful when he says No genera.-- thi
339d025 inks Temmink doubtful when he says No genera.-- THINKS there are some small divisions.-- does not see
364d104 owledge that sheep originally. black. & Yarrell THINKS the occasional production of black lambs is ow
366d112 ng the feathers along one toe of the Pouter one THINKS there is a law.,-- that there must have been a
368d114 ut on (like some waders) the bright plumage.-- «THINKS» Hence specific character most perfect in <mal
369d115 , is now universally admitted.""-- p. 483. Owen THINKS from climate of Australia, & from Ornithorhync

Page **(Key Word)**

```
374d134 able-- 28th. «I do not doubt, every one till he THINKS deeply has assumed that increase of animals ex
376d137 up trousers of men. Evidently knew <men> women, THINKS perhaps by smell.-- but monkeys examine sexes
377d139 the other when Dr. Smith more distant.-- But he THINKS other monkeys make st.-- noise In case of wom
377d139 phic naturalist pregnant with interest» Hyaena. THINKS, when pleased cocks his ears., when frighten d
385d163 alled everlasting layer--. or Polish breed. (he THINKS half pheasant, half fowls.-- eggs fertile, but
388d172 remarkable Athenaeum 1838. p 653. Ehrenberg<h> THINKS multiplication by division <only> is developem
401e015 aw a nonpareil sowed by Mr Tollet so produce.-- THINKS it probable that great part of those varieties
401e015 hat some effect from apple trees is produced.-- THINKS probably experiment was never tried of separat
401e015 so my experiment of strawberry not so absurd.-- THINKS-- that such variety as red cabbage produced fr
420e088 of geology.-- L'Institut 1838. p. 414; M. Guyon THINKS Monsters more common in Africa than in  Europe
431e121 her on coast of France.-- L. doubts.-- Lonsdale THINKS Ammonites would afford instance of such facts.
436e136 1 12th..> Cestracion, Port Jackson Shark-- Owen THINKS Australia part of Old World <If> It «may» be s
445e159 that any shell fish is really hermaphrodite. & <THINKS> even oyster may fecundate each other, by  the
449e173 World, has. written. Mastodon found at Timor.-- THINKS he has seen specimen at Paris Museum.-- Athena
462t051 ms-- yet how valueless this objection, when one THINKS of different kinds of cattle in every part of
477z002 ed to Mauritius. 18 Azara Voyage Vol. I. p. 279 THINKS the Moruffetes of Chile different from those o
508q016 nsmitted through females, like Hydrocele Dr. H. THINKS asthma in females takes place of gout.-- How a
520m001 ed Dec 16 1856 July 15th 1838 My father says he THINKS bodily complaints <& mental disposition> often
520m002 d he could not. remember his father.-- My father THINKS people of weak minds, below par in intellect
521m007 f hunting «-- habitual fits.--» which my Father THINKS is mentioned in the Zoonomia.-- Now if memory
525m024 le in Squib's case any direct fear.-- My father THINKS that selfishness, pride & kind of folly like (
526m028 rences between memory & imagination. «Catherine THINKS that children like looking at <ani> pictures,
534m061 r instinct. Hensleigh says to say. Brain per se THINKS is nonsense; yet who will venture to say  germ
544m102 phy of Living. p. 140-- Dreams good account of «THINKS» are recollected when intense, or when so near
544m102 d is kept up with waking thought.-- Ld Brougham THINKS no dreams except at this time. how does he acc
544m103 ideal, as all the other perceptions.-- The mind THINKS with extraordinary rapidity-- We may  conclude
545m106 quickly uncovering their teeth, this the Keeper THINKS is from pleasure, & may be compared to laughin
547m113 bell??)-- It may be deception to say the mind <THINKS> quicker in sleep, it may do less work & yet d
552m130 y whole imagination." Septembe. 3d Why when one THINKS of any object, (or having looked at any object
553m137 ulky «as» often as keeper spoke to it,-- but he THINKS not sulkiness-- this expression he believes is
554m138 ld male.-- A very green monkey (from Senegal he THINKS Callitrix Sebe??) he has seen place its head d
554m140 , which she had been told not to do.-- when she THINKS keeper will not see her.-- <but is> then knows
554m140 I do not know whether fear or shame.-- When she THINKS she is going to be whipped. will cover herself
577n052 ing» is connected with sexual, because each sex THINKS more of what another thinks of him, than of an
577n052 l, because each sex thinks more of what another THINKS of him, than of any one of his own sex.-- Henc
578n055 e blood rushed to her face,"-- One blush if one THINKS that any one suspects one of having done eithe
580n060 time, & perhaps of space-- in latter respect he THINKS he certainly has observed that some people of
581n065 ns long before they can speak-- or understand-- THINKS so it must have been in the dawn of civilizati
581n065 it must have been in the dawn of civilization-- THINKS many words, roar, scrape, crack, &c, imitative
603o012 origin of sacrifices. «common to many races»-- THINKS action towards <man> «a king» <changed into> i
604o017 struggle <between> when insanity is coming on «THINKS clearest analogy between dreams & insanity.» D
606o022 t., sculpture & painting, is beauty.-- which he THINKS is a better definition than Winkleman's. who s
608o026 iscover them: this is important explanation) he THINKS they have none.-- Effects.-- One must view a w
613o036 usness than our intestines have? [RHC] 5) Kirby THINKS that <all> there is one instinct to all  an
620o044 ally soon conquers, & the dog [RHC] 3) probably THINKS no more of it.-- Not so man, from his memory &
620o044 passion fades away, so that when man afterwards THINKS why was such an instinct not followed for a pl
621o047 en conquers it). It will be only rarely that it THINKS that nasty, which the natural tastes say is go
635j56r heme. would require constant miracles.-- p. 420 THINKS the great fecundity of germs is to afford supp
064r134 ays Elephant lives on very wretched cou[n]tries THINLY covered by vegetation. Rhinoceros quite in des
118a103 not formed volcanos. or become scoriform. as THINNED upwards & is now cut off by denudation it give
107a077 rd from effects of Elevation if not crust much THINNER beneath ocean than above it no because heat pr
107a078 hiefly beneath water & volcanos. crust must be THINNER «under water» but cause most difficult (better
125a121 kness of crust.-- {P} if crust were metal then THINNER if better conductor, then still thinner → The
125a121 1 then thinner if better conductor, then still THINNER → The problem is, you have temperature known a
229b233 say great tailed sheep aboriginal at Cape & a THINNER-tailed kind further inland.-- NB. There is div
136a147 n Lardner Vol II. p. 81. «&83» Some remarks on THINNESS of crust as implied by meeting with granite e
066r142 ot distinguish from stone Caradoc from lower of THIRD Silurian division--Together with same general c
099a049 re aggregated, would they not attract strong. a THIRD.-- & this would make layers.-- (Gravity can hav
202b130 tures in the two animals bearing relations to a THIRD body., or common end of structure A Race of dom
272c108 .-- Donacia.-- some orthopterous insects & some THIRD, have got thighs with same peculiar structure &
281c138 n of animals-- talking of men as related in the THIRD & fourth degree.-- a species must be compared t
298c191 of which the central parts become occupied by a THIRD best adapted kind.-- lower species would then r
387d169 he 3d tendency from first pair.-- Now if two of THIRD pair of same peculiarity breed they will have s
472s01v ler with less riged foliage & no black spot & a THIRD considerably paler, all rest very similar-- Jun
568n016 erience.-- Two sides of a triangle shorter than THIRD. is this necessary notion, ass has it.-- When o
369d115 compelled to traverse in order to quench their THIRST"-- But New Guinea.--!! S. America.-- Such diff
262c073 -- the fact of guavas having overrun-- Tahiti. THISTLE. Pampas. show how nicely things adapted--.-- T
470t177 ssum--Reseda wild very many Bees & Humbles--on THISTLES many (curious because a Composite) Asparagus
350d054 no way.?-- "Natura nihil agit frustra", as Sir THOMAS Browne says "is the only indisputable axiom in
448e167 nference I imagine that there are Baboons in St THOMAS on W. coast of Africa Owen Linn. Soc. April 2d
596n184 "Hartley" I should think well worth studying-- "THOMAS Brown" on Association worthy of close study.--
312c231 » new acquisition, we must [not located] Henry THOMPSON tells me best way to improve cattle is to cro
477z001 onnected with transform of Crust-- Westwood & THOMPSONS-- Part II.-- 35. Phil Trans Burrowing & borin
313c233 nguinal louse African & Europaean. different.-- THORAX & head differ Africa Australia Parasites die,
471tf07 lgoridae enumerates the strange forms which the THORAX & head displays.-- most fantastic & use unknow
401e017 ar may be Grafted on pear. Mountain-ash & white THORN! Species not being observed to change «is very
275c120 t end of chase would not run up hill-- he took THOROUGH bred <greyhound>. bull-dog. & crossed & recro
599o007 y, netted with edges & crowded with towns & THOROUGHFARES, I grant that man, from the effects of here
031r038 ases in Europe.-- Auvergne. very little Pumice, THOUGH Trachyte. same fact in Galapagos. Daubeny P 24
035r048 ny one country would produce equable effects.--<THOUGH so immense to short breathed traveller» Mounta
058r116 literate all land marks.--At the first it would THOUGH be easy to see on beach successive lines of se
109a084 Ehrenbergh on flints in chalk must be studied-- THOUGH I do not think good p. 411 When discussing con
119a104 um.) 3d. there are mountains in the moon, which THOUGH not very analogous (see Edinburgh. Phil. Journ
129a131 y strata of S America» read parts of this work, THOUGH it is but poor. Athenaeum. 1838 p. 791 -- Most
155g066 water in same «drainage» lines Mound of Gneiss THOUGH wonderful-- <that they are preserved> how much
171b003 ts sown in rich soil, many kinds, are produced, THOUGH new individuals produced by buds are constant,
173b011 ltered Hence the type would be of the continent THOUGH species all different. In cases as Galapagos &
188b068 seeds of white flower. all would come up white, THOUGH planted in same soil with blue. Now this is sa
189b073 ust like buds of plants, which die at one time, THOUGH produced either sooner or later.-- Prove anima
190b079 d, each fresh--> Mr Don remarked to me, that he THOUGH species became obscurer as knowledge increased
198b114 pression mean used by Cuvier, that all animals (THOUGH some may be) have not been created on the same
206b147 .-- We may conclude that there will be a period THOUGH long distant, when of the present men (of all
218b191 peculiar «constant» aspect. That is varieties, THOUGH of trifling order are formed by nature. Carmic
263c074 (if mankind destroyed) some intellectual being THOUGH not MAN.-- is as difficult to understand as Ly
265c087 y be of some use,-- Nature is never extravagant THOUGH clearly not of the use to which wings are gene
266c092 either coast not found here not even Antelopes, THOUGH common on coast of Arabia not even antelopes t
266c093 uch common on coast of Arabia not even antelopes THOUGH common on islets off Arabian Coast.-- Vol VI.
272c109 ould says most subgenera confined to continent, THOUGH we have seen species <of subgenera» scattered
273c112 ucture is not always applicable to same habits, THOUGH swallow hawk, milvulus, may catch insects on
273c113 yet the Tropic bird, has very different habits, THOUGH preeminently belonging to this type, the Humm
274c113 th claw like lark (NB The La jeune veuve parrot THOUGH so much on the ground has not this structure..
280c136 ong in blood, gets stronger & stronger, so that THOUGH by great effort <pr> one unlike can be produce
281c137 consider it as proved that they are varieties, (THOUGH that would be best).-- Argue the case theoreti
283c145 in same country with their own instinct & (even THOUGH <fertile> when compelled to breed hybrids prod
288c160 -- --Yellow Wagtail never seen in one district, THOUGH common on another, (golden creted wren so rare
293c173 individual objects as Ichneumon & caterpillar, THOUGH our ignorance, may make us think so, but only
304c206 this if series now existed from Man to Monads-- THOUGH physiology would profit. if the series were be
308c217 always for a week took of cover of right side, THOUGH my hand <vibrate> would sometimes vibrate-- «s
```

Page		(Key Word)	
315c242	ly in plants movements effects of irritability,	THOUGH	means injection of fluid different from contra
334d011	aft mare & another leader mare,-- this stallion	THOUGH	eager to all other mare had been entirely brok
334d011	are had been entirely broken from their mares,	(THOUGH	horsing every month) & worked in the same cart
343d038	ecies or varieties are becoming extinct. others	THOUGH	the negro of Africa is not loosing ground. Yet
344d042	n went back to either paret, & breed not fixed.	THOUGH	she resembled a harrier & her husband was pure
348d051	,-- I believe affinity may be taken literally,,	THOUGH	how far we can ever discover the real relation
349d052	(p. 53) of M. Edwards, thinking any group good,	THOUGH	not circular, if characters can be established
350d055	ve nothing to do with perfection of individual,	THOUGH	such relation seems common, but that perfectio
356d069	written I pretend to no originality of idea--	(THOUGH	I arrived at them quite independently & have u
360d090	ffspring did not go to heat. but parts swelled,	THOUGH	no fluid came from them.-- showing how gradual
360d090	l appearance of face & tail somewhat like dog--	THOUGH	it has full share of Jackall shape of body.--
360d090	l shape of body.-- disposition wild, & fearful.	THOUGH	not so much as in Jackall.-- In case where Jac
360d093	arder to cross than two less opposed in habits,	THOUGH	externally similar.-- this however is a sophis
363d102	& Nightingale which both sing their own songs,	THOUGH	imperfectly.-- Male birds always second their
366d111	g door outside, as often as she touched handle,	THOUGH	really fully aware she was not coming in, coul
371d128	cannot transmit through seeds these characters	THOUGH	transmitting them with such facility to bud.--
376d136	t, but in two or three days learn its comfort &	THOUGH	could not put it on, yet threw it over it, & m
379d152	d Moreton's Case.-- When cows have twins, <one>	THOUGH	capable of producing both <male> pair of male
380d153	of any case of old Male. becoming like female,	THOUGH	many of old female becoming like cocks.-- It i
380d155	e of hybrid birds) After animal has copulated.,	THOUGH	no offspring, Milk sometimes comes in Mammae &
380d155	n string will not.-- some of the tigers.-- cat,	THOUGH	caterwhalling. & put into female, when muzzled
382d158	hade of perfection --How came it nipples <are>	«THOUGH»	abortive, are so plain in Man, & yet no trace
386d167	in animals very different from that of plants.	(THOUGH	latter does sometimes occur in animals). latte
389d176	.-- it affects the subsequent offspring, <when>	THOUGH	other male may have copulated.-- two animals m
409e048	ere would be as many species, as individuals, &	THOUGH	we may not trace out all the ill effects. -- w
414e063	hem> such a variety far more easily than man,--	THOUGH	man's practiced judgment. even without time ca
414e064	fly intellect, in animals chiefly organization:	THOUGH	Cont of Africa & West Indies shows organizatio
419e087	with the Lamantin of Steller tells much less,	(THOUGH	<the> it also the effect of change) than a slo
423e097	t of the complexity of structure is adaptation.	THOUGH	perhaps difference between jaguar & tiger may
425e105	ps.-- many peculiar plants on single mountains,	THOUGH	these are connected with other mountains later
431e123	verage size: (surely this is very limited view,	THOUGH	perhaps a true element) «give examples, pigs,
440e148	of <growth> «generation» is only modification,	THOUGH	important one, of growth «Lamark. Vol II. p. 1
446e162	ed» by buds-- (the other three by buds & seeds	«THOUGH	by the latter very rarely») here is a case in
447e164	sometimes seeds All species of Lemna sometimes	THOUGH	very rarely flower [bu]t the one does on the c
460t015	f the swan-goose in appearance Bell at Hornsey	(THOUGH	only ¼ of blood). that it appears about half w
465t081	of other races, so it is with domestic breeds.	(THOUGH	in this case crossing has had somewhat to do w
466t093	w is it generally.-- In Azalea <do> «it is so»	<THOUGH	I saw no Bees «several» visiting it>.-- In yel
466t093	w day lily, the Bees visit base of upper petal,	THOUGH	not differently coloured-- & stamens bend up a
470t178	atile & other common kind--I think not on Phlox	THOUGH	they examine it.--«Little Dusty & Blue» Butter
494q004	adgers,-- How few wild animals are propagated,,	THOUGH	valuable as show. & curiosities!! What is pric
521m009	,-- could receive a new train through eyesight,	THOUGH,	not through hearing,-- Thus when dinner was a
524m021	f near Drayton & Ternhill, (where she was born)	THOUGH	she never naturally talked of these places.--
524m022	e in old age. exceedingly sharp in some things,	THOUGH	so confused in others.-- Mrs P. when in state
525m023	ket, bringing back game, or picking up a stone,	THOUGH	only acquired rules by art.-- like the law of
526m028	ery different from the inventive power,-- this,	THOUGH	very odd is perhaps true.-- mem Erasmus & mine
526m028	for music.-- Children like hearing a story told	THOUGH	they remember it so well that they can correct
526m031	(but what makes him fix!? <)>-- frame of mind,	THOUGH	perhaps he chooses wrongly,-- & what is frame
527m031	.-- now as we know birds learn from each other	«THOUGH	different species» when in confinement, so may
532m054	ht. being slightly unwell & felt so much afraid	THOUGH	my reason was laughing & told me there was not
532m056	Nina, when brought from Shrewsbury to Clayton,	(THOUGH	so fond of her & of servant of Richard & of Ma
533m060	I had a sort of consciousness I was not right;	THOUGH	I never realized the idea that I was tending t
535m064	us, watching few days till larva excluded, then	THOUGH	not feeding them «nor helping larva from egg»
536m072	& a polype (& a plant in some senses, perhaps,	THOUGH	from not having pain or pleasure actions unavo
538m079	r song, then the song was to her like one which	THOUGH	learnt in infancy, had often been repeated: No
542m093	teeth.-- He may feel satisfied with himself, &	THOUGH	dreading to say so, his step will grow erect &
548m114	reasoning-- hence delirium & sleep mental rest.	THOUGH.	most vivid & rapid thought.-- There may be so
548m115	reasoning besides these (which must be present,	THOUGH	one is not conscious of them, else one would n
548m115	eping one idea present is, perhaps, hard work--	THOUGH	dreams do that One Reflective Consciousness is
548m116	in one state almost as much as in the other,--	THOUGH	she when well did not recollect <it> «anything
548m117	insane idea; & ones expression of double self,	THOUGH	as in Dr Ashe's case, one here was conscious o
549m120	. A wise man will try to obtain this happiness.	THOUGH	he sees some «intellectual» good men, from ins
549m122	rules to mortify yourself do not tend to this--	THOUGH	believing it to be true, & then acting on it,
551m129	en pouting protrudes its lips into point-- man,	THOUGH	he does not pout. pushes out both lips in cont
555m143	ion with open mouths like the old world ones.--	THOUGH	the[y] move whole skin of head they do not mov
557m149	ive away rival.-- Fear is open mouthed to hear.	THOUGH	in individual case. nothing can be heard.-- Sh
565n007	ing involuntary.-- When one fear any bad news,	«THOUGH	in a letter> why is person painted with mouth
565n007	aises its head, & pricks its ears» when afraid,	THOUGH	not every time really wishing to smell its ene
567n014	Has an oyster necessary notion of space-- plant	THOUGH	it moves doubtless has not.-- Turkey cock in p
572n034	does not hurt the conscience of a Boy to swear,	THOUGH	reason may tell him not, but it does hurt his
582n067	it:-- there must, however, be a mental impulse	(THOUGH	unconscious of it) to move its legs so, as muc
586n078	feels inclined to stop at right number of house	THOUGH	one cannot remember it.]CD back, without consc
594n112	s would have fled equally if man had appeared--	THOUGH	instinct so firmly implanted, birds soon <dis>
602o011	intuition. when individual cannot give reason,	THOUGH	he feels he is right-- it is because each deci
611o034	rritability, which at least shows a local will,	THOUGH	perhaps not conscious sensation. [RHC] During
612o035	ulties into one common one always satisfactory,	THOUGH	not adding to positive knowledge. lessening am
614o037	existence of mankind requires these instincts,:	THOUGH	very weak so as to be overcome easily by reaso
620o044	uch an instinct not followed for a pleasure now	THOUGH	so trifling he feels remorse.-- He reasons on
624o051	-- Hence this becomes the law of right & wrong,	THOUGH,	that part, which is acquired by association f
628o53v	taught, but not cat, that is not act by gusto,	THOUGH	by fear it might be partly made.]CD p. 21. "Wh
629o55v	ene between this kind of conscience & the will,	THOUGH	«this» conscience does between the desires & w
635j56r	his does not explain death, but reproduction]CD	THOUGH	such a scheme. would require constant miracles
639j28v	s the long legged race will get the upper hand.	THOUGH	continually dragged back to old type by interm
055r106	est: the veins which follow this direction are	THOUGHT	by the <oldest> «most intelligent» miners to b
065r139	on: countenance so well preserved. that it was	THOUGHT	not to have belonged to an Englishman.--On 8th
089a019	arble requires polish to see structure,-- «He»	THOUGHT	of erecting machine to see if water fell. -- <
131a135	ng. L'Institut. 1838 p. 360. on orbicular trap	THOUGHT	to be bombs submarine L'Institut 1838 p. 400.
148g025	f between parents» (& not like dogs), but they	THOUGHT	the breed liable to vary-- I asked this questi
148g025	s question in many ways & received same answer	THOUGHT	lambs most like MOTHER!-- the cross not so har
165g126	ever obseved? Shows that <nervous> brain makes	THOUGHT	Glen Roy B C. Darwin All useful pages cut out
178b031	effected; a change whi[ch] the Fr. naturalists	THOUGHT	was species <Ascensi> Study Lesson Voyage of C
197b112	Cuvier's theory of Conditions of existence is	THOUGHT	to account resemblances &∴ quinary system, or
199b117	- Propagation explains this.---- Ancient Flora	THOUGHT	to more uniform than existing. Ed. N. Philos J
219b196	uted by such means as we can recognize, may be	THOUGHT	to explain nothing.-- it being as easy to prod
224b214	s not so great as between living thing without	THOUGHT	(plants) & living thing with thoughts (animal)
291c166	out importance of habits in classification.--	THOUGHT	(or desires more properly) being heredetary<)>
291c166	structure are facts full of meaning.-- Why is	THOUGHT.	being a secretion of brain, more wonderful th
310c222	I be asked by what power the creator has added	THOUGHT	to <an> so many animals of different types. I
310c223	es will turn out not nearly so confined as now	THOUGHT.	-- N. American, European & Chinese Genera & so
312c233	of different litters or of father & child are	THOUGHT	long breeding in. Must not trust him Hope says
315c242	y-- scenes in themselves accidental-- My first	THOUGHT	of sea side-- Study Bell on Expression & the Z
316c244	play.-- Hensleigh says the love of the deity &	THOUGHT	of him «or eternity», only difference between
331dIFC	lected Dec. 14-- 1856]CD Towards close I first	THOUGHT	of selection owing to struggle July 15th. 1838
345d042	se of avitism.++» Three gentlemen of party all	THOUGHT	with pigs &c, that hybrids were uncertain. Mr
345d042	&c, that hybrids were uncertain. Mr Drinkwater	THOUGHT	that a <pure blooded> «"first blood"» animal m
346d047	this in many ways, but received same answer.--	THOUGHT	lambs were more like father than Mother.-- The
360d092	& Herberts law of ideosyncrasy I have hitherto	THOUGHT	that a small difference <of any kind>, if very
364d104	ld Testament placed before sheep-- it has been	THOUGHT	that silver Pheasants about a house made other
365d106	so-- but very improbably, for it can hardly be	THOUGHT	that the cross would have adapted it to changi

Page **(Key Word)***
366d108 ns with imperfect ones.-- now Sir J. Sebright. THOUGHT if he had had a pair he could have produced fr
370d115 difficulties will always occur if animals are THOUGHT to have been created.-- it might as well be at
372d131 andth of inch should produce a Newton is often THOUGHT wonderful. it is part of same class of facts t
377d139 se facts may, be turned to ridicule, or may be THOUGHT disgusting, but to philosophic naturalist preg
382d157 plant & not in other), Hunter <asks> p. 36 is THOUGHT by Owen to ask. whether a Heautandrous animal
432e125 frowns connected with pain, as well as intense THOUGHT.--» No one but a practised geologist can reall
434e129 etween these varieties, so that they have been THOUGHT to be different species. Lychnis dioica, gener
436e137 e?? Seeing the beautiful seed of a Bull Rush I THOUGHT, surely no "fortuitous" growth could have prod
448e165 eriwinkle seldom produces seeds, because it is THOUGHT to require insects to impregnate it.-- it is a
464tf6r nting leopard, how strange, anyone, would have THOUGHT isolated species Mr Blyth, however, believes i
468t106 -bells-- wild-raspberry--leeks-- Flowers which THOUGHT very unattractive-- Found Rhubarb blossom swar
521m008 d not remembered, it was an habitual action of THOUGHT-secreting organs, brought into play by morbid
522m013 going against natural instincts.-- My Grand F. THOUGHT the feeling of anger, which rises almost invol
524m020 tual.-- a person twitching when a disagreeable THOUGHT occurs, is closely analogous to Epilepsy & con
524m021 antly do.-- In Mrs P. <...> of B. <talked of.> THOUGHT herself near Drayton & Ternhill, (where she wa
527m034 class.-- Now that I have a test of hardness of THOUGHT, from weakness of my stomach I observe a long
527m034 g required) as the closest train of geological THOUGHT.-- the capability of such trains of thought ma
527m034 al thought.-- the capability of such trains of THOUGHT makes a discoverer, & therefore (independent o
527m035 feel a severe disappointment «in real train of THOUGHT this does not happen. because papers, &c &c ro
529m040 mind supply of food was evasive & ill defined THOUGHT would receive pleasure from thinking of the fe
530m046 otion becomes habitual & involuntary.-- when a THOUGHT is thought very often it becomes habitual & in
530m046 es habitual & involuntary.-- when a thought is THOUGHT very often it becomes habitual & involuntary,-
530m046 at is involuntary memory, as in sleep.-- a new THOUGHT arises?? compounded of the involuntary thought
530m046 the tendency to habit of producing a train of THOUGHT.-- [not located] Fox believe cats discover bir
533m061 onscience, is obscure memory of having read or THOUGHT of some such remarke as now advanced; for I ca
535m069 e fixed laws of nature should be «universally» THOUGHT to be the will of a superior being; whose natu
538m078 ese things, & drunkedness, show what trains of THOUGHT depend on state of turn In drunkedness same di
538m081 explained. These facts showing what a train of THOUGH[T] action &c will arise from physical action on
539m083 plain what habit is-- In the habitual train of THOUGHT one idea. calls up other, & the consciousness
540m087 se of the intense labour of original inventive THOUGHT is that none of the ideas are habitual, nor re
541m090 & full of levity (¿ do I not confound action & THOUGHT here?) The opposite extreme of this desultory
541m090 here?) The opposite extreme of this desultory THOUGHT is following out such an idea, as effect of se
541m091 e.-- may explain excessive labour of inventive THOUGHT.-- Examine frame of mind in following changes
543m101 riginate.-- strong argument for brain bringing THOUGHT, & not merely instinct, a separate thing super
543m101 g superadded.-- we can thus trace causation of THOUGHT.-- it is brought within <our own> limits of ex
544m102 ing. that an associated is kept up with waking THOUGHT.-- Ld Brougham thinks no dreams except at this
547m111 cause to start at once to Shrewsbury., vaguely THOUGHT of packing up.-- was lying on my back fell to
547m111 haracterized» dream, in continuation of waking THOUGHT-- my servant was in the room. with my trunk ou
547m112 hether he had come & opened box, whether I had THOUGHT what clothes to take (how often one cannot tel
547m113 would so so!) Now all these parallel trains of THOUGHT necessary heirs of every action, & always runn
547m114 ng. effect of not having <all> other trains of THOUGHT, or memory from innumerable late events.-- the
548m114 sleep mental rest. though. most vivid & rapid THOUGHT.-- There may be some «two or three» trains of
548m114 .-- There may be some «two or three» trains of THOUGHT, therefore one may be imperfectly reason -- <I
548m115 ne would not stand) a crowd of other trains of THOUGHT are in progress-- In castle of air the trouble
555m143 1st Was witty in a dream in a confused manner. THOUGHT that a person was hung & came to life, & then
565n009 > lesser in degree, no smile, no frown showing THOUGHT, no compression of mouth showing action,-- sul
568n017 uated «in life time» to any line of action, or THOUGHT one feels pain, at not performing it, (either
572n035 n say expression was invented to conceal one's THOUGHT.-- Macculloch in his Chapter on the Existence
573n038 will turn in bed.-- in case spittle, effect of THOUGHT is to make saliva flow, & therefore thinking o
577n051 ed with thinking of ones appearance,--does the THOUGHT drive blood to surface exposed, face of man, f
578n055 blood gushed into his face,-- "as <she> «the» THOUGHT of his knowing «it», suddenly came across her,
583n069 e,, but relation of imitation & reason must be THOUGHT of.-- p. 19. animals capable of education; (th
584n073 an perceive sigh, commences as soon as painful THOUGHT crosses mind, before it can have affected resp
594n113 nausea, (when stomach a little disordered) at THOUGHT of almost anything ugly. baby-- association--
608o027 lly convinced of its truth. except man who has THOUGHT very much, & he will know his happiness lays i
614o037 s I have said, no soul superadded, so [RHC] 6) THOUGHT, however unintelligible it may be, seems as mu
615o038 t of acquired memory. a permanent secretion of THOUGHT, (or under contingencies of stimulants of cert
616o038 en heredetary memory & individual secretion of THOUGHT, may be no more «difference» than sexual inter
616o039 ry in others. [(1)] Why may it not be said that THOUGHT perceptions will, consciousness memory &c. hav
616o039 r is by a metaphor said to attract; & hence if THOUGHT &c bore the same relation to the brain that at
616o039 riety be said that the living brain perceived, THOUGHT, remembered &c. Well the heart is said to feel
616o39v undwork <of this> is entirely wanting by which THOUGHT or memory. might be in like manner attributed
617o39v ects as different phases of the same object of THOUGHT is a question which ought to be clearly compre
618o041 s nothing analogous to this in the relation of THOUGHT, perceptions, memory &c. either to our bodily
618o041 consciousness, & have no objective aspect. If THOUGHT bore the same relation to the brain that force
618o041 h other. A person might be quite familiar with THOUGHT & yet be ignorant of the existence of the brai
618o041 existence of the brain. We cannot perceive the THOUGHT attraction of sulphuric acid for metal of anot
618o041 , we can only infer it from his its behaviour. THOUGHT is only known subjectively? -- ? the brain onl
618o41v know attraction objectively 6) The reason why THOUGHT &c. should imply «X» the existence of somethin
618o41v e insufficient to account for the phenomena of THOUGHT. (The objects of thought have no reference to
618o41v for the phenomena of thought. (The objects of THOUGHT have no reference to place. [We see a particle
618o41v us as blue & weight; all that can be said that THOUGHT & organization run in a parallel series: if bl
618o41v tween <action> «things» so different as action THOUGHT & organization: But if the weight never came u
629o55v ry it.-- p. 241 (1) Any action by habit may be THOUGHT wrong.-- & conscience will imperiously say so,
632j53r probably from ignorance]CD Who would ever have THOUGHT that the intestines of a thrush were means suf
634j54v hat we become satisfied respecting an original THOUGHT, or design, pursued to its utmost exhaustion,
165g125 rs at head of which hill is round & not merely THOUGHTS laying dormant-- Man from Glen Turret said he
224b214 g without thought (plants) & living thing with THOUGHTS (animal). «∴ my theory very distinct from Lam
310c223 passions acquired & heredetary & such definite THOUGHTS, I will never allow that because there is a c
315c242 o sympathetic nerves-- The vividness of first <THOUGHTS> «memory» in children or rather their memory.
523m016 father was nearly drowned at High Ercall, the THOUGHTS of it, for some years after, was far more pai
527m031 sand together & one will be uppermost:-- so in THOUGHTS, one will rise according to law. How strange
527m033 mind to create short vivid flashes of images & THOUGHTS.-- Poetry. the latter thoughts are in same ma
527m033 hes of images & thoughts.-- Poetry. the latter THOUGHTS are in same manner vivid & grand. the frame o
527m035 y advantageous, before real train of inventive THOUGHTS are brought into play & then perhaps the soon
529m039 happiness.-- recall scraps of poetry;-- former THOUGHTS, & in experienced people-- recall pictures &
529m040 re of imitation come into play.-- the train of THOUGHTS vary no doubt in different people., an agricu
530m046 thought arises?? compounded of the involuntary THOUGHTS.-- An intentionally recollection of anything
530m046 physical effect of brain the «similar remark» THOUGHTS, being functions of same part of brain, or th
538m080 possibility of the brain having whole train of THOUGHTS, feeling & perception separate, from the ordi
544m103 when conscious connect past, present & future THOUGHTS are broken-- Sir J. Franklin when starved, al
546m110 ra (by mistake) have lower animals these vivid THOUGHTS In same book (p. 143) wonderful case of perfe
548m117 ntensity to degree of <happi> pleasure of such THOUGHTS We give no credit to instinctive feelings.--
549m118 peasant.-- To make greatest number of pleasant THOUGHTS, he must have contingency of good food, no pa
549m119 in themselves, or if recollected, such part of THOUGHTS innumerable, which past through mind.-- These
549m119 innumerable, which past through mind.-- These THOUGHTS are most pleasant. when the conscience tells
578n055 bad action, it always bear some references to THOUGHTS of other person Decemb. 27th.-- Fear loose th
599o006 of two ideas in nature; a developement of the THOUGHTS expressed in Fingals cave, & in the arched &
613o036 dices about souls, we see particular trains of THOUGHTS as fear of man,-- crows fear gun,-- pointers
614o037 lace out of bodies) really less wonderful than THOUGHTS-- One organic body likes one <m> kind more th
618o041 our bodily frame or the cerebral portion of it THOUGHTS, perception &c. are modes of subjective actio
317c251 gth of ears & length of limbs, so that he first THOUGT only one species. & all hares on East side hav
344d042 Now Jones, before this happened from her looks THOUGT she was halfbred Beagle Staghound. «++»∴the gr
409e047 ence between mind of a dog & a porpoise was not THOUGT overwhelming.-- yet I will not shirk difficult
430e120 rd Mr Lonsdale says Trigonia costata & elongata THOUGT considerably different, in proportional dimens
447e165 s & yet can be raised in gardens.-- Poa alpina, THOUGT generally vivaparous sometimes seeds.-- [There
055r108 f the most solid rocks.--The grand cliffs of a THOUSAND feet in height, of the solid lavas.--proporti
090a024 00> x 100 = 50, 0,0,000 = 2500 = tons in fifty THOUSAND years]CD If world increased a tenth; would th

Page
(Key Word)

```
146g011 -- enormous mass thunder storm, many <hundred> THOUSAND tuns. Black faced sheep, sometimes mottled wi
181b041 st any one «of them» having progeny living ten THOUSAND years hence; because at present day many  are
193b090 er every animal produces in course of ages ten THOUSAND varieties, (influenced itself perhaps by circ
195b103 & Galapagos orpheus.= Put this strong so many THOUSAND miles distant.-- Absolute knowledge that spec
199b116 stand excepting by propagation that out of the THOUSAND of new insects all belong to same types alrea
213b171 varieties  do the same, May you not breed, ten THOUSAND grey hounds & will they not be  greyhounds?--
225b216 being made otter-- why to be sure there were a THOUSAND intermediate forms.-- Opponent will say. show
285c152 being.-- one species May have passed through a THOUSAND changes, keeping distinct from other & if a f
375d135 -- One may say there is a force like a hundred THOUSAND wedges trying force <into> every kind of adap
397e004 change of level of a few feet during last two THOUSAND years in Italy, but what «changes» would such
441e151 line, or others. which only generate once in a THOUSAND generations.-- any amount of generation may t
527m031 free-will & chance are synonymous.-- Shake ten THOUSAND grains of sand together & one will be uppermo
638j58v every adaptation, as the surviving one of ten, THOUSAND trials.-- each step being perfect «or  nearly
639j28v er oftener than any other one. survive. in ten THOUSAND years the long legged race will get the upper
126a123 ow comes it in volcanos that have gone on for THOUSANDS of years, that surface does not become hot?--
199b117 ame types already established. why out of the THOUSANDS of forms, should they all be classified.-- Pr
280c136 one  case it changes one, in other it changes THOUSANDS in futurity.-- This is right way of viewing i
301c200 ntomologist going into a country & collecting THOUSANDS & tens of thousands New insects, perhaps scar
301c200 to a country & collecting thousands & tens of THOUSANDS New insects, perhaps scarcely one new  family
418e084 ixed, as if added to old individuals,-- during THOUSANDS of centuries,-- each of us, then <is as  old,
372d131 ld produce another man.-- That the embryo the THOUSANDTH of inch should produce a Newton is often tho
335d014 red to animal with amputated limb.-- Heredetary <THR> Six fingered people, <Hill> <Lord Berwick> fami
261c070 on of species from distant countries,. may give THREAD to conduct to laws of change of  organization!
452e181 instinct,  how could it have originated-- spins THREAD of cotton.-- do p. 583, It APPEARS probable,«?
458t002 «Bengal Journal» The Taylor Bird uses pieces of THREAD, picked «up» instead of spinning-- better cas
350d056 Blackwell (Abercrmbies) comparison of sight to THREADS.-» Hence the Pecten, which move imperfectly h
376d136 n its comfort & though could not put it on, yet THREW it over it, & made it meet in front.-- Dr Smith
523m017 ally come on suddenly.-- Case of Mrs. C. O. who THREW herself out of the window to kill herself  from
545m107 e. The ourang outang, under same circumstances, THREW itself down on its back & kicked & cryed like n
599o007 d & leafy forests" Very good!. I grant that the THRILL, which runs throug every fibre, when one behol
539m082 re. association with much pleasure immediately THRILLED across me, bringing up old indistinct ideas o
468t105 er large & rough-- flowers common-- many winged THRIPS, covered with pollen-- «Thrips» about as large
468t105 on-- many winged thrips, covered with pollen-- «THRIPS» about as large as bit of chopped horse hair w
225b217 turns a Bullfinch black, or iodine on glands of THROAT, (or colour of plumage altered during passage
415e068 been, (with what attendant organization, Hand & THROAT) has made a man.-- CD [any monkey probabl
573n038 ary & involuntary movements.-- Person with sore-THROAT told not swallow spittle. will have involuntar
287c157 & Annals. 1830 (?)." if she has put man on the THRONE (of reason), she has also placed a series of a
170b001 is a longer process, the new individual passing THROUG several stages (¿typical, <of the> or shortene
599o007 Very good!. I grant that the thrill, which runs THROUG every fibre, when one behold the last rays  of
190b078 er of adaptation) is given by true generation, THROUGHE means of every step of progressive increase o
035r047 me beds» should have been kept; it shows that THROUGHOUT all England, whole surface oscillated equabl
348d050 no  matter of what importance, which prevails THROUGHOUT the group & serves to insulate <them> it".--
370d117 es &c &c The line of argument «often» pursued THROUGHOUT my theory is to establish a point as a proba
452e181 ot exist. there & it is a singular thing that THROUGHOUT the Moluccas Archipelago they are only to be
120a107 considered.  study Hopkins. theory of dikes may THROW some light.-- thin dikes not cooling if they ha
342d034 species. Yarrell says the bird fanciers say the THROW of any two species crossed is uncertain-- Yarre
356d071 orld. (Major <I> Mitchell p. 244. vol I) spit & THROW dust <according to my theory of generation  (p.
385d163 g to mongrel, & all subsequent litters having a THROW of this mongrel.-- I did not ask the question.-
389d176 life, like tendency to budding, which wishes to THROW itself off.--» as might be inferred from annual
393dIBC ertile heterogenous? When bird fanciers say the THROW of two varieties is uncertain do they mean they
442e152 heory «without seeds are freshly transported»-- THROW over this theory, & the sexual reproduction  of
484z014 egs-- habits-- Does the Secretary, make noise & THROW head back M Edwards,--on polypi of Tubulipores
512q019 Do most Hawks eat stomach. of finches-- do they THROW up pellets-- (4) About hybrid pheasants treadin
538m078 utely two people. Consider this profoundly, may THROW light on consciousness, explained by Dr Dewar o
546m110 ally in her senses in either state.- does this THROW light on instinct, showing what trains of actio
570n024 eplies nothing. if he did go to reply. he would THROW back his shoulders. he wishes to show, he is de
526m027 escent of a farthing, free will determines our THROWING it up.-- equall true the two statements.-- Ca
155g065 ust have been so carefully preserved as to have THROWN water in same «drainage» lines Mound of Gneiss
523m016 ntal <was> delay of money, he was «only» NEARLY THROWN into a hospital.-- My father was nearly drowne
280c135 y animals not breeding at all in domestication. THROWS great difficulty in way of ascertaining  about
442e152 - The fact just alluded to of Northern flowers, THROWS enormous difficulty in the way of Mr  Knights.
446e163 reds of years, without fresh seeds arriving.»-- THROWS a very great difficulty in my theory, here  we
590n093 p. 44.-- Jealousy. causes spasm in bile duct, & THROWS bile in circulation p. 75. Haller says tooth a
124a119 rass a piece of copper not melted absorb, zinc THRUGOUT its thickness.-- this most curious with respe
271c105 hose Southern range is? <One> The black & white THRUSH of Azara builds its nest in <same country> som
297c189 hout doom. Vol 2. Mag of Z. & B. p. 431. Missel THRUSH lately increased in numbers over whole of Engl
319c255 lycatcher.-- Bachman says he thinks the Mocking THRUSH beats all English birds in song-- one of thei
319c256 s of N. America. singing so well. & the mocking THRUSH being so very beautiful gret contrast with Sou
341d031 st closely represented.-- Thus the red breasted THRUSH is represented by one not differing except  by
461tf03 ers. Athenaeum p. 605 Mr. Macgillivray says "<A THRUSH &> Blackbird have been known in <their> «its»
461tf03 n in <their> «its» natural state to mate with a THRUSH"-- Athenaeum 1839. p. 708.-- Shrew, found by M
482z011 Turdus  falklandii & then 9. passeres! Says the THRUSH & another species! birds of passage!! sylvia m
632j53r ould ever have thought that the intestines of a THRUSH were means sufficient to ensure propagation of
219b193 ight be transported some blown--floating trees THRUSHES & bunting & coots-- «(Turdus Guyanensis?) (Em
305c209 explained.-- this is fact analogous to mocking THRUSHES of Galapagos having tone of voice like S. Ame
319c255 eats all English birds in song.-- one of their THRUSHES exceeds our blackbird, but our blackbird exce
319c255 ackbird, but our blackbird exceeds their other THRUSHES-- yet they have one with very sweet  notes.--
319c256 few of N. America are quite as beautiful. The THRUSHES of N. America. singing so well. & the mocking
375d135 oeconomy of Nature, or rather forming gaps by THRUSTING out weaker ones. «The final cause of all this
023r009 h, 2 of them, 8 inches long, & as big as a mans THUMB, the rest not above half so long; The maw was f
066r140 . -- «(Think some 60 fathoms, none thicker than THUMB» Sea weed said at Kerguelen Isd. to grow on sho
403e022 articularly common.-- without arm, <skin> hands THUMB.-- one leg, hare lip &c. &c. in Vol II p 363 acc
146g011 undred feet in vertical height-- enormous mass THUNDER storm, many <hundred> thousand tuns. Black fac
535m069 ue, «&c» (Very true, no doubt savage attribute THUNDER & lightening to Gods anger.-- ∴ more poetry i
553m135 ) that savages (mem York Minster) consider the THUNDER & lightning the direct will of the God <thus>
566n012 mmenced, for instance probably such a thing as THUNDER. would be placed to the will of God. Zoology i
021rIFC & Lyell Sailed, 27th «Friday gale 29th> Friday THURSDAY 29th gale Lyell's Geology The living atoms ha
146g013 certain there were none on surface of any hill THURSDAY On side of Hill South of upper end of Loch Do
147g019 or sea could not be placed in present position THURSDAY Evening ½ past 8 Tyndrum 29.<625> «636» Temp.
159g085 pply.-- Consider profoundly Boulder hypothesis THURSDAY, from Glen Turrit to Fort Augustus Barom on u
472s02r aler, all rest very similar-- June 2. 42 Maer <THURSDAY> Thursday After watching 14 days. many  times
472s02r rest very similar-- June 2. 42 Maer <Thursday> THURSDAY After watching 14 days. many times every day.
044r074 Hyd:  Carb: A. Mur: A. = (& this effect of water THUS holding matter in solution must be great: &  in
048r084 each under a cascade, one can understand pebbles THUS coated.--The motion is most wonderful, from che
049r088 rom conical form I incline to <latter> former; & THUS occurring in groups.--As these greenstone rocks
089a019 t des Gonaïves it is difficult to find stone not THUS composed on the NE part more like marble requir
098a047 spar) within other concretion.-- state last page THUS. point of attempted crystallization, & therefor
104a067 ranitic areas &c &c volcanos {P} fissure dike.-- THUS dikes terminated Solubility of fluids varies wi
105a069 f cliffs. but then how spread abroad?-- There is THUS wide difference between erosive power of  river
126a122 onductor is {P} against my views-- if we had rod THUS & judged by increments at, how wrong, would our
171b004 full  grown individual «with fixed organization» THUS being modified,-- therefore generation to adapt
172b008 x-- Chiloe, fox,-- Inglish & Irish Hare.-- As we THUS believe species vary, <in> changing climate  we
177b026 nly makes it excessively complicated.» {P} Is it THUS fish can be traced right down to simple organiz
179b032 ing to <first> having once borne Mongrels he has THUS seen the black blood come out from the grandfat
180b036 cies in same genus (as is). REQUIRES extinction. THUS between A. & B. immens gap of relation. C & B.
180b036 nest gradation, B & D rather greater distinction THUS genera would be formed.-- bearing relation to a
180b037 ty of ostrich, Petise may not be well adapted, & THUS perish out, or on other hand like Orpheus. bein
181b040 als in a country not rapidly increasing.-- If we THUS go very far back to look to the source of the M
```

Page **(Key Word)**
183b051 i. in present states. <they> it might continue & THUS two species be created) & live in same country.
216b180 f hybrid lillies &c &c &, (V Herbert on hybrids) THUS act.-- Now the point will be to find whether kn
226b221 few representative species. this must happen. & THUS acquired will explain representative system Of
227b226 (or between extremities of any great divisions) THUS a knowledge of possible changes is discovered,
235b262 ed <p> having great tendency to vary? Is not man THUS circumstanced; varieties of dogs in different c
239c001 est variety, takes greatest effect on offspring. THUS presuming those varieties to be oldest which ha
247c029 in England"-- Black & Grey varieties of rabbits THUS handed down for nearly 70 year. Galapagos Mouse
248c030 ll not readily breed together: The argument must THUS be taken, as «in» wild state (where instinct no
255c052 ion of ancestor of all birds, & so for birds. We THUS obtain an abstract idea of a bird-- An animal w
265c086 t family, having it with very different habits-- THUS bill & nostril of Puffinuria I think we may cle
272c107 cture (probably accompanied by habits) of other, THUS in Chalcididous insect, which I brought from Au
272c109 ngst the same species & subgenera in families.-- THUS the banded tarsi is common to all the Laniadae
273c111 e latter is obscure because nearly all are so.-- THUS in Hawks, there is a swallow, both in structure
280c135 strange.-- This even might be said.-- My theory THUS explains a grand apparent anomaly in nature. <t
282c142 shing men-- but might hire the preexisting race, THUS the analogy would not in all cases be produced,
284c149 irds is visible, when young, but not when old.-- THUS speckled form of young blackbird. good remark i
286c154 ht into the world same way» they may convey much THUS, Man has expression.-- animals signals. (rabbit
298c190 onate.-- the Key to the affections might perhaps THUS be found-- a person who is habitually kind to c
305c211 the endless forms of the living beings.-- We see THUS Unity in thinking and acting principle in the v
305c211 es of <dif> separation between those individuals THUS endowed, & the community of mind, even in the t
309c219 arent, races of animals-- argue «opening» case. «THUS» Educate all classes-- avoid the contamination
312c231 be> the provincial breed, & that first offspring THUS produced are better, than those bred in & in.--
316c245 in land & F W shells there is more confinement. THUS the Naiads (study De Ferrussac) are confined to
336d019 of organization] CD» false.-- The creator would THUS contradict his own law. So far is there any app
341d029 t he did not see where the line could be drawn-- THUS the most remarkable character in Apteryx, small
341d031 -- which however are most closely represented.-- THUS the red breasted thrush is represented by one n
342d035 uliarity overbears the crossing with females not THUS characterized.-- 16th Aug.-- What a magnificent
343d038 hanging from <hot>. Warm to cold, damp to dry.-- THUS Tierra del Fuego has <not> «only» one «Guanaco»
350d056 ore organ fitted to animals place in creation.-- THUS senses, especially sight connected with locomot
358d085 is very common by hybrids, that are infertile.-- THUS the common. pheasant & fowl when crossed never
358d086 igure, & some degree of whiteness like a Male.-- THUS castration, hybridity, & breeding in & in tend
363d102 nging birds, part instinctive & part acquired,-- THUS Yarrel has Lark & Nightingale which both sing t
364d103 ase they exchange birds with some other fancier, THUS getting fresh blood, without fresh feather, & c
370d117 gement, of the perfection of their separation.-- THUS Vertebrate blend with Annelidae by some fish.--
370d117 ficulty of growing plants in all parts of world, THUS tea tree in Brazil must have degenerated. as mu
372d130 always different tail? An Individual bud may be THUS produced from the growth of one part, (not stri
376d138 evince more lewdness for bitch than dog: Monkey THUS examine each other sexes,-- «by taking up tail»
378d147 169) has the colour on its back bright blue.-- «THUS young of» Many of the pies assume the metallic
384d162 o that Molluscous hermaphroditism takes place.-- THUS one organ in each becomes obliterated, & sexes
386d165 of plants, does annual give buds.-- life may be THUS prolonged bud being formed & one part dying for
400e011 c having relation to geographical distribution-- THUS Hattica is great genus.-- because found in all
404e025 ucture would become obscure & therefore it might THUS have arisen, & M. Edwards p. 330 distinctly sta
405e031 ame Manner as they would vary, if in wild state; THUS mark on ear of cats, colour can be brownish do
430e118 . & poutter Pidgeons «race-horse». have not been THUS produced, but by training, & crossing & keeping
433e127 he deposition of the beds-- The argument must be THUS put, shall we give up whole system, of transmut
440e146 cause. either in present, or past generation.-- THUS cabbages growing like Nepenthes.-- cases of pid
441e148 onstitution. but not structure of the parents.-- THUS would a Crab tree vary if planted in rich soil,
442e153 ly similar; [all the gorze in Norway ought to be THUS characterized study Von Buch.]CD Now Mr Knights
443e156 to our lives-- Being myself a geologist, I have THUS argued to myself, till I can honestly reject su
449e169 female common goose took after the common goose THUS contradicting (probably) Yarrells law & Walkers
520m001 ftener go with colour, than with form of body.-- THUS the late Colonel Leigton resembled his father i
521m007 oonomia.-- Now if memory «of a tune & words» can THUS lie dormant, during a whole life time, quite un
521m009 through eyesight, though, not through hearing,-- THUS when dinner was announced he could not understa
521m010 why do you not know, that is A. B my gardener.-- THUS was he in every respect, no communication could
522m011 es before he maintained he had never heard of.-- THUS in many things if he began at one end, he knew
526m030 l exercise of the mind, altering form of head, & THUS these qualities become heredetary.-- When a man
526m030 actions, which are modified by circumstances, & THUS the appetites themselves become changed.-- appe
527m032 any years]CD.-- Beauty is instinctive feeling, & THUS cuts the Knot:-- Sir J. Reynolds explanation ma
534m063 re attacked by insects & snails of this country (THUS Dahlias by snails)-- <The> Apion radiolum under
535m064 n birds blind storge-- They continue till death, THUS acting 4 to 6 weeks. The deserted broods appear
536m070 - such efforts prevent anger, but observing eyes THUS unconsciously discover struggle of feeling.-- I
536m074 ind, because he would tend to be an atheist. Man THUS believing, <yet> would more earnestly pray "del
538m080 ey would make one's father & self one person-- & THUS eternal punishment explained. These facts showi
541m090 right in saying delirium rest-- therefore dreams THUS act.-- ∴ weak minded people are fickle & full o
543m097 ility of understanding language is considerable, THUS carthorse & dog.-- birds many cries. monkeys co
543m101 instinct, a separate thing superadded.-- we can THUS trace causation of thought.-- it is brought wit
553m135 thunder & lightning the direct will of the God (<THUS> & hence arises the theological age of science
553m135 n according to M. le Comte).-- Those savages who THUS argue, make the same mistake, more apparent how
553m137 the expression & noises of monkeys go in groups. THUS the pig-tailed baboon, shoved out its lip, look
558m152 guggling noise. Puma did same & & some others-- THUS <sudden> «forcible prolonged» expulsion of air
568n017 fulness is connected with some disturbed habit» [THUS shepherd dog. has pleasure in following its ins
571n028 character???) Why may not our heredetary nature THUS acquire some general notions, which are taste?
574n041 an has great tendency. to exert all senses, when THUS stimulated, smell, as Sir. Ch. Bell says, & hea
584n071 p. 22. instincts & structure always go together: THUS woodpecker: but this is not so,, the instincts
603o013 & whole classes of words «are abbreviations» he THUS derives from nouns & verbs-- so that much of EV
605o18v ollection the original cause of these feelings & THUS we apply to them the metaphorical term sublime
617o040 o other matter in the course of its DIRECTION, & THUS when we apprehend force in inanimate matter we
618o41v nvariable» priority of one to other: no not only THUS, for if day was first, we should not think nigh
621o047 ally acquired from some trifling circumstance.-- THUS a child may be taught to think almost anything
622o048 elings which do not lead to action are repressed THUS avarice. &c &c.-- [RHC] 8) in the beginning I m
623o050 e is much truth in doctrine, for [RHC] 9) We can THUS explain love of place.-- although here we have
624o051 e instincts, for they are in accordance with it. THUS a dog may be trained to hunt one pig sooner tha
629o55v imperiously say so, & produce shame & remorse-- [THUS pungency of one's feelings for indecency-- prep
633j53v is hard on my theory of gain of small advantages THUS to explain the curling of the valves of the bro
640j167 , from adjoining coast. Admirable explanation is THUS offered.-- From these views, one would infer th
558m150 giving rise "do unto others as yourself". "love THY neighbour as thyself". Analyse this out.-- beari
466t099 ividual. June 1st 1841. Maer Examined the Lemon-THYME.-- equally abortive as it was in autumn: filame
467t099 ly doubled, brown, but with no pollen.-- Common THYME growing close by is equally abortive--and both
500q010 st Indies-- Asiatic Researches) (23) Talk about THYME. Horned Oranges. Spallanzani Essay-- Figs 2 kin
501q011 ters-- in double flower. do Henslow Speaking of THYME doubts about stigma in similar manner ever fail
504q013 ages given (3) ______ in Heartease (4) Does the THYME bear abortive stamens every year & Spring. & wi
505q014 te it (7) History of Potato field= (8) Abortive THYME seeds weather wet--? Linum flavum put in Spirit
505q014 = failed to germinate 16 Will plant some of the THYME with abortive stamens by Terrace to see, whethe
558m150 to others as yourself". "love thy neighbour as THYSELF". Analyse this out.-- bearing in mind many new
463t057 t, <be> (as vertebrae into skull., two bones of TIBIA into one.--) because if the animals were taken
509q017 Remark. Horse or cow.-- degree of soldering of TIBIA & fibula: in Man any abortive bones??? do. Wing
416e072 d races L'Institut 1838. p. 394. Rhinoceros «TICHORHINUS» in Paris basin.-- its relation to African S
451e177 . -- p. 555. Lieut. Hutton counted, the ova of a TICK «in India» & found there were 5,283 attached to
592n103 express contempt. p. 76.-- children have been TICKLED into excessive laughter & so into convulsions.
581n064 silk-handkerchief-- cats & dogs fond of slight TICKLING sensation.-- in savages other tastes few. Mar
031r038 base of great lava cliffs [Fig. I] line of high TIDAL action {P} NB. patches of modern Conglomerates
032r039 . will be to eat in the land in line of highest TIDAL action. this will at length be checked by incre
032r040 f coast of Chile.-- Must first explain «top of» TIDAL band of action. This case differs. I think. fro
051r094 ments by mere force of waves: & action on upper TIDAL band, I do not see how to account for oceans po
154g057 Glen Guoy flat peat plain divortium aquarium-- TIDAL channel-- 12ft obscure obscure NB In Glen Colla
154g057 nnel-- 12ft obscure obscure NB In Glen Collarig TIDAL channel, sides <alm> 15 ft above bank or terrac
158g081 plain about 60 ft beneath shelf peat on pebbles TIDAL plain as sea gradually retired, hard to explain
163g107 hells. My informant saw them himself-- Sand with TIDE ripple Near Fort Augustus hill & fringe as if i

Page **(Key Word)**
306c212 gument) & therefore down the stream followed ebb TIDE, therefore got into habit of going down stream
064r134 rom shore. rock of oysters quite above reach of TIDES.--thinks them same as recent species. -- May I
066r141 rts which break through the N & South lines the TIDES form eddies with its extreme force. Yet, no out
117a102 s of lakes. to see effects of degradation, «no» TIDES, water always falling or at least not rising ar
151g039 ed.-- die away also, without any cause, must be TIDES. &c. {P} <notch> roads very much this character
355d066 N; as long as opponents will «are» not «able to» TIE themselves down, they can find loopholes) "It is
217b187 - A genus with species in Van Diemen's land and TIERRA del Fuego.-- Araucaria, species. Brazil {P} Ch
226b221 tely large ones.-- Is the flora of <S. America> TIERRA del Fuego like that of North Europe, many gene
268c099 g how many species common to Patagonia desert & TIERRA del Fuego. & forest «Parrots in Macquarrie Isd
343d038 g from <hot>. Warm to cold, damp to dry.-- Thus TIERRA del Fuego has <not> «only» one «Guanaco» of th
409e047 some difficulty in conceiving how inhabitant of TIERRA del Fuego is to be converted into civilized ma
481z010 do stellata. Meyen p. 92.-- great Kingfisher of TIERRA del Fuego killed in Chile. Dobrizhoffer., Vol
497q007 .-- Henslow.-- (1) Character of alpine Flora of TIERRA del Fuego and Entomology of.-- most important,
527m051 ge <all> «so many» birds singing in England, in TIERRA del Fuego not one.-- now as we know birds lear
553m137 superstitions of an Australian savage or one of TIERRA del Fuego.-- Mr Miller (superintendent of the
219b196 creator» two quadrupeds at S. America Jaguar & TIGER & Europe, as to produce same one. NAlthough in
283c145 th such sentences as "oh turn a Buccinum into a TIGER."-- but perhaps I feel the impossibility of thi
331dIFC ese?-- Are the number of kittens between Lion & TIGER at litter as numerous as in common lion? Are th
423e097 ion. though perhaps difference between jaguar & TIGER may not be so.-- Considering the Kingdom of nat
439e144 t absurd. that the presence of of the Leopard & TIGER together depended on some nice qualifications e
439e144 some nice qualifications each possess., & that TIGER springing an inch further would determine his p
450e174 1. Elephant. Rhinoceros Leopard (but not Royal TIGER), &c are found but only in one part the norther
529m041 ry view. if one were admiring one in India. & a TIGER stalked across the plains, how ones feelings wo
558m152 - I do not think they arch their backs-- Bengal TIGER. when slightly angry. curls tip of tail.-- do t
312c227 5» will have long tail.-- in raptorial birds, & TIGERS & sharks, being spotted, & colours of little v
312c228 tempt to keep up offspring, are not half lion & TIGERS ditto. (see Griffith) & half Muscovy ducks, «b
333d008 ding question was put.-- Fox thinks half Lion & TIGERS are exactly intermediate in character & Kitten
360d091 to Jackall.-- This Keeper has seen when sickly TIGERS have first come over, insects somewhat like «b
380d155 a dog if led in string will not.-- some of the TIGERS.-- cat, though caterwhalling. & put into femal
290c163 y. Transact. Vol V. Birds bones-- in strata of TILGATE forest Seeing common gull in garden at Zoology
231b244 ains-- ¿agrees with non-blending of languages?-- TILL man acquired reason, he would be limited animal
257c056 a miracle, but induction leads to other view.-- TILL we know uses of organs clearly, we cannot guess
275c120 ed <greyhound>. bull-dog. & crossed & recrossed, TILL there was a dash of blood, with whole form of g
275c120 that old variety» & then recrossing off spring. TILL size diminished, but feathers continued by pick
277c126 ss tells us that ¿O. Modulator & O. Patagonicus. TILL neutral ground ascertained, call them varieties
288c158 el parts of his series, ie, cannot be discovered TILL circles completed Major Mitchell, does not know
347d050 to genus or group of any kind not being perfect TILL circular. p. 5 Most clearly shows that genus ex
348d051 iscover the real relationship is doubtful.-- not TILL much knowledge is elicited.-- It will rest upon
355d067 towards the end of science. namely prediction.-- TILL facts are grouped. & called. there can be no pr
364d103 they in evident rivalry sing against each other, TILL it has been known one has killed itself.-- Q Si
374d134 unfavourable-- 28th. «I do not doubt, every one TILL he thinks deeply has assumed that increase of a
389d176 ht be inferred from annual plant being prolonged TILL it has bred.-- Offspring like both father & mot
436e138 d near, but not in sight of each other will sing TILL they drop off their perch.-- p. 101-- Kingfishe
443e156 yself a geologist, I have thus argued to myself, TILL I can honestly reject such false reasoning Bell
469t135 d many flowers-- Saw Bees frequent these flowers TILL late in evening-- On rough calc. 280 flowers--
530m049 x believe cats discover birds nests & watch them TILL the young are big enough to eat.-- There was bl
535m064 eggs on leaves of Eucalyptus, watching few days TILL larva excluded, then though not feeding them «n
535m064 oup.-- (as in birds blind storge-- They continue TILL death, thus acting 4 to 6 weeks. The deserted b
551m127 old man, he had just seen mind went on RAMBLING TILL excited by question.» Sept. 4th. Lyell in his P
555m142 l associated animals?) I doubted it in Fuegians, TILL I remembered Bynoes story of the women.-- The C
564n004 ar more sensitive. ulitmate effects of actions.→ TILL at last he face «instinct of» hunger, «of» deat
575n045 of Newfoundland dogs. who will not enter water, TILL he sees. whether birds badly wounded, or only w
615o36v l about in the evening «seldom leave their perch TILL evening» crow different.-- Heredetary effect of
634j54v , or design, pursued to its utmost exhaustion, & TILL it must be abandoned for another".-- What bosch
038r059 tions? & not sources of lava streams.--Urge not TILTED strata.-- It will be well to urge the case of
042r068 hili & delta of Indus», my belief in submarine TILTING alone, must be modified. «Moreover, the Volcan
217b184 strongly after either parent. about <half & half TIM> as often one way as other.-- He has known case
029r033 on ripple on the water. It was quite calm at the TIME. Latitude 8 deg. 47 min. N: longitude 61 deg. 2
032r040 as at first. & according to the greater or less TIME of rest. so would the size of the triangular ma
043r070 e than» one orifice <...> does not occur at same TIME: this is contrasted to contemporaneous action o
045r075 Ulloa states that Volcanos!! were in eruption at TIME of great Lima earthquake» In the Chili earthqua
055r107 historical records of eruptions how immense the TIME!! How well agrees with number of Craters!--At S
058r116 to a distance of more than 500 leagues. A little TIME after a bad earthquake in Chili; Arequipa in 82
062r129 empted to believe animals created for a definite TIME:--not extinguished by change of circumstances:
063r130 co to recent: in former case position, in latter TIME. (or changes consequent on lapse) being the rel
104a067 h superincumbent mass (varying hardness,-- takes TIME to trace) from few dikes which have given rise
113a091 gree. being little we may confidently infer that TIME has not been allowed for lower beds to cool dow
127a125 urther N. when soil frozen for greater length of TIME depth of ice ought to be less.-- Memoir of the
132a137 urious investigation it would be to compare, the TIME of the earthquake of Chile, with that of the pa
138a153 erve for comparison with the moon at some future TIME [blank] Sir. J. Halls Paper on the consolidatio
173b009 ion made perfect.-- truer even than in Lamarck's TIME. Gray's remark, best known species. (as some co
174b015 esentatives-- «as in Australia» This presupposes TIME when no Mammalia existed; Australian; Mamm were
175b016 lt to prove cats &c from Egypt no answer because TIME short & no great change has happened I look at
175b017 nge,-- as we see them in space, so might they in TIME As I have before said isolate species <& give e
187b063 Horse, Elephant & Mastodon dying out about same TIME in such different quarters.-- Will Mr Lyell say
187b065 d but different, because country separated since TIME of extinct quadrupeds:-- same argument applies
189b070 d tell, I do not doubt even colour hereditary in TIME as in space. (Mem: Galapagos). Little wings of
189b073 ether just like buds of plants, which die at one TIME, though produced either sooner or later.-- Prov
203b136 ess not heredetary?, having been gained in short TIME. Milvulus forficetus <has a wide range> is a ty
210b158 it not present analogy to what takes place from TIME? Von Buch distinctly states that permanent vari
212b165 - a sort of internal bark. would remain for long TIME together in tub of water with only nose project
214b175 scended from a breed known to be there since the TIME of Charles.,-- and now in the possession of Mr
217b188 ting, because Iceland, must have been all ice in TIME of ice transported.-- This gives room to fine s
228b229 have near genera far back, as well as at present TIME, we might expect confusion of species.-- Import
231b244 Dromedaries & Camels breed?-- As man has not had TIME to form good species, so cannot the domesticate
232b246 ng sea.-- For change of species does not measure TIME but physical changes (we assume like weather on
240c003 rk of Apteryx having feathers.-- It is possible, TIME being an element in the transmission of form, m
248c034 king» become more & more impressed in blood with TIME, then generation will «only» produce an offspri
252c045 America» «some doubt, from want of knowledge of TIME-- Analogy from three first will give one almost
252c045 rst will give one almost certain guide ∵ because TIME required too separate isld very long» America &
266c091 lants from Cape of Good Hope continuing for some TIME to flower at their own periods.-- Arcana of Sci
275c119 geography species tell of Physical relations in TIME «& forms» distribution tells of horizontal barr
282c140 se heredetary ones.-- which would elapse; during TIME such changes had elapsed.-- let these families
294c175 Lonsdale do. says Sheep could not live for some TIME at New York «instance of the fine relations of
294c177 s its proof, «its limiting, the allowing at same TIME true species.» & its adaption to classification
296c185 ogy p. 56. Peregrine Falcon holds birds for some TIME alive ¿therefore other species mice & only kill
299c195 cated kind with large root by sowing it at wrong TIME of year & manuring it.-- Epigonous. Perigonous
301c199 g grains. acquirable through hoarding from short TIME.-- My theory must encounter all these difficult
302c201 me to say that there can be no animal at present TIME having an intermediate affinity between two cla
303c204 essed a structure in these points for <t> a less TIME than other. points. female genital organs «in s
304c208 ant> but, where much death, may be inferred much TIME elapsed & therefore descended from branch high
308c216 nstration that all animals have never at any one TIME formed chain, since if cretceous period assumed
308c219 l show state of knowledge «Negroes existed since TIME of earliest Egyptian drawings & Old Testament»
312c232 ns non-heredetary & variations produced in short TIME in some extent counterpart, mutilation being «v
312c232 being «variation» produced in shortest possible TIME. Mr Willis Long eared little dogs, I am told, g
313c236 ssary to generation (lateral with no relation to TIME) as in buds.-- I can scarcely doubt final cause
328cIBC bo. Cerithium Jardin du Roi Java fossils at same TIME Study Botanical work on Buds & Gemmae. C D Char
332d001 not caused by a wound, when suddenly during one TIME he had three patients at very distant quarters

Page	**	(Key Word)	**
332d002	cases of virulent cancer in women, & since that	TIME	it has been rare disease.-- but now (July 1838)
336d017	ore persistent.-- «any amount of change» shorter	TIME	less [s]o.-- the result of this is that animal
337d023	n different countries in exact proportion to the	TIME	they have been separated; together with physica
338d023	gether with physical differences of country: the	TIME	of separation depends on facility of transport
342d033	on must be considered as distant species?? or is	TIME	the varying element). Then do those SPECIES whi
343d036	» from a period short of eternity to the present	TIME,	to the future-- How far grander than idea from
343d037	the Rhinoceros of Java & Sumatra, that since the	TIME	of the Silurian, he has made a long succession
355d069	bility of only small amount of change at any one	TIME	Seeing what Von Buch (Humboldt). G. St. Hilaire
360d091	iation) which impresses offspring most.-- «& not	TIME»	thinking of the Penguin duck & Herberts law of
360d092	ce <of any kind>, if very firmly fixed from long	TIME,	made no difference what its kind was.-- but if
362d100	rfect-- also to trace the laws of change in this	TIME.	-- the impossibility of discovering their origi
365d106	en greater puzzle, than none, for the «e»normous	TIME	which it must have taken to separate. Van Dieme
366d112	te, ie with the chance of two being born at same	TIME,	& make breed, one would doubt any law.-- Yet s
375d135	in increase at geometrical ratio in FAR SHORTER	TIME	than 25 years-- yet until the one sentence of M
377d139	hen present animals-- lived.-- we know the great	TIME,	necessary to form channel & (& Basses St) yet
377d139	(& Basses St) yet no change in English species--	TIME	no element in making change, only in fixing it:
377d139	fixing it: only circumstances. a contingency of	TIME.	When we multiply the effects of <earthquakes>,
386d165	eing formed & one part dying for great length of	TIME.	-- There is probably law of nature that any org
387d168	several ova in mother. (with only difference of	TIME)	is the above law anyways connected with the ca
387d169	wo animals which have some peculiarity for first	TIME	& if their <D & E> «all their offspring» inheri
391d174	a necessity that there should be something «each	TIME»	added to that kind of generation, which typifi
392d175	es. shows that difference need not be added EACH	TIME.	but after some time]CD What kind of plants are
392d175	ence need not be added EACH TIME. but after some	TIME]	CD What kind of plants are Monooecious or dioec
398e006	theory requires each form to have lasted for its	TIME:	but we ought in same bed if very thick to find
400e013	ations before they began to double.-- at present	TIME	Uncle J. does not suppose one aboriginal variet
406e037	s. of same two great orders destroyed about same	TIME	in North & South. America.-- Whole wor[l]d, for
414e063	-- though man's practiced judgment. even without	TIME	can do much.-- (yet one cross, & the permanence
417e079	than others.-- does this more refer to length of	TIME	that the resemblance is permanent, or the simil
418e084	tions of structure being added to the germ, at a	TIME,	(as even in childhood) when the organization i
419e087	ly secondarily,, by assumption well grounded, on	TIME;	-- therefore the mere loss of species, which ma
423e095	- It may be said, why should there not be at any	TIME	as many species tending to dis-developement (so
432e125	ession of organisms tell nothing about length of	TIME,	only order of succession.-- Splendid Pamplet.
433e127	e up whole system, of transmut., or believe that	TIME	has been much greater, & that systems, are only
436e135	on is far more efficient in making species, than	TIME	(as cause of change) which can hardly be believ
438e141	rmed.-- » if it shall be difficult to show that	<TIME>	the fixity of characters «from antiquity» prev
438e142	the amount of may depend on many circumstances,	TIME	of domestication [see Wikinson on dogs of Egypt
438e142	nson on dogs of Egypt & Cuvier on Mummies]CD [NB	TIME	is element in change, as in Dahlias]CD all much
443e155	a being hermaphrodite is, that intercourse every	TIME	is of no consequence in that degree of develope
455eIBC	ate them, «as by an insect coming always at same	TIME»	see if by so doing can be made sensitive The f
460t019	pes, it must not be supposed that this refers to	TIME.	-- Marsupial in Oolite.-- insects, of do orders
473s07r	arities shd be born,-- may come in corresponding	TIME	of life of offspring-- No peculiarity in extern
489qIFC	n &c. July. 1842.-- Shrewsbury p. 14 Henslow (2d	TIME)	p. 14.-- Father. And. Smith Dr. Holland p. 16
498q008	Willows grow in same situation & flower at same	TIME.	Has H. seen group of different species growing
513q21.	anded-- in reference to Lobelia & Clarkia-- Peas	TIME	of impregnation.-- says many flowers are dichog
513q21.	llen & stigma generally not being mature at same	TIME	on same plant --Flora of Australian Mountains.-
515q021	to Primroses. (4) Do apples "sport" in fruit, or	TIME	of leafing (5) Do the most cultivated show Hear
521m007	words» can thus lie dormant, during a whole life	TIME,	quite unconsciously of it, surely memory from
521m009	watch upon which he exclaimed, why it is dinner	TIME.	-- » My father asked him whether he had gardene
522m012	a thing it often does not take any effect at the	TIME,	but some time afterwards it calls up pain. or
522m012	n does not take any effect at the time, but some	TIME	afterwards it calls up pain. or pleasure. & is
522m012	which had just taken place.-- as if the idea of	TIME	had been disturbed.-- These foregoing cases of
522m013	and insane.-- that everybody is insane. at some	TIME.	Mania is quite distinct, different also from d
523m017	she was not insane, answered she had known it at	TIME	& had bought arsenic for that purpose.-- this f
523m017	these two years that she had been insane all the	TIME.	-- There are numberless people insane of partic
528m037	likewise is novelty of view even old one. every	TIME	one looks at it.-- these two causes very weak.-
530m044	f everything in <he a [...] Mr B> journey. short	TIME	previous,-- because, pain prevents repetition o
534m063	leap it would have been instinctive, seeing that	TIME	is lost & endeavours made must be experience &
536m073	umstances & education, & by choice which at that	TIME	organization gave me to will-- Verily the fault
541m091	ly; for that one does with novel for a length of	TIME.	-- Then if one endeavur to keep any simple idea
543m099	e chess Players-- A man at Cambridge, during his	TIME,	almost an absolute fool used to play regularly
544m102	t.-- Ld Brougham thinks no dreams except at this	TIME.	how does he account for dogs & men speaking in
544m102	se, at the violation of all <rules> relations of	TIME,	<identity,> place & personal connections-- ide
546m111	e years before.-- was greatly astonished, at the	TIME.	& could trace no chain of association» Mayo Ph
548m117	of pleasant ideas passing through mind in given	TIME.	-- intensity to degree of <happi> pleasure of s
550m124	ant to unpleasant mental sensations in any given	TIME	«-- compared to what other people experience.--
550m124	y be more or less pleasant & unpleasant, in same	TIME,	-- therefore degrees of happiness-- Entire happ
550m125	in,-- compared to what others experience in same	TIME.	-- Pleasure more usually refers to the sensatio
550m125	ccupy greater proportion of <each> «every» man's	TIME.	-- Begin discussion-- by saying what is Happine
559m156	nsitive plant if irritated very regularly at one	TIME	every day.-- naturally close at that time after
559m156	at one time every day.-- naturally close at that	TIME	after long period.-- My Father about double con
565n007	& pricks its ears» when afraid, though not every	TIME	really wishing to smell its enemy.-- Man & dogs
568n017	s it.-- When one is «simply» habituated «in life	TIME»	to any line of action, or thought one feels pa
571n029	we know many tastes become acquired during life	TIME:	-- the latter correspond to fashions in ideal t
572n034	d had dwealt on each word separately, neglecting	TIME,	& general sense, anymore than connected with g
579n059	ed by» other emotions? When a man keeps perfect.	TIME	in walking, to chronometer, is seen to be muscu
580n060	My father says old people first fail in ideas of	TIME,	& perhaps of space-- in latter respect he thin
580n061	very loose ideas.-- Have children loose ideas of	TIME?	-- Characteristic of one kind of intellect is t
582n069	nest; it knows its object but not result (first	TIME	of building?), but not the means of performing
587n089	ction.-- p. 248. Theory of Association. owing to	TIME	when entered brain, try contiguity of parts of
600o008	r.-- avarice a compounded passion gained in life	TIME]	CD 3. The Infinite, -- lives by hopes, looks to
601o009	onscious.-- Consciousness bears some relation to	TIME	& memory Reynolds X discourse very curious as s
607o025	e) circumstances. As man hearing Bible for first	TIME,	& great effect being produced.-- the wax was s
616o038	ill: there must be cases of secretion being some	TIME	governed by will in some animals, involuntary i
620o044	sons on it & determines to act more wisely other	TIME,	for he knows that the instinct. (or conscience
620o044	ays present (which is indeed, often felt at very	TIME	it is disobeyed) & is sure guide.-- Hence consc
620o045	ncts are best to be followed.-- Yet even at this	TIME,	malevolence,, when not urged to it by passion,
624o051	correctly «in which it» has been so in some past	TIME,	hence passions]CD «although perhaps useful at
151g040	(one of Gneiss remarkably water worn) are often	TIMES	of rock not in immediate neighbourhood, (as gra
227b224	I) a horizontal history of earth «within recent	TIMES.	». & many curious points of speculation; for ha
255c051	f structure.-- shape of wings have altered many	TIMES,	but all have had feathers.-- if wing totally o
259c065	principle that, cut a sheeps tail off plenty of	TIMES	& you will have no tail (example probably not t
289c162	produce young, capable of producing young many	TIMES	& lay two sorts of eggs-- one remaining through
298c190	e from simplest reasoning in lower animals many	TIMES	produced, a general tendency produced. Such as
337d020	probably the effect of <sev> local variety many	TIMES	changed together with some training in the earl
338d023	the local circumstances of the two countries in	TIMES	present & past. The effect of physical conditio
362d099	arrell says the «male» Axis of India, breeds at	TIMES	when horns not perfect-- (is not this so in S.
362d100	ted cattle have tendency to breed at particular	TIMES.	Mr Yarrell has old book 1765? Treatise on Dome
387d168	rom parental stock having been repeated several	TIMES,	that it becomes fixed in blood.-- Looking at o
397e004	ry, that the amount of change within historical	TIMES	has been small-- because change in forms is <al
412e057	shall be perfectly adapted to circumstances of	TIMES.	& from persistency «owing to their slow format
424e101	ere any cases of union of two regions in modern	TIMES.	-- this would depend on negative evidence of fo
436e135	halopods, seem to have decreased since earliest	TIMES	-- Apterix has a most perfect Struthio head pull
467t104	Keel of Lupine-- Seen Bees on Potato & several	TIMES	on Beans Rough.--green-cabbage «in flower»-- sw
469t135	28 minutes-- say then each flower is visited 30	TIMES	a day is considerably under mark, & this has no
472s02r	Thursday> Thursday After watching 14 days. many	TIMES	every day. many clumps of heartseases, never sa
531m050	s are placed) Memory cannot solely be number of	TIMES	repeated, because some people can remember poet
563n001	s building near houses. yet so shy at all other	TIMES.	-- Birth Hill shows it is evergreens they seek
567n013	not knowing what to do with them, came several	TIMES	& opened my hand, & put them in-- like child. T

Page **(Key Word)**
571n031 & vice versa.-- almost proves that at earliest TIMES there must have been intimate connection betwee
202b132 Melville Isd.-- "The buffaloes, introduced from TIMOR, herded separate from the English cattle, nor c
233b249 rk Australia Fauna so far. Indian all the rest. TIMOR according to Mountain chain ought to be Austral
264c082 owing to variety of station inhabited by them-- TIMOR. Australian forms amongst birds Java. not so mu
306c213 not for vertebrata, but mammalia & reptiles &c TIMOR is connected with Australia «map to King's Aust
306c213 es distant).-- directly beyond produced line of TIMOR 215 degree. What productions Sandal. Wood Isd.?
314c239 ot observe scarcely any Australian character in TIMOR plants, yet it seems there may be Eucalyptus!--
400e011 ia from approach to Asiatic [...]t in part near TIMOR, & to Europaean in Van Diemens land, where ther
449e170 of Japan-- that the «animals of» islands N. of TIMOR are allied to the «type of genera in» islas de
449e173 p of the World, has. written. Mastodon found at TIMOR.-- thinks he has seen specimen at Paris Museum.
451e180 t.-- V.-- : do. Chat XXI. Wild cattle & Hogs on TIMOR-land-- monkeys do not exist. there & it is a si
260c068 ds or animals are placed in white rooms to give TINGE to offspring.-- Darkness effect on human offspr
493q003 coloured = I have grey-cat «wh was female» with TINGE of tortoise-shell «on back.--» = Length of inte
378d147 young of» Many of the pies assume the metallic TINTS, such as Magpie, Jay, & perhaps all the rollers
383d160 each feather is fine metallic green. <from> with TIP & part of shaft metallic green.-- This green dou
402e019 culiarity in the cats «p 10» the joints near the TIP of the tail are generally crooked, as if they ha
453e182 ies «the» cats are born with the joints near the TIP crooked.-- is the form [...] Dampier. Vol I. p.
467t104 Humble press down wings which ejects pollen from TIP of sheath.-- «Also in Lathyrus pratensis yellow
558m152 backs-- Bengal tiger. when slightly angry. curls TIP of tail.-- do two cats arch their back when figh
290c164 ash grey back, like a black bird washed, Whilst TIPS of primaries black, by examining series I canno
400e012 ups.-- ears varying so much,-- kind of fur-- (do TIPS of ears take any colour?)-- length of tail vari
522m014 ich rises almost involuntarily when a person is TIRED is akin to insanity.-- «I know the feeling also
531m051 h came over me, when listening one evening when TIRED «-- how true the heart the scene of anger.--» t
552m131 ce to muscles, that in my grandfather remark, a TIRED man. involuntarily feels angry, when brain is p
611o034 nscious sensation. [RHC] During growth <extres> TISSUE <[...]> unites matter into certain form; invar
611o034 bear fixed relation to such accidents. But such TISSUE <must> bears relation to whole, that is enough
612o035 to external contingencies, as much as growth of TISSUE and are subject to accident; the sexual willin
268c096 tural History. «vol I» p. 159 curious account of TIT mouse feeding young of redstart & actually drivi
298c189 Annals of Natural History Vol. I. p. 185 case of TIT lark placing withered grass over nest, when ofte
218b191 sactions and a distinct work on Hybridity under TITLE of Amaryllidae & Narcissus. Mr Donn considers M
353d061 to Van Diemen's land & Australia well developed <TITS> «Mammae» in male ourang-outang. other point of
242c018 > D'Orbigny is said to have brought a tortoise & TOAD from S. America & identical with those from S.
242c018 se from S. Africa. «M. Bibron doubts fact.--» My TOAD is same species Coquille Voyage p. 25 Mais il n
496q05a so do with honey-- What is use of Bee Larkspur= =TOAD Orchis= How many flowers in minute do they visi
053r100 inoco mentioned by Humboldt under name of Rothe-TODTE-liegende is perhaps same with that of Pernambuc
366d112 ubt any law.-- Yet seeing the feathers along one TOE of the Pouter one thinks there is a law.,-- that
469t119 examined 7 says one specimen had on one foot, a TOE-nail & two joints-- as it is on one foot probabl
539m081 is of a man wagging his foot & working with his TOE to perform some difficult task.-- Aug. 12th. Whe
043r071 n formations & durability of similar causes go TOGETHER. add. <">> <from> "in the same line" to "from
046r078 le» surface; how comes it they do not flow out TOGETHER? How are they eliminated.--«Sulphur last.--»
049r089 ny. P 95. Glassy & Stony Pearlstones alternate TOGETHER in contorted layers: Mem: Phillips Mineralogy
056r111 canos separate.> Volcanos blend all substances TOGETHER; & products being similar over whole world, g
063r131 from. Cordillera. rocks.--When beneath water.--TOGETHER with hypothetical case of Brazil.-- Propagati
066r142 Caradoc from lower of third Silurian division--TOGETHER with same general character of fossils decept
098a047 ggregated (I assume the same force which draws TOGETHER two particles of Carb. of Lime, tends to crys
179b033 lants hybrids to go back?-- If so Men & plants TOGETHER would establish law. ≠as above stated:-- no o
189b073 ca All animals <are> of same species are bound TOGETHER just like buds of plants, which die at one ti
191b082 y forming species.-- Man & wife being constant TOGETHER for life, is in accordance with The Male anim
193b093 all the remarkable Australia genera, collected TOGETHER Man has no heredetray prejudices «or instinc»
193b093 ay prejudices «or instinc» to conquer or breed TOGETHER:-- Man has no limits to desire, in proportion
194b096 Do not plants, which have male & female organs TOGETHER, yet receive influence from other plants.-- D
200b122 ant, which of course most rare, or when placed TOGETHER they will breed.-- But what a character is th
202b132 ish cattle, nor could we get them to associate TOGETHER" There is long rigmarole article by S Hilaire
212b165 t of internal bark. would remain for long time TOGETHER in tub of water with only nose projecting.--
224b213 t remains «at large» with constant characters, TOGETHER with other <animals> beings of very near stru
228b229 nstance take Voluta & Conus (??) which now run TOGETHER, were not both genera formerly abundant. Seed
229b232 here> one common ancestor we may be all netted TOGETHER.-- Hermaphrodite animals couple: argument for
231b241 ones all different.-- Join Sumatra & Java <to> TOGETHER, by elevations now in Progress, & you will ha
248c030 of many ages standing, will not readily breed TOGETHER: The argument must thus be taken, as «in» wil
248c030 h plants) no animals VERY different will breed TOGETHER, so when we grant «(which can be shown probab
248c030 will be presumption that they would not. breed TOGETHER.-- We see even in domesticated varieties a te
254c049 digestion in Acrite Kingdom all organs blended TOGETHER, & same organ «where eliminated is» often rep
260c068 n (p. 5). describes many kinds of bird uniting TOGETHER in pursuit of Blue-Jay, when <one> birds hear
264c081 d seems to doubt how far structure & habits go TOGETHER. This must be profoundly considered.-- Struct
267c094 dogs of Guayana always hunt in packs «30 or 40 TOGETHER» colour reddish <brown>. ears long.-- like bu
274c115 c & In relations of affinity all organs change TOGETHER, in analogy certain parts perfect of typical
276c125 habits & structure similar would have blended TOGETHER Mem Mr Herberts law; habits determining ferti
277c127 as much as those (naming them) which are found TOGETHER.-- If two species come over to this country,
284c149 on non-variation, & not on extension ¿these go TOGETHER? Therefore value of organs vary in different
285c152 other & if a first & last individual were put TOGETHER, they would not according to all analogy bree
285c152 they would not according to all analogy breed TOGETHER.-- The bottom of the tree of life is utterly
305c211 , laught &c&c Man & Man may have some relation TOGETHER, as well as Man & child, polypus & polypus, b
337d021 fect of <sev> local variety many times changed TOGETHER with some training in the earlier branches «a
337d023 oportion to the time they have been separated; TOGETHER with physical differences of country: the tim
354d065 struck looking at the Indian cattle with Bump. TOGETHER with Bison, at some resemblance as if the "va
377d138 hynx makes is also made by the C. porcarious., TOGETHER with a grunting noise, the former signifies r
413e060 ephant, Rhinoceros, Hippot, Haena &c are found TOGETHER.-- Read this Work-- Decb. 4th.-- Why has the
431e121 onch. it is however, said they have been found TOGETHER on coast of France.-- L. doubts.-- Lonsdale t
433e127 The fact of tumbling pidgeons; flying high all TOGETHER & then tumbling, far more wonderful than here
439e144 d. that the presence of of the Leopard & Tiger TOGETHER depended on some nice qualifications each pos
466t091 /, observed 3 plants of Caltha Palustris alone TOGETHER. one had seed-pods turning brown, whilst both
469t151 hered same year came up true «in 1840»: All in TOGETHER blossomed together The seeds of these plants
469t151 e up true «in 1840»: All in together blossomed TOGETHER The seeds of these plants will be collected &
473s004 ion then all continents of cretaceous periods, TOGETHER with their littoral deposits are probably bur
485z017 of Mephites 4 distinct Camelidae. do not breed TOGETHER Mag: of Zoolog & B. Vol. II. p. 127. List of
492q002 oses impregnate each other. when close planted TOGETHER: <do> Can Holyoak be raised distinct by seed-
497q06v to cross.-- are there races-- if so plant them TOGETHER. & raise. seed.-- In letter Mr Herbert says d
507q15v 50) Any cases of wild varieties plants growing TOGETHER. under same conditions.-- like cowslip & prim
527m031 ynonymous.-- Shake ten thousand grains of sand TOGETHER & one will be uppermost:-- so in thoughts, on
544m102 lace & personal connections-- ideas are strung TOGETHER in manner <they> quite different from when aw
570n024 mined not to say anything. he presses his lips TOGETHER & shrugs his shoulders & walks off,-- I think
584n071 e cell; p. 22. instincts & structure always go TOGETHER: thus woodpecker: but this is not so,, the in
618o41v allel series: if blueness & weight always went TOGETHER. & as a thing grew blue it «uniquely» grew he
623o050 t of sociability & sociability, doubtless grow TOGETHER [RHC] This feeling seems to vary in races of
640j167 hat they are not species, until their breeding TOGETHER has been tried.-- With respect to the six pup
081rIBC Constant log always additive to convert French TOISE into English ft. 0.8058372 French metre into En
055r106 the richest Vol II 147 Shells at Concepecion 50 TOISES above the sea. = talks of them being packed cl
055r106 levation of sea's bottom. beds of shells. 2 - 3 TOISES thick.--Vol II. p. 252 Urge cliff form of land
081rIBC 372 French metre into English ft. 0.5159929 {T} TOISES Pieds Myriametre = 5130., 4. 5 inch Kilometre
130a133 f hot bottom.-- Study Bishoofs Paper.-- Weelsted TOLD me of some large fresh Water springs off coast
156g068 ts which had fallen before lake drained could be TOLD from «some of» those since fallen. «on the 3 sh
162g102 . enters by old tower called Glengarry (Nead Roy TOLD me) it is impossible to see my new shelf, from
212b165 latter not to former-- Mr Martens of Zoolog Soc TOLD me an Australian dog he had, used to burrow lik
217b184 ther, puppies differ, & like both parents.-- Fox TOLD me of case of mare covered by blood horse & Car
231b242 s case of Himalayan, plants ¿migratory birds, he TOLD me some story of crane from Holland!!! in stoma
289c162 terhouse says of affinity of many insects may be TOLD by their larvae) but the acts of condensing mus
302c202 fore more easily modified = = this is not easily TOLD, for any small family. having analogous charact
304c207 k &c without these trifles. it would not then be TOLD whether not descended from long way back.-- abe

```
312c232 ble time. Mr Willis Long eared little dogs, I am TOLD, go to heat, take dog. but do not become impreg
317c249 of Congo Expedition, NB. I met an old man--, who TOLD me that the mules between canary birds & goldfi
317c251 ed» of N. America, & of the shrews.-- Dr Bachman TOLD me. that near Charlestown ?three species, near
356d072 eory of generation (p. 175) if> 8th Sept Yarrell TOLD me he had just heard of Black game & Ptarmigan
363d102 very different.-- the two redpoles can hardly be TOLD apart, so that after differences were pointed o
363d102 ed out Selby confounded them, yet can readily be TOLD by incubation & other peculiarities.-- (Mem.--
363d103 Moschus &c & -- like young blackbirds Dr Bachman TOLD me that 1/2 Muscovy & common duck were often ca
364d104 wing to old <story> return.-- The Revd. R. Jones TOLD me precisely same story about some Southern «se
366d108 an a mere monstrosity propagated by art. Yarrell TOLD me of a cat & of a dog, born without front legs
376d136 onkeys». The Baboon of which anecdotes have been TOLD is Cyanocephalus Porcarius.-- this Monkey did n
389d173 held Hunters Eoeconomy So with inter-breeding as TOLD me by Willis Q Proved facts relating to Generation
410e053 numbers of species to genera &c &c can never be TOLD, without species being described.-- but then pe
521m007 o [not located] There is a case of Mr Anson. who TOLD a story of hunting «-- habitual fits.--» which
526m028 taste for music.-- Children like hearing a story TOLD though they remember it so well that they can c
532m054 t so much afraid though my reason was laughing & TOLD me there was nothing, & tried to seize hold of
546m110 as the ordinary state is concerned.?-- Mr. Mayo TOLD me the case of a lady, (whose name was told me,
546m110 Mayo told me the case of a lady, (whose name was TOLD me, who told the fact to Mr Mayo himself. she w
546m110 the case of a lady, (whose name was told me, who TOLD the fact to Mr Mayo himself. she was one day ne
554m140 Jenny will often do a thing, which she had been TOLD not to do.-- when she thinks keeper will not se
564n004 hich he wills to give his child.-- Octob 3d. Was TOLD by W of Downing. Coll. that he had seen chicken
573n038 ith mice.-- A person with St Vitus' dance badly, TOLD should have shilling to walk to door without to
573n038 involuntary movements.-- Person with sore-throat TOLD not swallow spittle. will have involuntary flow
573n039 wishing not to flow-- flow it will.-- My father TOLD Miss. C. of the bad conduct of Mrs C. (her brot
580n061 «Weak people say I know it because I was always TOLD so in childhood.-- hence the belief in the many
591n097 - July 20th Intelligent Keeper... Zoolog. Garden TOLD me. he has often watche tame young wolf & it ne
613o036 ple is the same in all animals. [LHC] «3]» Eyton TOLD me that his retriever Sailor he has seen push a
614o037 oy takes delight in mammae before any reason had TOLD him this distinctive mark, it is downright inst
232b246 anges (we assume like weather on long average TOLERABLY> uniform).-- Comparing fossils with whole wor
425e103 a.)-- «:CD(Does the Power of, «easily» making TOLERABLY fertile hybrids, bears relation to capability
401e015 up from it. lately saw a nonpareil sowed by Mr TOLLET so produce.-- thinks it probable that great pa
489qIPC Gray ---- 17 Yarrell---- 18 Blyth---- 19-- Mr. TOLLETT {T] Zoolog Soc «Gardens» ---- . 20 & Breeder
512q019 vidual case: subject to disease in youth.-- Mr TOLLETT-- about selection for milking-- loss of early
451e176 .--» A Malayan albino described "To this day the TOMB of his grandfather, who was also an albino is h
108a081 n page excised, not located] Bull:. Soc: Geolog. TOME IX 1837-8. p. 24. rocks of Chimborazo., & Pichi
247c028 e Pelew Islds.-- killed a woman. Chamisso p. 189 TOME III: Kotzebue.-- p 22. a Gecko on St Helena.--
376d138 s,-- «by taking up tail» Mem: Ourang Jenny with TOMMY.-- Good evidence of knowledge of Woman-- ʌ The
554m138 ook up womens petticoats-- just like Jenny with TOMMY ourang.-- Very curious.-- Mr Yarrell has seen J
554m139 -- will do anything-- will take & give food to TOMMY, or anything of any sort.-- I saw Tommy picking
554m139 food to Tommy, or anything of any sort.-- I saw TOMMY picking his nose with «a» straw.-- Jenny will o
567n013 & opened my hand, & put them in-- like child. TOMMY'S face, now ill, has expression of languor & suf
195b098 apagos it so actcd that bi[r]ds with plumage «&» TONE of voice partly American North & South.-- (& ge
288c159 alf way in appearance.-- bark about half way «in TONE»-- the native dogs howl most dismally, very rar
305c209 nalogous to mocking thrushes of Galapagos having TONE of voice like S. American.-- Have not Ruffs & R
333d005 fierce to dogs «otherwise habits not different; TONE of voice. perhaps rather different».-- crossed
241c015 ?). p. 159. & 160 «162» list of some birds of TONGATABOU. & New Ireland.-- Gould will hereafter know
148g029 n may often be due to rivers-- By Roy Bridge, a TONGUE of flat land, with terraces of each side of th
556m145 made by <ret> inspiration & quickly retracting TONGUE from behind upper & little between incisors.--
581n065 explanation it offers of radical diversity of TONGUES.-- [Emotions are the heredetary effects on the
090a024 0 x <50 = 250 0000» x 100 = 50, 0,0,000 = 2500 = TONS in fifty thousand years]CD If world increased a
603o013 etely men must have personified the deity.-- H. TOOKE has shown one chief object of language is promp
581n065 essively formed. (--names like sounds)--. Horne TOOKES tenses, &c &c -- <also g> if so & seeing how s
194b094 pear to represent the semnopitheque of India.-- TOOTH of <Spi> of Sapajou-- NB Sapajou is S. American
307c216 ra & classes. p. 479. fragment of tusk «& Molar TOOTH» of Hippotamus from Madagascar!!!!!! Proceeding
403e023 other Macleay says it is nonsense to say take a TOOTH of any animal (as Toxodon) & say its relations.
463t057 alking fish? difficult-- yet suggested. (vipers TOOTH also a difficult, the whole mind is constitute
507q016 eculiarity in the males of a family-- Where one TOOTH aborts, do you know whether any trace in germ.
577n052 ans most curiously shown in the sudden cures of TOOTH ache before being drawn,-- My father «even» bel
590n093 & throws bile in circulation p. 75. Haller says TOOTH ache, even from carious tooth cured by sight of
590n093 . 75. Haller says tooth ache, even from carious TOOTH cured by sight of instrument.-- Bennett's Wande
086a003 being found in America, when Mammoth & narrow TOOTHED Mastodon.-- argue against the prejudice of not
585n077 chmaker has faculty by his instruments to make TOOTHED wheel. he might by instinct make watch, but he
493q003 of intestine in Persian Cat-- , in Brazilian «TOOTHLESS» dog-- I. St. Hilaire says length differs in
032r040 & parts of coast of Chile.-- Must first explain «TOP of» tidal band of action. This case differs. I t
108a081 ll said probably much Carbonic Acid gaz here.-- [TOP portion page excised, not located] Bull:. Soc: G
108a082 und Vol XIV. (My Edition) p 500. Well described [TOP portion page excised, not located] -- do-- [...]
115a098 st regular slope-- {P} if not course enough flat TOP. ended by abrupt slope {P} each stratum would th
138a155 e heated sand red hot & brine was boiling on the TOP-- [blank] Would rotting wood by yielding Carboni
146g010 ath East End near Holyrood Palace In same way at TOP the trap could be traced Grey in front on wall p
159g084 ome way; several large ones (one 6 ft across) on TOP of spit between river & dry Corry Scarcely conce
207b151 . 325. Vol. IV. Ducks on rivers in Guiana. build TOP of trees carry duckling to the water in their be
443e154 quotes from Pliny, that is bad to graft from TOP shoots-- If prolongation of life by gemmation <
505q014 Flowers not seeding= Put pot of boiled earth on TOP of House =Aristolochia, plant wh require insects
147g015 in> space is thickly studded with ridges & flat TOPPED hill/ do alluvium. NB In one part pure sand in
150g038 e Hills in this neighbourhood appear very round-TOPPED with much drainage & far more earthy than what
086a004 ot believing recent elevation, yet sea shells at TOPS of mountains we ought to sympathize with. old d
086a007 streams of warm pole; in name of Heaven why are TOPS of Equatorial mountains so cold.-- Siberia no p
078r175 the most elevated summits, where mountains most TORN.--(¿anticlinal line?). -- Mines of Catorce «(Pr
352d060 f geology, of each formation being merely a page TORN out of a history, & the geologist being obliged
553m132 angaroos than emu, although young dogs get sadly TORN in conflicts with the former. But it is one thi
485z018 cause during winter they can bear the cold when TORPID.-- On this principle tropical forms in N. Amer
055r108 sediment «daily» yearly brought down by every TORRENT proves the decay atmospheric of the most solid
105a071 subsidence.-- removal downwards by successive TORRENT spread out. by sea-- beach action-- no one wi
030r035 2 & 3 miles from shore. V. Chart) Every winter TORRENTS must bring much vegetable matter from thickly
037r055 us> a herbivorous lizard.-- from the action of TORRENTS. «marine» Tortoise & other species of large l
156g070 ge to those now formed in same spot by present TORRENTS Maculloch wrong in saying no transported mate
284c148 velled by <dead> «each» trees dying & mountain TORRENTS.-- but to crawl up an hill, then by deaths?!)
229b234 in India.-- Range of Zebra?-- The Crocodile & TORTISE former inhabitants of Mauritius Freycinet Voya
037r055 arine remains» may contains <shells few corals TORTOISE» «remains of Amphibia, exclusively.» & Turtle
037r055 izard.-- from the action of torrents. «marine» TORTOISE & other species of large lizard.-- There woul
233b250 Waterhouse thinks two main divisions of cats. TORTOISE shell--& grey-banded. ¿species?-- thinks offs
236b279 malayan land emys, decidedly an Indian form of TORTOISE.-- On other hand. fresh water tortoise from G
236b279 form of Tortoise.-- On other hand. fresh water TORTOISE from Germany. (where Mr Murchison fox was fou
242c018 .-- <Says> D'Orbigny is said to have brought a TORTOISE & toad from S. America & identical with those
287c157 +. between Mammalia & fishes, one penguin, one TORTOISE shows hiatus-- but not saltus-- when Linnaeus
493q003 ke after male parent & vice versâ = History of TORTOISE-shell Cats. as only one sex so coloured = I h
493q003 I have grey-cat «wh was female» with tinge of TORTOISE-shell «on back.--» = Length of intestine in P
172b007 htly differen circumstances.--» Now Galapagos TORTOISES, Mocking birds; Falkland Fox-- Chiloe, fox,--
400e012 do Vol 10. p. 373, «& 374» Spaniards says no TORTOISES in other «places» besides Galapagos do. p. 37
458t001 Falconer on Sub. Him. fossils-- Ruminants. & TORTOISES gigantic-- hyaena-- bear & ruminants all of 1
480z006 bularia. p 52. May 1836 dimensions of immense TORTOISES. p. 81 & p 113 of 1834 On the passeres of S. A
482z012 writing with Dumeril says that two species of TORTOISES come from Galapagos!!! Azara. Voyage dans l'A
048r084 tate at St Helena. pebbles entirely coated with TOSCA. which implies motion in the «loose» bed of peb
123a117 0½ quadras from river; 20 varas from surface in TOSCA.-- remnant of Megetherium in interior.-- <The
134a142 p. 277. on whale bones in Falklands Some of the TOSCA nodules at Bahia Blanca Mr. Malcolmson says are
090a022 e weight «or size» is given of 25 stones.-- The TOTAL weight recorded is 473. pounds (taking about av
523m015 .-- his manner of curing it. by keeping the sum TOTAL of his accounts in his pocket, & studying mathe
549m120 ery one who has tasted them, will think the sum TOTAL of happiness greater. even if mixed with some p
640j167 med eighth part.-- or if other prey diminished, TOTAL number of dogs. would diminish, whilst the long
```

Page **(Key Word)**
075r166 stone" = "The wealth of the veins in most part TOTALLY independent of the nature of the beds they int
255c051 ny times, but all have had feathers.-- if wing TOTALLY obliterated. This may account for permanence i
359d087 one.-- It is singular pheasant & fowl being so TOTALLY infertile whereas animals further apart have b
567n016 y from all ordinary lines of association.-- is TOTALLY distinct from learning it by heart. Do not our
615o36v in any individual.-- His Malay breed «of fowl» TOTALLY different habits from Europaean. begin to prow
263c076 n other-- if this be granted!!) & whole fabric TOTTERS & falls.-- look abroad, study gradation. study
289c160 390,. young <grebes> «ring ouzels» dive instant TOUCH the water.-- capital instances of typical land
570n023 ed behind, without ones chest, being distended. TOUCH a person on the ribs & how he gulps in air.-- A
581n064 ens» admire peacock's tail, as much as we do.-- TOUCH apparently. ourang outang very fond of soft, si
585n074 rsley's story of chicken) to know that which we TOUCH & what [...] the same.--(this Hensleigh therefo
611o034 nt can be shewn to be direct physical effect of TOUCH & not irritability, which at least shows a loca
615o037 tive mark, it is downright instinct, leading to TOUCH a particular organ.-- I think Pincher shows sur
314c237 stamens united into long organ, moved on being TOUCHED, so as to protect itself, one segment of the c
366d111 servant cleaning door outside, as often as she TOUCHED handle, though really fully aware she was not
535m064 mies-- would not fly away, but bit pencil when TOUCHED with it-- do not know their own larvae, but on
540m085 ot disagree.-- Ourang outang at Zoolog Gardens TOUCHED pud. of young male & smelt its fingers. Seeing
545m107 ulatory angry look of black spider monkey when TOUCHED, also another monkey to dog. I showed nut & th
444e158 ence teach distances in air, in which it never TOUCHES objects.-- far better case than chicken peckin
585n075 oblem is how we know that thing is same, which TOUCHES two parts of our bodies, «or touches one part.
585n075 me, which touches two parts of our bodies, «or TOUCHES one part. very quickly successively.--» [& we
496q05a .-- Are Bees guided by smell-- or sight.-- --. TOUCHING Mr Brown theory of insect-like Orchis-- & fin
573n038 d should have shilling to walk to door without TOUCHING table.-- cannot avoid it.-- curious mixture o
022r008 s maw was like a leathern sack, very thick & so TOUGH that a sharp knife could not cut it: in which w
089a018 - Les grands tremblemens de terre sont presque TOUJOURS precedes et suivis, queque temps avant et apr
448e166 h in species and subgenera) between the Crag & TOURAINE beds, the one with neighbouring & Arctic sea,
448e167 shells in common «& are not far distant» with TOURAINE «which as L. says is strong argument for thei
452e182 - Vol do. p. 634, alludes to fact stated by M. TOURNAL that skulls found near Vienna appoximat to Neg
135a145 abundance Bengal Journal. Vol 4. 1835. p. 437. TOURS by Benza Neilgherries-- Much inform. on. decomp
257c057 espèces tres-analogues,-- quand ce nétaient pas TOUT à fait les mêmes." This good case. of replaceme
207b152 sente elle seule plus d'especes polymorphes que TOUTE la terre ferme de lancien monde".-- Considers f
058r118 ustrals Vol. I. p. 54. M. Bailly says."en effet TOUTES les montagnes de cette île se developpent auto
058r118 d'elle comme une ceinture d'immenses remparts; TOUTES affectent une pente plus ou moins inclinée ver
058r118 tent une coupe abrupte et souvent tailée a pic. TOUTES ces montagnes sont formées de couches parallel
603o012 rds <man> «a king» <changed into> is carried on TOWARD deity.-- & as king might like cruel pleasure,
021rIFC have undergone the greatest number of changes TOWARDS perfection (namely mammalia) must have a short
064r136 matter from CREVICE of Andes--therefore flowed TOWARDS it. a mass on each side 3000 ft thick & 150 br
071r157 h informs me that town near Tucuman and Salta. TOWARDS the Vermejo was utterly overthrown by earthqua
072r160 analogous to septa.-- would particle attracted TOWARDS space tend to form ring. [Fig. 10] {P} motion
125a121 uid, then cooling process must go from surface TOWARDS the interior -- who knows how far that may hav
148g023 e 40-- ‖ traces of them all along <Glencoe>.-- TOWARDS Fort William yet in Glencoe in parts no trace
148g027 dge many flat terraces one above much inclined TOWARDS river all these composed-- where side ravine e
149g031 culiarities? The slope of A & B regular & even TOWARDS {P} Spean Roy double terrace river «& to West
157g078 e 75 degree? A little below Divortium on slope TOWARDS Loch Spey 29.297 A 79.½ 29.316 divortium aquar
189b072 reat groups.-- Speculate on land being grouped TOWARDS centres near Equator in former periods & then
190b076 East & West. ends of New Holland. diminishing TOWARDS centre (p. 586)-- Parallel 33 degree-35 degree
190b076 l 33 degree-35 degree, source of forms. reduce TOWARDS Northern Eastern end & die away, & partake of
206b148 consequent on climate &c-- the whole races act TOWARDS each other, and are acted on, just like the tw
232b245 older climate whilst others died out, or moved TOWARDS equator.-- «or some species might then have be
232b247 of Europe»-- As Europaean forms have travelled TOWARDS Equator, <th> so would the plants from extreme
234b255 lyle, saw with his own eyes. new gate. opening TOWARDS pig.-- latch on other side.-- Pigs put legs ov
243c020 rk with regard to shells.-- But he says shells TOWARDS extremities of the continents peculiar to the
331dIFC of animals:]CD [All Selected Dec. 14-- 1856]CD TOWARDS close I first thought of selection owing to st
355d067 partly true they are of the greatest service, TOWARDS the end of science. namely prediction.-- till
408e043 When a species becomes rarer, as it progresses TOWARDS extermination. some other species must increas
410e051 er.-- (I from looking at all facts as inducing TOWARDS law of transmutation, cannot see the deduction
415e067 to the rearing up of every heredetary tendency TOWARDS fatal diseases, & such constitutions only bein
420e088 » account of Hyaenodon, a fossil dog-- leading TOWARDS Hyaena.-- see Comte Rendu.-- I suspect good ca
437e138 id to have been seen carrying a lamb two miles TOWARDS the Morne Mountains, it then dropped it & was
539m084 He who understands baboon <will> would do more TOWARDS metaphysics than Locke A dog whines, & so does
567n013 the willing of the simplelst animals, as hydra TOWARDS light. being direct effect of some law.-- have
582n007 legs so, as much as in the young salmon to go TOWARDS the sea. or down the stream; which it does unc
596n184 e» ears but <not> Chimpaze. does not gradation TOWARDS man.» Macacus especially pulls back skin of wh
603o012 ifices. «common to many races»-- thinks action TOWARDS <man> «a king» <changed into> is carried on to
614o037 just same argument. without indeed we are step TOWARDS some final end.-- production of higher animals
617o040 anifested in every particle of matter directed TOWARDS every other particle; but FORCE, <objectively>
619o042 man we should expect that acts of benevolence TOWARDS fellow <living> creatures, or of kindness to w
622o049 e social instinct may be combined with feeling TOWARDS one as a leader,-- the conjugal feeling may be
622o049 leader,-- the conjugal feeling may be directed TOWARDS one or more.-- It will be hard to discover thi
162g101 Head of which is so interesting. enters by old TOWER called Glengarry (Nead Roy told me) it is impos
165g124 ng-- Inverness & waters of the Tarf-- Kilfinnan TOWER where stream enters at head of which hill is ro
026r021 the Cleavage be There is a resemblance at Hobart TOWN between the older strata & the bottom of sea ne
066r142 f which dried up <all> a lake in neigbourhood of TOWN Mr Murchison insisted strongly. that taking up
071r157 log. Society Sir Woodbine Parish informs me that TOWN near Tucuman and Salta. towards the Vermejo was
152g047 tes new terrace Ballivard 2 miles North of Grant TOWN to Forrest road comminuted shells Important con
207b152 nks Van Diemen's land long separated from Hobart TOWN-- {from difference of races of men and animals}
252c045 en of arrangement, 3 <5> Species Rhinoceros Cape TOWN good species Indian species so distinct that al
341d030 loped to aid in breathing.-- Animals from Hobart TOWN mentioned, it seems most of species from there
041r065 s Africa it is mentioned that the Elephant came TOWNS driving by the want of water.--I believe in all
599o007 se, of county, netted with edges & crowded with TOWNS & thoroughfares, I grant that man, from the eff
086a005 camels in deserts & rodentia In Plata Mastodon TOXODON Is the general saline tendency of America conn
278c132 ared to Faunas of these countries, greter than TOXODON, Macrauchenia, &c compared to America.-- the w
278c132 e to escape to some more fitting country,-- if TOXODON had been found in Africa, the wonder have been
291c167 . Elephant has left no descendant in Europe<)> TOXODON in S. America) is absolutely necessary to expl
403e023 nonsense to say take a tooth of any animal (as TOXODON) & say its relations.-- if we know its congene
448e168 pril 2d. 1839 The Lepidosiren-- Amblyrhyncus & TOXODON, <all> equally aberrant-- the two former conne
448e168 rrant-- the two former connecting classes like TOXODON «In orders»-- Fish & reptiles in former case--
229b234 bility of seeds of Keeling standing transport.-- <TR but> Get him to discuss those mention[ed] by Les
360d092 & other very bright blue, it might be harder <to TR> for both parents to transmit there peculiaritie
595n125 g, «smiling», just as much instinctive as a bull <TR> calf, just born butting, or young crocodile sna
069r152 erious!-- Mem. SUBSIDENCE Uspallata of which no TRACE except by trees The structure of ice in columns
070r154 verlooked that except by trees, I could not see TRACE of Subsidence at Uspallata.-- ¿If crust very th
104a067 cumbent mass (varying hardness,-- takes time to TRACE) from few dikes which have given rise to erupti
120a107 the structure of Andes. where we believe we can TRACE the outlines of what were fluid undulations-- t
147g016 deposited it. Barrier of lake very lofty, & no TRACE of it; to the Sea more probable I did not look
148g023 towards Fort William yet in Glencoe in parts no TRACE of them-- Mem Coast of Chile--¿ is not Mica Sla
158g080 och Spey Forms terrace about 60 feet above Loch TRACE of this terrace «on <will> Granite ridge or a m
162g103 m with, left of Bright.-- like bed of lake with TRACE of terraces on each side High up the Tarf (a Gr
189b073 sooner or later.-- Prove animal like plants:-- TRACE gradation between associated & non associated a
191b081 t Indian Archipelago it would be interesting to TRACE limits of large animals-- Owls. transport mice
208b154 ermediate between every species.-- Who can find TRACE or history of species between Indian cow with h
274c114 peckers?-- In each division Gould thinks he can TRACE structure for insects & structure for vegetatio
293c175 agreeable to Musquitoes We never may be able to TRACE the steps by which the organization of the eye,
298c190 - [+] This might be mentioned in note.-- try to TRACE from simplest reasoning in lower animals many t
325c267 Cavernes d'Ossements 3d. Edit. Octav. (good to TRACE Europèan forms compared with African <A> Annals
353d060 ame with the <Zoologist> «philosopher», who has TRACE the structure of animals & plants.-- he get mer
362d100 peculiarities have been kept perfect-- also to TRACE the laws of change in this time.-- the impossib
363d101 tarmigan & in Bantam.--]CD CD[In the Pidgeons, TRACE the washing out of the forked band, like in plu
382d158 though» abortive, are so plain in Man, & yet no TRACE of abortive womb, or ovarium.-- or testicles in

Page **(Key Word)**
405e031 niform <dusky> blackish brown.-- yet retained a TRACE of horizontal mark on flank.; & tail. & kind of
409e048 ny species, as individuals, & though we may not TRACE out all the ill effects. -- we see it is not th
429e113 eve this Nature's plan.-- Whether we can or not TRACE history of first appearance of varieties of dom
470t176 nglish> But these mules <in our garden> show no TRACE of palmation!!? Bees at Wild St Johns Wort--Sca
507q016 Where one tooth aborts, do you know whether any TRACE in germ. (2) Any more cases of diseases, genera
542m094 nvoluntary, are common to animals.-- Curious to TRACE, which of these actions are habitual, & which n
543m101 nct, a separate thing superadded.-- we can thus TRACE causation of thought.-- it is brought within <o
547m111 -- was greatly astonished, at the time. & could TRACE no chain of association» Mayo Philos. seems cer
605o19v sublime.-- It appears to me, that we may often TRACE the source of this "inward glorying" to the gre
617o40v rections. We are satisfied therefore, if we can TRACE any force in inanimate matter up to the action
636j56v y shows life equally simple series, & therefore TRACE of beginning in organic world.-- Macculloch. At
146g010 ood Palace In same way at top the trap could be TRACED Grey in front on wall perhaps wall oblique The
155g062 plain <now> formed on shelf 4th 2d & 3d can be TRACED some way up, but most faintly on East side of
177b026 sively complicated.» {P} Is it thus fish can be TRACED right down to simple organization.-- birds-- n
186b060 dermata, less so in Miocen & so on.-- As I have TRACED the <type> great quadrupeds to Siberia; we mus
197b111 hest» as typical of <sa>. changes, which can be TRACED in same organ in different animals in scale.--
230b239 ut crossing.-- Now a gradual change can only be TRACED geologically (& then monuments imperfect) or h
281c138 er I cannot help thinking good analogy might be TRACED between relationship of all men now living & t
304c208 cumstances. Spines on Echidna.-- when it can be TRACED through series then probably heredetary & not
418e083 tebrates. [Müller's Physiolog. p.24.]CD» can be TRACED to a germ, endowed with the vital principle, w
535m069 uperior being; whose natures can only be rudely TRACED out. When one sees this, one suspects that our
555m142 inct, when origin of heredetary habit cannot be TRACED) V. D. p. 111, case of Association. Sept. 16th
617o040 tation of force i.e. movement? capable of being TRACED to the body of the individual to whom the acti
148g023 ces perhaps upper one 100 ft & other one 40-- TRACES of them all along <Glencoe>.-- towards Fort Wi
286c156 irds having web-feet, where we see scarcely any TRACES of passage a difficulty, but after all a sligh
381d157 Mollusca.-- «or Cephalopoda» are there abortive TRACES of other «sexual» organs; for if so, separtion
410e053 permanent varieties not in same country.-- The TRACES of changes in forms of organs, will care littl
462tf05 cidentally said that a mongrel man may lose all TRACES of his parentage in «about> seven «7» generati
472s01v ttle variation in the 60 one brighter with mere TRACES of black spot at base, one paler with less rig
484z015 ia, the Awks.-- What structure do the auks bear TRACES of.-- like Puffinuria does of Petrel?-- Study
599o05v original formation.-- declension &c often show TRACES of origin.-- Mayo Philosophy of Living. p. 264
603o013 & verbs-- so that much of EVERY language shows TRACES of anterior state?? Edinburgh Review Vol 18. (
633j54v ws.-- 4 «Study Cuviers Anatomie Comparé» p 308. TRACES the gradation of skeleton in Vertebrates & con
634j55r lly terrestrial.-- for we find even in Cetaceae TRACES of hind extremities.-- How are we to explain t
639j28v t once, but by steps. of which we have manifold TRACES in the several genera of Grallae Suppose six p
250c040 by Dillwyn, on replacement of Cephalopods & TRACHILIDOUS Molluscs. by each other in secondary & Tert
031r038 Europe.-- Auvergne. very little Pumice, though TRACHYTE. same fact in Galapagos. Daubeny P 24 V. back
034r044 n P 361. Daubeny Von Buch is very strong about TRACHYTE being the most inferior rocks--The stream at
046r078 l retorts.--neglecting the first production of TRACHYTE. look at Sulphur. salt. lime, are spread over
069r150 !! Analogous to Von Buch. Basalt where Basalt. TRACHYTE where trachyte. There must have been as much
069r150 Von Buch. Basalt where Basalt. trachyte where TRACHYTE. There must have been as much conglomerate on
090a021 his bear upon the sorting of matter. in making TRACHYTE come out before.-- What must be the effect of
091a026 mboldt on Quito Volcanoes & another on Mexican TRACHYTE <roc> lava called Andesite. Red Coral in the
097a043 elspathic veins?-- Mr Poulett Scrope. talks of TRACHYTE, "superficially coated by a thin pellicle of
074r165 es «Study Hoffmans account of steam acting on TRACHYTES. also Azores. We here have case of such vapou
094a035 ect to my theory of changes. of granites into TRACHYTES.-- Mention Osorno in lake. few Volcanos now i
033r043 any parts, in James Isd.--Mem St Helena-- All TRACHYTIC.--Daubeny P. 171. Vol I. Humboldt There is lo
181b041 at present day many are relatives, so that by TRACING back. the <descen> fathers would be reduced to
256c055 gradation between Man & animal, small point in TRACING history of Man.-- granted.--but if all other a
259c064 s all died certain year. Extreme difficulty of TRACING change of species to species «although we see
535m069 natures. accounting for fossils). & lastly the TRACING facts to laws. without any attempt to know the
285c150 e not born with any capacity for observation of TRACKS &c &c Dr. S. has some remarkable crochets abou
563n001 ibes. pheasant springing from nest & leaving no TRACKS.-- My Father says pea-hens do Wood pidgeons bu
187b063 ell say that some circumstance killed it over a TRACT from Spain to S. America.-- Never They die; wit
226b222 longing to its own genera Therefore if in small TRACT we have many species, we may insure mass contin
266c091 much fairer race than the Hindu's, in the same TRACTS;. & that in their appearance & manners they ar
343d038 interior are pushing into each other from slave TRADE, & colonization of S. Africa, so must the tribe
239cIFC d have any close species ever yet failed. About TRADES affecting form of man. Could you get racehorse
282c142 e relation to elements & not minding particular TRADES.--» then the second race would not obtain a ca
473s07r mals can adapt itself to course of life, «as in TRADES» there is no reason, why the peculiarities shd
530m045 of gentlefolks. & that peculiarities of form in TRADES (,as sailor tailor blacksmiths?) are likewise
530m045 ir children have some little advantage in these TRADES.-- Delirium seems to rest the sensorium.-- --
539m082 eness or power of intellect.-- Some complicated TRADES can hardly be considered as actions otherwise
558m151 nnexion.-- look at faces of people in different TRADES &c &c &c I observed the Asiatic Leopard. quarr
602o11v ything applicable to the high idea «p. 131.» in TRAGIC acting-- CD [My idea. would make the mind have
263c078 to have made man! Seclusion want &c & perhaps a TRAIN of animals of hundred generations of species to
398e005 rapid, I should say there was some link in our TRAIN of geological reasoning, extremely faulty» The
521m009 ubject when once started,-- could receive a new TRAIN through eyesight, though, not through hearing,-
522m010 , however, in conversation could catch up a new TRAIN if early association were called up.-- My F. as
527m034 organs of sense being required) as the closest TRAIN of geological thought.-- the capability of such
527m034 in the air are highly advantageous, before real TRAIN of inventive thoughts are brought into play & t
527m035 so as to feel a severe disappointment «in real TRAIN of thought this does not happen. because papers
528m035 castle by going to beginning of castle» because TRAIN cannot be discovered-- is closely analogous to
529m040 ing pleasure of imitation come into play.-- the TRAIN of thoughts vary no doubt in different people.,
530m046 brain, or the tendency to habit of producing a TRAIN of thought.-- [not located] Fox believe cats di
538m080 oking The possibility of the brain having whole TRAIN of thoughts, feeling & perception separate, fro
538m081 unishment explained. These facts showing what a TRAIN of though[t] action &c will arise from physical
539m083 really explain what habit is-- In the habitual TRAIN of thought one idea. calls up other, & the cons
555m144 e feeling of banter & joking» because the whole TRAIN of Dr Monro experiment about hanging came befor
371d127 o altered in disposition, as to be more easily TRAINED up to the (required) offices" &c &c Owen illus
466t095 horses having colts which can canter-- & DOGS TRAINED to pursuit having PUPPIES with the same powers
624o051 y are in accordance with it. thus a dog may be TRAINED to hunt one pig sooner than other, rather than
337d021 variety many times changed together with some TRAINING in the earlier branches «as in common greyhou
430e118 ce-horse». have not been thus produced, but by TRAINING, & crossing & keeping breed pure.-- «& so in
525m026 in families.-- My fathers does not know whether TRAINS of insanity are heredetary in any one family.-
527m034 of geological thought.-- the capability of such TRAINS of thought makes a discoverer, & therefore (in
538m078 enses.-- These things, & drunkedness, show what TRAINS of thought depend on state of turn In drunkedn
539m083 stances?? The possibility of two quite separate TRAINS going on in the mind as in double consciousnes
546m110 does this throw light on instinct, showing what TRAINS of action may be done unconsciously as far as
547m112 rent vividness) & there being no other parallel TRAINS of ideas connected with past circumstances.--
547m113 naturally would so so!) Now all these parallel TRAINS of thought necessary heirs of every action, &
547m113 refore could not doubt or believe.-- When I say TRAINS, it may be instantaneous changes in order <to
547m114 not reasoning. effect of not having <all> other TRAINS of thought, or memory from innumerable late ev
547m114 .-- the fatigue of thinking is keeping up these TRAINS,-- especially if they are invented as in imagi
548m114 pid thought.-- There may be some «two or three» TRAINS of thought, therefore one may be imperfectly r
548m115 hem, else one would not stand) a crowd of other TRAINS of thought are in progress-- In castle of air
613o036 ls no prejudices about souls, we see particular TRAINS of thoughts as fear of man,-- crows fear gun,-
079r177 s d'oeuvre de l[']industrie humaine, comme ils TRAITENT les loix de la nature & ses phenomenes."-- Ul
183b053 eory of the Earth attentively Cuvier objects to <TRAN> propagation of species, by saying, why not hav
477z002 uced in Paraguay in 1769 introduce in Governor's TRAN?? Azara Las Vinchuca or Benchuca. "Les individu
541m093 he will find it far more difficult to to look TRANQUIL.-- He may despise a man & say nothing, but wi
542m094 sigh of discomfort & weariness. & meditative TRANQUILITY. «whine of children. puppies do so dogs near
071r158 ountains falling in.-- Earthquakes at Quito. TRANQUILLITY «at Mendoza» exception.--«formerly perhaps
026r020 n metallic veins. 1830 P. 399.-- Carne. Geolog. TRANS: Cornwall «Vol II» It is a fact worth noticing
094a036 ie Transact Wern. Soc. Vol. 2. p. 35 Sir J Hall TRANS. Phils Royal Ed. Vol 7 Dr Buckland Reliquiae Di
094a036 uckland Reliquiae Diluvianae p. 201. & seq Murc TRANS Geolog Soc Vol 2. p 257 {T} The Pota: labiata c
471tf07 chis & Listera? Bryony saw common Bee on: Linn. TRANS 18. p. 133 Westwood on the Fulgoridae enumerate
471tf08 their use difficult to conceive any idea" Linn. TRANS. 18. p. 163. "D. Dod on two new genera of conif

441e148 wer of the lately acquired peculiarities are TRANSMITTED it is doubtful whether any are transmitted,
441e148 e transmitted it is doubtful whether any are TRANSMITTED, for the changes in fruit trees. mentioned b
508q016 f diseases, generally occurring in man being TRANSMITTED through females, like Hydrocele Dr. H. think
564n005 s supposed in man to guide to knowledge, was TRANSMITTED perfectly to chicken so as to seize small mo
574n042 ay affect the brain in a manner which can be TRANSMITTED.-- this is analogous to a blacksmith having
576n048 alizing of the means by which an instinct is TRANSMITTED.-- Arguing from man to animals is philosophi
614o037 d is it more wonderful that memory should be TRANSMITTED from generation.; than from hour to hour in
371d128 ansmit through seeds these characters though TRANSMITTING them with such facility to bud.-- this must
378d148 m quite remarkable". instance of old Species TRANSMITTING so much longer its Mental peculiarities. Wi
433e127 be thus put, shall we give up whole system, of TRANSMUT., or believe that time has been much greater,
462t051 generations.-- so many!! Hensleigh objects to TRANSMUT, theory, on the grounds of similarity in cond
463t057 ily been formed.-- With belief of <change.> TRANSMUTATION & geographical grouping we are led to endea
227b227 se) Waterhouse remarked, that any argument for TRANSMUT, from one organ graduating into other is lost
366d111 & aunts) & therefore does not tell against TRANSMUTATION of species-- Will it against genera.-- How
410e051 ing at all facts as inducing towards law of TRANSMUTATION, cannot see the deductions which are possib
419e087 hronology» «CONSIDER ALL THIS» Extinction & TRANSMUTATION, two foundations, hitherto confounded,. of
431e122 am, that extinction plays greater part then TRANSMUTATION.-- Do species migrate & die out.?-- In the
427e108 unds surprisingly well-- In early stages of TRANSMUTATIONS, the relations of animals & plants to each
407e039 een able to have survived a change, (& become TRANSMUTED), although other parallel species in other c
461t039 complex the series.-- Ch 6 Upland geese would TRANSPLANT seeds very far.-- Sept 31. The identity <of>
371d128 same atmosphere?-- may they produced not be TRANSPLANTED?, & yet year after year, successive roses &
372d129 g to their unity in one stem.-- a bud may be TRANSPLANTED & carry all these peculiarities not so a se
372d129 by being buds-- , more than if whole branch TRANSPLANTED? +.simplest forms of budding. Why does Geck
088a014 - Are any Fish seed-eaters. This important in TRANSPORT of Fish Let a Hawk fly at Heron.-- Ceratophyt
191b082 ting to trace limits of large animals-- Owls. TRANSPORT mice alive? Species formed by subsidence. Jav
196b105 t are there any genera, mundine, which cannot TRANSPORT easily.-- it would have been wonderful if the
210b158 nd species few in proportion to difficulty of TRANSPORT. For instance the temperate parts of Teneriff
219b194 l Fuego & C. of Good Hope show possibility of TRANSPORT. If some cannot be explained more philosophic
227b224 speculation; for having ascertained means of TRANSPORT, we should then know whether former lands int
229b234 sses possibility of seeds of Keeling standing TRANSPORT.-- <tr but> Get him to discuss those mention[
268c100 der it owing to one of each, being fitted for TRANSPORT ¿may it not be explained by mere chance?-- or
315c240 place on[ly] when creator sees. the means of TRANSPORT fail.-- otherwise no relation between means o
315c240 ail.-- otherwise no relation between means of TRANSPORT & creation exists..-- pooh. May have been Cre
316c245 is remarkable.-- Fish & drift sea weed-- may TRANSPORT ova of shells.-- Conchifera. hermaphrodites--
338d023 the time of separation depends on facility of TRANSPORT in the species itself, & in the local circums
041r067 nce of travelling may be owing to successive TRANSPORTAL from prevailing swell, (as Shingle travels o
052r097 --[Fig. 5] {P} * Slope necessary for seaward TRANSPORTAL of drift matter.-- Give various cases. [Fig.
09r152 ke a grand analogy between Wealden & Bolivia TRANSPORTAL of conglomerate between two ranges mysteriou
091a025 dust blown far out to sea valuable; because TRANSPORTAL of Minute seeds-- L. Institut. p. 209. May.
173b010 --<(> Then (remembering Lyells arguments of TRANSPORTAL) <continents> island near continents might h
195b102 ill be their successors.-- let the powers of TRANSPORTAL be such & so will be the form of one country
196b104 s are birds that are apt to wander & of easy TRANSPORTAL.-- Waders & Waterfowl.-- scrutinize genera,
198b115 not adopted to each country.-- Provision for TRANSPORTAL otherwise not so numerous: quote from Lyell:
499q010 eat percentage of seeds have contrivance for TRANSPORTAL, does he include seeds good to eat. (even Nu
357d074 chanism by which are adapted for long TRANSPORTATION, seems «?» to imply knowledge of whole wor
411e056 , & yet another means for individuals (Mem: TRANSPORTATION will be answered) one looks to analogy for
449e170 me as that which gives genera.-- <it is not TRANSPORTATION> now in case of large [not located] Mr Gre
632j53r acts p. 280. adduces provision of seeds for TRANSPORTATION through the air.-- cocoa nut by water «fuc
632j53r plants, if there was not some provision for TRANSPORTATION:-- But I do not want to deny laws.-- The w
632j53r (all have not it) was DIRECTLY created. for TRANSPORTATION. it follows from some more general law.--
074r165 us iron ore.» N.B. To show how metals may be TRANSPORTED by complicated chemical law & steam of salts
094a036 ner. as bits of Patagonian boulders might be TRANSPORTED.-- {T} On grooved rocks. Specimen of rock fr
156g070 resent torrents Maculloch wrong in saying no TRANSPORTED materials <into> on upper shelves granite &
191b081 valuable than any other, because less easily TRANSPORTED-- Mem plants on Coral islets.-- Next to anim
217b188 eland, must have been all ice in time of ice TRANSPORTED.-- This gives room to fine speculation.-- Ar
218b190 d good with facts about changes when animals TRANSPORTED.) Mr Herbert's papers are in the Horticultur
218b192 Graham isld.-- we know many seeds, might be TRANSPORTED some blown--floating trees Thrushes & buntin
219b194 re philosophical to state we do not know how TRANSPORTED.-- (Glaciers might have acted at Tristan d'A
226b221 rders being able to survive or chance having TRANSPORTED them to new station.-- When the new island s
248c029 mice. It is wonderful how it could have been TRANSPORTED. ¿whence does the New Zealand Rat belo
300c197 me the fathers of many species.-- a few eggs TRANSPORTED to the Str of Magellan.-- Change of habits i
306c214 ine-haired goats. which it is said cannot be TRANSPORTED from their country.-- the long-haired cats a
327c1BC ia Vol I. p. 71. account of Europaean plants TRANSPORTED.-- Crawford. Eastern Archipelago. probably s
354d065 or Ascension.-- The Galapagos mouse probably TRANSPORTED like the New Zealand one-- It should be obse
406e037 -- Addenda to Journal. I show erratic blocks TRANSPORTED far S. in Northern. Hemisphere-- likewise f
414e065 e range of man is not unlike that of animals TRANSPORTED by floating ice.-- I agree with Mr Lyell., m
442e152 r Knights. theory «without seeds are freshly TRANSPORTED»-- throw over this theory, & the sexual repr
497q006 ill spring up which will show, how seeds are TRANSPORTED, or how long they remain dormant. if kinds c
066r141 Yet, no outlet at head. Important in forming TRANSVERSE valleys Ice Sir W. Parish says they have Ear
100a053 r stalactites.-- agate rings, crystallization TRANSVERSE.-- or rather radiating to central point. can
114a094 rbank-- Dr. A. Smith's curious specimens of «TRANSVERSELY fibrous» quartz. & iron stone alternating.
073r163 rphyry. greenstone[,] amygdaloid. basalt & other TRAP cover it to great thickness. = Coast of Acapulc
112a089 of not discovering dike one end granite & other TRAP.-- It is in the mountain masses we must look fo
119a106 ock into gneiss &c judging from what we see when TRAP in dike & approach other rocks. & trap at least
119a106 see when trap in dike & approach other rocks. & TRAP at least as hot as lava-- of which temperature
131a135 h reading. L'Institut. 1838 p. 360. on orbicular TRAP thought to be bombs submarine L'Institut 1838 p
146g010 End near Holyrood Palace In same way at top the TRAP could be traced Grey in front on wall perhaps w
099a050 g direction.--> Therefore in PILE of mud from TRAPICHES. inclined layer!!!.-- The separation in the P
133a140 many points have been penetrated by volcanic & TRAPPEAN rocks, within say the Tertiary period. one is
193b092 in 4 degree.-- I have abstracted Mr Swainson's TRASH.-- at beginning of Volume on Geographical distr
235b272 of Zoolog. 842 White regular gradat in Man poor TRASH Lyell 1024 Flemings Philosophy of Zoolog Royle
321c275 Selbourne References at end Dr. Lang Australia «TRASH» skimmed Macleay's Horae Entomologica Ray's Wis
637j58r «young».-- good God & yet Mails have them. What TRASH p. 237. Gives as Summary of adaptations Horny p
042r067 e swell, generally & during gales would tend to TRAVEL a <me> central line of Patagonia. «NB. Mr L
047r082 oscillation, body of water manifestly does not TRAVEL up.--) If these view are right the coincidenta
058r116 a. next year Quito. considers these earthquakes TRAVEL in order.-- If we look at Elevations as consta
234b255 à France-- Par un Officier du Roi.-- Mackenzie TRAVEL. p. 280. says cattle in Iceland. «"are very» l
322c269 an. Mind Evelyns Sylva. skimmed, stupid Brownes TRAVEL in Africa; «well skimmed.» 1839 Jan 10t.-- All
510q017 says North of Siberia, no sea-current, icebergs TRAVEL by wind. Aug. St. Hilaire Bot. p. 787. positio
070r155 P. states that in Helm's travels accounts of TRAVELLED boulders. from the Cordovise range. Signor Ro
120a108 & light.-- thin dikes not cooling if they had TRAVELLED some hundred miles through nearly cold rock.--
232b247 «& north of Europe»-- As Europaean forms have TRAVELLED towards Equator, <th> so would the plants fro
246c025 not bark p. 433. birds & bats have certainly TRAVELLED from East Indies. isld, as far as Oualan.-- W
269c101 ., therefore 34 degree degrees of change have TRAVELLED that thickness in that period & no ways assis
278c129 ith, «Richardson» a Vaillant, a D'orbigny has TRAVELLED this will be most difficult. Sub-genera so fa
284c148 ere trees of such Nature far apart. Must have TRAVELLED by <dead> «each» trees dying & mountain tonre
427e109 imate in far north regions? Arctic forms have TRAVELLED S. From the analogy of the animal kingdom I s
035r048 ffects.--«though so immense to short breathed TRAVELLER» Mountains, which in size are grains of sand,
277c127 to observe.-- will aid us in physiology, tell TRAVELLER what to observe.-- if he knows he has done le
231b242 a & East Indian islds-- Monkeys different not TRAVELLERS?? Royles case of Himalayan, stands ¿migrator
041r067 . the blending of pebbles & the appearance of TRAVELLING may be owing to successive transportal from
047r081 ation, from violence. Is it not same as swell TRAVELLING across Pacifick.--excepting in number of wav
222b202 .-- Speculate on multiplication of species by TRAVELLING of climates & the backward & forward introdu
577n049 esume-- see p. 188 of Herschel's Treatise) a "TRAVELLING instance" a-- "frontier instance".-- for it
022r008 he discussion of winds & storms:--«in Volney's TRAVELS also» Dampier's last voyage to New Holland P 1
041r067 transportal from prevailing swell, (as Shingle TRAVELS on the Chesil bank. V. De la Beche). Ask Capt.
070r155 endoza smooth. Sir W. P. states that in Helm's TRAVELS accounts of travelled boulders. from the Cordo
080r181 action of sea on coast. 27. Bahama Isd De Lucs TRAVELS Beauforts Karamania Capt. Ross. & Scoresby dee

081r181 resby deep soundings Gilbert Farquhar Mathison TRAVELS Brazil. Peru. Sandwich Isd Mawes travels down
081r181 hison travels Brazil. Peru. Sandwich Isd Mawes TRAVELS down the Brazil.-- Did Melaspena publish his t
081r181 s down the Brazil.-- Did Melaspena publish his TRAVELS? Bellinghausen in 1819 Kotzebue 1816 Constant
115a097 en writing on Falkland Islds p. 94. Von Buch's TRAVELS account of Norway chain being broken through l
116a100 . p. 26. Geology of Arica <Schit> Schmidtmeyer TRAVELS into Chile p 29. gold is not sought for in Chi
133a100 fool.-- Feb 25 All facts show how slowly heat TRAVELS; & therefore the abysses where fluid rock has
252c042 at Leeward Isles. p. 388 Reference to Rüppel. TRAVELS (what language?) Hyena «venatica» <of> Cape fo
294c178 ation & affinities, its extension.-- Von Buch. TRAVELS, p. 306. account of trees ceasing to grow far
306c214 ies.-- -- Hamilton will give an account in his TRAVELS in Asia Minor of the domestic animals-- At Ang
309c221 euter into Queen, more wonderful case Dwights' TRAVELS in America, speaks of short legged sheep. here
320c275 n «Annals of Natural History" Skimmed Von Buch TRAVELS Whites Natural History of Selbourne References
323c269 -, except the V Volume.-- -- 19t. Mungo Park-- TRAVELS Feb 12. Sir. H. Davy Consolations in Travels -
323c269 - travels Feb 12. Sir. H. Davy Consolations in TRAVELS -- Observations on morals by Eugenius Feb 14th
323c269 ence of Physical causes: well skimmed Bartrams TRAVELS in N America May 18th Stanley familiar History
327cIBC peculiar plants-- in high points Read Volney's TRAVELS in Syria Vol I. p. 71. account of Europaean pl
327cIBC eful Knowledge on Horse & Cow & Sheep Clarke's TRAVELS.-- Temminck Hist. Nat. des Pigeons et des Gall
446e160 aneous generation.-- Introduction to Bartram's TRAVELS p. XXIII. <Some birds> Both sexes of some bird
483z012 us Cahirimus. of Geof.-- reference from Rüppel TRAVELS All Owens papers on Intestinal worms must be s
590n094 l has given an account of an Ourang.-- see his TRAVELS.-- When one sees in Cowper, whole sentences sp
046r078 that the dikes which so commonly (state facts) TRAVERS> "May have reference to the Great distances wh
369d115 alia of N. S. Wales are generally compelled to TRAVERSE granites, are granitic materials simply alter
077r171 estone & <many> «other secondary» rocks. Vein TRAVERSE in order to quench their thirst"-- But New Gu
512q019 ts-- (4) About hybrid pheasants treading-- any TRAVERSES both Clay slate, Porphyry North 52 W, & is ne
512q019 throw up pellets-- (4) About hybrid pheasants TREADEE?-- Difference in lambs of different breeds Is
079r177 shells Bougainville says P 291.-- The Fuegians TREADING-- any treadèe?-- Difference in lambs of diffe
136a147 drawer with important letters-- When I come to TREAT the "chefs d'oeuvre de l[']industrie humaine, c
570n026 yet would love mountains, & a negro, similarly TREAT of the age of the Pampas Deposit, I may properl
411e054 of Natural History, will be approximated to.-- TREATED would think negress beautiful,-- [male glow wo
508q016 ever heredetary (9) Are the works of Berhave (TREATING of the formal laws of corelation of parts & o
600o008 me" &c-- -- Confesses these faculties of soul, (TREATING of heredetary diseases) translated. (10) Abou
136a147 f Phillips as given p. 13. Vol II. Lardner's-- (TREATING of infinites not definable.-- Has little Chap
205b142 ains Asiatic types discoverable.-- Bridgewater TREATISE Phillips in Lardner Vol II p. 73.: some remar
205b143 re intermixture precluded.-- Kirby Bridgewater TREATISE p 85. Parasite of Negroes different from Euro
205b145 n Southern hemisphere. does this rule apply? A TREATISE There are some good accounts of passages of l
266c091 ran and the sword" [quote Whewells Bridgewater TREATISE on Form of Animals by Mr Cline "The character
323c269 Spallanzani's Essays on Animal Reproduction -- TREATISE (p..26). about plants from Cape of Good Hope
323c269 oshs' Ethical Philospohy -- Bell's Bridgewater TREATISE on Domestic pidgeons 30th Lives of Hayd & Moz
362d100 articular times. Mr Yarrell has old book 1765? TREATISE -- Wilkinsons Egyptian remains skimmed -- Pli
443e157 reject such false reasoning Bell Bridgewater's TREATISE on Domestic Pidgeon, in which it appears that
577n049 e: it is (I presume-- see p. 188 of Herschel's TREATISE on the Hand.-- p. 94.-- "The resemblance of t
637j57v eans-- but not in the other. [All Bridgewater TREATISE) a "travelling instance" a-- "frontier instan
446e162 think very close.-- «A fruit tree by certain TREATISES. are reduced simply statement of productivisme
498q007 character in grt degree from charcoal & good TREATMENT will suddenly send forth quantities of blosso
286c155 ht one It will be necessary from manner Fleming TREATMENT (8) Do bees frequent Cabbages «& Cowcumber's
071r156 coniferous must formerly have been most abundant TREATS subject to put in alternative of Man created b
176b021 ch as acid & alkali organized beings represent a TREE-- Metamorphic action: <most> coming so near sur
176b023 .-- Would there not be a triple branching in the TREE. irregularly branched some branches far more br
176b024 , double arrangement.-- if each Main stem of the TREE of life owing to three elements air, land & wat
177b025 sh & penguins really pass into each other.-- The TREE is adapted for these three elements, there will
192b083 l go back.-- Get instances of a variety of fruit TREE of life should perhaps be called the coral of l
228b229 genera formerly abundant. Seed of Ribston Pippin TREE or plant run wild in foreign country.-- When on
235b263 re so many, imported animals?-- At what. part of TREE <go> producing crab. is the offspring of a male
271c105 ild a nest lined with mud, in forest where not a TREE in which it builds, a berry on which it feeds.
278c130 urus being found fossil in Australia, & only one TREE of life, can orders like birds & animals separa
285c151 bly upset it-- The space which one branch of the TREE species (Mitchell's authority) in Australia, &
285c152 all analogy breed together.-- The bottom of the TREE if live occupied after its decay, will be occup
286c155 > relationship; latter word meaning descent.-- A TREE of life is utterly rotten & obliterated in the
352d059 ry of analogies-- <be> when it is considered the TREE is taken by Fleming as emblem of dichotomous ar
361d095 mon sparrow young & female similar plumage.-- in TREE of life must be erect not pressed on paper, to
370d117 f growing plants in all parts of world, thus tea TREE sparrow, (if I understand rightly) young cock &
384d161 ion alway put female first Will not even a fruit TREE in Brazil must have degenerated. as must spices
390d178 ncy to propagate the whole difference of parent, TREE or rose degenerate during its life so that succ
391d174 s do not procreate> & all alike in one parent or TREE, but it fails.-- therefore «each» seedling of o
401e015 y experiment was never tried of separating apple TREE, (but not in other trees.-- -- Why should there
440e145 t seems absurd proposition, that every «budding» TREE, & every buzzing insect & grazing animal owes i
441e149 ot structure of the parents.-- Thus would a Crab TREE vary if planted in rich soil, I presume not, bu
470t153 e collected & resown.-- Humble 22 flowers of Egg TREE by certain treatment will suddenly send forth q
470t178 es;»-- on Monk's Hood, brushing over stamen «Egg TREE in one minute Great Humble 17 flowers of Larksp
491q01v «above Exper» explains apples on side near other TREE»--"I think never on the Galeum saxatile & other
491q01v relized affect other branches-- The French Apple TREE being affected.-- does one branch of Cabbage be
497q06v (15). What is History of Viburnum, or snow-ball- TREE. what would result from seeds being sown= See i
499q009) Place pin's heads with Bird lime near male yew TREE & see whether they catch pollen-- <Ne> In Oenot
504q013 - No (6) There is apple with branch in middle of TREE with flowers near end of orchard.= At Shrewsbur
528m038 sea weed, &c &c-- this gives beauty to a single TREE,-- & the leaves of the foreground either owe th
546m111 le cause, remembered she had put it in branch of TREE, & apologising to party, went out & found it th
589n092 to exist, is clearly seen. in the absurdity of a TREE having reason: or dog, having high powers witho
038r058 d Islds from which proceeded pebbles & on which TREES grew.--? Are not the dikes in upper strata. qui
041r064 aps agrees with formation of pebbles & vertical TREES Grand Seco at B. Ayres; mention about the deer
061r128 aracter: Contrast low limit of Palms, evergreen TREES, arborescent grasses, parasitic plants, Cacti:
069r152 UBSIDENCE Uspallata of which no trace except by TREES, The structure of ice in columns. show that gran
070r154 : it is Let it not be overlooked that except by TREES, I could not see trace of Subsidence at Uspalla
085a001 p. 288. Salt deposited on windows of houses. & TREES all injured on Eastern side, far inland.-- even
088a014 ir nests; & then they may picked up beneath the TREES---- are any Fish seed-eaters. This important in
111a088 hile Geograph. Journal Vol. VII p. 216.-- Guava TREES, introduced about twenty years since (1835) fro
125a120 o subject do p. 447 & 449. «& 450». On Vertical TREES. Uspallata.-- do p. 473. on great Iceland strea
134a141 n road from Valparaiso to Santiago p. 328. dead TREES. on Isthmus of Pen. Tres Montes.-- as by subside
170b051 ividuals absolutely similar; for instance fruit TREES, probably polypi, gemmiparous propagation. bise
186b057 esentatives agrees with breeding.. in irregular TREES. & extinction of forms.?? It is in simplest cas
202b131 . Demerara. 10 «12» feet beneath surface forest TREES fallen «kind well known, carbonized»--; clay fi
206b149 h extinct. or of Pachydermata, or of coniferous TREES; or in certain shell cephalopoda.-- «Read Buckl
207b151 ol. IV. Ducks on rivers in Guiana. build top of TREES carry duckling to the water in their beaks, & t
207b151 lves in the grass.-- Beatson St. Helena says no TREES succeed so well at St. Helena. as. Pineaster &
218b192 eeds, might be transported some blown--floating TREES Thrushes & bunting & coots-- «(Turdus Guyanensi
219b193 iffe, St. Helena. J. Fernandez. Galapagos. Many TREES Compositae, because seeds first arrived «Ferns
219b193 s first arrived «Ferns ditto.--» & hence formed TREES]CD & would creator <on volcanic island.> make p
228b229 going back may not be owing to cross from other TREES.???? Do the seeds of Ribston Pippin & Golden Pi
259c064 of extermination of species-- Epidemic amongst TREES. Plane trees all died certain year. Extreme dif
259c064 tion of species-- Epidemic amongst trees. Plane TREES all died certain year. Extreme difficulty of TR
284c158 st salt water.-- Nuts prodigiously heavy (where TREES of such Nature far apart. Must have travelled b
284c168 far apart. Must have travelled by <dead> «each» TREES dying & mountain torrents.-- but to crawl up an
294c178 ension.-- Von Buch. Travels, p. 306. account of TREES ceasing to grow far N. becomes stunted, altered
317c249 40 leagues from land several patches of reed & TREES p. 259 120 ft in length, some branches of Justi
327c265 .--» Mr Coxe "Views of the Cultivation of Fruit TREES in. N. America" in Lib. of Hort. Soc Mr Neil. h
356d070 producing occasionally (as Fox says) same fruit TREES is analogous to some hybrids breedings-- there
374d133 supposed to be intermediate between Coniferous TREES & Lycopodiums.-- p. 437. Many. existing genera
386d165 tains same length of life.-- like Golden Pippen TREES! How is this with buds of plants, does annual g

Page **(Key Word)**
390d178 neration.-- therefore passions fail.-- In fruit TREES no doubt there is tendency to propagate the who
391d174 alike in one parent or tree, (but not in other TREES.-- -- Why should there be a necessity that ther
391d174 hermaphrodite being simple (Are not Coniferous TREES generally dioecious oldest forms) why are twin
401e015 ies may be due to impregnation from other apple TREES.-- now seeds of crab produce crab, so that some
401e015 ab produce crab, so that some effect from apple TREES is produced.-- Thinks probably experiment was n
417e077 een the fluids of the two as in the grafting of TREES.--]CD CD[The similarity of child to parent app
429e115 ith ease near London.-- what makes the line, as TREES in Beagle Channel.-- it is not elements.-- we c
429e115 it is other plants.-- a broad border of Killed TREES would form fringe.-- but there is a contest. &
435e130 genera <are> have this structure.-- Some willow TREES have been observed to change their sex,-- this
441e148 r any are transmitted, for the changes in fruit TREES. mentioned by Mr K may be caused by the diversi
442e153 Buch.]CD Now Mr Knights statements about fruit TREES. grafted. altering is hostile to this: but on o
442e153 ng is hostile to this: but on other hand, fruit TREES are propagated by means, which wild plants neve
443e153 aft from the roots is the best way to get young TREES, from worn-out kinds, & quotes from Pliny, that
446e162 s case of change analogous to change in grafted TREES «:so is not effect of different stocks in this
455eIBC ve buds, or tubers. Yes-- but these are same as TREES.-- Shake some sleeping mimosa-- do stamina of C
492q003 is referred to with doubt by Herbert Do forest-TREES sport much in nursery gardens? <are the> is the
503q013 ith cow: is not reverse possible?? Maer (1) Yew TREES near Boat House «ANY male branch.» --¿number of
503q013 umber of seeds in beginning of November 1841.-- TREES above male? (2) Result of Edwards experiment in
504q014 ment (30/p.11) (2) Yew Berries germinate?-- Yew TREES sexes-- (3) Get Holyhoaks. races planted & Linu
507q15v female flowers ever productive) Smith says many TREES in Tropics are of this class.-- (48) .Where «pu
515q021 ially Papilionaceous order (2) History of fruit TREES far north in Scotland-- do they flower-- do the
529m041 e poetical) the botanist might so view plants & TREES.-- I am sure I remember my pleasure in Kensingt
529m041 ns has often been greatly excited by looking at TREES at [i.e., as] great compound animals united by
473s05v roznerat, «on road to Bethgellert» wh flows by TREMADOC. but can tell them from lake S. of Moel Siabo
381d156 te female (as in plants) Cirrhipeds rotifers, TREMATODE & cestoid Entozoa Allotriandrous <or M> Mollu
557m150 can be heard.-- Shame would never make person TREMBLE, like fear.-- Why does any great mental affect
557m150 Why does any great mental affection make body TREMBLE. Why much laughter tears.-- & shaking body.--
089a018 e, C«o»urrejolles. 11th Observ.-- Les grands TREMBLEMENS de terre sont presque toujours precedes et s
532m057 panied by «troubled» beating of heart, sweat, TREMBLING of muscles, are not these effects of violent
539m084 f Queen) «O» of Hell Cant IV or V.) as pale & TREMBLING. & not as flushing & with muscles rigid.-- Ho
596nIBC of the face first affected?-- Can shivering & TREMBLING be considered convulsive.-- is convulsion. ar
596nIBC movement of voluntary muscles-- if so what is TREMBLING palsy? Expressions N [Old & USELESS notes abo
051r093 ean.--at Ascension. (where occassionally most TREMENDOUS surf & loose sandy beach) deposits «calcareo
509q017 y-- Teeth in foetal state: Mr. Horner. On Mr TREMENHERES Scottish Colliers, when men & women have lon
593n111 mind, & Dr. Hollands story of man in Delirium TREMENS hearing other man speaks. shows, that consciou
022r007 minated through the great Plas Newydd dike.--Mem TRES Montes. ((Henslow Anglesea)) great variety in n
049r088 tact varies in nature, -- Does not granite at C. TRES Montes become more siliceous in close contact?
134a141 o Santiago p. 328. dead trees on Isthmus of Pen. TRES Montes.-- as by subsidence ‖ Fitz Roy refers to
257c056 à observés en Patagonie. ou au moins des espèces TRES-analogues,-- quand ce nétaient pas tout à fait
400e013 ther «places» besides Galapagos do. p. 376. Isle TRES Marias off Mexico with small Hares & raccoons.-
270c104 Agassiz says Infusoria «are» insecta.-- G. R. TREVIRANUS, Biologie referred to,. as compilation of ac
638j58v aptation, as the surviving one of ten, thousand TRIALS.-- each step being perfect «or nearly so (exce
567n014 in balancing a body & an ass knows one side of TRIANGLE shorter than two. V. Whewell. Induct. Science
568n016 ur senses, or our experience.-- Two sides of a TRIANGLE shorter than third. is this necessary notion,
032r040 r less time of rest. so would the size of the TRIANGULAR mass removed vary.--The gradual rising conti
080r178 ward partaking of the character of a Araucarian TRIBE, with point affin of yew & intermediate Punctur
193b090 h are well adapted, This would account for each TRIBE <being> «acting as» in vacuum to each other p.
198b113 ree elements p 66]CD With unknown limits, every TRIBE appears fitted for as many situations as possib
206b149 there many anomalous lizards living; or of the TRIBE fish extinct. or of Pachydermata, or of conifer
264c081 most important element in considering to which TRIBE,-- structure without corresponding habits clear
271c107 are many plenty of instances of insects of one TRIBE taking on structure (probably accompanied by ha
179b033 wild men do not cross readily, distinctness of TRIBES in T. del Fuego. the existence of whiter tribe
179b033 tribes in T. del Fuego. the existence of whiter TRIBES in centre of S. America shows this.-- <If> Is
181b045 during subsequent ages.-- The Creator has made TRIBES of animals adapted preeminently for each eleme
181b045 ly for each element, but it seems law that such TRIBES, as far as compatible with such structure are
185b057 romera same. thing occurs with regards to other TRIBES in that same family.-- (NB I see Waterhouse th
206b148 mals first consider species of cats.-- <& other TRIBES>.-- &c &c Exclude mothers & then try this as s
250c038 re these great animals had not spread then such TRIBES as Marsupial & Edentata increased most. Certai
308c218 reverse habits Insects & birds are the only two TRIBES fitted for water, air, & land, (Macleay has th
343d038 ro of Africa is not loosing ground. Yet, as the TRIBES of the interior are pushing into each other fr
343d038 trade, & colonization of S. Africa, so must the TRIBES become blended & prevent that strong separatio
381d157 s-- Seeing sexes separate in some of the lowest TRIBES, leads one to suppose still more that they mus
451e178 ch «all» mankind «(& yet afterwards says native TRIBES can live there)» flee during 8 months out of 1
592n102 ed for Deaf & dumb school & used between Indian TRIBES are Many the same.-- Philosoph. Transactions V
539m082 interval. Augt. 15th. As child gains habit «or TRICK» so much more easily than man, so may animal ob
570n024 - mechanic apt to sigh.-- & hence carried on as TRICK) «Shrugging aroused acting» Octob 25. Why is mo
556m146 orses.-- 1838 good instance of useless muscular TRICKS accompanying emotion.-- when horses fighting,
557m149 (shame of ridicule) are singularly apt to catch TRICKS.-- so are people in passion my F. rubbing hand
560m156 r what they did in former one. about heredetary TRICKS & gestures, other cases like D. Corbet; «do» i
257c056 c &c D'orbigny (p 108) says having observed B. TRICOLOR in Patagonia. then in Chile & lastly 12,000 f
139a180 d [blank] Many interesting experiments might be TRIED by comparing Zoophite to plants.-- grafting len
344d041 om Ireland to Brighton Park--first rate bitch-- TRIED to breed from her, but her offspring came out o
401e015 roduced.-- Thinks probably experiment was never TRIED of separating apple tree entirely from all othe
472s01r um; all supposed to have been hybridised == Has TRIED several year to obtain seed, but the pods have
532m054 son was laughing & told me there was nothing, & TRIED to seize hold of objects to be frightened at.--
640j167 species, until their breeding together has been TRIED.-- With respect to the six puppies, if a hare w
280c136 tion unlike would go against the tendency.-- it TRIES to go back to grandfather, but if too unlike it
532m054 bly the system is affected, & by habit the mind TRIES to fix upon some object:-- When a man, child or
580n061 y the piece, she plays <f> better than when she TRIES is not this precisely the same, as the double-c
068r146 be too thick to be penetrated by the repeted TRIFELING injections.--Old vents would keep open long a
200b123 haracter is this? Race permanent, because every TRIFLE heredetary, without some cause of change,, yet
304c207 parent.-- same forms of beak &c without these TRIFLES. it would not then be told whether not descend
304c208 gh up.-- Such probabilities only guides.-- Yet TRIFLES are produced by circumstances. Spines on Echid
051r095 Davy experiment on the copper bottom. we see a TRIFLING circumstance determines whether an animal wil
056r109 at its age was limited? Who could suppose such TRIFLING means could efface & obliterate so grand a wo
179b033 s above stated:-- no one can doubt that lesser TRIFLING differences are blended «by» by intermarriage
218b191 constant» aspect. That is varieties, though of TRIFLING order are formed by nature. Carmichael. Trist
222b203 may look at it so--?? It holds good even with TRIFLING differences of expression -- one child like f
255c051 rated. This may account for permanence in many TRIFLING marks.-- such as the bands on pidgeons back.-
282c141 , we must suppose men instead of mere colour & TRIFLING form & head &c to become greatly changed. in
304c207 ween apterix & other birds.-- will having many TRIFLING characters, in common with other birds reveal
304c207 .-- Now all the different forms of Synallaxis. TRIFLING characters as red band on wing show to be fro
332d001 h. 1838 Finished. October 2d As a proof. what <TRIFLING> «unknown» causes act upon people. My Father
421e091 tatement of the tame ones.-- ‖an instance of a TRIFLING peculiarity not to be eradicated.-- ‖.do. p.
481z010 corals.-- nevertheless the details appear very TRIFLING Also Berre «p. 8» (I think Planariae) Sagitte
620o044 inct not followed for a pleasure now though so TRIFLING he feels remorse.-- He reasons on it & determ
621o047 rson. or it is accidentally acquired from some TRIFLING circumstance.-- Thus a child may be taught to
403e021 (Mariannes), "for example the prickly Limonia TRIFOLIATA, which cannot now be checked".-- Mar[s]den p
316c244 ut then they appears always very small ones as TRIGONIA in Australia or Concholepas in America,-- yet
430e120 nk it Hybrid.!!! Ask Woodward Mr Lonsdale says TRIGONIA costata & elongata thougt considerably differ
350d055 roderip were talking of some Crustacean, like TRILOBITE. (Polirus??) female blind & of quite differen
374d134 bratula, orbiculas, with many extinct forms & TRILOBITES Sept 25th. In considering infertility of hyb
152g048 sses, to upper edge of which they cut near Loch TRING-- Tuesday Bridge of Roy Level of «bed of» River
185b055 woodpecker-- a desert. Kingfisher.-- mountain TRINGAS.-- Upland goose.-- water chionis water rat wit
538m079 ness same disposition recurs, such as -- -- of TRINITY always thinking people were calling him a bast
507q15v ica Henslow presumes females produce. Polygam. TRIOECIA. (are female flowers ever productive) Smith s
176b023 ess of life of Mammalia.-- Would there not be a TRIPLE branching in the tree of life owing to three e
391d175 n in man more like «each other» than twins «or TRIPLETS &c &c in» in litter. Why is there some law ab

078r174 snumbering, no page 173] Under name of Sagitta TRIPTERA D'Orbigny has figured animal with setae like
479z005 451.-- many molluscs Under the name of Sagitta TRIPTERA M. D'Orbigny has described my animal with tee
065r138 ate with greater care. vegetation & climate of TRISTAN D. Acuneha. Kerguelen Land. Prince Edwards Isl
102a059 ay from coast is driven on to it.-- rollers at TRISTAN d'.Acunha.-- silting up. channels on coast of
218b192 ifling order are formed by nature. Carmichael. TRISTAN D'Acunha, a list of its Flora. is given Mr Don
219b195 w transported.-- (Glaciers might have acted at TRISTAN d'Acunha-- Carmichael Linn. Transacts. Vol XII
225b218 merican & African form of plant being found in TRISTAN D'Acunha. may be said to deceive man. as likel
225b220 ?-- I believe none.-- Canary islds.? Madeira? «TRISTAN d'Acunha?» «Iceland?--» The Connection between
226b222 ith Fernando Po.). with plants of St. Helena & TRISTAN D'Acunha, resolves itself into question of pro
269c100 (NB.-- Examine Abrolhos Flora with this view) TRISTAN D'Acunha, St Helena &c &c. Juan Fernandez A co
358d074 With respect to ancient geography of Atlantic TRISTAN D'Acunha ditto Juan Fernandez do Mitchell. Aus
480z008 chael Linn. Transacts Vol XII. p 496. Birds at TRISTAN d'Acunha.-- (Turdus Guayanensis?)) Emberiza Br
030r036 ff & low or sloping land What are the "palatal TRITORES" found in the coraliferous mountain Limestone
048r086 ia, are all beds same age? is white substance TRITURATED Porphyritic rock. s (mem white tufas with pu
556m144 shing]CD «Does Elephant know shame-- dog knows TRIUMPH.--» Sept. 23rd. Horses in Omnibus instantly st
565n007 wishing to smell its enemy.-- Man & dogs show TRIUMPH (& pride) same way walk erect & stiff, with he
570n025 acting»   Octob 25. Why is modesty, mixed with TRIUMPHANT feeling so similar to shame after asinine.--
365d105 e female in wood with Pheasants would sure to be TROD, & in many parts of Scandinavia these birds are
531m052 accompanies violent attack,-- Even the worm when TROD upon turneth,-- here probably there is no feelin
482z011 Embriza melanodera-- a linnet not caught.-- TROGLODYTIS Furnarius.-- Sturnus Magellanicus.-- p. 210
251c040 ia & Africa.-- NB. Any monograph like Gould on TROGONS worth studying.-- «do» Zoolog Journal Vol 2. p
272c109 cock peacock,. widowbird.-- Birds of Paradise. TROGONS.-- the one feather in wing the curious feather
273c110 f plumage (as «black & white» bars on wings of TROGONS are lengthened rump feathers.--; <then> & one
273c113 pratencole (¿connecte[d] with Chionis), yet the TROPIC bird, has very different habits, though preemi
450e175 ed horse is found East of y Burramposter & S of TROPIC-- By J. H. Moore after quitting Bengal this fa
028r029 t come that all Lime is not accumulated in the TROPICAL oceans detained by Organic powers. We know th
244c021 portant to show wide range of fish & shells in TROPICAL sea, it. would demonstrate.; not distance, ma
296c184 in relation to surrounding countries & present TROPICAL countries. p. 564. an abstract of Mr Swainson
310c224 capable of growing!! & propagating itself. In TROPICAL countries (as St Jago Cape de Verds) the shel
313c233 Australia countries die, when brought over on TROPICAL animals which accounts for the species changi
332d004 s are not healthy to worms, (like parasites of TROPICAL countries cannot endure this climate-- .) --
407e037 . America & Europe.-- S. America favourable to TROPICAL productions. The world formerly much more so.
407e038 e, than any part, (except Europe. in which all TROPICAL forms have been obliterated) of the world. fr
407e038 have been obliterated) of the world. from the <TROPICAL> Equable kind of climate to the extreme.-- Th
407e039 »-- formerly were not so «very» EQUABLE. or so TROPICAL, & therefore present state of world is not so
408e046 Looking over Lamark surprised to see how many TROPICAL genera come from New Holland, ¿Sydney? The do
413e059 s me Beck considers the characteristics of the TROPICAL Forms in shells. are numerous species, numero
485z018 Stizbergen wherever there is extreme heat, the TROPICAL forms extend further north, because during wi
486z018 bear the cold when torpid.-- On this principle TROPICAL forms in N. America extend much further N. in
486z018 ce it is odd that Amber insects of Europe have TROPICAL Forms See p. 256 of Note Book (C) for compari
615o36v crow different.-- Heredetary effect of former TROPICAL climate analogous to inflorescence of Tropica
615o36v tropical climate analogous to inflorescence of TROPICAL plants when imported & plants sleeping good s
231b243 r in feathers--seeds.-- Two inhabitants of the TROPICS, (whether one fossil or not) are related by re
231b243 emperature.-- now those of temperate regions & TROPICS are only related by one connection.-- viz desc
249c036 also with Temminks fact of forms being within TROPICS.-- Europaean birds at Japan. connected with Ea
484z014 75 A detailed comparison of production without TROPICS in Northern & Southern America-- valuable & pr
507q15v wers ever productive? Smith says many trees in TROPICS are of this class.-- (48) .Where «published» 1
360d089 «determined» by the sex? Individual instances TROUBLE Yarrels law. chief trust must be in general kn
364d103 esh blood, without fresh feather, & consequent TROUBLE in obliterating the fresh feather, by crossing
548m115 thought are in progress-- In castle of air the TROUBLE «I well recollect» is in making things somewha
035r046 h of the proposition."-- If such can happen in TROUBLED England; the more minute equalities of elevat
532m057 an.-- The sensation of fear is accompanied by «TROUBLED> beating of heart, sweat, trembling of muscle
564n002 ave been relieved, he would be sorry or have a TROUBLED conscience.-- Therefore I say grant reason to
508q018 translated. (10) About Daltonism in the MALE TROUGHTONS.-- paper in Taylors Scientific Memoirs (11)
376d137 s.-- These Monkeys had no curiosity to pull up TROUSERS of men. Evidently knew <men> women, thinks pe
473s05r e/42/ June/42/-- Mr. Bunbury says has heard the TROUT from different lakes of N. Wales can be disting
473s05r e (Capel-Curig) says that he can certainly tell TROUT from Ogwen, Capel Curig & some other lakes, (di
511q018 ether <Yar> knows whether Shaws hybrids between TROUT & Salmon were fertile & whether homogeneous {Ab
599o008 facultes, q'il possede seul sur la terre. J'ai TROUVE son âme" &c-- -- Confesses these faculties of
059r119 s une correspondance exacte, et lorsquelles se TROUVENT interrompues par quelque vallées ou par quelq
039r060 ce at the famous eruption of Rialeja, & the more TRUE analogy from the Galapagos-- Mr Lyell. P. 111 &
073r161 ns (as Hutton says) always come from without.-- "TRUE native iron that to which we cannot attribute a
110a087 t distinct on dike side.-- oozed from one of the TRUE rocks, most probably from the gneiss beds in th
157g078 um «about 12 ft higher than last station» 29.316 TRUE terrace «2d» near divortium aquarum is a lip wi
181b042 y. after that distant period.-- Hence if this is TRUE, that the greater the groups the greater the ga
182b046 ter. If my idea of origin of Quinarian system is TRUE, it will not occur in plants which are in far l
185b056 ilineata (Gray) new <liza> species, belonging to TRUE. American genus Waterhouse says he is certain,
189b071 f real repugnance.-- Waterhouse says there is no TRUE connection between great groups.-- Speculate on
190b078 ity is given (& power of adaptation) is given by TRUE generation, throughe means of every step of pro
227b224 present carnivora than Pachydermata If my theory TRUE, we get (I) a horizontal history of earth «with
227b228 e.--may look to first germ-- --led to comprehend TRUE affinities. My theory would give zest to recent
229b232 er.-- Hermaphrodite animals couple: argument for TRUE molluscs coupling.-- Dr. Smith's Information Lo
258c060 uld not become adapted to climate.-- Descent. or TRUE relationship, tends to keep to species to one f
259c065 es & you will have no tail (example probably not TRUE).-- or again healthy parents have healthy child
264c081 ure without corresponding habits clearly showing TRUE affinity, for instance tail of ground woodpecke
273c110 rally fill up genera & not species, which is not TRUE, with shells?.?.?» It looks as if animals peris
277c126 or abolishing specific names & giving subgenera. TRUE value.-- as in Opetiorhyncus. fulginosus. (a) f
278c129 ecies of each might not breed:-- Genus must be a TRUE cleft-- putting out of case the Analogys.-- If
280c135 n if more different might do so.-- <whi> is this TRUE?? My views, which would even lead to anticipate
281c137 ies, they would go back to grandfather, which is TRUE) & infertility is consequence.-- The simple exp
281c139 meaning of circular arrangement was only so far TRUE as avoided linear arrangement the central twigs
285c151 grow into affinity.--but whether ever arrive at TRUE affinity doubtful A species is only fixed thing
294c177 proof, «its limiting, the allowing at same time TRUE species.» & its adaption to classification & af
296c185 bstract of Mr Swainsons views. which if abstract TRUE are wonderfully absurd.-- p. 565 <breed> Scotch
300c196 nterposition of a deity, more humble & I believe TRUE to consider him created from animals.-- Insects
303c205 ing but experience. will, tell us. when group is TRUE,» there are no genera. if Mammalia are adduced.
341d030 oung of this animal, which is so anomalous among TRUE deer, yet is spotted like so many deer.-- very
355d067 be elicited by these speculation even if partly TRUE they are of the greatest service, towards the e
370d116 ales-- how has this been arranged-- Neuters are TRUE females, but with parts little developed.-- Sep
372d131 uld produce arm if supported., <so> & in making «TRUE» bud some such process is effected.-- a child m
372d131 be so born. but it would be very different from TRUE generation.-- there is no caterpillar state; th
373d132 changes from the placentates, Having Hair. like TRUE Mammalia, no more wonderful. than Echidna. & He
374d133 es have been found. p. 426 Sauroid fish in Coal, TRUE fish, & not intermediate between fish & reptile
378d147 & hen will be alike-- I presume converse is not TRUE for he says Hen & cock Starling alike, yet youn
379d151 ceived from his father so that case ceases to be TRUE avitism Annals of. Natural. History. p. 135. Na
409e049 oon probably is uninhabited]CD & if my theory be TRUE then the formation of sexes rigidly necessary.-
412e057 given milk.-- s: secretion from Pidgeons stomach TRUE milk.-- ╢ <Species. are innumerable variations>
416e071 Worms? [Barnacles, aquatic., <yet> Crustacean, & TRUE hermaphrodites] CD «It may be said that true he
416e071 , & true hermaphrodites] CD «It may be said that TRUE hermaphroditism is a consequence of non-locomot
416e071 s there are no fixed. land animals. so there are TRUE hermaphrodites.-- I suspect this rather effect
417e077 sexual differences) a variety. The offspring of TRUE <pare> hermaphrodite, would of course be like e
421e091 arded as affording very clear indications of its TRUE affinities. We are least likely in the modifica
422e092 dapted or transformed. & hence not indicative of TRUE affinity.-- -- Owen says Dugong connected wi
422e092 ifers> p. 306, the Dugongs cannot be united with TRUE Cetacea or whales.-- but are aquatic Pachyderms
423e096 w of developement in partial classes is far from TRUE.-- I doubt not if the simplest animals could be
426e105 tiary «epoch» like Sicily not having species, if TRUE important on my view.-- March 9th-- Is there an
426e107 other great series quite otherwise & make on[ly] TRUE hybrids.-- but this is false, [give instance of
428e112 & weak March 11th. Yarrell's law must be partly TRUE, as enuntiated by him to me, for otherwise bree

Page **(Key Word)**
429e116 ia, or even of polar regions of N. America.-- if TRUE curious on my view-- because these points were
430e118 - if &c &c.-- Then give my theory.-- excellently TRUE theory Examine list of St. Helena Plants & see
431e122 ndividuals) the deeper one goes-- surely is this TRUE?-- most strange.-- Does not spermatic animalcul
431c123 rely this is very limited view, though perhaps a TRUE element) «give examples, pigs, with small chine
432e124 my theory no land animal with fluid seeds can be TRUE hermaphrodite Man probably assumes the hairy ch
443e155 e of developement.-- [It is singular there is no TRUE hermaphrodite in beings with which have fluid s
447e164 , «with facility»-- such as cryptogamic plants & TRUE hermaphrodite Mollusca, & probably corals.-- th
469t151 ch other» & seeds gathered «all» came up in 1840 TRUE. Shrewsbury.-- Abberley-- Early Magazine-- &c.
469t151 nted in rows, & seeds gathered same year came up TRUE «in 1840»: All in together blossomed together T
472s01r me garden. & out of 60 seedlings not one came up TRUE.-- colour of flower & foliage not «being» like
472s01r colour of flower & foliage not «being» like the TRUE P. bracteatum; all supposed to have been hybrid
473s03v ndstone of. N. America is Red Sandstone. & Birds TRUE! Plants in Devonian-- How strange no plants in
479z006 p. 311 D'.Orbigny considers Dasypus villosus is TRUE Peludo Cavia Australis. Dorbigny Vol II, p 24 P
495q005 t seeds, & see how great a proportion springs up TRUE.-- This in fact always takes place in natural H
495q05a ass-- sow these seeds & see if they will come up TRUE-- whilst others are crossed.-- Are Bees guided
497q06v as, IMPORTANT, for if so, as these can be raised TRUE, there is no crossing by Bees.-- Henslow.-- (1)
504q014 s alone remain to be compared-- Cabbages.-- kept TRUE Try experiment (30/p.11) (2) Yew Berries germin
505q014 Bees July/42/ Mark has six day's puppy of small TRUE Bull-Dog-- length from nose over head to root o
523m017 ght arsenic for that purpose.-- this found to be TRUE.-- Her Husband never suspected during these two
523m018 - My father quite believe my grand F doctrine is TRUE, that the only cure for madness is forgetfulnes
526m027 ree will determines our throwing it up.-- equall TRUE the two statements.-- Catherine remarks that pl
526m028 entive power,-- this, though very odd is perhaps TRUE.-- mem Erasmus & mine taste for music.-- Childr
531m051 e, when listening one evening when tired «-- how TRUE the heart the scene of anger.--» to the pianofo
535m069 abstractions, such as plastic virtue, «&c» (Very TRUE, no doubt savage attribute thunder & lightening
535m069 eir nature.-- Reviewer considers this profoundly TRUE.-- How is it with children.-- Now it is not a l
549m122 do not tend to this-- though believing it to be TRUE, & then acting on it, will add to happiness.--
566n011 of obtaining <that> «a» faculty in the form of a TRUE instinct, which is a real instinct in the chick
571n029 spond to fashions in ideal taste & the former to TRUE taste.-- Everything that is habitual, if herede
577n051 unctions &c &c &c.-- bears. the same relation to TRUE memory, that the formation of a hinge «in a biv
578n056 ve action to remove disagreeable impression like TRUE convulsion. (Hence pass into convulsions?)-- sq
582n068 the result they will effect,.--" this not wholly TRUE, for we must grant a bird knows what is about w
583n069 ffer best cases of instincts].CD all this may be TRUE,, but relation of imitation & reason must be th
602o11b , & one often replies "what you say is perfectly TRUE, but you do not convince me.--" Belief allied t
611o034 (contingencies as heat light &c). [LHC] This is TRUE as long as movement of sensitive plant can be s
614o037 species change.-- Must be so if Lamarck's theory TRUE [RHC] Acquired instincts analogous «(& replace)
627o053 s young pointer to point-- clearly shows this is TRUE. p. 13. Affections cannot be analysed into "pow
635j56r of germs is to afford support to other beings.-- TRUE, (& the doctrine of checks & my theory) Maccull
640j167 rom their own Faunas, which in this case is only TRUE criterion.-- Hence it is highly unphilosophical
640j167 ry slowly changed & vertebrata much so.-- so far TRUE, but do not fish offer a most striking anomaly
173b009 Lamarck disappear as collection made perfect.-- TRUER even than in Lamarck's time. Gray's remark, bes
265c087 rcumstance of aberrant groups being small it is TRUISM. for it not so not aberrant.-- Tenioptera rufi
413e060 rom their alliance to Articulata, which are all TRULY bisexual. Buckland's Reliqu: Diluv. says Africa
524m019 s to those feelings. which may be considered as TRULY spritual.-- a person twitching when a disagreea
529m040 former animals, slow force cracking surface &c TRULY poetical. (V. Wordsworth about science being su
531m051 n, the violent passions they go into, shows how TRULY an instinctive feeling, <may not pa.> In reflec
552m130 more vivid? Surely the image in a dream cannot TRULY be <more> «as» vivid, «a reality» as in Spectra
582n068 what are their characteristics;-- they are more TRULY sensations??. a kind of mental pain & pleasure.
151g043 upper slope of which corresponds to shelf the TRUNCATION & the upper shores may correspond with some
547m111 g thought-- my servant was in the room. with my TRUNK out & I was engaged in hurriedly giving orders.
312c233 & child are thought long breeding in. Must not TRUST him Hope says that genus of parasite to genus o
360d089 Individual instances trouble Yarrels law. chief TRUST must be in general knowledge of breeders, where
132a139 at subterranean springs give result less to be TRUSTED than any others-- may not the cold «bottom of»
241c016 o see if Lessons' remarks on the Floras can be TRUSTED The changes in species must be very slow, owin
385d165 he mother, or sometimes of both." If L. can be TRUSTED, this is Lord. Moretons law.-- "How often do w
424e101 dence of fossil remains, & therefore not to be TRUSTED.-- -- Lyell tells me, on authority of Beck, th
035r046 little ground for skepticism, as to the general TRUTH of the proposition."-- If such can happen in tr
052r099 e terraces in Valleys of Chili may be with much TRUTH compared to the step = formed streams of lava a
186b057 esembles each other, (at least in one point, in TRUTH in all excepting specific character); and in pa
199b116 ise not so numerous: quote from Lyell: assuming TRUTH of quadrupeds being created on small spots of l
276c123 ey believe have endeavoured to advance cause of TRUTH It is of the utmost importance to show that hab
282c142 ground cuckoos. would not be affinity, but the TRUTH would never be discovered When one reads in Ehr
608o027 use no one can be really fully convinced of its TRUTH. except man who has thought very much, & he wil
622o049 ceived any benefit.-- Yet I think there is much TRUTH in doctrine, for [RHC] 9) We can thus explain l
035r048 cement, consequent of injection of fluid rock.-- TRY on globe. with slip paper a gradually curved enl
206b148 <& other tribes>.-- &c &c Exclude mothers & then TRY this as simile In a decreasing population at any
250c037 misphere fully characterized, of each continent. TRY amongst Europaean quadrupeds if Africa destroyed
298c190 estion.-- [+] This might be mentioned in note.-- TRY to trace from simplest reasoning in lower animal
376d137 he saw knew women.-- he has repeatedly seen them TRY to pull up petticoats., & if woman not afraid cl
455eIBC on sensitive plants; Physiology Get Habberley to TRY experiments. about raising plants. where they ca
472s01v ent three years since gave her some She means to TRY this year. Little variation in the 60 one bright
491q01v 1 Repeat the French experiment of Carrot 2 {also TRY Primrose & Cowslip in rich soil & propagate from
491q01v scarlet Lychnis can be propagated by cuttings.-- TRY.-- Important as discovering function of seeds--
496q006 tendency to double; what would be effect-- (10) TRY in how many generations. daisy. Fever-fuge Groun
496q006 ere is a double Crows-foot. or Ranunculus.= (11) TRY.. Nitrate of Soda-- Salt. Gypsum. Magnesium Iron
498q08v pecies growing White Mullein good plant to sow & TRY to get other species <near> close to each other.
504q014 ne remain to be compared-- Cabbages.-- kept true TRY experiment (30/p.11) (2) Yew Berries germinate?-
505q014 that some Bees are smaller & more vicious. Will TRY to get me some to look at:-- Was once offered a
549m120 of daily <happiness> «pleasure». A wise man will TRY to obtain this happiness. though he sees some «i
557m149 lousy probably originally entirely sexual; first TRY «to» attract female, (or object of attachment) &
587n089 f Association. owing to time when entered brain, TRY contiguity of parts of Brain.-- Mackintosh first
608o026 come into play.-- †It may be urged how often one TRY to persuade person to change line of conduct. as
620o046 dog really feels ashamed?) not so puppy, we <do> TRY to teach him & strengthen his instincts.-- so ma
629o54v atification,-- see what cases Mackintosh gives & TRY it.-- p. 241 (1) Any action by habit may be thou
228b230 ar or mere mongrels.-- It really would be worth TRYING to isolate some plants, under glass bells & se
375d135 there is a force like a hundred thousand wedges TRYING force <into> every kind of adapted structure i
541m089 risen? 19th. When I went down to Woollich I was TRYING to unbend my mind as much as possible (testing
581n063 even when you cannot remember it. as my father TRYING to remember the man's Christian name, writing
620o045 s-- + us the action was superfluous, as one man TRYING to save another in desperation.-- This shows,
638j059 nervous system is endowed with the knowledge of TRYING a hundred schemes of structure, in the course
565n007 n to hear well «as one will perceive if in night TRYS to listen to growl of hounds». <when> as fear t
270c103 imals than in vegetables. p. 243 radiate animals <TU> plants turned inside out. have position of orga
212b165 nal bark. would remain for long time together in TUB of water with only nose projecting.-- would pull
233b250 times heterogenous.-- Australian dog jumped into TUB leaving only nose above it-- pulled bell.-- -- I
314c237 d me Bauer's drawings of a curious plant where a TUBE consisting of pistils & stamens united into lon
502q11v oss. it is not possible to see orifice of poison-TUBE-- so put carmine in spirits & then experimentis
501q011 ual or Biennial be grafted or cuttings taken or TUBER-- talk about Mr Knights theory with Henslow.--
455eIBC es of children CD[Does any annual give buds, or TUBERS. Yes-- but these are same as trees.-- Shake so
435e133 ery different) that the granules exserted their TUBES: now Mr Herbert has shown that stigma swells, w
107a079 better conductor) Fitz Roy's Case of S. Maria & TUBUL applicable to Andes & Patagonia-- On Lyells ide
480z006 of Zoolog Soc. Important account of habits of TUBULARIA. p 52. May 1836 dimensions of immense Tortois
484z014 e & throw head back M Edwards,--on polypi of TUBULIPORES L'Institut-- 1838 p. 75 A detailed compariso
070r155 ordillera extend to near Salta. & not far from TUCAMA[N]. & at Chuquisaca. half across the continent.
317c249 s its offspring short tails /one born at Maer. TUCKEYS voyage-- p. 36 "Cercopithecus saboeus" said to
317c251 ng, with respect to forms.-- Study Appendix to TUCKEY'S Expedition Journal of the Academy of Natural
321c275 The British Aviary.--do--- & Lisle's Husbandry TUCKEYS voyage reread Appendix Ovington Voyage to Suri
071r157 Sir Woodbine Parish informs me that town near TUCUMAN and Salta. towards the Vermejo was utterly ove
481z010 pones-- says «the Condor» <it> is found in the TUCUMAN mountains The fourth Vol. «in Lyell's possessi
152g049 upper edge of which they cut near Loch Tring-- TUESDAY Bridge of Roy Level of «bed of» River 30.221/6

044r074 olution must be great: & in the fact of bombs in TUFA there is proof of such gaz) steam condensed.--P
048r086 tance triturated Porphyritic rock. s (mem white TUFAS with purple Claystones of P. Desire). = Where t
383d160 ver male-- Head like silver except in not having TUFT,-- back like do.-- but the black lines on each
507q15v t.= Any bored with drawing of Seed. Anemone with, TUFT-- Bull Rush-- Dandelion-- Sycamore. & seeds wit
440e146 rowing like Nepenthes.-- cases of pidgeons with TUFTS &c &c here there is no final cause yet it must
446e162 uring several years & then they break-- -- each TULIP is the <of> product of fresh bud-- here then is
442e151 mation «My theory will not admit this, now that TULIPS break by cultivation can a form become permane
446e162 n in its character.?» No--well characterized.-- TULIPS are cultivated during several years & then the
147g019 degree? Friday. Inverorum about 20 ft above Loch TULLA 29.804 Temp 62 degree Air 60 degree Below Loch
147g020 29.804 Temp 62 degree Air 60 degree Below Loch TULLA whole wide valley scattered with few very small
127a126 t bear on central fluidity.-- do p. 137. Lord TULLAMORE found Sulph of Soda in peat ashes in Ireland
462tf4v woodlouse-- -- a good analogy-- sea-Crustacea-- TULLUS. Athenaeum 1839 p. 772-- A curious theoretical
433e127 nly leaves out of whole volumes.-- The fact of TUMBLING pidgeons; flying high all together & then tum
433e127 ling pidgeons; flying high all together & then the TUMBLING, far more wonderful than heredetary ambling h
494q004 e of fox. otter. Badger &c &c &c.-- (11) Keep. TUMBLING pigeons. cross them with other breed.-- (12)
513q020 ck in very young state for skeleton-- Does the TUMBLING of pigeons vary in manner & perfection &c &c
105a070 istribution of blocks, that there has been no TUMULTUOUS rush.-- besides general improbability. strat
521m007 mentioned in the Zoonomia.-- How if memory «of a TUNE & words» can thus lie dormant, during a whole l
521m008 e of the muscles.???> Miss Cogan's memory of the TUNE, might be compared to birds singing, or some in
589n091 rain forming the instincts,-- could brain make a TUNE on the pianoforte, yes if every individual play
146g011 mous mass thunder storm, many <hundred> thousand TUNS. Black faced sheep, sometimes mottled with whit
051r093 ahia ferruginous.--At Pernambuco (great swell & TURBID water) organic bodies protect like peat reef o
509q017 ?? do. Wing in Apteryx ⫼ no as Os Coccygis-- TURBINATED bones? False ribs Wings of Apterix: clavicle
328cIBC ells of Genera Corlula Cham. Cardium. Porcellus TURBO. Cerithium Jardin du Roi Java fossils at same t
374d134 er Silurian-- several existing genera. Nautilus TURBO. buccinum. turritela. terebratura, orbiculas,
093a033 form salts of America. -- The number of minute TURBO in red earth with volutas. prove regular mud b
105a069 esides more surface exposed. bay more open to TURBULENCE. Bull. Soc. Geolog «1837» p. 320. paper on s
051r094 ed with living beings; In smooth seas (& even TURBULENT as at St Helena) I have mentioned point of gr
219b193 --floating trees Thrushes & bunting & coots-- «(TURDUS Guyanensis?) (Emberiza Brasiliensis?) (Fulica
271c105 upon life. Namely Carus.-- How remarkable that TURDUS Magellanicus. in the. S. Hemisphere. (replaced
271c105 should so strictly <f> agree in habits with the TURDUS Musicus «not found in N. America» whose Southe
480z008 s Vol XII. p 496. Birds at Tristan d'Acunha.-- (TURDUS Guayanensis?)) Emberiza Brasiliensis (?)) Fuli
482z011 ouette a huppe courte talks of nine terrestrial TURDUS falklandii & then 9. passeres! Says the thrush
482z017 not being replaced by Brazilian Species.-- Mem TURDUS Magellanicus.-- C, <Chingolo> Chimango- Diuca
160g090 A.75 degree 75 degree? Boulder, much covered by TURF 2ft. 8- long of syenite with pinkish felspar;--
502q11v nt: said not to sit on own eggs Flowers in short TURF. for abortion. or for sterility Land Birds Made
612o035 . was direct «physical» effects of mere or less TURGID vessels; effect of heat, light or shade.) Join
470t176 eem slow= Maer 1840 My Father formerly planted TURKEY or» Palmated and English, planted within few y
523m018 e of Shrewsbury gentleman, unnatural union with TURKEY cock,-- was restrained by remonstrances on him
542m093 , his step will grow erect & stiff like that of TURKEY.-- he may be amused, he need not express it, h
557m147 cies of swans, in parrots &c &c -- -- peacock & TURKEY cock in passion.-- Cat when pleased, erect its
567n014 ce-- plant though it moves doubtless has not.-- TURKEY cock in passion & sends blood to its breast &c
587n087 even taste of senna. might be acquired. as the TURKS have of Rhubarb: again on other hand, it is sai
022r007 7 Veins of quartz exceedingly rare Mem C. [Cape] TURN P. 434 & 419 As Limestone passes into schist sc
035r045 to preservation of bones &c &c--Yet <silicified> TURN over silicified wood. Cordilleras, Chiloe. &c s
044r072 lia.. Oceanic Isles. Geology of whole world will TURN out simple.-- Fortunate for this science. that
283c145 & fortify their minds with such sentences as "oh TURN a Buccinum into a Tiger."-- but perhaps I feel
283c145 the impossibility of this more than anyone.-- no TURN the Zebra into the Quagga.-- «let them be wild
310c223 Dr Royle seems to think Botanical Provinces will TURN out not nearly so confined as now thought.-- N.
332d004 . Eyton, a stone blind horse, seemed to perceive TURN on road where No houses to Eaton Mascott, where
332d004 o Eaton Mascott, where he had been accustomed to TURN down.-- -- applicable to birds migrations & Aus
410e052 auses of change".-- A philosopher, would as soon TURN tailor, as, mere describer of species, from its
468t112 } as stamens grow old «& shed some pollen». they TURN upwards & bend over stigma:-- but stigma «is» a
536m071 y do bulls & horses, animals of different orders TURN up their nostrils when excited by love? Stallio
538m078 . show what trains of thought depend on state of TURN In drunkedness same disposition recurs, such as
545m105 f forehead over eyes, at every emotion & <look> «TURN» of the head. I could not perceive «any» distin
573n037 t child of H. W. constantly. when refusing food, TURN his head first to one side & then to other. & h
573n038 flow & desire to swallow.-- tells himself not to TURN in bed. will turn in bed.-- in case spittle, ef
573n038 wallow.-- tells himself not to turn in bed. will TURN in bed.-- in case spittle, effect of thought is
586n082 others in building comb-- My faculty often will TURN out to be instincts, & so in some senses, is si
615o038 st ages; seems to indicate that when we <return> TURN into angels. this imperfect memory may become p
088a013 t several attempts at elevation From the lost & TURNED about position of strata, prooff thickness not
088a013 a, prooff thickness not very great; where piece TURNED over axis or hinge no doubt fluid.-- analogy &
131a136 son.-- From strata being not only vertical, but TURNED over in many parts of the world.-- argument st
204b141 «W. Fox. knew of case of male widgeon, winged & TURNED on pool, first season bred readily with common
264c080 of course they might be blended, if archipelago TURNED into continent &c &.-- There is beautiful grad
270c103 vegetables. p. 243 radiate animals <tu> plants TURNED inside out. have position of organs of generat
371d118 several fully developed tails, & one with beak TURNED up like Avocette. here is what [not located] t
377d139 rge them to these actions. «These facts may, be TURNED to ridicule, or may be thought disgusting, but
468t112 traightened pollen profusely shed; lengthened & TURNED up «more than stamens», so that all were brush
547m112 odily sensation & yet the Castle would not have TURNED into dream.-- It appears to me, that the mind
596nIBC ll into convulsions A carrier pidgeon carried & TURNED round & round in fainting state would it then
109a084 a. formed by Ca. of L. & Mur. of Soda mixed.-- TURNER'S Chemistry p. 206 Both Beck & Deshayes saw fos
326c266 . otherwise old whores would not have children TURNERS embassy to Thibet, perhaps worth reading quote
531m052 violent attack,-- Even the worm when trod upon TURNETH,-- here probably there is no feeling of passion
113a092 part of Australia.-- Probably a case of rivers TURNING round & penetrating their own range in Austral
307c216 ale», it will be splendid argument old female, TURNING into cock, abortive spurs. growing.-- Are ther
309c221 orm knowing female good case of instinct. bees TURNING neuter into Queen, more wonderful case Dwights
318c252 tation.-- (case of Squirrel from extreme north TURNING white like Hares??--) I never saw more beautif
466t091 ha Palustris alone together. one had seed-pods TURNING brown, whilst both others were in nearly full
506q15v 5. Charlsworth. vol II. p. 670-- oats cut down TURNING into Rye.-- 46). Book describing amount of Hor
512q019 st. would its offspring have ugly calves. also TURNING into fine beasts.-- For comparison with hybrid
556m146 hen horses fighting, they put down ears, when «TURNING round to kick» kicking they do the same. altho
593n105 & anything disagreeable being pursued.-- A dog TURNING round & round is some old instinct <perverted>
423e098 auliflower &c.-- Uncle John believes one single TURNIP in a garden is sufficient to spoil a bed of Cau
401e014 st care be taken to prevent fertilization from TURNIPS or other stocks. Says if any variety of apple
161g097 m. 29.264 A 82 75 degree? This last measurement TURNS out too low, (NB .260 would have been more corr
225b217 of which we are as ignorant. as why millet seed TURNS a Bullfinch black, or iodine on glands of throa
429e115 ge.-- but there is a contest. & a grain of sand TURNS the balance.-- Hort. Transact Vol I. M. Ramond.
512q019 n of coots-- variation in hounds= An ugly calf <TURNS> sometimes tuurns into fine beast. would its off
512q019 ation in hounds= An ugly calf <turns> sometimes TURNS into fine beast. would its offspring have ugly
439e145 l breeds of dogs.-- like greyhound-- fox-dog-- TURNSPIT & two other kinds It seems absurd proposition
447e164 wants insects to impregnate allied to Asclepias TURPIN cell is individual May 29th.-- -- Henslow says
448e165 out every part of plant producing buds, so that TURPIN says each cell of plant is individual.-- Most
310c224 rts of world Athenaeum June 3d 1838. quotes M. TURPINS assertion that globules of milk produce a plan
165g125 merely thoughts laying dormant-- Man from Glen TURRET said he learnt to know the lambs because most
345d043 n american form (if so)?.-- A Sphepherd of Glen TURRET. said he learnt to know lambs, because in thei
155g062 Shelf 3d dies away almost imperceptibly on Glen TURRIT side 2nd shelf very broad «& cut out, produced
155g062 e way up, but most faintly on East side of Glen TURRIT, where I believe they end in upwards inclined
159g085 ofoundly Boulder hypothesis Thursday, from Glen TURRIT to Fort Augustus Barom on upper (rather above)
374d134 al existing genera. Nautilus turbo. buccinum. TURRITELLA. terebratura, orbiculas, with many extinct f
037r055 Tortoise> «remains of Amphibia, exclusively.» TURTLE bones. & the bones of <two graniniverous> a he
261c070 t to laws of change of organization! The little TURTLE, without its parent running to the water, is a
437e138 s & chaffinches sometimes migratory.-- p. 103. TURTLES finding their way to the Caymans from Honduras
584n071 use of senses.-- knowledge of location ducks & TURTLES running to water.-- young crocodile snapping--
307c216 species to genera & classes. p. 479. fragment of TUSK «& Molar tooth» of Hippotamus from Madagascar!!
387d169 C. & theirs again in lesser degree.--now if the <TW> second race both have this peculiarity strongly
180b037 t (I)» is capable of making, 13 recent forms.-- TWELVE of the contemporarys must have left no offspri

(Key Word)
025r016 ast to Olinda [8 degree S.] 9-10 = 30-40 {T} at TWICE or 18-20 <60>--80 120 parallel of Olinda Shoale
231b243 escendants left in cooling climate might change TWICE over, whereas those which migrated a little to
360d089 ee little ones.-- Keeper said in <two> crosses «TWICE made» between terrier & hairless dogs of Africa
520m002 k with this fact.-- the resemblance was in odd TWICHING of muscles, & general manner of holding hands
495q05a aces of Cabbages near each other-- & enclose one TWIG of each in bell-glass-- sow these seeds & see i
281c139 true as avoided linear arrangement the central TWIGS dying, affinities would be <circular>, in broke
372d131 the skin grows over a wound.-- Does likeness of TWIN bear on this subject? A mans arm would produce
391d175 trees generally dioecious oldest forms) why are TWIN in man more like «each other» than twins «or tr
335d014 what these monsters are:-- are they «abortive» TWINS.-- ¶ The fertility of first cross, as stated by
379d152 iorate.-- Lord Moreton's Case.-- When cows have TWINS, <one> though capable of producing both <male>
391d175 why are twin in man more like «each other» than TWINS «or triplets &c &c in» in litter. Why is there
391d175 in litter. Why is there some law about sexes of TWINS in former case.-- (many monster are really twin
391d175 wins in former case.-- (many monster are really TWINS.)-- It is absolutely necessary that some «but n
541m092 n play has his mouth open ready to bark, & lip TWISTED up, in that peculiar manner they do, even more
524m020 y be considered as truly spritual.-- a person TWITCHING when a disagreeable thought occurs, is closel
524m020 f pain, emits its power on the muscles in the TWITCHING. Pride & suspicion are qualities, which my F
578n057 nvulsive wrinkling of some of the muscles «or TWITCHING».-- But why does joy & OTHER EMOTION make gro
595n117 g «whilst» dreaming, growling. & yelpings. «& TWITCHING paws» which they only do when <great> conside
091a026 he Mediterranean 700 feet deep in some of. the TWOPENNY periodical said so. «Campbell the Poet» Accra
503q012 Mr Riley to experimentise on hybridising ferns, TYNDRUM them back to back 37 Col. Sykes fertility of me
146g014 point could be followed up to neighbourhood of TYNDRUM where a large sort of <plain> space is thickly
147g019 in present position Thursday Evening ¼ past 8 TYNDRUM 29.<625> <636> Temp. 62 Friday morning ¼ past
173b011 rval arrived) might have grown altered Hence the TYPE would be of the continent though species all di
174b014 es. Propagation explains why modern animals same TYPE as extinct which is law almost proved.-- We can
174b015 separated from immens ages possibly two distinct TYPE, but each having its representatives-- «as in A
175b020 s & sloths as all offsprings of some still older TYPE some of the branches dying out.-- with this ten
178b030 ranches being dead from which they bifurcated.-- TYPE of Eocene with respect to Miocene of Europe? Lo
179b034 rding to all common language) will keep to their TYPE: in animals so far removed, with instinct in li
181b039 many» contemporary, would have left scarcely any TYPE of their existence in the present world.-- or w
181b040 far back to source of the Mammalian TYPE of organization; it is extremely improbable tha
181b044 aly & <the g> bearing stamp of <some> great main TYPE, & the gradation will be sudden-- Heaven know w
182b046 any in the Cryptogamic flora but not atmospheric TYPE. Hence probably only four, is not this Fries ru
183b052 conds. dès les premières générations" go back to TYPE of either animal when crossed with it.-- ¿Wheth
186b059 ll intermediate Lyell Vol III. p. 379. Mammalian TYPE of organization same from one period to another
186b060 ss so in Miocen & so on.-- As I have traced the <TYPE> great quadrupeds to Siberia; we must look to t
186b060 pe> great quadrupeds to Siberia; we must look to TYPE of organization.-- extinct species of that coun
190b075 ess would when the instincts were.-- relation of TYPE in two countries direct relation to facilities
195b097 w exist such strange forms as ornithorhyncus The TYPE of organization constant in the shells.-- The q
195b099 e same remarks applicable to fossil animals same TYPE, armadillo like covering created.-- passage for
199b116 eing created on small spots of land, of the same TYPE with the great continents we get a means of kno
202b133 ster of Paris.-- Now this is exception to law of TYPE. like horse in S. America. or like living Edent
208b154 son in Report about each genus having its parent TYPE in hotter parts of world Is monkey. peculiar to
209b156 xact proportion to distance (?). & similarity of TYPE (?).-- [«Mem:» Juan Fernandez]CD. From study of
222b206 improvement in system of articulation. ¿whether TYPE of each order may not be supposed that form, wh
236b278 not per saltum.-- Isld bordering continents same TYPE collect cases.-- African isld.-- «How is Juan F
256c055 rtiary formation amongst Graecian isles, ¿See if TYPE continued?-- See to Boblaye & Virlet.-- Whewell
263c076 .-- look abroad, study gradation. study unity of TYPE-- Study geographical distribution study relatio
266c088 ntries.-- Gould insist much upon knowing to what TYPE a bird belongs.-- I conceive without knowing fr
268c100 t class of animals having its aquatic, aerial &c TYPE?-- This of consequence, because applicable to N
274c113 nt habits, though preeminently belonging to this TYPE, ¿the Humming bird? the woodpeckers Gould says,
274c113 believes does. but also on fruit.-- The Rasorial TYPE is wonderfully shown in long legged cuckoos wit
274c114 ry bird.-- & Millisuga. Kingii very rasorial for TYPE.-- Now here I must observe these characters var
283c143 ps. where you have representations.-- The aerial TYPE in each family is relation to elements & not ha
352d058 ecome EXTINCT, & hence the IMPROVEMENTS of every TYPE of organization. such law would explain every t
360d094 would be different.-- Or if one species left its TYPE in having very long legs, & another in having v
370d115 e cause of N. Zealand not having any Mammalia.-- TYPE of geographical organization. no more can be sa
378d147 ange, & is the offspring brought back to earlier TYPE by Mother?-- do these differences indicate, spe
409e047 eys:-- Having proved mens & brutes bodies on one TYPE: almost superfluous to consider minds.-- as dif
436e137 c. The greatest difficulty to my theory, is same TYPE of shells in oldest formations:-- The Cambrian
437e141 4th.-- The Brussels Sprout returning suddenly to TYPE when brought back to home. (& yet all the varie
439e143 it determined by a facility in returning to old TYPE Mr Herbert showing the extreme facility of cros
449e170 imals of» islands N. of Timor are allied to the «TYPE of genera in» islas de Sonda as well by those w
566n010 y indicate mind & body retrograding to ancestral TYPE of consciousness &c &c.-- Lavater. (Holcroft Tr
640j28v per hand. though continually dragged back to old TYPE by intermarrying with ordinary race.-- «There i
175b019 uccessive animal is branching upwards different TYPES of organization improving as Owen says simplest
180b037 would be formed.-- bearing relation to ancient TYPES.-- with several extinct forms, for if each spec
197b112 ade for itself does not agree with old & modern TYPES being constant. Cuvier's theory of Conditions o
199b117 the thousand of new insects all belong to same TYPES already established. why out of the thousands o
205b142 te information) West of Rocky Mountains Asiatic TYPES discoverable.-- Bridgewater Treatise p 85. Para
205b142 t. Humboldt? Vol V. P II. p 565. Consult-- Says TYPES most subject to vary where intermixture preclud
236b278 isld.-- «How is Juan Fernandez-- Humming Birds» TYPES of former dogs. character of Miocene Mammalia--
250c039 estruction of large quadrupeds.-- common to two TYPES of animals What reptiles coexisted with Palaeot
258c062 developed & then another-- (according as parent TYPES are present) What must follow after there is proof o
294c177 killed in year 1000. reference to succession of TYPES ¿different species;-- Horse-- &c <Lonsdale says
304c206 to each other-- Different classes Keep to their TYPES. with different degrees of closeness. -- look h
310c222 ed thought to <an> so many animals of different TYPES. I will confess my profound ignorance.-- but se
324c268 164. on unfixed forms. Dr. Royle on Himalayan. TYPES. Smellie. Philosophy of Zoology Flemming. ditto
337d021 ock of all animals, but merely to classes where TYPES exist for if so. it will be necessary to show h
411e055 arts & organs it may serve perfectly to specify TYPES, & limits of variation., & hence indicate gaps.
427e108 med. would remain stationary, hence all present TYPES are ancient. According to my views of <Dioeciou
460t019 rld of infinite & growing complexity from a few TYPES, it must not be supposed that this refers to ti
460t019 t if it be asked how this complexity from a few TYPES originated, we must go to the first origin of t
634j55r found in Paris Basin?.-- NB) The explanation of TYPES of structure in classes-- as resulting from the
170b001 new individual passing throug several stages (¿TYPICAL, <of the> or shortened repetition of what the
176b023 r, land & water, & the endeavour of each <one> TYPICAL class to extend his domain into the other doma
197b111 animal» passes from worm to man; «highest» as TYPICAL of <sa>. changes, which can be traced in same
222b206 d least from ancestral form. If so are present TYPICAL species most near in form to ancient; in shell
227b225 lains the blending of two generas-- It explains TYPICAL structure.-- Every species is due to adaptatio
254c048 each class successively present modifications, TYPICAL of succeeding classes & likewise those much hi
254c049 not surprising to find many forms in Acrita,-- TYPICAL of other, (surely rather parents). (NB These v
261c072 adaptation to animals wants & not as change in TYPICAL structure?!! Whewell «in Comment/ few will dis
262c073 ere fact of division of lesser & more power (2. TYPICAL 3.subtypical) where power arbitrary. leaves do
274c115 together, in analogy certain parts perfect of TYPICAL structures certain <imper> parts changed Have
289c160 nstant touch the water.-- capital instances of TYPICAL land bird, having habits of a Grebe, structure
304c208 not isolated, we must suppose the changes from TYPICAL structure have either been more rapid than in
543m098 og & Elephant most intellectual.-- Hymenoptera TYPICAL insects. ie have all parts. Waterhouse Study w
391d174 time» added to that kind of generation, which TYPIFIES the whole course of change from simplest form
273c111 ird there is strongly marked variety,:.in the TYRANNIDAE.-- Milvulus-- Even flying woodpeckers., with
268c096 me district.-- <The same> <Mem> Tennioptera & TYRANNULA (NB work out how many forms Tyrannula can we
268c096 ptera & Tyrannula (NB work out how many forms TYRANNULA can we worked out into. Milvulus [not located
484z015 ntatives of this class--. Pyrocephalus & many TYRANNULAE-- replaces warblers of Europe-- Study profou
276c124 a peculiar structure, then Milvulus forficatus TYRANNUS Sulphureus if compelled solely to fish. struc
203b137 me. Milvulus forficetus <has a wide range> is a TYRANT flycatcher doing the service of a swallow I th
485z016 rather indefinite letter Mem Orpheus»--becoming TYRANT-- flycatcher-- shown by habits & plumage so ve
536m071 nostrils when excited by love? Stallion licking UDDERS of mare strictly analogous to men's affect for
424e100 «fossil» shells from North America, Scotland, UDDEVALLA. Many species same. & Northern forms-- & Amer
512q019 eep migration of coots-- variation in hounds= An UGLY calf <turns> sometimes turns into fine beast. w
512q019 turns into fine beast. would its offspring have UGLY calves. also turning into fine beasts.-- For co

```
****************************************(Key Word)****************************************
528m037 cannot be doubted if we look at buildings, even UGLY ones.-- the pleasure from perspective is derive
568n019 ly» says "he knew no more what was pretty & what UGLY than a cow--" «so it is with all uneducated.--»
571n030 ather altered. The Reason why New Buildings look UGLY is because there is some connection between the
594n113 little disordered) at thought of almost anything UGLY. baby-- association-- pouting child same as ang
595n121 art of same feeling which make us think anything UGLY-- a beau-ideal feeling. Same effect as acting o
603c014 in not explaining the possibility of <handsome> «UGLY healthy» young woman, with good expression-- st
564n003 ions, makes his conscience far more sensitive. ULTIMATE effects of actions.+ till at last he face «in
045r075 rian earthquake things pitched off the ground. «ULLOA states that Volcanos!! were in eruption at time
060r125 t. 1817.--p 371. Webster Antarctic veg:-- Study ULLOA to see if Indian habitation above regions of ve
079r176 [,]galena[,]quartz, Carb. of Lime. accompany.-- ULLOA has said silver in the highest & gold in the lo
479z005 ith teeth {P} p. 140. Flèche of Quoy et Gaimard ULLOA shell fish Purple die Marvellous stories Ulloa'
054r105 f America not reached. Juan. Galapagos. Cocos-- ULLOAS voyage North of Callao, the country, to the di
079r177 ent les loix de la nature & ses phenomenes."-- ULLOA'S Voyage, Shell fish purple die, marevellous sta
479z005 Ulloa shell fish Purple die Marvellous stories ULLOA'S Voyage Vol I, p. 168 Ceratophytes common in No
206b149 wer closely related;∴ (few species of genera) ULTIMATELY few genera (for otherwise the relationship w
571n028 dren acquire it soon. & why do all men. agree ULTIMATELY?-- We acquire many notions unconsciously, on
351d057 man undergoes metamorphosis., heart altered & UMBILICAL cord,-- Broderip alluded to Hunter's views on
106a076 temperature at great depths. All Earthquake UNACCOMPANIED by Volcanos must be sought after proofs of
569n022 -- Expression common to Savage & Frenchman, UNACCOMPANIED by dignity-- "no mon dieu," with a shrug--
138a153 bservaciones sobre El Clima del Lima par Dr. H. UNANUE says he believes the sea has formerly stood th
468t106 aspberry--leeks-- Flowers which thought very UNATTRACTIVE-- Found Rhubarb blossom swarming with small
536m072 ugh from not having pain or pleasure actions UNAVOIDABLE only to be changed by habits). now free wi
612c035 s are common to every animal instinctive and UNAVOIDABLE.-- +Can the word willing be used without con
541m089 h. When I went down to Woollich I was trying to UNBEND my mind as much as possible (testing success b
342d034 s say the throw of any two species crossed is UNCERTAIN-- Yarrell remarks he has somewhere met conjec
345d042 y all thought with pigs &c, that hybrids were UNCERTAIN. Mr Drinkwater thought that a <pure blooded>
393dIBC rd fanciers say the throw of two varieties is UNCERTAIN do they mean they cannot tell first result.,
393dIBC tell first result., or that «hybrid» breed is UNCERTAIN Is there any peculiarity or variation common
366d108 east show repugnance to breeding if instincts UNCHANGED, & if their characteristic qualities were all
264c079 being fit, for such an animal.--man, (rude, UNCIVILIZED man) might not have lived when certain other
184b054 armadilloes might be brother to Megatherium.-- UNCLE now dead. Bulletin Geologique April 1837. p. 21
400e013 » off province of Guadalaxura-- October 11th.-- UNCLE John-- says Decandoelle, distributed seeds of D
400e013 before they began to double.-- at present time UNCLE J. does not suppose one aboriginal variety.-- f
401e014 ll made by fertilizing one plant with another-- UNCLE John says he has no doubt bees fertilize enormo
401e017 o old stock, but not primrose producing cowslip UNCLE J. says common belief. that female plant impres
423e098 tionary or more probably increases.-- Jan 29th. UNCLE John says he feels sure, that the reason people
423e098 s have whole fields, some for cauliflower &c.-- UNCLE John believes one single turnip in a garden is
366d111 many good species; & therefore to genera (& the UNCLES & aunts) & therefore does not tell against tra
582n067 d to will to perform an instinct that it is UNCOMFORTABLE if it does not do it.-- My theory explains
029r033 lthough it lasted about a minute, there was no UNCOMMON ripple on the water. It was quite calm at the
075r167 «argentiferous lead»; sulfated Barytes very «UN»COMMON in Mexico. Fluor spar only in certain mines.
333d005 e. perhaps rather different».-- crossed with <UN>COMMON cat, exact variety unknown., three kittens, a
582n066 ersation of several people.-- Children have an UNCOMMON pleasure in hiding themselves & skulking abou
531m050 f great importance are easily forgotten, (if UNCONNECTED with fear &c) because people think that the
539m081 I in the other, & read so intently as to be UNCONSCIOUS of all around, yet there was no strain on th
545m104 gust 24th. As some impressions «Hume» become UNCONSCIOUS. so may some ideas.-- ie habits, which must
545m105 t the <becom> impression becoming very often UNCONSCIOUS, which makes the idea unconscious, if so (th
545m105 very often unconscious, which makes the idea UNCONSCIOUS, if so (think of this). study what impressio
545m105 hink of this). study what impressions become UNCONSCIOUS those which are viewed with little interest,
582m067 e must, however, be a mental impulse (though UNCONSCIOUS of it) to move its legs so, as much as in th
610o031 n name was Wilson!!-- How curious an inward. UNCONSCIOUS memory.-- Jan 14th. 1839.-- My father says h
308c217 g tea in pot made me go to tea chest almost UNCONSCIOUSLY.-- why do absent «Dr. Black. tea & sugar» p
521m007 ie dormant, during a whole life time, quite UNCONSCIOUSLY of it, surely memory from one generation to
536m070 orts prevent anger, but observing eyes thus UNCONSCIOUSLY discover struggle of feeling.-- It is as mu
538m080 dividuality implied by habit, when one acts UNCONSCIOUSLY with respect to more energetic self, & like
546m110 , showing what trains of action may be done UNCONSCIOUSLY as far as the ordinary state is concerned.?
571n028 agree ultimately?-- We acquire many notions UNCONSCIOUSLY, without abstracting them & reasoning on th
582n067 the sea. or down the stream; which it does UNCONSCIOUSLY of any end.-- N B. There is wide difference
100a054 hollow concretions & concretion filled with UNCONSOLIDATED matter-- Phillips Lardner p. 197. refers t
315c243 capital argument if man does move muscles for UNCOVERING canines) --. Blend this argument with his ha
432e125 yth's doctrine of young birds retrogressing-- UNCOVERING the canine teeth, or sneering, has no more r
542m096 - <&> therefore it is here continued when the UNCOVERING the canine useless.-- The distinction «as of
545m106 t like pish, but like chit-chit-chit, quickly UNCOVERING their teeth, this the Keeper thinks is from
565n006 learly analogous to a panther I saw in garden UNCOVERING its teeth to bite.-- the senseless grin of p
572n032 er by bowing the body, kneeling, prostration «UNCOVERING body» &c &c is matter of custom."-- this all
232b245 he <one> species which survives any change may UNDERGO indefinite change., (marking in their history
443e154 anic beings propagated by gemmation do not now UNDERGO metamorphoses, but to arrive at their present
461t025 ich are marsupial & the other have young which UNDERGO metamorphosis & are provided with fins, & henc
461t041 rmous changes of conditions, some species will UNDERGO & yet remain adapted.-- it does away with diff
635j56r of the world? & the physical changes it was to UNDERGO «animals feeding on each other &c &c».-- «(Cau
335d015 ocks «relative to changes which every species UNDERGOES») & hybrids between very near species (that i
351d057 um-- yet does nothing in vain!! Foetus of man UNDERGOES metamorphosis., heart altered & umbilical cor
534m063 hus Dahlias by snails)-- <The> Apion radiolum UNDERGOES transformation in the stem of Hollyhock, alth
606o021 erceive the various operations which the mind UNDERGOES in gaining the result.-- Lessings Laocoon. 2d
354d062 is not more wonderful than the body of a man UNDERGOING a constant round,--each particle is placed i
021rIFC ms having definite existence, those that have UNDERGONE the greatest number of changes towards perfec
022r006 dote seems commonly to occur where rocks have UNDERGONE action of heat. it is so found in Anglesea, a
040r063 ised, as in Brazil ferruginous sandy ones have UNDERGONE the same process.-- Neither lakes or Avalanch
065r139 consigned to the earth. yet body had scarcely UNDERGONE any decomposition: countenance so well preser
221b200 Plants Composites.--!!«good» those which have UNDERGONE most metamorphosis Islds X Is this applicable
296c184 her original. owing to being oldest. & having UNDERGONE changes «no near lofty country»?? p. 475 NB.
372d130 new individual), or he may produced by having UNDERGONE the endless changes, which its parents have.
407e038 ts of blocks both North & South, has probably UNDERGONE a greater change, than any part, (except Euro
459tflv & has been occupied [...] species, which has UNDERGONE all the changes. [im]portant view, copie[d] G
047r082 vement.--The swelling first on beach I connot UNDERSTAND, without (cs <[...]> raised above as). -- In
048r084 les. (On a sea beach under a cascade, one can UNDERSTAND pebbles thus coated.--The motion is most won
132a138 does not. (see resumé p. 536)-- «NB. I cannot UNDERSTAND the argument, that cold <oceans> <lakes> bot
186b058 n ancestor of Crysom. & Heterom, but I cannot UNDERSTAND the universality of such law.-- It would be
198b113 ch on the confines? Balanidae?-- --- I cannot UNDERSTAND whether. G. H. thinks developent in quite st
199b116 et a means of knowing movements.-- How can we UNDERSTAND excepting by propagation that out of the tho
210b158 neriffe, the proportion of genera I: I. I can UNDERSTAND in «one» small island species would not be m
263c074 al being though not MAN.-- is as difficult to UNDERSTAND as Lyells doctrine of slow movements &c &c.
286c154 abbit stamping ground) Man signals.-- animals UNDERSTAND the language, they know the crys of pain, as
293c172 t coat & this he put on & we afterwards could UNDERSTAND «(language better instance)» he had done thi
315c241 & Sandwich species & some of Japan. I do not UNDERSTAND any new ones.-- Menoir will be published St.
338d023 e> inter-breeding, for otherwise we could not UNDERSTAND the vast number of domesticated races.-- Ath
361d095 e resembled that of Cross-Beak-- In lark if I UNDERSTAND right, all species have same character which
361d095 ale similar plumage.-- in tree sparrow, (if I UNDERSTAND rightly) young cock & hen, all nearly simila
369d115 ly their place in nature, as well as fully to UNDERSTAND their oeconomy, is now universally admitted.
408e042 rnal. for those of Cuba.-- It is important to UNDERSTAND well the relation of passage from N. to S. A
417e077 body.-- or in two bodies, <th> we can as well UNDERSTAND the necessity of a relation between the flui
418e080 emi-hermaphrodite insects is it not easier to UNDERSTAND ;perfect?? developement of one sex on one si
510q017 in-inhabited spaces: has he published? does he UNDERSTAND English.-- Miguel to collect facts for me--
521m009 - Thus when dinner was announced he could not UNDERSTAND it, but the watch was <seen> shown him.-- «<
533m058 hink I have seen same thing before they could UNDERSTAND. what frowning means) if so this is precisel
542m097 cows, & many of insects-- they likewise must UNDERSTAND each other expressions, sounds, & signal mov
542m097 sounds, & signal movements.-- some say dogs UNDERSTAND expression of man's face.-- <That> How far t
551m128 iples talks of it as wonderful that Elephants UNDERSTAND contracts.-- but W. Fox's dog that shut the
```

Page **(Key Word)**
553m138 ous. as they depart in structure» The monkeys UNDERSTAND the affinities of man, better than the boast
554m139 f & liked the taste of Peppermint.--» Perfect UNDERSTAND voice.-- will do anything.-- will take & giv
567n013 e» when does such notion commence?-- Children UNDERSTAND before they can talk, so do many animals.--
580n062 -- says animals have abstraction because they UNDERSTAND signs.-- very profound.-- concludes that dif
581n065 ure, or pains long before they can speak-- or UNDERSTAND-- thinks so it must have been in the dawn of
617o39v ly comprehended by anyone who wishes to fully UNDERSTAND this subject, but the answer to it would req
641j29r decide their limits.-- To show how little we UNDERSTAND of the Physiological relations of animals. e
264c079 e to boast of his proud preeminence.-- «not UNDERSTANDING language of Fuegian, puts on par with Monke
325c267 strib: of British Plants. Humes Essay on H. UNDERSTANDING (sometime) Du Stewart works. & lives of Rei
411e054 of variation of races, may be important in UNDERSTANDING laws of specific change‖.-- When the laws o
526m029 th my Fathers case of Mr Corbet of the Hall UNDERSTANDING. (on hearing old association brought up) by
543m097 not easy to know,-- but this capability of UNDERSTANDING language is considerable, thus carthorse &
545m108 as in blindest memory-- also low faculty of UNDERSTANDING. Adam Smith (.D. Stewart life of. p. 27), s
559m155 y curious book.-- Hume's essay on the Human UNDERSTANDING well worth reading Copied <Smith> «D. Stewa
569n020 ike tastes of mouth & smell. Descent of Man UNDERSTANDING languages seem simplits case of Association
570n023 ion, without violence, without assigning or UNDERSTANDING reason.-- surprise with negation.-- like sh
539m084 proved.-- Metaphysic must flourish.-- He who UNDERSTANDS baboon <will> would do more towards metaphys
264c079 e, see its intelligence when spoken; as if it UNDERSTOOD every word said-- see its affection.-- to th
522m012 en accompanied by extreme anger, at not being UNDERSTOOD.-- My F. says there is perfect gradation bet
524m020 considering, if pride & suspicion can be well UNDERSTOOD. In insanity, the ideas do not go back to ch
047r083 rom local form of coast (as seen in swell) the UNDERTOW & overfall must vary proportionally Partial s
102a059 es would be lifted up & down. on coast itself, UNDERTOW would draw it outwards.-- form of breaker aff
281c138 eness, which as in absolute human family is UNDESCRIBABLE, yet holds good, so does it in real classif
078r174 rbigny has figured animal with setae like my UNDESCRIBED[.] p. 140. Flèche of Quoy et Gaimard.--D'Orb
207b150 soria found of unknown forms, a circumstance UNDISCOVERED by Ehrenbergh.-- <Marcel Serres p. 331. L'I
511q018 ?? (8) Get Sir. R. Heron to give me Pigs foot UNDIVIDED, & more particulars regarding effects of cros
537m077 answered effects of education, may be opposed UNDOUBTED cases of heredetary pride & in single familie
182b050 ew Zealand Mr Gould says in sub-genera, they UNDOUBTEDLY come from same countries.-- In mundine gener
399e009 crush of population.-- Octob 10th. Saw. two UNDOUBTEDLY rabbits in poulterer shops., of same colour
623o050 ly in «species of» animals, in which case it UNDOUBTEDLY is instinctive. But does not Hartley explana
070r154 allata.-- ¿If crust very thick would there be UNDULATION? would it not be mere vibration? but walls &
070r154 be mere vibration? but walls & feeling shows UNDULATION∴ crust thin.--Concepcion earthquake Draw clo
102a059 ach near the surface. that strata yield.-- In UNDULATION in open ocean. as pebbles would be lifted up
073r161 flexure <fr> in the fragmentary jasper.--do UNDULATIONS (as Hutton says) always come from without.--
120a107 we can trace the outlines of what were fluid UNDULATIONS-- the equal movements of Glen Roy road. (¿ m
625o052 t doing them gives pleasure & being prevented UNEASINESS, & that this is the feeling of right & wrong
568n019 & what ugly than a cow--" «so it is with all UNEDUCATED.--»-- Old man at Cambridge observed the igno
042r068 o elevated strata in eocene lakes of France, & UNEQUAL action of Earthquakes. «on Chili & delta of In
044r073 s of the earth;--Scarcity of Organic remains.--UNEQUAL distribution of Volcanic action, Australia S.
076r169 limates distribution of silver «in veins» very UNEQUAL sometimes disseminated <[...]> sometimes conce
121a112 ict could have been elevated without fissure & UNEQUAL.-- where were cracks?--? How came there ever t
276c125 de, which often enough repeated would cause an UNEQUAL number of vertebrae-- ¿Where two very close sp
283c145 ng been long in blood?-- My theory agrees with UNEQUAL distances between species. some fine & some wi
286c154 of countenance. «[s]hare of sickness,-- death, UNEQUAL life,-- stimulated by same passions-- brought
352d058 all united . (links in circle must be granted UNEQUAL, because fossil) Now what is group without cen
461t041 if so) with those of Spain & such facts-- This UNEQUAL duration is exactly same as some species exten
042r069 . America: in France we have freshwater lakes UNEQUALLY elevated, which movements if present in the A
285c153 Lyell has show such Physical changes will be UNEQUALLY rapid, with respect to their effects The AEgy
299c196 again in Mans mind, in different races, being UNEQUALLY developed.-- ¿is not Elephant intellectually
382d158 , that both are present in every animals, but UNEQUALLY developed.-- surely analogy of molluscs. & ne
271c106 is this halo.-- Continents are not stationary «UNERRING proofs not always continents».-- it is a plas
583n069 ustry of bee &c &c-- p. 15."instincts act with UNERRING precision".-- no p. 17. Contrast the invariab
148g024 Coast of Chile--¿ is not Mica Slate too hard & UNEVEN to be impressed Case of Birch Wood by Inveroru
303c205 place all that ever lived <on> into one list is UNFAIR, [moreover what will become of the future crea
035r045 r surface: Mem Patagonian pebbles beds, most UNFAVOURABLE to preservation of bones &c &c--Yet <silici
051r095 es not decompose. or vice versâ. Clay slates UNFAVOURABLE to attachment of many bodies [blank] Beeche
196b106 in S. America-- it will not do to say period UNFAVOURABLE to large quadrupeds--horse not large-- Ed.
201b126 o or three these form genera-- this is «from UNFAVOURABLE conditions» there are many gaps. & those fo
259c065 he best attempt of nature under certain very UNFAVOURABLE conditions.-- as an adaptation, but adaptat
285c151 decay in that species <shows> is effects of UNFAVOURABLE conditions. (hence rise & depression of imp
343d037 e latter will take place when Conditions are UNFAVOURABLE to numbers of animals. as in changing from
371d128 t of these roses, without circumstances very UNFAVOURABLE, will <deteriorated> continue of same varie
374d134 lly brothers & sisters,‖& therefore somewhat UNFAVOURABLE-- 28th. «I do not doubt, every one till he
404e026 mysterious such is case‖. therefore chance & UNFAVOURABLE conditions to parent may be become favourab
496q05a -- Magnolias. «Azaleas» & plants grown under UNFAVOURABLE circumstances, as Hyacinths in glasses &c &
285c153 fixed, certain physical changes at last become UNFIT, the animal cannot change quick enough & perish
365d106 re probably during known changes climate became UNFIT for. subalpina, or some Northern species, & bei
587n087 disapproval. were not something more than the UNFITNESS of the objects then viewed. to organs adapted
324c268 Lawrence Bory St. Vincent. Vol III p. 164. on UNFIXED forms. Dr. Royle on Himalayan. types. Smellie.
629o055 &c.-- <]> the "secondary passion" of Hutcheson UNFOLDED by D. Hartley.-- Darwin's Abstract of John Ma
576n048 a modification of <intellect> «instinct»-- an UNFOLDING & generalizing of the means by which an insti
532m056 bury) yet for a fortnight continued wretchedly UNHAPPY, constantly whined, would not remain quiet in
548m115 of the Botanical Somnambulist. (if he had been UNHAPPY)-- it is because in this state, the consciousn
549m120 some «intellectual» good men, from insanity &c UNHAPPY-- perhaps not so much as they appear & perhaps
359d086 aid, that breeding in & in tends to produce UNHEALTHINESS,-- «or» to perpetuate some organic differen
365d107 country, (& no monstrosity, or adaptation to UNHEALTHY state of womb).-- One can perceive that Natur
385d163 hybrids & inter se offspring in latter being UNHEALTHY.--» males «bred in & in» never lose passion.
409e048 . -- we see it is not the order in this perfect <UNI> world, either at the present, or many anterior
032r039 es be exposed to it, & the result would [be] a UNIFORM slope to base of cliff (Z). to which point the
156g071 degree? From this point plain appears like one UNIFORM slope slightly bending up each main valley.--
175b019 er similar climates & as far as world has been UNIFORM, at former epoch-- How is this Ehrenberg? ever
199b117 plains this.---- Ancient Flora thought to more UNIFORM than existing. Ed. N. Philos J. p. <191> «p 19
232b246 assume like weather on long average tolerably> UNIFORM).-- Comparing fossils with whole world. would
346d048 imals which breed, in & in are-- colour white, UNIFORM.--crafty, go in file, hide their young., bold.
405e031 k.-- the «Fallow» Deer. which were of a nearly UNIFORM <dusky> blackish brown.-- yet retained a trace
492q003 s exactly alike be interbred will offspring be UNIFORM.-- Mr Ford Has M. Sageret WRITTEN on crossing
223b211 rmediate character of offsprings accounts for UNIFORMITY of species & we Must confess. that we canot
299c193 h Canary Islds. Galapagos.-- Iceland has same UNIFORMITY Primrose & Cowslip, quite wild, but they aff
436e135 f change) which can hardly be believed, then, UNIFORMITY in «geological» formation. intelligible.-- «
491q002 in negative.-- Questions Regarding Plants. 1. UNIFORMITY of hybrid & Mongrel offspring 2. How have la
622o048 s.-- Those instructions, which the child sees UNIFORMLY performed by the teachers & all around him, w
299c195 dioecious plants many of the Female flowers UNIMPREGNATED Babington We see gradation to mans mind in
505q014 , & two had plenty of seed & these Seeds of UNIMPREGNATED Cowcumbers will they seed.?-- (11) Abberley
409e049 rom the Astronomer that the moon probably is UNINHABITED]CD & if my theory be true then the formation
632j53r are so.-- but the coral rock might have been UNINHABITED as the Alpine pinnacles.-- One thing must be
195b099 opagation gives. <ro> hiding place for many UNINTELLIGIBLE structures. it might have been of use in p
592n105 in of expression-- There are some instincts UNINTELLIGIBLE, <both> in the end gained «& therefore the
614o037 ul superadded, so [RHC] 6) thought, however UNINTELLIGIBLE it may be, seems as much function of organ
626o052 y: the remarks about "contact with will" is UNINTELLIGIBLE to me.-- conscience regulates feelings, as
165g126 ao scoticus & Tetrao-- not an American form The UNION of two instincts crossing most remarkable ever
186b059 vism) Probably this is first step in dislike to UNION, offspring not well intermediate Lyell Vol III.
335d015 fferent because not long in blood.= The case of UNION of perfect animals is distinct case,-- gradatio
336d018 imals cross. each sends his own likeness, & the UNION makes hybrid, in fact the parents beget child l
391d174 179, continued from Is a flower bud produced by UNION of two common buds??? Amongst buds each one exa
391d174 ammalia must long have so existed.» with double UNION.-- At present I can only say the whole object b
417e077 ht makes this analogy between grafting & sexual UNION-- Looking at simple generation as being the act
424e101 ting of two regions-- -- are there any cases of UNION of two regions in modern times.-- this would de
523m018 ideas, «Case of Shrewsbury gentleman, unnatural UNION with turkey cock,-- was restrained by remonstra

(Key Word)

568n018 o monkeys howl in harmony-- frogs chirp in do-- UNION of birds voice & taste for singing with Mammali
611o034 y vegetable or animal matter been formed by the UNION of simple non-organic matter, without action of
040r062 w Zealand, which agrees with St Helena in being UNIQUE, yet no quadrupeds. -- Is the white matter ben
618o41v ways went together. & as a thing grew blue it «UNIQUELY» grew heavier yet it could not be said that t
558m151 ore than mere love.-- fear for others acting in UNISON.-- active assistance. &c &c. it comes to Miss
377d140 , & instincts in the animal kingdom.-- It is the UNIT of our calendar.-- epochs & creations, reduce t
080c178 ts, as with man Does Indian rubber & black lead UNITE chemically like grease & mercury [blank] NB. P.
099a051 ratio in proportion to proximity would they not UNITE in B. K.> {P} on the diagonal of BK.-- ░ This i
103a063 izontal elevation excepting fissures from above UNITE with those from below. would always thin out ab
115a096 olarity to atoms in slates that cleave, & which UNITE the homogenious crystals., must aid in adding e
138a176 k] Would rotting wood by yielding Carbonic Acid UNITE with «piece of cabbage» alklali & precipitate s
156g072 helf at head where «granite &» «veined» gneiss <UNITE> «occurs» abundantly with perfectly rounded peb
157g077 ight Hand Cascade has <cut> «where two branches UNITE in upper Glen Roy» very little back from line 2
158g080 head of Glen Roy on same side where two rivers UNITE in Upper Glen Roy great plain about 60 ft benea
200b121 lity to marriage.-- ¿whether those genera which UNITE very different structure as. petrel & alk. do n
248c033 of hybridity, namely the [not located] animals UNITE, all the change that has been accumulated canno
261c071 n wings of Furnarius, Synallaxis &c &c. sure to UNITE the birds into group.-- it is same as Yarrell's
286c156 Linnaean genera.-- Now are the characters which UNITE these of older standing than constant number of
389d176 ther male may have copulated.-- two animals may UNITE & each have offspring by same mother.-- one ani
426e107 perfect series, from physical impossibility to UNITE to perfect prolifickness.-- (<a series might be
027r027 ut 500 Isd. & great banks. effect of Elevation. UNITED service Journal In the Iron sand formation <wo
063r132 ophite producing distinct animals. still partly UNITED. & eggs which become quite separate.--Consider
151g039 on steep earthy slope, two circumstances rarely UNITED.-- die away also, without any cause, must be t
203b135 poses world divided into Zoological provinces-- UNITED-- & now divided again-- Weakest part of theory
225b219 ince days of Didelphis in Stonefield. all lands UNITED (Falkland Fox. ice). . Mauritius what a diffic
292c169 » sexes <being in one normal-- when so the» are UNITED (which probably is first stage) the tendency t
314c237 nt where a tube consisting of pistils & stamens UNITED into long organ, moved on being touched, so as
352d058 hence there is no central radiating point, all UNITED . (links in circle must be granted unequal, be
377d139 s., when frighten depresses them.-- England was UNITED to Continent, when elephants lived. & when pre
422e092 quatic Mammifers> p. 306, the Dugongs cannot be UNITED with true Cetacea or whales.-- but are aquatic
440e147 ence of final cause mammae in man & wings under UNITED elytra The law of <growth> «generation» is onl
466t099 equally abortive as it was in autumn: filaments UNITED in whole length to corolla--anthers minute, di
468t112 over stigma:-- but stigma «is» almost roofed by UNITED filaments.-- This flower hostile to intermarri
502q11v ut Falklands Isds.-- Snipe Migratory-- probably UNITED by Land to S. America (33) Ornithologum common
529m041 g at trees at [i.e., as] great compound animals UNITED by wonderful & mysterious manner.-- There is m
568n017 of its own emotions. (which must be intimately UNITED with reason) it would feel «subsequent» sorrow
601o009 idence. when consciousness is absent) in fibres UNITED with nervous filaments.-- ¿plants? yes by dist
610o034 C] 1) Effects of Life in the abstract is matter UNITED by certain laws different from those., that go
611o034 he individual forms of living beings, matter is UNITED in different modification, peculiarities of ex
612o035 two kinds of life vegetable and animal strictly UNITED? [RHC] It is easy to conceive such movements &
387d168 that in (C) «in lesser degree» -- Then when (C) UNITES with Male (X) «assume that every peculiarity h
609o030 ays we have a moral sense.-- But my view <says> UNITES both «& shows them to be almost identical» + W
610o034 on. [RHC] During growth <extres> tissue <[...]> UNITES matter into certain form; invariable, as long
098a045 TED structure (∴ separation of ingredients) as UNITING with cretionary.-- it may <of> come of use in
260c068 ay Wilson (p. 5). describes many kinds of bird UNITING together in pursuit of Blue-Jay, when <one> bi
303c205 raordinary genus. Mesites bird from Madagascar UNITING pidgeons & gallinaceous birds & parrots.-- leg
607o025 conclusive remarks p. 205 & 206.]CD Motives are UNITS in the universe. [Effect of heredetary constitu
263c076 & falls.-- look abroad, study gradation. study UNITY of type-- Study geographical distribution study
269c103 d of with respect to life generally.-- where <">UNITY constantly develops multiplicity<"> [(his defin
269c103 <"> [(his definition "constant manifestation of UNITY through multiplicity" this unity,-- this distin
270c103 nifestation of unity through multiplicity" this UNITY,-- this distinctness of laws from rest of]CD un
305c211 dless forms of the living beings.-- We see thus UNITY in thinking and acting principle in the various
371d128 facility to bud.-- this must be owing to their UNITY in one stem.-- a bud may be transplanted & carr
292c169 ency is greater in Mammalia, than in shells ¿ UNIVALVES or bivalves.-- Anyman No VI. Magazine of Zool
033r043 glassy scoriae.--could walk round base:--not UNIVERSAL: could not climb up many parts, in James Isd.
537m075 rgues «with examples» very justly there is no UNIVERSAL moral sense.-- «from difference of action of
537m076 taces of mankind.-- p. 27. Mart. allows some UNIVERSAL feelings of right & wrong «(& therefore in fa
572m035 on lawless, whilst they are the only steady & UNIVERSAL means. recognized-- no one can say expression
186b058 ysom. & Heterom, but I cannot understand the UNIVERSALITY of such law.-- It would be curious to know
228b230 ariety.-- Strawberry produced by seed«s»??-- UNIVERSALITY of generation strongly shown by hybridity o
269c102 ife with laws of Chemical combination, & the UNIVERSALITY of latter render-- spontaneous generation n
261c070 Marsupial animals of N. America Hence it is UNIVERSALLY allowed that the discrimination of species i
369d115 s fully to understand their oeconomy, is now UNIVERSALLY admitted.""-- p. 483. Owen thinks from clima
535m069 ble that the fixed laws of nature should be «UNIVERSALLY» thought to be the will of a superior being;
171b005 ess (that is with our present system of body & UNIVERSE therefore final cause of life With this tende
270c103 y,-- this distinctness of laws from rest of]CD UNIVERSE «which Carus considers big animal» becomes mo
347d049 adapted to duration of sleep of man.!!! whole UNIVERSE so adapted!!! & not man to Planets.-- instanc
414e065 ated on principles. strictly applicable to the UNIVERSE.-- The range of man is not unlike that of ani
573n036 eity We can allow «satellites», planets, suns. UNIVERSE, nay whole systems of universe <of man> can b
573n036 planets, suns. universe, nay whole systems of UNIVERSE <of man> to be governed by laws., but the sma
573n036 ize the wonderful structure of a beetle than a UNIVERSE.-- November 20th Saw the youngest child of H.
607o025 arks p. 205 & 206.]CD Motives are units in the UNIVERSE. [Effect of heredetary constitution,-- educat
632j53r -- But I do not want to deny laws.-- The whole UNIVERSE is full of adaptations.-- but these are, I be
028r029 he actions of Springs or more probably by some UNKNOWN Volcanic process? How does it come that all Li
077r172 n & deposition under great pressure. (? heat!) UNKNOWN to us. ░ M. Chladni.--on meteoric Mexican ston
171b004 quire ideas ditto. V. Zoonomia.-- There may be UNKNOWN difficulty with full grown individual «with fi
175b017 ith some change probably <change> vary quicker UNKNOWN causes of change. Volcanic isld.-- Electricity
198b113 quinary system, or three elements p 66]CD With UNKNOWN limits, every tribe appears fitted for as many
207b150 rtiary epoch p. 330. Fossil Infusoria found of UNKNOWN forms, a circumstance undiscovered by Ehrenber
332d001 ished. October 2d As a proof. what «trifling» «UNKNOWN» causes act upon people. My Father mention, th
333d005 .-- crossed with <un>common cat, exact variety UNKNOWN., three kittens, alike each other, partaking <
333d005 isim Fox has half Persian cat. which bred with UNKNOWN common house cat.-- had four Kittens. two appe
342d036 view one can take of the world Astronomical <& UNKNOWN> causes, modified by unknown ones. cause chang
342d036 d Astronomical <& unknown> causes, modified by UNKNOWN ones. cause changes in geography & changes of
399e010 abound in the Navigators at & Manguia, but are UNKNOWN Eastward of the Navigators. Snakes occur there
399e010 f the Navigators. Snakes occur there, but are UNKNOWN in Hervey or Society isles. Hope says positive
463t059 perhaps animals of Fernando Noronha are found UNKNOWN coast in front of it.-- Cuvier has grand sente
471tf07 thorax & head displays.-- most fantastic & use UNKNOWN.-- "<when we find such an endless variety of f
534m062 wrong in comparing these cases, when agency is UNKNOWN, with simple exertion of intellectual faculty)
611o034 ble.-- And the Organic laws probably have some UNKNOWN relation to them-- [RHC] In the simplest forms
623o051 h is repeated under many generations. (& under UNKNOWN conditions) (for pig will not so readily attai
308c217 Two savages, two species.-- «discussion util, UNLESS it were fixed what a species means» civilized
232b247 h according to all analogy would have been very UNLIKE Southern Europaean ones.-- "a variation played
280c136 ss repugnant to nature to produce one offspring UNLIKE itself, than to produce that capable of produc
280c136 ronger, so that though by great effort <pr> one UNLIKE can be produced, yet to produce whole generati
280c136 an be produced, yet to produce whole generation UNLIKE would go against the tendency.-- it tries to g
280c136 it tries to go back to grandfather, but if too UNLIKE its own parent this impossible-- (Hence we mig
294c178 ere sickness)? fertility ¿because offspring too UNLIKE.--?? Memoire by Charles D'orbigny on Plastic C
414e065 able to the universe.-- The range of man is not UNLIKE that of animals transported by floating ice.--
594n112 -- flew to bed of flags. hernes are common. not UNLIKE in size in the air at a distance.-- How can su
292c169 change cannot be great, otherwise it would be UNLIMITED. We absolutely know that tendency is greater
086a003 pe, that the simplicity of Ventana's «Quartz.» UNMIXED is very pleasing; owing to the movements being
430e120 - Both species are found at Folkstone.-- it is UNNAMED this intermediate one.-- Mr Lonsdale evidently
588n090 useful to itself, how gained. reason? or some UNNAMED faculty-- Lamarck. Philosop. Zoolog. «p. 284.
275c120 exactly what I state.-- &c picking varieties. UNNATURAL circumstance Ld Orfords had breed of greyhoun
337d020 riety.-- our Europaean varieties must be very UNNATURAL-- Italian Greyhound is probably the effect of
523m018 ticular ideas, «Case of Shrewsbury gentleman, UNNATURAL union with turkey cock,-- was restrained by r
621o046 other natural appetites.-- he is monster, or UNNATURAL if malevolent, or hates his children without

350d054 anything framed to fill up empty cantons, & UNNECESSARY spaces" p 23. "for Nature is the art of God"
609o29v ng these instinctive passions--, which being UNNECESSARY we call vicious.-- (jealousy in a dog no one
516q024 icot grafted on what, & see what comes up.-- [UNNUMBERED blank] Experiment Cover patch of ground, wit
513q21. Does impregnation ever regularly take place in UNOPENED flower-- <doubt> disbelieve this in Bauers ca
108a080 ctly after great shock -- It appears to me UNPHILOSOPHICAL to think calcareous springs near coral ree
243c020 lizing without tables or references highly UNPHILOSOPHICAL xx Says same remark with regard to shells.
303c204 .-- He-will be bold. I will venture to say UNPHILOSOPHICAL. L'Institut 1838. p. 128. Extraordinary ge
640j167 only true criterion.-- Hence it is highly UNPHILOSOPHICAL to assert, that they are not species, unti
550m124 of child» is large proportion of pleasant to UNPLEASANT mental sensations in any given time «-- comp
550m124 then sensation may be more or less pleasant & UNPLEASANT, in same time,-- therefore degrees of happin
182b049 ds anterior to Man.. difficult for man to be UNPREJUDICED about self, but considering power, extendin
498q007 r, but not fruit-- -- Do not orchards become UNPRODUCTIVE from poorness of soil.-- yet crabs probably
506q015 is not produced or small in quantity -- Any UNPRODUCTIVE, where germen does not swell, although ther
196b109 ants live without carbonic acid gaz.)-- Yet UNQUESTIONABLY animals most dependent on vegetables. of t
538o080 when he struggled as it were with a second & UNREASONABLE man.-- If one could remember all ones farth
069r149 possibility of such change having taken place UNRECORDED must be insensible. Quantity of matter from
546r108 en we feel like them--. hence sympathy very UNSATISFACTORY because does not like Burke explain pleasu
337d022 hat difference would not be discovered by an UNSCIENTIFIC observer. -- <Transactions of the Entomologi
258c060 preventing adaptation owing to crossing, with UNSEASONED people. would cause destruction.-- simile Ma
371d127 d offspring.-- but surely in all these cases an UNSEEN change is produced in parents--colour is a dou
022c006 Australia.. Fitton's appendix Would Slate. & UNSTRATIFIED rocks show any difference in facility of co
106a074 herefore stream has no tendency to widen course UNTIL inclination is become comparatively small, & wh
277c127 e, or habits ascertained-- put them as (a). (b) UNTIL data be given.-- This will aid in preventing th
277c127 that he will not have brought home new species. UNTIL, he can show range & habits-- Take instances of
289c161 keep separate, which no doubt be overcome, but UNTIL it is the animals are distinct species If any o
375d135 l ratio in FAR SHORTER time than 25 years-- yet UNTIL the one sentence of Malthus no one clearly perc
415e066 species» of the Ornithorhyncus be found.;-- yet UNTIL man became cosmopolite, he would probably be co
595n115 d a most lamentable howls & & was not comforted UNTIL the Keeper took it <her> in his arms & carried
616o039 living bodies without these faculties & indeed UNTIL we know what answer they would give in support
617o040 force in inanimate matter we feel dissatisfied UNTIL we can point out How can force be recognized by
624o051 beneficial to race, will have reoccurred'. NB. UNTIL, it can be shewn, what things easiest become in
640j167 osophical to assert, that they are not species, UNTIL their breeding together has been tried.-- With
349d051 from fewness of forms-- Cannot be discovered «UN»TILL <in> «we ascend to» subgenera & families, <eve
618o41v ht & organization: But if the weight never came UNTILL the blueness had a certain intensity (& the ex
558m150 al, parental & social instincts, giving rise "do UNTO others as yourself". "love thy neighbour as thy
418e083 opement is spread over [not located] it utterly UNTOLD,-- what is added to the composition of the ato
446e160 ve developing power.-- My theory leaves quite UNTOUCHED the question of spontaneous generation.-- Int
523m015 his struggles against it, his knowledge of the UNTRUTH of the idea, namely his poverty.-- his manner
554m139 succeed of herself.-- <The male» "I saw» Jenny UNTYING a very difficult knot-- the sailor on board th
113a090 being 400 ft. shows that the strata have very UNUSUAL conducting power of heat from centre.-- But is
359d088 er says that breeds larger numbers, & rears an UNUSUAL number out of any one nest. even more than com
538o079 accurate & even profound knowledge or other & UNUSUAL line-- both odd appearance about eyes.-- one b
497q007 flowers appear difficult to cross, are there UNUSUALLY many species in genera of Leguminosae.-- Herb
532m054 g: I have awakened in the night. being slightly UNWELL & felt so much afraid though my reason was lau
559m154 s much in analogy, we never find out." This UNWILLINGNESS to consider Creator as governing by laws is
055r108 lainly» indicates former boundary. (as in other UNWORN islands) we take in at once the stupendous mas
021rIFC   R. N UP to 1 degree / July 1835. the excess of harbor --
033r043 walk round base--not universal: could not climb UP many parts, in James Isd.--Mem St Helena-- All T
036r049 e dilatation, which dilated cracks must be filled UP by dikes & mountain chains.-- Introduce part of
038r058 pearance leads me to believe mere fissures filled UP.--the appearance will here be the strongest argu
044r072 e-- Pisolitic balls occur in the Ashes which fill UP theatre of Pompeii (?). -- Such have been seen t
046r080 owing to water keeping its level whilst land rose UP & down.--But from above reasons, do not think so
047r082 Does the sea fall on banks as a Bore wave rushes UP? (NB. Earthquake wave is an oscillation, body of
047r082 llation, body of water manifestly does not travel UP.--) If these view are right the coincidental ret
057r114 egree 39 N. living worms in the mud which he drew UP from 1,000 f[athoms], & the temp of which was be
066r142 y have Earthquakes in Cordoba. one of which dried UP <all> a lake in neigbourhood of town Mr Murchiso
066r142 town Mr Murchison insisted strongly. that taking UP a piece of Falkland Sandstone. he could not dist
068r146 uidity of rock ∴ «in earliest stage» when covered UP beneath ocean).--The first dislocations & erupti
072r159 no mountains Mackenzie has talked of lava flowing UP Hill; ¿what does he mean?) Consult Dr Holland ab
075r166 o) no veins discovered. Humboldt suggests covered UP by volcanic rocks. //St Helena has been slightly
075r167 canic rocks. //St Helena has been slightly broken UP, & has there not been vein «of iron» discovered?
087a008 Encyclopaedia-- Lately elevated When Siberia went UP. Arctic land went down.-- Probably more Arctic l
088a014 eels alive to their nests; & then they may picked UP beneath the trees---- Are any Fish seed-eaters.
092a030 xplain their solution. Athenaeum M. 516 1837 High UP the Essequibo, granite & quartz, after passing s
102a059 ulation in open ocean. as pebbles would be lifted UP & down. on coast itself, undertow would draw it
102a059 to it.-- rollers at Tristan d'.Acunha.-- silting UP. channels on coast of England-- Any one. who has
103a063 ng agents, because not wedge-formed.-- Hence fill UP fissures-- If dikes effect of horizontal elevati
105a070 dillera. a rush of water will account for filling UP of valleys-- subsequent opening a medial gorge b
107a078 portant with respect to thickness of crust broken UP.----My view of Volcanos &c &c This view will bea
109a083 is lessened.-- Coral flats. argument for Heaping UP, -- very good this will show effects.-- analogous
117a102 ere obtained 100 miles E of Staten land. bringing UP pebbles 2 inches long?-- L'Institut. 1838 p. 151
125a121 etrated.-- lower down the temperature may be kept UP far higher from circulation of heated fluid or g
127a125 ranean isothermal line must be creeping «pushing» UP to «the» line of ice. Hence further N. when so
128a128 e sea mark.-- If mountain chains are matter piled UP. over crevice from effect of general elevation,-
146g014 t above Loch.-- From this point could be followed UP to neighbourhood of Tyndrum where a large spot o
150g035 N. side of Spean most clear & upper line running UP to great bight just as Dick shows NB. Lake graduall
150g037 onger than 3a Sunday In Glen Collarig, when water UP to shelf very shallow channel 50 ft wide & river
151g042 ve been waterworn after 3d lake.-- 4th shelf runs UP some way on great sloping plain of alluvium (muc
152g050 per terrace The butresses of Alluvium rise nearly UP to Glen Collarig up within 200 ft of level of 4t
152g050 esses of Alluvium rise nearly up to Glen Collarig UP within 200 ft of level of 4th shelf-- argument ag
153g053 below House of Glen Roy, seems to be which higher UP on is corroded {P} 4th [shelf] river 4th [shelf]
155g062 ormed on shelf 4th 2d & 3d can be traced some way UP, but most faintly on East side of Glen Turrit, w
156g071 n appears like one uniform slope slightly bending UP each main valley.-- & the river alone had modif
158g082 hill where Boulder lies. buttresses «occur» high UP on Shelf 2d «in Upper Glen Roy» In this upper pa
158g083 shelfs ground strewed with pebbles Shelf 3d runs UP with buttresses on each side «very little way» i
162g103 of lake with trace of terraces on each side High UP the Tarf (a Granite (boulder), sloping buttress
163g107 t Augustus hill & fringe as if it has been filled UP «at» 30 ft. higher with pebbles now worn away--
188b068 ver so many seeds of white flower. all would come UP white, though planted in same soil with blue. No
191b080 aphy.-- The motion of the earth must be excessive UP & down.-- Elephants in Ceylon-- East Indian arch
196b105 - Waders & Waterfowl.-- scrutinize genera, & draw UP tables-- Instinct may confine certain birds whic
199b119 settling a country.-- people very apt to be split UP into many isolated races. ¿are there any instanc
202b131 gree about) Vol IV P. I. Geograp. Journal. Voyage UP the Massaroony by W. Hillhouse.-- Demerara. In n
211b163 tches never being killed by them, whilst they eat UP the dogs.-- L.' Institut. Curious paper by M. Se
227b227 require fresh creation!-- If continent had sprung UP round Galapagos on Pacific side. the Oolite orde
231b241 amount of varying in wild state.-- When breaking UP «the primeval.» continent. Indian Rhinoceros. Ja
234b255 ide.-- Pigs put legs over, & then with snout lift UP latch & back.-- Frogs attempted to be introduced
240c004 ar them, eggs hatched under other birds & brought UP by hand.-- These facts all account for [not loca
245c025 only Cuscus & Barbyroussa «NB» [islds. Springing UP more likely to <M> have different species than t
249c036 supial forms been chiefly preserved,-- where shut UP by themselves without other animals? but they we
249c036 ves without other animals? but they were not shut UP!! Extreme southern points of S. Hemisphere fully
262c073 in any kingdom of nature, where scheme not filled UP, (most false to say no passages; nature is full
263c075 any kind of prejudices) <without> who just takes UP & lay down the subject without long meditation--
273c110 to expect that fossil forms would generally fill UP genera & not species, which is not true, with sh
273c111 erished) hawks & Milvulus &c would instantly fill UP their place.-- Humming bird there is strongly ma
275c120 o scent!) Mr Wynne) at end of chase would not run UP hill-- he took thorough bred <greyhound>. bull-d
281c137 simple expression of such a naturalist "splitting UP his species & genera very finely" show how arbit
284c148 » trees dying & mountain torrents.-- but to crawl UP an hill, then by deaths?!)-- looks like subsiden
287c157 placed a series of animals on the steps that lead UP to it" p. 20 ╫ +++hiatus & saltus not syn.-- Lin

Page
(Key Word)
293c172 ones   might yet be transmitted.= Memory springing UP after long intervals of forgetfulness.-- after s
293c172 versation before. is strong association recalling UP image which had been past-- so great an anomaly
299c194 in dry station it will for some generations come UP so.-- there are not Many intermediate shades in
304c208 me elapsed & therefore descended from branch high UP.-- Such probabilities only guides.-- Yet trifles
304c208 other birds, or that it sprung from a branch high UP.-- this argument not applicable to apterix.-- bu
312c228 hinese & 3 perfectly like spaniel even when grown UP.-- Are mules homogenious owing to no attempt to
312c228 Are mules homogenious owing to no attempt to keep UP offspring, are not half lion & tigers ditto. (se
312c276 ge de la Coquille Zoological Transactions. <done> UP to parts published March 1838 Whole of Geographi
350d054 grotesques in nature; not anything framed to fill UP empty cantons, & unnecessary spaces" p 23. "for
352d060 a history, & the geologist being obliged to fill UP the gaps.-- is possibly the same with the <Zoolo
358d085 were possible, the organs doubtless would shrivel UP.-- Yet odd they should have so much sexual chara
362d099 op off., replaced by hairy ones. which never «dry UP &» peel off their skin (not being wanted for war
370d116 in same work. (it is said that some kind lay <pu> UP honey even for single rainy day-- & from case of
371d118 wonderful that childs nervous system should build UP its body, like its parent, than that it should b
371d118 ral fully developed tails, & one with beak turned UP like Avocette. here is what [not located] that i
371d127 ered in disposition, as to be more easily trained UP to the (required) offices" &c &c Owen illustrate
376d137 women.-- he has repeatedly seen them try to pull UP petticoats., & if woman not afraid clasp them ro
376d137 do this.-- These Monkeys had no curiosity to pull UP trousers of men. Evidently knew <men> women, thi
376d138 xes of every Has repeatedly seen one he kept pull UP feathers of tail of Hen; which lived with it.--
376d138 onkey thus examine each other sexes,-- «by taking UP tail» Mem: Ourang Jenny with Tommy.-- Good evide
400e013 , showing effects of cultivation gradually adding UP. & four more generations before they began to do
401e015 s if any variety of apple be sown, all sorts come UP from it. lately saw a nonpareil sowed by Mr Toll
403e024 ch> not, cows hornless, (horses not) If they give UP infertility in largest sense. <es> «as» test of
411e055 come under this head» 27th November When summing UP argument against my theory, doubtless, the prese
415e067 ticated animals» no doubt is owing to the rearing UP of every heredetary tendency towards fatal disea
416e072 fference in external conditions.-- -- as in plant UP a mountain-- In races the differences depend upo
420e088 te Rendu.-- I suspect good case of fossil filling UP blank.-- CD[not between existing series of speci
427e109 & there only: as stated by Capt. Graah») & break UP. N. American Conchology from Europaean., & the c
433e127 ds-- The argument must be thus put, shall we give UP whole system, of transmut., or believe that time
458t002 l» The Taylor Bird uses pieces of thread, picked «UP» instead of spinning-- better case than English
459t009 deer, which are altogether immaculate when grown ". Saw at Mr Bell's at Hornsey the offspring of a
466t033 though not differently coloured-- & stamens bend UP a little In a wild purple Geranium, I see Bees v
466t094 of anthers & pistils, but these <do> do not bend UP-- In Lark-spur, if Bees put proboscis within nec
468t112 htemed pollen profusely shed; lengthened & turned UP «more than stamens», so that all were brushed by
469t151 «close to each other» & seeds gathered «all» came UP in 1840 true. Shrewsbury.-- Abberley-- Early Mag
469t151 planted in rows, & seeds gathered same year came UP true «in 1840»: All in together blossomed togeth
470t177 site) Asparagus very small flowers & as much shut UP, frequented by «many» Bees & Humbles-- «Humbles
472s01r n same garden. & out of 60 seedlings not one came UP true.-- colour of flower & foliage not «being» l
472s01v ocured with the P. orientale in garden & all came UP hybridised. It is possible to raise them pure fo
495q05a count seeds, & see how great a proportion springs UP true.-- This in fact always takes place in natur
495q05a rden» Coronella also sleeps on ditto-- Cover them UP periodically & see effect-- such dioecious indiv
495q05a l-glass-- sow these seeds & see if they will come UP true-- whilst others are crossed.-- Are Bees gui
497q006 & other odd places & see what plants will spring UP which will show, how seeds are transported, or h
497q006 d, or how long they remain dormant. if kinds come UP, not found in wood.-- but seeds continually drop
512q019 ost Hawks eat stomach. of finches-- do they throw UP pellets-- (4) About hybrid pheasants treading--
512q019 hereford cow similar to reverse cross.-- Sow cast- UP-balls of Hawks or even owls.-- How long do seeds
515q022 king after male & female parent.-- Will they grow UP in other respects different?--. Important.-- Oct
516q023 tandard Apricot grafted on what, & see what comes UP.-- [Unnumbered blank] Experiment Cover patch of
522m010 - Mr Corbet, however, in conversation could catch UP a new train if early association were called up.
522m010 h up a new train if early association were called UP.-- My F. asked him, did he know whom Mr Child «o
522m012 ct at the time, but some time afterwards it calls UP pain. or pleasure. & is often recurred to & ment
525m023 carrying a basket, bringing back game, or picking UP a stone, though only acquired rules by art.-- li
526m027 a farthing, free will determines our throwing it UP.-- equall true the two statements.-- Catherine r
526m028 detail, yet they have not imagination enough to <UP> recall up the image in their own mind,-- this m
526m028 t they have not imagination enough to <up> recall UP the image in their own mind,-- this may be worth
526m029 nderstanding. (on hearing old association brought UP) by sight & not by hearing One is tempted to bel
527m033 vivid & grand. the frame of mind being just kept UP by the music of the poetry. --(therefore singing
536m071 bulls & horses, animals of different orders turn UP their nostrils when excited by love? Stallion li
537m077 , & therefore he has these strong, & does not act UP to them, no doubt disobeys & hurts conscience mo
539m082 pleasure immediately thrilled across me, bringing UP old indistinct ideas of FitzWilliam Musm. I was
539m083 In the habitual train of thought one idea. calls UP other, & the consciousness of double individual
541m092 y has his mouth open ready to bark, & lip twisted UP, in that peculiar manner they do, even more than
542m095 of straining vision, as savages without hats put UP their hands, & as attention would amongst lowest
542m095 f the skin contract iris?-- same way as one lifts UP eyebrows to see things in dark. & hence is this
544m102 r when so near waking. that an associated is kept UP with waking thought.-- Ld Brougham thinks no dre
547m111 t once to Shrewsbury., vaguely thought of packing UP.-- was lying on my back fell to sleep for second
547m113 instantaneous changes in order <to every> calling UP ideas of every late impression.-- (do the ideas,
547m113 ss work & yet do so, from the exertion of keeping UP the memory of every late impression. & likewise
547m113 ew ones from senses. & <comparing their> «calling UP» old ones, to be sure of ones consciousness.-- M
547m114 late events.-- the fatigue of thinking is keeping UP these trains,-- especially if they are invented
550m127 re pleasant, every concomitant circumstance calls UP pleasure. or pleasure or pain of association.--
551m129 ping when word Jump said-- I saw the ourang. take UP a stone & pound the earth. Lockarts life of W. S
554m138 e??) he has seen place its head downwards to look UP womens petticoats-- just like Jenny with Tommy o
554m139 from a visitor, & before eating «everytime», look UP to «keeper» see whether, this was permitted & ea
559m156 as giving abstract of Smith's views «Take & pound UP inflorescent parts of mosses & see if Hybrid can
565m007 (& pride) same way walk erect & stiff, with head UP.-- Why does suspicion look obliquely.-- who can
578n057 g».-- But why does joy & OTHER EMOTION make grown UP people cry.-- What is emotion? At end of Burke's
588n089 hat they perceive the difference on being carried UP or downstairs, or dangled up & down-- in latter
588n089 nce on being carried up or downstairs, or dangled UP & down-- in latter case they struggle their arms
592n103 sions.-- «Paper» must be referred to, if I follow UP this subject & a reference to Brun's work.-- Shu
599o007 I cannot <admit> think reason sufficient to give UP my theory-- Viewing from eminence. the wide expa
602o011 is right-- it is because each decision &c is made UP of many partial results, & the impressions on th
609o29v sness jealousy, & every one being married to keep UP population. with the existences of so many posit
613o035 ther senses come into play, when relation is kept UP with distant object. where many such objects are
617o40v re, if we can trace any force in inanimate matter UP to the action of some animated agent Now the phe
620o044 rom his memory & <pow> mental capacity of calling UP past sensations, he will be forced to reflect on
621o047 to peculiarity of organ of taste, for when grown UP often conquers it). It will be only rarely that
623o050 place, but merely in the place. & yet place calls UP pleasure.-- [LHC] the instinct of sociability &
627o053 e have conception of moral obligation «when grown UP???» & the question is, whether this can be resol
641j29r ere will be a balance, continents have been split UP.-- who can decide their limits.-- To show how li
431e121 in the Inferior Oolite, & the T. elongata in the UPER formations Portland Stone &c &c.--if? so «it is
023r011 earance «arising from the manner of horizontal UPHEAVAL» of the shore of the Pacifick is 60 miles dis
031r037 extent; if lowered again & covered no sign of UPHEAVAL To Cleavage added other instances in old world
040r064 ius.-- There may have been oscillations in the UPHEAVAL of Andes.--but as long as all below water no
042r068 it is probable that point of Porphyry has been UPHEAVED in a dry form It is clear the forces have act
633j54r rm rain to wash down earth, from the mountains UPHEAVED by volcanic force, for these Marsh plants. Al
185b055 r-- a desert. Kingfisher.-- mountain tringas.-- UPLAND goose.-- water chionis water rat with land str
461t039 would tend to render complex the series.-- Ch 6 UPLAND geese would transplant seeds very far.-- Sept
063r133 ect success.--showing non Creation does not bear UPON solely adaptation of animals.--extinction in sa
090a021 ed to those of falling stones.--¿ does this bear UPON the sorting of matter. in making trachyte come
177b029 determinate life dependent on genus,, that genus UPON another, whole class would die out, therefore n
198b115 l about monsters worth reading NB well to insist UPON <different animals> large Mammalia not being fo
235b262 oint.-- All cases like Irish & English Hare bear UPON this.-- Why do Van Diemens land people require
266c088 loured in sterile countries.-- Gould insist much UPON knowing to what type a bird belongs.-- I concei
270c104 if any Metaphysical speculations are entered in UPON life. Namely Carus.-- How remarkable that Turdu
282c142 d not in all cases be produced, but would depend UPON exclusion.-- The same characters which are anal
294c176 how to show species-- I fear argument must rest UPON analogy & absence of varieties in a wild state-
306c212 0 miles long.-- a monkey. (Baboon) at Z. Gardens UPON being beaten behaved very differently from a do

Page **(Key Word)**
319c256 inferior to ours, & our lark ranks very high.-- UPON the whole thinks <many> more birds sing in Engl
332d001 As a proof. what <trifling> «unknown» causes act UPON people. My Father mention, than for ten years h
345d043 hers believes this resemblance general. ¿depends UPON mother bein[g] oldest breed?.-- -- Quarterly Jo
349d051 till much knowledge is elicited.-- It will rest UPON the discovery what characters VARY most easily:
380d154 Fowl cannot be distinguished.-- A capon will sit UPON eggs, as well as, & often better than a female.
413e059 s. the mystery of mysteries. & has grand passage UPON problem.! Hurrah.-- "intermediate causes" The S
416e072 up a mountain-- In races the differences depend UPON inheritance & in species are only ancient & per
419e087 overlooked that the chronology of geology rests UPON amount of physical change <affecting whole bodi
521m009 er is ready, what, what.-- then showed the watch UPON which he exclaimed, why it is dinner time.-- »
522m013 delirium, a peculiar complaint stomach not acted UPON by Emetics.-- people recognized,-- sudden chang
531m052 panies violent attack,-- Even the worm when trod UPON turneth,, here probably there is no feeling of
532m054 em is affected, & by habit the mind tries to fix UPON some object:-- When a man, child or colt has on
536m073 s of the fathers, corporeal & bodily are visited UPON the children.-- The above views would make a ma
541m091 cloth &c & with all this difficult EXPERIMENTIZE UPON this effort.-- it looks so analogous to muscle
601o009 , in order to avaoid it-- beetles feigning death UPON seeing an object.-- are Planariae conscious.--
602o011 gotten. Our happiness &c, our well-being depends UPON the "habitual reason".-- This power of the mind
602o11b s independent of the senses & experience p. 142 "UPON the whole it seems."-- "that the object of «all
038r058 & on which trees grew.--? Are not the dikes in UPPER strata. quite different from the Porphyries: ce
051r094 e fragments by mere force of waves: & action on UPPER tidal band, I do not see how to account for oce
057r114 e fact of Boulders not in lower strata. only in UPPER. in accordance in Europe with ice theory.-- Cap
074r164 uwacke: Silver appears far more abundant in the UPPER limestone, which H. calls by several secondary
092a028 .-- in Syria Geolog. Proc. p. 541. year 1837 In UPPER Assam. Geolog Proc p. 566 1837.-- Tertiary <bea
097a044 eries. p. 221.-- Mr Bollaert tells me, that the UPPER strata alone at Guantajaya contains salt see Ge
102a061 e greenstone dike in Granite residual matter of UPPER quartzose ones & felspar.?? Are the great cryst
113a090 erly much higher, «so» that we must look at the UPPER four hundred feet of strata having conducted aw
126a124 ording to this latter view the rod is reversed, UPPER part metal «conveying heat in one direction onl
145g006 is I believe about greatest dip of sandstone in UPPER part «of Salisbury Craigs» 25 degree perhaps mo
146g013 e of any hill Thursday On side of Hill South of UPPER end of Loch Dochart buttresses of Alluvium or r
147g015 her irregular horizontal strata I suppose these UPPER patches if prolonged would intersect alley abov
148g023 Canal opposite Loch Leven two terraces perhaps UPPER one 100 ft & other one 40-- ‖ traces of them al
150g035 red talus line on N. side of Spean most clear & UPPER line running up great bight just as Dick shows
150g037 In Glen Collarig, on side of Hill of Bohunthine UPPER road (2) extends as far nearly as house, the 3d
151g043 rivulet enters two great projecting butresses, UPPER slope of which corresponds to shelf the truncat
151g044 which corresponds to shelf the truncation & the UPPER shores may correspond with some line subsequent
152g048 y long at 4th shelf from size of buttresses, to UPPER edge of which they cut near Loch Tring-- Tuesda
152g050 . 30.127 A 72 degree Air 65 degree? at level of UPPER terrace The butresses of Alluvium rise nearly u
153g051 th shelf: others «lines not so level because of UPPER edge of cliff» Others below it--argument for la
155g066 hree» lines «should be» EQUALLY preserved 2d or UPPER one more perfect in this <part> «glen» than 3d.
156g067 s <part> «glen» than 3d. 3(a) less perfect than UPPER & lower but quite as perfect as those lines in
156g070 ng in saying no transported materials <into> on UPPER shelves granite & some other rocks at head of s
157g077 Cascade has <cut> «where two branches unite in UPPER Glen Roy» very little back from line 2d; little
157g077 ine 2d; little action since «that shelf» formed UPPER terrace near Loch Spey <29.35161> 29.360? A 79
158g079 r is horizontal line apparently continuation of UPPER terrace -- on right hand {P} rounded waterworn
158g080 Glen Roy on same side where two rivers unite in UPPER Glen Roy great plain about 60 ft beneath shelf
158g082 ies. buttresses «occur» high up on Shelf 2d «in UPPER Glen Roy» In this upper part «about junction of
158g083 high up on Shelf 2d «in Upper Glen Roy» In this UPPER part «about junction of Upper & from Glenroy» n
158g083 Glen Roy» In this upper part «about junction of UPPER & from Glenroy» near the upper shelfs ground st
158g083 bout junction of Upper & from Glenroy» near the UPPER shelfs ground strewed with pebbles Shelf 3d run
158g083 th buttresses on each side «very little way» in UPPER Glen Roy at pass {P} River Gorge 4th Sh side of
159g086 day, from Glen Turrit to Fort Augustus Barom on UPPER (rather above)? shelf 29.290 A. 69 degree Air 6
298c191 &c &c --Climate altering as island increases.-- UPPER parts attracting all the moisture.-- Henslow th
303c204 h of extremities, how are races in This respect UPPER & lower, which I do not know whether it <would
317c250 many common to Canary. isld., Europe, & St Jago UPPER region, & some to Cape.-- some proper well-wort
374d133 a depart more widely from living forms.-- p 458 UPPER Silurian, fishes oldest formation highly organi
391d175 of those «forms» slightly favoured, getting the UPPER hand. <]CD> & forming species)-- [Aphides havin
399e006 n same bed if very thick to find some change in UPPER & lower layers.-- good objection to my theory:
423e099 s in my fossils-- CD[If cases of one variety in UPPER part of bed-- & another in lower is very rare,
466t093 drum-- nectary marked by orange freckles on {a} UPPER petal; bees & flies seen directed to it-- The H
466t093 >.-- In yellow day lily, the Bees visit base of UPPER petal, though not differently coloured-- & stam
466t094 e Geranium, I see Bees visit always base {a} of UPPER petal from facility of alighting? which is not
467t103 ke on Kidney Bean, they go to nectar at foot of UPPER petal standing on «I saw Bee go to two species
467t104 {a} In all these nectar seems to be at base of UPPER petal & the curvature of <an> pistil, etc lies
472s02r this Heartease withered on Monday.--» alight on UPPER petals & insert proboscis, under sigma & draw i
556m145 iration & quickly retracting tongue from behind UPPER & little between incisors.-- like <W[...] what>
556m145 awings of Voltaire why is under lip curled over UPPER with mouth shut. expressing cool irony, not bit
564n006 e skin or muscles are contracted between eyes & UPPER lip., is most clearly analogous to a panther I
565n009 lies in wrinkles about the nose, & arrogance in UPPER lip. <The> Children having peculiar expression
577n051 to surface exposed, face of man, face, neck-- «UPPER» bosom in woman: like erection shyness is certa
639j28v housand years the long legged race will get the UPPER hand. though continually dragged back to old ty
527m031 housand grains of sand together & one will be UPPERMOST:-- so in thoughts, one will rise according to
191b080 lia From the consideration of these archipelagos UPS & downs in full conformity with Europaean format
285c151 close species, «on these isld» &c will probably UPSET it-- The space which one branch of the tree if
390d178 pple ought to differ from those of other.-- The UPSHOT of all this is that effect of Male is to impre
164g119 m mud.-- [blank] {P} muddy nodular strata coral UPSIDE down strata coarse agglomerate [...] shells fr
107a077 Does the isothermal {P} subterranean line moves UPWARD from effects of Elevation if not crust much th
025r019 mination of soundings, from about 80 fathoms & UPWARDS. that life is exceedingly rare, at the bottom
060r124 ts in the land to lift itself higher & to grow UPWARDS; for the land is constantly pushing the sea (w
060r124 his phenomena in connection with "the shooting UPWARDS" of the <ground> land in the W Indies.--p. 200
118a103 med volcanos. or become scoriform. has thinned UPWARDS & is now cut off by denudation it gives one gr
126a123 surface, would not the fluid matter be driven UPWARDS & so conduct heat?-- How comes it in volcanos
155g063 de of Glen Turrit, where I believe they end in UPWARDS inclined plains, as in Corry. & as «as I belie
175b019 hrenberg? every successive animal is branching UPWARDS different types of organization improving as O
350d053 Propagation, best rule for genera, & so mount UPWARDS.-- «judged by analogy»-- Consider all this NB.
466t093 » was dusted over. {P} Stamens & pistils curve UPWARDS, so that anthers & stigma lie in fairway to ne
468t112 amens grow old «& shed some pollen». they turn UPWARDS & bend over stigma:-- but stigma «is» almost r
588n090 their arms.-- do. p. 306 "the eyes are rolled UPWARDS during mental agony, & whilst strong emotions
090a022 fallen on the globe since the Cambrian system In URES dictionary between 1768 & 1818. that is fifty y
025r018 have heard of In a preface, it might be well to URGE, geologists to compare whole history of Europe,
037r052 ical ranges of plants. V. Lyell. Chap XI Vol II. URGE the entire absence of any rock situated beneath
038r059 r to elevations? & not sources of lava streams.--URGE not tilted strata.-- It will be well to urge th
039r059 s.--Urge not tilted strata.-- It will be well to URGE the case of St Helena, where dikes certainly ha
055r107 s of shells. 2 - 3 toises thick.--Vol II. p. 252 URGE cliff form of land, in St Helena. Ascension. Az
057r113 shed because too cold:--With discussion of camel URGE S. Africa productions.-- I think in Patagonia w
057r114 s. at Port Desire on plain. & interstratified.-- URGE fact of Boulders not in lower strata. only in u
060r125 nd nothing.» Mem Carolines quotation from Temple URGE the mineralogical difference of formations of S
062r129 rats, in the antipodes a parallel case.-- Should URGE that extinct Llama owed its death not to change
064r136 eat but local elevations of the land in Europe-- URGE difference of plutonic rocks & Volcanic metalli
064r136 nce of plutonic rocks & Volcanic metalliferous-- URGE enormous quantity of matter from CREVICE of And
069r150 ata may be older than (B).-- Most important view URGE curious fact felspar melted gneiss/// QUARTZ!!!
148g028 ide-streamlet had cut them out-- In all cases «I URGE» deposition marine-- because if not chain of la
377d139 definite than with bitch, for some feeling must URGE them to these actions. «These facts may, be tur
526m031 appetites themselves become changed.-- appetites URGE the man, but indefinitely, he chooses (but what
064r137 Lyell « <p 419> p 428» states that Von Buch has URGED that Java volcanos differ from all others in qu
070r153 different. not insensible change.-- Yet one is URGED to look to common parent? why should two of the
074r165 haps makes intersections richest-- Humboldt has URGED phenomena in veins, chemical affinities like in
297c185 efore other species mice & only kills them when URGED by hunger.-- p. 65. Aberrant groups few in numb
608o026 ll motives do not. come into play.-- †It may be URGED how often one try to persuade person to change
620o045 - Yet even at this time, malevolence,, when not URGED to it by passion, shows a bad child.-- Hence th

Page
(Key Word)

179b033 es «--:Humboldt. New Spain:--» Dr. Smith always URGES the distinct locality or Metropolis of every sp
390d179 m their stock, for to breed (as Sir J. Sebright URGES?) one with opposed characters is by impliance t
526m030 the faculty according to usual method, but what URGES him,-- absolute free will, motive may be anythi
526m030 &c &c An animal improves because its appetites URGES it to certain actions, which are modified by ci
496q006 . Magnesium Iron Rust Carb. of Ammonia.-- Horse URINE &c &c on associated plants. when proportional n
531m053 want of muscular exertion, palpitation, voiding URINE because done by some animals in defence, &c Sta
499q09v ry &c &c (a) Mercurialis-- Frog Bit, Valerian-- URTICA Dioica Sorrell. Lychnis. Butchers Broom-- «als
255c050 -- p. 56: Ornithological part of Voyage of??? A URUBU, (with one leg) attended the distribution of fo
241c016 .. Bull: Soc. Geolog. 1837-8. Tom: IX.-- M. D'.URVILLE on the Distrib of Ferns in South Sea (Indio Po
077r172 sition under great pressure. (? heat!) unknown to US. ⫷ M. Chladni.--on meteoric Mexican stone. Journ
277c126 istinct. analogy from every country & class tells US that ¿O. Modulator & O. Patagonicus. till neutra
277c127 aos.-- will point out what to observe.-- will aid US in physiology, tell traveller what to observe.--
293c173 mon & caterpillar, though our ignorance, may make US think so, but only between laws--]CD. Many disea
303c205 tain extent,-- nothing but experience. will, tell US. when group is true,» there are no genera. if Ma
317c249 some branches of Justicia still growing,) passed US. do. p. 243 (, Professor Smith's Journal) on the
352d059 not perform that function which experience shows US it was for.-- Most important law.-- Penguins win
358d075 he crows were amazingly bold, always accompanying US from camp to camp; it was absolutely necessary t
402e020 . p 367. "The Fauna of the Sunda islands presents US, for the most part, with the same families and g
403e023 ecause it has been so pronounced ex cathedrâ. let US look at facts. considering few domestic animals
418e084 iduals,, during thousands of centuries,-- each of US, then <is as old, as the oldest animal>, have pa
434e130 uppression of one organ. (here language forces on US the change, which <to> seems to have taken place
441e149 ye,-- if Australian Dog, could bud, analogy tells US, <be> «offspring» would be similar to <f> first
443e154 have some most important difference is forced on US.-- My theory only requires that organic beings p
536m074 lieving, <yet> would more earnestly pray "deliver US from temptation,. he would be most humble, he wo
544m103 else it is only our consciousness, & senses tell US it is not real. = = dreaming appears clearly rest
553m136 , make the same mistake, more apparent however to US, as does that philosopher who says the innate kn
553m136 knowledge of creator <is> «has been» implanted in US (<by> ¿ individually or in race?) by a separate
553m136 oduce every effect, of every kind which surrounds US. Moreover «it would be difficult to prove that»
553m136 goodness of God knowledge has been communicat to US».-- & that it does exist in different degrees in
571n027 general ideas of our ancestors being impressed on US.-- Surely we have taste naturally all has not be
573n036 we placidly believe the Astronomer, when he tells US satellites &c &c «The Savage admires not a steam
579n059 ed, with the habitual expressemotions, which make US love him, or her.-- it is blind feeling, somethi
589n092 ocomotion-- or that our faculties have been given US to exist, is clearly seen. in the absurdity of a
595n121 s Snow.-- Is this part of same feeling which make US think anything ugly-- a beau-ideal feeling. Same
595n121 -- a beau-ideal feeling. Same effect as acting on US-- <The Baby» «Effie Wedgwood» April 28th 1840 wa
616o39v ve action by which bodily action is made known to US, revealing respectively what are called it's sub
617o39v subjective aspect of bodily action is revealed to US by the effort it costs us to exert force or by i
617o39v y action is revealed to us by the effort it costs US to exert force or by internal consciousness; the
617o39v ble degree of attention. How do the senses affect US, except by internal consciousness 3) We must end
620o045 ngthened instinct, even when our reason tells-- + US the action was superfluous, as one man trying to
625o052 that certain feelings & actions are implanted in US. & that doing them gives pleasure & being preven
626o052 oes not say anything about any principles born in US.-- Great difference with my theory.-- see p. 349
633j53v imental accidents, & domesticated variations show US accidents may become heredetary [produce some pe
056r112 --which mingling & separating is well adapted to USE of mankind.--<Hutton show> Earthquakes part of n
071r156 ivine. Volcanic product.=> <Did Peruvian Indians USE arrows or Araucanians?--> If wood now preserved
087a012 o Europaean risings. Pacific great land. -- Will USE argument of proof of slow corrosion of valley of
098a045 s uniting with cretionary.-- it may <off> come of USE in discussion on Cleavage &c Geolog Transacts. V
121a111 seen clay stiff enough <to form> for potters to USE. in which great Knife formed crystals of ice wer
192b084 es nipple on man's breast. one does not say some USE, but <no.> sex not having been determined.-- so
195b099 unintelligible structures. it might have been of USE in progenitor-- or it may be of use-- like Mamm
195b099 have been of use in progenitor-- or it may be of USE.-- like Mammae on mens' breasts.-- How does it c
259c063 ng excessively abundant & tempting the Jaguar to USE its feet much in swimming, & every developement
265c086 ered wing-cases-- Yet these wings may be of some USE,-- Nature is never extravagant though clearly no
265c087 e is never extravagant though clearly not of the USE to which wings are generally applied.-- Therefor
265c087 se shrivelled wings could be shown to be of some USE. If we only had Puffinuria Garrottii & no other
275c119 s Since geograph distribution of animals, <as> I USE (new step in induction) as keystone of ancient g
282c140 n character,-- we here suppose these changes of <USE> adaptation greater than those heredetary ones.-
293c172 ness of reasoning to tell back from front. &c or USE of button holes it would be instinctive.-- My vi
296c184 ces of hybrids between pheasants & Black fowl.-- USE as argument possibly some few hybrids in nature.
352d057 f foetus.-- As Larvae may be more perfect (as we USE the word) than parent, so may species retrograde
352d059 cially of Mammalia As every organ is modified by USE, every abortive organ must have been once change
384d161 being developed, when the first <are> become of USE⫰ <Great characteristic of male greater strength,
471tf07 h the thorax & head displays.-- most fantastic & USE⫰ unknown.-- "<when we find such an endless variet
471tf07 gan "manifestation of divine power"?.--"of their USE difficult to conceive any idea" Linn. Trans. 18.
496q05a rtificial flowers-- also do with honey-- What is USE of Bee Larkspur-- =Toad Orchis= How many flowers
513q21. in Bauers case of orchidiae Where does J. Hunter USE expression of "male principle of arrangement."--
555m142 inct to destroy contagious disease.-- (Useful to USE term instinct, when origin of heredetary habit c
558m153 he immense difference. between man,-- forget the USE of language, & judge only by what you see. compa
584n071 to,» self preservation, (knowledge of enemies). USE of muscles, progression.-- use of senses.-- know
584n071 edge of enemies). use of muscles, progression.-- USE of senses.-- knowledge of location ducks & turtl
585n077 means of measuring cells, which is faculty, they USE this faculty instinctively; watchmaker has facul
586n082 the knowledge of size is merely judged by eye, & USE of limbs &c, or it result from mere impulse to s
588n091 s have reasons, because they have memory.-- what USE this faculty if not reason.-- or does this reaso
616o039 y be a startling expression, & so foreign to the USE of ordinary language that the onus probandi migh
632j53r ill be better always to refer to the author if I USE these facts p. 280. adduces provision of seeds f
639j28v - Macculloch p. 260 intimates canines no special USE to Man. Applicable to Bell's sneering-theory.--
127a125 > simply clinal lines. dipping so & so may be USED East-clinal lines & c & .-- But Siberia was onc
198b114 or branching S. H What does the expression mean USED by Cuvier, that all animals (though some may be
212b165 of Zoolog Soc told me an Australian dog he had, USED to burrow like fox.-- a sort of internal bark.
275c120 eestest in England lost courage-- (Bull-dogs are USED because they have no scent!) Mr Wynne) at end o
276c122 The expression hybrid & fertile Hybrids, may be USED to varieties, as well as species-- as formation
278c129 his it means nothing.--There should be some term USED, when there is series. Could I not give Catalog
284c149 such occur?» now some such «characters» rule are USED by Naturalists in their test of value of charac
309c219 to mules in their genitals & even to a limb not USED The only cause of similarity in individuals «we
311c226 a & Van Diemen's land diff.-- Habits can only be USED «in classification» as indication of structure
312c231 t, in the new cross.-- In the Bantam clubs, they USED to fix on the kind wanted, colouring of each fe
325c267 sticated plants; where function has ceased to be USED as tendril into stump Library of useful knowled
332d003 e.-- My Father Water-in the hair a century since USED to be called Worm Fever, as used much more latl
332d003 a century since used to be called Worm Fever, as USED much more latley diseased Mesenteric glands.--
334d009 s.-- Fox says a cousin «one of Mr Strutt» of his USED to breed to Common &⫰Muscovy Ducks.-- English.
347d050 5 Most clearly shows that genus expresses as now USED almost any group.-- ⫰all groups natural (p 6) a
356d069 ugh I arrived at them quite independently & have USED⫰ them since) the line of proof & reducing fact t
366d112 he NW. Coast blinking to keep out flies might be USED⫰» The wild ass has no cross. how comes it that t
375d135 crop. causes a dearth then in Spring. like food USED for other purposes as wheat for making brandy.-
386d166 bably law of nature that any organ. which is not USED is absorbed.-- this law acting against heredeta
416e071 ertain forms & not others.-- Term variety may be USED to gradation of changes which gradation shows i
525m024 , looking ashamed of himself.-- Squib at Maer, USED to betray himself by looking ashamed before it
543m099 bridge, during his time, almost an absolute fool USED to play regularly with D'Arblay of Christ of gr
543m099 blay of Christ of great genius, & yet invariably USED to beat him-- The son of a Fruiterer in Bond St
558m153 n grins & stamps with passion. can expression be USED more correctly than this for C. Sphynx.-- In th
565n008 er of lower lip are depressed & opposite muscles USED to when angry sneering is in progress.<--> the
574n039 subject see <A> ⫰ I think this argument might be USED to show language had a beginning, which my theo
583n070 these qualities of imitation & education may be USED as argument.-- for instinctive knowledge is not
587n089 ion very different from emotion.-- The former is USED with regard to the senses. Reason does not lead
592n102 that the signs invented for Deaf & dumb school & USED between Indian tribes are Many the same.-- Phil
593n105 inct <perverted> handed down & down.-- mem. Nina USED to get into hay & make a nest for herself.-- th
607o025 ommambulism or insanity. free will (as generally USED) is not there present, but he acts from motives
612o035 tive and unavoidable.-- +Can the word willing be USED without consciousness, for it is not evident, w

Page **(Key Word)**
613o036 ey result from organization of brain; «[LHC] not USED by Kirby» (:analogy:-- as races are formed or m
056r112 ess of terrestrial renovation & so is Volcano a USEFUL chemical instrument.--Yet neglecting these fin
170bIFC s> brain makes thought Glen Roy B C. Darwin All USEFUL pages cut out Dec. 7th. /1856/ (& again looked
188b069 ghtly different localities, so that they become USEFUL to know what is species.-- In proof that struc
309c221 ceeding from an accident. New England farmer,-- USEFUL could not leap fences:-- Dr Lang on Polynesian
325c267 sed to be used as tendril into stump Library of USEFUL knowledge. Horse, Cow, Sheep.-- Verey. Philoso
327cIBC pt. transitu et analogia commentatia Library of USEFUL Knowledge on Horse & Cow & Sheep Clarke's Trav
364d104 sh feather, by crossing-- It seems from Lib. of USEFUL. Knowledge that sheep originally. black. & Yar
390d179 d Buffon Suites de.-- Horse & Cattle Library of USEFUL Knowledge Bell's Quadrupeds the effects of bre
455eIBC made sensitive The function of sleeping someway USEFUL.-- it is only the association which is useless
458t002 using cotton &c instead of natural substances-- USEFUL perversion of instincts-- Beechey's Voyage Vol
467t100 on in young flowers. The abortive stamen are of USEFUL height.-- In Lupine, Bees «frequent» & seem to
491q01v is in these that male organs (not being always USEFUL). fail-- Really good subject for experiment.--
555m142 case instinct to destroy contagious disease.-- (USEFUL to use term instinct, when origin of heredetar
566n010 Chapt I on passions of mankind, as being really USEFUL to them: this must be studied. before my view
588n090 ter is invariability.-- if explained by habits, USEFUL to itself, how gained. reason? or some unnamed
595n125 think are better instances of instincts (highly USEFUL as only means of communication) in man, than s
624o051 past time, hence passions]CD «although perhaps USEFUL at present to some extent.» Hence this is the
192b084 no.> sex not having been determined.-- so with USELESS wings under elytra of beetles.-- born from bee
229b233 son channels, but not having them, instance of USELESS structure-- Smith thinks several species of Rh
257c058 in Old Red Sandstone.-- Nautili in----. it is USELESS to speculate «not only» about beginning of ani
278c128 laced in different subgenera, then it would be USELESS, but the formation of subgenera is empirical,
288c159 most dismally, very rarely bark.-- are almost USELESS not the least notion of hunting, or keeping wa
307c215 ed abortive in the womb.-- if these apparently USELESS organs do indicate such origin, then we are bo
386d166 ake a head; the other part may surely absorb a USELESS member.-- in fact they do it in disease & inju
410e052 ng mere «single specimens in» skins worse than USELESS.-- yet there is no cure «I may say all this, h
455eIBC useful.-- it is only the association which is USELESS. Granfather's Handwriting, to compare with my
542m096 of an habitual movement is continuing it when USELESS.-- <&> therefore it is here continued when the
542m096 here continued when the uncovering the canine USELESS.-- The distinction «as often said» of language
544m104 who does not produce children. or after he has USELESS. does not affect race. argument for early educ
556m146 , like bulls & horses.-- 1838 good instance of USELESS muscular tricks accompanying emotion.-- when h
556m146 ng they do the same. although it is then quite USELESS-- Cats kneeding when old, like kittens at the
560mIBC tural History of Babies-- Do babies start, (ie USELESS sudden movement of muscle) very early in life
585n076 owe in birds, seeing the sun &c are absolutely USELESS when applied to birds, which have been carried
599o004 what is trembling palsy? Expressions N [Old & USELESS notes about the moral sense & some metaphysica
635j55r ter of a physical law, «& is therefore utterly USELESS-- it foretells nothing» because we know nothin
216b179 17. Java Fossils 10 out of twenty have ANALOGUES USES this word «for» similar in the Indian sea.-- De
257c056 ut induction leads to other view.-- Till we know USES of organs clearly, we cannot guess causes of ch
349d052 ety & species!! (given in note.)-- Macleay <met> USES term genus when it is so many steps from a head
458t002 Vol II p. 502. «Bengal Journal» The Taylor Bird USES pieces of thread, picked «up-» instead of spinn
114a094 n -- I see Lyell talks of different composition USING difference in metamorphic action which I give a
275c119 cience--. the highest endowment of lofty genius USING geograph distribution of animals, <as> I use (n
301c199 t.-- p. 7. Mr Blyths arguments against squirrel USING reason in hiding its food is applicable to any
458t002 d of spinning-- better case than English birds, USING cotton &c instead of natural substances-- usefu
554m140 w, or with a blanket.-- these cases of commonly USING, foreign bodies, for end. most important step i
069r152 ween two ranges mysterious!-- Mem. SUBSIDENCE USPALLATA of which no trace except by trees The structu
070r154 trees, I could not see trace of Subsidence at USPALLATA.-- ¿If crust very thick would there be undula
125a120 do p. 447 & 449. «& 450». On Vertical trees. USPALLATA.-- do p. 473. on great Iceland stream. the 90
308c217 a &c &c Two savages, two species.-- «discussion USTIL, unless it were fixed what a species means» civ
150g038 th much drainage & far more earthy than what is USUAL-- Lines die away where slope less., best develo
451e176 & vow made at it. Both his parents were of the USUAL colour. His sister is an albino like himself sa
526m030 ee will,-- he improves the faculty according to USUAL method, but what urges him,-- absolute free wil
532m057 way, & must not <this> «running away» have been USUAL effects of fear.-- the state of collapse may be
607o025 re present, but he acts from motives, nearly as USUAL (a) one well feels how many actions are not det
610o033 our instincts-- surely in animals according to USUAL definition, there is much knowledge without exp
026r022 ncepcion Says Echinites. Encrinites. Asteriae, USUALLY petrified into a peculiar cream-coloured Limes
156g068 nite, all on these three shelves soil is <the> USUALLY slaty Point of rounded not scooped rock on <be
157g073 buttresses Yet certainly shelf 4th <near> only USUALLY contains many pebbles, but I believe this is c
174b013 ment applies only to hybridity.-- genera being USUALLY peculiar to same country, different genera dif
550m125 thers experience in same time.-- Pleasure more USUALLY refers to the sensations <it> when excited by
440e147 to bird as to allow any other kind of animal to USURP its place.-- & therefore the degree of injuriou
242c017 Malacca «& Cochin China» are said to have orang-UTANG & Pongo in common-- Galiopithecus common to Mo
388d172 bserved that transmission bears no relation to UTILITY of change-- hence harelips heredetary, disease
623o050 ne> doubts) « <I cannot> » [LHC] On the Law of UTILITY Nothing but that which has beneficial tendency
625o50v ks, on analogy of pleasure of imagination «the UTILITY part being blended & lost» & moral sense.-- My
629o55v sirable to be taught,-- all are agreed general UTILITY (3) It is other question whether any thing is
276c124 eavoured to advance cause of truth It is of the UTMOST importance to show that habits sometimes go be
634j54v an original thought, or design, pursued to its UTMOST exhaustion, & till it must be abandoned for an
086a003 ar character of Vegetation. -- So accustomed to UTTER confusion in Europe, that the simplicity of Ven
313c234 ficult case» Does this law of duration apply to UTTER extinction or rapidity of specific change.? <On
568n018 with Mammalian structure. «-- American monkeys UTTER pleasant plaintive cry--» The taste of recurrin
578n054 rdice, or cheating.-- one does not blush before UTTER stranger,-- or habitual friends.-- but half & h
637j57v as bee & butterfly» inconvenience.! extinction, UTTER extinction! let him study Malthus & Decandoelle
071r157 ear Tucuman and Salta. towards the Vermejo was UTTERLY overthrown by earthquake with great destructio
073r162 ly rare in all parts of the globe". p. 113 How UTTERLY incomprehensible that if meteoric stones simpl
189b072 es generate «other species», their race is not UTTERLY cut off:-- like golden pippen. if produced by
285c152 together.-- The bottom of the tree of life is UTTERLY rotten & obliterated in the course of ages.--
293c172 in brain people in fevers recollecting things UTTERLY forgotten» --it is scarcely more wonderful, th
418e083 l developement is spread over [not located] it UTTERLY untold,-- what is added to the composition of
443e155 in beings with which have fluid sperm.--]CD I UTTERLY deny the right to argue against my theory, bec
527m035 ith which a castle in the air is interrupted & UTTERLY forgotten--, so as to feel a severe disappoint
599o005 originates slowly-- if these speculations are UTTERLY valueless-- then argument fails-- if they have
599o007 ehold the last rays of & & or grand chorus are UTTERLY inexplicable-- I cannot <admit> think reason s
635j55r e character of a physical law, «& is therefore UTTERLY useless-- it foretells nothing» because we kno
123a116 eatly aided by considering space formed-- great VACUUM-- by dike.-- Mem. however. veins of segregatio
193b090 d account for each tribe <being> «acting as» in VACUUM to each other p. 306.--. Chamisso on Kamschatk
540m085 feel so much surprise at male animals smelling VAGINAE of females.-- when it is recollected that smel
483z013 oyage. Coquille's Voyage p 302 Vol II p. 302. VAGINULUS of Lima described "Arion" of Ascension. p. do
548m117 It is singular when looking at a table one has VAGUE idea something is not there, & then when one be
551m127 illusion & insanity the connexion appears to me VAGUE-- Delirium of every degree of intensity-- «in o
547m111 aginary cause to start at once to Shrewsbury., VAGUELY thought of packing up.-- was lying on my back
278c129 xcepting where an Andrew Smith, «Richardson» a VAILLANT, a D'orbigny has travelled this will be most
037r056 no other organic remains.-- On Pampas looked in VAIN for a pebble of any sort; not one was found.--
350d056 is some error-- Observed, nature does nothing in VAIN, therfore organ fitted to animals place in crea
351d056 - does not move per saltum-- yet does nothing in VAIN!! Foetus of man undergoes metamorphosis., heart
546m110 er return could not find paper cutter, hunted in VAIN for it-- ten years afterwards whilst at a meal,
569n022 be grieved-- "il leva les epaules et dit qu'il VALAIT mieux rester a Farroïlap quelque mal qu'on y f
354d062 s not Didelphis «Answered satisfactorily by. VALENCIENNES.» The change from caterpillar to butterfly-
499q09v ts necessary &c &c (a) Mercurialis-- Frog Bit, VALERIAN-- Urtica Dioica Sorrell. Lychnis. Butchers Br
042r068 as did those aerial Volcanos in Germany» In the VALLE del Yeso it is probable that point of Porphyry
059r119 rsquelles se trouvent interrompues par quelque VALLÉES ou par quelque scissures profondes, on les voi
059r119 evers de chacune des montagnes qui forment les VALLÉES ou les scissures.--M. B. thinks these parts in
032r040 long line of coast.--[Fig. 2] Mem San. Lorenzo; VALLEY of Copiapò & parts of coast of Chile.-- Must f
039r060 conduits to volcanoes.-- Talking of the cricket VALLEY «the most remarkable feature in the structure
043r070 ell. Vol I. P. 316. Earthquake of 1812 affected VALLEY of Missisippi & New Madrid & Caraccas.-- Is th
053r101 & Granite districts to be separated by profound VALLEY [.] Sydney. -- Lesson Zoologie Grand tertiary
077r171 ret.-- p. 189. "The small ravins into which the VALLEY of Marfil is divided, appear to have a decided
087a012 Will use argument of proof of slow corrosion of VALLEY of <Patagonia.> S Cruz -- from terrace like st

Page
**************************************(Key Word)**************************************
```
105a072 ----- Coquimbo on. other hand?-- The widening a VALLEY depends on serpentine course.-- the latter (it
105a073 endency of running water to deepen not to widen VALLEY.-- Why is serpentine course result of little i
106a074 . therefore rivers very ineffectual in widening VALLEY.-- it is essentially a deepening agent {P} The
147g020 gree Air 60 degree Below Loch Tulla whole wide VALLEY scattered with few very small & irregular hill
147g020 iver-- No exact terraces but appearances, as if VALLEY had been filled with sloping bed of rubbish Fr
149d030 deposit this Remember however the great Chilian VALLEY Acongua, must there have deposited much-- On o
150g038 to where side ravine enters On opposite side of VALLEY both extend below the Houses The Hills in this
152g046 pid descent of a terrace except at very head of VALLEY indicates new terrace Ballivard 2 miles North
156g068 shelves:  generally angular except near head of VALLEY fragments which had fallen before lake drained
156g071 one uniform slope slightly bending up each main VALLEY.-- & that river alone had modified it-- perhap
157g076 drifted»  here: on very summit no granite-- (in VALLEY «there are» granite) «boulders» {P} cory strea
158g083 Glen Roy at pass {P} River Gorge 4th Sh side of VALLEY Granite blocks on this side (return) between 2
159g087 r fringes of alluvium (?) still higher Slope of VALLEY much more gentle than in Glen Roy, & partly sh
161g095 utted Having crossed the mouth, (deep) of above VALLEY this road level with Peat moss most distinct t
161g095 , & with piece of excised rock lost at point of VALLEY chiefly from rockiness When on other side {P}
040r063 anches (Glaciers very rare) to cause floods in VALLEYS, which must aid in preserving the terraces <..
042r069 s, would have destroyed regularity of slope of VALLEYS.--All my observations of period «& manner» of
044r072 t see any number of dikes in the cliffs.--wide VALLEYS.--central peak small; yet great body of lavas
052r099 equal distance?) 1st cases. -- The terraces in VALLEYS of Chili may be with much truth compared to th
056r109 could efface & obliterate so grand a work?--In VALLEYS one is not sure whether fissures may not have
066r141 utlet at head. Important in forming transverse VALLEYS Ice Sir W. Parish says they have Earthquakes i
105a070 a rush of water will account for filling up of VALLEYS-- subsequent opening a medial gorge by slow er
105a071 sea harmonizes well with character of mouth of VALLEYS &c; Pampas.-- If blocks above their parent roc
105a071 introduce  other causes to explain «alluvi» in VALLEYS Lowe in his paper says land shells found with
106a075 y a deepening agent {P} Therefore when we have VALLEYS of this structure. as the inclination in all p
111a087 sis of Voyage: many observations on heights of VALLEYS in Chile Geograph. Journal Vol. VII p.  216.--
120a109 - The case of the shingle in the great Chilian VALLEYS must be profoundly considered. if elevation ne
121a112 August.   1838 Near Woolwich there are plains & VALLEYS just like Patagonia, & many shells in parts on
129a131 block  of rock moved by gale-- When writing on VALLEYS. «Tertiary strata of S America» read parts of
148g029 at land, with terraces of each side of the two VALLEYS corresponding as in Andes, composed of sand  &
429e116 any plants skirt each side of the great N & S. VALLEYS, which penetrate Pyrenees but are found no whe
429e116 but are found no where else not even in branch VALLEYS-- M. Ramond offers no explanation.-- Poet Cowp
106a075 hannel instead of wider.-- This applies to all VALLIES (except mere talus «over cliffs edge» of which
021r005 leavage. laminae fold round them; Quote this. VALPARAISO Granitic nodules in Gneiss. Epidote seems co
043r071 f Earthquakes in Chili:-- Chiloe. Concepcion. VALPARAISO (Copiapò & Guasco). yet whole territory vibr
047r081 do  not think so also elevating Earthquake of VALPARAISO. (1822) no great wave on record. -- «also ne
049r088 close contact? -- «Cordillera???» Porphyry at VALPARAISO; Epidote -- Must we look at regular greensto
095a037 I. Geograph. Journ. Analysis of Poenig Voyage VALPARAISO Dr. Gillies in MS. letter in Sir. W.  Parish
121a113 of small rise at Stockholm.-- analogous to my VALPARAISO case. Consider profoundly all consequences o
134a141 ast p. 204 do. do p. 210. Height on road from VALPARAISO to Santiago p. 328. dead trees on Isthmus of
149g030 her hand remember modelling power of sea N of VALPARAISO are those animals subject to much  variation
091a025 1. J. p 194. Fact of dust blown far out to sea VALUABLE; because transportal of Minute seeds-- L. Ins
191b081 ttle) Geographic distribution of Mammalia more VALUABLE than any other, because less easily transport
260c069 t!!?-- Wilson's American ornithology a mine of VALUABLE facts,. regarding habits range & all kinds of
268c095 o plants not being adapted to Air!! p. 11--&c. VALUABLE paper on quadrupeds of Van Diemen's land, whi
315c241 -- Paper read in 1837. semestre I suspect some VALUABLE analogies might be drawn between habitual act
344d041 des Developement of sexes in Caterpillar. very VALUABLE facts-- they are eating foetuses, as young of
385d163 ) <Think> Last litters are considered the most VALUABLE. because smallest sized dogs.-- one litter bi
415e067 umbers of fatal diseases in mankind, «the more VALUABLE domesticated animals» no doubt is owing to th
415e069 Whewells Inductive History. Contains many most VALUABLE references See if any law can be made out, th
461t041 f mammfers) in old beds & existing species is VALUABLE because it shows no innate power of change  &
484z014 thout Tropics in Northern & Southern America-- VALUABLE & practicable deed Caricaridae wanderers.--??
494q004 - How few wild animals are propagated,, though VALUABLE as show. & curiosities!! What is price of fox
543m101 gs-- Inherited Habits: Have Effect in Bones is VALUABLE soil in its seaward course,-- we sink into su
633j54r o say plants created to <arrest> «prevent» the VALUABLE soil in its seaward course,-- we sink into su
274c115 <water>  land birds,,-- Grups of very different VALUE have their represetatives;, the rasorial may be
277c126 lishing specific names & giving subgenera. true VALUE.-- as in Opetiorhyncus. fulginous, (a) Falklan
284c149 » rule are used by Naturalists in their test of VALUE of character-- Macleys rule is converse, «when»
284c149 of character-- Macleys rule is converse, «when» VALUE of character depends on non-variation, & not on
284c149 not  on extension ¿these go together? Therefore VALUE of organs vary in different group. & Not  known
287c158 gans in same state. in different animals, & the VALUE of those organs, when changed in different anim
302c202 > osculant groups between two circles «of equal VALUE» must be so from characters of analogy.-- see m
312c227 rs & sharks, being spotted, & colours of little VALUE Dr Smith if black & white Man crosses; children
348d051 lance,-- how can we estimate this amount, when <VALUE> no scale of value of difference is or can be s
348d051 estimate  this amount, when <value> no scale of VALUE of difference is or can be settled,-- I believe
349d053 on stock.-- all genera, common stock.-- so that VALUE can only be judged of in each «separate» line o
398e004 ns, not affecting organic forms, that the whole VALUE of the geological chronology depends, that most
415e068 only being cleared off by fatal diseases.-- The VALUE of a group does not depend on the number of the
558m151 ore than any other point of structure takes its VALUE. from its connexion with mind, (to show  hiatus
546m110 ding a book, with ivory paper cutter, which she VALUED, & she was suddenly called to go on the lawn t
462t051 & Sumatra & dissimilarity of forms-- yet how VALUELESS this objection, when one thinks of  different
464t063 his wonderful discoveries= negative facts are VALUELESS= monkeys= Owen has described a greatt Strutho
465t079 - any negative argument against-- monkey-man, VALUELESS.-- May not several generations have been conf
599o005 es slowly-- if these speculations are utterly VALUELESS-- then argument fails-- if they have, then la
633j53v l advantages thus to explain the curling of the VALVES of the broom.-- or the springing of other seed
399e010 lliams. Narrative of Miss. Enterprise, p. 497. VAMPIRE bat abound in the Navigators & at Manguia, but
030r036 s. & natural position» position at N. S. Wales & VAN Diemen's land.-- Whole coast S. of Concepcion wh
059r120 Bailly talks of much granite on all East side of VAN Diemen Land. All the Calcareous rocks which hard
182b050 t certainly distinct species between Australia & VAN Diemen's land. & Austral & New Zealand Mr  Gould
191b080 same as on continent-- 3 Paradupasi in common to VAN Diemen's land & Australia From the consideration
207b152 t volcanic islets not well fixed.-- Peron thinks VAN Diemen's land long separated from Hobart  Town--
217b187 us in Sandwich islands-- A genus with species in VAN Diemen's land and Tierra del Fuego.-- Araucaria,
225b219 ence New is only hope.-- New Zealand «compare to VAN Diemen's land.» glorious fact. of absence of qua
226b220 os mouse (?) brought by canoes Ceylon & India.-- VAN Diemen's land & Australia. England & Europe.-- T
235b263 ke Irish & English Hare bear upon this.-- Why do VAN Diemens Land people require so many, imported an
235b264 stem according to three elements How is Fauna of VAN Diemens Land & Australia (blank) Falconer's rema
242c017 s. Horsfield. Diard. Duvaucel. Leschenault Kuhl. VAN-Hasselt, Reinwardt «Forrest» authors on E. I«ndi
246c027 lland.-- New Guinea scarcely differs more from, <VAN Diemen's land.> Australia more than Van Diemen'
246c027 from,  <Van Diemen's land.> Australia more than VAN Diemen's land.-- Vol II p. 8 no snakes on isles
247c028 rctica «also Taeniatole austral» «caught» Chile, VAN Diemen's land & Cape of Good Hope V. p. 44 of th
268c095 Air!! p. 11--&c. valuable paper on quadrupeds of VAN Diemen's land, which appear diff. from Australia
278c130 Mitchell's authority) in Australia, & several in VAN Diemen's land is most important as showing forme
278c131 or Mitchell does not think that dog was found in VAN Diemen's land.-- V. 1s. Number of Geographical J
300c197 ed to the Str of Magellan.-- Change of habits in VAN Diemen's land. Study Mr Blyth's papers on Instin
311c225 80 degree & great diff of Latd. p 355 Echidna of VAN Diemen's land & Australia different Temminck Fau
311c225 (?!)  82 mammalia 293 Phalangista of Australia & VAN Diemen's land diff.-- Habits can only be made «i
314c239 fact)  Be cautious about Goulds case of birds of VAN Diemens Land & Australia.-- The wombat (Brown) i
319c257 Australian fossils intermediate between those of VAN Diemen's land & Australia proper.-- Irish Elk ca
353d061 use knows three species of Paradoxurus common to VAN Diemen's land & Australia well developed <tits>
365d107 rmous time which it must have taken to separate. VAN Diemen's land from Australia &c &c Sept. 14th. W
400e011 tic [...]t in part near Timor, & to Europaean in VAN Diemens land, where there is close species of el
414e064 on the plains of Bolivia-- strictly fossil «& in VAN Diemen's land»-- they have been exterminated  on
436e137 rld.-- Quartz of Falkland.-- Old Red Sandstone-- VAN Diemen's land.-- Porphyries of Andes. A familiar
500q10a cies?-- Are there any fine doubtful species from VAN Diemen's Land? or New Zealand? Babington about d
504q013 o see if pollen naturally carried, on account of VAN Mons views-- Also PEAS-- N.B. I think very likel
511q018 of colour «& length of ears» & skeleton, & skin= VAN. Voorst often writes to Lowe (7) In breeding. po
540m086 negros of Africa, (or again the black man of <B> VAN Diemens land & the energetic copper coloured nat
173b009 ficult to separate Every character continues to VANISH, bones instinct &c &c &c non fertility of hybr
```

Page
549m123 , & are now, like all other structures slowly VANISHING-- the mind of man is no more perfect, than in
086a005 ndency of America connected with its elevation. VAPOUR from below-- Malte Brun «Salt Lakes» Siberia m
090a024 tion be serious? if so other cause besides thin VAPOUR bringing planets to an end? Fragmentary granit
103a065 Plutonic rocks argument against great bodies of VAPOUR. according to Hopkins theory.-- general presen
119a104 continents, if globe be considered as condensed VAPOUR.-- inequlities are required to start with (& d
034r044 ments P. 386 Mem. Lyell's fact about sulphuric VAPOURS in East Indian Volcanos Gypsum Andes Mem. Beec
074r165 chytes. also Azores. We here have case of such VAPOURS washing a rock» Veins concretionary; concretio
137a149 ank] Ed. New. Phil J. 1838. p. 72. on metallic VAPOURS condensed from furnaces do/p. 84 on the effect
505q014 eds of Geranium pyrenaicum. small white-flowered VAR. with abortive stamens.-- show crossing & ¿hered
123a117 x.-- at Buenos Ayres 20½ quadras from river; 20 VARAS from surface in tosca.-- remnant of Megatherium
348d050 o be of importance in inverse ratio to their VARIABILITY.-- (Now caeteris paribus these will be the o
583n069 wers in individuals of the same species with VARIABILITY of reasoning power in one species man.-- fal
285c153 nder care of Man. & external circumstances not VARIABLE.-- Animals have voice, so has man. Not saltus
302c202 ed.-- we must argue reversely. WHERE CHARACTER VARIABLE it is (one of analogy or) LATELY ACQUIRED. In
539m082 obtain it far more easily, in proportion to VARIABLENESS or power of intellect.-- Some complicated t
032r041 cted from same point. & changes. «(changes in VARIATION?)» as in Cordillera.-- From poles to Equator
110a085 ternal changes are in process. connected with VARIATION of compass & these may cause «or be effect of
132a139 ars to me necessary) as M. Parrots shows from VARIATION in strata earth a very bad conductor.-- shows
149g031 Valparaiso are those animals subject to much VARIATION which have lately acquired their peculiaritie
230b236 producing more fertile offspring.-- 1st. All VARIATION of animal is either effect or adaptation,∴ an
232b247 een very unlike Southern Europaean ones.-- "a VARIATION played on secular refrigeration".-- <The Phen
284c149 ose it is oldest, & therefore lest subject to VARIATION.-- + <being> good for generic divisions [+] o
284c149 rse, <when> value of character depends on non- VARIATION, & not on extension ¿these go together? There
284c149 hose groups, where it remains less subject to VARIATION" Dr. A. Smith. knows lots of instances of rep
292c169 re complicated the animal the more subject to VARIATION. therefore sexes or two animals:-- «When» sex
298c192 <whole cross» «all organs» vary in plant. The VARIATION in character of leaf of plants is remarkable
302c202 IRED. In pigs number of vertebrae. subject to VARIATION. therefore lately acquired.-- I fear «great e
312c232 in some extent counterpart, mutilation being «VARIATION» produced in shortest possible time. Mr Willi
354d065 with Bison, at some resemblance as if the "VARIATION in one, was analogous to specific character o
360d091 ut Yarrell's law, is it local (not artificial VARIATION) which impresses offspring most.-- «& not tim
364d104 s in domesticated animals, that the amount of VARIATION is soon reached-- as in pidgeons no new races
366d112 erred to <other> Book M.» Is there any law of VARIATION. -- «(as Hunter supposes with Monsters)» if a
371d118 varieties of roses!!! proof of capability of VARIATION.-- Saw his collection of «Humming» birds, saw
388d171 f Mother.-- We see in a litter every possible VARIATION from being very near mother, & some very near
393dIBC reed is uncertain Is there any peculiarity or VARIATION common to any zoophyte «born in succession» w
400e012 m sure a very good case, might be made out of VARIATION analogous to specific variations.-- Kerr's Co
403e023 rong passage might be made-- why seeing great VARIATION in external form of varieties, do we suppose
404e024 my mice is good, because it is an involuntary VARIATION made by man, common to every individual & the
410e051 ed. of co relation of parts, from the laws of VARIATION of one part affecting another.-- (I from look
411e054 sible contingent circumstance.-- \the laws of VARIATION of races, may be important in understanding l
411e055 serve perfectly to specify types, & limits of VARIATION, & hence indicate gaps-- by this means the
416e071 dapted to external conditions.-- (hence great VARIATION in each birth) from man arbitrarily destroyin
425e103 tile hybrids, bears relation to capability of VARIATION?? my theory says so.--» March 6th. Mr Bentham
436e136 gainst this-- -- analogy will certainly allow VARIATION as much as «the» difference between <pi> spec
438e141 of characters «from antiquity» prevents their VARIATION, which is not improbable as Mr Herbert does n
446e162 nywhere?-- Yes on the continent is there more VARIATION in its character.?» No--well characterized.--
460t017 ization (ie those laws which prevent infinite VARIATION in every possible way.-- the laws which deter
460t017 kinds of monstrosity, & determine the kind of VARIATION «& sporting» in flowers & domestication of an
472s01v e her some She means to try this year. Little VARIATION in the 60 one brighter with mere traces of bl
498q007) It is most important to ascertain amount of VARIATION in plants raised by Scions, as Elms. &c &c--
501q10a this very important for it wd show that such VARIATION is not a generic or specific character, but
502q11v f different Cabbages most carefully to see if VARIATION equal in flower with leaves.-- strawberries G
507q15v 46). Book describing amount of Horticultural VARIATION? Henslow knows only on Citrons 47. Ficus cari
510q017 greatly-- how very interesting to see if any VARIATION in varieties. G. St. Hilaires law of Balancem
512q019 its in Dorsetshire sheep migration of coots-- VARIATION in hounds= An ugly calf <turns> sometimes tur
591n098 - the other one nothing will frighten-- hence VARIATION in character in different animals of same spe
636j56v luded that Plants would be subject to extreme VARIATION as long as crossing with other varieties was
055r108 s over immense areas. (& the counterbalancing VARIATIONS) of rain. = The Bulk of sediment «daily» sea
181b043 ties from three elements, from the «infinite» VARIATIONS, & all coming from one stock & obeying one l
206b148 es of man: these races may be overlooked mere VARIATIONS consequent on climate &c-- the whole races a
287c158 hen changed in different animals.-- + whether VARIATIONS in eye of vertebrate afford better character
287c158 e of vertebrate afford better character, than VARIATIONS in eye of mollusca. [+] These questions ma
312c232 stant. ¿ whether mutilations non-heredetary & VARIATIONS produced in short time in some extent counte
355d066 es) "It is well worthy of examination whether VARIATIONS are produced only in those character which a
400e012 e made out of variation analogous to specific VARIATIONS.-- Kerr's Collect of Voyages Vol 8 «p. 46» C
412e057 much true milk.-- \ <Species. are innumerable VARIATIONS>. Every structure is capable of innumerable
412e057 s>. Every structure is capable of innumerable VARIATIONS, as long as each shall be perfectly adapted
412e057 istency «owing to their slow formation» these VARIATIONS tend to accumulate. «on any structure.» L'In
430e117 positions: this is important as showing small VARIATIONS in offspring of wild animals.-- grateful & i
633j53v not be detrimental accidents, & domesticated VARIATIONS show us accidents may become heredetary [pro
390d179 but. «on> » breeding animals that have neither VARIED from their stock, for to breed (as Sir J. Sebr
390d179 rs is by impliance to breed two which have each VARIED from parent stock.-- The very theory of genera
422e095 umber of animals in the world depends, of their VARIED structure & complexity.-- hence as the forms b
438e142 is element in change, as in Dahlias]CD all much VARIED breeds both plants & animals have long been su
607o025 on,-- education under the influence of others-- VARIED capability of receiving impressions-- accident
618o41v s had a certain intensity (& the experiment was VARIED) then might it now be said, that blueness caus
049r088 f veins, is there said granite in close contact VARIES in nature, -- Does not granite at C. Tres Mont
099a050 of consequence.-- Therefore < S of inclination «VARIES with chemical attraction &c.» becomes measure
104a068 e.-- thus dikes resemble Solubility of fluids VARIES with temperature ¿ with pressure? Salt on surf
284c149 to Fleming p. 32 "where it (mode of generation) VARIES according to the species, is manifestly of
298c192 the moisture.-- Henslow thinks if leaf of plant VARIES, <whole cross» «all organs» vary in plant. The
354d065 Paradoxurus Phillippensis. Philippines Man have VARIES the range-- Argue the case of Probability. has
400e012 tips of ears take any colour?)-- length of tail VARIES, & character of fur-- I am sure a very good ca
537m075 e Mackintosh.-- Must grant, that the conscience VARIES in different races.-- no more wonderful than d
583n069 n one species man.-- false instinctive pointing VARIES.-- p. 18. Animals possess strong imitative fac
586n078 ay do all this instinctively «yes because power VARIES in breeds,» something of kind oneself knows in
032r041 n of Coral limestone in intertropical)» hence VARIETIES of substances ejected from same point. & chan
180b038 -- This requires principle that the permanent VARIETIES produced by <inter> confined breeding & chang
187b064 efore adaptation), & to obliterate accidental VARIETIES, & to accomodate itself to change, (for of co
187b064 self to change, (for of course change even in VARIETIES is accomodation). Now this argument applies t
192b083 ppins, & Golden Pippen, goldens-- hence-- sub- VARIETIES & hence possibility of reproducing any variet
192b085 le, that each is not more change than we know VARIETIES can produce.-- Therefore all genera MAY have
193b090 nimal produces in course of ages ten thousand VARIETIES, (influenced itself perhaps by circumstances)
194b098 ts.-- Does not Lyell give some argument about VARIETIES being difficult to keep on account of pollen
196b107 is not generally known that Ireland possesses VARIETIES of the furze, broom, & yew very different fro
196b107 rent from any found in great Britain, British VARIETIES are also found in Ireland-- There must be pro
200b125 eds in salt water & has any tendency to form VARIETIES? Ed. N. Phil. J. Morse found in Virginia p. 3
208b154 edition We need not expect to find <species>, VARIETIES, intermediate between every species.-- Who ca
209b155 no interest in perpetuating these particular VARIETIES. If species made by isolation; then their dis
210b158 me? Von Buch distinctly states that permanent VARIETIES. become species. p.147 «p. 150»-- not being c
210b159 lants not cryptogamic but I . 2,53.-- In know VARIETIES. there is analogy to species & genera.-- for
210b159 reyhound.-- In plants. do the seeds of marked VARIETIES produce no difference. if they do.--there pro
213b169 tion--» Tabulate Mammalia on this principle. <VARIETIES> < «Races» > Man in savage state may be calle
213b171 species is their exactness,' but do not know VARIETIES do the same, May you not breed, ten thousand
213b171 t be greyhounds?-- Yarrell's remark about old VARIETIES affecting the cross most well worthy of obser
216b180 -- Now the point will be to find whether know VARIETIES in plants do so.-- As in cacti &c.-- as in
218b191 ll have a peculiar «constant» aspect. That is VARIETIES, though of trifling order are formed by natur
222b203 less strongly marked) between what are called VARIETIES.-- NB. one mother bringing forth young having

Page	**	(Key Word)	**
223b210	mbs & new ones are formed) but yet propagates	VARIETIES	according to same law with animals?? Why are
224b215	her species., man already has produced marked	VARIETIES	& may someday produce something else., but no
234b256	Waterhouse says he is certain there are local	VARIETIES	«of colour & size, but not for[m]» of animals
234b256	rom Swansea.-- Again Waterhouse finds certain	VARIETIES	of a Harpalus. common at South end, but <rare
235b262	dency to vary? Is not man thus circumstanced;	VARIETIES	of dogs in different countries a case in poin
236b278	lax.-- Read Swainson [blank] In production of	VARIETIES	is it not per saltum.-- Isld bordering contin
239cIFC	fects of colour on parent, white room How are	VARIETIES	produced, by picking offspring? Instances of
239c001	est effect on offspring. Thus presuming those	VARIETIES	to be oldest which have long been known in an
239c001	appen with Australian dog & any of our common	VARIETIES	. He has no doubt that Chesnut, for many gener
239c002	, because oldest variety.-- -- He says of two	VARIETIES	of pidgeon, although having skulls so differe
240c004	ll these facts clearly point out two kinds of	VARIETIES	.-- One approaching to nature of Monster, here
242c017	ffspring not picked.-- as men do. when making	VARIETIES	.-- Voyage of Coquille. Zoolog. p 19.. Tapir,
245c024	found in all Malasia-- & oceania, offers many	VARIETIES	in each place to puzzle naturalists.-- p. 372
246c026	ut I do not see how it is known that they are	VARIETIES	& not species.-- Vol :694. King-fisher of Eur
246c026	htly different. Who can say whether species &	VARIETIES	P. 708. Columba Oceanica (Less) inhabits Caro
247c029	coloured as a herd in England"-- Black & Grey	VARIETIES	of rabbits thus handed down for nearly 70 yea
248c030	he most hypoth: part of my theory, that «two»	VARIETIES	of many ages standing, will not readily breed
248c030	hen we grant «(which can be shown probable,)»	VARIETIES	may be made in wild state, there will be pres
248c030	breed together.-- We see even in domesticated	VARIETIES	a tendency to go back to oldest race, which e
248c034	se therefore will be chiefly heredetary.-- If	VARIETIES	«produced by slow causes, without picking» be
249c034	wo> such as itself.-- therefore two different	VARIETIES	will produce hybrids but not varieties. which
249c034	ferent varieties will produce hybrids but not	VARIETIES	. which are not deeply impressed on blood., wi
249c035	or such as not capable of producing again The	VARIETIES	of Cardoon are cases <sp> like those of Primr
252c045	IS & Mice of America «good case on account of	VARIETIES	in N America» «some doubt, from want of knowl
258c059	Does not atavism relate to this law.?-- Local	VARIETIES	formed with extreme slowness, even where isol
262c073	w nicely things adapted--.-- These «aberrant»	VARIETIES	will be formed in any kingdom of nature, wher
265c083	counteracted by nature by crossing with other	VARIETIES	»-- but <accidental> changes after birth do no
271c106	expression for ignorance Two grand classes of	VARIETIES	.; one where offspring picked, one where not.-
275c120	ntal barriers-- Mr Yarrell.-- says my view of	VARIETIES	is exactly what I state.-- &c picking varieti
275c120	rieties is exactly what I state.-- &c picking	VARIETIES	. unnatural circumstance Ld Orfords had breed
275c121	& got some yellow & others yellowish & white	VARIETIES	by picking the yellow one & crossing with duc
275c121	sorry to find Mr Yarrell's evidence about old	VARIETIES	is reduced to scarcely anything.-- almost all
276c122	sion hybrid & fertile Hybrids, may be used to	VARIETIES	, as well as species-- as formation of species
276c122	ns of generation?! By profound study of local	VARIETIES	laws of change-- whether beak (as it appears
277c126	s. till neutral ground ascertained, call them	VARIETIES	. but two ostriches good species because inter
279c133	t propagated by nature.-- Whole art of making	VARIETIES	may be inferred from <this> facts stated.-- +
280c136	expect even if two mules bred or two certain	VARIETIES	, they would go back to grandfather, which is
281c137	eries is graeduated.-- Dr Beck doubt if local	VARIETIES	should be remembered, therefore do not consid
281c137	re do not consider it as proved that they are	VARIETIES	, (though that would be best).-- Argue the cas
285c153	itution, which other marked difference in the	VARIETIES	«made by» of Nature & Man.-- The constitution
294c176	manent.-- It will be easy to prove persistent	VARIETIES	in wild animals-- but how to show species-- I
294c176	argument must rest upon analogy & absence of	VARIETIES	in a wild state-- it may be said argument wil
299c194	lis perhaps, offers another case of permanent	VARIETIES	in wild state-- The two. former produced by d
299c195	oduced by difference of <locality>. station &	VARIETIES	chiefly produced by cultivating parent in ric
308c219	ving same idiosyncrasy, cause of fertility.--	VARIETIES	not produced as by nature. if so. the habits
324c268	zine (paper on Geograp. range Study Buffon on	VARIETIES	of Domesticated animals see if law's cannot b
335d015	primitive stock) are heredetary: «Hybrids of»	VARIETIES	is different because not long in blood.= The
337d020	ack, or not being perfect would perish.-- The	VARIETIES	of the domesticated animals must be most comp
337d020	d, not one is picked out, & few even of local	VARIETIES	approaches quite to wild local variety.-- our
337d020	quite to wild local variety.-- our Europaean	VARIETIES	must be very unnatural-- Italian Greyhound is
339d025	world into Zoological Provinces» according to	VARIETIES	of Man.? «In Australia. plants E & W very dif
342d033	cies-- Mr Blyth remarked only near species or	VARIETIES	produce heterogenous offsprings.-- «are not t
343d038	ture destinies of mankind, some of species or	VARIETIES	are becoming extinct. others though the negro
345d044	d of seclusion in breeding. how easy races or	VARIETIES	are made.-- The Highland Shepherd dogs, colou
349d053	eparate» line of descent.-- <& here limits of	VARIETIES	being constant. it would be exceedingl wrong
354d063	changes caterpillar to Butterfly.-- When two	VARIETIES	of dogs cross, Erasmus says it look lik[e] In
354d065	in genus."-- Is there any law of this. Do any	VARIETIES	of sheep «evidently artificial» approach in c
362d100	geon, in which it appears that all the <bird>	VARIETIES	«,now know» were then <pr> existing.-- he has
363d101	ion of cooing.--» & compare them with all the	VARIETIES	.-- Habits of rock pidgeon. (I suspect Pennant
365d107	their is no difference between <t> "permanent	VARIETIES	" & species. he overlooks-- <restric> those <r
365d107	ate of womb).-- One can perceive that Natural	VARIETIES	or species., all the structure of which is ad
371d118	y.-- Sept. 23rd. Saw in Loddiges Garden. 1279	VARIETIES	of roses!!! proof of capability of variation.
393dIBC	nous? When bird fanciers say the throw of two	VARIETIES	is uncertain do they mean they cannot tell fi
401e014	a great many will not return to all sorts of	VARIETIES	, which he attributes to crossing.-- Cape Broc
401e015	- thinks it probable that great part of those	VARIETIES	may be due to impregnation from other apple t
401e016	s red cabbage produced from passage from many	VARIETIES	, & probably would take long before all the st
403e023	hy seeing great variation in external form of	VARIETIES	, do we suppose bones will not change in numbe
406e036	od. with regard to marriage of individuals, &	VARIETIES	of same species & to different species-- some
410e053	species being described.-- but then permanent	VARIETIES	in same country, must be distinguished, from
410e053	ountry, must be distinguished, from permanent	VARIETIES	not in same country.-- The traces of changes
415e070	ferences See if any law can be made out, that	VARIETIES	are generally additive, & not abortive: with
417e077	o of «the» same <species>, variety, as in two	VARIETIES	, & this we might expect, as the difference be
424e102	n Edinburgh.-- Why does Fleming consider them	VARIETIES	& what says Jenyns to it?-- -- In argument of
426e106	dantly infertile hybrids, & the fact that old	VARIETIES	do not so much affect first race, as it does
426e107	: but the intervention of domesticated ie new	VARIETIES	destroys the appearance of this series & make
426e107	ries & makes one think that one large body of	VARIETIES	are fertile & make mongrel, & other great ser
429e113	n or not trace history of first appearance of	VARIETIES	of domesticated animals, yet as we know how m
430e118	t for all species. even if it will for all.--	VARIETIES	are made in two ways-- local varieties, when
430e118	all.-- Varieties are made in two ways-- local	VARIETIES	, when whole mass of species are subjected to
430e120	imensions, must <almost> be considered merely	VARIETIES	. & even Mr Sowerby is coming to this conclusi
434e129	ere is some difference of habit between these	VARIETIES	, so that they have been thought to be differe
437e141	ype when brought back to home. (& yet all the	VARIETIES	of Brassica certainly not becoming Brussels S
438e141	recognize any difference in crossing between	VARIETIES	& species, yet the amount of may depend on ma
439e144	t so much crossing as I think,, or that these	VARIETIES	have become as fixed as species, & prefer the
441e150	owers, Hydrangea -- black bullfinches-- & all	VARIETIES	must be presumed to be result of such laws.--
442e153	d plants never are, namely on stocks of other	VARIETIES	& we know that the kind of stock greatly affe
453e183	islds.-- <Cocks> The possibility of different	VARIETIES	being raised by seed is highly odd-- as it is
468t111	spur to Lupine two species of Larkspur -- two	VARIETIES	of Cistus Speedwell to Rhododendron-- <Loasa>
491q002	f hybrid & Mongrel offspring 2. How have late	VARIETIES	of Peas &c been obtained? 3.. Whether the viv
492q002	bert in vol IV. Hort. Transact.-- 4.. Are any	VARIETIES	of Cabbages not attacked in bad years from Ca
492q002	wild Bananas of Otaheite seedless;-- are all	VARIETIES	seedless-- if so. how have varieties been for
492q002	are all varieties seedless-- if so. how have	VARIETIES	been formed?-- 8. Can any annuals be budded.
493q003	species-- answered «by Henslow» see notes In	VARIETIES	is there any difference in off spring from A.
494q004	ecies similar.-- if not how is it in <allied>	VARIETIES	(9) Cross largest Malay with Bantam-- will eg
495q005	ross with cowslip pollen.-- as these are wild	VARIETIES	. Is any intermediate form found wild a The Le
498q008	s.-- (9) In the nurseries, when «seed of» the	VARIETIES	of Cabbages, peas, beans, as raised, do the S
498q008	,-- as the Bees will mingle the infinitesimal	VARIETIES	which must occur.-- ¿is «it» not these infini
498q008	ust occur.-- ¿is «it» not these infinitesimal	VARIETIES	, which counterbalance each other? (10) Is num
500q010	(24) Do Bees distinguish species, they do not	VARIETIES	.-- (25) Does the yellow white Butterfly depos
500q010	he yellow white Butterfly deposit eggs in all	VARIETIES	of Cabbage. (26) Do deer Keepers cross the br
500q010	in Cattle in Chillingham Park-- What Book on	VARIETIES	&c of deer. Contests of sexes.-- Q.30) March
503q012	Book no information & Hope about Silk worms.	VARIETIES	effects of domestication-- said to require Se
507q15v	oldts spinning seed.-- (50) Any cases of wild	VARIETIES	plants growing together. under same condition
510q017	w very interesting to see if any variation in	VARIETIES	. G. St. Hilaires law of Balancement Wm Yarrel
513q020	r «sometimes» all on one side, as in crossing	VARIETIES	Amongst varieties cross one with abortive tai
513q020	on one side, as in crossing varieties Amongst	VARIETIES	cross one with abortive tail or horn, with an
514q21.	lpine Australia Flora= Banana's seedless-- 20	VARIETIES	in mountains of Tahiti. Dr. Boott-- says cari
543m101	nces.-- The facts of half instincts. when two	VARIETIES	are crossed as in Shepherd dogs-- Inherited H

Page   ***(Key Word)***
559m155 out early production of foreign seeds.-- many VARIETIES.-- Rev R. Jones has it.-- very curious book.-
636j56v reme variation as long as crossing with other VARIETIES was prevented Do races of peas become intermi
637j57v of perfection Get instances of adaptations in VARIETIES.-- greyhound to hare.-- waterdog hair to wate
022r007 .--Mem tres Montes. ((Henslow Anglesea)) great VARIETY in nature of a dike.--Mem. at Chonos & Concepc
078r176 W & SE). «Vol III» Mexican Cordillera "immense VARIETY of Porphyries which are destitute of quartz, &
171b003 ings, become permanently changed or subject to VARIETY, according «to» circumstance,-- seeds of plant
172b008 hange is effected at once, -- something like a VARIETY produced-- --[every grade in that case surely
180b037 h respect to extinction we can easily see that VARIETY of ostrich, Petise may not be well adapted, &
192b083 rieties & hence possibility of reproducing any VARIETY, although many of the seeds will go back.-- Ge
192b083 f the seeds will go back.-- Get instances of a VARIETY of fruit tree or plant run wild in foreign cou
203b138 that old races when mingled with newer, hybrid VARIETY partakes chiefly of the former Eyton's paper o
204b140 does Wolf & Dog cross? Mr Yarrel thinks oldest VARIETY impresses the offspring most forcibly-- Esquim
218b189 s coexisting. proves this-- but when Man makes VARIETY these are vitiated.-- This barely applies to p
228b229 the offspring of a male & female animal of one VARIETY going back ¿whether this going back may not be
228b230 again, not a monstrous plant, but any marked. VARIETY.-- Strawberry produced by seed«s»??-- Universa
239c001 theory» tells me. he has no doubt that oldest VARIETY, takes greatest effect on offspring. Thus pres
239c002 on, offspring most like latter, because oldest VARIETY.-- -- He says of two varieties of pidgeon, alt
246c026 Centropus (Coucal) of Java & Phillippines, has VARIETY at Madagascar, Calcutta & Sumatra, but I do n
253c047 of Guzerat by Capt. W. Shee. considered merely VARIETY.-- red form of skull very slightly different.-
264c082 ds Birds vary much (more than shells) owing to VARIETY of station inhabited by them-- Timor. Australi
273c111 place.-- Humming bird there is strongly marked VARIETY,:.in the Tyrannidae.-- Milvulus-- Even flying
275c120 ssing with common Polish cock «is not that old VARIETY» & then recrossing off spring. till size dimin
275c121 one & crossing with duck bantams procured old VARIETY.-- The pidgeons which have such different skul
280c136 futurity.-- This is right way of viewing it.-- VARIETY when long in blood, gets stronger & stronger,
289c162 winter. «It would be curious to know whether a VARIETY could be transmitted more easily in those born
299c195 n rich soils & their seeds produce «offspring» VARIETY. wild carrot. made into biennial domesticated
333d005 fferent».-- crossed with «un»common cat, exact VARIETY unknown., three kittens, alike each other, par
334d012 ite-tailed squirrels, which form a marked wild VARIETY.-- doubtful whether all are white. Fox says th
337d020 local varieties approaches quite to wild local VARIETY.-- our Europaean varieties must be very unnatu
337d020 reyhound is probably the effect of <sev> local VARIETY many times changed together with some training
341d032 olog Gardens there is hybrid of Penguin duck a VARIETY of Muscovy) with goose!!) Dr. Bachman regulari
347d049 es, or lower than in old days: «for a very old VARIETY will be harder to vary, & therefore more apt t
349d052 does not know any difference between permanent VARIETY & species!! (given in note.)-- Macleay <met> u
355d066 lian dog) or donkeys to Zebras.-- «Mr Herberts VARIETY of horse, dun-coloured with stripe approaches
359d088 ers. (as if they were a good species, or local VARIETY, & not effect of breeding in & in like our pid
366d108 duced from these.-- this instance of monstrous VARIETY. which could not have been persistent in natur
371d128 vourable, will <deteriorated> continue of same VARIETY as long as life lasts, yet they cannot transmi
384d161 s life so that successive buds do differ-- any VARIETY. is not handed down. but is handed down for som
400e013 time Uncle J. does not suppose one aboriginal VARIETY.-- for they are all made by fertilizing one pl
401e014 tion from turnips or other stocks. Says if any VARIETY of apple be sown, all sorts come up from it. l
401e015 strawberry not so absurd.-- Thinks-- that such VARIETY as red cabbage produced from passage from many
410e053 tle, <in> whether the individual be species or VARIETY, but to discover physical laws of such correlat
414e063 d had the picking she would make <them> such a VARIETY far more easily than man,-- though man's pract
416e071 destroying certain forms & not others.-- Term VARIETY may be used to gradation of changes which grad
416e075 d & perfected Hence difference between races & VARIETY? «Man picks the Male, instead of allowing stre
417e077 ollow same law in two of «the» same <species>, VARIETY, as in two varieties, & this we might expect,
417e077 «indeed» (independent of sexual differences) a VARIETY. The offspring of «true <pare» hermaphrodite, w
423e099 ral species in my fossils.-- CD[If cases of one VARIETY in upper part of bed-- & another in lower is v
426e106 ee» p. 103» would lead me to think that a <do> VARIETY of one species would cross easier with 2d spec
427e110 ct produced would be different.-- same way one VARIETY of «animal» dog does not prefer other. but pro
439e144 es, & prefer their own pollen to that of other VARIETY.-- «Elizabeth & Hensleigh. seemed to think it
466r095 perial Lily & many other flowers-- My view of <VARIETY acquired» « <character> » of characters being
471tf07 use unknown.-- "<when we find such an endless VARIETY of form in the same» organ "manifestation of d
513q020 manner & perfection &c &c &c--if so probably a VARIETY, not specific character= Cross Rumpless fowls
640j167 f dogs. would diminish, whilst the long legged VARIETY would prevail.-- Not separately; NB. These vie
047r082 also the similar fact at Concepcion? Read the VARIOUS accounts & see if fall is not the first very e
052r098 or seaward transportal of drift matter.-- Give VARIOUS cases. [Fig. 6] {P} A advancing coast to Seawa
093a033 mber 8th 1877 (Memoranda so far distributed to VARIOUS subjects) Dr. A. Smith informs me that in the
305c211 Unity in thinking and acting principle in the VARIOUS shades of <dif> separation between those indiv
317c251 delphia Vol VII. Part II/. 1837 account of the VARIOUS hares «some since discovered» of N. America, a
411e056 in habits before form.-- I have already given VARIOUS examples The Pipe-fish is instance of part of
494q005 riments-- Plants Raise seedlings surrounded by VARIOUS bright colours, any effect? and silk caterpill
526m027 her animal & probably the only one affected by VARIOUS knowledge which is not heredetary & instinctiv
605o019 tempt «by one common principle» to explain the VARIOUS causes of those sensations, which we call meta
606o021 nstantaneous. that we cannot ever perceive the VARIOUS operations which the mind undergoes in gaining
247c029 ours, like the cattle, <">which I say "are as VARIOUSLY coloured as a herd in England"-- Black & Grey
097a043 llicle of a blackish colour like a dull & poor VARNISH, which I conceive to be analogous to the black
459t013 ll more.-- So Penguin impresses its form both on VARS of Lion: Annales des Sciences‖ (4) Prolifickne
508q016 wild animals breeding at the Cape.-- ‖About two VARS: of Lion: Annales des Sciences‖ (4) Prolifickne
032r040 so would the size of the triangular mass removed VARY.--The gradual rising continuing. a another slop
047r083 (as seen in swell) the undertow & overfall must VARY proportionally Partial shrinking after elevatio
100a052 effect of gravity, versus some fault explaining VARY dip & inclination.-- which last is strong chara
148g025 like dogs), but they thought the breed liable to VARY-- I asked this question in many ways & received
171b003 t, hence we see generation here seems a means to VARY. or adaptation.-- Again we <believe> <know> in
172b008 refore final cause of life With this tendency to VARY by generation, why are. species are constant ov
175b017 e» especially with some change probably <change> VARY quicker Unknown causes of change. Volcanic isld
195b100 ut not coronata.-- We know that domestic animals VARY in countries, without any assignable reason.--
205b142 II. , p 565. Consult-- Says types most subject to VARY where intermixture precluded.-- Kirby Bridgewat
206b147 first start must be greater, & this number would VARY at each lustrum, & the calculation of chance of
235b262 d. & long interbred <p> having great tendency to VARY? Is not man thus circumstanced; varieties of do
236b280 ometimes we might expect disseminated species to VARY a little, but such should not be general circum
256c053 e sophism. in Lyells statement that some species VARY more,, than what makes species in other animals
264c082 bits in making Out classification of birds Birds VARY much (more than shells) owing to variety of sta
274c114 type.-- Now here I must observe these characters VARY in degree in last instance hardly at all develo
284c149 enus to be founded on such characters, as do not VARY in the species of it.-- «where does such occur?
284c149 on ¿these go together? Therefore value of organs VARY in different group. & Not known in single ones-
297c185 unger.-- p. 65. Aberrant groups few in numbers & VARY much in character, divided into many small gene
298c192 leaf of plant varies, <whole cross» «all organs» VARY in plant. The variation in character of leaf of
305c209 lose to European species, and the sexes of which VARY in colour of plumage in same remarkable manner
306c214 Minor.-- tail like setters. long ears-- colours VARY, but form constant.-- The females of some moths
336d019 ordinary laws, the character of offspring would VARY, or rather they would not have offspring-- On t
342d033 se or with parent species.-- The hybrids do not VARY (ie the hens all alike & Cocks all alike) More
346d047 ve breedin[g] [not located] half breed liable to VARY. I asked this in many ways, but received same a
347d049 days: «for a very old variety will be harder to VARY, & therefore more apt to be extinguished.--???»
349d051 It will rest upon the discovery what characters VARY most easily:-- those which do not vary being fo
349d051 haracters VARY most easily:-- those which do not VARY being foundation for chief divisions.≠ p. 7. «I
355d066 duced only in those character which are seen to <VARY among> be different in species of same genus."
355d067 in different SPECIES of genus Sus. do vertebrae VARY? «See Cuvier Ossemens Fossiles» Although no new
405e031 haracters are partly retained, therefore colours VARY in same Manner as they would vary, if in wild s
405e031 refore colours vary in same Manner as they would VARY, if in wild state; thus mark on ear of cats, do
417e075 .-- Now in the ass-- there is little tendency to VARY. & hence offspring are hybrids,-- Mr G. B. Sow
424e103 ve done so in hot countries.-- CD[Camel does not VARY more than others» & horse in lesser degree,-- h
424e103 ridise the Camel» like ass «same way some plants VARY more than others» & horse in lesser degree,-- h
431e122 ion. the gardener separates a plant he wishes to VARY-- domesticated animals tend to vary. March 20th
431e122 he wishes to vary-- domesticated animals tend to VARY. March 20th. Phillips in Lecture in Royal Insti
436e136 ld <If> It «may» be said, that wild animals will VARY, according to my Malthusian views, within certa
441e149 ructure of the parents.-- Thus would a Crab tree VARY if planted in rich soil, I presume not, but its

Page ***(Key Word)***
441e151 sion, that every organ is become fixed. & cannot VARY.-- which all facts show to be absurd.-- As ther
447e163 st persistent-- if form exceedingly difficult to VARY.-- the run of chances, would prevent it varying
465t081 y immigration which tends to make the destroyers VARY; so that we here see reasons-- why no perfect g
490q001 ologist Does individual Shell or insect or group VARY more in one country or district than in another
490q001 her species Hooker says the species of Aquilegia VARY much in their spurs & Ranunculus in the nectari
495q005 & see what the effect will be.-- will seedlings VARY much more than cuttings &c (4) Raise annuals or
496q006 er appear equal-- & see whether proportions will VARY, which will show that such proportions not effe
497q06v tstemon, which have abortive parts, whether such VARY.-- Do Bees go to Sweet Peas, IMPORTANT, for if
501q10a nd, doubtful species-- Does any genus of Plants. VARY & hard to separate specifically in one country
501q10a rs or Helix. Or does any «one» species of plant, VARY in one region of Europe & less in another regio
509q017 rocess developed into ribs.) & does its abortion VARY, according to Bentham's Remark. Horse or cow.--
513q020 tate for skeleton== Does the tumbling of pigeons VARY in manner & perfection &c &c &c--if so probably
515q021 are reduced from normal number, they are apt to VARY in number in individuals of same species Eyton
529m040 mitation come into play.-- the train of thoughts VARY no doubt in different people., an agriculturist
584n071 odpecker: but this is not so,, the instincts may VARY before the structure does; & hence we get over
623o050 btless grow together [RHC] This feeling seems to VARY in races of man. & certainly in «species of» an
022r006 heat. it is so found in Anglesea, amongst the VARYING & dubious granites.--Wide limits of this miner
099a050 . < ∴ where little inclination, little force & VARYING direction.--> Therefore in PILE of mud from Tr
104a067 ssure going right through superincumbent mass (VARYING hardness,-- takes time to trace) from few dike
206b148 tance one man killing another.-- So is it with VARYING races of man: these races may be overlooked me
230b241 irds. Mr Goulds' case of Willow wren) & others VARYING in wild state to show that we do not know what
231b241 ding;--or as others would exjudge it amount of VARYING in wild state.-- When breaking up «the primeva
232b248 e me an account of parasitic animals of beasts VARYING in different climates Those will not object to
305c209 erican.-- Have not Ruffs & Reeves a remarkably VARYING plumage for wild birds-- At Zoolog Gardens the
342d033 considered as distant species?? or is time the VARYING element). Then do those SPECIES which breed mo
400e012 difficulty in establishing good groups.-- ears VARYING so much,-- kind of fur-- (do tips of ears take
447e163 o vary.-- the run of chances, would prevent it VARYING. A plant <producing> propagating itself by bud
515q022 stumps of tailess dogs & cats.-- (3) Hounds-- VARYING-- (4) About blended instincts of the geese whi
207b150 Says Coniferous structure intermediate between VASCULAR or cryptogam. (original Flora) and Dicotylede
048r086 urable; In what part of the globe are there such VAST numbers of wild animals. both species & individ
069r151 gone to? Intermediate space protected.-- Oh the VAST power of the ocean! Make a grand analogy betwee
154g060 d all well & good, but how came river to do this VAST quantity when during repose of lake it did but
302c202 fore lately acquired.-- I fear «great evil» from VAST opposition in opinion on all subjects of classi
338d023 eding, for otherwise we could not understand the VAST number of domesticated races.-- Athenaeum. p. 5
372d131 generation.-- there is no caterpillar state; the VAST difference of two kinds of generation shown by
408e045 rious facts.-- The Himmalaya. case, bears on the VAST changes even in that quarter of the world-- --
605o18v t a little to the effect: as when we look at the VAST ocean from any height.-- 6 That the superiority
614o037 t life) instincts.-- this loss is compensated by VAST power of memory, reason &. & many general insti
605o19v associated with it. as the idea of Deity. with VASTNESS of Eternity. which superiority we transfer to
196b108 ve development; for instance none--?, of the <VEBTETRATA> «vertebrates» could exist without plants &
060r125 780, Sept. 21st. 1817.--p 371. Webster Antarctic VEG:-- Study Ulloa to see if Indian habitation above
030r035 Chart) Every winter torrents must bring much VEGETABLE matter from thickly wooded mountains, probabl
117a101 ata according to composition thinks sand with VEGETABLE remains formed near coast, limestone deep wat
367d113 s of cock & Neck of Bull.-- is most common in VEGETABLE feeders. because males always armed in carniv
423e096 life.-- for otherwise a frost if killing the VEGETABLE in one quarter of the world would kill all of
571n029 s knowing poisonous «herbs:» & man not.-- ¿no VEGETABLE good «for man» to eat poisonous?-- How did an
611o034 is definition given in full.-- [LHC] ¿Has any VEGETABLE or animal matter been formed by the union of
612o035 LHC] ¿ in Corallina are not two kinds of life VEGETABLE and animal strictly united? [RHC] It is easy
196b108 nsects & worms.-- In abstract we may say that VEGETABLES & mass of insects could live without animals
196b109 Yet unquestionably animals most dependent on VEGETABLES. of the two great Kingdoms.-- Principes de Z
270c103 omes more developed in higher animals than in VEGETABLES. p. 243 radiate animals <tu> plants turned i
453e183 ighly odd-- as it is not so with the esculent VEGETABLES-- how is it with hollyoaks, flaxes &c &c? Mr
480z008 erman.-- Stuttgart ranks these bodies amongst VEGETABLES in Linn. Soc.-- Mr. Donn Carmichael Linn. Tr
502q11v > «early» do characters of races of different VEGETABLES & animals come on.-- Compare calves.: Compar
611o034 orms of living beings namely «one individual» VEGETABLES, the vital laws act definitely (<like> «as»
453e183 s first appeared.-- "Storia della Riproduzione VEGETALE". by Gallesio. Pisa 1816 p. 27. Dr. Holland.
046r079 al on the afflux of the former. -- Ascension. VEGETATION? Rats & Mices. At St Helena there is a nativ
060r125 to see if Indian habitation above regions of VEGETATION.--«I can find nothing.» Mem Carolines quotat
062r128 parasitic plants, Cacti: & with limits of no VEGETATION at S. Shetland = Great contrast of two sides
064r134 n very wretched cou[n]tries thinly covered by VEGETATION. Rhinoceros quite in deserts.--Much struck w
065r138 oli & Vesuvius Investigate with greater care. VEGETATION & climate of Tristan D. Acuneha. Kerguelen L
085a002 ent. which agrees. with peculiar character of VEGETATION. -- So accustomed to utter confusion in Euro
108a080 calcareous springs near coral reefs.-- Where VEGETATION luxuriant it might be almost as well said pr
241c016 rib of Ferns in South Sea (Indio Polynes: <>) VEGETATION far East) Ann: des Sciences. Semptemb. 1825
260c069 offspring.-- white, snow.-- the fine green of VEGETATION,-- ¿account for colour of bird in district.
274c114 n trace structure for insects & structure for VEGETATION.-- In conversation in Museum-- I could not d
296c184 roper dampness seeds can arrive quick enough» VEGETATION of peak-- altogether original. owing to bein
363d102 in the former place breeds in «flags> «thick VEGETATION» in swamps-- (owing to barns, perhaps, not b
398e004 anges» would such a change produce in climate VEGETATION &c.-- It is the circumstance of small physic
407e041 hat period) Analyse this,-- consider state of VEGETATION, & conchology,-- shells of Africa ought most
446e161 that the climate is on the decline, as far as VEGETATION is concerned, in parts of the Northern «Fren
324c268 Bevan on Honey Bee Dutrochet Memoires sur les VEGETAUX et animaux.-- on sleep & movements of Plant/
041r066 ircle,.{P}, had in its middle a short <fissure> «VEIN» terminated each way, which little vein was lik
041r066 issure> «vein» terminated each way, which little VEIN was like the rest of these thin veins which pro
074r165 (& Chiloe case, at least corelation)--Galapagos VEIN. vein of secretion.--metallic veins follow moun
074r165 iloe case, at least corelation)--Galapagos vein. VEIN of secretion.--metallic veins follow mountain c
075r167 as been slightly broken up, & has there not been VEIN «of iron» discovered?-- Klaproth analysed silve
077r171 one, limestone & <many> «other secondary» rocks. VEIN traverses both Clay slate, Porphyry North 52 W,
077r171 SW. with respect to latter doubts whether bed or VEIN (very like that of Spital of Schemnitz in Hunga
078r175 n parallel to the direction & inclination of the VEIN".-- at Zacatecas the veta grande has same direc
078r175 (Principal veins)» 25 degree to 30 degree to NE. VEIN of Moran 84 degree NE. of Real del Monte 85 deg
078r176 vitreous felspar".-- p. 215 Same metal in Tasco VEIN in Mica Slate & overlying Limestone Balls of Si
097a042 ra Can Greenstone dikes. be residue of quartzose VEIN in higher parts? & felspathic veins?-- Mr Poule
156g072 th» shelf This shelf at head where <granite &> «VEINED» gneiss <unite> «occurs» abundantly with perfe
022r007 of a dike.--Mem. at Chonos & Concepcion. P. 417 VEINS of quartz exceedingly rare Mem C. [Cape] Turn P
026r020 nded.-- Fox Philosoph. Transactions on metallic VEINS. 1830 P. 399.-- Carne. Geolog. Trans: Cornwall
026r021 elted rocks The frequent coincidence of line of VEINS & cleavage is importants; veins appearing a gal
026r021 ence of line of veins & cleavage is importants; VEINS appearing a galvanic phenomenon, so probably wi
041r066 ch fossils. lived in groups or not. Ferruginous VEINS of this figure {P} in sandstone: evidently depe
041r066 «quantity of iron» being there in excess.-- If VEINS {P} are secretionary, so are all those plates i
041r066 ich little vein was like the rest of these thin VEINS which project outwards.-- In Patagonia. the ble
045r074 on the surface bespeak the changes. -- metallic VEINS solution of silex & many other phenomena I do n
046r078 re they eliminated.--«Sulphur last.--» Metallic VEINS likewise must separate ingredients if we look t
049r087 facts of Passages marked by do.» discuss quartz VEINS, there contemp--yet similar ones in Clay. Slate
049r088 sides. V. Lyell P 355 Vol III. constitution of VEINS, is there said granite in close contact varies
055r106 c &c &c Vol II. Chapt VIII. p. 97 at Potosi the VEINS run from North <inclining> to South. inclining
055r106 > to South. inclining a little to the West: the VEINS which follow this direction are thought by the
068r147 a. & it is nothing odd to find them injected by VEINS & mass[es] [Fig. 8] {P} (A. B. C, now grown sol
073r163 .--in parts of table granits & gneiss with gold VEINS visible:--"Porphyries of Mexico may be consider
073r164 hibole frequently only vitreous felspar: = gold VEINS in a phonolitic porphyry. = several parts of N.
073r164 l parts of N. Spain great analogy to Hungary. = VEINS of Zimapan offer zeolite. stilbite. grammalite.
074r164 hrom. of lead. orpiment. chrysop[r]ase. opal:-- VEINS in Limestone & Grauwacke: Silver appears far mo
074r165 here have case of such vapours washing a rock» VEINS concretionary; concretions <dt> determined by f
074r165)--Galapagos vein. vein of secretion.--metallic VEINS follow mountain chain. there after NW <W>.-- «s
074r165 tions richest-- Humboldt has urged phenomena in VEINS, chemical affinities like in composed rock. gra
074r165 posed rock. granites syenite» «strangling &c of VEINS can only be accounted for by concretionary acti
075r166 asco in "alpine limestone" = "The wealth of the VEINS in most part totally independent of the nature
075r166 ect". = In the Guatemala part. (& Chiloe do) no VEINS discovered. Humboldt suggests covered up by vol

076r168 contain silver are peculiar to that part of the VEINS, nearest to the surface of the earth."--p. 156.
076r168 same minerals <as> there, that are found in the VEINS of Kongsberg in Norway.--namely dendritic silve
076r169 ica: In all climates distribution of silver «in VEINS» very unequal sometimes disseminated <[...]> so
077r171 me with that of the veta grande of Zacatecas, & VEINS of Tasco & Moran--of Guanaxuato to SW. with res
078r175 as same direction as Guanax.--the other E & W.--VEINS richest not in ravins or along gentle slopes. b
078r175 iclinal line?). -- Mines of Catorce «(Principal VEINS)» 25 degree to 30 degree to NE. vein of Moran 8
079r176 lying Limestone Balls of Silver ore occur in do VEINS. At Huantajaia. Humboldt says, mur of Silv.[,]
088a013 - from terrace like structure-- Intersection of VEINS prove, that there are at least several attempts
090a025 gmentary granite showing schistose structure (& VEINS appearing): mem. Henslows Anglesea solution of
097a042 of quartzose vein in higher parts? & felspathic VEINS?-- Mr Poulett Scrope. talks of Trachyte, "super
098a047 hence the thick wedges of feldspar in gneiss.-- VEINS in septaria. a kind of concretionary process (a
102a060 - the branching cracks-- only bear relations to VEINS in primitive rocks-- Are substances soluble und
110a086 par, (some crystals being red) «with» cleavage, VEINS of pyrites, few curious fissures; base in part.
120a110 . Scotland at least 2200. Jura 4000 feet.-- The VEINS of segregation in Greenstone of Salisbury Craig
123a116 volcanos.-- if so why not metals. The theory of VEINS will, I suspect be greatly aided by considering
123a116 ormed-- great vacuum-- by dike.-- Mem. however. VEINS of segregation in Salisbury Craigs Letter from
136a147 llips in Lardner Vol II p. 73.: some remarks on VEINS: Phillips in Ladner Vol. II p. 80-- some remark
136a147 adner Vol II p. 125. Good discussion on mineral VEINS p. 125 to 129 & p. 135--160 & 162 [blank] Ed. N
137a149 densed from furnaces do/p. 84 on the effects of VEINS of slag in iron Furnaces affecting to some dist
145a003 nic fox., an instance of Provincial breeds. [3] VEINS of Segregation in Salisbury Craigs [4] Salisbur
145a004 Craigs [4] Salisbury Craigs V. Specimens-- {P} VEINS, amygdaloidal-- as well as base not always para
157a075 composed of remarkable gneiss with red granite VEINS & quartz, & garnets.-- Boulders as before certa
483z013 ons & my own: & Capt. King's p 453-- Planariae VELELLAE (Less) parasitical on Vellellae in Atlantic O
483z013 53-- Planariae velellae (Less) parasitical on VELLELLAE in Atlantic Ocean Gould agrees with D'Orbigny
252c042 nce to Rüppel. travels (what language?) Hyena «VENATICA» <of> Cape found in Desert of Korto & Steppes
093a032 been sublimed into the tertiary limestones of VENDARQUES. Mem sublimation of sulphur to form salts of
247c028 e.-- p 22. a Gecko on St Helena.-- <in 1813, a VENEMOUS snake was> one Gecko on Isle of France Scincu
146g013 hepherd dogs The Patches of Conglomerate on S. VENTANA, excellent instance, how accidental is the pre
086a003 r confusion in Europe, that the simplicity of VENTANA'S «Quartz.» unmixed is very pleasing; owing to
404e025 curious mechanism of respiration, or rather VENTILATION peculiar to <the class> «some orders» of cru
068r146 rated by the repeated trifeling injections.--Old VENTS would keep open long after emersion, but improb
303c204 of present animals.-- He·will be bold. I will VENTURE to say unphilosophical. L'Institut 1838. p. 12
534m061 Brain per se thinks is nonsense; yet who will VENTURE to say germ within egg, cannot think-- as well
633j54r mistake they are created for them. If we once VENTURE to say plants created to <arrest> «prevent» th
305c210 insects, so is there none of reason in order of <VER> Mammals.-- Mem Elephants & dog.-- There is one
502q11v wers of wild & tame carrot-- Parsley & Fennel. VERBENA Compare flower of different Cabbages most care
554m139 her own mouth.-- seemed to relish the smell of VERBENA & Pocket Handerchief & liked the taste of Pepp
603o013 are abbreviations» he thus derives from nouns & VERBS-- so that much of EVERY language shows traces o
089a020 l. L Institut p. 192.-- (1837. Peninsula of Cape VERD. volcanic.-- Isle of Gory. rocks encrusted with
317c249 ecus saboeus" said to be monkey of St Jago C. de VERD; same as on coast of Africa.-- Maclay tells me
183b052 avec le chardonneret, avec la linotte, avec le VERDIER" &, <fr> silver gold & common pheasants & fowl
052r099 step = formed streams of lava at St Jago. C. de VERDS Quartz pebbles in the Cordilleras look as if so
208b154 ter parts of world Is monkey. peculiar to C. de VERD'S.--? NO Macleay Name given in Congo Expedition
310c224 self. In Tropical countries (as St Jago Cape de VERDS) the shells in equal periods with Europe would
325c267 brary of useful knowledge. Horse, Cow, Sheep.-- VEREY. Philosophie d'Histoire Naturelle Marcel de Ser
527m031 y.-- & what is frame of mind owing to.--<>-- I VERILY believe free-will & chance are synonymous.-- S
536m073 ich at that time organization gave me to will-- VERILY the faults of the fathers, corporeal & bodily
071r157 that town near Tucuman and Salta. towards the VERMEJO was utterly overthrown by earthquake with grea
470t178 -«Little Dusty & Blue» Butterflies at Clover.--VERONICA--, Ranunculus in numbers =what insect can get
058r118 outes affectent une pente plus ou moins inclinée VERS le rivage de la mer, tandis, au contraire, que
058r118 s le rivage de la mer, tandis, au contraire, que VERS le centre de' l île, elles presentent une coupe
059r119 ouches parallèles et inclinées du centre d'lile, VERS la mer; ces couches ont entre elles une corresp
047r083 n Patagonia plains. long periods of rest & vice VERSA more likely to be coincidental than single elev
051r095 to rock because it does not decompose. or vice VERSA. Clay slates unfavourable to attachment of many
155g064 nging more «detritus» than they corrode or vice VERSA Same inclination when serpentine might remove,
196b109 insects could live without animals but not vice VERSA. ¿could plants live without carbonic acid gaz.)
493q003 es male offspring take after male parent & vice VERSA = History of Tortoise-shell Cats. as only one s
525m026 mon remark that fat men are goodnatured, & vice VERSA Walter Scotts remark how odious an illtempered
547m112 senses are closed probably by sleep & not vica VERSA. anyhow I might have been quite still, & not at
568m017 ral sense, or instinct,) one feels pain, & vice VERSA pleasure in performing it.-- As soon as memory
571n031 ribing gentle things in gentle language, & vice VERSA.-- almost proves that at earliest times there m
100a052 th consideration. especially effect of gravity, VERSUS some fault explaining vary dip & inclination.-
263c077 falls! But Man-- -- wonderful Man. "divino ore VERSUS coelum attentus" is an exception.-- He is Mamm
242c017 olog. p 19.. Tapir, «des» couroucous et rupicole VERT instances of American forms in East. Ind: Archi
417e076 linear descendant of <Mammferus> «Mammiferous» VERT animal, which would find its place in the Syst
214b175 ossession of Mr Howard Galton, have one of the VERTEBRA, about 2/3 from base of tail, enlarged two ve
420e089 like some molluscous «bisexual» animal with a VERTEBRA only & no head-- !! Handwriting is determined
195b099 rmadillo like covering created.-- passage for VERTEBRAE in neck same cause, such beautiful adaptation
276c124 It is a difficulty how a different number of VERTEBRAE are produced, where, (& in all such structure
276c124 be gradation. See what Eytons young pigs-- if VERTEBRAE much lengthened &c there may be tendency to d
276c125 ugh repeated would cause an unequal number of VERTEBRAE-- ¿Where two very close species inhabit same
302c202 nalogy or) LATELY ACQUIRED. In pigs number of VERTEBRAE. subject to variation. therefore lately acqui
312c228 gs always go against this, without <number of VERTEBRAE» new acquisition, we must [not located] Henry
353d061 s to Eyton's discovery of different number of VERTEBRAE in Irish & English Hare.-- good case these ha
355d067 on) Now in different SPECIES of genus Sus. do VERTEBRAE vary? «See Cuvier Ossemens Fossiles» Although
420e089 ove his ears The head being six metamorphosed VERTEBRAE, the parent of all vertebrate animals.-- must
463t055 a Nov 15th Waterhouse showed me the component VERTEBRAE of the head of Snake wonderful!! distinct!!--
463t057 organ graduating into other is lost, <be> (as VERTEBRAE into skull., two bones of tibia into one.--)
509q017 ive bone. (ask more about the lowest cervical VERTEBRAE process developed into ribs.) & does its abor
515q022 on's statement of English Horses having fewer VERTEBRAE in tail, than Continental horses. {About the
181b043 ome habits may approach animals, & some of the VERTEBRATA invertebrates-- Such on few on each side wil
196b110 iculty of hypothesis of connecting Mollusca & VERTEBRATA, that there must be very great gaps.-- yet s
211b163 n Molluscous animals representing foetuses of VERTEBRATA, &c 1837 p. 370 Owen says Nonsense The distr
290c162 amount of change only & not kind» insects-- & VERTEBRATA & plants. At first classification on generat
306c213 ons. but for ones of very high order. not for VERTEBRATA, but mammalia & reptiles &c Timor is connect
640j167 w species, or rather be very slowly changed & VERTEBRATA much so.-- so far true, but do not fish offe
641j29v e eye being formed in Mollusca, Articulata, & VERTEBRATA, & Planaria, & light affecting plants. in in
181b043 tween birds & mammalia, Still greater between VERTEBRATE and Articulata. still greater between animal
233b252 cilities of change in the articulate than <M> VERTEBRATE. But how does this agree with longevity of s
287c158 ent animals.-- + whether variations in eye of VERTEBRATE afford better character, than variations in
299c196 ed Babington We see gradation to mans mind in VERTEBRATE Kingdom in more instincts in rodents than i
370d117 of the perfection of their separation.-- thus VERTEBRATE blend with Annelidae by some fish.-- But bir
384d162 an in each becomes obliterated, & sexes as in VERTEBRATE tak place.-- ∴ Every man & woman is hermaphr
420e089 ix metamorphosed vertebrae, the parent of all VERTEBRATE animals.-- must have been like some mollusco
196b108 ; for instance none--?, of the <vebtetrata> «VERTEBRATES» could exist without plants & insects had be
347d049 - Articulate animals must articulate. <i> in VERTEBRATES tendency to improve in intellect,-- if gener
418e083 re impressed on it.-- Seeing that <Man> «all VERTEBRATES. [Müller's Physiolog. p.24.]CD» can be trace
633j54v » p 308. Traces the gradation of skeleton in VERTEBRATES & constantly alludes «(& at p. 312)» to the
215b178 iority refers to introduction to Animaux Sans VERTEBRES as latest authority. The case of the tailess
032r039 n. this will at length be checked by increased VERTICAL «height» thickness (DZ) of mass to be removed
038r059 ks to those parts. Are not the dikes generally VERTICAL? if so posterior to elevations? & not sources
041r064 a.--Perhaps agrees with formation of pebbles & VERTICAL trees Grand Seco at B. Ayres; mention about t
044r073 present. day.-- applied by me geologically to VERTICAL movements. In Cord: after seeing small Bombs.
046r080 movement communicated to it as well as by the VERTICAL as lateral movement.--At first one would thin
091a027 direction of transitions clay slate &c nearly VERTICAL Linear earthquake 500 by 90.-- in Syria Geolo
204a049 cretionary action all fluid at once, the films VERTICAL. Ascertain law of attraction of particles of
099a049 on particles of equal weight.--)¿ cleavage not VERTICAL ∵ combined with gravity.-- hence changes in d
120a107 & facility with which the earth is cracking by VERTICAL planes into small pieces-- mem coal-field.--

125a120 181 on do subject do p. 447 & 449. «& 450». On VERTICAL trees. Uspallata.-- do p. 473. on great Icela
131a136 erica Richardson.-- From strata being not only VERTICAL, but turned over in many parts of the world.
146g011 eat Slip, 10 years since three hundred feet in VERTICAL height-- enormous mass thunder storm, many <h
399e006 does not Lonsdale know some case of change in VERTICAL series: Look at whole Glacial period? [not lo
099a048 icles in such cases moved more laterally than VERTICALLY, in concretions more vertically than lateral
099a048 aterally than vertically, in concretions more VERTICALLY than laterally. -- <In Area of this> {P} If
044r074 . In Cord: after seeing small Bombs. without a VESICLE. we may consider appearances of eruption at bo
244c022 ilio bonar«i»ensis (from Buenos Ayres) replaces <VESP.> holds same relation with equator--that Vesp.
244c022 s <Vesp.> holds same relation with equator--that VESP. lasiurus does in North. Hemisphere.-- p. 158 C
244c022 some species which escaped there.-- p. 139. VESPERTILIO bonar«i»ensis (from Buenos Ayres) replaces <
023r012 g fossil remains: Sharks followed Capt. Henry's VESSEL from the Friendly Isles. to Sydney; know by ha
633j53v me heredetary [produce some peculiarity in seed VESSEL]CD if man takes care they are not detrimental.
570n025 both accompanied by depending head., & active VESSELS of skin.-- What difference is there between Sq
612o035 rect «physical» effects of more or less turgid VESSELS; effect of heat, light or shade.) Joining two
146g014 aceous or sandy soil-- These Buttresses formed VESTIGE of irregular terrace perhaps near 300 ft above
465t079 species in S. America. -- so see what a «mere» VESTIGE, is preserved in this country-- same argument
351d056 - Macleay then answered, because nature leaves VESTIGES of what she does-- does not move per saltum--
040r063 preserving the terraces <..> Molina's Case At VESUVIUS. Vol III P. 124. Lyell. dikes have a parting
040r063 obably cold & not warm as sides of a crater as VESUVIUS.-- There may have been oscillations in the up
064r137 eruption [.] give instance of Etna Stromboli & VESUVIUS Investigate with greater care. vegetation & c
077r171 orth 52 W, & is nearly the same with that of the VETA grande of Zacatecas, & veins of Tasco & Moran--
077r171 have a decided influence on the richness of the VETA madre of [misnumbered page] Dr D. remarks. bad
078r175 n & inclination of the vein".-- at Zacatecas the VETA grande has same direction as Guanax.--the other
274c113 -, parrots with claw like lark (NB The La jeune VEUVE parrot though so much on the ground has not thi
308o217 k took of cover of right side, though my hand <VIBRATE> would sometimes vibrate-- «seeing no tea brou
308o217 side, though my hand <vibrate> would sometimes VIBRATE-- «seeing no tea brought back memory» old habi
043r071 araiso (Copiapo & Guasco). yet whole territory VIBRATES from any one shock-- In S. America--continuit
070r154 uld there be undulation? would it not be mere VIBRATION? but walls & feeling shows undulation'. crust
337d021 parts of creation) & another nerve to finest VIBRATION of sound.-- which is impossible.-- Mr Spence
497q06v ids have been formed. (15). What is History of VIBURNUM. or snow-ball-tree. what would result from se
547m112 -- The senses are closed probably by sleep & not VICA versa. anyhow I might have been quite still. &
553m135 apparent, one fixes on imaginary beings, many VICARIOUS, like ourselves) that savages (mem York Minst
303c204 es of modern causes. & considering over the VICCISSIUDES of present animals.-- He-will. be bold. I wi
047r083 rock In Patagonia plains. long periods of rest & VICE versâ more likely to be coincidental than singl
051r095 adhere to rock because it does not decompose. or VICE versâ. Clay slates unfavourable to attachment o
155g064 er bringing more «detritus» than they corrode or VICE versa Same inclination when serpentine might re
196b109 ss of insects could live without animals but not VICE versâ. ¿could plants live without carbonic acid
493q003 ses does male offspring take after male parent & VICE versâ = History of Tortoise-shell Cats. as only
525m026 he common remark that fat men are goodnatured, & VICE versa Walter Scotts remark how odious an illtem
568n017 (or moral sense, or instinct,) one feels pain, & VICE versa pleasure in performing it.-- As soon as m
571n031 s describing gentle things in gentle language, & VICE versa.-- almost proves that at earliest times t
609o029 e was justice.-- No checks were necessary to the VICE of intemperance, circumstances made the check.-
609o029v call vicious.-- (jealousy in a dog no one calls VICE). on same principle that Malthus had shown inco
609o029v iple that Malthus had shown incontinence to be a VICE & especially in the female October q. 1838 perh
406e036 sheep-- all cases: d degree p. 354-- The most VICIOUS dog. will not attack any animal except, dog wh
505q014 bberley says that some Bees are smaller & more VICIOUS. Will try to get me some to look at:-- Was onc
609o029v ve passions--, which being unnecessary we call VICIOUS.-- (jealousy in a dog no one calls vice). on s
258c061 t peculiarity of structure. «hence seals take VICTORIOUS seals, hence deer victorious deer, hence mal
258c061 hence seals take victorious seals, hence deer VICTORIOUS deer, hence males armed & pugnacious (all or
362d099 September 13th The passion of the doe to the VICTORIOUS stag. who rubs the skin off horns to fight--
024r015 a Peyrouse. South of Mocha; 19 miles. 65 Fathoms VIDE facts in Beechey. on NW coast of America off Ca
027r024 Mem.; rapidity of germination in young corals.--VIDE L. Jackson's paper. Philosoph Transact: at R. d
180b039 m America) of non adaptation of circumstances.-- VIDE two. pages back. Diagram The largeness of prese
200b123 t such causes are most obscure. without doubt:-- VIDE cattle: The grand fact is to establish whether
405e035 le cosmopolite man.-- L'Institut. 1838. p. 338.«V[IDE]» Important account of cross of sheep & Moufflo
440e145 & reproduction» is common to all living beings. VIDE Lamarck Vol II. p. 115. 4 four laws Who can say
544m103 rly rest of the mind, with all other faculties: «VIDE page 110, by mistake.» N B. Everything which ha
546m110 how many happy days he spent in such a place.-- «VIDE page 103, supra (by mistake) have lower animals
037r056 s in Granitic countries, enumerate cases. -- M. VIDEO exception, but even there, hills of Basalt & ot
049r087 Porphyry specimens, must be well examined At M. VIDEO «facts of Passages marked by do.» discuss quart
067r144 herefore so wonderful that volcanic rocks at M. VIDEO «Volcano in Pampas» Pasto Earthquake. Happened
251c042 on Capromys. 4 species probably in Cuba (p 271 VIEDO says American dogs silent. Mem contrary asserti
326c266 her has published in first volume of Annales of VIENNA. sketch of south Sea. Botany R. Brown. has cur
452e182 act stated by M. Tournal that skulls found near VIENNA appoximat to Negro form; those from Rhine to t
021rIFC a shorter duration, than the more constant: This VIEW supposes the simplest infusoria same since comm
023r010 re arriving at the Abrolhos shoals ￼ -- N.B. The VIEW of the Volcanos of the chain of the Cordilleras
026r019 allow water: so have the Conglomerates: Yet this VIEW is directly opposed to common opinion The Terti
035r048 tains, which in size are grains of sand, in this VIEW sink into their proper insignificance; as fract
047r082 water manifestly does not travel up.--) If these VIEW are right the coincidental retreat at Portugal
069r150 s strata may be older than (B).-- Most important VIEW Urge curious fact felspar melted gneiss/// QUAR
092a029 rents tend to deposit metals, if in solution. My VIEW of metamorphic in contradistinct to Volcanic wi
103a065 lly intersect metallic dikes: It is an important VIEW being subsequent to dislocation of strata. A ca
107a078 h respect to thickness of crust broken up.----My VIEW of Volcanos &c &c This view will bear much refl
107a078 ust broken up.----My view of Volcanos &c &c This VIEW will bear much reflection on method of cooling-
112a089 d volcanos within Cordillera-- allude to Lyell's VIEW of not discovering dike one end granite & other
118a104 reached the surface Arguments against Herschel's VIEW of cause of continental elevations (I) the alte
122a114 obey that Law. & lie over the platform:-- On my VIEW the degrading action must prevent internal flui
125a121 ated fluid or gases under pressure.-- {P} Lyells VIEW of transmission of heat by gases-- does not app
126a124 springs beneath sea-- → According to this latter VIEW the rod is reversed, upper part metal «conveyin
126a124 ce.-- ￼ the depth of frozen soil is against this VIEW.-- however it is said in some of the papers tha
172b007 who will dare say what result According to this VIEW animals, on separate islands, ought to become d
175b016 different set, as the rest of the world.-- This VIEW supposes that in course of ages. & therefore ch
182b047 xamine «good» collection of insects with this in VIEW.-- Geogr. Journal Vol VI. P II. p 89.-- Lieut.
183b051 species) if they do not breed readily. point in VIEW.--¿whether highly domesticated animals like rac
184b054 & the species now living.-- Now according to my VIEW. in S. America parent of all armadilloes might
195b099 aptations yet other animals live so well.-- This VIEW of propagation gives. <ro> hiding place for man
204b141 g.-- There is some much higher generalization in VIEW. In Marsupial division <do> we not see a splitt
216b182 as hybrid once. Is not this contradiction to his VIEW of races not mingling?-- In Foxes case of Blood
224b216 with same general structure.-- miserable limited VIEW.-- With respect to how species are. Lamaks "wil
227b226 structure may pass into each other: now on this VIEW no one need look for intermediate structures <b
227b227 mes full of speculation & line of observation.-- VIEW of generation. being condensation, test of high
254c048 gher in scale. So Owen actually believes in this VIEW!!! p. 392.-- except generation & digestion in A
254c049 epeated, as mouths in Polypi, Surely not correct VIEW of Flustra or Ascidia spicule in sponge. stomac
257c055 n may be a miracle, but induction leads to other VIEW.-- Till we know uses of organs clearly, we cann
259c066 o more than offspring (like atavism) & shows my «VIEW of» generation right?-- If puppy born with thic
264c081 whilst habits slightly preceed them-- From this VIEW habits must form most important element in cons
269c100 misphere (NB.-- Examine Abrolhos Flora with this VIEW») Tristan D'Acunha, St Helena &c &c. Juan Fernan
269c102 ds of Australia with plants, with this object in VIEW» The intimate relation of Life with laws of Che
275c120 s of horizontal barriers-- Mr Yarrell.-- says my VIEW of varieties is exactly what I state.-- &c pick
292c171 ch conviction to my mind.-- Reflect much over my VIEW of particular instinct being memory transmitted
293c173 se of button holes it would be instinctive.-- My VIEW of instinct explains its loss ¿ if it explains
312c232 ple transferable., not wonderful According to my VIEW <ins> beccause actions are constant they are in
315c243 nce.-- crying is a puzzler-- Under this point of VIEW. expression «of all animals» becomes very curio
327c265 ntly worthy of studying in Metaphysical point of VIEW Henslow has list of plants of Mauritius with lo
338d025 sil Elephant of Africa Most important under this VIEW, & Hippotamus of Madagascar: because. contempor
342d036 characterized.-- 16th Aug.-- What a magnificent VIEW one can take of the world Astronomical <& unkno
352d060 rtive???. Apterix certainly.-- Lyell's excellent VIEW of geology, of each formation being merely a pa
366d108 ave been persistent in nature.-- According to my VIEW, the domesticated animals would cease being fer

```
Page      ***********************************************(Key Word)***********************************************
367d112 key has. CD[old Buffon should be read on Mare My VIEW, why hybrids are infertile. supposes that  when
372d130 ider gemmation as artificial division.-- On this VIEW each particle of animal must have structure of
373d132 aring??-- Have Marsupiata abortive Mammae?.-- My VIEW would make every individual a spontaneous gener
382d157 t in two-- keeping sexes separate. Owen say such VIEW worthy of a Lamarckian.-- Mine is much simpler.
388d171 shade of fathers character.-- according to this VIEW more semen to one child. more like father.-- st
388d172 ting different puppies out of same mother.-- The VIEW that man & «or cock» pheasant &c is abortive he
389d173 s received into bud matured by female;<]CD> such VIEW no ways explains Ld. Moretons case: without the
391d174 stances «alone» make changes or species!! CD[The VIEW of <In> each Man or mammalia being abortive her
391d175 ual is different). (All this agrees well with my VIEW of those «forms» slightly favoured, getting the
418e080 moths, which can be impregnated externally-- My VIEW of every animal being Hermaphrodite-- probably
426e105 cily not having species, if true important on my VIEW.-- March 9th-- Is there any relation between th
429e116 r regions of N. America.-- if true curious on my VIEW-- because these points were last connected with
431e123 es not spermatic animalcule in Mosses, render my VIEW of the crossing of mosses & all others by actio
431e123 than  average size: (surely this is very limited VIEW, though perhaps a true element) «give examples,
439e143 pollen of same flower,-- as it tends to show my VIEW of <i>nfertility of hybrids «with parent specie
459tflv which has undergone all the changes. [im]portant VIEW, copie[d] Gleanings of Sciences. Vol. III p. 83
466t095 in Crown-Imperial Lily & many other flowers-- My VIEW of <variety acquired> « <character> » of charac
497q007 of Cape Heath by facility. ¿Knight take opposite VIEW. Gaertner talks of the several great & natural
528m037 organ of sight, which is common to every kind of VIEW-- as likewise is novelty of view even old one.
528m037 o every kind of view-- as likewise is novelty of VIEW even old one. every time one looks at it.-- the
528m039 gle cause) this symmetry & rhythm applies to the VIEW as a whole.-- Colour «& light» has very much to
529m041 bitual to become poetical) the botanist might so VIEW plants & trees.-- I am sure I remember my pleas
529m041 ous manner.-- There is much imagination in every VIEW. if one were admiring one in India. & a tiger s
537m075 different instincts.-- Fact most opposed to this VIEW, where the moral sense seems to have changed su
543m098 ore» than in other orders (study Kirby with this VIEW) therefore there is Instinctual development in
553m135 -- The history of Metaphysicks shows that such a VIEW cannot be, anyhow, easily overturned.-- so read
564n003 ssion» he must have conscience-- this is capital VIEW.-- Dogs conscience would not have been same wit
566n011 useful  to them: this must be studied. before my VIEW of origin of evil passions.-- Man getting sight
577n049 s great probability against free action.-- on my VIEW of free will, no one could discover he had  not
579n060 ng of Camelion & Octopus; strong analogy with my VIEW of blushing-- in former irritation on a piece o
581n063 with ordinary habits that is my new part of the VIEW.-- let the proof of heredetariness in habits. b
582n067 remains of savages state.-- N B. According to my VIEW marrying late, will make average of life longer
593n109 e. This [blank] will not do for insects. if this VIEW holds good-- then man, a socialist, does not kn
593n109 er animals is hostile «is subversive of» to this VIEW, & fowls hatching stones. in some degree is so.
602o11v e so little to do with art (p 128) R. compares a VIEW taken by a camera obscura &c a Poussin.-- How a
603o11b chitecture does not come under imitative art [my VIEW says yes. <old> mass of rock--]CD or poetry, CD
608o026 he  thinks they have none.-- Effects.-- One must VIEW a wrecked man, like a sickly one(P)-- We cannot
608o026 help loathing a diseased offensive object, so we VIEW wickedness.-- it would however be more proper t
608o027 ould be necessary. wickedness than disease. This VIEW should teach one profound humility, one deserve
608o027 d temper), nor ought one to blame others.-- This VIEW will not do harm, because no one can be  really
609o030 - The other says we have a moral sense.-- But my VIEW <says> unites both «& shows them to be almost i
616o039 what  answer they would give in support of their VIEW it is impossible to show satisfactorily it's er
626o052 on  this point.-- [LHC] p. 194. «&c &c» Butler's VIEW given on conscience: I cannot admit it.-- see n
638j059 .-- New theory of instinct, returning to Kirby's VIEW.-- Macculloch. Attributes of Deity Vol I. p. 25
289c162 xcept as distinct creation.-- Generation may be VIEWED as condensor, «+++ must (on my theory) =suppor
300c198 & instinct very just, but these faculties being VIEWED as replacing each other it is hiatus & not sal
545m105 impressions  become unconscious those which are VIEWED with little interest, & those which are viewed
545m105 viewed  with little interest, & those which are VIEWED very often.-- former do not give rise to ideas
550m126 friend, whose who family can draw-- says friend VIEWED him as Newfoundland dog would Greyhound  about
587n087 ing more than the unfitness of the objects then VIEWED. to organs adapted to other objects. (as  that
280c136 thousands  in futurity.-- This is right way of VIEWING it.-- Variety when long in blood, gets stronge
315c243 his  having canine teeth at all.-- This way of VIEWING the subject important.-- Laughing modified bar
571n030 s of rock.-- I was much struck with this, when VIEWING Windsor Castle which rises naturally & hence s
599o007 think reason sufficient to give up my theory-- VIEWING from eminence. the wide expanse, of county, ne
117a101 during elevation & depression. C. Prevost.-- My VIEWS of insensible oscillations of level will  alone
125a121 Mem.  S. Cruz. Assuming from Sir. W. Herschel's VIEWS earth originally fluid, then cooling process mu
126a122 a dumpling being bad conductor is {P} against my VIEWS-- if we had rod thus & judged by increments at,
242c018 s of Great ocean says in conformity with Bory's VIEWS.-- <Says> D'Orbigny is said to have brought a t
252c042 acleay to Bicheno much excellent detail & fine, VIEWS about Species-- MUST BE STUDIED: genera founded
254c049 al of other, (surely rather parents). (NB These VIEWS must lead to spontaneous generation??) This who
257c057 t-- all Nature answers to the possibility.-- My VIEWS will explain no Mammalia in secondary-epocks, &
280c135 ifferent might do so.-- <whi> is this true?? My VIEWS, which would even lead to anticipate mules is v
292c168 of  change in Scicily.-- Splendid Harmony these VIEWS-- did Lamarck connect extermination of some for
292c168 ck connect extermination of some forms with his VIEWS.-- as genera are large probably only few of ext
292c170 r of the demonstrations offered of the singular VIEWS there offered, & he must be a zealous man in th
292c170 d <">= I confess. no dissertation against these VIEWS, could possibly have had <to so convin> brought
296c185 countries.  p. 564. an abstract of Mr Swainsons VIEWS. which if abstract true are wonderfully absurd.
301c199 h squirrel to kill ears of corn according to my VIEWS, habits give structure,.. habits precedes struc
304c207 f range « <far more prob> » tends to alteration VIEWS.-- ostriches do-- but then there may have exist
309c219 erefore less fertile. according to Mr Herbert's VIEWS very ancient? Study with profound care, abortiv
325c267 rnithology Read Aristotle to see whether any my VIEWS-- Quoted by Owen.-- Hunter has written quarto.
325c267 y. St. Hilaires. 1832. contains all his fathers VIEWS Mackintosh Ethical Philos: Prostitution of Dan
325c267 lives of Reid, Smith & giving abstract of their VIEWS of the Cultivation of Fruit trees in. N. Americ
327c265 Oats,  &c «Horticultural Transacts.--» Mr Coxe "VIEWS-- Fox says a cousin «one of Mr Strutt» of his
334d009 y is Lord Moreton's case opposed to this fact & VIEWS.-- Seeing means separate in some of the lowest t
351d057 umbilical  cord,-- Broderip alluded to Hunter's VIEWS on this subject.-- Monstrosities, kind of deter
381d157 hybrids--  this is most important support to my VIEWS-- Seeing means separate in some of the lowest t
388d172 ng given milk.-- testis & ovaria The following VIEWS show that transmission of mutilation impossible
407e038 as that of S. America.-- (Explained by profound VIEWS of Lyell) Now «Equatorial» America from the <lo
426e106 it  does indelibly the many subsequent ones. My VIEWS, «V <see» p. 103» would lead me to think that a
427e109 all  present types are ancient. According to my VIEWS of <Dioecious p> all plants, being occasionally
436e136 d animals will vary, according to my Malthusian VIEWS, within certain limits, but beyond these not.--
486z018 this other order is comparatively rare.-- These VIEWS clearly explain rarity of insects in T. del Fue
504q013 ollen naturally carried, on account of Van Mons VIEWS-- Also PEAS-- N.B. I think very likely the Peas
535m070 argues  against all contrivance-- it is what my VIEWS tend to.-- When a man is in a passion he puts h
536m074 dily are visited upon the children.-- The above VIEWS would make a man a predestinarian of a new kind
553m137 Ancient Greeks, with their mystical but sublime VIEWS, or the wretched fears & strange  superstitions
559m155 &c worth reading. as giving abstract of Smith's VIEWS «Take & pound up inflorescent parts of mosses &
608o027 dent of himself to do harm.-- Believer in these VIEWS will pay great attention to Education.-- 4) The
608o028 ll pay great attention to Education.-- 4) These VIEWS are directly opposed & inexplicable if we suppo
609o29v many positive checks.-- (This is encroaching on VIEWS in second volume of Malthus). Adam Smith also t
640j167 iety would prevail.-- Not separately: NB. These VIEWS quite exclude the idea of domesticated  animals
640j167 of  domesticated animals changing.-- From these VIEWS we can deduce why small islands. should possess
640j167 rable explanation is thus offered.-- From these VIEWS, one would infer that Mollusca would offer  few
641j29v of antiquity & extinction of such forms-- these VIEWS will bear on geology-- There is an analogy betw
358d075 fire, & on one occasion, not withstanding our VIGILANCE A piece of pork 3 lb was taken from a boiling
285c151 upied after its decay, will be occupied by the VIGOROUS shoots from each branch No: because decay  in
368d114 most  attracted). -- singing best sign of most VIGOROUS males.-- «(NB. most strange cocks & hens. bei
428e112 ought to be no weeding or encouragement, but a VIGOROUS battle between strong & weak March 11th. Yarr
258c061 Capt. King do. p. 434. Table of birds from Cuba VIGORS.-- nothing of much interest XX. Hence relation
344d040 distribution of reptiles in Suites de Buffon.-- VIGORS has given list in Linnaean Transactions of bir
258c061 hether species may not be made by a little more VIGOUR being given to the chance offspring who have a
259c063 n swimming, & every developement giving greater VIGOUR to the parent so tending to produce effect  on
359d088 Yarrells law.-- it probably is explained by the VIGOUR of their propagating powers. (as if they  were
343d037 f the Silurian, it has made a long succession of VILE Molluscous animals-- How beneath the dignity of
251c041 (Bos  Gazoeus) does not mix with the Gobbah or VILLAGE Gazal.-- ¿is latter same species domesticated,
479z006 a Marsden. p. 311 D'.Orbigny considers Dasypus VILLOSUS is true Peludo Cavia Australis. Dorbigny Vol
499q09v ioica Sorrell. Lychnis. Butchers Broom-- «also, VINCA,» Examine all these, are they much frequented b
```

207b152 ctory remarks to Himalaya Mountains-- Bory St. VINCENT Vol. III. p. 164. Lile de la Reunion presente
324c268 of Nat. History Prichard.-- Lawrence Bory St. VINCENT. Vol III p. 164. on unfixed forms. Dr. Royle o
477z003 1769 introduce in Governor's tran?? Azara Las VINCHUCA or Benchuca. "Les individus ailes peuvent avo
572n034 as been cowardly, or has injured another bad, VINDICTIVE.-- or lied &c &c Are the facts (about commun
286c154 t the white Man, who has debased his Nature «& VIOLATES every best instinctive feeling» by making sla
544m102 p.-- Characters of dreams no surprise, at the VIOLATION of all <rules> relations of time, <identity,>
033r043 ble little table showing long PERIODS of great VIOLENCE volcanic. from Humboldt: Comparison P 361. Da
038r059 d. As argument in favor of lines of anticlinal VIOLENCE crossing lines of crater, <arg> state that al
047r081 -It must be considered as an oscillation, from VIOLENCE. Is it not same as swell travelling across Pa
470t178 by pulling back Wings, pollen is ejected with VIOLENCE in shower On many Papilionaceous; all wh. are
570n023 s & went away."-- he implies negation, without VIOLENCE, without assigning or understanding reason.--
029r033 e shock of earthquake shook the ship in a most VIOLENT manner. Although it lasted about a minute, the
051r093 one.--Corals, & Corallina survive, in the most VIOLENT surfs: in both latter cases become petrified;
522m013 sudden changes of disposition, like people in VIOLENT intoxication, often ends in insanity or deliri
531m051 isfortunes of others.-- In young children, the VIOLENT passions they go into, shows how truly an inst
531m052 about heart as of excited action, accompanying VIOLENT movement; may not passion be the feeling «cons
531m052 not passion be the feeling «consequent on the VIOLENT muscular exertion" which accompanies violent a
531m052 e violent muscular exertion» which accompanies VIOLENT attack,-- Even the worm when trod upon turneth
532m057 trembling of muscles, are not these effects of VIOLENT running away, & must not <this> «running away»
532m057 of sweat is the <state> effect of short -- but VIOLENT action.-- To avoid stating how far, I believe,
070r155 e at seven oclock Novem <5th> Concepcion most VIOLENTLY shaken, by earthquake. but no serious injury.
346d048 its peculiar.-- young one 203 days old butted VIOLENTLY. & fell.-- gore to death the old & wounded,--
467t103 {a} how the Humbles force down the wings most VIOLENTLY: in Beans the wings seem beautifully to prote
529m042 18 years old eating white lead. who was most VIOLENTLY purged «believe worms were passed off.» & vom
121a111 cture superinduced Lyell on Sweden p. 5. «& 7.» VIOLET strata from decomposed muscles.. Smith of Jord
514q21. Where is Boerhaave's paper on impregnation of VIOLETS.= Zostera= Are dwarf plants on Wellington Moun
463t057 & mud-walking fish? difficult-- yet suggested. (VIPERS tooth also a difficult), the whole mind is con
134a141 s. on coast of do p 8.-- soft Clay beds near C. VIRGIN p. 59. dip of Clay slate in T del Fuego Admira
200b125 form varieties? Ed. N. Phil. J. Morse found in VIRGINIA p. 325. July 1828. Animal now confined to ext
637j58r ses: consider this!--]CD consider these barren VIRGINS p. 235. talks of the long spinous processes in
256c055 les, ¿See if type continued?-- See to Boblaye & VIRLET.-- Whewell thinks (p 642) anniversary Speech.
262c072 civilization heredetary; ie instincts of wisdom VIRTUE? «like senses of savages» (How come its some c
271c106 roofs not always continents».-- it is a plastic VIRTUE.-- it is expression for ignorance Two grand cl
535m069 d by metaphysical abstractions, such as plastic VIRTUE, «&c» (Very true, no doubt savage attribute th
535m069 ains are as God made them,-- next step plastic <VIRTUE> natures. accounting for fossils). & lastly th
568n017 h strangers» «as in case of temperance, or real VIRTUE, that is action which experience shows will be
586n078 faculty, whether by sun, & heavens, or magnetic VIRTUE,-- the most probably supposition. with respect
614o037 reason &. & many general instincts, as love of VIRTUE, of association, parental affection-- The very
615o038 specially «the» general kind taking pleasure in VIRTUE because acquired in past ages; seems to indica
616o038 ed by memory of what has been heard; so love of VIRTUE enhanced by this heredetary kind of memory.--
628o54v ification will check the consciences desire for VIRTUE.-- [I expect there is some fallacy here.-- at
528m039 ith poetry, abundance, fertility, rustic life, VIRTUOUS happiness.-- recall scraps of poetry;-- forme
332d002 mber during a year or two he saw many cases of VIRULENT cancer in women, & since that time it has bee
380d153 ds young her growth is immediately checked-- the VIS formativa goes entirely to the offspring-- this
380d153 le which have» larvae which have bred before the VIS formativa had completed them-- (but this argumen
477z002 yaguaré «the zorilla-arskink» le quiyá (Coipu) VISCACHA.-- A. Patagonicus les tatous (.4 pichye, pelu
313c236 f, not from instruction Even the action of the VISCERA. under sympathetic nerve may be instinct or ha
035r048 at the action as a deep & extensive movement of VISCID nucleus, which in any one country would produc
073r163 arts of table granits & gneiss with gold veins VISIBLE:--"Porphyries of Mexico may be considered for
150g036r rier {P} great waterworn frame terrace 4 4 not VISIBLE 3a 3a 3 Bouthoner 3 2 Terrace 3 Alluvials 3a 3
284c149 hat a resemblanc between some form in birds is VISIBLE, when young, but not when old.-- thus speckled
381d156 maphrodites. Cryptandrous. (only female organs VISIBLE). Oyster. cystic Entozoa. Echinoderms. Acaleph
385d163 it was said little cock «yet very odd loosing VISIBLE powers» in Zoolog Gardens. & Kings at Otaheite
451e178 & the Dhangar who can live there & do not pine VISIBLY. p. 337. it would appear as if p. 345. The Cey
542m095 arisome man.-- Is frowning, result of straining VISION, as savages without hats put up their hands, &
542m095 vages clearly be directed chiefly by objects of VISION.-- Does the contraction & wrinkling of the ski
606o20v s of experiments & observations. & yet, like in VISION, it becomes so instantaneous. that we cannot e
264c079 mals were alive, which have perished.-- Let man VISIT Ourang-outang in domestication, hear expressive
466t093 l» visiting it>.-- In yellow day lily, the Bees VISIT base of upper petal, though not differently col
466t094 a little in a wild purple Geranium, I see Bees VISIT always base {a} of upper petal from facility of
496q05a Toad Orchis= How many flowers in minute do they VISIT?? good=!! Examine pollen of double flowers. com
515q022 nd-walk, on which I think I have never seen Bee VISIT. Experiments in Garden Sow stones of Standard A
531m049 ewsbury, which the cat was seen by Hubberley to VISIT daily to see how the young got on. this nest th
469t135 th seven flower stalks for ten minutes. it was VISITED by 13 Bees-- & each examined very many flowers
469t135 al succeeding days «many» «most numerous» bees VISITED this same bunch & on this day in five minutes
469t135 day in five minutes eleven Humbles came & each VISITED many flowers-- Saw Bees frequent these flowers
469t135 its 10 flowers in «minute» each flower will be VISITED in 28 minutes-- say then each flower is visite
469t135 isited in 28 minutes-- say then each flower is VISITED 30 times a day is considerably under mark, & t
536m073 faults of the fathers, corporeal & bodily are VISITED upon the children.-- The above views would mak
584n072 : the facts of memory of roads long after once VISITED by horse & dogs. (even blind horses & dogs) sh
139aIBC led R. N.-- Massac[h]usset would be well worth VISITING really good account of ice.-- C. Darwin A. Gl
466t093 do> «it is so» <Though I saw no Bees «several» VISITING it.-- In yellow day lily, the Bees visit 'bas
466t094 elf resembles a Bee, but does not prevent bees VISITING it.» In Columbine nectaries are placed all ro
554m199 was very curious to see her take bread from a VISITOR, & before eating «everytime», look up to «keep
469t135 On rough calc. 280 flowers-- allowing each Bee VISITS 10 flowers in «minute» each flower will be vis
139a180 <buds> galls.-- is it not effect of superadded VITAL influence?-- See End of Note Book. called R. N.
282c143 & <in» merely determined to such points by the VITAL laws.-- so that all character originally may «m
388d173 o such need.-- one would <one> suppose that the VITAL portion ¿nerves? passed through transformation,
418e083 .]CD» can be traced to a germ, endowed with the VITAL principle, which gives rise to the sexual organ
591n097 [not located] <& other cows--> Mr. Hamilton on VITAL laws (in the Athaenaeum Library) describes effe
611o034 of simple non-organic matter, without action of VITAL laws-- According to the individual forms of liv
611o034 beings namely «one individual» vegetables, the VITAL laws act definitely (<like> «as» chemical laws,
419e085 culated might be owing to absolute quantity of VITALITY «in the World»,-- the production of vitality,
419e085 f vitality «in the World»,-- the production of VITALITY, as argued by Müller from propagation of infi
218b189 es this-- but when Man makes variety these are VITIATED.-- This barely applies to plants Female pig a
571n030 eredetary habits. & perhaps even latter may be VITIATED. or rather altered. The Reason why New Buildi
073r164 rized by no quartz & amphibole frequently only VITREOUS felspar: = gold veins in a phonolitic porphyr
078r176 ute of quartz, & wh abound both in hornblend & VITREOUS felspar".-- p. 215 Same metal in Tasco vein i
242c018 the Scincus with golden streaks-- the lacerta VITTATA extends <to> from Amboina to New Ireland p. 23
573n030 rs.-- like Kitten with mice.-- A person with St VITUS' dance badly, told should have shilling to walk
447e164 alogy shows some most important end.-- Festuca VIVAPARA F ovina-- propagated like oni[on] Poa alpina
447e165 ys he has not the slightest doubt that Festuca VIVAPARA is the same species with F. ovina, <& this> r
446e163 f blossoms--» The case of the Lemma, «and the VIVAPAROUS grasses, which no doubt are propagated durin
447e164 -- propagated like oni[on] Poa alpina because VIVAPAROUS. Henslow has seen this-- (Poa alpina vivapar
447e164 vaparous. Henslow has seen this-- (Poa alpina VIVAPAROUS sometimes seeds All species of Lemna sometim
447e165 same species with F. ovina, <& this> rendered VIVAPAROUS by growing on heights.-- yet he has seen it
447e165 ed in Hort. Transact. Aira caespitosa becomes VIVAPAROUS on mountains & yet can be raised in gardens.
447e165 ed in gardens.-- Poa alpina, thougt generally VIVAPAROUS sometimes seeds.-- There are endless curiou
240c013 he Moluccas «Matchian» & Celebes.-- Amboina; VIVERRA Zibetha.-- All the Moluccas, Waggious New Gu
527m033 sound per se} & causes the mind to create short VIVID flashes of images & thoughts.-- Poetry. the lat
527m033 Poetry. the latter thoughts are in same manner VIVID & grand. the frame of mind being just kept up b
528m035 forgetfulness.-- & the approach to believing a VIVID castle in the air, or dreams real again explain
546m110 03, supra (by mistake) have lower animals these VIVID thoughts in same book (p. 143) wonderful case o
547m111 for second & wakened.-- had very clear & pretty VIVID «& perfectly characterized» dream, in continuat
548m114 ence delirium & sleep mental rest. though. most VIVID & rapid thought.-- There may be some «two or th
552m130 object<>» one Shuts ones eyes) is the image not VIVID as in sleep-- (one can dream of intense scarlet
552m130 erceptions, & that on[e] fancies the image more VIVID? Surely the image in a dream cannot truly be <m
552m130 he image in a dream cannot truly be <more> «as» VIVID, «a reality» as in Spectral images-- Mem Chiloe

Page
(Key Word)

```
569n021 mparison by senses of any two objects-- they by VIVID power of conception between one or two absent t
577n052 ensitive people apt to blush.-- -- The power of VIVID mental affection, on separate organs most curio
315c242 ntestines subject to sympathetic nerves-- The VIVIDNESS of first <thoughts> «memory» in children or r
544m103 pidity-- We may conclude that neither number, VIVIDNESS, rapidity, novelty of separate ideas cause fa
547m112 wholly absorbed with one idea (hence apparent VIVIDNESS) & there being no other parallel trains of id
569n021 things.-- reason probably mere consequence of VIVIDNESS & multiplicity of things remembered & the ass
384d162 sexes (woman makes, bud, man puts primordial VIVIFYING principle) one individual secretes two substa
491q002 ies of Peas &c been obtained? 3.. Whether the VIVIPAROUS grasses & onion, produce flowers, like the O
160g089 d Loch, making <several> two divortiums aquarum, VIZ two branches of River Bought & between one of th
199b118 ds & say life is short for this object & others, VIZ not too much change In Number 6'.? of E.d. N. Ph
231b243 & tropics are only related by one connection.-- VIZ descent.-- Hence far greater discordance in latt
284c149 n different group. & Not known in single ones--. VIZ. Macleay letter to Fleming p. 32 "where it (mode
542m095 ce is this the cause of expression of surprise-- VIZ seeing something obscurely with the wish to make
576n049 -- Arguing from man to animals is philosophical. VIZ. man is not a cause like a deity, as M. Cousin s
324c268 tom  IV. p 273 Latreille Geographie des insectes 8VO p. 181.-- See (p. 17) for references to authors
473s03r ore of the flowers withered.-- Sillimans Journal <VO> 1842. p. 142-- Sus americana & Hippotamus «with
601o08b - & Le Parfait Chasseur, par Desgraviers, un Vol 8VO Keratry-- Inductions morales et physiologiques T
195b098 so  acted that bi[r]ds with plumage «&» tone of VOICE partly American North & South.-- (& geographica
286c154 rnal circumstances not variable.-- Animals have VOICE, so has man. Not saltus. but hiatus animals exp
305c209 to mocking thrushes of Galapagos having tone of VOICE like S. American.-- Have not Ruffs & Reeves a r
333d005 o dogs «otherwise habits not different; tone of VOICE. perhaps rather different».-- crossed with <un>
341d032 last  associated with the ducks.-- most strange VOICE often in the night, like peacock.-- tail as lon
426e108 parentage?-- Wonderful as is the possession of VOICE by Man. we should remember, that even birds can
554m139 the  taste of Peppermint.--» Perfect understand VOICE.-- will do anything.-- will take & give food to
568n018 in harmony-- frogs chirp in do-- union of birds VOICE & taste for singing with Mammalian structure. «
569n020 ed in some necessary connexion between things & VOICE, as roaring for lion &c &c. (in same way alphab
589n092 son: or dog, having high powers without hand or VOICE.-- there is some great puzzle in what Sir. J. M
589n092 an abortive groan.-- more power over muscles of VOICE than respiration.-- like sigh before false snee
531m053 d with want of muscular exertion, palpitation, VOIDING urine because done by some animals in defence,
599o008 oralité, raison, beau ideal, infini conscience; VOILA l'homme separe de la matiere et du temps! voila
599o008 voila l'homme separe de la matiere et du temps! VOILA les facultes, q'il possede seul sur la terre. J
059v119 llées ou par quelque scissures profondes, on les VOIT se reproduire a des hauteurs communes sur le re
445e159 akes series of flying mammifers-- says lemur.-- VOLANS, has skin between its legs.-- -- strangely con
089a017 no  fossil shells --¿ action of Heat bubbles VOLATILIZED at bottom, condensed before rising?-- Mem. g
036r051 he absence of Second form, except near submarine VOLC: in harmony with the prevailing movement  being
060r125 ions of S. America & Europe.-- If great chain of VOLC. had been in action during secondary period how
061r127 f one species altered: <altered> Mem: my idea of VOLC: islands. elevated. then peculiar plants create
061r127 at new creation for large.--Australia's = if for VOLC. isld. then for any spot of land. = Yet new cre
023r011 ick is 60 miles distant from the grand ancient VOLCANIC axis «of the Andes».-- «Has this fault determ
023r011 e Andes».-- «Has this fault determined side of VOLCANIC activity.» That axis was produced, from a fis
025r017 n Ascencion known to be inactive 300 years? No VOLCANIC Earthquakes or Hot Springs in T. del Fuego =
028r029 ns of Springs or more probably by some unknown VOLCANIC process? How does it come that all Lime is no
028r032 f fleat prevented by sedimentary rocks, & hence VOLCANIC action, contradicted by Cordillera, where tha
033r043 e table showing long PERIODS of great violence VOLCANIC. from Humboldt: Comparison P 361. Daubeny Von
037r056 ption, but even there, hills of Basalt & other VOLCANIC rocks. Bahia, Rio de Jan: B. Oriental? level
042r069 observations of period «& manner» of elevation VOLCANIC action, must be more exclusively confined to
044r073 y of Organic remains.--Unequal distribution of VOLCANIC action, Australia S. Africa-- on one side. S.
046r079 ortional» particles altered.-- With respect to VOLCANIC theory. I want to ground, that the first phen
046r079 that  Plutonic rocks are generated as often as VOLCANIC. I consider latter as accidental on the afflu
049r087 ] {P} In Cordill: should basal lavas be called VOLCANIC or Plutonic The cellular state of all the Por
050r091 den Sumatra. M. De. Jonnes seems to think that VOLCANIC eruptions form foundations for Coral reefs.--
054r102 tion of Payta: N. part of New Zeeland entirely VOLCANIC!! New Zeeland rich in particular genera of pl
056r111 rculation of fluid nucleus,--the similarity of VOLCANIC products «over whole world» argument, as well
056r111 lar over whole world, general circulation. But VOLCANIC action separates some sulphur (perhaps  lime)
064r135 important effect.--? Capt. FitzRoy. -- Limited VOLCANIC action & limited earthquakes & great but loca
064r136 n Europe-- Urge difference of plutonic rocks & VOLCANIC metalliferous-- Urge enormous quantity of mat
067r144 dillera: It is not therefore so wonderful that VOLCANIC rocks at M. Video «Volcano in Pampas» Pasto E
068r145 ed to me that no metals in Polynesian Islds--. VOLCANIC plenty in S. America!! Metamorphic Volcanos o
068r146 hat to be surrounded by continent.-- change of VOLCANIC focus.-- <it is certain, if strata can be> Pr
071r156 nburgh. Phisoph. Transactions. = Mem: Olivine. VOLCANIC product.=> <Did Peruvian Indians use arrows o
072r159 mean?) Consult Dr Holland about bubbles.-- No VOLCANIC action on coast line of Old Greenland,  close
074r165 a fossil» Insist strongly on the grand fact of VOLCANIC & non Volcanic. Then Solfataras. «Mem: Volcan
074r165 t strongly on the grand fact of Volcanic & non VOLCANIC. Then Solfataras. «Mem: Micaceous iron ore.»
075r166 ns discovered. Humboldt suggests covered up by VOLCANIC rocks. //St Helena has been slightly broken u
085aIFC tracted as far as concerns "Geolog Observat on VOLCANIC islands & Coral Formation Lyell's Salband p.
085o002 that Himalayas penetrated like Bolivian Chain. VOLCANIC islands. from number of craters very ancient.
088a015 boldt. Fragmens Asiatiques account of American VOLCANIC action. -- Fragments of slate converted into
089a017 tamorphic action in red sandstone.-- Certainly VOLCANIC-- CD[Might not bottom of ocean boil; yet heat
089a020 titut p. 192.-- (1837. Peninsula of Cape Verd. VOLCANIC.-- Isle of Gory. rocks encrusted with serpula
092a029 n. My view of metamorphic in contradistinct to VOLCANIC will explain their solution. Athenaeum M. 516
096a039 tities of shells at Iquique. <Ceylon>. Band of VOLCANIC action in Iceland parellel to Greenland: Mem.
102a062 with  magnetism &c counteracting gravity.-- As VOLCANIC eruptions are accompanied by horizontal eleva
124a119 orbed.-- if so exactly parallel to limestone & VOLCANIC rock containing magnesia Lyell. Elements p.11
132a138 ing point.-- accounts for increase on earth by VOLCANIC action.-- <Why> now as we know volcanic actio
132a138 rth by volcanic action.-- <Why> now as we know VOLCANIC action prevails more beneath the sea,  <than>
133a140 we see how many points have been penetrated by VOLCANIC & trappean rocks, within say the Tertiary per
175b017 change> vary quicker Unknown causes of change. VOLCANIC isld.-- Electricity Each species changes. doe
195b103 on. or when country changes. Will it said that VOLCANIC soil of Galapagos under equator that external
207b152 de lancien monde".-- Considers forms in recent VOLCANIC islets not well fixed.-- Peron thinks Van Die
211b164 elago--, very good in connection with Von Buch VOLCANIC chart & my idea of double line of intersectio
218b192 y.-- Now when we hear that the whole island is VOLCANIC surmounted by water & studded with  others.--
219b193 -» & hence formed trees]CD & would creator <on VOLCANIC island.> make plants «grow closely» When this
219b193 island.> make plants «grow closely» When this VOLCANIC point appeared in the great ocean, have  made
219b194 -- We cannot consider it as adaptation because VOLCANIC isld. whilst <neig> Africa, sandstone, & gran
221b201 ater genera? The absence of Lime in Plutonic & VOLCANIC rocks. most remarkable.--¿ Have the changes b
408e044 y result of elevation,-- «all» modern & wholly VOLCANIC-- Azores might be prophecied to have this cha
453e182 he theory of polymorphous plants, abounding in VOLCANIC islds.-- <Cocks> The possibility of different
500q10a ing species of confined genera By my theory in VOLCANIC or rising isld, there ought to be a good many
514q21. s &-- carnosa?-- good-- Norfolk Isd-- geology. VOLCANIC? Applies to my geology & Species theory-- pec
633j54r ash down earth, from the mountains upheaved by VOLCANIC force, for these Marsh plants. All flow  from
132d138 eat.--» «(does M. Parrot suppose there is no VOLCANICITY beneath lakes)?» Suppose ocean represents pr
132a138 rrot does conjecture that in Scoresby's case VOLCANICITY has warmed it. Is not cold of ocean accounte
049r088 stone cones at S. T. del Fuego as nucleus of a VOLCANO or as an injected mass.--From conical form I i
050r090 s of Sumatra no connection with a neighbouring VOLCANO of Priamang.--Marsden Sumatra. M. De. Jonnes s
056r112 sary process of terrestrial renovation & so is VOLCANO a useful chemical instrument.--Yet  neglecting
057r112 --What more awful scourges to mankind than the VOLCANO & Earthquake.--Earthquakes act as ploughs  [,]
067r144 so wonderful that volcanic rocks at M. Video «VOLCANO in Pampas» Pasto Earthquake. Happened on Janua
071r158 ferent case from shore of Pacific.--Isabelle's VOLCANO, many amygdaloids.--Boussinquault «(Lyell)» cra
120a108 e hundred miles through nearly cold rock.-- in VOLCANO the pool is not deep. --Hot springs &c &c--the
134a142 r South of Part Luconia-- Phillipines there is VOLCANO on isld in large lake-- Berghaus Chart of do J
038c059 nes of crater, <arg> state that all the great VOLCANOES. have been elevated considerably. which shows
039c059 or in the coal measures have been conduits to VOLCANOES.-- Talking of the cricket valley «the most re
062c128 . = Perpetual snow.--subterranean lakes, near VOLCANOES. lakes of brine all inhabited: Go steadily th
091a026 p.  209. May. 1837 Paper by Humboldt on Quito VOLCANOES & another on Mexican Trachyte <roc> lava call
023r010 the  Abrolhos shoals ░ -- N.B. The view of the VOLCANOS of the chain of the Cordilleras as arising fr
023r010 y have happened from incipient elevation.» The VOLCANOS originated in the bottom of the ocean. & the
023r011 ated in the bottom of the ocean. & the present VOLCANOS have been said to be merely accidental apertu
034r044 's fact about sulphuric vapours in East Indian VOLCANOS Gypsum Andes Mem. Beechey. account of regular
```

042r068 ilting alone, must be modified. «Moreover, the VOLCANOS from sea there burst out, after rise from sea
042r068 er rise from sea: <As did> as did those aerial VOLCANOS in Germany» In the Valle del Yeso it is proba
043r070 s. -- P. 322 In any archipelago. & neigbouring VOLCANOS. eruption from «more than» one orifice <...>
045r075 e on difficulty of evidence about eruptions of VOLCANOS. (where there are no country newspapers)--At
045r075 ngs pitched off the ground. «Ulloa states that VOLCANOS!! were in eruption at time of great Lima eart
046r078 ussing connection of Pacifick & S. America. -- VOLCANOS must be considered as chemical retorts.--negl
046r079 enomem. is an inward afflux of melted matter.--VOLCANOS perhaps may be admittance of water, through t
056r111 es.--(intertropics at present fix lime). <Also VOLCANOS separate.> Volcanos blend all substances toge
056r111 t present fix lime). <Also Volcanos separate.> VOLCANOS blend all substances together; & products bei
057r112 & Earthquake.--Earthquakes act as ploughs [,] VOLCANOS as Marl-pits: Consider well age of Bones. = s
058r117 s constantly going on we shall see a cause for VOLCANOS part of same phenomena lasting so long.-- The
064r137 428» states that Von Buch has urged that Java VOLCANOS differ from all others in quantity of Sulph.
068r146 -. Volcanic plenty in S. America!! Metamorphic VOLCANOS only burst out where strata in act of disloca
094a035 into Trachytes.-- Mention Osorno in lake. few VOLCANOS now in lakes.-- Mr Murchison. M.S. Chapter on
096a040 NE & SW.-- Von Buch. Can. Ile p. 406. List of VOLCANOS Salomon Isld,-- New Britain-- &c &c In Ascens
104a067 suppose everywhere--, in granitic areas &c &c VOLCANOS {P} fissure dike.-- thus dikes terminated Sol
106a076 great depths. All Earthquake unaccompanied by VOLCANOS must be sought after proofs of sinking.-- No
107a078 to thickness of crust broken up.----My view of VOLCANOS &c &c This view will bear much reflection on
107a078 slocation taking place chiefly beneath water & VOLCANOS. crust must be thinner «under water» but caus
108a080 Earthquake if Subsidence we should not expect VOLCANOS.-- not so much horizontal oscillation. or so
112a089 nsideration. When discussing nucleuse's of old VOLCANOS within Cordillera-- allude to Lyell's view of
112a089 sses we must look for that.-- how few isolated VOLCANOS there are. where one alone has been formed--L
112a089 alone has been formed--Look at the now active VOLCANOS & see what high they are «See Athenaeum. 1838
118a103 pkins) & that every dike. which has not formed VOLCANOS. or become scoriform. has thinned upwards & i
122a113 .-- Will this not explain littoral mountains & VOLCANOS.-- Why on one coast? How can Herschel conside
123a116 must go round of dissemination & separation in VOLCANOS.-- if so why not metals. The theory of veins
126a123 n upwards & so conduct heat?-- How comes it in VOLCANOS that have gone on for thousands of years, tha
130a134 Phy-- Nat. t. I, 1831. sur le temp du globe on VOLCANOS &c worth reading. L'Institut. 1838 p. 360. on
477z003 s peuvent avoir <quatre> cinq lignes de long et VOLENT. p. 208 Fleas only appear in winter in Paragua
376d135 ows, is the final effect, (by means however of VOLITION) of this populousness, on the energy of Man»
586n081 - heredetary habit, is a part never subject to VOLITION.-- like plants going to sleep.-- "A bird has
602o10v tson's Physiology much about sleep-- Nerves.-- VOLITION &c Reynold XIII Discourse (p 115) a very good
036r052 nce: hence difference; action on land different VOLNEY, P 351. Vol I. woody bushes, «gazelles» hares,
022r008 pect to the discussion of winds & storms:--«in VOLNEY'S travels also» Dampier's last voyage to New Ho
327cIBC whether peculiar plants-- in high points Read VOLNEY'S travels in Syria Vol I. p. 71. account of Eur
067r143 thenaeum April 1836 (p302) Coleccion de obras. 2 VOLS fol: Buenos Ayres 1836: W. Parish?? «by Pedro d
320c276 . voyage a l isle Malouines Zoological Journal 5 VOLS Voyage de la Coquille Zoological Transactions.
323c269 genius Feb 14th. Bo«s»well's life of Johnson. 4. VOLS 25th Phillips. Geology. Larder 2d vol.-- March
323c269 l notes. -- 20th. Carlyle's French Revolution 3? VOLS. oct: -- 26th Blumenbach's Essay on Generation.
556m145 in men. than in animals.-- In the drawings of VOLTAIRE why is under lip curled over upper with mouth
193b092 stracted Mr Swainson's trash.-- at beginning of VOLUME on Geographical distribution of animals Brown
322c270 of France.= on Etna. Almost reread the previous VOLUME. & C. Prevost on L'Ile Julie Waterton's Essays
323c269 an 10t.-- All life of W. Scott.--, except the V VOLUME.-- -- 19t. Mungo Park-- travels Feb 12. Sir. H
326c266 .-- very good. Endlicher has published in first VOLUME of Annales of Vienna. sketch of south Sea. Bot
358d074 rast with otaheite in relation «See Gaudichauds VOLUME on the Botany of the Pacific.--» to nearest co
415e069 pecies, & in Man December 16th. The end of each VOLUME of Whewells Inductive History. Contains many m
559m155 n Geographical distribution of British Plants A VOLUME published by Colonel in army on "Wheat." in Je
609o29v ecks.-- (This is encroaching on views in second VOLUME of Malthus). Adam Smith also talks of the nece
021rIFC the excess of harbor = 180 See Daubisson both VOLUMES, and Molina 1st Vol & Lyell Sailed, 27th <Frid
296c184 like neighbouring Continent. This fact speaks VOLUMES. 2 Chapters. translated by Hooker.-- my theory
322c270 ll's Australia Walter Scotts life I & 2d & 3rd VOLUMES Abercrombie on the Intellectual powers. Hunter
326c266 Mind.-- Audubons. Ornithological Biography. 4. VOLUMES well worth reading Bevans work on Bees, new Ed
433e127 , & that systems, are only leaves out of whole VOLUMES.-- The fact of tumbling pidgeons; flying high
573n038 able.-- cannot avoid it.-- curious mixture of VOLUNTARY & involuntary movements.-- Person with sore-t
578n056 power of mind by habit gets more perfect over VOLUNTARY muscles, these convulsive actions-- (except i
578n057 s).-- But, the lachyrmal gland is «not» under VOLUNTARY power, (or only very little so) & hence by as
596nIBC -- is convulsion. are involuntary movement of VOLUNTARY muscles-- if so what is trembling palsy? Expr
616o038 intercourse in plants is involuntary, in man VOLUNTARY: ¿ False,-- secretion in both involuntary, <a
228b229 sion of species.-- Important. For instance take VOLUTA & Conus (??) which now run together, were not
481z008 rnal habits of Crustaceae Mr Broderip says that VOLUTA found in not less than 7 fathoms water. Mem Ba
093a033 The number of minute turbos in red earth with VOLUTAS. prove regular mud bank at Bahia Blanca. <fl>
499q010 l, does he include seeds good to eat. (even Nux VOMICA is eaten by a Buceros in East Indies-- Asiatic
529m042 ntly purged «believe worms were passed off.» & VOMITED, but who when he recovered. was found to be ig
034r044 lcanic. from Humboldt: Comparison P 361. Daubeny VON Buch is very strong about Trachyte being the mos
064r137 w remaining-- Lyell « <p 419> p 428» states that VON Buch has urged that Java volcanos differ from al
069r150 felspar melted gneiss//| QUARTZ!!! Analogous to VON Buch. Basalt where Basalt. trachyte where trachy
096a039 rellel to Greenland: Mem.¿ Greenland subsiding.) VON Buch Canary Isd. p. 351.. NB. Mackenzie talks of
096a040 ravel on basalt of Heckla-- All the Azores Isld. VON Buch p 359 stretched out NE & SW.-- Von Buch. Ca
096a040 es Isld. Von Buch p 359 stretched out NE & SW.-- VON Buch. Can. Ile p. 406. List of Volcanos Salomon
097a042 tut No degree 221 Lamellar dikes like Mica Slate VON. Buch. Canary Isd. p. 170.-- Mem. Cordillera Can
115a097 my rocks. when writing on Falkland Islds p. 94. VON Buch's Travels account of Norway chain being bro
123a115 - SPLENDID PAPER on fossil shells of S. America. VON Buch Lyell. (under head of Delta) describes near
209b156 ld Linnaean doctrine & Lyells. to certain extent VON Buch,.-- Canary Isles: French Edit. Flora of Isl
210b158 t present analogy to what takes place from time? VON Buch distinctly states that permanent varieties.
210b159 ut how do plants cross?-- = admirable discussion VON Buch says from Humboldt, in Laponia. genera to s
211b164 dian Archipelago--, very good in connection with VON Buch Volcanic chart & my idea of double line of
294c178 to classification & affinities, its extension.-- VON Buch. Travels, p. 306. account of trees ceasing
320c275 Continuation «Annals of Natural History» Skimmed VON Buch Travels Whites Natural History of Selbourne
355d068 h Secondary Species distinct-- but close.-- Mem. VON Buch on Cordillera fossils same remark. ¿was the
356d069 all amount of change at any one time Seeing what VON Buch (Humboldt). G. St. Hilaire, & Lamarck have
442e153 e in Norway ought to be thus characterized study VON Buch.]CD Now Mr Knights statements about fruit t
511q018 our «& length of ears» & skeleton, & skin= Van. VOORST often writes to Lowe (7) In breeding. pointers
451e176 lbino is held sacred by the credulous natives, & VOW made at it. Both his parents were of the usual c
022r008 of the dikes.--P 432. as in Andes. In Dampier's VOYAGE there is a mine of metereology with respect to
022r008 rms:--«in Volney's travels also» Dampier's last VOYAGE to New Holland P 127.--Caught a shark 11 ft lo
026r022 el Fuego.-- Is there account of Baron Roussin's VOYAGE.-- In Europe proofs of many oscillations of le
029r033 composition of Granitic rocks Mem. Chanticleers VOYAGE at <[...] Maranh> Pernambuco. EARTHQUAKE AT SE
029r033 he James Cruikshank, Captain John Young, on her VOYAGE from Demerara to London:-- "Feb. 12, 1835. At
054r105 ca not reached. Juan. Galapagos. Cocos-- Ulloas VOYAGE North of Callao, the country, to the distance
058r118 by Coral hypoth. agree with great continents). VOYAGE aux terres Australs Vol. I. p. 54. M. Bailly s
060r124 nd> land in the W Indies.--p. 200. Bollingbroke VOYAGE to the Demerary Earthquakes at St Helena. 1756
065r140 oast. -- Capt. Cook found soundings. (end of 2d VOYAGE outside coast of T. del Fuego. off. Christmas
079r177 loix de la nature & ses phenomenes."-- Ulloa's VOYAGE, Shell fish purple die, marevellous statements
095a037 ies Vol VI. Geograph. Journ. Analysis of Poenig VOYAGE Valparaiso Dr. Gillies in MS. letter in Sir. W
111a087 365. Meyen on Chile must be studied Analysis of VOYAGE: many observations on heights of valleys in Ch
116a099 gh Philosophical Journal Rapport on D'Orbigny's VOYAGE. good section of Rio Negro beds.-- -- refers t
178b031 ists thought was species <Ascensi> Study Lesson VOYAGE of Coquille.-- Dr. Smith says he is certain th
193b091 misso on Kamschatka quadrupeds Kotezebues first VOYAGE Copied into list Entomological Magazine paper
202b131 37 degree about) Vol IV P. I. Geograp. Journal. VOYAGE up the Massaroony by W. Hillhouse.-- Demerara.
218b190 France-- -- Madagascar oxen with hump.-- p 173. VOYAGE par un Officier du Roi Mem. Capt. Owen's story
226b220 ll isld off New Guinea same fact see Coquille's VOYAGE.-- Galapagos mouse (?) brought by canoes Ceylo
229b234 rtise former inhabitants of Mauritius Freycinet VOYAGE, agrees. with several mammalia being peculiar
233b249 sld. (as far East as New Ireland. see Coquilles VOYAGE), Waterhous remark Australia Fauna so far. Ind
234b255 ed race of cows from Madagascar-- p 173. Vol I. VOYAGE à France-- Par un Officier du Roi.-- Mackenzie
240c013 f wild animals, nor in Dobrizhoffer Abipones.-- VOYAGE. de L'Astrolabe Zoologie. p. 60. Vol I. Cynoce
241c016 es from about 8 to 20 of Zoologie of Coquille's VOYAGE to see if Lessons' remarks on the Floras can b
242c017 t picked.-- as men do. when making varieties.-- VOYAGE of Coquille. Zoolog. p 19., Tapir, «des» couro
242c018 extends <to> from Amboina to New Ireland p. 23 VOYAGE of Coquille Lesson No (p. 24) batrachian in is

243c019 oubts fact.--» My toad is same species Coquille VOYAGE p. 25 Mais il n'y a pas jusqu'aux îles Macquar
243c021 nts peculiar to the different points.-- Consult VOYAGE aux terres Australes Chap XXXIX tom IV p. 273
245c023 ecies, with different number of teats-- Coquille VOYAGE Durville has written Flora of Falkland Islds,
246c025 of America. would account for this.-- Coquille VOYAGE Says no reptiles. p 460 & very doubtful whethe
246c027 New Guinea.-- (Case of replacement)-- Coquille VOYAGE The caswary, inhabits Ceram, Bourou & especial
255c050 of such forms.-- p. 56: Ornithological part of VOYAGE of??? A Urubu, (with one leg) attended the dis
267c093 ts" Geograp. Journ. Vol VII. p. 216. Mr Bennett VOYAGE round world, 20 years have scarcely elapsed si
269c102 Holland, supplementary to Appendix to Flinders VOYAGE by Brown.-- great space seems to act per se as
311c227 phical range very good.-- Blainville Ovington's VOYAGE to Surat, floating isld. off coast of Africa .
317c249 ffspring short tails /one born at Maer. Tuckeys VOYAGE-- p. 36 "Cercopithecus saboeus" said to be mon
320c276 . Phil Journal. about 13 numbers have been read VOYAGE a l'isle de Frances Voyage de l'Astrolabe Part
320c276 mbers have been read Voyage a l'isle de Frances VOYAGE de l'Astrolabe Partie Zoologique Pernety. voya
320c276 oyage de l'Astrolabe Partie Zoologique Pernety. VOYAGE a l isle Malouines Zoological Journal 5 Vols V
320c276 ge a l isle Malouines Zoological Journal 5 Vols VOYAGE de la Coquille Zoological Transactions. <done>
321c275 tish Aviary.--do--- & Lisle's Husbandry Tuckeys VOYAGE reread Appendix Ovington Voyage to Surinam. To
321c275 sbandry Tuckeys voyage reread Appendix Ovington VOYAGE to Surinam. Voyage Congo expedition: Zaire exc
321c275 age reread Appendix Ovington Voyage to Surinam. VOYAGE Congo expedition: Zaire except Brown's Appendi
321c270 out [not located] <Rays Wisdom of> Lisiansky's VOYAGE round world. 1803-6 Nothing Lyells Elements of
322c269 otzebue's two voyages, skimmed well. do Lutke's VOYAGE. carefully read.-- Reynolds Discourses Lessing
324c268 Brown at end of Flinders & at end of the Congo VOYAGE Decandoelle. Philosophie. or Geographical dist
324c268 animaux.-- on sleep & movements of Plant/ I£:4s VOYAGE aux terres australes. Chapt. XIX. tom IV. p 27
402e019 Seas. p 233 Octob 12th Kotzebue's second «1st» VOYAGE. Vol II p. 344. account of insects of St. Pete
402e020 this archipelago Octob. 13th.-- Kotzebues first VOYAGE. Vol II p 367. "The Fauna of the Sunda island
403e021 the number is very limited.-- Kotzebue's Second VOYAGE do Vol III p. 77. Many foreign plants have bee
448e166 e this is always the case according to Brown.-- VOYAGE of Adventure & Beagle Vol I. p. 306 Shells, as
450e173 y stand India. better than the latter-- Forrest VOYAGE p. 323. Sooloo. imported elephants. wild hogs-
451e180 bolder & more generous temper-- Hodgson Koloff. VOYAGE through the Moluccas 1825--"No wild animals in
458t002 es-- useful perversion of instincts-- Beechey's VOYAGE Vol I. p. 499. «4to. Edit»-- Horses in Lao Cho
459tf1r case as showing gradations, Boteler's Narrative VOYAGE East coast of Africa-- Vol II. p. 256-- wild c
477z001 ns on Planariae by Johnson CXII. & CXV do Azara VOYAGE Vol I. p. 196. According to Charpentier de Coss
477z002 no> snail was introduced to Mauritius. 18 Azara VOYAGE Vol. I. p. 279 Thinks the Moruffes of Chile
478z003 s Mentions stinging Millepora. Quoy. Freycinets VOYAGE Vol p. 597 Many descriptions about lower anima
479z005 i Rapid growth of Coral-- RN. p 24 Bougainville VOYAGE round world no land animal besides Wolf at Fal
479z005 ∴ black rabbits not indigenous p 112 M Lesson--VOYAGE of Coquille wide limits of Nullipora Discussio
479z005 hell fish Purple die Marvellous stories Ulloa's VOYAGE Vol I, p. 168 Ceratophytes common in Northern
481z010 urth Vol. «in Lyell's possession» of Zoolog. of VOYAGE of Astrolabe must be studied for anatomy. of.
481z010 s. Madrepores p. 26 Nullipora p. 29-- In Meyen. VOYAGE round World German a reference to a luminous S
482z012 cies of Tortoises come from Galapagos!!! Azara. VOYAGE dans l'Amerique Merid. Tatu noir. abundant fro
483z013 p. 387. "on Classification of such animals.."-- VOYAGE. Coquille's Voyage p 302 Vol II p. 302. Vaginu
483z013 ication of such animals.."-- Voyage. Coquille's VOYAGE p 302 Vol II p. 302. Vaginulus of Lima describ
486z019 hiloe. Amblyrhyncus de marlin James Isd-- Lutke VOYAGE Vol III p 322 Dr Martens says only one Reptile
569n022 ite ends of series or harmonious prose.-- Lutké VOYAGE in Carolinas Vol II p. 132. offered to take a
569n022 » out not in Library no good There is a Lutké's VOYAGE autour du Monde (1826-9) Paris. 1835 Quoted re
133a141 at globe as resting on film of molten rock.-- VOYAGES of Adventure & Beagle vol I. p. 2 & 3. Porphyr
320c276 celand Molinas Chile Falkners Patagonia Azaras VOYAGES & Quadrupeds of Paraguay Dobrizhoffer. Abipom<
322c269 on's Embassy to China. Oct. 12t Kotzebue's two VOYAGES, skimmed well. do Lutke's Voyage. carefully re
400e012 us to specific variations.-- Kerr's Collect of VOYAGES Vol 8 «p. 46» Capt Davis in 1598 found cattle
452e181 Batchian near SE. end of Gilolo.-- "-- Forrest VOYAGES. p. 39-- deer but no wild animals in Gilolo.--
260c067 there acquired» then adaptation.-- No Carrion VULTURES in Australia!! Wilsons Ornithology, Vol III.
482z011 alco poliosoma -- novozelandiae -- histrionicus VULTUM aura Excessively inaccurate Saw a Chouette a h
522m010 eard of such a man.-- (My Father explained who he WA & all about him, but still maintained he had nev
111a087 gree. E. dip to NW to 80 degree faults with red WACKE contorted evidently dike. V. VII. p. 316 & 328
459t019 of its body & in the manner of walking but not WADDLING; its colour was darker than the penguin & th
534m062 efessa placed table in cups of water which they WADED. or swam across.-- they then stretched themselv
346d047 Albatross, on some Gulls. Flamingo-- (Spoonbill WADER. Ibis)-- laws of plumage might possibly be made
062r129 animals in S. America. Zorilla: wide limits of WADERS: Ascension. Keeling: at sea so commonly seen.
196b104 that are apt to wander & of easy transportal.-- WADERS & Waterfowl.-- scrutinize genera, & draw up ta
368d114 this with those females which put on (like some WADERS) the bright plumage.-- «thinks» Hence specific
482z011 5. only 9. Terrestrial birds at Falklands Isd 8 WADERS. 22 palmipedes: out of the first 9.:4 raptores
262c073 to say no passages; nature is full off them.-- WADING birds partially webbed; &c &c.--)-- & in round
135a144 g-- Burnetts. vol 4. p. 193 in Lat 26 degree S. WAFER looking for Copiapo. found inland a great many
537m074 t may be doubted whether a man intentionally can WAG his finger from real caprice. it is chance, whic
557m147 s (I believe) Hunter says. neither fox. nor wolf WAG their tails, &c. it is very curious, recurrence
591n097 thing or excited.-- so do young dingos, as I saw WAG tail when watching anything-- Keeper does not th
025r017 rthquakes or Hot Springs in T. del Fuego = The WAGER'S Earthquake the most Southern one I have heard
591n097 ung wolf & it never dropped its ears like dog-- WAGGED its tail «a little» when attending to anything
305c210 ars droop like dog older character & manner of WAGGING tail.-- habitual movements connected with mind
539m081 tellectual powers-- the difference is of a man WAGGING his foot & working with his toe to perform som
557m147 ched. just contrary. when pleased tail loose & WAGGING-- if as (I believe) Hunter says. neither fox.
606o025 (-- must be so, analyse (a) ones feelings when WAGGING one's finger-- one feels it in passion, love--
232b249 Holland form) is found in many island Celebes «WAGGIOU» &c &c. (See Lyell. Vol III p. 30) different s
240c013 mboina; Viverra Zibetha. ▌-- All the Moluccas, WAGGIOOS New Guinea. New Ireland, have phalangista, wh
288c160 st banished the Grasshopper Warbler-- --Yellow WAGTAIL never seen in one district, though common on a
363d102 Goulds Willow Wren.--) (Goulds story of Water-WAGTAILS mistake both species scattered over Europe)--
244c023 58 Cuscus albus. New Ireland ---- maculatus -- WAIGIOU Speaking of Lepus Magellanicus says; «after» "
376d137 icoats., & if woman not afraid clasp them round WAIST & look in their faces & Mak the st. st noise.--
569n022 Monde (1826-9) Paris. 1835 Quoted repeatedly by WAITZ (In Theil. V) in describing Caroline Archipelag
547m111 as lying on my back fell to sleep for second & WAKENED-- had very clear & pretty vivid «& perfectly
544m102 » are recollected when intense, or when so near WAKING. that an associated is kept up with waking tho
544m102 near waking. that an associated is kept up with WAKING thought.-- Ld Brougham thinks no dreams except
547m111 fectly characterized» dream, in continuation of WAKING thought.-- my servant was in the room. with my
030r036 m rivers. & natural position» position at N. S. WALES & Van Diemen's land.-- Whole coast S. of Concep
050c093 is nature of strip of Mountain Limestone in N. WALES. was it reef. -- I remember many Corals?? Brecc
242c017 Britain same kind of dog, with those of New S. WALES. <V.> p. 123 Crocodile at New Guinea. All the i
247c028 Scincus multilineatus (p 45) Moluccas & New S. WALES. Scincus Cyanurus «p 8 &». p 49 on all the Moluc
369d115 the Great distances which the Mammalia of N. S. WALES are generally compelled to traverse in order to
434e129 lieves that «only» red Lychnis grows in <south> WALES & certainly <old> only white in Cambridge, in s
473s05r has heard the Trout from different lakes of N. WALES can be distinguished-- & Jackson here (Capel-Cu
033r043 em Galapagos. chiefly red glassy scoriae.--could WALK round base:--not universal: could not climb up
484z016 comparison to find how many of the small finches WALK at Maldonado & Patagonia compared with those of
495q05a nd, on «dry» windy day, «flower garden on gravel WALK» will drift many seeds= Necessary to answer Wie
515q022 if get sterile-- Cover that little Ervum in Sand-WALK, on which I think I have never seen Bee visit.
536m070 r struggle of feeling.-- It is as much effort to WALK then lightly as to endeavur to stop heart beati
565n007 my.-- Man & dogs show triumph (& pride) same way WALK erect & stiff, with head up.-- Why does suspici
573n038 Vitus' dance badly, told should have shilling to WALK to door without touching table.-- cannot avoid
160g190 ntorted gneiss «narrow sharp ridge with peak» I WALKED all round hill. Boulder about 20 ft. below sum
334d012 ce a great many miles.-- yet one day <th> a cow WALKED in, then disappeared, & these days afterwards
327c265 Medical Review No XIV. April 1839.-- Review on "WALKER on intermarriage" price 14s. Marh. 20t. 1839.
428e113 d race-horses, which Eclipse? has begotten <?> "WALKER attributes this to effect of male sex on locom
493q004 le of old breed & see result.-- According to Mr WALKER the form of male ought to preponderate; accord
449e169 e thus contradicting (probably) Yarrells law & WALKERS of the male giving form-- they interbred. & th
459tf02 e)-- -- this may be, perhaps. squeezed into Mr WALKER'S law Gleanings of Science Vol III. p 320. Mr H
036r051 ly known. the acute chirping sound produced in WALKING over the sand: I am nearly sure, it is necessa
258c061 ked for in the aberrant groups.-- It is having WALKING fly catcher, woodpecker &c & which causes the
274c115 restial.-- How is it in water birds, there are WALKING forms in water birds,-- but no web forms in <w
292c171 consciousness «a most possible thing. see men WALKING in sleep».-- an action becomes habitual is pro
305c211 al Magnetism--" principles of irritations sleep WALKING. fits, laught &c&c Man & Man may have some rel
459t013 uin in the form of its body & in the manner of WALKING but not waddling; its colour was darker than t

463t055 anyone. have foreseen, sailing, climbing & mud-WALKING fish? difficult-- yet suggested. (vipers tooth
484z014 entative of Caracaras of Americas.-- manner of WALKING-- foot bill crest feathering on legs-- habits-
533m059 ch full grown men can experience-- Instinctive WALKING of animals. that is the ready movement & co-re
579n059 er emotions? When a man keeps perfect. time in WALKING, to chronometer, is seen to be muscular moveme
586n078 in breeds,» something of kind oneself knows in WALKING [one feels inclined to stop at right number of
586n081 o certain quarter"-- "An animal has faculty of WALKING. which in man is learnt by experience is in ot
615o36v cular organ.-- I think Pincher shows surprise, WALKING home one day met him, with Mark riding instant
536m070 a man is in a passion he puts himself stiff, & WALKS hard.-- «He cannot avoid sending will of action
570n024 sses his lips together & shrugs his shoulders & WALKS off,-- I think shrugging connected with many em
033r043 laced on the rim of conical crater: at Teneriffe WALL of Porph. Lava with base of Pitchstone; Mem Gal
056r110 . Helena.-- [Fig. 7] {P} effect of heat on inner WALL, hence resists degradation longer than outer pa
146g010 at top the trap could be traced Grey in front on WALL perhaps wall oblique The hill has been well-- d
146g010 ap could be traced Grey in front on wall perhaps WALL oblique The hill has been well-- denuded.-- «of
491q01v ilosop. of Zoolog. vol 1. p. 427-- says biennial-WALL-flowers & scarlet Lychnis can be propagated by
534m062 am across.-- they then stretched themselves from WALL to table.-- table being removed a little furthe
473s006 intermediate between Orchis & other plants-- & WALLICH has described Indian Plant.-- June /42/-- June
070r154 undulation? would it not be mere vibration? but WALLS & feeling shows undulation:. crust thin.--Concep
099a051 aration in the Ponza case of Scrope parallel to WALLS of dykes-- Mem. laminated dikes in Cordillera.}
501q10a country & not in other: Rosa is hard in Europe, WALNUT in America.-- Heaths in Africa; Hooker? are th
422e092 acea or whales.-- but are aquatic Pachyderms. & WALRUS-- aquatic seal.-- (Consult this passage, when
639j28r pus & Chamaelion.-- C. D. Sucking feet in Frog. WALRUS. Fly. Gecko &c. Prehensile tail. in Monkeys &
641j29r arctic regions-- Whales. «Narwhal» Polar bear. WALRUS, great Seals of Antarctic seas. (on other hand
322c270 arginal notes <Rengger &c> Mitchell's Australia WALTER Scotts life I & 2d & 3rd Volumes Abercrombie o
525m026 mark that fat men are goodnatured, & vice versa WALTER Scotts remark how odious an illtempered fat ma
576n046 er faculty than the acquirement of new ideas.-- WALTER Scott «(Antiquary)» Vol II p. 126 says seals k
195b100 did it only create three kinds not so likely to WANDER. Did it create two species closely allied to M
195b104 era, Bats .Foxes. Mus are birds that are apt to WANDER & of easy transportal.-- Waders & Waterfowl.--
222b206 order may not be supposed that form, which has WANDERED least from ancestral form. If so are present
203b135 e does not seem(?) to consider the monkey as a WANDERER, but as produced by climate?-- M. Baer (think
318c255 t migrats every year,; and probably a <chance» WANDERER like the first pair of Pipra flycatcher.-- Ba
202b134 rld more perfectly continental. we might have WANDERERS. (as Peccari in N. America) then if it is doo
202b134 family has offspring the chance is that these WANDERERS would not, but where original forms most nume
202b134 e original forms most numerous there would be WANDERERS.-- Some however might have offspring, & then
217b188 how that Mr D. wonders at these species being WANDERERS.-- Iceland no species to itself, a remark com
232b245 ator.-- «or some species might then have been WANDERERS.--» There ought to be fewer species in propor
274c116 se Northern regions as the receptacles of the WANDERERS out of the rest of the world?-- Will this not
484z014 ica-- valuable & practicable deed Caricaridae WANDERERS.--?? in N. America?? Wilson N. American Ornit
195b100 e Mammae on mens' breasts.-- How does it come WANDERING birds. such sandpipers. not new at Galapagos.
436e135 l» formation. intelligible.-- «-- No. but the WANDERING & separation of a few, probably would be most
448e167 the Eocene beds.-- see Lyells tables Bennetts WANDERING Vol II. p 155. By inference I imagine that th
500q10a t parts or same district.-- About <endemic &> WANDERING species of confined genera By my theory in vo
590n094 oth cured by sight of instrument.-- Bennett's WANDERINGS, Australian Dog does not Bark-- quotes Gardn
403e021 p. 189. The gaut, kind of crocodile, sometimes WANDERS from Pellew to Eap.-- There is another great l
041r065 oned that the Elephant came towns driving by the WANT of water.--I believe in all flat countries. yea
046r079 es altered.-- With respect to Volcanic theory. I WANT to ground, that the first phenomem. is an inwar
113a092 of force in that part.-- Important as explaining WANT of levelness Major Mitchell showed me a river <
199b118 modification, into a durable form, of a fugitive WANT into a fundamental propenity, of an accidental
252c045 unt of varieties in N America» «some doubt, from WANT of knowledge of time-- Analogy from three first
263c078 have been necessary to have made man! Seclusion WANT &c & perhaps a train of animals of hundred gene
272c108 similar.-- The question which I more immediately WANT are there Heteromera, which have habits & part
283c145 to breed hybrids produced--)» & then all that I WANT is granted.-- For. at Galapagos. make ten speci
292c171 obably first stage, & an habitual action implies WANT of consciousness & will & therefore may be call
315c243 news. discovery of prey.-- arising no doubt from WANT of assistance.-- crying is a puzzler-- Under th
498q007 retrograde into leaves-- is this ever effect of WANT of nutrition.-- Horned oranges so? --Yes, my Fa
531m053 hamming death, or running away. accompanied with WANT of muscular exertion, palpitation, voiding urin
565n008 ess.<--> the hypothesis of opposite muscles will WANT much confirmation. A grave person close those m
568n019 ny common to t[he] whole kingdom of nature. If I WANT some good passages against, opposition of divin
577n090 very day, & closing of the leaves, comes on from WANT of stimulus, after certain other actions, & hen
581n064 children have no difficulty in expressing their WANT, pleasure, or pains long before they can speak-
624o051 ation & imitation, has often been perverted from WANT of reason.-- Hence as Eugenius says, slow growt
632j53r ome provision for transportation:-- But I do not WANT to deny laws.-- The whole universe is full of a
312c231 the Bantam clubs, they used to fix on the kind WANTED, colouring of each feather weight & size & the
362d099 never «dry up &» peel off their skin (not being WANTED for war) & hence never fall off.| Curious the
410e053 will care little for species, except so far as WANTING names to refer to, to those forms, where the t
444e157 ns.-- In the Batracian Order the «32» ribs are WANTING. p. 144 in the Icthyosaurus 60 or 70 bones in
536m070 oticed this by perceiving myself skipping when WANTING not to feel angry-- such efforts prevent anger
616o39v shew that the groundwork <of this> is entirely WANTING by which thought or memory. might be in like m
261c072 ly be explained by direct adaptation to animals WANTS & not as change in typical structure?!! Whewell
292c170 yman No VI. Magazine of Zoology & Botany p. 566 WANTS to see absurdity of Quinary arrangement let him
388d170 nearly independently of its parent & therefore WANTS independent supply of food.-- is real. differen
432e125 r sneering, has no more relation to our present WANTS. or structure, than the muscles of the ears to
447e164 ntinent-- well characterised species periwinkle WANTS insects to impregnate allied to Asclepias Turpi
295c178 by song. do they by beauty, analogy of man if so WAR not [not located] Erasmus says he has seen old S
362d099 up &» peel off their skin (not being wanted for WAR) & hence never fall off.| Curious the rapidity o
429e114 difficult to believe in the dreadful «but quiet» WAR of organic beings. going on the peaceful woods.
288c160 t of-- to have almost banished the Grasshopper WARBLER-- --Yellow Wagtail never seen in one district,
484z015 s--. Pyrocephalus & many Tyrannulae-- replaces WARBLERS of Europe-- Study profoundly shells of Bahia
232b245 epoch), whilst others may die out or move South WARD.-- species must be compared to neighboring sea
291c168 tter, then Brazilian species would migrate south WARD being ready made.-- & so destroy individuals, w
148g028 formed successive bays but plains sloped centre-WARDS which would not have happened if the side-strea
378d148 o the windows to be fed, but still they have a WARINESS about them quite remarkable". instance of old
258c061 males armed & pugnacious (all order; cocks all WARLIKE)» «thiis wars against in any class:, those poi
040r063 ase from intersecting a mass probably cold & not WARM as sides of a crater as Vesuvius.-- There may h
086a007 rly by sun's position = If equatorial streams of WARM pole; in name of Heaven why are tops of Equator
132a138 g to M. Parrots argument against central heat to WARM the ocean).-- and M. Parrot does conjecture tha
343d037 o numbers of animals. as in changing from <hot>. WARM to cold, damp to dry.-- Thus Tierra del Fuego h
132a138 jecture that in Scoresby's case volcanicity has WARMED it. Is not cold of ocean accounted for, by the
220b199 ecies in each other, are migratory species from WARMER countries. When will this paper be published i
328m039 autumn, on clear day.-- 3d pleasure association WARMTH, exercise, birds singings.-- 4th. Pleasure of
343d037 ea from cramped imagination that God created. (WARRING against those very laws he established in all
375d134 of <Malthus>. «Decandoelle» does not convey the WARRING of the species as inference from Malthus.-- «i
258c061 ugnacious (all order; cocks all warlike)» «thiis WARS against in any class:, those points which are d
262c073 ially the powers of reasoning &c &c.-- Study the WARS of organic being.-- the fact of guavas having o
345d044 f males impressing most is in harmony with their WARS & rivalry.--» The very many breeds of animals i
633j54r e earth, then the earth revolves to form rain to WASH down earth, from the mountains upheaved by volc
028r029 forms slates: How is the Lime separated; is it WASHED from the solid rock by the actions of Springs
054r105 f this bay are perforated & smoothed like those WASHED by the waves, a sufficient proof, that the sea
057r113 as the Eocene. = Should Mr Owen consider bones WASHED about much at Coll. of. Surgeon's? I really sh
247c028 .--» Vol II. p. 10. it seems that Crocodile was WASHED on shore at one of the Pelew Islds.-- killed a
290c164 Soc. it's pale ash grey back, like a black bird WASHED, Whilst tips of primaries black, by examining
074r165 also Azores. We here have case of such vapours WASHING a rock» Veins concretionary; concretions <dt>
282c142 hen the second race would not obtain a cast of WASHING men-- but might hire the preexisting race, thu
363d101 in Bantam.--]CD CD[In the Pidgeons, trace the WASHING out of the forked band, like in plumage of duc
030r036 ases. «.Mytilus.--» «at Guacho» «on N. Chile» WASHINGTON.--» Mem: Micaceous formation of Chonos. inte
534m063 Newport says Dr Darwin mistaken in saying common WASP cuts off wings of flies from intellect. but it
535m063 habitually.-- good Heavens is it disputed that a WASP has this much intellect. yet habit may make it
370d116 oney even for single rainy day-- & from case of WASPS, is supposed cells properly are made for larvae
370d116 properly are made for larvae.-- CD[(p. 451.)-- WASPS breed many females, but almost all die.-- bees

Page
(Key Word)
```
212b165 pull the garden bell, & then run into Kennel to WATCH who would come to the door-- would constantly d
288c159 ess not the least notion of hunting, or keeping WATCH. how completely «nature & instinct» modified.--
358d075 om camp to camp; it was absolutely necessary to WATCH our meat, while in kettles on the fire, & on on
521m009 s announced he could not understand it, but the WATCH was <seen> shown him.-- «<Mr Corb> the  servant
521m009 shown him.-- «<Mr Corb> the servant showed him WATCH & said dinner is ready, what, what.-- then show
521m009 dinner is ready, what, what.-- then showed the WATCH upon which he exclaimed, why it is dinner time.
530m049 ocated] Fox believe cats discover birds nests & WATCH them till the young are big enough to eat.-- Th
585n077 o make toothed wheel. he might by instinct make WATCH, but he does it by reason & experience, or habi
620o046 t to stand a <spaniels> «housedog's» duty is to WATCH the house.-- it is part of <duty> their nature.
591n097 Keeper...   Zoolog. Garden told me. he has often WATCHE tame young wolf & it never dropped its ears li
363d102 very close species, of birds, habits when well WATCHED always very different.-- the two redpoles  can
469t135 a  second species.-- <Saw> Maer. June 15./41/.  WATCHED plants of Fraxinella, with seven flower stalks
047r081 a's bottom being in motion what difference? In WATCHING heavy swell, sea retreats & then breaks: i  e
437e140 not been shot by <some> «a» shepherds, who was WATCHING the scene.-- «In Shiant Isld. it is said, tha
472s02r ar-- June 2. 42 Maer <Thursday> Thursday After WATCHING 14 days. many times every day. many clumps of
535m064 ebrantia, laying eggs on leaves of Eucalyptus, WATCHING few days till larva excluded, then though not
535m064 not  feeding them «nor helping larva from egg» WATCHING them, brooding over them, preserving them fro
591n097 .-- so do young dingos, as I saw wag tail when WATCHING anything-- Keeper does not think they drop th
585n077 faculty, they use this faculty instinctively; WATCHMAKER has faculty by his instruments to make tooth
024r013 ing said about K. Georges Sound The idea of the WATER at Cauquenes. coming from the [...] Cordillera
024r014 ordillera & flowing The gradual shoaling of the WATER to more than 100 fathoms. proves the  existence
026r019 ey «the limestones» have been formed in shallow WATER: so have the Conglomerates: Yet this view is di
028r031 Chalk ever been dissolved? Singularity of fresh WATER at Iquiqui. not from rain, because alluvium sal
028r031 erefore such «rain» is cause, hence at least no WATER is absorbed into the earth <I did not see one d
029r033 t a minute, there was no uncommon ripple on the WATER. It was quite calm at the time. Latitude 8 deg.
037r052 entire absence of any rock situated beneath low WATER in the Southern ocean not being buoyed with Kel
038r058 er not so much; or no rapilli; & from action of WATER probably not so much aluminated. As argument in
040r064 he upheaval of Andes.--but as long as all below WATER no evidence--The depth of shells (which being p
041r064 eer approaching the wells.-- the effect of Salt WATER of the Salado.--Mem. in Owens Africa it is ment
041r065 the  Elephant came towns driving by the want of WATER.--I believe in all flat countries. years of dro
041r065 . Gay's fact about shells: Hibernation of fresh WATER Shells. multitudes.-- The question of shell's c
044r074 Sulp. Hyd: Carb: A. Mur: A. = (& this effect of WATER thus holding matter in solution must be  great:
046r079 matter.--Volcanos  perhaps may be admittance of WATER, through the rent strata: «Mr Lyell considers t
046r080 1746).--At  great Lisbon Earthquake Loch Lomond WATER oscillated between 2 & 3 ft. (as in Chili lake)
046r080 t.--At first some would think movement. owing to WATER keeping its level whilst land rose up & down.--
047r082 (NB. Earthquake wave is an oscillation, body of WATER manifestly does not travel up.--) If these view
050r091 sand less & gravel more common. the shoaler the WATER & nearer the Banks Is there not a sudden deepen
051r093 rruginous.--At Pernambuco (great swell & turbid WATER) organic bodies protect like peat reef of sands
056r111 orld» argument, as well as separating causes by WATER.--Or rather begin & explain how water separates
056r111 causes by water.--Or rather begin & explain how WATER separates.--(intertropics at present fix lime).
057r115 a to explain preserved animals.--Mem: stream of WATER in the country.-- Sir J. Herschel. says. precip
063r131 e drawn. from. Cordillera. rocks.--When beneath WATER.--together with hypothetical case of  Brazil.--
085a001 ern side, far inland.-- even 70 miles from salt WATER. Mr. Arrowsmith tells me, that Himalayas penetr
086a004 to such terms "fixed as the land, stable as the WATER"-- It may be worth noticing edentates &  camels
089a019 e,-- «He» Thought of erecting machine to see if WATER fell. -- <Keys off extreme point of Flori[da]>
100a055 ced by local heat, Ask Capt. Beaufort, whether, WATER flashing into steam, would Babbage.-- Webster P
102a061 with little pressure? An important question! If WATER yields substances from impact, «it» would  look
104a068 tion therefore capillary attraction would bring WATER with salt to surface Lyell remarked to me  that
105a070 leavage. C. Prevost.-- In Cordillera. a rush of WATER will account for filling up of valleys-- subseq
105a073 ll erosive power. therefore tendency of running WATER to deepen not to widen valley.-- Why is serpent
106a073 ion at first angle owing to momentum. which the WATER has obtained.-- If {P} inclination be great whe
107a078 k from dislocation taking place chiefly beneath WATER & volcanos. crust must be thinner «under water»
107a078 water & volcanos. crust must be thinner «under WATER» but cause most difficult (better conductor) Fi
113a092 lid rock.-- 4 & 5 fathoms deep. perfectly still WATER. Major Mitchell inferred subsidence; Mem my rem
114a094 distinct formation deep «& therefore extensive» WATER ∴ not formed in modern formation & not ever  in
115a098 very atom in rock On a coast, the shallower the WATER, the greater power of oscillations & currents.-
117a101 table remains formed near coast, limestone deep WATER. will bear on formations. during elevation & de
117a102 kes. to see effects of degradation, «no» tides, WATER always falling or at least not rising are there
120a109 common salt more soluble in <hot> cold than ho WATER with <-- especially if very hot under high pres
126a123 at better than plain?-- Mem 1000 {P} how easily WATER percolates rocks,-- when pressure increased  or
126a123 III   But equilibrium is not attained, & if cold WATER did not percolate surface, would become hotter.
126a123 s spring requires connected column.--» of cold WATER show, that water does percolate, & springs bene
126a123 s connected column.--» of cold water show, that WATER does percolate, & springs beneath sea-- → Accor
126a124 tal «conveying heat in one direction only, like WATER below 39 degree» & lower part glass.-- then the
126a124 that  there are springs even in siberia.-- from WATER {P} thawed at + in isothermal curve. East-clina
127a126 ss.-- Memoir of the Irish Academy Vol 8. p. 118 WATER no--. oil will freeze if cooled in a closed glo
128a128 nal vol II. p 89. at Madras. surrounded by salt WATER. purest fresh water must be sought for below th
128a128 Madras.  surrounded by salt water. purest fresh WATER must be sought for below the sea mark.-- If mou
128a129 sion of land on one side. produce subsidence of WATER on other. from tendency to regain statical equi
128a129 be only a modifying cause. {P} land protuberant WATER to counterbalance How strongly the Glen Roy cas
130a133 continent  to fall across a hot.--spring.-- Hot WATER would not lie. at bottom.-- Surely we here have
130a133 s Paper.-- Weelsted told me of some large fresh WATER springs off coast of Persia In Glen Roy paper I
131a136 9. p. 52. On Frozen soil of Siberia.-- facts of WATER flowing from beneath frozen crust in America Ri
132a139 to bottom) be caused, by absence of circulating WATER.-- & therefore that temperature of earth beneat
147g021 Temp of Air 65 degree? Glenoe, 6 ft above high WATER mark 30.380. 68 degree 65 degree? For compariso
150g037 er longer than 3a Sunday In Glen Collarig, when WATER up to shelf very shallow channel 50 ft wide & r
151g040 acter.-- The boulders (one of Gneiss remarkably WATER worn) are often times of rock not in  immediate
155g065 e been so carefully preserved as to have thrown WATER in same «drainage» lines Mound of Gneiss though
155g066 preserved? how much more so, these lines & even WATER-scooped rock «only decay from fragment falling»
156g069 aculloch's supposition;-- the old ravine, where WATER entered are not proportionately large to  those
161g098 .200 minus  .008 ---- .192 Loch Lochy 4 ft above WATER Barom: 30.372 A 76 degree 75 degree? The River
162g104 bles-- dip sideward, & inwards-- deposited when WATER stood at higher Loch Keeper tells me, that Loch
176b023 ree of life owing to three elements air, land & WATER, & the endeavour of each <one> typical class to
181b045 f birds were fitted solely for air & fishes for WATER. If my idea of origin of Quinarian system is tr
184b054 216 Deshayes on change in shells from salt & F. WATER-- on what is species. very good Has not Maccull
185b055 ngfisher.-- mountain tringas.-- Upland goose.-- WATER chionis water rat with land structures; :-law o
185b055 untain tringas.-- Upland goose.-- water chionis WATER rat with land structures; :-law of chance would
185b055 cause this to have happened in all. but less in WATER birds-- carrion eagles.-- This is but  carrying
200b125 xperiment to know whether soaking seeds in salt WATER &c has any tendency to form varieties? Ed. N. P
205b143 l I.-- compare with my planariae Leaches out of WATER Does the odd Petrel of F. del F. take form of a
207b151 uiana. build top of trees carry duckling to the WATER in their beaks, & the young one <inland> direct
212b165 .  would remain for long time together in tub of WATER with only nose projecting.-- would pull the gar
212b166 lish. Waterhouse has information respecting the WATER Rat.-- ¡Consult Dr Smith History of S.  African
217b184 er.-- He has known case of good pointer & rough WATER spaniel «produce litter like both parents» & Mr
218b192 that the whole island is volcanic surrounded by WATER & studded with others.-- we see a beginning  to
221b201 . Dela Beche. Geolog. Researches. facts of salt-WATER shells living in absolutely fresh water.-- orig
221b201 of salt-water shells living in absolutely fresh WATER.-- origin of Fresh-water genera? The absence of
221b201 ng in absolutely fresh water.-- origin of Fresh-WATER genera? The absence of lime in Plutonic & Volca
227b224 co & Europe. one gret sea (Coral reefs∴ shallow WATER at Melville Isd. (3d) We know that structure of
232b248 at gained? Experimentise on land shells in salt WATER & lizards do.-- Ask Eyton to procure me some Ge
233b251 guin-- Logger-headed Duck-- Large proportion of WATER & small of land or few quadrupeds-- Study Produ
233b251 quadrupeds-- Study Productions. of great Fresh WATER lakes of North America If Parasite different, w
236b279 Indian form of Tortoise.-- On other hand. fresh WATER tortoise from Germany. (where Mr Murchison  fox
255c051 ucture (habits of ducklings and chickens) Young WATER ouzels hence aversion to generation, before gre
259c064 lecting that the ground woodpecker &c.--, fresh WATER animals of great lakes are American form,  that
261c070 ittle turtle, without its parent running to the WATER, is a good instance of connate instinct, better
274c115 ns besides aerial, & terrestial,-- How is it in WATER birds, there are walking forms in water birds,-
274c115 s it in water birds, there are walking forms in WATER birds,-- but no web forms in <water> land birds
```

(Key Word)

274c115 ng forms in water birds,-- but no web forms in <WATER> land birds,,-- Grups of very different value h
284c148 & only on those, & the islets separated at high WATER.-- not other islands, nor any any other part of
284c148 se isld.-- ¿Brown can «not» bear the least salt WATER.-- Nuts prodigiously heavy (where trees of such
289c160 g <grebes> «ring ouzels» dive instant touch the WATER.-- capital instances of typical land bird, havi
301c199 instincts precede structure.--duckling runs to WATER. before it is conscious of web. feet.-- p. 7. M
306c212 ries its taste. taught it to go to <sea> salter WATER (& its necessities teach it taste, but that is
308c218 ects & birds are the only two tribes fitted for WATER, air, & land, (Macleay has this remark) Mem. nu
321c270 rve Mayo Philosophy of Art of Living Several of WATER Savage Landons Imaginary Conversations-- very ɔ
332d003 He believes all pretty much alike.-- My Father WATER-in the hair a century since used to be called A/
342d034 s he has somewhere met conjecture that all salt-WATER fish were once salt water (as they almost must
342d034 jecture that all salt-water fish were once salt WATER (as they almost must have been on elevation of
361d096 male??)-- Yarrell observed that female of some WATER birds, (as Phalarope) assume for breeding a mor
363d102 (Mem.-- Goulds Willow Wren.--) (Goulds story of WATER-Wagtails mistake both species scattered over Eu
369d115 eing Marsupial. («but» also mice) & these being WATER animals <that this> structure <connected with a
377d147 d] :Hence, also structure not really fitted for WATER, only habits & instincts-- The young of the <p>
379d151 . p. 135. Natural History of the Caspian. Fresh WATER Fish!! ¡adapted to salt water?-- peculiar speci
379d151 he Caspian. Fresh Water Fish!! ¡adapted to salt WATER?-- peculiar species, crabs & molluscs few.-- ¿a
381d157 n effected in latter; are <it> «organs» open to WATER? Would not Ferns according to this doctrine be
409e048 argument. of change.-- now take greater area of WATER & snow line descent. My theory gives great fina
414e063 ession, otherwise [not located] Are the feet of WATER-dogs all more webbed than those of other dog
416e071 d semen: therefore animal life commenced in the WATER!» It is a beautiful part of my theory, that «do
419e085 has published Fauna of Caspian.-- fishes fresh WATER kinds. (yet living in the salt?.)-- very few an
432e125 E. frowns prodigiously when drinking very cold WATER «frowns connected with pain, as well as intense
444e158 ions of species.!!-- p. 203 Chaetodon squirting WATER at fly.-- instinct, for how could experience te
481z008 ys that Voluta found in not less than 7 fathoms WATER. Mem Bahia Blanca. De la Beche theoretical rese
485z017 ine from T. del Fuego p. 141. How comes it salt WATER so soon putrifies?? p. 319. on Hydra-- polypi--
495q05a all kinds of seeds for week in Salt. artificial WATER.-- 12. Plant two races of Cabbages near each ot
524m019 -- It is an argument for materialism. that cold WATER brings on suddenly in head, a frame of mind, an
534m062 es on Formica indefessa placed table in cups of WATER which they waded. or swam across.-- then then s
550m126 Newfoundland dog would Greyhound about dread of WATER-- innate Septemb 1-- If one performs some actio
568n019 see Lyell on Scrope, Quarterly Review. 1827? In WATER Scotts life.. Tom Purdie, (beginning of Vol V)
575n045 . case of Newfoundland dogs. who will not enter WATER, till he sees. whether birds badly wounded, or
584n071 nowledge of location ducks & turtles running to WATER,-- young crocodile snapping-- p. 28. how curiou
585n074 Entry Madagascar Lemur seemed to like Lavendar WATER «very much» Henslow. N.. Necker has remarks on
593n111 waterhens feeding on grassy bank some way from WATER, suddenly, as if by word of command, they all t
632j53r transportation through the air.-- cocoa nut by WATER «fucus for adhesion».-- as examples of design.-
637j57v ieties.-- greyhound to hare.-- waterdog hair to WATER-- bull dog to bulls.-- primrose to <open fields
584n072 eans of guiding themselves through the air,-- WATERBIRDS, the bee to its nest,-- cats when carried in
637j57v aptations in varieties.-- greyhound to hare.-- WATERDOG hair to water-- bull dog to bulls.-- primrose
051r094 oral reef--or confervae in the breakers or in WATERFALL: Excepting by removal of large fragments by m
196b103 t to wander & of easy transportal.-- Waders & WATERFOWL.-- scrutinize genera, & draw up tables-- Inst
495q05a ot tame duck on pond with Duck-weed-- coots-- WATERHENS-- examine dog, which has swum-- on pools & ri
593n111 n's mind.-- At Maer. Pool. I saw many coots & WATERHENS feeding on grassy bank some way from water, s
233b249 r East as New Ireland. see Coquilles Voyage), WATERHOUSE remark Australia Fauna so far. Indian all the
088a014 o would line of mountain chain be Mr <Lyell> «WATERHOUSE» has frequently heard that Herons bring eels
185b056 a> species, belonging to true. American genus WATERHOUSE says he is certain, that in insects, each fa
185b057 other tribes in that same family.-- (NB I see WATERHOUSE thinks Quinary only three elements) How far
189b071 our of negroes, a point of real repugnance.-- WATERHOUSE says there is no TRUE connection between gre
190b079 nowledge increased, but genera stronger.-- Mr WATERHOUSE says no real <separ> passage between good ge
211b162 instance probably of rudimentary bones.-- As WATERHOUSE remarked Mere length of bill does not <indic
212b166 . 53. an Irish Rat.-- different from English. WATERHOUSE has information respecting the Water. Rat.--
232b249 of man over animals, without such resorts Mr WATERHOUSE has most curious facts about the distributio
233b250 with similar extension of form in birds.-- | WATERHOUSE thinks two main divisions of cats. Tortoise
234b256 d. --M. Gaimard, however, will settle this.-' WATERHOUSE says he is certain there are local varieties
234b256 evonshire, from another from Swansea.-- Again WATERHOUSE finds certain varieties of a Harpalus. commo
236bIBC oc. Geoffry-St. Hilaire Philosophy of Zoology WATERHOUSE B C N[a]me of two pigeons,-- «with specks» w
244c022 -- Procured two makis alive from there.-- Mem WATERHOUSE knows of some species which escaped there.--
268c095 en's land, which appear diff. from Australia WATERHOUSE <disputes this]> says differently) do. do. o
271c107 intermediate country-- ie. mundine groups.-- WATERHOUSE tells me in insects there are many plenty of
274c116 rest of the world?-- Will this not agree with WATERHOUSE & <birds> Mammalia.-- We have clear indicati
289c161 e to doubt :Scales into Teeth in Bering Pike (WATERHOUSE) Magazine of Zooly & Bot-- Vol II p. Dr John
289c162 ported by foetal lower developed forms.=» (NB (WATERHOUSE) says of affinity of many insects may be told
328cIBC e et de L'empire du Japon Wowett on Cattle-- (WATERHOUSE) has it) shells from Barrier isld many relati
353d061 ng to same section with with those of India-- (WATERHOUSE) knows three species of Paradoxurus common to
449e169 s-- Magazine of Nat. History. 1839. p. 106.-- WATERHOUSE refers to fossil remains of the Hamster.-- i
449e170 rbred. & the young kept constant. & all alike WATERHOUSE says some of the Galapagos Heteromerous inse
463t055 cases such as that of Java & Sumatra Nov 15th WATERHOUSE showed me the component vertebrae of the hea
463t057 posing much extinction. give a parallel case) WATERHOUSE remarked, that any argument for transmut, fr
463t059 be intermediate, but this is not required..-- WATERHOUSE says perhaps animals of Fernando Noronha are
481z009 e land shells of Galapagos different islds.-- WATERHOUSE remarks that no insectivore in S. America er
543m098 es. monkeys communicate much to each other.-- WATERHOUSE says far more instincts in all of the Hymeno
543m098 menoptera typical insects. ie have all parts. WATERHOUSE Study well the greater number of insects in
186b057 ks Quinary only three elements) How far Does WATERHOUSE'S representatives agrees with breeding.. in i
249c036 e not species, generally wide range? Mice.-- WATERHOUSE'S remarkable fact of no forms peculiar to wor
028r030 oceans detained by Organic powers. We know the WATERS of the ocean all are mingled. These reflection
122a114 f fluid, but of solid. because if of fluid, the WATERS of the ocean would obey that Law. & lie over t
137a151 2. «& 134» Bischoff. On the effects of meteoric WATERS on the temperature of the interior & p. 142 /
160g092 f Glen Bright flowing into E. end of L. Oich, & WATERS flowing into west end with obscure terraces on
165g124 chy would be well worth examining-- Inverness & WATERS of the Tarf-- Kilfinnan Tower where stream ent
473s05r wen, Capel Curig & some other lakes, (different WATERS) He cannot, however, tell them from L. Grozner
541m090 ut such an idea, as effect of sea on coves when WATERS had fallen, as in my Glen Roy paper.-- this gr
378d148 mage.-- How is this in Pidgeons & fowls.--??? WATE[R]TON «p. 197» put 12 wild duck's eggs under commo
563n001 -- October 2d.. 1838 Essays on Natural History WATERTON describes. pheasant springing from nest & lea
322c270 previous volume. & C. Prevost on L'Ile Julie WATERTON'S Essays on Natural History. Octob 2d Transact
095a039 iscovery side, found clear proofs of shells & WATERWORN rocks «at Cobija.» At Iquique of elevation to
147g018 must once have been very considerable mass of WATERWORN pebbles in Alluvium which without lake or sea
150g036 gradually drained, where is barrier {P} great WATERWORN frame terrace 4 4 not visible 3a 3a 3 Bouthon
151g042 r gravel plain of other, which must have been WATERWORN after 3d lake.-- 4th shelf runs up some way o
158g079 of upper terrace -- on right hand {P} rounded WATERWORN buttresses of granite obscure terrace 15 ft d
159g086 terraces (& conical hills on same) of «semi» WATERWORN & some partly well worn pebbles-- «which rive
163g108 t above sea-- soon decayed on exposure Mr H. C. WATSON Geographical distribution of British Plants Sh
325c267 ousness in Brutes in Blackwood. June 1838 WATSON on Geograph. Distrib: of British Plants. Humes
559m155 lackwood's Magazine June. 1838. Copied Mr H. C. WATSON on Geographical distribution of British Plants
046r080 Mices. At St Helena there is a native mouse Did WAVE first retreat at Juan Fernandez: the first grea
047r081 vating Earthquake of Valparaiso. (1822) no great WAVE on record. -- «also neighbouring sea must parta
047r081 sea must partake in absolute movement» Moreover WAVE «with same general character» reaches far beyon
047r081 swell, sea retreats & then breaks: i e to form a WAVE in ocean. is not this [Fig. 3] {P} form present
047r082 er part. -- Does the sea fall on banks as a Bore WAVE rushes up? (NB. Earthquake wave is an oscillati
047r082 banks as a Bore wave rushes up? (NB. Earthquake WAVE is an oscillation, body of water manifestly doe
047r083 <[...]> raised above as). -- In great Calabrian WAVE did not sea break first? I can imagine from loc
032r039 offered to the greater lateral extension of the WAVES. by the part beneath the band of greatest actio
032r039 slope to base of cliff (Z). to which point the WAVES would not reach. If now the ocean should sudden
047r081 elling across Pacifick.--excepting in number of WAVES & in wind, instead of sea's bottom being in mot
051r094 by removal of large fragments by mere force of WAVES: & action on upper tidal band, I do not see how
054r105 perforated & smoothed like those washed by the WAVES, a sufficient proof, that the sea formed these
056r109 er fissures may well have helped it, or diluvial WAVES. but when we see an entire island so encircled,
058r115 cion [.] no sign of elevation. Effects of great WAVES to obliterate all land marks.--At the first it
105a069 r) whereas sea. on coast, as long as exposed to WAVES of sea, cutting power increased with width. for

Page ***(Key Word)***
109a082 from beach action probably modified by form of WAVES & currents.-- but this must be continued. no cu
114a095 mple of Serapis. (now we have banished diluvial WAVES). & likewise <tells,> «offers a presumption» it
377d140 ins, when we estimate the matter removed by the WAVES of the sea, on beaches-- we really, measure the
528m037 s narrow in the distance.-- & even on paper two WAVING perfectly parallel lines are elegant.-- Again
293c174 urious instance of difference in races of men.-- WAX of Ear, bitter perhaps to prevent insects lodgin
586n082 limbs &c, or it result from mere impulse to save WAX.]CD which it instinctively exerts in concert wit
607o025 first time, & great effect being produced.-- the WAX was soft,-- the condition of mind which leads to
041r066 middle a short <fissure> «vein» terminated each WAY, which little vein was like the rest of these th
066r141 fathoms deep 24» under 50. Kerguelen Land, = the WAY it stands gales = very strong. Stones as bigger
102a059 raw it outwards.-- form of breaker affected some WAY out to sea.--¿ effects on bottom a thing floatin
102a059 sea.--¿ effects on bottom a thing floating some WAY from coast is driven on to it.-- rollers at Tris
109a082 land.-- Sea must always on actual beach act same WAY.-- a little further from beach action probably m
146g010] {P} Path East End near Holyrood Palace In same WAY at top the trap could be traced Grey in front on
151g042 aterworn after 3d lake.-- 4th shelf runs up some WAY on great sloping plain of alluvium (much corrode
153g052 isappear The normal condition of 4th shelf, some WAY below House of Glen Roy, seems to be which highe
155g062 > formed on shelf 4th 2d & 3d can be traced some WAY up, but most faintly on East side of Glen Turrit
158g083 uns up with buttresses on each side «very little WAY» in Upper Glen Roy at pass {P} River Gorge 4th S
159g084 sed of» Gneiss Block on 2d shelf & below it some WAY; several large ones (one 6 ft across) on top of
159g086 . 75 degree Air 70 degree? This station a little WAY down slope of obscure terraces (& conical hills
204b139 cidedly exceedingly prolific & hybrid about half WAY. Eyton says Hybrid about half aways & result the
217b184 her parent. about <half & half tim> as often one WAY as other.-- He has known case of good pointer &
240c003 ssion of form, may explain mule & pig being half WAY. Yet dogs sometimes like father, sometimes like
260c067 n hearing crys of hunger of little bird, in same WAY Wilson (p. 5). describes many kinds of bird unit
265c085 seed, one adaptation, other monster.-- The only WAY of judging whether structure is owing to habits,
280c135 all in domestication. throws great difficulty in WAY of ascertaining about hybrids.-- & is a very rem
280c136 t changes thousands in futurity.-- This is right WAY of viewing it.-- Variety when long in blood, get
281c138 m death or slow propagation of forms.--just same WAY as <we> «all men» not all equally related to eac
286c154 d by same passions-- brought into the world same WAY» they may convey much thus, Man has expression.-
286c154 shed to consider him as other animal-- it is the WAY of mankind. & I believe those who soar above Suc
288c159 dogs & English cross readily-- think about half WAY in appearance.-- bark about half way «in tone»--
288c159 about half way in appearance.-- bark about half WAY «in tone»-- the native dogs howl most dismally,
292c171 uccessive generations are impelled to do in same WAY-- The improvement of reason implies diversity &
301c198 t in animals. « & what reason» in precisely same WAY not possible to say what habitual in men & what
304c207 not then be told whether not descended from long WAY back.-- aberrant forms produced where many speci
308c217 of one father.-- yet differences carried a long WAY. --Case of Habit I kept my tea in right hand sid
312c231 must [not located] Henry Thompson tells me best WAY to improve cattle is to cross between a good bul
315c243 ment with his having canine teeth at all.-- This WAY of viewing the subject important.-- Laughing mod
318c255 lly separated the birds might yet remember which WAY to fly.-- There is a kind of Wren (Bebyk??) whic
350d053 rom temporal species as in two formations? by no WAY.?-- "Natura nihil agit frustra", as Sir Thomas B
357d072 to extinction of larger forms.-- From observing WAY the Marsupials of Australia have branched out in
391d174 is hermaphrodite (hence monstrosities tend that WAY «& from frequency of this tendency all mammalia
406e036 like one parent & sometimes other & sometimes ½ WAY. Ed. New-Phil. Transact. Rabies, common to men,
415e066 ima.-- caves.-- There being no fossils, the only WAY, that I can see to discover whether the parent o
417e078 her or mother, independently of its sex, or half WAY between, or someway different from either: & or
424e103 o be able to hybridise the Camel» like ass «same WAY some plants vary more than others» & horse in le
427e110 only effect produced would be different.-- same WAY one variety of <animal> dog does not prefer othe
437e138 times migratory.-- p. 103. Turtles finding their WAY to the Caymans from Honduras. good case of migra
440e147 n exceedingly small.-- This is far more probable WAY of explaining, much structure, than attempting a
442e152 thern flowers, throws enormous difficulty in the WAY of Mr Knights. theory «without seeds are freshly
443e153 bserved that to graft from the roots is the best WAY to get young trees, from worn-out kinds, & quote
448e167 ases highly improbable.-- yet I can see no other WAY of accounting for them.-- Think over this-- The
460t015 ugh only ¼ of blood). that it appears about half WAY between swan-goose & common goose.-- the stripe
460t017 ich prevent infinite variation in every possible WAY.-- the laws which determine the kinds of monstro
484z016 & Southern Hemisphere It is most interesting the WAY Synallaxis leads into Furnarius. by Patagonian F
495q05a les test by suspending magnet within & see which WAY they fly.-- (9) I have noticed leaves covered wi
506q014 7½/. The Greyhound. was in length (measured same WAY) 47½-- in heigt 30 inches Examine Keel of Common
533m059 xtreme difficulty of moving muscles in different WAY from what they have been accustomed to, in certa
534m062 Sykes compares this with pidgeons finding their WAY home-- there is something wrong in comparing the
537m074 is finger from real caprice. it is chance, which WAY it will be, but yet it is settled by reason.--
542m095 on & wrinkling of the skin contract iris?-- same WAY as one lifts up eyebrows to see things in dark.
542m096 lips in peculiar position, & he holds them this WAY, when opening mouth between interval of barking,
543m099 people of very limited intellects, & in the same WAY are chess Players-- A man at Cambridge, during h
557m149 rom shooting. runs away. is not afraid the whole WAY. but ashamed of himself.-- Jealousy probably ori
558m152 nged» expulsion of air «dogs snarl much the same WAY» generic manifestation of great passion.-- I do
565n007 enemy.-- Man & dogs show triumph (& pride) same WAY walk erect & stiff, with head up.-- Why does sus
569n020 ngs & voice, as roaring for lion &c &c. (in same WAY alphabet. arose from letters, symbol of word beg
574n042 theory intelligible An habitual action must some WAY affect the brain in a manner which can be transm
578n053 ordinarily to have one» you wont. ==)== No surer WAY to blush, than particularly to wish not to do so
583n071 , it must have impulse to make a cell in certain WAY, which way its organs are sufficient for hence i
583n071 ave impulse to make a cell in certain way, which WAY its organs are sufficient for hence it must some
583n071 its organs are sufficient for hence it must some WAY be able to measure the cell; p. 22. instincts &
584n072 ass of so called instincts to which my theory no WAY applies.-- it is the acquirement of a new sense,
585n076 return.-- Hence I conclude. pidgeon taken little WAY, whirled, & then taken other way-- would not fin
585n076 on taken little way, whirled, & then taken other WAY-- would not find its way back.-- ?? this is not
585n077 led, & then taken other way-- would not find its WAY back.-- ?? this is not instinct, but a faculty,
586n078 Heavens, points of compass, & they do know which WAY they go; & so return.-- «but does not apply to d
586n081 sleep.-- "A bird has the faculty of finding its WAY, which in certain species is instinctively «not
593n111 ny coots & waterhens feeding on grassy bank some WAY from water, suddenly, as if by word of command,
607o025 f mind which leads to motion being inclined that WAY]CD one sees this law in man in somnambulism or i
608o026 than to hate & be †A man may put himself in the WAY of Contingencies.--but his desire to do arises f
608o026 assist & educate by putting contingencies in the WAY to aid motive power.--if incorrigably bad nothin
615o36v t think this only pleasure; for it was different WAY of showing it, nor was there any cause, & if sur
617o39v he objective, by our external what senses in the WAY in which we apprehend the force of inanimate bod
640j28v intermarrying with ordinary race.-- «There is no WAY of eliminating the evils of old age, after breed
148g025 d liable to vary-- I asked this question in many WAYS & received same answer Thought lambs most like
200b122 en that> or Falkland rabbit.-- There is only two WAYS of proving to them it is not; one when they can
269c101 ave travelled that thickness in that period & no WAYS assisted by fluid currents which, may take plac
346d047 half breed liable to vary. I asked this in many WAYS, but received same answer.-- Thought lambs were
389d173 ed into bud matured by female;<]CD> such view no WAYS explains Ld. Moretons case: without the nervous
430e118 if it will for all.-- Varieties are made in two WAYS-- local varieties, when whole mass of species a
529m042 as found to be ignorant, but quite sensible & no WAYS an ideot.-- «in this case must have been functi
532m056 Shrewsbury,, then immediately fell into her old WAYS & became fat! What remarkable affection to a pl
336d019 eing created. it is probable if created at once. <WD> according to ordinary laws, the character of of
469t135 has now gone on 14 days. (except some wet ones/ & WD go on longer-- Woodfords Marrow fat, Early frame
471tf08 heir similarity «in structure» he says "indeed it WD be difficult to point out a family so completely
501q10a few are they constant: this very important for it WD show that such variation is not a generic or spe
022r008 ugh that a sharp knife could not cut it: in which WE found the Head & Boans of a Hippotomus; the hair
023r009 etrified, and the jaw was also firm, out of which WE pluckt a great many teeth, 2 of them, 8 inches l
025r018 be supported from that part of the globe: & when WE see conclusions substantiated over S. America &
025r018 nclusions substantiated over S. America & Europe. WE may believe them applicable to the world.-- My g
028r029 n the Tropical oceans detained by Organic powers. WE know the waters of the ocean all are mingled. Th
035r048 oscillated equably.-- These facts become easy if WE look at the action as a deep & extensive movemen
039r061 ulsion of fluid rock, which elevates a continent> WE are more abound to take analogy of movements of
040r063 ve to animal life.--Patagonia In the Chonos Islds WE must imagine bituminous shales have been metamor
041r065 bottom is important; because in this latter case. WE cannot judge whether such fossils. lived in grou
042r069 with far more regularity in S. America: in France WE have freshwater lakes unequally elevated, which
044r074 ord: after seeing small Bombs. without a vesicle. WE may consider appearances of eruption at bottom.-
045r077 xplanation of low Barometer? In a subsiding area. WE may believe the fluid matter instead of afflux (

```
*********************************************(Key Word)*********************************************
046r078 allic veins likewise must separate ingredients if WE look to a constant revolution.--Are we to consid
046r078 edients if we look to a constant revolution.--Are WE to consider that the dikes which so commonly (st
048r085 lump of hard clay. -- In the History of S America WE cannot dive into the causes of the losses of the
048r086 he same caution is applicable to the Siberia case WE must not think alluvial plains «always» most fav
049r088 llera???» Porphyry at Valparaiso; Epidote -- Must WE look at regular greenstone cones at S. T. del Fu
051r095 From Sir. H Davy experiment on the copper bottom. WE see a trifling circumstance determines whether a
055r108 of the solid lavas.--proportionally high to age. (WE do not wonder to see tertiary plains consumed) W
056r109 tes former boundary. (as in other unworn islands) WE take in at once the stupendous mass which has be
056r109 y not have helped it, or diluvial waves. but when WE see an entire island so encircled, the one slow
058r117 considers these earthquakes travel in order.-- If WE look at Elevations as constantly going on we sha
058r117 - If we look at Elevations as constantly going on WE shall see a cause for Volcanos part of same phen
063r130 first cases distinct species inosculate, so must WE believe ancient ones: «.» not gradual change or
063r132 mal in two. (gemmiparous. by nature or accident). WE see an individual divided either at one moment o
063r132 t one moment or through lapse of ages.--Therefore WE are not so much surprised at seeing Zoophite pro
070r153 to balls. must exhibit orbicular structure.--When WE recollect connection of columnar & orbicular in
070r153 nection of columnar & orbicular in basalt.-- When WE see Avestruz two species. certainly different. n
073r161 e from without.-- "True native iron that to which WE cannot attribute a meteoric origin & which is co
074r165 ccount of steam acting on trachytes. also Azores. WE here have case of such vapours washing a rock» V
086a004 nt elevation, yet sea shells at tops of mountains WE ought to sympathize with. old doubters of what a
090a023 nes of large numbers (& how few cases recorded if WE say «100» <5>0 lbs a year too little.-- How come
104a067 m few dikes which have given rise to eruptions.-- WE must suppose everywhere--, in granitic areas &c
105a070 quent opening a medial gorge by slow erosion. but WE have evidence in distribution of blocks, that th
105a071 succession of flights of steps; if one lake then WE must suppose barrier in the very part, where bar
106a075 essentially a deepening agent {P} Therefore when WE have valleys of this structure. as the inclinati
108a080 ing & rising areas.-- In Earthquake if Subsidence WE should not expect volcanos.-- not so much horizo
112a089 nite & other trap.-- It is in the mountain masses WE must look for that.-- how few isolated volcanos
113a090 wer of heat from centre.-- But is this not wrong? WE know mean of surface formerly much higher, «so»
113a090 w mean of surface formerly much higher, «so» that WE must look at the upper four hundred feet of stra
113a091 ter then 32 degree would have been found lower.-- WE have no right to consider the conducting powers
113a091 r or worse & the depth of 32 degree. being little WE may confidently infer that time has not been all
114a095 C. of Good Hope.-- A bare hill of greenstone. if WE know origin of greenstone tells subsidence as pl
114a095 subsidence as plainly as Temple of Serapis. (now WE have banished diluvial waves). & likewise <tells
118a103 Pentlands Fossils & Meyens --<Jura &> Chalk When WE consider parallelism of dikes (Hopkins) & that e
119a104 eat circular mountains, yet so analogous, that as WE see mountains formed (& mountains are effect of
119a104 & mountains are effect of continental elevations) WE may conclude that elevation is independent of sp
119a106 to convert rock into gneiss &c judging from what WE see when trap in dike & approach other rocks. &
120a107 - mem coal-field.-- the structure of Andes. where WE believe we can trace the outlines of what were f
120a107 field.-- the structure of Andes. where we believe WE can trace the outlines of what were fluid undula
126a122 them, this is ascertained by conducting powers-- WE judge from the surface, & say 60 ft to degree.--
126a122 being bad conductor is {P} against my views-- if WE had rod thus & judged by increments at, how wron
130a133 g.-- Hot water would not lie. at bottom.-- Surely WE here have proofs of hot bottom.-- Study Bishoofs
132a138 rease on earth by volcanic action.-- <Why> now as WE know volcanic action prevails more beneath the s
133a140 ust remain fluid for an enormous period: now when WE see how many points have been penetrated by volc
159g087 higher parts?? Bought Glen name of Glen by which WE descended, it is to the west of Glen Tarf What I
171b002 is life short, Why such high object generation.-- WE know world subject to cycle of change, temperatu
171b003 ll circumstances which influence living beings.-- WE see <living beings>. the young of living beings,
171b003 individuals produced by buds are constant, hence WE see generation here seems a means to vary. or ad
171b003 ere seems a means to vary. or adaptation.-- Again WE <believe> «know» in course of generations even m
172b008 d Fox-- Chiloe, fox,-- Inglish & Irish Hare.-- As WE thus believe species vary, <in> changing climate
172b008 thus believe species vary, <in> changing climate WE ought to find representative species; this we do
172b008 ate we ought to find representative species; this WE do in South America closely approaching.-- but a
172b008 ca closely approaching.-- but as they inosculate, WE must suppose the change is effected at once, --
173b012 parated.-- On this idea of propagation of species WE can see why a form peculiar to continents; all b
174b014 ame type as extinct which is law almost proved.-- WE can see «why» structure is common in certain cou
174b014 hy» structure is common in certain countries when WE can hardly believe necessary, but if it was nece
175b017 rong argument of possibility of such change,-- as WE see them in space, so might they in time As I ha
175b018 cannot help..-- becoming more complicated,; & if WE look to first origin there must be progress. if
175b018 e look to first origin there must be progress. if WE suppose monads are «constantly» formed ¿would th
175b020 ated recent terebratula, but Megatherium nothing. WE may look at Megatheria, armadillos & sloths as a
176b022 stranger in death of species, than individuals If WE suppose monad definite existence, as we may supp
176b022 iduals If we suppose monad definite existence, as WE may suppose is the case. their creation being de
177b025 nce to intermarriage <increases it--> settles it ¿WE need not think that fish & penguins really pass
177b027 t down to simple organization.-- birds-- not. (P} WE may fancy, according to shortness of life of spe
179b035 ar very probable:-- Mem; Horse, Llama. &c &c-- If WE <suppose> «grant» similarity of animals in one c
180b037 of species constant.-- With respect to extinction WE can easily see that variety of ostrich, Petise m
181b039 ype of their existence in the present world.-- or WE may suppose only each species in each generation
181b040 viduals in a country not rapidly increasing.-- If WE thus go very far back to look to the source of t
181b040 s relatives shall now exist,-- In same manner, if WE take «a man from.» any large family of 12 brothe
185b055 f to as many situations as possible.-- Why should WE have in open country a ground «do. <w> parrot.--
186b060 ve traced the <type> great quadrupeds to Siberia; WE must look to type of organization.-- extinct spe
189b070 & Megatherium. each with same kind of coat.-- If WE could tell, I do not doubt even colour hereditar
189b074 o talk of one animal being higher than another.-- WE consider those, where the {cerebral structure in
191b083 female insures often mixing of individuals. Here WE have avitism the ordinary event. & succession th
192b085 so insensible, that each is not more change than WE know varieties can produce.-- Therefore all gene
192b087 ganization. [Spines in Echidna & Hedgehog]CD-- As WE have one Marsupial animal in Stonefied slate, th
192b088 » with fishes & reptiles.-- In mere eocine rocks. WE can only expect some steps.-- I may ask whether
194b097 xture.-- But how with «hermaphrodite» shells.!!!? WE have not the slightest right to say there never
195b098 not permanent. this might be made very strong. if WE believe the Creator creates by any laws. which,
195b098 ed.-- Such an influence Must exist in such spots. WE know birds do arrive & seeds.-- The same remarks
195b100 sely allied to Mus. coronata, but not coronata.-- WE know that domestic animals vary in countries, wi
196b108 asitical nature of insects & worms.-- In abstract WE may say that vegetables & mass of insects could
199b116 land, of the same type with the great continents WE get a means of knowing movements.-- How can we u
199b116 ts we get a means of knowing movements.-- How can WE understand excepting by propagation that out of
199b118 <191> «p 191» No. 5. Ap 1837» F. Cuvier says. "But WE could only produce domestic individuals & not ra
202b132 erded separate from the English cattle, nor could WE get them to associate together" There is long ri
202b133 Now if suppose world more perfectly continental. WE might have wanderers. (as Peccari in N. America)
203b135 ve offspring, & then «V. L. Institut p 245. 1837» WE should have anomalies. as Cape Anteater,.-- This
203b137 flycatcher doing the service of a swallow I think WE may conclude from Australia & S. America. that o
204b141 eneralization in view. In Marsupial division «do» WE not see a splitting in orders, Carnivora, rodent
206b147 would have different formula for each lustrum.-- WE may conclude that there will be a period though
206b149 w species which stand between great groups, which WE are bound to consider the increasing ones.-- NB
208b154 d's.--? NO Macleay Name given in Congo Expedition WE need not expect to find <species>, varieties, in
209b157 k Islds = ∴ Islands & Artic are in same relation. WE find species few in proportion to difficulty of
218b189 and. & fresh species made. parents do not cross-- WE see it even in men}; the possibility of Caffers
218b192 ms» forms would have some peculiarity.-- Now when WE hear that the whole island is volcanic surmounte
218b192 anic surmounted by water & studded with others.-- WE see a beginning to isld. Graham isld.-- we know
218b192 ers.-- we see a beginning to isld. Graham isld.-- WE know many seeds, might be transported some blown
219b194 can form, merely because intermediate position.-- WE cannot consider it as adaptation because volcani
219b194 e cannot be explained more philosophical to state WE do not know how transported.-- (Glaciers might h
219b196 m one stock, & since distributed by such means as WE can recognize, may be thought to explain nothing
220b198 k hybrid plants; analogous to Men. & dogs. Now if WE take structure as criterion of species Hogs diff
220b198 species Hogs different species, dogs not, but if WE take character of offspring. Hogs not different.
222b203 urn to one parent, this is only character. & yet WE find this same tendency (only less strongly mark
222b203 s attempt at returning to parent stock.-- I think WE may look at it so--?? It holds good even with tr
222b204 ious-- My idea of propagation almost infers, what WE call improvement, --All Mammalia from one stock,
223b211 f offsprings accounts for uniformity of species & WE Must confess. that we canot tell, what is the am
223b211 for uniformity of species & we Must confess. that WE canot tell, what is the amount of difference, wh
224b214 e remain distinct. Where country changes rapidly, WE should expect most species.-- The difference int
```

```
Page   ********************************************(Key Word)********************************************
225b217 changes were effects of external causes, of which WE are as ignorant. as why millet seed turns a Bull
226b221 quired will explain representative system Of this WE see example in English & Irish Hare.-- Galapagos
226b222 ing to its own genera Therefore if in small tract WE have many species, we may insure mass continenta
226b222 Therefore if in small tract we have many species, WE may insure mass continental or many large island
227b224 nt carnivora than Pachydermata If my theory true, WE get (I) a horizontal history of earth «within re
227b224 ation; for having ascertained means of transport, WE should then know whether former lands intervened
227b224 .-- (2d) By character of any «two» ancient fauna, WE may form some idea of <origin under> connection
227b224 (Coral reefs∴ shallow water at Melville Isd. (3d) WE know that structure of every organ in A. B. C. t
227b224 «of one genus» can pass into each other «by steps WE see»: but this cannot be predicated of <genus.>
227b224 edicated of <genus.> structures in two genera.-- <WE then cease to know the steps.> although D E F. f
227b225 steps.> although D E F. follow close to A. B. C. WE cannot be sure that structure (C) could pass int
227b225 be sure that structure (C) could pass into (D).-- WE may foretell species. limits of good species bei
227b225 fact  that they are not far the most serviceable. WE may speculate of durability of succession from w
227b225 y speculate of durability of succession from what WE have seen. in old world, & on amount changes whi
227b227 f <change.> transmutation & geographical grouping WE are led to endeavour to discover causes of chang
227b228 eration, causes of change «in order» to know what WE have come from & to what we tend.-- this & «dire
228b228 n order» to know what we have come from & to what WE tend.-- this & «direct» examination of direct pa
228b229 fusorian. is "What are the laws of life".-- Where WE have near genera far back, as well as at present
228b229 near genera far back, as well as at present time, WE might expect confusion of species.-- Important.
228b231 y showing connexion of two plants. Animals-- whom WE have made our slaves we do not like to  consider
228b231 wo plants. Animals-- whom we have made our slaves WE do not like to consider our equals.-- «Do not sl
228b231 <of  death>. pain. sorrow for the dead.-- respect WE have no more reason to expect the father of man
228b231 an kind. than Macrauchenia yet he may be found:-- WE must not compare «chances of embedment in» man i
228b232 eradded, animals not got it, not look forward» if WE choose to let conjecture run wild then <our> ani
229b232 e, from our origin in <there> one common ancestor WE may be all netted together.-- Hermaphrodite anim
230b240 evidence  of two races of plants run wild.-- (for WE know that such can take place without impregnati
231b241 wren) & others varying in wild state to show that WE do not know what amount of difference prevents b
231b243 Now this is difficult to explain by creation-- or WE must suppose a multitude of small creations.-- W
232b245 ability of starting from one point.-- In the crag WE see the process of change of those forms, which
232b246 ecies does not measure time but physical changes (WE assume like weather on  long average tolerably» u
232b246 o countries» finding a very hot day, «in one», oh WE will take a day from the equator to add to the m
233b252 ree with longevity of species in Molluscs!!! When WE talk of higher orders, we should always say, int
233b252 ies in Molluscs!!! When we talk of higher orders, WE should always say, intellectually higher.-- But
236b280 s, as Decandoelle says, no he only says sometimes WE might expect disseminated species to vary a litt
236b280 n England» surely it is not-- intermediate genera WE might expect.-- Lindley Introduct Dict. Science.
248c029 each point, or migrated from those quarter, where WE know quadrupeds have existed for ages.-- ∴ The m
248c030 imals VERY different will breed together, so when WE grant «(which can be shown probable,)» varieties
248c030 resumption that they would not. breed together.-- WE see even in domesticated varieties a tendency to
248c033 id & parent stock, than between two hybrids.-- As WE see external influences first affect external [f
249c036 rms on Himalaya??-- This is very remarkable, when WE consider number of quadrupeds in Eocene  period.
254c048 cceed each other with the greatest rapidity"-- so WE find species each class successively present mod
255c051 ty in propagation.-- Feathers on, Apteryx because WE may suppose longest part of structure.-- shape o
255c052 ription of ancestor of all birds, & so for birds. WE thus obtain an abstract idea of a bird-- An anim
257c056 iracle, but induction leads to other view.-- Till WE know uses of organs clearly, we cannot guess cau
257c056 ther view.-- Till we know uses of organs clearly, WE cannot guess causes of change.-- hump on back of
257c058 secondary-epocks, & developement of lizards.-- As WE have birds impressions in Red Sandstone. great l
259c064 of TRACING change of species to species «although WE see it affected» tempts one to bring one back to
263c077 «under present form» will come, (or how dredfully WE are deceived) then he is no exception.-- he poss
263c078 ements in mental machinery-- so analogous to what WE see in bodily. that <I> it does not stagger me.-
265c086 abits-- Thus bill & nostril of Puffinuria I think WE may clearly attribute to heredetary origin & not
265c087 velled wings could be shown to be of some use. If WE only had Puffinuria Garrottii & no other species
265c087 had  Puffinuria Garrottii & no other species-- as WE have only ornithorhyncus, then we should never k
265c087 er species-- as we have only ornithorhyncus, then WE should never know how much structure was connect
266c091 manners  they are as opposite as day & night: yet WE know how remote the periods at which both left t
268c096 rannula (NB work out how many forms Tyrannula can WE worked out into. Milvulus [not located]  element
268c099 tribution is.-- ¿Pelagic forms similar--birds??-- WE must always bear in mind proofs of most  equable
272c109 says most subgenera confined to continent, though WE have seen species «of subgenera» scattered  over
272c109 seen  species «of subgenera» scattered over it.-- WE have abundant instances of remarkable  structure
274c116 Europe & northern Asia & Northern America.-- may WE not look to these Northern regions as the recept
274c116 s not agree with Waterhouse & <birds» Mammalia.-- WE have clear indication [not located] alone, but o
276c122 es-- as formation of species <of> gradual, so may WE suppose., that something intermediate, between n
278c129 ose sections & subgenera; are analogical, because WE do not know, whether nearest species of each mig
280c136 oo unlike its own parent this impossible- (Hence WE might expect even if two mules bred or two certa
281c138 or  slow propagation of forms.--just same way as <WE> «all men» not all equally related to each other
282c140 - Yet each family might have its own character,-- WE here suppose these changes of <use> adaptation g
282c141 ltogether different.-- To make this case perfect, WE must suppose men instead of mere colour & trifli
282c141 structure & even to certain degree in habits, yet WE might have these analogies.-- We must two  races
282c141 e in habits, yet we might have these analogies.-- WE must two races of such men living in same countr
282c142 ty with respect to species of each other, because WE suppose all descended from same.-- but if two or
283c146 destroyed far remote genera. will be produced. As WE know from Ehrenbergh, there are fossil (see Scie
283c146 iary fossil Infusoria, of same forms with recent, WE have nothing to do with CREATION.-- <On> The end
284c149 any  structure is general in all species in group WE may suppose it is oldest, & therefore lest subje
286c154 language,  they know the crys of pain, as well as WE.-- It is our arrogance, to raise on the same she
286c155 ke, at least every nation has. done so as. yet.-- WE now know what is the natural arrangement, it  is
286c156 ught they to hold,-- Birds having web-feet, where WE see scarcely any traces of passage a difficulty,
290c163 t be owing to heredetary power of Muscles.-- then WE SEE structure gained by habit Talent &c in man n
291c167 oditisms. one step in <scale>. Series-- in plants WE have a step between <mon> monoeecious & dioeciou
291c167 to  imagine how sexes were separated.-- in plants WE have some flowers monoecious & others dioecious.
292c169 cannot be great, otherwise it would be unlimited. WE absolutely know that tendency is greater in Mamm
292c171 s become heredetary & instinctive & not others.-- WE even see they must be done often «to be habitual
293c172 put note. Sir W. Scott has written about it]CD If WE saw a child do some action-- which its father. h
293c172 me action-- which its father. had done habitually WE should exclaim it was instinct.-- Even if savage
293c172 akes. & was given a Great coat & this he put on & WE afterwards could understand «(language better in
293c175 smell  of Man would be disagreeable to Musquitoes WE never may be able to trace the steps by which th
294c176 very close Species in islds. near continent, Must WE resort to quite different origin when species ra
297c186 death, makes the group aberrant When species rare WE infer extermination, when group few in number of
298c191 lower country during gradual elevation of isld.-- WE must imagine a considerable range of one species
299c195 onous &c-- very important in classification. here WE have generative organs. first character.-- In di
299c196 any of the Female flowers unimpregnated Babington WE see gradation to mans mind in Vertebrate Kindgdo
302c201 anterior to giving off of these two families, but WE see analogies between fish.-- Birds same remarks
302c202 ving analogous characters, might be multiplied.-- WE must argue reversely. WHERE CHARACTER VARIABLE i
304c204 bject from Lawrence. Blumenbach & Prichard -- Now WE might expect that animal halfway between man & m
304c208 y circumstances In ostrich which is not isolated, WE must suppose the changes from typical  structure
305c211 kind to the endless forms of the living beings.-- WE see thus Unity in thinking and acting  principle
307c215 ntly useless organs do indicate such origin, then WE are bound to consider abortive organs of same te
308c217 means»  civilized Man, May exclaim with Christian «WE are all» Brothers in spirit-- all children of on
309c219 used The only cause of similarity in individuals «WE know of:», is relationship, children of one pare
312c228 s, without «number of vertebrae» new acquisition, WE must [not located] Henry Thompson tells me  best
313c235 s men-- when habits much more firmly impressed.-- WE see in the Entomostraca. The sexual curiosity of
333d006 ail, fur &c to the half bred Persian.-- Here then WE have clear case of heterogenous offspring from o
335d016 considerations  are here combined). In last page, WE have seen mules could have no offspring, & this
338d023 separation  on <be> inter-breeding, for otherwise WE could not understand the vast number of domestic
343d039 10,000  years Negro probably a distinct species-- WE know how long a Mammal may go on as one  species
348d050 n species & become worthless-- Mammalia Edentata\ WE do (p 6) say such is group. because it has  such
348d050 ecause it has such characters of importance, "but WE say such happens to be the character, of no matt
348d051 s, what is that, amount of resemblance,-- how can WE estimate this amount, when <value> no scale of v
348d051 affinity  may be taken literally,, though how far WE can ever discover the real relationship is doubt
349d051 ss of forms-- Cannot be discovered «un»till <in> «WE ascend to» subgenera & families, <even in Cetion
```

Page **(Key Word)**
352d057 ge of foetus.-- As Larvae may be more perfect (as WE use the word) than parent, so may species retrog
355d068 rate parts of same plant<s>)-- now in some Polypi WE see young bud changing into ovules.-- Captain Gr
358d085 on on females.-- [not located] hen freely.-- here WE have beautiful proof of the breeding in & in (li
374d134 ly proportiona[l] to the number that can live.--» WE ought to be far from wondering of changes in num
377d139 lephants lived. & when present animals-- lived.-- WE know the great time, necessary to form channel &
377d140 : only circumstances. a contingency of time. When WE multiply the effects of <earthquakes>, elevating
377d140 ising continents, & forming mountain-chains, when WE estimate the matter removed by the waves of the
377d140 ter removed by the waves of the sea, on beaches-- WE really, measure the rapidity of change of form,
378d147 ection If masculine character. added to species,. WE can see why young & female alike Good Ch 6 Keep
383d160 le?-- legs pale coloured.-- In the back feathers, WE have character very different from either parent
385d164 eir bodies "It is a fact equally well known, that WE observe in the temper, especially of the younges
385d165 sted, this is Lord. Moretons law.-- "How often do WE find in the son, the character, constitution, &
388d171 ring being determined by imagination of Mother.-- WE see in a litter every possible variation from be
397e003 e, without the immediate agency of the deity. But WE know from experience! that these operations of w
397e003 ow from experience! that these operations of what WE call nature, have been conducted almost! invaria
397e003 constant as any of the laws of nature with which WE are acquainted."-- this applies to one species--
397e004 of one race to some change of circumstances; now WE know how slowly & insensibly such changes are in
397e004 owly & insensibly such changes are in progress.-- WE feel interest in discovering a change of level o
398e006 ».-- the latter pages in the history are perfect, WE obtain a glimpse only of the changes which the g
398e006 hich the government is subject to.-- further back WE obtain here & there in order a scattered page; w
398e006 we obtain here & there in order a scattered page; WE find <great> sensible change in the institutions
398e006 nd <great> sensible change in the institutions. & WE suppose not only revolutions, but entire obliter
398e006 quires each form to have lasted for its time: but WE ought in same bed if very thick to find some cha
402e020 end of Borneo to the adjacent island-- In Sooloo WE find the elephant-- in Magindanao several kinds
403e023 any animal (as Toxodon) & say its relations.-- if WE know its congeners then we can.-- now on my theo
403e023 ay its relations.-- if we know its congeners then WE can.-- now on my theory this «certainly» can be
403e023 great variation in external form of varieties, do WE suppose bones will not change in number. (even s
408e043 ss so, that either Americas.-- If species change, WE see external conditions have great effect on the
408e043 re extermination becomes part of same law.-- When WE know what a great effect. light has in colouring
409e048 ould be as many species, as individuals, & though WE may not trace out all the ill effects. -- we see
409e048 ough we may not trace out all the ill effects. -- WE see it is not the order in this perfect <uni> wo
409e049 er at the present, or many anterior epochs.-- but WE can see if all species, there would not be socia
415e066 ty like Ornithorhyncus,: since being cosmopolite, WE do find his remains.-- Lima.-- caves.-- There be
417e077 two organs in one body.-- or in two bodies, <th> WE can as well understand the necessity of a relati
417e077 e <species>, variety, as in two varieties, & this WE might expect, as the difference between man & wo
419e086 e as those of present seas.-- now in this country WE have better means of judging of slowness of phys
421e091 ng very clear indications of its true affinities. WE are least likely in the modifications of these o
423e095 as the simplest fish &), my answer is because, if WE begin with the simplest forms & suppose them to
423e099 y to have most slightly modified organic forms.-- WE know not rate of deposition has been equal even
424e103 s, which brought its puppies to be fondled.-- and WE see in the Australian dog an instance of a half
426e108 - Wonderful as is the possession of voice by Man. WE should remember, that even birds can imitate the
429e113 -- I do not believe this Nature's plan.-- Whether WE can or not trace history of first appearance of
429e113 ance of varieties of domesticated animals, yet as WE know how many plants have been produced (look at
429e113 y plants have been produced (look at the Dahlias. WE may infer it in animals.)-- Azara gives account
429e114 going on the peaceful woods. & smiling fields.-- WE must recollect the multitudes of plants introduc
429e114 ht spread themselves, as well as our wild plants, WE see how full nature. how firmly each holds its p
429e114 l nature. how firmly each holds its place.-- When WE hear from authors (Ramond. Hort. Transact Vol I.
429e115 trees in Beagle Channel.-- it is not elements.-- WE cannot believe in such a line., it is other plan
431e122 -- In the place where any species is most common, WE need not look for change, because its number sho
432e124 efore the children do not, (& in hairless kittens WE see same fact) go back, & this is argument again
433e127 f the beds-- The argument must be thus put, shall WE give up whole system, of transmut., or believe t
434e128 king about the controversy on Didelphys says. "If WE cannot reason from the analogies of the existing
434e128 of the existing to the events of the past world, WE have no foundation for our science".-- <it is on
434e128 -- <it is only analogy.> but experience has shown WE can & that analogy is sure guide & my theory exp
440e146 . results of complicated laws of organization: as WE see these strange plumage in pidgeons yet no cha
440e147 esponding change in Birds of Paradise.-- All that WE can say in such cases, is that the plumage has n
441e151 cross with another.-- to escape it «in any case» WE must draw such a monstrous conclusion, that ever
442e153 never are, namely on stocks of other varieties & WE know that the kind of stock greatly affects the
446e163 throws a very great difficulty in my theory, here WE have a plant remaining constant, without crossin
459t009 k Deer-- young spotted <like in> "prettty much as WE see in the young of the wild hog & of several sp
460t019 how this complexity from a few types originated, WE must go to the first origin of the world.-- our
462tf4v rew, found by M. Lartet same as existing species. WE see the same object gained by the Mataco-armadil
465t079 rom Lund more Mammals, than at present «in Europe WE know there has been several successions of Mamma
465t080 eds of La Plata. (except close to B. Ayres).-- If WE may take this as guide, the shells preserved mus
465t081 which tends to make the destroyers vary; so that WE here see reasons-- why no perfect gradation can
467t099 ive--and both growing within Kitchen Garden.-- As WE see in Hybrids that although anther «nor filamen
467t099 «nor filaments» shrivel, yet stigma does not, so WE may feel somewhat «but little» less surprised at
468t105 chopped horse hair with legs & take flight-- Yet WE have crosses-- I see Bees almost every flower--
471tf07 isplays.-- most fantastic & use unknown.-- "<when WE find such an endless variety of form in the same
473s07r s--how is this in young pigeons--dogs--cattle? As WE see the frame of animals can adapt itself to cou
491q01v oned by Mr Knight. Vol IV Hort. Transact.-- 4 May WE no suppose, that certain plants, like Aphides pr
498q008 s.» much-- or the minute Orthopt.-- important, as WE know how readily they cross.-- (9) In the nurser
498q009 e dioecious, if no hybrids were produced by seed, WE might feel sure, that pollen of own kind is much
506q15v the Nettle. at Galapagos-- Dioecious.-- Carex.-- WE may presume Nettle spreads by seeds= (44) Zoster
526m027 e same means & therefore properly no free will.-- WE may easily fancy there is, as we fancy there is
526m027 no free will.-- we may easily fancy there is, as WE fancy there is such a thing as chance.-- chance
527m031 in England, in Tierra del Fuego not one.-- now as WE know birds learn from each other «though differe
528m037 easure of perspective. which cannot be doubted if WE look at buildings, even ugly ones.-- the pleasur
540m085 merely coorganic as connexion of mammae & womb.-- WE need not feel so much surprise at male animals s
543m101 t merely instinct, a separate thing superadded.-- WE can thus trace causation of thought.-- it is bro
544m103 .-- The mind thinks with extraordinary rapidity-- WE may conclude that neither number, vividness, rap
546m108 ith (.D. Stewart life of. p. 27), says <sympathy> WE can only know what others think by putting ourse
546m108 k by putting ourselves in their situation, & then WE feel like them--. hence sympathy very unsatisfac
549m118 ty to degree of <happi> pleasure of such thoughts WE give no credit to instinctive feelings.-- for ma
549m121 appiness is the object of living.-- or whether if WE obey literally New Testament future life is almo
550m125 discussion-- by saying what is Happiness?-- When WE look back to happy days, are they not those of w
553m136 grant part of his most magnificent laws. of which WE profane «degnen» in thinking not capable to <do>
556m145 hysiognomy translated by Holcroft «Vol I» .p. 86 "WE ought never to forget-- --; that every man is bo
559m153 nce so great... "Ay Sir there is much in analogy, WE never find out." This unwillingness to consider
559m154 as governing by laws is probably that as long as WE consider each object an act of separate creation
559m154 consider each object an act of separate creation. WE admire it more. because we can compare it to the
559m154 of separate creation. we admire it more. because WE can compare it to the standard of our own minds.
559m154 f our own minds. which ceases to be the case when WE consider the formation of laws invoking laws. &
560m156 of eyes.-- Do female monkeys care for men.-- Have WE any ferns in the hothouse at home Natural Histor
564n005 citadel itself.-- the mind is function of body.-- WE must bring some stable foundation to argue from.
570n026 J. Reynolds.-- Is our idea of beauty, that which WE have been most generally accustomed to:-- analog
571n027 of our ancestors being impressed on us.-- Surely WE have taste naturally all has not been acquired b
571n028 re it soon. & why do all men. agree ultimately?-- WE acquire many notions unconsciously, without abst
571n029 al-- & hence become heredetary; on same principle WE know many tastes become acquired during life tim
572n032 applies to bodily weakness & inferiority, but now WE carry it on to mental inferiority-- when we do n
572n032 t now we carry it on to mental inferiority-- when WE do not expect any bodily harm-- case of habitual
573n036 e as mine about our origin of a notion of a Deity WE can allow «satellites», planets, suns. universe,
573n036 to be governed by laws,, but the smallest insect, WE wish to be created at once by special act, provi
573n036 c:--must be a special act, or result of laws. yet WE placidly believe the Astronomer, when he tells u
574n040 e end.-- Children & old people get into habits.-- WE probably can hardly form an idea of a mind so li
575n043 ids» with <the> it the provision for death.-- can WE deny that brain would be intermediate like rest
575n043 rain would be intermediate like rest of body? Can WE deny relation of mind & brain. «Do we deny the m
575n043 f body? Can we deny relation of mind & brain. «Do WE deny the mind of a greyhound & spaniel. differs

Page		(Key Word)	
575n043	nd & spaniel. differs from their brains» then can	WE	deny that the grand child dug for mice from some
575n043	ouble consciousness? What other explanation-- can	WE	suppose some essence. The facts about crossing r
575n044	to skulk about & hunt mice-- Jenners Jackall Have	WE	somewhat right to deny identity of instinct.-- H
576n048	ession, excepting from favourable circumstances!»	WE	must believe, that it require a far higher & far
577n049	an antagonist quality to life & mind of man.-- &	WE	do not suppose an hydatid to be a cause of itsel
581n064	- "food, sm<e>ll. (ourang-outang), music, colours	WE	must suppose <we> «Pea-hens» admire peacock's ta
581n064	(ourang-outang), music, colours we must suppose	<WE>	«Pea-hens» admire peacock's tail, as much as we
581n064	<we> «Pea-hens» admire peacock's tail, as much as	WE	do.-- touch apparently. ourang outang very fond
582n068	t they will effect,.--" this not wholly true, for	WE	must grant a bird knows what is about when build
583n070	its cells, by means of ordinary senses & muscles,	WE	cannot look at him, as machine to make cell of c
584n071	incts may vary before the structure does; & hence	WE	get over an apparent anomaly,, for if anyone has
585n074	Mr Worsley's story of chicken) to know that which	WE	touch & what [...] the same.--(this Hensleigh th
585n075	e same.--(this Hensleigh therefore problem is how	WE	know that thing is same, which touches two parts
585n075	ouches one part. very quickly successively.--» [&	WE	know from experiment of crossing fingers, that w
585n075	we know from experiment of crossing fingers, that	WE	only do know that it is one, when applied in pec
585n077	this is not instinct, but a faculty, or sense-- "	WE	know not how, stonge henge raised, yet not insti
586n080	difference, which clearly ought to be separated--	WE	apply instinct to one part. or another-- but (an
588n090	ion.-- Lamarck. Phil. Zoolog.-- Vol II p. 445. If	WE	compare the judgments & actions of a young anima
588n090	of a young animal with an old.-- (dog horse, sow)	WE	perceive great difference.-- «(& is not this dif
599o05v	-- if they have, then language was progressive.--	WE	cannot doubt that language is an altering elemen
599o05v	annot doubt that language is an altering element,	WE	see words invented-- we see their origin in name
599o05v	e is an altering element, we see words invented--	WE	see their origin in names of People.-- Sound of
600o08v	reat difference between moral sense & conscience?	WE	admire what is right by one & are ordered to do
604o016	s,, it choosing food-- crawling from light.-- Yet	WE	can split Planaria into three animals, & this co
604o018	Sublimity is height. & with the idea of ascension	WE	associate something extraordinary & of great pow
604o018	of great power-- -- 2 From these & other reasons	WE	apply to God the notion of living in lofty regio
604o018	power. being associated with God. these phenomena	WE	(feel & ?) call sublime.-- 4 From the associatio
604o018	From the association of power &c &c with height,	WE	often apply the term sublime, where there is no
605o18v	sublime. adds not a little to the effect: as when	WE	look at the vast ocean from any height.-- 6 That
605o18v	ction the original cause of these feelings & thus	WE	apply to them the metaphorical term sublime 7 So
605o019	ain the various causes of those sensations, which	WE	call metaphorically sublime, but that it is thro
605o019	through a complicated series of associations that	WE	apply to such emotions. this same term.-- Hence
605o19v	emotion, from the associations before mentioned.	WE	call sublime.-- It appears to me, that we may of
605o19v	tioned. we call sublime.-- It appears to me, that	WE	may often trace the source of this "inward glory
605o19v	ity. with vastness of Eternity. which superiority	WE	transfer to ourselves in the same manner as we a
605o19v	ty we transfer to ourselves in the same manner as	WE	are acted on by sympathy. D. Stewart on taste Th
605o020	pleasures from beauties of nature & art." But as	WE	often see people who are susceptible of pleasure
606o021	like in vision, it becomes so instantaneous. that	WE	cannot ever perceive the various operations whic
606o024	r painter or sculpture, but rather subjects which	WE	know, it is therefore the embodying of a floatin
608o026	e must view a wrecked man, like a sickly one(P)--	WE	cannot help loathing a diseased offensive object
608o026	not help loathing a diseased offensive object, so	WE	view wickedness.-- it would however be more prop
608o026	es.-- '(P) Animals do attack the weak & sickly as	WE	do the wicked.--we ought to pity & assist & educ
608o026	do attack the weak & sickly as we do the wicked.--	WE	ought to pity & assist & educate by putting cont
608o028	hese views are directly opposed & inexplicable if	WE	suppose that the sins of a man, are under his co
609o029	ing feel between conscience & impulse) but shame «	WE	alas know» is far easier conquered than the deep
609o29v	e instinctive passions--, which being unnecessary	WE	call vicious.-- (jealousy in a dog no one calls
609o030	produce the greatest happiness.-- The other says	WE	have a moral sense.-- But my view <says> unites
609o030	. In judging of <our ha> of the rule of happiness	WE	must look far forward «& to the general action»
609o030	best for our good far back.-- (much further than	WE	can look forward: hence our <[...]> rule may som
610o033	is regarding the sources of knowledge.-- whether <	WE	th there> "anything can be <any> «the» object of
610o033	erience".-- is this not almost a question whether	WE	have any instincts, or rather the amount of our
613o036	f brain) As in animals no prejudices about souls,	WE	see particular trains of thoughts as fear of man
614o037	ut further end just same argument. without indeed	WE	are step towards some final end.-- production of
615o038	cquired in past ages; seems to indicate that when	WE	<return> turn into angels. this imperfect memory
616o038	ngels. this imperfect memory may become perfect &	WE	may look back to definite action or to our consc
616o039	ing bodies without these faculties & indeed until	WE	know what answer they would give in support of t
616o39v	of indifference 2) In the absence of such a guide	WE	can only <shew> point out the mode «of perceptiv
616o39v	oint out the mode «of perceptive action» by which	WE	come to conceive of matter as attracting & shew
617o39v	, by our external what senses in the way in which	WE	apprehend the force of inanimate bodies. How we
617o39v	h we apprehend the force of inanimate bodies. How	WE	identify the two aspects as different phases of
617o040	es affect us, except by internal consciousness 3)	WE	must endeavour to do without it as well as we ca
617o040	3) We must endeavour to do without it as well as	WE	can. <The objective aspect of> bodily action as
617o040	atter in the course of its DIRECTION, & thus when	WE	apprehend force in inanimate matter we feel diss
617o040	thus when we apprehend force in inanimate matter	WE	feel dissatisfied until we can point out How can
617o040	ce in inanimate matter we feel dissatisfied until	WE	can point out How can force be recognized by our
617o40v	e exertion of our own power & consciousness of it	WE	are conscious that we ourselves can originate in
617o40v	power & consciousness of it we are conscious that	WE	ourselves can originate in any point an oppositi
617o40v	ncing each other & moving in opposite directions.	WE	are satisfied therefore, if we can trace any for
617o40v	posite directions. We are satisfied therefore, if	WE	can trace any force in inanimate matter up to th
617o40v	ested would be <sat> fundamentally accounted for,	WE	prefer this metaphorical mode of stating the fac
618o041	& yet be ignorant of the existence of the brain.	WE	cannot perceive the thought attraction of sulphu
618o041	ulphuric acid for metal of another person at all,	WE	can only infer it from his its behaviour. Though
618o041	wn subjectively? -- ? the brain only objectively.	WE	do not know attraction objectively 6) The reason
618o41v	e objects of thought have no reference to place. [	WE	see a particle move one to another, & <the> (or
618o41v	o another, & <the> (or conceive it) & that is all	WE	know of attraction, but we cannot see an atom th
618o41v	eive it) & that is all we know of attraction, but	WE	cannot see an atom think: they are as incongruou
618o41v	to other: no not only thus, for if day was first,	WE	should not think night an effect.]CD Cause and e
619o042	] The history of every race of man shows this, if	WE	judge him by his habits, as <if> another animal.
619o042	ject in question. Without regarding their origin,	WE	see in other animals they consist in such active
619o042	tecting sheep or hurting them.-- Therefore in man	WE	should expect that acts of benevolence towards f
619o043	n or appetite, what would be the result? In a dog	WE	see a struggle between its appetite, or love of
620o044	asure. passion in its nature is only temporary, &	WE	do not afterwards think of it.-- Whatever the ca
620o046	n ought to follow-- it is his duty to do so.-- So	WE	say a pointer ought to stand a <spaniels> «house
620o046	he «old» dog really feels ashamed?) not so puppy,	WE	<do> try to teach him & strengthen his instincts
621o047	Hence he has the right & wrong in his mind.-- Now	WE	know it is easy by association to give «almost»
622o049	ent races of man may have different instincts, as	WE	see in dogs & pidgeons.-- But as man is animal a
622o049	ere very few & general in their nature.-- So that	WE	have some, it is sufficient to give rise to the
623o050	ink there is much truth in doctrine, for [RHC] 9)	WE	can thus explain love of place.-- although here
623o050	e can thus explain love of place.-- although here	WE	have not received pleasure from the place, but m
623o050	y through <all> «many» ages. could be acquired, &	WE	are certain from our reason, that all which <has
623o050	certain from our reason, that all which <has> (as	WE	must admit) has been acquired, does possess the
627o053	sed as "disinterested" p. 14. It is allowed, that	WE	have conception of moral obligation «when grown
633j53v	oom.-- or the springing of other seeds.-- But are	WE	certain that these are necessary adaptations.--
633j53v	essary adaptations.-- May they not be accidental?	WE	have good reason to know that they would not be
633j54r	- Hence the mistake they are created for them. If	WE	once venture to say plants created to <arrest> «
633j54r	event» the valuable soil in its seaward course,--	WE	sink into such contemptible queries, as why shou
633j54r	h, why not created to live on alpine pinnacle? if	WE	once to presume that God «created plants to» arr
633j54r	e a Dutchman plants them to stop the moving sand)	WE	<do> lower the creator to the standard of one hi
634j54v	pted; & it is from these very circumstances, that	WE	become satisfied respecting an original thought,
634j55r	Mammiferous animals originally terrestrial.-- for	WE	find even in Cetaceae traces of hind extremities
634j55r	in Cetaceae traces of hind extremities.-- How are	WE	to explain this.-- Did reptiles first inhabit se
635j55r	e utterly useless-- it foretells nothing» because	WE	know nothing of the will of the Deity. how it ac
635j55r	or inconstant like that of Man.-- the cause given	WE	know not the effect [blank] 6 p. 412. Macculloch
637j57v	re adaptations just as much as Woodpecker. --only	WE	here see means-- but not in the other. All Brid
638j58v	making structure, instead of parts of body.-- Now	WE	know what instinct is-- consider this I look at
639j28v	become long. not at once, but by steps. of which	WE	have manifold traces in the several genera of Gr
640j167	domesticated animals changing.-- From these views	WE	can deduce why small islands. should possess man
641j29r	who can decide their limits.-- To show how little	WE	understand of the Physiological relations of ani

Page **(Key Word)**

641j29v nculated eye of Chamelion. crabs Crabs & Mollusca WE have analogues The stillness p. 276) of flight o
023r011 s produced, from a fissure in a deep & therefore WEAK part of the ocean's bottom. With respect to Sha
032r041 must   there not be a circulation «however slow & WEAK.»; «(cause of not accumulation of Coral limesto
116a099 er of levelling their shores where currents very WEAK??-- too great an abundance of matter would have
380d153 had   completed them-- (but this argument is VERY WEAK without knowing whether if kept they would have
428e112 ragement, but a vigorous battle between strong & WEAK March 11th. Yarrell's law must be partly true,
520m002 member his father.-- My father thinks. people of WEAK minds, below par in intellect frequently <are>
528m037 ry time one looks at it.-- these two causes very WEAK.-- (2d) form. some forms seem instinctively bea
541m090 g delirium rest-- therefore dreams thus act.-- ∴ WEAK minded people are fickle & full of levity (¿ do
578n057 y muscles, these convulsive actions-- (except in WEAK people & hysterical people inclined to convulsi
580n061 certainly has observed that some people of very WEAK intellect (As Miss Clive) have only possessed v
580n061 ld of the mind, no subsequent ones modify it.-- «WEAK people say I know it because I was always told
596nIBC in <body> face in pashion.?-- cry? Do people of WEAK intellects easily fall into habits Get facts ab
608o025 trong invariable passions-- when these passions, WEAK, opposed & complicated one calls them free will
608o026 ental capabilities.-- '(P) Animals do attack the WEAK & sickly as we do the wicked.--we ought to pity
610o032 & comforts.-- Both ideots, old People & those of WEAK intellects.-- Westminster Review. March 1840 p.
614o037 f mankind requires these instincts,: though very WEAK so as to be overcome easily by reason.-- Consci
621o046 assion.-- If his passions strong & his instincts WEAK. he will have many struggles, & experience only
624o051 ctive, this part of argument fails, or rather is WEAK.-- [RHC] Better simply put it, beneficial tende
633j54r do> lower the creator to the standard of one his WEAK creations.-- All such facts are merely relation
637j58r - let egg shells grow harder. so must those with WEAK beaks be sifted away.-- 4 & the species, like 1
103a064 itching against each other, is, I suspect much WEAKENED by <vi> considering how close the dislocation
437e139 hares, rabbits, rats & not being sufficiently WEAKENED by wounds got off from the young ones while t
615o038 n so perhaps general ones.-- Parental feelings WEAKENED in Otahiati; fear of death in Hindoo populati
375d135 Nature, or rather forming gaps by thrusting out WEAKER ones. «The final cause of all this weddings, m
428e112 ch are confessed to by Herbert) to sift out the WEAKER ones: there ought to be no weeding or encourag
574n042 hance? produced with strong arms, outliving the WEAKER ones, may be applicable to the formation of in
203b135 cal provinces-- united-- & now divided again-- WEAKEST part of theory death of species without appare
441e150 it makes fourth cause or law of change.-- The WEAKEST part of my theory is, the absolute necessity,
527m034 hat I have a test of hardness of thought, from WEAKNESS of my stomach I observe a long castle in the
540m088 ind is rendered ductile by grief, or by bodily WEAKNESS, melts into tears, with sensations of sorrow f
572n032 atter of custom."-- this all applies to bodily WEAKNESS & inferiority, but now we carry it on to ment
590n093 fear is the same, as that which attends great WEAKNESS.-- <Diarrhaea> & syncope p. 42. Sighing from
069r152 wer of the ocean! Make a grand analogy between WEALDEN & Bolivia Transportal of conglomerate between
100a055 ation kept to sea coast ∴ curious exception in WEALDEN.-- Would crystals arrange themselves in that d
213b172 y sea, from Equatorial to extreme poles.-- Oh. WEALDEN.-- Wealden. Do the N. American. Tertiary depos
213b172 Equatorial to extreme poles.-- Oh. Wealden.-- WEALDEN. Do the N. American. Tertiary deposits present
373d133 n in shells in Arctic Ocean. p. 350 Grallae in WEALDEN. oldest birds. p. 411 -- Decapod Crust in Musc
411e055 discovered before them in any part of World.-- WEALDEN to boot.-- When one sees in Coralline powers o
641j29r s bear on, the absence of their remains in the WEALDEN? In the strongly separated Arctic genera, ther
075r166 c or Chota & Pasco in "alpine limestone" = "The WEALTH of the veins in most part totally independent
542m094 nimals) scream of agony, sigh of discomfort & WEARINESS. & meditative tranquility. «whine of children
626o052 ill lead to action.-- the passion rising from WEARINESS leads to striking blows.--' p. 224.-- Hume's
542m095 ransferred, (also how often) to the tale of a WEARISOME man.-- Is frowning, result of straining visio
232b244 sure time but physical changes (we assume like WEATHER on long average tolerably) uniform).-- Compari
505q014 tory of Potato field= (8) Abortive Thyme seeds WEATHER wet--? Linum flavum put in Spirits which plant
635j56v aken that the anthers should not be exposed to WEATHER.-- this is against my theory of frequent inter
037r053 remains protect a rock, or that the rock not WEATHERING allows such Compare the elevated estuary of
070r153 ure of ice in columns. show that granite when WEATHERING into balls. must exhibit orbicular structure
257c057 n asked how did the otter live before it had its WEB-feet-- all Nature answers to the possibility.--
274c115 there are walking forms in water birds,-- but no WEB forms in <water> land birds,.-- Grups of very di
286c156 ion in books, ought they to hold,-- Birds having WEB-feet, where we see scarcely any traces of passag
293c173 rement.-- Analogy. a bird can swim without being WEB footed yet with much practice & led on by circum
293c173 much practice & led on by circumstance it becomes WEB footed, now Man by effort of Memory can remember
301c199 uckling runs to water. before it is conscious of WEB. feet.-- p. 7. Mr Blyths arguments against squir
296c184 possibly some few hybrids in nature.-- «p. 473» WEBB &. Berthelot. must be studied on Canary islands
262c073 ture is full off them.-- wading birds partially WEBBED. &c.--)-- & in round of chances every famil
318c252 spect.. to the Cordillera,---- Bachman has seen WEBBED Shrews. case of adaptation.-- (case of Squirre
414e063 located] Are the feet of water-dogs at all more WEBBED than those of other dogs.-- if nature had had
293c173 a regular contingency the brain would become WEBFOOTED & there would be no act of memory.-- [There i
060r125 na. 1756. June 1780, Sept. 21st. 1817.--p 371. WEBSTER Antarctic veg:-- Study Ulloa to see if Indian
060r126 ogs. Azores. although kept in numbers. p. 124. WEBSTER Consult W. Parish. & Azara about dry season[.]
100a055 r, water flashing into steam, would Babbage.-- WEBSTER Phillips insists of analogy between Australia
103a063 es have not been the moving agents, because not WEDGE-formed.-- Hence fill up fissures-- If dikes eff
353d061 rds-- all Eagles. of Australia characterized by WEDGE tails.-- many of the hawks <to> are analogues t
098a046 ncretions in this case laminar. hence the thick WEDGES of feldspar in gneiss.-- Veins in septaria. a
375d135 ay say there is a force like a hundred thousand WEDGES trying force <into> every kind of adapted stru
375d135 out weaker ones. «The final cause of all this WEDGINGS, must be to sort out proper structure & adapt
486z020 k of birds singing in the forests of Brazil H. WEDGWOOD says in <14th> «13th.» Vol of Archaeologia ar
539m083 fter my Father in heraldic principle. & Eras a WEDGWOOD in many respects & some of Aunt Sarahs. crank
595n121 ame effect as acting on us-- <The Baby> «Effie WEDGWOOD» April 28th 1840 was frightened at wild beast
155g062 terraces if the largest has hollowed out most WEDNESDAY Shelf 3d dies away almost imperceptibly on Gl
165g126 n face-- asked stated this generally the case WEDNESDAY 12/ & 3/ Why is the Tetrao scoticus & Tetrao-
058r116 be easy to see on beach successive lines of sea WEED-- Histoire Naturelle des Indes Acosta. p. 125.
066r140 nk some 60 fathoms, none thicker than thumb» Sea WEED said at Kerguelen Isd. to grow on shoals like F
316c245 desmas.-- this is remarkable.-- Fish & drift sea WEED-- may transport ova of shells.-- Conchifera. he
478z003 ape Pidgeon's stomach small shells (patella) sea WEED & many pebbles Mentions stinging Millepora. Quo
495q03a on hills.-- 10 Shoot tame duck on pond with Duck-WEED-- coots-- waterhens-- examine dog, which has sw
528m038 ty of some as Norfolk Isd fir shows this, or sea WEED, &c &c-- this gives beauty to a single tree,--
428e112 sift out the weaker ones: there ought to be no WEEDING or encouragement, but a vigorous battle betwee
392d180 ok out for instances of Avitism Examine English WEEDS in Hot. Houses will they flower Make Hybrids wi
495q005 enerations to accustom them to such soil.-- Sow WEEDS in such soil.-- 7 (a) Experimentise on Primrose
308c217 ed kept it in-- <right> left, but I always for a WEEK took of cover of right side, though my hand <vi
495q05a ond for seeds.-- 11. Soak all kinds of seeds for WEEK in Salt. artificial water.-- 12. Plant two race
543m100 a, fool that his Father only left him a guinea a WEEK. yet. he was inimitable chess player.-- Peacock
362d099 .¦ Curious the rapidity of the change in 5 or 6 WEEKS after castration, fresh horns begin to grow.--
380d155 does not menstruate in the month, she will in 5 WEEKS.-- A Bull is never taken from his own field to
532m056 o grow thin & did not seem quite happy. in five WEEKS was so thin, that she was sent back to Shrewsbu
535m064 -- They continue till death, thus acting 4 to 6 WEEKS. The deserted broods appeared healthy-- This re
130a133 roofs of hot bottom.-- Study Bishoofs Paper.-- WEELSTED told me of some large fresh Water springs off
493q003 mine semen of Hybrid animal. in comparison with WEIGH skeleton of Tame Duck & Wild Duck, & then weigh
493q003 Weigh skeleton of Tame Duck & Wild Duck, & then WEIGH their wing bones & see if relation is same good
090a023 ds average for each stone. that fell, that was WEIGHED,; <but> carrying on this ratio I can count 90
090a022 his ninety includes all actually counted.-- The WEIGHT «or size» is given of 25 stones.-- The total w
090a022 ht «or size» is given of 25 stones.-- The total WEIGHT recorded is 473. pounds (taking about average
099a049 avity can have no effect, on particles of equal WEIGHT.--)¿ cleavage not vertical ∴ combined with gra
103a065 kes & «axis of» mountain-chain in proportion to WEIGHT of super [..] mass.-- Absence of Caverns, in
312c231 x on the kind wanted, colouring of each feather WEIGHT & size & they would produce number agreeing al
437e139 agles nest, which shows power of carrying great WEIGHT. p. 125 is said that Eagles bring rabbits & ha
618o41v n atom think: they are as incongruous as blue & WEIGHT; all that can be said that thought & organizat
618o41v ization run in a parallel series: if blueness & WEIGHT always went together. & as a thing grew blue i
618o41v could not be said that the blueness caused the WEIGHT, anymore that weight, the blueness, still less
618o41v at the blueness caused the weight, anymore that WEIGHT, the blueness, still less between <action> «th
618o41v nt as action thought & organization: But if the WEIGHT never came untill the blueness had a certain i
618o41v then might it now be said, that blueness caused WEIGHT, because both due to some common cause:-- The
128a128 fall-- the problem will be falling of an arch WEIGHTED in its centre.-- Will not abrasion of land on
514q21. ion of violets.= Zostera= Are dwarf plants on WELLINGTON Mountain described in Flinders= Alpine Austr
041r064 . Ayres; mention about the deer approaching the WELLS.-- the effect of Salt water of the Salado.--Mem

043r076 er whole world) Yet eruptions <both> at sea (as WELLS as in the Cordillera), they may be considered a
323c269 y on Generation. Englis Transla -- The Revd. A. WELLS. Lecture on instinct -- Cline on the Breeding o
582n068 of mental pain & pleasure.-- The Revd. Algernon WELLS Lecture on animal instinct. 1834: p. 15. "To ac
583n070 med as more allied to reason than instinct.) Mr WELLS I can see mentally refers by reason knowledge g
182b048 -- Geogr. Journal Vol VI. P II. p 89.-- Lieut. WELLSTED obtained many sheep from Arabian coast. "Thes
266c092 low doubts? GEOGRAPHICAL JOURNAL. Vol V. p 201 WELLSTED. Memoir on isld of Socotra. Cattle generally
266c093 s off Arabian Coast.-- [Vol VI. p. 89.-- Lieut WELLSTED "on coast of Arabia between Ras Mohammed & Je
164g119 ata coarse agglomerate [...] shells from [...] WENLOCK Edge [blank] L. Lochy 12 ft 96 L. Oich 12 84 {
087a008 lls Encyclopaedia-- Lately elevated When Siberia WENT up. Arctic land went down.-- Probably more Arct
087a008 ately elevated When Siberia went up. Arctic land WENT down.-- Probably more Arctic land would be requ
120a109 heet of matter over surface.-- if elevation then WENT on at greater rate, not only river would carry
217b183 produced a very large litter.-- never afterwards WENT in heat.-- This is good instance of same fact i
336d019 nge, a bud could not be taken, without it either WENT back, or not being perfect would perish.-- The
344d042 alfbred Beagle Staghound. «++».;the grandchildren WENT back to either paret, & breed not fixed. though
364d104 white heads; which years afterwards occasionally WENT back-- (Effect of imagination on mother. white
541m089 w has this instinctive fear arisen? 19th. When I WENT down to Woollich I was trying to unbend my mind
546m111 ut it in branch of tree, & apologising to party, WENT out & found it there!!! Lady in perfect «mental
547m111 ervous influence replace each other August 29th. WENT to Bed. & built «common» Castle in the air, of
551m127 f intensity-- «in old man, he had just seen mind WENT on RAMBLING till excited by question.» Sept. 4t
569n023 "-- "made no reply, but shrugged his shoulders & WENT away."-- he implies negation, without violence,
585n076 emarks that Chimpanze pouted & whined, when, man WENT out of room.-- all theories of magnetic powe in
618o41v n a parallel series: if blueness & weight always WENT together. & as a thing grew blue it «uniquely»
094a036 storphine at Geolog. Soc: Colonel Imrie Transact WERN. Soc. Vol. 2. p. 35 Sir J Hall Trans. Phils Roy
032r041 .-- From poles to Equator current downwards & to WEST.--From Equator to poles. nearer the surface & t
037r056 B. Oriental? level surface not disturbed.--Whole WEST coast. Chonos to Copiapo.--Sydney. K. G. Sound.
055r106 <inclining> to South. inclining a little to the WEST: the veins which follow this direction are thou
069r151 te. There must have been as much conglomerate on WEST of Peuquenes as on East. Where gone to.?-- Ther
089a018 ses."-- Tom 54. p. 106 do-- p. 110. Mountains on WEST side of Domingo formed of coral limestone, with
110a085 206 Both Beck & Deshayes saw fossil shells from WEST Indies & declare them to be recent species-- Ly
127a125 P} thawed at + in isothermal curve. East-clinal. WEST clinal. S.-clinal. N-clinal & anticlinal «syncl
149g032 towards (P} Spean Roy double terrace river «& to WEST of Spean» difficult to explain on <formation> d
160g088 name of Glen by which we descended, is to the WEST of Glen Tarf What I called Alluvium shows the a
160g092 ng into E. end of L. Oich, & waters flowing into WEST end with obscure terraces on one side Barom 29.
179b032 white) in the two first children How is this in WEST Indies «--:.Humboldt. New Spain:--» Dr. Smith al
190b076 r Brown, about peculiarities of Flora. on East & WEST. ends of New Holland. diminishing towards centr
191b080 Elephants in Ceylon-- East Indian archipelago.-- WEST Indies = Opossum & Agouti same as on continent-
205b142 OMMENCING. Kirby says (not definite information) WEST of Rocky Mountains Asiatic types discoverable.-
218b190 ficier du Roi Mem. Capt. Owen's story of cats on WEST coast of Africa.-- changing hair-- The Edinbur
226b223 in S. America.-- <East.> «W» Coast of Africa & <WEST.> «E» of America, ought to present great contra
250c040 ? Zoological Journal.--- Vol I. p. 81. Capromys, WEST Indian isld. p. 120. «ref.» Philosop. Transacts
268c099 e Isd.--» very good. Study D'Orbigny. & range on WEST Coast «Guayaquil & Peru» Henslow in talking of
298c192 says in most plants, even those on Guernsey & on WEST coast of Ireland, are absolutely (& who better
310c224 ied» species Himalayas, 13,000 & Melville Isd.-- WEST Africa & India some plants same. --America.-- S
316c246 Mr Sowerby says there are some shells common to WEST coast of Africa & E. S. America.-- get instance
317c251 ced by three other species.-- Says all the hares WEST of <all> Rocky Mountains have peculiar characte
414e064 ls chiefly organization: though Cont of Africa & WEST Indies shows organization in Black Race there g
485z017 ?-- The Birds seem to move much further North on WEST coast of S. America. than on East.-- not being
510q017 urrent & winds in Antarctic ocean: are they from WEST, like as between Australia & S. America? Sabine
032r041 great depth. from axis to surface must gain a WESTERLY current:--If great changes of climate have ha
072r159 n Isld.--Mr Barrow thinks N & S. line connects WESTERN isles of Scotland & Iceland.--Bosh nor on Norw
134a141 Admiralty Sound. SE dip. much p. 136. Rocks on WESTERN Coast p. 204 do. do p. 210. Height on road fro
134a141 er coast of T. del. Fuego.-- p 385 Rocks of S. WESTERN Coast Vol II p. 277. on whale bones in Falklan
610o033 ts, old People & those of weak intellects.-- WESTMINSTER Review. March 1840 p. 267--- says the great
204b139 t & England» Loudon Mag: Septemb or Octob 1837 WESTWOOD has written paper on affinity and analogy in
471tf07 yony saw common Bee on: Linn. Trans 18. p. 133 WESTWOOD on the Fulgoridae enumerates the strange form
477z001 3. papers connected with transform of Crust-- WESTWOOD & Thompsons-- Part II.-- 35. Phil Trans Burro
534m063 must be experience & intellect.-- do. p. 157. WESTWOOD remarks that some imported plants are attacke
469k135 rk, & this has now gone on 14 days. (except some WET ones/ & wd go on longer-- Woodfords Marrow fat,
505q014 f Potato field= (8) Abortive Thyme seeds weather WET--? Linum flavum put in Spirits which plant seeds
078r176 ty of Porphyries which are destitute of quartz, & WH abound both in hornbland & vitreous felspar".--
162g105 tells me, that Loch Lochy is 8 ft below Loch Oich WH is 92 ft above sea-- Loch Ness 40 ft above do. W
466t094 nectary «they do» they must disturb all anthers, WH otherwise lie protected by the hairy black lip o
466t094 he hairy black lip of lower division of nectary: «WH. itself resembles a Bee, but does not prevent be
466t095 rs & colours breaking only hereditary characters, WH. come on in after life of Plants-- also goodness
470t177 es-- «Humbles & common» On silene, many plants of WH. have abortive stamens= Many Humbles on hedge Li
470t178 th violence in shower On many Papilionaceous; all WH. are in flower «I saw Bees;»-- on Monk's Hood, fl
472s02r many flowers, on one of which pollen was routed. WH. was not case, on several flowers I examined som
473s05r them from L. Groznerat, «on road to Bethgellert» WH flows by Tremadoc. but can tell them from lake S
473s05v c. but can tell them from lake S. of Moel Siabod. WH. flows into Conway by Bettws & there joins strea
493q003 s. as only one sex so coloured = I have grey-cat «WH was female» with tinge of tortoise-shell «on bac
505q014 boiled earth on top of House =Aristolochia, plant WH require insects to impregnate it (7) History of
505q014 ssum. never germinated 12 Does the horned orange. WH. never has seeds produced good pollen? Yes «From
134a141 385 Rocks of S. Western Coast Vol II p. 277. on WHALE bones in Falklands Some of the Tosca nodules a
228b229 ist ought to have before him, when dissecting a WHALE, or classifyng a mite, a fungus, or an infusori
287c157 ows hiatus-- but not saltus-- when Linnaeus put WHALE between cow & hawk a frolicsome saltus. «p. 19»
509q017 young» teeth, more plain in young Rhinoceros or WHALE, than in old?? Falconer says all in cases. Owen
639j28r o & Duck, Ornithorhyncus «externally». petrel & WHALE in some respects Chamaelion like power in Octop
641j29r ls of Antarctic seas. (on other hand Spermaceti WHALE & Manatee.-- Naturalists must be cautious.-- <s
211b162 ationship express, a real affinity & affinity-- WHALES & fish.-- Progeny of Manks cats without tails:
422e092 e Dugongs cannot be united with true Cetacea or WHALES.-- but are aquatic Pachyderms. & Walrus-- aqua
458t001 ient gigantic salamanders-- Every animal (except WHALES) have great prototype!!.-- Copied Vol II p. 50
641j29r ze» <to> are best nourished by arctic regions-- WHALES. «Narwhal» Polar bear. Walrus, great Seals of
250c037 ome forms be peculiar to it, so on, & so on.-- WHATEVER destroyed great <quadrupeds>; Pachyderm in S.
303c203 S. existing species becomes father of genera-- WHATEVER the cause is. <the> any osculant species. whi
312c231 icking opposite qualities, with no other means WHATEVER.--» Individual Men «& animals» could only exi
347d049 .-- 27th. August. There must be some law, that WHATEVER organization an animal has, it tends to multi
534m062 of such sensations, & memory is repetition of WHATEVER takes place in brain. when sensation is perce
568n017 ith reason) it would feel «subsequent» sorrow, WHATEVER the cause had been]CD-- «Also» When one is pr
582n067 - [the means must be present on any hypothesis WHATEVER]CD an animal may so far be said to will to pe
606o025 impression 1) September 6th. 1838 Every action WHATEVER is the effect of a motive.-- [-- must be so,
620o044 mporary, & we do not afterwards think of it.-- WHATEVER the cause of this may be, everyone must know,
334d010 who placed his stallion as second horse between <WHE> shaft mare & another leader mare,-- this stalli
375d135 in Spring, like food used for other purposes as WHEAT for making brandy.--» take Europe on an average
501q01l Henslow ask question of Col. Le. Couteur about WHEAT-- Change of Soil-- crossing-- when seeds raised
502q11v ng-- when seeds raised.-- His Book.-- 32. Would WHEAT from AEgypt ripen in Scotland?-- to show acclim
506q015 iculturists. whom he know.-- Col. le Couteur on WHEAT.-- (41) Have any monooecious or dioecious plant
559m155 lants A Volume published by Colonel in army on "WHEAT." in Jersey.-- very curious facts about early p
552m131 erms) much intellect.-- mem: Yarrell's story of WHEEL horse in drays, scraping against cornice stone
585n077 has faculty by his instruments to make toothed WHEEL. he might by instinct make watch, but he does i
177b028 mstances, to which> is it an index of the point WHENCE, two favourable points of organization commenc
420e088 species of dogs & Hyaena.-- but a common point, WHENCE both may have descended.-- Jan. 6th The rudime
285c150 has some remarkable crochets about instincts. WHENEVER instinct is mentioned some definition must be
378d147 ie, Jay, & perhaps all the rollers-- «He says» WHENEVER metallic brilliancy is present in Young birds
291c168 d being ready made.-- & so destroy individuals, WHERAS in Falkland Isd they would change & make new s
077r170 a, chiefly owing to destruction of porphyries. WHEREAS other to ancient rock.--this N degree 2. super
105a069 nation becomes less & ∴ tends to finite power) WHEREAS sea. on coast, as long as exposed to waves of
211b161 lings against other «for sexual ends» species, WHEREAS Man has such instincts very little. in Zoolog.
231b243 ft in cooling climate might change twice over, WHEREAS those which migrated a little to the southward

```
280c135 excessively  old, they would not make hybrids, WHEREAS two newer ones, even if more different might d
359d087 lar pheasant & fowl being so totally infertile WHEREAS animals further apart have bred inter se.-- Th
531m050 event by itself will make it to be remembered. WHEREAS it is the importance.-- people very often forg
305c211 a  more extended relations of the individuals, WHEREBY choice with memory. or reason? is necessary.--
261c071 es, perhaps not on now existing» Mr Gould says  WHEREVER any mark like red patch on wings of Furnarius
357d073 entata & Marsupials have been almost destroyed  WHEREVER other animals existed.-- Athenaeum 1838. p. 6
485z018 tomolog. insects swarm in Lapland & Stizbergen  WHEREVER there is extreme heat, the tropical forms ext
037r053 to degradation of rocks--It may be a question.  WHETHER organic remains protect a rock, or that the ro
041r064 . in beds) lived there, makes it very doubtful  WHETHER they could have lived in so deep a sea.--Perha
041r065 ; because in this latter case. we cannot judge  WHETHER such fossils. lived in groups or not. Ferrugin
051r095 tom. we see a trifling circumstance determines  WHETHER an animal will adhere to a certain part. Aprop
056r109 e so grand a work?--In valleys one is not sure  WHETHER fissures may not have helped it, or diluvial w
058r115 ves in planes. «Mem silky lustre» ask Erasmus.  WHETHER electricity would affect this. -- State the ci
063r132 th hypothetical case of Brazil.-- Propagation.  WHETHER ordinary. hermaphrodite. or by cutting an anim
077r171 uanaxuato to SW. with respect to latter doubts  WHETHER bed or vein (very like that of Spital of Schem
100a055 ng produced by local heat, Ask Capt. Beaufort,  WHETHER water flashing into steam, would Babbage.-- W
119a105 with  fossils,  therefore formed near surface.  WHETHER they can have been plunged so many miles  deep
124a119 f lead. it would be still more curious to know  WHETHER it would be absorbed.-- if so exactly parallel
172b006 place  them on fresh isld. it is very doubtful  WHETHER they would remain constant; is it not said tha
181b044 , & the gradation will be sudden-- Heaven know  WHETHER this agrees with Nature: Cuidado The above spe
183b051 if they do not breed readily. point in view.--¿WHETHER highly domesticated animals like races of man.
183b053 type of either animal when crossed with it.--  ¿WHETHER extinction of great S. American quadrupeds. pa
186b059 uld be curious to know in plants, (or animals)  WHETHER, <in> races have tendency to keep to <each> ei
192b085 some good instances. But it is other question,  WHETHER there have existed all those intermediate step
192b088 ks. we can only expect some steps.-- I may ask  WHETHER the series is not more perfect by the discover
193b090 er being climatized, climatizes the child. ¿--  WHETHER every animal produces in course of ages ten th
198b113 confines? Balanidae?-- ---- I cannot understand WHETHER. G. H. thinks development in quite straight lin
200b113 spring: physical impossibility to marriage.--  ¿WHETHER those genera which unite very different struct
200b123 -- Vide cattle: The grand fact is to establish  WHETHER in crossing very opposite races, whether you w
200b123 blish whether in crossing very opposite races,  WHETHER you would expect equal fertility-- ditto in Pl
200b125 ing.--- It would be curious experiment to know  WHETHER soaking seeds in salt water &c has any tendenc
216b180 ids) thus act.-- Now the point will be to find  WHETHER know varieties in plants do so.-- As in  cacti
220b197 to  coition & not production. But who can say,  WHETHER offspring does not depend on mind or  instinct
222b205 d in same predicament. It is another question,  WHETHER whole scale of Zoology may not be perfecting b
222b206 t have improvement in system of articulation.  ¿WHETHER type of each order may not be supposed that fo
223b208 s between perfect insect & former hard to tell  WHETHER articulate or intestinal, or even a mite.--  a
224b212 on, «(but experience according to each group)» WHETHER good species, & hence the importance Naturalis
225b218 pigs  foot with cloven hoof) Ask Entomologists  WHETHER they know of any case of introduced plant, whi
227b224 tained means of transport, we should then know  WHETHER former lands intervened.-- (2d) By character o
228b229 ale & female animal of one variety going back  ¿WHETHER this going back may not be owing to cross from
231b243 ers--seeds.-- Two inhabitants of the Tropics,  (WHETHER one fossil or not) are related by real relatio
241c014 140,  calls it Bos. leucoprymnus. does not say  WHETHER wild or not p. 156.-- Parroket with stiff tail
246c025 Voyage Says no reptiles. p 460 & very doubtful  WHETHER any birds Except. Dodo!!-- in Mauritius Lesson
246c026 ortion, colour slightly different. Who can say  WHETHER species & varieties P. 708. Columba Oceanica (
257c059 able of producing young ones like itself, but  ¿WHETHER great assumption? not solely producing like it
258c061 auses the confusion in this system of nature--  WHETHER species may not be made by a little more vigou
265c085 tion, other monster.-- The only way of judging  WHETHER structure is owing to habits, or heredetary is
265c085 e is owing to habits, or heredetary is to see,  WHETHER a large family has it, & one member of that fa
271c106 has  extended to Juan Fernandez in birds. but  ¿WHETHER to same island in plants?-- What is this halo.
273c112 in swallow??? which is most wonderful of all.  ¿WHETHER in most aerial of swallows.» Milvulus, & still
276c122 ound study of local varieties laws of change--  WHETHER beak (as it appears to me). colour of  plumage
278c129 enera; are analogical, because we do not know,  WHETHER nearest species of each might not breed:-- Gen
278c131 e breccia accumulated-- surely ask Owen to see  WHETHER species same, excessive improbability. Mem in
278c131 ls. Number of Geographical Journal to discover  WHETHER dog found at Swan river. The change in England
281c137 ally if animals did change excessively slowly.  WHETHER geologists would not find fossils such as they
285c157 of  analogy, but will grow into affinity.--but  WHETHER ever arrive at true affinity doubtful A specie
287c158 organs, when changed in different animals.-- +  WHETHER variations in eye of vertebrate afford  better
288c159 ircles completed Major Mitchell, does not know  WHETHER breeds of oxen have deteriorated, or  altered,
289c162 g through winter. «It would be curious to know  WHETHER a variety could be transmitted more easily  in
296c184 died on Canary islands-- Endeavour to find out  WHETHER African forms. (anyhow not Australian) on Peak
298c189 st, when often looked at.-- this most puzzling  WHETHER instinct, or reason?? Gould says he believes t
300c197 s instinctive dread it is exceedingly doubtful  WHETHER animals have any fear of death, only of  pain.
303c204 his respect upper & lower, which I do not know  WHETHER it <would have> differs in present races, & fo
304c207 gences??? A Question of immense difficulty is,  WHETHER Apteryx descends from same parent with other b
304c207 thout these trifles. it would not then be told  WHETHER not descended from long way back.-- aberrant f
312c232 are  instincts, & not ∴ instincts, constant.  ¿  WHETHER mutilations non-heredetary & variations produc
325c267 n's American Ornithology Read Aristotle to see  WHETHER any my views very ancient? Study with profound
327c265 y in which each one is found. very good to see  WHETHER peculiar plants-- in high points Read Volney's
334d012 ls, which form a marked wild variety. doubtful  WHETHER all are white. Fox says the Half Muscovy Fox s
341d029 - (& two Rheas still closer).-- Mr Blyth asked  WHETHER structure of pelvis & was not adaptive structu
341d029 ich does not make that bird a Penguin.-- (i.e.  WHETHER relation in one point or many) Owen answered t
348d051 -- i.e what characters chance to be heredetary  WHETHER important or not,). p. 7. "The Natural arrange
353d060 nimal, if there be many others somewhat allied  WHETHER «like» parent stock, or not. Now wings for fli
355d066 d loopholes) "It is well worthy of examination  WHETHER variations are produced only in those characte
363d101 ut.-- Study Temminks work on Pidgeons--, & see  WHETHER feathered legs.-- «Carruncles on beak & in Mus
370d117 apply  it as hypothesis to other points. & see  WHETHER it will solve them.-- It is less wonderful tha
380d153 but this argument is VERY WEAK without knowing  WHETHER if kept they would have wings.--).-- Seep p. 8
382d157 Hunter <asks> p. 36 is thought by Owen to ask.  WHETHER a Heautandrous animal is <evidently>  actually
383d159 , one like cock & other like Hen.-- one doubts  WHETHER they are not Hermaphrodites, like J.  Hunters.
386d165 me animal with definite life & split it, & see  WHETHER it retains same length of life.-- like  Golden
405e032 o hemispheres.-- It might be worth investigate  WHETHER. Megatherium & Mastodon are coembedded in N. A
406e035 fter> which parent impresses offspring most is  WHETHER mother has had any offspring before.-- -- now
409e049 h the world was brought into present state. --  WHETHER he was or not. He is present a social  animal»
410e053 f parts in individuals, will care little, <in>  WHETHER the individual be species or variety, but to d
411e054 ons, & changes of individual organs, must know  WHETHER the individuals «forms» are permanent, all step
414e060 wonderfully  changed, since Cretaceous period,  WHETHER progressive I know not.» (& insects.-- Stonest
415e066 sils, the only way, that I can see to discover  WHETHER the parent of man was quadruped or bimanous,,
424e102 phical distribution of Crustaceae.-- (I forget  WHETHER I have already referred to it.-- also on sperm
429e113 orns.-- I do not believe this Nature's plan.--  WHETHER we can or not trace history of first appearanc
430e119 theory Examine list of St. Helena Plants & see  WHETHER those which grow in low grounds are those, whi
434e127 more wonderful than heredetary ambling horses.  WHETHER the body of parent be altered, that is the Nis
434e128 it) succeeds in altering <or> form of body, or  WHETHER it merely has tendency (as effects of cultivat
441e148 d peculiarities are transmitted. it is doubtful WHETHER any are transmitted, for the changes in  fruit
442e152 which never flowers!!-- <How did it get there?  WHETHER> According to the above suggestion my theory w
471tf11 nd to end-- so that he is almost led to doubt.  WHETHER there is such a thing as a species-- Jun 1. 18
491q01v - (6) To hybridise EVERY flower on melon & see  WHETHER fruit affected. Mr. B. seemed to say impregnat
491q002 e late varieties of Peas &c be constant? 3..    WHETHER the viviparous grasses & onion, produce flower
492q002 ot attacked in bad years from Caterpillars. 5.  WHETHER Roses impregnate each other. when close plante
493q004 s of different heredetary constitution, to see  WHETHER offspring infertile.-- (4) Does the number of
494q005 a) between sensitive & sleeping species, & see  WHETHER association can be given (2) do the stamina of
495q005 own pollen» on flowers of other cabbages & see  WHETHER there will result hybrids-- (5) Dust flowers o
495q05a ed with some sticky stuff in flat places & see  WHETHER wind, on «dry» windy day, «flower garden on gr
496q05a of  double flowers. compared with single & see  WHETHER grains flaccid, as Koelreuter describes Kill S
496q006 when proportional number appear equal-- & see  WHETHER proportions will vary, which will show that su
497q06v nts, as Pentstemon, which have abortive parts,  WHETHER such vary.-- Do Bees go to Sweet Peas, IMPORTA
499q009 Bees, Butterflies-- Syrphus-- Meligethes & see  WHETHER they are dusted with pollen-- in what state (w
499q009 heads  with Bird lime near male yew tree & see  WHETHER they catch pollen-- <Ne> In Oenothera  bush.--
500q10a yone, where there are very close species & see  WHETHER they come from islds. or different parts or sa
```

```
501q011 perimentize on by sowing near each other & see WHETHER cross can be obtained-- I name these three pla
505q014 Thyme with abortive stamens by Terrace to see, WHETHER stamens will be produced in individual plants
507q15v ublished» list of spontaneous Hybrids-- to see WHETHER any Papilionaceous plants,-- whether many mono
507q15v -- to see whether any Papilionaceous plants,-- WHETHER many mono or Dioeious plants, & any with pecul
507q016 a family-- Where one tooth aborts, do you know WHETHER any trace in germ. (2) Any more cases of disea
509q017 nothing [<...>] It is very important to know, WHETHER Gould's observation holds good, that in the mu
509q017 h Colliers, when men & women have long worked, WHETHER children, who have not worked have any peculia
511q018 ifferent looking animal be formed-- not caring WHETHER good or bad.-- are any actually rejected?? (8)
511q018 rent appearance.-- {will introduce it in work} WHETHER <Yar> knows whether Shaws hybrids between Trou
511q018 will introduce it in work} Whether <Yar> knows WHETHER Shaws hybrids between Trout & Salmon were fert
511q018 hybrids between Trout & Salmon were fertile & WHETHER homogeneous {About German ornithologists, Bhem
513q020 orking fowls,-- or tailless dogs & fox, to see WHETHER the characters are then intermediate or «somet
515q021 merely not ripen.-- The point to attend to is WHETHER good & plenty of pollen is produced. & 2d if s
515q021 od & plenty of pollen is produced. & 2d if so, WHETHER concepcion takes place,-- the mere fact of see
521m009 why it is dinner time.-- » My father asked him WHETHER he had gardener of name A. B., &c &c. & he mai
522m013 ese emotions» acquired.-- this may be doubted, WHETHER rather not going against natural instincts.--
523m017 ore painful than the thing itself. Asked my F. WHETHER insanity is not distinguished from whims passi
525m026 ations in families.-- My fathers does not know WHETHER trains of insanity are heredetary in any one f
532m053 cannot help doing it.-- Fanny Hensleigh doubts WHETHER young babies start.-- ¶ If children wink. it i
532m055 ongs «& tales» of infancy, it is very doubtful WHETHER they could recollect these same things from an
532m074 t of his example on others. ¶ It may be doubted WHETHER a man intentionally can wag his finger from re
540m085 ational character let him compare the American WHETHER in the cold regions of the North,-- the elevat
547m112 ideas connected with past circumstances.-- as WHETHER i really was going to Shrewsbury, whether I ha
547m112 - as whether I really was going to Shrewsbury, WHETHER I had rung for Covington. whether he had come
547m112 Shrewsbury, whether I had rung for Covington. WHETHER he had come & opened box, whether I had though
547m112 r Covington. whether he had come & opened box, WHETHER I had thought what clothes to take (how often
547m113 hat clothes to take (how often one cannot tell WHETHER one has rung the bell,. when one recollects ci
547m113 effect of perception by senses fail first, as WHETHER I had pulled the bell??)-- It may be deception
549m119 e happiness.-- but as they are not recollected WHETHER from frequency, or inherent structure of mind.
549m120 as they appear & perhaps partly their fault.-- WHETHER this rule of happiness agrees with that of New
549m121 nse happiness.-- it is again another question, WHETHER this happiness is the object of living.-- or w
549m121 r this happiness is the object of living.-- or WHETHER if we obey literally New Testament future life
549m121 e life is almost the sole object--. -- I doubt WHETHER the last be right. The two rules come very nea
553m136 it does exist in different degrees in races.-- WHETHER in Ancient Greeks, with their mystical but sub
554m139 re eating «everytime», look up to «keeper» see WHETHER, this was permitted & eat it.-- good case of a
554m140 one wrong & will hide herself.-- I do not know WHETHER fear or shame.-- When she thinks she is going
564n055 s head, & picked it-- Here then, that faculty, WHETHER for position of axe of eyes, state of surface,
568n018 mparison of sensations would first take place, WHETHER to pursue immediate inclination or some future
572n032 ct is by making yourself less, but the manner, WHETHER by bowing the body, kneeling, prostration «unc
575n045 t having any smell.-- 27th. November.-- Think, WHETHER there is any analogy between grief & pain-- ce
575n045 dogs. who will not enter water, till he sees. WHETHER birds badly wounded, or only winged.-- fetches
576n047 egret «having acquired» this sense of right (& WHETHER wholly instinctive as in the dog, or chiefly h
580n061 playing by memory. she does not think at all, WHETHER she can or can not play the piece, she plays <
581n063 arly, all I must do is to generalize it, & see WHETHER applicable to all cases.-- & analogize it with
586n078 ut his knowledge of that quarter,. is faculty, WHETHER by sun, & heavens, or magnetic virtue,-- the m
610o033 idge, is regarding the sources of knowledge.-- WHETHER <we th' there> "anything can be <any> «the» obj
610o033 r experience".-- is this not almost a question WHETHER we have any instincts, or rather the amount of
622o048 ts into instincts. [RHC] Feelings of the mind, WHETHER leading to action or not, are the parts of our
627o053 ligation «when grown up???» & the question is, WHETHER this can be resolved into some operation of in
629o55v greed general utility (3) It is other question WHETHER any thing is taught instinctively; I say yes,
635j55r othing of the will of the Deity. how it acts & WHETHER constant or inconstant like that of Man.-- the
640j167 either sex determines life:.» «With respect to WHETHER Galapagos beings are species. it should be rem
256c055 f type continued?-- See to Boblaye & Virlet.-- WHEWELL thinks (p 642) anniversary Speech. Feb 1838 th
262c072 wants & not as change in typical structure?!! WHEWELL «in Comment/ few will dispute-- says civiliza
340d026 (!) according to affinities). confounds, like WHEWELL affinity with analogy-- Good table at end of d
347d049 uished.--???» Mayo (.Philosop of Living) quote WHEWELL as profound. because he says length of days ad
567n014 nows one side of triangle shorter than two. V. WHEWELL. Induct. Sciences-- Vol I p. 334 Does a negres
266c091 med alike with the Koran and the sword" quote WHEWELLS Bridgewater treatise, (p.,26). about plants f
322c269 read.-- Reynolds Discourses Lessing's Laocoon WHEWELLS-- inductive History.-- References at end of e
415e069 n Man December 16th. The end of each volume of WHEWELLS Inductive History. Contains many most valuab
434e128 he effects are equally handed to offspring.-- WHEWELL'S anniversary address 1839, p. 9.-- talks abou
625o052 y explains both, perhaps, by habit-- [LHC] 11) WHEWELLS preface. [RHC] It appears that Sir. J. & othe
625o052 -- only so far do I admit its supremacy p. 37. WHEWELLS gives Mackintosh's theory: the remarks about
280c135 wer omen, even if more different might do so.-- <WHI> is this true?? My views, which would even lead
281c139 may account for certain organs not being fixed, <WHI> in some genera, which are most fixed in others.
528m036 is probably arises from (1) harmony of colours, <WHI> & their absolute beauty. (which is as real a ca
226b220 alia. England & Europe.-- It will be well worth WHILE to study profoundly the origin & history of eve
358d075 it was absolutely necessary to watch our meat, WHILE in kettles on the fire, & on one occasion, not
437e139 weakened by wounds got off from the young ones WHILE they were amusing themselves with them, and one
582n068 tions, which will bring about a certain result, WHILE the creature performing those actions neither k
046r080 hink movement. owing to water keeping its level WHILST land rose up & down.--But from above reasons,
211b163 dogs.-- the bitches never being killed by them, WHILST they eat up the dogs.-- L.' Institut. Curious
219b194 onsider it as adaptation because volcanic isld. WHILST <neig> Africa, sandstone, & granite, (that is
222b202 ome species being recent agreeing with Senegal. WHILST Crag «agrees with» according to Beck has none
232b245 ceeded in becoming habituated to colder climate WHILST others died out, or moved towards equator.-- «
232b245 ir history an eocene miocene & pliocene epoch), WHILST others may die out or move South ward.... spec
233b252 r lakes of North America If Parasite different, WHILST man & his domesticated quadrupeds are not so.
264c081 ly considered.-- Structure may be obliterating, WHILST habits are changing-- or structure may be obta
264c081 s are changing-- or structure may be obtaining, WHILST habits slightly preceed them-- From this view
271c105 much mud.-- These facts show, habits heredetary WHILST species have changed Argumentum ad absurdum. T
290c164 s pale ash grey back, like a black bird washed, WHILST tips of primaries black, by examining series I
375d134 ire.--» in Nature production does not increase, WHILST no checks prevail, but the positive check of f
379d151 e children of Sir J. H. was so very like Sir W. WHILST Sir J. himself is not like-- now this is clear
388d173 transformations, should there need two organs; WHILST in common bud there is no such need.-- one wou
445e158 -- far better case than chicken pecking fly.-- "WHILST the shell stuck to its tail" as mentioned by S
466t091 lone together. one had seed-pods turning brown. WHILST both others were in nearly full flower Maer Ju
495q05a w these seeds & see if they will come up true-- WHILST others are crossed.-- Are Bees guided by smell
532m055 llect these same things from any effort of will WHILST their minds were sound. Caroline tells me that
546m111 r, hunted in vain for it-- ten years afterwards WHILST at a meal, she suddenly like a flash without a
565n006 re done in extreme.-- Looking at ones face <&> «WHILST» laughing in glass. & then as one ceases, or s
569n019 stinct from sexual beauty) is acquired taste.-- WHILST music extremely primitive.-- almost like taste
572m035 munication of ideas, &c) of expression lawless, WHILST they are the only steady & universal means. re
588n090 eyes are rolled upwards during mental agony, & WHILST strong emotions of reverence & piety are felt.
595n117 in his arms & carried to see.-- [blank] A Dog «WHILST» dreaming, growling. & yelpings. «& twitching
600o045 eing part of our nature, & its effects lasting, WHILST passions although equally natural leave effect
628o54v ecks the wish to <other> outward gratification, WHILST <the> no desire of gratification will check th
628o54v of conscience is stop to wishes of passion &c. WHILST the passions have no relation I think this «bo
636j56v h says, life, forms a broken, recurrent series, WHILST the habitation «or world» simple series.-- My
640j167 minished, total number of dogs. would diminish, WHILST the long legged variety would prevail.-- Not s
523m017 y F. whether insanity is not distinguished from WHIMS passion &c by coming on suddenly. Ans no.-- bec
264c079 Ourang-outang in domestication, hear expressive WHINE, see its intelligence when spoken; as if it und
542m094 comfort & weariness. & meditative tranquility. «WHINE of children. puppies do so dogs nearly silent,
545m107 d like naughty child.-- Do monkeys cry?-- «they WHINE like children.--» Expression, is an heredetary
532m056 tnight continued wretchedly unhappy, constantly WHINED, would not remain quiet in any room, would not
585n076 bite.-- Henslow remarks that Chimpanze pouted & WHINED, when, man went out of room.-- all theories of
540m084 ld do more towards metaphysics than Locke A dog WHINES, & so does man.-- dogs laughs for joy, so does
587n088 men-- orang-outang & chimpanze. pout.-- Former, WHINES just like a child. Get a Dictionary & make a l
554m140 or shame.-- When she thinks she is going to WHIPPED. will cover herself with straw, or with a blan
```

585n076 -- Hence I conclude. pidgeon taken little way, WHIRLED, & then taken other way-- would not find its w
489qIFC Cuming. -- p 1 Owen p 17 Hooker p. 17 {T} Mrs. WHITBY. Newlands Lymington Hants. Habits of different
032r042 d.-- I feel no doubt. respecting the brecciated WHITE stone of Chiloe, after having examined the chan
040r062 a in being unique, yet no quadrupeds. -- Is the WHITE matter beneath pebbles. the degraded matter of
048r086 led. -- In Patagonia, are all beds same age? is WHITE substance triturated Porphyritic rock. s (mem w
048r086 e substance triturated Porphyritic rock. s (mem WHITE tufas with purple Claystones of P. Desire). = W
049r087 n increased dip is not parallel case to Isle of WHITE. but rather to one out of a series of faults. [
057r114 e S. Africa productions.-- I think in Patagonia WHITE beds having proceeded from gravel proved.-- cur
146g011 tuns. Black faced sheep, sometimes mottled with WHITE black legs & tail like species in colouring Str
178b032 uille.-- Dr. Smith says he is certain that when WHITE Men & Hottentots or Negros cross at C. of Good.
179b032 grandfather, (when the mother was nearly quite WHITE) in the two first children How is this in West
179b034 lended «by» by intermarriages, then the black & WHITE is so far gone, that the species (for species t
182b048 bian coast. "These were of two kinds one <with> WHITE with a black face, & similar to those brought f
188b068 root out.-- For instance ever so many seeds of WHITE flower. all would come up white, though planted
188b068 o many seeds of white flower. all would come up WHITE, though planted in same soil with blue. Now thi
216b179 is inclined to think that offspring of Negro & WHITE will return to native stock (the cross often wh
216b179 to native stock (the cross often whiter<>> than WHITE parent) the mulattos themselves explain it by i
216b180 es with people. either a little nearer black or WHITE as it may happen.-- Dr Smith says he is sure of
216b180 e case at Cape.-- McClay argues from it Black & WHITE species.-- For, says he Seeds of hybrid lillies
216b181 eons. fowls. rabbits cats &c &c.-- When black & WHITE men cross some offspring black others white whi
216b181 k & white men cross some offspring black others WHITE which is more closely allied to case of cross o
235b272 s &c on 242. Hook Smellie Philos of Zoolog. 842 WHITE regulars gradat in Man poor trash Lyell 1024 Fle
239cIFC eep colour on wing Effects of colour on parent, WHITE room How are varieties produced, by picking off
246c025 arge size, resemble, chien-loup.--long, black & WHITE, ears short & straight-- do not bark p. 433. bi
251c041 biting Borneo & Sumatra. differ only in form of WHITE mark on breast: p. 234.-- good case.-- p. 526.
260c068 says-- that some birds or animals are placed in WHITE rooms to give tinge to offspring.-- Darkness ef
260c069 spring.-- Darkness effect on human offspring.-- WHITE, snow.-- the fine green of vegetation.-- ¿accou
265c084 s adaptation.-- Ermine, ptarmigan hare becoming WHITE in winter of Arctic countries few will say it i
267c093 ned & Jeddah". sheep numerous "of two kinds one WHITE with «a» black face, & similar to those brought
271c105 ica» whose Southern range is? <One> The black & WHITE thrush of Azara builds its nest in <same countr
273c110 is any marked colouring of plumage (as «black & WHITE» bars on wings of trogons are lengthened rump f
275c121 se bantam feathers at last got ducky, then took WHITE Chinese Bantam crossed & got some yellow & othe
275c121 crossed & got some yellow & others yellowish & WHITE varieties by picking the yellow one & crossing
286c154 scarcely[)] conceivable in savages) Has not the WHITE Man, who has debased his Nature «& violates eve
290c164 bred.-- (hence hybrids in this case have bred). WHITE & common pheasants. have crossed.-- </I saw> .-
303c204 robably was first black at base of nails & over WHITE of eyes,-- + + +. Will he say creation is at en
312c228 , & colours of little value Dr Smith if black & WHITE Man crosses; children heterogenous, he feels su
318c252 -- (case of Squirrel from extreme north turning WHITE like Hares??--) I never saw more beautiful adap
318c252 like snow shoes. than feet & hind legs of these WHITE hares, fitted for regions of snow.-- Acclimatis
318c254 rtnight in one particular part of country, like WHITE of Selbournes Rock Ouzels.-- ..If the line or b
334d012 in the Duke of Marlborough" there is a breed of WHITE-tailed squirrels, which form a marked wild vari
334d012 a marked wild variety. doubtful whether all are WHITE. Fox says the Half Muscovy Fox says a settler n
341d031 Bunting by one only differing by some permanent WHITE streaks.-- &c &c Dr. Bachman has crossed cock G
345d043 ixed» breed,: Jones says Sussex cattle were all WHITE headed, but this was bred out & now all are pur
345d043 e pure red, yet calf every now & then born with WHITE head (,or «short-horned with» black lip) & then
346d048 s all animals which breed, in & in are-- colour WHITE, uniform.--crafty, go in file, hide their young
363d101 of the Rock Pidgeon,-- several have a group of WHITE speckles on elbow joint-- in Bewick drawing the
363d103 d their songs, the ++ Cervus Campestris spotted WHITE when a fawn compare with fallow? deer. & Moschu
364d104 Southern «see p. 43 supra» breed of cattle with WHITE heads; which years afterwards occasionally went
364d104 y went back-- (Effect of imagination on mother. WHITE peeled rods mentioned in old Testament placed b
364d104 easants about a house made other pheasants have WHITE feathers).-- It certainly appears in domesticat
376d136 oss between the Guaranis & Spaniards are almost WHITE from first generation., that with Quichuas the
401e017 a medlar may be Grafted on pear. Mountain-ash & WHITE Thorn! Species not being observed to change «is
414e064 em preponderance. intellect in Australia to the WHITE.-- The peculiar skulls of the men on the plains
434e129 s grows in <south> Wales & certainly <old> only WHITE in Cambridge, in some counties sometimes one &
459t013 Mr Bell's at Hornsey the offspring of a black & WHITE <duck of pecu> «drake» with the penguin duck. i
467t105 ll-- Saw one small Bee; saw another on Cabbage--WHITE Butterflies suck nectar: «Maer June 41» Rhubarb
478z003 t of the genera-- Cyclops p. 134. and p. 115 In WHITE Cape Pidgeon's stomach small shells (patella) s
498q08v Has H. seen group of different species growing WHITE Mullein good plant to sow & try to get other sp
500q010 , they do not varieties.-- (25) Does the yellow WHITE Butterfly deposit eggs in all varieties of Cabb
505q014 Has planted seeds of Geranium pyrenaicum. small WHITE-flowered var. with abortive stamens.-- show cro
506q015 guin Henslow &c (36) Has not H. raised races of WHITE & Blue Linum-- did parent plants grow near each
511q018 nsmit to male more of his features-- in negro & WHITE (3) About the Bantams at Zoolog Soc.-- did Sir.
529m042 oph. Transactions, of ideot 18 years old eating WHITE lead. who was most violently purged «believe wo
595n117 ty in mind.-- think of Eyton's horses becoming <WHITE> with <lather> <foame> & sweat, when hearing me
358d086 bird, has long tail figure, & some degree of WHITENESS like a Male.-- Thus castration, hybridity, &
179b033 ess of tribes in T. del Fuego. the existence of WHITER tribes in centre of S. America shows this.-- <
216b179 te will return to native stock (the cross often WHITER<>> than white parent) the mulattos themselves
317c248 ten on fossils of N. America.-- At the end of "WHITE'S Selbourne." many references very good. also "R
321c275 ls of Natural History» Skimmed Von Buch Travels WHITES Natural History of Selbourne References at end
324c268 Falconers remarks on the influence of climate WHITE'S regular gradation in Man. Lindlys introduction
024r015 power. / Submarine currents Find instances; The WHOLE coast of New Holland shoals much: Dampier remar
025r016 R. San Francisco [10 degree 32 minute S.] 10 50 WHOLE coast to Olinda [8 degree S.] 9-10 = 30-40 {T}
025r018 it might be well to urge, geologists to compare WHOLE history of Europe, with America; I might add {I
028r031 d into the earth <I did not see one dike in the WHOLE Galapagos Arch; because no sections> same cause
030r036 position at N. S. Wales & Van Diemen's land.-- WHOLE coast S. of Concepcion where there are Tertiary
035r047 een kept; it shows that throughout all England, WHOLE surface oscillated equably.-- These facts becom
037r056 Jan: B. Oriental? level surface not disturbed.--WHOLE West coast. Chonos to Copiapo.--Sydney. K. G. S
038r057 e of geologist. Lyell considers (P 84 Vol III.) WHOLE of Etna series of coatings; hence it will be ne
038r057 ideration of the state at a grand eruption when WHOLE summit of mountain is blown off; & again when i
041r065 only preserved in that part. having lived over WHOLE bottom is important; because in this latter cas
043r071 Concepcion. Valparaiso (Copiapo & Guasco). yet WHOLE territory vibrates from any one shock-- In S. A
044r072 . Africa. Australia.. Oceanic Isles. Geology of WHOLE world will turn out simple.-- Fortunate for thi
045r075 l slope of the country; (perhaps generally over «WHOLE» surface; now comes it they do not flow out tog
046r078 . look at Sulphur. salt. lime, are spread over WHOLE line of coast Darby mentions beds of marine she
047r083 to be coincidental than single elevations along WHOLE line of coast Darby mentions beds of marine she
051r094 ct surface; On «hard» exposed rocks near Bahia, WHOLE surface to where highest spray (there pale gree
051r097 nky edition» Shores of Pacifick, as compared to WHOLE E. America. <East> Africa. Australia. profoundl
056r111 eus,--the similarity of Volcanic products «over WHOLE world» argument, as well as separating causes b
056r111 stances together; & products being similar over WHOLE world, general circulation. But Volcanic action
059r119 ese parts incontestably formed the parts of one WHOLE burning mountain, & that the central part fell
061r126 bout dry season[.] 1791. seen commonly bad over WHOLE world. «(Was it so in Sydney, consult history?
067r143 . Parish says. that beds of shells are found on WHOLE coast from P. Indio to Quilmes. & at least seve
104a068 with pressure? Salt on surface of plains due to WHOLE moisture being lost by evaporation therefore ca
104a068 nished with him that <th> gneiss, mica-slate of WHOLE kingdom of Norway was contorted yet no mountain
107a079 icable to Andes & Patagonia-- On Lyells idea of WHOLE centre of earth same heat, then change in form
147g017 t appeared as if they belonged to double series WHOLE very obscure but it is certain there must once
147g020 4 Temp 62 degree Air 60 degree Below Loch Tulla WHOLE wide valley scattered with few very small & irr
157g075 which shelf 2d «almost» forms it into island-- WHOLE hill composed of remarkable gneiss with red gra
160g090 2ft. 8- long of syenite with pinkish felspar;-- WHOLE hill dark grey fine grained. Much contorted gne
172b005 generation, why are. species are constant over WHOLE country; beautiful law of intermarriages <separ
174b015 - Marsupials. at Australia-- Will this apply to WHOLE organic kingdom, when our planet first cooled.-
175b018 » formed ¿would they not be pretty similar over WHOLE world under similar climates & as far as world
177b029 e dependent on genus,. that genus upon another, WHOLE class would die out, therefore not.-- Monad has
183b053 adrupeds. part of some great system acting over WHOLE world, the period of great quadrupeds declining
203b137 me mundine cause has destroyed animals over the WHOLE world-- For instance gradual reduction of tempe
206b148 mere variations consequent on climate &c-- the WHOLE races act towards each other, and are acted on,
218b192 e some peculiarity.-- Now when we hear that the WHOLE island is volcanic surmounted by water & studde

Page **(Key Word)**
222b205 me predicament. It is another question, whether WHOLE scale of Zoology may not be perfecting by chang
227b228 dy of instincts, heredetary. & mind heredetary, WHOLE metaphysics.-- it would lead to closest examina
228b231 es. His arts would not then have taken him over WHOLE world.-- «--the soul by consent of all is super
230b240 nce, as so many plants produce hybrids, or else WHOLE fabric will be overturned.-- Hence extreme diff
232b246 e tolerably> uniform).-- Comparing fossils with WHOLE world. would be like in a Meteorologic table «i
250c038 at is supposed to have been condition of former WHOLE world. America might have been string of island
254c049 ews must lead to spontaneous generation??) This WHOLE Paper Must be studied.-- <Three p. 7. Am.> D'or
259c063 so tending to produce effect on offspring-- but WHOLE race of that species must take to that particul
259c065 n, but adaptation during earliest existence; if WHOLE life then real adaptation The case of heredetar
263c076 reasoning than other-- if this be granted!!) & WHOLE fabric totters & falls.-- look abroad, study gr
273c112 the Humming bird, which is one instance of its WHOLE family where female is not dull.-- I must obser
275c120 recrossed, till there was a dash of blood, with WHOLE form of grey hound.-- picking out finest of eac
276c124 rd practising imperfectly some habit, which the WHOLE rest of other family practise with a peculiar s
279c133 g cut off-- so that not propagated by nature.-- WHOLE art of making varieties may be inferred from <t
280c136 <pr> one unlike can be produced, yet to produce WHOLE generation unlike would go against the tendency
282c140 ower of change yet, as external conditions over WHOLE world. similar-- & constitution of man original
288c158 ans,-- habits, range. &c &c-- Macleay rests his WHOLE groundwork of analogy on its concurrence in par
292c170 f <classi> real affinities. ie structure of the WHOLE animal let him read Mr Swainson's on the Classi
293c174 reeding there must be some corelation. but. the WHOLE Mechanism is so beautiful.-- The corelations ar
293c175 n.-- This really perhaps greatest difficulty to WHOLE theory.-- There is breed of tailless cats near
294c177 he opinion of many people in conversation.» the WHOLE object of the Work is its proof, «its limiting,
297c189 Missel thrush lately increased in numbers over WHOLE of England & Ireland.-- curious in so wild a bi
298c192 ure.-- Henslow thinks if leaf of plant varies, <WHOLE cross» «all organs» vary in plant. The variatio
298c192 & who better authority) similar with those over WHOLE of country.-- some species are larger &c in dif
306c213 appears to be> is not more than 60F. & «in» the WHOLE area,. 120 is greatest (about 200 miles distant
316c246 f new species some few have been scattered over WHOLE world Many shells at present day same (or accor
319c256 to ours, & our lark ranks very high.-- Upon the WHOLE thinks <many> more birds sing in England than i
320c276 ctions. <done> up to parts published March 1838 WHOLE of Geographical Journal Asiatic Journal to end
333d009 els & setters are produced. one would argue the WHOLE effect of race was determined by male: & How co
336d017 een gained slowly, now all the mules have their WHOLE <body> form of body gained in one generation, s
336d019 generation being a <slip> «bud» from parent. if WHOLE parent not entirely embued with the change, a b
337d021 ow one nerve becomes sensitive to light.-- (Mem WHOLE plant may be considered as one large eye-- have
342d035 1) accidentally says "--is distinctly marked as WHOLE dynasties have been featured by the Austrian li
347d049 of days adapted to duration of sleep of man.!!! WHOLE universe so adapted!!! & not man to Planets.--
356d071 ive solutions & linking of facts-- Savages over WHOLE world. (Major <I> Mitchell p. 244. vol I) spit
357d072 rule? The destruction of the great Mammals over WHOLE world shows there is rule.-- S. America & Austr
357d074 transportation, seems «?» to imply knowledge of WHOLE world-- if so doubtless «part of» system of gre
372d129 bably is like cutting off tail of Planaria, the WHOLE grown to that part.-- claw added to crab, tail
372d129 to another part of body.-- [in plants does not WHOLE individual change into generative organs?]CD it
372d129 se any qualities by being buds-- , more than if WHOLE branch transplanted? +.simplest forms of buddin
372d130 each particle of animal must have structure of WHOLE comprehended in itself.-- it must have the know
383d159 only dimidiate, but quarter-grown seems to show WHOLE body imbued with possibility of becoming either
386d166 knowledge of the part, of what is good for the WHOLE.-- if cut off nerves in snail. (Encyclop of Ana
389d176 rd to theory, showing generation connected with WHOLE system, «as if there was, a superabundance of l
390d178 ees no doubt there is tendency to propagate the WHOLE difference of parent, tree, but it fails.-- the
390d179 theory of generation being the passing through WHOLE series of forms to acquire differences: if none
391d174 to that kind of generation, which typifies the WHOLE course of change from simplest form.-- (Because
391d174 h double union.-- At present I can only say the WHOLE object being to acquire differences «indifferen
397e004 se change in forms is <al> solely adaptation of WHOLE of one race to some change of circumstances; no
398e004 llations, not affecting organic forms, that the WHOLE value of the geological chronology depends, tha
399e006 some case of change in vertical series: Look at WHOLE Glacial period? [not located] Study introductio
399e009 tructure will last. without it is adaptation to WHOLE life of animal, & not if it be solely to womb,
402e017 f one river did pour sediment in one spot, for <WHOLE> many epochs-- such changes would be observed.-
404e024 from which plain inference might be drawn that WHOLE infertility «of hybrids receive no explanation»
404e025 mere simple modification of an organ present in WHOLE class. Case of Mexican greyhounds.-- young bein
405e031 emilunar {P} mark on each side darker,, so that WHOLE colour is changed, these best marked characters
406e037 ed about same time in North & South. America.-- WHOLE wor[l]d, formerly possessed a climate compared
408e043 ours». acting. by a most delicate organ, on the WHOLE system may. produce-- ? When a species becomes
410e050 e changes should be slow & bear relation to the WHOLE changes of country, & not to the local changes.
419e087 rests upon amount of physical change <affecting WHOLE bodies of species>, & only secondarily,, by ass
423e096 enormous complexity, it is impossible to cover WHOLE surface of world with life.-- for otherwise a f
423e098 on is that people in the southern Counties have WHOLE fields, some for cauliflower &c.-- Uncle John b
430e118 es are made in two ways-- local varieties, when WHOLE mass of species are subjected to some influence
431e123 f foetus in proportion to male parent p. 8. his WHOLE doctrine of the advantage of crossing consists
433e127 The argument must be thus put, shall we give up WHOLE system, of transmut., or believe that time has
433e127 greater, & that systems, are only leaves out of WHOLE volumes.-- The fact of tumbling pidgeons; flyin
461t025 rious facts & this paper deserves fresh study & WHOLE order of the fish.-- Embryology p. 97. for Man
463t057 suggested. (vipers tooth also a difficult), the WHOLE mind is constituted that a difficulty makes gre
466t099 ortive as it was in autumn: filaments united in WHOLE length to corolla--anthers minute, distinctly d
467t104 covered with pollen was pressed & rubbed along WHOLE breast-- {b} pressing either one or both of Pea
472s02v pe off pollen. (as a needle becomes covered) so WHOLE sides of flower & stigma dusted.-- <I think> Wh
498q008 ision of germen <?>-- (11) Must pollen grain be WHOLE, to impregnate?-- I presume only stigma impregn
499q009 er they are dusted with pollen-- in what state (WHOLE or broken) is ball of pollen on Bees thighs (18
521m007 a tune & words» can thus lie dormant, during a WHOLE life time, quite unconsciously of it, surely me
522m011 many things if he began at one end, he knew the WHOLE subject.-- if at the other nothing.-- He could
528m039 this symmetry & rhythm applies to the view as a WHOLE.-- Colour «& light» has very much to do, as may
530m043 e asked what bond he could have had. yet during WHOLE illness, he had been able to direct about his o
538m080 illy joking The possibility of the brain having WHOLE train of thoughts, feeling & perception separat
541m090 I am capable-- I suspect from these facts that WHOLE effort consists in keeping one idea before your
551m129 ey flag & something like a snow-haze. covers my WHOLE imagination." Septembe. 3d Why when one thinks
552m131 ing force to legs & body, & especially, when to WHOLE body, being failed, & not to any particular mus
555m143 s like the old world ones.-- Though the[y] move WHOLE skin of head they do not move eyebrows.-- (I se
555m144 was the feeling of banter & joking» because the WHOLE train of Dr Monro experiment about hanging came
557m149 ome from shooting. runs away. is not afraid the WHOLE way. but ashamed of himself.-- Jealousy probabl
558m151 ce to this "ought," as well as the works of the WHOLE world.-- Read Mackintosh on Moral sense & emoti
558m151 ead Mackintosh on Moral sense & emotions.-- The WHOLE argument of expression more than any other poin
564n006 exaggerated habitual sneer» the manner in which WHOLE skin or muscles are contracted between eyes & u
567n015 me sensation of heat shows blood is pumped over WHOLE body.-- is it connected with surprise.-- heart
568n018 of recurring sounds in Harmony common to t[he] WHOLE kingdom of nature. If I want some good passages
573n036 llow «satellites», planets, suns. universe, nay WHOLE systems of universe <of man> to be governed by
586n079 could be formed or afterwards.-- child sucking WHOLE wonder instinctive.-- carrier pidgeon just as w
590n094 .-- see his Travels.-- When one sees in Cowper, WHOLE sentences spoken & believed to be audible. one
596n184 rds man.» Macacus especially pulls back skin of WHOLE forehead & 2 ears.-- emotions of every kind.--
599o005 .-- quotes Ld Mondobbo.-- language commenced in WHOLE sentences.-- signs-- ¿ were signs originally mu
601o009 her order. where the sensation is conveyed over WHOLE body (which it may be in first case. as when th
602o11b ent of the senses & experience p. 142 "Upon the WHOLE it seems."-- "that the object of «all» art is t
603o013 consequence» hence languages become corrupt, & WHOLE classes of words «are abbreviations» he thus de
611o034 dents. But such tissue <must> bears relation to WHOLE, that is enough must be present to be able to e
614o037 m> kind more than another-- What is matter? the WHOLE a mystery.-- [LHC] This Materialism does not te
623o050 tendency, (not to any one individual but to the WHOLE past race).-- <no one> doubts) « <I cannot> » [
626o052 ence regulates feelings, as of cowardice.-- the WHOLE appears to me rather rigmarole.-- He does not s
632j53r tation:-- But I do not want to deny laws.-- The WHOLE universe is full of adaptations.-- but these ar
638j58v infringement on the wisdom or Providence. when WHOLE rocks nay very mountains are formed of such dea
132a137 turelles. Tom I. p 501.-- shows first that data WHOLLY insufficient to calculate rate of increase of
408e044 en purely result of elevation,-- «all» modern & WHOLLY volcanic-- Azores might be prophecied to have
448e167 , numerically the same with recent & yet almost WHOLLY different, is same, as if Isthmus of Panama.--
547m112 nto dream.-- It appears to me, that the mind is WHOLLY absorbed with one idea (hence apparent vividne
576n047 having acquired» this sense of right (& Whether WHOLLY instinctive as in the dog, or chiefly habitual

**************************************(Key Word)**************************************

582n068 tends the result they will effect,.--" this not WHOLLY true, for we must grant a bird knows what is a
325c266 sted, of organ being worn out as. otherwise old WHORES would not have children Turners embassy to Thi
417e076 he common species: from one locality, all left WHORLED.-- He kept two to see if they would breed, It
049r089 conical mass. will this conical mass be granite? WHY not more probably greenstone? What probable orig
070r153 ge.-- Yet one is urged to look to common parent? WHY should two of the most closely allied species oc
086a007 uatorial streams of warm pole; in name of Heaven WHY are tops of Equatorial mountains so cold.-- Sibe
098a048 (Hence endless passages from gneiss to granite): WHY not horizontal? Why have particles in such cases
098a048 ges from gneiss to granite): Why not horizontal? WHY have particles in such cases moved more laterall
105a073 f running water to deepen not to widen valley.-- WHY is serpentine course result of little inclinatio
119a105 What causes that of tendency to irregularity,--. WHY does Sir John assume it to be constant.-- It is
122a113 his not explain littoral mountains & volcanos.-- WHY on one coast? How can Herschel consider figure o
123a116 dissemination & separation in volcanos.-- if so WHY not metals. The theory of veins will, I suspect
132a138 nts for increase on earth by volcanic action.-- <WHY> how as we know volcanic action prevails more be
154g060 P) now that it has got to the rock of cols if--. WHY should it deposits River terraces often descend
165g126 tated this generally the case Wednesday 12/ & 3/ WHY is the Tetrao scoticus & Tetrao-- not an America
171b002 hs nor «cut» Stallions nor nuns are longer lived WHY is life short, Why such high object generation.-
171b002 ons nor nuns are longer lived Why is life short, WHY such high object generation.-- We know world sub
171b005 f life With this tendency to vary by generation, WHY are. species are constant over whole country; be
173b012 n this idea of propagation of species we can see WHY a form peculiar to continents; all bred in from
173b012 uliar to continents; all bred in from one parent WHY. Myothera several species in S. America why, 2 o
174b013 rent why. Myothera several species in S. America WHY, 2 of ostriches in S. America-- This is answer
174b014 genera different countries. Propagation explains WHY modern animals same type as extinct which is law
174b014 xtinct which is law almost proved.-- We can see «WHY» structure is common in certain countries when w
183b053 cts to <tran> propagation of species, by saying, WHY not have some intermediate forms been discovered
185b055 date itself to as many situations as possible.-- WHY should we have in open country a ground «do. <w>
186b058 etains some one character of all its family; but WHY so? I can see no reason for these. analogies; CD
187b064 ion of species like generation of individuals.-- WHY does individual die, to perpetuate certain pecul
196b106 ants I believe not..-- It is a very great puzzle WHY Marsupials & Edentata should only have left off
196b106 y produced in several places & died off in some? WHY did not fossil horse breed in S. America-- it wi
198b115 ing found on all isld, (if act of fresh creation WHY not produced on New Zealand; if generated <No> «
199b117 ss all belong to same types already established. WHY out of the thousands of forms, should they all b
210b158 l island species would not be manufactured, <but WHY they should be manu> Does it not present analogy
223b209 37 Contrast New Zealand with Tasmania The reason WHY there is not perfect gradation of change in spec
223b210 s varieties according to same law with animals?? WHY are species not formed. during ascent of mountai
225b216 ely how did otter live before being made otter-- WHY to be sure there were a thousand intermediate fo
225b217 external causes, of which we are as ignorant. as WHY millet seed turns a Bullfinch black, or iodine o
235b263 ases like Irish & English Hare bear upon this.-- WHY do Van Diemens land people require so many, impo
235b274 es introduced would not change be superinduced-- WHY is every one so anxious to cross. animals from d
265c088 tempted to suppose from beholding the ground.-- WHY do beetles & birds & <f> become dull coloured in
291c166 corporeal structure are facts full of meaning.-- WHY is thought. being a secretion of brain, more won
292c171 ill & therefore may be called instinctive.-- But WHY do some actions become heredetary & instinctive
308c217 made me go to tea chest almost unconsciously.-- WHY do absent «Dr. Black. tea & sugar» people. rever
313c236 rts, & scission cannot effect the process.-- but WHY if louse created should not new genus have been
316c244 als.-- yet how faint in a Fuegian or Australian! WHY two sexes scission in all cases probably gemmati
345d043 ses» is killed. Notes from Glen Roy Note Book.-- WHY not gradation.-- no greater difficulty for Deity
359d087 s animals are healthy which is often the case, & WHY is not Tetrao Scoticus. an american form (if so)
367d112 s. CD[old Buffon should be read on Mare My view, WHY should organic affections always influence the s
372d129 ranch transplanted? +.simplest forms of budding. WHY hybrids are infertile. supposes that when foetus
372d130 -- but this has nothing to do with generation.-- WHY does Gecko produce always different tail? An Ind
373d132 t this argument, for Mammalia recent creation.-- WHY crab can produce claw. but man not arm. hard to
378d147 sculine character. added to species,. we can see WHY. what tendency can there be for abortive organ e
383d160 s effect of Metallic hue of silver pheasant. yet WHY young & female alike Good Ch 6 Keep Is it Male t
388d173 ows that semen. must actually reach the ovum.-- [WHY green? & not purple?-- legs pale coloured.-- In
391d174 ne parent or tree, (but not in other trees.-- -- WHY in making a bud, which is to pass through all tr
391d174 revious changes, which mutilations are not). but WHY should there be a necessity that there should be
391d175 niferous trees generally dioecious oldest forms) WHY should it demand some further change?-- Man prop
391d175 er» than twins «or triplets &c &c in» in litter. WHY are twin in man more like «each other» than twin
403e023 tob. 16th. A very strong passage might be made-- WHY is there some law about sexes of twins in former
407e041 ctivore being in S. America & Australia. reason, WHY seeing great variation in external form of varie
414e060 found together.-- Read this Work-- Decb. 4th.-- WHY: Marsupiata, when first introduced live & multip
423e095 dvancing complexity of others.-- It may be said, WHY has the organization of fishes & Mollusca (& pla
423e096 ery changes <len> tend to give rise to others.-- WHY should there not be at any time as many species
424e102 I presume Carrion Crow is found in Edinburgh.-- WHY then has there been a retrograde movement in Cep
434e128 that analogy is sure guide & my theory explains WHY does Fleming consider them varieties & what says
465t081 e destroyers vary; so that we here see reasons-- WHY it is sure guide.-- Lychnis April 3d.-- Henslow
473s07r urse of life, «as in trades» there is no reason, WHY no perfect gradation can be expected in any one
521m009 - then showed the watch upon which he exclaimed, WHY the peculiarities shd be born,-- may come in cor
521m010 the Gardener & said, who is tha? Mr C. answered WHY it is dinner time.-- » My father asked him wheth
533m058 t it sees it.-- it is frightened without knowing WHY do you not know, that is A. B my gardener.-- Thu
533m058 y-- the child dislikes the frown without knowing WHY-- the child dislikes the frown without knowing w
533m059 eing the scenes of his childhood without knowing WHY-- a man as in Guy. Mannering. feels, pleasure. i
536m071 that makes one dog smell posterior at another.-- WHY-- had not conscious of recollecting it-- this ma
540m086 ident,-- children with other children naughty.-- WHY do bulls & horses, animals of different orders t
540m089 haze. covers my whole imagination." Septembe. 3d WHY does person cry for joy? 17th. August Montaigne
556m145 . than in animals.-- In the drawings of Voltaire WHY when one thinks of any object, (or having looked
556m146 of donkey, horse & zebra. when going to kick.-- WHY is under lip curled over upper with mouth shut.
557m150 me would never make person tremble, like fear.-- WHY does dog put down ears, when pleased.-- is it op
557m150 es any great mental affection make body tremble. WHY does any great mental affection make body trembl
565n007 When one fear any bad news, «though in a letter» WHY much laughter tears.-- & shaking body.-- Are tho
565n007 letter» why is person painted with mouth open.-- WHY is person painted with mouth open.-- why when pe
565n007 de) same way walk erect & stiff, with head up.-- WHY when person is listening is mouth open to hear w
565n009 ant object, brightened & moistened by emotion,-- WHY does suspicion look obliquely.-- who can analyse
570n023 rom inspiration, which accompanies surprise.-- & WHY does emotion make tears fall?? Lavater says deri
570n025 n as trick) «Shrugging aroused acting» Octob 25. WHY does one inspire, when surprise, can one resist
571n028 lly all has not been acquired by education. else WHY is modesty, mixed with triumphant feeling so sim
571n028 on. else why do some children acquire it soon. & WHY do some children acquire it soon. & why do all m
571n028 d did high forehead sign of exalted character??) WHY do all men. agree ultimately?-- We acquire many
571n030 r may be vitiated. or rather altered. The Reason WHY may not our heredetary nature thus acquire some
578n057 ing of some of the muscles «or twitching».-- But WHY New Buildings look ugly is because there is some
587n088 with idea of scarlet, as well as remember it.-- WHY does joy & OTHER EMOTION make grown up people cr
591n099 is disobeyed-- I often have «as a boy» wondered WHY do children pout & not men-- orang-outang & chim
591n100 followed; good case of instinctive conscience.-- WHY all abnormal sexual actions or even impulses. (w
594n113 he room. nearly 3 months old. What is absurdity, WHY does not man eating cause disgust, because he do
603o014 tatues not painted-- «music» very good article-- WHY does one laugh at it-- sensation of disgust with
609o030 nstinctive» moral senses: (& this alone explains WHY flower beautiful? ¿even to children S. Jenyn's I
612o035 ude irritability for that require will in part. ¿WHY our moral sense points <is> to revenge). In judg
616o039 ill in some animals, involuntary in others. [1]] WHY more so than movement of sap. or sunflower to su
618o41v do not know attraction objectively 6) The reason WHY may it not be said that thought perceptions will
620o044 n fades away, so that when man afterwards thinks WHY thought &c. should imply «X» the existence of so
628o53v ough by fear it might be partly made.]CD p. 21. "WHY was such an instinct not followed for a pleasure
633j54r se,-- we sink into such contemptible queries, as WHY ought I to keep my word"-- gives the problem, of
633j54r e queries, as why should the earth have drifted; WHY should the earth have drifted; why should plants
633j54r h have drifted; why should plants require earth, WHY should plants require earth, why not created to
637j58r es in Giraffe &c, as adaptations to long necks-- WHY not created to live on alpine pinnacle? if we on
637j58r y point to chickens beak, to break egg. shells-- WHY they may as well say, «long» neck is adapted to
638j059 ss.) <th> form these schemes.-- I see no reason, WHY chicken could not have lived had it not been so.
638j059 ss.) <th> form these schemes.-- I see no reason, WHY structure of brain should not be born. with tend

```
640j167  imals changing.-- From these views we can deduce WHY small islands. should possess many peculiar spec
608o026  nimals do attack the weak & sickly as we do the WICKED.--we ought to pity & assist & educate by putti
608o026  thing a diseased offensive object, so we view WICKEDNESS.-- it would however be more proper to pity t
608o027  more  strange that there should be necessary. WICKEDNESS than disease. This view should teach one pro
022r006  glesea, amongst the varying & dubious granites.--WIDE limits of this mineral in Australia. Fitton's a
044r072  did  not see any number of dikes in the cliffs.--WIDE valleys.--central peak small; yet great body of
062r129  imits of birds & animals in S. America. Zorilla: WIDE limits of Waders: Ascension. Keeling: at sea so
087a011  Mississippi  -- No Vol. I. p212. Cuvier Oss Foss WIDE range of Mammalia really very important. harmon
105a069  ffs. but then how spread abroad?-- There is thus WIDE difference between erosive power of river & sea
120a109  ould carry further its own matter. but would cut WIDE gorge. leaving cliffs, on each side, such as no
147g020  p 62 degree Ait 60 degree Below Loch Tulla whole WIDE valley scattered with few very small & irregula
150g037  hen water up to shelf very shallow channel 50 ft WIDE & river get formed in centre In Glen  Collarig,
196b105  -- Instinct may confine certain birds which have WIDE powers of flight; but are there any genera, mun
198b113  I G. St Hilaire Insects & Molluscs allowed to be WIDE hiatus: states in one the sanguineous system, i
203b137  gained in short time. Milvulus forficetus <has a WIDE range> is a tyrant flycatcher doing the service
213b170  have long remained are those ¿Lyell?, which have WIDE range and therefore cross & keep similar. But t
223b208  ht well say how> The difference is that there is WIDE gap between Man & next animals in mind, more th
226b223  ing Molluscs, like Carnivorous Mammalia in their WIDE range & in their duration of species. ¿are car
243c020  ti found at <New> Isle of France: xx instance of WIDE range, where means of wide range work this out-
243c020  rance: xx instance of wide range, where means of WIDE range work this out-- L. Jenyns, about my  fish
244c021  e. Plants.--» It would be very important to show WIDE range of fish & shells in tropical sea, it. wou
246c025  lled from East Indies, isld, as far as Oualan.-- WIDE space of sea, to East of America. would account
249c035  not form two sections is this not connected with WIDE range of animals. Follow this out, where specie
249c035  genera in two words. have not species, generally WIDE range? Mice.-- Waterhouse's remarkable fact  of
253c046  of character. Hence Elephas primigenious over so WIDE a range, & Mastodon angustidens.-- Ogleby has f
258c060  ces effecting the area equably.-- Animals having WIDE range, by preventing adaptation owing to crossi
283c145  qual distances between species. some fine & some WIDE. which is strange if creator had so created the
304c206  s. -- look how close birds! look at Mammals: how WIDE.-- therefore birds younger???? or «have» not «b
479z005  ot indigenous p 112 M Lesson--Voyage of Coquille WIDE limits of Nullipora Discussion good on Falkland
509q017  at in the mundane genera, the species <are> have WIDE range-- How is this in «Plants??» Are abortive
546m108  st 26th. I cannot help. thinking horses admire a WIDE prospect.-- The very superiority of man perhaps
558m152  observed the Asiatic Leopard. quarrelling. mouth WIDE open, each [lip] drawn back & driving air out o
558m152  driving  air out of mouth «hairs erect on back» «WIDE open» with prodigious force.-- making growling,
582n067  t does unconsciously of any end.-- N B. There is WIDE difference, between the means by which an anima
599o007  o give up my theory-- Viewing from eminence. the WIDE expanse, of county, netted with edges & crowded
640j167  offer a most striking anomaly to this. Have they WIDE ranges? Agassiz has shewn that they most widely
641j29r  has shewn that they most widely differ» 3 A very WIDE range must be destructive to species, when phys
374d133  t from plants!) But the Cephalopoda depart more WIDELY from living forms.-- p 458 Upper Silurian, fis
640j167  y wide ranges? Agassiz has shewn that they most WIDELY differ» 3 A very wide range must be destructiv
105a073  fore tendency of running water to deepen not to WIDEN valley.-- Why is serpentine course result of li
106a074  stration.-- Therefore stream has no tendency to WIDEN course until inclination is become comparativel
105a072  recent)  ------- Coquimbo on. other hand?-- The WIDENING a valley depends on serpentine course.--  the
106a074  een case. if inclination small.-- The power of WIDENING channel depends on power of deflection with s
106a074  lessened. therefore rivers very ineffectual in WIDENING valley.-- it is essentially a deepening agent
105a069  river & sea.; the former as its channel becomes WIDER. its cutting power. (as does it when  the
106a073  d <cut> be to cut a narrower channel instead of WIDER.-- This applies to all vallies (except mere tal
288c159  n greater migrations, if American & intersected WIDER & wider at Rio Plata. birds which had originall
288c159  r migrations, if American & intersected wider & WIDER at Rio Plata. birds which had originally crosse
204b141  individual force> Mr Wynne has crossed Ducks & WIDGEON & offspring either amongst themselves or  with
204b141  parent birds.-- «W. Fox. knew of case of male WIDGEON, winged & turned on pool. first season bred re
339d025  reenfinch bred, & surely wild Duck & «pintail» WIDGEON!-- Divides animals «world into Zoological Prov
283c144  lying eagle.-- Hawk Gould seemed to think, that WIDOW bird. replaced Birds of Paradise-- if such fant
272c109  d superabundant,-- the tail in cock peacock,.. WIDOWBIRD.-- Birds of Paradise. Trogons.-- the one feat
524m022  most. intently.-- criminals before execution.-- WIDOWS not of their husbands-- My father's test of si
065r139  of  Deception Isl is six Geographical miles and WIDTH 2 & ½ miles S. Shetland. Lat. 62 degree 55 minu
105a069  d to waves of sea, cutting power increased with WIDTH. for besides more surface exposed. bay more ope
118a103  52)  fringe of sublittoral deposit always equal WIDTH --subject of fine paper this would make.-- L'In
268c096  ifferently) do. do. on the genus Procyon.-- by WIEGMAN Classified catalogue of animals of Nepal  read
326c265  p. 348. gives reference to Kolkreuter's Papers WIEGMAN has published German Pamphlet on crossing Oats
504q13v  ght to be placed far from all other Peas, from WIEGMAN Shrewsbury (1) Peas.-- Beans seeds alone remai
495q05a  » will drift many seeds= Necessary to answer WIESSENBORNS doctrine of Equivocal Generation Charlworth
191b082  subsidence  continually forming species.-- Man & WIFE being constant together for life, is in accorda
566n010  rationally explained.-- on the wish to support a WIFE a ruling motive.-- Book IV, Chapt I on passions
569n022  ol II p. 132. offered to take a savage, said his WIFE would be grieved-- "il leva les epaules et  dit
573n039  . C. of the bad conduct of Mrs C. (her brother's WIFE). & she said nothing but shrugged her shoulders
576n047  n, might not <t> do so even to save a friend, or WIFE.-- yet he would ever repent, & wished he had lo
619o042  rds fellow <living> creatures, or of kindness to WIFE [RHC] 2) and children would give him  pleasure,
438e142  many circumstances, time of domestication [see WIKINSON on dogs of Egypt & Cuvier on Mummies]CD [NB I
048o086  part of the globe are there such vast numbers of WILD animals. both species & individuals as in the h
063r133  Cats.  Horses. Cattle. Goat. Asses. have all run WILD & bred. no doubt with perfect success.--showing
171b004  ild of savage not civilized man.--birds rendered WILD <through> generations, acquire ideas ditto.  V.
179b033  believes in repugnance in crossing of species in WILD state.-- No doubt «C. D.» wild men do not cross
179b033  ing of species in wild state.-- No doubt «C. D.» WILD men do not cross readily, distinctness of tribe
179b034  d to make marriage.-- as Dr Smith remarked Man & WILD animals in this respect are differently circums
192b083  nstances of a variety of fruit tree or plant run WILD in foreign country.-- When one sees nipple on m
218b191  idae & Narcissus. Mr Donn considers Mr H. rather WILD Mr Donn remarks to me. that give him a  species
228b232  look forwards if we choose to let conjecture run WILD then <our> animals our fellow brethren in pain,
230b240  sible to get evidence of two races of plants run WILD.-- (for we know that such can take place withou
230b241  Goulds' case of Willow wren} & others varying in WILD state to show that we do not know what amount o
231b241  as  others would exjudge it amount of varying in WILD state.-- When breaking up «the primeval.» conti
235b275  The GRAND QUESTION Are there races of plants run WILD or nearly so, which <breed> do not intermix,--a
240c003  genera.,  yet retains markings of wings like the WILD rock pidgeon-- fact analogous to Owen's  Phil:
240c013  not located} Falkner Patagonia no description of WILD animals, nor in Dobrizhoffer Abipones.-- Voyage
241c014  calls it Bos. leucoprymnus. does not say whether WILD or not p. 156.-- Parroket with stiff tail  like
244c021  y is» «has» taken place with quadrupeds» p. 118. WILD pigs of Falklands, generally "red of brick" hai
245c023  considered  good species from dental characters, WILD pigs. said by Forrest to swim from one isld  to
248c030  gether: The argument must thus be taken, as «in» WILD state (where instinct not interfered with, or g
248c030  an be shown probable,)» varieties may be made in WILD state, there will be presumption that they woul
249c035  cases  <sp> like those of Primrose & Cowslip run WILD, The two species of Clenomys. case of replacing
251c041  ngely contradictory to Azaras fact of conduct of WILD & tame horses.-- p. 246-- Gmnura-- new genus of
266c092  gure those of Arabia & Egypt.-- CIVETS CATS only WILD animals on isld.-- Niether Hyenas; jackals monk
267c094  s of birds drifting out to sea Vol VII. p. 325-- WILD dogs of Guayana always hunt in packs «30 or  40
280c134  ir J. Sebright admirable essay} heredetary Young WILD ducks.-- lose as well as gain instincts. Wild &
280c134  ng wild ducks.-- lose as well as gain instincts. WILD & tame rabbits good instance-- instincts of man
280c134  clearly  applicable to formation of instincts in WILD animals, many species in one genus-- external c
283c145  o turn the Zebra into the Quagga.-- «let them be WILD in same country with their own instinct & (even
289c161  s bird & get account of habits My definition <in WILD> of species. has nothing to do with hydridity,,
294c176  s ready to admit, permanent small alterations in WILD animals, & thinks Lyell has overlooked argument
294c176  It will be easy to prove persistent Varieties in WILD animals-- but how to show species-- I fear argu
294c176  st rest upon analogy & absence of varieties in a WILD state-- it may be said argument will explain ve
296c185  are  wonderfully absurd.-- p. 565 <breed> Scotch WILD Cattle. breed freely with the tame Vol II. Maga
298c189  over whole of England & Ireland.-- curious in so WILD a bird.-- Annals of Natural History Vol. I.  p.
299c194  nd has same uniformity Primrose & Cowslip, quite WILD, but they affect different localities,-- latter
299c194  s, offers another case of permanent varieties in WILD state-- The two. former produced by difference
299c195  soils & their seeds produce <offspring> variety. WILD carrot. made into biennial domesticated kind wi
305c209  Ruffs & Reeves a remarkably varying plumage for WILD birds-- At Zoolog Gardens there is half  Jackal
308c219  rent climates &c. Do I mean that ideosyncracy of WILD animals is generally different, because their d
```

311c225 . «replacing each other». good to consult p. 326 WILD ass extending over 90 degree of Long. & Col. Sy
313c235 by association (case of Elephant, which had run WILD in India. in Heber?) is analogous to dormant in
325c267 rnal Rengger on Mammalia of Paraguay. account of WILD cattle & Montagu on birds (facts about close sp
333d006 the effect is the same.-- Fox thinks that when a WILD animal is crossed with tame, offspring always t
333d007 ssed with tame, offspring always take most after WILD.-- i.e that <alw> «no» domesticated ones have b
333d007 alw> «no» domesticated ones have been so long as WILD one under present form.-- Fox has seen several
334d012 d of white-tailed squirrels, which form a marked WILD variety. doubtful whether all are white. Fox sa
337d020 few even of local varieties approaches quite to WILD local variety.-- our Europaean varieties must b
339d025 »! Has not Goldfinch & Greenfinch bred, & surely WILD Duck & «pintail» Widgeon!-- Divides animals «wo
342d033 that hybrid Muscovy & Common duck have been shot WILD (escaped from Carolina?) off New York. therefor
346d048 enaeum (1838) p. 611. Ld. Tankerville account of WILD cattle of Chillingham,-- habits peculiar,-- you
356d072 eard of Black game & Ptarmigan having crossed in WILD state-- & the English & Some African dove.-- Th
359d087 art have bred inter se.-- These hybrids are very WILD & take <very little> in disposition after their
359d089 ike each other,-- (& not very like either either WILD or Pintail duck) from which they were descended
360d090 ll share of Jackall shape of body.-- disposition WILD, & fearful. though not so much as in Jackall.--
363d101 awing the the rock Pidgeon has not: now how many WILD pidgeons have spangles on this part: this will
363d101 ct Pennant has described them)-- [Study horns of WILD cattle.-- plumage of fowls-- long ears of rabbi
363d103 that 1/2 Muscovy & common duck were often caught WILD off coast of America.-- showing hybrids can far
364d105 ter has crossed with the Ptarmigan. subalpina in WILD state.-- Neilson has given figure of it.-- In E
365d105 dicament, probably, alone would species cross in WILD state.-- Is English red Grouse. a cross between
366d112 st blinking to keep out flies might be used» The WILD ass has no cross. how comes it. that the tame do
371d127 rates case of Dingo (he alludes to the dholes or WILD dogs of India) in Zoolog. Garden having colours
378d148 idgeons & fowls.--??? Wate[r]ton «p. 197» put 12 WILD duck's eggs under common ducks, the young cross
378d148 for they were of all colours.-- they were "half WILD half tame, they came to the windows to be fed,
392d180 erimentis on crossing of the several species of WILD fowl <in ⌐> of India «with our common ones» in
404e024 urd.-- their only escape is that rule applies to WILD animals only. from which plain inference might
405e031 rs vary in same Manner as they would vary, if in WILD state; thus mark on ear of cats, colour can be
421e091 to Ireland. ▯do p. 283. on the dark ears of the WILD Chillingham Cattle, with reference to Mr Bell's
424e103 alf reclaimed animal.-- The dogs, which have run WILD have, have done so in hot countries.-- CD[Camel
426e106 , & pheasant, & hooded crow goes against this. & WILD hybrid plants. If many wild animals were crosse
426e107 goes against this. & wild hybrid plants. If many WILD animals were crossed, the many would probably be p
426e107 but this is false, [give instance of series from WILD animals & plants]CD.-- Mr Marsh has some nephew
428e113 Cowslip is great difficulty. «I should doubt if WILD species ever formed like short-tailed cat or do
429e114 & which might spread themselves, as well as our WILD plants, we see how full nature. how firmly each
430e117 tant as showing small variations in offspring of WILD animals.-- grateful & intelligent.-- The theory
436e136 ia part of Old World <If> It «may» be said, that WILD animals will vary, according to my Malthusian v
439e143 on instincts in animals.-- yet the existence of WILD close species of plants shows there is tendency
442e153 hand, fruit trees are propagated by means, which WILD plants never are, namely on stocks of other var
450e173 rrest Voyage p. 323. Sooloo. imported elephants. WILD hogs-- spotted deer, no loonies, but cocatores
450e174 e» strictly monogamous-- geese polygamous (¿when WILD) but only some birds are so when wild-- wild du
450e174 ous (¿when wild) but only some birds are so when WILD-- wild ducks monogamous; tame ones highly polyg
450e174 hen wild) but only some birds are so when wild-- WILD ducks monogamous; tame ones highly polygamous--
450e175 re shades of difference in all the isld, like in WILD animals).-- There are prevailing colours in th
450e175 the different islands.-- The horse is only found WILD in the plains of Celebes. (but language shows t
450e175 not fit for horse. Forrest--. (p. 270) says many WILD horses, bullocks, & deer South part of Mildanao
451e180 on Koloff. voyage through the Moluccas 1825--"No WILD animals in Moa.--" Chapt.-- V.-- : do. Chat XXI
451e180 animals in Moa.--" Chapt.-- V.-- : do. Chat XXI. WILD cattle & Hogs on Timor-land-- monkeys do not ex
452e181 lolo.-- "-- Forrest Voyages. p. 39-- deer but no WILD animals in Gilolo.-- p. 134: Birds of Paradise
453e182 s the form [...] Dampier. Vol I. p. 320. says no WILD (carnivora) beasts on Phillipines. Forrest some
453e182 Forrest somewhere says same.-- do p. 393. <">The WILD, small fowls at Pulo Condore "crow like ours, b
454e184 as to become <all> monooecious.-- Are there not WILD plants, some partly dioecious? Mushroom Hybrids
454e184 s, some partly dioecious? Mushroom Hybrids? Any «WILD» plants in England, which do not perfect their
459t011r e Voyage East coast of Africa-- Vol II, p. 256-- WILD cattle at Madagascar-- «p. 121» No beasts of Pr
459t009 in> "pretty much as we see in the young of the WILD hog & of several species of deer, which are alt
466b094 ently coloured-- & stamens bend up a little In a WILD purple Geranium, I see Bees visit always base {
468t106 -- I see Bees almost every flower-- Blue-bells-- WILD-raspberry--leeks-- Flowers which thought very u
470t177 ur garden> show no trace of palmation!!? Bees at WILD St Johns Wort--Scabies, Cyanoglossum--Reseda wi
470t177 ild St Johns Wort--Scabies, Cyanoglossum--Reseda WILD very many Bees & Humbles--on Thistles many (cur
478z004 g it to be one animal In Australia I was assured WILD dog copulates <.> freely with tame: comes to ho
492q002 nct by seed-- Heartease. 6. -- Do not species of WILD Roses run into each other very much.-- Has not
492q002 - Has not some one written on them? 7... Are the WILD Bananas of Otaheite seedless;-- are all varieti
493q003 in comparison with Weigh skeleton of Tame Duck & WILD Duck, & then weigh their wing bones & see if re
494q004 al healthiness-- Fox's, Bears Badgers,-- How few WILD animals are propagated, though valuable as sho
495q005 ral Hybrids of Cabbages (7) Sow <daisy> seeds of WILD cabbage in VERY rich soil, will plants abort?,
495q005 case-- cross with cowslip pollen.-- as these are WILD varieties. Is any intermediate form found wild
495q005 e wild varieties. Is any intermediate form found WILD a The Leptosiphon densifolium «an annual» <slee
497q06v .-- (14) Examine pollen of those genera of which WILD hybrids have been formed. (15). What is History
502q11v e: for gradation in structure Compare flowers of WILD & tame carrot-- Parsley & Fennel. Verbena Compa
506q014 7¼-- in heigt 30 inches Examine Keel of Common & WILD Duck-- Black Duck & Penguin Henslow &c (36) Has
507q15v - & Humboldts spinning seed.-- (50) Any cases of WILD varieties plants growing together. under same c
508q016 organ. no answer?-- 3 Andrew Smith, about tamed WILD animals breeding at the Cape.-- ▯About two vars
512q019 eeder who raises many English birds-- will young WILD ones breed as well as,, as those already bred i
558m153 more correctly than this for C. Sphynx.-- In the WILD ass there is a curious drawing out of the side
592n105 e, and origin being so is not odd.; for instance WILD cattle & deer pursuing a wounded one.-- porpois
594n112 at Maer have learned that he is not dangerous-- WILD-bullocks would have fled equally if man had appear
595n121 ffie Wedgwood» April 28th 1840 was frightened at WILD beasts in Zoolog. Garden [blank] [not located]
615o038 lies to particular instincts of animals. even in WILD state; certainly to the domesticated.-- [LHC] N
203b136 1837, Mem. Sir F. Darwin cross breed boars were WILDER than parents. which is same as Indian Cattle..
356d071 mine▯ The facts about half breed animals being WILDER than parents is very curious as pointing out d
316c244 glimpses bursting on mind & giving rise to the WILDEST imagination & superstitions.-- +York's Minster
291c165 n but master.-- heredetary tameness as well as WILDNESS-- cf Sir J. Sebright.-- love. of man gained &
378d148 tting so much longer its Mental peculiarities. WILDNESS Reversion Q [not located] The present age is
279c133 s> facts stated.-- + +. Fully supported by Mr WILKINSON. = Milking heredetary, developemen of importa
320c275 tter in Quarterly Sir J. Sebright's Pamphlets WILKINSON on Cattle not abstracted Scientific Memoirs.
323c269 Philosophy -- Bell's Bridgewater Treatise -- WILKINSONS Egyptian remains skimmed -- Pliny Nat. Hist
439e145 ch saved him, would be the one encouraged-- » WILKINSONS Manners & Customs of the Ancient Egyptians V
025r019 cient, yet good» «(I suspect fragments of shells WILL generally be found to be old & dead)» «(I have
026r021 ins appearing a galvanic phenomenon, so probably WILL the Cleavage be There is a resemblance at Hobar
032r039 nglomerates [Fig. 2] {P} The action of sea A. B. WILL be to eat in the land in line of highest tidal
032r039 n the land in line of highest tidal action. this WILL at length be checked by increased vertical <hei
036r049 rved enlargement see its increased length. which WILL represent the dilatation, which dilated cracks
038r057 III.) whole of Etna series of coatings; hence it WILL be necessary to state all arguments for believi
038r057 low a field of fluid rock.--In the discussion it WILL be better not to refer to Lyell. but merely to
038r058 believe mere fissures filled up.--the appearance WILL here be the strongest argument:--¿ Consider cau
039r059 s of lava streams.--Urge not tilted strata.-- It WILL be well to urge the case of St Helena, where di
044r072 ustralia.. Oceanic Isles. Geology of whole world WILL turn out simple.-- Fortunate for this science.
045r077 ly oscillating as that of a spring) moves away.--WILL geology ever succeed in showing a direct relati
049r089 deed when do these dikes lead to a conical mass. WILL this conical mass be granite? Why not more prob
051r095 ifling circumstance determines whether an animal WILL adhere to a certain part. Apropos to question d
087a009 resembling S. America in Europaean latitudes.-- WILL it be supposed that the armadilloes have eaten
087a012 ing to Europaean risings. Pacific great land. -- WILL use argument of proof of slow corrosion of vall
090a023 king about average when several are given), this WILL give nearly 19 pounds average for each stone. t
092a029 iew of metamorphic in contradistinict to Volcanic WILL explain their solution. Athenaeum M. 516 1837 H
102a060 ous rocks of Ascension, each particle coated. &c WILL be aware how little common Gravity has to do wi
105a070 ge. C. Prevost.-- In Cordillera. a rush of water WILL account for filling up of valleys-- subsequent
105a071 rent spread out. by sea-- beach action -- no one WILL dispute. sea. once came to Mendoza--. Will they
105a071 o one will dispute. sea. once came to Mendoza--. WILL they introduce other causes to explain «alluvi»
106a074 flection with stream retaining its force, now it WILL be evident that deflected stream cannot retain

Page ***(Key Word)***

107a078 roken up.----My view of Volcanos &c &c This view WILL bear much reflection on method of cooling--Very
109a083 flats. argument for Heaping up.-- very good this WILL show effects.-- analogous to broad flat sand be
112a090 h they are «See Athenaeum. 1838. p 274. probably WILL be published in the Geograph. Journal.--» A mee
113a091 ds to cool down. & then in 50000 years the depth WILL be greater than <5000.> 400.-- These facts of S
117a101 remains formed near coast, limestone deep water. WILL bear on formations. during elevation & depressi
117a101 .-- My views of insensible oscillations of level WILL alone explain the immense amount of change whic
122a113 hels astronomy with oscillations of level.-- {P} WILL point {P} be the one which generally yields.--
122a113 l point {P} be the one which generally yields.-- WILL this not explain littoral mountains & volcanos.
123a116 nos.-- if so why not metals. The theory of veins WILL, I suspect be greatly aided by considering spac
127a126 the Irish Academy Vol 8. p. 118 water no--. oil WILL freeze if cooled in a closed globule of glass.
127a126 ssure in change of form as the result of heat.-- WILL it bear on central fluidity.-- do p. 137. Lord
128a128 ation,-- when subsidence takes place.-- Mountain WILL first fall-- the problem will be falling of an
128a128 place.-- Mountain will first fall-- the problem WILL be falling of an arch weighted in its centre.--
128a129 be falling of an arch weighted in its centre.-- WILL not abrasion of land on one side. produce subsi
128a129 rom tendency to regain statical equilibrium This WILL be only a modifying cause. {P} land protuberant
138a153 s table of every earthquake, during two years.-- WILL serve for comparison with the moon at some futu
139a180 phite to plants.-- grafting length of life &c &c WILL any inorganic substance cause such monstrous gr
146g007 alisbury Craigs» 25 degree perhaps most common-- WILL not curved form of hill be explained by my idea
158g080 ut 60 feet above Loch trace of this terrace «on <WILL> Granite ridge or a modified Granite ridge» at
172b007 from many enemies. so as often to intermarry who WILL dare say what result According to this view ani
174b015 at C. of Good Hope-- Marsupials. at Australia-- WILL this apply to whole organic kingdom, when our p
176b021 be about equally numerous. changes not result of WILL of animal, but law of adaptation as much as aci
176b024 tree is adapted for these three elements, there WILL be certainly points of affinity in each branch
178b030 other species have <come> «"air" of that place. WILL it be said, those have been there created there
179b034 certainly are according to all common language) WILL keep to their type: in animals so far removed,
181b041 sisters «in a state which does not increase» it WILL be chances against any one «of them» having pro
181b043 tebrata invertebrates-- Such on few on each side WILL yet present some anomaly & <the g> bearing stam
181b044 stamp of <some> great main type, & the gradation WILL be sudden-- Heaven know whether this agrees wit
182b046 y idea of origin of Quinarian system is true, it WILL not occur in plants which are in far larger Pro
182b047 rica. &c &c &c The new system of Natural History WILL be to describe limits of form. (& where possibl
187b063 ut about same time in such different quarters.-- WILL Mr Lyell say that some circumstance killed it o
188b067 ds-- (I) Gnu reeaches Orange river & says so far WILL I go and no further:-- Prof. Henslow says. that
189b073 sociated & non associated animals.-- & the story WILL be complete.-- It is absurd to talk of one anim
192b083 roducing any variety, although many of the seeds WILL go back.-- Get instances of a variety of fruit
194b093 re, in proportion instinct more. reason less. so WILL aversion be L. Institut «1837. No 246» a sectio
195b101 ated, then by the fixed laws of generation, such WILL be their successors.-- let the powers of transp
195b102 rs.-- let the powers of transportal be such & so WILL be the form of one country to another.-- let ge
195b102 .-- let geological changes go at such a rate, so WILL be the numbers & distribution of the species!!
195b103 ption of communication. or when country changes. WILL it said that Volcanic soil of Galapagos under e
196b106 hy did not fossil horse breed in S. America-- it WILL not do to say period unfavourable to large quad
198b113 e take birds animals, reptiles fish-- Conditions WILL not explain status (Perhaps consideration of ra
200b122 f course most rare, or when placed together they WILL breed.-- But what a character is this? Race per
200b124 expect equal fertility-- ditto in Plants.<==> It WILL be well to refer to Chamisso Vol III p. 155. ab
205b145 ss ram be put to horned ewe almost all the lambs WILL be hornless.--" does this apply to where same a
206b147 a for each lustrum.-- We may conclude that there WILL be a period though long distant, when of the pr
206b147 e present men (of all races) not more than a few WILL have successors. at present day. in looking at
206b148 s one with successors «for» centuries, the other WILL become extinct.-- Who can analyze causes, disli
206b149 sooner) & lastly perhaps some one single one.-- WILL not this account for the odd genera with few sp
210b159 oduce no difference. if they do.--there probably WILL be this relation also.‖Yes « Fox» ‖ The creativ
213b171 e, May you not breed, ten thousand grey hounds & WILL they not be greyhounds?-- Yarrell's remark abou
216b179 nclined to think that offspring of Negro & white WILL return to native stock (the cross often whiter<
216b180 (V Herbert on hybrids) thus act.-- Now the point WILL be to find whether know varieties in plants do
218b191 England, Scotland & other localities & each one WILL have a peculiar «constant» aspect. That is vari
220b199 re migratory species from warmer countries. When WILL this paper be published it will be curious.-- S
220b199 countries. When will this paper be published it WILL be curious.-- Some general statements about mun
222b205 ut intermediate in character, the same reasoning WILL allow of decrease in character. (which perhaps
223b209 s &c the species have not been much altered they WILL cross (perhaps more fertility & so make that su
223b211 ted to A. B. C. D.-- Destroy plants B. C. D. & A WILL soon form good species! The increased fertility
224b214 very distinct from Lamarcks» Without two species WILL generate common kind, which is not probable; th
224b214 common kind, which is not probable; then monkeys WILL never produce man. but both monkeys & man may p
225b217 e were a thousand intermediate forms.-- Opponent WILL say. show them me, I will answer yes, if you wi
225b217 iate forms.-- Opponent will say. show them me, I WILL answer yes, if you will show me every step betw
225b217 ill say. show them me, I will answer yes, if you WILL show me every step between bull Dog & Greyhound
225b218 g called <Phitophagous> omniphitophagous. But it WILL be said there are latent insects.-- as crows ag
225b219 termediate.-- Reference to Pig & Dogs. My theory WILL make me deny the creation of any new quadruped
226b220 Diemen's land Australia. England & Europe.-- It WILL be well worth while to study profoundly the ori
226b221 ative species. this must happen. & thus acquired WILL explain representative system Of this we see ex
230b239 s with difficulty permanently transmitted.-- <It WILL admit> a plant will admit of a certain quantity
230b239 rmanently transmitted.-- <It will admit> a plant WILL admit of a certain quantity of change at once.
230b239 rtain quantity of change at once. but afterwards WILL not alter. This need not apply to very slow cha
230b240 ach other), for if they are different then, they WILL be called species & mere producing fertile hybr
230b240 called species & mere producing fertile hybrids WILL not destroy that evidence, as so many plants pr
230b240 any plants produce hybrids, or else whole fabric WILL be overturned.-- Hence extreme difficulty, argu
231b241 > together, by elevations now in Progress, & you WILL have two. Tapir existing in East Indian Seas. M
231b244 e must suppose a multitude of small creations.-- WILL Dromedaries & Camels breed?-- As man has not ha
232b246 untries» finding a very hot day, «in one», oh we WILL take a day from the equator to add to the mean
232b248 ls of beasts varying in different climates Those WILL not object to my theory, those the philosophers
234b256 tion of birds of Iceland. --M. Gaimard, however, WILL settle this.-- Waterhouse says he is certain th
241c015 some birds of Tongatabou. & New Ireland.-- Gould WILL hereafter know about birds of N. Zealand L'Inst
248c030 ory, that «two» varieties of many ages standing, WILL not readily breed together: The argument must t
248c030 fected as with plants) no animals VERY different WILL breed together, so when we grant «(which can be
248c030 le,)» varieties may be made in wild state, there WILL be presumption that they would not. breed toget
248c033 rnal influences first affect external [for]m, so WILL the internal parts be of longest [consta]nt & t
248c033 ected by external circumstances> these therefore WILL be chiefly heredetary.-- If varieties «produced
248c034 re impressed in blood with time, then generation WILL «only» produce an offspring capable of producin
249c034 h as itself.-- therefore two different varieties WILL produce hybrids but not varieties. which are no
249c034 eties. which are not deeply impressed on blood., WILL cross & produce fertile offspring in the first
249c034 & produce fertile offspring in the first case it WILL either produce no of[fspri]ng or such as not ca
249c035 f Clenomys. case of replacing species. Dr. Smith WILL give me some capital information ¿Carnivora of
251c041 end to Denham Clapperton &c on Mammalia no doubt WILL all be included in Smiths work «do» Vol. IV p 2
252c045 of knowledge of time-- Analogy from three first WILL give one almost certain guide ∵ because time re
256c054 ecies in other animals.--? Forster on South Sea, WILL probably contain descriptions of domesticated a
257c057 ll Nature answers to the possibility.-- My views WILL explain no Mammalia in secondary-epocks, & deve
259c065 hat, cut a sheeps tail off plenty of times & you WILL have no tail (example probably not true).-- or
262c072 in typical structure?!! Whewell «in Comment/ few WILL dispute--» says civilization heredetary; ie ins
262c073 y things adapted--.-- These «aberrant» varieties WILL be formed in any kingdom of nature, where schem
262c073 . &c &c.--)-- & in round of chances every family WILL have some aberrant groups.-- but as for number
263c075 nd (although this may appear an absurd saying) & WILL never be conquered by anyone (if has any kind o
263c075 s of species & certainty of destruction; then he WILL choose & firmly believe in his new faith of the
263c077 he is not a deity, his end «under present form» WILL come, (or how dredfully we are deceived) then h
265c084 becoming white in winter of Arctic countries few WILL say it is direct effect, <of> according to Phys
265c086 ry origin & not adaptation. to its habits.-- Few WILL dispute that it is possible to have structure w
272c108 with similar habits. But the Horae Entomologicae WILL tell this.-- What peculiar conditions the Staph
274c116 of the wanderers out of the rest of the world?-- WILL this not agree with Waterhouse & <birds> Mammal
275c121 s are Blue Pouters & small Bald Heads Mr Yarrell WILL mention in his work I am sorry to find Mr Yarre
276c123 n. did for SPECIES-- between England & France. I WILL do with forms.-- Mention persecution of early A
277c127 put them as (a). (b) until data be given.-- This WILL aid in preventing the chaos.-- will point out w
277c127 iven.-- This will aid in preventing the chaos.-- WILL point out what to observe.-- will aid us in phy

Page ***(Key Word)***
277c127 g the chaos.-- will point out what to observe.-- WILL aid us in physiology, tell traveller what to ob
277c127 -- if he knows he has done least part.-- that he WILL not have brought home new species. until, he ca
277c128 ls, such as Cyrena This is reform which probably WILL be slow. but must take place-- such a classific
278c128 arison with other genera in other families.-- it WILL however be much <shu> surer, when false species
278c129 dson» a Vaillant, a D'orbigny has travelled this WILL be most difficult. Sub-genera so far may be eli
283c145 strange if creator had so created them.-- People WILL argue & fortify their minds with such sentences
283c145 let these only have progeny with species & there WILL be two genera.-- let short billed one, be exagg
283c145 gerated, & all rest destroyed far remote genera. WILL be produced. As we know from Ehrenbergh, there
285c151 eft out of case, that difference occurring.-- It WILL be necessary to show hybridity from few forms,
285c151 ons to islands close species, «on these isld» &c WILL probably upset it-- The space which one branch
285c151 ch of the tree if live occupied after its decay, WILL be occupied by the vigorous shoots from each br
285c151 connection of even distant ones) the characters WILL be first those of analogy, but will grow into a
285c151 e characters will be first those of analogy, but WILL grow into affinity.--but whether ever arrive at
285c153 perishes.-- Lyell has show such Physical changes WILL be unequally rapid, with respect to their effec
286c156 sage a difficulty, but after all a slight one It WILL be necessary from manner Fleming treats subject
289c161 him look at wings & orbits of penguin & then he WILL cease to doubt :Scales into Teeth in Bering Pik
290c164 examining series I cannot doubt laws of change, WILL be known.-- It appeared to me that half between
292c171 habitual action implies want of consciousness & WILL & therefore may be called instinctive.-- But wh
294c176 so would the change in animal be permanent.-- It WILL be easy to prove persistent Varieties in wild a
294c176 ieties in a wild state-- it may be said argument WILL explain very close Species in islds. near conti
294c177 ent. or two of the willow wrens &c &c. & analogy WILL necessarily explain the rest.-- Lonsdale says h
296c184 Hooker.-- my theory explains this. but no other WILL.-- St. Helena (& flora of Galapagos?) same cond
296c184 ant in form, others altered much.-- these others WILL be plants & land animals. & land shells.-- all
299c194 y; yet if T. palustris be sown in dry station it WILL for some generations come up so.-- there are no
300c197 of death acquired?. The S. American dung beetles WILL each become the fathers of many species.-- a fe
302c201 .-- the only connection between two such classes WILL be those of analogy, which when sufficiently Mu
303c204 at base of nails & over white of eyes,-- + + +, WILL he say creation is at end, seeing that Tertiary
303c204 over the viccissitudes of present animals.-- He-WILL be bold. I will venture to say unphilosophical.
303c204 situdes of present animals.-- He-will be bold. I WILL venture to say unphilosophical. L'Institut 1838
303c205 y to a certain extent,-- nothing but experience. WILL, tell us. when group is true,» there are no gen
303c205 ved <on> into one list is unfair, [moreover what WILL become of the future creations, if the list is
303c205 ect.--]CD. the creator so creates animals, «, it WILL be said» that although at any one there. are ga
304c206 s, which would render our knowledge a chaos: who WILL doubt this if series now existed from Man to Mo
304c207 existed series between apterix & other birds.-- WILL having many trifling characters, in common with
306c213 ifference with dogs of La Plata & Guyana people WILL say. not species.-- organs of generation a capt
306c214 f colours dependent on localities.-- -- Hamilton WILL give an account in his Travels in Asia Minor of
307c216 if so as she can be converted «into female», it WILL be splendid argument old female, turning into c
308c219 h glancing at, as she has no original ideas., it WILL show state of knowledge «Negroes existed since
310c222 ht to <an> so many animals of different types. I WILL confess my profound ignorance.-- but seeing suc
310c223 cquired & heredetary & such definite thoughts, I WILL never allow that because there is a chasm betwe
310c223 ion» Dr Royle seems to think Botanical Provinces WILL turn out not nearly so confined as now thought.
311c226 uding brain & other organs difficult to analyse) WILL not this separate facts about abortive organs &
312c227 irds in many families, «+very often in number 5» WILL have long tail.-- in raptorial birds, & tigers
314c238 us Florae Norfolkicae. 1833 Steph. Endlicker (He WILL give sketch of botany of islands of south seas
315c241 apan. I do not understand any new ones.-- Menoir WILL be published St. Petersburgh Academy Imperial--
335d013 to common goose.-- What has long been in blood, WILL remain in blood.-- --converse, what has not bee
335d013 emain in blood.-- --converse, what has not been, WILL not remain,-- yet offspring must be somewhat li
335d013 be somewhat like parents,-- therefore offspring WILL tend to go back, or have none-- the argument do
335d015 hat in proportion as things are long in blood so WILL they remain, a mule «being new species» will ha
335d015 so will they remain, a mule «being new species» WILL have no tendency to have offspring like parent,
335d015 ring like parent, but as they must like or there WILL be none, therefore a mule can have no offspring
335d015 quickly)» sometimes have similar offsprings, so WILL the worst mules (as real mule) have offspring,-
337d021 erely to classes where types exist for if so. it WILL be necessary to show how the first eye is forme
337d022 ars, who is curious &c &c &c who imitates.-- who WILL say there is distinct Creation required if he b
339d025 +++ Caledonia. two races of Men, but not plants» WILL it hold good.-- Thinks Temmink doubtful when he
343d037 ion of prominent ones in <latte> one: The latter WILL take place when Conditions are unfavourable to
347d049 lower than in old days: «for a very old variety WILL be harder to vary, & therefore more apt to be e
348d050 their variability.-- (Now caeteris paribus these WILL be the oldest) "The most important characters
349d051 ful.-- not till much knowledge is elicited.-- It WILL rest upon the discovery what characters VARY mo
349d051 <even in Cetionidae> «in the Cetoniadae»,-- when WILL ornithorhynchus come in circle?!!! p. 8-- Anoma
349d052 ently artificial, as interlopement of Marsupials WILL change all.-- & so on no one will settle number
349d052 of Marsupials will change all.-- & so on no one WILL settle number of primary divisions.-- Complains
349d052 ing to my theory, every species in any sub-genus WILL be. descended from one stock, & that stock with
349d053 rom one stock, & that stock with other subgenera WILL come from. common stock.-- all genera, common s
352d058 of> a group of species is made. father probably WILL be dead-- hence there is no central radiating p
355d066 then species are fertile; as long as opponents WILL «are» not «able to» tie themselves down, they c
355d067 dvantage of discovering laws is to foretell what WILL happen & to see bearing of scattered facts..--
357d073 t in New Ireland & continent since grown.-- This WILL explain. S. American case & Didelphis being Mun
363d101 y wild pidgeons have spangles on this part: this WILL be well worth working out.-- Study Temminks wor
364d103 The bird fanciers match their birds to see which WILL sing longest, & they in evident rivalry sing ag
366d111 does not tell against transmutation of species-- WILL it against genera.-- How long will the wretched
366d111 of species-- Will it against genera.-- How long WILL the wretched inhabitants of NW. Australia, go o
366d111 out extermination, & change of structure.-- When WILL the musquitoes of S. America take an effect.--
366d112 ncy for feathers to grow there «That Mutilations WILL not alter form may be inferred from Australians
367d112 ed, genetal organs by that co-relation of parts, WILL not be produced.-- Sept. 17th. Saw mule. appare
370d115 New Guinea.--!! S. America.-- Such difficulties WILL always occur if animals are thought to have bee
370d117 as hypothesis to other points. & see whether it WILL solve them.-- It is less wonderful that childs
371d128 roses, without circumstances very unfavourable, WILL <deteriorated> continue of same variety as long
378d147 esent in Young birds, one may be sure cock & hen WILL be alike-- I presume converse is not true for h
379d152 ver be put to any other breed else all the lambs WILL deteriorate.-- Lord Moreton's Case.-- When cows
379d152 of male & female.-- if there be one female, she WILL be free Marten. <Owen.> See Hunter's Owen-- In
380d154 & Guinea Fowl cannot be distinguished.-- A capon WILL sit upon eggs, as well as, & often better than
380d155 - if woman does not menstruate in the month, she WILL in 5 weeks.-- A Bull is never taken from his ow
380d155 field to bull a cow.-- -- a dog if led in string WILL not.-- some of the tigers.-- cat, though caterw
384d161 In speaking of generation alway put female first WILL not even a fruit tree or rose degenerate during
385d163 ngrel.-- I did not ask the question.-- His bitch WILL not take «& if she did take, probably would not
385d164 ant, half fowls.-- eggs fertile, but parent bird WILL never sit on them.-- May be just worth remember
387d169 two of third pair of same peculiarity breed they WILL have same influence as first pair + tendency th
389d176 each have offspring by same mother.-- one animal WILL fecundate female for several births, & even pro
390d179 n, it is not merely the too close animals, which WILL not breed, but the female at least (¿male?) loo
392d180 ous..) only simple form of life are monooecious. WILL ova of fishes & Mollusca «& Frogs» pass through
392d180 of Avitism Examine English weeds in Hot. Houses WILL they flower Make Hybrids with moths, where fecu
398e005 y, & know the amounts of change now in progress, WILL be the last to object to this theory on the sco
399e009 ntroduction to Cuviers Regne Animal No structure WILL last. without it is adaptation to whole life of
399e009 solely to childhood, or solely to manhood,-- it WILL decrease & be driven outwards in the grand crus
401e014 urchase seeds of any cabbage, where a great many WILL not return to all sorts of varieties, which he
401e017 ce <often> sometimes graft pears on apples. they WILL live but not flourish-- a medlar may be Grafted
403e023 ly» can be accounted for, on any other it is the WILL of God.-- Octob. 16th. A very strong passage mi
403e023 external form of varieties, do we suppose bones WILL not change in number. (even species do not this
406e035 against my modification of it-- Goat & Moufflon WILL not breed.-- p. do.-- Fish of Teneriffe. St. He
406e036 l cases: d degree p. 354-- The most vicious dog. WILL not attack any animal except, dog when absent f
407e042 ly & individually.-- I see clearly from F. R. it WILL be highly necessary to show that if species fal
409e047 a porpoise was not thougt overwhelming.-- yet I WILL not shirk difficulty-- I have felt some difficu
409e049 r not. He is present a social animal» a fact few WILL dispute, [although, that it was the sole object
409e049 spute, [although, that it was the sole object, I WILL dispute, when I hear from the geologist the his
410e051 Thinking of effects of my theory, laws probably WILL be discovered. of co relation of parts, from th
410e052 external conditions of country, most important & WILL be done to all countries,-- but naming mere «si
410e053 try.-- The traces of changes in forms of organs, WILL care little for species, except so far as wanti

Page
(Key Word)
410e053 laws of the corelation of parts in individuals, WILL care little, <in> whether the individual be spe
411e054 n, & laws of life,-- the end of Natural History, WILL be approximated to.-- Treating of the formal la
411e056 other means for individuals (Mem: transportation WILL be answered) one looks to analogy for cause in
412e058 ossil Mamm: from Poland. &c.-- Three principles, WILL account for all (1) Grandchildren. like. grandf
418e080 w of every animal being Hermaphrodite-- probably WILL recieve illustration from domestication of Mono
422e095 imals to become complicated although all perhaps WILL have done so from the new relations caused by t
423e099 & another in lower is very rare, the conclusion WILL be that our greatest formations <are> have been
430e117 ny of ours,-- even if extinction is denied.-- it WILL not account for all species. even if it will fo
430e117 it will not account for all species. even if it WILL for all.-- Varieties are made in two ways-- loc
434e129 y very slightly abortive & bed of female flowers WILL sometimes produce a few seeds.-- -- Ruscus acul
436e136 d World <If> It «may» be said, that wild animals WILL vary, according to my Malthusian views, within
436e136 ond these not.-- argue against this-- -- analogy WILL certainly allow variation as much as «the» diff
436e138 ches placed near, but not in sight of each other WILL sing till they drop off their perch.-- p. 101--
442e151 eneration may take place by gemmation «My theory WILL not admit this, now that tulips break by cultiv
444e157 ecies are the result of circumstances,;-- or the WILL of the Animal. p. 145. Seems to argue, that as
446e162 very close.-- «A fruit tree by certain treatment WILL suddenly send forth quantities of blossoms--» T
450e176 Sooloo.-- said to have been imported: shows they WILL propagate get dimensions-- do App. p 73 State o
451e178 ike man almost (this looks inaccurate C. D) they WILL catch the Malaria & die.-- On the other hand th
461t037 (or requiring nutrition)» to a certain amount it WILL be so handed down«(».. as mammae of men «callos
461t041 hat enormous changes of conditions, some species WILL undergo & yet remain adapted.-- it does away wi
463t057 ng of «many» facts with laws & their explanation WILL probably reject this theory-- (I must answer it
469t135 ch Bee visits 10 flowers in «minute» each flower WILL be visited in 28 minutes-- say then each flower
469t151 her blossomed together The seeds of these plants WILL be collected & resown.-- Humble 22 flowers of E
492q003 two half bred animals exactly alike be interbred WILL offspring be uniform.-- Mr Ford Has M. Sageret
494q004 varieties (9) Cross largest Malay with Bantam-- WILL egg kill Hen Bantam.-- Cross common Fowl with D
495q005 erent soils & temperatures & see what the effect WILL be.-- will seedlings vary much more than cuttin
495q005 & temperatures & see what the effect will be.-- WILL seedlings vary much more than cuttings &c (4) R
495q005 on flowers of other cabbages & see whether there WILL result hybrids-- (6) Dust flowers of one branch
495q005 <daisy> seeds of wild cabbage in VERY rich soil, WILL plants abort?, does it require successive gener
495q05a «dry» windy day, «flower garden on gravel walk» WILL drift many seeds= Necessary to answer Wiessenbo
495q05a ch in bell-glass-- sow these seeds & see if they WILL come up true-- whilst others are crossed.-- Are
496q006 ons. daisy. Fever-fuge Groundsil.-- gilly flower WILL break & become double.-- There is a double Crow
496q006 number appear equal-- & see whether proportions WILL vary, which will show that such proportions not
496q006 ual-- & see whether proportions will vary, which WILL show that such proportions not effect of Chance
497q006 d be known» & other odd places & see what plants WILL spring up which will show, how seeds are transp
497q006 dd places & see what plants will spring up which WILL show, how seeds are transported, or how long th
498q007 s.-- & Orchidacaeous plants no other case.-- (6) WILL plant accustomed to rich soil, when placed in v
498q008 an> such plants do not degenerate,-- as the Bees WILL mingle the infinitesimal varieties which must o
499q009 wman female branch At What distances from males, WILL female (a) Willows or Yews some poplar's produc
505q014 f seed & these Seeds of unimpregnated Cowcumbers WILL they seed.?-- (11) Abberley has planted seeds o
505q014 arge Asparagus: result? = failed to germinate 16 WILL plant some of the Thyme with abortive stamens b
505q014 rtive stamens by Terrace to see, whether stamens WILL be produced in individual plants 17 A dead-nett
505q014 individual plants 17 A dead-nettle in Hot-house. WILL it seed?-- (Skim through Penny Cyclopaedia) Abb
505q014 says that some Bees are smaller & more vicious. WILL try to get me some to look at;-- Was once offer
506q015 remember at all. (37) Any cases of plants. which WILL not produce seed in this country-- where cause
511q018 ies, having somewhat of different appearance.-- {WILL introduce it in work} Whether <Yar> knows wheth
512q019 ons some breeder who raises many English birds-- WILL young wild ones breed as well as,, as those alr
513q21. es, as dioecious plants, when crossed R. BROWN-- WILL pollen act on any flower before stigmas expande
514q21. ntains of N. America similar to Lapland Plants-- WILL get answer= Is pollen of cultivated Orchis & As
515q022 ur of beak, taking after male & female parent.-- WILL they grow up in other respects different?--. Im
516q23v xperiments not connected with Species Theory (1) WILL an extract of peat do to preserve fungi or anim
516q23v aying wood absorbs oxygen & forms Carbonic Acid. WILL this bear on Petrifaction?-- [blank] Questions
526m027 over these things, one doubts existence of free WILL every action determined by heredetary constitut
526m027 h by the same means & therefore properly no free WILL.-- we may easily fancy there is, as we fancy th
526m027 - chance governs the descent of a farthing, free WILL determines our throwing it up.-- equall true th
526m028 n mind,-- this may be worth thinking over.-- it. WILL perhaps show differences between memory & imagi
526m030 qualities become heredetary.-- When a man says I WILL improve my powers of imagination, & does so,--
526m030 rs of imagination, & does so,-- is not this free WILL,-- he improves the faculty according to usual m
526m030 sual method, but what urges him,-- absolute free WILL, motive may be anything ambition, avarice, &c &
527m031 e of mind owing to.--<)>-- I verily believe free-WILL & chance are synonymous.-- Shake ten thousand g
527m031 Shake ten thousand grains of sand together & one WILL be uppermost:-- so in thoughts, one will rise a
527m031 r & one will be uppermost:-- so in thoughts, one WILL rise according to law. How strange <all> «so ma
529m049 he young got on. this nest the cat could If cats WILL «ever» eat little birds, this most curious inst
531m050 think that the importance of the event by itself WILL make it to be remembered. whereas it is the imp
532m055 d recollect these same things from any effort of WILL whilst their minds were sound. Caroline tells m
534m061 to say. Brain per se thinks is nonsense; yet who WILL venture to say germ within egg. cannot think--
535m069 risea described [not located] as first caused by WILL of Gods. «or God» secondly that these are repla
535m069 nature should be «universally» thought to be the WILL of a superior being; whose natures can only be
535m070 d out. When one sees this, one suspects that our WILL may <be> «arise from» as fixed laws of organiza
536m070 stiff, & walks hard.-- «He cannot avoid sending WILL of action to muscles, any more than «prevent» h
536m072 idea of beautiful forms.-- With respect to free WILL, seeing a puppy playing cannot doubt that they
536m072 a puppy playing cannot doubt that they have free WILL, if so all animals.. then an oyster has & a pol
536m072 idable & only to be changed by habits). now free WILL of oyster, one can fancy to be direct effect of
536m072 s senses give it of pain or pleasure, if so free WILL is to mind, what chance is to matter «(M. Le Co
536m072 chance is to matter «(M. Le Compte)»-- the free WILL (if so called) makes change in bodily organizat
536m073 ge in bodily organization of oyster. so may free WILL make change in man.-- the real argument fixes o
536m073 hoice which at that time organization gave me to WILL-- Verily the faults of the fathers, corporeal &
537m074 er from real caprice. it is chance, which way it WILL be, but yet it is settled by reason.--¶ How slo
537m077 nce or instinct may be most firmly fixed, but it WILL not prevent other beings engrafted.-- No one dou
537m077 hurts conscience more than other.-- A Scotchman WILL his country or Swis.-- it may be answered effec
538m078 other cases somewhat analogous, & which I think WILL lead to fact of old people singing songs of the
538m081 acts showing what a train of though[t] action &c WILL arise from physical action on the brain, render
539m083 ings done in the other habitual state because it WILL (without direct consciousness?) change its habi
539m084 ysic must flourish.-- He who understands baboon <WILL> would do more towards metaphysics than Locke A
541m093 rgive his enemy & not wish to strike him, but he WILL find it far more difficult to to look tranquil.
541m093 a man & say nothing, but without a most distinct WILL, he will find it hard to keep his lip from stif
541m093 ay nothing, but without a most distinct will, he WILL find it hard to keep his lip from stiffening ov
542m093 h himself, & though dreading to say so, his step WILL grow erect & stiff like that of turkey.-- he ma
542m093 ly wish [not] to do it, but an involuntary laugh WILL burst forth, this & yawning. (common to other a
549m118 to dog losing his puppies-- This looks like free WILL.-- V. last page. A healthy child is «more» enti
549m119 very great, that every one who has tasted them, WILL think the sum total of happiness greater. even
549m120 tion of daily <happiness> «pleasure». A wise man WILL try to obtain this happiness. though he sees so
549m122 gh believing it to be true, & then acting on it, WILL add to happiness.-- Men having some instincts a
550m126 24. Curious passages showing how easily chance & WILL of Deity are confounded.-- well applicable to f
550m126 Deity are confounded.-- well applicable to free WILL. Mayo. Philosop. of Living p. 293. Animals "hav
552m132 , & not as Paleys rule is those that on long run WILL do good.-- alter will in all cases to have & or
552m132 is those that on long run will do good.-- alter WILL in all cases to have & origin as well as rule w
552m132 ll in all cases to have & origin as well as rule WILL be given.-- Descent of Man Moral Sense Mitchell
553m135 ter) consider the thunder & lightning the direct WILL of the God (<thus> & hence arises the theologic
553m139 te of Peppermint.--» Perfect understand voice.-- WILL do anything.-- will take & give food to Tommy,
553m139 Perfect understand voice.-- will do anything.-- WILL take & give food to Tommy, or anything of any s
554m140 w Tommy picking his nose with «a» straw.-- Jenny WILL often do a thing, which she had been told not t
554m140 ad been told not to do.-- when she thinks keeper WILL not see her.-- <but is> then knows she has done
554m140 her.-- <but is> then knows she has done wrong & WILL hide herself.-- I do not know whether fear or s
554m140 e.-- When she thinks she is going to be whipped, WILL cover herself with straw, or with a blanket.--
554m141 to go in with a stick, if he drops it, the bird WILL fly at him-- Knowledge.-- Sept. 13th It will be
554m141 ird will fly at him-- Knowledge.-- Sept. 13th It WILL be good to give Abercrombie's definition of "re
558m150 c nerve. most subject to habit, as being less so WILL.-- May not moral sense arise from our enlarged

558m153 tril. when passion commences.-- <All> Nearly all WILL exclaim, your arguments are good but look at th
565n007 is listening is mouth open to hear well «as one WILL perceive if in night trys to listen to growl of
565n007 does not recognise look of suspicion, even child WILL do so.-- Contempt look obliquely so does dog. w
565n008 progress.<--> the hypothesis of opposite muscles WILL want much confirmation. A grave person close th
566n012 such a thing as thunder. would be placed to the WILL of God. Zoology itself is now purely theologica
568n017 al virtue, that. is action which experience shows WILL be for general good, or in case of any fantasti
569n020 uphing being necessary sounds... not produced by WILL <by> but by corporeal structure.-- Devotional f
569n022 I am very sorry so it is"-- does not accompany I WILL not. I am sorry I cannot.-- Expression leave «t
570n024 s & how he gulps in air.-- Again a master says I WILL see you damned first." the man shrugs his shoul
573n038 erson with sore-throat told not swallow spittle. WILL have involuntary flow & desire to swallow.-- te
573n038 to swallow.-- tells himself not to turn in bed. WILL turn in bed.-- in case spittle, effect of thoug
573n039 subject, even when wishing not to flow-- flow it WILL.-- My father told Miss. C. of the bad conduct o
575n045 of Ornith. Biog. case of Newfoundland dogs. who WILL not enter water, till he sees. whether birds ba
575n046 77) remembering things of youth, when new ideas WILL not enter. is something analogous. to instinct,
577n049 instance".-- for it can be shown that the life & WILL of a conferva is not an antagonist quality to l
577n049 bility against free action.-- on my view of free WILL, no one could discover he had not it.-- The mem
577n050 e establishment of this principle of Association WILL help my theory of sensitive Plants Habitual act
578n053 sh not to do so.= = How directly personal remark WILL make any one blush.-- Is there not some saying
578n054 e carries on, by association, the question, "one WILL anyone, especially a women think of my face,"?
579n058 squinny-- when they consider profoundly,-- this WILL be curious if it is so.-- frown with grief,¿ bo
579n060 come.-- it is an excitement of surface under the WILL? of the animal.(-- Jan 21. 1839. Herchel's Disc
582n067 tate.-- N B. According to my view marrying late, WILL make average of life longer.-- for short-lived
582n067 of life longer.-- for short-lived constitutions WILL then be cut off.-- <Horses> Colts cantering in
582n067 esis whatever]CD an animal may so far be said to WILL to perform an instinct that it is uncomfortable
582n068 rmance of a number of prearranged actions, which WILL bring about a certain result, while the creatur
582n068 ctions neither knows nor intends the result they WILL effect,.--" this not wholly true, for we must g
586n082 with others in building comb-- My faculty often WILL turn out to be instincts, & so in some senses,
590n094 faculty, a power, quite distinct from self. «or WILL» [not located] <& other cows--> Mr. Hamilton on
591n098 xtraordinarily cowardly.-- the other one nothing WILL frighten-- hence variation in character in diff
593n109 o by association receives pleasure. This [blank] WILL not do for insects. if this view holds good-- t
595n117 cited, shows their power of imagination-- for it WILL not be allowed they can dream, & not have day-d
604o017 on insanity.-- Prevailing idea. owing to loss of WILL.-- chiefly excited by passive emotions.-- Canno
606o023 e imagination through the medium of the eye"; he WILL allow the secondary pleasures of harmonious col
607o025 his law in man in somnambulism or insanity. free WILL (as generally used) is not there present, but h
607o025 ctions are not determined by what is called free WILL, but by strong invariable passions-- when these
608o025 weak, opposed & complicated one calls them free WILL--the chance of mechanical phenomena.-- (Mem: M.
608o026 ith organization The general delusion about free WILL obvious.-- because man has power of action, & h
608o026 o aid motive power.--if incorrigably bad nothing WILL cure him' 3) disgusted. with them. Yet it is ri
608o027 per). nor ought one to blame others.-- This view WILL not do harm, because no one can be really fully
608o027 lays in doing good & being perfect, & therefore WILL know his happiness lays in doing good & being p
608o027 ruth. except man who has thought very much, & he WILL not be tempted, from knowing every thing he doe
608o027 of himself to do harm.-- Believer in these views WILL pay great attention to Education.-- 4) These vi
609o030 of moralists: one says our rule of life is what WILL produce the greatest happiness.-- The other say
610o034 & not irritability, which at least shows a local WILL, though perhaps not conscious sensation. [RHC]
612o035 ts. (But I include irritability for that require WILL in part. ¿Why more so than movement of sap. or
612o035 s. arms of polypus, show either local or general WILL, & stomach likewise «does». [LHC] ¿ in Corallin
613o035 ct. where many such objects are present, & where WILL directs <to> other parts of body. to do such.--
613o036 ed The meaning of Words, must be made out Reason WILL Consciousness Definite instincts being acquired
616o038 nvoluntary, <application in> «ejection only has» WILL: there must be cases of secretion being some ti
616o038 e cases of secretion being some time governed by WILL in some animals, involuntary in others. [1)] Wh
616o039 Why may it not be said that thought perceptions WILL, consciousness memory &c. have the same relatio
619o043 t with respect to himself it would be remorse as WILL be presently shown.-- This then is moral approb
620o044 ental capacity of calling up past sensations, he WILL be forced to reflect on his choice: an appetite
621o046 If his passions strong & his instincts weak. he WILL have many struggles, & experience only will tea
621o046 . he will have many struggles, & experience only WILL teach him, that the instinctive feeling in its
621o047 ir J. M. talks too much about the continguity to WILL. (a) The origin of passions too strong for our
621o047 taste, for when grown up often conquers it). It WILL be only rarely that it thinks that nasty, which
621o047 & wrong in action, so a child may be taught, or WILL acquire from seeing conduct of others, the feel
621o048 t or wrong.-- [RHC] 7) Hence, what parents think WILL be good for the child on the long run, & for th
621o048 s the parents are instinctively benevolent) they WILL teach to be wrong or right; this teaching may b
622o048 riously modified by circumstances of country, so WILL the conscience in these cases.-- Those instruct
622o048 rmly performed by the teachers & all around him, WILL be paramount,-- hence the law of honour. & the
622o049 on of law of honour from Paley [RHC] Anyone, who WILL reflect must feel, how like to injured conscien
622o049 eeling may be directed towards one or more.-- It WILL be hard to discover this, for the different rac
623o050 g under certain conditions, in this world. they <WILL conform to the law,> «can only be such, as are
623o051 nerations. (& under unknown conditions) (for pig WILL not so readily attain instinct of pointing as a
623o051 nce.)-- & only that which is beneficial to race, WILL have reoccurred'. NB. Until, it can be shewn, w
624o051 > changes «in accordance to beneficial tendency» WILL most readily affect. the instincts, for they ar
626o052 intosh's theory: the remarks about "contact with WILL" is unintelligible to me.-- conscience regulate
626o052 s. shewing that instinct cannot be said to guide WILL. as bird building nest, but supplies it-- insti
626o052 ing nest, but supplies it-- instinctive feelings WILL doubtless lead to similar actions which in prio
626o052 action on body.-- This feeling, when instinctive WILL lead to action.-- the passion rising from weari
627o053 into some operation of intellectual faculties-- WILL Eugenius allow this moral obligation? [2] [The
628o054 th that line of conduct, & if taught rightly, it WILL be for the general good, that is, the same caus
628o054v ication, whilst <the> no desire of gratification WILL check the consciences desire for virtue.-- (I e
628o054v fallacy here.-- at least point of «false» honour WILL stop all wish to gratify <it>-- anything contra
628o054v s <boshes> «nonsense»-- My theory of durableness WILL explain it.-- Would not the maternal affections
629o55v on by habit may be thought wrong.-- & conscience WILL imperiously say so, & produce shame & remorse--
629o55v intervene between this kind of conscience & the WILL, though «this» conscience does between the desi
629o55v ugh «this» conscience does between the desires & WILL?)) (2) It is other question what it is desirabl
632j53r s of God Macculloch. Attribs of Deity. Vol: I it WILL be better always to refer to the author if I us
633j53v One limit to the transmission of abortive organs WILL be as long as they are not detrimental.-- p. 28
634j55r of structure in classes-- as resulting from the WILL of the deity, to create animals on certain plan
635j55r oretells nothing» because we know nothing of the WILL of the Deity. how it acts & whether constant or
638j58v & the species, like 10,000 others, perish. & who WILL dare to say that this is an infringement on the
639j28v vive. in ten thousand years the long legged race WILL get the upper hand. though continually dragged
641j29r continents   <bri> abounding with species-- there WILL be a balance, continents have been split up.--
641j29v tiquity & extinction of such forms-- these views WILL bear on geology-- There is an analogy between f
148g023 es of them all along <Glencoe>.-- towards Fort WILLIAM yet in Glencoe in parts no trace of them- Mem
225b220 ery good on opposite tendency.-- Study Ellis & WILLIAMS. zoology of South Sea islds. any animals?-- I
399e010 y bred, they have not lost power of producing WILLIAMS. Narrative of Miss. Enterprise, p. 497. Vampi
471tf10 amm. of Europe-- Large Lizards in Navigatores. WILLIAMS. Narrative of Missionary enterprises Dr Andre
506q015 es he think of Dr. Flemings statement of Sweet WILLIAMS & Stocks, being propagated many years, by cut
225b216 ew.-- With respect to how species are. Lamaks "WILLING" doctrine absurd. (as equally are arguments ag
259c063 y peculiarity. or any race of plants--Lamark's WILLING absurd, ∵ not applicable to plant, Epidemics o
544m103 ars that, but yet scarcely really moves.-- the WILLING therefore so. it is absurd, as all the other perceptio
567n012 ecessary notion it is connected with <our> the WILLING of the simplest animals, as hydra towards lig
612o035 cation from without, & gives wondrous power of WILLING. These +willings are common to every animal in
612o035 l instinctive and unavoidable. -- +Can the word WILLING be used without consciousness, for it is not e
612o035 tissue and are subject to accident; the sexual WILLING comes on period of year as much as inflorescen
613o036 instinct to plants. but surely instincts imply WILLING, therefore word misplaced The meaning of Words
612o035 out, & gives wondrous power of willing. These +WILLINGS are common to every animal instinctive and un
612o035 vident, what animals have consciousness. These WILLINGS have relation to external contingencies, as m
312c232 riation" produced in shortest possible time. Mr WILLIS Long eared little dogs, I am told, go to heat,
338d024 - -- Says Negro-- thick skinned My hairdresser (WILLIS) says «black» «that strength of» hair goes wit
385d163 .-- «Men giving milk--»» Sept. 25th Young man at WILLIS «Grt. Marlborough Str, Hair dresser, assures m
389d173 ers Eoeconomy So with inter-breeding as told by WILLIS Q Proved facts relating to Generation One copu
437e139 dropped having wounded the bird. p. 124-- Mr WILLOUGHBY found a dead lamb «& hare» by the side of Ea

```
207b151 elena. as. Pineaster & Mimosa called Botany Bay WILLOW V. Dr Royle introductory remarks to Himalaya M
230b241 close (take Europaean birds. Mr Goulds' case of WILLOW wren) & others varying in wild state to show t
276c125 country are not habits different, (Mem: Gould's WILLOW Wren) but where close species inhabit differen
294c177 on crow & rook formed by descent. or two of the WILLOW wrens &c &c. & analogy will necessarily explai
363d102 ubation & other peculiarities.-- (Mem.-- Goulds WILLOW Wren.--) (Goulds story of Water-Wagtails mista
435e130 nly in genera <are> have this structure.-- Some WILLOW trees have been observed to change their sex,--
286c156 ane Fleming Quarterly review says nat: fam: of WILLOWS contains many Linnaean genera.-- Now are the c
498q008 Ants-- Enquire (13) Do any of same species of WILLOWS grow in same situation & flower at same  time.
499q009 ve than of foreign-- Eyton has such a grove of WILLOWS.-- (14) Bowman female branch At What distances
499q009 At  What distances from males, will female (a) WILLOWS or Yews some poplar's produce.-- (15) Would Ye
544m103 dreamt  of <goo> feasts of good food-- The mind WILLS to do this & hears that, but yet scarcely reall
564n004 nce, feeling in his heart those rules, which he WILLS to give his child.-- Octob 3d. Was told by W of
260c067 ring crys of hunger of little bird, in same way WILSON (p. 5). describes many kinds of bird uniting t
484z014 deed Caricaridae wanderers.--?? in N. America?? WILSON N. American Ornitholog must be studied  before
610o031 irected his letter, & I observed he had written WILSON & pointed it out; he was astonished, & said ho
610o031 how  very odd.-- --could not think what had put WILSON into his head.-- remembered, that he had. look
610o031 he  had. looked in direction book under head of WILSON, referred to Robert & found his Christian name
610o031 ferred to Robert & found his Christian name was WILSON!!-- How curious an inward. unconscious memory.
260c067 aptation.-- No Carrion Vultures in Australia!! WILSONS Ornithology, Vol III. p. 226 Wilson's Ornithol
260c067 stralia!! Wilsons Ornithology, Vol III. p. 226 WILSON'S Ornithology, D'orbigny, Spix, &c might compar
260c069 of  bird in district. which they frequent!!?-- WILSON'S American ornithology a mine of valuable facts
325c267 & Montagu on birds (facts about close species) WILSON'S American Ornithology Read Aristotle to see wh
047r081 oss Pacifick.--excepting in number of waves & in WIND, instead of sea's bottom being in motion what d
408e044 character.-- worth going there for.-- «Gales of WIND would blend species» Buckland. Reliquiae Diluvi
431e123 the crossing of mosses & all others by action of WIND difficult.-- Cline on the breeding of  Animals,
495q05a in some sticky stuff in flat places & see whether WIND, on «dry» windy day, «flower garden on gravel w
510q017 h of Siberia, no sea-current, icebergs travel by WIND. Aug. St. Hilaire Bot. p. 787. position of embr
521m009 rdener.-- My F. then asked Mr C. to come to the WINDOW & pointed out the Gardener & said, who is tha?
523m017 Case of Mrs. C. O. who threw herself out of the WINDOW to kill herself from jealousy of husband conne
085a001 n: Transact. Vol. 8. p. 288. Salt deposited on WINDOWS of houses. & trees all injured on Eastern side
378d148 ey were "half wild-half tame, they came to the WINDOWS to be fed, but still they have a wariness abou
022r008 f metereology with respect to the discussion of WINDS & storms:--«in Volney's travels also» Dampier's
510q017 on  double creations & where? How are current & WINDS in Antarctic ocean: are they from West, like as
405e031 probably did not.-- Octob. 25th. I observed in WINDSOR Park.-- the «Fallow» Deer. which were of a nea
453e182 g attested account of fall of fish in India.-- WINDSOR Earl-- Eastern Seas p. 229. Believes the <Rhin
571n030 k.-- I was much struck with this, when viewing WINDSOR Castle which rises naturally & hence sublimely
096a041 e??? The fact of Galapagos Isld. steep side to WINDWARD in allusion to St. Helena discussion. Mr Bray
495q05a uff in flat places & see whether wind, on «dry» WINDY day, «flower garden on gravel walk» will  drift
239cIFC ns,-- «with specks» which cross & keep colour on WING Effects of colour on parent, white room How are
255c051 ered many times, but all have had feathers.-- if WING totally obliterated. This may account for perma
265c086 after  seeing beetle with wings beneath soldered WING-cases-- Yet these wings may be of some use,-- N
272c109 Birds of Paradise. Trogons.-- the one feather in WING the curious feathers in tail of Edolius.-- Rema
273c112 allow hawk, .milvulus,, may catch insects on the WING & pratencole (¿connecte[d] with Chionis), yet t
304c207 f Synallaxis. trifling characters as red band on WING show to be from one parent.-- same forms of bea
352d059 us  it was for.-- Most important law.-- Penguins WING perhaps not abortive???. Apterix certainly.-- L
459t013 er than the penguin & the bright feathers on its WING resembled the drake.-- another of same half bre
493q003 ton of Tame Duck & Wild Duck, & then weigh their WING bones & see,,if relation is same good, avoids ef
509q017 tibia & fibula: in Man any abortive bones??? do. WING in Apteryx ||| no as Os Coccygis-- Turbinated bo
558m153 arch raises neck & depresses chin-- strikes with WING arches wings-- as does black Swan.-- Goose do a
204b141 birds.-- «W. Fox. knew of case of male widgeon, WINGED & turned on pool, first season bred readily wi
468t105 ma rather large & rough-- flowers common-- many WINGED thrips, covered with pollen-- «Thrips» about a
575n045 l he sees. whether birds badly wounded, or only WINGED.-- fetches two birds out at once.-- Old People
233b251 of Hope.-- His book Probably worth studying.-- WINGLESS birds S. Continents-- Ostriches. Dodo. Aptery
189b070 y in time as in space. (Mem: Galapagos). Little WINGS of Apteryx Dacelo & Kingfisher same colours Str
192b084 x not having been determined.-- so with useless WINGS under elytra of beetles.-- born from beetles wi
192b084 der elytra of beetles.-- born from beetles with WINGS.& modified.-- if simple creation, surely would
201b129 on  external influences.-- For instance he says WINGS of bat, are from external influence.--...... He
240c003 ould be called genera., yet retains markings of WINGS like the wild rock pidgeon.-- fact analogous to
255c051 y  suppose longest part of structure.-- shape of WINGS have altered many times, but all have had feath
261c017 Gould  says wherever any mark like red patch on WINGS of Furnarius, Synallaxis &c &c. sure to unite t
265c086 cture without habits-- after seeing beetle with WINGS beneath soldered wing-cases-- Yet these wings m
265c086 h wings beneath soldered wing-cases-- Yet these WINGS may be of some use.-- Nature is never extravaga
265c087 ravagant though clearly not of the use to which WINGS are generally applied.-- Therefore argument not
265c087 argument not destroyed even if these shrivelled WINGS could be shown to be of some use. If we only ha
273c110 olouring of plumage (as «black & white» bars on WINGS of trogons are lengthened rump feathers.--:-<th
273c111 vulus-- Even flying woodpeckers, with powerful WINGS, but tail stiff.-- Swallow & goatsuckers likewi
275c121 h have such different skulls, but same marks on WINGS are Blue Pouters & small Bald Heads Mr  Yarrell
289c161 scales. passing into each other let him look at WINGS & orbits of penguin & then he will cease to dou
307c215 ae in Men, capable of giving milk!)» rudimentary WINGS. so nature can produce in sex, what she does in
307c215 Apterix This is important for if these abortive WINGS in the female are allowed to the fully organize
307c215 n the female are allowed to the fully organized WINGS of the male rendered abortive in the womb.-- if
341d029 elvis & was not adaptive structure, like little WINGS of Auks which does not make that bird a Penguin
341d030 o little Movement.-- nocturnal crawling bird.-- WINGS reduced to rudiment.-- clavicle scapula &c stro
353d060 allied whether «like» parent stock, or not. Now WINGS for flight-- therefore ostrich not. The peculia
380d153 without knowing whether if kept they would have WINGS.--).-- Seep p. 84. Hens «like»-- Cocks from eff
440e147 ocked at absence of final cause mammae in man & WINGS under united elytra The law of <growth» «genera
464t065 here have been at least other birds, with small WINGS, & surely the Apteryx is more closely allied to
467t103 on «I saw Bee go to two species of Lupine,» two WINGS. & when the Lupine flower is perfectly ripe & p
467t103 is wonderful {a} how the Humbles force down the WINGS most violently: in Beans the wings seem beautif
467t103 rce down the wings most violently: in Beans the WINGS seem beautifully to protect sheath {a} In all t
467t104 ay= In Lotus corniculatus saw Humble press down WINGS which ejects pollen from tip of sheath.-- «Also
467t104 east-- {b} pressing either one or both of Pea's WINGS, stigma & mass of yellow pollen protrudes at sh
470t178 , 1854, Sept.) In Spanish Broom by pulling back WINGS, pollen is ejected with violence in shower On m
484z016 creeper  of same plumage.-- general red mark on WINGS all-- Spix has described Philedon. allied to
484z016 s strictly American. Colouring on under side of WINGS It would be interesting comparison to find  how
509q017 o as Os Coccygis-- Turbinated bones? False ribs WINGS of Apterix: clavicle in-? Combs in combless Po
534m063 Darwin  mistaken in saying common wasp cuts off WINGS of flies from intellect. but it does it  always
558m153 eck & depresses chin-- strikes with wing arches WINGS-- as does black Swan.-- Goose do all species pu
563n001 is evergreens they seek Cock Pheasant claps his WINGS before? crowing & only in breeding season &  on
532m053 ubts whether young babies start.-- ▌ If children WINK. it is instinct Fear must be simple instinctive
560mIBC n movement of muscle) very early in life Do they WINK, when anything placed before their eyes, very y
606o022 which  he thinks is a better definition than WINKLEMAN'S. who says it is simplicity with grandeur  of
557m147 ooth bone contrary to wrinkled: (a horse when WINNOWING & pleased pricks his ears?--).-- How is expre
030r035 fathoms  2 & 3 miles from shore. V. Chart) Every WINTER torrents must bring much vegetable matter from
265c084 ion.-- Ermine, ptarmigan hare becoming white in WINTER of Arctic countries few will say it is  direct
289c162 & lay two sorts of eggs-- one remaining through WINTER.-- «It would be curious to know whether a variet
437e138 t of England stationary, in southern stays only WINTER.-- Jays & chaffinches sometimes migratory.-- p
478z003 de  long et volent. p. 208 Fleas only appear in WINTER in Paraguay p 207 Slight notice on habit of Ig
485z018 ical forms extend further north, because during WINTER they can bear the cold when torpid.-- On  this
472s02r s, under sigma & draw it out over & over again & WIPE off pollen. (as a needle becomes covered) so wh
262c072 » says civilization heredetary; ie instincts of WISDOM virtue? «like senses of savages» (How come its
317c248 lbourne." many references very good. also "Rays WISDOM of God." Often refer to these.-- Also some few
321c275 ash» skimmed Macleay's Horae Entomologica Ray's WISDOM of God references at end-- The British Aviary.
321c270 . (read remainder) when out [not located] <Rays WISDOM of> Lisiansky's Voyage round world. 1803-6 Not
638j58v dare to say that this is an infringement on the WISDOM or Providence. when whole rocks nay very mount
549m120 arts» portion of daily <happiness» «pleasure». A WISE man will try to obtain this happiness. though h
620o044 se.-- He reasons on it & determines to act more WISELY other time, for he knows that the instinct. (o
227b227 er causes of change.-- the manner of adaptation (WISH of parents??) instinct & structure becomes full
```

228b231 to consider our equals.-- «Do not slave holders WISH to make the black man other kind?» Animals with
409e048 ent. My theory gives great final cause «I do not WISH to say only cause, but one great final cause,--
511q018 cter believe No or did result appear without his WISH Has since recrossed this breed.-- Have secondar
531m053 ing must be habitual «involuntary» movement from WISH to avoid some danger-- but it is instinctive be
536m073 should be grateful if it were pointed out.-- My WISH to improve my temper, what does it arise from b
541m093 .-- a man «insulted» may forgive his enemy & not WISH to strike him, but he will find it far more dif
542m093 d, he need not express it, he may most earnestly WISH [not] to do it, but an involuntary laugh will b
542m095 rprise-- viz seeing something obscurely with the WISH to make it out?-- Seeing a Baby (like Hensleigh
563n002 so & so for his interest, & found he disobeyed a WISH which was part of his system, & constant, for a
563n002 which was part of his system, & constant, for a WISH which was only short & might otherwise have bee
566n010 stity in women.-- rationally explained.-- on the WISH to support a wife a ruling motive.-- Book IV, C
573n036 e governed by laws,, but the smallest insect, we WISH to be created at once by special act, provided
578n053 you <think> «fear» you shall not have e---n, «or WISH extraordinarily to have one» you wont. ==)== No
578n053 =)== No surer way to blush, than particularly to WISH not to do so.= = How directly personal remark w
628o54v change produced-- 4) p 38 Conscience checks the WISH to <other> outward gratification, whilst <the>
628o54v - at least point of «false» honour will stop all WISH to gratify <it>-- anything contrary to it]CD NB
629o54v tinct> «conscience») equally <prefe> destroy all WISH of outward gratification,-- see what cases Mack
286c154 ing» by making slave of his fellow black, often WISHED to consider him as other animal-- it is the wa
563n002 that he constantly compared his impressions, & WISHED he had done so & so for his interest, & found
576n047 friend, or wife.-- yet he would ever repent, & WISHED he had lost his life in doing so.-- nor would
389d176 ndance of life, like tendency to budding, which WISHES to throw itself off.--» as might be inferred f
431e122 esh Creation. the gardener separates a plant he WISHES to vary-- domesticated animals tend to vary. M
549m118 rely happy (contentmt is different it refers to WISHES for future) than perhaps well «regulated» phil
570n024 to reply. he would throw back his shoulders. he WISHES to show, he is determined not to say anything.
606o025 ct of bodily organisms-- one knows it, when one WISHES to do some action (as jump off a bridge to sav
617o39v ought to be clearly comprehended by anyone who WISHES to fully understand this subject, but the answ
628o54v it]CD NB. the very end of conscience is stop to WISHES of passion &c. whilst the passions have no rel
225b219 ssils in old rocks for same purpose.!! Can the WISHING of the Parent produce any character on offspri
565n007 ars» when afraid, though not every time really WISHING to smell. its enemy.-- Man & dogs show triumph
573n039 ow, & therefore thinking of subject, even when WISHING not to flow-- flow it will.-- My father told M
555m144 hanged hanging into his head cut off) as kind of WIT, showing he had honourable wounds.-- all this wa
555m144 he had honourable wounds.-- all this was kind of WIT.-- I changed I believe from hanging to head cut
472s02v d?! opens & shuts end of sucker, after having WITHDRAWN it.-- Saw 4 more Bees at work-- another odd g
298c189 istory Vol. I. p. 185 case of tit lark placing WITHERED grass over nest, when often looked at.-- this
467t103 do in Broom & certainly when over-ripe & half WITHERED-- I saw Bees going to clover & once this happ
472s02r terday remarked that many flowers had suddenly WITHERED, & to day saw very odd dusky humble (with pol
472s02r such a one before-- Saw Fly 21 «this Heartease WITHERED on Monday.--» alight on upper petals & insert
472s02v e-- & more of same fly Two more of the flowers WITHERED.-- Sillimans Journal <vo> 1842. p. 142-- Sus
021r005 s the floating marine confervae, is very common WITHIN E. Indian Archipelago, no minute description,
030r034 ween Mocha & main land). <[...]> At Carelmapu.--WITHIN Chiloe:-- On open coast, near where Challenger
072r160 ce tend to form ring. [Fig. 10] {P} motion from WITHIN and without H. Kingdom N. Spa. Vol III p. 113
098a047 cess (analogous to layers of quartz & feldspar) WITHIN other concretion.-- state last page thus. poin
112a089 ion. When discussing nucleuse's of old volcanos WITHIN Cordillera-- allude to Lyell's view of not dis
133a140 e been penetrated by volcanic & trappean rocks, WITHIN say the Tertiary period. one is led, to look a
152g050 of Alluvium rise nearly up to Glen Collarig up WITHIN 200 ft of level of 4th shelf= argument against
206b146 e chance of 200 people <might be> being related WITHIN 200 years backward might be calculated & this
227b224 true, we get (I) a horizontal history of earth =WITHIN recent times.=. & many curious points of specu
249c036 gratory; also with Temminks fact of forms being WITHIN Tropics.-- Europaean birds at Japan. connected
335d015 animals» be like either parent, or intermediate WITHIN certain small limits (within which limits they
335d015 t, or intermediate within certain small limits (WITHIN which limits they might return to either paren
367d112 . supposes that when foetus is forming the ovum WITHIN it, is forming «& this must be so, else avitis
397e004 bjected to my theory, that the amount of change WITHIN historical times has been small-- because chan
436e136 ls will vary, according to my Malthusian views, WITHIN certain limits, but beyond these not.-- argue
466t094 t bend up-- In Lark-spur, if Bees put proboscis WITHIN nectary «they do» they must disturb all anther
467t099 close by is equally abortive--and both growing WITHIN Kitchen Garden.-- As we see in Hybrids that al
470t176 anted «Turkey or» Palmated and English, planted WITHIN few yards of each other actually produced hybr
495q05a , reversing the poles test by suspending magnet WITHIN & see which way they fly.-- (3) I have noticed
504q013 me bear abortive stamens every year & Spring. & WITHIN garden //Yes// (5) Examine the Parnassia whose
534m061 is nonsense; yet who will venture to say germ WITHIN egg, cannot think-- as well as animal born wit
543m101 hus trace causation of thought.-- it is brought WITHIN <our own> limits of examination.-- obeys same
358d073 ketless on the fire, & on one occasion, not WITHSTANDING our vigilance a piece of pork 3 lb was take
555m143 .-- -- S. American group sneer.-- Sept 21st Was WITTY in a dream in a confused manner. thought that a
510g018 in varieties. G. St. Hilaires law of Balancement WM Yarrell (1) About non-breeding of animals in con
025q016 rallel of Olinda Shoaler N. of Olinda.--a little WNW of C. Rock. [5 degree 29 minute S.] still shoale
572n033 he first page & pronounced each word distinctly. WOKE instantly but could not gather general sense of
183b051 d) & live in same country. How is propagation of WOLF & Dog. (because being believed same species) if
204b140 intermediate between parents.-- How easily does WOLF & Dog cross? Mr Yarrel thinks oldest variety im
333d007 all in Zoolog Garden) he has seen in a show half WOLF & «half Esquimaux» dog which <likewise more res
333d007 squimaux» dog which <likewise more resembled the WOLF than dog.--> appeared to be intermediate betwee
385d163 then is actually obliged to be held.-- like she WOLF of Hunter.-- young take distemper very readily
389d173 only the more complicated animals.]CD p. 310 She WOLF took dog. but had such aversion to it, that she
409e046 he dog being so much more intellectual than fox, WOLF &c-- is precisely analogous case to man, exc
409e046 s done so.-- Show a savage a dog, & ask him, how WOLF was so changed. When discussing extinction of m
424e102 says Jenyns to it?-- -- In argument of origin of WOLF, difference of mind is most relied on, but Bell
424e103 is most relied on, but Bell has some account of WOLF in Zoolog. Gardens, which brought its puppies t
459tf02 to common mule & hinnus-- in one case bastard of WOLF & dog had more form of male, & another of both
479z005 nville Voyage round world no land animal besides WOLF at Falkland ∴ black rabbits not indigenous p 11
557m147 if as (I believe) Hunter says. neither fox. nor WOLF wag their tails, &c. it is very curious, recurr
575n044 he same laws, as the crossing of jackall & Fox & WOLF & dog.-- the only test this is most important:
591n097 . Garden told me. he has often watche tame young WOLF & it never dropped its ears like dog-- wagged i
211b163 some short: therefore like dogs.-- Ogleby says, WOLVES at Hudson bay breed with dogs.-- the bitches n
247c028 on shore at one of the Pelew Islds.-- killed a WOMAN. Chamisso p. 189 Tome III: Kotzebue.-- p 22. a
295c178 p. 15 (Lyell's Pamphlet) Is man more hairy than WOMAN. because ancestors so, or has he assumed that c
355d068 ery same must take place in copulation-- (Man & WOMAN separate parts of same plant<s>)-- now in some
362d099 ff horns to fight-- is analogous to the love of WOMAN (as Mitchell remarks seen in savages) to brave
376d137 edly seen them try to pull up petticoats., & if WOMAN not afraid clasp them round waist & look in the
376d138 nny with Tommy.-- Good evidence of knowledge of WOMAN-- ¶ The noise st st. which the C. Sphynx makes
377d139 inks other monkeys make st.-- noise¶ In case of WOMAN instinctive desire may be said to be more defin
380d155 Gestation is always some multiple of seven-- if WOMAN does not menstruate in the month, she will in 5
384d162 nded down for some generations Theory of sexes (WOMAN makes, bud, man puts primordial vivifying princ
384d162 xes as in Vertebrate tak place.-- ∴ Every man & WOMAN is hermaphrodite:-- ∴ developed instincts of Ca
390d178 impress some difference: to make the bud of the WOMAN, not a bud in every respect.-- [Is this connect
417e077 e might expect, as the difference between man & WOMAN is «indeed» (independent of sexual differences)
560m156 emale monkeys not show signs of impatience when WOMAN present? Do they pout, or spit, or cry.-- <fe>
577n051 sed, face of man, face, neck-- «upper» bosom in WOMAN: like erection shyness is certainly very much c
591n099 e held in abhorrence it is because instincts to WOMAN is not followed; good case of instinctive consc
603o014 possibility of <handsome> «UGLY healthy» young WOMAN, with good expression-- statues not painted-- <
640j28v ecessary: the fertility of Man in old age keeps WOMAN alive: for Man & woman are same: fertility of e
640j28v of Man in old age keeps woman alive: for Man & WOMAN are same: fertility of either sex determines li
190b078 e increase of organization being imitated in the WOMB, which has been passed through to form that spe
307c215 nized wings of the male rendered abortive in the WOMB.-- if these apparently useless organs do indica
365d107 monstrosity, or adaptation to unhealthy state of WOMB).-- One can perceive that Natural varieties or
382d158 are so plain in Man, & yet no trace of abortive WOMB, or ovarium.-- or testicles in female.-- the <a
388d171 near father.-- now if one of these staid in the WOMB, when it came out. it might partake of shade of
399e009 o whole life of animal, & not if it be solely to WOMB, as in monster. or solely to childhood, or sole
540m085 obably merely coorganic as connexion of mammae & WOMB.-- We need not feel so much surprise at male an
314c239 of birds of Van Diemens land & Australia.-- The WOMBAT (Brown) is found in Isd of Bass' Straits The c
512q020 oung ones, bred in captivity --Mr Miller says WOMBWALLS were (4) About fertility of ass-zebra-horse=

Page
(Key Word)
290c163 &c in man not heredetary, because crossed with WOMEN with pretty faces When horse goes a round, the
309c220 d the contamination of <c1> castes. improve the WOMEN. (double influence) & mankind must improve-- Th
319c257 History of Man at Maer, it is said the Samoyed WOMEN (; north end of the Oural mountains) have black
332d002 or two he saw many cases of virulent cancer in WOMEN, & since that time it has been rare disease.--
354d062 ccording to age of individuals-- (see Mammae of WOMEN) in different parts when age changes caterpilla
358d076 e female differs from the male?)».-- children & WOMEN = "women recognized inferior intellectually"= O
358d076 differs from the male?)».-- children & women = "WOMEN recognized inferior intellectually"= Opposed to
376d137 on & monkey, big & little that ever he saw knew WOMEN.-- he has repeatedly seen them try to pull up p
376d137 o pull up trousers of men. Evidently knew <men> WOMEN, thinks perhaps by smell.-- but monkeys examine
385d164 - May be just worth remembering that ovarium of WOMEN (Paper in Vol I of Irish Royal Academy) have co
388d172 suming plumage of cock, & beards growing on old WOMEN = Stags horns & testes curious instances of cor
509q017 On Mr Tremenheres Scottish Colliers, when men & WOMEN have long worked, whether children, who have no
545m106 emain still.-- I do not doubt this Baboon. knew WOMEN.-- Another little old American monkey «(Mycelis
553m138 any of the American Monkey shew any desire for WOMEN-- «very curious. as they depart in structure» T
555m142 Fuegians, till I remembered Bynoes story of the WOMEN.-- The Chillingham cattle (& Porpoises) have no
566n010 t\| Malthus on Pop. p. 32, origin of Chastity in WOMEN.-- rationally explained.-- on the wish to suppo
574n041 hence with action of mouth & jaws.-- Lascivious WOMEN. are described as biting: so do stallions alway
578n054 n, the question, "one will anyone, especially a WOMEN think of my face,"? to one moral conduct.-- eit
536m071 of mare strictly analogous to men's affect for WOMENS breasts.∴ Dr Darwin's theory probably wrong, o
554m138 he has seen place its head downwards to look up WOMENS petticoats-- just like Jenny with Tommy ourang
538m081 ysical action on the brain, renders much less WONDEFFUL the instincts of animals-- Aug. 12th. 38. At
055r107 f Craters!--At S. Cruz. there is no occasion to WONDER what has become of the Basalt. Gone into fine
055r108 lavas.--proportionally high to age. (we do not WONDER to see tertiary plains consumed) Where slope «
063r133 n same manner may not depend.--There is no more WONDER in extinction of species than of individual.--
155g066 fragment falling» of no particular hardness no WONDER that all «three» lines «should be» EQUALLY pre
202b133 There is long rigmarole article by S Hilaire on WONDER of finding Monkey in France.-- of genus peculi
208b153 allied species. p. 127. p. 132 There is no more WONDER in extinction of individuals than of species P
278c132 on, Macrauchenia, &c compared to America.-- the WONDER is that the Europaean forms, were able to esca
278c132 try.-- if Toxodon had been found in Africa, the WONDER have been same for S. America & Europe.-- The
386d167 head & tail, & the belly both head & tail,--no WONDER there should be sympathy in human frame.-- «on
541m093 ing how ancient these expressions are, it is no WONDER that they are so difficult to conceal.-- a man
586m079 stinctive. it is a test to know how much of the WONDER consists in the action being performed or emot
586m079 be formed or afterwards.-- child sucking whole WONDER instinctive.-- carrier pidgeon just as wonderf
605o18v is no real sublimity 5 The emotions of terror & WONDER so often concomitant with sublime. adds not a
591n099 rvation is disobeyed-- I often have «as a boy» WONDERED why all abnormal sexual actions or even impul
035r047 mineralogical nature & dependent: & then how WONDERFUL level «of same beds» should have been kept; i
048r084 tand pebbles thus coated.--The motion is most WONDERFUL, from chemical attraction, as a blade of gras
051r094 iffs evident); the action is anomalous; It is WONDERFUL to see Coral reef--or confervae in the breake
067r144 nce inside Cordillera: It is not therefore so WONDERFUL that volcanic rocks at M. Video «Volcano in P
076r169 disseminated <[...]> sometimes concentrated: WONDERFUL quantity of pure silver in S. America. Geolog
085a001 of plains: Lias in Shropshire. or some other WONDERFUL outlyer.-- Linn: Transact. Vol. 8. p. 288. Sa
129a131 cock Report on Massacuhssets. p. 133 The most WONDERFUL case of great block of rock moved by gale-- W
155g066 same «drainage» lines Mound of Gneiss though WONDERFUL-- <that they are preserved» how much more so,
187b062 ins how Zebras reached South Africa-- It is a WONDERFUL fact Horse, Elephant & Mastodon dying out abo
196b105 cannot transport easily.-- it would have been WONDERFUL if the two Rheas had existed in different Con
201b126 nt See Geolog. Proc. p. 569. 1837. Account of WONDERFUL fossils of India.-- & p. 545 «great monkey» M
222b207 rison be instituted. People often talk of the WONDERFUL event of intellectual Man appearing..-- the a
223b207 pearance of insects with other senses is more WONDERFUL. its mind more different probably & introduct
223b208 -- a bee «compared with cheese mite» with its WONDERFUL intincts «might well say how» The difference
248c029 not the same section, with house mice. It is WONDERFUL how it could have been transported. ¿What sec
259c064 applicable to plant, Epidemics of South Sea, WONDERFUL case of extermination of species-- Epidemic a
263c077 l with recent. the fabric falls! But Man-- -- WONDERFUL Man. "divino or versus coelum attentus" is a
273c111 s if animals perished by errors.-- It is most WONDERFUL how in every family of bird, even the most «
273c112 this not the case in swallow??? which is most WONDERFUL of all. ¿whether in most aerial of swallows.»
291c166 is thought. being a secretion of brain, more WONDERFUL than gravity a property of matter? It is our
291c167 one «both» sex«es». is strongly supported by WONDERFUL fact of bees changing the sex by feeding.-- n
291c167 no it is developing an hybrid female it is a WONDERFUL relation going through all Nature.-- Makes he
293c172 ings utterly forgotten» --it is scarcely more WONDERFUL, that it should be remembered in next generat
293c175 more perfect. preserving its relations.-- the WONDERFUL power of adaptation given to organization.--
302c200 ps scarcely one new family & no new orders,-- WONDERFUL, partly explained on my theory, = otherwise m
309c221 nstinct. bees turning neuter into Queen, more WONDERFUL case Dwights' Travels in America, speaks of s
312c232 - therefore same principle transferable. not WONDERFUL According to my view <ins> beccause actions a
313c235 er?) is analogous to dormant instinct.-- (How WONDERFUL a case bees developing sex of neuters) specie
354d062 e from caterpillar to butterfly-- is not more WONDERFUL than the body of a man undergoing a constant
369d114 shes have no secondary characters--)p. 49. WONDERFUL case of Pea hen. taking feathers of Peacock &
371d118 see whether it will solve them.-- It is less WONDERFUL that childs nervous system should build up it
372d131 inch should produce a Newton is often thought WONDERFUL. it is part of same class of facts that the s
373d132 tes, Having Hair. like true Mammalia, no more WONDERFUL. than Echidna. & Hedgehog having spines.-- Do
408e044 n is the gap, for the new one to enter?-- The WONDERFUL species of Galapagos, must be owing to these
426e108 ed grandmother; what is Mr S. S. parentage?-- WONDERFUL as is the possession of voice by Man. we shou
433e127 g high all together & then tumbling, far more WONDERFUL than heredetary ambling horses. Whether the b
452e181 ournal of [Asiatic Soc] [...] p [...] -- most WONDERFUL instinct, how could it have originated-- spin
463t055 the component vertebrae of the head of Snake WONDERFUL!! distinct!!-- He would not allow such series
464t063 e the conclusion of his work-- Lund makes his WONDERFUL discoveries= negative facts are valueless= mo
467t103 ce this happened.-- And in common Beans it is WONDERFUL {a} how the Humbles force down the wings most
521m007 nscience, as instincts are, is not so very WONDERFUL.-- <Now is not epilepsy an habitual disease o
529m041 t [i.e., as] great compound animals united by WONDERFUL & mysterious manner.-- There is much imaginat
537m075 nscience varies in different races.-- no more WONDERFUL than dogs should have different instincts.--
546m110 ls these vivid thoughts In same book (p. 143) WONDERFUL case of perfect double consciousness Mayo com
551m128 . 4th. Lyell in his Principles talks of it as WONDERFUL that Elephants understand contracts.-- but W.
573m036 ur faculties are more fitted to recognize the WONDERFUL structure of a beetle than a Universe.-- Nove
575m043 iarity of structure of brain.?-- is this more WONDERFUL than memory. affected by diseases. &c &c, dou
586m079 wonder instinctive.-- carrier pidgeon just as WONDERFUL in old bird as new.-- migration, <only> «only
586m079 bird as new.-- migration, <only> «only» more WONDERFUL in young, because can not have been taught, w
586m080 rk night & not loosing its direction, equally WONDERFUL in young & old.-- These facts point out some
594n115 en to the Bee its instinct is not <more> less WONDERFUL than man his intellect Lyell has seen a littl
600o08a because most imperious.-- It would indeed be WONDERFUL, if, mind of animal was not closely allied to
610o032 knowing things, which are often repeated in a WONDERFUL manner.-- as the hour of the day &c-- All hab
614o037 m what takes place out of bodies) really less WONDERFUL than thoughts-- One organic body likes one <m
614o037 xperience for child sucking.-- And is it more WONDERFUL that memory should be transmitted from genera
614o037 d by man in lifetime Heredetary memory not so WONDERFUL as at first appears, & no too great advantage
636j56v ous Plants by foreign agency-- as insects, as WONDERFUL case of adaptation.! There would not have bee
054t102 P. 199; Falkland account of cleavage differs WONDERFULLY from mine: phyllade covered by quartzose san
273c112 aerial of swallows.» Milvulus, & still more WONDERFULLY to the Humming bird, which is one instance o
274c113 s. but also on fruit.-- The Rasorial type is WONDERFULLY shown in long legged cuckoos with claw like
296c185 Swainsons views. which if abstract true are WONDERFULLY absurd.-- p. 565 "breed> Scotch wild Cattle.
302c202 nalogy.-- see my notes on p. 37. of Macleay. WONDERFULLY accordant. with fact there stated, only in m
305c204 ale genital organs «in some monkeys clitoris WONDERFULLY produced».-- make abstract on this subject f
414o060 een so little progressive «!Agassiz makes it WONDERFULLY changed, since Cretaceous period, whether pr
374d134 ber that can live.--» We ought to be far from WONDERING of changes in number of species, from small c
605o19v ing. (often however accompanied with terror & WONDERMENT) «which> «this» emotion, from the associatio
217b187 ms-- The above facts evidently show that Mr D. WONDERS at these species being wanderers.-- Iceland no
612o035 h receives communication from without, & gives WONDROUS power of willing. These +wilings are common
578n053 e---n, «or wish extraordinarily to have one» you WONT. ==»== No surer way to blush, than particularly
027r028 rvice Journal In the Iron sand formation «would> WOOD converted into siliceous pyritous & coaly matte
035r046 nes &c &c--Yet <silicified> turn over silicified WOOD. Cordilleras, Chiloe. &c seems the organic stru
071r156 eruvian Indians use arrows or Araucanians?--> If WOOD now preserved over world Dicotyledones far prep
138a176 e was boiling on the top-- [blank] Would rotting WOOD by yielding Carbonic Acid unite with «piece of

Page   **(Key Word)***
148g024 too hard & uneven to be impressed Case of Birch WOOD by Inverorum being determined by sheep & not de
257c058 ssions in Red Sandstone. great lizards in do.-- <WOOD> <Dicot wood> Coniferous wood in Coal Measure.-
257c058 Sandstone. great lizards in do.-- <Wood> <Dicot WOOD> Coniferous wood in Coal Measure.-- highest fis
257c058 lizards in do.-- <Wood> <Dicot wood> Coniferous WOOD in Coal Measure.-- highest fish in Old Red Sand
259c062 lakes, hot spring &c &c inhabited therefore mud WOOD be inhabited, then how is this effected by-- fo
265c085 to Physical laws, as sulphuric acid disorganizes WOOD, but adaptation.-- albino however is monster. y
299c193 s a plant like Paris quadrifolium growing in one WOOD far from any other plants of same species Chann
306c213 ne of Timor 215 degree. What productions Sandal. WOOD Isd.? ought to agree with Java?? Terrestrial Pl
365d105 me is owing to their rarity., a single female in WOOD with Pheasants would sure to be trod, & in many
424e102 going North of the Elbe.-- yet they meet in one WOOD in Anhault. & there every year produce hybrids-
497q006 soil from centre of woods «especially if date of WOOD be known» & other odd places & see what plants
497q006 y remain dormant. it kinds come up, not found in WOOD.-- but seeds continually dropping in woods. by
516g23v ysis of Peat (2) Athenaeum 1840 p. 777. Decaying WOOD absorbs oxygen & forms Carbonic Acid. will this
563n001 leaving no tracks.-- My Father says pea-hens do WOOD pidgeons building near houses. yet so shy at al
071r157 Cordillera by Humboldt in Geolog. Society Sir WOODBINE Parish informs me that town near Tucuman and
288c160 s of knowing directions: mysterious. Were the WOODCOCKS, which came Madeira & <seized> ceased their
430e119 all the lower part rayed longitudinally (give WOODCUT) like I. sulcatus.-- Both species are found at
507q15v e (.49) List of seeds Gaertner de fruct:-- for WOODCUT-- 1 double hook-- -- Geum. Galium Burrh ≡ sing
639j28r in Scorpion & Crust in Squilla. & Mantis. C D WOODCUTS stones swallowed by birds & by Aphysia. C. D
030r035 s must bring much vegetable matter from thickly WOODED mountains, probably chiefly leaves.--This posi
202b131 adapted. Near the Caspian «Province of Ghilan» WOODED district cattle with humps as in India. Geogra
469k151 s. (except some wet ones/ & wd go on longer-- WOODFORDS Marrow fat, Early frame, Groom's Dwarf. plant
267c094 ery attempt to check its increase. The <bush> WOODLANDS for miles in extent are composed solely of th
462tf4v e object gained by the Mataco-armadillo & the WOODLOUSE-- -- a good analogy-- sea-Crustacea-- Tullus.
185b055 in open country a ground «do. <w> parrot.--» WOODPECKER-- a desert. Kingfisher.-- mountain tringas.-
241c014 r not p. 156.-- Parroket with stiff tail like WOODPECKER-- Birds of Australia. Many in common ¿speci
258c061 t groups.-- It is having walking fly catcher, WOODPECKER &c & which causes the confusion in this syst
259c064 -- it is only be recollecting that the ground WOODPECKER &c.--, fresh water animals of great lakes ar
264c082 ng true affinity, for instance tail of ground WOODPECKER-- -- but tail of some ducks aberrant from--
274c114 r-- Gould has seen with long tarsi.-- «Ground WOODPECKER» Secretary bird.-- & Millisuga. Kingii very
584n071 nstincts & structure always go together: thus WOODPECKER: but this is not so,, the instincts may vary
584n071 pparent anomaly,, for if anyone has taken the WOODPECKER as an example fitted for climbing, his argum
636j57r ulloch. Attrib. of Deity. Vol I p. 232. gives WOODPECKER as instance of beautiful adaptation.-- & the
636j57r sil forms show such losses.-- Consider ground WOODPECKER stiff tailed cormorant: pain & disease in wo
637j57v ields-- these are adaptations just as much as WOODPECKER. --only we here see means-- but not in the o
257c057 distant. <Study> The circumstance of ground WOODPECKERS.-- birds that cannot fly &c &c. seem clearly
273c111 .in the Tyrannidae.-- Milvulus-- Even flying WOODPECKERS Gould says, he believes does. but also on fr
274c113 longing to this type, ¿the Humming bird? the WOODPECKERS?-- In each division Gould thinks he can trac
274c114 ny ground parrots? are there not many ground WOODPECKERS?-- In the prevalence of Coniferous
491q01v st pollen produces any effect, as in case of WOODPIDGEON & Hen. mentioned by Mr Knight. Vol IV Hort.
221b202 trine removed?? Is the prevalence of Coniferous WOODS before Dicotyledenous a fact analogous to repti
377d140 s Lion. because inhabitant of plain & Jaguar of WOODS &c like ground birds [not located] :Hence, also
429e114 t» war of organic beings. going on the peaceful WOODS. & smiling fields.-- we must recollect the mult
497q006 nce Maer.-- (12) Take Bag of soil from centre of WOODS «especially if date of wood be known» & other o
497q006 nd in wood.-- but seeds continually dropping in WOODS. by birds 13. Mr. Herbert says Crocuses are ver
430e120 evidently inclines to think it Hybrid.!!! Ask WOODWARD Mr Lonsdale says Trigonia costata & elongata
036r052 action on land different Volney, P. 351. Vol I. WOODY bushes, «gazelles» hares, grasshoppers & Rats.
085aIFC al Formation Lyell's Salband p. 86 Shells near WOOLLICH p. 112 Speculate on the extension of Patagoni
541m089 inctive fear arisen? 19th. When I went down to WOOLLICH I was trying to unbend my mind as much as pos
085a FC rks Nothing For any Purpose A. Geology Note on WOOLWICH Nothing on any Subject As far as p. 33. distr
121a112 there ever to be cracks 11th August. 1838 Near WOOLWICH there are plains & valleys just like Patagoni
216b179 ossils 10 out of twenty have ANALOGUES uses this WORD «for» similar in the Indian sea.-- Deshayes.--
249c035 some capital information ¿Carnivora of New & Old WORD. do not form two sections is this not connected
249c036 rhouse's remarkable fact of no forms peculiar to WORD to special districts???? north of 30 degree.--,
264c079 elligence when spoken; as if it understood every WORD said-- see its affection.-- to those it knew.--
286c155 sification of <arrangement> relationship; latter WORD meaning descent.-- A tree is taken by Fleming a
305c210 There is one living spirit, prevalent over this WORD, (subject to certain contingencies of organic m
352d057 .-- As Larvae may be more perfect (as we use the WORD) than parent, so may species retrograde, but th
541m091 re one effort less» one is obliged to repeat the WORD, & think of qualities as flowers, cloth &c & wi
551m129 out "right." or Drinkwater's horse jumping when WORD Jump said-- I saw the ourang. take up a stone &
569n020 se of Association.-- Elephant often given food & WORD open your mouth said, recognizes that sound as
569n020 same way alphabet. arose from letters, symbol of WORD beginning with the sound of letter)-- crying ya
572n033 n French I read the first page & pronounced each WORD distinctly. woke instantly but could not gather
572n034 ds: it appears as if the mind had dwelt on each WORD separately, neglecting time, & general sense, a
587n088 a child. Get a Dictionary & make a list of every WORD, expressing a mental <desire> «quality» &c &c M
593n111 ssy bank some way from water, suddenly, as if by WORD of command, they all took flight & flappered ac
612o035 y animal instinctive and unavoidable.-- +Can the WORD willing be used without consciousness, for it i
613o036 s, but surely instincts imply willing, therefore WORD misplaced The meaning of Words, must be made ou
628o53v e partly made.]CD p. 21. "Why ought I to keep my WORD"-- gives the problem. of ethics-- [my answer wo
249c035 w this out, where species of same genera in two WORDS. have not species, generally wide range? Mice.-
521m007 d in the Zoonomia.-- Now if memory «of a tune & WORDS» can thus lie dormant, during a whole life time
524m021 n shows that repetition is not necessary)-- the WORDS second childhood full of meaning:-- Dreams do n
555m141 ning"-- only rather mere steps.-- dispute about WORDS.-- Miss Martineau (How to Observe p. 213) says
572n033 k quickly, & <running> «running» over imaginary WORDS: it appears as if the mind had dwelt on each w
581n065 been in the dawn of civilization-- thinks many WORDS, roar, scrape, crack, &c, imitative of the thin
595n127 [blank] Goldsmiths Essays No XV,, on sounds of WORDS being expressive, (Vol. 4 of Works) [blank] "Ad
599o05v bt that language is an altering element, we see WORDS invented-- we see their origin in names of Peop
599o05v see their origin in names of People.-- Sound of WORDS-- argument of original formation.-- declension
603o013 ce languages become corrupt, & whole classes of WORDS «are abbreviations» he thus derives from nouns
613o036 illing, therefore word misplaced The meaning of WORDS, must be made out Reason Will Consciousness Def
529m040 force cracking surface & truly poetical. (V. WORDSWORTH about science being sufficiently habitual to
579n057 eautiful there are some notes. & likewise on WORDSWORTH'S dissertation on Poetry.-- The expression of
056r109 fling means could efface & obliterate so grand a WORK?--In valleys one is not sure whether fissures m
067r143 es 163E: W. Parish?? «by Pedro de Angelis.» This WORK is reviewed in present Edinburgh March 1835 Sir
129a131 Tertiary strata of S America» read parts of this WORK, though it is but poor. Athenaeum. 1838 p. 791
182b050 t does as yet appear clim In Mr Gould Australian WORK some most curious cases. of close but certainly
198b114 cannot make out his ideas about propagation His WORK. Philosophie Anatomique. 2d Vol about monsters
209b156 nctes et permanentes." p. 145. In Humboldt great WORK de distribut. Plantarum. relation of genera to
218b191 in the Horticultural Transactions and a distinct WORK on Hybridity under title of Amaryllidae & Narci
228b232 ing «& famine»; our slaves in the most laborious WORK, our companion in our amusements. they may part
233b251 es. Anoplotherium.-- <M. Jerrod> & Dumeril great WORK on Reptiles. M. J. says some reptiles same from
235b263 , can orders like birds & animals separate &c &c WORK out Quinary system according to three elements
235b273 on Himalaya Plants-- Would it not be possible to WORK through all genera, & see how many confined to
243c020 nstance of wide range, where means of wide range WORK this out-- L. Jenyns, about my fish New Zealand
251c041 Mammalia no doubt will all be included in Smiths WORK «do» Vol. IV p 273. Macleay on Capromys. 4 spec
268c096 .-- <The same> <Mem> Tennioptera & Tyrannula (NB WORK out how many forms Tyrannula can we worked out
275c121 small Bald Heads Mr Yarrell will mention in his WORK I am sorry to find Mr Yarrell's evidence about
294c177 people in conversation.» the whole object of the WORK is its proof, «its limiting, the allowing at sa
300c196 a.-- Man in his arrogance thinks himself a great WORK. worthy the interposition of a deity, more humb
302c202 pinion on all subjects of classification, I must WORK out hypothesis, & compare it with resuts. if I
304c206 s & not all organs blending away.-- +++ Hopeless WORK to systematist, who believed that all his divis
325c267 -- Quoted by Owen.-- Hunter has written quarto. WORK on Physiology besides the papers collected by O
326c266 Biography. 4. Volumes well worth reading Bevans WORK on Bees, new Edit 1838 Harlaam. Physical & Medi
328cIBC du Roi Java fossils at same time Study Botanical WORK on Buds & Gemmae. C D Charles Darwin 36 Great M
338d024) of plants of Nova Zenbla -- in review of Baers WORK Edinburgh. Royal. Transact.-- p. 297. Vol 9. Dr
356d069 to law only merit if merit there be in following WORK.-- The history of Medicine, the extraordinary w
363d101 will be well worth working out.-- Study Temmincks WORK on Pidgeons--, & see whether feathered legs.--
370d116 no more can be said.... In paper on bees in same WORK. (it is said that some kind lay <pu> up honey e

```
413e060 Hippot, Haena &c are found together.-- Read this WORK-- Decb. 4th.-- Why has the organization of fish
419e087 efore the mere loss of species, which may be the WORK of a few years as with the Lamantin of Steller
425e104 pecies peculiar to the separate islands-- In his WORK on the Labiatae, some of these species are desc
463t063 same   such remark & before the conclusion of his WORK-- Lund makes his wonderful discoveries= negativ
464tf6v There   is ibex of Alp Pyrenees &c-- (see Blyth's WORK on Ruminants,-- these species must have migrate
472s02v after   having withdrawn it.-- Saw 4 more Bees at WORK-- another odd genus-- & a small common Humble--
482z011 -- p. 210. Scolopax very close to ours Rengger's WORK of Mammali: of Paraguay must be most  important
511q018 of different appearance.-- {will introduce it in WORK} Whether <Yar> knows whether Shaws hybrids betw
527m034 h I observe a long castle in the air, is as hard WORK (abstracting it being done in open air, with ex
539m081 struck with an intense headache «after good days WORK» which came on from reading «review of» M. Comt
547m113 e mind <thinks> quicker in sleep, it may do less WORK & yet do so, from the exertion of keeping up th
548m115 new means,-- therefore works of imagination hard WORK,-- Keeping one idea present is, perhaps, hard w
548m115 rk,-- Keeping one idea present is, perhaps, hard WORK-- though dreams do that One Reflective Consciou
580n062 ls & men only in Kind.-- probably very important WORK.-- Feb. 12. 1839. Sir. H. Davy -- Consolats: "t
582n066 r pleasure these are sensations" <Gardner in his WORK> In the life of Hayd & Mozart. fine music is ev
592n103 I follow up this subject & a reference to Brun's WORK.-- Shutting eyes in contempt opposite action to
606o023 here? Laocoon p. 75 "The beauties developed in a WORK of art are not approved by the eye itself, but
268c096 la (NB work out how many forms Tyrannula can we WORKED out into. Milvulus [not located] element geogr
334d011 rom their mares, (though horsing every month) & WORKED in the same cart in loose chains, by being at
483z013 The North & S. Range of shells might perhaps be WORKED out with advantage. with Cumms collections & m
509q017 s Scottish Colliers, when men & women have long WORKED, whether children, who have not worked have an
509q017 ave long worked, whether children, who have not WORKED have any peculiar configuration.-- Hooker <Met
363d101 spangles on this part: this will be well worth WORKING out.-- Study Temminks work on Pidgeons--, & se
539m081 the  difference is of a man wagging his foot & WORKING with his toe to perform some difficult task.--
048r085 at it was so arranged from the foresight of the WORKS of man Feeling surprise at Mastodon  inhabiting
325c267 Essay on H. Understanding (sometime) Du Stewart WORKS. & lives of Reid, Smith & giving abstract of th
350d054 sophy<"> Religio Medici. Vol II. Sir T Browne's WORKS. p. 20 There are no grotesques in nature; not an
508q016 -- (8) In Hump-back ever heredetary (9) Are the WORKS of Berhave (treating of heredetary diseases) tr
526m028 - Catherine remarks that pleasure received from WORKS of imagination very different from the inventiv
548m115 every  step, & inventing new means,-- therefore WORKS of imagination hard work,-- Keeping one idea pr
558m151 ", in reference to this "ought," as well as the WORKS of the whole world.-- Read Mackintosh on  Moral
569n019 erved the ignorant. merely looked at picture as WORKS of imitation.-- Hence pleasure in the beautiful
572n032 ture?? which would naturally happen.-- Reynolds WORKS. Vol I, p. 226-- "The general idea of showing r
595n127 on sounds of words being expressive, (Vol. 4 of WORKS) [blank] "Adam Smith Moral Sentiments" much  on
638j58v e of the dead & extinct The analogy between the WORKS of art «or intellect» such as hinge, & hinge of
638j58v «or intellect» such as hinge, & hinge of shell, WORKS of laws of organization is remarkable-- what is
021rIFC e simplest infusoria same since commencement of WORLD.-- [not located] La. billardiere mentions the f
025r018 & Europe. we may believe them applicable to the WORLD-- My general opinion from the examination of s
031r038 upheaval To Cleavage add other instances in old WORLD of symetrical structure. East India Archipelago
032r041 in cleavage lines./ possibly general symetry of WORLD.-- I feel no doubt. respecting the brecciated w
044r072 ca. Australia.. Oceanic Isles. Geology of whole WORLD will turn out simple.-- Fortunate for this scie
045r075 e of the country; (perhaps generally over whole WORLD) Yet eruptions <both> at sea (as wells as in th
056r111 the similarity of Volcanic products «over whole WORLD» argument, as well as separating causes by wate
056r111 s together; & products being similar over whole WORLD, general circulation. But Volcanic action separ
061r126 ry season[.] 1791. seen commonly bad over whole WORLD. «(Was it so in Sydney, consult history? Philli
071r156 s or Araucanians?--> If wood now preserved over WORLD Dicotyledones far preponderant, if so coniferou
080r180 . 73. General reflections on the geology of the WORLD P. 14-91. gradual shoaling of coasts 93  action
086a007 a great edentata. -- How much is temperature of WORLD regulated by atmospheric currents?-- chiefly cl
090a024 000 = 2500 = tons in fifty thousand years]CD If WORLD increased a tenth; would the perturbation be se
117a101 ange which must have taken place, otherwise the WORLD would daily be scene of ruin in late Natical Ma
128a129 the  Glen Roy case shows that the figure of the WORLD has just that form which forces dilemma. Transa
131a136 vertical,  but turned over in many parts of the WORLD -- argument strong in favour of thin crust theo
171b002 ort, Why such high object generation.-- We know WORLD subject to cycle of change, temperature & all c
171b004 eneration to adapt & alter the race to changing WORLD.-- On other hand, generation destroys the effec
174b015 pagation from different set, as the rest of the WORLD.-- This view supposes that in course of ages. &
175b018 ed ¿would they not be pretty similar over whole WORLD under similar climates & as far as world has be
175b019 whole  world under similar climates & as far as WORLD has been uniform, at former epoch-- How is this
181b039 cely any type of their existence in the present WORLD.-- or we may suppose only each species in each
183b053 ds. part of some great system acting over whole WORLD, the period of great quadrupeds declining as gr
194b095 y think a very strong case might be made out of WORLD before zoological divisions.-- Mem. species dou
202b133 ving Edentata in Africa &c &c.-- Now if suppose WORLD more perfectly continental. we might  have wande
203b135 e anomalies. as Cape Anteater,-- This supposes WORLD divided into Zoological provinces-- united--  &
203b137 dine cause has destroyed animals over the whole WORLD-- For instance gradual reduction of tempereture
208b154 genus having its parent type in hotter parts of WORLD Is monkey. peculiar to C. de Verd's.--? NO Macl
222b205 tiles, which can only be adaptation to changing WORLD,-- I cannot for a moment doubt, but what cetace
227b225 ty of succession from what we have seen. in old WORLD, & on amount changes which may happen-- ‖ It le
227b225 hich may happen-- ‖ It leads you to believe the WORLD older than geologists think. it agrees with exc
228b231 s arts would not then have taken him over whole WORLD.-- «--the soul by consent of all is superadded,
232b246 rably> uniform).-- Comparing fossils with whole WORLD. would be like in a Meteorologic table «in comp
232b247 r to add to the mean of the other.-- If the the WORLD had cooled by secular refrigeration in chief pa
233b250 t of fossils of Sewalick «India» Monkeys of old WORLD. Crocodiles. Anoplotherium.-- <M. Jerrod> & Dum
233b252 to say that intellectuality is only aim in this WORLD [not located] T. Carlyle, saw with his own eyes
239cIFC taking greatest effect Account of the [...] of WORLD.-- Charles Darwin written between («beginning o
250c038 Animals in Europe & America Some portion of the WORLD «Africa» being left more equable. yet America p
250c038 supposed to have been condition of former whole WORLD. America might have been string of islands.-- ¿
250c039 ny species but not genera distinct from rest of WORLD??? Lyells Principles, must be abstracted & answ
263c078 monkeys  might not,-- but probably would.-- the WORLD now being fit, for such an animal.--man, (rude,
267c093 Journ. Vol VII. p. 216. Mr Bennett Voyage round WORLD, 20 years have scarcely elapsed since the Guava
272c109 ommon to all the Laniadae & Muscicapidae of new WORLD, but not found in Old-World--. + + If in any «w
272c109 Muscicapidae of new World, but not found in Old-WORLD--. + + If in any «well developed» family (Gould
274c116 ptacles of the wanderers out of the rest of the WORLD?-- Will this not agree with Waterhouse & <birds
282c140 f change yet, as external conditions over whole WORLD. similar-- & constitution of man originally sim
284c148 -- not other islands, nor any any other part of WORLD.-- no other plants peculiar to these isld.-- ¿B
286c154 stimulated  by same passions-- brought into the WORLD same way» they may convey much thus, Man has ex
310c224 : 400 Australian plants found in other parts of WORLD Athenaeum June 3d 1838. quotes M. Turpins asser
316c246 species some few have been scattered over whole WORLD Many shells at present day same (or according t
321c270 ated] <Rays Wisdom of> Lisiansky's Voyage round WORLD. 1803-6 Nothing Lyells Elements of Geology Gibb
323c269 Egyptian  remains skimmed -- Pliny Nat. Hist of WORLD do -- Lamarck. II Vol. Philo. Zoology «referenc
339d025 ld Duck & «pintail» Widgeon!-- Divides animals «WORLD into Zoological Provinces» according to varieti
342d036 .-- What a magnificent view one can take of the WORLD Astronomical <& unknown> causes, modified by un
343d036 hese superindoce changes of form in the organic WORLD, as adaptation & these changing affect each ot
343d036 ves.-- instincts alter, reason is formed, & the WORLD peopled «with Myriads of distinct forms» from a
356d071 lutions & linking of facts-- Savages over whole WORLD, (Major <I> Mitchell p. 244. vol I) spit & thro
357d072 destroy without rule-- But what does he in this WORLD without rule? The destruction of the great Mamm
357d072 The destruction of the great Mammals over whole WORLD shows there is rule.-- S. America & Australia a
357d074 ortation, seems «?» to imply knowledge of whole WORLD, -- if so doubtless «part of» system of great har
370d117 of difficulty of growing plants in all parts of WORLD, thus tea tree in Brazil must have degenerated.
397e003 variably according to fixed laws: And since the WORLD began, the causes of population & depopulation
398e003 us of man Those who have studied history of the WORLD most closely, & know the amounts of change now
405e031 s to be so.-- [not located] Did man spread over WORLD as early as Elephants &c &.-- if in next 20 yea
406e037 t same time in North & South. America.-- Whole WOR[L]D, formerly possessed a climate compared to S. A
407e038 America favourable to Tropical productions. The WORLD formerly much more so. yet climate of same orde
407e038 ll Tropical forms have been obliterated) of the WORLD. from the <Tropical> Equable kind of climate to
407e039 E. or so tropical, & therefore present state of WORLD is not so different, with <d> regard to their p
408e042 able & low-- more so than any other part of the WORLD. Europe perhaps less so, that either Americas
408e045 on the vast changes even in that quarter of the WORLD-- -- Mem. elevation & subsidence of East Indian
409e048 e see it is not the order in this perfect <uni> WORLD, either at the present, or many anterior epochs
409e049 &c-- If man is one great object, for which the WORLD was brought into present state.-- «whether he w
```

```
411e054 s in the series, their relation to the external WORLD, & every possible contingent circumstance.-- ‖t
411e055 le & none discovered before them in any part of WORLD.-- Wealden to boot.-- When one sees in Corallin
419e085 owing to absolute quantity of vitality «in the WORLD»,-- the production of vitality, as argued by Mü
422e095 located]  The enormous number of animals in the WORLD depends, of their varied structure & complexity
423e096 ity, it is impossible to cover whole surface of WORLD with life.-- for otherwise a frost if killing t
423e096 if  killing the vegetable in one quarter of the WORLD would kill all of the one herbivorous. & its on
430e119 on & nearest being common to other parts of the WORLD-- March 16th. Mr Lonsdale showed me two specime
432e125 sed geologist can really comprehend how old the WORLD is, as the measurements refer not to revolution
434e128 ogies of the existing to the events of the past WORLD, we have no foundation for our science".--  <it
436e136 ckson Shark-- Owen thinks Australia part of Old WORLD <If> It «may» be said, that animals will v
436e137 ambrian formations do not however, extend round WORLD.-- Quartz of Falkland.-- Old Red Sandstone-- Va
443e155 o argue against my theory, because it makes the WORLD far older than what Geologists, think: it would
447e163 her.--»  -- these simple forms perhaps oldest in WORLD & hence most persistent-- if form exceedingly d
449e173 ge [not located] Mr Greenough on his Map of the WORLD, has. written. Mastodon found at Timor.-- think
451e178 onths out of 12.-- the largest mammifers in the WORLD consistently reside & are bred. "take tame anim
460t019 t is said that there is evidence in the organic WORLD of infinite & growing complexity from a few typ
460t019 iginated, we must go to the first origin of the WORLD.-- our present organic beings are the descendan
479z005 h of Coral-- RN. p 24 Bougainville Voyage round WORLD no land animal besides Wolf at Falkland ∴ black
481z010 p.  26 Nullipora p. 29-- In Meyen. Voyage round WORLD German a reference to a luminous Sertularia Les
555m143 ession of passion with open mouths like the old WORLD ones.-- Though the[y] move whole skin of head t
555m143 do  not move eyebrows.-- (I see some of the old WORLD ones move skin of head & ears,-- ∴ some men hav
558m151 this "ought," as well as the works of the whole WORLD.-- Read Mackintosh on Moral sense & emotions.--
610o034 erent from those,, that govern in the inorganic WORLD; life itself being, the capability of such matt
610o034 so great <worlds, inor> systems of laws «in the WORLD» the organic & inorganic-- The inorganic are pr
623o050 imals, living under certain conditions, in this WORLD, they <will conform to the law,» «can only be s
624o051 h himself» & parents, & hence to nearly all the WORLD.-- As conditions change, from civilization, edu
635j56r each  species) owing to the growing size of the WORLD? & the physical changes it was to undergo «anim
636j56v en, recurrent series, whilst the habitation «or WORLD» simple series.-- My theory shows life  equally
636j56v ries, & therefore trace of beginning in organic WORLD.-- Macculloch. Attrib. of Deity. Vol I p.  232.
636j57r ecker stiff tailed cormorant: pain & disease in WORLD & yet talk of perfection Get instances of adapt
639j28r ons.-- [There must have been deserts in the old WORLD!]CD p. 252 analogy of hand in mole, & Mole cric
640j167 ss species to themselves.-- Probably no case in WORLD like Galapagos. no hurricanes.-- islds never jo
611o034 of  movements. [LHC] Hence there are two great <WORLDS, inor> systems of laws «in the world» the orga
197b111 t process, by which man «one animal» passes from WORM to man; «highest» as typical of <sa>. changes,
254o049 in infusoria, generation in each joint of taenia WORM.-- formative energies easily expended & no  one
309c221 on in two classes however different.-- Male glow-WORM knowing female good case of instinct. bees turn
332d003 er-in the hair a century since used to be called WORM Fever, as used much more latley diseased Mesent
478z004 r J. Murray has given paper to Royal Soc on glow WORM. luminous property-- Curious arrangement of ani
531m052 on» which accompanies violent attack,-- Even the WORM when trod upon turneth,, here probably there is
570n027 ated would think negress beautiful,-- [male glow WORM doubtless admires female. showing. no connectio
057r114 und in Possession Bay in 73 degree 39 N. living WORMS in the mud which he drew up from 1,000 f[athoms
196b108 e on since from parasitical nature of insects & WORMS.-- In abstract we may say that vegetables & mas
332d003 attended by the transmission of large number of WORMS the child not having passed them before.-- Henc
332d004 Hence  disordered intestines are not healthy to WORMS, (like parasites of Tropical countries cannot e
416e071 animals.-- in shells?-- insects?.-- all!??!?-- WORMS? [Barnacles, aquatic., <yet> Crustacean, & true
483z013 m Rüppel travels All Owens papers on Intestinal WORMS must be studied in Vol I, Zoolog: Transact. bef
503q012 Resources Book no information & Hope about Silk WORMS. Varieties effects of domestication-- said to r
529m042 te lead. who was most violently purged «believe WORMS were passed off.» & vomited, but who when he re
032r039 eath the band of greatest action not having been WORN away.--If the level of the sea was to sink by v
048r086 esire). = Where talking of such substances being WORN into channels. mention submarine channels. such
051r094 ocks show signs of degradation; (soft substances WORN into bare cliffs evident); the action is anomal
053r100 ess interstratified with sediment.--& escarpment WORN away like english escarpment The great conglome
055r107 andstone first gives» half demolished craters).--WORN into mud & dust.--connection with age, & agreem
151g040 .--  The boulders (one of Gneiss remarkably water WORN) are often times of rock not in immediate neigh
151g040 between  Inn & Bouhunthine the summit «doubtless WORN into coincidence» has beach or band of  pebbles
159g086 on  same) of «semi» waterworn & some partly well WORN pebbles-- «which river could not have deposited
163g107 en filled up «at» 30 ft. higher with pebbles now WORN away-- The above shells must have been about 60
325c266 s not effect, as Lyell suggested, of organ being WORN out as. otherwise old whores would not have chi
343d039 sting animals found fossil when Europe must have WORN a quite different figure 19th. With respect  to
443e153 e roots is the best way to get young trees, from WORN-out kinds, & quotes from Pliny, that it is bad
113a091 consider the conducting powers either better or WORSE & the depth of 32 degree. being little we may c
205b145 roportion to their increase of size they become WORSE in form, less hardy, & more liable to  disease"
410e052 ,-- but naming mere «single specimens in» skins WORSE than useless.-- yet there is no cure «I may say
609o029 know» is far easier conquered than the deeper & WORSER feelings. These bad feelings no doubt orginall
585n074 ng> «which is only present in youth» (Mem. Mr WORSLEY'S story of chicken) to know that which we touch
335d015 sometimes  have similar offsprings, so will the WORST mules (as real mule) have offspring,-- slight d
508q016 election of Males in «cattle» or in Killing the WORST =or in dogs= (12) Do Hottentots generally resem
539m084 same  as Grandfather. Aug.  16th Anger «Rage» in WORST form is described by Spenser (Faery Queene. CD
470t177 w no trace of palmation!!? Bees at Wild St Johns WORT--Scabies, Cyanoglossum--Reseda wild very many B
026r021 . Geolog. Trans: Cornwall «Vol II» It is a fact WORTH noticing that cryst of glassy felspar in Phonol
050r090 at  Galapagos. <|> Sir George Mackenzie must be WORTH reading Some earthquakes of Sumatra no connecti
086a005 d as the land, stable as the water"-- It may be WORTH noticing edentates & camels in deserts & rodent
100a052 rticles of granite in Henslow's Grit, yet it is WORTH consideration. especially effect of gravity, ve
100a055 ic period.-- comparison rather loose.-- perhaps WORTH Says from Lardner's (p. 213) form of escarpment
121a112 ce, but I saw none embedded this point would be WORTH examining. to support. shells on surface of Pat
130a134 t. I, 1831. sur le temp du globe on Volcanos &c WORTH reading. L'Institut. 1838 p. 360. on  orbicular
134a144 parently good geological paper. by Malcolmson-- WORTH reading-- Burnetts. vol 4. p. 193 in Lat 26 deg
139aIBC ok. called R. N.-- Massac[h]usset would be well WORTH visiting really good account of ice.-- C. Darwi
165g124 hic conglomerates near Loch Lochy would be well WORTH examining-- Inverness & waters of the Tarf-- Ki
198b114 n created on the same plan. [Second resumé well WORTH studying] CD says grand idea god giving laws  &
198b114 . Philosophie Anatomique. 2d Vol about monsters WORTH reading NB well to insist upon <different anima
213b168 numerous  genera in fossil & recent state, well WORTH consideration--» Tabulate Mammalia on this prin
226b220 Australia.  England & Europe.-- It will be well WORTH while to study profoundly the origin & history
228b230 similar  or mere mongrels.-- It really would be WORTH trying to isolate some plants, under glass bell
233b251 dagascar & C. of Good Hope.-- His book Probably WORTH studying.-- Wingless birds S. Continents-- Ostr
251c040 rica.-- NB. Any monograph like Gould on Trogons WORTH studying.-- «do» Zoolog Journal Vol 2. p 221. H
270c104 matter already organised.-- This paper might be WORTH consulting, if any Metaphysical speculations ar
288c159 etch very <go> large price. as is evident to be WORTH introducing, instead of breeding from  original
308c219 e, Connection of Physical Sciences p 276 May be WORTH glancing at, as she has no original ideas., it
317c250 pper region, & some to Cape.-- some proper well-WORTH studying, with respect to forms.-- Study Append
325c267 culture of Dahlias Mrs. Gore on Roses might be WORTH consult. Paper on Consciousness in Brutes in Bl
326c266 ave children Turners embassy to Thibet, perhaps WORTH reading quoted by Malthus.-- Heberdens Observat
326c266 lop. of Anat. & Physiology.-- Dampier. probably WORTH reading Lessings. Laoccaon.-- (translated in 18
326c266 bons. Ornithological Biography. 4. Volumes well WORTH reading Bevans work on Bees, new Edit 1838 Harl
343d039 gure 19th. With respect to the Deluge it may be WORTH adding in note than amongst the Mammalia of Eur
349d052 n be established-- clearly so.-- NB. This paper WORTH referring to again.-- According to my theory, a
363d101 s have spangles on this part: this will be well WORTH working out.-- Study Temminks work on Pidgeons-
379d151 an, what is land to continent-- Original Paper, WORTH studying. Archiv. fur. Naturgeschichte. Septemb
385d164 rent bird will never sit on them.-- May be just WORTH remembering that ovarium of women (Paper in Vol
402e019 monkeys, buffaloes &c &c-- Malte Brun. would be WORTH skimming over with regard to this archipelago O
405e032 ewhat similar in two hemispheres.-- It might be WORTH investigate whether. Megatherium & Mastodon are
408e044 s might be prophecied to have this character.-- WORTH going there for.-- «Gales of wind would blend s
478z004 c &c Bennett on Chinchillidae Zoolog Transacts. WORTH reading Cuvier's Memoire 133 1803. on Pennatula
524m020 le, likely to become insane.-- now this is well WORTH considering, if pride & suspicion can be well u
526m028 l up the image in their own mind,-- this may be WORTH thinking over.-- it. will perhaps show differen
559m155 -- Hume's essay on the Human Understanding well WORTH reading Copied <Smith> «D. Stewart» lives of Ad
559m155 mith> «D. Stewart» lives of Adam Smith Reid, &c WORTH reading. as giving abstract of Smith's views «T
596n184 on the Passions." "Hartley" I should think well WORTH studying-- "Thomas Brown" on Association worthy
```

Page **(Key Word)**
348d050 acters break down in certain species & become WORTHLESS-- Mammalia Edentata‖ We do (p 6) say such is
112a089 IV p. 36. on subsidence of the land in Guiana, WORTHY of consideration. When discussing nucleuse's o
120a110 regation in Greenstone of Salisbury Craigs well WORTHY of attention-- rear Glen Roy Notebook-- & scra
213b171 out old varieties affecting the cross most well WORTHY of observation.-- I think it is certain strata
300c196 n in his arrogance thinks himself a great work. WORTHY the interposition of a deity, more humble & I
301c199 me reason, & actions habitual. it surely is not WORTHY interposition of deity to teach squirrel to ki
311c226 ans &c The doctrine of monsters is preeminently WORTHY of study on the idea of those parts being most
311c226 nsane men in civilized countries-- this is well WORTHY of investigation.-- Institut 1838. p. 174. Ape
327c265 tised. /6s Mrs Necker on Education preeminently WORTHY of studying in Metaphysical point of view Hens
355d066 lves down, they can find loopholes) "It is well WORTHY of examination whether variations are produced
358d076 e respects as this Caracara.-- Sept. 9th. It is WORTHY of observation that in insects where one of th
382d157 wo-- keeping sexes separate. Owen say such view WORTHY of a Lamarckian.-- Mine is much simpler.-- Hun
592n103 es by Parsons.» following pages contain remarks WORTHY of attention p. 15, 25. 40. 61. CD[a person is
596n184 worth studying-- "Thomas Brown" on Association WORTHY of close study.-- full of practical observatio
332d001 ipelas spreading over the head, not caused by a WOUND, when suddenly during one time he had three pat
372d129 law added to crab, tail to lizard,-- healing of WOUND.-- reproductive faculty + in the separated part
372d131 same class of facts that the skin grows over a WOUND.-- Does likeness of twin bear on this subject?
575n045 ef & pain-- certain ideas hurting brain, like a WOUND hurts body-- tears flow from both, as when one
346d048 d violently. & fell.-- gore to death the old & WOUNDED,-- see Annals. vol. 2. 1839.-- are bad breeder
437e139 stoats being carried (p. 121) & dropped having WOUNDED the bird. p. 124-- Mr Willoughby found a dead
575n045 enter water, till he sees. whether birds badly WOUNDED, or only winged.-- fetches two birds out at on
592n105 d.; for instance wild cattle & deer pursuing a WOUNDED one.-- porpoises a ditto-- it is probably some
372d130 he knowledge how to grow, & therefore to repair WOUNDS-- but this has nothing to do with generation.-
437e139 bits, rats & not being sufficiently weakened by WOUNDS got off from the young ones while they were am
555m144 off) as kind of wit, showing he had honourable WOUNDS.-- all this was kind of wit.-- I changed I bel
328cIBC ne des iles de la sonde et de L'empire du Japon WOWETT on Cattle-- (Waterhouse has it) shells from Ba
608o026 s they have none.-- Effects.-- One must view a WRECKED man, like a sickly one(P)-- We cannot help loa
230b241 (take Europaean birds. Mr Goulds' case of Willow WREN) & others varying in wild state to show that we
276c125 y are not habits different, (Mem: Gould's Willow WREN) but where close species inhabit different coun
289c160 strict, though common on another, (golden creted WREN so rare in some countries-- nightingale do.-- a
318c255 remember which way to fly.-- There is a kind of WREN (Bebyk??) which seems common in Rocky Mountains
363d102 n & other peculiarities.-- (Mem.-- Goulds Willow WREN.--) (Goulds story of Water-Wagtails mistake bot
224b213 y in any external character:-- For instance two WRENS forced to haunt two islands one with one kind o
294c177 & rook formed by descent. or two of the willow WRENS &c &c. & analogy will necessarily explain the r
064r134 ual.-- Mr Birchell says Elephant lives on very WRETCHED cou[n]tries thinly covered by vegetation. Rhi
366d111 -- Will it against genera.-- How long will the WRETCHED inhabitants of NW. Australia, go on blinking
553m137 with their mystical but sublime views, or the WRETCHED fears & strange superstitions of an Australia
640j167 ow anomalous, that the smallest newest, & most WRETCHED isld should possess species to themselves.--
532m056 rom Shrewsbury) yet for a fortnight continued WRETCHEDLY unhappy, constantly whined, would not remain
545m106 the head. I could not perceive «any» distinct WRINKLE, but such movements in skin of eyebrow importa
565n008 ion. A grave person close those muscles, which WRINKLE when smile.-- Hope is the expectant eye. looki
332d005 om coast of Guinea, -- ears bare. skin black & WRINKLED-- fur short. (tail cut off in progeny peculia
557m147 - Man when at ease has smooth brow contrary to WRINKLED: (a horse when winnowing & pleased pricks his
565n009 ake tears fall?? Lavater says derision lies in WRINKLES about the nose, & arrogance in upper lip. <Th
542m095 y objects of vision.-- Does the contraction & WRINKLING of the skin contract iris?-- same way as one
578n057 pour out tears, & there is slight convulsive WRINKLING of some of the muscles «or twitching».-- But
512q019 , as those already bred in cages. Get direction WRITE to-- (2) Does he believe. Stanley's fact of Haw
627o52v beneficial tendency of affections.-- If ever I WRITE on these subjects consult <following> pages. <p
229b233 oast. Elephant he believes is mentioned by old WRITERS on extreme Northern Coast. Hippopotamus do.--
511q018 of ears» & skeleton, & skin= Van. Voorst often WRITES to Lowe (7) In breeding. pointers. Bull-Dogs.
115a097 . if numbered compare them with my rocks. when WRITING on Falkland Islds p. 94. Von Buch's Travels ac
129a131 se of great block of rock moved by gale-- When WRITING on Valleys. «Tertiary strata of S America» rea
482z012 Jerrold?> «Bibrons coworker of Dumeril» who is WRITING with Dumeril says that two species of Tortoise
483z013 be studied in Vol I, Zoolog: Transact. before WRITING on Planariae or Polypi & is especially grand p
484z014 N. American Ornitholog must be studied before WRITING my general account-- ¿ Do not the Penguins rep
533m060 h) of Froude's life. that author remarks, that WRITING down his confessions of sins. did not make him
581n063 r trying to remember the man's Christian name, WRITING for the surname,, analogous to instinctive mem
422e092 hern Cetaceae).-- -- ‖.do. p. 318 M. Pictet of WRITINGS of Goethe.-- who maintains, that «Alludes to
086a006 st be read as well as Pallas before Geology is WRITTEN Cuvier. Europe possessed a great edentata. --
170bIFC ook was commenced about July. 1837 p. 235. was WRITTEN in January 183[8]: probably ended in beginning
184b054 what is species. very good Has not Macculloch WRITTEN on same changes in Fish Mem. Rabbit of Falklan
193b092 sld not having large quadrupeds-- Humboldt has WRITTEN on the geography of plants. Essai sur la Geogr
204b139 Loudon Mag: Septemb or Octob 1837 Westwood has WRITTEN paper on affinity and analogy in Linnaean Tran
215b178 sique. Tom 59. p 467. Peron G. St. Hilaire has WRITTEN "opuscule" entitled "Paleontographie" developi
222b204 child like father another like mother Has Lowe WRITTEN any other papers besides one in Latin one <of>
239cIFC count of the [...] the world.-- Charles Darwin WRITTEN between («beginning of» February & July 1838)
245c024 number of teats» Coquille Voyage Durville has WRITTEN Flora of Falkland Islds, where is it? All the
280c134 - Sir J.. Sebright excellent authority because WRITTEN on dog-Breaking.-- applies it to national char
293c172 brain not probable) put note. Sir W. Scott has WRITTEN about it]CD If we saw a child do some action--
316c247 nt between N. America & Europe?-- ‖.Norton has WRITTEN on fossils of N. America.-- At the end of "Whi
325c267 fathers views.-- Quoted by Owen.-- Hunter has WRITTEN quarto. work on Physiology besides the papers
327c265 N. America" in Lib. of Hort. Soc Mr Neil. has WRITTEN good article on Horticulture in Edinburgh. Enc
328cIBC by Bachman on migration of birds Temminck has WRITTEN "Coup d'oeil sur la Faune des iles de la sonde
354d062 er L'Institut p. 275. (1838) M. Blainville has WRITTEN paper to show Stonesfield Didelphis not Didelp
356d069 uch (Humboldt). G. St. Hilaire, & Lamarck have WRITTEN I pretend to no originality of idea-- (though
449e173 ed] Mr Greenough on his Map of the World, has. WRITTEN. Mastodon found at Timor.-- thinks he has seen
480z006 ondor by Humboldt Zoologie Recuiel-- Meyen has WRITTEN account of Guanaca. In transaction of Bonn Soc
480z008 own Caracara Krauss on Corallinae from S. Seas WRITTEN in German.-- Stuttgart ranks these bodies amon
492q002 into each other very much.-- Has not some one WRITTEN on them? 7... Are the wild Bananas of Otaheite
492q003 offspring be uniform.-- Mr Ford Has M. Sageret WRITTEN on crossing of Cabbages, quoted by (as if oral
510q017 s Blume say on alpine Flora of Java? Has Schow WRITTEN on double creations & where? How are current &
592n101 ed. see Hume on Sceptical Philosophy. Hume has WRITTEN "Natural Hist. of Religion" on its origin in H
599o004 out the moral sense & some metaphysical points WRITTEN about the year 1837 & Earlier--]CD in Athenae
610o031 out-- Directed his letter, & I observed he had WRITTEN Wilson & pointed it out; he was astonished, &
113a090 ng power of heat from centre.-- But is this not WRONG? we know mean of surface formerly much higher,
124a118 long period Subsidence in Demarara p. 131 (B.) WRONG Entrance. Book C. p. 101. On Frozen Soil of Sib
124a118 Frozen Soil of Siberia (with refer to Metamor) WRONG entrance Athenaeum. 1838. p. 652. Dr. Daubeny o
126a122 , & say 60 ft to degree.-- but this may be very WRONG,-- The fact of a dumplin being bad conductor is
126a122 we had rod thus & judged by increments at, how WRONG, would our judgement be-- Does condensed metal,
131a135 uces the case to show Sir. J. Herschel's theory WRONG.-- Geograph. Journal Vol. 8. p. 402.-- ground i
156g070 rmed in same spot by present torrents Macculloch WRONG in saying no transported materials <into> on up
205b145 al breeds often with same female p. 28 "It <is> WRONG to enlarge a Native breed of animals. for in <t
260c068 lcyon bird of passage.-- M. Coronata of Latham, WRONG,-- Mr Yarrell says-- that some birds or animals
299c195 mesticated kind with large root by sowing it at WRONG time of year & manuring it.-- Epigonous. Perigo
349d053 arieties being constant. it would be exceedingl WRONG to call,, one group genus & other subgenus,,-->
524m019 aware that they are insane & that their idea is WRONG.-- (Dr Ashe, the Birmingham Doctor), in this pr
525m023 - they feel shame, when doing anything which is WRONG.-- as eating meat., doing their dirt, running h
533m060 ncy in these cases? How did my mind feel it was WRONG (& it was not merely morally wrong, but hurting
533m061 feel it was wrong (& it was not merely morally WRONG, but hurting my character I felt it)-- this is
534m062 ons finding their way home-- there is something WRONG in comparing these cases, when agency is unknow
535m063 this much intellect. yet habit may make it act WRONG, as I have done when taking lid off <tea> side
536m071 or womens breasts.∴ Dr Darwin's theory probably WRONG, otherwise horses would have idea of beautiful
537m076 Mart. allows some universal feelings of right & WRONG «(& therefore in fact only limits moral sense)»
537m078 she seems to think «are» to make others happy & WRONG to injure them without temptation.-- This proba
538m079 emark about his slippers bad for fires, what is WRONG in his head. & Babington's silly joking The pos
554m140 not see her.-- <but is> then knows she has done WRONG & will hide herself.-- I do not know whether fe
584n073 re it can have affected respiration V E. p. 125 WRONG Entry Madagascar Lemur seemed to like Lavendar
603o012 ruel pleasure, so sacrifices cruel.-- Something WRONG here.-- Origin is certainly curious. Chinese, S

```
620o045 ced to the instincts.-- -- One does not feel it WRONG in very young child to be in passion, any  more
621o047 past  recollections.-- Hence he has the right & WRONG in his mind.-- Now we know it is easy by associ
621o047 that  as there nice & nasty in taste, & right & WRONG in action, so a child may be taught, or will ac
621o047 atural instincts) any action is either right or WRONG.-- [RHC] 7) Hence, what parents think will be g
621o048 instinctively benevolent) they will teach to be WRONG or right; this teaching may be curiously modifi
622o048 ense» instantaneous so declaring it is right or WRONG.-- «[just as in tastes of the mouth]CD» [LHC] M
622o049 fficient to give rise to the feeling of right & WRONG -- on which «almost» any other might be grafted
624o051 the  law of our instinctive feelings of right & WRONG,-- education, of parents strives* to same end.--
624o051 me end.-- Hence this becomes the law of right & WRONG, though, that part, which is acquired by associ
625o050v on comes into play that anger can be said to be WRONG.-- for then only is it perceived, that our pas
625o052 easiness, & that this is the feeling of right & WRONG, so far it has independent existence. & is su
628o053v - causing many actions to be considered right & WRONG,-- to be associated with the approving or disap
629o055v - p. 241 (1) Any action by habit may be thought WRONG.-- & conscience will imperiously say so, & prod
629o055v stion, how the feeling of ought, shame. right & WRONG comes into mind in first case-- seeing how sham
629o055 perhaps deviation from the instinctive. right & WRONG.-- (animals excepting domesticated ones have no
629o055 als excepting domesticated ones have no right & WRONG except instinctive ones) Perhaps my theory of g
629o055 ocial instincts explains the feeling of right & WRONG.-- arrived at first <rationally> by feeling-- r
526m031 <)>-- frame of mind, though perhaps he chooses WRONGLY,-- & what is frame of mind owing to.--<)>--  I
204b139 ffinity and analogy in Linnaean Transactions Mr WYNNE distinctly says that the mixture between Chines
204b141 e courage independently of individual force» Mr WYNNE has crossed Ducks & Widgeon & offspring either
275c120 l-dogs are used because they have no scent!) Mr WYNNE) at end of chase would not run up hill-- he too
511q018 there  ever any degeneration?? HOUNDS. Eyton Mr WYNNE, &c Could by selection a different looking anim
525m025 heredetary.-- My father says on authority of Mr WYNNE that bitch's offspring is affected by  previous
615o36v & if surprise was felt.-- analyse feelings. Mr WYNNE says, that beyond doubt courage is heredetary i
271c106 s & long bred in-- Mem, <an> a statement in Mr WYNNE'S BOOK, about not altering breed of animals in c
324c268 Plant/ If:4s Voyage aux terres australes. Chapt. XIX. tom IV. p 273 Latreille Geographie des insectes
027r023 movements  take place in semiconsolidated rocks P XV. mentions in what formations Conglomerates are f
591n101 f religion or polytheism, at p. 424 Vol. II «Sect XV. Dialogue on Natural Religion.» however, he seem
595n127 seen  others pout]CD [blank] Goldsmiths Essays No XV., on sounds of words being expressive, (Vol. 4 o
129a131 ter into this case.-- Ed. New. Phil. Journal Vol XXI. p. 213. Beyond the limits of Alps size of bould
451e180 wild animals in Moa.--" Chapt.-- V.-- : do. Chat XXI. Wild cattle & Hogs on Timor-land-- monkeys do n
243c021 nts.-- Consult Voyage aux terres Australes Chap XXXIX tom IV p. 273 2d Edit Consult Latreille. Geogra
482z011 log. Coq: p. 120 Coati Roux. Tatous & perhaps YAGOURUNDI near Concepcion!!!.-- (no species mentioned)
477z002 3  cats (mbara caya. le negro, et le pajero) l'YAGUARÉ «the zorilla-arskink» le quiyá (Coipu) viscach
503q012 . & other Hybrids-- Dogs &c &c 38 Does only male YAK cross with cow: is not reverse possible?? Maer (
112a090 er from M. Erhman stating that the mean temp at YAKOUS in Siberia being -8 Reaumur.-- there ought  to
503q018 eeding when tame in India?-- does not know About YAKS. & other Hybrids-- Dogs &c &c 38 Does only male
034r045 g the coast of NW. America P. 209--13 P & 444 «(YANKY Edit)» <I think>* At Ascension, the laminae <...
051r097 to  sand &c &c. <Vol II> P. 209. 211. 213. 444 «YANKY edition» Shores of Pacifick, from my theory, to
511q018 pearance.-- {will introduce it in work} Whether <YAR> knows whether Shaws hybrids between Trout & Sal
157g074 st measurement of shelf of 3d:-- granite block a YARD across. On side of «that» hill, in front of whi
470t176 ey or» Palmated and English, planted within few YARDS of each other actually produced hybrids-- My Fa
204b140 parents.-- How easily does Wolf & Dog cross? Mr YARREL thinks oldest variety impresses the  offspring
363d102 birds, part instinctive & part acquired.-- thus YARREL has Lark & Nightingale which both sing their o
203b138 phical or central heat.-- But then shells-- Mr YARRELL says that old races when mingled with newer, h
239c001 t horse by picking without change of habits Mr YARRELL «Give it as his theory» tells me. he has no do
239c002 ck to either parent is lucidly explained.-- Mr YARRELL states that if any odd pidgeon crossed with co
240c004 of Monster, heredetary. other adaptation.-- Mr YARRELL says, that after breeding in pidgeons with var
260c068 passage.-- M. Coronata of Latham, wrong.-- Mr YARRELL says-- that some birds or animals are placed i
275c120 distribution tells of horizontal barriers-- Mr YARRELL.-- says my view of varieties is exactly what I
275c121 n wings are Blue Pouters & small Bald Heads Mr YARRELL will mention in his work I am sorry to find Mr
307c216 rtive organs in neuter bee, (There is paper by YARRELL «in Zoolog Transactions» & Hunter on this subj
333d008 after  Esquimaux.-- this agrees perfectly with YARRELL & no leading question was put.-- Fox thinks ha
342d034 serve Bachman calls these Hybrids new species. YARRELL says the bird fanciers say the throw of any tw
342d034 hrow of any two species crossed is uncertain-- YARRELL remarks he has somewhere met conjecture that a
356d072 my  theory of generation (p. 175) if> 8th Sept YARRELL told me he had just heard of Black game & Ptar
361d096 t the change is effected through the male??)-- YARRELL observed that female of some water birds, (as
362d099 er castration, fresh horns begin to grow.-- Mr YARRELL says the «male» Axis of India, breeds at times
362d099 with  C. Campestris<)> refer to my notes) & Mr YARRELL supposes this a consequence of the female bree
362d100 have tendency to breed at particular times. Mr YARRELL has old book 1765? Treatise on Domestic Pidgeo
363d102 he forked band, like in plumage of ducks.-- Mr YARRELL says in very close species, of birds, habits w
364d103 most lost his Owl-Pidgeons from infertility,-- Mr YARRELL says in such case they exchange birds with som
364d104 ful. Knowledge that sheep originally. black. & YARRELL thinks the occasional production of black lamb
366d108 es. than are more monstrosity propagated by art. YARRELL told me of a cat & of a dog, born without fron
379d152 .  Naturgeschichte. September 11' Generation Mr YARRELL says it is well known that in breeding very pu
380d153 -- Cocks from effect of breeding in & in.-- Mr YARRELL does not know of any case of old Male. becomin
380d155 omes in Mammae & even when bitch is in heat.-- YARRELL believes Gestation is always some multiple of
449e168 this  right?-- June 18th. Eyton tells me, that YARRELL knows of a Gull, which has laid in domesticati
450e174 t cocatores & small green parrots. June 26th-- YARRELL.:-- Black Swan «in domestication & nature» str
461t028 he fish.-- Embryology p. 97. for Man Chapt see YARRELL Syngnathus Ch 6 I presume, from my theory, as
489qIFC Babington-- Gould ---- 10.(a) J. Gray ---- 17 YARRELL.-- 18 Blyth-- 19-- Mr. Tollett (T) Zoolog S
493q004 of male ought to preponderate; according to Mr YARRELL the latter ought: either in first breed or per
510q018 rieties. G. St. Hilaires law of Balancement Wm YARRELL (1) About non-breeding of animals in confineme
554m138 Jenny  with Tommy ourang.-- Very curious.-- Mr YARRELL has seen Jenny, when Keeper was away, take her
213b171 grey  hounds & will they not be greyhounds?-- YARRELL'S remark about old varieties affecting the cros
261c071 o unite the birds into group.-- it is same as YARRELL'S remark about rock Pidgeons.-- & the latter mo
275c121 ill mention in his work I am sorry to find Mr YARRELL'S evidence about old varieties is reduced to sc
325c267 the  papers collected by Owen. (at Shrewsbury) YARRELLS Paper on change of plumage in Hen Pheasants <
359d088 ke Penguin duck.-- which is strange anomaly in YARRELLS law.-- it probably is explained by the vigour
360d091 h they are kept?-- Is there any mistake about YARRELL'S law, is it local (not artificial variation) w
406e035 ufflon of Corsica. <would not>, sadly against YARRELL'S law.-- not so much against my modification of
428e112 rous battle between strong & weak March 11th. YARRELL'S law must be partly true, as enuntiated by him
449e169 the  common goose thus contradicting (probably) YARRELLS law & Walkers of the male giving form-- they
552m131 nts, (both Pachyderms) much intellect.-- mem: YARRELLS story of wheel horse in drays, scraping again
360d089 ined» by the sex? Individual instances trouble YARRELS law. chief trust must be in general  knowledge
540m084 s, <so> he smiles. Many of actions as hiccough & YAWN are probably merely coorganic as connexion of m
540m085 & smelt its fingers. Seeing a dog & horse & man YAWN, makes me feel how <much> all animals <are> bui
542m093 an  involuntary laugh will burst forth, this & YAWNING. (common to other animals) scream of agony, si
542m095 might yet remain just like sneering does.-- is YAWNING habitual from awaking from sleep see how a dog
542m095 how  a dog yawns when he awakes. & streching & YAWNING can be explained from too long rest of muscles
569n020 d beginning with the sound of letter)-- crying YAWNING laughing being necessary sounds... not produce
542m095 habitual  from awaking from sleep see how a dog YAWNS when he awakes. & streching & yawning can be ex
565n010 emember how Lavater ran away with new Lavaters,-- YE Gods!:-- says fleshy lips denote sensuality (p 1
058r116 Arequipa  in 82 was overthrown, & 86. Lima. next YEAR Quito. considers these earthquakes travel in or
067r142 f fossils deception complete.= Silliman Journal. YEAR 1835 excellent account of N. American geology.
090a023 50 years. ∴ 90 x 19 = 1710 ÷ 50 = 34 pounds each YEAR.-- but instead of 90 stones in many cases there
090a023 ow few cases recorded if we say «100» <5>0 lbs a YEAR too little.-- How comes it none in fossil state
092a028 uake 500 by 90.-- in Syria Geolog. Proc. p. 541. YEAR 1837 In Upper Assam. Geolog Proc p. 566 1837.--
093a034 us subjects) Dr. A. Smith informs me that in the YEAR a Rhinoceros was found <emb> in the mud, of the
206b146 ving souls on average are related to the (200dth YEAR) degree. Then 200 years ago, there were 200 peo
247c029 ieties of rabbits thus handed down for nearly 70 YEAR. Galapagos Mouse not the same section, with hou
259c064 emic amongst trees. Plane trees all died certain YEAR. Extreme difficulty of TRACING change of specie
294c177 seen  in old Book last Bear in England killed in YEAR 1000. reference to succession of types ¿differe
299c195 nd with large root by sowing it at wrong time of YEAR & manuring it.-- Epigonous. Perigonous &c-- ver
318c255 pot on the Alleghanies to which it migrats every YEAR,; and probably a «chance» wanderer like the fir
332d002 ods When he began practice, he remember during a YEAR or two he saw many cases of virulent cancer  in
362d099 his a consequence of the female breeding all the YEAR round. ask Colonel Sykes.-- Even our domesticat
364d103 wing hybrids can fare for themselves.‖ + + first YEAR.-- The bird fanciers match their birds to see w
```

Page ***(Key Word)***
371d128 -- may they produced not be transplanted?, & yet YEAR after year, successive roses & bud are produced
371d128 produced not be transplanted?, & yet year after YEAR, successive roses & bud are produced, like pare
375d135 age, every species must have same number killed, YEAR with year, by hawks. by. cold &c--.. even one s
375d135 species must have same number killed, year with YEAR, by hawks. by. cold &c--.. even one species of
400e013 distributed seeds of Dahlia all over Europe same YEAR.-- he sowed them for four generation before the
424e102 they meet in one wood in Anhault. & there every YEAR produce hybrids-- now this is independent good
469t119 ile to intermarriage!!xx In Phil Transact. about YEAR 1778. Paper by Camper on Ourang-outang, has exa
469t151 ean, were planted in rows, & seeds gathered same YEAR came up true «in 1840»: All in together blossom
472s01r sed to have been hybridised == Has tried several YEAR to obtain seed, but the pods have (except this
472s01r obtain seed, but the pods have (except this one YEAR (1827), always been empty.-- See separate note-
472s01v years since gave her some She means to try this YEAR. Little variation in the 60 one brighter with m
500q10a r. Contests of sexes.-- Q.30) March 1842. <Last> YEAR «before last» beans & peas were planted in rows
500q10a oining & seeds gathered there were planted «last YEAR» pell mell, without sticks & seeds gathered & t
500q10a eeds gathered & these are now to be planted this YEAR copied Gould.-- Number of species of Birds in N
504q013 e (4) Does the Thyme bear abortive stamens every YEAR & Spring. & within garden //Yes// (5) Examine t
599o004 nse & some metaphysical points written about the YEAR 1837 & Earlier--]CD in Athenaeum "Smart-- Begi
612o035 accident; the sexual willing comes on period of YEAR as much as inflorescence.-- [LHC] I here omit t
055r108 ts which strike the mind with force.--the exact YEARLY rise of the great rivers prove better than any
055r108 ations) of rain. = The Bulk of sediment «daily» YEARLY brought down by every torrent proves the decay
025r017 s of Lava in Ascencion known to be inactive 300 YEARS? No Volcanic Earthquakes or Hot Springs in T. d
041r065 ant of water.--I believe in all flat countries. YEARS of drought are common.--Mr Lyell has mentioned
054r105 lled Marques, where in all appearances not many YEARS since, the sea covered above half a league of w
090a022 s dictionary between 1768 & 1818. that is fifty YEARS-- 90 <showers of> stones are recorded as fallin
090a023 can count 90 stones which have fallen in the 50 YEARS. ∴ 90 x 19 = 1710 ÷ 50 = 34 pounds each year.--
090a024 0 = 50, 0,0,000 = 2500 = tons in fifty thousand YEARS]CD If world increased a tenth; would the pertur
111a088 p. 216.-- Guava trees, introduced about twenty YEARS since (1835) from Norfolk Isd into Geograph Jou
113a091 ed for lower beds to cool down. & then in 50000 YEARS the depth will be greater than <5000.> 400.-- T
126a123 in volcanos that have gone on for thousands of YEARS, that surface does not become hot?-- this looks
138a153 7. do has table of every earthquake, during two YEARS.-- will serve for comparison with the moon at s
146g011 rd metamorph» path only covering Great Slip, 10 YEARS since three hundred feet in vertical height-- e
181b041 ne «of them» having progeny living ten thousand YEARS hence; because at present day many are relative
206b146 e related to the (200dth year) degree. Then 200 YEARS ago, there were 200 people living who now have
206b146 200 people <might be> being related within 200 YEARS backward might be calculated & this number elim
206b146 s number elimanated say 150 people four hundred YEARS since were progenitors of present people, and s
251c040 ter from <Tarton> Barton.-- swifts return after YEARS to nest Vol II. p. 49. on the localities of cer
266c092 account of Bulbous root from Mummy: after 2000 YEARS, germinating.--!! Henslow doubts? GEOGRAPHICAL
267c093 VII. p. 216. Mr Bennett Voyage round world, 20 YEARS have scarcely elapsed since the Guava introduce
276c123 l scientific men is to push their science a few YEARS in advance only of their age. (differently from
332d001 ct upon people. My Father mention, than for ten YEARS he never saw one case of malignant erysipelas s
332d001 f communication, were seized with it, & for ten YEARS afterwards, he then did not see other cases.--
332d002 more case in a month, than in several previous YEARS, two having consulted him on one day.-- Mark at
341d031 ng female, yet so infertile never even in seven YEARS produced even an egg.-- a most curious bird, di
343d039 ise would have taken place. otherwise in 10,000 YEARS Negro probably a distinct species-- We know how
345d042 first blood"» animal must have gone on for many YEARS, before deserves <name> «to be so called»,-- th
345d042 short horned cattle have gone on for 50 or 70? YEARS-- now «well fixed» breed,: Jones says Sussex ca
362d100 his would be most curious to show that in sixty YEARS-- (how many generations) the strangest peculiar
364d104 supra» breed of cattle with white heads; which YEARS afterwards occasionally went back-- (Effect of
375d135 t geometrical ratio in FAR SHORTER time than 25 YEARS-- yet until the one sentence of Malthus no one
375d135 ived the great check amongst men.-- «Even a few YEARS plenty, makes population in Men increase, & an
398e004 of level of a few feet during last two thousand YEARS in Italy, but what «changes» would such a chang
405e031 rld as early as Elephants &c &.-- if in next 20 YEARS none of his remain found in the Americas probab
419e087 loss of species, which may be the work of a few YEARS as with the Lamantin of Steller tells much less
423e099 re> have been deposited in a period (say 10,000 YEARS) which is sufficient only to have most slightly
443e156 think: it would be doing, what others but fifty YEARS since to geologists.-- & what is older-- what r
446e162 terized.-- Tulips are cultivated during several YEARS & then they break-- -- each tulip is the <of> p
446e163 hich no doubt are propagated during hundreds of YEARS, without fresh seeds arriving.»-- throws a very
470t176 English,--the Palmated was introduced about '65 YEARS ago--& soon after mules abounded--so that palma
472s01r as a species-- Jun 1. 1842 Allen W. sowed some YEARS since gathered the seeds of Papaver bracteatum,
472s01v y.-- See separate note-- Elizabeth says several YEARS ago seeds were procured with the P. orientale i
472s01v possible to raise them pure for Miss Bent three YEARS since gave her some She means to try this year.
477z001 . According to Charpentier de Cossigny. only 10 YEARS ago <no> snail was introduced to Mauritius. 18
492q002 e any varieties of Cabbages not attacked in bad YEARS from Caterpillars. 5. Whether Roses impregnate
494q004 ow pulse being heredetary. (6) In the last 1000 YEARS how many generations of man have there been.--
506q015 Sweet Williams & Stocks, being propagated many YEARS, by cuttings.-- (40) Ask Henslow to distribute
520m002 stance as he is very deficient, he was nearly 9 YEARS old. when his father died.-- The omnipotence of
523m016 ed at High Ercall, the thoughts of it, for some YEARS after, was far more painful than the thing itse
523m017 alousy of husband connection with housemaid two YEARS before, to prove she was not insane, answered s
523m017 -- Her Husband never suspected during these two YEARS that she had been insane all the time.-- There
527m032 Migratory birds return to same quarter for many YEARS]CD.-- Beauty is instinctive feeling, & thus cut
529m042 elieves in Philosoph. Transactions, of ideot 18 YEARS old eating white lead. who was most violently p
539m082 zWilliam Musm. I was amused at this after seven YEARS interval. Augt. 15th. As child gains habit «or
546m111 find paper cutter, hunted in vain for it-- ten YEARS afterwards whilst at a meal, she suddenly like
546m111 pen to him about a knife. which he had hid some YEARS before.-- was greatly astonished, at the time.
566n011 -- Man getting sight slowly,, but when in grown YEARS, thinking he instinctively knows distances:, is
580n062 recollections of the infant likewise before two YEARS are soon lost; yet many of the habits acquired
595n121 t located] Ernest W. playing with Snow. when 2½ YEARS old. was frightened when Snow put a guaze over
639j28v er than any other one. survive. in ten thousand YEARS the long legged race will get the upper hand. t
264c080 ent localities, as my Furnarii.-- some genus of YELLOW & brown-breasted bird in Australia &c &c-- but
275c121 en took white Chinese Bantam crossed & got some YELLOW & others yellowish & white varieties by pickin
275c121 hers yellowish & white varieties by picking the YELLOW one & crossing with duck bantams procured old
288c160 ave almost banished the Grasshopper Warbler-- --YELLOW Wagtail never seen in one district, though com
466t093 ough I saw no Bees «several» visiting it>.-- In YELLOW day lily, the Bees visit base of upper petal,
467t104 om tip of sheath.-- «Also in Lathyrus pratensis YELLOW saw stigma project» In common Pea saw Humble s
467t104 er one or both of Pea's wings, stigma & mass of YELLOW pollen protrudes at sheath.-- At last I saw Be
472s02r s I examined some days ago-- This Bee flew from YELLOW to yellow & purple heartease without doubt.--
472s02r ed some days ago-- This Bee flew from yellow to YELLOW & purple heartease without doubt.-- Bee, not l
500q010 species, they do not varieties.-- (25) Does the YELLOW white Butterfly deposit eggs in all varieties
146g014 ium or rather mass of well rounded pebbles in YELLOWISH argillaceous or sandy soil-- These Buttresses
275c121 ese Bantam crossed & got some yellow & others YELLOWISH & white varieties by picking the yellow one &
595n117 - [blank] A Dog «whilst» dreaming, growling. & YELPINGS. «& twitching paws» which they only do when <
210b159 do.--there probably will be this relation also.‖YES « Fox» ‖ The creative power seems to be checked
225b217 - Opponent will say. show them me, I will answer YES, if you will show me every step between bull Dog
355d066 oach in character to goats.-- or dogs to foxes. (YES Australian dog) or donkeys to Zebras.-- «Mr Herb
373d132 Hedgehog having spines.-- Does not male Pidgeon (YES) surely) secrete milk? from stomach. analogous t
446e162 ase like Corallina-- «Does it flower anywhere?-- YES on the continent is there more variation in its
455eIBC hildren CD[Does any annual give buds, or tubers. YES-- but these are same as trees.-- Shake some slee
494q004 of another Duck. ¿in Pidgeon?-- Mr. Miller said YES with regard to former (8) Is form of globule of
498q007 ect of want of nutrition.-- Horned oranges so? --YES, my Father lost this character in grt degree fro
501q011 Knights theory with Henslow.-- Dr. Fleming says YES. (29) Are there RACES of Lupine, Stocks Clover,
504q013 e stamens every year & Spring. & within garden //YES// (5) Examine the Parnassia whose stamens move o
505q014 range. wh. never has seeds produced good pollen? YES «From cultivation lost their horns» is impregnat
512q020 there, sometimes couple but not conceive :Bears /YES/ (2) Foxes & English animals & birds breed (3) I
574n041 mber 27th.-- Sexual desire makes saliva to flow «YES, certainly»-- curious association: I have seen N
586n078 to dogs.--» they may do all this instinctively «YES because power varies in breeds,» something of ki
589n091 ts,-- could brain make a tune on the pianoforte, YES if every individual played a little, & something
601o009 fibres united with nervous filaments.-- ¿plants? YES by distinct mechanism 2. Sensation of higher ord
603o11b does not come under imitative art [my view says YES. <old> mass of rock--]CD or poetry, CD[my thery
603o11b d> mass of rock--]CD or poetry, CD[my thery says YES. imitating song -- two primary sources, sight, &

```
628o054 icial, as avarice love of gold.-- love of fame-- YES Hartley explains this & Mackintosh shows the cha
629o55v whether any thing is taught instinctively; I say YES, & my explanation agrees. with last head.-- (4)
042r068 ose aerial Volcanos in Germany» In the Valle del YESO it is probable that point of Porphyry has been
472s02r of heartseases, never saw any Bee go to them.     YESTERDAY remarked that many flowers had suddenly withe
025r019 ottom of the sea.--«certainly data insufficient, YET  good» «(I suspect fragments of shells will gener
026r019 med in shallow water: so have the Conglomerates:  YET  this view is directly opposed to common opinion
035r045 ost unfavourable to preservation of bones &c &c-- YET  <silicified> turn over silicified wood. Cordille
040r062 nd, which agrees with St Helena in being unique,  YET  no quadrupeds. -- Is the white matter beneath pe
042r068 distance?-- Fossil bones black as if from peat.--YET  cetaceous bones so likewise «of miocene period».
043r071 iloe. Concepcion. Valparaiso (Copiapò & Guasco).  YET  whole territory vibrates from any one shock-- In
044r072 the cliffs--wide valleys.--central peak small;    YET  great body of lavas have flowed from centre-- Pi
045r076 he country; (perhaps generally over whole world)  YET  eruptions <both> at sea (as wells as in the Cord
046r079 subterranean  fluid mass itself changed.--No. --  YET  the fluid granitic mass under <[...]> less press
049r087 ked by do.» discuss quartz veins, there contemp-- YET  similar ones in Clay. Slates contemporaneous oth
051r094 is owing to protection of Organic productions. -- YET  everywhere on coast (Il Defonsos «Kelp») rocks s
055r107 liffs at Ascension (or modern streams of St Jgo)  YET  no historical records of eruptions how immense t
056r112 n & so is Volcano a useful chemical instrument. = YET  neglecting these final causes.--What more  awful
061r127 if  for volc. isld. then for any spot of land. =  YET  new creation affected by Halo of neighbouring co
065r139 «buried in a mound» long consigned to the earth.  YET  body had scarcely undergone any decomposition: c
066r141 es the tides form eddies with its extreme force.  YET, no outlet at head. Important in forming transve
070r153 s. certainly different. not insensible change.--  YET  one is urged to look to common parent? why shoul
086a004 the prejudice of not believing recent elevation, YET  sea shells at tops of mountains we ought to symp
089a017 ly Volcanic-- CD[Might not bottom of ocean boil;  YET  heat never reach surface.-- Journal de Physique,
089a018 ingo formed of coral limestone, with interstices  YET  emptty.-- In all the mountains of Saint Marc  et
100a052 ement of particles of granite in Henslow's Grit,  YET  it is worth consideration. especially effect  of
104a068 a-slate of whole kingdom of Norway was contorted  YET  no mountain chain case parallel to Banda Oriente
119a104 ndes or Himalayas, but great circular mountains,  YET  so analogous, that as we see mountains formed (&
121a111 nts of rock, which retained their angles sharp--  YET  with character completely altered, & a crystalli
121a112 ing. to support. shells on surface of Patagonia,  YET  none in shingle beds. Lyell on Sweden. p. 12. pr
132a137 te rate of increase of heats in earth's crust.--  YET  heat does increase,-- but in Ocean does not. (se
145g001 emale p 367 Quarterly Journal of Agricl Dec 1837  YET  instances given against it-- Mere fact of many r
147g020 egular hills of alluvium-- nothing very striking  YET  possibly sea more probably than river-- No exact
148g023 them all along <Glencoe>.-- towards Fort William  YET  in Glencoe in parts no trace of them-- Mem Coast
157g073 ebbles of granite & forming «sloping» buttresses  YET  certainly shelf 4th <near> only usually contains
171b002 to breed.-- annuals rendered perennial. &c &c.--  YET  Eunuchs nor «cut» Stallions nor nuns are  longer
181b043 lata. still greater between animals & Plants But  YET  besides affinities from three elements, from the
181b043 ta invertebrates-- Such on few on each side will  YET  present some anomaly & <the g> bearing stamp  of
182b049 , extending range, reason & futurity. it does as  YET  appear clim In Mr Gould Australian work some mos
192b086 e never may have been grade between pig & tapir,  YET  from some common progenitor,-- Now if the interm
194b096 lants, which have male & female organs together,  YET  receive influence from other plants.-- Does  not
195b099 e in neck same cause, such beautiful adaptations  YET  other animals live so well.-- This view of propa
196b109 ¿could plants live without carbonic acid gaz.)--  YET  unquestionably animals most dependent on vegetab
196b110 ertebrata, that there must be very great gaps.--  YET  some analogy The existence of plants, & their pa
197b112 onsters «also» organs of lower animals appear.--  YET  nothing about propagation= I see nothing like gr
200b122 nimals as species, for instance Australian dog: <YET  when that? or Falkland rabbit.-- There is only t
200b123 rifle heredetary, without some cause of change,,  YET  such causes are most obscure. without doubt:-- V
206b148 tary disease, effects of contagions & accidents:  YET  some causes are evident, as for instance one man
208b154 uction of those regions, & consequently not many YET  multiplied: NB How does this bear with law refer
209b155 mp & Common;-- between Esquimaux & European dog?  YET  man has had no interest in perpetuating these pa
220b197 plants,  you cannot say that instinct perverted,  YET  organization «especially» connected with generat
222b202 «agrees with» according to Beck has none recent,  YET  genera same.-- Speculate on multiplication of sp
222b203 return to one parent, this is only character., &  YET  we find this same tendency (only less strongly m
223b210 hers, & cut off limbs & new ones are formed) but  YET  propagates varieties according to same law  with
228b231 expect the father of man kind. than Macrauchenia  YET  he may be found:-- We must not compare  «chances
239cIFC ere hybrids produced have any close species ever  YET  failed. About trades affecting form of man. Coul
240c003 so different, that they would be called genera.,  YET  retains markings of wings like the wild rock pid
240c003 of  form, may explain mule & pig being half way.  YET  dogs sometimes like father, sometimes like mothe
240c013 a. a Kangaroo d'Aroe (Didelphis Brunii) which as  YET  had only been found in isle of Aroe & Solor), «V
246c027 l II p. 8 no snakes on isles of central Pacific,  YET  there appears to be one at Botouma from  account
250c038 n of the world «Africa» being left more equable.  YET  America preeminently equable. might have allowed
256c054 nguishing which were species, (theory admirably)  YET  a glance would tell from which country,-- I ofte
264c079 asting his parent, naked, artless, not improving YET  improvable» & then let him dare to boast of  his
264c080 ide into shrikes & at the other into into Crows.  YET  all forming, according to Gould, good genus Goul
265c083 six  fingered people are sometimes heredetary,-- YET  these not adaptations «they are counteracted  by
265c083 and must have been lopped off & sheeps tails cut YET  there is no record of any effect.-- New Hollande
265c085 od, but adaptation.-- albino however is monster. YET  albino may so far be considered an adaptation as
265c086 beetle  with wings beneath soldered wing-cases-- YET  these wings may be of some use.-- Nature is neve
266c091 e & manners they are as opposite as day & night: YET  we know how remote the periods at which both lef
271c105 it feeds. or insects it devours is same species. YET  that it should so strictly <f> agree in habits w
273c113 e wing & pratencole (¿connecte[d] with Chionis), YET  the Tropic bird, has very different habits.  thou
278c132 t animals not preserved, in central S. America & YET  Africa & India???-- & Indian Islds.-- Sir J. Seb
280c133 ks on gaps.)-- Crosses of diff: breeds succeed,  YET  seems to grant, that difficult & other go back t
280c136 by great effort <pr> one unlike can be produced, YET  to produce whole generation unlike would go agai
281c138 ch in an absolute human family is undescribable, YET  holds good, so does it in real classification Th
282c140 imagine  the men to have greater power of change YET, as external conditions over whole world. simila
282c140 inally similar limits of change, would be same-- YET  each family might have its own character,-- we h
282c141 ould be possessed by the different races of man, YET  altogether different.-- To make this case perfec
286c154 I  believe those who soar above Such prejudices, YET  we might have these analogies.-- We must two rac
286c155 godlike,   at least every nation has. done so as.YET.-- We now know what is the natural  arrangement,
292c171 would  banish individual, but general ones might YET  be transmitted.= Memory springing up after  long
293c173 nalogy. a bird can swim without being web footed YET  with much practice & led on by circumstanc it be
296c184 t Australian) on Peaks. Did Creator make all new YET  forms like neighbouring Continent. This fact spe
299c194 wn to be same.-- one grows in marsh & other dry; YET  if T. palustris be sown in dry station it will f
302c201 ch when sufficiently Multipliied become affinity YET  often retaining a family likeness, & this I beli
303c205 said»   that although at any one there. are gaps.YET  <what> altogether «he» has created a perfect cha
304c208 nch high up.-- Such probabilities only guides.-- YET  trifles are produced by circumstances. Spines on
308c217 others in spirit-- all children of one father.-- YET  differences carried a long way. --Case of Habit
314c239 arcely any Australian character in Timor plants, YET  it seems there may be Eucalyptus!-- (Hostile fac
316c244 difference  between the mind of man & animals.-- YET  how faint in a Fuegian or Australian! why not gr
316c244 igonia in Australia or Concholepas in America,-- YET  many countries have far more species than  other
316c245 oma in Phillippines & Anphidesma in S. America-- YET  there are a few Cyclostomes & a few Anphidesmas.
318c255 f Migration) gradually separated the birds might YET  remember which way to fly.-- There is a kind of
319c255 but our blackbird exceeds their other thrushes-- YET  they have one with very sweet notes.-- Their sof
334d012 ely, changed his residence a great many miles.-- YET  one day <th> a cow walked in, then  disappeared,
335d013 -converse, what has not been, will not remain,-- YET  offspring must be somewhat like parents,-- there
341d030 s animal, which is so anomalous among true deer, YET  is spotted like so many deer.--- very curious lik
341d031 ea Fowl with Pea <cock> Hen.-- offspring female, YET  so infertile never even in seven years  produced
343d038 hough the negro of Africa is not loosing ground. YET, as the tribes of the interior are pushing  into
345d043 d, but this was bred out & now all are pure red, YET  calf every now & then born with white head  (,or
345d044 that  male impresses offspring more than female, YET  instances given on opposite side,-- «The  theory
347d049 of changes. then animals must tend to improve.-- YET  fish same as, or lower than in old days: «for  a
351d056 es of what she does.-- does not move per saltum-- YET  does nothing in vain!! Foetus of man undergoes m
358d085 ssible, the organs doubtless would shrivel up.-- YET  odd they should have so much sexual character as
368d102 ferences were pointed out Selby confounded them, YET  can readily be told by incubation & other peculi
366d112 e time, & make breed, one would doubt any law.-- YET  seeing the feathers along one toe of the  Pouter
371d128 ere?-- may they produced not be transplanted?, & YET  year after year, successive roses & bud are prod
371d128 continue of same variety as long as life lasts, YET  they cannot transmit through seeds these charact
```

374d133 sh, & not intermediate between fish & reptiles-- YET osteology closely resembles reptiles.-- p. 432 s
375d135 trical ratio in FAR SHORTER time than 25 years-- YET until the one sentence of Malthus no one clearly
376d136 learn its comfort & though could not put it on, YET threw it over it, & made it meet in front.-- Dr
377d139 time, necessary to form channel & (& Basses St) YET no change in English species-- time no element i
378d147 not true for he says Hen & cock Starling alike, YET young ones brown.-- Sexual Selection If masculin
380d154 eous birds have cock & hen plumage so different, YET the Cassowary & Guinea Fowl cannot be distinguis
382d158 <are> «though» are, so plain in Man, & YET no trace of abortive womb, or ovarium.-- or test
383d160 ss is effect of Metallic hue of silver pheasant. YET why green? & not purple?-- legs pale coloured.--
384d162 for the double purpose are not distinguished. --YET may be presumed from hybridity of ferns) afterwa
385d163 lose   passion. (Mem: so it was said little cock «YET very odd loosing visible powers» in Zoolog Garde
397e003 Epidemics-- seem intimately related to famines.. YET very inexplicable.-- do p. 529. "It accords with
398e006 ut entire obliterations & fresh laws created., & YET with <gov> symmetry «& regular laws» that baffle
399e006 y: a modern bed at present might be very thick & YET have same fossils.. does not Lonsdale know some c
402e019 sects of St. Peter & St. Pauls in Lat' 53 degree YET fauna like that 60 degree & 70 degree of Europe.
405e031 re of a nearly uniform <dusky> blackish brown.-- YET retained a trace of horizontal mark on flank.; &
407e038 al productions. The world formerly much more so. YET climate of same order as that of S. America.-- (
409e047 dog & a porpoise was not thougt overwhelming.-- YET I will not shirk difficulty-- I have felt some d
409e047 ed man.-- ask the missionaries about Australians YET slow progress has done so.-- Show a savage a dog
410e052 single specimens in» skins worse than useless.-- YET there is no cure «I may say all this, having mys
411e056 lline powers of multiplication of individuals, & YET another means for individuals (Mem: transportati
411e056 here innumerable individuals can be produced. & YET sexual apparatus.-- My account of Circus cinereu
414e063 iced judgment. even without time can do much.-- (YET one cross, & the permanence of his breed is dest
415e066 inct species» of the Ornithorhyncus be found.;-- YET until man became cosmopolite, he would probably
416e071 cts?.-- all!??!?-- Worms? [Barnacles, aquatic., <YET> Crustacean, & true hermaphrodites] CD «It may b
419e085 d Fauna of Caspian.-- fishes fresh water kinds. (YET living in the salt?.)-- very few animals of any
419e086 owness of physical changes, than in any other, & YET 200-300 ft elevation & no change & even no loss
420e089 ircumstances, as shown by difficulty in forging, YET handwriting said to be heredatary. shows well wh
422e095 resh, means of adding to their complexity.-- but YET there is no «NECESSARY» tendency in the simple a
424e102 t ranges-- latter not going North of the Elbe.., YET they meet in one wood in Anhault. & there every
429e113 appearance of varieties of domesticated animals, YET as we know how many plants have been produced (1
429e115 gins at 1600 metres precisely & stops at 2600. & YET know that plant can be cultivated with ease near
434e129 ent species. Lychnis dioica, generally dioicous. YET parts only very slightly abortive & bed of femal
436e135 rix has a most perfect Struthio head pulled out. YET feathers retain character? If separation in hori
436e137 th could have produced these innumerable seeds-- YET if a seed were produced with infinitesimal advan
437e141 g suddenly to type when brought back to home. (& YET all the varieties of Brassica certainly not beco
437e141 & depends on character of antecedent races.-- «& YET in all probability the Brussels Sprout was slowl
438e141 ference in crossing between varieties & species, YET the amount of may depend on many circumstances,
439e143 oves how much depends on instincts in animals.-- YET the existence of wild close species of plants sh
440e146 ns with tufts &c &c here there is no final cause YET it must be effect of some condition of external
440e146 ion: as we see these strange plumage in pidgeons YET no change of habits, so no <cause> corresponding
441e149 Australian Dog is not affected by domestication, YET offspring are,-- if Australian Dog, could bud, a
443e155 ese «sexual» functions to complexity is evident, YET the inference from some plants & some mollusca b
445e157 n the Icthyosaurus 60 or 70 bones in the paddle, YET all in the arm are perfect.-- p. 144.-- Alludes
445e158 282. Allows this instinctive power in chicken, YET says it is evidently acquired by experience in b
446e162 of Lemna only reproduces itself «in England, as YET observed» by buds-- (the other three by buds & s
447e165 is> rendered vivaparous by growing on heights.-- YET he has seen it propagated in a garden, which is
447e165 ira caespitosa becomes vivaparous on mountains & YET can be raised in gardens.-- Poa alpina, althougt g
448e167 erga & Paris, numerically the same with recent & YET almost wholly different, is same, as if Isthmus
448e167 f Panama.-- These two cases highly improbable.-- YET I can see no other way of accounting for them.--
451e178 ntioned by Heber) «from» which «all» mankind «(& YET afterwards says native tribes can live there)» f
453e183 istinctly, that Hollyoak reproduce each other. & YET I presume seed raised in same garden.-- now this
461t041 anges of conditions, some species will undergo & YET remain adapted.-- it does away with difficulty o
462t051 ion in Java & Sumatra & dissimilarity of forms-- YET how valueless this objection, when one thinks of
463t055 He would not allow such series showed passages-- YET in talking, constantly said as the <brain> spina
463t057 ailing, climbing & mud-walking fish? difficult-- YET suggested. (vipers tooth also a difficult), the
464t065 ous Bird from Nзw Zealand-- <so> not an Apteryx, YET it shows the Apteryx is not «quite» isolated in
465t079 here has been several successions of Mammals.--» YET only two monkeys, <there are now> have been foun
465t079 shells; where land broken, rivers entering.-- & YET no shells-- now look at Scotland-- coasts of Chi
467t099 ds that although anther «nor filaments» shrivel, YET stigma does not, so we may feel somewhat «but li
468t105 of chopped horse hair with legs & take flight-- YET we have crosses-- I see Bees almost every flower
473s07r ne/42/ You can select cattle & sheep for horns & YET no difference in calves--how is this in young pi
498q007 rds become unproductive from poorness of soil.-- YET crabs probably would grow there (7) Where parts
503q012 le & Horsfield (35) Talk about races of Banana & YET seedless-- no light Henslow or Royle, latter say
522m014 perhaps, slightly injured me, plotting speeches, YET with a sort of consciousness not just.-- From ha
523m014 m Doctor praising his sister who confined him. & YET disinheriting her.-- This «N B. I have read pape
524m022 escribed, (forgetting that her husband was dead) YET they have not imagination enough to <up> recall
526m028 r it so well that they can correct every detail, YET in some inflammatory diseases, where there has b
529m043 book of Sheriffs.-- July 22d. 1838 No Deliriums, YET when he paid the Attorneys bill, he asked what b
530m043 neys bill, he asked what bond he could have had. YET during whole illness, he had been able to direct
530m043 bsolutely forgotten.-- My father signed a bond, YET for a fortnight continued wretchedly unhappy, co
532m056 ard & of Mary & her bed brought from Shrewsbury) YET who will venture to say germ within egg, cannot
534m061 al born with instinctive knowledge.-- but if so, YET this knowledge acquired by senses,-- then thinki
536m061 gh says to say. Brain per se thinks is nonsense; YET habit may make it act wrong, as I have done when
536m074 it disputed that a wasp has this much intellect. <YET> would more earnestly pray, «deliver us from temp
537m074 ould tend to be an atheist. Man thus believing.- YET it is settled by reason.-- How slow habits are
537m074 caprice. it is chance, which way it will be, but YET as I think, the opposite side has been shown--
537m075 sense.-- «from difference of action of approved» YET perhaps if they had murdered their children, thi
538m079 ian mothers ceased to destroy their offspring--¿ YET Allen. W. remark about his slippers bad for fire
538m080 r when drunk they would not be more different, & YET they would make one's father & self one person--
539m081 so intently as to be unconscious of all around, YET there was no strain on the intellectual powers--
541m092 mbers in their head must have this high faculty, YET not clever people. Aug. 21st. 38 When a dog in p
542m094 ger nor have monkeys & many other animals,-- but YET when angry it is hard not to growl out some soun
542m095 anization were changed, I conceive sighing might YET remain just like sneering does.-- is yawning hab
543m099 larly with D'Arblay of Christ of great genius, & YET invariably used to beat him-- The son of a Fruit
543m100 l that his Father only left him a guinea week.-- YET. he was inimitable chess player.-- Peacocks rema
544m103 od-- The mind wills to do this & hears that, but YET scarcely really moves.-- the willing therefore i
547m112 ite still, & not attending to bodily sensation & YET the Castle would not have turned into dream.-- I
547m113 <thinks> quicker in sleep, it may do less work & YET do so, from the exertion of keeping up the memor
549m118 ure) than perhaps well <regulated> philosopher-- YET the philosopher has a much more intense happines
556m146 t now if horns were to grow on horses, they must YET continue to put down ears, when kicking.-- -- so
557m148 ause less discoverable in animals than latter.-- YET I think one can remonstrate with a dog, & make h
557m148 ation to jump away,-- it is, ill-defined fear.-- YET one knows oneself it is quite different from the
558m150 sense arise from our enlarged capacity <acting> «YET being obscurely guided» or strong instinctive se
563n001 pea-hens do Wood pidgeons building near houses. YET so shy at all other times.-- Birth Hill shows it
564n003 n to any animal with social & sexual instinct «& YET with passion» he must have conscience-- this is
565n007 on look obliquely.-- who can analyse suspicion-- YET who does not recognise look of suspicion, even c
567n015 one may like. dislike, or be indifferent about, YET feel shy.-- not if quite stranger.-- or less so.
570n026 e that a mountaineer <takes> born out of country YET would love mountains, & a negro, similarly treat
573n036 c &c:--must be a special act, or result of laws. YET we placidly believe the Astronomer, when he tell
576n047 not <t> do so even to save a friend, or wife.-- YET he would ever repent, & wished he had lost his l
580n062 infant   likewise before two years are soon lost; YET many of the habits acquired in that age are reta
585n075 icks with front legs & knocks with back of Head, YET never puts down its ear. good to contrast with h
585n077 r sense-- "We know not how, stonge henge raised, YET not instinct, but if all men placed stones in sa
588n090 es this reasoning apply chiefly to recollection. YET a dog hunting for a bone shows he has recollecti
589n092 . M. says of pure reason not leading to action & YET our emotions being only bodily actions associate
594n112 re,, that probably few had ever before seen one, YET all-- flew to bed of flags. hernes are common. n
603o11b inted must not a actors, or a scene in garden.-- YET both beautiful! p. 136. Says Architecture does n

```
604o016 ness,, it choosing food-- crawling from light.--  YET we can split Planaria into three animals, & this
606o20v y a long series of experiments & observations. &  YET, like in vision, it becomes so instantaneous. th
606o025 action  (as jump off a bridge to save another) &  YET dare not -- one could do it, but other motives p
608o026 being better & making him happier.-- he agrees &  YET does not.-- because motive power not in proper s
608o027 nothing  will cure him' 3) disgusted. with them.  YET it is right to punish criminals; but solely to d
608o027 humility,  one deserves no credit for anything.  (YET one takes it for beauty & good temper), nor ough
616o038 me of mind which makes music pleasant, a memory;  YET that frame is enhanced by memory of what has bee
618o041 A  person might be quite familiar with thought &  YET be ignorant of the existence of the brain. We ca
618o41v as  a thing grew blue it «uniquely» grew heavier  YET it could not be said that the blueness caused th
620o045 hich of its instincts are best to be followed.--  YET even at this time, malevolence,  when not urged
621o047 hat nasty, which the natural tastes say is good.  YET horseflesh show that even this is possible.-- So
622o049 from whom, she has never received any benefit.--  YET I think there is much truth in doctrine, for [RH
623o050 asure from the place, but merely in the place. &  YET place calls up pleasure.-- [LHC] the instinct of
623o050 to  say how far & minutely our instincts extend,  YET as they are acquired by social animals, living u
625o50v stincts. to gain long-lived good, ie happiness--  YET this system not selfish.-- explained by principl
636j57r tiff tailed cormorant: pain & disease in world &  YET talk of perfection Get instances of adaptations
637j58r es especial adaptation, to «young».-- good God &  YET Mails have them. What trash p. 237. Gives as Sum
638j58v made by intellect this process is shortened, but  YET analogous, no savage ever made a perfect hinge.-
641j28r because it must take so long to change species--  YET this is contradicted by continents <bri> aboundi
080c178 acter of a Araucarian tribe, with point. affin of  YEW & intermediate Puncture one animal with recent d
196b107 eland possesses varieties of the furze, broom, &  YEW very different from any found in great  Britain,
499q009 lows or Yews some poplar's produce.-- (15) Would  YEW fruit without impregnation.-- (16) Any calculati
499q009 (18)  Place pin's heads with Bird lime near male  YEW tree & see whether they catch pollen-- <Ne> In O
503q013 oss with cow: is not reverse possible?? Maer (1)  YEW Trees near Boat House «ANY male branch.» --¿numb
504q014 bbages.-- kept true Try experiment (30/p.11) (2)  YEW Berries germinate?-- Yew trees sexes-- (3) Get H
504q014 xperiment (30/p.11) (2) Yew Berries germinate?--  YEW trees sexes-- (3) Get Holyhoaks. races planted &
499q009 distances from males, will female (a) Willows or  YEWS some poplar's produce.-- (15) Would Yew fruit w
102a059 when  dikes reach near the surface. that strata  YIELD.-- In Undulation in open ocean. as pebbles woul
130a133 coast  of Persia In Glen Roy paper I show crust  YIELD easily. & if easily must be thin: <beside  mere
078r175 additional information ⫙ Guanaxuato, which has  YIELDED the most metal, where the direction of ravins,
101a058 ze with superincumbent strata. where they have  YIELDED conical axis of mountain.-- only when dikes re
138a176 ing on the top-- [blank] Would rotting wood by  YIELDING Carbonic Acid unite with «piece of cabbage» a
102a061 ittle pressure? An important question! If water  YIELDS substances from impact, «it» would look like i
122a113 - {P} will point {P} be the one which generally  YIELDS.-- Will this not explain littoral mountains  &
294c175 . says Sheep could not live for some time at New  YORK «instance of the fine relations of adaptation o
317c251 . that near Charlestown ?three species, near New  YORK. (600 miles N.?) replaced by three other specie
342d033 been  shot wild (escaped from Carolina?) off New  YORK. therefore instincts not imperfect.-- Are Pheas
553m135 any vicarious, like ourselves) that savages (mem  YORK Minster) consider the thunder & lightning the d
316c244 to  the wildest imagination & superstitions.-- +YORK'S Minster story of storm of snow after his broth
126a122 conductor,  then still thinner → The problem is,  YOU have temperature known at surface,-- you have te
126a122 lem is, you have temperature known at surface,--  YOU have temperature known far below surface, say 10
185b056 present every other; for instance in Heteromera,  YOU have representatives (which at first would be mi
200b123 whether in crossing very opposite races, whether  YOU would expect equal fertility-- ditto in Plants <
213b171 ss,' but do not known varieties do the same, May  YOU not breed, ten thousand grey hounds & will  they
220b197 pe, as to produce same one. ⫙Although in plants,  YOU cannot say that instinct perverted, yet organiza
225b217 nt will say. show them are, I will answer yes, if  YOU will show me every step between bull Dog & Greyh
227b225 on  amount changes which may happen-- ‖ It leads  YOU to believe the world older than geologists think
230b235 approaching desert country or ascending mountain  YOU ought to have a gradation of species, now this n
230b236 f species, now this notoriously is not the case,  YOU have stunted species, but not such as would make
231b241 <to> together, by elevations now in Progress, &  YOU will have two. Tapir existing in East Indian Sea
239cIFC ailed. About trades affecting form of man. Could  YOU get racehorse from Cart horse by picking without
259c065 le that, cut a sheeps tail off plenty of times &  YOU will have no tail (example probably not true).--
271c107 aser remarked to me at Zoological Society,, that  YOU never find two «similar» groups of birds in  two
273c110 others large, then he says from long experience,  YOU may be almost sure, that there exist intermediat
283c143 .-- Gould says it is only in large groups. where  YOU have representations.-- The aerial type in  each
291c166 .-- love of the deity effect of organization. oh  YOU Materialist!-- Read Barclay on organization!! Av
293c172 ation. [NB what are those Marvellous cases, when  YOU feel sure you have heard conversation before. is
293c172 t are those Marvellous cases, when you feel sure  YOU have heard conversation before. is strong associ
473s07r s described Indian Plant.-- June /42/-- June/42/  YOU can select cattle & sheep for horns & yet no dif
505q014 shaped by care 13 Arum before pollen is shed can  YOU find flys dusted with pollen from other flowers?
507q016 e males of a family-- Where one tooth aborts, do  YOU know whether any trace in germ. (2) Any more cas
521m010 rdener & said, who is tha? Mr C. answered why do  YOU not know, that is A. B my gardener.-- Thus was h
522m010 ed he had never heard of him).-- My F. then said  YOU remember Jack Baldwin at school.-- Answered To b
559m153 forget the use of language, & judge only by what  YOU see. compare, the Fuegian & Ourang & outang, & d
570n024 e gulps in air.-- Again a master says I will see  YOU damned first." the man shrugs his shoulders & re
578n053 ffect of mind on individual parts of body.== (if  YOU <think> «fear» you shall not have e---n, «or wis
578n053 dividual parts of body.== (if you <think> «fear»  YOU shall not have e---n, «or wish extraordinarily t
578n053 ave e---n, «or wish extraordinarily to have one»  YOU wont. ==)== No surer way to blush, than particul
581n063 16th.-- Is not that kind of memory. which makes  YOU do a thing properly, even when you cannot rememb
581n063 . which makes you do a thing properly, even when  YOU cannot remember it. as my father trying to rememb
602o11b Vol I. p. 115. Attributes of Deity. on Belief.--  YOU belief things you can give no proof for, & one o
602o11b ributes of Deity. on Belief.-- you belief things  YOU can give no proof for, & one often replies "what
602o11b can give no proof for, & one often replies "what  YOU say is perfectly true, but you do not convince m
602o11b en replies "what you say is perfectly true, but  YOU do not convince me.--" Belief allied to instinct
616o038 eward! X Perhaps should hardly be called memory;  YOU cannot call the frame of mind which makes  music
027r024 scending steps Mem.; rapidity of germination in  YOUNG corals.--vide L. Jackson's paper. Philosoph Tra
029r033 log-book  of the James Cruikshank, Captain John  YOUNG, on her voyage from Demerara to London:-- "Feb.
148g026 fatten,  This man confirmed the account of the «YOUNG» Shepherd dogs Saturday. Before coming to Bridg
171b003 ce living beings.-- We see <living beings>. the  YOUNG of living beings, become permanently changed or
207b151 rry duckling to the water in their beaks, & the  YOUNG one <inland> directly by instinct. can dive & c
222b203 lled varieties.-- NB. one mother bringing forth  YOUNG having very different characters is attempt  at
240c014 liance with New Holland. The Barbaroussa, (when  YOUNG very like the Siam race with long nozzle &  few
235c051 re structure (habits of ducklings and chickens)  YOUNG water ouzels hence aversion to generation, befo
257c059 .-- It is ASSUMPTION to say generation produces  YOUNG ones capable of producing young ones like itsel
257c059 ration produces young ones capable of producing  YOUNG ones like itself, but ¿whether great assumption
260c067 ferent kind have been know to assist in feeding  YOUNG cuckoo; as if there was storge, which could not
268c096 I~ p. 159 curious account of Tit mouse feeding  YOUNG of redstart & actually driving away parent bird
276c124 ure) there cannot be gradation. See what Eytons  YOUNG pigs-- if vertebrae much lengthened & there ma
280c134 ct (Sir J. Sebright admirable essay) heredetary  YOUNG wild ducks.-- lose as well as gain instincts. W
284c149 anc between some form in birds is visible, when  YOUNG, but not when old.-- thus speckled form of youn
284c149 oung, but not when old.-- thus speckled form of  YOUNG blackbird. good remark if general.-- Where  any
285c150 & grandfather <Might> Must be introduced & made  YOUNG.--⫙ father must be left out of case, that diffe
289c160 icely adapted species to localities⫙-- p. 390,.  YOUNG <grebes> «ring ouzels» dive instant touch the w
289c162 Dr  Johnston <on> Entomostraca Daphnia, produce  YOUNG, capable of producing young many times & lay tw
289c162 ca Daphnia, produce young, capable of producing  YOUNG many times & lay two sorts of eggs-- one remain
295c178 or has he assumed that character -- -- female &  YOUNG seem most like mean characters the others assum
295c183 llion tempted to cover old mare by being shown,  YOUNG one.-- Many African monkeys in Fernando Po-- no
306c212 ud, polypus & germ plant & seed.-- instincts in  YOUNG animals, well developed, just like, habits easi
306c212 ust like, habits easily gained in child hood.--  YOUNG salmons. first a species which lived in estuari
313c235 osity of the orang outang of (in June 1838 when  YOUNG male was added good instance of instinct showin
334d010 se of Stallion, according to Erasmus preferring  YOUNG mare to old, explained by Stallions, (according
341d030 characterized  by Ogleby, who observed that the  YOUNG of this animal, which is so anomalous among tru
342d035 logy of birds to Reptiles shown in osteology of  YOUNG Ostrich. 16th. D Israeli (Cur of Literat. Vol I
344d041 y valuable facts-- they are eating foetuses, as  YOUNG of Marsup. is sucking foetus.-- August 23d The
346d048 ild cattle of Chillingham,-- habits peculiar,--  YOUNG one 203 days old butted violently. & fell.-- go
346d048 white, uniform.--crafty, go in file, hide their  YOUNG., bold.-- a Mr W: Hall remarked that it was aga
350d055 ces in sex confined to annulosa?) Remarked that  YOUNG of Cirrhipedes can move & see, parent  fixed,--
350d055 of  Cirrhipedes can move & see, parent fixed,--  YOUNG of sponges move.-- young of Cochineal insects m
```

```
350d055 & see, parent fixed,-- young of sponges move.-- YOUNG of Cochineal insects move about & see, parent «
352d057 ave larvae more perfect-- this is applicable to YOUNG of Cochineal?? Is there some law in nature an a
355d068 s of same plant<s>)-- now in some Polypi we see YOUNG bud changing into ovules.-- Captain Grants. Him
358d076 larva, or less developed state.-- the female & YOUNG of all birds resemble each other in plumage «(t
359d088 y animal certainly seems chiefly to impress the YOUNG most with its form & disposition Saw three youn
359d089 oung most with its form & disposition Saw three YOUNG ducks, like each other,-- (& not very like eith
361d095 age, other alters entirely]CD In common sparrow YOUNG & female similar plumage.-- in tree sparrow, (i
361d095 e.-- in tree sparrow, (if I understand rightly) YOUNG cock & hen, all nearly similar.-- in blackbird
361d096 & hen, all nearly similar.-- in blackbird group YOUNG like some of the species-- (¿do these facts ind
363d103 mpare with fallow? deer. & Moschus &c & -- like YOUNG blackbirds Dr Bachman told me that 1/2 Muscovy
371d127 here is what [not located] that it shall beget YOUNG different in colour, form, & so altered in disp
373d132 from stomach. analogous to other males feeding YOUNG, & to abortive <organs> «mammae» in male Mammal
377d147 fitted for water, only habits & instincts-- The YOUNG of the <p> Kingfisher (.p. 169) has the colour
378d147 has the colour on its back bright blue.-- «thus YOUNG of» Many of the pies assume the metallic tints,
378d147 ays» whenever metallic brilliancy is present in YOUNG birds, one may be sure cock & hen will be alike
378d147 true for he says Hen & cock Starling alike, yet YOUNG ones brown.-- Sexual Selection If masculine cha
378d147 ne character. added to species,. we can see why YOUNG & female alike Good Ch 6 Keep Is it Male that a
378d148 child is like parent, so is species old: Hence <YOUNG> Kingfisher & pies, have long had their present
378d148 put 12 wild duck's eggs under common ducks, the YOUNG crossed amongst themselves, & I presume with co
380d153 f sexes, at birth & causes. If an animal breeds YOUNG her growth is immediately checked-- the vis for
385d163 astrated male.-- «Men giving milk--» Sept. 25th YOUNG man at Willis «Grt. Marlborough Str, Hair dress
385d163 bliged to be held.-- like she wolf of Hunter.-- YOUNG take distemper very readily & are subject to fi
387d169 athus, or Pipe fish the male of which receives <YOUNG> «eggs» in belly.-- analogous to men having mam
392d180 ough birds stomachs & live? In Muscovy ducks do YOUNG take most after father or Mother according as t
404e026 t in whole class. Case of Mexican greyhounds.-- YOUNG being habituated. instance such as Hunter, or s
421e090 (p 225. 1838.) account of metamorphosis in the YOUNG of Syngnathus-- curious as showing generality o
432e125 & this is argument against Blyth's doctrine of YOUNG birds retrogressing-- Uncovering the canine tee
437e139 s said that Eagles bring rabbits & hares to the YOUNG ones to exercise them in killing them.-- "Somet
437e139 ufficiently weakened by wounds got off from the YOUNG ones while they were amusing themselves with th
437e140 ..-- The parent bird another day brought to her YOUNG ones the cub of a fox, which after it had fough
437e140 ter it had fought well & desperately bitten the YOUNG ones, would in all probability have escaped"--
443e153 to graft from the roots is the best way to get YOUNG trees, from worn-out kinds, & quotes from Pliny
446e160 t, sitting on eggs. & feeding & defending their YOUNG..-- The oriolus (icterus Cat.) is an instance o
449e169 not this Siberian animal?-- Eyton says that the YOUNG of two hatches «all alike» between the male Chi
449e169 of the male giving form-- they interbred. & the YOUNG kept constant. & all alike Waterhouse says some
459t009 ience Vol III. p 320. Mr Hodgson on Musk Deer-- YOUNG spotted <like in> "prettty much as we see in th
459t009 potted <like in> "prettty much as we see in the YOUNG of the wild hog & of several species of deer, w
461t025 ns, one of which are marsupial & the other have YOUNG which undergo metamorphosis & are provided with
461t025 not require sac.-- but the male in these hatch YOUNG-- are there not some. Marsup. Mammalia, which <
467t100 d «Have dried some».-- some with no division in YOUNG flowers. The abortive stamen are of useful heig
473s07v s & see no difference in calves--how is this in YOUNG pigeons--dogs--cattle? As we see the frame of a
480z007 ol I p. 358. D'.Orbigny <considers> states that YOUNG birds of prey have longer tails than old ones--
491q01v ertain plants, like Aphides produce impregnated YOUNG ones; & that it is in these that male organs (n
502q11v s & animals come on.-- Compare calves.: Compare YOUNG. beans. cabbages.-- History of Pheasant-fowl. H
509q019 w is this in «Plants??» Are abortive organs as <YOUNG> teeth, more plain in young Rhinoceros or Whale
509q017 abortive organs as <young> teeth, more plain in YOUNG Rhinoceros or Whale, than in old?? Falconer say
512q019 me breeder who raises many English birds-- will YOUNG wild ones breed as well as,, as those already b
512q020 ere Lions have bred, have they been raised from YOUNG ones, bred in captivity --Mr Miller says Wombwa
513q020 om different countries= Do the Peacocks cross.= YOUNG Chinese or Penguin Duck in very young state for
513q020 s cross.= Young Chinese or Penguin Duck in very YOUNG state for skeleton== Does the tumbling of pigeo
530m049 cats discover birds nests & watch them till the YOUNG are big enough to eat.-- There was blackbirds n
531m049 seen by Hubberley to visit daily to see how the YOUNG got on. this nest the cat could If cats will «e
531m051 re in beholding the misfortunes of others.-- In YOUNG children, the violent passions they go into, sh
531m053 t it is instinctive because Nancy tells me very YOUNG babies start at anything they hear or see. whic
532m053 help doing it.-- Fanny Hensleigh doubts whether YOUNG babies start.-- If children wink. it is insti
540m085 Ourang outang at Zoolog Gardens touched put. of YOUNG male & smelt its fingers. Seeing a dog & horse
542m096 instinctive-- child does not sneer. because no YOUNG animal has canine teeth.-- A dog when he barks
546m110 sness Mayo compares it with Somnambulism.-- the YOUNG lady almost equally in her senses in either sta
551m129 ence.-- read monkeys for preexistence <">-- The YOUNG Ourang in <Zoolog> Gardens pouts. partly out di
553m132 arn sooner to take kangaroos than emu, although YOUNG dogs get sadly torn in conflicts with the forme
560mIBC k, when anything placed before their eyes, very YOUNG, before experience can have taught them to avoi
564n005 as to seize small moving object like fly.-- YOUNG partridge can run even with its shell on back.-
573n037 ays. that she has constantly observed that very YOUNG children. express the greatest surprise at emot
580n062 re of glory, immortal fame, &c so common in the YOUNG are symptoms of the infinite & progressive natu
582n066 en other people are about: this is analogous to YOUNG pigs hiding themselves; & heredetary remains of
582n051 s of it) to move its legs so, as much as in the YOUNG salmon to go towards the sea. on down the strea
584n071 of location ducks & turtles running to water,-- YOUNG crocodile snapping-- p. 28. how curious the mea
586n079 w.-- migration, <only> «only» more wonderful in YOUNG, because can not have been taught, where to go-
586n080 not loosing its direction, equally wonderful in YOUNG & old.-- These facts point out some essential d
588n090 445. If we compare the judgments & actions of a YOUNG animal with an old.-- (dog horse, sow) we perce
591n097 oolog. Garden told me. he has often watche tame YOUNG wolf & it never dropped its ears like dog-- wag
591n097 when attending to anything or excited.-- so do YOUNG dingos, as I saw wag tail when watching anythin
593n107 -- no doubt it may be attempted to be said that YOUNG animal learns parent smell & look so by associa
595n125 tive as a bull <tr> calf, just born butting, or YOUNG crocodile snapping.-- these I think are better
603o014 ng the possibility of <handsome> «UGLY healthy» YOUNG woman, with good expression-- statues not paint
620o045 tincts.-- -- One does not feel it wrong in very YOUNG child to be in passion, any more than in an ani
621o047 by association to give «almost» any taste to a YOUNG person. or it is accidentally acquired from som
627o053 leasure a dog has in obeying its instinct,-- as YOUNG pointer to point-- clearly shows this is true.
637j58r . 236. Marsupial bones especial adaptation, to «YOUNG».-- good God & yet Mails have them. What trash
067r144 ke. Happened on January 20th. 1834 Mr Sowerby. YOUNGER.. says that Falkland fossils decidedly belong t
304c206 ! look at Mammals: how wide.-- therefore birds YOUNGER???? or «have» not «been» exposed to so many co
334d012 ards came again, bringing with her the other & YOUNG cow.-- {P} Fox says when common & China goose
385d164 at we observe in the temper, especially of the YOUNGEST children, a striking <resemblance> similarity
573n037 eetle than a Universe.-- November 20th Saw the YOUNGEST child of H. W. constantly. when refusing food
255c053 l kinds of plan to insure stability; but isolate YOUR species her plan is frustrated or rather a new
303c205 genera. if Mammalia are adduced. say oh look to YOUR fossils, now if extinction had gone, without cr
534m061 thinking consists of sensation of images before YOUR eyes, or ears (language mere means of exciting
541m090 whole effort consists in keeping one idea before YOUR mind steadily, & not merely thinking intently;
558m153 sion commences.-- <All> Nearly all will exclaim, YOUR arguments are good but look at the immense diff
569n020 ciation.-- Elephant often given food & word open YOUR mouth said, recognizes that sound as perfectly
578n054 F. A. said to Mrs. B. A. how nice it would be if YOUR son would marry Miss. O. B.-- Mrs. B. A. blushe
615o038 n of pleasures with certain actions performed by YOUR parents, conscience This «X» memory especially
549m122 me very near each other.+ The rules to mortify YOUR do not tend to this-- though believing it to
558m150 cial instincts, giving rise "do unto others as YOURSELF. "love thy neighbour as thyself". Analyse th
572n032 e general idea of showing respect is by making YOURSELF less, but the manner, whether by bowing the b
473s07v a tendency to alter or assume some form late in YOUTH.-- only facts can decide-- some peculiarities m
512q019 ?? or in individual case: subject to disease in YOUTH.-- Mr Tollett-- about selection for milking-- 1
571n030 difficult to distinguish. between prejudices of YOUTH from <here> habits. & heredetary habits. & perh
575n046 remembering things of YOUTH, when new ideas will not enter. is something an
585n074 «by an» instinct<ing> «which is only present in YOUTH» (Mem. Mr Worsley's story of chicken) to know t
640j28v reeding season, or gaining adaptations, but for YOUTH most necessary: the fertility of Man in old age
077r171 arly the same with that of the veta grande of ZACATECAS, & veins of Tasco & Moran--of Guanaxuato to S
078r175 he direction & inclination of the vein".-- at ZACATECAS the veta grande has same direction as Guanax.
321c275 ton Voyage to Surinam. Voyage Congo expedition: ZAIRE except Brown's Appendix & excellent table of Ca
540m088 " &c &c &c so is conscience &c &c Coleridge.-- ZAPOYLA p. 117, Galignani Edition Fine poetry, or a st
031r038 lapagos. Daubeny P 24 V. back of page 1 of New ZEALAND Geological Notes. at St. Helena. This structur
040r062 Add from M. Lesson. character of Flora to New ZEALAND, which agrees with St Helena in being unique,
062r129 long distances; generally first arrives:-- New ZEALAND rats offering in the history of rats, in the a
```

182b050 Australia & Van Diemen's land. & Austral & New ZEALAND Mr Gould says in sub-genera, they undoubtedly
198b115 act of fresh creation why not produced on New ZEALAND; if generated <No> «an» answer <could> «can» b
223b209 log. Proceedings October (?) 1837 Contrast New ZEALAND with Tasmania The reason why there is not perf
225b219 e elevation Subsidence New is only hope.-- New ZEALAND «compare to Van Diemen's land.» glorious fact.
241c015 Guinea & rest of isle in E. Indi: Arch: In New ZEALAND. a sturnus of American form-- a Synallaxis. ¿A
241c015 -- Gould will hereafter know about birds of N. ZEALAND L'Institut. 1838. A Dipus. & other rongeur in
243c019 ances given) in East Ind. Arch:-- Birds of New ZEALAND absolutely different.-- --Philedon circinnatus
243c019 on circinnatus not found in Australia only New ZEALAND-- Norfolk. Isd. & New Caledonia peculiar speci
243c020 e work this out-- L. Jenyns, about my fish New ZEALAND & New Holland fish very similar.-- NB. Lesson
246c025 y one isid different]CD.-- p. 414. dogs of New ZEALAND of large size, resemble, chien-loup.--long, bl
248c029 e been transported. ¿What section does the New ZEALAND, a division of nature of Apterix, many genera
263c074 terix-- split, depress & elevate & enlarge New ZEALAND; a division of nature of Apterix, many genera
314c238 reface.-- Mr Brown says character of Flora, N. ZEALAND & N. Caledonia with a dash of New Holland. As
314c238 Caledonia with a dash of New Holland. As in N. ZEALAND-- Some species of Australian Genera Some speci
314c239 species same (Palm & Phormium tenax) as in New ZEALAND & Australia, some SPECIES of Australian GENERA
339d025 very different.-- Man not so, but N. & S. New ZEALAND & New +++ Caledonia. two races of Men, but not
354d065 apagos mouse probably transported like the New ZEALAND one-- It should be observed with what facility
370d115 hown from peculiarities of climate cause of N. ZEALAND not having any Mammalia.-- Type of geographica
464k065 described a greatt Struthonidous Bird from New ZEALAND-- <so> not an Apteryx, yet it shows the Aptery
500q10a ied Gould.-- Number of species of Birds in New ZEALAND, plants so few-- Range of mundane genera, «in
500q10a outbful species from Van Diemen's Land? or New ZEALAND? Babington about differences of Irish & Britis
502q11v d Birds Madeira Migratory-- ask Gould about N. ZEALAND, as Cuculus lucidus is.-- Ask Sulivan about Fa
540m086 & the energetic copper coloured natives of New ZEALAND)-- the American in Brazil is under same condit
292c170 singular views there offered, & he must be a ZEALOUS man in the cause if his faith is not staggered
189b072 die.== The fossil horse, generated in S. Africa ZEBRA.-- & continued.-- perished in America All anima
229b234 Institute on range of Bos in India.-- Range of ZEBRA?-- The Crocodile & Tortise former inhabitants o
283c145 ibility of this more than anyone.-- no turn the ZEBRA into the Quagga.-- «let them be wild in same co
367d113 pe on back also.-- legs reminded me strongly of ZEBRA.-- Mem. Quagga & Ld Moreton Mare ringed Owen sa
458t002 Gardens. informs me that a hybrid between ass & ZEBRA, crossed with pony mare & produced a very prett
512q020 says Wombwalls were (4) About fertility of ass-ZEBRA-horse-- (4) About fertility of ass-zebra-horse-
513q020 of ass-zebra-horse- (4) About fertility of ass-ZEBRA-horse- (5) About callosities on Camels-horses.
556m146 howing real affinity in face of donkey, horse & ZEBRA. when going to kick.-- Why does dog put down ea
187b062 e & Elephant reached S. America.-- explains how ZEBRAS reached South Africa-- It is a wonderful fact
355d066 gs to foxes. (yes Australian dog) or donkeys to ZEBRAS.-- «Mr Herberts variety of horse, dun-coloured
585n075 ear. good to content with horses, asses, <mi> ZEBRAS &c &c.-- Here there is kicker but not bite.--
374d133 eptiles.-- <M> p. 417. Magnesian Limestones & ZECHSTEIN oldest rock in which reptiles have been found
054r102 nd tertiary formation of Payta: N. part of New ZEALAND entirely volcanic!! New Zealand rich in partic
054r102 N. part of New Zeeland entirely volcanic!! New ZEELAND rich in particular genera of plants: All St. C
361d096 plumage than male.-- «My case of Caracara. N. ZELANDIAE--» Mr Blyth stated «that there are» two duck
338d024 505. some (very poor account) of plants of Nova ZENBLA -- in review of Baers work Edinburgh. Royal. T
073r164 t analogy to Hungary. = Veins of Zimapan offer ZEOLITE. stilbite. grammalite. pyenite. native sulphur
227b228 comprehend true affinities. My theory would give ZEST to recent & Fossil Comparative Anatomy, & it wo
240c013 cas «Matchian» & Celebes.-- ¶ Amboina; Viverra ZIBETHA. ¶-- All the Moluccas, Waggious New Guinea. Ne
073r164 N. Spain great analogy to Hungary. = Veins of ZIMAPAN offer zeolite. stilbite. grammalite. pyenite.
124a119 aking brass a piece of copper not melted absorb, ZINC thrugout its thickness.-- this most curious wit
124a119 urious with respect to epigmous action.-- if the ZINC were mixed with 90 percent of lead. it would be
427e111 ardier constitution.-- Now Sir. J. Banks. says ZIZANIA in 16 generations did become, acclimatized. &
428e111 red artificially, & only very partially to the ZIZANIAS in in Sir. J's ponds-- my principle being the
255c050 merica like other forms, but those inhabiting 3d ZONE of <latit> height & 3d of latitude more commonl
510q017 rphology. ¶-- Schelgel is he serpent man? about ZONES separated by non-inhabited spaces: has he publi
196b110 ables. of the two great Kingdoms.-- Principes de ZOOL: Philosop:-- I deduce from extreme difficulty o
296c184 Mauritius. p. 112. & paper on genus Magazine of ZOOL. & Bot.-- Vol I. p. 450. 4 instances of hybrids
594n115 escent --Affection & [...] Monkeys «Ogleby» seen ZOOL. Soc-- 1838 remember with distress their compan
343d037 was light".-- » August 17th Two regions may be ZOOLO-geographically divided either by developement o
095a037 s found with the Mactra. at Buenos Ayres at the ZOOLOG: Soc: Terebratula from Hudson's Bay. 2. specie
210b160 eculiar. to Mauritius are not found at Bourbond ZOOLOG. Proceedings¶. <p> 1832.p. III Mr Owen suggest
211b162 whereas Man has such instincts very little. in ZOOLOG. Proceedings. Jan 1837, «by Eyton» Account of
212b165 culiar to. latter not to former-- Mr Martens of ZOOLOG Soc told me an Australian dog he had, used to
212b166 mmon to Mauritius & Madagascar.? Proceedings of ZOOLOG. Soc June 1837 p. 53. an Irish Rat.-- differen
223b209 - Small «new» animal mentioned from Fernando Po ZOOLOG. Proceedings October (?) 1837 Contrast New Zea
235b272 s, situations &c on 242. Hook Smellie Philos of ZOOLOG. 842 White regular gradat in Man poor trash Ly
235b272 an poor trash Lyell 1024 Flemings Philosophy of ZOOLOG Royle on Himalaya Range-- Would it not be pos
240c014 o. when making varieties.-- Voyage of Coquille. ZOOLOG. p 19.. Tapir, «des» couroucous et rupicole ve
251c041 ph like Gould on Trogons worth studying.-- «do» ZOOLOG Journal Vol 2. p 221. Horsfield on two bears v
253c047 ntains some facts about close species of Birds. ZOOLOG. Transact. Vol I p. 165.-- <a> "an account of
254c047 red form of skull very slightly different.-- Q ZOOLOG. T. V. I. p 389. Owen remarks on Entozoa, the
258c061 ll changes may be considered in this light.--XX ZOOLOG. Journal-- Parrots in Macquarrie isld. vol III
297c186 hey are remnants.-- Cephalopoda ditto.-- Mag of ZOOLOG. & Bot. Vol. II p. 125 Allusion to abortive sp
305c210 remarkably varying plumage for wild birds-- At ZOOLOG Gardens there is half Jackal & Scotch Terrier.
307c216 s in neuter bee, (There is paper by Yarrell «in ZOOLOG Transactions» & Hunter on this subject) becaus
333d007 more resembled foxes than dogs (Mem Jackall in ZOOLOG Garden) He has seen in a show half Wolf & «hal
340d026 explicable every instinct in animals. Heard at ZOOLOG Soc their Pintail & Common Ducks, breed one wi
341d032 as long as Pea hen.-- about intermediate.-- (In ZOOLOG. Gardens there is hybrid of Penguin duck a vari
347d050 ogance!! August-- 29th.-- Macleay in A. Smith's ZOOLOG.-- of Africa -- p. 4. sticks to genus or group
361d095 easier than two last. Sept. 11. N Mr. Blyth, at ZOOLOG. Meeting stated, that Green-finch, all linnets
371d127 alludes to the dholes or wild dogs of India) in ZOOLOG. Garden having coloured offspring.-- but surel
383d160 the common mule must often have been dissected ZOOLOG. Garden. Sept 16." Hybrid between Silver & Com
385d163 e cock «yet very odd loosing visible powers» in ZOOLOG Gardens. & Kings at Otaheite) <Think> Last lit
392d180 fowl <in Z> of India «with our common ones» in ZOOLOG. Gardens; <Buffalo & common cattle-- Esquimaux
393d1BC brids pintail & common ducks. similar inter se? ZOOLOG. Gardens Are the hybrids of those species. whi
407e042 353, on animals of Antilles.-- (see Macleay in ZOOLOG. Journal. for those of Cuba.-- It is important
424e103 relied on, but Bell has some account of wolf in ZOOLOG. Gardens, which brought its puppies to be fond
478z004 als of Falklands &c &c Bennett on Chinchillidae ZOOLOG Transacts. worth reading Cuvier's Memoire 133
480z006 Australia. Dorbigny Vol II, p 24 Proceedings of ZOOLOG Soc. Important account of habits of Tubularia.
480z007 America translated from D Orbigny no IV Mag. of ZOOLOG & Botany p. 356 Lesson on Berre. do-- Magazine
480z007 Botany p. 356 Lesson on Berre. do-- Magazine of ZOOLOG & Botany. Vol I p. 358. D'.Orbigny <considers>
481z010 ains The fourth Vol. «in Lyell's possession» of ZOOLOG. of Voyage of Astrolabe must be studied for an
482z011 man a reference to a luminous Sertularia Lesson ZOOLOG. Coq: p. 120 Coati Roux. Tatous & perhaps Yago
483z012 not fossil at Isle of France: <Jerrold?> Bibron ZOOLOG. Journ Vol I. p. 125, owls seen crossing Atlan
483z013 s on Intestinal worms must be studied in Vol I, ZOOLOG: Transact. before writing on Planariae or Poly
485z017 stinct Camelidae. do not breed together Mag: of ZOOLOG & B. Vol. II. p. 127. List of submarine insect
489qIFC 7 Yarrell---- 18 Blyth--- 19-- Mr. Tollett {T} ZOOLOG Soc «Gardensa ----- 20 & Breeders Dr. Boott
491q01v w <few> one flower (5) Dr Fleming. Philosop. of ZOOLOG. vol 1. p. 427-- says biennial-wall-flowers &
494q004 owl with Dorking (10) Statistics of breeding in ZOOLOG. Gardens-- with respect to conditions of anima
511q018 res-- in negro & white (3) About the Bantams at ZOOLOG Soc.-- did Sir. J. Sebright select to destroy
540m085 ones own pud. not disagree.-- Ourang outang at ZOOLOG Gardens touched pud. of young male & smelt its
558m129 eys for preexistence <"»-- The young Ourang in <ZOOLOG> Gardens pouts. partly out displeasure (& part
588m090 one shows he has recollection.-- Lamarck. Phil. ZOOLOG.-- Vol II p. 445. If we compare the judgments
588n091 n? or some unnamed faculty-- Lamarck. Philosop. ZOOLOG. «p. 284. Vol. II» -- gives explanation & inst
591m097 nding on end.-- July 20th Intelligent Keeper... ZOOLOG. Garden told me. he has often watche tame youn
595n121 pril 28th 1840 was frightened at wild beasts in ZOOLOG. Garden [blank] [not located] A child crying.
194b095 strong case might be made out of world before ZOOLOGICAL divisions.-- Mem. species doubtful when know
195b098 ch, I think is shown by the very facts of the ZOOLOGICAL character of these islands so permanent a br
203b135 Anteater,.-- This supposes world divided into ZOOLOGICAL provinces-- united-- & now divided again-- W
250c040 erium in India--: connection with Latitudes!? ZOOLOGICAL Journal.--- Vol I. p. 81. Capromys, West Ind
260c067 s in N. America¶ Study Bonapartes list In the ZOOLOGICAL Journal I read a curious account to show tha
271c107 certain countries.-- Fraser remarked to me at ZOOLOGICAL Society,, that you never find two «similar»

Page ***(Key Word)***
320c276 Zoologique Pernety. voyage a l isle Malouines ZOOLOGICAL Journal 5 Vols Voyage de la Coquille Zoologi
320c276 ological Journal 5 Vols Voyage de la Coquille ZOOLOGICAL Transactions. <done> up to parts published M
325c267 Paper on change of plumage in Hen Pheasants <ZOOLOGICAL> Philosop. Transactions. 1827 Paxton on the
339d025 ntail» Widgeon!-- Divides animals «world into ZOOLOGICAL Provinces» according to varieties of Man.? «
389d115 e living habits of animals, with anatomical & ZOOLOGICAL research, in order to establish entirely the
458t002 legs can hardly ride on them. Mr Miller-- in ZOOLOGICAL Gardens. informs me that a hybrid between as
512q020 s-- Mem: how many miles they fly in few hours ZOOLOGICAL Soc (1) Do the animals there, sometimes coup
553m137 del Fuego.-- Mr Miller (superintendent of the ZOOLOGICAL Gardens) remarked that <exp> the   expression
555m142 V. D. p. 111, case of Association. Sept. 16th ZOOLOGICAL Gardens-- Endeavoured to classify expression
054r102 rated by profound valley [.] Sydney. -- Lesson ZOOLOGIE Grand tertiary formation of Payta; N. part of
240c013 obrizhoffer Abipones.-- Voyage. de L'Astrolabe ZOOLOGIE. p. 60. Vol I. Cynocephalus. niger. comes fro
241c016 w to read over the pages from about 8 to 20 of ZOOLOGIE of Coquille's Voyage to see if Lessons' remar
445e159 in baby Lamarck. Vol II p. 152.-- Philosophie ZOOLOGIE. says it is not sufficiently proved that any
480z006 . No.-- 221 Good account of Condor by Humboldt ZOOLOGIE Recuiel-- Meyen has written account of Guanac
320c276 'isle de Frances Voyage de l'Astrolabe Partie ZOOLOGIQUE Pernety. voyage a l isle Malouines Zoologica
353d060 up the gaps.-- is possibly the same with the <ZOOLOGIST> «philosopher», who has trace the structure o
214b174 ogers report to Brit Assoc <to> on N. American ZOOLOGY-- A breed of Blood Hounds from Aston Hall clos
217b189 hybrids between grouse & pheasant-- Magazine. ZOOLOGY & Botany Vol I p. 450 There is in nature a «re
222b205 It is another question, whether whole scale of ZOOLOGY may not be perfecting by change of Mammalia fo
225b220 n opposite tendency.-- Study Ellis & Williams. ZOOLOGY of South Sea islds. any animals?-- I believe n
236bIBC e Linn: Soc. Geoffry-St. Hilaire Philosophy of ZOOLOGY Waterhouse B C N[a]me of two pigeons,-- «with
288c160 ng been formerly nearer.-- «Selby» Magazine of ZOOLOGY & Botany No XI p. 390. a slight change in encl
290c164 Tilgate forest Seeing common gull in garden at ZOOLOGY Soc. it's pale ash grey back, like a black bir
292c170 alves or bivalves.-- Anyman No VI. Magazine of ZOOLOGY & Botany p. 566 wants to see absurdity of Quin
296c185 breed freely with the tame Vol II. Magazine of ZOOLOGY p. 56. Peregrine Falcon holds birds for some t
313c235 extinction-- In the Entomostraca (Magazine of ZOOLOGY & Botany) where several generations are produc
320c275 ific Memoirs. published by Taylor Magazine of. ZOOLOGY & Botany & Continuation «Annals of Natural His
323c269 t. Hist of World do -- Lamarck. II Vol. Philo. ZOOLOGY «references at end of each Chapter» Crabbes Li
324c268 le on Himalayan. types. Smellie. Philosophy of ZOOLOGY Flemming. ditto Falconers remarks on the influ
477z FC s later-- All poultry with same down-feathers. ZOOLOGY 1856 Skimmed through & abstracted Zoology Some
477zIFC ers. Zoology 1856 Skimmed through & abstracted ZOOLOGY Some excellent references in L. Jenyn's introd
477z001 t references in L. Jenyn's introduct to Mag of ZOOLOGY and Botany. Philosoph. Transacts. 3. papers co
481z009 or Journal.-- 18 Admirable engravings in Meyen ZOOLOGY on animal of Campanularia Alcedo stellata. Mey
566n012 s thunder. would be placed to the will of God. ZOOLOGY itself is now purely theological.-- Origin of
201b126 & p. 545 «great monkey» Mr Johnston says Mag of ZOOLY & Bot. p 65 Vol II talking of annelidae.-- <">T
289c162 o Teeth in Bering Pike (Waterhouse) Magazine of ZOOLY & Bot-- Vol II p. Dr Johnston <on> Entomostraca
170b001 83[8]: probably ended in beginning of February ZOONOMIA Two kinds of generation the coeval kind, all
171b002 ons to other life has not come into play)--See ZOONOMIA arguements, fails in hybrids where every thin
171b004 <through> generations, acquire ideas ditto. V. ZOONOMIA.-- There may be unknown difficulty with full
315c243 t of sea side-- Study Bell on Expression & the ZOONOMIA, for if the former shows that a man grinning
521m007 --» which my Father thinks is mentioned in the ZOONOMIA.-- Now if memory «of a tune & words» can thus
212b167 fossils & recent shells between herbivorous & ZOOPHAGOUS Mollusca according to periods.-- NB. Was Eur
063r132 erefore we are not so much surprised at seeing ZOOPHITE producing distinct animals. still partly unit
139a180 esting experiments might be tried by comparing ZOOPHITE to plants.-- grafting length of life &c &c Wi
393dIBC ere any peculiarity or variation common to any ZOOPHYTE «born in succession» which is not transmitted
576n049 says. because if so ourang outang.-- oyster & ZOOPHYTE: it is (I presume-- see p. 188 of Herschel's
526m029 ing over the scenes which I first recollect, «at ZOOS» they are all things, which are brought to mind
062r128 l the limits of birds & animals in S. America. ZORILLA: wide limits of Waders: Ascension. Keeling: at
477z002 a caya. le negro, et le pajero) l'yaguaré «the ZORILLA-arskink» le quiyá (Coipu) viscacha.-- A. Patag
506q15v - We may presume Nettle spreads by seeds= (44) ZOSTERA. Has he seen it in flower? does he know Botani
513q21. regnation.-- says many flowers are dichogamous ZOSTERA-- Knights notion of pollen & stigma generally
514q21. Boerhaave's paper on impregnation of violets.= ZOSTERA= Are dwarf plants on Wellington Mountain descr

Appendix 1: List of Suppressed Words and Their Frequencies

Total number of words in text = 132614
Number of different suppressed words = 213
Total number of suppressed words in text = 56854

1378	a	856	from	7	o	282	than
231	about	16	g	6389	of	998	that
61	above	6	gave	1092	on	4005	the
75	also	66	give	717	one	210	their
18	am	57	given	103	ones	157	them
241	an	38	gives	273	only	223	then
88	and	22	giving	676	or	500	there
78	another	28	h	438	other	279	these
338	any	158	had	92	others	357	they
808	are	452	has	5	ou	764	this
1171	as	717	have	127	our	179	those
580	at	124	having	159	out	45	three
66	b	327	he	110	over	53	through
1148	be	50	her	42	own	2948	to
281	been	48	here	955	p	7	tom
22	below	75	him	11	par	28	too
204	between	21	himself	142	part	11	took
88	both	203	his	83	parts	3	twenty
585	but	28	ie	4	plus	245	two
856	by	67	ii	2	pp	5	un
475	c	38	iii	9	q	66	under
276	can	3381	in	6	que	5	une
111	cannot	178	into	3	qui	44	v
1	cii	1883	is	1	quil	425	very
1	cxii	1132	it	1	quon	10	vi
1	cxv	207	its	28	r	9	vii
1	cxvi	49	itself	3	rd	2	viii
160	d	19	iv	196	s	238	vol
64	de	5	ix	110	said	55	w
27	des	60	j	54	saw	262	was
67	did	44	just	102	say	133	well
443	do	5	k	14	saying	140	were
293	does	80	l	289	says	281	what
13	doing	30	la	27	se	466	when
28	done	20	le	255	see	208	where
87	dr	31	least	34	seeing	717	which
15	du	15	les	80	seen	105	who
49	during	72	less	8	seven	8	whom
34	e	78	m	53	she	16	whose
215	each	295	many	62	sir	1007	with
9	edit	343	more	10	six	138	without
71	either	260	most	570	so	470	would
1	est	210	mr	527	some	16	x
30	et	13	mrs	6	son	2	xi
2	etc	3	ms	93	st	4	xii
139	even	215	much	234	such	1	xiii
165	every	123	n	7	sur	3	xiv
37	f	1	nd	52	t	8	xx
105	far	105	near	80	take	1	xxiii
91	few	8	neither	29	taken	6	y
4	fifty	1	nine	24	takes	7	z
6	five	436	no	22	taking		
489	for	1396	not	12	ten		
11	four	200	now	141	th		

Appendix 2: List of Keywords and Their Frequencies

Total number of different keywords = 11503

1 ab	7 accident	1 aculeatus	12 affinities
2 abandoned	16 accidental	1 acuneha	23 affinity
6 abberley	3 accidentally	9 acunha	4 afflux
1 abbreviations	9 accidents	2 acute	4 afford
1 abdominals	2 acclimatisation	1 ad	1 afforded
1 abel	1 acclimatised	4 adam	2 affording
1 aberant	1 acclimatized	4 adapt	5 afraid
1 abercrmbies	1 acclimatizing	55 adaptation	63 africa
3 abercrombie	2 accomodate	12 adaptations	21 african
2 abercrombies	1 accomodated	36 adapted	92 after
13 aberrant	1 accomodation	1 adaption	25 afterwards
2 abhorrence	14 accompanied	5 adaptive	35 again
1 abipomenes	4 accompanies	13 add	54 against
2 abipones	2 accompany	23 added	3 agassiz
14 able	6 accompanying	1 addenda	1 agate
1 abnormal	9 accordance	5 adding	1 agates
1 abo	1 accordant	5 addition	1 agaziz
1 abolishing	79 according	3 additional	32 age
4 aboriginal	1 accords	1 additions	6 agency
1 abort	1 accouchers	2 additive	4 agent
8 abortion	82 account	1 address	4 agents
60 abortive	6 accounted	2 adds	15 ages
1 aborts	2 accounting	2 adduced	1 agglomerate
5 abound	8 accounts	2 adduces	2 aggregated
1 abounded	2 accra	3 adhere	1 aggregates
3 abounding	3 accumulate	1 adhesion	1 agit
1 abrasion	5 accumulated	1 adjacent	5 ago
2 abroad	2 accumulating	4 adjoining	2 agony
1 abrolhos	4 accumulation	1 admettre	1 agouti
1 abrupt	1 accumulations	10 admirable	12 agree
1 abrupte	1 accurate	1 admirably	1 agreed
13 absence	1 accustom	1 admiralty	2 agreeing
10 absent	8 accustomed	2 admiration	4 agreement
17 absolute	2 ache	5 admire	27 agrees
1 absoluteley	11 acid	3 admires	1 agricl
18 absolutely	1 acongua	1 admiring	1 agriculture
3 absorb	1 acorn	1 admission	1 agriculturist
6 absorbed	1 acosta	10 admit	1 agriculturists
1 absorbing	1 acquaintances	1 admittance	1 ahead
1 absorbs	1 acquainted	3 admitted	1 ai
1 abstinence	2 acquirable	1 admonition	8 aid
16 abstract	10 acquire	2 adopted	3 aided
5 abstracted	48 acquired	2 adult	1 aids
2 abstracting	6 acquirement	2 advance	1 aient
2 abstraction	1 acquires	4 advanced	1 ailes
1 abstractions	2 acquiring	2 advancing	3 aim
12 absurd	1 acquisition	7 advantage	2 aimé
3 absurdity	1 acrid	1 advantageous	41 air
2 absurdly	1 acrita	2 advantages	1 aira
1 absurdum	2 acrite	2 adventure	1 aisément
4 abundance	12 across	1 adverse	1 aisle
15 abundant	33 act	1 advertised	1 akin
2 abundantly	10 acted	1 aegypt	1 al
1 abutted	1 acter	2 aegyptian	1 alacrity
1 abysses	21 acting	7 aerial	1 alas
3 abyssinia	138 action	15 affect	1 alauda
2 acad	54 actions	23 affected	1 albatross
6 academy	5 active	1 affectent	6 albino
2 acalepha	3 activity	11 affecting	3 albite
1 acalephes	1 actors	9 affection	1 albus
1 acanthosoma	7 acts	13 affections	2 alcedo
2 acapulco	3 actual	3 affects	1 alcyone
1 acceleration	12 actually	1 affin	4 alderney

1 aleutian	1 analogie	1 antilles	6 areas
1 algernon	11 analogies	1 antilope	1 arequipa
1 alight	1 analogize	1 antimony	1 arg
1 alighted	63 analogous	1 antipodes	2 argentiferous
1 alighting	2 analogue	3 antiquary	1 argillaceous
1 alights	5 analogues	1 antiquities	20 argue
17 alike	83 analogy	4 antiquity	8 argued
10 alive	1 analogys	3 ants	1 arguements
1 alk	11 analyse	1 antuco	5 argues
1 alkali	5 analysed	1 anxious	1 arguing
1 alklali	6 analysis	5 anyhow	1 argumen
506 all	1 analyze	1 anyman	82 argument
1 alleghanies	1 anapsis	3 anymore	11 arguments
3 allen	1 anastatica	11 anyone	1 argumentum
1 alley	4 anat	28 anything	1 arica
4 alliance	2 anatidae	1 anyways	1 arion
27 allied	1 anatifs	1 anywhere	10 arise
1 allotriandrous	2 anatomical	2 ap	4 arisen
14 allow	1 anatomie	6 apart	5 arises
1 allowance	1 anatomique	1 apercu	4 arising
8 allowed	1 anatomist	1 apertures	1 aristolochia
4 allowing	1 anatomists	1 apes	1 aristotle
4 allows	3 anatomy	5 aphides	9 arm
2 allude	4 ancestor	1 aphrodites	2 armadillo
4 alluded	5 ancestors	1 aphysia	3 armadilloes
12 alludes	2 ancestral	1 apion	1 armadillos
2 allusion	3 anchusa	1 apollo	3 armed
1 alluvi	26 ancient	1 apologising	3 armless
2 alluvial	1 anciently	1 apoplexy	5 arms
1 alluvials	1 ancients	1 app	2 army
16 alluvium	1 andaman	1 apparatus	2 aroe
1 alm	1 andersons	10 apparent	1 arose
67 almost	18 andes	9 apparently	3 around
23 alone	2 andesite	39 appear	1 aroused
8 along	2 andite	26 appearance	4 arrange
1 alp	8 andrew	4 appearances	5 arranged
2 alphabet	1 anecdote	14 appeared	21 arrangement
3 alpina	1 anecdotes	4 appearing	1 arranging
14 alpine	1 anemone	48 appears	3 arrest
9 alps	2 angelis	2 append	1 arrests
6 already	3 angels	10 appendix	1 arrival
11 alter	19 anger	4 appetite	2 arrivals
1 alteration	1 angle	6 appetites	6 arrive
3 alterations	1 angles	11 apple	7 arrived
22 altered	5 anglesea	4 apples	1 arrivers
7 altering	1 anglles	1 applicability	1 arrives
2 alternate	1 angora	35 applicable	3 arriving
5 alternating	9 angry	1 application	5 arrogance
1 alternation	1 anguiformis	6 applied	3 arrow
1 alternative	1 anguish	16 applies	1 arrows
4 alters	7 angular	23 apply	1 arrowsmith
34 although	2 angustidens	1 appoximat	1 arsenic
6 altogether	1 anhault	2 apprehend	1 arsenical
2 alum	3 ani	1 apprehended	1 arskink
1 alumen	163 animal	1 apprehensible	20 art
1 aluminated	1 animalcular	11 approach	1 artic
1 alw	3 animalcule	7 approaches	7 article
2 alway	1 animalcules	7 approaching	5 articulata
68 always	420 animals	2 approbation	4 articulate
1 amaryllidae	3 animated	1 appropriately	1 articulation
1 amazingly	5 animaux	2 approved	1 artifical
1 amazon	1 ann	1 approving	1 artificer
3 amazons	4 annales	1 approximated	1 artifices
1 amber	16 annals	2 apres	14 artificial
1 ambient	1 années	1 apri	4 artificially
2 ambition	1 annelida	1 apricot	1 artless
2 ambling	2 annelidae	16 april	2 arts
2 amblyrhyncus	1 annelidous	1 apropos	1 arum
4 amboina	2 anniversary	10 apt	1 asbestos
1 âme	1 announced	10 apterix	1 ascaris
167 america	6 annual	9 apteryx	1 ascencion
75 american	1 annually	1 aquarium	3 ascend
1 americana	6 annuals	6 aquarum	1 ascended
1 americans	1 annulosa	9 aquatic	2 ascending
4 americas	2 anomalies	1 aquatica	1 ascensi
1 amerique	6 anomalous	1 aqueo	18 ascension
1 ammocoetus	9 anomaly	1 aquilegia	1 ascent
1 ammonia	2 anoplotherium	4 arabia	2 ascertain
1 ammonite	1 anphidesma	2 arabian	5 ascertained
1 ammonites	1 anphidesmas	1 arabs	1 ascertaining
4 among	1 ans	1 araucanians	1 ascertainment
25 amongst	1 anson	1 araucaria	1 ascidia
26 amount	1 answ	1 araucarian	1 asclepiadae
1 amounting	18 answer	1 arbitrarily	5 asclepias
1 amounts	16 answered	4 arbitrary	2 ash
1 amph	5 answers	1 arblay	5 ashamed
1 amphibia	3 antagonist	1 arborescent	1 ashe
3 amphibole	4 antarctic	1 arc	4 ashes
1 amplification	1 antarctica	1 arcana	1 ashs
1 amputated	2 anteater	12 arch	8 asia
2 amused	2 antecedent	1 archaeologia	11 asiatic
1 amusements	1 antelope	3 arched	1 asiatiques
2 amusing	3 antelopes	1 arches	1 aside
1 amygdaloid	6 anterior	19 archipelago	1 asinine
1 amygdaloidal	1 anteriorly	1 archipelagos	36 ask
1 amygdaloids	2 anther	2 architecture	12 asked
1 amyyralidae	8 anthers	1 archiv	1 asks
1 anagallis	1 anticipate	12 arctic	1 asleep
1 analogia	1 anticipated	1 ardour	2 asparagus
4 analogical	5 anticlinal	7 area	6 aspect

705

1 aspects	1 auk	7 balls	154 because
15 ass	2 auks	1 banana	1 beccause
1 assam	4 aunt	2 bananas	5 beche
1 asseel	1 aunts	10 band	2 beches
1 assent	1 aura	1 banda	10 beck
1 assert	1 aurock	2 banded	1 becom
1 asserted	2 austral	5 bands	81 become
3 assertion	2 australes	1 bang	37 becomes
1 asserts	74 australia	1 banish	19 becoming
2 asses	38 australian	6 banished	24 bed
2 assignable	3 australians	7 bank	1 bedrock
1 assigned	1 australias	11 banks	32 beds
1 assigning	1 australis	6 bantam	30 bee
4 assist	1 australs	5 bantams	1 beech
3 assistance	1 austrian	1 banter	3 beechey
1 assisted	2 author	2 bar	1 beecheys
4 assoc	7 authority	1 barbaroussa	1 beechy
3 associate	4 authors	1 barbs	45 bees
21 associated	2 autour	1 barbyrousa	1 beet
1 associatical	2 autumn	1 barbyroussa	2 beetle
49 association	3 auvergne	1 barclay	7 beetles
5 associations	3 aux	5 bare	1 beetsons
8 assume	2 ava	1 barely	102 before
6 assumed	1 avalanches	8 bark	5 began
3 assumes	1 avant	4 barking	2 beget
6 assuming	1 avaoid	1 barks	1 begetting
5 assumption	5 avarice	1 barlyroussa	1 beggar
1 assured	4 avec	1 barn	6 begin
1 assures	1 avenue	1 barnacles	15 beginning
1 asteriae	10 average	1 barns	4 begins
1 asterias	3 aversion	10 barom	1 begotten
1 asthma	1 avestruz	2 barometer	1 behaved
1 aston	2 aviary	2 baron	1 behaviour
4 astonished	1 avila	4 barren	3 behind
1 astonishing	1 avions	8 barrier	1 behold
1 astonishingly	8 avitism	3 barriers	2 beholding
2 astonishment	1 avocette	1 barrington	1 bein
3 astrolabe	8 avoid	1 barrow	309 being
3 astronomer	2 avoided	2 bars	22 beings
2 astronomers	1 avoiding	1 barton	13 belief
1 astronomical	1 avoids	2 bartrams	1 belier
2 astronomy	1 avoir	3 barytes	66 believe
1 atacama	1 avow	2 basal	7 believed
7 atavism	5 awake	10 basalt	1 believer
1 athaenaeum	4 awakened	19 base	21 believes
1 atheism	2 awakes	1 based	10 believing
1 atheist	1 awaking	1 bashfulness	19 bell
27 athenaeum	6 aware	5 basin	1 bellinghausen
4 atlantic	31 away	2 basins	1 bellows
2 atmosphere	1 aways	1 basis	8 bells
3 atmospheric	1 awful	2 basket	2 belly
2 atom	2 awk	1 bass	6 belong
2 atoms	2 awks	1 basses	2 belonged
2 attach	1 axe	1 bastard	6 belonging
2 attached	2 axiom	3 bat	1 belongs
2 attachment	15 axis	1 batchian	4 ben
4 attack	1 axolotl	1 bath	1 benchuca
2 attacked	1 ay	1 batopilas	5 bend
3 attacking	7 ayres	1 batrachian	1 bending
1 attacks	3 azalea	1 batracian	1 bends
1 attain	1 azaleas	7 bats	25 beneath
3 attained	7 azara	1 battle	10 beneficial
9 attempt	2 azaras	2 bauers	1 benefit
4 attempted	7 azores	14 bay	1 benefits
1 attempting	2 babbage	1 bayfields	1 benevelo
4 attempts	7 babies	1 bays	3 benevolence
3 attend	5 babington	1 bayte	2 benevolent
1 attendant	1 babingtons	1 bea	7 bengal
3 attended	9 baboon	17 beach	3 bennett
4 attending	2 baboons	4 beaches	2 bennetts
2 attends	4 baby	10 beagle	1 bent
1 attentif	1 baccalao	11 beak	2 bentham
8 attention	11 bachman	2 beaks	2 benthams
1 attentively	83 back	6 bean	1 benza
1 attentus	1 backed	7 beans	1 berberis
1 attest	2 backs	42 bear	1 berbice
2 attested	2 backward	1 beards	1 berenica
1 attitude	3 backwards	8 bearing	1 berghaus
1 attorneys	32 bad	22 bears	1 berhave
6 attract	1 badger	1 beast	1 bering
2 attracted	2 badgers	7 beasts	1 berlin
2 attracting	3 badly	3 beat	2 berre
14 attraction	1 badness	2 beaten	1 berries
1 attractive	1 baer	3 beating	1 berry
2 attrib	1 baers	1 beats	1 berthelot
1 attribs	1 baffles	1 beatson	1 berwick
6 attribute	1 bag	8 beau	1 berwicks
2 attributed	2 bahama	1 beaucoup	1 berzelius
6 attributes	12 bahia	1 beaufort	13 beside
3 au	2 bailly	1 beauforts	1 bespeak
1 auckland	3 balance	1 beaum	1 bessy
1 audible	1 balancement	1 beaumonts	26 best
3 audubon	3 balancing	2 beauties	1 bethgellert
1 audubons	1 balançons	29 beautiful	1 betray
11 aug	2 balanidae	1 beautifully	34 better
1 augment	1 bald	23 beauty	1 bettws
1 augt	1 baldwin	1 beaver	1 bevan
1 augus	2 ball	1 beavers	1 bevans
13 august	1 ballivard	1 bebyk	1 bewick
2 augustus	1 balloon	12 became	

706

10 beyond	1 blushed	16 branch	25 bud
1 bhem	1 blushes	4 branched	2 budded
1 bible	9 blushing	11 branches	5 budding
3 bibron	12 blyth	7 branching	23 buds
1 bibrons	5 blyths	1 brandy	4 buenos
1 bicheno	1 boans	2 brasiliensis	2 buffalo
1 bichoffs	3 board	1 brass	2 buffaloes
2 bien	2 boars	1 brassica	1 buffer
4 biennial	1 boast	1 brave	7 buffon
1 bifurcated	1 boasted	1 brayley	1 bug
13 big	1 boat	19 brazil	3 build
3 bigger	1 boblaye	5 brazilian	8 building
1 bight	19 bodies	1 bread	2 buildings
1 bilary	23 bodily	10 break	4 builds
3 bile	60 body	1 breaker	2 built
1 bilineata	1 boerhaaves	1 breakers	1 bulbous
4 bill	1 boethius	3 breaking	1 bulbs
1 billardiere	2 bogota	1 breaks	1 bulk
2 billed	1 bohunthine	11 breast	26 bull
1 billin	1 boil	2 breasted	2 bulletin
1 bills	1 boiled	4 breasts	1 bullfinch
2 bimanous	2 boiling	2 breath	1 bullfinches
1 binstead	5 bold	1 breathed	1 bullocks
1 biog	1 bolder	2 breathing	7 bulls
1 biography	3 bolivia	6 breccia	1 bulwark
1 biologie	2 bolivian	2 brecciated	2 bump
1 birch	2 bollaert	49 bred	3 bunbury
1 birchell	1 bollingbroke	93 breed	2 bunch
1 birchels	1 bombay	1 breeder	2 bunting
60 bird	3 bombs	4 breeders	1 bunwood
213 birds	2 bonapartes	47 breeding	2 buoyed
2 birgos	1 bonariensis	1 breedings	1 burchell
3 birmingham	3 bond	26 breeds	1 buried
8 birth	5 bone	1 brent	3 burke
5 births	34 bones	1 brethren	2 burkes
1 biscay	1 bonin	1 brevis	1 burkhardt
1 bischoff	1 bonite	1 brewster	1 burnetts
1 bisection	1 bonn	1 bri	1 burning
3 bisexual	1 bonnet	1 brick	1 burns
1 bishoofs	41 book	6 bridge	1 burramposter
1 bison	8 books	6 bridgewater	1 burrh
4 bit	2 boot	1 bridgewaters	1 burrow
12 bitch	1 borabhum	9 bright	1 burrowing
3 bitches	2 border	1 brightened	4 burst
1 bitchs	1 bordering	1 brighter	2 bursting
6 bite	1 borders	1 brighton	1 bursts
2 biting	3 bore	2 brilliancy	2 bush
2 bits	3 borealis	1 brilliant	1 bushes
1 bitten	3 boring	3 brine	1 bustard
2 bitter	32 born	12 bring	2 bustards
1 bituminous	1 borne	9 bringing	1 butchers
1 bivalve	11 borneo	3 brings	1 butler
1 bivalves	3 bory	1 brit	1 butlers
1 bk	1 borys	9 britain	2 butresses
72 black	3 bos	1 britannicus	1 butted
5 blackbird	1 bosch	12 british	1 butter
2 blackbirds	1 bosh	8 broad	4 butterflies
1 blacker	1 boshes	2 broader	5 butterfly
3 blackish	1 bosom	3 broadly	1 butting
1 blacksmith	1 bosoms	1 broccoli	1 button
1 blacksmiths	1 boswells	1 broccolli	3 buttress
1 blackwell	5 bot	4 broderip	9 buttresses
1 blackwood	5 botanical	1 broke	1 buzzard
1 blackwoods	1 botanically	15 broken	1 buzzing
2 bladder	1 botanique	1 brongniart	1 bynoe
1 bladders	4 botanist	1 brood	1 bynoes
1 blade	16 botany	1 brooding	1 ca
2 blainville	1 botelers	1 broods	10 cabbage
1 blair	1 botouma	2 brook	12 cabbages
1 blakeway	27 bottom	5 broom	2 cacti
1 blakeways	1 bottomed	3 brother	2 cad
1 blame	2 bougainville	8 brothers	1 caermarthen
8 blanca	5 bought	3 brougham	1 caespitosa
28 blank	1 bouhunthine	37 brought	1 caetacea
1 blanket	10 boulder	1 brow	1 caeteris
1 blattae	15 boulders	33 brown	2 caffers
5 blend	4 bound	2 browne	1 caffres
9 blended	1 boundary	3 brownes	1 cages
6 blending	1 bourbon	1 brownish	1 cahirimus
8 blind	1 bourbond	2 browns	2 cairn
1 blindest	2 bourou	1 brows	1 calabria
2 blinking	1 bourous	1 bruce	2 calabrian
7 block	1 boussingualt	4 brun	1 calandria
11 blocks	1 bouthoner	1 brunii	4 calc
37 blood	2 bowels	1 bruns	8 calcareous
1 blooded	1 bowen	2 brush	1 calceolaria
1 bloodhound	1 bowerbank	2 brushed	2 calculate
1 blossom	1 bowing	1 brushing	3 calculated
2 blossomed	1 bowman	3 brussels	2 calculation
1 blossoms	2 box	6 brutes	1 calculators
4 blow	4 boy	1 bryony	1 calcutta
4 blown	2 boys	1 bu	1 caldcleughs
1 blows	1 boz	3 bubbles	3 caledonia
3 blubbering	1 bra	1 buccinum	2 caledonian
14 blue	2 bracteatum	1 buceros	1 calendar
5 blueness	1 brahmin	17 buch	4 calf
1 blume	52 brain	1 buchs	1 calidonia
1 blumenbach	2 brains	1 buck	2 california
1 blumenbachs	1 bramidor	3 buckland	13 call
9 blush		1 bucklands	3 callao

41 called	1 caring	1 centropus	1 chemically
5 calling	1 carious	1 centrum	1 chemistry
1 callitrix	2 carlyle	4 centuries	1 chert
2 callosities	1 carlyles	2 century	1 chesil
11 calls	3 carmichael	6 cephalopoda	2 chesnut
2 calm	2 carmine	4 cephalopods	3 chess
1 calosoma	1 carnatic	1 ceram	4 chest
1 caltha	1 carne	2 ceratophytes	1 chetah
3 calves	1 carniola	5 cercopithecus	1 chevre
4 cambrian	7 carnivora	4 cerebral	1 chick
3 cambridge	3 carnivorous	1 cerithium	7 chicken
33 came	1 carnosa	2 cerro	3 chickens
5 camel	2 carolina	104 certain	1 chico
1 camelidae	1 carolinas	55 certainly	9 chief
1 camelion	5 caroline	1 certainment	29 chiefly
8 camels	2 carolines	2 certainty	2 chien
1 camera	18 carried	1 cervical	60 child
2 camp	3 carrier	3 cervus	10 childhood
2 campanula	2 carries	3 ces	1 childr
1 campanularia	5 carrion	2 cessation	68 children
1 campbell	4 carrot	1 cessent	1 childrens
1 camper	1 carruncles	1 cestoid	1 childs
2 campestris	7 carry	1 cestracion	20 chile
3 canada	6 carrying	3 cetacea	1 chileno
1 canadense	3 cart	7 cetaceae	12 chili
1 canadian	1 carter	1 cetaceans	3 chilian
1 canal	2 carthorse	1 cetaceous	5 chillingham
1 canaries	4 carus	1 cetionidae	17 chiloe
13 canary	1 cas	1 cetoniadae	1 chimango
1 canarys	2 cascade	1 cette	1 chimborazo
2 cancer	271 case	4 ceylon	1 chimera
1 candoelle	86 cases	1 ceylonese	1 chimney
1 candour	1 casoars	2 cf	2 chimpanze
1 cane	3 caspian	8 ch	1 chimpaze
1 canelones	1 cassay	1 chacune	2 chin
7 canine	2 cassicans	1 chaetodon	6 china
2 canines	2 cassowary	1 chaffinches	1 chinchillidae
1 cannabalism	4 cast	1 chaffy	1 chinense
1 cannt	3 castes	15 chain	1 chinensis
1 canoes	12 castle	10 chains	13 chinese
1 canot	3 castles	1 chair	1 chineses
1 cant	1 castrate	1 chalcididous	1 chingolo
1 cantal	1 castrated	6 chalk	2 chionis
1 canter	5 castration	1 challenger	1 chirp
3 cantering	2 casts	1 cham	1 chirping
1 canthairides	1 casualty	1 chamaeleon	3 chit
1 canton	1 casuistical	3 chamaelion	1 chladni
1 cantons	1 caswary	2 chamelion	1 chlorite
1 cantors	23 cat	7 chamisso	1 chlorites
2 capabilities	4 catalogue	32 chance	1 chloritic
7 capability	1 catasetums	7 chances	2 chloropus
17 capable	1 catastrophes	208 change	4 choice
1 capacities	7 catch	31 changed	4 chonos
3 capacity	1 catcher	115 changes	1 choo
22 cape	7 caterpillar	18 changing	3 choose
3 capel	4 caterpillars	16 channel	3 chooses
1 caperailkie	1 caterwhalling	6 channels	1 choosing
1 capercailzie	1 cathedrâ	1 chanticleers	1 chopped
1 capillary	6 catherine	2 chaos	1 chops
13 capital	1 catorce	2 chap	1 chorus
3 capon	28 cats	6 chapt	1 chose
1 capped	45 cattle	5 chapter	1 chota
1 caprice	1 caubul	1 chapters	1 chouette
1 capricious	6 caught	2 chara	1 christ
2 capromys	2 cauliflower	106 character	5 christian
1 capsules	1 cauquenes	1 characterised	2 christmas
17 capt	5 causation	5 characteristic	1 chrom
2 captain	101 cause	2 characteristics	4 chronology
1 captial	13 caused	1 characterize	1 chronometer
1 captivity	44 causes	9 characterized	1 chrysoprase
2 cara	3 causing	40 characters	1 chupat
2 carabidae	3 caution	2 charcoal	1 chuquisaca
1 carabids	5 cautious	1 chardonneret	1 churche
4 caracara	1 cave	1 charged	1 cinbermere
1 caracaras	1 cavernes	4 charity	1 cinereus
2 caraccas	2 caverns	7 charles	1 cinq
1 caradoc	4 caves	1 charlestown	1 circinnatus
7 carb	1 cavia	1 charlesworth	6 circle
1 carbbuncl	1 cavities	1 charlsworth	5 circles
2 carbon	1 caya	1 charlworth	8 circular
1 carbonaceous	1 cayenne	1 charpentier	1 circularity
3 carbonate	1 caymans	4 chart	1 circulating
5 carbonic	2 cd	1 charts	10 circulation
3 carboniferous	2 ce	1 chase	1 circulations
1 carbonized	3 cease	1 chasm	1 circum
3 carcases	7 ceased	1 chasms	1 circumference
2 cardium	3 ceases	1 chasseur	14 circumstance
1 cardoon	2 ceasing	1 chastity	3 circumstanced
18 care	1 ceinture	2 chat	60 circumstances
3 careful	6 celebes	1 cheating	1 circus
5 carefully	6 cell	7 check	2 cirrhipedes
1 carelmapu	8 cells	4 checked	1 cirrhipeds
1 cares	1 cellular	1 checking	1 cisalpina
1 carex	1 celui	7 checks	1 cisticola
1 caribs	1 cementation	1 cheek	1 cistus
1 carica	2 cent	1 cheerful	1 citadel
1 caricaridae	1 centimetre	1 cheese	1 citrons
1 caricas	21 central	1 chefs	1 civet
1 caricature	19 centre	1 cheiroptera	1 civets
1 carimon	1 centres	15 chemical	

4 civilization	1 coelum	1 complaints	4 confused
8 civilized	1 coembedded	1 complanata	4 confusion
1 cl	1 coeval	4 complete	1 congeners
1 claims	1 coexisted	2 completed	3 congenital
1 clapperton	1 coexisting	6 completely	6 conglomerate
1 claps	1 coffin	1 completion	6 conglomerates
1 clarkes	3 cogan	4 complex	5 congo
1 clarkia	1 cogans	8 complexity	6 conical
1 clasp	3 coincidence	17 complicated	1 coniferae
1 clasping	2 coincidental	1 component	7 coniferous
20 class	1 coipu	8 composed	1 conjectural
27 classes	3 coition	1 compositae	4 conjecture
1 classfication	1 coitus	1 composite	2 conjoined
1 classi	8 col	2 composites	2 conjugal
21 classification	32 cold	5 composition	1 connate
2 classified	2 colder	1 compound	4 connect
1 classifies	1 coleccion	2 compounded	59 connected
2 classify	1 coleoptera	2 comprehend	6 connecting
1 classifying	2 coleridge	2 comprehended	30 connection
1 classifyng	2 coll	1 compressed	1 connections
2 clavicle	3 collapse	1 compression	2 connects
4 claw	9 collarig	1 compte	7 connexion
21 clay	5 collect	5 comte	1 conquer
1 clayey	5 collected	2 comtes	2 conquered
1 claystones	3 collecting	3 conceal	2 conquers
1 clayton	9 collection	2 concealed	49 conscience
1 clean	2 collections	2 conceivable	1 consciences
1 cleaned	2 collector	15 conceive	14 conscious
1 cleaning	1 collects	1 conceiving	53 consciousness
25 clear	1 colliers	2 concentrated	1 conse
1 cleared	1 colobes	1 concentric	1 consent
1 clearest	5 colonel	1 concentricus	1 consequen
31 clearly	1 colonization	16 concepcion	22 consequence
1 clearness	43 colour	1 concepcional	4 consequences
1 clears	20 coloured	1 concepecion	9 consequent
24 cleavage	10 colouring	4 conception	5 consequently
1 cleave	21 colours	7 concerned	56 consider
1 cleaving	1 cols	1 concerning	7 considerable
1 cleft	1 colt	1 concerns	7 considerably
1 clenomys	2 colts	1 concert	8 consideration
1 clever	2 columba	1 conch	2 considerations
5 cliff	1 columbia	1 conchifera	34 considered
10 cliffs	1 columbine	1 concholepas	15 considering
1 clifts	1 column	2 conchology	23 considers
1 clim	1 columnar	5 conclude	1 consigned
1 clima	1 columns	5 concluded	6 consist
1 climat	1 comb	1 concludes	2 consisted
41 climate	1 combination	1 conclure	1 consistent
7 climates	6 combined	9 conclusion	1 consistently
1 climatized	1 combining	1 conclusions	3 consisting
1 climatizes	1 combless	2 conclusive	11 consists
2 climb	1 combs	3 concomitant	1 consolations
3 climbing	62 come	1 concourse	1 consolats
6 clinal	30 comes	1 concret	1 consolidated
4 cline	2 comfort	4 concretion	1 consolidation
1 clinging	1 comforted	6 concretionary	1 constance
2 clitoris	1 comforts	16 concretions	2 constancy
1 clive	15 coming	1 concurrence	36 constant
1 clock	1 command	1 condamaine	18 constantly
55 close	3 comme	2 condensation	1 constituents
3 closed	4 commence	5 condensed	2 constituted
26 closely	9 commenced	1 condensing	14 constitution
2 closeness	1 commencement	1 condensor	1 constitutional
3 closer	3 commences	12 condition	2 constitutions
1 closes	1 commencing	1 conditioned	2 constructive
7 closest	1 comment	36 conditions	1 consul
3 closing	1 commentatia	2 condor	20 consult
1 cloth	1 commerce	1 condore	1 consulted
1 clothes	1 comminuted	1 conducct	1 consulting
2 clotted	1 committed	1 conduce	1 consume
1 cloud	188 common	16 conduct	1 consumed
1 cloven	1 commoner	2 conducted	2 cont
3 clover	9 commonly	6 conducting	1 conta
1 club	1 communes	6 conductor	4 contact
1 clubs	1 communicat	1 conduits	1 contagions
3 clump	3 communicate	1 cone	2 contagious
3 clumps	3 communicated	2 cones	4 contain
4 co	1 communicates	1 conferva	3 contained
14 coal	1 communicating	4 confervae	5 containing
1 coaly	10 communication	4 confess	11 contains
4 coarse	3 community	1 confessed	1 contamination
99 coast	1 comp	1 confesses	1 contemp
4 coasts	1 compact	1 confession	1 contemplation
4 coat	7 companion	1 confessions	3 contemporaneous
6 coated	1 companions	1 confidence	2 contemporaries
2 coati	1 company	1 confidently	2 contemporary
1 coating	3 comparative	3 configuration	1 contemporarys
1 coatings	2 comparatively	1 confine	7 contempt
1 cobalt	32 compare	20 confined	1 contemptible
2 cobija	26 compared	6 confinement	1 contend
1 cocatores	5 compares	1 confines	1 contentmt
1 coccygis	8 comparing	1 confirmation	1 contents
1 cochin	28 comparison	1 confirmed	1 contest
2 cochineal	3 compass	1 conflicts	1 contests
30 cock	1 compatible	1 conform	1 contiguity
1 cocked	5 compelled	2 conformable	33 continent
1 cockles	1 compensated	3 conformity	6 continental
8 cocks	1 compilation	1 confound	26 continents
3 cocoa	1 complains	4 confounded	1 continet
5 cocos	2 complaint	2 confounds	1 contingences

12 contingencies
7 contingency
2 contingent
1 contingents
1 contiguity
4 continually
3 continuation
5 continue
13 continued
1 continues
3 continuing
1 continuity
3 continuous
4 contorted
4 contract
1 contracted
4 contraction
1 contracts
1 contradict
4 contradicted
1 contradicting
2 contradiction
1 contradictory
1 contradistinct
1 contradistinction
1 contraire
7 contrary
10 contrast
2 contrasted
2 contrivance
1 control
1 controversy
1 conus
1 converge
6 conversation
1 conversations
6 converse
2 convert
6 converted
2 convey
1 conveyed
2 conveying
2 conviction
1 convin
1 convince
2 convinced
3 convulsion
4 convulsions
6 convulsive
1 conway
3 conybeare
1 cooing
2 cook
3 cool
7 cooled
6 cooling
1 coorganic
4 coots
1 copenhagen
7 copiapò
6 copied
5 copper
1 copris
1 copulate
2 copulated
1 copulates
5 copulation
1 coq
9 coquille
4 coquilles
5 coquimbo
19 coral
1 coraliferous
4 corallina
1 corallinae
2 coralline
1 corallines
6 corals
1 corb
6 corbet
1 corbets
1 corbules
2 cord
2 cordill
37 cordillera
3 cordilleras
2 cordoba
3 cordova
1 cordovise
1 core
8 corelation
5 corelations
1 corlula
1 cormorant
2 corn
1 corner
1 cornice
1 corniculatus
2 cornwall
2 corolla

1 coromandel
3 coronata
1 coronella
3 corporeal
4 correct
2 correctly
5 correspond
1 correspondance
9 corresponding
1 corresponds
1 corresponing
1 corroborates
2 corrode
3 corroded
2 corrosion
1 corrupt
3 corry
1 corsica
1 cory
3 cosmopolite
1 cosmopolites
1 cossigny
1 costata
1 costatus
2 costorphine
1 costs
1 cotopaxi
3 cotton
1 coucal
2 couches
114 could
1 coulter
2 count
3 counted
4 countenance
3 countenances
2 counteracted
1 counteracting
2 counterbalance
1 counterbalancing
1 counterpart
3 counties
56 countries
86 country
2 county
1 coup
1 coupe
4 couple
1 coupling
5 courage
1 couroucus
1 courrejolles
22 course
1 courses
1 court
1 courte
2 cousin
1 cousins
2 couteur
1 cove
10 cover
21 covered
2 covering
1 covers
1 coves
1 covington
15 cow
3 cowardice
2 cowardly
3 cowcumbers
1 coworker
2 cowper
1 cowpox
8 cows
11 cowslip
1 coxe
7 crab
1 crabbes
6 crabs
1 crack
2 cracking
5 cracks
1 crafty
5 crag
7 craigs
1 cramped
1 crane
1 cranks
7 crater
10 craters
2 crawford
1 crawl
3 crawling
3 cream
1 creat
6 create
35 created
2 creates
1 creating
28 creation

7 creations
4 creative
18 creator
1 creature
4 creatures
2 credit
1 credulous
1 creeks
2 creeper
1 creeping
1 crepitando
1 cresselly
1 crest
2 crested
2 cretaceous
1 cretceous
1 creted
1 cretionary
2 crevice
2 cricket
4 cries
1 criminal
2 criminals
2 criterion
1 crochets
6 crocodile
2 crocodiles
1 crocuses
2 crooked
1 croonian
1 crop
1 crops
89 cross
1 crosse
46 crossed
8 crosses
56 crossing
7 crow
2 crowd
1 crowded
1 crowing
1 crown
7 crows
1 crozet
2 cruel
1 cruikshank
1 crush
26 crust
5 crustacea
2 crustaceae
2 crustacean
1 crustaceous
7 cruz
6 cry
1 cryed
4 crying
1 crypt
3 cryptandrous
2 cryptogam
1 cryptogamia
6 cryptogamic
3 crys
1 crysom
2 crysomela
1 crysomelidae
6 cryst
2 crystalline
3 crystallization
1 crystallize
1 crystallized
1 crystallizes
1 crystalls
9 crystals
1 cs
1 cub
4 cuba
1 cubic
1 cuckoo
3 cuckoos
1 cuculus
1 cuidado
7 cultivated
1 cultivating
8 cultivation
2 culture
1 cuming
1 cumms
1 cunningness
1 cups
1 cur
1 curculionidae
3 cure
4 cured
1 cures
3 curig
1 curing
1 curiosities
4 curiosity
90 curious
3 curiously

2 curled
1 curling
1 curls
6 current
13 currents
3 curvature
1 curvatures
1 curve
6 curved
1 curvilinear
3 cuscus
3 custom
1 customary
1 customs
32 cut
3 cuts
2 cutter
5 cutting
5 cuttings
2 cuttle
17 cuvier
5 cuviers
6 cyanocephalus
1 cyanoglossum
1 cyanurus
2 cycle
1 cyclopaedia
1 cyclops
1 cyclostoma
1 cyclostomes
1 cylinder
1 cynocephalus
1 cynoglossum
1 cyrena
1 cystic
2 dacelo
1 dahlia
4 dahlias
5 daily
1 daines
2 daisy
1 daltonism
1 dam
1 damned
3 damp
4 dampier
2 dampiers
1 dampness
2 dance
1 dandelion
1 dandelions
3 danger
1 dangerous
1 dangled
1 dangling
1 dans
1 daphnia
1 darby
6 dare
1 daresay
1 daresays
13 dark
2 darker
2 darkness
1 darnleys
1 darwar
11 darwin
3 darwins
4 dash
2 dasypus
1 dasyurus
3 data
1 date
10 daubeny
1 daubisson
2 daughters
1 davis
3 davy
1 dawn
39 day
21 days
20 dead
1 deaden
1 deadly
2 deaf
2 deal
1 dealt
2 dear
1 dearth
38 death
2 deaths
1 debased
9 dec
8 decandoelle
1 decandolles
1 decapod
4 decay
1 decayed
2 decaying
1 decb

2 deceive
1 deceived
4 decemb
3 december
1 decency
4 deception
2 decide
2 decided
6 decidedly
1 decimetre
1 decision
1 decisions
1 declare
1 declared
1 declaring
1 declension
1 decline
1 declined
1 declining
2 decompose
2 decomposed
3 decomposition
5 decrease
2 decreased
3 decreasing
1 decrite
1 decumanus
2 deduce
1 deduces
1 deduction
1 deductions
1 deed
23 deep
1 deepen
2 deepening
1 deepens
2 deeper
2 deepest
8 deeply
25 deer
1 defect
2 defective
1 defence
3 defending
1 defends
1 defiance
2 deficiency
2 deficient
1 definable
1 define
2 defined
18 definite
1 definitely
11 definition
2 deflected
2 deflection
1 defonsos
1 deformations
2 deformed
2 deformities
3 deformity
1 defunct
2 deg
2 degenerate
1 degenerated
1 degeneration
1 degnen
9 degradation
1 degraded
2 degrading
142 degree
7 degrees
1 deinotherium
22 deity
1 déjà
31 del
1 dela
1 delabechs
2 delaware
1 delay
1 deletereous
4 delicate
2 delight
8 delirium
1 deliriums
1 deliver
1 delivery
1 della
2 delta
1 deltas
1 deluge
1 delusion
2 demand
1 demarara
4 demerara
1 demerary
1 demolished
1 demonstrate
1 demonstration
1 demonstrations

1 dendritic
1 denham
1 denied
1 denies
2 denmark
1 denote
1 densifolium
1 dental
3 denudation
1 denuded
9 deny
2 depart
11 depend
1 depended
5 dependent
2 depending
19 depends
2 depopulation
8 deposit
8 deposited
6 deposition
9 deposits
2 depress
1 depressed
2 depresses
5 depression
14 depth
6 depths
1 derision
1 derivable
3 derived
1 derives
2 derry
1 desaguadero
1 descen
4 descend
4 descendant
3 descendants
11 descended
3 descending
1 descends
14 descent
2 describe
33 described
2 describer
2 describers
10 describes
5 describing
2 descript
11 description
2 descriptions
15 desert
1 deserted
4 deserts
1 deserve
3 deserves
1 desgraviers
4 deshayes
3 design
1 designs
3 desirable
22 desire
6 desires
1 despair
1 desperately
1 desperation
1 despise
1 destined
1 destinies
1 destiny
1 destitute
8 destroy
17 destroyed
1 destroyer
1 destroyers
2 destroying
2 destroys
9 destruction
2 destructive
1 desultory
6 detail
1 detailed
1 details
1 detained
1 deter
3 deteriorate
2 deteriorated
1 deteriorates
2 determinate
2 determination
5 determine
17 determined
5 determines
1 determining
3 detrimental
2 detritus
30 developed
1 developemen
21 developement
1 developent

4 developing
5 development
1 developpent
1 develops
1 deviate
1 deviation
1 devil
2 devonian
1 devonshire
1 devotional
1 devourer
1 devours
1 dew
1 dewar
1 dhangar
1 dholbum
1 dholes
1 di
1 diagonal
1 diagram
1 dialogue
1 diamante
1 diameter
1 diametrically
1 diard
1 diarrhaea
1 dichogamous
1 dichotomous
3 dick
2 dicks
2 dicot
3 dicotyledenous
1 dicotyledones
2 dict
1 dictionary
1 didelphidae
6 didelphis
1 didelphys
20 die
7 died
1 diemen
28 diemens
4 dies
1 dieteriorating
1 dieu
1 dif
1 diferent
11 diff
19 differ
2 differed
1 differen
90 difference
30 differences
248 different
8 differently
2 differing
14 differs
44 difficult
5 difficulties
55 difficulty
1 digested
1 digestion
1 digestive
2 dignity
16 dike
36 dikes
1 dilatation
1 dilated
1 dilemma
1 dillwyn
1 diluv
2 diluvial
2 diluvianae
5 dimensions
1 dimidiate
1 diminish
2 diminished
1 diminishing
1 dingo
1 dingos
4 dinner
1 dinners
2 dioecia
30 dioecious
1 dioeecous
3 dioica
1 dioicous
2 diorite
13 dip
2 dipping
1 diptera
3 dipus
24 direct
12 directed
20 direction
6 directions
12 directly
1 directory
1 directs

1 dirt
3 dis
1 disabled
1 disagree
7 disagreeable
3 disappear
2 disappeared
1 disappearing
1 disappointment
1 disapproval
1 disapproving
1 disbelieve
2 discomfort
1 discordance
1 discordant
5 discourse
1 discourses
15 discover
2 discoverable
20 discovered
2 discoverer
2 discoveries
6 discovering
6 discovery
1 discriminating
1 discrimination
2 discuss
1 discusses
8 discussing
25 discussion
17 disease
5 diseased
17 diseases
1 disfigure
4 disgust
1 disgusted
1 disgusting
1 disinheriting
1 disintegrated
1 disinterested
8 dislike
1 dislikes
1 dislocate
4 dislocation
2 dislocations
1 dismally
2 disobey
3 disobeyed
1 disobeys
2 disordered
1 disorders
1 disorganized
1 disorganizes
1 dispersion
1 displacement
1 display
1 displays
1 displeasure
14 disposition
1 dispositions
1 disputable
6 dispute
5 disputed
1 disputes
1 dissatisfied
1 dissected
1 dissecting
5 disseminated
1 dissemination
1 dissenblances
1 dissert
4 dissertation
1 dissertations
1 dissimilarity
2 dissolved
14 distance
7 distances
22 distant
1 distaste
1 distemper
2 distended
50 distinct
1 distincte
6 distinction
1 distinctive
8 distinctly
2 distinctness
4 distinguish
2 distinguishable
7 distinguished
4 distinguishing
1 distract
1 distracted
3 distress
1 distri
6 distrib
2 distribut
1 distribute
8 distributed

711

712

17 enormous
13 enough
1 enquire
1 enquiries
1 enquiry
1 ensd
1 ensue
1 ensure
7 enter
5 entered
3 entering
1 enterprise
1 enterprises
4 enters
2 enthusiasm
4 entire
14 entirely
1 entitled
1 entomolgicae
2 entomolog
1 entomologica
1 entomologicae
6 entomological
2 entomologist
1 entomologists
1 entomology
1 entomost
5 entomostraca
5 entozoa
3 entrance
1 entre
1 entry
1 enumerate
1 enumerates
1 enunciated
1 enuntiated
1 envelope
2 envelopes
1 envy
11 eocene
1 eocine
1 eoeconomy
1 epaules
1 epic
1 epidemic
2 epidemics
2 epidote
1 epigmous
1 epigonous
4 epilepsy
1 epizoa
6 epoch
6 epochs
1 epock
1 epocks
11 equable
2 equably
19 equal
1 equalities
1 equall
22 equally
2 equals
8 equator
1 equatorial
9 equilibrium
1 equivocal
1 eradicate
1 eradicated
1 eras
11 erasmus
1 ercall
5 erect
1 erecting
1 erection
1 erhman
4 erin
1 erman
1 ermans
1 ermine
1 ernest
2 erosion
2 erosive
3 erratic
1 erroneousness
5 error
1 errors
1 errs
7 eruption
9 eruptions
1 ervum
1 erysipelas
1 es
7 escape
5 escaped
3 escarpment
1 esculent
1 ese
1 espèce
4 especes
1 especial

2 especiall
38 especially
1 esq
10 esquimaux
1 essai
9 essay
7 essays
4 essence
4 essential
1 essentially
1 essequibo
4 establish
6 established
1 establishes
1 establishing
1 establishment
2 estimate
1 estuaries
3 estuary
1 été
1 eternal
7 eternity
3 ethical
2 ethics
1 ethnical
1 etiquettes
1 etn
3 etna
1 être
2 eucalyptus
4 eugenius
1 eunuchs
1 euphorbia
30 europaean
1 europaeans
79 europe
5 european
1 eussent
1 eût
1 evaporation
1 evasive
1 evelyns
6 evening
3 event
3 events
39 ever
1 evergreen
1 evergreens
1 everlasting
1 everybody
1 everyone
4 everything
1 everytime
3 everywhere
15 evidence
10 evident
21 evidently
4 evil
1 evils
1 evince
2 ewe
1 ex
6 exact
1 exacte
11 exactly
1 exactness
3 exaggerated
1 exaggeration
1 exalted
1 examen
7 examination
25 examine
10 examined
3 examining
11 example
6 examples
1 excalted
1 excavated
1 exceeding
1 exceedingl
7 exceedingly
2 exceeds
1 excellences
14 excellent
1 excellently
39 except
15 excepting
11 exception
2 exceptions
5 excess
5 excessive
11 excessively
1 exchange
4 excised
1 excite
15 excited
1 excitement
1 excites
2 exciting
5 exclaim

1 exclaimed
2 exclude
1 excluded
1 exclusion
2 exclusively
1 excuses
1 execution
6 exercise
2 exert
6 exertion
1 exerts
1 exeters
1 exhausted
1 exhaustion
2 exhibit
3 exhibited
1 exhibits
17 exist
17 existed
30 existence
1 existences
20 existing
3 exists
1 exjudge
1 exosmic
1 exotics
2 exp
1 expand
1 expanded
1 expands
1 expanse
1 expansion
1 expansions
1 expatiate
19 expect
1 expectant
3 expected
1 expecting
8 expedition
1 expended
2 expense
1 exper
34 experience
2 experienced
15 experiment
5 experimentise
3 experimentize
11 experiments
39 explain
26 explained
5 explaining
29 explains
19 explanation
1 explanations
1 explicable
1 explosion
7 exposed
1 exposes
1 expostulatory
1 exposure
1 expr
5 express
1 expressed
1 expressemotions
1 expresses
5 expressing
59 expression
4 expressions
3 expressive
2 expulsion
2 exquisite
1 exserted
1 extant
11 extend
7 extended
5 extending
3 extends
6 extension
3 extensive
8 extent
2 exterminated
8 extermination
51 external
5 externally
22 extinct
26 extinction
2 extinguished
1 extra
3 extract
2 extraordinarily
8 extraordinary
1 extravagant
1 extreamly
37 extreme
6 extremely
5 extremities
2 extremity
1 extres
1 exudation
1 exudes

1 exuviae
15 eye
1 eyebrow
3 eyebrows
1 eyed
1 eyelids
20 eyes
1 eyesight
15 eyton
8 eytons
1 fable
3 fabric
25 face
5 faced
1 facedness
6 faces
1 facilitate
2 facilities
13 facility
98 fact
91 facts
1 facultes
9 faculties
29 faculty
1 fades
1 faery
1 fah
9 fail
5 failed
4 failing
9 fails
1 failure
1 faint
2 fainting
3 faintly
1 fair
1 fairer
1 fairly
1 fairway
1 fait
3 faith
1 falco
1 falcon
2 falconer
2 falconers
21 falkland
1 falklandii
12 falklands
2 falkner
1 falkners
19 fall
1 fallacy
1 falle
7 fallen
9 falling
2 fallow
4 falls
14 false
1 falsely
1 fam
2 fame
6 familiar
18 families
33 family
4 famine
1 famines
1 famous
2 fan
2 fancied
1 fancier
3 fanciers
1 fancies
7 fancy
1 fang
1 fanny
3 fantastic
3 fare
1 farmer
1 farquhar
1 farroilap
1 farthers
1 farthing
1 fashions
2 fast
4 fat
4 fatal
87 father
1 fathered
13 fathers
12 fathoms
3 fatigue
1 fatiguing
1 fatness
2 fatten
1 fatty
5 fault
10 faults
2 faulty
1 faun
13 fauna

3 faunas	1 finch	1 flowering	1 fortunately
1 faune	4 finches	57 flowers	1 fortune
1 favor	36 find	6 flowing	5 forward
5 favour	5 finding	3 flows	1 forwards
12 favourable	1 finds	34 fluid	1 foss
2 favoured	25 fine	3 fluidity	50 fossil
1 favouring	2 finely	2 fluids	2 fossiles
1 fawn	1 fineness	2 fluor	32 fossils
2 fe	2 finer	1 fluoric	1 fought
48 fear	4 finest	2 flush	1 fouled
1 fearful	1 fingals	1 flushing	123 found
3 fears	3 finger	2 flustra	6 foundation
1 feasts	3 fingered	1 flute	2 foundations
8 feather	6 fingers	1 fluvicolae	2 founded
1 feathered	4 finished	15 fly	1 fourmillier
1 feathering	1 finite	4 flycatcher	3 fourth
19 feathers	1 finlay	9 flying	11 fowl
2 feature	1 fins	2 flys	17 fowls
1 featured	1 fintec	1 foame	51 fox
6 features	1 fir	1 focus	8 foxes
11 feb	1 fire	2 foetal	3 foxs
3 february	1 fires	15 foetus	5 fr
1 feconds	1 firm	2 foetuses	1 fracture
3 fecundate	1 firma	1 fol	1 fractured
1 fecundated	9 firmly	1 fold	1 fractures
2 fecundation	1 firola	1 folds	2 fragmens
1 fecundity	153 first	2 foliage	7 fragment
1 fed	63 fish	2 folkstone	2 fragmentary
1 feeders	1 fisher	21 follow	14 fragments
10 feeding	13 fishes	10 followed	14 frame
1 feeds	6 fissure	13 following	2 framed
33 feel	9 fissures	2 follows	1 frames
1 feelin	5 fit	1 folly	15 france
51 feeling	3 fits	3 fond	1 frances
35 feelings	13 fitted	1 fondled	1 francisco
14 feels	1 fitting	1 fondling	1 franklin
25 feet	1 fitton	22 food	1 fraser
1 feigning	2 fittons	1 foods	5 fraxinella
1 feis	3 fitz	3 fool	1 fre
4 feldspar	3 fitzroy	14 foot	1 freckles
1 feldspathes	1 fitzwilliam	2 footed	23 free
6 fell	4 fix	1 footsteps	7 freely
6 fellow	30 fixed	1 forbes	2 freestone
12 felspar	2 fixes	45 force	2 freeze
1 felspathic	2 fixing	3 forced	2 freezing
13 felt	2 fixity	9 forces	16 french
76 female	1 fl	1 forchammers	2 frenchman
21 females	2 flaccid	1 forcible	7 frequency
1 fences	1 flag	1 forcibly	7 frequent
1 fennel	3 flags	1 ford	2 frequented
1 ferguson	2 flamingo	1 forefather	6 frequently
1 ferme	1 flank	2 forefathers	1 frequet
12 fernandez	1 flanks	2 foregoing	34 fresh
6 fernando	1 flappered	1 foreground	1 freshly
12 ferns	1 flaqueza	3 forehead	1 freshness
1 ferrugineous	2 flash	10 foreign	2 freshwater
1 ferrugineum	1 flashes	1 foreknowledge	1 freycinet
2 ferruginous	1 flashing	1 forequarters	1 freycinets
1 ferrussac	13 flat	1 foreseen	2 friction
32 fertile	2 flats	6 forest	6 friday
1 fertilise	1 flavour	7 forests	4 friend
22 fertility	1 flavum	2 foretell	4 friendly
1 fertilization	1 flax	1 foretells	1 friends
3 fertilize	1 flaxes	1 forficatus	2 friendship
1 fertilized	3 fleas	1 forficetus	3 fries
2 fertilizing	3 flèche	6 forget	1 frigate
1 feruginous	1 fled	4 forgetfulness	2 fright
2 festuca	1 fledge	4 forgets	2 frighten
1 fetch	1 flee	2 forgetting	8 frightened
1 fetches	1 fleestest	1 forging	1 frightens
1 fetid	9 fleming	1 forgive	3 fringe
3 fever	2 flemings	11 forgotten	2 fringes
1 fevers	1 flemming	2 forked	1 fringilla
6 fewer	1 fleshy	168 form	2 frio
1 fewness	2 flew	2 formal	3 frog
2 fibre	1 flexure	36 formation	5 frogs
1 fibres	3 flies	23 formations	1 frolicsome
2 fibrous	6 flight	2 formativa	1 frondosi
1 fibula	4 flights	1 formative	9 front
1 fickle	3 flinders	1 formativus	1 frontier
1 ficus	1 flint	79 formed	1 frost
7 field	2 flints	1 formées	1 froudes
5 fields	1 flipper	1 forment	8 frown
2 fierce	2 floated	51 former	6 frowning
12 fig	7 floating	20 formerly	2 frowns
4 fight	2 flock	1 formica	8 frozen
3 fighting	1 flocks	10 forming	1 fruct
2 figs	1 flood	160 forms	1 fructification
8 figure	1 floods	1 formula	22 fruit
1 figured	25 flora	7 forrest	1 fruiterer
1 figures	1 florae	1 forseeing	1 frustra
6 filaments	1 floras	1 forseight	2 frustrated
1 file	1 flore	2 forster	52 ft
7 fill	1 florida	3 fort	3 fucus
6 filled	1 flourens	4 forth	1 fuegia
2 filling	2 flourish	1 fortify	3 fuegian
2 film	9 flow	4 fortnight	2 fuegians
1 films	2 flowed	1 forts	25 fuego
19 final	55 flower	2 fortuitous	1 fuge
2 finally	2 flowered	1 fortunate	1 fugitive

1 fulfil	124 genera	1 glengarry	1 grandson
1 fulginosus	77 general	1 glenoe	1 granfathers
1 fulgoridae	1 generality	2 glenroy	1 graniniverous
2 fulica	3 generalization	1 glimpse	56 granite
26 full	3 generalize	1 glimpses	5 granites
1 fullers	2 generalized	14 globe	10 granitic
1 fullfil	2 generalizing	1 globes	1 granits
2 fullfilled	1 generalle	2 globule	1 granny
14 fully	59 generally	2 globules	13 grant
1 fulva	1 generat	1 glöger	5 granted
7 function	3 generate	3 gloomy	1 granting
1 functional	7 generated	1 glorious	1 grants
3 functions	86 generation	1 glorrying	1 granules
2 fundamental	36 generations	1 glory	1 grapple
1 fundamentally	6 generative	2 glorying	5 grass
1 fungi	6 generic	1 gloss	4 grasses
1 fungus	1 generl	1 glove	1 grasshopper
7 fur	1 generosity	3 glow	1 grasshoppers
2 furious	1 generous	2 glowing	1 grassy
2 furnaces	1 genetal	2 glowworm	3 grateful
1 furnarii	1 geneva	1 glowworms	5 gratification
5 furnarius	1 genital	1 gmnura	2 gratified
1 furor	1 genitals	26 gneiss	1 gratify
23 further	3 genius	1 gnu	1 gratitude
1 furthest	1 genres	72 go	1 grauwacke
1 fury	5 gentle	4 goat	2 grave
1 furze	1 gentlefolks	4 goats	11 gravel
1 fusibility	1 gentleman	1 goatsucker	1 gravities
2 fusible	1 gentlemen	1 goatsuckers	12 gravity
1 fût	1 gentler	1 gobbah	4 gray
13 future	1 gently	23 god	1 grays
2 futurity	56 genus	1 godlike	3 grazing
3 gaertner	1 geo	1 gods	1 grease
4 gaimard	1 geof	1 godwits	264 great
6 gain	1 geoffroy	9 goes	59 greater
24 gained	1 geoffry	1 goethe	26 greatest
5 gaining	1 geogr	28 going	8 greatly
4 gains	6 geograp	10 gold	1 greatness
1 gait	21 geograph	8 golden	1 greatt
51 galapagos	1 geographic	1 goldens	2 grebe
3 gale	26 geographical	1 goldfinch	1 grebes
1 galena	2 geographically	2 goldfinches	1 greek
3 gales	1 geographico	1 goldsmiths	2 greeks
1 galeum	4 geographie	1 gonaïves	12 green
1 galignani	7 geography	17 gone	1 greenfinch
2 galiopithecus	2 geol	1 goo	6 greenland
1 galium	27 geolog	214 good	1 greenough
1 gallesio	13 geological	1 goodhumoured	13 greenstone
3 gallinaceous	3 geologically	1 goodnatured	1 gregory
1 gallinaces	1 geologico	2 goodness	1 grenats
1 gallinules	1 geologique	15 goose	1 grenish
1 gallionella	6 geologist	2 gore	3 gret
2 galls	5 geologists	4 gorge	1 greter
1 galton	25 geology	1 gory	7 grew
1 galtons	1 geometrical	4 gorze	11 grey
1 galvanic	2 george	11 got	15 greyhound
1 galvanize	1 georges	29 gould	6 greyhounds
6 game	1 georgia	6 goulds	1 greywacke
2 gander	2 geranium	2 gout	7 grief
1 ganders	7 germ	1 gov	1 grieved
1 ganges	4 german	2 governed	1 griffith
1 ganglionic	4 germany	1 governing	3 grin
1 gangway	2 germen	1 government	2 grinding
4 gap	1 germinate	1 governors	1 grinning
12 gaps	1 germinated	1 governs	1 grins
1 garcia	1 germinating	2 gowen	1 grisea
23 garden	1 germination	1 graah	1 grit
5 gardener	3 germs	1 gradat	1 groan
1 gardeners	2 gestation	29 gradation	1 grooms
22 gardens	1 gesture	3 gradations	1 grooved
1 gardiners	2 gestures	2 grade	1 grooves
2 gardner	44 get	1 grades	1 grotesques
2 gardners	3 gets	11 gradual	1 grotius
1 garments	8 getting	11 gradually	37 ground
1 garnets	1 geum	1 graduate	2 grounded
1 garrottii	1 ghilan	1 graduating	3 grounds
1 garsipa	1 gibbon	2 graecian	1 groundsil
2 gases	1 gibbons	1 graecians	2 groundwork
1 gastrobranchus	1 gift	1 graeduated	35 group
3 gate	1 giganteus	4 graft	2 grouped
1 gather	2 gigantic	7 grafted	2 grouping
5 gathered	1 gilbert	3 grafting	26 groups
1 gatherers	1 gillies	1 graham	8 grouse
1 gaudichauds	1 gilly	2 grain	1 grove
1 gaudy	1 gilolo	2 grained	22 grow
1 gault	3 giraffe	6 grains	1 growers
1 gaut	1 giraffes	3 grallae	16 growing
1 gay	2 glacial	1 grammalite	2 growl
1 gays	2 glaciers	35 grand	2 growling
3 gaz	1 glance	1 grandchild	15 grown
1 gazal	1 glancing	2 grandchildren	4 grows
1 gazelles	1 gland	3 grande	15 growth
1 gazes	2 glands	1 grander	1 groznerat
1 gazoeus	7 glass	1 grandest	5 grt
4 gecko	1 glasses	1 grandeur	1 grunting
10 geese	6 glassy	13 grandfather	1 grups
1 gemma	1 glazing	3 grandfathers	1 guacho
1 gemmae	2 gleanings	1 grandmother	1 guadalaxura
9 gemmation	44 glen	1 grands	1 guahon
2 gemmiparous	3 glencoe		2 gualgayoc

1 guanaca	1 harbors	1 helps	1 hit
4 guanaco	23 hard	1 hemi	1 hitchcock
1 guanax	2 harden	2 hemiptera	2 hitherto
2 guanaxuato	4 harder	1 hemipterous	2 hive
1 guantajaya	1 hardier	16 hemisphere	3 hoarding
1 guanuaxuato	1 hardinge	1 hemispheres	3 hobart
1 guaranis	16 hardly	19 hen	1 hodgkins
1 guasco	4 hardness	150 hence	3 hodgson
1 guatemala	5 hardy	1 henge	1 hoffmans
2 guava	15 hare	1 henry	3 hog
1 guavas	1 harelips	1 henrys	6 hogs
2 guayana	14 hares	6 hens	3 holcroft
1 guayanensis	1 harlaam	9 hensleigh	6 hold
3 guayaquil	2 harlan	2 hensleighs	1 holders
1 guaze	3 harm	35 henslow	4 holding
1 guernsey	1 harmonicon	3 henslows	10 holds
2 guess	2 harmonious	1 hepatici	1 hole
1 guggling	2 harmonizes	1 heraldic	1 holes
2 guiana	9 harmony	1 herbage	24 holland
12 guide	1 harpalus	18 herbert	1 hollanders
5 guided	3 harrier	7 herberts	1 hollands
1 guides	1 harrison	3 herbivorous	2 hollow
1 guiding	6 hartley	1 herbs	1 hollowed
1 guillemost	1 harvest	1 herchels	1 hollowness
1 guilty	1 hasselt	1 herd	1 holly
16 guinea	1 hat	1 herded	1 hollyhock
2 gulf	2 hatch	2 hereafter	1 hollyoak
2 gull	2 hatched	1 heredetariness	1 hollyoaks
1 gullies	1 hatches	108 heredetary	1 holman
1 gulls	1 hatching	1 heredetray	1 holocentrus
1 gully	2 hate	4 hereditary	1 holyhoaks
1 gulps	2 hates	1 hereford	1 holyoak
1 gum	1 hatred	22 hermaphrodite	1 holyoaks
2 gun	1 hatreds	9 hermaphrodites	1 holyrood
1 guns	1 hats	4 hermaphroditism	8 home
3 guoy	1 hattica	1 hermaphroditisms	2 homes
1 gushed	1 haunt	2 hermoso	1 homme
1 gusto	1 haustellata	1 hernes	1 hommes
1 guy	1 hautes	1 hero	2 homogeneous
2 guyana	1 hauteurs	2 heron	3 homogenious
1 guyanensis	1 hawfinch	3 herons	1 honduras
1 guyon	10 hawk	1 hersche	2 honestly
1 guzerat	8 hawks	3 herschel	5 honey
1 gypseous	1 hay	8 herschels	4 honour
6 gypsum	2 hayd	8 herself	1 honourable
1 ha	1 haze	1 hershel	4 hood
1 habberley	71 head	1 hervey	2 hooded
1 haberclador	2 headache	2 hesitate	1 hoof
51 habit	2 headed	1 heterodox	3 hook
1 habitat	7 heads	7 heterogenous	1 hooked
3 habitation	1 healing	1 heterom	6 hooker
1 habitations	1 heals	4 heteromera	2 hop
110 habits	3 health	1 heteromerous	24 hope
49 habitual	1 healthily	6 hiatus	2 hopeless
7 habitually	2 healthiness	1 hibbert	1 hopes
4 habituated	9 healthy	1 hibernation	1 hopkins
1 hackles	1 heaping	1 hiccough	4 horae
1 hackneyed	1 heaps	2 hid	14 horizontal
1 haemorragic	12 hear	2 hide	1 horizontally
1 haena	19 heard	4 hiding	1 horn
16 hair	14 hearing	34 high	1 hornblend
1 hairdresser	2 hears	31 higher	4 hornblende
3 haired	20 heart	20 highest	1 horne
6 hairless	7 heartease	5 highland	10 horned
3 hairs	1 heartseases	1 highlands	1 horner
11 hairy	42 heat	14 highly	5 hornless
1 hal	5 heated	1 highness	14 horns
1 halcyon	1 heath	8 hilaire	2 hornsey
59 half	1 heathen	2 hilaires	1 horny
1 halfbred	1 heather	1 hilianthemum	1 horsbrugh
1 halfway	1 heaths	40 hill	1 horsburgs
1 halimeda	1 heats	2 hillhouse	52 horse
6 hall	4 heautandrous	1 hillock	1 horseflesh
1 haller	4 heaven	11 hills	34 horses
1 halls	3 heavens	1 hilly	6 horsfield
3 halo	1 heaverns	7 himalaya	1 horsing
3 hamilton	1 heavier	3 himalayan	8 hort
1 hampers	1 heavily	3 himalayas	3 horticultural
1 hamster	3 heavy	2 himmalaya	2 horticulture
27 hand	2 heber	5 hind	2 horticulturists
12 handed	1 heberdens	1 hinder	1 hortus
1 handerchief	1 hebrides	1 hindoism	1 hospital
1 handkerchief	1 heckla	1 hindoo	4 hostile
1 handle	1 hectometre	1 hindus	26 hot
1 handling	2 hedge	5 hinge	2 hothouse
7 hands	3 hedgehog	2 hinnus	4 hottentots
1 handsome	17 height	1 hints	1 hotter
4 handwriting	3 heights	2 hippopotamus	4 hound
1 hang	1 heigt	1 hippot	10 hounds
4 hanging	1 heigth	1 hippotami	4 hour
1 hants	1 heirs	6 hippotamus	3 hours
12 happen	9 held	1 hippotomus	17 house
11 happened	29 helena	1 hire	1 housedogs
3 happening	1 helix	1 hirundo	1 housemaid
3 happens	2 hell	1 hiss	8 houses
1 happi	2 helms	6 hist	313 how
1 happier	11 help	4 histoire	1 howard
27 happiness	1 helped	2 historical	36 however
6 happy	1 helping	55 history	2 howl
1 harbor	1 helplessness	1 histrionicus	1 howled

716

2 howling
1 howls
1 huantajaia
1 hubberley
1 hudson
1 hudsons
1 hue
1 humaine
14 human
2 humb
14 humble
7 humbles
1 humbold
27 humboldt
3 humboldts
4 hume
5 humes
2 humility
5 humming
1 humour
7 hump
3 humps
1 hunbt
13 hundred
1 hundreds
1 hung
2 hungary
3 hunger
7 hunt
1 hunted
18 hunter
8 hunters
6 hunting
1 huppe
1 hurrah
1 hurricane
1 hurricanes
1 hurriedly
2 hurt
4 hurting
2 hurts
4 husband
1 husbandry
1 husbands
1 husks
1 hutcheson
4 hutton
1 hyacinths
10 hyaena
1 hyaenodon
41 hybrid
2 hybridise
3 hybridised
1 hybridisim
1 hybridising
12 hybridity
71 hybrids
1 hyd
1 hydatid
2 hydra
2 hydrangea
1 hydras
1 hydridity
1 hydrocele
1 hydrogen
1 hydromys
2 hydrophobia
1 hyena
1 hyenas
3 hymenoptera
1 hypnum
3 hypoth
1 hypotheses
12 hypothesis
2 hypothetical
1 hysterical
567 i
1 ibex
2 ibis
15 ice
2 icebergs
16 iceland
1 ichneumon
1 ichthyosaurus
2 icterus
1 icthiology
1 icthyo
1 icthyosaurus
82 idea
11 ideal
49 ideas
10 identical
1 identically
1 identify
5 identity
3 ideosyncracy
2 ideosyncracy
2 ideot
2 ideotcy
2 ideotic

4 ideots
1 idiosyncrasy
548 if
3 igneous
8 ignorance
4 ignorant
1 iguana
1 iimagine
5 il
4 ile
3 iles
5 ill
1 illdefined
3 illness
1 illnesses
1 illtempered
1 illusion
2 illustrated
1 illustrates
3 illustration
3 illustrations
1 ils
5 image
3 images
4 imaginary
25 imagination
8 imagine
2 imagining
2 imago
2 imbued
2 imitate
2 imitated
1 imitates
1 imitating
9 imitation
1 imitations
4 imitative
1 immaculate
5 immediate
6 immediately
2 immens
13 immense
1 immensely
1 immenses
1 immersed
2 immigration
1 immortal
1 immortality
1 immutability
1 imp
1 impact
1 impaired
1 impatience
1 impediment
1 impelled
1 imper
1 imperative
1 imperceptibly
11 imperfect
4 imperfectly
2 imperial
1 imperious
1 imperiously
1 imperiousness
1 imperishable
4 implanted
1 impliance
1 implicit
3 implied
5 implies
4 imply
2 import
18 importance
83 important
1 importants
1 importation
6 imported
9 impossibility
12 impossible
1 impregnable
10 impregnate
8 impregnated
1 impregnating
15 impregnation
1 impregnations
2 impress
10 impressed
9 impresses
1 impressing
9 impression
9 impressions
1 improbabilities
3 improbability
5 improbable
3 improbably
1 improperly
1 improvable
10 improve
4 improved
7 improvement

1 improvements
3 improves
4 improving
1 impudence
7 impulse
1 impulses
1 impunity
1 impure
1 imrie
3 inaccurate
1 inactive
1 inadmissible
1 inamimate
2 inanimate
1 inarticulate
1 incensed
1 incessant
3 incestuous
4 inch
6 inches
1 incidentally
1 incipient
1 incisors
19 inclination
1 incline
10 inclined
1 inclin
1 inclinées
2 inclines
2 inclining
4 include
1 included
2 includes
1 including
1 incompatible
1 incomprehensible
1 incongruous
1 inconstant
1 incontestably
1 incontinence
2 inconvenience
1 incorrect
1 incorrigably
26 increase
14 increased
5 increases
5 increasing
1 increments
1 incubation
1 inculcate
2 ind
1 indecency
12 indeed
1 indefessa
3 indefinite
1 indefinitely
1 indelibleness
3 indelibly
1 indelicate
1 independency
17 independent
5 independently
1 indes
1 index
1 indexed
1 indi
42 india
48 indian
2 indians
9 indicate
1 indicated
3 indicates
6 indication
1 indications
1 indicative
1 indicus
7 indies
1 indifference
1 indifferent
1 indifferently
1 indigenous
1 indignant
3 indio
1 indisputable
1 indissoluble
1 indistinct
1 individ
56 individual
6 individuality
2 individually
42 individuals
1 individus
2 induce
1 inducing
1 induct
4 induction
1 inductions
3 inductive
1 indus
1 industrie

1 industry
1 ineffectual
1 inequality
1 inequlities
1 inevitable
3 inexplicable
1 inf
1 infancies
5 infancy
1 infant
5 infer
5 inference
6 inferior
2 inferiority
7 inferred
1 infers
11 infertile
7 infertility
1 infiltering
1 infini
10 infinite
2 infinitely
1 infinites
4 infinitesimal
1 infinity
1 inflamed
1 inflammatory
1 inflating
3 inflorescence
1 inflorescent
27 influence
4 influenced
3 influences
1 inform
2 informant
8 information
4 informs
1 infra
1 infrequent
1 infringement
1 infusible
9 infusoria
1 infusorian
1 inglish
2 ingredients
1 inguinal
5 inhabit
5 inhabitant
5 inhabitants
5 inhabited
3 inhabiting
5 inhabits
1 inhaling
1 inherent
3 inherit
2 inheritable
1 inheritance
3 inherited
1 inherits
1 inimitable
5 injected
1 injecting
4 injection
1 injections
1 injure
6 injured
1 injuries
2 injurious
1 injuriousness
4 injury
1 injustness
9 inland
2 inn
5 innate
1 inner
1 innumberable
10 innumerable
1 ino
1 inoceranus
1 inor
8 inorganic
2 inosculate
1 inosculating
2 inosculation
2 inquiry
2 ins
13 insane
21 insanity
13 insect
3 insecta
2 insectes
1 insectiferous
2 insectivore
74 insects
4 insensible
2 insensibly
1 insert
1 inshore
4 inside
1 insignificance

2 insignificant
6 insist
2 insisted
3 insists
3 inspiration
1 inspire
1 instability
72 instance
1 instanced
33 instances
1 instant
4 instantaneous
1 instantaneously
1 instantaneousness
5 instantly
22 instead
1 instinc
144 instinct
1 instincting
69 instinctive
1 instinctivedly
10 instinctively
109 instincts
1 instinctual
1 instinctus
47 institut
3 institute
1 instituted
3 institution
1 institutions
2 instruction
1 instructions
3 instrument
2 instruments
4 insufficient
1 insular
1 insulate
1 insulted
1 insulting
1 insure
1 insures
1 integrant
30 intellect
3 intellects
17 intellectual
1 intellectuality
3 intellectually
1 intelligble
4 intelligence
4 intelligent
3 intelligible
4 intemperance
3 intends
9 intense
3 intensity
1 intent
1 intention
2 intentionally
3 intently
14 inter
4 interbred
1 intercalated
1 intercepting
4 intercourse
17 interest
1 interested
10 interesting
1 interfere
1 interfered
1 interference
7 interior
1 interlineal
2 interlock
1 interlopement
1 interm
5 intermarriage
4 intermarriages
1 intermarry
1 intermarrying
55 intermediate
1 intermittent
1 intermix
1 intermixed
2 intermixture
9 internal
1 internally
2 interposition
1 interrompues
1 interrupted
1 interruption
3 intersect
1 intersected
2 intersecting
2 intersection
2 intersections
1 interstices
2 interstratified
2 intertropical
1 intertropics
1 intertwined

4 interval
1 intervals
1 intervene
1 intervened
1 intervening
1 intervention
3 intestinal
1 intestine
5 intestines
3 intimate
6 intimately
1 intimates
1 intincts
1 intolerable
1 intoxication
8 introduce
22 introduced
1 introducing
4 introduct
8 introduction
1 introductory
1 intruder
1 intuition
1 inutility
2 invariability
4 invariable
4 invariably
8 invented
1 inventing
1 invention
5 inventive
1 inverness
3 inverorum
1 inverse
1 invertebrates
3 investigate
3 investigation
2 invisible
1 invoking
2 involuntarily
17 involuntary
6 inward
2 inwards
1 iodine
1 ipecacuhan
1 iquique
1 iquiqui
1 ira
22 ireland
3 iris
14 irish
19 iron
1 irony
8 irregular
1 irregularities
1 irregularity
2 irregularly
3 irritability
1 irritate
1 irritated
2 irritation
1 irritations
1 isabelles
1 ischia
27 isd
4 isds
1 isid
2 isl
21 island
36 islands
1 islas
64 isld
28 islds
18 isle
15 isles
11 islets
3 isolate
10 isolated
1 isolately
4 isolation
5 isothermal
1 ispida
1 israeli
1 issue
8 isthmus
3 italian
3 italy
1 itapicuru
1 ivory
1 jack
2 jackal
11 jackall
1 jackals
2 jackson
1 jacksons
8 jago
6 jaguar
1 jamaica
3 james
13 jan

3 janeiro
3 january
4 japan
2 japon
1 japonica
1 jar
1 jardin
1 jardines
1 jasper
26 java
2 jaw
3 jaws
2 jay
1 jays
9 jealousy
1 jeddah
2 jeffrey
1 jelly
1 jemmy
2 jenner
2 jenners
7 jenny
7 jenyns
1 jerrod
2 jerrold
1 jersey
1 jet
1 jeune
1 jews
1 jgo
1 joatingua
8 john
1 johns
3 johnson
2 johnston
2 join
3 joined
2 joining
1 joins
2 joint
3 joints
1 jokers
1 jokes
2 joking
8 jones
1 jonnes
4 jordan
1 jordanhill
1 joun
10 journ
62 journal
2 journey
5 joy
12 juan
5 judge
7 judged
1 judgement
1 judges
5 judging
7 judgment
1 judgments
1 juices
3 julian
1 julians
1 julie
15 july
6 jump
1 jumped
1 jumping
1 jumps
1 jun
3 junction
21 june
1 jungles
1 jura
1 jusquaux
2 justice
1 justicia
1 justly
1 kaluz
1 kames
1 kamschatka
1 kamtchatka
1 kamtschatka
2 kangaroo
1 kangaroos
1 kankaer
1 kant
1 karamania
2 keel
7 keeling
23 keep
16 keeper
2 keepers
10 keeping
2 keeps
4 kelp
1 kennel
1 kensington
1 kenyon

23 kept
1 keratry
5 kerguelen
1 kerrs
1 kettles
4 key
2 keys
1 keystone
2 kick
1 kicked
1 kicker
2 kicking
1 kicks
3 kidney
1 kilfinnan
6 kill
16 killed
5 killing
1 kills
1 kilometre
82 kind
1 kindgdom
1 kindness
42 kinds
6 king
9 kingdom
3 kingdoms
7 kingfisher
1 kingii
6 kings
1 kinlet
1 kinlett
8 kirby
1 kirbys
1 kiss
1 kitchen
1 kitten
9 kittens
1 klaproth
1 knats
1 kneeding
1 kneeling
10 knew
3 knife
5 knight
5 knights
1 knit
1 knobs
1 knocking
1 knocks
2 knot
121 know
20 knowing
50 knowledge
50 known
24 knows
1 koelreuter
1 kolkreuters
1 koloff
1 kolreuter
1 kongsberg
1 koran
1 kordofan
1 korto
1 kosir
1 kotezebues
4 kotzebue
4 kotzebues
1 krauss
1 kuhl
1 kurdish
1 kurukuru
1 kylau
1 kylow
1 labial
2 labiata
2 labiatae
1 labillardiere
2 laborious
2 labour
1 labouring
1 laburnum
1 lacépède
1 lacerta
1 lachyrmal
1 lacrustine
2 ladner
3 lady
1 lagoon
2 laid
25 lake
17 lakes
1 lamaks
1 lamantin
15 lamarck
1 lamarckian
3 lamarcks
2 lamark
1 lamarks
2 lamb

9 lambs	11 leads	1 lifts	2 lockarts
1 lamellar	4 leaf	21 light	2 locke
1 lament	1 leafing	1 lightening	1 locks
1 lamentable	1 leafy	1 lighter	4 locomotion
3 laminae	3 league	1 lighthouse	1 locomotive
1 laminar	9 leagues	1 lightly	2 loddiges
2 laminated	2 leap	1 lightning	1 lodging
2 lamination	1 leaping	1 lights	6 lofty
1 lancien	1 leapt	2 lignes	2 log
107 land	9 learn	1 lignite	1 logger
1 landed	2 learned	2 lik	1 loins
1 landing	2 learning	373 like	1 loix
1 landlocked	1 learns	1 liked	1 lomond
1 landons	7 learnt	7 likely	5 london
5 lands	1 leathern	1 liken	157 long
2 lang	5 leave	4 likeness	16 longer
36 language	13 leaves	3 likenesses	5 longest
5 languages	4 leaving	5 likes	1 longevity
1 languid	1 lect	30 likewise	1 longicornes
1 languor	3 lecture	1 liking	1 longirostris
1 laniadae	1 lectures	2 lile	2 longitude
2 lao	10 led	1 lillies	1 longitudinally
1 laoccaon	1 ledges	2 lily	9 lonsdale
4 laocoon	1 leeks	7 lima	1 lonsdales
3 lapland	1 leer	3 limb	82 look
2 laponia	1 leet	7 limbs	17 looked
2 lapse	1 leeward	16 lime	23 looking
1 lardens	19 left	19 limestone	18 looks
1 larder	4 leg	5 limestones	1 loonies
6 lardner	6 legged	6 limit	1 loopholes
4 lardners	1 legrand	11 limited	10 loose
65 large	28 legs	1 limiting	1 loosed
1 largeness	1 leguminosae	26 limits	2 looses
14 larger	1 leighton	1 limonia	8 loosing
6 largest	1 leigton	1 lin	1 lopped
7 lark	1 lemagne	2 linaria	11 lord
1 larks	4 lemna	1 lincolnshire	1 lorenzo
4 larkspur	1 lemon	3 lindley	1 lorsquelles
1 lartet	2 lemur	1 lindlys	6 lose
4 larva	1 lemurs	62 line	2 losing
8 larvae	1 len	6 linear	9 loss
1 las	26 length	3 lined	3 losses
1 lasch	3 lengthened	36 lines	22 lost
1 lascivious	1 leone	2 link	1 lots
1 lasiurus	4 leopard	1 linking	1 lotus
55 last	1 lepidosiren	3 links	2 loudon
2 lasted	1 leptosiphon	11 linn	2 loudons
3 lasting	1 leptuse	6 linnaean	1 louis
4 lastly	1 lepus	6 linnaeus	1 louisiana
1 lasts	1 leroy	1 linnet	2 loup
11 lat	1 leschenault	1 linnets	2 louse
2 latch	2 lessened	1 linotte	29 love
1 latd	1 lessening	1 linseed	3 loves
13 late	10 lesser	3 linum	1 loving
11 lately	5 lessings	9 lion	20 low
3 latent	13 lesson	1 lions	3 lowe
4 later	1 lessonia	17 lip	45 lower
5 lateral	1 lessons	9 lips	2 lowered
3 laterally	1 lest	2 liquid	9 lowest
1 latest	31 let	1 lisbon	1 lr
1 latham	18 letter	1 lisianskys	1 lucanidae
1 lather	3 letters	1 lisles	1 lucanus
2 lathyrus	1 lettres	24 list	1 lucidly
1 latin	1 leuconotes	1 listen	1 lucidus
1 latit	1 leucoprymnus	1 listened	1 lucky
3 latitude	1 leur	2 listening	2 lugon
4 latitudes	1 leurs	1 listera	1 luconia
1 latley	1 leva	1 lists	1 lucs
2 latreille	29 level	1 literal	2 luminous
1 latte	1 levelling	2 literally	1 lump
56 latter	1 levelness	2 literary	2 lund
1 lauder	1 levels	1 literat	1 lunds
6 laugh	1 leven	1 lithospernum	6 lupine
8 laughing	1 levity	9 litter	1 lustre
1 laughs	1 lewd	4 litters	3 lustrum
1 laught	1 lewdness	104 little	2 lutke
3 laughter	1 lewis	3 littoral	2 lutkes
1 laurels	1 leyden	25 live	1 luxuriant
13 lava	28 lhc	20 lived	1 lx
3 lavas	1 liability	1 liver	7 lychnis
4 lavater	3 liable	1 livers	1 lycopodiums
4 lavaters	1 lias	11 lives	59 lyell
1 lavendar	2 lib	55 living	28 lyells
89 law	1 liberal	1 liza	3 lying
1 lawless	5 library	4 lizard	1 lymington
1 lawn	2 lice	8 lizards	1 ma
2 lawrence	2 licentiousness	2 llama	5 mac
88 laws	1 lichen	2 loasa	1 macaco
7 lay	2 licking	1 loathing	3 macacus
7 layer	1 lid	1 lobe	14 macculloch
9 layers	6 lie	1 lobelia	1 maccullochs
2 laying	1 lied	1 lobsters	1 macdonald
1 lays	1 liegende	22 local	1 macgillivray
1 lb	6 lies	10 localities	3 machine
1 lbs	1 lieu	8 locality	1 machinery
11 ld	4 lieut	1 locally	6 mackenzie
2 leaches	105 life	48 located	1 mackenzies
28 lead	2 lifetime	1 location	14 mackintosh
2 leader	3 lift	28 loch	3 mackintoshs
8 leading	1 lifted	8 lochy	1 maclay

720

1 minutely	1 monoeecious	1 multiplications	1 nautili
5 minutes	2 monogamous	4 multiplicity	1 nautilus
1 miocen	2 monograph	5 multiplied	1 navigatores
7 miocene	2 monooecia	1 multipliied	2 navigators
2 miracle	12 monooecious	3 multiply	2 nay
1 miracles	1 monotrematous	1 multiplying	65 nb
1 miserable	1 monro	2 multitude	7 ne
3 misery	1 mons	2 multitudes	1 nea
1 misfortunes	1 monsoons	2 mummies	1 nead
1 misgivings	6 monster	1 mummy	8 nearer
1 mishaped	11 monsters	2 mundane	12 nearest
1 misnumbered	5 monstrosities	10 mundine	45 nearly
2 misnumbering	7 monstrosity	1 mungo	1 nebria
1 misplaced	8 monstrous	4 mur	8 necessarily
17 miss	1 monstruous	1 murc	58 necessary
1 missel	3 montagnes	4 murchison	2 necessities
1 misseltoe	2 montagu	1 murchisons	11 necessity
2 mission	2 montaigne	1 murder	8 neck
1 missionaries	1 monte	1 murdered	2 necker
1 missionary	3 montes	1 muriated	4 necks
1 missisippi	4 month	1 murray	5 nectar
1 mississippi	4 months	4 mus	3 nectaries
9 mistake	1 monucle	1 musalmans	2 nectarines
3 mistaken	1 monucule	1 muscadivora	7 nectary
1 mistakes	1 monuments	1 muschelkalk	11 need
1 mistook	6 moon	1 musci	1 needle
9 mitchell	1 moor	1 musical	4 negation
2 mitchells	1 moore	1 muscicapa	7 negative
1 mitchill	36 moral	1 muscicapidae	1 neglecting
3 mite	1 morales	5 muscle	2 negress
1 mitterschlich	1 moralists	35 muscles	18 negro
1 mix	1 moralité	9 muscovy	4 negroes
10 mixed	1 morally	9 muscular	2 negros
1 mixing	2 morals	2 museum	1 neig
3 mixture	2 moran	1 mush	1 neigbourhood
1 mixtures	1 morbid	1 mushroom	1 neigbouring
1 mn	8 moreover	22 music	2 neighboring
1 moa	2 moreton	2 musical	1 neighbour
2 mocha	8 moretons	1 musicus	3 neighbourhood
1 mock	1 mormodes	1 musk	12 neighbouring
4 mocking	1 morne	1 muskitoes	1 neil
3 mode	2 morning	1 musm	1 neilgherries
1 modelling	1 morocco	3 musquitoes	1 neilson
2 moderately	1 morphology	329 must	1 nematoid
15 modern	1 morro	1 mutations	2 nepal
2 modes	1 morse	1 mutilated	1 nepenthes
2 modest	1 mortify	2 mutilation	1 nephews
2 modesty	1 moruffetes	5 mutilations	6 nerve
9 modification	2 moschus	1 mutually	6 nerves
6 modifications	3 moss	1 muzzled	11 nervous
25 modified	4 mosses	294 my	3 ness
1 modify	1 mossy	1 mycelis	17 nest
1 modifying	1 mostly	1 mylodon	2 nests
1 modulator	1 mostrified	1 myothera	1 net
2 moel	1 motacilla	1 myriads	1 nétaient
1 mohamned	1 moth	1 myriametre	1 netted
2 moins	34 mother	7 myself	6 nettle
1 moist	5 mothers	1 mysteries	4 neuter
1 moistened	1 mothes	6 mysterious	3 neuters
2 moisture	4 moths	1 mystery	2 neutral
1 mojos	8 motion	1 mystical	82 never
1 molar	1 motionless	1 mytilus	3 nevertheless
3 mole	6 motive	1 naiads	181 new
1 molecular	6 motives	1 nail	1 newbold
1 molecule	1 mottle	2 nails	1 newcastle
3 molina	5 mottled	3 naked	2 newer
3 molinas	2 moufflon	24 name	1 newest
1 molluccas	1 mould	14 namely	3 newfoundland
19 mollusca	4 mound	7 names	1 newlands
4 molluscous	1 mount	3 naming	1 newly
11 molluscs	35 mountain	1 nancy	1 newman
1 molten	1 mountaineer	1 napoleon	1 newport
1 molting	1 mountainous	1 narcissus	2 news
1 molucca	46 mountains	3 narrative	1 newspapers
8 moluccas	9 mouse	6 narrow	1 newton
1 moluccensis	30 mouth	1 narrower	1 newydd
1 molybdated	1 mouthed	1 narrowness	10 next
6 moment	3 mouths	1 narwhal	3 nice
1 momentarily	20 move	3 nasty	2 nicely
1 momentary	8 moved	13 nat	1 nichol
1 moments	39 movement	1 natical	1 nickel
1 momentum	30 movements	2 nation	1 niether
2 mon	5 moves	1 national	2 niger
3 monad	7 moving	3 nations	15 night
2 monads	2 mozart	13 native	2 nightingale
2 monday	1 muar	5 natives	1 nightingales
2 monde	15 mud	1 natura	1 nihil
1 mondobbo	1 muddy	1 naturae	5 nina
3 money	1 muhanmad	54 natural	1 ninety
6 mongrel	1 mulatto	5 naturalist	1 nipple
1 mongrelized	1 mulattos	9 naturalists	3 nipples
3 mongrels	11 mule	8 naturally	1 nisus
29 monkey	19 mules	105 nature	2 nitrate
47 monkeys	2 mulita	4 naturelle	1 nitrous
2 monks	1 mullein	1 naturelles	1 nne
1 mono	1 müller	3 natures	1 nobody
1 monoceros	1 müllers	1 naturgeschichte	3 nocturnal
1 monocot	1 multilineatus	1 naughtiness	1 nodding
2 monocotyledenous	1 multiple	2 naughty	1 nodular
1 monoecious	4 multiplication	1 nausea	1 nodule

3 nodules
1 noir
12 noise
1 noises
22 non
22 none
1 nonpareil
6 nonsense
28 nor
7 norfolk
1 norfolkicae
6 normal
2 noronha
42 north
29 northern
1 northwards
1 norton
9 norway
7 nose
3 noses
1 nostril
2 nostrils
1 notch
1 notched
20 note
1 notebook
14 notes
49 nothing
3 notice
4 noticed
4 notices
2 noticing
19 notion
7 notions
1 notoriously
1 nouns
1 nourished
2 nous
2 nov
1 nova
2 novel
3 novelty
1 novem
8 november
1 novozelandiae
1 nozzle
5 nucleus
1 nucleuses
2 nullipora
89 number
1 numbered
1 numberless
20 numbers
2 numerical
1 numerically
19 numerous
1 nuns
1 nurse
1 nursed
1 nurseries
1 nursery
2 nursling
6 nut
3 nutrition
2 nuts
1 nux
14 nw
2 oak
3 oats
1 obediance
1 obedience
5 obey
1 obeyed
7 obeying
2 obeys
40 object
1 objected
3 objection
4 objective
3 objectively
14 objects
2 obligation
7 obliged
2 oblique
2 obliquely
4 obliterate
4 obliterated
2 obliterating
1 obliteration
1 obliterations
1 obras
1 obscura
13 obscure
3 obscurely
1 obscurer
1 observ
1 observable
1 observaciones
3 observat
9 observation

12 observations
15 observe
30 observed
2 observer
5 observés
5 observing
1 obseved
3 obsidian
1 obstacle
2 obstruction
10 obtain
12 obtained
2 obtaining
1 obviating
2 obvious
1 obviously
2 occasion
1 occasional
5 occasionally
1 occassionally
5 occupied
4 occupy
19 occur
1 occurence
3 occurred
3 occurrence
4 occurring
8 occurs
35 ocean
2 oceania
1 oceanic
1 oceanica
7 oceans
1 oclock
5 oct
2 octav
19 octob
9 october
2 octopus
21 odd
1 oddity
1 odious
2 odour
6 oeconomy
1 oeil
2 oenothera
1 oeuvre
64 off
2 offence
1 offended
1 offensive
5 offer
7 offered
1 offering
7 offers
1 office
1 officer
1 officers
1 offices
3 officier
136 offspring
4 offsprings
79 often
2 oftener
5 ogleby
1 ogwen
6 oh
5 oich
3 oil
2 oiseaux
125 old
12 older
19 oldest
3 olinda
2 olivine
1 omit
1 omnibus
1 omnibuss
1 omniphitophagous
1 omnipotence
1 omnipotent
44 once
1 oncitate
4 oneself
3 onion
1 ont
1 onus
3 oolite
2 oolitic
2 oozed
2 op
1 opal
31 open
3 opened
7 opening
1 openly
1 opens
2 operation
2 operations
1 opetiorhyncus

1 ophrys
1 ophyressa
10 opinion
2 opossum
1 opossums
4 opponent
1 opponents
1 opportunities
15 opposed
27 opposite
1 opposites
3 opposition
1 optional
1 opuscule
1 oral
1 oran
3 orang
3 orange
3 oranges
1 orangs
3 orbicular
1 orbiculas
20 orbigny
1 orbignys
1 orbits
1 orchard
1 orchards
1 orchidacaeous
2 orchidiae
6 orchis
35 order
3 ordered
3 ordering
22 orders
22 ordinary
3 ore
2 ores
1 orfords
31 organ
35 organic
1 organics
4 organisms
60 organization
1 organizaton
8 organized
84 organs
1 orginally
1 oriental
1 orientale
1 orientel
3 orifice
1 orifices
63 origin
14 original
2 originality
14 originally
4 originate
3 originated
2 originates
2 orinoco
1 oriolus
1 ornaments
1 ornith
1 ornitho
1 ornitholog
2 ornithological
1 ornithologists
1 ornithologum
5 ornithology
1 ornithorhynchus
7 ornithorhyncus
1 ornityhyrhycus
5 orpheus
1 orpiment
1 orthopt
1 orthopterous
1 os
1 oscillate
2 oscillated
1 oscillating
1 oscillation
8 oscillations
4 osculant
1 osorno
1 oss
1 ossemens
1 ossements
3 osteology
2 osteopora
7 ostrich
8 ostriches
4 otaheite
1 otahiati
25 otherwise
5 otter
2 oualan
48 ought
1 oural
16 ourang
4 ours

5 ourselves
12 outang
3 outer
1 outlet
1 outlines
1 outliving
1 outlyer
1 outlying
5 outside
2 outward
4 outwards
1 ouzel
3 ouzels
5 ova
1 ovals
3 ovaria
2 ovarium
1 overbears
2 overcome
1 overfall
1 overlooked
1 overlooks
1 overlying
1 overrate
1 overrun
1 overtake
1 overtempted
3 overthrown
1 overturned
3 overwhelming
2 ovina
1 ovington
1 ovingtons
1 ovules
8 ovum
2 owe
1 owed
38 owen
6 owens
2 owes
54 owing
3 owl
6 owls
1 owners
5 ox
1 oxalis
4 oxen
1 oxford
1 oxidated
1 oxidation
2 oxide
1 oxided
1 oxygen
1 oxyurus
7 oyster
2 oysters
1 pa
3 pachyderm
9 pachydermata
2 pachyderms
8 pacific
5 pacifick
2 packed
1 packing
1 packs
1 paddle
21 page
9 pages
1 paid
36 pain
3 painful
3 pains
3 painted
1 painter
2 painting
11 pair
1 pajero
1 palace
3 palaeotherium
1 palatal
1 palates
7 pale
1 paleontographie
3 paler
1 paley
1 paleys
1 pallas
1 palliates
1 palm
3 palmated
1 palmation
1 palmipedes
3 palms
1 palpitation
1 palsy
2 palustris
5 pampas
3 pamphlet
1 pamphlets
1 pamplet

1 panama	1 pause	1 perfume	1 phitophagous
1 panther	1 paws	1 perga	2 phlox
1 pantomimic	1 paxton	104 perhaps	1 phocae
2 papaver	1 pay	1 perigonous	1 phonolite
83 paper	4 payta	36 period	1 phonolites
15 papers	1 pe	1 periodical	1 phonolitic
1 papil	9 pea	1 periodically	1 phormium
5 papilionaceous	1 peacable	18 periods	1 phrase
1 pappus	1 peaccocks	5 perish	2 phrenologists
1 papuensis	1 peaceful	8 perished	1 phryganea
4 paradise	2 peaches	1 perishes	1 phsiognominical
2 paradoxurus	4 peacock	3 periwinkle	2 phy
1 paradupasi	3 peacocks	5 permanence	1 phyllade
1 paragua	6 peak	21 permanent	3 phys
7 paraguay	2 peaks	1 permanentes	42 physical
28 parallel	2 pear	3 permanently	2 physiognomy
1 paralleles	1 pearlstones	1 permanet	1 physiolog
1 parallelism	1 pears	1 permitted	1 physiological
1 paralytic	12 peas	4 pernambuco	1 physiologiques
1 paralyzes	2 peasant	1 pernetty	6 physiology
1 parameles	12 peat	1 pernety	2 physique
1 paramount	1 peaty	3 peron	2 pi
1 paranagua	2 pebble	1 perpetual	1 pianoforte
4 parasite	38 pebbles	3 perpetuate	1 pic
3 parasites	2 peccari	1 perpetuating	1 pichincha
2 parasitic	1 pecking	1 perplexing	1 pichye
2 parasitical	1 pecten	1 perroquets	9 picked
1 pare	1 pectinibr	1 pers	13 picking
1 parellel	2 pectinibranchiate	1 persecution	1 picks
78 parent	1 pecu	1 persia	1 pictet
2 parentage	71 peculiar	6 persian	7 picture
6 parental	23 peculiarities	1 persians	3 pictures
49 parents	22 peculiarity	2 persistency	1 picturesquely
1 paret	1 ped	6 persistent	14 pidgeon
1 parfait	1 pedro	35 person	30 pidgeons
1 paribus	1 pedunculated	2 personal	10 piece
17 paris	1 peel	1 personified	3 pieces
6 parish	1 peeled	1 personnal	1 pieds
6 park	2 peevish	2 perspective	2 pies
1 parkers	2 pegu	1 persuade	1 piety
1 parks	1 pegue	1 perturbation	9 pig
1 parnassia	1 pelagic	8 peru	6 pigeons
1 parroket	2 pelew	1 peruvian	21 pigs
6 parrot	1 pelican	3 perversion	1 pike
13 parrots	1 pell	3 perverted	1 pile
1 parrott	1 pellet	1 pestiferous	1 piled
1 parsley	1 pellets	5 petal	1 pincers
1 parsons	2 pellew	1 petals	3 pincher
5 partake	1 pellicle	1 peter	1 pine
2 partakes	1 pellucid	1 peters	1 pineaple
5 partaking	1 pelud	1 petersburgh	1 pineaster
4 partial	1 peludo	1 petise	1 pines
3 partially	1 pelvis	1 petisse	2 pink
10 particle	1 pen	2 petites	1 pinkish
10 particles	1 pencil	5 petrel	1 pinks
2 particls	1 penetrate	2 petrels	1 pinna
20 particular	4 penetrated	1 petrifaction	1 pinnacle
5 particularly	3 penetrating	4 petrified	1 pinnacles
2 particulars	17 penguin	2 petticoats	1 pins
1 particuliere	3 penguins	1 peu	6 pintail
1 partie	3 peninsula	1 peuquenes	3 pipe
2 parting	1 pennant	1 peuvent	4 pippen
1 partition	1 pennatula	1 peyrouse	1 pippens
24 partly	1 penny	1 phaedo	4 pippin
3 partridge	1 pente	1 phalanges	1 pippins
1 parturition	2 pentlands	3 phalangista	2 pipra
3 party	1 pentstemon	1 phalarope	1 pisa
3 pas	84 people	1 phases	1 piscatoria
1 pasco	2 peopled	23 pheasant	1 pish
1 pashion	2 peppermint	10 pheasants	1 pisolitic
17 pass	9 per	1 phenomem	1 pisses
21 passage	13 perceive	12 phenomena	4 pistil
12 passages	8 perceived	1 phenomenes	4 pistils
11 passed	1 perceives	4 phenomenon	3 pitched
2 passeres	1 perceiving	1 phi	1 pitching
5 passes	1 percent	22 phil	3 pitchstone
1 passess	2 percentage	1 philadelphia	1 pits
7 passing	6 perception	2 philedon	4 pity
57 passion	4 perceptions	1 philip	89 place
1 passionate	2 perceptive	1 philippines	24 placed
24 passions	2 perch	3 phillipines	1 placent
1 passive	1 perching	1 phillippensis	1 placentates
25 past	1 percieved	3 phillippines	7 places
1 pasto	1 percisely	22 phillips	1 placidly
30 patagonia	1 percolate	1 philo	1 placing
9 patagonian	1 percolates	6 philos	1 plagued
2 patagonicus	1 peregrine	9 philosop	26 plain
1 patagonie	1 perenne	6 philosoph	2 plainly
1 pataz	1 perennial	6 philosopher	22 plains
2 patch	72 perfect	1 philosophers	1 plaintive
6 patches	1 perfected	1 philosophic	9 plan
1 patella	1 perfecting	4 philosophical	5 planaria
2 path	16 perfection	4 philosophie	9 planariae
3 pathology	25 perfectly	1 philosophique	1 planchon
1 patient	2 perforated	16 philosophy	2 plane
1 patients	6 perform	1 philosphically	4 planes
1 patriotic	1 performance	1 philospohy	5 planet
1 patriotism	3 performed	1 philotis	3 planets
1 paulo	5 performing	2 phils	1 plans
3 pauls	4 performs	1 phisoph	63 plant

723

2 plantarum	5 pool	1 precision	33 principle
13 planted	1 pools	1 precluded	13 principles
1 plantigrade	9 poor	1 precludes	3 prior
262 plants	2 poorness	1 precursor	2 priority
1 plas	1 pop	1 predestinarian	1 pristine
1 plaster	1 poplars	4 predicament	1 pritchards
4 plastic	2 poppy	1 predicated	3 private
8 plata	12 population	2 prediction	1 prize
1 plate	1 populousness	1 predominant	1 prizes
2 plates	1 porcarious	1 predominantly	1 prob
3 platform	1 porcarius	1 predominates	2 probabilities
2 plato	1 porcellus	1 preeminence	8 probability
2 platycephalus	1 porcupine	1 preeminent	23 probable
14 play	1 pork	11 preeminently	167 probably
2 played	1 porph	1 preeminiently	1 probandi
1 player	6 porphyries	2 preexistence	9 problem
1 players	1 porphyritic	3 preexisting	4 proboscis
1 playfair	10 porphyry	4 preface	3 proc
5 playing	1 porphyrys	1 prefe	1 proceed
2 plays	1 porpoise	4 prefer	2 proceeded
1 ple	3 porpoises	1 preferred	1 proceeding
12 pleasant	6 port	1 preferring	8 proceedings
7 pleased	2 portezuelo	1 pregnancy	2 proceeds
1 pleases	3 portillo	2 pregnant	21 process
1 pleasing	8 portion	1 prehensile	1 processes
78 pleasure	2 portions	1 prejudice	3 procreate
7 pleasures	2 portland	1 prejudices	1 procure
1 plentiful	2 porto	1 preliminary	6 procured
1 plentifully	1 portsmouth	1 premières	1 procyon
7 plenty	1 portugal	1 premises	1 prodigious
1 plesiosaurus	1 portuguese	1 prepared	3 prodigiously
1 plesiossaurus	18 position	1 preparis	1 prodromus
1 pleurisy	2 positions	1 preponderance	76 produce
1 pliable	5 positive	1 preponderant	96 produced
2 pliny	1 positively	1 preponderate	8 produces
3 pliocene	1 possede	1 preponderated	23 producing
1 plodding	5 possess	1 preposterously	2 product
1 plotting	1 possesse	1 prepuce	19 production
1 ploughed	8 possessed	1 prescribing	7 productions
1 ploughs	3 possesses	7 presence	3 productive
1 pluckt	6 possession	105 present	1 productiveness
35 plumage	20 possibility	1 presente	2 products
1 plunged	31 possible	1 presentent	1 proesentum
5 plutonic	11 possibly	1 presently	3 prof
4 po	2 post	1 presents	1 profane
3 poa	3 posterior	11 preservation	1 profession
2 pocket	3 pot	1 preservative	1 professor
1 pod	1 pota	1 preserve	1 profit
2 pods	3 potato	18 preserved	1 profondes
1 poem	2 potosi	4 preserving	16 profound
1 poenig	1 potters	1 presque	15 profoundly
1 poenis	1 pouches	2 press	1 profusely
3 poet	1 poulett	2 pressed	7 progenitor
2 poetical	1 poulterer	1 presses	4 progenitors
13 poetry	2 poultry	1 pressing	9 progeny
1 poets	2 pound	13 pressure	13 progress
66 point	3 pounds	15 presume	1 progresses
4 pointed	2 pour	2 presumed	1 progressife
7 pointer	1 pourrait	1 presumes	6 progression
2 pointers	1 poussin	1 presuming	12 progressive
5 pointing	5 pout	2 presumption	1 progressively
24 points	1 pouted	2 presupposes	4 project
3 poison	1 pouter	3 pretend	3 projecting
2 poisonous	2 pouters	1 prettty	1 projects
1 poisons	6 pouting	10 pretty	1 prolific
1 poked	2 pouts	3 prevail	2 prolifickness
1 poland	1 poutter	4 prevailing	1 prolongation
5 polar	1 poverty	3 prevails	7 prolonged
1 polarity	1 pow	1 prevalence	1 prominent
2 pole	1 powder	2 prevalent	1 promptness
5 poles	1 powdered	16 prevent	1 prone
1 poliosoma	1 powe	9 prevented	2 pronounced
1 polirus	75 power	2 preventing	15 proof
3 polish	2 powerful	10 prevents	1 prooff
2 politics	34 powers	6 previous	10 proofs
1 polity	1 powis	1 previously	1 prop
1 poll	1 pox	3 prevost	9 propagate
63 pollen	2 pr	7 prey	18 propagated
1 polygam	1 practicable	1 priamang	1 propagates
1 polygamia	1 practical	3 price	6 propagating
2 polygamous	1 practically	2 prichard	22 propagation
1 polymorphes	3 practice	1 prick	1 propelling
1 polymorphous	1 practiced	1 pricked	1 propenity
1 polynes	1 practise	2 prickly	13 proper
1 polynesia	2 practised	2 pricks	6 properly
3 polynesian	1 practising	10 pride	1 properties
1 polynesians	1 praeternatural	1 priest	5 property
1 polype	1 praising	1 prietly	1 prophecied
9 polypi	1 pratencole	1 primaries	1 prophetic
1 polyps	1 pratensis	7 primary	29 proportion
4 polypus	2 pray	1 primeval	6 proportional
1 polytheism	1 prayer	1 primigenious	2 proportionally
1 pompeei	1 prearranged	8 primitive	1 proportionately
2 pond	1 precede	1 primordial	4 proportions
1 ponder	3 precedes	11 primrose	3 proposition
1 ponds	1 preceed	1 primroses	1 propositions
1 pongo	1 precip	1 prin	2 propriety
3 pony	1 precipitate	1 prince	1 propulsion
2 ponza	2 precise	1 principal	1 prose
1 pooh	13 precisely	1 principes	1 prospect

1 prospective
1 prostitution
1 prostration
1 proteaceae
5 protect
3 protected
1 protecting
1 protection
2 proteus
1 prototype
1 protrude
1 protruded
2 protrudes
3 protrusion
1 protuberant
1 proud
16 prove
13 proved
2 proverbially
6 proves
4 provided
2 providence
2 province
3 provinces
3 provincial
1 proving
4 provision
1 prowl
1 prowlers
1 proximity
5 ptarmigan
1 pu
1 publish
14 published
2 pud
1 pudu
4 puffinuria
1 pugnacious
5 pull
3 pulled
2 pulling
1 pulls
1 pulo
6 pulse
2 puma
7 pumice
1 pumiceous
1 pumped
1 pumping
1 punaise
1 puncture
1 pungency
1 punish
1 punishment
1 punjab
1 punning
15 puppies
5 puppy
1 purchase
1 purdie
17 pure
3 purely
1 purest
1 purged
7 purple
8 purpose
1 purposely
1 purposes
1 pursue
3 pursued
1 pursuing
2 pursuit
1 push
1 pushed
1 pushes
3 pushing
1 pustular
44 put
1 putrid
1 putrifies
1 puts
4 putting
6 puzzle
2 puzzler
2 puzzling
1 pyenite
1 pyrenaicum
3 pyrenees
1 pyrennees
1 pyrite
3 pyrites
1 pyritous
1 pyrocephalus
1 pyrrho
1 quadras
1 quadrifolium
2 quadrumanous
3 quadruped
29 quadrupeds
4 quagga

1 quakes
1 qualifications
11 qualities
4 quality
1 quand
4 quantities
15 quantity
1 quantum
1 quarrelling
1 quarrelsome
1 quarries
9 quarter
5 quarterly
6 quarters
1 quarto
19 quartz
4 quartzose
2 quaternary
1 quatre
2 queen
1 queene
1 queens
3 quelque
1 quels
1 quench
1 queque
1 queries
42 question
1 questioning
8 questions
2 quetelet
1 quichuas
6 quick
2 quicker
6 quickly
2 quiet
2 quillota
1 quilmes
1 quinarian
1 quinarians
5 quinary
57 quite
5 quito
1 quitting
1 quiyá
2 quotation
8 quote
9 quoted
7 quotes
1 quoting
4 quoy
7 rabbit
12 rabbits
1 rabies
1 raccoons
32 race
1 racehorse
75 races
3 radack
1 radiate
2 radiating
1 radiation
1 radical
1 radicle
1 radiolum
2 raffles
1 rafflesia
2 rage
5 rain
2 rains
1 rainy
9 raise
20 raised
1 raisers
3 raises
2 raising
1 raison
1 rakhekna
1 rakkelhan
1 ralix
1 ram
1 rambling
3 ramond
1 rams
2 ran
51 range
1 ranged
8 ranges
3 ranks
3 ranunculus
11 rapid
7 rapidity
5 rapidly
1 rapilli
2 rapport
2 raptores
2 raptorial
22 rare
10 rarely
2 rarer

2 rarity
1 ras
4 rasorial
1 raspberry
6 rat
6 rate
52 rather
4 ratio
1 rational
1 rationally
6 rats
5 ravine
1 ravines
3 ravins
1 rayed
6 rays
1 razor
1 re
7 reach
7 reached
2 reaches
2 reaction
44 read
15 readily
2 readiness
17 reading
1 reads
10 ready
37 real
4 reality
1 realized
1 realizing
40 really
2 reappear
1 reappearance
2 rear
3 reared
2 rearing
1 rears
89 reason
2 reasonable
1 reasoned
1 reasoners
22 reasoning
1 reasonings
10 reasons
1 reaumur
1 rebukes
4 recall
2 recalled
1 recalling
1 recalls
1 recede
1 recedes
9 receive
15 received
3 receives
2 receiving
36 recent
1 recently
1 receptacle
1 receptacles
1 recieve
1 reciprocally
1 reclaimed
1 recognise
1 recognised
1 recognising
1 recognition
5 recognize
2 recognized
2 recognizes
8 recollect
6 recollected
4 recollecting
5 recollection
4 recollections
3 recollects
1 reconsult
5 record
4 recorded
1 records
1 recovered
1 recovering
3 recrossed
1 recrossing
1 recrystallized
1 rectangular
1 recuiel
1 recurred
1 recurrence
3 recurrent
1 recurring
2 recurs
40 red
5 reddish
1 redissolved
1 redpoles
1 redstart
2 reduce

9 reduced
1 reduces
2 reducing
1 reduction
1 reeaches
1 reed
4 reef
4 reefs
1 reeves
4 ref
11 refer
28 reference
14 references
7 referred
2 referring
13 refers
1 refiltered
3 reflect
2 reflecting
2 reflection
3 reflections
1 reflective
1 reflex
1 reform
2 refrigeration
1 refusing
1 regain
13 regard
1 regarded
1 regarder
8 regarding
2 regards
8 region
18 regions
1 regne
1 regrafted
1 regressed
1 regret
13 regular
2 regularity
5 regularly
2 regulated
2 regulates
2 reid
1 reinwardt
2 reject
1 rejected
1 rejects
1 relate
15 related
1 relates
1 relating
91 relation
27 relations
13 relationship
6 relative
4 relatives
1 relic
1 relics
1 relict
1 relied
2 relieve
1 relieved
1 relieving
3 religio
3 religion
1 religions
1 religious
1 reliqu
2 reliquiae
1 relish
20 remain
1 remainder
4 remained
5 remaining
25 remains
28 remark
43 remarkable
4 remarkably
1 remarke
21 remarked
40 remarks
34 remember
16 remembered
5 remembering
2 remembers
1 remind
1 reminded
1 remnant
1 remnants
1 remodelling
1 remonstrances
1 remonstrate
4 remorse
7 remote
1 remotely
1 remotest
4 removal
3 remove
12 removed

1 sc	1 sects	1 setae	102 show
1 scabies	2 secular	2 setters	12 showed
11 scale	1 secured	1 setting	1 shower
4 scales	1 security	4 settle	1 showers
4 scandinavia	2 sedgwick	3 settled	51 showing
1 scapula	1 sedgwicks	1 settler	36 shown
1 scar	6 sediment	1 settles	102 shows
1 scarabadae	2 sedimentary	2 settling	1 shrew
26 scarcely	45 seed	1 seul	4 shrews
1 scarcity	1 seeding	1 seule	17 shrewsbury
9 scarlet	5 seedless	1 sev	1 shrikes
8 scattered	2 seedling	64 several	1 shrill
4 scene	6 seedlings	4 severe	1 shrink
2 scenery	90 seeds	1 sewalick	2 shrinking
5 scenes	2 seedsmen	28 sex	2 shrivel
1 scent	1 seek	34 sexes	1 shrivelled
1 sceptical	25 seem	38 sexual	3 shropshire
1 scepticisms	8 seemed	1 sexually	1 shrub
1 schelgel	60 seems	1 sh	1 shrubbery
4 scheme	1 seep	4 shade	1 shrug
2 schemes	11 sees	3 shades	2 shrugged
1 schemnitz	2 segment	2 shaft	3 shrugging
1 schiede	4 segregation	3 shake	2 shrugs
1 schist	2 seize	1 shaken	1 shu
1 schistose	2 seized	2 shaking	1 shudder
1 schit	2 selbourne	1 shale	7 shut
1 schmidtmeyer	1 selbournes	2 shales	2 shuts
2 schomburgk	2 selby	11 shall	1 shutting
5 school	6 seldom	4 shallow	3 shy
1 schow	4 select	1 shallower	4 shyness
1 sci	5 selected	1 shallowest	1 si
2 scicily	7 selection	20 shame	1 siabod
1 scienc	1 selenite	3 shamming	2 siam
16 science	10 self	4 shape	1 siamese
9 sciences	2 selfish	2 shaped	17 siberia
8 scientific	1 selfishness	1 shapes	2 siberian
3 scincus	1 sellow	2 share	1 siciences
1 scions	1 selves	4 shark	2 sicily
3 scission	5 semen	6 sharks	1 sick
2 scissures	1 semestre	5 sharp	4 sickly
1 scolopax	3 semi	1 sharper	2 sickness
2 scooped	1 semiconsolidated	1 shatsbury	75 side
1 scopes	1 semilunar	1 shaws	8 sides
1 score	2 seminal	1 shd	1 sideward
1 scoresby	1 semnopitheque	6 sheath	1 sierra
1 scoresbys	1 semptemb	3 shed	1 sift
1 scoriaceous	4 send	2 shee	1 sifted
1 scoriae	3 sending	29 sheep	5 sigh
1 scoriform	3 sends	2 sheeps	5 sighing
1 scorpion	4 senegal	1 sheet	14 sight
4 scotch	2 senna	1 sheets	2 sighted
1 scotchman	19 sensation	46 shelf	1 sightedness
2 scoticus	11 sensations	1 shelfs	1 sigma
7 scotland	32 sense	15 shell	6 sign
4 scott	1 senseless	112 shells	1 signal
1 scottish	38 senses	7 shelves	2 signals
4 scotts	3 sensible	1 shelving	1 signed
1 scourges	11 sensitive	9 shepherd	1 signifies
1 scrape	2 sensorium	3 shepherds	1 signor
1 scraped	2 sensual	1 sheriffs	8 signs
1 scrapes	1 sensuality	4 shetland	1 siicily
1 scraping	3 sent	5 shew	1 silene
2 scraps	3 sentence	9 shewn	2 silent
1 scratch	3 sentences	1 shiant	1 silex
1 scream	1 sentimental	1 shilling	2 silica
4 scrope	3 sentiments	4 shingle	2 siliceous
1 scrutinize	2 sep	1 shinglle	3 silicified
2 sculpture	1 separ	3 ship	3 silk
1 scum	36 separate	1 shirk	1 silky
114 sea	26 separated	1 shivering	3 silliman
3 seal	3 separately	1 shivers	5 sillimans
6 seals	5 separates	1 shoal	1 silly
1 seams	4 separating	3 shoaler	1 silting
17 seas	17 separation	2 shoaling	7 silurian
4 season	1 separe	3 shoals	1 siluris
1 seasons	3 separtion	4 shock	1 silv
1 seat	23 sept	1 shocked	32 silver
1 seated	1 septa	1 shocks	63 similar
7 seaward	3 septaria	1 shoes	17 similarity
1 sebastian	5 septemb	1 shook	2 similarly
1 sebe	1 septembe	1 shoot	2 simile
10 sebright	5 september	2 shooting	31 simple
1 sebrights	1 seq	2 shoots	1 simplelst
2 seclusion	1 serapis	1 shops	2 simpler
1 seco	1 sergipe	9 shore	14 simplest
19 second	48 series	6 shores	1 simplets
1 secondarily	1 serin	48 short	2 simplicity
26 secondary	2 serious	2 shortened	1 simplified
1 secondly	2 serpent	8 shorter	2 simplify
1 secousses	6 serpentine	2 shortest	1 simplifys
2 secret	1 serpula	1 shortlived	1 simplits
3 secretary	4 serres	5 shortness	12 simply
1 secrete	1 sertularia	1 shortsighted	36 since
1 secretes	5 servant	3 shot	1 sincere
1 secreting	3 serve	109 should	1 sincerity
11 secretion	1 serves	2 shoulder	8 sing
1 secretionary	4 service	6 shoulders	2 singapore
1 sect	1 serviceable	1 shout	1 singe
7 section	1 ses	1 shouts	1 singer
3 sections	5 set	1 shoved	14 singing

727

1	singings	1	snort	15	space	11	spread
20	single	1	snorts	4	spaces	2	spreading
2	singly	2	snout	10	spain	1	spreads
10	singular	16	snow	2	spallanzani	10	spring
1	singularity	1	snuffs	3	spallanzanis	6	springing
2	singularly	1	soak	1	spangles	14	springs
3	sink	1	soaking	2	spaniards	1	spritual
3	sinking	2	soar	4	spaniel	2	sprout
2	sinks	1	sobbing	3	spaniels	1	sprouting
3	sins	1	sobre	1	spanish	1	sprouts
1	siren	37	soc	2	spar	4	sprung
4	sister	1	soci	1	spark	2	spur
4	sisters	2	sociability	4	sparrow	4	spurs
3	sit	25	social	1	spasm	1	squares
1	site	1	socialist	1	spath	1	squatting
2	sitting	1	societies	2	spawn	1	squeeze
2	situ	17	society	2	spe	2	squeezed
1	situated	1	socotra	2	speak	2	squib
3	situation	1	socrates	10	speaking	1	squibs
5	situations	6	soda	6	speaks	1	squilla
1	sivatherium	10	soft	6	spean	1	squinny
1	sixty	1	softness	1	spec	5	squirrel
30	size	24	soil	6	special	3	squirrels
2	sized	2	soils	658	species	1	squirting
6	skeleton	1	solanum	12	specific	2	ssw
1	skeletons	1	sold	3	specifically	1	stability
1	skepticism	1	soldered	1	specify	2	stable
3	sketch	1	soldering	4	specimen	1	stag
1	sketches	3	sole	8	specimens	6	stage
1	skim	21	solely	1	speciosissimus	3	stages
8	skimmed	1	soley	1	speciosus	1	stagger
1	skimming	2	solfataras	1	speckled	2	staggered
20	skin	10	solid	1	speckles	1	staghound
1	skinned	1	solidifying	2	specks	1	stags
2	skins	1	solitaria	1	spectral	1	staid
1	skip	3	solitary	9	speculate	2	stain
1	skipping	1	solor	2	speculated	1	stained
1	skirt	1	solubility	2	speculates	1	stalactiform
1	skulk	2	soluble	2	speculating	1	stalactite
1	skulking	7	solution	6	speculation	1	stalactites
2	skull	2	solutions	9	speculations	1	stalked
4	skulls	1	solve	1	speculats	1	stalks
1	skunk	4	solved	1	speech	5	stallion
1	slag	1	solves	1	speeches	3	stallions
24	slate	4	somebody	1	speedwell	3	stamen
6	slates	1	someday	1	spence	18	stamens
1	slaty	1	somehow	1	spencer	2	stamina
3	slave	1	someone	1	spenser	1	stammering
1	slavery	1	somersetshire	1	spent	2	stamp
2	slaves	1	somerville	1	sperm	2	stamping
25	sleep	34	something	1	spermaceti	1	stamps
6	sleeping	1	sometime	4	spermatic	6	stand
1	sleeps	47	sometimes	3	spey	4	standard
1	sleepy	2	someway	1	sphepherd	7	standing
19	slight	1	someways	1	sphincter	2	stands
2	slightest	22	somewhat	1	sphincters	1	stank
17	slightly	5	somewhere	5	sphynx	2	stanley
1	sliminess	1	somnabulism	1	spi	2	stanleys
3	slip	3	somnambulism	1	spices	1	staphylini
1	slippers	1	somnambulist	1	spicule	4	staphylinidae
1	slit	1	sonda	6	spider	1	starch
21	slope	1	sonde	1	spiders	1	starling
1	sloped	7	song	1	spinal	8	start
1	slopes	4	songs	1	spine	3	started
7	sloping	3	sont	8	spines	5	starting
1	sloths	3	sooloo	2	spinning	1	startled
1	slovering	28	soon	1	spinous	1	startling
32	slow	6	sooner	1	spins	1	starvation
2	slower	3	sophism	1	spiny	1	starved
16	slowly	1	sore	1	spiracles	81	state
3	slowness	1	sorex	1	spiral	22	stated
90	small	2	sorrell	5	spirit	13	statement
10	smaller	3	sorrow	3	spirits	5	statements
5	smallest	1	sorrowful	4	spit	1	staten
1	smart	4	sorry	1	spital	23	states
26	smell	14	sort	1	spite	3	statical
2	smellie	1	sorted	2	spittle	4	stating
1	smelling	1	sorting	2	spitzbergen	9	station
1	smells	3	sorts	3	spix	2	stationary
2	smelt	3	sought	5	splendid	2	stations
8	smile	7	soul	1	splendour	1	statisics
2	smiles	3	souls	6	split	1	statistical
3	smiling	22	sound	2	splits	1	statistics
42	smith	1	sounding	4	splitting	1	statue
7	smiths	10	soundings	1	spoil	2	statues
4	smooth	10	sounds	1	spoke	2	stature
1	smoothed	10	source	2	spoken	1	status
3	snail	4	sources	1	spondylus	1	staunchness
3	snails	1	sous	1	sponge	1	staunton
3	snake	42	south	2	sponges	1	stauntons
4	snakes	24	southern	6	spontaneous	1	stay
1	snap	1	southward	1	spoonbill	1	stays
2	snapping	1	souvent	4	sport	3	steadily
2	snarl	10	sow	1	sporting	2	steady
1	snarling	3	sowed	1	sports	5	steam
4	sneer	12	sowerby	1	sportsmans	1	steel
6	sneering	2	sowing	9	spot	3	steep
1	sneeze	4	sown	5	spots	1	stellata
1	snipe	1	sp	5	spotted	1	steller
1	snipes	1	spa	1	spray	4	stem

20 step
1 steph
1 stephens
1 stephenson
1 steppes
22 steps
4 sterile
1 sterility
6 stewart
1 stewarts
1 stick
1 sticking
3 sticks
1 sticky
11 stiff
1 stiffening
20 stigma
2 stigmas
1 stikingly
1 stilbite
26 still
2 stillness
2 stimulants
2 stimulated
2 stimulus
2 sting
1 stinging
1 stizbergen
1 stoats
28 stock
1 stockholm
8 stocks
1 stokes
1 stolen
19 stomach
2 stomachs
20 stone
1 stonefied
1 stonefield
16 stones
3 stonesfield
1 stonge
2 stony
2 stood
1 stooping
8 stop
1 stopped
1 stopping
3 stops
1 store
3 storge
2 storia
1 stories
1 storm
1 storms
19 story
3 str
6 straight
1 straightened
2 strain
1 straining
1 strait
1 straits
1 stranded
26 strange
2 strangely
4 stranger
1 strangers
1 strangest
1 strangling
52 strata
4 stratification
1 stratified
2 stratum
4 straw
1 strawberries
2 strawberry
1 stray
1 strayed
2 streaks
15 stream
1 streamlet
7 streams
1 streched
1 streching
8 strength
1 strengthen
1 strengthened
2 stretched
1 stretches
1 strewed
1 striae
1 strickland
1 strict
12 strictly
6 strike
1 strikes
7 striking
2 string
2 strings

1 strip
3 stripe
1 strive
2 strives
2 stroke
1 stromboli
39 strong
6 stronger
6 strongest
19 strongly
11 struck
160 structure
17 structures
7 struggle
1 struggled
2 struggles
1 struggling
1 strung
1 struthio
1 struthionidae
1 struthios
1 struthonidae
1 struthonidous
1 strutt
1 sts
1 stuck
2 studded
1 student
15 studied
57 study
9 studying
2 stuff
2 stump
1 stumps
3 stunted
1 stupendous
1 stupid
2 sturnus
1 sturt
1 stuttgart
9 sub
1 subaerial
4 subalpina
3 subaqueous
1 subclassed
3 subdivision
1 subdivisions
10 subgenera
1 subgenus
50 subject
3 subjected
4 subjective
1 subjectively
8 subjects
1 subkingdom
1 sublimation
12 sublime
1 sublimed
1 sublimely
4 sublimity
2 sublittoral
9 submarine
1 submersion
1 submits
3 subordinate
12 subsequent
1 subsequently
2 subsided
33 subsidence
1 subsidiary
2 subsiding
1 subsist
6 substance
10 substances
1 substantiated
1 substitution
1 substracted
7 subterranean
1 subtypical
1 subularia
1 subversive
5 succeed
1 succeeded
2 succeeding
2 succeeds
1 succesive
2 success
16 succession
1 successions
19 successive
3 successively
5 successors
3 suck
2 sucked
2 sucker
8 sucking
13 sudden
20 suddenly
1 suffered
4 suffering

1 sufficent
10 sufficient
4 sufficiently
1 sufficing
2 suffolk
3 sugar
6 suggested
1 suggestion
1 suggests
1 suicide
3 suites
1 suivis
2 sulcatus
1 sulfated
1 sulivan
4 sulkiness
1 sulky
1 sullen
1 sulleness
3 sulp
4 sulph
7 sulphur
1 sulphuretted
1 sulphureus
3 sulphuric
1 sulpuret
3 sum
22 sumatra
1 summagi
1 summary
2 summer
1 summing
9 summit
1 summits
6 sun
1 sunda
1 sunday
1 sundorne
1 sunflower
2 suns
1 sup
1 super
1 superabundance
1 superabundant
7 superadded
1 superficial
1 superficially
1 superfluities
2 superfluous
1 superfoetation
2 superga
1 superimposed
2 superincumbent
2 superinduce
4 superinduced
1 superintendent
3 superior
7 superiority
2 supernumerary
1 superstition
2 superstitions
1 supplementary
1 supplies
1 supply
9 support
6 supported
43 suppose
18 supposed
9 supposes
1 supposing
5 supposition
1 suppression
5 supra
3 supremacy
1 supreme
1 surat
24 sure
29 surely
2 surer
1 surf
49 surface
1 surfs
1 surgeons
1 surinam
1 surmounted
1 surname
15 surprise
5 surprised
1 surprising
1 surprisingly
3 surrounded
2 surrounding
1 surrounds
1 surtout
1 survive
3 survived
1 survives
1 surviving
5 sus
1 susceptibility

1 susceptible
15 suspect
1 suspected
1 suspecting
2 suspects
1 suspending
5 suspicion
1 sussex
4 sw
3 swainson
4 swainsons
9 swallow
1 swallowed
3 swallows
1 swam
1 swamp
1 swamps
11 swan
1 swans
1 swansea
2 swarm
1 swarmed
1 swarming
1 swear
4 sweat
6 sweden
6 sweet
7 swell
1 swelled
2 swelling
1 swells
1 swift
1 swifts
3 swim
2 swimming
1 swine
1 swis
1 sword
1 swum
1 sycamore
8 sydney
5 syenite
8 sykes
1 sylva
1 sylvestris
2 sylvia
2 symbol
1 symetrical
1 symetry
1 symmetrically
5 symmetry
5 sympathetic
1 sympathetics
1 sympathize
13 sympathy
1 symptoms
1 syn
1 synallaxis
2 synclinal
1 syncope
3 syngnathus
1 synonymous
3 syria
1 syrphus
39 system
1 systema
1 systematic
1 systematist
1 systematizing
3 systems
23 table
5 tables
1 tabulate
1 tadpole
1 tadpoles
2 taenia
1 taeniatole
5 tahiti
1 tahitian
51 tail
9 tailed
1 tailée
2 tailless
2 tailless
1 tailor
1 tailors
13 tails
1 taint
1 taiti
1 tak
2 talcose
1 tale
1 talent
1 tales
10 talk
6 talked
13 talking
20 talks
1 tall
2 talus

20 tame	1 testis	1 tilted	1 transat
1 tamed	1 testudo	1 tilting	1 transcendental
3 tameness	4 tetrao	1 tim	1 transfer
1 tanagra	1 teyde	138 time	1 transferable
1 tandis	2 tha	22 times	1 transferred
1 tankerville	1 thârû	10 timor	1 transform
7 tapir	2 thawed	2 tinge	4 transformation
1 tapirus	1 theatre	1 tints	2 transformations
6 tarf	1 theil	5 tip	1 transformed
2 tarsi	1 theirs	2 tips	2 transition
1 tartary	34 themselves	3 tired	1 transitions
1 tarton	3 theological	3 tissue	1 transitu
3 tasco	2 theology	2 tit	1 transl
1 task	5 theoretical	1 title	1 transla
1 tasmania	1 theoretically	1 tits	1 translat
31 taste	2 theories	3 toad	1 translate
1 tasted	137 theory	1 todte	7 translated
8 tastes	164 therefore	3 toe	2 translation
3 tatous	1 therfore	57 together	10 transmission
1 tatu	2 thermometer	1 toise	7 transmit
10 taught	1 thermometrical	3 toises	14 transmitted
10 taw	1 thery	36 told	2 transmitting
3 taylor	1 thes	2 tolerably	3 transmut
3 taylors	1 thibet	1 tollet	5 transmutation
9 tea	16 thick	2 tollett	1 transmutations
10 teach	3 thicker	1 tomb	1 transmuted
1 teachers	2 thickly	2 tome	1 transplant
3 teaching	10 thickness	4 tommy	3 transplanted
1 teal	3 thighs	1 tommys	12 transport
6 tears	1 thiis	4 tone	9 transportal
1 teats	13 thin	1 tongatabou	6 transportation
1 teazle	1 thincks	2 tongue	18 transported
24 teeth	39 thing	1 tongues	2 transverse
1 teethless	33 things	1 tons	1 transversely
1 tel	122 think	1 tooke	6 trap
28 tell	32 thinking	1 tookes	1 trapiches
30 tells	83 thinks	8 tooth	1 trappean
4 temminck	1 thinly	2 toothed	4 trash
1 temmincks	1 thinned	1 toothless	6 travel
1 temmink	5 thinner	10 top	8 travelled
2 temminks	1 thinness	2 topped	2 traveller
11 temp	9 third	2 tops	1 travellers
5 temper	1 thirst	3 torn	4 travelling
2 temperance	1 thistle	1 torpid	23 travels
4 temperate	1 thistles	2 torrent	1 travers
20 temperature	3 thomas	4 torrents	2 traverse
1 temperatures	1 thompson	1 tortise	1 traverses
1 tempereture	1 thompsons	9 tortoise	1 treadèe
3 temple	2 thorax	5 tortoises	1 treading
1 temporal	1 thorn	1 tosca	2 treat
2 temporary	1 thorough	4 total	1 treated
2 temps	1 thoroughfares	1 totally	3 treating
5 temptation	112 though	1 totters	10 treatise
15 tempted	102 thought	6 touch	1 treatises
1 tempting	25 thoughts	5 touched	2 treatment
1 tempts	5 thougt	2 touches	1 treats
1 tenacious	16 thousand	2 touching	33 tree
1 tenacity	6 thousands	1 tough	53 trees
1 tenax	1 thousandth	1 toujours	1 tremadoc
19 tend	1 thr	1 touraine	1 trematode
1 tendancy	3 thread	1 tournal	2 tremble
1 tendencies	1 threads	1 tours	1 tremblemens
85 tendency	3 threw	1 tout	4 trembling
4 tending	1 thrill	1 toute	1 tremendous
2 tendril	1 thrilled	3 toutes	1 tremenheres
10 tends	2 thrips	1 toward	1 tremens
6 teneriffe	3 throat	34 towards	5 tres
1 tenioptera	1 throne	2 tower	1 treviranus
1 tennioptera	2 throug	7 town	1 trials
1 tenrec	1 throughe	2 towns	2 triangle
1 tenrecs	4 throughout	7 toxodon	1 triangular
1 tens	12 throw	3 tr	6 tribe
1 tenses	1 throwing	30 trace	13 tribes
1 tenth	2 thrown	12 traced	2 trick
1 ter	4 throws	12 traces	3 tricks
1 terebrantia	1 thrugout	1 trachilidous	1 tricolor
4 terebratula	9 thrush	8 trachyte	6 tried
11 term	5 thrushes	2 trachytes	3 tries
1 termed	1 thrusting	1 trachytic	1 trifeling
1 terminal	3 thumb	4 tracing	1 trifle
1 terminated	4 thunder	2 tracks	2 trifles
2 termination	6 thursday	1 tract	14 trifling
1 terms	101 thus	1 tracts	1 trifoliata
1 ternhill	1 thy	1 trade	2 trigonia
1 terra	7 thyme	7 trades	1 trilobite
22 terrace	1 thyself	1 tragic	1 trilobites
15 terraces	2 tibia	14 train	1 tring
3 terre	1 tichorhinus	3 trained	1 tringas
3 terres	1 tick	2 training	1 trinity
1 terrestial	1 tickled	13 trains	1 trioecia
12 terrestrial	1 tickling	1 traitent	1 triple
3 terrier	7 tidal	2 tran	1 triplets
2 territory	2 tide	1 tranquil	2 triptera
2 terror	4 tides	1 tranquility	10 tristan
24 tertiary	1 tie	1 tranquillity	1 tritores
12 test	9 tierra	6 trans	1 triturated
4 testament	9 tiger	20 transact	2 triumph
3 testes	5 tigers	2 transaction	1 triumphant
1 testicles	1 tilgate	17 transactions	2 trod
1 testing	24 till	14 transacts	1 troglogdytis

5 visible	3 war	1 whereby	1 wondered
1 visibly	1 warbler	3 wherever	57 wonderful
3 vision	1 warblers	153 whether	7 wonderfully
6 visit	2 ward	5 whewell	1 wondering
7 visited	1 wards	6 whewells	1 wonderment
3 visiting	1 wariness	3 whi	1 wonders
1 visitor	1 warlike	4 while	1 wondrous
1 visits	4 warm	29 whilst	1 wont
7 vital	1 warmed	1 whims	18 wood
2 vitality	1 warmer	3 whine	1 woodbine
2 vitiated	1 warmth	2 whined	1 woodcocks
2 vitreous	2 warring	2 whines	2 woodcut
1 vittata	3 wars	1 whipped	1 woodcuts
1 vitus	1 wash	1 whirled	2 wooded
2 vivapara	5 washed	1 whitby	1 woodfords
6 vivaparous	3 washing	63 white	1 woodlands
1 viverra	1 washington	1 whiteness	1 woodlouse
11 vivid	2 wasp	2 whiter	11 woodpecker
4 vividness	2 wasps	3 whites	4 woodpeckers
1 vivifying	9 watch	140 whole	1 woodpidgeon
1 viviparous	1 watche	6 wholly	5 woods
6 viz	2 watched	1 whores	1 woodward
3 vo	6 watching	1 whorled	1 woody
11 voice	1 watchmaker	106 why	2 woollich
1 voiding	126 water	1 wicked	2 woolwich
2 voila	1 waterbirds	2 wickedness	18 word
1 voit	1 waterdog	33 wide	11 words
1 volans	1 waterfall	2 widely	1 wordsworth
1 volatilized	1 waterfowl	1 widen	1 wordsworths
4 volc	2 waterhens	3 widening	43 work
52 volcanic	1 waterhous	4 wider	5 worked
2 volcanicity	28 waterhouse	3 widgeon	2 working
8 volcano	2 waterhouses	1 widow	12 works
4 volcanoes	7 waters	1 widowbird	104 world
32 volcanos	2 waterton	1 widows	1 worlds
1 volent	1 watertons	1 width	7 worm
3 volition	6 waterworn	3 wiegman	8 worms
1 volney	3 watson	1 wiessenborns	12 worn
2 volneys	7 wave	6 wife	3 worse
4 vols	11 waves	1 wikinson	1 worser
1 voltaire	1 waving	117 wild	1 worsleys
8 volume	3 wax	2 wilder	3 worst
5 volumes	77 way	1 wildest	1 wort
5 voluntary	8 ways	2 wildness	39 worth
2 voluta	4 wd	2 wilkinson	1 worthless
1 volutas	445 we	2 wilkinsons	13 worthy
1 vomica	20 weak	434 will	4 wound
1 vomited	3 weakened	1 william	4 wounded
18 von	3 weaker	4 williams	3 wounds
1 voorst	2 weakest	8 willing	1 wowett
1 vow	4 weakness	2 willings	1 wrecked
68 voyage	7 wealden	4 willis	5 wren
5 voyages	1 wealth	1 willoughby	2 wrens
1 vultures	2 weariness	6 willow	4 wretched
1 vultus	1 wearisome	4 willows	1 wretchedly
1 wa	3 weather	2 wills	1 wrinkle
1 wacke	2 weathering	6 wilson	2 wrinkled
1 waddling	6 web	4 wilsons	1 wrinkles
1 waded	1 webb	5 wind	2 wrinkling
1 wader	3 webbed	2 window	2 write
4 waders	1 webfooted	2 windows	1 writers
1 wading	3 webster	2 winds	1 writes
1 wafer	2 wedge	3 windsor	7 writing
3 wag	2 wedges	1 windward	1 writings
1 wagers	1 wedgings	1 windy	26 written
1 wagged	3 wedgwood	11 wing	41 wrong
1 wagging	2 wednesday	3 winged	1 wrongly
1 waggiou	6 weed	1 wingless	6 wynne
1 waggious	1 weeding	36 wings	1 wynnes
1 wagtail	2 weeds	2 wink	1 xix
1 wagtails	3 week	1 winklemans	3 xv
1 waigiou	4 weeks	1 winnowing	2 xxi
1 waist	1 weelsted	6 winter	1 xxxix
1 waitz	2 weigh	1 wipe	1 yagourundi
1 wakened	1 weighed	5 wisdom	1 yaguaré
3 waking	12 weight	1 wise	1 yak
7 wales	1 weighted	1 wisely	1 yakous
7 walk	1 wellington	18 wish	1 yaks
2 walked	5 wells	3 wished	2 yanky
3 walker	3 wellsted	7 wishes	1 yar
2 walkers	1 wenlock	3 wishing	1 yard
13 walking	14 went	2 wit	1 yards
2 walks	1 wern	1 withdrawn	2 yarrel
6 wall	25 west	5 withered	30 yarrell
1 wallich	1 westerly	22 within	10 yarrells
2 walls	3 western	1 withstanding	1 yarrels
1 walnut	1 westminster	1 witty	2 yawn
3 walrus	4 westwood	1 wm	4 yawning
3 walter	2 wet	1 wnw	1 yawns
2 wander	13 wh	1 woke	1 ye
1 wandered	6 whale	15 wolf	32 year
2 wanderer	4 whales	1 wolves	2 yearly
7 wanderers	9 whatever	18 woman	58 years
4 wandering	1 whe	7 womb	10 yellow
1 wanderings	5 wheat	1 wombat	2 yellowish
1 wanders	2 wheel	1 wombwalls	1 yelpings
18 want	2 whence	18 women	20 yes
2 wanted	2 whenever	2 womens	1 yeso
4 wanting	1 wheras	1 wondefful	1 yesterday
5 wants	7 whereas	13 wonder	231 yet

```
  7 yew
  1 yews
  2 yield
  2 yielded
  1 yielding
  2 yields
  4 york
  1 yorks
 36 you
112 young
  3 younger
  2 youngest
  8 your
  3 yourself
  6 youth
  2 zacatecas
  1 zaire
  1 zapoyla
 26 zealand
  1 zealous
  8 zebra
  3 zebras
  1 zechstein
  2 zeeland
  1 zelandiae
  1 zenbla
  1 zeolite
  1 zest
  1 zibetha
  1 zimapan
  2 zinc
  1 zizania
  1 zizanias
  1 zone
  1 zones
  3 zool
  1 zoolo
 48 zoolog
 15 zoological
  5 zoologie
  1 zoologique
  1 zoologist
 18 zoology
  2 zooly
  5 zoonomia
  1 zoophagous
  2 zoophite
  2 zoophyte
  1 zoos
  2 zorilla
  3 zostera
```

Appendix 3: List of Variant Spellings of Keywords

Darwin's variant spellings, foreign words, proper names, word fragments or
abbreviations, and British spellings.

Total number of words in dictionary appendix = 2773

abberley	amphibia	arica	ballivard	birgos
abdominals	amygdaloid	arion	banda	biscay
aberant	amygdaloidal	aristolochia	barbaroussa	bischoff
abercrmbies	amygdaloids	armadillo	barbyrousa	bishoofs
abercrombie	amyyralidae	armadilloes	barbyroussa	blackwell
abercrombies	anagallis	aroe	barclay	blackwood
abipomenes	analogia	arrowsmith	barlyroussa	blackwoods
abipones	analogie	arskink	barom	blainville
abo	analogys	artic	barrington	blair
abrolhos	anapsis	articulata	bartrams	blakeway
abrupte	anastatica	artifical	batchian	blakeways
absoluteley	anat	ascaris	batopilas	blanca
absurdum	anatidae	ascencion	batracian	blattae
acad	anatifs	ascensi	bauers	blume
acalepha	anatomie	asclepiadae	bayfields	blumenbach
acalephes	anatomique	asclepias	bayte	blumenbachs
acanthosoma	anchusa	ashe	bea	blyth
acapulco	andaman	asiatiques	beatson	blyths
acclimatisation	andesite	asseel	beaucoup	boans
acclimatised	andite	assoc	beaufort	boblaye
accomodate	angelis	associatical	beaum	boerhaaves
accomodated	anglesea	asteriae	beaumonts	boethius
accomodation	anglles	asterias	bebyk	bohunthine
accouchers	anguiformis	aston	beccause	bollaert
acongua	angustidens	astrolabe	beche	bollingbroke
acosta	anhault	atacama	beches	bonapartes
acrita	animaux	athaenaeum	becom	bonariensis
acrite	annales	athenaeum	beechey	bonin
acter	années	attentif	beecheys	bonite
aculeatus	annelida	attentus	beechy	boott
acuneha	annelidae	attribs	beetsons	borabhum
acunha	annelidous	audubons	bein	borealis
admettre	annulosa	aug	bellinghausen	borneo
aegypt	anoplotherium	augt	benchuca	bory
aegyptian	anphidesma	augus	benevelo	borys
affectent	anphidesmas	aurock	bennetts	bos
affin	ans	australes	benthams	boshes
afflux	anson	australis	benza	boswell
agaziz	answ	australs	berberis	botanique
agit	antilope	autour	berbice	botelers
agricl	antuco	auvergne	berenica	botouma
ai	anyman	aux	berghaus	bougainville
aient	anyways	ava	berhave	bouhunthine
ailes	ap	avaoid	bering	bourbond
aimé	apercu	avec	berre	bourou
aira	aphrodites	avestruz	berthelot	bourous
aisément	aphysia	avila	berwick	boussingualt
alauda	apion	avions	berwicks	bouthoner
albite	apologising	avitism	berzelius	bowen
albus	appoximat	avocette	bessy	bowerbank
alcedo	après	avoir	bethgellert	boz
alcyone	apri	aways	bettws	bracteatum
alk	apterix	awk	bevan	bramidor
alklali	aquarium	awks	bevans	brasiliensis
alleghanies	aquarum	ayres	bewick	brassica
allotriandrous	aquatica	azara	bhem	brayley
alluvi	aqueo	azaras	bibron	brecciated
alm	aquilegia	babbage	bibrons	brent
alpina	araucanians	babington	bicheno	brevis
alumen	araucaria	babingtons	bichoffs	brewster
aluminated	araucarian	baccalao	bien	bri
alw	arblay	bachman	bilary	bridgewater
alway	arch	baer	bilineata	bridgewaters
amaryllidae	archaeologia	baers	billardiere	britannicus
amblyrhyncus	archiv	bahama	billin	broccolli
amboina	arequipa	bahia	bimanous	broderip
âme	arg	bailly	binstead	brongniart
amerique	arguements	balancement	biologie	brownes
ammocoetus	argumen	balançons	birchell	brun
amph	argumentum	balanidae	birchels	brunii

734

bruns
bu
buccinum
buceros
buch
buchs
buckland
bucklands
buffon
bunbury
bunwood
burchell
burkhardt
burnetts
burramposter
burrh
butresses
bynoe
bynoes
caermarthen
caespitosa
caetacea
caeteris
caffers
caffres
cahirimus
calabria
calabrian
calandria
calc
calceolaria
caldcleughs
calidonia
callao
callitrix
calosoma
caltha
camelidae
camelion
campanularia
campbell
campestris
canadense
canarys
candoelle
canelones
cannabalism
cannt
canot
cantal
canthairides
capel
caperailkie
capercailzie
capromys
capt
captial
cara
carabidae
carabids
caraccas
caradoc
carb
carbbuncl
carcases
cardium
cardoon
carelmapu
carex
carica
caricaridae
caricas
carimon
carmichael
carnatic
carne
carniola
carnivora
carnosa
carolines
carruncles
carthorse
carus
cas
casoars
cassay
cassicans
caswary
catasetums
caterwhalling
cathedrâ
catorce
caubul
cauquenes
cavernes
cavia
caya
ceinture
celui
centropus

centrum
cephalopoda
ceram
ceratophytes
cercopithecus
cerithium
cerro
certainment
cervus
ces
cessent
cestoid
cestracion
cetacea
cetaceae
cetionidae
cetoniadae
cette
ch
chacune
chaetodon
chalcididous
cham
chamaeleon
chamaelion
chamelion
chamisso
chara
characterised
chardonneret
charlestown
charlesworth
charlsworth
charlworth
charpentier
cheiroptera
chesil
chesnut
chetah
chevre
chico
chien
childr
childs
chileno
chilian
chillingham
chiloe
chimango
chimborazo
chimpanze
chimpaze
chinchillidae
chinense
chinensis
chineses
chingolo
chionis
chladni
chloritic
chloropus
chonos
choo
chota
chouette
chrom
chrysoprase
chupat
chuquisaca
churche
cinbermere
cinereus
cinq
circinnatus
circum
circumstanc
cirrhipedes
cirrhipeds
cisalpina
cisticola
cistus
clapperton
clarkes
clarkia
classfication
classi
classifyng
claystones
clayton
clenomys
clifts
clim
clima
climat
climatized
climatizes
clinal
cline
cobija
cocatores

coccygis
coelum
coembedded
cogan
cogans
coipu
col
coleccion
coleoptera
collarig
colobes
cols
columba
combless
comme
commentatia
communicat
comparé
complanata
compositae
compte
concentricus
concepcional
concepecion
conchifera
concholepas
conclure
concret
concretionary
condamaine
condensor
condore
conducct
conferva
confervae
coniferae
connate
connexion
conse
consequen
consolats
cont
conta
contemp
contemporarys
contentmt
continet
contingences
continguity
contradistinct
contraire
conus
convin
conway
conybeare
coorganic
copiapo
copiapò
copiápò
copris
coq
coquille
coquilles
coquimbo
coraliferous
corallina
corallinae
coralline
corallines
corb
corbet
corbets
corbules
cordill
cordova
cordovise
corelation
corelations
corlula
corniculatus
coromandel
coronata
coronella
correspondance
corresponing
corry
cory
cossigny
costata
costatus
costorphine
cotopaxi
coucal
couroucous
courrejolles
courte
couteur
cowcumbers
coworker
coxe

crabbes
craigs
crawford
creat
crepitando
cresselly
cretceous
creted
cretionary
croonian
crozet
cruikshank
crustacea
crustaceae
cryed
cryptandrous
cryptogamia
crys
crysom
crysomela
crysomelidae
cryst
crystalls
cuculus
cuidado
cuming
cumms
curculionidae
curig
cuscus
cuvier
cuviers
cyanocephalus
cyanoglossum
cyanurus
cyclopaedia
cyclostoma
cyclostomes
cynocephalus
cynoglossum
cyrena
dacelo
daines
daltonism
dampier
dampiers
dans
daphnia
darby
daresay
daresays
darnleys
darwar
dasypus
dasyurus
daubeny
daubisson
decandoelle
decandolles
decb
decemb
decrite
decumanus
defonsos
degnen
deinotherium
déjà
dela
delabechs
deletereous
demarara
demerara
demerary
denham
densifolium
derry
desaguadero
descen
descript
desgraviers
deshayes
developemen
developement
developent
developpent
devonshire
dewar
dhangar
dholbum
dholes
di
diard
diarrhaea
dichogamous
dicot
dicotyledenous
dicotyledones
didelphidae
didelphis
didelphys

diemen
diemens
dieteriorating
dieu
dif
diferent
diff
differen
dillwyn
diluv
diluvianae
dimidiate
dingos
dioecia
dioecious
dioeecous
dioeious
dioica
dioicous
diorite
diptera
dipus
dissenblances
dissert
distincte
distinctes
distri
distrib
distribut
dit
diuca
divatium
divellent
divino
divisent
divortium
divortiums
dn
dobrizhoffer
dochart
dod
dodecatheon
domestica
domingo
donacia
donn
dorbigny
dorking
dorsetshire
doubl
douglas
draco
drayton
dredfully
drinkwater
drinkwaters
drunkedness
dt
dth
duchesne
dumeril
dumplin
durableness
durville
dutrochet
duvaucel
dwealt
dwights
dz
eap
eaton
echinites
echinodermata
echnida
ed
edentata
edin
edolius
education
effet
egalement
ehrenberg
ehrenbergh
ehrenbergs
eichwald
elater
elena
elephas
elevatory
elie
elimanated
elle
elles
elliotsons
éloigné
elongata
elspeths
elytra
emb
emberiza

735

embriza	ferguson	gaudichauds	happi	huppe
embued	ferme	gault	hardinge	hutcheson
emeu	fernandez	gaut	harlaam	hutton
empire	ferrugineous	gaz	harmonicon	hyaenodon
emptty	ferrugineum	gazal	harpalus	hybridise
emys	ferrussac	gazoeus	hartley	hybridised
encrinites	fertilise	gemmiparous	hasselt	hybridisim
encyclop	feruginous	generalle	hattica	hybridising
endeavoured	festuca	generat	haustellata	hyd
endeavours	fever	generl	hautes	hydatid
endeavur	ficus	genetal	hauteurs	hydridity
endlicher	fingals	geo	hawfinch	hydrocele
endlicker	finlay	geof	hayd	hydromys
endosmic	fintec	geoffroy	heartease	hymenoptera
englis	firma	geoffry	heautandrous	hypnum
ensd	firola	geogr	heaverns	hypoth
entomolgicae	fitton	geograp	heber	ichthyosaurus
entomolog	fittons	geograph	heberdens	icterus
entomologica	fitz	geographically	heckla	icthiology
entomologicae	fitzroy	geographico	hectometre	icthyo
entomost	fitzwilliam	geographie	heigt	icthyosaurus
entomostraca	flappered	geolog	heigth	ideosyncracy
entozoa	flaqueza	geologico	hemi	ideosyncrasy
entre	flavum	geologique	hemiptera	ideot
enuntiated	fleche	germen	henge	ideotcy
eocine	flèche	geum	hensleigh	ideotic
eoeconomy	fléche	ghilan	hensleighs	ideots
epaules	fleestest	giganteus	henslow	iimagine
epidote	flemming	gilolo	henslows	il
epigmous	flore	glencoe	hepatici	ile
epigonous	flourens	glenoe	herberts	ile
epizoa	fluoric	glenroy	herchels	iles
epock	flustra	glöger	heredetariness	iles
epocks	fluvicolae	glorrying	heredetary	illdefined
equall	flys	gmnura	heredetray	illtempered
ercall	foame	gobbah	hermaphrodite	immens
erhman	foetal	goldens	hermoso	immenses
erman	fol	gonaïves	hernes	imper
ermans	folkstone	goodhumoured	hersche	impliance
ervum	forbes	goodnatured	herschels	importants
espèce	forchammers	gorze	hershel	imrie
especes	forficatus	goulds	hervey	inamimate
espèces	forficetus	gov	heterogenous	inclinêe
especiall	formativa	gowen	heterom	inclinées
esq	formativus	graah	heteromera	incorrigably
essai	formées	gradat	heteromerous	ind
essequibo	forment	graecian	hibbert	indefessa
été	forrest	graecians	hilaire	indelibleness
etn	forseeing	graeduated	hilaires	indes
etna	forseight	grallae	hilianthemum	indi
être	fossiles	grammalite	hillhouse	indicus
eugenius	fourmillier	grande	himmalaya	indio
europaean	fragmens	granfathers	hindoism	individ
europaeans	fraser	graniniverous	hinnus	individus
europèan	fraxinella	granits	hippot	industrie
eussent	fre	grauwacke	hippotami	inequlities
eût	frequet	greatt	hippotamus	inf
evelyns	freycinet	greenfinch	hippotomus	infiltering
everytime	freycinets	greenough	hirundo	infini
exacte	fringilla	greenstone	histoire	infinites
examen	frio	grenats	histrionicus	inflorescence
excalted	frondosi	grenish	hitchcock	infra
exceedingl	froudes	gret	hodgkins	infusoria
exeters	fruct	greter	hodgson	inglish
exjudge	fruiterer	greywacke	hoffmans	injustness
exosmic	frustra	grisea	holcroft	innumberable
exp	ft	grotius	hollands	ino
exper	fucus	groundsil	hollyoak	inoceranus
experimentise	fuegia	groznerat	hollyoaks	inor
experimentize	fuegian	grt	holocentrus	inosculate
expr	fuegians	grups	holyhoaks	inosculating
expressemotions	fuego	guacho	holyoak	inosculation
exserted	fuge	guadalaxura	holyoaks	insecta
extreamly	fulginosus	guahon	holyrood	insectes
extres	fulgoridae	gualgayoc	homme	insectiferous
exuviae	fulica	guanaca	hommes	instinc
eyton	fullfil	guanax	homogenious	instincting
eytons	fullfilled	guanaxuato	honourable	instinctivedly
facedness	fulva	guantajaya	hooker	instinctus
facultes	furnarii	guanuaxuato	hopkins	institut
fah	furnarius	guasco	horae	integrant
fait	fût	guayana	hornblend	intelligble
falco	gaertner	guayanensis	horne	intelligence
falkland	gaimard	guaze	horner	interlopement
falklandii	galapagos	guggling	hornsey	interm
falklands	galeum	guillemost	horsbrugh	interrompues
falkner	galignani	guoy	horsburgs	interstratified
falkners	galiopithecus	guyanensis	horsfield	intertropical
falle	galium	guyon	hort	intertropics
fam	gallesio	guzerat	hortus	intincts
farroïlap	gallinaces	gypseous	housedogs	introduct
farthers	gallionella	habberley	huantajaia	inutility
faune	galtons	haberclador	hubberley	inverorum
favourable	garcia	haemorragic	humaine	ipecacuhan
feb	gardiners	haena	humb	iquique
feconds	gardner	halfbred	humbold	iquiqui
feelin	gardners	halimeda	humboldt	isabelles
feis	garrottii	haller	humboldts	isd
feldspathes	garsipa	handerchief	humes	isds
felspathic	gastrobranchus	hants	hunbt	isid

736

isl	lathyrus	macleay	mephites	muscicapa
islas	latit	macleays	mer	muscicapidae
isld	latley	macleys	mercu	muscovy
islds	latreille	macloviana	mercurialis	musicus
isle	latte	macphee	mères	muskitoes
isolately	laught	macquarie	merid	musm
ispida	lavas	macquarrie	mesites	musquitoes
itapicuru	lavater	macqueries	meta	mycelis
jackall	lavaters	macrauchenia	metals	mylodon
jago	lavendar	macrotherium	metamor	myothera
japon	lb	mactra	metamorph	myriametre
jardin	lbs	maculatus	metamorphised	mytilus
jardines	ld	maculloch	metamorphs	natical
jeddah	lect	macullochs	metaphysic	natura
jemmy	leet	macúsie	metaphysicks	naturae
jenners	legrand	maed	meteorolite	naturelle
jenyns	leguminosae	maer	metereology	naturelles
jerrod	leighton	magellanic	meterological	naturgeschichte
jerrold	leigton	magellanicus	métis	navigatores
jeune	lemagne	magindanao	meyen	n.b
jgo	lemna	mahé	meyens	nead
joatingua	len	mais	micaceous	nebria
johnston	lepidosiren	mak	mices	neig
jonnes	leptosiphon	makis	michell	neigbourhood
jordanhill	leptuse	mal	miers	neigbouring
joun	lepus	malasia	mieux	neighbourhood
journ	leschenault	malcolmson	migrats	neighbouring
julians	lessings	maldonado	mildanao	neilgherries
jun	lessonia	malouines	millepora	neilson
jura	lettres	malte	millisuga	nematoid
jusquaux	leuconotes	malthus	milvulus	ness
justicia	leucoprymnus	malva	miocen	nétaient
kaluz	leur	mam	mishaped	newbold
kames	leurs	mamalogy	misnumbered	newlands
kamschatka	leven	mamals	misnumbering	newydd
kamtchatka	leyden	mamm	missel	nichol
kamtschatka	lhc	mammali	misseltoe	niether
kankaer	lias	mammalia	missisippi	nihil
karamania	liegende	mammalidae	mitchill	nisus
kensington	lieut	mammferus	mitterschlich	norfolkicae
kenyon	lignes	mammiferous	moel	noronha
keratry	lik	mammifers	mohamned	norton
kerguelen	lile	mandibulata	moins	nous
kerrs	lile	manguia	mojos	nov
kilfinnan	lillies	manks	molina	novem
kindgdom	limonia	mannering	molinas	novozelandiae
kingii	lin	mans	molluccas	nullipora
kinlet	linaria	manu	mollusca	nux
kinlett	lincolnshire	maranh	molluscous	obediance
kirby	lindley	marchantia	molucca	obras
kirbys	lindlys	marevellous	moluccensis	obscura
klaproth	linn	marfil	molybdated	observ
knats	linnaeus	marh	mondobbo	observaciones
kneeding	linotte	mariana	monoceros	observat
koelreuter	linum	marianna	monocot	observés
kolkreuters	lisianskys	marianne	monocotyledenous	obseved
koloff	listera	mariannes	monoecious	occassionally
kolreuter	literat	marianus	monoeecious	occurence
kongsberg	lithospernum	marias	monooecia	oclock
kordofan	liza	marquesans	monooecious	oct
korto	loasa	marsden	monro	octav
kosir	lochy	marsup	mons	octob
kotezebues	lockarts	marsupiata	monstruous	oeconomy
kotzebue	loddiges	martineau	montagnes	oeil
kotzebues	loix	martineaus	montagu	oenothera
krauss	lomond	martius	montes	officier
kuhl	longicornes	mascott	monucle	ogleby
kurukuru	longirostris	massachusset	monucule	ogwen
kylau	lonsdale	massacuhssets	moralité	oich
kylow	lonsdales	massaroony	moran	oiseaux
labiata	lorsquelles	mataco	moreton	olinda
labiatae	loudon	matchian	moretons	omnibuss
labillardiere	loudons	mathison	mormodes	omniphitophagous
lacépède	loup	matica	morne	oncitate
lacerta	lowe	matiere	moruffetes	ont
lachyrmal	lucanidae	mawes	moschus	op
lacrustine	lucanus	maypo	mostrified	opetiorhyncus
ladner	lucidus	mbara	motacilla	ophrys
lamaks	luçon	mcclay	mothes	ophyressa
lamantin	luconia	meckel	moufflon	opinion
lamarck	lucs	meditteranean	muar	opuscule
lamarcks	lund	megalonyx	muhanmad	orang
lamark	lunds	megatheria	mulattos	orangs
lamarks	lutke	megatherium	mulita	orbiculas
lancien	lutké	megetherium	müller	orbigny
landons	lutkes	melanodera	müllers	orbignys
lang	lx	melaphyre	multilineatus	orchidacaeous
laniadae	lychnis	melaspena	multipliied	orchidiae
lao	lycopodiums	melegethes	mundine	orfords
laoccaon	lyell	meligethes	mungo	organization
laponia	lyells	mell	mur	organizaton
lardens	lymington	melolonittha	murc	orginally
lardner	macaco	mem	murchison	orientale
lardners	macacus	mêmes	murchisons	orientel
lartet	macculloch	memoire	muriated	oriolus
lasch	maccullochs	memoires	musalmans	ornith
lasiurus	macdonald	mendoza	muscadivora	ornitho
lat	macgillivray	menoir	muschelkalk	ornithholog
latd	mackintoshs	mens	musci	ornithologum
latham	maclay	menyanthes	muscical	ornithorhynchus

ornithorhyncus	pernetty	portillo	reid	scoresby
ornityhyrhycus	pernety	possede	reinwardt	scoresbys
orpiment	peron	possesse	religio	scoriform
orthopt	perroquets	pota	reliqu	scoticus
osorno	pers	potosi	reliquiae	scotts
oss	personnal	poulett	remarke	scrope
ossemens	petersburgh	pourrait	remparts	sebe
ossements	petise	poussin	rendu	sebright
osteopora	petisse	poutter	rendus	sebrights
otaheite	petites	powe	rengger	seco
otahiati	peu	powis	renggers	secondary
oualan	peuquenes	praeternatural	repesentative	secousses
oural	peuvent	pratencole	repeted	secretionary
ourang	peyrouse	pratensis	représentants	sedgwick
outang	phaedo	preceed	represetatives	sedgwicks
outlyer	phalangista	precip	reproduire	selbourne
ovaria	phalarope	predestinarian	reproduzione	selbournes
ovarium	phenomem	preeminiently	reseda	selby
overfall	phenomenes	prefe	resemblanc	sellow
overtempted	phil	premières	resown	semestre
ovina	philedon	preparis	restr	semi
ovington	phillipines	presente	restric	semiconsolidated
ovingtons	phillippensis	presentent	resumè	semilunar
owen	phillippines	presque	resumé	semnopitheque
owens	philo	prettty	resuts	semptemb
oxalis	philos	prevost	revd	sensorium
oxidated	philosop	priamang	reynold	sep
oxided	philosoph	prichard	rhc	separ
oxyurus	philosophie	prietly	rhinocerose	separe
pachydermata	philosophique	primigenious	rhod	separtion
pacifick	philosphically	prin	rhodendron	sept
pajero	philotis	principes	rhododendrum	septaria
palaeotherium	phils	pritchards	rhyncus	septemb
paleontographie	phisoph	prob	rialeja	septembe
paley	phitophagous	probandi	ribston	seq
paleys	phocae	prodromus	ribstone	serapis
palmipedes	phonolite	proesentum	riged	sergipe
palustris	phonolites	profondes	riproduzione	serin
pamplet	phonolitic	progressife	rivage	serpula
papaver	phormium	prolifickness	ro	serres
papil	phryganea	prooff	roberts	sertularia
papilionaceous	phsiognominical	propenity	rodentia	ses
papuensis	phy	prophecied	rogers	seul
paradoxurus	phyllade	proteaceae	roi	seule
paradupasi	phys	pud	rongeur	sev
paragua	physiolog	pudu	rosa	sewalick
paralleles	physiologiques	puffinuria	rothe	shame
parameles	pic	pulo	roussette	shatsbury
paranagua	pichincha	punaise	roussin	shaws
parellel	pichye	purdie	roussins	shd
paret	pictet	putrifies	roxburgh	shee
paribus	pidgeon	pyenite	royle	sheeps
parnassia	pidgeons	pyrenaicum	royles	shelfs
parroket	pieds	pyrennees	roys	shiant
parrott	pineaple	pyritous	rozales	shinglle
particls	pineaster	pyrocephalus	ruber	short
particuliere	pippen	pyrrho	rufiventris	shortlived
partie	pippens	quadras	rumpless	shrewsbury
pasco	pipra	quadrifolium	rupestris	shropshire
pashion	piscatoria	quadrumanous	rupicole	shu
passeres	pish	quand	rüppel	siabod
passess	pisolitic	quatre	ruppia	siciences
pasto	pisses	queene	ruscus	sightedness
patagonicus	pitchstone	quelque	saboeus	siicily
patagonie	placent	quels	sageret	silene
pataz	placentates	queque	sagitta	silicified
paulo	planaria	quetelet	sagittella	silliman
pauls	planariae	quichuas	sait	sillimans
paxton	planchon	quillota	salado	siluris
payta	plantarum	quilmes	salamandra	silv
pe	plas	quinarian	salband	simplelst
peacable	platycephalus	quinarians	salomon	simplets
peaccocks	playfair	quinary	salpa	simplifys
pearlstones	ple	quiyá	salsisbury	simplits
peccari	plesiosaurus	quoy	salta	singings
pecten	plesiossaurus	radack	saltum	situ
pectinibr	pluckt	radiolum	saltus	sivatherium
pectinibranchiate	pm	rafflesia	sambawa	slovering
pecu	poa	rakhekna	sammon	smellie
ped	poenig	rakkelhan	santo	sobre
pedunculated	poenis	ralix	santos	soc
pegu	poliosoma	ramond	sapajou	soci
pegue	polirus	ranunculus	sarahs	socotra
pelew	polygam	rapilli	sauroid	solanum
pellew	polygamia	raptores	savigny	soley
pellicle	polymorphes	ras	saxatile	solfataras
pelud	polynes	rasorial	scarabadae	solitaria
peludo	polype	ravins	schelgel	solor
pennatula	polypi	reaumur	schemnitz	somersetshire
pente	polypus	recieve	schiede	somnabulism
pentlands	pompeei	recognise	schit	sonda
pentstemon	pongo	recognised	schmidtmeyer	sont
percieved	ponza	recognising	schomburgk	sooloo
percisely	porcarious	reconsult	schow	sorex
perenne	porcarius	recuiel	sci	sorrell
perga	porcellus	redissolved	scicily	sort
perigonous	porph	redpoles	scienc	souvent
periwinkle	porphyrys	reeaches	scincus	sowerby
permanentes	portezuelo	regne	scissures	spallanzani
permanet		regrafted	scolopax	spallanzanis

spath
spe
spean
speciosissimus
speciosus
speculats
spence
spey
sphepherd
sphynx
spi
spital
spitzbergen
spix
spondylus
spritual
squilla
squinny
staghound
stalactiform
stanleys
staphylini
staphylinidae
staten
statisics
staunton
stauntons
stellata
steller
steph
stephens
stephenson
stewarts
stikingly
stilbite
stizbergen
stonefied
stonefield
stonesfield
stonge
storge
storia
str
streched
streching
strickland
stromboli
struthio
struthionidae
struthios
struthonidae
struthonidous
strutt
sts
sturnus
sturt
subaerial
subalpina
subaqueous
subclassed
subgenera
subkingdom
sublittoral
substracted
subtypical
subularia
succesive
sufficent
suffolk
suivis
sulcatus
sulivan
sulleness
sulp
sulph
sulphuretted
sulphureus
sulphuric
sulpuret
summagi
sunda
sundorne
superadded
superfoetation
superga
superincumbent
surat
sus
sussex
swainson
swainsons
swis
syenite
sykes
sylva
sylvestris
symetrical
symetry
sympathetics
synallaxis

syncope
syngnathus
syrphus
systema
systematist
taenia
taeniatole
tailée
tailess
taiti
tak
talcose
tanagra
tandis
tankerville
tapirus
tarf
tarton
tasco
tatous
tatu
taylors
teethless
tel
temminck
temmincks
temmink
temminks
tempereture
tenax
tendancy
teneriffe
tenioptera
tennioptera
tenrec
tenrecs
ter
terebrantia
terebratula
ternhill
terre
terres
terrestial
tetrao
teyde
tha
thârû
theil
therfore
thermometrical
thery
thes
thiis
thincks
thompsons
thougt
thr
thrips
throug
throughe
thrugout
tichorhinus
tierra
tilgate
todte
toise
toises
tollet
tollett
tongatabou
tooke
tookes
tortise
tosca
toujours
touraine
tournal
toute
toutes
toxodon
tr
trachilidous
trachyte
trachytes
trachytic
traitent
tran
trans
transat
transitu
transl
transla
translat
transmut
transportal
trapiches
trappean
travers
treadèe
tremadoc

tremblemens
tremenheres
tremens
tres
treviranus
trifeling
trifoliata
trigonia
tring
tringas
trioecia
triptera
tritores
troglogdytis
troughtons
trouve
trouvent
trys
tu
tubul
tubularia
tubulipores
tucaman
tuckeys
tucuman
tufas
tulla
tullamore
tullus
turbinated
turbo
turbos
turdus
turneth
turnspit
turpin
turpins
turrit
turritella
tw
twiching
tyndrum
tyrannidae
tyrannula
tyrannulae
tyrannus
uddevalla
ulitmate
ulloa
ulloas
unanùe
undescribable
undescribed
unfavourable
uni
until
untill
uper
ures
urtica
urubu
urville
uspallata
ustil
utang
vaginulus
vaillant
valait
valenciennes
valle
vallées
vallies
van
varas
vars
vebtetrata
veg
vegetale
vegetaux
velellae
vellellae
venatica
vendarques
venemous
ventana
ventanas
ver
verd
verdier
verds
verey
vermejo
vers
versâ
vert
vertebrae
vertebrata
vertèbres
vesp
vespertilio

veta
veuve
vica
viccissitudes
viedo
villosus
vinca
vinchuca
virlet
viscacha
vittata
vitus
vivapara
vivaparous
viverra
vo
voit
volans
volc
volcanicity
volent
volney
volneys
vols
voluta
volutas
vomica
voorst
vultus
wacke
waggiou
waggious
waigiou
waitz
wallich
watche
waterbirds
waterdog
waterhens
waterhous
waterhouse
waterhouses
waterton
watertons
watson
wd
wealden
webb
webfooted
weelsted
wellsted
wenlock
wern
westminster
westwood
wh
whe
wheras
whewell
whewells
whi
whitby
widowbird
wiegman
wiessenborns
wikinson
wilkinson
wilkinsons
willings
willis
willoughby
wilsons
winklemans
wm
wombwalls
womens
wondefful
woodfords
woodlouse
woodpidgeon
woodward
woollich
woolwich
worser
worsleys
wowett
wynne
wynnes
xxi
xxxix
yagourundi
yaguaré
yakous
yanky
yar
yarrel
yarrell
yarrells
yarrels
yeso

zacatecas
zaire
zapoyla
zechstein
zeeland
zelandiae
zenbla
zeolite
zibetha
zimapan
zizania
zizanias
zool
zoolo
zoolog
zoologie
zoologique
zooly
zoonomia
zoophagous
zoophite
zorilla
zostera

739